DIRECTORY OF BIOMEDICAL AND HEALTH CARE GRANTS 2012

23rd Edition

Schoolhouse Partners LLC
West Lafayette, Indiana

Why 'Partners'? Partners work together as a team, with each person contributing different skills and expressing his or her individual interests and opinions to the unity and efficiency of the group in order to achieve common goals. Our aim is to make worthwhile resources available to you so that you and your institution or organization can and will achieve those goals.

Schoolhouse Partners, 1281 Win Hentschel Blvd, IN 47906
www.schoolhousepartners.net
Published 1985, 23rd edition 2012

Printed and bound in the United States of America

ISBN 978-0-9837622-7-0
0-9837622-7-9

Table of Contents

Introduction

For over three decades the GRANTS Database and its print complements, including the annual Directory of Biomedical and Health Care Grants, have provided the research community with current, accurate information regarding funding for research-related programs and projects, scholarships, fellowships, conferences, and internships. To meet the increased information needs of professionals, the GRANTS Database, researched and edited by members of the Schoolhouse Partners editorial staff, has grown significantly. This 23rd edition of the annual Directory of Biomedical and Health Care Grants offers factual and concise descriptions of nearly 4,800 funding programs. With this new edition both experienced and novice grantseekers will be better able to target only the most applicable programs for their biomedical and healthcare needs.

Directory of Biomedical and Health Care Grants 2012

The *Directory of Biomedical and Healthcare Grants 2012* features listings of programs that offer non-repayable funding for projects ranging from laboratory investigations to those designed to study the needs of society in healthcare delivery. Listings in the main section of the *Directory* contain annotations describing each program's focuses and goals, program requirements explaining eligibility, funding amounts, deadlines, *Catalog of Federal Domestic Assistance* program number (for U.S. government programs only), sponsor name and address, contact information, and the sponsor's Internet Web address. Grantseekers with access to the Internet can use the addresses to locate further information about the organizations and their application procedures. Internet addresses are also provided, when available, within listings. Some of the grant programs listed in the main section have geographic restrictions for applicants.

The following people have devoted a great deal of patience, hard work and thought to keep this project and the database an informative service for the user. It is especially important to mention the GrantSelect team – Jeanne Winstead, Susan Bales, Louis S. Schafer, Ed.S., and Anita Schafer – for their diligent research, editorial work, development and assignment of index terms, and production – as well as all Schoolhouse Partners staff who contribute to the GRANTS effort.

Indexes. The Subject Index of this *Directory* lists all program titles—with accession numbers—under their applicable subject terms. Other indexes follow, including the Grants by Program Type Index, which lists 38 program categories, such as Basic Research, Fellowships, Travel Grants, etc., with the grants that fall within their scope; and the Geographic Index, which lists programs that have state, regional, or international focus. See How to Use This Directory on pages v-vii for sample index entries.

Using the *Directory* for Grantseeking

By using the *Directory of Biomedical and Health Care Grants*, grantseekers can match the needs of their particular programs with those sponsors offering funding in the researchers' area of interest. The information listed here is meant to eliminate the costs incurred by both grantseekers and grantmakers when inappropriate proposals are submitted for a sponsor's funding program. However, because the GRANTS Database is updated daily basis, with program listings continually added, deleted, and revised, grantseekers using this *Directory* may also search the GRANTS database online through *GrantSelect* (www.grantselect.com).

All new and revised information has been extracted from (1) the sponsor's updated of previously published program statements included in earlier editions of GRANTS publications, (2) questionnaires sent to sponsors whose programs were not listed in previous editions, or (3) other materials published by the sponsor and furnished to Schoolhouse Partners. Updated information for U.S. government programs includes new and revised program information published in the *Federal Register*; the latest edition of the *Catalog of Federal Domestic Assistance*; the *NIH Guide*, published weekly by the National Institutes of Health; and the *NSF E-Bulletin*, a monthly publication of the National Science Foundation. Included in this edition are identifying document numbers from the NIH and NSF publications. Located at the ends of the program descriptions in certain entries, the numbers indicated the ongoing NIH program number (PA) or the request for applications number (RFA). For programs of the National Science Foundation, the *NSF Bulletin* number appears. This information will help users identify the programs when seeking additional information from program staff.

 Anyone having questions regarding the content, format, or any other aspect of the *Directory of Biomedical and Health Care Grants 2012*, GrantSelect, or the GRANTS Database should contact the Editors, GRANTS, Schoolhouse Partners or editor@schoolhousepartners.net.

How to Use This Directory

The *Directory of Biomedical and Health Care Grants 2012* is designed to allow the user quick and easy access to information regarding funding programs in a researcher's specific area of interest. This *Directory* is composed of a main section, Grant Programs, which lists grant programs in alphabetical order, and three indexes: the Subject Index, the Grants by Program Type Index, and the Geographic Index.

GRANT PROGRAMS

Each listing in this section consists of the following elements: an annotation describing each program's focus and goals, requirements explaining eligibility, funding amounts, application and renewal dates, the *Catalog of Federal Domestic Assistance* program number (for U.S. government programs only), sponsor information, contact information, and Internet address.

GRANT TITLE — **Paul G. Allen Family Foundation Grants** **4298** — ACCESSION NUMBER

The single foundation, created through the consolidation of Allen's six previous foundations (The Allen Foundation for the Arts, The Paul G. Allen Charitable Foundation, The Paul G. Allen Foundation for Medical Research, The Paul G. Allen Forest Protection Foundation, The Allen Foundation for Music, and The Paul G. Allen Virtual Education Foundation), will continue to focus on the Allen family's philanthropic interests in the areas of arts and culture, youth engagement, community development and social change, and scientific and technological innovation. The Arts and Culture Program fosters creativity and promotes critical thinking by helping strong arts organizations become sustainable and supporting projects that feature innovative and diverse artistic forms. — GRANT DESCRIPTION The Youth Engagement Program improves the way young people learn by supporting organizations that use innovative teaching strategies and provide opportunities for children to address issues relevant to their lives. The Community Development and Social Change Program promotes individual and community development by supporting initiatives and organizations that provide access to resources and opportunities. The Scientific and Technological Innovation Program advances promising scientific and technology research that has the potential to enhance understanding and stewardship of the world in which we live. Organizations may only receive one grant per year. Organizations must not have any delinquent final reports due to any of the Paul G. Allen Foundations for previous grants. Grantseekers are encouraged to apply through the online application process, where basic organizational and project information will be requested. Guidelines are available online.

REQUIREMENTS — *Requirements* 501(c)3 tax-exempt organizations, status from the Internal Revenue government entities, and IRS-recognized tribes are eligible. Eligible organizations must be located in, or serving populations of, the Pacific Northwest, which includes Alaska, Idaho, Montana, Oregon, and Washington.

RESTRICTIONS — *Restrictions* In general, the foundation will not consider requests for general fund drives, annual appeals, or federated campaigns; special events or sponsorships; direct grants, scholarships, or loans for the benefit of specific individuals; projects of organizations whose policies or practices discriminate on the basis of race, ethnic origin, sex, creed, or sexual orientation; contributions to sectarian or religious organizations whose principle activity is for the benefit of their own members or adherents; loans or debt retirement; projects that will benefit the students of a single school; general operating support for ongoing activities; or projects not aligned with the foundation's specified program areas. 509(a) private foundations are ineligible.

APPLICATION/ DUE DATE — *Date(s) Application Is Due* Mar 31; Sep 30.

Contact Grants Administrator, (206) 342-2030; fax: (206) 342-3030; — CONTACT
email: info@pgafamilyfoundation.org

INTERNET ADDRESS — *Internet* http://www.pgafamilyfoundation.org

Sponsor Paul G. Allen Family Foundation — SPONSOR INFORMATION
505 Fifth Ave S, Ste 900
Seattle, WA 98104

SUBJECT INDEX

The most effective way to access specific funding programs is through the Subject Index. This index lists the subject terms with applicable grants program titles – and their accession numbers – alphabetically under each term. Terms were assigned to target the specific area of research designated in the description of each program. Cross-references are used to link subjects and assist the user in finding specific grant information.

Following are general guidelines that can make your search of this index more successful. First, check under the specific topic of interest rather than a more general term. For instance, if you are interested in chemical engineering, look under "Chemical Engineering" rather than "Engineering." Items indexed under "Engineering" indicate funding in broad areas of engineering.

Use general headings when you want grants covering broader areas or if you can't find a specific topic. For example, many grants list funding for humanities research, health programs, or science and technology. To find these grants use such headings as "Humanities," "Medical Programs," "Science," or "Technology." For additional grant information on more specific humanities research opportunities, such as in American History or Cultural Anthropology, also check under the topics "United States History" and "Anthropology, Cultural."

Many of the grant programs provide funding for research-related scholarships, faculty fellowships, dissertations, undergraduate education, conferences, or internships. If grant funds are designated for specific disciplines, you will find the items under the specific subject. Scholarships and fellowships are also listed under the terms "Native American Education," "African Americans (Student Support)," "Hispanic Education," "Minority Education," and "Women's Education."

Grants concerning study of a particular country are listed under the name of the country. Grants concerning the history, literature, art and language of a country are listed under the name of the country also, e.g., "Chinese Art" and "Chinese Language/Literature."

SUBJECT TERM ———— **Education**

A.L. Mailman Family Foundation Grants, 5
AAAS Science & Technology Policy Fellowships - Health, Education & Human Services, 22
AARP Andrus Foundation Grants, 104
Abbott Laboratories Fund Grants, 125
ACIE Host University Edmund S. Muskie/Freedom Support Act Graduate Fellowships, 156
ACT Awards, 189
Akonadi Foundation Anti-Racism Grants, 351 ———— **PROGRAM TITLE**
Akron Community Foundation Grants, 352
Albert and Margaret Alkek Foundation Grants, 370
Albuquerque Community Foundation Grants, 379
Alcoa Foundation Grants, 380
Alcon Foundation Grants Program, 382

GEOGRAPHIC INDEX

This index lists programs that have state, regional, or international geographic focus. The Geographic Index is arranged by state, followed by Canadian programs, then by international programs by country, and lists grant program titles and their corresponding accession numbers.

COUNTRY ———— **United States**

Alabama ———— **STATE**

3M Fndn Grants, 2
Alabama Humanities Fndn Grants Program, 368
Arkema Inc. Fndn Science Teachers Program, 705
CDC Injury Control Research Centers Grants, 1198
PROGRAM TITLE ———— DOE Experimental Program to Stimulate Competitive Research (EPSCoR), 1653
Hill Crest Fndn Grants, 2293
Linn-Henley Charitable Trust Grants, 2777
NOAA Community-Based Restoration Program (CRP) Grants, 3843
Southern Company's Longleaf Pine Reforestation Fund, 4726

PROGRAM TYPE INDEX

This index is broken into 43 categories according to the type of program funded:

- Adult Basic Education
- Adult/Family Literacy Training
- Awards/Prizes
- Basic Research
- Building Construction and/or Renovation
- Capital Campaigns
- Centers: Research/Demonstration/Service
- Citizenship Instruction
- Community Development
- Consulting/Visiting Personnel
- Cultural Outreach
- Curriculum Development/Teacher Training
- Demonstration Grants
- Development (Institutional/Departmental)
- Dissertation/Thesis Research Support
- Educational Programs
- Emergency Programs
- Endowments
- Environmental Programs
- Exchange Programs
- Exhibitions, Collections, Performances, Video/ Film Production
- Faculty/Professional Development
- Fellowships
- General Operating Support
- Graduate Assistantships
- Grants to Individuals
- International Exchange Programs
- International Grants
- Land Acquisition
- Matching/Challenge Funds
- Materials/Equipment Acquisition (Computers, Books, Tapes, etc.)
- Preservation/Restoration
- Professorships
- Publishing/Editing/Translating
- Religious Programs
- Scholarships
- Seed Grants
- Service Delivery Programs
- Symposia, Conferences, Workshops, Seminars
- Technical Assistance
- Training Programs/Internships
- Travel Grants
- Vocational Education

Government and Organization Acronyms

AAAAI American Academy of Allergy Asthma and Immunology
AAAS American Association for the Advancement of Science
AACAP American Academy of Child and Adolescent Psychiatry
AACN American Association of Critical Care Nurses
AACR American Association of Cancer Research
AAFCS American Association of Family and Consumer Sciences
AAFP American Academy of Family Physicians Foundation
AAFPRS American Academy of Facial Plastic and Reconstructive Surgery
AAR American Academy in Rome
AAS American Antiquarian Society
AASL American Association of School Libraries
AAUW American Association of University Women
ACC Asian Cultural Council
ACLS American Council of Learned Societies
ACM Association for Computing Machinery
ACS American Cancer Society
ADA American Diabetes Association
ADHF American Digestive Health Foundation
AF Arthritis Foundation
AFAR American Federation for Aging Research
AFOSR Air Force Office of Scientific Research
AFUD American Foundation for Urologic Disease
AFUW Australian Federation of University Women
AGS American Geriatrics Society
AHA American Heart Association
AHAF American Health Assistance Foundation
AFHMR Alberta Heritage Foundation for Medical Research
AHRQ Agency for Healthcare Research and Quality
AICR American Institute for Cancer Research
AIIS American Institute for Indian Studies
AJA American Jewish Archives
ALA American Library Association
ALISE Association for Library and Information Science Education
AMNH American Museum of Natural History
AMS American Musicological Society
ANL Argonne National Library
ANS American Numismatic Society
AOA American Osteopathic Association
APA American Psychological Association
APSA American Political Science Association
ARIT American Research Institute in Turkey
ARO Army Research Office
ASA American Statistical Association
ASCSA American School of Classical Studies at Athens
ASECS American Society for Eighteenth-Century Studies
ASF American-Scandinavian Foundation
ASHRAE American Society of Heating, Refrigerating, and Air Conditioning Engineers
ASME American Society of Mechanical Engineers
ASNS American Society for Nutritional Sciences
ASPRS American Society of Photogrammetry and Remote Sensing
AWU Associated Western Universities
AWWA American Water Works Association
BA British Academy
BWF Burroughs Wellcome Fund
CBIE Canadian Bureau for International Education
CCFF Canadian Cystic Fibrosis Foundation
CDC Centers for Disease Control and Prevention
CES Council for European Studies
CF The Commonwealth Fund
CFF Cystic Fibrosis Foundation
CFPC College of Family Physicians of Canada
CFUW Canadian Federation of University Women
CHEA Canadian Home Economics Association
CIES Council for International Exchange of Scholars
CIUS Canadian Institute of Ukrainian Studies
CLA Canadian Lung Association
CLF Canadian Liver Foundation
CRI Cancer Research Institute
DAAD Deutscher Akademische Austauschdienst (German Academic Exchange Service)
DHHS Department of Health and Human Services
DOA Department of Agriculture
DOC Department of Commerce
DOD Department of Defense
DOE Department of Energy
DOI Department of the Interior
DOJ Department of Justice
DOL Department of Labor
DOS Department of State
DOT Department of Transportation
EFA Epilepsy Foundation of America
EPA Environmental Protection Agency
ESF European Science Foundation
ETS Educational Testing Service
FCAR Formation de Chercheurs et L'Aide a la Recherche
FDA Food and Drug Administration
FIC Fogarty International Center
GAAC German American Academic Council

GCA	Garden Club of America
HHMI	Howard Hughes Medical Institute
HRSA	Health Resources and Services Administration
HUD	Department of Housing and Urban Development
IRA	International Reading Association
IREX	International Research and Exchanges Board
IUCP	Indiana University Center on Philanthropy
JDF	Juvenile Diabetes Foundation International
JMO	John M. Olin Foundation
JSPS	Japan Society for the Promotion of Science
KFC	Kidney Foundation of Canada
LSA	Leukemia Society of America
MHRC	Manitoba Health Research Council
MLA	Medical Library Association
MMA	Metropolitan Museum of Art
MSSC	Multiple Sclerosis Society of Canada
NAACP	National Association for the Advancement of Colored People
NARSAD	National Alliance for Research on Schizophrenia and Depression
NASM	National Air and Space Museum
NATO	North Atlantic Treaty Organization
NCCAM	National Center for Complementary and Alternative Medicine
NCI	National Cancer Institute
NCIC	National Cancer Institute of Canada
NCRR	National Center for Research Resources
NEH	National Endowment for the Humanities
NEI	National Eye Institute
NFID	National Foundation for Infectious Diseases
NFWF	National Fish and Wildlife Foundation
NHGRI	National Human Genome Research Institute
NHLBI	National Heart, Lung and Blood Institute
NIA	National Institute on Aging
NIAAA	National Institute on Alcohol Abuse and Alcoholism
NIAF	National Italian American Foundation
NIAID	National Institute of Allergy and Infectious Diseases
NIAMS	National Institute of Arthritis and Musculoskeletal Skin Diseases
NICHD	National Institute of Child Health and Human Development
NIDA	National Institute on Drug Abuse
NIDCD	National Institute on Deafness and Other Communication Disorders
NIDCR	National Institute of Dental and Craniofacial Research
NIDDK	National Institute of Diabetes, and Digestive and Kidney Diseases
NIDRR	National Institute on Disability and Rehabilitation Research
NIEHS	National Institute of Environmental Health Sciences
NIGMS	National Institute of General Medical Sciences
NIH	National Institutes of Health
NIJ	National Institute of Justice
NIMH	National Institute of Mental Health
NINDS	National Institute of Neurological Disorders and Stroke
NINR	National Institute of Nursing Research
NIOSH	National Institute for Occupational Safety and Health
NIST	National Institute of Standards and Technology
NKF	National Kidney Foundation
NL	Newberry Library
NLM	National Library of Medicine
NMF	National Medical Fellowships, Inc.
NMSS	National Multiple Sclerosis Society
NOAA	National Oceanic and Atmospheric Administration
NRC	National Research Council
NSERC	Natural Sciences and Engineering Research Council of Canada
NSF	National Science Foundation
OAH	Organization of American Historians
ONF	Oncology Nursing Foundation
ONR	Office of Naval Research
OREF	Orthopaedic Research and Education Foundation
ORISE	Oak Ridge Institute for Science and Education
OSI	Open Society Institute
PDF	Parkinson's Disease Foundation
PhRMA	Pharmaceutical Research and Manufacturers of American Foundation
PHSC	The Photographic Historical Society of Canada
RCPSC	Royal College of Physicians and Surgeons of Canada
RSC	Royal Society of Canada
RWJF	Robert Wood Johnson Foundation
SLA	Special Libraries Association
SSHRC	Social Sciences and Humanities Research Council of Canada
SSRC	Social Science Research Council
STTI	Sigma Theta Tau International
USHMM	United States Holocaust Memorial Museum Research Institute
USIA	United States Information Agency
USIP	United States Institute of Peace
WAWH	Western Association of Woman Historians

1 in 9: Long Island Breast Cancer Action Coalition Grants 1

The Coalition's mission is to promote awareness of the breast cancer epidemic through education, outreach, advocacy, and direct support of research which is being done to find the causes of and cures for breast cancer and other related cancers. The Coalition's major goals are to: raise awareness of the epidemic of breast cancer and keep the disease in the forefront; obtain funding for research; investigate the link between pesticides and breast cancer since Long Island is a farm area; increase detection of breast cancer; and find the causes, a cure and how to prevent breast cancer.
Geographic Focus: New Jersey, New York, Pennsylvania
Contact: Geri Barish, President; (516) 374-3190; fax (516) 569-1894; info@1in9.org
Internet: http://www.1in9.org/
Sponsor: Long Island Breast Cancer Action Coalition
86 East Rockaway Road
Hewlett, NY 11557

1st Source Foundation Grants 2

Established in 1952 and administered by the 1st Source Bank, the Foundation supports community foundations, youth clubs and organizations involved with television, education, health, and human services. The Foundation provides support to organizations working in the following areas: social welfare and human services; education; culture and the arts; and community, civic and neighborhood involvement. Giving is primarily centered in Indiana, and the major type of funding given is for general operating support. Since there are no specific applications forms required or deadlines with which to adhere, applicants should send a letter of request detailing the project and the amount of funding needed.
Requirements: Any 501(c)3 serving the residents of Indiana communities where 1st Source Banks are located are eligible to apply.
Geographic Focus: Indiana
Amount of Grant: 500 - 5,000 U.S.D
Samples: South Bend Heritage Foundation, South Bend, Indiana, $5,000 - general operations support (2010); Boys and Girls Club of St. Joseph County, Mishawaka, Indiana, $5,000 - general operations support (2010).
Contact: Lee Morton, Vice President and Trustee; (574) 235-2790 or (574) 235-2254
Internet: https://www.1stsource.com/about-us/community-involvement
Sponsor: 1st Source Foundation
100 N Michigan Street, P.O. Box 1602
South Bend, IN 46601-1630

1st Touch Foundation Grants 3

Established by Derek L. Lee in 2006 in an effort to help his daughter, Jada, giving is centered around the testing and researching of Leber's Congenital Amaurosis (LCA), an extremely rare disease which causes severe vision loss and blindness. The Foundation's primary long-range mission is to help researchers find a cure and eradicate this disease. It will also continue to provide grants for higher education, and help children and youth organizations within its local communities. There are no specific guidelines, application forms, or deadlines with which to adhere, so applicants should contact the Foundation office directly.
Contact: Mark Martella, (415) 421-0535 or (818) 501-4421; Mmartella@1sttouch.org
Internet: http://1sttouch.org/
Sponsor: 1st Touch Foundation
5098 Foothills BOulevard, Suite 3, #492
Roseville, CA 95747

2 Depot Square Ipswich Charitable Foundation Grants 4

In December of 2005, Ipswich Co-operative Bank established the 2 Depot Square Ipswich Charitable Foundation with an initial contribution of $200,000. Charitable giving has been a cornerstone of the Bank's business philosophy for many years. The Foundation plays a vital role in supporting economic development and improving the quality of life in the communities that it serves. The Foundation focuses its giving in the following areas: economic and community empowerment—includes programs which focus on the promotion and development of access to safe and affordable housing, and programs which support community revitalization efforts; youth development—initiatives that encourage youth through social, educational, athletic or cultural programs; arts and culture—programs and organizations which provide art and cultural programs that enrich communities; and health and human services—organizations which strive to enhance the health and well-being of children and families in its communities. The Grant Committee meets in April and November. Completed applications are due April 1st and November 1st.
Requirements: The Foundation awards funds to non-profit organizations based in Ipswich and the surrounding communities. Applications for grants will only be accepted from qualified 501(c)3 or 501(c)1 organizations.
Geographic Focus: Massachusetts
Date(s) Application is Due: Apr 1; Nov 1
Contact: Tammy Roeger; (978) 462-3106; fax (978) 462-1980; tammy.roeger@ifs-nbpt.com
Internet: http://www.institutionforsavings.com/site/charitable_2depot_about.html
Sponsor: 2 Depot Square Ipswich Charitable Foundation
2 Depot Square
Ipswich, MA 01938-1914

2COBS Private Charitable Foundation Grants 5

The Foundation, established in Colorado in 2005, supports a variety of causes, which include: health care; cancer services; community services; and local community foundations. Although giving is centered around the Durango, Colorado, region, it is not uncommon for support to be offered to 501(c)3 programs across the country. There are no specific application formats or deadlines with which to adhere, and applicants should send a letter of request outlining the program need and overall budget.
Amount of Grant: 250 - 1,000 U.S.D
Contact: Christopher J. O'Brien, Chairperson; (970) 247-7828 or (970) 385-1740
Sponsor: 2COBS Private Charitable Foundation
10 Town Plaza, Suite 100
Durango, CO 81301-6910

3M Company Health and Human Services Grants 6

Parallel to its Foundation, the 3M Company makes charitable contributions to nonprofit health and human services organizations directly. Support is given on a national basis. Most giving is initiated through a Request for Proposal process that allows the company to focus our giving and maximize results. Areas of interest for Health and Human Services include: to increase resiliency in youth through prevention efforts from early childhood to 12th grade; and to build and sustain healthy communities. Types of support include: building and renovation; capital campaigns; donated equipment; donated land; donated products; employee volunteer services; general operating support; in-kind gifts; internship funds; program development; seed money; technical assistance; and use of facilities. The Company also reaches out to bring assistance and help communities prepare for disaster.
Restrictions: No support is offered for: political, fraternal, social, veterans, or military organizations; propaganda or lobbying organizations; religious organizations not of direct benefit to the entire community; animal-related organizations; or disease-specific organizations. No grants are given to individuals, or for electronic media promotion or sponsorships, athletic events, non-3M equipment, endowments, emergency needs, conferences, seminars, workshops, symposia, fund raising or testimonial events, travel, or film or video production; no cause-related marketing.
Geographic Focus: Alabama, Alaska, Arkansas, California, Connecticut, Georgia, Hawaii, Illinois, Indiana, Iowa, Kentucky, Massachusetts, Michigan, Minnesota, Missouri, Nebraska, New Jersey, New York, Ohio, South Carolina, South Dakota, Texas, Utah, Wisconsin
Contact: Cynthia F. Kleven; (651) 733-0144; fax (651) 737-3061; cfkleven@mmm.com
Internet: http://solutions.3m.com/wps/portal/3M/en_U.S./CommunityAffairs/CommunityGiving/U.S./HHS/
Sponsor: 3M Company Contributions
3M Center, Building 225-01-S-23
Saint Paul, MN 55144-1000

3M Foundation Community Giving Grants 7

The foundation supports organizations involved with arts and culture, K-12 education, higher education, the environment, and health and human services. Special emphasis is directed toward programs designed to help prepare individuals and families for success. Fields of interest are: arts; arts education; business education; disaster relief, preparedness, and services; economics; elementary and secondary education; employment training; engineering; environmental causes; family services; federated giving programs; health care; higher education; human services; mathematics; minorities; science programs; youth development; and youth, services. Types of support include: building construction and renovation, capital campaigns curriculum development, employee matching gifts, general operating support, in-kind gifts, program development, and scholarship funds. The foundation utilizes an invitational Request For Proposal (RFP) process for organizations located in Minneapolis and St. Paul, Minnesota, and Austin, Texas. Application forms are not required.
Requirements: Established 501(c)3 tax-exempt organizations in all 3M communities are eligible.
Restrictions: The 3M Foundation does not accept unsolicited proposals in St. Paul/Minneapolis, Minnesota, and Austin, Texas. No support for religious organizations, conduit agencies, political groups, fraternal organizations, social groups, or veteran organizations. No funding for hospitals, K-12 schools, military organizations, animal-related organizations, or disease-specific organizations. No grants to individuals, or for endowments, emergency operating support, advocacy and lobbying efforts, fundraising events and associated advertising, travel, publications, start-up needs, non -equipment, debt reduction, conferences, athletic events, or film or video production; no loans or investments.
Geographic Focus: Alabama, Alaska, Arkansas, California, Connecticut, Georgia, Hawaii, Illinois, Indiana, Iowa, Kentucky, Massachusetts, Michigan, Minnesota, Missouri, Nebraska, New York, Ohio, South Carolina, South Dakota, Texas, Utah, Wisconsin
Samples: Nature Conservancy (Arlington, VA)—to conserve and restore grasslands in northwestern and west-central Minnesota, and for conservation work along the lower Rio Grande River in Texas, $5.15 million; U of Minnesota-Twin Cities (Minneapolis, MN)—to endow the 3M Science and Technology Fellowships; for the Carlson School of Management's new center, which will evaluate the business potential of new technologies; and for scholarships for undergraduates, $8.6 million and $1 million matching grant.

Contact: Cynthia F. Kleven; (651) 733-0144; fax (651) 737-3061; cfkleven@mmm.com
Internet: http://solutions.3m.com/wps/portal/3M/en_U.S./3M-Sustainability/Global/Stakeholders/Giving/
Sponsor: 3M Foundation
3M Center, Building 225-01-S-23
Saint Paul, MN 55144-1000

25th Anniversary Foundation Grants 8
Established by Arista Records, which was founded by Clive Jay Davis in the late 1970s, the 25th Anniversary Foundation was named to honor a quarter-century of company number one hits. The Foundation's primary field of interest is supporting AIDS research at nationally recognized organizations. There are no particular application formats or deadlines with which to adhere, and potential applicants should first approach the Foundation in writing, outlining their overall need.
Amount of Grant: 10,000 U.S.D
Samples: City of Hope, Los Angeles, California, $10,000—for AIDS research; T.J. Martel Foundation, New York, New York, $10,000—for AIDS research.
Contact: Robert J. Sorrentino; (212) 782-1137 or (212) 782-1000; (212)782-1010
Sponsor: 25th Anniversary Foundation (formerly Arista Records Foundation)
1745 Broadway
New York, NY 10019

49ers Foundation Grants 9
The San Francisco 49ers foundation is designed to enhance the educational, health, social, and cultural needs of the San Francisco Bay Area community. The current funding priority concentrates on supporting organizations and programs that focus on tackling violence and that help establish life goals; encourage positive character building choices; teach life skills; and demonstrate respect, tolerance and appreciation of diversity. Grant proposals are accepted January through April each year. Submit proposals of three or less pages.
Requirements: 49ers Foundation grants will be awarded only to 501(c)3 organizations in the greater San Francisco Bay Area that address violence prevention.
Geographic Focus: California
Amount of Grant: Up to 5,000 U.S.D
Contact: Shauna Standart, Director; (408) 562-4949; fax (408) 727-4937; 49ersfoundation@niners.nfl.com
Internet: http://www.49ers.com/community/foundation.html
Sponsor: 49ers Foundation
4949 Centennial Boulevard
Santa Clara, CA 95054-1229

100 Mile Man Foundation Grants 10
Established in 2004 and formerly known as the Itzler Family Foundation, the 100 Mile Man Foundation currently offers support primarily to New York organizations, including: health organizations and agencies, social services groups, Jewish agencies and synagogues, and medical research groups. There are no specific guidelines, applications, or deadlines with which to adhere. Generally, grants have ranged from $100 to $160,000.
Requirements: Giving is limited to 501(c)3 organizations serving the residents of the greater New York metropolitan area.
Geographic Focus: New York
Contact: Jesse Itzler; (212) 867-0790 or (888) 843-1006; info@the100mileman.com
Internet: http://the100mileman.com/?page_id=276
Sponsor: 100 Mile Man Foundation
230 Park Avenue, Suite 850
New York, NY 10169

118 Foundation Grants 11
The mission of The 118 Foundation is to support organizations that play a key role in making communities better in the areas of housing, education, and health care. The foundation has also expanded its outreach internationally to support organizations that also fight poverty, create jobs, and transform lives. Giving is primarily to public charities that match the foundation's mission in the areas of child and youth development, affordable housing, and micro-finance and micro-enterprise development, in disadvantaged communities in both the U.S. and abroad.
Restrictions: No support is available for organizations lacking 501(c)3 status. No grants are given to individuals, or for endowments, capital campaigns, scholarships, or medical research.
Amount of Grant: Up to 135,000 U.S.D
Contact: Diane Tollefson, Executive Director; (703) 338-5948; the118foundation@mac.com or grants@webmethods.org
Sponsor: 118 Foundation
1350 Beverly Road, #115-325
McLean, VA 22101

1104 Foundation Grants 12
The Foundation gives primarily to support religious programs, health, education, children and youth, and community service organizations. Amounts range from $100 to $50,000. There are no application forms or deadlines to which applicants must adhere. Though giving is primarily restricted to the Charlotte, North Carolina, region, grants are also awarded to national organizations which are aligned with the foundation's interests.
Samples: Women's Ministries of Mount Gilead, Gilead, North Carolina, $1,000; Providence Day School, Charlotte, North Carolina, $2,500; The Fletcher School, Charlotte, North Carolina, $3,000.
Contact: Charles V. Ricks, Treasurer; (704) 566-3349 or (704) 568-5550
Sponsor: 1104 Foundation
6000 Monroe Road, Suite 100
Charlotte, NC 28212-6119

1675 Foundation Grants 13
The foundation is a private family foundation dedicated to improving the quality of life for individuals and families through the support of non-profit organizations working in the areas of health, human services, education, the environment, and history. Priority is given to organizations serving Chester County, southeastern Pennsylvania, the Greater Boston area, and other geographic areas of interest to the Trustees. Grants are made for operating support, special projects, endowment, and capital. Grants range from $2,000 to $50,000 and are made twice a year at the discretion of the Trustees. Requests should be in writing and mailed to the executive director.
Requirements: 501(c)3 nonprofit organizations and public charities under IRS Code 509(a) are eligible.
Restrictions: The foundation does not make grants to individuals, nor are they made for political purposes.
Geographic Focus: Massachusetts, Pennsylvania
Date(s) Application is Due: Mar 1; Oct 1
Amount of Grant: 2,000 - 50,000 U.S.D
Contact: Marge Brennan, Grants Manager; (610) 896-3868; fax (610) 896-3869; mbrennan@1675foundation.org
Internet: http://www.1675foundation.org/guidelines.htm
Sponsor: 1675 Foundation
16 East Lancaster Avenue, Suite 102
Ardmore, PA 19003-2228

1976 Foundation Grants 14
Established in Pennsylvania in 1976, the 1976 Foundation gives primarily to organizations that support services for the blind, pediatric medicine, health care, and research. The average grant ranges up to $10,000, though larger grants will be considered. Though there are no specific guidelines or application forms, the annual deadline is June 30. Applicants should contact the office in writing, explaining the project and amount needed.
Geographic Focus: Pennsylvania
Date(s) Application is Due: Jun 30
Amount of Grant: Up to 10,000 U.S.D
Samples: Children's Hospital of Philadelphia, Philadelphia, Pennsylvania, $10,000; Purnell School 51, Pottersville, New Jersey, $10,000.
Contact: Nathaniel Peter Hamilton, President; (610) 254-9401; fax (610) 254-9404
Sponsor: 1976 Foundation
200 Eagle Road, Suite 308
Wayne, PA 19087-3115

A-T Children's Project Grants 15
The A-T Children's Project strives to assist respected scientists in developing a clearer understanding of ataxia-telangiectasia. The Project is determined to find a timely cure, or life-improving treatments, for this serious disease. Grant awards are made through a careful and detailed selection process. The members of the Scientific Advisory Board examine each proposal and make their independent recommendations to its Board of Directors which then votes on each proposed project. Proposals from junior investigators, from scientists in related disciplines, and from individuals with innovative new ideas for A-T research are particularly encouraged, as are laboratories and teams working together from industry as well as teaching universities. The Project provides competitive grant awards for basic and translational research grants related to A-T. One- and two-year projects are funded up to a maximum total direct cost of $75,000 per year. Grants of $75,000 per year, however, are rare; grants in the $25,000 to 50,000 per year range are much more common. Budgets for up to $150,000 for a two-year project are acceptable.
Requirements: Proposals from junior investigators, from scientists in related disciplines, and from individuals with innovative new ideas for A-T research are particularly encouraged, as are laboratories and teams working together from industry as well as teaching universities. A Letter of Intent is not required. However, prior to submission of a full-length proposal, applicants seeking Scientific Advisory Board input may submit a Letter of Intent directly to the A-TCP Science Coordinator. This letter (not to exceed two pages) should include a brief abstract describing the proposed research, specific aims and an estimated budget. Applicants must submit an electronic copy of their Proposal in either MSWord or PDF formats to grants@atcp.org.
Restrictions: The sponsor does not pay for administrative overhead and indirect costs. The sponsor does not pay for institutional construction or renovation; purchase of major capital equipment other than directly needed for proposed experiments; office equipment or furniture; travel (except as required to perform the project); tuition fees; journal subscriptions; dues or memberships; and printing or publishing costs.
Date(s) Application is Due: Mar 1; Sep 1
Amount of Grant: Up to 75,000 U.S.D
Contact: Cynthia Rothblum, Science Coordinator; (954) 481-6611 or (800) 543-5728; fax (954) 725-1153; grants@atcp.org or cynthia@atcp.org
Internet: http://www.communityatcp.org//Page.aspx?pid=625&srcid=633
Sponsor: A-T Children's Project
5300 W. Hillsboro Boulevard, #105
Coconut Creek, FL 33073

A-T Medical Research Foundation Grants 16
The Foundation is a national 501(c) tax-exempt organizations offering a comprehensive program of information and support services for patients and their families on ataxia-

telangiectasia, a lethal genetic disease that attacks children, causing progressive loss of muscle control, immune system problems, and a strikingly high rate of cancer, especially leukemia and lymphoma. The Foundation also offers patient and professional information and education materials, sponsors meetings and scientific workshops, funds research, and provides referrals to chapters and support groups. Seed grants and research grants are available.
Amount of Grant: 35,000 - 255,000 U.S.D
Contact: Pamela Smith, President; (818) 906-2861; fax (818) 906-2870; becca4435@aol.com or atmrf@aol.com
Sponsor: Ataxia Telangiectasia Medical Research Foundation
16224 Elisa Place
Encino, CA 91436-3320

A. Gary Anderson Family Foundation Grants 17
The foundation supports nonprofits in its areas of interest, including arts and culture, higher education, children and youth, and social services. Requests from California receive priority. There are no application deadlines with which to adhere, but application forms are required. Applicants should begin by initially forwarding a letter outlining the proposed project.
Requirements: Funding is primarily given for higher education, and arts and culture in California.
Restrictions: Grants are not made to individuals.
Geographic Focus: California
Amount of Grant: 1,000 - 25,000 U.S.D
Samples: Hoag Memorial Hospital Presbyterian (Newport Beach, CA)—for its capital campaign for the Hoag Women's Pavilion, $5 million.
Contact: Erin J. Lastinger, President; (714) 685-3990; fax (714) 685-3994
Sponsor: A. Gary Anderson Family Foundation
300 S Harbor Boulevard, Suite 1010
Anaheim, CA 92805-3721

A.O. Smith Community Grants 18
The foundation supports nonprofit organizations in communities where A.O. Smith Corporation has facilities. The foundation supports elementary and secondary school projects that focus on the quality of educational programs and curriculum development; civic, cultural, and social welfare of communities; and medical research and improved local health services. Types of support include continuing support, annual campaigns, building/renovation, scholarship funds, and employee matching gifts. Proposals should describe the project's benefits and constituency, budget information including other sources of funding, and how results will be reported. There is no deadline for applications, although proposals to be considered for the following year's foundation budget must be received by October 30.
Requirements: Nonprofit organizations in company areas may apply.
Restrictions: Grants do not support political organizations or organizations whose chief purpose is to influence legislation.
Date(s) Application is Due: Oct 30
Amount of Grant: 500 - 20,000 U.S.D
Contact: Edward J. O'Connor, (414) 359-4000; fax (414) 359-4064
Internet: http://www.aosmith.com/About/Detail.aspx?id=132
Sponsor: A.O. Smith Foundation
P.O. Box 245008
Milwaukee, WI 53224-9510

A/H Foundation Grants 19
The A/H Foundation was established in California in 1996, and was named after the initials of its two founders, Albert Friedman and Harvey Friedman. Its primary mission is to support: cancer research, community and economic development; law enforcement and police agencies; community health organizations; public and private universities; human services; and Jewish agencies and synagogues. There are no specific deadlines with which to adhere, and initial approach should be by letter.
Requirements: Giving is primarily in the State of California to 501(c)3 organizations (although there is some giving outside of California).
Geographic Focus: California
Amount of Grant: Up to 25,000 U.S.D
Samples: Concern Foundation, Beverly Hills, California, $24,500; Loyola Marymount University, Los Angeles, California, $12,500; Calicinto Ranch, San Jacinto, California, $400.
Contact: Keith J. Rosen, (818) 920-9888; fax (818)920-9388; info@wantacpa.com
Sponsor: A/H Foundation
15545 Devonshire Street, Suite 210
Mission Hills, CA 91345

AAAAI / Schering Allergy/Immunology Fellowship of Excellence Training Awards 20
The Award will support the two years of training (MDs and DOs) for a fellow at an ACGME-accredited allergy/immunology training program who is a member of an underrepresented minority community, a graduate of a U.S. medical school certified in Internal Medicine or Pediatrics, and a U.S. citizen. The application must be received no later than April 1, and should be completed by the fellow who will start training July 1.
Requirements: Receipt of the Award does not preclude the applicant from obtaining grant support from other sources
Restrictions: Funds may not be used for institutional indirect costs
Date(s) Application is Due: Apr 1
Amount of Grant: 50,000 U.S.D
Contact: Jerome Schultz, (414) 272-6071; fax (414) 276-6070; jschultz@aaaai.org or info@aaaai.org
Internet: http://www.aaaai.org/members/grants%5Fawards/immunology_fellowship/
Sponsor: American Academy of Allergy, Asthma, and Immunology
555 E Wells Street, Suite 1100
Milwaukee, WI 53202-3823

AAAAI Allergy, Asthma and Immunology Education and Research Trust (ART) Allied Health Travel Grants 21
The AAAAI abstract submission and travel grant application provides a new process for growth and learning. The Program includes: presentation of original research; attendance at AAAAI Annual Meeting postgraduate program and symposia; participation in a special program that includes basic research presentations by AAAAI members who are NIH funded researchers in allergy/immunology; information about career opportunities in allergy, asthma and immunology research; and funds for travel (including a stipend for ground transportation), hotel accommodation, and meeting registration.
Requirements: Eligible candidates include first-, second-, and third-year RNs in training.
Restrictions: Individuals with a faculty appointment will not be considered.
Amount of Grant: Up to 750 U.S.D
Samples: Consuelo M. Carcel, RN, Children's Hospital of Orange County—for The Impact of Home Visits on Retention in a Low-Income School Based Mobile Asthma Clinic, $750; Patricia S. Hutcheson, Saint Louis University Medical School—for Is Allergic Fungal Sinusitis A Distinct Form Of Chronic Rhinosinusitis?, $750,; Carrie L. Vibbert, RN, University of Maryland School of Medicine—for Does Use of Asthma Controllers Really Indicate Asthma Control?, $750.
Contact: Mikelle Johnson, Program Manager; (414) 272-6071; fax (414) 276-3349; mjohnson@aaaai.org
Internet: http://www.aaaai.org/members/allied_health/abstractstravelgrants.stm
Sponsor: American Academy of Allergy, Asthma, and Immunology
555 E Wells Street, Suite 1100
Milwaukee, WI 53202-3823

AAAAI Allergy, Asthma and Immunology Education and Research Trust (ART) Clinical Research Grants 22
Formerly the Education and Research Trust (ERT), training program directors, deans, department chairpersons, and sponsors are invited to nominate promising investigators involved in research dealing with asthma for this grant. The award is a grant in aid provided to the program sponsoring the applicant and is intended to provide salary support. Application should include a biographical sketch, including a curriculum vita, an outline of past accomplishments, a summary of a current research project, and plans for future research.
Requirements: Applicants must be MDs or PhDs who are fellows in training or residents associated with an approved allergy and immunology program and are AAAAI members (or individuals who have submitted applications for membership by September 1).
Restrictions: Individuals with a faculty appointment and past AAAAI research award recipients will not be considered.
Samples: Seema Sharma Aceves, MD, PhD, Children's Hospital San Diego—for Evaluating Disease Severity and Tissue Remodeling in Pediatric Patients with Eosinophilic Esophagitis, $10,000; Kevin P. O'Brien, MD, Parker, Colorado—for Prevention of Atopy by Combination of Lactobacillus Supplements (pro-TH1) and Omega-3 Supplements (anti-TH2) During Pregnancy and Breastfeeding: A Randomized, Placebo-controlled Trial, $10,000.
Contact: Angela Gade, Development Senior Project Manager; (414) 272-6071; fax (414) 276-6070; agade@aaaai.org or info@aaaai.org
Internet: http://www.aaaai.org/members/grants_awards
Sponsor: American Academy of Allergy, Asthma, and Immunology
555 E Wells Street, Suite 1100
Milwaukee, WI 53202-3823

AAAAI Allergy, Asthma and Immunology Education and Research Trust (ART) Faculty Development Award 23
The award offers an individual three years of financial support to concentrate his/her research efforts in either basic or clinical allergy and immunology. The objectives of this award are to recognize and support young researchers and promote the specialty of allergy/immunology. The recipient is expected to participate in a seminar at the AAAAI annual meeting, where awardees present their research.
Requirements: The award is open to researchers of junior faculty status—either instructor, assistant professor, or the equivalent grade. This award is directed toward MDs, but in unusual circumstances PhDs will be considered. Applicants must be U.S. or Canadian citizens or permanent residents who are AAAAI members (or have applied for membership by September 1).
Restrictions: Past recipients are ineligible.
Date(s) Application is Due: Nov 15
Amount of Grant: 100,000 U.S.D
Samples: Christine Seroogy, MD, University of Wisconsin School of Medicine and Public Health—for The role of toll-like receptors (TLRs) in CD25+ T regulatory cell development and function, $300,000 over three years; Jordan S. Orange, MD, PhD, Children's Hospital of Philadelphia—for Directed Secretion Through the Immunological Synapse, $300,000 over three years; Sanjeev Jain, MD, PhD, University of Washington—for Development of Structure Based Inhibitors of Nitric Oxide Synthase for the Treatment of Asthma, $300,000 over three years.
Contact: Angela Gade, Development Senior Project Manager; (414) 272-6071; fax (414) 276-6070; agade@aaaai.org or info@aaaai.org

Internet: http://www.aaaai.org/art/
Sponsor: American Academy of Allergy, Asthma, and Immunology
555 E Wells Street, Suite 1100
Milwaukee, WI 53202-3823

AAAAI Phoenix Program Award 24
Funded by the ERT, the AAAAI Allergy Educational Forum and the AAAAI, the grant is a critical element in the development of new allergy/immunology training programs. This grant award can be utilized to effectively increase the size of programs, increase research staff, increase the number of fellows being trained and to develop future faculty members. It is hoped that such additions will increase research quantity and quality, teaching and training mission, and the delivery of health care. Awards average $50,000 to $100,000 per year for three successive years.
Requirements: U.S. college and university medical centers are eligible to apply.
Amount of Grant: Up to 25,000 U.S.D
Samples: Penn State University College of Medicine, $150,000 over three years; University of Mississippi Medical Center, $250,000 over three years; University of North Carolina-Chapel Hill, $250,000 over three years.
Contact: Angela Gade, Development Senior Project Manager; (414) 272-6071; fax (414) 276-6070; agade@aaaai.org or info@aaaai.org
Internet: http://www.aaaai.org/art/
Sponsor: American Academy of Allergy, Asthma, and Immunology
555 E Wells Street, Suite 1100
Milwaukee, WI 53202-3823

AAAAI Research Award Grants 25
The organization promotes and stimulates research and study in allergy, asthma and immunology. The goal of this program is to support the research by a fellow-in-training, resident, and post-doctoral scholar at an allergy and immunology or a clinical and laboratory immunology training program.
Requirements: Fellow-in-training, resident, or post-doctoral scholar affiliated with an allergy and immunology or clinical and laboratory immunology training program that is in good standing with Residency Review Committee and is fully-accredited by the Accreditation Council of Graduate Medical Education may apply. Applicants must be an AAAAI member or must submit a membership application with the award application.
Restrictions: Applicants with faculty status will not be considered. Past AAAAI research award recipients are not eligible.
Amount of Grant: 25,000 U.S.D
Contact: Angela Gade, Development Senior Project Manager; (414) 272-6071; fax (414) 276-6070; agade@aaaai.org or info@aaaai.org
Internet: http://www.aaaai.org/members/grants_awards
Sponsor: American Academy of Allergy, Asthma, and Immunology
555 E Wells Street, Suite 1100
Milwaukee, WI 53202-3823

AAAAI STAR Program Grants 26
The goal of the Program is to provide U.S. and Canadian Ph.D. candidates and Ph.D. postdocs with an opportunity to consider research opportunities in the field of allergy and immunology. The AAAAI Annual Meeting provides a forum for those interested in translating basic science advances into clinical practice to network with basic scientists, translational clinical researchers, and clinical experts in allergy/ immunology U.S. and Canadian Ph.D. candidates and Ph.D. postdocs are invited to submit abstracts for presentation at the AAAAI Annual Meeting. Up to 40 selected participants will be invited to participate in the STAR Program, which will include: presentation of original research at the STAR Program; attendance at AAAAI Annual Meeting postgraduate program and symposia; participation in a special STAR Program that includes basic research presentations by AAAAI members who are NIH funded researchers in allergy/ immunology; information about career opportunities in allergy, asthma and immunology research; and funds for travel (including a stipend for ground transportation), hotel accommodation, and meeting registration.
Requirements: Eligible candidates include first-, second-, and third-year MD fellows in training in either an allergy and immunology or a clinical and laboratory immunology training program approved by the Accreditation Council of Graduate Medical Education or a clinical immunology and allergy training program approved by the Royal College of Physicians and Surgeons of Canada.
Restrictions: Individuals with a faculty appointment will not be considered.
Contact: Mikelle Johnson; (414) 272-6071; fax (414) 276-3349; mjohnson@aaaai.org
Internet: http://www.aaaai.org/members/allied_health/abstractstravelgrants.stm
Sponsor: American Academy of Allergy, Asthma, and Immunology
555 E Wells Street, Suite 1100
Milwaukee, WI 53202-3823

AAAAI Summer Clinical Research Fellowship Medical Student Grants 27
The Program goal is to award grants to medical students wishing to conduct clinical research in allergy/immunology during their summer recess in order to encourage their consideration of careers as a clinician in allergy/immunology. Grants will be awarded to outstanding medical students who wish to pursue clinical research. Clinical topics for research examples include optimizing health care of asthma, collecting a series of cases with an interesting diagnosis and preparing a literature review of the topic, a case report with a literature review, reviewing practice parameters related to management of a problem in allergy/immunology and applying to care in a clinic or office, as well as other topics of interest in the practice of allergy/immunology.
Requirements: Applicant must be a full-time medical student residing in the U.S. or Canada and will have successfully completed at least 8 months of medical school by May 15.
Restrictions: Past Summer Fellowship Grant recipients are not eligible.
Date(s) Application is Due: Mar 3
Amount of Grant: 2,000 U.S.D
Contact: Program Administrator; (414) 272-6071; fax (414) 276-6070; info@aaaai.org
Internet: http://www.aaaai.org/members/grants%5Fawards/
Sponsor: American Academy of Allergy, Asthma, and Immunology
555 E Wells Street, Suite 1100
Milwaukee, WI 53202-3823

AAAAI Summer Research Fellowship Medical Student Grants 28
Fellowships are awarded on an annual basis to assist medical students interested in pursuing research in allergy and immunology during their summer recess. The fellowships will be awarded to outstanding individuals who wish to pursue research in the following areas: physiology of allergic diseases, pharmacology of allergy and inflammation, basic cellular and molecular immunology, AIDS, and other topics pertinent to the understanding of allergic and immune mechanisms of disease.
Requirements: Applicant must be a full time medical student residing in the U.S. or Canada and must have successfully completed at least eight months of medical school by May 15.
Restrictions: Past recipients are ineligible.
Date(s) Application is Due: Mar 3
Amount of Grant: 2,000 U.S.D
Samples: Jason Konner, Aaron Diamond AIDS Research Ctr (New York, NY)—summer fellowship recipient, $1500; Goron Lai, Johns Hopkins Asthma and Allergy Ctr (Baltimore, MD)—summer fellowship recipient, $1500.
Contact: Program Administrator; (414) 272-6071; fax (414) 276-6070; info@aaaai.org
Internet: http://www.aaaai.org/members/grants%5Fawards/
Sponsor: American Academy of Allergy, Asthma, and Immunology
555 E Wells Street, Suite 1100
Milwaukee, WI 53202-3823

AAAS Science & Technology Policy Fellowships - National Defense & Global Security 29
Fellows spend one year contributing external perspective and expertise on technology in defense policy, national and international security and systems analysis, and assisting with project and program oversight. Fellows will be placed in one of the following: Federal Bureau of Investigation (FBI), Nuclear Regulatory Commission, U.S. Department of Defense or U.S. Department of Homeland Security. Fellows will work on a variety of issues related to defense policy, national security, weapons of mass destruction, technology applications, defense systems analysis and support, and program oversight and management. Assignments may involve significant interagency, congressional or international activity. The program includes an orientation on international affairs and executive branch and congressional operations, and a year-long seminar series on issues involving science, technology and public policy. Fellows should not expect to work specifically on technical issues related to their dissertations or previous post-doctoral appointments, but rather to apply their technical, organizational, and communication skills to technical and policy issues in the placement agency. Additionally, Fellows help increase the awareness of the placement agency and its departments as a challenging and rewarding career environment for scientists and engineers.
Requirements: All applicants must have a doctoral-level degree (PhD, MD, DVM, DSc, PharmD, and other terminal degrees), in any physical, biological, health/medical or social science, any field of engineering, or any relevant interdisciplinary field (individuals with a master's degree in engineering and at least three years of post-degree professional experience also may apply). Note: All requirements for the degree must be completed by the application deadline. A prospective fellow must demonstrate exceptional competence in their specialty appropriate to their career stage, and have the strong endorsement of three references. Show an understanding of the opportunities for science and engineering to support a broad range of non-scientific issues, and display a commitment to apply their scientific or technical expertise to serve society; Exhibit awareness and sensitivity to the political, economic and social issues that influence policy; Be articulate communicators, both verbally and in writing, to decision-makers and non-scientific audiences, and have the ability to work effectively with individuals and groups outside the scientific community; Demonstrate integrity and good judgment, and the flexibility and willingness to address policy issues outside their scientific realm.
Restrictions: Federal employees are ineligible. All applicants must be U.S. citizens (dual citizenship from the United States and another country is acceptable). Some program areas and host agencies seek additional qualifications. Holding a Title 42 position is considered federal employment and therefore renders an applicant ineligible for a AAAS Science & Technology Policy Fellowship. Individuals working in postdoctoral positions at government labs or agencies, for private contractors, and in state government posts are not federal employees and therefore are eligible for the AAAS Fellowships.
Date(s) Application is Due: Dec 15
Amount of Grant: 70,000 - 90,000 U.S.D
Contact: Rick Kempinski, Senior Program Manager; (202) 326-6700; fax (202) 289-4950; rkempinski@aaas.org or fellowships@aaas.org
Internet: http://fellowships.aaas.org/02_Areas/02_index.shtml
Sponsor: American Association for the Advancement of Science
1200 New York Avenue NW
Washington, DC 20005-3920

AACAP-NIDA K12 Career Development Awards 30

The grant provides up to five years of salary support and mentored addiction research training for qualified child and adolescent psychiatrists (CAPs) who intend to establish careers as independent investigators in mental health and addiction research. Letters of Intent are required and due on or before January 5. Full applications are due March 30, with new grants awarded on June 1. Applicants are strongly encouraged to contact program staff regarding the program and candidate qualifications before submitting Letters of Intent.

Requirements: Eligible applicants must be a child and adolescent psychiatrist with a commitment and goal of becoming an independent investigator in addiction-related research focused on children and adolescents. Applicants must also be U.S. citizens, non-citizen nationals, and/or individuals with permanent residence status. Women and minority candidates are especially encouraged to apply.

Date(s) Application is Due: Jan 5

Amount of Grant: Up to 140,000 U.S.D

Contact: Stacia Fleisher, (202) 966-7300; fax (202) 966-2891; sfleisher@aacap.org or research@aacap.org

Internet: http://www.aacap.org/cs/root/research_and_training_awards/aacap_nida12

Sponsor: American Academy of Child and Adolescent Psychiatry
3615 Wisconsin Avenue NW
Washington, DC 20016-3007

AACAP Beatrix A. Hamburg Award for the Best New Research Poster by a Child and Adolescent Psychiatry Resident 31

This award recognizes the author of the best new research poster presented at the AACAP Annual Meeting. The award provides a $1,000 honorarium. The recipient will receive a plaque at the Young Leader Awards Ceremony during the AACAP Annual Meeting. There is no application process for this award. The AACAP Program Committee will review all new research poster submissions by child and adolescent psychiatry residents to determine the recipient of this award. The recipient will be notified through an acceptance letter. Non-recipients will not be notified.

Requirements: The author must be a child and adolescent psychiatry resident at an accredited institution and AACAP member. Applicants must: be a child and adolescent psychiatry resident; attend an accredited institution; be an AACAP member; and attend and present a new research poster at the AACAP Annual Meeting.

Date(s) Application is Due: Jun 14

Amount of Grant: 1,000 U.S.D

Contact: Stacie Hall; (202) 966-7300, ext. 113; fax (202) 966-2891; shall@aacap.org

Internet: http://www.aacap.org/cs/root/research_and_training_awards/aacap_beatrix_a_hamburg_award_for_the_best_new_research_poster_by_a_child_and_adolescent_psychiatry_resident_at_the_aacap_annual_meeting

Sponsor: American Academy of Child and Adolescent Psychiatry
3615 Wisconsin Avenue NW
Washington, DC 20016-3007

AACAP Child and Adolescent Psychiatry (CAP) Teaching Scholarships 32

The program is offered through the Harvard Macy Institute for Physician Educators Program, and is designed to fashion master teachers who will inspire medical students and residents to become child and adolescent psychiatrists. To facilitate this objective, the field of child and adolescent psychiatry needs educators trained in state-of-the-art teaching principles, methodologies, and technologies of adult education and learning. The AACAP Teaching Scholars Program addresses this challenge by creating a group of highly trained medical educators. This program is an opportunity for CAP faculty who are committed to teaching careers to train at the prestigious Harvard Medical School, gain research and teaching experience, and to give back to future generations of CAP faculty.

Requirements: Applicants must: have a mid-level academic appointment (Senior Assistant Professor, Associate Professor or early Professor); have a commitment to an active career in medical student/resident teaching; have been accepted by the joint AACAP-Harvard Macy Institute admissions group; attend at two meetings in Boston, MA (12 days in January and 6 days in May); have active participation in a collaborative CAP project spanning January through May; be potential participants as program faculty during a subsequent second year; participate in AACAP Annual Meetings and Mid-Year Institutes to bring modern teaching and educational methods to the AACAP membership; participate in an AACAP College of Teaching Scholars; and possess strong support from the applicant's School of Medicine, Department of Psychiatry and Division of Child and Adolescent Psychiatry (letter of institutional commitment and some release time required).

Date(s) Application is Due: Jul 15

Amount of Grant: Up to 8,500 U.S.D

Contact: Gabe Robbins, (202) 966-7300, ext. 117; fax (202) 966-2891; grobbins@aacap.org

Internet: http://www.aacap.org/cs/root/research_and_training_awards/child_and_adolescent_psychiatry_teaching_scholars_program

Sponsor: American Academy of Child and Adolescent Psychiatry
3615 Wisconsin Avenue NW
Washington, DC 20016-3007

AACAP Congressional Fellowships 33

The Program is designed to educate policy makers and Congressional staff about child and adolescent psychiatry and to foster awareness of children's mental health issues. This experience will allow the Fellow to develop a keen understanding of how public policy affects patient care, education, and health insurance issues. The AACAP Congressional Fellow will be placed with a Congressional office or Committee where they will gain invaluable experience as they assist in the development of legislative and public policy initiatives. The award includes an $85,000 stipend, up to $3,000 in relocation or moving expenses, and health insurance for the length of the fellowship program.

Requirements: Successful applicants must: be an AACAP member; be a PGY-4 psychiatry resident or beyond; have completed training in or be scheduled to complete training in one of the child and adolescent psychiatry programs (CAP training program, Combined General Psychiatry/CAP Program, or Triple Board Program); have a demonstrated interest in public policy issues; exhibit an interest in applying medical and scientific knowledge to the resolution of public policy issues; and be a U.S. citizen or permanent resident.

Date(s) Application is Due: Feb 1

Amount of Grant: 85,000 - 88,000 U.S.D

Contact: Gabe Robbins, (202) 966-7300, ext. 117; fax (202) 966-2891; grobbins@aacap.org

Internet: http://www.aacap.org/cs/root/legislative_action/2008_congressional_fellowship_program

Sponsor: American Academy of Child and Adolescent Psychiatry
3615 Wisconsin Avenue NW
Washington, DC 20016-3007

AACAP Educational Outreach Program for Child and Adolescent Psychiatry Residents 34

The Educational Outreach Program (EOP) funding provides the opportunity for up to 50 child and adolescent psychiatry residents to receive a formal overview to the field of child and adolescent psychiatry, establish child and adolescent psychiatrists as mentors and experience the AACAP Annual Meeting. Participants will be exposed to the breadth and depth of the field of child and adolescent psychiatry, including research opportunities, access to mentors, and various networking opportunities. Participation in this program provides up to $750 for travel expenses.

Requirements: Applicants must: be child and adolescent psychiatry residents at the time of the AACAP Annual Meeting; be currently enrolled in a residency program in the United States; be residents in their first or second year of child fellowship training are eligible (Triple Boarders in their fourth or fifth year of training in their triple board programs are eligible); either be members of the AACAP or have a membership application pending at the time of application; and attend all AACAP Annual Meeting events specified by AACAP.

Date(s) Application is Due: Jul 12

Amount of Grant: Up to 750 U.S.D

Contact: Gabe Robbins, (202) 966-7300, ext. 117; fax (202) 966-2891; grobbins@aacap.org

Internet: http://www.aacap.org/cs/residents/eop-capfellows

Sponsor: American Academy of Child and Adolescent Psychiatry
3615 Wisconsin Avenue NW
Washington, DC 20016-3007

AACAP Educational Outreach Program for General Psychiatry Residents 35

The Educational Outreach Program (EOP) for General Psychiatry residents provides the opportunity for up to 20 general psychiatry residents to receive a formal overview to the field of child and adolescent psychiatry, establish child and adolescent psychiatrists as mentors and experience the AACAP Annual Meeting. Participants will be exposed to the breadth and depth of the field of child and adolescent psychiatry, including research opportunities, access to mentors, and various networking opportunities. Participation in this program provides up to $1,500 for travel expenses to the AACAP Annual Meeting.

Requirements: Applicants must: be general psychiatry residents at the time of the AACAP Annual Meeting; be currently enrolled in a residency program in the United States; be residents in their first, second or third year of general psychiatry training (Triple Boarders in their first, second or third year of training in their triple board programs are eligible); either be members of the AACAP or have a membership application pending at the time of application; and attend all AACAP Annual Meeting events specified by AACAP.

Date(s) Application is Due: Jul 12

Amount of Grant: Up to 1,500 U.S.D

Contact: Gabe Robbins, (202) 966-7300, ext. 117; fax (202) 966-2891; grobbins@aacap.org

Internet: http://www.aacap.org/cs/residents/eop-generalresidents

Sponsor: American Academy of Child and Adolescent Psychiatry
3615 Wisconsin Avenue NW
Washington, DC 20016-3007

AACAP Elaine Schlosser Lewis Award for Research on Attention-Deficit Disorder 36

This annual award acknowledges outstanding leadership and continuous contributions in the field of research by giving $5,000 for the best paper published in the Journal on attention-deficit disorder written by a child and adolescent psychiatrist. The award winner will be recognized at a Distinguished Awards Luncheon and make an Honors Presentation about his or her work during the AACAP Annual Meeting.

Requirements: Nomination letters must be accompanied by a CV for the individual nominated.

Date(s) Application is Due: May 7

Amount of Grant: 5,000 U.S.D

Samples: Margaret Weiss, M.D., Children's Hospital, Vancouver, BC—Sleep Hygiene and Melatonin Treatment for Children and Adolescents With ADHD and Initial Insomnia.

Contact: Alyssa Sommer, Research Coordinator; (202) 966-7300, ext. 157; fax (202) 364-5925; asommer@aacap.org

Internet: http://www.aacap.org/cs/root/research_and_training_awards/aacap_journal_awards

Sponsor: American Academy of Child and Adolescent Psychiatry
3615 Wisconsin Avenue NW
Washington, DC 20016-3007

AACAP George Tarjan Award for Contributions in Developmental Disabilities 37
This award recognizes a child and adolescent psychiatrist and Academy member who has made significant contributions in a lifetime career or single seminal work to the understanding or care of those with mental retardation and developmental disabilities. These contributions must have national and/or international stature and clearly demonstrate lasting effects. The contributions may be in areas of teaching, research, program development, direct clinical service, advocacy or administrative commitment. A cash prize of up to $1,000 will be awarded. The award winner will be recognized at a Distinguished Awards Luncheon and make an Honors Presentation about his or her work during the AACAP Annual Meeting.
Requirements: Nomination letters must be accompanied by a CV for the individual nominated.
Date(s) Application is Due: Apr 30
Amount of Grant: 1,000 U.S.D
Samples: Peter Tanguay, M.D., University of Louisville, Louisville, KY—Autism: Fixing the DSM-IV.
Contact: Edwin Cook; (202) 966-7300; fax (202) 966-2891; training@aacap.org
Internet: http://www.aacap.org/cs/distinguishedmembers
Sponsor: American Academy of Child and Adolescent Psychiatry
3615 Wisconsin Avenue NW
Washington, DC 20016-3007

AACAP Irving Philips Award for Prevention 38
This award recognizes a child and adolescent psychiatrist and Academy member who has made significant contributions in a lifetime career or single seminal work to the prevention of mental illness in children and adolescents. These contributions must have national and/or international stature and clearly demonstrate lasting effects. The contributions may be in the areas of teaching, research, program development, direct clinical service, advocacy or administrative commitment. The award pays $2,500 to the winner and a $2,000 donation to a prevention program or center of the awardee's choice. The award winner will be recognized at a Distinguished Awards Luncheon and make an Honors Presentation about his or her work during the AACAP Annual Meeting.
Requirements: Nomination letters must be accompanied by a CV for the individual nominated.
Date(s) Application is Due: Apr 30
Amount of Grant: 4,500 U.S.D
Samples: Charles Zeanah, M.D., Tulane University New Orleans, LA—Preventing Adverse Outcomes for Abandoned Children: The Bucharest Early Intervention Project.
Contact: Harry H. Wright; (202) 966-7300; fax (202) 966-2891; training@aacap.org
Internet: http://www.aacap.org/cs/distinguishedmembers
Sponsor: American Academy of Child and Adolescent Psychiatry
3615 Wisconsin Avenue NW
Washington, DC 20016-3007

AACAP Jeanne Spurlock Lecture and Award on Diversity and Culture 39
The lecture and award reflects the spirit of the nomination for outstanding contributions to diversity and culture and in some way encourages individuals from diverse cultural backgrounds to become child and adolescent psychiatrists. Nominees should be individuals who have made contributions in the areas of social awareness including: civil rights; spirituality and/or religion; social welfare; public information; scientific research; education and mentoring; and the arts (literature, theatre, music, painting, sculpture or photography). The award includes an honorarium of $2,500, with the award winner being recognized at a Distinguished Awards Luncheon where he or she will make an honors award presentation about his or her work during the AACAP Annual Meeting.
Date(s) Application is Due: May 1
Amount of Grant: 2,500 U.S.D
Contact: Jennifer Mecidus; (202) 966-7300, ext. 137; fax (202) 966-9518; jmedicus@aacap.org
Internet: http://www.aacap.org/cs/awards
Sponsor: American Academy of Child and Adolescent Psychiatry
3615 Wisconsin Avenue NW
Washington, DC 20016-3007

AACAP Jeanne Spurlock Minority Medical Student Clinical Fellowship in Child and Adolescent Psychiatry 40
Up to fourteen (14) fellowships are available each year for minority medical students to explore a career in child and adolescent psychiatry, gain valuable work experience, and meet leaders in the child and adolescent psychiatry field. The fellowship opportunity provides up to $3,500 for 12 weeks of clinical training under a child and adolescent psychiatrist mentor. Research assignments may include responsibility for part of the observation or evaluation, developing specific aspects of the research mechanisms, conducting interviews or tests, use of rating scales, and psychological or cognitive testing of subjects. Contact AACAP for applications.
Requirements: African American, Asian American, Native American, Alaskan Native, Mexican American, Hispanic, and Pacific Islander students in accredited U.S. medical schools are eligible.
Date(s) Application is Due: Mar 1
Amount of Grant: 3,500 U.S.D
Contact: Gabe Robbins, (202) 966-7300, ext. 117; fax (202) 966-2891; grobbins@aacap.org
Internet: http://www.aacap.org/cs/students/opportunities/SpurlockClinical
Sponsor: American Academy of Child and Adolescent Psychiatry
3615 Wisconsin Avenue NW
Washington, DC 20016-3007

AACAP Jeanne Spurlock Research Fellowship in Substance Abuse and Addiction for Minority Medical Students 41
The Fellowship offers a unique opportunity for minority medical students to explore a research career in substance abuse in relation to child and adolescent psychiatry, gain valuable work experience, and meet leaders in the child and adolescent psychiatry field. The fellowship opportunity provides up to $3,500 for 12 weeks of summer research under a child and adolescent psychiatrist researcher/mentor. The research training plan must provide for significant contact between the student and the mentor and for exposure to state-of-the-art drug abuse and addiction research. The plan should include program planning discussions, instruction in research planning and implementation, regular meetings with the mentor, laboratory director, and the research group, and assigned readings. Research assignments may include responsibility for part of the observation or evaluation, developing specific aspects of the research mechanisms, conducting interviews or tests, use of rating scales, and psychological or cognitive testing of subjects.
Requirements: All fellowship participants must attend the AACAP Annual Meeting. Applications are considered from African-American, Native American, Alaskan Native, Mexican American, Hispanic, Asian, and Pacific Islander students in accredited U.S. medical schools.
Date(s) Application is Due: Mar 1
Amount of Grant: Up to 3,500 U.S.D
Contact: Gabe Robbins, (202) 966-7300, ext. 117; fax (202) 966-2891; grobbins@aacap.org or apartner@aacap.org
Internet: http://www.aacap.org/cs/students/opportunities/SpurlockResearch
Sponsor: American Academy of Child and Adolescent Psychiatry
3615 Wisconsin Avenue NW
Washington, DC 20016-3007

AACAP Junior Investigator Awards 42
The AACAP Junior Investigator Award offers two awards of up to $30,000 a year for two years for child and adolescent psychiatry junior faculty (assistant professor level or equivalent). The program is intended to facilitate innovative research. The research may be basic or clinical in nature but must be relevant to our understanding, treatment and prevention of child and adolescent mental health disorders. The award also includes the cost of attending the AACAP Annual Meeting for five days.
Requirements: Recipients are required to submit a poster or oral presentation on his or her research for the AACAP's Annual Meeting. Applicants must: be board eligible or certified in child and adolescent psychiatry; have a doctoral level degree and be in a faculty or independent research position; and have an on-site mentor who has had experience in the type of research that is being proposed that will normally include work with children and adolescents. Candidates must either be AACAP members or have a membership application pending (not paid by the award).
Restrictions: Applicants who have served as Principal Investigator on an NIH R01 grant are not eligible.
Date(s) Application is Due: Mar 1
Amount of Grant: Up to 30,000 U.S.D
Contact: Alyssa Sommer; (202) 966-7300, ext. 157; fax (202) 364-5925; asommer@aacap.org
Internet: http://www.aacap.org/cs/root/research_and_training_awards/aacap_junior_investigator_awards
Sponsor: American Academy of Child and Adolescent Psychiatry
3615 Wisconsin Avenue NW
Washington, DC 20016-3007

AACAP Klingenstein Third Generation Foundation Award for Research in Depression or Suicide 43
This annual award acknowledges outstanding leadership and continuous contributions in the field of research by giving $5,000 for the best paper on suicide and/or depression published in the Journal during the past year. The award winner will be recognized at a Distinguished Awards Luncheon and make an Honors Presentation about his or her work during the AACAP Annual Meeting.
Requirements: Nomination letters must be accompanied by a CV for the individual nominated.
Date(s) Application is Due: May 7
Amount of Grant: 5,000 U.S.D
Contact: Gabe Robbins, (202) 966-7300, ext. 117; fax (202) 966-2891; grobbins@aacap.org
Internet: http://www.aacap.org/cs/root/research_and_training_awards/aacap_journal_awards
Sponsor: American Academy of Child and Adolescent Psychiatry
3615 Wisconsin Avenue NW
Washington, DC 20016-3007

AACAP Norbert and Charlotte Rieger Award for Scientific Achievement 44
This award is given annually for the most significant article published in the Journal of the American Academy of Child and Adolescent Psychiatry (JAACAP) in the past year (July-May). The article must be written by a psychiatrist specializing in child and/or adolescent psychiatry and cannot be a review.
Requirements: The only submissions accepted are papers from child and adolescent psychiatrists that have been published in the Journal during the past year.
Date(s) Application is Due: May 7
Amount of Grant: Up to 4,500 U.S.D
Samples: Gail Bernstein, M.D., University of Minnesota, Minneapolis, MN—School-Based Interventions for Anxious Children.
Contact: Gabe Robbins, (202) 966-7300, ext. 117; fax (202) 966-2891; grobbins@aacap.org

Internet: http://www.aacap.org/cs/root/research_and_training_awards/aacap_journal_awards
Sponsor: American Academy of Child and Adolescent Psychiatry
3615 Wisconsin Avenue NW
Washington, DC 20016-3007

AACAP Norbert and Charlotte Rieger Psychodynamic Psychotherapy Award 45
The Award recognizes the best published or unpublished paper, written by an AACAP member that addresses the use of psychodynamic psychotherapy in clinical practice and fosters development, teaching, and practice of psychodynamic psychotherapy within child and adolescent psychiatry. Papers that express a novel hypothesis, raise questions about existing theory, or integrate new neuroscience and developmental psychotherapy research with psychodynamic principles may be nominated. Unpublished, new papers and papers published within the last three years may be submitted by their authors. Published papers may be nominated by any member of the AACAP. Authors may be senior or junior faculty members or residents. Delivery of the winning paper will be made at the AACAP Annual Meeting Honors Presentation.
Date(s) Application is Due: May 7
Amount of Grant: 4,500 U.S.D
Samples: Theodore Gaensbauer, M.D., University of Colorado Health Sciences Center, Denver, CO—Traumatized Young Children: The Treatment Process.
Contact: Gabe Robbins, (202) 966-7300, ext. 117; fax (202) 966-2891; grobbins@aacap.org
Internet: http://www.aacap.org/cs/2007_press_releases/rachel_seidel_md_wins_aacap_2007_norbert_and_charlotte_rieger_psychodynamic_psychotherapy_award
Sponsor: American Academy of Child and Adolescent Psychiatry
3615 Wisconsin Avenue NW
Washington, DC 20016-3007

AACAP Pilot Research Award for Attention-Deficit Disorder, Supported by the Elaine Schlosser Lewis Fund 46
The Award for Attention-Deficit Disorder is available for a junior faculty or child psychiatry resident, and is supported by the Elaine Schlosser Lewis Fund. The award recipient will be encouraged to work with a child and adolescent psychiatric investigator with expertise in his or her particular area of interest. Work must be completed within one year of receipt of the award.
Requirements: Candidates must be board eligible, certified in child and adolescent psychiatry, or enrolled in a child psychiatry residency or fellowship program. Candidates must also have a faculty appointment in an accredited medical school or be in a fully accredited child and adolescent psychiatry clinical research or training program.
Restrictions: At the time of application, candidates may not have more than two years experience following graduation from residency/fellowship training. Candidates must not have any previous significant, individual research funding in the field of child and adolescent mental health. These include the following: NIMH/NIH Funding (Small Grants, T Award or R-01) or similar foundation or industry research funding.
Date(s) Application is Due: May 7
Amount of Grant: 15,000 U.S.D
Contact: Alyssa Sommer; (202) 966-7300, ext. 157; asommer@aacap.org
Internet: http://www.aacap.org/cs/root/research_and_training_awards/pilot_research_award_for_attention_disorder_for_a_junior_faculty_or_child_psychiatry_resident_supported_by_the_elaine_schlosser_lewis_fund
Sponsor: American Academy of Child and Adolescent Psychiatry
3615 Wisconsin Avenue NW
Washington, DC 20016-3007

AACAP Pilot Research Award for Learning Disabilities, Supported by the Elaine Schlosser Lewis Fund 47
The Award for Learning Disabilities and Psychiatric Disorders is available for a junior faculty or child psychiatry resident, and is supported by the Elaine Schlosser Lewis Fund. The award encourages work with a child and adolescent psychiatric investigator with expertise in his or her particular area of interest. Work must be completed within one year of receipt of the award. Please note that travel to both the Annual Meeting and the Research Update Luncheon is not included in the award and is the responsibility of the recipient.
Requirements: Candidates must be board eligible, certified in child and adolescent psychiatry, or enrolled in a child psychiatry residency or fellowship program. Candidates must also have a faculty appointment in an accredited medical school or be in a fully accredited child and adolescent psychiatry clinical research or training program.
Restrictions: Candidates must not have any previous significant, individual research funding in the field of child and adolescent mental health. These include the following: NIMH/NIH Funding (Small Grants, T Award or R-01) or similar foundation or industry research funding. The recipient must be an AACAP member at the time of application and agree to present his or her research.
Date(s) Application is Due: May 7
Amount of Grant: 15,000 U.S.D
Contact: Alyssa Sommer, Research Coordinator; (202) 966-7300, ext. 157; fax (202) 364-5925; asommer@aacap.org
Internet: http://www.aacap.org/cs/root/research_and_training_awards/pilot_research_award_for_learning_disabilities_for_a_junior_faculty_or_child_psychiatry_resident_supported_by_the_elaine_schlosser_lewis_fund
Sponsor: American Academy of Child and Adolescent Psychiatry
3615 Wisconsin Avenue NW
Washington, DC 20016-3007

AACAP Pilot Research Awards, Supported by Eli Lilly and Company 48
Eight awards are available each year to junior faculty and child psychiatry fellows for pilot research with child and adolescent psychiatric researchers. Award recipients will be matched with child and adolescent psychiatric investigators working in their particular area of interest to act as consultants or mentors during the course of the award. Work must be completed within one year of receipt of the award.
Requirements: Candidates must be board eligible, certified in child and adolescent psychiatry, or enrolled in a child psychiatry residency or fellowship program. Candidates must have a faculty appointment in an accredited medical school or be in a fully accredited child and adolescent psychiatry clinical research or training program.
Restrictions: At the time of application, candidates may not have more than two years experience following graduation from residency/fellowship training. Candidates must not have any previous significant, individual research funding in the field of child and adolescent mental health. These include the following: NIMH/NIH Funding (Small Grants, T Award or R-01) or similar foundation or industry research funding.
Date(s) Application is Due: May 7
Amount of Grant: 15,000 U.S.D
Contact: Alyssa Sommer, Research Coordinator; (202) 966-7300, ext. 157; fax (202) 364-5925; asommer@aacap.org
Internet: http://www.aacap.org/cs/root/research_and_training_awards/aacap_pilot_research_award_for_junior_faculty_and_child_psychiatry_fellows_supported_by_lilly_usa_llc
Sponsor: American Academy of Child and Adolescent Psychiatry
3615 Wisconsin Avenue NW
Washington, DC 20016-3007

AACAP Quest for the Test Bipolar Disorder Pilot Research Award 49
The AACAP Quest for the Test Bipolar Disorder Pilot Research Award, supported by The Ryan Licht Sang Bipolar Foundation, offers $15,000 for child and adolescent psychiatry residents and junior faculty who have an interest in beginning a career in child and adolescent mental health research. Proposed research projects should address questions concerning the pathology or diagnosis of children and adolescents with bipolar disorder. The award also includes the cost of attending the AACAP Annual Meeting for five days.
Requirements: The recipient is required to submit a poster presentation on his or her research for the AACAP's Annual Meeting. Candidates must be board eligible, certified in child and adolescent psychiatry, or enrolled in a child psychiatry residency or fellowship program. Candidates must also have a faculty appointment in an accredited medical school or be in a fully accredited child and adolescent psychiatry clinical research or training program.
Restrictions: At the time of application, candidates may not have more than two years experience following graduation from residency/fellowship training. Candidates must not have any previous significant, individual research funding in the field of child and adolescent mental health. These include the following: NIH Funding (K or R level grants) or similar foundation or industry research funding.
Date(s) Application is Due: May 7
Amount of Grant: 15,000 U.S.D
Contact: Alyssa Sommer; (202) 966-7300, ext. 157; asommer@aacap.org
Internet: http://www.aacap.org/cs/root/research_and_training_awards/aacap_quest_for_the_test_bipolar_disorder_pilot_research_award
Sponsor: American Academy of Child and Adolescent Psychiatry
3615 Wisconsin Avenue NW
Washington, DC 20016-3007

AACAP Rieger Service Program Award for Excellence 50
This award recognizes innovative programs that address prevention, diagnosis, or treatment of mental illnesses in children and adolescents, and serve as model programs to the community. This award of $4,500 is shared among the awardee and his or her service program. The award winner will be recognized at a Distinguished Awards Luncheon and make an Honors Presentation about his or her work during the AACAP Annual Meeting.
Requirements: Nomination letters must be accompanied by a CV and any support materials for the individual or organization nominated.
Date(s) Application is Due: Apr 30
Amount of Grant: 4,500 U.S.D
Samples: Bradley Stein, M.D., Ph.D., Rand Corporation, Pittsburgh, PA—The Development, Implementation, and Dissemination of an Effective School-based Program for Children Exposed to Violence.
Contact: Kaye McGinty, (202) 966-7300; fax (202) 966-2891; aboccanelli@aacap.org
Internet: http://www.aacap.org/cs/distinguishedmembers
Sponsor: American Academy of Child and Adolescent Psychiatry
3615 Wisconsin Avenue NW
Washington, DC 20016-3007

AACAP Robert Cancro Academic Leadership Award 51
This award recognizes a currently serving General Psychiatry Training Director, Medical School Dean, CEO of a Training Institution, Chair of a Department of Pediatrics or Chair of a Department of Psychiatry for his or her contributions to the promotion of child and adolescent psychiatry. Named in honor of Robert Cancro, M.D., Chairman at New York University, this award offers $500 to the awardee. Nominations for the award may be made by Child and Adolescent Training Directors or Division Directors. The recipient of this award will receive a plaque and be recognized at the AACAP Annual Meeting. The award recipient will be honored at the Distinguished Awards Luncheon, Training Directors Dinner, and will provide an Honors Presentation on his or her work during the Annual Meeting.
Requirements: Nominations must include a CV for the individual nominated
Date(s) Application is Due: Apr 30

Amount of Grant: 500 U.S.D
Samples: Susan McLeer, M.D., Drexel University Philadelphia, PA—Money, Mergers and Hospital Closures: Building and Rebuilding Programs in the 21st Century.
Contact: Jeffrey Hunt, Co-Chair; (202) 966-7300; fax (202) 966-2891
Internet: http://www.aacap.org/cs/distinguishedmembers
Sponsor: American Academy of Child and Adolescent Psychiatry
3615 Wisconsin Avenue NW
Washington, DC 20016-3007

AACAP Sidney Berman Award for the School-Based Study and Intervention for Learning Disorders and Mental Ilness 52
This award recognizes an individual or program that has shown outstanding achievement in the school-based study or delivery of intervention for learning disorders and mental illness. A cash prize of $4,500 will be awarded. The award winner will be recognized at a Distinguished Awards Luncheon and make an Honors Presentation about his or her work during the AACAP Annual Meeting.
Requirements: Nomination letters must be accompanied by a CV for the individual nominated or program information.
Date(s) Application is Due: Apr 30
Amount of Grant: 4,500 U.S.D
Contact: Steven Adelsheim, (202) 966-7300; fax (202) 966-2891; aboccanelli@aacap.org
Internet: http://www.aacap.org/cs/distinguishedmembers
Sponsor: American Academy of Child and Adolescent Psychiatry
3615 Wisconsin Avenue NW
Washington, DC 20016-3007

AACAP Simon Wile Leadership in Consultation Award 53
This annual award acknowledges outstanding leadership and continuous contributions in the field of liaison child and adolescent psychiatry. Nomination letters must be accompanied by a CV for the individual nominated. The award winner will be recognized at a Distinguished Awards Luncheon and make an Honors Presentation about his or her work during the AACAP Annual Meeting.
Requirements: Practicing child and adolescent psychiatrists are eligible for nomination.
Date(s) Application is Due: Apr 30
Amount of Grant: 500 U.S.D
Contact: Richard Martini; (202) 966-7300; fax (202) 966-2891; aboccanelli@aacap.org
Internet: http://www.aacap.org/cs/distinguishedmembers
Sponsor: American Academy of Child and Adolescent Psychiatry
3615 Wisconsin Avenue NW
Washington, DC 20016-3007

AACAP Summer Medical Student Fellowships 54
The American Academy of Child and Adolescent Psychiatry (AACAP) is offering summer fellowships for work with a child and adolescent psychiatrist mentor in a clinical or research setting. Sponsored by the AACAP Campaign for America's Kids, the Fellowships offer a chance for up to 6 medical students to explore a career in child and adolescent psychiatry, gain valuable work experience, and meet leaders in the child and adolescent psychiatry field. The fellowship opportunity provides up to $3,500 for 12 weeks of clinical or research training under a child and adolescent psychiatrist mentor. The anticipated application deadline for this fellowship is April.
Requirements: Applicants must be students enrolled in accredited U.S. medical schools.
Amount of Grant: 3,500 U.S.D
Contact: Gabe Robbins, (202) 966-7300, ext. 117; fax (202) 966-2891; grobbins@aacap.org
Internet: http://www.aacap.org/cs/students/summerfellowship
Sponsor: American Academy of Child and Adolescent Psychiatry
3615 Wisconsin Avenue NW
Washington, DC 20016-3007

AACAP Systems of Care Special Program Scholarships 55
The scholarship includes up to $750 for travel expenses to the AACAP's Annual Meeting. Residents may apply to both the Special Program Scholarship and the Educational Outreach Program (EOP); however, individuals cannot receive both awards at the same time. This includes airfare, hotel, and meals, maximum $75 per day.
Date(s) Application is Due: Jul 13
Amount of Grant: 750 U.S.D
Contact: Adriano Boccanelli, Clinical Practice Assistant; (202) 966-7300, ext. 133; fax (202) 966-2891; aboccanelli@aacap.org
Internet: http://www.aacap.org/cs/root/medical_students_and_residents/residents/2009_systems_of_care_special_program_scholarship
Sponsor: American Academy of Child and Adolescent Psychiatry
3615 Wisconsin Avenue NW
Washington, DC 20016-3007

AACN-Edwards Lifesciences Nurse-Driven Clinical Practice Outcomes Grants 56
Co-sponsored by Edwards Lifesciences and the American Association of Critical-Care Nurses, each grant supports a nurse experienced in research who is conducting a clearly articulated research study that relates to use of protocol-based care driven by nurses. A commitment is required to produce results that will be available and disseminated in a short timeframe. Proposals submitted by an interdisciplinary team or an experienced researcher mentoring a novice will also be considered. Funds may be used for original research or replication of existing research. Especially desirable are proposals that focus on any of the following at any point in the continuum of acute or critical care: implementation of nurse-driven protocols using technology, including minimally invasive devices, with evaluation of patient outcomes and nurses' role in implementing the protocols; or development and clinical evaluation of nurse-driven protocols. Two awards are available each year. Each proposal can be funded up to of $5,000.
Requirements: Principal Investigators must be nurses holding current AACN membership and must remain AACN members throughout the life of the grant funding.
Restrictions: Funds may not be used to pay salaries or institutional overhead or augment funding from formal grants. Principal investigators who have received funding from AACN are ineligible to receive additional funding from AACN during the lifetime of the original award.
Date(s) Application is Due: Oct 1
Amount of Grant: Up to 5,000 U.S.D
Contact: Pamela Shellner, Clinical Practice Specialist; (949) 362-2050, ext. 321 or (800) 394-5995, ext. 321; fax (949) 362-2020; research@aacn.org or info@aacn.org
Internet: http://www.aacn.org/WD/Practice/Content/grant-Edwards.pcms?pid=1@menu=practice
Sponsor: American Association of Critical-Care Nurses
101 Columbia
Aliso Viejo, CA 92656-4109

AACN-Philips Medical Systems Clinical Outcomes Grants 57
Co-sponsored by Philips Medical Systems and the American Association of Critical-Care Nurses, each grant supports experienced nurses in conducting clearly articulated research studies. The research must be relevant to clinical nursing practice in acute or critical care. Proposals submitted by an interdisciplinary team or an experienced researcher mentoring a novice will also be considered. Funds may be used for original research or replication of existing research. Qualified proposals should seek to achieve improved outcomes and/or system efficiencies in care of acute or critically ill individuals of any age. Especially desirable are proposals for hospital-based inquiry that focuses on any of the following: intervention strategies addressing technology integration into patient care; use of computerized medical record systems in assessing patient outcomes and managing care; improving specific patient care outcomes (e.g., clinical, safety, financial); use of simulation in clinical education of nurses; and one or more of the 2006 National Patient Safety Goals issued by the Joint Commission on Accreditation of Healthcare Organizations.
Requirements: Principal Investigators must be nurses holding current AACN membership and must remain AACN members throughout the life of the grant funding.
Restrictions: Funds may not be used to pay salaries or institutional overhead or augment funding from formal grants. Principal investigators who have received funding from AACN are ineligible to receive additional funding from AACN during the lifetime of the original award.
Date(s) Application is Due: Oct 1
Amount of Grant: Up to 10,000 U.S.D
Contact: Pamela Shellner, Clinical Practice Specialist; (949) 362-2050, ext. 321 or (800) 394-5995, ext. 321; fax (949) 362-2020; research@aacn.org or info@aacn.org
Internet: http://www.aacn.org/WD/Practice/Content/grant-Philips.pcms?pid=1@menu=practice
Sponsor: American Association of Critical-Care Nurses
101 Columbia
Aliso Viejo, CA 92656-4109

AACN-Sigma Theta Tau Critical Care Grant 58
The grant, cosponsored by AACN and Sigma Theta Tau International, supports critical care nursing research and is awarded annually for a study relevant to critical care nursing practice. The proposed study may be used to meet requirements of an academic degree.
Requirements: Principal investigators must be a member of AACN and/or of Sigma Theta Tau International. The PI must have a minimum Masters Degree.
Restrictions: Funds may not be used to pay salaries or institutional overhead or augment funding from formal grants. Principal investigators who have received funding from AACN are ineligible to receive additional funding from AACN during the lifetime of the original award.
Date(s) Application is Due: Oct 1
Amount of Grant: Up to 10,000 U.S.D
Contact: Pamela Shellner, Clinical Practice Specialist; (949) 362-2050, ext. 321 or (800) 394-5995, ext. 321; fax (949) 362-2020; research@aacn.org or info@aacn.org
Internet: http://www.aacn.org/WD/Practice/Content/grant-STT.pcms?pid=1@menu=practice
Sponsor: American Association of Critical-Care Nurses
101 Columbia
Aliso Viejo, CA 92656-4109

AACN Clinical Inquiry Fund Grants 59
The fund provides small awards to qualified individuals carrying out clinical research projects that directly benefit patients and/or families. Interdisciplinary projects are encouraged. Funds may be awarded for new projects, projects in progress, and projects required for an academic degree as long as all other project criteria are met. The funds may be used to cover direct project expenses—i.e., printed materials, small equipment, and supplies including computer software.
Requirements: Principal Investigators must be nurses holding current AACN membership and must remain AACN members throughout the life of the grant funding.
Restrictions: Funds may not be used to pay salaries or institutional overhead or augment funding from formal grants.
Date(s) Application is Due: Oct 1
Amount of Grant: 500 U.S.D

Contact: Pamela Shellner, Clinical Practice Specialist; (949) 362-2050, ext. 321 or (800) 394-5995, ext. 321; fax (949) 362-2020; research@aacn.org or info@aacn.org
Internet: http://www.aacn.org/WD/Practice/Docs/Grant_Table09-10.pdf
Sponsor: American Association of Critical-Care Nurses
101 Columbia
Aliso Viejo, CA 92656-4109

AACN Clinical Practice Grants **60**
This $6,000 grant supports research focused on one or more AACN Research priorities. Research conducted in fulfillment of an academic degree is acceptable. Funds may be awarded for new projects, projects in progress and projects required for an academic degree as long as all other project criteria are met. Eligible projects may include research utilization studies, CQI projects or outcomes evaluation projects. Interdisciplinary and collaborative projects are encouraged and may involve interdisciplinary teams, multiple nursing units, home health, subacute and transitional care, other institutions or community agencies. Funds may be used to cover direct project expenses, such as printed materials, small equipment or supplies including computer software. Six awards are available each year. One award is funded up to $6,000 each year.
Requirements: Principal Investigators must be nurses holding current AACN membership and must remain AACN members throughout the life of the grant funding.
Restrictions: Funds may not be used to pay salaries or institutional overhead or augment funding from formal grants. Principal investigators who have received funding from AACN are ineligible to receive additional funding from AACN during the lifetime of the original award.
Date(s) Application is Due: Oct 1
Amount of Grant: Up to 6,000 U.S.D
Contact: Pamela Shellner; (949) 362-2050, ext. 321; research@aacn.org
Internet: http://www.aacn.org/WD/Practice/Content/grant-CP.pcms?pid=1@menu=practice
Sponsor: American Association of Critical-Care Nurses
101 Columbia
Aliso Viejo, CA 92656-4109

AACN Critical Care Grants **61**
AACN's research agenda promotes the creation of cultures of inquiry, broad sharing and evidence-based practice. This $15,000 grant supports research focused on one or more of AACN Research priorities. Research priority areas include: effective and appropriate use of technology to achieve optimal patient assessment, management, and/or outcomes; creating a healing, humane environment; processes and systems that foster the optimal contribution of critical care nurses; effective approaches to symptom management; and prevention and management of complication.
Requirements: Principal Investigators must be nurses holding current AACN membership and must remain AACN members throughout the life of the grant funding.
Restrictions: The proposed research may not be used to meet the requirements of an academic degree. Funds may not be used to pay salaries or institutional overhead or augment funding from formal grants. Principal investigators who have received funding from AACN are ineligible to receive additional funding from AACN during the lifetime of the original award.
Date(s) Application is Due: Oct 1
Amount of Grant: 15,000 U.S.D
Contact: Pamela Shellner; (949) 362-2050, ext. 321; research@aacn.org
Internet: http://www.aacn.org/WD/Practice/Content/grant-CC.pcms?pid=1@menu=practice
Sponsor: American Association of Critical-Care Nurses
101 Columbia
Aliso Viejo, CA 92656-4109

AACN End of Life/Palliative Care Small Projects Grants **62**
The end of life/palliative care grant provides an award to a qualified individual(s) carrying out a project focusing on end of life and/or palliative care outcomes in acute and/or critical care settings. Eligible projects may focus on any age group (neonatal to elderly) in the acute or critical care arena. They may include patient education programs, staff development programs, competency-based educational programs, CQI/outcomes evaluation projects, or small clinical research studies. Funds may be awarded for new projects, projects in progress and projects required for an academic degree as long as all other project criteria are met. Collaborative projects are encouraged and may involve interdisciplinary teams, multiple nursing units, home health, subacute and transitional care, other institutions or community agencies. Funds may be used to cover direct project expenses, such as printed materials, small equipment, supplies including computer software. Two awards are available each year. Each proposal can be funded for up to $500.
Requirements: Principal Investigators must be nurses holding current AACN membership and must remain AACN members throughout the life of the grant funding.
Restrictions: Funds may not be used to pay for salaries, speaker honorarium or costs, institutional overhead nor augment funding from formal large grants.
Date(s) Application is Due: Oct 1
Amount of Grant: Up to 500 U.S.D
Contact: Pamela Shellner; (949) 362-2050, ext. 321; research@aacn.org
Internet: http://www.aacn.org/WD/Practice/Content/grant-EOL.pcms?pid=1@menu=Practice
Sponsor: American Association of Critical-Care Nurses
101 Columbia
Aliso Viejo, CA 92656-4109

AACN Evidence-Based Clinical Practice Grants **63**
The Evidence-Based Clinical Practice Grant provides awards to grants that stimulate the use of patient-focused data and/or previously generated research findings to develop, implement and evaluate changes critical and acute care nursing practice. Funds may be awarded for new projects, projects in progress and projects required for an academic degree as long as all other project criteria are met. Eligible projects may include research utilization studies, CQI projects or outcomes evaluation projects. Interdisciplinary and collaborative projects are encouraged and may involve interdisciplinary teams, multiple nursing units, home health, subacute and transitional care, other institutions or community agencies. Funds may be used to cover direct project expenses, such as printed materials, small equipment or supplies including computer software. Six awards are available each year. Each proposal can be funded for up to $1,000.
Requirements: Principal Investigators must be nurses holding current AACN membership and must remain AACN members throughout the life of the grant funding.
Restrictions: Funds may not be used to pay salaries or institutional overhead or augment funding from formal grants. Principal investigators who have received funding from AACN are ineligible to receive additional funding from AACN during the lifetime of the original award.
Date(s) Application is Due: Oct 1
Amount of Grant: Up to 1,000 U.S.D
Contact: Pamela Shellner; (949) 362-2050, ext. 321 research@aacn.org
Internet: http://www.aacn.org/WD/Practice/Content/grant-EBP.pcms?pid=1@menu=Practice
Sponsor: American Association of Critical-Care Nurses
101 Columbia
Aliso Viejo, CA 92656-4109

AACN Mentorship Grant **64**
This award, cosponsored by Mallinckrodt Inc. and AACN, facilitates critical care nursing practice research between a novice and an experienced researcher. The novice researcher is a beginning researcher with limited or no research experience in the area of the proposed investigation. The novice researcher is also a registered nurse with current AACN membership. The grant may be used to fund research for an academic degree. The mentor must show strong evidence of research expertise in the proposed area of research to be pursued by the novice investigator.
Requirements: Principal Investigators must be nurses holding current AACN membership and must remain AACN members throughout the life of the grant funding.
Restrictions: The mentor may not be designated mentor in two consecutive years and may not be conducting research as part of an academic degree.
Date(s) Application is Due: Oct 1
Amount of Grant: Up to 10,000 U.S.D
Contact: Pamela Shellner; (949) 362-2050, ext. 321; research@aacn.org
Internet: http://www.aacn.org/WD/Practice/Content/grant-Mentor.pcms?pid=1@menu=practice
Sponsor: American Association of Critical-Care Nurses
101 Columbia
Aliso Viejo, CA 92656-4109

AACN Physio-Control Small Projects Grants **65**
The program provides awards to qualified individuals carrying out projects focusing on aspects of acute myocardial infarction, resuscitation, or sudden cardiac death. Such projects may include the use of defibrillation, synchronized cardioversion, noninvasive pacing, or interpretive 12-lead electrocardiogram. Eligible projects may include patient education programs, staff development programs, competency-based educational programs, continuous quality improvement projects, outcomes evaluation projects, or small clinical research studies. Funds may be awarded for new projects, projects in progress, and projects required for an academic degree as long as all other project criteria are met. Collaborative projects are encouraged and may involve interdisciplinary teams, multiple nursing units, home health, subacute and transitional care, other institutions, or community agencies. Funds may be used to cover direct project expenses, such as printed materials, small equipment, or supplies including computer software.
Requirements: Principal Investigators must be nurses holding current AACN membership and must remain AACN members throughout the life of the grant funding.
Restrictions: Funds may not be used to pay for salaries, speaker honorarium or costs, institutional overhead nor augment funding from formal large grants.
Date(s) Application is Due: Oct 1
Amount of Grant: Up to 1,500 U.S.D
Contact: Pamela Shellner; (949) 362-2050, ext. 321; research@aacn.org
Internet: http://www.aacn.org/WD/Practice/Docs/Grant_Table09-10.pdf
Sponsor: American Association of Critical-Care Nurses
101 Columbia
Aliso Viejo, CA 92656-4109

AACR-American Cancer Society Award for Research Excellence in Cancer Epidemiology and Prevention **66**
The AACR and the American Cancer Society established this Award in 1992 to honor outstanding research accomplishments in the fields of cancer epidemiology, biomarkers, and prevention. Nominations may be made via letter from any scientist, whether an AACR member or nonmember, who is now or has been affiliated with any institution involved in cancer research, cancer medicine, or cancer-related biomedical science. The winner will receive an honorarium of $5,000 and give a 50-minute lecture during the AACR Annual Meeting.
Requirements: Candidacy is open to all cancer researchers who are affiliated with any institution involved in cancer research, cancer medicine, or cancer-related biomedical

science anywhere in the world. Such institutions include those in academia, industry, or government.
Restrictions: Institutions or organizations are not eligible for the Award. Candidates may not nominate themselves.
Date(s) Application is Due: Oct 15
Amount of Grant: 5,000 U.S.D
Contact: Monique P. Eversley, Program Associate; (267) 646-0576 or (215) 440-9300, ext. 1400; fax (267) 646-0576; monique.eversley@aacr.org
Internet: http://www.aacr.org/home/scientists/scientific-achievement-awards/acs-award.aspx
Sponsor: American Association for Cancer Research
615 Chestnut Street, 17th Floor
Philadelphia, PA 19106-4404

AACR-Colorectal Cancer Coalition Fellows Grant 67
The Fellowship program supports innovative research by a meritorious young investigator by presenting the Fellow with research funds to pursue an independent line of investigation within the context of his/her current placement. This is a one-year grant of $30,000. Research projects are restricted to translational or clinical cancer research that has an ultimate goal of developing or improving therapeutic interventions for patients with metastatic colorectal cancer. One grant will be awarded this cycle.
Requirements: Candidates must be, at the start of the grant term, in their third, fourth, or fifth year of their fellowship status.
Date(s) Application is Due: Dec 18
Amount of Grant: 30,000 U.S.D
Contact: Hanna Hopfinger, Grants Program Assistant; (267) 646-0665; fax (215) 440-9372; hanna.hopfinger@aacr.org
Internet: http://www.aacr.org/home/scientists/research-funding—fellowships/fellows-grants.aspx
Sponsor: American Association for Cancer Research
615 Chestnut Street, 17th Floor
Philadelphia, PA 19106-4404

AACR-FNAB Fellows Grant for Translational Pancreatic Cancer Research 68
The Fellows Grants support innovative research by a meritorious young investigator by presenting the Fellow with research funds to pursue an independent line of investigation within the context of his/her current placement. This is a one-year grant of $35,000. Research projects are restricted to translational pancreatic cancer research that includes the use of human tissue and has implications for therapeutic application and individualized medicine.
Requirements: Candidates must be, at the start of the grant term, in their third, fourth, or fifth year of their fellowship status.
Date(s) Application is Due: Dec 18
Amount of Grant: 35,000 U.S.D
Contact: Julia Laurence, Program Assistant; (215) 440-9300, ext. 102 or (267) 646-0655; fax (215) 440-9372; julia.laurence@aacr.org or laurence@aacr.org
Internet: http://www.aacr.org/home/scientists/research-funding—fellowships/fellows-grants.aspx
Sponsor: American Association for Cancer Research
615 Chestnut Street, 17th Floor
Philadelphia, PA 19106-4404

AACR-FNAB Foundation Career Development Award for Translational Cancer Research 69
Career development awards are two-year awards that support research by scientists who are engaged in meritorious cancer research at academic institutions. These awards provide important transitional support for direct research expenses as researchers move from the ranks of early career scientists to faculty status. The Awards are open to junior faculty who completed postdoctoral studies or clinical fellowships between specified dates and who are at an academic or medical institution. Research projects in this area should have direct applicability to translational cancer and be focused on any individualized therapeutic area. Projects should have implications for individualized cancer treatment and make use of human biopsies or samples, such as needle biopsies or circulating cancer cells. One grant will be presented annually. The submission deadline is December 3, with the decision date March 31.
Requirements: Candidates must have acquired a doctoral degree in a related field and may not currently be a candidate for a further doctoral or professional degree. If a candidate has obtained an equivalent degree at a foreign institution, information on the nature of the degree must be provided at the time of application.
Restrictions: Awards are restricted to investigators or institutions within the U.S. Employees of a national government and employees of private industry are ineligible.
Date(s) Application is Due: Dec 3
Amount of Grant: 50,000 U.S.D
Samples: Tara A. Young, M.D., Ph.D., University of California, Los Angeles, Los Angeles, CA, $50,000 - Project: High Resolution Cytogenetic Study of Archival Metastatic Choroidal Melanoma (2008-2010).
Contact: Julia Laurence, Program Assistant; (215) 440-9300, ext. 102 or (267) 646-0655; fax (215) 440-9372; julia.laurence@aacr.org or awards@aacr.org
Internet: http://www.aacr.org/home/scientists/research-funding—fellowships/research-funding/research-funding—fellowships/career-development-award-recipients.aspx
Sponsor: American Association for Cancer Research
615 Chestnut Street, 17th Floor
Philadelphia, PA 19106-4404

AACR-Genentech BioOncology Career Development Award for Cancer Research on the HER Family Pathway 70
Career development awards are two-year awards that support research by scientists who are engaged in meritorious cancer research at academic institutions. These awards provide important transitional support for direct research expenses as researchers move from the ranks of early career scientists to faculty status. The Awards are open to junior faculty who completed postdoctoral studies or clinical fellowships between specified dates and who are at an academic or medical institution. Research projects in this area should have direct applicability to basic, translational, or clinical cancer research focused on the HER Family Pathway. One grant will be presented annually. The submission deadline is December 3, with the decision date March 31.
Requirements: Candidates must be affiliated with an institution in the U.S. and must hold an M.D. or combined M.D./Ph.D. If a candidate has obtained an equivalent degree at a foreign institution, information on the nature of the degree must be provided at the time of application.
Restrictions: Employees of a national government and employees of private industry are ineligible.
Date(s) Application is Due: Dec 3
Amount of Grant: 50,000 U.S.D
Contact: Julia Laurence, Program Assistant; (215) 440-9300, ext. 102 or (267) 646-0655; fax (215) 440-9372; julia.laurence@aacr.org or awards@aacr.org
Internet: http://www.aacr.org/
Sponsor: American Association for Cancer Research
615 Chestnut Street, 17th Floor
Philadelphia, PA 19106-4404

AACR-Genentech BioOncology Fellowship for Cancer Research in Angiogenesis 71
AACR Research Fellowships are open to Postdoctoral Fellows and Clinical Research Fellows at an academic facility, teaching hospital, or research institution who will be in the 1st, 2nd, or 3rd year of their postdoctoral training at the start of the fellowship term on July 1. This is a two-year term with a grant of $35,000 per year. One Fellowship will be awarded.
Requirements: The recipients of the Fellowships must attend the AACR Annual Meeting to accept the award.
Restrictions: Research projects are restricted to cancer research with direct applicability to angiogenesis.
Date(s) Application is Due: Dec 3
Amount of Grant: 35,000 U.S.D
Contact: Hanna Hopfinger, Program Assistant; (267) 646-0665 or (215) 440-9300; fax (215) 440-9372; hanna.hopfinger@aacr.org or awards@aacr.org
Internet: http://www.aacr.org/
Sponsor: American Association for Cancer Research
615 Chestnut Street, 17th Floor
Philadelphia, PA 19106-4404

AACR-GlaxoSmithKline Clinical Cancer Research Scholar Awards 72
These awards provide support of $4,000 which may be used by the recipient to support attendance at the Annual Meeting and any other AACR meeting within a two-year period. Up to $2,000 per meeting may be used, and reimbursement checks are provided after the meeting upon submission of proper documentation of travel, housing, registration, and/or subsistence expenses. Award recipients are still required to pay the Annual Meeting registration fee and make their own travel and housing arrangements. Selection is made by the Program Committee based upon the novelty, quality, and significance of the abstract submitted. Award recipients will receive notification separate from abstract acceptance and scheduling information.
Requirements: Graduate students, medical students and residents, clinical fellows or equivalent, and postdoctoral fellows (AACR members and nonmembers) who are presenters of a meritorious abstract in clinical cancer research are eligible.
Restrictions: Employees or subcontractors of private industry are not eligible.
Date(s) Application is Due: Nov 28
Amount of Grant: 2,000 - 4,000 U.S.D
Samples: Manisha Bhutani, M.D., UT M.D. Anderson Cancer Center; David W. Cescon, M.D., Princess Margaret Hospital, University of Toronto; Zeshaan A. Rasheed, M.D., Ph.D., Sidney Kimmel Comprehensive Cancer Center, Johns Hopkins University School of Medicine.
Contact: Mona E. Shater, Staff Assistant; (215) 440-9300 or (267) 646-0654; fax (215) 440-9372; mona.shater@aacr.org
Internet: http://www.aacr.org/home/scientists/meetings—workshops/travel-grants/aacr-glaxosmithkline-clinical-cancer-research-scholar-awards.aspx
Sponsor: American Association for Cancer Research
615 Chestnut Street, 17th Floor
Philadelphia, PA 19106-4404

AACR-Minorities in Cancer Research Jane Cooke Wright Lectureship Awards 73
The AACR-Minorities in Cancer Research Jane Cooke Wright Lectureship was first presented in 2006. The Lectureship is intended to give recognition to an outstanding scientist who has made meritorious contributions to the field of cancer research and who has, through leadership or by example, furthered the advancement of minority investigators in cancer research. The recipient will present the AACR-Minorities in Cancer Research Jane Cooke Wright Lectureship during the AACR Annual Meeting, receive an honorarium and commemorative plaque, and receive support for the winner and a guest to attend the Annual Meeting.

Requirements: Candidacy is open to all cancer researchers who are affiliated with any institution involved in cancer research, cancer medicine, or cancer-related biomedical science anywhere in the world. Such institutions include those in academia, industry, or government. Nominations may be made by any scientist, whether an AACR member or nonmember, who is now or has been affiliated with any institution involved in cancer research, cancer medicine, or cancer-related biomedical science. Candidates may not nominate themselves. Nominations must be submitted online at https://proposalcentral.altum.com, no later than 4:00 p.m. United States Eastern Time on October 15th. Paper nominations will not be accepted.
Restrictions: Institutions or organizations are not eligible for the Lectureship.
Date(s) Application is Due: Oct 15
Contact: Monique P. Eversley, Staff Associate; (267) 646-0576 or (215) 440-9300, ext. 1400; fax (215) 440-9372; monique.eversley@aacr.org or awards@aacr.org
Internet: http://www.aacr.org/home/scientists/scientific-achievement-awards/m.cr-wright-lectureship.aspx
Sponsor: American Association for Cancer Research
615 Chestnut Street, 17th Floor
Philadelphia, PA 19106-4404

AACR-National Brain Tumor Foundation Fellows Grant 74

The AACR-National Brain Tumor Society Fellowship, in memory of Bonnie Brooks, is open to Postdoctoral Fellows and Clinical Research Fellows at an academic facility, teaching hospital, or research institution. This fellowship provides a one-year grant of $40,000 to support the salary and benefits of Fellows engaged in Glioblastoma multiforme (GBM) research. A partial amount of funds may be designated for research support. It is anticipated that two fellowships will be awarded. The AACR requires applicants to submit both an online and a paper application. The online application must be submitted by 12:00 noon (United States Eastern Time) on August 10, the application must be submitted using the proposalCENTRAL website at https://proposalcentral.altum.com. The paper application with original signatures and all supporting documents must be postmarked and sent no later then August 12.
Requirements: Candidates must be, at the start of the grant term, in their third, fourth, or fifth year of their fellowship status.
Restrictions: Investigators holding the rank of Instructor, Adjunct Professor, Assistant Professor, Research Assistant Professor, the equivalent or higher are not eligible.
Date(s) Application is Due: Aug 10; Aug 12
Amount of Grant: 40,000 U.S.D
Contact: Hanna Hopfinger, Grants Program Assistant; (267) 646-0665; fax (215) 440-9372; hanna.hopfinger@aacr.org
Internet: http://www.aacr.org/home/scientists/research-funding—fellowships/postdoctoral-fellowships.aspx
Sponsor: American Association for Cancer Research
615 Chestnut Street, 17th Floor
Philadelphia, PA 19106-4404

AACR-NCI International Investigator Opportunity Grants 75

Recognizing that there is no substitute for the collegial interaction and scholarly discussion that is a cornerstone of scientific advancement, these grants provide significant financial support for meeting registration, housing, travel, and incidentals to attend the annual American Association for Cancer Research Meeting. Grants are available for ten cancer researchers from countries where such opportunities are limited.
Requirements: Applicants must be nationals or permanent residents of, and must reside and conduct cancer research within, low and middle-income countries as defined by the World Bank. Applicants should also hold an doctorate or equivalent degree and be actively engaged in a program of cancer research with a record of cancer-focused publications in peer-reviewed scientific journals.
Restrictions: Funds may not be used for entertainment expenses (such as for tours or in-room movies), meals for other individuals, souvenirs, and other non-meeting expenses.
Date(s) Application is Due: Sep 28
Samples: Hakan Akbulut, M.D., Professor, Medical Oncology, Ankara University, Ankara, Turkey; Soo H. Teo, D.Phil., Chief Executive, Cancer Research Initiatives Foundation, Subang Jaya, Malaysia; Jozsef Tovari, Ph.D., Lab Head, Department of Tumor Progression, National Institute of Oncology, Budapest, Hungary.
Contact: Susan Waskey, Program Assistant; (215) 440-9300 or (866) 423-3965; fax (215) 440-9313; susan.waskey@aacr.org
Internet: http://www.aacr.org/home/scientists/meetings—workshops/travel-grants.aspx
Sponsor: American Association for Cancer Research
615 Chestnut Street, 17th Floor
Philadelphia, PA 19106-4404

AACR-Pancreatic Cancer Action Network Career Development Award for Pancreatic Cancer Research 76

The Pancreatic Cancer Action Network-AACR Career Development Award represents a joint effort to encourage and support early career scientist who are in the first four years of a faculty appointment to conduct pancreatic cancer research and establish successful career paths in this field. The research proposed for funding may be basic, translational, clinical or epidemiological in nature and must have direct applicability and relevance to pancreatic cancer. The Award provides a two year grant of $200,000 ($100,000 per year) for direct research expenses, which may include salary and benefits of the grant recipient, postdoctoral or clinical research fellows, and/or research assistants, research/laboratory supplies and equipment. It is anticipated that three Career Development Awards will be funded.
Requirements: Candidates must have acquired a doctoral degree in a related field and may not currently be a candidate for a further doctoral or professional degree. If a candidate has obtained an equivalent degree at a foreign institution, information on the nature of the degree must be provided at the time of application.
Restrictions: Awards are restricted to investigators or institutions within the U.S. Employees of a national government and employees of private industry are ineligible.
Date(s) Application is Due: Oct 28
Contact: Julia Laurence, Program Assistant; (215) 440-9300, ext. 102 or (267) 646-0655; fax (215) 440-9372; julia.laurence@aacr.org or awards@aacr.org
Internet: http://www.aacr.org/home/scientists/research-funding—fellowships/career-development-awards.aspx
Sponsor: American Association for Cancer Research
615 Chestnut Street, 17th Floor
Philadelphia, PA 19106-4404

AACR-Pancreatic Cancer Action Network Innovative Grants 77

The AACR-PanCAN Pilot Grants represent a joint effort of the AACR and the Pancreatic Cancer Action Network to support innovative research focusing on the early detection or treatment of pancreatic cancer. Research projects may have direct application to pancreatic cancer or may demonstrate relevance to pancreatic cancer in order to encourage investigators with experience in other areas of research to apply their ideas to this challenging field. Proposals will be accepted for innovative research projects that have a direct application or demonstrate relevance to the early detection or treatment of pancreatic cancer; projects may be in any discipline of basic, translational, clinical, or epidemiological cancer research. Proposals that will develop preliminary data necessary to prepare and submit a competitive research grant application to a major federal funding agency will also be accepted. Special emphasis will be placed on research that is not duplicative of other efforts and has the potential for national application. Researchers who are affiliated with any institution involved in cancer research, cancer medicine, or cancer-related biomedical science within the United States may apply. There are no residency status restrictions. Each grantee will be awarded a total of $200,000, paid over the two-year term.
Requirements: Applicants must have acquired a doctoral degree in a related field. If an applicant has obtained an equivalent degree at a foreign institution, information on the nature of the degree must be provided at the time of application.
Restrictions: Employees or subcontractors of a national government or the for-profit private industry are not eligible to apply. Members of the Scientific Review Committee are not eligible for these grants. Current or past recipients of a PanCAN Pilot Grant are not eligible.
Date(s) Application is Due: Oct 5; Nov 30
Amount of Grant: 200,000 U.S.D
Contact: Julia Laurence, Program Assistant; (215) 440-9300, ext. 102 or (267) 646-0655; fax (215) 440-9372; julia.laurence@aacr.org or laurence@aacr.org
Internet: http://www.aacr.org/home/scientists/research-funding—fellowships/pancreatic-cancer-action-network-aacr-innovative-grants.aspx
Sponsor: American Association for Cancer Research
615 Chestnut Street, 17th Floor
Philadelphia, PA 19106-4404

AACR-Prevent Cancer Foundation Award for Excellence in Cancer Prevention Research 78

he Award for Excellence in Cancer Prevention Research will be given to a scientist residing in any country in the world for his or her seminal contributions to the field of cancer prevention. Such investigations must have been conducted in basic, translational, clinical, epidemiological, or behavioral science in cancer prevention research. Further, these studies must have had not only a major impact on the field, but must also have stimulated new directions in this important area. The recipient of the Award will receive a $5,000 honorarium and will present a lecture at the Annual AACR International Conference. Nominations for the Award must be submitted electronically no later than 4:00 p.m. United States Eastern Time on Monday, July 31.
Requirements: All cancer researchers who are affiliated with any institution involved in cancer research, cancer medicine, or cancer-related biomedical science anywhere in the world. Such institutions include those in academia, industry, or government may be nominated for the Award. Candidates must currently maintain an active research program, have a record of recent publications, and be able to present the Award lecture at the Conference.
Restrictions: Institutions or organizations are not eligible for the Award. Paper nominations will not be accepted.
Date(s) Application is Due: Jul 31
Amount of Grant: 5,000 U.S.D
Contact: Monique P. Eversley, Staff Associate; (267) 646-0576 or (215) 440-9300, ext. 1400; fax (215) 440-9372; monique.eversley@aacr.org or awards@aacr.org
Internet: http://www.aacr.org/home/scientists/scientific-achievement-awards/prevent-cancer-award.aspx
Sponsor: American Association for Cancer Research
615 Chestnut Street, 17th Floor
Philadelphia, PA 19106-4404

AACR-WICR Scholar Awards 79

AACR-WICR Scholar Awards are awarded to ten members of Women in Cancer Research who are scientists-in-training and presenters of meritorious scientific papers at AACR Annual Meetings.
Requirements: Eligible candidates for these awards must fit the following criteria, candidates must be: members of Women in Cancer Research (WICR); full time scientists-in-training who are graduate students, medical students, residents, clinical fellows or equivalent, or postdoctoral fellows; travelling from within the United States or abroad; first authors on abstracts submitted for consideration for presentation at the AACR Annual Meeting.

Date(s) Application is Due: Dec 1
Contact: Lauren Medvetz, Program Coordinator; (267) 646-0689; fax (215) 440-9372; lauren.medvets@aacr.org
Internet: http://www.aacr.org/home/scientists/meetings—workshops/travel-grants/wicr-scholar-awards.aspx
Sponsor: American Association for Cancer Research
615 Chestnut Street, 17th Floor
Philadelphia, PA 19106-4404

AACR-Women in Cancer Research Charlotte Friend Memorial Lectureship Awards 80
The AACR-Women in Cancer Research Charlotte Friend Memorial Lectureship was established in 1998 in honor of renowned virologist and discoverer of the Friend virus, Dr. Charlotte Friend, for her pioneering research on viruses, cell differentiation, and cancer. The lecture is intended to give recognition to an outstanding female or male scientist who has made meritorious contributions to the field of cancer research and who has, through leadership or by example, furthered the advancement of women in science. The winner of the Lectureship will present a 50-minute lecture during the AACR Annual Meeting, receive an honorarium and commemorative plaque, and receive support for the winner and a guest to attend the Annual Meeting.
Requirements: Candidacy is open to all cancer researchers who are affiliated with any institution involved in cancer research, cancer medicine, or cancer-related biomedical science anywhere in the world. Such institutions include those in academia, industry, or government.
Restrictions: Nominations may be made by any scientist, whether an AACR member or nonmember, who is now or has been affiliated with any institution involved in cancer research, cancer medicine, or cancer-related biomedical science. Candidates may not nominate themselves. Nominations must be submitted online at https://proposalcentral.altum.com, no later than 4:00 p.m. United States Eastern Time on October 15th. Paper nominations will not be accepted. Institutions or organizations are not eligible for the Lectureship.
Date(s) Application is Due: Oct 15
Contact: Lauren Medvetz, Program Coordinator; (267) 646-0689; fax (215) 440-9372; lauren.medvets@aacr.org
Internet: http://www.aacr.org/home/scientists/scientific-achievement-awards/wicr-friend-lectureship.aspx
Sponsor: American Association for Cancer Research
615 Chestnut Street, 17th Floor
Philadelphia, PA 19106-4404

AACR Award for Lifetime Achievement in Cancer Research 81
The AACR Award for Lifetime Achievement in Cancer Research was established and first given in 2004 to honor an individual who has made significant fundamental contributions to cancer research, either through a single scientific discovery or a body of work. These contributions, whether they have been in research, leadership, or mentorship, must have had a lasting impact on the cancer field and must have demonstrated a lifetime commitment to progress against cancer. Nominations may be made by any scientist, whether an AACR member or nonmember, who is now or has been affiliated with any institution involved in cancer research, cancer medicine, or cancer-related biomedical science. Nominations for the Award must be submitted electronically to the AACR no later than 4:00 p.m. United States Eastern Time on November 16.
Requirements: Nominations may be made on behalf of individuals who are living at the time of the nomination. Candidates need not be members of the AACR. Candidates need not be currently engaged in cancer research.
Restrictions: Institutions or organizations are not eligible for the Award. Paper nominations will not be accepted.
Date(s) Application is Due: Nov 16
Contact: Monique P. Eversley, Staff Associate; (267) 646-0576 or (215) 440-9300, ext. 1400; fax (215) 440-9372; monique.eversley@aacr.org or awards@aacr.org
Internet: http://www.aacr.org/home/scientists/scientific-achievement-awards/lifetime-achievement-award.aspx
Sponsor: American Association for Cancer Research
615 Chestnut Street, 17th Floor
Philadelphia, PA 19106-4404

AACR Award for Outstanding Achievement in Cancer Research 82
This award is given in recognition of a young investigator on the basis of meritorious achievement in cancer research. The winner receives an honorarium of $5,000, presents a 50-minute lecture, and is given full support for the winner and a guest to attend the AACR Annual Meeting.
Requirements: Candidacy is open to all cancer researchers who are affiliated with any institution involved in cancer research, cancer medicine, or cancer-related biomedical science anywhere in the world. Such institutions include those in academia, industry, or government. In accordance with the wishes of the donor, the recipient must be no more than 40 years of age by the time the award is received.
Restrictions: Institutions or organizations are not eligible for the Award.
Date(s) Application is Due: Oct 15
Amount of Grant: 5,000 U.S.D
Contact: Monique P. Eversley, Program Associate; (267) 646-0576 or (215) 440-9300, ext. 1400; fax (267) 646-0576; monique.eversley@aacr.org
Internet: http://www.aacr.org/home/scientists/scientific-achievement-awards/outstanding-achievement-award.aspx
Sponsor: American Association for Cancer Research
615 Chestnut Street, 17th Floor
Philadelphia, PA 19106-4404

AACR Award for Outstanding Achievement in Chemistry in Cancer Research 83
The Award will be given for outstanding, novel, and significant chemistry research, which has led to important contributions to the fields of basic cancer research; translational cancer research; cancer diagnosis; the prevention of cancer; or the treatment of patients with cancer. Such research may include, but is not limited to, drug discovery and design; structural biology; proteomics, metabolomics and biological mass spectrometry; chemical aspects of carcinogenesis; imaging agents and radiotherapeutics; and chemical biology. The winner of the award will give a 50-minute lecture during the AACR Annual Meeting, will receive a commemorative plaque and a $10,000 honorarium, and receive support for the winner and a guest to attend the Annual Meeting. Nominations for the Award must be submitted electronically no later than 4:00 p.m. United States Eastern Time on October 15.
Requirements: Candidacy is open to all researchers who are affiliated with any institution involved in cancer research, cancer medicine, or cancer-related biomedical science anywhere in the world. Such institutions include those in academia, industry, or government.
Restrictions: Institutions or organizations are not eligible for the Award. Paper nominations will not be accepted.
Date(s) Application is Due: Oct 15
Amount of Grant: 10,000 U.S.D
Contact: Lauren Medvetz, (267) 646-0689; fax (215) 440-9372; lauren.medvets@aacr.org
Internet: http://www.aacr.org/home/scientists/scientific-achievement-awards/chemistry-in-cancer-research-award.aspx
Sponsor: American Association for Cancer Research
615 Chestnut Street, 17th Floor
Philadelphia, PA 19106-4404

AACR Basic Cancer Research Fellowships 84
The AACR Basic Cancer Research Fellowships consist of two Fellowships, each one-year grants. The AACR Anna D, Barker Fellowship in Basic Cancer Research, funds $40,000, and the AACR-Astellas U.S.A Foundation Fellowships in Basic Cancer Research provides funding in the amount of $30,000. These grants are for salary and benefits to the Fellows. A partial amount of funds may be designated for direct research support. The recipients must attend the AACR Annual Meeting to accept the grants. The AACR requires applicants to submit both an online and a paper application. The online application must be completed by 12:00 noon (United States Eastern Time) on December 10, using the proposalCENTRAL website at http://propasalcentral.altum.com. The paper application with original signatures and all supporting documents must be postmarked and sent no later then December 15.
Requirements: Applications are open to Postdoctoral Fellows and Clinical Research Fellows at an academic facility, teaching hospital, or research institution who will be in the 1st, 2nd, or 3rd year of their postdoctoral training at the start of the fellowship term.
Restrictions: Academic faculty holding the rank of assistant professor or higher, graduate or medical students, government employees, and employees of private industry are not eligible.
Date(s) Application is Due: Dec 10; Dec 15
Amount of Grant: 30,000 - 40,000 U.S.D
Contact: Hanna Hopfinger, Grants Program Assistant; (267) 646-0665; fax (215) 440-9372; hanna.hopfinger@aacr.org
Internet: http://www.aacr.org/home/scientists/research-funding—fellowships/postdoctoral-fellowships.aspx
Sponsor: American Association for Cancer Research
615 Chestnut Street, 17th Floor
Philadelphia, PA 19106-4404

AACR Brigid G. Leventhal Scholar in Cancer Research Awards 85
The purpose of this award program is to enhance the education and training of early career scientists by providing financial support for their participation in AACR Annual Meetings and Special Conferences. The AACR-WICR Brigid G. Leventhal Scholar Award Committee makes selection for these competitive awards after careful consideration of the candidate's application and accompanying materials.
Requirements: Candidates must be: Women in Cancer Research members; full time scientists in training who are graduate students, medical students, residents, clinical fellows or equivalent, or postdoctoral fellows; and first authors on abstracts submitted for consideration for presentation at the AACR Annual Meeting or Special Conference which the applicant wants to attend.
Date(s) Application is Due: Sep 10
Contact: Mona E. Shater, Staff Assistant; (215) 440-9300 or (267) 646-0654; fax (215) 440-9372; mona.shater@aacr.org or wicr@aacr.org
Internet: http://www.aacr.org/home/membership-/association-groups/women-in-cancer-research.aspx
Sponsor: American Association for Cancer Research
615 Chestnut Street, 17th Floor
Philadelphia, PA 19106-4404

AACR Career Development Awards for Pediatric Cancer Research 86
Two Career Development Awards will be presented, the AACR-Aflac Incorporated Career Development Award for Pediatric Cancer Research and the AACR Centennial Career Development Award for Childhood Cancer Research. These Awards are open to junior faculty who are in their first full-year, faculty appointment. The Awards provide two-year grants of $100,000 ($50,000 per year) for direct research expenses and/or salary support. Research projects must have direct applicability to pediatric cancer.
Requirements: Candidates must have acquired a doctoral degree and have completed postdoctoral studies or clinical fellowships on or after July 2 (two years prior to the start of the grant term). Candidates must be in their first full-time, faculty appointment and

hold the title of Instructor, Research Assistant Professor, Assistant Professor, or an equivalent full-time faculty appointment at the start of the grant term (July 1).
Restrictions: Non AACR members must submit a satisfactory application for Active Membership with the grant application by December 10th. Employees or subcontractors of a government or a for-profit private industry are not eligible.
Date(s) Application is Due: Dec 18
Amount of Grant: 100,000 U.S.D
Contact: Julia Laurence, Program Assistant; (215) 440-9300, ext. 102 or (267) 646-0655; fax (215) 440-9372; julia.laurence@aacr.org or awards@aacr.org
Internet: http://www.aacr.org/home/scientists/research-funding—fellowships/career-development-awards.aspx
Sponsor: American Association for Cancer Research
615 Chestnut Street, 17th Floor
Philadelphia, PA 19106-4404

AACR Centennial Postdoctoral Fellowships in Cancer Research **87**
AACR Centennial Postdoctoral Research Fellowships in Cancer Research will provide $60,000 per year for up to three years to provide salary support and research funding for clinical and postdoctoral fellows in the first, second, or third year of their fellowship status. In addition to financial flexibility these fellowships also provide career flexibility. Fellowship funds may be transferred to move with the grantee if the grantee's fellowship is completed during the grant term - a valuable commodity for a newly-independent investigator. Under this award mechanism, the postdoctoral trainee is considered the Lead Investigator and should write the proposal independently with appropriate direction from the mentor. Proposals may be submitted in any area of cancer research whether laboratory-based, translational, clinical, epidemiological, behavioral, or other. Applications must be completed online using the proposalCENTRAL website, with one paper copy submitted to the AACR office. Application instructions and program guidelines are available online.
Requirements: Applicants must: be in the 1st, 2nd, or 3rd year of their postdoctoral training at the start of the fellowship term; be members of the AACR or must submit an application for membership prior to the fellowship application deadline; have acquired a doctoral degree in a biological sciences-related field; and hold the title of Postdoctoral Fellow or Clinical Research Fellow at an academic facility, teaching hospital, or research institution.
Restrictions: Employees, trainees, or subcontractors of a national government or private industry are not eligible.
Date(s) Application is Due: Feb 1
Amount of Grant: 60,000 - 180,000 U.S.D
Contact: Hanna Hopfinger, Program Assistant; (267) 646-0665 or (215) 440-9300; fax (215) 440-9372; hanna.hopfinger@aacr.org or awards@aacr.org
Internet: http://www.aacr.org/home/scientists/research-funding—fellowships/research-funding.aspx
Sponsor: American Association for Cancer Research
615 Chestnut Street, 17th Floor
Philadelphia, PA 19106-4404

AACR Centennial Pre-doctoral Fellowships in Cancer Research **88**
The AACR Centennial Pre-doctoral Research Fellowships will provide full-time graduate students a rare opportunity to invest in their future cancer research careers. The grant terms are flexible, allowing applicants to tailor their grant funds to fit their individual research funding and salary needs. Grantees may also transfer their funds to a postdoctoral fellowship position, should they advance in their careers during the grant term. Under this award mechanism, the pre-doctoral trainee is considered the Lead Investigator and should write the proposal independently with appropriate direction from the mentor. Proposals may be submitted in any area of cancer research whether laboratory-based, translational, clinical, epidemiological, behavioral, or other. A grant of $30,000 per year over a maximum period of three years will be awarded for salary support and research funding.
Requirements: Applicants must: be full-time graduate students who are pursuing a doctoral degree and engaged in a program of cancer research; be members of the AACR or must submit an application for membership prior to the fellowship application deadline; have completed at least one year of research training at the graduate level or have equivalent research experience; and be engaged in basic, clinical, translational or epidemiological cancer research. At the time of award acceptance the awardee must have advanced to degree candidacy.
Restrictions: Summer student positions and undergraduate summer studentships do not qualify. Employees, trainees, or subcontractors of a national government or private industry are not eligible.
Date(s) Application is Due: Feb 1
Amount of Grant: 30,000 - 90,000 U.S.D
Contact: Hanna Hopfinger, Program Assistant; (267) 646-0665 or (215) 440-9300; fax (215) 440-9372; hanna.hopfinger@aacr.org or awards@aacr.org
Internet: http://www.aacr.org/home/scientists/research-funding—fellowships/research-funding.aspx
Sponsor: American Association for Cancer Research
615 Chestnut Street, 17th Floor
Philadelphia, PA 19106-4404

AACR Clinical and Translational Research Fellowships **89**
The AACR Clinical and Translational Research Fellowships consist of three different Fellowships: AACR-Astellas U.S.A Foundation Fellowship in Clinical/Translational Cancer Research, which is a one-year grant of $30,000; AACR-AstraZeneca Fellowship for Translational Lung Cancer Research, which is a two-year grant of $90,000; and, AACR-Bristol-Myers Squibb Oncology Fellowship in Clinical Cancer Research, which is a one-year grant of $40,000. These fellowships support the salary and benefits of the Fellow. A partial amount of funds may be designated for direct research support. The AACR requires applicants to submit both an online and a paper application. The online application deadline is December 10th, the application must be submitted using the proposalCENTRAL website at https://proposalcentral.altum.com. The paper application with original signatures and all supporting documents must be postmarked and sent no later then December 15.
Requirements: Applications are open to Postdoctoral Fellows and Clinical Research Fellows at an academic facility, teaching hospital, or research institution who will be in the 1st, 2nd, or 3rd year of their postdoctoral training at the start of the fellowship term.
Restrictions: Academic faculty holding the rank of assistant professor or higher, graduate or medical students, government employees, and employees of private industry are not eligible.
Date(s) Application is Due: Dec 10; Dec 15
Amount of Grant: 30,000 - 90,000 U.S.D
Contact: Hanna Hopfinger, Program Assistant; (267) 646-0665 or (215) 440-9300; fax (215) 440-9372; hanna.hopfinger@aacr.org or awards@aacr.org
Internet: http://www.aacr.org/home/scientists/research-funding—fellowships/postdoctoral-fellowships.aspx
Sponsor: American Association for Cancer Research
615 Chestnut Street, 17th Floor
Philadelphia, PA 19106-4404

AACR Fellows Grants **90**
The Fellowship program supports innovative research by a meritorious young investigator by presenting the Fellow with research funds to pursue an independent line of investigation within the context of his/her current Fellowship placement. By allowing a Fellow to acquire the equipment and supplies needed to pursue a new direction in his/her research program, the Fellows Grant assists the Fellow in developing preliminary data to support a future project or investigating a new technique that otherwise would not be possible in the absence of this funding. The Selection Committee favors proposals that provided evidence of an applicant's research initiative and creativity, and the support of a scientific mentor. The Committee also weighed the potential of the candidate to make meaningful contributions to the field, and promise of success as a cancer researcher in future years.
Requirements: Candidates must be, at the start of the grant term, in their third, fourth, or fifth year of their fellowship status.
Date(s) Application is Due: Dec 18
Amount of Grant: 30,000 - 40,000 U.S.D
Contact: Julia Laurence, Program Assistant; (215) 440-9300, ext. 102 or (267) 646-0655; fax (215) 440-9372; julia.laurence@aacr.org or awards@aacr.org
Internet: http://www.aacr.org/home/scientists/research-funding—fellowships/fellows-grants.aspx
Sponsor: American Association for Cancer Research
615 Chestnut Street, 17th Floor
Philadelphia, PA 19106-4404

AACR G.H.A. Clowes Memorial Award **91**
The AACR and Eli Lilly and Company established this Award in 1961 to honor Dr. G.H.A. Clowes, who was a founding member of the AACR and a research director of Eli Lilly. The Clowes Award recognizes an individual with outstanding recent accomplishments in basic cancer research. Nominations for the Award must be submitted electronically to the AACR no later than 4:00 p.m. United States Eastern Time on October 15. Paper nominations will not be accepted.
Requirements: Candidacy is open to all cancer researchers who are affiliated with any institution involved in cancer research, cancer medicine, or cancer-related biomedical science anywhere in the world. Such institutions include those in academia, industry, or government.
Restrictions: Institutions or organizations are not eligible for the Award. Paper nominations will not be accepted.
Date(s) Application is Due: Oct 15
Amount of Grant: 10,000 U.S.D
Contact: Monique P. Eversley, Staff Associate; (267) 646-0576 or (215) 440-9300, ext. 1400; fax (215) 440-9372; monique.eversley@aacr.org or awards@aacr.org
Internet: http://www.aacr.org/home/scientists/scientific-achievement-awards/clowes-award.aspx
Sponsor: American Association for Cancer Research
615 Chestnut Street, 17th Floor
Philadelphia, PA 19106-4404

AACR Gertrude Elion Cancer Research Award **92**
This prestigious award provides a one-year research grant to an assistant professor at an academic or nonprofit research institute worldwide for salary and benefits, laboratory supplies, and limited domestic travel to support research in cancer etiology, diagnosis, treatment, or prevention (basic, translational, or clinical cancer research). The AACR requires applicants to submit both an online and a paper application. The online application must be submitted by 12:00 noon (United States Eastern Daylight Time) on February 27, using the proposalCENTRAL website at https://proposalcentral.altum.com. The paper application with original signatures and all supporting documents must be postmarked and sent no later then March 3.
Requirements: Candidates must be nominated by an AACR member and submit a detailed application. Candidates must have completed postdoctoral studies or clinical fellowships no later than July 1 of the award year and ordinarily not more than five years earlier. Candidates must be tenure-tracked scientists at the level of assistant professor at an academic institution anywhere in the world.
Restrictions: Tenured faculty in academia, employees of a national government, and employees of private industry are not eligible.
Date(s) Application is Due: Feb 27; Mar 3

Amount of Grant: 50,000 U.S.D
Contact: Julia Laurence, Program Assistant; (215) 440-9300, ext. 102 or (267) 646-0655; fax (215) 440-9372; julia.laurence@aacr.org or laurence@aacr.org
Internet: http://www.aacr.org/home/scientists/research-funding—fellowships/elion-award.aspx
Sponsor: American Association for Cancer Research
615 Chestnut Street, 17th Floor
Philadelphia, PA 19106-4404

AACR Henry Shepard Bladder Cancer Research Grants 93

These two-year Grants will provide up to $250,000 in total support for innovative cancer research projects designed to accelerate the discovery, development, and application of new agents to treat bladder cancer and/or for pre-clinical research with direct therapeutic intent. Laboratory-based projects must present plans with clinical collaborators indicating how the work will be translated into the clinic. Similarly, clinical studies must show how the work was derived from basic preclinical work and how results will be channeled back to laboratory-based collaborators. Four grants will be awarded. The submission deadline is July 25, at 12:00 noon, U.S. Eastern Time.
Requirements: Independent investigators who are affiliated with any institution involved in cancer research, cancer medicine, or cancer-related biomedical science anywhere in the world may apply. There are no geographic, national, or residency status restrictions. Applicants must have acquired a doctoral degree in a related field.
Restrictions: Pre- and Postdoctoral fellows are not eligible. Employees or subcontractors of a national government or the for-profit private industry are not eligible to serve as the Principal Investigator for the purpose of these Grants. However, collaborations with such individuals are encouraged. Neither the members of the Scientific Review Committee nor the members of their individual laboratories are eligible for these Grants.
Date(s) Application is Due: Jul 25
Amount of Grant: Up to 250,000 U.S.D
Contact: Hanna Hopfinger, Program Assistant; (267) 646-0665 or (215) 440-9300; fax (215) 440-9372; hanna.hopfinger@aacr.org or awards@aacr.org
Internet: http://www.aacr.org/home/scientists/research-funding—training-grants/research-funding/research-funding—fellowships/bladder-cancer-research-grants.aspx
Sponsor: American Association for Cancer Research
615 Chestnut Street, 17th Floor
Philadelphia, PA 19106-4404

AACR Joseph H. Burchenal Memorial Award 94

This Award was established in 1996 to recognize outstanding achievements in clinical cancer research. It is named for the late Dr. Joseph H. Burchenal, Honorary Member and Past President of the AACR, and a major figure in clinical cancer research and chemotherapy. The winner of the Award for Outstanding Achievement in Clinical Cancer Research will give a 50-minute lecture during the AACR Annual Meeting, and receive an honorarium and support for attendance at the meeting.
Requirements: Candidacy is open to all cancer researchers who are affiliated with any institution involved in cancer research, cancer medicine, or cancer-related biomedical science anywhere in the world. Such institutions include those in academia, industry, or government.
Restrictions: Institutions or organizations are not eligible for the Award. Candidates may not nominate themselves.
Date(s) Application is Due: Oct 15
Samples: Joseph R. Bertino, M.D., Interim Director & Chief Scientific Officer, The Cancer Institute of New Jersey.
Contact: Monique P. Eversley, Program Associate; (267) 646-0576 or (215) 440-9300, ext. 1400; fax (267) 646-0576; monique.eversley@aacr.org
Internet: http://www.aacr.org/home/scientists/scientific-achievement-awards/burchenal-award.aspx
Sponsor: American Association for Cancer Research
615 Chestnut Street, 17th Floor
Philadelphia, PA 19106-4404

AACR Judah Folkman Career Development Award for Anti-Angiogenesis Research 95

The AACR Judah Folkman Career Development Award for Angiogenesis Research is open to junior faculty who completed postdoctoral studies or clinical fellowships no more then three years prior to the start of the grant term (Dec 1); who are in their first full-time, faculty appointment and hold the title of Instructor, Research Assistant Professor, Assistant Professor, or an equivalent full-time faculty appointment; and who are at an academic, medical, or research Institution. The Career Development Award provides a two-year grant of $100,000 ($50,000 per year) for direct research expenses and/or salary support. One Award will be presented.
Requirements: Applicants must have acquired a medical degree or hold a combined M.D./Ph.D and may not currently be a candidate for a further doctoral or professional degree. The Award is open to AACR members. Non-members must submit a satisfactory application for Active Membership and additional Curriculum Vitae with the grant application.
Restrictions: Employees, trainees, or subcontractors of a national government or private industry are not eligible. Exceptions may apply if an applicant holds a full-time position at a Veterans' Hospital or national laboratory in the U.S. Current recipients of any AACR grant are also not eligible.
Date(s) Application is Due: Aug 10
Amount of Grant: 100,000 U.S.D
Contact: Julia Laurence, Program Assistant; (215) 440-9300, ext. 102 or (267) 646-0655; fax (215) 440-9372; julia.laurence@aacr.org or awards@aacr.org
Internet: http://www.aacr.org/home/scientists/research-funding—fellowships/career-development-awards.aspx
Sponsor: American Association for Cancer Research
615 Chestnut Street, 17th Floor
Philadelphia, PA 19106-4404

AACR Kirk A. Landon and Dorothy P. Landon Foundation Prizes 96

These two major international awards recognize outstanding scientists who have made seminal basic and translational cancer research discoveries at the cutting edge of scientific novelty and significance, which have accelerated progress against cancer and have implications for future discoveries and contributions to cancer research. The Kirk A. Landon-AACR Prize for Basic Cancer Research recognizes significant, fundamental contributions to laboratory research. The Dorothy P. Landon-AACR Prize for Translational Cancer Research recognizes extraordinary achievement in translational cancer research—the interface between basic research and its application to the clinic in the areas of diagnosis, treatment, or the prevention of cancer. Both prizes bring heightened public attention to landmark achievements in the continuing effort to prevent and cure cancer through the presentation of dynamic lectures during the AACR annual meeting; and promote and reward continued, productive cancer research.
Requirements: The prizes are open to all cancer researchers who are affiliated with any institution involved in cancer research, cancer medicine, or cancer-related biomedical science anywhere in the world. Such institutions include those in academia, industry, or government. Candidates must be active researchers and have a record of recent publications.
Restrictions: Institutions or organizations are ineligible.
Date(s) Application is Due: Aug 25
Amount of Grant: 100,000 U.S.D
Samples: Arnold J. Levine, Ph.D., Professor, The Cancer Institute of New Jersey—Kirk A. Landon-AACR Prize for Basic Cancer Research; John Mendelsohn, M.D., President, UT M.D. Anderson Cancer Center, Houston, Texas—Dorothy P. Landon-AACR Prize for Translational Cancer Research.
Contact: Monique P. Eversley, Staff Associate; (267) 646-0576 or (215) 440-9300, ext. 1400; fax (215) 440-9372; monique.eversley@aacr.org or awards@aacr.org
Internet: http://www.aacr.org/home/scientists/scientific-achievement-awards/landon-prizes.aspx
Sponsor: American Association for Cancer Research
615 Chestnut Street, 17th Floor
Philadelphia, PA 19106-4404

AACR Margaret Foti Award for Leadership and Extraordinary Achievements in Cancer Research 97

This Award will recognize a true champion of cancer research, who embodies the sustained commitment of Margaret Foti to the prevention and cure of cancer. The Award will be given to an individual whose leadership and extraordinary achievements in cancer research have made a major impact in the field. Such achievements may include contributions to the acceleration of progress in cancer research; raising national or international awareness of cancer research; or in other ways demonstrating a sustained commitment to cancer research. Nominations for the Award must be submitted electronically no later than 4:00 p.m. U.S. Eastern Time on November 16. The recipient will receive an honorarium, a commemorative plaque, and support for the winner to attend the AACR Annual Meeting.
Requirements: Candidacy is open to all individuals who are affiliated with any organization whose mission supports cancer research or any institution involved in cancer research, cancer medicine, or cancer-related biomedical science anywhere in the world. Such institutions include those in academia, industry, or government.
Restrictions: Institutions or organizations are not eligible for the Award. Paper nominations will not be accepted.
Date(s) Application is Due: Nov 16
Contact: Monique P. Eversley, Program Associate; (267) 646-0576 or (215) 440-9300, ext. 1400; fax (215) 440-9372; monique.eversley@aacr.org or awards@aacr.org
Internet: http://www.aacr.org/home/scientists/scientific-achievement-awards/margaret-foti-award.aspx
Sponsor: American Association for Cancer Research
615 Chestnut Street, 17th Floor
Philadelphia, PA 19106-4404

AACR Minority-Serving Institution Faculty Scholar in Cancer Research Awards 98

The AACR is pleased to announce the availability of Scholar Awards in Cancer Research for full-time faculty members of Minority-Serving Institutions [Historically Black Colleges and Universities (HBCUs), Hispanic Serving Institutions (HSIs), and Tribal Colleges and Universities and other post-secondary institutions as defined by the U.S. Department of Education]. The purposes of this Award program is to increase the scientific knowledge base of faculty members at Minority-Serving Institutions, and to encourage them and their students to pursue careers in cancer research. AACR Minority-Serving Institution Faculty Scholar in Cancer Research Awards are presented by the American Association for Cancer Research to scientists at the level of Assistant Professor or above at a Minority-Serving Institution, who are engaged in meritorious basic, clinical, or translational cancer research.
Requirements: Candidates must have completed doctoral studies or clinical fellowships and hold full-time faculty status at an institution designated as a Minority-Serving Institution. Candidates must have acquired doctoral degrees in fields relevant to cancer research. Candidates must be citizens or permanent residents of the United States or Canada.
Date(s) Application is Due: Jul 20
Samples: Joseph Agyin, Ph.D., UT Health Science Center, San Antonio, Texas; Wellington Ayensu, M.D., Jackson State University, Jackson, Mississippi; Rebecca Hartley, Ph.D., University of New Mexico, Albuquerque, New Mexico.

Contact: Mona E. Shater, Staff Assistant; (215) 440-9300 or (267) 646-0654; fax (215) 440-9372; mona.shater@aacr.org or micr@aacr.org
Internet: http://www.aacr.org/home/scientists/meetings—workshops/special-conferences/metabolism-and-cancer/financial-support-for-attendance.aspx
Sponsor: American Association for Cancer Research
615 Chestnut Street, 17th Floor
Philadelphia, PA 19106-4404

AACR Minority Scholar in Cancer Research Awards **99**
The AACR administers this award program that is supported by a grant from the Comprehensive Minority Biomedical Branch of the National Cancer Institute to provide funds for participation of meritorious minority scientists in the AACR Special Conferences Series (U.S. domestic conferences only). These awards are intended to enhance the education and training of minority researchers and to increase the visibility and recognition of minorities involved in cancer research.
Requirements: Candidates must be full-time graduate students, medical students, residents, clinical or postdoctoral fellows, or junior faculty members who are either engaged in cancer research or who have the training and potential to make contributions to this field. Candidates must also be citizens or permanent residents of the United States or Canada.
Restrictions: This program applies only to racial/ethnic minority groups identified by the National Cancer Institute as being traditionally underrepresented in cancer and biomedical research. These groups include African Americans/Blacks, Alaskan Natives, Hispanic Americans, Native Americans, and Native Pacific Islanders. Only citizens of the United States or Canada or scientists who are permanent residents of these countries may receive one of these awards.
Date(s) Application is Due: Jul 20
Samples: Irene Alvarez, B.S., University of Arizona, Tucson, Arizona; Cristina M. Contreras, B.S., UT Southwestern Medical Center, Dallas, Texas; Vivian A. Galacia, B.S., University of Southern California, Los Angeles, California.
Contact: Mona E. Shater, Staff Assistant; (215) 440-9300 or (267) 646-0654; fax (215) 440-9372; mona.shater@aacr.org or micr@aacr.org
Internet: http://www.aacr.org/home/scientists/meetings—workshops/special-conferences/metabolism-and-cancer/financial-support-for-attendance.aspx
Sponsor: American Association for Cancer Research
615 Chestnut Street, 17th Floor
Philadelphia, PA 19106-4404

AACR Outstanding Investigator Award for Breast Cancer Research **100**
The AACR Outstanding Investigator Award for Breast Cancer Research, funded by Susan G. Komen for the Cure, will recognize an investigator of no more than 50 years of age whose novel and significant work has had or may have a far-reaching impact on the etiology, detection, diagnosis, treatment, or prevention of breast cancer. Such work may involve any discipline across the continuum of biomedical research, including basic, translational, clinical, and epidemiological studies. The recipient of the Award will receive a $10,000 honorarium and present a 25-minute lecture at the Annual San Antonio Breast Cancer Symposium. Nominations may be made by any scientist, whether an AACR member or nonmember, who is now or has been affiliated with any institution involved in cancer research, cancer medicine, or cancer-related biomedical science. Candidates may not nominate themselves. Nominations must be submitted online at https://proposalcentral.altum.com, no later than 4:00 p.m. United States Eastern Time on May 15th.
Requirements: All cancer researchers who are affiliated with any institution involved in cancer research, cancer medicine, or cancer-related biomedical science anywhere in the world may be nominated. Such institutions include those in academia, industry, or government. Candidates must be no more than 50 years of age at the time the Award is received.
Restrictions: Institutions or organizations are not eligible for the Award. Paper nominations will not be accepted.
Date(s) Application is Due: May 15
Amount of Grant: 10,000 U.S.D
Contact: Monique P. Eversley, Program Associate; (267) 646-0576 or (215) 440-9300, ext. 1400; fax (215) 440-9372; monique.eversley@aacr.org or awards@aacr.org
Internet: http://www.aacr.org/home/scientists/scientific-achievement-awards/outstanding-investigator-award.aspx
Sponsor: American Association for Cancer Research
615 Chestnut Street, 17th Floor
Philadelphia, PA 19106-4404

AACR Princess Takamatsu Memorial Lectureship **101**
This Lectureship will recognize an individual scientist whose novel and significant work has had or may have a far-reaching impact on the detection, diagnosis, treatment, or prevention of cancer, and who embodies the dedication of the Princess to multinational collaborations. The Lecturer will receive an unrestricted cash award of $10,000, support to attend the AACR Annual Meeting, and a commemorative item serving as tangible witness to the singular honor of his/her selection. Nominations must be submitted electronically no later than 4:00 p.m. U.S. Eastern Time on November 16.
Requirements: Candidacy is open to all cancer researchers who are affiliated with any institution involved in cancer research, cancer medicine, or cancer-related biomedical science anywhere in the world. Such institutions include those in academia, industry, or government.
Restrictions: Institutions or organizations are not eligible for the Lectureship. Paper nominations will not be accepted.
Date(s) Application is Due: Nov 16
Amount of Grant: 10,000 U.S.D
Contact: Monique P. Eversley, Staff Associate; (267) 646-0576 or (215) 440-9300, ext. 1400; fax (215) 440-9372; monique.eversley@aacr.org or awards@aacr.org
Internet: http://www.aacr.org/home/scientists/scientific-achievement-awards/takamatsu-lectureship.aspx
Sponsor: American Association for Cancer Research
615 Chestnut Street, 17th Floor
Philadelphia, PA 19106-4404

AACR Richard and Hinda Rosenthal Memorial Awards **102**
AACR and the Rosenthal Family Foundation established this Award in 1977 to recognize research that has made, or promises to soon make, a notable contribution to improved clinical care in the field of cancer. In its desire to honor and provide incentive to investigators relatively early in their careers, the Foundation has stipulated that recipients not be more than 50 years of age at the time the Award is received. The winner of the AACR Richard and Hinda Rosenthal Memorial Award will receive an honorarium of $10,000, give a 50-minute lecture during the AACR Annual Meeting and be given support for the winner and a guest to attend the Annual Meeting.
Requirements: Candidacy is open to cancer researchers, engaged in the practice of medicine, who are affiliated with any institution involved in cancer research, cancer medicine, or cancer-related biomedical science. Such institutions include those in academia, industry, or government.
Restrictions: Candidates may not nominate themselves. Nominations must be submitted online at https://proposalcentral.altum.com, no later than 4:00 p.m. United States Eastern Time on October 15th. Paper nominations will not be accepted. Institutions or organizations are not eligible for the Award.
Date(s) Application is Due: Oct 15
Amount of Grant: 10,000 U.S.D
Contact: Monique P. Eversley, Staff Associate; (267) 646-0576 or (215) 440-9300, ext. 1400; fax (215) 440-9372; monique.eversley@aacr.org or awards@aacr.org
Internet: http://www.aacr.org/home/scientists/scientific-achievement-awards/clowes-award.aspx
Sponsor: American Association for Cancer Research
615 Chestnut Street, 17th Floor
Philadelphia, PA 19106-4404

AACR Scholar-in-Training Awards **103**
Scholar-in-Training Awards provide financial support for early-career scientists to attend the AACR Annual Meeting. Since its inception in 1986, the program has provided over 3,200 grants and has received support from more than 30 cancer research foundations, corporations, individuals, and other organizations dedicated to the fight against cancer. A stipend (from $400-$2,000 depending on the geographic location of the recipient or the type of award) will be presented to recipients on-site. Recipients may pick up their a check and welcome package at the Associate Member Resource and Career Center.
Requirements: Eligible candidates are graduate students, medical students and residents, clinical fellows or equivalent, and postdoctoral fellows. Eligible candidates must be the presenter of a proffered paper. Eligible candidates may be traveling within the U.S. or from abroad. Some awards are specifically designated for those traveling from Asia, Europe, and countries with emerging economies.
Restrictions: Employees or subcontractors of private industry are not eligible.
Date(s) Application is Due: Jul 20
Amount of Grant: 400 - 2,000 U.S.D
Contact: Mona E. Shater, Staff Assistant; (215) 440-9300 or (267) 646-0654; fax (215) 440-9372; mona.shater@aacr.org
Internet: http://www.aacr.org/home/scientists/meetings—workshops/special-conferences/metabolism-and-cancer/financial-support-for-attendance.aspx
Sponsor: American Association for Cancer Research
615 Chestnut Street, 17th Floor
Philadelphia, PA 19106-4404

AACR Scholar-in-Training Awards - Special Conferences **104**
The AACR offers Scholar-in-Training Awards to enhance the education and training of early career scientists by providing financial support for their attendance at AACR Special Conferences. These awards are provided to offset a portion of the registration, travel, and subsistence expenses incurred in attending these conferences. A stipend (typically $400-$1,400) will be presented to award recipients at the conference.
Requirements: All graduate and medical students, postdoctoral fellows, and physicians-in-training who submit abstracts may be considered for a Scholar-in-Training Award.
Restrictions: Employees or subcontractors of private industry are not eligible.
Amount of Grant: 400 - 1,400 U.S.D
Contact: Mona E. Shater, Staff Assistant; (215) 440-9300 or (267) 646-0654; fax (215) 440-9372; mona.shater@aacr.org or programs@aacr.org
Internet: http://www.aacr.org/home/scientists/meetings—workshops/travel-grants/scholar-in-training-awards—-special-conferences.aspx
Sponsor: American Association for Cancer Research
615 Chestnut Street, 17th Floor
Philadelphia, PA 19106-4404

AACR Team Science Award **105**
The Award has been established by the AACR and Eli Lilly and Company to acknowledge and catalyze the growing importance of interdisciplinary teams to the understanding of cancer and/or the translation of research discoveries into clinical cancer applications. In addition, through the presentation of this Award, the AACR and Eli Lilly seek to effect change within the traditional cancer research culture by recognizing those institutions that value and foster interdisciplinary team science. These institutions will have demonstrated their support of a team science environment by creating mechanisms

to enhance the required infrastructure, such as through pilot funding, technology transfer offices, shared resources, etc., and by presenting awards, honors, appointments, and promotions to those who participate in interdisciplinary teams. Nominations for the 2009 Award will be accepted until November 16.
Requirements: For the purpose of this Award, a team is defined as a group of individuals representing interdisciplinary expertise, each of whom have made substantive and quantifiable contributions to the research being recognized. Team members may be working within the same institution or at several institutions. Candidacy is open to all cancer researchers who are affiliated with any institution involved in cancer research, cancer medicine, or cancer-related biomedical science anywhere in the world. Such institutions include those in academia, industry, or government.
Date(s) Application is Due: Nov 16
Amount of Grant: 50,000 U.S.D
Contact: Monique P. Eversley, Program Associate; (267) 646-0576 or (215) 440-9300, ext. 1400; fax (215) 440-9372; monique.eversley@aacr.org or awards@aacr.org
Internet: http://www.aacr.org/home/scientists/scientific-achievement-awards/team-science-award.aspx
Sponsor: American Association of Cancer Research
615 Chestnut Street, 17th Floor
Philadelphia, PA 19106-4404

AACR Thomas J. Bardos Science Education Awards 106
AACR-Thomas J. Bardos Science Education Awards for Undergraduate Students are available to full-time third-year undergraduate students majoring in science. The purpose of these awards is to inspire young science students to enter the field of cancer research. The award consists of a waiver of registration fees for two annual AACR meetings and $1500/year for research and travel expenses. The award is for a two-year period.
Requirements: Candidates must be full-time, third year undergraduate students majoring in science. Applications from students who are not yet committed to cancer research are welcome.
Date(s) Application is Due: Dec 7
Amount of Grant: 1,500 U.S.D
Contact: Mona E. Shater, Staff Assistant; (215) 440-9300 or (267) 646-0654; fax (215) 440-9412; mona.shater@aacr.org or scienceeducation@aacr.org
Internet: http://www.aacr.org/home/scientists/meetings—workshops/travel-grants/aacr-thomas-j-bardos-science-education-awards-for-undergraduate-students-.aspx
Sponsor: American Association for Cancer Research
615 Chestnut Street, 17th Floor
Philadelphia, PA 19106-4404

AAFA Investigator Research Grants 107
AAFA is dedicated to finding the causes, new treatments and cures for asthma and allergic diseases. AAFA offers a program that sponsors seed grants for investigators wishing to explore new areas of scientific merit related to asthma and allergic diseases. Support has enabled 80% of the researchers who have received the AAFA research grant to win additional research funding from other sources totaling more than $18 million. The Asthma and Allergy of Foundation of America is the only patient organization in the United States that funds research grants for both asthma and allergies. At this time, the application process for grants is through an NIH referral system.
Samples: Stephen C. Dreskin, MD, PhD, University of Colorado, Boulder, Colorado; Anthony A. Horner, MD, University of California, San Diego, CA—Analysis of airway; David Lewis, MD, Stanford University School of Medicine—Mechanisms of viral induced asthma.
Contact: Mary Brasler, Program Director; (202) 466-7643, ext. 238; mbrasler@aafa.org
Internet: http://www.aafa.org/display.cfm?id=6&sub=64
Sponsor: Asthma and Allergy Foundation of America
8201 Corporate Drive, Suite 100
Landover, MD 20785

AAFCS International Graduate Fellowships 108
The Association encourages graduate study in family and consumer sciences and its subspecialties, such as textiles and clothing, nutrition, institutional management/food service systems administration, consumer studies, Cooperative Extension, and family and consumer sciences communication, education, and administration. International graduate fellowships are awarded to individuals who have exhibited the potential to make contributions to the family and consumer sciences profession.
Requirements: Applicants must be citizens or permanent residents of the United States. Applicant must show clearly defined plans for full-time graduate or undergraduate study during the time for which the fellowship is awarded. Each request for appropriate application form must be accompanied by an application fee of $40.
Contact: Fellowships and Awards Committee; (703) 706-4600; fax (703) 706-4663; staff@aafcs.org
Internet: http://www.aafcs.org/programs/scholarships.htm
Sponsor: American Association of Family and Consumer Sciences
400 N Columbus Street, Suite 202
Alexandria, VA 22314

AAFCS National Graduate Fellowships 109
AAFCS annually awards national graduate fellowships to support graduate study in areas such as family and consumer sciences, nutrition, textiles, and home economics. The fellowships are awarded to individuals who have exhibited the potential for making contributions to these professions. It is recommended that applicants first complete at least one year of professional family and consumer science experience by the time of application.
Requirements: Applicants must be citizens or permanent residents of the United States. Applicant must show clearly defined plans for full-time graduate or undergraduate study during the time for which the fellowship is awarded. Each request for appropriate application form must be accompanied by an application fee of $40.
Date(s) Application is Due: Jan 7
Amount of Grant: Up to 5,000 U.S.D
Contact: Fellowships and Awards Committee; (703) 706-4600; fax (703) 706-4663; staff@aafcs.org
Internet: http://www.aafcs.org/programs/fellowshipseven.html
Sponsor: American Association of Family and Consumer Sciences
400 N Columbus Street, Suite 202
Alexandria, VA 22314

AAFCS National Undergraduate Scholarships 110
The Association encourages undergraduate study in family and consumer sciences and its subspecialties, such as textiles and clothing, nutrition, institutional management/food service systems administration, consumer studies, Cooperative Extension, and family and consumer sciences communication, education, and administration. Scholarships are awarded to individuals who have exhibited the potential to make contributions to the family and consumer sciences profession.
Requirements: Applicants must be citizens or permanent residents of the United States. Applicant must show clearly defined plans for full-time graduate or undergraduate study during the time for which the fellowship is awarded. Each request for appropriate application form must be accompanied by an application fee of $40.
Date(s) Application is Due: Jan 7
Amount of Grant: 5,000 U.S.D
Contact: Scholarships and Awards Committee; (703) 706-4600; fax (703) 706-4663; staff@aafcs.org
Internet: http://www.aafcs.org/programs/scholarships.htm
Sponsor: American Association of Family and Consumer Sciences
400 N Columbus Street, Suite 202
Alexandria, VA 22314

AAFP Foundation Health Literacy State Grants 111
The AAFP Foundation recognizes that nearly half of all American adults (90 million people) have difficulty reading, understanding and acting upon all types of health information. They are often ashamed and devise methods to mask their difficulty or may be reluctant to ask questions for fear of being perceived as ignorant. To help address literacy concerns, the AAFP Foundation has developed the Health Literacy State Grant Awards Program. The purpose of the program is to support collaborative projects that address health literacy concerns at the state and local level. Five awards in the amount $10,000 each will be offered.
Date(s) Application is Due: Mar 26
Amount of Grant: 10,000 U.S.D
Contact: Susie Morantz, Senior Program Manager; (913) 906-6000, ext. 4470 or (800) 274-2237; fax (913) 906-6095; smorantz@aafp.org
Internet: http://www.aafpfoundation.org/x821.xml
Sponsor: American Academy of Family Physicians Foundation
11400 Tomahawk Creek Parkway, Suite 440
Leawood, KS 66211-2672

AAFP Foundation Joint Grants 112
The foundation supports clinical research projects within family practice that ultimately will result in improved patient care. This program includes a research grant stimulation program that is designed to stimulate research in family practice by supporting small and first-time research projects. Longer research proposals must be prepared in accordance with standard guidelines for grant proposals.
Requirements: Eligible to apply are individual family physicians, residents in family practice, and medical students planning to pursue the specialty of family practice; educational and health care institutions or organizations that will use the grant exclusively for programs or projects directly involved with family practice or family medicine; departments of family practice; family practice residency programs; and family practice organizations or associations. Research stimulation proposals may be reviewed at any time during the year depending on availability of funds.
Restrictions: Proposals seeking support for the cost of instituting programs or support for such activities as videotape production, designing curriculum, or implementing a project, etc., will be denied.
Amount of Grant: Up to 30,000 U.S.D
Contact: Susie Morantz, Senior Program Manager; (913) 906-6000, ext. 4470 or (800) 274-2237; fax (913) 906-6095; smorantz@aafp.org
Internet: http://www.aafpfoundation.org/x270.xml
Sponsor: American Academy of Family Physicians Foundation
11400 Tomahawk Creek Parkway, Suite 440
Leawood, KS 66211-2672

AAFP Foundation Pfizer Teacher Development Awards 113
The purpose of the awards is to honor community-based physicians who give up time from their practice to teach family medicine students and/or residents part time. Eligible candidates may serve as preceptors or as volunteer teachers at other sites, including family medicine teaching centers. Guidelines are available online.
Requirements: Applicant must have graduated from an ACGME-approved family practice residency program within the last seven years; have entered family medicine; be a member of the American Academy of Family Physicians; have entered or plan to enter

part-time family medicine teaching (not less than four hours per month and no more than 32 hours per month—averaged over a year); and teach voluntarily or receive no more than $18,000 compensation for the educational time devoted to residents and/or students.
Amount of Grant: 2,000 U.S.D
Contact: Sondra Goodman, Grants Program Manager; (913) 906-6000, ext. 4457 or (800) 274-2237; fax (913) 906-6095; sgoodman@aafp.org
Internet: http://www.aafpfoundation.org/x269.xml
Sponsor: American Academy of Family Physicians Foundation
11400 Tomahawk Creek Parkway, Suite 440
Leawood, KS 66211-2672

AAFP Foundation Wyeth Immunization Awards **114**
This program offers family medicine residency programs the opportunity to receive recognition for identifying and developing creative solutions to overcoming barriers to childhood immunizations, thus increasing immunization rates, and to share the best practices that have benefited the communities in which they live. Applications are accepted annually for one of the following award track categories: Best Practices Award—overall achievement with systems already in place to overcome immunization barriers and achieve high rates within a certain time frame; Most Improved Award—overcame barriers and other challenges to greatly enhance immunization rates; and Implement New System Grant—residency programs seeking to implement a new system to help increase immunization rates in under-served children. Grant winners will receive: one of 10 monetary awards (four at $5,000 and six at $10,000) determined by rank of application as scored by review panel; an education scholarship for $1,000 to send a resident to the annual AAFP National Conference of Family Medicine Residents and Medical Students; an additional travel scholarship to send a resident to the AAFP Scientific Assembly (top six); and a plaque of recognition.
Requirements: Family medicine residency programs that have increased immunization rates within a defined time frame and can demonstrate the best practices implemented to achieve the higher compliance are eligible.
Date(s) Application is Due: Apr 21
Amount of Grant: 1,000 - 10,000 U.S.D
Contact: Sondra Goodman, Grants Program Manager; (913) 906-6000, ext. 4457 or (800) 274-2237; fax (913) 906-6095; sgoodman@aafp.org
Internet: http://www.aafpfoundation.org/x655.xml
Sponsor: American Academy of Family Physicians Foundation
11400 Tomahawk Creek Parkway, Suite 440
Leawood, KS 66211-2672

AAFP Research Stimulation Grants **115**
This Program provides relatively small grants to stimulate research in family practice in the areas of preliminary efforts leading to a larger research project, pilot projects, and data collection for a larger research project. The applicant must include a discussion of how the proposed project is anticipated to lead to a larger project. Grant applications will be accepted twice a year: by March 1 and September 1.
Requirements: Eligible to apply are individual family physicians, family practice/medicine organizations or associations, residents in family practice, departments of family medicine, and educational and health care institutions directly involved in family practice/medicine.
Date(s) Application is Due: Mar 1; Sep 1
Amount of Grant: 7,500 U.S.D
Contact: Susie Morantz, Senior Program Manager; (913) 906-6000, ext. 4470 or (800) 274-2237; fax (913) 906-6095; smorantz@aafp.org
Internet: http://www.aafpfoundation.org/x446.xml
Sponsor: American Academy of Family Physicians Foundation
11400 Tomahawk Creek Parkway, Suite 440
Leawood, KS 66211-2672

AAFPRS Ben Shuster Memorial Award **116**
This award is given annually for the best paper based on clinical work or research in the field of facial plastic surgery by a resident or fellow in training. The paper must have been delivered at a national meeting (or its equivalent) within the preceding two years prior to the February 1 date of submission. Studies prepared during the first year after completion of residency training will be considered provided that the research was conducted during the author's residency training program or fellowship. A certificate and an award of $1,000 are presented.
Requirements: The competition is open to U.S. or Canadian residents or fellows in otolaryngology who are members of AAFPRS. Each entrant must be the sole or senior author and an AAFPRS member.
Date(s) Application is Due: Feb 1
Amount of Grant: 1,000 U.S.D
Contact: Awards Coordinator; (703) 299-9291; fax (703) 299-8898; info@aafprs.org
Internet: http://www.aafprs.org/physician/awards_grants/awards.html
Sponsor: American Academy of Facial, Plastic, and Reconstructive Surgery
310 S Henry Street
Alexandria, VA 22314

AAFPRS Bernstein Grant **117**
The purpose of the grant is to encourage original research projects that will advance facial plastic and reconstructive surgery. The grant may be used as seed money for research projects. Applicants should submit an application, a detailed proposal, the investigator's curriculum vita, and letters from the appropriate institutional review committees certifying institutional conformity to the U.S. government guidelines for human and animal experiments. The recipient should complete his or her work within three (3) years with interim reports due each year. All research grant monies awarded are intended to fund direct costs related to the research as described in the grant proposal.
Requirements: Applicants should be U.S. or Canadian AAFPRS fellow members. The primary criteria are that the research be original and have direct application to facial plastic and reconstructive surgery.
Date(s) Application is Due: Jan 15
Amount of Grant: 25,000 U.S.D
Contact: Grant Coordinator; (703) 299-9291; fax (703) 299-8898; info@aafprs.org
Internet: http://www.aafprs.org/physician/awards_grants/bernstein_grant.html
Sponsor: American Academy of Facial, Plastic, and Reconstructive Surgery
310 S Henry Street
Alexandria, VA 22314

AAFPRS Fellowships **118**
The Educational and Research Foundation for the American Academy of Facial Plastic and Reconstructive Surgery fellowship program provides postgraduate training in facial plastic surgery. The objectives of the fellowship program are to: provide an outstanding academic opportunity for the acquisition of specialized knowledge and skills in facial plastic surgery; develop trained specialists who will contribute to the ongoing development of facial plastic and reconstructive surgery; foster development of facial plastic and reconstructive surgery educators, especially in residency programs; and encourage the development of new skills and knowledge in facial plastic and reconstructive surgery through basic research and clinical trials. Eligible physicians may apply for 40 positions available each year.
Requirements: Applicants to the Fellowship Program must be physicians who are in or have completed an otolaryngology or plastic surgery residency program accredited by the Accreditation Council for Graduate Medical Education (ACGME) or Royal College of Physicians and Surgeons of Canada (RCPSC) or board-certified in otolaryngology-head and neck surgery or plastic surgery. Applicants must be members of the AAFPRS before submitting an application.
Restrictions: Applicants should not be full-time faculty members holding the rank of assistant professor or higher at the institution where the fellowship will take place.
Date(s) Application is Due: Feb 1
Contact: Fellowship Coordinator; (703) 299-9291; fax (703) 299-8898; info@aafprs.org or fsanders@aafprs.org
Internet: http://www.aafprs.org/physician/benefits/ben_fellow.html
Sponsor: American Academy of Facial, Plastic, and Reconstructive Surgery
310 S Henry Street
Alexandria, VA 22314

AAFPRS Investigator Development Grant **119**
The purpose of the Investigator Development Grant is to support the work of a young faculty member in facial plastic surgery conducting significant clinical or laboratory research and involved in the training of resident surgeons in research. One $15,000 grant may be awarded each year upon receipt of the signed agreement. Recipients should complete their work within three (3) years with interim reports due each year. An extension is possible upon written request to the Research Committee. The original proposal and all other requirements should be submitted to C.O.R.E. no later than January 15 or the next business day thereafter.
Requirements: An applicant must be an AAFPRS member at any level involved in the training of resident surgeons.
Date(s) Application is Due: Jan 15
Amount of Grant: Up to 15,000 U.S.D
Contact: Grant Coordinator; (703) 299-9291; fax (703) 299-8898; info@aafprs.org
Internet: http://www.aafprs.org/physician/awards_grants/investigator_grant.html
Sponsor: American Academy of Facial, Plastic, and Reconstructive Surgery
310 S Henry Street
Alexandria, VA 22314

AAFPRS Ira J. Tresley Research Award **120**
This annual award is given to a U.S. or Canadian academy member for the best paper based on any research study in the field of facial plastic surgery. The competition is open to any U.S. or Canadian physician who is an AAFPRS member and who has been board certified for at least three years. The paper must have been presented at a national meeting or its equivalent within the preceding two years prior to the February 1 date of submission. A certificate and an award of $1,000 are presented.
Date(s) Application is Due: Feb 1
Amount of Grant: 1,000 U.S.D
Contact: Awards Coordinator; (703) 299-9291; fax (703) 299-8898; info@aafprs.org
Internet: http://www.aafprs.org/physician/awards_grants/awards.html
Sponsor: American Academy of Facial, Plastic, and Reconstructive Surgery
310 S Henry Street
Alexandria, VA 22314

AAFPRS John Orlando Roe Award **121**
This award, named for John Orlando Roe, the surgeon who accomplished the first rhinoplasty in 1887, includes a certificate and an award of $1,000 to be presented each year to an AAFPRS Fellow in an AAFPRS Foundation Fellowship Program who submits the best clinical research paper written during the current fellowship year.
Date(s) Application is Due: Feb 1
Amount of Grant: 1,000 U.S.D
Contact: Awards Coordinator; (703) 299-9291; fax (703) 299-8898; info@aafprs.org

Internet: http://www.aafprs.org/physician/awards_grants/awards.html
Sponsor: American Academy of Facial, Plastic, and Reconstructive Surgery
310 S Henry Street
Alexandria, VA 22314

AAFPRS Leslie Bernstein Resident Research Grants 122
Two resident research grants, tenable in the United States or Canada, may be awarded each year to stimulate resident research in facial plastic surgery on projects that are well conceived and scientifically valid. Residents are encouraged to enter early in their training so that their applications may be revised and resubmitted if not accepted the first time. This award may be integrated with other funding to complete a project. Interested residents should write for entry guidelines.
Requirements: U.S. and Canadian residents at any level who are AAFPRS members are eligible to apply, even if the research work will be done during their fellowship year. Grant recipients must complete their work in one (1) to two (2) years with interim reports due each year.
Date(s) Application is Due: Jan 15; Dec 15
Amount of Grant: 5,000 U.S.D
Contact: Grant Coordinator; (703) 299-9291; fax (703) 299-8898; info@aafprs.org
Internet: http://www.aafprs.org/physician/awards_grants/grants.html
Sponsor: American Academy of Facial, Plastic, and Reconstructive Surgery
310 S Henry Street
Alexandria, VA 22314

AAFPRS Residency Travel Awards 123
Two Residency Travel Awards may be given each year for two outstanding papers in facial plastic and reconstructive surgery primarily authored by a resident or medical student in training. The paper must be submitted by February 1 for consideration, which will be presented at the Annual Fall Meeting. Entries should conform to the guidelines of the Archives of Facial Plastic Surgery, the official journal of the AAFPRS. Papers submitted will be judged anonymously. Only the cover page should contain the applicant's name or institutional identification.
Date(s) Application is Due: Feb 1
Contact: Awards Coordinator; (703) 299-9291; fax (703) 299-8898; info@aafprs.org
Internet: http://www.aafprs.org/physician/awards_grants/awards.html
Sponsor: American Academy of Facial, Plastic, and Reconstructive Surgery
310 S Henry Street
Alexandria, VA 22314

AAFPRS Sir Harold Delf Gillies Award 124
This award is named for Sir Harold Delf Gillies, a British otolaryngologist who in September 1917 described the tubed pedicle flap. Dr. Gillies frequently visited the U.S. and lectured widely to surgeons of various specialties and was given the title Father of Plastic Surgery. A certificate and an award of $1,000 will be presented each year to an AAFPRS Fellow in an AAFPRS Foundation Fellowship Program who submits the best basic science research paper written during the current fellowship year.
Date(s) Application is Due: Feb 1
Amount of Grant: 1,000 U.S.D
Contact: Awards Coordinator; (703) 299-9291; fax (703) 299-8898; info@aafprs.org
Internet: http://www.aafprs.org/physician/awards_grants/awards.html
Sponsor: American Academy of Facial, Plastic, and Reconstructive Surgery
310 S Henry Street
Alexandria, VA 22314

AAHN Grant for Historical Research 125
The American Association for the History of Nursing is offering a research grant of $3,000 for new researchers. (Indirect costs of 8% are also available). It is expected that the research and new materials produced by the award winner will help ensure the growth of scholarly work focused on the history of nursing. The deadline for submission of applications is April 1. The AAHN Panel will make its decision about the award by May 15 and the recipient will be notified by June 1 of each year. Funding will start July 1 of the grant year and last for one year. A no-cost extension may be granted on request.
Requirements: Applicants must be members of AAHN and hold the doctorate. They may be faculty members or independent researchers.
Date(s) Application is Due: Apr 1
Amount of Grant: 3,000 U.S.D
Contact: David Stumph; (303) 422-2685; fax (303) 422-8894; aahn@aahn.org
Internet: http://www.aahn.org/grants.html
Sponsor: American Association of the History of Nursing
10200 W 44th Avenue, Suite 304
Wheat Ridge, CO 80033

AAHN Pre-Doctoral Research Grant 126
This grant is designed to encourage and support graduate training and historical research at the Masters and Doctoral levels. The grant will be in the amount of $2,000. Proposals will focus on a significant question in the history of nursing. Selection criteria include the scholarly merit of the proposal, consideration of the student's preparation for this study, the advisor's qualifications for guiding the study and the project's potential for contributing to scholarship in the field of nursing history.
Requirements: The student must be: enrolled in an accredited masters program or doctoral program; and a member of AAHN. The research advisor must be doctorally prepared with scholarly activity in the field of nursing history and prior experience in guidance of research training.
Date(s) Application is Due: Apr 1
Amount of Grant: 2,000 U.S.D
Contact: David Stumph; (303) 422-2685; fax (303) 422-8894; aahn@aahn.org
Internet: http://www.aahn.org/grants.html
Sponsor: American Association of the History of Nursing
10200 W 44th Avenue, Suite 304
Wheat Ridge, CO 80033

AAHPERD-AAHE Barbara A. Cooley Scholarships 127
The award will be $1,000 per scholarship. Recipients may spend the award in any manner they desire. The awarded scholarship will be recognized at the annual Awards Luncheon at the AAHE/AAHPERD annual meeting. Recipients will also receive a one-year complimentary student membership in AAHE. Additional guidelines and application are available at the AAHPERD website. Application materials must be received by November 15.
Requirements: The award is open to a master's level student who is currently enrolled in a health education program at an accredited college/university in the United States or a U.S. territory. To be eligible, the applicant must have a minimum current overall grade point average of 3.0 on a 4.0 scale.
Restrictions: Prior AAHE scholarship recipients are not eligible.
Date(s) Application is Due: Nov 15
Amount of Grant: 1,000 U.S.D
Contact: Linda Moore; (703) 476-3437; fax (703) 476-6638; aahe@aahperd.org
Internet: http://www.aahperd.org/aahe/events/Barbara-Cooley.cfm
Sponsor: American Alliance for Health, Physical Education, Recreation and Dance
1900 Association Drive
Reston, VA 20191-1599

AAHPERD-AAHE Delbert Oberteuffer Scholarships 128
The award will be $1,500, which may be used at the discretion of the recipient. The scholarship recipient will be recognized at the annual Awards Luncheon at the AAHE/AAHPERD convention. Additional guidelines and application are available at the AAHPERD website. Application materials must be received by November 15.
Requirements: The award is open to a doctoral level student who is currently enrolled in a health education program in the United States or a U.S. territory. To be eligible to apply, you must be an AAHE member and a graduate student at an accredited college/university in a doctoral level course of study leading to preparing you to develop, implement and/or evaluate health education programs for children and youth. The applicant must also have a minimum cumulative grade point average of 3.5 in graduate courses.
Date(s) Application is Due: Nov 15
Amount of Grant: 1,500 U.S.D
Contact: Linda Moore; (703) 476-3437; fax (703) 476-6638; aahe@aahperd.org
Internet: http://www.aahperd.org/aahe/events/awards/Delbert-Oberteuffer.cfm
Sponsor: American Alliance for Health, Physical Education, Recreation and Dance
1900 Association Drive
Reston, VA 20191-1599

AAHPERD-AAHE Undergraduate Scholarships 129
AAHE Undergraduate Scholarships are available to an outstanding student officially recognized as an undergraduate health education major at any four-year university/college. These are one year scholarships.
Requirements: Open to undergraduate students who are enrolled full time in a health education program at a 4-year college or university. Applicants must a GPA of 3.25 or higher and be active in health education professional activities and organizations at their school and/or in their community. They must submit a list of extracurricular service activities; an official transcript; 3 letters of recommendation; and an essay that includes what they hope to accomplish as a health educator (during training and in the future) and attributes and aspirations brought to the field of health education. Financial need is not considered in the selection process.
Contact: Linda Moore; (703) 476-3437; fax (703) 476-6638; aahe@aahperd.org
Internet: http://www.aahperd.org/whatwedo/scholarships/Abernathy_Scholarship.cfm
Sponsor: American Alliance for Health, Physical Education, Recreation and Dance
1900 Association Drive
Reston, VA 20191-1599

AAHPERD Ruth Abernathy Presidential Scholarships 130
The Ruth Abernathy Presidential Scholarship, developed by the past presidents of AAHPERD to honor deserving students, is awarded to three undergraduate students and two graduate students in January of each year. All scholarships are presented at the AAHPERD National Convention & Exposition held in the spring. Undergraduate awards are $1,250 each and graduate awards are $1,750 each. Recipients also receive a complimentary three-year AAHPERD membership.
Requirements: Applicants must be enrolled in a full-time program at the time that the scholarship recipients are selected and must be majoring in a field related to one or more of the disciplines represented by AAHPERD and its associations (i.e., health, physical education, recreation, dance, sport - see the list of interest areas at the website). Applicants will be asked to describe the field of study/degree area and its relationship to the HPERDS-related fields. Application forms containing detailed application criteria and procedures for submission are available at the website.
Restrictions: Individuals in the following medical programs are not eligible: Nursing, Pharmacology, Physical Therapy, Occupational Therapy, Athletic Training, and Acupuncture and Oriental Medicine. Students pursuing teaching licensure, national board certification, or completing prerequisite coursework in preparation for admission to a master's degree program are are also not eligible to apply.

Date(s) Application is Due: Oct 15
Amount of Grant: 1,750 - 1,250 U.S.D
Contact: Deb Callis, Director; (703) 476-3400; dcallis@aahperd.org
Internet: http://www.aahperd.org/whatwedo/scholarships/Abernathy_Scholarship.cfm
Sponsor: American Alliance for Health, Physical Education, Recreation and Dance
1900 Association Drive
Reston, VA 20191-1599

AAHPM Hospice and Palliative Medicine Fellowships 131
The program offers training programs to prepare physicians for careers in academic or community-based palliative medicine. Fellowships provide clinical experience and supervision for palliative medicine in hospitals, homes, hospices, chronic care institutions, and outpatient settings, and include consultational and longitudinal care exposure, as well as training in bereavement care. Some programs also provide opportunities for mentored research or teaching. Most programs begin in July, and clinical training generally lasts one year, allowing for board eligibility upon successful completion. For a complete list of U.S. palliative medicine fellowship training programs, members should visit the web site.
Requirements: Trainees must have already completed residency training in a primary specialty, such as internal or family medicine, or have similar clinical training and experience.
Contact: Fellowships Director; (847) 375-4712; fax (847) 375-6475; info@aahpm.org
Internet: http://www.aahpm.org/fellowship/index.html
Sponsor: American Academy of Hospice and Palliative Medicine
4700 W Lake Avenue
Glenview, IL 60025-1485

AAMC Abraham Flexner Award 132
The Abraham Flexner Award for Distinguished Service to Medical Education is granted annually to an individual for extraordinary contributions to the medical education community. The award was first presented by the Association in 1958 and is the AAMC's most prestigious honor. The Flexner Award recognizes the highest standards in medical education and honors individuals whose impact on medical education is national in scope. Nominations may be made by any faculty or staff member of a medical school or teaching hospital or by any member of an academic society. The Award Review Committee may also solicit nominations.
Requirements: Any individual who has made a significant contribution to academic medicine is eligible for nomination.
Date(s) Application is Due: May 1
Amount of Grant: 10,000 U.S.D
Contact: Sandra D. Gordon, (202) 828-0472; fax (202) 828-1125; sgordon@aamc.org
Internet: http://www.aamc.org/about/awards/flexner.htm
Sponsor: Association of American Medical Colleges
2450 N Street NW
Washington, DC 20037-1127

AAMC Alpha Omega Alpha Robert J. Glaser Distinguished Teacher Awards 133
The award recognize the significant contributions to medical education made by gifted teachers. Up to four awards will be granted each year. Each awardee will receive a $10,000 grant. The awardee's nominating institution will receive $2,500 for teaching activities, and, if the nominating institution has an AOA chapter, the chapter will receive a stipend of $1,000 toward its activities. The awards will be presented during the AAMC Annual Meeting in Washington, D.C.
Requirements: The nomination must be made by the Dean of the nominating institution.
Date(s) Application is Due: May 1
Amount of Grant: 12,500 U.S.D
Contact: Henry M. Sondheimer, (202) 828-0680; fax (202) 828-1125; hsondheimer@aamc.org
Internet: http://www.aamc.org/about/awards/aoa.htm
Sponsor: Association of American Medical Colleges
2450 N Street NW
Washington, DC 20037-1127

AAMC Award for Distinguished Research 134
The Award for Distinguished Research in the Biomedical Sciences honors outstanding biomedical research related to health and disease. The research recognized should have contributed to the substance of medicine. Nominations may be made by anyone on the faculty or staff of a medical school or teaching hospital or by a member of an academic society.
Requirements: The nominee must be an individual who serves on the faculty of an AAMC member medical school or teaching hospital.
Date(s) Application is Due: May 1
Amount of Grant: 5,000 U.S.D
Contact: Sandra D. Gordon, (202) 828-0472; fax (202) 828-1125; sgordon@aamc.org
Internet: http://www.aamc.org/about/awards/research.htm
Sponsor: Association of American Medical Colleges
2450 N Street NW
Washington, DC 20037-1127

AAMC Caring for Community Grants 135
The association, with the support of the Pfizer Medical Humanities Initiative, conducts an institutional grant program to encourage the development of student-initiated services and programs to the community. Eligible programs may range from those that promote awareness about sexually transmitted diseases, to vaccination and literacy programs, to any program that fulfills an unmet need within the community. Grant awards will also be offered to eligible service programs that are currently underway. Up to 10 grant awards are made each year. Program requirements and applications are available online.
Requirements: Medical schools are eligible to receive support for community service-oriented projects in which they explore new ways to serve their local communities.
Date(s) Application is Due: Mar 12
Amount of Grant: 2,000 - 15,000 U.S.D
Contact: Ally Anderson, (202) 828-0682; fax (202) 828-1125; aanderson@aamc.org
Internet: http://www.aamc.org/about/awards/cfc/start.htm
Sponsor: Association of American Medical Colleges
2450 N Street NW
Washington, DC 20037-1127

AAMC David E. Rogers Award 136
The Award honors Dr. David Rogers, a former president of the Foundation and an exemplar of academic medicine's commitment to meeting the health care needs of our country. The award is granted annually to a member of a medical school faculty who has made major contributions to improving the health and health care of the American people. Presentation of the award and a will be made at the association's annual meeting. Nomination guidelines are available online.
Requirements: Nominations may be made by any faculty or staff member of a medical school or teaching hospital or by any member of an academic society. The award is limited to an individual who has spent the majority of his or her career in academic medicine in the United States.
Date(s) Application is Due: May 1
Amount of Grant: 10,000 U.S.D
Contact: Sandra D. Gordon, (202) 828-0472; fax (202) 828-1125; sgordon@aamc.org
Internet: http://www.aamc.org/about/awards/rogers.htm
Sponsor: Association of American Medical Colleges
2450 N Street NW
Washington, DC 20037-1127

AAMC Herbert W. Nickens Award 137
The annual award is given to an individual who has made outstanding contributions to promoting justice in medical education and health care for people in the United States. The recipient receives a monetary award and presents the Nickens Memorial Lecture at the association's annual meeting. Nomination guidelines are available online.
Requirements: Nominees may come from a wide range of fields including medicine, public health, education, law, nursing, and the social sciences.
Date(s) Application is Due: May 1
Amount of Grant: 10,000 U.S.D
Contact: Juan Amador, (202) 862-6149; fax (202) 828-1125; jamador@aamc.org
Internet: http://www.aamc.org/about/awards/nickensaward.htm
Sponsor: Association of American Medical Colleges
2450 N Street NW
Washington, DC 20037-1127

AAMC Herbert W. Nickens Faculty Fellowship 138
The award recognizes an outstanding junior faculty member who has demonstrated leadership in the United States in addressing inequities in medical education and health care; demonstrated efforts in addressing educational, societal, and health care needs of minorities; and is committed to a career in academic medicine. Funding for the fellowship begins on the year after the fellowship is awarded, and it can be used over a two year period. The recipient is required to submit final narrative and financial reports. Nomination guidelines are available onnline.
Requirements: A medical school may nominate one faculty member for this award. A candidate must: be a U.S. citizen or permanent resident; hold the rank of full-time assistant professor in a LCME-accredited U.S. medical school department; have held a faculty position for no more than three years; have received only one appointment as assistant professor; and hold a M.D., Ph.D., or have earned another doctoral degree.
Date(s) Application is Due: May 1
Amount of Grant: 15,000 U.S.D
Contact: Juan Amador, (202) 862-6149; fax (202) 828-1125; jamador@aamc.org or nickensawards@aamc.org
Internet: http://www.aamc.org/about/awards/nickensfellowships.htm
Sponsor: Association of American Medical Colleges
2450 N Street NW
Washington, DC 20037-1127

AAMC Herbert W. Nickens Medical Student Scholarships 139
The association makes scholarship awards that are given to outstanding third-year medical students who have demonstrated leadership in efforts to eliminate inequities in medical education and health care, and who have demonstrated leadership efforts in addressing educational, societal, and health care needs of minorities in the United States. Each recipient receives a scholarship and an award certificate. Institutions that have a pass/fail grading system and do not assign honor grades should fully discuss the applicant's academic accomplishments in the nomination letter. Nomination guidelines are available online.
Requirements: A medical school may nominate one student. A candidate must be: a U.S. citizen or permanent resident; and entering the third year of study in a LCME-accredited U.S. medical school in the fall semester. Students enrolled in combined degree programs (such as M.D./Ph.D.) are eligible when they are entering their third year of medical school.
Date(s) Application is Due: May 1
Amount of Grant: 5,000 U.S.D

Samples: Nehkonti Adams, Eastern Virginia Medical School; Dora Cristina Castaneda, Stanford University School of Medicine; Katherine Leila Neuhausen, Emory University School of Medicine.
Contact: Juan Amador, (202) 862-6149 or (202) 828-0400; fax (202) 828-1125; jamador@aamc.org or nickensawards@aamc.org
Internet: http://www.aamc.org/about/awards/nickensscholarships.htm
Sponsor: Association of American Medical Colleges
2450 N Street NW
Washington, DC 20037-1127

AAMC Humanism in Medicine Award 140
The award recognizes a medical school faculty physician who exemplifies the qualities of a caring and compassionate mentor in the teaching and advising of medical students. The goal of the award is to emphasize and reinforce among medical school faculty and students the importance of humanistic qualities in physicians, and to enhance interactions between medical school faculty and students. Nominations and award information were sent to medical school's organization of student representatives and student affairs deans. The award will be presented at the association's annual meeting.
Requirements: Each medical school in the United States may nominate one physician faculty member.
Date(s) Application is Due: Apr 16
Amount of Grant: Up to 5,000 U.S.D
Samples: Yasmin S. Meah, M.D., Assistant Professor of Medical Education and Medicine, The Mount Sinai Visting Doctors Program, Mount Sinai School of Medicine; Robert J. Paeglow, M.D., Albany Medical College; Melissa A. Warfield, M.D., Eastern Virginia Medical School (EVMS); Sharad Jain, M.D., University of California, San Francisco School of Medicine.
Contact: Denine Hales, (202) 828-0681; fax (202) 828-1125; dhales@aamc.org
Internet: http://www.aamc.org/about/awards/humanism.htm
Sponsor: Association of American Medical Colleges
2450 N Street NW
Washington, DC 20037-1127

AANS Neurosurgery Research and Education Foundation/Spine Section Young Clinician Investigator Award 141
Established by the Neurosurgery Research and Education Foundation (NREF) of the American Association of Neurological Surgeons (AANS), the Joint Section on Disorders of the Spine and Peripheral Nerves Award grants support to young faculty who are pursuing careers in the area of spinal cord, vertebral column or peripheral nerve disorders. The purpose of the award is to fund pilot studies that provide preliminary data used to strengthen applications for more permanent funding from other sources. The one-year award totals $40,000.
Requirements: Applicants must be physicians or neurosurgeons who are full-time faculty in North American teaching institutions and in the early years of their careers.
Restrictions: Applicant may not accept another award for the same project during the same time period. No more than one award per year will be awarded to the same institution.
Date(s) Application is Due: Oct 31
Contact: Michele S. Gregory, Director of Development; (847) 378-0540 or (847) 378-0500; fax (847) 378-0600; msg@aans.org or info@aans.org
Internet: http://www.aans.org/research/fellowship/nref_y.asp
Sponsor: Research Foundation of the American Association of Neurological Surgeons
5550 Meadowbrook Drive
Rolling Meadows, IL 60008-3852

AANS Neurosurgery Research and Education Foundation Young Clinician Investigator Award 142
This one-year, renewable award is available to North American neurosurgeons who are full-time faculty in teaching institutions and are in the early years of their careers. The purpose of the program is to fund pilot studies that could provide preliminary data to be used for strengthening applications for more permanent funding from other sources. Applications related to any field of neurosurgery are encouraged. The NREF encourages both patient-oriented clinical research and basic science research projects. The one-year award totals $40,000.
Restrictions: Applicant may not accept another award for the same project during the same time period. No more than one award per year will be awarded to the same institution.
Date(s) Application is Due: Oct 31
Amount of Grant: 40,000 U.S.D
Contact: Michele S. Gregory, Director of Development; (847) 378-0540 or (847) 378-0500; fax (847) 378-0600; msg@aans.org or info@aans.org
Internet: http://www.aans.org/research/fellowship/nref_y.asp
Sponsor: Research Foundation of the American Association of Neurological Surgeons
5550 Meadowbrook Drive
Rolling Meadows, IL 60008-3852

AAO-HNSF CORE Research Grants 143
CORE Research Grants are available in three different categories: Research Project Grants—an independent or assisted project using existing skills and resources, and is hypothesis-based, with specific aims and timetable; Research Training Grants—a research project that is designed to help you refine existing research skill set or acquire new skills, and requires mentor and evidence of training process; and Career Development Grants—a formal, mentored period of research combined with coursework, designed to help researcher become independent researcher. Applicants should complete a Letter of Intent by December 15th, using the online application system, and should then submit the completed application before 12 o'clock midnight EST on January 15.
Date(s) Application is Due: Jan 15; Dec 15
Contact: Stephanie L. Jones; (703) 519-1586 or (703) 836-4444; sljones@entnet.org
Internet: http://www.entnet.org/EducationAndResearch/coreGrants.cfm
Sponsor: American Academy of Otolaryngology-Head and Neck Surgery Foundation
1650 Diagonal Road
Alexandria, VA 22314-2857

AAO-HNSF Health Services Research Grants 144
The purpose of the Health Services Research Grant is to foster research that will improve the effectiveness and appropriateness of medical practice. Projects supported under this program will develop and disseminate scientific information on the effects of otolaryngology services and procedures on patients' survival, health status, functional capacity, and quality of life. The award is intended to promote increased participation by otolaryngologists in the rapidly expanding area of health services research. Proposed projects may be related to any area of otolaryngology-head and neck surgery, but must have direct or potential clinical significance for patients seen by otolaryngologist-head and neck surgeons. Grants range up to $10,000 maximum, are for one year, and are not renewable. Projects must be completed within two years of the award date; no-cost extensions are available upon written request. Letters of intent must be received by December 15, and completed applications must be submitted by January 15.
Requirements: Any otolaryngologist in the United States or Canada is eligible to apply for the Health Services Research Grant. Applicants may be independent practitioners, residents or fellows in an approved training program, or practitioners affiliated with academic or similar institutions. All applicants must be members in good standing of AAO-HNSF; Associate and Corresponding Members are not eligible to apply as Principal Investigator, but may participate actively in the proposed project.
Date(s) Application is Due: Jan 15; Dec 15
Amount of Grant: Up to 10,000 U.S.D
Contact: Stephanie L. Jones; (703) 519-1586 or (703) 836-4444; sljones@entnet.org
Internet: http://www.entnet.org/EducationAndResearch/coreGrants.cfm
Sponsor: American Academy of Otolaryngology-Head and Neck Surgery Foundation
1650 Diagonal Road
Alexandria, VA 22314-2857

AAO-HNSF Maureen Hannley Research Training Awards 145
The purpose of the Maureen Hannley Research Training Award program is to foster the acquisition of contemporary basic or clinical research skills among new full-time academic otolaryngologist-head and neck surgeons. The award is intended as a preliminary step in clinical investigator career development and is expected to facilitate the recipient's preparation of a more comprehensive individualized research training plan suitable for submission to one of the National Institutes of Health Clinical Investigator Career Development Award (CIA) programs (K series). Grants range up to $15,000 maximum. The award is one year, and is non-renewable. Projects must be completed within two years of the award date; no-cost extensions are available upon written request. Letters of intent must be received by December 15, and completed applications must be submitted by January 15.
Requirements: Applicants in the United States or Canada and members in good standing of the AAO-HNSF are welcome. Applicants must have demonstrated potential for excellence in research and teaching and a serious commitment to an academic research career in otolaryngology-head and neck surgery. Priority will be given to senior residents, fellows or faculty who have completed residencies or fellowships within two years of the application receipt date. All candidates must be sponsored by the Chairperson of his/her Division or Department and by an official representative of the institution which would administer the Award and in whose name the application is formally submitted.
Date(s) Application is Due: Jan 15; Dec 15
Amount of Grant: Up to 15,000 U.S.D
Contact: Stephanie L. Jones; (703) 519-1586 or (703) 836-4444; sljones@entnet.org
Internet: http://www.entnet.org/EducationAndResearch/coreGrants.cfm
Sponsor: American Academy of Otolaryngology-Head and Neck Surgery Foundation
1650 Diagonal Road
Alexandria, VA 22314-2857

AAO-HNSF Percy Memorial Research Award 146
The Percy Memorial Research Award is an annual grant-in-aid of a worthy research project proposed in any area within the scope of otolaryngology-head and neck surgery. The Award was established in memory of A. Edward Percy, Jr., M.D. and his parents. Projects must have direct or potential clinical significance for patients seen by otolaryngologists-head and neck surgeons. They must be designed so as to yield useful information within the period of award, but priority will be given to projects which are also innovative and promise to develop into new long range research programs which will attract funding from other sources. Grants range up to $10,000 maximum, are for one year, and are not renewable. Letters of intent must be received by December 15, and completed applications must be submitted by January 15.
Requirements: Applicants in the United States or Canada and members in good standing of AAO-HNSF are welcome to apply. Applicants should be experienced independent otolaryngologist investigators affiliated with academic or similar institutions eligible to apply for and administer Federal research awards. Fellows with substantial research experience are eligible to apply, as are individuals who have already competed successfully for independent research grant support from a private or Federal funding agency.
Restrictions: Funds may not be requested to pay any portions of the salaries of the principal investigator or of any support personnel with strictly secretarial or clerical

responsibilities. Funds may also not be used for the purchase of any item of equipment costing more than $500.
Date(s) Application is Due: Jan 15; Dec 15
Amount of Grant: Up to 25,000 U.S.D
Contact: Stephanie L. Jones; (703) 519-1586 or (703) 836-4444; sljones@entnet.org
Internet: http://www.entnet.org/EducationAndResearch/coreGrants.cfm
Sponsor: American Academy of Otolaryngology-Head and Neck Surgery Foundation
1650 Diagonal Road
Alexandria, VA 22314-2857

AAO-HNSF Rande H. Lazar Health Services Research Grant 147
The purpose of the Rande H. Lazar Health Services Research Grant is to support the gathering of socioeconomic data for otolaryngology. The award is intended to promote increased participation by otolaryngologists in the rapidly expanding area of health services research. Proposed projects may be related to any area of otolaryngology-head and neck surgery, but must have direct or potential clinical significance for patients seen by otolaryngologist-head and neck surgeons. Grants range up to $10,000 maximum, are for one year, and are not renewable. Letters of intent must be received by December 15, and completed applications must be submitted by January 15.
Requirements: Any otolaryngologist in the United States or Canada is eligible to apply for the Rande H. Lazar Health Services Research Grant. Applicants may be independent practitioners, residents or fellows in an approved training program, or practitioners affiliated with academic or similar institutions. All applicants must be members in good standing of AAO-HNSF; Associate and Corresponding Members are not eligible to apply as Principal Investigator, but may participate actively in the proposed project.
Date(s) Application is Due: Jan 15; Dec 15
Amount of Grant: Up to 10,000 U.S.D
Contact: Stephanie L. Jones; (703) 519-1586 or (703) 836-4444; sljones@entnet.org
Internet: http://www.entnet.org/EducationAndResearch/coreGrants.cfm
Sponsor: American Academy of Otolaryngology-Head and Neck Surgery Foundation
1650 Diagonal Road
Alexandria, VA 22314-2857

AAO-HNSF Resident Research Awards 148
The purpose of this grant is to stimulate original resident research in otolaryngology projects that are well-conceived and scientifically valid, with the potential to advance otolaryngology. Proposed projects may be related to any area of otolaryngology-head and neck surgery, and should be designed in collaboration with a preceptor investigator and approved by the candidate's department chairperson and institution. Grants range up to $10,000 maximum. Projects must be completed within one year of the award date; no-cost extensions are available upon written request. Letters of intent must be received by December 15, and completed applications must be submitted by January 15. Up to eight awards are available annually.
Requirements: The award is open to any AAO-HNS member in good standing who is a resident in an accredited otolaryngology - head and neck surgery training program in the U.S. and Canada.
Date(s) Application is Due: Jan 15; Dec 15
Amount of Grant: Up to 10,000 U.S.D
Contact: Stephanie L. Jones; (703) 519-1586 or (703) 836-4444; sljones@entnet.org
Internet: http://www.entnet.org/EducationAndResearch/coreGrants.cfm
Sponsor: American Academy of Otolaryngology-Head and Neck Surgery Foundation
1650 Diagonal Road
Alexandria, VA 22314-2857

AAO-HNS Medical Student Research Prize 149
The Medical Student Research Prize is given annually for a manuscript submitted to the Editor of Otolaryngology-Head and Neck Surgery for review and consideration for publication in the journal. An award winning manuscript will be automatically accepted as an oral or poster presentation at the Translational and Basic Research Presentations Program (previously called the Research Forum). First prize is $1,000, and the award is available in both clinical and basic science categories through the generosity of GYRU. S. ACMI ENT Division.
Requirements: All manuscripts submitted for this competition must have a medical student as senior author and must describe work that has not been placed under active consideration for publication other than in Otolaryngology-Head and Neck Surgery.
Date(s) Application is Due: Mar 9
Amount of Grant: 1,000 U.S.D
Contact: Stephanie L. Jones; (703) 519-1586 or (703) 836-4444; sljones@entnet.org
Internet: http://www.entnet.org/educationandresearch/grantsandfellowships.cfm
Sponsor: American Academy of Otolaryngology-Head and Neck Surgery Foundation
1650 Diagonal Road
Alexandria, VA 22314-2857

AAO-HNS Resident Research Prizes 150
Resident Research Prizes are given annually for manuscripts submitted to the Editor of Otolaryngology-Head and Neck Surgery for review and consideration for publication in the journal. All award winning manuscripts will be automatically accepted as oral or poster presentations at the Translational and Basic Research Presentations Program (previously called the Research Forum). First prize is $2,000, second prize is $1,500, and third prize is $1,000. All awards are available in both clinical and basic science categories through the generosity of GYRU.S. ACMI ENT Division.
Requirements: All manuscripts submitted for this competition must have a resident as senior author and must describe work that has not been placed under active consideration for publication other than in Otolaryngology-Head and Neck Surgery.
Date(s) Application is Due: Mar 9
Amount of Grant: 1,000 - 2,000 U.S.D
Contact: Stephanie L. Jones; (703) 519-1586 or (703) 836-4444; sljones@entnet.org
Internet: http://www.entnet.org/educationandresearch/grantsandfellowships.cfm
Sponsor: American Academy of Otolaryngology-Head and Neck Surgery Foundation
1650 Diagonal Road
Alexandria, VA 22314-2857

AAOA Foundation/AAO-HNSF Combined Research Grants 151
The purpose of this award is to support a collaborative AAOA Foundation research project by fostering basic or clinical research related to otolaryngic allergy, rhinology, or related immunology by otolaryngologists-head and neck surgeons. Grants range up to $10,000 maximum. Projects must be completed within two years of the award date; no-cost extensions are available upon written request. Letters of intent must be received by December 15, and completed applications must be submitted by January 15.
Requirements: Applicants must have demonstrated potential for excellence in research. There are no restrictions on career stage, education, or country of residence. AAOA members are preferred but not required. All candidates must be sponsored by the Chairperson of his/her Division or Department and by an official representative of the institution that would administer the Award and in whose name the application is formally submitted.
Date(s) Application is Due: Jan 15; Dec 15
Amount of Grant: Up to 10,000 U.S.D
Contact: Stephanie L. Jones; (703) 519-1586 or (703) 836-4444; sljones@entnet.org
Internet: http://www.entnet.org/EducationAndResearch/coreGrants.cfm
Sponsor: American Academy of Otolaryngic Allergy Foundation
1990 M. Street NW, Suite 680
Washington, DC 20036

AAOHN Foundation Experienced Researcher Grants 152
The foundation funds research focused on improving the health and safety of the nation's workforce. Research grants are awarded to support research in the field of occupational and environmental health by an experienced occupational and environmental health registered nurse. Proposals may be submitted that are in the early stages of development (prior to data collection). Two $10,000 grants are awarded annually.
Requirements: Experienced occupational and environmental health registered nurse researchers are eligible to apply.
Restrictions: Completed research projects will not be accepted for consideration.
Date(s) Application is Due: Dec 1
Amount of Grant: 10,000 U.S.D
Contact: Ann Cox, Executive; (770) 455-7757; fax (770) 455-7271; ann@aaohn.org
Internet: http://www.aaohn.org/foundation/grants/index.cfm
Sponsor: American Association of Occupational Health Nurses Foundation
2920 Brandywine Road, Suite 100
Atlanta, GA 30341

AAOHN Foundation New Investigator Researcher Grants 153
The foundation funds research, scholarship, and leadership development grants focused on improving the health and safety of the nation's workforce. Grants are awarded to encourage research in the field of occupational and environmental health by a new or novice registered nurse principal investigator. Proposals may be submitted that are in the early stages of development (prior to data collection).
Restrictions: Completed research projects are ineligible.
Date(s) Application is Due: Dec 1
Amount of Grant: 3,000 U.S.D
Contact: Ann Cox; (770) 455-7757; fax (770) 455-7271; ann@aaohn.org
Internet: http://www.aaohn.org/foundation/grants/index.cfm
Sponsor: American Association of Occupational Health Nurses Foundation
2920 Brandywine Road, Suite 100
Atlanta, GA 30341

AAOHN Foundation Professional Development Scholarships 154
Annually, the AAOHN Foundation awards grants to support professional development (academic and continuing education) and leadership development in occupational and environmental health. Leadership development scholarship applications, due August 1, support volunteer leadership development in occupational and environmental health nursing. Scholarship applications for academic study, due December 1, are offered to provide opportunities to further professional education for occupational and environmental health professionals. To support occupational and environmental health professionals in attending and successfully completing continuing education activities that will further their professional development and continued competence, the AAOHN Foundation offers continuing education scholarships, with applications due December 1.
Restrictions: These grants are not intended to supplement tuition in an academic program.
Date(s) Application is Due: Dec 1
Amount of Grant: 3,000 U.S.D
Contact: Ann Cox; (770) 455-7757; fax (770) 455-7271; ann@aaohn.org
Internet: http://www.aaohn.org/foundation/scholarships/index.cfm
Sponsor: American Association of Occupational Health Nurses Foundation
2920 Brandywine Road, Suite 100
Atlanta, GA 30341

AAP Anne E. Dyson Child Advocacy Awards 155

The AAP Resident Section Anne E. Dyson Child Advocacy Award, supported by the Dyson Foundation, celebrates the efforts of pediatricians-in-training as they work in their communities to improve the health of children. This award seeks to showcase projects that are designed and implemented by residents, which aim to improve the lives of children. The Award includes: $300 in funds to advance the winning projects' goals; travel and lodging expenses for up to 2 residents per project to the AAP National Conference and Exhibition (NCE); and presentation of the Advocacy Award plaque during the Resident Section assembly meeting.
Requirements: Any resident-sponsored and/or resident led project that seeks to advocate on behalf of children is eligible for this award.
Date(s) Application is Due: Jul 15
Amount of Grant: 300 U.S.D
Contact: Kimberley VandenBrook, Grants Administrator; (800) 433-9016, ext. 7134; kvandenbrook@core.com
Internet: http://www.aap.org/sections/ypn/r/funding_awards/anne_dyson.html
Sponsor: American Academy of Pediatrics
141 Northwest Point Boulevard
Elk Grove Village, IL 60007-1098

AAP Community Access To Child Health (CATCH) Advocacy Training Grants 156

The Program supports five pediatric faculty-resident pairs (10 people) to attend the AAP Legislative Conference in Washington, DC each year. Each faculty-resident pair is required to implement an educational activity on child advocacy in coordination with their local AAP chapter following the conference. Local chapters will receive up to $1,000 toward completing the educational activity in conjunction with the faculty-resident pairs. The next call for applications will be available each fall.
Date(s) Application is Due: Nov 9
Amount of Grant: Up to 1,000 U.S.D
Contact: Kathy Kocvara, Program Coordinator; (847) 434-7085 or (800) 433-9016, ext. 7632; kkocvara@aap.org or catch@aap.org
Internet: http://www.aap.org/commpeds/cpti/Opportunities.htm
Sponsor: American Academy of Pediatrics
141 Northwest Point Boulevard
Elk Grove Village, IL 60007-1098

AAP Community Access To Child Health (CATCH) Implementation Grants 157

This program supports pediatricians in the initial and/or pilot stage of developing and implementing a community-based child health initiative. Grants of up to $12,000 are awarded each year on a competitive basis to pediatricians who want to address the local needs of children in the community. Priority will be given to projects serving communities with the greatest demonstrated health care access needs and health disparities. Strong collaborative community partnerships and future sustainability of the project are encouraged.
Requirements: A pediatrician must lead the project and be significantly involved in proposal development and project activities.
Restrictions: Only applicants from the United States and its territories are eligible to apply.
Date(s) Application is Due: Jan 29
Amount of Grant: Up to 12,000 U.S.D
Contact: Kathy Kocvara, Program Coordinator; (847) 434-7085 or (800) 433-9016, ext. 7632; kkocvara@aap.org or catch@aap.org
Internet: http://www.aap.org/catch/implementgrants.htm
Sponsor: American Academy of Pediatrics
141 Northwest Point Boulevard
Elk Grove Village, IL 60007-1098

AAP Community Access to Child Health (CATCH) Planning Grants 158

The program awards competitive grants to pediatricians who want to plan community-based initiatives to increase access to children's health care. Proposed initiatives must be broad-based community partnerships. Priority will be given to proposals promoting medical homes for under-served children and those with special health care needs; collaborating with SCHIP or Medicaid, and representing new initiatives within the community. Grant funds must be used for planning, not implementation. Planning activities may include needs assessments and community asset mapping, feasibility studies, community meetings, focus groups, and development of grant proposals. If an applicant is not an AAP member, a letter of support from the AAP chapter president in his/her area must be obtained. Applicants must contact their chapter CATCH facilitators for approval of applications prior to submission. Technical assistance for applicants is available from CATCH staff and facilitators. Annual deadline dates may vary; contact program staff for exact dates.
Requirements: U.S. pediatricians and those in U.S. territories are eligible. Every program must be led by, facilitated by, or have the significant involvement of a pediatrician. During the planning phase, the involvement of other community members should be secured in order to ensure local support for the program. Applicants are encouraged to demonstrate collaboration with state child health insurance programs and/or state Medicaid programs.
Restrictions: Only applicants from the United States and its territories are eligible to apply.
Date(s) Application is Due: Jul 30
Amount of Grant: 2,500 - 12,000 U.S.D
Contact: Kathy Kocvara, Program Coordinator, CATCH; (847) 434-7632; fax (847) 228-6432; kkocvara@aap.org
Internet: http://www.aap.org/catch/planninggrants.htm
Sponsor: American Academy of Pediatrics
141 Northwest Point Boulevard
Elk Grove Village, IL 60007-1098

AAP Community Access To Child Health (CATCH) Residency Training Grants 159

The Community Pediatrics Training Initiative (CPTI) partners with the Community Access to Child Health (CATCH) Program to provide this grant opportunity targeting pediatric residency training programs. The mission of the program is to provide support to residency programs to build sustainable opportunities for residents to gain experience working on community-based child health initiatives that increase access to medical homes or specific health services not otherwise available. Grants of up to $15,000 will be awarded on a competitive basis to pediatric residency programs which submit proposals to plan and implement community-based child health initiatives as part of the training curriculum over the course of 16 months. The next Call for Proposals will be available in September.
Requirements: A pediatric faculty member must oversee the project and provide mentorship for residents participating in project activities.
Date(s) Application is Due: Nov 18
Amount of Grant: Up to 15,000 U.S.D
Contact: Kathy Kocvara, Program Coordinator; (847) 434-7085 or (800) 433-9016, ext. 7632; kkocvara@aap.org or catch@aap.org
Internet: http://www.aap.org/commpeds/cpti/Opportunities.htm
Sponsor: American Academy of Pediatrics
141 Northwest Point Boulevard
Elk Grove Village, IL 60007-1098

AAP Community Access To Child Health (CATCH) Resident Grants 160

The Community Access to Child Health (CATCH) Resident Funds program supports pediatric residents in the planning of community-based child health initiatives. Grants of up to $3,000 are awarded twice each year on a competitive basis for pediatric residents to address the needs of children in their communities. CATCH Resident Funds grant projects must include planning activities and also may include some implementation activities. Resident grants are available twice a year - May to July, during the CATCH Planning Funds grant cycle, and November to January, during the CATCH Implementation Funds grant cycle.
Requirements: A pediatric resident must lead the project and be significantly involved in proposal development and project activities.
Restrictions: Only pediatric residents from the United States and its territories are eligible to apply for CATCH Resident Funds grants.
Date(s) Application is Due: Jan 31; Jul 31
Amount of Grant: Up to 3,000 U.S.D
Contact: Kathy Kocvara, Program Coordinator; (847) 434-7085 or (800) 433-9016, ext. 7632; kkocvara@aap.org or catch@aap.org
Internet: http://www.aap.org/catch/residentgrants.htm
Sponsor: American Academy of Pediatrics
141 Northwest Point Boulevard
Elk Grove Village, IL 60007-1098

AAPD Henry B. Betts Award 161

Each year the award honors an individual whose work and scope of influence has significantly improved the quality of life for people with disabilities in the past, and will be a force for change in the future. It is named for Henry B. Betts, MD, in recognition of his pioneering leadership in the field of physical medicine and rehabilitation and decades of dedicated service to the Rehabilitation Institute of Chicago. The awardee receives an unrestricted cash award and a commemorative crystal piece, which is presented publicly at AAPD's Leadership Gala in Washington, DC.
Requirements: Successful nominees will have demonstrated a strong vision and understanding of how to improve the quality of life for Americans with disabilities; possess a record of efforts and accomplishments that have affected a wide disability population; and have served as a powerful force for change, enhancing the opportunities for people with disabilities to participate fully in all aspects of society.
Date(s) Application is Due: Oct 9
Amount of Grant: 50,000 U.S.D
Contact: Tracey Murray, Leadership Coordinator; (800) 840-8844 or (770) 232-9001; murr9001@bellsouth.net
Internet: http://www.aapd-dc.org
Sponsor: American Association of People with Disabilities
1629 K Street NW, Suite 503
Washington, DC 20006

AAPD Paul G. Hearne Leadership Award 162

The awards are presented to people with disabilities who are emerging as leaders in their respective fields to help them continue their progress as leaders. They will also have an opportunity to meet and network with national disability leaders at the AAPD Leadership Gala in Washington, DC. Winners of the Award must demonstrate all of the following: leadership achievements that show a positive impact on the community of people with disabilities or within their area of disability interest; connections they have made between individuals with disabilities and others in their communities; a positive vision for the disability community and a continuing commitment to their leadership activities; and potential to contribute at a national level. AAPD encourages emerging leaders with disabilities of any age to apply. Previous awardees represent a diverse group of people with disabilities aged 11 to 56.
Requirements: U.S. residents with any type of disability are eligible to apply.
Date(s) Application is Due: Oct 1
Amount of Grant: 10,000 U.S.D
Samples: Claudia Gordon, a consultant to the National Council on Disability (Washington, DC)—award winner, the first black deaf female lawyer in the United States and a long-time advocate for the rights of deaf people, $10,000; Peter Cody

Hunt—award winner, a graduate student in rehabilitation science and technology in Pittsburgh, who has developed outreach programs for Asian-Americans with disabilities, $10,000; Sarah Louise Triano—award winner, director of the Yield the Power to the Youth program at Access Living (Chicago), who works to mobilize and educate disabled people and increase their ability to influence public and social policy, $10,000.
Contact: David Hale, Program Manager; (202) 521-4306; dHale@aapd.com or awards@aapd.com
Internet: http://www.aapd.com/DMD/PaulHearneAward.html
Sponsor: American Association of People with Disabilities
1629 K Street NW, Suite 503
Washington, DC 20006

AAP International Travel Grants 163

The AAP has set aside a minimum of twelve (12) five-hundred dollar ($500.00) grants to be awarded to categorical pediatric or combined-training pediatric residents who wish to complete a clinical pediatric elective in the developing world during residency. The selection committee is composed of members from the AAP Section on International Child Health and the Section on Residents. After the selection committee reviews the applications the recipients will be notified mid-May and mid-November.
Requirements: Must be a categorical pediatric or combined-training pediatric resident at a U.S. or Canadian university.
Date(s) Application is Due: Mar 15; Sep 15
Amount of Grant: 500 U.S.D
Contact: Kimberley VandenBrook, Grants Administrator; (800) 433-9016, ext. 7134; kvandenbrook@core.com
Internet: http://www.aap.org/sections/ypn/r/funding_awards/international_travel.html
Sponsor: American Academy of Pediatrics
141 Northwest Point Boulevard
Elk Grove Village, IL 60007-1098

AAP Legislative Conference Scholarships 164

The American Academy of Pediatrics Legislative Conference Scholarships is specifically designed to help allay financial difficulties for members interested in attending the annual Legislative Conference, typically help each March. The Academy grants annually a number of scholarships varying from $1,000 to $2,000 on average. Each applicant will be reviewed carefully and awards will be granted to the most deserving.
Contact: Kimberley VandenBrook, Grants Administrator; (800) 433-9016, ext. 7134; kvandenbrook@core.com
Internet: http://www.aap.org/sections/ypn/r/funding_awards/
Sponsor: American Academy of Pediatrics
601 13th Street NW, Suite 400 North
Washington, DC 20005

AAP Leonard P. Rome Community Access to Child Health (CATCH) Visiting Professorships 165

The purpose of the Leonard P. Rome Community Access to Child Health (CATCH) Visiting Professorships Program is to promote advocacy for children and advance the field of community pediatrics. Beginning with the 2008 grant cycle, CATCH will be collaborating with the Community Pediatrics Training Initiative (CPTI) to implement this program. Four accredited pediatric residency programs will receive up to $4,500 each to fund a 2-or 3-day educational program focusing on the field of community pediatrics. District CATCH Facilitators (DCFs) and Chapter CATCH Facilitators (CCFs) are available to provide technical assistance.
Amount of Grant: Up to 4,500 U.S.D
Contact: Alanna Bailey Whybrew, (847) 434-7085 or (800) 433-9016, ext. 7397; awhybrew@aap.org or catch@aap.org
Internet: http://www.aap.org/catch/vp.htm
Sponsor: American Academy of Pediatrics
141 Northwest Point Boulevard
Elk Grove Village, IL 60007-1098

AAP Nutrition Award 166

One award is given annually for outstanding achievements in research relating to nutrition of infants and children. The award is made to an individual or for one project. There is no age restriction for the award; however, it is hoped that younger persons will be considered. The award is made possible by a grant from the Infant Formula Council. In addition to the honorarium, the award includes round-trip airfare and two days' lodging for the recipient and a guest to attend the annual meeting and receive the award.
Requirements: The competition is open to residents of the United States or Canada whose research has been completed and publicly reported. Separate letters should be written for each individual nominated; a letter should contain a description of the nominee's achievements, stating clearly the basis for the recommendation, including references to the literature that describes her/his work. Nominee's bibliography is to be submitted with the nominating letter together with copies of available reprints. Letters supporting the nomination are to be screened by the nominator and forwarded with the nomination.
Restrictions: An individual may not submit more than five letters supporting the nomination. Current members of the AAP Committee on Nutrition are not eligible for this award.
Date(s) Application is Due: Mar 1
Amount of Grant: 3,000 U.S.D
Internet: http://www.aap.org/visit/nutrannouncemts.htm
Sponsor: American Academy of Pediatrics
141 Northwest Point Boulevard
Elk Grove Village, IL 60007-1098

AAP Program Delegate Awards 167

The travel grants are available to residents who serve as program delegates to the AAP. This grant serves to support the program delegate's travel to the AAP Resident Section meeting at the annual National Conference and Exhibition (NCE). The travel grant is available to both categorical and combined-training pediatric residents. However, there is only one grant available per institution. The AAP Program Delegate will automatically receive the $300 travel grant.
Date(s) Application is Due: Sep 14
Amount of Grant: 300 U.S.D
Contact: Kimberley VandenBrook, Grants Administrator; (800) 433-9016, ext. 7134; kvandenbrook@core.com
Internet: http://www.aap.org/sections/ypn/r/funding_awards/nce.html
Sponsor: American Academy of Pediatrics
141 Northwest Point Boulevard
Elk Grove Village, IL 60007-1098

AAP Resident Initiative Fund Grants 168

The Program is offering twenty-five (25) grants of up to $1,000 each. The funding will enable residents to educate fellow residents and/or parents on a specific aspect of one of the Academy's national child health priorities (Special Health Care Needs, Foster Care, Oral Health, Disaster Preparedness, Mental Health, Obesity, and Immunizations). The Academy hopes that this program will: increase the number of new, creative and innovative opportunities for residents to apply their leadership and advocacy skills within the AAP structure; increase residents' knowledge of the Academy's child health priorities; and increase resident collaboration with national AAP and its chapters.
Date(s) Application is Due: Mar 31
Amount of Grant: 1,000 U.S.D
Contact: Kimberley VandenBrook, Grants Administrator; (800) 433-9016, ext. 7134; kvandenbrook@core.com
Internet: http://www.aap.org/sections/ypn/r/funding_awards/res_initiative.html
Sponsor: American Academy of Pediatrics
141 Northwest Point Boulevard
Elk Grove Village, IL 60007-1098

AAP Resident Research Grants 169

The Program is designed to give pediatric residents with limited research experience an opportunity to initiate and complete research projects related to their professional interests. Projects may be related to the full spectrum of child health research, such as behavioral sciences, biomedical sciences, epidemiology, health services, perinatal/neonatal health, prevention, public health, quality improvement, quality measurement, or basic laboratory-based science. Research projects can be conducted for a maximum of 2 years and should be completed during the residency program.
Requirements: Applicants must be legal residents of the United States or Canada. Although awards are primarily intended for support of first- and second-year pediatric residents, individuals who have secured a position for a third or fourth year of residency beginning in July may also apply.
Date(s) Application is Due: Feb 27
Amount of Grant: Up to 2,000 U.S.D
Contact: Jeannine Hess; (800) 433-9016, ext. 7876; fax (847) 434-8000; jhess@aap.org
Internet: http://www.aap.org/sections/ypn/r/funding_awards/research_grants.html
Sponsor: American Academy of Pediatrics
141 Northwest Point Boulevard
Elk Grove Village, IL 60007-1098

Aaron & Cecile Goldman Family Foundation Grants 170

The Aaron and Cecile Goldman Family Foundation, established in 1962 in the District of Columbia, offers its primary support to Jewish and local agencies, including education and human services, and to local children's programs and camps. Fields of interest include: arts and humanities; children and youth services; education; human services; Jewish agencies and synagogues; Jewish federated giving programs; recreational programs; and camps. Types of support include building and renovation, capital campaigns, emergency funding, endowments, equipment purchase, fellowships, matching/challenge grants, support for professions, program development, and scholarships. Applicants should forward a letter of proposal to the Foundation office, whose board members meet in the spring and fall of each year.
Requirements: One copy of the proposal letter is required. 501(c)3 organizations supporting the residents of Washington, DC, and its surrounding area are eligible to apply.
Restrictions: No grants are offered to individuals.
Geographic Focus: District of Columbia
Amount of Grant: Up to 20,000 U.S.D
Contact: Jennifer Margolius Fisher, (202) 332-6500; fax (202) 332-1800; jennifer@themargoliusfirm.com
Sponsor: Aaron and Cecile Goldman Family Foundation
3000 Connecticut Avenue NW, Suite 100
Washington, DC 20008-2509

Aaron & Freda Glickman Foundation Grants 171

The Aaron and Freda Glickman Foundation, established in New York, offers grant support primarily in the State of New York. Funding typically comes in the form of general operating support for health organizations and associations, as well as Jewish agencies and synagogues. There are no specific deadlines or application forms with which to adhere, and applicants should contact the Foundation in writing, outlining the need and a description of the program.
Requirements: 501(c)3 organizations and Jewish agencies serving the residents of New York are eligible to apply.

Geographic Focus: New York
Amount of Grant: Up to 2,500 U.S.D
Contact: Edwin J. Glickman, Secretary; (561) 470-1434
Sponsor: Aaron and Freda Glickman Foundation
3080 N 36th Avenue, V-61
Hollywood, FL 33021

Aaron Foundation (California) Grants 172
The Aaron Foundation, established in California in 1987 by the Aaron Mortgage Corporation, supports programs for children and youth, health care facilities, and human services agencies. Giving is primarily centered in the Bakersfield, California, region. Applicants should submit: a copy of current year's organizational budget and/or project budget; a copy of IRS Determination Letter; a detailed description of project and amount of funding requested; a statement of the problem that the project will address; and a brief history of organization and description of its mission.
Requirements: The project must benefit the residents of Bakersfield, California.
Geographic Focus: California
Date(s) Application is Due: Nov 30
Contact: Hal E. Aaron, Trustee; (661) 322-6353; fax (661) 322-6120
Sponsor: Aaron Foundation
651 H Street, Suite 100
Bakersfield, CA 93304-1305

Aaron Foundation (Massachusetts) Grants 173
Established in Massachusetts in 1951, the Foundation awards grants to eligible nonprofit organizations in its areas of interest, including: arts and culture; health care; higher education; Jewish services and temples; health care; social services; and youth. There are no application deadlines or forms with which to adhere. Applicants should submit: results expected from proposed grant; a copy of IRS Determination Letter; copy of most recent annual report/audited financial statement or 990; a listing of board of directors, trustees, officers and other key people and their affiliations; and a copy of current year's organizational budget and/or project budget.
Requirements: Connecticut, Massachusetts, and Rhode Island nonprofit organizations are eligible to apply.
Restrictions: Individuals are ineligible to apply.
Geographic Focus: Connecticut, Massachusetts, Rhode Island
Amount of Grant: 1,000 - 100,000 U.S.D
Contact: Avram J. Goldberg, Trustee; (617) 695-1300 or (617) 695-1946
Sponsor: Aaron Foundation
225 Franklin Street, Suite 1450
Boston, MA 02110

AARP Andrus Foundation Grants 174
The foundation actively seeks new approaches to maximizing the independence of older persons. Grants are awarded under two initiatives: living with chronic health conditions, and aging and living environments. Three types of grants are awarded. Research grants support cutting-edge, applied psychosocial, behavioral, health, and policy research and the dissemination of research findings. Dissemination grants support translating research information into language and materials useful to older adults and the practitioners, service providers, and others who work on their behalf. Grants generally are awarded for projects with the potential for national impact and high visibility, and may include public education programs, major media events, and the production of consumer or practitioner-oriented materials. Funding will also be allocated for the packaging and dissemination of research results and products related to aging and financial security, but not for conferences or the publication of conference proceedings. Education and training grants support graduate students engaged in aging-related research. National programs that provide leadership development, professional development, and research training to students and young professionals receive preference. Grants are awarded only to national organizations, not to individuals or individual educational institutions. Submit a letter of inquiry by the listed deadline dates; full proposals are by invitation.
Requirements: Grants are awarded to colleges, universities, and 501(c)3 non-profits.
Restrictions: Grants do not fund capital campaigns, organizational infrastructure (bricks and mortar), or vehicle purchases; individuals; biomedical or basic (bench) research; programs and services for older adults; education and training programs for professionals; pilot or preliminary research studies; or organizational overhead expenses.
Date(s) Application is Due: Mar 1; Jul 1; Nov 1
Samples: Lighthouse International (New York, NY)—to develop, implement, and evaluate a public-awareness campaign about key research findings on vision, vision impairment, and vision rehabilitation among elderly people, $150,000; American Society on Aging (San Francisco, CA)—to sponsor four minority practitioners who are conducting research projects as part of their participation in a leadership and professional-development program, $33,400; International Longevity Ctr (New York, NY)—to produce five consumer publications that address various trends in aging, $150,700; National Hospice and Palliative Care Organization (Alexandria, VA)—for research needed to develop a more comprehensive set of performance measures for end-of-life professionals, $100,000.
Contact: Grants Administrator; (800) 775-6776 or (202) 434-6190; fax (202) 434-6458; andrus@aarp.org
Internet: http://www.aarp.org/about_aarp/nrta/nrta_past_present_and_future.html#top
Sponsor: American Association of Retired Persons
601 E Street NW
Washington, DC 20049

Abbot and Dorothy H. Stevens Foundation Grants 175
The Abbot and Dorothy H. Stevens Foundation funds Massachusetts non-profits with a emphasis on the greater Lawrence/Merrimack Valley area. Giving primarily for the arts, education, conservation, and health and human services. Fields of interest include: arts; children/youth, services; crime/violence prevention, domestic violence; education; elderly; environment, natural resources; health care; health organizations, association; historic preservation/historical societies; humanities; human services; immigrants; medical school/education; museums. Types of support include: building/renovation; capital campaigns; continuing support; endowments; equipment; general/operating support; management development/capacity building; matching/challenge support; program-related investments/loans; program development; technical assistance.
Requirements: Massachusetts 501(c)3 tax-exempt organizations serving the greater Lawrence and Merrimack Valley are eligible. There is no deadline date when applying for funding, nor is there a application form required. The Foundation will however accept the Associated Grantmakers Common Proposal Form. Applicants should submit a copy of their IRS Determination Letter and one copy of their proposal when applying for funding. The board meets and reviews proposals monthly except for the months of July and August.
Restrictions: Grants do not support national organizations, state or federal agencies, individuals, annual campaigns, deficit financing, exchange programs, internships, professorships, scholarships, or fellowships.
Geographic Focus: Massachusetts
Amount of Grant: 1,000 - 20,000 U.S.D
Samples: Boston Childrens Museum, Boston, MA, $15,000; Boston Symphony Orchestra, Boston, MA, $10,000; Boston Ballet, Boston, MA, $5,000.
Contact: Josh Miner, Executive Director; (978) 688-7211; fax (978) 686-1620; grantprocess@stevensfoundation.com
Sponsor: Abbot and Dorothy H. Suitevens Foundation
P.O. Box 111
North Andover, MA 01845-7211

Abbott-ASM Lifetime Achievement Award 176
The award, sponsored by Abbott Laboratories, is given to honor a distinguished scientist for a lifetime of outstanding contributions in fundamental biomedical research in any of the microbiological sciences. The award consists of a cash prize, framed certificate, medallion, and national or international travel expenses incidental to receiving the award at the ASM general meeting. Guidelines are available online. All nomination materials must be submitted electronically.
Requirements: Mature scientists, both active and retired, from all areas of microbiology are eligible.
Restrictions: Self-nominations are not be accepted.
Date(s) Application is Due: Oct 1
Amount of Grant: Up to 20,000 U.S.D
Samples: Norman R. Pace, $20,000—award recipient; R. John Collier, $20,000—award recipient; Jonathan Beckwith, $20,000—award recipient.
Contact: Awards Committee; (202) 942-9226; fax (202) 942-9353; awards@asmusa.org
Internet: http://www.asm.org/Academy/index.asp?bid=2587
Sponsor: American Society for Microbiology
1752 N Street, N.W.
Washington, DC 20036-2904

Abbott Fund Access to Health Care Grants 177
Abbott's Access to Health Care programs seek innovative solutions to improve and expand access to health care services for disadvantaged populations. Specific areas of focus include cardiovascular health, diabetes, nutrition, maternal and child health and neonatal care. Many Abbott Fund programs are engaged in closing gaps in ethnic and minority communities, promoting health and nutrition education for families, training health workers, and improving delivery of health services.
Requirements: Grants are made to tax-exempt organizations supporting access to health care.
Restrictions: Contributions will not be made to individuals; for-profit entities; purely social organizations; political parties or candidates; sectarian religious organizations; advertising; symposia, conferences, and meetings; ticket purchases; memberships; business-related purposes; volunteer efforts of non-Abbott employees; or marketing sponsorships.
Geographic Focus: Arizona, California, Illinois, Kansas, Massachusetts, Michigan, New Jersey, New York, North Carolina, Ohio, Puerto Rico, Texas, Utah, Virginia
Amount of Grant: 20,000 - 10,000 U.S.D
Contact: Cindy Schwab, Vice President; (847) 937-7075; fax (847) 935-5051; cindy.schwab@abbott.com
Internet: http://www.abbottfund.org/sections/what.html
Sponsor: Abbott Fund
100 Abbott Park Road, Department 379, Building 6D
Abbott Park, IL 60064-3500

Abbott Fund Community Vitality Grants 178
The Abbott Fund Community Vitality Grant program is active in communities around the world where Abbott has a significant presence. It pursues local partnerships and creative programs that address unmet needs of a community. Emphasis is placed on improving access to health care and promoting science education. The program also supports major civic, arts and other cultural institution programming, primarily in the Chicago metropolitan area where Abbott is headquartered.
Requirements: Grants are made to tax-exempt organizations supporting company operating areas in Arizona, California, Illinois, Kansas, Massachusetts, Michigan, New Jersey, New York, North Carolina, Ohio, Puerto Rico, Texas, and Virginia, and Utah. It also supports the communities of: Abingdon, England; Brockville, Canada;

Campoverde, Italy; Clonmel, Ireland; Cootehill, Ireland; Delkenheim, Germany; Kanata, Canada; Katsuyama, Japan; Ludwigshafen, Germany; Queenborough, England; Rio de Janeiro, Brazil; Sligo, Ireland; and Zwolle, the Netherlands.
Restrictions: Contributions will not be made to individuals; for-profit entities; purely social organizations; political parties or candidates; sectarian religious organizations; advertising; symposia, conferences, and meetings; ticket purchases; memberships; business-related purposes; volunteer efforts of non-Abbott employees; or marketing sponsorships.
Geographic Focus: Arizona, California, Illinois, Kansas, Massachusetts, Michigan, New Jersey, New York, North Carolina, Ohio, Puerto Rico, Texas, Utah, Virginia
Amount of Grant: 10,000 U.S.D
Contact: Cindy Schwab, Vice President; (847) 937-7075; fax (847) 935-5051; cindy.schwab@abbott.com
Internet: http://www.abbottfund.org/sections/what.html
Sponsor: Abbott Fund
100 Abbott Park Road, Department 379, Building 6D
Abbott Park, IL 60064-3500

Abbott Fund Global AIDS Care Grants 179
Abbott has been a significant contributor to the fight against HIV/AIDS for more than two decades. Since 2000, Abbott and the Abbott Fund have invested $225 million in grants and product donations targeted to resource poor countries most impacted by HIV/AIDS. The focus of these efforts includes expanding access to care, testing and treatment; strengthening HIV/AIDS health care systems; preventing mother-to-child transmission; and supporting children and families affected by HIV/AIDS. In addition, Abbott and the Abbott Fund have helped pioneer innovative model programs to combat the disease. For example, we helped build the Baylor International Pediatric AIDS Initiative's first pediatric outpatient clinic in Romania in 2001 that has served an average of 600 patients per year. The clinic has reduced pediatric HIV mortality rates by more than 90 percent. Today Baylor has replicated the Romania model in additional clinics throughout Africa and now serves nearly 60,000 children and young people with HIV – including those at two clinics built by and supported by the Abbott Fund in Malawi and Tanzania.
Requirements: Grants are made to tax-exempt organizations supporting global HIV/AIDS programs.
Restrictions: Contributions will not be made to individuals; for-profit entities; purely social organizations; political parties or candidates; sectarian religious organizations; advertising; symposia, conferences, and meetings; ticket purchases; memberships; business-related purposes; volunteer efforts of non-Abbott employees; or marketing sponsorships.
Geographic Focus: Arizona, California, Illinois, Kansas, Massachusetts, Michigan, New Jersey, New York, North Carolina, Ohio, Puerto Rico, Texas, Utah, Virginia
Amount of Grant: 10,000 - 100,000 U.S.D
Contact: Cindy Schwab, Vice President; (847) 937-7075; fax (847) 935-5051; cindy.schwab@abbott.com
Internet: http://www.abbottfund.org/project/20/30/Helping-Children-and-Young-People-Living-with-HIV-AIDS
Sponsor: Abbott Fund
100 Abbott Park Road, Department 379, Building 6D
Abbott Park, IL 60064-3500

Abbott Fund Science and Medical Innovation Grants 180
The Abbott Fund Science and Medical Innovation program supports new approaches to learning that are designed to foster a better understanding of science and medical innovation and the value they bring to improving human health. Partnerships support K-12 education through postgraduate level research including funding and volunteer assistance for student and educator workshops and programs; career development; and educational grants for college and university science facilities, scholarships and fellowships.
Requirements: Grants are made to tax-exempt organizations supporting company operating areas in Arizona, California, Illinois, Kansas, Massachusetts, Michigan, New Jersey, New York, North Carolina, Ohio, Puerto Rico, Texas, and Virginia, and Utah. It also supports the communities of: Abingdon, England; Brockville, Canada; Campoverde, Italy; Clonmel, Ireland; Cootehill, Ireland; Delkenheim, Germany; Kanata, Canada; Katsuyama, Japan; Ludwigshafen, Germany; Queenborough, England; Rio de Janeiro, Brazil; Sligo, Ireland; and Zwolle, the Netherlands.
Restrictions: Contributions will not be made to individuals; for-profit entities; purely social organizations; political parties or candidates; sectarian religious organizations; advertising; symposia, conferences, and meetings; ticket purchases; memberships; business-related purposes; volunteer efforts of non-Abbott employees; or marketing sponsorships.
Geographic Focus: Arizona, California, Illinois, Kansas, Massachusetts, Michigan, New Jersey, New York, North Carolina, Ohio, Puerto Rico, Texas, Utah, Virginia
Amount of Grant: 20,000 - 10,000 U.S.D
Contact: Cindy Schwab, Vice President; (847) 937-7075; fax (847) 935-5051; cindy.schwab@abbott.com
Internet: http://www.abbottfund.org/sections/what.html
Sponsor: Abbott Fund
100 Abbott Park Road, Department 379, Building 6D
Abbott Park, IL 60064-3500

Abdus Salam ICTP Federation Arrangement Scheme Grants 181
Federation Arrangements are agreements signed by the ICTP and a scientific institute (Federated Institute) in developing countries. This enables the Institute to send junior scientists - up to the age of 40 - to the ICTP for a total of 60 to 150 days, during the validity of the contract (i.e. three years), on a cost-sharing basis. While living expenses are always covered by the Center, travel expenses are either covered by the Federated Institute or shared with the Center. The Federation Arrangement Scheme gives the opportunity to junior scientists from developing countries to participate in the Center's scientific programs. Their participation is proposed by the Federated Institute and approved by the ICTP. Presently there are five different standard types of Federation Arrangements depending upon geographical areas. While the junior scientists' living expenses are always covered by the Centre, travel expenses are either covered by the Federated Institute or shared with the Center.
Requirements: Scientific institutes from developing countries (to send junior scientists to ICTP) are eligible to apply.
Geographic Focus: All Countries
Contact: Patrizia Passarella, Program Assistant; (+39) 040 2240 389; fax (+39) 040 2240 388; itlabs@ictp.trieste.it or diploma@ictp.it
Internet: http://assoc.ictp.it/federation-scheme/description.html
Sponsor: Abdus Salam International Centre for Theoretical Physics
Strada Costiera 11
Trieste, I - 34014 Italy

Abdus Salam ICTP Training and Research in Italian Laboratories Grants 182
The ICTP Programme gives scientists from Developing Countries, Central and Eastern Europe the opportunity to spend periods ranging from a few months to one year at Italian research laboratories of universities, governmental and private institutions, which operate in different branches of Physics. This includes academic studies as well as practical applications and industrial projects. Follow up visits and collaboration programs are also envisaged. Agreements of collaboration have been established with Italian research institutions, i.e. CNR (National Research Council), INFN (National Institute for Nuclear Physics), ENEA (Italian Commission for New Technologies, Energy and Environment) and INFM (National Institute for the Physics of Matter), APAT (Agency for Environmental Protection and Technical Services), Sincrotrone Trieste (Elettra Synchrotron Light Source), National Institute of Oceanography and Experimental Geophysics (OGS) and others. They offer hospitality and a special financial contribution. The fields covered, which reflect activities held at the ICTP, are presently: physics of condensed matter - solid state physics and materials science, computational methods, atomic and molecular physics, laser spectroscopy, semiconductor physics and technology; physics and energy - plasma physics and fusion, non-conventional energy sources, applied nuclear physics; physics and technology - optical physics and lasers, communication physics and technology, optical fibers, microprocessors and informatics, synchrotron radiation applications; earth and environmental sciences - geophysics and seismology, oceanography, physics of the atmosphere and aeronomy, meteorology and climatology, soil physics, environmental control, mathematical ecology, modelling of environmental systems; physics of the living state - biophysics, medical physics, neurophysics; and miscellaneous - instrumentation for nuclear and subnuclear physics, space physics, topics at the interface with chemistry, engineering, or biology. The minimum amount of these grants will be euro 1,300 per month. Limited funds for travel expenses will be available in case these are not covered by other sources. Health insurance coverage will be provided for visitors and their accompanying family members. A monthly family allowance for those having dependent family members will be granted. The average length of a visit is 6 to 12 months.
Requirements: Less developed countries in Central and Eastern Europe may apply.
Geographic Focus: All Countries
Amount of Grant: Up to 15,600 EUR
Contact: Patrizia Passarella, Program Assistant; (+39) 040 2240 389; fax (+39) 040 2240 388; itlabs@ictp.trieste.it or diploma@ictp.it
Internet: http://users.ictp.it/www_users/ItaLab/index.html
Sponsor: Abdus Salam International Centre for Theoretical Physics
Strada Costiera 11
Trieste, I - 34014 Italy

Abel Foundation Grants 183
The foundation awards grants to eligible Nebraska nonprofit organizations in its areas of interest, including environmental programs and natural resource conservation, health care, higher education, Protestant religion and churches, and social services. Types of support include building construction/renovation, capital campaigns, general operating support, and program grants. Contact the office for application forms. Applications qualifying for review will be considered within 1 to 3 months of receipt.
Requirements: Nebraska nonprofit organizations are eligible. Preference is given to requests from Licoln, Nebraska, and the state's southeastern area.
Restrictions: The Foundation does not accept applications from organizations: that have had requests approved or declined in the past 12 months; or that are currently received payments for a multi-year grant.
Geographic Focus: Nebraska
Date(s) Application is Due: Apr 15; Jul 31; Oct 31
Amount of Grant: 50 - 40,000 U.S.D
Contact: J. Ross McCown, Vice President; (402) 434-1212; fax (402) 434-1799; rossm@nebcoinc.com
Internet: http://www.abelfoundation.org/grant.htm
Sponsor: Abel Foundation
1815 Y Street, P.O. Box 80268
Lincoln, NE 68501-0268

Abell-Hanger Foundation Grants 184
The foundation makes grants to nonprofit Texas organizations, other than private foundations, that are involved in such undertakings for the public/society benefit including arts, cultural, and humanities; education; health; human services; and religion. Types of support include general operating support, continuing support, annual campaigns, capital campaigns, building construction/renovation, equipment acquisition, endowment funds, program development, seed funds, scholarship funds, research

grants, and matching funds. Block scholarship grants are made only to institutions of higher education located in Texas. Recipient colleges and universities are free to administer the grants. Education grants are limited generally to institutions of higher education, including religious institutions (Baptist, Christian, Lutheran, Methodist, and Presbyterian). Applicants must seek funding for the same proposal from various sources because sole sponsorship of programs is rarely undertaken. Grant requests are considered and awarded throughout each year. The trustees prefer to consider only one request per applicant each fiscal year. Unsuccessful proposals may not be resubmitted for at least 12 months. Applicant organizations that have never received funding from the foundation should request a preproposal questionnaire; the trustees will review the request to determine whether it warrants a complete proposal.
Requirements: Applicant organizations must be located in Texas and be 501(c)3 tax-exempt. National organizations with significant operations in, or providing material benefits to the citizens of, Texas will be considered based on the degree of operations/benefits within the state.
Restrictions: The foundation does not fund grants, scholarships, or fellowships for individuals.
Geographic Focus: Texas
Date(s) Application is Due: Feb 28; May 31; Sep 30; Nov 11
Amount of Grant: 20,000 - 50,000 U.S.D
Samples: Church on Wheels, Inc. (Midland, TX), $36,000 for operating expenses; Rock the Desert Ministries, Inc. (Midland, TX), $15,000 for operating expenses for the 2006 Christian music rock concert to reach teenagers; St. Andrew's Presbyterian Mission (Midland, TX), $37,500 for the utility bill assistance program.
Contact: David L. Smith, Executive Director; (432) 684-6655; fax (432) 684-4474; ahf@abell-hanger.org
Internet: http://www.abell-hanger.org/GrantCriteria.htm
Sponsor: Abell-Hanger Foundation
P.O. Box 430
Midland, TX 79702-0430

Abell Foundation Health and Human Services Grants 185

Through grants awarded in this area, the Foundation seeks to address societal issues associated with family disintegration, family planning, child support, teenage parenting, domestic violence, children's health and well-being, child abuse and neglect, hunger, food self-sufficiency and homelessness. The Foundation also supports advocacy programs for better health care and social services for children and youth as well as for a comprehensive system of universal health care. Of particular concern is the support of efforts to combat childhood lead paint poisoning and mental health disorders. Furthermore, the Foundation continues to provide opportunities for low-income families to live in quality housing in good neighborhoods in the region. While the Foundation's primary focus is on the development of permanent housing, it also will consider emergency and transitional housing.
Requirements: 501(c)3 organizations serving Maryland communities, especially in the Baltimore area, may apply. The foundation prefers grantees that show strong fiscal management, their project's benefit to the community, ability to achieve goals, unique work, and other sources of financial support.
Restrictions: Grants are not given to individuals, or for sponsorships, deficit financing, operating budgets, annual sustaining funds, travel, or loans. The foundation currently does not fund housing projects, hospitals, or medical research.
Geographic Focus: Maryland
Date(s) Application is Due: Jan 1; Mar 1; May 1; Aug 1; Sep 1; Nov 1
Amount of Grant: 5,000 - 50,000 U.S.D
Contact: Melanie Styles, Program Officer; (410) 547-1300; fax (410) 539-6579; styles@abell.org or abell@abell.org
Internet: http://www.abell.org/programareas/health.html
Sponsor: Abell Foundation
111 S Calvert Street, Suite 2300
Baltimore, MD 21202-6174

Abington Foundation Grants 186

David Knight Ford was born in Cleveland in 1894. Descended from a New England family, his great-grandfather established a farm on a site that is now part of University Circle. The Abington Foundation, named for the place in New England where Mr. Ford's ancestors first settled, was created in 1983 to support the values which the senior Fords championed and which their descendents now uphold. The Foundation awards grants to Ohio nonprofit organizations serving Cuyahoga County in its areas of interest, including pre-primary and higher education, geriatric healthcare and nursing, the promotion or sustenance of individual and family economic independence, and cultural activities. Current priority for the Foundations resources is given to funding requests for specific programs or projects that represent critical periods in a childs development from birth to age 5 and to early adolescence, ages 10-15. Requests for endowment or general operating support are discouraged. Foundation staff is available for consultation during the proposal preparation process.
Requirements: Ohio nonprofit organizations serving Cuyahoga County are eligible.
Geographic Focus: Ohio
Date(s) Application is Due: May 1; Sep 1; Dec 1
Samples: Cleveland Play House, Cleveland, Ohio, $50,000 - to build the Allen Theatre Complex, a three theatre center for performing arts and arts education, in partnership with Playhouse Square and Cleveland State University (2011); North Coast Health Ministry, Cleveland, Ohio, $33,000 - to hire a chronic care nurse case manager (2011); St. Martin de Porres High School, Cleveland, Ohio, $23,000 - for summer programming and tutoring for incoming freshmen (2011).
Contact: Janet E. Narten, Executive Director; (216) 621-2901 or (216) 621-2632; fax (216) 621-8198
Internet: http://www.fmscleveland.com/abington
Sponsor: Abington Foundation
1422 Euclid Avenue, Suite 627
Cleveland, OH 44115-1952

Able To Serve Grants 187

Able to Serve Inc. is a public charity, providing services and opportunities for persons with mental and physical disabilities. Able primarily serves Wake and Johnston counties in North Carolina with some services extended to surrounding counties when resources are available. There are currently four programs that cover the services Able fund within the local community; Local Ministry Network, Community Partnerships, Van Transportation Services, and Computer Learning/Donations.
Requirements: Able to Serve provides each service and program at no charge to the individual or family receiving services within the Wake and Johnston counties of North Carolina.
Geographic Focus: North Carolina
Contact: Carlton S. McDaniel, President; (919) 779-5545; carlton@abletoserve.com
Internet: www.abletoserve.com
Sponsor: Able To Serve
P.O. Box 334
Garner, NC 27529-0334

Able Trust Vocational Rehabilitation Grants for Individuals 188

The trust supports individuals and non-profit vocational rehabilitation programs throughout Florida with fund-raising, grant making and public awareness of disability issues. Created by the Florida Legislature in 1990, the Florida Endowment Foundation for Vocational Rehabilitation, parent organization of the Able Trust, is 501(c)3 non-profit public/private partnership with a goal of assisting Floridians with disabilities in achieving employment. The trust provides grant funds to Florida not-for-profit agencies and Floridians with disabilities for a wide array of projects leading to the employment of individuals with disabilities. To be considered for funding a proposal must address the employment individuals with disabilities and priority is given to those projects with direct employment placement outcomes during the grant time period. Grants to individuals are typically around $2,500. Historically grants have been made between $500-$4,000 for job accommodations.
Requirements: Individuals currently residing in the state of Florida that need emergency on-the-job accommodations to accept an employment offer, retain or receive a promotion at their current employment and are not currently open clients with a state agency provider may apply to be considered for assistance. There are no deadlines; guidelines, helpful tips and forms can be found at the sponsor's website.
Restrictions: Grants funds may not be used to purchase: vehicles, property, building improvements, capital campaigns, endowments, fellowships, scholarships, travel grants, tuition where state and federal aid is available, lobbying, medical items, incurred debt, and proposed expenses prior to grant approval.
Geographic Focus: Florida
Amount of Grant: 500 - 4,000 U.S.D
Contact: Guenevere Crum, Senior Vice President; (850) 224-4493; fax (850) 224-4496; guenevere@abletrust.org or info@abletrust.org
Internet: http://www.abletrust.org/grant/booklet.shtml
Sponsor: Able Trust
3320 Thomasville Road, Suite 200
Tallahassee, FL 32308

Abney Foundation Grants 189

The foundation makes grants for innovative and creative projects, and to programs that are responsive to changing community needs in the areas of health, social service, education, and cultural affairs in South Carolina. The foundation's primary focus is on higher education. All requests must be in writing and in accordance with foundation guidelines, which are available upon request, and from the website. Applicants may submit a Letter of Intent (LOI) briefly describing the project before submitting a proposal in order to find out if their ideas are potentially supported by the foundation.
Requirements: Agencies applying for funds should be serving the citizens of South Carolina.
Restrictions: The foundation does not generally fund requests for operating expenses.
Geographic Focus: South Carolina
Date(s) Application is Due: Nov 15
Amount of Grant: 5,000 - 100,000 U.S.D
Samples: Anderson College (Anderson, SC)—for scholarships, $100,000; Anderson College (Anderson, SC)—to establish a scholarship fund, $100,000.
Contact: David C. King, Executive Director; (864) 964-9201; fax (864) 964-9209; info@abneyfoundation.org
Internet: http://www.abneyfoundation.org/guideline.htm
Sponsor: Abney Foundation
100 Vine Street
Anderson, SC 29621-3265

Aboudane Family Foundation Grants 190

Established in Michigan in 2007, the Aboudane Family Foundation offers awards for operating support to Islamic organizations throughout the United States. Primary fields of interest are stated to be: health; education; and community services. There are no specific application forms or annual deadlines, and applicants should contact the Foundation directly before beginning the writing process. Grants generally range from $1,000 to $15,000.
Amount of Grant: 1,000 - 15,000 U.S.D

Samples: Grand Blanc Islamin Center, Grand Blanc, Michigan, $8,802 - for operating support (2010); Al-Fajr Institute, Flint, Michigan, $5,000 - for operating support (2010); Islamic Relief, Alexandria, Virginia, $12,000 - for operating support (2010).
Contact: Zakwan A. Aboudane, (810) 742-8770; fax (810) 742-8772
Sponsor: Aboudane Family Foundation
5032 Parkwood Court
Flushing, MI 48433-1390

Abramson Family Foundation Grants 191
The foundation awards grants nationwide to Jewish organizations, educational institutions, and health associations in its areas of interest, including arts, education, higher education; hospitals (general); Israel; Jewish agencies, temples, and charitable giving; and social services delivery. The foundation also sponsors a scholarship program. Contact the office for application forms.
Contact: Judith Abramson Felgoise
Sponsor: Abramson Family Foundation
376 Regatta Drive
Jupiter, FL 33477

Abundance Foundation International Grants 192
The Abundance Foundation makes grants to organizations aligned with its mission to improve global health through education, economic empowerment and health systems strengthening. It is focused on programs that unlock the potential of local communities particularly in Africa, Central America and Haiti. Its grantee partners train, support and empower local leaders to create new capabilities that result in lasting improvement in quality of life. In addition to direct grant funding that allows for upscaling of successful existing programs, the Foundation raise awareness about local leaders whose vision and heroism are creating positive change in their communities.
Geographic Focus: All States, Haiti
Contact: Stephen Kahn, President; (510) 841-4123; fax (510) 841-4093; info@abundancefound.org
Internet: http://www.abundancefound.org/grants/
Sponsor: Abundance Foundation
127 University Avenue
Berkeley, CA 94710-1616

ACAAI Foundation Research Grants 193
The Foundation develops, promotes and funds clinical research and educational programs related to allergy, asthma & immunology. The Foundation is dedicated to: strengthening, supporting and funding allergy/immunology training programs; heightening awareness of the critical role of the allergist in cost-effective treatment of asthma and other allergic and immunologic diseases; and supporting asthma camp programs to teach children and their parents how to live with asthma. Programs include clinical fellowship stipends, fellow-in-training research grants, young faculty support awards, scholar's return awards, and support for asthma camps.
Amount of Grant: Up to 250,000 U.S.D
Contact: Program Administrator; (847) 427-1200; fax (847) 427-1294; mail@acaai.org
Internet: http://www.acaai.org/Member/ACAAI_Foundation/ACAAI_Foundation.htm
Sponsor: American College of Allergy, Asthma and Immunology Foundation
85 West Algonquin Road, Suite 550
Arlington Heights, IL 60005

AcademyHealth Alice S. Hersh New Investigator Awards 194
The Alice S. Hersh New Investigator Award recognizes the contribution of new scholars to the field of health services research. Eligible nominees are those whose first research appointments were no earlier than January 1, 2002. Post-doctoral positions will be considered research appointments unless these years provided training necessary for entry into the field. In addition, the eligibility period may be extended for a nominee who has taken a formal leave of absence from their employer. The awards will be presented at the Annual Research Meeting in June. All awardees receive complimentary registration and lodging to attend the meeting and a $2,500 prize from the Alice S. Hersh Memorial Fund.
Requirements: Nominators must be members of AcademyHealth although nominees need not be. Nominees from academic and non-academic institutions are encouraged. Nominations must include one letter of nomination (no more than two pages) that reviews the nominee's contribution to the field and specifies the date of the nominee's first research appointment, and must include curriculum vitae.
Date(s) Application is Due: Mar 2
Amount of Grant: 2,500 U.S.D
Contact: Jennifer Muldoon, Director; (202) 292-6770; fax (202) 292-6870; jennifer.muldoon@academyhealth.org
Internet: http://www.academyhealth.org/content.cfm?ItemNumber=954@navItemNumber=2342
Sponsor: AcademyHealth
1150 17th Street NW, Suite 600
Washington, DC 20036

AcademyHealth Alice S. Hersh Student Scholarships 195
AcademyHealth is pleased to announce a new scholarship program, The Alice S. Hersh Student Scholarship, which provides two free registrations to the National Health Policy Conference (NHPC) and four free registrations to the Annual Research Meeting (ARM). The scholarship is designed to encourage professional and educational development in health services research and policy among student members. This scholarship commemorates the dedication of Alice S. Hersh, the founding executive director of the Association for Health Services Research (AHSR), to supporting the next generation of health services researchers. Application deadline dates are: November 15, for the National Health Policy Conference Scholarships; and March 15, for the Annual Research Meeting Scholarships.
Requirements: Applicants must be current student members of AcademyHealth and must be enrolled in a masters or doctoral program at the time of application. To apply, students must submit the following: a completed scholarship application form (available on the AcademyHealth website); essay outlining how attendance at the ARM or NHPC will assist in your educational and professional pursuits in health services research or policy; current resume; letter from faculty advisor or project coordinator (optional).
Date(s) Application is Due: Mar 15; Nov 15
Contact: Jennifer Muldoon, Director; (202) 292-6770; fax (202) 292-6870; jennifer.muldoon@academyhealth.org
Internet: http://www.academyhealth.org/Training/content.cfm?ItemNumber=2064&navItemNumber=2341
Sponsor: AcademyHealth
1150 17th Street NW, Suite 600
Washington, DC 20036

AcademyHealth Awards 196
AcademyHealth's awards programs recognize individuals who have made significant contributions to the fields of health services research and health policy. Nominations are requested for four awards: Distinguished Investigator Award; Article-of-the-Year; Alice S. Hersh New Investigator Award; and the Dissertation Award. These awards will be presented at the Annual Research Meeting in June. To obtain a description of each award and guidelines, see AcademyHealth website at: http://www.academyhealth.org/content.cfm?ItemNumber=954@navItemNumber=2342.
Requirements: Nominators must be members of AcademyHealth although nominees need not be. Self-nominations are acceptable. All nominations must include the name and contact information for the nominator and nominee, as well as any additional information specified under the description of each award. Complete nominations must be received at AcademyHealth by March 2. Nominations may be submitted by email or postal mail. All awardees receive complimentary registration and lodging to attend the Annual Research Meeting. The Alice S. Hersh New Investigator Awardee receives a $2,500 prize from the Alice S. Hersh Memorial Fund.
Date(s) Application is Due: Mar 2
Contact: Jennifer Muldoon, Director; (202) 292-6770; fax (202) 292-6870; jennifer.muldoon@academyhealth.org
Internet: http://www.academyhealth.org/content.cfm?ItemNumber=954@navItemNumber=2342
Sponsor: AcademyHealth
1150 17th Street NW, Suite 600
Washington, DC 20036

AcademyHealth HSR Impact Awards 197
The program requests nominations of health services research that has made a positive impact on health policy and/or practice. The winning impacts will be disseminated widely as part of AcademyHealth's ongoing efforts to promote the field of health services research and communicate its value for health care decision-making. Nominated research may be published or unpublished, a single study or a body of work, the work of an individual or a team. The time frame for when the research was conducted is open, but the impact should have occurred recently. Nominators and nominees are not required to be AcademyHealth members. Self-nominations are accepted. Award winners are announced at the National Health Policy Conference, February 6 and 7, and the winners receive complimentary registration, travel, and lodging to the conference.
Requirements: Each nomination must include a summary describing the research used, the method of translation, and the impact on policy and/or practice (no more than 500 words); a letter of recommendation from the user of this research discussing the impact the research has had; contact information for the lead researcher; and contact information for the nominator (may be the lead researcher).
Date(s) Application is Due: Jul 29
Amount of Grant: 2,000 U.S.D
Contact: Jennifer Muldoon, Director; (202) 292-6770; fax (202) 292-6870; jennifer.muldoon@academyhealth.org
Internet: http://www.academyhealth.org/awards/hsrimpacts.htm
Sponsor: AcademyHealth
1150 17th Street NW, Suite 600
Washington, DC 20036

AcademyHealth Nemours Child Health Services Research Awards 198
The Nemours Child Health Services Research Award recognizes the scientific work of emerging scholars in the field of child health services research, particularly research on quality improvement of pediatric health services. Recipients must be within seven years of entry into the field of child health services research as of December 31, 2007. Year of entry will be judged by the nominee's first publication in child health services research. The winner will receive $1,000 in recognition of his/her contribution to child health services research, and the award will be presented at the Child Health Services Research Meeting, held in in Washington, D.C., in conjunction with AcademyHealth's Annual Research Meeting in June.
Requirements: Nominations should include: Letter of nomination that identifies when the candidate first entered the field (i.e., title and year of first publication in child health services research); Curriculum vitae that includes a detailed list of accomplishments that merit this recognition. Self-nominations will be accepted.
Date(s) Application is Due: Apr 15

Amount of Grant: 1,000 U.S.D
Contact: Jennifer Muldoon, Director; (202) 292-6770; fax (202) 292-6870; jennifer.muldoon@academyhealth.org
Internet: http://www.academyhealth.org/content.cfm?ItemNumber=982&navItemNumber=2342
Sponsor: AcademyHealth
1150 17th Street NW, Suite 600
Washington, DC 20036

AcademyHealth PHSR Interest Group Student Scholarships 199
Public Health Systems Research (PHSR) is a field of inquiry examining the organization, financing, performance, and impact of health systems, defined as the constellation of governmental and non-governmental actors that influence population health, including health care providers, insurers, purchasers, public health agencies, community-based organizations, and entities that operate outside the traditional sphere of health care. The Student Scholarships recognize graduate students who demonstrate potential to contribute to the field of PHSR through promising research. Scholarship recipients will receive $1,000 to be put toward registration and travel to attend AcademyHealth's Annual Research Meeting (ARM) and the PHSR IG Annual Meeting. Additionally, students will have an opportunity to present their research during the PHSR IG Meeting's poster session.
Requirements: Masters or doctoral students engaged in research that shows potential to advance the understanding of public health systems are encouraged to apply. The deadline for applications is April 15. Applications may be submitted by email or postal mail. To apply, students must submit the following: completed Student Scholarship Form (available at the AcademyHealth website); an abstract for a poster to be presented at the PHSR Interest Group Meeting (this research maybe in-progress); a letter of recommendation from a faculty member or research project director.
Date(s) Application is Due: Apr 15
Contact: Jennifer Muldoon, Director; (202) 292-6770; fax (202) 292-6870; jennifer.muldoon@academyhealth.org
Internet: http://www.academyhealth.org/Communities/content.cfm?ItemNumber=2535&navItemNumber=2035
Sponsor: AcademyHealth
1150 17th Street NW, Suite 600
Washington, DC 20036

AcademyHealth PHSR Research Article of the Year Awards 200
Public Health Systems Research (PHSR) is a field of inquiry examining the organization, financing, performance, and impact of health systems defined as the constellation of governmental and non-governmental actors that influence population health, including health care providers, insurers, purchasers, public health agencies, community-based organizations and entities that operate outside the traditional sphere of health care. The Article of the Year Award recognizes the best scientific work that the field of PHSR has produced and published in the previous year. The awardee will receive complimentary registration for both the PHSR Interest Group Annual Meeting and AcademyHealth's Annual Research Meeting, travel assistance, and a $1,000 cash award. Nominated articles must have been published in a peer-reviewed journal between January 1 and December 31 of the previous year, using either qualitative or quantitative techniques, articles must present, analyze, and comment on new data or synthesize and analyze data that have already been collected. The selection will be based on the article's contribution to the understanding of public health systems; provision of new insights to the field of PHSR; and potential to advance the field and/or challenge current thinking. Nominators must be members of AcademyHealth, although nominees need not be. Self-nominations are acceptable.
Requirements: Nominations must: be received at AcademyHealth by April 15; include the name and contact information for both the nominator and nominee and a copy of the article(s). Nominations may be submitted by email or postal mail. email to: PHSR@academyhealth.org. Mail to: AcademyHealth, Attention: PHSR Article of the Year Committee, 1150 17th Street, NW, Suite 600, Washington, DC 20036.
Date(s) Application is Due: Apr 15
Amount of Grant: 1,000 U.S.D
Contact: Jennifer Muldoon, Director; (202) 292-6770; fax (202) 292-6870; jennifer.muldoon@academyhealth.org
Internet: http://www.academyhealth.org/Communities/content.cfm?ItemNumber=2536
Sponsor: AcademyHealth
1150 17th Street NW, Suite 600
Washington, DC 20036

ACCP Anticoagulation Training Program Grants 201
The Anticoagulation Traineeship is a minimum of a four-week intensive training program for pharmacy students, residents, and fellows that includes a structured didactic component; extensive clinical experience and participation in ongoing clinical research. The training period can be extended 1 or 2 weeks if needed to meet academic requirements of the trainee's academic institution. Under the current system, the trainee spends five half-days per week actively involved in clinics providing patient care under the direct supervision of one of the ACNA clinicians. Trainees whose usual residence is outside the metropolitan San Antonio area will receive a grant-in-aid of $1000 to partially offset travel and living expenses incurred in conjunction with the traineeship.
Requirements: The program is targeted to doctoral-level pharmacy students enrolled in their final year of professional study, residents, and fellows. The necessary arrangements will be made with the student's home institution so the traineeship qualifies for academic and clerkship credit. Students actively enrolled in a Pharm.D. degree program need not be members of ACCP to be eligible. Residents and fellows must be current members of ACCP to be eligible.
Date(s) Application is Due: Nov 1
Amount of Grant: 1,000 U.S.D
Contact: Cathy Englund, Executive Secretary; (913) 492-3311; fax (913) 492-0088; cenglund@accp.com
Internet: http://www.accp.com/frontiers/research.php#resfel
Sponsor: American College of Clinical Pharmacy
13000 W 87th Street Parkway
Lenexa, KS 66215-4530

ACCP Heart Failure Training Program Grants 202
Funded through an educational grant from Scios this heart failure training program is conducted at six different sites within the U.S; the University of Illinois Medical Center, the University of Michigan Health System, the University of North Carolina Heart Failure Program, the Ohio State University, the University of Southern California Medical Center, and the University of Utah Medical Center UTAH Affiliated Heart Failure Program. Depending on the specific site, the Traineeship is a two to four week, intensive training program that includes extensive clinical experience in either the ambulatory care and/or inpatient setting(s), a structured didactic component, and exposure to ongoing clinical research. The primary goals of the heart failure traineeship are to provide pharmacy practitioners, fellows, and residents with specific knowledge and skills central to the management of patients with heart failure. For practitioners, the traineeship will provide sufficient knowledge and experience such that they should have the basis to establish a heart failure clinic or disease management program within their own practices.
Requirements: Residents, fellows, or practicing clinical pharmacists must be current members of ACCP at the time of application to be eligible.
Date(s) Application is Due: May 1; Sep 1
Amount of Grant: 1,000 U.S.D
Contact: Kathy Lemons, Administrative Assistant, Professional Development; (913) 492-3311; fax (913) 492-0088; klemons@accp.com
Internet: http://www.accp.com/
Sponsor: American College of Clinical Pharmacy
13000 W 87th Street Parkway
Lenexa, KS 66215-4530

ACE Charitable Foundation Grants 203
ACE supports the communities around the world in which its employees live and work through its established ACE Foundations and through company-sponsored volunteer initiatives. They focus their philanthropic support in the areas of education, poverty and health, and the environment. Particular consideration is given to opportunities where ACE employees' time and expertise can be utilized in addition to financial support. ACE encourages the development of local and regional initiatives, which reflect their employees' commitment to the needs of the communities in which they live and work. ACE employees direct a significant portion of the company's charitable giving.
Requirements: The mission of the ACE Charitable Foundation is to assist less fortunate individuals and communities in achieving and sustaining productive and healthy lives in geographic areas where ACE employees live and work. The ACE Charitable Foundation strives to accomplish this by focusing the majority of its funds on clearly defined projects that have measurable objectives and outcomes and solve problems in the areas of education, the environment and poverty and health. Submit a brief letter of introduction to the Executive Director for consideration.
Geographic Focus: Arizona, Arkansas, California, Colorado, Connecticut, Delaware, District of Columbia, Florida, Georgia, Hawaii, Illinois, Indiana, Kansas, Louisiana, Maryland, Massachusetts, Michigan, Minnesota, Nevada, New Jersey, New York, North Carolina, Ohio, Oregon, Pennsylvania, Puerto Rico, South Carolina, Texas, Virginia, Washington
Samples: The following is a partial list of organizations with which the ACE Charitable Foundation is associated: American Red Cross; City of Hope Cancer Center; Fairmount Park Conservancy; Kids' Chance; Need in Deed; Philadelphia Cares; Philadelphia Museum of Art; Philadelphia Academies, Inc; United Way
Contact: Eden Kratchman, Executive Director; acefoundation@acegroup.com
Internet: http://www.acelimited.com/AceLimitedRoot/About+ACE/ACE+Philanthropy+Worldwide/The+ACE+INA+Foundation+One+Column.htm
Sponsor: ACE Charitable Foundation
436 Walnut Street, WA 08G
Philadelphia, PA 19106

ACF Ethnic Community Self-Help Grants 204
The Office of Refugee Resettlement (ORR) invites the submission of applications for funding, on a competitive basis, to connect newcomer refugees and their communities with community resources. The objective of this program is to strengthen organized ethnic communities comprised and representative of refugee populations to ensure ongoing support and services to refugees after initial resettlement. Ethnic Community Self-Help applications are for project periods of up to three years (36 months). Awards, on a competitive basis, will be for twelve (12)-month budget periods although project periods may be up to thirty-six (36) months. Applications for continuation grants funded under these awards, beyond the twelve (12)-month budget period but within the thirty-six (36) month project period, will be entertained in subsequent years on a non-competitive basis subject to availability of funds, satisfactory progress of the grantee and determination that continued funding would be in the best interest of the government.
Requirements: Public or private nonprofit agencies are eligible applicants. Faith-based and community organizations that meet the statutory eligibility requirements are also eligible to apply. ORR is interested in applications from national, regional (multi-state) or local refugee community organizations that address community building, facilitate cultural adjustment and integration of refugees, and deliver mutually supportive functions such as

information exchange, civic participation, resource enhancement, orientation and support to newly arriving refugees (and other refugees that maybe in need of such assistance regardless of their resettlement date) and public education to the larger community on the background, needs and potential of refugees. Applicants to this program will be of two general types: (1) National or Multi-site ethnic organizations that propose to develop or strengthen local ethnic groups and/or a national network of ethnic entities for purposes of linking refugees to community resources and promoting and strengthening community participation; or (2) Emerging local ethnic communities that seek to function as bridges between newly arrived refugees and mainstream local resources and organizations.
Restrictions: Foreign entities are not eligible under this announcement.
Date(s) Application is Due: Feb 24
Amount of Grant: 50,000 - 200,000 U.S.D
Contact: Thomas Giossi, Program Manager, Division of Community Resettlement; (202) 401-5720; Thomas.Giossi@acf.hhs.gov
Internet: http://www.acf.hhs.gov/grants/open/HHS-2008-ACF-ORR-RE-0117.html
Sponsor: Administration for Children and Families
370 L'Enfant Promenade SW, 5th Floor
Washington, DC 20447

ACF Foundation Grants **205**
The Foundation supports organizations involved with arts and culture, children and youth services, community and economic development, education, health care, higher education, and human services. Giving is limited to areas of company operations, with emphasis on St. Louis and St. Charles, Missouri. There are no specific deadlines or forms, and applicants should forward the entire proposal to the office.
Requirements: 501(c)3 nonprofit organizations, based in or serving the the communities of St. Louis or St. Charles, Missouri, are eligible to apply.
Restrictions: No grants are given to individuals.
Geographic Focus: Missouri
Contact: Nancy Collins; (636) 940-5101 or (636) 940-5000; ncollins@arleasing.com
Sponsor: ACF Foundation
101 Clark Street, Suite 201
St. Charles, MO 63301-2081

ACF Preferred Communities Grants **206**
The purpose and objectives of the Preferred Communities program are to support resettlement of newly arriving refugees with the best opportunities for their self-sufficiency and integration into new communities, and to support refugees with special needs that require more intensive case management. Applicants may apply to support resettlement in new communities targeted to geographic sites or special populations agreed to in consultation with the Department of State/Bureau for Population, Refugees and Migration (BPRM) and ORR. There are two types of Preferred Communities programs for the purposes of this grant. The first type of Preferred Communities program should expect to receive a minimum of 100 new refugees annually. ORR will consider exceptions to this standard where the applicant provides substantial justification for the request and documents the community's history of arrivals, the period of time needed to reach a level of 100 new refugees and the record of outcomes for achieving self-sufficiency soon after arrival. The second type of Preferred Communities program will expect to receive a proposed number of cases that will need intensive case management. If the Preferred Community plans to focus resources on special needs cases, a history of its qualifications and experience with serving special needs cases should be provided.
Requirements: Eligible applicants are ten national voluntary agencies that currently resettle refugees under a Reception and Placement Cooperative Agreement with the Department of State or with the Department of Homeland Security. The Preferred Communities program is restricted to these agencies because placements of new arrivals occur under the terms of the cooperative agreements, and no other agencies place new arrivals or participate in determining their resettlement sites. The ten eligible applicants are the following: Church World Service, Inc; Domestic and Foreign Missionary Society; Ethiopian Community Development Council, Inc; Hebrew Immigrant Aid Society, Inc; International Rescue Committee; Lutheran Immigration and Refugee Service; State of Iowa; U.S. Conference of Catholic Bishops; U.S. Committee for Refugees and Immigrants; and World Relief Corporation. Faith-based and community organizations that meet the statutory eligibility requirements are eligible to apply under this announcement. Preferred Communities are those localities that support populations where refugees have excellent opportunities to achieve early employment and sustained economic independence without public assistance. Preferred Communities should have a history of low welfare utilization by refugees. In addition, refugees should have the potential for earned income at a favorable level relative to the cost of living and to public assistance benefits. Characteristics of these communities include: (1) A moderate cost of living; (2) excellent employment opportunities in a strong, entry-level labor market; (3) affordable housing and transportation accessible for employment; (4) low secondary out-migration rates for refugees; (5) communities that meet the religious needs of arriving populations; (6) local community support and positive reception for the refugees; (7) receptive school environments; and (8) other related community features that contribute to a favorable quality of life for arriving refugees, such as excellent medical facilities.
Restrictions: Foreign entities are not eligible under this announcement.
Date(s) Application is Due: Feb 24
Amount of Grant: 100,000 - 300,000 U.S.D
Contact: Susan S. Benjamin, (202) 401-4851; Susan.Benjamin@acf.hhs.gov
Internet: http://www.acf.hhs.gov/grants/open/HHS-2008-ACF-ORR-RP-0112.html
Sponsor: Administration for Children and Families
370 L'Enfant Promenade SW, 5th Floor
Washington, DC 20447

ACF Standing Announcement for Unaccompanied Alien Children Grants **207**
One of the functions of the Office of Refugee Resettlement (ORR) Division of Unaccompanied Children's Services (DUCS) is to provide temporary shelter care and other related services to children in ORR custody. This program seeks to fund providers for basic shelter and/or group homes, staff secure, secure and more specialized types of care. There are two priority areas for funding and applicants can apply to both priority areas under separate applications.
Requirements: Eligibility is open to all types of domestic applicants other than individuals. Non-profit organizations (including faith-based and community organizations) and for-profit organizations are eligible to apply. Organizations must be appropriately licensed (at the time of submission of the application) as facilities for the provision of shelter care and other related services to dependent children. For-profit organizations must clearly demonstrate that they are only charging the program actual costs incurred. Program Services: The provision of services will include a structured, safe and productive environment that meets or exceeds respective State guidelines, the Flores Settlement Agreement, and ORR service requirements. Applicants shall provide or arrange for the services listed below for each child in their care. Service delivery shall be accomplished in a manner that is sensitive to the age, culture, religion, dietary needs, native language, and the individual needs of each child. These critical services and assessments must be administered even during the shortest length of stay. Staffing *Requirements:* Applicants shall hire and retain direct care staff that has experience in child welfare, youth work, child care or a related field. A majority of staff that are responsible for direct care delivery shall be bilingual in English and the major language of the children under their care. While Spanish is the primary language of the majority of UAC, access to other languages should be available, if necessary. Applicants are encouraged to identify direct care staff with a minimum level of education equivalent to a high school diploma or higher. Applicants shall screen all employees for criminal background including past history of child sexual abuse. Additionally, applicants shall demonstrate a commitment to comply with state mandatory reporting requirements for child sexual abuse. This requirement includes but is not limited to children who have been victims of commercial sexual exploitation. Program Location: Applicants must plan for the delivery of the minimum services as described in Program Services (above). The shelter/s should be located in areas easily accessible to immigration courts, pro bono legal services, national airports, and community mental health and medical services. Geographic priority consideration and discretion will be given based on overall program capacity and services delivery considerations. Applicants must designate their application in one of the following priority areas: Priority 1 Mega Grants (National Network Applicant)— Priority will be given to applicants who have large national network capability to assist ORR with building capacity and who will be able to accommodate volume fluctuations in the UAC population. Applicants will also be required to provide oversight and monitoring of the facilities within their networks. Clarity of plan for maintaining minimum capacity, as well as their ability to expand capacity in response to program fluctuations and serve specialized types of care (such as staff secure and/or therapeutic care) in addition to primary shelter care. All identified capacity must be appropriately licensed to ensure immediate accessibility as the program necessitates. Priority 2 Micro Grants (Individual Applicants)— ORR may also seek applications from individual care providers or partnerships between two entities to provide specialized services.
Restrictions: Foreign entities are not eligible under this announcement. The U.S. Government is opposed to prostitution and related activities, which are inherently harmful and dehumanizing, and contribute to the phenomenon of trafficking in persons. U.S. non-governmental organizations, and their sub-grantees, cannot use U.S. Government funds to lobby for, promote, or advocate the legalization or regulation of prostitution as a legitimate form of work. It is the responsibility of the primary grantee to ensure these criteria are met by its sub-grantees. Accordingly, the grant application must ensure that no monies, if awarded, will be used for these unallowable purposes.
Date(s) Application is Due: Mar 24
Amount of Grant: 2,000,000 - 40,000,000 U.S.D
Contact: Belinda Rochelle, Division of Unaccompanied Children Services; (202) 401-5269; belinda.rochelle@acf.hhs.gov
Internet: http://www.acf.hhs.gov/grants/open/HHS-2008-ACF-ORR-ZU-0119.html
Sponsor: Administration for Children and Families
370 L'Enfant Promenade SW, 5th Floor
Washington, DC 20447

ACF Supplemental Services for Recently Arrived Refugees Grants **208**
The purpose and objectives ORR seeks to achieve through Supplemental Services for Recently Arrived Refugees are to provide additional resources to communities where the refugee services are insufficient. Under these circumstances, resources are needed to provide additional service capacity to accommodate an increase of refugees. Through Supplemental Services for Recently Arrived Refugees, ORR intends to offer to communities the resources to respond to the needs of newly arrived refugees with adequate and culturally and linguistically appropriate social services. This grant program is intended to provide services to arriving refugees or sudden and unexpected large secondary migration of refugees where communities are not sufficiently prepared in terms of linguistic or culturally appropriate services. In providing services to refugees who have arrived in country less than three years, applicants should view these resources as a temporary solution to insufficient services currently available to meet emerging needs of recently arrived refugees in a specific community. ORR formula social service funds are awarded to States to provide services proportionate to the number of refugee arrivals during the previous three years. ORR invites applications that propose seventeen (17) month projects for a minimum of 100 refugees annually. Examples of situations for which applicants may request funds for grants under Supplemental Services for Recently Arrived Refugees are as follows: (1) Refugee services do not presently exist or the service capacity is not sufficient to accommodate significant increases in arrivals; and (2) the existing service system does not have culturally and linguistically compatible staff.

Requirements: Public or private nonprofit agencies are eligible applicants. Faith-based and community organizations that meet the statutory eligibility requirements are eligible to apply. Allowable activities in the Supplemental Services for Recently Arrived Refugees program are social services for refugees that are appropriate and accessible in language and culture. Examples of allowable social or emergency services to recently arrived refugees include: information and referral services, outreach services, social adjustment services, emergency services, health-related services, home management services, day care for children, transportation, translation and interpretation, case management, citizenship and naturalization preparation services, employment services, employability assessment, on-the job training, English language instruction, vocational training, skills recertification and any additional services approved by the Director of ORR as supporting the objectives of this program.
Restrictions: Foreign entities are not eligible under this announcement.
Date(s) Application is Due: Feb 24
Amount of Grant: 75,000 - 500,000 U.S.D
Contact: Susan S. Benjamin, (202) 401-4851; Susan.Benjamin@acf.hhs.gov
Internet: http://www.acf.hhs.gov/grants/open/HHS-2008-ACF-ORR-RU-0113.html
Sponsor: Administration for Children and Families
370 L'Enfant Promenade SW, 5th Floor
Washington, DC 20447

ACGT Investigators Grants **209**

The Investigators Award in Clinical Translation of Gene Therapy for Cancer distributes funds over 3-5 years, inclusive of a maximum of 10% indirect costs. Funds may be used at the recipient's discretion for salary, technical assistance, supplies, animals or capital equipment, but may not support staff not directly related to the project, e.g. secretaries or administrative assistants. Purchase of equipment is not allowed in the final year of the grant. ACGT is accepting grant applications to produce and release-testing of the clinical trial agents under cGMP, conduct the necessary pre-clinical pharmacological and toxicological studies in appropriate animal models, and/or conducting the clinical translational trials in patients in support of an Investigative New Drug application to the FDA.
Requirements: The candidate must hold an MD, PhD, or equivalent degree and be a tenure-track or tenured faculty. The investigator must be conducting original research as an independent faculty member. ACGT has no citizenship restrictions, and research supported by the award must be conducted at medical schools and research centers only in the United States. While the unambiguous demonstration of preclinical efficacy in cancer treatment by gene therapy is a pre-requisite, entering into the clinical trial during the funding period is also a requirement.
Amount of Grant: Up to 1,000,000 U.S.D
Contact: Grace Pedersen, Foundation Administrator; (203) 358-8000, ext. 495; aneslage@acgtfoundation.org
Internet: http://www.acgtfoundation.org/grants.html
Sponsor: Alliance for Cancer Gene Therapy
96 Cummings Point Road
Stamford, CT 06902

ACGT Young Investigator Grants **210**

The alliance funds research aimed at furthering the development of gene therapy approaches to the treatment of cancer. The overall objectives of this grant are to advance gene therapy into the causes, treatment, and prevention of all types of cancer by promoting development of novel and innovative studies by young investigators. The emphasis of this initiative is to promote basic, and preclinical research approaches utilizing cells and genes as medicine. The six main areas of research ACGT will support are: tumor-specific replicating viruses and bacteria, anti-angiogenesis, immune-mediated gene therapy and cancer vaccines, oncogenes/suppressor oncogenes/apoptosis, tumor targeting and vector development, and other cancer gene therapy research. The three-year grant may be used at the recipient's discretion for salary, technical assistance, supplies, animals or capital equipment, but may not support staff not directly related to the project, e.g. secretaries or administrative assistants. Purchase of equipment is not allowed in the third year of the grant. Continued support is contingent upon submission and approval of a noncompetitive renewal application each year. Guideline and applications are available online.
Requirements: Candidates must hold an MD, MPH, PhD, or equivalent degree and be a tenure-track assistant professor within five years of their initial appointment to this rank, at the time of award activation. The investigator must be conducting original research as an independent faculty member. ACGT has no citizenship restrictions, and research supported by the award must be conducted at medical schools and research centers only in the United States.
Date(s) Application is Due: Sep 21
Amount of Grant: Up to 500,000 U.S.D
Contact: Grace Pedersen, Foundation Administrator; (203) 358-8000, ext. 495; aneslage@acgtfoundation.org
Internet: http://www.acgtfoundation.org/grants.html
Sponsor: Alliance for Cancer Gene Therapy
96 Cummings Point Road
Stamford, CT 06902

ACHE Foster G. McGaw Student Scholarship **211**

The scholarship is designated for students enrolled in their final year of a healthcare management graduate program. One scholarship is awarded per candidate. Application is available online in the Fall.
Requirements: Applicant must be a student associate in good standing of the American College of Healthcare Executives; enrolled in full-time study for the upcoming fall term, which is the student's final year of didactic work in a healthcare management graduate program; demonstrate financial need; and a U.S. or Canadian citizen.
Restrictions: Previous recipients of this scholarship and the Albert W. Dent Graduate Student Scholarship are ineligible.
Date(s) Application is Due: Mar 31
Amount of Grant: 3,500 U.S.D
Contact: Thomas Killam, (312) 424-9400; fax (312) 424-0023; tkillam@ache.org
Internet: http://www.ache.org/Faculty_Students/mcgaw_scholarship.cfm
Sponsor: American College of Healthcare Executives
1 N Franklin Street, Suite 1700
Chicago, IL 60606-3491

Achelis Foundations Grants **212**

Elisabeth Achelis was born in Brooklyn Heights in 1880. She used her inheritance from her father Fritz Achelis, who was President of the American Hard Rubber Company, to establish the Achelis Foundation in 1940 to "aid and contribute to charitable, benevolent, educational and religious uses and purposes for the moral, ethical, physical, mental and intellectual well-being and progress of mankind; to aid and contribute to methods for the peaceful settlement of international differences; to aid and contribute to the furtherance of the objects and purposes of any charitable, benevolent, educational or religious institution or agency; and to establish and maintain charitable, benevolent and educational institutions and agencies." The Achelis Foundation shares trustees, staff, office space, and even a website with the Bodman Foundation which has a similar mission and geographic area of concentration (both foundations give in New York City, while the Bodman Foundation also gives in New Jersey). Funding is concentrated in six program areas: Arts and Culture, Education, Employment, Health, Public Policy, and Youth and Families.
Requirements: 501(c)3 organizations based in New York City that fall within the foundation's areas of interest are welcome to submit an inquiry or proposal letter by regular mail (initial inquiries by email or fax are not accepted, nor are CDs, DVDs, computer discs, or video tapes). An initial inquiry to the foundation should include only the following items: a proposal letter that briefly summarizes the history of the project, need, objectives, time period, key staff, project budget, and evaluation plan; the applicant's latest annual report and complete set of audited financial statements; and the applicant's IRS 501(c)3 tax-exemption letter. Applications may be submitted at any time during the year. Each request is reviewed by staff and will usually receive a written response within thirty days. Those requests deemed consistent with the interests and resources of the foundation will be evaluated further and more information will be requested. Foundation staff may request a site visit, conference call, or meeting. All grants are reviewed and approved by the Trustees at one of their three board meetings in May, September, or December.
Restrictions: The foundation generally does not make grants for the following purposes or program areas: nonprofit organizations outside of New York; annual appeals, dinner functions, and fundraising events; endowments and capital campaigns; loans and deficit financing; direct grants to individuals; individual day-care and after-school programs; housing; organizations or projects based outside the U.S; films or video projects; small art, dance, music, and theater groups; individual K-12 schools (except charter schools); national health and mental health organizations; and government agencies or nonprofit organizations significantly funded or reimbursed by government agencies. Limited resources prevent the foundations from funding the same organization on an ongoing annual basis.
Geographic Focus: New York
Samples: American Ballet Theatre, New York, New York, $25,000 - general operating support (2010); Achievement First, Brooklyn, New York, $100,000 - for start-up and first-year expenses of two new charter schools (2010); Andrew Glover Youth Program, New York, New York, $25,000 - general operating support (2010).
Contact: John B. Krieger; (212) 644-0322; main@achelis-bodman-fnds.org
Internet: http://foundationcenter.org/grantmaker/achelis-bodman/guidelines.html
Sponsor: Achelis Foundation
767 Third Avenue, 4th Floor
New York, NY 10017-2023

Ackerman Foundation Grants **213**

The foundation was established as a charitable trust in 1992 by James F. Ackerman, a local entrepreneur and philanthropist. As an Indianapolis based organization, grants are made predominately to central Indiana organizations as well as a few national medical research institutions. Specifically, the foundation focuses on Indiana cultural institutions and organizations benefiting health and human services, community development, and education. Grant requests will be considered for both operating fund purposes as well as capital campaigns. The foundation does not have grant application forms. To be considered for assistance, an organization should write a brief one or two page letter describing its proposal. The Trustees of the foundation meet semi-annually on the business day that falls on or closest to June 15 and December 15.
Requirements: Established under the laws of the State of Indiana, the foundation considers grant proposals from eligible organizations which are tax exempt under the United States Internal Revenue Service Code section 501(c)3.
Restrictions: The foundation does not make grants to individuals.
Date(s) Application is Due: May 15; Nov 15
Amount of Grant: 500 - 100,000 U.S.D
Samples: Conner Prairie Museum, Fishers, Indiana, $5,000 - general operating support (2010); Brebeuf Jesuit Prep School, Indianapolis, Indiana, $10,000 - general operating support (2010); Providence Cristo Rey High School, Indianapolis, Indiana, $25,000 - general operating support (2010).
Contact: Grants Director; (317) 663-0205; fax (317) 663-0215; jdisbro@cardinalep.com
Internet: http://ackermanfoundation.com/
Sponsor: Ackerman Foundation of Indiana
280 E. 96th Street, Suite 350
Indianapolis, IN 46240-3858

ACOG History of American Obstetrics and Gynecology Fellowship 214
ACOG and Ortho-McNeil Pharmaceutical Corporation jointly sponsor the fellowship in the History of American Obstetrics and Gynecology each year. The recipient of the fellowship spends one month in the Washington, DC, area working full-time to complete their specific historical research project. Although the fellowship will be based in the ACOG History Library, the fellow is encouraged to use other national, historical, and medical collections in the Washington, DC, area. The results of this research must be disseminated through either publication or presentation at a professional meeting. Application and guidelines are available online.
Requirements: ACOG junior fellows and fellows are encouraged to apply.
Date(s) Application is Due: Oct 1
Amount of Grant: 5,000 U.S.D
Contact: Debra Scarborough, (202) 863-2578; fax (202) 484-1595; history@acog.org
Internet: http://www.acog.org/from_home/departments/dept_notice.cfm?recno=21&bulletin=254
Sponsor: American College of Obstetricians and Gynecologists
409 12th Street SW
Washington, DC 20024-2588

ACS Cancer Control Career Development Awards for Primary Care Physicians 215
ACS annually awards three three-year career development awards to physicians specializing in primary care (e.g., family practice, internal medicine, pediatrics, and obstetrics and gynecology). These awards are intended to encourage and assist in the development of promising candidates who will pursue academic careers in primary care specialties. It is anticipated that physicians trained under these awards will improve cancer control through involvement in primary care practice, education, and research activities related to cancer control. Awards are made for three years with progressive stipends.
Requirements: Candidates must be U.S. citizens or permanent residents; must hold an MD, DO, or equivalent degree; and must have completed residency requirements of the appropriate primary care specialty board. Candidate must hold academic rank from instructor to assistant professor and must not be more than 10 years out of training.
Restrictions: Applicants may not have academic rank above that of assistant professor and must not be tenured or be the section head (or equivalent) in his/her discipline. Additionally, the applicant may not be training as a fellow at the time of the award.
Date(s) Application is Due: Oct 15
Amount of Grant: Up to 100,000 U.S.D
Contact: Extramural Grants Department; (404) 329-7558 or (800) 875-2562; fax (404) 321-4669; grants@cancer.org
Internet: http://www.cancer.org/docroot/RES/content/RES_5_2x_Cancer_Control_Career_Development_Awards_for_Primary_Care_Physicians.asp?sitearea=RES
Sponsor: American Cancer Society
3100 Wyman Park Drive, Suite W400
Baltimore, MD 21211

ACS Clinical Research Professor Grants 216
The American Cancer Society offers a limited number of grants to established investigators in mid-career who have made seminal contributions that have changed the direction of clinical, psychosocial, behavioral, health policy or epidemiologic cancer research. Furthermore, it is expected that these investigators will continue to provide leadership in their research area. Up to two awards are made annually for a five-year term that can be renewed once. The award of up to $80,000 per year (direct costs only) may be used for salary or research project support. Interested individuals should submit their curriculum vitae with a complete bibliography and a letter of intent that briefly describes their seminal contributions to research. This information must be submitted via the Research Professor letter of intent on proposalCENTRAL at: https://proposalcentral.altum.com. The Letter of Intent for the Clinical Research Professor Award must be submitted no later than August 1. Upon receipt, the ACS will provide appropriate guidance either by telephone or email. Clinical Research Professor candidates whose Letter of Intent has been approved must submit their application for the award before the October 15 deadline.
Date(s) Application is Due: Aug 1; Oct 15
Amount of Grant: Up to 80,000 U.S.D
Contact: John J. Stevens, Vice President; (404) 329-7550; john.stevens@cancer.org
Internet: http://www.cancer.org/docroot/RES/content/RES_5_2x_Clinical_Research_Professorships.asp?sitearea=RES
Sponsor: American Cancer Society
3100 Wyman Park Drive, Suite W400
Baltimore, MD 21211

ACS Doctoral Degree Scholarships in Cancer Nursing 217
Awarded to graduate students pursuing doctoral study in the field of cancer nursing research, and preparing for careers as nurse scientists. Initial awards are made for up to two years with the possibility of a two-year renewal. There is a stipend of $15,000 per year.
Requirements: The applicant must: be currently enrolled in or applying to a doctoral degree program in nursing or a related field of research; meet requirements for doctoral study and must have been accepted by the institution to which s/he has applied at the time of funding; have a current license to practice as a registered nurse; project a program of study that integrates cancer nursing and provides evidence of faculty support for the program of study. Scholarship recipients must take a minimum of 18 credit hours or 6 courses per year (unless coursework has been completed and the student accepted to candidacy).
Restrictions: Students in programs that award the doctorate of nursing practice (DNP) are not eligible for this program.
Date(s) Application is Due: Oct 15
Amount of Grant: 15,000 U.S.D
Contact: Virginia Krawiec, MPA Director; (404) 329-7612; fax (404) 321-4669; ginger.krawiec@cancer.org
Internet: http://www.cancer.org/docroot/RES/content/RES_5_2x_Doctoral_Degree_Scholarships_in_Cancer_Nursing.asp?sitearea=RES
Sponsor: American Cancer Society
3100 Wyman Park Drive, Suite W400
Baltimore, MD 21211

ACS Doctoral Scholarships in Cancer Nursing 218
Doctoral degree scholarships in cancer nursing are awarded to graduate students pursuing doctoral study in the following cancer nursing fields: research, education, administration, or clinical practice. Awards are made for up to four years.
Requirements: Applicants must be currently enrolled in or applying to a doctoral degree program and have current licensure to practice as a registered nurse. American citizens or permanent residents are eligible.
Date(s) Application is Due: Oct 15
Amount of Grant: 15,000 U.S.D
Contact: Grants Department; (404) 329-7558; fax (404) 321-4669; grants@cancer.org
Internet: http://www.cancer.org/docroot/RES/content/RES_5_2x_Doctoral_Degree_Scholarships_in_Cancer_Nursing.asp?sitearea=RES
Sponsor: American Cancer Society
3100 Wyman Park Drive, Suite W400
Baltimore, MD 21211

ACS Doctoral Training Grants in Oncology Social Work 219
This grant is awarded to qualifying doctoral students at schools of social work that train individuals to conduct research related to the psychosocial needs of persons with cancer and their families. Initial 2-year grant providing a stipend of $20,000 per year with possibility of a 2-year competitive renewal.
Requirements: The following eligibility requirements must be met: the applicant must have master's degree in social work; at least one year of social work experience in a health care setting (oncology/cancer control experience preferred, but not required); a demonstrated commitment to a career in oncology social work as evidenced by recent experience, education, and/or research; the applicant must be currently enrolled in or applying to a doctoral degree program in social work; if an applicant is not yet enrolled in a Ph.D. program, the applicant must meet requirements for doctoral study and must have been accepted by the institution to which s/he has applied at the time of funding; the applicant must project a full-time program of study that integrates oncology social work and provides evidence of faculty support for the program of study; to qualify for funding for the dissertation phase of the doctoral program, the applicant must conduct a dissertation relevant to Oncology Social Work. Successful applicants must also agree to attend the American Cancer Society activities at the annual Society for Social Work Research meeting. (Recipients at the dissertation phase will be expected to present their research at this meeting.)
Date(s) Application is Due: Oct 15
Amount of Grant: 20,000 U.S.D
Contact: Virginia Krawiec, MPA Director; (404) 329-7612; fax (404) 321-4669; ginger.krawiec@cancer.org
Internet: http://www.cancer.org/docroot/RES/content/RES_5_2x_Post-Masters_Training_Grants_in_Clinical_Oncology_Social_Work.asp?sitearea=RES
Sponsor: American Cancer Society
3100 Wyman Park Drive, Suite W400
Baltimore, MD 21211

ACS Graduate Scholarships in Cancer Nursing Practice 220
This award supports graduate students pursuing a master's degree in cancer nursing or doctorate of nursing practice (DNP). Awards may be for two years, with stipend of $10,000 per year. The goal of this program is to strengthen nursing practice by providing assistance for advanced preparation in the following fields of cancer nursing: clinical practice, education, and administration.
Requirements: The applicant must: be currently enrolled in or applying to a master's or DNP degree graduate program with demonstrated integration of cancer content; meet requirements for graduate study and at the time of funding have been accepted by the institution to which s/he has applied. The institution must verify the applicant's acceptance into the master's or DNP program; be pursuing an advanced degree and not solely a post master's certificate. All applicants must have a current license to practice as a registered nurse. Students in bridge programs must have passed the N-CLEX examination by the time the award begins.
Date(s) Application is Due: Feb 1
Amount of Grant: 10,000 U.S.D
Contact: Virginia Krawiec, MPA Director; (404) 329-7612; fax (404) 321-4669; ginger.krawiec@cancer.org
Internet: http://www.cancer.org/docroot/RES/content/RES_5_2x_Masters_Degree_Scholarships_in_Cancer_Nursing.asp?sitearea=RES
Sponsor: American Cancer Society
3100 Wyman Park Drive, Suite W400
Baltimore, MD 21211

ACS Institutional Research Grants 221
The IRG program provides seed money for the initiation of promising new projects by independent junior investigators so they may obtain preliminary results that will enable them to compete successfully for national research grants.
Requirements: Only full-time, tenure-track JHU faculty members at the ranks of Instructor or Assistant Professor or equivalent who are within the first six years of their

first independent research or faculty appointment may apply. Relevance to the cancer problem is important and must be well documented. Within this research scope, a wide breadth of research approaches will be considered; these include basic laboratory studies and clinical research.
Restrictions: Senior investigators, postdoctoral fellows and junior investigators who have competitive national research grants or who have received prior support form the IRG are not eligible.
Date(s) Application is Due: Apr 1
Amount of Grant: 90,000 U.S.D
Contact: Grace Bigelow, Research Projects Administration; (410) 516-5256; fax (410) 516-7775; graceb@jhu.edu
Internet: http://www.cancer.org/docroot/RES/content/RES_5_2x_Institutional_Research_Grants.asp?sitearea=RES
Sponsor: American Cancer Society
3100 Wyman Park Drive, Suite W400
Baltimore, MD 21211

ACS Master's Training Grants in Clinical Oncology Social Work **222**
These grants are awarded annually to qualifying institutions that train clinical oncology social workers to provide cancer patients and their families with psychosocial services. Grants are available to second-year students in a master's program and post-master's social workers within five years of graduation. The master's training must introduce social workers to the special needs of cancer patients and their families. Contact the society for application forms.
Requirements: Applications will be considered from institutions that identify clinical oncology social work training activities, cancer programs with defined psychosocial support services, and relationships with schools accredited by the Council on Social Work Education.
Date(s) Application is Due: Oct 15
Amount of Grant: Up to 12,000 U.S.D
Contact: Grants Department; (404) 329-7558; fax (404) 321-4669; grants@cancer.org
Internet: http://www.cancer.org/docroot/RES/content/RES_5_2x_Masters_Training_Grants_in_Clinical_Oncology_Social_Work.asp?sitearea=RES
Sponsor: American Cancer Society
3100 Wyman Park Drive, Suite W400
Baltimore, MD 21211

ACS MEN2 Thyroid Cancer Professorship Grants **223**
The Professorship award is intended for an outstanding senior investigator who is well known for their contributions to the understanding and/or management of MEN2. Appropriate areas of investigation include, but are not limited to: understanding consequences of RET mutations, molecular events underlying the development of MEN2-related tumors, improved animal models of MEN2, new screening and monitoring tools, new imaging approaches, and new pharmacologic and other strategies to blunt the effects of mutations in RET and other genes associated with medullary thyroid cancer. This person will serve as the leader for a consortium consisting of up to seven (7) investigators with Research Scholar and/or Mentored Research Scholar grants and up to five (5) Postdoctoral Fellows. Responsibilities primarily involve serving as the chairman of an annual research meeting of MEN2 investigators. The amount of the award is $80,000 per year for five years and can be budgeted at the recipient's discretion. Only one award will be made and a letter of intent must be submitted. The awardee must be willing to be an occasional spokesperson for the American Cancer Society and for the Consortium. Interested individuals must electronically submit their curriculum vitae with a complete bibliography and a letter of intent that briefly describes their seminal contributions to thyroid cancer research. This information must be submitted via Access Electronic Grant application process at proposalCENTRAL no later than December 31, by applying for the American Cancer Society â€“ MEN2 Thyroid Cancer Professorship. Qualified candidates will be invited no later than January 15 to submit full applications by April 1.
Requirements: Candidates must be American citizens or permanent residents with at least 10 years of experience beyond receipt of their terminal degree and in general are expected to hold the rank of full professor. Department chairs or individuals with equivalent administrative positions are eligible.
Date(s) Application is Due: Dec 31
Amount of Grant: 80,000 U.S.D
Contact: Charles Saxe, Program Director; (404) 929-6919; charles.saxe@cancer.org
Internet: http://www.cancer.org/docroot/RES/content/RES_5_2x_RFA_for_The_American_Cancer_Society_MEN2_Thyroid_Cancer_Consortium_Professorship.asp?sitearea=RES
Sponsor: American Cancer Society
3100 Wyman Park Drive, Suite W400
Baltimore, MD 21211

ACS Mentored Research Scholar Grant in Applied and Clinical Research **224**
This funding opportunity supports mentored research by full-time faculty, typically within the first four years of their appointment, with the goal of becoming independent investigators in clinical, cancer control and prevention, epidemiologic, psychosocial, behavioral, health services and health policy research. Awards are made up to five years and for up to $135,000 per year (direct costs) plus 8% allowable indirect costs. A maximum of $10,000 per year for the mentor(s) (regardless of the number of mentors) is included in the $135,000. These grants are not renewable.
Date(s) Application is Due: Apr 1; Oct 15
Amount of Grant: 135,000 U.S.D
Contact: Grants Department; (404) 329-7558; fax (404) 417-5974; grants@cancer.org
Internet: http://www.cancer.org/docroot/RES/content/RES_5_2x_Mentored_Research_Scholar_Grant_in_Applied_and_Clinical_Research.asp?sitearea=RES
Sponsor: American Cancer Society
3100 Wyman Park Drive, Suite W400
Baltimore, MD 21211

ACS Physician Training Awards in Preventive Medicine **225**
The Physician Training Award in Preventive Medicine (PTAPM) is intended to encourage and assist the development of promising individuals who will pursue careers in preventive medicine. This program is designed to create a cadre of preventive medicine specialists who are expert in cancer prevention and control, and with the potential to become leaders in research, education, and intervention in this area. Through the Physician Training Award in Preventive Medicine, the Society seeks to support physicians in accredited residency programs that will lead to eligibility for certification in preventive medicine; such programs must provide cancer prevention and control research and practice opportunities. Awards are for four years in the total amount of $300,000, based on an average of $50,000 per resident training year. These grants are renewable.
Requirements: Requirements for the institution are available on the ACS website. Eligibility requirements for the residents are: nominated residents must be citizens or non-citizen nationals of the United States or its possessions and territories, or must have been lawfully admitted to the United States for permanent residence at the time of application; must have an MD, DO, or equivalent degree; must have completed the clinical year of a residency program in preventive medicine, or have at least one year of postgraduate clinical training; must be accepted by or applying to the sponsoring residency program. It is preferred that the candidates not have completed more than half of the required academic work at the beginning of the award period. Residents that have completed the MPH degree will be considered on a case-by-case basis, and must propose course work in cancer prevention and control.
Date(s) Application is Due: Apr 1
Amount of Grant: Up to 300,000 U.S.D
Contact: Grants Department; (404) 329-7558; fax (404) 417-5974; grants@cancer.org
Internet: http://www.cancer.org/docroot/RES/content/RES_5_2x_Physician_Training_Awards_in_Preventive_Medicine.asp?sitearea=RES
Sponsor: American Cancer Society
3100 Wyman Park Drive, Suite W400
Baltimore, MD 21211

ACS Pilot and Exploratory Projects in Palliative Care of Cancer Patients and Their Families Grants **226**
This request for application (RFA) provides funding for investigators performing pilot and exploratory research studies whose purpose is to test interventions, develop research methodologies, and explore novel areas of research in palliative care of cancer patients and their families. A condition of funding is a clearly defined plan as to how the investigator will use the results of the project to develop larger, extramurally funded research projects. This RFA is limited to applications that focus on palliative care research projects for seriously ill cancer patients and their families in three specific areas: exploring the relationship of pain and other distressing symptoms on quality and quantity of life, independence, function, and disability and developing interventions directed at their treatment in patients with advanced and chronic illnesses; studying methods of improving communication between adults living with serious illness, their families and their health care providers; evaluating models and systems of care for patients living with advanced illness and their families. As a condition of accepting the award, each recipient of this grant, will agree to: attend the required meetings of the NPCRC in Fall during the award period; present results of the funded research at this required NPCRC meeting; prepare annual progress reports for each year of funding and a final report at the conclusion of the award period; list the American Cancer Society as funding this study on all publications and presentations.
Requirements: Applications may be submitted by not-for-profit institutions located within the United States, its territories and the Commonwealth of Puerto Rico. Applicants must: be United States citizens, non-citizen nationals or permanent residents of the United States; hold a doctorate degree (MD, PhD, or equivalent) and have a full-time faculty position or equivalent at a college, university, medical school, or other fiscally responsible not-for-profit organization within the United States. Independent investigators at all stages of their career are eligible to apply. Applicants must submit their full application electronically and in paper format per guidelines on cancer.org no later than 5pm (Eastern) on October 15th.
Restrictions: Awards may not exceed $60,000 per year (direct costs) plus 20% indirect costs. Salary support for the Principal Investigator may not exceed 20% of the direct costs. Awards may not exceed a period of one to two years duration.
Date(s) Application is Due: Oct 15
Amount of Grant: Up to 60,000 U.S.D
Contact: Ronit Elk; (404) 417-5957; fax (404) 321-4669; ronit.elk@cancer.org
Internet: http://www.cancer.org/docroot/RES/content/RES_5_2x_RFA__Pilot_and_Exploratory_Projects_in_Palliative_Care_of_Cancer_Patients_and_Their_Families.asp?sitearea=RES
Sponsor: American Cancer Society
3100 Wyman Park Drive, Suite W400
Baltimore, MD 21211

ACS Postdoctoral Fellowships **227**
Fellowships are designed to enable a new investigator to qualify for an independent career in cancer research. Postdoctoral fellowships may be made for one, two, or three years. Stipend payments are made directly to the individual, or the institution if requested,

at the beginning of each month. Travel funds are paid to the individual. Institutional allowances are paid annually at the start of the grant and on the anniversary date thereafter. Awards are made for one to three years with progressive stipends.
Requirements: Applicants must be citizens or permanent residents of the United States (the latter must provide notarized evidence of their legal resident alien status) and shall have been awarded a doctoral degree prior to the activation date of the grant. Application must be endorsed by applicant's mentor and the head of the department in which the training will be received. A plan of training must be formulated and agreed upon by the mentor and the applicant and described in detail in the application.
Restrictions: Awards will not be made to applicants who have completed five or more years of postdoctoral training prior to the start date of the fellowship.
Date(s) Application is Due: Apr 1; Oct 15
Amount of Grant: 142,000 U.S.D
Contact: Grants Department; (404) 329-7558; fax (404) 321-4669; grants@cancer.org
Internet: http://www.cancer.org/docroot/RES/content/RES_5_2x_Postdoctoral_Fellowships.asp?sitearea=RES
Sponsor: American Cancer Society
3100 Wyman Park Drive, Suite W400
Baltimore, MD 21211

ACS Research Professor Grants 228
The American Cancer Society offers a limited number of grants to investigators in mid-career who have made seminal contributions that have changed the direction of cancer research. Furthermore, it is expected that these investigators will continue to provide leadership in their research area. Up to two awards are made annually for a five-year term that can be renewed once. The award of up to $80,000 per year (direct costs only) may be used for salary or research project support. Interested individuals should submit their curriculum vitae with a complete bibliography and a letter of intent that briefly describes their seminal contributions to research. This information must be submitted no later then February 1, via the Research Professor letter of intent on proposalCENTRAL at: https://proposalcentral.altum.com. Upon receipt, ACS will provide appropriate guidance either by telephone or email.
Date(s) Application is Due: Feb 1; Apr 1
Amount of Grant: Up to 80,000 U.S.D
Contact: John J. Stevens, Vice President; (404) 329-7550; john.stevens@cancer.org
Internet: http://www.cancer.org/docroot/RES/content/RES_5_2x_Research_Professorships.asp?sitearea=RES
Sponsor: American Cancer Society
3100 Wyman Park Drive, Suite W400
Baltimore, MD 21211

ACS Research Scholar Grants for Health Services and Health Policy Research 229
The grants support research projects centered on health services and health policy initiated by investigators at any stage in their careers. The initial award includes 20 percent for indirect costs for up to four years, and may be renewed once for up to four years.
Date(s) Application is Due: Apr 1; Oct 15
Amount of Grant: Up to 200,000 U.S.D
Contact: Grants Department; (404) 329-7558; fax (404) 321-4669; grants@cancer.org
Internet: http://www.cancer.org/docroot/RES/content/RES_5_2x_Research_Scholar_Grants_For_Health_Services_and_Health_Policy_and_Outcomes_Research.asp?sitearea=RES
Sponsor: American Cancer Society
3100 Wyman Park Drive, Suite W400
Baltimore, MD 21211

ACS Research Scholar Grants in Basic, Preclinical, Clinical and Epidemiology Research 230
This grant supports investigator-initiated research projects in basic, preclinical, clinical and epidemiologic research. Awards are for up to four years and for up to $200,000 per year (direct costs), plus 20% allowable indirect costs.
Date(s) Application is Due: Apr 1; Oct 15
Amount of Grant: 200,000 U.S.D
Contact: Grants Department; (404) 329-7558; fax (404) 417-5974; grants@cancer.org
Internet: http://www.cancer.org/docroot/RES/content/RES_5_2x_Research_Scholar_Grants_for_Beginning_Investigators.asp?sitearea=RES
Sponsor: American Cancer Society
3100 Wyman Park Drive, Suite W400
Baltimore, MD 21211

ACS Research Scholar Grants in Psychosocial and Behavioral and Cancer Control Research 231
The grants support research projects focusing on the psychosocial and behavioral aspects of cancer by independent investigators at any stage in their careers. Applications are encouraged in which an individual at an early career stage is coprincipal investigator with an established researcher. The initial award supports four years of research and may be renewed once for four additional years.
Requirements: Individuals must not have held an independent position for more than six years at the time the application is submitted.
Date(s) Application is Due: Apr 1; Oct 15
Amount of Grant: Up to 200,000 U.S.D
Contact: Grants Department; (404) 329-7558; fax (404) 417-5974; grants@cancer.org
Internet: http://www.cancer.org/docroot/RES/content/RES_5_2x_Research_Scholar_Grants_in_Psychological_and_Behavioral_Research_for_Beginning_and_Senior_Investigators.asp?sitearea=RES
Sponsor: American Cancer Society
3100 Wyman Park Drive, Suite W400
Baltimore, MD 21211

Active for Life Program Grants 232
This is a four-year initiative supported by the Robert Wood Johnson Foundation that seeks to increase the number of American adults age 50 and older who engage in regular physical activity (30 minutes a day on most days). The program will pursue a multipronged strategy that includes the replication and expansion of programs already developed that have demonstrated efficacy in increasing physical activity levels among midlife and older adults, and will sustain such programs through existing community institutions including, but not limited to, community or senior centers, recreation centers, public health departments, housing authorities, and religious institutions. Active for Life has selected two previously tested program models: group-based problem solving to integrate physical activity into everyday living; and individually selected activities that are facilitated with ongoing, brief telephone and mail follow-up delivered to the home.
Requirements: Applications are accepted from local, state, or regional entities with experience meeting the health and functional needs of populations of adults age 50 and older. Preference will be given to organizations with management teams with experience in interagency collaboration, as well as strong administration, training, and supervision practices.
Date(s) Application is Due: Mar 15
Amount of Grant: Up to 87,000,000 U.S.D
Contact: Dr. Diane Dowdy, (979) 458-4202; activeforlife@srph.tamu.edu
Internet: http://www.activeforlife.info/resources/funding_opportunities.html
Sponsor: Texas AandM University System Health Science Center
1103 University Drive E, Suite 100
College Station, TX 77840

Actors Fund Social Services and Financial Assistance 233
The Actors Fund Social Services offer comprehensive programs designed to meet the critical needs of entertainment professionals throughout their lives. Social workers provide crisis intervention, individual and family needs assessments, and develop long-term plans include ongoing support, education, information and referrals. In addition, financial assistance can be provided for essential living expenses such as rent, utilities or medical costs.
Requirements: Eligibility for financial assistance varies by program. You are encouraged to call and speak with an Intake Social Worker who can best determine how the program may be of assistance. Overall eligibility for financial assistance is based on: a minimum of five years of industry employment with earnings of at least $6,500 for three out of the last five years and financial need. To apply for financial assistance, an application with supporting documentation and an interview is required. There are no specific deadlines.
Contact: Western Region; (323) 933-9244, ext. 55; intakela@actorsfund.org
Internet: http://new.actorsfund.org/services/Social_Services_and_Financial_Assistance/index_html
Sponsor: Actors Fund
729 7th Avenue, 10th Floor
New York, NY 10019

Acumen East Africa Fellowship 234
The East Africa Fellowships are designed to develop leadership skills. The fellowship is structured like an executive MBA, so the awardee remains on the job for the fellowship period. The fellowship creates a unique opportunity for East Africans to receive world-class business training, leadership development, and mentoring. Highlights of the fellowship include: support to refine the applicant's own projects for a large social impact; leadership skills development; and continual guidance from experienced mentors and trainers in strategy and problem solving. Awardees are most likely be successful in their fellowship if they have a deep passion and commitment to the East African region; a proven track record of leadership and management responsibilities; and an unrelenting perseverance, personal integrity and critical thinking skills.
Requirements: Applicants must be at least 18 to apply. There is no educational requirement for the program. The most important requirement is that the applicant have been driving a social change project for at least three years. A "social change project" is defined as any project, initiative, or program that addresses a pressing social issue with an innovative solution that has a positive impact in the lives of the community. The fellowship includes five seminars, including one regional trip. A conference at the end of the fellowship is designed and implemented by the fellowship awardees so they may implement the skills they have learned. The selection process is rigorous and takes place over three months. Online applications are reviewed from March 16 to April 15; phone interviews are conducted from April 17 to May 7; selection conference takes place on May 21 in Nairobi; and finalists are announced on May 31.
Restrictions: Because awardees hold jobs during their fellowship, they do not receive stipends or health insurance.
Geographic Focus: Burundi, Djibouti, Eritrea, Ethiopia, Kenya, Rwanda, Somalia, Sudan, Tanzania, Uganda
Contact: Suraj Sudhakar, East Africa Fellows Manager; +254-20-386-1559 or +254-20-386-1561; ssudhakar@acumenfund.org
Internet: http://www.acumenfund.org/fellows/east-africa-fellows-program.html
Sponsor: Acumen Fund
76 Ninth Avenue, Suite 315
New York, NY 10011

Acumen Global Fellowships 235

The Global Fellowship is a full-time one-year fellowship focused on leadership development and operational experience. Fellows come from all over the world and undergo two months of leadership training in New York and are then placed in one of Acumen Fund's global investments for nine months. Applicants are encouraged to review the website videos of past fellowship recipients. Applicants should also refer to the website for current application information.

Requirements: Applicants should submit a resume/curriculum vitae; personal profile information; two professional references; a letter of recommendation; and four short essay questions. Applicants should review the website's key factors as indicators of a successful fellowship: a proven record of leadership and management responsibilities; experience working in emerging markets; unrelenting perseverance, personal integrity, and critical thinking skills; strong passion and commitment; and 3-7 years of work experience, with a graduate degree preferred.

Restrictions: Fellows are expected to travel frequently and on short notice, sometimes to difficult environments. The program is intensive, time-consuming and requires significant commitment and flexibility. When in the field, access to Internet and phone systems will be available wherever possible, but applicants should expect times when frequent communication with family and friends is not possible or affordable.

Geographic Focus: All States, All Countries

Contact: Blair Miller, Fellows Manager; (646) 747-3961; bmiller@acumenfund.org

Internet: http://www.acumenfund.org/fellows/fellows-programs.html

Sponsor: Acumen Fund
76 Ninth Avenue, Suite 315
New York, NY 10011

Acushnet Foundation Grants 236

The foundation awards one-year, renewable grants to eligible Massachusetts nonprofit organizations in its areas of interest, including community, hospitals and health care, family, and youth and at-risk youths. Types of support include capital grants (building funds and materials), seed money grants, annual campaigns, and emergency funds. Applicants should submit a letter of inquiry.

Requirements: Massachusetts 501(c)3 tax-exempt organizations serving the New Bedford area are eligible.

Restrictions: Grants are not made to individuals.

Geographic Focus: Massachusetts

Amount of Grant: 1,000 - 95,000 U.S.D

Contact: William Blasdale, Seamark Financial Services Contact; (508) 758-6159

Sponsor: Acushnet Foundation
P.O. Box 1498
Mattapoisett, MA 02739

ACVIM Foundation Clinical Investigation Grants 237

The Foundation improves the health and well-being of animals by funding humane studies and communicating information about this work to the veterinary community and the general public. It supports leading veterinary scientists as they take on a range of critical animal health issues within the specialties of small and large animal internal medicine, cardiology, neurology, and oncology. This work, in turn, holds promise for better preventive care, better nutrition, and more sensitive screening and improved treatment for such devastating diseases as cancer, epilepsy, cardiomyopathy, and kidney failure. The Foundation's main goal is to support clinical investigations that lead to new diagnostic, treatment, and prevention techniques. Resident applicants may request up to $12,500 per year, while diplomate applicants may request up to $15,000. One $20,000 grant will be awarded to a collaborative/multi-center proposal, with preference toward private practice/academia partnerships. Funding for multi-year projects up to 3 years will be considered, subject to annual non-competitive renewal based on availability of funds and demonstration of adequate progress on the project. The foundation will consider funding for multi-year projects up to three years.

Date(s) Application is Due: Sep 15

Amount of Grant: Up to 20,000 U.S.D

Samples: Dr. Jill Beech, University of Pennsylvania School of Veterinary Medicine—Evaluation of Endogenous Alpha MSH Concentrations at Different Seasons and in Response to TRH in Normal Horses and Those with Pituitary Hyperplasia; Dr. Stephanie Kottler, University of Missouri-Columbia College of Veterinary Medicine—Prevalence of Staphylococcus aureus and MRSA carriage in three populations.

Contact: Angela E. Frimberger, Executive Director; (303) 231-9933 or (800) 245-9081; info@acvimfoundation.org

Internet: http://www.acvimfoundation.org/grants/info.html

Sponsor: American College of Veterinary Internal Medicine
997 Wadsworth Boulevard, Suite A
Lakewood, CO 80214

ADA Career Development Awards 238

The awards assist exceptionally promising new investigators in their transition to the level of established investigators. The awards support such individuals' salary and research program for five years to enable the investigators to initiate an independent research effort that would result in sufficient accomplishments that in turn would qualify the investigators for long-term research funding. Application request forms are available on the Web site.

Requirements: Applicants must hold full-time or clinical faculty positions or the equivalent at university-affiliated institutions within the United States or one of its possessions; have an MD or PhD degree, or in the case of other health professions, an appropriate advanced degree; and have only two to five years postdoctoral/postfellowship research experience. Applicants must be U.S. citizens or permanent residents or have applied for permanent resident status and must be members of ADA's professional section.

Restrictions: Time spent in residency training is not considered equivalent to research experience.

Date(s) Application is Due: Jan 15

Amount of Grant: Up to 200,000 U.S.D

Contact: Research Department; (703) 299-2071; fax (703) 549-1715; research@diabetes.org

Internet: http://www.diabetes.org/diabetes-research/research-grant-application-forms/development-awards.jsp

Sponsor: American Diabetes Association
1701 N Beauregard Street
Alexandria, VA 22311

ADA Clinical Research Awards 239

The grants provide a stipend for three years for studies undertaken that involve humans directly. In vitro research on human blood or tissue samples does not qualify under this program unless there has been a major in vivo intervention (i.e., diet, drugs, exercise, etc.) and the study protocol is designed specifically to quantitate the effect of this manipulation on the tissue being examined in vitro. The funds may be used for equipment, supplies, and/or technician support. Up to $20,000 of the funds may be used as salary support for the principal investigator. Awards are made for up to three years. Application request forms are available on the Web site.

Requirements: Applicants must be U.S. citizens or permanent residents or have applied for permanent resident status, hold an MD or PhD degree (for other health professionals, an appropriate health- or science-related degree), and hold a faculty-level appointment at a university-affiliated institution within the United States or U.S. possessions. Award recipients must be members of the ADA's professional section.

Date(s) Application is Due: Jan 15; Jul 15

Amount of Grant: 20,000 - 100,000 U.S.D

Contact: Research Department; (703) 299-2071 or (703) 549-1500; fax (703) 549-1715; research@diabetes.org

Internet: http://www.diabetes.org/diabetes-research/research-grant-application-forms/nationwide-research-awards.jsp

Sponsor: American Diabetes Association
1701 N Beauregard Street
Alexandria, VA 22311

ADA Foundation Bud Tarrson Dental School Student Community Leadership Awards 240

American Dental Association Foundation's Bud Tarrson Dental School Student Community Leadership Awards recognize one exemplary volunteer community service project that is organized and/or conducted by a group of dental students enrolled in a pre-doctoral dental education program accredited by the Commission on Dental Accreditation. One application from each dental school is accepted. One award will be made to one dental school on behalf of the student's or mentor's winning application. Current program information including deadlines can be found at the Foundation's website.

Requirements: Award funds must be expressly used to continue to enhance student service learning and outreach to underserved populations.

Restrictions: The Foundation does not accept unsolicited proposals outside of its request for proposal programs announced via their website.

Amount of Grant: 5,000 - 5,000 U.S.D

Contact: Jane Jasek, Administrator; (312) 440-2547; fax (312) 440-3526; JasekJ@ada.org

Internet: http://www.ada.org/applyforassistance.aspx#tarrson

Sponsor: American Dental Association Foundation
211 E Chicago Avenue
Chicago, IL 60611

ADA Foundation Dentsply International Research Fellowships 241

American Dental Association Foundation's Dentsply International Research Fellowships may be applied for by finalists in the Annual Dental Student Research Competition conducted by Dentsply International. Fellowship recipients receive a ten-week research opportunity, a $4,000 stipend, and room and board expenses in order to work in the Foundation's Paffenbarger Research Center located in Gaithersburg, Maryland.

Requirements: Applications are available at the dental school research advisor and/or dean's office.

Contact: Dr. Clifton M. Carey, Director of Administration; (301) 975-6806; fax (301) 926-9143; clif.carey@nist.gov

Internet: http://www.ada.org/applyforassistance.aspx#research

Sponsor: American Dental Association Foundation
211 E Chicago Avenue
Chicago, IL 60611

ADA Foundation Disaster Assistance Grants 242

The ADA Foundation's Disaster Assistance Grants provide assistance to members of the dental profession who have lost property in a declared disaster and to support emergency dental care in areas affected by disasters. There are two types of assistance. Grants to dentists provide immediate assistance to dentists affected by disaster. Grants to organizations provide needed dental services to communities affected by disaster. The maximum award made to dentist is $2,500. The award made to organizations is determined at the Foundation's discretion based on need.

Requirements: To be eligible for funding, a "disaster" must be declared by a governmental agency, or an event meeting the Foundation's definition must have occurred. The Foundation define disaster as "a sudden occurrence which inflicts widespread catastrophic damage to a large geographic area and/or which generally affects a large number of individuals." Disasters can be both natural and caused by human conduct. Examples include, but are not

limited to: civil disorders (excluding acts of war), explosions, fires, tornadoes, earthquakes, floods, tidal waves, forest fires and hurricanes. Eligible dentists are those who are a victim of a disaster. Requests must generally be made within two months of the occurrence of the disaster, unless the time is extended at the discretion of the Board. Dentist applying must show that he or she suffered property damages. Eligible organizations are dental-related 501(c)3 organizations that can provide dental services to victims of a disaster in a community where emergency dental care is needed by victims of a disaster.
Restrictions: Applicants will be offered assistance only once per disaster. The Foundation does not accept unsolicited proposals outside of its request for proposal programs announced via their website.
Contact: Grants Administrator; (312) 440-2547; fax (312) 440-3526; adaf@ada.org
Internet: http://www.ada.org/applyforassistance.aspx
Sponsor: American Dental Association Foundation
211 E Chicago Avenue
Chicago, IL 60611

ADA Foundation George C. Paffenbarger Student Research Awards 243
The ADA Foundation George C. Paffenbarger Student Research Awards are offered in collaboration with the Academy of Operative Dentistry. Awards honors George C. Paffenbarger, an internationally recognized authority on dental materials and a pioneer in the development of specifications and standards for testing materials. The Foundation's Paffenbarger Research Center is named in his honor. Awards are open to dental students at all levels who wish to undertake novel research relevant to contemporary operative dentistry. The budget for the proposed research may not exceed $6,000.
Requirements: The application should take the form of a protocol outlining the background, aims and hypothesis of the proposed research, the methodology to be employed, and the anticipated work on the clinical practice of operative dentistry. Application information, including deadline and submission format, can be found at the Foundation's website.
Restrictions: Applications must be submitted electronically.
Contact: Grants Administrator; (312) 440-2547; fax (312) 440-3526; adaf@ada.org
Internet: http://www.ada.org/applyforassistance.aspx#tarrson
Sponsor: American Dental Association Foundation
211 E Chicago Avenue
Chicago, IL 60611

ADA Foundation Relief Grants 244
ADA Foundation Relief Grant Programs provide financial assistance to dentists and their dependents who, because of accidental injury, a medical condition, disability or advanced age, are unable to meet basic living expenses. Grant rules and criteria as well as an application can be found at the Foundation's website.
Requirements: Eligible applicants are dentists, their dependents and former dependents of deceased dentists if an accidental injury, advanced age, physically debilitating illness or medically-related condition prevents them from gainful employment and results in an inability to be wholly self-sustaining. Dependents are limited to a dentist's current spouse and blood-related or legally adopted children under the age of eighteen. A deceased dentist's spouse, if married at the time of the death, is also considered a dependent.
Restrictions: If a deceased dentist's spouse remarries, he/she is no longer eligible.
Contact: Grants Administrator; (312) 440-2547; fax (312) 440-3526; adaf@ada.org
Internet: http://www.ada.org/applyforassistance.aspx#research
Sponsor: American Dental Association Foundation
211 E Chicago Avenue
Chicago, IL 60611

ADA Foundation Samuel Harris Fund for Children's Dental Health Grants 245
ADA Foundation Samuel Harris Fund for Children's Dental Health Grants provides funding to programs that are designed to improve and maintain children's oral health through outreach, primary prevention, and education. The program currently focuses on funding parent and caregiver education programs to prevent early childhood caries, also known as baby bottle tooth decay, by circumventing primary oral bacterial infection before it can take hold. See the website for the complete request for proposal, application materials, and deadlines.
Requirements: Proposals are accepted from nonprofit 501(c)3 organizations and government agencies for community-based oral health and prevention education programs for parents and caregivers of infants in the United States and its territories.
Restrictions: The Foundation does not accept unsolicited proposals outside of its request for proposal programs announced via their website.
Date(s) Application is Due: Jul 8
Contact: Grants Administrator; (312) 440-2547; fax (312) 440-3526; adaf@ada.org
Internet: http://www.ada.org/applyforassistance.aspx#harris
Sponsor: American Dental Association Foundation
211 E Chicago Avenue
Chicago, IL 60611-2616

ADA Foundation Scholarships 246
ADA Foundation Scholarships are for dental students, underrepresented minority dental students and allied dental students including dental hygiene, dental assisting and dental laboratory technology. Submission deadlines vary and application materials must be obtained from the applicant's school.
Requirements: Eligible students must: be enrolled full-time student at an educational institution; be taking a minimum of twelve credit hours; and have maintained the minimum grade-point average (3.50) required for application to the ADAF Scholarship Program.
Restrictions: Scholarship funds may be used exclusively to defray educational costs such as tuition, fees, books, and supplies.
Amount of Grant: 1,000 - 1,000 U.S.D
Contact: Grants Administrator; (312) 440-2547; fax (312) 440-3526; adaf@ada.org
Internet: http://www.ada.org/applyforassistance.aspx#adaf
Sponsor: American Dental Association Foundation
211 E Chicago Avenue
Chicago, IL 60611

ADA Foundation Scientist in Training Fellowship 247
American Dental Association Foundation's Scientist in Training Fellowships are conducted at the Foundation's Paffenbarger Research Center (PRC). The program includes a full-time fellow working in conjunction with the PRC's scientific research staff. Fellowships have been made possible by the Great-West Life and Annuity Insurance Company.
Requirements: Individuals interested in learning more about the program, application process or deadlines are invited to contact Paffenbarger Research Center.
Contact: Dr. Clifton M. Carey, Director of Administration; (301) 975-6806; fax (301) 926-9143; clif.carey@nist.gov
Internet: http://www.ada.org/applyforassistance.aspx#adaf
Sponsor: American Dental Association Foundation
211 E Chicago Avenue
Chicago, IL 60611

ADA Foundation Summer Scholars Fellowships 248
The ADA Foundation's Paffenbarger Research Center annually appoints two young investigators to its summer scholars program. The program, begun in 1940, brings industrial and dental researchers together in an environment outside the dental school. This program is made possible by Dentsply International.
Requirements: Individuals interested in learning more about the program, including the application process and deadlines, are invited to contact Paffenbarger Research Center.
Contact: Gretchen Duppins, Summer Scholars Program Coordinator; (301) 975-6806; fax (301) 926-9143; gretchen@dupins.nist.gov
Internet: http://www.ada.org/applyforassistance.aspx
Sponsor: American Dental Association Foundation
211 E Chicago Avenue
Chicago, IL 60611

ADA Medical Scholars Program Awards and Physician-Scientist Training Awards 249
The goal of the program is to produce new leaders in diabetes research. The medical scholars program provides one year of research support to students in medical school; the physician-scientist award provides three years of support for the doctoral portion of an MD/PhD degree. Applications may be requested from the Web site.
Date(s) Application is Due: Jan 15
Amount of Grant: 30,000 U.S.D
Contact: Research Department; (703) 299-2071 or (703) 549-1500; fax (703) 549-1715; research@diabetes.org
Internet: http://www.diabetes.org/diabetes-research/research-grant-application-forms/ADA-grant-opportunities/ADA-current-grant-opportunities.jsp
Sponsor: American Diabetes Association
1701 N Beauregard Street
Alexandria, VA 22311

ADA Mentor-Based Postdoctoral Fellowships 250
The purpose of this program is to support the training of scientists in an environment most conducive to beginning a career in diabetes research. An award is given to an established and active investigator in diabetes re-search for the annual stipend support of a postdoctoral fellow to work closely with the mentor. The applicant investigator will be responsible for the selection of the qualified fellow. The term of the award is 4 years. Funding for each year will be contingent upon submission of the name and CV of a fellow by June 1 each year.
Requirements: The postdoctoral fellow must have an MD or PhD degree, no more than three years of postdoctoral research experience, and cannot serve an internship or residency during the award period. There are no citizenship requirements for the fellow. The investigator must meet citizenship and employment eligibility requirements. Award recipients must be members of ADA's professional section.
Date(s) Application is Due: Jul 15
Amount of Grant: Up to 49,000 U.S.D
Contact: Research Department; (703) 299-2071 or (703) 549-1500; fax (703) 549-1715; research@diabetes.org
Internet: http://www.diabetes.org/diabetes-research/research-grant-application-forms/training-awards.jsp#medical
Sponsor: American Diabetes Association
1701 N Beauregard Street
Alexandria, VA 22311

ADA Minority Dental Student Scholarships 251
Scholarships are awarded to minority students who are entering their second year of dental school. The number of scholarships awarded is dependent on funds available. Application forms are available at accredited dental schools.
Requirements: Applicants must be citizens of the United States from one of the following minority groups currently underrepresented in the dental profession: Native American, African American, or Hispanic. Students selected must have been accepted by a dental school in the United States that is accredited by the Commission on Dental Accreditation. Applicants must demonstrate a minimum need of $2500 and maintain a 3.0 grade point average on a 4.0 scale.

Date(s) Application is Due: Jul 31
Amount of Grant: Up to 2,500 U.S.D
Contact: Health Foundation; (312) 440-2547; fax (312) 440-3526; adaf@ada.org
Internet: http://www.ada.org/ada/prod/adaf/prog_scholarship_prog.asp
Sponsor: ADA Endowment and Assistance Fund
211 E Chicago Avenue
Chicago, IL 60611

Adam Richter Charitable Trust Grants 252
The Trust, which is administered by the Bank of America in Dallas, Texas, was established in California in 1994. Offering grants on a national basis, its primary fields of interest include: the arts; animal welfare; Christian agencies and churches; the environment; health organizations; higher education; hospitals; and human services. There are no deadlines, though a formal application is required. Interested applicants should begin by sending a letter of inquiry to the Trust office.
Samples: Habitat for Humanity International, San Francisco, California, $13,082; Greenpeace, Washington, DC, $2,000; Sonoma Land Trust, Santa Rosa, California, $12,000.
Contact: Michael Schlebach, Specialty Asset Manager; (214) 922-9290
Sponsor: Adam Richter Charitable Trust
411 N Akard Street
Dallas, TX 75201-3307

Adams and Reese LLP Corporate Giving Grants 253
At Adams and Reese, the corporation takes pride in giving back to its communities and believes success is directly related to the prosperity and the quality of life within the communities it serves. Its corporate philanthropy program, HUGS (Hope, Understanding, Giving, and Support) was founded in 1988 by Partner Mark Surprenant. Since its inception, the firm has devoted financial resources and thousands of volunteer hours to offer assistance to those in need. A fundamental commitment to volunteerism is the deep-rooted characteristic of the corporation. Primary activities include grants in support of health, youth development, and human services. Fields of interest are: general charitable giving, operating support, health care organizations, legal services, social services, and youth programs.
Requirements: Regions of corporate operation include the states of: Alabama, Florida, Louisiana, Mississippi, Tennessee, Texas, and Washington, DC
Geographic Focus: Alabama, District of Columbia, Florida, Louisiana, Mississippi, Tennessee, Texas
Contact: Mark Surprenant, Liaison Partner; (504) 581-3234; fax (504) 566-0210; Mark.Surprenant@arlaw.com
Internet: http://www.adamsandreese.com/community/
Sponsor: Adams and Reese LLP Corporation
701 Poydras Street, Suite 4500
New Orleans, LA 70139-7755

Adams County Community Foundation Grants 254
The Adams County Community Foundation was established in October of 2007 as a Pennsylvania corporation to succeed the Adams County Foundation which for 22 years had operated as a trust-based community foundation. The Foundation is a public charity that the IRS has determined to be a 501(c)3 organization. The purpose of this foundation is to inspire people and communities to build and distribute charitable funds for Good, for Adams County, forever. After one year of operation, the foundation had acquired nine funds established by individual people or families, members of the Foundation's Board of Directors, groups of business people, a local charity and by a visionary patriot from the 1700s. The Foundation's assets are managed by a professional investment company hired by the Board of Directors and overseen by the Foundation's Investment & Finance Committee. From these funds, distribution of grants is made to qualified, local charities who demonstrate that they are meeting community needs. With a 16 member Board of Directors, Adams County Community Foundation serves as a good steward of the monies of our donors and demonstrates accountability, transparency, confidentiality, compassion, inclusiveness and excellence in its work. The foundation gives to organizations that assist, promote and improve the moral, mental, social and physical well-being of area residents. Fields of interest include: arts; Christian agencies and churches; community and economic development; education; and health organizations.
Requirements: Applicants must serve residents living in Adams County, Pennsylvania.
Geographic Focus: Pennsylvania
Contact: Barbara Ernico, (717) 337-0060 or (717) 337-3353; fax (717) 337-1080; info@adamscountycf.org
Internet: http://www.adamscountycf.org/funds.html
Sponsor: Adams County Community Foundation
101 W Middle Street, P.O. Box 4565
Gettysburg, PA 17325-2109

Adams Rotary Memorial Fund A Grants 255
Established in Indiana, the Adams Rotary Memorial Fund A offers funding in Howard County, Indiana, with a primary goal to support handicapped children throughout the county. With this in mind, its primary fields of interest include: children and youth services; alleviation of disabilities; and money to support families who have a family member with a disability. There are no application deadlines or specific formats, and those in need should begin by contacted the Fund trustee directly.
Geographic Focus: Indiana
Contact: Glenn Grunmann, Trustee; (317) 464-8212 or (574) 282-8839
Sponsor: Adams Rotary Memorial Fund
202 S. Michigan Street
South Bend, IN 46601-2021

Adams Rotary Memorial Fund B Scholarships 256
Established in Indiana, the Adams Rotary Memorial Fund B offers scholarships in Howard County, Indiana, with a primary goal of supporting the continued education of graduates of high schools throughout the county. Financial support is offered for those students pursuing a degree in medicine or nursing. Applications are available upon request, and the annual deadline is the 3rd Friday of March. Awards range from $1,000 to $3,000, and can be used for tuition, books, and room and board.
Requirements: Any high school graduate of a Howard County, Indiana, school is eligible.
Date(s) Application is Due: Mar 15
Amount of Grant: 1,000 - 3,000 U.S.D
Samples: Theodore T. Brown, Indianapolis, Indiana, $3,000 - scholarship award to attend Indiana University School of Medicine (2010); Kristin Antosik, Kokomo, Indiana, $1,000 - scholarship award to attend Purdue University (2010); Erik Pattison, Indianapolis, Indiana, $2,000 - scholarship award to attend Indiana University School of Medicine (2010).
Contact: Glenn Grunmann, Trustee; (317) 464-8212 or (574) 282-8839
Sponsor: Adams Rotary Memorial Fund
202 S. Michigan Street
South Bend, IN 46601-2021

Adaptec Foundation Grants 257
The corporate contributions program is primarily directed, although not exclusively, to organizations and programs serving the communities where its employees work and live, with special emphasis in the Silicon Valley area. Funding priorities include education and research—programs that lower the drop-out rate of youth in accredited schools and colleges, address career development and preparation, job training and placement, educational scholarship assistance, and investment in higher education (primarily in engineering-related university programs); health and human services—research cures for devastating diseases, emergency shelter and subsistence for the homeless, treatment and assistance for abused spouses and children, drug and alcohol rehabilitation, and supporting organizations that strengthen youths and families; helping people enjoy life more fully through music, dance, art museums, opera, theater, and performing arts groups; and $25-$500 employee matching gifts. The Corporate Contributions Committee meets quarterly to consider requests.
Restrictions: In general, Adaptec does not make contributions to any organization that does not have 501(c)3 or comparable status; individuals, except through the Adaptec Scholarship program; churches, synagogues, and other religious groups. (Requests are considered from religious-sponsored and other groups whose activities do not support any specific religious doctrine or ethnic/cultural group.); political or fraternal organizations; underwrite or assist with the development of films, television, or video/radio productions, whether for commercial, independent, or nonprofit ventures, unless such activity is in support of the computer I/O software/hardware industry and is directly related to the business of the company; tickets for raffles, contests, or other fundraising activities (except requests from major fundraising organizations that meet the defined objectives of the contribution program, such as United Way, the Silicon Valley Charity Ball and the San Jose Symphony); organizations whose primary objective is to assist, benefit, and address animal welfare or environmental issues, such as the preservation of endangered species and plants or conservation efforts to preserve the quality of wildlife, water, soil, or air; reduce debts, fund general operating expenses or capital improvement projects, or retroactively fund activities or programs that are already completed; and school sports leagues and events, and other non-educational related activities such as band, debate teams, and goodwill trips.
Amount of Grant: 250 - 2,500 U.S.D
Contact: Contribution Program Manager; (408) 945-8600; fax (408) 262-2533
Internet: http://cms.adaptec.com/ko-KR/company/about/_corporate/adaptec_corpcontributions.htm?nc=/ko-KR/company/about/_corporate/adaptec_corpcontributions.htm
Sponsor: Adaptec Foundation
691 South Milpitas Boulevard, MS-15
Milpitas, CA 95035-5473

ADA Research Awards 258
This program provides grant support to both new and established investigators. Applications will be considered in any area that is relevant to the etiology or pathophysiology of diabetes and its complications. Awards are made for up to three years for the purchase of equipment, supplies, and salary for technical assistance. Up to $20,000 may be used for the investigator's salary. Up to 15 percent of the award may be used for indirect costs.
Requirements: Applicants must hold MD or PhD degrees or, in the case of other health professions, an appropriate advanced degree; must be U.S. citizens or permanent residents or have applied for permanent resident status; and must hold full-time or clinical faculty positions or the equivalent at university-affiliated institutions within the United States or U.S. possessions. Award recipients must be members of the ADA's professional section.
Date(s) Application is Due: Jan 15; Jul 15
Amount of Grant: 20,000 - 100,000 U.S.D
Contact: Magda Galindo, Research Program Manager; (703) 299-2071 or (703) 549-1500; fax (703) 549-1715; mgalindo@diabetes.org. or research@diabetes.org
Internet: http://www.diabetes.org/diabetes-research/research-grant-application-forms/ADA-grant-opportunities/ADA-current-grant-opportunities.jsp
Sponsor: American Diabetes Association
1701 N Beauregard Street
Alexandria, VA 22311

Addison H. Gibson Foundation Medical Grants 259

The Foundation's principal philanthropic interest is to provide medical care; specifically, to enable working people to continue their employment without the burden of untreated illness, or the financial hardship resulting from loss of income associated with untreated illness. The funds provide such medical treatment to needy, self-supporting, long-term western Pennsylvania residents who otherwise cannot afford the medical aid which they require. Physicians, social service and hospital administrators may apply on behalf of their patients.
Requirements: Applicants for medical grants must be referred by their attending physicians, or via social service or hospital administrators. Medical grants are applied for in advance of prescribed treatment. Previously incurred medical expenses are not eligible for consideration. Patients with a correctable condition only will be considered.
Restrictions: Recipients who benefit from these grants must be long-term residents of western Pennsylvania. Grant monies are paid directly to the providers, never to the patients.
Geographic Focus: Pennsylvania
Contact: Rebecca Wallace, Director; (412) 261-1611; fax (412) 261-5733
Internet: http://www.gibson-fnd.org/Docs/Medical_Grants.htm
Sponsor: Addison H. Gibson Foundation
One PPG Place, Suite 2230
Pittsburgh, PA 15222-5401

Adelaide Breed Bayrd Foundation Grants 260

The foundation supports programs and projects of nonprofit organizations serving the Boston, Massachusetts area with emphasis on the Malden community. Areas of interest include adult/continuing education, aging, centers and services, arts, children and youth services, community and economic development, education, family services, health care, health organizations, hospitals (general), human services, libraries and library science, residential and custodial care, and hospices. The foundation funds annual campaigns, building and renovation, capital campaigns, emergency requests, equipment, program development, and scholarship funds. Grants have been awarded in the past to support public libraries, hospitals, and the Girl Scouts. The foundation accepts the Associated Grant Makers (AGM) Common Proposal Form available from the AGM website; however an application form is not required. Applicants should submit the following: a copy of their IRS Determination Letter; copies of their most recent annual report, audited financial statements, and 990; a detailed description of their project and the amount of funding requested; and a copy of their organizational or project budget for the current year. Proposals should be submitted before the second Tuesday in February. Awardees will be notified in April or May.
Requirements: 501(c)3 organizations serving the Boston, Massachusetts area are eligible. Preference will be given to organizations serving the community of Malden, Massachusetts.
Restrictions: The foundation does not support the following categories: requests from individuals; the performing arts (except for certain educational programs); matching or challenge grants; demonstration projects; conferences; publications; research or endowment funds; or loans.
Geographic Focus: Massachusetts
Amount of Grant: 1,000 - 150,000 U.S.D
Contact: C. Henry Kezer, President; (781) 324-1231
Sponsor: Adelaide Breed Bayrd Foundation
350 Main Street, Suite 13
Malden, MA 02148-5023

ADHF/Elsevier Research Initiative Grant 261

This research initiative grant is offered to investigators for support of pilot research projects in gastroenterology-related areas. The objectives of the program are to provide funds for new investigators to help them establish their research careers and to support pilot projects that represent new research directions for established investigators. The intent is to stimulate research in gastroenterology-related areas by permitting investigators to obtain new data that can ultimately provide the basis for subsequent grant applications of more substantial funding and duration. The awards may be used for salary, supplies, or equipment.
Requirements: Investigators must possess an MD or PhD degree (or equivalent) and must hold faculty positions at an accredited North American institution.
Restrictions: Applicants may not hold awards on a similar topic from other agencies.
Date(s) Application is Due: Jan 14
Amount of Grant: 25,000 U.S.D
Contact: Grants Administrator; (301) 222-4005; fax (301) 222-4010; awards@fdhn.org
Internet: http://www.fdhn.org/html/awards/elect_app.html
Sponsor: American Digestive Health Foundation
4930 Del Ray Avenue
Bethesda, MD 20814

ADHF Miles and Shirley Fiterman Foundation Basic Research Awards 262

Two awards are offered: Hugh R. Butt Award in Hepatology or Nutrition, and the Joseph B. Kirsner Award in Gastroenterology. The awards are funded by the Miles and Shirley Fiterman Foundation and recognize excellence in clinical research in hepatology or nutrition and enterology and help support the clinical research efforts of the recipients. Each award, given annually, is provided to the awardee's institution to support his or her research. There are no application forms. A nominating letter, curriculum vita, bibliography, and five reprints of the candidate's most important work should be received by the listed deadline date for applications.
Requirements: Young to midlevel investigators will be ranked above senior people whose contributions, although substantive, may have been accomplished in the distant past and whose major current activities do not include active, cutting-edge research activities.
Date(s) Application is Due: Jan 14
Amount of Grant: 35,000 U.S.D
Contact: Grants Administrator; (301) 222-4005; fax (301) 222-4010; info@fdhn.org
Internet: http://www.fdhn.org/html/awards/elect_app.html
Sponsor: American Digestive Health Foundation
4930 Del Ray Avenue
Bethesda, MD 20814

Administaff Community Affairs Grants 263

Five areas qualify for Administaff corporate contribution consideration: the elderly, education, social service, environment, and health. Administaff will consider four types of requests: volunteers and grants; grants; event sponsorships; and in-kind donations—equipment such as computers, copiers and furniture. All contribution requests must be made in writing. Guidelines and sample cover letter are available online.
Requirements: Charitable organizations must have 501(c)3 status, according to IRS standards; operate in one of Administaff's district markets; and submit a written request. Qualifying organizations also must submit a financial statement to the Better Business Bureau before consideration.
Restrictions: Ineligible entities include: religious organizations (i.e., churches, temples, and other houses of worship, or those whose main purpose is to promote a particular faith or creed); political organizations; individuals; athletic groups; or school clubs.
Amount of Grant: 5,000 U.S.D
Contact: Betty Collins, Program Director, Community Involvement; (281) 358-8986; fax (281) 312-3559; community_involvement@administaff.com
Internet: http://www.administaff.com/about_asf/grants.asp
Sponsor: Administaff Corporation
19001 Crescent Springs Drive
Kingwood, TX 77339-3802

Administration on Aging Alzheimer's Disease Demonstration Grants to States 264

The purpose of this program is to develop models of assistance for persons with Alzheimer's disease and their families, and improve the responsiveness of existing home- and community-based care systems for persons with Alzheimer's disease and related disorders and their families. HHS-2005-AoA-AZ-0502
Requirements: Eligibility for grant awards is limited to state agencies. Only one application per state will be accepted.
Date(s) Application is Due: May 27
Amount of Grant: 250,000 - 325,000 U.S.D
Contact: Margaret Tolson, Program Officer; (202) 357-3440; fax (202) 401-7620; Margaret.Tolson@aoa.hhs.gov
Internet: http://www.aoa.dhhs.gov/doingbus/fundopp/fundopp.asp
Sponsor: Administration on Aging
1 Massachusetts Avenue NW
Washington, DC 20001

Administration on Aging Senior Medicare Patrol Project Grants 265

The goal of SMP Projects is to empower beneficiaries/consumers to prevent health care fraud through outreach and education. Program coverage must target vulnerable, hard-to-reach population beneficiaries, their families and other consumers. The purpose of this competition is to provide the opportunity to fund one (1) Project in each of 26 eligible states including the District of Columbia for a project period of up to three (3) years.
Requirements: Domestic public or private and non-profit entities including state, local and Indian tribal governments (American Indian/Alaskan Native/Native American), faith-based organizations, community-based organizations, hospitals, and institutions of higher education are eligible to apply. Through this competition, AoA plans to fund one project in each of the following 26 states and the District of Columbia: Alaska, Arizona, Arkansas, Colorado, Delaware, the District of Columbia, Florida, Georgia, Idaho, Kansas, Kentucky, Maine, Massachusetts, Michigan, Montana, New Jersey, New Mexico, Ohio, Oklahoma, Oregon, Puerto Rico, Tennessee, Texas, Virginia, Washington, and West Virginia.
Date(s) Application is Due: Apr 11
Amount of Grant: Up to 345,000 U.S.D
Contact: Barbara Lewis, Project Officer; (202) 357-3532; Barbara.Lewis@aoa.hhs.gov
Internet: http://www.aoa.dhhs.gov/doingbus/fundopp/fundopp.asp
Sponsor: Administration on Aging
1 Massachusetts Avenue NW
Washington, DC 20001

Administrative and Health Policy Dental Externships 266

The administrative externships give selected applicants with leadership ability the opportunity to be directly involved in organized dentistry. Students spend four consecutive weeks during the summer at the ASDA and ADA offices in Chicago. Only those students available for four consecutive weeks between June 1 and August 31 should apply. The award consists of a small stipend, transportation to and from Chicago or Washington, DC, and housing.
Requirements: Applicants must be members of the American Student Dental Association and students in good standing.
Contact: Program Contact; (800) 621-8099, ext. 2846; fax (312) 440-2820; ASDA@ASDAnet.org
Internet: http://www.asdanet.org
Sponsor: American Student Dental Association
211 E Chicago Avenue, Suite 1160
Chicago, IL 60611

Adobe Community Investment Grants 267

Adobe supports strategic programs and partnerships that help make its communities better, stronger, and more vibrant places to live, work and do business. Adobe's focus areas for giving and grants programs are designed to: increase Adobe's impact in the community through support of more organizations; and strengthen Adobe's role as a corporate partner by creating deeper, stronger, and richer partnerships. The Adobe Community Investment Grant program provides multi-year (with annual review), comprehensive support, including cash, software, volunteers, and facilities use through an Adobe-initiated, RFP application process. Grant amounts are at least $20,000 and can be for up to three years. Organizations selected become Adobe Community Investment Partners and are required to sit out one year after the grant ends before reapplying.

Requirements: Adobe is currently accepting grant proposals in the following communities: San Jose/Silicon Valley, California (southern San Mateo County, Santa Clara County, southern Alameda County); San Francisco, California; Seattle/King County, Washington; and Ottawa, Ontario, Canada.

Geographic Focus: California, Washington, Canada

Contact: Lesley Dierks; (408) 536-6000 or (408) 536-3993; ldierks@adobe.com

Internet: http://www.adobe.com/aboutadobe/philanthropy/commgivingprgrm.html#Community%20giving%20programs

Sponsor: Adobe Systems
345 Park Avenue
San Jose, CA 95110-2704

Adolph Coors Foundation Grants 268

The Coors Foundation supports organizations that promote the western values of self-reliance, personal responsibility, and integrity. The foundation believes these values foster an environment where entrepreneurial spirits flourish and help Coloradans reach their full potential. High priority is placed on programs that help youth to prosper, that encourage economic opportunities for adults, and that advance public policies that uphold traditional American values. Traditional areas of support include one-on-one mentoring programs, job training, and a variety of self-help initiatives. The foundation also has an interest in bringing integrative medicine into the medical mainstream. In each of its giving areas, the foundation seeks evidenced-based results. Civic and cultural programs attracting the Foundation's attention are typically those that enhance our culture and heritage, that demonstrate our creativity as a people and that are likely to be of economic benefit to and broadly used by the communities they serve. Past grants have supported boys and girls clubs and inner-city health programs. Types of support include building funds, general operating budgets, seed money, and special projects. The foundation has moved to an online screening and application system which is accessible from the foundation website. Application deadlines are March 1, July 1, and November 1.

Requirements: All applicants must be classified as 501(c)3 organizations by the Internal Revenue Service and must operate within the United States.

Restrictions: The foundation does not provide support for the following expenses or entities: organizations primarily supported by tax-derived funds; conduit organizations that pass funds to non-exempt organizations; organizations with two consecutive years of operating loss; K-12 schools or the ancillary programs and projects of those schools; individuals; research projects; production of films or other media-related projects; historic renovation; churches or church projects; museums or museum projects; animals or animal-related projects; preschools, day-care centers, nursing homes, extended-care facilities, or respite care; deficit funding or retirement of debt; special events, meetings, or seminars; purchase of computer equipment; adaptive sports programs; and national health organizations. Organizations applying for start-up funding must have been in operation for at least one full year.

Date(s) Application is Due: Mar 1; Jul 1; Nov 1

Samples: A Christian Ministry in the National Parks, Denver, Colorado, $20,000 - general operating support (2010); Accion New Mexico, Albuquerque, New Mexico, $10,000 - general operating support (2010); Cato Institute, Washington, D.C., $150,000 - general operating support (2010).

Contact: Jeanne Bistranin, Program Officer; (303) 388-1636; fax (303) 388-1684

Internet: http://www.coorsfoundation.org/Process/index.html

Sponsor: Adolph Coors Foundation
4100 E Mississippi Avenue, Suite 1850
Denver, CO 80246

Advance Auto Parts Corporate Giving Grants 269

The Advance Auto Parts Corporate Giving program was founded on the belief that good business is more than just selling merchandise. Since the company was founded in 1932, it has been guided by the following principles, known as the Advance Values: inspire and build the self-confidence and success of every Team Member; serve customers better than anyone else and help them succeed; and grow the business and profitability with integrity. While the corporation understands that there are a variety of worthy causes in every community, to maximize its giving and make the greatest possible impact, Advance has chosen to focus its charitable efforts on serving those in need through support of the following four impact areas: health – improving the health and well-being of others; education – helping people reach their potential by providing educational opportunities; at-risk children and families – helping children and families to assure their critical needs are met through comprehensive, community-based programs; and disaster relief – providing timely support to those impacted by disasters such as earthquakes, hurricanes, floods and other natural disasters.

Contact: Grants Manager; (540) 362-4911; fax (540) 561-1448

Internet: http://corp.advanceautoparts.com/english/about/public.asp

Sponsor: Advance Auto Parts Corporate Giving
5008 Airport Road
Roanoke, VA 24012

Advanced Micro Devices Community Affairs Grants 270

AMD awards grants to nonprofit organizations and schools that serve the communities in which it operates: Sunnyvale, CA, and Austin, TX, and in Europe and Asia. AMD grants funds to agencies that serve the community's health and human service, education and civic needs. AMD K-12 initiatives target programs that increase student interest and/or proficiency in literacy, math, science, and computer technology. AMD also funds programs aimed at developing and supporting effective classroom instruction. AMD also funds basic needs programs to local health and human services organizations. In addition, AMD matches employee gifts to organizations whose activities fall within one of the approved funding areas. AMD also supports engineering education at universities throughout the United States. Generally, funds are granted for specific projects, not for general operating expenses. Requests must be submitted in writing, using appropriate forms, and must include a description of the applicant organization with a statement of its purposes and objectives, a statement regarding the need the grant will address and the geographic area and population served, the names and qualifications of the person(s) who will administer the grant, and amount requested with an explanation of how the funds will be used. Awards are made on a quarterly basis.

Requirements: IRS 501(c)3 tax-exempt organizations and schools based in Sunnyvale, CA, and Austin, TX, are eligible.

Restrictions: AMD will not fund political activities, national programs, advocacy programs, religious or fraternal organizations, individuals, fund-raising events, conferences, seminars, door prizes, registration fees, advertising, or sports leagues.

Geographic Focus: California, Texas, Albania, Andorra, Armenia, Austria, Azerbaijan, Belarus, Belgium, Bosnia & Herzegovina, Bulgaria, Croatia, Cyprus, Czech Republic, Denmark, Estonia, Finland, France, Georgia, Germany, Greece, Hungary, Iceland, Ireland, Italy, Kosovo, Latvia, Liechtenstein, Lithuania, Luxembourg, Macedonia, Malta, Moldova, Monaco, Montenegro, Norway, Poland, Portugal, Romania, Russia, San Marino, Serbia, Slovakia, Slovenia, Spain, Sweden, Switzerland, The Netherlands, Turkey, Ukraine, United Kingdom, Vatican City

Date(s) Application is Due: May 1

Samples: Children's Hospital of Austin (TX)—to develop an outdoor garden for patients and their families, $100,000; El Buen Samaritano (Austin, TX)—to provide prenatal healthcare to uninsured women, and to construct a nurses' station in its new facility, $85,000; Elizabeth Ann Seton Fund (Austin, TX)—to support pharmacy programs at Seton Clinics throughout the city that provide inexpensive medicines to low-income patients, $50,000; United Way Capital Area (Austin, TX)—for its volunteer resource and referral program, as well as a new program that provides mini grants to youth volunteers, $90,000.

Contact: Community Affairs Manager; (408) 749-5373

Internet: http://www.amd.com/us-en/Corporate/AboutAMD/0,,51_52_7697_7702,00.html

Sponsor: Advanced Micro Devices
P.O. Box 3453
Sunnyvale, CA 94088-3453

Advancing Colorado's Mental Health Care Project Grants 271

This project, a joint venture of the foundation, the Colorado Trust, the Denver Foundation, and HealthOne Alliance, focuses on improving the integration and coordination of mental health services for adults with serious mental illness and children with severe emotional disturbances. The project will support the integration of mental health services in five to eight Colorado communities so that people with significant mental health needs can be treated across agencies, regardless of funding sources, organizational structures, or policy and practice differences. Applicant communities propose the target population(s), local approaches for accomplishing improved system integration, and specific strategies and structures for integrating services. Complete program information is available online.

Requirements: Applicants must have the following qualifications and characteristics: a fiscal agent that has 501(c)3 tax-exempt status or is a governmental entity; and an existing collaborative identified to oversee project policy and funding that includes members from agencies and provider organizations serving persons with mental health needs, as well as recipients of mental health services and their families.

Restrictions: Applicants may not propose the use of grant or match funds for new construction or renovation; grants to individuals; retirement of debt; institutional indirect cost recovery; lobbying or political purposes; religious purposes; activities other than charitable activities; replacement of existing or lost funds; or payment for individual services that could be covered by existing funding sources or that falls outside the intent of the grant (i.e., hospital or medication expenses).

Geographic Focus: Colorado

Date(s) Application is Due: May 31

Amount of Grant: 375,000 - 650,000 U.S.D

Contact: Rachel Mondragon, (303) 837-1200 or (888) 847-9140; rachel@coloradotrust.org

Internet: http://www.coloradotrust.org/index.cfm?fuseAction=Grantmaking.Funding Opportunities

Sponsor: Caring for Colorado Foundation
4100 E Mississippi Avenue, #605
Denver, CO 80246

Advancing Newborn Medicine Fellowship Grants 272

The program's objectives are to promote research in the area of neonatal cardiorespiratory medicine, support the professional growth of neonatology fellows, and contribute to improvements in the medical care of hospitalized newborns. Fellows are invited to submit proposals stating a specific question to be answered, methods to be used, and type of analysis to be performed.

Requirements: Proposals are accepted from individuals in accredited neonatal-perinatal medicine training programs in the United States.

Date(s) Application is Due: Nov 30

Amount of Grant: 7,500 U.S.D
Contact: Glen Santiago, (888) 706-3899
Internet: http://www.forestpharm.com
Sponsor: Forest Pharmaceuticals
909 Third Avenue
New York, NY 10022

Advocate HealthCare Post Graduate Administrative Fellowship 273
The program is open to recent graduates from programs in Healthcare Administration or Business Administration, and is designed to provide its fellows with an educational working environment that enables them to use their analytical, project management and communication skills in team-based setting. Under the supervision of senior leaders, fellows complete projects in operational and/or strategic areas. Project assignments are adapted to the individual's needs and interests. The postgraduate fellowship is a one-year program designed to enables the graduate to develop leadership skills and management skills. Fellows are given the opportunity for exposure across Advocate's eight hospitals, three physician practice organizations, home health care agency and corporate offices (Chicago area).
Requirements: In order to be considered, the following must be submitted no later than November 1, the year previous to entry in to the program: (1) Personal statement of interest from explaining qualifications and career objectives; (2) Statement of community involvement; (3) A current resume; (4) Letters of recommendation from applicant's program director and at least one other professional recommendation; and, (5) Copy of unofficial undergraduate transcripts and most recent graduate school transcripts.
Date(s) Application is Due: Nov 1
Contact: Melissa O'Neill, Director; (630) 572-9393
Internet: http://www.advocatehealth.com/body.cfm?id=888
Sponsor: Advocate HealthCare
2025 Windsor Drive
Oak Brook, IL 60523

AEC Trust Grants 274
The AEC Trust is a private foundation established in 1980 as a philanthropic, grantmaking organization. The trust awards grants to eligible nonprofits in Colorado, Florida, Georgia, Massachusetts, and Wisconsin in its areas of interest, including AIDS, arts and culture, community development, elementary education, the environment, health care, higher education, social services, and women. Types of support include building construction/renovation, capital campaigns, challenge/matching grants, equipment acquisition, general operating support, land acquisition, project support, and publications. Request guidelines in writing.
Requirements: 501(c)3 nonprofit organizations in the communities of Boulder, Colorado, Gainesville, Florida, Atlanta, Georgia, Amherst, Massachusetts, and Green Bay, Wisconsin, are eligible.
Restrictions: Grants do not support individuals, international organizations, political organizations, religious organizations, school districts, special events/benefit dinners, state and local government agencies, United Way agencies, national public charities, endowments, sponsorships, or annual fund campaigns.
Geographic Focus: Colorado, Florida, Georgia, Massachusetts, Wisconsin
Date(s) Application is Due: Apr 1; Sep 1
Amount of Grant: 10,000 - 50,000 U.S.D
Contact: Julia Beisel, (800) 839-1754 or (401) 881-3409; requests@foundationsource.com
Internet: https://online.foundationsource.com/public/home/aec
Sponsor: AEC Trust
191 Peachtree Street, Suite 2300
Atlanta, GA 30309-7676

AEGON Transamerica Foundation Grants 275
The foundation supports organizations involved with arts and culture, education, cancer, ALS research, hunger, human services, and the disabled. Giving on a national basis in areas of company operations, with some emphasis on Little Rock, Arkansas, California, Atlanta, Georgia, Cedar Rapids, Iowa, Louisville, Kentucky, Baltimore, Maryland, Charlotte, North Carolina, New York, Pennsylvania, and Fort Worth, Texas.
Requirements: Grant recipient must be a 501(c)3 organization.
Restrictions: Grants are not awarded to individuals. Unsolicited proposals not accepted.
Geographic Focus: Arkansas, California, Georgia, Iowa, Kentucky, Maryland, New York, North Carolina, Pennsylvania, Texas
Amount of Grant: 50 - 2,500,000 U.S.D
Samples: Johns Hopkins University, Baltimore, MD, $2,500,000; United Way of Central Carolinas, Charlotte, NC, $46,676; Theater Cedar Rapids, Cedar Rapids, IA, $150,000.
Contact: Grants Administrator; (319) 398-8935
Sponsor: AEGON Transamerica Foundation
4333 Edgewood Road, NE
Cedar Rapids, IA 52499-3210

Aetna Foundation Health Grants in Connecticut 276
The Aetna Foundation is the independent charitable and philanthropic arm of Aetna Inc. Founded in 1972, the Foundation helps build healthy communities by promoting volunteerism, forming partnerships and funding initiatives that improve the quality of life where our employees and customers live and work. With its Health Grants in Connecticut, the foundation supports programs designed to enhance the quality of health care, and address disabilities and chronic diseases. Support is limited to Hartford and Middletown, Connecticut. Grants of up to $35,000 are awarded.
Requirements: 501(c)3 tax-exempt organizations serving the residents of Hartford and Middletown, Connecticut, are eligible.
Restrictions: The Foundation does not support grants or scholarships to individuals; religious organizations in support of their sacramental or theological functions; advertising; political causes and events; multi-year requests; or fund raising activities, including tables, tickets, dinners, walks, sporting events, etc.
Geographic Focus: Connecticut
Date(s) Application is Due: Apr 10
Amount of Grant: Up to 35,000 U.S.D
Contact: Marilda L. Gandara, President; (860) 273-6382 or (860) 273-0123; fax (860) 273-4764; aetnafoundation@aetna.com
Internet: http://www.aetna.com/foundation/grants_ct/
Sponsor: Aetna Foundation
151 Farmington Avenue
Hartford, CT 06156

Aetna Foundation Quality of Care Grants 277
Aetna is committed to enhancing the quality of health care for all Americans through its business practices and charitable giving. Aetna will support philanthropic initiatives focused on: disparities in health—programs to improve the quality of health care delivered to racial and ethnic populations; and end-of-life care—programs to enhance care and services for both caregivers and patients at the end of life. Priority consideration will be given to programs that address both areas of interest (i.e., end-of-life care for racially and ethnically diverse populations). The listed application deadline is for proposal abstracts; full proposals are by invitation.
Requirements: Grants will be made only to nonprofit organizations with evidence of IRS 501(c)3 designation or de facto tax-exempt status.
Restrictions: The program does not support grants and scholarships to individuals, fundraising activities, advertising, and political or religious activities.
Amount of Grant: 25,000 - 100,000 U.S.D
Contact: Marilda L. Gandara, President; (860) 273-6382 or (860) 273-0123; fax (860) 273-4764; aetnafoundation@aetna.com
Internet: http://www.aetna.com/foundation/grants_qc/index.html
Sponsor: Aetna Foundation
151 Farmington Avenue
Hartford, CT 06156

Aetna Foundation Quality of Care Grants for Depression 278
With its Quality of Care Grants for Depression, the foundation supports programs designed to provide screening and outreach efforts to enhance early identification, diagnosis, and treatment of depression, with a particular emphasis on children and women; integrate mental health services with general medical and preventive health services; provide public education and awareness programs to enhance understanding and treatment of depression and reduce the stigma associated with mental illness; and provide training and education that enhances the ability of health care professionals to communicate and deliver culturally sensitive mental health care and services. Special emphasis is directed toward programs designed to focus on depression for racially and ethnically diverse populations. Grants range from $50,000 to $300,000. Depression Grant abstracts are due on October 15, with full proposals (by invitation from abstracts) due on February 15 of the following year.
Requirements: 501(c)3 tax-exempt organizations are eligible.
Restrictions: The Foundation does not support grants or scholarships to individuals; religious organizations in support of their sacramental or theological functions; advertising; political causes and events; multi-year requests; or fund raising activities, including tables, tickets, dinners, walks, sporting events, etc.
Date(s) Application is Due: Feb 15; Oct 15
Amount of Grant: 50,000 - 300,000 U.S.D
Contact: Marilda L. Gandara; (860) 273-6382; aetnafoundation@aetna.com
Internet: http://www.aetna.com/foundation/grants_qc/overview.html
Sponsor: Aetna Foundation
151 Farmington Avenue
Hartford, CT 06156

Aetna Foundation Quality of Care Grants for End of Life Care 279
With its Quality of Care Grants for End of Life Care, the foundation supports programs designed to provide training and education, including cultural competency, for professionals across the health care continuum to enhance communication and care at the end of life; provide support services and resources for caregivers to help them deal more effectively with complex and emotionally charged end of life care issues; and reduce barriers and increase use of palliative and home care services for patients at the end of life. Special emphasis is directed toward programs designed to focus on end of life care for racially and ethnically diverse populations. Grants range from $50,000 to $300,000. End of Life Care Grant abstracts are due on September 15, with full proposals (by invitation from abstracts) due on February 15 of the following year.
Requirements: 501(c)3 tax-exempt organizations are eligible.
Restrictions: The Foundation does not support grants or scholarships to individuals; religious organizations in support of their sacramental or theological functions; advertising; political causes and events; multi-year requests; or fund raising activities, including tables, tickets, dinners, walks, sporting events, etc.
Date(s) Application is Due: Feb 15; Sep 15
Amount of Grant: 50,000 - 300,000 U.S.D
Contact: Marilda L. Gandara; (860) 273-6382; aetnafoundation@aetna.com
Internet: http://www.aetna.com/foundation/grants_qc/overview.html
Sponsor: Aetna Foundation
151 Farmington Avenue
Hartford, CT 06156

Aetna Foundation Regional Community Health General Support Grants 280

The Aetna Foundation is the independent charitable and philanthropic arm of Aetna Inc. Founded in 1972, the Foundation helps build healthy communities by promoting volunteerism, forming partnerships and funding initiatives that improve the quality of life where our employees and customers live and work. The foundation awards Regional Community Health General Support Grants to nonprofit organizations that have discussed possibilities with a local Aetna Representative, received an invitation to apply, and are involved with arts and culture, education, health, human services, and economic development.

Requirements: 501(c)3 tax-exempt organizations in regionally designated communities are eligible (list available online).

Restrictions: Restrictions include: grants to individuals; scholarships; tables or tickets for sporting events; fund-raising activities, e.g., dinners, walks, golf tournaments or similar events; religious organizations in support of their sacramental or theological functions; advertising; multiyear requests; or political causes and events.

Date(s) Application is Due: Mar 15; Apr 15; May 15

Amount of Grant: 25,000 - 50,000 U.S.D

Contact: Marilda L. Gandara, President; (860) 273-6382 or (860) 273-0123; fax (860) 273-4764; aetnafoundation@aetna.com

Internet: http://www.aetna.com/foundation/grants_reg/overview.html#2

Sponsor: Aetna Foundation
151 Farmington Avenue
Hartford, CT 06156

Aetna Foundation Regional Community Health Grants 281

The foundation's grants focus on reducing disparities in health care among racial and ethnic populations. Areas of interest include cultural competency—training and education for health care professionals, including culturally appropriate end-of-life care training; disease prevention, awareness, and delivery of culturally sensitive care and services related to children's oral health—community-based screening, treatment, and/or family education initiatives regarding the importance of dental care; diabetes— initiatives targeting individuals and families with prevention and healthy behavior modification messages to help combat the growing incidence of type 2 diabetes in children and adults; and depression—screening, education, and outreach efforts to enhance early identification, diagnosis, and treatment of depression in youth and adults. Applicants may submit only one application per regional program. Guidelines are available online.

Requirements: 501(c)3 tax-exempt organizations in regionally designated communities are eligible (list available online).

Restrictions: Restrictions include: grants to individuals; scholarships; tables or tickets for sporting events; fund-raising activities, e.g., dinners, walks, golf tournaments or similar events; religious organizations in support of their sacramental or theological functions; advertising; multiyear requests; or political causes and events.

Date(s) Application is Due: Mar 31

Amount of Grant: 25,000 - 50,000 U.S.D

Contact: Marilda L. Gandara, President; (860) 273-6382 or (860) 273-0123; fax (860) 273-4764; aetnafoundation@aetna.com

Internet: http://www.aetna.com/foundation/grants_reg/

Sponsor: Aetna Foundation
151 Farmington Avenue
Hartford, CT 06156

Aetna Foundation Regional Community Health Outreach Grants 282

Awards will be granted to nonprofit organizations to support grassroots efforts that increase awareness of critical health initiatives through health walks, health fairs and health education outreach. Grants of up to $25,000 will be considered. Applicants should provide all levels of event sponsorships on the application. Funding decisions and notifications will be made throughout the year while all notifications will be made by November 1.

Requirements: 501(c)3 tax-exempt organizations in regionally designated communities are eligible (list available online).

Restrictions: Restrictions include: grants to individuals; scholarships; tables or tickets for sporting events; fund-raising activities, e.g., dinners, walks, golf tournaments or similar events; religious organizations in support of their sacramental or theological functions; advertising; multiyear requests; or political causes and events.

Amount of Grant: Up to 25,000 U.S.D

Contact: Marilda L. Gandara, President; (860) 273-6382 or (860) 273-0123; fax (860) 273-4764; aetnafoundation@aetna.com

Internet: http://www.aetna.com/foundation/grants_reg/overview.html#2

Sponsor: Aetna Foundation
151 Farmington Avenue
Hartford, CT 06156

Aetna Foundation Sponsorships Grants in Connecticut 283

The Aetna Foundation is the independent charitable and philanthropic arm of Aetna Inc. Founded in 1972, the Foundation helps build healthy communities by promoting volunteerism, forming partnerships and funding initiatives that improve the quality of life where our employees and customers live and work. With its Sponsorships Grants in Connecticut, the foundation awards grants to nonprofit organizations to sponsor fundraising and other events, including walks, dinners, health fairs, or similar events, with an emphasis on health-related issues. Support is limited to Hartford and Middletown, Connecticut. Grants of up to $10,000 are awarded on a quarterly basis.

Requirements: 501(c)3 tax-exempt organizations serving the residents of Hartford and Middletown, Connecticut, are eligible.

Restrictions: The Foundation does not support grants or scholarships to individuals; religious organizations in support of their sacramental or theological functions; advertising; political causes and events; multi-year requests; or fund raising activities, including tables, tickets, dinners, walks, sporting events, etc.

Geographic Focus: Connecticut

Date(s) Application is Due: Feb 27; Jul 24; Oct 30

Amount of Grant: Up to 10,000 U.S.D

Contact: Marilda L. Gandara, President; (860) 273-6382 or (860) 273-0123; fax (860) 273-4764; aetnafoundation@aetna.com

Internet: http://www.aetna.com/foundation/grants_ct/

Sponsor: Aetna Foundation
151 Farmington Avenue
Hartford, CT 06156

Aetna Foundation Voice Of Conscience Award 284

The program was created in memory of Arthur Ashe Jr., tennis great, citizen of the world, member of Aetna's board of directors for more than 10 years, and chairman of the Aetna Foundation at the time of his death. The award is made to a citizen/leader who emulates the selfless commitment to education, civil rights, health and human services that Arthur had. The award winner will receive a grant to the nonprofit(s) organizations of his/her choice. Characteristics of the award recipient include: advancing human rights, opportunity, and dignity; commitment to serve other, regardless of personal benefit; devotion to achieving humanitarian ideals; accomplishments in civil rights, education, health care, and social equity; willingness to take risks for challenging causes; and breaking boundaries across race, culture, class, faith, and ethnicity. Guidelines and nomination form are available online.

Date(s) Application is Due: Jun 15

Amount of Grant: 50,000 U.S.D

Samples: Marie Alfano, case management nurse consultant, Cranbury, New Jersey.

Contact: Marilda L. Gandara, President; (860) 273-6382 or (860) 273-0123; fax (860) 273-4764; aetnafoundation@aetna.com

Internet: http://www.aetna.com/foundation/award/index.html

Sponsor: Aetna Foundation
151 Farmington Avenue
Hartford, CT 06156

AFAR/Merck Junior Investigators In Geriatric Clinical Pharmacology 285

The training program provides the opportunity for two fellows to acquire competency in clinical pharmacology, geriatrics/gerontology, and research in geriatric clinical pharmacology. Applications must be submitted by an institution on behalf of an individual candidate for the fellowship.

Requirements: The candidate must be board certified or eligible in a primary specialty by July 1. Proof of board eligibility must be submitted. At the time of application, the candidate must be within three years of having completed postdoctoral or fellowship training. Previous training in geriatrics or clinical pharmacology is not required, but one or the other is highly desirable. Candidate must be a citizen or a permanent resident of the United States.

Date(s) Application is Due: Nov 3

Amount of Grant: 60,000 U.S.D

Samples: Dr. Daniel Kaufer, Assistant Professor of Psychiatry and Neurology, U of Pittsburgh (Pittsburgh, PA)—for research to identify risk factors for anticoagulant-related bleeding, $100,000.

Contact: Grants Manager; (212) 703-9977; fax (212) 997-0330; grants@afar.org

Internet: http://www.afar.org/grants.html

Sponsor: American Federation for Aging Research
55 West 39th Street, 16th Floor
New York, NY 10018

AFAR CART Fund Grants 286

The goal of the CART Fund is to encourage exploratory and developmental AD research projects within the United States by providing support for the early and conceptual plans of those projects that may not yet be supported by extensive preliminary data but have the potential to substantially advance biomedical research. This proposal should be distinct from those projects designed to increase knowledge in a well established area unless it is intended to extend previous discoveries toward new directions or applications. Applications may encompass a project period of up to two years with a combined budget for direct cost up to $250,000. No indirect costs are allowed. Domestic public and private institutions are eligible, such as universities, colleges, hospitals and laboratories. This is for NEW projects only. Up to two $250,000 awards will be available this year. Dr. John Trojanowski will chair a scientific review group that will triage the Letters-of-Intent and select a maximum of fifteen deemed to have the highest merit. Those selected will be invited to submit a subsequent standard grant application from which the final recommendation will be made by the review group. The final selection will be made by the CART Fund selection committee. Notifications of the finalists (maximum of 15) will be mailed by January 7th. Investigators whose proposals are accepted for further consideration will be sent a full application form which must be completed and returned by February 16th.

Requirements: The CART Fund is inviting interested applicants from within the United States to submit a Letter-of-Intent that includes sufficient detail to communicate the importance of your study as well as information on its feasibility. Submit email questions and Letters-of-Intent to Dr. James B. Puryear, Chairman, CART Grants Program: jimpuryear@comcast.net.

Restrictions: Applications will be deemed ineligible from for-profit organizations as well as those already supported by regular or program grants.

Date(s) Application is Due: Feb 16
Amount of Grant: 250,000 U.S.D
Samples: Sanjay W. Pimplikar, Ph.D. - The Cleveland Clinic, OH; Todd E. Golde, M.D., Ph.D. - Mayo Clinic, Jacksonville, FL; Claudio Soto, Ph.D. - University of Texas Medical Branch, TX;
Contact: Grants Manager; (212) 703-9977 or (888) 582-2327; fax (212) 997-0330; grants@afar.org or info@afar.org
Internet: http://www.afar.org/CART.html
Sponsor: American Federation for Aging Research
55 West 39th Street, 16th Floor
New York, NY 10018

AFAR Medical Student Summer Research Training in Aging Program **287**

The program provides eight- to 12-week scholarships for clinical geriatrics and aging research. Scholars receive training from top experts in geriatrics and/or gerontology and other disciplines such as physiology, molecular biology, neurology, and epidemiology at one of four national training centers or, for a limited number of medical schools, at their own institutions. Scholars also participate in research initiatives in basic, clinical, or health services research relevant to care of the elderly. Following the scholarship, scholars participate in aging research and/or related activities under the supervision of their home-site sponsor. Most scholars will conduct training and research during the summer months. Application and guidelines are available online.
Requirements: Any allopathic or osteopathic medical student in good standing who has completed at least one year of medical school and is a U.S. citizen or permanent resident may apply. Each student submits an application, which must be accompanied by a letter of support from an individual who knows the candidate and by a biographical sketch from the home school faculty member who has agreed to nurture the work of the scholar's geriatric career development.
Date(s) Application is Due: Feb 7
Amount of Grant: 2,596 - 5,193 U.S.D
Contact: Grants Manager; (212) 703-9977; fax (212) 997-0330; amfedaging@aol.com
Internet: http://www.afar.org/grants.html
Sponsor: American Federation for Aging Research
55 West 39th Street, 16th Floor
New York, NY 10018

AFAR Medical Student Training in Aging Research Program **288**

The Medical Student Training in Aging Research Program (MSTAR) provides medical students, early in their training, with an enriching experience in aging related research and geriatrics, under the mentorship of top experts in the field. This program introduces students to research and academic experiences that they might not otherwise have during medical school. Students participate in an eight to twelve week structured research, clinical, and didactic program in geriatrics, appropriate to their level of training and interests. Students may train at a National Training Center supported by the National Institute on Aging or, for a limited number of medical schools, at their own institution. Refer to the AFAR website for a complete listing of the participating institutions.
Requirements: Any allopathic or osteopathic medical student in good standing, who will have successfully completed one year of medical school at a U.S. institution by June, may apply. Applicants must be citizens or non-citizen nationals of the United States, or must have been lawfully admitted for permanent residence (i.e., in possession of a currently valid Alien Registration Receipt Card I-551, or some other legal verification of such status.)
Restrictions: Individuals on temporary or student visas and individuals holding PhD, MD, DVM, or equivalent doctoral degrees in the health sciences are not eligible.
Date(s) Application is Due: Feb 6
Contact: Grants Manager; (212) 703-9977 or (888) 582-2327; fax (212) 997-0330; grants@afar.org or info@afar.org
Internet: http://www.afar.org/medstu.html
Sponsor: American Federation for Aging Research
55 West 39th Street, 16th Floor
New York, NY 10018

AFAR Paul Beeson Career Development Awards in Aging Research for the Island of Ireland **289**

The Paul Beeson Career Development Awards In Aging Research Program offers faculty development awards to outstanding junior physician faculty committed to academic careers in aging-related research, teaching, and practice. The goals of the program are: to encourage and assist the development of future leaders in the field of aging by supporting faculty members early in their careers to gain additional research training as needed and to establish independent programs in aging research; to deepen the commitment of research institutions to academic research in aging and to translating research outcomes to geriatric medicine by involving mentor and recipient in establishing and advancing the recipient's career in aging research; to expand medical research on aging broadly defined as including the biology of aging, maintenance of health and independence in old age, diseases and disabilities of old age and issues in their clinical management, and systems of care for the elderly. To maximize the educational/training opportunities of this fellowship program, Beeson Ireland Scholars are encouraged to spend three to six months abroad with a research team in the U.S. in the second or the third year of the award. The goals of this rotation are varied and can be for purposes of training in techniques, data collection, data analysis, etc. to enhance the Scholars' transition to independence. This is unique to the Beeson Ireland Program. To date, the program has provided awards to 149 very promising junior faculty at institutions in the United States as well as 3 junior faculty on the Island of Ireland. The program was established in 1994 by The Atlantic Philanthropies, The John A. Hartford Foundation, The Commonwealth Fund and The Starr Foundation, and administered by the American Federation for Aging Research (AFAR.) In 2004, The Atlantic Philanthropies, The John A. Hartford Foundation, The Starr Foundation and AFAR entered a partnership with The National Institute on Aging (NIA.) Until 2006, the program soley funded Scholars at U.S.-based institutions but beginning in 2007, with support from The Atlantic Philanthropies, Scholars are also funded in Ireland (The Republic of Ireland and Northern Ireland.) The scholar will receive a grant for three years. The salary will be at specialty registrar/clinical lecturer level to protect a minimum of 75% of the scholar's time for research, with the remainder available annually for research support, including an institutional overhead on the research support component.
Requirements: Nominations for the Paul Beeson Career Development Awards In Aging Research Program are to be made by the deans of medical schools (or equivalent) in Ireland (the Republic of Ireland and Northern Ireland.) Institutions may submit as many applications as they wish. To be eligible for nomination, a candidate must: be a U.S. physician or an Irish citizen physician of the Republic of Ireland, a UK citizen of Northern Ireland, an EU citizen living and working in the Island of Ireland or a non-EU citizen resident in the island of Ireland with a valid work permit; commit at least 75% of his/her full-time professional effort to the goals of this award; be a medical graduate who has recently completed or is about to complete a doctoral degree (PhD/MD) and be undertaking or have recently completed higher clinical training; have at least one research publication in a high-impact journal. If the candidate is not the first author, he/she must provide a short description of his/her contribution to the research and the writing of the paper; this should be substantial. For each scholar, a senior faculty member at the scholar's institution must be selected to serve as a mentor to help guide the scholar's research and career planning and provide access to organizations, programs, and colleagues helpful to the scholar's efforts. More than one mentor may be selected. Letters of endorsement including specific information on institutional support for the scholar should be provided by the dean, the relevant department chairperson (or equivalent - may be one person), and the mentor. In addition, three letters of reference should be provided by other faculty members and/or senior professionals with whom the scholar has worked and who are well acquainted with his/her capabilities, potential, accomplishments, and commitment. All candidates must submit applications endorsed by the Dean of School of Medicine (or equivalent).
Date(s) Application is Due: Jan 22
Contact: Grants Manager; (212) 703-9977 or (888) 582-2327; fax (212) 997-0330; beeson@afar.org
Internet: http://afar.org/BeesonIreland.html
Sponsor: American Federation for Aging Research
55 West 39th Street, 16th Floor
New York, NY 10018

AFAR Research Grants **290**

The major goal of this program is to assist in the development of the careers of junior investigators committed to pursuing careers in the field of aging research. Of particular interest are research projects concerned with understanding the basic mechanisms of aging; projects investigating age-related diseases, especially if approached from the point of view of how basic aging processes may lead to these outcomes; and projects concerning mechanisms underlying common geriatric functional disorders, as long as these include connections to fundamental problems in the biology of aging. Examples of promising areas of research include: aging and immune function; genetic control of longevity; neurobiology and neuropathology of aging; invertebrate or vertebrate animal models; cardiovascular aging; aging and cellular stress resistance; metabolic and endocrine changes; age-related changes in cell proliferation; caloric restriction and aging; DNA repair and control of gene expression; biology of the menopause; and aging and apoptosis. Application and instruction sheet are available online.
Requirements: The applicant must be an independent investigator with assigned independent space and must be within the first four years of a junior faculty appointment (instructor, assistant professor or equivalent) by July 1st.
Restrictions: The program does not provide support for postdoctoral fellows in the laboratory of a senior investigator; investigators who have already received major independent funding for research on aging, such as an R01 grant or a grant of equal to or greater than $100,000 from another private funding source; senior faculty, i.e., at the rank of associate professor level or higher; and projects that deal strictly with clinical problems, such as the diagnosis and treatment of disease, health outcomes, or the social context of aging. Former AFAR grant recipients are not eligible to reapply. Applicants for the Glenn/AFAR Breakthroughs in Gerontology (BIG) program cannot also submit an application for this research grant.
Date(s) Application is Due: Dec 16
Amount of Grant: 75,000 U.S.D
Contact: Grants Manager; (212) 703-9977 or (888) 582-2327; fax (212) 997-0330; grants@afar.org or info@afar.org
Internet: http://afar.org/afar99.html
Sponsor: American Federation for Aging Research
55 West 39th Street, 16th Floor
New York, NY 10018

AF Doctoral Dissertation Award for Arthritis Health Professionals **291**

The award provides one or two years of salary and/or research support to projects related to arthritis management and/or comprehensive patient care in rheumatology practice, research, or education. A dissertation project is preferred. Suitable studies include, but are not limited to, functional, behavioral, nutritional, occupational, or epidemiological aspects of patient care and management. Drug studies and laboratory in vitro studies are not appropriate. Application and guidelines are available online. Annual deadline dates may vary; contact program staff for exact dates.
Requirements: The awards are designed for predoctoral students entering the research phase of their programs. The doctoral chairperson must approve the project. A candidate

must have membership or eligibility for membership in his/her professional organization. Applicants must be U.S. citizens or have permanent residence.
Restrictions: The award does not support laboratory research.
Date(s) Application is Due: Sep 1
Contact: Research Department; (404) 965-7537; grantsupport@arthritis.org
Internet: http://www.arthritis.org/research/ProposalCentral.asp
Sponsor: Arthritis Foundation
1330 W Peachtree Street, Suite 100
Atlanta, GA 30309

Affymetrix Corporate Contributions Grants **292**
Affymetrix makes charitable donations to nonprofit organizations through its corporate philanthropy program. The company focuses its giving in three main areas: education—programs that support science and math education, focusing on K-12 students and their teachers; ethics—organizations that help foster an ongoing public dialog about genetic-related ethics; and cancer research and advocacy—nonprofit organizations working in the areas of disease research and advocacy, with a specific emphasis on cancer. Requests are reviewed four times per year. The average grant size is approximately $2500.
Requirements: 501(c)3 nonprofit organizations are eligible.
Restrictions: The corporation does not provide funding for advertising journals or booklets; fundraising events such as telethons, walkathons, and races; specific performances or concerts; sporting events; endowment campaigns; film, video, television, or radio projects; grants to individuals; political causes or candidates; or organizations that practice discrimination or limit membership on the basis of race, creed, gender, age, sexual orientation, or national origin.
Contact: Contributions Manager, outreach@affymetrix.com
Internet: http://www.affymetrix.com/corporate/outreach/corporate.affx
Sponsor: Affymetrix
3420 Central Expressway
Santa Clara, CA 95051

AFG Industries Grants **293**
The corporation awards general operating grants to nonprofits in its headquarters area in the categories of arts and humanities, civic and public affairs, education at all levels, health care, and social services. There are no application deadlines. Submit a brief letter of inquiry.
Requirements: Tennessee nonprofits are eligible.
Geographic Focus: Tennessee
Amount of Grant: 100,000 - 250,000 U.S.D
Contact: Human Resources; (800) 251-0441 or (423) 229-7200; fax (423) 229-7459
Internet: http://www.afgglass.com
Sponsor: AFG Industries
P.O. Box 929
Kingsport, TN 37662

AF Investigator Awards **294**
The purpose of the award is to provide support to physicians and scientists in research fields broadly related to arthritis for the period between completion of postdoctoral fellowship training and establishment as an independent investigator. A senior scientist familiar with the applicant's area of research should be designated as the sponsor. The sponsor and the chairperson of the relevant academic department are responsible for stating the role of the applicant within the department, promising protection of time for research activities related to the award, guaranteeing space for the investigative work, and outlining future opportunities for the applicant. Each investigator is expected to devote at least 80 percent of his/her professional time to laboratory research. Permission for U.S. applicants to work abroad may be granted but the award may not be used for sabbatical leave. Awards will begin July 1 and will be renewed each year for a period of three years pending receipt of satisfactory evidence of progress. The awards will be paid to the sponsoring institution for stipend only, stipend plus research expenses, or research expenses only; the budget should indicate how the grant will be allocated. Application and guidelines are available online.
Requirements: Applicants must be U.S. citizens or have permanent residence. Applicants must have completed a minimum of three and a maximum of six years postdoctoral research experience, and must hold an MD, PhD, or equivalent degree, and have demonstrated distinction and productivity in research.
Restrictions: The applicant may not hold an NIH R01, NSF Grant, FIRST, Howard Hughes, Pew, VA Merit, Wellcome, Searle, Arthritis Foundation Biomedical Science Grant or Clinical Science Grant or equivalent award at the time of application. If such an award is made for a different project after the start date of the Arthritis Investigator Award, the individual may retain his/her Arthritis Foundation award. Individuals at the NIH and CDC are not eligible to apply.
Date(s) Application is Due: Sep 1
Amount of Grant: 76,000 U.S.D
Contact: Research Department; (404) 965-7537; grantsupport@arthritis.org
Internet: http://www.arthritis.org/research/ProposalCentral.asp#career
Sponsor: Arthritis Foundation
1330 W Peachtree Street, Suite 100
Atlanta, GA 30309

AFLAC Inc Grants **295**
The company awards grants to nonprofit organizations for cancer research, health-related projects, and community services. International contributions are determined by the company's senior officers. Submit a letter of inquiry that includes an organization description, amount requested, purpose of request, recently audited financial statement, and proof of tax-exempt status.
Requirements: Nonprofit organizations are eligible.
Restrictions: Grants are not made to individuals.
Samples: National Museum of African American History and Culture (Washington, DC)—to develop this new museum, whose collections will cover such topics as slavery, the Harlem Renaissance, and the civil-rights movement, $1 million.
Contact: Patsy Thomas, (706) 317-6109; fax (706) 320-2288; pthomas3@aflac.com
Internet: http://www.aflac.com
Sponsor: AFLAC
1932 Wynnton Road
Columbus, GA 31999

AF New Investigator Research Grants for Arthritis Health Professionals **296**
These research grants are to encourage PhD level health professionals who have research expertise to design and carry out innovative research projects related to the rheumatic diseases. The grant is intended to provide support for the period between completion of doctorate work and establishment as an independent investigator. The research project must be related to arthritis management and/or comprehensive patient care in rheumatology practice, research, or education. Suitable studies include, but are not limited to, functional, behavioral, nutritional, occupational, or epidemiological aspects of patient care and management. Individuals with limited research experience must apply in conjunction with a supervisor or co-investigator with demonstrated research expertise in the applicant's area of study. Awards are made for three years. Application and guidelines are available online.
Requirements: Grants are designed for applicants with a PhD or equivalent doctoral degree and demonstrated research experience. These awards are meant to encourage investigators who have received a doctoral degree within the last five years. A candidate must have membership or eligibility for membership in her/his professional organization.
Restrictions: Drug studies and laboratory in vitro studies are not appropriate. Individuals at the NIH and CDC are not eligible to apply. Not for laboratory research. MDs are not eligible.
Date(s) Application is Due: Sep 1
Amount of Grant: 50,000 U.S.D
Contact: Research Department; (404) 965-7636; grantsupport@arthritis.org
Internet: http://www.arthritis.org/research/ProposalCentral.asp#grants
Sponsor: Arthritis Foundation
1330 W Peachtree Street, Suite 100
Atlanta, GA 30309

AF Physician Scientist Development Award **297**
The award, which is cosponsored by the Scleroderma Foundation, is designed to enable MDs and those with an equivalent medical degree with no more than 1-1/4 years of research experience at the time of application to embark on careers in biomedical and/or clinical research related to the understanding of arthritis and the rheumatic diseases. Candidates with more than 1-1/2 years of postdegree laboratory experience must apply to the foundation's postdoctoral fellowship program. This is a two-year award that may be renewed for a third year. The stipend includes salary and fringe benefits. Application and guidelines are available online.
Requirements: Only physicians with no more than 1-1/4 years of research training at the time of application are eligible. Individuals must have received an MD, DO, or equivalent medical degree from an accredited institution within the past seven years; completed training in internal medicine or pediatrics; and will have completed at least one year of specialty training as of the start date of the award. For those awards funded by the Scleroderma Foundation, applicants will have completed at least one year of specialty training in rheumatology. Applicants must be U.S. citizens or have permanent residence.
Restrictions: MD/PhDs, DO/PhDs, and PhDs are not eligible.
Date(s) Application is Due: Sep 1
Amount of Grant: 50,000 U.S.D
Contact: Research Department; (404) 965-7537; grantsupport@arthritis.org
Internet: http://www.arthritisfoundation.org/research/ProposalCentral.asp
Sponsor: Arthritis Foundation
1330 W Peachtree Street, Suite 100
Atlanta, GA 30309

AF Postdoctoral Research Fellowships **298**
The grant provides a salary stipend for MDs, DOs, PhDs, or equivalent for a three-year period. Ninety percent time must be devoted to arthritis-related research. A stipend and institutional grant are awarded. There are no citizenship requirements for this award. Application and guidelines are available online.
Requirements: MDs are not eligible after six years of postdoctoral research training (or seven years in the case of a clinical training program, which includes one year in the laboratory).
Restrictions: MDs are not eligible after six years of laboratory training or seven years in the case of a clinical training program that includes one year in the laboratory. PhDs are not eligible after four years of postdegree laboratory experience. Individuals at or above the assistant professor level, or those who have tenured positions, are ineligible; however, if one is promoted or establishes tenure after receiving an award, he/she may continue to receive the award for the remainder of the fellowship period.
Date(s) Application is Due: Sep 1
Amount of Grant: 50,500 U.S.D
Contact: Research Department; (404) 965-7537; grantsupport@arthritis.org
Internet: http://www.arthritis.org/research/proposalcentral.asp
Sponsor: Arthritis Foundation
1330 W Peachtree Street, Suite 100
Atlanta, GA 30309

African American Fund of New Jersey Grants 299
The fund concentrates on providing significant funding to New Jersey's African American communities to promote and enhance human services delivery and community empowerment. Program areas that will receive primary funding consideration are in the areas of pre-school through high school education initiatives, technology-oriented initiatives, health-related programs, community issues, and building wealth. Programs that emphasize positive family values, traditions, social customs, and family practices that have contributed to the survival of the African American family are encouraged. Types of support include general operating grants, program development grants, and technical assistance. Proposals must demonstrate the intention to commit the necessary financial and human resources to a program that is well coordinated with related human services within the respective communities. All applicants will be notified of approval or disapproval of their proposals. Annual deadline dates may vary; contact program staff for exact dates.
Requirements: New Jersey 501(c)3 nonprofit, tax-exempt agencies, community-based organizations, and schools are eligible.
Geographic Focus: New Jersey
Amount of Grant: 1,000 - 5,000 U.S.D
Samples: Hudson County Child Abuse Prevention Center - grant recipient; Link Community School - grant recipient; Newark Literacy Campaign, Inc. - grant recipient; Young Science Achievers, Inc. - grant recipient.
Contact: Sondra Clark, President; (973) 676-5283; fax (973) 672-5030; sclark@aafnj.org
Internet: http://www.aafnj.org/HBW_Grants.htm
Sponsor: African American Fund of New Jersey
132 S Harrison Street
East Orange, NJ 07018

A Friend's Foundation Trust Grants 300
The Foundation, founded as the Hubbard Foundation in 1959 by philanthropist Frank M. Hubbard, serves central Florida with its support of health research and health care, education at all levels, children and youth activities, the arts, and religious agencies. Grants typically range from $5,000 to $35,000, though some higher amounts are given. There are no specific application formats or deadlines with which to adhere, and applicants should send a letter of request to the Foundation address listed.
Requirements: Applicants must be 501(c)3 organizations serving residents of central Florida.
Geographic Focus: Florida
Amount of Grant: 5,000 - 35,000 U.S.D
Samples: Tales and Tails Gala, Orlando, Florida—to sponsor the event; Orlando Opera Company, Orlando, Florida—for general operating costs.
Contact: L. Evans Hubbard, (407) 876-3122; ehubbard@cfl.rr.com
Sponsor: A Friend's Foundation Trust
9000 Hubbard Place
Orlando, FL 32819

AF Target Identifications in Lupus Grants 301
The purpose of the grants program is to fund highly meritorious and innovative research directed to the identification of targets for new therapies in the treatment of lupus. This award is open to nonprofit and industry applications from anywhere in the world. Grants are made for two years with a possibility of renewal. Application and guidelines are available online.
Requirements: Individuals with doctoral degrees (MD, PhD, DO, or equivalent) are eligible to apply. Scientific independence as evidenced by direction of a research program or a publication record or other experience that establishes scientific leadership is necessary to apply.
Date(s) Application is Due: Sep 1
Amount of Grant: 250,000 U.S.D
Contact: Research Department; (404) 965-7537; grantsupport@arthritis.org
Internet: http://www.arthritis.org/research/ProposalCentral.asp
Sponsor: Arthritis Foundation
1330 W Peachtree Street, Suite 100
Atlanta, GA 30309

A Fund for Women Grants 302
The fund provides grants to women and girls in the community that enhance education, employment and self-esteem. All grants are driven by the overall goal of helping women and children learn self-reliance and reach self-sufficiency. Under that umbrella, the Fund focuses on the following four key areas of need: keeping elderly or disabled women in their homes, or in community settings; providing services to victims of domestic abuse; helping women achieve economic self sufficiency and increase their earning potential; and reducing homelessness for women and girls. AFFW is particularly interested in innovative programs and services that will reach women and girls from diverse backgrounds in urban, suburban or rural Dane County. Proposals should indicate how the project or program will help connect women or girls to a support network or community that will help them to overcome emotional, social, intellectual, spiritual, occupational and/or physical barriers to self-sufficiency. Applications are due annually by July 1, at 4:30 p.m.
Requirements: Applicants must be non-profit organizations (exempt from Federal income taxes under section 501c(3) of the Internal Revenue Code), schools, governmental bodies, or under the supervision of such a group. Projects must focus on women and girls in Dane County.
Geographic Focus: Wisconsin
Date(s) Application is Due: Jul 1
Amount of Grant: 2,000 - 25,000 U.S.D
Samples: The African American Ethnic Academy, Madison, Wisconsin, $25,000; Domestic Abuse Intervention Services, Madison, Wisconsin, $25,000.
Contact: Kathleen Woit, President; (608) 232-1763, ext. 232; fax (608) 232-1772; kwoit@madisoncommunityfoundation.org or AFFW@madisoncommunityfoundation.org
Internet: http://www.affw.org/grants/apply.php
Sponsor: A Fund for Women
2 Science Court, P.O. Box 5010
Madison, WI 53711-1088

AGHE Graduate Scholarships and Fellowships 303
The association awards graduate scholarships and fellowships for research on aging. Gerontology and aging studies faculty are invited to nominate qualified students. This program is funded by the AARP Andrus Foundation. Faculty nomination and student application forms are available online. Annual deadline dates may vary; contact program staff for exact dates.
Requirements: Gerontology and aging studies faculty are invited to nominate qualified students for these awards.
Amount of Grant: 2,000 - 15,000 U.S.D
Contact: M. Angela Baker; (202) 289-9806, ext. 125; abaker@aghe.org
Internet: http://www.aghe.org/677858
Sponsor: Association for Gerontology in Higher Education
1220 L Street, NW, Suite 901
Washington, DC 20005-4015

Agnes B. Hunt Trust Grants 304
The trust awards grants to eligible Georgia nonprofit organizations in the general areas of health, welfare, and education for the poor and needy in its local community. Apply online; complete guidelines are provided.
Requirements: 501(c)3 nonprofit organizations in Griffin and Spalding County, GA, are eligible.
Geographic Focus: Georgia
Date(s) Application is Due: Nov 1
Amount of Grant: 500 - 5,000 U.S.D
Contact: Joseph Walker
Internet: http://www.agnesbhunttrust.org
Sponsor: Agnes B. Hunt Trust
P.O. Box 1610
Griffin, GA 30224-1610

Agnes Gund Foundation Grants 305
The foundation awards general operating grants in its areas of interest, including health organizations, higher education, and performing arts (dance, music). There are no application deadlines or forms. Submit a letter of inquiry.
Requirements: Although there are no funding restrictions there is a focus on New York City.
Restrictions: Applicants must be a tax-exempt organization.
Geographic Focus: New York
Amount of Grant: 5,000 - 100,000 U.S.D
Samples: Museum of Modern Art (New York, NY)—for general operating support, $1 million; Cleveland Museum of Art (OH)—for general operating support, $200,000; Virginia Museum of Fine Arts (Richmond, VA)—for general operating support, $100,000; Creative Capital Foundation (New York, NY)—for general operating support, $75,000.
Contact: Program Director; (330) 385-3400
Sponsor: Agnes Gund Foundation
517 Broadway, 3rd Floor
East Liverpool, OH 43920

Agnes M. Lindsay Trust Grants 306
The Trust makes grants to nonprofit organizations in the states of Maine, Massachusetts, New Hampshire and, Vermont to improve the quality of life. Areas of interest include: health & welfare organizations; recreation, camp scholarships (camperships); education; educational scholarships. The Trust awards grants for capital needs; capital campaigns, building renovations, equipment; computers, furniture, etc. Grant proposals are reviewed on a monthly basis, grants average $1,000 - $15,000. Campership applicantions must be submitted by March 1st and range from $1,000 - $4,000.
Requirements: 501(c)3 Maine, Massachusetts, New Hampshire, and Vermont tax-exempt organizations are eligible to apply for funding. To begin the application process, submit a letter of inquiry prior to submitting a full proposal for funding. In addition to basic information about your organization, include a brief outline of your statement of need and a project budget. You may complete this as a word document and attach it to an email addressed to: adminatlindsaytrustdotorg. Grant proposals should contain: proposal summary document (available at Trust website); narrative of the organization and description of need; financial statements for last two years or Form 990; budget for capital expenditures; estimates or quotes obtained; IRS Determination letter. Submit your proposal and all attachments electronically in either a PDF format and/or Microsoft Word and EXCEL to: proposalsatlindsaytrustdotorg. Applicants requesting a camp scholarships (camperships) provide the following: proposal summary document (available at Trust website); most recent audited financial statements (one calender year) or Form 990; and IRS Determination Letter. Submit your proposal electronically in a PDF file to: proposalsatlindsaytrustdotorg. The trust will accept Microsoft Word documents and Excel documents as attachments. If you are unable to send your proposal or any portion thereof electronically, you may submit it via U.S. mail. A status report for the grant awarded is required after the grant has been expended.
Restrictions: The Trust does not fund: endowments, public entities, awarded to individuals, municipalities, libraries, museums, sectarian organizations, or capital grants to private schools. The Trust very rarely provides funding for operating /program support. If your organization received a grant, wait one year before reapplying for additional funding.

Geographic Focus: Maine, Massachusetts, New Hampshire, Vermont
Date(s) Application is Due: Mar 1
Amount of Grant: 1,000 - 15,000 U.S.D
Samples: Child and Family Services of New Hampshire, Manchester, NH, $50,000—challenge grant to support the Capital Expansion Project to purchase land and furnishings for the new home on Union Street to serve homeless youth; Massachusetts Board of Higher Education, Boston, MA, $30,000—annual scholarship grant; Big Brothers Big Sisters of Bath Brunswick, Brunswick, ME, $5,000—for a new server and computer equipment/software providing the capacity to interface with the national office.
Contact: Susan Bouchard, Administrative Director; (603) 669-1366 or (866) 669-1366; fax (603) 665-8114; admin@lindsaytrust.org
Internet: http://www.lindsaytrust.org/index.html
Sponsor: Agnes M. Lindsay Trust
660 Chestnut Street
Manchester, NH 03104

A Good Neighbor Foundation Grants 307
Established in Cincinnati, Ohio, in 2002, A Good Neighbor Foundation offers support primarily to individuals and families that are in need of immediate aid and services. Its major fields of interest include: cancer; children's services; human services; and emergency aid. Funding comes in the form of general operating support for residents of Cincinnati, Ohio, and northern Kentucky. There are no specific deadlines or application forms, and applicants should contact the office in writing or via telephone.
Geographic Focus: Kentucky, Ohio
Amount of Grant: Up to 400,000 U.S.D
Contact: Michele Kelley, Secretary-Treasurer; (513) 651-9333
Sponsor: A Good Neighbor Foundation
414 Walnut Street, Suite 1014
Cincinnati, OH 45202-3913

AGS/Merck New Investigator Awards 308
The awards, funded through an educational grant from Merck U.S. Human Health, are given to individuals whose original research, as presented in a submitted abstract, reflects new and relevant research in geriatrics. Topics are invited in basic research, clinical investigation, clinical medicine, public health, and research in the fundamental neurosciences. Awards will be chosen based on originality, scientific merit, relevance of the research, and the applicant's overall academic accomplishments. The awards are intended, in part, to cover awardees' travel expenses to attend the AGS annual meeting. Annual deadline dates may vary; contact program staff for exact dates.
Requirements: The awards are restricted to fellows in training and new and junior investigators holding an academic appointment not longer than five years postfellowship. The work reported must not have been published prior to the date of application.
Restrictions: Materials presented at other national meetings will not be accepted.
Amount of Grant: 1,500 U.S.D
Contact: Dennise McAlpin, Program Contact; (212) 308-1414, ext. 321; fax (212) 832-8646; dmcalpin@americangeriatrics.org or info.amger@americangeriatrics.org
Internet: http://www.americangeriatrics.org/news/meeting/am2004_newInvestAward.shtml
Sponsor: American Geriatrics Society
350 Fifth Avenue, Suite 801
New York, NY 10118

AGS Dennis W. Jahnigen Career Development Scholars Awards 309
The awards are designed to address the urgent need to create a structure for developing leaders in geriatrics in academic surgery and such related specialties as anesthesiology, emergency medicine, general surgery, gynecology, ophthalmology, orthopedic surgery, otolaryngology, physical medicine and rehabilitation, thoracic surgery, and urology. Awards are for two years. Nomination may be made online.
Requirements: Nominations may be made by department chairs in any of the specialty areas listed. Departments may submit one application; institutions may submit more than one application. To be eligible, a candidate must be a physician who is a U.S. citizen or permanent resident, and be certified or board eligible to practice in one of the targeted specialty areas; have a primary academic appointment in a U.S. institution in one of the targeted specialty departments; and have completed his/her training (residency and/or fellowship) on or after June 30, 1995. (Exceptions to this limit will be considered for compelling reasons and must be reviewed and approved prior to application submission.)
Date(s) Application is Due: Dec 6
Amount of Grant: 100,000 U.S.D
Contact: Amy Tam-Liao, (212) 308-1414; atamliao@americangeriatrics.org
Internet: http://www.americangeriatrics.org/hartford/jahnigen.shtml
Sponsor: American Geriatrics Society
350 Fifth Avenue, Suite 801
New York, NY 10118

AGS Edward Henderson Student Award 310
The award is presented to a medical student interested in pursuing a career in geriatrics, who has demonstrated excellence in the field. The student must be nominated by one faculty member and have at least two supporting letters of nomination from other faculty. Annual deadline dates may vary; contact program staff for exact dates.
Requirements: A student must have demonstrated a commitment to the field of geriatrics through leadership in areas pertinent to geriatrics; initiation of new information or programs in geriatrics; or scholarship in geriatrics through original research or reviews.
Amount of Grant: 500 U.S.D
Contact: Dennise McAlpin, Program Contact; (212) 308-1414, ext. 321; fax (212) 832-8646; dmcalpin@americangeriatrics.org or info.amger@americangeriatrics.org
Internet: http://www.americangeriatrics.org/news/meeting/awrdbroc.shtml#Henderson
Sponsor: American Geriatrics Society
350 Fifth Avenue, Suite 801
New York, NY 10118

AGS Foundation T. Franklin Williams Scholars Award 311
This program supports physician-scientists committed to improving the health care of older adults. The award is intended to allow individuals to initiate and ultimately sustain a career in research and education. Junior faculty members devoting 75 percent time to research are eligible to receive project support. The award must be matched. Guidelines and request for application are available online.
Requirements: To be eligible, the applicant must have an MD or DO degree; and hold a full-time faculty appointment at a U.S. academic medical institution at the level of assistant professor for no longer than four years at the time the grant becomes effective. The applicant must have at least two sponsors who are committed to providing guidance and collaboration throughout the course of the proposed project.
Date(s) Application is Due: Feb 28
Amount of Grant: 37,500 U.S.D
Contact: Award Administrator; (800) 247-8779 or (212) 308-1414; fax (212) 832-8646; sreinthaler@americangeriatrics.org
Internet: http://www.frycomm.com/ags/Franklin_Williams/index.asp
Sponsor: AGS Foundation for Health in Aging
350 Fifth Avenue, Suite 801
New York, NY 10118

AGS John A. Hartford Foundation Geriatrics for Specialists Initiative 312
AGS, through a program funded by the John A. Hartford Foundation, seeks proposals for specialty-specific initiatives from academic training centers to develop, initiate, and evaluate programs designed to increase education for residents in the geriatrics aspect of their disciplines. The overall project addresses the urgent need to create a structure for developing leaders in geriatrics in academic surgery and related medical specialties. The program was created to assure that specialty residency programs adopt specific learning objectives and curricula in geriatric care as part of a targeted effort to enhance residents' knowledge, skills, and attitudes relevant to care of the older patient, in order to improve dissemination of new clinical research findings in geriatrics into the surgical and medical specialties. Eight two-year grants will be awarded. Application and guidelines are available online.
Requirements: Proposals must be from institutions that have well-developed academic programs in the applicant disciplines and in geriatrics. The proposal should be submitted by a member of the specialty department or division.
Date(s) Application is Due: Apr 8
Amount of Grant: 16,000 U.S.D
Contact: Grants Administrator; (212) 308-1414; fax (212) 832-8646
Internet: http://www.americangeriatrics.org/2005_GESR.shtml
Sponsor: American Geriatrics Society
350 Fifth Avenue, Suite 801
New York, NY 10118

AGS Student Research Award 313
An award will be given to the student presenting the most outstanding paper or poster at the AGS annual meeting. The awardee will be chosen based on originality, scientific merit, and relevance of the research. The awardee will present his/her research at the annual meeting. Annual deadline dates may vary; contact program staff for exact dates.
Requirements: Students must submit an abstract of research or research in progress on the official AGS abstract form. In addition, the student's curriculum vita and a letter from the student's advisor verifying the student's contribution to the work must be provided.
Amount of Grant: 500 U.S.D
Contact: Dennise McAlpin, Program Contact; (212) 308-1414, ext. 321; fax (212) 832-8646; dmcalpin@americangeriatrics.org or info@americangeriatrics.org
Internet: http://www.americangeriatrics.org/education/geristudents/Scholarships_Awards
Sponsor: American Geriatrics Society
350 Fifth Avenue, Suite 801
New York, NY 10118

Agway Foundation Grants 314
The primary goal of the foundation is to support organizations dedicated to serving the interests of farmers and rural communities in the Northeast, including health care, children and youth services, rural youth organizations, and agriculture. The foundation also supports organizations that contribute to the quality of life in the Agway headquarters area and it actively promotes employee volunteer involvement in community service.
Requirements: 501(c)3 nonprofit organizations are eligible for grant support.
Restrictions: The foundation funds are not used for individuals; political, religious, or labor organizations; scholarships; matching gift programs or capital campaigns of educational institutions; memberships in professional societies or trade associations; or the general operating funds of health care facilities.
Geographic Focus: Connecticut, Maine, Massachusetts, New Hampshire, Rhode Island, Vermont
Amount of Grant: 1,000 - 10,000 U.S.D
Contact: Stephen Hoefer; (315) 449-6474; info@agway.com or chairman@agway.com
Internet: http://www.colebrook-nh.com/Public_Documents/ColebrookNH_BBoard/I008D1C75

Sponsor: Agway Foundation
P.O. Box 4933
Syracuse, NY 13221

AHA Beginning Grants-in-Aid 315

Grants-in-aid promote the independent status of promising beginning scientists. Research areas broadly related to cardiovascular function and disease or stroke, or to related clinical, basic science, and public health problems are funded. All basic disciplines as well as epidemiological, community, and clinical investigations that bear on cardiovascular and stroke problems are eligible for funding. Awards are made for two years.
Requirements: Applicants must have an MD, PhD, DO, or equivalent and be initiating an independent research career. Faculty up to and including assistant professor may apply. Applicants must have a full-time faculty/staff appointment at activation and must be U.S. citizens, permanent residents, exchange visitors, temporary workers in a specialty occupation, Canadian or Mexican citizens engaging in professional activities, temporary workers with extraordinary abilities in the sciences, or students with student visas.
Geographic Focus: Arizona
Date(s) Application is Due: Jan 5
Amount of Grant: Up to 62,000 U.S.D
Contact: Program Director; (214) 706-1158; fax (214) 706-1341; affil@heart.org
Internet: http://www.americanheart.org/presenter.jhtml?identifier=2483
Sponsor: American Heart Association
7272 Greenville Avenue
Dallas, TX 75231-4596

AHA Established Investigator Grants 316

The program supports the career development of highly promising clinician-scientists and PhDs who have recently acquired independent status by funding high quality, innovative research projects for which no previous financial support has been obtained from other granting agencies. Grants provide four-year project and salary support. At the time of award activation, the time period since applicant's first faculty appointment must be four to nine years. Approximately 50 grants will be awarded annually. Award amount includes salary/fringe and 10 percent indirect costs.
Requirements: U.S. citizens or permanent residents with the MD, PhD, DO, or equivalent doctoral degrees are eligible. Eligible applicants must have full-time faculty/staff appointment at application, up to and including Associate Professor (or equivalant). Applicant must have at least four years but no more than nine years since first full-time faculty/staff appointment.
Restrictions: Current and past principal investigators are ineligible.
Amount of Grant: 1,000,000 - 400,000 U.S.D
Contact: National Center; (214) 706-1158; fax (214) 706-1341; ncrp@heart.org
Internet: http://www.americanheart.org/presenter.jhtml?identifier=9713
Sponsor: American Heart Association
7272 Greenville Avenue
Dallas, TX 75231-4596

AHAF Alzheimer's Disease Research Grants 317

AHAF funds outstanding scientists and physicians in neurobiology, physiology, pathology, molecular and developmental biology, chemistry, pharmacology, epidemiology, and surgery who are conducting research to better understand and/or treat Alzheimer's disease. Junior or senior investigators will be considered. Grants are awarded for up to two years and are renewable. Grant applications are reviewed by a scientific review committee on a competitive peer-review system.
Requirements: Grants are awarded to universities, medical centers, and independent research institutions. AHAF funds grants for research at nonprofit organizations only. The principal investigator must hold the academic rank of assistant professor (or equivalent) or higher.
Restrictions: Funding is not provided for overhead costs, construction, or building expenses.
Date(s) Application is Due: Oct 19
Amount of Grant: Up to 300,000 U.S.D
Contact: Kara Hurst, Grants Coordinator; (800) 437-2423 or (301) 948-3244; fax (301) 258-9454; khurst@ahaf.org
Internet: http://www.ahaf.org/alzdis/research/grants.htm
Sponsor: American Health Assistance Foundation
22512 Gateway Center Drive
Clarksburg, MD 20871

AHAF Macular Degeneration Research Grants 318

The foundation awards research grants to advance study of macular degeneration. Applications are evaluated based on the scientific merit of the proposal, the feasibility of the proposed research, the potential of the research to lead to better understanding and treatment of eye diseases, and the demonstrated ability of the investigator to complete the research. Application materials are available on the Web site.
Date(s) Application is Due: Jul 10; Oct 29
Amount of Grant: Up to 50,000 U.S.D
Contact: Kara Hurst, Grants Coordinator; (800) 437-2423 or (301) 948-3244; fax (301) 258-9454; khurst@ahaf.org
Internet: http://www.ahaf.org/macular/research/grants.htm
Sponsor: American Health Assistance Foundation
22512 Gateway Center Drive
Clarksburg, MD 20871

AHAF National Glaucoma Research Grants 319

AHAF funds outstanding scientists and physicians with expertise in cell and molecular biology, physiology, biochemistry, endocrinology, and pharmacology. Grants are awarded on the basis of scientific merit of the proposal, the relevance of the research, and the potential impact of the proposed study on better understanding and/or treatment of glaucoma. Applications are reviewed by a scientific review committee on a competitive peer-review system. AHAF is interested in receiving focused research grant applications from investigators at all stages of their careers. AHAF is particularly interested in new investigators with little or no previous grant support and established investigators with new ideas or directions for their research.
Requirements: AHAF grants are awarded to universities, medical centers, and independent research institutions. AHAF provides grants for research at nonprofit organizations only.
Restrictions: Grants are not made to individuals. Funding is not provided for overhead costs, construction, or building expenses.
Date(s) Application is Due: Oct 27
Amount of Grant: Up to 100,000 U.S.D
Samples: Teri Belecky-Adams, Ph.D., Trustees of Indiana University, Indianapolis, IN, $100,000; David Calkins, Ph.D., Vanderbilt University Medical Center, Nashville, TN, $100,000; Mortimer Civan, M.D., University of Pennsylvania, Philadelphia, PA, $100,000.
Contact: Kara Hurst, Grants Coordinator; (800) 437-2423 or (301) 948-3244; fax (301) 258-9454; khurst@ahaf.org
Internet: http://www.ahaf.org/glaucoma/research/glresrch.htm
Sponsor: American Health Assistance Foundation
22512 Gateway Center Drive
Clarksburg, MD 20871

AHA New York State Affiliate Postdoctoral Fellowships 320

Research fellowships are awarded for a two- to three-year period to help a trainee initiate a career in cardiovascular research while obtaining significant research results.
Requirements: Applicants must be MDs, CVMs, PDs, or PhDs, no more than three years postdoctoral, who agree to commit full-time effort to research. U.S. citizens, permanent residents, exchange visitors, temporary workers in specialty occupations, Canadian or Mexican citizens engaging in professional activities, temporary workers with extraordinary ability in the sciences, and those with student visas are eligible. Postdoctoral applicants who are outside the United States at time of application must provide visa documentation prior to award activation.
Restrictions: The program is not intended for individuals of faculty rank.
Date(s) Application is Due: Jan 13; Nov 15
Amount of Grant: Up to 38,000 U.S.D
Contact: Grants Administrator; (214) 706-1158 or (214) 706-1744; fax (214) 706-1341; ncrp@heart.org
Internet: http://www.americanheart.org/presenter.jhtml?identifier=2402
Sponsor: American Heart Association
7272 Greenville Avenue
Dallas, TX 75231-4596

AHA New York State Affiliate Scientist Development Grants 321

The purpose of the program is to help promising beginning scientists to move from completion of research training to the status of independent investigators. The focus of the research should be broadly related to CV function and disease, stroke, or to related clinical, basic science, and public health problems. The duration of the award is three years.
Requirements: Eligible to apply are U.S. citizens or permanent residents; individuals of up to and including assistant professor (or equivalent) rank at application; individuals holding MD, PhD, DO, or equivalent at application. At activation, no more than four years should have elapsed since first full-time faculty/staff appointment. At activation, applicant must have full-time faculty/staff appointment.
Restrictions: At activation, applicants shall have received no prior national level grant. Awards are non-renewable.
Date(s) Application is Due: Jan 13
Amount of Grant: Up to 66,000 U.S.D
Contact: Grant Administrator; (214) 706-1158 or (214) 706-1744; fax (214) 706-1341; ncrp@heart.org
Internet: http://www.americanheart.org/presenter.jhtml?identifier=2402#sdg
Sponsor: American Heart Association
7272 Greenville Avenue
Dallas, TX 75231-4596

AHA Postdoctoral Fellowship 322

The Pacific Mountain Affiliate focuses on services in its region, including Alaska, Arizona, Colorado, Hawaii, Idaho, Montana, New Mexico, Oregon, Washington and Wyoming. Fellowships assist postdoctoral trainees to initiate a career in cardiovascular research. Research funded is broadly related to cardiovascular function and disease, stroke, or to related clinical, basic science, and public health problems. All basic disciplines as well as epidemiological, community, and clinical investigations that bear on cardiovascular and stroke problems are funded. Awards are made for two to three years.
Requirements: Applicants must be U.S. citizens, permanent residents, exchange visitors, temporary workers in a specialty occupation, Canadian or Mexican citizens engaging in professional activities, temporary workers with extraordinary abilities in the sciences, or students with student visas. Applicants must have an MD, PhD, DO, or equivalent at the time the fellowship is activated.
Date(s) Application is Due: Jan 10
Amount of Grant: 40,568 - 48,428 U.S.D

Contact: Program Director; (214) 706-1457; fax (214) 706-1341; ncrp@heart.org
Internet: http://www.americanheart.org/presenter.jhtml?identifier=2418
Sponsor: American Heart Association
7272 Greenville Avenue
Dallas, TX 75231-4596

AHA Predoctoral Fellowship **323**

Fellowships help students initiate careers in cardiovascular research by providing research assistance and training. Research broadly related to cardiovascular function and disease, stroke or to related clinical, basic science, and public health problems will be funded. The fellowships are awarded in basic disciplines as well as epidemiological, community and clinical investigations that bear on cardiovascular and stroke problems. Awards are made for one or two years.
Requirements: Predoctoral PhD, MD, DO (or equivalent) students seeking research training with a sponsor/mentor prior to embarking on a research career may apply. Applicants must be U.S. citizens, permanent residents, exchange visitors, temporary workers in a specialty occupation, Canadian or Mexican citizens engaging in professional activities, temporary workers with extraordinary abilities in the sciences, or students with student visas.
Restrictions: An awardee may not hold another association award concurrently. No more than one fellow per sponsor will be funded. Funds may not be used to pay tuition.
Geographic Focus: Arizona, Colorado, New Mexico, Wyoming
Date(s) Application is Due: Jan 10
Amount of Grant: Up to 24,000 U.S.D
Contact: Program Director; (214) 706-1744; fax (214) 706-1341; ncrp@heart.org
Internet: http://www.americanheart.org/presenter.jhtml?identifier=2418
Sponsor: American Heart Association
7272 Greenville Avenue
Dallas, TX 75231-4596

AHA Scientist Development Grants **324**

The program supports highly promising beginning scientists in their progress toward independence and bridges the gap between completion of research training and readiness for competition as independent investigators. Grants provide four-year project and salary support. Applications may be submitted for review in the final year of a postdoctoral research fellowship or in the first four years of a faculty appointment. Peer review criteria will include originality and scientific merit of the proposed project (which cannot overlap with other funded work), prior productivity of the applicant, and evidence that the award will promote independent status for the applicant. Approximately 70 grants will be made annually. The award includes salary/fringe, project support, and 10 percent for indirect costs. The award is not renewable.
Requirements: U.S. citizens or permanent residents with the MD, PhD, DO, or equivalent doctoral degrees are eligible. Eligible applicants must have full-time faculty/staff appointment at activation, up to and including associate professor or equivalent.
Restrictions: Applicants cannot hold or have held any other national award. Applicant must have no more than four years since first full-time faculty/staff appointment at activation.
Date(s) Application is Due: Jan 9
Amount of Grant: Up to 65,000 U.S.D
Contact: National Center; (214) 706-1158 or (214) 706-1744; fax (214) 706-1341; ncrp@heart.org
Internet: http://www.americanheart.org/presenter.jhtml?identifier=3004142
Sponsor: American Heart Association
7272 Greenville Avenue
Dallas, TX 75231-4596

AHFMR Clinical Investigator Awards **325**

The Award allows highly qualified clinicians in the early stages of their careers in health research to commit the majority of their time (greater than 75%) to research. This award includes a $110,000 annual contribution toward salary and benefits, a $10,000 annual research prize, a $35,000 annual research allowance, and additional funds for renovations, operations, relocation, and equipment. In all cases, the relevance of the applicant's proposed research to human health should be apparent. All applicants must name a mentor or group of mentors who will foster their early career development.
Requirements: Canadian citizens or permanent residents of Canada who hold an MD or DDS degree, have completed all requirements for clinical specialty recognition, and qualify to hold a full-time position in a clinical department of the sponsoring institution are eligible to apply.
Geographic Focus: All States, Canada
Date(s) Application is Due: Sep 15
Amount of Grant: 155,000 U.S.D
Contact: Pamela Valentine, Director of Grants and Awards; (780) 423-5727; fax (780) 429-3509; pamela.valentine@ahfmr.ab.ca
Internet: http://www.ahfmr.ab.ca/grants/ClinInvest.php
Sponsor: Alberta Innovates - Health Solutions
10104-103 Avenue, Suite 1500
Edmonton, AB T5J 4A7 Canada

AHFMR Community Support Program Grants **326**

Community support program grants will provide two types of awards: Community Health Research Visiting Lecturer Awards and Community Health Research Travel Grants. The former (Community Health Research Visiting Lecturer Awards) are intended to assist Alberta's non-profit organizations in attracting outstanding health researchers to the province. The latter (Community Health Research Travel Grants) facilitate contact among health researchers in Alberta, as well as with national and international colleagues. Candidates from a wide variety of disciplines may apply as long as the purpose of the travel is to promote community health research.
Requirements: The Community Health Research Visiting Lecturer Awards program is open to community organizations that are directly involved in health research. The Universities of Alberta, Calgary and Lethbridge are not eligible to apply to this program but must submit requests to the AHFMR Visiting Lecturer Program. The Community Health Research Travel Grants program is open to individuals (independent investigators) actively engaged in health research in a non-profit organization in Alberta are eligible to apply.
Restrictions: University faculty are not eligible to apply for these awards.
Geographic Focus: All States, Canada
Date(s) Application is Due: Jan 9; Apr 14; Jul 10; Oct 9
Contact: Pamela Valentine, Director of Grants and Awards; (780) 423-5727; fax (780) 429-3509; pamela.valentine@ahfmr.ab.ca
Internet: http://www.ahfmr.ab.ca/grants/commsupp.php
Sponsor: Alberta Innovates - Health Solutions
10104-103 Avenue, Suite 1500
Edmonton, AB T5J 4A7 Canada

AHFMR Dr. Lionel E. McLeod Health Research Scholarship **327**

The award enables academically superior graduate students to undertake full-time training in medical or health research. The Scholarship is tenable through the Faculty of Medicine in any area related to health including: medicine; health services research; social and behavioral sciences (sociology, anthropology, psychology, economics, or political science); quantitative and population sciences (epidemiology and biostatistics, operations research, decision sciences, or computer sciences); and the humanities (philosophy, history, or theology).
Requirements: Candidates must be students at the University of Alberta, the University of Calgary, or the University of British Columbia and sponsored by the Faculty of Medicine.
Geographic Focus: All States, Canada
Amount of Grant: 21,500 U.S.D
Contact: Pamela Valentine, Director of Grants and Awards; (780) 423-5727; fax (780) 429-3509; pamela.valentine@ahfmr.ab.ca
Internet: http://www.ahfmr.ab.ca/grants/mcleod.php
Sponsor: Alberta Innovates - Health Solutions
10104-103 Avenue, Suite 1500
Edmonton, AB T5J 4A7 Canada

AHFMR Establishment and Independent Establishment Grants **328**

The two types of grants available include: Establishment Grants—AHFMR provides funds to assist reseachers (newly recruited to Alberta) in the start-up of their laboratories and/or research projects; and Independent Establishment Grants—open to new Alberta Investigators who are not candidates for Heritage personnel awards (it is understood that these investigators will commit much of their time to medical or health research). Those sponsoring investigators for these awards must notify the Foundation of their intent to do so at least four months in advance of the application deadline, and request a determination of the eligibility of the candidate. The Foundation will not accept applications for which advanced notice has not been received. There are two categories of grant funding: Operating Funds and Major Equipment Funds.
Requirements: Eligibility to apply for these grants is restricted to applicants recruited to leadership positions (such as Department Head or Director of a centre, institute or division), where the recruitment is expected to stimulate the development of research in a field or discipline of interest to the Foundation.
Geographic Focus: All States, Canada
Date(s) Application is Due: Sep 15
Contact: Pamela Valentine, Director of Grants and Awards; (780) 423-5727; fax (780) 429-3509; pamela.valentine@ahfmr.ab.ca
Internet: http://www.ahfmr.ab.ca/grants/EstabIndep.php
Sponsor: Alberta Innovates - Health Solutions
10104-103 Avenue, Suite 1500
Edmonton, AB T5J 4A7 Canada

AHFMR Forefront Internship Program Grants **329**

The Internship Program addresses the need for highly trained staff and management of the Alberta-based health, medical products and biotechnology industries. This Program is intended to support technology commercialization training and experience for individuals with an appropriate background in science and/or business. Although the work plan should identify the specific area of focus, the intern should receive a broad exposure to all of the following areas: product development; intellectual property strategy; regulatory requirements; market research and strategy; and business planning and development. Interns are awarded a $5,000 training allowance, which may be used for expenses related to educational workshops/seminars/courses necessary for relevant commercialization training, plus a $45,000 to $50,000 stipend.
Requirements: The applicant must hold, or be in the final year of a degree in business, management, or science. The applicants should have some relevant experience in the business or private sector. Applicants must be sponsored by an Alberta organization actively engaged in the commercialization of medical or health related technology. The organization must have the experience and resources to provide an appropriate training environment, and designate one person within the organization who will be the primary supervisor for the intern. The organization must agree to provide direct and continuous supervision.
Geographic Focus: All States, Canada
Date(s) Application is Due: Feb 15

Amount of Grant: 45,000 - 50,000 U.S.D
Contact: Carla Weyland, ForeFront Officer; (780) 423-5727; fax (780) 429-3509; carla.weyland@ahfmr.ab.ca
Internet: http://www.ahfmr.ab.ca/forefront/internship.php
Sponsor: Alberta Innovates - Health Solutions
10104-103 Avenue, Suite 1500
Edmonton, AB T5J 4A7 Canada

AHFMR Health Research Career Renewal Awards 330
Research Career Renewal Awards allow qualified Alberta-based faculty members to obtain rigorous training in any of the following disciplines: epidemiology; biostatistics; psychosocial sciences; and clinical experimental method and design. Training must be undertaken in a center that is renowned for the quality of research and training in the particular area of study. The Career Renewal award will provide a contribution to stipend and/or a research allowance (which can be applied toward offsetting tuition costs, for example).
Requirements: An Alberta-based institution must propose and sponsor the candidate (who must be a full-time faculty member) in addition to committing to provide an adequate research environment upon the trainee's return.
Geographic Focus: All States, Canada
Date(s) Application is Due: Mar 1; Oct 1
Contact: Pamela Valentine, Director of Grants and Awards; (780) 423-5727; fax (780) 429-3509; pamela.valentine@ahfmr.ab.ca
Internet: http://www.ahfmr.ab.ca/grants/CareerAward.php
Sponsor: Alberta Innovates - Health Solutions
10104-103 Avenue, Suite 1500
Edmonton, AB T5J 4A7 Canada

AHFMR Health Research Fund Grants 331
The Health Research Fund (HRF), an operating grant program that has been in existence for over ten years, is a collaborative program with the funding originating from Alberta Health and Wellness and with AHFMR doing the program administration. The Fund supports relevant, high quality health research studies in the areas of: health services; population health; health technology assessment. There are no specific deadlines.
Requirements: The HRF competition is aimed at investigative teams which include both researchers and decision-makers or users of health research information working together on issues relating to the appropriate research areas. Proposals that do not include a relevant decision-maker and /or user of health research information as part of the investigative team are not eligible for funding. Applications may be sponsored by either non-profit decision-making organizations (health agencies, associations, service provider organizations, regional or provincial health authorities, Ministry of Health and Wellness) or non-profit research organizations (universities, hospitals, research institutions or centers). The sponsoring organization must be Alberta-based and have been in existence for at least one year.
Restrictions: Funding will not be provided for the following activities/items: projects consisting strictly of basic biomedical or clinical research (such as clinical trials); projects that are limited to program development or evaluation; projects that involve issues not related to human health; projects that duplicate services already provided; staff training, unless such training is required for the proposed project; purchase of furniture, or renting/leasing space and office equipment, (unless such items are not provided for by existing organizations).
Geographic Focus: All States, Canada
Amount of Grant: 50,000 - 75,000 U.S.D
Contact: Pamela Valentine, Director of Grants and Awards; (780) 423-5727; fax (780) 429-3509; pamela.valentine@ahfmr.ab.ca
Internet: http://www.ahfmr.ab.ca/grants/HRF/hrfguide.php
Sponsor: Alberta Innovates - Health Solutions
10104-103 Avenue, Suite 1500
Edmonton, AB T5J 4A7 Canada

AHFMR Heritage Summer Studentships 332
AHFMR Summer Studentships offer motivated students with exceptional academic records an opportunity to participate in medical or health research in Alberta during the summer months. The award is meant to encourage students to consider pursuing formal training and a career in health research. The award is tenable for a minimum of 2 months, and a maximum of 4 months during the period of May to August. The award consists of a $1300 per month stipend only.
Requirements: Those candidates who meet one of the following criteria are eligible to apply: registered in an Alberta-based undergraduate degree program in a medical or health-related field; registered in an undergraduate degree program outside of Alberta, and desiring to engage in research during the summer at an Alberta institution; registered in an M.D. program, and who may also hold an undergraduate or graduate degree; exceptional high school students with records of participation in the health care system, and a clear interest in pursuing a health research career; or currently in the last term of their undergraduate degrees and who have applied either to Medical School or to a graduate program that would start in the coming Fall.
Geographic Focus: All States, Canada
Date(s) Application is Due: Feb 1
Amount of Grant: 2,600 - 5,200 U.S.D
Contact: Christina Van Gelderen, (780) 423-5727; fax (780) 429-3509; christina.van.gelderen@ahfmr.ab.ca
Internet: http://www.ahfmr.ab.ca/grants/Sum-student.php
Sponsor: Alberta Innovates - Health Solutions
10104-103 Avenue, Suite 1500
Edmonton, AB T5J 4A7 Canada

AHFMR Infrastructure Funding Grants 333
AHFMR has developed a program to support infrastructure costs associated with the Canada Research Chairs (CRC) program. Infrastructure support for the Chairs comes from the Canada Foundation for Innovation (CFI), which covers 40% of total eligible costs. The remaining costs come from other funding partners. AHFMR has developed a program that will provide a contribution to the 60% of the cost of the CFI-eligible projects that normally comes from the universities and their funding partners.
Requirements: The Program provides support for successful CRC holders who are: faculty members at an Alberta institution, who are ineligible to apply for AHFMR Establishment awards and are principal investigators on active Canadian Institute for Health Research (CIHR) Operating Grants: or are currently recipients of unconditional AHFMR Personnel Awards and are principal investigators on non-AHFMR peer-reviewed operating grants.
Restrictions: Investigators being recruited to the province, and who are eligible to apply for, or have received AHFMR Establishment grants are not eligible for consideration under this program.
Geographic Focus: All States, Canada
Amount of Grant: 62,500 U.S.D
Contact: Pamela Valentine, Director of Grants and Awards; (780) 423-5727; fax (780) 429-3509; pamela.valentine@ahfmr.ab.ca
Internet: http://www.ahfmr.ab.ca/grants/crcinfo.php#1
Sponsor: Alberta Innovates - Health Solutions
10104-103 Avenue, Suite 1500
Edmonton, AB T5J 4A7 Canada

AHFMR Interdisciplinary Team Grants 334
The Alberta Heritage Foundation for Medical Research (AHFMR) Interdisciplinary Team Grants Program provides opportunities for high-quality, internationally recognized teams of investigators to complete research initiatives with defined health outcomes. Funds available in this competition are to support collaborative, interdisciplinary and multi-institutional teams that address important research questions, health problems or issues in defined areas of research that are aligned with strategic research priorities of the Foundation and Alberta. The Program will provide up to $1 million per year per team for up to five years. Funding can be used for a broad range of research-related costs including research infrastructure, research operating costs, core administrative/management costs, scientific support for the team (including the costs associated with the recruitment of new investigators to the team), collaborative/linkage activity and knowledge exchange/translation activity.
Requirements: Each eligible AHFMR Interdisciplinary Team Grant application will include: A Team Leader—the Team Leader must be an established researcher with proven leadership skills and experience who will act as research program director and who will assume administrative responsibility for the grant. It is expected that the Team Leader will devote a significant and appropriate portion of her/his time to these tasks. The Team Leader will have their primary academic appointment at an Alberta-based university; at least two additional independent investigators with established research track records. Teams with a nucleus of experienced investigators are encouraged to include promising new investigators as part of their group; team members who collectively have an extensive record of success, are creative and original in their approach to research and its translation, and who have experience working in research teams. The specific contribution of each team member and end-user partner, where applicable, must be described; representation from more than one research discipline and from more than one "research pillar" (i.e. biomedical; clinical science; health systems and services; and the social, cultural and other factors that affect the health of populations.); and representation from more than one Alberta-based university. Multi-institutional collaboration is strongly encouraged in this Program.
Geographic Focus: All States, Canada
Date(s) Application is Due: Nov 14
Amount of Grant: Up to 1,000,000 U.S.D
Contact: Pamela Valentine, Director of Grants and Awards; (780) 423-5727; fax (780) 429-3509; pamela.valentine@ahfmr.ab.ca
Internet: http://www.ahfmr.ab.ca/grants/team_guidelines.php
Sponsor: Alberta Innovates - Health Solutions
10104-103 Avenue, Suite 1500
Edmonton, AB T5J 4A7 Canada

AHFMR Major Equipment Grants 335
AHFMR holds an annual competition to assist Alberta's medical or health researchers in obtaining major scientific equipment (defined as equipment costing more than $10,000 per unit or setup). It is AHFMR's expectation that large requests (e.g. greater than $100,000) will include contribution from other sources. Major equipment funds may be applied toward the purchase of the equipment, as well as toward the following expenses: installation; freight; transportation; duty; sales tax; and GST.
Requirements: An Alberta-based institution must sponsor applications on behalf of full-time, competitively-funded, productive faculty members.
Geographic Focus: All States, Canada
Date(s) Application is Due: Mar 1
Amount of Grant: 10,000 - 100,000 U.S.D
Contact: Pamela Valentine, Director of Grants and Awards; (780) 423-5727; fax (780) 429-3509; pamela.valentine@ahfmr.ab.ca
Internet: http://www.ahfmr.ab.ca/grants/equipment.php
Sponsor: Alberta Innovates - Health Solutions
10104-103 Avenue, Suite 1500
Edmonton, AB T5J 4A7 Canada

AHFMR Medical or Health Scholars Investigator Grants 336
Heritage Medical or Health Scholars are investigators who have recently completed their postdoctoral research training, and are currently seeking their first faculty-level appointments, or who are at the Heritage Population Heath Investigator or Heritage Clinical Investigator levels. Candidates must demonstrate an ability to initiate and conduct independent and collaborative research, as well as an interest and ability, or potential ability, to train future investigators. The Scholarship award includes a contribution toward salary and benefits, and may also include a research allowance and funds for renovations, and relocation.
Requirements: Candidates for the Heritage Scholar award must possess a M.D., D.D.S., D.V.M., Ph.D. or equivalent qualification. Furthermore, they must be eligible to hold a full-time appointment with the sponsoring institution.
Geographic Focus: All States, Canada
Date(s) Application is Due: Sep 15
Amount of Grant: 20,000 U.S.D
Contact: Pamela Valentine, Director of Grants and Awards; (780) 423-5727; fax (780) 429-3509; pamela.valentine@ahfmr.ab.ca
Internet: http://www.ahfmr.ab.ca/grants/scholars.php
Sponsor: Alberta Innovates - Health Solutions
10104-103 Avenue, Suite 1500
Edmonton, AB T5J 4A7 Canada

AHFMR Medical or Health Scientist Awards 337
The Heritage Medical or Health Scientist Award is the most senior personnel award offered by AHFMR. Heritage Scientists must be established, internationally recognized scientific leaders in their fields of study. Their stature should be apparent by: the quality of their publication records; the originality and vision of their research proposals; the quality and number of their trainees; their ability to attract and retain research grants; and the leadership qualities they exhibit locally, nationally, and internationally. The Scientist award includes a contribution toward salary and benefits, and may also include a research allowance and funds for renovations, and relocation.
Requirements: Candidates for the Heritage Scientist award must possess a M.D., D.D.S., D.V.M., Ph.D. or equivalent qualification. Furthermore, they must be eligible to hold a full-time appointment with the sponsoring institution. AHFMR expects a time commitment of not less than 75% to health research activities. The department chair and the dean will designate the time expected for teaching, clinical duties, examining, and other activities.
Geographic Focus: All States, Canada
Date(s) Application is Due: Sep 15
Amount of Grant: 10,000 - 40,000 U.S.D
Contact: Pamela Valentine, Director of Grants and Awards; (780) 423-5727; fax (780) 429-3509; pamela.valentine@ahfmr.ab.ca
Internet: http://www.ahfmr.ab.ca/grants/scientists.php
Sponsor: Alberta Innovates - Health Solutions
10104-103 Avenue, Suite 1500
Edmonton, AB T5J 4A7 Canada

AHFMR Part-Time Studentships 338
Alberta Innovates - Health Solutions (AIHS) Part-Time Studentships enable students enrolled in a full-time professional degree program (usually the M.D. program) in Alberta to engage in part-time research training during the regular academic year. The award consists of a stipend only. Part-time students may elect to engage in full-time research during the summer; these students should contact AIHS to request that full-time summer trainee status be implemented.
Requirements: Candidates must normally have been accepted into, or be currently engaged in, a full-time graduate program at an Alberta-based university in a health-related discipline leading to a Master's or doctoral degree. It is expected that the amount of time the candidate will commit to research will not be less than the equivalent of one day per week, taken either as one full day or two half days.
Restrictions: The award is tenable at an Alberta-based university only.
Geographic Focus: Canada
Date(s) Application is Due: Mar 1; Oct 1
Contact: Pamela Valentine, Vice President of Research & Innovation; (780) 423-5727, ext. 230; fax (780) 429-3509; pamela.valentine@albertainnovates.ca or grants.health@albertainnovates.ca
Internet: http://www.ahfmr.ab.ca/grants/PT-student.php
Sponsor: Alberta Innovates - Health Solutions
10104-103 Avenue, Suite 1500
Edmonton, AB T5J 4A7 Canada

AHFMR Personnel Funding Grants 339
AHFMR has developed a program to support personnel costs associated with the Canada Research Chairs (CRC) program. In general, support will be given to Heritage personnel who receive CRC awards consistent with AHFMR guidelines concerning competitive personnel funding contributions from other agencies. Though it does not duplicate personnel funding, the Foundation has approved stipend incentives to partner with other personnel awards received. Currently the stipend incentive program provides funding to about 50 AHFMR researchers who have been successful in both AHFMR Personnel Award competitions and the personnel award competitions of other agencies.
Requirements: Researchers recruited to Alberta with a CRC award are eligible to apply for AHFMR Personnel Awards. If they are successful, the latter may be used as a contribution to matching the CFI allocation for CRC awards.
Geographic Focus: All States, Canada
Contact: Pamela Valentine, Director of Grants and Awards; (780) 423-5727; fax (780) 429-3509; pamela.valentine@ahfmr.ab.ca
Internet: http://www.ahfmr.ab.ca/grants/crcinfo.php#1
Sponsor: Alberta Innovates - Health Solutions
10104-103 Avenue, Suite 1500
Edmonton, AB T5J 4A7 Canada

AHFMR Polaris Awards 340
The Alberta Heritage Foundation for Medical Research (AHFMR) Polaris Award was established as a means to help Alberta universities recruit outstanding mid-career health researchers of exceptional international calibre to the province. Individuals supported via this program will be known as AHFMR Polaris Investigators. The goal of this Award is to accelerate research activity in key areas that are priorities for Albertans. AHFMR will provide up to $1 million per year per award for up to ten years. This funding is to be at least matched by the institution nominating the AHFMR Polaris Investigator. The institutional support can include funding provided from other partners including government, the private sector and non-profit agencies. It generally cannot include funding originating from standard grants and awards programs of research organizations such as CFI, CIHR, and NSERC. The first AHFMR Polaris Award will be made in 2008. It is anticipated that initially up to three awards will be available.
Requirements: AHFMR Polaris Investigators must meet a number of eligibility criteria, including being: mid-career investigators with outstanding records of accomplishments in health research. Investigators active in any one (or more) of the four pillars of health research (basic, clinical, health services research, population health) are eligible for consideration; leaders who have histories of accomplishment in science and research demonstrated via executive positions in scientific societies, editorship of a prestigious journal, and invitations to speak at international conferences (they would be recognized as candidates for, or would have recently received, a major international recognition award); nominated by the University of Alberta, the University of Calgary, or the University of Lethbridge. Nominations may also be considered for individuals to be based within an Alberta Regional or Provincial Health Authority, but it is anticipated that they will have academic appointments at one of the universities as well; currently based outside Alberta and prepared to relocate to Alberta full-time for the duration of the AHFMR Polaris Award; and prepared to commit a majority of their time to research.
Geographic Focus: All States, Canada
Amount of Grant: 1,000,000 U.S.D
Contact: Pamela Valentine; (780) 423-5727; pamela.valentine@ahfmr.ab.ca
Internet: http://www.ahfmr.ab.ca/grants/polaris.php
Sponsor: Alberta Innovates - Health Solutions
10104-103 Avenue, Suite 1500
Edmonton, AB T5J 4A7 Canada

AHFMR Population Health Investigator (PHI) Awards 341
The Award allows independent investigators in the early stages of their careers in health research to commit the majority of their time (greater than 75%) to research. This award includes a $110,000 annual contribution toward salary and benefits, a $10,000 annual research prize, a $15,000 annual research allowance, and additional funds for renovations, operations, relocation, and equipment. Potential areas of research are: quantitative and population sciences (epidemiology and biostatistics, operations research, decision sciences, and computer sciences); health services research; social and behavioural sciences (sociology, anthropology, psychology, economics, and political science); and the humanities (law, philosophy, history, and theology). In all cases, the relevance of the applicant's proposed research to human health should be apparent.
Requirements: Canadian citizens or permanent residents of Canada who hold an MD or DDS degree, have completed all requirements for clinical specialty recognition, and qualify to hold a full-time position in a clinical department of the sponsoring institution are eligible to apply.
Geographic Focus: All States, Canada
Date(s) Application is Due: Sep 15
Amount of Grant: 135,000 U.S.D
Contact: Pamela Valentine; (780) 423-5727; pamela.valentine@ahfmr.ab.ca
Internet: http://www.ahfmr.ab.ca/grants/PopHealthInvest.php
Sponsor: Alberta Innovates - Health Solutions
10104-103 Avenue, Suite 1500
Edmonton, AB T5J 4A7 Canada

AHFMR Proposals for Special Initiatives 342
AHFMR welcomes proposals for innovative initiatives that accelerate the achievement of its overall objectives. Institutions recruiting internationally-recognized researchers to lead new initiatives should consult AHFMR, if current programs are inappropriate. AHFMR will also consider proposals for pilot projects in patient or population-based research. These projects must have the potential to develop information, and the investigators must have the expertise to attract grants from other agencies. The review process and financial support for special initiatives may be shared by AHFMR and other agencies.
Restrictions: Applications that are normally eligible for regular operating grants from other agencies will NOT be considered under this program.
Geographic Focus: All States, Canada
Contact: Pamela Valentine, Director of Grants and Awards; (780) 423-5727; fax (780) 429-3509; pamela.valentine@ahfmr.ab.ca
Internet: http://www.ahfmr.ab.ca/grants/initiatives.php
Sponsor: Alberta Innovates - Health Solutions
10104-103 Avenue, Suite 1500
Edmonton, AB T5J 4A7 Canada

AHFMR Research Prize 343

The AHFMR Research Prize is intended to maintain and improve AHFMR Personnel Awards as a potent vehicle for the recruitment and retention of highly qualified, internationally competitive investigators at Alberta institutions. The Prize will be made to every eligible investigator supported by AHFMR through the AHFMR Personnel Support Programs (Population Health Investigator, Clinical Investigator, Scholar, Senior Scholar, Scientist).
Requirements: To be eligible, the investigator must be currently supported by an unconditional AHFMR investigator award. Investigators receiving AHFMR Terminal support or those on Leaves of Absence are not eligible to receive the Research Prize. The Research Prize is not determined by institutional rank and salary scale. It is provided in recognition of the individual investigator's research accomplishments, and of his/her success in obtaining an AHFMR Independent Investigator award. The Research Prize is a personal prize to the recipient.
Restrictions: Successful applicants for an Independent Establishment Grant are not eligible to receive the AHFMR Research Prize.
Geographic Focus: All States, Canada
Amount of Grant: 10,000 - 20,000 U.S.D
Contact: Pamela Valentine; (780) 423-5727; pamela.valentine@ahfmr.ab.ca
Internet: http://www.ahfmr.ab.ca/grants/resprize.php
Sponsor: Alberta Innovates - Health Solutions
10104-103 Avenue, Suite 1500
Edmonton, AB T5J 4A7 Canada

AHFMR Scholar Awards 344

The Scholars grant allows independent researchers in the early stages of their careers in health research to commit the majority of their time (greater than 75%) to research. Scholars are defined as investigators who have recently completed their postdoctoral research training, and are currently seeking their first faculty-level appointments, or who are completing their term as AHFMR Population Heath Investigator or Clinical Investigator awards. This award includes a $110,000 annual contribution toward salary and benefits, a $10,000 to $20,000 annual research prize, a $35,000 annual research allowance, and additional funds for renovations, operations, relocation, and equipment. Potential areas of research are: quantitative and population sciences (epidemiology and biostatistics, operations research, decision sciences, and computer sciences); health services research; social and behavioral sciences (sociology, anthropology, psychology, economics, and political science); and the humanities (law, philosophy, history, and theology). In all cases, the relevance of the applicant's proposed research to human health should be apparent. All new awards are for seven years.
Requirements: Candidates must possess a M.D., D.D.S., D.V.M., Ph.D. or equivalent qualification. Furthermore, they must be eligible to hold full-time faculty appointments with the sponsoring institutions.
Geographic Focus: All States, Canada
Date(s) Application is Due: Sep 15
Amount of Grant: 770,000 - 1,155,000 U.S.D
Contact: Pamela Valentine; (780) 423-5727; pamela.valentine@ahfmr.ab.ca
Internet: http://www.ahfmr.ab.ca/grants/scholars.php
Sponsor: Alberta Innovates - Health Solutions
10104-103 Avenue, Suite 1500
Edmonton, AB T5J 4A7 Canada

AHFMR Scientist Awards 345

The Scientist award is the most senior investigator award offered by AHFMR. Scientists must be established, internationally recognized scientific leaders in their fields of study. Their stature should be demonstrated through: the quality of their publications; the originality and vision of their research proposals; the quality and number of their trainees; their ability to attract and retain research grants; and the leadership qualities they exhibit locally, nationally, and internationally. This award includes a $160,000 annual contribution toward salary and benefits, a $20,000 annual research prize, a $50,000 annual research allowance, and additional funds for renovations, operations, relocation, and equipment. In all cases, the relevance of the applicant's proposed research to human health should be apparent. All new awards are for seven years.
Requirements: Candidates for the award must possess a M.D., D.D.S., D.V.M., Ph.D. or equivalent qualification. Furthermore, they must be eligible to hold full-time faculty appointments with the sponsoring institutions. At time of implementation, candidates will typically have ten to fifteen years experience as independent investigators and are expected to have attained full professor level faculty appointments.
Geographic Focus: All States, Canada
Date(s) Application is Due: Sep 15
Amount of Grant: 230,000 U.S.D
Contact: Pamela Valentine; (780) 423-5727; pamela.valentine@ahfmr.ab.ca
Internet: http://www.ahfmr.ab.ca/grants/scientists.php
Sponsor: Alberta Innovates - Health Solutions
10104-103 Avenue, Suite 1500
Edmonton, AB T5J 4A7 Canada

AHFMR Senior Investigator Awards 346

Like the Scientist Award, this is the most senior investigator award offered by AHFMR. Applicants must be established, internationally recognized scientific leaders in their fields of study. Their stature should be demonstrated through: the quality of their publications; the originality and vision of their research proposals; the quality and number of their trainees; their ability to attract and retain research grants; and the leadership qualities they exhibit locally, nationally, and internationally. This award includes a $160,000 annual contribution toward salary and benefits, a $40,000 annual research prize, a $20,000 annual research allowance, and additional funds for renovations, operations, relocation, and equipment. In all cases, the relevance of the applicant's proposed research to human health should be apparent. All new awards are for seven years.
Requirements: Candidates for the award must possess a M.D., D.D.S., D.V.M., Ph.D. or equivalent qualification. Furthermore, they must be eligible to hold full-time faculty appointments with the sponsoring institutions. At time of implementation, candidates will typically have ten to fifteen years experience as independent investigators and are expected to have attained full professor level faculty appointments.
Geographic Focus: All States, Canada
Amount of Grant: 220,000 U.S.D
Contact: Pamela Valentine; (780) 423-5727; pamela.valentine@ahfmr.ab.ca
Internet: http://www.ahfmr.ab.ca/grants/independent-investigator-revised.php
Sponsor: Alberta Innovates - Health Solutions
10104-103 Avenue, Suite 1500
Edmonton, AB T5J 4A7 Canada

AHFMR Senior Medical or Health Scholars Investigator Grants 347

Heritage Senior Medical or Health Scholars must have an excellent track record of independent research over several years. They are more senior than Heritage Scholars and have usually been previously supported via an external salary support award (e.g. as an AHFMR Scholar, CIHR New Investigator, or equivalent). At the time of application, candidates for a Senior Scholar Award will typically have four to six years as an independent investigator and will be appointed at either the Assistant Professor or Associate Professor level. The Senior Scholarship award includes a contribution toward salary and benefits, and may also include a research allowance and funds for renovations, and relocation.
Requirements: Candidates for the Heritage Senior Scholar award must possess a M.D., D.D.S., D.V.M., Ph.D. or equivalent qualification. Furthermore, they must be eligible to hold a full-time appointment with the sponsoring institution. AHFMR expects a time commitment of not less than 75% to health research activities. The department chair and the dean will designate the time expected for teaching, clinical duties, examining, and other activities.
Geographic Focus: All States, Canada
Date(s) Application is Due: Sep 15
Amount of Grant: Up to 30,000 U.S.D
Contact: Pamela Valentine; (780) 423-5727; pamela.valentine@ahfmr.ab.ca
Internet: http://www.ahfmr.ab.ca/grants/seniorscholars.php
Sponsor: Alberta Innovates - Health Solutions
10104-103 Avenue, Suite 1500
Edmonton, AB T5J 4A7 Canada

AHFMR Senior Scholar Awards 348

The Senior Scholars award allows independent researchers in the later stages of their careers in health research to commit the majority of their time (greater than 75%) to research. Applicants are more senior than Scholars and have usually been previously supported via salary support awards (e.g. as an AHFMR Scholar, CIHR New Investigator, or equivalent). At time of implementation, candidates for a Senior Scholar award will typically have five to ten years experience as independent investigators and are expected to have attained an associate professor level faculty appointment. This award includes a $135,000 annual contribution toward salary and benefits, a $20,000 annual research prize, a $50,000 annual research allowance, and additional funds for renovations, operations, relocation, and equipment. In all cases, the relevance of the applicant's proposed research to human health should be apparent. All new awards are for seven years.
Requirements: Candidates must possess a M.D., D.D.S., D.V.M., Ph.D. or equivalent qualification. Furthermore, they must be eligible to hold a full-time faculty appointment with the sponsoring institution.
Geographic Focus: All States, Canada
Amount of Grant: 205,000 U.S.D
Contact: Pamela Valentine; (780) 423-5727; pamela.valentine@ahfmr.ab.ca
Internet: http://www.ahfmr.ab.ca/grants/seniorscholars.php
Sponsor: Alberta Innovates - Health Solutions
10104-103 Avenue, Suite 1500
Edmonton, AB T5J 4A7 Canada

AHIMA Dissertation Assistance Grants 349

The Dissertation Assistance Award program supports research undertaken as part of an academic program to qualify for a doctorate in areas relevant to health information management (HIM). Each recipient shall be limited to one funded grant per year. Ordinarily, the scope of the proposal shall be such that it can be completed within 18 months from the date of funding. There are no minimum or maximum grant request expectations, but the range of grants historically has been between $5,000 and $10,000. Submissions that address one or more of the AHIMA research priorities will receive priority consideration for funding.
Requirements: To qualify for an award under this program the student (principal investigator) must be enrolled in an accredited doctoral degree program in an area related to HIM (computer science, business management, education, public health, and so forth) and must be an active, associate, or student member of AHIMA. All requirements for the doctoral degree, other than the dissertation, must be completed by the award date.
Date(s) Application is Due: Oct 31
Amount of Grant: 5,000 - 10,000 U.S.D
Samples: Ellen Jacobs, MEd, RHIA—titled In Search of a Message to Promote the Personal Health Record; Melinda Wilkins, MEd, RHIA—titled Adoption of Electronic Health Records: Perceptions of Health Information Managers; Rebecca Reynolds,

MHA, RHIA—titled The Intentions of First Year Medical Students to use Electronic Health Records.
Contact: Carol Nielsen, Director; (312) 233-1175; carol.nielsen@ahimafoundation.org
Internet: http://www.ahimafoundation.org/Scholarships/dissertation.aspx
Sponsor: Foundation of Research and Education of the American Health Information Management Association
233 N Michigan Avenue, 21st Floor
Chicago, IL 60601

AHIMA Faculty Development Stipends 350

FORE Faculty Development Stipends are intended to encourage excellence in effective teaching and leadership in areas related to e-HIM. Program goals are to: expedite curriculum innovation and development in e-HIM content areas; research and develop new methods of teaching, evaluation, and assessment in the HIM learning process; support faculty in maintaining the highest standards of knowledge and practice in e-HIM; integrate new technologies effectively into the HIM learning process; and strengthen the capacity of HIM academic programs to address emerging issues in design and implementation of the electronic health record at all levels of health care delivery, in support of an electronic health information infrastructure.
Requirements: Application should be for activity planned during the four (4) month period following the expected notification date.
Date(s) Application is Due: Mar 14; Jul 11; Oct 3
Contact: Susan H. Fenton, Director of Research; (312) 233-1532 or (312) 233-1100; fax (210) 479-1043; susan.fenton@ahima.org or fore@ahima.org
Internet: http://www.ahimafoundation.org/Scholarships/Faculty.aspx
Sponsor: Foundation of Research and Education of the American Health Information Management Association
233 N Michigan Avenue, 21st Floor
Chicago, IL 60601

AHIMA Grant-In-Aid Research Grants 351

The FORE Grant-in-Aid program is directed toward supporting the development of HIM professionals as leaders in defining and validating the unique body of knowledge encompassed by HIM. The results of these studies not only provide information for the HIM professional to apply in meeting current and future challenges, but also support policy initiatives and the redefinition of the roles of HIM practitioners. Ordinarily, the scope of the proposal shall be such that it can be completed within 18 months from the date of funding. There are no minimum or maximum grant request expectations, but the range of grants historically has been between $15,000 and $40,000. Submissions that address one or more of the AHIMA research priorities will receive priority consideration for funding.
Requirements: The primary or secondary investigator must be an active, associate, or student member of AHIMA. Each recipient shall be limited to one funded grant per year.
Date(s) Application is Due: Sep 19
Amount of Grant: 15,000 - 40,000 U.S.D
Contact: Susan H. Fenton, Director of Research; (312) 233-1532 or (312) 233-1100; fax (210) 479-1043; susan.fenton@ahima.org or fore@ahima.org
Internet: http://www.ahima.org/fore/research/grantinaid.asp
Sponsor: Foundation of Research and Education of the American Health Information Management Association
233 N Michigan Avenue, 21st Floor
Chicago, IL 60601

Ahmanson Foundation Grants 352

The Foundation reviews grant requests from 501(c)3 organizations that are based in and serving Los Angeles County in the areas of education (elementary and secondary education, higher education, nursing school education, and adult literacy and basic skills), the arts and humanities (including libraries and cultural programs), disadvantaged, domestic violence, health and medicine, human services including youth, and religion (Christian, Episcopal, interdenominational, Jewish, Lutheran, Methodist, Presbyterian, Roman Catholic, and Salvation Army). Types of grant support include capital campaigns, challenge, building/renovations, equipment, endowment funds, matching funds, program development, scholarship funds, and seed money. There are no application forms; the board meets four times annually to consider requests. Full proposals will be invited. Letters of inquiry are to be sent through the postal service following the guidelines listed.
Requirements: 501(c)3 nonprofit organizations that serve or are based in Los Angeles County, California, may apply.
Restrictions: Grants are not awarded to individuals or for continuing support, annual campaigns, professorships or internships, fellowships, or film production.
Geographic Focus: California
Amount of Grant: 10,000 - 50,000 U.S.D
Samples: Optimist Youth Homes and Family Services (Los Angeles, CA)—to help construct its new facility that will expand and enhance academic and therapeutic programs for at-risk youngsters, $100,000; California Lutheran U (Thousand Oaks, CA)—to equip the Exercise Physiology/ Biomechanics Laboratory in the new Gilbert Sports and Fitness Center, $500,000.
Contact: Leonard Walcott Jr., Vice President & Managing Director; (310) 278-0770; info@theahmansonfoundation.org
Internet: http://www.theahmansonfoundation.org/fund.html
Sponsor: Ahmanson Foundation
9215 Wilshire Boulevard
Beverly Hills, CA 90210

Ahn Family Foundation Grants 353

The Ahn Family Foundation, established by Sangwoo and Laura Ahn in 1997, has as its primary purposes to support aging, religion, human and social services, art, culture, education, and the environment. Giving is primarily limited to Connecticut, Massachusetts, and New York, although the Foundation has also supported projects in Maryland and Washington. There are no particular guidelines or deadlines with which to adhere, and applicants should begin by contacting the Foundation in writing. This letter of inquiry should outline the purpose for the request, the mission of the applicant organization, and the amount requested.
Geographic Focus: Connecticut, Massachusetts, New York
Amount of Grant: 1,000 - 20,000 U.S.D
Samples: Holy Apostles Soup Kitchen, New York, New York, $20,000; Saint Lukes LifeWorks, Stamford, Connecticut, $20,000; All Saints Church, Chevy Chase, Maryland, $5,000.
Contact: Alison D. Ahn, Treasurer; fax (203) 869-4875
Sponsor: Ahn Family Foundation
901 Hillsboro Mile
Hillsboro Beach, FL 33062-2801

AHNS/AAO-HNSF Surgeon Scientist Combined Award 354

The Grant is open to surgeons beginning a clinician-scientist career track to support research in the pathogenesis, pathophysiology, diagnosis, prevention, or treatment of head and neck neoplastic disease. This grant is only available during odd numbered years. Awards are two year, nonrenewable, with a possible $70,000 maximum total costs ($35,000 per year). One award available per year. Letters of intent must be received by December 15, and completed applications must be submitted by January 15.
Requirements: Applicants must be members or candidate members of the American Academy of Otolaryngology-Head and Neck Surgery and/or the American Head and Neck Society.
Date(s) Application is Due: Jan 15; Dec 15
Amount of Grant: Up to 70,000 U.S.D
Contact: Stephanie L. Jones, Assistant Director; (703) 519-1586 or (703) 836-4444; sljones@entnet.org
Internet: http://www.headandneckcancer.org/research/grants.php
Sponsor: American Head and Neck Society
11300 W Olympic Boulevard, Suite 600
Los Angeles, CA 90064

AHNS/AAO-HNSF Young Investigator Combined Award 355

The purpose of this award is to support a collaborative AHNS/AAO-HNSF research project by fostering the development of contemporary basic or clinical research skills focused on neoplastic disease of the head and neck among new full-time academic surgeons. The award is intended as a preliminary step in clinical investigator career development and is expected to facilitate the recipient's preparation of a more comprehensive individualized research plan suitable for submission to the National Institutes of Health or comparable funding agency. This is a two year, non-renewable award, offering $40,000 maximum ($20,000 per year). One award is available annually. Letters of intent must be received by December 15, and completed applications must be submitted by January 15.
Requirements: Applicants must be physicians with demonstrated potential for excellence in research and teaching and serious commitment to an academic research career in head and neck surgery. Applicants must be members or candidate members of the American Academy of Otolaryngology-Head and Neck Surgery and/or the American Head and Neck Society. Priority will be given to fellows or junior faculty who have completed residencies or fellowships within four years of the application receipt date. All candidates must be sponsored by the Chair of his/her Division or Department and by an official representative of the institution which would administer the Award and in whose name the application is formally submitted.
Date(s) Application is Due: Jan 15; Dec 15
Amount of Grant: Up to 40,000 U.S.D
Contact: Stephanie L. Jones, Assistant Director; (703) 519-1586 or (703) 836-4444; sljones@entnet.org
Internet: http://www.headandneckcancer.org/research/grants.php
Sponsor: American Head and Neck Society
11300 W Olympic Boulevard, Suite 600
Los Angeles, CA 90064

AHNS Alando J. Ballantyne Resident Research Pilot Grant 356

The purpose of this award is to support basic, translational, or clinical research projects in head and neck oncology. Clinical or translational research studies are strongly encouraged and should be specifically related to the prevention, diagnosis, treatment, outcomes, or pathophysiology of head and neck neoplastic disease. Research supported by this award should be specifically directed toward the pathogenesis, pathophysiology, diagnosis, prevention, or treatment of head and neck neoplastic disease, and may be either basic or clinical/translational in approach. While not specifically required, proposals which aim to introduce new knowledge and methodology from other disciplines to research in head and neck disease, or which demonstrate collaborative effort with members of other related disciplines are encouraged. Grants range up to $10,000 maximum, are for one year, and are not renewable. Letters of intent must be received by December 15, and completed applications must be submitted by January 15.
Requirements: This grant is open to resident in U.S. or Canadian training programs. Previous AHNS or AAO-HNS Foundation research grant recipients are eligible to compete for this grant. However, candidates who have successfully obtained funding from a private or federal funding agency for the same research are ineligible. Candidates who have applied for support of the same research from other funding sources, and who

are notified of an award from both another agency and from AHNS must choose only one of the awards.
Date(s) Application is Due: Jan 15; Dec 15
Amount of Grant: Up to 10,000 U.S.D
Contact: Stephanie L. Jones, Assistant Director; (703) 519-1586 or (703) 836-4444; sljones@entnet.org
Internet: http://www.headandneckcancer.org/research/grants.php
Sponsor: American Head and Neck Society
11300 W Olympic Boulevard, Suite 600
Los Angeles, CA 90064

AHNS Pilot Grant **357**
The purpose of this award is to support basic, translational, or clinical research projects in head and neck oncology. Clinical or translational research studies are strongly encouraged and should be specifically related to the prevention, diagnosis, treatment, outcomes, or pathophysiology of head and neck neoplastic disease. Research supported by this award should be specifically directed toward the pathogenesis, pathophysiology, diagnosis, prevention, or treatment of head and neck neoplastic disease, and may be either basic or clinical/translational in approach. While not specifically required, proposals which aim to introduce new knowledge and methodology from other disciplines to research in head and neck disease, or which demonstrate collaborative effort with members of other related disciplines are encouraged. Grants range up to $10,000 maximum, are for one year, and are not renewable. Letters of intent must be received by December 15, and completed applications must be submitted by January 15.
Requirements: Candidates for this award should reside in the U.S. or Canada, be medical students, residents, Ph.D.s or faculty members at the rank of associate professor or below.
Date(s) Application is Due: Jan 15; Dec 15
Amount of Grant: Up to 10,000 U.S.D
Contact: Stephanie L. Jones, Assistant Director; (703) 519-1586 or (703) 836-4444; sljones@entnet.org
Internet: http://www.headandneckcancer.org/research/grants.php
Sponsor: American Head and Neck Society
11300 W Olympic Boulevard, Suite 600
Los Angeles, CA 90064

AHRF Eugene L. Derlacki, M.D. Research Grants **358**
The Eugene L. Derlacki, M.D. Grant is awarded for excellence in the field of hearing research. This grant provides $25,000 per year for two years, for a total award of $50,000. It is dedicated to hearing research and is intended for more significant research projects requiring funding that extends beyond the normal AHRF grant of $25,000. Researchers are invited to apply for this grant using the Research Grant Application Guidelines. According to the rules of the grant, the recipient will receive the first year funding of $25,000. After the first year, a progress report must be submitted to the Research Committee. If the report is approved, the second year funding of $25,000 will then be awarded.
Date(s) Application is Due: Aug 3
Amount of Grant: 25,000 U.S.D
Samples: Steven H. Green, Ph.D., University of Iowa—for Role of JNK Signaling in the Death of Spiral Ganglion Neurons After Hair Cell Loss (2005-06); Keiko Hirose, M.D., The Cleveland Clinic Foundation, Ohio—for Cellular Repair of the Murine Cochlea After Acoustic Injury.
Contact: Kristen Madhuizen, Development and Communication Associate; (312) 726-9670; fax (312) 726-9695; kristen.madhuizen@american-hearing.org or ahrf@american-hearing.org
Internet: http://www.american-hearing.org/research/derlacki_grant.html
Sponsor: American Hearing Research Foundation
8 South Michigan Avenue, Suite 814
Chicago, IL 60603-4539

AHRF Georgia Birtman Grant **359**
The American Hearing Research Foundation (AHRF), together with the Northwestern Memorial Foundation (NMF) give the Georgia Birtman Grant, named in honor of long-time supporter of the AHRF. The one-year $75,000 grant ($25,000 each from the Birtman Fund, the AHRF and Northwestern Memorial Foundation) supports the advancement of research and education in otology and neurotology. The grant is awarded to an exceptional researcher in audiology, otology or neurotology who will work in a lab at Northwestern University. The research topic involves some aspect of the diagnosis, treatment, and rehabilitation of hearing and balance disorders related to the inner ear. The research has the potential to generate clinical care innovations, facilitate translational clinical studies, and develop creative educational programs.
Date(s) Application is Due: Aug 3
Amount of Grant: Up to 75,000 U.S.D
Samples: Timothy C. Hain, M.D., Northwestern University, Feinberg School of Medicine, Chicago, Illinois—for Vestibular Evoked Myogenic Potentials; Claus-Peter Richter, M.D., Ph.D., Northwestern University, Feinberg School of Medicine, Chicago, Illinois—for Electrical Stimulation of Spiral Ganglion Cells.
Contact: Kristen Madhuizen, Development and Communication Associate; (312) 726-9670; fax (312) 726-9695; kristen.madhuizen@american-hearing.org or ahrf@american-hearing.org
Internet: http://www.american-hearing.org/research/birtman_grant.html
Sponsor: American Hearing Research Foundation
8 South Michigan Avenue, Suite 814
Chicago, IL 60603-4539

AHRF Regular Research Grants **360**
The American Hearing Research Foundation funds five to ten $20,000 research grants each year. Research Grants should relate to the hearing or balance functions of the ear. Both basic and clinical studies may be proposed. Priority is given to providing start-up funds for new projects. Applications are reviewed by a Research Committee and awards begin in January. Applications are due no later that noon on August 3 of the previous year.
Date(s) Application is Due: Aug 3
Amount of Grant: Up to 20,000 U.S.D
Contact: Kristen Madhuizen, Development and Communication; (312) 726-9670; fax (312) 726-9695; kristen.madhuizen@american-hearing.org or ahrf@american-hearing.org
Internet: http://www.american-hearing.org/research/grant_guidelines.html
Sponsor: American Hearing Research Foundation
8 South Michigan Avenue, Suite 814
Chicago, IL 60603-4539

AHRF Wiley H. Harrison Memorial Research Award **361**
The purpose of this award is to support clinical research projects in otology or neurotology designed to increase understanding of hearing disorders. Research supported by this award should be specifically directed toward the clinical identification, diagnosis, prevention, or treatment of diseases, disorders, or conditions of the ear. While not specifically required, proposals which aim to introduce new knowledge and methodology from other disciplines to research in otology or neurotology, or which demonstrate collaborative effort with members of other related disciplines are encouraged. This is a one year, non-renewable grant of $25,000 maximum, with only one award available annually. Completed applications must be submitted by August 3.
Requirements: Candidates for this award should be physicians (M.D.) at the resident, fellow, or junior faculty stage, or PhD scientists. Previous AHRF or AAO-HNS Foundation research grant recipients are eligible to compete for this grant.
Date(s) Application is Due: Aug 3
Amount of Grant: Up to 25,000 U.S.D
Contact: Kristen Madhuizen, Development and Communication Associate; (312) 726-9670; fax (312) 726-9695; kristen.madhuizen@american-hearing.org or ahrf@american-hearing.org
Internet: http://www.american-hearing.org/research/harrisongrant.html
Sponsor: American Hearing Research Foundation
8 South Michigan Avenue, Suite 814
Chicago, IL 60603-4539

AHRQ Assessment of Quality Improvement Strategies in Health Care **362**
The projects undertaken as a result of this RFA will analyze the relative utility and costs of various approaches to health care quality improvement. The fundamental long-term goal of this effort is to strengthen the evidence base underlying the choice of strategies to employ when attempting to improve the quality of clinical care. Studies should focus on comparing improvement efforts that target those areas where the greatest improvements in health and functional status can occur, reliable and valid quality measures exist, and a variety of strategies are being employed. Partnerships between academic and other research organizations with existing health care quality improvement efforts through established mechanisms such as Peer Review Organizations (PROs), Quality Improvement Organizations (QIOs), purchaser groups, health plans, and accrediting bodies are required under this RFA. Special preference will be accorded to applications from investigators not recently or currently funded as principal investigators of an AHCPR grant for research on quality improvement strategies. RFA: HS-99-002
Requirements: Applications may be submitted by public or private nonprofit organizations. For-profit entities may participate as members of consortia or subcontractors if the applicant is nonprofit.
Restrictions: Organizations described in section 501(c)4 of the Internal Revenue Code that engage in lobbying are not eligible.
Amount of Grant: 5,000 - 28,000,000 U.S.D
Contact: Mable Lam, Grants Management Officer; (301) 427-1448
Internet: http://www.grants.nih.gov/grants/guide/rfa-files/RFA-HS-99-002.html
Sponsor: Agency for Healthcare Research and Quality
2101 E Jefferson Street, Executive Office Center
Rockville, MD 20852

AHRQ Centers for Education and Research on Therapeutics (CERTS) Grants **363**
CERTs is a three-year program that will support demonstration centers. These centers will evaluate and develop options and methods, and conduct and perform pilot studies. Studies will consist of state-of-the-art clinical, health services, or laboratory research to increase awareness of the benefits, risks and effectiveness of new uses, existing uses, or combined uses of therapeutics. This demonstration program seeks new and more effective ways to develop, translate, and disseminate objective information on therapeutics to health care providers and other decision makers to improve practice. In addition, CERTs may selectively develop protocols and possibly undertake pilot studies on the comparative cost effectiveness and safety of medical products. This will be accomplished with data on appropriate therapeutic usage and outcomes; and the identification and prevention of medical errors and adverse effects. The long-term goal of the program will be to improve the quality of care while reducing costs. RFA: HS-99-004
Requirements: Applications may be submitted by public or private nonprofit organizations. For-profit organizations may participate as members of consortia or subcontractors if the applicant is nonprofit.
Restrictions: Organizations described in section 501(c)4 of the Internal Revenue Code that engage in lobbying are not eligible.
Amount of Grant: 5,000 - 28,000,000 U.S.D

Contact: Mable Lam, Grants Management Officer; (301) 427-1447
Internet: http://www.grants.nih.gov/grants/guide/rfa-files/RFA-HS-99-004.html
Sponsor: Agency for Healthcare Research and Quality
2101 E Jefferson Street, Executive Office Center
Rockville, MD 20852

AHRQ Centers of Excellence for Patient Safety Research and Practice 364
The purpose of this program is to support the development of multidisciplinary research teams to build the knowledge base on the scope and impact of medical errors; identify the root causes of threats to patient safety and effective system approaches to prevent the occurrence of errors; study the effectiveness of various interventions to capture information on medical errors; and evaluate the outcomes of promising interventions in a variety of healthcare settings. A letter of intent is requested by January 3; full application is due January 24. Annual deadlines may vary; contact program staff for exact dates.
Requirements: Applications may be submitted by domestic, both public and private, nonprofit organizations; units of state and local governments; and eligible agencies of the federal government.
Restrictions: Foreign institutions are not eligible to apply.
Date(s) Application is Due: Jan 3; Jan 24
Amount of Grant: 5,000 - 28,000,000 U.S.D
Contact: Mable Lam, Grants Management Officer; (301) 427-1447
Internet: http://grants.nih.gov/grants/guide/rfa-files/RFA-HS-01-002.html
Sponsor: Agency for Healthcare Research and Quality
2101 E Jefferson Street, Executive Office Center
Rockville, MD 20852

AHRQ Development and Coordination of Rural Health Services Grants 365
This program provides funds to develop and disseminate information to assist rural communities and rural health care organizations in developing and coordinating rural health care services. Applications will be reviewed and rated on the applicant's ability to address the following areas: developing and disseminating knowledge; organizing, publicizing, and managing meetings to enhance the dissemination of information on rural health care issues; providing technical and other nonfinancial assistance to state offices of rural health to increase their capability to serve rural residents; developing a network to promote recruitment and retention of health professionals in rural areas; and providing updated information to the public about rural health clinics. Contact the office for deadline dates.
Requirements: Nonprofit private organizations that represent national, state, and local rural health constituencies that are interested in and committed to improvements in rural health care. may apply.
Samples: Examples of projects previously funded include the following: two national meetings of state offices; two workshops on rural minority health; meetings and support for the National Recruitment Network; technical assistance to states and local organizations; and a paper on rural oral health.
Contact: Lawrence Poole, Grants Management; (301) 443-2385
Internet: http://www.ruralhealth.hrsa.gov
Sponsor: Agency for Healthcare Research and Quality
5600 Fishers Lane, 9A-55
Rockville, MD 20857

AHRQ Health Care Access, Quality, and Insurance for Low-Income Children Grants 366
AHCPR and the David and Lucile Packard Foundation invite applications for cooperative agreements to conduct research that will generate information useful to purchasers and designers of health insurance and health care delivery systems for low-income children. Specifically, answers are sought to the policy question: How do features of U.S. health care insurance and delivery systems improve health care access and quality for low-income children, particularly racial and ethnic minority children or those with special health care needs? Studies will examine how the features of U.S. insurance programs and the organization of the health care delivery systems associated with these programs affect access to services and the quality of care received by low-income children. Studies will fall into one of two categories: those examining the impact on low-income children enrolled in these insurance programs (enrollee impact), and those examining the impact on a low-income community's health care delivery system and all the children that it serves (community impact). Applicants may propose research that falls into one or both of these categories. Care should be taken not to replicate the work of other major health care studies. Application kits are available at most institutional offices of sponsored research and may be obtained from the Division of Extramural Outreach and Information Resources, Office of Extramural Research, National Institutes of Health, 6701 Rockledge Drive, MSC 7910, Bethesda, MD 20892-7910, telephone (301) 435-0714, email: Grantsinfo@nih.gov. RFA: HS-99-005
Requirements: Applications may be submitted by domestic and foreign, public and private nonprofit organizations.
Restrictions: For-profit organizations are not eligible as applicants, but may participate as members of consortia or as subcontractors. Organizations described in section 501(c)4 of the Internal Revenue Code that engage in lobbying are not eligible to receive grant/cooperative agreement awards.
Amount of Grant: 5,000 - 28,000,000 U.S.D
Contact: Mable Lam, Grants Management Officer; (301) 427-1447
Internet: http://www.grants.nih.gov/grants/guide/rfa-files/RFA-HS-99-005.html
Sponsor: Agency for Healthcare Research and Quality
2101 E Jefferson Street, Executive Office Center
Rockville, MD 20852

AHRQ Independent Scientist Award 367
The award is a special salary-only grant designed to provide protected time for newly independent scientists who currently have non-research obligations such as heavy teaching loads, clinical work, committee assignments, service, and administrative duties that prevent them from having a period of intensive research focus. The award is targeted to persons with doctoral degrees who have completed their research training, have independent peer-reviewed research support, and who need a period of protected research time in order to foster their research career development. New applications are due February 12, June 12, and October 12, while renewal applications are due March 12, July 12, and November 12. Up to $90,000 per year of base salary plus fringe benefits is available. No other research development support funds are provided.
Requirements: The following organizations and institutions are eligible to apply: public/state controlled institutions of higher education; private institutions of higher education; nonprofits with 501(c)3 IRS status; nonprofits without 501(c)3 IRS status; small businesses; for-profit organizations; State governments; U.S. territories or possessions; Indian/Native American tribal governments (Federally recognized and other than Federally recognized); Indian/Native American tribally designated organizations; Hispanic-serving institutions; historically Black colleges and universities (HBCUs); tribally controlled colleges and universities (TCCUs); Alaska Native and Native Hawaiian serving institutions; regional organizations; and faith-based or community based organizations. The candidate must have a doctoral degree and peer-reviewed, independent research support at the time the award is made. The candidate must spend a minimum of 75 percent effort conducting research during the period of the award.
Restrictions: The award is not intended for investigators who already have full time to perform research, or have substantial publication records or considerable research support indicating that they are well established in their fields. Foreign institutions are not eligible to apply.
Date(s) Application is Due: Feb 12; Mar 12; Jun 12; Jul 12; Oct 12; Nov 12
Contact: Kay Anderson, (301) 427-1555; fax (301) 427-1562; Kay.Anderson@ahrq.hhs.gov
Internet: http://grants.nih.gov/grants/guide/pa-files/PAR-07-444.html
Sponsor: Agency for Healthcare Research and Quality
540 Gaither Road
Rockville, MD 20850

AHRQ Individual Awards for Postdoctoral Fellows Ruth L. Kirschstein National Research Service Awards (NRSA) 368
The purpose of the postdoctoral fellowship award is to provide support to promising postdoctoral applicants who have the potential to become productive and successful independent research investigators. The proposed postdoctoral training must offer an opportunity to enhance the applicant's understanding of health services research and must be responsive to AHRQ's mission, which is to improve the quality, safety, efficiency, and effectiveness of health care for all Americans. The research sponsored and conducted by AHRQ develops and presents scientific evidence regarding all aspects of health care. It addresses issues of organization, delivery, financing, utilization, patient and provider behavior, outcomes, effectiveness and cost. It evaluates both clinical services and the system in which these services are provided. These scientific results improve the evidence base to enable better decisions about health care, including such areas as disease prevention, appropriate use of medical technologies, improving diagnosis and treatment utilizing comparative effectiveness research, and reducing racial and ethnic disparities. In addition, AHRQ is interested in the application of health information technology (health IT), as well as reducing medical errors and improving patient safety.
Requirements: The following organizations and institutions are eligible to apply: public or non-profit private institutions, such as a university, college, or a faith-based or community-based organization; units of local or State government; eligible agencies of the Federal government; Indian/Native American tribal governments (Federally recognized); Indian/Native American tribal governments (other than Federally recognized); and Indian/Native American tribally designated organizations.
Restrictions: Awards for fellowship training may only be made to domestic institutions.
Date(s) Application is Due: Apr 8; Aug 8; Dec 8
Contact: Shelley M. Benjamin, (301) 427-1528; fax (301) 427-1562; Shelley.Benjamin@ahrq.hhs.gov
Internet: http://grants.nih.gov/grants/guide/pa-files/PA-09-229.html
Sponsor: Agency for Healthcare Research and Quality
540 Gaither Road
Rockville, MD 20850

AHRQ Translating Research into Practice Grants 369
AHCPR invites applications to conduct research related to implementing evidence-based tools and information in diverse health care settings among practitioners caring for diverse populations. Applications are sought for studies that apply innovative strategies for implementing evidence-based tools and information and demonstrate improved clinical practice and sustained practitioner behavior change. Evidence-based tools and information include findings from rigorously conducted research and from clinical practice guidelines, algorithms, treatment protocols, practice parameters, quality indicators, and continuous quality improvement initiatives that are developed using a systematic approach to evidence synthesis. AHCPR is especially interested in studies that implement AHCPR-supported, evidence-based tools and information, including Patient Outcome Research Team (P.O.RT) and other research findings; AHCPR-supported clinical practice guideline recommendations; and evidence reports and technology assessments produced by the AHCPR Evidence-based Practice Centers. The goal of this RFA is to improve the translation and use of research findings and evidence-based tools by developing and validating innovative principles, methods, and tools that work in diverse settings, populations, and payment systems. Applicants are encouraged to form public-private

partnerships or consortia, such as between academic and other research organizations and health plans and purchasers, to perform this research. Such partnerships may help to more quickly translate research findings into actual practice settings. Applicants may also seek collaborative funding for projects funded under this RFA. Roles of collaborators should be clearly defined in the application. RFA: HS-99-003
Requirements: Applications may be submitted by public or private nonprofit organizations. For-profit organizations may participate as members of consortia or subcontractors if the applicant is nonprofit.
Restrictions: Organizations described in section 501(c)4 of the Internal Revenue Code that engage in lobbying are not eligible.
Amount of Grant: Up to 2,000,000 U.S.D
Contact: Joan Metcalfe; (301) 594-1841; fax (301) 594-3210; jmetcalf@ahcpr.gov
Internet: http://www.grants.nih.gov/grants/guide/rfa-files/RFA-HS-99-003.html
Sponsor: Agency for Healthcare Research and Quality
2101 E Jefferson Street, Executive Office Center
Rockville, MD 20852

AHRQ Understanding and Eliminating Minority Health Disparities **370**
The purpose of this program is to conduct research on racial and ethnic disparities in health that are amenable to improvements in health services. Projects funded under this RFA will build on previous research that has identified disparities in access to, and utilization, quality and outcomes of health care services and the excess burdens of illness and death for African Americans, Hispanic Americans, American Indians, Alaska Natives, Asian Americans and Pacific Islanders compared to the United States nonminority population. Projects funded by AHCPR will analyze causes and contributing factors for the inequalities that are related to the delivery and practice of health care, and identify and implement strategies to eliminate them. A letter of intent is requested by December 22; full application is due January 21. Annual deadline dates may vary; contact program staff for exact dates. RFA: HS-00-003
Requirements: Applications may be submitted by domestic or foreign, public or private nonprofit organizations, including American Indian/Alaska Native organizations, universities, clinics, units of state, tribal and local governments, and eligible agencies of the federal government. AHCPR, by statute, can make grants only to nonprofit organizations; however, for-profit organizations may participate in grant projects as members of consortia or as subcontractors.
Restrictions: 501(c)4 organizations that engage in lobbying are not eligible.
Date(s) Application is Due: Jan 21; Dec 22
Amount of Grant: Up to 100,000,000 U.S.D
Contact: Joan Metcalfe; (301) 594-1841; fax (301) 594-3210; jmetcalf@ahcpr.gov
Internet: http://grants.nih.gov/grants/guide/rfa-files/RFA-HS-00-003.html
Sponsor: Agency for Healthcare Research and Quality
2101 E Jefferson Street, Executive Office Center
Rockville, MD 20852

AICR Investigator-Initiated Research Grants **371**
Grants are awarded to nonprofit institutions for research relevant to understanding the effects of dietary and nutritional practices on the etiology, pathogenesis, and treatment of cancer. Applications for this program should fall within the scope of diet and prevention of cancer and diet and treatment of cancer. New ideas in research that focus on dietary and nutritional means of preventing and treating cancer or improving the quality of life of the cancer patient are encouraged. Funded organizations receive two years of support with possible renewal for two additional years. Preproposal letters of intent are encouraged but are not required. Application forms, instructions, and policy information may be obtained from the institute.
Restrictions: Indirect costs cannot exceed 10 percent of direct cost budget.
Date(s) Application is Due: Jul 1; Dec 16
Amount of Grant: 75,000 U.S.D
Contact: Research Department; (202) 328-7744 or (800) 843-8114; fax (202) 328-7226; research@aicr.org
Internet: http://www.aicr.org/research/investigator.lasso
Sponsor: American Institute for Cancer Research
1759 R Street NW
Washington, DC 20009

AICR Postdoctoral Award **372**
These international fellowships are awarded to beginning investigators to stimulate innovative and new research on the prevention, etiology, or treatment of cancer by dietary or nutritional methods. Applications should propose relevant feasibility studies to obtain data in support of a new hypothesis that then could be expanded to increase understanding of the role of dietary and nutritional factors in the etiology, pathogenesis, or treatment of cancer.
Requirements: The principal investigator must have a PhD or MD degree that was awarded no more than three years prior to the date of the application and must hold an academic appointment no higher than assistant professor. The applicant must be sponsored by a professor in whose laboratory the applicant is to perform his or her research.
Date(s) Application is Due: Jul 1; Dec 16
Amount of Grant: 25,010 U.S.D
Contact: Research Department; (202) 328-7744 or (800) 843-8114; fax (202) 328-7226; research@aicr.org
Internet: http://www.aicr.org/research/post_doc.lasso
Sponsor: American Institute for Cancer Research
1759 R Street NW
Washington, DC 20009

AIDS Vaccine Advocacy Coalition (AVAC) Fund Grants **373**
The Fund functions as a small-scale emergency fund to assist needy clinical sites that require immediate help with purchases such as additional medical or lab supplies not covered by grants or contracts for vaccine research. All AIDS vaccine clinical trial sites in resource-limited countries or needy communities, regardless of sponsor, can apply for grants up to $2,000. A committee consisting of HIV vaccine investigators from around the world reviews proposals and makes recommendations to a special Fund Committee of the AVAC Board.
Amount of Grant: 2,000 U.S.D
Samples: Kenya AIDS Vaccine Initiative, Department of Medical Microbiology, Nairobi University, $2,000—for HIV Rapid Test Kits and Emergency Medicines; Brooke Bond Tea Plantation Hospital, $2,000—for general health screenings.
Contact: Marie Semmelbeck, (212) 367-1188 or (212) 367-1279; fax (646) 365-3452; fund@avac.org
Internet: http://www.avac.org/ht/d/sp/i/962/pid/962
Sponsor: AIDS Vaccine Advocacy Coalition
101 West 23rd Street, #2227
New York, NY 10011

AIG American General Corporation Grants **374**
Corporate contributions support organizations that make the greatest impact in its operating communities in the areas of education, health and human services, arts and culture, and community development. The program also recognizes employee volunteer efforts with its Community Spirit Awards program. Award recipients each designate a grant to the organizations they served. Contact the office for application procedures.
Requirements: 501(c)3 nonprofits, particularly those located in the corporation's operating communities, may apply.
Restrictions: Contributions will generally not be made to organizations without 501(c)3 status; United Way agencies (except through an annual employee-giving campaign and corporate gift); veterans groups, along with labor, religious, political, fraternal or external athletic groups, except in specific cases in which such groups provide needed benefits or services to the community at large; national organizations; private foundations; individuals; hospitals, unless it is a teaching institution; and international organizations.
Date(s) Application is Due: Mar 31; Jun 30; Sep 30; Dec 31
Amount of Grant: 25,000 U.S.D
Samples: Anti-Defamation League—to support its No Place for Hate campaign, $115,000; America's Promise—for youth programs, $1.5 million.
Contact: Christina Minkley, Christina_Minkley@aigag.com
Internet: http://www.aigag.com/life/life.nsf/contents/aboutus_community_givingguidelines
Sponsor: AIG American General Corporation
2929 Allen Parkway, A35-00
Houston, TX 77019

AIHP Sonnedecker Visiting Scholar Grants **375**
The Sonnedecker Visiting Scholar Program offers assistance for short-term historical research related to the history of pharmacy, including the history of drugs, at the University of Wisconsin-Madison. The program provides assistance for travel, maintaining temporary residence in Madison, and meeting research expenses associated with utilizing the collection. A brochure is available on request that describes the pharmaco-historical collections, which have been developed in Madison during more than a century by the University of Wisconsin-Madison, the State Historical Society of Wisconsin, and the American Institute of the History of Pharmacy. At least $1,000 becomes available annually to defray part of the expenses of a recipient, for whatever period of residence is appropriate. Grants are made throughout the year on the basis of the merit of previous historical work and on the appropriateness of historical resources on the University of Wisconsin campus to the research proposed.
Requirements: Eligible applicants are: Historians, Pharmacists, and other scholars working in the Pharmacy field.
Amount of Grant: 1,000 U.S.D
Contact: Gregory Higby, Director; (608) 262-5378; grants@aihp.org
Internet: http://cms.pharmacy.wisc.edu/aihp/programs/sonnedecker
Sponsor: American Institute of the History of Pharmacy
777 Highland Avenue
Madison, WI 53705-2222

AIHP Thesis Support Grants **376**
The Program offers a grant-in-aid totaling $2,000 or more annually to a graduate student to reinforce historical investigations of some aspect of pharmacy, whether ancient or modern, to pay research expenses not normally met by the university granting the degree. Any thesis project devoted to the history of pharmacy, history of drugs, or other humanistic study utilizing a pharmaco-historical approach, is eligible if based in an institution of higher learning of the U.S.A. The maximum grant in this program of the Institute will be $2,500.
Requirements: Any graduate student in good standing at an institution of the United States may apply, regardless of the department through which the Doctor of Philosophy degree will be granted. The graduate student need not be an American citizen; nor does the research topic need to be in the field of American history.
Restrictions: Examples of ineligible expenses would be: living expenses of the applicant at the home university; routine typing to produce research notes or the thesis manuscript; routine illustrations for the manuscript; and publication of research results. Indirect expenses, such as overhead and other institution-related costs, may not be included in a grant application.
Date(s) Application is Due: Feb 1
Amount of Grant: Up to 2,500 U.S.D

Contact: Gregory Higby, Director; (608) 262-5378; grants@aihp.org
Internet: http://cms.pharmacy.wisc.edu/aihp/programs/thesisgrant
Sponsor: American Institute of the History of Pharmacy
777 Highland Avenue
Madison, WI 53705-2222

AIHS Clinical Fellowships **377**
The AIHS Clinical Fellowships program is designed for highly qualified individuals who hold an M.D. or D.D.S. degree, and who anticipate undertaking a career in health related or clinical research in Alberta. The awards consist of a stipend and a research allowance. A Clinical Fellowship award is normally tenable for a maximum of 3 years. However, if the trainee is registered in a graduate program, the term may be extended. A Clinical Fellowship award is normally tenable for a maximum of 3 years. However, if the trainee is registered in a graduate program, the term may be extended.
Requirements: Normally, an award will be held within the Province of Alberta; however, candidates who are Canadian citizens or permanent residents with records of outstanding performance in postgraduate training may seek research training elsewhere, if sponsored by an Alberta faculty with an expressed interest in future recruitment of the candidate. Candidates must hold an MD or DDS degree and have received a significant portion of their postgraduate training in Alberta.
Geographic Focus: Canada
Date(s) Application is Due: Mar 1; Oct 1
Amount of Grant: 20,000 - 50,000 U.S.D
Contact: Pamela Valentine, Vice President of Research & Innovation; (780) 423-5727, ext. 230; fax (780) 429-3509; pamela.valentine@albertainnovates.ca or grants.health@albertainnovates.ca
Internet: http://www.ahfmr.ab.ca/grants/Clin-fellow.php
Sponsor: Alberta Innovates - Health Solutions
10104-103 Avenue, Suite 1500
Edmonton, AB T5J 4A7 Canada

AIHS Fast-Track Fellowships **378**
AIHS will accept, at any time, applications from Alberta-based institutions for full-time Fellowship awards for outstanding candidates currently training outside of Alberta. The fast-track program is intended to assist these institutions in recruiting highly qualified trainees to the province. Up to a maximum of 10 awards per calendar year will be awarded through the fast-track review process in each category. Applications to the program will be reviewed on a first come-first serve basis. Fast-track awards are tenable for up to 12 months.
Requirements: Applications must include a letter from the Dean/designate of the faculty indicating that the proposal for a fast-track award has been reviewed and stating why a fast-track application should be considered for the candidate in question. Please refer to the sections on Full-Time Fellowships or Full-Time Studentships for further details regarding eligibility for training awards.
Geographic Focus: Canada
Amount of Grant: 20,000 - 50,000 U.S.D
Contact: Pamela Valentine, Vice President of Research & Innovation; (780) 423-5727, ext. 230; fax (780) 429-3509; pamela.valentine@albertainnovates.ca or grants.health@albertainnovates.ca
Internet: http://www.ahfmr.ab.ca/grants/fasttrack.php
Sponsor: Alberta Innovates - Health Solutions
10104-103 Avenue, Suite 1500
Edmonton, AB T5J 4A7 Canada

AIHS Full-Time Fellowships **379**
The AIHS Full-Time Fellowships program is designed to enable highly qualified doctoral graduates to prepare for careers in medical or health research as independent investigators. A fellowship award will provide to the host institution funding for one year's stipend, its associated benefits, and a research allowance. Fellows may engage in teaching activities related to their research discipline a maximum of 20 percent of their time. The maximum term is three years
Requirements: Candidates must have a Ph.D., M.D., D.D.S., D.V.M. or D.Pharm. degree. Normally, support will not be provided beyond 6 years after receipt of the Ph.D. degree, or beyond 8 years after receipt of the M.D., D.D.S., D.V.M. or D.Pharm. degrees.
Geographic Focus: Canada
Date(s) Application is Due: Mar 1; Oct 1
Amount of Grant: 35,000 - 50,000 U.S.D
Contact: Pamela Valentine, Vice President of Research & Innovation; (780) 423-5727, ext. 230; fax (780) 429-3509; pamela.valentine@albertainnovates.ca or grants.health@albertainnovates.ca
Internet: http://www.ahfmr.ab.ca/grants/FT-fellow.php
Sponsor: Alberta Innovates - Health Solutions
10104-103 Avenue, Suite 1500
Edmonton, AB T5J 4A7 Canada

AIHS Full-Time Health Research Studentships **380**
The Health Research Studentship program enables academically superior students to undertake full-time training in health research. The award supports training in: research on the organization and delivery of health care; technology assessment; community health; health promotion; disease prevention; and related disciplines. The Health Research Studentship consists of a stipend and a research allowance. The term of the award is a maximum of five years (three years at the graduate level).
Requirements: Candidates must have been accepted into, or be currently studying in, a full-time, thesis-based, graduate program at an Alberta-based university in a health-related discipline leading to a Master's or doctoral degree. This award is not available to students registered in a course-based program. Normally, support will not be provided beyond 6 years of enrollment in graduate school. Candidates who have interrupted their training for parenting or other reasons and who have consequently exceeded this time limit, may apply for studentship support. In such cases, however, the candidate is advised to clearly explain the nature of his/her particular circumstances at the time of application.
Geographic Focus: Canada
Date(s) Application is Due: Mar 1; Oct 1
Amount of Grant: 20,000 U.S.D
Contact: Pamela Valentine, Vice President of Research & Innovation; (780) 423-5727, ext. 230; fax (780) 429-3509; pamela.valentine@albertainnovates.ca or grants.health@albertainnovates.ca
Internet: http://www.ahfmr.ab.ca/grants/FT-student.php#HRS
Sponsor: Alberta Innovates - Health Solutions
10104-103 Avenue, Suite 1500
Edmonton, AB T5J 4A7 Canada

AIHS Full-Time M.D./Ph.D. Studentships **381**
AIHS Full-Time M.D./Ph.D. Studentships are intended to provide an opportunity for exceptional candidates, who wish to pursue careers as Clinical Investigators, to study for the M.D. and the Ph.D. degrees simultaneously. Support is complementary to the formal M.D./Ph.D. programs at the University of Alberta and the University of Calgary. M.D./Ph.D. studentship awards are tenable only at an Alberta-based university. The term of the award is a maximum of six years.
Requirements: Applicants are limited to those students that formally enter the MD/PhD program not later than the start of their second year of PhD studies at either the University of Alberta or the University of Calgary. AIHS's offer of support is contingent on the granting of complementary stipend support from the Faculty of not less than 15% of the value of the AIHS award, for the duration of the award. Furthermore, it is expected that the universities will undertake to provide administrative support to the offices of the coordinators of the M.D./Ph.D. programs.
Geographic Focus: Canada
Date(s) Application is Due: Mar 1; Oct 1
Amount of Grant: 20,000 U.S.D
Contact: Pamela Valentine, Vice President of Research & Innovation; (780) 423-5727, ext. 230; fax (780) 429-3509; pamela.valentine@albertainnovates.ca or grants.health@albertainnovates.ca
Internet: http://www.ahfmr.ab.ca/grants/FT-student.php#MPS
Sponsor: Alberta Innovates - Health Solutions
10104-103 Avenue, Suite 1500
Edmonton, AB T5J 4A7 Canada

AIHS Full-Time Studentships **382**
AIHS Full-Time Studentships enable academically superior students to undertake full-time research training in the basic biomedical sciences or in clinical research. The award consists of a stipend and a research allowance. Approved uses of the research allowance include: the purchase of scientific materials, supplies and expendables; the purchase of minor equipment; computer software programs; costs for the use of libraries, or computers; costs associated with the publication of research results; travel expenses to attend scientific meetings; and purchase of books, periodicals and journals. The term of the award is a maximum of five years (three years at the graduate level).
Requirements: Candidates must have been accepted into, or be currently engaged in, a full-time, thesis-based, graduate program at an Alberta-based university in a health-related discipline leading to a Master's or doctoral degree.
Restrictions: This award is not available to students registered in a course-based program. Normally, support will not be provided beyond 6 years of enrollment in graduate school. Candidates who have interrupted their training for parenting or other reasons and who have consequently exceeded this time limit, may apply for studentship support. In such cases, however, the candidate is advised to clearly explain the nature of his/her particular circumstances at the time of application.
Geographic Focus: Canada
Date(s) Application is Due: Mar 1; Oct 1
Amount of Grant: 20,000 U.S.D
Contact: Pamela Valentine, Vice President of Research & Innovation; (780) 423-5727, ext. 230; fax (780) 429-3509; pamela.valentine@albertainnovates.ca or grants.health@albertainnovates.ca
Internet: http://www.ahfmr.ab.ca/grants/FT-student.php#fts_description
Sponsor: Alberta Innovates - Health Solutions
10104-103 Avenue, Suite 1500
Edmonton, AB T5J 4A7 Canada

AIHS Part-Time Fellowships **383**
AIHS Part-Time Fellowships enable highly qualified doctoral graduates to prepare for careers in medical or health research as independent investigators doing part-time research training during the regular academic year while enrolled in a full-time professional degree program in Alberta. Part-time awards will consist of a stipend only and will be pro-rated to the amount of time the candidate is prepared to commit to research. The full-time fellowship rate will be used as the basis for this calculation. The maximum term for the award is three years.
Requirements: Candidates must hold a PhD and must be engaged in a professional degree program (usually the MD program).
Restrictions: Part-time fellowship awards are tenable only at an Alberta-based university.

Geographic Focus: Canada
Date(s) Application is Due: Mar 1; Oct 1
Amount of Grant: 5,000 - 30,000 U.S.D
Contact: Pamela Valentine, Vice President of Research & Innovation; (780) 423-5727, ext. 230; fax (780) 429-3509; pamela.valentine@albertainnovates.ca or grants.health@albertainnovates.ca
Internet: http://www.ahfmr.ab.ca/grants/PT-fellow.php
Sponsor: Alberta Innovates - Health Solutions
10104-103 Avenue, Suite 1500
Edmonton, AB T5J 4A7 Canada

Air Products and Chemicals Grants **384**
The foundation supports nonprofit organizations in company-operating areas in the fields of precollege and higher education; fitness, health, and welfare; community and economic development, arts and culture; and the environment and safety. Types of support include capital grants, employee matching gifts, general operating support, multiyear-continuing grants, project grants, seed grants, fellowships, employee-related scholarships, and donated equipment. There are no application forms; requests must be in writing. Guidelines are available online.
Requirements: 501(c)3 nonprofit organizations in company-operating areas are eligible.
Restrictions: Grants are not made to/for individuals, sectarian or denominational organizations, political candidates or activities, veterans organizations, organizations receiving United Way support, labor groups, elementary or secondary schools, capital campaigns of national organizations, hospital operating expenses, national health organizations, or goodwill advertising.
Contact: Kassie Hilgert, Program Manager; hilgerk@airproducts.com
Internet: http://www.airproducts.com/Responsibility/SocialResponsibility
Sponsor: Air Products and Chemicals Corporation
7201 Hamilton Boulevard
Allentown, PA 18195-1501

AJCC Cancer Staging Grants **385**
AJCC seeks proposals to develop and evaluate improved staging algorithms for specific cancer sites and types. Successful proposals will present new approaches and utilize data sets available to the applicant to test and validate current or revised staging algorithms. Applicants may use elements currently used for AJCC/UICC TNM staging, additional or modified anatomic and non-anatomic factors to define prognosis, and incorporate validated factors that identify response to treatment. Proposals that address and validate the use of non-anatomic factors to improve the utility of staging and define the response to therapy are encouraged. The AJCC will provide funding to a number of groups for a variety of disease sites to promote innovative strategies for enhancing the clinical utility of staging while maintaining compatibility with current staging data collection systems. The listed application deadline is for letters of intent. Guidelines are available online.
Requirements: Eligible institutions include for-profit and nonprofit organizations; public or private institutions, such as universities, colleges, laboratories, and hospitals; units of state and local government; eligible agencies of the federal government; and domestic or foreign institutions/organizations.
Date(s) Application is Due: Dec 15
Amount of Grant: Up to 150,000 U.S.D
Contact: Valerie Vesich, (312) 202-5420; fax (312) 202-5009; AJCC@facs.org
Internet: http://www.cancerstaging.org/initiatives/rfp.html#
Sponsor: American Joint Committee on Cancer
633 N Saint Clair Street
Chicago, IL 60611

Alabama Power Foundation Grants **386**
The foundation's mission is to improve the lives and circumstances of Alabama residents and to strengthen the communities in which they live. It supports programs that will improve education (by supporting innovative programs and assisting teachers in their crucial responsibilities), strengthen communities, promote arts and culture or restore and enhance the environment. The project must meet several of the guidelines to be considered. The majority of grants are made to support targeted efforts with specific objectives. The foundation also provides general operating assistance to organizations, capital grants for endowment and building, and scholarship funds. Applications should be sent to the local Alabama Power Company office for review and recommendation to the foundation. The amount of the request determines the review frequency.
Requirements: Applications are accepted from Alabama nonprofit organizations whose programs fall within foundation guidelines.
Restrictions: Grants are not made to individuals, for sectarian religious purposes, or for political activities.
Geographic Focus: Alabama
Date(s) Application is Due: Feb 1; May 1; Aug 1; Nov 1
Amount of Grant: 1,000 - 100,000 U.S.D
Samples: Stillman College (Tuscaloosa, AL)—to construct the Cordell Wynn Humanities, Communications, and Fine Arts Ctr, $65,000.
Contact: William Johnson, President; (205) 257-2508; fax (205) 257-1860
Internet: http://www.alabamapower.com/foundation/grantsandinitiatives.asp
Sponsor: Alabama Power Foundation
600 N 18th Street, P.O. Box 2641
Birmingham, AL 35291-0011

Aladdin Industries Foundation Grants **387**
The foundation awards grants, primarily in Tennessee, in its areas of interest, including arts and culture, youth services, health and human services, education, business, drug abuse, and the elderly. Types of support include general operating support, seed grants, and employee-related scholarships. Proposals are reviewed quarterly.
Requirements: Most grants are awarded in Tennessee. Proposals are reviewed quarterly.
Geographic Focus: Tennessee
Contact: L.B. Jenkins, Secretary & Treasurer; (615) 748-3360
Sponsor: Aladdin Industries Foundation
703 Murfreesboro Road
Nashville, TN 37210-4521

Alaska Airlines Corporate Giving Grants **388**
Alaska Airlines and Horizon Air Corporate Giving support health and human services, arts and cultural programs, as well as education, environmental, and civic organizations. The program focuses on communities served now or in the near future and where a significant number of its employees live or work. When considering requests, the corporation favors organizations and efforts that are most likely to enhance a community's cultural and economic vitality and improve the quality of life for its citizens. Objectives of the Corporate Giving Program are to: provide support to charitable and cultural organizations within our communities; and maintain a reasonable balance of contributions supporting a wide range of organizations.
Requirements: Giving is on a national and international basis in areas of company operations, with emphasis on Alaska and Washington; some further giving occurs in Canada and Mexico.
Restrictions: Alaska Airlines or Horizon Air do not provide support for: loans and grants to individuals or private business; groups that discriminate; endowments; religious organizations (if for sacramental purposes); pageants; capital projects; multi-year commitment or automatic renewal grants; general operating expenses; publicly or privately funded educational institutions; or organizations whose prime purpose is to influence legislation.
Geographic Focus: Alaska, Washington, Canada, Mexico
Contact: Susan Bramstedt, Director of Public Affairs (Alaska); (907) 266-7230; fax (907) 266-7229; susan.bramstedt@alaskaair.com
Internet: http://www.alaskaair.com/as/www2/company/csr/social-responsibility.asp
Sponsor: Alaska Airlines
4750 Old International Airport Road
Anchorage, AK 99502

Alaska Airlines Foundation Grants **389**
A small number of cash grants ranging on average from $5,000 to $15,000 are given to 501(c)3 non-profit organizations classified as public charities in Alaska and Washington. These grants should focus on educational efforts that address a unique need or value to a community. The Foundation supports health and human services, arts and cultural programs, as well as education, environmental, and civic organizations.
Requirements: 501(c)3 organizations based in Alaska and Washington are eligible.
Restrictions: Alaska Airlines or Horizon Air do not provide support for: loans and grants to individuals or private business; groups that discriminate; endowments; religious organizations (if for sacramental purposes); pageants; capital projects; multi-year commitment or automatic renewal grants; general operating expenses; publicly or privately funded educational institutions; or organizations whose prime purpose is to influence legislation.
Geographic Focus: Alaska, Washington
Amount of Grant: 5,000 - 15,000 U.S.D
Contact: Susan Bramstedt, Director of Public Affairs (Alaska); (907) 266-7230; fax (907) 266-7229; susan.bramstedt@alaskaair.com
Internet: http://www.alaskaair.com/as/www2/company/csr/as-foundation.asp
Sponsor: Alaska Airlines Foundation
4750 Old International Airport Road
Anchorage, AK 99502

Albany Medical Center Prize in Medicine and Biomedical Research **390**
The prize recognizes a physician or biomedical scientist (or group of physicians or scientists) who has made extraordinary and sustained leadership contributions to improving healthcare and patient care; or who has successfully pursued innovative biomedical research with demonstrated translational benefits applied to improved patient care. Each year's prize winner will have demonstrated significant outcomes that offer medical value of national or international importance. Prize winner activities will include but not be limited to disease and injury management, clinical research, and basic science investigations of diseases and injuries, leading to new discoveries and improved clinical outcomes. Nomination guidelines are available online.
Requirements: Any physician or scientist or group whose work has led to significant advances in the fields of health care and scientific research with demonstrated translational benefits applied to improved patient care may be nominated. Those honored will be practitioners and/or scientists whose accomplishments and outcomes have been demonstrated in the past quarter century, with preference to demonstrated accomplishments in the past decade.
Date(s) Application is Due: Jan 8
Amount of Grant: 500,000 U.S.D
Contact: Teri A. Cerveny, Secretary; (518) 262-8043; fax (518) 262-4769; AMCprize@mail.amc.edu
Internet: http://www.amc.edu/Academic/AlbanyPrize/prize_criteria.html
Sponsor: Albany Medical Center
628 Madison Avenue, Center Building, 1st Floor
Albany, NY 12208

Albert and Margaret Alkek Foundation Grants 391

The foundation awards grants to Texas nonprofit organizations to support charitable, religious, scientific (primarily medical), literary, cultural and educational organizations and programs. Preference will be given to research and education-related projects that will pay lasting dividends in terms of new discoveries and improved quality of life. One application per 12-month period will be considered. There are no application deadlines or forms. Applicants should submit a one- to two-page letter of inquiry that includes a brief description of the organization and the project for which funds are being considered, and the amount of funding need in total as well as the amount being requested. Inquiries should be sent via U.S. postal mail; fax or email applications are not accepted.
Requirements: Texas nonprofit organizations are eligible.
Restrictions: The foundation does not make grants to individuals or loans of any type. The foundation does not make direct scholarships to students. All scholarship programs are administered through educational institutions. The foundation prefers not to fund: organizations that in turn make grants to others; grants intended to influence legislation or to support candidates for political office; fund-raising events such as luncheons, dinners, galas, advertising in programs, or other similar activities; charities operated by service clubs; memorials for individuals; student organizations; or purchase of uniforms, equipment, or trips for school-related organizations or sports teams.
Geographic Focus: Texas
Samples: Baylor College of Medicine (Houston, TX)—for research on cancer, cardiovascular science, diabetes, pharmacogenomics, and other biomedical topics, $31.25 million.
Contact: Grants Administrator; (713) 951-0019; fax (713) 951-0043; info@alkek.org
Internet: http://www.alkek.org/grantguidelines.htm
Sponsor: Albert and Margaret Alkek Foundation
1221 McKinney, Suite 4525
Houston, TX 77010-2023

Albert and Mary Lasker Foundation Awards 392

The major purpose of the awards is to honor the individual(s) who have made significant contributions in basic or clinical research in diseases that are the main cause of death and disability. Three categories of awards are made. The Basic Medical Research Award honors the scientist or scientists who have made fundamental investigations that open new areas of biomedical science. The Clinical Medical Research Award honors the scientist(s) whose contributions, directly or indirectly, have led to the improvement of the clinical management or treatment of patients and to the alleviation or elimination or one of the major medical causes of disability or death. The Special Achievement Award in Medical Science honors a scientist whose contributions to research are of unique magnitude and immeasurable influence on the course of science, health, or medicine, and whose professional career has engendered extreme respect within the biomedical community. Nomination forms are available online. Nominations must be received by the listed application deadline.
Date(s) Application is Due: Feb 1
Amount of Grant: 25,000 - 50,000 U.S.D
Samples: Pierre Chambon, Institute of Genetics and Molecular and Cellular Biology (Strasbourg, France)—for the 2004 Lasker Award for Basic Medical Research, $50,000; Charles Kelman, International Retinal Research Foundation (Birmingham, AL)—for the 2004 Lasker Award for Clinical Medical Research, $50,000; Matthew Meselson, Harvard U (Cambridge, MA)—for the 2004 Lasker Award for Special Achievement in Medical Science, $25,000.
Contact: David Keegan, (212) 286-0222; fax (212) 286-0924; dkeegan@laskerfoundation.org
Internet: http://www.laskerfoundation.org/awards/awards.html
Sponsor: Albert and Mary Lasker Foundation
110 East 42nd Street, Suite 1300
New York, NY 10017

Albert Ellis Institute Clinical Fellowships 393

This fellowship program is a unique one- to two-year part-time course of study intended to help the therapist develop clinical proficiency in an active psychotherapy approach emphasizing rational emotive behavior psychotherapy and allied cognitive behavioral techniques. Fellows receive highly diversified training in clinical experience, supervision, workshop experience, and research. By special arrangement, persons on sabbatical or from out of town may complete the program in one year. Contact the institute for applications and deadline dates.
Requirements: The fellowship is open to those who hold doctorates in psychology or counseling (or near completion of PhD); to MSWs, RNs, and MDs; and to licensed mental health counselors. All candidates must be licensed or license-eligible. Predoctoral interns pursue the same program as postdoctoral fellows and receive the same stipend.
Amount of Grant: 6,000 U.S.D
Contact: Dr. Kristene Doyle, Program Contact; (212) 535-0822 or (800) 323-4738; fax (212) 249-3582; krisdoyle@albertellis.com or info@albertellis.org
Internet: http://www.rebt.org/titlepages/professionals.asp
Sponsor: Albert Ellis Institute
45 E 65th Street
New York, NY 10021

Alberto Culver Corporate Contributions Grants 394

The company prefers to make small donations to a large number of organizations rather than substantial contributions to a few institutions. Major areas of support are civic and community programs, health and welfare, education, culture and art, and youth activities. Special consideration is given to proposals that benefit a large number of people; directly or indirectly assist women's groups or other purchasers or potential customers of Alberto-Culver products; are involved in programs that rehabilitate, train, teach skills or employ the underprivileged, handicapped or minorities; have received support from Alberto-Culver in the past; and received matching grants for donations. Grants are awarded for one year and may be renewed.
Requirements: Only 501(c)3 organizations are eligible. Requests should be in writing on organization letterhead and should state amount requested, and whether the donation will be used for operating funds, capital expansion, supplies or special projects. The objectives and programs of the organization should be clearly defined.
Restrictions: Grants are not awarded to support United Way affiliates, religious groups, preschools, K-12 schools, tax-supported colleges, projects that duplicate other efforts, and multiyear commitments.
Contact: V. James Marino, Director; (708) 450-3000; fax (708) 450-3435
Internet: http://www.alberto.com
Sponsor: Alberto Culver
2525 Armitage Avenue
Melrose Park, IL 60160

Albert Pick Jr. Fund Grants 395

The fund contributes to organizations in the following categories: culture, education, health and human services, civic and community organizations. The fund will consider assistance to new and creative programs within these areas and to operating support. Support will be given to requests from organizations that conduct their programs in Chicago, IL, only. The fund functions on a calendar year; applications are considered four times a year based on dates of board meetings. The fund prefers that prospective grantees call or write for guidelines prior to sending their proposals.
Requirements: Only Illinois organizations may apply.
Restrictions: Funds will not be provided for reduction or liquidation of debts, religious purposes, endowments, long-term commitments, building programs, individuals, political purposes, or advertising/program books.
Geographic Focus: Illinois
Date(s) Application is Due: Jan 21; Apr 1; Jul 1; Oct 1
Amount of Grant: 3,000 U.S.D
Samples: Children First Fund/CPS Foundation (Gale Academy Links to Literacy Program) $6900; Art Resources in Teaching (Chicago, IL)—for general operating support, $7500; Chicago Academy of Sciences/Peggy Notebaert Nature Museum (Chicago, IL)—for general operating support, $10,000; Chicago Opera Theater (Chicago, IL)—for general operating support, $10,000.
Contact: Cleopatra Alexander, Executive Director; (312) 236-1192; cleopatra@albertpickjrfund.org
Internet: http://www.albertpickjrfund.org
Sponsor: Albert Pick Jr. Fund
30 N Michigan Avenue, Suite 1002
Chicago, IL 60602

Albertson's Charitable Giving Grants 396

Albertson's Inc invests in its operating communities and makes corporate contributions to support nonprofits. Areas of charitable giving include health and hunger relief—food banks, churches, and other community-based relief groups; health and nutrition—medical services, flu shots, health screening for diseases such as diabetes and heart disease; and education and development of youth—academic excellence or nurturing efforts. Submit requests in writing and include information about the organization's goals, accomplishments, evaluation plans, leadership, and finances. Requests are accepted throughout the year.
Requirements: Tax-exempt organizations in Arizona, Arkansas, California, Colorado, Delaware, Florida, Georgia, Idaho, Illinois, Indiana, Iowa, Kansas, Louisiana, Maine, Maryland, Massachusetts, Michigan, Minnesota, Mississippi, Missouri, Montana, Nebraska, Nevada, New Hampshire, New Jersey, New Mexico, North Dakota, Oklahoma, Oregon, Pennsylvania, South Dakota, Tennessee, Texas, Utah, Vermont, Washington, Wisconsin, and Wyoming are eligible. Applicants must pass an online eligibility test. Preference will be given to requests that offer volunteerism opportunities.
Restrictions: Contributions cannot be made to churches or religious organizations for purposes of religious advocacy.
Geographic Focus: Arizona, Arkansas, California, Colorado, Delaware, Florida, Georgia, Idaho, Illinois, Indiana, Iowa, Kansas, Louisiana, Maine, Maryland, Massachusetts, Michigan, Minnesota, Mississippi, Missouri, Montana, Nebraska, Nevada, New Hampshire, New Jersey, New Mexico, North Dakota, Oklahoma, Oregon, Pennsylvania, South Dakota, Tennessee, Texas, Utah, Vermont, Washington, Wisconsin, Wyoming
Contact: Community Relations Manager; (877) 932-7948 or (208) 395-6200; fax (208) 395-4382; albertsonscustomercare@albertsons.com
Internet: http://www.albertsons.com/abs_inthecommunity
Sponsor: Albertson's
250 E Parkcenter Boulevard
Boise, ID 83706

Albert Spencer Andrew Charitable Foundation Grants 397

The foundation awards grants in California to nonprofits in its areas of interest, including eye health and diseases, AIDS, and museums.
Requirements: Nonprofits in California are eligible to apply.
Geographic Focus: California
Amount of Grant: 1,000 - 110,000 U.S.D
Contact: Daniel Renberg, c/o Gibson, Dunn and Crutcher LLP
Sponsor: Albert Spencer Andrew Charitable Foundation
2029 Century Park E, Suite 40
Los Angeles, CA 90067

Albert W. Rice Charitable Foundation Grants 398

The Albert W. Rice Charitable Foundation was established in 1959 to support and promote quality educational, human-services, and health-care programming for underserved populations. In the area of education, the foundation supports academic access, enrichment, and remedial programming for children, youth, adults, and senior citizens that focuses on preparing individuals to achieve while in school and beyond. In the area of health care, the foundation supports programming that improves access to primary care for traditionally underserved individuals, health education initiatives and programming that impact at-risk populations, and medical research. In the area of human services the foundation tries to meet evolving needs of communities. Currently the foundation's focus is on (but is not limited to) youth development, violence prevention, employment, life-skills attainment, and food programs. Grant requests for general operating support are strongly encouraged. Program support and occasional capital support will also be considered. Special consideration is given to charitable organizations that serve the people of Worcester, Massachusetts, and its surrounding communities. The majority of grants from the Rice Foundation are one year in duration; on occasion, multi-year support is awarded. Applicants must apply online at the grant website. Applicants are strongly encouraged to do the following before applying: review the downloadable state application procedures for additional helpful information and clarifications; review the downloadable online-application guidelines at the grant website; review the foundation's funding history (link is available from the grant website); review the online application questions in advance; and review the list of required attachments. These will generally include: a list of board members, financial statements (audited, reviewed, or compiled by independent auditor); an organization summary; a list of other funding sources; an IRS Determination letter; and other required documents. All attachments must be uploaded in the online application as PDF, Word, or Excel files. The application deadline for the Albert W. Rice Charitable Foundation is 11:59 p.m. on July 1. Applicants will be notified of grant decisions before September 30.
Requirements: Applicants must have 501(c)3 tax-exempt status.
Restrictions: The foundation does not support requests from individuals, organizations attempting to influence policy through direct lobbying, or any political campaigns.
Geographic Focus: Massachusetts
Date(s) Application is Due: Jul 1
Samples: Big Brothers, Big Sisters of Central Massacusetts Metrowest, Worcester, Massachusetts, $30,000, program development (2010); Youth Opportunities Upheld, Worcester, Massachusetts, $20,000, Dynamic Youth Academy (2010); Legal Assistance Corporation of Central Massachusetts, Worcester, Massachusetts, $20,000 (2010).
Contact: Michealle Larkins; (866) 778-6859; michealle.larkins@baml.com
Internet: https://www.bankofamerica.com/philanthropic/fn_search.action
Sponsor: Albert W. Rice Charitable Foundation
225 Franklin Street, 4th Floor, MA1-225-04-02
Boston, MA 02110

Albuquerque Community Foundation Grants 399

The foundation seeks to improve the quality of life in the greater Albuquerque, New Mexico, area by providing support for projects and organizations that serve the community in arts and culture, education, environmental and historic preservation, children and youth, and health and human services. Through its grant program, the foundation supports projects that are innovative, meet the needs of underserved segments of the community, encourage matching funds or additional gifts, promote cooperation among agencies, empower the disadvantaged and disabled, and enhance the effectiveness of local charitable organizations. Types of support include continuing support, general operating support, program development, publication, seed grants, scholarships funds and scholarships to individuals, and technical assistance.
Requirements: IRS 501(c)3 organizations based in Albuquerque, NM, are eligible. Proposals are reviewed on the basis of the following priorities: impact, innovation, leverage, management, and nonduplication.
Restrictions: Grants are generally not made to or for individuals, political or religious purposes, debt retirement, payment of interest or taxes, annual campaigns, endowments, emergency funding, to influence legislation or elections, scholarships, awards, or to private foundations and other grantmaking organizations.
Geographic Focus: New Mexico
Date(s) Application is Due: Apr 16; Aug 15
Amount of Grant: Up to 10,000 U.S.D
Contact: R. Randall Royster, Executive Director; (505) 883-6240; fax (505) 883-3629; rroyster@albuquerquefoundation.org
Internet: http://www.albuquerquefoundation.org/grants/grant-home.htm
Sponsor: Albuquerque Community Foundation
P.O. Box 36960
Albuquerque, NM 87176-6960

Alcatel-Lucent Technologies Corporate Contributions Grants 400

Alcatel-Lucent is committed to enhancing and serving the communities where its employees and customers live and work. The corporation demonstrates that commitment through employee volunteerism in community service projects around the world and Foundation grant making. It is building on that heritage as it develops plans for future Alcatel-Lucent corporate sponsorship and philanthropic endeavors. The company's dedication supports programs to improve education and youth development. And Alcatel-Lucent continues to support photography worldwide as well as citizenship initiatives, especially those that use information and communication technologies which enable social inclusion.
Requirements: Government recognized public charities with appropriate tax-exempt identification numbers, such as 501(c)3 status or the international equivalent, are eligible.
Restrictions: The program does not support fund raisers or events; individuals, including awards, sponsorships such as walk-a-thons, or any form of personal financial assistance; endowments; conferences, seminars, trips, tours, etc; multiyear pledges; capital campaigns, including construction/renovation; ads in fundraising publications or journals; religious programs; veterans, labor, or political organizations; fraternal, athletic, and social clubs or programs; operating expenses of organizations supported by United Way; national health organizations other than United Way; or local chapters of organizations receiving Lucent support at the national level.
Contact: Michele Donato, Manager; (908) 582-7906; fax (908) 508-2576; foundation@alcatel-lucent.com
Internet: http://www.alcatel-lucent.com/wps/portal/!ut/p/kcxml/04_Sj9SPykssy0xPLMnMz0vM0Y_QjzKLd4y3NA0FSYGYRq6m-pEoYgbxjgiRIH1vfV-P_NxU_QD9gtzQiHJHR0UAvqLryQ!!/delta/base64xml/L3dJdyEvd0ZNQUFzQUMvNElVRS82X0FfNDZP
Sponsor: Alcatel-Lucent Technologies Corporation
600 Mountain Avenue, Room 6F430C
Murray Hill, NJ 07974-2008

Alcatel-Lucent Technologies Foundation Grants 401

The Alcatel-Lucent Foundation is the philanthropic arm of Alcatel-Lucent and it leads the company's charitable activities. With a focus on volunteerism, the Foundation's mission is to support the commitment of Alcatel-Lucent to social responsibility by serving and enhancing the communities where its employees and customers live and work. To accomplish its mission, the Foundation manages grants and employee volunteerism on a global level. It receives its income from the corporation - Alcatel-Lucent - whose name it bears. However, legally the Foundation is an independent, charitable, non-profit and private entity and is governed by its own board of trustees that is separate from the corporate board of directors. Global Foundation grants are dedicated to the main focus areas of the Foundation and are managed by the Foundation.
Requirements: Giving is on an international basis.
Amount of Grant: 500 - 5,000 U.S.D
Contact: Michele Donato; (908) 582-7906; foundation@alcatel-lucent.com
Internet: http://www.alcatel-lucent.com/wps/portal/foundation
Sponsor: Alcatel-Lucent Technologies Foundation
600 Mountain Avenue, Room 6F4
Murray Hill, NJ 07974-2008

Alchemy Foundation Grants 402

The foundation awards small grants worldwide to individuals and organizations to help persons (in particular babies and children) suffering from mental or physical illness or disability; assist in the relief of famine; provide for the mental and spiritual welfare of the elderly and dying; and fund medical research.
Amount of Grant: 2,000 - 5,000 GBP
Contact: Richard Stilgoe, 44-1883-730800
Internet: http://www.dundeecity.gov.uk/funding/alchemy.htm
Sponsor: Alchemy Foundation
Trevereux Manor
Limpsfield Chart, RH8 0TL England

Alcoa Foundation Grants 403

General priorities of the foundation include safe and healthy children and families—ensuring that children and their families have the tools, the knowledge and the services to remain healthy and safe at home, in the community and in the workplace; conservation and sustainability—educating young leaders on conservation issues, protecting forests, promoting sound public policy research, and understanding the linkages between business and the environment; skills today for tomorrow—providing individuals with critical skills and services to be economically connected, workplace-ready, and productive in a changing economy; business and community partnerships—strengthening the nonprofit sector and developing meaningful partnerships among nonprofits, the private sector, and local government; and global education in science, engineering, technology and business—broadening student participation in areas central to Alcoa to prepare a diverse cross-section of our communities for a global workplace. Types of support include capital grants, building funds, challenge grants, matching gifts, general support, research grants, scholarships, and seed money. Initial contact should be a letter of inquiry.
Requirements: The foundation awards grants to nonprofit public charities in communities where Alcoa has a presence. Local Alcoans work within their communities to evaluate organizations and make recommendations for funding to Alcoa Foundation. Nonprofit organizations that serve localized communities should find the Alcoa facility nearest to them and write a one-page letter describing their mission, nature of request, connection to the areas of excellence and offering contact information. If interested, the Alcoa location contact will notify the requesting organization and invite them to submit more information. Areas of operation include western Pennsylvania; Davenport, Iowa; Evansville, Indiana; Massena, New York; New Jersey; Cleveland, Ohio; Knoxville, Tennessee; and Rockdale, Texas.
Restrictions: The foundation does not make gifts to local projects other than those near Alcoa plant or office locations; endowment funds, deficit reduction, or operating reserves; hospital capital campaign programs unless the hospital presents a comprehensive area analysis that justifies, on a regional rather than an individual institutional basis, the need for the capital improvement; individuals, except for the scholarship program for children of Alcoa employees; tickets and other promotional activities; trips, tours, or student exchange programs; or documentaries and videos.
Amount of Grant: 1,000 - 50,000 U.S.D
Contact: Meg McDonald, President and Treasurer; (412) 553-2348; fax (412) 553-4498; alcoa.foundation@alcoa.com

Internet: http://www.alcoa.com/global/en/community/foundation.asp
Sponsor: Alcoa Foundation
201 Isabella Street
Pittsburgh, PA 15212-5858

Alcohol Misuse and Alcoholism Research Grants **404**

The foundation's research interests include factors influencing transitions in drinking patterns and behavior, effects of moderate use of alcohol on health and well-being, mechanisms underlying the behavioral and biomedical effects of alcohol, biobehavioral/interdisciplinary research on the etiology of alcohol misuse. Applications may be obtained from the Web site or by contacting the office. Grantees are usually notified within two weeks following the advisory council meetings, which are held in April and November.
Requirements: Applications may be submitted by public or private nonprofit organizations such as universities, colleges, hospitals, research institutes and organizations, and governmental research agencies and laboratories in the United States and Canada.
Restrictions: Non-research activities such as education projects, public awareness efforts and treatment or referral services are not eligible for support. The Foundation also does not support the training of pre- and post-doctoral fellows, undergraduates, graduate students, medical students, interns or residents. It does not fund thesis or dissertation research. The Foundation does not encourage applications on treatment of the complications of advanced alcoholism. However, research involving treatment intended to elucidate the pathogenesis of alcohol-related problems will be considered.
Date(s) Application is Due: Feb 1; Sep 1
Amount of Grant: Up to 100,000 U.S.D
Contact: Research Grants Administrator; (410) 821-7066; fax (410) 821-7065; info@abmrf.org
Internet: http://www.abmrf.org/grants.htm
Sponsor: Alcoholic Beverage Medical Research Foundation (ABMRF)
1122 Kenilworth Drive, Suite 407
Baltimore, MD 21204

Alcon Foundation Grants **405**

The foundation supports organizations in the fields of health care, leadership programs, research, education and community responsibility. Programs that advance the education and skill levels of eye care professionals are given special consideration. General operating grants to organizations and institutions improving education and research in the areas of specialization of Alcon Laboratories—ophthalmology and vision care. Grants are also awarded to community activities that benefit company employees. Applications may be submitted at any time.
Requirements: Grants are not made for building programs.
Restrictions: Non 501(c)3organizations, individuals and scholarship programs, religious, veterans or fraternal organizations, political causes, capital campaigns, matching gifts, trips, tournaments and tours, and endowments are not supported.
Amount of Grant: 100 - 50,000 U.S.D
Contact: Mary Dulle, Chair; (817) 293-0450; Mary.Dulle@Alconlabs.com
Internet: http://www.alconlabs.com/corporate-responsibility/alcon-foundation.asp
Sponsor: Alcon Foundation
6201 S Freeway
Fort Worth, TX 76134

Alexander & Baldwin Foundation Hawaiian and Pacific Island Grants **406**

The foundation awards grants to eligible Hawaii nonprofit organizations in its areas of interest, including health and human services, education, the community, culture and arts, the maritime arena and the environment. Grant preferences are given to organizations and projects that address significant community needs, have the active support of A&B employees, are preventive in nature, and have demonstrated support of the community. Start-up, general operating and special project needs, as well as major and minor capital requests are considered. Deadlines listed are for organizations in Hawaii and the Pacific Islands. Although the majority of grants range between $1,000 and $5,000, the Foundation considers request upward of $20,000.
Requirements: The foundation gives funding to community-based projects and organizations that qualify with 501(c)3 status.
Restrictions: Grants are not awarded to support United Way agencies for operating support, individuals, events, travel expenses, or scholarships.
Geographic Focus: Hawaii
Date(s) Application is Due: Feb 1; Apr 1; Jun 1; Aug 1; Oct 1; Dec 1
Amount of Grant: 1,000 - 5,000 U.S.D
Contact: Nami Nielipinski, (808) 525-6641; fax (808) 525-6677; nnielipinski@abinc.com
Internet: http://www.alexanderbaldwinfoundation.org/appguide.htm
Sponsor: Alexander and Baldwin Foundation
P.O. Box 3440
Honolulu, HI 96801-3440

Alexander & Baldwin Foundation Mainland Grants **407**

The foundation awards grants to eligible U.S. nonprofit organizations in its areas of interest, including health and human services, education, the community, culture and arts, the maritime arena and the environment. Grant preferences are given to organizations and projects that address significant community needs, have the active support of A&B employees, are preventive in nature, and have demonstrated support of the community. Start-up, general operating and special project needs, as well as major and minor capital requests are considered. Although the majority of grants range between $1,000 and $5,000, the Foundation considers request upward of $20,000.
Requirements: The foundation gives funding to community-based projects and organizations that qualify with 501(c)3 status.
Restrictions: Grants are not awarded to support United Way agencies for operating support, individuals, events, travel expenses, or scholarships.
Date(s) Application is Due: Jan 1; Feb 1; Mar 1; Apr 1; May 1; Jun 1; Jul 1; Aug 1; Sep 1; Oct 1; Nov 1; Dec 1
Amount of Grant: 1,000 - 5,000 U.S.D
Contact: Paul Merwin, (707) 421-8121; fax (707) 421-1835; plmifm@aol.com
Internet: http://www.alexanderbaldwinfoundation.org/appguide.htm
Sponsor: Alexander and Baldwin Foundation
555 12th Street
Oakland, CA 94607

Alexander and Margaret Stewart Trust Grants **408**

The trust awards grants in the greater Washington, DC, area for cancer treatment, especially equipment used in diagnosis and treatment; and caring for children who are physically ill, mentally ill, or disabled. Grants also support research, education, and prevention of common childhood diseases, including negative societal behavioral patterns that impact children. Proposals for projects aiding the economically deprived receive preference. The trust awards start-up funding. Applications may be submitted at any time; requests received by the listed deadline date are reviewed by the end of the year.
Requirements: Nonprofits in the greater District of Columbia area are eligible.
Restrictions: Requests for endowments, buildings, or capital campaigns are denied.
Geographic Focus: District of Columbia
Date(s) Application is Due: Sep 15
Amount of Grant: 50,000 - 150,000 U.S.D
Samples: Childrens National Medical Center, Washington, DC, $420,000—for Center for Cancer and Blood Disorders; DC Campaign to Prevent Teen Pregnancy, Washington, DC, $75,000—for general support.
Contact: William J. Bierbower, (202) 785-9892; fax (202) 785-0918; wbierbower@stewart-trust.org
Internet: http://www.stewart-trust.org/guidelines.htm
Sponsor: Alexander and Margaret Stewart Trust
888 17th Street NW, Brawner Building, Suite 610
Washington, DC 20006-3321

Alexander Eastman Foundation Grants **409**

The purposes of the Alexander Eastman Foundation are served through grant support to organizations awarded in response to proposals and special initiative commitments planned collaboratively by the Foundation with local service providers. Grants are awarded to support the capital, special projects and operations needs of qualifying organizations. In considering proposals, priority is given to funding activity which serves the Foundation's priority interests. The Alexander Eastman Foundation supports: Education - provide information and community education to improve the health and well-being of residents of the greater Derry area; address goals for healthy individuals and families through a long-term commitment to prevention, health promotion and education of consumers and providers; foster individual responsibility, independence, self-care and healthy life-style choices; Family Systems - strengthen families as the critical unit for community health and well-being; recognize the changing nature of families and provide resources and assistance to reduce stress on families and improve family function; Access - expand access to quality health care and prevention services for people with financial need. Applications are available at the Foundations website.
Requirements: Nonprofit organizations serving Derry, Londonderry, Windham, Chester, Hampstead, and Sandown, NH, are eligible.
Restrictions: Grants are made neither to individuals nor to qualifying organizations to support the cost of services to particular individuals, except through the Alexander Eastman Scholarship Program.
Geographic Focus: New Hampshire
Date(s) Application is Due: Apr 1; Oct 1
Samples: Community Caregivers of Greater Derry, Derry, NH, $10,000 - to support the volunteer-driven program of quality client care in the greater Derry community; The Upper Room, A Family Resource Center, Derry, NH, $28,000 - to support prevention and intervention for families at risk of having a child placed in foster care; Child and Family Services of New Hampshire, Manchester, NH, $19,760 - to support services to twelve non-Medicaid-eligible teens in the Healthy Family Home Visiting program, to support tuition assistance for low-income, at risk greater Derry area youth to attend a session of residential summer camp at Camp Spaulding.
Contact: Amy Lockwood, Director of Grants; (888) 228-1821, ext. 80; alockwood@alexandereastman.org
Internet: http://www.alexandereastman.org/02grants.html
Sponsor: Alexander Eastman Foundation
26 South Main Street, PMB 250
Concord, NH 03301

Alexander Foundation Cancer, Catastrophic Illness and Injury Grants **410**

The program is intended to provide support to lesbian, gay, bisexual and transgendered people who are at risk of losing their ability to provide basic life needs due to treatment or complications related to cancer, a catastrophic illness, or a serious illness. Some examples of basic life needs are medical care, housing, food, clothing, phones, utilities, and so on. Unlike the one grant payment provided through the Emergency Grant Program, this program provides a fixed monthly grant for up to 12 months. The amount of the grant is determined by the recipients' needs but not to exceed $2,400.

Requirements: Applicants must have diagnosis and treatment of any form of cancer, catastrophic illness, or serious injury except those occurring simultaneously with HIV/AIDS. Recipients must be gay, lesbian, bisexual or transgendered, diagnosed with AIDS or HIV positive, demonstrate a perceived financial need and have been recommended by a recognized referral agency. Additionally, recipients must reside in Colorado.
Restrictions: Recipients must demonstrate financial need and cannot have received financial assistance from the Foundation during the previous 6 months. The Alexander Foundation will only accept applications from an approved referring agency or source.
Geographic Focus: Colorado
Amount of Grant: Up to 2,400 U.S.D
Contact: Jack Heruska, President; (303) 331-7733; fax (303) 331-1953; egrants@thealexanderfoundation.org
Internet: http://www.thealexanderfoundation.org/cancer.html
Sponsor: Alexander Foundation
P.O. Box 1995
Denver, CO 80201-1995

Alexander Foundation Insurance Continuation Grants 411

The program provides funds on a monthly basis to pay medical insurance premiums for those individuals who have been diagnosed with AIDS or HIV positive. After a recipient has been accepted into the program, they remain in the program as long as they can demonstrate a continued need. New applicants may be added to a waiting list until such time as funds become available.
Requirements: Recipients must be gay, lesbian, bisexual or transgendered, diagnosed with AIDS or HIV positive, demonstrate a perceived financial need and have been recommended by a recognized referral agency. Additionally, recipients must reside in Colorado.
Geographic Focus: Colorado
Contact: Jack Heruska, President; (303) 331-7733; fax (303) 331-1953; egrants@thealexanderfoundation.org
Internet: http://www.thealexanderfoundation.org/insurance.html
Sponsor: Alexander Foundation
P.O. Box 1995
Denver, CO 80201-1995

Alexander von Humboldt Foundation Georg Forster Fellowships for Experienced Researchers 412

Fellowships are available to scholars from developing countries in humanities and social sciences; political science and economics; projects in the public health sector and in the fields of agriculture, forestry, and geosciences; and interdisciplinary projects relating to environmental and resource protection. Fellowships are designed to promote the transfer of knowledge and methods and to contribute to further development in fellows' home countries. Applications may be submitted for long-term research stays of between six and 18 months. The fellowship is flexible and can be divided up into as many as three stays within three years. The Foundation grants up to 60 Fellowships for postdoctoral researchers and experienced researchers annually.
Requirements: Scholars from developing countries (excluding Turkey, India, and the Peoples' Republic of China) may apply. Applicants must have completed their doctorates less than twelve years ago. Scholars in the humanities or social sciences and physicians must have a good knowledge of German if it is necessary to carry out the project successfully; otherwise a good knowledge of English; scientists and engineers must have a good knowledge of German or English.
Restrictions: Short-term study tours, participation in conferences, or educational visits cannot be funded. Applications for extension of research stays already commenced in Germany cannot be considered.
Amount of Grant: 2,450 EUR
Contact: Program Contact; (+49) 0228-833-0; fax (+49) 0228-833-199; info@avh.de
Internet: http://www.humboldt-foundation.de/web/georg-forster-research-fellowship-experienced.html
Sponsor: Alexander von Humboldt Foundation
Jean-Paul-Street 12
Bonn, D-53173 Germany

Alexander von Humboldt Foundation Georg Forster Fellowships for Postdoctoral Researchers 413

Fellowships for postdoctoral researchers are the instrument with which the Foundation enables highly-qualified scientists and scholars who are just embarking on their academic careers and who completed their doctorates less than four years agol to spend extended periods of research (6-24 months) in Germany. Scientists and scholars from all disciplines from developing and threshold countries (excluding People's Republic of China, India and Turkey) may apply. The research proposal must deal with issues of major relevance to the future development of the candidate's country of origin and, in this context, particularly suited to transferring knowledge and methods to developing and threshold countries. Research projects are carried out in cooperation with academic hosts at research institutions in Germany. Candidates choose their own research projects and their host in Germany and prepare their own research plan.
Amount of Grant: 2,250 EUR
Contact: Program Contact; (+49) 0228-833-0; fax (+49) 0228-833-199; info@avh.de
Internet: http://www.humboldt-foundation.de/web/georg-forster-research-fellowship-postdoc.html
Sponsor: Alexander von Humboldt Foundation
Jean-Paul-Street 12
Bonn, D-53173 Germany

Alexander von Humboldt Foundation Institutional Academic Coop Grants 414

Under the follow-up program of the Foundation, institutional partnerships for long-term research collaboration can be created by combining various sponsorship opportunities. Special emphasis is placed on the exchange of younger scientists. The aim is to give these scientists access to the world-wide network of scholars sponsored by the Foundation and to alert them to the Humboldt Research Fellowship Program. The program includes: reciprocal research visits by applicants and their co-workers at the partner institutes (including travel and living expenses); jointly hosted academic conferences and workshops (including travel and living expenses for participants from the partner institutes, as well as expenses for organizing and realizing the events); and funds for the purchase of scientific equipment and additional material (for the foreign institutes).
Requirements: Young scientists and scholars who are involved in the co-operation should preferably have completed a PhD or should at least be doctoral candidates and must be under 35 years of age.
Restrictions: Funds up to a maximum of 50,000 Euro. Funds for material costs (scientific equipment and additional material) are limited to a maximum of 50% of the total funds requested. The purchase price of scientific equipment may not exceed a total amount of 20,000 Euro.
Amount of Grant: Up to 55,000 EUR
Contact: Program Coordinator; (+49) 0228-833-0; fax (+49) 0228-833-199; info@avh.de
Internet: http://www.humboldt-foundation.de/web/31390.html
Sponsor: Alexander von Humboldt Foundation
Jean-Paul-Street 12
Bonn, D-53173 Germany

Alexander von Humboldt Foundation Research Fellowships for Experienced Researchers 415

These Fellowships for experienced researchers are the instrument with which the Foundation enables highly-qualified scientists and scholars from abroad, who completed their doctorates less than twelve years ago to spend extended periods of research (6-18 months; may be divided up into a maximum of three blocks) in Germany. Candidates are expected to have their own, clearly defined research profile. This means they should usually be working at least at the level of Assistant Professor or Junior Research Group Leader or be able to document independent research work over a number of years. Scientists and scholars from all disciplines and countries may apply. The fellowship is worth 2,450 EUR per month. This includes a mobility lump sum and a contribution towards health and liability insurance.
Requirements: Application requirements include: a doctorate or comparable academic degree (Ph.D., C.Sc. or equivalent), completed less than twelve years prior to the date of application; the candidate's own research profile documented by a comprehensive list of academic publications reviewed according to international standards and printed in journals and/or by publishing houses; and a good knowledge of German if it is necessary to carry out the project successfully (otherwise a good knowledge of English; scientists and engineers must have a good knowledge of German or English).
Restrictions: Short-term visits for study and training purposes or for attending conferences are not eligible for sponsorship.
Amount of Grant: 14,700 - 44,100 EUR
Contact: Program Coordinator; (+49) 0228-833-0; fax (+49) 0228-833-199; info@avh.de
Internet: http://www.humboldt-foundation.de/web/humboldt-research-fellowship-experienced.html
Sponsor: Alexander von Humboldt Foundation
Jean-Paul-Street 12
Bonn, D-53173 Germany

Alexander von Humboldt Foundation Research Fellowships for Postdoctoral Researchers 416

These Fellowships for postdoctoral researchers are the instrument with which the Foundation enables highly-qualified scientists and scholars from abroad who are just embarking on their academic careers and who completed their doctorates less than four years ago to spend extended periods of research (6-24 months) in Germany. Scientists and scholars from all disciplines and countries may apply. Research projects are carried out in cooperation with academic hosts at research institutions in Germany. Candidates choose their own research projects and their host in Germany and prepare their own research plan. Details of the research project and the time schedule must be agreed upon with the prospective host in advance. Candidates are selected solely on the basis of their academic record. There are no quotas for individual disciplines or countries. The fellowship is worth 2,250 EUR per month. This includes a mobility lump sum and a contribution towards health and liability insurance.
Requirements: Application requirements include: a doctorate or comparable academic degree (Ph.D., C.Sc. or equivalent), completed less than twelve years prior to the date of application; the candidate's own research profile documented by a comprehensive list of academic publications reviewed according to international standards and printed in journals and/or by publishing houses; and a good knowledge of German if it is necessary to carry out the project successfully (otherwise a good knowledge of English; scientists and engineers must have a good knowledge of German or English).
Restrictions: Short-term visits for study and training purposes or for attending conferences are not eligible for sponsorship.
Amount of Grant: 13,500 - 54,000 EUR
Contact: Program Coordinator; (+49) 0228-833-0; fax (+49) 0228-833-199; info@avh.de
Internet: http://www.humboldt-foundation.de/web/humboldt-research-fellowship-postdoc.html
Sponsor: Alexander von Humboldt Foundation
Jean-Paul-Street 12
Bonn, D-53173 Germany

Alex Brown and Sons Charitable Foundation Grants 417
The foundation awards one-year renewable grants to U.S. nonprofit organizations in its areas of interest, including arts and culture, community affairs, education, medicine, and science. Types of support include annual campaigns, capital grants, endowment funds, general operating grants, program development grants, and scholarship funds. Applicants should submit a letter of intent that briefly outlines the purpose of the grant.
Requirements: 501(c)3 tax-exempt organizations in Maryland are eligible.
Restrictions: No support for private schools or churches. Grants are not made to individuals.
Geographic Focus: Maryland
Amount of Grant: 1,000 - 100,000 U.S.D
Samples: Baltimore Symphony Orchestra (MD)—for discretionary use, $1 million; Johns Hopkins Hospital (Baltimore, MD)—for discretionary use, $1 million; Baltimore Zoo (MD)—for discretionary use, $1 million.
Contact: Margaret Preston, Secretary, c/o Deutsche Bank Alex Brown Inc
Sponsor: Alex Brown and Sons Charitable Foundation
P.O. Box 2257
Baltimore, MD 21203

Alexis Gregory Foundation Grants 418
The Foundation, established in 1986, gives primarily for education purposes. Interests include: AIDS and AIDS research; arts; France; health organizations; human services; Italy; Jewish agencies and temples; libraries and library science; museums; performing arts; and the United Kingdom. Types of support include program development and scholarships to individuals. The primary region of giving is New York, though grants are sometimes approved for out-of-state organizations. Though there is no specific deadline date, the board generally reviews applications in January. An application form is not required. Applicants should initially send a letter or full proposal.
Geographic Focus: New York
Amount of Grant: Up to 15,000 U.S.D
Contact: Alexis Gregory, President; (212) 737-5297; fax (212) 737-5340
Sponsor: Alexis Gregory Foundation
1334 York Avenue, 3rd Floor
New York, NY 10021

ALFJ Astraea U.S. and International Movement Fund 419
Astraea's Movement Resource Fund provides grants to enhance the capacity and effectiveness of our grantee partners and ally organizations to engage in movement building work. Grants are generally provided in three areas: Technical Assistance, Travel/Peer-to-Peer Learning and Historic Convenings. Letters of inquiry are accepted year round.
Requirements: One page letters of inquiry should be sent to the program officer. Information should include information about the organization, the purpose of the request, and a budget summary of how the funds would be used.
Restrictions: Although any organization that fits the funding criteria may apply, Astraea prioritizes current Astraea grantee partners.
Geographic Focus: All States, All Countries
Contact: Namita Chad, Program Officer; (212) 529-8021; fax (212) 982-3321; nchad@astraeafoundation.org
Internet: http://www.astraeafoundation.org/grants/grant-applications-and-deadlines
Sponsor: Astraea Lesbian Foundation for Justice
116 E 16th Street, 7th Floor
New York, NY 10003

Alfred and Tillie Shemanski Testamentary Trust Grants 420
Alfred Shemanski was an immigrant from Poland with little formal education. He understood the struggles of being a stranger in a strange land and became a champion for both Jewish and secular education. His successes never blinded him to the plight of those less fortunate. Mr. Shemanski exercised philanthropy in the purest definition of the word — love of mankind. The Alfred & Tillie Shemanski Trust was established in 1974 to: improve the capacity of and cooperation among Jewish congregations in the City of Seattle, Washington; support interfaith tolerance and understanding; provide scholarship assistance, primarily to the University of Washington and Seattle University; and support and promote quality educational, human-services, and health-care programming for economically disadvantaged individuals and families. Grant requests for general operating support, start-up funding, and prizes or awards are encouraged. Grants from the Shemanski Trust are one year in duration. Application materials are available for download at the grant website. Applicants are strongly encouraged to review the state application guidelines for additional helpful information and clarifications before applying. Applicants are also encouraged to review the trust's funding history (link is available from the grant website). The application deadline for the Shemanski Trust is October 15. Applicants will be notified of grant decisions before November 30.
Requirements: Applicants must serve residents of the Seattle and Puget Sound area.
Restrictions: Requests for fundraising events or sponsorship opportunities will not be considered. The trust does not support requests from individuals, organizations attempting to influence policy through direct lobbying, or any political campaigns.
Geographic Focus: Washington
Date(s) Application is Due: Oct 15
Samples: Seattle University, Seattle, Washington, $45,000, for Schemanski Scholarship Fund (2010); College Success Foundation, Issaquah, Washington, $10,000, scholarship funds (2010); Temple De Hirsch Sinai, Seattle, Washington, $35,000, education initiative (2010); Mockingbird Society, Seattle, Washington, $10,000, Mockingbird Network (foster-youth-led program)(2010).
Contact: Nancy Atkinson, Vice President; (800) 848-7177; nancy.l.atkinson@baml.com
Internet: https://www.bankofamerica.com/philanthropic/fn_search.action
Sponsor: Alfred and Tillie Shemanski Testamentary Trust
800 5th Avenue, WA1-501-33-23
Seattle, WA 98104

Alfred Bader Award in BioInorganic or BioOrganic Chemistry 421
This award is given annually to recognize outstanding contributions to bioorganic or bioinorganic chemistry. The award will be granted for outstanding research accomplishments without regard to age or nationality. The award is intended to recognize significant accomplishments that are at the interface between biology and organic or inorganic chemistry. Special consideration will be given to applications of the fundamental principles and experimental methodology of chemistry to areas of biological significance. The award consists of $5,000 and a certificate.
Requirements: Any individual, except a member of the awards committee, may submit one nomination or seconding letter for the award in any given year. The nominating documents consist of a letter of not more than 1000 words containing an evaluation of the nominee's accomplishments and a specific identification of the work to be recognized, a biographical sketch including date of birth, and a list of publications and patents authored by the nominee.
Restrictions: Self-nominations are not accepted.
Date(s) Application is Due: Nov 1
Amount of Grant: 5,000 U.S.D
Samples: Kevin M. Smith—award recipient, $5,000; Lawrence Que, Jr.—award recipient, $5,000; Eckard Munck—award recipient, $5,000.
Contact: Felicia Dixon, Awards Administrator; (800) 227-5558 or (202) 872-4408; fax (202) 776-8008; f_dixon@acs.org or awards@acs.org
Internet: http://portal.acs.org/portal/acs/corg/content?_nfpb=true&_pageLabel=PP_ARTICLEMAIN&node_id=1319&content_id=CTP_004490&use_sec=true&sec_url_var=region1
Sponsor: American Chemical Society
1155 Sixteenth Street, NW
Washington, DC 20036-4801

Alfred Bersted Foundation Grants 422
The Alfred Bersted Foundation was established in 1972 to support and promote quality educational, human services, and health care programming for underserved populations. The Foundation specifically serves the people of DeKalb, DuPage, Kane, and McHenry counties in Illinois. Application materials are available for download at the grant website. Applicants are strongly encouraged to review the state application guidelines for additional helpful information and clarifications before applying. Applicants are also encouraged to review the foundation's funding history (link is available from the grant website). The foundation has a rolling application deadline. In general, applicants will be notified of grant decisions three to four months after proposal submission.
Requirements: Applicant organizations must have 501(c)3 tax-exempt status and a physical presence in one of the following counties: DeKalb, DuPage, Kane, or McHenry.
Restrictions: The Alfred Bersted Foundation does not make grants to degree-conferring institutions of higher education, religious houses of worship, or organizations testing for public safety. In general, grant requests for endowment campaigns will not be considered.
Geographic Focus: Illinois
Contact: Debra Grand, Senior Vice President; (312) 828-4154; ilgrantmaking@bankofamerica.com
Internet: https://www.bankofamerica.com/philanthropic/fn_search.action
Sponsor: Alfred Bersted Foundation
231 South LaSalle Street, IL1-231-13-32
Chicago, IL 60604

Alfred Burger Award in Medicinal Chemistry 423
This award, supported by GlaxoSmithKline in 1978, is awarded biennially in even-numbered years to recognize outstanding contributions to research in medicinal chemistry. The award is granted without regard to age or nationality, and the recipient presents an award address at the spring meeting of the ACS Division of Medicinal Chemistry. Applications are accepted in odd-numbered years. The award consists of $3,000 and a certificate.
Requirements: Any individual, except a member of the award committee, may submit one nomination or seconding letter for the award in any given year. The nominating documents consist of a letter of not more than 1000 words containing an evaluation of the nominee's accomplishments and a specific identification of the work to be recognized, a biographical sketch including date of birth, and a list of publications and patents authored by the nominee.
Restrictions: Self-nominations are not accepted.
Date(s) Application is Due: Nov 1
Amount of Grant: 3,000 U.S.D
Samples: Magid Abou-Gharbia—award recipient, $3,000; Joel R. Huff—award recipient, $3,000; William Greenlee—award recipient, $3,000.
Contact: Felicia Dixon, Awards Administrator; (800) 227-5558 or (202) 872-4408; fax (202) 776-8008; f_dixon@acs.org or awards@acs.org
Internet: http://portal.acs.org/portal/acs/corg/content?_nfpb=true&_pageLabel=PP_ARTICLEMAIN&node_id=1319&content_id=CTP_004491&use_sec=true&sec_url_var=region1
Sponsor: American Chemical Society
1155 Sixteenth Street, NW
Washington, DC 20036-4801

Alfred E. Chase Charitable Foundation Grants 424

The Alfred E. Chase Charitable Foundation was established in 1956 to support and promote quality educational, human-services, and health-care programming for underserved populations. Special consideration is given to charitable organizations that serve the people of the city of Lynn and the North Shore of Massachusetts. The foundation is a generous supporter of the Associated Grant Makers (AGM) Summer Fund which provides operating support for summer camps serving low-income urban youth from Boston, Cambridge, Chelsea, and Somerville. Excluding the grant to the AGM Summer Fund, the typical grant range is $10,000 to $30,000. In the area of education the foundation supports academic access, enrichment, and remedial programming for children, youth, adults, and senior citizens that focuses on preparing individuals to achieve while in school and beyond. In the area of human services the foundation supports organizations meeting the basic needs of all individuals to include but not limited to youth development, violence prevention, employment, life skills attainment, and food programs. In the area of health care, the foundation supports programming that improves access to primary care for traditionally underserved individuals, as well as supporting health education initiatives and programs that impact at-risk populations. Grant requests for general operating support are strongly encouraged. Program support will also be considered. Small, program-related capital expenses may be included in general operating or program requests. The majority of grants from the Chase Charitable Foundation are one year in duration. On occasion, multi-year support is awarded. Applicants must apply online at the grant website. Applicants are strongly encouraged to do the following before applying: review the downloadable Massachusetts state application procedures for additional helpful information and clarifications; review the downloadable online-application guidelines at the grant website; review the foundation's funding history (link is available from the grant website); review the online application questions in advance; and review the list of required attachments. These will generally include: a list of board members, financial statements (audited, reviewed, or compiled by independent auditor); an organization summary; a list of other funding sources; an IRS Determination letter; and other required documents. All attachments must be uploaded in the online application as PDF, Word, or Excel files. The application deadline is 11:59 p.m. on April 1. Applicants will be notified of grant decisions before June 30.
Requirements: Applicants must have 501(c)3 tax-exempt status.
Restrictions: The foundation does not support requests from individuals, organizations attempting to influence policy through direct lobbying, or any political campaigns.
Geographic Focus: Massachusetts
Date(s) Application is Due: Apr 1
Amount of Grant: 10,000 - 30,000 U.S.D
Samples: Uphams Corner Community Center, Dorchester, Massachusetts, $15,000 (2010); Massachusetts Advocates for Children, Boston, Massachusetts, $20,000 (2010); Union Social Action Foundation, Cambridge, Massachusetts, $15,000.
Contact: Miki C. Akimoto, Vice President; (866) 778-6859; miki.akimoto@bam..com
Internet: https://www.bankofamerica.com/philanthropic/fn_search.action
Sponsor: Alfred E. Chase Charitable Foundation
225 Franklin Street, 4th Floor, MA1-225-04-02
Boston, MA 02110

Alfred I. DuPont Foundation Grants 425

Grants are awarded primarily to elderly adults requiring health, economic, or educational assistance. Support is also given for higher education and medical research. All grants to the elderly are made to individuals in the southeastern United States. General operating support and grants to individuals are awarded. An application form is required. Applications are accepted at any time and are dealt with promptly.
Requirements: Nonprofit organizations in the southeastern U.S. are eligible. Preference is given to those in Florida.
Geographic Focus: Florida
Amount of Grant: 1,000 - 35,000 U.S.D
Samples: Gulf Coast Community College Fdn (Panama City, FL)—for operating support, $50,000; Dreams Come True of Jacksonville (Jacksonville, FL)—for operating support, $10,000.
Contact: Rosemary Cusimano Wills, Secretary; (904) 232-4123
Sponsor: Alfred I. DuPont Foundation
4600 Touchton Road E, Building 200, Suite 120
Jacksonville, FL 32246

Alfred P. Sloan Foundation History of Science and Technology Grants 426

The goal of this program is to preserve the raw material of history by supporting archival projects now centered on Charles Darwin, Thomas A. Edison, and Kurt Gödel, and via new projects based on the World Wide Web. Two specific areas are of interest: the archival program; and recent history of science and engineering on the web. Grant requests can be made at any time for support of activities related to Foundation program areas and interests. The Foundation is generally limited to supporting tax-exempt organizations. The Foundation has no deadlines or standard forms. The Foundation accepts proposals sent by email. A brief letter of inquiry, rather than a fully developed proposal, is an advisable first step for an applicant, conserving his or her time and allowing for a preliminary response regarding the possibility of support.
Requirements: Concise, well-organized proposals are preferred. In no case should the body of the proposal exceed 20 double-spaced pages.
Restrictions: The Foundation's activities do not normally extend to religion, the creative or performing arts, elementary or secondary education, medical research or health care, the humanities or to activities outside the United States. Grants are not made for endowments or for buildings or equipment.
Contact: Jesse H. Ausubel; (212) 649-1649; fax (212) 757-5117; ausubel@sloan.org
Internet: http://sloan.org/programs/scitech_historysci.shtml
Sponsor: Alfred P. Sloan Foundation
630 Fifth Avenue, Suite 2550
New York, NY 10111-0242

Alfred P. Sloan Foundation Research Fellowships 427

Fellowships stimulate fundamental research by young scholars of outstanding promise at a time in their careers when their creative abilities are especially high and when government or other support is difficult to obtain. Fellows are free to pursue whatever lines of inquiry that are of the most compelling interest to them. Funds are awarded directly to the fellow's institution and may be used by the fellow for such purposes as equipment, technical assistance, professional travel, trainee support, or activities directly related to the fellow's research. The foundation welcomes nominations of all candidates who meet the traditional high standards of this program and strongly encourages the participation of women and members of underrepresented minority groups. Awards are made in the fields of physics, chemistry, neuroscience, economics, pure mathematics, applied mathematics, computer science, and computational and evolutionary molecular biology. Fellowships are awarded for a two-year period, with possible extension for another two years. Each year 118 fellows are selected from the nominations received. Nominations are due September 15 for awards to begin the following September. Nomination forms are available upon request.
Requirements: Candidates must hold a PhD or equivalent in chemistry, physics, mathematics, computer science, economics, neuroscience or computational and evolutionary molecular biology, or in a related interdisciplinary field, and must also be a faculty member at a U.S. or Canadian university. Candidates must be no more than six years from the completion of the most recent PhD or equivalent as of the year of their nomination.
Restrictions: Direct applications are not accepted Candidates must be nominated by department heads or other senior scholars.
Geographic Focus: All States, Canada
Date(s) Application is Due: Sep 15
Amount of Grant: 45,000 U.S.D
Contact: Michael Teitelbaum, Program Director; (212) 649-1649; fax (212) 757-5117; teitelbaum@sloan.org
Internet: http://sloan.org/programs/scitech_fellowships.shtml
Sponsor: Alfred P. Sloan Foundation
630 Fifth Avenue, Suite 2550
New York, NY 10111-0242

Alfred W. Bressler Prize in Vision Science 428

The Prize recognizes a professional in the field of vision science whose leadership, research and service have resulted in important advancements in the treatment of eye disease or rehabilitation of persons with vision loss. A panel of distinguished vision science professionals will select the winner who receives a prize of $40,000.
Requirements: The application process is open to established professionals in the field of vision science whose contributions have advanced vision care, the treatment of eye disease, or the rehabilitation of persons with visual disabilities or blindness and whose further work is expected to contribute significantly. Candidates from the United States and countries around the world are eligible for the award.
Date(s) Application is Due: Dec 31
Amount of Grant: 40,000 U.S.D
Contact: Program Administrator; (212) 769-7801; bressler@jgb.org
Internet: http://www.jgb.org/programs_bressler.asp
Sponsor: Jewish Guild for the Blind
15 W 65th Street
New York, NY 10023

Alice C. A. Sibley Fund Grants 429

The purpose of the Alice C.A. Sibley Fund is to provide medical eye care to vulnerable populations of all ages in Worcester including: hospital and surgical fees; medical equipment; and eye glasses and corrective treatment. Approximately $24,000 in grants will be awarded annually to one or more organizations that directly support people's eye health.
Requirements: Applicants must be tax-exempt, nonprofit organizations as recognized by IRS code 501(c)3, and must be located in or providing services to residents of Worcester, Massachusetts.
Geographic Focus: Massachusetts
Date(s) Application is Due: Apr 15
Amount of Grant: Up to 24,000 U.S.D
Contact: Lois Smith, Senior Program Officer; (508) 755-0980, ext. 107; lsmith@greaterworcester.org
Internet: http://www.greaterworcester.org/grants/Sibley.htm
Sponsor: Greater Worcester Community Foundation
370 Main Street, Suite 650
Worcester, MA 01608-1738

Alice Fisher Society Fellowships 430

The Barbara Bates Center for the Study of the History of Nursing offers a fellowship of $2,500 to support two weeks in residence at the Center and ongoing collaboration with Center historians. Selection of Alice Fisher Society scholars will be based on evidence of interest in and aptitude for historical research related to nursing. It is expected that the research and new materials produced by Alice Fisher Society scholars will help ensure the growth of scholarly work focused on the history of nursing. Fisher scholars will participate in Center activities and will present their research at a Center seminar.

Requirements: The scholarships are open to individuals with master's and doctoral level preparation. The application should be sent via email to either Patricia D'Antonio, PhD, RN, FAAN (dantonio@nursing.upenn.edu) or Barbra Mann Wall, PhD, RN (wallbm@nursing.upenn.edu).
Date(s) Application is Due: Dec 31
Amount of Grant: 2,500 U.S.D
Contact: Dr. Karen Buhler-Wilkerson, Director; (215) 898-4725; fax (215) 573-2168; karenwil@nursing.upenn.edu
Internet: http://www.nursing.upenn.edu/history/Pages/Alice_Fisher_Fellowship.aspx
Sponsor: University of Pennsylvania
School of Nursing, 418 Curie Boulevard
Philadelphia, PA 19104-6020

Alice Tweed Tuohy Foundation Grants 431

The foundation promotes organizations that promote young people; that provide outstanding opportunities for performance, growth, and creativity; that nurture personal integrity and ambition; and that reward high achievement. The foundation assists organizations offering services to children whose choices might otherwise be unfairly restricted by need; supported are activities both academic and extracurricular that challenge young people while encouraging the growth of responsibility and personal integrity. Organizations dedicated to improving the quality of life by meeting the vital needs of the community are also supported. Types of support include building construction/ renovation, scholarship funds, and matching funds. Financing priority is accorded those organizations with the least in-house capacity to raise capital, assisting these groups to surmount critical monetary obstacles and continue productive service to the community. Proposals may be submitted annually between July 1 and September 15. However, the foundation has announced that a three-year partial moratorium period on awarding grants will commence July 1, 2009.
Requirements: Applications are considered only from Santa Barbara, CA public, tax-exempt organizations. Priority consideration is given to applications from organizations serving young people, education, health and medicine, community affairs, and the arts.
Restrictions: Excluded from consideration are applications for the benefit of specific individuals, organizations outside the Santa Barbara area, organizations in overpopulated nonprofit areas, national campaigns, fund-raising normally carried out by the organization, operating expenses, or budgetary support.
Geographic Focus: California
Date(s) Application is Due: Sep 15
Amount of Grant: 750 - 98,000 U.S.D
Contact: Program Contact; (805) 962-6430; fax (805) 962-7135; atuohyfdn@aol.com
Sponsor: Alice Tweed Tuohy Foundation
205 E Carrillo Street, Room 219, P.O. Box 1328
Santa Barbara, CA 93102-1328

Allan C. and Leila J. Garden Foundation Grants 432

The mission of the Allan C. and Leila J. Garden Foundation is to support charitable organizations that maintain, care for and educate orphan or underprivileged children. It is also the Foundation's intent to support organizations that provide medical, dental, hospital care, nursing and treatment of crippled or physically handicapped children. Grants from Allan C. and Leila J. Garden Foundation are primarily one year in duration. On occasion, multi-year support is awarded. Applicants must apply online at the grant website. Applicants are strongly encouraged to do the following before applying: review the downloadable state application procedures for additional helpful information and clarifications; review the downloadable online-application guidelines at the grant website; review the foundation's funding history (link is available from the grant website); review the online application questions in advance; and review the list of required attachments. These will generally include: a list of board members, financial statements (audited, reviewed, or compiled by independent auditor); an organization summary; a list of other funding sources; an IRS Determination letter; and other required documents. All attachments must be uploaded in the online application as PDF, Word, or Excel files. The Allan C. & Leila J. Garden Foundation application deadline is 11:59 p.m. on June 1. Applicants will be notified of grant decisions by letter within three to four months after the deadline.
Requirements: Applicants must have 501(c)3 tax-exempt status and serve residents of Ben Hill, Irwin, and Wilcox Counties in Georgia. A breakdown of number/percentage of people served by specific counties is required on the online application.
Restrictions: The foundation does not support requests from individuals, organizations attempting to influence policy through direct lobbying, or any political campaigns.
Geographic Focus: Georgia
Date(s) Application is Due: Jun 1
Contact: Quanda Allen, Vice President; (404) 264-1377; quanda.allen@baml.com
Internet: https://www.bankofamerica.com/philanthropic/fn_search.action
Sponsor: Allan C. and Leila J. Garden Foundation
3414 Peachtree Road, N.E., Suite 1475, GA7-813-14-04
Atlanta, GA 30326-1113

Allegan County Community Foundation Grants 433

The foundation awards one-year grants to eligible Michigan charitable organizations to improve the quality of life in Allegan County. Areas of interest include education, health, and social services. Types of support include youth grants, general grants, matching/challenge grants, building construction/renovation, program development, and emergency funds. A copy of the IRS tax-exemption letter must be submitted with application. Any youth group, club, school or class can apply for the T.A.G. awards but must have an adult leader who oversees the project and any funds awarded.
Requirements: 501(c)3 tax-exempt organizations serving Allegan County, Michigan, residents are eligible. All applicants are required to schedule an appointment with foundation staff prior to submitting an application.
Restrictions: Individuals are ineligible. All T.A.G. mini-grants must be handwritten by youth 18 or younger.
Geographic Focus: Michigan
Date(s) Application is Due: Dec 4
Contact: Theresa Bray, Executive Director; (269) 673-8344; fax (269) 673-8745
Internet: http://www.alleganfoundation.org/grants.htm
Sponsor: Allegan County Community Foundation
524 Marshall Street
Allegan, MI 49010

Allegheny Technologies Charitable Trust 434

The corporation awards one-year renewable grants to nonprofit organizations to enhance the quality of life for people in company operating locations. Areas of interest include arts and culture, civic and public affairs, education, health, and social services. Types of support include capital grants, general operating grants, program development grants, and employee matching gifts. Applicants should submit a letter of inquiry that describes the organization, purpose of grants, and funds sought. No particular form or information is required. Potential applicants are requested to provide proof of exempt public charity status. There are no deadlines.
Requirements: 501(c)3 organizations in company-operating areas are eligible.
Restrictions: Contributions are made only to public charities. Individuals and private foundations are excluded.
Geographic Focus: Pennsylvania
Amount of Grant: 1,000 - 35,000 U.S.D
Contact: Jon D. Walton, Jr., Trustee; (412) 394-2800; fax (412) 394-3034
Internet: http://www.alleghenytechnologies.com
Sponsor: Allegheny Technologies
1000 Six PPG Place
Pittsburgh, PA 15222

Allegis Group Foundation Grants 435

The foundation awards grants to support organizations providing services or programs in the areas of education, health, and underprivileged children. Giving is national in scope, with an emphasis on Baltimore. There are no application deadlines or forms. Send a letter of inquiry as the initial approach.
Samples: Leukemia and Lymphoma Society (San Jose, CA)—$1000; Odyssey School (Baltimore, MD)—$4000; Heart to Heart International (Olathe, KS)—$15,000.
Contact: Hillary Murray, (410) 579-3509
Internet: http://www.allegisgroup.com
Sponsor: Allegis Group Foundation
7301 Parkway Drive
Hanover, MD 21076

Allen P. and Josephine B. Green Foundation Grants 436

The mission of the foundation is to improve the quality of life in Missouri. Grants support programs to bring health care services to people; counseling and educational services for children and adults with physical, mental, or behavioral problems; innovative developmental and educational programs for children; cultural and preservation projects to safeguard our heritage and to promote broader awareness and understanding of such programs and activities on the part of Missourians; and environmental and conservation projects based in Missouri. Grants will be made for one year only. Priority will be given to new projects and to those that have not been funded for at least one year. The board meets in May and November. Guidelines are available online.
Requirements: Grants are awarded to 501(c)3 nonprofit organizations in Missouri.
Restrictions: Grants are not awarded to individuals; to projects and programs located outside of the United States; to charities that are not publicly supported; for social causes or for social activism; for lobbying, propagandizing, or for political campaigns; or to requests deemed as inappropriate for a variety of other reasons. Grant requests that probably will not be approved include projects and programs located outside of Missouri; commitment of funds for more than one year; funding to the same organization for two years in a row; grants for operating funds and annual budgets; unspecified general funds; other than to specific projects or programs; projects with a perceived small chance of success; programs that would seem unlikely to continue without further funding from the Green Foundation; large capital fund drives; large building fund drives; large endowments; to an organization whose primary mission is to raise and distribute funds; or to federated giving programs.
Geographic Focus: Missouri
Date(s) Application is Due: Mar 15; Jun 15; Sep 15; Dec 15
Samples: Childhaven (Saint Louis, MO)—for a computer system, $5000; Cardinal Ritter Institute Adult Day Care Ctr (Saint Louis, MO)—toward purchase of bus to transport elderly, $20,000; Ctr for Environmental Studies, Mineral Area College (Park Hills, MO)—to support the environmental action plan, $10,000.
Contact: Walter Staley, Jr. , Secretary-Treasurer; (573) 581-5568; fax (573) 581-1714; wstaley@greenfdn.org
Internet: http://www.greenfdn.org
Sponsor: Allen P. and Josephine B. Green Foundation
P.O. Box 523
Mexico, MO 65265

Alliance Healthcare Foundation Grants 437

The Foundation provides a public voice for the critical health care needs of its communities. Areas of interest include: Access to Health, Mental Health, Community Health, and HIV Health Care programs. Grants to local organizations support a wide range of programs and services. Priority is given to program and financing strategies that can produce discernible outcomes. All grant applicants should start by submitting a Letter of Intent (LOI), written and addressed according to the Foundation's guidelines. The LOI is the only form of application the Foundation accepts. Letters are reviewed on a monthly basis. Each is carefully evaluated by program staff in relation to other funding requests. Applicants will be notified about the results of the review within four to six weeks.
Restrictions: The Foundation does not fund projects and programs outside San Diego and Imperial Counties. In addition, the Foundation does not fund: research; lobbying; underwriting of medical expenses; general operating expenses we deem to be excessive; construction or renovation; the purchase of costly equipment; development activities, such as fundraising events, capital campaigns or annual fund drives; projects or proposals from individuals; or organizations that do not have 501(c)3 status.
Geographic Focus: California
Samples: Imperial County Office of Education (El Centro, CA)—to establish a family-resource center on the grounds of Seeley Elementary School, $450,000; Point Loma Nazarene U (San Diego, CA)—for its free clinic that provides health services to uninsured residents of San Diego's City Heights neighborhood, $122,880; Shakti Rising (San Diego, CA)—for strategic planning at this organization that uses holistic health practices and counseling to assist women struggling with abuse, addiction, and unhealthy relationships and behavior patterns, $159,415 over two yars; Whittier Institute for Diabetes (La Jolla, CA)—to establish the Child Obesity and Diabetes Prevention Program at Horton Elementary School, in San Diego, $62,657.
Contact: Karen Romero, (858) 614-4888; fax (858) 874 3656; kromero@alliancehf.org
Internet: http://www.alliancehf.org/grants_prog/what_we_fund/access_healthcare.html
Sponsor: Alliance Healthcare Foundation
9325 Sky Park Court, Suite 350
San Diego, CA 92123

All Stars Helping Kids Grants 438

The program focuses on helping disadvantaged children in the San Francisco Bay area. Areas of interest include children facing health problems, disabilities, family struggles, crime, and economic hardship. Types of support include grants for programs and projects, celebrity endorsements, special events, and corporate sponsorships.
Requirements: Nonprofits in California's San Francisco Bay area are eligible.
Geographic Focus: California
Samples: Stanford U (CA)—to help establish the Martin Luther King Jr. Research Institute, $1 million challenge grant.
Contact: Lindsey Ford, Executive Director; (650) 363-1395; fax (650) 363-1194; lindsey@allstarshelpingkids.com
Internet: http://www.allstarshelpingkids.com
Sponsor: All Stars Helping Kids
970 Main Street
Redwood City, CA 94063

Allstate Corporate Giving Grants 439

Allstate is a company of energized people with great ideas. The Corporate Giving program is committed to supporting the communities where company employees live and work by contributing to programs where its experience, partnership and leadership will have the greatest impact. The company offers financial support to a variety of programs and organizations throughout the country that help create strong and vital communities.
Requirements: The Allstate Corporation makes grants to nonprofit, tax-exempt groups under Section 501(c)3 of the Internal Revenue Code.
Contact: Executive Director; (847) 402-5000 or (847) 402-5502; fax (847) 326-7517; allfound@allstate.com
Internet: http://www.allstate.com/social-responsibility/corporate/corporate-giving.aspx
Sponsor: Allstate Corporation
2775 Sanders Road, Suite F4
Northbrook, IL 60062

Allstate Corporate Hometown Commitment Grants 440

Allstate takes a special interest in the greater Chicagoland community, the company's hometown for more than 75 years. The corporation is particularly invested in this community because it recognizes that a thriving hometown is critical to Allstate's success. The company recruits local talent, relies on local infrastructure, and depends on the city's vibrancy to ensure that its associates have a rich quality of life. By supporting organizations that build strong Chicagoland communities, the company contributes to the city's position as a center of global culture, education and business.
Requirements: The Allstate Corporation makes grants to Chicago area nonprofit, tax-exempt groups under Section 501(c)3 of the Internal Revenue Code.
Geographic Focus: Illinois, Indiana
Samples: Chicago Children's Museum, Chicago, Illinois, $15,000,000.
Contact: Executive Director; (847) 402-5000 or (847) 402-5502; fax (847) 326-7517; allfound@allstate.com
Internet: http://www.allstate.com/social-responsibility/corporate/corporate-giving.aspx
Sponsor: Allstate Corporation
2775 Sanders Road, Suite F4
Northbrook, IL 60062

Allyn Foundation Grants 441

The mission of the Allyn Foundation is to improve the quality of life in Central New York. To accomplish this, the foundation focuses its grant making in the following areas: health facilities and services: promote the delivery of quality health care in our communities; higher education in the Onondaga and Cayuga counties: expand access and success in education beyond high school, particularly for students of low income or other underrepresented backgrounds; basic human services: work to ensure that all people have basic daily needs of food, clothing, and shelter and enable the means to provide for themselves; family planning medical and educational services: promote education in family planning and improved access to service; quality of life for youth: improve quality of life for children and families through organizations that are focused on at risk youth and families.
Requirements: Nonprofit organizations in Onondaga and Cayuga Counties, NY, may submit grant applications.
Restrictions: Grants are not awarded for religious purposes, endowment funds, loans, or to individuals.
Geographic Focus: New York
Contact: Margaret O'Connell, Director; (315) 685-5059; info@allynfoundation.org
Internet: http://www.allynfoundation.org/apply.html
Sponsor: Allyn Foundation
14 West Genesee Street, P.O. Box 22
Skaneateles, NY 13152

ALO U.S.-Mexico Training, Internships, Exchanges, and Scholarships (TIES) Partnership Initiative 442

ALO, in cooperation with U.S.AID/Mexico, is issuing this request for applications for the United States-Mexico Training, Internships, Exchanges, and Scholarships (TIES) Initiative. ALO anticipates recommending approximately 10 applications for awards (for completion by September 2008), in the areas of rural and micro-finance; democracy; environment; rural prosperity for small-scale producers; and health. Successful applicant institutions will have significant private sector involvement and propose at least 10 long-term (two academic semesters) scholarships for Mexicans to be embedded within a partnership, in addition to other training, internships, and exchanges. Application and guidelines are available online.
Requirements: The lead institution must be a U.S. college, university, or community college.
Date(s) Application is Due: Apr 15
Amount of Grant: Up to 500,000 U.S.D
Contact: Grants Administrator; (202) 478-4700; fax (202) 478-4715; alo@aascu.org
Internet: http://www.aascu.org/alo/RFPs/TIES05/TIES05.htm
Sponsor: Association Liaison Office for University Cooperation in Development
1307 New York Avenue NW, Suite 500
Washington, DC 20005-4701

Alpha Omega Foundation Grants 443

The Foundation was established to advance the dental profession through research, grants, and scholarships. Fields of interest include: dental care, dental education, and medical research. Grants are given to individuals, program development, and higher education institutions. The application, which can be found at the web site, must be typed and submitted via email by September 1 each year. Notification will be completed by the following December.
Date(s) Application is Due: Sep 1
Amount of Grant: 1,000 - 36,000 U.S.D
Contact: Ben Williamowsky, (877) 368-6326 or (301) 294-2773; fax (301) 738-6403; bawilly@verizon.net or foundation@ao.org
Internet: http://www.ao.org/index.php?option=com_content&task=view&id=80&Itemid=147
Sponsor: Alpha Omega Foundation
50 W Edmonston Drive, Suite 404
Rockville, MD 20852-1274

Alpha Research Foundation Grants 444

The primary purpose of the Foundation is to encourage and support biomedical research for the welfare of the general public. There are no specific deadlines or application formats with which to adhere. Applicants should contact the foundation prior to submission, in order to discuss their proposed project. If approved, applicants should then submit an outline of the proposed research along with a proposed budget.
Geographic Focus: Maryland
Amount of Grant: Up to 2,000 U.S.D
Contact: Barry Rueben Fierst, (301) 762-8872; fax (301) 762-8874; bfierst@aol.com
Sponsor: Alpha Research Foundation
7118 Glenbrook Road
Bethesda, MD 20814-1238

Alpine Winter Foundation Grants 445

Established in California in 1963, the Alpine Winter Foundation's primary purpose is to support alpine safety, health programs, and education within the Tahoe-Donner, California, region. Applications should be in the form of a written request, and should include the specific purpose and history of the requesting organization. There are no specific deadlines with which to adhere.
Requirements: Giving is limited to organizations described in Section 170(b)(1)(a) and Section 501(c)3(a) that promote alpine safety, heath, and education within the Tahoe-Donner area of California,
Geographic Focus: California

Amount of Grant: Up to 40,000 U.S.D
Samples: Donner Trail School, Soda Springs, California, $1,500—for winter survival training programs; Sugar Bowl Ski Team Foundation, Norden, California, $37,500—for need-based scholarships and a building fund *2008).
Contact: Mary S. Tilden, President; (415) 221-7762
Sponsor: Alpine Winter Foundation
3863 Jackson Street, P.O. Box 591659
San Francisco, CA 94118-1610

ALSAM Foundation Grants 446
The foundation awards grants in its areas of interest, including agriculture, Christian agencies and churches, higher education, health care and medical research, human services, minorities, and the economically disadvantaged. Types of support include building construction/renovation, general operating costs, and scholarships. The board meets in January and October. Contact the office for application forms.
Requirements: Higher education institutions, nonprofit organizations, religious organizations, and research institutions are eligible.
Restrictions: No grants to individuals.
Amount of Grant: 5,000 - 50,000 U.S.D
Samples: University of Utah, Moran Eye Center (Salt Lake City, UT)—for building fund, $2.5 million; Saint George Catholic Church (Saint George, UT)—for building fund, $200,000; Harvard University (Cambridge, MA)—for research of Neurofibromatosis and Allied Disorders, $600,000; Verbum Dei High School (Los Angeles, CA)—for conversion, $2,374,378.
Contact: Ron Cutshall, Chair; (801) 266-4950
Sponsor: ALSAM Foundation
6190 Moffat Farm Lane
Salt Lake City, UT 84121

ALS Clinical Management Research Grants 447
The mission of the association is to find a cure for and improve living with amyotrophic lateral sclerosis (often called Lou Gehrig's disease). Through this program, the association is encouraging new research, the results of which will build an evidence base and demonstrate measurable, positive effects on the clinical management and lives of patients with ALS. Clinical problems relating to the following priority areas for clinical management research include breathing; swallowing/ nutrition and PEG issues/ choking/excess saliva; speech/communication; musculoskeletal symptoms/treatment; mobility/ activities of daily living; and psychosocial/mental health. The association also will consider abstracts on other clinical management topics.
Requirements: Applicants must be faculty members or scientists at reputable scientific institutions.
Date(s) Application is Due: Jan 7
Amount of Grant: Up to 30,000 U.S.D
Contact: Dr. Mary Lyon, Grants Administrator; (818) 880-9007, ext. 217; fax (818) 880-9006; mary@alsa-national.org
Internet: http://www.alsa.org/patient/research.cfm?CFID=102432&CFTOKEN=86382360
Sponsor: Amyotrophic Lateral Sclerosis Association
27001 Agoura Road, Suite 150
Calabasas Hills, CA 91301

ALS Research Grants 448
Research into the cause, means of prevention and possible cure of ALS is the driving priority of the ALS Association. The ALS Association annually awards $2 million or more in new research grants and currently holds two calls for abstracts each year. After review of the abstracts, selected investigators are asked to submit full applications which are also peer reviewed by ALSA's scientific review committee. Annual deadline dates may vary; contact program staff for exact dates.
Requirements: Applicants must be faculty members or scientists at reputable scientific institutions.
Amount of Grant: 40,000 - 80,000 U.S.D
Contact: Grants Administrator; (888) 949-2577 or (818) 880-9007; researchgrants@alsa-national.org
Internet: http://www.alsa.org/research/process.cfm?CFID=102432&CFTOKEN=86382360
Sponsor: Amyotrophic Lateral Sclerosis Association
27001 Agoura Road, Suite 150
Calabasas Hills, CA 91301

Alternatives Research and Development Foundation Alternatives in Education Grants 449
The mission of the Alternatives Research and Development Foundation is to fund and promote the development, validation and adoption of non-animal methods in biomedical research, product testing and education. ARDF has announced a special grant initiative seeking proposals to replace traditional use of animals for education and training purposes. Up to $10,000 in funding is available to support individual projects. the deadline for Letters of Intent is January 8, while invited full proposals are due March 12.
Requirements: Grants from the Foundation are made only to individuals affiliated with a non-profit, tax-exempt, institution, organization, or foreign equivalent.
Date(s) Application is Due: Jan 8; Mar 12
Amount of Grant: Up to 10,000 U.S.D
Contact: Sue A. Leary, President and Executive Director; (215) 887-8076; fax (215) 887-0771; grants@ardf-online.org or info@ardf-online.org
Internet: http://www.ardf-online.org/
Sponsor: Alternatives Research and Development Foundation
801 Old York Road, Suite 316
Jenkintown, PA 19046

Alternatives Research and Development Foundation Grants 450
The foundation awards grants to research centers, educational institutions, and other nonprofit organizations exploring non-animal testing. Grants will be awarded for research projects that use human, rather than nonhuman, vertebrae tissue; do not use intact, vertebrae animals; and can be completed within one year. Multiyear projects are considered on a case-by-case basis. The foundation often provides partial grants to initiate projects, with continuation funding available at a later date. The foundation will not consider proposals if program staff uses animals acquired from a shelter, or if individuals are employed who use such animals in their personal research programs. Winners are announced on July 15. Telephone inquiries are discouraged.
Requirements: Individuals attending or employed by U.S. universities and research institutions may apply. Applications from non-U.S. institutions or investigators may be considered.
Restrictions: The foundation does not make grants for in vitro projects that use nonhuman animal serum, indirect costs, purchase of personal computers, salary supplements, fringe benefits, travel expenses, or publication costs for the principal investigators. Phone calls to the foundation are discouraged. Guidelines and application are available online.
Date(s) Application is Due: Apr 30
Amount of Grant: Up to 40,000 U.S.D
Contact: Sue A. Leary, President and Executive Director; (215) 887-8076; fax (215) 887-0771; grants@ardf-online.org or info@ardf-online.org
Internet: http://www.ardf-online.org
Sponsor: Alternatives Research and Development Foundation
801 Old York Road, Suite 316
Jenkintown, PA 19046

Altman Foundation Health Care Grants 451
The Altman Foundation's mission is to support programs, and institutions that enrich the quality of life in New York City, with a particular focus on initiatives that help individuals, families, and communities benefit from the services and opportunities that will enable them to achieve their full potential.
Requirements: IRS 501(c)3 organizations in New York are eligible.
Restrictions: Grants are not awarded to individuals. As a general rule, the foundation does not consider requests for bricks and mortar funds or the purchase of capital equipment.
Geographic Focus: New York
Amount of Grant: 10,000 - 100,000 U.S.D
Samples: Literacy Assistance Venter, Inc., N.Y., $200,000 - to provide support for the Health Literacy Resource Center; The Boys' Club of New York, N.Y., $50,000 - to renew support for the After School Academy; The New York Community Trust, N.Y., to help support the New York City Workforce Development Funders Group $50,000.
Contact: Karen L. Rosa, Executive Director; (212) 682-0970; krosa@altman.org
Internet: http://www.altmanfoundation.org/guide.html
Sponsor: Altman Foundation
521 Fifth Avenue, 35th Floor
New York, NY 10175

Alton Ochsner Award 452
The Award is presented to one or more clinical or basic science investigators, without regard to age, race, gender, or nationality, for outstanding and exemplary original scientific investigations that relate tobacco consumption and health. This scientific work may be clinical, fundamental, epidemiological or preventive in scope. The prime criterion for award selection is its scientific impact on this major health threat. The $15,000 award is presented at the Annual Convocation of the American College of Chest Physicians.
Requirements: All nominations, whatever the category of scientific inquiry, must be supported by letters and copies of peer-reviewed scientific publications.
Date(s) Application is Due: Mar 31
Amount of Grant: 15,000 U.S.D
Contact: Edward D. Frohlich, Alton Ochsner Distinguished Scientist; (504) 842-3000; fax (504) 842-3258
Internet: http://www.ochsner.org/homepage.cfm
Sponsor: Ochsner Clinic Foundation
1514 Jefferson Highway
New Orleans, LA 70121

ALVRE Casselberry Award 453
The Casselberry Award has been established to encourage the advancement of the art and science of Laryngology and Rhinology. The award is given for outstanding manuscripts or accomplishments in Laryngology and Rhinology and consists of a $1,000 award and a certificate from the Association. Competition for this award will be limited to those persons whose abstracts are selected in consideration for inclusion in the Annual Scientific Program.
Date(s) Application is Due: Nov 30
Amount of Grant: 1,000 U.S.D
Contact: Maxine Cunningham, Administrator; (615) 322-6326; fax (615) 322-9102; Maxine@alahns.org or ala-hns@comcast.net
Sponsor: American Laryngological Association / American Laryngological Voice and Research Education Foundation
1215 21st Avenue South, 7302 MCE South
Nashville, TN 37232-8783

ALVRE Grants 454

The purpose of this award is to support basic, translational, or clinical research projects in laryngology, voice, outcomes, and related subjects. Research supported by this award should be specifically directed toward the pathogenesis, pathophysiology, diagnosis, prevention, or treatment of diseases, disorders, or conditions of the larynx and may be either basic or clinical/translational in approach. While not specifically required, proposals which aim to introduce new knowledge and methodology from other disciplines to research in laryngology or neurolaryngology, or which demonstrate collaborative effort with members of other related disciplines are encouraged. Projects must be designed so as to yield useful information within the period of award, but priority will be given to projects that are also innovative with promise to develop into new long-range or expanded research programs capable of attracting funding from other sources. A single, one-year, non-renewable award of $25,000 maximum is available annually. The foundation will consider requests to cover travel expenses up to $1,000 for the principal investigator to present his/her results at the ALA annual meeting.

Requirements: Candidates for this award should be otolaryngologists in the U.S. or Canada who have completed their training at an ACGME accredited program in otolaryngology—head and neck surgery. The prinicipal investigator should be a physician faculty member of a recognized department, division, or section of otolaryngology-head and neck surgery.

Amount of Grant: Up to 25,000 U.S.D

Contact: Maxine Cunningham, Administrator; (615) 322-6326; fax (615) 322-9102; Maxine@alahns.org or ala-hns@comcast.net

Internet: http://www.entnet.org/EducationAndResearch/coreGrants.cfm

Sponsor: American Laryngological Association / American Laryngological Voice and Research Education Foundation
1215 21st Avenue South, 7302 MCE South
Nashville, TN 37232-8783

ALVRE Seymour R. Cohen Award 455

The Seymour R. Cohen Award was established with a bequest by the Cohen family during Dr. Cohen's presidency in 1989. In order to qualify for the Award, the candidate must perform basic science research in the area of pediatric laryngology and/or pediatric neurolaryngology and be a citizen of the United States or Canada. The material must be approved and accepted for presentation at the Annual Meeting. Should multiple candidates collaborate in the research, all authors must comply with the citizenship requirements. Dr. Cohen wishes this Award only to be used for the funding of research in pediatric laryngology and/or pediatric neurolaryngology. The Award is to be presented bi-annually.

Requirements: The scope of the Award is limited to citizens of the United States of America and Canada.

Contact: Maxine Cunningham, Administrator; (615) 322-6326; fax (615) 322-9102; Maxine@alahns.org or ala-hns@comcast.net

Internet: http://www.alahns.org/i4a/pages/index.cfm?pageid=3332

Sponsor: American Laryngological Association / American Laryngological Voice and Research Education Foundation
1215 21st Avenue South, 7302 MCE South
Nashville, TN 37232-8783

Alzheimer's Association Conference Grants 456

The Alzheimer's Association has a long history of supporting scientific conferences that advance research on Alzheimer's disease. One of the principal goals of the Association from its inception has been to increase public awareness and to facilitate the exchange of information through the scientific and clinical communities. The support of conferences, workshops and meetings has been a key vehicle in achieving this goal. The objectives for conference support are to: facilitate and speed the exchange of information relevant to Alzheimer's disease research; convene experts to address emerging issues in Alzheimer's disease research; offer opportunities for new investigators and graduate students to participate in scientific meetings; facilitate the creation of networks among investigators in related areas; and increase visibility of the research interests and programs of the Alzheimer's Association. At this time, support requests must be limited to no more than $10,000 per conference. Most awarded conference support requests have been in the range of $2000 to $5000.

Requirements: Requests for conference support may be submitted at any time. It is recommended that requests be submitted at least three months before the conference.

Date(s) Application is Due: Jan 4; Mar 6; May 1; Jun 7; Aug 7; Oct 2; Nov 1

Amount of Grant: Up to 10,000 U.S.D

Contact: Angela Worlds, (312) 335-5807; fax (312) 335-4034; angela.worlds@alz.org

Internet: http://www.alz.org/research/alzheimers_grants/types_of_grants.asp

Sponsor: Alzheimer's Association
225 North Michigan Avenue, Suite 1700
Chicago, IL 60601-7633

Alzheimer's Association Development of New Cognitive and Functional Instruments Grants 457

The Alzheimer's Association is launching a new initiative to stimulate the scholarly investigation and development of cognitive or functional evaluation instruments that can capture the earliest changes in the disease, are sensitive to change over time, and/or could be used in clinical trials. This RFA is designed to enable pilot research or proof-of-principle studies that can provide preliminary data for subsequent inquiry. Allowable costs under this award include: purchase and care of laboratory animals; small pieces of laboratory equipment and laboratory supplies; computer equipment if used strictly for data collection; travel (up to $1,000 per year); and salary for the principal investigator, scientific (including post-doctoral fellows) and technical staff (including laboratory technicians and administrative support staff whose work is directly related to the funded project). The Association anticipates funding up to 2 approaches addressing needs in cognitive and functional instrument development. Each award is limited to $400,000 (direct and indirect costs) for two to three years. Requests in any given year may not exceed $200,000 (direct and indirect costs). Indirect costs are capped at 10 percent (rent for laboratory or office space is expected to be covered by indirect costs paid to the institution). Letters of Intent (LOIs) must be received by December 20, with completed applications due by February 7.

Requirements: Researchers with full-time staff or faculty appointments are encouraged to apply.

Restrictions: Applications from post-doctoral candidates will not be accepted. Costs not allowed under this award include: tuition; computer hardware or software for investigators; rent for laboratory or office spaces; or construction or renovation costs.

Date(s) Application is Due: Feb 7

Amount of Grant: Up to 400,000 U.S.D

Contact: Nico Stanculescu, Program Administrator; (312) 335-5747 or (312) 335-5862; fax (312) 335-4034; nico.stanculescu@alz.org or grantsapp@alz.org

Internet: http://www.alz.org/research/alzheimers_grants/types_of_grants.asp

Sponsor: Alzheimer's Association
225 North Michigan Avenue, Suite 1700
Chicago, IL 60601-7633

Alzheimer's Association Everyday Technologies for Alzheimer Care Grants 458

Everyday Technologies for Alzheimer Care (ETAC) is a cooperative research funding initiative sponsored by the Alzheimer's Association and Intel Corporation. ETAC seeks proposals on personalized diagnostics, preventive tools and interventions for adults coping with the spectrum of cognitive aging and neurodegenerative disease, particularly Alzheimer's disease. ETAC is designed to support exploratory multidisciplinary research that would not typically be funded by national health and science granting foundations. Minor iterations in testing plans or populations will not be considered for funding. Collaboration between social science/medical/public health and computer science/engineering researchers is valued. Mobile computing, high bandwidth sensing, robotics, imaging, face recognition, natural language processing, statistical modeling and a host of other technology advances allow unprecedented opportunities to study disease progression and therapeutic strategies in the context of everyday life. ETAC supports research that integrates such emerging technology capabilities with leading directions in behavioral science and biomedical research. Grants that merely create Internet-based versions of existing services or paper tools will not be considered. Submissions must be original ideas, not continuations of previously funded ETAC projects. Please see links provided below for examples of studies that have been funded by ETAC. Letters of Intent (LOIs) must be received by December 20, with completed applications due by February 7. The Association anticipates funding two awards under this program.

Requirements: Researchers with full-time staff or faculty appointments are encouraged to apply.

Restrictions: The Alzheimer's Association will not accept new research grant applications from currently funded Alzheimer's disease investigators who are delinquent in submitting interim/final scientific or interim/final financial reports on active grants. ETAC applications from post-doctoral candidates will not be accepted. Each total award is limited to $200,000 (direct and indirect costs) for up to three years. Requests in any given year may not exceed $90,000 (direct and indirect costs). Indirect costs are capped at 10 percent (rent for laboratory/office space is expected to be covered by indirect costs paid to the institution).

Date(s) Application is Due: Feb 7

Amount of Grant: Up to 200,000 U.S.D

Contact: Nico Stanculescu, Program Administrator; (312) 335-5747 or (312) 335-5862; fax (312) 335-4034; nico.stanculescu@alz.org or grantsapp@alz.org

Internet: http://www.alz.org/research/alzheimers_grants/types_of_grants.asp

Sponsor: Alzheimer's Association
225 North Michigan Avenue, Suite 1700
Chicago, IL 60601-7633

Alzheimer's Association Investigator-Initiated Research Grants 459

The program is structured to provide one to three years of sustained project support for independent, ongoing research. Proposals are solicited for basic, clinical, and social/behavioral research relevant to degenerative brain diseases such as Alzheimer's disease. Allowable costs for this award include the purchase and care of laboratory animals; small pieces of laboratory equipment and laboratory supplies; and salary for the principal investigator, scientific (including postdoctoral fellows) and technical staff (including laboratory technicians and modest secretarial support). It is required that most of the funds awarded under this program be used for direct research support. Allowable costs under this award include: purchase and care of laboratory animals; small pieces of laboratory equipment and laboratory supplies; computer software if used strictly for data collection; salary for the principal investigator, scientific (including post-doctoral fellows) and technical staff (including laboratory technicians and administrative support related directly to the funded project); and travel to scientific and professional meetings, not to exceed $1,000 per year. Letters of Intent (LOIs) must be received by December 20, with completed applications due by February 7.

Requirements: Public, private, domestic and foreign research laboratories, medical centers and hospitals, and universities are eligible to apply. Investigators from all stages of their research career development are encouraged to apply.

Restrictions: Your budget must not exceed the maximum amount of the award, $240,000 or $100,000 per year. Costs not allowed under this award include: tuition; computer hardware or standard software (e.g., Microsoft Office) for investigators; rent for laboratory or office space; or construction or renovation costs.

Date(s) Application is Due: Feb 7

Amount of Grant: 100,000 - 240,000 U.S.D
Contact: Nico Stanculescu, Program Administrator; (312) 335-5747 or (312) 335-5862; fax (312) 335-4034; nico.stanculescu@alz.org or grantsapp@alz.org
Internet: http://www.alz.org/research/alzheimers_grants/types_of_grants.asp
Sponsor: Alzheimer's Association
225 North Michigan Avenue, Suite 1700
Chicago, IL 60601-7633

Alzheimer's Association Mentored New Investigator Research Grants to Promote Diversity 460

The Mentored New Investigator Research Grant to Promote Diversity (MNIGD) is a three-year award intended to be a research-based and mentoring investment in the process of closing the health disparities gap between diverse and non-diverse investigator populations. The Alzheimer's Association feels strongly that the mentoring and involvement of diverse researchers in independently funded Alzheimer's research is a pressing need. The MNIRGD is intended to enhance the capacity of diverse and non-diverse scientists to conduct basic, clinical and social/behavioral research. Each MNIRGD award is limited to $170,000. A total of $150,000 will be awarded for costs related to the proposed research for up to three years (direct and indirect costs). Allowable costs under this award include: purchase and care of laboratory animals; small pieces of laboratory equipment and laboratory supplies; computer equipment if used strictly for data collection; travel (up to $1,000 per year); and salary for the principal investigator, scientific (including post-doctoral fellows) and technical staff (including laboratory technicians and administrative support related directly to the funded project) Letters of Intent (LOIs) must be received by December 20, with completed applications due by February 7.
Requirements: Eligibility for this grant competition is restricted to investigators who have less than 10 years of research experience after receipt of their terminal degree.
Restrictions: Each MNIRGD award is limited to $170,000. A total of $150,000 will be awarded for costs related to the proposed research for up to three years (direct and indirect costs). Requests in any given year may not exceed $60,000 (direct and indirect costs). Indirect costs are capped at 10 percent (rent for laboratory and/or office space is expected to be covered by indirect costs paid to the institution). Costs not allowed under this award include: tuition; computer hardware or software for investigators; rent for laboratory/office space; or construction or renovation costs.
Date(s) Application is Due: Feb 7
Amount of Grant: Up to 170,000 U.S.D
Contact: Nico Stanculescu, Program Administrator; (312) 335-5747 or (312) 335-5862; fax (312) 335-4034; nico.stanculescu@alz.org or grantsapp@alz.org
Internet: http://www.alz.org/research/alzheimers_grants/types_of_grants.asp
Sponsor: Alzheimer's Association
225 North Michigan Avenue, Suite 1700
Chicago, IL 60601-7633

Alzheimer's Association Neuronal Hyper Excitability and Seizures in Alzheimer's Disease Grants 461

The Alzheimer's Association is launching a new initiative to stimulate the development of new pharmacological strategies to prevent or treat seizures and abnormal neural network activity in Alzheimer's disease (AD). The Association's Request for Applications (RFAs) is aimed at the identification, screening and development of therapeutic strategies to reduce seizures and other types of abnormal neural network activity and at the evaluation of drug safety and efficacy at the preclinical and clinical levels. The RFA is designed to enable preliminary pilot research or proof-of-principle studies that can provide data for further research support by other funding agencies. Allowable costs under this award include: purchase and care of laboratory animals; small pieces of laboratory equipment and laboratory supplies; computer equipment if used strictly for data collection; travel (up to $1,000 per year); and salary for the principal investigator, scientific (including post-doctoral fellows) and technical staff (including laboratory technicians and administrative support staff whose work is directly related to the funded project). The Association anticipates funding up to two (2) Neuronal Hyper Excitability and Seizures in AD awards. Each award is limited to $400,000 (direct and indirect costs) for two to three years. Requests in any given year may not exceed $200,000 (direct and indirect costs). Indirect costs are capped at 10 percent (rent for laboratory/office space is expected to be covered by indirect costs paid to the institution). Letters of Intent (LOIs) must be received by December 20, with completed applications due by February 7.
Requirements: Researchers with full-time staff or faculty appointments are encouraged to apply.
Restrictions: Applications from post-doctoral candidates will not be accepted. Costs not allowed under this award include: tuition; computer hardware or software for investigators; rent for laboratory/office spaces; or construction or renovation costs.
Date(s) Application is Due: Feb 7
Amount of Grant: Up to 400,000 U.S.D
Contact: Nico Stanculescu, Program Administrator; (312) 335-5747 or (312) 335-5862; fax (312) 335-4034; nico.stanculescu@alz.org or grantsapp@alz.org
Internet: http://www.alz.org/research/alzheimers_grants/types_of_grants.asp
Sponsor: Alzheimer's Association
225 North Michigan Avenue, Suite 1700
Chicago, IL 60601-7633

Alzheimer's Association New Investigator Research Grants 462

The purpose of this program is to provide new investigators with funding that will allow them to develop preliminary or pilot data, to test procedures, and develop hypotheses which will then underpin the preparation of research grant applications to NIH, NSF, and other funding agencies and groups, including the Alzheimer's Association. All applications submitted to the program must focus on a question or questions in interventions for Alzheimer's disease to be considered responsive to the program announcement. Thirty awards will be made under this program. Letters of Intent (LOIs) must be received by December 20, with completed applications due by February 7.
Requirements: Public, private, domestic and foreign research laboratories, medical centers and hospitals, and universities are eligible to apply. Eligibility is restricted to investigators who have less than 10 years of research experience, including postdoctoral fellowships or residencies, after receipt of the doctoral degree. Applications from graduate and doctoral students for research projects, which will be used for the thesis or dissertation, will be accepted.
Restrictions: The Association anticipates funding 45 awards under this competition. Each total award is limited to $100,000 (direct and indirect costs) for up to two years. Requests in any given year may not exceed $60,000 (direct and indirect costs). Indirect costs are capped at 10 percent (rent for laboratory/office space is expected to be covered by indirect costs paid to the institution).
Date(s) Application is Due: Feb 7
Amount of Grant: Up to 100,000 U.S.D
Contact: Nico Stanculescu, Program Administrator; (312) 335-5747 or (312) 335-5862; fax (312) 335-4034; nico.stanculescu@alz.org or grantsapp@alz.org
Internet: http://www.alz.org/research/alzheimers_grants/types_of_grants.asp
Sponsor: Alzheimer's Association
225 North Michigan Avenue, Suite 1700
Chicago, IL 60601-7633

Alzheimer's Association New Investigator Research Grants to Promote Diversity 463

The New Investigator Research Grant to Promote Diversity (NIRGD) in Alzheimer's research is a two-year award to investigators who are currently underrepresented at academic institutions in Alzheimer's or related dementias research. The objective of this award is to increase the number of highly trained investigators from diverse backgrounds whose basic, clinical and social/behavioral research interests are grounded in the advanced methods and experimental approaches needed to solve problems related to Alzheimer's and related dementias in general and in health disparities populations. Each NIRGD award is limited to $100,000 (direct and indirect costs) for up to two years. Requests in any given year may not exceed $60,000 (direct and indirect costs). Letters of Intent (LOIs) must be received by December 20, with completed applications due by February 7.
Requirements: Eligibility to apply for this grant competition is restricted to investigators who have less than 10 years of research experience after receipt of their terminal degree.
Restrictions: Costs not allowed under this award include: tuition; computer hardware or software for investigators; rent for laboratory/office space; or construction or renovation costs. Your budget must not exceed the maximum amount of the award, $100,000 or $60,000 per year.
Date(s) Application is Due: Feb 7
Amount of Grant: Up to 100,000 U.S.D
Contact: Nico Stanculescu, Program Administrator; (312) 335-5747 or (312) 335-5862; fax (312) 335-4034; nico.stanculescu@alz.org or grantsapp@alz.org
Internet: http://www.alz.org/research/alzheimers_grants/types_of_grants.asp
Sponsor: Alzheimer's Association
225 North Michigan Avenue, Suite 1700
Chicago, IL 60601-7633

Alzheimer's Association U.S.-U.K. Young Investigator Exchange Fellowships 464

The U.S.-U.K. Young Investigator Exchange Fellowship provides a three-year grant to fund quality scientific research into the causes, diagnosis and treatment of Alzheimer's disease. The fellowship aims to address important research questions as well as to help and encourage promising scientists as they establish their careers within Alzheimer's research internationally. It is also hoped that by supporting meaningful scientific collaboration between scientists in the United Kingdom and the United States, there will be mutual benefit to the research output of both countries. In addition to an exchange of ideas, an important aim is to promote the learning of new experimental techniques and methodologies. The purpose of this fellowship is to provide new investigators with funding that will allow them to develop preliminary or pilot data, to test procedures and to develop hypotheses on an international level. The intent is to support international early-career development that will lay the groundwork for future research grant applications to other international funding agencies and groups. Allowable costs under this award include: purchase and care of laboratory animals; small pieces of laboratory equipment and laboratory supplies; computer equipment if used strictly for data collection; travel (up to $1,000 per year); and salary for the principal investigator, scientific (including post-doctoral fellows) and technical staff (including laboratory technicians and administrative support staff whose work is directly related to the funded project). The Alzheimer's Association and the Alzheimer's Reearch United Kingdom anticipate funding up to four awards (two originating from each country) under this competition. Each total award is limited to $300,000 U.S.D. A total of $260,000 U.S.D will be awarded for costs related to the proposed research for up to three years (direct and indirect costs). Requests in any given year may not exceed $90,000 U.S.D (direct and indirect). Indirect costs are capped at 10 percent (rent for laboratory/office space is expected to be covered by indirect costs paid to the institution). The PI must commit to a 75 percent effort toward the proposed project over the funding period. Letters of Intent (LOIs) must be received by December 20, with completed applications due by February 7.
Requirements: Researchers with full-time staff or faculty appointments are encouraged to apply.
Restrictions: Applications from post-doctoral candidates will not be accepted. Costs not allowed under this award include: tuition; computer hardware or software for investigators; rent for laboratory/office spaces; or construction or renovation costs.
Geographic Focus: All States, United Kingdom

Date(s) Application is Due: Feb 7
Amount of Grant: Up to 300,000 U.S.D
Contact: Nico Stanculescu, Program Director; (312) 335-5747 or (312) 335-5862; fax (312) 335-4034; nico.stanculescu@alz.org or grantsapp@alz.org
Internet: http://www.alz.org/research/alzheimers_grants/types_of_grants.asp
Sponsor: Alzheimer's Association
225 North Michigan Avenue, Suite 1700
Chicago, IL 60601-7633

Alzheimer's Association Zenith Fellows Awards 465

The Zenith Fellows award was initiated in 1991 to provide a vehicle for research support for donors with a substantial personal commitment to the advancement of Alzheimer's disease research. The objective of the Zenith Fellows Awards competition is to provide major support for investigators who have: contributed significantly to the field of Alzheimer's disease research or made significant contributions to other areas of science and are now beginning to focus more directly on problems related to Alzheimer's disease; and are likely to make substantial contributions in the future. We anticipate funding up to four awards under this competition. Each award is limited to $450,000 (direct and indirect costs) for three years. Requests in any given year may not exceed $250,000 (direct and indirect costs). Indirect costs are capped at 10 percent (rent for laboratory/office space is expected to be covered by indirect costs paid to the institution).
Requirements: Only established independent investigators are eligible as evidenced by: academic appointment; major, peer-reviewed, external multi-year grant support on which the applicant is the principal investigator (PI); independent laboratory operation; and quality and independence of publication record.
Restrictions: Previous recipients of Zenith Awards, Alzheimer's Disease Center Directors (P50 and P30), Medical and Scientific Advisory Council, and members of the National Board of the Alzheimer's Association are not eligible to apply.
Date(s) Application is Due: Feb 7
Amount of Grant: Up to 450,000 U.S.D
Contact: Nico Stanculescu, Program Administrator; (312) 335-5747 or (312) 335-5862; fax (312) 335-4034; nico.stanculescu@alz.org or grantsapp@alz.org
Internet: http://www.alz.org/research/alzheimers_grants/types_of_grants.asp
Sponsor: Alzheimer's Association
225 North Michigan Avenue, Suite 1700
Chicago, IL 60601-7633

Alzheimer's Foundation of America Brodsky Grant 466

The foundation awards this grant each year for an innovative program or service that improves the lives of individuals with Alzheimer's disease and related illnesses, and their families. Major criteria for the award are innovativeness, greatest need, and replication potential. Programs can involve, but are not limited to, hands- on care, delivery of social services, cognitive stimulation, the arts, safety, and training. The grant will be awarded to one of AFA's member organizations. Organizations that become members prior to the grant deadline are eligible for consideration. The Brodsky Grant is in addition to AFA's regular grants made to member organizations biannually based on a competitive review process. Membership applications are available online.
Requirements: AFA welcomes nonprofit organizations as members. Membership is open to organizations dedicated to aiding individuals with Alzheimer's disease and related illnesses, and their caregivers and families. Entities may be of any size budget and staff.
Restrictions: Nonprofit organizations that are chapters of national organizations are not eligible to become member organizations, but may apply as associate members.
Date(s) Application is Due: May 15
Amount of Grant: 20,000 U.S.D
Contact: Grants Administrator; (866) 232-8484; fax (646) 638-1546
Internet: http://www.alzfdn.org/join/member.shtml
Sponsor: Alzheimer's Foundation of America
322 Eighth Avenue, 6th Floor
New York, NY 10001

AMA-MSS Chapter Involvement Grants 467

The AMA-MSS Chapter Involvement Grant (CIG) Program provides each MSS chapter with up to $1,000 per academic year for chapter activities including recruitment and retention, chapter development, education, and community service events. A maximum of $250 or $500 is available for each event, depending on the type of event. Applications are due at least 30 days prior to the event.
Amount of Grant: 250 - 500 U.S.D
Contact: Rebecca Gierhahn, Director; (800) 262-3211, ext. 4753; rebecca.gierhahn@ama-assn.org or mss@ama-assn.org
Internet: http://www.ama-assn.org/ama/pub/about-ama/our-people/member-groups-sections/medical-student-section/opportunities/grants-awards-scholarships.shtml
Sponsor: American Medical Association
515 N State Street
Chicago, IL 60654

AMA-MSS Chapter of the Year (COTY) Award 468

The AMA-MSS Chapter of the Year (COTY) Award recognizes the true strength of the AMA-MSS organization, the local chapters. Applicant chapters are judged in a number of areas, including membership, community service, advocacy, innovation, and collaboration. The winning chapter is awarded a $500 grant to be used toward chapter activities and is recognized at the MSS Annual Meeting. All MSS chapters are encouraged to apply.
Date(s) Application is Due: Apr 30
Amount of Grant: 500 U.S.D
Samples: University of Texas Medical School at Houston; University of Arkansas for Medical Sciences College of Medicine.
Contact: Rebecca Gierhahn, Director; (800) 262-3211, ext. 4753; rebecca.gierhahn@ama-assn.org or mss@ama-assn.org
Internet: http://www.ama-assn.org/ama/pub/about-ama/our-people/member-groups-sections/medical-student-section/opportunities/grants-awards-scholarships.shtml
Sponsor: American Medical Association
515 N State Street
Chicago, IL 60654

AMA-MSS Government Relations Advocacy Fellowship 469

The AMA-MSS Government Relations Advocacy Fellowship (GRAF) for medical students is currently in its fifth year. One Fellow is selected each spring to work in Washington, D.C., as a full-time paid member of the AMA's Federal advocacy team for one year. The Fellow is responsible for working with the AMA's federal advocacy team to advance the Association's legislative agenda and policies on behalf of physicians, patients and medical students. A key goal for the fellowship is to educate medical student, resident and young physician AMA members about issues in public health and health policy in order to encourage activism and leadership in local communities. Applicants are not expected to be proficient advocates and managers at the start of the fellowship. It is meant to be a learning experience where proficiencies will be developed or enhanced. Prior health policy, legislative, or political experience is helpful. For additional information, please contact the Department of Medical Student Services at mss@ama-assn.org.
Requirements: Any AMA-MSS member in good standing at a medical school in the United States is eligible to apply.
Date(s) Application is Due: Jan 31
Contact: Rebecca Gierhahn, Director; (800) 262-3211, ext. 4753; rebecca.gierhahn@ama-assn.org or mss@ama-assn.org
Internet: http://www.ama-assn.org/ama/pub/about-ama/our-people/member-groups-sections/medical-student-section/opportunities/internships-fellowships.shtml
Sponsor: American Medical Association
515 N State Street
Chicago, IL 60654

AMA-MSS Government Relations Internship Programs 470

The Department of Medical Student Services, in conjunction with the Washington, D.C., office of the American Medical Association, offer assistance to students seeking to increase their involvement and education in national health policy and in the national legislative activities of organized medicine. The Government Relations Internship Program (GRIP) provides stipends to assist selected students who are completing summer health policy internships in the Washington, D.C., area. Medical students report that their credibility as AMA-sponsored interns allows them to pursue issues in depth, and in many cases, become experts on a specific topic. Similarly, internship sponsors welcome the enthusiasm and unique perspective that medical students bring to discussions about issues that require medicine's point of view. All students selected to participate in GRIP will receive a stipend. Additionally, GRIP participants completing internships with non-governmental organizations will benefit from attendance at weekly seminars conducted at the AMA Washington Office. These seminars are designed to increase the continuity of the internship experience, promote camaraderie among medical students working in health policy, and facilitate continuing education on important political issues. Upon completion of the program, all GRIP participants will be required to complete and submit a report on their internship experiences. The GRIP application deadline is February 15. However, late applications (through April 15) will be considered on a rolling basis if space remains. If you have any questions, please contact Keith Voogd, MSS Policy Analyst, in the Department of Medical Student Services at (312) 464-4745.
Requirements: To be eligible for the GRIP program, students must be AMA members who have secured policy internships in the Washington, DC, area during the summer. Eligible internship sites include: National specialty societies seated in the AMA House of Delegates; Public health advocacy groups; Non-clinical international health policy programs (limited spots); Congressional offices*; Federal agencies*.
Restrictions: Please note that internships with narrowly focused lobbying organizations or with specialty organizations that are not seated in the AMA House of Delegates are strongly discouraged. *Due to new Congressional lobbying disclosure rules, interns in Congressional offices or federal agencies will not be able to participate in the weekly educational seminars at the AMA's Washington, D.C., office.
Date(s) Application is Due: Feb 15
Contact: Rebecca Gierhahn, Director; (800) 262-3211, ext. 4753; rebecca.gierhahn@ama-assn.org or mss@ama-assn.org
Internet: http://www.ama-assn.org/ama/pub/about-ama/our-people/member-groups-sections/medical-student-section/opportunities/internships-fellowships.shtml
Sponsor: American Medical Association
515 N State Street
Chicago, IL 60654

AMA-MSS Research Poster Award 471

The MSS/RFS Joint Research Poster Symposium is held annually at the MSS Interim Meeting (November). Research is presented in eight categories (biochemistry/cell biology, cancer biology, cardiology/vascular biology, clinical/epidemiological/health care, immunology/microbiology, neurobiology/neuroscience, radiology/imaging, surgery). The overall winner receives a free trip to the MSS Annual Meeting in Chicago.
Date(s) Application is Due: Sep 1

Contact: Rebecca Gierhahn, Director; (800) 262-3211, ext. 4753; rebecca.gierhahn@ama-assn.org or mss@ama-assn.org
Internet: http://www.ama-assn.org/ama/pub/about-ama/our-people/member-groups-sections/medical-student-section/opportunities/grants-awards-scholarships.shtml
Sponsor: American Medical Association
515 N State Street
Chicago, IL 60654

AMA-RFS and AMA Foundation Medical Student & Resident/Fellow Elective Medicine & the Media 472

Discovery Communications, the leading global real-world media company, is offering an exclusive four to six week elective for American Medical Association (AMA) medical student and resident/fellow members. As part of Discovery's ongoing dedication to education, Discovery Channel produces continuing medical education (CME) programs that air on the Discovery Channel weekly. During the elective, one AMA medical student and one AMA resident/fellow will each work closely with all members of the Discovery team to develop a medical education program that will air on the network and be viewed by over 2 million people. This exciting opportunity gives the student and resident/fellow hands-on experience in the translation of rigorous scientific data into an entertaining and informational program. The student and resident/fellow will work primarily at Discovery Communications headquarters in Silver Spring, MD, but will also travel to relevant on-site and studio shoots for the program. A stipend will be provided to cover living expenses based on length of rotation ($3,000 for four-week rotation). Dates and length of rotation are flexible. Applications for the Winter elective are due September 30. Applicants not selected for the Winter elective must resubmit for the Summer elective if they wish to be considered for it. Summer applications are due March 15.
Requirements: All medical student and resident/fellow members of the American Medical Association (AMA) are eligible to apply for an exclusive elective in health communications with Discovery Channel in Silver Spring, MD. One AMA medical student and one AMA resident/fellow will be selected.
Date(s) Application is Due: Mar 15; Sep 30
Amount of Grant: 3,000 - 4,500 U.S.D
Contact: Jon Fanning, Director; (800) 262-3211, ext. 4978 or (312) 464-4978; fax (312) 464-5845; Jon.Fanning@ama-assn.org or rfs@ama-assn.org
Internet: http://www.ama-assn.org/ama/pub/about-ama/our-people/member-groups-sections/resident-fellow-section/awards-grants.shtml
Sponsor: American Medical Association Resident and Fellow Services
515 N State Street, 14th Floor
Chicago, IL 60654

AMA-RFS Legislative Awareness Internships 473

The Department of Resident and Fellow Services, in conjunction with the Washington, DC office of the American Medical Association, is sponsoring a two-week legislative internship program in the spring. Two residents and/or fellows will be selected. The program will afford residents and fellows the unique opportunity to participate in the political process of organized medicine at the national level. Each resident or fellow will receive a $1,000 stipend to help defray program-related expenses (e.g., airfare, hotel, cabs); all other expenses are the responsibility of the resident or fellow. The selected resident or fellow is also responsible for securing his or her own housing accommodations.
Requirements: Applicants must be members of the AMA.
Restrictions: The selected resident or fellow must obtain permission to participate in the program from their program director prior to the selection process.
Amount of Grant: 1,000 U.S.D
Contact: Jon Fanning, Director; (800) 262-3211, ext. 4978 or (312) 464-4978; fax (312) 464-5845; Jon.Fanning@ama-assn.org or rfs@ama-assn.org
Internet: http://www.ama-assn.org/ama/pub/about-ama/our-people/member-groups-sections/resident-fellow-section/awards-grants.shtml
Sponsor: American Medical Association Resident and Fellow Services
515 N State Street, 14th Floor
Chicago, IL 60654

AMA-WPC Joan F. Giambalvo Memorial Scholarships 474

The American Medical Association (AMA) Foundation in association with the AMA Women Physicians Congress (WPC) has established the Joan F. Giambalvo Memorial Scholarship Fund with the goal of advancing the progress of women in the medical profession and strengthening the ability of the AMA to identify and address the needs of women physicians and medical students. Proposals for the Joan F. Giambalvo Memorial Scholarship Fund will be accepted between Nov. 1 and Feb. 15. The AMA-WPC Joan F. Giambalvo Memorial Scholarship seeks innovative research proposals focusing on professional work/practice issues that affect women physicians, including, but not limited to: part-time working strategies; leadership training protocols; gender-based physician practice patterns; physician satisfaction or burnout; retention incentives; practice re-entry issues. Proposals for projects with concurrent/complementary funding and/or plans to use this grant as seed money for larger studies to follow will be received favorably, as will proposals for independent new projects. This award is for a maximum of $10,000. A budget submitted with each applicant's proposal should reflect the anticipated use of funds for the amount requested. Proposals will be evaluated and awardee(s) selected based on a variety of factors, including, but not limited to: the innovation, quality and/or feasibility of the idea; project/research methodology; potential to produce action, change or more comprehensive studies; and career goals of the applicant.
Requirements: Applicants/awardees shall be: female or male; working alone or collaboratively on the specific research project; physicians, medical students, other health professionals, or individuals working or doing graduate work in an applicable profession such as public health, sociology, psychology, etc.
Restrictions: Members of the AMA-WPC Governing Council, AMA Foundation Board of Directors, AMA Board of Trustees and Joan F. Giambalvo Memorial Scholarship Fund selection committee, and AMA and AMA Foundation staff are not eligible for this award.
Date(s) Application is Due: Feb 15
Amount of Grant: Up to 10,000 U.S.D
Contact: Alice Reed, Scholarship Contact; (312) 464-5523; alice.reed@ama-assn.org
Internet: http://www.ama-assn.org/ama/pub/about-ama/our-people/member-groups-sections/women-physicians-congress/about-wpc/joan-f-giambalvo-memorial-scholarship.shtml
Sponsor: American Medical Association
515 N State Street
Chicago, IL 60654

Amador Community Foundation Grants 475

The foundation supports organizations that enhance the quality of life for the people of Amador County by: encouraging private giving for the public good by providing a flexible, cost-effective and tax-exempt vehicle for donors with varied charitable interests and abilities to give; building and maintaining a permanent endowment fund in order to provide a continuing source of income for grants; making grants that are innovative, strategic and relevant in the support of nonprofit sectors; and serving as a catalyst to address changing and challenging community issues. There are no set project or program areas for funding.
Restrictions: The Foundation does not award grants for political or religious purposes, to retire long-term indebtedness, to influence legislation or elections, to private foundations and other grant-making organizations.
Geographic Focus: California
Contact: Grants Administrator; (209) 223-2148; fax (209) 223-4569; acf@amadorcommunityfoundation.org
Internet: http://www.amadorcommunityfoundation.org/grant.html
Sponsor: Amador Community Foundation
21-B Main Street, P.O. Box 1154
Jackson, CA 95642

AMA Foundation Arthur N. Wilson, MD, Scholarship 476

The Arthur N. Wilson, MD Scholarship provides a $5,000 tuition assistance scholarship to a currently enrolled medical student who graduated from a high school in Southeast Alaska. This scholarship highlights the importance of supporting future physicians in rural communities. The majority of Southeast Alaska is part of the Tongass National Forest, the United States' largest national forest. Major cities are Juneau, Sitka, and Ketchikan. Due to the rural nature of Southeast Alaska, many communities have no road connections outside of their locale, so aircraft and boats are a major means of transport.
Geographic Focus: Alaska
Date(s) Application is Due: Jun 15
Amount of Grant: 5,000 U.S.D
Samples: 2008 - Jodie Totten, University of Washington School of Medicine; 2007 - Mackenzie Slater - University of Washington School of Medicine.
Contact: Steven W. Churchill, Executive Director; (312) 464-4200; fax (312) 464-4142; Steven.Churchill@ama-assn.org
Internet: http://www.ama-assn.org/ama/pub/about-ama/ama-foundation/our-programs/medical-education/arthur-n-wilson.shtml
Sponsor: American Medical Association Foundation
515 N State Street
Chicago, IL 60654

AMA Foundation Dr. Nathan Davis International Awards in Medicine 477

Named for the founder of the AMA, the Dr. Nathan Davis International Award in Medicine recognizes physicians whose influence reach the international patient population and change the future of their medical care. By treating, educating and counseling patients beyond the U.S. border, the physician's work is having a positive impact on health care in the global arena. A $2,500 grant will be given to the institution or organization with which the recipient works. The recipient will also receive travel expenses and accommodations to the Excellence in Medicine Awards Banquet, and the AMA National Advocacy Conference in Washington D.C.
Requirements: To qualify, nominees must: have improved dramatically medical practice, medical education or medical research outside of the United States; embody the values of the medical profession through leadership, service, excellence, integrity and ethical behavior; during a lifetime of service or in a current initiative, has benefited the health and well-being of a specific patient population.
Date(s) Application is Due: Nov 16
Amount of Grant: 2,500 U.S.D
Contact: Steven W. Churchill, Executive Director; (312) 464-2593 or (312) 464-4200; fax (312) 464-4142; Steven.Churchill@ama-assn.org or amafoundation@ama-assn.org
Internet: http://www.ama-assn.org/ama/pub/about-ama/ama-foundation/our-programs/public-health/excellence-medicine-awards.shtml
Sponsor: American Medical Association Foundation
515 N State Street
Chicago, IL 60654

AMA Foundation Fund for Better Health Grants 478

The philosophy of the AMA Foundation Fund for Better Health begins with the idea that local communities and organizations have great knowledge and insight into their community's health care issues. Based on this thought, the AMA Foundation, with

support from the AMA Alliance, created the Fund for Better Health. Through this program, the AMA Foundation provides seed grants for grassroots, public health projects in communities throughout America. Over the years, the fund has provided over 200 grants totaling nearly $300,000 to projects that address healthy lifestyles, domestic violence prevention, substance abuse prevention, health literacy, patient safety and care for the uninsured. A maximum of $5,000 will be distributed to each grant recipient. The number of grant recipients will be determined by the AMA Foundation after all applications have been received. Typically, the number of grants awarded does not exceed twenty.
Requirements: Organizations are eligible to apply for grants which further the charitable and educational purposes of the AMA Foundation. Grants made in 2009 support programs addressing the issue of healthy lifestyles in the areas of nutrition and physical fitness, alcohol, substance abuse and smoking prevention (and cessation), and violence prevention. The three types of organizations eligible to apply are organizations with annual operating budgets of $1 million or less; new organizations begun in the last 5 years; or established organizations starting a new service or expanding a current service to an underserved population.
Restrictions: None of the funds awarded are to pay for staff salary or overhead expenses.
Date(s) Application is Due: Jul 15
Amount of Grant: Up to 5,000 U.S.D
Contact: Dina Lindenberg, Program Officer; (312) 464-4193; fax (312) 464-5973; dina.lindenberg@ama-assn.org or amafoundation@ama-assn.org
Internet: http://www.ama-assn.org/ama/pub/about-ama/ama-foundation/our-programs/public-health/fund-better-health.shtml
Sponsor: American Medical Association Foundation
515 N State Street
Chicago, IL 60654

AMA Foundation Health Literacy Grants 479
The foundation awards grants to programs that promote clearer communication between patients and their physicians and other health care providers. They are given to four groups: medical students; residents and fellows; physicians, hospital staffs and medical societies; and AMA Alliance groups and community organizations.
Amount of Grant: 500 - 4,000 U.S.D
Contact: Steven W. Churchill, Executive Director; (312) 464-2593 or (312) 464-4200; fax (312) 464-4142; Steven.Churchill@ama-assn.org or healthliteracy@ama-assn.org
Internet: http://www.ama-assn.org/ama/pub/about-ama/ama-foundation/our-programs/public-health/health-literacy-program.shtml
Sponsor: American Medical Association Foundation
515 N State Street
Chicago, IL 60654

AMA Foundation Healthy Communities/Healthy America Grants 480
Through the Healthy Communities/Healthy America program, the AMA Foundation awards $10,000-$25,000 grants to physician-led free clinics. Grants will be awarded to free clinics that: are requesting funds for specific projects, not for activities such as routine operations, maintenance or facility repairs; have regular and considerable operating hours; have significant physician involvement; and provide medical services. Preference will be given to applicants who demonstrate how grant dollars will be leveraged to provide the greatest amount of care.
Restrictions: Clinics that provide both medical and dental services, but are requesting funds for a project that is dental care-specific will not be considered.
Amount of Grant: 10,000 - 25,000 U.S.D
Contact: Steven W. Churchill, Executive Director; (312) 464-2593 or (312) 464-4200; fax (312) 464-4142; Steven.Churchill@ama-assn.org or healthliteracy@ama-assn.org
Internet: http://www.ama-assn.org/ama/pub/about-ama/ama-foundation/our-programs/public-health/healthy-communities-healthy.shtml
Sponsor: American Medical Association Foundation
515 N State Street
Chicago, IL 60654

AMA Foundation Jack B. McConnell, MD Awards for Excellence in Volunteerism 481
The Jack B. McConnell, MD, Award for Excellence in Volunteerism recognizes the work of senior physicians who provides treatment to U.S. patients who lack access to health care. After a full career of practice, these physicians remain dedicated to the future of medicine through the spirit of volunteerism. A $2,500 grant will be given to the institution or organization with which the recipient works. The recipient will also receive travel expenses and accommodations to the Excellence in Medicine Awards Banquet, and the AMA National Advocacy Conference in Washington D.C.
Requirements: To qualify, nominees must: have volunteered a significant portion of their medical services while over the age of 55; demonstrate their commitment to health care access by assisting underserved U.S. patients.
Date(s) Application is Due: Nov 16
Amount of Grant: 2,500 U.S.D
Contact: Steven W. Churchill, Executive Director; (312) 464-2593 or (312) 464-4200; fax (312) 464-4142; Steven.Churchill@ama-assn.org or amafoundation@ama-assn.org
Internet: http://www.ama-assn.org/ama/pub/about-ama/ama-foundation/our-programs/public-health/excellence-medicine-awards.shtml
Sponsor: American Medical Association Foundation
515 N State Street
Chicago, IL 60654

AMA Foundation Joan F. Giambalvo Memorial Scholarships 482
The American Medical Association (AMA) Women Physicians Congress (WPC) has established the Joan F. Giambalvo Memorial Scholarship Fund and partners with the AMA Foundation with the goal of advancing the progress of women in the medical profession and strengthening the AMA's ability to identify and address the needs and interests of women physicians and medical students.
Requirements: Applicants shall be: Female or male; Working alone or collaboratively on the specific research project; Physicians, medical students, other health professionals, or individuals working or doing graduate work in an applicable profession such as public health, sociology, psychology, etc. Proposals will be evaluated and awardee(s) selected based on a variety of factors, including, but not limited to: the innovation, quality and/or feasibility of the idea; project/research methodology; potential to produce action, change or more comprehensive studies; and career goals of the applicant. AMA membership is not required in order to be eligible for this award.
Restrictions: Members of the AMA-WPC Governing Council, AMA Foundation Board of Directors, AMA Board of Trustees and Joan F. Giambalvo Memorial Scholarship Fund selection committee, and AMA and AMA Foundation staff are not eligible for this award.
Date(s) Application is Due: Feb 15
Amount of Grant: Up to 10,000 U.S.D
Samples: 2008 Recipients: Sue Yom, MD, PhD, Assistant Professor of Radiation Oncology at the University of California San Francisco; Maurice Clifton, MD, MSEd, Associate Dean for Admissions and Student Affairs with the Mercer University School of Medicine.
Contact: Alice Reed, (312) 464-5523; alice.reed@ama-assn.org
Internet: http://www.ama-assn.org/ama/pub/category/15566.html
Sponsor: American Medical Association Foundation
515 N State Street
Chicago, IL 60654

AMA Foundation Jordan Fieldman, MD, Awards 483
The Jordan Fieldman, MD Award was established in the name of Dr. Jordan Fieldman and is sponsored by the AMA Resident and Fellow Section in association with the AMA Foundation. Dr. Jordan Fieldman was an outstanding physician and deeply concerned with helping his patients and making the world of medicine a better place. Unfortunately, Jordan lost his battle with cancer in June 2004. In establishing this award, the AMA-RFS desires to continue to create physicians like Dr. Fieldman and give them the skills and means to do so. The award money will cover resident travel to both the Annual and Interim meeting during the year of the award and necessary expenses while attending these meetings. This award will give a young doctor an opportunity that he or she may otherwise never be able to experience.
Requirements: The Awardee must be: a first-time delegate or attendee to the AMA-RFS meetings; from a state or district that does not have funding available to support resident or fellow travel to attend the AMA-RFS Meetings; interested and active in patient advocacy efforts; involved in organized medicine. Additional guidelines, as well as, an online application are available at, http://www.ama-assn.org/ama/pub/about-ama/ama-foundation/our-programs/public-health/jordan-fieldman-md.shtml.
Contact: Sharyn Grose, Program Administrator; (312) 464-4978 or (800) 621-8335; fax (312) 464-5845; sharyn.grose@ama-assn.org or rfs@ama-assn.org
Internet: http://www.ama-assn.org/ama/pub/about-ama/ama-foundation/our-programs/public-health/jordan-fieldman-md.shtml
Sponsor: American Medical Association Foundation
515 N State Street
Chicago, IL 60654

AMA Foundation Leadership Awards 484
An exceptional medical professional goes beyond the medical practice to positively influence health care. Through organized medicine and community activities, individuals from each stage of the physician lifecycle can, and do, make a difference in the quality of health care and the medical environment. The Leadership Awards are presented to 15 medical students, 10 residents/fellows and 5 early career physicians to recognize their strong, nonclinical leadership skills in advocacy, community service and/or education. Award recipients will be invited to attend leadership development training in Washington, D.C. This training will strengthen leadership skills and result in a greater effort to advance health care in America. Directly following the training, recipients will attend the Excellence in Medicine Awards Banquet and stay for the AMA National Advocacy Conference. Airfare and accommodations are provided, in addition to a nominal reimbursement to help defray additional travel-related costs.
Requirements: All Leadership Awards are self-nominated, with applicants responsible for submitting all appropriate materials and documentation. To qualify, applicants must demonstrate outstanding leadership in the areas of advocacy, community service and/or education and be either: a medical student enrolled in an accredited medical school; a resident physician enrolled in an accredited residency program; a fellow physician enrolled in a fellowship program; an early-career physician under the age of 40 or in his or her first eight years of practice following residency/fellowship.
Date(s) Application is Due: Nov 16
Contact: Steven W. Churchill, Executive Director; (312) 464-2593 or (312) 464-4200; fax (312) 464-4142; Steven.Churchill@ama-assn.org or amafoundation@ama-assn.org
Internet: http://www.ama-assn.org/ama/pub/about-ama/ama-foundation/our-programs/public-health/excellence-medicine-awards.shtml
Sponsor: American Medical Association Foundation
515 N State Street
Chicago, IL 60654

AMA Foundation Minority Scholars Awards 485
The AMA Foundation, in collaboration with the AMA Minority Affairs Consortium (MAC), with support from Pfizer Inc, offers the Minority Scholars Award. Approximately twelve Minority Scholars Awards are awarded annually, each in the amount of a $10,000 scholarship.

Requirements: You must be a current first or second-year student and a permanent resident or citizen of the U.S. Eligible students of minority background include African American/ Black, American Indian, Native Hawaiian, Alaska Native and Hispanic/Latino. Each medical school is invited to submit up to two nominees. Contact your medical school if you are interested in being nominated for the Minority Scholars Award.
Date(s) Application is Due: Apr 15
Amount of Grant: 10,000 U.S.D
Samples: 2007 Recipients: Chuck Branch, University of Minnesota Medical School, Duluth; Michael Nevarez, Harvard Medical School.
Contact: Steven W. Churchill, Executive Director; (312) 464-4200; fax (312) 464-4142; Steven.Churchill@ama-assn.org
Internet: http://www.ama-assn.org/ama/pub/category/20116.html
Sponsor: American Medical Association Foundation
515 N State Street
Chicago, IL 60654

AMA Foundation Physicians of Tomorrow Scholarships **486**
These $10,000 scholarships reward current third-year medical students, who are entering their fourth-year of study. There will be 12 Physicians of Tomorrow scholarships funded by the AMA Foundation. Multiple scholarships, funded by the AMA Foundation, the Audio-Digest Foundation and the Johnson F. Hammond, MD Fund will be awarded.
Requirements: The selection of the recipients will be based on academic achievement and financial need. The recipient of the one Physicians of Tomorrow Scholarship funded by the Audio-Digest Foundation should have an interest in 'the communication of science.' Activities such as mentoring and/or teaching are examples of 'communication of science.' The recipient of the one Physicians of Tomorrow Scholarship funded by the Johnson F. Hammond, MD Fund should have an interest in and commitment to a career in medical journalism. Contact your medical school if you are interested in being nominated for the Physicians of Tomorrow Scholarships. Each medical school may submit one nomination for each of these scholarship opportunities. Thus, each school may submit up to three nominations in total.
Date(s) Application is Due: May 29
Amount of Grant: 10,000 U.S.D
Samples: 2008: Puya Alikhani, Texas A&M Health Science Center College of Medicine; Diana Badillo, Stanford University School of Medicine; Andrew Barina, Saint Louis University School of Medicine; Karl Bezak, Vanderbilt University School of Medicine; Rozalina Grubina, Johns Hopkins University School of Medicine; Nadia Hernandez, University of Texas Health Science Center at Houston; Stephanie Hu, Harvard Medical School; Arman Kilic, University of Pittsburgh School of Medicine; Youssra Marjoua, Yale University School of Medicine; Monica Patton, University of Vermont College of Medicine; Janae Phelps, Howard University College of Medicine; Dominic Sanford, University of Missouri Columbia; Javay Ross, Charles Drew University/UCLA Medical Education Program; Matthew Bivens, George Washington University School of Medicine and Health Sciences; Helena Hart, University of California, San Francisco School of Medicine.
Contact: Steven W. Churchill, Executive Director; (312) 464-4200; fax (312) 464-4142; Steven.Churchill@ama-assn.org
Internet: http://www.ama-assn.org/ama/pub/category/20119.html
Sponsor: American Medical Association Foundation
515 N State Street
Chicago, IL 60654

AMA Foundation Pride in the Profession Grants **487**
The Pride in the Profession Awards honor physicians whose lives encompass the true spirit of being a medical professional: caring for people. By practicing medicine in areas of challenge or crisis, or by devoting their time to volunteerism or public service, these physicians serve as the voice of patients in the United States who otherwise might not be heard. A $2,500 grant will be given to the institution or organization with which the recipient works. The recipient will also receive travel expenses and accommodations to the Excellence in Medicine Awards banquet, and the AMA National Advocacy Conference in Washington D.C.
Requirements: To qualify, nominees must: promote the art and science of medicine and the betterment of the public health; embody the values of the medical profession through leadership, service, excellence, integrity and ethical behavior; enrich patients, colleagues and the community through dedicated medical practice or service; offer better access to quality health care for an underserved patient population in the U.S.
Date(s) Application is Due: Nov 16
Amount of Grant: 2,500 U.S.D
Contact: Steven W. Churchill, Executive Director; (312) 464-2593 or (312) 464-4200; fax (312) 464-4142; Steven.Churchill@ama-assn.org or amafoundation@ama-assn.org
Internet: http://www.ama-assn.org/ama/pub/about-ama/ama-foundation/our-programs/public-health/excellence-medicine-awards.shtml
Sponsor: American Medical Association Foundation
515 N State Street
Chicago, IL 60654

AMA Foundation Seed Grants for Research **488**
The AMA Foundation established the Seed Grant Research Program to encourage medical students, physician residents and fellows to enter the research field. The program provides $2,500-$5,000 grants to help them conduct small basic science, applied, or clinical research projects. These funds will round out new project budgets, rather than sustain current initiatives. One-year grants will be awarded in the following research categories: Cardiovascular/pulmonary diseases; HIV/AIDS; Leukemia; Neoplastic diseases; Secondhand smoke. Grants in Cardiovascular/Pulmonary Diseases, HIV/AIDS, Leukemia, and Neoplastic Diseases will be $2,500. Grants in the Secondhand smoke category will be $5,000.
Requirements: Applicants must be a medical student, physician resident or fellow of an accredited U.S. medical school or institution; they must also be either a U.S. citizen or a permanent resident of the U.S.. Projects must be applicant-conceived, rather than ongoing research of their mentor or Principal Investigator.
Restrictions: Seed grant funds cannot be used for salary or stipend, indirect/administrative costs, to hire a consultant or contractor, and solely for travel expenses. Seed grants will not be awarded to any applicant who has previously received an AMA Foundation seed grant in the research category in which they are applying.
Date(s) Application is Due: Dec 15
Amount of Grant: 2,500 - 5,000 U.S.D
Contact: Seed Grants Program Supervisor; (312) 464-4200; fax (312) 464-5973; seedgrants@ama-assn.org
Internet: http://www.ama-assn.org/ama/pub/category/7785.html
Sponsor: American Medical Association Foundation
515 N State Street
Chicago, IL 60654

AMA Foundation Worldscopes Program **489**
WorldScopes collects stethoscopes from U.S. physicians and distributes them, with the help of humanitarian organizations, to communities around the world where medical supplies are scarce. A stethoscope donation will help a colleague in diagnosing and treating diseases that deprive so many of health, life, and human dignity.
Contact: Ethical Force Program Office; (312) 464-4075; fax (312) 464-4799; WorldScopes@ama-assn.org
Internet: http://www.ama-assn.org/ama/pub/category/12768.html
Sponsor: American Medical Association Foundation
515 N State Street
Chicago, IL 60654

Amarillo Area/Harrington Foundations Grants **490**
The community foundation seeks to improve the quality of life in the 26 northernmost counties of the Texas Panhandle. Grants are awarded to nonprofits in support of arts and culture, education, health care, and social services. Types of support include research, building construction/renovation, equipment acquisition, program development, seed grants, scholarship funds, and matching funds.
Requirements: Nonprofit 501(c)3 organizations located in the northernmost 26 counties of the Texas Panhandle, including Dallam, Sherman, Hansford, Ochiltree, Lipscomb, Hartley, Moore, Hutchinson, Roberts, Hemphill, Oldham, Potter, Carson, Gray, Wheeler, Deaf Smith, Randall, Armstrong, Donley, Collingsworth, Parmer, Castro, Swisher, Briscoe, Hall, and Childress, are eligible.
Restrictions: The foundation does not make grants to or for religious activities; political lobbying or legislative activities; endowments; debt retirement; deficit financing, reduction of operating deficit, etc; private or parochial schools; national, state, or local fund-raising activities; general operating expenses for United Way agencies; or umbrella funding organizations that would distribute requested funds at their own discretion.
Geographic Focus: Texas
Date(s) Application is Due: Jan 5; Jul 6
Samples: U of Texas at Austin (TX)—for an international fellowship program for faculty members and graduate students, $30 million; U of Texas at Austin (TX)—to endow an international fellowship program that will provide salary, research, and travel grants to approximately five faculty members and 16 graduate students annually, $30 million.
Contact: Kathy Grant, (806) 376-4521; kathie@aaf-hf.org
Internet: http://aaf-hf.org/grants/guidelines.htm
Sponsor: Amarillo Area/Harrington Foundations
801 S Fillmore Street, Suite 700
Amarillo, TX 79101

AMA Virtual Mentor Theme Issue Editor Grants **491**
Virtual Mentor [www.virtualmentor.org] is the American Medical Association's online ethics journal. Its mission is to promote the ethical and professional development of tomorrow's physicians, and its primary audiences are medical students, residents, other physicians, and medical educators. VM serves as an educational resource for a rapidly growing number of readers. VM invites medical students and residents to apply to serve as theme issue editors. Each theme issue editor undertakes, with help from VM staff editors, responsibility for: (1) selecting a theme for one issue and defining the ethical and professionalism concerns inherent in that theme, (2) generating case narratives that provide opportunities for examining those concerns in clinical and educational contexts, (3) identifying, securing, and corresponding with contributors to the issue, (4) editing copy and reviewing page proofs before the issue goes live on the first working day of the month. These tasks will be accomplished at each editor's home location. Theme issue editors, selected in late November, will be flown to Chicago to meet with Virtual Mentor staff some time in February. Each theme issue editor receives a $1,000-stipend.
Requirements: Experience in editing or journalism is desirable. The abilities to critique an argument or article, work well with authors, and meet deadlines are essential. To apply, send a short letter stating your current medical student or resident program status and your reason for wishing to be a theme issue editor for VM. Attach to the letter, the following: a curriculum vitae; a writing sample of not more than 1,000 words; your response to this exercise in 500 words or fewer. Additional guidelines available at: http://virtualmentor.ama-assn.org/site/issue-edscall.html. Send your letter of application and attachments by email to: Faith.Lagay@ama-assn.org. The deadline for applications is midnight CST on November 13.

Date(s) Application is Due: Nov 13
Amount of Grant: 1,000 U.S.D
Contact: Faith Lagay, Managing Editor; (312) 464-5438 or (312) 464-5260; fax (312) 464-4799; Faith.Lagay@ama-assn.org or virtualmentor@ama-assn.org
Internet: http://www.ama-assn.org/ama/pub/about-ama/our-people/member-groups-sections/medical-student-section/opportunities/internships-fellowships.shtml
Sponsor: American Medical Association
515 N State Street
Chicago, IL 60654

AMD Corporate Contributions Grants 492
AMD has established two global focus areas: strengthening community and strengthening education. Priority is given to basic needs (food, shelter, and basic health care), education (math and science, teacher development, and college and career awareness), and university education. The k-12 initiatives target programs that increase student interest and/or proficiency in literacy, match, science, and computer technology. AMD also funds programs aimed at developing and supporting effective classroom instruction.
Requirements: Most contributions are made to accredited schools and 501(c)3 nonprofit agencies operating in Austin, Texas (Travis Couunty) or Sunnyvale, California (Santa Clara County) with which AMD has a strong established relationship.
Restrictions: AMD does not consider unsolicited applications for programs outside of the communities in which they operate or outside their focus areas. Also excluded are individuals, medical research, religious, political, service or fraternal organizations, arts or cultural programs, advocacy groups, athletic teams, recreational programs, or individual scouting troops. Organization must be non-discriminatory.
Date(s) Application is Due: May 1
Amount of Grant: Up to 25,000,000 U.S.D
Contact: Community Affairs Manager; (800) 538-8450, ext. 45373; fax (408) 749-5373
Internet: http://www.amd.com/us-en/Corporate/AboutAMD/0,,51_52_7697_7702,00.html
Sponsor: AMD Corporation
P.O. Box 3453, M/S 42, 1 AMD Pl
Sunnyvale, CA 94088

Amelia Peabody Charitable Fund 493
This fund gives grants for higher education, hospitals and health services, the environment, and culture primarily in Massachusetts and New England. Types of support include capital campaigns, building construction/renovation, equipment acquisition, endowment funds, and research. Initial approach should be a proposal giving pertinent information concerning the structure of the requesting organization, amount requested, purpose of the project, budget, personnel involved, and other foundations with whom proposals have been filed for this project. Application guidelines are available upon request.
Requirements: Nonprofit organizations operating in Massachusetts are eligible to apply.
Restrictions: Awards will not be made to individuals, for films, or for periods of more than one year, nor will grants be awarded to religious or political organizations or organizations maintained by local, state, or federal governments. Salaries will not be supported and very seldom will operating grants be awarded.
Geographic Focus: Massachusetts
Date(s) Application is Due: Feb 1; Jun 1; Oct 1
Amount of Grant: 10,000 - 100,000 U.S.D
Samples: Boston Public Library Foundation (Boston, MA)—program support, $100,000; YMCA of Cambridge Family (Cambridge, MA)—program support, $100,000.
Contact: Cheryl Gideon, Program Contact; (617) 451-6178
Sponsor: Amelia Peabody Charitable Fund
10 Post Office Square, North Suite 995
Boston, MA 02109-4603

Amelia Sillman Rockwell and Carlos Perry Rockwell Charities Fund Grants 494
The Amelia Sillman Rockwell and Carlos Perry Rockwell Charities Fund was established in 1962 to support and promote quality educational, human-services, and health-care programming for underserved populations. Special consideration is given to charitable organizations that serve children or the elderly. Grant requests for general operating support are strongly encouraged. Program support will also be considered. Small, program-related capital expenses may be included in general operating or program requests. The majority of grants from the Rockwell Charities Fund are one year in duration; on occasion, multi-year support is awarded. Applicants must apply online at the grant website. Applicants are strongly encouraged to do the following before applying: review the downloadable state application procedures for additional helpful information and clarifications; review the downloadable online-application guidelines at the grant website; review the foundation's funding history (link is available from the grant website); review the online application questions in advance; and review the list of required attachments. These will generally include: a list of board members, financial statements (audited, reviewed, or compiled by independent auditor); an organization summary; a list of other funding sources; an IRS Determination letter; and other required documents. All attachments must be uploaded in the online application as PDF, Word, or Excel files. The application deadline for the Rockwell Charities Fund is 11:59 p.m. on February 1. Applicants will be notified of grant decisions before May 31.
Requirements: Applicants must have 501(c)3 tax-exempt status.
Restrictions: The trust does not support requests from individuals, organizations attempting to influence policy through direct lobbying, or any political campaigns.
Geographic Focus: Massachusetts
Date(s) Application is Due: Feb 1
Contact: Miki C. Akimoto, Vice President; (866) 778-6859; miki.akimoto@baml.com
Internet: https://www.bankofamerica.com/philanthropic/fn_search.action
Sponsor: Amelia Sillman Rockwell and Carlos Perry Rockwell Charities Fund
225 Franklin Street, 4th Floor, MA1-225-04-02
Boston, MA 02110

American-Italian Cancer Foundation Fellowships 495
The mission of the program is to further the advancement of cancer research by sponsoring the research and training of promising young scientists in Italy and in the United States. A major activity is the award of fellowships to Italian and American scientists so that they might spend a period of research and advanced training in preclinical or clinical research centers in the United States or Italy. The fellowships are intended to augment the training of motivated scientists who have the potential to conduct path-breaking original research upon the return to their respective countries. The one-year fellowships are renewable for an additional year.
Requirements: Individuals in the United States and Italy who have been awarded the degrees of PhD, MD, DSc, or DVM may apply in conjunction with a host institution. Candidates must have received their degree no more than three years prior to the date of application.
Date(s) Application is Due: Mar 1
Amount of Grant: 25,000 - 30,000 U.S.D
Contact: Franca Gaudio, (212) 628-9090; fax (212) 517-6089; fgaudio@aicfonline.org
Internet: http://www.aicfonline.org
Sponsor: American-Italian Cancer Foundation
112 E 71st Street, Suite 2B
New York, NY 10021

American-Italian Cancer Foundation Prize for Scientific Excellence 496
The prize recognizes a major scientific discovery in basic or translational cancer research advances that emanate from outstanding scientific work that has reached the early stages of clinical trials and shows real promise. The foundation is keenly interested in research that promises to make a difference in the quality of life for cancer patients. Prize winners will have made a significant contribution to understanding cancer, continue to be active in research, and their work holds promise for future outstanding contributions to the field. Each year two awards will be presented to one U.S. and one European scientist. Nomination guidelines are available online.
Requirements: Nominations may be made by any scientist currently or previously affiliated with a cancer research institution or an advocacy organization.
Restrictions: Institutions or organizations are not eligible for this award, and candidates may not nominate themselves.
Amount of Grant: 50,000 U.S.D
Contact: Dr. Esther Dyer, Executive Director; (212) 628-9090; fax (212) 517-6089; aicf@aicfonline.org
Internet: http://www.aicfonline.org
Sponsor: American-Italian Cancer Foundation
112 E 71st Street, Suite 2B
New York, NY 10021

American-Scandinavian Foundation Visiting Lectureship Grants 497
The ASF invites U.S. colleges and universities to apply for funding to host a visiting lecturer from Norway or Sweden. The $20,000 awards are for appointments of one semester. Lectureships should be in the areas of: public policy, conflict resolution, health care, environmental studies or multiculturalism. Letters of Intent are due by November 6, with full applications postmarked by February 5.
Requirements: The competition is open to all American colleges and universities. The award is appropriate not just for Scandinavian studies departments, but for any department or inter-disciplinary program with an interest in incorporating a Scandinavian focus into its course offerings. The lecturer must be a Norwegian or Swedish citizen, and a scholar or expert in a field appropriate to the host department or program.
Date(s) Application is Due: Feb 5; Nov 6
Amount of Grant: 20,000 U.S.D
Contact: Ellen McKey, Director of Fellowships and Grants; (212) 879-9779; fax (212) 249-3444; grants@amscan.org
Sponsor: American-Scandinavian Foundation
58 Park Avenue
New York, NY 10016

American Academy of Dermatology Camp Discovery Scholarships 498
Every year, the academy sponsors a week of fishing, boating, swimming, water skiing, arts and crafts for young people with a serious skin condition. Under the expert care of dermatologists and nurses, the camp offers the opportunity of spending a week among other young people who have similar skin conditions. Many of the counselors have serious skin conditions as well, and can provide support and advice to campers. The camp fee is covered by full scholarships, including transportation.
Requirements: Campers may attend at any one of four locations selected each year. Campers must meet the age criteria at time of camp (generally between the ages of 8 and 16). Campers must also get a recommendation from a dermatologist who is an academy member. Registration can be completed online at the website.
Contact: Jill Mueller, (847) 240-1737; fax (847) 330-8907; jmueller@aad.org
Internet: http://www.campdiscovery.org/
Sponsor: American Academy of Dermatology
930 E Woodfield Road
Schaumburg, IL 60173

American Academy of Dermatology Shade Structure Grants 499

The program awards grants to support the purchase of shade structures designed to provide shade and ultraviolet radiation protection for outdoor areas. Locations can include any area where children and adults gather and are exposed to the harmful rays of the sun, such as playgrounds, pools, bleachers, and eating or recreation areas.

Requirements: The program is open to non-profit organizations or educational institutions that serve children and teenagers, ages 18 and younger. Applicants will be reviewed based on the following: demonstrated commitment to sun safety within the organization and community; sponsorship of application by an academy member dermatologist; and ability to meet the build timeline outlined in grant criteria. Application and guidelines are available online.

Restrictions: Faxed nor emailed applications will not be accepted.

Date(s) Application is Due: Apr 12

Amount of Grant: Up to 8,000 U.S.D

Samples: Amelia Earhart Elementary, Goddard, Kan.; Anthony Wayne Rehabilitation Center for Handicapped and Blind Inc., Fort Wayne, Ind.; Bear Exploration Center, Montgomery, Ala..

Contact: Jennifer Allyn, Program Director; (847) 240-1730; jallyn@aad.org

Internet: http://www.aad.org/public/sun/grants.html

Sponsor: American Academy of Dermatology
930 E Woodfield Road
Schaumburg, IL 60173

American Academy of Neurology Clinical Research Training Fellowship 500

This fellowship is a two-year award to support research in clinical neurosciences. The award supplies a stipend plus tuition coverage of formal education in clinical research methodology at the institution or elsewhere. Supplementation of the stipend by other grants is allowable.

Requirements: The fellowship is open to non-U.S. neurologists with interest in an academic career in clinical research who have completed residency training less than five years prior to application. Preference will be given to applicants who have completed one or two years subspecialty training in an area of clinical research interest. Only one application per department will be considered. The applicant must supply a letter of nomination from the chair of the institution's department of neurology; a letter of intent to pursue a two-year program indicating specific clinical focus, proposed institution and preceptor, and future goals; a three-page research proposal; a copy of current curriculum vita and three letters of reference; letter from proposed preceptor detailing any ongoing clinical research in the area of interest, manner the applicant will interact in the service and research components of the fellowship, assurance of restricted clinical responsibilities, and assumption of fellow's research costs; and documentation of formal coursework to include quantitative clinical epidemiology, biostatistics, study design, data analysis, and ethics. Ten copies of all materials must be submitted.

Date(s) Application is Due: Oct 3

Amount of Grant: 57,000 U.S.D

Contact: Martin Schaefer; (651) 695-2759; fax (651) 695-2791; foundation@aan.com or mschaefer@aan.com

Internet: http://www.neurofoundation.org/research/research.cfm

Sponsor: American Academy of Neurology
1080 Montreal Avenue
Saint Paul, MN 55116

American Academy of Neurology International Scholarship Program 501

The purpose of this scholarship is to enable a young investigator to attend the academy's annual meeting to deliver a platform or poster presentation at the scientific sessions. A stipend, travel and lodging expenses, and program admission fees are provided. Application forms are available on the Web site. Annual deadline dates may vary; contact program staff for exact dates.

Requirements: This scholarship is available to residents of countries other than the United States or Canada who are investigators in the field of neurology. Applications are especially welcome from qualified individuals in underdeveloped countries.

Restrictions: Foreign residents living in the United States or Canada are not eligible.

Date(s) Application is Due: Dec 1

Contact: Cheryl Alementi, (651) 695-2737; calementi@aan.com

Internet: http://www.aan.com/professionals/awards/award/awa_res_scho.cfm

Sponsor: American Academy of Neurology
1080 Montreal Avenue
Saint Paul, MN 55116

American Academy of Nursing Building Academic Geriatric Nursing Capacity (BAGNC) Scholarships 502

The Foundation's overall goal is to increase the nation's capacity to provide effective and affordable care to its rapidly increasing older population. Specifically, the Foundation seeks to enhance the training of physicians, nurses, social workers and other health professionals who care for older adults, and promote innovations in the integration and delivery of services. The goal of the Scholarship program is to increase academic geriatric nursing capacity in the United States. BAGNC focuses on the development of academic leadership in gerontological nursing through strong mentorship in the components of academic geriatric nursing (research, teaching and community service); leadership development, a national network of scholars and academic geriatric nurses; and exposure to a wide range of experts in gerontology and geriatrics. Scholars in collaboration with their identified mentor will design and implement a tailored professional development plan designed to support development of new competencies and enhanced effectiveness as an academic leader.

Requirements: Registered nurses who are U.S. citizens or U.S. permanent residents and who hold a degree(s) in nursing are eligible. Predoctoral applicants must: be registered nurses; hold degree(s) in nursing; be United States citizens or permanent U.S. residents; plan an academic and research career; and demonstrate potential for long-term contributions to geriatric nursing.

Date(s) Application is Due: Jan 15

Amount of Grant: Up to 50,000 U.S.D

Contact: Patricia Archbold; (202) 777-1172; fax (202) 777-0107; parchbold@aannet.org

Internet: http://www.aannet.org/i4a/pages/Index.cfm?pageID=3295

Sponsor: American Academy of Nursing
888 17th Street NW, Suite 800
Washington, DC 20006

American Academy of Nursing Claire M. Fagin Fellowships 503

The Fellowship supports two years of full time advanced research and leadership training for doctorally prepared faculty committed to careers in academic geriatric nursing by providing $120,000 for the 2-year fellowship ($60,000 per annum). Program focuses on the development of academic leadership in gerontological nursing through such activities as: research; focused study; networking among scholars, mentors and colleagues in other fields; demonstration of growth in ability to transform self and organizations by moving outside of traditional modes of success; completion and write-up of a significant research project; and by success in achieving funding from other sources. Selected fellows, in collaboration with their mentor, will design and implement an individual professional development plan that will support them in developing new competencies and enhanced effectiveness as an academic leader and researcher. Award programs must begin between July 1st and September 1st of the award year. The program is committed to advancing well-qualified applicants from under-represented minority groups to improve the nation's ability to provide culturally competent care to its increasingly diverse aging population.

Requirements: Applicants must: be doctorally-prepared registered nurses; hold degree(s) in nursing; be United States citizens or permanent U.S. residents; be doctorally-prepared registered nurses,; have the potential to develop into independent investigators; and demonstrate potential for long-term contributions to geriatric nursing. Applications will be accepted from doctoral students who will complete their doctoral program prior to the award. Faculty members in accredited Schools of Nursing who hold the rank of assistant professor or associate professor may apply for fellowships.

Date(s) Application is Due: Jan 13

Amount of Grant: 120,000 U.S.D

Contact: Patricia Archbold, Program Director; (202) 777-1172; fax (202) 777-0107; parchbold@aannet.org

Internet: http://www.geriatricnursing.org/applications/cmf-fellowship.asp

Sponsor: American Academy of Nursing
888 17th Street NW, Suite 800
Washington, DC 20006

American Academy of Nursing Mayday Fund Grants 504

The Foundation's overall goal is to increase the nation's capacity to provide effective and affordable care to its rapidly increasing older population. Specifically, the Foundation seeks to enhance the training of physicians, nurses, social workers and other health professionals who care for older adults, and promote innovations in the integration and delivery of services. This program is aimed at candidates whose research includes the study of pain in the elderly. Award programs must begin between July 1st and September 1st.

Requirements: Predoctoral applicants must: be registered nurses; hold degree(s) in nursing; be United States citizens or permanent U.S. residents; plan an academic and research career; and demonstrate potential for long-term contributions to geriatric nursing.

Date(s) Application is Due: Jan 9

Amount of Grant: Up to 5,000 U.S.D

Contact: Patricia Archbold, Program Director; (202) 777-1172; fax (202) 777-0107; parchbold@aannet.org

Internet: http://www.geriatricnursing.org/applications/applications.asp

Sponsor: American Academy of Nursing
888 17th Street NW, Suite 800
Washington, DC 20006

American Academy of Nursing MBA Scholarships 505

The Program seeks nurse applicants pursuing a Masters of Business Administration (including Executive MBA) at a highly ranked school of business, and committed to a career focus on the management/leadership of institutions serving older persons. The Scholarship offers grant-funded scholarship support of $50,000 over the period of one or two years to successful candidates who will commence or continue full-time in an MBA program. The Scholarship program includes additional leadership development experiences.

Requirements: Scholarship applicants must: be registered nurses; hold degree(s) in nursing; and be United States citizens or permanent U.S. residents. Applicants must be accepted to, or be currently enrolled in, a highly ranked business school. Applicants must provide specific examples of ability to understand organizational and management challenges as well as strategic approaches to such challenges and provide evidence of previous commitment to gerontology.

Date(s) Application is Due: Jan 15

Amount of Grant: Up to 50,000 U.S.D

Contact: Patricia Archbold, Program Director; (202) 777-1172; fax (202) 777-0107; parchbold@aannet.org

Internet: http://www.geriatricnursing.org/applications/mba-scholarship.asp

Sponsor: American Academy of Nursing
888 17th Street NW, Suite 800
Washington, DC 20006

American Academy of Nursing Media Awards 506

The Award is presented to the individual(s) or organization(s) whose use of media has highlighted the unique contribution of nursing and: increased public awareness of the value nursing plays in promoting health and providing health care to the public and/or the need to address the growing nursing shortage; reported the impact of specific public policies on the health status of individuals, communities or the general population; been responsible for motivating specific actions to improve healthcare for diverse groups of people; disseminated nursing research findings nationally (and possibly internationally) using both interdisciplinary and public media; and depicted specific examples of health-enhancing interactions, culturally sensitive health care, health care addressing health disparity issues of under-served and vulnerable populations, health promoting activities or healing actions. Examples of items that may be submitted for the Academy's Media Awards Program include, but are not limited to: newspaper articles, radio programs, motion pictures, public service announcements, books, feature films, television shows, magazine articles, novels, documentaries, popular literature, creative endeavors, and multimedia. Entries are encouraged that highlight nursing practice and/or nursing issues.
Requirements: A non-refundable entry fee of $50.00 is required.
Date(s) Application is Due: Jul 27
Contact: Patricia Archbold, Program Director; (202) 777-1172; fax (202) 777-0107; parchbold@aannet.org
Internet: http://www.aannet.org
Sponsor: American Academy of Nursing
888 17th Street NW, Suite 800
Washington, DC 20006

American Chemical Society Award for Creative Advances in Environmental Science and Technology 507

The purpose of the Award is to encourage creativity in research and technology or methods of analysis to provide a scientific basis for informed environmental control decision-making processes, or to provide practical technologies that will reduce health risk factors. The award consists of $5,000 and a certificate. Up to $2,500 for travel expenses to the meeting at which the award will be presented will be reimbursed.
Requirements: Any individual, except a member of the award committee, may submit one nomination or seconding letter for the award in any given year. The nominating documents consist of a letter of not more than 1000 words containing an evaluation of the nominee's accomplishments and a specific identification of the work to be recognized, a biographical sketch including date of birth, and a list of publications and patents authored by the nominee.
Restrictions: The award will be granted without regard to age or nationality. Self-nominations are not accepted.
Date(s) Application is Due: Nov 1
Amount of Grant: 5,000 - 7,500 U.S.D
Contact: Felicia Dixon, Awards Administrator; (800) 227-5558, ext. 6283 or (202) 872-4408; fax (202) 776-8008; f_dixon@acs.org or awards@acs.org
Internet: http://portal.acs.org/portal/acs/corg/content?_nfpb=true&_pageLabel=PP_ARTICLEMAIN&node_id=1319&content_id=CTP_004504&use_sec=true&sec_url_var=region1&__uuid=ed529e7a-2f63-4c81-8359-960558b43759
Sponsor: American Chemical Society
1155 Sixteenth Street, NW
Washington, DC 20036-4801

American Chemical Society Award in Separations Science and Technology 508

This award is given annually to recognize outstanding accomplishments in fundamental or applied research directed to separations science and technology. The award shall be granted to an individual without regard to age or nationality. The scope of the award is to be as broad as possible, covering all fields where separations science and technology is practiced, including but not limited to biology, chemistry, engineering, geology, and medicine. The award consists of $5,000 and a certificate. Up to $2,500 for travel expenses to the meeting at which the award will be presented will be reimbursed. The recipient will deliver a lecture at the annual Division of Industrial and Engineering Chemistry Separations Science and Technology Symposium.
Requirements: Any individual, except a member of the award committee, may submit one nomination or seconding letter for the award in any given year. The nominating documents consist of a letter of not more than 1000 words containing an evaluation of the nominee's accomplishments and a specific identification of the work to be recognized, a biographical sketch including date of birth, and a list of publications and patents authored by the nominee.
Restrictions: Self-nominations are not accepted.
Date(s) Application is Due: Nov 1
Amount of Grant: 5,000 - 7,500 U.S.D
Samples: Abraham M. Lenhoff—award recipient, $5,000; Allan S. Myerson—award winner, $5,000; William H. Pirkle—award winner, $5,000.
Contact: Felicia Dixon, Awards Administrator; (800) 227-4575 or (202) 872-4408; fax (202) 776-8008; f_dixon@acs.org or awards@acs.org
Internet: http://portal.acs.org/portal/acs/corg/content?_nfpb=true&_pageLabel=PP_ARTICLEMAIN&node_id=1319&content_id=CTP_004552&use_sec=true&sec_url_var=region1
Sponsor: American Chemical Society
1155 Sixteenth Street, NW
Washington, DC 20036-4801

American Chemical Society Claude S. Hudson Awards 509

This award, supported by the National Starch and Chemical Company, is given biennially in odd-numbered years to recognize outstanding contributions to carbohydrate chemistry, whether in education, research, or application. The award is granted without regard to age or nationality. Applications are accepted in even-numbered years. The award consists of $5,000 and a certificate. Up to $1,000 for travel expenses to the meeting at which the award will be presented will be reimbursed.
Requirements: Any individual, except a member of the award committee, may submit one nomination or seconding letter for the award in any given year. Nominating documents consist of a letter of not more than 1000 words containing an evaluation of the nominee's accomplishments and a specific identification of the work to be recognized, a biographical sketch including date of birth, and a list of publications and patents authored by the nominee.
Restrictions: Self-nominations are not accepted.
Date(s) Application is Due: Nov 1
Amount of Grant: 6,000 U.S.D
Samples: Peter H. Seeberger—award recipient,; Pierre Sinay—award recipient,; David Bundle—award recipient,.
Contact: Felicia Dixon, Awards Administrator; (202) 872-4408 or (202) 872-4408; fax (202) 776-8008; f_dixon@acs.org or awards@acs.org
Internet: http://portal.acs.org/portal/acs/corg/content?_nfpb=true&_pageLabel=PP_ARTICLEMAIN&node_id=1319&content_id=CTP_004501&use_sec=true&sec_url_var=region1
Sponsor: American Chemical Society
1155 Sixteenth Street, NW
Washington, DC 20036-4801

American Chemical Society GCI Pharmaceutical Roundtable Research Grants 510

The Roundtable is a partnership with the ACS GCI (Green Chemistry Institute) and pharmaceutical corporations united by a shared commitment to integrate the principles of green chemistry and engineering into the business of drug discovery and production. The program was created to support research in the key green chemistry research areas. (Ten chemical transformations and two process related operations were identified and subsequently published - download the list from the sponsor's website). The primary purpose of this grant is to publish research to make information publicly available. One grant is planned to be awarded. The total award is limited to $150,000 for a grant period of 12 to 24 months.
Requirements: Proposals will only be accepted from public and private institutions of higher education. The grant is not limited to institutions in the United States. Proposals must be submitted through the appropriate institutional office for external funding. Proposed research must focus on the key targeted areas identified by the Roundtable. Proposals must clearly describe how the proposed research will provide improvements over existing technology that are consistent with the goals of Green Chemistry including improved atom economy, use of less hazardous reagents, more energy efficient processes, elimination of by-products, reduced waste generation, improved environmental, health, and safety and life cycle impacts, catalytic versus stoichiometric methods, and development of inherently safer transformations. Illustrations of the relevance of the proposed technology to synthetic and process related problems within the pharmaceutical industry are encouraged. Complete guidelines can be downloaded at the sponsor's website.
Restrictions: Applicants may have only one research grant with the ACS GCI Pharmaceutical Roundtable at a time. Current research grant holders may not apply for a new grant until their active grant or no-cost time extension officially expires.
Date(s) Application is Due: Sep 14
Amount of Grant: Up to 150,000 U.S.D
Contact: Gayle Peterman, Finance and Grants Manager; (202) 872-6092 or (202) 872-4481; fax (202) 872-6319; g_peterman@acs.org or gcipr@acs.org
Internet: http://portal.acs.org/portal/acs/corg/content?_nfpb=true&_pageLabel=PP_TRANSITIONMAIN&node_id=1456&use_sec=false&sec_url_var=region1
Sponsor: American Chemical Society
1155 Sixteenth Street, NW
Washington, DC 20036-4801

American College of Rheumatology Fellows-in-Training Travel Scholarships 511

The American College of Rheumatology Fellows-in-Training (FIT) Travel Scholarship is designed to provide funding to assist Fellows- In-Training at U.S. and Canadian programs in rheumatology attend the Annual Meeting. Preference will be given to those applicants who are currently a first or second year adult or first, second or third year pediatric fellows in a training program approved by the Accreditation Council of Graduate Medical Education or the Royal College of Physicians and Surgeons of Canada. Applicants in a third year of an adult or fourth year of a pediatric fellowship will be given consideration after the registration deadline, space permitting on a first come first serve basis.
Requirements: Applicant must be first or second year adult or first, second or third year pediatric fellows in a training program approved by the Accreditation Council of Graduate Medical Education or the Royal College of Physicians and Surgeons of Canada. Applicants must be in an accredited fellowship position at the time of the Annual Meeting.
Restrictions: Individuals with a faculty appointment will not be considered.
Date(s) Application is Due: Jul 24
Contact: Fellowship Coordinator; (404) 633-3777; fax (404) 633-1870; fittravel@rheumatology.org
Internet: http://www.rheumatology.org/fellows/fitscholarship.asp
Sponsor: American College of Rheumatology
1800 Century Place, Suite 250
Atlanta, GA 30345-4300

American College of Surgeons and The Triological Society Clinical Scientist Development Awards 512

The Triological Society (TRIO) and the American College of Surgeons (ACS) announce a competitive research career development program that will provide supplemental funding to otolaryngologist-head and neck surgeons who receive NIH Mentored Clinical Scientist Development Awards (K08) or Mentored Patient-Oriented Research Development Awards (K23). The TRIO and ACS are offering these awards as a means to facilitate the research career development of otolaryngologists-head and neck surgeons, with the expectation that awardees will have sufficient pilot data to submit a competitive R01 proposal prior to the conclusion of the K awards. Awards will use the K08 or K23 mechanisms. Planning, direction, and execution of the program will be the responsibility of the candidate and her/his mentor on behalf of the applicant institution. The project period will be for a period of up to five years with a minimum of three years. Awards are not renewable. TRIO and ACS will provide salary and fringe benefits for the NIH K08/K23 recipient. A combined total of $80,000 per year in direct costs will be awarded to the institution of the awardee.

Requirements: Applicants must be MD otolaryngologist-head and neck surgeons and members of the American College of Surgeons. Applicants must meet all requirements as set forth in the K08 and K23 application criteria and must be recipients of a new K08 or K23 award or have an existing K08/K23 award with a minimum of three years remaining in the funding period as of June.

Restrictions: Pre-existing applications and awards are not eligible for consideration.

Date(s) Application is Due: Apr 15

Amount of Grant: 80,000 U.S.D

Contact: Kate Early, Administrative Assistant; (312) 202-5281; fax (312) 202-5021; kearly@facs.org

Internet: http://www.facs.org/memberservices/acstriol.html

Sponsor: American College of Surgeons
633 N Saint Clair Street
Chicago, IL 60611-3211

American College of Surgeons Australia and New Zealand Chapter Travelling Fellowships 513

The International Relations Committee of the American College of Surgeons announces the availability of a traveling fellowship, the Australia and New Zealand Chapter of the American College of Surgeons Travelling Fellowship. The purpose of this fellowship is to encourage international exchange of information concerning surgical science, practice, and education and to establish professional and academic collaborations and friendships. The Chapter and the College will provide the sum of $8,000 to the successful applicant, who will also be exempted from registration fees for the Annual Scientific Congress. He/she must meet all travel and living expenses.

Requirements: The scholarship is available to a Fellow of the American College of Surgeons in any of the surgical specialties who meets the following *Requirements:* has a major interest and accomplishment in clinical and basic science related to surgery; holds a current full-time academic appointment in Canada or the United States; is under 45 years of age on the date the application is filed; and is enthusiastic and personable and possesses good communication skills. The Fellow is required to spend a minimum of two weeks in Germany. The Fellow is required to spend a minimum of two or three weeks in Australia and New Zealand.

Date(s) Application is Due: Nov 16

Amount of Grant: 8,000 U.S.D

Contact: Kate Early, Administrative Assistant; (312) 202-5281; fax (312) 202-5021; kearly@facs.org

Internet: http://www.facs.org/memberservices/traveling.html

Sponsor: American College of Surgeons
633 N Saint Clair Street
Chicago, IL 60611-3211

American College of Surgeons Co-Sponsored K08/K23 NIH Supplement Awards 514

The American College of Surgeons announces a program that will provide supplemental funding to up to five individuals who receive a Mentored Clinical Scientist Development Award (K08/K23). It is directed at surgeon-scientists working in the early stages of their research careers. The award requires co-sponsorship with an approved surgical society of a three, four or five year period of supervised research experience that may integrate didactic studies with laboratory or clinically based research. This award program will offer a means to facilitate the career development of individuals pursuing careers in surgical research by enhancing salary support over and above that offered by the K08/K23 mechanism. The application deadline is June 12, with funding to begin July 1.

Requirements: Awardees must be members in good standing of both the College and a co-sponsoring surgical society. Participating surgical societies include American Association of Plastic Surgeons (AAPS), American Head and Neck Society (AHNS), American Society of Transplant Surgeons (ASTS), American Vascular Association (AVA), Society of Gynecologic Oncologists (SGO), Society of University Surgeons (SU. S.), and Thoracic Surgery Foundation for Research and Education (TSFRE).

Date(s) Application is Due: Jun 12

Contact: Kate Early, Administrative Assistant; (312) 202-5281; fax (312) 202-5021; kearly@facs.org

Internet: http://www.facs.org/memberservices/acs-nih.html

Sponsor: American College of Surgeons
633 N Saint Clair Street
Chicago, IL 60611-3211

American College of Surgeons Faculty Career Development Award for Neurological Surgeons 515

The American College of Surgeons (ACS) and the Neurosurgery Research and Education Foundation of the American Association of Neurological Surgeons (NREF-AANS) are offering a two-year faculty career development award to neurological surgeons. The award is to support the establishment of a new and independent research program in an area of neurological surgery. The award is $40,000 per year for each of two years, to support the research, and is not renewable thereafter.

Requirements: The award is open to surgeons who are members or candidate members in good standing of both the American College of Surgeons (ACS) and the American Association of Neurological Surgeons (AANS); and have completed specialty training within the preceding five years and have received a full-time faculty appointment at a medical school accredited by the Liaison Committee on Medical Education in the United States or by the Committee for Accreditation of Canadian Medical Schools in Canada.

Restrictions: Applicants may not be current recipients of major research grants.

Date(s) Application is Due: Dec 1

Amount of Grant: 40,000 U.S.D

Contact: Michele S. Gregory, Director of Development; (847) 378-0540 or (847) 378-0500; fax (847) 378-0600; msg@aans.org or info@aans.org

Internet: http://www.facs.org/memberservices/acsnrefaans07.pdf

Sponsor: American College of Surgeons
633 N Saint Clair Street
Chicago, IL 60611-3211

American College of Surgeons Faculty Research Fellowships 516

The American College of Surgeons is offering two-year faculty research fellowships to surgeons entering academic careers in surgery or a surgical specialty. The fellowship is to assist a surgeon in the establishment of a new and independent research program. Applicants should have demonstrated their potential to work as independent investigators. The fellowship award is $40,000 per year for each of two years, to support the research.

Requirements: The fellowship is restricted to surgeons who have recently (usually within three years) completed formal surgical education and entered the field of full-time academic surgery and received a faculty appointment in a department of surgery, or one of the surgical specialties, at an approved medical school in the United States or Canada. Preference will be given to applicants who directly enter academic surgery following residency or fellowships. Approval of the application is required from the administration (dean or fiscal officer) and the head of the department under whom the recipient will be studying during the fellowship.

Date(s) Application is Due: Nov 2

Amount of Grant: 40,000 U.S.D

Contact: Kate Early, Administrative Assistant; (312) 202-5281; fax (312) 202-5021; kearly@facs.org

Internet: http://www.facs.org/memberservices/acsfaculty.html

Sponsor: American College of Surgeons
633 N Saint Clair Street
Chicago, IL 60611-3211

American College of Surgeons Health Policy Scholarships 517

The American College of Surgeons, in partnership with a number of other organizations, is offering annual scholarships to subsidize attendance and participation in the Executive Leadership Program in Health Policy and Management at Brandeis University. The awards are in the amount of $8,000, to be used toward the cost of tuition, travel, housing, and subsistence during the period of the course. The closing date for receipt of all applications is February 2. All applicants will be notified of the outcome of the selection process by March 31.

Requirements: The award is open to surgeons who are members in good standing of both the American College of Surgeons (ACS) and the specific partnering organization for each award. These include: the American Association of Neurological Surgeons (AANS); the American Academy of Ophthalmology (AAO); the American Academy of Otolaryngology, Head and Neck Surgery Foundation (AAO-HNS); the American Association for the Surgery of Trauma (AAST); the American Pediatric Surgical Association (APSA); the American Surgical Association (ASA); the American Society of Breast Surgeons (ASBrS); the American Society of Colon and Rectal Surgeons (ASCRS); the American Society of Plastic Surgeons (ASPS); the American Urogynecologic Society (AUGS); the Society of Thoracic Surgeons (STS); and the Society for Vascular Surgery (SVS). Applicants must be at least 30 years old, but under 55, on the date that the completed application is filed.

Restrictions: Indirect costs are not paid to the recipient or to the recipient's institution.

Date(s) Application is Due: Feb 2

Amount of Grant: 8,000 U.S.D

Contact: Michele S. Gregory, Director of Development; (847) 378-0540 or (847) 378-0500; fax (847) 378-0600; msg@aans.org or info@aans.org

Internet: http://www.facs.org/memberservices/acsaans.html

Sponsor: American College of Surgeons
633 N Saint Clair Street
Chicago, IL 60611-3211

American College of Surgeons Health Policy Scholarships for General Surgeons 518

The American College of Surgeons is offering an annual scholarship to subsidize attendance and participation in the Executive Leadership Program in Health Policy and Management at Brandeis University. The award is in the amount of $8,000, to be used toward the cost of tuition, travel, housing, and subsistence during the period of the course. The closing date for receipt of applications is February 2. All applicants will be notified of the outcome of the selection process by March 31

Requirements: The award is open to general surgeons who are members in good standing of the American College of Surgeons (ACS). Applicants must be at least 30 years old,

but under 55, on the date that the completed application is filed. The scholarship must be used in the year for which it is designated.
Restrictions: Indirect costs are not paid to the recipient or to the recipient's institution.
Date(s) Application is Due: Feb 2
Amount of Grant: 8,000 U.S.D
Contact: Administrator, Scholarships Division; (800) 621-4111 or (312) 202-5000; fax (312) 202-5001
Internet: http://www.facs.org/memberservices/acshealthpolicy.html
Sponsor: American College of Surgeons
633 N Saint Clair Street
Chicago, IL 60611-3211

American College of Surgeons International Guest Scholarships 519
The American College of Surgeons offers International Guest Scholarships to competent young surgeons from countries other than the United States or Canada who have demonstrated strong interests in teaching and research. The scholarships, in the amount of $8,000 each, provide the Scholars with an opportunity to visit clinical, teaching, and research activities in North America and to attend and participate fully in the educational opportunities and activities of the American College of Surgeons Clinical Congress.
Requirements: Applicants must: be graduates of schools of medicine; be at least 35 years old, but under 45, on the date that the completed application is filed; submit their applications from their intended permanent location; and have demonstrated a commitment to teaching and/or research in accordance with the standards of the applicant's country.
Date(s) Application is Due: Jul 1
Amount of Grant: 8,000 U.S.D
Contact: Kate Early, Administrative Assistant; (312) 202-5281; fax (312) 202-5021; kearly@facs.org
Internet: http://www.facs.org/memberservices/igs.html
Sponsor: American College of Surgeons
633 N Saint Clair Street
Chicago, IL 60611-3211

American College of Surgeons Nizar N. Oweida, MD, FACS, Scholarships 520
The Board of Governors of the American College of Surgeons announces the availability of a scholarship for young rural surgeons, the Nizar N. Oweida, MD, FACS, Scholarship of the American College of Surgeons. The Scholarship provides an award of $5,000 to subsidize the participation of a young rural-based Fellow or Associate Fellow in attendance at the annual Clinical Congress of the American College of Surgeons. This amount is to be used to help defray travel expenses for the Clinical Congress, postgraduate course fees, hotel costs, and per diem expenses during the Clinical Congress.
Requirements: The Oweida Scholarship is available to a member of the American College of Surgeons in any of the surgical specialties who meets the following *Requirements:* serves a rural community in the U.S. or Canada; is a Fellow or Associate Fellow in good standing; and is under 45 years of age on the date the application is filed.
Date(s) Application is Due: Dec 15
Amount of Grant: 5,000 U.S.D
Contact: Kate Early, Administrative Assistant; (312) 202-5281; fax (312) 202-5021; kearly@facs.org
Internet: http://www.facs.org/memberservices/oweida.html
Sponsor: American College of Surgeons
633 N Saint Clair Street
Chicago, IL 60611-3211

American College of Surgeons Resident and Associate Society Leadership Scholarships 521
The Resident and Associate Society of the American College of Surgeons (RAS-ACS) is accepting applications for its leadership scholarship for Resident Members and Associate Fellows of the College. The RAS Scholarship will be awarded competitively to young surgeons who exemplify an important goal the RAS has set for itself—developing future leaders in the field of surgery. The award is for the purpose of allowing the young surgeon awardee to attend one of several courses that will be offered for this scholarship. The courses are intended to engage young surgeons and expose them to opportunities to improve their abilities in areas such as leadership, communications skills, and research, skills that are necessary to succeed as leaders in surgery.
Date(s) Application is Due: Jan 30
Contact: Peg Haar, Division of Member Services; (312) 202-5312; phaar@facs.org
Internet: http://www.facs.org/ras-acs/resources/scholarship.html
Sponsor: American College of Surgeons
633 N Saint Clair Street
Chicago, IL 60611-3211

American College of Surgeons Resident Research Scholarships 522
ACS scholarships will be awarded each year to encourage residents to pursue careers in academic surgery. Priority will be given to residents beginning full-time investigative activities, with no clinical responsibilities, for the two-year period of the scholarship. Study outside the United States or Canada is permissible. The scholarship is for the personal use of the recipient and is not to diminish or replace the usual or expected compensation; award is made directly to the scholar and not to the institution. Renewal for the second year is contingent upon acceptable progress and study protocol for the second year.
Requirements: The applicant must have completed two postdoctoral years in an accredited surgical training program in the United States or Canada at the time the scholarship is awarded and shall not complete formal residency training before the end of the scholarship. Approval of the application is required from the administration (dean or fiscal officer) and the head of the department under whom the recipient will be studying.
Restrictions: Only in exceptional circumstances will more than one ACS scholarship be granted in a single year to applicants from the same institution.
Date(s) Application is Due: Sep 1
Amount of Grant: 30,000 U.S.D
Contact: Administrator, Scholarships Division; (800) 621-4111 or (312) 202-5000; fax (312) 202-5001
Internet: http://www.facs.org/memberservices/acsresident.html
Sponsor: American College of Surgeons
633 N Saint Clair Street
Chicago, IL 60611-3211

American College of Surgeons Wound Care Management Award 523
The American College of Surgeons is offering a two-year faculty research award through KCI U.S.A, to a general surgeon engaged in a research project addressing wound care management. The purpose of this fellowship will be to acquire knowledge leading to new clinical applications or projects that will provide the medical community with a better understanding of the use of advanced wound healing therapies. The fellowship award is $85,000 per year, with the possibility of an extension for a second year if satisfactory progress is made. Specifically, translational projects such as new methods to mechanically stimulate wound healing, design of clinically applicable skin substitutes, methods to enhance specific, difficult-to-heal wounds related to general or extremity trauma, various soft-tissue injuries, quality limb salvage, or methods to reduce surgical intervention to heal a variety of wounds, are being sought. This award may be used by the recipient for support of his/her research or academic enrichment in any fashion that the recipient deems maximally supportive of his/her investigations. This may include faculty salary replacement, support for clinical research personnel, supplies directly related to the research activity, consumable equipment, and research related travel costs up to $2,000 per year.
Requirements: The fellowship is open to Fellows or Associate Fellows of the College who have: (1) completed the chief residency year or accredited fellowship training within the preceding 10 years; and (2) received a fulltime faculty appointment in a department of surgery or a surgical specialty at a medical school accredited by the Liaison Committee on Medical Education in the United States or by the Committee for Accreditation of Canadian Medical Schools in Canada. Preference will be given to applicants who directly enter academic surgery following residency or fellowship.
Date(s) Application is Due: May 1
Amount of Grant: 85,000 U.S.D
Contact: Administrator, Scholarships Division; (800) 621-4111 or (312) 202-5000; fax (312) 202-5001
Internet: http://www.facs.org/memberservices/woundcareposter2008.pdf
Sponsor: American College of Surgeons
633 N Saint Clair Street
Chicago, IL 60611-3211

American Dietetic Association Foundation Scholarships 524
As the largest provider of dietetic scholarships, the program awards funds to deserving students at all levels of study. Additional funds are granted in the form of continuing education stipends, research grants, and recognition awards to practicing dietitians and nutritionists. Additional program information is available online. Application deadlines vary, and applications are typically made available four to six months before the deadline. To receive an application for a particular award or for more information contact the office.
Requirements: Applicant must be a U.S. citizen. Scholarships are awarded in the field of dietetics at all levels: undergraduate, graduate, internship, technician, and for preprofessional practice programs.
Amount of Grant: 500 - 5,000 U.S.D
Contact: Grants Administrator; (800) 877-1600, ext. 5400; education@eatright.org
Internet: http://www.eatright.org/Public/7772.cfm
Sponsor: American Dietetic Association Foundation
120 S Riverside Plaza, Suite 2000
Chicago, IL 60606

American Electric Power Grants 525
The corporate contributions program awards grants to nonprofit organizations in company-operating areas. Contributions are made principally in the areas of education, the environment, and human services (i.e., hunger, housing, health, safety). Priority is based on the perceived overall benefit to communities in the company's service area. In the area of education, preference is given to grades pre-K through 12 in the fields of science, technology and math. Multiyear commitments to capital campaigns generally do not exceed five years. AEP and its employees provide strong support to many annual United Way campaigns within its service territory; therefore, additional support to United Way agencies is extremely limited. There is no formal application form or deadline. Each request should include evidence of tax-exempt status; amount of funds requested; description of intended use of requested funds, who will benefit, and how they will benefit; overall budget of related program; and description of how the organization will measure results and report outcome of the use of funds being requested. Written requests for local or statewide projects may be submitted to the appropriate AEP state office. Electronic requests in MS Word format may be submitted to the corresponding email address. Guidelines are available online.
Requirements: 501(c)3 nonprofit organizations in Arkansas, Indiana, Kentucky, Louisiana, Michigan, Ohio, Oklahoma, Tennessee, Texas, Virginia and West Virginia are eligible. Grant-seekers should first approach their local AEP operating company.
Restrictions: Grants are not awarded to religious, fraternal, service, and veteran organizations, except for nonsectarian social service activities available to the broader

community; organizations with a purpose that is solely athletic in nature; or to individuals.
Geographic Focus: Arkansas, Indiana, Kentucky, Louisiana, Michigan, Ohio, Oklahoma, Tennessee, Texas, Virginia, West Virginia
Contact: Contributions Administrator; (614) 716-1000; mkwalsh@aep.com or mkwalsh@aep.com
Internet: http://www.aep.com/about/community/corpGive/Default.htm
Sponsor: American Electric Power
1 Riverside Plaza, 19th Floor
Columbus, OH 43215

American Express Foundation Community Service Grants 526
Whether it is feeding the hungry, mentoring students, building homes for the homeless or cleaning up the environment, tens of thousands of American Express employees serve their communities through volunteerism and personal financial contributions, and the Foundation views this activity as an extension of the service ethic that lies at the heart of its business. It encourages good citizenship by supporting organizations that cultivate meaningful opportunities for civic engagement by employees and members of the community. The Foundation also serves its communities by supporting immediate and long-term relief and recovery efforts to help victims of natural disasters. Funding also goes to support preparedness programs that allow relief agencies to be better equipped in responding to emergencies as they occur.
Requirements: Eligible organizations must: certify tax-exempt status under Section 501(c)3 and 509(a)1, 2 or 3 of the U.S. Internal Revenue Code. Organizations outside the U.S. must be able to document not-for-profit status.
Restrictions: The program does not fund: individual needs, including scholarships, sponsorships and other forms of financial aid; fund-raising activities, such as galas, benefits, dinners and sporting events; goodwill advertising, souvenir journals or dinner programs; travel for individuals or groups; sectarian activities of religious organizations; political causes, candidates, organizations or campaigns; or books, magazines or articles in professional journals.
Geographic Focus: All States, All Countries
Samples: Association for Persons with Special Needs, Singapore - volunteers were engaged in the development of the center and as mentors to its clients (2011); Feeding America, Chicago, Illinois - funds the expansion of its Store Donation Program which recovers perishable products that do not meet retailers' marketing standards (2011).
Contact: Timothy McClimon, President; (212) 640-5661; fax (212) 693-1033
Internet: http://about.americanexpress.com/csr/comm_serv.aspx
Sponsor: American Express Foundation
200 Vesey Street, 48th Floor
New York, NY 10285-4804

American Foodservice Charitable Trust Grants 527
American Foodservice Charitable Trust provides funding in the following areas of interest: Christian agencies & churches; community/economic development; higher education; hospitals (general) and; youth development. The types of support available are: general/operating support; program development; scholarship funds. Giving is primarily available in the areas of: King of Prussia, Pennsylvania; Thomasville, Georgia; and Fort Worth, Texas.
Requirements: Qualifying IRA 501(c)3 nonprofit organizations are eligible to apply. There's no application form nor is there a deadline date to adhere to when submitting a proposal to the Foundation.
Restrictions: No grants to individuals.
Geographic Focus: Georgia, Pennsylvania, Texas
Amount of Grant: 200 - 8,000 U.S.D
Contact: Richard S. Downs, Trustee; (610) 933-9792
Sponsor: American Foodservice Charitable Trust
860 First Avenue, Suite 9A
King of Prussia, PA 19406-1404

American Foundation for Aging Research Fellowhips 528
AFAR awards fellowships to undergraduate, graduate and pre-doctoral (MD, PhD, DDS, etc.) students who are actively involved or are planning active involvement in a biomedical age-related research project. Awards are to alleviate the costs of tuition and fees at a United States degree granting institution. If such costs are already covered by another source of funds, other than personal funds, the awards are to be used for research purposes. Awards are made each semester and are renewable once. Preapplications may be submitted at any time; renewal requests should be made at least 30 days before the next semester.
Requirements: Applicants must be undergraduate, graduate and pre-doctoral (MD, PhD, DDS, etc.) students who are enrolled at degree programs at colleges or universitites in the United States.
Restrictions: Sociology, psychology, and health-professional related research such as physical therapy and exercise physiology are not currently being funded.
Amount of Grant: 500 - 1,000 U.S.D
Samples: Dr. Ralph Snyderman, Duke U Medical Ctr (Durham, NC)—for aging-related research, $1000.
Contact: Executive Secretary; (919) 515-5679; fax (919) 515-2047; afar_office@ncsu.edu
Internet: http://www4.ncsu.edu/unity/users/a/agris/afar/afar.htm
Sponsor: American Foundation for Aging Research
Campus Box 7622, North Carolina State University
Raleigh, NC 27695-7622

American Foundation for Suicide Prevention Grants 529
The research grants are awarded to investigators conducting clinical, biological, or psychosocial research on the problem of suicide, for one- or two-year periods, and for up to three years for research fellowships. Grant applicants compete in five categories: Pilot Grants provide seed money for new projects and are awarded to individual investigators without regard to academic rank or pervious research experience with suicide; Young Investigator Grants are awarded for two years of support to investigators at the level of assistant professor or lower; Standard Research Grants are awarded to individual investigators for a two-year period; Distinguished Researcher Awards are awarded to investigators at the level of associate professor or higher with a proven history of research in the area of suicide (to fund new directions and initiatives in suicidology); Postdoctoral Research Fellowships are awarded for full-time training projects by investigators who have received a PhD degree within the preceding three years and have not had more than three years of fellowship support. Application deadlines are December 15 for Young Investigator Grants, Standard Research Grants, Distinguished Researcher Awards, and Postdoctoral Fellowships; and April 15, August 15, and December 15 for Pilot Grants. Guidelines and application forms are available online.
Date(s) Application is Due: Apr 15; Aug 15; Dec 15
Amount of Grant: 20,000 - 100,000 U.S.D
Contact: Grants Administrator; (888) 333-2377 or (212) 363-3500; fax (212) 363-6237; bkoestner@afsp.org
Internet: http://www.afsp.org/research/grants.htm
Sponsor: American Foundation for Suicide Prevention
120 Wall Street, 22nd Floor
New York, NY 10005

American Indian Science and Engineering Scholarships 530
The A.T. Anderson Memorial Scholarship is awarded to Native American/Alaskan Native graduate or undergraduate students pursuing programs in the sciences, engineering, health-related fields, business, natural resources, and math and science secondary education. Awards are made for one academic year. Deadlines, guidelines, and application forms are available on the Web site.
Requirements: Applicants must be society members who are at least one-quarter Native American or recognized as a member of a tribe and full-time students at the undergraduate or graduate level.
Date(s) Application is Due: Jun 15
Amount of Grant: 5,000 U.S.D
Contact: Scholarship Department; (505) 765-1052, ext. 105; fax (505) 765-5608; scholarship@aises.org
Internet: http://www.aises.org/highered/scholarships/index.html
Sponsor: American Indian Science and Engineering Society
P.O. Box 9828
Albuquerque, NM 87119-9828

American Jewish World Service Grants 531
AJWS' grantmaking supports community-based organizations in the developing world that are undertaking holistic community development programs. These groups design and implement projects that creatively and effectively address economic development, education, healthcare, and sustainable agriculture. All of the initiatives also have strong components of strengthening civil society and/or promoting women's empowerment. Grantmaking links human rights and sustainable development.
Requirements: Grassroots organizations are eligible.
Restrictions: Grants are not awarded to individuals.
Amount of Grant: 3,000 - 30,000 U.S.D
Samples: Cross-Cultural Solutions (New Rochelle, NY)—to provide food to rescue workers and to coordinate volunteer efforts at the site of the World Trade Center, $5000; Independent Press Assoc (New York, NY)—to place advertisements in New York ethnic publications on the availability of disaster resources in the wake of the September 11 attacks, $25,500; Mount Sinai Hospital, Disaster Psychiatry Outreach (New York, NY)—to work with the New York City Department of Mental Health in its outreach efforts to counsel rescue workers and victims of the World Trade Center collapse, $10,000; Jewish Board of Family and Children Services (New York, NY)—to provide mental-health services, including drop-in centers, telephone hotlines, and on-site counseling in the wake of the September 11 attacks, $10,000.
Contact: Grants Administrator; (800) 889-7146 or (212) 792-2100; fax (212) 792-2930; grants@ajws.org
Internet: http://www.ajws.org/index.cfm?section_id=3
Sponsor: American Jewish World Service
45 W 36th Street, 10th Floor
New York, NY 10018-7904

American Legacy Foundation National Calls for Proposals Grants 532
The American Legacy Foundation provides grants to support innovations in tobacco control. National calls for proposals, such as the Priority Populations Initiative, are issued annually to address a variety of tobacco prevention and control issues. Comprehensive assistance and training are provided to the foundation's grantees as well as to state and local tobacco programs. Application forms, guidelines, and procedures are available online. There are no application deadlines.
Requirements: Funding is available only to state or local political subdivisions and legally constituted tax-exempt 501(c)3 organizations based in the 46 states, the District of Columbia, and five territories (American Samoa, Guam, Northern Mariana Islands, Puerto Rico, and the Virgin Islands) identified in the MSA with tobacco product manufacturers. An Indian reservation, Indian tribe, or tribal organization located within

the 46 settling states or a non-governmental entity that serves such a reservation may also apply for funding.
Restrictions: The foundation will not award grants to applicants that are in current receipt of grant monies or in-kind contributions from any tobacco manufacturer, distributor, or other tobacco-related entity.
Contact: Katherine Wilson, Assistant Vice President of Grants; (202) 454-5555; fax (202) 454-5599; grantsinfo@americanlegacy.org
Internet: http://www.americanlegacy.org/64.aspx
Sponsor: American Legacy Foundation
1724 Massachusetts Avenue NW
Washington, DC 20036

American Legacy Foundation Small Innovative Grants 533
The program supports projects that advance creative, promising solutions based on sound principles of tobacco control to remedy the harm caused by tobacco use in America. The program was created to seed new projects or enable an organization to pilot a new idea or approach. The proposed project must demonstrate an element of creativity, ingenuity or innovation and must distinguish itself from the large number of solid programs proposed to the foundation in each grant round.
Requirements: Funding is available only to state or local political subdivisions and legally constituted tax-exempt 501(c)3 organizations based in the 46 states, the District of Columbia, and five territories (American Samoa, Guam, Northern Mariana Islands, Puerto Rico, and the Virgin Islands) identified in the MSA with tobacco product manufacturers. An Indian reservation, Indian tribe, or tribal organization located within the 46 settling states or a nongovernmental entity that serves such a reservation may also apply for funding. Successful applications submitted under the Small Innovative Grants Program must: address one or both of Legacy's goals; demonstrate innovative or new tobacco prevention or cessation efforts; demonstrate a strong likelihood for a sustainable effort after the grant period; demonstrate that the project may be replicated; address the Healthy People 2010 risk reduction objectives related to tobacco use, and; incorporate the CDC's Best Practices for Comprehensive Tobacco Control Programs as appropriate. The foundation will give special consideration to applications addressing these current areas of interest.
Restrictions: The foundation will not award a grant to any applicant that is in current receipt of any grant monies or in-kind contribution from any tobacco manufacturer, distributor, or other tobacco-related entity. In addition, the foundation expects that a grantee will not accept any grant monies or in-kind contribution from any tobacco manufacturer, distributor, or other tobacco-related entity over the duration of the grant. Additionally, will not consider applications for: Projects focusing on youth prevention, cessation, activism, or education (up to 18 years old); Nicotine replacement therapy (NRT) and pharmaceuticals (as the sole or primary focus of the grant); Conference support (as the sole or primary focus of the grant); Media or marketing campaigns (as the sole or primary focus of the grant); Research projects EXCEPT community-based participatory research, which is allowed; Projects focusing on substances other than tobacco; Replication of an existing program; Expansion of an existing program; and Replacement funds; Grants to individuals, for religious activities, to build endowments, to support operating deficits, to retire debt, for capital purchases for building improvements, for construction, for lobbying, or for real estate purchase or development.
Date(s) Application is Due: Feb 27; Aug 8
Amount of Grant: 20,000 - 100,000 U.S.D
Contact: Katherine Wilson, Assistant Vice President of Grants; (202) 454-5555; fax (202) 454-5599; grantsinfo@americanlegacy.org
Internet: http://www.americanlegacy.org/2530.aspx
Sponsor: American Legacy Foundation
1724 Massachusetts Avenue NW
Washington, DC 20036

American Liver Foundation Innovative Hepatology Seed Grant 534
The goal of this program is to foster development of an imaginative research program in Alpha 1 Antitrypsin Deficiency and/or directly related areas of scientific investigation.
Requirements: To be eligible candidates must be a faculty member in an academic institution in the United States or its possessions. Applicants who hold research awards directed at salary support, such as Veterans Administration, Research Associates, Clinical Investigator, AGA Industry Research Awards, Glaxo Institute Awards, and ALF Liver Scholar Awards are also eligible.
Restrictions: Individuals with an RO1, Merit Review, or NIH FIRST Award are ineligible if their grant overlaps with the present application.
Date(s) Application is Due: Oct 1
Amount of Grant: 100,000 U.S.D
Contact: Joan Gallagher, Program Director; (973) 256-2550, ext. 224 or (800) 465-4837; jgallagher@liverfoundation
Internet: http://www.liverfoundation.org/db/grants/114
Sponsor: American Liver Foundation
1425 Pompton Avenue, Suite 3
Cedar Grove, NJ 07009

American Liver Foundation Liver Scholar Awards Program 535
The goal of the program is to permit scientists with liver research training to bridge the gap between completion of research training and attainment of status as independent research scientists. This additional research experience should enable scientists to compete for research grants from national sources, particularly NIH. The program aims to attract well-trained, basic scientists, who hold MD, PhD, or MD/PhD degrees, to a career in liver disease research. These awards are for the purpose of encouraging research into liver physiology and disease research and developing the potential of young, outstanding scientists; therefore, individuals who already are well established in the fields are not considered eligible. The awards will be made to eligible institutions and are for salary only. Each awardee must continue to hold a full-time faculty appointment in the sponsoring institution; devote at least 75 percent of time to their research project; attend the annual meeting of the American Association for the Study of Liver Diseases; acknowledge support by the ALF in any publications resulting from research performed during the tenure of the award; and submit an annual progress report describing results, their significance, and pending or planned applications for research support.
Requirements: Candidates must be sponsored by a nonfederal public or private nonprofit institution engaged in health care and health-related research within the United States and its possessions. Applicants must have had three to four years of relevant postdoctoral experience prior to the beginning date of the award; have obtained their doctorate degree within the last 10 years; and have institutional confirmation of a faculty appointment at the time the award commences and throughout its duration. Applicants who hold or have applied for nonfederal research awards must provide the title and specific aims of the awards with their application.
Restrictions: At the time of application, the applicant cannot hold or have held any of the following awards: NIH R01, R29, K11, K08; Veterans Administration Merit Review; Associate Investigation; Research Associate; Clinical Investigator; or AGA Industry Award.
Date(s) Application is Due: Oct 1
Amount of Grant: 75,000 U.S.D
Contact: Joan Gallagher, Program Director; (973) 256-2550, ext. 224; fax (973) 256-3214; jgallagher@liverfoundation.org
Internet: http://www.liverfoundation.org/db/grants/113
Sponsor: American Liver Foundation
1425 Pompton Avenue, Suite 3
Cedar Grove, NJ 07009

American Liver Foundation Physician Research Development Award 536
The purpose of this award is to support and develop clinical research training of fellows seeking junior faculty appointment (or in their first years as a junior faculty member). Preference is given to translational research in pathobiology and treatment of viral hepatitis toward developing preclinical and clinical applications as well as addressing natural history, prevention, and epidemiology of viral hepatitis. Funding is available from $225,000 - $270,000 over three or four years.
Date(s) Application is Due: Oct 1
Amount of Grant: 225,000 - 270,000 U.S.D
Contact: Joan Gallagher, Program Director; (973) 256-2550, ext. 224; fax (973) 256-3214; jgallagher@liverfoundation.org
Internet: http://www.liverfoundation.org/db-list/grants/0/descend/ID/Validated
Sponsor: American Liver Foundation
1425 Pompton Avenue, Suite 3
Cedar Grove, NJ 07009

American Liver Foundation Postdoctoral Research Fellowships 537
The foundation offers postdoctoral research fellowships to encourage the development of individuals with research potential who require additional research training and experience, specifically in investigational work relating to liver physiology and disease, in preparation for careers of independent research in this field. The goal of the program is to encourage MD, PhD, and MD/PhD postdoctoral fellows to enter an academic career in liver disease research. Fellows are expected to devote at least 75 percent of time to their specific research project, attend the annual meeting of the American Association for the Study of Liver Diseases, acknowledge support by the ALF in any publications resulting from the research performed during the tenure of the award, and submit a progress report describing results obtained and their significance. The application must be written by the applicant.
Requirements: Candidates must be sponsored by a nonfederal public or private nonprofit institute engaged in health care and health-related research within the United States and its possessions. Physician applicants (MD and MD/PhD) must submit documentation that all clinical training has been completed. All applicants must be in the first year of appointment as a postdoctoral research fellow or trainee.
Restrictions: Individuals with more than two years of postdoctoral research training or those already well established in the field of hepatology are ineligible. In addition, the fellowship is designed as a supplement to augment NIH or nonfederal fellowship stipends and will not be awarded to an applicant who has no other source of research salary support.
Date(s) Application is Due: Oct 1
Amount of Grant: 12,500 U.S.D
Contact: Joan Gallagher, Program Director; (973) 256-2550, ext. 224; fax (973) 256-3214; jgallagher@liverfoundation.org
Internet: http://www.liverfoundation.org/db/grants/131
Sponsor: American Liver Foundation
1425 Pompton Avenue, Suite 3
Cedar Grove, NJ 07009

American Liver Foundation Student Research Fellowship Award 538
The foundation offers awards to MD and PhD students to encourage them to gain exposure to the research laboratory, and possibly consider liver research as a career option. Each fellowship is for a three-month period. The candidate must be in full-time research for a period of 10-12 weeks under the supervision of a preceptor in hepatic physiology or disease.
Requirements: Full-time students of U.S. undergraduate, graduate, or medical school programs are eligible to apply.
Date(s) Application is Due: Oct 1
Amount of Grant: 2,500 U.S.D

Contact: Joan Gallagher, Program Director; (973) 256-2550, ext. 224; fax (973) 256-3214; jgallagher@liverfoundation.org
Internet: http://www.liverfoundation.org/db-list/grants/0/descend/ID/Validated
Sponsor: American Liver Foundation
1425 Pompton Avenue, Suite 3
Cedar Grove, NJ 07009

American Lung Association Career Investigator Award **539**
The award supports individuals who have demonstrated success in research and who show great promise for a career in investigation. Awards are subject to annual review and may be granted for up to three years. Awardees are expected to devote full time, and in no case less than 75 percent of their time, to research. Awardees may undertake limited administrative, teaching, and clinical responsibilities if they are directly related to the nature of the research supported by this award.
Requirements: U.S. citizens and foreign nationals with appropriate visa immigration status are eligible. At the time of application, an applicant must hold a doctoral degree and faculty appointment at the level of assistant or associate professor and be undertaking a project related to lung disease.
Restrictions: Individuals who have attained the rank of full professor or who have more than eight years of faculty experience at the time of application are ineligible.
Date(s) Application is Due: Sep 1
Amount of Grant: Up to 60,000 U.S.D
Contact: Evita Mendoza; (212) 315-8793; emendoza@lungusa.org or info@lungusa.org
Internet: http://www.lungusa.org/site/apps/s/content.asp?c=dvLUK9O0E&b=34706&ct=67676
Sponsor: American Lung Association
1301 Pennsylvania Avenue, NW, Suite 800
Washington, DC 20004

American Lung Association Clinical Patient Care Research Grants **540**
The objective of the grant is to support investigators working in patient-focused research. The areas of research may include the disciplines of health services research, behavioral research, epidemiology, pharmacology, and medical ethics. The outcome of these research activities is to develop new insights into clinical, social, behavioral, and biological factors that affect the manifestation of lung health and disease. Grants are subject to annual review and may be granted for up to two years.
Requirements: U.S. citizens and foreign nationals with appropriate visa immigration status are eligible. At the time of application, an applicant must hold a doctoral degree and faculty appointment with an academic institution and have completed two years of postdoctoral research training.
Restrictions: Residents, interns, postdoctoral fellows, graduate students, and established investigators are ineligible.
Date(s) Application is Due: Sep 1
Amount of Grant: Up to 40,000 U.S.D
Contact: Evita Mendoza; (212) 315-8793; emendoza@lungusa.org or info@lungusa.org
Internet: http://www.lungusa.org/site/apps/s/content.asp?c=dvLUK9O0E&b=34706&ct=67676
Sponsor: American Lung Association
1301 Pennsylvania Avenue, NW, Suite 800
Washington, DC 20004

American Lung Association Dalsemar Research Grant **541**
The grant supports up to two years of research in interstitial lung disease. Grants are subject to annual review and may be granted for up to two years. The second year of support is based on demonstrating satisfactory progress as well as the availability of American Lung Association funding.
Requirements: U.S. citizens and foreign nationals with appropriate immigration visa status are eligible. At the time of application, an applicant must hold a doctoral degree and a faculty appointment with an academic institution and have completed two years of research training.
Restrictions: Residents, interns, postdoctoral fellows, students enrolled in degree-granting programs, and established investigators are not eligible.
Date(s) Application is Due: Sep 1
Amount of Grant: Up to 40,000 U.S.D
Contact: Evita Mendoza; (212) 315-8793; emendoza@lungusa.org or info@lungusa.org
Internet: http://www.lungusa.org/site/apps/s/content.asp?c=dvLUK9O0E&b=34706&ct=67676
Sponsor: American Lung Association
1301 Pennsylvania Avenue, NW, Suite 800
Washington, DC 20004

American Lung Association Research Grants **542**
The grants provide starter or seed money to new investigators working in areas relevant to the prevention of lung disease and the promotion of lung health. The research supported may be clinical, laboratory, epidemiologic, social, environmental, or any other kind as long as it is relevant to lung biology. Six individuals will be designated Edward Livingston Trudeau Scholars. Awards are for one year and may be renewed for an additional year, depending on the availability of funds.
Requirements: At the time of application, an applicant must hold a doctoral degree, faculty appointment with an academic institution, and have completed two years of research training. U.S. citizens, Canadian citizens, and permanent residents of the United States training in U.S. institutions are eligible.
Restrictions: An American Lung Association grantee cannot simultaneously hold any other national ALA award. An investigator who has previously received a research grant or a Dalsemer Award is not eligible for additional national research grant support. An ALA research grant awardee may not hold an award of established investigatorship or large grant during the tenure of her or his ALA award.
Date(s) Application is Due: Sep 1
Amount of Grant: Up to 35,000 U.S.D
Contact: Evita Mendoza; (212) 315-8793; emendoza@lungusa.org or info@lungusa.org
Internet: http://www.lungusa.org/site/apps/s/content.asp?c=dvLUK9O0E&b=34706&ct=67676
Sponsor: American Lung Association
1301 Pennsylvania Avenue, NW, Suite 800
Washington, DC 20004

American Lung Association Senior Research Training Fellowship **543**
The fellowship supports the training of scientific investigators in the fields of adult and pediatric pulmonary medicine and lung biology. Preference is given to applicants who demonstrate a program of training that will enable them to pursue academic careers. Fellows are expected to devote full time or at least 75 percent of their time to research training. Fellowships may be renewed for an additional year.
Requirements: U.S. citizens and foreign nationals with appropriate immigration visa status are eligible. At the time of application, an applicant must hold a doctoral degree or have comparable qualifications. MD applicants must have completed their clinical training, have some research experience, and be entering the third or fourth year of research fellowship training. PhD applicants must not be beyond the third postdoctoral year at the time of application.
Date(s) Application is Due: Sep 1
Amount of Grant: Up to 32,500 U.S.D
Contact: Evita Mendoza; (212) 315-8793; emendoza@lungusa.org or info@lungusa.org
Internet: http://www.lungusa.org/site/apps/s/content.asp?c=dvLUK9O0E&b=34706&ct=67676
Sponsor: American Lung Association
1301 Pennsylvania Avenue, NW, Suite 800
Washington, DC 20004

American Medical Association Awards **544**
The American Medical Association recognizes excellence through the following AMA Awards: Distinguished Service Award; Citation for Distinguished Service; AMA Medal of Valor; Scientific Achievement Award; Benjamin Rush Award for Citizenship and Community Service; President's Citation for Service to the Public; Joseph B. Goldberger Award in Clinical Nutrition; Dr. William Beaumont Award in Medicine; AMA Foundation Award for Health Education; Isaac Hays, MD and John Bell, MD Award for Leadership in Medical Ethics and Professionalism; Medical Executive Lifetime Achievement Award; and Medical Executive Meritorious Achievement Award. Candidates may be nominated for more than one award. Nominations should be emailed by February 23 each year.
Date(s) Application is Due: Feb 23
Contact: Roger Brown, (312) 464-4344; fax (312) 464-4505; roger.brown@ama-assn.org
Internet: http://www.ama-assn.org/ama/pub/about-ama/awards/about-ama-awards-program.shtml
Sponsor: American Medical Association
515 N State Street
Chicago, IL 60654

American Nurses Foundation Nursing Research Grants **545**
The program is designed to award grants for nursing research conducted by registered nurses who are either beginning nurse researchers or experienced nurse researchers entering a new field of investigation. Priority is based upon the scientific merit of the proposal, with consideration given to the investigator's ability to conduct the study.
Requirements: Principal investigators must be baccalaureate-prepared registered nurses; however, the coinvestigator does not have to be a nurse as long as the proposal is for a nursing research project.
Restrictions: Some ANF grants are restricted as to their use by corporations and organizations that have contributed the funds.
Date(s) Application is Due: May 3
Amount of Grant: 2,700 U.S.D
Contact: Leo Schargorodski, Executive Director; (301) 628-5230; anf@ana.org or lschargo@ana.org
Internet: http://www.nursingworld.org/anf/nrggrant.htm
Sponsor: American Nurses Foundation
8515 Georgia Avenue, Suite 400 W
Silver Spring, MD 20910

American Otological Society Research Fund Fellowhips and Grants **546**
Research grants and fellowships are awarded to U.S. and Canadian physicians and doctoral-level investigators in all aspects of otosclerosis, Meniere's disease, and related ear disorders. The one-year grants, which begin July 1, are renewable for a second year and are tenable at U.S. and Canadian institutions. Awards are made to an institution on behalf of a candidate. Appropriate areas of research include diagnosis, management, and pathogenesis of otosclerosis and Meniere's disease as well as underlying processes. The applicant should describe correlations between proposed research with the clinical pathological entities of otosclerosis and Meniere's disease.
Requirements: Physician and nonphysician investigators are eligible for research grants. Fellowships are open to physicians (residents and medical students).

Restrictions: Research grants do not support funding for investigator's salary.
Date(s) Application is Due: Jan 31
Amount of Grant: 40,000 U.S.D
Samples: Timothy Doerr, Wayne State U (Detroit, MI)—for fellowship proposal, Regulatory Mechanisms of Vestibular Blood Flow, $39,621; Dr. Bradley William Kesser, U of Virginia (Charlottesville, VA)—for fellowship proposal, The Pathophysiology and Pathogenesis of Changes in the Inner Ear During Meningitis, $39,750.
Contact: Dr. Jeffrey Harris, (619) 543-7896; fax (619) 543-5521; jpharris@ucsd.edu
Internet: http://itsa.ucsf.edu/~ajo/AOS/SchGra.html
Sponsor: American Otological Society
200 W Arbor Drive, 8895
San Diego, CA 92103-8895

American Parkinson's Disease Association Cotzias Fellowships 547

This three-year fellowship is intended to recruit and assist young neurologists in establishing careers in research, teaching, and patient service relevant to the problems, causes, prevention, diagnosis, and treatment of Parkinson's disease and other neurological movement disorders.
Requirements: An applicant must be a U.S. citizen, have an MD degree obtained within 10 years of application, and must have training in a clinical discipline concerned with disorders of the nervous system. Rank of assistant professor is required. The applicant must be sponsored by a nonprofit institution in the United States or its territories.
Date(s) Application is Due: Mar 1
Amount of Grant: 80,000 U.S.D
Samples: Dr. Russell H. Swerdlow, U of Virginia Health System (Charlottesville, VA)—fellowship recipient, for research entitled Genetic Studies in Parkinson's Disease. Dr. Mel Feany, Brigham and Women's Hospital (Boston, MA)—fellowship recipient, for research entitled Genetic Model of Parkinsonian Movement Disorders. Dr. Serge Przedborski, Columbia U (New York, NY)—fellowship recipient, for research entitled Transplanted Dopamine Cell Survival in Animal Model of Parkinson's Disease. Dr. David Standaert, Massachusetts General Hospital (Boston, MA)—fellowship recipient, for research entitled NMDA Receptors: Regulation of Basal Ganglia Function and Pathogenesis of Parkinsonism.
Contact: Grants Administrator; (718) 981-8001 or (800) 223-2732; fax (718) 981-4399; apda@apdaparkinson.org
Internet: http://www.apdaparkinson.org/user/ViewFellowshipAndGrants.asp
Sponsor: American Parkinson's Disease Association
1250 Hylan Boulevard, Suite 4B
Staten Island, NY 10305-1946

American Parkinson's Disease Association Research Grants 548

Qualified scientists may apply for grants for research in the area of Parkinson's disease. Grants are awarded for one year for direct costs of the research project; indirect costs, salary for the principal investigator, travel expenses, and institutional overhead are not covered. Grants may be renewed for a second year on deadline date, but reapplication must be made. Description of the research proposal should not exceed two pages and should include a statement of how research is related to Parkinson's disease.
Requirements: Scientists affiliated with U.S. institutions are eligible.
Restrictions: Publication costs may not exceed $300, and equipment budget may not exceed $5000.
Date(s) Application is Due: Mar 1
Amount of Grant: 50,000 U.S.D
Contact: Grants Administrator; (718) 981-8001 or (800) 223-2732; fax (718) 981-4399; apda@apdaparkinson.org
Internet: http://www.apdaparkinson.org/user/ViewFellowshipAndGrants.asp
Sponsor: American Parkinson's Disease Association
1250 Hylan Boulevard, Suite 4B
Staten Island, NY 10305-1946

American Philosophical Society Daland Fellowships in Clinical Investigation 549

A limited number of fellowships are awarded annually for research in clinical medicine including the fields of internal medicine, neurology, psychiatry, pediatrics, and surgery. Patient-oriented research is emphasized. Essentially 100 percent of the fellow's time will be devoted to research. Teaching or clinical service of a limited amount is permitted. Additional salary may be granted by the institution at which the fellow is located, from another fellowship, or from a similar award during the tenure of the fellowship. The term of the fellowship is one year, with possible renewal for another year. Applications are available on the Web site, or may be obtained by written request, stating when the MD or MD/PhD degree was awarded; include a self-addressed mailing label.
Requirements: These fellowships are designed for qualified persons who have held an MD, or MD/PhD degree for less than eight years. The fellowship is generally intended to be the first post-clinical fellowship; but each case will be decided on its merits. Preference is generally given to candidates who have not more than two years of post-doctoral training and research. Applicants must expect to perform their research at an institution in the United States, under the supervision of a scientific adviser. Candidates are to be nominated by their department chairman, in a letter providing assurance that the nominee will work with the guidance of a scientific adviser of established reputation who has guaranteed adequate space, supplies, etc. for the Fellow. The adviser need not be a member of the department nominating the Fellow, nor need the activities of the Fellow be limited to the nominating department. As a general rule, no more than one fellowship will be awarded to a given institution in the same year of competition.
Restrictions: The society does not provide funds for institutional overhead.
Date(s) Application is Due: Sep 1
Amount of Grant: Up to 40,000 U.S.D
Contact: Linda Musumeci, Research Administrator, (215) 440-3429; LMusumeci@amphilsoc.org
Internet: http://www.amphilsoc.org/grants/daland.htm
Sponsor: American Philosophical Society
104 South Fifth Street
Philadelphia, PA 19106-3387

American Psychiatric Association/AstraZeneca Young Minds in Psychiatry International Awards 550

The program recognizes and supports promising international young psychiatrists within five years of completing a psychiatric residency. Four unrestricted career development awards of $45,000 (U.S.D) will be available. Awards will be made to two promising physicians from the U.S. with one in Bipolar Disorder research and one on research in Schizophrenia. An additional two awards will be made to promising physicians from countries outside the U.S., with one in Bipolar Disorder research and one on research in Schizophrenia. Three other awards of $30,000 (U.S.D) in either Bipolar Disorder or Schizophrenia research will specifically focus on applicants from developing countries whose economies are classified by the World Bank as low income or lower middle income.
Requirements: The U.S. applicants must be citizens or permanent residents of the United States. Applications are evaluated based on evidence of academic promise; how the proposal will advance the applicant's career; and innovative or original concepts, approaches, or methods for developing the applicant's career.
Date(s) Application is Due: Nov 30
Amount of Grant: 45,000 U.S.D
Contact: Ernesto Guerra, (703) 907-7300; fax (703) 907-1085; eguerra@psych.org or apa@psych.org
Internet: http://psych.org/MainMenu/Research/ResearchTrainingandFunding/ResearchFellowships/APAAstraZenecaYoungMindsinPsychiatryInternationalAward.aspx
Sponsor: American Psychiatric Association
1000 Wilson Boulevard, Suite 1825
Arlington, VA 22209-3901

American Psychiatric Association/Bristol-Myers Squibb Fellowships in Public Psychiatry 551

This important collaborative venture between the American Psychiatric Association and the Bristol-Myers Squibb Company provides for the selection of ten residents in psychiatry, based on recommendations from their Departments and evidence of a track record of commitment to and accomplishment in public psychiatry. The purposes of the APA/Bristol-Myers Squibb Fellowship are twofold: (1) to heighten psychiatric residents' awareness of the many activities of psychiatry in the public sector and of the career opportunities in this work, and (2) to provide experiences which will contribute to the professional development of those residents who will play leadership roles within the public sector in future years.
Requirements: Nomination Procedure and Eligibility Criteria: One outstanding resident who has a substantial interest in any area of public sector psychiatry and a significant potential for a leadership role in such work may be nominated by a psychiatric residency training program that is accredited by the Accreditation Council for Graduate Medical Education or the Royal College of Physicians and Surgeons of Canada; submission of a nomination represents an explicit commitment by the residency program to permit the candidate, if selected, time off to attend APA meetings in their entirety during the fellowship term, as well as the candidate's willingness to do so; Psychiatric residents who will be in the PG-3 year of training during the fellowship term are eligible to participate in the program; residents who will be in their PG-4 year will also be considered if they are in a 5-year psychiatric training program; nominee must be an APA member or in the process of becoming a member. To be considered, the following must be submitted: completed application form (submitted electronically or via mail); two-page essay; letter of nomination from the Residency Training Director or Chair, Department of Psychiatry.
Restrictions: Residents who have received (or are scheduled to receive) other APA fellowships or comparable outside fellowships (e.g., Ginsberg, Laughlin) are not eligible. Those residents who currently serve on APA's Committee of Residents and Fellows are not eligible for the fellowship.
Date(s) Application is Due: Mar 31
Contact: Nancy Delanoche, APA Office of Graduate and Undergraduate Education; (703) 907-8635; ndelanoche@psych.org
Internet: http://www.psych.org/Departments/EDU/ResidentMIT/fellowshipinpublicpsychiatry.aspx
Sponsor: American Psychiatric Association
1000 Wilson Boulevard, Suite 1825
Arlington, VA 22209-3901

American Psychiatric Association/Janssen Resident Psychiatric Research Scholars 552

The program is intended to identify promising PGY-1, PGY-2, and PGY-3 psychiatric residents with the potential to become leaders in clinical and health services research in all areas of psychiatric research. Emphasis will be placed on special mentoring and career enrichment programs both at the APA Annual Meeting and throughout the year. An individual research mentor will be assigned to oversee the resident's fellowship. The mentors will be chosen among nationally known leaders in clinical and health services research. The mentor will advise and encourage the Scholar during the two-year fellowship. The scholar will receive $2,500 during the second year of the fellowship, to assist in their research career development (e.g., for use in developing a pilot research project, obtaining statistical consultation, or visiting potential research training programs). The program also provides funding for travel to the American Psychiatric Association (APA) Annual Meeting during both years of the fellowship.

Date(s) Application is Due: Jan 15
Amount of Grant: 2,500 U.S.D
Contact: Ernesto Guerra, (703) 907-7300; fax (703) 907-1085; eguerra@psych.org or apa@psych.org
Internet: http://psych.org/MainMenu/Research/ResearchTrainingandFunding/ResearchFellowships/APIREJanssenResidentPsychiatricResearchScholars.aspx
Sponsor: American Psychiatric Association
1000 Wilson Boulevard, Suite 1825
Arlington, VA 22209-3901

American Psychiatric Association/Merck Co. Early Academic Career Research Award 553
The Award is intended to help support the research of a junior faculty member with an interest in sleep disorders or schizophrenia. Two separate awards will be made to candidates who have completed a psychiatry residency, at least one year of a psychiatry research fellowship, and are seeking to make a commitment to a research career with the end goal of successfully transitioning to that of an independent investigator. The Award is intended to assist in this key transition period by providing one year of funding of $45,000. Salary support provided by this award will allow junior faculty in departments of psychiatry at U.S. academic institutions to devote more time to research.
Requirements: Candidates must be citizens or permanent residents of the U.S. and should also be APA members. Eligible candidates will be trained psychiatrists with an MD, MD/Ph.D. or DO degree who have completed residency training in general psychiatry or child psychiatry in the U.S. Candidates also should have completed at least one year of a psychiatry research fellowship but are not 3 or more years post fellowship completion, and do not currently hold an academic rank higher than Assistant Professor.
Restrictions: Individuals who have obtained K-awards or are considered an independent Principal Investigator on R-01 or R-21 research grants from NIH are not eligible.
Date(s) Application is Due: Oct 14
Amount of Grant: 45,000 U.S.D
Contact: Ernesto Guerra, (703) 907-7300; fax (703) 907-1085; eguerra@psych.org or apa@psych.org
Internet: http://www.psych.org/research/apire/res_careerdev/apamerck.cfm
Sponsor: American Psychiatric Association
1000 Wilson Boulevard, Suite 1825
Arlington, VA 22209-3901

American Psychiatric Association Award for Research in Psychiatry 554
The Award is given in recognition of a single distinguished contribution, a body of work, or a lifetime contribution that has had a major impact on the field and/or altered the practice of psychiatry. The Award is intended to cover the full spectrum of psychiatric research. The Award consists of a $5,000 prize and an honorary plaque to be presented at APA's Annual Meeting in May. The Award also includes an honorary lecture by the awardee.
Requirements: Candidates for the Award must be citizens of the United States or Canada and be nominated by a sponsor. Sponsors must be members of the American Psychiatric Association.
Restrictions: Members of the Award Committee are excluded from submitting nominations.
Date(s) Application is Due: Aug 28
Amount of Grant: 5,000 U.S.D
Contact: Harold Goldstein, (703) 907-8623; goharold@psych.org
Internet: http://www.psych.org/research/apire/training_fund/psychaward.cfm
Sponsor: American Psychiatric Association
1000 Wilson Boulevard, Suite 1825
Arlington, VA 22209-3901

American Psychiatric Association Minority Fellowships Program 555
The APA offers psychiatric residents one-year fellowships in the APA organization to increase the knowledge of cultural factors influencing psychiatric diagnosis and treatment, provide opportunities for these residents to participate in the deliberations and decision-making processes, and provide role models for the fellows. Each fellow is appointed to a component of the APA's organizational structure and attends the association's annual meeting as an observer and active participant. Fellows are selected on the basis of their commitment to serving underrepresented populations, demonstrated leadership abilities, and interest in the interrelationship between mental health/illness and transcultural factors.
Requirements: Psychiatric residents who are starting their second year of psychiatric training, or third year if they are in a four-year residency program, are eligible to apply.
Contact: Marilyn King, Fellowship Coordinator; (703) 907-8653; fax (703) 907-7849; mking@psych.org
Internet: http://www.psych.org/Resources/OMNA/MFP.aspx
Sponsor: American Psychiatric Association
1000 Wilson Boulevard, Suite 1825
Arlington, VA 22209-3901

American Psychiatric Association Minority Medical Student Fellowship in HIV Psychiatry 556
The APA invites ethnic minority medical students who have an interest in psychiatric issues to apply. The program is intended to identify minority medical students who have primary interests in services related to HIV/AIDS and substance abuse and its relationship to the mental health or psychological well being of ethnic minorities.
Requirements: These programs are open to ethnic minority medical students who are U.S. citizens or permanent residents currently enrolled in a U.S. medical school.
Date(s) Application is Due: Mar 31
Contact: Diane Pennessi, Senior Project Manager; (703)907-8668; fax (703)907-8668; dpennessi@psych.org
Internet: http://www.psych.org/Resources/OMNA/MFP/HIV-Information-10.aspx
Sponsor: American Psychiatric Association
1000 Wilson Boulevard, Suite 1825
Arlington, VA 22209-3901

American Psychiatric Association Minority Medical Student Summer Externship in Addiction Psychiatry 557
The APA invites ethnic minority medical students who have an interest in psychiatric issues to apply. This clinical shadowing program identifies minority medical students who may have a specific interest in services related to substance abuse treatment/prevention and provide a setting where the student can work closely with a mentor who specializes in addiction psychiatry for one month.
Requirements: These programs are open to ethnic minority medical students who are U.S. citizens or permanent residents currently enrolled in a U.S. medical school.
Date(s) Application is Due: Feb 28
Amount of Grant: 1,500 U.S.D
Contact: Marilyn King, Assistant Program Director; (703) 907-8653; fax (703) 907-7852; mking@psych.org
Internet: http://www.psych.org/Resources/OMNA/MFP/minoritymedicalstudentsummerexternshipinaddictionpsychiatry.aspx
Sponsor: American Psychiatric Association
1000 Wilson Boulevard, Suite 1825
Arlington, VA 22209-3901

American Psychiatric Association Minority Medical Student Summer Mentoring Program 558
The APA invites ethnic minority medical students who have an interest in psychiatric issues to apply. This program is intended to identify ethnic minority medical students who have an interest in psychiatric issues and expose students to a setting where they can work closely with a psychiatrist mentor for one month.
Requirements: These programs are open to ethnic minority medical students who are U.S. citizens or permanent residents currently enrolled in a U.S. medical school.
Date(s) Application is Due: Feb 28
Amount of Grant: 1,500 U.S.D
Contact: Marilyn King, Assistant Program Director; (703) 907-8653; fax (703) 907-7852; mking@psych.org
Internet: http://www.psych.org/Resources/OMNA/MFP/minoritymedicalstudentsummermentoringprogram.aspx
Sponsor: American Psychiatric Association
1000 Wilson Boulevard, Suite 1825
Arlington, VA 22209-3901

American Psychiatric Association Program for Minority Research Training in Psychiatry (PMRTP) 559
The program is designed to increase the number of underrepresented minority men and women in the field of psychiatric research. Research training offers the opportunity to engage in scientific investigation across the full array of disciplines, from basic neuroscience, genetics, and pharmacology to the cognitive behavioral, and social sciences, clinical psychiatry, and mental health services research. Research exposure can help students and trainees develop sound skills for clinical assessment and treatment planning. The program provides funding for short and long-term training opportunities at three levels: Medical School, Residency and Post-residency. National competitions also enable qualified mini-fellows to attend research-oriented meetings of psychiatric organizations. For medical students and residents, the duration can be two months to one year. For post-residency fellows, the duration is generally two years. A third year of fellowship support may be available if appropriate to a trainee's career development. Support from PMRTP falls into three categories: stipends, travel, and tuition and fees. Stipends are based on the trainee's years of relevant experience and the length of the research training experience.
Requirements: Preference in selection is given to underrepresented minorities such as American Indians, Asian-Americans, Blacks/African-Americans, Hispanics, Pacific Islanders, or other ethnic or racial group members found to be underrepresented in biomedical or behavioral research.
Date(s) Application is Due: Apr 1; Dec 1
Amount of Grant: 20,772 - 51,036 U.S.D
Contact: Ernesto Guerra, (703) 907-7300; fax (703) 907-1085; eguerra&@psych.org or apa@psych.org
Internet: http://www.psych.org/MainMenu/Research/ResearchTrainingandFunding/ResearchFellowships/ProgramforMinorityResearchTraininginPsychiatry.aspx
Sponsor: American Psychiatric Association
1000 Wilson Boulevard, Suite 1825
Arlington, VA 22209-3901

American Psychiatric Association Research Colloquium for Junior Investigators 560
The purpose of the colloquium is to provide guidance, mentorship and encouragement to young investigators in the early phases of their training. Junior investigators will have an opportunity to obtain feedback about their past, present, and future research interests from mentors who are tops in their field in a small group setting as well as general information about research career development and grantsmanship. Candidates whose research interests are similar to those listed below are also encouraged to apply. An all-day workshop for junior psychiatric investigators will focus on these three areas: childhood disorders including autism, ADHD, conduct, eating, and mood disorders;

genomics, epigenetics, and proteomics; and treatment of major psychiatric disorders: from substance abuse to schizophrenia.
Requirements: Psychiatrists who are senior residents, fellows, or junior faculty, and who have an interest and potential in developing research careers in the areas of research listed above. Participants should hold a medical degree or be a member of the APA; or be eligible to become members of the APA.
Restrictions: Those with individual federal research awards are not eligible.
Date(s) Application is Due: Nov 16
Amount of Grant: 1,000 U.S.D
Contact: Ernesto Guerra, (703) 907-7300; fax (703) 907-1085; eguerra@psych.org or apa@psych.org
Internet: http://psych.org/MainMenu/Research/ResearchTrainingandFunding/ResearchColloquiumforJuniorInvestigators.aspx
Sponsor: American Psychiatric Association
1000 Wilson Boulevard, Suite 1825
Arlington, VA 22209-3901

American Psychiatric Association Travel Scholarships for Minority Medical Students 561

The APA invites ethnic minority medical students who have an interest in psychiatric issues to apply. The program supports travel and related costs for approximately 10 minority medical students interested in psychiatric to attend either the APA annual meeting in May or the Institute on Psychiatric Services (IPS) meeting in October. This program is a way for medical students to witness organized psychiatry at work and to learn more about the field. Not only will students attend sessions for experts and trainees alike, but they will be assigned to a mentor who will help them maximize their annual meeting or IPS experience and discuss career plans and resident training programs.
Requirements: These programs are open to U.S. ethnic minority medical students who are U.S. citizens or permanent residents currently enrolled in a U.S. medical school. Ethnic minorities are: American-Indian, Alaska Native, Native Hawaiian, Asian-American, African-American, and Hispanic/Latino.
Date(s) Application is Due: Jan 29
Contact: Marilyn King, Assistant Program Director; (703) 907-8653; fax (703) 907-7852; mking@psych.org
Internet: http://www.psych.org/Resources/OMNA/MFP/MedicalStudentTravelScholarships.aspx
Sponsor: American Psychiatric Association
1000 Wilson Boulevard, Suite 1825
Arlington, VA 22209-3901

American Psychiatric Foundation Alexander Gralnick, MD, Award for Research in Schizophrenia 562

The award will annually acknowledge research achievements in the treatment of schizophrenia, emphasizing early diagnosis and treatment and psychosocial aspects of the disease process. Additional preference will be given to researchers working in a psychiatric facility. The applicant should submit a statement summarizing the nature of the research relevant to the award, emphasizing its internal consistency and scientific implications; a list of publications (or articles accepted for publication) relevant to the award; a current curriculum vita; and a current bibliography. Four copies and the original application should be submitted. The $4,000 award will be presented at the annual meeting.
Date(s) Application is Due: Nov 2
Amount of Grant: 4,000 U.S.D
Contact: Paul T. Burke, Executive Director; (703) 907-8518 or (703) 907-8512; fax (703) 907-7851; pburke@psych.org or apf@psych.org
Internet: http://www.psychfoundation.org/GrantAndAwards/AwardsandFellowships/AlexanderGralnickMDAward.aspx
Sponsor: American Psychiatric Foundation
1000 Wilson Boulevard, Suite 1825
Arlington, VA 22209

American Psychiatric Foundation Awards for Advancing Minority Mental Health 563

The program was established by the foundation, with support of Otsuka America Pharmaceutical, to recognize psychiatrists, other health professionals, mental health programs, and other organizations that have undertaken innovative and supportive efforts to raise awareness of mental illness in minority communities, the need for early recognition, the availability of treatment and how to access it, and the cultural barriers to treatment; increase access to quality mental health services for minorities; and improve the quality of care for minorities, particularly those in the public health system or with severe mental illness. The awards, although given to mental health professionals, will focus on patient issues such as raising awareness of mental illness, the need for treatment, and how to access treatment. Programs or efforts that reduce language and cultural barriers, enhance direct community understanding of mental illness, expand efforts to reach out to their community, and support community reintegration are just a few examples of the types of activities that could be recognized by these awards. Winning programs will be featured in publications such as Psychiatric News, and on American Psychiatric Association and American Psychiatric Foundation Web sites. In addition, award winners and their programs will be recognized at the APA Annual Meetings each May. Guidelines are available online.
Requirements: Psychiatrists, other health professionals, mental health programs, and other organizations are eligible. Psychiatrists and other health professionals must be licensed and/or accredited in the state in which you practice. Mental health programs and other organizations must have been in operation for at least two years before the application deadline.
Amount of Grant: 5,000 U.S.D
Samples: Family Service of El Paso, El Paso, Texas, $5,000 - for its unique and innovative mental health services provided to individuals and families who have fled the violence in Mexico (2011); Northside Center for Child Development in New York, New York, $5,000 - for the youth mental health programs it runs (2011); Student Stress and Anger Management Program, Los Angeles, California, $5,000 - for its unique services to address the emotional components that contribute to the mental health of middle and high school students (2011).
Contact: Paul T. Burke, Executive Director; (703) 907-8518 or (703) 907-8512; fax (703) 907-7851; pburke@psych.org or apf@psych.org
Internet: http://www.psychfoundation.org/GrantAndAwards/AwardsandFellowships/AAMMH.aspx
Sponsor: American Psychiatric Foundation
1000 Wilson Boulevard, Suite 1825
Arlington, VA 22209

American Psychiatric Foundation Call for Proposals 564

APF support allows organizations across the country to make a difference through unique educational, informational and outreach initiatives that promote the early recognition and treatment of mental illness and encourage leadership in the field of psychiatry. The Call for Proposals program is intended to fund public education programs, as well as information and outreach initiatives that promote the early recognition and treatment of mental illness. Up to $750,000 in grant funds are available over the course of three years. Average grants are in the $50,000 range. Grant making under this program has been temporarily suspended while a new Call for Proposals is being developed. Contact the office for further updates.
Requirements: Organizations that have been in existence for at least two years and currently maintain a 501(c)3 charitable status and American Psychiatric Association District Branches and subsidiaries. Organizations need not be mental health programs.
Amount of Grant: Up to 100,000 U.S.D
Contact: Paul T. Burke, Executive Director; (703) 907-8518 or (703) 907-8512; fax (703) 907-7851; pburke@psych.org or apf@psych.org
Internet: http://www.psychfoundation.org/GrantAndAwards/Grants/CallforProposals.aspx
Sponsor: American Psychiatric Foundation
1000 Wilson Boulevard, Suite 1825
Arlington, VA 22209

American Psychiatric Foundation Disaster Recovery Fund for Psychiatrists 565

The Disaster Recovery Fund for Psychiatrists grant program is intended to assist psychiatrists and psychiatric residents who have been adversely affected by disasters such as hurricanes Katrina, Rita and Wilma. Psychiatrists and psychiatric residents who have been affected by disasters are eligible to apply for grant funding up to $2,500. Grants will be made to allow psychiatric residents to continue their medical training and to help psychiatrists rebuild their practices in order to serve the mental health needs of the people in their communities.
Amount of Grant: Up to 2,500 U.S.D
Contact: Paul T. Burke, Executive Director; (703) 907-8518 or (703) 907-8512; fax (703) 907-7851; pburke@psych.org or apf@psych.org
Internet: http://www.psychfoundation.org/GrantAndAwards/Grants.aspx
Sponsor: American Psychiatric Foundation
1000 Wilson Boulevard, Suite 1825
Arlington, VA 22209

American Psychiatric Foundation Helping Hands Grants 566

The program was established to encourage medical students to participate in community mental health service activities, particularly those focused on underserved populations. The program seeks to raise awareness of mental illness and the importance of early recognition and builds an interest among medical students in psychiatry and working in underserved communities. The foundation makes grants up to $5,000 to medical schools for mental health service projects that are created and managed by medical students. Projects can be new initiatives conducted in partnership with community agencies or in conjunction with ongoing medical school outreach activities. Examples of fundable activities include screening in community health centers or homeless shelters; outreach and community education about mental health; and health literacy programs.
Requirements: Medical schools are eligible to apply. All projects must be conducted under the supervision of medical faculty. Medical students who participate in the program must be in their second, third, or fourth year of medical school at the time they are engaged in community service. Funds must be expended within one year. Unused funds must be returned to the foundation.
Restrictions: Grants will not fund services beyond basic mental health screenings and referrals.
Date(s) Application is Due: May 26
Amount of Grant: Up to 5,000 U.S.D
Samples: Wake Forest University, Winston-Salem, North Caroline, $5,000 - to provide mental health literacy, basic mental health screening, and referrals at the student-run Delivering Equal Access to Care free clinic (2011); University of Cincinnati College of Medicine, Cincinnati, Ohio, $5,000 - to implement a partnership with a local non-profit, Off the Streets, to educate and empower women coming out of a cycle of prostitution and addiction (2011); Vanderbilt University School of Medicine, Nashville, Tennessee, $5,000 - to raise awareness in the community about mental illness, improve referral of patients to community resources, and educate all medical student volunteers at Shade Tree Clinic (STC) about the burden of mental illness in underserved populations (2011).
Contact: Paul T. Burke, Executive Director; (703) 907-8518 or (703) 907-8512; fax (703) 907-7851; pburke@psych.org or apf@psych.org
Internet: http://www.psychfoundation.org/GrantAndAwards/Grants/HelpingHands.aspx

Sponsor: American Psychiatric Foundation
1000 Wilson Boulevard, Suite 1825
Arlington, VA 22209

American Psychiatric Foundation James H. Scully Jr., M.D., Educational Fund Grants 567

The fund supports education and research within the American Psychiatric Association (APA), its subsidiary organizations, or affiliated organizations, including district branches, state associations, or other psychiatric organizations that are in partnership with the APA. The fund can be used to support activities that improve the quality of care for patients with mental illness, advance the prevention of mental illness, raise awareness of mental illness and its treatments, and increase access to mental health services. Requests for funding should be sent in writing to the foundation and include the following: identification and significance of need; description of program/activity to address that need; expected outcomes and evaluation process; budget and time line; and background of key project personnel. Funding decisions are made by the foundations board of directors.
Restrictions: The fund will not support public policy or lobbying activities.
Contact: Paul T. Burke, Executive Director; (703) 907-8518 or (703) 907-8512; fax (703) 907-7851; pburke@psych.org or apf@psych.org
Internet: http://www.psychfoundation.org/GrantAndAwards/Grants/JamesHScullyEducationalFund.aspx
Sponsor: American Psychiatric Foundation
1000 Wilson Boulevard, Suite 1825
Arlington, VA 22209

American Psychiatric Foundation Jeanne Spurlock Congressional Fellowship 568

The Jeanne Spurlock Congressional Fellowship provides general psychiatry and child psychiatry residents an opportunity to work in a congressional office or committee on federal health policy, particularly policy related to child and minority issues. The recipient will serve a ten-month fellowship in Washington, DC. The fellow will be introduced to the structure and development of federal and congressional health policy procedures, with a focus on mental health issues affecting minorities and under-served populations, including children. Fellows traditionally help develop legislative proposals, track and analyze legislative initiatives, arrange hearings, and brief Members of Congress and their staff, and interact with their constituents. During the fellowship, recipients have opportunities to interact with health policymakers and advocacy/professional groups, including the APA.
Requirements: PGY-II and III general psychiatry residents, child psychiatry residents, or child psychiatrists who will be out of training for less than one year at the time of the fellowship may apply. Applicants must be U.S. citizens or permanent residents. The recipient will be required to submit a written summary of the Fellowship experience at the end of the fellowship and may make recommendations or suggestions for improving the fellowship.
Date(s) Application is Due: Mar 14
Amount of Grant: 35,000 - 37,500 U.S.D
Contact: Paul T. Burke, Executive Director; (703) 907-8518 or (703) 907-8512; fax (703) 837-5451; pburke@psych.org or apf@psych.org
Internet: http://www.psychfoundation.org/GrantAndAwards/Fellowships/SpurlockFellowship.aspx
Sponsor: American Psychiatric Foundation
1000 Wilson Boulevard, Suite 1825
Arlington, VA 22209

American Psychiatric Foundation Minority Fellowships 569

The Minority Fellowships Program (MFP) endeavors to eliminate racial and ethnic disparities in mental health and substance abuse care by providing specialized training to psychiatry residents and medical students interested in serving minority communities. The MFP is designed to: provide fellowship recipients with enriching training experiences through participation in the APA September and Annual Meetings; provide recipients with resources to support activities that enhance culturally relevant aspects of their training program; stimulate their interest in pursuing training in areas of psychiatry where minority groups are underrepresented, such as research, child psychiatry, and addiction psychiatry; and develop leadership to improve the quality of mental health care for the following federally recognized ethnic minority groups: American Indians, Native Alaskans, Asian Americans, Native Hawaiians, Native Pacific Islanders, African Americans and Hispanics/Latinos.
Contact: Paul T. Burke, Executive Director; (703) 907-8518 or (703) 907-8512; fax (703) 837-5451; pburke@psych.org or apf@psych.org
Internet: http://www.psychfoundation.org/GrantAndAwards/Fellowships/MFP.aspx
Sponsor: American Psychiatric Foundation
1000 Wilson Boulevard, Suite 1825
Arlington, VA 22209

American Psychiatric Foundation Typical or Troubled School Mental Health Education Grants 570

The Grant Program (also known as Típico o Problematico in Spanish) provides funding to implement the Typical or Troubled School Mental Health educational model in communities nationwide. Community organizations, high schools and school districts are eligible to receive funding. The educational program is designed for school personnel (teachers, coaches guidance counselors, etc.) to raise their awareness of mental disorders in teens. The program focuses on promoting the importance of early recognition and treatment, recognizing the early warning signs of mental health problems, and encouraging action and appropriate referral to a mental health professional. Grant support is as follows: for implementation in two to four high schools, $1,000; for implementation in five or more high schools, $2,000.
Date(s) Application is Due: Mar 30
Amount of Grant: Up to 2,000 U.S.D
Contact: Paul T. Burke, Executive Director; (703) 907-8518 or (703) 907-8512; fax (703) 907-7851; pburke@psych.org or apf@psych.org
Internet: http://www.psychfoundation.org/GrantAndAwards/Grants/TypicalorTroubled.aspx
Sponsor: American Psychiatric Foundation
1000 Wilson Boulevard, Suite 1825
Arlington, VA 22209

American Psychological Foundation F. J. McGuigan Young Investigator Prize 571

The American Psychological Foundation awards this biennial prize to recognize the efforts of a young psychological science investigator in the areas of research consistent with those pursued by Frank Joseph McGuigan, PhD. The prize is focused to support research to explicate the concept of the human mind. The approach must be a materialistic one fostering both empirical and theoretical research. Empirical research would primarily be psychophysiological, but physiological and behavioral research may also qualify for support. The recipient will be selected based on the excellence of the full breadth of research conducted and published to date, as well as the promise of research planned for five years. The prize will be awarded to the recipient's institution for the benefit of his or her research. Nomination guidelines are available online.
Requirements: Nominees must have earned a doctoral degree in psychology or a related field, and be nine or fewer years post doctoral degree at the time of the nomination deadline. Nominees must have an affiliation with an accredited college, university, or other research institution.
Restrictions: Faculty salaries and indirect costs (i.e., overhead) may not be requested. Dualistic approaches such as espoused by many contemporary cognitive psychologists do not qualify for support.
Date(s) Application is Due: Mar 1
Amount of Grant: 25,000 U.S.D
Contact: Science Directorate; (202) 336-6000; fax (202) 336-5953; science@apa.org
Internet: http://www.apa.org/science/mcguigan.html
Sponsor: American Psychological Association
750 First Street NE
Washington, DC 20002-4242

American Psychological Foundation Todd E. Husted Memorial Award 572

This award supports dissertation research that indicates the most potential to contribute toward the development and improvement of mental illness services for those with severe and persistent mental illness. Relevant topics include those that foster the development of a more comprehensive, humane, and responsive system of mental health care; develop a protective and humane sequencing of interventions that prevents the deterioration, homelessness, and premature deaths of those with serious mental illness; develop effective methods of improving patient compliance with medication and treatment for those having impaired insight as a result of schizophrenia and bipolar affective disorder; demonstrate practical methods of improved identification, diversion, and treatment of persons with mental illness who, as a result of that illness, enter the criminal justice system; foster methods to improve training and social attitudes of professionals in the criminal justice system (attorneys, public defenders, judges) regarding the role of serious mental illness in the behaviors of mentally ill offenders; and increase access to and utilization of appropriate services and supports for the most treatment-resistant and severely mentally ill persons. Application and guidelines are available online.
Requirements: APA student affiliates enrolled in a psychology graduate program are eligible to apply.
Date(s) Application is Due: Sep 15
Amount of Grant: 1,000 U.S.D
Contact: Science Directorate; (202) 336-6000; fax (202) 336-5953; science@apa.org
Internet: http://www.apa.org/science/dissinfo.html
Sponsor: American Psychological Association
750 First Street NE
Washington, DC 20002-4242

American Psychosomatic Society Herbert Weiner Early Career Awards 573

These annual awards honor individuals early in their career who have contributed significantly to the field of psychosomatic medicine and show substantial promise of continued meritorious academic accomplishments in the field. Recipients have the opportunity to present their research findings during the annual meeting of the society. A plaque is included with the award.
Requirements: Nominees of any nationality must be fewer than 10 years past their final academic degree and must be members of the society.
Date(s) Application is Due: Nov 1
Amount of Grant: 1,000 U.S.D
Contact: Program Contact; (703) 556-9222; fax (703) 556-8729; info@psychosomatic.org
Internet: http://www.psychosomatic.org/awards/index.htm
Sponsor: American Psychosomatic Society
6728 Old McLean Village Drive
McLean, VA 22101

American Schlafhorst Foundation Grants 574

The foundation awards grants to eligible North Carolina nonprofit organizations in its areas of interest, including arts, children and youth, elderly, education, health care, science, and social services delivery. Types of support include building construction/renovation, equipment acquisition, general operating grants, research grants, scholarships, and seed money grants.

Requirements: 501(c)3 tax-exempt organizations serving the greater Charlotte, NC, area are eligible.
Restrictions: Individuals are not eligible.
Geographic Focus: North Carolina
Amount of Grant: 1,000 - 40,000 U.S.D
Contact: Grants Administrator; (704) 554-0800; info@schlafhorst.com
Sponsor: American Schlafhorst Foundation
8801 South Boulevard
Charlotte, NC 28224

American Society for Dermatologic Surgery Cutting Edge Research Grants 575
The goal of the program is to encourage research in areas of specific relevance to dermatologic surgery and/or cutaneous oncology. It is the hope of the ASDS Research Grant Committee to encourage well-conceived basic and clinical research projects to stimulate the transfer of new technologies from the clinical setting or laboratory to dermatologic surgery practice.
Restrictions: Only one application for an ASDS grant award will be considered from an institution or practice.
Date(s) Application is Due: Jul 15
Amount of Grant: 5,000 - 15,000 U.S.D
Contact: Medical and Scientific Committee Contact; (847) 956-0900; fax (847) 956-0999
Internet: http://www.asds-net.org/education/ResearchGrant/education-research.html
Sponsor: American Society for Dermatologic Surgery
5550 Meadowbrook Drive Suite 120
Rolling Meadows, IL 60008

American Society for Surgery of the Hand Outcome Studies Grants 576
Outcome studies grants are awarded to encourage young investigators to initiate projects that will lead to additional funding from other sources. There are no specific eligibility requirements, although the society encourages applications from surgeons and other allied health personnel who have a primary interest in hand surgery. All proposals must make use of a recognized and well-documented outcomes instrument, such as the SF-36, Sickness Impact Profile, Arthritis Impact Measurement Scale, etc., and must address an issue of importance to hand surgery. The society encourages applications from surgeons and other allied health personnel who have a primary interest in hand surgery.
Restrictions: Grants may not be used for the personal compensation of investigators, travel or conferences, publication costs, the payment of hospital care for patients under study, or the purchase of major pieces of equipment.
Date(s) Application is Due: Jul 11
Amount of Grant: 10,000 U.S.D
Samples: Dr. Thomas Trumble, Department of Orthopedic Surgery, U of Washington Medical Ctr (Seattle, WA)—for clinical trial of hand therapy following carpal tunnel surgery, $10,000; Dr. Jeffrey Katz, Department of Rheumatology, Brigham and Women's Hospital (Boston, MA)—for research on carpal tunnel syndrome and shoulder disorders, $10,000.
Contact: Julie Quinn, (847) 384-9250; fax (847) 384-1435; jquinn@assh.org
Internet: http://www.assh.org/Content/NavigationMenu/Hand_Surgery_Professionals/Outcome_Studies/Translational_and_Clinical_Research.htm
Sponsor: American Society for Surgery of the Hand
6300 N River Road, Suite 600
Rosemont, IL 60018-4256

American Society for Surgery of the Hand Research Grants 577
The society has established grant funding of research projects to encourage young investigators. It is anticipated that the funding will enable investigators to initiate projects that will lead to additional funding from other sources. Submission of proposals intended for NIH or OREF is discouraged. There are no specific guidelines within the framework of hand surgery; in the past, funding has been provided for a range of subjects, including microvascular, tendons, nerves, artificial joints, biochemical, biomechanical, rehabilitation, and other subjects.
Requirements: Awards are made to universities, medical schools, research institutions, or individuals. Grants may be used for salaries of technical and professional assistants, purchase of supplies, or minor pieces of equipment.
Restrictions: Grants may not be used for the personal compensation of investigators, travel or conferences, publication costs, the payment of hospital care for patients under study, or the purchase of major pieces of equipment. No overhead or indirect costs may be imposed by the recipient institution.
Date(s) Application is Due: May 2
Amount of Grant: Up to 16,000 U.S.D
Samples: Dr. Steven Moran—for Effects of S-Fluoracil on Flexor Tendon Repairs, $6000; Dr. Jack Choucka and Dr. Daniel Mass—for Cyclical Failure of Cadaveric Zone II Human Flexor Tendon Repairs, $8000.
Contact: Julie Quinn, (847) 384-9250; fax (847) 384-1435; jquinn@assh.org
Internet: http://www.assh.org/Content/NavigationMenu/Hand_Surgery_Professionals/Research/Basic_Science_Research.htm
Sponsor: American Society for Surgery of the Hand
6300 N River Road, Suite 600
Rosemont, IL 60018-4256

American Society on Aging Graduate Student Research Award 578
The award is presented annually to a graduate student for research relevant to aging and applicable to practice. The winner will be expected to attend the joint conference of the ASA and the National Council on the Aging to present the winning paper. The award comprises a certificate, complimentary registration and one-night's lodging for the conference, presentation of research findings at a highlighted session at the conference, and complimentary one-year membership in ASA. Submission guidelines and application are available online.
Requirements: Only graduate-level research will be considered. Findings must be from completed research; submission of the conceptual framework alone is not sufficient. Applicants must either be currently enrolled in a graduate degree program or must have completed their studies no more than one year before the time of submission. Applicants must be sponsored by a faculty member. Applicants and faculty sponsors need not be ASA members.
Restrictions: Papers that have been published are not eligible for submission.
Date(s) Application is Due: Oct 1
Contact: Nancy Decia, Coordinator, Education & Training; (415) 974-9610; fax (415) 974-0300; awards@asaging.org
Internet: http://www.asaging.org/asav2/awards/description_grad.cfm?submenu1=grad
Sponsor: American Society on Aging
833 Market Street, Suite 511
San Francisco, CA 94103-1824

American Society on Aging Mental Health and Aging Awards 579
The awards are made to organizations that have demonstrated high-quality, innovative programs that enhance the health-related quality of life in older adults. The program recognizes outstanding intervention programs developed in recent years to improve the health and health care of the aging. Model programs can include, but are not limited to, wellness/health promotion programs, community-based care, dementia care, mental health, caregiver supports, medication management, chronic illness care, end-of-life care, and programs that expand access to underserved groups of older adults. Winning programs will be honored at the Conference of The National Council on the Aging (NCOA) and the American Society on Aging (ASA). Up to six programs will receive cash awards, one-night's lodging for one person at the conference, and a one-year complimentary membership to ASA and the Healthcare and Aging Network.
Requirements: Applicants must be ASA members. The program must have at least one year's worth of program results and cannot have been in existence longer than five years (or must represent a substantially new program model in the past five years) by the date of submission.
Restrictions: Organizations that have won this award within the past two years are not eligible to apply.
Date(s) Application is Due: Oct 31
Amount of Grant: 2,500 U.S.D
Contact: Tobi Abramson, (516) 596-0073; tabramson@excite.com
Internet: http://www.asaging.org/asav2/awards/description_mhan.cfm?submenu1=mhan
Sponsor: American Society on Aging
833 Market Street, Suite 511
San Francisco, CA 94103-1824

American Society on Aging NOMA Award for Excellence in Multicultural Aging 580
Awards will be given to organizations that have demonstrated high-quality, innovative programs that enhance the lives of a multicultural aging population. Using the framework that supports the broader vision of diversity at the American Society on Aging, the NOMA Award for Excellence in Multicultural Aging seeks to identify and recognize best practices in developing and implementing programs and/or services that meet the needs of a multicultural aging population. Up to three programs will receive cash awards of $1,500, one complimentary conference registration, one night's lodging for one person at the conference and a one-year complimentary membership to ASA and the Network on Multicultural Aging (NOMA).
Requirements: Eligible applicants must: be an organization or have an affiliation with an organization that focuses on providing information and services to an aging population; have a program and/or service targeted to a multicultural aging population that has been tested, has a proven and successful track record, and has been in existence for a minimum of one year; and have a membership with ASA.
Date(s) Application is Due: Dec 15
Amount of Grant: 1,500 U.S.D
Contact: Mahi Sadeghi, Awards Coordinator; (415) 974-9602; fax (415) 974-0300; awards@asaging.org
Internet: http://www.asaging.org/asav2/awards/description_noma.cfm?submenu1=noma
Sponsor: American Society on Aging
833 Market Street, Suite 511
San Francisco, CA 94103-1824

American Sociological Association Minority Fellowships 581
Through its Minority Fellowship Program (MFP), the American Sociological Association (ASA) supports the development and training of sociologists of color in mental health and drug abuse research. Funded by a training grant sponsored by the National Institute of Mental Health (NIMH) and co-funded by the National Institute of Drug Abuse (NIDA), MFP seeks to attract talented doctoral students to ensure a diverse and highly trained workforce is available to assume leadership roles in research related to the nation's mental health and drug abuse research agendas.
Requirements: American citizens and permanent visa residents including, but not limited to, persons who are African American, Latino/Hispanic (e.g., Chicano, Cuban, Puerto Rican), Native American, and Asian American (e.g., Chinese, Japanese, Korean), and Pacific Islanders (e.g., Hawaiian, Guamanian, Samoan, Filipino) are eligible. The program is open to students beginning or continuing study in graduate sociology departments. New students must qualify for acceptance at accredited institutions of higher learning and express a commitment to sociological research on mental health. Upon completion of their support, recipients are expected to engage in behavioral research or teaching, or a combination thereof, for a period equal to the period of support beyond 12 months.

Date(s) Application is Due: Jan 31
Amount of Grant: 20,000 - 25,000 U.S.D
Contact: Margaret Weigers Vitullo, Director; (202) 383-9005, ext. 323; fax (202) 638-0882; spivack@asanet.org or apap@asanet.org
Internet: http://www.asanet.org/cs/root/leftnav/funding/minority_fellowship_program
Sponsor: American Sociological Association
1430 K Street NW, Suite 600
Washington, DC 20005

Amerigroup Foundation Grants 582
Helping to create healthy communities is the cornerstone of the Amerigroup Foundation's mission. The objective is to serve as a national resource that fosters an environment where there is a continuum of education, access and care, all of which improve the health and well-being of the financially vulnerable and uninsured Americans. Support is available in the states of: Delaware; District of Columbia; Florida; Georgia; Nevada; New Mexico; New York; Ohio; South Carolina; Tennessee; Texas and; Virginia. The Foundation primarily provides grants is the form of general support but program development and, sponsorships are also available to qualified non-profit organizations.
Requirements: 501(c)3 non-profits are eligible to apply from the following states: Delaware; District of Columbia; Florida; Georgia; Nevada; New Mexico; New York; Ohio; South Carolina; Tennessee; Texas and; Virginia.
Restrictions: Funding is unavailable for: projects or organizations that offer a direct benefit to the trustees of the Foundation or to employees or directors of Amerigroup; projects or organizations that might in any way pose a conflict with Amerigroup's mission, goals, programs, products or employees; projects or organizations that do not benefit a broad cross section of the community; individuals; political parties, candidates or lobbying activities; benefits, raffles, souvenir programs, trips, tours or similar events; for-profit entities, including start-up businesses.
Geographic Focus: Delaware, District of Columbia, Florida, Georgia, Nevada, New Mexico, New York, Ohio, South Carolina, Tennessee, Texas, Virginia
Amount of Grant: 500 - 10,000 U.S.D
Contact: Amy Sheyer, Grants Manager; fax (757) 490-6900; (757) 222-2360
Internet: http://www.realsolutions.com/company/pages/Foundation.aspx
Sponsor: Amerigroup Foundation
4425 Corporation Lane
Virginia Beach, VA 23462-3103

Amerisure Insurance Community Service Grants 583
The company awards grants nationwide to nonprofit organizations in the areas of education and health. There are no application forms. Send proposals to the headquarters office or the nearest facility. Proposals should contain a detailed description of the project and the amount requested. Junior achievement is supported especially in K-12 classrooms in the Farmington Hills, MI school system.
Requirements: Nonprofit organizations are eligible.
Date(s) Application is Due: Dec 31
Contact: Human Resources; (248) 615-9000
Internet: http://www.amerisure.com/au_1d_community.cfm
Sponsor: Amerisure Insurance Company
26777 Halsted Road
Farmington Hills, MI 48331-3586

Ametek Foundation Grants 584
Corporate contributions are made through the foundation to nonprofit organizations in company-operating areas. The foundation supports programs and projects in the categories of health, education, and social services, with education being the largest area of support. Under the category of health, grants are awarded to hospitals, health care facilities, and for medical research. Support for education is given to colleges and universities and technical schools, as well as scholarship funds. Welfare funding supports philanthropic organizations, including the United Way. Arts groups, museums, and civic groups also receive support. Types of support include general operating support, annual campaigns, building construction/renovation, equipment acquisition, endowment funds, scholarship funds, research, technical assistance, and matching funds. Annual application deadline dates may vary; contact the office for specific dates. The board meets in November and May to consider requests.
Requirements: IRS 501(c)3 organizations are eligible.
Restrictions: Grants are not made to individuals or to political, fraternal, or veterans organizations.
Date(s) Application is Due: Feb 28; Sep 1
Amount of Grant: 1,000 - 25,000 U.S.D
Samples: Rochester City School District (Rochester, NY)—grant recipient, $116, 231; Gnaden Huetten Memorial Hospital (Lehighton, PA)—grant recipient, $1000; Abilities, Inc of Florida (Clearwater, FL)—grant recipient, $15,000.
Contact: Kathryn Londra, (610) 647-2121; fax (610) 296-3412
Sponsor: Ametek Foundation
37 N Valley Road, Building 4, P.O. Box 1764
Paoli, PA 19301-0801

amfAR Fellowships 585
An amfAR fellowship is a grant that encourages the postdoctoral (M.D., Ph.D., or equivalent) investigator with limited experience in the field to advance a career in HIV/AIDS research. Fellowships are awarded for two years and may not be renewed for additional funding. amfAR fellows and mentors must be affiliated with the same nonprofit institution. The applicant's interest in a career in HIV/AIDS will be demonstrated by previous relevant work at the postdoctoral fellow or instructor level and will be carefully evaluated. Each fellowship is funded at a total of up to $125,000: a maximum of $110,000 is allowed for personnel (salary and fringe benefits) and other research-related direct costs. It is expected that a fellow will devote the decided majority of his or her time to the approved fellowship project. Personnel costs supported by the fellowship must represent a minimum of 85% effort and be consistent with institution policy for other institution personnel of similar rank and title, regardless of source(s) of support. An additional $3,636 is provided to support attendance at amfAR-approved professional development activities, for a direct cost maximum of $113,636. The period of performance for fellowships awarded under this RFP will be for two years starting January 1.
Requirements: The fellowship applicant must be mentored by an experienced investigator who: is qualified to oversee the proposed research; has successfully supervised postdoctoral fellows; and is at the associate professor level or higher.
Restrictions: Institutional indirect costs may not exceed 10% of direct costs.
Date(s) Application is Due: Jul 7
Amount of Grant: Up to 125,000 U.S.D
Contact: Grant Administrator; (212) 806-1696 or (212) 806-1600; fax (212) 806-1601; grants@amfar.org
Internet: http://www.amfar.org/lab/grants/default.aspx?id=7455&terms=Fellowships
Sponsor: American Foundation for AIDS Research
120 Wall Street, 13th Floor
New York, NY 10005-3908

amfAR Global Initiatives Grants 586
Since awarding its first international grant in 1986, amfAR has supported HIV/AIDS research, prevention, education, and training programs in dozens of countries across the developing world. Today, amfAR's international grant making supports two important initiatives. Formed in 2001, TREAT Asia is a network of clinics, hospitals, and research centers working with civil society to ensure the safe and effective delivery of HIV/AIDS treatments across Asia and the Pacific. And through a newly established MSM Initiative, amfAR awards grants to grassroots groups in support of innovative HIV/AIDS services for men who have sex with men in resource-limited countries.
Date(s) Application is Due: Jul 21
Amount of Grant: Up to 15,000 U.S.D
Samples: Matthew G. Law, Ph.D., University of New South Wales, Sydney, Australia, $157,465—for Asia Pacific HIV Observational Database (APHOD); Shivananda Khan, Naz Foundation International, London, United Kingdom, $10,000—for Risks and Responsibilities: Male Sexual Health and HIV.
Contact: Grant Administrator; (212) 806-1696 or (212) 806-1600; fax (212) 806-1601; grants@amfar.org
Internet: http://www.amfar.org/world/grants/default.aspx?id=192
Sponsor: American Foundation for AIDS Research
120 Wall Street, 13th Floor
New York, NY 10005-3908

amfAR Mathilde Krim Fellowships in Basic Biomedical Research 587
amfAR, The Foundation for AIDS Research, is pleased to announce the availability of support for Mathilde Krim Fellowships in Basic Biomedical Research. Each fellowship is funded at a total of up to $125,000 (phase I): a maximum of $110,000 is allowed for personnel (salary and fringe benefits) and other research-related direct costs. It is expected that a Krim fellow will devote the decided majority of his or her time to the approved fellowship project. Personnel costs supported by the fellowship must represent a minimum of 85% effort and be consistent with institution policy for other institution personnel of similar rank and title, regardless of source(s) of support. An additional $3,636 is provided to support attendance at amfAR-approved professional development activities, for a direct cost maximum of $113,636. Contingent upon subsequent application and peer review, phase II funding for an additional consecutive year may be approved to support basic biomedical HIV/AIDS research costs (up to $50,000 total) during the first year of a tenure-track position at any U.S. or international nonprofit research institution. The period of performance for Mathilde Krim Fellowships awarded under this RFP will be from January 1 to December 31 of the following year.
Restrictions: Institutional indirect costs may not exceed 10% of direct costs.
Date(s) Application is Due: Jul 7
Amount of Grant: Up to 125,000 U.S.D
Contact: Grant Administrator; (212) 806-1696 or (212) 806-1600; fax (212) 806-1601; grants@amfar.org
Internet: http://www.amfar.org/lab/grants/default.aspx?id=7455&terms=Fellowships
Sponsor: American Foundation for AIDS Research
120 Wall Street, 13th Floor
New York, NY 10005-3908

amfAR Public Policy Grants 588
AmfAR's public policy grants support the advancement of evidence-based AIDS-related public policies. Many of these awards facilitate the study and implementation of HIV prevention methods, such as harm reduction programs, that are based on sound science and help save lives. Public policy grants also help fund policy-related educational programs, including conferences.
Samples: Ronald Bayer, Ph.D., The Evolving Ethics and Politics for HIV Testing, Columbia University, Joseph L. Mailman School of Public Health, New York, NY, $66,000; Paula Santiago, 7th National Harm Reduction Conference, Harm Reduction Coalition, New York, NY, $10,000.
Contact: Grant Administrator; (212) 806-1696 or (212) 806-1600; fax (212) 806-1601; grants@amfar.org

Internet: http://www.amfar.org/hill/article.aspx?id=3920
Sponsor: American Foundation for AIDS Research
120 Wall Street, 13th Floor
New York, NY 10005-3908

amfAR Research Grants 589

Grants are awarded to support basic and clinical research projects in biomedical, humanistic, and social sciences research relevant to AIDS. In general, funds are applied to direct costs of salaries and fringe benefits for professional and technical personnel, laboratory supplies and equipment, travel, and the publication of findings. Research grants are awarded for one year without assurance of continued funding. $100,000 in direct costs plus up to 20% for indirect costs is available. Investigators are required to submit a pre-application letter of intent (LOI). Any postdoctoral investigator who is affiliated with a nonprofit institution may submit an LOI.
Requirements: Principal investigators for research grants must be faculty-level researchers affiliated with a nonprofit institution. Research grants are given to nonprofit institutions worldwide to support investigator-led projects approved by the Foundation.
Amount of Grant: 100,000 U.S.D
Samples: Columbia U (New York, NY)—to support the work of Susana Valente, for studies on microbicide development and AIDS-related research, $90,000.
Contact: Grant Administrator; (212) 806-1696 or (212) 806-1600; fax (212) 806-1601; grants@amfar.org
Internet: http://www.amfar.org/rfp/
Sponsor: American Foundation for AIDS Research
120 Wall Street, 13th Floor
New York, NY 10005-3908

Amgen Foundation Grants 590

Amgen seeks to: advance science education, improve quality of care and access for patients, and support resources that create sound communities where Amgen staff members live and work. Requests must be received at least 90 days in advance of the desired contribution date. Guidelines are available online.
Requirements: 501(c)3 tax-exempt organizations located in Amgen communities are eligible. Eligible grantees may include public elementary and secondary schools, as well as public colleges and universities, public libraries and public hospitals.
Restrictions: In general, Amgen does not consider requests for the following: support to individuals, fundraising or sports-related events, corporate sponsorship requests, religious organizations unless the program is secular in nature and benefits a broad range of the community, political organization or lobbying activity, labor unions, fraternal, service or veterans' organizations, private foundations, or organizations that are discriminatory.
Contact: Program Contact; (805) 447-4056 or (805) 447-1000; fax (805) 447-1010
Internet: http://wwwext.amgen.com/citizenship/apply_for_grant.html
Sponsor: Amgen Foundation
1 Amgen Center Drive, MS 38-3-B
Thousand Oaks, CA 91320-1799

AMHPS Dr. James A. Ferguson Emerging Infectious Diseases Fellowships 591

The Dr. James A. Ferguson Emerging Infectious Diseases Fellowship is a paid, eight-week, professional-development program (June and July) for racial- and ethnic-minority graduate students with an interest in public health. The fellowship exposes students to research opportunities at the Centers for Disease Control and Prevention (CDC). Students gain research experience in infectious-diseases research areas such as antimicrobial resistance, emerging infectious diseases, animal-related diseases, hospital-related infectious issues, etc. In addition to laboratory research, students are exposed to issues concerning public health via bi-weekly seminars (Brown Bag Sessions) and networking activities (Rap Sessions). Fellows are required to submit a progress report, literature review, abstract and final report throughout the course of the fellowship. Upon completion of all research, students are required to deliver a formal presentation for CDC researchers and Association of Minority Health Professions Schools, Inc. (AMHPS) staff. The fellowship provides a $4,000 stipend which is released in two installments – at the beginning and the end of the fellowship. Additionally the fellowship provides roundtrip travel or mileage reimbursement for students who attend school outside of the Atlanta, Georgia area (housing is provided to non-resident students.) The application form with instructions is available for download from the fellowship website and should be received by the Program Officer no later than the deadline date. Applicants are encouraged to start the application process early to give references enough time to fill out and return the recommendation section. The application must be notarized. The deadline date may vary from year to year. Interested applicants are encouraged to check the website or contact the Program Officer to verify current deadline dates.
Requirements: Applicants must be currently enrolled as full-time graduate level students at an AMHPS member institution or in a Non-AMHPS Master of Public Health Program. Applicants must commit to the length of the fellowship.
Restrictions: This internship program is restricted to members of an under-represented minority group as defined by the CDC Office of Minority Health & Health Disparities (OMHD).
Date(s) Application is Due: Feb 25
Amount of Grant: 4,000 U.S.D
Contact: Jamel Slaughter, Program Officer; (678) 904-4219; fax (678) 904-4518; jamel.slaughter@amhps.org
Internet: http://www.minorityhealth.org/p-student-drjames.php
Sponsor: Association Of Minority Health Professions Schools, Inc.
3525 Piedmont Road, 7 Piedmont Center, Suite 300
Atlanta, GA 30305

Amica Company and Foundation Grants 592

The program financially supplements employee volunteer efforts. Philanthropic programs include the Citizenship Grant Program—annual grants ($1000 maximum) to community groups in which Amica employees are involved; Matching Gift Funds—employee-matching grants ($1000 maximum) to elementary and secondary schools, two-year colleges and four-year colleges, and universities and college level graduate schools; and contributions to United Way. The company makes corporate donations to health organizations, hospitals, libraries, schools, museums, and other deserving institutions. Recent recipients include the American Heart Association, the Miami Project to Cure Paralysis, and The Nature Conservancy.
Contact: Grants Administrator; (800) 622-6422, ext. 2100; fax (401) 334-4241
Internet: http://www.amica.com/aboutUs/community/inYourComm.html#Amica Co. Foundation
Sponsor: Amica Insurance Company
100 Amica Way
Lincoln, RI 02865-1167

AMI Semiconductors Corporate Grants 593

The company makes grants to nonprofit organizations in support of the performing arts, economic development, business education, health cost containment, and social services for senior citizens. Types of support include conferences and seminars, general operating support, matching grants, multiyear grants, professorships, research, and scholarships. There are no application deadlines. Submit a letter of inquiry that includes a description of the organization and program, amount of funds requested, purpose of the request, recently audited financial statement, and proof of tax-exempt status. Contact office for grant availability.
Restrictions: The company does not support political or lobbying groups.
Amount of Grant: 1,000 - 2,500 U.S.D
Contact: Tamera Drake, (208) 234-6890; fax (208) 234-6795; tamera_Drake@amis.com
Internet: http://www.amis.com/about
Sponsor: AMI Semiconductors
2300 Buckskin Road
Pocatello, ID 83201

AMSSM Foundation Research Grants 594

The purpose of the AMSSM Foundation Research Grant Awards is to foster original scientific investigations by members of AMSSM. in this program, the Foundation welcomes research grant proposals that investigate issues within the broad discipline of sports medicine, including clinical practice, injury prevention and rehabilitation, basic science, epidemiology, and education. Grant awards are designed to provide partial support of research projects. Grantee institutions are expected to provide all necessary basic facilities and services that normally would be expected to exist in any institution qualified to undertake such research. Overhead or indirect costs will be supported to a maximum of 10% of direct costs. It is anticipated that the requested support will fall into one of three ranges. 1) $500 to $2,500 total costs annually for mailings, supplies, equipment, and technical support of small or pilot projects; 2) $2,500 to $10,000 total costs annually for a project larger in scope than the previous category and may provide partial support of research assistants or study coordinators; and 3) $10,000 to $25,000 total costs annually for a significant project that includes well documented pilot or previous studies and may provide partial salary support for key investigators. The maximum grant award is $25,000.
Requirements: The primary investigator of the grant must be an AMSSM member at the time of grant submission. Resident and fellow AMSSM members may apply as the principal investigator but must have at least one full AMSSM member listed as a co-investigator at the time of application.
Date(s) Application is Due: Oct 1
Amount of Grant: Up to 25,000 U.S.D
Contact: Grants Administrator; (913) 327-1415; fax (913) 327-1491; office@amssm.org
Internet: http://www.newamssm.org/Research%20Grant%20Information.html
Sponsor: American Medical Society for Sports Medicine
11639 Earnshaw
Overland Park, KS 66210

AMWA Anne C. Carter Student Leadership Award 595

AMWA honors the memory of Anne C. Carter, MD, with an annual award for outstanding student leadership. Nominees must meet the following criteria: national AMWA medical student member; demonstrated exceptional leadership skills through vision, inspiration, innovation, and coordination of local projects that further the mission of AMWA by improving women's health and/or supporting women in medicine; and must be nominated by an AMWA student chapter. Awardees will be honored during AMWA's annual meeting. Nomination guidelines are available online.
Date(s) Application is Due: Oct 31
Amount of Grant: 1,500 U.S.D
Contact: Award Administrator; (703) 838-0500; fax (703) 549-3864; info@amwa-doc.org
Internet: http://www.amwa-doc.org/index.cfm?objectId=855A0811-D567-0B25-5094E9176D8FFF58
Sponsor: American Medical Women's Association Foundation
801 N Fairfax Street, Suite 400
Alexandria, VA 22314

AMWA Carroll L. Birch Award 596

The annual award, sponsored by the Chicago Branch of AMWA, is presented for the best original research paper written by an AMWA student member. Award criteria include: original research paper written by a national AMWA medical student member; and applicant must attend an accredited U.S. allopathic or osteopathic medical school.

The paper may have been previously published. The recipient receives a cash award, a plaque presented at AMWA's annual meeting, and an article noting the award winner will appear in one of the association's publications. The award recipient is strongly encouraged to attend AMWA's annual meeting. Submit the following materials to the office: cover letter stating applicant's medical school, graduation date, and permanent mailing address; an abstract of 250 words or less; four copies of the manuscript; and a letter from the faculty research sponsor.
Date(s) Application is Due: Jun 30
Amount of Grant: 1,000 U.S.D
Contact: Award Administrator; (703) 838-0500; fax (703) 549-3864; info@amwa-doc.org
Internet: http://www.amwa-doc.org/index.cfm?objectId=E5F39C87-D567-0B25-5377ABBB15148B2E
Sponsor: American Medical Women's Association Foundation
801 N Fairfax Street, Suite 400
Alexandria, VA 22314

AMWA Glasgow-Rubin Essay Award **597**
The annual award is presented to a medical student for the best essay about a mentor. Award criteria include: AMWA national student member; attends an accredited U.S. allopathic or osteopathic medical school; original research paper written about a personal relationship with a woman physician mentor; and essay must be approximately 1000 words. Submitted essays should be of publishable quality; typed, double-spaced; and title page to include title, author's name, address and phone number, email address, and name medical school.
Date(s) Application is Due: May 31
Amount of Grant: 1,000 U.S.D
Contact: Award Administrator; (703) 838-0500; fax (703) 549-3864; info@amwa-doc.org
Internet: http://www.amwa-doc.org/index.cfm?objectId=288C83B1-D567-0B25-539772C31D5C35EC
Sponsor: American Medical Women's Association Foundation
801 N Fairfax Street, Suite 400
Alexandria, VA 22314

Andersen Corporate Foundation **598**
The Foundation contributes to: organizations that enhance self-sufficiency for people living in poverty, senior citizens, and people with disabilities; organizations that promote safe and healthy environments, as well as organizations that seek to improve health through prevention and education programs, primarily for young people, senior citizens, and people in vulnerable situations; organizations that offer intellectual and social opportunities with a focus primarily on young people, senior citizens, and people with disabilities; and support that builds, promotes, and preserves communities. Complete guidelines are available online.
Requirements: All grant recipients must meet these giving guidelines: the organization's programs and services are consistent with Andersen Corporate Foundation's mission and values; the organization's purpose and programs fit within Andersen Corporate Foundation's defined program focus areas; the organization can demonstrate sound fiscal management and effective delivery of services; nonprofit corporation is registered under section 501(c)3 of the IRS Code in the United States or for Canadian Charities with the Canadian Revenue Agency.
Restrictions: The Foundation avoids making grants in organizations that are in competition with each other to provide the same service to the community. The Foundation does not support endowments or make grants to individuals. Funding is not granted to national research organizations.
Geographic Focus: Iowa, Minnesota, Virginia, Wisconsin, Canada
Date(s) Application is Due: Apr 15; Jul 15; Oct 15; Dec 15
Contact: Program Director; (888) 439-9508 or (651) 439-1557; fax (651) 439-9480; andersencorpfdn@srinc.biz
Internet: https://www.srinc.biz/bp/index.html
Sponsor: Andersen Corporate Foundation
342 5th Avenue N
Bayport, MN 55003

Andre Agassi Charitable Foundation Grants **599**
The foundation awards grants to at-risk youth programs targeting education and recreation in the Las Vegas area. There are no application forms or deadlines. Submitted requests will be considered for funding in the following calendar year. Detailed guidelines are available online.
Requirements: Nevada 501(c)3 nonprofits serving the Las Vegas area are eligible.
Restrictions: Grants do not support organizations or projects outside the Las Vegas community, organizations that discriminate, individuals, advertising, religious or sectarian organizations for religious purposes, or political organizations and programs designed to influence legislation or elect candidates to public office.
Geographic Focus: Nevada
Samples: Las Vegas Philharmonic, Las Vegas, NV, $10,000—art program for underprivileged youths; Boys and Girls Club of Nevada, Las Vegas, NV, $334,403—after school program; Center for Independent Living, Las Vegas., NV, $5,000—residential treatment for adolescents.
Contact: Julie Pippenger, Chief Operating Officer; (702) 227-5700; fax (702) 866-2928; info@agassi.net
Internet: http://www.agassifoundation.org
Sponsor: Andre Agassi Charitable Foundation
3960 Howard Hughes Parkway, Suite 750
Las Vegas, NV 89169

Andrew Balint Charitable Trust Grants **600**
The trust operates in the United Kingdom and internationally and awards grants to nonprofit charities in the fields of education, medicine and health, and social welfare.
Contact: J.K. Oliver, 020 7624 2098
Sponsor: Andrew Balint Charitable Trust
4-6 Canfield Place, Suite A
London, NW6 3BT England

Andrew Family Foundation Grants **601**
The Andrew Family Foundation is a private, philanthropic organization that will consider proposals from public, non-profit organizations under IRS Section 501(c)3 to support projects and organizations that foster individual growth and enhance communities through education, humanitarian efforts, and the arts. Funding primarily in the Illinois with a special interest in the Cook County region. The types of support available include: annual campaigns; building/renovation; capital campaigns; general/operating support; scholarship funds. There is no deadline date when applying for funding. Qualified grant proposals will be reviewed by the Andrew Family Foundation Grant Making Committee prior to quarterly board meetings. The committee will make a recommendation to the Board of Directors of the Foundation.
Requirements: 501(c)3 tax-exempt organizations are eligible to apply for funding. To begin the application process, take the Eligibility Quiz to confirm that you qualify for a grant from the foundation. Upon successful completion of the Eligibility Quiz, you will be invited to complete an online Letter of Inquiry. The Board will review your Letter of Inquiry and may invite you to submit a Full Application for review. Generally, the Board meets in February, May, August and November of each year.
Restrictions: The foundation does not provide funds: to individuals; taxable corporations; religious programs; political organizations; and other private foundations.
Geographic Focus: Illinois
Contact: Connor Humphrey, Grants Administrator; (708) 460-1288 or (602) 828-8471; fax (602) 385-3267; aff@inlignwealth.com or Connor.Humphrey@GenSpring.com
Internet: https://online.foundationsource.com/andrew/board2.htm
Sponsor: Andrew Family Foundation
14628 John Humphrey Drive
Orland Park, IL 60462

Anesthesia Clinical Research Starter Grants **602**
The grants offer start-up money to institutions on behalf of faculty members who seek to obtain further funding for continuation of their anesthesiology-related projects that focus on scholarly, hypothesis-based clinical studies. Institutional matching funds are required.
Requirements: Faculty members may be at the associate, instructor, or assistant professor level.
Date(s) Application is Due: Feb 15; Aug 15
Amount of Grant: 35,000 - 50,000 U.S.D
Contact: Thomas Bruckman, Executive Director; (507) 266-6866; fax (507) 284-0120; grunewald.nathan@mayo.edu
Internet: http://www.faer.org/grants.php
Sponsor: Foundation for Anesthesia Education and Research
200 First Street SW, Charlton Building, Mayo Clinic
Rochester, MN 55905

Angel Kiss Foundation Grants **603**
Angel Kiss Foundation (AKF) is a nonprofit organization dedicated to helping families with children who are being treated for cancer. The Foundation provides financial assistance to families to help with the astronomical expenses associated with the disease. The Foundation works with families on an individual basis and reviews the needs of each family. It provides financial assistance for a variety of expenses including: medical expenses, insurance deductibles, uncovered procedures, household living expenses, mortgage payments, rent, food, utilities, travel expenses related to treatment and other family expenses causing hardship during the child's cancer treatment. Families can work with their social worker and medical doctor for a referral to Angel Kiss Foundation, call Angel Kiss Foundation for guidance in the application procedure, or access the Web site for an online application.
Contact: Donna Breen; (775) 323-7721 or (888) 589-5477; donna@dlbreen.com
Internet: http://angelkissfoundation.org/About/FAQ.html
Sponsor: Angel Kiss Foundation
150 Ridge Street, Suite 3
Reno, NV 89501

Angels Baseball Foundation Grants **604**
Since its inception in 2004, the Angels Baseball Foundation has awarded many grants to worthy organizations throughout the community, ranging from music programs to at-risk youth shelters. Currently, it focuses on initiatives aimed to create and improve education, health care, arts and sciences, and community related youth programs, in addition to providing children the opportunity to experience the game of baseball. Giving is limited to the greater Los Angeles, California, area. Fields of interest include: the arts, athletics/sports activities, children and youth, cancer detection, cancer research, community and economic development, education (at all levels), health care, and community service programs. Applicants should begin by contacting the Foundation directly.
Requirements: Applicants must serve the residents of the greater Los Angeles area.
Geographic Focus: California
Contact: Anne Bafus, Community Relations; (714) 940-2174 or (714) 940-2244
Internet: http://losangeles.angels.mlb.com/ana/community/baseball_foundation.jsp
Sponsor: Angels Baseball Foundation
2000 E Gene Autry Way
Anaheim, CA 92806-6100

Anheuser-Busch Foundation Grants 605

Support is provided almost exclusively to causes that are located in communities in which the company has manufacturing facilities. Contributions are made for education, health, social services, minorities and youth, cultural enrichment, and environmental protection programs. Types of support include capital grants, employee matching gifts, equipment and material acquisition, general support, and donated products. Full proposals are accepted throughout the year. At the Web site, click on Guidelines for Charitable Giving.
Requirements: 501(c)3 tax-exempt organizations in St. Louis, MO; Baldwinsville, N.Y; Cartersvile, Ga; Columbus, Oh; Fairfield, Ca; Fort Collins, Co; Houston, Tx; Jacksonville, Fl; Los Angeles, Ca; Merrimack, N.H; Newark, N.J; Williamsburg, Va., Orlando, Fl; Tampa, Fl; San Antonio, Tx; San Diego, Tx.
Restrictions: Grants are not made to individuals; political, social, fraternal, religious, or athletic organizations; or hospitals for operating funds.
Geographic Focus: California, Colorado, Florida, Georgia, Missouri, New Hampshire, New Jersey, New York, Ohio, Texas, Virginia
Amount of Grant: 25,000 - 100,000 U.S.D
Contact: Assistant Manager; (314) 577-2453; fax (314) 557-3251
Internet: http://www.anheuser-busch.com/CharitableGivingIndex.html
Sponsor: Anheuser-Busch Foundation
One Busch Place
Saint Louis, MO 63118-1852

Anna Fitch Ardenghi Trust Grants 606

The Anna Fitch Ardenghi Trust was established in 1981 to support and promote quality educational, cultural, human services, and health care programming for underserved populations living in New Haven, Connecticut. Special preference is given to charitable organizations that focus on the arts or youth-related programming. The deadline for applications is July 1, and applicants will be notified of grant decisions by letter within 2 to 3 months after the proposal deadline. Grants from the Ardenghi Trust are 1 year in duration.
Requirements: 501(c)3 organizations serving the residents of New Haven, Connecticut, are eligible to apply.
Restrictions: Grant requests for capital projects will not be considered. Applicants will not be awarded a grant for more than 3 consecutive years.
Geographic Focus: Connecticut
Date(s) Application is Due: Jul 1
Contact: Carmen Britt, Senior Foundation Officer; (860) 657-7019; carmen.britt@baml.com
Internet: https://www.bankofamerica.com/philanthropic/grantmaking.action
Sponsor: Anna Fitch Ardenghi Trust
200 Glastonbury Boulevard, Suite #200, CT2-545-02-05
Glastonbury, CT O6033-4056

Anna Louise Hoffman Award for Outstanding Achievement in Graduate Chemistry Research 607

At the time of nomination, the candidate must be a full-time female graduate student who is a candidate for a graduate degree at an accredited institution. Research presented by the candidate must be original and of one of the main chemical divisions—analytical, biochemical, inorganic, organic, physical, and/or ancillary. The nominee may be, but need not be, a member of Iota Sigma Pi.
Requirements: Nomination must be made by members of the institution's graduate faculty. Dossier must contain an academic history, including all transcripts; candidate's permanent and school addresses; two letters of recommendation; the nomination; a brief description of the candidate's research; and a list of publications and talks or papers presented outside of degree requirements. Description of the research should be no more than 1000 words and prepared by the candidate.
Date(s) Application is Due: Feb 15
Amount of Grant: 500 U.S.D
Contact: Vicki Grassian, (319) 335-1392; fax (319) 353-1115; vicki-grassian@uiowa.edu
Internet: http://www.iotasigmapi.info//studentAwards.htm
Sponsor: Iota Sigma Pi National Honor Society for Women in Chemistry
Euclid Avenue at East 24th Street
Cleveland, OH 44115

Ann and Robert H. Lurie Family Foundation Grants 608

Founded in 1986, the foundation initially had set up six categories: medical services and research; child-related medical organizations; basic services including food and shelter; education; the arts; and so-called wild things. When Ann Lurie's husband, Robert, died in 1990, she was left with a fortune worth hundreds of millions and six kids — ages 5 to 15. She has given away well over $100-million since Robert H. Lurie died at age 48, generally spurning requests from nonprofit groups with undistinguished records and concentrating her gifts on some of the world's biggest problems, including hunger, cancer, and inadequate health care. Today, the foundation supports educating and providing resources to children so they are better prepared for the future.
Requirements: There are no specific deadlines or application forms to fill out. An organization requesting funding should have finances in good order, demonstrating that it is making good use of its resources. Note: Ann Lurie has a preference for identifying her own projects, rather than responding to solicitations.
Geographic Focus: Illinois, Michigan
Samples: University of Michigan College of Engineering (Ann Arbor, Michigan) - $2,500,000; University of Michigan Entrepreneurial Institute (Ann Arbor, Michigan) - $1,000,000; Africa Infectious Disease Village Clinic (Chicago, Illinois) - $403,429 Ancient Egypt Research Associates (Cambridge, Massachusetts) - $400,000; Infant Welfare Society of Chicago (Chicago, Illinois) - $154,457; K-idea Trust (Chicago, Illinois) - $111,160; Riders For Health II (Rockville, Maryland) - $100,000; Greater Chicago Food Depository (Chicago, Illinois) - $65,000; Robert H. Lurie Comprehensive Cancer Center (Chicago, Illinois) - $63,870; The Synergos Institute, Inc. (New York, New York) - $25,000; The Chicago Public Library Foundation (Chicago, Illinois) - $25,000; Marwen Foundation (Chicago, Illinois) - $25,000; The Smile Train (New York, New York) - $25,000; Choate Rosemary Hall (Wallingford, Connecticut) - $25,000; The Chicago Public Library Foundation (Chicago, Illinois) - $25,000; Children Affected By AIDS Foundation (Chicago, Illinois) - $25,000; Children's Memorial Foundation (Chicago, Illinois) - $23,750; Chicago Cultural Center Foundation (Chicago, Illinois) - $10,000
Contact: Ann Lurie, President; (312) 466-3750; fax (312) 466-3700
Sponsor: Ann and Robert H. Lurie Family Foundation
2 N Riverside Plaza, Suite 1500
Chicago, IL 60606-2600

Ann Arbor Area Community Foundation Grants 609

The foundation is interested in funding projects that will improve the quality of life for citizens of the Ann Arbor, Michigan, area. Eligible projects generally fall within the categories of education, culture, social service, community development, environmental awareness, health and wellness, and youth and senior citizens. Types of support include: emergency funds, program development, conferences and seminars, publication, seed grants, scholarship funds, research, and matching funds. Higher priority is given to programs that are preventive rather than remedial, increase individual access to community resources, examine and address the underlying causes of local problems, promote independence and personal achievement, attract volunteer resources and support, strengthen the private nonprofit sector, encourage collaboration with other organizations, and build the capacity of the applying organization. Organizations interested in applying are strongly encouraged to discuss their project with the program director prior to submitting an application.
Requirements: 501(c)3 nonprofit organizations in the Ann Arbor, MI, area, which is the area that falls within the boundaries of the Ann Arbor public schools district, are eligible.
Restrictions: The foundation usually does not make grants for construction projects, annual giving campaigns or capital campaigns, normal operating expenses (except for start-up purposes), religious or sectarian purposes, computer hardware equipment, individuals, advocacy or political purposes, multiyear funding, or regranting.
Geographic Focus: Michigan
Date(s) Application is Due: Feb 11; Oct 7
Samples: Jazzistry, $9,139 for Finding Your Place in the Jazz Story; Community Respite Center, Inc., $14,079 for development of a Lifespan Respite Center for Washtenaw County; Jewish Family Services, $9,000 for a database implementation project; Washtenaw United Way, $25,000 for a 2-1-1 Call Center; Vineyard Church of Ann Arbor, $8,000 or a mentoring ministry for single parent families; University of Michigan-Turner Geriatric Clinic, $3,500 for the Senior Unmet Needs Fund (all 2006).
Contact: Phil D'Anieri, Program Director; (734) 663-0401; fax (734) 663-3514; pdanieri@aaacf.org
Internet: http://www.aaacf.org/grants.asp
Sponsor: Ann Arbor Area Community Foundation
301 N Main Street, Suite 300
Ann Arbor, MI 48104

Anne J. Caudal Foundation Grants 610

The Anne J. Caudal Foundation was established in 2007 to benefit disabled veterans of any time or of any branch of the United States armed forces and to perpetuate the recognition or memory of their accomplishments or sacrifice in time of war or otherwise. Special consideration is given to organizations that serve disabled veterans in New Jersey. Grant requests for general operating support are strongly encouraged. Program support will also be considered. Small, program-related capital expenses may be included in general operating or program requests. The majority of grants from the Caudal Foundation are one year in duration. On occasion, multi-year support is awarded. Application materials are available for download from the grant website. The application deadline for the Anne J. Caudal Foundation is July 1. Applicants are encouraged to review the state application guidelines for additional helpful information and clarification before applying. Applicants are also encouraged to view the foundation's funding history (link is available at the grant website). Applicants will be notified of grant decisions before August 15.
Requirements: Applicants must have 501(c)3 tax-exempt status.
Restrictions: The foundation does not support requests from individuals, organizations attempting to influence policy through direct lobbying, or any political campaigns.
Date(s) Application is Due: Jul 1
Samples: United States Wounded Soldiers Foundation, Dallas, Texas, $12,500 (2010); Our Military Kids, McLean, Virginia, $12,500 (2010); Jewish Family Service of New Jersey, Wayne, New Jersey, $12,500 (2010).
Contact: Maryann Clemente; (646) 855-0786; maryann.clemente@baml.com
Internet: https://www.bankofamerica.com/philanthropic/fn_search.action
Sponsor: Anne J. Caudal Foundation
One Bryant Park, NY1-100-28-05
New York, NY 10036

Annenberg Foundation Grants 611

The foundation provides support for projects within its grantmaking interests of education and youth development, arts and culture, civic, community and the environment, and health and human services. It encourages the development of more effective ways to share ideas and knowledge. Letters of inquiry may be submitted at all times during the year and there are no deadlines. Please review the grants database for additional types of grants given.
Requirements: 501(c)3 tax-exempt organizations are eligible.

Restrictions: Full proposals are not accepted unless requested by a Foundation representative. The foundation is not presently considering inquiries for: individuals, individual K-12 schools, for-profit organizations, political activities or attempts to influence specific legislation, individual scholarships, projects focused exclusively on research, or programs outside of its grant-making interests.
Samples: Centro Latino De Educacion Popular, Los Angeles, CA, $50,000 - for the Spanish Literacy Campaign to mobilize, train, equip, and support Spanish speakers to teach their non-literate friends and family members to read and write using an online literacy program; American Patrons of the Tate Gallery Foundation, Inc., New York, NY, $50,000 - to support special exhibitions and education programs at the Tate Modern; National Public Radio, Inc., Washington, DC, $500,000 - for the programs of NPR West to broaden the creation and distribution of high-quality public radio programming.
Contact: Leonard Aube, Executive Director; (310) 209-4560; fax (310) 209-1631; info@annenbergfoundation.org
Internet: http://www.annenbergfoundation.org/grants
Sponsor: Annenberg Foundation
2000 Avenue of the Stars, Suite 1000
Los Angeles, CA 90067

Annie Sinclair Knudsen Memorial Fund/Kaua'i Community Grants 612
The priorities of the Fund are projects and services that benefit the Kaua'i community including: culture and arts; education; environment; and health and human services. Priority will be given to programs that are well-defined and likely to be successful, address a community need, demonstrate an ability to deliver, and have an adequate budget. Application information is available online.
Requirements: Any nonprofit, tax-exempt 501(c)3 organization, neighborhood group or project is eligible.
Geographic Focus: Hawaii
Date(s) Application is Due: Jul 16
Amount of Grant: 2,000 - 15,000 U.S.D
Contact: Deborah Rice, (808) 245-4585; drice@hcf-hawaii.org
Internet: http://www.hawaiicommunityfoundation.org/index.php?id=71&categoryID=22
Sponsor: Hawai'i Community Foundation
1164 Bishop Street, Suite 800
Honolulu, HI 96813

Ann Jackson Family Foundation Grants 613
The foundation awards grants to eligible nonprofit organizations in its areas of interest, including secondary education, healthcare, child welfare, animal welfare, and the disabled. Types of support include building construction/renovation, capital campaigns, and general operating grants. Most grants are awarded to nonprofits in Santa Barbara, California. There are no application forms or deadlines. Contact the foundation for document attachments required.
Geographic Focus: California
Amount of Grant: 5,000 - 50,000 U.S.D
Samples: All Saints By the Sea Episcopal Church, Santa Barbara, CA, $14,000—general purposes grant; Bowl Foundation of Santa Barbara, Santa Barbara, CA, $20,000—building project grant; Dunn School, Los Olivos, CA, $50,000—faculty housing grant.
Contact: Palmer Jackson, President; (805) 969-2258
Sponsor: Ann Jackson Family Foundation
P.O. Box 5580
Santa Barbara, CA 93150-5580

Ann Peppers Foundation Grants 614
The foundation awards grants to eligible California nonprofit organizations in its areas of interest, including arts and cultural programs, disabled, elderly, health care, private education, and social services. Types of support include capital grants, general operating grants on a temporary basis, matching grants, scholarships, and research grants. Grants are initiated by the foundation manager. There are no application deadlines; the board meets quarterly to consider requests.
Requirements: Southern California 501(c)3 nonprofits, colleges, and universities are eligible. Preference is given to requests from Los Angeles County.
Geographic Focus: California
Amount of Grant: 2,000 - 50,000 U.S.D
Samples: Boys Republic (Chino Hills, CA)—for kitchen renovation at residential facility , $5,000; Pepperdine University (Malibu, CA)—for scholarships, $25,000; Huntington Medical Research Institute (Pasadena, CA)—to purchase imaging technology for preclinical breast cancer, $5,000; Marymount College (Rancho Palos Verdes, CA)—for scholarships, $10,000.
Contact: Jack Alexander, Secretary; (626) 449-0793
Sponsor: Ann Peppers Foundation
625 S Fair Oaks Avenue
South Pasadena, CA 91030

Annunziata Sanguinetti Foundation Grants 615
The foundation supports charitable organizations in San Francisco, California that are devoted wholly or partially to the care, treatment, rehabilitation, and education of children with physical or mental defects or illnesses. The distribution committee meets annually, usually in November or early December. Proposals are accepted each year between July 1 and October 31.
Requirements: Nonprofit organizations in the San Francisco, California area are eligible to apply. Initial approach should be through a letter of inquiry. There is no application form required when submitting a proposal. The Proposal should include the following items: copy of IRS Determination Letter; copy of most recent annual report/audited financial statement/990; detailed description of project and amount of funding requested; 1 copy of the proposal.
Restrictions: Restricted to organizations which benefit the children residing in San Francisco County, California.
Geographic Focus: California
Date(s) Application is Due: Oct 31
Amount of Grant: 5,000 - 50,000 U.S.D
Samples: Lighthouse for the Blind and Visually Impaired, San Francisco, CA, $20,000; Edgewood Children's Center, San Francisco, CA, $20,000; Lukemia & Lymphona Society, San Francisco, CA, $12,000.
Contact: Eugene Ranghiasci, Vice President; (415) 396-3215; fax (415) 834-0604
Sponsor: Annunziata Sanguinetti Foundation
420 Montgomery Street, 5th Floor
San Francisco, CA 94106

ANS/AAO-HNSF Herbert Silverstein Otology and Neurotology Research Award 616
The purpose of this award, which is jointly sponsored by the AAO-HNS Foundation and the American Neurotology Society, is to support a clinical or translational research project focused on diseases, disorders, or conditions of the peripheral or central auditory and/or vestibular system among new full-time academic surgeons. The award is intended as a preliminary step in clinical investigator career development and is expected to facilitate the recipient's preparation of a more comprehensive individualized research plan suitable for submission to the National Institutes of Health or comparable funding agency.
Requirements: Applicants must be physicians in the United States or Canada with demonstrated potential for excellence in research and teaching and serious commitment to an academic research career in otology or neurotology. Priority will be given to fellows or junior faculty, who have completed residencies or fellowships within four years of the application receipt date, although otolaryngology-head and neck surgery residents are eligible. All candidates must be sponsored by the Chair of his/her Division or Department and by an official representative of the institution which would administer the Award and in whose name the application is formally submitted.
Amount of Grant: Up to 25,000 U.S.D
Contact: Stephanie L. Jones; (703) 519-1586 or (703) 836-4444; sljones@entnet.org
Internet: http://www.americanneurotologysociety.com/funding.html
Sponsor: American Academy of Otolaryngology-Head and Neck Surgery Foundation
1650 Diagonal Road
Alexandria, VA 22314-2857

Anschutz Family Foundation Grants 617
The foundation makes grants in Colorado's rural and urban communities, to assist the elderly, the young, and the economically disadvantaged. Support is given to programs and projects that strengthen families and enable individuals to become productive and responsible citizens of society. Religious organizations, including Christian, Lutheran, Presbyterian, Roman Catholic, and the Salvation Army, also are eligible for funding. Types of support include special projects, general operating budgets, continuing support, seed money, emergency funds, technical assistance, and publications.
Requirements: Nonprofit organizations in Colorado may apply.
Restrictions: Grants are not awarded to individuals, programs outside of Colorado, graduate and post-graduate research, religious organizations for religious purposes, special events, promotions or conferences, candidates for political office, endowments, debt reduction, multi-year grants, and capital campaigns.
Geographic Focus: Colorado
Date(s) Application is Due: Jan 15; Aug 1
Amount of Grant: 2,500 - 10,000 U.S.D
Samples: The Center for Hearing, Speech, and Language (Denver, CO)—to support Kidscreen, a vision, hearing, speech and language developmental screening service for low-income families, $5,000; The Learning Source for Adults and Families (Denver, CO)—general operating support for neighborhood literacy programs for adults and families, $7,500; Seniors Resource Center (Denver, CO)—for general operating support for programs that maximize seniors' independence and dignity, $5000; Women's Resource Center (Durango, CO)—To support the resource and referral program, $5,000.
Contact: Sue Anschutz-Rodgers, President & Executive Director; (303) 293-2338; fax (303) 299-1235; info@anschutzfamilyfoundation.org
Internet: http://www.anschutzfamilyfoundation.org/info.htm
Sponsor: Anschutz Family Foundation
555 17th Street, Suite 2400
Denver, CO 80202

Ansell, Zaro, Grimm & Aaron Foundation Grants 618
The Ansell, Zaro, Grimm and Aaron Foundation was established in New Jersey in 2007 by a leading law firm of the same name. Funding is offered primarily in the State of New Jersey to Jewish agencies and synagogues, health research, and social service agencies in support of the economically disadvantaged. There are no specific deadlines or application forms with which to adhere, and applicants should initially approach the Foundation with a letter of request detailing the purpose of the grant and verification of tax exempt status. The maximum amount given is $5,000.
Requirements: 501(c)3 organizations supporting the residents of New Jersey are eligible.
Geographic Focus: New Jersey
Amount of Grant: Up to 6,000 U.S.D
Samples: 180 Turning Lives Around, Hazlet, New Jersey, $3,500; American Recreational Military Service, Red Bank, New Jersey, $100; Asbury Park 4th of July Celebration, Asbury Park, New Jersey, $5,400.

Contact: Jerold L. Zaro, Trustee; (732) 922-1000; fax (732) 922-6161; jlz@ansellzaro.com
Internet: http://www.ansellzaro.com/index_contact.html
Sponsor: Ansell, Zaro, Grimm and Aaron Foundation
1500 Lawrence Avenue
Ocean, NJ 07712-4023

ANS Neurotology Fellows Award 619

The Neurotology Fellows Award was established by the American Neurotology Society in 1996 to reward scientific excellence. The award has been generously endowed through royalties received for the textbook Neurotology, Mosby Publishers, D. Brackmann and R. Jackler, editors. One or more awards of $500 each is intended to subsidize travel expenses incurred while giving a scientific presentation at the Annual Spring Meeting of the American Neurotology Society. The material presented need not have been performed during the Fellowship year; it may derive from earlier work performed during the residency or extracurricular research experience. Applicants should submit electronically a cover letter, a supporting letter from the Fellowship director to verify qualifications, and electronic submission of the abstract by October 15.
Requirements: Applicants must be full-time participants in a post-residency Neurotology Fellowship in the United States or Canada. In addition, the Fellow must be both podium presenter and first author of the paper submitted for publication.
Date(s) Application is Due: Oct 15
Amount of Grant: 500 U.S.D
Contact: Shirley Gossard, Administrator; (352) 751-0932 or (217) 414-4868; fax (352) 751-0696; segossard@aol.com
Internet: http://www.americanneurotologysociety.com/funding.html
Sponsor: American Neurotology Society
3096 Riverdale Road
The Villages, FL 32162

ANS Nicholas Torok Vestibular Award 620

The Nicholas Torok Vestibular Award, established under the aegis of the American Neurotology Society instituting a yearly lecture, is to be presented at the Annual Scientific Spring Meeting of the Society. The subject should be an innovative observation, experience or technique in the field of vestibular basic science or clinical science. The candidate shall be selected preferably from the membership of the Society or upon specific merit of any M.D. or Ph.D. A yearly award of $1,500 will be given for the selected manuscript.
Requirements: The manuscript must be presented electronically in the publication format recommended by Otology & Neurotology, the official journal of the American Neurotology Society.
Date(s) Application is Due: Oct 15
Amount of Grant: 1,500 U.S.D
Contact: Shirley Gossard, Administrator; (352) 751-0932 or (217) 414-4868; fax (352) 751-0696; segossard@aol.com
Internet: http://www.americanneurotologysociety.com/funding.html
Sponsor: American Neurotology Society
3096 Riverdale Road
The Villages, FL 32162

ANS Trainee Award 621

The American Neurotology Society offers an award for the best clinical or basic science paper in Neurotology submitted by a Resident or Fellow in training in the field of Otolaryngology-Head and Neck Surgery. A clinical study or basic research study is acceptable. The Society has established a cash award of $1,000 for the winning paper. The author will be expected to attend the Annual ANS Spring Meeting and present the paper personally.
Requirements: Residents from any approved residency program or Fellows in Neurotology in the United States or Canada are eligible to compete. The Resident or Fellow must be the primary author. He or she must provide the main inspiration for the paper and do the literature search, data collection, original and final drafts of the article, the discussion, and the conclusions.
Restrictions: The paper must be an original contribution, not previously published.
Date(s) Application is Due: Oct 15
Amount of Grant: 1,000 U.S.D
Contact: Shirley Gossard, Administrator; (352) 751-0932 or (217) 414-4868; fax (352) 751-0696; segossard@aol.com
Internet: http://www.americanneurotologysociety.com/funding.html
Sponsor: American Neurotology Society
3096 Riverdale Road
The Villages, FL 32162

Anthem Blue Cross and Blue Shield Grants 622

As a multistate entity, the program supports charitable, civic, and nonprofit groups in company operating areas. Categories of giving include healthy minds (educational), healthy bodies (health oriented), and healthy communities (community projects and human services initiatives). Joint sponsorships with hospitals and providers for programs that benefit the community are encouraged. Requests must be submitted in writing. Contact the office for guidelines.
Requirements: 501(c)3 tax-exempt organizations in corporate communities are eligible.
Restrictions: Grants do not support individuals; political organizations/candidates; religious organizations; fraternal, military, or labor organizations; or providers/hospitals.
Geographic Focus: Colorado, Connecticut, Indiana, Kentucky, Maine, Missouri, Nevada, New Hampshire, Ohio, Virginia, Wisconsin
Contact: Vicki Perkins, (317) 488-6216
Internet: http://www.anthem.com/wps/portal/ahpculdesac?content_path=shared/noapplication/f4/s0/t0/pw_018991.htm&na=aboutanthem&rootLevel=3&label=Charity%20Guidelines
Sponsor: Anthem Blue Cross and Blue Shield
120 Monument Circle
Indianapolis, IN 46204

Anthem Blue Cross and Blue Shield School Nurse Grants 623

The mission of the project is to promote cardiovascular health among school-age children. Anthem's goals are to assist school nurses in initiating programs that may encourage a healthy lifestyle, reduce cardiovascular risk factors, promote physical activity, and encourage healthy eating behaviors; and facilitate the development of innovative school health programs that are based on research. No awards will be made to individuals, but will be awarded to the school. To apply, the school nurse representing the school must submit a completed application form with the appropriate attachments by the listed application deadline. Guidelines, including program contacts in Kentucky and Indiana, and application are available online.
Requirements: Eligible applicants include school nurses who are licensed in the state in which they practice and who represent a public or private school (elementary, junior high, middle, or high school) in Ohio, Kentucky, or Indiana.
Restrictions: Funds cannot be used for the salary or wages of any staff or employee.
Geographic Focus: Indiana, Kentucky, Ohio
Date(s) Application is Due: Feb 1
Amount of Grant: 500 - 1,000 U.S.D
Contact: Gabrielle Karpowicz, (614) 529-5068; gkarpowicz@aol.com
Internet: http://oasn.org/Awards_Grants/awards.htm
Sponsor: Ohio Association of School Nurses
6167 Heritage Point Court
Hilliard, OH 43026

Anthony R. Abraham Foundation Grants 624

For more than 30 years, the Anthony R. Abraham Foundation mission has been to help non-profit organizations worldwide. The foundations strives to: provide programs and services to help people around the world become self-productive and give back to their communities; ensure that no child is denied medical treatment due to a lack of insurance; provide education that breaks barriers; guarantee that research into the cure of catastrophic diseases continues; help raise the quality of life. It's been a privilege for the Foundation to be able to help ease poverty, raise hospitals, build orphanages and further medical research and the Foundation looks forward to doing even more. Some of the organizations helped include: Domestic organizations -St. Jude Children's Research Hospital; Camillus House; Miami Rescue Mission; Habitat for Humanity of Greater Miami; America's Second Harvest; Miami Children's Hospital; Jackson Memorial Hospital; Florida Heart Research Institute; Big Brothers/Big Sisters; Alonzo Mourning Charities; Honey Shine Mentoring Program; Cancer Link; The Miami Lighthouse for the Blind; Overtown Youth Center. International organizations: Brothers of the Good Shepherd; School for the Blind, Lebanon; Rene Moawad Foundation; Haitian Foundation; Children's International Network; Doctors without Borders; Little Sisters of Nazareth; Maronite Order of the Holy Family.
Requirements: Non-profit 501 (c) 3 organization requesting funding should fill out a funds-request form and submit it at the Foundation's website. The Foundation will contact you, if your organization qualifies under the Foundation's guidelines and the law.
Restrictions: Grants are not made to individuals.
Amount of Grant: 100 - 50,000 U.S.D
Contact: Anthony R. Abraham, Chairman; (305) 665-2222
Internet: http://www.abrahamfoundation.com/about
Sponsor: Anthony R. Abraham Foundation
1320 S Dixie Highway, Suite 241
Coral Gables, FL 33146-2937

Antone & Edene Vidinha Charitable Trust Grants 625

The trust provides partial support to programs and projects of tax-exempt, public charities in Hawaii to improve the quality of life in the state, particularly the island of Kauai. Grants of one year's duration are awarded in categories of interest to the trust, including: churches on Kauai; hospitals; health organizations which benefit the people of Kauai; educational scholarships to colleges and universities in the State of Hawaii for deserving students from the Island of Kauai. Types of support include building/renovation; equipment; general/operating support; program development; scholarship funds. Grants average from $2,000 - $80,000.
Requirements: 501 (c)3 nonprofit organizations in Hawaii are eligible to apply.The Trust places a special emphasis on the island of Kauai. Contact Paula Boyce to acquire the cover sheet/application forms and any additional guidelines required to begin the application process. Proposals must be submitted by December 1st.
Restrictions: No grants to/for: individuals; endowments; multi-year pledges.
Geographic Focus: Hawaii
Date(s) Application is Due: Dec 1
Amount of Grant: 2,000 - 80,000 U.S.D
Contact: Paula Boyce, c/o Bank of Hawaii; (808) 538-4944; fax (808) 538-4647; pboyce@boh.com
Internet: http://www.hawaiicommunityfoundation.org/index.php?id=290
Sponsor: Antone and Edene Vidinha Charitable Trust
Bank of Hawai'i, Foundation Administration Department 758
Honolulu, HI 96802-3170

AOA-Burnett Osteopathic Student Research Award 626

AOA offers this annual award in recognition of a student of osteopathic medicine for the most outstanding concept paper pertaining to an osteopathic-oriented research proposal. The award consists of an all-expense paid trip to the association's research conference plus a cash award and submittal of the paper to the Journal of the American Osteopathic Association for possible publication.

Requirements: Eligibility is limited to students enrolled in AOA-accredited colleges of osteopathic medicine.

Date(s) Application is Due: Apr 1

Amount of Grant: 100 U.S.D

Contact: Elizabeth Freeman, Division of Research Development; (800) 621-1773, ext. 8006; fax (312) 202-8200; research@osteopathic.org

Internet: http://www.do-online.osteotech.org/index.cfm?PageID=res_grant

Sponsor: American Osteopathic Association
142 E Ontario Street
Chicago, IL 60611-2864

AOA Clinical Investigator Development Award 627

This program recruits and supports osteopathic physicians for research and teaching careers in the medical sciences. Support is given for three years. Contact program staff for availability.

Amount of Grant: 50,000 U.S.D

Contact: Elizabeth Freeman, Division of Research Develoopment; (800) 621-1773, ext. 8006; research@osteopathic.org

Internet: http://do-online.osteotech.org/index.cfm?PageID=res_main

Sponsor: American Osteopathic Association
142 E Ontario Street
Chicago, IL 60611-2864

AOA Gutensohn-Denslow Award 628

The award recognizes and rewards a person who has made an outstanding contribution to the osteopathic profession in the areas of research and education. AOA and the National Osteopathic Foundation administer this program, which is funded by Glaxo Wellcome Inc.

Requirements: A nominee must be either an osteopathic physician or hold a doctoral level degree and be on the faculty or staff of an osteopathic institution. Nominations may come from anyone in the osteopathic profession and from colleagues of osteopathic physicians. Individuals nominated previously must be renominated for consideration. Initial nominations must include a current curriculum vita of the nominee and a detailed account of his/her contributions to research.

Date(s) Application is Due: Mar 15

Contact: Elizabeth Freeman, Division of Research Development; (800) 621-1773, ext. 8006; fax (312) 280-3860; research@osteopathic.org

Internet: http://www.osteopathic.org/index.cfm?PageID=ost_resoverview

Sponsor: American Osteopathic Association
142 E Ontario Street
Chicago, IL 60611-2864

AOA Osteopathic Research Fellowships 629

These research fellowships are designed to enable the applicant to conduct a basic science or clinical research project that will make a significant contribution to osteopathic medicine. The fellowship also serves as seed funding to encourage an osteopathic physician to contribute to research throughout his/her career in osteopathic medicine.

Requirements: Applicant must be either a postdoctoral osteopathic medical student, possessing an earned DO degree awarded by an AOA-accredited institution and be enrolled in an internship, residency, or research fellowship; or an undergraduate osteopathic medical student enrolled in an AOA-accredited institution.

Date(s) Application is Due: Mar 15

Amount of Grant: 5,000 U.S.D

Contact: Elizabeth Freeman, Division of Research Develoopment; (800) 621-1773, ext. 8006; fax (312) 202-8200; research@osteopathic.org

Internet: http://www.do-online.osteotech.org/index.cfm?PageID=res_grant

Sponsor: American Osteopathic Association
142 E Ontario Street
Chicago, IL 60611-2864

AOA Osteopathic Research Grants 630

Grants support projects investigating problems of a biological nature that will lead to a better understanding and a more effective application of the principles and concepts of osteopathic medicine. Grants are supported by AOA and the A.T. Still Osteopathic Foundation and Research Institute and administered by the foundation at the same address.

Requirements: Grants are made to qualified AOA-accredited, -affiliated, or -approved osteopathic institutions, not to individuals. Osteopathic physicians having the DO degree who are at non-osteopathic-accredited, -affiliated, or -approved institutions are also eligible to apply. Grantee must operate within an established institutional research program.

Date(s) Application is Due: Dec 1

Amount of Grant: 25,000 - 50,000 U.S.D

Contact: Elizabeth Freeman, Division of Research Development; (800) 621-1773, ext. 8006; research@osteopathic.org

Internet: http://www.do-online.osteotech.org/index.cfm?PageID=res_grant

Sponsor: American Osteopathic Association
142 E Ontario Street
Chicago, IL 60611-2864

AON Foundation Grants 631

AON funds community-based initiatives that focus on youths' development through health, education, the arts and cultural activities to schools, school districts and other nonprofit organizations in communities where the company operates, especially the Chicago area. Supported initiatives include community-based projects that focus on youths' development mentally, physically, and academically, and demonstrate positive intervention by including an outcome-evaluation component. The foundation supports higher and other education, social services, community funds, and hospitals and health associations. The board meets three times each year to consider requests. There is no specific application. A letter is acceptable.

Requirements: Nonprofit organizations in AON Corporation domestic and international operating communities, especially organizations in the Chicago area, may apply.

Restrictions: Awards are restricted or limited to charitable, educational (excluding the operation of a secondary educational institution or vocational school) and scientific organizations that qualify as 501(c)3 or 509(a)(1),(2), or (3).

Date(s) Application is Due: Apr 1; Jul 1; Oct 1

Amount of Grant: 1,000 - 10,000 U.S.D

Contact: Carolyn Labutka, VP/Executive Director; (312) 381-3549; fax (312) 701-4533

Internet: http://edreform.com/info/grant.htm

Sponsor: AON Foundation
200 East Randolph Street
Chicago, IL 60601

AO North America Grants 632

Support is given to trauma research projects, training in the management and care of teachers, and to visiting professors in the teaching of orthopedic, maxillofacial, spine, or veterinary trauma treatment.

Samples: Orthopaedic Trauma Assoc (Rosemont, IL)—for research, $75,000. Vanderbilt U, School of Medicine (Nashville, TN)—for research, $25,000. Foothills Hospital (Calgary, Canada)—for research, $7255.

Contact: Grants Administrator; (610) 251-9007; fax (610) 251-9059; ellisa@aona.com

Internet: http://www.aona.org/fellow_preceptor.asp

Sponsor: AO North America
P.O. Box 308
Devon, PA 19333-0308

AORN Foundation Scholarships 633

AORN scholarships provide funds for tuition and registration fees to members enrolled in baccalaureate, master's, and doctoral degree programs. Payment is made directly to the educational institution. The scholarships are granted for a one-year period; reapplication for funds is required annually through a new application and all supporting documents. An applicant must submit evidence of acceptance or enrollment in the particular program with specific information in support of accreditation status and a school catalog description of course of study; a personal statement reflecting contributions in nursing practice, research, and/or education; an official AORN scholarship application; official school transcripts; and three personal reference letters.

Requirements: Applicants must be RNs and currently active or associate members of AORN in good standing; membership must have been maintained continuously for 12 months prior to application deadline. Applicants must maintain a cumulative GPA of 3.0.

Restrictions: Members of the national board of directors during their elected terms, the scholarship board, and AORN headquarters personnel are ineligible. It is not possible to combine an AORN scholarship with a federal nurse traineeship or other total funding awards; other nonconflicting one-time only scholarships received by an applicant will not affect the funds of an AORN scholarship.

Date(s) Application is Due: May 1

Amount of Grant: 500 U.S.D

Contact: Ingrid Bendzsa, (800) 755-2676, ext. 328; ibendzsa@aorn.org

Internet: http://www.aorn.org/foundation/scholarship.htm

Sponsor: Association of Operating Room Nurses Foundation
2170 S Parker Road, Suite 300
Denver, CO 80231-5711

APA Congressional Fellowships 634

The fellowships afford an opportunity for the recipient psychologists to participate in the policy-making process of the U.S. Congress. Trained scientists and practitioners work with the staff members of Congress or congressional committees and are involved in the process of relating social and behavioral science to public policy. The fellowships are sponsored by APA in cooperation with American Association for the Advancement of Science (AAAS). The fellowship is for a period of one year beginning in September. The Fellowship stipend ranges from $60,000 to $75,000, depending upon years of post-doctoral experience. In addition, APA provides $375 per month for health insurance and funding may also be available for professional development and relocation expenses during the Fellowship year.

Requirements: Applicants must have a doctorate in psychology and be members of APA. Persons who are receiving sabbatical funds are also eligible to apply.

Date(s) Application is Due: Jan 7

Amount of Grant: 60,000 - 75,000 U.S.D

Contact: Micah Haskell-Hoehl, Program Administrator; (202) 336-5935; fax (202) 336-5935; mhaskell-hoehl@apa.org

Internet: http://www.apa.org/ppo/fellows/

Sponsor: American Psychological Association
750 First Street NE
Washington, DC 20002-4242

APA Culture of Service in the Psychological Sciences Award 635

The Award recognizes departments that demonstrate a commitment to service in the psychological sciences. Departments selected for this award will show a pattern of support for service from faculty at all levels, including a demonstration that service to the discipline is rewarded in faculty tenure and promotion. Successful Departments will also demonstrate that service to the profession is an integral part of training and mentoring. Service to the discipline includes such activities as departmental release time for serving on boards and committees of psychological associations; editing journals; serving on a review panel; or chairing an IRB. Other culture of service activities that a department would encourage include mentoring students and colleagues; advocating for psychological science's best interests with state and federal lawmakers; and promoting the value of psychological science in the public eye. The focus of this award is a department's faculty service to the discipline and not their scholarly achievements. Both Undergraduate and Graduate Departments of Psychology are eligible. Self-nominations are encouraged.

Amount of Grant: 5,000 U.S.D

Samples: Davidson College, Departments of Psycholog, Davidson, North Carolina; University of Minnesota, Department of Psychology, Minneapolis, Minnesota.

Contact: Suzanne Wandersman, Director for Governance Affairs; (202) 336-6000; fax (202) 336-5953; swandersman@apa.org

Internet: http://www.apa.org/science/dept_award.html

Sponsor: American Psychological Association
750 First Street NE
Washington, DC 20002-4242

APA Dissertation Research Awards 636

Dissertation awards are granted to graduate students in psychology programs to help offset the costs associated with dissertation research. The research may be done in any area of psychology. Applicants who have already defended their dissertations are eligible to apply for these funds, as long as they have not yet received doctoral degrees as of the application deadline. APA provides dissertation research awards to approximately 50 students annually. Dissertations must be approved by the students' committees before application. The application deadline falls in mid-September of each year. Awards are announced in late December.

Requirements: APA student affiliates enrolled in a psychology graduate program are eligible to apply.

Restrictions: Departments may not nominate more than three students per year for dissertation awards.

Date(s) Application is Due: Sep 15

Amount of Grant: 1,000 - 5,000 U.S.D

Contact: APA Science Directorate, (800) 374-2721 or (202) 336-5500; fax (202) 336-5953; science@apa.org

Internet: http://www.apa.org/science/dissinfo.html

Sponsor: American Psychological Association
750 First Street NE
Washington, DC 20002-4242

APA Distinguished Scientific Award for Early Career Contribution to Psychology 637

The Distinguished Scientific Award for Early Career Contribution to Psychology recognizes excellent young psychologists. Nominations of persons who received doctoral degrees during and since 1998 are being sought in these five areas: animal learning and behavior, comparative; cognition/human learning; developmental psychology; health; and psychopathology. These categories should be interpreted broadly and are not meant to be exclusive; all of psychology is of sufficient merit to be considered for awards. The Award consists of a citation and a cash prize, which are presented at the APA Annual Convention.

Samples: Robert D. Gray, PhD (applied research), Department of Applied Psychology, Arizona State University; Patrik O. Vuilleumier, MD (behavioral /cognitive neuroscience), Department of Neurology, University of California, Davis; R. Chris Fraley, PhD (individual differences), Department of Psychology at University of Illinois at Urbana-Champaign; Jorn Diedrichsen, PhD (perception/motor performance), School of Psychology, Adeilad Brigantia University, Wales; Matthew D. Lieberman, PhD (social), University of California, Los Angeles.

Contact: Suzanne Wandersman, Director for Governance Affairs; (202) 336-6000, fax (202) 336-5953; swandersman@apa.org

Internet: http://www.apa.org/science/sciaward.html

Sponsor: American Psychological Association
750 First Street NE
Washington, DC 20002-4242

APA Distinguished Scientific Award for the Applications of Psychology 638

The Distinguished Scientific Award for the Applications of Psychology honors psychologists who have made distinguished theoretical or empirical advances in psychology leading to the understanding or amelioration of important practical problems. The Award consists of a citation and a cash prize, which are presented at the APA Annual Convention.

Samples: Karl G. Joreskog, PhD, Emeritus Professor, University of Upssala, Sweden,; Peter M. Bentler, PhD, UCLA Psychology Department, Los. Angeles, CA; John P. Campbell, PhD, Professor and Chair of Psychology and Industrial Relations, University of Minnesota, MN.

Contact: Suzanne Wandersman, Director for Governance Affairs; (202) 336-6000; fax (202) 336-5953; swandersman@apa.org

Internet: http://www.apa.org/science/sciaward.html

Sponsor: American Psychological Association
750 First Street NE
Washington, DC 20002-4242

APA Distinguished Scientific Contribution Award 639

The Distinguished Scientific Contribution Award honors psychologists who have made distinguished theoretical or empirical contributions to basic research in psychology. The Award consists of a citation and a cash prize, which are presented at the APA Annual Convention.

Samples: Marilynn B. Brewer, PhD, Department of Psychology, Ohio State University, Columbus, OH; Jean M. Mandler, PhD, Research Professor, Department of Cognitive Science, University of California, San Diego; Paul Rozin, PhD, Professor of Psychology, University of Pennsylvania.

Contact: Suzanne Wandersman, Director for Governance Affairs; (202) 336-6000; fax (202) 336-5953; swandersman@apa.org

Internet: http://www.apa.org/science/sciaward.html

Sponsor: American Psychological Association
750 First Street NE
Washington, DC 20002-4242

APA Distinguished Service to Psychological Science Award 640

This Award recognizes individuals who have made outstanding contributions to psychological science through their commitment to a culture of service. Award recipients will receive an honorarium of $1,000. Nominees will have demonstrated their service to the discipline by aiding in association governance; serving on boards, committees and various psychological associations; editing journals; reviewing grant proposals; mentoring students and colleagues; advocating for psychological science's best interests with state and federal lawmakers; and promoting the value of psychological science in the public eye. Nominees may be involved in one service area, many of the areas, or all of the service areas noted above. An individual's service to the discipline and not a person's scholarly achievements are the focus of this award.

Amount of Grant: 1,000 U.S.D

Samples: Robert L. Balster, Ph.D., Professor of Pharmacology, University of Michigan Medical School; Nora Newcombe, Professor of Psychology, Temple University; Robert A. Bjork, Chair of the Psychology Department, University of California, Los Angeles; J. Bruce Overmier, Ph.D., Department of Psychology, University of Minnesota.

Contact: Suzanne Wandersman, Director for Governance Affairs; (202) 336-6000; fax (202) 336-5953; swandersman@apa.org

Internet: http://www.apa.org/science/serv_award.html

Sponsor: American Psychological Association
750 First Street NE
Washington, DC 20002-4242

APA Meritorious Research Service Commendation 641

The Commendation recognizes individuals who have made outstanding contributions to psychological science through their service as employees of the federal government or other organizations. Contributions are defined according to service to the field that directly or indirectly advances opportunities and resources for psychological science. This may include staff at federal or non-federal research funding, regulatory or other agencies. Nominees may be active or retired but ordinarily will have a minimum of 10 years of such service. The individual's personal scholarly achievements (i.e., research, teaching, and writing) are not considered in the selection process independent of their service contributions.

Samples: Robert B. Huebner, National Institute on Alcohol Abuse and Alcoholism; Jack D. Maser, University of California, San Diego; Robert S. Ruskin, Consortium Research Fellows Program, Universities of the Washington.

Contact: Suzanne Wandersman, Director for Governance Affairs; (202) 336-6000; fax (202) 336-5953; swandersman@apa.org

Internet: http://www.apa.org/science/meritorious.html

Sponsor: American Psychological Association
750 First Street NE
Washington, DC 20002-4242

APA Scientific Conferences Grants 642

The program seeks proposals for research conferences in psychology. The purpose of this program is to promote the exchange of important new contributions and approaches in scientific psychology. Conference formats include festschrifts—conferences organized as tributes to distinguished scholars; stand-alone conferences—usually two days long, involves psychologists as the primary organizers; and add-a-day conference—meetings that occur at the beginning or end of a scientific conference other than APA (e.g., Society for Ingestive Behavior or Psychonomics Society). The conference must also be supported by the host institution with direct funds, in-kind support, or a combination of the two.

Requirements: One of the primary organizers must be a member of APA. Only academic institutions accredited by a regional body may apply. Independent research institutions must provide evidence of affiliation with such an accredited institution. Joint proposals from cooperating institutions are encouraged. Conferences may be held only in the United States, its possessions, or Canada.

Restrictions: APA governance groups, APA divisions, and related entities are not eligible for funding under this program.

Date(s) Application is Due: Jun 1; Dec 1

Amount of Grant: 500 - 20,000 U.S.D

Contact: Deborah McCall, Science Directorate; (202) 336-6000; fax (202) 336-5953; dmccall@apa.org

Internet: http://www.apa.org/science/psa/sep07ann.html

Sponsor: American Psychological Association
750 First Street NE
Washington, DC 20002-4242

APA Travel Awards 643

The purpose of the travel award program is to help psychology graduate students travel to the annual APA convention to present their research. The APA awards up to 100 travel awards annually. Multiple authors are accepted, but the student applicant must be the designated presenter of the research. The application deadline falls in mid-April of each year. Awards will be announced in mid-June. Applications are available on the Web site.
Requirements: Applicants must be student affiliates of the APA; students who are not affiliates may apply for affiliation at the time of application.
Restrictions: Departments may not endorse more than three students per year for travel awards.
Date(s) Application is Due: Apr 12
Amount of Grant: 300 U.S.D
Contact: Science Directorate; (800) 374-2721 or (202) 336-5500; fax (202) 336-5963; science@apa.org
Internet: http://www.apa.org/science/funding.html
Sponsor: American Psychological Association
750 First Street NE
Washington, DC 20002-4242

APA Young Investigator Grants 644

The program provides financial support to teaching, research, and health care delivery projects in general pediatrics. The number of awards is dependent on available funds and the size of grant requests of selected projects. Proposals should address the goal of the association: to encourage, promote, and facilitate improved patient care, teaching, and research in general pediatrics.
Requirements: The principal investigator must be a member of the APA or have submitted an application for membership. Preference will be given to new investigators, including those in training.
Restrictions: Multiple year funding requests will not be considered.
Date(s) Application is Due: Oct 25
Amount of Grant: Up to 10,000 U.S.D
Contact: Grants Administrator; (703) 556-9222; fax (703) 556-8729; info@academicpeds.org
Sponsor: Ambulatory Pediatric Association
6728 Old McLean Village Drive
McLean, VA 22101

APF/COGDOP Graduate Research Scholarships in Psychology 645

The American Psychological Foundation (APF) and the Council of Graduate Departments of Psychology (COGDOP) jointly offer graduate research scholarships, including the Clarence J. Rosecrans Scholarship ($2000), the Ruth G. and Joseph D. Matarazzo Scholarship ($3000), as well as a number of $1,000 scholarships. The scholarships will be given directly to the individual graduate students enrolled in an interim master's program or doctoral program. If a student is currently enrolled in a terminal master's program, the student must intend to enroll in a PhD program. Several fellowships have been reserved for students who, at the time of application, are within the first two years of graduate study in psychology. The purpose of the scholarship program is to assist graduate students of psychology with research costs. The scholarships are administered by APA. Application and guidelines are available online.
Requirements: Each graduate department of psychology that is a member of COGDOP may submit nominations. The number of candidates that each member department is allowed to nominate depends upon the total number of students enrolled in the graduate program. Departments that have 100 or fewer students enrolled in their graduate programs may nominate one candidate; departments that have 101-200 graduate students enrolled may nominate up to two candidates; and departments that have more than 200 graduate students enrolled may nominate up to three candidates.
Date(s) Application is Due: Jun 15
Amount of Grant: 1,000 - 3,000 U.S.D
Samples: Arathi Sethumadhavan, Texas Tech University —$3,000; Kristen M. Culbert, Michigan State University—$2,000; Jennifer M. Brielmaier, George Mason University—$1,000.
Contact: Science Directorate; (202) 336-6000; fax (202) 336-5953; science@apa.org
Internet: http://www.apa.org/science/apf-cogdop.html
Sponsor: American Psychological Association
750 First Street NE
Washington, DC 20002-4242

APhA Foundation Incentive Grants 646

The incentive grants provide seed funding to help pharmacists in all practice settings explore new methods and services that enhance their role as pharmaceutical care providers, and to encourage them to share their experiences with other pharmacists. This year, the incentive grants program will continue to expand to include project proposals specifically for pharmacists interested in establishing or supporting an already existing practice model in the area of pain management. Projects should concentrate on new, innovative patient care service that is of significance to pharmacy care settings and that can be evaluated for its relevance. Application and guidelines are available online.
Requirements: Applicants must be members of APhA, currently licensed, and actively engaged in ambulatory pharmacy practice. Recipients are encouraged to attend the APhA annual meeting and submit present findings, if applicable.
Restrictions: Indirect costs will not be funded. Salaries for recipients are not funded; however, consultant fees, fees for reseach/technician services, and expenses for adminsitrative services may be allowable if essential if essential to development/ implementation of the project.
Date(s) Application is Due: Oct 3
Amount of Grant: 1,000 U.S.D
Contact: Grants Administrator; info@aphafoundation.org
Internet: http://www.aphafoundation.org/IncentiveGrants/incentive.htm
Sponsor: American Pharmacists Association
2215 Constitution Avenue NW
Washington, DC 20037-2985

APHL Emerging Infectious Diseases Fellowships 647

The Emerging Infectious Diseases (EID) Laboratory Fellowship Program, sponsored by the Association of Public Health Laboratories (APHL) and the Centers for Disease Control and Prevention (CDC), trains and prepares scientists for careers in public health laboratories and supports public health initiatives related to infectious disease research. Two types of fellowships are offered: the EID Advanced Laboratory Training Fellowship is a one-year program designed for bachelor's or master's level scientists, with emphasis on the practical application of technologies, methodologies and practices related to emerging infectious diseases; the EID Laboratory Research Fellowship is a two-year program designed for doctoral level (Ph.D., M.D. or D.V.M.) scientists to conduct high-priority research in infectious diseases. Up to 30 pre-doctoral one-year and 15 post-doctoral two-year fellowships will be awarded. Fellows are placed in local, state or federal (CDC) public health laboratories throughout the U.S.. Fellows are provided with a stipend, medical insurance, travel to the host laboratory and a professional development allowance. The fellowship program offers a wide variety of training and research experiences. Examples of projects include research in molecular genetics, pathogenesis, epidemiology and cell biology. Fellows may be trained in specific laboratory techniques such as real-time PCR, DNA sequencing and mass spectrometry. Fellows may receive specialized training and experience with a specific pathogen, or generalized training in influenza or STD surveillance, newborn screening or chemical and biological warfare agents. Other work may include diagnostic virology, bioanalytical chemistry, foodborne disease research including Pulsed Field Gel Electrophoresis (PFGE), clinical and environmental diagnostics and epidemiologic studies of disease outbreaks. Fellows also have opportunities to participate in research seminars, teleconferences, regional and national meetings and other continuing education opportunities. All fellows participate in an orientation session at CDC in Atlanta to gain a general understanding of the public health laboratory system and how it relates to infectious disease surveillance, prevention, research and control. All applicants are required to submit one original single-sided application to APHL by the application deadline. The application may be downloaded from the EID fellowship website in writeable pdf format. All required supporting documents (listed in the application) should be collected by the applicant and forwarded with the application to APHL. Deadline dates may vary from year to year. Applicants are encouraged to check the EID fellowship website or contact the Fellowship Manager to verify current the deadline date. APHL member laboratories may apply to host one or more EID Fellows. The link to the Host Laboratory Instructions and Application page is available from the EID Fellowship website.
Requirements: EID Advanced Laboratory Training Fellowship applicants must have completed a bachelor's or master's degree in microbiology, virology or a related discipline and/or completed an accredited medical technologist program prior to the start of the fellowship. EID Laboratory Research Fellowship applicants must have received a Ph.D., M.D., or D.V.M., or have completed all requirements for such a degree prior to the start of the fellowship. Applicants without a degree in a physical science will be considered if they have significant laboratory science and/or public health coursework or experience. All bachelor's and master's level candidates must have completed their degree by the time the fellowship begins (mid-September). Postdoctoral candidates may delay their start date to February 2012. Degree requirements must be completed by by this time. All appointees must attend the August Orientation session in Atlanta.
Restrictions: All applicants must be U.S. citizens.
Date(s) Application is Due: Feb 11
Contact: Heather Roney, Manager, Fellowship Program; (240) 485-2778 or (240) 485-2745; fax (240) 485-2700; fellowships@aphl.org or heather.roney@aphl.org
Internet: http://www.aphl.org/mycareer/fellowships/eid/pages/default.aspx
Sponsor: Association of Public Health Laboratories
8515 Georgia Avenue, Suite 700
Silver Spring, MD 20910

Appalachian Regional Commission Health Grants 648

Access to comprehensive, affordable health care is vital to social and economic growth in the Appalachian Region. ARC's health projects focus on community-based efforts to encourage health-promotion and disease-prevention activities. Strategies include: using best practices in public health to develop targeted approaches to wellness and disease prevention; supporting partnerships that educate children and families about basic health risks; using telecommunications and other technology to reduce the high cost of health-care services; and encouraging the development and expansion of health professional education services within the Region. ARC health grants have helped provide equipment for hospitals and rural clinics, training for health care professionals, and support for community-based health education activities. ARC also works with other organizations to address the high incidence of life-threatening diseases in the Region, as in its ongoing partnership with the Centers for Disease Control and Prevention in diabetes and cancer education, prevention, and treatment programs in the Region's distressed counties.
Requirements: States, and through states, public bodies and private nonprofit organizations are eligible to apply.
Restrictions: Generally, ARC grants are limited to 50% of project costs.
Geographic Focus: Alabama, Georgia, Kentucky, Maryland, Mississippi, New York, North Carolina, Ohio, Pennsylvania, South Carolina, Tennessee, Virginia, West Virginia
Contact: Jill Wilmoth, Budget and Program Specialist; (202) 884-7668; fax (202) 884-7691; jwilmoth@arc.gov

Internet: http://www.arc.gov/health
Sponsor: Appalachian Regional Commission
1666 Connecticut Avenue NW, Suite 700
Washington, DC 20009-1068

APPEAL Social Justice and Tobacco Control Grants **649**
This national initiative will provide resources for organizations to promote policy change in tobacco control among Asian American and Pacific Islander communities through community organizing and advocacy policy. Previous experience with tobacco-related work is not necessary. Application and guidelines are available online.
Requirements: Qualified applicants must be experienced in grassroots community organizing and have 501(c)3 nonprofit status.
Date(s) Application is Due: Apr 25
Amount of Grant: 20,000 - 50,000 U.S.D
Contact: Lisa Fu, (510) 272-9536, ext. 104; fax (510) 272-0817; lfu@aapcho.org
Internet: http://www.appealforcommunities.org/pages/Call_for_Proposals44.php?project_id=44
Sponsor: Asian Pacific Partners for Empowerment and Leadership
300 Frank H Ogawa Plaza, Suite 620
Oakland, CA 94612

APSAA Fellowships **650**
The mission of the Fellowship Program is to encourage interest and involvement in psychoanalysis among the future leaders, researchers and educators of mental health and academia. Early-career psychiatrists, psychologists, social workers and academics are eligible for to apply for the Fellowship. The Fellowship offers the opportunity to attend the biannual meetings of APSAA, to meet analysts and Fellows from across the country, to have a Mentor, and to present their clinical work or research at the meetings. The application for the Fellowship Program is available online September 15.
Requirements: Psychiatrists, psychologists, social workers, and academics who meet the specific eligibility requirements (available online) are eligible. Applicants may be nominated by their department chairs or program directors if applicable. When not applicable, self-nominations are encouraged. Applicants must be training or working in the United States during the fellowship year. All applicants should have demonstrated leadership ability in their discipline, or have special aptitude in research, teaching, and/or clinical endeavors; and have special interest in psychodynamics, psychoanalysis, or applied psychoanalysis.
Contact: James Guimaraes, Fellowship Coordinator; (212) 752-0450; fax (212) 593-0571; jguimaraes@apsa.org or info@apsa.org
Internet: http://www.apsa.org/FELLOWSHIPPROGRAM/GENERALINFORMATION/tabid/298/Default.aspx
Sponsor: American Psychoanalytic Association
309 East 49th Street
New York, NY 10007-1601

APSAA Foundation Grants **651**
The mission of the APF Committee is to raise funds and sponsor programs promoting a better understanding of psychoanalysis and encouraging effective and innovative dissemination of psychoanalytic ideas and services to the public. The APF Committee's objective is to educate the public, the community of mental health workers, and allied disciplines about the relevance of psychoanalysis as a powerful therapeutic and research instrument whose applications span a wide range of understanding of individual behavior and cultural phenomena. One or several of the following elements should be a part of the proposal in order to be considered: community outreach; national focus—not a program that is generally put on as a regular activity by a society or institute; fund raising element—seeks co-funding, matching or other sources of support for project (grant money may be used as seed money for helping Grantee to get other sources of funding); transportable—broad applicability to other psychoanalytic institutes, societies and foundations (the program can serve as a model for other programs); and creative and original.
Amount of Grant: 1,000 - 5,000 U.S.D
Samples: Washington Center for Psychoanalysis—to enable the design and implementation of a system of psychodynamic support for Jubilee JumpStart, a new child care and education facility for low-income families, $5,000; Allen Creek Preschool, Ann Arbor, Michigan—to support the Early Childhood Training Initiative, a new outreach program designed to nurture the growth of early childhood educators and day care workers through professional training, workshops, and consultations, $4,000.
Contact: Dean K. Stein, Executive Director; (212) 752-0450; fax (212) 593-0571; deankstein@apsa.org or info@apsa.org
Internet: http://www.apsa.org/AmericanPsychoanalyticbrFoundationCommittee/tabid/70/Default.aspx
Sponsor: American Psychoanalytic Association
309 East 49th Street
New York, NY 10007-1601

APSAA Mini-Career Grants **652**
The Fund conceives of psychoanalytic research along the broadest lines, including: scholarly and empirical investigative contributions that can advance knowledge of psychoanalytic theory, practice, and links between psychoanalysis and neighboring disciplines such as developmental psychology or neuroscience. Awards of up to $15,000 annually for up to two years are intended to support the beginning career of a psychoanalytic investigator. These mini-career awards are intended, for example, to buy time for a junior faculty member or clinician just starting a practice so that they may consult with other investigators, join an investigative team, or attend year long seminars on research methodology or specific methods relevant to their research. Applicants for these awards must provide a year-long career development plan with letters from appropriate faculty and mentors describing the candidate's career development plan and how the award will facilitate the candidate's goals.
Requirements: Applicants should meet the following criteria: proposal relevant to psychoanalysis; applicant is either a new applicant to the Fund or resubmitting a previously reviewed application; applicant has a masters degree or higher; and human subject protection and consent procedures must be specified for research studies involving ongoing data collection.
Restrictions: Applicants who have received $80,000 total from the Fund are ordinarily not eligible to apply again.
Date(s) Application is Due: Apr 1; Nov 1
Amount of Grant: Up to 15,000 U.S.D
Contact: Dean K. Stein, Executive Director; (212) 752-0450; fax (212) 593-0571; deankstein@apsa.org or info@apsa.org
Internet: http://www.apsa.org/RESEARCH/FUNDFORPSYCHOANALYTICRESEARCH/GRANTS/tabid/137/Default.aspx
Sponsor: American Psychoanalytic Association
309 East 49th Street
New York, NY 10007-1601

APSAA Research Grants **653**
The Fund conceives of psychoanalytic research along the broadest lines, including: scholarly and empirical investigative contributions that can advance knowledge of psychoanalytic theory, practice, and links between psychoanalysis and neighboring disciplines such as developmental psychology or neuroscience. Grants of one or two years duration at a maximum of $20,000 yearly are intended for a specific project building upon psychoanalytic principles or directly investigating the process and/or outcome of psychoanalytically informed treatments. The Board recognizes that this amount of money is not sufficient usually to cover the costs of a large research study, particularly one that may involve a large number of subjects or a longitudinal design. Hence, the Board encourages applicants to consider feasibility of their request, that is, what they can achieve for this amount of money, and also to think about this type of award as supporting the beginning stages of a research project, the gathering of pilot data, or the refinement of design and methods.
Requirements: Applicants should meet the following criteria: proposal relevant to psychoanalysis; applicant is either a new applicant to the Fund or resubmitting a previously reviewed application; applicant has a masters degree or higher; and human subject protection and consent procedures must be specified for research studies involving ongoing data collection.
Restrictions: Applicants who have received $80,000 total from the Fund are ordinarily not eligible to apply again.
Date(s) Application is Due: Apr 1; Nov 1
Amount of Grant: Up to 20,000 U.S.D
Contact: Dean K. Stein, Executive Director; (212) 752-0450; fax (212) 593-0571; deankstein@apsa.org or info@apsa.org
Internet: http://www.apsa.org/RESEARCH/FUNDFORPSYCHOANALYTICRESEARCH/GRANTS/tabid/137/Default.aspx
Sponsor: American Psychoanalytic Association
309 East 49th Street
New York, NY 10007-1601

APSAA Small Beginning Scholar Pre-Investigation Grants **654**
The Fund conceives of psychoanalytic research along the broadest lines, including: scholarly and empirical investigative contributions that can advance knowledge of psychoanalytic theory, practice, and links between psychoanalysis and neighboring disciplines such as developmental psychology or neuroscience. These small grants of less than $5,000 for one year are intended to permit a beginning scholar to gather pilot data in preparation for the submission of a full grant to the Fund or to another agency. The Fund accepts revisions of a previously submitted application up to four years after the previous submission date. Up to two revised submissions may be submitted within four years of the original submission date.
Requirements: Applicants should meet the following criteria: proposal relevant to psychoanalysis; applicant is either a new applicant to the Fund or resubmitting a previously reviewed application; applicant has a masters degree or higher; and human subject protection and consent procedures must be specified for research studies involving ongoing data collection.
Restrictions: Applicants who have received $80,000 total from the Fund are ordinarily not eligible to apply again.
Date(s) Application is Due: Apr 1; Nov 1
Amount of Grant: Up to 5,000 U.S.D
Contact: Dean K. Stein, Executive Director; (212) 752-0450; fax (212) 593-0571; deankstein@apsa.org or info@apsa.org
Internet: http://www.apsa.org/RESEARCH/FUNDFORPSYCHOANALYTICRESEARCH/GRANTS/tabid/137/Default.aspx
Sponsor: American Psychoanalytic Association
309 East 49th Street
New York, NY 10007-1601

APSAA Small Beginning Scholar Visiting and Consulting Grants **655**
The Fund conceives of psychoanalytic research along the broadest lines, including: scholarly and empirical investigative contributions that can advance knowledge of psychoanalytic theory, practice, and links between psychoanalysis and neighboring disciplines such as developmental psychology or neuroscience. Small grants of $3,000 are

intended to permit a beginning scholar to spend time visiting and consulting with a more experienced, senior investigator who has agreed to help the junior investigator begin their investigative work. Typically these awards are made to permit the beginning scholar the funds to travel and/or to reimburse the senior investigator for travel and consultation. These grants may also be used to support travel to a research training seminar in the U.S. or abroad. The Fund accepts revisions of a previously submitted application up to four years after the previous submission date. Up to two revised submissions may be submitted within four years of the original submission date.
Requirements: Applicants should meet the following criteria: proposal relevant to psychoanalysis; applicant is either a new applicant to the Fund or resubmitting a previously reviewed application; applicant has a masters degree or higher; and human subject protection and consent procedures must be specified for research studies involving ongoing data collection.
Restrictions: Applicants who have received $80,000 total from the Fund are ordinarily not eligible to apply again.
Date(s) Application is Due: Apr 1; Nov 1
Amount of Grant: Up to 3,000 U.S.D
Contact: Dean K. Stein; (212) 752-0450; deankstein@apsa.org or info@apsa.org
Internet: http://www.apsa.org/RESEARCH/FUNDFORPSYCHOANALYTIC RESEARCH/GRANTS/tabid/137/Default.aspx
Sponsor: American Psychoanalytic Association
309 East 49th Street
New York, NY 10007-1601

APSA Congressional Health and Aging Policy Fellowships **656**
Supported by the Atlantic Philanthropies and administered by Columbia University, this national program seeks to provide professions in health and aging with the experience and skills necessary to make a positive contribution to the development and implementation of health policies that affect older Americans. The program will offer two different tracks for individual placement: a residential track that includes a nine-to-12-month placement in Washington, D.C. or at a state agency (as a legislative assistant in Congress, a professional staff member in an executive agency or in a policy organization; and a non-residential track that includes a health policy project and brief placement(s) throughout the year at relevant sites. Core program components focused on career development and professional enrichment are provided for fellows in both tracks.
Requirements: The program is open to physicians, nurses and social workers at all career stages (early, mid, and late) with demonstrated commitment to health and aging issues and a desire to be involved in health policy at the federal, state or local level. Other professionals with clinical backgrounds (e.g. pharmacists, dentists, clinical psychologists) working in the field of health and aging are also eligible to apply.
Contact: Jeffrey R. Biggs, Program Director; (202) 483-2512, ext. 521; fax (202) 483-2657; jbiggs@apsanet.org or cfp@apsanet.org
Internet: http://www.apsanet.org/content_50835.cfm
Sponsor: American Political Science Association
1527 New Hampshire Avenue, NW
Washington, DC 20036-1206

APSA Robert Wood Johnson Foundation Health Policy Fellowships **657**
The purpose of this fellowship is to give outstanding mid-career health professionals in academic and community-based settings an opportunity to learn more about the legislative process through direct participation. Fellows actively contribute to the formulation of national health policies and accelerate their careers as leaders in health policy. Initiated in 1973, the program is funded by The Robert Wood Johnson Foundation (RWJF)and conducted by the Institute of Medicine (IOM) of the National Academy of Sciences. The application process typically opens in September and the application deadline typically occurs in mid-November of each year. Fellows must commit 100 percent of their time to program activities during the first 12 months. Up to six grants of up to $165,000 each will be made.
Date(s) Application is Due: Nov 14
Amount of Grant: Up to 165,000 U.S.D
Contact: Jeffrey R. Biggs, Program Director; (202) 334-1506 or (202) 483-2512, ext. 521; fax (202) 483-2657; jbiggs@apsanet.org or info@healthpolicyfellows.org
Internet: http://www.apsanet.org/content_3544.cfm
Sponsor: American Political Science Association
1527 New Hampshire Avenue, NW
Washington, DC 20036-1206

APS Foundation Grants **658**
Through its Corporate Giving and Foundation, APS continues to be the leading Arizona corporate citizen. It supports nonprofit organizations with a 501(c)3 Internal Revenue Service tax-exempt status through cash and/or in kind services. The Foundation support Arizona communities in five strategic areas: health and human services, community development, education, arts and culture, and environment.
Restrictions: APS does not fund individual requests, charter or private schools, religious, political, fraternal, legislative or lobbying efforts or organizations, travel-related or hotel expenses, private or family foundations, private non-profit organizations, salaries and/or debt reduction.
Geographic Focus: Arizona
Amount of Grant: Up to 100,000 U.S.D
Samples: Phoenix Art Museum, Phoenix, Arizona, $100,000; Phoenix Zoo, Phoenix, Arizona, $40,000; Phoenix Indian Center, Phoenix, Arizona, $25,000.
Contact: Teresa DeValle, Contributions Coordinator; (602) 250-2259; fax (602) 250-3066; Teresa.DeValle@aps.com
Internet: http://www.aps.com/general_info/AboutAPS_14_Archive.html
Sponsor: Arizona Public Service Foundation
P.O. Box 53999, MS 8010
Phoenix, AZ 85072-3999

AptarGroup Foundation Grants **659**
The foundation awards grants to Illinois nonprofit organizations in its areas of interest, including in order of priority, health and social services, cultural activities and programs, and higher education.
Requirements: Illinois nonprofit organizations are eligible.
Restrictions: Grants do not support nursing homes, animal welfare groups, national or international relief organizations, primary, secondary, or theological schools, or religious, civic, fraternal, veterans', social, or political organizations. Individuals, testimonial dinners, fundraising events, courtesy advertising, or trips or tours will not be supported.
Geographic Focus: Illinois
Date(s) Application is Due: Mar 15; Oct 15
Contact: Lawrence Lowrimore, Secretary; (815) 477-0424; fax (815) 477-0481; info@aptargroup.com
Sponsor: AptarGroup Charitable Foundation
475 W Terra Cotta Avenue, Suite E
Crystal Lake, IL 60014-9695

Aragona Family Foundation Grants **660**
Giving primarily in Austin, Texas, the Aragona Family Foundation provides general operating support grants in the following areas of interest: animals/wildlife, preservation/protection; cancer; children/youth, services; community/economic development; education; food banks; hospitals (general); philanthropy/voluntarism; protestant agencies & churches. Grants range from $2,000 - $100,000.
Requirements: Qualifying 501 (c)3 organizations are eligible to apply for funding. There is no: application form required; deadline date to adhere to.
Geographic Focus: Texas
Amount of Grant: 2,000 - 100,000 U.S.D
Samples: St. Stephens Episcopal School, Austin, TX, $100,000—to further the organizations charitable mission; Harvard University, Cambridge, MA, $533,300—to further the organizations charitable mission.
Contact: Joseph C. Aragona, President; (512) 328-2178
Sponsor: Aragona Family Foundation
3311 Westlake Drive
Austin, TX 78746-1901

Aratani Foundation Grants **661**
The foundation awards grants to nonprofits in its areas of interest, including education, health care, museums, recreation, and religion. Preference is given to Japanese-American cultural organizations. Types of support include annual campaigns, building construction/renovation, capital campaigns, conferences and seminars, continuing support, curriculum development, endowments, exchange programs, fellowships, general operating support, program development, scholarship funds, and seed grants.
Requirements: The Foundation gives primarily in the state of California but funding is also available in the states of: New York; District of Columbia; Washington; Florida; Rhode Island; Oregon. Application forms are not required, but application outlines are available. Contact the Foundation directly for additional guidelines. Applications maybe submitted in English and Japanese. There are no application deadline dates.
Restrictions: Grants are not made to individuals.
Geographic Focus: California, District of Columbia, Florida, New York, Oregon, Rhode Island, Washington
Amount of Grant: 1,000 - 150,000 U.S.D
Samples: Asian Pacific American Legal Center of South California, Los Angeles, CA, $50,000; Asian Pacific American Institute for Congressional Studies, Washington, DC, $10,000; Buddhist Churches of America, San Francisco, CA, $200,000.
Contact: George Aratani, President; (310) 530-9900
Sponsor: Aratani Foundation
23505 Crenshaw Boulevard, North 230
Hollywood, CA 90505

Arcadia Foundation Grants **662**
The foundation awards grants to Pennsylvania nonprofit organizations to improve the quality of life. Areas of interest include hospitals and hospital building funds, health agencies and services, nursing, hospices, early childhood, adult and higher education, libraries, child development and welfare agencies, youth organizations, and social service and general welfare agencies, including care of the handicapped, aged, and hungry. Also supported are family services, environment and conservation, wildlife and animal welfare, religious organizations, historical preservation, and music organizations. Types of support include general operating support, continuing support, annual campaigns, capital campaigns, building construction/renovation, equipment acquisition, endowment funds, program development, scholarship funds, and research. Applications are accepted between September 1 and November 1.
Requirements: Eastern Pennsylvania organizations whose addresses have zip codes of 18000-19000 are eligible. Application form not required. The initial approach should be a letter or proposal—not exceeding two pages.
Restrictions: Grants are not awarded to support individuals, deficit financing, land acquisition, fellowships, demonstration projects, publications, or conferences.
Geographic Focus: Pennsylvania

Date(s) Application is Due: Nov 1
Samples: Thomas Jefferson U (Philadelphia, PA)—for a new medical education building, $1 million.
Contact: Marilyn Lee Steinbright, President; (610) 275-8460
Sponsor: Arcadia Foundation
105 E Logan Street
Norristown, PA 19401

Archer Daniels Midland Foundation Grants 663
The foundation prefers to fund programs that will directly impact the communities where operating units are located. Nearly a third of the funding is directed toward educational institutions, including elementary, secondary, and higher education. The advocacy category covers world affairs and foreign relations groups that deal with international trade; also supported are projects stressing free enterprise and assistance for women. Support of social services goes to minority group development, cultural activities, and hospital and youth agencies. Health service funding supports hospital and disease association programs. Conservation funding supports programs for protecting the environment and beautification as well as conservation. Religion grants support Christian, Jewish, and Roman Catholic organizations, such as churches, colleges and universities, international ministries/missions, Jewish welfare, and the Salvation Army. Applications are accepted at any time; initial contact by letter of inquiry is encouraged.
Requirements: Tax-exempt organizations are eligible. Current United Way recipients and national organizations are also eligible.
Geographic Focus: Arkansas, California, Georgia, Illinois, Indiana, Iowa, Kansas, Kentucky, Massachusetts, Michigan, Minnesota, Mississippi, Missouri, Montana, Nebraska, New Jersey, New York, North Carolina, North Dakota, Ohio, Oklahoma, Pennsylvania, Puerto Rico, South Carolina, Tennessee, Texas, Washington, Wisconsin
Amount of Grant: 10,000 U.S.D
Contact: Brian Peterson, Program Contact; (217) 424-5413 or (217) 424-5200; corpaffairs@admworld.com
Internet: http://www.adm.com/en-U.S./responsibility/social_investing/Pages/default.aspx
Sponsor: Archer Daniels Midland Foundation
4666 Faries Parkway, P.O. Box 1470
Decatur, IL 62526-5630

Archstone Foundation Grants 664
The foundation's goal is to support projects that meet the needs of the aging population. Grants are awarded for efforts to educate, provide services, alter behavior, raise awareness, convene diverse stakeholders, or promote public policy. Top priorities include fall prevention, elder abuse, dealing with end-of-life issues, and offering delivery services to noninstitutionalized seniors. Types of support include conferences and seminars, demonstration grants, program development grants, and technical assistance grants. Grants are awarded for one year and may be renewed for up to three years. There are no application deadlines. The board meets in March, June, September, and December. Contact office for availability.
Requirements: U.S. nonprofits, with emphasis on Southern California, are eligible.
Restrictions: The foundation does not fund capital expenditures, biomedical research, brick and mortar projects, building campaigns, or programs that solely provide or distribute food.
Geographic Focus: California
Date(s) Application is Due: Jun 25
Samples: U of California (Los Angeles)—for a meeting at which more than 150 participants—including representatives of public health departments, hospitals and health systems, health insurance plans, and social service agencies—created a statewide plan for reducing the incidence of falls among older Californians, $135,000; Zen Hospice Project (San Francisco, CA)—for a one-year counseling curriculum for practitioners involved in the direct care of dying people, $50,000 over two years.
Contact: Mary Ellen Courtwright, Program Officer; (562) 590-8655; fax (562) 495-0317; archstone@archstone.org
Internet: http://www.archstone.org/info-url_nocat2293/info-url_nocat_show.htm?doc_id=34504
Sponsor: Archstone Foundation
401 E Ocean Boulevard, Suite 1000
Long Beach, CA 90802

ARCO Foundation Education Grants 665
The foundation will concentrate its aid to education in support of the following: precollege programs to improve the quality of teaching and learning in urban public education; programs aimed at decreasing attrition rates among low-income and minority students; programs to motivate low-income and minority students to succeed in college, especially in mathematics-based careers of engineering, science, and business; support for laboratory renovation and scientific equipment in academic disciplines of interest at major research universities; programs to retain the most talented young faculty in academic careers in selected disciplines; selected liberal arts programs at colleges and universities of interest; state associations of private colleges in the states where the company has interests; academic programs relevant to energy interests at regional universities and colleges; and national education associations and organizations that seek to improve education in public high schools and at higher academic levels. Grants are awarded for operating budgets, seed money, equipment, land acquisition, matching funds, employee matching gifts, employee related scholarships, special projects, and technical assistance. Applications are accepted at any time; annual report should be obtained prior to submitting a formal proposal.
Requirements: The foundation is a regional organization funding nonprofit organizations in states where ARCO has facilities and personnel, including Alaska, Arizona, California, Colorado, Nevada, Texas, and Washington. Requests from those states and those nearby should be addressed to the local community affairs managers.
Restrictions: The foundation discourages applications from the following: programs not focused on promoting self-sufficiency and economic development of minority populations; historic preservation or urban development projects not tied to neighborhood economic revitalization; proposals from religious organizations or funding requests from federal, state, county, and municipal agencies, including school districts. The foundation does not generally consider support of hospital building or endowment campaigns, medical equipment, medical research programs, single-issue health organizations, or health services not directed at low-income people.
Geographic Focus: Alaska, Arizona, California, Colorado, Nevada, Texas, Washington
Amount of Grant: 1,500 - 360,000 U.S.D
Samples: U of Montana (Butte, MT)—to help retain and graduate minority students pursuing engineering degrees, $16,500; U of California (Davis, CA)—to help retain and graduate minority students pursuing engineering degrees, $47,000.
Contact: Virginia Victorin, Grants Administrator; (213) 486-3342; fax (213) 486-0113
Internet: http://www.ntlf.com/html/grants/5977.htm
Sponsor: ARCO Foundation
151 South Flower Street
Los Angeles, CA 90071

Argyros Foundation Grants 666
The foundation annually awards grants for cultural activities, higher education, religious projects, social services, recreation, and health services. Types of support include general operating support and program development. Proposals must give a detailed outline of the project for which funds are requested.
Requirements: Giving is primarily made to organizations in Orange County, CA.
Geographic Focus: California
Date(s) Application is Due: Jun 1
Amount of Grant: 5,000 - 50,000 U.S.D
Samples: Estell Doheny Eye Institute (Los Angeles, CA)—for operating support, $110,000; South Coast Repertory Theater (Costa Mesa, CA)—for program support, $20,000; Boy Scouts of America, Orange County (Santa Ana, CA)—for operating support, $7000.
Contact: Daniel Russo, Argyros Charitable Trusts; (714) 481-5000
Sponsor: Argyros Foundation
949 S Coast Drive, #600
Costa Mesa, CA 92626

Arie and Ida Crown Memorial Grants 667
The program supports programs that offer opportunities to the disadvantaged, strengthens the bond of families, and improves the quality of people's lives. As a general rule, the Foundation funds organizations that serve the greater Chicago area as well as organizations that serve the broader Jewish community. Most grants are awarded to organizations within the city of Chicago. Organizations are supported in the areas of arts and culture (concentrating on educational and enrichment programs for youth), civic affairs, education, health (stressing access to services, hospice and health promotion), and human service (focusing on programs which offer assistance for children and families).
Requirements: Nonprofit organizations in Chicago and Cook County, IL, may apply for grant support.
Restrictions: Grants are not made to support individuals, conference expenses, film projects, government programs (50 percent government funded), or research projects.
Geographic Focus: Illinois
Date(s) Application is Due: Jan 31; Jul 31
Amount of Grant: 1,000 - 200,000 U.S.D
Contact: Susan Crown, President; (312) 236-6300; fax (312) 984-1499; AICM@crown-Chicago.com
Internet: http://www.crownmemorial.org/
Sponsor: Arie and Ida Crown Memorial
222 N LaSalle Street, Suite 2000
Chicago, IL 60601

Arizona Cardinals Grants 668
The National Football League franchise supports programs designed to improve the quality of life and enhance opportunities for children, women, and minorities in the state of Arizona. Specific areas of interest include arts and culture, civic affairs, education, health, science, and social services. The foundation is interested in expanding its giving and looks for new charities to fund. First-time applicant organizations generally will receive grants of $5000 or less.
Requirements: Applicants must be exempt under 501(c)3 of the Internal Revenue Service code.
Geographic Focus: Arizona
Date(s) Application is Due: Aug 1
Amount of Grant: 2,000 - 5,000 U.S.D
Samples: AZCenters for Comprehensive Education and Life Skills; Chrysalis Shelter for Victims of Domestic Violence; Child crisis Center-East Valley; Jewish Family and Children's Services; Juvenile Diabetes Research Foundation; United Food Bank.
Contact: Pat Tankersley, (602) 379-0101; fax (480) 785-7327
Internet: http://www.azcardinals.com/community/charities.php
Sponsor: Arizona Cardinals
P.O. Box 888
Phoenix, AZ 85001-0888

Arizona Community Foundation Scholarships 669

The Arizona Community Foundation sponsors the Dorrance Family Foundation scholarships to provide academic and financial support to Arizona's first generation college students. The foundation awards up to 25 scholarships annually to incoming freshmen who will attend one of three Arizona public universities. The award is renewable for up to three years for a total of eight semesters of full-time undergraduate study. Scholarships are renewed based on academic standing, participation in program events and activities, and community service projects. Any academic major is eligible.
Requirements: Applicants must be Arizona residents; seniors in good standing at an accredited Arizona high school; first generation applicants to attend college; hold a minimum cumulative 3.0 GPA; an SAT score of 1040 or composite ACT score of 22 (excluding writing score); demonstrate financial need; accepted by one of Arizona's three residential public universities (Arizona State, Northern Arizona University, or University of Arizona); and demonstrated leadership and community service. Visit the website for a current application and deadline information.
Restrictions: Applicants must be Arizona residents.
Geographic Focus: Arizona
Amount of Grant: 10,000 U.S.D
Contact: Grant Administrator; (602) 381-1400 or (800) 222-8221; fax (602) 381-1575; nstaylor@azfoundation.org
Internet: http://www.dorrancescholarship.org/index.php?hs=1
Sponsor: Arizona Community Foundation
2201 E Camelback Road, Suite 202
Phoenix, AZ 85016

Arizona Diamondbacks Charities Grants 670

The major-league baseball franchise awards grants in Arizona to support as wide as possible a variety of charitable causes. Priority will to be given to organizations that fall under the foundation's focus areas of health care for the indigent, homelessness and youth education. Types of support include general support, project grants, donated equipment, employee matching gifts, corporate sponsorships, and speakers. Organizations wishing to apply for a grant can fax a request to the community affairs department.
Requirements: Applicants must be 501(c)3 Arizona-based organizations committed to spending grant proceeds in Arizona.
Geographic Focus: Arizona
Contact: Program Contact; (602) 462-6500; fax (602) 462-6575
Internet: http://arizona.diamondbacks.mlb.com/ari/community/foundation.jsp
Sponsor: Arizona Diamondbacks
P.O. Box 2095
Phoenix, AZ 85001

Arizona Public Service Corporate Giving Program Grants 671

Grants are awarded to support Arizona nonprofit organizations in the areas of health and human services, community development, arts and culture, education, and environment. The foundation awards support for project grants, capital building funds, research, employee matching gifts, in-kind services, conferences and seminars, and operating support. Applications are accepted on an ongoing basis.
Requirements: Arizona 501(c)3 nonprofits are eligible.
Restrictions: APS Corporate Giving does not fund individual request, charter or private schools, religious, political fraternal, legislative or lobbying efforts to organizations, travel-related or hotel expenses, private or family foundation, private non-profit organizations, salaries and/or debt reduction.
Geographic Focus: Arizona
Contact: Cindy Slick, Manager; (602) 250-4707; fax (602) 250-2113; Cindy.Slick@aps.com
Internet: http://www.aps.com/main/community/dev/default.html
Sponsor: Arizona Public Service Corporation
P.O. Box 53999, MS 8010
Phoenix, AZ 85072-3999

Arkansas Community Foundation Arkansas Black Hall of Fame Grants 672

Grants support programs and projects that address problems, challenges, and opportunities in black communities in Arkansas. Categories of interest include education; health and wellness; youth development; and small business and/or economic development. Priority goes to proposals that show multiple sponsoring agencies/organizations; include evidence of local financial support; demonstrate collaborative ventures among community organizations; and have promise for sustainability beyond the period of the grant.
Requirements: Tax-exempt entities such as schools, churches, or government entities are eligible. Preference is given to applications submitted by 501(c)3 nonprofit organizations.
Restrictions: Funds cannot be allocated for adult salary support in carrying out the duties of the project; used to support general operating budgets outside of the specific proposal or project; or for scholarships for formal education at any level.
Geographic Focus: Arkansas
Date(s) Application is Due: Apr 10
Amount of Grant: Up to 5,000 U.S.D
Contact: Cecilia Patterson, Program Director; (501) 372-1116; fax (501) 372-1166; cpatterson@arcf.org or arcf@arcf.org
Internet: http://www.arcf.org/page27973.cfm
Sponsor: Arkansas Community Foundation
1400 West Markham, Suite 206
Little Rock, AR 72201

Arkansas Community Foundation Grants 673

The Arkansas Community Foundation (ARCF) serves philanthropic donors and supports non-profits in Arkansas. ARCF makes grants from charitable funds established by individuals, families, corporations, and non-profit organizations. From among over 1100 funds, its donors make grants in the following areas of interest: animal welfare, arts and humanities, community development, education, environment, health, human services, and religion. While the Foundation's discretionary grant cycles are few, grants are usually modest and, generally, fund capital or building campaigns are not funded.
Requirements: 501(c)3 Arkansas nonprofit organizations are eligible.
Geographic Focus: Arkansas
Amount of Grant: 500 - 2,500 U.S.D
Contact: Cecilia Patterson, Program Director; (501) 372-1116; fax (501) 372-1166; cpatterson@arcf.org or arcf@arcf.org
Internet: http://www.arcf.org/page12714.cfm
Sponsor: Arkansas Community Foundation
1400 West Markham, Suite 206
Little Rock, AR 72201

Arkell Hall Foundation Grants 674

The primary mission of the foundation is the operation and maintenance of a home for elderly women. Funds that may become available above the needs of the home may be distributed annually to tax-exempt organizations providing services in the target community—Western Montgomery County, NY—with preference given to those active in service to senior citizens, education (higher education, medical education, adult basic education and literacy), religion (Christian, Christian Reformed Church, Lutheran, Methodist, Roman Catholic, the Salvation Army, and United Methodist), and health care. Types of support include capital, challenge, endowment, general support, matching, and seed money grants. Grants are made on a single-year basis and are awarded annually in October or November. It is recommended that requests be submitted between July 1 and September 15. Initial review of requests is performed upon receipt. Results are forwarded to the applicant within one month.
Requirements: IRS 501(c)3 organizations directly impacting the Western Montgomery County, NY, community are eligible.
Restrictions: Requests that do not include written proof of 501(c)3 status will not be considered. Projects or organizations with large service areas, such as national or regional, will not qualify for funding, nor will projects in which the target community is not the primary area of focus.
Geographic Focus: New York
Date(s) Application is Due: Oct 1
Amount of Grant: 1,000 - 50,000 U.S.D
Samples: Cornell University, College of Human Ecology (Ithaca, NY)—for general support, $20,000; College of Saint Rose (Albany, NY)—to support scholarships for single mothers, $3000.
Contact: Joseph Santangelo, Vice President; (518) 673-5417; fax (518) 673-5493
Sponsor: Arkell Hall Foundation
68 Front Street, P.O. Box 240
Canajoharie, NY 13317-0240

Arkema Foundation Science Teachers Program 675

This program is an intensive week-long session for elementary and secondary school teachers. Armed with innovative science experiment kits and the guidance of chemical engineers and scientists, teachers learn new and fascinating ways to illustrate scientific concepts. Scientific topics explored include life, earth and physical science and technology. School principals are asked to nominate two teachers in grades three through six, who are then chosen by the corporate committee to participate in the program. Application forms and contact information are available on the Web site.
Requirements: Teachers in the following geographic areas may participate: Alabama (Mobile county); Kentucky (Graves, Livingston, Lyon, Marshall, McCracken, Carroll, Gallatin, Trimble, Jefferson counties); Michigan (Wayne county); Minnesota (Didge, Mower, Steele counties); New Jersey (Burlington, Camden, Gloucester, and Salem counties); New York (Genesee, Livingston counties); Pennsylvania (Berks, Bucks, Chester, Delaware, Montgomery, and Philadelphia counties); Tennessee (Shelby county); Texas (Jasper, Newton, Jefferson, Orange, Harris, and Brazos counties).
Geographic Focus: Alabama, Kentucky, Michigan, Minnesota, New Jersey, New York, Pennsylvania, Tennessee, Texas
Amount of Grant: 500 U.S.D
Contact: Jane Crawford, (215) 419-7614; jane.crawford@arkemagroup.com
Internet: http://www.products.arkemagroup.com/index.cfm?pag=190
Sponsor: Arkema Foundation
2000 Market Street
Philadelphia, PA 19103-3222

Arlington Community Foundation Grants 676

The foundation awards grants in Arlington to educators and nonprofit organizations for innovative projects that supplement and enrich the learning environment for preschool to adult students. The focus is curriculum enrichment, the arts (musical, dramatic, visual), pursuit of higher education, vocational education, after school and summer programs, life-long learning, environmental issues, parent involvement, and community involvement. The community enhancement grants support arts and humanities, children & families, community improvement, health, housing/homeless & hunger, legal, social services, and senior enrichment. The purpose of the Prompt Response Fund is to enable nonprofit groups in Arlington respond quickly to unanticipated opportunities or unexpected, urgent community needs. Deadlines may vary. Guidelines and applications are available online.

Requirements: Organizations and individuals with projects designed to meet educational needs of Arlington residents are encouraged to apply.
Restrictions: The foundation does not make grants for endowments, capital campaigns, religious purposes, individual debts, or political lobbying.
Geographic Focus: Virginia
Contact: Wanda L. Pierce, Executive Director; (703) 243-4785; fax (703) 243-4796; info@arlcf.org
Internet: http://www.arlcf.org/grants.html
Sponsor: Arlington Community Foundation
2525 Wilson Boulevard
Arlington, VA 22201

Armstrong McDonald Foundation Grants 677
The foundation awards grants to eligible nonprofit organizations in its areas of interest, including animal welfare, children and youth, education, special needs, health, and relief and social.
Requirements: 510(c)3 nonprofit organizations, colleges and universities, schools, and school districts are eligible. Application must be downloaded from website.
Restrictions: Grants do not support advocacy organizations, individuals, international organizations, political organizations, or state and local government agencies. The foundation does not fund capital campaigns, salaries/stipends, and multi-year projects. Organization must have received a grant from the foundation within the last five years unless it is located within the states of Arizona or Nebraska. No organization east of the Mississippi is eligible for a grant.
Date(s) Application is Due: Sep 30
Amount of Grant: 1,000 - 80,000 U.S.D
Samples: San Diego Zoo (San Diego, CA) $40,000; Tucson Children's Museum (Tucson, AZ) $20,000; Omaha Street School (Omaha, NE) $17,500; UofA Found for Steele Children (Tucson, AZ) $72,900; Saddlebrook Community Outreach (Tucson, AZ) $10,000; Canine Co-Pilots (Flagstaff, AZ) $10,000; NW School for Hearing Impaired (Seattle, WA) $7,500.
Contact: Laurie Bouchard, Vice President; (520) 878-9627; info@ArmstrongMcDonaldFoundation.org
Internet: http://www.armstrongmcdonaldfoundation.org/
Sponsor: Armstrong McDonald Foundation
P.O. Box 70110
Tucson, AZ 85737.01

Arnold Fund Grants 678
The primary focus of the fund is to support charitable organizations located in Newton County, GA. Fields of interest include performing arts, music, higher education, library, education, biomedicine, and medical research. Grants are also made to institutions of higher education in which the founders had a special interest. Proposers are requested to write a brief letter explaining the project for which the funds are being requested. A formal application form is not required.
Restrictions: Grants are not made to individuals.
Geographic Focus: Georgia
Amount of Grant: 500 - 279,000 U.S.D
Contact: John Sawyer, Executive Director; (404) 881-7886
Sponsor: Arnold Fund
1201 W Peachtree Street, Suite 4200
Atlanta, GA 30309-3400

Arnold P. Gold Foundation Grants 679
The foundation's goal is to nurture and perpetuate the tradition of the caring doctor. Toward this end, grants support innovative medical education programs that emphasize humanism in medicine. The foundation is interested in the development, implementation, evaluation, and replication of innovative medical educational programs and projects that influence the way physicians are trained. A grant request form is available on the Web site.
Requirements: Medical schools and nonprofit organizations are eligible.
Amount of Grant: Up to 25,000 U.S.D
Contact: Grants Administrator; (201) 567-7999; fax (201) 567-7880; goldfdtn@gold-foundation.org
Internet: http://www.humanism-in-medicine.org
Sponsor: Arnold P. Gold Foundation
619 Palisade Avenue
Englewood, NJ 07632

ARS Foundation Grants 680
The mission of the foundation is to make a positive difference in the lives of others. They support human and social services, the arts, children and the elderly, and health care but will consider other organizations. Giving is focused in New York City and Fairfield County, CT, areas. Applicants should submit a cover letter containing a brief history and description of the organization, the program(s), the needs/goals to be addressed, and a description of the project; an itemized budget, current organizational annual budget, and recent audited financial statement; annual report; other funding sources; and 501(c)3 determination letter. There are no application deadlines or forms.
Requirements: Qualified 501(c)3 charitable organizations in the New York City and Fairfield County, CT, areas are eligible.
Restrictions: The foundation does not give grants to individuals, private foundations, or organizations not qualified as charitable organizations.
Geographic Focus: Connecticut, New York
Contact: Administrator; (212) 986-1533; fax (212) 972-2303; info@arsfoundation.com
Internet: http://www.arsfoundation.com
Sponsor: Adolph and Ruth Schnurmacher Foundation
551 Fifth Avenue, Suite 1210
New York, NY 10176

ARS New Investigator Award 681
The purpose of this award is to support basic, translational, or clinical research projects in rhinology. Proposed projects may be related to any area of rhinology. Proposed project shall be designed in collaboration with a preceptor investigator and approved by the candidate's department chairperson and institution. One award of up to $25,000 over a two-year period will be given annually. Letters of Interest should be received no later than December 15, with full proposals due on January 15.
Requirements: Any member of the American Rhinologic Society who has not received previous, significant outside funding is eligible. The applicant must have, as a mentor, an established researcher who will provide a letter of support stating the extent of involvement in the project and provide a summary of his/her research experience.
Date(s) Application is Due: Jan 15; Dec 15
Amount of Grant: Up to 25,000 U.S.D
Contact: Stephanie L. Jones, Assistant Director; (703) 519-1586 or (703) 836-4444; sljones@entnet.org
Internet: http://www.entnet.org/EducationAndResearch/coreGrants.cfm
Sponsor: American Rhinology Society
One Prince Street
Alexandria, VA 22314-3357

ARS Resident Research Grants 682
The purpose of this award is to support basic, translational, or clinical research projects in rhinology. Proposed projects may be related to any area of rhinology. Proposed project shall be designed in collaboration with a preceptor investigator and approved by the candidate's department chairperson and institution. Two awards of up to $8,000 over a one-year period will be given annually. Letters of Interest should be received no later than December 15, with full proposals due on January 15.
Requirements: Any resident in training in an approved program in the U.S. or Canada is eligible to apply for the American Rhinologic Society Research Grant. Resident applicants must have as a co-investigator a supervising faculty who is a member in good standing in the American Rhinologic Society (ARS).
Date(s) Application is Due: Jan 15; Dec 15
Amount of Grant: Up to 8,000 U.S.D
Contact: Stephanie L. Jones, Assistant Director; (703) 519-1586 or (703) 836-4444; sljones@entnet.org
Internet: http://www.entnet.org/EducationAndResearch/upload/2009-ARS-Resideint-Research-Grant.pdf
Sponsor: American Rhinology Society
One Prince Street
Alexandria, VA 22314-3357

Artemas W. Stearns Trust Grants 683
The trust awards grants to eligible Massachusetts area nonprofit organizations in its areas of interest, including education, health, youth development, and the elderly and poor. Grants are awarded for one year and may be renewed. Guidelines will be sent upon request.
Requirements: Nonprofits serving the greater Lawrence, MA, area are eligible.
Geographic Focus: Massachusetts
Date(s) Application is Due: Jan 31
Amount of Grant: 1,500 - 35,000 U.S.D
Contact: Clifford Elias, Managing Trustee; (978) 687-0151; fax (978) 685-9132
Sponsor: Artemas W. Suitearns Trust
70 East Street
Methuen, MA 01844

Arthritis Foundation of Australia Grants 684
The foundation awards grants in the fields of medicine and health, with emphasis on arthritis research and education. Types of awards include grants-in-aid, fellowships, and scholarships. Grants are awarded in Australia, the United Kingdom, and the United States. Annual deadline dates may vary; contact program staff for exact dates.
Requirements: Awards are available to clinical, scientific, and allied health professionals who are Australian citizens or permanent residents. Scholarships, fellowships, and grants may only be held in universities, hospitals, or recognized research institutes.
Date(s) Application is Due: Jun 30
Contact: Grant Coordinator; 61 (0)2 9552 6085; fax 61 (0)2 9552 6078; info@arthritisaustralia.com.au
Internet: http://arthritisaustralia.com.au/Research%20Grants
Sponsor: Arthritis Foundation of Australia
P.O. Box 121
Sydney, NSW 2001 Australia

Arthritis Medical Research Grants 685
The foundation's primary purpose is to provide financial support to research studies aimed at discovering new knowledge for the prevention, treatment, and cure of arthritis and other rheumatic diseases. Both clinical and basic studies will be considered. The focus of proposals should be on high-incidence diseases. Grants are awarded for one year; projects requiring multiyear funding must reapply each year. Partial funding may be awarded if the grantee has documentation that the remainder of the funding for proposed

research has been acquired. Grants support salaries, supplies, and equipment directly related to the proposed studies. Overhead or indirect costs will not be funded.
Requirements: Priority will be given to newer investigators possessing either a MD and/or PhD degree who are associated with qualified institutions.
Date(s) Application is Due: Jan 15
Amount of Grant: 20,000 - 50,000 U.S.D
Samples: Arunan Kaliyaperuman, Northwestern U (Evanston, IL)—for autoimmune diseases research, $38,000; Morteza Setareh, Scripps Research Institute (La Jolla, CA)—for research on mechanisms in the human body that trigger autoimmune diseases, $38,000.
Contact: Helene Belisle, Executive Director; (800) 588-2873; fax (562) 983-1410; anrf@ix.netcom.com
Internet: http://www.curearthritis.org/grant.htm
Sponsor: Arthritis National Research Foundation
200 Oceangate, Suite 400
Long Beach, CA 90802

Arthritis Society Investigator Awards **686**
A limited number of grants are offered to individuals who have demonstrated their ability as independent research scientists with international impact and are pursuing a full-time academic career in medical science. Candidates will only be considered between three and eight years after receiving their first academic appointment. Registration deadline is August 15; full application is due by September 15.
Requirements: Application is made on behalf of candidates by the chair of the department or director of the research institute in which the scholar will be appointed. Application should be submitted on form NF-109A.
Restrictions: Grants will not be made to replace faculty salary support for well-established investigators who have held faculty appointment for longer than five years at the applying institution.
Geographic Focus: All States, Canada
Amount of Grant: 70,000 CAD
Contact: Medical and Scientific Program Supervisor; (416) 967-3353, ext. 311 or (416) 979-7228; fax (416) 979-8366; can@mtsinai.on.ca
Internet: http://www.arthritis.ca/look%20at%20research/awards/investigators/default.asp?s=1
Sponsor: Arthritis Society (Canada)
393 University Avenue, Suite 1700
Toronto, ON M5G 1E6 Canada

Arthritis Society Multicenter Grants **687**
Multicenter grants are designed to foster collaboration of individuals, working at different geographic locations, on a research project deemed relevant to the rheumatic diseases. Grants are usually made for a period of three years and are intended to cover the costs of the research for which they are provided. Appropriate space and basic facilities at the institutions concerned are prerequisite to an application. Registration deadline is November 15; full application is due by December 15.
Requirements: Completed applications for grants or renewals must be submitted on form NF-107A; obtain detailed instructions from the society.
Geographic Focus: All States, Canada
Date(s) Application is Due: Nov 15; Dec 15
Amount of Grant: 60,000 CAD
Contact: Medical and Scientific Program Supervisor; (416) 967-3353, ext. 311 or (416) 979-7228; fax (416) 979-8366; can@mtsinai.on.ca
Internet: http://www.arthritis.ca/look%20at%20research/awards/default.asp?s=1
Sponsor: Arthritis Society (Canada)
393 University Avenue, Suite 1700
Toronto, ON M5G 1E6 Canada

Arthritis Society New Investigator Award **688**
The awards are intended to support newly appointed junior faculty members (within three years of appointment) who are planning to pursue a career in basic or clinical research related to arthritis. The scholarships assist individuals with the PhD or MD degree who have received extensive postdoctoral research training in their chosen area and who have clearly demonstrated their potential as independent research scientists to establish themselves. Scholarships are not renewable; prior recipients may apply for research scientist grants. Registration deadline is November 15; full application is due by December 15.
Requirements: Application is made on behalf of candidates by the chair of the department or director of the research institute in which the scholar will be appointed. Application should be submitted on form NF-110A.
Restrictions: The awards are not intended to replace faculty salary support for well-established investigators who have been on faculty for a number of years at the applying institution.
Geographic Focus: All States, Canada
Date(s) Application is Due: Nov 15; Dec 15
Amount of Grant: 50,000 CAD
Contact: Medical and Scientific Program Supervisor; (416) 967-3353, ext. 311 or (416) 979-7228; fax (416) 979-8366; can@mtsinai.on.ca
Internet: http://www.arthritis.ca/look%20at%20research/awards/new%20investigators/default.asp?s=1
Sponsor: Arthritis Society (Canada)
393 University Avenue, Suite 1700
Toronto, ON M5G 1E6 Canada

Arthritis Society Research Grants **689**
Grants in aid of research are designed to promote and support research proposed by investigators holding staff appointments at Canadian universities or other recognized Canadian institutions where the research is deemed relevant to the rheumatic diseases. Grants are usually made for a period of three years and are intended to cover the costs of the research for which they are provided. Appropriate space and basic facilities at the institution concerned are prerequisite to an application. Requests for all funds on grant applications must be carefully and fully justified.
Requirements: Completed applications for grants or renewals must be submitted on form NF-104A; obtain detailed instructions from the society.
Geographic Focus: All States, Canada
Date(s) Application is Due: Dec 15
Amount of Grant: 60,000 CAD
Contact: Medical and Scientific Program Supervisor; (416) 967-3353, ext. 311 or (416) 979-7228; fax (416) 979-8366; can@mtsinai.on.ca
Internet: http://www.arthritis.ca/look%20at%20research/awards/d9/default.asp?s=1
Sponsor: Arthritis Society (Canada)
393 University Avenue, Suite 1700
Toronto, ON M5G 1E6 Canada

Arthur and Rochelle Belfer Foundation Grants **690**
The foundation awards grants to nonprofit organizations of the Jewish faith, with a focus on New York. Grants are targeted toward programs supporting the elderly and women; education/higher education; institutions such as seminaries, synagogues, and temples; hospitals; Jewish welfare; and medical centers. Types of support include general support grants and fellowships. There are no application deadlines. Applicants should send a brief letter of inquiry describing the program.
Restrictions: Grants are not made to individuals.
Geographic Focus: New York
Amount of Grant: 1,000 - 100,000 U.S.D
Samples: Joan S. Brugge, PhD, chair, Department of Cell Biology, arvard Medical School—for breast cancer studies; Dana-Farber Cancer Institute (Boston, MA)—for general support, $1 million; Anti-Defamation League of B'nai B'rith (New York, NY)—for operating support, $22,000; American Friends of Israel Museum (New York, NY)—for operating support, $5000.
Contact: Robert Belfer, President; (212) 508-6020
Sponsor: Arthur and Rochelle Belfer Foundation
767 Fifth Avenue, 46th Floor
New York, NY 10153-0002

Arthur Ashley Williams Foundation Grants **691**
AAW Foundation awards grants to charitable organizations to help improve quality of life for those seeking assistance from these charitable organizations. Grant requests will be considered for program support, seed money, challenge grants, and capital improvements.
Requirements: To apply for a grant, applicants must be non-profit 501(c)3 organizations or public schools.
Restrictions: The Foundation will not fund political or sectarian activities.
Amount of Grant: 1,000 - 15,000 U.S.D
Samples: The Barry R. Kirschner Wildlife Foundation, Durham, CA, $11,500; Framingham Historical and Natural Historic Society, Framingham, MA, $15,000.
Contact: Program Contact; (508) 893-0757; clambert@rcn.com
Internet: http://www.aawfoundation.org/requests.php
Sponsor: Arthur Ashley Williams Foundation
P.O. Box 6280
Holliston, MA 01746

ArvinMeritor Foundation Health Grants **692**
The foundation provides grants primarily in company-operating locations in the areas of education and training, civic and health, youth organizations, and arts and culture. In the area of Health, the foundation supports programs designed to promote health research, treatment, education, and awareness; and reduce the cost of illness. Types of support include general operating budgets, building funds, continuing support, equipment, and projects. The committee foundation meets every six to eight weeks to review grant requests. Contact program staff for current guidelines.
Requirements: Nonprofit 501(c)3 organizations should submit a one- to two-page letter outlining the purpose and needs of the program, its budget, duration, goals, leadership, and amount requested.
Restrictions: Ineligibility applies to individuals; organizations that limit participation or services based on race, gender, religion, color, creed, age, or national origin; projects without ties to a community that is home to an ArvinMeritor facility; organizations that pose any conflict with the goals and mission of ArvinMeritor, its employees, communities, or products; operating expenses for United Way local agencies, except through the foundation's support of annual United Way campaigns; sponsorships of fund raising activities by individuals (i.e., walk-a-thons); requests for loans or debt retirement; religious or sectarian programs for religious purposes; labor, political, or veterans organizations; fraternal, athletic, or social clubs; or seminars, conferences, trips, and tours.
Contact: Jerry Rush, Senior Director; (248) 435-7907; fax (248) 245-1031; jerry.rush@arvinmeritor.com
Internet: http://www.arvinmeritor.com/community/community.asp
Sponsor: ArvinMeritor Foundation
2135 W Maple Road
Troy, MI 48084

ArvinMeritor Grants 693

ArvinMeritor???s philanthropic programs focus on education and the communities where its employees live and work. Grants support engineering and technical schools worldwide to increase the number of engineers and scientists in the field of automotive engineering, and to stimulate technological development and innovation in this field. The corporation also partners with local, private, and public agencies in order to improve the quality of life in the areas of education, cultural programs, civic responsibility, and health and human services. There are no application deadlines.
Requirements: Nonprofit organizations in ArvinMeritor communities are eligible.
Restrictions: Grants do not support organizations without nonprofit status; individuals; organizations that limit participation or services based on race, gender, religion, color, creed, age, or national origin; projects without ties to a community that is home to an ArvinMeritor facility; organizations that pose any conflict with the goals and mission of ArvinMeritor, its employees, communities, or products; operating expenses for United Way local agencies, except through support of annual United Way campaigns; sponsorships of fund-raising activities by individuals; requests for loans or debt retirement; religious or sectarian programs for religious purposes; labor, political, or veterans organizations; fraternal, athletic, or social clubs; or seminars, conferences, trips, and tours.
Amount of Grant: 1,000 - 100,000 U.S.D
Contact: Jerry Rush, Senior Director; (248) 435-7907; fax (248) 245-1031; jerry.rush@arvinmeritor.com
Internet: http://www.arvinmeritor.com/community/guidelines.asp
Sponsor: ArvinMeritor
2135 W Maple Road
Troy, MI 48084-7186

ASA Metlife Foundation MindAlert Awards 694

The awards were established to recognize innovations in mental fitness programming for older adults. Awards recognize programs, products, or tools that promote cognitive fitness in later life. One award will be made to a program in each of three categories: lifelong learning and third age learning programs where mental fitness is implicit; educational programs focused on enhancing mental fitness for the general population of older adults; and programs specifically designed to enhance mental fitness for cognitively-impaired older adults. Applications are judged for their innovation, their basis in research, demonstration of their effectiveness, potential for replicability and the extent to which the programs are accessible to diverse populations of elders. Guidelines and application are available online.
Requirements: Programs and products or tools that promote cognitive function in later life are eligible.
Date(s) Application is Due: Oct 1
Amount of Grant: 1,500 U.S.D
Contact: Nancy Decia, Coordinator, Education & Training; (415) 974-9610; fax (415) 974-0300; awards@asaging.org
Internet: http://www.asaging.org/asav2/mindalertaward/index.cfm
Sponsor: American Society on Aging
833 Market Street, Suite 511
San Francisco, CA 94103-1824

ASCO/UICC International Cancer Technology Transfer Fellowships 695

The scholarships are intended primarily for researchers in the early stages of their career as well as for experienced clinical oncologists. The ICRETT objective is to facilitate rapid international transfer of cancer research and clinical technology, and it allows fellows to exchange knowledge and enhance skills in basic, clinical, behavioral and epidemiological areas of cancer research, and cancer control and prevention. Through the program fellows also acquire appropriate clinical management, diagnostic and therapeutic expertise for effective application and use in home organizations upon return. The UICC has also developed the ICRETT for teaching faculty which funds experts to teach at host institutions throughout the world for a duration of 1 week to 1 month.
Requirements: The candidate must possess the appropriate professional qualifications and experience according to the specifications of each fellowship described on the website and must currently be engaged in cancer research, clinical oncology practice, or cancer society work. To permit effective communication at the host institute, the candidate must have adequate fluency in a common language. The candidate must also be on the staff payroll of a university, research laboratory or institute, hospital, oncology unit, or voluntary cancer society (or be accredited volunteers of such societies) to where they will return at the end of a fellowship.
Restrictions: Candidates attached to commercial entities, going to a profit organization or have associations with the tobacco industry not eligible.
Amount of Grant: 3,400 U.S.D
Contact: Nancy R. Daly, Executive Director; (571) 483-1432; dalyn@asco.org
Internet: http://www.ascocancerfoundation.org/TACF/Awards/Award+Opportunities/ASCO+%26+UICC+ICRETT+Fellowship
Sponsor: American Society of Clinical Oncology
1900 Duke Street, Suite 200
Alexandria, VA 22314

ASCO Advanced Clinical Research Award in Colorectal Cancer 696

The Advanced Clinical Research Award (ACRA) in Colorectal Cancer is designed to fund investigators who are committed to clinical cancer research and who wish to conduct original research not currently funded. The grant totals $450,000 and is paid in 3 annual increments of $150,000 made on or about July 1 of each year of the grant term. The grant funds will be directed to the sponsoring institution and should be used towards salary support, supplies, equipment, travel, etc. necessary for the pursuit of the recipient's research project.
Requirements: Appplicants must meet the following criteria: be a physician (MD, DO, or international equivalent) who is in the fourth to ninth year of a full-time, primary faculty appointment in a clinical department at an academic medical institution at the time of grant submission; have completed productive post doctoral/post fellowship research and demonstrated the ability to undertake independent investigator-initiated clinical research; be an active member of ASCO or have submitted a membership application with the grant application; be able to commit to 75% of full-time effort in research (applies to total research, not just the proposed project) during the award period.
Date(s) Application is Due: Nov 3
Amount of Grant: 450,000 U.S.D
Contact: Nancy R. Daly, Executive Director; (571) 483-1432; dalyn@asco.org
Internet: http://www.ascocancerfoundation.org/TACF/Grants/Grant+Opportunities/Advanced+Clinical+Research+Award+in+Colorectal-Cancer
Sponsor: American Society of Clinical Oncology
1900 Duke Street, Suite 200
Alexandria, VA 22314

ASCO Advanced Clinical Research Awards in Breast Cancer 697

The ASCO Foundation and Genentech BioOncology invite physicians with full-time faculty appointments to apply for a clinical oncology research award in Breast Cancer. The program funds clinical investigators who are committed to clinical cancer research. Their research must have a direct patient-oriented focus including clinical trials and/or translational research involving human subjects. The Advanced Clinical Research Award is designed to provide funding beyond the Clinical Research Career Development Award (CDA). By continuing to support proven clinical researchers who are post-CDA but at a critical stage in their early career, the foundation hopes to expand the cadre of expert clinical oncology researchers who are developing promising research initiatives.
Requirements: The applicant must meet the following criteria at the time of grant award: be a physician (MD or DO) who is 5-10 years post final sub-specialty training; have a full-time faculty appointment in a clinical department at an academic medical center; have completed productive postdoctoral/post-fellowship research and demonstrated the ability to undertake independent investigator-initiated clinical research; be an active member of the American Society of Clinical Oncology (ASCO); and expect to spend 75% of time during the award period dedicated to research.
Date(s) Application is Due: Oct 20
Amount of Grant: 450,000 U.S.D
Contact: Nancy R. Daly, Executive Director; (571) 483-1432; dalyn@asco.org
Internet: http://www.ascocancerfoundation.org/TACF/Grants/Grant+Opportunities/Advanced+Clinical+Research+Award+in+Breast+Cancer
Sponsor: American Society of Clinical Oncology
1900 Duke Street, Suite 200
Alexandria, VA 22314

ASCO Advanced Clinical Research Awards in Sarcoma 698

The Advanced Clinical Research Award (ACRA) in Sarcoma is designed to fund investigators who are committed to clinical cancer research and who wish to conduct original research not currently funded. The award is intended to support proposals with a patient-oriented focus, including a clinical research study and/or translational research involving human subjects. The grant totals $450,000 and is paid in three annual increments of $150,000 made on or about January 1 of each year of the grant term. Preference will be given to proposals involving therapeutic vaccine or immunotherapeutic approaches for sarcoma.
Requirements: Applicants must meet the following criteria: be a physician (MD, DO, or international equivalent) who is in the fourth to ninth year of a full-time, primary faculty appointment in a clinical department at an academic medical institution at the time of grant submission; have completed productive post doctoral/post fellowship research and demonstrated the ability to undertake independent investigator-initiated clinical research; be an active member of ASCO or have submitted a membership application with the grant application; be able to commit to 75% of full-time effort in research (applies to total research, not just the proposed project) during the award period.
Restrictions: Proposals with a predominant focus on in vitro or animal studies (even if clinically relevant) are not allowed.
Date(s) Application is Due: Aug 6
Amount of Grant: 450,000 U.S.D
Contact: Nancy R. Daly, Executive Director; (571) 483-1432; dalyn@asco.org
Internet: http://www.ascocancerfoundation.org/TACF/Grants/Grant+Opportunities/Advanced+Clinical+Research+Award+in+Sarcoma
Sponsor: American Society of Clinical Oncology
1900 Duke Street, Suite 200
Alexandria, VA 22314

ASCO Career Development Award 699

The program provides funds to clinical investigators who have received their initial faculty appointment to establish an independent clinical cancer research program. Since many awards are supported with restricted grants from outside organizations, the ASCO is particularly interested in identifying young researchers working in the following sub-specialties and emerging disciplines: breast cancer; cancer survivorship; geriatric oncology; health disparities; kidney cancer; multiple myeloma; ovarian cancer; pancreatic cancer; sarcoma; survivorship; and young adult cancer. However, The ASCO Foundation welcomes application submissions in all oncology sub-specialties.
Requirements: Clinical investigators who have received their initial faculty appointment to establish an independent clinical cancer research program are eligible.
Date(s) Application is Due: Sep 9
Amount of Grant: 200,000 U.S.D

Contact: Nancy R. Daly, Executive Director; (571) 483-1432; dalyn@asco.org
Internet: http://www.ascocancerfoundation.org/TACF/Grants/Grant+Opportunities/Career+Development+Award/Career+Development+Award
Sponsor: American Society of Clinical Oncology
1900 Duke Street, Suite 200
Alexandria, VA 22314

ASCO Community Oncology Research Grants **700**
The Community Oncology Research Grant is open to all eligible, community-based practices. The award is intended to support community-based practices, in their efforts to enhance their clinical trials program. Areas of improvement are based on the American Society of Clinical Oncology Statement on Minimum Standards and Exemplary Attributes of Clinical Trial Sites (Journal of Clinical Oncology, May 20 2008). The grant period begins July 1 and concludes June 30, in addition to a six-month follow-up period for evaluation. Progress reports are required at six and twelve months. A final report is due at eighteen months. Eligible practices are encouraged to apply for the award by completing an online application. Grant proposals will be entertained up to $30,000, including no more than $26,000 for grant support, $1,500 for travel to the ASCO Annual Meeting, and $2,500 for overhead costs. The grant funds will be directed to the community practice, not to an individual. A total of three awards will be presented at the ASCO's Annual Meeting in Chicago, IL.
Requirements: CORG eligibility is based on the following criteria, practices must: have at least one member of ASCO involved in the research program; have at least one member of an ASCO State/Regional Affiliate Program involved in the research program; meet basic standards for a quality clinical trials site (based on Good Clinical Practice standards, the accepted international ethical and scientific quality standard for designing, conducting, recording, and reporting trials); be community-based, with high-quality audit reports and investigators in good standing.
Amount of Grant: 30,000 U.S.D
Contact: Nancy R. Daly, Executive Director; (571) 483-1432; dalyn@asco.org
Internet: http://www.ascocancerfoundation.org/TACF/Grants/Grant+Opportunities/Community+Oncology+Research+Grant
Sponsor: American Society of Clinical Oncology
1900 Duke Street, Suite 200
Alexandria, VA 22314

ASCO Comparative Effectiveness Research Professorship in Breast Cancer **701**
The Comparative Effectiveness Research Professorship (CERP) in Breast Cancer is intended to provide support to qualified individuals to investigate comparative effectiveness in breast cancer and to train future generations of researchers in this field. The CERP will provide funding to outstanding researchers who have made and are continuing to make significant contributions that have changed the direction of breast cancer research and who will provide mentorship to junior researchers. The award totals $500,000 and is paid in five annual increments of $100,000 made on or about July 1 of each year of the grant term. The award is not designed to cover the total cost of a research project or the investigator's entire compensation, but can be used at the discretion of the awardee to support their overall research activities. See website for complete budget guidelines.
Requirements: Applications focusing on all aspects of breast cancer including prevention, diagnosis, treatment, delivery of care, and health economics will be accepted. Applicants must: have an MD, DO, PhD or equivalent degree; have the rank of full professor (or equivalent). If current rank is not that of a full professor but meets all other eligibility criteria, a written explanation must be provided for equivalency of current rank; have a full-time faculty appointment at an academic medical center; have made significant contributions that have changed the direction of breast cancer research; be serving as a research mentor to one or more researcher(s) in training, and must be planning to continue to provide leadership in this area throughout the award period; lead a research team in conducting research on comparative effectiveness in breast cancer; be an active member of ASCO or have submitted a membership application with the grant application; commit to spending 75% of time during the award period dedicated to research (applies to total research, not just the proposed project) including leading a team of researchers (in the lab and/or the clinic) and mentoring physician-scientists. Individuals with an endowed professorship from their own institution are eligible to apply for this award. Individuals with administrative responsibilities (such as a Division or Department Chair) must clearly document their time commitment for performing research activities and mentoring activities. The sponsoring institution must be a not-for-profit institution.
Date(s) Application is Due: Nov 3
Amount of Grant: 500,000 U.S.D
Contact: Nancy R. Daly, Executive Director; (571) 483-1432; dalyn@asco.org
Internet: http://www.ascocancerfoundation.org/TACF/Grants/Grant+Opportunities/Comparative+Effectiveness+Research+Professorship
Sponsor: American Society of Clinical Oncology
1900 Duke Street, Suite 200
Alexandria, VA 22314

ASCO Improving Cancer Care Grants **702**
The ASCO Cancer Foundation Improving Cancer Care Grant, funded by Susan G. Komen for the Cure, will provide extramural research funding to address important issues regarding access to healthcare, quality of care, and delivery of care, with general applicability to breast cancer. The goal of this program is to encourage multi-disciplinary research that will have a major impact on cancer care, with general applicability in the breast cancer arena. This grant opportunity is part of the Susan G. Komen for the Cure/ASCO Cancer Foundation Research Initiative. Susan G. Komen for the Cure and The ASCO Cancer Foundation expect to award up to three research grants, each totaling 1.35 million dollars. This money will be distributed as $450,000 total costs per year over three years. Applicants must submit their letter of intent in the online application system by 5:00 PM EDT on September 1st. Only online applications will be accepted. After review by the Research Initiative Grant Review Subcommittee, applicants will be notified about the status of their letter of intent. Only applicants who have submitted an approved letter of intent will be eligible to submit the full proposal. Applicants can expect to be notified on October 1st.
Requirements: Eligible research teams: will focus on implementing and/or evaluating new solutions to existing problems in quality of, access to, and delivery of care, with general applicability to breast cancer; will be led by a single Principal Investigator, who must be an active ASCO member (or have submitted a membership application) with an MD, DO, PhD or equivalent degree. All categories of active ASCO members are eligible; will have a multidisciplinary team of investigators that may include clinicians, nurses, pharmacists, statisticians, epidemiologists, information technologists, and other research experts; will be allowed to obtain expertise not represented in the core team through consultants and/or sub-contracts.
Date(s) Application is Due: Sep 1; Dec 1
Amount of Grant: 450,000 - 1,350,000 U.S.D
Contact: Nancy R. Daly, Executive Director; (571) 483-1432; dalyn@asco.org
Internet: http://www.ascocancerfoundation.org/TACF/Grants/Grant+Opportunities/Improving+Cancer+Care+Grant
Sponsor: American Society of Clinical Oncology
1900 Duke Street, Suite 200
Alexandria, VA 22314

ASCO International Development and Education Awards (IDEA) **703**
The award provides support for oncologists in developing countries to attend the ASCO annual meeting and spend additional time at a comprehensive cancer center. The grants are designed to provide continuing medical education, assist in career development, and help establish strong relationships with leading ASCO members who serve as scientific mentors to each recipient. The strongest consideration will be given to candidates who: have submitted an abstract for the most recent Annual Meeting; and have less than 10 years of experience in the field of oncology. Guidelines are available online.
Requirements: Applicants must: complete an application, which includes a 250-word essay, letter of recommendation, and submission of the applicant's CV; be able to travel to the United States and have no visa restrictions or anticipate having any difficulty obtaining a U.S. visa; be citizens of a country identified by ASCO as one of greatest need or with limited resources; and either be or meet the requirements of being either an ASCO Active Member, Active-Junior Member, or an Associate Member.
Restrictions: Applicants cannot: have received one year or more of formal training (for instance, attended medical school, participated in internships, fellowships, or residency programs) in the United States, Western Europe, Australia, New Zealand, Canada, or the United Kingdom; be a previous recipient of the IDEA (formally known as ITG) program; or have previously attended more that one (1) ASCO Annual Meeting.
Date(s) Application is Due: Jan 13
Contact: Amanda Woodell, International Affairs; (571) 483-1501; idea@asco.org
Internet: http://www.ascocancerfoundation.org/TACF/Awards/Award+Opportunities/International+Development+and+Education+Award+%28IDEA%29
Sponsor: American Society of Clinical Oncology
1900 Duke Street, Suite 200
Alexandria, VA 22314

ASCO Loan Repayment Grants **704**
The ASCO Diversity in Oncology Initiative, funded by Susan G. Komen for the Cure, is designed to facilitate the recruitment and retention of individuals from populations underrepresented in medicine to cancer careers, with particular attention to the development of clinical practitioners and investigators. The ASCO Loan Repayment Program, funded by Susan G. Komen for the Cure (LRP), provides repayment of qualifying educational debt to oncologists, or to oncology fellows upon completion of training, who commit to practicing oncology in a medically underserved region of the United States. The LRP will repay up to $35,000 per year for two years (up to $70,000 total) of qualifying education loans. The application period opened September 16, and will close December 16. Applications are available online. Awardees will be announced in early April.
Requirements: To qualify to receive LRP loan assistance, applicants must have completed an approved oncology fellowship training program in the United States by July 1. Candidates must be U.S. citizens, U.S. nationals, or permanent residents. Candidates must have an M.D., D.O., or equivalent medical doctoral degree from an accredited institution. An applicant must be an Associate or Active ASCO member or submit an ASCO membership application with the LRP application.
Date(s) Application is Due: Sep 16
Amount of Grant: 35,000 U.S.D
Contact: Nancy R. Daly, Executive Director; (571) 483-1432; dalyn@asco.org
Internet: http://www.ascocancerfoundation.org/TACF/Awards/Award+Opportunities/ASCO+Diversity+in+Oncology+Initiative+funded+by+Susan+G.+Komen+for+the+Cure/The+ASCO+Loan+Repayment+Program%2C+funded+by+Susan+G.+Komen+for+the+Cure
Sponsor: American Society of Clinical Oncology
1900 Duke Street, Suite 200
Alexandria, VA 22314

ASCO Long-Term International Fellowships **705**
The ASCO Cancer Foundation Long Term International Fellowship (LIFe) provides young oncologists in developing nations the support and resources needed to advance their training by deepening their relationship with a U.S. or Canadian colleague and his or her institution. Through a one or two year fellowship the recipient will earn valuable training and experience with which they can affect change in cancer care in their home country.

Recipients must return to their home institutions after completing their fellowship and are expected to disseminate the knowledge they have gained. The ASCO Cancer Foundation (TACF) will administer fundraising and grant distribution for the fellowships. Funds will be sent from TACF to the host institution in January of the award year. The host institution will then be responsible for financial management of the fellowship. Any funds not spent on the fellowship will be returned to TACF. Recipients must attend the ASCO Annual Meeting that falls during the first year of their award term where they will be honored at The ASCO Cancer Foundation Grants Brunch. Recipients will also be expected to submit their research to ASCO via an abstract or publication submission following their fellowship. Applications are due to ASCO by October 15 for the fellowship to begin anytime in the next calendar year. Applications should be completed and returned jointly by the prospective host and fellow. A decision will be distributed by December 15.
Requirements: Hosts must: be an ASCO member in good standing; be employed at a U.S. or Canadian institution with sufficient infrastructure to provide for a 1 year fellowship; independently identify fellowship applicants; agree to cover any costs in excess of ASCO's contribution; complete the host section of the LIFe Application; complete fellowship evaluations as requested by ASCO. Applicants must: be an ASCO member in good standing; reside in a country identified by the World Bank as income level low, lower middle, and upper middle; have the support of their home institution including a letter of support from their supervisor; be a medical doctor who has finished a fellowship in clinical oncology or the equivalent; have less than 10 years experience in the field of oncology; complete the applicant section of the LIFe application including submission of a CV of up to 3 pages in length; commit to return to country of origin at the completion of their fellowship.
Date(s) Application is Due: Oct 15
Contact: Nancy R. Daly, Executive Director; (571) 483-1432; dalyn@asco.org
Internet: http://www.ascocancerfoundation.org/TACF/Awards/Award+Opportunities/Long-term+International+Fellowship+%28LIFe%29
Sponsor: American Society of Clinical Oncology
1900 Duke Street, Suite 200
Alexandria, VA 22314

ASCO Medical Student Rotation Grants 706
The ASCO Diversity in Oncology Initiative, funded by Susan G. Komen for the Cure, is designed to facilitate the recruitment and retention of individuals from populations underrepresented in medicine to cancer careers, with particular attention to the development of clinical practitioners and investigators. The ASCO Medical Student Rotation, funded by Susan G. Komen for the Cure, provides 8 - 10 week clinical or clinical research oncology rotations for U.S. medical students from populations underrepresented in medicine who are interested in pursuing oncology as a career. Recipients receive a $5,000 stipend for the rotation plus $1,500 for future travel to the ASCO Annual Meeting. An additional $2,000 will be provided to support the student's mentor. The application period opened October 7 and will close January 7. Awardees will be announced in early April.
Requirements: To qualify, applicants must be enrolled in an MD or DO, U.S. medical school program and be of an underrepresented population as defined by the program eligibility criteria. Applicants self-identify and recipients are drawn from this pool. Candidates must be U.S. citizens, U.S. nationals, or permanent residents. Candidates also must demonstrate an interest in pursuing oncology as a career and have a record of good academic standing.
Date(s) Application is Due: Jan 7
Amount of Grant: Up to 8,500 U.S.D
Contact: Nancy R. Daly, Executive Director; (571) 483-1432; dalyn@asco.org
Internet: http://www.ascocancerfoundation.org/TACF/Awards/Award+Opportunities/ASCO+Diversity+in+Oncology+Initiative+funded+by+Susan+G.+Komen+for+the+Cure/The+ASCO+Medical+Student+Rotation%2C+funded+by+Susan+G.+Komen+for+the+Cure
Sponsor: American Society of Clinical Oncology
1900 Duke Street, Suite 200
Alexandria, VA 22314

ASCO Merit Awards 707
The Merit Awards are designed to further promote clinical research by young scientists and to provide fellows with an opportunity to present their research and interact with other clinical cancer investigators at ASCO scientific meetings. A select number of Merit Awards are given annually to recognize outstanding abstracts submitted for consideration for presentation at an ASCO scientific meeting. The awards are given to oncology fellows who are first authors on selected abstracts.
Requirements: The first author of an abstract must meet the following requirements to be considered for an Award: be a physician (MD or DO) in an oncology fellowship training program or a doctoral degree candidate (such as PharmD or PhD) enrolled in an approved oncology specialty training program at the time of abstract submission; work in an oncology laboratory or clinical research setting; agree to present the abstract at the ASCO scientific meeting; check the Merit Award consideration box on the online abstract submitter; provide a letter of support from the Training Program Director indicating eligibility for the award; and submit a two-page curriculum vitae.
Contact: Nancy R. Daly, Executive Director; (571) 483-1432; dalyn@asco.org
Internet: http://www.ascocancerfoundation.org/TACF/Awards/Award+Opportunities/Merit+Award
Sponsor: American Society of Clinical Oncology
1900 Duke Street, Suite 200
Alexandria, VA 22314

ASCO Resident Travel Awards 708
The ASCO Diversity in Oncology Initiative, funded by Susan G. Komen for the Cure, is designed to facilitate the recruitment and retention of individuals from populations underrepresented in medicine to cancer careers, with particular attention to the development of clinical practitioners and investigators. The ASCO Resident Travel Award, funded by Susan G. Komen for the Cure, provides financial support for residents from underrepresented populations to attend ASCO's Annual Meeting. The intention of this award is to attract residents from under-represented populations to a possible career in one of the oncology specialties. This travel award will allow the recipients to travel to the ASCO Annual Scientific Meeting, where they will have an opportunity to meet oncologists and to understand the career possibilities in this area. The Resident Travel Award includes a $1,500 travel advance, complimentary Annual Meeting registration, and access to the Annual Meeting housing block and ASCO's travel agent. The application period opened October 7 and close January 7. Awardees will be announced in early April. Applications are available online.
Requirements: To qualify, applicants must be enrolled in an ACGME-accredited residency program required for future training in a cancer related subspecialty (i.e., Internal Medicine considering Med/Onc, Surgery considering Surg/Onc) and be of an underrepresented population as defined by the program eligibility criteria. Candidates must be U.S. citizens, U.S. nationals or permanent residents. Candidates must demonstrate an interest in pursuing oncology as a career as well as a record of good academic standing.
Date(s) Application is Due: Jan 7
Amount of Grant: 1,500 U.S.D
Contact: Nancy R. Daly, Executive Director; (571) 483-1432; dalyn@asco.org
Internet: http://www.ascocancerfoundation.org/TACF/Awards/Award+Opportunities/ASCO+Diversity+in+Oncology+Initiative+funded+by+Susan+G.+Komen+for+the+Cure/The+ASCO+Resident+Travel+Award%2C+funded+by+Susan+G.+Komen+for+the+Cure
Sponsor: American Society of Clinical Oncology
1900 Duke Street, Suite 200
Alexandria, VA 22314

ASCO State Affiliate Grants 709
The Program is designed to enable State/Regional Affiliates to engage in activities that will have an overall benefit to their members, ASCO members, other state societies, and the oncology community at-large. The purpose of the Program is to provide financial support to State/Regional Affiliates, which will assist them in implementing specific activities. The grant Program allows ASCO to utilize its strength and expertise in raising funds, taking advantage of resources not available on the state level, and demonstrates ASCO's commitment to strengthening state societies. Possible projects include, but are not limited to, conducting an educational meeting, developing an informational brochure or newsletter, creating a state clinical trials network, taking part in a quality cancer care initiative, or collaborating with a patient advocacy group on an awareness campaign.
Requirements: Grant applicants must be either the current President of Administrator/Executive Director of a state society that is formally affiliated with ASCO. Applications must illustrate a commitment by society leadership in carrying out the requested activities.
Restrictions: Funds should not be used for general operating costs of the state society. Grants will also not be provided for State/Regional Affiliates to carry out lobbying activities.
Date(s) Application is Due: Nov 17
Amount of Grant: 10,000 U.S.D
Samples: Denali (Alaska) Oncology Group, $10,000; Georgia Society of Clinical Oncology, $10,000; Maryland/DC Society of Clinical Oncology, $10,000; New York State Society of Medical Oncologists & Hematologists, $10,000; Northern New England Society of Clinical Oncology, $10,000.
Contact: Nancy R. Daly, Executive Director; (571) 483-1432; dalyn@asco.org
Internet: http://www.ascocancerfoundation.org/TACF/Professional+Education+and+Support/Cancer+Policy
Sponsor: American Society of Clinical Oncology
1900 Duke Street, Suite 200
Alexandria, VA 22314

ASCO Translational Research Professorships 710
The Translational Research Professorship is designed to provide flexible funding to outstanding translational researchers who have made and are continuing to make significant contributions that have changed the direction of cancer research and who provide mentorship to future translational researchers. The award is intended to support qualified individuals who are dedicated to bringing advances in basic sciences into the clinical arena and to mentoring other translational researchers. The award totals $500,000 and is paid in 5 annual increments of $100,000 throughout the award period. Only online applications will be accepted. In order to complete an application, a Letter of Intent must first be submitted by November 20th. Completed applications must be submitted by January 22nd.
Requirements: The applicant must meet the following criteria at the time of grant award (July 1): be a physician (MD, DO or international equivalent) with a full-time faculty appointment in a clinical department at an academic medical center; currently hold the rank of full professor (or equivalent) at an academic medical center. If applicant does not hold the rank of full professor, but meets all other eligibility criteria, he/she must provide written explanation of why his/her current rank at their institution should be considered equivalent; made significant contributions that have changed the direction of cancer research; be serving as a research mentor to one or more translational researcher(s) in training, and must be planning to continue to provide leadership in this area throughout the award period Lead a research team in the conduct of translational research; be a member of the American Society of Clinical Oncology (ASCO); expect to spend 75% of time during the award period dedicated to translational research activities, leading a team of translational researchers (both in the lab and in the clinic) and mentoring physician-scientists. Individuals with an endowed professorship from their own institution are eligible to apply for this award. Individuals with administrative responsibilities (such

as a Division or Department Chair) must clearly document their time commitment for performing research activities and mentoring activities. The application must originate jointly from the sponsoring institution and the applicant. The sponsoring institution must be a not-for-profit institution.
Date(s) Application is Due: Jan 22; Nov 20
Amount of Grant: 500,000 U.S.D
Samples: Merrill Egorin, MD, The Hillman Cancer Institute, Pittsburgh, PA—Preclinical and Clinical Studies of PARP Inhibitors as Cancer Therapy.
Contact: Nancy R. Daly, Executive Director; (571) 483-1432; dalyn@asco.org
Internet: http://www.ascocancerfoundation.org/TACF/Grants/Grant+Opportunities/Translational+Research+Professorship
Sponsor: American Society of Clinical Oncology
1900 Duke Street, Suite 200
Alexandria, VA 22314

ASCO Young Investigator Award **711**
This grant provides funding to promising investigators to encourage and promote quality research in clinical oncology. The purpose of this award is to fund physicians during the transition from a fellowship program to a faculty appointment. Priority consideration will be given to proposals that include patient-oriented and, ultimately, clinical research. Since many awards are supported with restricted grants from outside organizations, the ASCO is particularly interested in identifying young researchers working in the following sub-specialties and emerging disciplines: breast cancer; cancer survivorship; geriatric oncology; health disparities; kidney cancer; multiple myeloma; ovarian cancer; pancreatic cancer; sarcoma; survivorship; and young adult cancer. However, The ASCO Foundation welcomes application submissions in all oncology subspecialties.
Requirements: The recipient must be a physician (MD or DO) who, at the time of grant award is in the final year of his/her final sub-specialty training program or in the first year post his/her final sub-specialty training. The sponsoring facility must be an academic medical institution. The primary mentor must be in the candidate's proposed research field, must assume responsibility, and provide guidance for the research. The applicant must either be a member of ASCO or submit a membership application with the grant application. The applicant should spend at least 60 to 75% of his or her time in research during the award period.
Date(s) Application is Due: Sep 9
Amount of Grant: 50,000 U.S.D
Contact: Nancy R. Daly, Executive Director; (571) 483-1432; dalyn@asco.org
Internet: http://www.ascocancerfoundation.org/TACF/Grants/Grant+Opportunities/Young+Investigator+Award
Sponsor: American Society of Clinical Oncology
1900 Duke Street, Suite 200
Alexandria, VA 22314

ASGE / ConMed Award for Outstanding Manuscript by a Fellow/Resident **712**
The American Society for Gastrointestinal Endoscopy and ConMed Endoscopic Technologies sponsor the annual ASGE/ConMed Outstanding Manuscript Award. The prize consists of $10,000 cash award to the Fellow or Resident in training in a gastroenterology or a surgical gastrointestinal endoscopic program whose manuscript is accepted in for publication in Gastrointestinal Endoscopy. The award is presented during Digestive Disease Week, the largest and most prestigious meeting in the world for the GI professional.
Amount of Grant: 10,000 U.S.D
Contact: Chair, ASGE Research Committee; (630) 573-0600; fax (630) 573-0691; grants@asge.org
Internet: http://www.asge.org/index.aspx?id=4892&terms=ConMed+Award
Sponsor: American Society for Gastrointestinal Endoscopy
1520 Kensington Road, Suite 202
Oak Brook, IL 60523

ASGE Don Wilson Award **713**
The Award provides Advanced Fellows or Junior Faculty with the opportunity to train outside of their home country with a premier GI endoscopist or group in order to advance their training. The award assists in underwriting reasonable and customary travel and living expenses for a period of one to three months. The award includes a $7,500 cash stipend prior to the recipient's travel. In addition, a 20% disbursement will be made to the host institution. A total of three awards will be available annually for North American and International ASGE members. Note: One training must take place in the United States.
Requirements: Applicants must be the equivalent of Junior Faculty or Advanced Fellows and be proficient in the English Language. Applicants must also: be a current ASGE Member; be a Junior Faculty or Advanced Fellow (3rd or 4th year or international equivalent); have permission from their own institution to undertake the travel; and have permission from an ASGE member at the host institution to accept the application for training.
Restrictions: Individuals applying for 1-2 year advanced fellowships do not qualify for this award.
Date(s) Application is Due: Sep 1
Amount of Grant: 7,500 U.S.D
Contact: Chair, ASGE International Committee; (630) 573-0600; fax (630) 573-0691; membership@asge.org
Internet: http://www.asge.org/nspages/research/applications/donWilson.cfm
Sponsor: American Society for Gastrointestinal Endoscopy
1520 Kensington Road, Suite 202
Oak Brook, IL 60523

ASGE Endoscopic Research Awards **714**
These research awards are offered to physicians for projects in basic and clinical endoscopic technology research, outcomes and effectiveness of endoscopy research. The primary objective is to foster research in gastrointestinal endoscopy both within and outside of academic centers. Two categories of grants may be used for up to two years of study: grants of $1-15,000 (category A) and grants of $15,001 - 50,000 (category B). Requests for funding seed projects that will lead to further research as well as larger requests for definitive clinical trials will be considered. Funding requests may also include: personnel expenses for research assistant and/or faculty salary support (percentage of time for study should be specified and appropriately justified); study supplies; and equipment essential for study.
Requirements: Candidate must be: an ASGE member; an MD (or have equivalent degree); and current in a gastroenterology-related and endoscopic practice in academic institutions or private practice in North America.
Restrictions: Funding will not be provided for: salary support for trainees; computer purchases (unless a unique application is proposed); standard equipment and supplies needed for appropriate patient care (for example, sclerotherapy needles); travel to meetings; or indirect costs.
Date(s) Application is Due: Sep 12
Amount of Grant: Up to 50,000 U.S.D
Contact: Chair, ASGE Research Committee; (630) 573-0600; fax (630) 573-0691; grants@asge.org
Internet: http://www.asge.org/index.aspx?id=4892&terms=ConMed+Award
Sponsor: American Society for Gastrointestinal Endoscopy
1520 Kensington Road, Suite 202
Oak Brook, IL 60523

ASGE Endoscopic Research Career Development Awards **715**
These awards provide the salary and/or research support necessary for the investigator to enhance his/her career development. The award must be used to acquire new skills for furthering a career in endoscopic research. Examples of such skills include advanced training in endoscopic procedures, training in outcomes research relevant to endoscopy, training in use of large databases such as CORI, and new endoscopic research techniques, including animal models.
Requirements: Applicants must be ASGE members and hold full-time faculty positions at North American (U.S., Canada, or Mexico) universities or professional institutions at the time of application. The award is intended for faculty who have demonstrated promise and have some record of accomplishment in research. Candidates must devote at least 30 percent of their effort to research related to gastrointestinal endoscopy during the period of the award.
Restrictions: The award is not available for fellows. Faculty with principle investigator current federal funding or other concurrent career development support still active at the time of this award are not eligible. Recipients of the ASGE Career Development Awards may not be granted an ASGE Research & Outcomes & Effectiveness Award during their two-year award period.
Date(s) Application is Due: Dec 12
Amount of Grant: 75,000 U.S.D
Contact: Chair, ASGE Research Committee; (630) 573-0600; grants@asge.org
Internet: http://www.asge.org/index.aspx?id=4892&terms=ConMed+Award
Sponsor: American Society for Gastrointestinal Endoscopy
1520 Kensington Road, Suite 202
Oak Brook, IL 60523

ASGE Given Capsule Endoscopy Research Award **716**
These research awards are offered to physicians for projects specific to capsule endoscopy both within and outside of academic centers. The award must be used for projects directly relating to capsule endoscopy. Requests for funding seed projects that will lead to further research as well as requests for definitive clinical trials will be considered. Funding requests may include: personnel (research assistant and/or faculty salary support—percentage of time for study should be specified and appropriately justified); study supplies; and equipment essential for study.
Requirements: Candidate must be an ASGE member, be an MD (or have equivalent degree), and be current in a gastroenterology-related and endoscopic practice in academic institutions or private practice in North America.
Restrictions: Funding will not be provided for: salary support for trainees; computer purchases (unless a unique application is proposed); standard equipment and supplies needed for appropriate patient care (for example, sclerotherapy needles); travel to meetings; publication costs; or indirect costs.
Date(s) Application is Due: Dec 7
Amount of Grant: Up to 25,000 U.S.D
Contact: Chair, ASGE Research Committee; (630) 573-0600; fax (630) 573-0691; grants@asge.org
Internet: http://www.asge.org/nspages/research/applications/ceapp.cfm
Sponsor: American Society for Gastrointestinal Endoscopy
1520 Kensington Road, Suite 202
Oak Brook, IL 60523

ASHA Advancing Academic-Research Careers (AARC) Award **717**
This award is given to new faculty in higher education to support their academic and research endeavors in the field of communication sciences and disorders (CSD). Up to five awards of $5,000 each will be granted to support activities such as: Improving teaching skills; Mentoring graduate students; Participating in research activities; Preparing a grant application; Preparing a manuscript or publication; Presenting at a professional meeting (e.g., the ASHA Convention).

Requirements: Candidates must possess: A research doctoral degree (PhD, ScD, CScD, or equivalent, e.g., EdD), which must be conferred by August 1; A full-time faculty appointment; Primary responsibilities of teaching and conducting research; Less than six years of teaching/research experience or are below the level of associate professor; A tenure or non-tenure track appointment. A list of required documents and a downloadable application form are available at the website.
Restrictions: Postdoctoral fellows are not eligible. Only students attending a university in the U.S. are eligible to apply for the award.
Date(s) Application is Due: May 15
Amount of Grant: 5,000 U.S.D
Contact: Margaret Rogers, Chief Staff Officer for Science & Research; (800) 498-2071 or (301) 296-5706; fax (301) 296-8580; mrogers@asha.org
Internet: http://www.asha.org/students/awards.htm#aarc
Sponsor: American Speech-Language-Hearing Association
2200 Research Boulevard, #245
Rockville, MD 20850-3289

ASHA Minority Student Leadership Program Awards **718**
The MSLP is a leadership development program established for undergraduate seniors and graduate students who are enrolled in communication sciences and disorders programs. Aditionally, ASHA's Focused Initiative on the PhD Shortage in Higher Education will provide funding for 10 eligible candidates enrolled in a research doctoral program. The award package includes complimentary registration to the ASHA Convention held annually; up to five nights lodging; meal stipend; and, program materials. Forty students are chosen to participate - 5 undergraduate seniors, 20 master's students, 5 entry level clinical doctoral students and 10 PhD students.
Requirements: Undergraduate seniors and graduate students who are or will be enrolled in the fall communication sciences and disorders programs and are not members of ASHA are eligible. Preference will be given to students who are members of racial/ethnic minority groups historically under-represented within ASHA, including American Indian or Alaska Native, Asian, Black or African American, Native Hawaiian or other Pacific Islander, and/or Hispanic/Latino. For Graduate Study - The student must be enrolled in, or accepted for, graduate study in a communication sciences and disorders program in the United States. Master's degree candidates and entry level clinical doctoral candidates must be in a program accredited by the Council of Academic Accreditation (CAA). PhD Track - You must be enrolled in a research doctoral degree program. ASHA members may apply. The required application packet is available for download at the website (http://www.asha.org/students/awards.htm#MSLP).
Restrictions: Program participants are responsible for transportation expenses to and from the convention city and personal incidentals.
Date(s) Application is Due: May 18
Contact: Melanie Johnson, Program Manager; (301) 296-8681; mjohnson@asha.org
Internet: http://www.asha.org/students/awards.htm#MSLP
Sponsor: American Speech-Language-Hearing Association
2200 Research Boulevard, #245
Rockville, MD 20850-3289

ASHA Multicultural Activities Projects Grants **719**
The program is open to clinical and/or school-based speech, language, and hearing programs; university programs, state associations, allied and related professional organizations, and ASHA's Special Interest Divisions. ASHA has allocated funding for projects on multicultural activities with a maximum of $18,000 awarded for a single grant. It is expected that the average individual award given will be between $7,000 and $10,000. Applications must be postmarked by April 4. Proposals receiving funding will be announced in July.
Requirements: Proposals must have a multicultural focus. Multicultural is defined to include issues dealing with race, ethnicity, language, gender or gender identification, age, sexual orientation, and disability. There is particular interest in, but not limited to, proposals that respond to one or more of ASHA's Strategic Initiatives from ASHA's Strategic Pathway to Excellence. Projects must: be compatible with ASHA's mission and vision; have a clear high-quality plan for meeting its objectives; be completed, including evaluation, within 12 months of initiation of the project; be adaptable by other clinical and/or school-based programs, university programs, state associations, and associated and related professional organizations for their own use; and describe what will happen to the project once ASHA funding has ended. Criteria for evaluating proposals and the required downloadable forms are available at the website.
Restrictions: An ASHA member must serve as project director. Grants will not be awarded for fund raising, governmental lobbying, or awards ceremonies.
Amount of Grant: Up to 18,000 U.S.D
Contact: Karen Beverly-Ducker, (800) 638-8255; fax (301) 296-8580; OMAGrants@asha.org
Internet: http://www.asha.org/about/leadership-projects/multicultural/funding/rfp.htm
Sponsor: American Speech-Language-Hearing Association
2200 Research Boulevard, #245
Rockville, MD 20850-3289

ASHA Research Mentoring-Pair Travel Award **720**
This award is designed to foster the professional development of students, clinicians, and emerging scientists who have expressed an interest in research in communication sciences and disorders. Up to 10 travel awards of $1,000 each will be granted to eligible mentor mentee pairs (mentee= $750; mentor= $250) to help defray the cost of attending the annual research symposium. Special consideration will be given to individuals whose research or research interests are (a) relevant to the symposium topic, (b) interdisciplinary in nature, (c) translational in nature, or (d) related to issues of cultural or linguistic diversity.
Requirements: Mentees must be (a) a current student at the bachelor's, master's, AuD, or PhD level, (b) a postdoctoral fellow, (c) a junior level faculty member (less than 6 years teaching/research experience in a tenure-track position or below the level of associate professor), or (d) a clinician. Mentors must be a seasoned investigator in an academic or health care environment. Download the Mentee and Mentory Application Forms at the website.
Date(s) Application is Due: Jun 22
Amount of Grant: 1,000 U.S.D
Contact: Margaret Rogers, Chief Staff Officer for Science & Research; (800) 498-2071 or (301) 296-5706; fax (301) 296-8580; mrogers@asha.org
Internet: http://www.asha.org/students/awards.htm#2008%20Research%20Mentoring-Pair%20Travel%20Award
Sponsor: American Speech-Language-Hearing Association
2200 Research Boulevard, #245
Rockville, MD 20850-3289

ASHA Student Research Travel Award **721**
The awards are intended to highlight the research activities of budding scientists and encourage careers in science and research. Approximately 40 student research travel awards of $500 each are available for undergraduate, Master's, and doctoral students who are first authors on a paper (all session types are acceptable: technical, poster, one hour seminar, two hour seminar, or short course). Identification of prospective awardees will be based on the recommendation of Convention co-chairs and topic coordinators. Awardees will receive a $500 travel award; have their exceptional paper highlighted in the convention program; and, be recognized at the Researcher-Academic Town Meeting and other designated events.
Requirements: Each award recipient will be required to: register for the Convention at his or her own expense; present his or her paper at the assigned time (as noted in the Convention program booklet); secure a mentor to shadow throughout the convention; and, attend selected research-related sessions during the convention. In order to apply, students must: submit an abstract summarizing their research through the normal Call for Papers process; check the appropriate box (when prompted) indicating that their paper is 'student-authored'; and, check the appropriate box (when prompted) indicating that they desire to have this abstract considered for the Student Research Travel Award.
Restrictions: Papers must be submitted online by the deadline date and must be designated as a student-authored paper.
Date(s) Application is Due: Mar 31
Amount of Grant: 500 U.S.D
Contact: Margaret Rogers, Chief Staff Officer for Science & Research; (800) 498-2071 or (301) 296-5706; fax (301) 296-8580; mrogers@asha.org
Internet: http://www.asha.org/students/awards.htm#research
Sponsor: American Speech-Language-Hearing Association
2200 Research Boulevard, #245
Rockville, MD 20850-3289

ASHA Students Preparing for Academic & Research Careers (SPARC) Award **722**
The goal of SPARC is to foster students' interest in the pursuit of PhD education and careers in academia in order to fill faculty/researcher vacancies in communication sciences and disorders (CSD). Students will identify a primary faculty mentor and propose a one-year CSD Career Mentoring Plan. As many as ten students will be awarded up to $1,500 each to be used for teaching and research enhancement activities, such as travel to a research, pedagogy conference, or meeting; travel for a visit to an off-campus site that provides learning opportunities in a research lab and/or a college classroom setting; or course registration to support the mentoring plans outlined in the application.
Requirements: Eligible applicants include junior or senior undergraduates, 1st year master's students, and 1st and 2nd year entry-level clinical doctoral (e.g., AuD) students enrolled part-time or full-time in a CSD program in the United States. At the time of the award, eligible recipients' education status for the upcoming academic year must be undergraduate senior, 1st or 2nd year master's, or 1st, 2nd or 3rd year entry-level clinical doctoral (i.e., AuD) in a full-time or part-time CSD program in the United States. Master's degree and entry-level clinical doctoral students must be in a program accredited by the Council for Academic Accreditation in Audiology and Speech-Language Pathology (CAA). Full requirements (essay question, documentation, etc.) are listed at the website, along with the downloadable application form.
Date(s) Application is Due: May 15
Amount of Grant: 5,000 U.S.D
Contact: Arlene A. Pietranton, Executive Director; apietranton@asha.org or academicaffairs@asha.org
Internet: http://www.asha.org/students/awards.htm#sparc
Sponsor: American Speech-Language-Hearing Association
2200 Research Boulevard, #245
Rockville, MD 20850-3289

ASHFoundation Clinical Research Grants **723**
These $50,000-$75,000 grants support scientists with a research doctorate within the discipline of communication sciences and disorders to support investigations that will advance knowledge of the efficacy of treatment and assessment practices. Project funding is available for mentored treatment research, independent treatment research, or collaborative treatment research as specified in grant guidelines.
Amount of Grant: 50,000 - 75,000 U.S.D
Samples: Swathi Kiran, Assistant Professor, University of Texas at Austin—Semantic Feature Analysis in the Treatment of Lexical Retrieval Deficits in Spanish-English and French-English Bilingual Aphasia, $50,000; Tiffany Johnson, Assistant Professor, University of Kansas Center for Research, Inc.—New Directions in Clinical Applications of Otoacoustic Emissions. $50,000; Andrea Pittman, Assistant Professor, Arizona

State University—Evaluating Advanced Signal Processing in Children with Hearing Loss, $75,000; Nancy Pearl Solomon, Research Speech Pathologist, Walter Reed Army Medical Center—Treating Dysphagia in Blast-Injured Patients. $50,000.
Contact: Emily Diaz, Project Assistant; fax (301) 296-8703; ediaz@asha.org
Internet: http://www.ashfoundation.org/Foundation/grants/research_grants.htm
Sponsor: American Speech-Language-Hearing Foundation
2200 Research Boulevard
Rockville, MD 20850-3289

ASHFoundation Graduate Student International Scholarship 724
One award is available for a full-time international graduate student who has demonstrated outstanding academic achievement. Applicants must be studying communications sciences and disorders at a U.S. institution accredited by ASHA's Council of Academic Accreditation. All scholarship recipients will be announced and recognized at the ASHA convention.
Requirements: Applicants must be full-time students.
Amount of Grant: Up to 4,000 U.S.D
Samples: Beverly A. Collisson, PhD Candidate, University of Connecticut, $5,000; Beula M. Magimairaj, PhD Candidate, Ohio University, $5,000; Maria V. Ivanova, PhD candidate, Ohio University, $4,000; Victoria Lee, AuD candidate, Northwestern University, $4,000.
Contact: Emily Diaz, Project Assistant; fax (301) 296-8703; ediaz@asha.org
Internet: http://www.ashfoundation.org/Foundation/grants/GraduateScholarships.htm
Sponsor: American Speech-Language-Hearing Foundation
2200 Research Boulevard
Rockville, MD 20850-3289

ASHFoundation Graduate Student Scholarships 725
The foundation invites full-time graduate students to submit applications in competition for one of seven scholarships. Full-time master's or doctoral students in communication sciences and disorders programs demonstrating outstanding academic achievement are eligible to compete for these $4,000 awards. Supported in part by Psi Iota Xi National Philanthropic Organization and the Marni Reisberg Memorial Fund. All scholarship recipients will be announced and recognized at the annual convention held in November.
Requirements: Applicants must be accepted for graduate study in a communication sciences and disorders program (master's degree candidates must be in an ASHA Educational Standards Board accredited program; this is not mandatory for doctoral degree candidates); be enrolled for full-time study (12 or more credit hours or the full-time standard); submit official university transcripts of academic coursework, credits, and grades; and be recommended by a committee of two or more past or present college faculty and colleagues (at least one supervisor) at the student's current place of employment. Annual deadline dates may vary; contact program staff for exact dates.
Restrictions: Applicants must not have received a prior scholarship from the foundation.
Amount of Grant: 4,000 U.S.D
Contact: Emily Diaz, Project Assistant; fax (301) 296-8703; ediaz@asha.org
Internet: http://www.ashfoundation.org/Foundation/grants/GraduateScholarships.htm
Sponsor: American Speech-Language-Hearing Foundation
2200 Research Boulevard
Rockville, MD 20850-3289

ASHFoundation Graduate Student Scholarships for Minority Students 726
Racial/ethnic minority students who are U.S. citizens, who are accepted for graduate study in speech-language pathology or audiology, and who demonstrate outstanding academic achievement are eligible to compete for this scholarship. Applicants should submit a formal paper, such as a term paper, in competition for this award. The paper must be either an integrative literature review or an opinion paper on a current professional issue. All scholarship recipients will be announced and recognized at the ASHA convention.
Requirements: Candidates must be U.S. citizens and members of a racial/ethnic minority, including Native American and Alaska Native, Asian and Pacific Islander, African American, and Hispanic. Applicants must be accepted for master's or doctoral study in a speech-language pathology or audiology program for the upcoming academic year. The student must be enrolled for full-time (12 or more credit hours) study.
Restrictions: Applicants may not have received a prior scholarship from the foundation.
Amount of Grant: 2,000 - 4,000 U.S.D
Samples: Tracy Conner, MA candidate, University of Massachusetts-Amherst, $5,000; Rhona Galera, CScD candidate, University of Pittsburgh, $5,000; Derek E. Daniels, PhD candidate, Bowling Green State University, $4,000.
Contact: Emily Diaz, Project Assistant; fax (301) 296-8703; ediaz@asha.org
Internet: http://www.ashfoundation.org/Foundation/grants/GraduateScholarships.htm
Sponsor: American Speech-Language-Hearing Foundation
2200 Research Boulevard
Rockville, MD 20850-3289

ASHFoundation Graduate Student with a Disability Scholarship 727
Full-time graduate students with a disability who are enrolled in a communication sciences and disorders program and demonstrate outstanding academic achievement are eligible to compete for a $4,000 scholarship. Applicants must demonstrate superior academic achievement and submit transcripts, essay and references with official application. The Award is intended for the blind, hearing impaired, physically handicapped, or learning impaired. Master's, but not doctoral, students must be enrolled in an ASHA Council on Academic Accreditation approved program.
Requirements: Candidates must be U.S. citizens and either blind, hearing impaired, physically handicapped, or learning impaired. Applicants must be accepted for master's study in a speech-language pathology or audiology program for the upcoming academic year. The student must be enrolled for full-time (12 or more credit hours) study.
Amount of Grant: 4,000 U.S.D
Samples: Stephanie Cochran, MA Candidate, Duquesne University; Stephanie P. Hirsh, MA candidate, Northwestern University; Alexandra Eng, MS candidate, University of the Pacific.
Contact: Emily Diaz, Project Assistant; fax (301) 296-8703; ediaz@asha.org
Internet: http://www.ashfoundation.org/Foundation/grants/GraduateScholarships.htm
Sponsor: American Speech-Language-Hearing Foundation
2200 Research Boulevard
Rockville, MD 20850-3289

ASHFoundation New Century Scholars Program Doctoral Scholarships 728
The Program is a special funding initiative resulting from the ASHFoundation's Dreams and Possibilities Campaign. The Award is for students accepted or currently enrolled in a research doctoral program in communications sciences and disorders. Full-time study will be given priority; part-time status will be considered. Students should be committed to a teacher-investigator career in communication sciences and disorders in the United States.
Requirements: The Award is available to U.S. citizens only. Applicants must be accepted or currently enrolled in a research doctoral program.
Amount of Grant: 10,000 U.S.D
Samples: Sophie Ambrose, PhD Candidate, University of Kansas; Megha Bahl, PhD Candidate, University of Arizona; Angela Yarnell Bonino, PhD candidate, University of North Carolina at Chapel Hill; Deanna Britton, PhD Candidate, University of Washington; Jamie L. Desjardins, PhD Candidate, Syracuse University; Kerry Danahy Ebert, PhD Candidate, University of Minnesota, Twin Cities; Aaron M. Johnson, PhD Candidate, University of Wisconsin, Madison; Karen Le, PhD Candidate, University of Connecticut; Jimin Lee, PhD Candidate, University of Wisconsin, Madison; Megan K. McPherson, PhD Candidate, Purdue University; Kimberly M. Meigh, PhD Candidate, University of Pittsburgh.
Contact: Emily Diaz, Project Assistant; fax (301) 296-8703; ediaz@asha.org
Internet: http://www.ashfoundation.org/Foundation/grants/GraduateScholarships.htm
Sponsor: American Speech-Language-Hearing Foundation
2200 Research Boulevard
Rockville, MD 20850-3289

ASHFoundation New Century Scholars Research Grant 729
The Program is a special funding initiative resulting from the ASHFoundation's Dreams and Possibilities Campaign. The Grant is a one-time award intended to advance the knowledge base in communication sciences and disorders. Applicants must have a PhD or equivalent research doctorate within the discipline. Priority will be given to studies that are innovative, groundbreaking, or that meet research needs not yet addressed.
Restrictions: Students enrolled in a degree program or working on dissertation research are not eligible.
Amount of Grant: 10,000 U.S.D
Samples: Jungmee Lee, Assistant Professor, University of Arizona—Proposal: Correlation Between Cochlear Tuning and Otoacoustic Emissions: Exploring Scientific and Clinical Implications, $10,000; Chang Liu, Assistant Professor, University of Texas at Austin—Auditory Processing: Comparing Phonologically Disordered and Typically Developing Children, $10,000; John McCarthy, Assistant Professor, Ohio University—Improving Auditory Scanning Interfaces in AAC Devices, $10,000; Valeriy Shafiro, Assistant Professor, Rush University Medical Center—Effects of Environmental Sound Training on the Perception of Environmental Sounds and Speech in Cochlear Implant Patients, $10,000.
Contact: Emily Diaz, Project Assistant; fax (301) 296-8703; ediaz@asha.org
Internet: http://www.ashfoundation.org/Foundation/grants/research_grants.htm
Sponsor: American Speech-Language-Hearing Foundation
2200 Research Boulevard
Rockville, MD 20850-3289

ASHFoundation Research Grant for New Investigators 730
The grants are designed to help further research activities of new investigators and should have particular clinical relevance to speech-language pathology and audiology. The awards are designed to encourage research activities of new scientists who earned their latest degree within the last five years. Proposals, while not limited in topic, are encouraged in the area of treatment research, particularly efficacy and outcome studies. Include abstract, research plan, bibliography, management plan, and budget with application.
Requirements: The individual must have received the master's or doctoral level degree in the last five years; must not have received prior funding for research, with the exception of internal university funding; and the proposal must be for research to be initiated.
Restrictions: Open to individuals with a Master's degree or doctoral degree who are not currently enrolled in a degree program.
Samples: Nina C. Capone, Associate Professor, Seton Hall University—The Effects of Gesture Cues on Object Word Learning by Children with Language Impairments; Gayle L. DeDe Assistant Professor, University of Arizona—On-Line Sentence Comprehension in Aphasia: Is Reading Different than Listening?; Ciara Leydon, Assistant Professor, Brooklyn College—Construction and Characterization of a Novel Model of Vocal Fold Mucosa; Rita R. Patel, Assistant Professor, University of Kentucky—High Speed Digital Analysis of Vocal Fold Vibration in Children; Yasmeen Faroqi Shah, Assistant Professor, University of Maryland, College Park—Retreival of Action Names in Aphasia: An Investigation of the Embodied Cognition Framework; Yana Yunosova, Assistant Professor, University of Toronto—Visual Feedback Systems in Speech Rehabilitation: Defining Vocal Tract Targets.

Contact: Emily Diaz, Project Assistant; fax (301) 296-8703; ediaz@asha.org
Internet: http://www.ashfoundation.org/Foundation/grants/research_grants.htm
Sponsor: American Speech-Language-Hearing Foundation
2200 Research Boulevard
Rockville, MD 20850-3289

ASHFoundation Research Grant in Speech Science 731

ASLH, in conjunction with the Acoustical Society of America, invites new researchers to submit proposals in competition for one grant every other year. The grant is designed to further research activities of new investigators and promulgate the work of the late Dennis Klatt, a noted researcher and scientist in the area of speech communication. Priority will be given to areas reflecting Dr. Klatt's broad interests, such as proposals studying speech perception, synthesis, and acoustics, with an emphasis on an interdisciplinary research approach. The grant can be used to initiate new research or supplement an existing research project. Funds may be requested for a variety of purposes, i.e., equipment, subjects, research assistants, or research-related travel. The grant recipient will be announced at the ASHA convention.
Requirements: Individuals having received a doctoral degree within the last five years and who wish to further research activities in the areas of speech communication are eligible to compete for the grant.
Amount of Grant: 5,000 U.S.D
Samples: Mary K. Fagan, Indiana University School of Medicine—Vocalization and Sound Exploration in Hearing, Deaf, and Cochlear-Implanted Infants, $5,000; Rajka Smiljanic, Northwestern University—Effect of Clear Speech on Production and Perception of Croatian Rhythm, $4,000; Lori L. Holt, Carnegie Mellon University—An Investigation of the Perceptual and Learning Influences on Phonetic Category Formation, $5,000.
Contact: Emily Diaz, Project Assistant; fax (301) 296-8703; ediaz@asha.org
Internet: http://www.ashfoundation.org/Foundation/grants/research_grants.htm
Sponsor: American Speech-Language-Hearing Foundation
2200 Research Boulevard
Rockville, MD 20850-3289

ASHFoundation Student Research Grants in Audiology 732

Each year the foundation awards one grant for support of research to be initiated in the area of clinical and/or rehabilitative audiology. One $2,000 award to a master's or doctoral student for research to be initiated in the area of audiology. Applicants must submit up to a ten-page research plan, one-page abstract, two-page management plan and budget, and letter of support. All study must be done in the United States.
Requirements: The competition open to master's or doctoral degree students enrolled in, or accepted for, graduate study in a communication sciences and disorders program in the United States. Master's degree candidates must be in a program accredited by the Council on Academic Accreditation in Audiology and Speech-Language Pathology (CAA); this is not mandatory for doctoral degree candidates. Applicants must be enrolled for full-time study for the full academic year.
Amount of Grant: 2,000 U.S.D
Samples: Tara D. Reed, University of Texas at Dallas, Mentor: James F. Jerger—Effects of Aging on Interaural Asymmetry in a Competing Speech; Faith M. Parker, Montclair State University, Mentor: Janet Koehnke—The Effects of Age and Reverberation on Cortical Audiotry Processing; TaskYu-Hsiang Wu, University of Iowa, Mentor: Ruth Bentler—Impact of Visual Cues on Microphone Mode Preference.
Contact: Emily Diaz, Project Assistant; fax (301) 296-8703; ediaz@asha.org
Internet: http://www.ashfoundation.org/Foundation/grants/research_grants.htm
Sponsor: American Speech-Language-Hearing Foundation
2200 Research Boulevard
Rockville, MD 20850-3289

ASHFoundation Student Research Grants in Early Childhood Language Development 733

One $2,000 grant per year is awarded to a graduate or postgraduate student to support research in early childhood language development. The applicant must submit up to a ten-page research plan, one-page abstract, two-page management plan and budget, and letter of support. All study must be done in the United States. The recipient and the recipient's mentor will be announced and recognized at the ASHA Convention.
Requirements: The competition is open to graduate or postgraduate students in the area of communication sciences and disorders. Applicants must submit a proposal according to foundation guidelines, to be received by the indicated deadline date for applications.
Amount of Grant: 2,000 U.S.D
Samples: Kathryn Wright Brady, University of Missouri, Mentor: Judith C. Goodman—Clues to Meaning: Exploring Potential Effects of Paired, Congruent Cues on Toddlers' Word Learning; Dawn Vogler-Elias, University of Buffalo, Mentor: Geralyn Timler—A Shared Storybook Reading Intervention for Preschoolers with Autism; Jonathan L. Preston, Syracuse University, Mentor: Mary Louise Edwards—Preliminary Investigation of a Weighted Measure of Speech Sound Accuracy.
Contact: Emily Diaz, Project Assistant; fax (301) 296-8703; ediaz@asha.org
Internet: http://www.ashfoundation.org/Foundation/grants/research_grants.htm
Sponsor: American Speech-Language-Hearing Foundation
2200 Research Boulevard
Rockville, MD 20850-3289

ASHG Public Health Genetics Fellowship 734

The objectives of this co-sponsored Fellowship are to enhance the Center for Disease Control (CDC)'s current activities and develop new approaches in integrating genetics and genomics into public health policy and programs at the national, state, and local levels. A wide range of challenges and opportunities associated with the integration of genetics and genomics are emerging for health policy makers and program staff within these health agencies. Key issues of planning, workforce training, expansion of surveillance systems, health policy development, and changes in disease prevention programs are important program activities. The Fellow is expected to work independently and in conjunction with epidemiologists, geneticists, public health specialists, and communications specialists across CDC. Activities may include: planning and development of current and future projects in genetics and genomics with CDC; providing technical assistance to CDC programs, and collaboratively with CDC programs providing assistance to state and local health agencies on the integration of genomics into policy and practice around specific diseases; collaborating with national public health organizations and professional organizations, including the American Society of Human Genetics (ASHG), in the planning and development of national public health strategies associated with genomics and disease prevention; participating in ASHG activities related to the work at CDC to promote timely transfer of information about policy and program development; participating in writing reports about the impact of genomics and public health policy and practice; enhancing awareness about the importance of genomics in public health by providing content for the ASHG website and materials for the Mentor network; assisting with the development and delivery of training and educational activities related to genomics and public health. This career development opportunity will offer a broad experience in the emerging role of genomics in public health practice. The Fellow will learn strategies for and practical skills in integrating genomics into specific health policies and programs associated with disease prevention and health promotion. Benefits include access to and cooperation of genetics and public health professionals to enhance collaboration between the fields. The Fellow will work closely with CDC staff and state and local health agencies to assess and develop the role of genomics in public health activities. The Fellow will gain insight and expertise through the training of, and with, public health professionals as well as development and delivery of workshops and case studies appropriate for a variety of audiences. The length of the fellowship is two years for early career candidates and one year for mid-career candidates. The Fellowship starts in July. Deadlines for application may vary from year to year. Interested applicants are encouraged to check the fellowship website and contact program staff for information on how to apply for the fellowship.
Requirements: Successful applicants must have either an M.D. or Ph.D. degree or a master's degree in genetics, genetic counseling, or a related field and experience or interest in public health genetics. Applicants must qualify as early career (3 years experience) or mid-career professionals.
Restrictions: Interested applicants are encouraged to call program staff with any questions on requirements and restrictions.
Geographic Focus: All States, All Countries
Date(s) Application is Due: Apr 23
Contact: Joann Boughman; (866) 486-4363 or (301) 634-7300; jboughman@ashg.org
Internet: http://www.ashg.org/pages/education_cdc.shtml
Sponsor: American Society of Human Genetics
9650 Rockville Pike
Bethesda, MD 20814

Ashland Corporate Contributions Grants 735

The foundation awards grants to eligible Kentucky, Ohio, and West Virginia nonprofit organizations in its areas of interest, including education, arts, communities/civic, disaster relief, environment, and health and human services. Types of support include seed money grants, project grants, endowments, matching gifts, and employee volunteers. The foundation does not provide an application. Most giving is centered on programs that best address the needs of Ashland's employees, stockholders, customers, and other constituencies, and the communities in which they live. Funding requests are not solicited but will be considered. The primary focus is on education, with an emphasis on mentoring, literacy and/or diversity.
Requirements: 501(c)3 tax-exempt organizations in Boyd and Greenup Counties, KY, Lawrence County, OH, and Cabell and Wayne Counties, WV, are eligible. Charitable groups within church or religious organizations are eligible.
Restrictions: The foundation does not support individuals, capital campaigns for building or equipment, endowments, travel, film or video production, tickets, religious or political activities, or goodwill advertising.
Geographic Focus: Kentucky, Ohio, West Virginia
Contact: Program Contact; (859) 815-3333; community@ashland.com
Internet: http://www.ashland.com/commitments/contributions.asp
Sponsor: Ashland
50 East Rivercenter Boulevard, P.O. Box 391
Covington, KY 41012-0391

ASHP Research and Education Foundation Research Grants 736

The ASHP offers grants in three different categories. Medication Safety Research grants give priority to projects that study medication-use design in hospitals and health systems and result in promotion of safety initiatives in medication use. Junior Investigator Research Grants support health services research in medication use that is conducted by pharmacist junior investigators; and strengthen the skills of newer pharmacist investigators while fostering development of mentoring relationships with more experienced senior investigators. The Pharmacy Resident Medication Safety Grant program is intended to support a worthy research project devoted to improving medication safety. Projects that demonstrate the role of the pharmacist in safe medication use, enhance medication-use systems, and can be duplicated in other health systems will be given high priority. Annual deadline dates may vary; contact program staff for exact dates.
Requirements: Eligibility requirements vary by program; contact staff for eligibility.
Contact: Daniel Cobaugh; (301) 664-8767; fax (301) 664-8872; foundation@ashp.org

Internet: http://www.ashpfoundation.org/Research/fundingOpps.cfm
Sponsor: American Society of Health-System Pharmacists Research and Education Foundation
7272 Wisconsin Avenue
Bethesda, MD 20814

ASM-PAHO Infectious Diseases Epidemiology and Surveillance Fellowships 737
Funded by the Pan American Health Organization and ASM, this award offers fellowships to young scientists from Bolivia, Colombia, Costa Rica, Dominican Republic, Ecuador, El Salvador, Guatemala, Honduras, Nicaragua, Panama, Paraguay, or Peru, who have obtained their Masters, PhD, or other equivalent academic degree, and who have at least five years of laboratory experience in the area of antimicrobial resistance, to visit a host scientist in the U.S.. The ASM-PAHO Infectious Diseases Epidemiology and Surveillance Fellowship will award a maximum grant of $4,000. The award is not intended to provide travel to obtain a degree at the host institution.
Requirements: The application must be made jointly between a host institution and the visiting fellow. The Investigator must be: a member of ASM or any other national microbiological society; actively involved in research in the microbiological sciences; have obtained, or be in the process of obtaining, their masters, PhD, or other equivalent academic degree within the last five years; a national of a resource-limited Africa, Asian or Latin American country submitting the application from an institution in their home region. Investigators from developed nations, as defined by BOTH the UN high HDI group and the World Bank high income economy group, are ineligible. Investigators submitting an application from the U.S. where they are spending a training period are ineligible; sufficiently proficient in the use of the English language. The Host Scientist must be: an ASM member; actively involved in research in the microbiological sciences in the U.S; actively involved in teaching in the microbiological sciences in the U.S; interested in sustaining international collaborations Preference will be given to: applicants who can prove three years of membership in ASM or any other national microbiological society, and; applicants who have not previously had the opportunity to travel to a facility in another country. The application should demonstrate to the Review Committee: academic excellence of the applicant - Honors & Awards, CV, and Letters of Recommendation; depth of the applicant's research experience - Statement of prior Research Experience; quality and originality of work proposed during the Fellowship - Potential Collaboration and Proposed Research Plan; relevance of the work proposed to the applicant's locale - Potential Collaboration and Proposed Research Plan; excellence of the host, and the research and training environment
Geographic Focus: All States, Argentina, Bolivia, Brazil, Chile, Colombia, Ecuador, Guyana, Paraguay, Peru
Date(s) Application is Due: Apr 15; Oct 15
Amount of Grant: 4,000 U.S.D
Contact: Supervisor; (202) 942-9368; fax (202) 942-9328; international@asmusa.org
Internet: http://www.asm.org/International/index.asp?bid=2778
Sponsor: American Society for Microbiology
1752 N Street, N.W.
Washington, DC 20036-2904

ASM-UNESCO Leadership Grant for International Educators 738
This new program, sponsored jointly by ASM and UNESCO, has been developed to enable a select group of educators from resource-limited countries to attend the ASM Conference for Undergraduate Educators (ASMCUE) and a pre-conference workshop to provide leaders in education with the resources to build innovative teaching modules that engage students and lead to enduring understandings in microbiology. Successful applicants will receive financial support to cover airfare (economy class), room, board and registration to the pre-conference workshop and ASMCUE. Participants will be reimbursed for airfare via wire transfer upon submission of a confirmed invoice. Room, board and registration will be arranged by ASM.
Requirements: Preference will be given to ASM members. Applicants must be a citizen of, and residing in, a resource-limited country in Latin America, Asia, Africa, and Central and Eastern Europe. Resource-limited countries are classified as upper-middle income and below by the World Bank Development Group. (To find out if your country fits this criteria, visit http://tinyurl.com/7he3u.) Participants must have a minimum of eight years teaching in microbiology/biology at a university level and must also have regular and sustained access to the Internet.
Restrictions: Members of the ASM International Board, International Education Committee, International Membership Committee, International Laboratory Capacity Building Committee, and the ASM International Ambassador Network are not eligible for this program.
Date(s) Application is Due: Dec 1
Contact: Michelle Godinez, Coordinator, Education Programs and Events; (202) 942-9317; fax (202) 942-9328; mgodinez@asmusa.org
Internet: http://www.asm.org/International/index.asp?bid=61306
Sponsor: American Society for Microbiology
1752 N Street, N.W.
Washington, DC 20036-2904

ASM/CDC Fellowships in Infectious Disease and Public Health Microbiology 739
The ASM/CDC Postdoctoral Research Fellowship Program is sponsored by the American Society for Microbiology (ASM) and the Centers of Disease Control and Prevention (CDC). Founded in 1899, the American Society for Microbiology (ASM) promotes the microbiological sciences and their application for the common good. The Society accomplishes its mission by disseminating information, stimulating research, promoting education, and advancing the profession. ASM's headquarters office is located in Washington, D.C. The Centers for Disease Control and Prevention (CDC) is the agency of the United States Public Health Service (PHS) responsible for developing and applying disease control and prevention methods and improving the health of the citizens of the United States. The CDC has been a leader in efforts to prevent and control infectious diseases for nearly half a century. CDC supports surveillance, research, prevention efforts, and training in the area of infectious diseases. The goal of the two-year ASM/CDC Fellowships is to support the development of new approaches, methodologies and knowledge in infectious disease prevention and control in areas within the public health mission of the CDC. The program awards up to ten fellowships annually. The fellowships are similar to those at the postdoctoral level in universities, industry, and other government laboratories and allow bright, highly motivated, recent doctoral graduates to perform research-in-residence while headquartered at a CDC location. Participating CDC organizational units are located in: Atlanta, Georgia; Fort Collins, Colorado; and San Juan, Puerto Rico and are comprised of: the Division of Bacterial Diseases; the Division of Foodborne, Bacterial and Mycotic Diseases; the Division of Healthcare Quality Promotion; the Division of HIV/AIDS Prevention; the Division of Influenza; the Division of Viral Hepatitis; the Division of Parasitic Diseases; the Division of Tuberculosis Elimination; the Divisions of Vector-Borne Infectious Diseases; the Division of Viral and Rickettsial Diseases; the Division of Viral Diseases; and the Division of Viral Hepatitus. The fellowship appointments are for two years assuming satisfactory progress. A stipend (up to $45,243 annually), health benefits (up to $3000 annually), relocation benefits (up to $500), and professional development funds (up to $2000 annually) are provided. Fellows are classified as fellows and are not employees of either CDC or ASM but rather have the status of a visiting scientist. Applicants must use the ASM online application system to apply for the fellowship. The link to the online system is available at the program website on the top right corner of the screen. Applicants are encouraged to review the "Full Program Catalog" to obtain complete information about the program, research opportunities, and the application process. The link to the catalog is available at the program website. The ASM/CDC Postdoctoral Research Fellowship Program is sponsored by the U.S. Department of Health and Human Services.
Requirements: The program is intended for individuals who either earned their doctorate degree or completed a primary residency within three years of their proposed start date (considerations will be given to individuals with more experience if there are compelling reasons). Non-U.S. nationals whose primary language is not English must demonstrate command of the English language through the Test of English as a Foreign Language (TOEFL). These applicants may also be subject to an oral interview. Non-U.S. nationals who are offered awards must have valid visas throughout their tenure. Fellows will be responsible for managing all their own visa matters. Fellows must devote their full-time effort to the approved research project and must be in residence at the sponsoring CDC laboratory for the entire two-year period. Fellows must begin the program between October 1 and December 31 during the year in which they are accepted. All fellows must undergo security clearance to work at CDC. Support after the first year of the Program is contingent upon satisfactory progress in the proposed research. Interim and final reports and/or presentations are required. Fellows are expected to publish results of their studies in the scientific literature and to present findings at national meetings. Fellows are subject to the general regulations of the laboratory and are expected to conduct research according to the highest scientific and ethical standards and in compliance with all applicable laws, regulations, and policies regarding protection of human research subjects, humane care and use of laboratory animals, and laboratory safety. Participation in CDC seminar programs is also expected. After completion of the two-year program, fellows are required to keep ASM informed of their current position, affiliation, and continued professional development for a minimum of ten years. This arrangement will ensure proper assessment and longitudinal studies regarding the effectiveness of such training.
Restrictions: Applicants may not have a faculty position or be enrolled in a graduate degree program during the fellowship. Monies from other appointments, fellowships, private employment, or consulting or contract work are not permitted during the two-year appointment.
Geographic Focus: All States, All Countries
Date(s) Application is Due: Jan 15
Amount of Grant: Up to 100,986 U.S.D
Contact: Tiffani Fonseca, Coordinator, Fellowship Programs; (202) 942-9295 or (202) 942-9283; fax (202) 942-9329; fellowships@asmusa.org or tfonseca@asmusa.org
Internet: http://www.asm.org/asm/index.php/education/asmcdc-program-in-infectious-disease-and-public-health-microbiology.html
Sponsor: American Society for Microbiology
1752 N Street, N.W.
Washington, DC 20036-2904

ASM Abbott Laboratories Award in Clinical and Diagnostic Immunology 740
The ASM Abbott Laboratories Award in Clinical and Diagnostic Immunology honors a distinguished scientist in clinical or diagnostic immunology for outstanding contributions to those fields.
Requirements: The nominee must demonstrate significant contributions to the understanding of the functioning of the host immune system in human disease, clinical approaches to diseases involving the immune system or development or clinical application of immunodiagnostic procedures.
Restrictions: Self-nominations will not be accepted.
Date(s) Application is Due: Oct 1
Amount of Grant: 2,000 U.S.D
Contact: Awards Committee; (202) 942-9226; fax (202) 942-9353; awards@asmusa.org
Internet: http://www.asm.org/Academy/index.asp?bid=2301
Sponsor: American Society for Microbiology
1752 N Street, N.W.
Washington, DC 20036-2904

ASM BD Award for Research in Clinical Microbiology 741

The ASM BD Award for Research in Clinical Microbiology honors a a distinguished scientist for research accomplishments that form the foundation for important applications in clinical microbiology.

Requirements: The nominee must be a distinguished clinical microbiologist. All nomination components must be emailed to awards@asmusa.org. Nominations must consist of the following: curriculum vitae; letter of nomination; letters of support.

Restrictions: Self-nominations will not be accepted.

Date(s) Application is Due: Oct 1

Amount of Grant: 2,000 U.S.D

Contact: Awards Committee; (202) 942-9226; fax (202) 942-9353; awards@asmusa.org

Internet: http://www.asm.org/Academy/index.asp?bid=2304

Sponsor: American Society for Microbiology
1752 N Street, N.W.
Washington, DC 20036-2904

ASM bioMerieux Sonnenwirth Award for Leadership in Clinical Microbiology 742

Given in memory of Alexander Sonnenwirth, Ph.D., this award recognizes a distinguished microbiologist for the promotion of innovation in clinical laboratory science, dedication to ASM, and the advancement of clinical microbiology as a profession.

Requirements: Nominations will be considered without updating for three years. email (see contact info.) all nomination components. Nominations must consist of the following: curriculum vitae, including a list of nominee's publications; letter of nomination, describe the nominee's innovation in clinical laboratory science, dedication to ASM, role as a leader in the field of diagnostic microbiology, and the advancement of clinical microbiology as a profession; letters of support, two letters of support must come from persons, other than the nominator, who are familiar with the nominee's qualifications and accomplishments. No more than one of the three letters may be from the nominee's institution or the same institution.

Restrictions: Self-nominations will not be accepted.

Date(s) Application is Due: Oct 1

Amount of Grant: 2,000 U.S.D

Contact: Awards Committee; (202) 942-9226; fax (202) 942-9353; awards@asmusa.org

Internet: http://www.asm.org/Academy/index.asp?bid=36368

Sponsor: American Society for Microbiology
1752 N Street, N.W.
Washington, DC 20036-2904

ASM Congressional Science Fellowships 743

The program will select a postdoctoral to mid-career microbiologist to spend one year on the staff of an individual congressman, congressional committee, or with some other appropriate organizational unit of Congress. The purpose of the program is to make practical contributions to more effective use of scientific knowledge in government, to educate the scientific communities regarding public policy, and to broaden the perspective of both the scientific and governmental communities regarding the value of such science-government interaction. The ASM Fellow will function as special legislative assistant within the congressional staff. The American Association for the Advancement of Science will arrange a carefully structured orientation program, guide the placement process, and coordinate weekly seminars throughout the year for the ASM Fellow, as well as other Congressional Fellows. The period of appointment is from September 1 and extends for one year. The award will include a $60,000 stipend plus health care benefits. The stipend is supported by the Martin Frobisher Fund.

Requirements: Prospective Fellows must be citizens of the United States, be members of ASM for at least one year and must have completed their Ph.D. by the time the fellowship begins in September. Candidates are expected to show competence in some aspect of microbiology, have a broad background in science and technology, and have interest and some experience in applying scientific knowledge toward the solution of social problems. Candidates are expected to be articulate, literate, adaptable, interested in work on a range of public policy problems, and able to work with a variety of people from diverse professional backgrounds. A Complete Application for the Fellowship must include: a letter from the candidate indicating a desire to apply and listing three references; three letters of references; a statement from the candidate about his or her qualifications and career goals.

Restrictions: Candidates selected for interviews must provide travel to and from Washington at their own expense. The candidate is responsible for soliciting the required references, providing the six guidelines for the reference response and seeing that the references are forwarded before the deadline.

Date(s) Application is Due: Feb 20

Amount of Grant: 60,000 U.S.D

Contact: Heather Garvey, Program Assistant; (202) 942-9209; fax (202) 942-9335; hgarvey@usmusa.org

Internet: http://www.asm.org/Policy/index.asp?bid=12335

Sponsor: American Society for Microbiology
1752 N Street, N.W.
Washington, DC 20036-2904

ASM Corporate Activities Program Student Travel Grants 744

The Corporate Activities Program Student Travel Grant is a $500 grant given a maximum of 160 students who are ASM members and who are the presenting author of a poster at the General Meeting. Students are selected based on a review of their abstract submission for the meeting. The awards are highly competitive and only about 25-30% of those seeking grants will receive them. If a grant recipient cannot present the accepted poster, the grant money must be returned to ASM.

Requirements: To qualify for consideration the student must: be an ASM Member and the presenting author; be accepted to present a poster at the General Meeting; be an undergraduate/pre-doctoral student, or; be a Ph.D. or M.D. whose degree has been or will be awarded after May of the current year and before May of the meeting year; have indicated a desire to be considered for the grant on the abstract submission, and submit a letter of nomination from a faculty member or department head on official letterhead. The letter must state that the student is expected to be active in the training or degree program at the time of the meeting. If the student will complete their terminal degree between the date of submission and the meeting, the letter must indicate the expected date of completion. The abstract tracking number must be included in the body of the letter. The letter should be submitted to ASM not later than December of each year. Exact date will be published in the Call for Abstracts for the General Meeting.

Amount of Grant: 500 U.S.D

Contact: Awards Committee; (202) 942-9226; fax (202) 942-9353; awards@asmusa.org

Internet: http://www.asm.org/Awards/index.asp?bid=14968

Sponsor: American Society for Microbiology
1752 N Street, N.W.
Washington, DC 20036-2904

ASMCUE/GM Travel Assistance Grants 745

ASM offers travel awards to attend the 2009 ASMCUE to ASM Members who have been accepted to present a poster at ASM General Meeting and ASMCUE.

Requirements: Eligibility: accepted to present a poster at both ASM General Meeting in any division and ASMCUE; be an active ASM member; teach an undergraduate microbiology course or participates in outreach activities related to microbiology; have an interest in pursuing a teaching career at an undergraduate institution; not be a recipient of substantial federal funding or any other ASM faculty enhancement or travel award.

Date(s) Application is Due: Mar 20

Amount of Grant: 500 U.S.D

Contact: Awards Committee; (202) 942-9226; fax (202) 942-9353; awards@asmusa.org

Internet: http://www.asm.org/Education/index.asp

Sponsor: American Society for Microbiology
1752 N Street, N.W.
Washington, DC 20036-2904

ASM D.C. White Research and Mentoring Award 746

The ASM D.C. White Research and Mentoring Award recognizing distinguished accomplishments in interdisciplinary research and mentoring in microbiology, this award honors D.C. White, who was known for his interdisciplinary scientific approach and for being a dedicated and inspiring mentor.

Requirements: Consideration will be given to the breadth of the nominee's contributions, as well as their originality and overall impact. There are no age restrictions, but the nominee must have a distinguished record of accomplishments in microbiological research. Nominees in all areas of microbiology will be considered. Nominations must consist of the following: curriculum vitae, including a list of publications; list of those mentored; letter of nomination, describe the nominee's accomplishments of unusual merit in interdisciplinary research in the field of microbiology; letters of support, two letters of support must come from persons, other than the nominator, who are familiar with the nominee's qualifications and accomplishments. One letter should be from someone who can comment specifically on the nominee's mentoring. No more than one of the three letters may be from the nominee's institution or the same institution.

Restrictions: Self nominations will not be accepted.

Date(s) Application is Due: Oct 1

Amount of Grant: 5,000 U.S.D

Contact: Awards Committee; (202) 942-9226; fax (202) 942-9353; awards@asmusa.org

Internet: http://www.asm.org/Academy/index.asp?bid=50931

Sponsor: American Society for Microbiology
1752 N Street, N.W.
Washington, DC 20036-2904

ASM Early-Career Faculty Travel Award 747

SM offers travel awards to attend the ASM Conference for Undergraduate Educators (ASMCUE) to increase the participation of early-career undergraduate faculty, post-doctoral scientists or senior level graduate students interested in teaching careers.

Requirements: Eligibility: accepted to present a poster at both ASM General Meeting in any division and ASMCUE; be an active ASM member; teach an undergraduate microbiology course or participates in outreach activities related to microbiology; have an interest in pursuing a teaching career at an undergraduate institution; not be a recipient of substantial federal funding or any other ASM faculty enhancement or travel award.

Date(s) Application is Due: Mar 20

Amount of Grant: 500 U.S.D

Contact: Awards Committee; (202) 942-9226; fax (202) 942-9353; awards@asmusa.org

Internet: http://www.asm.org/Education/index.asp

Sponsor: American Society for Microbiology
1752 N Street, N.W.
Washington, DC 20036-2904

ASME H.R. Lissner Award 748

The ASME H.R. Lissner Award was established in honor of H.R. Lissner for his pioneering contributions in biomechanics. The award is bestowed annually for outstanding accomplishments in the area of bioengineering. These achievements may be in the form of (1) significant research contributions in bioengineering; (2) development of new methods of measuring in bioengineering; (3) design of new equipment and instrumentation in bioengineering; (4) educational impact in the training of bioengineers; and/or (5) service to the bioengineering community, in general, and to the Bioengineering Division of

ASME, in particular. The award includes a bronze medal, certificate, $1,000, and travel expenses to attend the award presentation. Nomination forms and instructions are available at the website. Nomination packets will be held and considered for three years; a new nomination packet must be submitted after that time.
Requirements: Candidate must be an active member of the bioengineering division.
Date(s) Application is Due: Sep 1
Amount of Grant: 1,000 U.S.D
Contact: T.P. Andriacchi, Committee Chair; (650) 559-9170; tandriac@stanford.edu
Internet: http://www.asme.org/about-asme/honors-awards/achievement-awards/h-r—lissner-medal
Sponsor: American Society of Mechanical Engineers
3 Park Avenue
New York, NY 10016

ASM Eli Lilly and Company Research Award 749
ASM's oldest and most prestigious prize, it rewards fundamental research of unusual merit in microbiology or immunology by an individual on the threshold of his or her career.
Requirements: The nominee's work is not judged by comparison with the research of more experienced scientists, and special consideration is given to originality and independence of thought. The nominee must not have reached his or her 45th birthday by April 30 of the year the award is given. The nominee must be working in the United States or Canada and be actively engaged in the line of research for which the award is to be made.
Restrictions: Nominations must consist of the following: curriculum vitae, including a list of publications; verification of date of birth. Photocopy or scanned file of driver's license, passport or birth certificate; letter of nomination, describe the nominee's fundamental research of unusual merit in microbiology or immunology; letters of support, two letters of support must come from persons, other than the nominator, who are familiar with the nominee's qualifications and accomplishments. No more than one of the three letters may be from the nominee's institution or the same institution.
Date(s) Application is Due: Oct 1
Amount of Grant: 5,000 U.S.D
Contact: Awards Committee; (202) 942-9226; fax (202) 942-9353; awards@asmusa.org
Internet: http://www.asm.org/academy/index.asp?bid=2281
Sponsor: American Society for Microbiology
1752 N Street, N.W.
Washington, DC 20036-2904

ASM Faculty Ehancement Program -Travel Awards 750
ASM offers travel awards to attend the ASMCUE to faculty who teach microbiology at 2- or 4-year institutions with a large percentage of historically excluded and underrepresented students.
Requirements: Eligibility: accepted to present a poster at both ASM General Meeting in any division and ASMCUE; be an active ASM member; teach an undergraduate microbiology course or participates in outreach activities related to microbiology; have an interest in pursuing a teaching career at an undergraduate institution; not be a recipient of substantial federal funding or any other ASM faculty enhancement or travel award.
Date(s) Application is Due: Mar 20
Amount of Grant: 500 U.S.D
Contact: Awards Committee; (202) 942-9226; fax (202) 942-9353; awards@asmusa.org
Internet: http://www.asm.org/Education/index.asp
Sponsor: American Society for Microbiology
1752 N Street, N.W.
Washington, DC 20036-2904

ASM Founders Distinguished Service Award 751
The ASM Founders Distinguished Service Award honors a member of the ASM for outstanding contributions and commitment to the ASM as a volunteer at the national level.
Requirements: Selection is based on commitment to furthering the goals of the ASM, ability to inspire commitment from others, and significance of contributions to the membership of ASM and its various audiences. The nominee must be an ASM member in good standing who has served in a volunteer capacity for ASM at the national level (e.g., as a member of committees or editorial boards or as a workshop leader) for a minimum of five years, and has not held office as ASM President, Secretary, Treasurer, or Chair of a CPC Board or the American Academy of Microbiology. Nominations must consist of the following: curriculum vitae; letter of nomination, describe the nominee's ASM service and contributions on which the nomination is based; letters of support, two letters of support must come from persons, other than the nominator, who are familiar with the nominee's service and contributions to the ASM. No more than one of the three letters may be from the nominee's institution or the same institution.
Date(s) Application is Due: Oct 1
Samples: Self-nominations will not be accepted.
Contact: Awards Committee; (202) 942-9226; fax (202) 942-9353; awards@asmusa.org
Internet: http://www.asm.org/Academy/index.asp?bid=2324
Sponsor: American Society for Microbiology
1752 N Street, N.W.
Washington, DC 20036-2904

ASM Gen-Probe Joseph Public Health Award 752
The ASM Gen-Probe Joseph Public Health Award honors a distinguished microbiologist who has exhibited exemplary leadership and service in the field of public health. This award has been established in memory of J. Mehsen Joseph, Ph.D., who dedicated his life toward the advancement of both microbiology and public health.
Requirements: The nominee must be a microbiologist identified with public health. The recipient will be recognized for significant achievements in integrating the science of microbiology into the practice of public health and for promoting the importance of linking these two disciplines. Nominations must consist of the following: curriculum vitae, including a list of nominee's publications; letter of nomination, describe the nominee's leadership and service in the field of public health; letters of support, two letters of support must come from persons, other than the nominator, who are familiar with the nominee's qualifications and accomplishments, citing in particular the nominee's leadership and service in the field of public health. No more than one of the three letters may be from the nominee's institution or from the same institution.
Date(s) Application is Due: Oct 1
Amount of Grant: 2,000 U.S.D
Contact: Awards Committee; (202) 942-9226; fax (202) 942-9353; awards@asmusa.org
Internet: http://www.asm.org/Academy/index.asp?bid=36040
Sponsor: American Society for Microbiology
1752 N Street, N.W.
Washington, DC 20036-2904

ASM General Meeting Minority Travel Grants 753
The American Society for Microbiology (ASM) will offer travel grants to increase the participation of underrepresented minority (URM) groups in the ASM General Meeting. The ASM will select post-doctoral scholars from URM groups in the microbiological sciences or faculty from Minority Serving Institutions. Each grantee will be offered up to $1500 to defray expenses associated with travel to the ASM General Meeting.
Requirements: Eligibility: ASM Member; U.S. citizen or permanent resident; faculty member from a Minority Serving Institution (MSI), Historically Black Colleges and Universities (HBCU), Hispanic Serving institutions (HSI), Tribal Colleges and Universities (TCU) and other institutions with substantial enrollments of minority students are considered to be Minority Serving Institutions (MSI); post-doctoral Scholars from an URM Group
Date(s) Application is Due: Jan 30
Amount of Grant: Up to 1,500 U.S.D
Contact: Awards Committee; (202) 942-9226; fax (202) 942-9353; awards@asmusa.org
Internet: http://www.asm.org/Awards/index.asp?bid=38173
Sponsor: American Society for Microbiology
1752 N Street, N.W.
Washington, DC 20036-2904

ASM GlaxoSmithKline International Member of the Year Award 754
The ASM GlaxoSmithKline International Member of the Year Award honors a distinguished microbiologist who exhibited exemplary leadership in the international microbiological community. It recognizes an international ASM member for education, communication, research, and advancement of the profession to the international microbiology community while demonstrating a commitment to the ASM.
Requirements: The nominee can be any international member of ASM who has made major contributions toward the advancement of the microbiological sciences within the international community through education, research, communication, and leadership. The nominee must not have served on the International Board, the International Education Committee or the International Membership Committee within the past two years. Nominations must consist of the following: curriculum vitae, including a list of publications; letter of nomination, describe the nominee's contributions towards the advancement of the microbiological sciences within the international community through education, research, communication, and leadership. Emphasis should be on the nominee's creativity and originality, as well as the nominee's role in ASM-sponsored activities; letters of support, two letters of support must come from persons, other than the nominator, who are familiar with the nominee's contributions and accomplishments. No more than one of the three letters may be from the nominee's institution or the same institution.
Restrictions: Self-nominations will not be accepted.
Date(s) Application is Due: Oct 1
Contact: Awards Committee; (202) 942-9226; fax (202) 942-9353; awards@asmusa.org
Internet: http://www.asm.org/Academy/index.asp?bid=36033
Sponsor: American Society for Microbiology
1752 N Street, N.W.
Washington, DC 20036-2904

ASM ICAAC Young Investigator Awards 755
The ASM ICAAC Young Investigator Awards recognize and reward early career scientists for research excellence and potential in microbiology and infectious disease.
Requirements: Nominees must be no more than three years beyond completion of postdoctoral research training in microbiology or infectious diseases at the time of the nomination deadline. Nominations must consist of the following: curriculum vitae, including a list of publications; letter of nomination, describe the nominee's research excellence and potential in microbiology and infectious diseases; letters of support,two letters of support must come from persons, other than the nominator, who are familiar with the nominee's research excellence and potential. No more than one of the three letters may be from the nominee's institution or the same institution.
Restrictions: Self-nominations will not be accepted.
Date(s) Application is Due: Apr 1
Contact: Awards Committee; (202) 942-9226; fax (202) 942-9353; awards@asmusa.org
Internet: http://www.asm.org/academy/index.asp?bid=2278
Sponsor: American Society for Microbiology
1752 N Street, N.W.
Washington, DC 20036-2904

ASM Intel Awards 756
ASM sponsors a special prize in microbiology at the Intel International Science and Engineering Fair. The fair recognizes excellence in science and technology by high school students. Ten cash prizes are given and all laureates receive a one-year student membership in ASM and access to members-only web resources. The first place laureate receives a cash prize of $2,000, the second place laureate receives a cash prize of $1,250, the thrid place laureate receives a cash prize of $750, and the fourth place laureate receives a cash prize of $500. Six honorable mentions of $250 are also awarded.
Amount of Grant: 250 - 2,000 U.S.D
Contact: Awards Committee; (202) 942-9226; fax (202) 942-9353; awards@asmusa.org
Internet: http://www.asm.org/Academy/index.asp?bid=36845
Sponsor: American Society for Microbiology
1752 N Street, N.W.
Washington, DC 20036-2904

ASM International Fellowship for Latin America and the Caribbean 757
Funded by ASM this award offers fellowships to promising young investigators in South and Central America and the Caribbean who are within five years of obtaining, or are in the process of obtaining, their Masters, PhD, or other equivalent academic degree, and who are working in any of the microbiology disciplines, to visit a host scientist in the U.S.. The International Fellowship Program provides funding to meet the costs of a visit to one institution for between 6 weeks to 6 months. The International Fellowship Program provides funding to meet the costs of a visit to one institution for between 6 weeks to 6 months.
Requirements: The application must be made jointly between a host institution and the visiting fellow. The Investigator must be:a member of ASM or any other national microbiological society; actively involved in research in the microbiological sciences; have obtained, or be in the process of obtaining, their masters, PhD, or other equivalent academic degree within the last five years; a national of a resource-limited Africa, Asian or Latin American country submitting the application from an institution in their home region. Investigators from developed nations, as defined by BOTH the UN high HDI group and the World Bank high income economy group, are ineligible. Investigators submitting an application from the U.S. where they are spending a training period are ineligible; sufficiently proficient in the use of the English language. The Host Scientist must be: an ASM member; actively involved in research in the microbiological sciences in the U.S; actively involved in teaching in the microbiological sciences in the U.S; interested in sustaining international collaborations Preference will be given to: applicants who can prove three years of membership in ASM or any other national microbiological society, and; applicants who have not previously had the opportunity to travel to a facility in another country. The application should demonstrate to the Review Committee: academic excellence of the applicant - Honors & Awards, CV, and Letters of Recommendation; depth of the applicant's research experience - Statement of prior Research Experience; quality and originality of work proposed during the Fellowship - Potential Collaboration and Proposed Research Plan; relevance of the work proposed to the applicant's locale - Potential Collaboration and Proposed Research Plan; excellence of the host, and the research and training environment
Geographic Focus: All States, Antigua & Barbuda, Argentina, Bahamas, Barbados, Belize, Bolivia, Brazil, Chile, Colombia, Costa Rica, Cuba, Ecuador, Guyana, Paraguay, Peru
Date(s) Application is Due: Apr 15; Oct 15
Amount of Grant: 4,000 U.S.D
Contact: Supervisor; (202) 942-9368; fax (202) 942-9328; international@asmusa.org
Internet: http://www.asm.org/International/index.asp?bid=2778
Sponsor: American Society for Microbiology
1752 N Street, N.W.
Washington, DC 20036-2904

ASM International Fellowships for Asia and Africa 758
Funded by ASM this award offers a maximum of one fellowship to a promising young investigator in Asia, as well as to one young investigator in Africa who is within five years of obtaining, or in the process of obtaining, his/her Masters, PhD, or other equivalent academic degree, and who is working in any of the microbiology disciplines, to visit a host scientist in the U.S.. Both International Fellowships will award a maximum grant of $5,500. The award is not intended to provide travel to obtain a degree at the host institution.
Requirements: The application must be made jointly between a host institution and the visiting fellow. The Investigator must be: a member of ASM or any other national microbiological society; actively involved in research in the microbiological sciences; have obtained, or be in the process of obtaining, their masters, PhD, or other equivalent academic degree within the last five years; a national of a resource-limited Africa, Asian or Latin American country submitting the application from an institution in their home region. Investigators from developed nations, as defined by BOTH the UN high HDI group and the World Bank high income economy group, are ineligible. Investigators submitting an application from the U.S. where they are spending a training period are ineligible; sufficiently proficient in the use of the English language. The Host Scientist must be: an ASM member; actively involved in research in the microbiological sciences in the U.S; actively involved in teaching in the microbiological sciences in the U.S; interested in sustaining international collaborations Preference will be given to: applicants who can prove three years of membership in ASM or any other national microbiological society, and; applicants who have not previously had the opportunity to travel to a facility in another country.
Date(s) Application is Due: Apr 15; Oct 15
Contact: Supervisor; (202) 942-9368; fax (202) 942-9328; international@asmusa.org
Internet: http://www.asm.org/International/index.asp?bid=2778
Sponsor: American Society for Microbiology
1752 N Street, N.W.
Washington, DC 20036-2904

ASM International Professorship for Latin America 759
The program provides funding support to an ASM member from the U.S. who is scientifically recognized in his/her area to teach a hands-on, highly interactive short course on a single topic in the microbiological sciences at a Latin American institution of higher learning. Up to five Teaching Professorships offered per program year.
Requirements: The application must be made jointly between a hosting institution and the visiting professor. Eligibility for Host Institution: institution with graduate students enrolled in a masters, doctoral or equivalent program, post-doctoral fellows or residents, and teaching faculty; institution where at least 12 students will be enrolled full-time in the short course; commitment to maximizing use of the course, as demonstrated by the applicability of the course's contents to existing programs at the institution or to the field of research of the Research Professor. demonstrated commitment to international collaborations and partnerships; demonstrated commitment to advancing the microbiological sciences through interactions with professional organizations; institutions in developed nations, as defined by the UN high HDI group, are ineligible. Eligibility for Visiting Teaching Professor: scientifically recognized their area of microbiological expertise; actively engaged in teaching at the post-secondary level; commitment to international collaborations and partnerships; must be an ASM member during the term of the Professorship. Preference will be given to: applicants who will be able to make maximal use of the course, as demonstrated either by the applicability of the course's contents to existing programs at the host institution or to the field of research of the visiting Professor; applications that demonstrate a commitment to building an ongoing institutional relationship. It is strongly recommended that the course include a hands-on component, such as a wet lab or other practical activity.
Date(s) Application is Due: Apr 15; Oct 15
Amount of Grant: 4,000 U.S.D
Contact: Supervisor; (202) 942-9368; fax (202) 942-9328; international@asmusa.org
Internet: http://www.asm.org/International/index.asp?bid=57785
Sponsor: American Society for Microbiology
1752 N Street, N.W.
Washington, DC 20036-2904

ASM International Professorships for Asia and Africa 760
Each program provides funding support to an ASM member from the U.S. who is scientifically recognized in his/her area to teach a hands-on, highly interactive short course on a single topic in the microbiological sciences at an institution of higher learning in Asia and in Africa. The Teaching Professorship for both programs is limited to one per program year.
Requirements: The application must be made jointly between a hosting institution and the visiting professor. Eligibility for Host Institution: institution with graduate students enrolled in a masters, doctoral or equivalent program, post-doctoral fellows or residents, and teaching faculty; institution where at least 12 students will be enrolled full-time in the short course; commitment to maximizing use of the course, as demonstrated by the applicability of the course's contents to existing programs at the institution or to the field of research of the Research Professor. demonstrated commitment to international collaborations and partnerships; demonstrated commitment to advancing the microbiological sciences through interactions with professional organizations; institutions in developed nations, as defined by the UN high HDI group, are ineligible. Eligibility for Visiting Teaching Professor: scientifically recognized their area of microbiological expertise; actively engaged in teaching at the post-secondary level; commitment to international collaborations and partnerships; must be an ASM member during the term of the Professorship. Preference will be given to: applicants who will be able to make maximal use of the course, as demonstrated either by the applicability of the course's contents to existing programs at the host institution or to the field of research of the visiting Professor; applications that demonstrate a commitment to building an ongoing institutional relationship. It is strongly recommended that the course include a hands-on component, such as a wet lab or other practical activity.
Date(s) Application is Due: Apr 15; Oct 15
Contact: Supervisor; (202) 942-9368; fax (202) 942-9328; international@asmusa.org
Internet: http://www.asm.org/International/index asp?bid=57785
Sponsor: American Society for Microbiology
1752 N Street, N.W.
Washington, DC 20036-2904

ASM Maurice Hilleman/Merck Award 761
The award is presented in memory of Maurice R. Hilleman, whose work in the development of vaccines has saved the lives of many throughout the world.
Requirements: The nominee must have made outstanding achievements in pathogenesis, vaccine discovery, vaccine development, and/or control of vaccine-preventable diseases. Nominations must consist of the following: curriculum vitae, including a full list of publications; letter of nomination, describe the nominee's remarkable contributions to pathogenesis, vaccine discovery, vaccine development, and/or the control of vaccine-preventable diseases, including a list of the ten publications most relevant to the award; letters of support, two letters of support must come from persons, other than the nominator, who are familiar with the nominee's achievements. No more than one of the three letters may be from the nominee's institution or the same institution.
Restrictions: Self-nominations will not be accepted.
Date(s) Application is Due: Oct 1
Amount of Grant: 20,000 U.S.D
Contact: Awards Committee; (202) 942-9226; fax (202) 942-9353; awards@asmusa.org
Internet: http://www.asm.org/Academy/index.asp?bid=57948
Sponsor: American Society for Microbiology
1752 N Street, N.W.
Washington, DC 20036-2904

ASM Merck Irving S. Sigal Memorial Awards 762

The ASM Merck Irving S. Sigal Memorial Awards recognize and award excellence in basic research in medical microbiology and infectious diseases. The awards are presented in memory of Irving S. Sigal, who was instrumental in the early discovery of therapies to treat HIV/AIDS.

Requirements: The nominee must be no more than five years beyond completion of postdoctoral research training in microbiology or infectious diseases at the time of the nomination deadline. Nominations must consist of the following: curriculum vitae, including a list of publications; letter of nomination, describe the nominee's excellence in basic research in medical microbiology and infectious diseases; letters of support, two letters of support must come from persons, other than the nomination, who are familiar with the nominee's qualifications and accomplishments. No more than one of the three letters may be from the nominee's institution or the same institution.

Date(s) Application is Due: Oct 1

Amount of Grant: 2,500 U.S.D

Contact: Awards Committee; (202) 942-9226; fax (202) 942-9353; awards@asmusa.org

Internet: http://www.asm.org/academy/index.asp?bid=15056

Sponsor: American Society for Microbiology

1752 N Street, N.W.

Washington, DC 20036-2904

ASM Microbiology Undergraduate Research Fellowship 763

The goal of the Microbiology Undergraduate Research Fellowship (MURF) program is to increase the number of underrepresented undergraduate students who wish to, and have demonstrated the ability to pursue graduate careers (Ph.D. or M.D./Ph.D.) in microbiology. Students will have the opportunity to conduct full time summer research with an ASM member at a host institution, and present research results at the Annual Biomedical Research Conference for Minority Students and the ASM General Meeting. Students will: agree to participate in an undergraduate summer research program at a U.S. based institution; conduct a research project for a minimum of 10 weeks; work with a faculty mentor who is an ASM member; submit a research abstract to the Annual Biomedical Research Conference for Minority Students (ABRCMS); submit a research abstract to ASM for presentation at the ASM General Meeting. Students decide the institution, research area, and level of activity for the summer. Based on interests, independence, and ability, students will be placed at a host U.S. Institution of the student's choice to conduct basic science research. From a list provided on the application, students interested in conducting research at a host institution will select three institutions where they would like to conduct their summer research. Every effort will be made to place fellows at their first choice.

Requirements: Eligible student candidates for the fellowship must be from groups that have been determined by the applicant's institution to be underrepresented in the microbiological sciences. The ASM encourages institutions to identify individuals that have been historically underrepresented, and remain underrepresented today in the microbiological sciences nationally. These groups include African-Americans, Hispanics, Native Americans, Alaskan Natives, and Pacific Islanders. In addition, applicants must also: be U.S. citizen or permanent U.S. resident; be enrolled as full-time matriculating undergraduate students during the academic year at an accredited U.S. institution; be either freshmen with college level research experience, sophomores, juniors, or seniors who will not graduate before the completion date of the summer program; be members of an underrepresented group in microbiology; have taken introductory courses in biology, chemistry, and preferably microbiology prior to submission of the application; have strong interests in obtaining a Ph.D., or M.D./Ph.D. in the microbiological sciences, and have lab research experience.

Restrictions: Travel funds are contingent upon acceptance of an abstract for the General Meeting.

Date(s) Application is Due: Feb 1

Amount of Grant: 5,850 U.S.D

Contact: Fellowship Committee; (202) 942-9283; fax (202) 942-9329; fellowships-careerinformation@asmusa.org

Internet: http://www.asm.org/Education/index.asp?bid=4322

Sponsor: American Society for Microbiology

1752 N Street, N.W.

Washington, DC 20036-2904

ASM Millis-Colwell Postgraduate Travel Grant 764

This new grant, negotiated between the American Society for Microbiology (ASM U.S.A) and the Australian Society for Microbiology (ASM Australia), allows for the exchange of one student member from each society to present an abstract at the annual General Meeting of the other society and to spend a week at nearby research laboratory. The grant aims to create a long lasting bond between ASM U.S.A and ASM Australia and is designed to benefit PhD students in both countries by giving them the opportunity to travel overseas to present their work and experience the best of microbiology in the partner country. ASM U.S.A will submit the successful applicant's abstract for an oral or poster presentation to ASM Australia for their Annual Scientific Meeting & Exhibition. The applicant will also have negotiated an agreement to visit the research laboratory of an Australian scientist in the period either immediately before or immediately after the ASM Australia annual meeting. The award will be conditional upon ASM Australia's acceptance of the submitted abstract.

Requirements: The successful applicant will be: a U.S. citizen or permanent resident of the U.S; a student member of the American Society for Microbiology; and enrolled in a PhD program at an American University. The award will cover roundtrip economy airfare between the U.S. and Australia, and any necessary internal flights; per diem living allowance for up to 14 days; and, registration fee for the Australian Society for Microbiology's Annual Scientific Meeting & Exhibition.

Date(s) Application is Due: Jan 1

Amount of Grant: Up to 5,000 U.S.D

Contact: Program Coordinator; (202) 942-9282; fax (202) 942-9329

Internet: http://www.asm.org/International/index.asp?bid=53729

Sponsor: American Society for Microbiology

1752 N Street, N.W.

Washington, DC 20036-2904

ASM Procter & Gamble Award in Applied and Environmental Microbiology 765

The ASM Procter & Gamble Award in Applied and Environmental Microbiology recognizes distinguished achievement in research and development in applied (non-clinical) and environmental microbiology.

Requirements: The nominee must demonstrate outstanding accomplishment in research or development in an appropriate field and be actively engaged in research or development at the time that the award is presented. Nominations must consist of the following: curriculum vitae, including a list of publications; letter of nomination, describe the nominee's outstanding accomplishments in research or development in the appropriate field, including a list of the ten most relevant publications to the award; letters of support, two letters of support must come from persons, other than the nominator, who are familiar with the nominee's outstanding accomplishments. No more than one of the three letters may be from the nominee's institution or the same institution.

Restrictions: Self-nominations will not be accepted.

Date(s) Application is Due: Oct 1

Amount of Grant: 2,000 U.S.D

Contact: Awards Committee; (202) 942-9226; fax (202) 942-9353; awards@asmusa.org

Internet: http://www.asm.org/academy/index.asp?bid=2288

Sponsor: American Society for Microbiology

1752 N Street, N.W.

Washington, DC 20036-2904

ASM Promega Biotechnology Research Award 766

The ASM Promega Biotechnology Research Award honors outstanding contributions to the application of biotechnology through fundamental microbiological research and development.

Requirements: A nomination can be a single exceptionally significant achievement or the aggregate of a number of exemplary achievements. Nominations must consist of the following: curriculum vitae, including a list of publications; letter of nomination, describe the nominee's outstanding accomplishments in microbiological research and development, including a list of the ten most relevant publications to the award; letters of support, two letters of support must come from persons, other than the nominator, who are familiar with the nominee's accomplishments. No more than one of the three letters may be from the nominee's institution or the same institution.

Restrictions: Neither self-nominations nor co-nominations will be accepted.

Date(s) Application is Due: Oct 1

Amount of Grant: 5,000 U.S.D

Contact: Awards Committee; (202) 942-9226; fax (202) 942-9353; awards@asmusa.org

Internet: http://www.asm.org/academy/index.asp?bid=2291

Sponsor: American Society for Microbiology

1752 N Street, N.W.

Washington, DC 20036-2904

ASM Public Communications Award 767

The ASM Public Communications Award, sponsored by ASM, recognizes outstanding achievement in increasing public awareness, knowledge and understanding of microbiology. Microbiology is concerned with issues such as the environment, prevention and treatment of infectious diseases, laboratory and diagnostic medicine, and food and water safety, among others.

Requirements: Eligibility is limited to primary authors of information concerning the microbiological sciences appearing in mass media, including print (newspapers and periodicals only) and broadcast media readily available to the general public. Individual items and series are both eligible. Applicants may nominate themselves. International entries are accepted but must be translated into English prior to submission.

Restrictions: Books, websites, and institutionally sponsored publications are not eligible.

Date(s) Application is Due: Jan 31

Amount of Grant: 2,500 U.S.D

Contact: Garth Hogan, Office of Communications; (202) 942-9226; fax (202) 942-9353; ghogan@asmusa.org

Internet: http://www.asm.org/Media/index.asp?bid=2676

Sponsor: American Society for Microbiology

1752 N Street, N.W.

Washington, DC 20036-2904

ASM Raymond W. Sarber Awards 768

The ASM Raymond W. Sarber Awards recognize students at the undergraduate and predoctoral levels for research excellence and potential. The awards honor Raymond W. Sarber for his contributions to the growth and advancement of ASM.

Requirements: Nominees must be a student at the undergraduate or predoctoral level, attending an accredited institution in the United States, in an academic program involving microbiology. Two awards are presented at the ASM General Meeting, one to an undergraduate and one to a graduate student. They will be judged separately.

Restrictions: Nominations must consist of the following: curriculum vitae; statement, no longer than two pages, from the student outlining academic achievements, research accomplishments, and future career goals; letter of nomination, describe the nominee's

research excellence and potential; letters of support, two letters of support: one from a supervisor or mentor, and the other letter may be from an academic advisor or colleague.
Date(s) Application is Due: Oct 1
Amount of Grant: 1,500 U.S.D
Samples: Self-nominations will not be accepted.
Contact: Awards Committee; (202) 942-9226; fax (202) 942-9353; awards@asmusa.org
Internet: http://www.asm.org/academy/index.asp?bid=2294
Sponsor: American Society for Microbiology
1752 N Street, N.W.
Washington, DC 20036-2904

ASM Richard and Mary Finkelstein Travel Grant 769
The Richard and Mary Finkelstein Travel Grant is a $500 grant given to 6 students who are ASM members and who are the presenting author of a poster at the General Meeting. Selection criteria for this award are the same as the Corporate Activities Program Student Travel Grant but the research presented must be in the area of microbial pathogenesis. Potential nominees must first be selected to receive a Corporate Activities Program Student Travel Grant and then the abstracts are reviewed again to establish the quality of the research in the additional area of microbial pathogenesis.
Requirements: To qualify for consideration the student must: be presenting research in the area of microbial pathogenesisbe; be an ASM Member and the presenting author; be accepted to present a poster at the General Meeting; be an undergraduate/pre-doctoral student, or; be a Ph.D. or M.D. whose degree has been or will be awarded after May of the current year and before May of the meeting year; have indicated a desire to be considered for the grant on the abstract submission, and submit a letter of nomination from a faculty member or department head on official letterhead. The letter must state that the student is expected to be active in the training or degree program at the time of the meeting. If the student will complete their terminal degree between the date of submission and the meeting, the letter must indicate the expected date of completion. The abstract tracking number must be included in the body of the letter. The letter should be submitted to ASM not later than December of each year. Exact date will be published in the Call for Abstracts for the General Meeting.
Amount of Grant: 500 U.S.D
Contact: Awards Committee; (202) 942-9226; fax (202) 942-9353; awards@asmusa.org
Internet: http://www.asm.org/Awards/index.asp?bid=14968h
Sponsor: American Society for Microbiology
1752 N Street, N.W.
Washington, DC 20036-2904

ASM Robert D. Watkins Graduate Research Fellowship 770
The goal of the fellowship is to increase the number of underrepresented groups completing doctoral degrees in the microbiological sciences. The ASM Robert D. Watkins Graduate Research Fellowship is aimed at highly competitive graduate students who are enrolled in a Ph.D. program and who have completed their graduate course work in the microbiological sciences. The fellowship encourages students to continue and complete their research project in the microbiological sciences. Students will be required to submit an abstract each year to ASM for presentation at the annual ASM General Meeting and attend the ASM Kadner Institute one time during the three-year tenure of the fellowship.
Requirements: Eligible candidates must be from groups that have been determined by the applicant's institution to be underrepresented in the microbiological sciences. The ASM encourages institutions to identify individuals that have been historically underrepresented, and remain underrepresented today in the microbiological sciences nationally. These groups include African-Americans, Hispanics, Native Americans, Alaskan Natives, and Pacific Islanders. In addition, applicants must: be formally admitted to a doctoral program in the microbiological sciences in an accredited U.S. institution; have successfully completed the first year of the graduate program (first year graduate students cannot apply); have successfully completed all graduate coursework requirements for the doctoral degree by the date of activation of the fellowship; be a student member of ASM; be mentored by an ASM member; be a U.S. citizen or a permanent resident; not have funding or have funding that will expire by the start date of the fellowship(fellowship cannot run concurrently with other national fellowships from NIH, NSF, HHMI, etc.); three letters of recommendations must be submitted with your application. One letter must be from your research advisor/mentor; Official transcripts from all colleges and universities attended.
Date(s) Application is Due: May 1
Amount of Grant: 63,000 U.S.D
Contact: Education Board; (202) 942-9283; fax (202) 942-9329; fellowships-careerinformation@asmusa.org
Internet: http://www.asm.org/Education/index.asp?bid=6278
Sponsor: American Society for Microbiology
1752 N Street, N.W.
Washington, DC 20036-2904

ASM Roche Diagnostics Alice C. Evans Award 771
The ASM Roche Diagnostics Alice C. Evans Award recognizes contributions toward the full participation and advancement of women in microbiology. This award was established by the ASM's Committee on the Status of Women in Microbiology, and is given in memory of Alice C. Evans, the first woman to be elected ASM President in 1928.
Requirements: The nominee can be any member of ASM who has made major contributions toward fostering the inclusion, development, and advancement of women in careers in microbiology. The nominee must demonstrate commitment to women in science through mentorship and advocacy and by setting an example through scientific and professional achievement. Nominations must consist of the following: curriculum vitae, including a list of nominee's publications; letter of Nomination, describe the nominee's major contributions toward fostering the inclusion, development, and advancement of women in careers in microbiology; letters of support, two letters of support must come from persons, other than the nominator, who are familiar with the nominee's qualifications and accomplishments. No more than one of the three letters may be from the nominee's institution or from the same institution.
Restrictions: Self-nominations will not be accepted.
Date(s) Application is Due: Oct 1
Contact: Awards Committee; (202) 942-9226; fax (202) 942-9353; awards@asmusa.org
Internet: http://www.asm.org/Academy/index.asp?bid=2332
Sponsor: American Society for Microbiology
1752 N Street, N.W.
Washington, DC 20036-2904

ASM sanofi-aventis ICAAC Award 772
ASM's premier award in antimicrobial chemotherapy research; it stimulates research and honors outstanding accomplishment in antimicrobial chemotherapy.
Requirements: The nominee must be actively engaged in research involving development of new agents, investigation of antimicrobial action or resistance to antimicrobial agents, and/or the pharmacology, toxicology or clinical use of those agents. The nominee must not have served on the ICAAC Program Committee within the past two years. Nominations must consist of the following: curriculum vitae, including a list of publications; letter of nomination, describe the nominee's outstanding accomplishment in antimicrobial chemotherapy, including a list of the ten most relevant publications to the award; letters of support, two letters of support should come from two people, other than the nominator, who are familiar with the nominee's accomplishments. No more than one of the three letters may be from the nominee's institution or the same institution.
Restrictions: Self-nominations will not be accepted.
Date(s) Application is Due: Apr 1
Amount of Grant: 20,000 U.S.D
Contact: Awards Committee; (202) 942-9226; fax (202) 942-9353; awards@asmusa.org
Internet: http://www.asm.org/academy/index.asp?bid=2268
Sponsor: American Society for Microbiology
1752 N Street, N.W.
Washington, DC 20036-2904

ASM Scherago-Rubin Award 773
The ASM Scherago-Rubin Award recognizes an outstanding, bench-level clinical microbiologist. The award was established by the late Sally Jo Rubin, an active member of ASM's Clinical Microbiology Division, in honor of her grandfather, Professor Morris Scherago.
Requirements: The nominee must be a non-doctoral-level clinical microbiologist involved primarily in routine diagnostic work, rather than in research, who has distinguished himself or herself with excellent performance in the clinical laboratory. Nominations must consist of the following: curriculum vitae, including a list of the nominee's publications; letter of nomination, a letter from the supervisor describing how the nominee's routine diagnostic work, rather than research, has distinguished himself or herself in the clinical laboratory; letters of support, two letters of support preferably from individuals from different institutions who are familiar with the nominee's laboratory accomplishments.
Restrictions: Self-nominations will not be accepted.
Date(s) Application is Due: Oct 1
Amount of Grant: 1,500 U.S.D
Contact: Awards Committee; (202) 942-9226; fax (202) 942-9353; awards@asmusa.org
Internet: http://www.asm.org/academy/index.asp?bid=2335
Sponsor: American Society for Microbiology
1752 N Street, N.W.
Washington, DC 20036-2904

ASM Siemens Healthcare Diagnostics Young Investigator Award 774
The ASM Siemens Healthcare Diagnostics Young Investigator Award honors outstanding laboratory research in clinical microbiology or antimicrobial agents and is intended to further the career development of a young clinical scientist and promote awareness of clinical microbiology as a career.
Requirements: The nominee should be conducting outstanding research in clinical microbiology, automation in clinical laboratories, development of novel antimicrobial agents, mechanisms of action of antimicrobial agents or mechanisms of resistance to antimicrobial agents. The nominee must be no more than five years beyond completion of postdoctoral research training in microbiology, infectious diseases or related disciplines at the time of the nomination deadline. Nominations must consist of the following: curriculum vitae, including a list of publications, abstracts, and manuscripts in preparation; letter of nomination, describe the nominee's outstanding clinical research and potential; letters of support, two letters of support must come from persons, other than the nominator, who are familiar with the nominee's qualifications and accomplishments. No more than one of the three letters may be from the nominee's institution or from the same institution.
Restrictions: Self-nominations will not be accepted.
Date(s) Application is Due: Oct 1
Amount of Grant: 2,000 U.S.D
Contact: Awards Committee; (202) 942-9226; fax (202) 942-9353; awards@asmusa.org
Internet: http://www.asm.org/Academy/index.asp?bid=2307
Sponsor: American Society for Microbiology
1752 N Street, N.W.
Washington, DC 20036-2904

ASM Student and Post Doctoral Fellow Travel Grants 775

The ICAAC Program Committee will award travel grants worth at least $500, to support students' attendance at the 43rd ICAAC. Travel grants are selected on the basis of the scientific quality of the submitted abstracts. Preference is given to students who have not received a travel grant in the past two years, and the Committee strives to award only one travel grant per laboratory. They are sponsored by the ASM Corporate Partners.

Requirements: Criteria: ASM membership at time of application; undergraduate or pre-doctoral student; Ph.D. or M.D. within two years of graduation at the time of the meeting; or an M.D. in a residency or fellowship program within six years of receipt of the M.D. degree at the time of the meeting; cover letter from the applicant, including their contact information, accompanied by letter of nomination from a faculty member on department letterhead, stating that the applicant is expected to be active in the training or degree program at the time of the meeting. If the applicant will have attained their terminal degree at the time of the meeting, the date of graduation must be included in this letter; submission of an abstract for the ICAAC, which is accepted for presentation; letter of nomination, copy of the abstract, and submission control number.

Date(s) Application is Due: May 30

Amount of Grant: 500 U.S.D

Contact: Awards Committee; (202) 942-9226; fax (202) 942-9353; awards@asmusa.org

Internet: http://www.asm.org/Meetings/index.asp?bid=15264

Sponsor: American Society for Microbiology

1752 N Street, N.W.

Washington, DC 20036-2904

ASM Student Travel Grants 776

ASM has provided funds for the awarding travel grants of $500 each to predoctoral students.

Requirements: To qualify, applicants must be: a member of ASM (student membership in ASM is only $15); a presenter on an accepted abstract; and must be at the predoctoral level at the time of abstract submission. Applicants submitting an abstract via the ASM web site should check the box on the screen when filling out the electronic submission form where indicated. In addition, selected recipients must return a confirmation form sent with notice of award selection, including proof of predoctoral status, either in the form of a copy of the validated student identification card, or a letter on departmental letterhead from their departmental chair which verifies their student status.

Amount of Grant: 500 U.S.D

Contact: Awards Committee; (202) 942-9226; fax (202) 942-9353; awards@asmusa.org

Internet: http://www.asm.org/Meetings/index.asp?bid=15264

Sponsor: American Society for Microbiology

1752 N Street, N.W.

Washington, DC 20036-2904

ASM TREK Diagnostic ABMM/ABMLI Professional Recognition Award 777

The ASM TREK Diagnostic ABMM/ABMLI Professional Recognition Award recognizes a Diplomate of the American Board of Medical Microbiology (ABMM) or the American Board of Medical Laboratory Immunology (ABMLI) for outstanding contributions to the professional recognition of certified microbiologists and/or immunologists and the work they do. Primary consideration will be given to ABMM or ABMLI Diplomates who have made significant contributions to the advancement and public recognition of the profession over and above scientific achievements or board-related activities.

Requirements: Nominations must consist of the following: curriculum vitae, including a list of nominee's publications; letter of nomination, describe the nominee's outstanding contributions to the professional recognition of clinical microbiologists and/or immunologists; letters of support, two letters of support must come from persons, other than the nominator, who are familiar with the nominee's qualifications and accomplishments. No more than one of the three letters may be from the nominee's institution or from the same institution.

Restrictions: Self-nominations will not be accepted.

Date(s) Application is Due: Oct 1

Contact: Awards Committee; (202) 942-9226; fax (202) 942-9353; awards@asmusa.org

Internet: http://www.asm.org/Academy/index.asp?bid=2321

Sponsor: American Society for Microbiology

1752 N Street, N.W.

Washington, DC 20036-2904

ASM Undergraduate Research Fellowship 778

The ASM Undergraduate Research Fellowship (URF) is aimed at highly competitive students who wish to pursue graduate careers (Ph.D. or MD/Ph.D) in microbiology. Students will have the opportunity to conduct full time research at their home institutions with an ASM member and present research results at the ASM General Meeting the following year. Students will: conduct a research project for a minimum of 10 weeks beginning in the summer; work with faculty mentors who are ASM members and who are employed at the students home institutions, and submit a research abstract for presentation at the ASM General Meeting. The Fellowship provides: up to $4000 for student stipend; two-year ASM student membership, and up to $1000 in travel support for students to present the results of the research project at the ASM General Meeting. Travel funds are contingent upon acceptance of an abstract for the General Meeting.

Requirements: Eligible student candidates for the fellowship must: be enrolled as full-time matriculating undergraduate students during the academic year at an accredited U.S. Institution; be involved in a research project; have an ASM member at their home institutions willing to serve as a mentor, and not receive financial support for research (i.e., Council for Undergraduate Research, Minority Access to Research Careers, Sigma Xi) during the fellowship. Applicants who do not meet all eligibility requirements will not be considered.

Restrictions: The program requires a joint application from both the student and a faculty mentor. It is student's responsibility to submit a completed application.

Date(s) Application is Due: Feb 1

Amount of Grant: 4,000 - 5,000 U.S.D

Contact: Fellowship Committee; (202) 942-9283; fax (202) 942-9329; fellowships-careerinformation@asmusa.org

Internet: http://www.asm.org/Education/index.asp?bid=4319

Sponsor: American Society for Microbiology

1752 N Street, N.W.

Washington, DC 20036-2904

ASM U.S.FCC/J. Roger Porter Award 779

This award recognizes outstanding efforts by a scientist who has demonstrated the importance of microbial biodiversity through sustained curatorial or stewardship activities for a major resource used by the scientific community. It honors the memory of Dr. Porter and his remarkable contributions to science.

Requirements: The nominee will have greatly aided other scientists by demonstrating the fundamentals of culture collections and related resources and the rich biodiversity that such collections preserve. These resources may include collections of cells or microorganisms, databases or tools, or major reference works or services as an essential aspect of either one or more major publications or an extensive work effort.

Restrictions: Nominations must consist of the following: curriculum vitae, including a list of publications; letter of nomination, describe the nominee's outstanding efforts demonstrating the importance of microbial biodiversity through sustained curatorial or stewardship activities for a major resource used by the scientific community; letters of support, two letters of support must come from persons, other than the nominator, who are familiar with the nominee's outstanding efforts. No more than one of the three letters may be from the nominee's institution or the same institution.

Date(s) Application is Due: Oct 1

Amount of Grant: 2,000 U.S.D

Samples: Self-nominations will not be accepted.

Contact: Awards Committee; (202) 942-9226; fax (202) 942-9353; awards@asmusa.org

Internet: http://www.asm.org/academy/index.asp?bid=2297

Sponsor: American Society for Microbiology

1752 N Street, N.W.

Washington, DC 20036-2904

ASM William A Hinton Research Training Award 780

The ASM William A Hinton Research Training Award honors outstanding contributions toward fostering the research training of underrepresented minorities in microbiology. It is given in memory of William A. Hinton, a physician-research scientist and one of the first African-Americans to join the ASM.

Requirements: The nominee must contribute to the research training of undergraduate students, graduate students, postdoctoral fellows or health professional students and efforts leading to the increased participation of underrepresented minorities in microbiology. Nominations must consist of the following: curriculum vitae, including a list of nominee's publications; letter of nomination, describe the nominee's training of undergraduate students, graduate students, postdoctoral fellows, or health professional students. Detail specific examples of excellence in training and mentoring of underrepresented minorities. Describe the duration, extent, and impact of the nominee's training of underrepresented minority students, graduate students, postdoctoral fellows, or health professional students; letters of support, two letters of support must come from persons, other than the nominator, who are familiar with the nominee's outstanding contributions to the training of underrepresented minorities. No more than one of the three letters may be from the nominee's institution or from the same institution.

Date(s) Application is Due: Oct 1

Contact: Awards Committee; (202) 942-9226; fax (202) 942-9353; awards@asmusa.org

Internet: http://www.asm.org/Academy/index.asp?bid=2317

Sponsor: American Society for Microbiology

1752 N Street, N.W.

Washington, DC 20036-2904

ASNS Centrum Center for Nutrition Science Award 781

The award is given for recent investigative contributions of contemporary significance to the understanding of human nutrition. The contributions need not be restricted to investigative work with humans as long as they have relevance to human nutrition and health. Preference is usually given to scientists in the Western Hemisphere. The award and an engraved plaque are made available by Wyeth Consumer Healthcare. Nomination should include a letter stating the significance of the work, a selected bibliography that supports the nomination, and a reprint or series of reprints reporting such research.

Requirements: Nominations may be made by anyone, including members of the respective nominating committees and nonmembers of the institute. Nominations will be retained for two years. Candidates may be renominated. Candidates need not be members of the institute. However, the awards are usually given to professionally active nutrition scientists.

Restrictions: An individual who has received one ASNS award is not eligible to receive another award unless it is for accomplishments not covered by the first award.

Date(s) Application is Due: Sep 1

Amount of Grant: 1,500 U.S.D

Contact: Program Officer; (301) 530-7050; fax (301) 634-7892; sec@asns.org

Internet: http://www.asns.org/awards.html

Sponsor: American Society for Nutritional Sciences

9650 Rockville Pike, Suite L-4500

Bethesda, MD 20814

ASNS Conrad A. Elvehjem Award for Public Service in Nutrition 782

The award is given for specific and distinguished service to the public through the science of nutrition. Such service could be rendered through governmental, industrial, private, or international institutions, but contributions of an investigative character are not excluded. The award and an engraved plaque are made available by Kraft Foods. Nominations should include a letter stating the basis for the nomination, a selected bibliography indicating the candidate's contributions to public service, and the candidate's curriculum vita.
Requirements: Nominations may be made by anyone, including members of the respective nominating committees and nonmembers of the institute. Nominations will be retained for two years. Candidates may be renominated. Candidates need not be members of the institute. However, the awards are usually given to professionally active nutrition scientists.
Restrictions: An individual who has received one ASNS award is not eligible to receive another award unless it is for accomplishments not covered by the first award.
Date(s) Application is Due: Oct 1
Amount of Grant: 1,500 U.S.D
Contact: Secretariat; (301) 530-7050; fax (301) 637-7892; sec@asns.org
Internet: http://www.asns.org/awards.html
Sponsor: American Society for Nutritional Sciences
9650 Rockville Pike, Suite L-4500
Bethesda, MD 20814

ASNS Mead Johnson Nutrition Award 783

The award is given to an investigator for a single outstanding piece of nutrition research or a series of papers on the same subject completed within 10 years of postgraduate training. The award and an inscribed scroll are made available by Mead Johnson Nutritionals. Nominations should include a letter stating the significance of the work, a selected bibliography that supports the nomination, and a reprint or series of reprints reporting such research.
Requirements: Nominations may be made by anyone, including members of the respective nominating committees and nonmembers of the institute. Nominations will be retained for two years. Candidates may be renominated. Candidates need not be members of the institute. However, the awards are usually given to professionally active nutrition scientists.
Restrictions: An individual who has received one ASNS award is not eligible to receive another award unless it is for accomplishments not covered by the first award.
Date(s) Application is Due: Sep 1
Amount of Grant: 2,500 U.S.D
Contact: Program Officer; (301) 530-7050; fax (301) 634-7892; sec@asns.org
Internet: http://www.asns.org/awards.html
Sponsor: American Society for Nutritional Sciences
9650 Rockville Pike, Suite L-4500
Bethesda, MD 20814

ASNS Osborne and Mendel Nutrition Award 784

The award is given for outstanding recent basic research accomplishments in nutrition. Nominations should include a letter stating the significance of the work, a selected bibliography of all papers relating to the research on which the nomination is based, and a reprint or series of reprints reporting this research.
Requirements: Nominations may be made by anyone, including members of the respective nominating committees and nonmembers of the institute. Nominations will be retained for two years. Candidates may be renominated. Candidates need not be members of the institute. However, the awards are usually given to professionally active nutrition scientists.
Restrictions: An individual who has received one ASNS award is not eligible to receive another award unless it is for accomplishments not covered by the first award.
Date(s) Application is Due: Sep 1
Contact: Program Officer; (301) 530-7050; fax (301) 634-7892; sec@asns.org
Internet: http://www.asns.org/awards.html
Sponsor: American Society for Nutritional Sciences
9650 Rockville Pike, Suite L-4500
Bethesda, MD 20814

ASNS Predoctoral Fellowship Program 785

This fellowship is designed to support research in human and clinical nutrition. Preference will be given to work of immediate relevance to human nutrition information and human nutrition status. Contact the institute for application materials.
Requirements: Applicants must be enrolled in a graduate program registered in the AIN Directory of Graduate Programs in Nutritional Sciences.
Date(s) Application is Due: Dec 1
Amount of Grant: 5,000 U.S.D
Contact: Secretariat; (301) 530-7052; fax (301) 364-7892; sec@asns.org
Internet: http://www.asns.org/winners.html
Sponsor: American Society for Nutritional Sciences
9650 Rockville Pike, Suite L-4500
Bethesda, MD 20814

A Social Corporation Grants 786

The Fund, established in Wisconsin in 1987, has as its primary purpose to support research in medical physics within the Madison, Wisconsin, region. Specific application guidelines are available upon request. There are no particular deadlines with which to adhere, and applicant organizations should contact the fund administrator directly for further details.
Geographic Focus: Wisconsin
Contact: Charles Lescrenier; (608) 831-1188 or (800) 426-6391; fax (608) 836-9201
Sponsor: A Social Corporation
2500 West Beitline Highway at University Avenue, P.O. Box 620327
Middleton, WI 53562-0327

ASPEN C. Richard Fleming Grant 787

The Foundation provides annual grant support to nutrition researchers at all stages of their careers. This grant, funded by the donations of ASPEN members, is intended to assist a nutrition investigator by providing preliminary funding for promising new research in the field of nutrition and metabolic support and related areas of clinical nutrition. Applicants still in training positions may apply. Corporate employees are eligible to receive grants if the research project is not part of their normal duties and they have at least a part-time academic appointment or are associated with a professional institute that conducts nutrition research. In making choices among otherwise high-quality applicants, the reviewers will take into account the geographic distribution of grant winners; their ethnic background and gender, should the applicant choose to supply such information; their professional discipline; and other criteria as appropriate to ensure an equitable system. Funds may be used only for technician salary, equipment, supplies, animals, clinical research costs or other expenses directly related to the conduct of the proposed research.
Requirements: Applicants must submit a letter from their supervisor or department head at the institution confirming his/her commitment to the project. If the project involves human subjects, a letter pledging support in recruiting patients from the primary care provider and the institutional review board overseeing human studies is required. For junior faculty and applicants who do not hold faculty positions, three letters of reference are required.
Restrictions: This grant is not intended to support pursuit of an additional degree. Grant funds may not be used for: comparison of commercial products; indirect costs or overhead; costs of patient care; constructing or renovating facilities; furniture or office equipment; secretarial services; honoraria or membership dues; textbooks or periodicals; repair or service contract costs on institutional equipment; entertainment; travel; or salary support for the Principal Investigator.
Date(s) Application is Due: Sep 8
Amount of Grant: 3,500 U.S.D
Contact: Paula Bowen, Research Program Administrator; (301) 587-6315, ext. 132 or (301) 920-9132; fax (301) 587-2365; paulab@aspen.nutr.org
Internet: http://www.nutritioncare.org/wcontent.aspx?id=784
Sponsor: American Society for Parenteral and Enteral Nutrition Rhoads Research Foundation
8630 Fenton Street, Suite 412
Silver Spring, MD 20910

ASPEN Douglas Wilmore Grant 788

The Foundation provides annual grant support to nutrition researchers at all stages of their careers. The grant is intended to assist a nutrition investigator by providing preliminary funding for promising new research in the field of nutrition and metabolic support and related areas of clinical nutrition. Priority consideration is given to applications that involve a multidisciplinary team or that investigate the efficacy of parenteral or enteral nutrition. Applicants still in training positions may apply for the Fleming Grant. Corporate employees are eligible to receive grants if the research project is not part of their normal duties and they have at least a part-time academic appointment or are associated with a professional institute that conducts nutrition research. In making choices among otherwise high-quality applicants, the reviewers will take into account the geographic distribution of grant winners; their ethnic background and gender, should the applicant choose to supply such information; their professional discipline; and other criteria as appropriate to ensure an equitable system. Funds may be used only for technician salary, equipment, supplies, animals, clinical research costs or other expenses directly related to the conduct of the proposed research. The grant can be renewed for a second year of funding, assuming satisfactory progress.
Requirements: Applicants must submit a letter from their supervisor or department head at the institution confirming his/her commitment to the project. If the project involves human subjects, a letter pledging support in recruiting patients from the primary care provider and the institutional review board overseeing human studies is required. For junior faculty and applicants who do not hold faculty positions, three letters of reference are required.
Restrictions: Individuals who have received other research grants in excess of $25,000 are ineligible to apply, as are those who completed training more than 5 years prior to the start of the grant period. This grant is not intended to support pursuit of an additional degree. Grant funds may not be used for: comparison of commercial products; indirect costs or overhead; costs of patient care; constructing or renovating facilities; furniture or office equipment; secretarial services; honoraria or membership dues; textbooks or periodicals; repair or service contract costs on institutional equipment; entertainment; travel; or salary support for the Principal Investigator.
Date(s) Application is Due: Sep 8
Amount of Grant: Up to 25,000 U.S.D
Contact: Paula Bowen, Research Program Administrator; (301) 587-6315, ext. 132 or (301) 920-9132; fax (301) 587-2365; paulab@aspen.nutr.org
Internet: http://www.nutritioncare.org/wcontent.aspx?id=784
Sponsor: American Society for Parenteral and Enteral Nutrition Rhoads Research Foundation
8630 Fenton Street, Suite 412
Silver Spring, MD 20910

ASPEN Dudrick Research Scholar Award 789

A.S.P.E.N. members are invited to nominate individuals to be considered by the Dudrick Award Committee for the Stanley Dudrick Research Scholar Award. A.S.P.E.N. presents this annual award to recognize and support a mid-career investigator who has shown significant achievements in nutrition support, is an A.S.P.E.N. member, and demonstrates exceptional research productivity and the potential to continue to make contributions in the field of nutrition therapy. The award includes $5,000, plus recognition during the Rhoads Lecture and Awards Symposium at Clinical Nutrition Week.

Requirements: The applicant should: be an A.S.P.E.N. member; have an advanced degree (MD, PharmD, MS, PhD, or equivalent); currently be in a training program or have completed formal training in nutritional research/therapy within the past 10-15 years, e.g. mid-career; have demonstrated exceptional research productivity during and/or following formal training.
Date(s) Application is Due: Oct 15
Amount of Grant: 5,000 U.S.D
Contact: Paula Bowen, Research Program Administrator; (301) 587-6315, ext. 132 or (301) 920-9132; fax (301) 587-2365; paulab@aspen.nutr.org
Internet: http://www.nutritioncare.org/wcontent.aspx?id=878
Sponsor: American Society for Parenteral and Enteral Nutrition
8630 Fenton Street, Suite 412
Silver Spring, MD 20910

ASPEN Harry M. Vars Award **790**
The Award is made annually for the best research presentation by an investigator at Nutrition Week. The recipient of the award is selected by the Vars Award Selection Subcommittee based upon a review of manuscripts submitted by qualified candidates. The author of the best paper presents his or her paper at the Premier Paper Session and is given a $1,000 cash prize, a travel grant and a plaque during the Research Awards Ceremony at Nutrition Week. The winning manuscript is published in the Journal of Parenteral and Enteral Nutrition (JPEN). The Award is open to all disciplines.
Amount of Grant: 1,000 U.S.D
Samples: Zhi Yong Peng, Natalie J. Serkova, Douglas J. Kominsky, Jaimi L. Brown, and Paul E. Wischmeyer—Glutamine-Mediated Attenuation of Cellular Metabolic Dysfunction and Cell Death After Injury Is Dependent on Heat Shock Factor-1 Expression.
Contact: Michelle Spangenburg, Program Director for Education; (301) 587-6315, ext. 127; fax (301) 587-2365; michelles@aspen.nutr.org
Internet: http://www.nutritioncare.org/wcontent.aspx?id=878
Sponsor: American Society for Parenteral and Enteral Nutrition
8630 Fenton Street, Suite 412
Silver Spring, MD 20910

ASPEN Maurice Shils Grant **791**
The Foundation provides annual grant support to nutrition researchers at all stages of their careers. This grant, funded by a donation from Baxter Health Care and Nestle Clinical Nutrition, is intended to assist a nutrition investigator by providing preliminary funding for promising new research in the field of nutrition and metabolic support and related areas of clinical nutrition. Persons applying for the grant must be in a post-training position and commit at least 20% of their time to research. Corporate employees are eligible to receive grants if the research project is not part of their normal duties and they have at least a part-time academic appointment or are associated with a professional institute that conducts nutrition research. In making choices among otherwise high-quality applicants, the reviewers will take into account the geographic distribution of grant winners; their ethnic background and gender, should the applicant choose to supply such information; their professional discipline; and other criteria as appropriate to ensure an equitable system. Funds may be used only for technician salary, equipment, supplies, animals, clinical research costs or other expenses directly related to the conduct of the proposed research. Grants can be renewed for a second year of funding, assuming satisfactory progress.
Requirements: Applicants must submit a letter from their supervisor or department head at the institution confirming his/her commitment to the project. If the project involves human subjects, a letter pledging support in recruiting patients from the primary care provider and the institutional review board overseeing human studies is required. For junior faculty and applicants who do not hold faculty positions, three letters of reference are required.
Restrictions: Individuals who have received other research grants in excess of $25,000 are ineligible to apply, as are those who completed training more than 5 years prior to the start of the grant period. This grant is not intended to support pursuit of an additional degree. Grant funds may not be used for: comparison of commercial products; indirect costs or overhead; costs of patient care; constructing or renovating facilities; furniture or office equipment; secretarial services; honoraria or membership dues; textbooks or periodicals; repair or service contract costs on institutional equipment; entertainment; travel; or salary support for the Principal Investigator.
Date(s) Application is Due: Sep 8
Amount of Grant: 18,000 U.S.D
Contact: Paula Bowen, Research Program Administrator; (301) 587-6315, ext. 132 or (301) 920-9132; fax (301) 587-2365; paulab@aspen.nutr.org
Internet: http://www.nutritioncare.org/wcontent.aspx?id=784
Sponsor: American Society for Parenteral and Enteral Nutrition Rhoads Research Foundation
8630 Fenton Street, Suite 412
Silver Spring, MD 20910

ASPEN Norman Yoshimura Grant **792**
The Foundation provides annual grant support to nutrition researchers at all stages of their careers. This grant, supported by B. Braun Medical, Inc. funding, is intended to assist a nutrition investigator by providing preliminary funding for promising new research in the field of nutrition and metabolic support and related areas of clinical nutrition. Applicants still in training positions may apply. Corporate employees are eligible to receive grants if the research project is not part of their normal duties and they have at least a part-time academic appointment or are associated with a professional institute that conducts nutrition research. In making choices among otherwise high-quality applicants, the reviewers will take into account the geographic distribution of grant winners; their ethnic background and gender, should the applicant choose to supply such information; their professional discipline; and other criteria as appropriate to ensure an equitable system. Funds may be used only for technician salary, equipment, supplies, animals, clinical research costs or other expenses directly related to the conduct of the proposed research.
Requirements: Applicants must submit a letter from their supervisor or department head at the institution confirming his/her commitment to the project. If the project involves human subjects, a letter pledging support in recruiting patients from the primary care provider and the institutional review board overseeing human studies is required. For junior faculty and applicants who do not hold faculty positions, three letters of reference are required.
Restrictions: This grant is not intended to support pursuit of an additional degree. Grant funds may not be used for: comparison of commercial products; indirect costs or overhead; costs of patient care; constructing or renovating facilities; furniture or office equipment; secretarial services; honoraria or membership dues; textbooks or periodicals; repair or service contract costs on institutional equipment; entertainment; travel; or salary support for the Principal Investigator.
Date(s) Application is Due: Sep 8
Amount of Grant: 15,000 U.S.D
Contact: Paula Bowen, Research Program Administrator; (301) 587-6315, ext. 132 or (301) 920-9132; fax (301) 587-2365; paulab@aspen.nutr.org
Internet: http://www.nutritioncare.org/wcontent.aspx?id=784
Sponsor: American Society for Parenteral and Enteral Nutrition Rhoads Research Foundation
8630 Fenton Street, Suite 412
Silver Spring, MD 20910

ASPEN Nutrition Research Grants **793**
The Foundation provides annual grant support to nutrition researchers at all stages of their careers. This grant, funded by a donation from the Abbott Foundation, is intended to assist a nutrition investigator by providing preliminary funding for promising new research in the field of nutrition and metabolic support and related areas of clinical nutrition. Persons applying for the grant must be in a post-training position and commit at least 20% of their time to research. Corporate employees are eligible to receive grants if the research project is not part of their normal duties and they have at least a part-time academic appointment or are associated with a professional institute that conducts nutrition research. In making choices among otherwise high-quality applicants, the reviewers will take into account the geographic distribution of grant winners; their ethnic background and gender, should the applicant choose to supply such information; their professional discipline; and other criteria as appropriate to ensure an equitable system. Funds may be used only for technician salary, equipment, supplies, animals, clinical research costs or other expenses directly related to the conduct of the proposed research. Grants can be renewed for a second year of funding, assuming satisfactory progress.
Requirements: Applicants must submit a letter from their supervisor or department head at the institution confirming his/her commitment to the project. If the project involves human subjects, a letter pledging support in recruiting patients from the primary care provider and the institutional review board overseeing human studies is required. For junior faculty and applicants who do not hold faculty positions, three letters of reference are required.
Restrictions: Individuals who have received other research grants in excess of $25,000 are ineligible to apply, as are those who completed training more than 5 years prior to the start of the grant period. This grant is not intended to support pursuit of an additional degree. Grant funds may not be used for: comparison of commercial products; indirect costs or overhead; costs of patient care; constructing or renovating facilities; furniture or office equipment; secretarial services; honoraria or membership dues; textbooks or periodicals; repair or service contract costs on institutional equipment; entertainment; travel; or salary support for the Principal Investigator.
Date(s) Application is Due: Sep 8
Amount of Grant: Up to 25,000 U.S.D
Contact: Paula Bowen, Research Program Administrator; (301) 587-6315, ext. 132 or (301) 920-9132; fax (301) 587-2365; paulab@aspen.nutr.org
Internet: http://www.nutritioncare.org/wcontent.aspx?id=784
Sponsor: American Society for Parenteral and Enteral Nutrition Rhoads Research Foundation
8630 Fenton Street, Suite 412
Silver Spring, MD 20910

ASPEN Scientific Abstracts Awards for Papers or Posters **794**
ASPEN will consider all Scientific Abstract submissions for awards. Applicants will be notified if their abstract is a candidate. The first authors of the three highest ranked abstracts from this category (paper or poster) will receive travel grants to attend the Research Workshop, and to present their work at Clinical Nutrition Week. In addition, cash awards in the amounts of $750, $500 and $250 will be offered to the presenters of the three best papers. Awards are presented during the Rhoads Lecture and Research Awards Ceremony at Clinical Nutrition Week.
Amount of Grant: 250 - 750 U.S.D
Contact: Paula Bowen, Research Program Administrator; (301) 587-6315, ext. 132 or (301) 920-9132; fax (301) 587-2365; paulab@aspen.nutr.org
Internet: http://www.nutritioncare.org/Index.aspx?id=724
Sponsor: American Society for Parenteral and Enteral Nutrition
8630 Fenton Street, Suite 412
Silver Spring, MD 20910

ASPEN Scientific Abstracts Promising Investigator Awards **795**
ASPEN will consider all Scientific Abstract submissions for awards. Applicants will be notified if their abstract is a candidate. These awards are presented to the top three Scientific Abstract authors, who are also in the early stages of their careers. Winners receive $750 in

travel funds to offset their travel expenses to and from Clinical Nutrition Week. Vars Award winners are eligible for these awards if they are in the early stages of their careers.
Amount of Grant: 750 U.S.D
Contact: Paula Bowen, Research Program Administrator; (301) 587-6315, ext. 132 or (301) 920-9132; fax (301) 587-2365; paulab@aspen.nutr.org
Internet: http://www.nutritioncare.org/wcontent.aspx?id=878
Sponsor: American Society for Parenteral and Enteral Nutrition
8630 Fenton Street, Suite 412
Silver Spring, MD 20910

ASPET-Astellas Awards in Translational Pharmacology **796**
The ASPET-Astellas Awards in Translational Pharmacology are intended to extend fundamental research closer to applications directed towards improving human health. The awards will be given to: recognize those individuals whose research has the potential to lead to the introduction of novel pharmacologic approaches or technologies that may offer significant advances in clinical medicine in the future; and to facilitate that translational process. Three (3) awards of $30,000 each will be made to individuals. The money may be used for supplemental research funding, travel, training, or in any way that furthers the goals of the organization. Nominations should be made electronically.
Requirements: Any ASPET member in good standing may nominate an individual for this award.
Restrictions: Self-nominations will not be accepted.
Date(s) Application is Due: Sep 15
Amount of Grant: 30,000 U.S.D
Contact: Christine Carrico, Executive Officer; (301) 634-7060; fax (301) 634-7061; ccarrico@aspet.org or info@aspet.org
Internet: http://www.aspet.org/public/awards/Astellas_awards.html
Sponsor: American Society for Pharmacology and Experimental Therapeutics
9650 Rockville Pike
Bethesda, MD 20814-3995

ASPET Benedict R. Lucchesi Award in Cardiac Pharmacology **797**
The Benedict R. Lucchesi Award in Cardiac Pharmacology was established to honor Dr. Lucchesi's lifelong scientific contributions to our better understanding and appreciation of pharmacological treatment and prevention of cardiovascular disease and for his mentoring of countless prominent functional (in vivo) cardiovascular pharmacologists. The Benedict R. Lucchesi Award is a biennial award, consisting an honorarium of $1,000, a custom-designed crystal bowl depicting the named Lectureship, and up to $2,000 travel expenses including registration to the annual spring ASPET meeting. A recipient will be selected and invited to deliver a state-of-the-art lecture on recent advances in the field of cardiac and electropharmacology at the spring ASPET meeting (Division's programming session). The presentation of his/her research should be of broad interest and contribute to the growth of the Cardiovascular Pharmacology Division.
Requirements: There are no restrictions on institutional affiliation, nationality, or age of the candidate, but the recipient must be a member of the ASPET. Nominations must be made by a member of the ASPET, and no member may nominate more than one candidate per year.
Date(s) Application is Due: Sep 15
Amount of Grant: 1,000 - 2,000 U.S.D
Contact: Christine Carrico, Executive Officer; (301) 634-7060; fax (301) 634-7061; ccarrico@aspet.org or info@aspet.org
Internet: http://www.aspet.org/public/divisions/cardiovascular/lucchesi_award.htm
Sponsor: American Society for Pharmacology and Experimental Therapeutics
9650 Rockville Pike
Bethesda, MD 20814-3995

ASPET Bernard B. Brodie Award in Drug Metabolism **798**
This biennial award (even-numbered years), established by CIBA-GEIGY Corporation to honor the fundamental contributions of Bernard B. Brodie, is presented to recognize outstanding original research contributions in drug metabolism and disposition, particularly those having a major impact on future research in the field. Only one nominator is necessary although more are acceptable. The award consists of a $2,000 honorarium, a commemorative medal, hotel and economy airfare to the award ceremony at the annual meeting. Nominations must be received by the deadline prior to the award year.
Requirements: Nominees must be members of ASPET; nominators need not be members of ASPET.
Restrictions: Supporting research accomplishments must not have been used to win any other major award.
Date(s) Application is Due: Sep 15
Amount of Grant: 2,000 U.S.D
Contact: Christine Carrico, Executive Officer; (301) 634-7060; fax (301) 634-7061; ccarrico@aspet.org or info@aspet.org
Internet: http://www.aspet.org/public/awards/brodie_award.html
Sponsor: American Society for Pharmacology and Experimental Therapeutics
9650 Rockville Pike
Bethesda, MD 20814-3995

ASPET Division for Drug Metabolism Early Career Achievement Award **799**
The ASPET Division for Drug Metabolism Early Career Achievement Award has been established to recognize excellent original research by early career investigators in the area of drug metabolism and disposition. The award is presented biennially in odd-numbered years. The award consists of $1,000, a plaque, and complimentary registration plus travel expenses (to a maximum of $1,000) for the winner to attend the award ceremony at the annual meeting. The awardee will deliver a lecture at the annual meeting describing their relevant research accomplishments. The awardee will be invited to publish a review article on the subject matter of the award lecture in Drug Metabolism and Disposition.
Requirements: Nominees for this award must have a doctoral degree (e.g. Ph.D., M.D., Pharm. D., D.V.M.) and must be within 15 years of having received their final degree, as of December 31 of the year of the award. There are no restrictions on institutional affiliation and a candidate need not be a member of ASPET.
Date(s) Application is Due: Sep 15
Amount of Grant: 1,000 U.S.D
Contact: Christine Carrico, Executive Officer; (301) 634-7060; fax (301) 634-7061; ccarrico@aspet.org or info@aspet.org
Internet: http://www.aspet.org/public/divisions/drugmetab/early_achievement_award.htm
Sponsor: American Society for Pharmacology and Experimental Therapeutics
9650 Rockville Pike
Bethesda, MD 20814-3995

ASPET Epilepsy Research Award for Outstanding Contributions to the Pharmacology of Antiepileptic Drugs **800**
The International League Against Epilepsy (ILAE) has sponsored this biennial award of $2,000 and a Certification of Citation to be awarded by the American Society for Pharmacology and Experimental Therapeutics, for the purpose of recognizing and stimulating outstanding research leading to better clinical control of epileptic seizures. This research may include the basic screening and testing of new therapeutic agents, studies on mechanisms of action, metabolic disposition, pharmacokinetics, and clinical pharmacology studies. The recipient will be selected by the Epilepsy Award Committee appointed by the President of ASPET, with representation of ILAE.
Requirements: Nominations for the Award may be submitted by members of any recognized scientific association, domestic or foreign.
Date(s) Application is Due: Sep 15
Amount of Grant: 2,000 U.S.D
Contact: Christine Carrico, Executive Officer; (301) 634-7060; fax (301) 634-7061; ccarrico@aspet.org or info@aspet.org
Internet: http://www.aspet.org/public/awards/epilepsy_award.html
Sponsor: American Society for Pharmacology and Experimental Therapeutics
9650 Rockville Pike
Bethesda, MD 20814-3995

ASPET Goodman and Gilman Award in Drug Receptor Pharmacology **801**
This award is given biennially (even-numbered years) to recognize and stimulate outstanding research in the pharmacology of biological receptors. It is the hope that such research might provide a better understanding of the mechanism of biological processes and potentially provide the basis for the discovery of drugs useful in the treatment of diseases. The award includes a monetary prize and travel support for winner and spouse to attend the award ceremony. Nominations must be received September 15 of the year prior to the year in which the award will be made.
Requirements: Nominations must be made by ASPET members; however, there are no restrictions as to age, sex, nationality, or institutional affiliation. Nominee need not be a member of ASPET.
Date(s) Application is Due: Sep 15
Amount of Grant: 2,500 U.S.D
Contact: Christine Carrico, Executive Officer; (301) 634-7060; fax (301) 634-7061; ccarrico@aspet.org or info@aspet.org
Internet: http://www.aspet.org/awards/goodman_and_gilman/
Sponsor: American Society for Pharmacology and Experimental Therapeutics
9650 Rockville Pike
Bethesda, MD 20814-3995

ASPET John J. Abel Award in Pharmacology **802**
This award, sponsored by Eli Lilly and Company and administered by ASPET, is given annually to stimulate original and outstanding fundamental research in pharmacology and experimental therapeutics by young investigators. Candidates may not have passed their 39th birthday as of April 30 of the year of the award, or previously received an award from the sponsor for the same technical accomplishment.
Requirements: Candidates need not be members of the society; however, nominations must be made by ASPET members.
Date(s) Application is Due: Sep 15
Amount of Grant: 2,500 U.S.D
Contact: Christine Carrico, Executive Officer; (301) 634-7060; fax (301) 634-7061; ccarrico@aspet.org or info@aspet.org
Internet: http://www.aspet.org/public/awards/abel_award.html
Sponsor: American Society for Pharmacology and Experimental Therapeutics
9650 Rockville Pike
Bethesda, MD 20814-3995

ASPET Julius Axelrod Award in Pharmacology **803**
The Julius Axelrod Award is presented annually for significant contributions to understanding the biochemical mechanisms underlying the pharmacological actions of drugs and for contributions to mentoring other pharmacologists. The award consists of an honorarium of $5,000, a medal, hotel and economy airfare for the winner and spouse to the annual meeting. The formal presentation of this award and medal will be made at the annual meeting of ASPET. The recipient will be invited by the President of the Society to deliver the Julius Axelrod Lecture and organize the Julius Axelrod Symposium at the annual meeting a year hence. The recipient will also be invited by the Catecholamine Club to give a less formal presentation at its annual dinner meeting that year of the award.

Date(s) Application is Due: Sep 15
Amount of Grant: 5,000 U.S.D
Contact: Christine Carrico, Executive Officer; (301) 634-7060; fax (301) 634-7061; ccarrico@aspet.org or info@aspet.org
Internet: http://www.aspet.org/public/awards/axelrod_award.html
Sponsor: American Society for Pharmacology and Experimental Therapeutics
9650 Rockville Pike
Bethesda, MD 20814-3995

ASPET P. B. Dews Lifetime Achievement Award for Research in Behavioral Pharmacology 804

ASPET's Division of Behavioral Pharmacology sponsors the P. B. Dews Award for Research in Behavioral Pharmacology to recognize outstanding lifetime achievements in research, teaching and professional service in the field of Behavioral Pharmacology and to honor Peter Dews for his seminal contributions to the development of behavioral pharmacology as a discipline. The Award consists of $750, a plaque, and partial travel expenses to the award ceremony at the ASPET annual meeting. The recipient will be invited by the Chair of the Division of Behavioral Pharmacology to deliver a special lecture on this occasion. The lecture will be published subsequently in an appropriate ASPET-sponsored publication.
Requirements: Nominations may be made by members of ASPET or of any relevant scientific society.
Restrictions: There are no restrictions on nominees for this award.
Date(s) Application is Due: Sep 15
Amount of Grant: 750 U.S.D
Contact: Christine Carrico, Executive Officer; (301) 634-7060; fax (301) 634-7061; ccarrico@aspet.org or info@aspet.org
Internet: http://www.aspet.org/public/awards/dews_award.html
Sponsor: American Society for Pharmacology and Experimental Therapeutics
9650 Rockville Pike
Bethesda, MD 20814-3995

ASPET Paul M. Vanhoutte Award 805

The Paul M. Vanhoutte Award in Vascular Pharmacology was established to honor Dr. Vanhoutte's lifelong scientific contributions to our better understanding and appreciation of the importance of endothelial cells and vascular smooth muscle function in health and disease and for his mentoring of countless prominent endothelial and vascular biologists and pharmacologists. This is a biennial award, consisting an honorarium of $1,000, a custom-designed crystal bowl depicting the named Lectureship, and up to $2,000 travel expenses including registration to the annual spring ASPET meeting.
Requirements: There are no restrictions on institutional affiliation, nationality, or age of the candidate, but the recipient must be a member of the ASPET.
Date(s) Application is Due: Sep 15
Amount of Grant: 1,000 - 2,000 U.S.D
Contact: Christine Carrico, Executive Officer; (301) 634-7060; fax (301) 634-7061; ccarrico@aspet.org or info@aspet.org
Internet: http://www.aspet.org/public/divisions/cardiovascular/vanhoutte_award.htm
Sponsor: American Society for Pharmacology and Experimental Therapeutics
9650 Rockville Pike
Bethesda, MD 20814-3995

ASPET Torald Sollmann Award in Pharmacology 806

This award, established by Wyeth-Herst Labs, is given in odd-numbered years to commemorate the pioneer work in the United States of Dr. Torald Sollmann in the fields of pharmacological investigation and education. The award is made to a nominee for significant contributions over many years to the advancement and extension of knowledge in the field of pharmacology. The award consists of an honorarium of $3,500, a medal, hotel and economy airfare for the winner and spouse to the annual meeting. The formal presentation of this biennial award and medal will be made at the annual meeting of ASPET. The recipient will be invited by the President of the Society to deliver a lecture to the membership that may be published in an appropriate ASPET journal.
Requirements: Nominations must be made by ASPET members; however, the nominee need not be a member. There are no restrictions as to age or institutional affiliation.
Date(s) Application is Due: Sep 15
Amount of Grant: 3,500 U.S.D
Contact: Christine Carrico, Executive Officer; (301) 634-7060; fax (301) 634-7061; ccarrico@aspet.org or info@aspet.org
Internet: http://www.aspet.org/public/awards/sollmann_award.html
Sponsor: American Society for Pharmacology and Experimental Therapeutics
9650 Rockville Pike
Bethesda, MD 20814-3995

ASPET Travel Awards 807

Graduate Student Travel Awards, Minority Graduate Student Travel Awards, and Summer Undergraduate Student Travel Awards are a fixed sum of $600 plus registration. Young Scientist Travel Awards and Minority Young Scientist Travel Awards are a fixed sum of $600 plus registration. All travel awards are inclusive of registration and are given to partially defray the costs of travel and housing. The check for the award and an award certificate will be presented at the annual ASPET Opening Awards Ceremony. Applications will be available in late September.
Requirements: Applications must be submitted electronically online or as email attachments.
Date(s) Application is Due: Dec 1
Amount of Grant: 600 U.S.D
Contact: Christine Carrico, Executive Officer; (301) 634-7060; fax (301) 634-7061; ccarrico@aspet.org or travelawards@aspet.org
Internet: http://www.aspet.org/public/awards/awards_fellowships.html#Travel_Awards
Sponsor: American Society for Pharmacology and Experimental Therapeutics
9650 Rockville Pike
Bethesda, MD 20814-3995

ASPH/CDC Allan Rosenfield Global Health Fellowships 808

In recognition of a career dedicated to improving the health of the world's most vulnerable populations, the Association of Schools of Public Health (ASPH), with support from the Centers for Disease Control and Prevention (CDC), established a fellowship in honor of Dr. Allan Rosenfield, former Dean of the Columbia University Mailman School of Public Health. As a global health pacesetter Dr. Rosenfield addressed many issues, ranging from women's health to HIV/AIDS to advocating healthcare for the poor and disadvantaged. His commitment to improving lives in neighborhoods not only in America, but in far away places such as rural sub-Saharan Africa and Asia, is an example ASPH hopes students will learn to follow during the period of their fellowships and throughout their public health careers. Students selected as Allan Rosenfield Global Health Fellows are expected to involve themselves, as Dr. Rosenfield did throughout his career, in all aspects of global health. The training offered through this program will expand the global health protection workforce and provide fellows an opportunity to gain practical, first-hand experience on the front lines of international public health and to learn from leading global health experts. One- or two-year international fellowships will be placed within a global health program at CDC headquarters in Atlanta or in one of the fifty foreign countries where CDC is working with the Ministry of Health and other Public Health partners. Additional travel during the fellowship period may be required depending on program activities and priorities. Exact geographical locations will not be determined until the final placement process. At that time, successful applicants will be offered positions and notified of all details for that particular position. Applicants must apply through ASPH's online grant application system (click on the APPLICATION tab at the top of the ASPH fellowship website) and mail required hard copy and supporting documents to ASPH. Required application materials include an essay, a resume or curriculum vitae, two letters of recommendation (from the references provided in the online application), official graduate transcripts, and proof of residency. Applicants are encouraged to review requirements for submitting these documents at the fellowship website as well as to read the FAQ section of the website for more information about the fellowship. Deadlines may vary from year to year; however, requests for application are announced in ASPH's weekly electronic newsletter "The Friday Letter" which applicants may register to receive at no cost. The subscription link is available at http://fellowships.asph.org/ from the bottom right hand corner of the screen. Additionally, applicants may check the same website for updates on available fellowships.
Requirements: Applicants must have received their Masters or Doctorate degree no later than the August prior to the beginning of the fellowship or no earlier than May within the last five years. Graduate degrees must come from an ASPH-member graduate school of public health accredited by the Council on Education for Public Health (CEPH). (A list of accredited schools with full ASPH membership can be accessed via the sponsor website.) Prior to leaving for the fellowship assignment locations, fellows will be required to travel to CDC in Atlanta, Georgia for a ten-day orientation in mid-September. Fellowship assignments will begin following orientation.
Restrictions: All applicants must be U.S. citizens or hold a visa permitting permanent residence ("Green Card") in the U.S. to be eligible for the fellowship program. Applicants may apply to any number of ASPH fellowship programs; however applicants may apply to a maximum of only three positions within each program.
Geographic Focus: All States, All Countries
Date(s) Application is Due: Feb 16
Amount of Grant: 43,000 - 52,000 U.S.D
Contact: Binley G. Taylor, Coordinator; (202) 296-1099; fax (202) 296-1252; btaylor@asph.org or trainingprograms@asph.org
Internet: https://fellowships.asph.org/programs/details.cfm?programID=!%260%20%20%0A
Sponsor: Association of Schools of Public Health
1900 M Street NW, Suite 710
Washington, DC 20036

ASPH/CDC Public Health Fellowships 809

The ASPH/CDC Public Health Fellowships were established in 1995 to address emerging needs of public health and to provide leadership and professional opportunities at the Centers for Disease Control and Prevention (CDC) for graduate students of members of the Association of Schools of Public Health (ASPH). The fellowships are sponsored cooperatively by ASPH and CDC. Positions are full-time opportunities that last one to two years depending on the needs of CDC and the fellow. Fellowships typically commence in August and are located at CDC headquarters in Atlanta, Georgia. The types of fellowships will vary according to specific areas of research or training within CDC's Centers/Institutes/Offices (CIOs). Selected fellows will have the opportunity to apply graduate education in the field, develop professional experience and expertise, and contribute to public health research and practice. Fellows will receive a stipend payment for the duration of the fellowship on a biweekly basis based on the education level/degree requirement of the fellowship position. In addition to the fellowship stipend, the fellow will have access to a sufficient "fellowship allowance" that is intended to cover health/dental/vision insurance premiums during the fellowship period. In addition, some positions may offer an additional "travel/training allowance" that may be used to cover project-related travel, tuition, journal subscriptions, association dues, etc. (not all positions will have a travel/training allowance). The fellowship stipend is intended to cover all living expenses including housing and project-related travel. Fellows are

responsible for their own move to and from the fellowship site (relocation expenses are not covered). Typically, the ASPH application season opens in November and runs through March. Requests for application are announced in ASPH's weekly electronic newsletter "The Friday Letter" which applicants may register to receive at no cost. The subscription link is available at http://fellowships.asph.org/ from the bottom right hand corner of the screen. Additionally, applicants may check the same website for updates on available fellowships. Interested applicants must apply through ASPH's online grant application system (click on the APPLICATION tab at the top of the ASPH fellowship website) and mail required hard copy and supporting documents to ASPH. Required application materials include an essay, a resume or curriculum vitae, two letters of recommendation (from the references provided in the online application), official graduate transcripts, and proof of residency. Applicants are encouraged to review requirements for submitting these documents at the fellowship website.
Requirements: Students must have received an M.P.H. or Doctorate degree from an ASPH-member, CEPH-accredited graduate school of public health prior to the beginning of the fellowship. A list of accredited schools with full ASPH membership can be accessed via the sponsor website. Early career professionals with M.P.H. or Doctorate degrees (within 5 years of graduation) may also apply for the fellowship program. Selected fellows are required to relocate to Atlanta. Fellows are also required to maintain their own health insurance while in the fellowship program.
Restrictions: The ASPH/CDC Public Health Fellowships are available only to U.S. citizens or students holding a visa permitting permanent residence in the U.S. Applicants may apply to any number of ASPH fellowship programs; however applicants may apply to a maximum of only three positions within each program.
Amount of Grant: 43,000 - 52,000 U.S.D
Contact: Binley G. Taylor, Coordinator; (202) 296-1099; fax (202) 296-1252; btaylor@asph.org or trainingprograms@asph.org
Internet: https://fellowships.asph.org/Programs/details.cfm?programID=%22%26M8%20%0A
Sponsor: Association of Schools of Public Health
1900 M Street NW, Suite 710
Washington, DC 20036

ASPH/CDC Public Health Internships 810
The ASPH/CDC Public Health Internship Program is a collaborative effort between the Association of Schools of Public Health (ASPH) and the Centers for Disease Control and Prevention (CDC). The program enables schools of public health to assign graduate students in public health to the CDC for full-time internships in which "graduate candidates and early career professionals with graduate degrees can practice applying skills and knowledge learned in the classroom and field." The internships take place for ten weeks over the summer. Internship positions will be located either at CDC headquarters in Atlanta, Georgia, or at CDC geographical locations throughout the United States. Each intern will receive a training stipend for the ten-week assignment, paid in four equal installments. In addition to possibly fulfilling an internship requirement, students will benefit from: exposure to state-of-the-art information; acquisition of skills and knowledge useful to their careers; and interaction with technical experts in their chosen fields. Upon completion of the internship, the selected applicant will have made a useful contribution to a project of public health importance that is related to the mission of CDC. In addition, the selected applicant will have carried out scientific work that, if applicable, may be acceptable to the sponsoring academic institution for degree credit. Typically, the ASPH application season opens in November and runs through March. Requests for application are announced in ASPH's weekly electronic newsletter "The Friday Letter" which applicants may register to receive at no cost. The subscription link is available at http://internships.asph.org/. Additionally, applicants may check the ASPH Internship website for updates on available internships (http://internships.asph.org/). Prospective interns must apply through ASPH's online grant application system (available at the internship website) and mail required hard copy and supporting documents to ASPH. Required application materials include an essay, a resume or curriculum vitae, two letters of recommendation (from the references provided in the online application), official graduate transcripts, and proof of residency. Applicants are encouraged to review requirements for submitting these documents at the internship website.
Requirements: Applicants must be currently enrolled in a Masters or Doctoral program at an ASPH-member graduate school of public health accredited by the Council on Education for Public Health (CEPH.) A list of ASPH members is available from the sponsor website. Applicants must be students at the time of application, although they may have graduated prior to entering the program. Applicants must be U.S. citizens or hold visas permitting permanent residence in the U.S. Selected interns are required to relocate to the internship location. Relocation costs of any kind will not be covered by ASPH or CDC.
Restrictions: Applicants may apply to as many ASPH internships programs as they like (that are accepting applications). Additionally applicants may apply to up to three positions per program.
Contact: Binley G. Taylor, Coordinator; (202) 296-1099; fax (202) 296-1252; trainingprograms@asph.org or btaylor@asph.org
Internet: https://internships.asph.org/programs/details.cfm?programID=!%230%20%20%0A
Sponsor: Association of Schools of Public Health
1900 M Street NW, Suite 710
Washington, DC 20036

ASP.O. Daiichi Innovative Technology Grant 811
The Daiichi Innovative Technology Grant will fund exploratory or hypothesis-generating projects that are not well-suited to a formal grant application (eg, based on statistical analysis and sample size specification) such as: development of new surgical or diagnostic instruments; survey or quality of life measures; new use of internet technology or computer software such as CDs or DVDs; educational brochures, materials, software for patients or physicians; and other applications of innovative technology for education or research in pediatric otolaryngology. Full patent and copyright control must be retained by the applicant and the applicant's institution. If patented innovations funded by this award generate more than $5,000, the applicant may be required to return funds to the Society. One award of up to $5,000, to be used over a one-year period, will be given annually. Letters of Intent should be received no later than December 15, with full proposals due on January 15.
Requirements: Researchers (MD, PhD, DMD, DC) in disciplines who will conduct research directly relevant to innovative technology in pediatric otolaryngology are eligible to apply. Applications submitted by otolaryngologists or demonstrating collaborations with otolaryngologists are preferred. Participation of an ASP.O. member is not required, but is preferred.
Date(s) Application is Due: Jan 15; Dec 15
Amount of Grant: 5,000 U.S.D
Contact: Stephanie L. Jones, Assistant Director; (703) 519-1586 or (703) 836-4444; sljones@entnet.org
Internet: http://www.aspo.us/information.php?info_id=12
Sponsor: American Society for Pediatric Otolaryngology
One Prince Street
Alexandria, VA 22314-3357

ASP.O. Fellowships 812
ASP.O. exists to foster excellence in the care of children with otorhinolaryngologic disorders through education and research and thereby enhance the profession of Pediatric Otolaryngology. Its fellowship program intends to: facilitate the creation and dissemination of knowledge about the care of infants and children with ORL disorders; serve as an advocate for infants and children with ORL disorders, ASP.O. members and others with shared goals; and preserve and promote dedication to excellent and humane care for infants and children with ORL disorders. See the web site for details and all available fellowships.
Contact: Stephanie L. Jones, Assistant Director; (703) 519-1586 or (703) 836-4444; sljones@entnet.org
Internet: http://www.aspo.us/fellowships.php
Sponsor: American Society for Pediatric Otolaryngology
One Prince Street
Alexandria, VA 22314-3357

ASP.O. Research Grants 813
The American Society of Pediatric Otolaryngology (ASP.O.) awards funds annually to support innovative research in pediatric otolaryngology. ASP.O. will consider applications from both individuals and institutions. Preference is given to proposed projects that are to be completed within one year, although exceptional proposals that have duration in excess of one year will be considered. Two awards of up to $15,000, to be used over a one-year period, will be given annually. Letters of Intent should be received no later than December 15, with full proposals due on January 15.
Requirements: Researchers (MD, PhD, DMD) in disciplines who will conduct research directly relevant to pediatric otolaryngology are eligible to apply.
Restrictions: No portion of any grant may be used for travel expenses or for principal investigator salaries.
Date(s) Application is Due: Jan 15; Dec 15
Amount of Grant: 15,000 U.S.D
Contact: Stephanie L. Jones, Assistant Director; (703) 535-3747 or (703) 836-4444; sljones@entnet.org
Internet: http://www.aspo.us/information.php?info_id=12
Sponsor: American Society for Pediatric Otolaryngology
One Prince Street
Alexandria, VA 22314-3357

ASSE Safety Research Grants 814
The foundation supports safety and health research projects, particularly in the following areas: effective interventions for prevention of traumatic injuries; effectiveness of incentive programs; developing management commitment to safety/risk control; criteria to measure risk control effectiveness; effective safety programs for temporary workers; proactive incorporation of risk control principles in early design of facilities, processes, and management systems; study cultures within organizations with outstanding safety records; and safety in public places.
Date(s) Application is Due: Oct 1
Amount of Grant: 20,000 U.S.D
Samples: Dr. Daniel Della-Giustina, West Virginia U—for the project How to Cope with Terrorism in Schools, $3000; N. Kumar Kittusamy, Marshall U—for the project Assessment of Ergonomic Risk Factors among Operators of Heavy Earth Moving Machinery, $3000.
Contact: Mary Goranson; (847) 768-3412; fax (847) 296-3769; mgoranson@asse.org
Internet: http://www.asse.org/foundat.htm
Sponsor: American Society of Safety Engineers
1800 E Oakton Street
Des Plaines, IL 60018-2187

Assisi Foundation of Memphis Capital Project Grants 815
The foundation supports organizations in its areas of interest, including health and human services—promote the health and well-being of the Mid-South community and help the health care system respond more effectively to community needs; education and literacy—projects/programs that build organizational capacity of provider agencies, provide professional development to service providers, promote collaboration among

provider agencies, and leverage resources (local, state, and federal); social justice/ethics—projects/programs that strengthen ethical values among Mid-South citizens and promote social justice leading to a better understanding of and a more effective response to economic or social threats to the community; and cultural enrichment and the arts—projects/programs that foster an appreciation of the arts in the Greater Memphis community. Religious organizations seeking funding for religious programs also are eligible. Typical capital projects include: building - new construction, addition to existing facility, or renovation; technology - Information Management System installation/upgrade, computer hardware, audio-visual equipment/systems; and furnishings/equipment. Typically, payment of capital project grants is made when the organization begins the construction/renovation. Specific terms and schedule of payments will be based upon the scope of the project, amount of award, duration of the project, and completion of required reporting at appropriate intervals.
Requirements: The Foundation makes grants only to organizations that are classified as tax-exempt under Section 501(c)3 of the Internal Revenue Code and as public charities under Section 509(a) of that Code. The Foundation uses its resources for charitable endeavors that advance the well-being of people and institutions located in Shelby, Fayette, and Tipton Counties in Tennessee; Crittenden County, Arkansas; and Desoto County, Mississippi.
Restrictions: Grants are not made for individuals, national fundraising drives, projects that address the needs of only one congregation, tickets for benefits, political organizations or candidates for public office, lobbying activities, recurring budget deficits, or tournament fees and/or travel for athletic competitions.
Geographic Focus: Arkansas, Mississippi, Tennessee
Date(s) Application is Due: Feb 15; May 17; Aug 16; Nov 15
Contact: Jan Young, Executive Director; (901) 684-1564; jyoung@assisifoundation.org
Internet: http://www.assisifoundation.org/capitalproject.html
Sponsor: Assisi Foundation of Memphis
515 Erin Drive
Memphis, TN 38117

Assisi Foundation of Memphis General Grants **816**
The foundation supports organizations in its areas of interest, including health and human services—promote the health and well-being of the Mid-South community and help the health care system respond more effectively to community needs; education and literacy—projects/programs that build organizational capacity of provider agencies, provide professional development to service providers, promote collaboration among provider agencies, and leverage resources (local, state, and federal); social justice/ethics—projects/programs that strengthen ethical values among Mid-South citizens and promote social justice leading to a better understanding of and a more effective response to economic or social threats to the community; and cultural enrichment and the arts—projects/programs that foster an appreciation of the arts in the Greater Memphis community. Religious organizations seeking funding for religious programs also are eligible. Deadlines are set to coordinate with quarterly meetings of the Board of Directors of the Foundation.
Requirements: The Foundation makes grants only to organizations that are classified as tax-exempt under Section 501(c)3 of the Internal Revenue Code and as public charities under Section 509(a) of that Code. The Foundation uses its resources for charitable endeavors that advance the well-being of people and institutions located in Shelby, Fayette, and Tipton Counties in Tennessee; Crittenden County, Arkansas; and Desoto County, Mississippi.
Restrictions: Grants are not made for individuals, national fundraising drives, projects that address the needs of only one congregation, tickets for benefits, political organizations or candidates for public office, lobbying activities, recurring budget deficits, or tournament fees and/or travel for athletic competitions.
Geographic Focus: Arkansas, Mississippi, Tennessee
Date(s) Application is Due: Feb 15; May 17; Aug 16; Nov 15
Amount of Grant: Up to 20,000 U.S.D
Contact: Jan Young, Executive Director; (901) 684-1564; jyoung@assisifoundation.org
Internet: http://www.assisifoundation.org/generalgrants.html
Sponsor: Assisi Foundation of Memphis
515 Erin Drive
Memphis, TN 38117

Assurant Foundation Grants **817**
The Foundation awards approximately 15 grants per year to improve the quality of life in the New York metropolitan community. The Foundation is committed to fulfilling the basic needs of the community, be they educational, health, nutrition, or housing.
Requirements: 501(c)3 nonprofits in the New York City metropolitan area are eligible.
Restrictions: As a general policy the foundation does not make grants to individuals; political parties, candidates or lobbying activities; religious organizations; or fundraising events.
Geographic Focus: New York
Amount of Grant: Up to 5,000 U.S.D
Contact: Kristy Hall, (212) 859-7000; Kristy.Hall@assurant.com
Internet: http://www.assurant.com/inc/assurant/community/new-york.html
Sponsor: Assurant Foundation
1 Chase Manhattan Plaza, 41st Floor
New York, NY 10005

Assurant Health Foundation Grants **818**
The foundation awards grants in Wisconsin in its areas of interest, including health care access and promotion, disease prevention, the arts and education, as well as the foundation's national public policy initiatives. Types of support include capital campaigns, consulting services, continuing support, employee matching gifts, employee-related scholarships, in-kind gifts, matching/challenge support, and program development.
Requirements: Wisconsin nonprofit organizations are eligible. Preference is given to requests serving southeastern Wisconsin.
Restrictions: Unsolicited requests for funds not accepted. Requests for funds only accepted after the foundation sends out requests for proposals.
Geographic Focus: Wisconsin
Samples: Gerald Ignace Clinic (WI)—$10,000; Task Force on Family Violence (WI)—$10,000; Latino Arts Organization (WI)—$5000.
Contact: Grants Administrator; (414) 299-1348; megan.hindman@assurant.com
Internet: http://www.newsroom.eassuranthealth.com/corp/news/newsroom/articles/press_06092005.htm
Sponsor: Assurant Health Foundation
501 W Michigan Street
Milwaukee, WI 53203

Astellas Foundation Research Grants **819**
The Foundation was established with the objectives of contributing to the improvement of national health and welfare, and to the progress of therapeutic medicines. In fulfilling the objectives, the Foundation aims to conduct pioneering research in new fields of diseases and metabolisms of medicines, by elucidating the mechanism of diseases and their treatment, in particular the relationship between therapeutic medicines and in-situ metabolisms. Major activities include: subsidizing research on metabolic disorders related to the diagnosis of diseases and therapeutic medicine; publication and editing of periodicals for research achievements; supporting and production of educational movies; and other projects to achieve the objectives of the Foundation.
Contact: Grants Manager; 81 3 3244 3397; fax 81 3 5201 8512
Sponsor: Astellas Foundation
2-3-11 Nihonbashihoncho, Chuo-ku
Tokyo, 103-8411 Japan

AT&T Foundation Civic and Community Service Program Grants **820**
The foundation supports programs that enhance education by integrating new technologies and increasing learning opportunities, improve economic development through technology and local initiatives, provide vital assistance to key community-based organizations, support cultural institutions that make a community unique, and advance the goals and meet the needs of diverse populations.
Requirements: The foundation makes 501(c)3 grants to tax-exempt, nonprofit organizations in certain states. Requirements are available online.
Samples: Leader Dogs for the Blind (Rochester, Mich.) received a $25,000 grant for blind and visually impaired individuals. Communities in Schools of San Antonio - received a grant to enhance the technology infrastructure to reach more young people at risk of dropping out of school.
Contact: Marilyn Reznick, (212) 387-6555; fax (212) 387-5097; reznick@att.com
Internet: http://www.att.com/gen/corporate-citizenship?pid=7736&DCMP=att_foundation
Sponsor: AT&T Foundation
32 Avenue of the Americas, 24th Floor
New York, NY 10013

Atherton Family Foundation Grants **821**
The foundation supports the establishment, maintenance, or promotion of such religious, charitable, educational, benevolent, scientific, or other like purposes in Hawaii. Types of support include general operating grants, equipment acquisition, capital fund projects, seed grants, and theology scholarships to children of ministers in the state. The grant period generally is one year, though some multiyear grants will be awarded. Applications are due one month prior to board meetings. Contact the office for application procedures.
Requirements: Hawaii nonprofit organizations are eligible.
Restrictions: Requests for annual operating support or mature programs, an organization that has already received a grant in the same calendar year (with some exceptions), individuals or for the benefit of specific individuals, loans, grants to endowment funds, private foundations, purpose of lobbying, conferences, festivals and one-time events, re-granting, or individual Department of Education schools not typically funded.
Geographic Focus: Hawaii
Date(s) Application is Due: Feb 1; Apr 1; Aug 1; Oct 1; Dec 1
Amount of Grant: 2,000 - 75,000 U.S.D
Samples: La Pietra-Hawaii School for Girls (Honolulu, HI)—for capital campaign, $112,000; Le Jardin Academy (HI)—to purchase computer network equipment, $5000; Hawaii State Committee on Family Violence (HI)—for general operating support, $4000.
Contact: Grant Coordinator; (808) 566-5524 or (888) 731-3863, ext. 524; fax (808) 521-6286; foundations@hcf-hawaii.org
Internet: http://www.Athertonfamilyfoundation.org
Sponsor: Atherton Family Foundation
1164 Bishop Street, Suite 800
Honolulu, HI 96813

Athwin Foundation Grants **822**
Established in 1956, the Athwin Foundation serves the Minneapolis, St. Paul, Bloomington, Minnesota, Wisconsin area. Primary areas of interest include: arts and culture, church related projects, education, and social services. Providing support in the form of: capital campaigns, general/operating support and, program development. Contact the Foundation by submitting a Letter of Inquiry, if they are interested in your project, a formal request will be made & then a proposal maybe submitted.
Requirements: IRS 501 (c)3 non-profit organizations are eligible to apply. The foundation prefers applicants to use the Minnesota Common Grant Application Form that can be

obtained by calling the Minnesota Council on foundations (612) 338-1989 or by visiting their web site.
Restrictions: Grants do not support: individuals, scholarships, fellowships, or loans.
Geographic Focus: Minnesota, Wisconsin
Date(s) Application is Due: Mar 1
Amount of Grant: 2,000 - 100,000 U.S.D
Samples: Foundation of Childrens Hospitals and Clinics of Minnesota, Roseville, MN, $100,000— for general operating support; Neighborhood Involvement Program, Minneapolis, MN, $15,000—for general operating support; Jungle Theater, Minneapolis, MN, $2,000—for general operating support.
Contact: Bruce Bean, Trustee; (952) 915-6165
Sponsor: Athwin Foundation
5200 Wilson Road, Suite 307
Minneapolis, MN 55424-1344

Atlanta Foundation Grants 823

Grants are awarded to assist Georgia charitable and educational institutions to improve the quality of life in Fulton and DeKalb Counties. Primary areas of interest include education, cultural programs, housing, and other charitable giving. Additional areas of support include adult basic education and literacy training; higher education; health care organizations; recreation; and youth, family, and human services. The board meets in August to consider requests.
Requirements: Nonprofit organizations in Georgia's DeKalb and Fulton Counties may apply for grant support.
Restrictions: Grants are not awarded to individuals or for scholarships, fellowships, or loans.
Geographic Focus: Georgia
Date(s) Application is Due: Mar 1; Sep 1
Amount of Grant: 2,500 - 50,000 U.S.D
Contact: Trustee, c/o Wachovia Bank of Georgia NA; grantinquiriesga@wachovia.com
Internet: http://www.wachovia.com/corp_inst/charitable_services/0,,4269_3296,00.html
Sponsor: Atlanta Foundation
3414 Peachtree Road, 5th Floor, MC GA8023
Atlanta, GA 30326

Atran Foundation Grants 824

The foundation supports Jewish nonprofits with a focus on New York and Israel in the areas of higher education, Jewish education, religious education, community services, Jewish welfare, temples, international ministries and missions, temples, medical centers, and women's affairs. Types of support include conferences and seminars, endowment funds, general support, matching funds, multiyear/continuing support, project support, research, scholarships, and seed money grants. The foundation requires that proposals be in writing. Proposals should include the nature of the project, its objectives and significance, time estimate, and budget.
Requirements: 501(c)3 tax-exempt Jewish organizations are eligible.
Restrictions: Grants are not made to individuals.
Date(s) Application is Due: Sep 30
Amount of Grant: 250 - 100,000 U.S.D
Samples: Yivo Institute for Jewish Research (New York, NY)—for project support, $90,000; Brandeis University (Waltham, MA)—for operating support, $28,000; Folksbiene Yiddish Theater (New York, NY)—for operating support, $15,000.
Contact: Diane Fischer, President; (212) 505-9677
Sponsor: Atran Foundation
23-25 E 21st Street, 3rd Floor
New York, NY 10010

AUAF/NIH Intramural Urology Research Training Grants 825

This intramural program, developed by NIDDK and NCI in collaboration with AFUD, provides an opportunity for selected individuals to complete a research project under the direction of a senior investigator at NIDDK or NCI. Two positions each are available in NIDDK and NCI for laboratory-based research. Up to two positions are available in the surgery branch of NCI, of which six months are dedicated to clinical training in urologic oncology. The program consists of a two-year fellowship with possible extension for a third year. Applications are available on the Web site. Annual deadline dates may vary; contact program staff for exact dates.
Requirements: Eligibility is open to physicians in urology residency programs, or those who have recently completed a urology residency, who desire intensive exposure to research experience utilizing basic molecular and cellular biologic techniques in projects relevant to urology. Applicants must have less than five years of postdoctoral research experience (clinical residency training is not counted as postdoctoral research).
Date(s) Application is Due: Sep 1
Amount of Grant: 40,000 - 44,000 U.S.D
Contact: Kym Liddick, Director; (410) 468-1812; fax (410) 468-1808; kym@afud.org
Internet: http://www.afud.org/research/index.asp
Sponsor: American Urologic Association Foundation
1128 N Charles Street
Baltimore, MD 21201

AUAF Health Services Research Scholar Grants in Urology 826

The program supports urology residents or trained urologists (MD/DO) within five years of residency who aspire to conduct research in health care services as it relates to urology. The candidate must be committed to a career in academic urology and able to provide evidence of current interest and/or accomplishments in matters relating to the delivery of health services and health economics and policy. The project should include didactic experience and may include plans for obtaining an advanced degree in public health, health policy, health financing and economics, or other relevant areas. In view of the multidisciplinary nature of health services research and health care policy, the proposal should include an explanation of how researchers from different clinical and nonclinical disciplines will be involved in the research project. The candidate must dedicate a minimum of 80 percent of his/her time to the research project. This experience should prepare the scholar for a subsequent career in academic urology as it relates to health care policy, health financing, and health economics. Applications are available on the Web site. Annual deadline dates may vary; contact program staff for exact dates.
Requirements: An accredited medical education research institution and/or department within such an institution must sponsor the candidate A preceptor, an established health services researcher in a urology department or a nonurology clinician/researcher in a school of public health/health policy, must have prior experience in health services research with funded grant support and must make a two-year commitment to sponsor the scholar.
Date(s) Application is Due: Sep 1
Amount of Grant: 44,000 U.S.D
Contact: Anthony Caputi, Manager, Research Scholar Program; (800) 828-7866; fax (410) 468-1808; anthony@afud.org
Internet: http://www.afud.org/research/index.asp
Sponsor: American Urologic Association Foundation
1128 N Charles Street
Baltimore, MD 21201

AUAF PhD Research Scholars Program 827

These fellowships are offered to postdoctoral basic scientists with a research interest in urologic or related diseases and dysfunctions. Research projects focus on many different diseases including, but not limited to: prostate cancer, bladder cancer, kidney stones, incontinence, impotence, BPH, prostatitis, urinary tract infections, interstitial cystitis, and pediatric dysfunctions. A commitment to dedicate two years in the program as a full-time researcher is required. The sponsoring urology department is required to provide space, laboratory equipment, and supplies to enable the recipient to perform the proposed research. Continued support is subject to yearly review and evaluation by the scientific and education committee. Application forms are available on the Web site. Annual deadline dates may vary; contact program staff for exact dates.
Requirements: An accredited medical education research institution or department within such an institution must sponsor the candidate by guaranteeing research time commitment and adequate financial support, including responsibility for the adequacy of the scientific environment, laboratory equipment, and supplies to perform the proposed research.
Date(s) Application is Due: Sep 1
Amount of Grant: 10,000 - 60,000 U.S.D
Contact: Anthony Caputi, Manager, Research Scholar Program; (410) 468-1803; fax (410) 468-1808; anthony@afud.org
Internet: http://www.afud.org/research/index.asp
Sponsor: American Urologic Association Foundation
1128 N Charles Street
Baltimore, MD 21201

AUAF Practicing Urologist Research Awards 828

To aid and encourage practicing urologists with research ideas to undertake collaborative investigations at a urological research laboratory, the AUAF grants research awards to be matched by the sponsoring institution. Applications are available on the Web site. Annual deadline dates may vary; contact program staff for exact dates.
Requirements: An accredited medical research institution or department must sponsor the candidate by guaranteeing adequate support, including responsibility for the adequacy of the environment, laboratory equipment, and supplies needed to perform the proposed research.
Date(s) Application is Due: Sep 1
Amount of Grant: 5,000 U.S.D
Contact: Anthony Caputi, Manager, Research Scholar Program; (410) 468-1803; fax (410) 468-1808; anthony@afud.org
Internet: http://www.afud.org/research/index.asp
Sponsor: American Urologic Association Foundation
1128 N Charles Street
Baltimore, MD 21201

AUAF Research Scholar Program 829

The fellowship program supports urology residents in training or post residency (not more than five years) for one year of full-time investigation in laboratory research of urologic diseases. An established basic science academician in either a basic science or clinical department with substantial experience in grant support and mentoring of scientific trainees must make a one-year commitment to sponsor the scholar. One-year awards can be renewed for a second year. Annual deadline dates may vary; contact program staff for exact dates.
Requirements: An accredited medical education research institution or department within such an institution must sponsor the candidate by guaranteeing research time commitment and adequate financial support, including responsibility for the adequacy of the scientific environment, laboratory equipment, and supplies to perform the proposed research.
Date(s) Application is Due: Sep 1
Amount of Grant: 25,000 U.S.D
Contact: Anthony Caputi, Manager, Research Scholar Program; (410) 468-1803; fax (410) 468-1808; anthony@afud.org
Internet: http://www.afud.org/research/scholars/current.asp
Sponsor: American Urologic Association Foundation
1128 N Charles Street
Baltimore, MD 21201

Auburn Foundation Grants 830

The purpose of the Foundation is to stimulate giving and cooperative leadership among the citizens of Auburn; help improve the lives of all community residents, especially those who are most vulnerable; and enrich the cultural environment and community life. Of special interest are projects that bring together all ages and sections of the town, or that contribute to healthy, active living. Application is available online.

Requirements: Any nonprofit organization that serves residents of Auburn is invited to apply.

Restrictions: Grants will not be awarded to for-profit businesses or expenses already incurred by the applicant.

Geographic Focus: Massachusetts

Date(s) Application is Due: Apr 15

Amount of Grant: 5,000 U.S.D

Samples: Bancroft School (Worcester, MA)—for general support, $45,000; New England Science Ctr, Rutland House (Worcester, MA)—for repairs, $20,000; Tower Hill Botanic Garden (Boylston, MA)—for general support, $30,000.

Contact: Lois Smith, Senior Program Officer; (508) 755-0980, ext. 107; lsmith@greaterworcester.org

Internet: http://www.greaterworcester.org/grants/Auburn.htm

Sponsor: Auburn Foundation
370 Main Street, Suite 650
Worcester, MA 01608-1738

Audrey and Sydney Irmas Charitable Foundation Grants 831

The foundation awards grants to eligible nonprofit organizations in its areas of interest, including arts and culture, higher education, homeless and urban issues, hospitals, and Jewish welfare. Grants support long-term, nonrenewable pledges and general operating grants. Most grants are made to Los Angeles County, CA, nonprofits.

Requirements: 501(c)3 nonprofit organizations are eligible.

Geographic Focus: California

Amount of Grant: 1,000 - 75,000 U.S.D

Samples: Childrens Hospital of Los Angeles (Los Angeles, CA)—$2810.65; Jewish Family Services (Los Angeles, CA)—$500; Downtown Womens Center (Los Angeles, CA)—$200; Carnegie Museum of Art (Pittsburg, PA)—$25,000.

Contact: Robert Irmas, Administrator; (818) 382-3313; fax (818) 382-3315; robirm@aol.com

Sponsor: Audrey and Sydney Irmas Charitable Foundation
16830 Ventura Boulevard, Suite 364
Encino, CA 91436-2797

AUPHA Baxter International Foundation Prize for Health Services Research 832

The prize acknowledges national or international contributions of health services research, defined as a product of the application of analytic methods to the organization, financing, and/or delivery of health services. An individual's specific contribution or a career-long achievement may be recognized. The prize recognizes a person who has made significant and demonstrable contributions to the health of the public in three primary focus areas: health services management, health policy development, and health care delivery.

Requirements: Nominations must include a letter of nomination, two letters of recommendation, and a current curriculum vita. Nominations should focus on the nominee's contribution in the areas of health services management, health policy development, and health care delivery. Nominations should be sent to and further information can be obtained from the Secretary, HSR Prize Committee, AUPHA, 2000 N. 14th Street, Suite 780, Arlington, VA 22201.

Date(s) Application is Due: Dec 15

Contact: Lydia Reed, (703) 894-0940; lydia.reed@aupha.org

Internet: http://www.e-guana.net/organizations.php3?action=printContentItem&orgid=75&typeID=506&itemID=3744

Sponsor: Association of University Programs in Health Administration
2000 14th Street N, Suite 780
Arlington, VA 22201

Aurora Foundation Grants 833

The foundation provides grants to eligible nonprofit organizations in the following categories: education, social services, health care, and the arts and humanities. In general, grants are made for capital purposes only, not for operating expenses. Evidence of tax-exempt status must be submitted with all applications. Preference will be given to requests that demonstrate the most urgent and immediate need for funding. Grants are ordinarily made for one year only.

Requirements: 501(c)3 and 170(b)1a nonprofit agencies located in the foundation's service area, which includes the City of Aurora, southern Kane County, and Kendall County in Illinois, are eligible.

Restrictions: The foundation rarely provides the entire support of a project. Grants are not generally available for those agencies and institutions that are funded primarily through tax support. Telephone the office prior to submitting a formal proposal.

Geographic Focus: Illinois

Date(s) Application is Due: Aug 1

Samples: Association for Individual Development—van, $10,000. Aurora U (IL)—to renovate a facility and create the Center for Community Enrichment, which will house cultural programs, lectures, and other activities for students and community members, $500,000 over five years.

Contact: Grants Administrator; (630) 896-7800; grant@aurorafdn.org

Internet: http://www.aurorafdn.org/grantmaking.html

Sponsor: Aurora Foundation
111 W Downer Place, Suite 312
Aurora, IL 60506

Austin-Bailey Health and Wellness Foundation Grants 834

The foundation awards grants to Ohio nonprofit organizations to support programs that promote the physical and mental well-being of the citizens of Holmes, Stark, Tuscarawas and Wayne Counties in the state of Ohio. The Foundation emphasizes healthcare affordability concerns of the uninsured adn underinsured, the poor, children, single parents and the aging. It also advocates programs that speak to the mental health needs of the individuals and families. Opportunities to work with other foundations and organizations to promote the principles of health and wellness are welcome.

Requirements: Applicant organizations must provide a current copy of IRS determination indicating 501(c)3 and 509(a) tax-exempt status; serve part or all of the four county area of Holmes, Stark, Tuscarawas, and Wayne; provide a description of the organization and its activities; state the need to be addressed; and provide a brief description of the project or program and how it will address the health and wellness needs of the community.

Restrictions: The foundation generally does not support annual appeals or membership drives; fund-raising events, program advertising, lobbying activities, endowment funds or organizations not classified as tax exempt by the IRS.

Geographic Focus: Ohio

Date(s) Application is Due: Jun 30; Dec 17

Amount of Grant: 1,000 - 15,000 U.S.D

Samples: Western Stark Free Clinic, Massillon, OH, $20,000— provide health care for the homeless delivered via their 40 foot mobile Care-A-Van; Habitat for Humanity, Canton, OH, $12,500—livability costs associated with one home to enable a family with physical disabilities to live independently; Stark County Health Department, Canton, OH, $15,000—operating support for their dental clinic.

Contact: Grants Administrator; (330) 580-2380; fax (330) 580-2381; abfdn@sbcglobal.net

Internet: http://fdncenter.org/grantmaker/austinbailey/guide.html

Sponsor: Austin-Bailey Health and Wellness Foundation
2719 Fulton Road NW, Suite D
Canton, OH 44718

Austin College Leadership Award 835

The award honors an outstanding individual who through his/her life's work has demonstrated the principles of servant leadership by: taking a courageous stand on a public policy issue that advances a humanitarian or educational purpose; serving the youth of a state, nation, or international community to improve the quality of health, educational, or community services; or creating opportunities for young people that help them enhance their educational experience and move to a new level of service to society. The recipient also will visit Austin College to speak to community leaders and interact with students and faculty. Nomination form and guidelines are available online.

Requirements: Nominees will be leaders from all levels whose leadership position will have placed him/her in a unique and special position among leaders throughout the world.

Date(s) Application is Due: Jun 1

Amount of Grant: 100,000 U.S.D

Contact: Award Administrator; (903) 813-2000 or (800) 526-4276; fax (903) 813-3199

Internet: http://www.austincollege.edu/category.asp?3523

Sponsor: Austin College
900 North Grand Avenue
Sherman, TX 75090-4400

Australian Academy of Science Grants 836

The objectives of the Academy are to promote science through a range of activities. It has defined four major program areas as their focus. The Academy focus is on recognition of outstanding contributions to science, education and public awareness, science policy, and international relations. Guidelines are available online.

Geographic Focus: All States, Australia

Contact: Executive Secretary; 61-2-6201-9400; fax 61-2-6201-9494; eb@science.org.au

Internet: http://www.science.org.au

Sponsor: Australian Academy of Science
GP.O. Box 783
Canberra, ACT 2601 Australia

Australian Cancer Foundation for Medical Research Grants 837

The foundation awards grants in Australia in the field of cancer research. This includes, but is not limited to, cell or molecular biology, epidemiology, prevention or drug development. Types of support include capital projects and equipment.

Geographic Focus: All States, Australia

Date(s) Application is Due: May 12

Contact: Lorraine McNuff, (02) 9223 7833; fax (02) 9223 1800; lmcnuff@acrf.com.au

Internet: http://www.acrf.com.au/page/research_grants.html

Sponsor: Australian Cancer Foundation for Medical Research
Strand Arcade, Suite 403
Sydney, NSW 2000 Australia

Australian Institute of Aboriginal and Torres Strait Islander Studies—Research Grants 838

Grants cover costs associated with conduct of research for projects concerned with Australian Aboriginal and Torres Strait Island cultures in fields such as health, human biology, social anthropology, linguistics, ethnomusicology, material culture, rock art, archaeology, ethnobotany, psychology, education, and Aboriginal history including oral history. Annual deadline dates may vary; contact program staff for exact dates.

Requirements: Nationals of any country are eligible for assistance.

Date(s) Application is Due: Dec 21

Samples: Glynn Barratt—for Russian Texts Translation and Annotation project, $9800; Andree Grau—for Tiwi Dance and the Embodiment of Knowledge project, $11,747.
Contact: Research Grants Administrator; 61 2 6246 1145; fax 61 2 6249 7714; grants@aiatsis.gov.au
Internet: http://www.aiatsis.gov.au/rsrch/rsrch_grnts/rg_abt.htm
Sponsor: Australian Institute of Aboriginal and Torres Strait Islander Studies
GP.O. Box 553
Canberra, ACT 2601 Australia

Australian Spinal Research Foundation Grants 839
The foundation invites applications for grants to support research concerning chiropractic. Priority will be given to projects that investigate the vertebral subluxation complex and how chiropractic contributes to wellness. In addition, priority will be given to chiropractic-based research projects that cover topics as diverse as those comparing the clinical outcome of various chiropractic techniques, or the effect of the chiropractic adjustment (e.g. sleep patterns, colic or immune function), or an analysis of how ergonomics may benefit posture, whether this be lying, sitting, or standing.
Geographic Focus: All States, Australia
Date(s) Application is Due: May 31
Amount of Grant: 2,000 U.S.D
Contact: Grants Administrator; 61 (7) 3808 4098; fax 61 (7) 3808 8109; asrf@spinalresearch.com.au
Internet: http://www.spinalresearch.com.au/Research/research.html
Sponsor: Australian Spinal Research Foundation
P.O. Box 1047
Springwood, Queensland QLD 4127 Australia

Autodesk Community Relations Grants 840
Autodesk plays an active role in the communities where employees live and work. Grants are awarded to local nonprofit organizations in the program's areas of interest, including community development, arts, the disabled, education, environment, health and human services, science and technology, and civic affairs. Grants are awarded on a national basis. Preference is given to organizations where Autodesk does business. There are no application deadlines. Written proposal requirements are available online.
Requirements: 501(c)3 tax-exempt organizations are eligible.
Restrictions: Grants do not support sporting events or athletic teams, churches or religious organizations, political parties or organizations, general advertising, video, film or television productions, individuals, subscription fees or admission tickets, fraternal, veteran or sectarian groups or direct funds to Autodesk employees participating in fund-raising events.
Amount of Grant: 1,000 - 3,000 U.S.D
Contact: Community Relations Manager; (415) 507-6138
Internet: http://usa.autodesk.com/adsk/servlet/index?siteID=123112&id=1064603
Sponsor: Autodesk
111 McInnis Parkway
San Rafael, CA 94903

Autzen Foundation Grants 841
The foundation awards grants to Oregon nonprofit organizations in its areas of interest of arts, children/youth, services, environment, health care, higher education, human services, and performing arts. Types of support include building construction/renovation, continuing support, matching/challenge grants, program development, and seed grants.
Requirements: Nonprofit organizations in the Pacific Northwest are eligible with an emphasis on those in Oregon.
Restrictions: Grants do not support individuals, scholarships, fellowships, or loans.
Geographic Focus: Oregon, Washington
Date(s) Application is Due: Mar 15; Aug 15; Nov 15
Amount of Grant: 2,000 - 27,000 U.S.D
Samples: Clackamas Womens Services, Oregon City, OR, $4,000; Christie School, Marylhurst, OR, $15,000; Salvation Army, Cascade Division, Portland, OR, $10,000.
Contact: Kim Freed, Administrator; (503) 226-6051; autzen@europa.com
Sponsor: Autzen Foundation
P.O. Box 3709
Portland, OR 97208-3709

AVDF Health Care Grants 842
This program supports efforts to encourage caring attitudes in the delivery of patient care. Projects should have potential for widespread practical application and should be of interest to other groups. New ideas are encouraged, especially if they facilitate communication between patients (and their families) and doctors, nurses and other caregivers; ameliorate patient anxieties; strengthen trust and cooperation; and foster caring attitudes. Proposals, must have potential for wide application and not merely local improvement of requesting institutions or their communities.
Requirements: Proposals should be submitted by the head of the institution under whose auspices the project would be accomplished.
Amount of Grant: 100,000 - 200,000 U.S.D
Samples: Southern Illinois U School of Medicine (IL)—for a program designed to foster caring attitudes toward aging patients, $195,523; U of Chicago, Pritzker School of Medicine (IL)—to expand the community-based teaching of cross-cultural patient-doctor communication, $199,986; U of Massachusetts Medical School (Worcester, MA)—for efforts to identify the unique attributes of a caring attitude in health-care settings, $198,314.
Contact: Dr. Jonathan Howe, Executive Director; (904) 359-0670; arthurvining@msn.com
Internet: http://jvm.com/davis/PROGRAMS.HTM#health
Sponsor: Arthur Vining Davis Foundations
225 Water Street, Suite 1510
Jacksonville, FL 32202-5185

Avon Foundation Breast Care Fund Grants 843
The fund awards grants for the development of breast cancer education and outreach activities, particularly breast health education and early detection services for under-served women. AFBCF supports programs that: recruit women for both first time screening and annual screening; develop partnerships between community-based outreach providers and local medical providers; work with health care providers to ensure proper clinical follow-up of abnormal screening results; and educate older women (65+ years old) about Medicare coverage of annual screening mammograms and assist them in obtaining the service from providers who accept Medicare.
Requirements: Private, non-government, nonprofit organizations in the United States and Puerto Rico are eligible. Grants are awarded to community-based programs and/or health care agencies that provide medically under-served women aged 40 and older with direct access to breast cancer education, annual clinical screening services, and prompt follow-up care. All programs must utilize the three-part approach to breast cancer early detection including regular screening mammography, clinical breast examination, and monthly breast self-examination.
Date(s) Application is Due: Aug 28
Amount of Grant: 30,000 - 50,000 U.S.D
Samples: Family Services Center of Coffee County, Enterprise, AL; The Witness Project, University of Arkansas for Medical Sciences, Little Rock, AR; Latino Community Development Agency, Oklahoma City, OK.
Contact: Coordinating Center; (212) 244-5368; fax (212) 695-3081; admin@avonbreastcare.org
Internet: http://www.avonbreastcare.org/fundinginfo.htm
Sponsor: Avon Foundation
505 Eighth Avenue, Suite 1601
New York, NY 10018-6505

Avon Products Foundation Grants 844
The foundation's two-fold focus is to support education, community and social services, and arts organizations and programs that provide economic opportunities for women and girls; and to support breast cancer and other women's health organizations and programs. The foundation awards grants in cities and regions with a large concentration of representatives and business operations, with the majority of funds going to U.S.-based institutions. National, international, and New York metropolitan area programs are administered through the foundation's headquarters in New York. For regional funding support, refer to the Avon Foundation website.
Requirements: Applying organizations must be tax-exempt; national and municipal organizations are eligible. Request the guidelines brochure prior to submitting a formal proposal.
Restrictions: Grants do not support individuals; memberships; lobbying organizations; political activities and organizations; religious, veteran, or fraternal organizations; fundraising events; and journal advertisements.
Amount of Grant: Up to 23,000,000 U.S.D
Contact: Grants Administrator; (866) 505-2866; info@avonfoundation.org
Internet: http://www.avoncompany.com/women/avonfoundation
Sponsor: Avon Products Foundation
1345 Avenue of the Americas
New York, NY 10105

AWHONN Research Grants 845
The AWHONN small grants program is designed primarily for novice researchers, serving first-time efforts or beginning development of a program of research. The purpose of the funding is to provide seed money, pilot funding, or total funding for small projects with promising contributions to nursing knowledge in clinical practice. The focus of the research is women's health, obstetric, or neonatal nursing phenomena.
Requirements: Applicants must be members of the association. An application for membership may accompany the proposal.
Restrictions: Primary investigators of current federally funded research and of previous AWHONN research are not eligible. Grant funds may not be used for indirect costs, tuition, computer hardware/printers.
Date(s) Application is Due: Nov 1
Amount of Grant: 2,500 - 10,000 U.S.D
Contact: Daniella McCarthy, Research Program Manager; (202) 261-2434; fax (202) 728-0575; ResearchPrograms@awhonn.org or daniellam@awhonn.org
Internet: http://www.awhonn.org/awhonn/?pg=0-874-2240
Sponsor: Association of Women's Health, Obstetric, and Neonatal Nurses
2000 L Street NW, Suite 740
Washington, DC 20036

Axe-Houghton Foundation Grants 846
The foundation operates in the New York City metropolitan area exclusively for charitable, educational, and scientific purposes to foster and encourage an appreciation of the English language, with major emphasis on the spoken language. Priority is given to projects for the improvement of speech and its uses in the areas of public affairs, education, theater, poetry, debate, and the oral interpretation of literature. A portion of available funds is devoted to speech remediation and to scientific research pertaining to speech. Types of support include program grants and seed money grants.

Requirements: Initial inquiries should be by full proposal and should contain a brief statement of objective, duration of time required to complete this objective, qualifications of personnel responsible for conducting the project, and the general amounts of funds needed. The proposal should also indicate the degree of internal support that would be committed.
Restrictions: Grants are made only to tax-exempt institutions and not to individuals, private foundations or organizations outside of the United States. Grants are not made for scholarships, fellowships, capital, or general support programs.
Geographic Focus: New York
Date(s) Application is Due: Sep 1
Amount of Grant: 1,000 - 7,500 U.S.D
Contact: Claire Brook, President; (212) 909-8304
Internet: http://www.foundationcenter.org/grantmaker/axehoughton/index.html
Sponsor: Axe-Houghton Foundation
919 3rd Avenue, 2nd Floor
New York, NY 10022

B.C. McCabe Foundation Grants **847**
The foundation awards grants to California and Nevada nonprofit organizations in its areas of interest, including education, medical care, human services, and youth development. Types of support include matching gifts, program grants, and seed money grants. Applicant should make an initial phone call and submit a letter of inquiry, including the IRS tax-exemption letter.
Requirements: 501(c)3 nonprofit organizations serving the greater Los Angeles, CA, and Reno, NV, areas are eligible.
Geographic Focus: California, Nevada
Amount of Grant: 10,000 - 100,000 U.S.D
Contact: Jim Shepard, Trustee; (562) 696-1433; fax (562) 698-5508
Sponsor: B.C. McCabe Foundation
8152 Painter Avenue, Suite 201
Whittier, CA 90602

Bacon Family Foundation Grants **848**
The foundation awards grants to Colorado nonprofit organizations in its areas of interest, including arts and culture, community development, disabilities, economic development, education, food distribution, environment, health care, historic preservation, homelessness, housing, literacy, recreation/parks, religion, social services, and youth. Types of support include capital campaigns, challenge/matching grants, equipment acquisition, general operating support, project grants, and seed grants. A letter should outline the project or program in brief but sufficient detail for the Foundation to initially evaluate the proposal. There are no application deadlines. The Bacon Family Foundation meets quarterly to consider applications.
Requirements: Colorado nonprofit organizations are eligible. Preference will be given to requests from western Colorado.
Restrictions: No grants are made to individuals.
Geographic Focus: Colorado
Contact: Linda Simpson, (970) 243-3767; lsimpson@wc-cf.org
Internet: http://www.wc-cf.org/bacon.htm
Sponsor: Bacon Family Foundation
P.O. Box 4570
Grand Junction, CO 81502-4570

Bailey Foundation Grants **849**
The foundation supports South Carolina nonprofits, primarily in Laurens County, its areas of interest, include: health clinics and organizations involved with theater, education, human services, business promotion, Christianity and awards college scholarships to students graduating from public high schools in Laurens County, South Carolina. Support is offered in the form of : capital campaigns; employee matching gifts; endowments; matching/challenge support; scholarship funds, including individual scholarships; annual campaigns; building/renovation grants. Initial contact may be by telephone or letter. Deadline for scholarship applications is April 15, contact Foundation to request an application form.
Requirements: Nonprofit organizations in Laurens County, South Carolina are eligible.
Restrictions: No grants to individuals (except for scholarships), or for general operating support.
Geographic Focus: South Carolina
Date(s) Application is Due: Apr 15; Oct 1
Amount of Grant: 2,500 - 50,000 U.S.D
Samples: YMCA of Greater Clinton South Carolina, Clinton, SC, $50,000; South Carolina Independent Colleges and Universities, Columbia, SC, $20,000; Thornwell Home and School for Children, Clinton, SC, $25,500.
Contact: Robert S. Link, Jr. , Administration, Carolina First Bank; (864) 938-2632; fax (864) 938-2669
Sponsor: Bailey Foundation
P.O. Box 494
Clinton, SC 29325-0494

Balance Bar Sports Individual/Team Grants **850**
The program awards grants to enable amateur athletes and teams to pursue a sport or recreational activity that enhances physical health. Past grants have supported equipment costs, training, and other activities related to archery, cycling, hiking, the martial arts, rock climbing, running, yoga, and other pursuits. Complete guidelines and application, which must be submitted online, are available on the Web site.
Requirements: U.S. citizens who are at least 18 years old and are amateur athletes/enthusiasts/hobbyists who have not competed in athletic activities as their main source of income are eligible. Amateur teams also may apply, provided that all members meet the above qualifications.
Date(s) Application is Due: Mar 15; Aug 30
Amount of Grant: 500 - 10,000 U.S.D
Contact: Hunter PR; (212) 679-6600 or (212) 481-6076; atucker@hunterpr.com
Internet: http://www.balance.com/grants/default.aspx
Sponsor: Balance Bar
41 Madison Avenue, 5th Floor
New York, NY 10010

Balfe Family Foundation Grants **851**
The Balfe Family Foundation, established in Florida in 1999, has as its primary fields of interest: education, health care and research, human services, and social services. Types of support are typically in the form of general operations, service delivery support, and seed funding. There are particular application forms or deadlines with which to adhere. Applicants should contact the Foundation in writing, outlining the program and funding need.
Amount of Grant: 50 - 50,000 U.S.D
Samples: Crystal Cathedral, Garden Grove, California, $1,200—religious contribution; Greater Lafayette Community Foundation, Lafayette, Indiana, $48,594—for a general contribution.
Contact: Opal M. Balfe, President; (954) 462-6300; fax (954) 462-4607
Sponsor: Balfe Family Foundation
2731 NE 14 Street Causeway, #829
Pomano Beach, FL 33062-3562

Ball Brothers Foundation General Grants **852**
Ball Brothers Foundation offers funding opportunities for initiatives that have one of the six broad subject areas: education; arts; culture and humanities; human services; public/society benefit; health; and environment. For those interested in school or community wellness, application must be submitted through application found under Ball Wellness Grants in the application materials.
Requirements: Ball Brothers Foundation is restricted by its charter to grants to nonprofit institutions and organizations within Indiana.
Restrictions: Applications are limited to organization and institutions within in the State of Indiana. The Foundation will not support: direct assistance to individuals or scholarships; applications coming from outside of Indiana; booster organizations; on-going salary requests of staff personnel to support an organization; services that the community-at-large should normally underwrite (i.e. roads, bus transportation, etc.0; capital building projects; research projects (except for philanthropic studies); or unsolicited proposals (all requests must begin with a preliminary proposal).
Geographic Focus: Indiana
Date(s) Application is Due: Feb 15; Jul 15
Amount of Grant: 5,000 - 100,000 U.S.D
Contact: Donna Munchel, (765) 741-5500; fax (765) 741-5518; donna.munchel@ballfdn.org
Internet: http://www.ballfdn.org/index/staff/generalinfo.asp
Sponsor: Ball Brothers Foundation
222 South Mulberry Street
Muncie, IN 47305

Ball Brothers Foundation Grants **853**
The foundation seeks to build and sustain a high quality of life in Indiana by awarding grants to nonprofit organizations in broad subject areas, including elementary, secondary, higher, and adult basic education and literacy skills; cultural activities; community betterment; and health and human services. Usually Muncie and Delaware Counties receive a higher priority for funding than requests from within East Central Indiana or across the state. Types of support include general operations, annual campaigns, capital campaigns, building construction/renovation, program development, conferences and seminars, professorships, publication, curriculum development, research, fellowships, matching funds, seed grants, and technical assistance. Preference will be given to catalytic grants that will stimulate others to participate in problem solving or in matching fund programs and to innovative approaches for addressing either traditional or emerging community needs. Applications are reviewed by the board of directors in January, May, and September of each calendar year. Proposals are encouraged to be submitted from February to May. Grant seekers may send a preliminary proposal, complete proposal, or ask for a personal visit to discuss a potential grant request.
Requirements: Indiana nonprofit institutions and organizations are eligible.
Restrictions: The Foundation will not support: direct assistance to individuals or scholarships; applications coming from outside of Indiana; booster organizations; on-going salary requests of staff personnel to support an organization; services that the community-at-large should normally underwrite (i.e. roads, bus transportation, etc.); capital building projects; research projects (except for philanthropic studies); or unsolicited proposals (all requests must begin with a preliminary proposal).
Geographic Focus: Indiana
Contact: Donna Munchel, (765) 741-5500; fax (765) 741-5518; donna.munchel@ballfdn.org
Internet: http://www.ballfdn.org
Sponsor: Ball Brothers Foundation
222 South Mulberry Street
Muncie, IN 47305

Baltimore Washington Center Adult Psychoanalysis Fellowships 854
The Adult Psychoanalysis Fellowship Program is a one-year program designed for individuals who have a significant interest in psychoanalysis as a body of knowledge and as a framework with which to understand and carry out therapeutic efforts. Intended for advanced graduate students and residents as well as recent graduates in Psychiatry, Psychology and Social Work, the program at its core consists of monthly seminars presented by psychoanalysts on clinically-related topics. Fellows are also welcome to attend Forums on Psychoanalysis and other conferences offered by the Institute and Society.
Date(s) Application is Due: Jul 15
Contact: Elizabeth Manne, Executive Director; (301) 470-3635 or (410) 792-8060; fax (410) 792-4912; admin@bwanalysis.org
Internet: http://www.bwanalysis.org/Fellowship%20Program-Adult.html
Sponsor: Baltimore Washington Center for Psychoanalysis
14900 Sweitzer Lane, Suite 102
Laurel, MD 20707

Baltimore Washington Center Child Psychotherapy Fellowships 855
The Child Psychotherapy Fellowship program is a one-year program specifically designed for child mental health professionals who are interested in deepening their understanding of working with children from a psychoanalytic or psychodynamic perspective. The program is tuition-free. Detailed discussions of actual clinical work with children and reviews of relevant psychoanalytic literature will be used as a framework to enhance understanding and ability to carry out therapeutic work. Fellowship meetings will be held monthly, September through May, on the second Wednesday of every month, at the homes of Baltimore and Washington faculty members.
Requirements: Professionals, including psychiatrists, psychologists, social workers, and counselors currently working with children in psychotherapy should apply. Recent graduates of training programs should also apply.
Date(s) Application is Due: Jul 15
Contact: Elizabeth Manne, Executive Director; (301) 470-3635 or (410) 792-8060; fax (410) 792-4912; admin@bwanalysis.org
Internet: http://www.bwanalysis.org/Fellowship%20Program-Child.html
Sponsor: Baltimore Washington Center for Psychoanalysis
14900 Sweitzer Lane, Suite 102
Laurel, MD 20707

Bamford-Lahey Children's Foundation Grants 856
THe foundation was created to improve the lives of children with developmental language disorders. One or two grants are awarded each year to a university, hospital, or public school for projects that are directly related to improving the lives of children with developmental language disorders and if indirect costs are not involved or are very low.
Requirements: 501(c)3 organizations are eligible.
Date(s) Application is Due: Nov 1
Amount of Grant: Up to 20,000 U.S.D
Contact: Grants Administrator; info@bamford-lahey.org
Internet: http://bamford-lahey.org/guidelines.html
Sponsor: Bamford-Lahey Children's Foundation
2995 Woodside Drive, Suite 400
Woodside, CA 94062

BancorpSouth Foundation Grants 857
The BancorpSouth Foundation supports organizations involved with orchestras, secondary and higher education, legal aid, housing, youth development, and human services. Support is available in the form of general/operating grants in areas of operation, that include Arkansas, Mississippi and Tennessee. Applications for funding are accepted on a rolling basis and reviewed quarterly.
Requirements: Arkansas, Mississippi and Tennessee 501(c)3 non-profit organizations are eligible to apply. There is no application deadline nor is there an application form required when applying for funding. Applicants should include a detailed description of project and amount of funding requested in the proposal.
Restrictions: No grants to individuals.
Geographic Focus: Arkansas, Mississippi, Tennessee
Amount of Grant: 5,000 - 15,000 U.S.D
Samples: Southern Arkansas University, Magnolia, AR, $15,000; Alpha House Home For Boys, Tupelo, MS, $5,000; Affordable Housing, Jackson, TN, $5,000.
Contact: Nash Allen, Grants Manager; (662) 680-2000
Sponsor: BancorpSouth Foundation
P.O. Box 789
Tupelo, MS 38802-0789

Bandai Foundation Grants 858
The foundation awards grants to nonprofits in its areas of interest, including theater, libraries, animal welfare, medical care, children and youth, international affairs, and science. There are no application forms or deadlines.
Requirements: Giving is primarily in California and New York.
Geographic Focus: California, New York
Amount of Grant: 10,000 - 150,000 U.S.D
Contact: Alison Miller
Sponsor: Bandai Foundation
5551 Katella Avenue
Cypress, CA 90630

Banfi Vintners Foundation Grants 859
The foundation awards general operating grants to nonprofits in its areas of interest, including higher education, civic and public affairs, arts and humanities, wildlife protection, health (hospitals and disease research/prevention), religion, science, social services, and international. Grants are made nationwide, with preference given to requests from Massachusetts and the New York, NY, area.
Geographic Focus: Massachusetts, New York
Amount of Grant: 100 - 465,000 U.S.D
Samples: Colgate U (Hamilton, NY)—for operating support, $75,000; Friends for Long Island's Heritage (Syosset, NY—for operating support, $25,000; Huntington Hospital Assoc (Huntington, NY)—for operating support, $25,000; Cornell U (Ithaca, NY)—for operating support, $465,000.
Contact: Philip Calderone, Executive Director; (516) 626-9200
Sponsor: Banfi Vintners Foundation
1111 Cedar Swamp Road
Glen Head, NY 11545

Bank of America Charitable Foundation Anchor Institutions Grants 860
Bank of America's local leaders, working within their communities, look for opportunities to support community stakeholders; opportunities that will further economic and community development, that will do more than put bricks and mortar on your block. When it does find those opportunities, the Foundation invites anchor institutions to apply for a grant. Such institutions might include: arts and cultural institutions such as museums, zoos, and performance halls; colleges, universities, and community colleges that attract and educate the workforce; and hospitals that provide excellent health care.
Contact: Anne Finucane, Foundation Chairperson/Chief Marketing Officer; (800) 218-9946 or (617) 434-9405; anne.m.finucane@bankofamerica.com
Internet: http://www.bankofamerica.com/foundation/index.cfm?template=fd_anchorinstitutions
Sponsor: Bank of America Charitable Foundation
100 North Tryon Street
Charlotte, NC 28255

Bank of America Charitable Foundation National Programs and Partners Grants 861
Part of the Foundation's philanthropic objective is to sustain community efforts and provide funding flexibility so nonprofits can distribute funds as they know will best serve their clients and build their organizations. Its national partners are an integral part of the neighborhoods they serve. These are organizations that complement what the Foundation is able to do in a local community and collectively impact hundreds of neighborhoods across the country. It's possible an organization needs funding for capacity building or operating support at the national level; it may be just as possible the funding needs to be directed to the local level to strengthen that office's reach. And when disaster strikes, the Foundation is prepared to respond nationally when need overwhelms local capabilities to respond.
Requirements: To be considered for a local grant, an organization must be: tax-exempt under section 501(c)3 of the Internal Revenue Code; must be based and serve communities in particular markets; and must be aligned with funding priorities for those markets. A map of markets is available online.
Restrictions: The following organizations are not eligible for funding: organizations that deny service to a potential client or beneficiary on the basis of race, religion, color, sex, sexual orientation, age, national origin, ancestry, citizenship, veteran, or disability status; individuals, including those seeking scholarships or fellowship assistance; political, labor, fraternal organizations, or civic clubs; religious organizations (for example, churches and synagogues). Some secular community service programs such as soup kitchens, shelters for the homeless, and job banks are eligible, even when sponsored by religious groups; individual pre-K-12 schools (public or private); sports, athletic events, or athletic programs; travel-related events, including student trips or tours; development or production of books, films, videos, or television programs; or memorial campaigns.
Contact: Program Administrator; (800) 218-9946
Internet: http://www.bankofamerica.com/foundation/index.cfm?template=fd_nationalprograms
Sponsor: Bank of America Charitable Foundation
100 North Tryon Street
Charlotte, NC 28255

Bank of America Charitable Gift Fund - General Fund Grants 862
The Bank of America Charitable Gift Fund - General Fund was established in 1955 to support and promote quality educational, human services, and health care programming for underserved populations. Grant requests for general operating support are strongly encouraged. Program support will also be considered. Small, program-related capital expenses may be included in general operating or program requests. The application deadline is September 1, and applicants will be notified of grant decisions before November 30.
Requirements: 501(c)3 organizations serving the residents of Massachusetts are eligible to apply.
Restrictions: The majority of grants from the Bank of America Charitable Gift Fund - General Fund are 1 year in duration. On occasion, multi-year support is awarded.
Geographic Focus: Massachusetts
Date(s) Application is Due: Sep 1
Contact: George Thorn, Vice President; (866) 778-6859; george.thorn@baml.com
Internet: https://www.bankofamerica.com/philanthropic/grantmaking.action
Sponsor: Bank of America Charitable Gift Fund
100 Federal Street, MA5-100-16-01
Boston, MA 02110

Bank of America Client Foundation Grants 863

The Foundation gives primarily to the arts, education, human services, natural science, youth and family organizations, and historic preservation. Fields of interest include: adult education; aging; AIDS; arts; children and youth services; community and economic development; crime and violence prevention; economically disadvantaged; education; the environment; family services; health care; historic preservation and historical societies; human services; marine science; media; mental health/crisis services; museums; performing arts; recreation; and urban/community development.
Requirements: Giving is limited to organizations operating in or providing services to residents of Sarasota County, Florida.
Restrictions: Generally the Foundation will not provide grants for: endowments, deficit financing, debt reduction, or ordinary operating expenses; conferences, seminars, workshops, travel, surveys, advertising, fund-raising costs or research; annual giving campaigns; individuals; projects that have already been completed; public or private schools; or child care facilities.
Geographic Focus: Florida
Date(s) Application is Due: Apr 15; Sep 15
Amount of Grant: 35,000 U.S.D
Samples: Jewish Family and Children's Service of Sarasota-Manatee, Inc., $50,000 - to purchase conference room equipment and Brain Fitness Center technology and furnishings; Pines of Sarasota Foundation, Inc., $45,000 - to renovate sections of The Garden - The Pines Alzheimer's/Dementia Facility; Forty Carrots of Sarasota, Inc., $50,000 - Capital Campaign to purchase building and property and create a reserve to cover long-term maintenance.
Contact: Renee Armbruster, CTFA Senior Vice President; (941) 952-2804; fax (941) 952-2768; renee.armbruster@ustrust.com
Sponsor: Bank of America Client Foundation
Bank of America Private Wealth Management, 1605 Main Street, Suite 800
Sarasota, FL 34236

Banks Family Foundation Grants 864

The foundation's mission is to empower youths to become responsible adults by supporting organizations that benefit the health, education, and welfare of the children of northern Alameda County, CA. Organizations are limited to two grants in three calendar years. Applications should include a brief description of the organization including names of officers and board members; a copy of the most recent financial statements; a copy of the IRS 501(c)3 ruling; and a description of and budget for the funding request. Grant disbursements are made twice yearly, on January 15 and July 15. Guidelines and application are available online.
Requirements: New grant requests are limited to 501(c)3 nonprofit organizations located in northern Alameda County, CA.
Geographic Focus: California
Date(s) Application is Due: May 15; Nov 15
Amount of Grant: Up to 10,000 U.S.D
Contact: Grants Administrator; (510) 763-3512; info@banksfamilyfoundation.net
Internet: http://www.banksfamilyfoundation.org
Sponsor: Banks Family Foundation
P.O. Box 10287
Oakland, CA 94610

Baptist-Trinity Lutheran Legacy Foundation Grants 865

The Foundation provides support in the greater Kansas City area for crisis related medical assistance, neighborhood school health grants, and health education programs and services in the service area of Baptist Medical Center and Trinity Lutheran Hospital. Funding priorities include: emergency medical assistance; health education; and health education in neighborhood schools. Highest consideration will be given for funding programs or services that demonstrate diversity, have measurable outcomes and can document continued existence after grant funding. Health Education Grant requests will be reviewed in January, March, May, July, September and November.
Requirements: All applicants must be non-profit, tax exempt organizations.
Restrictions: No grants will be made to individuals, political parties, candidates or political activities. Grants will not be given for special events, annual campaigns, endowments, or construction projects.
Geographic Focus: Missouri
Contact: Becky Schaid, Executive Director; (816) 276-7515 or (816) 276-7555; fax (816) 926-2261; becky@btllf.org
Internet: http://www.btllf.org/funding.html
Sponsor: Baptist-Trinity Lutheran Legacy Foundation
6601 Rockhill Road
Kansas City, MO 64131

Baptist Community Ministries Grants 866

BCM funds grants primarily in four major areas of interest: health; education; public safety; and governmental oversight. Applications not falling in these four major areas of interest are not typically considered. BCM funds transom grants and strategic grants. Transom grants are unsolicited grant proposals submitted by qualified, nonprofit organizations during our two semi-annual open transom cycles, and encourage the development of new ideas and nurture inventive solutions to community problems. Strategic grants generally target the long-range goals in each area of interest and are longer in duration. Strategic grants are usually generated through Request for Proposals or by invitation.
Requirements: Grants are made to eligible organizations in one or more of the five Louisiana parishes of Orleans, Jefferson, St. Bernard, St. Tammany and Plaquemines. Beneficiaries of grants should reside in one or more of the five parishes. It is preferred that the organization receiving a grant also be located in the same five-parish region and be overseen by unpaid volunteers living in the area.
Restrictions: Grants are not made to private foundations described in Section 509(a) of the Internal Revenue Code, private or publicly held corporations, limited liability corporations, partnerships, or sub-chapter S corporations. Funding requests not supported are: capital projects; ongoing general operating expenses; projects of national or statewide scope; single-disease charities; direct support to individuals or fraternal bodies.
Geographic Focus: Louisiana
Date(s) Application is Due: Mar 15; Sep 15
Amount of Grant: 50,000 U.S.D
Contact: Joanne M. Schmidt, Grants Manager; (504) 593-2345 or (504) 593-2323; fax (504) 593-2301; jschmidt@bcm.org or info@bcm.org
Internet: http://www.bcm.org/grantmaking/
Sponsor: Baptist Community Ministries
400 Poydras Street, Suite 2950
New Orleans, LA 70130-3245

Barberton Community Foundation Grants 867

The foundation awards grants to improve the quality of life for the citizens of Barberton. In addition to the regular grants program quarterly deadlines, the Foundation's Small Grant Program accepts applications monthly for grants up to $1,000. Areas of interest include charitable endeavors, education, public health, and public recreation.
Requirements: The eligibility requirements for small grants and large grants differ. See the Application Guidelines for specific requirements.
Restrictions: Grants are not given to: individuals; endowments housed at institutions other than the Barberton Community Foundation; religious organizations for religious purposes; projects that do not exclusively benefit the citizens of Barberton; fund debt reductions, deficits, or previous obligations; fund annual fund raising drives; fund ongoing operating expenses; fund political projects, sabbatical leaves or scholarly research; or fund venture capital for competitive profit-making activities.
Geographic Focus: Ohio
Date(s) Application is Due: Jan 2; Apr 1; Jul 1; Oct 1
Amount of Grant: Up to 1,000 U.S.D
Samples: Barberton All-Town Little League (OH)—to refurbish two ball fields and construct two fields at the U.L. Light Middle School, $14,075; City of Barberton (OH)—for summer concert programs at Lake Anna Park, $55,000; Metro Regional Transit Authority (Akron, OH)—for a transit center in Barberton, $102,000; Barberton Area United Way (OH)—to help needy residents make utility and rent payments, $60,000.
Contact: Chuck Sandstrom, Executive Director; (330) 745-5995; fax (330) 745-3990; csandstrom@bcfcharity.org
Internet: http://www.bcfcharity.org/bcf/grant/grants.shtml
Sponsor: Barberton Community Foundation
460 W Paige Avenue
Barberton, OH 44203

Barker Foundation Grants 868

The foundation offers funding to New Hampshire non-profit organizations in the areas of: children/youth services; education; health organizations, association; hospitals (general); and human services. Giving primarily for health associations, social services and youth. Grants are awarded in the following types: Annual campaigns building/renovation; capital campaigns; continuing support; equipment; general/operating support and; program development.
Requirements: New Hampshire nonprofit organizations are eligible for funding. The Foundation accepts written requests only. No formal application form is required. Send a 1 page concept paper and request for guidelines to the Foundation with a SASE for response.
Restrictions: Grants are not made to individuals.
Geographic Focus: New Hampshire
Amount of Grant: 2,000 - 13,000 U.S.D
Samples: Boys and Girls Club of Greater Nashua, Nashua, NH, $10,000; Senior Activity Center, Nashua, NH, $2,000; Nashua Soup Kitchen and Shelter, Nashua, NH, $13,000.
Contact: Allan Barker, Treasurer
Sponsor: Barker Foundation
P.O. Box 328
Nashua, NH 03061-0328

Barker Welfare Foundation Grants 869

The mission of the foundation is to make grants to qualified charitable organizations whose initiatives improve the quality of life, with an emphasis on strengthening youth and families and to reflect the philosophy of Catherine B. Hickox, the Founder. Consideration will be given to applications from institutions and agencies operating in the fields of health, welfare, education and literacy, cultural activities, and civic affairs, primarily serving the metropolitan area of New York, NY, and Michigan City, IN. The foundation board meets twice per year to consider requests.
Requirements: In advance of submitting a request, a brief letter or telephone call is suggested to determine if the organization seeking to apply for a grant falls within the current general policy of the Foundation. Before the Foundation sends out its application form, a brief 2-3 page letter describing the organization, the purpose and amount requested should be sent. A copy of the first page of the most recent 990 filed with the IRS and a current budget for the whole organization and a budget for program/project (if requesting program/project support) including income and expense should be sent with the letter of inquiry. Grants are made to tax-exempt organizations which have

received a ruling by the Internal Revenue Service that they are organizations described in Section 501(c)3 and classified in Section 509(a)(1),(2), or (3) of the Internal Revenue Code (publicly supported organizations and their affiliates).
Restrictions: Appeals for the following will be declined: organizations not located in nor directly serving the defined areas; national health; welfare; or education agencies; institutions or funds; scholarships; fellowships; loans; student aid; appeals from individuals; medical and scientific research; private elementary and secondary schools; colleges; universities; professional schools; trade organizations; films; program advertising; conferences; seminars; benefits and fund raising costs; start-up organizations; emergency funds; and deficit financing; lobbying-related or legislative activities; endowment funds; and intermediary organizations.
Geographic Focus: Indiana, New York
Date(s) Application is Due: Feb 1; Aug 1
Amount of Grant: 7,500 - 15,000 U.S.D
Contact: Sarane Ross, President; (516) 759-5592; BarkerSMD@aol.com
Internet: http://www.barkerwelfare.org
Sponsor: Barker Welfare Foundation
P.O. Box 2
Glen Head, NY 11545

Barnes Group Foundation Grants 870
The foundation awards grants in the following areas of interest: Education; Health and Welfare; Hospitals; Culture and Art and Civic and Youth Organizations. Highest priority is given to the support of organizations and projects in communities where the company has its executive office, group headquarters offices, plants and other facilities. There are no deadlines for submitting requests. Initial contact should be by letter.
Requirements: In order to be considered for support by the Foundation, a grant seeker must furnish a copy of its ruling from the Internal Revenue Service determining that it is an organization described in Section 501(c)3 and Section 509(a)(1), (2), (3), or (4) of the Internal Revenue Code.
Restrictions: No grants are made to organizations located outside the United States. The Foundation will consider only one grant request per organization during any one year period. The Foundation will not contribute directly or indirectly to any political activities or legislative lobbying efforts.
Geographic Focus: Connecticut, Maine, Massachusetts, New Hampshire, Rhode Island, Vermont
Amount of Grant: 500 - 3,500 U.S.D
Contact: Secretary; (860) 973-2112; fax (860) 589-7466
Internet: http://www.barnesgroupinc.com/about/foundation.html
Sponsor: Barnes Group Foundation
123 Main Street, P.O. Box 489
Bristol, CT 06011-0489

Barra Foundation Community Fund Grants 871
Community Fund grants provide unrestricted contributions to qualified organizations primarily in the Greater Philadelphia area. These grants are generally in amounts between $1,000 and $10,000 per year. The Foundation's categories of funding for these grants are: human services; arts and culture; health; and education. These grants are made at certain times of the year, depending on the category of grant. The categories and time frames for each are as follows: human services—March 1 with grants made in the Spring of each year; arts and culture—June 1 with grants made in the Summer of each year; and health and education—September 1 with grants made in the fall of each year.
Requirements: Applications are welcome from 501(c)3 organizations serving the greater Philadelphia region.
Geographic Focus: Pennsylvania
Date(s) Application is Due: Mar 1; Jun 1; Sep 1
Amount of Grant: 2,000 - 15,000 U.S.D
Contact: William Harral, III, President; (215) 233-5115; fax (215) 836-1033; william.harral@verizon.net
Internet: http://www.barrafoundation.org/grants/index.html
Sponsor: Barra Foundation
8200 Flourtown Avenue, Suite 12
Wyndmoor, PA 19038-7976

Barra Foundation Project Grants 872
Project grants are one-time grants generally for amounts above $10,000. The Foundation considers grants for innovative projects that aid research in advancing the frontiers of human services, arts and culture, health and education. Grants are not made for ongoing or expanding programs where substantial initial support was previously provided from other sources. Three principal criteria are strictly adhered to in judging the merits of a proposed project. They are: innovation; evaluation; and dissemination. Proposals may be submitted at any time of the year. Initial requests for project grants should be in the form of a preliminary concept paper, generally not to exceed two pages, summarizing: the principal focus and objectives of the proposed project; uniqueness of the concept; the overall methodology; estimated timetable; preliminary budget data; and other sources of support for the project.
Restrictions: The foundation does not provide grants for: ongoing operating budgets; staff salaries; budget deficits; endowments; capital campaigns and projects; international programs; environmental and religious organizations; scholarships and fellowships; or audio/video projects, publications, catalogs, and exhibitions.
Geographic Focus: Pennsylvania
Amount of Grant: 10,000 - 500,000 U.S.D
Contact: William Harral, III, President; (215) 233-5115; fax (215) 836-1033; william.harral@verizon.net
Internet: http://www.barrafoundation.org/grants/index.html
Sponsor: Barra Foundation
8200 Flourtown Avenue, Suite 12
Wyndmoor, PA 19038-7976

Barr Fund Grants 873
The fund awards grants to Illinois nonprofits in its areas of interest, including services for children and youth, orchestras, higher education, mental health/crisis services, Jewish temples and organizations, and social services. Grants are awarded for general operating support. There are no application forms or deadlines. Submit a letter of request.
Requirements: Illinois nonprofits are eligible.
Restrictions: Individuals are not eligible.
Geographic Focus: Illinois
Amount of Grant: 100 - 75,000 U.S.D
Samples: Rush University Medical Center, Chicago, IL, $10,000; New Israel Fund, Washington, DC, $1,000; Highland Park Community Foundation, Highland Park, IL, $1,000; The Barr Harris Center of the Institute For Psychoanalysis, Chicago, IL, $50,000.
Contact: Donald Lubin, President; (312) 782-4710; fax (312) 876-8000
Sponsor: Barr Fund
230 West Monroe Street, Suite 330
Chicago, IL 60606-4701

Barth Syndrome Foundation Research Grants 874
Barth syndrome is a serious X-linked recessive condition associated with cardiomyopathy, neutropenia, skeletal muscle weakness, exercise intolerance, growth delay, and diverse biochemical abnormalities (including defects in mitochondrial metabolism and phospholipid biosynthesis). Because many clinical and biochemical abnormalities of Barth syndrome remain poorly understood, the program is seeking proposals for research that will advance knowledge on any aspect of the syndrome. The foundation prefers to award seed grants to experienced investigators for testing of initial hypotheses and collection of preliminary data leading to successful long-term funding by NIH and other major granting institutions. The foundation also encourages investigators new to the field of Barth Syndrome research. Send an electronic version of the full application and all the attachments by the listed application deadline. The Foundation anticipates awarding several one- or two-year grants of up to $40,000 each. Funds will be available as soon as the successful grant applicants have been notified.
Requirements: Principal investigators who are affiliated with nonprofit institutions are eligible.
Date(s) Application is Due: Oct 31
Amount of Grant: 10,000 - 40,000 U.S.D
Contact: Matthew J. Toth, Science Director; (617) 469-6769; fax (617) 849-5695; mtoth@barthsyndrome.org
Internet: http://www.barthsyndrome.org/english/View.asp?x=1635
Sponsor: Barth Syndrome Foundation
675 VFW Parkway, #372
Chestnut Hill, MA 02467

Batchelor Foundation Grants 875
The foundation supports food banks and organizations involved with arts and culture, education, the environment, animals and wildlife, health, human services, and economically disadvantaged people. Special emphasis is directed toward programs designed to engage in medical research and provide care for childhood diseases; and promote study, preservation, and public awareness of the natural environment. Funding is available in the form of: capital campaigns; continuing support; endowments; general/operating support; and program development grants. There are no application forms. Initial approach should be a letter that details the grant proposal.
Requirements: Florida area nonprofits are eligible.
Restrictions: Individuals are not eligible.
Geographic Focus: Florida
Amount of Grant: 1,000 - 1,000,000 U.S.D
Samples: Community Partnership for Homeless, Miami, FL, $1,056,600—for general operating support and endowment; Fairchild Tropical Botanic Garden, Coral Gables, FL, $540,000— For general operating support; University of Miami, Miami, FL, $525,000— for general operating support.
Contact: Anne Batchelor-Robjohns, Co-C.E.O.; (305) 416-9066 or (305) 534-5004; jbatchelor@bellsouth.net
Sponsor: Batchelor Foundation
111 NE 1st Street, Suite 820
Miami, FL 33132

Baton Rouge Area Foundation Grants 876
The foundation seeks to enhance the quality of life for all citizens of Baton Rouge, Louisiana. The foundation concentrates on projects and programs in the areas of community development, education, environment, health and medical, and religion. Applicants should call the office to determine if their project or program is consistent with the foundation's goals before sending an application.
Requirements: Only nonprofits that are registered as 501(c)3 corporations working in the service region of East and West Baton Rouge Parish, East and West Feliciana, Ascension, Livingston, Iberville and Pointe Coupee can apply for grants.
Geographic Focus: Louisiana
Samples: Louisiana Art and Science Museum, Baton Rouge, LA., $125,835; Rhode Island School of Design, Providence, RI, $10,000; Friends of Hilltop Arboretum, Inc., Baton Rouge, LA., $5,450.

Contact: Angela Fitzgerald, Executive Assistant; (225) 387-6126 or (877) 387-6126; fax (225) 387-6153; afitzgerald@braf.org
Internet: https://www.braf.org/braf/GRANTS/tabid/103/Default.aspx
Sponsor: Baton Rouge Area Foundation
402 North Fourth Street
Baton Rouge, LA 70802

Battle Creek Community Foundation Grants **877**
The Battle Creek Community Foundation focuses on the greater Battle community, and is most interested in projects that focus on education, health, human services, arts and culture, public affairs or community development. In general, priority is given to those grant ideas that: increases the capacity of the community to participate in identifying needs and developing and implementing solutions; encourages cooperation; demonstrates a clean and convincing need; develops self-reliance; avoids unnecessary duplication of services; targets gaps in services; Mirrors the diversity of our community; promotes equity among its various segments; and has clean, defined goals and/or measurable outcomes. Grant applications and guidelines are available on the Web site.
Requirements: Grants will be considered only from 501(c)3 nonprofit agencies in Michigan offering services to the residents of Battle Creek.
Restrictions: Individuals are not eligible to receive grants.
Geographic Focus: Michigan
Amount of Grant: 5,000 - 10,000 U.S.D
Samples: Drug Free Workplace (Battle Creek, MI)—program support; Battle Creek Health System (Battle Creek, MI)—program support.
Contact: Kelly Boles Chapman, Vice President of Programs; (269) 962-2181; fax (269) 962-2182; bccf@bccfoundation.org
Internet: http://www.bccfoundation.org/grants/
Sponsor: Battle Creek Community Foundation
1 Riverwalk Centre, 34 W Jackson Street
Battle Creek, MI 49017-3505

Batts Foundation Grants **878**
Established in 1988, the Batts Foundation supports organizations involved with arts and culture, K-12 and higher education, disease, and human services. Types of support include: annual campaigns, building and renovation; capital campaigns; continuing support; operating support; endowments; matching grants; program development; and scholarship funding. Grants will be awarded primarily in the western Michigan area, particularly the communities of Huron, Zeeland, and Grand Rapids. There are no application forms or deadlines with which to adhere, and applicants should begin by submitting a one page letter summarizing the project.
Requirements: Michigan nonprofit organizations are eligible.
Restrictions: Individuals are ineligible.
Geographic Focus: Michigan
Amount of Grant: 250 - 25,000 U.S.D
Contact: Robert Batts, Director; (616) 956-3053; jsand@battsgroup.com
Sponsor: Batts Foundation
3855 Sparks Drive SE, Suite 222
Grand Rapids, MI 49546-2427

Baxter International Foundation Grants **879**
The foundation supports the development of more accessible and affordable healthcare. The foundation funds initiatives that improve the access, quality and cost-effectiveness of healthcare. Focusing on these priorities, the foundation focusing on communities where Baxter has a corporate presence. In Illinois, grants are restricted to Lake, McHenry and Cook counties. The foundation also funds in Asia, Australia, Canada, Europe, Latin America and Mexico.
Requirements: U.S. nonprofits in Lake, McHenry and Cook counties of Illinois are eligible to apply. Internationally, the following countries are eligible to apply: Asia, Australia, Canada, Europe, Latin America and Mexico.
Restrictions: In general, The Baxter International Foundation does not make grants to: capital and endowment campaigns (includes requests for infrastructure of any kind, equipment, vehicles, etc.); disease or condition-specific organizations or programs; educational grants/continuing professional education scholarships; educational institutions, except in instances where a grant would help achieve other goals, such as increasing community-based direct health services or the skills and availability of community health-care providers, in areas where there are Baxter facilities; general operating support or maintenence of effort; hospitals; individuals, including scholarships for individuals; lobbying and political organizations; magazines, professional journals, documentary, film, video, radio or website productions; medical missions; organizations seeking travel support for individuals or groups, medical missions or conferences; organizations soliciting contributions for advertising space, tickets to dinners, benefits, social and fund-raising events, sponsorships and promotional materials; organizations with a limited constituency, such as fraternal, veterans or religious organizations; research.
Geographic Focus: Illinois
Date(s) Application is Due: Jan 21; Apr 13; Jul 13; Sep 29
Samples: Abraham Low Self-Help Systems, Chicago, IL, $70,000—for staff to initiate veterans' mental health support groups at the North Chicago VA facility; Fundacao Julita, Sao Paulo, Brazil, $51,700—to expand and stabilize mental health services to highly disadvantaged children, youth and families in Jardim Sao Luis neighborhood of Sao Paulo, Brazil; Charities Aid Foundation for the Chronic Care Foundation, New Delhi, India, $240,000—to expand programs to prevent non-communicable chronic diseases in India; The Irish Hospice Foundation, Dublin, Ireland, $109,390—for a pilot program to train healthcare specialists in palliative care to expand access to non-cancer patients in Dublin, Ireland.
Contact: Foundation Contact; (847) 948-4605; fdninfo@baxter.com
Internet: http://www.baxter.com/about_baxter/foundation/index.html
Sponsor: Baxter International Foundation
1 Baxter Parkway, DF2-2E
Deerfield, IL 60015-4625

Bayer Clinical Scholarship Award **880**
This award is intended to facilitate the development of specific clinical expertise in the field of hemophilia for applicants who have completed medical training and have an interest in pursuing a career as a hemophilia treater/researcher. The award will support a mentored physician in training for two years. Clinical duties will encompass diagnosis, evaluation, and the planning of management strategies for patients with hereditary bleeding disorders. In addition to the clinical experience, the applicant may pursue a research project in the field of hemostasis. Clinical scholarships will provide funding for two years. Up to five new awards will be made each year. Guidelines and application are available online.
Requirements: The applicant should have earned his/her medical degree within the previous eight years. The commitment of the mentor to the grantee and the research project (if applicable) is a vital element of the application, as is the quality of the clinical environment at which the applicant will undertake the scholarship.
Amount of Grant: 70,000 U.S.D
Contact: Program Administrator; programadministrator@bayer-hemophilia-awards.com
Internet: http://www.bayer-hemophilia-awards.com/awards.cfm#clinical
Sponsor: Bayer HealthCare Corporation
100 Bayer Road
Pittsburgh, PA 15205-9741

Bayer Hemophilia Caregivers Education Award **881**
This award recognizes the essential role of caregivers and allied health professionals in the care of patients with hemophilia. It is designed to support their role by promoting continuing education. Awards will be provided for education activities in the field of hemophilia. This may include, but is not limited to, the development of educational seminars/symposia, mentored experiences, or training workshops. The Award is not designed for the development of materials for patients and parents. Applications will be judged by: Educational merit of the proposed initiative; Innovation of study rationale and techniques; Impact on the development of the professional and his/her ability to deliver care; Educational environment. The duration of the award is one year. Up to six Awards of up to $25,000 will be made annually. Part of the award may fund salary support. The candidate must spend at least 25% of his/her time on the project in order to request salary support.
Requirements: Requests for these grants must come from caregivers and allied health professionals involved in the care of patients with hemophilia. These may include nurses and nurse practitioners, physical therapists, pharmacists, psychiatrists, psychologists, genetic counselors, social workers, clinical dieticians, dentists and dental hygienists, among others. The Award is not designed for medically qualified hematologists who treat patients with bleeding disorders.
Restrictions: Unauthorized expenses are: Salaries, travel and/or housing related to sabbaticals; Purchase or rental of office equipment; Fees for tuition; Membership dues, congress/meeting registrations, subscriptions.
Amount of Grant: 25,000 U.S.D
Contact: Program Administrator; programadministrator@bayer-hemophilia-awards.com
Internet: http://www.bayer-hemophilia-awards.com/awards.cfm#caregivers
Sponsor: Bayer HealthCare Corporation
100 Bayer Road
Pittsburgh, PA 15205-9741

Bayer Hemophilia Early Career Investigator Award **882**
This award will fund salary support and research funds for a junior faculty member who wishes to undertake a mentored basic and/or clinical research project in the bleeding disorders field. The applicant would be expected to dedicate a significant amount of time to the project. Examples of topics for research projects that might be considered for these awards include, but are not limited to: clinical studies; properties and delivery of clotting factor proteins; assays and models; genetics and epidemiology; and molecular aspects and mechanisms of clotting factor inhibitor formation. Awards will provide funding for two years. Up to five new awards will be made each year. Part of the award may fund salary support. The candidate must spend at least 25 percent of his/her time on the project in order to request salary support. Applicants should submit a letter of intent via email describing the proposed project in 500 words or fewer; full proposals are by invitation. Guidelines are available online.
Requirements: The applicant should have an entry-level academic or clinical appointment within his or her institution. This award is open to applicants with a medical degree and/or PhD. Applicants should have earned their terminal degree within the previous 10 years.
Date(s) Application is Due: Oct 31
Amount of Grant: 100,000 U.S.D
Contact: Corporate Communications Coordinator; (412) 777-2000
Internet: http://www.bayer-hemophilia-awards.com/awards.cfm#early
Sponsor: Bayer HealthCare Corporation
100 Bayer Road
Pittsburgh, PA 15205-9741

Bayer Hemophilia Special Project Award **883**
The award is designed to support a wide range of research projects in the field of hemophilia. Examples of the types of projects that might be considered for these awards include, but are not limited to, those related to: clinical research; basic research; assessment and intervention in psychosocial issues facing patients and their families; assessment of quality of life and other health economic outcomes in patients with

bleeding disorders; and the effects of treatment modalities on such outcomes. The award is designed to encourage hypothesis-driven research, where the investigator is attempting to prove or disprove a set of assumptions. It is not designed to support studies such as the collection of epidemiological data. The funding will be awarded to allow the project to run for one or two years. Part of the award may fund salary support. The candidate must spend at least 25 percent of his/her time on the project in order to request salary support. Applicants should submit a letter of intent via email describing the proposed project in 500 words or fewer; full proposals are by invitation.
Requirements: Any individual affiliated with a facility that carries out research in inherited bleeding disorders, or provides care to patients with those disorders, may make a request for these grants. Such facilities may include medical universities, hospitals, treatment centers, blood centers, and other laboratories. All applications must be in English.
Date(s) Application is Due: Oct 31
Amount of Grant: Up to 200,000 U.S.D
Contact: Corporate Communications Coordinator; (412) 777-2000
Internet: http://www.bayer-hemophilia-awards.com/awards.cfm
Sponsor: Bayer HealthCare Corporation
100 Bayer Road
Pittsburgh, PA 15205-9741

BCBS Louisiana Angel Award 884
The program seeks nominations of individuals in Louisiana who have made extraordinary efforts on behalf of the state's children. Individuals, not groups, may be nominated. However, a person may be nominated for individual work done as a member of a group. A person's lifetime accomplishments will be considered. Nominated effort must have been done in Louisiana. Nominees can be paid for their work, but efforts going above and beyond job responsibilities will be considered. Work-release and course credit efforts are also eligible. The award includes a grant to one overall winner and grants to finalists to further their efforts of improving the quality of life for Louisiana's children. Nomination guidelines and form are available online.
Restrictions: Past Angel Award finalists, members of the award selection committee, elected political officials, and Blue Cross and Blue Shield of Louisiana employees and their immediate families are ineligible.
Geographic Focus: Louisiana
Date(s) Application is Due: Apr 15
Amount of Grant: 5,000 - 10,000 U.S.D
Contact: Award Administrator, c/o Louisiana Child Caring Foundation; (888) 219-2583; angel.award@bcbsla.co
Internet: http://www.angelaward.org/default.asp
Sponsor: Blue Cross Blue Shield of Louisiana
P.O. Box 98022
Baton Rouge, LA 70898-9022

BCBSM Building Healthy Communities Engaging Elementary Schools and Community Partners Grants 885
BCBSM (Blue Cross Blue Shield of Michigan) is offering this grant program to strengthen school and community efforts to reduce the risk and prevalence of childhood obesity through prevention and partnership. With this request for proposals Blue Cross invited Michigan elementary schools to collaborate with students, parents and community partners to address childhood obesity's root causes by promoting awareness of and education about physical activity and nutrition. See BCBSM website for all required forms.
Requirements: Electronic notice of intent is to be submitted no later then April 3. The application must be postmarked by April 24 to be eligible and be submitted with three copies of the proposal.
Geographic Focus: Michigan
Date(s) Application is Due: Apr 3; Apr 24
Amount of Grant: 40,000 - 60,000 U.S.D
Contact: BHC Director; (800) 658-6715; buildhealth@bcbsm.com
Internet: http://www.bcbsm.com/buildhealth/index.shtml
Sponsor: Blue Cross Blue Shield of Michigan Foundation
600 East Lafayette Boulevard
Detroit, MI 48226-2998

BCBSM Claude Pepper Award 886
The Claude Pepper Award is presented annually to two outstanding senior citizen advocates who demonstrate a strong concern for the special needs of the elderly. The BCBSM (Blue Cross Blue Shield of Michigan) named the award after the late U.S. Congressman who campaigned for senior citizens' rights. Blue Cross Blue Shield of Michigan's Senior Advisory Council selects two individuals, a retiree volunteer over age 55 and another person of any age who's employed in the senior advocacy field. Each award recipient receives a $1,000 donation for the nonprofit organization of his or her choice. Candidates for this award will have clearly demonstrated their dedication to Michigan's seniors through work that positively impacts older adults, particularly those who don't receive adequate health services due to mental, physical, financial or geographical limitations. To nominate someone for a Claude Pepper Award, submit the nomination online using the form provided at the BCBSM website.
Requirements: Nomination Criteria: awards are given to individuals, not groups or teams of people. However, you may nominate a person for their individual work on behalf of a group to which he or she belongs; volunteers are defined as people who freely choose to provide a service that contributes to the well-being of an individual or a community; nominees can be paid workers; however, the committee is looking for efforts that go above and beyond job responsibilities; the candidate must have performed the services in the State of Michigan; monetary contributions can only to made to 501(c)3 nonprofit organizations.
Restrictions: Employees of Blue Cross Blue Shield of Michigan, its subsidiaries and members of all selection committees are not eligible for this award, nor are their family members or elected political office holders.
Geographic Focus: Michigan
Contact: Community Affairs Hotline, (800) 733-2583
Internet: http://www.bcbsm.com/home/commitment/claude_pepper_award.shtml
Sponsor: Blue Cross Blue Shield of Michigan Foundation
600 East Lafayette Boulevard
Detroit, MI 48226-2998

BCBSM Corporate Contributions Grants 887
The BCBSM (Blue Cross Blue Shield of Michigan) corporate giving program supports nonprofit 501(3)(c) organizations throughout Michigan. BCBSM invests in local communities by supporting organizations that share a commitment to Michigan. BCBSM charitable contributions support programs that: address health issues specific to Michigan's youth and seniors; extend a nonprofit organization's reach through effective use of volunteers; emphasize collaboration across regions within Michigan. Contribution requests are accepted year round and are reviewed monthly by BCBSM's director of community affairs.
Requirements: To be considered for a charitable contribution from the BCBSM, an organization must be: nonprofit 501(3)(c) status; Michigan based and focused. Requests for contributions must be submitted in writing, and include: Corporate Contributions Cover Sheet (available on BCBSM website); IRS 501(c)3 tax exemption documentation; organization's annual report. Submit requests for contribution one of three ways: mail; fax and; email.
Restrictions: Contributions are not made to: individuals; political organizations, campaigns or candidates; extracurricular school activities; endowments; alumni associations; scholarship programs; research projects; multi-year pledges; operational expenses; group travel expenses; capital campaigns.
Geographic Focus: Michigan
Contact: Cathy Mozham, BCBSM Community Responsibility Director; (313) 225-0539; fax (313) 225-9693; cmozham@bcbsm.com
Internet: http://www.bcbsm.com/home/commitment/corporate_contributions.shtml
Sponsor: Blue Cross Blue Shield of Michigan Foundation
600 East Lafayette Boulevard
Detroit, MI 48226-2998

BCBSM Foundation Community Health Matching Grants 888
The program focuses on access to care for the uninsured and under insured. The program's purpose is to encourage nonprofit community based organizations to form partnerships with health care organizations, research organizations, or governmental agencies to develop and rigorously evaluate new ways of increasing access to care for the under and uninsured, in Michigan. The program offers up to $50,000 per year for two years, contingent on a 25% match.
Requirements: Nonprofit 501(c)3 organizations based in Michigan are eligible. Nonprofits do not need a firm commitment of matching support prior to the submission of the proposal [letters from potential funding partners are encouraged, indicating their support for the project and interest in possible funding partnership(s)]. Proposals must include a rigorous evaluation of the outcome of the initiative in accomplishing its objectives (preferably an independent or university-based evaluation).
Restrictions: In-kind contributions are not accepted as a match. Grants under this program do not pay for the cost of equipment (e.g., personal computers), hardware or software. The BCBSM Foundation does not provide support to for?profit organizations or individuals associated with organizations not located in Michigan.
Geographic Focus: Michigan
Amount of Grant: Up to 100,000 U.S.D
Samples: TahLeekah Partee, Pine Rest Christian Mental Health Services, $75,000 - Hope and Healing for Uninsured Youth in Grand Rapids (2010); Michigan League for Human Services, Ingham County, $45,000 - Kids Count in Michigan (2010); Community Church of Douglas, Allegan County, $99,692 - The Provision of Community Nursing (2010).
Contact: Nora Maloy, Dr.P.H., Senior Program Officer; (313) 225-8706; fax (313) 225-7730; foundation@bcbsm.com
Internet: http://www.bcbsm.com/foundation/grant.shtml
Sponsor: Blue Cross Blue Shield of Michigan Foundation
600 East Lafayette Boulevard
Detroit, MI 48226-2998

BCBSM Foundation Excellence in Research Awards for Students 889
The annual Excellence in Research Award for Students acknowledges doctoral candidates or medical students enrolled in Michigan universities who have published research papers that represent significant contributions to health policy or clinical care. Entries will be judged based on the subject matter's potential to make a contribution to health policy and medical care as well as the quality and originality of the research. There will be three awards: 1st place - $1,000; 2nd place - $750; 3rd place - $500.
Requirements: Doctoral candidates or medical students enrolled in Michigan universities are eligible. The nomination must be made by a faculty member at the student's university. If the paper has multiple authors, the student must be the first author.
Restrictions: Basic research papers are not eligible.
Geographic Focus: Michigan
Date(s) Application is Due: Jan 1
Amount of Grant: 500 - 1,000 U.S.D
Contact: Ira Strumwasser; (313) 225-8706; fax (313) 225-7730; foundation@bcbsm.com
Internet: http://www.bcbsm.com/foundation/grant.shtml
Sponsor: Blue Cross Blue Shield of Michigan Foundation
600 East Lafayette Boulevard
Detroit, MI 48226-2998

BCBSM Foundation Investigator Initiated Research Grants 890
The program provides grants for applied research that focuses on quality, cost and appropriate access to health care in Michigan. Grants average $75,000 for one year. Multi-year grants or grants in excess of $75,000 are awarded for exemplary projects. Projects that focus on the quality and cost of health care and appropriate access are considered priority and include: Organization and delivery of health care services; Evaluation of new methods or approaches to containing health care costs; Evaluation of new methods or approaches to providing access to high quality health care; Assessment and assurance of quality care; Identification and validation of clinical protocols and evidence-based practice guidelines. Applications are accepted at any time.
Requirements: Applications will be accepted from medical- and doctoral-level researchers based in hospitals, university settings, non-profit health care organizations, health systems, medical or nursing schools, schools of public health, or other relevant academic disciplines such as psychology, sociology and urban studies.
Restrictions: This program does not support basic science or biomedical research, including drug studies or studies using animals, costs of equipment (e.g., personal computers), hardware, software or on-going operating or developmental expenses. The BCBSM Foundation does not provide support to for-profit organizations or individuals associated with organizations not located in Michigan. Blue Cross and Blue Shield of Michigan employees, members of their immediate families, and employees and immediate family members of any Blue Cross and Blue Shield of Michigan affiliate or subsidiary are not eligible to receive BCBSM Foundation grants.
Geographic Focus: Michigan
Amount of Grant: Up to 100,000 U.S.D
Samples: Wei Du, Ph.D., Children's Hospital of Michigan, Department of Pediatrics, $74,994 - Diagnosing Adverse Drug Reactions in Pediatric ICU (2010); Teresa Wehrwein, Ph.D., R.N., Michigan State University – College of Nursing, $82,290 - Preparing Nurses for Roles in Patient-Centered Medical Homes: Case Management and Quality/Safety Management (2010); Faith Hopp, Ph.D., Hospice of Michigan, $89,646 - Evaluation of @HOMe Support Program (2010).
Contact: Nora Maloy, Dr. P.H., Senior Program Officer; (313) 225-8706; fax (313) 225-7730; foundation@bcbsm.com
Internet: http://www.bcbsm.com/foundation/grant.shtml
Sponsor: Blue Cross Blue Shield of Michigan Foundation
600 East Lafayette Boulevard
Detroit, MI 48226-2998

BCBSM Foundation Physician Investigator Research Awards 891
The program provides seed money to physicians to explore the merits of a particular research idea for further study. Grants are offered of up to $10,000 for projects that include pilot, feasibility or small research studies in clinical or health services research.
Requirements: Applicants must be a physician interested in research, who is licensed and domiciled in the state of Michigan. Applicants may include physicians working in research environments such as medical schools or university-affiliated hospitals, health care systems, or nonprofit agencies. An application may be submitted at any time.
Geographic Focus: Michigan
Amount of Grant: Up to 10,000 U.S.D
Contact: Nora Maloy, Dr.P.H., Senior Program Officer; (313) 225-8706; fax (313) 225-7730; foundation@bcbsm.com
Internet: http://www.bcbsm.com/foundation/grant.shtml
Sponsor: Blue Cross Blue Shield of Michigan Foundation
600 East Lafayette Boulevard
Detroit, MI 48226-2998

BCBSM Foundation Primary and Clinical Prevention Grants 892
The Blue Cross Blue Shield of Michigan Foundation requests letters of interest from Michigan physicians and members of the research community interested in preventing the three leading causes of death in Michigan: heart disease, cancer and stroke. The purpose of this grant initiative is to increase the use of evidence-based prevention efforts to improve the health of Michigan residents. The BCBSM Foundation has allocated $500,000 to fund several projects for up to two years. Funds will be available for salary support, program costs, supplies, office operations and other costs related to the proposed project. Computer equipment expenses, including PC hardware and software, will not be supported. In addition to the overall quality of the project, the feasibility and appropriateness of the budget is a factor in the funding decisions. The required application cover page and the terms and conditions may be downloaded for the BCBSM web site.
Requirements: Applicants must be doctoral level researchers (including physicians) based at Michigan universities, academic medical settings, community hospitals, health systems and community based nonprofit organizations. The principal investigator must have appropriate research and clinical credentials.
Geographic Focus: Michigan
Date(s) Application is Due: Sep 1
Contact: Nora Maloy; (313) 225-8706; fax (313) 225-7730; nmaloy@bcbsm.com
Internet: http://www.bcbsm.com/foundation/grant.shtml
Sponsor: Blue Cross Blue Shield of Michigan Foundation
600 East Lafayette Boulevard
Detroit, MI 48226-2998

BCBSM Foundation Proposal Development Awards 893
The program is intended to help Michigan-based nonprofit community organizations develop grant proposals for new healthcare initiatives. This award of up to $3,500 is available to help nonprofits obtain grant writing resources so they may secure the funding they need to bring creative community-based healthcare programs to life. Applications are accepted at any time.
Requirements: Applicants must be Michigan-based nonprofit community organizations that deliver health services. Michigan nonprofits are encouraged to apply for all available funding, including Federal grants. Nonprofits may also use the award funds to develop proposals for private, community or corporate foundations, either through local Michigan or national foundations, or from the state of Michigan. Expenditure of award funds is restricted to proposal development costs, such as technical assistance, freelance proposal writers and related proposal development costs.
Restrictions: Award funds are not intended to support any type of project activity or to supplement operational expenses. Proposals already completed are not eligible for funding.
Geographic Focus: Michigan
Amount of Grant: Up to 3,500 U.S.D
Samples: Warren Conner Development Coalition, Wayne County, $3,500 - Promise: 48213 (2010); Eastern Huron Ambulance Service, Huron County, $3,500 - Expanding Access to Quality Emergency Medical Services in Rural Michigan (2010); BRAINS Foundation, Kent County, $3,500 - BRAINS Foundation Neuropsychological Training Program (2010).
Contact: Nora Maloy, Dr.P.H., Senior Program Officer; (313) 225-8706; fax (313) 225-7730; foundation@bcbsm.com
Internet: http://www.bcbsm.com/foundation/grant.shtml
Sponsor: Blue Cross Blue Shield of Michigan Foundation
600 East Lafayette Boulevard
Detroit, MI 48226-2998

BCBSM Foundation Student Award Program 894
The program offers a one year stipend to fund a wide range of health care projects, including applied research, pilot programs, or demonstration and evaluation projects.
Requirements: All doctoral and medical students enrolled in Michigan universities are eligible. For consideration, the proposed project must focus geographically on the state of Michigan and address the BCBSM Foundation's objectives.
Restrictions: Completed or substantially completed dissertations or research projects are not eligible. Students who previously received this award are not eligible. Blue Cross and Blue Shield of Michigan affiliates and subsidiaries are not eligible. Investigation of pharmaceutical efficacy, basic research or research involving non-human subjects is not eligible. Grant monies are not intended to support field placements, practica or internships.
Geographic Focus: Michigan
Date(s) Application is Due: Apr 30
Amount of Grant: 3,000 U.S.D
Contact: Ira Strumwasser, Director; (313) 225-8706; fax (313) 225-7730; foundation@bcbsm.com
Internet: http://www.bcbsm.com/foundation/
Sponsor: Blue Cross Blue Shield of Michigan Foundation
600 East Lafayette Boulevard
Detroit, MI 48226-2998

BCBSM Frank J. McDevitt, DO, Excellence in Research Awards 895
The Frank J. McDevitt Excellence in Research Awards for Health Services, Policy and Clinical Care recognize Michigan-based researchers and physicians who have published research in research journals that contributes to improving health and medical care in Michigan. The BCBSM Foundation will award $10,000 for unrestricted research to physician and doctoral level researchers for research on: health policy; health services; clinical care.
Requirements: Applicants must be from Michigan based researchers with terminal research degrees (e.g. PhD, DrPH) and from Michigan based physicians (MD or DO). In the case of multiple authorship, the nominee must be the first author. All authors will be nominated and will constitute one team nomination. The nomination research must have been published, or accepted for publication in a refereed health or medical care journal, within the previous two years. The financial award will be made to the recipient's 501(c)3 non-profit or educational organization.
Restrictions: Only one article per nominee will be accepted for consideration each year.
Geographic Focus: Michigan
Date(s) Application is Due: Jan 1
Amount of Grant: 10,000 U.S.D
Contact: Nora Maloy (313) 225-8706; fax (313) 225-7730; nmaloy@bcbsm.com
Internet: http://www.bcbsm.com/foundation/grant.shtml
Sponsor: Blue Cross Blue Shield of Michigan Foundation
600 East Lafayette Boulevard
Detroit, MI 48226-2998

BCBSNC Foundation Fit Together Grants 896
The program provides funding and technical assistance to rural North Carolina communities that seek to improve community health by implementing innovative and integrated strategies to increase physical activity. Up to five grantees will be chosen to develop and maintain interdisciplinary community partnerships that seek to implement programs, policies, physical projects, and promotional strategies that increase access and reduce barriers to opportunities for active living.
Requirements: North Carolina nonprofit or government organizations are eligible.
Geographic Focus: North Carolina
Date(s) Application is Due: Dec 16
Amount of Grant: 40,000 U.S.D
Contact: Polly Leousis, (919) 765-4102
Internet: http://www.bcbsnc.com/foundation/fitogether_grants.html
Sponsor: Blue Cross Blue Shield of North Carolina Foundation
P.O. Box 2291
Durham, NC 27707-0718

BCBSNC Foundation Grants 897

The foundation's primary objective is to fund clearly defined, innovative grants that further the foundation mission of improving the health and well-being of North Carolinians. The foundation funds programs typically possessing the following characteristics: programs and/or services designed to produce measurable, long-term impact; programs that are sustainable and designed to be ongoing, rather than one-time or sporadic events; programs and/or services that are replicable. The three primary focus areas include health of vulnerable populations, healthy active communities and community impact through nonprofit excellence.

Requirements: Applicants must meet the following criteria to be eligible for funding: organization is located within North Carolina; organization is a 501(c)3 organization or an educational or governmental entity with tax-exempt status, that is not a private foundation or Type III supporting organization; organization must be able to provide its most recent IRS Form 990. Depending on the type of grant and the size of the organization, an audit may be required as part of the submitted proposal.

Restrictions: The foundation will not provide funding for annual campaigns, political campaigns, religious purposes, individuals, endowments, purchase of advertisements, or for the sole purpose of receiving goods or entitlements from a charitable organization.

Geographic Focus: North Carolina

Samples: Diabetes Management Solutions (Raleigh, NC)—to support a pilot program for a physician diabetes day clinic as a model for improved outcomes in diabetes management, $10,980; Interfaith Assistance Ministry (Henderson, NC)—funding for a computer in every office to expedite the access and exchange of information for the Medi-Find program, $8430; Housing for New Hope (Durham, NC)—to build capacity of a comprehensive and sustainable health program for residents of five housing programs, $10,000; Graham Children???s Health Services of Toe River (Burnsville, NC)—to support preventive care for Hispanic and other uninsured children at school-based health centers, $20,465.

Contact: Grant Review Committee; (919) 765-7347; foundation@bcbsnc.com

Internet: http://www.bcbsnc.com/foundation/grants.html?previouslyOver=true¤tlyOver=true

Sponsor: Blue Cross Blue Shield of North Carolina Foundation
P.O. Box 2291
Durham, NC 27707-0718

BCBS of Massachusetts Catalyst Fund Grants 898

The program provides small grants to support capacity building in community-based organizations. Grants will assist with one-time expenses needed to strengthen an organization's ability to fulfill its mission. Examples of activities that could be supported include community outreach for health insurance enrollment, health prevention activities, or health care education convening; hiring a grant writer; community meetings to engage stakeholders and constituents; leadership development opportunities; strategic planning activities; and board development and staff training.

Requirements: Grants will only be made to Massachusetts-based organizations that serve the health care needs of uninsured and low-income residents of the state.

Restrictions: Grants will not be made to individuals, for-profit entities, hospitals, or public agencies. Grants are not intended to cover recurring general operating funds, such as salaries, overhead, or membership dues.

Amount of Grant: 500 - 2,500 U.S.D

Contact: Grants Administrator; (617) 246-3744; bcbsmafoundation.org

Internet: http://www.bcbsmafoundation.org/foundationroot/en_U.S./documents/catalystFundguidelines.pdf

Sponsor: Blue Cross Blue Shield of Massachusetts Foundation
401 Park Drive
Boston, MA 02115-3326

BCBS of Massachusetts Foundation Grants 899

With a mission to expand access to health care, the foundation awards grants intended to make a significant impact on the health of Massachusetts's low income and uninsured residents. Grant programs include Innovation Fund for the Uninsured; Connecting Consumers with Care; Strengthening the Voice for Access; Pathways to Culturally Competent Health Care; Catalyst Fund; Building Bridges in Children's Mental Health; and Policy Research and Analysis. Grant awards in particular focus areas are made during specific grant cycles. Deadlines and additional information are available online.

Requirements: Massachusetts 501(c)3 nonprofit organizations or government agencies with a mission that includes improving access and removing barriers to healthcare for low-income or uninsured Massachusetts's residents are eligible.

Restrictions: The foundation will not fund individuals; for-profit organizations; capital campaigns, endowments, or building drives; fundraising drives and events; retiring debt or operating deficits; direct political or lobbying activity; projects outside of Massachusetts; or religious organizations for religious purposes (proposals for secular programs of faith-based organizations that meet funding criteria will be considered). Grants are rarely made for curriculum development, conferences, film or video production, or scholarships.

Geographic Focus: Massachusetts

Date(s) Application is Due: Jan 20; Sep 7

Samples: Beth Israel Deaconess Medical Ctr (Boston, MA)—to help managers select, retain, and promote highly qualified, diverse staff members, $40,000; Boston U, Goldman School of Dental Medicine (Boston, MA)—to develop policy recommendations designed to increase access to fluoridated water, with a focus on uninsured and low-income people and minorities, $46,716; John Snow Research and Training Institute (Boston, MA)—to study the extent to which a subsidized health-insurance program for people in the Massachusetts fishing industry can be duplicated in other industries, $50,000; Massachusetts Department of Mental Health (Boston, MA)—to integrate primary care and mental-health services for underserved Hispanics in the Worcester, MA, area, $50,000.

Contact: Program Coordinator; (617) 246-3744; fax (617) 246-3992; grantinfo@bcbsmafoundation.org

Internet: http://www.bcbsmafoundation.org/foundationroot/en_U.S./grants/focusArea.jsp#aInnovation%20Fund%20for%20the%20Uninsured

Sponsor: Blue Cross Blue Shield of Massachusetts Foundation
401 Park Drive
Boston, MA 02115-3326

BCBS of Massachusetts Health Coverage Fellowships 900

The fellowship, the first of its kind in the country, is designed to help New England media do a better job covering critical health care issues. Ten reporters and editors from newspapers, radio stations, and television outlets are chosen each year to participate in a nine-day intensive program to deepen their understanding of the complex and changing world of medicine and health care. Throughout the remainder of the year, fellows participate in half-day retreats and can consult on stories and projects with former Boston Globe medical reporter Larry Tye. Application is available online.

Requirements: New England health-care journalists are eligible. Reporters must have an editor's approval.

Date(s) Application is Due: Dec 1

Contact: Larry Tye, info@bcbsmafoundation.org

Internet: http://www.bcbsmafoundation.org/foundationroot/en_U.S./policyAndPrograms/policyPresent.jsp?repositoryId=/foundationroot/en_U.S./repositories/commonMainContent/policyAndPrograms/healthCoverageFellowship.xml

Sponsor: Blue Cross Blue Shield of Massachusetts Foundation
401 Park Drive
Boston, MA 02115-3326

BCBS of Minnesota Foundation Healthy Together Grants 901

The program, Healthy Together: Creating Community with New Americans, will explore, implement, and evaluate strategies at the intersection of health, social connectedness, and immigrant integration. Applicants may propose planning or implementation projects related to one or more of the initiative's three objectives: to promote the mental health and healthy social adjustment of new Americans; to strengthen the organizational capacity of immigrant-led institutions; and to promote exchanges among newcomers and the community-at-large that will lead to greater social connectedness and healthier communities. The foundation awards one-year planning grants and one- to three-year implementation grants for community, regional, or statewide efforts. The listed application deadlines are for letters of inquiry; full proposals are by invitation.

Requirements: Minnesota 501(c)3 tax-exempt organizations and nonprofit organizations serving Minnesotans are eligible.

Restrictions: As a general rule, the foundation does not award more than one grant to an organization at a time. In addition, the foundation does not provide grants to or make loans for the following: individuals; lobbying, political, or fraternal activities; sports events and athletic groups; religious purposes; clinical quality improvement activities; biomedical research, capital purposes (building, purchase, remodeling, or furnishing of facilities); equipment or travel, except as related to requests for program support; endowments, fundraising events, or development campaigns; retiring debt or covering deficits; payment of services or benefits reimbursable from other sources; supplanting funds already secured for budgeted staff and/or services; or long-term financial support.

Geographic Focus: Minnesota

Date(s) Application is Due: May 9; Aug 8

Amount of Grant: Up to 150,000 U.S.D

Contact: Grants Administrator; (866) 812-1593 or (651) 662-3950; fax (651) 662-1361; foundation@bluecrossmn.com

Internet: http://www.bluecrossmn.com/public/foundation/healthy_together.html

Sponsor: Blue Cross and Blue Shield of Minnesota Foundation
3535 Blue Cross Road, Route M459
Eagan, MN 55122

BCBS of Minnesota Growing Up Healthy Grants 902

Growing Up Healthy: Kids and Communities is a statewide initiative to improve the health of Minnesota's children through a focus on social and environmental determinants of health. The goal is to build strong and connected communities where children can thrive and grow up healthy by working at the intersection of health and two or more of the following health determinants: early childhood development; stable, affordable housing; and the physical environment.

Requirements: Eligible applicants include community- and faith-based organizations; state, county and municipal agencies; tribal governments and agencies; professional associations or collaboratives; and policy and research organizations. Applicants must be located in Minnesota or serve Minnesotans. Eligible applicants include units of government as well as organizations designated as nonprofit under section 501(c)3 of the IRS code.

Geographic Focus: Minnesota

Date(s) Application is Due: Apr 13

Contact: Lisa Simer, Program Officer; (651) 662-0138; Lisa_K_Simer@bluecrossmn.com.

Internet: http://www.bluecrossmn.com/bc/wcs/groups/bcbsmn/@mbc_foundation/documents/public/mbc2_index.hcsp

Sponsor: Blue Cross Blue Shield of Minnesota Foundation
3535 Blue Cross Road, Route M459
Eagan, MN 55122

BCBS of Minnesota Healthy Together Grants 903

Healthy Together: Creating Community with New Americans is a grantmaking initiative designed to reduce health disparities for immigrants and improve the health and vitality of the entire community. The foundation awards grants to projects that foster exchanges and interactions between newcomers and the receiving community, strengthen the

capacity of immigrant-led organizations and their attention to health, and address social adjustment and mental health.
Requirements: Eligible applicants include community- and faith-based organizations; state, county and municipal agencies; tribal governments and agencies; professional associations or collaboratives; and policy and research organizations. Applicants must be located in Minnesota or serve Minnesotans. Eligible applicants include units of government as well as organizations designated as nonprofit under section 501(c)3 of the IRS code.
Geographic Focus: Minnesota
Date(s) Application is Due: Feb 1
Amount of Grant: 1,200 - 250,000 U.S.D
Samples: Episcopal Community Services, Inc., $25,000 to engage Native Americans in the Bemidji community in early childhood development and stable housing issues; Folwell Neighborhood Association, $25,000 to improve school readiness of 500 children; National Center for Healthy Housing, $150,000 to demonstrate green building principles and tenant education in Worthington, Minnesota; Resources for Child Caring, $25,000 to improve home environments of children in the Payne-Phalen neighborhood.
Contact: Jocelyn Ancheta; (651) 662-2894; Jocelyn_L_Ancheta@bluecrossmn.com
Internet: http://www.bluecrossmn.com/public/foundation/index.html
Sponsor: Blue Cross Blue Shield of Minnesota Foundation
3535 Blue Cross Road, Route M459
Eagan, MN 55122

BCRF-AACR Grants for Translational Breast Cancer Research **904**
The grants will provide direct support for innovative cancer research projects designed to accelerate the discovery, development, and application of new agents to treat breast cancer and/or for pre-clinical research with direct therapeutic intent. Special emphasis will be placed on research that holds promise for leading to individualized therapeutic options for treatment in the near future.
Requirements: Researchers who are affiliated with any institution involved in cancer research, cancer medicine, or cancer-related biomedical science anywhere in the world may apply. There are no geographic, national, or residency status restrictions. Applicants must have acquired a doctoral degree in a related field. If an applicant has obtained an equivalent degree at a foreign institution, information on the nature of the degree must be provided at the time of application.
Restrictions: Employees or subcontractors of a national government or the for-profit private industry are not eligible to serve as the Principal Investigator for the purpose of these grants. However, collaborations with such individuals are encouraged. Neither the members of the Scientific Review Committee nor the members of their individual laboratories are eligible for these grants.
Date(s) Application is Due: May 20; Jul 20
Amount of Grant: 200,000 U.S.D
Contact: Hanna Hopfinger, Program Assistant; (267) 646-0665 or (215) 440-9300; fax (215) 440-9372; hanna.hopfinger@aacr.org or awards@aacr.org
Internet: http://www.aacr.org/home/scientists/research-funding—fellowships/bcrf-aacr-grants.aspx
Sponsor: Breast Cancer Research Foundation and the American Association for Cancer Research
615 Chestnut Street, 17th Floor
Philadelphia, PA 19106-4404

BCRF Research Grants **905**
The goal of the grant process is a serious peer review of all proposals. Proposals are invited by the Medical Advisory Board, rather than accepted as unsolicited requests. The MAB generally reviews proposals in the late summer. There is no set format, but the BCRF requests a brief narrative (no more than 4-5 pages), lay language summary, annual budget (no more than $225,000 initially, with a maximum of 20% in indirect costs) and budget narrative. Considerable latitude is given to the investigators in terms of the work proposed. Both the Board of Directors and the Medical Advisory Board concur that some of the most important advances in understanding the disease will most likely occur by enabling brilliant minds to pursue some of their most creative theories.
Restrictions: While most of the grants are unrestricted, the Foundation has donors requesting restricted grants for specific purposes. When this occurs, the Medical Advisory Board will meet to determine whether the area of interest merits further exploration. If so, they will identify various research projects meeting the donor's criteria.
Amount of Grant: Up to 225,000 U.S.D
Contact: Program Administrator; (646) 497-2600; fax (646) 497-0890; bcrf@bcrfcure.org
Internet: http://www.bcrfcure.org/action_grantguidelines.html
Sponsor: Breast Cancer Research Foundation
60 East 56th Street, 8th Floor
New York, NY 10022

Bearemy's Kennel Pals Grants **906**
The program provides direct support for animals in domestic pet programs including animal welfare organizations, pet rescue and rehabilitation organizations, and therapeutic and humane education pet programs. Grants will be a one-time contribution and generally range from $1,000 to $10,000, but the average grant tends to be $4,000. Programs that will be funded include: (1) Individual Project grants generally for one-time purchases or to fulfill a short-term need. Examples include equipment purchases or spay/neuter events, etc; (2) Organization Program grants - start-up or operational costs for ongoing programs. Examples include spay/neuter programs, humane education initiatives, and service dog training programs.
Requirements: While the geographic focus of the program is broad (United States and Canada), priority is given to organizations located near Build-A-Bear Workshop stores. United States applicants must be a tax-exempt organization under Section 501(c)3 of the IRS Code, and not a private foundation, within the meaning of Code Sections 509(a)(1) or 509(a)(2), or a state college or university within the meaning of Code Section 511(a)(2)(B) (a Public Charity). In addition, grant recipients must certify that they are not a supporting organization within the meaning of Code Section 509(a)(3). Canadian applicants must be a registered Canadian charity.
Restrictions: Grant types not funded: (1) Capital Campaigns; (2) Construction or new facility expenses; (3) Fundraising or Event Sponsorships; (4) Political Activities; (5) Religious organizations for religious purposes.
Geographic Focus: All States, Canada
Date(s) Application is Due: Mar 31; Jun 30; Sep 31; Dec 31
Amount of Grant: 1,000 - 10,000 U.S.D
Samples: Animal Advocates of South Carolina (Barnwell, SC); Bunny Castle Rabbit Rescue (Bishop, CA); EquiFriends (Snohomish, WA); Hearing and Service Dogs of Minnesota (Minneapolis, MN); Pet Adoption Fund (Canoga Park, CA).
Contact: Maxine Clark, President; (314) 423-8000, ext. 5366; giving@buildabear.com
Internet: http://www.buildabear.com/aboutus/community/bearhugs.aspx
Sponsor: Build-A-Bear Workshop Bear Hugs Foundation
1954 Innerbelt Business Center Drive
Saint Louis, MO 63114

Beatrice Laing Trust Grants **907**
The trust was founded for the relief of poverty and for the advancement of the evangelical faith internationally. Grants are awarded for general charitable purposes at the discretion of the Trustees, with an emphasis on social services and relief, and medical aid and research. Grant recipients typically are charities working in deprived sections of the community in the United Kingdom, to missionary societies and, less frequently, to individuals working in the field of missions in United Kingdom and abroad.
Geographic Focus: All States, All Countries
Samples: ActionAid, 10,000 pounds to improve access to education for pastoralist families in Somaliland; OxFam, 10,000 pounds to the You Are Here program; Sense International, 15,000 pounds for work with deaf-blind people in Romania.
Contact: Trust Administrator; 020 8238 8890; fax 020 8238 8897
Sponsor: Beatrice Laing Trust
33 Bunns Lane
London, NW7 2DX England

Beazley Foundation Grants **908**
The foundation makes grants to Virginia nonprofit organizations engaged in educational, charitable, and religious activities. The focus of support is within the South Hampton Roads cities and counties. Past awards include: music appreciation; educational outreach and mentoring programs; health and wellness; equipment and operating expenses; environmental education programs; renovations and repairs; literacy and community development. Grant applications can be found online.
Requirements: 501(c)3 nonprofit organizations in Virginia are eligible.
Restrictions: Individuals are ineligible.
Geographic Focus: Virginia
Date(s) Application is Due: Jan 4; Mar 1; Jun 1; Oct 1
Samples: The Tidewater Winds, Norfolk, VA, $7,500—operating expenses; The Union Mission Ministries, Norfolk, VA, $125,000—capital campaign; Virginia College Fund, Richmond, VA, $50,000—Beazley Scholarship Endowments; The Academy of Music, Norfolk, VA, $42,000— Park View Elementary Strings Program.
Contact: Donna Russell, Associate Director; (757) 393-1605; fax (757) 393-4708; info@beazleyfoundation.org
Sponsor: Beazley Foundation
3720 Brighton Street
Portsmouth, VA 23707-3902

Bechtel Group Foundation Building Positive Community Relationships Grants **909**
Bechtel Group Foundation was created in 1954 to respond to the needs of the communities around the world in which Bechtel has offices or major projects. These grants support educational, civic and cultural, and social service programs in the communities that host major Bechtel offices and projects. Bechtel offices are located in the following regions/countries: Australia; Brazil; Canada; Chile; Greater China; Egypt; France; India; Indonesia; Japan; Korea; Libya; Malaysia; Peru; Philippines; Poland; Qatar; Russia; Saudi Arabia; Singapore; Taiwan; Thailand; Turkey; United Arab Emirates; United Kingdom - London, England; United States - Maryland, Arizona; Texas; Virginia; New York; Tennessee; Washington; California; Washington, D.C. These grants are typically under U.S. $5,000.
Geographic Focus: Arizona, California, Maryland, Tennessee, Texas, Virginia, Washington, Australia, Brazil, Canada, China, Japan, Peru, Philippines, Poland, Russia, United Kingdom
Amount of Grant: 1,000 - 5,000 U.S.D
Contact: Susan Grisso, Foundation Manager; (415) 768-5444; becfoun@bechtel.com
Internet: http://www.bechtel.com/foundation.html
Sponsor: Bechtel Group Foundation
50 Beale Street
San Francisco, CA 94105

Beckley Area Foundation Grants **910**
The Foundation makes annual discretionary grants to charitable organizations in the areas of education, health and human services, the arts, public recreation and civic beautification. Using several general unrestricted endowments which generate income,

BAF makes these grants to address the community's current most pressing needs and promising opportunities.
Requirements: To be eligible for a grant from the Beckley Area Foundation, applicants must be a non-profit organizatiion tax-exempt under section 501(c)3 of the Internal Revenue Code, or they must be a public institution. Programs funded through this annual grant must be located in Raleigh County.
Restrictions: As a Community Foundation, BAF cannot accept requests on behalf of individuals or groups that do not have official recognition as a non-profit charity or public entity. Grants are for one year only. Generally grants are not made for: annual campaigns; endowments; sectarian religious programs; political purposes or lobbying; ongoing operating expenses; expenses outside the grant period.
Geographic Focus: West Virginia
Date(s) Application is Due: Dec 15
Amount of Grant: 10,000 U.S.D
Contact: Grants Administrator; (304) 253-3806; fax (304) 253-7304; funds@beckleyareafoundation.com
Internet: http://beckleyareafoundation.com/id7.html
Sponsor: Beckley Area Foundation
129 Main Street, Suite 203
Beckley, WV 25801

Beckman Coulter Community Relations Grants 911
The corporation funds nonprofit organizations, charities and educational institutions that support and promote progress and interest in the areas of science, research-related health care and science education in our plant and customer communities.
Requirements: Organizations seeking funding should submit a letter of request providing the following information: full legal name of the organization; brief description of the organization, including a mission statement and services provided; amount requested; description of how the funds will be used. Additional attachments should include: history of previous support by Beckman Coulter; description of any involvement by Beckman Coulter employees; statement as to why you consider Beckman Coulter an appropriate donor; list of governing board members.
Restrictions: Grants do not support golf tournaments, galas, or religious or political activities.
Contact: Cathy Doherty, Community Relations; (714) 961-4478
Internet: http://www.beckman.com/hr/ourcompany/oc_communityRelations.asp#
Sponsor: Beckman Coulter Corporation
Community Relations, 200 S Kraemer Boulevard
Brea, CA 92821

Beckman Scholars Program Grants 912
The Program is an invited program for accredited universities and four-year colleges in the United States. It provides scholarships that contribute significantly in advancing the education, research training and personal development of select students in chemistry, biochemistry, and the biological and medical sciences. The sustained, in-depth undergraduate research experiences and comprehensive faculty mentoring are unique in terms of program scope, content and level of scholarship awards. The amount given is $19,300 for two summers and one academic year.
Amount of Grant: 19,300 U.S.D
Contact: Program Administrator; (949) 721-2222; fax (949) 721-2225; beckmanscholars@beckman-foundation.com or k.williams@beckman-foundation.com
Internet: http://www.beckman-foundation.com/bsp.html
Sponsor: Arnold and Mabel Beckman Foundation
100 Academy Drive
Irvine, CA 92617

Beckman Young Investigators Program Grants 913
The foundation makes grants to nonprofit research institutions to promote research in chemistry and the life sciences, broadly interpreted, and particularly to foster the invention of methods, instruments, and materials that will open up new avenues of research in science. The program is intended to provide research support to the most promising young faculty members in the early stages of academic careers in the chemical and life sciences. The program is open to persons with tenure-track appointments in academic and nonprofit institutions that conduct fundamental research in the chemical and life sciences. The program is intended primarily for U.S. institutions. Only proposals of exceptional merit from foreign institutions will receive consideration. Projects are normally funded for a period of two years. When extraordinary circumstances warrant, support may be provided over a one-year or three-year period. Application materials are available on the Web site.
Requirements: To be eligible, an applicant should not have completed more than three full years in his or her tenure-track or other comparable independent research appointment on the anniversary date of initial appointment in the year in which application is to be made. Individuals may not apply more than three times.
Restrictions: Funding will not be considered for general institutional expenses or general fund-raising campaign expenses such as dinners and mass mailings. The foundation does not provide funds for overhead or indirect costs.
Date(s) Application is Due: Sep 28
Amount of Grant: 264,000 U.S.D
Contact: Program Administrator; (949) 721-2222; fax (949) 721-2225; younginvestigators@beckman-foundation.com or k.williams@beckman-foundation.com
Internet: http://www.beckman-foundation.com/byi_guides.html
Sponsor: Arnold and Mabel Beckman Foundation
100 Academy Drive
Irvine, CA 92617

Becton Dickinson and Company Grants 914
The company supports national and international activities in keeping with its commitment to advance the quality of medical practice and patient care in four areas: the global healthcare fund; the local initiatives fund the matching gifts program; and the BD product donation program. Call for guidelines.
Contact: Community Relations Manager; (201) 847-5800
Internet: http://www.bd.com/responsibility/contributions
Sponsor: Becton Dickinson and Company
1 Becton Drive
Franklin Lakes, NJ 07417

Bedford Community Health Foundation Grants 915
The foundation awards grants to nonprofit health associations. Types of support include general operating support, equipment acquisition, and scholarships to individuals. Application forms are not required. Grants may be submitted throughout the year, however the Board of Trustees will only review grants at the May and November meetings. Applicants should call the executive director to discuss the proposal and to arrange a meeting.
Requirements: Nonprofits serving Bedford and Bedford County are eligible.
Restrictions: Individuals are not eligible.
Geographic Focus: Virginia
Date(s) Application is Due: Apr 1; Oct 1
Amount of Grant: 250 - 50,000 U.S.D
Contact: Program Contact; (540) 586-5292; fax (540) 587-5819; bchf@bchf.org
Internet: http://www.bchf.org
Sponsor: Bedford Community Health Foundation
P.O. Box 1104
Bedford, VA 24523

Beez Foundation Grants 916
The foundation's mission is to raise money for pediatric brain cancer research and pediatric patient services. Research and/or support grants are awarded for brain cancer research projects or studies that are designed to advance the treatment, cure, outcome, and etiology of pediatric brain cancer; and any patient services that are focused on making the lives of children with cancer and their families a little easier. The foundation's particular interest is in the initial funding of projects that are likely to generate data that can be used in larger scale projects or studies. Preference is made to pediatric cancer-related requests. Applicants are requested to forward a one-page summary (in Microsoft Word or Adobe Acrobat format), the amount requested, and how the funds will be spent.
Date(s) Application is Due: Nov 20
Amount of Grant: Up to 40,000 U.S.D
Contact: Grants Administrator; (732) 563-1144; proposals@beezfoundation.org
Internet: http://www.beezfoundation.org/Pages/Grants.html
Sponsor: Beez Foundation
26 H Worlds Fair Drive
Somerset, NJ 08873

Beirne Carter Foundation Grants 917
The Foundation places an emphasis on health, education, local history, nature, ecology and youth. Except in rare instances, grants are limited to institutions and organizations located in the Commonwealth of Virginia. Requests that are not of a recurring nature are given preference. Any organization that has been awarded a grant is encouraged to wait at least three years before submitting a new grant request. Any organization seeking grant consideration must complete a Preliminary Proposal form which is online. Its content should be limited to the two page format provided online.
Requirements: The Foundation is organized exclusively for charitable purposes. Grants are made only to institutions and organizations which qualify under IRS regulations as tax exempt and which are not private operating foundations as defined by the IRS.
Restrictions: Grants will not normally be made: to endowment funds (including scholarship funds); to organizations supported primarily by government funds (such as public secondary schools and colleges and local municipalities); to churches and related organizations; for ongoing operating expenses (such as salaries, rent, office supplies, etc.); for existing deficits; or for debt reduction. Grants are not made to individuals.
Geographic Focus: Virginia
Date(s) Application is Due: Feb 1; Aug 3
Contact: Peter C. Toms; (804) 521-0272; fax (804) 521-0274; bcarterfn@aol.com
Internet: http://www.bcarterfdn.org
Sponsor: Beirne Carter Foundation
1802 Bayberry Court, Suite 401
Richmond, VA 23226-3773

Beldon Fund Grants 918
The fund focuses project and general support grants in two programs: human health and the environment, and key states. Human health and the environment—the fund seeks proposals that engage new constituencies in exposing the connection between toxic chemicals and human health and in promoting public policies that prevent or eliminate environmental risks to people's health. The program focuses grant making in three areas: new advocates, human exposure to toxic chemicals, and environmental justice. Key states—the fund believes that states hold the key to bringing about rapid, real change on environmental issues and policy in the United States. By strengthening public support for environmental protection in several of these key states, the fund hopes to transform the nation's approach to environmental protection. The fund is currently accepting proposals from Florida, Michigan, Minnesota, Wisconsin, and North Carolina for this program. Proposals do not need to be tied to any particular issue or set of issues, but targeted

issues must be those that will build active public support for the environment. From time to time, the fund will add and remove states from this program. Types of support include special projects, seed money, general operating budgets, and technical assistance. One-year and multiyear grants are awarded. Due date applies to the letter of intent; full proposals are by invitation.By supporting effective, nonprofit advocacy organizations, the Beldon Fund seeks to build a national consensus to achieve and sustain a healthy planet. The Fund plans to invest its entire principal and earnings by 2009 to attain this goal.
Requirements: 501(c)3 tax-exempt organizations are eligible.
Restrictions: Grants do not support international efforts, academic or university efforts, school-based environmental education, land acquisition, wildlife or habitat preservation, film or video production, deficit reduction, endowments, capital campaigns, acquisitions of museums, service delivery, scholarshp, publications, or arts/culture.
Geographic Focus: Florida, Michigan, Minnesota, North Carolina, Wisconsin
Date(s) Application is Due: Feb 28; Jun 13
Amount of Grant: 5,000 - 100,000 U.S.D
Samples: Ctr for Public Interest Research (Boston, MA)—for its Campaign Staff Training Institute Program to ameliorate the shortage of skilled activists, $75,000; Institute for Conservation Leadership (Takoma Park, MD)—for its outreach and hands-on support efforts, along with Environmental Support Ctr, $10,000; State Environmental Leadership Program (Madison, WI)—for general support, $175,000; Washington Environmental Alliance for Voter Education (Seattle, WA)—for general support, $50,000.
Contact: Holeri Faruolo, Grants Manager; (800) 591-9595 or (212) 616-5600; fax (212) 616-5656; info@beldon.org
Internet: http://www.beldon.org
Sponsor: Beldon Fund
99 Madison Avenue, 8th Floor
New York, NY 10016

Belk Foundation Grants 919
The Foundation makes grants to a wide variety of community-based nonprofit organizations and institutions whose missions and actions support the advancement of Christian causes and the up-building of mankind. The Foundation supports local and regional organizations by: assisting secondary schools, colleges and universities and their programs; assisting religious institutions and organizations and their programs; supporting area arts and other cultural organizations and their programs; supporting community-based human services organizations and their programs; and aiding hospitals and health care organizations and their programs.
Requirements: 501(c)3 nonprofits in communities in the 14 states where Belk stores are located. Preference is given to organizations in North Carolina.
Restrictions: Grants are not awarded to: individuals, including students; public, government or quasi-governmental programs, agencies or organizations (excluding certain public secondary schools, colleges and universities); or international programs and/or organizations. Additionally, the Foundation does not provide door prizes, gift certificates, merchandise or other giveaways.
Geographic Focus: Alabama, Arkansas, Florida, Georgia, Kentucky, Louisiana, Maryland, Mississippi, North Carolina, South Carolina, Tennessee, Texas, Virginia, West Virginia
Date(s) Application is Due: Apr 15; Oct 15
Contact: Susan C. Blount, Foundation Administrator; (704) 426-8396; susan_blount@belk.com
Internet: http://www.belk.com/AST/Misc/Belk_Stores/About_Us/Belk_Community/Belk_Foundation.jsp
Sponsor: Belk Foundation
2801 W Tyvola Road
Charlotte, NC 28217-4500

Bella Vista Foundation Grants 920
The foundation has two main focus areas: early childhood—healthy emotional development in children during the first years of their lives, and parents involvement to help their children deal with the internal issues of attachment and independence; and environment restoration—restoration of land, streams, wetlands, habitat, etc., and acquisition of land for the purpose of preservation, particularly requests that include restoration efforts. Deadlines are January 30 and June 15 for environmental restoration grants, and January 15 and June 15 for early childhood grants. Guidelines are available online.
Requirements: Early childhood grants support public charities located in California's San Francisco, Marin, San Mateo, and Santa Clara counties. Environmental restoration grants support projects in California or Oregon.
Restrictions: Grants will not be made to or for the arts, sectarian religious purposes, individuals, or benefit events, and will only be made for medical research, health care, publications, or video production under special circumstances and only in the early childhood development focus area. The foundation does not make multiyear grants.
Geographic Focus: California, Oregon
Date(s) Application is Due: Jan 15; Jan 30; Jun 15
Amount of Grant: Up to 100,000,000 U.S.D
Contact: Mary Gregory, Executive Director; (415) 561-6540; fax (415) 561-6477; mgregory@pfs-llc.net
Internet: http://www.pfs-llc.net/bellavista/index.html
Sponsor: Bella Vista Foundation
Presidio Building 1016, Suite 300, P.O. Box 29906
San Francisco, CA 94129-0906

Ben B. Cheney Foundation Grants 921
Ben B. Cheney Foundation supports projects in communities where the Cheney Lumber Company was active. These areas include: Tacoma, Pierce, southwestern Washington, southwestern Oregon with a focus on Medford, portions of Del Norte, Humboldt, Lassen, Shasta, Siskiyou, and Trinity counties in California. The Foundation's goal is to improve the quality of life in those communities by making grants to a wide range of activities including: Charity (programs providing for basic needs such as food, shelter, and clothing); Civic (programs improving the quality of life in a community as a whole such as museums and recreation facilities); Culture (programs encompassing the arts); Education (programs supporting capital projects and scholarships, primarily for fourteen pre-selected colleges and universities in the Pacific Northwest); Elderly (programs serving the social, health, recreational, and other needs of older people); Health (programs related to providing health care); Social Services (programs serving people with physical or mental disabilities or other special needs); Youth (programs helping young people to gain the skills needed to become responsible and productive adults). Ben B. Cheney Foundation prefers to fund projects that: develop new and innovative approaches to community problem; facilitate the improvement of services or programs; invest in equipment or facilities that will have a long-lasting impact on community needs. The Foundation's application process always begins with a two to three page proposal letter. The process is the same for past grantees and new grant seekers. There are no deadlines. The Foundation accepts proposal letters throughout the year. It may take six to nine months from the receipt of a proposal letter to consideration of a grant application by the Board of Directors.
Requirements: Ben B. Cheney Foundation makes grants to private, nonprofit organizations that have received their 501(c)3 status from the IRS and that qualify as public charities. In special circumstances proposals from governmental organizations are allowed.
Restrictions: Ben B. Cheney Foundation generally does not make grants for: general operating budgets, annual campaigns, projects which are primarily or normally financed by tax funds, religious organizations for sectarian purposes, basic research, endowment funds, individuals, produce books, films, videos, conferences, seminars, attendance, individual student, or student groups raising money for school related trips.
Geographic Focus: California, Oregon, Washington
Amount of Grant: Up to 300,000 U.S.D
Samples: Kitsap Mental Health Services, Bremerton,Washington, $50,000 - build Keller House for residential and outpatient programs; Ocean Shores Friends of the Library, Oceans Shores, Washington $4,000 - support the conference for Nonprofit Leaders in SW Washington to enhance fundraising; Hoover Elementary PTO, Medford, Oregon, $1,260 - provide the last monies needed to complete the new Boundless community playground at the school; Catholic Community Services, Tacoma, Washington, $40,000 - support operations of the downtown emergency shelter.
Contact: Bradbury F. Cheney; (253) 572-2442; info@benbcheneyfoundation.org
Internet: http://www.benbcheneyfoundation.org
Sponsor: Ben B. Cheney Foundation
3110 Ruston Way, Suite A
Tacoma, WA 98402-5307

Bender Foundation Grants 922
The foundation supports projects for higher education, health agencies, Jewish welfare funds and organizations, Christian youth organizations, and social welfare. Emphasis is support for programs to assist the elderly and aging, children and parenting, and environmental programs. Grants are awarded for challenge/matching grants, endowments, general operations, and scholarships.
Requirements: Grants are made to organizations and institutions in Maryland and Washington, DC. An applicant should initially send a brief letter of intent. Full proposals are by invitation.
Restrictions: Grants are not made to individuals.
Geographic Focus: District of Columbia, Maryland
Date(s) Application is Due: Nov 30
Amount of Grant: 100 - 100,000 U.S.D
Contact: Julie Bender Silver, President; (202) 828-9000; fax (202) 785-9347
Sponsor: Bender Foundation
1120 Connecticut Avenue NW, Suite 1200
Washington, DC 20036

Benign Essential Blepharospasm Research Grants 923
The foundation supports research to find the cause and cure for benign essential blepharospasm/meige and related disorders of the facial musculature. A number of grants are issued annually. Proposals detailing research projects to be considered for support are welcomed.
Date(s) Application is Due: Jul 1
Amount of Grant: Up to 150,000 U.S.D
Samples: Dr. John McCann, U of Utah (Salt Lake City, UT)—for the project Trial of Cervical Sympathetic Blockade for the Treatment of Benign Essential Blepharospasm, $13,000; Dr. Mark Hallett, NINDS, NIH (Bethesda, MD) and Dr. Robert Daroff, Case Western Reserve U (Cleveland, OH)—for the blepharospasm workshop, $25,763; Dr. Michael Hutchinson, NYU Medical Ctr (New York, NY)—for the project Neuroimaging of Blepharospasm, $62,325.
Contact: Office Manager; (409) 832-0788; fax (409) 832-0890; bebrf@blepharospasm.org
Internet: http://www.blepharospasm.org/res-prop.html
Sponsor: Benign Essential Blepharospasm Research Foundation
P.O. Box 12468
Beaumont, TX 77726-2468

Benton Community Foundation - The Cookie Jar Grant 924

The Benton Community Foundation - The Cookie Jar Grant was established by women in the Benton County area to fund programs that empower women. The Cookie Jar is a collaboration of women working together to create a stronger community where women and girls are empowered to reach their full potential. Priority is given to programs that improve the overall health or well-being of women and/or girls in Benton County. Priority is also given to programs that: reach as many people as possible; are run by a collaboration of non-profit organizations; and have multiple sources of funding for their project.
Requirements: Organizations should submit the grant application to the Foundation to include the following information: their organizational information and contacts; a summary of their program with amount requested; program details, such as how the program will benefit women or girls in Benton County; the program's anticipated timeline; the organization's top three goals for the project and how they plan to meet them; budget details; and a list of the organization's board of directors.
Restrictions: Foundation grant programs generally do not fund: ongoing operating expenses; individuals; special events such as parades, festivals and sporting events; debt or deficit reduction.
Geographic Focus: Indiana
Date(s) Application is Due: Sep 15
Amount of Grant: 4,000 U.S.D
Contact: Ashley Bice; (765) 884-8022; fax (765) 884-8023; ashley@bentoncf.org
Internet: http://www.bentoncf.org/cookiejar.html
Sponsor: Benton Community Foundation
P.O. Box 351
Fowler, IN 47944

Benton Community Foundation Grants 925

The Foundation grants address the broad needs of Benton County residents. Funding is allocated in the following categories: civic affairs; cultural affairs; education; health and safety; and social services. Priority is given to programs that: reach as many people as possible; improve the ability of the organization to serve the community over the long term; serve Benton County residents; are run by a collaboration of nonprofit organizations; and have multiple sources of funding for the project.
Requirements: Grant requests are accepted year-round. Applicants are encouraged to contact the Foundation prior to submitting a request to confirm that the project is appropriate for Foundation funding. To make a request of $750 or less, organizations should submit a letter of application which includes a description of the organization, statement of the problem or need being addressed, explanation of the project, estimated expenses, timeframe for completion, and the amount being requested. Grant decisions are made within 30 days of application. Organizations requesting more than $750 are encouraged to contact the Foundation to discuss the proposed project/program prior to submitting an application. They will then complete the Community Grant Application. Grant decisions are made within 90 days of application.
Restrictions: Benton Community Foundation generally does not fund: political organizations or candidates; endowments; ongoing operating expenses; individuals; special events such as parades, festivals, and sporting events; debt or deficit reduction; or projects funding in a previous year, unless invited to resubmit.
Geographic Focus: Indiana
Date(s) Application is Due: Aug 1
Contact: Ashley Bice; (765) 884-8022; fax (765) 884-8023; ashley@bentoncf.org
Internet: http://www.bentoncf.org/grants_community.html
Sponsor: Benton Community Foundation
P.O. Box 351
Fowler, IN 47944

Benton County Foundation - Fitzgerald Family Scholarships 926

The Benton Community Foundation Fitzgerald Family Scholarships provide funding for students pursuing post-secondary education. Priority is given to students who intend to study in a health-related profession and/or have overcome a specific life challenge.
Requirements: Students should contact their guidance office for the scholarship application and deadline information.
Geographic Focus: Indiana
Contact: Ashley Bice, Executive Director; (765) 884-8022; fax (765) 884-8023; ashley@bentoncf.org or info@bentoncf.org
Internet: http://www.bentoncf.org/scholarships_fitzgerald.html
Sponsor: Benton Community Foundation
P.O. Box 351
Fowler, IN 47944

Benton County Foundation Grants 927

The Foundation's investment philosophy is built on the precepts of diversification, long-term strategic focus, and prudent risk management. The mission is to establish endowments, manage the funds received, and distribute a portion of the earnings each year to benefit youth and community. The Foundation focuses on funding programs that enhance youth education and/or provide positive character building and life skills experience.
Requirements: Organizations must be located in Benton County, Oregon.
Geographic Focus: Oregon
Date(s) Application is Due: Mar 13
Contact: Dick Thompson; (541) 753-1603 or (541) 231-7604; BCF@peak.org
Internet: http://www.bentoncountyfoundation.org/grants/
Sponsor: Benton County Foundation
P.O. Box 911
Corvallis, OR 97330

Berks County Community Foundation Grants 928

The foundation supports a broad range of community projects in this Pennsylvania county, including the arts and culture, economic development, education, the environment, and health and human services. Types of support include general operating grants, capital campaigns, demonstration grants, seed grants, and program grants. Although applicants do not have to be located in Berks County, they must provide programs and services within the county.
Requirements: Tax exempt or public benefit organizations, individuals, associations and private or public agencies are eligible to apply. The grant must be used for charitable purposes only. Organizations are eligible to apply to more than one grant program in the same year.
Geographic Focus: Pennsylvania
Contact: Richard Mappin; (610) 685-2223; fax (610) 685-2240; info@bccf.org
Internet: http://www.bccf.org/pages/grants.html
Sponsor: Berks County Community Foundation
501 Washington Street, Suite 801, P.O. Box 212
Reading, PA 19603-0212

Berlex Foundation Reproductive Scientist Development Program Grants 929

The program supports the postdoctoral career development of young obstetrician-gynecologists who are committed to careers in academic medicine. Grants support two to three years of full-time research under the mentorship of internationally recognized senior scientists. Applicants should contact Dr. Robert Jaffee, University of California, San Francisco, Box 0922, San Francisco, CA 94145-0922, (415) 476-9047.
Amount of Grant: 25,000 U.S.D
Contact: Grants Administrator; (201) 342-4441; llisanti@berlex-foundation.org
Internet: http://www.berlex-foundation.org/aw04.asp
Sponsor: Berlex Foundation
433 Hackensack Avenue, 9th Floor
Hackensack, NJ 07628

Berlex Foundation Scholar Award in Basic Science Research 930

The award provides the opportunity for clinician-investigators who wish to initiate new studies or who are currently conducting promising basic research in the area of reproductive medicine to continue work on their projects for an additional year. One or more awards may be made annually.
Requirements: Candidates must have completed an approved residency or fellowship in the field of obstetrics/gynecology. The applicant should be junior faculty. If not yet faculty, the individual should have reasonable assurance of such a position within a department of obstetrics and gynecology as evidenced by a Letter of Endorsement from the department head.
Restrictions: Candidates who are in an active training program as of January 1 are ineligible.
Date(s) Application is Due: Oct 1
Amount of Grant: 10,000 - 40,000 U.S.D
Samples: Dr. Robert Barbieri—for research, The Role of Hyperinsulinemia in the Pathogenesis of PCO, $40,000; Dr. Jill Flood—for research, Identification of Factor(s) Required for Embryonic Developmental Competence, $40,000.
Contact: Grants Administrator; (201) 342-4441; llisanti@berlex-foundation.org
Internet: http://www.berlex-foundation.org/aw01.asp
Sponsor: Berlex Foundation
433 Hackensack Avenue, 9th Floor
Hackensack, NJ 07628

Berlex Foundation Scholar Award in Clinical Research 931

The award is intended to support studies that are specifically focused on clinical research in obstetrics and gynecology. Its goal is to promote excellent clinical research by young investigators in the areas of diagnosis, treatment, and prognosis. The foundation plans to give more weight to proposals that incorporate more rigorous methods such as randomized controlled trials. Analytic studies, including cohort and case-control studies, also may be considered. One or more awards may be presented annually. Support is awarded to the institution in which the recipient will conduct the research. The recipient must identify the total amount required to complete the study and designate the specific allocation of funds. It is expected that the candidate devote a minimum of 50 percent of his/her time to the research activity.
Requirements: Candidates must have full-time faculty appointments in schools of medicine or public health. Preference will be given to proposals by academicians who are early in their careers.
Restrictions: Applicants should not currently be in training programs. The foundation is not seeking proposals for descriptive studies. This award is not intended for support of basic science research or animal studies.
Date(s) Application is Due: Oct 1
Amount of Grant: Up to 50,000 U.S.D
Samples: Dr. Debra Guinn—for research, Multicenter, Randomized, Placebo-Controlled Trial of Serial Doses of Antenatal Corticosteroids, $50,000; Andrea Witlin, DO—for research, The Efficacy of Magnesium Sulfate in Women with Mild Preeclampsia, $50,000.
Contact: Svetlana Lisanti, Administrator; (201) 342-4441; llisanti@berlex-foundation.org
Internet: http://www.berlex-foundation.org/aw02.asp
Sponsor: Berlex Foundation
433 Hackensack Avenue, 9th Floor
Hackensack, NJ 07628

Bernard and Audre Rapoport Foundation Health Grants 932

Bernard and Audre Rapoport Foundation seeks to improve the quality and delivery of healthcare services to all citizens, especially to women, children, and those who do not have access to conventional medical resources. Community-based outreach initiatives such as immunization programs are of interest to the Foundation. The primary focus of the Foundation is on programs that benefit children and youth in Waco and McLennan County, Texas. Proposals that fall outside of this geographical focus are considered as long as they offer imaginative, and when possible, long-range solutions to the problems of the most needy members of society, and ideally solutions that can be replicated in other communities.

Requirements: Program seeking funding must be a catalyst for change and promote both individual competence and social capacity.

Restrictions: Bernard and Audre Rapoport Foundation only supports organizations that are nonprofit 501(c)3 tax-exempt.

Date(s) Application is Due: Jun 15; Aug 15

Amount of Grant: 4,000 - 250,000 U.S.D

Samples: University of Texas Health Science Center, Houston, Texas, $20,000 - stroke research; Potter's Vessel Ministry, Waco, Texas, $10,000 - new health care provider; Planned Parenthood of Central Texas, Waco, Texas, $60,000 - general support.

Contact: Carole Jones, Foundation Coordinator; (254) 741-0510; fax (254) 741-0092; carole@rapoportfdn.org

Internet: http://www.rapoport-fdn.org/

Sponsor: Bernard and Audre Rapoport Foundation
5400 Bosque Boulevard, Suite 245
Waco, TX 76710

Bernard F. and Alva B. Gimbel Foundation Grants 933

The foundation's areas of interest are currently education, workforce and economic development, civil legal services, criminal justice, reproductive rights, and the environment. In most program areas, the Foundation seeks to fund both direct services programs and advocacy efforts. The Foundation's support for direct services programs is limited to those operating in New York City.

Requirements: Grants are made only to tax-exempt 501(c)3 organizations.

Restrictions: The Foundation does not make grants to: individuals; direct service programs outside of New York City; individual schools; short-term educational programs and workshops; mentoring programs; after-school and summer programs; youth development programs. Unsolicited proposals for the Environment program are not accepted.

Geographic Focus: New York

Contact: Leslie Gimbel, Executive Director; (212) 895-8050; fax (212) 895-8052

Internet: http://www.gimbelfoundation.org/index.html

Sponsor: Bernard F. and Alva B. Gimbel Foundation
271 Madison Avenue, Suite 605
New York, NY 10016

Bernard Van Leer Foundation Grants 934

The foundation operates in 34 countries in Europe, the Far East, Australia, Central and South America, the United States, and Africa to further the holistic development of young children. Grantmaking focuses on thematic initiatives, which may include partners from countries where the foundation does not otherwise have a grantmaking program. Current themes are: children affected by HIV/AIDS; respect for diversity; and growing up in indigenous societies. The foundation funds projects that promote a holistic approach to early childhood development; the enhancement of parental capacity to support their children's development; a development strategy that is rooted in the local context and is culturally, socially, and economically appropriate; and the building of capacity, local ownership, and working in partnership. Unsolicited proposals are rarely funded. To be considered, submit information about the organization; the project's objectives, strategies, beneficiaries, scope, location, and duration; and the overall budget and the amount required.

Requirements: Grants are made to governmental and nongovernmental organizations located in the list of eligible countries, which include: Brazil, Colombia, the East Caribbean region, Germany, Greece, India, Indonesia, Israel, Kenya, Mexico, Morocco, the Netherlands, Peru, Poland, South Africa, Tanzania, Thailand, Turkey, Uganda, United States of America and Zimbabwe.

Restrictions: The foundation does not fund applications for support to individual children; projects that concentrate solely on one aspect of children's development or learning; projects that specifically focus on children with special needs such as mental and/or physical handicap; proposals for the construction and maintenance of buildings, or the purchase of equipment and materials; isolated requests for scholarships, conferences, media or theatre events; or general support to organisations, or requests to cover recurrent costs or deficits

Contact: Peter Laugharn, Executive Director; 31-70-351-20-40; fax 31-70-350-23-73; proposal.administration@bvleerf.nl

Internet: http://www.bernardvanleer.org/about/Applications_for_funding

Sponsor: Bernard Van Leer Foundation
Eisenhowerlaan 156, P.O. Box 82334
The Hague, 2508 EH Netherlands

Berrien Community Foundation Grants 935

The community foundation awards grants to nonprofits in Berrien County, Michigan, that address community needs. The Foundation is very interested in providing start-up funding for programs that address our focus areas of nurturing children, building community spirit/arts and culture and youth leadership and development. Higher priority is given to requests that demonstrate community-based collaborative solutions likely to stay in place after Foundation funding concludes. Low priority is given to requests for bricks and mortar, operational funds on a repetitive basis, annual fund drives, equipment, and ongoing programs where alternative funding is not planned to carry a program/project forward following a Foundation grant. Low priority is also given to advertising and capital campaigns and grants to cover deficits or other previously incurred obligations. Applicants must first call the Foundation's program director to discuss the proposed program/project.

Requirements: Grant applications will only be accepted from nonprofit 501(c)3 and grass roots organizations serving Berrien County residents.

Restrictions: Grants are not made for sectarian religious purposes, national fundraising efforts, political organizations or campaigns. Grants are not made to individuals, and form letters/emails are neither reviewed nor acknowledged.

Geographic Focus: Michigan

Date(s) Application is Due: Sep 1

Amount of Grant: Up to 10,000 U.S.D

Contact: Anne McCausland, Program Director; (269) 983-3304, ext. 2; AnneMcCausland@BerrienCommunity.org

Internet: http://www.berriencommunity.org

Sponsor: Berrien Community Foundation
2900 South State Street, Suite 2 East
Saint Joseph, MI 49085

Bertha Russ Lytel Foundation Grants 936

The foundation awards grants to eligible California nonprofit organizations in its areas of interest, including civic affairs; cultural programs, libraries and museums; elementary education; higher education, including agricultural and nursing scholarships; health organizations, including hospitals and hospice; and social services for the elderly and disabled. The primary focus of the Foundation is programs for seniors. Types of support include building construction/renovation, continuing support, equipment acquisition, general operating grants, matching grants, scholarship funds, and seed grants. Contact the Foundation Manager for guidelines and application information.

Requirements: California nonprofit organizations in Humboldt County are eligible.

Geographic Focus: California

Contact: Don Hindley, Manager; (707) 786-9236

Sponsor: Bertha Russ Lytel Foundation
P.O. Box 893
Ferndale, CA 95536

Bert W. Martin Foundation Grants 937

The foundation awards grants to eligible nonprofit organizations in its areas of interest, including higher education, healthcare, sports, and general charitable giving. There are no application forms or deadlines. Proposal letters are accepted by postal mail only. No email proposals accepted.

Requirements: Nonprofit organizations in Florida and Arizona.

Geographic Focus: Arizona, Florida

Samples: Columbia U (New York, NY)—to enable undergraduate students to attend the Earth Semester program, a series of studies in earth systems and policy designed to foster an understanding of critical global issues, at its Biosphere 2 Center, $800,000.

Contact: Grants Administrator, c/o Northern Trust Co.

Sponsor: Bert W. Martin Foundation
50 S LaSalle Street, B-3
Chicago, IL 60675-0002

Besser Foundation Grants 938

The foundation limits its giving to nonprofits in the Alpena, MI area. Areas of interest include education, social services, civic affairs, arts and culture, religion, and international. Types of support include scholarship funds, matching funds, operating budgets, and continuing support. The board meets quarterly to consider requests.

Requirements: Only nonprofits in Michigan may apply.

Restrictions: Unless specifically requested by a Trustee, the Foundation will not consider grant requests from organizations outside of Alpena; nor for endowment funds, to defray meeting or conference expenses, or to pay for travel of individuals or groups. We will not relieve organizations or the public of their responsibilities, nor make grants to individuals for any purpose.

Geographic Focus: Michigan

Contact: J. Richard Wilson, President; (989) 354-4722; fax (517) 354-8099; besserfoundation@verizon.net

Sponsor: Besser Foundation
123 North Second Avenue, Suite 3
Alpena, MI 49707-2801

Bethesda Foundation Grants 939

The foundation awards grants to nonprofits in Hornell, NY, in its areas of interest, including education, hospitals, health care, substance abuse services, nutrition, AIDS research, and social services. Types of support include general operating support, equipment acquisition, program development, and scholarship funds. Application forms must be obtained from the office. The board meets in March, June, September, and December.

Requirements: Nonprofit organizations in Hornell, NY, are eligible.

Geographic Focus: New York

Amount of Grant: 1,000 - 24,000 U.S.D

Contact: Grants Administrator; (513) 745-1616; fax (513) 745-1623; bethesdafoundation@trihealth.com

Internet: http://www.bethesdafoundation.com

Sponsor: Bethesda Foundation
10506 Montgomery Road, Suite 304
Cincinnati, OH 45242

BFHF Sapphire Award Recognizing Community Health Excellence **940**

The award is designed to recognize and promote programs that have demonstrated impact in improving the health-related outcomes of Florida's at-risk populations and communities. The award is meant to promote greater and more lasting support for improved services and systems that can be sustained over time. Strategies that build on community assets, enhance organizational capacity, foster systemic change, and/or lead to lasting policy changes that improve health will be especially appropriate for this award. Up to three nonprofit winners will be selected annually, each receiving a financial award toward its defined community health objectives. In addition limited promotional support will be made available to help educate the public and various stakeholders about winning projects. The organization can nominate itself, and seek nomination support from three key people so long as all three supporters complete the forms themselves. Guidelines and nomination form are available online.

Requirements: Florida 501(c)3 tax-exempt organizations that have a proven record of improving the health of their community through measurable outcomes; and have programs that are involved with health-related services are eligible for nomination. Nominations can be initiated by a health professional, volunteer, board member, or client who knows first-hand the organization's work; a representative of a partnering organization, such as another nonprofit organization or governmental agency; a representative of a foundation, corporation, or other funding institution; a program evaluator; or someone else who has direct knowledge of the organization's work and impact. Three persons familiar with the organization are required to nominate the organization.

Restrictions: Nominations should not be based on the idea of raising funds without regard to recognizing demonstrated success.

Geographic Focus: Florida

Date(s) Application is Due: Jun 29

Amount of Grant: Up to 120,000 U.S.D

Contact: Administrator; (800) 477-3736, ext. 63215, op; TheSapphireAward@bcbsfl.com

Internet: http://www.bcbsfl.com/index.cfm?section=&fuseaction=BlueFoundation.Sapphire#three

Sponsor: Blue Foundation for a Healthy Florida

P.O. Box 1798

Jacksonville, FL 32231-0014

BHHS Legacy Foundation Grants **941**

BHHS Legacy Foundation funds efforts that enhance and improve the quality of life and health of children, families and senior citizens that are primarily low-income and under-served in the greater Phoenix and Tri-State regions. The Foundation places a high priority on collaborative projects and programs with other foundations, government agencies and nonprofit entities seeking to strengthen communities and support systemic change in health care access, health improvement and delivery. It also places a strong emphasis on funding nonprofit community organizations to increase their capacity to serve. These may include projects and programs offered by established nonprofit entities as well as those provided by start-up nonprofit organizations. Types of grants include: program/project; education; capacity building; capital equipment; operating; transition; and : Community Assistance Relief for Emergencies (CARE) funding. The foundation recommends discussing project or program concept and proposal with a foundation representative prior to submitting a written request for funding.

Restrictions: Giving is primarily restricted to the greater Phoenix and Bullhead/Laughlin regions, Arizona.

Geographic Focus: Arizona

Date(s) Application is Due: Apr 3; Aug 7; Dec 4

Amount of Grant: 25,000 - 75,000 U.S.D

Contact: Karen E. Orth, Grants Manager; (602) 778-1200; fax (602) 778-1255; korth@bhhslegacy.org or info@bhhslegacy.org

Internet: http://bhhslegacy.org/access_healthcare.aspx

Sponsor: BHHS Legacy Foundation

2999 N 44th Street, Suite 530

Phoenix, AZ 85018

BIA Assistance for Native American Children with Severe Disabilities **942**

The objective of the program is to provide for the special education and related services of Native American children with severe disabilities. Funds may be used for physical and occupational therapy, speech therapy, direct academic services, and administration including staff development. Funds may only be used for students who are enrolled in a bureau-funded school. Applications are accepted at any time.

Requirements: Members or direct descendants of members of federally recognized tribes and enrolled in a BIA-funded school are eligible.

Amount of Grant: Up to 3,600,000 U.S.D

Contact: Chief, Center for School Improvement, Office of Indian Education Programs; (505) 248-6942

Internet: http://www.doi.gov/bureau-indian-affairs.html

Sponsor: U.S. Department of the Interior

1849 C Street NW, MS-3512 MIB

Washington, DC 20240-0001

BibleLands Grants **943**

The organization was founded to support Christian missions in the lands of the Bible. The program awards grants to hospitals and schools in the Middle East. BibleLands supports schools and colleges that offer good quality Christian-based education for children whose families could not otherwise afford it.

Requirements: Projects must be Christian led.

Restrictions: Grants are not made to individuals or to projects outside of the Middle East.

Contact: N. Edward-Few, 01494 897 950; fax 01494 897 951; info@biblelands.org.uk

Internet: http://www.biblelands.org.uk/index2.htm

Sponsor: BibleLands

P.O. Box 50

High Wycombe, HP15 7QU England

Bikes Belong Foundation Paul David Clark Bicycling Safety Grants **944**

The Bikes Belong Foundation, launched in 2006, is a separate, complementary organization to the Bikes Belong Coalition, an organization sponsored by the U.S. bicycle industry. The foundation's focus is on bicycle-safety projects and children's bicycle programs. The foundation offers three distinct grant programs: the REI Grant Program; Bikes Belong Research Grants; and the Paul David Clark Bicycling Safety Fund. The latter is geared toward advocacy efforts to improve bicycling safety. The fund is named for Paul David Clark, one of more than 600 U.S. cyclists killed in motor vehicle collisions and accidents in 2005. Paul was an avid cyclist who loved the outdoors. An attorney, he provided pro bono service to nonprofit conservation groups, including the Natural Resources Defense Council and the Trust for Public Land. In the wake of the accident, Bikes Belong teamed with Paul's brother, Blair, to create a fund to support projects that increase bicycle safety, particularly in northern California. The fund's mission is two-fold: to encourage motorists to be more aware of bicyclists; and to compel motorists and cyclists to respectfully share the road. Interested applicants are encouraged to contact the foundation for information on funding availability and how to apply.

Samples: Community Cycling Center, Portland, Oregon, $10,000; Freiker, Boulder, Colorado, $10,000; NorCal High School Mountain Bike League, Petaluma, California, $10,000.

Contact: Zoe Kircos, Grants Manager; (303) 449-4893 ext. 5; fax (303) 442-2936; zoe@bikesbelong.org

Sponsor: Bikes Belong Foundation

207 Canyon Boulevard Suite 202

Boulder, CO 80302

Bildner Family Foundation Grants **945**

The foundation awards grants to nonprofit organizations in its areas of interest, including arts and performing arts, health care and health organizations, Jewish and social service delivery. Types of support include continuing support, general operating support, and program support. The majority of grants are awarded in New Jersey and New York. There are no application forms or deadlines.

Restrictions: The foundation does not support private foundations or individuals.

Geographic Focus: New Jersey, New York

Amount of Grant: 2,000 - 300,000 U.S.D

Samples: Dartmouth College, $6 million for residence hall construction; Rutgers, (New Brunswick, NJ)—for general support, $162,197; Bergen Community College (Paramus, NJ)—to develop the Center for the Study of International Understanding, $225,000.

Contact: Allen Bildner, President

Sponsor: Bildner Family Foundation

293 Eisenhower Parkway, Suite 150

Livingston, NJ 07039

Bill and Melinda Gates Foundation Emergency Response Grants **946**

The foundation supports supports effective relief agencies and local organizations that respond quickly to people's most pressing needs in challenging conditions. The foundation is interested in proposals that: deliver food and clean water; improve sanitation; provide medical attention and shelter; prevent or minimize outbreaks of disease; and support livelihoods through cash-for-work programs. The foundation currently supports: people affected by the global food crisis; people in Sri Lanka and Pakistan displaced by political unrest and violence; victims of the earthquake in Haiti; communities affected by Typhoon Ketsana in the Philippines and Vietnam; and a consortium of leading humanitarian aid organizations.

Requirements: Relief agencies must: have extensive experience and local relationships and be able to deliver help within days, when needs are most crucial. The foundation also funds organizational capacity—building and explore learning opportunities to reinforce emergency response capabilities.

Restrictions: The majority of funding is made to organizations that are independently identified by our staff. Unsolicited proposals are not accepted. Proposals must be made through 501(c)3 or other tax-exempt organizations. The foundation is unable to make grants directly to individuals. The Bill and Melinda Gates Foundation will not fund: projects addressing health problems in developed countries; political campaigns and legislative lobbying efforts; building or capital campaigns; or projects that exclusively serve religious purposes.

Contact: Kate James, Chief Communications Officer; (206) 709-3140; info@gatesfoundation.org

Internet: http://www.gatesfoundation.org/topics/Pages/emergency-response.aspx

Sponsor: Bill and Melinda Gates Foundation

P.O. Box 23350

Seattle, WA 98102

Bill and Melinda Gates Foundation Grand Challenges in Global Health Grants **947**

A grand challenge is a call for a specific scientific or technological innovation that removes a critical barrier to solving an important health problem in the developing world with a high likelihood of global impact and feasibility. Grand challenges address the diseases and health conditions that cause the greatest morbidity and mortality in the developing world, thus accounting for the enormous health disparities between the developing and the developed world, and that receive disproportionately less attention from the scientific and technical community than their consequences demand. The goals are broad, encompassing prevention, detection, diagnosis, treatment, rehabilitation, and surveillance and control of diseases. Challenges include to improve childhood vaccines; create new vaccines; control insects that transmit agents of disease; improve nutrition

to promote health; improve drug treatment of infectious diseases; cure latent and chronic infections; and measure disease and health status accurately and economically in developing countries. Letters of intent are due by the listed application deadline; form and guidelines are available online.
Requirements: For-profit, nonprofit, academic, domestic, and foreign institutions may submit letters of intent.
Restrictions: Individuals not affiliated with an institution are ineligible.
Contact: Program Administrator; (301) 402-4968; fax (301) 480-1661; Grants@GrandChallengesGH.org
Internet: http://www.grandchallengesgh.org/challenges.aspx?SecID=258
Sponsor: Foundation for the National Institutes of Health
Natcher Building 45, Room 3AN-44
Bethesda, MD 20892-6300

Bill and Melinda Gates Foundation Urban Development Grants 948
The foundation seeks proposals that help the urban poor strengthen their voices and improve their economic, health, and social opportunities. More than 900 million people currently live in poor, densely populated urban areas in developing countries. By 2050, an additional 2.6 billion people are expected to live in cities—most of them in the developing world. Urban centers are often ill-prepared to meet the basic needs of rapidly expanding populations, and the poor are the most likely to suffer the consequences.
Requirements: The foundation seeks proposals that: are able to produce measurable results; use preventive approaches; promise significant and long-lasting change; leverage support from other sources; and accelerate or are in accordance with work the foundation already supports.
Restrictions: The majority of funding is made to organizations that are independently identified by our staff. Unsolicited proposals are not accepted. Proposals must be made through 501(c)3 or other tax-exempt organizations. The foundation is unable to make grants directly to individuals. The Bill and Melinda Gates Foundation will not fund: projects addressing health problems in developed countries; political campaigns and legislative lobbying efforts; building or capital campaigns; or projects that exclusively serve religious purposes.
Samples: CHF International, Silver Spring, Maryland, $9,037,927 (over four years) - building local intermediaries, community-based organizations and urban local body staff to implement inclusive pro-poor urban development interventions on global level, with a focus in Asia and Africa; International Institute for Environment and Development, London, United Kingdom, $10,000,000 (over three years) - develop Slumdwellers International Urban Poor Fund into a separate entity that will grow from a grantmaking fund into a capital supporting the work of national slumdweller programs; Affordable Housing Institute, Boston, Massachusetts, $2,500,000 (over five years and nine months) - to promote research and education on the socioeconomics of shelter in urbanizing communities outside the United States.
Contact: Kate James; (206) 709-3140; info@gatesfoundation.org
Internet: http://www.gatesfoundation.org/global-development/Pages/urban-poverty-special-initiative.aspx
Sponsor: Bill and Melinda Gates Foundation
P.O. Box 23350
Seattle, WA 98102

Bill and Melinda Gates Foundation Water, Sanitation and Hygiene Grants 949
Poor sanitation causes severe diarrhea, which kills 1.5 million children each year. Smart investments in sanitation can reduce disease, increase family incomes, keep girls in school, help preserve the environment, and enhance human dignity. The foundation is looking to work with partners in an effort to expand affordable access to sanitation.
Requirements: The Bill and Melinda Gates Foundation seeks proposals that: are able to produce measurable results; use preventive approaches; promise significant and long-lasting change; leverage support from other sources; and accelerate or are in accordance with work the foundation already supports.
Restrictions: The majority of funding is made to organizations that are independently identified by our staff. Unsolicited proposals are not accepted. Proposals must be made through 501(c)3 or other tax-exempt organizations. The foundation is unable to make grants directly to individuals. The Bill and Melinda Gates Foundation will not fund: projects addressing health problems in developed countries; political campaigns and legislative lobbying efforts; building or capital campaigns; or projects that exclusively serve religious purposes.
Samples: Swiss Federal Institute of Aquatic Science and Technology, Dubendorf, Switzerland, $399,899 (over one year and three months) - investigation of transport and treatment for suitable on-site toilets and on-site water recovery (2011); Deutsche Gesellschaft für Internationale Zusammenarbeit, Eschborn, Germany, $2,869,244 (over five years) - support technical cooperation to improve living conditions, and improve the health of the urban poor in particular, in low-income urban areas in Kenya by offering access to sustainable sanitation services (2011); Water Services Trust Fund; Nairobi, Kenya, $7,130,699 (four years and eleven months) - to improve living conditions, and improve children's health in particular, in low-income urban areas in Kenya (2011).
Contact: Kate James , Chief Communications Officer; (206) 709-3140; info@gatesfoundation.org
Internet: http://www.gatesfoundation.org/watersanitationhygiene/Pages/home.aspx
Sponsor: Bill and Melinda Gates Foundation
P.O. Box 23350
Seattle, WA 98102

Billie Jean King Foundation Grants 950
The mission of the foundation is to inspire humankind in the pursuit of excellence, ensure equal opportunity, and enhance the quality of life for all individuals--regardless of gender, race, religion, appearance, or sexual orientation. Areas of interest include health, education, and athletics. Programs eligible for support include public education, training, service provision, research, legal services, advocacy, and legislative reform. Priority will be given to projects focusing on women; multicultural, lesbian, gay, bisexual, and transgendered individuals; minorities; and youth. Innovative, entrepreneurial efforts where new funding is likely to have a broad and lasting impact receive preference. There are no application deadlines. Submit a brief letter of inquiry; full applications are by invitation.
Requirements: 501(c)3 nonprofit organizations are eligible.
Amount of Grant: 1,000 - 25,000 U.S.D
Contact: Billie Jean King, President; (623) 362-7208
Internet: http://vpr2.admin.arizona.edu/rso/02010411.htm
Sponsor: Billie Jean King Foundation
P.O. Box 10777
Phoenix, AZ 85064-0777

Binney and Smith Community Grants 951
Awards include cash and product donations to nonprofit organizations that address the needs of children and young adults through projects in the arts, education, health and welfare, and civic areas. Special consideration is given to nonprofit organizations with which employees serve in a volunteer capacity. Funding is concentrated on specific programs or capital campaign projects rather than general operational support, galas and other fundraising events and event sponsorship. Direct funding and product donations are limited to non-profit organizations in Lehigh and Northampton Counties along with the cities of Kutztown, PA and Phillipsburg, N.J.
Requirements: Eastern Pennylvania nonprofit organizations are eligible.
Restrictions: Grants do not support: nonprofit organizations outside the Lehigh Valley; individuals; religious organizations; fraternal organizations; political organizations; national or international organizations; individual child care centers; nursing and convalescent homes; prisoners; conventions; hospitals; endowment funds; labor groups; social clubs; veterans organizations; athletic organizations. In addition, organizations eligible for support generally will not be funded when the purpose is to cover any of the following: past operating deficits; travel; conferences; non-product goodwill advertising; undefined operational support; tickets for or tables to special events; advertising in programs.
Geographic Focus: Pennsylvania
Contact: Community Affairs Manager; (800) 272-9652
Internet: http://www.binney-smith.com/page.cfm?id=17&hasFlash=false
Sponsor: Binney and Smith Corporation
1100 Church Lane, P.O. Box 431
Easton, PA 18044-0431

Biogen Corporate Giving Grants 952
The corporate giving program gives back to Biogen communities through grants awarded to nonprofit, community-based organizations, most of which are dedicated to education, community service, healthcare, and culture and the arts. There are no application forms or deadlines; guidelines are available online. See submission information for proposal mailing instructions.
Requirements: To be eligible for funding, an organization must be: a non-profit agency with tax-exempt status under section 501(c)3 of the Internal Revenue Service code; serving residents of Cambridge or the Greater Boston area, San Diego or San Diego County, and Raleigh-Durham, NC.
Restrictions: Eligible organizations must not: advocate, support or practice discrimination based on race, religion, age, national origin, language, gender, sexual preference or physical limitation; include the promotion of any particular religious faith as an explicit and primary goal; have the nomination or election of candidates to political office as an explicit goal, be government agencies, political organizations or individuals. Grants should not cover fundraising or capital campaign activities.
Geographic Focus: California, Massachusetts, North Carolina
Amount of Grant: 500 - 2,500 U.S.D
Contact: Kathryn Bloom, Communications; (617) 679-2000; fax (617) 679-2617
Internet: http://www.biogen.com/site/014.html
Sponsor: Biogen
14 Cambridge Center
Cambridge, MA 02142

Birds Eye Vegetables Share the Wonder Grants 953
Birds Eye and Pinnacle Foods Group LLC will sponsor a video and essay competition on their Facebook site for organizations that inspire communities and children to unlock the wonder of vegetables. During the initial competition period, applicants may visit the Birds Eye and Pinnacle Foods Facebook page and follow the directions provided to write and submit an essay of three hundred words or less about how their organization plans to use the award money to inspire their community. Finalists will then be required to produce a video that articulates the story and themes set forth in their written entries. Finalists' essays and videos will both be featured on Facebook for a period of thirty days during which the public will vote on their favorite entries. The organization that receives the highest number of valid votes will receive a $20,000 check. The organization receiving the second highest number of valid votes will receive $15,000. The organization receiving the third highest number of valid votes will receive $10,000. The two organizations receiving the fourth and fifth highest number of valid votes will receive $5,000 each. Essay submissions are due by 11:59 p.m. on the deadline date. Submission dates may vary from year to year. Interested organizations are encouraged to check the website or Facebook page to verify current deadlines.
Requirements: The contest is open only to contestants who are legal residents of the fifty United States and the District of Columbia and who are eighteen years of age or older at the time of entry. Individuals may enter only on behalf of a 501(c)3 or equivalent organization (e.g. school or community based group).

Restrictions: Employees, officers, directors, members, agents, and representatives of Pinnacle Foods Group LLC ("Sponsor"), Weber Shandwick, and their respective parents, subsidiaries, affiliates, promotion and advertising agencies, and law firms are not eligible to enter or win. Immediate family members (spouse and parent, children and siblings and their respective spouses, regardless of where they reside) and persons living in the same household, whether or not related, of such employees, officers, directors, members, agents, and representatives are also not eligible to enter or win. The contest is subject to all applicable federal, state and local laws. Previous winners from any other Pinnacle Foods Group LLC contest are not eligible to enter or win.
Date(s) Application is Due: Aug 26
Amount of Grant: 5,000 - 20,000 U.S.D
Samples: The Food Trust, Philadelphia, Pennsylvania, $20,000 (2011).
Contact: Susan Riker, Senior Vice President, Administration; (973) 541-6620
Internet: http://www.birdseye.com/community
Sponsor: Pinnacle Foods Group, LLC
399 Jefferson Road
Parsippany, NJ 07054-3707

Birmingham Foundation Grants 954

The Birmingham Foundation is a private, independent foundation, dedicated to health-related and human services grant-making in the south neighborhoods of Pittsburgh, Pennsylvania, which are part of the 15203, 15210 and 15211 zip code areas. The Board of Directors meets three times a year to act on grant proposals, in March, June and November.
Requirements: 501(c)3 nonprofits in South Pittsburgh communities including Allentown, Arlington and Arlington Heights, Beltzhoover, Bon Air, Carrick, Duquesne Heights, Knoxville, Mount Oliver, Mount Washington, Saint Clair Village, and South Side are eligible.
Restrictions: Grants are not normally made for operating budgets, deficits, fund-raising campaigns, general research, overhead costs, scholarships, political campaigns, or loans. Grants are not made to individuals, to other private foundations, for sectarian religious activities, or for the efforts to influence legislation unless grant-related.
Geographic Focus: Pennsylvania
Amount of Grant: 2,500 - 50,000 U.S.D
Contact: Chris Mason, Administrative Assistant; (412) 481-2777; fax (412) 481-2727; info@birmfoundpgh.org
Internet: http://www.birminghamfoundation.org
Sponsor: Birmingham Foundation
2005 Sarah Street, 2nd Floor
Pittsburgh, PA 15203

BJ's Charitable Foundation Grants 955

Established with the goal of creating a positive, long-lasting impact on the communities BJ's serves, the mission of BJ's Charitable Foundation is to enhance and enrich community programs that primarily benefit children and families. The majority of the foundation's giving is focused on organizations that: (a) Promote the safety, security and well-being of children and families; (b) Support education and health programs; (c) Provide community service opportunities; or, (d) Aid in hunger and disaster relief.
Requirements: Organizations that are tax-exempt under 501(c)3 of the Internal Revenue Code and recognized as a 'public charity' by the IRS may apply. The program must align with the foundation's mission of supporting children and families in the specific areas of safety, security and well-being, education, health, community service, hunger/homelessness and disaster relief. And, the program must positively impact communities where BJ's Clubs are located. All organizations are limited to one (1) application per 12-month period. Additional applications will not be considered.
Restrictions: The foundation does not provide funding for: (1) Organizations that discriminate on the basis of race, color, gender, sexual orientation, age, religion, national or ethnic origin or physical disability. (2) Political organizations, fraternal groups or social clubs that engage in any kind of political activity. (3) Religious organizations, unless they serve the general public in a significant non-denominational way. (4) Programs that have been in place for less than one (1) year. (5) Individuals. (6) Organizations located outside BJ's markets. (7) Capital campaigns. (8) Private foundations. (9) Sponsorships for music, film and art festivals. (10) Business expositions/conferences. (11) Journal or program advertisements.
Geographic Focus: Connecticut, Delaware, Florida, Georgia, Maine, Maryland, Massachusetts, New Hampshire, New Jersey, New York, North Carolina, Ohio, Pennsylvania, Rhode Island, South Carolina, Virginia
Date(s) Application is Due: Mar 6; Jul 10; Dec 6; Dec 30
Contact: Community Relations Manager; (508) 651-7400; fax (508) 651-6623
Internet: http://www.bjs.com/about/community/charity.shtml
Sponsor: BJ's Charitable Foundation
1 Mercer Road, P.O. Box 9614
Natick, MA 01760

Blackford County Community Foundation - WOW Grants 956

The Blackford County Community Foundation - WOW Grants offer funding for projects that encourage, educate, and enlighten women and/or children. Applicants may be individuals, groups, or organizations.
Requirements: Applicants should submit the following items to the Foundation: cover page; project narrative with a detailed description of the project, including timeline and results expected; and a detailed budget. All grant recipients will be required to give a brief report about their project at the spring meeting.
Restrictions: The following are not eligible for funding: operating deficits, post-event or after the fact situations; special fundraising events, including endowment campaigns; political endeavors and propaganda; and profit-making enterprises and/or projects for personal gain.
Geographic Focus: Indiana
Date(s) Application is Due: Feb 1
Amount of Grant: 100 - 2,500 U.S.D
Contact: Patricia Poulson, Executive Director; (765) 348-3411; fax (765) 348-4945; ppoulson@blackfordcounty.org
Internet: http://www.blackfordcofoundation.org/pages.asp?Page=Women%20of%20Worth&PageIndex=411
Sponsor: Blackford County Community Foundation
121 North High Street
Hartford City, IN 47348

Blackford County Community Foundation Grants 957

The Blackford County Community Foundation and its supporting organizations award numerous grants annually. Primary areas of interest include: community and economic development; education; community services planning and coordination; and human services. The Foundation has quarterly deadlines for submitting applications in January, March, June, and September.
Requirements: Applicants must contact the Executive Director to determine if their project is suitable for the funding. If the organization is asked to submit a formal proposal, it must consider the following criteria: purpose and definition of the project or program; background of the request office; officers and staff personnel of requesting organization; financial information and budgets; evaluation results; and how the project will be affected if funding is not received. The Foundation will judge the proposal on its merit, priority, and substantive quality.
Restrictions: The Blackford County Community Foundation generally does not fund: profit-making enterprises; political activities; operating budgets of organizations, except for limited experimental or demonstration periods; sectarian or religious organizations operated primarily for the benefit of their own members; endowment purposes; capital grants to building campaigns will only be made when there is evidence that such support is vital to the success of a program meeting priority needs of the community. Grants are not awarded for endowment purposes.
Geographic Focus: Indiana
Contact: Patricia Poulson, Executive Director; (765) 348-3411; fax (765) 348-4945; ppoulson@blackfordcounty.org or foundation@blackfordcounty.org
Internet: http://www.blackfordcofoundation.org/pages.asp?PageIndex=377
Sponsor: Blackford County Community Foundation
121 North High Street
Hartford City, IN 47348

Black River Falls Area Foundation Grants 958

Grant applications are evaluated with consideration to the policies, funding objectives and the mission statement of the Foundation. General categories of support include: education, health services, cultural activities, social services and civic projects. Grants to be considered are those that meet charitable needs of: a new or innovative nature to fulfill unmet needs; one-time projects; capital improvements; equipment needs; start-up expenses; special programs; and emergency funding. Applications will be accepted April 1 through May 15.
Requirements: Any non-profit organization in Jackson County, Wisconsin, may apply for a grant. A copy of the organization's tax-exempt status is required.
Restrictions: Requests to support endowments, operating expenses, debt repayment, religious purposes, individuals, and lobbying expenses are usually discouraged.
Geographic Focus: Wisconsin
Date(s) Application is Due: May 15
Contact: Beth Overlien, Administrative Assistant; (715) 284-3113
Internet: http://www.brfareafoundation.org/application
Sponsor: Black River Falls Area Foundation
P.O. Box 99
Black River Falls, WI 54615

Blanche and Irving Laurie Foundation Grants 959

The Blanche & Irving Laurie Foundation was established in 1983 by New Brunswick philanthropist Irving Laurie. The foundation makes charitable gifts to institutions and nonprofits in broad areas of interest, including the arts, especially theater and music; education; health care; social services; and needs and concerns of the Jewish community. Capital grants, operating support grants, grants for programs/projects, and scholarships are awarded. Applicants should submit seven copies of a written proposal containing the following items: copies of the most recent annual report, audited financial statement, and 990; a detailed description of the project and amount of funding requested; and a copy of the current year's organization budget and/or project budget. The foundation's board meets quarterly to evaluate proposals. Final notification occurs within three to four months from submission. An informational brochure and application guidelines are available from the foundation.
Requirements: Nonprofit organizations in New Jersey are eligible.
Restrictions: Giving is primarily concentrated in New Jersey. The foundation does not support medical research.
Geographic Focus: New Jersey
Amount of Grant: 250 - 200,000 U.S.D
Samples: Atlantic Theatre Company, New York, New York, $109,500 - capital campaign and production support (2010); Neward Beth Israel Medical Center, Newark, New Jersey, $20,000 - improvements geriatrics emergency room (2010); Sesame Workshop, New York, New York, $20,000 - Shalom Sesame Project (2010).
Contact: Gene Korf, Executive Director; (973) 993-1583
Sponsor: Blanche and Irving Laurie Foundation
P.O. Box 53
Roseland, NJ 07068-5788

Blanche and Julian Robertson Family Foundation Grants 960

The Blanche and Julian Robertson Family Foundation is totally committed to the goal of improving the quality of life in Salisbury and Rowan County. The general direction of the Foundation's interest is: programs which address social problems and nurture positive social relationships; efforts aimed at enriching lives through exposure to the cultural arts; neighborhood revitalization programs, especially when such programs encourage development of transitional housing and enable first-time homeowners to purchase homes; programs that improve opportunities for youth at risk and families in crisis; efforts to improve broad-based educational, recreational, and athletic opportunities; efforts which address health and the environment. The Foundation is also interested in programs and projects that demonstrate the attributes of leverage (where a grant will attract matching gifts or other funding), as well as innovation, thoroughness, passion, and commitment. Contact the office for application and guidelines.

Requirements: North Carolina nonprofit organizations serving Salisbury and Rowan County.

Restrictions: Foundation does not make grants outside Salisbury and Rowan County.

Geographic Focus: North Carolina

Date(s) Application is Due: Mar 30

Samples: Rowan Regional Medical Ctr Foundation (Salisbury, NC)—for its capital campaign for a new emergency department, patient-care facility, and community-education center, $1 million.

Contact: David Setzer, Executive Director; (704) 637-0511; fax (704) 637-0177; bjrfoundation@aol.com

Sponsor: Blanche and Julian Robertson Family Foundation
141 East Council Street, P.O. Box 4242
Salisbury, NC 28145-4242

Blowitz-Ridgeway Foundation Early Childhood Development Research Award 961

The foundation supports nonprofit organizations and programs in the areas of medicine, psychology, residential care, and education, and for research in medicine, psychology, social science, and education. Types of support include capital grants, endowments, program development grants, research grants, and scholarships. The Foundation prefers prospective grantees whose programs or services benefit persons who have not yet reached their majority and/or are for the care of individuals or elderly persons who lack sufficient resources to provide for themselves. Although grants may be made to organizations outside the state of Illinois, preference will generally be given to applicants from Illinois. Application information is available on the Web site.

Requirements: Applicants must be classified as 501(c)3 by the IRS.

Restrictions: Grants will not be made for religious or political purposes, nor generally for the production or writing of audio-visual materials.

Contact: Laura Romero, Program Associate; (847) 330-1020; fax (847) 330-1028; laura@blowitzridgeway.org

Internet: http://www.blowitzridgeway.org/information/information.html

Sponsor: Blowitz-Ridgeway Foundation
1701 E Woodfield Road, Suite 201
Schaumburg, IL 60173

Blowitz-Ridgeway Foundation Grants 962

The foundation supports nonprofit agencies that provide medical, psychiatric, and psychological care to economically disadvantaged children and adolescents. Program and capital grants are awarded, primarily in Illinois, in support of medical, psychiatric, psychological, and/or residential care; and research programs in medicine, psychology, social science, and education. The foundation supports operating budgets, and applicants may request commitments that extend beyond one year, but requests for annual funding will not be considered. Applications are accepted throughout the year and are reviewed in the order in which they are received. Guidelines and applications are available online.

Requirements: 501(c)3 nonprofit organizations that offer services to people who lack resources to provide for themselves may apply.

Restrictions: Grants will not be awarded to government agencies or to organizations that subsist mainly on third-party funding and have demonstrated no ability or expended little effort to attract private funding. Grants will not be made for religious or political purposes or for the production or writing of audio-visual materials.

Geographic Focus: Illinois

Amount of Grant: 5,000 - 30,000 U.S.D

Samples: Elizabeth Ann Seton Program, $5,000 in general operating support of programs for pregnant and parenting mothers; The Enterprising Kitchen, $15,000 for programs improving self-sufficiency and employability of low-income women; Faith in Action of McHenry County, $7,000 to recruit volunteers; Gospel Rescue Mission, $10,000 for children's social and recreational activities; Guardian Angel Community Services, $5,000 for the Groundwork Domestic Violence Program.

Contact: Serena Moy, Administrator; (847) 330-1020; fax (847) 330-1028; serena@blowitzridgeway.org

Internet: http://www.blowitzridgeway.org/information/information1.html

Sponsor: Blowitz-Ridgeway Foundation
1701 E Woodfield Road, Suite 201
Schaumburg, IL 60173

Blue Cross and Blue Shield of Oklahoma Champions of Health Awards 963

The awards recognize individuals, health care professionals, legislators, and for-profit and nonprofit organizations that are working to make a difference in the health of their communities, and to facilitate community partnerships to work toward improving Oklahoma's health status. Award categories include champion of the uninsured, education and awareness champion, champion of health care in rural Oklahoma, minority health champion, legislative health champion, ageless hero, health and medical media champion, and corporate health champion. Winners will be presented with an award and monetary contribution at the awards banquet in October.

Requirements: Oklahoma individuals, health care professionals, legislators, for-profit organizations, and nonprofit organizations may be nominated.

Geographic Focus: Oklahoma

Date(s) Application is Due: Aug 12

Amount of Grant: 1,000 U.S.D

Contact: Champions of Health; (866) 876-4376

Internet: http://www.bcbsok.com/imenu_bcbsok_11.html

Sponsor: Blue Cross and Blue Shield of Oklahoma
P.O. Box 3283
Tulsa, OK 74102-3283

Blue Grass Community Foundation Grants 964

The foundation awards grants to nonprofits in central Kentucky in its areas of interest, including arts and culture, housing, employment support and education. Types of support include building construction/renovation, equipment acquisition, seed grants, matching grants, and scholarships to individuals. Grants will range from approximately $5,000 to $20,000. To apply, submit a letter of inquiry of no more than three pages by 4:00 pm, August 17. Applicants will then be invited to submit full proposals, which are due by October 14.

Requirements: Central Kentucky nonprofit organizations are eligible.

Geographic Focus: Kentucky

Date(s) Application is Due: Aug 17; Oct 14

Amount of Grant: 5,000 - 20,000 U.S.D

Samples: The Nest, a Center for Women, Children and Families, $7500 - to replenish the emergency food bank; Central Music Academy, $5000 - for operating funds and programs benefitting disadvantaged youth; Ruby Bailey Family Services Inc., $7000 - to expand free meal programs between school terms.

Contact: Barbara A. Fischer, Grants Officer; (859) 225-3343; fax (859) 243-0770; bfischer@bgcf.org

Internet: http://www.bgcf.org/page27526.cfm

Sponsor: Blue Grass Community Foundation
250 W Main Street, Suite 1220
Lexington, KY 40507-1714

Blue Mountain Community Foundation Grants 965

The Foundation administers charitable funds to benefit people of the Blue Mountain Area. Most of the money for discretionary grants is designated by donors for use by agencies serving Walla Walla County. The Foundation's grant making policies are generally directed toward the fields of social and community services, the arts and humanities, education and health. In reviewing grant applications, careful consideration will be given to: potential impact of the program/project on the community and the number of people who will benefit; local volunteer involvement and support; commitment of the organization's Board of Directors; degree to which the applicant works with or complements the services of other community organizations; organization's fiscal responsibility and management skills; possibility of using the grant as seed money for matching funds from other sources; ability of the organization to obtain additional funding and to provide ongoing funding after the term of the grant.

Requirements: Nonprofit organizations serving the Walla Walla Valley, from Dayton to Milton-Freewater are encouraged to submit proposals.

Restrictions: Grants usually will not be made for the following: programs outside the Blue Mountain Area, operating expenses, annual fund drives, field trips, travel to or in support of conferences. No grants will be made for sectarian religious purposes nor to influence legislation or elections.

Geographic Focus: Oregon, Washington

Date(s) Application is Due: Jul 1

Amount of Grant: 125 - 4,000 U.S.D

Samples: Blue Mountain Action Council, WA, $7,000 - Volunteer Adult Literacy Program; Walla Walla Community College Foundation, WA, $1,500 - Parent Education Program Scholarships; Carnegie Art Center, WA, $3,000 - general support.

Contact: Lawson F. Knight, Executive Director; (509) 529-4371; fax (509) 529-5284; BMCF@bluemountainfoundation.org

Internet: http://www.bluemountainfoundation.org/grant-making-programs.php

Sponsor: Blue Mountain Community Foundation
8 South Second, Suite 168, P.O. Box 603
Walla Walla, WA 99362-0015

Blue River Community Foundation Grants 966

The Blue River Community Foundation is a community-based philanthropic organization that identifies, promotes, supports, and manages programs that will enhance the quality of life in Shelby County, Indiana, for this generation and future generations. To this end, the Foundation has established five areas of interest for the competitive grant making program: community and civic—support for community programs designed to improve life in Shelby County; arts and culture—support for programs and facilities that offer wide-spread opportunities for participation and appreciation; education—support for programs at all levels of education; health—support for the promotion of health and well-being for Shelby County residents; and social services—support of human service organization programs. Organizations interested in submitting a grant request should first submit the grant interest form found on the website. If the grant request meets the Foundation's funding guidelines, the organization will be invited to submit our formal Grant Application Form. Grant applications may be submitted at any time, but will only be reviewed during the next upcoming grant cycle.

Requirements: Any 501(c)3 organization serving the residents of Shelby County, Indiana, may apply.
Geographic Focus: Indiana
Date(s) Application is Due: Feb 1; Jun 1; Oct 1
Contact: Lynne Ensminger, Program Administrator; (317) 392-7955; fax (317) 392-4545; lensminger@blueriverfoundation.com or brf@blueriverfoundation.com
Internet: http://blueriverfoundation.com/main.asp?SectionID=6&TM=34321.26
Sponsor: Blue River Community Foundation
54 W Broadway Street, Suite 1, P.O. Box 808
Shelbyville, IN 46176

Blue Shield of California Grants 967
Consideration for funding will be given exclusively to organizations that pursue activities directly related to the foundation's program goals, including domestic violence prevention through service provision, education, and outreach; research and education regarding medical best practices and health technologies; and direct or indirect provision of medical insurance or health care to those populations that are uninsured or underinsured, and related policy development.
Requirements: The Foundation funds organizations that are non-profit and tax-exempt under 501(c)3 of the Internal Revenue Service Code (IRC) and defined as a public charity under 509(a)1, 2, or 3 (types I, II, or a functionally integrated type III); accredited schools; units of government/public agencies; tribal governments. The foundation will only support projects that meet the following criteria: the mission of the grantee organization is consistent with the goals and mission of the foundation; the grant is used primarily to serve Californians; the grant seeking organization has a reputation for credibility and integrity; the grant seeking organization is pursuing activities directly related to one of the Foundation's three Program Areas: Health Care and Coverage, Health and Technology and Blue Shield Against Violence.
Restrictions: The foundation does not fund award dinners, athletic events, competitions, special events, or tournaments; conferences or seminars; capital construction; television/film/media production; religious organizations for religious purposes; political causes, candidates, organizations or campaigns; capital projects over $50,000; multi-year projects (generally); grants to individuals (with the exception of the regulated Blue Shield of California Employee Scholarship Program); grants to 509(a) 3, type III supporting organizations that are not functionally integrated.
Geographic Focus: California
Contact: Grants Administrator; (415) 229-5785; fax (415) 229-6268
Internet: http://blueshieldcafoundation.org/grant-center/index.cfm
Sponsor: Blue Shield of California
50 Beale Street
San Francisco, CA 94105-1808

Blum-Kovler Foundation Grants 968
The Blum-Kovler Foundation was established in 1985 after Everett Kovler retired from his position as President of James Beam Distilling Company. The foundation awards general operating grants to eligible nonprofit organizations in its areas of interest, including social services, Jewish welfare funds, higher education, health services and medical research, and cultural programs. The foundation also supports youth- and child-welfare agencies and public-interest and civic-affairs groups. Grants are awarded primarily in the Chicago metropolitan area and in the Washington, D.C. area. There are no application forms. Applicants should submit a one to two page written proposal with a copy of their IRS determination letter by mid-November to considered for the current year. Typical grant awarded is between $1,000-$5,000.
Requirements: Illinois and District of Columbia nonprofit organizations are eligible.
Geographic Focus: District of Columbia, Illinois
Amount of Grant: 1,000 - 1,000,000 U.S.D
Samples: Camp of Dreams, Chicago, Illinois, $3,000; Community Foundation for the National Capital Region, Washington, D.C., $1,316,527; Compassion and Choices, Denver, Colorado, $1,000.
Contact: Hymen Bregar, Secretary; (312) 664-5050
Peter Kovler, Chairperson and Vice President; (312) 664-5050
Sponsor: Blum-Kovler Foundation
875 N Michigan Avenue, Suite 3400
Chicago, IL 60611-1958

Blumenthal Foundation Grants 969
In 1924 Mr. I. D. Blumenthal was a traveling salesman in need of repair to his car's radiator. A local tinsmith in Charlotte, North Carolina repaired the radiator with a "magic powder". Impressed with the product, I.D. teamed with the tinsmith and SOLDER SEAL became the first product of Radiator Specialty Company. The Blumenthal Foundation was founded in 1953 and was endowed with the sucess of the Radiator Speciality Company. The foundation focuses the majority of its grants on programs and projects that have an impact on Charlotte, and the state of North Carolina. The philanthropic efforts of the Foundation are focused in nine areas of grant making: arts, science and culture; civic and community; education; environment; foundation affiliates; health; Jewish institutions and philanthropies; religious and interfaith; and social sciences. The foundation believes that basic operational funding for non-profits is just as important, if not more so, than support for special programs or projects; consequently, grants are provided for seed money, annual operating costs, capital campaigns, conferences and seminars, special projects, and endowments. Interested organizations may click the Grant Guidelines link at the website for detailed submission instructions. Applications must be mailed. There are no deadlines, and requests are accepted on an ongoing basis. The Board of Trustees meets quarterly to consider grant applications.
Requirements: 501(c)3 organizations and institutions that serve the city of Charlotte and the state of North Carolina in the foundation's areas of interest are eligible to apply.
Restrictions: Grants are not made to individuals for any purpose.
Geographic Focus: North Carolina
Samples: Arts & Science Council, Charlottte, North Carolina, $10,000 - a grant for support for the annual drive fund (2010); Community Building Initiative, Charlotte, North Carolina, $2,500 - a grant to support the organization's efforts to promote racial and ethnic inclusion; Friends of DuPont Forest, Brevard, North Carolina, $2,500 - a grant for support of the Muddy Sneakers "Joy of Learning Outside" program that helps awaken in school-age children a deeper connection with the natural world and that stimulates learning and brings new life to a child's classroom performance.
Contact: Philip Blumenthal, Director; (704) 688-2305; fax (704) 688-2301; foundation@gunk.com
Internet: http://www.blumenthalfoundation.org/BFGrantListings.htm
Sponsor: Blumenthal Foundation
P.O. Box 34689
Charlotte, NC 28234-4689

BMW of North America Charitable Contributions 970
The corporation funds charitable programs that benefit society in the areas of education, road-traffic safety, and the environment. The corporation supports education at all levels and specifically focuses on the following: intercultural learning for K-12 students and their teachers; automotive technology, mechanics, and career and repair programs in high schools, technical schools, and community colleges; and research in the areas of safety design, ergonomics, and new materials. In the area of road traffic safety, the corporation supports driver-education programs geared at teenagers and new drivers; basic auto-maintenance programs for women; consumer education on general road-safety issues; and programs to promote the safety of children and young people on the road. In the area of the environment, BMW is committed to sustainable development and focuses grant making on the following: conservation/preservation of natural resources, in particular park lands and waterways; research and promotion of alternative fuels; and environmental education for K-12 students. In general, grants are awarded for specific projects rather than for general operating support, although some operating and capital grants are given consideration. Interested organizations may download application instructions and guidelines at the grant website. Organizations wishing to be considered for a grant must submit an application; telephone solicitations will not considered.
Requirements: 501(c)3 charities or 501(c)9 organizations are eligible to apply.
Restrictions: The corporate giving program does not support non-tax-exempt organizations; individuals; religious organizations for religious purposes; political candidates or lobbying organizations; organizations with a limited constituency, such as fraternal, labor, or veterans groups; travel by groups or individuals; national or local chapters of disease-specific organizations; national conferences, sports events, and other one-time, short-term events; sponsorships or advertising; anti-business groups; team sponsorships or athletic scholarships; or organizations outside the United States or its territories.
Samples: Spelman College, Atlanta, Georgia - Spelman Scholars Program; Susan G. Komen for the Cure, Dallas, Texas - Drive for the Cure; The School, Home, and Office Products Association, Moraine, Ohio - Call Me MISTER Program.
Contact: Grants Coordinator; (201) 307-4000; fax (201) 307 3607
Internet: http://www.bmwgroupna.com/07_Philanth.htm
Sponsor: BMW of North America, LLC
300 Chestnut Ridge Road
Woodcliff Lake, NJ 07677-7731

Bob and Delores Hope Charitable Foundation Grants 971
The foundation is primarily focusing its efforts on the economically disadvantaged populations of California and Texas. Projects of strong interest will be programs designed to help those living at or below the poverty guidelines. Other fields of interest include: the arts; Catholic agencies and churches; education; higher education; human services; medical research; and treatments for substance abuse. There are no specific application forms or deadlines, and applicants should send a letter of inquiry.
Requirements: Proposals are only accepted upon request. Applicants should submit the following: listing of additional sources and amount of support; timetable for implementation and evaluation of project; results expected from proposed grant; brief history of organization and description of its mission; and a detailed description of project and amount of funding requested.
Restrictions: No grants are given to/for: individuals; capital construction; fund raising; deficit financing; conferences; seminars; media events or workshops (unless they are an integral part of a broader program); or loans.
Geographic Focus: California, Texas
Amount of Grant: Up to 100,000 U.S.D
Contact: Linda Hope, Vice-President; (818) 841-2020
Sponsor: Bob and Delores Hope Charitable Foundation
10346 Moorpark Street
North Hollywood, CA 91602-2407

Bob Evans Farms Grants 972
Grants support Ohio nonprofit organizations focusing on families and children issues. Priority is given to work that provides the skills youth and adults need to become self-sufficient. Additional areas of interest include education, health care, and community improvement. Preference is given to requests involving a large number of Evans' employees in communities where it maintains operations. Applicants may apply online.
Requirements: Ohio nonprofit organizations are eligible.
Geographic Focus: Ohio

Date(s) Application is Due: Mar 1
Contact: Sarah Morton; (614) 491-2225; fax (614) 497-4330
Internet: http://www.bobevans.com/website/homepage.nsf/pages/philanthropy?opendocument
Sponsor: Bob Evans Farms
3776 S High Street
Columbus, OH 43207-0863

Bodenwein Public Benevolent Foundation Grants 973

The Bodenwein Public Benevolent Foundation was established in 1938 under the will of Theodore Bodenwein, owner and publisher of The Day newspaper, to support and promote quality educational, cultural, human-services, and health-care programming for underserved populations. The Foundation specifically serves the people of Greater New London County where The Day has a substantial circulation. The majority of grants from the Bodenwein Public Benevolent Foundation are one year in duration; on occasion, multi-year support is awarded. Applicants must apply online at the grant website. Applicants are strongly encouraged to do the following before applying: review the downloadable state application procedures for additional helpful information, requirements, and restrictions; review the downloadable online-application guidelines at the grant website; review the foundation's funding history (link is available from the grant website); review the online application questions in advance; and review the list of required attachments. These will generally include: a list of board members, financial statements (audited, reviewed, or compiled by independent auditor); an organization summary; a list of other funding sources; an IRS Determination letter; and other required documents. All attachments must be uploaded in the online application as PDF, Word, or Excel files. The annual application deadline is 11:59 p.m. on November 15. Applicants will be notified of grant decisions by letter within three to four months after the proposal deadline.
Requirements: Nonprofit organizations serving Greater New London County (East Lyme, Groton, Ledyard, Lyme, Montville, Mystic, New London, North Stonington, Old Lyme, Salem, Stonington, and Waterford, Connecticut) are eligible to apply. A breakdown of number/percentage of people served by specific towns will be required in the online application.
Restrictions: The foundation does not support requests from individuals, organizations attempting to influence policy through direct lobbying, or any political campaigns.
Geographic Focus: Connecticut
Date(s) Application is Due: Nov 15
Samples: New London Public Library, New London, Connecticut, $60,000 (2010); Malta, Inc., Groton, Connecticut, $4,000; Eastern Connecticut Ballet, East Lyme, Connecticut, $2,000.
Contact: Amy Lynch, Senior Foundation Manager; (860) 657-7015; amy.r.lynch@baml.com
Internet: https://www.bankofamerica.com/philanthropic/fn_search.action
Sponsor: Bodenwein Public Benevolent Foundation
200 Glastonbury Boulevard, Suite # 200
Glastonbury, CT 06033-4056

Bodman Foundation Grants 974

The Bodman Foundation was established by George M. Bodman and his wife Louise Clarke Bodman in 1945. Mr. Bodman was born in Toledo, Ohio in 1882 and died in 1950. Mrs. Bodman was born in Chicago in 1893 and died in 1955. The Bodmans lived for much of their lives in Red Bank, New Jersey and in New York City where Mr. Bodman was a senior partner at the investment banking firm of Cyrus J. Lawrence and Sons. The Bodmans were generous supporters of numerous cultural, civic, and service organizations. During World War I Mr. Bodman headed the Intelligence Service of the War Trade Board. During World War II, he served as executive assistant to the Red Cross Commissioner for Great Britain and was regional director in charge of American Red Cross Club operations in England, Scotland, and Ireland. The Bodman Foundation's Certificate of Incorporation states that its funds are to be used for "the aid, support or benefit of religious, educational, charitable, and benevolent objects and purposes for the moral, ethical and physical well-being and progress of mankind." The Bodman Foundation shares trustees, staff, office space, and even a website with the Achelis Foundation which has a similar mission and geographic area of concentration (both foundations give in New York City, while the Bodman Foundation also gives in New Jersey). Funding is concentrated in six program areas: Arts and Culture, Education, Employment, Health, Public Policy, and Youth and Families.
Requirements: 501(c)3 organizations based in New York City that fall within the foundation's areas of interest are welcome to submit an inquiry or proposal letter by regular mail (initial inquiries by email or fax are not accepted, nor are CDs, DVDs, computer discs, or video tapes). An initial inquiry to the foundation should include only the following items: a proposal letter that briefly summarizes the history of the project, need, objectives, time period, key staff, project budget, and evaluation plan; the applicant's latest annual report and complete set of audited financial statements; and the applicant's IRS 501(c)3 tax-exemption letter. Applications may be submitted at any time during the year. Each request is reviewed by staff and will usually receive a written response within thirty days. Those requests deemed consistent with the interests and resources of the foundation will be evaluated further and more information will be requested. Foundation staff may request a site visit, conference call, or meeting. All grants are reviewed and approved by the Trustees at one of their three board meetings in May, September, or December.
Restrictions: The foundation generally does not make grants for the following purposes or program areas: nonprofit organizations outside of New York; annual appeals, dinner functions, and fundraising events; endowments and capital campaigns; loans and deficit financing; direct grants to individuals; individual day-care and after-school programs; housing; organizations or projects based outside the U.S; films or video projects; small art, dance, music, and theater groups; individual K-12 schools (except charter schools); national health and mental health organizations; and government agencies or nonprofit organizations significantly funded or reimbursed by government agencies. Limited resources prevent the foundations from funding the same organization on an ongoing annual basis.
Geographic Focus: New Jersey, New York
Amount of Grant: 10,000 - 100,000 U.S.D
Samples: Hispanic Society of America, New York, New York, $50,000 - general operating support (2010); Uncommon Schools, New York, New York, $100,000 - for the start-up and first-year expenses of two new charter schools in Newark, New Jersey (2010); Center for Alternative Sentencing and Employment Services, New York, New York, $25,000 - to support the Career Exploration program for court-involved youth (2010).
Contact: John B. Krieger, Executive Director; (212) 644-0322; fax (212) 759-6510; main@achelis-bodman-fnds.org
Internet: http://foundationcenter.org/grantmaker/achelis-bodman/guidelines.html
Sponsor: Bodman Foundation
767 Third Avenue, 4th Floor
New York, NY 10017-2023

Boeing Company Contributions Grants 975

The Boeing U.S. contributions program welcomes applications in five focus areas: education; health and human services; arts and culture; civic; and the environment. Primary fields of interest include: arts; elementary and secondary education; the environment; family services, prevention of domestic violence; health care; public affairs; public safety; substance abuse programs; and general human services. The largest single block of charitable contributions goes toward supporting programs and projects related to education. Boeing also looks for innovative initiatives that promote the economic well-being of the community and neighborhood revitalization. Boeing invests in programs that promote participation in arts and cultural activities and experiences, programs that increase public understanding of and engagement in the processes and issues that affect communities and programs that protect and conserve the natural environment. Boeing accepts applications for cash grants, in-kind donations, and services.
Requirements: To apply for support you must be a U.S. based IRS 501(c)3 qualified charitable or educational organization or an accredited K-12 educational institution. U.S. grant guidelines and applications are available online.
Restrictions: Grants do not support: an individual person or families; adoption services; political candidates or organizations; religious activities, in whole or in part, for the purpose of further religious doctrine; memorials and endowments; travel expenses; nonprofit and school sponsored walk-a-thons, athletic events and athletic group sponsorships other than Special Olympics; door prizes or raffles; U.S. hospitals and medical research; school-affiliated orchestras, bands, choirs, trips, athletic teams, drama groups, yearbooks and class parties; general operating expenses for programs within the United States; organizations that do not follow our application procedures; follow-on applications from past grantees that have not met our reporting requirements or satisfactorily completed the terms of past grants; fundraising events, annual funds, galas and other special-event fundraising activities; advertising, t-shirts, giveaways and promotional items; documentary films, books, etc; debt reduction; dissertations and student research projects; loans, scholarships, fellowships and grants to individuals; for-profit businesses; gifts, honoraria, gratuities; capital improvements to rental properties.
Geographic Focus: Alabama, Arizona, California, Colorado, District of Columbia, Florida, Georgia, Hawaii, Illinois, Kansas, Maryland, Missouri, Nevada, New Mexico, Ohio, Oklahoma, Oregon, Pennsylvania, South Carolina, Texas, Utah, Washington, Australia, Canada
Contact: Antoinette Bailey, (312) 544-2000; fax (312) 544 - 2082
Internet: http://www.boeing.com/companyoffices/aboutus/community/charitable.htm
Sponsor: Boeing Company Contributions
100 North Riverside
Chicago, IL 60606-1596

Boettcher Foundation Grants 976

Grant support is given to promote the general well-being of humanity. Grants are awarded for arts and culture, community and social service, education and healthcare. Organizations seeking support from the Foundation should send a preliminary letter, describing the organization that wishes to submit a proposal and the project for which funding is being requested. The letter should be signed by the head of the applicant agency and should include a statement related to the priority of the project within the organization's overall plans. Letters of inquiry should be mailed or emailed.
Requirements: Capital grants are made in the form of challenges, conditional on an applicant agency's ability to raise the balance of the funds needed for a project. Although no absolute guidelines have been established, 50 to 75 percent of the goal should already be in hand before the grant request will be considered.
Restrictions: The Foundation does not accept proposals, or provide grants, for the following giving interests: operations; gymnasiums/athletic fields; housing; purchase of tables or tickets for dinner/events; individuals; large urban hospitals; out-of-state projects; media presentations; small business start-ups; open space/parks; conferences, seminars, workshops; organizations that primarily serve animals; debt reduction; pilot programs; endowments; religious groups or organizations for their religious purposes; scholarships; travel.
Geographic Focus: Colorado
Contact: Grants Administrator; (800) 323-9640 or (303) 534-1937; grants@boettcherfoundation.org
Internet: http://www.boettcherfoundation.org/grants/index.html
Sponsor: Boettcher Foundation
600 Seventeenth Street, Suite 2210 South
Denver, CO 80202-5422

Bonfils-Stanton Foundation Grants 977

Colorado nonprofit organizations are eligible to apply, and funds must be used within the state for the benefit of Colorado citizens. The focus of the foundation is to advance excellence in the areas of arts and culture, community service, and science and medicine. Types of support include operating grants, capital campaigns, and capacity-building grants. Proposals will be reviewed at quarterly meetings. Guidelines and forms are available online.

Requirements: Colorado 501(c)3 organizations are eligible.

Restrictions: Areas generally not eligible for funding include loans, grants, or scholarships to individuals; events, media productions, seminars, conferences, or travel expenses related to meetings; activities or initiatives that have a religious purpose or objective; endowment funding, fellowships, endowed chairs; funding to retire operating debt; requests from organizations outside the State of Colorado or that are not for the benefit of Colorado citizens.

Geographic Focus: Colorado

Date(s) Application is Due: Jan 31; Apr 30; Jul 31; Oct 31

Samples: ArtReach (CO)—for a Web-based ticket reservation system; Capitol Hill Community Services (CO)—to fund meal sites for the homeless; Kids in Need of Dentistry (CO)—to develop a comprehensive technology plan.

Contact: Susan France, Vice President of Programs; (303) 825-3774; fax (303) 825-0802; susan@bonfils-stanton.org

Internet: http://www.bonfils-stantonfoundation.org

Sponsor: Bonfils-Stanton Foundation
1601 Arapahoe Street, Suite 500
Denver, CO 80202

Booth-Bricker Fund Grants 978

The Foundation makes contributions for the purposes of promoting, developing and fostering religious, charitable, scientific, literary and educational programs. Requests are welcomed for capital needs, special projects and other one-time requirements. Applications should be made by letter. There are no forms or deadlines. Requests should include complete information about the applicant organization, including its history, purpose, finances, current operations, governing board and tax status. A detailed explanation of the proposed use of the funds must be provided.

Requirements: Requests are accepted for the funding of projects within the state of Louisiana. Priority is given to the New Orleans area.

Restrictions: The Foundation generally does not provide sustaining (operations and maintenance) funding. No grants are made to individuals.

Geographic Focus: Louisiana

Amount of Grant: 5,000 - 50,000 U.S.D

Samples: YMCA (New Orleans, LA)—program support, $51,700; Tulane U (New Orleans, LA)—for the Medical Ctr Cerise Chair, $30,000; Saint Dominic School (New Orleans, LA)—program support, $15,000.

Contact: Gray S. Parker, Chairperson; (504) 581-2430

Sponsor: Booth-Bricker Fund
826 Union Street, Suite 300
New Orleans, LA 70112

Borkee-Hagley Foundation Grants 979

The foundation awards grants in a wide range of interests, including social services to children and families, religious organizations, and environmental programs. Delaware nonprofits receive preference. The board meets in December to consider requests.

Requirements: Delaware nonprofits are eligible to apply.

Restrictions: No support for specific churches or synagogues. Grants are not made to individuals.

Geographic Focus: Delaware

Date(s) Application is Due: Nov 1

Amount of Grant: 1,000 - 25,000 U.S.D

Samples: Artistic Productions Inc (Hockessin, DE)—$1000; Delaware Hospice (Wilmington, DE)—$13,000; Better Life Outreach Ministries (Newport, DE)—$5000; Children and Families First (Wilmington, DE)—$13,000.

Contact: Henry H. Silliman Jr., President; (302) 652-8616

Sponsor: Borkee-Hagley Foundation
P.O. Box 4590
Wilmington, DE 19807-4590

Bosque Foundation Grants 980

The foundation gives primarily for higher education, human services, Baptist & Protestant agencies/churches and, medical research in Texas. Types of support include capital campaigns, building/renovation and research grants. There are no application forms or deadlines. Applicants should submit a one-page letter of intent that describes the program and request.

Requirements: Texas nonprofit and for-profit organizations are eligible.

Restrictions: Individuals are not eligible.

Geographic Focus: Texas

Amount of Grant: 5,000 - 20,000 U.S.D

Samples: Midland Community Theater, Midland, TX, $1,000; Peoples Missionary Baptist Church, Detroit, MI, $20,000; Dallas Baptist University, Dallas, TX, $10,000.

Contact: Louis A. Beecherl Jr., Trustee; (214) 956-6732; fax (214) 956-6733

Sponsor: Bosque Foundation
5950 Cedar Springs Boulevard, Suite 210
Dallas, TX 75235-6803

Boston Foundation Grants 981

The Boston Foundation has a particular concern for low income and disenfranchised communities and residents and supports organizations and programs whose work helps advance the Foundation's high priorities in a variety of subject areas: Arts and Culture; Civic Engagement; Community Safety, Economic Development; Education/Out-of-School Time, Health and Human Services; Housing and Community Economic Development; the Nonprofit Sector, Urban Environment and Workforce Development. The Foundation generally makes the following types of grants: Project or program support for community-based efforts that improve the quality of life in the community, test new models, and promote collaborative and innovative ventures; advocacy and public policy research that is linked to specific action; support for planning to enable organizations and residents to assess community needs, respond to new challenges and opportunities, and provide for the inclusion of new populations; organizational support to develop and build the capacity of nonprofit organizations - support that helps organizations keep pace with the changing requirements and demands of their communities and broader environments; small grants awarded on a rolling basis for one-time organizational development needs through the Vision Fund. In addition, on a very limited basis, the Foundation will consider development grants and strategic alliances.

Requirements: Grants are made only to tax-exempt organizations in Massachusetts.

Restrictions: The committee does not consider more than one proposal from the same organization within a 12-month period. Discretionary grants are generally not made to the following applicants: city or state government agencies or departments; individuals; medical research; endowments; equipment; replacement of lost/expired government funding or gap funding to cover the full cost of providing services; scholarships and fellowships; video and film production; construction and renovation projects and capital campaigns; programs with religious content; travel; summer camps and lobbying. Activities that are generally lower priorities for the Foundation are conferences, lectures, one-time events, programs benefiting only a small number of participants or routine service delivery and/or operating expenses.

Geographic Focus: Massachusetts

Date(s) Application is Due: Jan 5; Jul 1

Samples: Raw Art Works in Lynn, MA, $75,000 - for arts-based youth development programs to young people in Lynn; Boston Private Industry Council, Boston, MA, $200,000 - to administer the Accuplacer assessment to all program participants to determine the need for remedial writing support, and provide college transition coaching to 75 participating students headed to Bunker Hill and Roxbury Community College;

Contact: Corey Davis; (617) 338-1700; fax (617) 338-1604; info@tbf.org

Internet: http://www.tbf.org

Sponsor: Boston Foundation
75 Arlington Street, 10th Floor
Boston, MA 02116

Boston Globe Foundation Grants 982

The foundation concentrates on three focus areas: strengthen the reading, writing, and critical thinking of young people, while fostering their inherent love of learning; strengthen the roads that link people to culture; and strengthen the civic fabric of the city. The foundation also sponsors the Neighbor to Neighbor Initiative, which funds exceptional Dorchester focused nonprofits.

Requirements: Massachusetts nonprofit organizations in the greater Boston area are eligible.

Restrictions: The Foundation will only review one proposal per year from any organization.

Geographic Focus: Massachusetts

Amount of Grant: 5,000 - 15,000 U.S.D

Contact: Leah P. Bailey, Director; (617) 929-2895; fax (617) 929-2041; foundation@globe.com

Internet: http://bostonglobe.com

Sponsor: Boston Globe Foundation
P.O. Box 55819
Boston, MA 02205-55819

Boston Jewish Community Women's Fund Grants 983

The fund, a project of Combined Jewish Philanthropies, invites letters of intent for projects that benefit women and girls. Grants are made in the areas of health, abuse, hunger, education, and empowerment. In Massachusetts, projects funded are sponsored by organizations both within the Jewish community and from the community at large. In Israel, proposals from programs that serve the Haifa community or have a documented history of funding from other North American organizations are invited to apply. Letters of intent should not be more than two pages long and must contain: your organization's mission; a statement of need for the project, how the project will address it, and how it meets BJCWF objectives; a budget narrative including the amount of funding requested, projected major expenditures, and use of BJCWF funds.

Requirements: Grants are made to 501(c)3 organizations in Massachusetts or comparable organizations in Israel.

Geographic Focus: Massachusetts, Israel

Date(s) Application is Due: Oct 15

Amount of Grant: Up to 25,000 U.S.D

Contact: Susan Ebert, (617) 457-8500; susane@cjp.org

Internet: http://www.cjp.org/section_display.html?ID=572

Sponsor: Boston Jewish Community Women's Fund
126 High Street
Boston, MA 02110

Boston Psychoanalytic Society and Institute Fellowship in Child Psychoanalytic Psychotherapy 984
The Child Fellowship Program offers participants an intensive, one year course of study in psychoanalytic approaches to child and adolescent psychotherapy. The curriculum includes both a theoretical course and a clinical course to be taught weekly. The theory courses will introduce the fundamentals of child analytic theories and will review historical as well as modern theoretical approaches. The clinical courses are organized around different developmental stages of childhood. Those who complete the program and wish to continue on in the Child and Adolescent Advanced Training Program will receive credit for their participation and can apply for advanced standing. A printable application form is available for download. Tuition will be $1677 for the academic year.
Requirements: The program is open to mental health clinicians who have completed or are in an advanced phase of their clinical training (residency, graduate internship) and who are interested in further didactic and clinical education in psychoanalytic work with children. Other professionals who work with children and their families such as pediatricians, teachers, and ancillary therapists are also encouraged to apply.
Date(s) Application is Due: May 15
Contact: Elizabeth Jordan, Program in Psychoanalytic Studies; (617) 266-0953; fax (617) 266-3466; office@bostonpsychoanalytic.org
Internet: http://www.bostonpsychoanalytic.org/child_fellowship
Sponsor: Boston Psychoanalytic Society and Institute
15 Commonwealth Avenue
Boston, MA 02116

Boston Psychoanalytic Society and Institute Fellowship in Psychoanalytic Psychotherapy 985
The Fellowship is a one-year program designed to enhance participants' grasp of the principles and practice of psychoanalytic psychotherapy. It is also well suited for clinicians considering more extensive training in psychoanalytic psychotherapy or training in psychoanalysis, but who are either uncertain or not yet ready for a multi-year training commitment. The Fellowship offers two seminars weekly, one theoretical and one clinical, led by outstanding teachers from the Boston Psychoanalytic Society and Institute. The theoretical seminar focuses on Fundamental Concepts of Psychoanalytic Psychotherapy for the first term, on Comparative Theories of Psychotherapeutic Technique for the second term and on Selected Topics in Psychoanalytic Psychotherapy for the third term. The clinical seminar consists of discussion of case material presented by instructors and class members, with senior faculty as discussants. The seminars meet on Thursday evenings for 30 weeks starting in September. Each Fellow is matched with a faculty advisor who is available to discuss individual training goals and options. There is also an opportunity for individual supervision at reduced fees with a wide choice of experienced supervisors from the Boston Psychoanalytic Society and Institute. Tuition will be $1,677 for the academic year. The application fee is $25.
Requirements: The Fellowship is designed especially for mental health clinicians who have completed or are in an advanced phase of their requisite clinical training (psychiatric residency, psychology internship, MSW, masters in psychiatric nursing, etc.) and who desire further didactic and clinical education in psychoanalytic psychotherapy.
Date(s) Application is Due: May 15
Amount of Grant: 1,677 U.S.D
Contact: Elizabeth Jordan, Program in Psychoanalytic Studies; (617) 266-0953; fax (617) 266-3466; office@bostonpsychoanalytic.org
Internet: http://www.bostonpsychoanalytic.org/fellowship
Sponsor: Boston Psychoanalytic Society and Institute
15 Commonwealth Avenue
Boston, MA 02116

Boston Women's Fund Grants 986
The fund gives grants to support community-based projects in the greater Boston, MA, area run by and for women/girls organized for social and economic change. Also funded are direct service projects that have an organizing component. Types of support include challenge/matching grants, general operating grants, grants in aid, project grants, seed money grants, and technical assistance. Groups just getting started are encouraged to apply, particularly those including women/girls who are most vulnerable and have the least access to other resources—minority women, low-income women, girls, disabled women, lesbians, and older women. To apply, send a letter of intent to the foundation by September ninth. Letter of intent is available on the website.
Requirements: Applicant must be a Boston-area group with an organizational budget of less than $150,000 per year.
Geographic Focus: Massachusetts
Date(s) Application is Due: Oct 1
Amount of Grant: Up to 15,000 U.S.D
Contact: Catherine Joseph, Program Director; (617) 725-0035, ext. 3002; fax (617) 725-0277; Catherine@bostonwomensfund.org
Internet: http://www.bostonwomensfund.org/grantinfo.html
Sponsor: Boston Women's Fund
14 Beacon Street, Suite 805
Boston, MA 02108

Bothin Foundation Grants 987
The foundation has traditionally elected to support charitable organizations promoting youth, the elderly, the physically and mentally disabled, health care, minorities, community social services, and the environment. Grants for the arts are made only to those groups predominantly serving youth or with heavy emphasis on youth participation. The foundation prefers to award grants for capital or building and equipment needs. It is also preferred that a full three years elapse between grants. The board of directors meets at least three times annually to consider requests, usually in February, May and October. Applications should be submitted twelve weeks prior to the board meeting. Annual deadline dates vary; contact program staff for exact dates.
Requirements: Applicants must submit a brief preliminary letter of request for a specific grant including the purpose and history of the agency, goals and objectives of the project, proposed budget for which funding is sought, and a copy of tax-exemption letters. The foundation supports public nonprofit organizations located in San Francisco, Marin, Sonoma, and San Mateo Counties in California.
Restrictions: Grants are not made to individuals, for endowment drives, medical research, general operating expenses, films and other media presentations, religious organizations, conferences, program support, or to educational institutions other than those directly serving the learning and developmentally disabled.
Geographic Focus: California
Amount of Grant: Up to 12,000,000 U.S.D
Contact: Mary Callender, Program Officer; (415) 561-6540; fax (415) 561-6477; mcallender@pfs-llc.net
Internet: http://www.pfs-llc.net/bothin/index.html
Sponsor: Bothin Foundation
Presidio Building 1016, Suite 300, P.O. Box 29906
San Francisco, CA 94129-0906

Boyd Gaming Corporation Contributions Program 988
From giving generously to a variety of worthy charitable organizations to continually enhancing the effectiveness of our diversity programs, Boyd Gaming Corp. has a long-standing commitment to responsible gaming, and concern for the plantet, Boyd strives to make a positive differences in the communities in which it operates. The type of support offered is in: employee volunteer services; general/operating support; in-kind gifts
Requirements: All charitable requests must originate from a 501(c)3 non-profit organization and be from a state in which the company operates.
Restrictions: Boyd Gaming is not unable to act favorably on any request: for an individual, team or school-sponsored endeavor; for programs that discriminate for any reason, including race, color, creed, religion, age, sex or national origin
Geographic Focus: Hawaii, Illinois, Indiana, Louisiana, Mississippi, Nevada, New Jersey
Contact: Corporate Office; (702) 792-7200
Internet: http://www.boydgaming.com/about-boyd/corporate-responsibility
Sponsor: Boyd Gaming Corporation
3883 Howard Hughes Parkway, 9th Floor
Las Vegas, NV 89169

Bradley-Turner Foundation Grants 989
Incorporated as the W.C. and Sarah H. Bradley Foundation in Georgia in 1943, the Bradley-Turner Foundation uses contributions from the company's success to support the community and region through many different programs and facilities funded in whole or in part by foundation donations. The foundation has a special interest in endeavors related to family and children services, education, religion (Baptist, Christian, interdenominational, Methodist, Presbyterian, Salvation Army, and United Methodist), health, and culture and the arts. Major focus is placed on the vitality and quality of life in Columbus, Georgia, though compelling programs beyond the city's boundaries will also be considered. The foundation is particularly interested in projects that have a broad base of community support. There are no application forms. Applicants are asked to submit a letter of three to five pages describing the project. Grants are reviewed quarterly, in February, May, August, and November, when the board meets.
Requirements: IRS 501(c)3 tax-exempt organizations in Georgia are eligible. Heavily focuses on the southern region of the United States, but the Foundation is not solely limited to this area. It has also donated in Massachusetts, Illinois, and Colorado.
Restrictions: Grants are not made to individuals or to for-profit businesses or corporations.
Geographic Focus: Georgia
Contact: Phyllis Wagner, Executive Secretary; (706) 571-6040; fax (706) 571-3408
Internet: http://www.wcbradley.com/divisions.asp
Sponsor: Bradley-Turner Foundation
1017 Front Avenue, P.O. Box 140
Columbus, GA 31902-0140

Brain Trust Translational Research Grants 990
The program, a collaborative effort of Accelerate Brain Cancer Cure, the Alzheimer's Association, the Michael J. Fox Foundation for Parkinson's Research, and the Robert Packard Center for ALS Research at Johns Hopkins University, seeks to identify common challenges, explore possible solutions, and fund development of new therapeutic approaches with application to diverse brain diseases. The program seeks applications focused on technologies that can achieve selective targeting and/or delivery of therapeutic agents to specific regions/cells in the brain, including overcoming the limitations imposed by the blood-brain barrier, which can be applied in the clinical setting in the next 24 months. Funding is for one year, and the funds can only be applied to direct costs of research.
Requirements: For-profit and nonprofit organizations, individuals affiliated with nonprofit and for-profit organizations, and unaffiliated individuals are eligible.
Date(s) Application is Due: Jun 1
Contact: John Reher, (650) 685-2202; info@brain-trust.org
Internet: http://www.brain-trust.org/html/rfa.html
Sponsor: Brain Trust
800 Airport Boulevard, Suite 508
Burlingame, CA 94010

Brain Tumor Research Grants 991

Grants support basic scientific research directed at finding a cure for brain tumors. Grants are awarded for up to a two-year period and may be used for start-up projects or supplementary funding. Research guidelines and application are available on the Web site.

Date(s) Application is Due: Mar 16

Amount of Grant: Up to 100,000 U.S.D

Contact: Carrie Treadwell, Research Manager; (800) 770-8287, ext. 10 or (480) 575-8388; grants@tbts.org

Internet: http://www.tbts.org/itemDetail.asp?categoryID=300&itemID=16402

Sponsor: Brain Tumor Society
124 Watertown Street
Watertown, MA 02472-2500

Breast Cancer Fund Grants 992

Traditionally, the fund has awarded three types of grants: innovative research grants support projects investigating cutting-edge scientific approaches to the detection, treatment, and prevention of breast cancer; community model grants (non-research grants) to encourage and support the start-up of truly creative, innovative, and replicable U.S. programs that are developing new methods for addressing the support, education, health care access, and advocacy needs of women affected by breast cancer; and discretionary grants that generally fund smaller education, support, and advocacy initiatives. The fund gives preference to projects that address the needs of an ethnically, economically, and geographically diverse representation of women, including persons who are underserved or have low incomes. Initial application for research and community model grants is by letter of intent. If interested, the fund will invite a full proposal. Applicants for discretionary grants may submit an abbreviated application form available from the fund. Contact program staff for project availability.

Requirements: The fund will consider applications from individuals or organizations, whether public or private, for-profit or nonprofit.

Restrictions: No support is available for the direct cost of medical services, and no funds may be used for any lobbying efforts or political expenditures.

Contact: Jeanne Rizzo, Executive Director; (415) 346-8223; fax (415) 543-2975; ed@breastcancerfund.org or info@breastcancerfund.org

Internet: http://www.breastcancerfund.org

Sponsor: Breast Cancer Fund
1388 Sutter Street, Suite 400
San Francisco, CA 94109-5400

Brian G. Dyson Foundation Grants 993

The Foundation, established in Atlanta in 1994 by former Coca-Cola Bottling executive, Brian G. Dyson, offers funding to community foundations, higher education, and federated programs in the Atlanta region and throughout Georgia. Grants typically range up to $20,000, and funding supports general operating costs. There are no specific guidelines, application formats, or deadlines with which to adhere, and initial contact should be made in writing.

Requirements: Applicants must be colleges, public or private schools, or other non-profit organizations located within the State of Georgia.

Geographic Focus: Georgia

Amount of Grant: Up to 20,000 U.S.D

Contact: Brian G. Dyson, Director; (404) 364-2940

Sponsor: Brian G. Dyson Foundation
3060 Peachtree Road NW, Suite 1465
Atlanta, GA 30305-2241

Bridgestone/Firestone Trust Fund Grants 994

The trust supports programs and projects of nonprofit organizations in the areas of education, child welfare, and environment and conservation in communities where the company has operations. While primary consideration is given to organizations and causes related to the three major focus points, the Fund recognizes the importance and value in supporting all types of civic, community and cultural activities. Assistance is regularly given for: community and neighborhood improvements; civil rights and equal opportunity; voter registration and education; job training; performing arts programs; public radio and television; cultural programs; non-academic libraries; and museums. Types of support include: annual campaigns; building/renovation; capital campaigns; continuing support; donated equipment; emergency funds; employee matching gifts; employee-related scholarships; endowments; exchange programs; fellowships; general/operating support; matching/challenge support; program development; research; scholarship funds; and sponsorships. Applications must be submitted in writing and should include a description of the organization (two-page maximum) and its record of accomplishment, objectives of the program, whom the program benefits, and proposed method to evaluate the program's success; amount sought from the trust in relation to the total need; exactly how trust fund money would be used; copy of IRS 501(c)3 confirmation letter; list of board of directors and their professional affiliations; previous year's financial report; current year's operating budget; Form 990; list of other contributors and the amount of their donations; and copy of recent audit if available. Proposals are reviewed upon receipt.

Requirements: IRS 501(c)3 nonprofit tax-exempt organizations in Alabama, Arkansas, Colorado, Connecticut, Florida, Kentucky, Illinois, Indiana, Iowa, Louisiana, Michigan, Minnesota, Mississippi, North Carolina, Ohio, Pennsylvania, South Carolina, Tennessee, Texas, Utah, and Wisconsin are eligible. Schools, governmental agencies or other nonprofit, civic organizations are included. Grant proposals should be sent directly to the management of the local Bridgestone Firestone facility.

Restrictions: It is essential that all organizations receiving grants be equal opportunity employers who will operate their programs in support of equal opportunity objectives. Contributions will not be made to groups that discriminate on the basis of race, color, religion, gender, mental or physical disabilities, sexual orientation, national origin, age, citizenship, veteran/reserve/national guard status, or other protected status; partisan political organizations; or groups limited to members of a single religious organization.

Geographic Focus: Alabama, Arkansas, Colorado, Connecticut, Florida, Illinois, Indiana, Iowa, Kentucky, Louisiana, Michigan, Minnesota, Mississippi, Nevada, North Carolina, Ohio, Pennsylvania, South Carolina, Tennessee, Texas, Utah, Wisconsin

Amount of Grant: 50 - 50,000 U.S.D

Contact: Bernice Csaszar, Administrator; (615) 937-1415 or (615) 937-1000; fax (615) 937-1414; CsaszarBernice@bfusa.com or bfstrustfund@bfusa.com

Internet: http://www.bridgestone-firestone.com/trustfund.asp

Sponsor: Bridgestone/Firestone Trust Fund
535 Marriott Drive, P.O. Box 140990
Nashville, TN 37214-0990

Bright Family Foundation Grants 995

Established in 1986, the foundation primarily serves the Stanislaus County area of California. Areas of interest include: religion; education; medical services; medical school/education; internships; human services; children/youth services; arts & culture. The Board meets once a year in December to review proposals. Deadline date for applications is November 1.

Requirements: Stanislaus County, CA, IRS 501(c)3 tax-exempt organizations within a 20 mile radius of Modesto, CA, are eligible to apply.

Geographic Focus: California

Date(s) Application is Due: Nov 1

Amount of Grant: 5,000 - 50,000 U.S.D

Samples: Beggs High School, Beggs, OK, $32,000—scholarship program; UCSF School of Medicine, San Francisco, CA, $45,000; Modesto Symphony Orchestra, Modesto, CA, $20,000.

Contact: Calvin Bright, President; (209) 526-8242

Sponsor: Bright Family Foundation
1620 North Carpenter Road, Building B
Modesto, CA 95351-1155

Bright Mountain Foundation Grants 996

The foundation awards grants in three areas of priority: children—parenting skills programs, bilingual education programs, after-school care programs, and educational programs for children (including those that teach respect for our planet, and programs that address the nutritional needs of children); senior citizens—senior volunteer programs, programs that address the nutritional needs of seniors, and programs that help seniors maintain independent living; and persons living with HIV/AIDS and/or Hepatitis C—healthcare programs for persons living with HIV/AIDS that use a psychosocial integrated approach, medication adherence programs using a multidisciplinary team approach, and programs that address the nutritional needs of persons affected by HIV/AIDS. The foundation is also interested in funding programs that serve people dually diagnosed Hepatitis C and/or HIV positive. Projects are preferred that measure positive impact on target populations and provide comprehensive bilingual services.

Requirements: Colorado nonprofit organizations are eligible.

Restrictions: The foundation does not fund administrative costs, general operating expenses, annual fund drives, scholarships, endowments, fundraising events, or debt reduction.

Geographic Focus: Colorado

Date(s) Application is Due: Jan 16; Mar 15; Aug 15; Oct 14

Amount of Grant: 3,000 - 5,000 U.S.D

Contact: Cathy Lopez Wessell, Co-Executive Director; (303) 381-2245; fax (303) 381-2246; cwessell@brightmtnfdtn.org

Internet: http://www.brightmtnfdtn.org

Sponsor: Bright Mountain Foundation
1470 Walnut Street, Suite 101
Boulder, CO 80302-5138

Bright Promises Foundation Grants 997

The Bright Promises Foundation's primary activities are identifying the most pressing unmet needs of disadvantaged children in Illinois; calling for individuals, foundations, agencies, legislators, parents and the media to join the foundation in supporting these needs; soliciting grant applications and making grants that support these needs; attracting volunteers and funds to the foundation; and recognizing important role models with awards. Currently, the Bright Promises Foundation's focus is promoting better health among low-income and other at-risk children between the ages of 8-12. The Bright Promises Foundation initiated a four-year grant program called Healthy Children/Healthy Adults: Promoting Health through Better Nutritional Choices. The program responds to the escalating problem of childhood obesity in the state of Illinois. Now in its third year, the Foundation has so far paid and pledged $481,646 to community-based multi-purpose agencies to promote better health among low-income and other at-risk children between the ages of 8-12. The foundation revisits its focus every four years to ensure relevancy.

Requirements: Grant applications are considered annually from a pool of invited applicants. Proposals are evaluated based on criteria including measurable goals and objectives, and sustainability after Bright Promises Foundation funding ends. Grantees are required to report at least twice each year on their measurable objectives, and project coordinators from the Bright Promises Foundation board of directors conduct site visits with staff during the application process and during the grant year.

Geographic Focus: Illinois

Samples: Centers for New Horizons, "Healthy Children/Healthy Adults Project," Chicago, Illinois - program includes nutrition classes delivered to children, parents and staff as well as menu planning, budgeting, food preparation and cooking, developing a working community garden and a peer education program where students will learn to be leaders in their school and to promote health and nutrition school-wide; Children's Home + Aid, "Community Schools Student Health Fitness Project," Austin, Illinois - the program serves 125 children at Howe Elementary School to provide them the foundation for a lifetime of healthy eating and exercise habits; Erie Neighborhood House, "Super H - Healthy Kids Make Happy Kids Project," Chicago, Illinois - the bi-monthly club for children and their parents includes education about the nutritional value of chosen "super foods" and how to prepare them, field trips, staff training and hands-on activities.
Contact: Iris Krieg, Executive Director; (312) 704-8260; info@brightpromises.org
Internet: http://www.brightpromises.org/OurPrograms/Grants/
Sponsor: Bright Promises Foundation
333 N. Michigan Avenue, Suite 510
Chicago, IL 60601

Brinson Foundation Grants **998**
The foundation supports education, public health, and scientific research programs that engage, inform, and inspire committed citizens to confront the challenges that face humanity. Grantmaking priorities are education—awareness and outreach, democracy and citizenship, economically disadvantaged, and libraries and literacy; public health—awareness and outreach, and economically disadvantaged; and scientific research—astrophysics, cosmology, geophysics, medical research (i.e., Alzheimer's disease, cancer, Lou Gehrig's Disease (ALS), and stroke). Types of support include general operating grants and project grants. The foundation does not accept unsolicited grant applications. Grantseekers are asked to review the foundation's mission, vision, beliefs, priorities (accessed from the Who We Are link), and guidelines. If a grantseeker believes the request would match one or more of the foundation's grantmaking priorities, they can make an inquiry by completing the online Grantseeker Information Form. The completed form should be emailed to the office. Further application is by invitation.
Requirements: The foundation will consider inviting grant applications from organizations: whose request matches one or more of the Foundation's grantmaking priorities; located in the United States of America that are exempt from tax under Section 501(c)3 of the Internal Revenue Code and are defined as charitable organizations as described in Section 509(a)1, 2 or 3 or 170(b)1A; located outside the United States of America provided they produce a written legal opinion stating that they are a charitable equivalency to a qualifying U.S. organization and/or a written affidavit containing sufficient information for the Foundation to make a reasonable judgment that the organization is charitable. The Foundation's education and public health grants are generally made to organizations that serve individuals and communities in the greater Chicago area. It also considers leading U.S.-based programs that reach broader populations across the U.S. and internationally or have the potential to have a meaningful impact on best practices at the national or international level. The Foundation's physical science research grants are made to leading organizations across the United States. In this priority area, the location of the program is less critical than the match with the Foundation's grantmaking priorities.
Restrictions: The Foundation will not consider grant inquiries from organizations that: discriminate on the basis of race, gender, religion, ethnicity or sexual orientation. The Foundation will not consider grant inquiries that request funding for: activities that attempt to influence public elections; voter registration; political activity; lobbying efforts; promotion of a specific religious faith; medical research involving human cloning. The Foundation discourages grant inquiries requesting funds for: capital improvements; endowments; fundraising events.
Contact: Cheryl A. Heads, Grants Manager; (312) 799-4500; fax (312) 799-4310; mail@brinsonfoundation.org
Internet: http://www.brinsonfoundation.org/grant_seekers/grantseekers.shtml
Sponsor: Brinson Foundation
737 N Michigan Avenue, Suite 1850
Chicago, IL 60611

Bristol-Myers Squibb / Meade Johnson Award for Distinguished Achievement in Nutrition Research **999**
The Award was first presented in 1978. The Award includes a $50,000 cash prize and a silver commemorative medallion. The winners are selected by an independent peer-review selection committee whose members are grant administrators of current Bristol-Myers Squibb/Mead Johnson Unrestricted Nutrition Research Grants. The annual application deadline is in late April.
Date(s) Application is Due: Apr 24
Amount of Grant: 50,000 U.S.D
Samples: Steven Zeisel, Kenan Distinguished Professor of Nutrition and Pediatrics and associate dean for research at the School of Public Health, $50,000.
Contact: John Damonti, President; (212) 546-4000; fax (212) 546-9574
Internet: http://www.bms.com/aboutbms/grants/data/nutrit.html
Sponsor: Bristol-Myers Squibb Foundation
345 Park Avenue, Suite 4364
New York, NY 10154

Bristol-Myers Squibb Foundation Global HIV/AIDS Initiative Grants **1000**
The intent of the Program is to develop new models in awareness, in medical care, in community development and in prevention and treatment in poor and resource-limited areas of the world, where the need for all such efforts is greatest. Health care infrastructures must be developed and enhanced, stigmatization must be overcome, health care worker capacity must be built and preserved and local people must be empowered to generate and sustain local solutions to this global problem.
Requirements: The Bristol-Myers Squibb Foundation considers requests for support only from tax-exempt organizations that satisfy the requirements of section 501(c)3 of the U.S. Internal Revenue Code.
Restrictions: The Foundation does not award funds to: individuals; political, fraternal, social or veterans' organizations; religious or sectarian organizations unless engaged in a significant project benefiting the entire community; organizations receiving support through United Way or other federated campaigns; endowments; courtesy advertising; or conferences/special events/videos.
Contact: John Damonti, President; (212) 546-4000; fax (212) 546-9574
Internet: http://www.bms.com/sr/philanthropy/data/globhiv.html
Sponsor: Bristol-Myers Squibb Foundation
345 Park Avenue, Suite 4364
New York, NY 10154

Bristol-Myers Squibb Foundation Health Disparities Grants **1001**
One mission of the Foundation is to reduce health disparities by strengthening community-based health care worker capacity, integrating medical care and community-based supportive services, and mobilizing communities to fight disease. To this end, this Program attempts to address health disparities in four strategic disease areas representing major public health burdens and in four highly affected geographies: hepatitis in Asia, HIV/AIDS in Africa, serious mental illness in the U.S., and cancer in Europe. Additional areas of concern include: metabolic diseases, infectious diseases; rheumatoid arthritis; cardiovascular diseases; substance abuse; women's health issues; and overal health care giving.
Requirements: Nonprofit organizations in communities where Bristol-Myers Squibb maintains a facility should submit their requests for company contributions directly to that location. Contact persons are listed at the company website.
Restrictions: The foundation does not support individuals; conferences, special events, or videos; political, fraternal, social, or veterans organizations; religious or sectarian activities, unless they benefit the entire community; organizations funded through federated campaigns; endowments; or courtesy advertising.
Geographic Focus: Connecticut, Indiana, Massachusetts, New Jersey, New York
Contact: John Damonti, President; (212) 546-4000; fax (212) 546-9574
Internet: http://www.bms.com/foundation/reducing_health_disparities/Pages/default.aspx
Sponsor: Bristol-Myers Squibb Foundation
345 Park Avenue, Suite 4364
New York, NY 10154

Bristol-Myers Squibb Foundation Health Education Grants **1002**
The Program supports the development and replication of novel approaches aimed at individuals in the community to learn more about their health and wellbeing so that they can become more informed decision makers about health care and they can actively participate in disease prevention and management. In many of the grants, the Foundation seeks to be a catalyzing force for changes in government or social policies, supporting programs that can demonstrate the value of new approaches to health care through educational efforts. Nearly all the education programs focus on taking on major diseases, like hepatitis B or cancer and creating model programs for either preventing them or helping treat aspects of them in community settings.
Requirements: The Bristol-Myers Squibb Foundation considers requests for support only from tax-exempt organizations that satisfy the requirements of section 501(c)3 of the U.S. Internal Revenue Code.
Restrictions: The Foundation does not award funds to: individuals; political, fraternal, social or veterans' organizations; religious or sectarian organizations unless engaged in a significant project benefiting the entire community; organizations receiving support through United Way or other federated campaigns; endowments; courtesy advertising; or conferences/special events/videos.
Samples: Children's Health Fund, $300,000 over two years—to establish a comprehensive health resource center at the Harlem Children's Zone's Promise Academy; World Walk Foundation, $300,000 over four years—to establish a model program to improve prevention and treatment of diabetic wounds by educating health care professionals, other health workers, family care givers and the patients themselves at rural health clinics in Central America and the Caribbean.
Contact: John Damonti, President; (212) 546-4000; fax (212) 546-9574
Internet: http://www.bms.com/sr/philanthropy/data/healthed.html
Sponsor: Bristol-Myers Squibb Foundation
345 Park Avenue, Suite 4364
New York, NY 10154

Bristol-Myers Squibb Foundation Product Donations Grants **1003**
The Foundation partners with designated non-governmental organizations, donating medical products to support long term health care programs in developing countries as well as addressing immediate needs to provide emergency health care disaster relief. Through this Program, it addresses health disparities and community infrastructure building alongside other Foundation and company charitable contributions programs. There are three main areas where the company's donations of products are used: the Medical Mission Box program; health care infrastructure support; and disaster relief.
Samples: Pillatre, Haiti, in support of a mission trip to serve the rural poor of Northern Haiti; Karigiri, India, to conduct primary care outreach in conjunction with Karigiri Hospital (Schieffeline Leprosy Research and Training Center).
Contact: John Damonti, President; (212) 546-4000; fax (212) 546-9574
Internet: http://www.bms.com/foundation/reducing_health_disparities/donations/Pages/default.aspx
Sponsor: Bristol-Myers Squibb Foundation
345 Park Avenue, Suite 4364
New York, NY 10154

Bristol-Myers Squibb Foundation Women's Health Grants 1004

Bristol-Myers Squibb Women's Health Program supports projects that enhance women's health with strategies that improve education, prevention, diagnosis, treatment and access to care for women worldwide. Support has been given to projects that test innovative outreach programs, cultivate multi-sectoral partnerships and add new information to the existing body of knowledge to help define and achieve improved health for women around the world. The goal of the program is to generate initiatives that will help enhance women's health through novel interdisciplinary strategies that improve education, prevention, diagnosis, treatment and access to care for women worldwide. Since its inception, significant resources have been invested in programs that educate women about diseases and conditions that particularly threaten them as women.
Requirements: The Bristol-Myers Squibb Foundation considers requests for support only from tax-exempt organizations that satisfy the requirements of section 501(c)3 of the U.S. Internal Revenue Code.
Restrictions: The Foundation does not award funds to: individuals; political, fraternal, social or veterans' organizations; religious or sectarian organizations unless engaged in a significant project benefiting the entire community; organizations receiving support through United Way or other federated campaigns; endowments; courtesy advertising; or conferences/special events/videos.
Contact: John Damonti, President; (212) 546-4000; fax (212) 546-9574
Internet: http://www.bms.com/Documents/foundation/women2002.pdf
Sponsor: Bristol-Myers Squibb Foundation
345 Park Avenue, Suite 4364
New York, NY 10154

Bristol-Myers Squibb Patient Assistance Program 1005

The Foundation was established to provide temporary assistance to qualifying patients with a financial hardship who generally have no private prescription drug insurance and are not enrolled in a prescription drug coverage plan through Medicaid or any other federal, state or local health program. The program provides free medications to indigent patients who qualify in all 50 states, Puerto Rico and the U.S. Virgin Islands. All patients and their health care providers must complete a program application to be considered for assistance. A printable version of the application is available online. Applications can also be obtained by calling the toll free telephone number.
Requirements: Applicants are required to provide information regarding household income, prescription insurance coverage, and citizenship, along with both patient and health care provider signatures.
Contact: Program Coordinator; (800) 736-0003
Internet: http://www.bmspaf.org/program0.html
Sponsor: Bristol-Myers Squibb Patient Assistance Foundation
345 Park Avenue, Suite 4364
New York, NY 10154-0037

Bristol-Myers Squibb Virology and Oncology M.D. Research Fellowships 1006

The Program provides support for fellows to gain experience in epidemiological and clinical research as it relates to the care of individuals infected with HIV/AIDS and/or HBV. Goals of the Program include: support of studies that will further strengthen the science and knowledge of HIV/AIDS and HBV; development of a foundation for future prospective and retrospective studies; to provide a forum to share research findings; and support of the development of future clinical researchers. Up to 18 fellows will be selected annually, with grant awards up to $20,000 to support research-related expenses for a one-year research period. Applications must be submitted online.
Requirements: Candidates must: be an active Fellow in good standing in an ACGME-accredited Fellows training program; be a senior Fellow not in the first years of the fellowship; desire to enhance knowledge and skill development in the area of HIV/AIDS, HBV, or Oncology research; and identify a faculty member to serve as the project mentor.
Restrictions: Total grant amount is inclusive of indirect costs and associated IRB fees and is not permitted for use towards travel to conferences or for durable equipment.
Date(s) Application is Due: Feb 26
Amount of Grant: Up to 20,000 U.S.D
Contact: John Damonti, President; (212) 546-4000; fax (212) 546-9574
Internet: http://www.bms.com/clinical_trials/investigator_sponsored_trials/fellows_research_program/Pages/default.aspx
Sponsor: Bristol-Myers Squibb Foundation
345 Park Avenue, Suite 4364
New York, NY 10154

Broad Foundation IBD Research Grants 1007

The program seeks to stimulate innovative research that will lead to both the prevention and successful therapy of inflammatory bowel disease (IBD), including Chrohn's disease and ulcerative colitis. The foundation's goal is to fund basic or clinical research projects that are in the early stages of exploration; propose new directions or ideas; are creative, novel, cutting edge, and imaginative; and are not ready for funding by other more traditional granting agencies. All proposals must be based on sound scientific evidence and careful evaluation of current knowledge in IBD research. Requests must be preceded by a brief (one to three pages)letter of interest, which may be submitted at any time. Funding will be granted for one year; continued funding is based on progress reports and the perceived value of the findings.
Requirements: Nonprofit institutions worldwide, such as universities, hospitals, and research institutes, are eligible.
Restrictions: It is anticipated that the foundation will not fund projects after sufficient progress and maturity have made them fundable by other agencies.
Amount of Grant: 100,000 U.S.D
Samples: Charles N. Bernstein, M.D., University of Manitoba, Winnipeg, Manitoba, Canada—Project Title: A population-based characterization of potential microbial etiologies of inflammatory bowel disease using geographically defined high and low rate prevalence/incidence areas in Manitoba; Cucchiara, Salvatore, M.D., Ph.D., Sapienza University of Rome, Rome, Italy—Project Title: Molecular characterization of mucosa-associated intestinal microbiota and intestinal innate immune response: searching for additional mechanisms in pediatric Crohn's disease; Huang, Emina, M.D., University of Florida, Gainesville, Florida—Project Title: Colitis derived tumorigenic stem cells and the inflammatory bowel.
Contact: Heather Kubinec, Senior Research Administrator; (310) 954-5091; fax (310) 954-5092; info@broadmedical.org
Internet: http://www.broadmedical.org/funding.htm
Sponsor: Broad Foundation
10900 Wilshire Boulevard, Suite 1200
Los Angeles, CA 90024-6532

Broadhurst Foundation Grants 1008

The foundation supports the arts and humanities, community development, and medical research. In addition to program support, funds are awarded for building programs and necessary equipment in these areas. Support for scholarship funds for students training for the Christian ministry at institutions selected by the foundation; grants also to educational and religious institutions, and to medical research institutions, especially those related to pediatric diseases.
Requirements: Organizations in Oklahoma are eligible to apply.
Restrictions: No grants to individuals or for scholarship funds.
Geographic Focus: Oklahoma
Amount of Grant: 100 - 18,000 U.S.D
Contact: Ann Cassidy Baker, Chair; (918) 584-0661; fax (918) 584-5831
Sponsor: Broadhurst Foundation
401 S Boston Avenue, Suite 100
Tulsa, OK 74103-4002

Broadway Cares/Equity Fights AIDS National Grants 1009

BC/EFA makes grants to AIDS service organizations across the country twice a year, usually in January and September. Annual grants are made to support programs that offer direct care and/or services to people with HIV/AIDS. Projects funded include those providing food, shelter, transportation, emergency financial aid, emotional/practical support systems, nonreimbursable medical expenses, and other services. BC/EFA prefers to direct grants to urban minority communities and to smaller and mid-sized service providers in rural and suburban communities where resources are scarce.
Requirements: U.S. and Puerto Rico AIDS community service organizations that have an annual budget under $15 million are eligible to apply.
Restrictions: BC/EFA does not fund publications, travel/registration for conferences, reimbursable medical expenses, sectarian ministry, capital investment, education services for either the general public or high-risk populations, or medical research.
Date(s) Application is Due: May 30
Amount of Grant: 2,500 - 10,000 U.S.D
Contact: Brian O'Donnell, National Grant Program Manager; (212) 840-0770, ext. 226; fax (212) 840-0551; odonnell@bcefa.org or info@bcefa.org
Internet: http://www.broadwaycares.org/index.cfm
Sponsor: Broadway Cares / Equity Fights AIDS
165 W 46th Street, Suite 1300
New York, NY 10036

Broadway Cares/Equity Fights AIDS National Grants for Community Based Organizations 1010

The purpose of the program is to provide funds to local community-based organizations across the country that provide direct services to people living with or at high risk of HIV/AIDS and their families. Areas of support typically include projects providing case management, food, housing, transportation, emergency financial assistance, emotional support, and nonreimbursable and emergency medical expenses. Guidelines are available online.
Requirements: 501(c)3 tax-exempt organizations with total budgets below $10 million are eligible.
Restrictions: Grants do not support publications, travel or registration for conferences, reimbursable or ongoing medical expenses, capital investment, educational services, or medical research.
Date(s) Application is Due: Apr 30
Amount of Grant: Up to 5,000 U.S.D
Contact: Brian O'Donnell, National Grant Program Manager; (212) 840-0770, ext. 226; fax (212) 840-0551; odonnell@bcefa.org
Internet: http://www.broadwaycares.org/index.cfm
Sponsor: Broadway Cares / Equity Fights AIDS
165 W 46th Street, Suite 1300
New York, NY 10036

Brookdale Foundation Leadership in Aging Fellowships 1011

The Leadership in Aging Fellowship provides two years of support to junior academics to focus on a project that will help establish them in an area of aging research. The fellowship is open to a broad range of disciplines including, but not limited to, medical, biological and basic sciences, nursing, social sciences, the arts and humanities. The award is a two-year grant paid to the candidate's sponsoring institution in support of the candidate's research project. The grant amount of up to $125,000 each year is intended to cover 75% of the fellow's time, base salary and fringe benefits. The award could also

be used to include the support of a graduate assistant if necessary as long as the total amount does not exceed $125,000.
Requirements: The fellowship is open to all professionals in the field of aging. Candidates must have: (1) leadership potential; (2) an ongoing commitment to a career in aging; (3) a mentor (or mentors); (4) and, a willingness to commit at least 75% of his or her time for career development during each of the two years of the fellowship.
Amount of Grant: Up to 125,000 U.S.D
Samples: Kate de Medeiros, Ph.D. - Johns Hopkins University School of Medicine; Jennifer Hagerty Lingler, PhD, MA, FNP - University of Pittsburgh Schools of Nursing and Medicine; Teresita M Hogan MD, FACEP - Department of Emergency Medicine at the University of Illinois Chicago; Brie A. Williams, MD, MS - Division of Geriatrics at the University of California, San Francisco.
Contact: Cara Kenien, Fellowship Contact; (212) 308-7355; fax (212) 750-0132; cjk@brookdalefoundation.org
Internet: http://www.brookdalefoundation.org/Leadership/Fellows/fellows.html
Sponsor: Brookdale Foundation
950 Third Avenue, 19th Floor
New York, NY 10022-3668

Brookdale Foundation National Group Respite and Early Memory Loss Grants 1012
The Group Respite Program and the specialized Early Memory Loss Program are both community-based, social model, day service programs that provide dementia-specific group activities for participants and respite from caregiving tasks for family caregivers. Programs are open a minimum of one day a week for at least four hours per session, offering people with AD and other memory disorders regularly scheduled opportunities for socialization and group activities in a supportive environment. At least one paid professional staff member and a core of trained volunteers, students and program assistants, provide staffing for sites. In addition, programs offer access to other services for participants and their families, such as individual counseling for caregivers, information and referral to other health and social services, and support group meetings for family members. Additional program guidelines are available at the Foundations website.
Requirements: Applicants must be non-profit organizations with tax-exempt status under Section 501(c)(30 of the Internal Revenue Code or public agencies as defined under Section 509(a).
Date(s) Application is Due: Jul 1
Samples: Alzheimer's Association, South Central, Madison, WI; First United Methodist Church of Magnolia, Magnolia, AR; Friends of Salem Senior Center, Salem, OR.
Contact: Carmen Mendieta, Director; (212) 308-7355; fax (212) 750-0132; cm@brookdalefoundation.org
Internet: http://www.brookdalefoundation.org/Respite/respiteprogram2008.html
Sponsor: Brookdale Foundation
950 Third Avenue, 19th Floor
New York, NY 10022-3668

Brook J. Lenfest Foundation Grants 1013
The foundation primarily supports programs and organizations operating in urban areas of southeastern and south central Pennsylvania, southern New Jersey, and northern Delaware. Nonprofit organizations in Philadelphia receive preference. The foundation is dedicated to making people aware of positive life choices and providing support and opportunities for those motivated to pursue them. The Foundation will focus mainly on education, job training, mentoring programs, wellness based healthcare and the arts. The Foundation considers grant requests twice annually. To apply, send a short (two to four page) letter of inquiry that includes: amount requested; description of the program; date by which funding should be received; short and long term outcomes information for existing programs; expected outcomes and how the results of the program will be evaluated for new programs; description of how the program will support the mission and interests of the Foundation; total program budget involved in the request; other support for the program received, requested or expected from foundations, corporation, government or other revenue sources.
Requirements: Nonprofit organizations operating in urban areas of southeastern and south central Pennsylvania, southern New Jersey, and northern Delaware with a particular interest in Philadelphia are eligible.
Geographic Focus: Delaware, New Jersey, Pennsylvania
Samples: Pennsylvania State U (University Park, PA)—to provide scholarships to low-income students from Philadelphia public high schools, $1.4 million over five years.
Contact: Grants Administrator; (610) 828-4510; fax (610) 828-0390; lenfestfoundation@lenfestfoundation.org
Internet: http://www.brookjlenfestfoundation.org
Sponsor: Brook J. Lenfest Foundation
Five Tower Bridge, 300 Barr Harbor Drive, Suite 450
West Conshohocken, PA 19428

Brooklyn Community Foundation Caring Neighbors Grants 1014
The Caring Neighbors Fund assists vulnerable Brooklyn families and individuals with immediate need for a social safety net and seeks to provide access to health and mental health services. Its goals are to: offer paths out of poverty by supporting the work of emergency food providers and human services agencies; provide access to care for unaddressed physical or mental health needs in accessible community settings; ensure that homeless individuals and families can access safe temporary shelter and support services. The Foundation typically has two grant cycles annually, Letter of Inquiry are accepted then. Contact the Foundation directly for current grant cycles. Organizations invited to submit a complete proposal should expect a final decision from the Foundation within eight to twelve weeks of receipt.
Requirements: Organizations applying for a grant from the Brooklyn Community Foundation must be classified as tax-exempt under Section 501(c)3 of the Internal Revenue Code and as public charities under Section 509(a) of that Code. Fiscally sponsored organizations may also apply. Your organization need not be based in Brooklyn; however grants from the Brooklyn Community Foundation must directly benefit Brooklyn neighborhoods and/or Brooklyn residents. Applying for funding is a two part process: Ist step—begins with a Letter of Inquiry (LOI). This is the opportunity for an organization to provide the Foundation with an overview of the group and its proposed activities, project or program; 2nd step—applicants who have been selected to proceed to the next stage will receive an email from the Foundation's staff indicating that the organization has been approved to submit a complete proposal which best describes in detail the activities, program or project. If selected, your organization will have 30 days to submit a complete a proposal online.
Restrictions: The Foundation does not: fund individuals; support for-profit organizations; purchase tickets for dinners, golf outings or similar fundraising events; make contributions to candidates for elective office or for partisan political purposes; provide funding for religious purposes.
Geographic Focus: New York
Samples: New York Asian Women's Center, $10,000; St. John's Bread & Life, $250,000; Brooklyn Bureau of Community Service, $25,000.
Contact: Diane John, (718) 722-5952 or (718) 722-2300; fax (718) 722-5757; info@BrooklynCommunityFoundation.org
Internet: http://www.brooklyncommunityfoundation.org/grants/caring-neighbors
Sponsor: Brooklyn Community Foundation
45 Main Street, Suite 409
Brooklyn, NY 11201

Brown County Community Foundation Grants 1015
The Brown County Community Foundation exists for the purpose of enhancing the lives of the citizens and organizations of the community. It is a non-profit enterprise that seeks to provide the mechanism through which those who desire to help others in the community may carry out their philanthropy. By supporting charitable organizations in broad areas of community need - education, social services, health care, arts and humanities, and environment - it helps build a stronger, healthier Brown County. In general, grants shall be made for capital purposes only, not for operating expenses. Grant applications are generally available in April. Applications are due late May, with decisions announced in late June, or July.
Requirements: Evidence of 501(c)3 and/or 170(b)(1)(A) tax status must be submitted with all applications. Proposals must serve the residents of Brown County, Indiana.
Geographic Focus: Indiana
Contact: Judy Bowling, (812) 988-4882; judy@browncountycommunityfoundation.org
Internet: http://www.browncountycommunityfoundation.org/grants.html
Sponsor: Brown County Community Foundation
96 S Jefferson Street, P.O. Box 191
Nashville, IN 47448

Bruce and Adele Greenfield Foundation Grants 1016
The Foundation offers funding primarily in the forms of ongoing support and capital campaigns. Focus areas include: the arts; community development; health care; higher education; environmental issues; human/social service programs that support families and seniors; and science. Of particular interest are the performing arts, including dance, music, and theater. Though the geographic focus is Florida and Pennsylvania, support is also occasionally given outside of these two states. Initial approach should be by Letter of Interest (LOI), with no specific application forms or deadlines.
Requirements: Grantees must be 501(c)3 organizations with a sound track record.
Restrictions: Generally the Foundation will not provide grants for individuals.
Geographic Focus: Florida, Pennsylvania
Amount of Grant: 100 - 3,000 U.S.D
Samples: Good Shepherd Home, Allentown, Pennsylvania, $500; Pennsylvania Hortocultural Society, Philadelphia, Pennsylvania, $100; Germantown Jewish Center, Philadelphia, Pennsylvania, $3,000.
Contact: Adele G. Greenfield, (508) 428-1762; fax (508) 428-0607
Sponsor: Bruce and Adele Greenfield Foundation
575 Mistic Drive
Marstons Mills, MA 02648-1405

Brush Foundation Grants 1017
The primary goal of the foundation is to ensure that family planning worldwide becomes acceptable, available, accessible, affordable, effective, and safe. It focuses on projects that protect and enhance people's ability to manage their reproductive health; carry out public policy analysis and public education in the areas related to reproductive behavior and its social implications; and advance the personal knowledge and purposeful behavior of young people with respect to sexuality within a broad social context. Letters on inquiry are requested by January 15 and June 15.
Requirements: Grants are limited to 501(c)3 organizations.
Restrictions: The foundation will not accept unsolicited proposals.
Date(s) Application is Due: Jan 15; Jun 15
Amount of Grant: 5,000 - 50,000 U.S.D
Samples: Zero Population Growth (Washington, DC)—$15,000; Women's Health Ctr Foundation (Duluth, MN)—$10,000; San Benito High School (Hollister CA)—$6037; Feminist Majority Foundation (Arlington, VA)—$5000.
Contact: Krystal Fletcher, Grants Administrator; (216) 881-5121; fax (216) 881-1834; brushfoundation@hotmail.com
Sponsor: Brush Foundation
3135 Euclid Avenue, Suite 102
Cleveland, OH 44115

Bullitt Foundation Grants 1018
The foundation functions to protect and restore the environment of the Pacific Northwest, including Washington, Oregon, Idaho, western Montana, coastal rainforests in Alaska, and British Columbia, Canada. Program priorities include aquatic ecosystems; terrestrial ecosystems; conservation and stewardship in agriculture; energy and climate change; growth management and transportation; toxic and radioactive substances; training, organizational development, and unique opportunities (including education and public outreach). Areas of interest include air pollution, climate change, endangered species, energy conservation, environmental education and justice, human health, transportation, and tribal communities. The foundation supports challenge/matching, general operating, project/program, seed money, demonstration, and development grants, as well as requests for conferences/seminars and technical assistance support. Grants will be awarded for one year with possible renewal.
Requirements: Nonprofit organizations in the Pacific Northwest, including Washington, Oregon, Idaho, western Montana, coastal rainforests in Alaska, and British Columbia, Canada are eligible.
Geographic Focus: Alaska, Idaho, Montana, Oregon, Washington, Canada
Date(s) Application is Due: May 1; Nov 1
Samples: North Seattle Community College Foundation (WA)—to support a model project for watershed educational activities, $10,000; Washington's National Park Fund (WA)—to support a program for youth from local school districts, $5000.
Contact: Program Officer; (206) 343-0807; fax (206) 343-0822; info@bullitt.org
Internet: http://www.bullitt.org
Sponsor: Bullitt Foundation
1212 Minor Avenue
Seattle, WA 98101-2825

Burden Trust Grants 1019
The trust supports international nonprofit organizations in the fields of medical research and hospitals; schools and training institutions; and care of and homes for the elderly, children, and other individuals in need. Preference will be given to organizations affiliated with the Anglican Church.
Date(s) Application is Due: Mar 31
Contact: Patrick O'Conor, 0117 9628611; p.oconor@netgates.co.uk
Sponsor: Burden Trust
51 Downs Park W
Bristol, BS6 7QL England

Burlington Industries Foundation Grants 1020
This is a company-sponsored foundation, giving primarily in areas of company operations in North Carolina, South Carolina, and Virginia. The foundation supports organizations involved with arts and culture, education, health, youth development, community development, and civic affairs.
Requirements: 501(c)3 organizations in Burlington communities (North Carolina, South Carolina, and Virginia), are eligible for grant support. Requests for funding must be accompanied by: proof of 501(c)3 tax-exempt status; description of the organization and its objective; justification for the project; evidence that the organization is well established; information about the organization's reputation, efficiency, management ability, financial status and sources of income.
Restrictions: No support for: sectarian or denominational religious organizations, national organizations, private secondary schools, historic preservation organizations, individuals (except for employees in distress), conferences, seminars, workshops, endowments, outdoor dramas, films, documentaries, medical research, loans.
Geographic Focus: North Carolina, South Carolina, Virginia
Amount of Grant: 1,000 - 50,000 U.S.D
Samples: Greensboro Day School, Greensboro, NC, $12,450; United Way of Alamance County, Burlington, NC, $3,000; Independent College Fund of North Carolina, Raleigh, NC, $15,000.
Contact: Delores Sides, Executive Director; (336) 379-2903; delores.sides@itg-global.com
Sponsor: Burlington Industries Foundation
P.O. Box 26540
Greensboro, NC 27415-6540

Burlington Northern Santa Fe Foundation Grants 1021
The foundation is focused on the communities where the company operates and areas where its railways pass. The foundation supports education, including scholarships for Native Americans and scholarships for children of employees in conjunction with the National Merit Scholarships program; the arts, including museums, performing arts, and libraries; and civic and public affairs. Support goes to the Nature Conservancy and for local fire departments and law enforcement. Types of support include general operating support, continuing support, annual campaigns, capital campaigns, and program development. Health and human services funding concentrates on the United Way. Awards are made for a single year and for continuing support. The company also matches employee funds given to public and private colleges and universities, cultural organizations, and hospitals in the United States. Requests for applications should describe the purpose for the grant. Requests are reviewed every six weeks.
Requirements: Any 501(c)3 organization located in Schaumburg, IL, and communities where the corporation operates, including 28 states and two Canadian provinces, are eligible to apply.
Samples: Texas Christian University, Fort Worth, TX, $600,000 - for Career Services Center and Neeley Schools Next Generation Leadership Program; Hastings Rural Fire District, Glenvil, NE, $20,000 - for replacement of rescue truck and grass firefighting rig; Foss Waterway Seaport, Tacoma, WA, $50,000 - for restoration and development of waterfront cultural center and maritime museum.
Contact: Deanna Dugas, Manager Corporate Contributions; (817) 867-6407; fax (817) 352-7924; Deanna.dugas@bnsf.com
Sponsor: Burlington Northern Santa Fe Foundation
2650 Lou Menk Drive, 2nd Floor, P.O. Box 961057
Fort Worth, TX 76131-2830

Burton G. Bettingen Grants 1022
The fields of activity of the corporation are education at all levels, mental health, crime and abuse victims and public protection programs, religion (Christian, Roman Catholic, and Salvation Army), environment, and welfare. Top funding priority is children and youth. The current focus is on child prostitutes, runaways, and abandoned children. Nonprofits servicing the economically disadvantaged also may apply. The corporation provides broad types of support, including operating, capital, research, challenge/matching grants, and endowments. A letter of inquiry stating the applicant's background, goals and objectives, and the specific need for funding is welcome. Unsolicited submissions are considered but receive low priority.
Requirements: IRS 501(c)3 organizations are eligible Giving primarily, but not limited to, Southern California.
Restrictions: The corporation does not award grants to individuals; for general fund-raising events, dinners, or mass mailings; or to grantmaking organizations.
Geographic Focus: California
Amount of Grant: 5,000 - 200,000 U.S.D
Samples: Childrens Hospital of Los Angeles (Los Angeles, CA)—for capital campaign, $700,000; Phoenix House (New York, NY)—for general support, $50,000; Children's Defense Fund (Washington, DC)—for general support, $750,000.
Contact: Patricia Brown, Executive Director; (323) 938-8478; fax (323) 938-8479; burtonbet@aol.com
Sponsor: Burton G. Bettingen Corporation
134 S Mansfield Avenue
Los Angeles, CA 90036-3019

Bush Foundation Health & Human Services Grants 1023
The foundation responds to a broad range of human services proposals. Proposals are reviewed on a case-by-case basis; applicant organizations take the lead in identifying promising solutions to the challenges faced by people who use their programs. In recent years, most grant dollars given to human services organizations have been for programs serving children, youth, and families. The foundation also considers program proposals that will improve the quality, accessibility, and efficiency of health care services in the region.
Requirements: The foundation is most interested in proposals that: (1) Promote opportunities for individuals and communities to become fully contributing members of society by supporting organizational projects that remove barriers to effective education, economic security and good health; (2) Improve the abilities of immigrant and refugee organizations, groups and individuals to obtain basic needs and rights, promote refugee and immigrant civic engagement and enhance their contribution to economic and cultural life; (3) The foundation will also consider proposals for comprehensive capital campaigns for building purchases, major building renovations and new construction to improve physical facilities. To be eligible for consideration, your organization must be a 501(c)3 nonprofit, tax-exempt organization, located in Minnesota, North Dakota or South Dakota, and able to demonstrate that you can take the lead in identifying promising solutions to challenges faced by people who use your programs. The two-step application process begins with a letter of inquiry. Guidelines for writing the letter of inquiry are on the information sheet which is available by contacting the sponsor or by download at the website.
Restrictions: The foundation does not make grants to: Individuals; Government agencies (except in special cases dictated by foundation priorities); Projects not benefiting the three-state region of Minnesota, North Dakota and South Dakota; Or, projects outside the United States. Download the Grant Restrictions file from the website for more details.
Geographic Focus: Minnesota, North Dakota, South Dakota
Date(s) Application is Due: Mar 1; Jul 1; Nov 1
Contact: Program Officer; (651) 227-0891; grants@bushfoundation.org
Internet: http://www.bushfoundation.org/grants/human_services.asp
Sponsor: Bush Foundation
332 Minnesota Street, Suite East 900
St. Paul, MN 55101-1315

Bush Foundation Medical Fellowships 1024
The program was established to enhance community health care in Minnesota, North Dakota, South Dakota through the professional and personal development of selected physician leaders. Each year, the program awards approximately 13 fellowships that enable physicians to take a leave of absence from their practices to pursue professional and personal goals that address the health care needs of their communities. Their programs are self designed and self managed; they may last from three to 12 months. During this time, the fellowship provides a monthly stipend, as well as other financial aid.
Requirements: Applicants to the program must be: Physicians currently practicing in Minnesota, North Dakota or South Dakota; At least 35 years old and at least 10 years out of medical or osteopathic college; Able to state clearly their needs and opportunities for application of new skills and knowledge, both for the communities they serve and their own career development; and, able to explain how their programs will benefit an underserved population or need. All applications and references should be submitted via postal or express service. emailed applications and references must be followed by hard copy.
Restrictions: Previous Bush Medical Fellows may not reapply. If a physician (or his or her spouse/partner) is an elected official or public policymaker, the applicant may receive a grant for a degree-granting program only.
Geographic Focus: Minnesota, North Dakota, South Dakota

Date(s) Application is Due: Dec 1
Contact: Michael R. Wilcox, Program Director; (952) 442-2420 or (952) 758-4144; fax (952) 442-5841; bushmed@bushfoundation.org
Internet: http://www.bushfellows.org/medical
Sponsor: Bush Foundation
332 Minnesota Street, Suite East 900
St. Paul, MN 55101-1315

Bushrod H. Campbell and Adah F. Hall Charity Fund Grants **1025**
Grants are awarded to organizations in the Boston area devoted to basic needs for the elderly, projects relating to medicine and medical research, health care, hospitals, the blind and deaf, and certain discretionary projects. Grants are also awarded countrywide for projects addressing population control. Types of support include capital grants, general operating grants, program grants, and research grants.
Requirements: Tax-exempt organizations located within Boston and neighboring communities and U.S. tax-exempt organizations devoted to population control are eligible.
Restrictions: Grants are not awarded to individuals.
Geographic Focus: Massachusetts
Date(s) Application is Due: Jan 15; Apr 15; Aug 15; Oct 15
Amount of Grant: 3,000 - 7,500 U.S.D
Contact: Brenda Taylor, c/o Palmer & Dodge; (617) 239-0556; fax (617) 227-4420
Sponsor: Bushrod H. Campbell and Adah F. Hall Charity Fund
111 Huntington Avenue at Prudential Center
Boston, MA 02199-7613

Business Bank of Nevada Community Grants **1026**
Business Bank of Nevada contributes financial assistance to nonprofit institutions and organizations that enhance the quality of life and promote public interest where the company conducts its business. The four main areas of giving include education, health and human services, community development, and arts and culture.
Requirements: Nevada 501(c)3 tax-exempt organizations in communities where the Bank has a presence. Community Grants are part of Business Bank of Nevada's grants program and must comply with the Bank's overall charitable funding guidelines. Requests must include the following information: a copy of the IRS letter of nonprofit tax-exempt 501C(3) status; a statement of purpose of the organization; the purpose and amount of the grant; the geographic area served; the project budget; a list of current board members with affiliations; a list of sources and amounts of other funding obtained, pledged, or requested; a copy of the latest audited or board approved financial statement.
Geographic Focus: Nevada
Amount of Grant: 250 - 2,500 U.S.D
Contact: Paul Stowell, SVP Public Relations Manager; (702) 952-4415; pstowell@bbnv.com
Internet: http://www.bbnv.com/charitable_giving.php
Sponsor: Business Bank of Nevada
6085 West Twain Avenue
Las Vegas, NV 89103

BWF Ad Hoc Grants **1027**
To complement the competitive award programs, BWF also makes modest grants on an ad hoc basis to nonprofit organizations conducting activities intended to improve the general environment for science. These noncompetitive grants are for activities closely related to the Foundation's focus areas. The Foundation places special priority on working with non-profit organizations, including government agencies, to leverage financial support for targeted areas of research, and on encouraging other foundations to support biomedical research. Proposals should be brief and include a description of the focus of the activity, the expected outcomes, qualifications of the organization or individuals involved; provide certification of the sponsor's Internal Revenue Service tax-exempt status; and give the total budget for the activity, including any financial support obtained or promised. Proposals are given careful preliminary review, and those deemed appropriate are presented for consideration by BWF's board of directors.
Contact: Nancy S. Sung; (919) 991-5100; fax (919) 991-5160; info@bwfund.org
Internet: http://204.85.36.154/page.php?mode=privateview&pageID=79&navID=110
Sponsor: Burroughs Wellcome Fund
21 TW Alexander Drive, P.O. Box 13901
Research Triangle Park, NC 27709-3901

BWF Career Awards at the Scientific Interface **1028**
This initiative supports the early career development of postdoctoral scientists in the physical and computational sciences, whose work addresses biological questions and who are dedicated to pursuing careers in academic research. Research methods may include any combination of experiment, computation, mathematical modeling, statistical analysis, or computer simulation. Areas of interest include: chemistry, computer sciences, engineering, mathematics, and physics. Awards provide a stipend over five years to support up to two years of advanced postdoctoral training and the first three years of a faculty appointment. During both the postdoctoral and the faculty periods, grants must be made to degree-granting institutions in the United States or Canada on behalf of the award recipient. Award recipients must complete at least one year of postdoctoral training during the period of the award, usually at the institution that nominates the candidate.
Requirements: U.S. and Canadian academic institutions, including medical schools, graduate schools, affiliated hospitals, and research institutes are eligible. Candidates must hold a PhD degree in the fields of mathematics, physics, chemistry (physical, theoretical, or computational), computer science, statistics, or engineering. Exceptions will be made only if the applicant can demonstrate significant expertise in one of these areas, evidenced by publications or advanced course work. Candidates must have completed at least six months but not more than 48 months of postdoctoral training at the time of application.
Restrictions: Candidates must not hold or have accepted a faculty appointment as a tenure-track assistant professor at the time of application.
Amount of Grant: Up to 500,000 U.S.D
Contact: Dr. Nancy S. Sung, Senior Program Officer; (919) 991-5100; fax (919) 991-5160; info@bwfund.org
Internet: http://204.85.36.154/page.php?mode=privateview&pageID=129&navID=160
Sponsor: Burroughs Wellcome Fund
21 TW Alexander Drive, P.O. Box 13901
Research Triangle Park, NC 27709-3901

BWF Career Awards for Medical Scientists (CAMS) **1029**
The program proved five-year, $700,000 awards for physician-scientists bridge advanced postdoctoral/fellowship training and the early years of faculty service. Proposals must be in the area of basic biomedical, disease-oriented, translational, or molecular, genetic, or pharmacological epidemiology research. Proposals in the area of epidemiology should contact BWF to determine their eligibility. Proposals in health services research or involving large-scale clinical trials are ineligible. Awards are made to degree granting institutions in the U.S. or Canada on behalf of the awardee. The postdoctoral/fellowship portion of the award may last up to two years, and allows the awardee the use of up to $95,000 per year ($65,000 per year salary support).
Requirements: Candidates must hold an M.D., D.D.S., D.V.M., Pharm.D., or equivalent clinical degree.
Amount of Grant: 700,000 U.S.D
Contact: Rolly L. Simpson Jr.; (919) 991-5100; fax (919) 991-5160; info@bwfund.org
Internet: http://204.85.36.154/page.php?mode=privateview&pageID=189&navID=19
1
Sponsor: Burroughs Wellcome Fund
21 TW Alexander Drive, P.O. Box 13901
Research Triangle Park, NC 27709-3901

BWF Clinical Scientist Awards in Translational Research **1030**
The Clinical Scientist Awards in Translational Research program supports established independent physician-scientists who are dedicated to translational research - the two-way transfer between work at the laboratory bench and patient care. The program is intended to help protect award recipients' time to pursue the vital link between basic and clinical research. Importantly, the program aims to identify and reward proven mentors and to increase their capacity to train the next generation of investigators skilled in translational research. In this way, BWF hopes to increase the ranks of experienced physician-scientists critically positioned to bridge the gap between bench and bedside. Selection criteria include the candidate's qualifications, quality of the proposed activities, the ability to conduct innovative research, originality of the proposed research, and the potential to advance research in the scientific fields of study. Applications must be submitted electronically.
Requirements: U.S. or Canadian nonprofit hospitals and educational, scientific, research, or health institutions may submit applications. Candidates must have an MD or MD/PhD degree; hold an appointment or a joint appointment in a sub-specialty of clinical medicine; present evidence of an established research career; and be investigators at the assistant-professor level, associate-professor level, or an equivalent tenure-track position. Institutions may nominate up to four candidates, as long as at least one of the candidates is female, and at least one of the candidates is a member of an underrepresented minority group (Hispanic-American, African-American, Native American).
Amount of Grant: 750,000 U.S.D
Samples: Gilbert Chu, Stanford U Medical Ctr (Palo Alto, CA)—to support researchers working to bridge the gap between basic research and patient treatment and care, $750,000 over five years.
Contact: Nancy S. Sung, Senior Program Officer; (919) 991-5100; fax (919) 991-5160; info@bwfund.org
Internet: http://204.85.36.154/page.php?mode=privateview&pageID=177&navID=180
Sponsor: Burroughs Wellcome Fund
21 TW Alexander Drive, P.O. Box 13901
Research Triangle Park, NC 27709-3901

BWF Institutional Unifying Population and Laboratory Based Sciences **1031**
The program offers five-year institutional training awards of $500,000 per year to bridge the gap between the population and computational sciences and the laboratory-based biological sciences. The award will support the training of researchers between existing concentrations of research strength in population approaches to human health and in basic biological sciences. The goal is to establish training programs by partnering researchers working in schools of medicine and schools (or academic divisions) of public health.
Requirements: Degree-granting institutions in the U.S. or Canada may submit applications.
Restrictions: For-profit companies may not participate in the application, but could be valuable partners in such training programs.
Amount of Grant: 500,000 U.S.D
Contact: Victoria McGovern, Senior Program Officer; (919) 991-5100; fax (919) 991-5160; info@bwfund.org
Internet: http://204.85.36.154/page.php?mode=privateview&pageID=159&navID=164
Sponsor: Burroughs Wellcome Fund
21 TW Alexander Drive, P.O. Box 13901
Research Triangle Park, NC 27709-3901

BWF Investigators in the Pathogenesis of Infectious Disease **1032**

These five-year awards provide $500,000 for opportunities for accomplished investigators at the assistant professor level to study pathogenesis, with a focus on the intersection of human and microbial biology. The program is intended to shed light on the overarching issues of how human hosts handle infectious challenge. The awards are intended to give recipients the freedom and flexibility to pursue new avenues of inquiry and higher-risk research projects that hold potential for advancing significantly the biochemical, pharmacological, immunological, and molecular biological understanding of how infectious agents and the human body interact.
Requirements: Candidates must: generally have an M.D., D.V.M., or Ph.D. degree; have an established record of independent research and hold a tenure-track position as an assistant professor or equivalent (at the time of application) at a degree-granting institution; and be nominated by accredited, degree-granting institutions in the United States or Canada.
Restrictions: Applications from non-tenure track investigators at tenure-offering, degree-granting institutions will not be accepted.
Geographic Focus: All States, Canada
Amount of Grant: 500,000 U.S.D
Contact: Victoria McGovern, Senior Program Officer; (919) 991-5100; fax (919) 991-5160; info@bwfund.org
Internet: http://204.85.36.154/page.php?mode=privateview&pageID=105&navID=135
Sponsor: Burroughs Wellcome Fund
21 TW Alexander Drive, P.O. Box 13901
Research Triangle Park, NC 27709-3901

Byron W. and Alice L. Lockwood Foundation Grants **1033**
The foundation awards grants to Washington nonprofit organizations in its areas of interest, including arts, biomedical research, higher education, hospitals and health care organizations, housing and homelessness, museums, religion and religious welfare programs, and social services. Types of support include capital improvements, continuing support, general operating grants, professorships, projects grants, and research grants. There are no application forms; submit a letter of request.
Requirements: Washington nonprofit organizations are eligible. Grants are awarded primarily in Washington's Seattle and Puget Sound areas.
Date(s) Application is Due: Oct 31
Amount of Grant: 900 - 56,000 U.S.D
Contact: Lee Kraft, Executive Director; (206) 230-8489
Sponsor: Byron W. and Alice L. Lockwood Foundation
P.O. Box 4
Mercer, WA 98040

C.R. Bard Foundation Grants **1034**
The foundation's areas of interest are health care; arts, cultural, and community life; and education and research in company-operating areas. In the area of health care, the foundation supports organizations and programs specializing in vascular medicine, urology, and oncology. In the arts/cultural/community category, the foundation considers requests from local civic, recreational, youth, and art and cultural organizations. In the area of education and research, the foundation considers requests from institutions of higher education that offer medical and managerial curricula, as well as organizations whose programs and services benefit the health care industry. Types of support include program development, scholarship funds, and employee matching gifts. Requests for funding must be in writing. Contact the office for application requirements.
Requirements: 501(c)3 organizations in primarily New Jersey, Massachusetts, and California may apply for grant support.
Geographic Focus: California, Massachusetts, New Jersey
Amount of Grant: Up to 14,000,000 U.S.D
Contact: Program Contact; (908) 277-8182 or (800) 367-2273; fax (908) 277-8240
Internet: http://www.crbard.com/about/com_relations/foundations/index.cfm
Sponsor: C.R. Bard Foundation
730 Central Avenue
Murray Hill, NJ 07974

C.W. Titus Foundation Grants **1035**
The foundation awards grants to eligible nonprofit organizations in Missouri and Oklahoma for programs in its areas of interest, including arts and culture, healthcare and health organizations, and social services. Types of support include building construction/renovation, general operating support, and materials/equipment acquisition. There are no application forms or deadlines.
Requirements: Missouri and Oklahoma nonprofit organizations are eligible.
Geographic Focus: Missouri, Oklahoma
Amount of Grant: 1,000 - 10,000 U.S.D
Samples: Southwest Missouri State U Foundation (Springfield, MO)—to construct and equip a new master-control facility for the university's Ozarks Public Television station, $1 million.
Contact: Grants Administrator
Sponsor: C.W. Titus Foundation
950 Philtower Building, Suite 950
Tulsa, OK 74103-4123

Cable Positive's Tony Cox Community Fund Grants **1036**
Cable Positive's Tony Cox Community Fund is a national grant program that exists to encourage community-based AIDS Service Organizations (ASOs) and cable outlets to partner in joint community outreach efforts, or to produce and distribute new, locally focused HIV/AIDS-related programs and Public Service Announcements (PSAs). Grants are available up to $7,000 for 501(c)3 organizations, with special consideration given to AIDS Service Organizations (ASOs) and cable systems and producers partnering with ASOs. Eligible local community outreach projects include, but are not limited to: World AIDS Day and National HIV Testing Day events, AIDS Rides/Walks, other joint efforts between AIDS organizations and local cable operators, etc. Funding is also available for production costs of HIV/AIDS-related programs and PSAs.
Requirements: In order to be considered, you must partner with a cable system. To find your local cable partner go to www.cableyellowpages.com. If you do not have an existing relationship with your local cable system, contact their Public Affairs or Community Relations department. Successful applicants usually demonstrate the following tactics: PSA or program production; Community partnering-third party participation; Local government involvement; Media outreach (press kits & promotion); and, Marketing campaign. The following three funding levels are available: 1. $3,000-Alpha Cable System (up to 150,000 television households); 2. $5,000-Beta Cable System (150,001 - 300,000 television households); 3. $7,000- Gamma Cable System more than 300,000 television households).
Geographic Focus: Oklahoma, Oregon, Pennsylvania, Rhode Island, South Carolina, South Dakota, Tennessee, Texas, Utah, Vermont, Virginia, Washington, West Virginia, Wisconsin, Wyoming
Date(s) Application is Due: Sep 12
Amount of Grant: Up to 7,000 U.S.D
Contact: Jennifer Medina, (212) 459-1504; Jennifer@cablepositive.org
Internet: http://www.cablepositive.org/programs-tonycox.html
Sponsor: Tony Cox Community Fund
1775 Broadway, Suite 443
New York, NY 10019

Cabot Corporation Foundation Grants **1037**
The goal of Cabot Corporation Foundation is to support community outreach objectives, with priority given to science and technology education and community and civic improvement efforts in the communities where the company has major facilities or operations. Types of support include capital grants, challenge grants, employee matching gifts, fellowships, general support, professorships, project support, research, scholarships, and seed money. The board meets in January, April, July, and October to consider requests. Applications must be received at least 30 days before a board meeting.
Requirements: The foundation supports only nonprofit 501(c)3 tax-exempt organizations in areas of company operation. Modest support is available for international organizations that qualify under U.S. tax regulations.
Restrictions: Contributions are not made to individuals; fraternal, political, athletic or veterans organizations; religious institutions; capital and endowment campaigns; sponsorships of local groups/individuals to participate in regional, national, or international competitions, conferences or events; advertising sponsorships; fund-raising events.
Geographic Focus: Georgia, Illinois, Louisiana, Massachusetts, New Mexico, Pennsylvania, Texas, West Virginia, Belgium, Canada, China, Switzerland, United Kingdom
Amount of Grant: 2,000 - 75,000 U.S.D
Samples: The Access for Education Foundation, Albuquerque, New Mexico, $2,500 - K-12 educational outreach activities focused on space exploration as a method to connect students to science, technology, engineering and mathematics; Hanson Memorial High School, Franklin, Louisiana, $13,650 - computer equipment and software for the school library; St. Joseph's Hospice Resource Centre of Sarnia Lambton, Sarnia, Ontario, Canada, $25,000 paid over five years ($5,000 each year) - build a residential hospice and establish an Operational Sustainability Fund to ensure sufficient financing for the first five years.
Contact: Jane A. Bell, Executive Director; (617) 345-0100; fax (617) 342-6312; jane_bell@cabot-corp.com
Internet: http://www.cabot-corp.com/About-Cabot/Corporate-Giving
Sponsor: Cabot Corporation Foundation
Two Seaport Lane, Suite 1300
Boston, MA 02210-2019

Caddock Foundation Grants **1038**
The foundation supports nonprofit national and international Evangelical Christian religious organizations including churches and religious institutions, community groups, hospitals, international missions and ministries, religious centers and facilities, and youth organizations. Types of support include conferences/seminars, fellowships, and general operating support. Application must be made in writing and include a description of the organization and its objectives and the purpose of the grant. There are no application deadlines.
Restrictions: Grants are made to Evangelical Christian organizations. Grants are not made to individuals.
Amount of Grant: 1,800 - 390,000 U.S.D
Samples: Set Free Prison Ministries (Riverside CA)—for operating support, $152,000; Grace Evangelical Society (Roanoke, TX)—for operating support, $40,000.
Contact: Richard E. Caddock, Jr., Treasurer; (951) 683-5361
Sponsor: Caddock Foundation
1717 Chicago Avenue
Riverside, CA 92507

Caesars Foundation Grants **1039**
Founded as Harrah's Foundation in Nevada in 2002, giving is in the area of company operations. The foundation supports programs designed to help older individuals live longer, healthier, and more fulfilling lives; promote a safe and clean environment; and improve the quality of life in communities where Caesars operates. Fields of interest include: aging centers and services; Alzheimer's disease; developmentally disabled services; the environment; food distribution programs; food services; health care; patient services; higher education; hospitals; human services; mental health services; public affairs; public safety; nutrition; and youth services. Types of support being offered include:

building and renovation; capital campaigns; continuing support; general operating support; program development; research; scholarship funding; and sponsorships. There are no specific deadlines or application forms. The foundation generally funds programs and projects of $10,000 or more.
Geographic Focus: Arizona, California, Illinois, Indiana, Iowa, Louisiana, Mississippi, Missouri, Nevada, New Jersey, North Carolina, Pennsylvania
Amount of Grant: Up to 250,000 U.S.D
Samples: Atlantic Cape Community College Foundation, Atlantic City, New Jersey, $75,000 - general operating support (2010); Calumet College of St. Joseph, Whiting, Indiana, $60,000 - construction of new student center building (2010); University of Chicago Medical Center, Chicago, Illinois, $50,000 - research and development (2010).
Contact: Lisa Mariani, Administrator; (702) 880-4728 or (702) 407-6358; fax (702) 407-6520; caesarsfoundation@caesars.com
Internet: http://www.caesarsfoundation.com/
Sponsor: Caesars Foundation
1 Caesars Palace Drive
Las Vegas, NV 89109-8969

Cailloux Foundation Grants **1040**
The foundation awards grants to nonprofit organizations in its areas of interest, including civic and cultural, education and youth, family and community service, and health and rehabilitation. Requests for specific projects or programs, technical assistance, and capital projects are considered; general operations are less common and usually small; challenge grants are awarded occasionally; and endowment grants are rare. The general policy of the foundation is not to make grants to an organization more than once within any 12-month period. Applicants must submit a letter of inquiry; instructions are available online.
Requirements: 501(c)3 tax-exempt organizations are eligible. Only grant proposals originating in Kerr and the surrounding communities (Gillespie, Bandera, Edwards, Real, and Kimble Counties) will be considered.
Restrictions: In general, grants are not made for fund raising events, professional conferences, membership drives, competition expenses, or programs/projects normally funded by governmental entities. The foundation does not fund church or seminary construction, or church related entities or activities other than ecumenically oriented projects/programs that otherwise meet foundation guidelines. Grants or loans to individuals are never made.
Contact: Grants Administrator; (830) 895-5222; info@caillouxfoundation.org
Internet: http://www.caillouxfoundation.org/grant_guidelines.htm
Sponsor: Cailloux Foundation
P.O. Box 291276
Kerrville, TX 78029-1276

Caleb C. and Julia W. Dula Educational and Charitable Foundation Grants **1041**
Grants are given in the areas of the arts and humanities (particularly museums and libraries), child welfare, the aged, community funds and appeals, health care, religion, and historical preservation. Most groups receiving foundation grants have an established reputation in their particular field. There are no set requirements. Applicants should submit a letter that describes the organization, project, and amount requested.
Requirements: Grants are given to support projects of tax-exempt organizations.
Restrictions: Support is not available to individuals.
Date(s) Application is Due: Apr 1; Oct 1
Amount of Grant: 5,000 - 50,000 U.S.D
Contact: James F. Mauze, (314) 726-2800; fax (314) 863-3821; jfmauze@msn.com
Sponsor: Caleb C. and Julia W. Dula Educational and Charitable Foundation
112 S Hanley Road
Saint Louis, MO 63105

California Breast Cancer Research Program Grants **1042**
The state-funded breast cancer research effort supports a partnership comprising community members (such as a breast cancer advocacy organization, a community clinic or organization serving women with breast cancer, or a member of a California community affected by breast cancer) and experienced research scientists. Two funding mechanisms are available. The 18-month pilot award supports the initial phase of the project, including feasibility of methods, strengthening collaborations, development of tools and methods, and/or collecting pilot data. The three-year full award is for projects with a fully developed research plan with supporting preliminary data, carried out by a well integrated team of scientific and community members with a previous work relationship. The partnership works together to identify the research question, develop the research plan, carry out the research, interpret the results, and disseminate information to the community. The listed application deadline is for concept papers. Guidelines are available online.
Requirements: A partnership consists of California community members (organization or individual) and experienced research scientists.
Geographic Focus: California
Date(s) Application is Due: Nov 3
Amount of Grant: Up to 600,000 U.S.D
Contact: Natalie Collins, (510) 987-0646; natalie.collins@ucop.edu
Internet: http://www.cbcrp.org/apply/crcCall
Sponsor: University of California
300 Lakeside Drive, 6th Floor
Oakland, CA 94612-3550

California Community Foundation Health Care Grants **1043**
The foundation strives to ensure that low-income adults and children have access to regular, sustainable, affordable and quality health care. Priorities include: efforts that serve federally designated Medically Underserved Areas or Health Professional Shortage Areas, and that deliver comprehensive health services (e.g., medical, dental, mental health, etc.) to low-income and underserved populations; efforts that enroll people in health care coverage and connect them with a source of regular care; efforts that integrate strong prevention and early intervention strategies into their health services; and requests that clearly demonstrate knowledge of the health conditions that affect the specific population to be served.
Requirements: Tax-exempt organizations (not private foundations) located within or serving primarily residents of Los Angeles County are eligible.
Restrictions: Grants will not be considered for annual campaigns or special fund-raising events; building campaigns, with the exception of community development grants; endowments; existing obligations; equipment, unless it is an integral part of an eligible project; incurring a debt liability; individuals, with the exception of arts and culture grants; routine operating expenses; sectarian purposes; or programs that will in turn make grants.
Geographic Focus: California
Date(s) Application is Due: Feb 1; Jun 1; Oct 1
Amount of Grant: 100,000 - 1,000,000 U.S.D
Samples: Northeast Valley Health Corporation, San Fernando and Santa Clarita Valleys, CA, $150,000; Planned Parenthood of Pasadena, Northern San Gabriel Valley, CA $125,000; American Heart Association Inc, Los Angeles County, CA, $80,633.
Contact: Tamu Jones, Program Officer; (213) 413-4130, ext. 250; fax (213) 383-2046; tjones@ccf-la.org
Internet: http://www.calfund.org/receive/health_care.php
Sponsor: California Community Foundation
445 S Figueroa Street, Suite 3400
Los Angeles, CA 90071

California Community Foundation Human Development Grants **1044**
The human development program aims to expand support services for special populations that enable them to acquire the resources and skills that will lead to self-sufficiency. Special populations include aging adults, developmentally and/or physically disabled children and adults, at-risk youth and foster youth. Priorities include: support services for aging adults, developmentally and/or physically disabled children and adults, at-risk youth and foster youth; and efforts that promote civic dialogue and advocate for critical issues affecting aging adults, developmentally and/or physically disabled children and adults, at-risk youth and foster youth.
Requirements: IRS 501(c)3 tax-exempt organizations (not private foundations) located within or serving primarily residents of Los Angeles County are eligible.
Restrictions: Grants will not be considered for annual campaigns or special fund-raising events; building campaigns, with the exception of community development grants; endowments; existing obligations; equipment, unless it is an integral part of an eligible project; incurring a debt liability; individuals, with the exception of arts and culture grants; routine operating expenses; sectarian purposes; or programs that will in turn make grants.
Geographic Focus: California
Date(s) Application is Due: Feb 1; Jun 1; Oct 1
Amount of Grant: 50,000 - 300,000 U.S.D
Samples: Alliance for Leadership and Education, Los Angeles County, CA, $102,000 - two-year core operating support for a statewide advocacy organization that provides public education and supports public policies that advance adult day services throughout California; Exceptional Children's Foundation, Los Angeles County, CA, $90,000 - one-year core operating support for a comprehensive education, rehabilitation, social service, and advocacy agency that assists children and adults with developmental disabilities throughout most of Los Angeles County; Valley Village, San Fernando Valley, CA, $100,000 - two-year capacity building support for staff and board development training and fund development activities for a multiservice organization serving individuals with developmental disabilities (including aging adults) in the San Fernando Valley.
Contact: Robert Lewis, Program Officer; (213) 413-4130, ext. 273; fax (213) 383-2046; rlewis@ccf-la.org
Internet: http://www.calfund.org/receive/human_development.php
Sponsor: California Community Foundation
445 S Figueroa Street, Suite 3400
Los Angeles, CA 90071

California Endowment Agricultural Worker Health Grants **1045**
The program seeks to effectively provide, enhance, and expand essential health services to California's more than one million agricultural workers and their families. Grants support organizations representing multiple geographic locations where agricultural workers are concentrated, as well as specific health and medical services. Complete guidelines and an application are available in English and in Spanish on the Web site.
Requirements: California nonprofit organizations, clinics, schools, and public health departments are eligible. Institutions such as local public health departments and schools that have demonstrated the active involvement of agricultural workers in the development of proposals also will be considered.
Geographic Focus: California
Amount of Grant: Up to 500,000 U.S.D
Samples: Central Coast Interfaith Sponsors (Watsonville, CA)—to support and expand resident-based organizing and leadership-development efforts, as part of the Agricultural Worker Health Initiative, $300,000; Community Bridges (Aptos, CA)—to strengthen its ability to advocate and provide access to occupational health care and workers' compensation benefits for farmworkers, $210,618; La Union Del Pueblo Entero (Keane,

CA)—for a leadership program that seeks to help agricultural workers, community leaders, and community-based organizations work together to improve the health of agricultural workers and their families, $450,000.
Contact: Grants Administrator; (800) 449-4149 or (818) 703-3311; fax (818) 703-4193; questions@calendow.org
Internet: http://www.calendow.org/opportunities/index.stm
Sponsor: California Endowment
1000 N Alameda Street
Los Angeles, CA 90012

California Endowment CommunitiesFirst Grants **1046**
The endowment supports California nonprofits in the provision of programs and services promoting the health of all California residents. Favorable consideration is given to programs serving women, children, the elderly, and minorities. Areas of support include community innovation, which encourages communities to explore new ways to upgrade access to care; health and well-being, which promotes prevention strategies for all populations; and multicultural health, which addresses sociocultural barriers to improve health. Preference will be given to letters of intent that show that the project is community-driven, takes an innovative approach, benefits underserved populations, demonstrates cultural competency, builds partnerships, and addresses persistent problems.
Requirements: California 501(c)3 nonprofits may apply.
Restrictions: Requests are denied for biomedical or nonapplied research, overhead costs, annual fund drives, endowments, operating deficits, general support, or scholarships.
Geographic Focus: California
Amount of Grant: 50 - 60,000,000 U.S.D
Contact: Grants Administrator; (818) 703-3311 or (800) 449-4149, ext. 3311; fax (818) 703-4193; CommunitiesFirst@calendow.org
Internet: http://www.calendow.org/opportunities/communitiesfirst.stm
Sponsor: California Endowment
1000 N Alameda Street
Los Angeles, CA 90012

California Endowment Local Opportunities Fund Grants **1047**
The fund will support a full spectrum of programs or activities to assist organizations in addressing the health needs of their local communities. The most common types of grants available include program grants—for specific projects or programs, including expansion of an existing project or program, pilot projects, or start-up of a new project or program; core operating support—day-to-day core operating costs for an existing program or for the general work of an organization; capacity building—to develop either the management or administrative capacity of an organization or to develop the organization's capacity to effectively implement a particular program; and planning—to develop a detailed plan and strategy for implementation of a project or program that will address a locally defined health need or health-related priority. Priority will be given to applications that address a locally defined health need or health-related priority in an underserved community; are from a grassroots, nontraditional and/or emerging organization, or that address an issue or community that traditionally does not benefit from mainstream funding resources; and utilize the talents, cultures, and assets of the local community to address the health priorities of that community.
Requirements: California nonprofit organizations are eligible. Requests for core operating support will only be considered from organizations with annual operating budgets of less than $500,000 (as determined by financial statements and the annual operating budget submitted with the application).
Date(s) Application is Due: Jan 10; May 15; Sep 15
Amount of Grant: Up to 50,000 U.S.D
Samples: Horizons Foundation (San Francisco, CA)—for organizational development and regranting assistance needed to expand programs that address disparities in health and well-being among lesbian, gay, bisexual, and transgender residents of the San Francisco Bay area, $150,000; Asian Pacific Fund (San Francisco, CA)—to strengthen community institutions and expand programs and services needed to address disparities among Asian residents of the San Francisco Bay area, $150,000; Los Angeles Brotherhood Crusade (CA)—for organizational development and regranting assistance needed to expand programs that address disparities in health and well-being among minorities in South Central Los Angeles, $150,000.
Contact: Fund Coordinator; (800) 449-4149 or (818) 703-3311; fax (818) 703-4193
Internet: http://www.calendow.org/apply/frm_apply.htm
Sponsor: California Endowment
1000 N Alameda Street
Los Angeles, CA 90012

California Wellness Foundation Sabbatical Program Grants **1048**
This program rewards the leaders of today and prepares them for the challenges of tomorrow with financial grants to subsidize well-deserved sabbaticals. Eight nonprofit health service organizations will receive grants to provide their executive directors with the rest and rejuvenation needed to return to their organizations replenished and prepared to continue to direct their organizations' missions. The program provides grants to organizations to cover the leaders' salaries for three- to six-month sabbaticals. In addition, awards will be given to each selected organization for the professional development of managers and staff required to take on additional responsibility in the absence of the sabbatical recipients. The program is designed not only to support nonprofit executives but also to provide resources for the organizations as a whole to better serve their clients, the community, and the health sector. Guidelines and application are available online.
Requirements: Candidates must hold an executive director or equivalent position at a nonprofit health organization; have served in an executive role (though not necessarily for one organization) for the past six consecutive years or more without a significant leave; have served in their current executive leadership role for a minimum of three years; be employed full time at the applicant organization; have approval from their organizations' boards for the proposed sabbatical; be willing to submit all reports as required; and not have any prohibited affiliation with the foundation. Candidates' organizations must be located in California and provide health services that directly benefit state residents; address the particular health needs of historically underserved populations, including low-income individuals, people of color, youth, and residents of rural areas; be a 501(c)3 nonprofit organization and not a 509(a) private foundation; be able to sustain regular operations during the candidate's absence; and guarantee maintenance of regular health and other benefits to the candidate during his/her period of leave.
Restrictions: No director, officer, or staff member of the foundation or any member of their immediate families are eligible.
Geographic Focus: California
Date(s) Application is Due: Apr 22
Amount of Grant: 5,000 - 30,000 U.S.D
Contact: Administrator; (818) 702-1970; fax (818) 593-6614; sabbaticalprogram@tcwf.org
Internet: http://www.tcwf.org/press_room/sabbatical_program/index.htm
Sponsor: California Wellness Foundation
6320 Canoga Avenue, Suite 1700
Woodland Hills, CA 91367-7111

California Wellness Foundation Work and Health Program Grants **1049**
The foundation focuses its activities on specific priority areas where it has a significant, long-term commitment. Within each priority area, the foundation allocates the majority of its funds toward initiatives. Initiatives are targeted grantmaking programs with distinct objectives and are generally announced through requests for proposals. The foundation awards general grants and project grants. Under general grants, requests for core operating support for organizations that provide direct services to Californians for disease prevention or health promotion are of primary interest. Priority areas under general grants are diversity in the health professions, environmental health, healthy aging, mental health, teenage pregnancy prevention,, violence prevention, women's health, and work and health. Special projects grants are awarded to areas that fall outside the priority areas. Of particular interest are proposals to help California communities respond to cutbacks in federally funded programs. Activities commonly supported under special projects include strengthening traditional safety-net providers, educating consumers about changes in health care systems, advocating for underserved communities in health policy debates, and informing public decision making through policy analysis.
Requirements: Eligible applicants are California 501(c)3 nonprofit organizations, or organizations with a preapproved fiscal sponsor. An organization should first write a succinct letter of interest (one to two pages) that describes the organization, its leadership, the region and population(s) served, and the activities for which funding is needed, including the amount requested.
Geographic Focus: California
Amount of Grant: 5,000 - 200,000 U.S.D
Contact: Grants Administrator; (818) 702-1900; fax (818) 593-6614
Internet: http://www.tcwf.org/grants_program/index.htm
Sponsor: California Wellness Foundation
6320 Canoga Avenue, Suite 1700
Woodland Hills, CA 91367-7111

Callaway Foundation Grants **1050**
The Foundation awards grants for the benefit of projects and people in LaGrange and Troup County, Georgia. Areas of interest, include: arts and entertainment, elementary, higher, and secondary education; libraries; health and hospitals; community funds; care for the aged; community development; historic preservation; and church support. Types of support include annual campaigns, building construction/renovation, capital campaigns, continuing support, equipment acquisition, general operating support, land acquisition, and matching/challenge support. Preference is given to enduring construction projects and capital equipment. The Foundation Board meets four times per year in January, April, July, and October. Grant requests and applications are due the last day of the month preceding the meetings.
Requirements: IRS 501(c)3 nonprofit organizations in LaGrange and Troup County, Georgia are eligible to apply. Letters of request should briefly cover all aspects of the project, including complete financial planning and costs involved. Copies of budgets and current financial statements should also be included. An application form is available at the Foundation's website.
Restrictions: Grants are usually not made for loans, debt retirement, endowment or operating expenses. Requests from churches located outside Troup County, Georgia, are not considered.
Geographic Focus: Georgia
Date(s) Application is Due: Mar 31; Jun 30; Sep 30; Dec 31
Amount of Grant: 1,000 - 4,000,000 U.S.D
Samples: Chattahoochee Valley Art Museum, LaGrange, GA, $59,000—for operating support; Georgia Cities Foundation, Atlanta, GA, $367,772—for Downtown Revitalization Fund; LaGrange College, LaGrange, GA, $4,000,000—for library building and renovation.
Contact: H. Speer Burdette, President; (706) 884-7348; fax (706) 884-0201; hsburdette@callaway-foundation.org
Internet: http://www.callawayfoundation.org/grant_policies.php
Sponsor: Callaway Foundation
209 Broome Street, P.O. Box 790
La Grange, GA 30241

Callaway Golf Company Foundation Grants 1051

The foundation strives to support initiatives in communities where Callaway Golf Company employees live and work. The geographic area of focus is primarily North San Diego County, California. The foundation offers support in the form of: matching funds, special projects, and general operating budgets. Areas of interest include but are not limited to: children and youth; biomedical research, with a special interest in the field of cancer; golf; education; boys & girls clubs of America; drug prevention; American Red Cross; housing; veterans; youth programs; food banks; social services; emergency programs; scholarship program for dependents of Callaway Golf employees; grants for training, competition and equipment needs. The foundation does not require the completion of a formal application document but requests a description of the organization and its history, the project at issue including goals and time lines, the qualifications of the leadership personnel involved in the project, and a detailed project budget. Grants are awarded semiannually. The Callaway Golf Company Foundation does not accept unsolicited requests for grants.
Requirements: IRS 501(c)3 nonprofit organizations in California are eligible.
Restrictions: The foundation will not fund applicants that illegally discriminate on the basis of gender, race, color, religion, national origin, ancestry, age, marital status, medical condition, or physical disability, either in the services they provide or in the hiring of staff; or promote political or particular religious doctrines.
Geographic Focus: California
Amount of Grant: 500 - 10,000 U.S.D
Samples: Saint Judes Children's Research Hospital, $300; Utah Food Bank Services, $100; Purdue Foundation, $1,000.
Contact: Paul Thompson, Executive Director; (760) 930-8686; fax (760) 930-5021; cgcfoundation@callawaygolf.com
Internet: http://www.callawaygolf.com/Global/en-U.S./Corporate/CallawayGolfFoundation.html
Sponsor: Callaway Golf Company Foundation
2180 Rutherford Road
Carlsbad, CA 92008-7328

Cambridge Community Foundation Grants 1052

Cambridge Community Foundation is dedicated to improving the quality of life for the residents of Cambridge, Massachusetts. The CCF serves Cambridge through our support of nonprofit community organizations, by making direct financial grants, providing technical assistance, and forming partnerships among organizations to coordinate services, address gaps, and highlight emerging issues. CCF primarily supports work in: early childhood services; youth service; senior services; community services; emergency outreach; arts; and the environment. See, the Foundations website http://www.cambridgecf.org/grant.html to download Proposal Summary Sheets, and additional guidelines.
Requirements: To be eligible to apply, the agency must be tax-exempt 501(c)3 under the IRS code), and the program must serve the people of Cambridge, Massachusetts.
Restrictions: Support is not provided to municipal, state, or federal agencies. Grants for individuals, scholarships, research studies, conferences, films, capital fund drives, or loans are not eligible.
Geographic Focus: Massachusetts
Date(s) Application is Due: Apr 1; Oct 1
Amount of Grant: 500 - 60,000 U.S.D
Samples: Family First, Cambridge, MA — funding for the Family First parenting education and support programs for low-income and homeless families living in Cambridge; Big Sister Association of Greater Boston, Boston, MA — funding to support Cambridge girls who participate in Big Sister's traditional one-to-one community-based mentoring program; YNA Care Network, Inc./Hospice of Cambridge, Cambridge, MA — funding to support the patient care needs and running of the Chilton House, Cambridge's hospice, as well as bereavement support groups and counseling services to patient family members.
Contact: Robert S. Hurlbut, Jr., Executive Director; (617) 576-9966; fax (617) 876-8187; RHurlbut@CambridgeCF.org or info@cambridgecf.org
Internet: http://www.cambridgecf.org
Sponsor: Cambridge Community Foundation
99 Bishop Richard Allen Drive
Cambridge, MA 02139

Camp-Younts Foundation Grants 1053

The foundation supports social services, higher and secondary education, youth organizations, Protestant religion, and hospitals and other health organizations in Florida, Georgia, North Carolina, and Virginia. Applicants should submit a letter describing the program, a copy of the 501(c)3 tax-determination letter, listing of board members, and an audited budget for the previous year.
Requirements: Nonprofit organizations in Florida, Georgia, North Carolina, and Virginia may request grant support.
Geographic Focus: Florida, Georgia, North Carolina, Virginia
Date(s) Application is Due: Sep 1
Amount of Grant: 1,000 - 55,000 U.S.D
Samples: Virginia Baptist General Board (Richmond, VA)—for operating support, $40,000; Saint Edwards School (Bon Air, VA)—for operating support, $25,000.
Contact: Bobby Worrell, Executive Director; (757) 562-3439
Sponsor: Camp-Younts Foundation
P.O. Box 4655
Atlanta, GA 30302

Campbell Foundation HIV/AIDS Research Grants 1054

The foundation awards grants for research projects on ways to prevent and treat HIV, AIDS, and related conditions or illnesses. Proposals focusing on alternative, nontraditional avenues of study receive preference. Most grants support clinical research, although other types of work are considered. After initial screening, requests are reviewed by a peer review board of noted HIV/AIDS physicians and then sent to the board for final decisions. There are no application deadlines, but as all grants are handled chronologically, the foundation suggests submitting grant requests as early as possible.
Requirements: U.S. nonprofit organizations are eligible.
Amount of Grant: 20,000 - 80,000 U.S.D
Contact: Program Contact; (954) 493-8822; fax (954) 493-8801; campfound@aol.com
Internet: http://members.aol.com/campfound
Sponsor: Campbell Foundation
5975 N Federal Highway, Suite 126
Fort Lauderdale, FL 33308

Campbell Hoffman Foundation Grants 1055

The mission of the Campbell Hoffman Foundation is to promote and fund efforts to increase access to comprehensive health care for underserved and uninsured populations in the Northern Virginia region. Northern Virginia is defined as the counties of Arlington, Fairfax, Loudoun and Prince William and the cities of Alexandria, Falls Church, Fairfax, Manassas and Manassas Park. A letter of Inquiry should be submitted, as the initial approach, when approaching the Foundation for funding. Upon review, if the letter of inquiry meets with Foundations criteria, the applicant will be asked to submit a proposal; guidelines will be provided to them when the proposal is requested.
Requirements: Eligible organizations include nonprofit 501(c)3 organizations, government agencies and faith-based organizations. Eligible organizations must be both located in and serve the target populations of Northern Virginia.
Restrictions: The Foundation will not provide funding for capital campaigns, endowment campaigns, special events and/or conferences (including travel to and participation in same) , emergency funding, loans, capital projects (including but not limited to building, construction or renovation), land purchases, lawsuits, films, video or publications. The Foundation may choose to provide general operating support.
Geographic Focus: Virginia
Samples: Northern Virginia Family Service, $12,200—expand Dentral Resource Recruitment Program; Arlington Free Clinic, $30,000—continue integration of mental health into primary care setting; Loudoun Community Health Center, $45,000—initiate mental heatlh program; integrate w/ primary care; create community mental health provider coalition.
Contact: Lyn S. Hainge, Executive Director; (703) 749-1794; fax (703) 442-0846; lhainge@campbellhoffman.org
Internet: http://www.campbellhoffman.org/applicants/default.aspx
Sponsor: Campbell Hoffman Foundation
1420 Spring Hill Road, Suite 600
McLean, VA 22102

Campbell Soup Foundation Grants 1056

Since 1953, the Campbell Soup Foundation has provided financial support to local champions that inspire positive change in communities throughout the United States where Campbell Soup Company employees live and work. The Foundation places particular emphasis on Camden, New Jersey, birthplace of Campbell's flagship soup business and world headquarters. The Campbell Soup Foundation focuses its giving on four key areas: hunger relief-supporting food bank organizations in the communities of operation; wellness-addressing the health of consumers in the communities where they live; education-leveraging the Campbell brand portfolio to support educational programs; community revitalization-enhancing the quality of life in the communities that Campbell operates in. The Foundation only considers applications that meet the following criteria: the proposal must fit one of the key focus areas; the organization must display strong and effective leadership; the proposed plan must be clear and compelling, with measurable and sustainable commitments expressed in terms of real results; the proposed activity must be sufficiently visible to leverage additional support from other funding sources. There is no formal deadline. Proposals are accepted and reviewed on a rolling basis.
Requirements: The Foundation limits grants to nonprofit organizations which are tax-exempt under Section 501(c)3 of the Internal Revenue Code. Grants are made to institutions that serve: Camden, New Jersey; Davis, California; Sacramento, California; Stockton, California; Bloomfield, Connecticut; Norwalk, Connecticut; Lakeland, Florida; Downers Grove, Illinois; Marshall, Michigan; Maxton, North Carolina; Camden, New Jersey; South Plainfield, New Jersey; Napoleon, Ohio; Wauseon, Ohio; Willard, Ohio; Denver, Pennsylvania; Downingtown, Pennsylvania; Aiken, South Carolina; Paris, Texas; Richmond, Utah; Everett, Washington; Milwaukee, Wisconsin. Organizations do not need to be located in these communities in order to qualify for funding. However, the programs to be funded must serve these communities. Proposals must be submitted electronically via email to community_relations@campbellsoup.com. Proposals should be prepared in a concise, narrative form, without extensive documentation.
Restrictions: Grants are not made to the following: organizations that are based outside the United States and its territories; individuals; organizations that limit their services to members of one religious group or whose services propagate religious faith or creed; political organizations and those having the primary purpose of influencing legislation of/or promoting a particular ideological point of view; units of government; events and sponsorships; sports related events, activities and sponsorships. Organizations may not submit the same or similar proposals more than once in a Foundation fiscal year (July 1 - June 30). Proposals submitted via regular mail will not be reviewed.

Geographic Focus: Connecticut, Florida, Illinois, Michigan, New Jersey, North Carolina, Ohio, Pennsylvania, South Carolina, Texas, Utah, Washington, Wisconsin
Samples: Coriell Institute for Medical Research, $12,500; Scotland County Literacy Council, $15,000; Delaware County Christian School, $500.
Contact: Grant Administrator; (856) 342-6423 or (800) 257-8443; fax (856) 541-8185; community_relations@campbellsoup.com
Internet: http://www.campbellsoupcompany.com/community_center.asp
Sponsor: Campbell Soup Foundation
1 Campbell Place
Camden, NJ 08103-1701

CampusRN/AACN Nursing Scholarship Fund 1057

The scholarship program supports students who are seeking a baccalaureate, master's, or doctoral degree in nursing. Special consideration will be given to students enrolled in a master's or doctoral program with the goal of pursuing a nursing faculty career; completing an RN to baccalaureate program (BSN); or enrolled in an accelerated baccalaureate or master's degree nursing program. Applicants must submit an online application and create a complete profile on CampusRN. All applicants must already be enrolled (not just accepted) at an AACN member institution. Applications are accepted only by email.
Requirements: Nursing students enrolled in a baccalaureate, master's or doctoral degree program in nursing are eligible.
Date(s) Application is Due: Jan 1; Mar 1; May 1; Jul 1; Sep 1; Nov 1
Amount of Grant: 2,500 U.S.D
Contact: Scholarship Administrator; (202) 463-6930; fax (202) 785-8320; info@campuscareercenter.com
Internet: http://aacn.campusrn.com/scholarships/scholarship_rn.asp
Sponsor: American Association of Colleges of Nursing
1 Dupont Circle NW, Suite 530
Washington, DC 20036

Canada-U.S. Fulbright New Century Scholars Program Grants 1058

The Fulbright New Century Scholars Program allows scholars who have attained a level of national or international recognition for demonstrated professional accomplishments pertaining to higher education and its role in national and global economic development. The deadline for this specific award is October 30 for an award to be taken up from May of the following year and conclude the next April. The New Century Scholar award carries a value of U.S.$30,000.
Requirements: Eligibility requirements include: Canadian or American citizenship at time of application (permanent resident status or landed immigrant status are not sufficient); a Ph.D. or equivalent professional/terminal degree by December 31 of the application year or equivalent professional experience; and English language proficiency.
Restrictions: Applicants are not eligible for a Canada-U.S. Fulbright Award if he or she: currently holds permanent residency status in the intended host country; is currently residing in or enrolled at a university in the intended host country; has resided abroad for five or more consecutive years in the six year period immediately preceding the date of application; has received a Fulbright Award within the last five years; has had recent substantial experience in the intended host country (recent substantial experience is defined as study, teaching, research or employment for a period of more than an academic year (nine months) during the past five years); or is a Canadian applicant and you are a dual citizen of Canada and the United States.
Geographic Focus: All States, Canada
Date(s) Application is Due: Oct 30
Amount of Grant: 30,000 U.S.D
Contact: Jennifer Regan, Senior Program Officer; (613) 688-5517 or (613) 688-5540; fax (613) 237-2029; jregan@fulbright.ca
Internet: http://www.fulbright.ca/en/award.asp
Sponsor: Foundation for Educational Exchange between Canada and the United States of America
350 Albert Street, Suite 2015
Ottawa, ON K1R 1A4 Canada

Canada-U.S. Fulbright Senior Specialists Program Grants 1059

The Fulbright Senior Specialists Program is designed to provide U.S. and Canadian faculty and professionals with opportunities to collaborate on curriculum and faculty development, institutional planning and a variety of other activities. During the course of their grant, Fulbright Senior Specialists may engage in any of the following activities at their Canadian host institution: conduct needs assessments, surveys, institutional or programmatic research; take part in specialized academic programs and conferences; consult with administrators and instructors of post-secondary institutions on faculty development; present lectures at graduate and undergraduate levels; participate in or lead seminars or workshops at overseas academic institutions; develop and/or assess academic curricula or educational materials; and conduct teacher-training programs at the tertiary level. These are short-term grants of two to six weeks to that are offered on a continuous basis.
Requirements: Eligibility requirements include: Canadian or American citizenship at time of application (permanent resident status or landed immigrant status are not sufficient); a Ph.D. or equivalent professional/terminal degree by December 31 of the application year or equivalent professional experience; and English language proficiency.
Restrictions: Applicants are not eligible for a Canada-U.S. Fulbright Award if he or she: currently holds permanent residency status in the intended host country; is currently residing in or enrolled at a university in the intended host country; has resided abroad for five or more consecutive years in the six year period immediately preceding the date of application; has received a Fulbright Award within the last five years; has had recent substantial experience in the intended host country (recent substantial experience is defined as study, teaching, research or employment for a period of more than an academic year (nine months) during the past five years); or is a Canadian applicant and you are a dual citizen of Canada and the United States.
Geographic Focus: All States, Canada
Contact: Jennifer Regan, Senior Program Officer; (613) 688-5517 or (613) 688-5540; fax (613) 237-2029; jregan@fulbright.ca
Internet: http://www.fulbright.ca/en/seniorspecialists.asp
Sponsor: Institute of Int'l Education, Council for International Exchange of Scholars
350 Albert Street, Suite 2015
Ottawa, ON K1R 1A4 Canada

Canada Graduate Scholarships and NSERC Postgraduate Scholarships 1060

Canada Graduate Scholarships and NSERC Postgraduate Scholarships provide financial support to high-calibre scholars who are engaged in master's or doctoral programs in the natural sciences or engineering. The Canada Graduate Scholarships will be offered to the top-ranked applicants at each level (master's and doctoral) and the next tier of meritorious applicants will be offered an NSERC Postgraduate Scholarship. This support allows these scholars to fully concentrate on their studies and to seek out the best research mentors in their chosen fields. NSERC encourages interested and qualified Aboriginal students to apply.
Requirements: To be considered eligible for support, applicants must: (1) be a Canadian citizen or a permanent resident of Canada; (2) hold, or expect to hold (at the time you take up the award), a degree in science or engineering from a university whose standing is acceptable to NSERC (if you have a degree in a field other than science or engineering, NSERC may accept your application at its discretion); (3) intend to pursue in the following year full-time graduate studies and research at the master's or doctoral level in an eligible program in one of the areas of the natural sciences and engineering supported by NSERC; and (4) have obtained a first-class average (a grade of 'A-') in each of the last two completed years of study (full-time equivalent). Any exceptions to this requirement must be accompanied by supporting comments that justify the submission.
Restrictions: Scholarships are not available to students entering graduate studies in a qualifying year. Students who, after receiving an NSERC scholarship, change their field of study and research to a field that is not supported by NSERC, cease to be eligible and their awards will be cancelled.
Geographic Focus: All States, Canada
Date(s) Application is Due: Oct 15
Amount of Grant: 17,300 - 35,000 CAD
Contact: PGS Program Officer; (613) 995-5521; fax (613) 996-2589; schol@nserc.ca
Internet: http://www.nserc-crsng.gc.ca/Students-Etudiants/PG-CS/BellandPostgrad-BelletSuperieures_eng.asp
Sponsor: Natural Sciences and Engineering Research Council of Canada
350 Albert Street, 14th Floor
Ottawa, ON K1A 1H5 Canada

Canadian Dental Research Foundation Award 1061

This research award is offered biennially to encourage research related to dentistry conducted by graduate or postgraduate students in Canada. The monetary prize will be presented to the student who will have submitted the best report on a research project at the CDA annual meeting and the paper will be presented at the ACFD/CADR biennial meeting.
Requirements: Research papers entered are limited to graduate or postgraduate students who have conducted their work in association with the dental faculty of a Canadian university.
Geographic Focus: All States, Canada
Contact: Richard Munro, Executive Director; (613) 236-4763; fax (613) 236-3935; Information@dcf-fdc.ca
Internet: http://www.dcf-fdc.ca
Sponsor: Dentistry Canada Fund
427 Gilmour Street
Ottawa, ON K2P OR5 Canada

Canadian Foundation for Dietetic Research Awards 1062

Within the broad category of practice-based dietetic research, the foundation prefers to support research that is of direct relevance to the nutritional well-being of populations within Canada. Priority research directions are outcomes of intervention, new roles for dietitians in meeting the health needs of all Canadians, identification of vulnerable groups and their nutritional needs, and determinants of food choice. An active member of Dietitians of Canada must be a member, if not the leader, of the research team. Applications for projects extending beyond one year will not be considered. The researchers must be associated with an institution or agency with charitable status (such as a teaching hospital or university) that will administer the funds awarded. Grants are intended to finance the direct operating costs of dietetic practice-based research projects. Calls for proposals are posted on the Web site.
Requirements: Applications will only be considered where a practicing dietitian who is an active member of Dietitians of Canada and who is delivering direct or indirect client/patient/public care of service is the principal investigator or a co-investigator.
Geographic Focus: All States, Canada
Date(s) Application is Due: Sep 30
Amount of Grant: Up to 10,000 CAD
Contact: Awards Manager; (416) 642-9307; fax (416) 596-0603; cfdr@dietitians.ca
Internet: http://www.dietitians.ca/cfdr/grants.htm
Sponsor: Canadian Foundation for Dietetic Research
480 University Avenue, Suite 604
Toronto, ON M5G 1V2 Canada

Canadian Friends of Hebrew University Research and Study Grants **1063**
Grants are awarded for one year of undergraduate or graduate study to Canadian citizens or residents of Canada who are able to fulfill the entrance requirements of the Hebrew University in Jerusalem in their chosen fields of study, which may include law, dentistry, social sciences and economics, humanities, arts, science, agriculture, pharmacy, and library science.
Geographic Focus: All States, Canada
Date(s) Application is Due: May 1
Amount of Grant: 1,000 - 4,000 CAD
Contact: Dina Wachtel, National Director; (416) 485-8000; fax (416) 485-8565; dwachtel@cfhu.org
Internet: http://www.cfhu.org
Sponsor: Canadian Friends of Hebrew University
3080 Yonge Street, Suite 5024
Toronto, ON M4N 3P4 Canada

Canadian Liver Foundation Grants **1064**
The foundation awards organization grants, fellowships, and studentships to support research into nephrology, hepatic function or disease, and other related disciplines.
Requirements: Canadian citizens and organizations are eligible.
Geographic Focus: All States, Canada
Date(s) Application is Due: Mar 31
Contact: Billie Potkonjak, (800) 563-5483; fax (416) 491-4952; bpotkonjak@liver.ca
Internet: http://www.liver.ca/Research/Grant_Program
Sponsor: Canadian Liver Foundation
2235 Sheppard Avenue E, Suite 1500
Toronto, ON M2J 5B5 Canada

Canadian Nurses Foundation Study Awards **1065**
Grants provide funding for nurses undertaking university studies in nursing and health care and for research related to nursing practice and health care. Awards are given for university study at the baccalaureate, master's, and doctoral levels. Priority will be given to nurses studying in programs with a nursing focus. Once recipients have completed their degrees, they must return to nursing and assist in the development of improved nursing and health care in Canada.
Requirements: Applicants must be members of the Canadian Nurses Foundation and must have gained acceptance into a university program of study; applicants must identify the practice area in which they plan to study and have definite career goals.
Geographic Focus: All States, Canada
Amount of Grant: 3,000 - 6,000 CAD
Contact: Linda Piazza, Executive Director; (613) 237-2159, ext. 250; fax (613) 237-3520; lpiazza@cna-aiic.ca or orcnf@cnursesfdn.ca
Internet: http://www.canadiannursesfoundation.com/welcome.htm
Sponsor: Canadian Nurses Foundation
50 Driveway
Ottawa, ON K2P 1E2 Canada

Canadian Optometric Education Trust Fund Grants **1066**
Since 1980, the COETF has received applications and awarded funding to projects covering a vast and varied scope of research. The fund supports quality vision and eye care services for Canadians of all ages. Projects have been funded for improving optometrical education, professional development, and research. The Awards Committee meets annually (generally in March) to consider applications. The deadline for applications is typically in early February. Interested applicants should secure forms from the fund.
Restrictions: The fund customarily does not provide support for travel to and from sites where the results of a project are intended to be presented, i.e., a symposium or continuing education seminar.
Geographic Focus: All States, Canada
Amount of Grant: 3,000 - 5,000 CAD
Contact: Glenn Campbell, Executive Director; (888) 263-4676 or (613) 235-7924; fax (613) 235-2025; gcampbell@opto.ca or info@opto.ca
Internet: http://www.opto.ca/en/public/03_optometry/03_06_coetf.asp#apply
Sponsor: Canadian Optometric Education Trust Fund
234 Argyle Avenue
Ottawa, ON K2P 1B9 Canada

Canadian Patient Safety Institute (CPSI) Patient Safety Studentships **1067**
The Canadian Patient Safety Institute (CPSI) is inviting applications from organizations/supervisors interested in enabling students to participate in patient safety work. Funding is to support the work of students in any healthcare profession, as well as other disciplines (including but not limited to social sciences, human kinetics, psychology and human factors engineering) interested in patient safety. The goal is to create new collaborative learning opportunities aligned with the mandate of CPSI. Students are required to be the major resource in carrying out the work for each patient safety project. Any person from organizations such as universities, health regions, and hospitals, non-governmental or not-for-profit private organizations who will provide supervision for a student project may apply for this award. The student selected can be a student of any health care profession as well as other disciplines (including but not limited to social sciences, human kinetics, psychology and human factors) with an interest in patient safety, who are registered in an undergraduate or graduate program (part-time or full-time), and who will be a student during the period of the award. The value of the award from CPSI is $6,000.
Requirements: Applying organizations/supervisors that are interested in accessing these resources must provide matching funding to augment studentship funding that will enable students to participate in patient safety work. Matched funding must be in actual dollars and cannot be in-kind contributions.
Restrictions: CPSI funding is to be used for student stipends only. CPSI will not make funds available for administrative or overhead purposes.
Geographic Focus: All States, Canada
Date(s) Application is Due: Dec 12
Amount of Grant: 6,000 U.S.D
Contact: Laurel Taylor, Director of Operations; (866) 421-6933 or (780) 409-8090; fax (780) 409-8098; ltaylor@cpsi-icsp.ca
Internet: http://www.patientsafetyinstitute.ca/education/Patientsafetystudentship2009.html
Sponsor: Canadian Patient Safety Institute
10235 - 101 Street
Edmonton, AB T5J 3G1 Canada

Canadian Patient Safety Institute (CPSI) Research Grants **1068**
The Canadian Patient Safety Institute (CPSI) announces its annual Research Competition to continue its strategic mandate of increasing the scope and scale of research and evaluation activities in patient safety. The primary goal for CPSI Research Competition is to develop knowledge about patient safety that can be helpful in a variety of settings and circumstances in organizations across Canada. It is the applicant team's responsibility to identify and secure matching funding from other contributing organization(s). Applicant teams may obtain matching funding from universities, foundations, voluntary health charities, provider associations, provincial government departments, regional health authorities, hospitals, research centers or the private sector.
Requirements: A 1:1 funding ratio is the acceptable minimum required for any project over $20,000.
Geographic Focus: All States, Canada
Date(s) Application is Due: Dec 2
Amount of Grant: 120,000 U.S.D
Contact: Laurel Taylor, Director of Operations; (866) 421-6933 or (780) 409-8090; fax (780) 409-8098; ltaylor@cpsi-icsp.ca
Internet: http://www.patientsafetyinstitute.ca/news/2008researchcompetition.html
Sponsor: Canadian Patient Safety Institute
10235 - 101 Street
Edmonton, AB T5J 3G1 Canada

CAN Autism Treatment-Related Study Awards **1069**
The organization recognizes the urgent need to develop effective therapies to treat both core and domain specific features of autism. Research proposals focused on all aspects of treatment, from basic research models, to clinical trials, to biomarker identification, are acceptable for submission. Due to the current lack of effective treatment and absence of known biomarkers, applications are reviewed as they are received. Funding amount and duration is determined per individual proposal. Indirect costs are limited to 10 percent of the total amount. Letters of intent are required prior to invitation of full proposals.
Date(s) Application is Due: Sep 1
Amount of Grant: 60,000 - 180,000 U.S.D
Contact: Therese Finazzo, Grants Officer; (888) 828-8476 or (323) 549-0500, ext. 18; fax (323) 549-0500; therese@cureautismnow.org
Internet: http://www.cureautismnow.org/research/funding/3523.jsp
Sponsor: Cure Autism Now
5455 Wilshire Boulevard, Suite 715
Los Angeles, CA 90036-4234

Cancer Research Postdoctoral Fellowships for Basic and Physician Scientists **1070**
This program grants initial postdoctoral fellowships to support promising young investigators pursuing careers in cancer research. The fund encourages all theoretical and experimental research that is relevant to the study of cancer and the search for cancer causes, mechanisms, therapies, and prevention. Candidates must apply for the fellowships under the guidance of a sponsor—a senior member of the scientific research community. The sponsor should be actively engaged in the planning, execution, and supervision of the proposed research and should encourage the fellow to report the results of the research in scientific journals and meetings. Awards are made to institutions for the support of the fellow under direct supervision of the sponsor.
Requirements: Candidates must have completed one or more of the following degrees or equivalent: MD, PhD, DDS, or DVM. For level one funding, basic and physician scientist applicants must have received their degrees no more than one year prior to the SAC meeting at which their applications are to be considered and DVM applicants no more than three years. For level two funding, physician scientists (MD, DDS, DVM, or the equivalent) must have completed their residencies and clinical training no more than three years prior to the SAC meeting at which their applications are to be considered.
Restrictions: Only one application will be accepted from a given sponsor or fellow per review session; there is no limit, however, on the number of applications from an institution. No more than two Runyon-Winchell Fellows will be funded to work with the same sponsor at any time.
Date(s) Application is Due: Mar 15; Aug 15
Amount of Grant: 128,000 - 168,000 U.S.D
Contact: Lorraine Egan, Executive Director; (212) 455-0541; fax (212) 697-4950; lorraine.egan@drcrf.org or fellowship@cancerresearchfund.org
Internet: http://www.drcrf.org/apFellowship.html
Sponsor: Cancer Research Fund of the Damon Runyon-Walter Winchell Foundation
675 Third Avenue, 25th Floor
New York, NY 10017

Cancer Research Society Fellowships and Grants 1071
Grants and fellowships are supported for projects researching the cause and cure of cancer. Recipients are selected annually by the society's medical advisory board on the basis of the scientific value of the research projects submitted.
Requirements: Research projects supported must be carried out in Canada.
Geographic Focus: All States, Canada
Date(s) Application is Due: Feb 15
Contact: Program Director; (514) 861-9227; fax (514) 861-9220; grants@cancerresearchsociety.ca
Internet: http://src-crs.ca/main.php?lang=2&id=305
Sponsor: Cancer Research Society
402-625 Av Du President-Kennedy
Montreal, QC H3A 3S5 Canada

Cancer Research UK Fellowships 1072
The fund operates in the United Kingdom and worldwide to investigate all matters connected with or bearing on the causes, prevention, treatment, and cure of cancer. Principal disciplines include cellular and molecular biology, clinical cancer research, and epidemiology. Types of support include research fellowships, visiting fellowships, clinical research fellowships, and graduate studentships.
Contact: Grants Administrator; 020 7242 0200; fax 020 7269 3100; research_enquiries@cancer.org.uk
Internet: http://science.cancerresearchuk.org/fandm/grantapplications/?version=1
Sponsor: Cancer Research UK
Lincoln's Inn Fields, P.O. Box 123
London, WC2A 3PX England

Cancer Research UK Grants 1073
The organization supports cancer research throughout the United Kingdom. Grants are awarded to universities, medical schools, and hospitals in support of research projects. Additional types of support include life fellowships, career development awards, clinical and nonclinical scientists fellowships, and capital grants for new research buildings and major items of equipment.
Geographic Focus: All States, United Kingdom
Contact: Administrator; 020 7242 0200; fax 020 7269 3100; research_enquiries@cancer.org.uk
Internet: http://science.cancerresearchuk.org/gapp/?version=3
Sponsor: Cancer Research UK
Lincoln's Inn Fields, P.O. Box 123
London, WC2A 3PX England

Cancer Society of New Zealand Grants 1074
The society supports cancer research through grants to individuals and organizations in New Zealand. Types of support include research grants and traveling fellowships to enable New Zealand graduates to travel abroad for study.
Geographic Focus: All States, New Zealand
Date(s) Application is Due: Nov 1
Contact: Administrator; (04) 494 7270; fax (04) 494 7271; admin@cancernz.org.nz
Internet: http://www.cancernz.org.nz
Sponsor: Cancer Society of New Zealand
P.O. Box 10847
Wellington, 6144
New Zealand

Cancer Treatment Research Foundation Clinical Investigation Grants 1075
The grants support new and pilot feasibility clinical projects in cancer therapy. Grants are intended to stimulate innovative research relevant to cancer therapy, such as new agents, innovative anticancer therapies, biological response modifiers, immunotherapy, and gene therapy; and quality of life, cancer education, and nutritional oncology. Funds may be used for salaries, supplies, and nonreimbursable research-related patient care costs, including extraordinary laboratory and imaging studies. There are no application deadlines. Preliminary grant request forms are available on the foundation's Web site. The foundation will invite a formal application.
Requirements: New and established investigators are encouraged to apply: young investigators working in established research programs relevant to clinical research who are without support from NIH or other cancer research agencies; and established, clinical researchers who wish to embark on innovative studies directly relevant to the foundation's mission with novel, new, or pilot projects distinctly removed from their currently funded research projects.
Date(s) Application is Due: Aug 15
Amount of Grant: 25,000 - 300,000 U.S.D
Contact: Kristine Nelson, Grants Administrator; (847) 342-7450; grants@ctrf.org
Internet: http://www.ctrf.org/grantapp.cfm
Sponsor: Cancer Treatment Research Foundation
1336 Basswood Road
Schaumburg, IL 60173

CAN Pilot Research Projects 1076
Pilot research programs will provide funding for investigators at any stage in their career to work on innovative pilot projects. The purpose of these awards is to encourage innovative approaches towards elucidating the causes, prevention, and appropriate therapy of autism. They are intended to provide support for innovative research that may not have reached a stage where funding can be obtained from federal sources. Annual deadline dates may vary; contact program staff for exact dates.
Requirements: Investigators whose focus has been outside the field of autism as well as innovative investigators from within the field are encouraged to apply.
Restrictions: Members of CAN's Board of Directors are not eligible as investigators in CAN supported research. Members of CAN's Scientific Advisory Committee are eligible, provided they comply with foundation policies regarding the avoidance of conflict of interest.
Date(s) Application is Due: Mar 3; Jun 9
Amount of Grant: Up to 120,000 U.S.D
Contact: Therese Finazzo, Grants Officer; (888) 828-8476 or (323) 549-0500, ext. 18; fax (323) 549-0547; therese@cureautismnow.org
Internet: http://www.cureautismnow.org/research/funding/3522.jsp
Sponsor: Cure Autism Now
5455 Wilshire Boulevard, Suite 715
Los Angeles, CA 90036-4234

CAN Young Investigator Awards 1077
The purpose of these awards is to encourage innovative approaches towards elucidating the causes, prevention, and appropriate therapy of autism. These awards are designed to support outstanding projects that involve innovative approaches and the application of cutting-edge technologies. All proposals must have direct and immediate relevance to autism and related disorders.
Requirements: Awards provide for the work of outstanding candidates to be carried out under the supervision of a mentor who is an established investigator. The investigator need not be directly involved in autism research but must provide an environment in which the CAN fellow can pursue a project with relevance to autism or related disorders.
Restrictions: Applicants should be no more than four years out of a MD or PhD program on the date the fellowship would begin.
Date(s) Application is Due: Mar 3; Jun 9
Amount of Grant: 41,000 U.S.D
Contact: Therese Finazzo, Grants Officer; (323) 549-0500, ext. 18 or (888) 828-8476; fax (323) 549-0500; therese@cureautismnow.org
Internet: http://www.cureautismnow.org/research/funding/3522.jsp
Sponsor: Cure Autism Now
5455 Wilshire Boulevard, Suite 715
Los Angeles, CA 90036-4234

Capital Region Community Foundation Grants 1078
The foundation awards grants to eligible Michigan nonprofit organizations in its areas of interest, including education, environment, health care, human services, humanities, and public benefit. Types of support include capital grants, program development grants, and seed money grants. Grants are awarded for one year and are nonrenewable. Guidelines and application are available on-line. Application must be made on-line. Program deadlines vary: Youth Fund grant deadline is January 31; Eaton County Community Foundation grant deadline is March 1; and Capital Region Community Foundation grant deadline is April 1.
Requirements: Michigan 501(c)3 tax-exempt organizations in Clinton, Eaton, and Ingham Counties are eligible. Churches in the tri-county area are eligible for programs including food banks, after-school programs, programs assisting the needy, and charitable work benefiting the community, there grant applications are due June 1st.
Restrictions: Grants do not support sectarian or religious programs; individuals; international organizations; endowment funds; administrative costs of fund raising campaigns; annual meetings; routine operating expenses; or existing obligations, debts, or liabilities.
Geographic Focus: Michigan
Date(s) Application is Due: Jan 31; Mar 1; Apr 1
Amount of Grant: 2,000 - 25,000 U.S.D
Contact: Brad Patterson, Vice President of Programs; (517) 272-2870; fax (517) 272-2871; bpatterson@crcfoundation.org
Internet: http://crcfoundation.org/grantmaking/grants
Sponsor: Capital Region Community Foundation
6035 Executive Drive, Suite 104
Lansing, MI 48901

Cardiac Arrhythmias Research and Education Foundation Research Grants 1079
The foundation's two primary goals are to raise funds for clinical research of cardiac arrhythmias and to educate the public and the medical community about sudden death. Awards will be granted to promising young investigators working on the genetics, mechanisms, therapy, and prevention of sudden death due to cardiac arrhythmias. Two research grants are awarded each year.
Amount of Grant: 50,000 U.S.D
Samples: Johns Hopkins U—to support the research of David Charles Johns on the mechanisms and treatment of cardiac arrhythmias and the prevention of sudden death caused by heart rhythm disorders, $100,000 over two years.
Contact: Grants Administrator; (800) 404-9500 or (425) 788-1987; fax (425) 788-1927; care@longqt.org
Internet: http://www.longqt.org/fundingmedical.html
Sponsor: Cardiac Arrhythmias Research and Education Foundation
26425 NE Allen Street, Suite 103, P.O. Box 369
Duvall, WA 98019

Cardinal Health Foundation Grants 1080

The foundation's mission is to support employees' interests and to advance and fund programs that improve access to and delivery of health care services in CardinalHealth communities. Additional areas of interest include arts and culture, education, and youth development. Application forms are not required; submit a project summary of no more than three pages. Proposals are by invitation.
Requirements: 501(c)3 organizations in CardinalHealth communities are eligible.
Restrictions: Capital campaigns, endowments, religious organizations or sectarian programs for religious purposes, veteran, labor, and political organizations or campaigns, fraternal, athletic or social clubs, requests for loans or debt retirements, individual endeavors or needs, organizations that discriminate on the basis of age, disability, religion, ethnic origin, gender, or sexual orientation, or organizations with divisive or litigious public agendas will not be supported.
Amount of Grant: 250 - 450,000 U.S.D
Contact: Debra Hadley, Executive Director; (614) 757-7450; cardinalfoundation@cardinal.com or communityrelations@cardinalhealth.com
Internet: http://www.cardinal.com/us/en/community/
Sponsor: Cardinal Health Foundation
7000 Cardinal Place
Dublin, OH 43017

Cargill Citizenship Fund-Corporate Giving Grants 1081

Cargill's purpose is to be the global leader in nourishing people. Cargill measures their performance through engaged employees, satisfied customers, profitable growth and enriched communities. Corporate giving is one important way Cargill works to enrich the 1,000 communities where they conduct business. With 149,000 employees in 63 countries, Cargill people are working everyday to nourish the lives of those around us. The Cargill Citizenship Fund provides strategic grants to organizations serving communities where Cargill has a presence. The Fund provides direct grants for regional, national and global partnerships and provides matching grants for selected local projects supported by our businesses. Cargill seeks to build sustainable communities by focusing our human and financial resources in three areas: Nutrition and Health-support for programs and projects that address long-term solutions to hunger, increase access to health education and/or basic health care in developing and emerging countries, and improve youth nutrition and wellness; Education-support for innovative programs that improve academic achievement, develop logic and thinking skills, promote leadership development, and/or increase access to education for socio-economically disadvantaged children. Cargill also supports mutually beneficial partnerships with selected higher education institutions; Environment-support for projects that protect and improve accessibility to water resources; promote biodiversity conservation in agricultural areas; and educate children about conservation and/or proper sanitation. Application and additional guidelines are available at: http://www.cargill.com/wcm/groups/public/@ccom/documents/document/doc-giving-funding-app.pdf
Requirements: Applicants must have 501(c)3 status or the equivalent; and they must be located in communities where Cargill has a business presence. Only under special circumstances will Cargill consider general operating or capital support. Organizations requesting capital or operating support should contact the Cargill Citizenship Fund staff before applying.
Restrictions: Cargil will not fund: organizations without 501(c)3 status or the equivalent; organizations that do not serve communities where Cargill has a business presence; individuals or groups seeking support for research, planning, personal needs or travel; public service or political campaigns; lobbying, political or fraternal activities; benefit dinners or tickets to the same; fundraising campaigns, walk-a-thons, or promotions to eliminate or control; specific diseases; athletic scholarships; advertising or event sponsorships; religious groups for religious purposes; publications, audio-visual productions or special broadcasts; endowments; medical equipment.
Geographic Focus: All States, All Countries
Amount of Grant: 500 - 100,000 U.S.D
Contact: Stacey Smida, Grants Administrator; (952) 742-4311; stacey_smida@cargill.com
Internet: http://www.cargill.com/wcm/groups/public/@ccom/documents/document/doc-giving-funding-app.pdf
Sponsor: Cargill Corporation
P.O. Box 5650
Minneapolis, MN 55440-5650

Caring for Colorado Foundation Grants 1082

The foundation aims to improve healthcare in Colorado. Grants are awarded in four areas: community health; enabling informed health decisions; health infrastructure; and oral health. Application deadlines are March 24 for community health, informed health decisions, and health infrastructure grants, and July 1 for oral health grants.
Requirements: Colorado nonprofit organizations are eligible.
Geographic Focus: Colorado
Date(s) Application is Due: Mar 24; Jul 1
Amount of Grant: Up to 150,000 U.S.D
Contact: Grants Administrator; (720) 524-0770; fax (720) 524-0787; health@caringforcolorado.org
Internet: http://www.caringforcolorado.org/grant.php
Sponsor: Caring for Colorado Foundation
4100 E Mississippi Avenue, #605
Denver, CO 80246

Caring Foundation Grants 1083

Blue Cross and Blue Shield of Alabama, Incorporated sponsors The Caring Foundation. The Foundation was established in 1990 in Alabama and focuses it's support primarily in the region. The foundation supports organizations involved with education, federated giving programs, health, hospitals, safety education, children and youth services. The Foundation offers support of a general/operating nature as well, as program development.
Requirements: There is no formal application form, submit a proposal containing the following: name, address and phone number of organization; brief history of organization and description of its mission; detailed description of project and amount of funding requested.
Restrictions: Grants aren't available to individuals, or for capital campaigns.
Geographic Focus: Alabama
Amount of Grant: 10,000 - 900,000 U.S.D
Samples: Alabama Child Caring Foundation, Birmingham, AL, $900,000; Ronald McDonald House Charities of Alabama, Birmingham, AL, $15,500; Auburn University Foundation, Auburn, AL, $105,000; Salvation Army of Birmingham, Birmingham, AL, $100,400;
Contact: James M. Brown, Senior Vice President; (205) 220-2500
Sponsor: Caring Foundation
450 Riverchase Parkway, East
Birmingham, AL 35244-2858

Carl and Eloise Pohlad Family Foundation Grants 1084

The mission of the foundation is to improve the lives of economically disadvantaged children and youth and participate in projects that positively impact the quality of life in the Minneapolis/St.Paul area. The foundation awards grants to Minnesota nonprofits in its areas of interest, including arts and culture, economic development, education, environment, health care, housing, and social services. Types of support include general operating support, continuing support, capital campaigns, building construction/renovation, endowments, emergency funds, scholarship funds, and research.
Requirements: Minnesota nonprofits are eligible.
Restrictions: Individuals are not eligible. Capital request are considered only for physical plant improvements or significant technology investments. Capital requests for housing construction, endowment, program start-up or expansion or to establish operating reserves are not considered.
Geographic Focus: Minnesota
Amount of Grant: Up to 32,000,000 U.S.D
Samples: Abbott Northwestern Hospital (Minneapolis, MN)—to construct a cardiology hospital, $3 million.
Contact: Josette Elstad, Grants Manager; (612) 661-3910; fax (612) 661-3715; info@pohladfamilygiving.org
Internet: http://www.pohladfamilyfoundation.org/pff/pff_default.aspx
Sponsor: Carl and Eloise Pohlad Family Foundation
60 South Sixth Street, Suite 3900
Minneapolis, MN 55402

Carl and Lily Pforzheimer Foundation Grants 1085

The foundation awards grants to U.S. nonprofit organizations in its areas of interest, including higher and secondary education, cultural activities and programs, public affairs, and healthcare. Types of support include capital campaigns, fellowships, internship funds, matching grants, professorships, program development, publication, scholarship funds, and seed grants. There are no application forms or deadlines. The board meets in April, June, October, and December to consider requests.
Restrictions: Grants are not awarded to individuals or for building construction/renovation.
Amount of Grant: 10,000 - 150,000 U.S.D
Samples: Bank Street College of Education (New York, NY)—for scholarships and unrestrictd support, $2 million.
Contact: Grants Administrator; (212) 223-6500
Sponsor: Carl and Lily Pforzheimer Foundation
950 Third Avenue, 30th Floor
New York, NY 10022

Carl B. and Florence E. King Foundation Grants 1086

The lives of Carl B. and Florence E. King were marked by warmth, compassion, and generosity. As they prospered, they believed in giving back. With gracious benevolence, they dedicated themselves to the betterment of individuals, communities, and society through informed giving. Today, the Carl B. and Florence E. King Foundation honors their memory, continues their tradition, and builds upon their vision. The Foundation is principally interested in the following areas: aging population; arts, culture, and history; children and youth; education; indigent; and to build non-profit capacity. The Foundation awards grants twice each year. Applicants must first submit a letter of inquiry, and then a full grant proposal only upon invitation.
Requirements: The King Foundation distributes grants only to entities that serve residents of Arkansas and Texas. Within Texas, the Foundation is principally interested in the Dallas-Fort Worth area and West Texas. Within Arkansas, the foundation focuses on the southern and eastern portions of the state. Applicants must also have a letter of determination from the Internal Revenue Service acknowledging tax-exempt status as described in Section 501(c)3 of the Internal Revenue Code.
Restrictions: The Foundation does not award grants: to individuals; to organizations or programs that do not serve residents of our geographic focus areas in Texas or Arkansas; to organizations that are not tax exempt; for general operating support, annual fund drives, or funds to offset operating losses (including retiring debt incurred to cover

operating losses); to create endowments; toward balls, events, or galas benefiting charitable organizations; to efforts to treat or cure a single disease or condition; to church or seminary construction, or religious programs (other than social service-based initiatives); toward the cost of hosting or attending professional conferences or symposia, or participating in amateur sports competitions or similar activities.
Geographic Focus: Arkansas, Texas
Date(s) Application is Due: Jun 15; Dec 15
Amount of Grant: Up to 100,000 U.S.D
Samples: The Dallas Foundation, Dallas, TX, $50,000 - support for the funding pool for the Zero-to-Five Funders' Collaborative project in the Bachman Lake area of Dallas; Amon Carter Museum, Fort Worth, TX, $25,000 - program support for professional development programs for North Texas educators; Learning Center Jonesboro, AR, $25,000 - capital support to purchase and install a playground for young children with physical and developmental disabilities.
Contact: Michelle D. Monse, President; (214) 750-1884; fax (214) 750-1651; michellemonse@kingfoundation.com
Internet: http://www.kingfoundation.com/Grants/King-Foundation-Grants.aspx
Sponsor: Carl B. and Florence E. King Foundation
2929 Carlisle Street, Suite 222
Dallas, TX 75204

Carl C. Icahn Foundation Grants 1087
The foundation awards grants to New York and New Jersey nonprofits in the areas of education, arts and culture, health care, child welfare, and Jewish temples and organizations. Types of support include general operating support, annual campaigns, building construction/renovation, and matching funds. There are no application deadlines or forms.
Requirements: New York and New Jersey nonprofits are eligible to apply.
Restrictions: No grants are provided to individuals.
Geographic Focus: New Jersey, New York
Amount of Grant: 500 - 1,600,000 U.S.D
Samples: Randall's Island Sports Foundation, New York, NY, $1,666,000—to support program/general purposes; Wildlife Conservation Society, Pelham, NY, $10,000—to support program/general purposes; The Ladies' Village Improvement Society, East Hampton, NY, $500—to support program/general purposes.
Contact: Gail Golden-Icahn, Vice Presicent; (212) 702-4300; fax (212) 750-5815
Sponsor: Carl C. Icahn Foundation
767 5th Avenue, 47th Floor
New York, NY 10153-0023

Carl Gellert and Celia Berta Gellert Foundation Grants 1088
The foundation funds religious, charitable, scientific, literary or educational purposes restricted in the nine counties of the greater San Francisco Bay Area (Alameda, Contra Costa, Marin, Napa, San Francisco, San Mateo, Santa Clara, Solano and Sonoma). No grants are made to individuals. Types of support include general operations, annual and capital campaigns, building construction/ renovation, equipment acquisition, debt reduction, program/project development, medical research, publication, and scholarships.
Requirements: California 501(c)3 tax-exempt nonprofit organizations that are not private foundations are eligible.
Restrictions: Grants are not awarded to individuals.
Geographic Focus: California
Date(s) Application is Due: Aug 15
Amount of Grant: 1,000 - 10,000 U.S.D
Samples: U of San Francisco, School of Business and Management (CA)—to renovate and add a new wing to the McLaren Center, home of the School of Business and Management, $500,000.
Contact: Jack Fitzpatrick, Executive Director; (415) 255-2829
Internet: http://home.earthlink.net/~cgcbg
Sponsor: Carl Gellert and Celia Berta Gellert Foundation
1169 Market Street, Suite 808
San Francisco, CA 94103

Carlisle Foundation Grants 1089
The foundation prefers to support proposals that are new, innovative, and/or demonstrate promise as models that might be replicated in other sites. In many instances, grants act as venture capital or seed money through which the applicant organization can demonstrate a new concept or expand its own capacity to deliver services. Although the foundation reviews a wide range of proposals, several areas receive high priority, including substance abuse; domestic and community violence; homelessness/housing; economic development; and services for children, youth, and families. The foundation prefers to make grants for restricted project support rather than general operating support. Applicant organizations should submit a one-page concept paper accompanied by a copy of the 501(c)3 determination letter. There is no time line or deadline for concept letters.
Requirements: Connecticut, Vermont, Rhode Island, Massachusetts, New Hampshire and Maine C501(c)3 organizations are eligible. Applicant organizations should submit a one-page concept paper accompanied by a copy of the 501(c)3 determination letter.
Restrictions: Requests for capital support are rarely considered; requests for support of endowments are never considered. The foundation does not usually support programs with a primary focus on day care, disabilities, education, health services, legal services, mental retardation, or older adults. The foundation also does not provide direct support for individuals.
Geographic Focus: Connecticut, Maine, Massachusetts, New Hampshire, Rhode Island, Vermont
Samples: South County Habitat for Humanity, Shannon, RI, $25,000 - to provide start up capital for the development of Re-Store, a retail outlet designed to distribute surplus building supplies at low cost; Wayside Youth & Family Network, Framingham, MA, $50,000 - community match to a major Robert Wood Johnson grant to develop a comprehensive center, Tempo Young Adult Resource Center, for disenfranchised and underserved youth and young adults; Communities for Restorative Justice, Concord, MA, $7,500 - to help sustain this Juvenile Justice program that provides a mediated settlement to a range of non-violent crimes and that allows the perpetrator to avoid the criminal justice system;
Contact: Richard Goldblatt, Executive Director; (401) 284-0368; fax (401) 284-0390; rag@carlislefoundation.org
Internet: http://www.carlislefoundation.org/annualgrants.htm
Sponsor: Carlisle Foundation
P.O. Box 5549
Wakefield, RI 02880-5549

Carl J. Herzog Foundation Grants 1090
The foundation awards grants, primarily in Connecticut, for research and general operating support in its areas of interest, including medical research, dermatology research, hospitals, and education. There are no application deadlines or forms.
Requirements: 501(c)3 tax-exempt organizations are eligible.
Geographic Focus: Connecticut
Amount of Grant: 1,000 - 200,000 U.S.D
Samples: Meharry Medical College (Nashville, TN)—for scholarships for minority students, $250,000.
Contact: David Babson, (203) 629-2424
Sponsor: Carl J. Herzog Foundation
321 Railroad Avenue
Greenwich, CT 06836-0788

Carl M. Freeman Foundation FACES Grants 1091
Founded in 2000 in Delaware, FACES stands for Freeman Assists Communities with Extra Support. The FACES program is designed to find and fund the smaller, overlooked projects in it's neighborhoods. The grants are limited to Montgomery County nonprofit organizations with operating budgets of $750,000 or less and Sussex County nonprofit organizations with operating budgets of $500,000 or less. Funding applications are available in five areas of interest: arts/culture; education/environment; health & human services; housing and; other-this may include anything you feel does not fit in the above categories, for example spaying cats/dogs. Additional guidelines and applications are available at: http://www.freemanfoundation.org/carl/CarlMFreemanFoundation/Grants/GrantGuidelines/ApplyForaGrant/tabid/185/Default.aspx.
Requirements: 501(c)3 tax-exempt organizations in Montgomery & Sussex County are eligible. Nonsectarian religious programs also are eligible.
Restrictions: Grants will not be distributed to: individuals; political associations or candidates; organizations that would disperse the funding to others; organizations that discriminate by race, creed, gender, sexual orientation, age, religion, disability or national origin.
Geographic Focus: Delaware, Maryland, West Virginia
Samples: Charles Town Health Right, Inc., Charles Town, WV—to help purchase needed medical and support supplies in order to meet daily medical needs; Olney Community Band, Silver Spring, MD—to support the 2009 Concert Series and help present affordable concerts to diverse crowds; Sussex Technical Adult Division, Georgetown, DE—to support the Even Start Family Literacy Program, which helps improve the literacy skills of low income and low literate families.
Contact: Trish Schechtman, Relationship Manager; (302) 436-3555; trish@freemanfoundation.org
Internet: http://www.freemanfoundation.org/carl/CarlMFreemanFoundation/Grants/FACES/tabid/204/Default.aspx
Sponsor: Carl M. Freeman Foundation
36097 Sand Cove Road
Selbyville, DE 19975

Carl M. Freeman Foundation Grants 1092
The Carl M. Freeman Foundation has historically emphasized it's support in the following communities: Montgomery County, Maryland; Sussex County, Delaware; and the Eastern Panhandle of West Virginia . Funding is available for a wide variety of community organizations, having supported everything from arts organizations and hunger centers to educational and health related organizations. To simplify the application process, funding applications are available in five areas of interest: arts/culture; education/environment; health & human services; housing and; other-this may include anything you feel does not fit in the above categories, for example spaying cats/dogs. Additional guidelines and applications are available at: http://www.freemanfoundation.org/carl/CarlMFreemanFoundation/Grants/GrantGuidelines/ApplyForaGrant/tabid/185/Default.aspx
Requirements: 501(c)3 tax-exempt organizations in Maryland, Delaware and West Virginia are eligible. Nonsectarian religious programs also are eligible.
Restrictions: Grants will not be distributed to: individuals; political associations or candidates; organizations that would disperse the funding to others; organizations that discriminate by race, creed, gender, sexual orientation, age, religion, disability or national origin.
Geographic Focus: Delaware, Maryland, West Virginia
Amount of Grant: 5,000 - 30,000 U.S.D
Contact: Trish Schechtman, Relationship Manager; (302) 436-3555; trish@freemanfoundation.org

Internet: http://www.freemanfoundation.org/carl/CarlMFreemanFoundation/Grants/GrantGuidelines/tabid/181/Default.aspx
Sponsor: Carl M. Freeman Foundation
36097 Sand Cove Road
Selbyville, DE 19975

Carlos and Marguerite Mason Trust Grants 1093

The trust awards grants in Georgia with a focus on organ transplants. The trust's interests are patient services—needs relating to the transplantation process for the patient and the immediate family members, as well as donor family members and living donors; donation and transplantation—related education (donor education, professional education, and technical education for transplantation providers; programs to improve consent rates; and programs to promote living organ donations); and research—medical research relating to the area of transplantation. Types of support include capital improvements, start-up grants for new projects, and challenge/matching grants. Grants will be awarded for up to three years. Applicants are encouraged to contact the trustee for an appointment prior to submitting a grant request.
Requirements: Georgia nonprofit organizations are eligible.
Restrictions: Requests for general operating or administrative funds are discouraged. The trust generally does not fund endowment, nor does it fund general goodwill advertising; make grants that would replace existing sources of funding; or pay indirect overhead expenses for projects at colleges, universities, governmental units, or other established organizations.
Geographic Focus: Georgia
Date(s) Application is Due: Jun 1
Amount of Grant: Up to 15,000,000 U.S.D
Samples: MCG Health System (Augusta, GA)—to educate the public about the importance of transplants for people with chronic kidney disease, and to help meet the needs of people in rural Georgia communities who are without transplant services, $1.1 million.
Contact: Randy Karesh, Grant Manager, Wachovia Bank of Georgia; (404) 332-6677; grantinquiriesga@wachovia.com
Internet: http://www.wachovia.com/corp_inst/charitable_services/0,,4269_3300,00.html
Sponsor: Carlos and Marguerite Mason Trust
3414 Peachtree Road, 5th Floor, MC GA8023
Atlanta, GA 30326

Carl R. Hendrickson Family Foundation Grants 1094

The Carl R. Hendrickson Family Foundation was established in 1991 to support and promote quality educational, human-services, and health-care programming for underserved populations. Carl R. Hendrickson was a Chicago entrepreneur who, along with his father and brothers, built the Hendrickson Trucking Company. Carl and his wife, Agnes, had one child, Virginia. Virginia followed in her father's footsteps by leading the family business and by serving as President of the Hendrickson Foundation. Virginia died in 1995, leaving no heirs. The Hendricksons prided themselves on their entrepreneurial spirit, having been in the forefront of the trucking business by inventing the tandem truck. Reflecting the Hendrickson family's strong Christian faith, special consideration is given to charitable organizations that help individuals meet their basic needs while also addressing their spiritual needs. Preference is given to organizations or programs that approach their mission from an entrepreneurial perspective. The majority of grants from the Hendrickson Family Foundation are one year in duration. Application guidelines as well as a link to the downloadable application are given at the grant website. Applicants are also encouraged to review the Illinois state application guidelines and the foundation's funding history before applying (links are available from the grant website). The deadline for application to the Carl R. Hendrickson Family Foundation is July 31. Grant decisions will be made by December 31.
Requirements: Applicants must have 501(c)3 status. Applications must be mailed.
Restrictions: In general, grant requests for individuals, endowment campaigns or capital projects will not be considered. The foundation does not support requests from individuals, organizations attempting to influence policy through direct lobbying, or any political campaigns.
Geographic Focus: All States, Illinois
Date(s) Application is Due: Jul 31
Samples: WorldVenture, Littleton, Colorado, $10,000 (2010); Chicago Lighthouse for People Who Are Blind or Visually Impaired, Chicago, Illinois, $10,000 (2010); Erie Family Health Network, Chicago, Illinois, $10,000 (2010).
Contact: Debra Grand, Senior Vice President; (312) 828-4154; ilgrantmaking@bankofamerica.com
Internet: https://www.bankofamerica.com/philanthropic/fn_search.action
Sponsor: Carl R. Hendrickson Family Foundation
231 South LaSalle Street, IL1-231-13-32
Chicago, IL 60604

Carlsbad Charitable Foundation Grants 1095

The mission of the Carlsbad Charitable Foundation is to advance philanthropy in its service region in order to: build community excellence; stimulate innovation; and enhance capacity of nonprofits. Its primary purpose is to: meet emerging needs by encouraging and increasing responsible and effective philanthropy by and for the benefit of all who live, work and play in Carlsbad; build a Carlsbad community endowment; provide funding annually to Carlsbad organizations and causes; and give Carlsbad community members a vehicle for legacy planning and gifts that will benefit Carlsbad now and forever. The field of grant giving focus changes annually, so applicants should contact the office directly for specific details. Most recently, the Foundation was interested in funding Senior Health and Human Services programs managed by non-profit or government agencies.
Requirements: Any 501(c)3 organization serving the residents of Carlsbad, California, are eligible to apply. Projects must specifically target seniors over the age of 55 who live in the City of Carlsbad.
Restrictions: The Carlsbad Charitable Foundation does not make grants for: annual campaigns and fund raising events for non-specific purposes; capital campaigns for buildings or facilities; stipends for attendance at conferences; endowments or chairs; for-profit organizations or enterprises; individuals unaffiliated with a qualified fiscal sponsor; projects that promote religious or political doctrine; research projects (medical or otherwise); scholarships; or existing obligations or debt.
Geographic Focus: California
Date(s) Application is Due: Jan 12
Contact: Trudy Amstrong; (619) 814-1384; trudy@sdfoundation.org
Internet: http://www.endowcarlsbad.org/grants.html
Sponsor: Carlsbad Charitable Foundation
2508 Historic Decatur Road, Suite 200
San Diego, CA 92106

Carls Foundation Grants 1096

The foundation has broadly defined charitable purposes, but the principal purpose and mission of the foundation is as follows: children's welfare (primarily in Michigan), including health care facilities and programs, with special emphasis on the prevention and treatment of hearing impairment; recreational, educational, and welfare programs especially for children who are disadvantaged for economic and/or health reasons; and preservation of natural areas, open space, and historic buildings and areas having special natural beauty or significance in maintaining America's heritage and historic ideals, through assistance to land trusts and land conservancies and directly related environmental educational programs. Types of grants include capital grants, limited budget support, start up/seed money, and multi-year grants. The foundation has no formal application for grant requests. The Trustees meet minimally three times per year. Requests are accepted at all times, but organizations are encouraged to submit requests well in advance of scheduled board meetings, in January, May, and September. A letter of inquiry is not required and phone calls are welcome. Guidelines are available online.
Requirements: 501(c)3 tax-exempt organizations are eligible.
Restrictions: Grants are not awarded to individuals or for endowments, publications, conferences, seminars, film, fellowships, educational loans, travel, research, playground structures or athletic facilities, or underwriting special events.
Geographic Focus: Michigan
Date(s) Application is Due: Mar 1; Jul 1; Nov 1
Amount of Grant: 5,000 - 50,000 U.S.D
Contact: Kathy Stenman, Program Officer; (313) 965-0990; fax (313) 965-0547
Internet: http://www.carlsfdn.org
Sponsor: Carls Foundation
333 W Fort Street, Suite 1940
Detroit, MI 48226

Carl W. and Carrie Mae Joslyn Trust Grants 1097

Grants support activities providing services to resident children, elderly, and the disabled in El Paso County, Colorado. Areas of interest include education, medical care, rehabilitation, children and youth services, and aging centers and services. Types of support include general operating support, annual campaigns, building construction and renovation, equipment acquisition, endowment funds, and program development. Application must be in writing and must specifically describe the use of the funds. Grants are not sustaining and new applications must be submitted semiannually for renewal.
Requirements: Nonprofit organizations located in, or serving the residents of, El Paso County, Colorado, are eligible.
Restrictions: Grants are not made to individuals or for research, scholarships, fellowships, loans, or matching gifts.
Geographic Focus: Colorado
Date(s) Application is Due: Apr 30; Oct 31
Amount of Grant: 500 - 15,000 U.S.D
Contact: Susan Bradt Laabs, Grants Administrator; (719) 227-6435 or (719) 227-6439; fax (719) 2276448
Sponsor: Carl W. and Carrie Mae Joslyn Charitable Trust
Trust Department, P.O. Box 1699
Colorado Springs, CO 80942

Carnahan-Jackson Foundation Grants 1098

The foundation's areas of interest include higher and other education, libraries, hospitals, youth, the disabled, drug abuse programs, ecology, housing, community development, dance and other performing arts groups, and churches. Types of support include general operating support, continuing support, capital campaigns, building construction/renovation, equipment acquisition, programs/projects, seed grants, curriculum development, scholarship funds, and matching funds.
Requirements: IRS 501(c)3 organizations serving western New York, particularly Chautauqua County, are eligible.
Geographic Focus: New York
Amount of Grant: 2,500 - 125,000 U.S.D
Contact: Stephen E. Sellstrom, (716) 483-1015
Sponsor: Carnahan-Jackson Foundation
13 East 4th Street, P.O. Box 3326
Jamestown, NY 14701-3326

Carnegie Corporation of New York Grants 1099

The foundation provides research, study, and support for projects to improve government at all levels, to increase public understanding of social policy issues, to equalize opportunities for minorities and women, and to increase participation in political and civic life. Also supported are projects that promote electoral reform; education reform from early childhood through higher education; early childhood development; and urban school reform. The foundation will also fund research on the increasing availability and success of after-school and extended service programs for children and teenagers, particularly those in urban areas, that promote high academic achievement. Dissemination of best practices in teacher education will also be emphasized. There is no formal procedure for submitting a proposal. To apply under any of the corporation's grantmaking programs, applicants should submit a full proposal that describes the project's aims, duration, methods, amount of financial support required, and key personnel. The board meets four times a year, in October, February, April, and June.
Restrictions: Grants are not made for construction or maintenance of facilities or endowments. The Corporation does not generally make grants to individuals except through the Carnegie Scholars Program, that supports the work of select scholars and experts conducting research in the foundation's fields of interest.
Amount of Grant: Up to 80,000,000 U.S.D
Samples: American Academy of Arts and Sciences, Cambridge, Massachusetts, $500,000—toward an initiative on minimizing the risks of nuclear energy expansion; George Washington University, Washington, DC, $349,900— toward research and writing on the effects of domestic foreign policy debates on U.S. global engagement.
Contact: Sarina Cipriano, Grants Manager; (212) 371-3200; fax (212) 754-4073
Internet: http://www.carnegie.org/sub/program/grant.html
Sponsor: Carnegie Corporation of New York
437 Madison Avenue
New York, NY 10022

Carolyn Foundation Grants 1100

Priorities for funding include community and environmental grantmaking. In the community focus area needs are addressed only in the communities of interest to the foundation: Minneapolis, Minnesota and, New Haven, Connecticut. There are two community focus areas for funding: economically disadvantaged children and youth; and community and cultural vitality. The foundation works to empower economically disadvantaged children and youth by supporting their families and others to inspire, nurture, educate and guide them to achieve long-term stability and well-being. In the environmental focus area, the Carolyn Foundation environmental committee is currently most interested in funding renewable energy programs. The Foundation will consider other environmental proposals if funds allow. All proposals submitted must: address root causes and create systemic and sustainable solutions and change; address global issues with local interventions that address local needs, as well as global needs; develop and implement solutions that can be replicated in other areas; collaborate effectively with others in the community: government, non-government, foundations and private parties. The Carolyn Foundation makes grants twice a year, in June and January. Applications must be submitted by January 15 for June grants, and July 15 for January grants. This is a postmark deadline. Grant applications will be reviewed by the Executive Director and a committee of foundation volunteers. Declinations will be sent at the time a decision is made to no longer consider a proposal, typically before the end of the review cycle. Successful applicants will be notified in June and January.
Requirements: IRS 501(c)3 nonprofit organizations in Minnesota and Connecticut may apply for the environmental grants. The Community grants program is limited to the cities of Minneapolis, Minnesota and New Haven, Connecticut. The foundation encourages use of Carolyn Foundation Application Form adapted from the Minnesota Common Grant Application Form. Applicants choosing not to use the common grant form must address the same information as required by the common grant. All proposals must use the Carolyn Foundation Cover Sheet, available at the Foundation's website. It is request that summary information and description of the project be no longer than six pages, printed on one side on 8 1/2 x 11-inch paper. Supporting documents, such as financial information, list of Officers, Directors, and Executive Staff, IRS determination letter may be in addition to the six pages. Do not send bound proposals, cassettes or VCR tapes.
Restrictions: Grants are not awarded to individuals, political organizations or candidates, veterans organizations, fraternal societies or orders, annual fund drives, umbrella organizations, or to deficits already incurred. The foundation does not generally make grants to religious organizations for religious purposes or to organizations in support of operations carried on in foreign countries.
Geographic Focus: Connecticut, Minnesota
Date(s) Application is Due: Jan 15; Jul 15
Amount of Grant: 5,000 - 50,000 U.S.D
Contact: Becky Erdahl, Executive Director; (612) 596-3279 or (612) 596-3266; fax (612) 339-1951; berdahl@carolynfoundation.org
Internet: http://www.carolynfoundation.org/guidelines.html
Sponsor: Carolyn Foundation
706 2nd Avenue South, Suite 760
Minneapolis, MN 55402

Carpenter Foundation Grants 1101

The foundation's primary areas of interest include the arts, education, public interest, and human services. The foundation is deeply concerned with the well-being of children and families and their relationship to their neighborhoods and communities. Also of concern is the health of the web of agencies and organizations which serve them. Grants are awarded for general operating support, program development, capital campaigns, equipment acquisition, scholarship funds, seed money, matching funds, and technical support. Deadlines are generally about six weeks before the quarterly board meetings, held in January, March, June, and September.
Requirements: Tax exempt agencies in the Jackson and Josephine Counties of Oregon may submit proposals.
Restrictions: Grants are not made to individuals. The foundation rarely makes grants for historical applications, hospital construction or equipment, group or individual trips, or activities for religious purposes.
Geographic Focus: Oregon
Amount of Grant: 250 - 25,000 U.S.D
Samples: Community Health Center (Medford, OR)—fir a program to provide discounted medications to low-income/uninsured patients, $15,000; Science Works Hands-On Museum, (Ashland, OR)—in support of a life science hands-on exhibition for local schools and students, $6,000; Southern Oregon State College (Ashland, OR)—for faculty development opportunities which improve teaching, $25,000; Oregon Water Trust, (Portland, OR)—for a community outreach and education project to locally distribute the General Elections Voter's Guide, $12 000.
Contact: Polly Williams, Program Officer; (541) 772-5732; fax (541) 773-3970; carpfdn@internetcds.com
Internet: http://www.carpenter-foundation.org
Sponsor: Carpenter Foundation
711 E Main Street, Suite 10
Medford, OR 97504

Carrie E. and Lena V. Glenn Foundation Grants 1102

Established in 1971, the Glenn Foundation provides annual grants in the following areas of interest: arts; children/youth, services; Christian agencies & churches; Elementary/secondary education; Environment; and Human services.
Requirements: Federally tax-exempt institutions and not-for-profit agencies that serve Gaston County, NC. agencies or out-of-county agencies whose projects have an impact on Gaston County citizens are eligible to apply.
Restrictions: Funding is not available for: planning grants; grants to individuals; scholarships; capital campaigns; umbrella campaigns; and multi-year grants.
Geographic Focus: North Carolina
Date(s) Application is Due: Mar 1
Amount of Grant: 3,000 - 25,000 U.S.D
Samples: Community Foundation of Gaston County, Gastonia, NC., $25,000 - for purchase and installment of This Little Light of Mine, a painting by John Biggers, recreated as mosaic time mural; Gastonia Potters House, Lowell, NC., $10,000 - for operating expenses for residential drug rehabiliation program for women; Alliance for Children and Youth, Gastonia, NC., $8,885 - for Strengthening Families program for parents and their children;
Contact: Barbara H. Voorhees, Executive Director; (704) 867-0296; fax (704) 867-4496; glennfnd@bellsouth.net
Sponsor: Carrie E. and Lena V. Glenn Foundation
1552 Union Road, Suite D
Gastonia, NC 28054

Carrie Estelle Doheny Foundation Grants 1103

The foundation primarily funds local, not-for profit organizations endeavoring to advance education, medicine and religion, to improve the health and welfare of the sick, aged, incapacitated, and to aid the needy. Educational funding includes support of inner city Catholic schools, and scholarship funds for Catholic high schools and universities. Adult education programs and religious education are also supported. Medical funding is focused in two areas: research and care of the disadvantaged. Religious funding is directed to support the gospel values as expressed in the Roman Catholic faith. Health and welfare funding is directed to organizations who assist individuals to lead independent satisfying lives. Specific areas of interest include adoption and foster care service groups, programs for the disabled, health education programs, and senior programs. Aiding the needy funding includes inner city youth clubs, summer camps, and food banks. Applications are accepted anytime, an application form is required for submission and may be downloaded from the foundation web site. The board meets on the last Friday of each month, except for the month of September. Requests should be submitted approximately 6 weeks in advance to allow sufficient time for processing and for distribution to the Board members prior to the meeting. Allow 2 or 3 months for notification of the Board's decision. This is done in writing following the Board of Directors Meeting each month.
Requirements: The Foundation limits its grants to programs located within the fifty states and certified as 501(c)3 non-profit public charities by the Internal Revenue Service. The vast majority of funding is done in the Greater Los Angeles area.
Restrictions: Grant requests are not considered from individuals or from tax-supported entities. Areas also excluded from consideration include support for individuals, endowment funds, publishing books, television or radio programs, travel funds, advertisement, scholarships, or political purposes in any form.
Amount of Grant: 5,000 - 150,000 U.S.D
Samples: Boys & Girls Club of Venice, Venice, CA, $7,000—collaborative mentoring program support; YMCA of Greater Los Angeles, CA, $10,000—camperships for disadvantaged youth; South Central Lamp, Los Angeles, CA, general operating support.
Contact: Shirley Bernard, Senior Grants Administrator; (213) 488-1122; fax (213) 488-1544; doheny@dohenyfoundation.org
Internet: http://www.dohenyfoundation.org/grant/grant.htm
Sponsor: Carrie Estelle Doheny Foundation
707 Wilshire Boulevard, Suite 4960
Los Angeles, CA 90017

Carrier Corporation Contributions Grants 1104

Carrier donates approximately $2 million around the world to registered nonprofit organizations. In the United States, Carrier funds only qualified 501(c)3 organizations that meet its eligibility criteria and operate in locations where the company has a significant employee base. Carrier believes in helping people in the communities where they live, work and do business. To better serve those communities and to better align its corporate contributions with mission and values, it focuses giving on the following these areas: environment and sustainability; civic & community; education; arts & culture; health & human services. All U.S. non-profits are required to complete an online grant application. Applications are accepted from March 1 through June 1 of each year, and are reviewed for funding to be paid the following year. Applicants will receive notification in the first quarter of the calendar year in which funding will occur.

Requirements: Carrier funds only qualified 501(c)3 organizations that meet eligibility criteria and operate in locations where it has a significant employee base.

Restrictions: Carrier will not fund: individuals; religious organizations; alumni groups, sororities or fraternities; booster clubs; political groups; any organization determined by Carrier to have a conflict of interest; any organization whose practices are inconsistent with the company's Code of Ethics

Geographic Focus: Alabama, Arizona, Connecticut, Georgia, Illinois, Indiana, Michigan, Nevada, New York, North Carolina, South Carolina, Tennessee, Texas

Date(s) Application is Due: Jun 1

Contact: Rajan Goel, Vice President; (860) 674-3420; fax (860) 622-0488

Internet: http://www.corp.carrier.com/vgn-ext-templating/v/index.jsp?vgnextoid=6afa80757d7e7010VgnVCM100000cb890b80RCRD

Sponsor: Carrier Corporation
One Carrier Place
Farmington, CT 06034-4015

Carroll County Community Foundation Grants 1105

The Carroll County Community Foundation funds initiatives that improve the quality of life for citizens of Carroll County, Indiana. The Foundation's grant program emphasizes change-oriented issues, with the following areas of interest: health and medical; social services; education; cultural affairs; civic affairs; and community beautification. Proposals: must strive to anticipate the changing needs of the community and be flexible in responding to them; must be change-oriented and problem-solving in nature with emphasis on "seed" money or pilot project support rather than for ongoing general operating support; support innovative efforts and projects that offer far-reaching gains and widespread community results; may coordinate with other funders and donors where possible, including using matching or challenge grant techniques; closely relate and coordinate with the programs of other sources for funding such as the government, other foundations, and associations; and achieve certain objectives such as become more efficient, increase fund-raising capabilities, and deliver better products. Grants will be made only to organizations: whose programs benefit the residents of the county, with preference given to those projects with high visibility in the community; which provide for a responsible fiscal agent and adequate accounting procedures, with preference given to those projects that generate revenue and/or have plans that sustain the project.

Requirements: A letter of inquiry to the Foundation is required as a pre-qualification. The letter should contain a brief statement of the applicant's needs for assistance, estimate of total cost of the project, and enough factual information to enable the Foundation to determine whether or not the application falls within the guidelines of its grants program. After the organization has received a response to apply for a grant, the grantee will then fill out the online application and submit ten copies of the application and all attachments to the Foundation for approval. Organizations should refer to the Foundation website for further information.

Restrictions: The Foundation typically does not award grants for: normal operating expenses and/or salaries; individuals; seminars or trips except where there are special circumstances which will benefit the larger community; sectarian religious purposes but can be made to religious organizations for general community programs; endowment purposes of recipient organizations; projects which have been proposed by individuals or organizations responsible to advisory bodies or persons; new projects and/or equipment purchased prior to the grant application being approved.

Geographic Focus: Indiana

Date(s) Application is Due: Sep 6

Contact: Ron Harper, President; (765) 454-7298 or (800) 964-0508; ron@cfhoward.org

Internet: http://cfcarroll.org/newsite/grantprogram.shtml

Sponsor: Carroll County Community Foundation
215 West Sycamore Street
Kokomo, IN 46901

Carylon Foundation Grants 1106

The foundation awards general support grants to nonprofits of the Christian, interdenominational, Jewish, and Presbyterian faiths. Higher education institutions, health care organizations, international missions/ministries, medical centers, religious organizations, and temples receive support. Application may be made by submitting a brief letter describing the organization and program. There are no application deadlines.

Amount of Grant: 50 - 50,000 U.S.D

Samples: Rush-Presbyterian-Saint Lukes Medical Center (Chicago, IL)—for educational programs, $25,000; Holocaust Museum (Chicago, IL)—for operating support, $220,500; Weizmann Institute of Science (Palm Beach, FL)—for operating support, $10,000.

Contact: Marcie Mervis, Trustee; (312) 666-7700

Sponsor: Carylon Foundation
2500 W Arthington
Chicago, IL 60612-4108

Cass County Community Foundation Grants 1107

The Cass County Community Foundation (CCCF) assists donors in building enduring sources of charitable assets to promote education, enhance humanity, and advance community development throughout Cass County. The Foundation has the following areas of interest: education; human services; and community development. All applications are reviewed by a committee comprised of CCCF Board members and other volunteers from the community. Non-profit organizations whose projects directly impact the lives of Cass County residents are eligible to apply.

Requirements: The grant application is available online and is also available at the Foundation office. Grant seekers should include their organizational, financial, and project information, and submit seven copies of the application packet to the Foundation. Applicants are encouraged to participate in the free grant writing workshop available each spring.

Restrictions: Public schools, while non-profit, are not 501(c)3 and therefore are not eligible. The Foundation will not consider grants for: existing obligations; services supported by tax dollars; individuals or travel expenses; repeat funding; on-going operating expenses; advocacy; religious purposes or affiliations; or loans or endowments.

Geographic Focus: Indiana

Date(s) Application is Due: Jul 1

Amount of Grant: 5,000 - 14,000 U.S.D

Samples: Galveston, IN, fire department equipment modernization; $5,000 (2010); Royal Center, IN, volunteer fire department emergency generator; $5,000 (2010); Salvation Army, security and alarm system, ADA compliant restrooms for new building; $14,250 (2010).

Contact: Deanna Crispen, Executive Director; (574) 722-2200; fax (574) 753-7501; dcrispen@casscountycf.org or info@casscountycf.org

Internet: http://casscountycf.org/page/Competitive-Grants-Cycle-id-24

Sponsor: Cass County Community Foundation
417 North Street, Suite 102, P.O. Box 441
Logansport, IN 46947

Castle Rock Foundation Grants 1108

The foundation's grants are awarded to promote a better understanding of the free-enterprise system; preserve principles on which democracy was founded, ensure a limited role for government, and protect individual rights; encourage personal leadership; and uphold traditional American values. Types of support include general operating grants and special project grants.

Requirements: U.S. 501(c)3 organizations are eligible.

Restrictions: Grants requests from human service agencies, museums, organizations primarily supported by tax-derived funding, endowments, scientific or medical research, publications or media projects, churches, debt retirement, special events, or individuals will be denied.

Date(s) Application is Due: Mar 15

Amount of Grant: 25,000 - 75,000 U.S.D

Samples: College Fund/United Negro Fund, Fairfax, VA, $40,000 for scholarships; Congressional Medal of Honor Society, Mt. Pleasant, SC, $20,000 for general operating expense; Foundation for Research on economics and the Environment, $50,000 for general operating expense; Fund for American Studies, Washington, DC, $25,000 for general operating expense.

Contact: Sally W. Rippey, Executive Director; (303) 388-1683; fax (303) 388-1684; generalinfo@castlerockfdn.org

Internet: http://www.castlerockfoundation.org

Sponsor: Castle Rock Foundation
4100 E Mississippi Avenue, Suite 1850
Denver, CO 80246

Catherine Holmes Wilkins Foundation Grants 1109

The foundation awards grants in Washington's Puget Sound region to nonprofit organizations in its areas of interest. Funding priorities include medical research and education (medical and academic centers conducting research and training in areas such as cancer, heart disease, and mental illness); physically handicapped and mentally ill (community nonprofit agencies providing direct social services to people with physical disabilities or mental illness); and services for the needy (community-based programs providing immediate support to the needy, with particular emphasis on services for abused women and children). Preference will be given to project support, rather than ongoing operating expenses. Proposals are accepted throughout the year. Grants are awarded quarterly.

Requirements: 501(c)3 nonprofit organizations that operate within or significantly affect the residents of the greater Seattle region (Tacoma to Everett, Seattle, and the Eastside) are eligible.

Restrictions: Grants are not made for multi-year projects, debt retirement, operational deficits, or to individuals or for scholarships.

Geographic Focus: Washington

Date(s) Application is Due: Mar 8; Oct 8

Amount of Grant: 3,000 - 10,000 U.S.D

Contact: Nancy Atkinson, Vice President; (206) 781-3472; OgleFounds@aol.com

Internet: http://fdncenter.org/grantmaker/wilkins

Sponsor: Catherine Holmes Wilkins Foundation
P.O. Box 24565, WA1-501-33-23
Seattle, WA 98124

Catherine Kennedy Home Foundation Grants 1110

The Catherine Kennedy Home Foundation established in 2001, supports nonprofit organizations involved with Aging, centers/services; Christian agencies & churches; developmentally disabled, centers & services; family services; domestic violence; food

distribution, meals on wheels; residential/custodial care, hospices; YM/YWCAs & YM/YWHAs. Giving primarily in Wilmington, North Carolina.
Requirements: Initial approach should be made through a letter to the Foundation requesting an application form. The Foundation will provide you with additional guidelines.
Restrictions: Funding not available outside of North Carolina.
Geographic Focus: North Carolina
Date(s) Application is Due: Feb 28
Amount of Grant: 1,000 - 30,000 U.S.D
Samples: Wilmington Area Rebuilding Ministry, Wilmington, NC, $30,000; Brunswick Senior Resources, Bolivia, NC, $25,000; YMCA of Wilmington, Wilmington, NC, $8,000;
Contact: Garry A. Garris, Director; (910) 452-0611
Sponsor: Catherine Kennedy Home
P.O. Box 4782
Wilmington, NC 28406-1782

CCFF Clinical Fellowships **1111**
Canadian Cystic Fibrosis Foundation (CCFF) Clinical Fellowships are intended for those physicians who have already obtained their residency training and who wish to pursue additional clinical training in Cystic Fibrosis (CF) care. The purpose is to train physicians to become CF specialists, so that they can provide ongoing clinical care to individuals with CF in Canada. This training would include developing competence to allow them to participate in clinical trials. This fellowship is not intended to train individuals from other countries who will not stay in Canada. Up to two highly-ranked competitive Clinical Fellowships will be offered by the Foundation each year. The value of a CCFF Clinical Fellowship is dependent upon academic qualifications, and/or research experience. Award levels are reviewed on an annual basis, and will correspond with prevailing Canadian rates. Additional guidelines are available at: http://www.cysticfibrosis.ca/en/treatment/TrainingAwards_ClinicalFellowships.php.
Requirements: Canadian citizens or permanent residents who have an M.D. degree, have recently completed their clinical training, are exam eligible, and have obtained medical licensure in Canada, are eligible to apply. Clinical Fellowships are tenable in Canada or abroad and it is intended that this experience will directly benefit the Canadian CF community.
Restrictions: Clinical Fellowships will not be awarded to individuals who have not completed residency training.
Geographic Focus: All States, Canada
Date(s) Application is Due: Oct 1
Contact: Grants Administrator; (416) 485-9149; fax (416) 485-0960; info@cysticfibrosis.ca
Internet: http://www.cysticfibrosis.ca/en/treatment/TrainingAwards_ClinicalFellowships.php
Sponsor: Canadian Cystic Fibrosis Foundation
2221 Yonge Street, Suite 601
Toronto, ON M4S 2B4 Canada

CCFF Clinical Project Grants **1112**
Clinical Project grants are intended to bolster the Canadian Cystic Fibrosis Foundation's (CCFF's) commitment to clinical research, and to promote the further development of the Clinical Studies Network. Awarded at the discretion of the leaders of the Medical/Scientific Advisory Committee, in consultation with the Clinical Studies Network, clinical project grants provide a mechanism whereby the most important ideas for clinical studies can be forged into viable protocols, and strategies of urgent clinical relevance can be pursued (i.e. seed money for small, start-up pilot projects). Applications should be made in the form of a letter in which a clear hypothesis is detailed, relevance to CF is demonstrated, and a brief budget is outlined. Applications may be submitted at any time, but the Foundation must be consulted in advance with respect to the availability of funds.
Geographic Focus: All States, Canada
Contact: Grants Administrator; (416) 485-9149; fax (416) 485-0960; info@cysticfibrosis.ca
Internet: http://www.cysticfibrosis.ca/en/research/clinicalprojectgrants.php
Sponsor: Canadian Cystic Fibrosis Foundation
2221 Yonge Street, Suite 601
Toronto, ON M4S 2B4 Canada

CCFF Clinic Incentive Grants **1113**
Clinic Incentive grants are intended to enhance the standard of clinical care available to Canadians with cystic fibrosis, by providing funds to initiate a comprehensive program for cystic fibrosis patient care, research, and teaching; or to strengthen an existing program.
Requirements: Canadian hospitals and/or medical schools are eligible to apply for Clinic Incentive grants.
Restrictions: Basic support for clinical care is the responsibility of provincial governments, and Canadian Cystic Fibrosis Foundation Clinic Incentive grants, which are under the jurisdiction of the cystic fibrosis clinic director, are not intended to supplant public funding.
Geographic Focus: All States, Canada
Date(s) Application is Due: Oct 1
Contact: Grants Administrator; (416) 485-9149; fax (416) 485-0960; info@cysticfibrosis.ca
Internet: http://www.cysticfibrosis.ca/en/treatment/ClinicIncentiveGrants.php
Sponsor: Canadian Cystic Fibrosis Foundation
2221 Yonge Street, Suite 601
Toronto, ON M4S 2B4 Canada

CCFF Fellowships **1114**
A limited number of competitive fellowships are offered by the foundation each year for basic or clinical research training in areas of the biomedical or behavioral sciences pertinent to cystic fibrosis. Fellowships are tenable at approved universities, hospitals, and research institutes in Canada. Initial fellowships are awarded for a two-year period. Fellows may apply for a one-year renewal. No one may receive more than three years of support under a CCFF fellowship. Canadian fellowship applicants of exceptional quality requesting funding to study abroad will be considered. Applicants are expected to demonstrate that comparable training is not available in Canada. Equitable consideration is given to applicants from outside of Canada who intend to return to their own country on completion of the fellowship.
Requirements: Individuals who hold MD or PhD degrees are eligible to apply.
Restrictions: Applicants who have already completed six or more years of postgraduate study or training are not eligible.
Date(s) Application is Due: Oct 1
Contact: Grants Administrator; (416) 485-9149; fax (416) 485-0960; info@cysticfibrosis.ca
Internet: http://www.cysticfibrosis.ca/en/research/TrainingAwards_Fellowships.php
Sponsor: Canadian Cystic Fibrosis Foundation
2221 Yonge Street, Suite 601
Toronto, ON M4S 2B4 Canada

CCFF Research Grants **1115**
Research grants are intended to facilitate the scientific investigation of all aspects of cystic fibrosis. Generally, the Foundation funds research which is carried out in Canada, and the principal investigator must be based at a Canadian institution. While the Foundation will fund a collaboration from outside of Canada, funds must be directed through a Canadian institution and Canadian investigators intending to collaborate with individuals from outside of Canada should contact the Foundation. Applications may be submitted by groups of individuals who plan to collaborate. Grants are awarded for up to three years. Additional guidelines are available at: http://www.cysticfibrosis.ca/en/research/researchgrants.php
Requirements: A principal investigator should hold a recognized, full-time faculty appointment in a relevant discipline at a Canadian university or hospital. Under exceptional circumstances, and at the discretion of the Research Subcommittee, research grant applications from other individuals may be evaluated on a case-by-case basis, with significant emphasis placed on the degree of independence of the applicant, and on the institutional commitments to this individual. Such applications must include a statement from the applicant addressing the issue of salary support during the term of the grant, and the availability of laboratory space; and letters from the Departmental Chair and Dean of Faculty, clarifying the nature and extent of the institutional commitment to the applicant. This same eligibility criterion regarding a faculty appointment applies to any named co-investigator.
Restrictions: Research grants do not provide support for graduate students, postdoctoral fellows, or summer students; construction costs; institutional overheads for laboratory facilities; or purchase of equipment in excess of $10,000.
Geographic Focus: All States, Canada
Date(s) Application is Due: Aug 4; Oct 1
Contact: Grants Administrator; (416) 485-9149; fax (416) 485-0960; info@cysticfibrosis.ca
Internet: http://www.cysticfibrosis.ca/en/research/researchgrants.php
Sponsor: Canadian Cystic Fibrosis Foundation
2221 Yonge Street, Suite 601
Toronto, ON M4S 2B4 Canada

CCFF Scholarships **1116**
Scholarships provide salary support to gifted investigators in Cystic Fibrosis research, offering them the opportunity to develop outstanding cystic fibrosis research programs. Scholarships are held in conjunction with a Canadian Cystic Fibrosis Foundation research grant. Applications are available online.
Requirements: Applications are accepted from candidates who have received their first faculty appointment within the preceding five calendar years.
Geographic Focus: All States, Canada
Date(s) Application is Due: Oct 3
Contact: Grants Administrator; (416) 485-9149; fax (416) 485-0960; info@cysticfibrosis.ca
Internet: http://www.cysticfibrosis.ca/en/research/TrainingAwards_Scholarships.php
Sponsor: Canadian Cystic Fibrosis Foundation
2221 Yonge Street, Suite 601
Toronto, ON M4S 2B4 Canada

CCFF Senior Scientist Research Training Awards **1117**
Senior Scientist Research Training awards provide support to a limited number of established Cystic Fibrosis (CF) investigators by offering them an opportunity to obtain additional training that will enhance their capacity to conduct research directly relevant to CF. This award can be used for sabbatical support for qualified individuals.
Requirements: Applicants must have held a recognized, full-time faculty appointment in a relevant discipline at a Canadian university or hospital for at least six years. In addition, applicants must hold major research grant funding from recognized granting agencies.
Restrictions: Applicants may hold only one Canadian Cystic Fibrosis Foundation personnel award concurrently.
Geographic Focus: All States, Canada
Date(s) Application is Due: Apr 1; Oct 0
Amount of Grant: Up to 30,000 U.S.D
Samples: Dr. Miguel Valvano, The University of Western Ontario, $18,600.
Contact: Grants Administrator; (416) 485-9149; fax (416) 485-0960; info@cysticfibrosis.ca

Internet: http://www.cysticfibrosis.ca/en/research/TrainingAwards_SeniorScientistResearch.php
Sponsor: Canadian Cystic Fibrosis Foundation
2221 Yonge Street, Suite 601
Toronto, ON M4S 2B4 Canada

CCFF Small Conference Grants 1118
Small conference grants are offered to support medical/scientific conferences which focus on subjects of direct relevance to cystic fibrosis. Such grants will be up to a maximum of $C2500 and are intended to supplement other funding sources. The small conference grants will also support inter-clinic exchanges which are used to facilitate the exchange of special expertise between larger, university-based CF clinics and smaller, more remote clinics. Requests of this type will involve fewer individuals, and will not normally exceed $C1000. Applications may be submitted at any time, but the foundation should be consulted in advance with respect to the availability of funds. Grants are on a first-come, first-served basis.
Requirements: CF clinic directors and CCFF-funded investigators are eligible to apply.
Geographic Focus: All States, Canada
Amount of Grant: 1,000 - 2,500 CAD
Samples: Dr. Christine Bear, University of Toronto, Toronto, ON, $2,500— 8th International Conference on AAA Proteins;
Contact: Grants Administrator; (416) 485-9149; fax (416) 485-0960; info@cysticfibrosis.ca
Internet: http://www.cysticfibrosis.ca/en/research/SmallConferenceGrants.php
Sponsor: Canadian Cystic Fibrosis Foundation
2221 Yonge Street, Suite 601
Toronto, ON M4S 2B4 Canada

CCFf Special Travel Grants For Fellows & Students 1119
The Foundation encourages Canadian Cystic Fibrosis Foundation (CCFF) supported Fellows and Students to attend and participate in scientific meetings related to cystic fibrosis. In addition, allowances may be utilized to acquire new knowledge (i.e. visit another lab). Special Travel allowances may be awarded to CCFF-supported Fellows and Students, for each year of their award, upon written application and pending the availability of funds. No CCFF-supported Fellow or Student may receive more than one grant per year. Subject to the availability of funds and under certain circumstances, CCFF-supported Scholars, physicians who are not CF clinic directors but who specialize in the clinical care of adults with cystic fibrosis at Canadian CF clinics, and recipients of the Skelly and Lackey Summer Studentship awards, may also be eligible. Special Travel allowances are intended to cover travel costs not to exceed the lowest applicable advance booking airfare, and reasonable accommodation and meals, up to a $1,800. Additional guidelines are available at: http://www.cysticfibrosis.ca/en/research/TravelAllowancesStudents.php.
Requirements: Applications should be made in the form of a letter, and must be submitted prior to the proposed travel. The letter should indicate the conference or meeting for which the allowance is requested and must include a proposed travel and accommodation budget, along with the supervisor's signed endorsement. Proof of active participation (i.e. abstract acceptance, etc.) is also required.
Restrictions: An itemized breakdown of expenses incurred and the original receipts must be submitted to the Foundation upon return from the conference or meeting, and any unexpended funds returned to the Foundation.
Geographic Focus: All States, Canada
Amount of Grant: Up to 1,800 U.S.D
Contact: Grants Administrator; (416) 485-9149; fax (416) 485-0960; ResearchPrograms@cysticfibrosis.ca or info@cysticfibrosis.ca
Internet: http://www.cysticfibrosis.ca/en/research/TravelAllowancesStudents.php
Sponsor: Canadian Cystic Fibrosis Foundation
2221 Yonge Street, Suite 601
Toronto, ON M4S 2B4 Canada

CCFF Studentships 1120
A limited number of competitive studentships are offered by the foundation each year to highly qualified graduate students who are registered for a higher degree and who are undertaking full-time research training in areas of the biomedical or behavioral sciences relevant to cystic fibrosis. Students are expected to spend at least 75 percent of their time on the research training described in their application.
Requirements: Awards are tenable only at Canadian universities.
Geographic Focus: All States, Canada
Date(s) Application is Due: Oct 1
Contact: Grants Administrator; (416) 485-9149; fax (416) 485-0960; info@cysticfibrosis.ca
Internet: http://www.cysticfibrosis.ca/en/research/TrainingAwards_Studentships.php
Sponsor: Canadian Cystic Fibrosis Foundation
2221 Yonge Street, Suite 601
Toronto, ON M4S 2B4 Canada

CCFF Summer Studentship 1121
Summer studentships are intended to provide support to students engaged in summer research projects in areas of the biomedical or behavioral sciences relevant to cystic fibrosis, under the direction of CF clinic directors or principal investigators. The research project should be attainable within the three-month term of the award.
Requirements: Full-time students pursuing an undergraduate degree in an appropriate discipline are eligible to receive this award. Under certain circumstances, other special academic situations, may be considered for a summer studentship.
Restrictions: Applications must be made by the proposed supervisor on behalf of a student. Applications directly from students cannot be accepted.
Geographic Focus: All States, Canada
Date(s) Application is Due: Feb 1
Contact: Grants Administrator; (416) 485-9149; fax (416) 485-0960; info@cysticfibrosis.ca
Internet: http://www.cysticfibrosis.ca/en/research/TrainingAwards_SummerStudentships.php
Sponsor: Canadian Cystic Fibrosis Foundation
2221 Yonge Street, Suite 601
Toronto, ON M4S 2B4 Canada

CCFF Transplant Center Incentive Grants 1122
The grants are intended to enhance the quality of care available to cystic fibrosis transplant candidates by providing eligible centers with supplementary funding for support of personnel directly involved in the provision of patient services; travel to the annual North American Cystic Fibrosis conference; and administrative costs associated with providing data to the Canadian Cystic Fibrosis Lung Transplant Registry. Grants are renewable on an annual basis.
Requirements: Any Canadian lung transplant center which currently has one or more individuals with cystic fibrosis listed for transplant may apply for a Transplant Center Incentive grant. (
Restrictions: Under no circumstances will funding be provided to more than one transplant centers in the same city.
Geographic Focus: All States, Canada
Date(s) Application is Due: Oct 1
Contact: Grants Administrator; (416) 485-9149; fax (416) 485-0960; info@cysticfibrosis.ca
Internet: http://www.cysticfibrosis.ca/en/treatment/TransplantCentreIncentiveGrants.php
Sponsor: Canadian Cystic Fibrosis Foundation
2221 Yonge Street, Suite 601
Toronto, ON M4S 2B4 Canada

CCFF Travel Supplement Grants For ECFS 1123
The Foundation has a special travel fund to assist a limited number of individuals who plan to play an active role in European Cystic Fibrosis Society (ECFS) conferences. Up to five travel supplement grants for the upcoming conference will be awarded on a competitive basis, the maximum award will be $2,000 per person.
Requirements: The travel supplements will be available to CCFF-funded investigators; Canadian CF clinic directors or clinicians; and Canadian CF clinic coordinators and allied health professionals (applications from clinicians, coordinators and health professionals must be endorsed by the clinic director); individuals requesting the supplement must show evidence of active participation at the conference, i.e. documentation indicating that you will be presenting a paper or abstract, or participating in a workshop; application is by letter to the Canadian CF Foundation, with a brief budget included; and frequency of application from any particular individual should be reasonable (i.e. once every four years).
Restrictions: An itemized breakdown of expenses incurred and the original receipts must be submitted to the Foundation upon return from the ECFS conference, and any unexpended funds returned to the Foundation.
Geographic Focus: All States, Canada
Date(s) Application is Due: Apr 1
Amount of Grant: Up to 2,000 CAD
Contact: Grants Administrator; (416) 485-9149; fax (416) 485-0960; ResearchPrograms@cysticfibrosis.ca or info@cysticfibrosis.ca
Internet: http://www.cysticfibrosis.ca/en/research/TravelSupplementECFs.php
Sponsor: Canadian Cystic Fibrosis Foundation
2221 Yonge Street, Suite 601
Toronto, ON M4S 2B4 Canada

CCFF Visiting Allied Health Professional Awards 1124
A limited number of awards may be made to allied health professionals from abroad who are invited to engage in Cystic Fibrosis clinical observation or activity at a Canadian institution; or Canadian allied health professionals who wish to visit another clinic in Canada or abroad. It is intended that this experience as a Canadian Cystic Fibrosis Foundation Visiting Allied Health Professional will, in some way, benefit Canadian Cystic Fibrosis clinical care. Applications may be submitted at any time, but the Foundation should be consulted in advance with respect to the availability of funds.
Requirements: To qualify, an allied health professional must be associated with a recognized Cystic Fibrosis clinic, and be an active member in Cystic Fibrosis clinical care. Applications should be made in the form of a letter, with a brief budget included and accompanied by a supporting letter from the head of the appropriate department of the applicant's institution. A supporting letter signed by the department head of the host institution should also be provided.
Contact: Grants Administrator; (416) 485-9149; fax (416) 485-0960; info@cysticfibrosis.ca
Internet: http://www.cysticfibrosis.ca/en/treatment/TrainingAwards.php
Sponsor: Canadian Cystic Fibrosis Foundation
2221 Yonge Street, Suite 601
Toronto, ON M4S 2B4 Canada

CCFF Visiting Clinician Awards 1125
Awards may be made to clinicians from abroad who are invited to engage in Cystic Fibrosis clinical observation or activity at a Canadian institution; or Canadian clinicians who wish to visit another clinic in Canada or abroad. Applications may be submitted at any time, but the Foundation should be consulted in advance with respect to the availability of funds.

Requirements: To qualify, a clinician must be associated with a recognized Cystic Fibrosis clinic, and be an active member in Cystic Fibrosis clinical care. It is intended that this experience as a Canadian Cystic Fibrosis Foundation Visiting Clinician will, in some way, benefit Canadian Cystic Fibrosis clinical care. Applications should be made in the form of a letter, with a brief budget included and accompanied by a supporting letter from the head of the appropriate department of the applicant's institution. A supporting letter signed by the department head of the host institution should also be provided.
Contact: Grants Administrator; (416) 485-9149; fax (416) 485-0960; info@cysticfibrosis.ca
Internet: http://www.cysticfibrosis.ca/en/treatment/TrainingAwards.php
Sponsor: Canadian Cystic Fibrosis Foundation
2221 Yonge Street, Suite 601
Toronto, ON M4S 2B4 Canada

CCFF Visiting Scientist Awards **1126**
Awards are made to senior investigators from abroad who are invited to engage in CF research at a Canadian institution; or junior and senior Canadian investigators who wish to work in another laboratory in Canada or abroad. It is intended that this experience as a CCFF visiting scientist will, in some way, benefit the Canadian CF research effort. Applications may be submitted at any time, but the foundation should be consulted in advance with respect to the availability of funds.
Requirements: Senior investigators from abroad or junior and senior Canadian investigators who wish to work in another laboratory in Canada or abroad are eligible to apply.
Contact: Grants Administrator; (416) 485-9149; fax (416) 485-0960; info@cysticfibrosis.ca
Internet: http://www.cysticfibrosis.ca/en/research/TrainingAwards_VisitingScientist.php
Sponsor: Canadian Cystic Fibrosis Foundation
2221 Yonge Street, Suite 601
Toronto, ON M4S 2B4 Canada

CCHD Community Development Grants **1127**
CCHD is committed to supporting groups of low-income individuals as they work to break the cycle of poverty and improve their communities. By helping the poor to participate in the decisions and actions that affect their lives, CCHD empowers them to move beyond poverty. The organization's efforts should directly benefit a relatively large number of people rather than a few individuals. The organization should generate cooperation among and within diverse groups in the interest of a more integrated and mutually understanding society. An applicant organization seeking seed or matching monies will also be considered. (If requesting these monies, applicants should present positive documentation that other public and/or private sources will commit their funds to support the organization's efforts.)
Requirements: Only organizations that are not now receiving an organizing grant are required to submit an Eligibility Quiz. Eligibility Quizzes are accepted on a rolling basis between September 1st and November 1st. The sponsor recommends submitting your Eligibility Quiz well in advance of the November 1st deadline, to help with processing and to give eligible applicants more time with the next step in application. To be eligible for CCHD funds, an organization must satisfy ALL the following criteria and guidelines: the activity for which funding is requested must conform to the moral and social teachings of the Catholic Church; the applicant organization must demonstrate both the intention and capacity to effectively work toward the elimination of the root causes of poverty and to enact institutional change; the organization's efforts must benefit people living in poverty. At least 50 percent of those benefiting from the organization's efforts must be people experiencing poverty; people living in poverty must have the dominant voice in the organization (at least 50 percent of those who plan, implement and make policy, hire and fire staff should be persons who are involuntarily poor; the organization should demonstrate ongoing leadership development because it is considered essential to the strength, depth and sustainability of the organization; the organization should demonstrate a clear vision for development of financial capacity that might include membership dues, grassroots fundraising, foundation and/or corporate support; and the organization must be fully nonpartisan when engaging in political activities. Organizations engaged in partisan political activity are not eligible. See the website for further explanations.
Restrictions: The following general classifications do not meet CCHD criteria and/or guidelines for community organizing grants: organizations with primary focus on direct service (e.g., daycare centers, recreation programs, community centers, scholarships, subsidies, counseling programs, referral services, cultural enrichment programs, direct clinical services, emergency shelters and other services, refugee resettlement programs, etc.); advocacy efforts where only staff, a few individuals or middle to upper-income people are speaking for a particular low income constituency without the direct involvement and leadership of low income individuals; organizations controlled by governmental (federal, state, local), educational, or ecclesiastical bodies; research projects, surveys, planning and feasibility studies, etc.; individually owned, for-profit businesses; or organizations that would use CCHD money for re-granting purposes or to fund other organizations.
Date(s) Application is Due: Nov 1
Amount of Grant: 25,000 - 50,000 U.S.D
Contact: Ralph McCloud, Director; (202) 541-3367; rmccloud@usccb.org
Internet: http://www.usccb.org/about/catholic-campaign-for-human-development/grants/community-development-grants-program/index.cfm
Sponsor: United States Conference of Catholic Bishops
3211 Fourth Street, NE
Washington, DC 20017-1194

CDC-Hubert Global Health Fellowship **1128**
The CDC-Hubert fellowship provides an opportunity for third- and fourth-year medical and veterinary students to gain public health experience in an international setting. Hubert fellows spend six to twelve weeks in a developing country working on a priority health problem in conjunction with Center of Disease Control (CDC) agency staff. Most schools award course credit for the fellowship. Hubert fellows receive a $4,000 stipend to help pay for travel and living expenses while in-country (any cost above $4,000 is the responsibility of the fellow). Deadline dates may vary from year to year. Interested applicants are encouraged to check the website for current information on how and when to apply for the fellowship. Further information on field assignment choices can be found at the grant website. Click on "More about the Program" on the left side of the screen. Examples of past assignments are: "Assessing Barriers to Use of Rapid HIV Tests" (China); "Assessing Health Information Systems for Border Populations" (Thailand); "China National Center for AIDS/STD Prevention and Control" (China); "Descriptive Analysis of HIV/AIDS" (Nigeria); "Field Development and Testing of a Manual to Measure the Burden of Severe Influenza-Related Lower Respiratory Infection" (Thailand); "Pathogen Discovery Focused on Lyssaviruses and Other Emerging Zoonotic Pathogens" and "Knowledge, Attitude and Practice (KAP) Surveys and Risk Assessment within Population at Higher Risk of Exposure to Bats" (Democratic Republic of the Congo, Guatemala or Peru); "Causes of Community-Acquired Pneumonia" (rural Thailand); "Surveillance System for Acute Respiratory Infection" (Egypt); "Hospital Acquired Infections" (Egypt); "Influenza Sentinel Surveillance" (Kenya); "Population-Based Surveillance for Emerging Infectious Diseases" (Kibera Informal Housing Settlement in Nairobi, Kenya); "Population-Based Surveillance for Emerging Infectious Diseases" (rural Lwak Area of Kisumu, Kenya); "A Study of Absenteeism in the Kenyan Health Workforce" (Kenya); "Validating Cortrimoxazole Coverage among People Living with HIV/AIDS" (Ethiopia); and "Surveillance for Influenza Viruses in Domestic Animals" (Kenya). The CDC-Hubert Global Health Fellowship is endowed by the O.C. Hubert Charitable Trust.
Requirements: Applicants must be: medical or veterinary students; in their 2nd or 3rd year when they apply; enrolled in a school accredited by the Liaison Committee on Medical Education, the American Osteopathic Association, or the American Veterinary Medical Association; covered by medical insurance during the fellowship; a U.S. citizen or permanent resident; and able to attend the fellowship orientation (unless out of the country on the field assignment). Orientation travel expenses are paid by the CDC-Hubert Global Health Fellowship.
Date(s) Application is Due: Jan 17
Amount of Grant: 4,000 U.S.D
Contact: Catherine Folowoshele, M.P.H., Program Coordinator; 404-498-6148; fax 404-498-6085; HubertFellowship@CDCFoundation.org
Internet: http://www.cdc.gov/HubertFellowship/
Sponsor: Centers for Disease Control and Prevention
1600 Clifton Road
Atlanta, GA 30333

CDC African American Communities Coalition Development Project Grants **1129**
The purpose of this program is to provide assistance to local, regional and, national nonprofit organizations to: develop and implement effective community-based Human Immunodeficiency Virus (HIV) prevention programs related to achieving national goals; promote coordination for primary and secondary HIV prevention efforts among community organizations, HIV education/prevention service agencies, and public organizations including local and state health departments and substance abuse agencies; and evaluate the HIV prevention programs for which support is provided
Requirements: Nongovernmental public and nonprofit private entities are eligible.
Amount of Grant: 62,000 - 100,000,000 U.S.D
Contact: Allyn Nakashima, National Center for HIV, STD, and TB Prevention, Centers for Disease Control and Prevention, Departm; (404) 639-0900
Internet: http://www.cdc.gov
Sponsor: Centers for Disease Control and Prevention
1600 Clifton Road
Atlanta, GA 30333

CDC Antimicrobial Resistance in Hospital-Acquired Infections Among Intensive Care Unit Patients **1130**
CDC offers funds to conduct studies on antimicrobial resistance in hospital-acquired infections among intensive care unit patients. Grants will be awarded to conduct both economic and outcome evaluations on the extent of the problem, as well as on factors that lead to different outcomes.
Requirements: Applications may be submitted by public and private nonprofit organizations and by governments and their agencies.
Amount of Grant: 60,000 - 100,000 U.S.D
Contact: Nealean Austin, Grants Management Branch; (770) 488-2700
Internet: http://edocket.access.gpo.gov/2004/04-9308.htm
Sponsor: Centers for Disease Control and Prevention
1600 Clifton Road
Atlanta, GA 30333

CDC Cancer Prevention and Control Programs **1131**
The purpose of the program is to improve and to promote health among at-risk cancer populations and to reduce cancer morbidity and mortality.
Requirements: Assistance will be provided only to the organizations listed. The Healthcare Association of New York to develop an integrated model for the delivery of comprehensive breast cancer services; the Health Choice Network, Miami/Dade County, Florida to administer the Jesse Trice Cancer Prevention Project; the East Tennessee State University, Cancer Prevention Research Center, James H. Quillen College of Medicine to address cancer care in the rural Appalachian region; the University of Rhode Island, Cancer Prevention Research Center to provide interactive interventions to at-risk populations; the Sisters of Charity Health Care System, to ensure that patients have access to early detection of gastrointestinal cancers; and Marin County, California to evaluate high incidence of breast cancer in the San Francisco Bay Area.

Geographic Focus: California, Florida, New York, Rhode Island, Tennessee
Amount of Grant: Up to 41,000,000 U.S.D
Contact: Grants Administrator; (770) 488-4751; cancerinfo@cdc.gov
Internet: http://www.cdc.gov/cancer
Sponsor: Centers for Disease Control and Prevention
1600 Clifton Road
Atlanta, GA 30333

CDC Collegiate Leaders in Environmental Health Internships **1132**

The Collegiate Leaders in Environmental Health (CLEH) is a 10-week summer environmental internship (June through August) for undergraduate students who are majoring in environmental studies or environmental, physical, biological, chemical, and/or social sciences; and who are passionate about the environment, interested in human health, and curious about how the two are linked. A healthy environment should be capable of sustaining a healthy population; however, with 6.7 billion people on the earth, there is a need for more sustainable interactions between humans and the environment. Environmental issues such as overpopulation, air pollution, food shortage, natural disasters, water contamination, and exposure to toxic substances provide challenges to human health. Human influence on the environment is the main focus of environmental studies; however there is a growing need to evaluate the effect that the environment has on human populations. Harmful environments can increase the risk of many health conditions: asthma, heart disease, cancer, neurological disease, infections, endocrine dysfunction, injuries, and more. Healthy environments, on the other hand, can promote good health in many ways — protecting people from toxic exposures, providing safe water and clean air, and encouraging healthy behaviors such as outdoor recreation. The link between the environment and health is aptly referred to as "environmental health." Environmental health is the discipline that focuses on the interrelationships between people and their environment, promotes human health and well-being, and fosters safe and healthy living. This branch of public health is concerned with all aspects of the natural and built environment that may affect human health. CLEH Interns will be placed in environmental health programs at the U.S. Centers for Disease Control and Prevention (CDC)'s National Center for Environmental Health and the Agency for Toxic Substances and Disease Registry (NCEH/ATSDR) at CDC's Chamblee Campus. Over the course of the summer, interns will be exposed to a broad overview of environmental public health issues at the federal level, participate in environmental health projects, interact with federal officials and scientists, and visit important environmental health sites in and around Atlanta. Other activities include "brown-bag" lunches with CDC staff, as well as attending lectures from prominent environmental health leaders in the Atlanta area. In addition, interns will be able to attend the many seminars offered by CDC during the summer. CLEH interns receive a weekly stipend of $500 for living expenses. All interns are responsible for the cost of travel to and from Atlanta and for housing during the internship. Information about student housing options at local universities can be found at the internship website. Links to application materials and guidelines are provided at the website along with specific instructions for submitting application materials via email, fax, or U.S. mail. Annual deadline dates may vary; applicants are encouraged to check the website or contact program staff to verify current deadline dates.
Requirements: Applicants must: be U.S. citizens or permanent residents with a green card; be enrolled full time at a college or university as rising juniors or seniors by the fall semester; have achieved a minimum cumulative GPA of 3.0 on a 4.0 scale; and major or have coursework concentration in environmental studies and/or environmental, physical, biological, chemical, or social sciences (if applicants are not majoring in one of these areas they can make the case for why their major or interests are applicable to this internship in essay question #2 of the application materials.) All participants are considered guest researchers and are subject to CDC regulations governing visiting scientists, engineers, and other professionals. As a guest researcher, each participant is responsible for payment of income taxes and is advised to become familiar with the relevant sections of the current tax codes. Health-insurance is required for all participants.
Restrictions: Participation in the program is contingent upon the individual's ability to obtain the proper security clearance. All applicants will be subject to a criminal records check and other background investigations conducted by the U.S. Government. These inquiries are conducted to develop information to assess various factors about the applicant, including reliability, trustworthiness, honesty, integrity, character, conduct and loyalty to the United States. Deferrals of the internship will not be allowed. Seniors graduating in Spring 2011 will not be accepted to this program.
Date(s) Application is Due: Feb 2
Amount of Grant: 5,000 U.S.D
Contact: Lt. Cory Moore, M.P.H., Environmental Health Specialist; (770) 488-0593; fax (404) 929-2820; CLEH@cdc.gov or cory.moore@cdc.hhs.gov
Internet: http://www.cdc.gov/nceh/cleh/
Sponsor: Centers for Disease Control and Prevention
1600 Clifton Road
Atlanta, GA 30333

CDC Cooperative Agreement for Continuing Enhanced National Surveillance for Prion Diseases in the United States **1133**

The purpose of the funding is to continue an active surveillance program similar to that conducted by the National Prion Disease Pathology Surveillance Center since 1997 to monitor the occurrence of potentially emerging human TSEs in the United States. This program addresses the Healthy People 2010 focus area(s) of Immunizations and Infectious Diseases. Eligible applicants should have experience in conducting prion disease surveillance and must be capable of fulfilling the ongoing needs for enhancing such surveillance at the national level.
Requirements: Eligible applicants include: public nonprofit organizations; private nonprofit organizations; small, minority, women-owned businesses; universities; colleges; research institutions; hospitals; community-based organizations; faith-based organizations; federally recognized Indian tribal governments; Indian tribes; Indian tribal organizations; state and local governments or their Bona Fide Agents (this includes the District of Columbia, the Commonwealth of Puerto Rico, the Virgin Islands, the Commonwealth of the Northern Marianna Islands, American Samoa, Guam, the Federated States of Micronesia, the Republic of the Marshall Islands, and the Republic of Palau); and political subdivisions of States (in consultation with States).
Restrictions: Recipients may: not use funds for research; not use funds for clinical care; only expend funds for reasonable program purposes, including personnel, travel, supplies, and services, such as contractual. Awardees may not generally use HHS/CDC/ATSDR funding for the purchase of furniture or equipment (any such proposed spending must be identified in the budget). The direct and primary recipient in a cooperative agreement program must perform a substantial role in carrying out project objectives and not merely serve as a conduit for an award to another party or provider who is ineligible.
Date(s) Application is Due: Jun 30
Amount of Grant: 100,000,000 U.S.D
Contact: Mattie Jackson, Grants Management Specialist; (770) 488-2696; MIJ3@cdc.gov
Internet: http://www.cdc.gov/od/pgo/funding/FOAs.htm
Sponsor: Centers for Disease Control and Prevention
1600 Clifton Road
Atlanta, GA 30333

CDC Cooperative Agreement for Partnership to Enhance Public Health Informatics **1134**

The purpose of the program is to strengthen the breadth and depth of the public health workforce by providing training in public health informatics and encouraging new explorations of innovations at the intersection of public health and health informatics. CDC wants to partner with a national organization that is established as a leader in clinical and health informatics and with established tutorials and other training in informatics. This program addresses the Healthy People 2010 focus area of Public Health Infrastructure and the CDC's Healthy Protection Goal of People Prepared for Emerging Health Threats to develop and measure how well public health systems perform in critical times.
Requirements: Assistance will be provided only to national non-profit organizations (a national non-profit, tax-exempt Section 501(c)3 agency. Eligible national organizations must have a web-based training and in person tutorials in clinical and public health informatics. These resources should have been established, fully operational, and deemed successful before application is submitted. Additionally, applicants are required to have an active national and international professional membership in informatics. Limiting assistance to national non-profit organizations is necessary to produce the maximum possible enhancement, given available resources.
Restrictions: The following restrictions apply: recipients may not use funds for research; recipients may not use funds for clinical care; recipients may only expend funds for reasonable program purposes, including personnel, travel, supplies, and services, such as contractual; awardees may not generally use HHS/CDC/ATSDR funding for the purchase of furniture or equipment; the direct and primary recipient in a cooperative agreement program must perform a substantial role in carrying out project objectives and not merely serve as a conduit for an award to another party or provider who is ineligible; reimbursement of pre-award costs is not allowed; and these funds may not be used to set up or support lobbying by interest/advocacy groups.
Date(s) Application is Due: Jul 2
Contact: Kaleema McLean, Grants Specialist; (770) 488-2742; kmclean@cdc.gov
Internet: http://www.cdc.gov/od/pgo/funding/FOAs.htm
Sponsor: Centers for Disease Control and Prevention
1600 Clifton Road
Atlanta, GA 30333

CDC Cooperative Agreement for the Development, Operation, and Evaluation of an Entertainment Education Program **1135**

The purpose of this program is raise awareness and behavioral change concerning public health issues through the use of the television media as means of accessing the television viewing public by providing: accurate public health information and public health issues, as well as accurate depictions of healthy living at all stages of life, to entertainment industry leadership for possible inclusion in television story lines; assistance with public health outreach, including providing additional information concerning public health topics depicted on television story lines, and assistance with public service announcements and informational short videos to be aired in conjunction with drama presentations; and evaluation of public health television story lines and the effect on the viewing public.
Requirements: Eligible applicants include: public nonprofit organizations; private nonprofit organizations; universities; colleges; research institutions; federally recognized Indian tribal governments; Indian tribes; Indian tribal organizations; state and local governments or their Bona Fide Agents (this includes the District of Columbia, the Commonwealth of Puerto Rico, the Virgin Islands, the Commonwealth of the Northern Marianna Islands, American Samoa, Guam, the Federated States of Micronesia, the Republic of the Marshall Islands, and the Republic of Palau); and political subdivisions of States (in consultation with States).
Date(s) Application is Due: Jun 7
Contact: Sharon Robertson, Grants Officer; (770) 488-2748; SRobertson1@cdc.gov
Internet: http://www.cdc.gov/od/pgo/funding/FOAs.htm
Sponsor: Centers for Disease Control and Prevention
1600 Clifton Road
Atlanta, GA 30333

CDC Core Capacity Program Grants 1136
This program seeks to develop basic cardiovascular health promotions and strategies at the state level. It should include statewide partnerships and program coordination among health agencies aimed at primary and secondary heart disease prevention.
Requirements: State health agencies, the District of Columbia, Puerto Rico, and U.S. territories are eligible to apply.
Restrictions: Funds may not be used to supplant state or local funds, to provide inpatient care or personal health services, nor to support construction or renovation of facilities.
Amount of Grant: 250,000 - 500,000 U.S.D
Contact: Nealean Austin, Team Lead, Branch B, Procurement and Grants Office, Centers for Disease Control and Prevention; (770) 488-2754
Internet: http://www.cdc.gov
Sponsor: Centers for Disease Control and Prevention
1600 Clifton Road
Atlanta, GA 30333

CDC David J. Sencer Museum Adult Group Tour 1137
The David J. Sencer CDC (Centers for Disease Control and Prevention) Museum's adult group tour (college, professional, other) is available for groups ranging from ten to thirty people (groups smaller than ten people are asked to visit the museum for a self-guided tour). Upon arrival to the Museum, the group will be greeted and oriented to CDC on the Global Symphony platform. The Global Symphony is a media installation featuring four, three–minute stories that describe in depth CDC's contributions to the elimination of polio, the investigation of Legionnaire's disease, the battle to stem the rise of obesity in the United States, and the study of how humans, animals, and the environment interact in the spread of Ebola. Following the orientation, the group will tour the current temporary exhibit and the History of CDC exhibit. The group's tour coordinator may also request a CDC speaker. Speakers are CDC employees who volunteer their time, so speaker availability is not guaranteed. Topics vary and depend on speaker availability. Following the guided tour, visitors will have the opportunity to explore the exhibit area independently. Tour requests may be submitted online at the program website (given above). A complete list of attendees' names must also be submitted via the given fax number or first listed email address. Please note the tour is not scheduled until the group's tour coordinator receives a confirmation email. Tour coordinators are encouraged to visit the program website for more information about current exhibits. The David J. Sencer CDC Museum is a Smithsonian Institution Affiliate, which provides access to unique programming, content and expertise, and provides a network and forum for the CDC Museum to showcase itself on a national stage and at a national level.
Requirements: Tour coordinators are expected to review driving directions and parking instructions sent in confirmation email. Each group member must bring a government-issued photo ID. Non-U.S. citizens must bring a passport as ID.
Contact: Judy Gantt, Curator; (404) 639-0830 or (404) 639-0831; fax (404) 639-0834; museum@cdc.gov or judy.gantt@cdc.hhs.gov
Internet: http://www.cdc.gov/museum/tours/adult.htm
Sponsor: Centers for Disease Control and Prevention
1600 Clifton Road
Atlanta, GA 30333

CDC David J. Sencer Museum Teacher Professional Development Workshops 1138
The David J. Sencer CDC (Centers of Disease Control and Prevention) Museum offers free professional-development workshops designed to support middle- and secondary-school science education and the incorporation of public health concepts into existing classroom curriculum. These workshops are a result of museum partnerships with external educational organizations, and divisions and offices within CDC. Currently being offered are: "Teach Epidemiology," a five-day workshop for high-school science, health, math, and social studies teachers on teaching epidemiology in the high-school classroom; and "Med Myst: Teaching About the World of Pathogens Through Web Adventures," a two-day workshop for middle-school science teachers who are looking for ways to teach the topics of infectious diseases, pathogens, and the scientific method in an exciting and relevant context. Professional Learning Unit credits are available for attending these workshops. Teachers are encouraged to visit the program website for detailed information on workshop content, presenter qualifications, and lunch arrangements. Application information will be posted at the website in December. The David J. Sencer CDC Museum is a Smithsonian Institution Affiliate, which provides access to unique programming, content and expertise, and provides a network and forum for the CDC Museum to showcase itself on a national stage and at a national level.
Requirements: All workshops require an application. Participants are expected to arrive on time each day, actively participate, complete workshop assignments and remain in the classroom for the duration of the workshop. PLU credits will only be awarded to participants who meet these expectations.
Restrictions: Teachers who fail to attend a workshop after accepting a slot forfeit their rights to attend future workshops.
Contact: Judy Gantt, Curator; (404) 639-0830 or (404) 639-0831; fax (404) 639-0834; museum@cdc.gov or judy.gantt@cdc.hhs.gov
Internet: http://www.cdc.gov/museum/teachers.htm
Sponsor: Centers for Disease Control and Prevention
1600 Clifton Road
Atlanta, GA 30333

CDC Disabilities Prevention Demonstration-Epidemiology Project Grants 1139
The objectives of this program are to provide a national focus for the prevention of disabilities in targeted disability groups: developmental disabilities, injury disabilities from health and spinal cord trauma, selected adult chronic conditions, and secondary disabilities in persons with physical disabilities; to build capacity at the state and community level to coordinate disabilities prevention activities and establish surveillance; to employ epidemiological methods to set priorities and target interventions; and to quantify and conduct programs to prevent secondary conditions in persons with primary disabilities. For demonstration/epidemiology projects, grant support may be used to implement and evaluate specified project activities to identify and quantify preventable secondary disabilities and measure the effectiveness and costs of preventive interventions. Funds may be used to support personnel services, equipment, supplies, travel, and services directly related to project activities. Projects will be supported for two to three years depending on availability of funds. Contact the office for deadline dates.
Requirements: Eligible applicants are public and private nonprofit entities, including disabilities services organizations (such as independent living centers), local health departments, other local governmental agencies including local organizations, units of state agencies, voluntary agencies, universities, colleges, medical facilities, research institutions, and federally recognized Native American tribal governments.
Restrictions: Project funds may not be used to supplant state or local funds available for disabilities prevention or for construction costs, or to lease or purchase facilities or space.
Amount of Grant: 130,000 - 460,000 U.S.D
Contact: Donald Betts, Program Contact; (404) 498-3957; dib3@cdc.gov
Internet: http://www.cdc.gov
Sponsor: Centers for Disease Control and Prevention
1600 Clifton Road
Atlanta, GA 30333

CDC Disease Detective Camp 1140
The Centers for Disease Control and Prevention (CDC)'s Disease Detective Camp (DDC) is an educational program started by CDC's David J. Sencer Museum in 2005 as a mechanism for developing an academic, public-health-day-camp curriculum for state and county health departments. The DDC camp is offered to upcoming high school juniors and seniors and is held at CDC's headquarters in Atlanta, Georgia at no charge. The museum will offer two one-week sessions each summer. The CDC Disease Detective Camp curriculum is based on contextual and situated-cognition learning principles. By learning through hands–on activities and seminars, high school juniors and seniors at the conclusion of the camp will be able to: identify five careers within public health; demonstrate an understanding of basic epidemiology terms; calculate basic epidemiologic rates given an outbreak scenario and data; recognize how infectious and chronic diseases are tracked in the United States; and understand the role of public health law in protecting the public's health in the United States. Over the course of five days, campers will take on the role of disease detectives and learn first–hand how the CDC safeguards the nation's health. Teams will probe a disease outbreak using epidemiologic and laboratory skills and report their findings to a group of CDC scientists. Activities may include short lectures by CDC experts, a mock press conference in the CDC press room, and a look behind the scenes of CDC. The application process for the current DDC is posted the preceding December at the DDC website. Future plans for DDC includes creating a curriculum toolkit for local health departments seeking to offer a similar camp experience. Due to the popularity of this camp, there are more interested students then the program can accommodate. For this reason, interested students must apply (parents are encouraged to help interested students with the application, but not apply for them). Applications will be made available at the DDC website in December or parents and students may call the second phone listed above to have an application mailed. Camp participants are selected in two phases. Applicants are first selected based on their answers to the essay questions on the application and on the recommendation form. The second phase consists of slots assigned through a lottery system. Twenty–seven high–school juniors and seniors will be selected for each camp session. The David J. Sencer CDC Museum is a Smithsonian Institution Affiliate, which provides access to unique programming, content and expertise, and provides a network and forum for the CDC Museum to showcase itself on a national stage and at a national level.
Requirements: The CDC Disease Detective Camp is open to motivated students who will be high-school juniors or seniors. Applicants must be 16 years old by the first day of the camp in order to comply with CDC's laboratory safety requirements. There is no cost associated with attending the CDC Disease Detective Camp, but campers will need to pay for their own lunches. Non–Atlanta residents may apply for the camp, but are responsible for providing their own accommodations and transportation. Campers in past years have stayed with family friends or relatives in Atlanta. Most attendees are from the Atlanta area, but out-of-state students also attend every year. This is a wonderful opportunity to make friends from other schools!
Contact: Judy Gantt, Curator; (404) 639-0830 or (404) 639-0831; fax (404) 639-0834; museum@cdc.gov or judy.gantt@cdc.hhs.gov
Internet: http://www.cdc.gov/museum/camp/index.htm
Sponsor: Centers for Disease Control and Prevention
1600 Clifton Road
Atlanta, GA 30333

CDC Epidemic Intelligence Service Training Grants 1141
The Epidemic Intelligence Service (EIS) is a unique two-year post-graduate training program of service and on-the-job learning for health professionals interested in the practice of applied epidemiology. About 75% of EIS graduates remain in public health at the Centers for Disease Control and Prevention (CDC) or in state or local health departments. Many become leaders in public health throughout the world. EIS officers work in state or local health departments or at CDC including the Center for Global Health (Atlanta, Georgia); the National Center for Chronic Disease Prevention and Health Promotion (Atlanta, Georgia); the National Center for Emerging and Zoonotic Infectious Diseases

(Atlanta, Georgia); the National Center for Environmental Health and Agency for Toxic Substances and Disease Registry (Atlanta, Georgia); the National Center for Health Statistics (Hyattsville, Maryland); the National Center for HIV/AIDS, Viral Hepatitis, STD and TB Prevention (Atlanta, Georgia); the National Center for Immunization and Respiratory Diseases (Atlanta, Georgia); the National Center for Injury Prevention and Control (Atlanta, Georgia); the National Center on Birth Defects and Developmental Disabilities (Atlanta, Georgia); the National Institute for Occupational Safety and Health (Anchorage, Alaska; Atlanta, Georgia; Cincinnati, Ohio; Denver, Colorado; Morgantown, West Virginia; Pittsburgh, Pennsylvania; and Spokane, Washington); the Office of Public Health Preparedness and Response (Atlanta, Georgia); the Office of State, Tribal, Local and Territorial Support (Atlanta, Georgia); and the Office of Surveillance, Epidemiology, and Laboratory Services (Atlanta, Georgia). Although international work may be part of any EIS assignment, no assignment currently is based outside the U.S. The Program is modeled after a traditional medical residency program where much of the education occurs through experiential learning. Thus, the EIS program not only organizes training for trainees (called officers), but a major contribution of the program (and work of the officers) is providing service to the CDC and its public health partners. EIS officers not only conduct routine research activities that are directly related to fulfilling CDC's mission; but they serve as one of CDC's primary resources for responding to urgent or emergent public health problems nationally and around the world. Recent EIS Investigations include: survey of adoptive parents and medical providers on the health status of Haitian orphans entering the United States after the Haiti earthquake (2010); investigation of potential hepatitis B virus and hepatitis C virus transmission associated with vessel conduits used for solid organ transplantation (2010); community-wide assessment of health impact and public health emergency response following extended disruption of drinking water service during an extreme winter freeze in Alabama (2010); investigation of cases of novel influenza A (H1N1) in multiple states; increased incidence of Haemophilus influenzae type b in Minnesota; investigation of infections with Staphylococcus aureus among patients undergoing procedures at a pain management clinic in West Virginia; outbreak of carbapenem-resistant Klebsiella pneumoniae at a long term care facility in Illinois; and mass exposures to a rabies positive bat carcass in Montana. Classroom instruction includes topics such as applied epidemiology, biostatistics, public health surveillance, scientific writing, and working with the media, as well as emerging public health issues. Each EIS class begins with a 1-month course, starting in July each year in Atlanta. As part of the on-the-job training, EIS officers are required to complete the following core activities of learning (CALs): conduct or participate in a field investigation of a potentially serious public health problem; design, conduct, and interpret an epidemiologic analysis on public health data; design, implement, or evaluate a public health surveillance system; write and submit a scientific manuscript for a peer-reviewed journal; write and submit a report to the Morbidity and Mortality Weekly Report (MMWR); present a paper or poster at the annual EIS Conference; give an oral presentation at CDC's Epidemiology Grand Rounds (Tuesday Morning Seminar) or at a national or international scientific meeting; and respond appropriately to written or oral public health inquiries. During the two-year training program, EIS officers are employees of the CDC and receive a salary and benefits. Salaries range from $55,000 to $75,000 per year, based on qualifications and experience. The application period opens in May of each year. Applicants apply online at the program website. The application deadline is midnight of September 1. Required supporting documents must be postmarked September 1. Further details about these may be found at the program website. Accepted applicants must attend an EIS Conference in Atlanta in April, travel expenses to be paid by the EIS. The two-year training program starts in July.
Requirements: The EIS seeks qualified health professionals with an interest in public health and a commitment to public service. Applicants must meet professional degree and licensing requirements, commit to a two-year full-time program starting in July, and be willing to relocate. Physicians (M.D., D.O., M.B.B.S., etc.) must have at least one year of clinical training. U.S citizens and U.S. permanent residents must have an active, unrestricted, U.S. license to practice their clinical specialty. Doctoral-level scientists (Ph.D., Dr.PH., Sc.D., etc.) should have a background in: epidemiology; biostatistics; biological, environmental, social, behavioral, or nutritional sciences; and/or other relevant health science. Other medical professionals such as scientists, nurses, physician assistants, and doctors of pharmacy must have a master of public health (M.P.H.) or equivalent degree. U.S. citizens and U.S. permanent residents must have an active, unrestricted U.S. license to practice their clinical specialty. Veterinarians (D.V.M., V.M.D., etc.) must have either a M.P.H. (or equivalent degree), or relevant public health experience. U.S. citizens and U.S. permanent residents must have an active, unrestricted, U.S. license to practice their clinical specialty. Non-U.S. citizens meeting the above degree requirements are eligible to apply. Non-U.S. citizens must be legal permanent residents or eligible for J-1 status prior to the program's start date. If selected for EIS, CDC will sponsor the J1 visa. Because of the domestic nature of the program, only a limited number of non-U.S. citizens are selected. Applicants are encouraged to contact the EIS office with any questions regarding eligibility.
Restrictions: Citizens of countries that have been determined by the U.S. Secretary of State to be State Sponsors of Terrorism will not be able to get security clearance to work at CDC and should not apply to the EIS Program. Countries currently on this list include: Cuba, Iran, Syria, and Sudan.
Geographic Focus: All States, All Countries
Date(s) Application is Due: Sep 1
Amount of Grant: 110,000 - 150,000 U.S.D
Contact: Douglas Hamilton, MD, PhD, Director; (404) 498-6110; fax (404) 498.6535; EIS@cdc.gov or douglas.hamilton@cdc.hhs.gov
Internet: http://www.cdc.gov/eis/index.html
Sponsor: Centers for Disease Control and Prevention
1600 Clifton Road
Atlanta, GA 30333

CDC Epidemiology and Laboratory Capacity Infectious Diseases Grants 1142
The program grants funds to states and eligible local public health agencies in strengthening basic epidemiologic and laboratory capacity to address infectious disease threats with a focus on notifiable diseases, food-, water-, and vector-borne diseases, vaccine-preventable diseases, and drug-resistant infections. Awards are intended to support activities that enhance the ability of a community to identify and monitor outbreaks of infectious diseases and other infectious disease emergencies, characterize disease determinants, use public health data to develop policy, and assess the effectiveness of these activities. Matching funds are not required. Funding preference will be given to competing continuation applications over new applications. Annual deadline dates may vary; contact program staff for exact dates.
Requirements: Eligible are state health departments, the District of Columbia, Puerto Rico and U.S. territories, and city and county public health agencies.
Restrictions: States that already have ELC projects are ineligible to apply.
Amount of Grant: Up to 300,000 U.S.D
Contact: Nealean Austin, Grants Management Branch; (770) 488-2700
Internet: http://www.cdc.gov/epo/index.htm
Sponsor: Centers for Disease Control and Prevention
1600 Clifton Road
Atlanta, GA 30333

CDC Epidemiology Elective Rotation 1143
The Epidemiology Elective Program (Epi Elective) is a 6 to 8 week rotation sponsored by the Centers of Disease Control and Prevention (CDC) for senior medical and veterinary students in their 3rd and 4th years. Epi Elective gives students an introduction to preventive medicine, public health, and the principles of applied epidemiology while working with CDC epidemiologists to solve real-world public health problems. Participants learn through hands-on experience working on a current public health project, and are mentored by experienced CDC staff. Most schools award course credit. During the elective, participants may: participate in the surveillance of public health problems; analyze public health data for new disease risk factors; become a coauthor on a publication of major health importance; and/or work in the field investigating an outbreak. Selected students are placed in one of the following subject areas within CDC or the Indian Health Service: National Center for Emerging and Zoonotic Infectious Diseases (NCEZID), Atlanta, Georgia; Arctic Investigations Program (part of NCEZID), Anchorage, Alaska; Division of Vector-Borne Diseases (part of NCEZID), Ft. Collins, Colorado; Indian Health Service (IHS), Albuquerque, New Mexico; National Center for Chronic Disease Prevention and Health Promotion (NCCDPHP), Atlanta, Georgia; National Center for Environmental Health / Agency for Toxic Substances and Disease Registry (NCEH/ATSDR), Atlanta, Georgia; National Center for Health Statistics (NCHS), Hyattsville, Maryland; National Center for HIV/AIDS, Viral Hepatitis, STD, and TB Prevention (NCHHSTP), Atlanta, Georgia; National Center for Immunization and Respiratory Diseases (NCIRD), Atlanta, Georgia; National Center for Injury Prevention and Control (NCIPC), Atlanta, Georgia; or the National Center on Birth Defects and Developmental Disabilities (NCBDDD), Atlanta, Georgia. Alternatively students may be placed at the National Institute for Occupational Safety and Health (NIOSH) in: Anchorage, Alaska; Atlanta, Georgia; Cincinnati, Ohio; Denver, Colorado; Morgantown, West Virginia; Pittsburgh, Pennsylvania; or Spokane, Washington. Every effort is made to assign students to their preferred public health subject area. Interested students apply online. Supporting documents are required and must be postmarked by the deadline date. Students should check the Epi Elective web site for further details. Application deadlines may vary from year to year. Students are encouraged to check the Epi Elective website or contact the sponsor to verify current deadline dates.
Requirements: Medical and veterinary students in their 3rd year apply for a 6 to 8 week rotation in the fall or spring semester of their 4th year. The rotations must occur between June and December or between January and May. Students must be U.S. citizens or permanent residents and be enrolled in a school accredited by one of the following: the Liaison Committee on Medical Education; the American Osteopathic Association; or the American Veterinary Medical Association. Applicants selected for the program must pay for round-trip transportation to their assigned CDC facility and for living expenses during the elective (travel and living expenses related to field investigations are paid by CDC.)
Date(s) Application is Due: Mar 30; May 30
Contact: Renee Amos, CNI; (404) 498-6152 or (800) 232-4636; fax (404) 498-6085; renee.douglas@cdc.hhs.gov or EpiElective@cdc.gov
Internet: http://www.cdc.gov/epielective/index.html
Sponsor: Centers for Disease Control and Prevention
1600 Clifton Road
Atlanta, GA 30333

CDC Evaluation of the Use of Rapid Testing For Influenza in Outpatient Medical Settings 1144
The purpose of this project is to evaluate how rapid tests for influenza are being implemented and used in clinical practice in outpatient medical settings such as community clinics, solo and group practice physician offices, and hospital emergency rooms across the United States. This evaluation will include a determination of the scope of rapid influenza test use, the types of tests in use and how they are selected, the personnel performing testing, the extent to which good laboratory practices and testing guidelines are being followed, how results are reported and interpreted, how results are used for patient care and antiviral and antibiotic prescribing practices, and the presence of linkages between these outpatient settings and the public health system. Additionally, this project may identify potential opportunities to provide guidance to assist sites in making decisions on the appropriate use of these tests and ways to enhance the connectivity with the public health system. Connectivity with public health is

especially important in light of the possibility of an influenza pandemic. This project will identify and evaluate practices used in outpatient settings related to processes such as specimen collection, testing, reporting and referral for influenza.
Requirements: Eligible applicants include: public nonprofit organizations; private nonprofit organizations; small, minority, women-owned businesses; universities; colleges; research institutions; hospitals; community-based organizations; faith-based organizations; federally recognized Indian tribal governments; Indian tribes; Indian tribal organizations; state and local governments or their Bona Fide Agents (this includes the District of Columbia, the Commonwealth of Puerto Rico, the Virgin Islands, the Commonwealth of the Northern Marianna Islands, American Samoa, Guam, the Federated States of Micronesia, the Republic of the Marshall Islands, and the Republic of Palau); and political subdivisions of States (in consultation with States).
Restrictions: Recipients may: not use funds for research; not use funds for clinical care; only expend funds for reasonable program purposes, including personnel, travel, supplies, and services, such as contractual. Awardees may not generally use HHS/CDC/ATSDR funding for the purchase of furniture or equipment (any such proposed spending must be identified in the budget). The direct and primary recipient in a cooperative agreement program must perform a substantial role in carrying out project objectives and not merely serve as a conduit for an award to another party or provider who is ineligible. Reimbursement of pre-award costs is not allowed. Reimbursement of construction costs is not allowed.
Date(s) Application is Due: Aug 6
Amount of Grant: 200,000 U.S.D
Contact: Yolanda Sledge, Grants Management Specialist; (770) 488-2787; yiso@cdc.gov
Internet: http://www.cdc.gov/od/pgo/funding/FOAs.htm
Sponsor: Centers for Disease Control and Prevention
1600 Clifton Road
Atlanta, GA 30333

CDC Evidence-Based Laboratory Medicine: Quality/Performance Measure Evaluation 1145
The purpose of the program is to evaluate clinical laboratory practice by identifying evidence-based laboratory medicine quality/performance measures associated with the pre- and post-analytic stages of the total testing process, and to identify and address gaps and opportunities for improvement consistent with national health care priorities to improve public health. The primary objectives are to: provide an evidence base which systematically identifies and defines important gaps in laboratory medicine quality related to individuals or populations and/or patient safety health outcomes with demonstrated impacts (i.e., clinical and/or economic); identify, develop, and define laboratory medicine performance/quality measures that can be broadly implemented to evaluate performance associated with patient outcomes; and identify interventions effective in improving performance.
Requirements: Eligible applicants include: public nonprofit organizations; private nonprofit organizations; small, minority, women-owned businesses; universities; colleges; research institutions; hospitals; community-based organizations; faith-based organizations; federally recognized Indian tribal governments; Indian tribes; Indian tribal organizations; state and local governments or their Bona Fide Agents (this includes the District of Columbia, the Commonwealth of Puerto Rico, the Virgin Islands, the Commonwealth of the Northern Marianna Islands, American Samoa, Guam, the Federated States of Micronesia, the Republic of the Marshall Islands, and the Republic of Palau); and political subdivisions of States (in consultation with States).
Restrictions: Recipients may: not use funds for research; not use funds for clinical care; only expend funds for reasonable program purposes, including personnel, travel, supplies, and services, such as contractual. Awardees may not generally use HHS/CDC/ATSDR funding for the purchase of furniture or equipment (any such proposed spending must be identified in the budget). The direct and primary recipient in a cooperative agreement program must perform a substantial role in carrying out project objectives and not merely serve as a conduit for an award to another party or provider who is ineligible. Reimbursement of pre-award costs is not allowed.
Date(s) Application is Due: Jul 17
Amount of Grant: 100,000 - 500,000 U.S.D
Contact: Yolanda Sledge, Grants Management Specialist; (770) 488-2787; yiso@cdc.gov
Internet: http://www.cdc.gov/od/pgo/funding/FOAs.htm
Sponsor: Centers for Disease Control and Prevention
1600 Clifton Road
Atlanta, GA 30333

CDC Experience Applied Epidemiology Fellowships 1146
The goal of the CDC Experience Applied Epidemiology Fellowship is to increase the number of physicians with a population health perspective by providing medical students with an understanding of applied epidemiology, the role of epidemiology in medicine and health, and the role of physicians in the public health system. The fellowship is held at the Centers of Disease Control and Prevention (CDC) headquarters in Atlanta, Georgia. Fellows start in August and work for ten or twelve months. While at CDC, fellows will: gain training and work experience in applied epidemiology and public health; perform epidemiologic analyses and research; participate in public health field experiences; report findings through written and oral presentations at scientific meetings; write scientific manuscripts for publication; participate in monthly seminars and journal clubs; and attend the annual Epidemic Intelligence Service conference. CDC Program areas available to applicants are chronic diseases, emergency preparedness, environmental health, global health, infectious diseases, injury prevention, and reproductive health. Fellows receive a stipend for living expenses. Applicants must apply online. Supporting documents are required and must be postmarked by the deadline date. Applicants should check the CDC Experience website for further details. Annual deadline dates may vary; applicants are encouraged to check the website or contact program staff to verify current deadline dates. The CDC Experience Applied Epidemiology Fellowship is made possible by a public/private partnership supported by a grant to the CDC Foundation from External Medical Affairs, Pfizer Inc., and is considered a valuable experience to those those seeking career advancement in the areas of clinical medicine, clinical epidemiology, health services research, preventive medicine, and public health.
Requirements: Applicants must: be medical students in their 2nd or 3rd year when they apply; be enrolled in a school accredited by the Liaison Committee on Medical Education or the American Osteopathic Association; be a U.S. citizen or permanent resident; and be willing to relocate to Atlanta, Georgia for the duration of the fellowship. Awardees may start the fellowship while in their 3rd or 4th year.
Date(s) Application is Due: Dec 2
Contact: Virginia Watson, MA, Program Contact; (404) 498-6151; VWatson1@cdc.gov or cdcexperience@cdcfoundation.org
Internet: http://www.cdc.gov/CDCExperienceFellowship/index.html
Sponsor: Centers for Disease Control and Prevention
1600 Clifton Road
Atlanta, GA 30333

CDC Foundation Atlanta International Health Fellowships 1147
The Atlanta International Health Fellowships (AIHF) were created in 1984 by Drs. Bob Chen and Katy Irwin to provide financial assistance to international participants in CDC's varied public health training courses. The two young physicians enlisted help from other CDC employees and retirees to raise funds and then formed a partnership with Emory University, which provided free tuition to one individual each year to attend Emory's International Course in Applied Epidemiology. Drs. Chen and Irwin also formed a partnership with Villa International to provide free housing to the AIHF recipient. In 1997, a grant from the Tull Charitable Foundation was added to existing funds, and the Atlanta International Health Fellowship became the first endowed fund at the CDC Foundation. Since the first fellowships were awarded in 1991, 25 individuals from over 22 countries have received stipends to cover costs of travel, tuition and lodging as they enrich their public health expertise at CDC. Interested applicants are encouraged to contact the CDC Foundation or Emory University for information on how to apply for the AIHF fellowship.
Geographic Focus: All States, All Countries
Contact: Verla S. Neslund, J.D., Vice President for Programs; (888) 880-4232 or (404) 653-0790; fax (404) 653-0330; vneslund@cdcfoundation.org
Pia Vaerliano, B.S.C., M.B.A, Co-Director of Special Programs; (404) 727-3485; fax (404) 727-4590; pvaleri@sph.emory.edu
Internet: http://www.cdcfoundation.org/what/program/atlanta-international-health-fellowship-endowment
Sponsor: Centers for Disease Control and Prevention (CDC) Foundation
55 Park Place NE, Suite 400
Atlanta, GA 30303

CDC Foundation Emergency Response Fund Grants 1148
The CDC Foundation's Emergency Response Fund provides immediate, flexible resources to CDC experts addressing public health emergencies in the U.S. - whether natural disasters, emerging disease outbreaks or bioterrorist threats. Following the events of September 11, 2001, and the anthrax attacks, the CDC Foundation established the Emergency Response Fund to give CDC what it needs most in an emergency: flexibility and access to immediate resources. Federal dollars, even during emergencies, are tied to restrictions and purchasing procedures that can limit CDC's ability to act quickly. The Foundation's Emergency Response Fund gives CDC a backup source of funding to fill critical gaps and meet immediate needs. The Fund was activated for the first time in 2005 to support the public health response to Hurricane Katrina in the Gulf Coast region. Donations to the Fund from Kaiser Permanente, the Robert Wood Johnson Foundation, and other organizations and individuals enabled the CDC Foundation to immediately respond to requests for help from CDC and their public health partners in the gulf coast region. The Foundation was also able to provide new facilities for two public health agencies on the Mississippi coast, replacing buildings that had been destroyed by the storm.
Contact: Verla S. Neslund, J.D., Vice President for Programs; (888) 880-4232 or (404) 653-0790; fax (404) 653-0330; vneslund@cdcfoundation.org
Internet: http://www.cdcfoundation.org/response
Sponsor: Centers for Disease Control and Prevention (CDC) Foundation
55 Park Place NE, Suite 400
Atlanta, GA 30303

CDC Foundation Tobacco Network Lab Fellowship 1149
According to a Centers for Disease Control and Prevention (CDC) morbidity and mortality report, tobacco use is the world's leading single preventable cause of death. Worldwide tobacco-related deaths currently exceed 5 million a year and are expected to exceed 8 million a year by 2030 if left unchecked. In the United States, tobacco use is the single leading preventable cause of disease, disability, and death. In 2003, the World Health Organization (WHO) created the Tobacco Laboratory Network (TobLabNet) of government, academic, and independent laboratories worldwide to strengthen national and regional capacity for testing and research of the contents and emissions of tobacco products. The Network's stated goals are: to establish global capacity to test tobacco products for regulatory compliance; to research and develop harmonized standards for contents and emissions testing; to share tobacco research and testing standards and results; to inform risk assessment activities related to the use of tobacco products; and to develop harmonized reporting of results so that data can be transformed into meaningful trend information that can be compared across countries and over time.

The Tobacco Network Lab Fellowship is a collaboration between CDC and WHO to place a scientist in CDC's Division of Laboratory Sciences to help TobLabNet address testing and research of tobacco products at the global level in accordance with the tobacco product regulation provisions outlined by the WHO Framework Convention on Tobacco Control. A link to brief information about the fellowship is at the grant website under Chronic Disease & Birth Defects Programs. Interested applicants are encouraged to contact the CDC Foundation or the World Health Organization for more information on how to apply for the fellowship.
Geographic Focus: All States, All Countries
Contact: Verla S. Neslund, J.D., Vice President for Programs; (888) 880-4232 or (404) 653-0790; fax (404) 653-0330; vneslund@cdcfoundation.org
Dr Haik Nikogosian, Convention Secretariat; 41 22 791 5043; fax 41 22 791 5830; fctcsecretariat@who.int
Internet: http://www.cdcfoundation.org/what/programs/list#program-518
Sponsor: Centers for Disease Control and Prevention (CDC) Foundation
55 Park Place NE, Suite 400
Atlanta, GA 30303

CDC Hepatitis Education for Inmates and Correctional Staff Program Grants 1150
CDC announces the availability of funds for a cooperative agreement with one or more national voluntary organizations involved with correctional facilities to develop and distribute materials to educate inmates and correctional staff about the risks of transmission and acquisition of viral hepatitis, as well as prevention, counseling, testing, and treatment of viral hepatitis, with an emphasis on hepatitis A, B, and C prevention and counseling.
Requirements: Applications may be provided only to organizations currently providing health education materials to correctional populations and health related training materials to staff of correctional institutions at a national or regional level.
Amount of Grant: 242,614 U.S.D
Contact: Nealean Austin, Grants Management Branch, Procurement and Grants Office, Centers for Disease Control and Prevention; (770) 488-2700
Internet: http://www.cdc.gov
Sponsor: Centers for Disease Control and Prevention
1600 Clifton Road
Atlanta, GA 30333

CDC HIV Demonstration, Research, Public and Professional Education Project Grants 1151
Grants are awarded to states, political subdivisions of states, and other public and nonprofit entities for research on the prevention of HIV infection at the community level. Funds may be used to develop, implement and evaluate new intervention, including those targeting people who are infected with HIV. Applicants are encouraged to have research groups participate in the program. Contact the program office for exact deadline dates.
Requirements: States, political subdivisions of states, other public and nonprofit private entities are eligible to apply.
Amount of Grant: 50,000 - 1,000,000 U.S.D
Contact: Cheryl Maddux, Grants Manager; (770) 498-1911
Internet: http://www.cdc.gov
Sponsor: Centers for Disease Control and Prevention
1600 Clifton Road
Atlanta, GA 30333

CDC HIV Prevention: Community-Based Organizations 1152
CDC invites comments under a planned request for applications to support community-based organizations (CBO) to develop, implement, and evaluate effective community-based HIV prevention programs for at-risk populations, especially racial and ethnic minority populations.
Requirements: Applicants must be tax-exempt. They may apply as minority organizations that intend to serve predominantly racial or ethnic minority populations at risk or as CBOs that will serve high-risk populations in general. Applicants must also have two years of service with the proposed target population.
Date(s) Application is Due: Oct 29
Amount of Grant: Up to 100,000,000 U.S.D
Contact: Program Contact; (404) 639-5230; fax (404) 639-2072; hivmail@cdc.gov
Internet: http://www.cdc.gov/od/pgo/funding/grantmain.htm
Sponsor: Centers for Disease Control and Prevention
1600 Clifton Road
Atlanta, GA 30333

CDC HIV Prevention among Ethnic Minority Populations 1153
The purpose of this program is to create community-based organizations aimed at HIV prevention among ethnic minority populations.
Requirements: State health departments and public and private institutions of higher education may apply, along with any community-based organization (CBO). Minority-oriented CBO's must have at least 50 percent of their governing board composed of the racial minority that will be served and at least 50 percent of the program's positions be held by persons who reflect the intended racial demographics.
Date(s) Application is Due: Mar 7
Amount of Grant: Up to 100,000,000 U.S.D
Contact: Grant Administrator; (404) 639-2072
Internet: http://www.cdc.gov/about/funding.htm
Sponsor: Centers for Disease Control and Prevention
1600 Clifton Road
Atlanta, GA 30333

CDC HIV Prevention Projects for Community-Based Organizations 1154
CDC announces the availability of funds to support community-based organizations (CBOs) to develop, implement, and evaluate state-of-the-art, model community-based HIV prevention programs for populations at risk for HIV infection, especially racial/ethnic minority populations at risk. The goals of this program are to reduce the disproportionate impact of the HIV epidemic on racial/ethnic minority populations and other at-risk populations; to improve and expand community-based HIV prevention services by supporting community-based HIV prevention programs that address priorities described in applicable state and local comprehensive HIV prevention plans (that is, the plans developed by the official HIV prevention community planning groups for the jurisdiction in which the CBO is located) or that adequately justify addressing other priorities; enhance CBOs' incorporation of scientific theory and data, and validated program experience into the design, implementation, and evaluation of HIV prevention services; and support collaboration and coordination of HIV prevention efforts among CBOs, community planning groups, other local organizations, local and state health departments, and managed care organizations serving populations at risk for HIV infection.
Requirements: Applicant organization must have current, valid 501(c)3 tax-exempt status; must be located in the community; and have an established record of at least two years of service to the proposed target population. Contact program staff for a complete list of requirements.
Amount of Grant: Up to 225,000 U.S.D
Contact: Cheryl Maddux, Grants Manager; (770) 498-1911
Internet: http://www.cdc.gov/about/funding.htm
Sponsor: Centers for Disease Control and Prevention
1600 Clifton Road
Atlanta, GA 30333

CDC Increasing Breast and Cervical Cancer Screening Services for Urban American Indian/Alaska Native Women 1155
The purpose of this program is to: increase the number of urban American Indian/ Alaska Native (AI/AN) women receiving BCCED screening; decrease time to access diagnosis and treatment; increase participation of Urban Indian Health Organizations (UIHOs) in state BCCEDP & Comprehensive Cancer Control (CCC) coalitions; and to provide a model for other state health department programs to work with UIHOs in a collaborative, culturally appropriate manner.
Requirements: Eligible applicants that can apply for this funding are non-profit organizations with experience and expertise in addressing the health care needs of urban Indian populations and providing culturally competent services to them. These include, but are not limited to organizations such as: Indian health boards; inter-tribal councils; American Indian/Alaska native health tribal organizations; inter-tribal councils; inter-tribal consortia; urban organizations; and other non-profit organizations, if incorporated for the primary purpose of improving AI/AN health and represent such interests for the tribes, or urban Indian communities throughout the United States.
Restrictions: Recipients may: not use funds for research; not use funds for clinical care; only expend funds for reasonable program purposes, including personnel, travel, supplies, and services, such as contractual. Awardees may not generally use HHS/CDC/ATSDR funding for the purchase of furniture or equipment (any such proposed spending must be identified in the budget). The direct and primary recipient in a cooperative agreement program must perform a substantial role in carrying out project objectives and not merely serve as a conduit for an award to another party or provider who is ineligible. Reimbursement of pre-award costs is not allowed.
Date(s) Application is Due: Jul 23
Amount of Grant: 350,000 - 1,050,000 U.S.D
Contact: Stephanie Lankford; (770) 488-2936; fzi8@cdc.gov
Internet: http://www.cdc.gov/od/pgo/funding/FOAs.htm
Sponsor: Centers for Disease Control and Prevention
1600 Clifton Road
Atlanta, GA 30333

CDC Injury Control Research Centers Grants 1156
Grants are intended to support injury control research and demonstrations on priority issues; to integrate aspects of engineering, public health, behavioral sciences, medicine, and other disciplines to prevent and control injuries more effectively; to rigorously apply and evaluate current and new interventions, methods, and strategies that focus on the prevention and control of injuries; to stimulate and support injury control research centers in academic institutions that will develop a comprehensive and integrated approach to injury control research and training; and to bring the knowledge and expertise of injury control research centers to bear on the development of effective public health programs for injury control. Refer to announcement number 01007. Annual deadline dates may vary; contact the office for exact dates.
Requirements: Any nonprofit or for-profit organization in eligible regions may apply. Eligible regions include 34 states as well as the District of Columbia, the Virgin Islands, and Puerto Rico.
Restrictions: Grantees may not award subgrants but may enter into contracts as necessary to achieve the aims of the program.
Geographic Focus: Alabama, Alaska, Arizona, Arkansas, California, Delaware, District of Columbia, Florida, Georgia, Hawaii, Idaho, Illinois, Indiana, Kentucky, Louisiana, Maryland, Michigan, Minnesota, Mississippi, Nevada, New Jersey, New Mexico, New York, North Carolina, Ohio, Oklahoma, Oregon, Pennsylvania, South Carolina, Tennessee, Texas, Virginia, Washington, West Virginia, Wisconsin
Amount of Grant: 905,500 U.S.D
Contact: Program Manager, National Center for Injury Prevention and Control; (770) 488-1506; OHCINFO@cdc.gov
Internet: http://www.cdc.gov/ncipc/profiles/icrcs/default.htm

Sponsor: Centers for Disease Control and Prevention
1600 Clifton Road
Atlanta, GA 30333

CDC Injury Prevention and Control Research and State and Community Based Programs 1157
Research grants support injury control research on priority issues; projects that integrate aspects of engineering, public health, behavioral sciences, medicine, and other disciplines in order to prevent and control injuries more effectively; projects that rigorously apply and evaluate current and new interventions, methods, and strategies that focus on the prevention and control of injuries; projects that stimulate and support Injury Control Research Centers (ICRC) in academic institutions which will develop a comprehensive and integrated approach to injury control research and training; and projects that bring the knowledge and expertise of ICRC's to bear on the development of effective public health programs for injury control. State and Community Program grants are to develop and evaluate new methods or to evaluate existing methods and techniques used in injury surveillance by public health agencies; and to develop, expand, or improve injury control programs to reduce morbidity, mortality, severity, disability, and cost from injuries.
Requirements: Applications may be submitted by all public and private nonprofit and for-profit organizations and by governments and their agencies.
Contact: Program Officer, Injury Prevention and Control Research Projects; (770) 488-1506; fax (770) 488-1667; OHCINFO@cdc.gov
Internet: http://www.cdc.gov/ncipc/profiles/icrcs/default.htm
Sponsor: Centers for Disease Control and Prevention
1600 Clifton Road
Atlanta, GA 30333

CDC Intervention Epidemiologic Research Studies of HIV/AIDS 1158
CDC announces the availability of funds for a cooperative agreement program to support intervention epidemiologic research studies of AIDS and HIV infection. These awards will help support researchers in two areas: the development and evaluation of innovative interventions for preventing and reducing the transmission of HIV infection in young and recently initiated injection drug users (IDUs); and the development and evaluation of an intervention study to improve access to antiretroviral therapy in HIV-infected disadvantaged populations.
Requirements: Applications may be submitted by public and private nonprofit organizations and by governments and their agencies.
Amount of Grant: 5,000 - 270,000 U.S.D
Contact: Allyn Nakashima, Epidemiology Branch, Center for Disease Control and Prevention; (404) 639-0900
Internet: http://www.cdc.gov/epo
Sponsor: Centers for Disease Control and Prevention
1600 Clifton Road
Atlanta, GA 30333

CDC Minority Health Statistics Dissertation Research Grants 1159
The purpose of the program is to make awards for the conduct of special surveys or studies on the health of racial and ethnic populations or subpopulations; analysis of data on ethnic and racial populations and subpopulations; and research on improving methods for developing statistics on ethnic and racial populations and subpopulations. The deadline date listed is for letters of intent. Annual dates may vary; contact the program office for exact dates.
Requirements: Eligible applicants may be public or private nonprofit institutions that will administer the grant on behalf of the proposed principal investigator. The proposed principal investigator must be a registered doctoral candidate in resident or nonresident status. All requirements for the doctoral degree other than the dissertation must be completed by the time of the award. Students seeking a doctorate in any relevant research discipline are eligible. An applicant institution may be either the degree-granting institution or another nonprofit institution with which the proposed principal investigator is professionally affiliated.
Amount of Grant: 15,000 - 30,000 U.S.D
Contact: Program Contact; (301) 458-4000; MGP@cdc.gov
Internet: http://www.cdc.gov/nchswww/about/grants/grants.htm
Sponsor: Centers for Disease Control and Prevention
1600 Clifton Road
Atlanta, GA 30333

CDC National Breast and Cervical Cancer Early Detection Program 1160
The purpose of this program is to apply a state, territorial, or tribal public health approach to increase access to and use of screening services. Funded programs will establish a comprehensive breast and cervical cancer early detection screening program that includes breast and cervical cancer screening, tracking, follow-up and case management; public education and outreach; professional education; quality assurance and improvement; surveillance and evaluation; coalitions and partnerships; and management.
Requirements: Assistance will be provided only to the official health departments of states or their bona fide agents, including the District of Columbia, the Commonwealth of Puerto Rico, the Virgin Islands, the Commonwealth of the Northern Mariana Islands, the Republic of Palau, and federally recognized Indian Tribal governments.
Restrictions: Alaska, Arizona, Arkansas, American Samoa, California, Colorado, Connecticut, Florida, Georgia, Illinois, Iowa, Kansas, Louisiana, Maine, Maryland, Massachusetts, Michigan, Minnesota, Missouri, Nebraska, New Jersey, New Mexico, New York, North Carolina, Ohio, Oklahoma, Oregon, Pennsylvania, Rhodes Island, South Carolina, Texas, Utah, Vermont, Washington, West Virginia, Wisconsin, Puerto Rico, and Guam are not eligible for funding. See announcement for complete listing.
Amount of Grant: 145,000 - 84,000,000 U.S.D
Contact: Carlos Smiley, (770) 488-2754
Internet: http://www.cdc.gov/cancer/nbccedp/index.htm
Sponsor: Centers for Disease Control and Prevention
1600 Clifton Road
Atlanta, GA 30333

CDC National Poison Prevention and Control Program Grants 1161
The purpose of the program is to support an integrated system of poison prevention and control services including: coordination of all poison control centers (PCCs) through development, implementation, and evaluation of standardized public education; and development of a plan to improve national toxicosurveillance and development of a single, nationwide toll-free telephone number and related public service media campaign.
Requirements: Applications may be submitted by public and private nonprofit organizations, governments and their agencies, other public and private nonprofit organizations, state and local governments or their bona fide agents, and federally recognized American Indian tribal governments, American Indian tribes, or American Indian tribal organizations.
Contact: Cheryl Maddux, Grants Management Officer, Centers for Disease Control and Prevention,; (770) 488-2645
Internet: http://www.cdc.gov
Sponsor: Centers for Disease Control and Prevention
1600 Clifton Road
Atlanta, GA 30333

CDC National Programs That Build the Capacity of Schools To Prevent Foodborne Illness Through Coordinated School Health Programs 1162
The purpose of this program is to develop a national program to build the capacity of state and local education and health agencies, and others to prevent foodborne illness and other important health problems as part of a coordinated school health program.
Requirements: Assistance will be provided to national organizations that are private health, education, or social service agencies (professional, or voluntary); qualify as a nonprofit 501(c)3 entity; have the capacity and experience to assist their local affiliates; and have affiliate offices or local, state, or regional membership constituencies in a minimum of ten states and territories.
Restrictions: National organizations that are funded currently by CDC/Division of Adolescent and School Health (DASH) under program announcements 99023, 97065, 00026, 00081, 00109, 00719, 98885, 99072, 00079 or 00618 are not eligible for this program announcement.
Amount of Grant: 299,000 U.S.D
Contact: Nealean Austin, Grants Management Officer; (770) 488-2754
Internet: http://www.cdc.gov/healthyyouth
Sponsor: Centers for Disease Control and Prevention
1600 Clifton Road
Atlanta, GA 30333

CDC National Training Information and Exchange Program 1163
The purpose of this program is to make data and information on trauma care in the United States more accessible to a broad spectrum of individuals and organizations, including trauma care professionals and professional associations, trauma centers and other acute care hospitals, trauma care systems, emergency medical services (EMS) systems, injury researchers and research organizations, public health agencies, health care payers, and the general public.
Requirements: Applications may be submitted by public and private nonprofit organizations and by governments and their agencies.
Samples: (1) Injury Control Research Centers (ICRC) have undertaken a broad range of work. For example, the John Hopkins University ICRC has sponsored Summer Training Institutes for injury control researchers and practitioners. Harvard has been a key planning, training, and program resource for injury control programs in the New England States. Work at the University of North Carolina ICRC has led to the creation of an injury control unit in the North Carolina State Health Department. Harborview ICRC serves as a State and regional resource in trauma and burn care and is a leader of efforts to reduce pedestrian injuries and injuries associated with motorcycles and bicycles. (2) Funded Injury Prevention and Control Projects address priority research concerns encompassing acute care, biomechanics, prevention, epidemiology, and rehabilitation. As examples, researchers are investigating the mechanism of traumatic brain injury due to impact, other investigators are defining risk factors for intimate partner violence and another group is evaluating the effectiveness of trauma systems. (3) Surveillance programs address E-coded hospital discharge data; model surveillance systems to address nonfatal injuries resulting from intentional and unintentional injuries.
Contact: Cheryl Maddux, Grants Management Officer, Centers for Disease Control and Prevention,; (770) 488-2645
Internet: http://www.cdc.gov
Sponsor: Centers for Disease Control and Prevention
1600 Clifton Road
Atlanta, GA 30333

CDC Oral Diseases Prevention and Oral Health Promotion Grants 1164
The purpose of the program is to develop initiatives related to oral disease prevention and related chronic disease and health promotion capacity; and, to coordinate the dissemination of comprehensive oral disease information and health promotion programmatic expertise

among state and local agencies, public and private sector organizations, and health care professionals in the United States. The purpose of this program includes conducting projects that promote the development of leadership and infrastructure to establish sustainable oral health programs at the state and local levels; advance tobacco control programs, especially smokeless tobacco; advance school oral health education and targeted oral disease and oral injury prevention efforts (e.g., athletic mouth guards); and develop and carry out oral health education and disease prevention programs for high-risk adults. Annual deadline dates may vary; contact program staff for exact dates.
Requirements: Assistance will be provided only to dental health organizations that are 501(c)3 private, nonprofit organizations. Eligible applicants must have established working relationships with public advocacy coalitions and national, state, and local organizations.
Amount of Grant: 70,000 U.S.D
Contact: Scott Presson; (770) 488-6056; skp4@cdc.gov
Internet: http://www.cdc.gov/oralhealth
Sponsor: Centers for Disease Control and Prevention
1600 Clifton Road
Atlanta, GA 30333

CDC Pregnancy Risk Assessment Monitoring Systems Program Grants 1165
The purpose of the program is to assist state public health agencies in establishing and maintaining population-based surveillance of pregnant women. The projects will focus on selected maternal behaviors and experiences that occur around the time of pregnancy and early infancy. Data generated will be used to guide state family-planning health programs and policies. Grant funds may be used to coordinate, develop, or implement data-gathering programs relating to pregnancy and family planning or to form partnerships with other states or territories for the purpose of pooling resources to gather information. Funds may also be used to disseminate the information gathered. CDC expects to make five awards. No matching grants are required, although recipients are expected to provide in-kind support for the programs.
Requirements: Eligible applicants are the health agencies of states and U.S. territories, with the exception of those that have previously received funds through the program.
Contact: Nealean Austin, Grants Management Officer; (770) 488-2700
Internet: http://www.cdc.gov
Sponsor: Centers for Disease Control and Prevention
1600 Clifton Road
Atlanta, GA 30333

CDC Preventive Health Services-Sexually Transmitted Diseases Control Grants 1166
The objective of the grants is to reduce morbidity and mortality by preventing cases and complications of sexually transmitted diseases. Grant funds may be used for sexually transmitted disease (STD) surveillance activities, including reporting, screening, and follow-up of diagnostic tests for, and diagnosed cases of, STD; STD case finding and case follow-up activities, including notification of sex partners to infectious cases of STD and follow-up systems; interstate epidemiologic referral and follow-up activities; program-specific professional (including appropriate allied health personnel) STD education, training, and clinical skills improvement activities as authorized under section 318(b) of the PHS Act; program-specific STD education and information activities as authorized under section 318(b) of the PHS Act; special studies or demonstrations to evaluate or test STD prevention activities; and control strategies and activities as may be prescribed by the program.
Requirements: Any state, state health authority, or political subdivision of a state may apply.
Amount of Grant: 19,000 - 51,000,000 U.S.D
Contact: Sevgi Aral, Director, Division of STD Prevention; (404) 639-8259; fax (404) 639-8608
Internet: http://www.cdc.gov
Sponsor: Centers for Disease Control and Prevention
1600 Clifton Road
Atlanta, GA 30333

CDC Preventive Medicine Residency and Fellowship 1167
The Preventive Medicine Residency and Fellowship (PMR/F) is an umbrella that includes two preventive medicine programs that provide hands-on experiences in public health agencies at the federal, state, and local levels. The programs consist of a residency (PMR) for physicians only and a fellowship (PMF) for physicians and other health professionals. The PMR is a 24-month nationally accredited graduate medical education program for physicians. It is accredited by the Accreditation Council for Graduate Medical Education (ACGME) and meets the residency requirement of the American Board of Preventive Medicine (ABPM) in the specialty of Public Health and General Preventive Medicine. Select candidates can be considered for one year of residency training. The PMF is a 12-month program open to veterinarians, dentists, registered nurses, physician assistants, and international medical graduates (physicians) who have no ACGME- or American Osteopathic Association (AOA)- accredited postgraduate clinical training. The residency (PMR) and the fellowship (PMF) both provide service, training, and supervised on-the-job learning in public health and preventive medicine practice. PMR/F activities include: designing, implementing, and evaluating public health programs; developing or analyzing health policy; managing public health and preventive medicine projects; bridging medical and public health sectors to improve wellness; and developing and applying leadership skills. The PMR/F program goals are to prepare clinicians for leadership roles in public health and general preventive medicine at the federal, state or local levels. Residents and fellows will receive mentoring by experienced public health and preventive medicine practitioners; hands-on experience in public health and preventive medicine practice; classroom training to enhance professional growth; and M.P.H. coursework for select physicians in the PMR program. Residents and fellows receive a salary and benefits, moving expenses, travel to PMR/F-sponsored training, and tuition support for an M.P.H. degree for residents (as needed). (Qualified applicants interested in the American Board of Preventive Medicine "complementary pathway to board exam certification" must have completed an ACGME accredited residency in another specialty and have an M.P.H. or equivalent degree per ACGME.) Interested applicants may apply online between July 1 and September 15. Additional supporting documents are required. Applications are due by midnight eastern daylight time on the due date. Supporting documents must also be received on the application due date. Applicants are encouraged to mail supporting documents early. All mail delivery to the CDC goes to a central mail facility before being delivered to the PMR/F office. Applicants are encouraged to click on the "Application Information" link at the PMR/F website for additional instructions, guidelines, and information on how to apply and submit the supporting documents.
Requirements: Applicants must: commit to a 1- or 2-year full-time training period (depending on the program) starting in mid-June; be willing to relocate for duration of training; meet the professional and licensing requirements for hiring by the U.S. government according to the U.S. Office of Personnel Management (link provided at the PMR/F website); and have trained in the Epidemic Intelligence Service (EIS) program or have comparable applied epidemiology experience. PMR applicants must be physicians who have completed at least 12 months of ACGME-accredited postgraduate clinical training involving at least 11 months of direct patient care; have a current, full, and unrestricted medical license from a U.S. licensing jurisdiction; and have a Master of Public Health (M.P.H.) or equivalent accredited degree, per ACGME requirements. See the PMR/F website for more information about these requirements. There are some exceptions to the M.P.H. degree requirement. Physicians may be sponsored during the residency to obtain an M.P.H. degree. The completion of necessary courses to earn the degree will be integrated into the training program. M.P.H. sponsored applicants must be Commissioned Corps officers in the U.S. Public Health Service at the time of application with a commitment to an additional 2-year service obligation in the Commissioned Corps following the residency and a willingness to attend the academic institution for the M.P.H. selected by the residency (be willing to relocate). PMF applicants must be veterinarians, dentists, registered nurses, physician assistants, or international medical graduates (physicians who have no ACGME- or American Osteopathic Association- accredited postgraduate clinical training) who have an M.P.H. or equivalent degree from an accredited institution; have a current, full, and unrestricted license from a U.S. licensing jurisdiction in their qualifying clinical specialty; and demonstrate ability in applied epidemiology, public health practice, and written and oral communication for scientific audiences.
Restrictions: Applicants must be U.S. citizens or permanent residents. Physicians who have graduated from a medical school outside the U.S. and do not have a current, full, and unrestricted license from a U.S. licensing jurisdiction may apply.
Date(s) Application is Due: Sep 15
Contact: PMR/F Coordinator; (404) 498-6140; fax (404) 498-6135; PrevMed@cdc.gov
Internet: http://www.cdc.gov/PrevMed/index.html
Sponsor: Centers for Disease Control and Prevention
1600 Clifton Road
Atlanta, GA 30333

CDC Promoting Prevention in Managed Care Cooperative Agreements 1168
CDC announces the availability of funds for a cooperative agreement program for promoting prevention in managed care. This announcement relates to all of the priority areas of Healthy People 2010. Its purpose is to promote the attainment of the objectives outlined in Healthy People 2010 through strengthening the infrastructure supporting the science and practice of prevention in managed care. Annual deadline dates may vary; contact program staff for exact dates.
Requirements: Applications may be submitted by private sector, nonprofit, managed care membership organizations who provide an array of services and products (e.g., technical support, publications, training and continuing education, communication and information sharing, etc.) to member plans in at least 20 states; and whose member plans and affiliated entities can demonstrate past and current experience conducting public domain, prevention research (i.e., research in the areas of health promotion, disease prevention, and chronic disease management) in the managed care environment.
Amount of Grant: 242,614 U.S.D
Samples: (1) Cooperative agreements to states for the Behavioral Risk Factor Surveillance system, (2) Public Health Conference support grants, (3) cooperative agreements to states for core capacity building for Tobacco Prevention and Control Programs, (4) cooperative agreements to states for the National Program of Cancer Registries, (5) cooperative agreements for Community Coalition Partnerships for the prevention of teen pregnancy, (6) cooperative agreements to states for the Pregnancy Risk Assessment Monitoring System, (7) cooperative agreement for epilepsy program activities, and (8) cooperative agreement for Oral Disease Prevention.
Contact: Nealean Austin, Grants Management Branch; (770) 488-2700
Internet: http://www.cdc.gov/nchs/hphome.htm
Sponsor: Centers for Disease Control and Prevention
1600 Clifton Road
Atlanta, GA 30333

CDC Public Health Associates 1169
The Public Health Associate Program (PHAP) is a Centers for Disease Control and Prevention (CDC) development program administered through CDC's Office for State, Tribal, Local and Territorial Support (OSTLTS). The program, which began in 1948, provides opportunities for promising future public health professionals to gain broad experience in the day-to-day operation of public health programs. PHAP is geared toward recent baccalaureate college graduates (BA/BS) who are beginning a career in public

health. Public Health Associates are employed by CDC on term-limited appointments for two years. As "assignees" to agencies in the governmental public health system, most associates hold host-agency positions that are functionally indistinguishable from their local co-workers. In addition to city, county and state assignments, associates may be assigned to territories, tribes and CDC quarantine stations in major international airports within the United States. Assignments are individually tailored to meet local needs in the delivery of public health services, such as case investigation, disease surveillance, health promotion, community outreach, and public health administration. Furthermore, each appointment consists of two different one-year assignments selected by the host agency in one of the following program areas: STD, TB and/or HIV; Other Communicable Diseases; Chronic Disease; Environmental Health; Public Health Preparedness; Global Migration and Quarantine; Immunization; and Injury Prevention. Associate job assignments in the program areas listed above include activities and responsibilities that will lead to the development of specific competency objectives. Both one-year assignments involve substantive activity in the following domains: patient/client interaction to elicit or provide information; investigation of public health problems and issues; data collection, analysis and reporting; surveillance activities; public health education; provision of public health services; community involvement and partnership; and liaison with other public or private providers and partners. While each associate has a different combination of assignments over their two-year employment with CDC, all associates participate in emergency response activities in support of federal, state or local host agencies. This may include participation in an exercise to prepare for major public health events or an actual response to an on-going event (e.g., hurricanes, H1N1 flu outbreak). Upon completion of the two-year appointment, associates are qualified to compete for entry-level career positions as CDC public health advisors and equivalent positions at state, tribal, local, territorial or non-governmental public health organizations. Host sites are responsible for providing the associates with a variety of training opportunities that will not only orientate them to the local or state health agency but also lead to required program competencies. These trainings include both local employment training (e.g., safety, security, information technology, standards of conduct) in accordance with state and local requirements and also program-specific training to enable and enhance work performance. In addition, CDC provides all associates with a program of core training in public health concepts and topics designed to support their experiential learning, reinforce the acquisition of the program competencies, and enhance professional growth and development. The core training curriculum involves: CDC employment orientation within thirty days of the program start date, a series of required one- to five-day training events hosted by CDC semi-annually, and monthly webinars and other e-learning. CDC funds annual travel and training expenses for associates to attend orientation and training required by CDC, many of which are located at CDC headquarters in Atlanta. In addition, host sites are obligated to spend local funds to train associates in program-specific areas. Associates are paid every two weeks at the equivalent of a GS-5, step 1 (adjusted for their location) following the federal pay calendar. Associates who successfully complete their first year assignments will then be promoted to the equivalent of a GS-7, step 1 for their final one-year assignment. All associates are given the opportunity to enroll in a choice of full health benefit options (that are partially compensated by the CDC) available to all federal employees. Prospective associates may apply for the PHAP online. The link to the online application system will be available at the PHAP website during mid-February through early March. The March deadline will be communicated when the online application link is posted. Additional supporting documents are required, including academic transcripts and letters of recommendation. Additional guidelines, instructions, and downloadable forms for supporting documents are provided at the PHAP website. Applications received after 11:59 p.m. (Eastern) on the due date will not be reviewed. Supporting documents must be postmarked by the closing date of the online application. Prospective associates are encouraged to visit the PHAP website for additional information on the PHAP experience.
Requirements: Applicants should have a keen interest in frontline service careers in public health and in developing strong public health programmatic and operations skills. Applicants must: be U.S. citizens; hold a bachelor's degree from an accredited academic institution by July 1 of the year of application; be willing to commit to full-time work for two years; be willing to work in a state, tribal, local or territorial health agency setting, or CDC quarantine station; and, be willing to relocate at their own expense. Note: Applicants with a master's degree will be considered but do not receive preference in hiring, nor a higher pay scale. Individuals holding a master's degree may be eligible for other CDC fellowship programs.
Restrictions: Applicants must be U.S. Citizens.
Amount of Grant: 76,284 U.S.D
Contact: Lynn Gibbs Scharf, Director; (404) 498-0030; fax (404) 639-9130; phap@cdc.gov
Internet: http://www.cdc.gov/phap/index.html
Sponsor: Centers for Disease Control and Prevention
1600 Clifton Road
Atlanta, GA 30333

CDC Public Health Associates Hosts 1170

The Center of Disease Control and Prevention (CDC)'s Office for State, Tribal, Local and Territorial Support (OSTLTS) seeks proposals for field-based assignments for its Public Health Associate Program (PHAP) on an annual basis. PHAP is a two-year workforce development program that provides opportunities for future public health professionals to gain broad experience in the day-to-day operation of public health programs. Each Public Health Associate is assigned to a designated host health agency for two one-year assignments beginning in July of each year. As a training and service program, PHAP offers many benefits to host agencies: a qualified associate for a two-year assignment (the class of 2010 had 1,231 eligible applicants competing for 50 positions in a rigorous screening process); salary and employment benefits for the associate; a core training program in public health concepts and topics designed to support associates' experiential learning, reinforce the acquisition of the program competencies and enhance professional growth and development; and ongoing technical support for host sites in managing the associate, including periodic site visits. The core training curriculum involves: CDC employment orientation within 30 days of the associate's start date; a series of required one- to five-day training events hosted by CDC semi-annually; and monthly webinars and other e-learning assignments. Costs related to attendance at PHAP core training will be supported by CDC. Host sites in their turn must provide all local or long-distance travel and expenses for trainings related to associate work assignments; all local or long distance travel and related expenses for conferences and meetings related to associate work assignments; and a suitable work site, including desk, computer, chair, and other equipment to successfully complete programmatic tasks as named in the application. Host site supervisors oversee the training and field activities of the associate, ensure the associate is familiar with relevant techniques in a given program specialty, and encourage the overall professional development of the associate. Supervisors receive an initial orientation conducted by the CDC PHAP office prior to the associate's arrival and participate in conference calls at various intervals during the course of the associateship. The PHAP Host Site Guidelines and Application are available as downloadable PDF documents at the PHAP Host website (click on the "Application Process" link on the left hand side of the screen to access the web page that has the download links). Applications must be submitted electronically as email attachments with the subject line "PHAP Host Site Application – (your health department name)" to the email address given. Applications are due at 11:59 p.m. on the application deadline date. Deadlines may vary from year to year. Prospective hosts are encouraged to check the PHAP Host website or use the email address given to verify current deadlines and to obtain complete information on the PHAP program and application submission requirements.
Requirements: Host sites must be state, tribal, local, and territorial public health agencies, or CDC quarantine stations. Host sites are selected based on the scope, quality, and diversity of the experience offered to the associate reflected in the initial host site application. Selection criteria include, but are not limited to, host-site year-one and year-two activities; ability of the host site to support the associate as an employee (e.g., resources, training needs for the associate, travel support for training, and other needs); and demonstrated capacity to provide associates with employment training (e.g., safety, security, information technology, standards of conduct) and program-specific training. Host health agencies and supervisors are strongly encouraged to provide financial support and opportunities for the associate to participate in public health activities which will expand the associate's scope and depth of public health knowledge and / or expand his / her job-related capabilities. Associates are expected to be integrated into the host site and treated as an entry level employee. If employee programs are offered to regular employees, host sites are expected to provide comparable programs and financial support to the associate. Host site supervisors are required to hold a bachelor's degree or above and have prior supervisory experience. In addition, the primary supervisor must devote four hours per week to spend with the associate through the duration of the assignment to provide direct guidance on weekly assignments.
Date(s) Application is Due: Feb 21
Contact: Lynn Gibbs Scharf, Director; (404) 498-0030; fax (404) 639-9130; phap@cdc.gov
Internet: http://www.cdc.gov/phap/host/index.html
Sponsor: Centers for Disease Control and Prevention
1600 Clifton Road
Atlanta, GA 30333

CDC Public Health Conference Support Grants 1171

The purpose of the grants is to provide partial support for specific non-federal conferences in the areas of health promotion and disease prevention information/education programs. Funds may be used for direct cost expenditures, such as salaries, speaker fees, rental of necessary equipment, registration fees, and transportation costs (not to exceed economy-class fare) for non-federal individuals. The program continues through March 8, 2009.
Requirements: Public and private organizations, including colleges and universities, and units of state and local government are eligible to apply. Applicants must provide a portion of the conference cost.
Restrictions: Funds may not be used for indirect costs, equipment purchases, honoraria, entertainment, or personal expenses.
Contact: Sharon Robertson, Grants Officer; (770) 488-2748; SRobertson1@cdc.gov
Internet: http://www.cdc.gov/od/pgo/funding/FOAs.htm
Sponsor: Centers for Disease Control and Prevention
1600 Clifton Road
Atlanta, GA 30333

CDC Public Health Informatics Fellowships 1172

The Public Health Informatics Fellowship Program (PHIFP) provides two-year paid fellowships in public health informatics. The competency-based and hands-on training will allow recipients to apply the disciplines of information/computer science and technology to solve real-world public health problems. Fellows will have the opportunities: to learn about informatics and public health; to work with teams involved in research and development of public health information systems; to lead an informatics project; and to design, develop, implement, evaluate, and manage public health information systems. Fellows will be temporary employees of the Centers of Disease Control and Prevention (CDC) and will receive employment benefits (e.g., medical insurance and vacation time). Salaries will be based on federal pay grades GS 11 or 12, depending on applicant qualifications. Applicants should apply online at the PHIFP website (the link to the online application system is available from the Eligibility and Application Information page.) Submissions are due on the deadline date before midnight. All required supporting documents must be mailed in one package and be postmarked by the second deadline date (deadlines may change from

year to year; applicants should check the PHIFP website to verify current submission deadlines for proposals and supporting documents). Further details and downloadable instructions for letters of recommendation are available from the PHIFP website.
Requirements: To apply for PHIFP, applicants must meet both the educational and professional requirements for the fellowship, be willing to commit to a two-year, full-time program, and be willing to relocate to Atlanta, Georgia. A doctoral (Ph.D., M.D.) or masters-level degree is required. Qualifying degrees must be from an accredited academic institution in one of the following areas: public health, medicine, health care, or health-services research; computer science, information science, information systems; statistics; epidemiology; or public-health informatics or related discipline. Documented one-year professional experience for doctoral-level candidates and three-year professional experience for masters-level candidates is required in one of the following fields: public-health informatics, health informatics or related field, information systems, information science, computer science, or information technology. Additionally, documented one-year professional experience for doctoral-level candidates and three-year professional experience for masters-level candidates is required in public health or a related healthcare profession (medicine, nursing, veterinary medicine, dentistry, and allied health professions). Applicants should have documented experience in performing research or evaluation during or after academic training (e.g., publication, thesis, poster presentation, or research/evaluation proposal). Non-U.S. citizens must be legal permanent residents or eligible for J-1 status prior to the program's start date.
Restrictions: Citizens of countries that have been determined by the U.S. Secretary of State to be State Sponsors of Terrorism will not be able to get security clearance to work at CDC and should not apply to the program (a current list is available at the U.S. Department of State website).
Geographic Focus: All States, All Countries
Date(s) Application is Due: Nov 14; Nov 28
Amount of Grant: 59,987 - 74,297 U.S.D
Contact: Herman Tolentino, M.D., Director, PHIF Program; (404) 498-6168 or (404) 498-6219; fax (404) 498-6135; phifp@cdc.gov or herman.tolentino@cdc.hhs.gov
Internet: http://www.cdc.gov/PHIFP/
Sponsor: Centers for Disease Control and Prevention
1600 Clifton Road
Atlanta, GA 30333

CDC Public Health Prevention Service Fellowships **1173**
The Public Health Prevention Service (PHPS) is a 3-year training and service fellowship that prepares master's level public health professionals for leadership positions in local, state, national, and international public health agencies. The fellowship focuses on public health program management and provides hands-on experience and mentorship in program planning, implementation, and evaluation at Centers for Disease Control and Prevention (CDC), and in state and local health organizations. PHPS fellows are placed in public health organizations throughout the U.S. after completing one year of training (two six-month rotations) at CDC. Examples of rotation activities include: leadership on policy and legislative activities related to nutrition, physical activity, and obesity in the Division of Nutrition, Physical Activity and Obesity; development of case management guidelines for the Division of HIV/AIDS Prevention; and evaluation of reproductive health activities for the World Health Organization. CDC rotations typically take place at CDC headquarters in Atlanta, Georgia; however, other possible locations include: Washington, D.C.; Morgantown, West Virginia; Hyattsville, Maryland; Cincinnati, Ohio; and Denver, Colorado. Examples of field assignments include: development and implementation of HIV/AIDS action plans for Global AIDS programs; creation of a website for physicians and nurses to increase knowledge, improve reporting, and decrease hospitalizations and deaths from pertussis (whooping cough); and coordination and implementation of a multi-cultural media campaign to highlight CDC's National Breast and Cervical Cancer Screening program. Field assignments typically take place at a public health organization in the U.S. or one of its territories. Fellows also have the opportunity to practice public health in an international setting through programs such as the Global AIDS Program (GAP) and Stop Transmission of Polio (STOP). Fellows are CDC employees for the duration of the fellowship. They initially earn a salary equivalent to a GS-9 pay grade and advance to GS-11 with geographic adjustments. They receive benefits such as medical insurance and vacation. Applicants must apply online at the PHPS website. Applications are due before midnight on the application deadline. All required supporting documents must be postmarked by the application deadline date. Deadlines may vary from year to year; applicants should check the PHPS website to verify current submission deadlines and to obtain other important submission details for proposals and supporting documents. Organizations wishing to host and mentor a PHPS fellow are encouraged to follow the "hosting and mentoring a PHPS fellow" link provided at the bottom of the screen at the PHPS website.
Requirements: Applicants should have a strong interest in a leadership and management career in public health and must: meet both the educational and professional requirements; be willing to commit to a 3-year full-time program; and be willing to relocate to Atlanta for year one and then again for an assignment at a state or local health organization. To satisfy educational requirements, applicants must have a master's degree in public health or management-related fields from an accredited college or university and attain academic achievement in areas of: epidemiology/biostatistics; biological, physical, and environmental sciences; behavioral and social sciences; health education and promotion; and management and administration. To satisfy professional requirements, applicants must have at least 2,080 hours of paid public health work experience. This is the equivalent of 1-year, full-time work. Multiple paid work experiences qualify. Work experience can include paid internships, externships, and fellowships. Examples of paid, public health work experience include, but are not limited to: special project coordinator in a health department; graduate student researcher or research assistant in a school of public health, assisting with a public health project; community disease investigator; health education specialist; program coordinator for a community-based public health organization; program consultant for a non-profit public health organization; and health educator in the Peace Corps.
Restrictions: Applicants must be U.S. Citizens.
Date(s) Application is Due: Feb 1
Amount of Grant: 170,954 U.S.D
Contact: Cindi Melanson, MPH, CHES, Director; (404) 498-6120 or (404) 498-6389; fax (404) 498-6125; PHPS@cdc.gov or cindi.melanson@cdc.hhs.gov
Internet: http://www.cdc.gov/PHPS/
Sponsor: Centers for Disease Control and Prevention
1600 Clifton Road
Atlanta, GA 30333

CDC Public Health Prevention Service Fellowship Sponsorships **1174**
The Public Health Prevention Service (PHPS) program offers three-year training and service fellowships for postgraduate masters-level professionals to prepare them for management positions in public health programs. PHPS fellows have advanced degrees in public health or other related fields including (but not limited to) the following areas: public policy; health sciences; health education; public administration; business administration; and social work. A strong interest in a public health leadership and management career, demonstrated achievement in the areas of: epidemiology/biostatistics; biological, physical, and environmental sciences; behavioral and social sciences; health education and promotion; and management and administration, and a prerequisite 2,080 hours of paid public health experience are also required. PHPS fellows complete one year of training at CDC and then spend two years in public health organizations throughout the U.S. and its territories. The purpose of the two-year field assignments is to provide PHPS fellows with high-quality work in the area of program management including (but not limited to) the following: development and implementation of management decisions; analysis and formulation of program and policy recommendations; involvement in budget design, preparation, and monitoring; staff and workforce planning and development; implementation and management of science-based interventions; provision of technical assistance and programmatic consultation on epidemiology-based projects and the role of data in program evaluation; evaluation of public health programs; development and implementation of planning processes; communication of public health information to audiences with diverse backgrounds; and development of partnerships to advance public health programs and priorities. All salaries, benefits, and PHPS-related travel expenses will be the responsibility of the PHPS training and service program. In addition, PHPS will provide PHPS fellows with limited assistance for relocation to the field assignment. The health organization must provide the fellow with training and learning opportunities; office space and equipment required for completing the assignment; periodic assessments of performance, portfolios, and professional development; travel expenses for assignment-related activities and professional development opportunities; ongoing mentoring and supervision; and meaningful assignments in program management. Two-year field assignments must provide opportunities for fellows to meet PHPS performance requirements, which are as follows: to conduct a public health assessment; to develop a plan for a public health program or initiative; to implement a public-health program plan or initiative; to evaluate a public health program or initiative; to analyze a public-health-policy-related issue and prepare a written response; to communicate public health information to a lay audience using a variety of media; to communicate public health information to professional audiences; and to participate in various aspects of the funding process. Mentoring is a critical aspect of all assignments. Assignment supervisors should be involved in the fellow's projects. Field supervisors are required to agree to the following: to commit to a minimum of four hours weekly for mentoring activities and performance feedback; to orient the PHPS fellow to program-specific areas, policies, and protocols; to provide technical guidance for all work assignments; to monitor the assignment, ensuring meaningful work-experiences and activities are provided; facilitate and assess successful completion of assignment-specific skill sets (based on the PHPS performance requirements); assist the PHPS fellow in identifying additional professional development opportunities; facilitate mentoring linkages with appropriate staff; submit periodic PHPS assessment and feedback reports to the PHPS office; and participate in ongoing communication with PHPS supervisors. The CDC encourages public health organizations to take advantage of the opportunity to provide a practical learning experience for PHPS fellows while addressing the organization's public health priorities. Organizations interested in sponsoring a PHPS fellow should download the updated "Guide for Health Organizations to Request a Fellow" booklet (Adobe PDF file; 21 pages; 265KB) for instructions, samples of a letter of intent, and a full application. A download link for this document is provided at the PHPS Fellows Sponsorship website. In brief, organizations must first submit a Letter of Intent (LOI) by January 10. Accepted applicants will be invited to submit a full application describing the proposed assignment by March 16. If the full application is approved, the organization will be invited to participate in open recruitment. During this two-week period, health organizations are encouraged to discuss their potential assignments with PHPS fellows. After open recruitment, selected health organizations are invited to formal interviews with PHPS fellows in Atlanta, Georgia. Primary supervisors are expected to participate. The health organization is responsible for all travel-related expenses. Application and LOI deadlines may vary from year to year. Applicants are encouraged to check the PHPS Fellows Sponsorship website to verify current deadline dates.
Requirements: Eligible health agencies or organizations (hereafter referred to as organizations) include the following: state and local health departments; U.S. territorial health departments; Indian Health Service area offices; federally-recognized tribes with established public health department structures (or their equivalent) providing public health services to their tribal members; and other health-related organizations (e.g., community-based organizations, foundations, and universities) that have an active collaboration with a state or local health department.
Date(s) Application is Due: Mar 16

Contact: Cindi Melanson, MPH, CHES, Director; (404) 498-6120 or (404) 498-6389; fax (404) 498-6125; PHPSAssignments@cdc.gov or cindi.melanson@cdc.hhs.gov
Internet: http://www.cdc.gov/PHPS/FieldAssignments/index.html
Sponsor: Centers for Disease Control and Prevention
1600 Clifton Road
Atlanta, GA 30333

CDC Research, Treatment, and Education Programs on Lyme Disease in the United States Cooperative Agreements 1175
The objective of the program is to develop and implement practical and effective measures for the primary and secondary prevention of Lyme disease. Funds will be available to develop disease surveillance, conduct ecological and epidemiological studies, develop prevention and control activities, develop better diagnostic tests, and develop and disseminate educational materials and programs. Cooperative agreements are usually awarded for a three-year project period. Initial awards are made for a one-year budget period with continuation awards for up to an additional two years. Contact the office listed for application deadline dates. PA-04-008
Requirements: Public and nonprofit organizations able to provide services to geographical areas where Lyme disease is endemic or found to be newly emerging are eligible to apply. Funding preference will be given to proposals that incorporate evaluation of effectiveness, cost, and acceptability of strategies, or combinations of strategies, for population-based control of tick-borne diseases.
Amount of Grant: 100,000 - 700,000 U.S.D
Contact: Dr. Brian Mahy, National Center for Infectious Diseases; (404) 639-0043; fax (404) 639-2469; bxm1@cdc.gov
Internet: http://www.cdc.gov/od/pgo/funding/Expired/04008.htm
Sponsor: Centers for Disease Control and Prevention
1600 Clifton Road
Atlanta, GA 30333

CDC School Health Programs to Prevent the Spread of HIV Cooperative Agreements 1176
The objective of the program is to support the development and implementation of effective health education for HIV and other important health problems for school-age populations (elementary- through college-age youth); parents; and relevant school, health, and education personnel. Cooperative agreement funds may be used to support personnel, their training and travel, and to purchase supplies and services for planning, organizing, and conducting activities directly related to the objectives of this program. These activities may include obtaining baseline data and establishing a system to monitor the availability and adequacy of HIV education in schools; increasing the number of schools providing HIV education; monitoring the levels of HIV-related knowledge, beliefs, and behaviors among students; collaborating with state and local health agencies in carrying out HIV education programs in the schools; establishing policies, guidelines, advisory committees, and/or standards to increase the number of schools providing HIV education; integrating HIV education within a more comprehensive program of school health education; disseminating information about accessibility, availability, and quality of educational strategies, materials, and curricula to local education agencies and schools; and providing technical assistance to local school districts and schools in implementing HIV education. The maximum project period is five years. Contact the office for deadline dates.
Requirements: Eligible state education agencies in U.S. states and territories, nonprofit organizations, and universities may apply.
Restrictions: Funds may not be spent for research activities, surveys, or questionnaires except as may be needed to collect basic evaluation requirements. In addition, funds may not be used for purchasing computer equipment, office equipment or furnishings, renting or leasing office space, or financing construction or renovation unless specifically approved.
Amount of Grant: 87,000 - 650,000 U.S.D
Contact: Mary Sue Lancaster, Program Development and Services Branch, Division of Adolescent and School Health; (770) 488-6130; fax (770) 488-6163; msl0@cdc.gov
Internet: http://www.cdc.gov
Sponsor: Centers for Disease Control and Prevention
1600 Clifton Road
Atlanta, GA 30333

CDC Sexually Transmitted Diseases Research, Demonstration, Public Information, and Education Grants 1177
The grants are meant to develop, improve, apply, and evaluate methods for the prevention and control of syphilis, gonorrhea, and other sexually transmitted diseases through demonstration and applied research. Also, the grants are meant to develop, improve, apply, and evaluate methods and strategies for public information and education about these diseases; to support professional (including appropriate allied health personnel) education, training, and clinical skills improvement activities; and to support particularly deserving public information and education programs that cannot be supported through other grant programs. Grant funds may be used for the costs associated with planning, organizing, and conducting applied research, demonstrations and education programs, and to reimburse individuals asked to be participants in the applied research. Contact the program office for deadline dates.
Requirements: Any state, political subdivision of a state, or any other public or private nonprofit institution may apply.
Contact: Sevgi Aral, Director; (404) 639-8259; fax (404) 639-8608
Internet: http://www.cdc.gov
Sponsor: Centers for Disease Control and Prevention
1600 Clifton Road
Atlanta, GA 30333

CDC Steven M. Teutsch Prevention Effectiveness Fellowships 1178
The Prevention Effectiveness Fellowship (PEF) is a two-year, highly-selective research fellowship for recent doctoral graduates with a background in economics, policy analysis, operations research, decision sciences, and other quantitative areas, and was established to address demand in the field of public health for economics-based inquiry, quantitative policy analysis, and integrative-health-services research. The PEF allows applicants to apply their academic training in quantitative methods to the science of health protection, health promotion, and disease prevention. Prevention effectiveness uses economic evaluation, policy analysis, and decision modeling to systematically assess the impact of public health interventions, policies, programs, and practices on health outcomes. Econometric, decision, simulation, and operations modeling and methods are employed to determine the efficacy, effectiveness, efficiency, and cost of public health services, policies, and interventions in order to facilitate better decision-making about the allocation of resources around programmatic priorities. At CDC, prevention effectiveness methods are used to explain: health-related burden and costs; effectiveness and efficiency of health protection, health promotion, and disease prevention programs; health program prioritization based on optimization modeling; disease transmission and intervention modeling; and health system performance. Fellows have the opportunity to work with leading researchers on issues of critical importance to create the expertise, information, and tools that enable people and communities to protect their health through: health promotion; prevention of disease, injury and disability; and preparedness for new health threats. Fellows are assigned to various CDC Centers, Institute, and Offices (CIOs) where they assess the public health impact of policies, programs, and practices by determining their cost, effectiveness, and overall impact. The PEF provides initial classroom training and hands-on experience to support the development of valuable research skills. After an introduction to basic public health sciences and prevention-effectiveness methods, fellows are assigned to a CDC program area where they take a lead role in prevention-effectiveness studies. Assignment activities include: designing and conducting studies; working closely with national and international experts in public health; presenting and publishing results; and providing technical assistance and service on specific projects or methods. In addition, professional support and career development are provided through didactic training, workshops, research seminars, and peer discussion groups. Fellows are full-time CDC employees during the fellowship and will receive the same benefits (such as medical insurance and vacation) as other full-time CDC employees. During the first year, fellows are paid the equivalent of a GS 12, Step 1 salary. In the second year, they are eligible for salary equivalent to a GS 12, Step 2 with geographic adjustments. Applicants must apply online at the PEF website (click on "Apply Now" on the left hand side of the screen). Applications are due on January 21 before midnight. All required supporting documents must be postmarked by January 19. Deadlines may vary from year to year; applicants should check the PEF website to verify current submission deadlines for proposals and supporting documents. Further essential details on application and supporting document requirements are available from the PEF website.
Requirements: Applicants must hold a doctoral degree in economics or applied economics, decision sciences, health services research or related health sciences, industrial engineering or operations research, public policy or policy analysis, or a related quantitatively-oriented field. A medical degree with additional relevant training is acceptable. Dissertations must be defended by June 17 and degree requirements must be met by July 15 prior to the start date of the fellowship. Written and spoken proficiency in English is essential. Relocation to Atlanta, Georgia is required. As stated in the FAQ at the PEF website, up to $3,000 is provided to cover relocation expenses. Non-U.S. citizens must be legal permanent residents or eligible for J-1 status prior to the program's start date.
Restrictions: Citizens of countries that have been determined by the U.S. Secretary of State to be State Sponsors of Terrorism will not be able to get security clearance to work at CDC and should not apply to the PE Fellowship Program. A list of these may be found at the U.S. Department of State website.
Geographic Focus: All States, All Countries
Date(s) Application is Due: Jan 11
Amount of Grant: 149,198 U.S.D
Contact: Adam G. Skelton, Ph.D., M.P.H., Fellowship Coordinator; (404) 498-6324 or (404) 498-6786; PEF@cdc.gov or adam.skelton@cdc.hhs.gov
Internet: http://www.cdc.gov/PEF/index.html
Sponsor: Centers for Disease Control and Prevention
1600 Clifton Road
Atlanta, GA 30333

CDC Suicide Prevention Research Center 1179
The purpose of this grant is to establish a Suicide Prevention Research Center. The recipient of the grant will conduct, evaluate, and publish suicide prevention research; design, implement, and evaluate suicide prevention programs; collaborate with outside agencies; develop a curricula and graduate training programs in disciplines relevant to suicide prevention; and disseminate findings.
Requirements: Eligible applicants must demonstrate expertise in some form of suicide prevention research. Applicant must provide a director/Principal Investigator.
Amount of Grant: 500,000 U.S.D
Contact: Dr. Thomas Shires; (702) 671-2297; fax (702) 385-9399; DVPINFO@cdc.gov
Internet: http://www.cdc.gov/ncipc/dvp/suinevada.htm
Sponsor: Centers for Disease Control and Prevention
1600 Clifton Road
Atlanta, GA 30333

CDC Summer Graduate Environmental Health Internships 1180
The Centers for Disease Control and Prevention (CDC)'s Graduate Environmental Health (GEH) Program provides paid ten week summer internships (June through August) for graduate students with majors or interests deemed to be in support of

environmental health missions at CDC's National Center for Environmental Health (NCEH) and Agency for Toxic Substances and Disease Registry (ATSDR). The mission of the internship is to offer selected students a broad overview of environmental health at the federal level and to foster an interest in environmental health as a career. Applicants should have a passion for the environment and an eagerness to learn about the environment's link to human health. In many cases interested students may receive academic credit for this internship. Academic credit will be considered on a case by case basis. During the program interns will be assigned to projects that utilize the skills they've acquired through graduate studies and personal experiences. Interns will also be able to take advantage of lecture series and other opportunities offered at the CDC. GEH interns receive a weekly stipend of $750 for living expenses. All interns are responsible for the cost of travel to and from Atlanta and for housing during the internship. Information about student housing options at local universities can be found at the internship website. Links to application materials and guidelines are provided at the website along with specific instructions for submitting applications and supporting documents via email, fax, or U.S. mail. Annual deadline dates may vary; applicants are encouraged to check the website or contact program staff to verify current deadline dates.
Requirements: Each GEH intern must: possess U.S. Citizenship or Permanent Resident status with a green card; be currently enrolled full-time in a college/university as a graduate student in a degree granting program; have interests and/or studies that are applicable to the field of environmental health; have a minimum cumulative GPA of 3.0 on a scale of 4.0; and have proof of health insurance. All participants are considered guest researchers and are subject to CDC regulations governing visiting scientists, engineers, and other professionals. As a guest researcher, each participant is responsible for payment of income taxes and is advised to become familiar with the relevant sections of the current tax codes.
Restrictions: Deferrals of the internship will not be allowed. Participation in the program is contingent upon the individual's ability to obtain the proper security clearance. All applicants will be subject to a criminal records check and other background investigations conducted by the U.S. Government. These inquiries are conducted to develop information to assess various factors about the applicant, including reliability, trustworthiness, honesty, integrity, character, conduct and loyalty to the United States.
Date(s) Application is Due: Feb 28
Amount of Grant: 7,500 U.S.D
Contact: Lt. Cory Moore, M.P.H., Environmental Health Specialist; (770) 488-0593; fax (404) 929-2820; geh@cdc.gov or cory.moore@cdc.hhs.gov
Internet: http://www.cdc.gov/nceh/geh/default.htm
Sponsor: Centers for Disease Control and Prevention
1600 Clifton Road
Atlanta, GA 30333

CDC Summer Program In Environmental Health Internships **1181**
The Centers for Disease Control and Prevention (CDC)'s Summer Program in Environmental Health (SUPEH) is a 10-week internship (June through August) for students majoring in environmental health. Interns participate in activities with the Environmental Health Services Branch of CDC's National Center for Environmental Health (NCEH). In many cases interested students may receive academic credit for this internship and are encouraged to check with their academic advisors or other appropriate school official. Over the summer, interns will explore many aspects of environmental public health and gain an understanding of environmental health work at the local, state, regional, tribal, and federal levels. Activities include: collaborating with federal employees working in the environmental health field and developing mentoring relationships; shadowing senior officials and scientists; spending time in local public health departments; traveling to important environmental health sites in the city of Atlanta and beyond; attending brown-bag lunches with CDC staff to discuss environmental health topics; and making presentations at CDC staff meetings. CDC's extensive lecture and seminar program is available to interns as well. SUPEH interns and interns in the Collegiate Leaders for Environmental Health (CLEH) program attend field trips and other activities together. SUPEH interns receive a weekly stipend of $500 for living expenses. All interns are responsible for the cost of travel to and from Atlanta and for housing during the internship. Information about student housing options at local universities can be found at the internship website. Applications and supporting materials are due by 5 p.m. on the due date. Applications are available for download at the internship website and may be submitted via email, fax, or U.S. mail. Applicants are advised to check the website for specific instructions for each method of submission. As actual deadlines may vary from year to year, applicants are advised to check the internship website to verify current deadline and other key dates. SUPEH internships are facilitated through the Oak Ridge Institute for Science and Education.
Requirements: Each SUPEH intern must: major in an environmental health program accredited by the National Environmental Health Science and Protection Accreditation Council (EHAC); have U.S. citizenship or Permanent Resident status with a green card; be enrolled full-time as a rising junior, senior, or graduate student at a college or university; have a minimum cumulative GPA of 3.0 on a 4.0 scale; have current health insurance coverage acceptable in the State of Georgia; and be fluent in the English language. Students must return to school to complete degree requirements at the end of their internship OR the internship must serve as needed academic credit to complete degree requirements. Participants in the Junior Commissioned Officer Student Training and Extern Program (JRCOSTEP) are also eligible for the internship. All summer interns are considered guest researchers. As such, interns are subject to CDC regulations governing visiting scientists, engineers, and other professionals. As a guest researcher, each intern is responsible for payment of income taxes and is advised to become familiar with the relevant sections of the Internal Revenue Code and other applicable state and federal laws and regulations.
Restrictions: The internship must be accepted in the summer it is offered; internships cannot be deferred. Participation in the program is contingent on the applicant's ability to obtain the proper security clearances. All applicants are subject to a criminal records check and other background investigations conducted by the U.S. government. These inquiries develop information to assess various factors about the applicant, including reliability, trustworthiness, honesty, integrity, character, conduct, and loyalty to the United States.
Date(s) Application is Due: Feb 11
Amount of Grant: 5,000 U.S.D
Contact: Lt. Cory Moore, M.P.H., Environmental Health Specialist; (770) 488-0593; fax (404) 929-2820; ehinternship@cdc.gov or cory.moore@cdc.hhs.gov
Internet: http://www.cdc.gov/nceh/ehs/supeh/
Sponsor: Centers for Disease Control and Prevention
1600 Clifton Road
Atlanta, GA 30333

CDC Technology Translation and Transfer of Effective HIV Behavioral Interventions **1182**
HIV/AIDS researchers have developed and tested prevention interventions that aim to reduce sex-related and drug-related risk behaviors. A number of these interventions have credible evidence of effectiveness, defined as reporting positive and significant results on an HIV-relevant behavioral or health outcome. The purposes of this cooperative agreement project are to: translate effective HIV prevention interventions whose original research was conducted with methodological rigor, preferably at multiple sites with multiple populations at risk; develop packages of materials in collaboration with health departments, community-based organizations, and/or other prevention providers and consumers, that prevention providers can use to replicate the interventions in nonresearch field situations; and study the process of technology transfer using the prevention packages in at least one field setting supported by training, quality assurance, and technical assistance.
Requirements: Applications may be submitted by public and private nonprofit organizations and governments and their agencies.
Amount of Grant: 50,000 - 1,000,000 U.S.D
Contact: Cheryl Maddux, Grants Manager; (770) 498-1911
Internet: http://www.cdc.gov/od/pgo/funding/grantmain.htm
Sponsor: Centers for Disease Control and Prevention
1600 Clifton Road
Atlanta, GA 30333

CDC Using Genetic Information to Prevent Disease and Improve Health Research Grants **1183**
The purpose of this program is to strengthen science for public health action, collaborate with health care partners for prevention, and promote healthy living at every stage of life. The program will provide funding for conducting population-based research to assess how risk for disease and disability in well-defined populations is influenced by the interaction of human genetic variation with modifiable risk factors, and ensure that genetic tests and services are incorporated in population-based interventions that promote health and prevent disease and disability. Annual deadline dates may vary; contact the program office for exact dates.
Requirements: Applications may be submitted by public and private nonprofit organizations and by governments and their agencies, other public and private nonprofit organizations, state and local governments or their bona fide agents, and federally recognized American Indian tribal governments, American Indian tribes, or American Indian tribal organizations.
Contact: Nealean Austin, Grants Management Branch, Centers for Disease Control and Prevention; (770) 488-2700
Internet: http://www.cdc.gov
Sponsor: Centers for Disease Control and Prevention
1600 Clifton Road
Atlanta, GA 30333

Cemala Foundation Grants **1184**
The Cemala Foundation is a private family foundation established in 1986 by Martha A. and Ceasar Cone II to continue the family tradition of commitment to enhancing the quality of life of the community through grants to qualified charitable organizations. Areas of interest are: arts/culture; education; health; homelessness; human services; the state of North Carolina and; public interest. Application and additional guidelines are available at the Foundation's website.
Requirements: Grants are made only to non-profit charitable organizations which are tax exempt under Section 501(c)3 of the Internal Revenue Code or to public governmental units. Generally, grants are limited to projects which benefit the citizens of Guilford County, North Carolina. Occasionally, projects which benefit the state of North Carolina as a whole are considered.
Restrictions: The Foundation does not consider support for annual campaigns, endowments, sectarian religious activities, or requests under $1,000. Grants are not made to individuals. Grants from the Cemala Foundation are usually awarded for one year only. Only one grant application may be submitted in any twelve-month period. Organizations receiving grants are required to complete an evaluation report within twelve months after receipt of the funds.
Geographic Focus: North Carolina
Date(s) Application is Due: Mar 1; Sep 1
Amount of Grant: 3,000 - 200,000 U.S.D
Samples: Eastern Music Festival for the MYEMF, $25,000—Community Outreach Project; Bennett College, $25,000—for general operating support; Women's Resource Center of Greensboro, $15,000—for the New Choices Program to provide pre-employment training and placement case management support to assist displaced homemakers; Salvation Army in Greensboro, $75,000—for emergency financial assistance program.
Contact: Susan S. Schwartz; (336) 274-3541; fax (336) 272-8153; cemala@cemala.org

Internet: http://www.cemala.org/grant/guidelines.php
Sponsor: Cemala Foundation
330 South Greene Street, Suite 101
Greensboro, NC 27401

Center for Advanced Study in the Behavioral Sciences Postdoctoral Fellowships 1185
The center offers nine- to 12-month residential postdoctoral fellowships for scientists and scholars of proven accomplishment or exceptional promise. Fellowships have been offered in psychology, sociology, anthropology, political science, history, economics, philosophy, psychiatry-psychoanalysis, linguistics, humanities, law, and education; and certain biomedical, mathematical, and statistical specialties. The selection process is started by nominations being received from well-known behavioral scientists, academic administrators, and former fellows. Nominees are screened by a review panel and the final selections are made by the center's board of trustees. The selection process normally takes several years, with the fellowship roster for any given year usually prepared more than a year in advance. Usually 45 to 50 fellowships are awarded annually. Nominations may be submitted at any time.
Requirements: Qualified scholars in behavioral and biological sciences, humanities, and certain areas of statistics and computer science from any country are eligible. Resident participants usually have completed university training, hold university appointments, are engaged in research and research training, and are in the top 5 to 10 percent of the scholars in their field.
Contact: Robert Scott, Associate Director; (650) 321-2052; fax (650) 321-1192; info@casbs.org
Internet: https://casbs.stanford.edu
Sponsor: Center for Advanced Study in the Behavioral Sciences
75 Alta Road
Stanford, CA 94305

Center for Alternatives to Animal Testing Grants 1186
The center seeks to develop innovative non-whole animal methods to evaluate fully commercial and/or therapeutic products to ensure the health and safety of the public. The program funds research that will lead to the refinement, replacement, or reduction of animals in toxicity testing and by disseminating scientifically correct information about these methods and their applications. The center encourages the development of in vitro approaches to toxicity evaluation including, but not limited to, methods using human cells/cell lines and studies in the areas of skin hypersensitivity/toxicity phototoxicity, target organ toxicity, and structure-activity relationships. Grants support one year of research with possible renewal for another year. Preproposals must meet the application deadline; full proposals are invited.
Requirements: Applicants must have an established mechanism for handling private funding.
Restrictions: Projects focusing on mutagenicity or carcinogenicity are not funded.
Amount of Grant: Up to 25,000 U.S.D
Contact: Alan Goldberg, Center Director; (410) 223-1692; fax (410) 223-1603; caat@jhsph.edu or goldberg@jhsph.edu
Internet: http://caat.jhsph.edu/programs/grants/grants.htm
Sponsor: Johns Hopkins University
111 Market Place, Suite 840
Baltimore, MD 21202-6709

Center for Black Music Research Rockefeller Fellowships 1187
Under the auspices of the Rockefeller Foundation, the Center for Black Music Research (CBMR) will offer a series of resident research fellowships under the theme of Researching the Circum-Caribbean. The CBMR will accept applications to pursue work on musico-cultural issues related to Latin-American and West Indian cultures in the circum-Caribbean region. Applications will be considered from scholars in the fields of music and music librarianship, history, Caribbean studies, and American cultural studies. Two fellowships will be offered. Fellows will spend half their time in residence at the AMRI in Saint Thomas, U.S. Virgin Islands, from which they will do field or archival research, and the other half at the CBMR's Chicago site preparing for or writing the results of the work they performed while at the Caribbean site or other sites of field research. The fellows will present work-in-progress at colloquia and will have opportunities to interact with faculty, students, and members of the Chicago and Saint Thomas arts, cultural, and educational communities. Each fellowship includes a stipend ($25,000), an allowance ($4000 maximum) for moving to and from the fellow's home and the CBMR (or the Virgin Islands); an allowance $2000 maximum) for inter-island or intra-Caribbean research travel; and free housing in both locations. Application guidelines are available online.
Requirements: Applicants must have completed work on their terminal degree (PhD degree or its equivalent) at the time of application and should be able to demonstrate their familiarity with the research of or related to the circum-Caribbean region.
Date(s) Application is Due: Feb 2
Amount of Grant: 37,750 U.S.D
Contact: Linda Hunter, Fellowship Administrator; (312) 344-7559; fax 312) 344-8029; lhunter@cbmr.colum.edu
Internet: http://www.cbmr.org/fellows0405.htm
Sponsor: Columbia College Chicago
600 S Michigan Avenue
Chicago, IL 60605-1996

Center for Science in the Public Interest Internships 1188
The center is a national consumer organization that focuses on health and nutrition issues. CSPI offers internships for a small number of qualified students in undergraduate, graduate, law, and medical schools each summer and during the school year. Generally, an internship is for 10 weeks. The specific dates of internship are flexible and depend on the center's needs and the applicant's schedule. The undergraduate interns earn $6 an hour and the graduate interns earn $7 an hour. Application materials should include a cover letter indicating issues of interest, future plans, available dates; resume; writing sample; two letters of recommendation; and official transcript of courses and grades. Application materials should be submitted as early as possible.
Contact: Human Resources Manager; (202) 332-9110; fax (202) 265-4954; cspi@cspinet.org
Internet: http://www.cspinet.org/about/jobs_internship_2005_2006.html
Sponsor: Center for Science in the Public Interest
1875 Connecticut Avenue NW, Suite 300
Washington, DC 20009-5728

Central Carolina Community Foundation Community Impact Grants 1189
The Foundation's Community Impact grants fund programmatic initiatives of nonprofit organizations. In order to support positive community change and help ensure a program's success, the Grants have a maximum request amount of $10,000. The grant making process is a bi-annual process. Proposals are accepted, April 1 - September 30, for the Fall cycle, October 1 - March 31 for the Spring cycle.
Requirements: Nonprofit organizations serving South Carolina communities are eligible.
Restrictions: Grants do not support routine operating expenses, fund raising projects, debt reduction, endowment development, medical research, conference travel, or conference underwriting or sponsorship. Letter of intent or proposal may not be submitted by email or fax.
Geographic Focus: South Carolina
Date(s) Application is Due: Mar 31; Sep 30
Amount of Grant: Up to 10,000 U.S.D
Samples: Cooperative Ministry, Columbia, SC, $10,000—for the Helping Hands of Hope program, which provides assistance to individuals and families for rent, utilities, food/household, hygiene, transportation and furniture; Project Life: Positeen, Orangeburg, SC, $10,000—to expand its services into two new sites, Rivelon and Bethune-Bowman Elementary Schools in Orangeburg County.
Contact: Veronica L. Pinkett-Barber, Program Officer; (803) 254-5601 ext. 331; fax (803) 799-6663; veronica@yourfoundation.org
Internet: http://www.yourfoundation.org/nonprofits/foundationgrants/impactgrants.aspx
Sponsor: Central Carolina Community Foundation
2711 Middleburg Drive, Suite 213
Columbia, SC 29204

Central Okanagan Foundation Grants 1190
The foundation seeks to enhance the quality of life in Canada and awards grants to nonprofit organizations in its areas of interest, including arts and culture, children, youth, families, education, health, community services, environmental conservation, and heritage preservation. Types of support include emergency funds, equipment acquisition, program development, scholarship funds, and seed money. Contact the foundation by phone to discuss the proposed project. Guidelines and application form will be sent if the project falls within the foundation's guidelines.
Requirements: Nonprofit British Columbia organizations in the Central Okanagan region as defined by the boundaries of BC School District 23 are eligible.
Restrictions: Grants are not awarded to individuals or religious organizations for religious purposes. Grants do not support capital campaigns, debt retirement, endowments, building construction/renovation, or operating expenses.
Geographic Focus: All States, Canada
Date(s) Application is Due: May 1; Oct 1
Contact: Cheryl Miller, Grants Manager; (250) 861-6160; fax (250) 861-6156; cheryl@centralokanaganfoundation.org
Internet: http://www.centralokanaganfoundation.org/pages/grants/cof-grants.php
Sponsor: Central Okanagan Foundation
217-1889 Springfield Road
Kelowna, BC V1Y 5V5 Canada

Cessna Foundation Grants 1191
Grants are made to organizations primarily in the areas where the company operates. Support is given for projects that take a creative approach to such fundamental issues as education, health and human services, youth enrichment, and arts and culture. The foundation values projects which take a creative approach to such fundamental issues as education, neighborhood improvement, youth development, community problem-solving, assistance to people who are disadvantaged, environmental conservation and cultural enrichment. Grants cover projects, building, equipment, capital campaigns, program development, employee matching gifts, and employee-related scholarships. All proposals must be in writing and should include the following information: needs statement and project objectives; purpose of organization; constituency; board members, community, and volunteer involvement; how results of the project will be measured; plans for continued funding of the project, if applicable; one-page project budget and an organizational chart; and primary source of funds. Proposals are accepted at any time.
Requirements: Giving limited to areas of company operations, with emphasis on Wichita, KS.
Restrictions: Grants are not awarded to individuals; national or regional organizations, unless their programs address specific local community needs; programs or initiatives where the primary purpose is the promotion of religious doctrine or tenets; elementary or secondary schools (except to provide special initiatives or programs not provided by regular school budgets); political action or legislative advocacy groups; operational funds; medical or other research organizations; organizations located in or benefiting nations other than the United States and its territories; or fraternal groups, athletic teams, bands, veterans organizations, volunteer firefighters, or similar groups.

Amount of Grant: 300 - 10,000 U.S.D
Contact: Rhonda Fullerton, Secretary-Treasurer; (316) 517-7810; fax (316) 517-7812
Internet: http://www.cessna.com/about/corporate-citizenship.html
Sponsor: Cessna Foundation
P.O. Box 7706
Wichita, KS 67277

Cetana Educational Foundation Scholarships **1192**
Cetana selects a limited number of scholars, simply because the foundation has limited resources. Scholars from all religious and ethnic groups in Myanmar are eligible for scholarships. Support for scholars covers the cost of education, room and board, travel to and from the university, and associated expenses for the duration of their study abroad. Cetana also provides mentors who are available to offer the scholars advice and practical support in bridging cultures. The Outreach Committee in Myanmar assists the graduate in finding rewarding employment upon return. Once they have finished their studies, scholars return to Myanmar for a period of at least three years so that they may share their new knowledge and experience. Most scholarships are for graduate studies in neighboring SE Asian countries. Only a few scholarships are available in the U.S. where colleges provide full tuition grants.
Requirements: Applicants must be living in Myanmar. Those living outside the country will not be considered. Scholars must be studying in majors that have application to work opportunities in country. Students must pledge to return home for a minimum of 3 years after completion of studies. If the recipient breaks the pledge, he or she must pay back full scholarship so that another student may take his or her place.
Contact: Tin May Thein, Director; (240) 242-3997; tinmaythein@cetana.org
Internet: http://www.cetana.org/Cetana/Scholarship_Program.html
Sponsor: Cetana Educational Foundation
12817 Twinbrook Parkway, #207
Rockville, MD 20851

CF Clinical Research Trainee Awards **1193**
This is a training program for physicians enrolled in a U.S. or Canadian subspecialty training program. Ten awards will be granted for clinical research in thrombosis, asthma, critical care, lung cancer, and women's health. Applicants may apply for only one type of clinical research award, and each program director may support only one applicant for an award in each area. Requests should be written to include goals and objectives that may be accomplished within one year of the starting date (July 1) of the research project. Research funding may be used as the research mentor and program director deem appropriate (e.g., salary support and supplies directly related to the research project).
Requirements: Applicants must be citizens of the United States or Canada or have a J-1 or H-1 visa and be working in a certified U.S. or Canadian institution and enrolled in a subspecialty training program at the time of application; hold an MD degree or its equivalent; be enrolled in a U.S. or Canadian subspecialty training program in allergy/immunology, cardiac electrophysiology, critical care anesthesiology, critical care intensive care, critical care medicine, infectious disease, cardiology, pediatric critical care, pediatric pulmonary disease, pulmonary disease, surgical critical care, or thoracic surgery; be affiliate members of the American College of Chest Physicians (contact Angela Popa-Padilla at 847/ 498-1400 for an application); show promise of significant contribution to his/her specialty; and apply for only one type of award and be the sole applicant from the subspecialty program to apply for that type of award.
Restrictions: Awards are to support clinical research and not basic or bench level research.
Date(s) Application is Due: Jun 30
Contact: Sue Ciezadlo, Grants Administrator; (847) 498-8363; fax (847) 498-5460; sciezadlo@chestnet.org
Internet: http://www.chestfoundation.org/researchAwards/index.php
Sponsor: Chest Foundation
3300 Dundee Road
Northbrook, IL 60062-2348

CFCR Health Care Grants **1194**
The foundation awards grants to eligible New York nonprofit organizations to improve health care for residents of the region. Types of support include capital grants and program development grants. Grants may be held for a maximum of three years. Applicants should submit a letter of inquiry that describes the organization, the program, and the amount sought. Full applications are by invitation.
Requirements: New York 501(c)3 tax-exempt organizations serving Albany, Rensselaer, Schenectady, or Saratoga Counties are eligible.
Geographic Focus: New York
Amount of Grant: 2,500 - 30,000 U.S.D
Contact: Jackie Mahoney, Program Director; (518) 446-9638; fax (518) 446-9708; Jmahoney@cfcr.org, or info@cfcr.org
Internet: http://www.cfcr.org/grantmaking/grantmaking.htm
Sponsor: Community Foundation for the Capital Region
6 Tower Place, Executive Park Drive
Albany, NY 12203

CFF-NIH Funding Grants **1195**
The objective of this award is to support excellent cystic fibrosis-related research projects that have been submitted to and approved by NIH but cannot be supported by available NIH funds. Applications must fall within the upper 40th percentile with a priority score of 200 or better. Investigators will be required to resubmit applications to NIH. Interested individuals are encouraged to contact the foundation for application procedures and/or to discuss the potential relevance of their work to the objectives of this program. Applications may be submitted on an ongoing basis.
Amount of Grant: 75,000 - 125,000 U.S.D
Contact: Office of Grants Management; (800) 344-4823 or (301) 951-4422; fax (301) 951-6378; grants@cff.org
Internet: http://www.cff.org/research
Sponsor: Cystic Fibrosis Foundation
6931 Arlington Road, 2nd Floor
Bethesda, MD 20814

CFF Clinical Research Grants **1196**
These awards offer support of clinical research projects directly related to cystic fibrosis. Projects may address diagnostic or therapeutic methods related to cystic fibrosis or the pathophysiology of cystic fibrosis. However, studies performed in animal models or cell/tissues removed from patients will not be supported under this award. Applicants must demonstrate access to a sufficient number of cystic fibrosis patients and appropriate controls. A letter of intent to apply must be submitted by June 1 with applications required by December 1.
Date(s) Application is Due: Jun 1; Dec 1
Amount of Grant: Up to 80,000 U.S.D
Contact: Office of Grants Management; (800) 344-4823 or (301) 951-4422; fax (301) 951-6378; grants@cff.org
Internet: http://www.cff.org/research/cystic_fibrosis_foundation_grants/research_grants
Sponsor: Cystic Fibrosis Foundation
6931 Arlington Road, 2nd Floor
Bethesda, MD 20814

CFF First- and Second-Year Clinical Fellowships **1197**
The foundation offers competitive clinical fellowships for up to three years for physicians interested in cystic fibrosis and other chronic pulmonary and gastrointestinal diseases of children and adolescents/adults. The intent of this award is to encourage specialized training early in a physician's career and to prepare well-qualified candidates for careers in academic medicine. Training must take place in one of the foundation's accredited centers and must provide thorough grounding in diagnostic and therapeutic procedures, comprehensive care, and cystic fibrosis-related research. All first- and second-year programs must commit a significant portion (at least 30 percent over two years) to research training. Fellows funded by other sources for their first year of training may apply to the foundation for subsequent training support. Third-year fellowships are available for additional basic and/or clinical research training. Research fellowships are also available from the foundation to support physicians beyond the fellowship experience; salary will be commensurate with experience.
Requirements: Applicants must be U.S. citizens or have permanent U.S. resident visas, and must have completed pediatric training and be eligible for board certification in pediatrics by the time the fellowship begins. Candidates with prior training in internal medicine may also apply but must have completed at least two years of an approved adult pulmonary or GI fellowship; be jointly sponsored by the departments of medicine and pediatrics; and be willing to commit at least 75 percent of their time to cystic fibrosis and related problems of young adults.
Date(s) Application is Due: Oct 1
Amount of Grant: 42,000 - 85,750 U.S.D
Contact: Office of Grants Management; (800) 344-4823 or (301) 951-4422; fax (301) 951-6378; grants@cff.org
Internet: http://www.cff.org/research/cystic_fibrosis_foundation_grants/training_grants
Sponsor: Cystic Fibrosis Foundation
6931 Arlington Road, 2nd Floor
Bethesda, MD 20814

CFF Leroy Matthews Physician-Scientist Awards **1198**
This award will provide up to six years of support for outstanding newly trained pediatricians and internists (MDs and MD/PhDs) to complete subspecialty training, develop into independent investigators, and initiate research programs. Institutional and individual grants are available.
Requirements: U.S. citizenship or permanent resident status is required.
Date(s) Application is Due: Sep 1
Amount of Grant: 52,000 - 345,000 U.S.D
Contact: Office of Grants Management; (800) 344-4823 or (301) 951-4422; fax (301) 951-6378; grants@cff.org
Internet: http://www.cff.org/research/cystic_fibrosis_foundation_grants/research_grants
Sponsor: Cystic Fibrosis Foundation
6931 Arlington Road, 2nd Floor
Bethesda, MD 20814

CFF Pilot and Feasibility Awards **1199**
These grants are for developing and testing new hypotheses and/or new methods, and to support promising new investigators as they establish themselves in research areas relevant to cystic fibrosis. Proposed work must be hypothesis driven and must reflect innovative approaches to critical questions in cystic fibrosis research. Two years of support may be requested.
Restrictions: The award is not meant to support continuation of programs begun under other granting mechanisms. Electronic submission is required.
Date(s) Application is Due: Sep 1
Amount of Grant: Up to 40,000 U.S.D
Contact: Office of Grants Management; (800) 344-4823 or (301) 951-4422; fax (301) 951-6378; grants@cff.org
Internet: http://www.cff.org/research/cystic_fibrosis_foundation_grants/research_grants

Sponsor: Cystic Fibrosis Foundation
6931 Arlington Road, 2nd Floor
Bethesda, MD 20814

CFF Postdoctoral Research Fellowships 1200
Competitive postdoctoral fellowships in basic or clinical research are made on an annual basis, renewable for a second year. The fellowships are awarded in areas of basic cellular and metabolic research or in other problem areas pertinent to cystic fibrosis and related chronic and recurrent pulmonary and gastrointestinal diseases of childhood. Preference is shown to recent graduates or those just beginning their investigative careers. A third year is offered on a limited basis to highly qualified candidates. Interested individuals are encouraged to contact the foundation for application procedures and/or to discuss the potential relevance of their work to the objectives of the foundation.
Requirements: These awards are offered to MDs, PhDs, and MD/PhDs interested in conducting basic or clinical research related to cystic fibrosis.
Restrictions: Electronic submission only.
Date(s) Application is Due: Sep 1
Amount of Grant: 33,000 - 103,400 U.S.D
Contact: Office of Grants Management; (800) 344-4823 or (301) 951-4422; fax (301) 951-6378; kcurley@cff.org
Internet: http://www.cff.org/research/cystic_fibrosis_foundation_grants/training_grants
Sponsor: Cystic Fibrosis Foundation
6931 Arlington Road, 2nd Floor
Bethesda, MD 20814

CFF Research Grants 1201
Grants are offered in support of high-quality research projects ranging from basic cellular and metabolic mechanisms to therapy of cystic fibrosis and related chronic and recurrent pulmonary and gastrointestinal diseases of childhood. These grants are broadly oriented to the support of projects for developing and initially testing new hypotheses and/or new methods, or those being applied to problems of cystic fibrosis for the first time. The intent of this award is to enable the investigator to collect sufficient data to compete successfully for long-term support from NIH. Interested individuals are encouraged to contact the foundation for guidelines and/or to discuss the potential relevance of their work to the objectives of this program.
Requirements: Applications may be submitted by established investigators or new investigators starting their independent research careers but may not represent support for continuation of a line of research already established by the applicant.
Date(s) Application is Due: Sep 1
Amount of Grant: Up to 90,000 U.S.D
Contact: Office of Grants Management; (800) 344-4823 or (301) 951-4422; fax (301) 951-6378; grants@cff.org
Internet: http://www.cff.org/research/cystic_fibrosis_foundation_grants/research_grants
Sponsor: Cystic Fibrosis Foundation
6931 Arlington Road, 2nd Floor
Bethesda, MD 20814

CFF Student Traineeships 1202
The traineeships are offered to qualified students to introduce them to cystic fibrosis research. Each recipient must work with a faculty sponsor on a research project related to cystic fibrosis. Interested individuals are encouraged to contact the foundation for application procedures and/or to discuss the potential relevance of their work to the objectives of the program. Applications are accepted throughout the year. A maximum of $300 of the award may be used for laboratory expenses with the remainder used as a stipend for the trainee.
Requirements: Applicants must be students in or about to enter a doctoral program (MD, PhD, or MD/PhD); senior-level undergraduates planning to pursue graduate training may also apply.
Amount of Grant: 1,500 U.S.D
Contact: Office of Grants Management; (800) 344-4823 or (301) 951-4422; fax (301) 951-6378; grants@cff.org
Internet: http://www.cff.org/research/cystic_fibrosis_foundation_grants/training_grants
Sponsor: Cystic Fibrosis Foundation
6931 Arlington Road, 2nd Floor
Bethesda, MD 20814

CFF Therapeutics Development Program 1203
The purpose of this program is to provide funds to businesses that will develop commercial products to benefit individuals with cystic fibrosis. Structured as a matching grants program, funds will be awarded only if they are matched by the recipient. First, the grantee will examine the scientific potential of new products under component 1. Component 2 studies will involve support for the continuation of component 1 developments and the initiation of patient clinical studies. Component 1 and component 2 awards each will receive a maximum of two years of support. Funds will be provided for research prior to, but not including, phase 3 clinical trials. Applications are accepted on an ongoing basis.
Amount of Grant: 200,000 - 1,500,000 U.S.D
Contact: Office of Grants Management; (800) 344-4823 or (301) 951-4422; fax (301) 951-6378; grants@cff.org
Internet: http://www.cff.org/research/cystic_fibrosis_foundation_therapeutics
Sponsor: Cystic Fibrosis Foundation
6931 Arlington Road, 2nd Floor
Bethesda, MD 20814

CFF Third-Year Clinical Fellowships 1204
The foundation offers competitive clinical fellowships for up to three years for physicians interested in cystic fibrosis and other chronic pulmonary and gastrointestinal diseases of children and adolescents/adults to encourage specialized training early in a physician's career and to prepare well-qualified candidates for careers in academic medicine. The third-year fellowship award offers support for additional intense basic and/or clinical research training related to cystic fibrosis.
Requirements: Applicants and sponsors must submit proposals of the research studies to be undertaken and other specialized training that will be offered during this year. Preference will be given to applicants whose training was supported by the foundation.
Restrictions: Recipients who do not enter a career of academic medicine will be subject to payback provisions.
Date(s) Application is Due: Oct 1
Amount of Grant: 52,600 - 62,600 U.S.D
Contact: Office of Grants Management; (800) 344-4823 or (301) 951-4422; fax (301) 951-6378; grants@cff.org
Internet: http://www.cff.org/research/cystic_fibrosis_foundation_grants/training_grants
Sponsor: Cystic Fibrosis Foundation
6931 Arlington Road, 2nd Floor
Bethesda, MD 20814

CFFVR Alcoholism and Drug Abuse Grants 1205
The Community Foundation for the Fox Valley Region (CFFVR) was established as a public, nonprofit organization in 1986 to enhance the quality of life for all people of the region. Since it was founded, funds within the Foundation have awarded more than $125 million in grants to nonprofit organizations, primarily in Wisconsin's Fox Valley region. The purpose of the Alcoholism and Drug Abuse Grants program is to: supports new or existing programs and projects that address the prevention and/or treatment of alcohol and other drug abuse in the Fox Valley and surrounding area.
Requirements: Organizations serving residents of Outagamie, Calumet, Waupaca, Shawano and northern Winnebago counties are eligible to apply. The grant application form is available at the CFFVR website.
Geographic Focus: Wisconsin
Amount of Grant: 2,500 U.S.D
Contact: Todd Sutton, Community Engagement Manager; (920) 830-1290, ext. 28; fax (920) 830-1293; lfilapek@cffoxvalley.org
Internet: https://www.cffoxvalley.org/Page.aspx?pid=652
Sponsor: Community Foundation for the Fox Valley Region
4455 West Lawrence Street, P.O. Box 563
Appleton, WI 54912-0563

CFFVR Basic Needs Giving Partnership Grants 1206
Supported by the U.S. Oil Open Fund for Basic Needs within the Community Foundation and the J.J. Keller Foundation, the partnership assists established charitable organizations with successful programs that address root causes of poverty. Available forms of support include: capacity building; general operating support; project support; project analysis & advocacy. A single organization may request up to $15,000 per year for three years, and collaborative proposals may request up to $100,000 per year for three years. Multiple years of support will be considered only if there is a compelling case for multi-year funding and the project clearly demonstrates how progression shall occur over time.
Requirements: Eligible applicants are well-established charitable organizations that are exempt from federal income taxes under the Internal Revenue Code and have been in operation for a minimum of three years. Wisconsin organizations must serve residents in Outagamie, Calumet, Waupaca, Shawano, or northern Winnebago counties.
Restrictions: Grants from the Basic Needs Giving Partnership will not support the following: technology projects; capital campaigns or building projects; organizational set-up costs; annual fund drives or endowments; lobbying for specific legislation; activities that occur before funding is awarded; organizations with past-due or incomplete grant reports.
Geographic Focus: Wisconsin
Date(s) Application is Due: Feb 15; Sep 15
Contact: Martha Hemwall, Community Engagement Officer; (920) 830-1290, ext. 26; mhemwall@cffoxvalley.org
Internet: https://www.cffoxvalley.org/Page.aspx?pid=400
Sponsor: Community Foundation for the Fox Valley Region
4455 West Lawrence Street, P.O. Box 563
Appleton, WI 54912-0563

CFFVR Capital Credit Union Charitable Giving Grants 1207
The Capital Credit Union Charitable Giving Fund was established to support projects and programs that provide for basic needs (food, shelter, clothing and medical care), making a positive impact on the lives of those in the communities served by Capital Credit Union. Grant applications are accepted annually with a submission deadline of December 31. The grant application is available online at the Community Foundation for the Fox Valley Region website.
Requirements: To be eligible to apply for a grant, organizations must: be a public charity, as determined by the IRS and described in 501(c)3 of the tax code (educational institutions and government programs/entities are not eligible for consideration); provide, or propose to develop, services that are focused directly on improving the availability of, or providing for, basic needs in our community; benefit specific communities served by Capital Credit Union. These include the counties of Calumet, Outagamie, and Winnebago in the state of Wisconsin.
Geographic Focus: Wisconsin
Date(s) Application is Due: Dec 31

Contact: Shelly Leadley; (920) 830-1290, ext. 34; sleadley@cffoxvalley.org
Internet: https://www.cffoxvalley.org/Page.aspx?pid=415
Sponsor: Community Foundation for the Fox Valley Region
4455 West Lawrence Street, P.O. Box 563
Appleton, WI 54912-0563

CFFVR Chilton Area Community Foundation Grants 1208

The Chilton Area Community Foundation works to enhance the quality of life for the people of the greater Chilton area. Grants support specific projects or new programs for which a moderate amount of grant money can make a significant impact on an area of need. Grants are made for a broad range of purposes to a wide variety of charitable organizations in the focus areas of health, arts/culture, education and community development. Grants support projects and programs with clear goals and financial accountability, including: creative new activities or services (new programs, one-time projects, events, exhibits, studies or surveys); enhancement or strengthening of existing activities (projects to enhance, expand or strengthen the range, quantity and/or quality of an organization's programs and services); and small capital investments (items that are directly related to program delivery or service to clients, such as a refrigerator for a food pantry or equipment to comply with Americans with Disabilities Act requirements). Applications are available on the website.
Requirements: Organizations serving residents of Outagamie, Calumet, Waupaca, Shawano and northern Winnebago counties are eligible to apply. The grant application form is available at the CFFVR website.
Geographic Focus: Wisconsin
Date(s) Application is Due: Dec 31
Contact: Terry Friederichs; (920) 849-4042; tjfriederichs@charter.net
Internet: https://www.cffoxvalley.org/Page.aspx?pid=415
Sponsor: Community Foundation for the Fox Valley Region
4455 West Lawrence Street, P.O. Box 563
Appleton, WI 54912-0563

CFFVR Clintonville Area Foundation Grants 1209

Clintonville Area Foundation (CAF) grants are awarded from unrestricted funds to support specific projects or new programs for which a moderate amount of grant money can make an impact on an area of need. Grants are made for a broad range of purposes to a wide variety of charitable organizations in the focus areas of health, education and community development. The Foundation is interesting in supporting: creative new activities or services—new programs, one-time projects, events, exhibits, studies or surveys; enhancement or strengthening of existing activities—projects to enhance, expand or strengthen the range, quantity and/or quality of an organization's programs and services; small capital investments—items that are directly related to program delivery or service to clients, such as a refrigerator for a food pantry or equipment to comply with ADA requirements. Grant applications are available online.
Requirements: Wisconsin 501(c)3 nonprofit organizations that serve the residents of the Clintonville area are eligible. General questions can also be directed to the CAF Grants Committee Chair or the Foundation.
Restrictions: The Clintonville Area Foundation will not generally fund the following: general operating expenses not related to the proposed project; annual fund drives or fundraising events; endowment funds; programs with a sectarian or religious purpose that promote a specific journey of faith; major capital projects such as the acquisition of land or buildings; medical research; travel for individuals or groups such as bands, sports teams or classes; activities that occur before funding is awarded; organizations with past-due or incomplete grant reports.
Geographic Focus: Wisconsin
Date(s) Application is Due: Dec 31
Contact: Jenny Goldschmidt; (715) 823-7125, ext. 2603; clintonvillefoundation@gmail.com
Internet: https://www.cffoxvalley.org/Page.aspx?pid=415
Sponsor: Community Foundation for the Fox Valley Region
4455 West Lawrence Street, P.O. Box 563
Appleton, WI 54912-0563

CFFVR Frank C. Shattuck Community Grants 1210

The Frank C. Shattuck Community Fund supports new or supplements existing services for youth and the elderly and benefits education, the arts and health care in Winnebago and Outagamie counties of Wisconsin. Funding is available for programs, capital expenses or operating expenses of qualifying charitable organizations. Grants to projects may be either one-time payments or multi-year commitments. Application forms are available online.
Requirements: Wisconsin 501(c)3 organizations that serve residents in Outagamie or northern Winnebago counties are eligible to apply. To begin the application process, submit the following prior to the application deadline: grant application form; list of the organization's governing board members, including their professional or community affiliation; current year (board-approved) operating budget.
Restrictions: The Shattuck Fund will not typically support the following: grants for religious or political purposes; grants to support endowment funds of organizations; travel for individuals or groups such as bands, sports teams or classes; reimbursement for previously incurred expenses.
Geographic Focus: Wisconsin
Date(s) Application is Due: Mar 1; Sep 1
Contact: Shelly Leadley; (920) 830-1290, ext. 34; sleadley@cffoxvalley.org
Internet: https://www.cffoxvalley.org/Page.aspx?pid=339
Sponsor: Community Foundation for the Fox Valley Region
4455 West Lawrence Street, P.O. Box 563
Appleton, WI 54912-0563

CFFVR Jewelers Mutual Charitable Giving Grants 1211

Through philanthropy, Jewelers Mutual Insurance Company aspires to achieve a lasting and positive impact on the Fox River Valley. Jewelers Mutual strives to be a responsible corporate citizen and a valued employer by supporting critical needs of the community. Priority areas for giving are: organizations in which Jewelers Mutual employees are actively involved; organizations that address needs in the following areas: basic needs (food, shelter, clothing), as well as programs that reduce/eliminate the root causes of poverty (such as literacy, affordable housing, job training); positive youth development and education; health and wellness, particularly diabetes, cancer, mental health, and affordable access for the disadvantaged; and the vitality of the Fox River area, including gifts to libraries, police or fire departments, parks, the arts, and preservation of the local environment. Letters of request may be sent to the Community Foundation for the Fox Valley Region.
Requirements: Nonprofit, 501(c)3 organizations that serve residents of the Fox River Valley are eligible.
Geographic Focus: Wisconsin
Contact: Shelly Leadley; (920) 830-1290; SLeadley@cffoxvalley.org
Internet: http://www.jewelersmutual.com/information.aspx?id=4305
Sponsor: Community Foundation for the Fox Valley Region
4455 West Lawrence Street, P.O. Box 563
Appleton, WI 54912-0563

CFFVR Myra M. and Robert L. Vandehey Foundation Grants 1212

The Myra M. and Robert L. Vandehey Foundation's mission is support the charitable interests of the Myra and Robert Vandehey family. Areas of interest include: education; children and youth; health care and; family services. Funding opportunities are limited to nonprofit organizations that serve residents in the Fox Cities or Keshena areas of Wisconsin. Contact the Vice President for CFFVR prior to submitting a request to verify that the need aligns with current priorities. Unsolicited grant requests are not accepted from organizations not previously awarded support. Applications may be submitted at any time, and no application form is required.
Requirements: Wisconsin 501(c)3 nonprofit organizations that serve residents in the Fox Cities or Keshena are eligible for funding.
Restrictions: No grants to individuals.
Geographic Focus: Wisconsin
Samples: Fox Valley Technical College Foundation, Appleton, WI, $20,000—educational grant; Boys & Girls Clubs of the Fox Valley, Appleton, WI, $25,000—youth development grant; Nami Fox Valley, Inc., $10,000—mental health and crisis intervention grant.
Contact: Cathy Mutschler, Vice President Community Engagement; (920) 830-1290, ext. 29; cmutschler@cffoxvalley.org
Internet: https://www.cffoxvalley.org/Page.aspx?pid=415
Sponsor: Community Foundation for the Fox Valley Region
4455 West Lawrence Street, P.O. Box 563
Appleton, WI 54912-0563

CFFVR Project Grants 1213

Project grants, support specific projects or new programs for which a moderate amount of grant money can make a significant impact on an area of need. Grants are made for a broad range of purposes to a wide variety of charitable organizations in the focus areas of arts and culture, community development, education, environment, health and human services. Organizations eligible to apply must serve residents in Outagamie, Calumet, Shawano, Waupaca or northern Winnebago counties. Grants for specific projects typically are for no more than $10,000 for one year.
Requirements: Wisconsin 501(c)3 tax-exempt organizations serving residents of Outagamie, Calumet, Waupaca, Shawano and northern Winnebago counties are eligible to apply. The grant application form is available at the CFFVR website.
Restrictions: Project grants typically will not fund the following: general operating expenses not related to the proposed project; annual fund drives or fund raising events; endowment funds; programs with a sectarian or religious purpose that promote a specific journey of faith; major capital projects such as the acquisition of land or buildings; medical research; travel for individuals or groups such as bands, sports teams or classes; activities that occur before funding is awarded; health and safety equipment; playground equipment; organizations with past-due or incomplete grant reports.
Geographic Focus: Wisconsin
Date(s) Application is Due: Feb 1; Aug 1
Amount of Grant: 10,000 U.S.D
Samples: Child Care Resource and Referral, $2,000—to provide at-risk families with young children increased access to supportive community services; Community Clothes Closet, $10,000—to purchase a used cargo van to provide a means of picking up donated clothing; Northeast Wisconsin Land Trust, $4,000—to acquire supplies and equipment necessary to fully implement a newly created land stewardship program.
Contact: Todd Sutton, Grants Officer; (920) 830-1290, ext. 28; fax (920) 830-1293; tsutton@cffoxvalley.org
Internet: https://www.cffoxvalley.org/Page.aspx?pid=343
Sponsor: Community Foundation for the Fox Valley Region
4455 West Lawrence Street, P.O. Box 563
Appleton, WI 54912-0563

CFFVR Robert and Patricia Endries Family Foundation Grants 1214

The Robert & Patricia Endries Family Foundation was established for the benefit of people in need, primarily in the Brillion area but with some consideration to the Fox Valley, Lakeshore and Northeastern areas of Wisconsin. Priority areas of giving include: the vitality of the Brillion area; the disadvantaged, particularly the disabled, homeless, low income, single parents, troubled youth, or the chronically or mentally ill; health and

human services, particularly diabetes, cancer, cerebral palsy, Alzheimer's disease, kidney disease, or mental health; religious causes or organizations with a spiritual purpose; sports or arts programming and, or sponsorships. Grants are considered for capital campaigns and, or specific capital improvements for the above priority organizations. Gifts will be directed to specific programs or opportunities, not to general operations (with exception for those organizations the foundation has had a long-established relationship with). Matching or challenge gifts are also encouraged, to motivate additional giving by others. Organizations that support needs outside of the Brillion area or that do not yet have an established relationship with the foundation should contact the foundation prior to submission of a formal request.
Requirements: Wisconsin 501(c)3 charitable organizations are eligible to apply. Contact Foundation directly be begin the application process.
Restrictions: The Foundation will not typically support the following: gifts to political organizations or causes; gifts to organizations that are not pro-life supporters or that lack sensitivity to promoting human life in any form (unborn or born); gifts to organizations affiliated with or in support of cloning or embryonic stem-cell research; grants to organizations that receive significant public/government funding; reimbursement for previously incurred expenses.
Geographic Focus: Wisconsin
Date(s) Application is Due: Jan 1; Apr 1; Oct 1
Contact: Shelly Leadley; (920) 830-1290, ext. 34; sleadley@cffoxvalley.org
Internet: https://www.cffoxvalley.org/Page.aspx?pid=415
Sponsor: Community Foundation for the Fox Valley Region
4455 West Lawrence Street, P.O. Box 563
Appleton, WI 54912-0563

CFFVR Schmidt Family G4 Grants **1215**
The Schmidt Family G4 grants provide funding to improve the quality of life of those most in need in the Fox Valley, Wisconsin region, with a focus on at-risk youth and self-sufficiency for women. This goal will be accomplished by seeking to address immediate needs and to affect meaningful change in the following areas: at-risk youth—especially those with a physical or mental illness, those who have experienced abuse or those who have significant financial need; adult self-sufficiency—with a priority on issues that affect the stability and independence of women, as well as literacy, job skills training and transitional living for all. The G4 Committee prefers: not to be the sole funder for most projects it considers, unless the amount requested is small and/or a one-time request; to support specific projects or new programs for which a moderate amount of grant money can make a significant impact on an area of need and sustainability. A broad array of requests will be considered, including capital campaigns, existing programs or recurring events as long as they fall within the other listed giving guidelines. Grant awards will typically not exceed $15,000. To assist with the educational aspect of this fund, a formal application is required (available online). Prior to submitting an application, organizations are strongly encouraged to contact Cathy Mutschlerto, discuss the potential proposal and process. Complete and submit the application prior to the March 1, October 1 deadlines.
Requirements: IRS 501(c)3 nonprofit organizations, as well as government agencies are eligible to apply for funding. Organizations that are not public charities may apply through a fiscal sponsor. Organizations must serve Fox Valley residents, particularly in Outagamie, Calumet or northern Winnebago counties of Wisconsin.
Restrictions: The G4 Fund typically will not support the following: organizations that have received funding from the G4 Committee in the most recent 20 months; multi-year requests; programs with a sectarian or religious purpose that promote a specific journey of faith; travel for individuals or groups such as bands, sports teams or classes; reimbursement for previously incurred expenses; endowment funds; fund-raising events; requests from organizations with past-due or incomplete grant reports; a program or need previously declined unless the organization is invited back by the committee; programs or needs that do not serve Fox Valley residents, particularly Outagamie, Calumet or northern Winnebago counties.
Geographic Focus: Wisconsin
Date(s) Application is Due: Mar 1; Sep 1
Amount of Grant: 15,000 U.S.D
Contact: Cathy Mutschler, VP Community Engagement; (920) 830-1290, ext. 29; cmutschler@cffoxvalley.org
Internet: https://www.cffoxvalley.org/Page.aspx?pid=340
Sponsor: Community Foundation for the Fox Valley Region
4455 West Lawrence Street, P.O. Box 563
Appleton, WI 54912-0563

CFFVR Shawano Area Community Foundation Grants **1216**
Shawano Area Community Foundation works to preserve and improve the quality of life in Shawano, Wisconsin and, the surrounding area, including communities having economic, educational, cultural and recreational ties with the area. Grant applications are available online.
Requirements: Wisconsin 501(c)3 non-profits in or serving surrounding area of Shawano are eligible to apply.
Geographic Focus: Wisconsin
Date(s) Application is Due: Oct 1
Amount of Grant: 5,000 U.S.D
Samples: City of Shawano/County Library, Shawano, WI, $3,000—reading machine for visually impaired; Shawano Oral Health Fund, Shawano, WI, $1,500—program support for ssecond grade children in Shawano County; St. John's Trinity Lutheran Church, Shawano, WI, $1,000—food purchase grant.
Contact: Susan Hanson, Grants Manager; (715) 253-2580; shawanofoundation@granitewave.com
Internet: https://www.cffoxvalley.org/Page.aspx?pid=412
Sponsor: Community Foundation for the Fox Valley Region
4455 West Lawrence Street, P.O. Box 563
Appleton, WI 54912-0563

CFFVR Waupaca Area Community Foundation Grants **1217**
Waupaca Area Community Foundation works to preserve and improve the quality of life for the people of the greater Waupaca area of Wisconsin. The Foundation awards grants to Wisconsin non-profits, primarily in the following areas of interest: human services; arts and culture; education; community development. Grant applications are available online.
Requirements: Wisconsin 501(c)3 non-profits in or serving the residents of the Waupaca area are eligible to apply.
Restrictions: The Foundation typically will not award grants to: support routine operating expenses of established organizations, to support annual fund drives of such organizations, or to eliminate their previously incurred deficits; support new or established endowment funds; fund specific research projects; support travel for individuals or groups (such as bands, sports teams, forensics competitors, or the like); support sectarian or religious purposes or causes; reimburse anyone for previously incurred expenses.
Geographic Focus: Wisconsin
Date(s) Application is Due: Jun 30
Amount of Grant: 5,000 U.S.D
Contact: Jack Rhodes, WACF President; (715) 256-1939
Internet: https://www.cffoxvalley.org/Page.aspx?pid=415
Sponsor: Community Foundation for the Fox Valley Region
4455 West Lawrence Street, P.O. Box 563
Appleton, WI 54912-0563

CFFVR Wisconsin King's Daughters and Sons Grants **1218**
The Wisconsin Branch of the International Order of King's Daughters and Sons, in alignment with the priorities of its national parent organization, will provide grants to Fox Valley charitable organizations who provide services related to autism and literacy. Application is available online.
Requirements: Wisconsin 501(c)3 non-profit organizations in the Fox Valley region may apply. Organizations that are not public charities may apply through a fiscal sponsor.
Geographic Focus: Wisconsin
Date(s) Application is Due: Oct 15
Amount of Grant: 3,000 U.S.D
Contact: Kathy Mutschler, Director, Donor Engagement; (920) 830-1290, ext. 27; fax (920) 830-1293; cmutschler@cffoxvalley.org
Internet: https://www.cffoxvalley.org/Page.aspx?pid=875
Sponsor: Community Foundation for the Fox Valley Region
4455 West Lawrence Street, P.O. Box 563
Appleton, WI 54912-0563

CFFVR Women's Fund for the Fox Valley Region Grants **1219**
The Women's Fund provides grants for programs that inspire women and girls to flourish personally, economically and professionally. Grants have been distributed to programs supporting the following areas: arts & culture; physical and mental health; economic, self-sufficency; education; parenting and child care; violence prevention. The Women's Fund believes that no project is too small or too new to be considered. Innovative approaches and projects with limited access to other funding are encouraged. Collaborative efforts are welcome. Grant applicants should address one or more of these funding priorities as they relate to women and girls: promotes economic self-sufficiency; improves safely from violence; provides opportunities to develop life skills; promotes physical and/or mental health; enhances dignity and self-worth; promotes leadership development; provides opportunities for artistic development and/or exposure to the arts; provides gender-specific solutions to problems facing women and girls; creates an environment that encourages social change. To apply submit a letter of interest by the deadline.
Requirements: To be eligible for a grant, the project must be consistent with the Women's Fund mission; benefits women and girls in the Wisconsin, Fox Valley region; organization must be a tax-exempt, not-for-profit organization under the Internal Revenue Code, section 501(c)3.
Restrictions: The Women's Fund will not fund: individuals, endowments, government agencies (however educational institutions may qualify), projects with a religious focus, and political parties, candidates or partisan activities.
Geographic Focus: Wisconsin
Date(s) Application is Due: Mar 15
Contact: Becky Boulanger, Program Director; (920) 830-1290, ext. 17; bboulanger@cffoxvalley.org or grants@womensfundfvr.org
Internet: https://www.cffoxvalley.org/Page.aspx?pid=415
Sponsor: Community Foundation for the Fox Valley Region
4455 West Lawrence Street, P.O. Box 563
Appleton, WI 54912-0563

CFPC Family Medicine Resident Leadership Awards **1220**
The CFPC Research and Education Foundation honors outstanding educational and research initiatives in the area of family medicine through conferral of Family Medicine Resident Awards. Canadian Research Awards for Family Medicine Residents provide national recognition for original research carried out by postgraduate trainees. The Murray Stalker Memorial Lecture Award is given to promote and recognize scholarly activities of family medicine residents. The Nadine St. Pierre Award recognizes the outstanding achievement of a francophone family practice resident. Award recipients

receive return air fare, complimentary registration for the Section of Teachers Annual Workshop, and a stipend for expenses. Deadline dates vary between programs; contact program staff for exact dates.
Geographic Focus: All States, Canada
Date(s) Application is Due: Jun 1
Amount of Grant: 250 - 1,000 CAD
Contact: Kiki Ziten; (905) 629-0900, ext. 432; fax (905) 629-0893; kziten@cfpc.ca
Internet: http://www.cfpc.ca/English/cfpc/programs/awards%20program/default.asp?s=1
Sponsor: College of Family Physicians of Canada
2630 Skymark Avenue
Mississauga, ON L4W 5A4 Canada

CFPC Family Physician of the Year Award **1221**
The annual award recognizes an individual for outstanding contribution to family medicine. Criteria for the award include excellence in family practice over time, community medical service, contribution to the community, and involvement with CFPC. The award is presented at the time of the CFPC Convocation in conjunction with the National Annual Scientific Assembly in Halifax, Nova Scotia. The recipient will present a short addresss at the time of convocation and will be involved as a spokesperson for Canadian family medicine throughout the year following receipt of the award. The recipient will receive return air fare, complimentary registration, and a stipend for expenses. The deadline for submission varies according to each Provincial Chapter. Please contact the Provincial Chapter of your nominee to confirm the deadline date.
Requirements: Each provincial chapter may select a nominee for the award. Names must be submitted to the National Honours and Awards Committee.
Geographic Focus: All States, Canada
Amount of Grant: 2,000 CAD
Contact: Kiki Ziten, (905) 629-0900, ext. 432; fax (905) 629-0893; kziten@cfpc.ca
Internet: http://www.cfpc.ca/English/cfpc/programs/awards%20program/members%20can%20apply%20for/default.asp?s=1#Perkinawards
Sponsor: College of Family Physicians of Canada
2630 Skymark Avenue
Mississauga, ON L4W 5A4 Canada

CFPC Fellowships **1222**
The program awards fellowships to CFPC members to recognize achievement in the discipline of family medicine. The conferral of fellowships will take into consideration the nominee's exemplary performance as a family physician and exceptional contributions to community activities that enhance health in its broadest sense; outstanding leadership in the academic discipline of family medicine and major administrative responsibilities within a university department of family medicine; and significant elected or appointed executive positions at the national or chapter level of the CFPC or involvement in major committee or other college activities.
Requirements: Nominees must hold certification in family medicine, have a minimum of 10 years continuous membership, and be members in good standing at the time of nomination.
Geographic Focus: All States, Canada
Date(s) Application is Due: Feb 15
Contact: Kiki Ziten; (905) 629-0900, ext. 432; fax (905) 629-0893; kziten@cfpc.ca
Internet: http://www.cfpc.ca/English/cfpc/research/section%20of%20researchers/home/default.asp?s=1
Sponsor: College of Family Physicians of Canada
2630 Skymark Avenue
Mississauga, ON L4W 5A4 Canada

CFPC Research and Education Foundation Grants **1223**
The program is supported by the CFPC Research and Education Foundation and industry partners and comprises research grants, scholarships, and training programs for educational and research initiatives in the area of family medicine. The Royal Canadian Legion One-Month Traineeship grants are awarded to improve the quality of medical education and the provision of care to elderly patients. The Hollister King grant provides an opportunity for a practicing family physician to attain or enhance special skills in the delivery of rural medical care. The Douglas M. Robb research grant is given to a community-based member who plans to conduct research on a topic relevant to the practice of family medicine. Research/development grants provide seed money to members to initiate or complete research/development projects in family medicine. Practice Enrichment grants enable members in active practice to participate in a course of study under the direction of a Canadian university for a minimum of three months in fields related to family medicine. Traveling Scholarship grants enable members to pursue clinical studies for a minimum of three weeks under the direction of a clinical department at a Canadian university. Graduate Study grants enable members to pursue a course of study at a Canadian university for a period of at least six months on a part-time basis.
Geographic Focus: All States, Canada
Amount of Grant: 1,000 - 5,000 CAD
Contact: Kiki Ziten; (905) 629-0900, ext. 432; fax (905) 629-0893; kziten@cfpc.ca
Internet: http://www.cfpc.ca/English/cfpc/programs/awards%20program/awardshome%20pdf/default.asp?s=1
Sponsor: College of Family Physicians of Canada
2630 Skymark Avenue
Mississauga, ON L4W 5A4 Canada

CFPC Special Recognition Awards **1224**
The CFPC Research and Education Foundation with support of industry partners honors outstanding educational and research initiatives in the area of family medicine through conferral of Special Recognition Awards. The W. Victor Johnston Oration award is presented each year at convocation by a renowned leader in family medicine; the orator is selected by the CFPC executive committee. The Donald I. Rice Merit Award allows a nationally or internationally renowned leader in family medicine the opportunity to travel and engage in educational activities. The Ian McWhinney Family Medicine Education Award is presented to an outstanding family medicine teacher. The Irwin Bean Award is presented annually to the practice-eligible candidate who achieves the highest standing in the Certification Examination in Family Medicine. The Janssen-Ortho Literary Awards are given annually to authors of articles published in Canadian Family Physician that are judged to be the best sumbission in English and in French. The Canadian Family Physician Research Award promotes research and publication by family physicians. Annual deadline dates may vary; contact program staff for exact dates.
Geographic Focus: All States, Canada
Amount of Grant: 500 - 5,000 CAD
Contact: Kiki Ziten; (905) 629-0900, ext. 432; fax (905) 629-0893; kziten@cfpc.ca
Internet: http://www.cfpc.ca/English/cfpc/programs/awards%20program/awardshome%20pdf/default.asp?s=1
Sponsor: College of Family Physicians of Canada
2630 Skymark Avenue
Mississauga, ON L4W 5A4 Canada

CF Roger C. Bone Award for Advances in End-of-Life Care **1225**
The award was created to annually recognize members of the American College of Chest Physicians who have demonstrated leadership in end-of-life care. Recipients of this award have shown substantial improvement in the care of patients as documented in the related sections of the application form. Application forms and guidelines are available online.
Requirements: Applicants must be members of the American College of Chest Physicians and have the endorsement of another ACCP member.
Date(s) Application is Due: Apr 29
Amount of Grant: 10,000 U.S.D
Contact: Sue Ciezadlo, Senior Project Coordinator; (847) 498-8363; fax (847) 498-5460; sciezadlo@chestnet.org
Internet: http://www.chestfoundation.org/criticalCare/index.php
Sponsor: Chest Foundation
3300 Dundee Road
Northbrook, IL 60062-2348

Challenged Athletes Foundation Grants **1226**
The foundation awards grants to individuals with physical disabilities who need funding for sport-specific training, competition, or athletic equipment needs. All sports are eligible. Applications are accepted year round with a December 1st postmarked deadline.
Requirements: People of all ages with permanent disabilities are eligible. Disabilities must be recognized by the International Paralympic Committee
Date(s) Application is Due: Dec 1
Contact: Virginia Tinley, Executive Director; (858) 793-9293; fax (858) 793-9291; execdir@challengedathletes.org
Internet: http://www.challengedathletes.org
Sponsor: Challenged Athletes Foundation
2148 Jimmy Durante Boulevard, Suite B
Del Mar, CA 92014

Chamberlain Foundation Grants **1227**
The foundation awards grants to Pueblo, CO, nonprofit organizations in its areas of interest, including arts and culture, education, religion, and science. Types of support include general operating support, equipment acquisition, and program development. Submit a brief proposal that includes the purpose of the request and its relevance to the foundation; amount requested; brief history of the organization and its achievements; names, titles, and qualifications of key personnel; name of agency that conducted the last annual audit; copy of the IRS tax-exemption letter; list of other sources of financial support during the past 12 months; and names and affiliations of board members, trustees, and officers of the organization by either of the two deadlines.
Requirements: 501(c)(30 Pueblo, CO, nonprofit organizations are eligible.
Restrictions: Grants do not support individuals; conferences; political activities; religious organizations whose services are limited to members; veteran, labor, fraternal, athletic, or social clubs; national health agencies concerned with specific diseases or health issues; operating expenses for United Way-supported organizations; publications, advertising campaigns, or travel expenses.
Geographic Focus: Colorado
Amount of Grant: 1,000 - 20,000 U.S.D
Contact: David Shaw, Foundation Chair; (719) 543-8596; fax (719) 543-8599
Sponsor: Chamberlain Foundation
501 North Main Street, Suite 222
Pueblo, CO 81003

Champ-A Champion Fur Kids Grants **1228**
Grants for children's health and wellness are awarded twice each year. The program provides direct support for children in the areas of health and wellness such as childhood disease research foundations, child safety organizations and organizations that serve children with special needs. The goal is to provide grants to help many programs that are

working hard to make the world a healthier and happier place for kids. Programs funded include: (a) Individual Project grants (generally for one-time purchases or to fulfill a short-term need, such as the purchase of materials or equipment); or, (b) Organization Program grants (start-up or operational costs for ongoing programs. Examples include funds for research, health and wellness educational programs, or financial assistance for children and families in-need.)
Requirements: United States applicants must be a tax-exempt organization under Section 501(c)3 of the IRS Code, and 'not a private foundation,' within the meaning of Code Sections 509(a)(1) or 509(a)(2), or a state college or university within the meaning of Code Section 511(a)(2)(B) (a 'Public Charity'). In addition, grant recipients must certify that they are not a supporting organization within the meaning of Code Section 509(a)(3). Canadian applicants must be a registered Canadian charity.
Restrictions: Programs that will not be funded include: (1) Annual Appeals or Capital Campaigns; (2) Construction or New Facility expenses; (3) Fundraising or Event Sponsorships; (4) Political Activities; (5) Religious organizations for religious purposes.
Geographic Focus: All States, Canada
Date(s) Application is Due: Feb 28; Aug 30; Nov 30
Amount of Grant: 1,000 - 10,000 U.S.D
Contact: Maxine Clark, President; (314) 423-8000, ext. 5366; giving@buildabear.com
Internet: http://www.buildabear.com/aboutus/community/bearhugs.aspx
Sponsor: Build-A-Bear Workshop Bear Hugs Foundation
1954 Innerbelt Business Center Drive
Saint Louis, MO 63114

Champlin Foundations Grants **1229**
The Champlin Foundations are comprised of three foundations, the first established in 1932 by George S. Champlin, Florence C. Hamilton and Hope C. Neaves, who also created The Second Champlin Foundation in 1947. Both of these trusts are administered by the same Distribution Committee. The Third Champlin Foundation was established by George S. Champlin in 1975 and is administered by a separate Distribution Committee. All three foundations share the same management and PNC Bank/Delaware is the trustee of each. The Champlin Foundations are private foundations as defined in Section 509(a) of the Internal Revenue Code and are exempt from Federal income tax under Section 501(c)3. The Foundations areas of interest include: Youth/Fitness; Hospitals/Healthcare; Open Space/Conservation/Environment; Education; Libraries; Social Services; Historic Preservation; Cultural/Artistic; Animal Humane Societies. The Foundations make direct grants to tax exempt organizations, substantially all in Rhode Island, mostly for capital needs. Capital needs may consist of equipment, construction, renovations, the purchase of real property and reduction of mortgage indebtedness. One important goal is to fund tax exempt organizations within Rhode Island that will have the greatest impact on the broadest possible segment of the population. Another important goal is to provide "hands on" equipment and facilities for those being served by these tax-exempt organizations. Grant applications are accepted March 1st through April 30th. No applications will be accepted via facsimile or email. Applications must be postmarked no later than April 30th, or in the case of April 30th falling on a weekend, then the first business day thereafter.
Requirements: 501(c)3 tax exempt organizations of Rhode Island are eligible to apply.
Geographic Focus: Rhode Island
Date(s) Application is Due: Apr 30
Amount of Grant: 25,000 - 65,000 U.S.D
Contact: Keith Lang, Executive Director; (401) 736-0370; fax (401) 736-7248; champlinfdns@worldnet.att.net
Internet: http://www.fdncenter.org/grantmaker/champlin
Sponsor: Champlin Foundations
300 Centerville Road, Suite 300S
Warwick, RI 02886-0226

Chapman Charitable Foundation Grants **1230**
At Chapman, the ongoing business mission includes supporting the non-profit community through innovation, service and charity. The corporation accomplishes this mission through its corporate endeavors by striving to provide non-profit agencies with comprehensive coverage at the most reasonable price. Likewise, since its inception in 2000, the Chapman Charitable Foundation has donated over $6.75 million dollars to more than 470 California based Social Service Agencies. Primarily, the foundation supports organizations involved with education, forest conservation, health, human services, and religion. Particular fields of interest include: children and youth services, foster care, Christian agencies and churches, education (all levels), the environment, health care access, health care clinics and centers, hospitals, human services, and religion. Applicants should begin by contacting the Foundation with a one-page letter of inquiry. The foundation utilizes a Recommendation Committee to select potential grantees. Application forms are not required.
Geographic Focus: California
Amount of Grant: Up to 150,000 U.S.D
Samples: AIDS Healthcare Foundation, Los Angeles, California, $10,000 - general operations (2010); Michael Pourson Ministries, San Rafael, California, $108,000 - support programs (2010); Pachamana Alliance, San Francisco, California, $36,000 - general operations (2010).
Contact: Mari Perez, Grants Coordinator; (626) 405-8031; fax (626) 405-0585; mperez@chapmanins.com or info@chapmanins.com
Internet: http://www.chapmanins.com/about/foundation
Sponsor: Chapman Charitable Foundation
265 North San Gabriel Boulevard
Pasadena, CA 91107-3423

CharityWorks Grants **1231**
Each year, CharityWorks seeks a partner whose programs improve the quality of life for children and families in the Washington metropolitan area. CharityWorks contributes manpower and money to qualified nonprofit organizations in the Washington Metropolitan area that make a significant impact on the area's most urgent societal and educational needs. Specific focus areas are education, health, and poverty reduction. The beneficiary will receive 80 percent of the proceeds of CharityWork's fundraising efforts. The remaining 20 percent of the proceeds will be divided among the other two finalists.
Geographic Focus: District of Columbia
Date(s) Application is Due: Apr 18
Contact: Grants Administrator; (703) 286-0758; fax (703) 286-0791; charityworks@aol.com
Internet: http://www.charityworksdc.org
Sponsor: CharityWorks
1616 Anderson Road, Suite 209
McLean, VA 22102

Charles A. Frueauff Foundation Grants **1232**
The foundation considers proposals 501(c)3 organizations that support private four-year colleges and universities, social service agencies, and health-related agencies and institutions. Types of support include building construction/renovation, capital campaigns, equipment acquisition, general operating support, matching/challenge grants, annual campaigns, and emergency funds. Consideration will be given to programs that support persons leaving welfare, preparing students for employment in non-profit agencies, tutoring at-risk youth, and revitalizing neighborhoods. Applicants are requested to send via postal service a one-page letter of inquiry and include the following information: very brief agency mission and purpose; agency location; brief purpose of request and amount requested; and email address if you wish to receive notification via email.
Requirements: Applicants must be private nonprofit corporations with 501(c)3 status.
Restrictions: The foundation funds nationwide except: Arizona, Alaska, California, Hawaii, Idaho, Iowa, Michigan, Minnesota, Montana, Nevada, New Mexico, North Dakota, Ohio, Oregon, Utah, Washington, Wisconsin, and Wyoming. K-12 schools are ineligible. Grants are not awarded to individuals, provide emergency funds, fund research, or for loans. Multi-year grants, international projects, state supported colleges or universities, primary and secondary schools, churches, or fund raising drives and special events are not supported.
Geographic Focus: Alabama, Arkansas, Colorado, Connecticut, Delaware, District of Columbia, Florida, Georgia, Illinois, Indiana, Kansas, Kentucky, Louisiana, Maine, Maryland, Massachusetts, Mississippi, Missouri, Nebraska, New Hampshire, New Jersey, New Mexico, New York, North Carolina, Oklahoma, Pennsylvania, Rhode Island, South Carolina, South Dakota, Tennessee, Texas, Vermont, Virginia, West Virginia
Date(s) Application is Due: Mar 15; Sep 15
Samples: Melmark Home (Berwyn, PA)—for classroom and office renovation, $50,000; Stetson U (Deland, FL)—to provide scholarships to students majoring in the natural sciences, $100,000.
Contact: Sue Frueauff, (501) 324-2233; fax (501) 324-2236
Internet: http://www.frueaufffoundation.com/application/default.asp
Sponsor: Charles A. Frueauff Foundation
200 South Commerce, Suite 100
Little Rock, AR 72201

Charles Delmar Foundation Grants **1233**
Established in 1957, the Foundation supports organizations involved with inter-American studies, higher, secondary, elementary, and other education, underprivileged youth, the disadvantaged, the aged, the homeless and housing issues, general welfare organizations, and fine and performing arts. Giving primarily in the Washington, DC area in the U.S., and in Europe and South America. There are no specific deadlines with which to adhere. Contact the Foundation for further application information and guidelines.
Restrictions: No grants to individuals, or for building or endowment funds, or matching gifts; no loans.
Geographic Focus: District of Columbia, Maryland, Virginia, West Virginia, Albania, Andorra, Argentina, Armenia, Austria, Azerbaijan, Belarus, Belgium, Bolivia, Bosnia & Herzegovina, Brazil, Bulgaria, Chile, Colombia, Croatia, Cyprus, Czech Republic, Denmark, Ecuador, Estonia, Finland, France, Georgia, Germany, Greece, Guyana, Hungary, Iceland, Ireland, Italy, Kosovo, Latvia, Liechtenstein, Lithuania, Luxembourg, Macedonia, Malta, Moldova, Monaco, Montenegro, Norway, Paraguay, Peru, Poland, Portugal, Romania, Russia, San Marino, Serbia, Slovakia, Slovenia, Spain, Sweden, Switzerland, The Netherlands, Turkey, Ukraine, United Kingdom, Vatican City
Amount of Grant: 500 - 10,000 U.S.D
Samples: James River Association, Mechanicsville, VA, $4,000 - for general support; Pan American Development Foundation, Washington, DC, $5,000 - for environmental operations; College of Wooster, Wooster, OH, $9,000 - for general support;
Contact: Mareen D. Hughes, President; (703) 534-9109
Sponsor: Charles Delmar Foundation
5205 Leesburg Pike, Suite 209
Falls Church, VA 22041-3858

Charles Edison Fund Grants **1234**
The Charles Edison Fund is an endowed philanthropic institution, incorporated in Delaware in 1948, dedicated to the support of worthwhile endeavors generally within the areas of medical research, science education and historic preservation. Funding is available on a national level but, the majority of institutions and, organizations assisted are based principally in the New York-New Jersey Metropolitan area. This concentration of interest

facilitates the efforts of the Trustees to evaluate the work of recipient groups, frequently accomplished by personal visitations. The Fund is a seed money organization, primarily starting projects which would not otherwise get off the ground. The Fund meets three times a year, usually in February or March, June and December at which time requests which have been submitted at least three weeks prior to the meeting will be considered.
Requirements: Public and private schools, universities and colleges, and nonprofits are eligible. The Fund recommends that a grant request be submitted on the requesting organization's letterhead and be signed by an official on behalf of the governing board. The Fund does not require or supply application forms. The request should be detailed, complete and include background information about the organization, a full explanation of the project and its costs, and a financial report, current budget and evidence of tax-exempt status of the requesting organization.
Geographic Focus: New Jersey, New York
Contact: John P. Keegan, President; (973) 648-0500; fax (973) 648-0400; info@charlesedisonfund.org
Internet: http://www.charlesedisonfund.org/thefund.html
Sponsor: Charles Edison Fund
One Riverfront Plaza, 4th Floor
Newark, NJ 07102

Charles F. Bacon Trust Grants **1235**
The Charles F. Bacon Trust was established in 1928 to support and promote quality educational, human services, and health care programming for underserved populations. Special consideration is given to charitable organizations that serve the needs of elderly women. Grant requests for general operating support are strongly encouraged. Program support will also be considered. Small, program-related capital expenses may be included in general operating or program requests. The application deadline is April 1, and applicants will be notified of grant decisions before June 30.
Requirements: 501(c)3 organizations serving residents of Massachusetts are eligible.
Restrictions: The majority of Bacon Trust grants are 1 year in duration. On occasion, multi-year support is awarded.
Geographic Focus: Massachusetts
Date(s) Application is Due: Apr 1
Contact: Michealle Larkins, Vice President; (866) 778-6859; michealle.larkins@baml.com
Internet: https://www.bankofamerica.com/philanthropic/fn_search.action
Sponsor: Charles F. Bacon Trust
225 Franklin Street, 4th Floor, MA1-225-04-02
Boston, MA 02110

Charles G. Koch Charitable Foundation Grants **1236**
The foundation provides funding for academic and public policy research directed at solving social problems through voluntary action and free enterprise. In the area of research, the foundation primarily funds organizations working with doctorate-level investigators in disciplines such as economics, history, philosophy, political science, and organizational behavior. Types of support include general operating, scholarship funds, conferences and seminars, research, special projects, and seed money. There are no application deadlines. Submit preproposal letters (three-page limit).
Amount of Grant: 25,000 - 300,000 U.S.D
Samples: George Mason U (Fairfax, VA)—to recruit faculty members, and for programs in experimental economics, $3 million.
Contact: Kelly Young, Vice President; (202) 393-2354; fax (202) 393-2355; email@cgkfoundation.org
Internet: http://www.cgkfoundation.org
Sponsor: Charles G. Koch Charitable Foundation
655 15th Street NW, Suite 445
Washington, DC 20005-2001

Charles H. Dater Foundation Grants **1237**
The foundation makes grants to private, nonprofit organizations and public agencies in Greater Cincinnati for programs that benefit children in the region in the areas of arts/culture, education, healthcare, social services, and other community needs. Grants are usually made for one year, and subsequent grants for an extended or ongoing program are based on an evaluation of annual results. Multiple grants to an organization in the foundation's same fiscal year (September through August) are possible, but rare. The foundation looks favorably on applications that leverage a grant to seek additional funding and resources. The foundation directors/officers meet monthly to review grant applications.
Requirements: Nonprofit organizations in the greater Cincinnati area are eligible. This area is defined as the eight-county metropolitan area made up of the counties of Hamilton, Butler, Warren and Clermont in Ohio; Boone, Kenton and Campbell in Northern Kentucky; and Dearborn in Indiana.
Restrictions: The foundation does not make grants to individuals, for scholarships for individuals, for debt reduction, and, with rare exception, for capital fund projects.
Geographic Focus: Indiana, Kentucky, Ohio
Amount of Grant: 10,000 - 20,000 U.S.D
Contact: Beth Broomall, Grants Coordinator; (513) 241-2658; fax (513) 274-2731; bb@DaterFoundation.org
Internet: http://www.daterfoundation.org/grants.php
Sponsor: Charles H. Dater Foundation
602 Main Street, Suite 302
Cincinnati, OH 45202

Charles H. Farnsworth Trust Grants **1238**
The Charles H. Farnsworth Trust was established in 1930 to assist older adults to live in dignity and with independence. In describing the purpose of his legacy, Mr. Farnsworth made clear his interest in supporting housing, particularly in developing affordable housing options, and in providing support services to older adults. Program interests include: development of housing, especially housing with support services; services for elderly persons (i.e. health care, homemaker assistance, and nutritional support to enable the elderly to continue living in the community); and research, planning, and communication to better inform individuals, institutions, and the community at large on ways to improve the quality and quantity of housing and support services for seniors. Capital grants related to construction or renovation of housing for older adults is of particular interest. While general operating grants are provided, requests for support for new or special projects/programs are preferred. Planning grants investigating strategies for the development of supportive housing for the elderly are encouraged. Capital grants generally range between $25,000 and $250,000. General operating grants are usually no more than $10,000. Applicants must apply online at the grant website. Applicants are strongly encouraged to do the following before applying: review the downloadable state application procedures for additional helpful information and clarifications; review the downloadable online-application guidelines at the grant website; review the trust's funding history (link is available from the grant website); review the online application questions in advance; and review the list of required attachments. These will generally include: a list of board members, financial statements (audited, reviewed, or compiled by independent auditor); an organization summary; a list of other funding sources; an IRS Determination letter; and other required documents. All attachments must be uploaded in the online application as PDF, Word, or Excel files. The Farnsworth Trust has biannual deadlines of February 1 and October 15. Applications must be submitted by 11:59 p.m. on the deadline dates. Grant applicants for the February deadline will be notified of grant decisions by May 31 and applicants for the October deadline will be notified of grant decisions by December 31.
Restrictions: Grant opportunities are restricted to the Greater Boston area.
Geographic Focus: Massachusetts
Date(s) Application is Due: Feb 1; Oct 15
Amount of Grant: 10,000 - 250,000 U.S.D
Samples: Hearth, Boston, Massachusetts, $100,000 (2010); Greater Boston Chinese Golden Age Center, Boston, Massachusetts, $289,346 (2010); Friends of Saint Josephs Food Pantry, Salem, Massachusetts, $10,000 (2010).
Contact: Michealle Larkins, Vice President; (866) 778-6859; michealle.larkins@baml.com
Internet: https://www.bankofamerica.com/philanthropic/fn_search.action
Sponsor: Charles H. Farnsworth Trust
225 Franklin Street, 4th Floor, MA1-225-04-02
Boston, MA 02110

Charles H. Hall Foundation **1239**
The Charles H. Hall Foundation was established in 2007 to support and promote educational, health- and human-services, religious, and arts and cultural programming for underserved populations. Special consideration is given to programs whose purpose is the prevention of cruelty to children or animals. Grants from the Charles H. Hall Foundation are one year in duration. Applicants must apply online at the grant website. Applicants are strongly encouraged to do the following before applying: review the downloadable state application procedures for additional helpful information and clarifications; review the downloadable online-application guidelines at the grant website; review the foundation's funding history (link is available from the grant website); review the online application questions in advance; and review the list of required attachments. These will generally include: a list of board members, financial statements (audited, reviewed, or compiled by independent auditor); an organization summary; a list of other funding sources; an IRS Determination letter; and other required documents. All attachments must be uploaded in the online application as PDF, Word, or Excel files. The Charles H. Hall Foundation application deadline is 11:59 p.m. on December 1. Applicants will be notified of grant decisions by letter within three to four months of the deadline.
Requirements: The foundation specifically serves organizations based in Berkshire, Hampden, Hampshire, or Franklin Counties, Massachusetts. Grants will be considered for specific programs or projects with preference given to organizations that provide direct services.
Restrictions: Applicants will not be awarded a grant for more than 3 consecutive years. The foundation does not support requests from individuals, organizations attempting to influence policy through direct lobbying, or any political campaigns.
Geographic Focus: Massachusetts
Date(s) Application is Due: Dec 1
Amount of Grant: 5,000 - 20,000 U.S.D
Samples: Friends of the Homeless, Springfield, Massachusetts, $50,000 (2010); Food Bank of Western Massachusetts, Hatfield, Massachusetts, $25,000 (2010); Boys and Girls Club, Springfield, Massachusetts, $50,000 (2010).
Contact: Amy Lynch, Senior Manager; (860) 657-7015; amy.r.lynch@baml.com
Internet: https://www.bankofamerica.com/philanthropic/fn_search.action
Sponsor: Charles H. Hall Foundation
200 Glastonbury Boulevard, Suite # 200, CT2-545-02-05
Glastonbury, CT 06033-4056

Charles H. Hood Foundation Child Health Research Grants **1240**
The grants assist in the promotion of child health in New England. Interest areas include pediatrics, pediatric surgery, pediatric medical and surgical subspecialties, and perinatal obstetrics. Emphasis is on medical research contributing to a reduction of the health problems and health needs of large numbers of children. The foundation seeks innovative,

start-up projects that are not yet candidates for government or large foundation support. Guidelines are published twice each year.
Requirements: Applicants must have completed postdoctoral training and must not be receiving salary support from training funds, including fellowships, at the time of the grant award. Giving is limited to Connecticut, Massachusetts, Maine, New Hampshire, Rhode Island, and Vermont.
Restrictions: Grants are not given for building funds, endowments, regular budgets, or public fund-raising campaigns. Grants are not awarded to individuals.
Geographic Focus: Connecticut, Maine, Massachusetts, New Hampshire, Rhode Island, Vermont
Amount of Grant: 75,000 U.S.D
Contact: Raymond Considine, Executive Director; (617) 695-9439; fax (617) 423-4619; chhoodfdn@tmfnet.org
Sponsor: Charles H. Hood Foundation
95 Berkeley Street, Suite 201
Boston, MA 02116

Charles H. Pearson Foundation Grants 1241

The Charles H. Pearson Foundation Fund was established in 1922 to support and promote quality educational, human-services, and health-care programming for underserved populations. In the area of education, the fund supports academic access, enrichment, and remedial programming for children, youth, adults, and senior citizens that focuses on preparing individuals to achieve while in school and beyond. In the area of health care, the fund supports programming that improves access to primary care for traditionally underserved individuals, health education initiatives and programming that impact at-risk populations, and medical research. In the area of human services the fund tries to meet evolving needs of communities. Currently the fund's focus is on (but is not limited to) youth development, violence prevention, employment, life-skills attainment, and food programs. Grant requests for general operating support are strongly encouraged. Program support will also be considered. Small, program-related capital expenses may be included in general operating or program requests. The majority of grants from the Pearson Fund are one year in duration; on occasion, multi-year support is awarded. Applicants must apply online at the grant website. Applicants are strongly encouraged to do the following before applying: review the downloadable state application procedures for additional helpful information and clarifications; review the downloadable online-application guidelines at the grant website; review the foundation's funding history (link is available from the grant website); review the online application questions in advance; and review the list of required attachments. These will generally include: a list of board members, financial statements (audited, reviewed, or compiled by independent auditor); an organization summary; a list of other funding sources; an IRS Determination letter; and other required documents. All attachments must be uploaded in the online application as PDF, Word, or Excel files. The application deadline for the Charles H. Pearson Foundation Fund is 11:59 p.m. on July 1. Applicants will be notified of grant decisions before September 30.
Requirements: Applicants must have 501(c)3 tax-exempt status.
Restrictions: In general, capital requests are not advised. The fund does not support endowment campaigns, events such as galas or award ceremonies, and costs of fundraising events. The fund does not support requests from individuals, organizations attempting to influence policy through direct lobbying, or any political campaigns.
Geographic Focus: Massachusetts
Date(s) Application is Due: Jul 1
Samples: United Way of Massachusetts Bay, Boston, Massachusetts, $36,000, (2010); Berklee College of Music, Boston, Massachusetts, $25,000, Berklee City Music Faculty Outreach Program to help underserved youth in the Boston Public Schools (2010); Community Legal Services and Counseling Center, Cambridge, Massachusetts, general operating support to provide free legal services and affordable mental health services to underserved families and individuals (2010).
Contact: Michealle Larkins, Vice President; (866) 778-6859; michealle.larkins@baml.com
Internet: https://www.bankofamerica.com/philanthropic/fn_search.action
Sponsor: Charles H. Pearson Foundation Fund
225 Franklin Street, 4th Floor, MA1-225-04-02
Boston, MA 02110

Charles H. Revson Foundation Grants 1242

The foundation awards grants nationwide in its areas of interest, including urban affairs and public policy, education and higher education, biomedical research policy, and Jewish education and philanthropy. Types of support include capital campaigns, continuing support, fellowships, internship funds, program development, and research. Preference is given to requests serving New York, NY. There are no application forms or deadlines. The board meets in April, June, October, and December.
Restrictions: Grants do not support local or national health appeals or direct service programs, individuals, building construction/renovation, book projects, charity events, travel expenses, or budgetary support.
Samples: The Legal Aid Society, New York, NY, $100,000—to support the Economic Security Initiative, which provides essential legal assistance to meet the current economic challenges; Jewish Theological Seminary of America, New York, NY, $500,000—to support the creation of the Center for Pastoral Education; Albert Einstein Collefe of Medicine of Yeshiva University, New York, NY, $170,619—to support a fellow in the Charles H. Revson Senior Fellowships in the Life Sciences Program; Fiscal Policy Institute, Somerville, MA, $200,000—to support research, public education, and advocacy work on budget, economic, and related policy issues that affect low- and moderate-income New Yorkers.
Contact: Maria Marcantonio, Grants Administrator; (212) 935-3340; fax (212) 688-0633; info@revsonfoundation.org
Internet: http://www.revsonfoundation.org/guidelines.html
Sponsor: Charles H. Revson Foundation
55 E 59th Street, 23rd Floor
New York, NY 10022

Charles Lafitte Foundation Grants 1243

The foundation is committed to helping groups and individuals foster lasting improvement on the human condition by providing support to education, children's advocacy, medical research, and the arts. Children's advocacy grants support organizations working to improve the quality of life for children, particularly in relation to child abuse, literacy, foster housing, hunger, and after-school programs. Education grants support innovative programs that work to resolve social service issues, address the needs of students with learning disabilities, provide technology and computer-based education, offer leadership skills education, and support at-risk students. Colleges and universities also receive support for research and conferences. The foundation's medical issues and research grants support healthcare studies, with emphasis on cancer research and treatment, children's health, health education, and promoting healthy living and disease prevention. Art grants support emerging artists and educational art programs.
Requirements: 501(c)3 tax-exempt organizations are eligible.
Samples: American Cancer Society, Eastern Division (North Brunswick, NJ)—for the Pain Initiative, an educational campaign that promotes chronic-pain relief as an integral part of comprehensive patient care, $500,000.
Contact: Jennifer Vertetis, President; jennifer@charleslafitte.org
Internet: http://www.charleslafitte.org
Sponsor: Charles Lafitte Foundation
29520 2nd Avenue SW
Federal Way, WA 98023

Charles M. and Mary D. Grant Foundation Grants 1244

The foundation awards grants primarily in the U.S. southeast in its areas of interest, including community and economic development, health and human services, education, and literacy. The foundation prefers to support projects, but may consider requests for operating support from organizations with budgets of less than $1 million. Grants are made in September. A minimum of three years must elapse between grant awards. Guidelines are available online.
Requirements: Southeast U.S. 501(c)3 tax-exempt organizations are eligible.
Restrictions: Individuals are ineligible, and loans are not awarded.
Date(s) Application is Due: Apr 30
Amount of Grant: 10,000 - 50,000 U.S.D
Samples: Free Clinic of Danville (VA)—for building renovations, $15,000; United Negro College Fund (New York, NY)—for the Faculty Advancement Program of the technology enhancement capital campaign, $30,000; Mountain Assoc for Community Economic Development (Berea, KY)—for general support, $35,000; Atlanta Legal Aid Society (GA)—for the Health Law Partnership, $25,000.
Contact: Jacqueline Elias, c/o JPMorgan Private Bank
Internet: http://fdncenter.org/grantmaker/grant
Sponsor: Charles M. and Mary D. Grant Foundation
345 Park Avenue, 4th Floor
New York, NY 10154

Charles M. Bair Family Trust Grants 1245

The trust awards general support grants to Montana nonprofits in its areas of interest, including museums, performing arts, arts and culture, higher education, youth, and social services. Application forms are required.
Requirements: Montana nonprofit organizations are eligible, with emphasis on Yellowstone, Meagher and Wheatland counties.
Restrictions: Funding is not provided for: churches, conventions, or associations of churches; individuals; conferences; symposiums; or for fundraising events.
Geographic Focus: Montana
Date(s) Application is Due: Jan 15; Aug 1
Amount of Grant: 2,000 - 300,000 U.S.D
Contact: Grants Administrator, c/o U.S. Bank, Tax Department
Sponsor: Charles M. Bair Family Trust
P.O. Box 20678
Billings, MT 59115

Charles Nelson Robinson Fund Grants 1246

The Charles Nelson Robinson Fund was established in 1970 to support and promote quality educational, human-services, and health-care programming for underserved populations in Hartford, Connecticut. Preference is given to organizations that provide human services programming to underserved adults. Grants from the Robinson Fund are one year in duration. Applicants must apply online at the grant website. Applicants are strongly encouraged to do the following before applying: review the downloadable state application procedures for additional helpful information and clarifications; review the downloadable online-application guidelines at the grant website; review the foundation's funding history (link is available from the grant website); review the online application questions in advance; and review the list of required attachments. These will generally include: a list of board members, financial statements (audited, reviewed, or compiled by independent auditor); an organization summary; a list of other funding sources; an IRS Determination letter; and other required documents. All attachments must be uploaded in the online application as PDF, Word, or Excel files. The Charles Nelson Robinson Fund has an annual deadline of 11:59 p.m. on February 15. Applicants will be notified of grant decisions by letter within two to three months of the proposal deadline.

Requirements: Applicant organizations must have 501(c)3 tax-exempt status and have a principal office located in the city of Hartford, Connecticut.
Restrictions: Grant requests for capital projects will not be considered. Applicants will not be awarded a grant for more than three consecutive years. The fund does not support requests from individuals, organizations attempting to influence policy through direct lobbying, or any political campaigns.
Geographic Focus: Connecticut
Date(s) Application is Due: Feb 15
Samples: Hartford Interval House, Hartford, Connecticut, $5,000, general operating support (2010); Loaves and Fishes Ministries, Hartford, Connecticut, $4,000, general operating support (2010); Bulkeley High School, Hartford, Connecticut, $3,000, VOAG Horse Care Program (2011).
Contact: Carmen Britt, Senior Officer; (860) 657-7019; carmen.britt@baml.com
Internet: https://www.bankofamerica.com/philanthropic/fn_search.action
Sponsor: Charles Nelson Robinson Fund
200 Glastonbury Boulevard, Suite # 200, CT2-545-02-05
Glastonbury, CT 06033-4056

Charlotte County (FL) Community Foundation Grants **1247**
The foundation awards grants from its unrestricted funds to support innovative solutions for the citizens of Charlotte County through the implementation of projects and programs that address specific identified needs of the community and demonstrate a wide-spread positive impact on its residents. The foundation places priority on funding requests from nonprofit organizations whose project can: Provide an unduplicated value and/or service; Leverage dollars and/or people power; Match foundation funds with funds from other sources; Establish cooperative efforts of two or more organizations where such synergy is likely to demonstrate superior results and/or lower costs; Enhance, expand and/or improve the organization's capabilities; Show visible end results within a specific timeline for completion; Demonstrate that the Foundation grant will play an important role; Provide a meaningful outcomes evaluations process with a project completion report; Reach a large number of Charlotte County residents; and, be completed within one year with a request of less than $7,500.
Requirements: Foundation grants are made to tax exempt organizations that have been in existence for at least two years and whose focus is in one of the following areas of interest: Animal Welfare; Arts and Culture; Community Development; Education; Environment; Health, Human, and Social Services; Historic Preservation; Nonprofit Organization Capacity Building. Additionally, organizations must also show that they have: a Board of Directors composed of individuals of diverse backgrounds, at least half of whom reside in Charlotte County; Strong management and leadership qualities; Demonstrated fiscal responsibility; Submitted its prior year IRS Form 990; Benefited Charlotte County residents. To apply, submit a Letter of Intent (LOI) summarizing your request. (See instructions at the foundation's website.) If the foundation feels that the proposed project has merit, you will be asked to complete a grant proposal package to be submitted as your formal request for funding.
Restrictions: The foundation is generally not interested in requests for: General operating support; Building or capital campaigns; Deficit financing and debt reduction; Endowment funds; Fraternal organizations, societies or orders; Loans or assistance to individuals; Religious organizations for sectarian purposes; Lobbying legislators or influencing elections; Political organizations or campaigns; Fund raising events; Basic scientific research; Start up organization funding; Travel expenses for individuals or groups; or, Re-granting.
Geographic Focus: Florida
Date(s) Application is Due: Apr 5; May 5; Jul 6; Sep 16
Amount of Grant: Up to 7,500 U.S.D
Contact: Gregory C. Bobonich, J.D.Chief Executive Officer; (941) 637-0077; gbobonich@charlottecommunityfoundation.org
Internet: http://www.charlottecommunityfoundation.org/index.php?subcat=20&articleid=68&page_num=1#
Sponsor: Charlotte County Community Foundation
1675 West Marion Avenue
Punta Gorda, FL 33950

Charlotte Geyer Foundation Cancer Research Grants **1248**
The foundation supports research into the cause, prevention, and treatment of cancer. The purpose of the awards is to provide one year's funding to exceptional proposals to give investigators the opportunity of advancing and improving projects to the point where they are able to successfully compete for an NCI RO1 or other award. Proposals are reviewed three times per year. Application guidelines and procedures are available online.
Requirements: The foundation will review recent proposals—either the original proposal, a revised proposal, or a portion of the original proposal deemed to have special merit—that were submitted to NCI; received a peer-review ranking in the 10 percentiles above the NCI payline; and did not receive funding.
Date(s) Application is Due: Feb 1; Jun 1; Oct 1
Amount of Grant: Up to 100,000 U.S.D
Contact: Nancy Falletta, Executive Director; (716) 632-6448; fax (716) 632-6098
Internet: http://www.charlottegeyer.org
Sponsor: Charlotte Geyer Foundation
P.O. Box 1276
Williamsville, NY 14231-1276

Chase Paymentech Corporate Giving Grants **1249**
The corporation awards grants to nonprofit organizations that provide needed services in the areas of education, health and human services. Preference for funding is given to organizations whose work impacts the citizens of those areas in which Paymentech offices are located, including Westerville, OH; Dallas, TX; Salem, NH; Silver Springs, MD; Tampa, FL; Tempe, AZ; and Toronto, Canada. The Corporate Giving Committee meets each quarter. Organizations are notified about funding status after each meeting.
Requirements: 501(c)3 non-profit organizations that provide needed services in the areas of education and health and human services are eligible. Applicants submit a written request on the organization's letterhead detailing the organization's mission, the amount requested and a detailed explanation of the purpose for which the funds would be used. Organizations should also submit proof of non-profit status.
Restrictions: Paymentech does not make grants to individuals; religious, fraternal, political, or veterans organizations; or colleges and universities. Nor is funding provided for the following purposes: or deficit spending or debt liquidation; annual capital campaigns of hospitals, colleges, universities, grade schools, or high schools.
Geographic Focus: Arizona, Florida, Maryland, New Hampshire, Ohio, Texas, Canada
Date(s) Application is Due: Mar 30; Jun 29; Sep 28; Dec 31
Amount of Grant: Up to 5,000 U.S.D
Contact: Corporate Giving Director
Internet: http://www.chasepaymentech.com/portal/community/chase_paymentech/public/public_website/about_us/company_information_pages/corporate_giving_program
Sponsor: Chase Paymentech
14221 Dallas Parkway, Building Two
Dallas, TX 75254

Chateaubriand Exact Sciences, Engineering, and Medicine Research Scholarships **1250**
Individuals currently pursuing the PhD or who have completed it in the last three years may qualify for a scholarship from the French government to conduct research at a French university, school of engineering, or state-funded laboratory. Candidates must contact the host institution to secure a position before applying. Scholarships are available for periods of six to 18 months and carry a stipend, health insurance, and round-trip travel expenses. Application forms are available on the Web site.
Requirements: Applicants must be U.S. citizens and registered in a U.S. university, and must be PhD candidates or recent PhD holders. Applicants must obtain written agreement from the French hosting institution before applying; an online registration form is available on the Web site.
Date(s) Application is Due: Jan 15
Amount of Grant: 52,200 - 156,600 EUR
Contact: Chateaubriand Fellowship Program Coordinator; (202) 944-6246; fax (202) 944-6244; Chateaubriand@amb-wash.fr
Internet: http://www.frenchculture.org/education/support/chateaubriand/
Sponsor: French Embassy Cultural Services
4101 Reservoir Road NW
Washington, DC 20007-2176

Chatlos Foundation Grants **1251**
The Chatlos Foundation supports nonprofit organizations in the U.S.A and around the globe. Support is provided to organizations currently exempt by the Internal Revenue Service of the United States. The Foundation's areas of interest are: Bible Colleges/Seminaries, Religious Causes, Medical Concerns, Liberal Arts Colleges and Social Concerns.
Requirements: Applicants must be U.S. tax-exempt, nonprofit organizations that provide services in the following areas: bible colleges, religious causes, medical concerns, liberal arts colleges, and social concerns. Proposals must include cover letter, specific request, tax-exemption letter, and budget. If proposal is to be considered at board level, additional information will be requested.
Restrictions: The foundation will not accept requests from individual church congregations, individuals, organizations in existence for less than two years as indicated by IRS tax-exempt letter of determination, for education below the college level, for medical research projects, or for support of the arts.
Amount of Grant: 10,000 - 25,000 U.S.D
Contact: C. J. Leff, Administrator; (407) 862-5077; cj@chatlos.org
Internet: http://www.chatlos.org/AppInfo.htm
Sponsor: Chatlos Foundation
P.O. Box 915048
Longwood, FL 32791-5048

Chazen Foundation Grants **1252**
The foundation awards grants nationwide to eligible nonprofit organizations in its areas of interest, including arts; business education; higher education; hospitals; human services; Israel; Jewish agencies, temples, and federated giving programs; museums; and music performance. Types of support include building construction/renovation, capital campaigns, general operating support, grants and scholarships to individuals, professorships, and scholarship funds. Education grants are awarded to students of the Rockland County, NY, area. There are no application deadlines.
Contact: Grants Administrator; (212) 750-6600
Internet: http://www.chazenscholar.com/project.php3
Sponsor: Chazen Foundation
767 5th Avenue, 26th Floor
New York, NY 10153-2696

CHCF Grants **1253**
The foundation's mission is to expand access to affordable, quality health care for underserved individuals and communities in California and to promote fundamental improvements in the health status of Californians. The foundation has five funding areas: improving care delivery, business of healthcare, healthcare quality, California health policy, and California's uninsured. The one-year, renewable grants will focus on areas where the foundation's

resources can initiate meaningful policy recommendations, innovative research, and the development of model programs. Philanthropic activities include foundation-initiated projects, requests for proposals, and unsolicited proposals. emailed letters of inquiry should include, in two to three pages, a brief description of the proposed project, along with an estimated time line and budget. A full proposal may be requested.
Requirements: California 501(c)3 nonprofit organizations are eligible.
Restrictions: The foundation does not generally support the cost of direct clinical care, ongoing general operating expenses, capital campaigns, annual appeals or other fund-raising events, construction, purchase or renovation of facilities, or purchase of equipment.
Geographic Focus: California
Date(s) Application is Due: Oct 14
Samples: California Society of Thoracic Surgeons (Torrance, CA)—to build the initial infrastructure to collect, analyze, and report data on CABG procedures performed in California and to identify quality improvements, $161,000; California Health Advocates (Sacramento, CA)—to improve the quality and availability of information for consumers on Medicare, Medicare+Choice, and Medicare supplemental insurance by increasing the technical capacity of the Health Insurance Counseling and Advocacy Program, $433,751 over 23 months; RAND (Santa Monica, CA)—to study the relationship between drug benefit design and the use of non-drug medical services to determine how drug benefit design affects key outcomes for chronic conditions, $92,000.
Contact: Grants Administrator; (510) 238-1040; fax (510) 238-1388; grants@chcf.org
Internet: http://www.chcf.org/grantinfo
Sponsor: California HealthCare Foundation
1438 Webster Street, Suite 400
Oakland, CA 94612

CHCF Health Care Affordability Grants 1254
The program awards grants to support initiatives that reduce rising health care costs in the short term. Instead of merely shifting costs from one part of the health care sector to another, the goal of these projects will be to achieve net cost savings for the overall health system. Projects funded under this RFP may include efforts to reduce health care costs by streamlining information exchange, improving efficiency, and decreasing costs; collaborating to reduce duplication and achieve economies of scale; aligning incentives among players in the health care system; and increasing transparency and promoting informed decision making. Initial funding will support grantees as they develop ideas, partnerships, and plans for implementing appropriate strategies. Based on the success of Phase I, CHCF anticipates a future funding phase that would support implementation of selected California-based interventions. Complete details, including submission instructions, are available online.
Requirements: Nonprofit and for-profit organizations are eligible. Proposals from outside of California need to clearly demonstrate applicability of the strategy to the California health care system; a partnership that would allow the cost-saving strategy to be introduced in California; and a specific plan for coordinating the project.
Date(s) Application is Due: Dec 1
Amount of Grant: 25,000 - 75,000 U.S.D
Contact: Grants Administrator; (510) 238-1040; fax (510) 238-1388
Internet: http://www.chcf.org/grantinfo/rfps/view.cfm?itemid=115045
Sponsor: California HealthCare Foundation
1438 Webster Street, Suite 400
Oakland, CA 94612

CHCF Health Care Leadership Fellowships 1255
The program is a two-year fellowship that offers clinically trained health care professionals the experiences, competencies, and skills necessary for effective vision and leadership of the health care system. Up to 30 physicians, nurses, dentists, pharmacists, and other health care professionals are selected annually to participate in this California HealthCare Foundation-sponsored program. The program covers tuition, the cost of seminars and other learning experiences, educational materials, electronic and print resources, lodging, and most meals related to attending the seminars. Application guidelines and forms are available online.
Requirements: Selection of fellows is based on the following criteria: experience (at least five years) in management or leadership positions; recognition and endorsement of the candidate as a future leader by his or her organization; demonstrated accomplishments in professional and/or civic settings; openness to change and understanding of the nature of the challenges confronting healthcare; and a commitment to a rigorous program of study.
Date(s) Application is Due: May 24
Contact: Tracy Scates, UCSF Center for the Health Professions Contact; (415) 502-1173 or (415) 476-8181; fax (415) 476-4113; tscates@itsa.ucsf.edu
Internet: http://futurehealth.ucsf.edu/futureleaders/index.html
Sponsor: California HealthCare Foundation
1438 Webster Street, Suite 400
Oakland, CA 94612

CHCF Local Coverage Expansion Initiative 1256
This initiative supports California counties and local coalitions seeking to expand coverage to uninsured children and adults that are ineligible for public insurance programs. This solicitation will fund two types of technical assistance grants: planning grants will assist coverage expansion efforts with activities such as coalition strengthening, better defining the target population's needs, and/or developing an implementation plan; and implementation grants will provide funding for specific unmet needs associated with launching a local coverage program. Grants will be awarded for period up to one year. Application and guidelines are available online.
Requirements: Any type of organization may apply, but the program will focus on public programs seeking to expand coverage, primarily through the use of public funds. It is anticipated that successful proposals will either come from public or quasi-public agencies, or include a plan for close collaboration with public sector entities.
Date(s) Application is Due: Oct 14
Amount of Grant: 100,000 - 200,000 U.S.D
Contact: Administrator; (510) 238-1040; fax (510) 238-1388; localcoverage@chcf.org
Internet: http://www.chcf.org/grantinfo/rfps/index.cfm?itemID=21263
Sponsor: California HealthCare Foundation
1438 Webster Street, Suite 400
Oakland, CA 94612

CHCF Tracking California's Health Insurance Markets Grants 1257
The foundation works to ensure that all Californians have access to affordable, quality health care. Foundation grants support two-year projects that monitor premiums and products available through California's health insurance markets, focusing in particular on individual, small groups, and Medicare segments; assess financial implications of new benefit designs for consumers by developing scenarios that involve specific health conditions and health-services-use patterns; and interpret and communicate findings in a timely, accessible format for policy, regulatory, consumer advocacy, media, research, and industry audiences. Guidelines are available online.
Requirements: California nonprofit organizations are eligible.
Geographic Focus: California
Contact: Grants Administrator; (510) 238-1040; fax (510) 238-1388; grants@chcf.org
Internet: http://www.chcf.org/grantinfo
Sponsor: California HealthCare Foundation
1438 Webster Street, Suite 400
Oakland, CA 94612

Cherokee Inspired Comfort Award 1258
The program invites nominations of nurses and other nonphysician health care professionals for exceptional service, sacrifice, and innovation. Candidates may be nominated in two of the following four categories, but they only may win in one: registered nurse, licensed practical nurse/licensed vocational nurse, student nurse, and other nonphysician health care professional. One dollar will be donated to Nurses House for each health care professional who is nominated for the award. Prizes are awarded in each nomination category. Electronic and paper nominations are accepted. Nomination form and guidelines are available online.
Date(s) Application is Due: Jul 31
Contact: Award Administrator
Internet: http://www.cherokeeuniforms.com/news/050622_rotation_01.htm
Sponsor: Cherokee Uniforms
13576 Desmond Street
Pacoima, CA 91331

Chesapeake Corporation Foundation Grants 1259
The foundation supports U.S. and international nonprofits in its areas of interest, including civic affairs, community development, cultural programs, higher education, and health. Types of support include endowments, capital grants, matching gifts, and scholarships. Grants are awarded for one year with the possibility of renewal.
Requirements: Nonprofits internationally are eligible.
Restrictions: Grants do not support athletic purposes or individuals, except for employee-related scholarships.
Amount of Grant: 1,000 - 10,000 U.S.D
Contact: J.P. Causey Jr., (804) 697-1000; fax (804) 697-1199
Internet: http://www.cskcorp.com
Sponsor: Chesapeake Corporation Foundation
P.O. Box 2350
Richmond, VA 23218-2350

Chest Foundation/LUNGevity Foundation Clinical Research in Lung Cancer Grants 1260
The goal of this joint clinical research program is, ultimately, to save the lives of those people afflicted with lung cancer by funding innovative research designed to treat and cure lung cancer. A clinical research award of $75,000 will be granted to a single award recipient for clinical research in lung cancer over a period of 2 years. The amount of $37,500 will be granted after July 1. Upon receipt of the Progress Report received before July 15, the remaining $37,500 will be granted.
Requirements: To be considered a candidate for The Chest Foundation/LUNGevity Foundation Clinical Research Award in Lung Cancer, an individual must: be an ACCP member who has completed at least 2 years of pulmonary or oncology fellowship and be within 5 years of fellowship or no more than 5 years in practice; hold a degree of MD, DO, MB,BCh, PharmD, PhD or its equivalent; can be a citizen of the United States, or have a J-1 or H-1 visa, and be working in a certified United States institution or organization, or be a Canadian or international member who is working in a certified institution or organization in his/her country; and be a clinician in lung cancer research.
Date(s) Application is Due: Apr 30
Amount of Grant: 75,000 U.S.D
Contact: Sue Ciezadlo, Senior Project Coordinator; (847) 498-8363; fax (847) 498-5460; sciezadlo@chestnet.org
Internet: http://www.chestfoundation.org/foundation/clinical/LUNGevityAward.php
Sponsor: Chest Foundation
3300 Dundee Road
Northbrook, IL 60062-2348

Chest Foundation Eli Lilly and Company Distinguished Scholar in Critical Care Medicine Award 1261

The Eli Lilly and Company Distinguished Scholar in Critical Care Medicine will have a 3-year opportunity to examine issues that relate to critically ill patients. The award is intended to permit the investigation of issues that are not easily supported by traditional funding, such as clinical trials or basic science. Rather, the award would support activities such as the development of public policy, patient education models, or economic analysis of treatment or care delivery in this patient group. A stipend of $50,000 annually over the 3-year term of the program will be provided. A fourth year stipend of $10,000 will also be awarded to the individual who will serve in a mentorship role to the fourth Distinguished Scholar position during his/her inaugural year of service.
Requirements: An applicant must be: an FCCP and hold the degree of MD, DO, MBChB, MBBCh, MBBS, DNSc, PharmD, PhD, or EdD; board-certified in critical care medicine; and a recognized clinician and/or scientist with a specialty in sepsis and/or critical care as evidenced by publications, presentations, and/or peers.
Date(s) Application is Due: Apr 30
Amount of Grant: Up to 160,000 U.S.D
Contact: Sue Ciezadlo, Senior Project Coordinator; (847) 498-8363; fax (847) 498-5460; sciezadlo@chestnet.org
Internet: http://www.chestfoundation.org/foundation/clinical/dsCriticalCare.php
Sponsor: Chest Foundation
3300 Dundee Road
Northbrook, IL 60062-2348

Chest Foundation Geriatric Development Research Awards 1262

The two-year award is intended to provide the impetus required for long-term career development focused on integrating geriatrics into the sub-specialties of internal medicine. The program awards the grant to an academic internist to develop and implement a basic, clinical, or health services research project focused on a geriatric aspect of chest medicine. The funding can support the salary of the award recipient and/or the purchase of supplies, the salaries of technical personnel, and other resources necessary for the completion of the research project. The award also includes a one-time travel grant to attend the meetings of the American Geriatrics Society and the American College of Chest Physicians in the second year of the award.
Requirements: To be eligible for the award, applicants must: have U.S. citizenship or permanent resident status; hold a degree of MD or its equivalent; have completed a sub-specialty internal medicine fellowship leading to certification in his/her sub-specialty by the American Board of Internal Medicine and be within the first three years of his/her faculty appointment; possess a faculty appointment by July 1 (this faculty appointment should be documented in the required letters of recommendation from the applicant's department chair and division director); be a member of the American College of Chest Physicians; commit 75 percent of his/her professional effort to research activities; develop and implement a basic, clinical, or health services research project focused on a geriatric aspect of chest medicine; and generate and implement a career development plan focused on the geriatrics aspects of chest medicine. This plan must include organizing and interacting with a mentorship team made up of a minimum of four members—the applicant's research mentor as the leader of the research team, a sub-specialist in the applicant's field of chest medicine, a geriatrician, and one other member at the applicant's discretion.
Restrictions: The funding cannot be used to acquire administrative or clerical support. Funding is to cover total costs; no indirect cost funds will be provided.
Date(s) Application is Due: Apr 30
Amount of Grant: 53,000 U.S.D
Contact: Sue Ciezadlo, Senior Project Coordinator; (847) 498-8363; fax (847) 498-5460; sciezadlo@chestnet.org
Internet: http://www.chestfoundation.org/foundation/clinical/geriatricAward.php
Sponsor: Chest Foundation
3300 Dundee Road
Northbrook, IL 60062-2348

Chest Foundation Grant in Venous Thromboembolism 1263

The Chest Foundation is issuing this Request for Proposal (RFP) that will fund two projects with up to a total grant amount of $180,000 each for a 1-year period. The goal of the grant process is to confer awards to projects that include evidence-based replicable programs and tools to address the current gaps in the management and prophylaxis in venous thromboembolism (VTE). The deadline is February 20.
Requirements: To be considered a candidate for The Chest Foundation Grant in Venous Thromboembolism, an individual must: be a current ACCP member and hold the degree of MD, DO, MBChB, MBBCh, MBBS, DNSc, PharmD, PhD, or EdD; and be a recognized scientist focused on an area of VTE or clinical decision making as evidenced by academic productivity and/or publications and presentations.
Date(s) Application is Due: Feb 20
Amount of Grant: Up to 180,000 U.S.D
Contact: Sue Ciezadlo, Senior Project Coordinator; (847) 498-8363; fax (847) 498-5460; sciezadlo@chestnet.org
Internet: http://www.chestfoundation.org/foundation/clinical/VT_Grant.php
Sponsor: Chest Foundation
3300 Dundee Road
Northbrook, IL 60062-2348

Chestnut Hill Charitable Foundation, Inc Grants 1264

The foundation awards grants to nonprofit organizations in the greater Boston, MA, area in support of medical research, children and youth, and arts and education. Specifically, the foundation's interests include visual arts, museums, humanities, arts and culture, early childhood education and development, education at all levels, hospitals and medical research (cancer, heart and lungs, AIDS, and diabetes), social services, and general charitable giving. Types of support include general operating support, annual campaigns, capital campaigns, building construction/renovations, programs and projects, seed grants, curriculum development, research, and matching funds. There are no application forms; submit a letter of request.
Requirements: Nonprofit organizations in the greater Boston, MA, area may submit letters of request.
Restrictions: Grants are not awarded to support individuals or political or religious organizations.
Geographic Focus: Massachusetts
Amount of Grant: 1,000 - 15,000 U.S.D
Contact: Kay Kilpatrick, Director Corporate Giving; (617) 630-2415
Sponsor: Chestnut Hill Charitable Foundation
27 Boylston Street
Chestnut Hill, MA 02167-1700

Chicago Board of Trade Foundation Grants 1265

The foundation provides grant support to organizations in the metropolitan Chicago, IL, area in its areas of interest: arts and culture, including libraries and museums; education, including higher education, adult basic education and literacy training, and science and technology; health care, including mental health services, cancer research, and rehabilitation; child and youth development; minorities; the economically disadvantaged; wildlife; and media and communication. Grants are awarded for general operating support, continuing support, annual campaigns, capital campaigns, and endowments. An application form is not required; the initial approach should be by letter, or via the CBOT Foundation form on the Web site. The board meets during the first quarter annually to consider requests.
Requirements: Nonprofit organizations in a 75-mile radius of the metropolitan Chicago, IL, area may apply for grant support.
Restrictions: Support is not given to hospitals or foundations. Loans and program-related investments will also be denied.
Geographic Focus: Illinois
Date(s) Application is Due: Oct 1
Contact: Grants Administrator; (312) 435-3456; fax (312) 341-3306
Internet: http://www.cbot.com/cbot/pub/page/0,3181,948,00.html
Sponsor: Chicago Board of Trade Foundation
141 W Jackson Boulevard, Suite 600-A
Chicago, IL 60604

Chicago Institute for Psychoanalysis Fellowships 1266

This is a program for advanced trainees in psychiatry, social work, and psychology who have significant interest in psychoanalysis as a body of knowledge and as a framework with which to understand and carry out therapeutic work. It is jointly sponsored by the Chicago Psychoanalytic Society and the Institute for Psychoanalysis. A mentor is assigned to each Fellow, who is available throughout the year-long program to discuss cases, papers, research, etc. In addition, Fellows are guests at Society meetings and conferences held locally. Opportunities are provided for the Fellows to discuss presented papers with their mentors. A monthly conference is held to acquaint Fellows with basic psychoanalytic clinical concepts through readings and discussion. Each Fellow receives a copy of the Annual of Psychoanalysis and other publications.
Date(s) Application is Due: Aug 15
Contact: Christine Susman, Director; (312) 922-7474, ext. 324; csusman@chicagoanalysis.org or admin@chicagoanalysis.org
Internet: http://www.chicagoanalysis.org/fellow.php
Sponsor: Chicago Institute for Psychoanalysis
122 S Michigan Avenue
Chicago, IL 60603

Chicago Tribune's Corporate Contributions Grants 1267

The foundation's mission is to promote public knowledge and strengthen the Chicago metropolitan community by encouraging journalistic excellence, diversity and liberty; supporting diverse cultural institutions; and promoting civic efforts. Program areas for the foundation are civic, culture, and journalism. Grants in the civic program area support business, nonprofit or educational initiatives. These grants are considered by invitation only and are accepted throughout the year. The range is $2,500 to $25,000. Grants in the culture area support cultural organizations within the Chicago metropolitan area that provide educational programs in the arts for children from low-income communities or that foster diverse art. The range for these grants is $5,000 to $7,500. Proposals are due February 1. Grants in the journalism are support organizations on a national level that offer programs for professional journalists that lead to journalistic excellence, promote diversity in the newspaper industry and protect freedom of the press. Support is considered for journalism programs in the metropolitan Chicago area that promote diversity. Support is not provided to high school or college newspapers. The grant range is $5,000 to $20,000. Proposals are due June 1. Types of support include general operating grants, matching gifts, and program development grants. A brief summary proposal should be sent; if interested, the foundation will request a full proposal.
Requirements: Grants are made to Chicago-area tax-exempt organizations.
Restrictions: Grants are not made to individuals nor for fund-raising events.
Geographic Focus: Illinois
Date(s) Application is Due: Feb 1; Jun 1
Amount of Grant: 2,500 - 25,000 U.S.D

Contact: Community Relations Manager; (312) 222-4300; fax (312) 222-3751; ctcommunityrelations@tribune.com
Internet: http://www.chicagotribune.com/about/custom/community/chi-foundationspage-htmlstory,0,7165967.htmlstory
Sponsor: Chicago Tribune Foundation
435 N Michigan Avenue, 2nd Floor
Chicago, IL 60611-4041

Chicago White Metal Charitable Foundation Grants 1268
The charitable foundation awards grants to Chicago-area nonprofits in its areas of interest, including community development, government/public administration, human services, performing arts, opera, recreation, scholarships/financial aid, and youth development. Grants are awarded for one year and are renewable for a second year.
Requirements: Chicago, IL, 501(c)3 organizations are eligible.
Restrictions: Religious organizations and individuals are ineligible.
Geographic Focus: Illinois
Contact: Grants Administrator; (630) 595-4424; fax (630) 595-4474
Sponsor: Chicago White Metal Charitable Foundation
Route 38 at Eastgate Drive
Bensenville, IL 60106-9806

Child Nutrition Foundation School Community Kitchens Grants 1269
The program awards seed grants for local school community kitchen projects. School community kitchens are school-community partnerships that help people move into foodservice jobs while providing meals for those in need.
Requirements: Applicants must represent a school foodservice or nutrition department at the district or school building level and be a current American School Food Service Association (ASFSA) member.
Date(s) Application is Due: Oct 15
Amount of Grant: 10,000 U.S.D
Contact: Administrator; (703) 739-3900; fax (703) 739-3915; cnf@schoolnutrition.org
Internet: http://www.schoolnutrition.org/CNF.aspx?id=1133
Sponsor: Child Nutrition Foundation
700 S Washington Street, Suite 300
Alexandria, VA 22314

Children's Advocacy Center Grants 1270
The program supports the establishment and expansion of children's advocacy centers, with the goal of helping to prevent child abuse. One-year grants are awarded in the following categories: program support, research, member training, prevention, associate member program development, nonmember training, and recognized chapters. Applicants must designate a steering committee that comprises child-welfare workers, law-enforcement officials, medical doctors, and mental-health professionals. Application and guidelines are available online.
Requirements: 501(c)3 tax-exempt organizations are eligible.
Date(s) Application is Due: Sep 13
Amount of Grant: 5,000 - 50,000 U.S.D
Contact: Grants Administrator; (800) 239-9950, ext. 116
Internet: http://www.nca-online.org
Sponsor: National Children's Alliance
516 C Street NE
Washington, DC 20002

Children's Brain Tumor Foundation Research Grants 1271
Science grants are awarded for basic laboratory research on pediatric brain and central nervous system tumors. Priority will be given to research that relates to: priorities I and II of the Research and Scientific Priorities for Pediatric Brain Tumors, as listed in the Report of the Brain Tumor Progress Review Group, which can be found at: http://prg.nci.nih.gov/brain/pediatrics.Submission of a two-page, preapplication form, equivalent to a letter of intent, is required. An original, signed form must be submitted along with five copies. The listed deadline is for preapplication forms; full proposals are by invitation.
Requirements: Funding is currently restricted to principal investigators at institutions within the United States.
Restrictions: The foundation does not award grants to individuals or to private foundations, and does not fund debt reduction, capital improvements, or travel expenses. Overhead expenses are not to exceed 10 percent of the total project cost.
Amount of Grant: 75,000 U.S.D
Samples: Princton U (NJ)—to support the work of Jonathan Eggenschwiler, for research on pediatric brain tumors, $150,000 over two years; U of California at San Francisco (CA)—to support the work of Daphne Haas-Kogan, for research on pediatric brain tumors, $150,000 over two years; U of Chicago (IL)—to support the work of Manuel Utset, for research on pediatric brain tumors, $150,000 over two years.
Contact: Judy Hurley; (212) 448-9494; fax (212) 448-1022; JHurley@cbtf.org
Internet: http://www.cbtf.org/grant_info.html
Sponsor: Children's Brain Tumor Foundation
274 Madison Avenue, Suite 1301
New York, NY 10016-0701

Children's Cardiomyopathy Foundation Research Grants 1272
The foundation awards funds to support research related to all forms of cardiomyopathy affecting children under the age of 18 years. The goal of CCF's research program is to advance medical knowledge on the disease and develop more accurate diagnostic methods, life-improving therapies and ultimately a cure. The grant program is designed to provide seed funding to investigators for the testing of initial hypotheses and collecting of preliminary data to help secure long-term funding by the National Institutes of Health and other major granting institutions.
Requirements: Proposals will be accepted on an annual basis for innovative basic, clinical or translational research relevant to the cause or treatment of cardiomyopathy in children. Principal investigators must hold a MD, PhD or equivalent degree and reside in the United States or Canada. The investigator must have a faculty appointment at an accredited U.S. or Canadian institution and have the proven ability to pursue independent research as evidenced by original research in peer-reviewed journals. Guidelines and application forms can be downloaded at the website.
Geographic Focus: All States, Canada
Date(s) Application is Due: Oct 5
Amount of Grant: 25,000 - 50,000 U.S.D
Contact: Lisa Yue, Executive Director; (866) 808-2873; fax (201) 227-7016; lyue@childrenscardiomyopathy.org or info@childrenscardiomyopathy.org
Internet: http://www.childrenscardiomyopathy.org/site/grants.php
Sponsor: Children's Cardiomyopathy Foundation (CCF)
P.O. Box 547
Tenafly, NJ 07670

Children's Leukemia Research Association Research Grants 1273
The association supports research efforts into the causes and cure of leukemia and gives patient aid to families in need while meeting the expenses incurred in leukemia treatment. A medical advisory committee consisting of prominent internationally known and respected hematologists reviews grant proposals submitted for consideration of projects that are not otherwise funded. Grants are available for start-up funding for laboratory or clinical investigations in leukemia. Although the association prefers to fund new investigators, applications from established investigators for new initiatives are also solicited. Renewal for a second year is considered if other funding for promising projects has not been obtained.
Restrictions: Previously funded projects are not usually considered for additional funding.
Date(s) Application is Due: Jun 30
Amount of Grant: Up to 20,000 U.S.D
Contact: Director; (516) 222-1944; fax (516) 222-0457; info@childrensleukemia.org
Internet: http://www.childrensleukemia.org/ResearchPast%20Pres.htm
Sponsor: Children's Leukemia Research Association
585 Stewart Avenue, Suite LL-18
Garden City, NY 11530

Children's Medical Research Institute Grants 1274
The program, which is designed to promote pediatric research as a career, supports summer scholarships for first- and second-year Oklahoma University College of Medicine students. Each year students are awarded grants to support their pediatric research interests as they assist OU Medical Center Department of Pediatrics physicians. The program also awards small research grants/short-term projects to individual scientists at the OU Medical Center, who often use the funds as seed money to finance pilot research projects.
Contact: Sana Rettig, (405) 271-8001, option 1, ext.; fax (405) 271-1175; sana-rettig@ouhsc.edu
Internet: http://www.cmri.net/scholarships.html
Sponsor: Children's Medical Research Institute
800 Research Parkway, Suite 150
Oklahoma City, OK 73104

Children's Trust Fund of Oregon Grants 1275
The trust's mission is to support the establishment of community-based direct service programs to provide primary prevention program to reduce the incidence of child abuse and neglect and to promote healthy families. Preference is given to programs that help children by increasing caretaker skills and the knowledge of child development. The trust prefers projects that focus on underserved minorities or geographic areas, and efforts that can be replicated as proven models. There are no deadline dates; proposals are reviewed twice each year.
Requirements: Oregon public and private nonprofit organizations are eligible.
Date(s) Application is Due: Apr 1
Amount of Grant: Up to 20,000 U.S.D
Contact: Cindy Thompson, Executive Director; (503) 222-7102; fax (503) 222-6975; cindy@ctfo.org
Internet: http://www.ctfo.org
Sponsor: Children's Trust Fund of Oregon
1410 SW Morrison Street, Suite 501
Portland, OR 97205

Children Affected by AIDS Foundation Camp Network Grants 1276
The mission of the CAAF Camp Network is to increase the number of children that attend camp every year by providing grants; to foster communication among agencies providing camp experiences for children and their families; and to support the development of new camps. Members of the network gain access to the competitive funding process from the CAAF's yearly financial allocation to the Camp Network.
Requirements: For consideration for membership, Camp Service Providers need to provide proof of the following: (a) Proof of 501(c)3 Not-for-Profit Status; (b) 990 IRS form if applicable; (c) Experience with children infected with or affected by HIV/AIDS and providing camp services. Camps seeking funding from CAAF need to provide proof of the following: (1) Proof of minimum Insurance Coverage's as recommended by the Camp

Network; (2) Indemnity for CAAF (Camp Network) from camp agencies, including any employment related claims (Contractual provision, prior to receiving a CAAF grant award.); (3) Compliance with applicable laws such as: employment, health and safety laws and the Americans with Disabilities Act. This will be a contractual provision of the grant agreement, before receiving a CAAF grant award; (4) Release of Liability from Parent or Guardians and Children, using unique identifiers; and, (5) Evidence that the Camp has policies and procedures in place for: (a) Children's safety and emergencies; (b) Pre-screening of paid and non-paid staff; (c) Other information as needed.
Date(s) Application is Due: May 16
Amount of Grant: 1,000 - 15,000 U.S.D
Contact: Rolla Bedford, (310) 258-0850, ext. 14; fax (310) 258-0851; rolla.bedford@caaf4kids.org
Internet: http://www.caaf4kids.org/
Sponsor: Children Affected by AIDS Foundation
6033 W Century Boulevard, Suite 603
Los Angeles, CA 90045

Children Affected by AIDS Foundation Domestic Grants 1277

The foundation awards grants to nonprofit organizations that provide for the medical, emotional, social, and basic needs of children who are infected with HIV or affected by AIDS. Organizations can request funds for their programs that directly impact the lives of children affected by AIDS by submitting a proposal during the foundation's once-a-year application period. Special needs requests are considered throughout the year, depending on funding availability. Funding categories include Basic Needs (clothing, formula, food, diapers, burial services, transportation to/from appointments, emergency child care, and respite services); Social & Recreational (social activities such as arts and crafts, sports, holiday parties, field expeditions, and other forms of recreation); Psychosocial Support (intended to benefit direct services related only to the provision of psychosocial services to children which may include individual and family therapy/counseling, support groups, and life skills classes); Guidelines and application are available online.
Requirements: 501(c)3 nonprofit organizations that provide direct care, support, and assistance to children, ages birth to 13 years, who are HIV-positive or otherwise affected by HIV/AIDS are eligible. This includes children who are HIV-infected, those who are or may become orphaned as a result of losing parents to HIV disease, and those who have siblings, parents, or caregivers who are infected with HIV.
Restrictions: Individuals, research activities, fraternal groups, political or religious activities, and organizations not in good standing with CAAF are not eligible. Additionally, CAAF will not grant funds to pass-through organizations.
Date(s) Application is Due: May 16
Amount of Grant: 1,000 - 25,000 U.S.D
Contact: Rolla Bedford, (310) 258-0850, ext. 14; fax (310) 258-0851; rolla.bedford@caaf4kids.org
Internet: http://www.caaf4kids.org
Sponsor: Children Affected by AIDS Foundation
6033 W Century Boulevard, Suite 603
Los Angeles, CA 90045

Children Affected by AIDS Foundation Family Assistance Emergency Fund Grants 1278

The foundation is making available its Emergency Fund to help HIV/AIDS Service Organizations in the U.S. to meet the needs of HIV-impacted children and their families. Funds can be accessed by qualified HIV/AIDS non-profits with IRS 501(c)3 status. The Foundation will only provide grants to organizations that do not discriminate on the basis of race, color, religion, gender, gender identity, gender expression, age, sexual orientation, marital status, national origin, disability, or other characteristics protected by law. The goal of the Emergency Fund is to meet the pressing, unanticipated needs in the day-to-day life of children, up to 13 years of age, and their families. Requests may be submitted at any time, and will be reviewed shortly after being received.
Requirements: Funds requested must directly improve the well-being of children in the household. Funds can be accessed by qualified HIV/AIDS non-profits with IRS 501(c)3 status. Eligible requests include: Food; Utility assistance; School uniforms, clothing; Transportation to medical/social service appointments; Medications and medical co-pays; Transitional housing assistance; Burial expenses. Other needs not listed may be considered-please contact the Foundation Program staff prior to sending a request. To download the Emergency Fund's guidelines, go to the sponsor's site, click on 'Initiatives' on the home page, then 'Family Assistance.'
Restrictions: The emergency fund cannot be accessed by or for the following purposes: political or fraternal organizations, or for research or religious purposes.
Amount of Grant: 50 - 2,000 U.S.D
Contact: Rolla Bedford, (310) 258-0850, ext. 14; fax (310) 258-0851; rolla.bedford@caaf4kids.org
Internet: http://www.caaf4kids.org/
Sponsor: Children Affected by AIDS Foundation
6033 W Century Boulevard, Suite 603
Los Angeles, CA 90045

Chiles Foundation Grants 1279

The foundation has a deep concern for and confidence in the future of Oregon and the Pacific Northwest. Although the foundation has made a steady commitment to the improvement of the quality of life for those who live and work in this area, it is not restricted in its grant making to the Pacific Northwest. The foundation has traditionally made grants to certain select institutions of higher education for business schools, scholarships, and athletics; supports basic research in certain select medical institutions; supports religion through divinity schools and religious education; and believes that the arts and cultural activities of a community are important and supports certain select, established institutions. Types of support include building construction/renovation, equipment acquisition, and scholarship funds. Annual deadline dates may vary; contact Foundation by a letter of Inquiry. The Foundation does not accept unsolicited proposals.
Requirements: The preferred initial method of contact is a phone call to the grants administration office to determine whether a prospective proposal is within guidelines; if so, an applicant will be invited to submit a one-page written preliminary proposal. An application form will be sent after approval of the preliminary proposal by the executive committee.
Restrictions: No support for projects involving litigation. Grants are not made to individuals, for deficit financing, mortgage retirement, or projects and conferences already completed.
Geographic Focus: California, Oregon, Germany
Amount of Grant: 1,000 - 270,000 U.S.D
Samples: Boston Symphony Orchestra, Boston, MA, $100,000— general operating support; Stanford University, Stanford, CA, $159,000—scholarship assistance; Ludwig-Maximilians University, Munich, Germany, $180,000—support continuing clinical and experimental research.
Contact: Earle M. Chiles; (503) 222-2143; fax (503) 228-7079; cf@uswest.net
Sponsor: Chiles Foundation
111 SW Fifth Avenue, Suite 4050
Portland, OR 97204-3643

Chiron Foundation Community Grants 1280

The foundation's focus areas are health and medicine, education, and community. Four imperatives guide health-care giving: accelerating progress toward the prevention and cure or successful management of cancer through research, education, early detection, and public-policy debate; combating infectious disease through prevention-related programs, educational efforts, and therapeutics targeting at-risk populations, with emphasis on the special needs of children and families; ensuring the availability and safety of the blood supply and promoting the highest standards of care for blood donors and recipients; and supporting initiatives in the international medical community to provide vaccines and immunization services to protect at-risk populations, especially children, against the devastation of crippling and lethal diseases. Four imperatives guide education giving: providing training and professional development opportunities to enable classroom teachers to teach science and math more effectively; increasing opportunities for economically disadvantaged students to continue on to postsecondary education; supporting community-based training programs to prepare underrepresented minorities and underprivileged groups for careers in the biosciences; and encouraging and assisting promising scholars to pursue advanced research in the biosciences and medicine. Four imperatives guide community giving: providing early diagnosis, intervention, and support for children with disabilities or other special physical or emotional needs; providing essential social services such as meals and housing to those in greatest need; empowering individuals to achieve self-sufficiency through employment training and job-skill development; and enriching the character and celebrating the distinctives of local communities. Requests are accepted at anytime, and proposals are reviewed quarterly.
Requirements: 501(c)3 tax-exempt organizations in Chiron communities (San Francisco East Bay Area, including Alameda, Contra Costa, and Solano Counties; Seattle; and Philadelphia) are eligible.
Restrictions: In general, Chiron does not support organizations that do not have 501(c)3 tax status; religious, fraternal, service, or veterans' organizations; civic or cultural organizations that do not serve the areas in which Chiron is located; alumni drives and teacher organizations; memorials; municipal and for-profit hospitals; labor unions; city, municipal, or federal government departments; organizations or causes that do not support the company's commitment to non-discrimination and diversity; projects of national scope or from national organizations not related to health care; matching gifts; individuals, including scholarships (other than those awarded as part of the company-sponsored college scholarship program); travel support; fund-raising activities related to individual sponsorship; and fund-raising dinners other than those for health care or medical research organizations aligned with company research and product interests.
Geographic Focus: California, Pennsylvania, Washington
Amount of Grant: 5,000 - 75,000 U.S.D
Contact: Community Relations; (510) 601-6952; Chiron_foundation@chiron.com
Internet: http://www.chiron.com/foundation/index.html
Sponsor: Chiron Foundation
4560 Horton Street
Emeryville, CA 94608

Christensen Fund Regional Grants 1281

The fund (TCF) focuses its grantmaking on maintaining the biological and cultural diversity of the world by focusing on four geographic regions: the greater South West (Southwest United States and Northwest Mexico); Central Asia and Turkey; the African Rift Valley (Ethiopia); and Northern Australia and Melanesia. Grants within these programs are generally directed to organizations based within those regions or, where appropriate, to internationally based organizations working in support of people and institutions on the ground. In general, grants are one year or less; currently grants up to two years are by invitation only.
Requirements: 501(c)3 nonprofit organizations and non-U.S.A institutions with nonprofit or equivalent status in their country of origin are eligible. Partnerships or associations with U.S.A-based nonprofit organizations are preferred.
Restrictions: The fund does not make grants directly to individuals but rather assists individuals through institutions qualified to receive nonprofit support with which such individuals are affiliated.

Amount of Grant: Up to 200,000 U.S.D
Contact: Grants Administrator; (415) 644-1600; fax (415) 644-1601; info@christensenfund.org
Internet: http://www.christensenfund.org/index.html
Sponsor: Christensen Fund
260 Townsend Street
San Francisco, CA 94107

Christine and Katharina Pauly Charitable Trust Grants 1282
The Christine and Katharina Pauly Charitable Trust was established in 1985 to support and promote quality educational, health, and human-services programming for underserved populations. Special consideration is given to charitable organizations that serve the needs of children or older adults. The majority of grants from the Pauly Trust are one year in duration. Applicants must apply online at the grant website. Applicants are strongly encouraged to do the following before applying: review the downloadable state application procedures for additional helpful information and clarifications; review the downloadable online-application guidelines at the grant website; review the trust's funding history (link is available from the grant website); review the online application questions in advance; and review the list of required attachments. These will generally include: a list of board members, financial statements (audited, reviewed, or compiled by independent auditor); an organization summary; a list of other funding sources; an IRS Determination letter; and other required documents. All attachments must be uploaded in the online application as PDF, Word, or Excel files. The application deadline for the Christine and Katharina Pauly Charitable Trust is 11:59 p.m. on September 1. Applicants will be notified of grant decisions by December 31. The Christine and Katharina Pauly Charitable Trust was created under the wills of Ms. Hazel Katharina Pauly and Ms. Frieda Christine Oleta Pauly.
Requirements: Applicants must have 501(c)3 tax-exempt status.
Restrictions: In general, grant requests for individuals, endowment campaigns, or capital projects will not be considered. The Fund will consider requests for general operating support only if the organization's operating budget is less than $1 million. The trust does not support requests from individuals, organizations attempting to influence policy through direct lobbying, or any political campaigns.
Geographic Focus: Missouri
Date(s) Application is Due: Sep 1
Contact: George Thorn; (312) 828-4154; ilgrantmaking@bankofamerica.com
Internet: https://www.bankofamerica.com/philanthropic/fn_search.action
Sponsor: Christine and Katharina Pauly Charitable Trust
231 South LaSalle Street, IL1-231-13-32
Chicago, IL 60604

Christine O. Gregoire Youth/Young Adult Award for Outstanding Use of Tobacco Industry Documents 1283
The award recognizes a person 24 years of age or younger who has made a contribution to the health of the public in the recent past through use of tobacco documents. The award also honors innovation in the use and application of tobacco industry documents to improve the public's health and, where applicable, to further the goals of tobacco prevention and control in order to help build a world where young people reject tobacco and anyone can quit. Those nominated should be individuals who have made a notable impact through innovative use of tobacco industry documents as applied to research, policy, or advocacy.
Requirements: At least one of the following two criteria must be met by the nominated individual: Nominees must have made a remarkable research, policy, or advocacy contribution with the use of tobacco industry documents, and/or; Nominees must have employed innovative, creative approaches to the employment of tobacco industry documents that result in an improvement in the health or public awareness of a community or nation. Nominations may be made by individuals such as colleagues, peers, coworkers, instructors/professors, governments, or CBOs that are qualified to make such a nomination because of their familiarity with the nominee's work and contribution.
Restrictions: Nominees must not have any affiliation with the tobacco industry. Members of the awards committee may not nominate potential recipients, although they may provide written support as part of a nomination package. Employees and directors of the American Legacy Foundation and their relatives are not eligible.
Date(s) Application is Due: Feb 6
Amount of Grant: 7,500 U.S.D
Contact: Virginia Lockmuller, Grant Coordinator; (202) 454-5555; fax (202) 454-5599; awards@americanlegacy.org
Internet: http://www.legacyforhealth.org/awards/
Sponsor: American Legacy Foundation
1724 Massachusetts Avenue NW
Washington, DC 20036

Christopher Columbus Fellowship Foundation-U.S. Chamber of Commerce Chairmen's Distinguished Life Sciences Awards 1284
The $25,000 Chairmen's Distinguished Life Sciences Award will be presented to a scientist or researcher who is making or has recently made a significant and positive contribution toward the development of a cutting edge innovation in the field of life sciences. The Chairmen's Award will provide incentive for continuing the research. Nominations are received online only at: www.ccolumbusfoundationawards.org. Individuals will compete and their accomplishments will be judged on how they exemplify excellence in life sciences. Winners must attend the awards ceremony in April at Washington, D.C., in order to receive the award.
Requirements: All Nominees must be living United States citizens. The Nomination must demonstrate a program of ongoing work with a specific outcome suggesting that an important achievement will result. The Nomination's proposed achievement must benefit human health in a unique and effective manner, demonstrating originality from other work in the same or related field(s). The proposed achievement must correlate to a tangible outcome to advance human health.
Restrictions: Nominees must be permitted, by their employer or any other relevant authority, to accept a monetary award bestowed by the Foundation, a Federal government agency. Anyone who cannot attend the award ceremony will automatically be disqualified.
Date(s) Application is Due: Mar 27
Amount of Grant: 25,000 U.S.D
Contact: Judith M. Shellenberger, Executive Director; (315) 258-0090; fax (315) 258-0093; judithmscolumbus@cs.com
Internet: http://www.columbusfdn.org/lifesciences/
Sponsor: Christopher Columbus Fellowship Foundation
110 Genesee Street, Suite 390
Auburn, NY 13021

Christopher D. Smithers Foundation Research Grants 1285
Activities of the foundation are concentrated in the field of alcoholism. The foundation supports research on treatment, prevention, and public education to create a public awareness of alcoholism as a treatable disease. Small groups, such as treatment groups, shelters, and halfway houses, may apply for one-year seed grants to help them get started. There are no deadlines; applications are accepted year-round. The foundation requests that organizations first write, call, or fax for an annual report and guidelines before applying.
Restrictions: Grants are not awarded to individuals.
Samples: Illinois Church Action on Alcohol Problems (Springfield, IL)—for the alcohol education program in public and private schools in Illinois, $10,000; Gratitude House (West Palm Beach, FL)—to help women suffering from alcoholism, $5000.
Contact: Grants Administrator; (516) 676-0067; fax (516) 676-0323; info@smithersfoundation.org
Internet: http://www.smithersfoundation.org
Sponsor: Christopher D. Smithers Foundation
Oyster Bay Road, P.O. Box 67
Mill Neck, NY 11765

Christopher Reeve Paralysis Foundation Individual Research Grants 1286
The mission of the foundation is to support cutting-edge research to develop effective treatments and cures for paralysis caused by spinal cord injury and other central nervous system disorders. International individual grants support research that seeks to promote neuronal growth and survival, encourage the formation of synapses, enhance the production of myelin, and restore conduction capabilities in the acutely and chronically injured spinal cord; evaluate drugs or other interventions that protect against secondary neuronal injury or provide insight into the mechanisms causing such damage; elucidate the biological mechanisms underlying approaches to improve concomitant function; and understand anatomical characteristics of injury in both animals and human spinal cord, documenting neuronal systems that are most vulnerable to injury and resultant losses in function. The program is designed to encourage promising new investigators to undertake research on spinal cord regeneration and recovery; encourage researchers who are well-established in other areas to transfer their efforts to spinal cord research; and enable researchers with novel ideas to test their ideas and develop pilot data for seeking larger awards from NIH and other funding sources.
Requirements: Senior scientists, young investigators, and postdoctoral fellows are eligible to serve as principal investigators.
Date(s) Application is Due: Jun 15; Dec 15
Amount of Grant: 60,000 - 75,000 U.S.D
Samples: 13 reseachers nationwide—for research on paralysis resulting from spinal-cord injuries, $1.976 million divided.
Contact: Dr. Douglas Landsman, (800) 225-0292; dlandsman@crpf.org
Internet: http://www.christopherreeve.org/research/researchmain.cfm
Sponsor: Christopher Reeve Paralysis Foundation
500 Morris Avenue
Springfield, NJ 07081

Christopher Reeve Paralysis Foundation Quality of Life and Health Promotions Grants 1287
The foundation awards grants to organizations nationwide that help improve opportunities, access, and day-to-day quality of life for individuals living with disabilities—primarily paralysis—and their families. This program recognizes the unique and numerous needs of these individuals and the importance of providing services and programs that enable them to participate in all areas of life. Quality of Life grants are given to programs or projects that improve the daily lives of people living with disabilities, particularly spinal cord injuries. Funding is awarded twice yearly to programs that provide assistance through access, advocacy, education, recreation, and technology, among others. Health Promotion Grants seek to remove societal and environmental barriers that limit the abilities of individuals with paralysis to participate in life activities. Participation in these activities improves physical and emotional health and prevents secondary conditions for persons living with paralysis. Priority funding goes to programs that focus on paralysis generally and are not specific to one condition or disease that results in paralysis. Complete guidelines are available online.
Requirements: 501(c)3 tax-exempt organizations are eligible.
Restrictions: No grants will be made to individuals, for benefit tickets, or for courtesy advertising.
Date(s) Application is Due: Mar 1; Sep 1
Amount of Grant: 5,000 - 25,000 U.S.D

Samples: To be distributed among 126 nonprofit groups nationwide—for awards through the Quality of Life program, which seeks to improve the well-being of people who are paralyzed, $779,321.
Contact: Donna Valente, Grants Coordinator; (800) 539-7309 or (973) 467-8270, ext. 211; qol@crpf.org
Internet: http://www.christopherreeve.org/qlgrants/qlgrantsmain.cfm
Sponsor: Christopher Reeve Paralysis Foundation
500 Morris Avenue
Springfield, NJ 07081

Christy-Houston Foundation Grants **1288**
The foundation awards grants to nonprofits in Rutherford County, TN, for health, education, arts and culture, and community development. Areas of interest include hospitals, nursing care, health care and health associations, nutrition, hospice care, and community development. Types of support include building construction/renovation, equipment acquisition, scholarship funds, and matching funds. Application forms are not required, and there are no application deadlines. The board meets in March, June, September, and December to consider requests.
Requirements: 501(c)3 nonprofits in Rutherford County, TN, are eligible.
Restrictions: Grants are not awarded to: religious organizations for religious purposes, veterans organizations, historical societies, individuals, operating expenses or endowments.
Geographic Focus: Tennessee
Amount of Grant: 50,000 - 200,000 U.S.D
Contact: Robert B. Mifflin, Executive Director; (615) 898-1140; fax (615) 895-9524
Sponsor: Christy-Houston Foundation
1296 Dow Street
Murfreesboro, TN 37130-2413

Chula Vista Charitable Foundation Grants **1289**
Community leaders throughout Chula Vista are partnering with the San Diego Foundation to create and sustain a local foundation to serve solely the community of Chula Vista. Endow Chula Vista is a community-specific effort to focus on endowment building for Chula Vista, now and for generations to come. It is part of The San Diego Foundation's region-wide initiative, Endow San Diego, to inform and inspire San Diegans regarding the benefits of endowment.
Requirements: Any 501(c)3 organization serving the residents of Chula Vista, California, are eligible to apply.
Restrictions: The Chula Vista Charitable Foundation does not make grants for: annual campaigns and fund raising events for non-specific purposes; capital campaigns for buildings or facilities; stipends for attendance at conferences; endowments or chairs; for-profit organizations or enterprises; individuals unaffiliated with a qualified fiscal sponsor; projects that promote religious or political doctrine; research projects (medical or otherwise); scholarships; or existing obligations or debt.
Geographic Focus: California
Contact: Kerry Helmer, (619) 814-1384; kerry@sdfoundation.org
Internet: http://www.endowchulavista.org/
Sponsor: Chula Vista Charitable Foundation / San Diego Foundation
2508 Historic Decatur Road, Suite 200
San Diego, CA 92106

CICF City of Noblesville Community Grant **1290**
The City of Noblesville Community Grants were established to support the charitable intentions of Noblesville. Other purposes of the grant include, but are not limited to, basic needs, economic stability, health and wellness, education, vitality of neighborhoods and communities, arts and culture, and environment. Priority is given to programs and projects most likely to have a positive effect on the city's residents. Proposals are accepted in February and July of each year. Applicants are strongly encouraged to view the Grantseeker's Guide posted on the website.
Requirements: The Foundation gives careful consideration to projects that; most benefit Noblesville residents; promote inclusiveness and diversity; respond to basic human needs; connect individuals and families to the community; complement other organizations to eliminate duplication of services; obtain additional funding and provide ongoing funding after the project. Organizations can apply through the online application.
Restrictions: The Foundation does not fund the following: organizations that are not tax exempt; multi-year grants; grants to individuals; projects aimed at promoting a particular religion or construction projects for religious institutions; operating, program, and construction costs at schools, universities, and private academies unless there is significant opportunity for community use or collaboration; organizations or projects that discriminate based on race, ethnicity, age, gender or sexual orientation; political campaigns or direct lobbying efforts by 501(c)3 organizations; post-event, after-the-fact situations or debt retirement; medical, scientific, or academic research; publications, films, audiovisual and media materials, programs produced for artistic purposes or produced for resale; travel for bands, sports teams, classes, and similar groups; annual appeals, galas, or membership contributions; fundraising events such as golf tournaments, walk-a-thons, and fashion shows.
Geographic Focus: Indiana
Amount of Grant: 1,000 - 10,000 U.S.D
Contact: Liz Tate; (317) 843-2479, ext. 302; fax (317) 848-5463; lizt@cicf.org
Internet: http://www.cicf.org/how-to-apply-grantmaking
Sponsor: Central Indiana Community Foundation
615 North Alabama Street, Suite 119
Indianapolis, IN 46204-1498

CICF Indianapolis Foundation Community Grants **1291**
CICF's mission is to inspire, support, and practice philanthropy, leadership, and service in the community. Proposed programs should align with any of the Foundation's Seven Elements of a Thriving Community: basic needs; economic stability; health and wellness; education; vitality and connectivity of neighborhoods and communities; arts and culture; and the environment. The application and grant request detail form are available online. Applications are accepted during the months of February and July.
Requirements: CICF welcomes grant applications from charitable organizations that are tax exempt under section 501(c)3 of the Internal Revenue Code, and from governmental agencies. New projects or organizations with pending 501(c)3 status may submit an application with the assistance of a fiscal sponsor. Grant inquiries and proposals will be prioritized using the following criteria: organizations that serve primarily Marion County residents; organizations with a demonstrable track record; programs serving populations disadvantaged due to income, age, ethnicity, language, education, disability, transportation or other adverse conditions; project/program ideas must be fully developed; and projects that strongly connect to existing community initiatives (e.g. the Blueprint to End Homelessness, Indianapolis Cultural Development Initiative, and Family Strengthening Coalition). Application information is available online.
Geographic Focus: Indiana
Contact: Liz Tate, Vice President for Grants; (317) 634-2423, ext. 175; fax (317) 684-0943; liz@cicf.org or program@cicf.org
Internet: http://www.cicf.org/the-indianapolis-foundation
Sponsor: Central Indiana Community Foundation
615 North Alabama Street, Suite 119
Indianapolis, IN 46204-1498

CICF James Proctor Grant for Aged Men and Women **1292**
The CICF - James Proctor Grant for Aged Men and Women funds projects that benefit elderly men and women in the fields of arts and culture, basic needs, and health and wellness. Applications are accepted in the months of February and July, and are available at the Foundation website.
Geographic Focus: Indiana
Amount of Grant: 10,000 - 60,000 U.S.D
Contact: Liz Tate, (317) 634-2423, ext. 175; fax (317) 684-0943; liz@cicf.org
Internet: http://www.cicf.org/examples-of-named-funds
Sponsor: Central Indiana Community Foundation
615 North Alabama Street, Suite 119
Indianapolis, IN 46204-1498

CICF Legacy Fund Grants **1293**
The CICF - Legacy Fund Grants fund Hamilton County, Indiana nonprofits in the following areas of interest: arts and culture; basic needs; economic stability; education; environment; health and wellness; and vitality of neighborhoods and communities. Applications are accepted during the months of February and July, and are available at the Foundation's website.
Requirements: In reviewing grant applications, the Foundation gives careful consideration to projects that: impact individuals served; promote inclusiveness and diversity and are responsive to basic human needs; connect individuals and families to the community; commit to carrying out the project; complement other organizations and eliminate duplication of services; manage their fiscal responsibilities; match funds; and obtain additional funding now and in the future.
Restrictions: Limited to not-for-profit organizations and for the benefit of the people of Hamilton County, Indiana.
Geographic Focus: Indiana
Amount of Grant: 5,000 - 20,000 U.S.D
Contact: Liz Tate; (317) 634-2423, ext. 175; fax (317) 848-5463; liz@cicf.org
Internet: http://www.cicf.org.php5-17.dfw1-1.websitetestlink.com/how-to-apply-grantmaking
Sponsor: Central Indiana Community Foundation
615 North Alabama Street, Suite 119
Indianapolis, IN 46204-1498

CIGNA Foundation Grants **1294**
CIGNA Foundation has identified four areas for grant consideration—health and human services, education, community and civic affairs, and culture and the arts. Health and education are of primary concern and receive priority. Under education, priority is placed on public secondary education, higher education for minorities, and adult basic education/literacy. The foundation also considers requests from U.S. cultural, educational, and public policy organizations that have international components. Requests are accepted and reviewed throughout the year. Consideration will be given to requests for general operating support, program development, annual campaigns, conferences and seminars, fellowships, scholarship funds and employee-related scholarships, and matching gifts and funds.
Requirements: Organizations with 501(c)3 tax-exempt status are eligible.
Restrictions: The foundation will not consider applications for grants to individuals, organizations operating to influence legislation or litigation, political organizations, or religious activities. In general, the foundation will not consider applications from organizations receiving substantial support through the United Way or other CIGNA-supported federated funding agencies; hospitals' capital improvements; or research, prevention, and treatment of specific diseases.
Contact: Program Contact; communityrelations@cigna.com
Internet: http://www.cigna.com/general/about/community/grant_information.html
Sponsor: CIGNA Foundation
1601 Chestnut Street, TL06B
Philadelphia, PA 19192-1540

CIIT Postdoctoral Fellowships 1295

Ten to 15 fellowships are offered annually to support persons holding recently earned PhD, MD, or DVM degrees during further training in genetic toxicology, immunotoxicology, biochemical toxicology, pathology, reproductive toxicology, epidemiology, teratology, and carcinogenesis. Fellowships are tenable at the institute for two to three years. The institute holds 501(c)6 status. Applications are available on the Web site.
Requirements: Applicants who recently were awarded PhD degrees in a discipline related to toxicology are eligible to apply; those holding recently awarded DVM or MD degrees are expected to have substantial research experience.
Restrictions: Traineeships and fellowships are awarded only for research conducted at the CIIT. Awardees may not use their stipends to supplement funds from other awards.
Amount of Grant: 27,000 - 28,500 U.S.D
Contact: Rusty Bramlage, Human Resources Manager; (919) 558-1331; fax (919) 558-1300; bramlage@ciit.org
Internet: http://www.ciit.org/careers/openings_postdoc.asp
Sponsor: Chemical Industry Institute of Toxicology
P.O. Box 12137
Research Triangle Park, NC 27709

Cinergy Foundation Grants 1296

The Cinergy corporate contributions program was developed to reinvest in southwest Ohio, northern Kentucky, and Indiana communities. The mission of the foundation is to be a national leader in community development by creating proactive, innovative partnerships that improve the quality of life in Indiana, Ohio, and Kentucky, especially within the electric utility's 69-county service area. Grant categories include arts and culture; lifelong learning—K-12 (BASICS) and higher education; and healthy communities—environment and youth. Applicants are requested to submit grant applications to the district office closest to them; contact program staff for office locations. The board meets in April, July, October, and December to consider requests.
Requirements: The foundation supports projects and programs of 501(c)3 tax-exempt organizations located in 10 southwestern Ohio counties, six northern Kentucky counties, and sixty nine Indiana counties.
Restrictions: Grants do not support advertising; capital and endowment campaigns; competitions (scholastic or athletic); construction/renovation projects; equipment purchases of any type (playground, computers, etc.); golf events; K-12 schools outside the BASICS program; membership dues; organizations without 501(c)3 status; programs that pose a potential conflict of interest; projects/organizations benefiting an individual or a few persons; requests outside the Cinergy service territory; scholarship programs; sponsorships for athletic teams; travel expenses for individuals or groups; uniforms; or veterans, labor, religious, political, or fraternal groups.
Geographic Focus: Indiana, Kentucky, Ohio
Amount of Grant: Up to 50,000 U.S.D
Contact: Karol King; (800) 262-3000, ext. 1251 or (513) 287-1251; kking@cinergy.com
Internet: http://www.cinergy.com/Community/Cinergy_Foundation/default.asp
Sponsor: Cinergy Foundation
139 E Fourth Street
Cincinnati, OH 45202

CIT Corporate Giving Grants 1297

The corporate giving program focuses its giving to programs that substantially strengthen the communities where CIT has a significant presence; provide services that benefit employees, their families, and communities; and support employee volunteerism. Funding priorities include organizations, institutions, or programs that work to foster excellence in education, particularly among at-risk and underprivileged youth; support local health and social welfare issues; help stabilize and improve neighborhoods by providing viable economic and educational growth opportunities to the community; and support the arts, and their accessibility, as a means of enhancing personal and social development. Proposals are reviewed on an ongoing basis and must be formally submitted by mail.
Requirements: 501(c)3 tax-exempt organizations and Canadian charities registered with the Canada Customs and Revenue Agency (CCRA) are eligible.
Restrictions: The following will not be considered for funding: political organizations, campaigns, or candidates; conferences, memberships, or sports competitions; organizations that discriminate in any way with national equal opportunity policies; organizations whose chief purpose is to influence legislation; veteran and fraternal organizations; religious organizations, unless engaged in a significant, nonsectarian project that benefits a broad base of the community; or individuals.
Contact: Stacy Papas, Community Affairs Manager; stacy.papas@cit.com
Internet: http://www.cit.com/about-cit/corporate-giving/index.htm
Sponsor: CIT Group
11 West 42nd Street
New York, NY 10036

Citizens Bank Mid-Atlantic Charitable Foundation Grants 1298

Charitable grants are made only to qualified 501(c)3 Rhode Island-based organizations serving Rhode Island residents. Citizens look for opportunities where moderate funding can affect significant results in the community. Priority consideration is given to programs that encourage the development of innovative responses to basic human needs; promote fair housing and focus on community issues of neighborhood development and economic self-sufficiency; support the availability of quality, cost-effective, community-based health care, particularly for low-income families and children who are at risk; promote new ways to provide a quality education to populations that are underserved, including job training; promote availability and accessibility in the area of culture and the arts; and promote citizen participation in the development of new and workable solutions for improving and maintaining a healthy environment. Charitable grants are usually for capital funding (to build or renovate a facility) or implementation of a specific program.
Requirements: 501(c)3 Rhode Island-based nonprofits are eligible.
Restrictions: Grants do not support individuals, single disease/issue research organizations, religious organizations for religious purposes, labor or fraternal or veterans groups, political organizations or projects, operating deficits, underwriting of conferences and seminars, governmental public agencies, endowments, annual operating support, trips and tours, payment on bank loans, advertising, or fund-raising events.
Geographic Focus: Rhode Island
Amount of Grant: 1,000 - 100,000 U.S.D
Contact: Jeanne Cola, Senior Vice President; (401) 456-7200; fax (401) 456-7366
Sponsor: Citizens Bank Mid-Atlantic Charitable Foundation
870 Westminster Street
Providence, RI 02903

Civil Justice Foundation Grants 1299

The foundation awards modest grants to tax-exempt grassroots advocacy organizations in three categories: (1) to groups of injured consumers, (2) for injury prevention advocacy work, and (3) for projects to improve the care and treatment of injured consumers. The foundation board prefers to fund advocacy, injury prevention, and service projects focused on model problems that can be replicated elsewhere. Letters of proposal must reach the office by the listed application deadline date. Call the foundation for specific deadlines. Organizations will be invited to submit full grant applications.
Requirements: 501(c)3 tax-exempt organizations may apply. Organizations chosen to receive support typically have operating budgets that do not exceed $500,000.
Restrictions: No grants are awarded to individuals.
Amount of Grant: 1,500 - 20,000 U.S.D
Samples: CALPRIG Toy Safety Education Project (Los Angeles, CA)—support for toy safety campaign, $6000. NYCOSH (New York, NY)—support for injured worker computer database, $10,000.
Contact: Director; (800) 424-2725, ext. 381; fax (202) 965-0355; cjf@atlahq.org
Internet: http://www.atla.org/foundations/civiljus/grants.aspx
Sponsor: Civil Justice Foundation
1050 31st Street NW
Washington, DC 20007

CJ Foundation for SIDS Express Grants 1300

The foundation addresses the special needs of the SIDS community. This program provides a simple manner for organizations to apply for funding. Areas considered for funding are: Programmatic: funding for specific programs/projects including support/bereavement services, education and public awareness, sponsorship of relevant meetings and conferences; Administrative/Management: includes general operating support directly associated with the effective delivery of services and/or efficient management of the organization; and Staffing Needs: salary support for staff positions (full-time or part-time) and/or project consultants.
Amount of Grant: 500 - 5,000 U.S.D
Contact: Program Director; (866) 314-7437; wendy@cjsids.com
Internet: http://www.cjsids.com/grant_programs.htm
Sponsor: CJ Foundation for SIDS
30 Prospect Avenue
Hackensack, NJ 07601

CJ Foundation for SIDS Support Services Grants 1301

The foundation addresses the special needs of the SIDS community. Support services grants comprise program support—funding for support services, education and public awareness, and sponsorship of relevant meetings and conferences; administrative/management support— general operating support directly associated with efficient management of the grantee's organization (may include the purchase of equipment); temporary staffing needs—grants for full- or part-time staff positions; and special projects—i.e., initial underwriting associated with new programs and innovative pilot or demonstration projects.
Date(s) Application is Due: Mar 15
Amount of Grant: Up to 5,000 U.S.D
Samples: Tomorrow's Child/Michigan SIDS (Lansing, MI)—to upgrade its systems in order to better support families who have lost a child because of sudden infant death syndrome or other causes, $15,000.
Contact: Program Director; (866) 314-7437; wendy@cjsids.com
Internet: http://www.cjsids.com/grant_programs.htm
Sponsor: CJ Foundation for SIDS
30 Prospect Avenue
Hackensack, NJ 07601

CLA Nurses Respiratory Society Fellowships 1302

The objective of the fellowships is to permit nurses to pursue graduate study so that they may be better able to contribute to the Canadian field of respiratory illness and health care. Applicants may register in any master's or doctoral program where knowledge, skill, and expertise may develop toward this objective. The program should show evidence of a plan to foster theory, research, and either a clinical or functional area of specialization. Application forms may be obtained from the association.
Requirements: Applicant must be a Canadian citizen or permanent Canadian resident, be a registered nurse, be enrolled full-time in a graduate program at the master's or doctoral level, and be a member of CNRS.
Geographic Focus: All States, Canada

Date(s) Application is Due: Feb 1
Amount of Grant: 3,750 - 7,500 CAD
Contact: Grants Administrator; (416) 864-9911; fax (416) 864-9916; orcs@on.lung.ca
Internet: http://www.on.lung.ca/orcs/fellowships.html
Sponsor: Canadian Lung Association
1900 City Park Drive, Suite 508
Gloucester, ON K1J 1A3 Canada

CLA Nurses Respiratory Society Nursing Research Grants **1303**
Research grants are offered to those studying respiratory nursing, chronic or acute lung disease, which will result in improved quality of patient care. Research may include clinical investigation of any nursing-related phenomenon; be pertinent to illness assessment, management, or responses; or be aimed at health promotion and prevention issues. Studies may use either quantitative or qualitative methodologies.
Requirements: Applicant must be a Canadian citizen or permanent Canadian resident; a registered nurse; hold an appointment in, or have an affiliation with, a health care agency, educational institution, or other organization in Canada that can administer the funds in an approved manner; and be a member of CNRS.
Geographic Focus: All States, Canada
Date(s) Application is Due: Nov 1
Amount of Grant: 3,000 - 30,000 CAD
Contact: Grant Administrator; (613) 747-6776; fax (613) 747-7430; dhogg@cha.ab.ca
Internet: http://www.lung.ca/resp/Fellowships/Fellow_Grants.htm
Sponsor: Canadian Lung Association
1900 City Park Drive, Suite 508
Gloucester, ON K1J 1A3 Canada

CLA Physiotherapy Cardio-Respiratory Society Fellowships **1304**
The purpose of the research program is to pursue increased scientific knowledge in the area of cardio-respiratory physiotherapy practice. Fellowships are offered to physiotherapists pursuing postgraduate training, with respiratory research as the major component. The award period is one year.
Requirements: Applicants must be Canadian citizens or permanent Canadian residents, be registered physiotherapists, be enrolled or accepted for full-time study in a graduate program at the master's or doctoral level, and be a member of the Canadian Physiotherapy Cardio-Respiratory Society.
Geographic Focus: All States, Canada
Date(s) Application is Due: Nov 1
Contact: Administrator, Societies & Special Projects; (613) 569-6411; nprt@lung.ca
Internet: http://www.lung.ca/crhp/index.html
Sponsor: Canadian Lung Association
1900 City Park Drive, Suite 508
Gloucester, ON K1J 1A3 Canada

Clara Blackford Smith and W. Aubrey Smith Charitable Foundation Grants **1305**
The Clara Blackford Smith & W. Aubrey Smith Charitable Foundation was established in 1978. Mrs. Smith was a well-known benefactor of health care. The foundation was established under her will to support and promote quality education, health-care, and human-services programming for underserved populations. Special consideration is given to charitable organizations that serve the people of Grayson County, Texas. The foundation makes an annual grant to Denison High School in Denison, Texas to provide college scholarship assistance for deserving graduates. The majority of grants from the Smith Charitable Foundation are one year in duration; on occasion, multi-year support is awarded. Applicants must apply online at the grant website. Applicants are strongly encouraged to do the following before applying: review the downloadable state application procedures for additional helpful information and clarifications; review the downloadable online-application guidelines at the grant website; review the foundation's funding history (link is available from the grant website); review the online application questions in advance; and review the list of required attachments. These will generally include: a list of board members, financial statements (audited, reviewed, or compiled by independent auditor); an organization summary; a list of other funding sources; an IRS Determination letter; and other required documents. All attachments must be uploaded in the online application as PDF, Word, or Excel files. The Clara Blackford Smith & W. Aubrey Smith Charitable Foundation has four deadlines annually: March 1, June 1, September 1, and December 1. Applications must be submitted by 11:59 p.m. on the deadline dates. Grant applicants are notified as follows: March deadline applicants will be notified of grant decisions by June 30; June applicants will be notified by September 30; September applicants will be notified by December 31; and December applicants will be notified by March 31 of the following year.
Requirements: Applicants must have 501(c)3 tax-exempt status.
Restrictions: The foundation does not support requests from individuals, organizations attempting to influence policy through direct lobbying, or any political campaigns.
Geographic Focus: Texas
Date(s) Application is Due: Mar 1; Jun 1; Sep 1; Dec 1
Contact: David Ross, Senior Vice President; tx.philanthropic@baml.com
Internet: https://www.bankofamerica.com/philanthropic/fn_search.action
Sponsor: Clara Blackford Smith and W. Aubrey Smith Charitable Foundation
901 Main Street, 19th Floor, TX1-492-19-11
Dallas, TX 75202-3714

Clarcor Foundation Grants **1306**
Clarcor, through its foundation, seeks to improve the quality of life in the communities in which the company operates. Grants support nonprofits in the categories of health, human services, education, culture and art, and civic activities. Types of support include general operating support, annual campaigns, capital campaigns, and employee matching gifts. There are no application deadlines. The board meets in February, May, August, and November.
Requirements: Tax-exempt organizations in company-operating areas in Rockford, IL; Louisville, KY; Kearney and Gothenburg, NE; Cincinnati, OH; Oklahoma, and Lancaster, PA; are eligible.
Restrictions: Grants are not made to support individuals, endowment funds, research, scholarships, fellowships, or loans.
Geographic Focus: Illinois, Kentucky, Nebraska, North Carolina, Ohio, Oklahoma, Pennsylvania
Amount of Grant: 500 - 50,250 U.S.D
Contact: Pete Nangel, (815) 962-8867; fax (815) 962-0417
Sponsor: Clarcor Foundation
840 Crescent Centre Drive, Suite 600
Franklin, TN 37067

Clarence E. Heller Charitable Foundation Grants **1307**
The charitable foundation supports nonprofit organizations, with priority given to proposals from California, in its areas of interest, including environment and health—to prevent serious risk to human health from toxic substances and other environmental hazards by supporting programs in research, education, and policy development; management of resources—to protect and preserve the earth's limited resources by assisting programs that demonstrate how natural resources can be managed on a sustainable and an ecologically sound basis, and supporting initiatives for sustainable agriculture, and for promoting the long-term viability of communities and regions; music—to encourage the playing, enjoyment, and accessibility of symphonic and chamber music by providing scholarship and program assistance at selected community music organizations and schools, and by helping community-based ensembles of demonstrated quality implement artistic initiatives, diversify and increase audiences, and improve fund-raising capacity; and education—to focus on support for programs that improve the teaching skills of educators and artists in environmental and arts education. Types of support include continuing support, general operating expenses, publications, research, seed money, and special projects. The foundation's board meets typically in March, June, and October. Letters of inquiry are accepted at any time.
Requirements: 501(c)3 tax-exempt organizations are eligible.
Restrictions: Grants are not made to individuals.
Geographic Focus: California
Amount of Grant: 5,000 - 600,000 U.S.D
Contact: Bruce Hirsch, Executive Director; (415) 989-9839; fax (415) 989-1909; info@cehcf.org
Internet: http://cehcf.org/app_info.html
Sponsor: Clarence E. Heller Charitable Foundation
44 Montgomery Street, Suite 1970
San Francisco, CA 94104

Clarence T.C. Ching Foundation Grants **1308**
The Clarence T. C. Ching Foundation provides funding opportunities to nonprofits primarily in Honolulu, Hawaii. The Foundation's areas of interest include education and health. Types of support available are: scholarship funds; general program support and; building/renovation. Grant applications are accepted year round with no application deadline date.
Requirements: Nonprofits in Honolulu, Hawaii are eligible for funding. When applying for a grant include the following in with your proposal: copy of IRS Determination Letter; detailed description of project and amount of funding requested; listing of additional sources and amount of support; five (5) copies of the proposal.
Restrictions: No grants to individuals.
Geographic Focus: Hawaii
Amount of Grant: 500 - 70,000 U.S.D
Samples: Contemporary Museum, Honolulu, HI, $500—general program support; St. Francis Healthcare of Hawaii, Honolulu, HI, $70,000—general program support; St. Louis Scholarship Fund, Hawaii, Honolulu, HI, $70,000—general program support.
Contact: R. Stevens Gilley, Executive Director; (808) 521-0344
Sponsor: Clarence T.C. Ching Foundation
1001 Bishop Street, Suite 960
Honolulu, HI 96813

Clarian Health Critical / Progressive Care Internships **1309**
The Clarian Health Critical / Progressive Care Internship Program is a 12 week individualized program. The Internship is offered year round, complete with salary. The program provides experience in the following specialized services: adult critical care; cardiac; comprehensive critical care; cardiovascular critical care; medical intensive care; neuro/trauma critical care; neuro-surgery intensive care; progressive care; surgical intensive care
Requirements: Contact Julie Ruschhaupt, Critical Care Recruiter, to begin application process
Geographic Focus: Indiana
Contact: Julie Ruschhaupt, Critical Care Recruiter; (317) 278-7082 or (877) 354-2996; jruschhaupt@clarian.org
Internet: http://www.clarian.org/clarianjobs/nursing/internships_externships.htm#top
Sponsor: Clarian Health Partners
950 N Meridian Street, Gateway Plaza
Indianapolis, IN 46204

Clarian Health Multi-Specialty Internships 1310
The Clarian Health Multi-Specialty Internship is a 7 to 8 week individualized program, that is offered year round.
Requirements: Contact Joy Fay for additional information to apply.
Geographic Focus: Indiana
Contact: Joy Fay, RN; (317) 962-3788; jfay@clarian.org
Internet: http://www.clarian.org/clarianjobs/nursing/internships_externships.htm#multi
Sponsor: Clarian Health Partners
950 N Meridian Street, Gateway Plaza
Indianapolis, IN 46204

Clarian Health OR Internships 1311
The Clarian Health OR Internship Program is offered twice a year, for a duration of 6 months. The program provides experience in the following specialized services: general; cardio thoracic; cardiovascular; endoscopies; ENT; GYN; laser; neurosurgery; ophthalmology; orthopedics; pediatrics; plastic / Reconstructive; transplants; trauma; urology
Geographic Focus: Indiana
Contact: Shalunda Tyler, Clarian Health Nurse Recruiter; (317) 962-9083; styler1@clarian.org
Internet: http://www.clarian.org/clarianjobs/nursing/internships_externships.htm#multi
Sponsor: Clarian Health Partners
950 N Meridian Street, Gateway Plaza
Indianapolis, IN 46204

Clarian Health Scholarships for LPNs 1312
The Clarian Health Scholarships for LPNs Program provides an award of $750 per semester. The scholarship does require a post graduate commitment of 6 months full-time employment for each semester of funding.
Requirements: Applicant must: be employed as an LPN at Clarian; be in good standing to receive the scholarship. A participant whose job performance does not meet the standards of Clarian's performance review program during any time of the employment period/year that the scholarship is in effect will jeopardize his/her opportunity for future funding; be actively involved in Clinical Unit Operations (i.e., attend staff meetings, keep mandatory education requirements up to date, etc.); must maintain a competitive GPA. To complete the application process, you must: complete the online application; schedule appointment to complete paperwork; submit proof of enrollment in an approved nursing school; submit 2 professional recommendations; one from current manager, one from RN co-worker; submit a one-page narrative on why you should be awarded the scholarship
Geographic Focus: Indiana
Date(s) Application is Due: Jul 1; Nov 1
Amount of Grant: 750 U.S.D
Contact: Clarian Nurse Recruitment Line; (877) 354-2996; dspilker@clarian.org
Internet: http://www.clarian.org/clarianjobs/nursing/lpnscholarship.htm
Sponsor: Clarian Health Partners
950 N Meridian Street, Gateway Plaza
Indianapolis, IN 46204

Clarian Health Student Nurse Extern Scholarships 1313
The Clarian Health Student Nurse Extern Scholarship Program offers scholarships for each semester worked as a student nurse extern (SNE) at one of the three downtown Clarian hospitals: Methodist Hospital, Indiana University Hospital and Riley Hospital for Children. The SNE Scholarship Program offers several options to meet scheduling and financial needs. See website for details.
Requirements: Any nursing student in good standing and actively enrolled in a program that results in a RN degree may apply. A nursing student must meet the requirements of the SNE job description by completing at least one semester of nursing clinicals and remain actively enrolled and a student in good standing in the school of nursing they are attending. To apply for a Student Nurse Extern position at Clarian health, please send resume to dspilker@clarian.org. In a cover letter or the objective section of your resume, indicate your interest in the Student Nurse Extern position. The number of sponsorship awards is limited.
Geographic Focus: Indiana
Amount of Grant: 250 - 1,000 U.S.D
Contact: Clarian Nurse Recruitment Line; (877) 354-2996; dspilker@clarian.org
Internet: http://www.clarian.org/portal/patients/education?clarianContentID=/education/nursing_education/SNE.xml
Sponsor: Clarian Health Partners
950 N Meridian Street, Gateway Plaza
Indianapolis, IN 46204

Clark-Winchcole Foundation Grants 1314
The foundation awards grants to nonprofit organizations in the District of Columbia, with an emphasis on higher education, hospitals and health care, cultural programs, youth, the disabled, and religion. Types of support include general operating support, scholarships, and building construction/renovation. Applicants should submit a letter of inquiry that includes a description of the project, amount requested, audited financial report, budget, and proof of tax-exempt status. Further information will be requested by the foundation if interested.
Requirements: Only nonprofit organizations in the District of Columbia are eligible to apply.
Restrictions: No support for individuals and private foundations.
Geographic Focus: District of Columbia
Amount of Grant: 5,000 - 225,000 U.S.D
Contact: Vincent Burke, President; (301) 654-3607
Sponsor: Clark-Winchcole Foundation
3 Bethesda Metro Center, Suite 550
Bethesda, MD 20814-5358

Clark and Ruby Baker Foundation Grants 1315
The Clark and Ruby Baker Foundation was established to address a number of charitable concerns. The Baker family cared deeply about the Methodist Church and founded the Foundation to support Methodist affiliated, higher educational institutions in rural or small towns; charitable organizations that serve infirm, deserving, and aged ministers; economically disadvantaged and deserving children; and orphans and orphanages. The Foundation also provides support to charitable organizations that extend financial aid to the sick and infirm receiving medical treatment in any hospital or clinic in the state of Georgia. Capital support may be considered for the following purposes: for construction of educational facilities at a college or university; for clinics and hospitals; for libraries; and for any building with a charitable use. Grants from the Clark and Ruby Baker Foundation are primarily one year in duration; on occasion, multi-year support is awarded. Applicants must apply online at the grant website. Applicants are strongly encouraged to do the following before applying: review the downloadable state application procedures for additional helpful information and clarifications; review the downloadable online-application guidelines at the grant website; review the foundation's funding history (link is available from the grant website); review the online application questions in advance; and review the list of required attachments. These will generally include: a list of board members, financial statements (audited, reviewed, or compiled by independent auditor); an organization summary; a list of other funding sources; an IRS Determination letter; and other required documents. All attachments must be uploaded in the online application as PDF, Word, or Excel files. The Clark and Ruby Baker Foundation application deadline is 11:59 p.m. on June 1. Applicants will be notified of grant decisions by letter within one to two months after the deadline.
Requirements: Applicants must have 501(c)3 tax-exempt status.
Restrictions: The foundation does not support requests from individuals, organizations attempting to influence policy through direct lobbying, or any political campaigns.
Geographic Focus: Georgia
Date(s) Application is Due: Jun 1
Contact: Quanda Allen, Vice President; (404) 264-1377; quanda.allen@baml.com
Internet: https://www.bankofamerica.com/philanthropic/fn_search.action
Sponsor: Clark and Ruby Baker Foundation
3414 Peachtree Road, N.E., Suite 1475, GA7-813-14-04
Atlanta, GA 30326-1113

Clark County Community Foundation Grants 1316
Clark County Community Foundation will award grants to 501(c)3 charitable organizations, schools, and exempt governmental units or their subdivisions serving Clark County, Kentucky. Areas of interest include arts and culture, housing, employment support and education. Grants will range from approximately $5,000 to $10,000. To apply, submit a letter of inquiry of no more than three pages by 4:00 pm, March 15. Applicants will then be invited to submit full proposals, which are due by May 15.
Requirements: Clark County, Kentucky, nonprofit organizations are eligible.
Geographic Focus: Kentucky
Date(s) Application is Due: Mar 15; May 15
Amount of Grant: 5,000 - 10,000 U.S.D
Contact: Barbara A. Fischer; (859) 225-3343; fax (859) 243-0770; bfischer@bgcf.org
Internet: http://www.bgcf.org/page27526.cfm
Sponsor: Clark County Community Foundation
250 W Main Street, Suite 1220
Lexington, KY 40507

Clark Foundation Grants 1317
The foundation awards grants in New York, NY, and upstate New York for charitable and educational purposes and in support of health care, youth, cultural, environmental, and community organizations. Types of support include general operating support, continuing support, annual campaigns, capital campaigns, building construction/renovation, equipment acquisition, program development, seed money, and scholarships to Cooperstown County residents.
Requirements: Nonprofits in upstate New York and New York, NY, are eligible.
Geographic Focus: New York
Date(s) Application is Due: Jan 1; Apr 1; Jul 15; Oct 1
Amount of Grant: 20,000 - 200,000 U.S.D
Samples: Harlem RBI (New York, NY)—for general operating support, $100,000; Partnership with Children (New York, NY)—for general support, $150,000.
Contact: Charles Hamilton, Executive Director; (212) 977-6900
Sponsor: Clark Foundation
1 Rockefeller Plaza, 31st Floor
New York, NY 10020

Claude Bennett Family Foundation Grants 1318
Established in Alabama in 1993, the Claude Bennett Family Foundation has as its primary interest areas: health care, health organizations, higher education, human services, and Protestant churches and agencies. The range of funding is up to $15,000, and the Foundation has supported agencies across the U.S. There are no specific applications or deadlines with which to adhere, and applicants should contact the Foundation in writing, stating the purpose of their request and offering a detailed budget.

Samples: Food Bank of Larimer County, Fort Collins, Colorado, $1,500; Poudre School District, Fort Collins, Colorado, $3,000; South Highland Presbyterian Church, Birmingham, Alabama, $11,025; Penguin Project, East Peoria, Illinois, $1,000.
Contact: Harold I. Apolinsky, Treasurer; (205) 930-5122; fax (205) 212-3888; hapolinsky@sirote.com
Sponsor: Claude Bennett Family Foundation
2311 Highland Avenue S, P.O. Box 130804
Birmingham, AL 35213-0804

Claude Pepper Foundation Grants **1319**
The foundation makes grants primarily to support the work of the Claude Pepper Center and the Pepper Institute on Aging and Public Policy, both located at Florida State University, Tallahassee, FL. The foundation makes limited grants to other organizations to continue the work and vision of Claude and Mildred Pepper. The foundation also supports a visiting scholars program and an oratory competition for Florida students. Grants are usually made for a period of one year and except in rare instances no grant will be made for longer than a three-year period of time. Guidelines are available online.
Requirements: 501(c)3 tax-exempt organizations are eligible.
Samples: Florida Council on Aging - To support Special Plenary Sessions at the Annual Aging Network Conferences; National Academy of Social Insurance - To sponsor the Academy's annual conference opening dinner in Washington, D.C., which addressed issues surrounding Social Security reform; Tallahassee Senior Center - To co-sponsor the Claude Pepper Senior Walk/Run, which consists of a competition for walkers over the age of 60, and one for runners, with prizes and refreshments.
Contact: John T. Herndon, Executive Director; 850) 644-9309; fax (850) 644-9301; herndon@claudepepperfoundation.org
Internet: http://www.claudepepperfoundation.org/programs_grants.cfm
Sponsor: Claude Pepper Foundation
636 West Call Street
Tallahassee, FL 32306-1122

Claude Worthington Benedum Foundation Grants **1320**
Grants are made in the areas of education, health and human services, community and economic development, environment, and the arts. Grants have been awarded to support education reform, teacher education, higher education, workforce development, rural health, professional developing in healthcare, human services, affordable housing, and economic development. Grants are awarded to organizations in West Virginia and southwestern Pennsylvania. Funds are provided for general operations for projects, sometimes including building and equipment, in West Virginia, and for projects in Pittsburgh that address regional problems and needs, that establish demonstration projects with strong potential for replication in West Virginia, or make outstanding contributions to the area. Additional types of support include matching funds, consulting services, technical assistance, capital campaigns, conferences and seminars, research, and seed grants. Organizations wishing to apply should request a copy of the annual report, which includes application guidelines. Applications may be submitted at any time; the board meets for review in March, June, September, and December.
Requirements: Southwestern Pennsylvania and West Virginia nonprofits may apply.
Restrictions: Support is not given for national health and welfare campaigns, medical research, religious activities, fellowships, scholarships, annual campaigns, or travel.
Geographic Focus: Pennsylvania, West Virginia
Contact: Margaret M. Martin, Grants Administrator; (800) 223-5948 or (412) 246-3636; fax (412) 288-0366; mmartin@benedum.org
Internet: http://www.benedum.org/pages.cfm?id=10
Sponsor: Claude Worthington Benedum Foundation
1400 Benedum-Trees Building, 223 Fourth Avenue
Pittsburgh, PA 15222

Clay Foundation Grants **1321**
The primary mission of the foundation is to promote and enhance the quality of life of the citizens of West Virginia. Initially, grants will be confined to organizations or projects within the greater Kanawha Valley that significantly affect its residents. The foundation considers a broad range of organizations, though special interest is given to programs in the field of aging; health care, research, and education; vocational education; and services to disadvantaged youth and their families. Applicants are asked to submit a preliminary letter. If foundation priorities and resources permit consideration of the request, a detailed formal proposal may be requested. There are no deadline dates for the submission of preliminary letters. The board meets four times annually, generally in January, April, July, and October, at which times applications are considered and grants made.
Requirements: IRS 501(c)3 tax-exempt organizations in West Virginia are eligible.
Restrictions: Grants are not made for ongoing normal operations, debt retirement or operational deficits, contributions to endowment or scholarship funds, annual appeals or other fund-raising events, national fund-raising campaigns, religious organizations for religious purposes, or grants to conduit organizations.
Geographic Focus: West Virginia
Amount of Grant: 5,000 - 75,000 U.S.D
Samples: U of Charleston (WV)—to construct the Buckner and Lyell Clay Ctr, $1.15 million; Charleston Catholic High School (Charleston, WV)—for the science wing, $60,000; Fund for the Arts (Charleston, WV)—challenge grant, $60,000. West Virginia State Museum—for renovation, $350,000.
Contact: Charles Avampato, President; (304) 344-8656; fax (304) 344-3805
Sponsor: Clay Foundation
1426 Kanawha Boulevard E
Charleston, WV 25301

Cleo Foundation Grants **1322**
The foundation supports public charities in San Francisco, San Benito, Mendocino, and southern Monterey counties, with primary interest in supporting organizations that benefit at-risk and/or low-income children, youth, and adults. The foundation is most interested in programs that can make good use of relatively small grants. Types of support include building construction/renovation, capital campaigns, continuing support, operating support, and program development. Applications are due 12 weeks before board meetings in March and September. Request guidelines in writing.
Requirements: 501(c)3 organizations in San Francisco, San Benito, Mendocino, and southern Monterey counties in California may apply.
Geographic Focus: California
Amount of Grant: 1,000 - 20,000 U.S.D
Contact: Mary Gregory, Executive Director; (415) 561-6540; fax (415) 561-6477; mgregory@pfs-llc.net
Internet: http://www.pfs-llc.net/cleo.html
Sponsor: Cleo Foundation
Presidio Building 1016, Suite 300, P.O. Box 29906
San Francisco, CA 94129-0906

Cleveland-Cliffs Foundation Grants **1323**
Contributions are made to nonprofit organizations to enhance the quality of life of Cleveland-Cliffs Inc employees and in recognition of a corporate responsibility toward educational, health, welfare, civic, and cultural matters within the communities where the company operates. The foundation was formed for the purpose of making contributions to groups organized and operated exclusively for religious, charitable, scientific, literary, or educational purposes and for the prevention of cruelty to children or animals. Types of support include general operating support, annual campaigns, capital campaigns, building/renovation, professorships, scholarship funds, research, and employee matching gifts. The foundation's major emphasis is on supporting education through a matching gift program and direct contributions to educational institutions. Requests for support must be in writing.
Requirements: Nonprofit organizations in the mining communities in which Cleveland-Cliffs Inc operates, including Alabama, Michigan, Minnesota, Ohio and, West Virginia are eligible.
Geographic Focus: Alabama, Michigan, Minnesota, Ohio, West Virginia
Amount of Grant: 250 - 50,000 U.S.D
Contact: Dana W. Byrne, Vice President; (216) 694-5700; fax (216) 694-4880; publicrelations@cleveland-cliffs.com
Internet: http://www.cliffsnaturalresources.com/Development/CommunityRelations/Pages/Cleveland-CliffsFoundation.aspx
Sponsor: Cleveland-Cliffs Foundation
1100 Superior Avenue, Suite 1500
Cleveland, OH 44114-2589

Cleveland Browns Foundation Grants **1324**
The foundation supports the northeast Ohio community by funding programs that improve the lives of disadvantaged children. The foundation's four focus areas are education, arts and culture, health, and career development as they relate to children. There are no application deadlines.
Requirements: Northern Ohio nonprofit organizations are eligible. Applications sold be submitted by mail
Restrictions: Organizations and causes that will not be considered for funding include fund-raising, sponsorship events, or donation requests; religious organizations for sectarian religious purposes; general or annual operation expenses; capital or building funds; or staff salaries or stipends.
Geographic Focus: Ohio
Contact: Grant Program Supervisor; (440) 891-5063; fax (440) 891-7529
Internet: http://www.clevelandbrowns.com/community/foundation
Sponsor: Cleveland Browns Foundation
Community Relations Department, 76 Lou Groza Boulevard
Berea, OH 44017

CLF Fellowships **1325**
The fellowships provide support for specialized clinical or experimental training in hepatic function or disease to individuals who have already completed the basic graduate program. The value of the fellowship is based on the qualifications and seniority of the applicant and will include, if applicable, a travel grant equivalent to return economy airfare for the grantee and spouse between the university of origin and the place of training. Guidelines and application forms are available from the office and in medical schools through the office of the Dean of the Faculty of Medicine.
Requirements: Candidates must hold an MD or PhD degree or equivalent. Canadian citizens and landed immigrants who are resident in Canada at the time of application or Canadian citizens who are normally resident in Canada are eligible. Applications must be sponsored by a Canadian Faculty of Medicine or Health Sciences and accompanied by a letter of support from the Dean (or designee).
Geographic Focus: All States, Canada
Date(s) Application is Due: Mar 31
Amount of Grant: Up to 30,000 CAD
Contact: Research Awards Administrator; (800) 563-5483 or (416) 491-3353; fax (416) 491-4952; clf@liver.ca
Internet: http://www.liver.ca/Research/Grant_Program
Sponsor: Canadian Liver Foundation
2235 Sheppard Avenue E, Suite 1500
Toronto, ON M2J 5B5 Canada

CLF Graduate Studentships 1326
Studentships enable academically superior students to undertake full-time studies in Canadian universities in disciplines relevant to liver function, structure, and disease. Awards are for one academic year and may be renewed three times for a total of four years of support. Registrants in an MS program are restricted to a maximum of three years of support. Guidelines are available from the office or at medical schools through the office of the Dean of the Faculty of Medicine.
Requirements: The candidate must be accepted into a full-time university graduate science program in a medically related discipline related to a master's or doctoral degree; hold a record of superior academic performance; and be sponsored by a faculty supervisor with a record of productive medical research and sufficient, competitively acquired research funding to ensure the satisfactory conduct of the student's research during the term of the award.
Geographic Focus: All States, Canada
Date(s) Application is Due: Mar 31
Amount of Grant: 16,500 CAD
Contact: Billie Potkonjak, (800) 563-5483; fax (416) 491-4952; bpotkonjak@liver.ca
Internet: http://www.liver.ca/Research/Grant_Program
Sponsor: Canadian Liver Foundation
2235 Sheppard Avenue E, Suite 1500
Toronto, ON M2J 5B5 Canada

CLF Operating Grants 1327
The grants provide funding to help fully qualified and trained hepatologists and basic scientists begin liver research projects in their own laboratories at Canadian universities. Candidates must be prepared to spend 80 percent of their time in research and to commit the remaining time to active participation in teaching and/or patient care.
Requirements: Candidates must hold an MD or PhD or equivalent and have a proven interest in liver structure, function, or disease. The applicant must have completed a minimum of two and preferably three years of formal research training (post medical specialty training in the case of MDs) and have obtained additional research experience as a clinical investigator or postdoctoral fellow under the supervision of a well-established, competitively funded, senior basic or clinical scientific faculty colleague who holds a faculty level position.
Geographic Focus: All States, Canada
Date(s) Application is Due: Mar 31
Amount of Grant: Up to 60,000 CAD
Contact: Research Awards Administrator; (416) 964-1953; fax (416) 964-0024; bpotkonjak@liver.ca
Internet: http://www.liver.ca/Research/Grant_Program
Sponsor: Canadian Liver Foundation
2235 Sheppard Avenue E, Suite 1500
Toronto, ON M2J 5B5 Canada

CLIF Bar Family Foundation Grants 1328
The foundation's mission is to support nonprofits - grassroots organizations in particular - working to promote environmental restoration and conservation, sustainable food and agriculture, people's health, and a wide range of social concerns. It focuses on organizations whose missions support environmental restoration and conservation, sustainable food and agriculture, people's health and youth.
Requirements: The foundation funds nonprofit organizations working on projects that are well informed, have clearly defined objectives, demonstrate strong community ties, and promote values such as compassion, inclusiveness, patience, and positive development both in the program and in its implementation. Emphasis is on grassroots organizations that have the ability to engage local groups, positively impact their communities, and focus most of their resources on useful and positive actions.
Restrictions: Applications that require funding in less then 12 weeks will not be accepted. The foundation will not fund: Deficit financing; Loans or grants to individuals; Capital construction; Research projects, conferences, seminars, media events or workshops, unless they are an integral part of a broader program; State agencies; Religious groups; Individual Sponsorships; Fundraising events for your organization such as fun runs, bike rides, etc. If you are a local/regional branch of a national organization they will not fund grants to local chapters if they fund the national organization. Otherwise, they only take into consideration local offices in Northern California.
Samples: The Alameda County Community Food Bank; American Farmland Trust Breast Cancer Fund; New Leaders Initiative/Earth Island Institute; Center for Food Safety; Center for Rural Affairs; Circle of Life; City Slicker Farms; Community Alliance with Family Farmers; Community Resources for Science; Diabetic Youth Foundation; Disability Rights Education & Defense Fund; Ecological Farming Association; Ecology Center; Focus the Nation: Global Warming Solutions for America; Food Alliance Green Belt Movement; Greentreks Grid Alternatives; Habitat for Humanity; Healthy Child Healthy World; Heifer International; The Land Institute; Leave No Trace; Leukemia & Lymphoma Society; Luna Kids Dance; Lungevity Foundation; KIPP Bay Area; Marin Agricultural Land Trust; NorCal High School Mountain Bike League; Organic Farming Research Foundation; Pedals For Progress; Project Open Hand; Rainforest Action Network; San Francisco AIDS Foundation; Save Mount Diablo; Save the Waves; Saving Teens in Crisis Collaborative; Trips for Kids; Wild Hope; Worldbike; World Neighbors
Contact: Kit Crawford, President; (510) 859-2283; fax (510) 588-5490; familyfoundation@clifbar.com
Internet: http://clifbarfamilyfoundation.org/subtemplate.php?s=1
Sponsor: CLIF Bar Family Foundation
1610 5th Street
Berkeley, CA 94710

Clinton County Community Foundation Grants 1329
The Clinton County Community Foundation is a catalyst for stimulating and funding initiatives that improve the quality of life for citizens of Clinton County. The Foundation Grants address needs that generally fall into the following categories: health and medical; social services; education; cultural affairs; civic affairs; and community beautification.
Requirements: All applicants must receive pre-qualification prior to submitting an application by submitting a letter of inquiry to the Program Director. The letter should contain a brief statement of the applicant's needs for assistance, estimate of total cost of project, and enough information to enable the Foundation to determine if the application falls within the guidelines of its grants program. If the grant application is decided, organizations must submit one original application, plus ten copies required for review by the grantmaking committee with the following information included: grant application with the cover page; project budget; board list; evidence of board approval; 501(c)3 letter; year-end audit or financial statement; current month and year-to-date financial statement; and when applicable, three estimates must be included, one from a Clinton County business. The organization may be contacted for additional information, an interview with the grant making committee, or a possible site visit, and the time period which a grant decision will likely be made.
Geographic Focus: Indiana
Date(s) Application is Due: May 6; Sep 7
Contact: Kim Abney; (765) 454-7298 or (800) 964-0508; kim@cfhoward.org
Internet: http://www.cfclinton.org/grant_seekers_cl.html
Sponsor: Clinton County Community Foundation
215 West Sycamore Street
Kokomo, IN 46901

Clowes ACS/AAST/NIGMS Jointly Sponsored Mentored Clinical Scientist Development Award 1330
The American College of Surgeons and American Association for the Surgery of Trauma announce a program that will provide supplemental salary funding of up to $75,000 per year to an individual who has received a Mentored Clinical Scientist Development Award (K08/K23) from the National Institute for General Medical Science (NIGMS). This award is directed at surgeon-scientists working in the early stages of their research careers. The award supports a three-, four-, or five-year period of supervised research experience that may integrate didactic studies with laboratory or clinical research. The award program offers a means to facilitate the career development of individuals pursuing careers in trauma surgery research by enhancing salary support over and above that offered by the K08/K23 mechanism.
Requirements: Awardees must be members in good standing of the College and eligible for membership in AAST.
Restrictions: Pre-existing applications and awards are not eligible for consideration.
Date(s) Application is Due: Oct 12
Amount of Grant: Up to 75,000 U.S.D
Contact: Kate Early, Administrative Assistant; (312) 202-5281; fax (312) 202-5021; kearly@facs.org
Internet: http://www.facs.org/memberservices/acs-aast-nigms.html
Sponsor: American College of Surgeons
633 N Saint Clair Street
Chicago, IL 60611-3211

CMA Foundation Grants 1331
The foundation seeks to improve the quality of life for central Ohio residents through the promotion of wellness, prevention of disease, and delivery of services and research. The foundation will review funding requests for matching or challenge grants, multiyear commitments, and capital expenditures on a project-by-project basis. The foundation is the leading local source of funding for health care projects benefiting the local community. Application may be made online.
Requirements: Nonprofit organizations in central Ohio, including Franklin, Delaware, Fairfield, Licking, Madison, Pickaway, and Union Counties, are eligible.
Geographic Focus: Ohio
Samples: Children's Hunger Alliance (Columbus, OH)—to promote school-breakfast programs, $30,000; Educational Council Foundation (Columbus, OH)—to train middle- and high-school educators to provide information about positive health behaviors, $50,000 maximum; Franklin County Health Dept (Columbus, OH)—to expand its anti-tobacco activities, $30,000 maximum; Mount Carmel Health Systems Foundation (Columbus, OH)—to provide education, information, and self-care skills training to people who have experienced heart failure, $90,000.
Contact: Program Contact; (614) 240-7410; yourthoughts@goodhealthcolumbus.org
Internet: http://www.cmaf-ohio.org/cmaf/grants.html
Sponsor: Columbus Medical Association Foundation
431 E Broad Street
Columbus, OH 43215

CMA Foundation Uninsured Population Grants 1332
The program targets the needs of the uninsured in Franklin County and identifies characteristics and barriers to obtaining health care. Grants are awarded to community organizations for programs that work toward eradication of the identified barriers faced by the uninsured in accessing health care. It is the intent of the foundation to grant direct service providers the financial means to target one or more these barriers.
Requirements: Nonprofit organizations in central Ohio, primarily those residing in Franklin County and the six contiguous counties of Delaware, Fairfield, Licking, Madison, Pickaway and Union. may submit requests for grant support.
Geographic Focus: Ohio

Samples: Prescription for Care Alliance (OH)—to provide financial and technical support to a consortium of grantees, including eight lead agencies and 21 community partners, to improve the availability of medications to low-income, uninsured Franklin County residents, $650,000.
Contact: Director of Programs; (614) 240-7410; fax (614) 240-7415; yourthoughts@goodhealthcolumbus.org
Internet: http://www.cmaf-ohio.org/cmaf/grants.html
Sponsor: Columbus Medical Association Foundation
431 E Broad Street
Columbus, OH 43215

CMAP Physician Advocacy Fellowship 1333

This program supports doctors to develop or enhance their advocacy skills by implementing a project in partnership with an advocacy organization. The fellowship seeks to make advocacy a core professional value for physicians by developing a cadre of advocates with expertise in achieving system or policy-level social change at the local, state, and national level. Projects must be focused within the United States and should identify system or policy level changes as the outcomes of the fellowship work. Application guidelines are available online. Deadline listed is for short proposals; full proposals are by invitation.
Requirements: U.S. physicians at all stages of their careers are eligible; however, the most competitive applicants are practicing physicians. Prospective applicants must secure the commitment of an advocacy organization that is prepared to house, mentor, and support them throughout the fellowship period. A list of advocacy organizations that are interested in participating in the fellowship is available online. Applicants also may apply with organizations other than those listed online.
Date(s) Application is Due: Oct 7
Amount of Grant: 2,000 - 80,000 U.S.D
Samples: Robert Goodman—for efforts to minimize the influence of commercial interests, particularly those of the pharmaceutical industry, in medical education and training, $111,804.
Contact: Claudia Calhoon; (212) 342-4769; fax cmc100@columbia.edu
Internet: http://www.cmap.columbia.edu/research_fellowship.shtml
Sponsor: Columbia University
630 W 168th Street, P&S Box 11
New York, NY 10032

CMS Hispanic Health Services Research Grants 1334

The purpose of the grant is to implement Hispanic American health services research activities to meet the needs of diverse CMS beneficiary populations. The program seeks competitive applications for small applied research projects that relate to identifying and evaluating solutions for eliminating health disparities among Hispanic Americans. Funding is available for grants to implement research related to health care delivery and health financing issues affecting Hispanic American communities, including issues of access to health care, utilization of health care services, health outcomes, quality of services, cost of care, health and racial disparities, socio-economic differences, cultural barriers, managed care systems, and active ties related to health screening, prevention, outreach, and education.
Requirements: Investigators should be associated with a university, college, community-based health organization, or a professional association that has a health services research component. Researchers are encouraged to use CMS data as part of their research projects. Researchers are also expected to become involved in the design, implementation, and operation of research projects that address health care issues such as financing, delivery, access, quality, and barriers affecting the Hispanic American community. Applicants must meet one of the following three requirements in order to qualify for funding under this grant program: 1) A health services/disparities researcher at an university or college offering a Ph.D. or Master's Degree Program in one or more of the following disciplines Allied Health, Gerontology, Health Care Administration, Health Education, Health Management, Nursing, Nutrition, Pharmacology, Public Health, Public Policy, Social Work; or 2) a member of a community-based health organization with a Hispanic health services research component; or 3) a member of a professional association focusing on Hispanic health services and health disparities issues. Applications must be submitted electronically (via grants.gov). CMS will accept a hard copy if the applicant is unable to access grants.gov or is having serious problems in sending the application electronically.
Restrictions: Grant funds may not be used for any of the following: to provide direct services to individuals except as explicitly permitted under the grant solicitation; to match any other Federal funds; or, to provide services, equipment, or supports that are already the legal responsibility of another party under Federal law.
Date(s) Application is Due: Jul 3
Amount of Grant: 200,000 - 250,000 U.S.D
Contact: Joi Grymes, (410) 786-7251; Joi.Grymes@cms.hhs.gov
Internet: http://www.cms.hhs.gov/ResearchDemoGrantsOpt/03_Hispanic_Health_Services.asp#TopOfPage
Sponsor: Centers for Medicare and Medicaid Services
7500 Security Boulevard, Room C2-21-15
Baltimore, MD 21244-1850

CMS Historically Black Colleges and Universities Health Services Research Grants 1335

The purpose of the grant program is to support researchers in implementing health services research activities to meet the needs of diverse CMS beneficiary populations. The goals of the grant program are to: 1) encourage HBCU health services researchers to pursue research issues which impact the Medicare, Medicaid, and SCHIP (State Children's Health Insurance Program) programs, 2) assist CMS in implementing its mission focusing on health care quality and improvement for its beneficiaries, 3) assist HBCU researchers by supporting extramural research in health care capacity development activities for the African American communities, 4) increase the pool of HBCU researchers capable of implementing the research, demonstration, and evaluation activities of CMS, and 5) assist in fostering interuniversity communication and collaboration regarding African American health disparity issues. Funding is available for grants to implement research related to health care delivery and health financing issues affecting African American communities, including issues of access to health care, utilization of health care services, health outcomes, quality of services, cost of care, health and racial disparities, socio-economic differences, cultural barriers, managed care systems, and activities related to health screening, prevention, outreach, and education.
Requirements: To be eligible for grants under this program, an organization must be an HBCU (Historically Black College or University) and meet one of the following three
Requirements: 1) offer a Ph.D. or Master's Degree Program in one or more of the following disciplines - Allied Health, Gerontology, Health Care Administration, Health Education, Health Management, Nursing, Nutrition, Pharmacology, Public Health, Public Policy, Social Work; or 2) have a School of Medicine; or 3) a member of the National HBCU Network for Health Services and Health Disparities. All proposals should describe research to be conducted with relevance to the CMS Medicare, Medicaid, and SCHIP programs and which area of Healthy People 2010 is served by this project. Applications must be submitted electronically (via grants.gov). CMS will accept a hard copy if the applicant is unable to access grants.gov or is having serious problems in sending the application electronically.
Restrictions: Grant funds may not be used for any of the following: to provide direct services to individuals except as explicitly permitted under the grant solicitation; to match any other Federal funds; to provide services, equipment, or supports that are already the legal responsibility of another party under Federal law.
Date(s) Application is Due: Jul 2
Amount of Grant: 200,000 - 250,000 U.S.D
Contact: Joi Grymes, (410) 786-7251; Joi.Grymes@cms.hhs.gov
Internet: http://www.cms.hhs.gov/ResearchDemoGrantsOpt/02_Historically_Black_Colleges_and_Universities.asp#TopOfPage
Sponsor: Centers for Medicare and Medicaid Services
7500 Security Boulevard, Room C2-21-15
Baltimore, MD 21244-1850

CMS Research and Demonstration Grants 1336

The general purpose of the Centers for Medicare & Medicaid Services' (CMS) research and demonstration program is to conduct and support projects to develop, test, and implement new health care financing and payment policies and to evaluate the impact of the agency's programs on its beneficiaries, providers, States, and other customers and partners. The scope of the agency's activities embraces all areas of health care: costs, access, quality, service delivery models, and financing and payment approaches.
Requirements: The following themes represent the agency's current priorities in research. Note that all projects must fall into the agency's statutory authorities to operate and improve Medicare, Medicaid, and other CMS programs and activities: (1) Monitoring and Evaluating CMS Programs; (2) Strengthening Medicaid, State Children's Health Insurance Program (SCHIP), and State Programs; (3) Expanding Beneficiaries' Choices and Availability of Managed Care Options; (4) Developing FFS Payment and Service Delivery Systems; (5) Improving Quality of Care and Performance Under CMS Programs; (6) Improving the Health of Our Beneficiary Population; (7) Prescription Drugs; (8) Building Research Capacity. Application packet is available online.
Restrictions: Applicants are expected to contribute towards the project costs. Generally 5 percent of the total project costs is considered acceptable. CMS rarely approves grants or cooperative agreements for research or demonstration projects in which the Federal Government covers 100 percent of the project's costs. The budget may not include costs for construction or remodeling or for project activities that take place before the applicant has received official notification of our approval of the project.
Amount of Grant: 25,000 - 1,000,000 U.S.D
Contact: Grant Officer; (410) 786-5130; Jnorris1@cms.hhs.gov
Internet: http://www.cms.hhs.gov/ResearchDemoGrantsOpt/04_Other_CMS-Grant_Opportunities.asp
Sponsor: Centers for Medicare and Medicaid Services
7500 Security Boulevard, Room C2-21-15
Baltimore, MD 21244-1850

CNA Foundation Grants 1337

The CNA Foundation concentrates its support primarily in programs designed to: meet the education needs of children, assist and support children, youth and adults in developing vocational skills. Support economically disadvantaged children and families. Requests for funding are accepted year-round. However, grants will be made only in accordance with the Foundation's budgetary guidelines. Proposals must be clear and brief. The Foundation will contact the organization if more information is needed.
Requirements: IRS 501(c)3 tax-exempt organizations are eligible.
Restrictions: Grants are not made to/for individuals; political causes, candidates, or organizations; veterans, labor, alumni, military, athletic clubs, or social clubs; sectarian organizations or denominational religious organizations; capital improvement or building projects; endowed chairs or professorships; United Way-affiliated agencies; or national groups whose local chapters have already received support.
Amount of Grant: 10,000 - 250,000 U.S.D
Samples: Chicago 2016 Exploratory Committee, Chicago, IL, $250,000; National Chamber Foundation, Washington, DC, $100,000; Starlight Starbright Childrens Foundation Midwest, Chicago, IL, $50,000.
Contact: Marlene Rotstein, Director; (312) 822-7065; marlene.rotstein@cna.com or cna_foundation@cna.com

Internet: http://www.cna.com/portal/site/cna/menuitem.7204aaf0316757e8715f09f6556631a0/?vgnextoid=b1e940fa11056010VgnVCM1000005566130aRCRD
Sponsor: CNA Foundation
333 South Wabash Avenue
Chicago, IL 60604

CNCS AmeriCorps VISTA Project Grants **1338**
AmeriCorps VISTA is a national-service program of the Corporation for National and Community Service (the Corporation) which oversees a variety of programs including other AmeriCorps programs, SeniorCorps programs, and Learn and Serve America programs. Designed specifically to fight poverty, VISTA was authorized in 1964 as Volunteers in Service to America. The program was incorporated into the AmeriCorps network of programs in 1993. VISTA supports efforts to alleviate poverty by encouraging volunteers (members), ages 18 years and older, from all walks of life, to engage in a year of full-time service with a sponsoring organization (sponsor) to create or expand programs designed to bring individuals and communities out of poverty. Under this arrangement, the Corporation places a team of VISTA members with a sponsor; the sponsor funds local operating and logistics costs of the project while the Corporation covers member and certain sponsor costs as follows: a biweekly living allowance for members; a Segal AmeriCorps Education Award or post-service stipend for members; health coverage for members; a moving allowance for members relocating to serve; liability coverage for members under the Federal Employees Compensation Act and the Federal Torts Claims Act; childcare (for income-eligible members); FICA; payroll services (members receive their paychecks directly from AmeriCorps VISTA); training in project management and leadership for VISTA members and project supervisors; and assistance for sponsors to recruit VISTA members. Applications for VISTA resources are handled by Corporation State Offices which are federal offices staffed by federal employees in the states. (A list of the offices along with their contact information is available at the AmeriCorps website.) Applying for VISTA resources is a two-step process. As step one, the organization must submit a VISTA Concept Paper. If the concept paper is accepted, the organization must, as step two, submit a VISTA Project Application. Applicants will receive their project-application materials when their concept paper has been approved. The length of the application process varies, but the average length of time from the initial contact to a final decision is three to five months. Both concept papers and project applications are usually submitted using eGrants, the Corporation's web-based system for applications. Organizations must visit the eGrant website (link available at the AmeriCorps website) to create an account prior to submitting concept papers. Organizations that cannot submit using eGrants may submit a paper copy. The forms are included in the downloadable Concept-Paper Instructions document at the AmeriCorps website. As of this writing, VISTA is giving priority to new projects that focus on the areas of housing, financial literacy, and employment.
Requirements: Public organizations such as nonprofit private organizations, Indian Tribes, state and local government organizations, and institutions of higher education can apply to be VISTA sponsors. Eligible nonprofit private organizations are not limited to those with IRS 501(c)3 status, but rather all organizations with IRS 501(c) status that focus on anti-poverty community development. Project sponsors are encouraged (but not required) to provide a financial match; however they must be able to direct the project, supervise the volunteers, and provide necessary administrative support to complete the goals and objectives of the project. Projects must be developed in accordance with all four of the VISTA Core Principles: Anti-Poverty Focus, Community Empowerment, Sustainable Solutions, and Capacity Building. All VISTA resources must be used to create, expand, or enhance projects that lift people out of poverty. Additionally, the Corporation has identified, in its strategic plan, six focus areas for funding: Economic Opportunity; Education; Healthy Futures; Veterans and Military Families; Disaster Services; and Environmental Stewardship. All new VISTA project development must fall within these six focus areas. As of this writing, the Corporation will direct most VISTA resources to the Economic Opportunity and Education focus areas; however, the Corporation will also address the other focus areas, according to the ability of Corporation State Offices to identify opportunities in those areas that can have a direct impact on breaking the cycle of poverty.
Restrictions: Organizations that focus solely on advocacy and lobbying are not eligible. Key legislation and regulations governing the VISTA program are as follows: the Domestic Volunteer Service Act (as amended by Public Law 113-13, April 2009); the National Service Trust Act (as amended by Public Law 113-13, April 2009); the Edward M. Kennedy Serve America Act (Public Law 113-13, April 2009); and the Code of Federal Regulations, Title 45, Parts 1206, 1210-1211, 1216-1220, 1222, and 1226.
Contact: Mary Strasser, Director, Office of AmeriCorps VISTA; (202) 606-6943 or (202) 606-5000
Kathy Little, Program Assistant; (202) 606-6852 or (202) 606-5000
Internet: http://www.americorps.gov/for_organizations/apply/vista.asp
Sponsor: Corporation for National and Community Service
1201 New York Avenue Northwest
Washington, DC 20525

CNCS Engaging Persons with Disabilities in National and Community Service Grants **1339**
The mission of the Corporation for National and Community Service (the Corporation) is to improve lives, strengthen communities, and foster civic participation through service and volunteering. The purpose of the Corporation's Engaging Persons with Disabilities program is to engage more persons with disabilities in national and community service. The program funds innovative national or regional (operating in at least three or more states) partnership models in which persons with disabilities are engaged in service that results in measurable impact on the community served. The following categories are of particular interest: using service as a strategy to assist in the transition of young persons with disabilities from school to adult life; and providing adults with disabilities, preferably disabled veterans, the opportunity to engage in service. Grants are awarded covering a period not to exceed three years, with continued funding beyond year one contingent on satisfactory performance. Proposals should be submitted through the eGrants system. However, if for a legitimate reason an applicant must submit a paper application, it must be received by 5 p.m. eastern standard time on the deadline date and include a cover letter explaining why the application could not be submitted through eGrants. Late applications may be accepted only if the applicant submits a letter via email to LateApplications@cns.gov within the 24-hour period following the deadline explaining the extenuating circumstance which caused the delay. Late applications are evaluated on a case-by-case basis. The process of competing for federal grants generally works as follows. Types of funding offered by federal programs are set forth in the program's enabling legislation. Grant competitions for a program's funding tend to follow congressional budget and appropriations cycles (as well as multi-year funding cycles in some cases) and are announced through various Notices which provide details, guidelines, and deadlines for workshops, letters of intent (if requested), and application submission. Notices are posted at the Corporation's and www.grants.gov websites. A subscription link is also provided at these sites for prospective applicants to be notified via email of upcoming grant competitions. In addition to statutory priorities, the Corporation determines priorities to focus each grant competition. Generally a series of technical assistance calls or workshops will be offered before the application is due. The dates of these will be posted at the Corporation's website. Application and other deadlines may vary from competition to competition. Prospective applicants are encouraged to verify current deadline dates. Notices of Intent are generally requested rather than required and are used to help the Corporation plan the review process.
Requirements: Public or private nonprofit organizations, including faith-based and other community-based organizations, are eligible. Applicant organizations must be national or regional (operating in at least three or more states) in scope.
Restrictions: Corporation grants are subject to the Cost Principles and Uniform Administrative Requirements under the applicable Office of Management and Budget (OMB) Circulars. Also, awards will be subject to the law(s) under which the award is made including the following pieces of legislation: the National and Community Service Act of 1990; National and Community Service Trust Act of 1993; and the Edward Kennedy Serve America Act of 2009.
Amount of Grant: 1,000,000 - 5,000,000 U.S.D
Samples: Ability Awareness, Costa Mesa, California in partnership with Habitat for Humanity, Atlanta, Georgia, $1.3 million - to launch the new Ability House Veterans and College Students with Disabilities Initiatives in California, District of Columbia, Florida, Hawaii, Massachusetts, and New Jersey; The University of California, Los Angeles, California, $1.4 million - for its Tarjan Center which will manage a collaborative program in three states to increase number of youth and adults with disabilitites, including veterans, in community service; Denver Options, Inc., Denver, Colorado, $1.2 million - to establish the capacity to provide service opportunities for young people with disabilities in Colorado, Montana, North Dakota, South Dakota, Utah, and Wyoming.
Contact: Grants Administrator; (202) 606-5000; disabilityoutreach@cns.gov
Internet: http://www.nationalservice.gov/about/initiatives/index.asp
Sponsor: Corporation for National and Community Service
1201 New York Avenue NW
Washington, DC 20525

CNCS Martin Luther King, Jr Day of Service Grants **1340**
In 1983 President Ronald Reagan signed a new national holiday into law to honor the life and work of Dr. Martin Luther King, Jr. Dr. King sought to forge the common ground on which people from all walks of life could join as equals to address important community issues. Service, he realized was the great equalizer. As he stated, "Everybody can be great, because everybody can serve." In 1994 Congress passed the King Holiday and Service Act which authorized the Corporation of National and Community Service (the Corporation) to partner with the Martin Luther King, Jr. Center for Nonviolent Social Change, Inc. in transforming the holiday into a national day of service for all Americans. The Martin Luther King, Jr. Day of Service grants are one means by which the Corporation has addressed this task. The grants support intermediary organizations who in turn must competitively subgrant 80% of awarded funds (in amounts of $1,000 per year or more) to eligible organizations to plan and carry out direct service activities on the Martin Luther King Holiday weekend. The Corporation especially seeks organizations with multi-state, national (ten or more states) or comprehensive regional service areas. Service activities may be part of a continuum of year-round service and contribute to advancing overall programmatic impact and organizational capacity of intermediaries and grantees. The award period is up to three years, with funding provided in annual increments, contingent upon the availability of appropriations, compliance with grant conditions, and satisfactory performance. The process of competing for federal grants generally works as follows. Types of funding offered by federal programs are set forth in the program's enabling legislation. Grant competitions for a program's funding tend to follow congressional budget and appropriations cycles (as well as multi-year funding cycles in some cases) and are announced through various Notices which provide details, guidelines, and deadlines for workshops, letters of intent (if requested), and application submission. Notices are posted at the Corporation's and www.grants.gov websites. A subscription link is also provided at these sites for prospective applicants to be notified via email of upcoming grant competitions. In addition to statutory priorities, the Corporation determines priorities to focus each grant competition. All grants must support the following types of service activities: activities that address one or more of the Corporation's strategic areas, which include education, health futures, environmental stewardship, veterans and military families, economic opportunity, and disaster preparedness (with special consideration given to activities that support and engage veterans and military families, and provide

opportunities to connect to service beyond the holiday); activities that reflect the life and teaching of Martin Luther King, Jr., and that expand the impact of Day of Service activities from one year to the next; and activities that strengthen communities engaged in service. Grant applications are submitted via eGrants, the Corporation's web-based system for grant application and management. Generally a series of technical assistance calls or workshops will be offered before the application is due. The dates of these will be posted at the Corporation's website. Applications are due no later than 5 p.m. eastern time on the deadline date. Prospective applicants must create an account at the eGrant website (link available at the Corporation website). Applicants are advised to do this at least three weeks prior to the application-submission deadline. Applicants are advised to draft the application as a word-processing document, then copy and paste the document into eGrants no later than ten days before the deadline. If extenuating circumstances make electronic submission impossible, applicants should send a hard copy of their application via overnight carrier (non-U.S. Postal Service because of security–related delays in receiving mail from the U.S. Postal Service). Hard-copy applications must include a cover letter detailing the circumstances that make it impossible to submit via e-Grants. Late applications may be accepted only if the applicant submits a letter via email to LateApplications@cns.gov within the 24-hour period following the deadline explaining the extenuating circumstance which caused the delay. Late applications are evaluated on a case-by-case basis. Application and other deadlines may vary from competition to competition. Prospective applicants are encouraged to verify current deadline dates. Notices of Intent are generally requested rather than required and are used to help the Corporation plan the review process.
Requirements: Eligible applicants include public and state controlled institutions of higher education; 501(c)3 nonprofits; private institutions of higher education; the Corporation's State Service Commissions; volunteer centers; national or regional education agencies; educational institutions; and faith-based institutions. Applicants must propose to provide at least 70% of the total project cost in cash or in-kind from non-federal sources. Applicants must include a Dun and Bradstreet Data Universal Numbering System (DUNS) number on their applications and register with the Central Contractor's Registry (CCR). The DUNS number does not replace an Employer Identification Number (EIN). DUNS numbers may be obtained at no cost by calling the DUNS number request line at (866) 705-5711 or by applying online: http://fedgov.dnb.com/webform. The website indicates a 24-hour email turnaround time; however, the Corporation suggests registering at least 30 days in advance of the application due date.
Restrictions: Corporation grants are subject to the Cost Principles and Uniform Administrative Requirements under the applicable Office of Management and Budget (OMB) Circulars. Also, awards will be subject to the law(s) under which the award is made.
Date(s) Application is Due: Jul 21
Amount of Grant: 50,000 - 300,000 U.S.D
Contact: Rochelle Barry, State Program Director; (404) 965-2102; MLKDay@cns.gov
Internet: http://mlkday.org
Sponsor: Corporation for National and Community Service
1201 New York Avenue NW
Washington, DC 20525

CNCS Next Generation Grants **1341**

The mission of the Corporation for National and Community Service (the Corporation) is to improve lives, strengthen communities, and foster civic participation through service and volunteering. The Corporation's Next Generation Grants foster the expansion of national service by providing seed money to help organizations plan and implement innovative service programs to meet community needs through the use of volunteers. Applicant's proposals must fall under at least one of three service areas: programs that engage individuals in an intensive commitment to service in communities (defined as serving at least 40 hours per week); volunteer programs for seniors (minimum age 55); and programs that connect service with education. Applications may address community needs in one of the following issues areas: education, environment, health and human services, homeland security, public safety, or another compelling community need. Proposals and/or Concept Papers are usually submitted through the eGrants system. If an applicant must submit these documents on paper, they must be received by 5 p.m. eastern standard time on the deadline date and include a cover letter explaining why the application could not be submitted through eGrants. Late submissions may be accepted if the applicant submits a letter via email to LateApplications@cns.gov within the 24-hour period following the deadline explaining the extenuating circumstance which caused the delay. Late submissions are evaluated on a case-by-case basis. The process of competing for federal grants generally works as follows. Types of funding offered by federal programs are set forth in the program's enabling legislation. Grant competitions for a program's funding tend to follow congressional budget and appropriations cycles (as well as multi-year funding cycles in some cases) and are announced through various Notices which provide details, guidelines, and deadlines for workshops, letters of intent (if requested), and application submission. Notices are posted at the Corporation's and www.grants.gov websites. A subscription link is also provided at these sites for prospective applicants to be notified via email of upcoming grant competitions. In addition to statutory priorities, the Corporation determines priorities to focus each grant competition. Generally a series of technical assistance calls or workshops will be offered before the application is due. The dates of these will be posted at the Corporation's website. Application and other deadlines may vary from competition to competition. Prospective applicants are encouraged to verify current deadline dates. Notices of Intent are generally requested rather than required and are used to help the Corporation plan the review process.
Requirements: Eligible applicants are nonprofit charitable organizations, such as public charities, community organizations (faith-based and secular), private foundations, and individual schools. Applicants other than individual schools generally will have an annual operating budget of $500,000 or less. Organizations that have opportunities to greatly expand community and volunteer service but that have little or no experience with federal grants are encouraged to apply.
Restrictions: Applicants cannot have received a previous grant from the Corporation. Corporation grants are subject to the Cost Principles and Uniform Administrative Requirements under the applicable Office of Management and Budget (OMB) Circulars. Also, awards will be subject to the law(s) under which the award is made including the following pieces of legislation: the National and Community Service Act of 1990; National and Community Service Trust Act of 1993; and the Edward Kennedy Serve America Act of 2009.
Amount of Grant: Up to 400,000 U.S.D
Contact: Grants Coordinator; (202) 606-5000; nextgeneration@cns.gov
Internet: http://www.nationalservice.gov/about/programs/initiatives.asp
Sponsor: Corporation for National and Community Service
1201 New York Avenue NW
Washington, DC 20525

CNCS Senior Companion Program Grants **1342**

The Senior Companions Program (SCP) was established in 1973 under Title II of the Domestic Volunteer Services Act (DVSA) to provide opportunities for older adult volunteers to assist other older adults and persons with disabilities, who, without support, might not be able to live independently. Eighteen model Senior Companion projects were funded initially. Today that number has grown to 223 projects with more than 15,000 volunteers who serve through nonprofit and public organizations (local sponsors) to help home-bound clients with chores such as light housekeeping, paying bills, buying groceries, and finding transportation to medical appointments. SCP volunteers serve from fifteen to forty hours a week and receive hourly stipends. They must be fifty-five or older and meet established income eligibility guidelines. In addition to the stipend, they receive accident, personal-liability, and excess-automobile insurance coverage; assistance with the cost of transportation; an annual physical examination; recognition; and, as feasible, meals during their assignments. Volunteers receive training in how to assist persons diagnosed with Alzheimer's disease, stroke, diabetes, mental illness, etc., as well as when to alert doctors and family members to potential health problems. Currently SCP is administered through the Corporation for National and Community Service (the Corporation)'s Senior Corps Program. The Corporation accepts SCP grant applications only when new funding is available or when it is necessary to replace an existing sponsor. In addition, eligible agencies or organizations may, under a Memorandum of Agreement with the Corporation, receive technical assistance and materials to aid in establishing and operating a non-federally-funded SCP project using local funds. Notices for nationwide competitions for new SCP grants are posted at www.grants.gov and at the Corporation and Senior Corps websites. (Subscription links for receiving RSS feeds on new funding opportunities are also available at the websites.) Notices to apply to replace a sponsor are advertised locally through Corporation State Offices. (A list of the offices along with their contact information is available at the Corporation website.) Grant applications are submitted through the Corporation's eGrants system. For more information, interested applicants may download the SCP Handbook from the Senior Corps and Corporation websites or contact their Corporation State Office.
Requirements: The Corporation awards grants to public agencies, Indian tribes, and secular or faith-based private non-profit organizations in the United States that have authority to accept and the capacity to administer an SCP project. The SCP requires a non-federal share of 10% of the total project cost. SCP projects are generally expected and required to be on-going. SCP Sponsors may apply for continued funding from the Corporation.
Restrictions: The total of cost reimbursements for Senior Companions, including stipends, insurance, transportation, meals, physical examinations, uniforms if appropriate, and recognition must be equal to at least 80 percent of the Corporation's Federal share of the grant. (Federal and non-Federal resources, including excess non-Corporation resources, can be used to make up this sum.) Key legislative pieces enabling and regulating the SCP have been the Domestic Volunteer Service Act (DVSA) of 1973, the National and Community Service Trust Act (1993), 45 C.F.R. § 1216 (non-displacement of contracts and employed workers), the Edward Kennedy Serve America Act, and 45 C.F.R. § 2551. SCP funding generally requires an Office of Management and Budget (OMB) audit.
Amount of Grant: 200,000 U.S.D
Contact: Wanda Carney; (202) 606-6934 or (202) 606-5000
Tan Erwin, Director, Senior Corps; (202) 606-6867
Internet: http://www.seniorcorps.gov/for_organizations/funding/funding.asp
Sponsor: Corporation for National and Community Service
1201 New York Avenue NW
Washington, DC 20525

CNCS September 11th National Day of Service and Remembrance Grantas **1343**

The mission of the Corporation for National and Community Service (the Corporation) is to improve lives, strengthen communities, and foster civic participation through service and volunteering. For more than 15 years, the Corporation—through its Senior Corps, AmeriCorps, Learn and Serve America, and other programs and activities—has helped to engage millions of citizens in meeting community and national challenges through service and volunteer action. In 2009, Congress passed the Edward M. Kennedy Serve America Act (SAA), which authorized the most sweeping expansion of national service in a generation. This law not only expands service opportunities, but focuses national service on key outcomes; builds the capacity of individuals, nonprofits, and communities to succeed; and encourages innovative approaches to solving problems. In alignment with the SAA the Corporation will award a limited number of grants to outstanding organizations to implement and support service and remembrance activities that support the agency's mission and priorities, as well as those that honor the lives lost on September 11, 2001, and individuals who rose in service as a result of the event. Service activities must be directly connected to the September 11th National Day of Service and Remembrance, occur on or near that day, and reflect its tone and tenor. Organizations are encouraged to apply as intermediaries (entities that award and manage sub-grants) and/or to apply for direct support to oversee and implement local service projects. In any case grantees are expected to expand participation in the September 11th National Day of Service and Remembrance as well as connect volunteers to ongoing opportunities to serve. The award period for the September 11th Day of Service grants is two years. Grantees are eligible for continuation funding in the second year contingent upon satisfactory performance, demonstrated capacity to manage the grant, compliance with grant requirements (including match), and availability of Congressional appropriations. The process of competing for federal grants generally works as follows. Types of funding offered by federal programs are set forth in the program's enabling legislation. Grant competitions for a program's funding tend to follow congressional budget and appropriations cycles (as well as multi-year funding cycles in some cases) and are announced through various Notices which provide details, guidelines, and deadlines for workshops, letters of intent (if requested), and application submission. Notices are posted at the Corporation's and www.grants.gov websites. A subscription link is also provided at these sites for prospective applicants to be notified via email of upcoming grant competitions. In addition to statutory priorities, the Corporation determines priorities to focus each grant competition. All grants must support service activities that address one or more of the Corporation's currently mandated areas of concentration: disaster services; economic opportunity; education; environmental stewardship; healthy futures; and veterans and military families. Additionally, proposals may honor those who lost their lives on September 11, 2001 and those who rose in service as a result of that tragedy, demonstrate opportunities to engage new and diverse groups in service, leverage new public-private partnerships, meet an identified community need, provide opportunities to connect to ongoing opportunities to serve, and emphasize community connections to veteran service organizations and military service organizations. Priority consideration will be given to proposals that address disaster services, veterans and military families, capacity building, or engaging veterans, military families, and youth (particularly disconnected youth) in service. Grant applications are submitted via eGrants, the Corporation's web-based system for grant application and management. Generally a series of technical assistance calls or workshops will be offered before the application is due. The dates of these will be posted at the Corporation's website. Applications are due no later than 5 p.m. eastern time on the deadline date. Prospective applicants must create an account at the eGrant website (link available at the Corporation website). Applicants are advised to do this at least three weeks prior to the application-submission deadline. Applicants are advised to draft the application as a word-processing document, then copy and paste the document into eGrants no later than ten days before the deadline. If extenuating circumstances make electronic submission impossible, applicants should send a hard copy of their application via overnight carrier (non-U.S. Postal Service because of security–related delays in receiving mail from the U.S. Postal Service). Hard-copy applications must include a cover letter detailing the circumstances that make it impossible to submit via e-Grants. Late applications may be accepted only if the applicant submits a letter via email to LateApplications@cns.gov within the 24-hour period following the deadline explaining the extenuating circumstance which caused the delay. Late applications are evaluated on a case-by-case basis. The Corporation anticipates announcing the results of the current competition by the end of June. Application and other deadlines may vary from competition to competition. Prospective applicants are encouraged to verify current deadline dates.

Requirements: This competition is open to public or private nonprofit organizations, including faith-based and other community organizations; institutions of higher education; government entities within states or territories (e.g., cities, counties); Indian Tribes; labor organizations; partnerships and consortia; state service commissions; state agencies; and intermediaries. The Corporation especially seeks the following entities: applicants with multi-state, national (10 or more states), or comprehensive regional focus; applicants that fund statewide projects; and applicants that have congressionally recognized status as Veteran Service Organizations. In addition, the Corporation seeks applicants with a demonstrated ability to increase participation in service to meet vital community needs, that will offer ongoing opportunities for individuals to serve throughout the year, and that engage youth (particularly disconnected youth) and/or veterans and military families in service. Applicants are required to provide a grantee share equal to 25 percent of the total program budget. The match can be cash or in-kind. Applicants must include a Dun and Bradstreet Data Universal Numbering System (DUNS) number on their applications and register with the Central Contractor's Registry (CCR). The DUNS number does not replace an Employer Identification Number (EIN). DUNS numbers may be obtained at no cost by calling the DUNS number request line at (866) 705-5711 or by applying online: http://fedgov.dnb.com/webform. The website indicates a 24-hour email turnaround time; however, the Corporation suggests registering at least 30 days in advance of the application due date. All programs are subject to national-service criminal-registry checks on all individuals receiving a salary through the grant, including those whose salary is included as match.

Restrictions: Organizations that have been convicted of a federal crime are not eligible to apply. Pursuant to the Lobbying Disclosure Act of 1995, an organization described in Section 501 (c)(4) of the Internal Revenue Code of 1986, 26 U.S.C. 501 (c)(4) that engages in lobbying activities is not eligible to apply. September 11th National Day of Service and Remembrance Challenge grantees who have have already received federal grants to implement service projects in 2012 – 2014 are not eligible to apply until 2014. Corporation grants are subject to the Cost Principles and Uniform Administrative Requirements under the applicable Office of Management and Budget (OMB) Circulars. Also, awards will be subject to the law(s) under which the award is made (e.g., the SSA), as well as specific terms and conditions established in discretionary grant or cooperative agreements or defined in the Provisions or Special Conditions attached. Applicants will have an opportunity to negotiate conditions prior to acceptance of an award, if necessary.

Geographic Focus: All States, Northern Mariana Islands, U.S. Virgin Islands, American Samoa

Date(s) Application is Due: Mar 13

Amount of Grant: 50,000 - 150,000 U.S.D

Contact: Vielka Garibaldi, Director for Grant Review Operations; (202) 606-3991 or (2020 606-5000; sept11@cns.gov or info@cns.gov

Internet: http://www.nationalservice.gov/for_organizations/funding/nofa_detail.asp?tbl_nofa_id=94

Sponsor: Corporation for National and Community Service
1201 New York Avenue NW
Washington, DC 20525

CNCS Social Innovation Grants for New Grantees **1344**

The Social Innovation Fund is authorized by the Edward M. Kennedy Serve America Act and is administered by the Corporation for National and Community Service (the Corporation), a federal agency that engages more than five million Americans as volunteers through well-known national-service programs like Senior Corps, AmeriCorps, and Learn and Serve America. The Social Innovation Fund is primarily concerned with advancing social innovation as a key strategy for solving critical social challenges. The program's goal is to identify and help spread those innovative and potentially transformative approaches that have been developed at the local level to solve community problems. An approach is considered transformative if it not only produces strong impact, but also if it (1) has the potential to affect how the same challenge is addressed in other communities, (2) addresses more than one critical community challenge concurrently, or (3) produces significant cost savings through gains in efficiency. The operating model of the Social Innovation Fund is distinguished by four key elements: (1) reliance on intermediaries with strong skills and track records of success in selecting, validating, and growing high-impact nonprofit organizations; (2) assuring participation from the non-federal stake-holders by requiring each federal dollar be matched 1:1 with money from non-federal sources not only by the intermediaries but also by their subgrantees; (3) requiring that all intermediaries engage each of their subgrantees in formal evaluations of program performance and impact; and (4) requiring each grantee to commit to knowledge sharing and other initiatives that advance social innovation more generally in the nonprofit sector. Program models funded by Social Innovation Fund intermediaries must produce rigorous evaluative evidence that not only assesses effectiveness, but also builds the existing evidence base for other programs addressing these issues. The Corporation expects and even strongly suggests that most intermediaries will choose to contract with a research partner and/or require their subgrantees to contract with a research partner to help with formal program evaluation. Both intermediary and subgrantee budgets should indicate estimated costs related to evaluation. The Social Innovation Fund award period is up to five years. Applicants' project and evaluation plans should represent the full five-year period. Selected applicants will receive an award for the first year of program expenses. During program years two through five, these intermediaries will participate in an annual continuation-funding review process during which time they will be eligible for additional annual funds. These additional funds are contingent on the availability of appropriations, compliance with grant conditions, and satisfactory performance, including having secured sufficient matching funds and making appropriate progress in advancing the evaluations and growth of subgrantees. Applying intermediaries may propose subgrant funding strategies that are of two distinct types. In both cases, the intermediary will have defined a specific social problem it intends to address through its own Social Innovation Fund. Under one strategy, the intermediary will be seeking and investing in subgrantee organizations that have themselves developed and/or implemented innovative solutions to the specified problem. Under the other strategy, the intermediary will have identified the innovative solution and will be seeking subgrantees that are able to effectively implement that solution. The process of competing for federal grants generally works as follows. Types of funding offered by federal programs are set forth in the program's enabling legislation. Competitions for this funding tend to follow congressional budget and appropriations cycles (as well as multi-year funding cycles in some cases) and are announced through various Notices which provide details, guidelines, and deadlines for workshops, letters of intent (if requested), and application submission. Notices are posted at the Corporation's and www.grants.gov websites. A subscription link is also provided at these sites for prospective applicants to be notified via email of upcoming grant competitions. In addition to statutory priorities, the Corporation determines priorities to focus each grant competition. The current competition is focused in the areas of youth development, economic opportunity, and healthy futures. Grant applications are submitted via eGrants, the Corporation's web-based system for grant application and management. Generally a series of technical assistance calls or workshops will be offered before the application is due. The dates of these will be posted at the Corporation's website. Applications are due no later than 5 p.m. eastern time on the deadline date. Prospective applicants must create an account at the eGrant website (link available at the Corporation website). Applicants are advised to do this at least three weeks prior to the application-submission deadline. Applicants are advised to draft the application as a word-processing document, then copy and paste the document into eGrants no later than ten days before the deadline. If extenuating circumstances make electronic submission impossible, applicants should send a hard copy of their application via overnight carrier (non-U.S. Postal Service because of security–related delays in receiving mail from the U.S. Postal Service). Hard-copy applications must include a cover letter detailing the circumstances that make it impossible to submit via e-Grants. Late applications may be accepted only if the applicant submits a letter via email to LateApplications@cns.gov within the 24-hour period following the deadline explaining the extenuating circumstance which caused the delay. Late applications are evaluated on a case-by-case basis. The review process for the Social Innovation fund consists of several stages. Initially applications are reviewed to determine compliance with match, deadline, and completeness requirements. The next stage is an eligibility review. The final stages are an expert review, a quality-control review, an internal review, request for clarification, selection for funding, and feedback to applicants. The Corporation anticipates announcing the results of the current competition

by August. Application and other deadlines may vary from competition to competition. Prospective applicants are encouraged to verify current deadline dates.
Requirements: Applicants must be an eligible grantmaking institution or partnership in existence at the time of the application. Providing grants to nonprofit community organizations should be central to the applicant's mission and should be clearly reflected in the organization's promotional materials and annual operating budget. Core operations must include conducting open competitive grant competitions, negotiating specific grant requirements with grant recipients; and overseeing and monitoring performance of grant recipients. By statute Social Innovation Fund intermediaries must operate either as geographically-based or as issue-based grantmakers. A geographically-based intermediary will address one or more priority issues withing a single geographic location. An issue-based intermediary will address a single priority issue in multiple geographic locations. At the time of submission, applicants must demonstrate through a letter or other form of documentation that they have either cash-on-hand or commitments (or a combination thereof) toward meeting 50 percent of their first year matching funds. Applicants must include a Dun and Bradstreet Data Universal Numbering System (DUNS) number on their applications and register with the Central Contractor's Registry (CCR). The DUNS number does not replace an Employer Identification Number (EIN). DUNS numbers may be obtained at no cost by calling the DUNS number request line at (866) 705-5711 or by applying online: http://fedgov.dnb.com/webform. The website indicates a 24-hour email turnaround time; however, the Corporation suggests registering at least 30 days in advance of the application due date.
Restrictions: Intermediaries must distribute at least 80 percent of awarded federal funds to subgrantees, run an open competition that is available to eligible nonprofit organizations beyond the intermediary's own existing grant portfolio or network, and provide sufficient public notice of the availability of Social Innovation Fund subgrants to eligible nonprofit community organizations. Given that innovation funds currently exist in the Departments of Education and Labor to invest specifically in evidence-based programs in education and job training, the Corporation does not intend to make Social Innovation Fund awards to programs in these areas unless they clearly propose a solution to an unmet need as identified in consultation with both Departments. The funding mechanism for Social Innovation Fund awards is a cooperative agreement that provides for substantial involvement by the Corporation with the intermediaries as they carry out approved activities. The assigned Corporation program officer will confer with the grantee on a regular and frequent basis to develop and/or review service delivery and project status, including work plans, budgets, periodic reports, evaluations, etc. In particular the Corporation anticipates having substantial involvement in developing and approving subgrantee selection plans; developing and approving subgrantee evaluation plans; documenting subgrantee growth plans; and documenting and sharing lessons learned through a Corporation-sponsored learning community. Grants under the Social Innovation program are subject to the Cost Principles and Uniform Administration Requirements under the applicable Office of Management and Budget (OMB) Circulars. Also, awards will be subject to the law(s) under which the award is made (e.g., the 1990 National Community Service Act and the 2009 Edward Kennedy Serve America Act), as well as specific terms and conditions established in the cooperative agreement or defined in the Terms and Conditions of the award. Applicants will have an opportunity to negotiate conditions prior to acceptance of an award, if necessary.
Date(s) Application is Due: Mar 27
Amount of Grant: 1,000,000 - 5,000,000 U.S.D
Samples: Foundation for a Healthy Kentucky, Louisville, Kentucky, $2 million - current subgrantees are testing innovative strategies to increase access to needed health services in underserved communities (2010); National Fund for Workforce Solutions, Boston, Massachusetts, $7.7 million - partnerships to provide targeted training and career support assistance to more than 20,000 low-income individuals (2010); New Profit Inc., Cambridge, Massachusetts, $5 million - collaboration with five to six innovative youth-focused organizations with existing evidence to yield significant improvements in helping young people navigate the increasingly complex path from high school to college and productive employment (2010).
Contact: Vielka Garibaldi, Director for Grant Review Operations; (202) 606-5000 or (202) 606-3223; info@cns.gov or innovation@cns.gov
Internet: http://www.nationalservice.gov/about/programs/innovation.asp
Sponsor: Corporation for National and Community Service
1201 New York Avenue NW
Washington, DC 20525

CNET Community Involvement Grants 1345
The program awards grants to San Francisco nonprofit organizations addressing critical community needs. Areas of interest include arts and culture, civic affairs and community development, education, and health and human services. Types of support include in-kind gifts and efforts to organize collections for goodwill drives. Submit requests anytime.
Requirements: San Francisco, CA, nonprofit organizations are eligible. Eligible nonprofit organizations must be located in the United States and must have received 501(c)3 tax-exempt status by the Internal Revenue Service.
Restrictions: Grants do not suppport multiyear projects, religious or political activities, sports programs without a charitable purpose, or endowments.
Geographic Focus: California
Contact: CNET Computer Network, (415) 364-8000; volunteer@cnet.com
Internet: http://www.cnet.com/aboutcnet/0-13620.html
Sponsor: CNET Community Involvement Program
235 Second Street
San Francisco, CA 94105

CNIB Baker Applied Research Fund Grants 1346
The CNIB Baker Applied Research Fund supports research that is focused on the social, educational, cultural and rehabilitative needs of Canadians living with vision loss, and the application of assistive devices to meet these needs. This program is intended both to advance the career development of new investigators, and to encourage senior researchers to develop programs of applied research. Proposals are invited from both. Grants are for a one-year period, with a limit of $35000.
Requirements: Projects must be completed within 12 months. Funds that are unused after 24 months after the start up of the grant must be returned unless a special request for continuance is approved. Applicants must be residents of Canada and research must be conducted primarily in Canada. The applicant must: have an academic appointment at a Canadian university or occupy a supervisory position at a health care facility, or vision rehab facility. All applications will be evaluated with regard to the following criteria: excellence in the health and social sciences as applied to vision care; feasibility of research results translating into improved interventions, programs, and services for Canadians living with vision loss; enhancement of Canada's outstanding research into the needs of Canadians living with vision loss; and creation of training opportunities for students and service providers.
Restrictions: Equipment costs will not generally be covered, but in certain circumstances equipment may be funded on the basis of matching funds from the host facility. Travel and publications costs should not exceed $2,000.
Geographic Focus: Canada
Date(s) Application is Due: Jan 15
Amount of Grant: Up to 35,000 CAD
Contact: Shampa Bose, Executive Assistant and Research Coordinator; (416) 486-2500, ext. 7622 or (800) 563-2642; fax (416) 480-7059; shampa.bose@cnib.ca
Internet: http://www.cnib.ca/en/research/funding/eabaker-applied/
Sponsor: Canadian National Institute for the Blind
1929 Bayview Avenue
Toronto, M4G 3E8 Canada

CNIB Baker Fellowships 1347
CNIB's Baker Fellowships are awarded annually for post-graduate training in ophthalmic sub-specialties. The program offers fellowships for one year in amounts up to $30,000. Applicants must apply online no later than January 15 to be eligible for consideration. Fellowship funds are provided quarterly commencing July 1 of the following year. Successful applicants are normally advised of their award in April.
Requirements: Recipients must begin or return to an academic posting in Canada within three years of completing their training, or return to an area of need as supported by a letter from the Medical Director of the community. An outline of future plans supported by a letter of acceptance to an academic posting or alternate should accompany the application where possible. This letter plays an important role in the eligibility of the candidate and strong efforts should be made to clarify that future position is available to the candidate. If the fellow does not return within a three-year time frame of completion of study to Canada, CNIB requests reimbursement for the funding provided.
Restrictions: Applicants who are expecting to receive funds or stipends from other sources amounting equal or more than $30,000 may not apply for CNIB Baker fellowship.
Geographic Focus: All States, Canada
Date(s) Application is Due: Jan 15
Amount of Grant: Up to 30,000 CAD
Contact: Shampa Bose, Executive Assistant and Research Coordinator; (416) 486-2500, ext. 7622 or (800) 563-2642; fax (416) 480-7059; shampa.bose@cnib.ca
Internet: http://www.cnib.ca/en/research/funding/eabaker-fellowship/
Sponsor: Canadian National Institute for the Blind
1929 Bayview Avenue
Toronto, M4G 3E8 Canada

CNIB Baker New Researcher Fund Grants 1348
The CNIB Baker New Researcher Fund provides one-year grants to encourage new investigations that may lead to the prevention of vision loss. It is intended to benefit new investigators (within 5 years after an academic faculty appointment) by giving them experience and results which can assist them in further grant applications and pilot investigations. A new investigator is someone who is affiliated with a research facility through an academic posting and who has been in the field for less than five years. Grants are for one year, in amounts up to $35,000 (CAD). The application must be submitted online no later than January 15 of each year, with grant decisions made by April 1.
Requirements: Projects should be completed within 12 months. Funds that are unused after 24 months of the start up of the grant must be returned unless a special request for continuance is approved. Applicants must be residents of Canada and research must be conducted primarily in Canada.
Restrictions: Equipment costs will not generally be covered, but in certain circumstances equipment may be funded on the basis of matching funds from the host facility. Travel and publication costs should not exceed $2,000. Applications must be focused and no longer than 10 pages in total.
Geographic Focus: Canada
Date(s) Application is Due: Jan 15
Amount of Grant: Up to 35,000 CAD
Contact: Shampa Bose, Executive Assistant and Research Coordinator; (416) 486-2500, ext. 7622 or (800) 563-2642; fax (416) 480-7059; shampa.bose@cnib.ca
Internet: http://www.cnib.ca/en/research/funding/eabaker-researcher/
Sponsor: Canadian National Institute for the Blind
1929 Bayview Avenue
Toronto, M4G 3E8 Canada

CNIB Barbara Tuck MacPhee Award 1349

The CNIB Barbara Tuck MacPhee Award supports researchers in the field of macular degeneration. The Tuck-MacPhee Award provides a one-year grant of up to $35,000. Projects should be completed within 12 months. The investigator will provide CNIB with a progress report six months following the start up of the grant. A second report including a copy of relevant findings of the research is required at the end of the research and prior to releasing the final quarterly payment. All publications and presentations of this research should acknowledge the support of the CNIB Baker Tuck MacPhee Fund. Applicants must apply online no later than January 15 of each year. The application will be reviewed by a multi-disciplinary review committee, and decisions will be finalized by April 1.
Requirements: Funds that are unused after 24 months must be returned unless a special request for continuance is approved. Equipment costs will not generally be covered, but in certain circumstances equipment may be funded on the basis of matching funds from the host facility. This competition is open to all researchers; however applicants must be residents of Canada and research must be conducted primarily in Canada.
Restrictions: Trainees (residents, fellows, post graduate students) are not eligible to apply.
Geographic Focus: All States, All Countries
Date(s) Application is Due: Jan 15
Amount of Grant: Up to 35,000 CAD
Contact: Shampa Bose, Executive Assistant and Research Coordinator; (416) 486-2500, ext. 7622 or (800) 563-2642; fax (416) 480-7059; shampa.bose@cnib.ca
Internet: http://www.cnib.ca/en/research/funding/Tuck-MacPhee/
Sponsor: Canadian National Institute for the Blind
1929 Bayview Avenue
Toronto, M4G 3E8 Canada

CNIB Canada Glaucoma Clinical Research Council Grants 1350

The CNIB-CGCRC Award represents a partnership of CNIB and the Canada Glaucoma Clinical Research Council designed to invest in clinical glaucoma research. This partnership contributes about $225,000 over a three year period. Funding of the CNIB-CGCRC award is via an unrestricted grant from Alcon Canada Inc. The application deadline for the CNIB-CGCRC Award is June 30. There are two levels of funding available: $5,000 or less for investigators with training or practice for less than 5 years and $5000 to $25,000 for investigators with training or practice over 5 years.
Amount of Grant: Up to 25,000 CAD
Contact: Shampa Bose, Executive Assistant and Research Coordinator; (416) 486-2500, ext. 7622 or (800) 563-2642; fax (416) 480-7059; shampa.bose@cnib.ca
Internet: http://www.cnib.ca/en/research/funding/cnib-cgrc/
Sponsor: Canadian National Institute for the Blind
1929 Bayview Avenue
Toronto, M4G 3E8 Canada

CNIB Chanchlani Global Vision Research Award 1351

The CNIB Chanchlani Global Vision Research Award was established in 2011 with the goal of encouraging world researchers in the area of vision science and vision rehabilitation. The award is consistent with the goals of CNIB (Canadian National Institute for the Blind) of conducting and funding world-class research to reduce the impact of sight loss in people's lives. The award will be given to vision scientists anywhere in the world who have made a major original contribution to the fields of vision science or vision rehabilitation. The nominations will be evaluated by the CNIB Research Committee, a group of eminent Canadian vision scientists, clinicians and vision rehabilitation practitioners. The evaluation process will depend heavily on the quality of the information provided in the nomination process. The CNIB Chanchlani prize will be valued at $25,000. The award will be to an individual recipient. There will be no shared prizes. The prize winner may be a resident of any country.
Requirements: Nominations may be made by anyone knowing the potential recipient well. This may include the potential recipient his/herself.
Geographic Focus: All States, All Countries
Date(s) Application is Due: Jul 15
Amount of Grant: 25,000 CAD
Contact: Shampa Bose, Executive Assistant and Research Coordinator; (416) 486-2500, ext. 7622 or (800) 563-2642; fax (416) 480-7059; shampa.bose@cnib.ca
Internet: http://www.cnib.ca/en/research/funding/Pages/Chanchlani-Award.aspx
Sponsor: Canadian National Institute for the Blind
1929 Bayview Avenue
Toronto, M4G 3E8 Canada

CNIB E. (Ben) & Mary Hochhausen Access Technology Research Grants 1352

The award was established by Ben and Mary Hochhausen to further their long-time volunteer and financial contribution to CNIB. This international award encourages research in the field of access technology for people living with vision loss. Grants up to $10,000 are made each year at the discretion of the E. (Ben) and Mary Hochhausen Fund trustees. Applications are accepted from any country in the world. Awards may be applied to: research projects; study at centres of excellence in technology; fellowships; development of prototypes; or development costs for bringing important new products to market. Initial application should be received by the E. (Ben) and Mary Hochhausen Fund trustees by September 30.
Requirements: Trustees will consider: students enrolled in a recognized College, University or equivalent program; or researchers from industry, academia or applied science.
Geographic Focus: All States, All Countries
Date(s) Application is Due: Sep 30
Amount of Grant: Up to 10,000 CAD
Contact: Shampa Bose, Executive Assistant and Research Coordinator; (416) 486-2500, ext. 7622 or (800) 563-2642; fax (416) 480-7059; shampa.bose@cnib.ca
Internet: http://www.cnib.ca/en/research/funding/hochhausen/
Sponsor: Canadian National Institute for the Blind
1929 Bayview Avenue
Toronto, M4G 3E8 Canada

CNIB Ross Purse Doctoral Fellowships 1353

The Ross C. Purse Doctoral Fellowship encourages and supports theoretical and practical research and studies at the postgraduate or doctoral level, in the field of vision loss in Canada. The fellowship is awarded for research in the social sciences, engineering, or other fields of study that are immediately relevant to the field of vision loss. One fellowship is awarded annually to a qualified applicant. Successful applicants will be considered for subsequent funding only in exceptional circumstances. Each fellowship is valued at up to CAD$12,500, to be paid in three equal installments. Initial payment is made at the time of the award, and the second installment is paid after receipt of an interim report outlining the development of your research. Final payment is made conditional upon receipt by the Secretariat of the dissertation from the candidate and his/her thesis supervisor or department head. Completed applications must be postmarked not later than April 2 of each year.
Requirements: Applications will be considered from persons studying at a Canadian university or college, or at a foreign university where a commitment to work in the field of vision loss in Canada for at least two years can be demonstrated. Preference will be given to graduates of a Canadian university or college. Applicants must have achieved a high academic standing and must have demonstrated superior intellectual ability and judgment. Recipients may undertake paid employment with the permission of their supervisor of studies.
Geographic Focus: All States, All Countries, Canada
Date(s) Application is Due: Apr 2
Amount of Grant: Up to 12,500 CAD
Contact: Shampa Bose, Executive Assistant and Research Coordinator; (416) 486-2500, ext. 7622 or (800) 563-2642; fax (416) 480-7059; shampa.bose@cnib.ca
Internet: http://www.cnib.ca/en/research/funding/Ross-Purse/
Sponsor: Canadian National Institute for the Blind
1929 Bayview Avenue
Toronto, M4G 3E8 Canada

Coastal Bend Community Foundation Grants 1354

The community foundation awards grants to nonprofit organizations in the Aransas, Bee, Jim Wells, Kleberg, Nueces, Refugio and San Patricio counties of Texas. Grants are usually made to provide seed money for innovative and start-up programs that will generate additional future funding or revenues. Capital projects will also receive favorable consideration. The board of directors approves grant recipients in early November after considering recommendations presented by the Grants Committee. Areas of interest include alcohol and drug abuse, libraries and literacy, arts and culture, higher and other education, adult basic education, child welfare, hospitals, community development, animal welfare, human services, and general charitable giving. Grants are awarded for general operating support, program and project development, equipment, seed money, scholarship funds, and fellowships. Application forms are not required.
Requirements: Nonprofit organizations in Texas in Aransas, Bee, Jim Wells, Kleberg, Nueces, Refugio, and San Patricio counties, may submit grant proposals.
Geographic Focus: Texas
Date(s) Application is Due: Sep 1
Amount of Grant: 250 - 50,000 U.S.D
Samples: Texas A&M U (Corpus Christi, TX)—for program support, $10,000;HIV/AIDS Housing Ctr (Corpus Christi, TX)—for operating support, $12,500.
Contact: Jim Moloney, Executive Vice President; (361) 882-9745; fax (361) 882-2865; jmoloney@cbcfoundation.org
Internet: http://www.cbcfoundation.org/grant.html
Sponsor: Coastal Bend Community Foundation
600 Building, Suite 1716
Corpus Christi, TX 78473

Coastal Community Foundation of South Carolina Grants 1355

The foundation awards grants to South Carolina nonprofit organizations in its areas of interest, including arts and culture, education, environment, health, religion, justice and equity, and social services. Types of support include general operating support, emergency funds, program development, publication, seed money, scholarship funds, and technical assistance. Deadlines vary per program; check website for exact dates.
Requirements: 501(c)3 South Carolina nonprofits located in the following coastal counties are eligible to apply: Beaufort, Berkeley, Charleston, Colleton, Dorchester, Georgetown, Hampton and Jasper.
Restrictions: Grants do not support individuals (except for designated scholarship funds), endowments, deficit financing, dinners, and rarely building funds.
Geographic Focus: South Carolina
Date(s) Application is Due: Jun 1
Amount of Grant: 500 - 10,000 U.S.D
Contact: George C. Stevens, President-CEO; (843) 723-3635; fax (843) 577-3671; gstevens@tcfgives.org or info@ccfgives.org
Internet: http://www.ccfgives.org
Sponsor: Coastal Community Foundation of South Carolina
90 Mary Street
Charleston, SC 29403

Coca-Cola Foundation Grants 1356

The foundation has established education as its philanthropic focus and set aside most of its funds to support educational initiatives that address pressing needs. To help prepare youth for life, the foundation gives in three areas: higher education—pipeline programs that connect various levels of education and help students stay in school, scholarships, and minority advancement; classroom teaching and learning—innovative K-12 projects, teacher development, and small projects that deal with classroom activities; and global education—projects that encourage international studies, global understanding, and student exchange. Grants are awarded to both public and private institutions at all levels of education: universities, colleges, and secondary and elementary schools. International educational institutions and health care organizations also receive consideration. Types of support include annual campaigns, donated equipment, employee matching gifts, operating budgets, special projects, capital campaigns, continuing support, fellowships, internships, endowment funds, matching funds, and scholarship funds. Proposals may be submitted at any time.

Requirements: The following are eligible to apply: IRS 501(c)3 non-profits; a foreign organization that has received a ruling from the IRS that it is a section 501(c)3 tax exempt organization; or a foreign organization that is the equivalent of a U.S. charity.

Restrictions: The foundation does not make grants to individuals, religious endeavors, political or fraternal organizations, or organizations without 501(c)3 status.

Amount of Grant: 10,000 - 6,000,000 U.S.D

Samples: Holyfield Foundation, Inc., Fairburn, GA, $75,000—contribution to the Giving Disadvantaged Youth a Fighting Chance scholarship program; United Way of Metropolitan Atlanta, Atlanta, GA, $15,000—support for Samaritan House, Impact Group and Atlanta Union Mission; Atlanta Children's Shelter, Inc., Atlanta, GA, $10,000—contribution to support shelter programs.

Contact: Helen Smith Price, Executive Director; (404) 676-2568; fax (404) 676-8804

Internet: http://www.thecoca-colacompany.com/citizenship/our_communities.html

Sponsor: Coca-Cola Foundation
P.O. Box 1734
Atlanta, GA 30301

Cockrell Foundation Grants 1357

The Cockrell Trust was established in 1957 with funds donated by Ernest and Virginia Cockrell. The Foundation's special emphasis is The University of Texas at Austin. Other fields of interest are: youth services, arts, health care, cultural programs, civic, religious and social services. The Foundation awards grants to support: annual campaigns, building funds, capital campaigns, endowment funds, general purposes, matching funds, and special projects. There are no deadlines for making a request to The Cockrell Foundation. The Foundation meetings are in the late spring and late Fall. An application form is not required. See website for recommended guidelines in making a proposal: http://www.cockrell.com/foundation/grant_guidelines.asp .

Requirements: Texas 501(c)3 nonprofit organizations in Houston are eligible.

Restrictions: The Foundation generally makes grants for only one-year periods. An organization should apply only once during any calendar year. If a grant request is denied, the applicant must wait until the following year before submitting a new request. The Foundation does not participate in feasibility studies and generally does not make grants for the following: individuals; mass appeal solicitations; medical or scientific research projects; organizations outside of Houston, Texas and the United States. Grant requests sent via email or fax will not be accepted.

Geographic Focus: Texas

Samples: The University of Texas, M.D. Anderson Cancer Center, $20,000—support for the Cord Blood Bank Program; Greater Houston Community Foundation, $100,000—support for the Mayor's Gulf Coast Ike Relief Fund; Houston Museum of Natural Science, $2,500—to support the HMNS student enrichment and teacher training programs for Houston schools.

Contact: M. Nancy Williams, Executive Vice President; (713) 209-7500; foundation@cockrell.com

Internet: http://www.cockrell.com/foundation/grant_guidelines.asp

Sponsor: Cockrell Foundation
1000 Main Street, Suite 3250
Houston, TX 77002

Coeta and Donald Barker Foundation Grants 1358

The foundation awards grants to California and Oregon nonprofit organizations in its areas of interest, including arts, children and youth, community development, disabled, environmental conservation, family services, federated giving, health care and health organizations, heart and circulatory research, higher education, hospitals, mental health, and secondary school education. Types of support include building construction/renovation, equipment acquisition, general operating support, program development, and scholarship funds.

Requirements: California and Oregon nonprofit organizations are eligible.

Restrictions: Grants do not support sectarian religious purposes, federal and tax-dependent organizations, individuals, or endowment funds.

Geographic Focus: California, Oregon

Date(s) Application is Due: Mar 1; Aug 1

Amount of Grant: 100 - 20,450 U.S.D

Samples: Santa Barbara Zoological Gardens (Santa Barbara, CA), $5000.

Contact: Nancy Harris, Executive Administrator; (760) 324-2656; fax (760) 321-8662

Sponsor: Coeta and Donald Barker Foundation
P.O. Box 936
Rancho Mirage, CA 92270

Coleman Foundation Cancer Care Grants 1359

Since the 1980's, the Foundation has been an active advocate for raising the standards of cancer care in the Midwest region. Cancer care including support services, treatment and research are primary initiatives of the Foundation. The Foundation has funded specific health care programs, community-based and psychosocial support services, cancer research, development of new treatments, and improvement of direct patient services. Additionally, the Foundation has supported both cancer-specific and general health patient programs.

Requirements: Nonprofits across the Midwest are eligible. Preference is given to requests from Illinois and Chicago.

Restrictions: The program does not fund for-profit businesses, individuals, individual scholarships, ad books, tickets, equipment purchases (including computer hardware or software), or advertising.

Geographic Focus: Illinois, Indiana, Iowa, Michigan, Ohio, Wisconsin

Samples: Sinai Health System, Illinois, $328,000—towards capital support for Oncology Infusion Center and OB Triage & Ante partum Testing, to improve oncologic and obstetric facilities, enhance direct patient service areas, and streamline delivery of health care; Wellness House , Illinois, $3,000—to support the 2008 Psychosocial Conference for medical and psychosocial oncology professionals to understand the implication of the Cancer Care for the Whole Patient report published by the Institute of Medicine and discuss ways to improve comprehensive care of cancer patients.

Contact: Michael Hennessy, President; (312) 902-7120; fax (312) 902-7124; info@colemanfoundation.org

Internet: http://www.colemanfoundation.org/cancer.html

Sponsor: Coleman Foundation
651 West Washington Boulevard, Suite 306
Chicago, IL 60661

Colin Higgins Foundation Grants 1360

The foundation supports organizations that help people respond to a variety of life challenges, with a focus on responding to the AIDS pandemic by supporting innovative programs in the areas of education, service, and advocacy for people infected, affected, and at risk for HIV; and empowering lesbian, gay, bisexual, and transgender peoples by supporting community-based organizations that combat homophobia and foster leadership. The foundation is particularly interested in organizations that have a significant impact working in underserved, rural areas with traditionally underserved constituencies. The process of applying consists of two stages: submission of a letter of inquiry, and, if requested, a full proposal.

Requirements: The foundation only considers organizations with overall budgets under $2 million, or project budgets under $500,000.

Restrictions: Grants do not support projects that had previously fallen within its funding priorities, including any organizations or programs based in urban areas (population over 1 million); film and video projects; or organizations with overall budgets of over $2 million, or project budgets over $500,000. The one exception to the exclusion of programs in urban areas is those organizations and projects working with underserved communities (i.e., communities of color, low-income communities, transgender people, etc.). The foundation does not accept letters of inquiry from universities, schools, individuals, or corporations. It does not support political or legislative activities, endowments, or deficit budgets. Additionally, the foundation does not support capital campaigns, although it will consider requests for capital improvements, and does not award grants to organizations outside the United States.

Date(s) Application is Due: Jul 1

Amount of Grant: 5,000 - 15,000 U.S.D

Samples: Iowa Pride Network, Des Moines, IA, $10,000—general support; Out Youth, Austin, TX, $15,000—Texas Gay-Straight Alliance Network; Equality Florida, St. Petersburg, FL, $15,000—Youth Voices for Equality.

Contact: Grants Administrator; (212) 509-4975; fax (212) 509-1059; info@colinhiggins.org

Internet: http://www.colinhiggins.org/grantmaking/index.cfm

Sponsor: Colin Higgins Foundation
55 Exchange Place, Suite 402
New York, NY 10005

Collective Brands Foundation Grants 1361

The Collective Brands Foundation invests financially in non-profit organizations that align with the Foundation's focus areas. Priority is given to organizations that provide involvement opportunities for team members and employees of Collective Brands, Inc. The Foundation may also consider sponsorships from charitable organizations in the following areas: Eastern Kansas, including Topeka, Lawrence and the Kansas City metropolitan area; New York City; Lexington, Massachusetts and the Greater Boston area; Denver, Colorado; Redlands, California; and Brookville, Ohio.

Requirements: The Collective Brands Foundation will consider requests for monetary grants from 501(c)3 non-profit organizations that manage programs in at least one of the following areas: women's preventative health; children's physical activity and fitness; improving the lives of children and youth in need; preserving the environment; and supporting industry in the United States. Applications must be submitted online at the Foundation's website.

Restrictions: Grants will not be awarded to religious organizations for projects that are sectarian and do not benefit a broad community base. Also, the Foundation will not award grants to: individuals; political causes, candidates or legislative lobbying efforts; or for capital campaigns, debt reduction, travel or conferences.

Date(s) Application is Due: Aug 15

Amount of Grant: Up to 3,000 U.S.D

Contact: Michele Gray, Director of Community Investments; (877) 902-4437; grants@greaterhorizons.org
Internet: http://www.collectivebrands.com/foundation
Sponsor: Collective Brands Foundation
3231 SE 6th Avenue
Topeka, KS 66607

Collins C. Diboll Private Foundation Grants 1362
The foundation awards grants to Louisiana nonprofit organizations in the areas of higher education, human services, and youth. Field of interest include: Catholic churches and agencies; education; higher education; human services; museums (art); and protestant agencies and churches. Types of support include building construction/renovation, capital campaigns, endowment funds, and general operating support. Applicants should submit a detailed description of the project and the amount of funding requested, along with a copy of an IRS determination letter.
Requirements: Louisiana nonprofit organizations are eligible.
Restrictions: Individuals are not eligible to apply.
Geographic Focus: Louisiana
Amount of Grant: 500 - 200,000 U.S.D
Samples: Alton Ochsner Medical Foundation (New Orleans, LA)—for Golden Jubilee/Initiative Campaign, $100,000; New Orleans Museum of Art (New Orleans, LA)—$100,000; Junior Achievement of New Orleans (New Orleans, LA)—$35,000; Preservation Resource Ctr of New Orleans (New Orleans, LA)—for renovations, $11,000.
Contact: Donald W. Diboll, Chair; (504) 582-8103
Sponsor: Collins C. Diboll Private Foundation
201 Saint Charles Avenue, 50th Floor
New Orleans, LA 70170-5100

Collins Foundation Grants 1363
Grants are awarded to organizations in Oregon with emphasis on higher education, youth and health agencies, social welfare, natural resource and wildlife conservation, religion, and arts and cultural programs. A special area of support includes building and medical equipment, and programs for those who are physically or mentally disabled. Challenge/matching grants, general operating support, scholarships, and internships also are awarded. Applications may be submitted at any time, additional guidelines are available at the Foundation's website.
Requirements: Grants are made to 501(c)3 nonprofit agencies domiciled in Oregon. The proposed project must directly benefit the citizens of Oregon.
Restrictions: Grants are not made to individuals or to organizations sponsoring requests intended to be used by or for the benefit of an individual. Grants normally are not made to elementary, secondary, or public higher education institutions; or to individual religious congregations. Grants normally are not made for development office personnel, annual fundraising activities, endowments, operational deficits, financial emergencies, or debt retirement.
Geographic Focus: Oregon
Samples: Portland Reading Foundation, Portland, OR, $9,000—support a literacy program in two inner-city elementary schools to advance students reading skills to grade level; Children's Community Clinic, Portland, OR, $25,000—provide healthcare services for low-income children; The Nature Conservancy, Portland, OR, $450,000 (3 year grant)—acquire and conserve six areas identified as the agency's highest priorities.
Contact: Cynthia Knowles, Director of Programs; (503) 227-7171; fax (503) 295-3794; information@collinsfoundation.org
Internet: http://www.collinsfoundation.org/submission-guidelines
Sponsor: Collins Foundation
1618 South West First Avenue, Suite 505
Portland, OR 97201

Colonel Stanley R. McNeil Foundation Grants 1364
Colonel Stanley R. McNeil and his wife Merna created the Colonel Stanley R. McNeil Foundation in 1993 to support and promote quality educational, human-services, and health-care programming for underserved populations. Special consideration is given to charitable organizations that serve the needs of children. During their lifetimes, the McNeils were actively involved in their local church and community. Colonel McNeil also served on a number of boards, including Lake Bluff Children's Home and Ravenswood Hospital. Although the McNeils had no children, they were strong supporters of children's causes. The foundation is particularly interested in funding programs or organizations that focus on children's causes, start-up initiatives within the human-services or arts and culture arenas, and healthcare. To better support the capacity of nonprofit organizations, multi-year funding requests are considered. Grant requests for naming opportunities that honor the donors are strongly encouraged. Applicants must apply online at the grant website. Applicants are strongly encouraged to do the following before applying: review the downloadable state application procedures for additional helpful information and clarifications; review the downloadable online-application guidelines at the grant website; review the foundation's funding history (link is available from the grant website); review the online application questions in advance; and review the list of required attachments. These will generally include: a list of board members, financial statements (audited, reviewed, or compiled by independent auditor); an organization summary; a list of other funding sources; an IRS Determination letter; and other required documents. All attachments must be uploaded in the online application as PDF, Word, or Excel files. The Colonel Stanley R. McNeil Foundation has biannual deadlines of February 1 and June 1. Applications must be submitted by 11:59 p.m. on the deadline dates. Applicants for the February deadline will be notified of grant decisions by June 30. Applicants for the June deadline will be notified by November 30.
Requirements: Illinois nonprofit organizations serving the Chicago metropolitan area are eligible.
Restrictions: Because requests for support usually exceed available resources, organizations can only apply to either the Lang Burk Fund or the Colonel Stanley McNeil Foundation in the same calendar year. Grant requests to both foundations during the same calendar year will no longer be accepted. The Foundation will consider requests for general operations only if the organization's operating budget is less than $1 million. In general, grant request for individuals, endowment campaigns or capital projects will not be considered.
Geographic Focus: Illinois
Date(s) Application is Due: Feb 1; Jun 1
Samples: Gads Hill Center, Chicago, Illinois, $35,000 (2010); Field Museum of Natural History, Chicago, Illinois, $50,000 (2010); Organic School Project, Chicago, Illinois, $35,000 (2010).
Contact: George Thorn, Vice President, Bank of America; (312) 828-4154; ilgrantmaking@bankofamerica.com
Internet: https://www.bankofamerica.com/philanthropic/fn_search.action
Sponsor: Colonel Stanley R. McNeil Foundation
231 South LaSalle Street, IL1-231-13-32
Chicago, IL 60604

Colorado Interstate Gas Grants 1365
Nonprofits in communities served by the company are eligible to apply for grants in the categories of education, including early child development, K-12, and higher education; civic; health and welfare; minorities; cultural programs; environmental programs, particularly those focusing on air quality; and youth groups that concentrate on improving areas where CIG employees live and work. The company also supports special event sponsorships, corporate memberships, joint projects with other organizations, and employee participation in volunteer activities.
Requirements: Colorado, Wyoming, Texas, Utah and Kansas 501(c)3 tax-exempt organizations where the company has pipelines are eligible.
Geographic Focus: Colorado
Date(s) Application is Due: Aug 1
Amount of Grant: 200 - 20,000 U.S.D
Contact: Richard Wheatley, Media Relations; (713) 420-6828
Internet: http://www.cigco.com
Sponsor: Colorado Interstate Gas Company
P.O. Box 1087
Colorado Springs, CO 80944

Colorado Public Health and Environment Grants 1366
Grants are awarded in Colorado to improve and protect the health of residents and the quality of the environment, to assure access to essential health and medical care, and to plan and regulate Colorado's medical care system. Types of support include block grants, capital grants, conferences and seminars, demonstration grants, development grants, formula grants, general operating grants, program development, technical assistance, training, and travel grants. Applicants must notify the department of interest for guidelines and procedures. Grants are awarded usually for one year, with possible renewal based on availability of funds.
Requirements: Most applicants are Colorado governmental or quasi-governmental agencies and corporations.
Geographic Focus: Colorado
Amount of Grant: 1,000 - 1,000,000 U.S.D
Contact: Howard Roitman, Program Director; (303) 692-3397; Howard.Roitman@state.co.us
Internet: http://www.cdphe.state.co.us
Sponsor: Colorado Department of Public Health and the Environment
4300 Cherry Creek Drive S
Denver, CO 80246-1530

Colorado Springs Community Trust Fund Grants 1367
Grants are awarded to El Paso and Teller Counties in Colorado. The community trust supports Colorado nonprofits with an emphasis on community funds, youth agencies, health, and child welfare. Fields of interest include the arts and cultural programs (visual arts, performing arts, museums), education (elementary education, higher education, adult basic education, literacy, and reading), health care and health organizations, substance abuse and mental health services, crime and law enforcement, human services, and community development. Continuing support for programs/projects will be awarded, as well as support for annual campaigns, building/renovations, equipment, emergency funds, and seed money. There are no application forms; initial approach should be by letter. The board meets four times each year to consider requests.
Requirements: Colorado nonprofits, with an emphasis on El Paso and Teller Counties, are eligible for grant support.
Restrictions: Grants are not awarded to individuals or for operating budgets, deficit financing, endowment funds, matching gifts, scholarships, fellowships, research, demonstration projects, publications, conferences, or loans.
Geographic Focus: Colorado
Date(s) Application is Due: Mar 1; Aug 1; Nov 1
Amount of Grant: 500 - 5,000 U.S.D
Contact: Michael Hannigan, Executive Director; (719) 389-1251; fax (719) 389-1252; mhannigan@PPCF.org
Sponsor: Colorado Springs Community Trust Fund
P.O. Box 1443
Colorado Springs, CO 80901

Colorado Trust Grants 1368

The Trust's strategic grantmaking supports the development of a coordinated system of policies, programs and services that, expand health coverage, improve and expand health care. The Trust issues Requests for Proposals (RFP) and welcomes responses from nonprofit organizations and governmental entities across Colorado. When a competitive funding opportunity is available, a detailed RFP with related instructions and specific application deadlines is posted to the Funds website. On occasion, The Trust also asks organizations that are focused on strategies specific to achieving access to health to submit individual, non-competitive proposals.

Requirements: The following types of organizations are eligible to apply for grants: nonprofit organizations that are exempt under Section 501(c)3 of the Internal Revenue Code and are classified as not a private foundation under Section 509(a); independent sponsored projects of a nonprofit 501(c)3 organization acting as a fiscal agent; government and public agencies.

Restrictions: The Trust asks for proposals through a Request for Proposal process, rather than accepting unsolicited proposals. Announcements of Requests for Proposals are posted at the website. Grant seekers also may register with the Trust to receive notification of new funding opportunities. The Colorado Trust does not make grants for the following: political campaigns or voter registration drives; capital funding for the purchase, construction or renovation of any facilities or other physical infrastructure; operating deficits or retirement of debt; indirect allocations (excluding fiscal agent fees); religious purposes.

Geographic Focus: Colorado

Contact: Ed Lucero, Senior Program Officer; (888) 847-9140 or (303) 837-1200; fax (303) 839-9034; ed@coloradotrust.org

Internet: http://www.coloradotrust.org

Sponsor: Colorado Trust
1600 Sherman Street
Denver, CO 80203-1200

Colorado Trust Partnerships for Health Initiative 1369

The trust supports this statewide initiative to improve the coordination of health services at the community level. The program will support new or existing community health partnerships, consisting of local health departments, county nursing services, community-based organizations, and other interested organizations, businesses, and community members. Each partnership will identify a Healthy People 2010 focus area important to its community and will develop and implement a coordinated plan to address that health issues. Grants will be awarded to help partnerships build, strengthen, and sustain the public health infrastructure of Colorado communities by proactively addressing public health issues. The complete RFP is available online.

Requirements: Partnerships must have a 501(c)3 fiscal agent or be a governmental entity; strong community support including representatives from the public, private, and nonprofit sectors; ability to develop strategic partnerships; knowledge of the the public health system and health promotion/disease prevention issues; ability to design a service approach to meet targeted Health People 2010 needs; cultural competency and a demonstrated commitment to diversity; sound fiscal management; and willingness to work in partnership with the Colorado Trust in the planning and implementation phases of the initiative.

Geographic Focus: Colorado

Date(s) Application is Due: Nov 17

Amount of Grant: 10,000 - 15,000 U.S.D

Contact: Susan Downs-Karkos; (888) 847-9140 or (303) 837-1200;; susan@coloradotrust.org

Internet: http://www.coloradotrust.org

Sponsor: Colorado Trust
1600 Sherman Street
Denver, CO 80203-1200

Columbus Foundation Allen Eiry Fund Grants 1370

The Fund provides support to organizations that serve older adults in Seneca County. Projects supported by the fund include transportation, recreation, homemaker services, hot meals, court ordered guardianships, information and referral programs, library programs for nursing homes, support for capital projects, and other projects that enhance the lives of needy elderly.

Requirements: The Fund welcomes requests from organizations having recognition under Section 501(c)3 of the Internal Revenue Code that serve older adults in Seneca County.

Restrictions: The Fund makes no grants to individuals. Requests for religious purposes, budget deficits, endowments, conferences, scholarly research, or projects that are normally the responsibility of a public agency are generally not funded.

Geographic Focus: Ohio

Date(s) Application is Due: Nov 5

Contact: Dottie Henderson, Executive Assistant; (614) 251-4000; fax (614) 251-4009; dhenderson@columbusfoundation.org

Internet: http://www.columbusfoundation.org/gogrants/targeted_needs/specialized_grants.aspx

Sponsor: Columbus Foundation
1234 E Broad Street
Columbus, OH 43205

Columbus Foundation Ann Ellis Fund Grants 1371

The Fund is used for eye research, with primary emphasis on sponsoring scientific research on the diagnosis, prevention, and treatment of glaucoma. Once this is accomplished, the fund will assist in the research of other hidden eye problems. For more information, please contact the Community Research and Grant Management Department at the Foundation

Restrictions: Individuals are ineligible. Requests for religious purposes, budget deficits, endowments, conferences, or projects that are normally the responsibility of a public agency are generally not funded.

Date(s) Application is Due: Oct 1

Contact: Dottie Henderson, Executive Assistant; (614) 251-4000; fax (614) 251-4009; dhenderson@columbusfoundation.org

Internet: http://www.columbusfoundation.org/gogrants/targeted_needs/specialized_grants.aspx

Sponsor: Columbus Foundation
1234 E Broad Street
Columbus, OH 43205

Columbus Foundation Central Benefits Health Care Foundation Grants 1372

The Foundation was established in 1997 with a focus on preventative health care for indigent children and adults; the Trustees have taken a step further by focusing grantmaking for the preventative health care needs of children, prenatal through age six. Proposals for funding are accepted at anytime. Funding is limited to the central Ohio area, giving primarily in Columbus, Ohio.

Requirements: Central Ohio, tax-exempt organizations are eligible to apply.

Restrictions: Individuals are ineligible.

Geographic Focus: Ohio

Contact: Tamera Durrence, Assistant Vice President and Director of Supporting Foundations; (614) 251-4000; fax (614) 251-4009; tdurrence@columbusfoundation.org

Internet: http://www.columbusfoundation.org/find/support/cenben_usa.aspx

Sponsor: Columbus Foundation
1234 E Broad Street
Columbus, OH 43205

Columbus Foundation Competitive Grants 1373

The foundation's competitive grants are made in the following fields: advancing philanthropy, arts and humanities, conservation, education, health, social services, and urban affairs. The Governing Committee approves distributions from unrestricted and field of interest funds through competitive grants. Competitive grants are the most common way nonprofit organizations request funding from the Foundation. Submit a Letter of Intent or a Full Proposal by accessing our online grant application system.

Requirements: Central Ohio, tax-exempt public charities under Section 501(c)3 of the Internal Revenue Service Code may submit grant requests.

Restrictions: Individuals are ineligible. Requests for religious purposes, budget deficits, endowments, conferences, scholarly research, or projects that are normally the responsibility of a public agency are generally not funded. Funding is not available for projects when funds are available elsewhere.

Geographic Focus: Ohio

Contact: Emily Savors, Director; (614) 251-4000; fax (614) 251-4009; esavors@columbusfoundation.org

Internet: http://www.columbusfoundation.org/gogrants/index.aspx

Sponsor: Columbus Foundation
1234 E Broad Street
Columbus, OH 43205

Columbus Foundation Estrich Fund Grants 1374

The Fund is designed to provide home health care and/or inpatient services to the terminally ill. Funds may be used to provide medical services (excluding doctor visits, in-hospital costs, or care facilities), medical supplies and equipment, transportation, and physical and occupational therapy. Funds may also be used to make available emotional, social, and spiritual support through the use of professionals and/or trained volunteers.

Requirements: Central Ohio, tax-exempt public charities under Section 501(c)3 of the Internal Revenue Service Code may submit grant requests.

Restrictions: Requests for religious purposes, budget deficits, endowments, conferences, scholarly research, or projects that are normally the responsibility of a public agency are generally not funded.

Geographic Focus: Ohio

Date(s) Application is Due: Apr 20

Samples: Fisrtlink, Columbus, OH, $18,725—to support the Dental Options Program, which provides free or low cost dental care to residents of central Ohio; Central Ohio Diabetes Association, Columbus, OH—to support outreach and education programs for the Hispanic community; Impact Safety Programs, Columbus, OH, $20,000—to provide a violence prevention program for residents of Amethyst; Columbus AIDS Task Force (CATF), Columbus, OH, $22,000—to support outreach offices at Neighborhood House and Just For Today.

Contact: Sandi Smith, (614) 251-4000; fax (614) 251-4009; ssmith@columbusfoundation.org

Internet: http://www.columbusfoundation.org/GD/Templates/Pages/TCF/TCFSecondaryaspx?page=66

Sponsor: Columbus Foundation
1234 E Broad Street
Columbus, OH 43205

Columbus Foundation J. Floyd Dixon Memorial Fund Grants 1375

The purpose of the Fund is to provide educational programs for children, health programs for the elderly, and social services programs in Jackson County. An individual grant seldom exceeds $10,000. Applications must be received by May 28, and decisions will be announced three to four months after the application deadline.

Requirements: The fund welcomes requests from organizations having recognition under Section 501(c)3 of the Internal Revenue Code that services the residents of Jackson County.

Restrictions: The Fund makes no grants to individuals, and generally does not fund governmental agencies. Requests for religious purposes, budget deficits, endowments,

conferences, scholarly research, or projects that are normally the responsibility of a public agency are generally not funded.
Geographic Focus: Ohio
Date(s) Application is Due: May 28
Amount of Grant: Up to 10,000 U.S.D
Contact: Dottie Henderson, Executive Assistant; (614) 251-4000; fax (614) 251-4009; dhenderson@columbusfoundation.org
Internet: http://www.columbusfoundation.org/gogrants/targeted_needs/specialized_grants.aspx
Sponsor: Columbus Foundation
1234 E Broad Street
Columbus, OH 43205

Columbus Foundation Mary Eleanor Morris Fund Grants **1376**
The Fund supports quality-of-life projects, with a focus on the arts, civic affairs, conservation, education, health, and social services that benefits the residents of Logan County. Application deadlines are the first Friday in January and the first Friday in July.
Requirements: The fund welcomes requests from organizations having recognition under Section 501(c)3 of the Internal Revenue Code that services the residents of Logan County, as well as public and private schools.
Restrictions: The Fund makes no grants to individuals. Requests for religious purposes, budget deficits, endowments, conferences, scholarly research, or projects that are normally the responsibility of a public agency are generally not funded.
Geographic Focus: Ohio
Date(s) Application is Due: Jan 1; Jul 2
Contact: Emily Savors, Director; (614) 251-4000; fax (614) 251-4009; esavors@columbusfoundation.org
Internet: http://www.columbusfoundation.org/gogrants/targeted_needs/small_grants.aspx
Sponsor: Columbus Foundation
1234 E Broad Street
Columbus, OH 43205

Columbus Foundation Paul G. Duke Grants **1377**
The Foundation is primarily interested in supporting projects for children, young adults, and the family. Grants are generally made in the fields of the arts, education, health, and social services. The Foundation will also consider making challenge grants for worthwhile projects to encourage matching gifts or additional funding from other donors. Requests may be for general, capital, or specific project support, including seed money for innovative programs. The Foundation accepts proposals twice per year, and support is generally given for one year.
Requirements: Nonprofit organizations having recognition under Section 501(c)3 of the Internal Revenue Code are eligible for grants.
Restrictions: The Foundation generally does not make grants for transportation, computer hardware or software, and research or treatment for specific diseases. The Foundation does not award grants to individuals. Generally, grants are not made for religious purposes, budget deficits, or projects that are normally the responsibility of a public agency.
Geographic Focus: Ohio
Date(s) Application is Due: Apr 1; Aug 1
Contact: Tami Durrence, (614) 251-4000; fax (614) 251-4009; mail@supportingfoundations.org
Internet: http://www.columbusfoundation.org/paul_g_duke.aspx
Sponsor: Paul G. Duke Foundation
1234 East Broad Street
Columbus, OH 43205-1453

Columbus Foundation Robert E. and Genevieve B. Schaefer Fund Grants **1378**
The purpose of the Fund is to benefit the residents of Chillicothe and/or Ross County. The funding supports programs that enhance the community in three broad areas: cultural development, health and human welfare, and economic development.
Requirements: The Fund welcomes applications from organizations recognized under Section 501(c)3 of the Internal Revenue Code that serves the residents of Chillicothe and/or Ross County. All organizations with budgets over $75,000 must provide an audit, and those with budgets less than $75,000 must provide their most recent form 990 and compiled financial statements.
Restrictions: The Fund makes no grants to individuals, churches, governmental agencies, or school districts.
Geographic Focus: Ohio
Date(s) Application is Due: Oct 29
Contact: Dottie Henderson, Executive Assistant; (614) 251-4000; fax (614) 251-4009; dhenderson@columbusfoundation.org
Internet: http://www.columbusfoundation.org/gogrants/targeted_needs/specialized_grants.aspx
Sponsor: Columbus Foundation
1234 E Broad Street
Columbus, OH 43205

Columbus Foundation Small Grants **1379**
The Small Grants provides support for several areas including: animals; arts; camber music; conservation; community improvement; education; health, special populations; Logan County; Hocking/Fairfield counties of Ohio. Applicants may apply for up to $25,000; however, the majority of the grants awarded will be $1,000 to $10,000. Applications must be submitted by the first Friday of April, July, and October.
Requirements: Central Ohio, tax-exempt public charities under Section 501(c)3 of the Internal Revenue Service Code may submit grant requests.
Geographic Focus: Ohio
Amount of Grant: 1,000 - 25,000 U.S.D
Contact: Emily Savors, Director; (614) 251-4000; fax (614) 251-4009; esavors@columbusfoundation.org
Internet: http://www.columbusfoundation.org/gogrants/targeted_needs/small_grants.aspx
Sponsor: Columbus Foundation
1234 E Broad Street
Columbus, OH 43205

Columbus Foundation Traditional Grants **1380**
The Traditional Grants Program creates quality opportunities and meets community need by focusing on three areas: Basic Needs funds programs and projects that provide food, shelter, and clothing. Priority will be given to programs and projects that meet current needs and simultaneously integrate services designed to support long-term stability; Disadvantaged Children funds programs and projects that meet the diverse needs of at-risk children. Priority will be given to programs and projects that are in the home, build relationships with the family, and create a support network around and for the families of disadvantaged children; Developmental Disabilities funds programs and projects that address the needs of children and adults with physical or cognitive disabilities that impair functions or behavior and that occurred before a person reaches the age of 22. (Blindness/visual impairments and deafness/hearing impairments are not considered within this category.) If you are implementing your project in a Columbus City School building, or collaborating with the district on a project, and asked to submit a full application, you must request a letter of endorsement from the Office of Development.
Requirements: Central Ohio, tax-exempt public charities under Section 501(c)3 of the Internal Revenue Service Code may submit grant requests .If you are implementing your project in a Columbus City School building, or collaborating with the district on a project, and asked to submit a full application, you must request a letter of endorsement from the Office of Development. Letter of Intent deadlines are twice a year on the first Friday in February and September. See Foundations website for further guidelines.
Restrictions: Operating support will only be considered when the applicant demonstrates continuous innovation that enhances services.
Geographic Focus: Ohio
Date(s) Application is Due: Feb 5; Sep 3
Contact: Emily Savors, Director; (614) 251-4000; fax (614) 251-4009; esavors@columbusfoundation.org
Internet: http://www.columbusfoundation.org/gogrants/targeted_needs/traditional_grants.aspx
Sponsor: Columbus Foundation
1234 E Broad Street
Columbus, OH 43205

Comer Foundation Grants **1381**
The foundation awards grants to U.S. nonprofits in its areas of interest, including needle-exchange programs grounded in harm-reduction practice and programs providing direct services addressing health related to drug use; education—AIDS education, job training, and career guidance for low-income groups; health, with an emphasis on AIDS; social services—temporary shelters for homeless people and organizations offering diverse child, youth, and family services; and arts, culture, and humanities—to make programs of established organizations accessible to the public at large. Types of support include operating budgets, special projects, and staff development. There are no application deadlines.
Restrictions: Individuals are ineligible for grant support.
Geographic Focus: Illinois
Date(s) Application is Due: Mar 1; Jul 1; Nov 1
Amount of Grant: 5,000 - 30,000 U.S.D
Contact: Stephanie Comer, (415) 256-9917; fax (415) 256-9918; scomer@comer-foundation.org
Internet: http://www.comer-foundation.com/home.html
Sponsor: Comer Foundation
P.O. Box 78154
San Francisco, CA 94107

Comerica Charitable Foundation Grants **1382**
The Comerica Charitable Foundation funding priorities support community needs in it's primary markets within Texas, Michigan, California, Arizona, and Florida. Applications are accepted for program support and capital expense of those non-profit organizations that support the foundation's priorities. Economic self-sufficiency for low and moderate income individuals and families will be supported in the areas of financial literacy, job readiness, job creation and retention, small business training and development, and transitional and supportive housing. Neighborhood revitalization areas to be supported are affordable housing and neighborhood business development. Financial literacy programs (K-12 and adult) and scholarships for students with income needs for studies in business, finance and growth industries will be supported. Also a priority is access to health care to include preventive care for the uninsured and under-insured. Also a priority are programs that support diversity and inclusion. Funding deadlines vary according to region. See the Foundation's website for regional offices.
Requirements: 501(c)3 nonprofits are eligible.
Restrictions: Grants are not awarded for United Way organizations, religious and fraternal groups, political parties, charitable golf events, athletic programs, multiyear pledges, or endowment funds.
Geographic Focus: Arizona, California, Florida, Michigan, Texas

Date(s) Application is Due: Mar 15; Jun 15; Sep 15; Nov 15
Contact: Program Contact; (313) 222-7356
Internet: http://www.comerica.com/vgn-ext-templating/v/index.jsp?vgnextoid=374970d75d994010VgnVCM1000004502a8c0RCRD
Sponsor: Comerica Charitable Foundation
P.O. Box 75000, MC 3390
Detroit, MI 48275-3390

Commission on Religion in Appalachia Grants 1383
The commission supports Christian and interdenominational nonprofits in U.S. Appalachian regions. Churches and ministries and nonprofits in the areas of parochial education, religious welfare, civil rights, health care, and social services (domestic violence, housing, women's affairs, and food distribution) are awarded grants. The commission strongly encourages applicants to obtain guidelines prior to submitting proposals.
Requirements: Nonprofits in the Appalachian regions of Mississippi, Alabama, Georgia, South Carolina, North Carolina, Tennessee, Kentucky, Virginia, West Virginia, Ohio, Pennsylvania, Maryland, and New York are eligible.
Geographic Focus: Mississippi
Date(s) Application is Due: Jan 31
Amount of Grant: 5,000 - 15,000 U.S.D
Contact: Administrator; (304) 720-2672; fax (304) 720-2673; corainappa@aol.com
Internet: http://www.geocities.com/appalcora/About.html
Sponsor: Commission on Religion in Appalachia
P.O. Box 11908
Charleston, WV 25339-1908

Commonweal Foundation Community Assistance Grants 1384
The foundation supports educational programs and projects assisting disadvantaged, at-risk youth. The foundation focuses on secondary and, to a lesser extent, elementary education. The foundation also considers grants for educational research and, to a limited extent, health care. Guidelines are available online.
Requirements: 501(c)3 tax-exempt organizations located in the District of Columbia, Maryland, or Northern Virginia are eligible. Applicant organizations should have an annual budget not exceeding $1 million.
Geographic Focus: District of Columbia, Maryland, Virginia
Date(s) Application is Due: Mar 1; Aug 1
Amount of Grant: Up to 25,000 U.S.D
Contact: Gloria Dairsow, (240) 450-0000; fax (240) 450-4115; gdairsow@cweal.org
Internet: http://www.cweal.org/cag.htm
Sponsor: Commonweal Foundation
10770 Columbia Pike, Suite 150
Silver Spring, MD 20901

Commonweal Foundation Grants 1385
The foundation awards grants to eligible nonprofit organizations serving disadvantaged youth in the area comprising the corridor between Baltimore, MD, and Washington, DC. It also offers some assistance to elementary education, educational research, and some health care causes. Programs include Partners in Learning (address low literacy skills of economically disadvantaged children), community assistance grants (i.e., after-school tutoring, parenting classes, and shelters for the homeless or abused women), and learning disabilities support program (special education services to economically disadvantaged children). Contact the office for application procedures.
Requirements: 501(c)3 tax-exempt organizations serving disadvantaged youth in the corridor between Baltimore, MD, and Washington, DC, are eligible.
Geographic Focus: District of Columbia, Maryland, Virginia
Date(s) Application is Due: Feb 1; May 1; Aug 1; Nov 1
Amount of Grant: 10,000 - 25,000 U.S.D
Contact: Renee Cottman-Reyes, (240) 450-0000; fax (240) 450-4115; gdairsow@cweal.org
Internet: http://www.commonweal-foundation.org/index.html
Sponsor: Commonweal Foundation
10770 Columbia Pike, Suite 150
Silver Spring, MD 20901

Commonwealth Edison Grants 1386
The company recognizes its social responsibility to the area it serves, comprising Chicago and 25 northern Illinois counties, and makes grants and contributions to nonprofit organizations that address the diverse needs of this region and that can most enhance the economic, cultural, educational, and social health of its communities. Financial support is given for general operations as well as for specific projects or purposes that are in the general interest of the company or support general community needs. Contributions to hospitals are generally limited to capital fund drives. Building endowment funds or requests to relieve operating deficits are not supported. Each of the company's six commercial divisions (addresses and phone numbers available upon request) maintains a small philanthropic budget to address local needs. The corporate budget addresses needs of a wider ranging nature, such as support to United Way programs, education, and employee volunteer programs. Applications are accepted at any time. The executive review committee meets and addresses proposals quarterly, normally in February, May, August, and November.
Requirements: Nonprofit organizations serving Chicago and 25 northern Illinois counties are eligible.
Restrictions: Support is not provided for individuals; fraternal or veteran organizations; sectarian religious organizations; non-tax-exempt organizations; municipal, state, or federal agencies; political organizations or campaigns; most United Way agencies; specific elementary or secondary schools or school systems; organizations requesting purchase of ads or sponsorship programs; or national or international organizations that do not have specific business relating to Commonwealth Edison and/or its customers.
Geographic Focus: Illinois
Amount of Grant: 1,000 - 5,000 U.S.D
Contact: Steve Solomon, (312) 394-4321; fax (312) 394-2231
Sponsor: Commonwealth Edison Company
440 S. Lasalle Street, P.O. Box 805379
Chicago, IL 60680-5379

Commonwealth Fund/Harvard University Fellowship in Minority Health Policy 1387
The Commonwealth Fund/Harvard University Fellowship in Minority Health Policy is a one-year, full-time, academic degree-granting program designed to create physician-leaders, particularly minority physician-leaders, who will pursue careers in health policy, public health practice, and academia. It is designed to incorporate the critical skills taught in schools of public health, government, business, and medicine with leadership forums and seminar series conducted by Harvard senior faculty and nationally recognized leaders in minority health and public policy; supervised practicums and shadowing opportunities; and site visits, conferences, and travel. Each fellowship provides: $50,000 stipend; full tuition; health insurance; books; travel; and related program expenses, including financial assistance for a practicum project.
Requirements: Applicants are required to complete applications to both the Commonwealth Fund/Harvard University Fellowship in Minority Health Policy and the Harvard School of Public Health. Applicants must have finished residency and be U.S. citizens.
Date(s) Application is Due: Jan 4
Amount of Grant: 50,000 U.S.D
Contact: Joan Y. Reede, Director; (617) 432-2922; mfdp_cfhuf@hms.harvard.edu
Internet: http://www.commonwealthfund.org/Fellowships/Minority-Health-Policy-Fellowship.aspx
Sponsor: Commonwealth Fund
164 Longwood Avenue, 2nd Floor
Boston, MA 02115-5818

Commonwealth Fund Affordable Health Insurance Grants 1388
The Program on Affordable Health Insurance envisions an efficient and equitable health insurance system that makes available to all Americans comprehensive, continuous, and affordable coverage. In support of that vision, the program seeks to: analyze changes in employer-based, private and public insurance coverage for people under age 65, and determine how those changes may affect the affordability and comprehensiveness of coverage, the number of uninsured, the number of under-insured, and churning in and out of coverage; document the consequences of being uninsured, under-insured, and unstably insured, with regard to access to care, health status, personal financial security, and economic productivity; and develop and evaluate federal and state policies to expand and stabilize health insurance, make it more affordable and aligned with incentives to access appropriate and high quality care, and enhance the efficiency with which it is administered.
Requirements: The Commonwealth Fund requests letters of inquiry to initiate the grant application process, and does not wish to review full proposals at this stage. Applicants are encouraged to submit letters of inquiry using our online form.
Restrictions: The Fund makes grants only to tax-exempt organizations and public agencies and does not support: general planning and ongoing activities or existing deficits; endowment or capital costs, including construction, renovation, or equipment; basic biomedical research; conferences, symposia, major media projects, or documentaries, unless they are an outgrowth of one of the Fund's programs; individuals; scholarships; churches or other religious organizations unless the project for which they seek funding is entirely secular in nature; and work for which achievements cannot be measured.
Contact: Sara R. Collins, Assistant Vice-President; (212) 606-3838 or (212) 606-3800; fax (212) 606-3500; src@cmwf.org
Internet: http://www.commonwealthfund.org/Content/Program-Areas/Affordable-Health-Insurance.aspx
Sponsor: Commonwealth Fund
1 E 75th Street
New York, NY 10021-2692

Commonwealth Fund Australian-American Health Policy Fellowships 1389
The Australian Government Department of Health and Aging hopes to enrich health policy thinking as Australian-American Health Policy Fellows study how Australia approaches health policy issues, share lessons learned from the United States, and develop an international perspective and network of contacts to facilitate policy exchange and collaboration that extends beyond the fellowship experience. Australian-American Health Policy Fellowships are open to accomplished, mid-career health policy researchers and practitioners, including, academics, physicians, decision makers in managed care and other private sector health care organizations, federal and state health officials, and journalists. The Fellowship provides up to $55,000 (AUD) for terms of six to ten months, with a minimum stay of six months in Australia required. Focused on issues of common concern to Australian and U.S. policymakers, the fellowships are structured around areas of mutual policy interest, for example: health care quality and safety, the private/public mix of insurance and providers, the fiscal sustainability of health systems, the health care workforce, management and efficiency of health care delivery, and investment in preventive care strategies.
Requirements: All applicants must also meet the following criteria: be a citizen of the United States; be a mid-career health services researcher or practitioner (e.g., a physician, decision maker in a managed care organization or other private health care organization, government official or policy analyst, or journalist); have a demonstrated expertise in health policy issues and track record of informing health policy through research, policy analysis,

health services, or clinical leadership; have completed a master's degree or doctorate (or the equivalent thereof) in health services research, health administration, health policy, or a related discipline, such as economics or political science; and if academically based, be at a mid-career level (e.g., research fellow to associate professor).
Restrictions: Fellowships are not awarded to support basic research or study for an academic degree.
Date(s) Application is Due: Aug 15
Amount of Grant: Up to 55,000 U.S.D
Contact: Robin Osborn, Director; (212) 606-3809 or (212) 606-3800; fax (212) 606-3875; ro@cmwf.org
Internet: http://www.commonwealthfund.org/Fellowships/Australian-American-Health-Policy-Fellowships.aspx
Sponsor: Commonwealth Fund
1 E 75th Street
New York, NY 10021-2692

Commonwealth Fund Harkness Fellowships in Health Care Policy and Practice 1390
The Harkness Fellowships in Health Care Policy and Practice provide an opportunity for professionals from the Australia, Germany, the Netherlands, New Zealand, Norway, Switzerland, and the United Kingdom to spend four to 12 months in the United States conducting a research study that is relevant to health care policy and practice in both the United States and the fellow's home country, and that is focused on the issues of greatest concern to the fund. Fellowship awards provide up to $107,000, which covers round trip airfare to the United States, a living allowance, funds for project-related travel, research, and conferences, travel to attend the Commonwealth Fund program of fellowship seminars, health insurance, and U.S. and state taxes. A family supplement - including airfare, living allowance, and health insurance - is also provided to Fellows accompanied by a partner and/or children up to age 18.
Requirements: Applicants must be citizens of Australia, Germany, the Netherlands, New Zealand, Norway, Switzerland, and the United Kingdom in their late 20s to early 40s with broad educational backgrounds, not just those in research or academic careers. In order to apply, applicants must be nominated by their institution and submit a formal application, which is available from The Commonwealth Fund in New York City or its representatives in their home country.
Geographic Focus: All States, United Kingdom
Date(s) Application is Due: Sep 15
Amount of Grant: Up to 107,000 U.S.D
Contact: Robin Osborn, Director; (212) 606-3809 or (212) 606-3800; fax (212) 606-3875; ro@cmwf.org
Internet: http://www.commonwealthfund.org/Fellowships/Minority-Health-Policy-Fellowship.aspx
Sponsor: Commonwealth Fund
1 E 75th Street
New York, NY 10021-2692

Commonwealth Fund Health Care Quality Improvement and Efficiency Grants 1391
The goal of the Fund's Program on Health Care Quality Improvement and Efficiency is to improve the quality and efficiency of health care in the United States. To that end, the program supports projects that: promote the development and widespread adoption of health care quality and efficiency measures; assess and enhance the capacity of health care organizations to provide better care more efficiently; and promote the development and adoption of payment and incentive models that encourage providers to improve quality and efficiency.
Requirements: The Commonwealth Fund requests letters of inquiry to initiate the grant application process, and does not wish to review full proposals at this stage. Applicants are encouraged to submit letters of inquiry using our online form.
Restrictions: The Fund makes grants only to tax-exempt organizations and public agencies and does not support: general planning and ongoing activities or existing deficits; endowment or capital costs, including construction, renovation, or equipment; basic biomedical research; conferences, symposia, major media projects, or documentaries, unless they are an outgrowth of one of the Fund's programs; individuals; scholarships; churches or other religious organizations unless the project for which they seek funding is entirely secular in nature; and work for which achievements cannot be measured.
Contact: Anne-Marie J. Audet; (212) 606-3800; fax (212) 606-3500; ama@cmwf.org
Internet: http://www.commonwealthfund.org/Content/Program-Areas/Health-Care-Quality-Improvement-and-Efficiency.aspx
Sponsor: Commonwealth Fund
1 E 75th Street
New York, NY 10021-2692

Commonwealth Fund Patient-Centered Coordinated Care Program Grants 1392
As defined by the Institute of Medicine, patient-centered care is health care that establishes a partnership among practitioners, patients, and their families (when appropriate), to ensure that decisions respect patients' needs and preferences, and that patients have the education and support they need to make decisions and participate in their own care. In primary care, such care is best provided in a medical home - a primary care practice or health center that ensures patients have enhanced access to their clinicians (for example, through the availability of evening or weekend appointments), coordinates care, and engages in continuous quality improvement. The goal of The Commonwealth Fund's Patient-Centered Coordinated Care Program, established in 2005, is to improve the quality of primary care by making it more patient- and family-centered. The initiative supports projects that: promote the collection of information on patient-centered care and the delivery of care to facilitate public reporting and quality improvement; stimulate adoption of effective practices, models, and tools to make primary care practices patient- and family-centered; and improve policy to encourage patient- and family-centered care in medical homes.
Requirements: The Commonwealth Fund requests letters of inquiry to initiate the grant application process, and does not wish to review full proposals at this stage. Applicants are encouraged to submit letters of inquiry using our online form.
Restrictions: The Fund makes grants only to tax-exempt organizations and public agencies and does not support: general planning and ongoing activities or existing deficits; endowment or capital costs, including construction, renovation, or equipment; basic biomedical research; conferences, symposia, major media projects, or documentaries, unless they are an outgrowth of one of the Fund's programs; individuals; scholarships; churches or other religious organizations unless the project for which they seek funding is entirely secular in nature; and work for which achievements cannot be measured.
Contact: Melinda Abrams, (212) 606-3800; fax (212) 606-3500; mka@cmwf.org
Internet: http://www.commonwealthfund.org/Content/Program-Areas/Patient-Centered-Coordinated-Care.aspx
Sponsor: Commonwealth Fund
1 E 75th Street
New York, NY 10021-2692

Commonwealth Fund Payment System Reform Grants 1393
To achieve a high performing health system, the U.S. must curb spending growth and improve the way health care is provided. Payment system reform is critical to accomplishing these objectives. As a nation, we need to align incentives so that health care providers are rewarded for high-value care rather than a high volume of services. Rewarding value over volume will also encourage the development of a more integrated health care delivery system. The Commonwealth Fund, through its Program on Payment System Reform, supports analysis and the development of policy options to accomplish these goals. Areas of interest include: reforming the existing payment structure to improve the alignment of incentives and to provide a base for more comprehensive payment reform; modeling and analyzing the potential impact of alternative options for payment reform in Medicare and throughout the health system; using payment reform to encourage the development of new models of health care delivery that provide better and more coordinated care; and using comparative effectiveness research to support better decision-making by providers, payers, and patients.
Requirements: The Commonwealth Fund requests letters of inquiry to initiate the grant application process, and does not wish to review full proposals at this stage. Applicants are encouraged to submit letters of inquiry using our online form.
Restrictions: The Fund makes grants only to tax-exempt organizations and public agencies and does not support: general planning and ongoing activities or existing deficits; endowment or capital costs, including construction, renovation, or equipment; basic biomedical research; conferences, symposia, major media projects, or documentaries, unless they are an outgrowth of one of the Fund's programs; individuals; scholarships; churches or other religious organizations unless the project for which they seek funding is entirely secular in nature; and work for which achievements cannot be measured.
Contact: Stuart Guterman, (202) 292-6735 or (212) 606-3800; fax (212) 606-3500; sxg@cmwf.org
Internet: http://www.commonwealthfund.org/Content/Program-Areas/Payment-System-Reform.aspx
Sponsor: Commonwealth Fund
1 E 75th Street
New York, NY 10021-2692

Commonwealth Fund Quality of Care for Frail Elders Grants 1394
The Commonwealth Fund Program on Quality of Care for Frail Elders aims to transform the nation's nursing homes and other long-term care facilities into resident-centered organizations that are good places to live and work, capable of providing the highest-quality care. The projects it supports aim to: identify, evaluate, and spread models of resident-centered care, or care delivered in accordance with the needs and desires of the people who live in nursing homes; equip nursing home operators to lead transformational change; and promote policy options that support resident-centered care.
Requirements: The Commonwealth Fund requests letters of inquiry to initiate the grant application process, and does not wish to review full proposals at this stage. Applicants are encouraged to submit letters of inquiry using our online form.
Restrictions: The Fund makes grants only to tax-exempt organizations and public agencies and does not support: general planning and ongoing activities or existing deficits; endowment or capital costs, including construction, renovation, or equipment; basic biomedical research; conferences, symposia, major media projects, or documentaries, unless they are an outgrowth of one of the Fund's programs; individuals; scholarships; churches or other religious organizations unless the project for which they seek funding is entirely secular in nature; and work for which achievements cannot be measured.
Contact: Mary Jane Koren, (212) 606-3800; fax (212) 606-3500; mjk@cmwf.org
Internet: http://www.commonwealthfund.org/Content/Program-Areas/Quality-of-Care-for-Frail-Elders.aspx
Sponsor: Commonwealth Fund
1 E 75th Street
New York, NY 10021-2692

Commonwealth Fund Small Grants 1395
The fund awards small grants for projects in its major program areas, which include: affordable health insurance; international health care policy and practice; improving the quality of health care services; and improving insurance coverage and access to care. Types of support include employee-matching gifts, program development, program evaluation, and research. Preference is given to projects that seek to solve problems,

especially those affecting vulnerable groups; that analyze the effects of policies and trends on well-defined health issues; and that develop and test practical solutions. Prospective grantees should submit a letter of inquiry via regular or electronic mail. Program staff will contact applicants if more detailed information is required. Proposals recommended by fund staff are reviewed and voted upon by the board of directors, which meets in July, April, and November. Letters of inquiry should be brief, no more than two pages.
Requirements: The fund makes grants only to tax-exempt organizations and public agencies.
Restrictions: The fund does not support general planning and ongoing activities or existing deficits; endowment or capital costs, including construction, renovation, or equipment; basic biomedical research; conferences, symposia, major media projects or documentaries, unless they are an outgrowth of one of the fund's programs; individuals; scholarships; churches or other religious organizations, unless the project is entirely secular in nature; or work for which achievements cannot be measured.
Contact: Andrea Landes, Director of Grants Management; (212) 606-3800; fax (212) 606-3508; acl@cmwf.org
Internet: http://www.commonwealthfund.org/Grants-And-Programs/Applicant-and-Grantee-Resources/Applying-for-a-Grant.aspx
Sponsor: Commonwealth Fund
1 E 75th Street
New York, NY 10021-2692

Commonwealth Fund State High Performance Health Systems Grants 1396
The Commonwealth Fund's State High Performance Health Systems program is designed to help states develop the infrastructure needed to improve health care quality and outcomes - and to draw out and share lessons of national import from the experience of states that are moving toward comprehensive health care reforms. To enable states to help local health care providers better meet the needs of their patient populations, the program will support approaches that offer providers shared access to the clinical support and practice-management services essential to achieving high performance. Such resources will often be developed though state-initiated, public-private partnerships. To inform and support state health care leaders, additional projects will facilitate information-sharing among states working toward health care reform.
Requirements: The Commonwealth Fund requests letters of inquiry to initiate the grant application process, and does not wish to review full proposals at this stage. Applicants are encouraged to submit letters of inquiry using our online form.
Restrictions: The Fund makes grants only to tax-exempt organizations and public agencies and does not support: general planning and ongoing activities or existing deficits; endowment or capital costs, including construction, renovation, or equipment; basic biomedical research; conferences, symposia, major media projects, or documentaries, unless they are an outgrowth of one of the Fund's programs; individuals; scholarships; churches or other religious organizations unless the project for which they seek funding is entirely secular in nature; and work for which achievements cannot be measured.
Contact: Edward L. Schor, Vice President; (212) 606-3800; fax (212) 606-3500; els@cmwf.org
Internet: http://www.commonwealthfund.org/Content/Program-Areas/State-High-Performance-Health-System.aspx
Sponsor: Commonwealth Fund
1 E 75th Street
New York, NY 10021-2692

Communities Foundation of Texas Grants 1397
The unrestricted funds of the foundation support programs and projects intended to improve the quality of life for the citizens of the Dallas, TX, area. The funds support education, health and hospitals, social services, youth, and cultural programs. The foundation encourages projects developed in consultation with other agencies and planning groups and that promote coordination, cooperation, and sharing among organizations. Types of support include seed grants, emergency funds, building funds, equipment acquisition, matching funds, technical assistance, research, capital campaigns, and operating budgets. Requests for operating funds generally are not granted.
Requirements: 501(c)3 organizations in the Dallas, TX, area may apply.
Restrictions: Grants are not made to or for individuals, endowments, sectarian religious purposes, political or lobbying efforts, deficit financing, media projects, or for operational expenses of established organizations.
Geographic Focus: Texas
Contact: Brent Christopher, President; (214) 750-4222; fax (214) 750-4210; bchristopher@cftexas.org
Internet: http://www.cftexas.org
Sponsor: Communities Foundation of Texas
5500 Caruth Haven Lane
Dallas, TX 75225-8146

Community-Campus Partnerships for Health Award 1398
The award recognizes exemplary partnerships between communities and health professional schools that build on each other's strengths to improve health professions education, civic responsibility, and the overall health of their communities. In addition, the award recognizes community-campus partnerships that reflect CCPH's principles of good practice. The intent of the award is to highlight the power and potential of community-campus partnerships. Award recipients receive complimentary travel, lodging, and registration to the CCPH annual conference for the awards for up to two of the partnership's representatives; complimentary space in the CCPH conference program for a description of the partnership; complimentary space in the CCPH conference exhibit/poster hall; and a commemorative plaque. Nominations from any country or nation are welcome, but submissions must be in English. Nomination guidelines and form are available online.
Requirements: At minimum, an eligible partnership must be an established partnership involving one or more health professional school partners and one or more community partners. Individuals are not eligible, only partnerships.
Contact: Award Administrator; (206) 543-8178; ccphuw@u.washington.edu.
Internet: http://www.ccph.info
Sponsor: Community-Campus Partnerships for Health
UW Box 354809
Seattle, WA 98195-4809

Community Campus Partnerships for Health Fellowships 1399
Community-based professionals, academic administrators, and faculty with significant knowledge and expertise in building and sustaining service-learning, community-based participatory research, and community-campus partnerships are encouraged to apply for a fellowship with CCPH. The expected outcome of the program is to advance and support these concepts in the context of health professions education and practice. Fellows will receive a stipend for a one-year fellowship period, with the potential for a second year of funding. Funding provides support for time spent participating in fellow-related activities. Fellows also receive a certificate of recognition upon completing the fellowship. Application procedures and additional information are available online.
Requirements: CCPH members and nonmembers are eligible.
Amount of Grant: 5,000 U.S.D
Contact: Sarena Seifer, Executive Director; (206) 616-4305; fax (206) 685-6747; sarena@u.washington.edu or ccphuw@u.washington.edu
Internet: http://depts.washington.edu/ccph
Sponsor: Community Campus Partnerships for Health
UW Box 354809
Seattle, WA 98195-4809

Community Foundation AIDS Endowment Awards 1400
The foundation administers an annual special grants program called the AIDS Endowment Awards. Grants of $1,000 will be awarded to one or two organizations that demonstrate a commitment to working towards the prevention, education and treatment of AIDS so that they may continue their efforts. Services provided could include education programs for the prevention of HIV/AIDS, medical and social services for those living with HIV/AIDS, housing services for those living with HIV/AIDS and medical research for the treatment and prevention of HIV/AIDS.
Requirements: Selected organizations in the metropolitan Richmond area interested in the field of AIDS education, research or community service and care are encouraged to apply. Organizations interested in applying should submit a brief proposal to foundation which should be no more than two single-spaced pages and include: [1] a description of the organization's mission; and [2] a description of the agency's work in the field of AIDS. Include a list of Board of Governors, a copy of the most recent IRS Form 990, an audited financial statement if available, and an IRS Tax Exempt Letter designating your organization a 501(c)3 non-profit.
Geographic Focus: Virginia
Date(s) Application is Due: Feb 15
Amount of Grant: 1,000 U.S.D
Contact: Susan Hallett; (804) 330-7400; fax (804) 330-5992; shallett@tcfrichmond.org
Internet: http://www.tcfrichmond.org/Page2954.cfm
Sponsor: Community Foundation Serving Richmond and Central Virginia
7501 Boulders View Drive, Suite 110
Richmond, VA 23225

Community Foundation Alliance City of Evansville Endowment Fund Grants 1401
The City of Evansville Endowment Fund originated in 1994, when former Mayor Frank McDonald II proposed that $5 million of revenue from the gaming boat be invested in a way that would serve the City of Evansville, Indiana, forever. The City of Evansville Endowment Fund's earnings allow for grant making year after year. During its most recent grant cycle, the Fund allowed for grants totaling more than $180,000 to 21 nonprofit organizations serving the City of Evansville. The Foundation's grant cycle runs from May through October each year. Funding requests are accepted only during the grant cycle. The CEEF grants committee will consider funding requests of any amount. Requests of at least $1,000 are preferred.
Requirements: CEEF serves to provide funds to organizations that qualify as tax-exempt organizations under sections 501(c)3 and 509(a) of the Internal Revenue Code. The organization must serve within the city limits of Evansville, Indiana. At least sixty percent (60%) of grant funding will be distributed for activities that will support or benefit the 4th and 6th wards of the City of Evansville.
Restrictions: Not more than thirty percent (30%) of any grant request may be for personnel costs, travel costs, office supplies and other program operating costs.
Geographic Focus: Indiana
Date(s) Application is Due: Oct 31
Samples: Ark Crisis Child Care Center—to purchase and install fencing to enclose the playground, $3,200; Evansville Junior Football League—to assist with the construction of a handicap accessible restroom and storage building on the football field grounds in Howell Park, $5,000; Little Lambs of Evansville—for the purchase and installation of shelving units and for a rolling cart, clothes racks, and storage containers, $1,700.
Contact: Melinda Waldroup, Program Director; (812) 429-1191 or (877) 429-1191; fax (812) 429-0840; mwaldroup@alliance9.org
Internet: http://www.alliance9.org/city-of-evansville-endowment-fund
Sponsor: Community Foundation Alliance
123 NW Fourth Street, Suite 322
Evansville, IN 47708-1712

Community Foundation for Greater Buffalo Grants 1402

The Community Foundation for Greater Buffalo (CFGB) is a public charity holding more than 800 different charitable funds, large and small, established by individuals, families, nonprofit agencies and businesses to benefit Western New York. Since 1919, the foundation has served the needs of it's community and the wishes of it's donors through personalized service, financial stewardship, local expertise, and community leadership. The Foundation focuse's on four main areas of interest: strengthen the region as a center for arts and culture; natural, historic, and architectural resources; reduce racial and ethnic disparities; increase economic self-sufficiency for low-income individuals and families. Special funding will also be available to support the following: AIDS research and its cure ($5,000); programs that serve the visual, speech or hearing impaired ($5,000). Preference will be given to requests that align with one or more of the four focus areas.

Requirements: Applicants must be 501(c)3 not-for-profit organizations located within the eight counties of Western New York: Allegany, Cattaraugus, Chautauqua, Erie, Genesee, Niagara, Orleans, and Wyoming.

Restrictions: The Foundation will not consider competitive funding for: endowments; religious purposes; schools not registered with the New York State Education Department; attendance at or sponsorship of fundraising events for organizations; annual events or festivals; any partisan political activity. Funds from the foundation cannot be used to support or oppose a candidate for political office. Projects and activities that have occurred. The Foundation will not, except in extraordinary cases, provide payment or reimbursement for expenses incurred prior to the funding decision being communicated to the applicant.

Geographic Focus: New York

Date(s) Application is Due: Mar 1

Amount of Grant: 1,000 - 25,000 U.S.D

Samples: Carnegie Art Center, $4,500—to support the continued restoration of the landmark Carnegie Art Center building; Grassroots Gardens, $25,000—to support new community gardens that provide beautification, food and employment; Youth Character Development, $9,100—to aid refugee students in the transition to formal education and encourage college enrollment.

Contact: Jean McKeown, Senior Program Officer; (716) 852-2857, ext. 204; fax (716) 852-2861; jeanm@cfgb.org

Internet: http://www.cfgb.org/page17000.cfm

Sponsor: Community Foundation for Greater Buffalo
712 Main Street
Buffalo, NY 14202

Community Foundation for Greater New Haven $5,000 and Under Grants 1403

The Community Foundation for Greater New Haven (CFFGNH) is a philanthropic institution that was established in 1928. The foundation's mission is to create positive and sustainable change in Connecticut's Greater New Haven region by increasing the amount of and enhancing the impact of community philanthropy. Funding through the CFFGNH $5,000 and Under Grants process are available to any organization with an operating budget of $2 million or less.

Requirements: IRS 501(c)3 nonprofit organizations are eligible to apply in the greater New Haven area, which includes: Ansonia; Bethany; Branford; Cheshire; Derby; East Haven; Guilford; Hamden; Madison; Milford; New Haven; North Branford; North Haven; Orange; Oxford; Seymour; Shelton; Wallingford; West Haven and; Woodbridge.

Restrictions: No organization shall be eligible to receive a grant under this process more often than once in any period of two calendar years. Grants are not made to support religious activities, lobbying, or travel.

Geographic Focus: Connecticut

Amount of Grant: 5,000 U.S.D

Samples: Christ Episcopal Church, $5,000—to support the Kathleen Samela Memorial Food Bank; Connecticut Alliance to Benefit Law Enforcement, Inc., $2,100—to support training for Hamden Fire and EMS personnel; Connecticut Pre-Engineering Program, $4,000—to provide operating support for science, technology, engineering and math (STEM) enrichment programming to students in High School in the Community, East Rock, Clinton Avenue, Micro-Society and Troupe schools.

Contact: Denise Canning, Grants Manager; (203) 777-2386 or (203) 777-7076; fax (203) 787-6584; dcanning@cfgnh.org

Internet: http://www.cfgnh.org/GrantsScholarships/TypesofGrants/tabid/199/Default.aspx

Sponsor: Community Foundation for Greater New Haven
70 Audubon Street
New Haven, CT 06510

Community Foundation for Greater New Haven Responsive New Grants 1404

The Community Foundation for Greater New Haven is a philanthropic institution that was established in 1928. The foundation's mission is to create positive and sustainable change in Connecticut's Greater New Haven region by increasing the amount of and enhancing the impact of community philanthropy. The Responsive New Grants Program is generally awarded to address an agency's general operating, programmatic, capital or technical assistance needs. This funding source is open to all requests for projects and organizational support.

Requirements: IRS 501(c)3 nonprofit organizations are eligible to apply in the greater New Haven area, which includes: Ansonia; Bethany; Branford; Cheshire; Derby; East Haven; Guilford; Hamden; Madison; Milford; New Haven; North Branford; North Haven; Orange; Oxford; Seymour; Shelton; Wallingford; West Haven and; Woodbridge.

Restrictions: Grants are not made to support religious activities, lobbying, or travel.

Geographic Focus: Connecticut

Date(s) Application is Due: Mar 5

Amount of Grant: 10,000 - 185,000 U.S.D

Samples: Arts Council of Greater New Haven, New Haven, CT, $50,000—to provide general operating support; Beth El Center Inc., Milford, CT, $25,000— general operating support for transitional shelter programs for homeless individuals and families and a soup kitchen/food pantry that feeds individuals and families from Greater New Haven and Lower Naugatuck Valley; Area Congregations Together Inc., Shelton, CT, $150,000—to establish a development program.

Contact: Denise Canning, Grants Manager; (203) 777-2386 or (203) 777-7076; fax (203) 787-6584; dcanning@cfgnh.org

Internet: http://www.cfgnh.org/GrantsScholarships/TypesofGrants/tabid/199/Default.aspx

Sponsor: Community Foundation for Greater New Haven
70 Audubon Street
New Haven, CT 06510

Community Foundation for Greater New Haven Sponsorship Grants 1405

The Community Foundation for Greater New Haven (CFFGNH) is a philanthropic institution that was established in 1928. The foundation's mission is to create positive and sustainable change in Connecticut's Greater New Haven region by increasing the amount of and enhancing the impact of community philanthropy. The available sponsorships are awarded throughout a calendar year, only for events and may not exceed $2,500 per event. It is recommended that nonprofits submit a sponsorship request at least 60 days in advance of their event. Applications are accepted year-round, application available online at: http://www.cfgnh.org/Grantmaking/TypesofGrants/tabid/199/Default.aspx.

Requirements: IRS 501(c)3 nonprofit organizations are eligible to apply in the greater New Haven area, which includes: Ansonia; Bethany; Branford; Cheshire; Derby; East Haven; Guilford; Hamden; Madison; Milford; New Haven; North Branford; North Haven; Orange; Oxford; Seymour; Shelton; Wallingford; West Haven and; Woodbridge.

Restrictions: Grants are not made to support religious activities, lobbying, or travel.

Geographic Focus: Connecticut

Amount of Grant: 2,500 U.S.D

Contact: Leigh Higgins, Assistant to CEO; (203) 777-2386 or (203) 777-7092; fax (203) 787-6584; lhiggins@cfgnh.org

Internet: http://www.cfgnh.org/Grantmaking/TypesofGrants/tabid/199/Default.aspx

Sponsor: Community Foundation for Greater New Haven
70 Audubon Street
New Haven, CT 06510

Community Foundation for Greater New Haven Valley Neighborhood Grants 1406

The Valley Community Foundation (VCF), a supporting organization of The Community Foundation for Greater New Haven (CFGNH), promotes investment in it's community leaders. The Valley Neighborhood Grant Program is rooted in CFGNH's Valley Neighborhood Small Grants Program but has the purpose of encouraging, identifying, engaging and supporting neighborhood leadership. It provides funding ($100 to $3,000), technical assistance and training to civic groups to assist them in carrying out their civic agenda, and to support development and implementation of projects that will improve the quality of Valley life.

Requirements: The following groups are eligible to apply: Art Groups; Blockwatches; Business Associations; Faith-based Organizations; Groups that serve the elderly; Individuals in partnership with a non-profit; Neighborhood Groups & Associations; Parent Groups; Tenant Associations; Service Clubs; Youth Groups; Partnerships between groups listed above are welcomed; Non-profits that are all-volunteer or have a staff of 4 or less are eligible to apply. If your group does not have legal nonprofit status, establish a relationship with a 501(c)3 organization to serve as your fiduciary agent. Please Note: If you choose to work with a 501(c)3, they must submit a letter of support.

Restrictions: The Neighborhood Program will not fund requests for: salaries or stipends; recreational trips; gas grills or coolers.

Geographic Focus: Connecticut

Date(s) Application is Due: Feb 1

Amount of Grant: 100 - 3,000 U.S.D

Samples: Ansonia Historical Commission, $2,000—to support the costs associated with the repairing and restoring of the monument located outside of the Ansonia Public Library; Ansonia High School Human Relations Club/Doyle Senior Center, $2,000—to support the cost of two intergenerational programs/dinners involving members of the Ansonia High School Human Relations Club and members of the Doyle Senior Center in Ansonia; Assumption Church Parish Nurse Program, $1,500—to support the costs associated with running Bereavement Group, CPR class and Children's Health & Safety Programs.

Contact: Stephanie Sutherland, Associate Philanthropic Officer; (203) 777-2386 or (203) 777-7077; fax (203) 787-6584; ssutherland@cfgnh.org

Internet: http://www.cfgnh.org/Grantmaking/TypesofGrants/tabid/199/Default.aspx

Sponsor: Community Foundation for Greater New Haven
70 Audubon Street
New Haven, CT 06510

Community Foundation for Greater New Haven Women & Girls Grants 1407

The Community Foundation for Greater New Haven (CFFGNH) is a philanthropic institution that was established in 1928. The foundation's mission is to create positive and sustainable change in Connecticut's Greater New Haven region by increasing the amount of and enhancing the impact of community philanthropy. The Community Fund for Women & Girls was created in 1995 by an anonymous woman to provide ongoing support for services important to women and girls in the Greater New Haven and Lower Naugatuck Valley area. Favorable grant proposals would: create and support opportunities for the economic, educational, physical, emotional, social, artistic, and personal growth of women and girls; meet special needs of women and girls and the diverse populations of women in our region; encourage the advancement and full participation of women and

girls in the community and in philanthropy; advance the status of women and girls in the core areas of economic security, health, violence, education and political participation.
Requirements: IRS 501(c)3 nonprofit organizations are eligible to apply in the greater New Haven area, which includes: Ansonia; Bethany; Branford; Cheshire; Derby; East Haven; Guilford; Hamden; Madison; Milford; New Haven; North Branford; North Haven; Orange; Oxford; Seymour; Shelton; Wallingford; West Haven and; Woodbridge.
Geographic Focus: Connecticut
Date(s) Application is Due: Jan 15
Amount of Grant: 1,000 - 10,000 U.S.D
Samples: The New Haven Diaper Bank, $10,000—to support a project manager for the Basic Human Needs Policy collaboration between the Liman Project, Wiggin and Dana attorneys and other interested parties. The goal is to secure a sustaining source of state and federal funding for diapers, serving over 1040 children regionally, and potentially all children living in poverty in the U.S.; All Our Kin, Inc., $3,600—to support the Family Child Care Tool kit licensing project. The project will provide six more licensing tool kits to the existing program, allowing six women to become licensed home care providers for 36 children; Hill Health Corporation—$10,000—to support the Prison to Community Peer Support Program, a 44 week peer support group serving 40 newly released formerly incarcerated women, led by women who have experienced and overcome substance abuse and or mental illness and who have made the transition from prison to the community.
Contact: Denise Canning, Grants Manager; (203) 777-2386 or (203) 777-7076; fax (203) 787-6584; dcanning@cfgnh.org
Internet: http://www.cfgnh.org/Grantmaking/TypesofGrants/tabid/199/Default.aspx
Sponsor: Community Foundation for Greater New Haven
70 Audubon Street
New Haven, CT 06510

Community Foundation for Monterey County Grants **1408**
The community foundation supports nonprofits in Monterey, CA, in the areas of social services, education, environment, arts, health, historic preservation, and general charitable giving. Collaboratives are highly encouraged. The foundation awards grants through its General Endowment. Technical Assistance grants and Neighborhood Grants also are awarded. Types of support include seed money, emergency funds, general operating support, building funds, equipment acquisition, land acquisition, matching funds, projects, consulting services, and technical assistance. Information on how to apply for each type of grant can be accessed online.
Requirements: 501(c)3 tax-exempt organizations in Monterey, California, may submit applications.
Restrictions: The foundation does not support individuals, religious activities, scholarships, fellowships, travel, research, salaries and other operating expenses of schools and public agencies, annual campaigns, or special events; create or add to endowment funds; or pay off debt.
Geographic Focus: California
Date(s) Application is Due: Jan 3; May 2; Aug 1
Amount of Grant: 5,000 - 40,000 U.S.D
Contact: Jackie Wendland, Grants Administrator; (831) 375-9712, ext. 11; fax (831) 375-4731; jackie@cfmco.org
Internet: http://www.cfmco.org/grantsOverview.php
Sponsor: Community Foundation for Monterey County
2354 Garden Road
Monterey, CA 93940

Community Foundation for Muskegon County Grants **1409**
The community foundation was established to serve the needs of the people of Muskegon County and nearby western Michigan. The foundation's strategic plan for grant making focuses on the prevention of problems rather than the cure; encourages programs that are collaborative, comprehensive, and have the potential to be continuous; encourages leveraging and matching grant opportunities from multiple funders; and supports seed money opportunities for innovative projects. Major grant interest areas include the arts, community development and urban revitalization, education, environment, health/human services, and needs of young children (ages 0-3). Types of support include seed money grants, special projects, matching funds, equipment, scholarship funds, loans, research, publications, conferences and seminars, endowment funds, consulting services, continuing support, emergency funds, internships, professorships, and renovation projects. Although the foundation generally makes one-year grant commitments, it will consider making longer term commitments for new efforts that show strong promise for positive impact. Applications are accepted on specific dates each year; it is recommended that the applicant contact the foundation to discuss their interests and to obtain the next grant application deadline.
Requirements: IRS 501(c)3 organizations and institutions in Muskegon County and nearby western Michigan are eligible.
Restrictions: Support will not be provided for routine operating expenses; capital equipment, computer hardware and software, and motor vehicles; conferences, publications, videos, films, television, or radio programs; endowment campaigns; special fund-raising events; religious programs that serve specific denominations; existing obligations or debts; individual schools or districts; or individuals.
Geographic Focus: Michigan
Amount of Grant: 250 - 50,000 U.S.D
Contact: Marcy Joy, Program Officer; (231) 722-4538 or (231) 332-4124; fax (231) 722-4616; mjoy@cffmc.org
Internet: http://www.cffmc.org/grantapply.php
Sponsor: Community Foundation for Muskegon County
425 West Western Avenue, Suite 200
Muskegon, MI 49440

Community Foundation for San Benito County Grants **1410**
The Foundation serves donors, advances philanthropy and achieves impact by supporting the work of nonprofit organizations. The Foundation continuously monitors the San Benito County community to understand the nature of need, the forces of change, available resources and the capacity for growth. The Foundation provides funding to support impactful programs within and across the following areas of interest: arts and culture; education and youth; health and social services; agriculture and environment; community enhancement; economic development. Application deadlines are announced on the Foundation's website.
Requirements: Selection criteria and priorities are as follows: projects or services which respond to a demonstrated need within San Benito County; effective use and greatest impact of grant funds; initiatives to solve significant community issues; collaboration and coordination of service delivery; demonstrate a level of cooperation with other organizations, including leveraging financial and in-kind support from other groups and individuals; strengthening organizational capacity; addressing diverse community interests; organizational or program sustainability; organizations with demonstrated financial need; and unduplicated services. Awards are generally up to $40,000. Major projects or initiatives over $40,000 must be preliminarily discussed with Foundation staff.
Restrictions: In general the following are not funded: organizations that discriminate on the basis of age, disability, ethnic origin, gender, race or religion; grants to individuals; fraternal or service organizations, unless in support of specific programs open to or benefiting the entire community; salaries and other operating expenses of schools and public agencies; fundraising events such as annual campaigns, walk-a-thons, tournaments, fashion shows, dinners and auctions; organizations and programs designed to support political activities; organizations located outside San Benito County unless for a specific program benefiting residents within San Benito County; pay off existing obligations or enable funding of reserve accounts; endowment funds; and scholarships, fellowships, travel grants and academic, technical or specialized research.
Geographic Focus: California
Samples: Chamberlain's Mental Health Services, Gilroy, California, $100,000 - behavioral management programs for functionally impaired preschool children (2011); SB Collaborative for Homeless Services, Santa Barbara, California, $5,000 - homeless information data program required to enable future HUD funding (2011); and Seniors Council (Foster Grandparents), San Benito, California, $6,159 - support of seniors mentoring and tutoring program for at-risk youth (2011).
Contact: Grants Manager; (831) 630-1924; fax (831) 630-1934; info@cffsbc.org
Internet: http://www.cffsbc.org/grantoverview.php
Sponsor: Community Foundation for San Benito County
829 San Benio Street, Suite 200
Hollister, CA 95023

Community Foundation for Southeast Michigan Grants **1411**
The Community Foundation for Southeast Michigan is always looking for effective program and project ideas that can improve life in southeast Michigan, specifically in the seven-county service area of Wayne, Oakland, Macomb, Monroe, Washtenaw, St. Clair and Livingston. Grants are provided considering various local needs and identifying those projects which promise the strongest long-term impact on the region. In general support is provided for projects and programs in the areas of arts and culture, civic affairs, health, human services, neighborhood and regional economic development, work force development, environment and land use. Interested applicants are encouraged to review the grantmaking guidelines at the website and call the Program Officer to discuss proposals. Grants range from $5,000 to $1 million, with the majority ranging from $35,000 to $100,000. Application may be made at any time. Proposals submitted prior to February 15, May 15, August 15 and November 15 will typically receive a response within three or four months.
Requirements: Eligible organizations: are a 501(c)3 tax-exempt organization, a government entity, a school district or a university; have headquarters (or a local partner) located in the seven-county service area; serve residents in the seven-county service area; have Board and/or the CEO approval to submit a proposal; and have a current certified financial audit.
Restrictions: All final reports due to the Foundation for previous grants must be submitted before applying. Requests for sectarian religious programs, individuals, and funding for deficits or other previously incurred obligations are not eligible.
Geographic Focus: Michigan
Contact: Katie Brisson, Senior Program Officer; (888) 933-6369 or (313) 961-6675; fax (313) 961-2886; kbrisson@cfsem.org
Internet: http://cfsem.org/apply-grant
Sponsor: Community Foundation for Southeast Michigan
333 W Fort Street, Suite 2010
Detroit, MI 48226-3134

Community Foundation for Southern Arizona Grants **1412**
The foundation awards grants in southern Arizona for a broad array of charitable purposes in the areas of arts and humanities, education, the environment, health, social services. Types of support include challenge/matching grants, conferences and seminars, scholarships, equipment acquisition, fellowships, general support, multiyear support, project development, publications, research, seed funding, and technical assistance. Priority is given to proposals that promote collaboration and build on the strengths of individuals and communities. Deadlines are determined on a yearly basis.
Requirements: Nonprofit and grassroots organizations in southern Arizona communities are eligible. Southern Arizona communities include all of Cochise, Santa Cruz, and Pima Counties; and the areas of Yuma, Mariposa, Pinal, Graham, and Greenlee Counties that lie south of the Gila River.

Restrictions: Funds are generally not available for ongoing operating or capital campaigns, debt retirement, endowments, individuals, individual schools, sectarian activities, or underwriting of fund-raising events.
Geographic Focus: Arizona
Amount of Grant: 1,000 - 10,000 U.S.D
Samples: Arizona's Children Association (AZ)—for KARE legal services, $7200; Tucson Clean and Beautiful (Tucson, AZ)—for Adopt-A-Park Program, $2000; Tucson Children's Museum, Inc (Tucson, AZ)—to cover admission costs for low-income children, $2250; Tucson Cinema Foundation Inc (Tucson, AZ)—funding for a ten-month program for high school students, $5600.
Contact: Barbara Brown, Executive Vice President; (520) 770-0800, ext. 107; fax (520) 770-1500; bbrown@CFSoAZ.org
Internet: http://www.cfsoaz.org/page17480.cfm
Sponsor: Community Foundation for Southern Arizona
2250 E Broadway Boulevard
Tucson, AZ 85719

Community Foundation for the National Capital Region Competitive Grants 1413
The Community Fund has a particular interest in supporting groups in the metropolitan Washington, DC, area working in the following issue areas: violence prevention; education; community building; cross-cultural or cultural partnership building; family literacy; and healthcare and dental services for underprivileged children, youth, and families. Types of support include general operating support, program development, technical assistance, and program evaluation. Approximately 20 to 30 grants per year will be awarded in two competitive cycles. Letters of inquiry (three-page maximum) must meet the listed application deadlines. Proposals will be invited.
Requirements: 501(c)3 nonprofits in the metropolitan Washington region, including the District of Columbia, northern Virginia, and suburban Maryland may submit letters of inquiry. Applicants must represent a neighborhood, citywide, or regional coalition effort, with one nonprofit organization serving as project sponsor.
Geographic Focus: District of Columbia, Maryland, Virginia
Samples: Nine nonprofit organizations (Washington, DC)—for awards through the Bridge Builders Fund, which supports projects that promote communication and understanding between the lesbian, gay, and bisexual populations and other city residents, $60,000 divided; George Washington U (Washington, DC)—for the Ctr for Excellence in Municipal Management, $150,000; Henry Ford Health System (Troy, MI)—to develop a set of standards for measuring the quality of health care provided to minority patients in managed-care plans, $179,464.
Contact: Alicia Reid; (202) 263-4774; fax (202) 955-8084; areid@cfncr.org
Internet: http://www.cfncr.org/page13847.cfm
Sponsor: Community Foundation for the National Capital Region
1201 15th Street NW, Suite 420
Washington, DC 20005

Community Foundation of Bartholomew County Heritage Fund Grants 1414
The goal of the Heritage Fund's grant program is to achieve the maximum impact with the available resources. The Fund will consider grant applications that: are change-orientated and problem-solving in nature; strive to anticipate the changing needs of the community and to be flexible in responding to them; address the needs of a significant number of community residents and provide the greatest benefit per dollar granted; encourage support from the community by using matching, challenge and other grant techniques; have a broad funding base, with additional support being sought from the government, foundations, associations and other funders; enable grant recipients to achieve certain objectives such as capacity building, and/or increasing efficiency, effectiveness and fund-raising capabilities; request technical assistance or specialized help with projects that respond to community needs; or positively impact the Heritage Fund's Areas of Initiative.
Requirements: Grants are made to not-for-profit organizations whose programs benefit the residents of Bartholomew County, Indiana.
Restrictions: The Heritage Fund provides seed money or support for the pilot phase of a program. The Heritage Fund does not generally provide operating support.
Geographic Focus: Indiana
Date(s) Application is Due: Mar 1; Jun 1; Sep 1; Dec 1
Amount of Grant: Up to 3,000 U.S.D
Contact: Lynda J. Morgan, Program Officer; (812) 376-7772; fax (812) 376-0051; lmorgan@heritagefundbc.org
Internet: http://www.heritagefundbc.com/grants/process_guidelines_deadlines.php
Sponsor: Community Foundation of Bartholomew County
538 Franklin Street, P.O. Box 1547
Columbus, IN 47202-1547

Community Foundation of Bartholomew County James A. Henderson Award for Fund-Raising 1415
This award recognizes the invaluable role volunteer fund-raisers play in advancing the quality of life within Bartholomew County, Indiana. Not-for-profit organizations in the county are invited to nominate volunteers who have performed outstanding fund-raising for their organizations. A committee will review the nominations and select the person to be honored. The committee will consider such criteria as innovation, sustainability, creativity, effectiveness, effort, ability to engage others, outreach to new donors, etc. Generally the amount raised will not be a major factor in the scoring. Efforts will be made to recognize unsung heroes in fund-raising. The successful nominee will be recognized at the Heritage Fund's Annual Report to the Community. He/she will receive a small gift and the nominating organization will receive a $2,500 grant in honor of the winner.
Geographic Focus: Indiana
Amount of Grant: 2,500 U.S.D
Contact: Lynda J. Morgan, Program Officer; (812) 376-7772; fax (812) 376-0051; lmorgan@heritagefundbc.org
Internet: http://www.heritagefundbc.com/grants/award_james_henderson.php
Sponsor: Community Foundation of Bartholomew County
538 Franklin Street, P.O. Box 1547
Columbus, IN 47202-1547

Community Foundation of Bloomington and Monroe County Grants 1416
Grants will be made for a wide variety of programs and purposes in support of the arts and cultural activities, social and health concerns, educational activities, beautification projects, and for other community development needs that will benefit the citizens of Bloomington and Monroe County, Indiana. Start-up funds for special projects reflecting incremental growth possibilities will be given particular attention. Additional types of support include building construction and renovation, equipment acquisition, program development, seed grants, and scholarship funds. Projects that provide leverage for generating other funds and community resources or are seed money for new programs will also receive special attention. It is expected that the majority of grants will be made for one time only; however, multi-year grants may be considered in unusual circumstances. Applications are reviewed in March, June, September, and December.
Requirements: Grants are made only to nonprofit organizations and programs in Bloomington or Monroe County, Indiana.
Restrictions: Grants will not be made for general operating support purposes; endowment campaigns or for previously incurred debts; or for political purposes.
Geographic Focus: Indiana
Samples: Camerata (IN)—to underwrite musical performances for children, $2000; Shelter Inc (IN)—to replace locks on temporary shelter homes, $1500; Bloomington Fire Department (IN)—to underwrite a fire safety program for children, $200; Bloomington Parks and Recreation (IN)—for historic park development, $9000.
Contact: Renee M. Chambers, Program Director; (812) 333-9016; fax (812) 333-1153; renee@cfbmc.org
Internet: http://www.communityfoundation.ws
Sponsor: Community Foundation of Bloomington and Monroe County
101 W Kirkwood Avenue, Suite 321
Bloomington, IN 47404

Community Foundation of Boone County Grants 1417
The Community Foundation of Boone County Grants provides funding in the following areas: arts/culture; community development; education; elderly; health; human services; youth; environment; and recreation. Applications for grants of $10,000 or less should be submitted on the Short Form located on the website, while applications for more than $10,000 should follow instructions in the Content and Format section of the website. Awards will be made to nonprofit organizations exempt from federal taxation under section 501(c)3 of the Internal Revenue Code. Grants may be allowed to individuals and to non-501(c)3 organizations if there is documented charitable activity benefiting or serving the residents of Boone County, or if the Community Foundation is acting as fiscal agent for the project or program.
Requirements: Grant proposals for over $10,000 must include the following format: abstract or executive summary; description of organization; statement of problem or need; objectives; proposed solution; materials/equipment; staff; facilities; evaluation; budget and its explanation; and appendices. Organizations should review the Foundation website for specific proposal submission instructions.
Restrictions: Non-allowable expenses include support of pre-award costs (i.e., project costs generated during the preparation of a proposal for the same project); existing general fund operating expenses; regular salaries of pre-award permanent staff (unless overload compensation is justified during the life of the funded project); international travel; first-class air fare; luxury accommodations; hospitality for purposes other than those directly related to meeting program objectives as defined in the proposal; alcohol; and indirect or regular existing administrative costs (e.g., telephone, utilities, general maintenance, etc.) of the applicant organization.
Geographic Focus: Indiana
Date(s) Application is Due: Jan 24; Mar 28; May 30; Aug 1; Oct 3; Nov 21
Samples: Lebanon Area Boys and Girls Club for the SCORE (Scholarship, Citizenship, Organization, Responsibility, Effort) Mentoring Program for high risk youth, $9,995 (2011); Youth As Resources Program of United Way of Central Indiana, $5,000 in support of youth-led volunteer community service projects, where young people are exposed to philanthropy and the needs of their community (2011); Otterbein United Methodist Church, $10,000 for the What's For Lunch Program, providing more than 400 lunches for needy families (2011).
Contact: Barbara J. Schroeder, Program Director; (317) 873-0210 or (765) 482-0024; fax (317) 873-0219; barb@communityfoundationbc.org
Internet: http://www.communityfoundationbc.org/grants.html
Sponsor: Community Foundation of Boone County
60 East Cedar Street
Zionsville, IN 46077

Community Foundation of Broward Grants 1418
Through its competitive grantmaking process, the foundation supports programs that strengthen families within low socioeconomic communities in Florida's Broward County. Grantmaking priorities include the arts, foster care, out-of-school time, and technology. Proposals also are accepted in specific areas of interest, including animal welfare—protect and educate; cancer and arthritis research—finding a cure; patient care—relief and assistance to individuals with cancer or other terminal diseases; literacy—opening

worlds through language, an AIDS/HIV—through issued RFP. Proposals are accepted anytime throughout the year. Application and guidelines are available online.
Requirements: 501(c)3 organizations, public charities as defined by the IRS code, and governmental agencies located in Broward County or directly benefiting the residents of Broward County are eligible.
Restrictions: Grants are not awarded in the following areas: annual fundraising, capital campaign/improvements, deficit financing, endowment efforts, grants to individuals, religious purposes, or routine operating needs.
Geographic Focus: Florida
Amount of Grant: 10,000 U.S.D
Samples: Covenant House Florida (Fort Lauderdale, FL)—for child-rearing education and childcare services for at-risk pregnant youths and young parents at its crisis shelter, $10,000; Eckerd Youth Alternatives (Clearwater, FL)—for after-school counseling and literacy services for youths recently released from residential delinquency-treatment programs, $10,000; Pines Youth Ctr (Pembroke Pines, FL)—for an extracurricular program for middle-school students that provides tutoring and computer and reading instruction, $12,000; First Call for Help of Broward (FL)—for the technological support of a crisis and referral hotline, $15,710.
Contact: Sheri Brown, Vice President of Strategic Community Initiatives; (954) 761-9503, ext. 103; fax (954) 761-7102; jbull@cfbroward.org
Internet: http://www.cfbroward.org/new/grant.html
Sponsor: Community Foundation of Broward
1401 E Broward Boulevard, Suite 100
Fort Lauderdale, FL 33301

Community Foundation of Central Illinois Grants 1419
The foundation funds programs in the fields of education, arts, human services, community service, or community development, and that advance one or more of the following objectives: address and help resolve important existing or emerging community issues; support new and creative projects and organizations offering the greatest opportunity for positive and significant change; promote cooperation and collaboration among organizations; identify, enhance, and develop leadership in the community through creative and innovative activities that empower individuals; and improve the quality or scope of charitable works in our community. Application guidelines are available online.
Requirements: Nonprofits within a 50-mile radius of Peoria, IL, are eligible.
Restrictions: The foundation will not fund annual campaigns, individuals, or endowments, or make grants for sectarian religious purposes.
Geographic Focus: Illinois
Date(s) Application is Due: Apr 15; Sep 15
Contact: Kristan Creek, Program Officer; (309) 674-8730; fax (309) 674-8754; Kristan@communityfoundationci.org
Internet: http://www.communityfoundationci.org/grant_guidelines.asp
Sponsor: Community Foundation of Central Illinois
331 Fulton Street, Suite 310
Peoria, IL 61602

Community Foundation of Crawford County Grants and Scholarships 1420
The CFCC's community grant program reflects the needs of the community. The CFCC maintains an effective relationship as grantor, and community organizations (as grantees), which is based on mutual respect and partnership. Since 1998 the CFCC has awarded over $5.5million in programming/grants and scholarships. Fields of interest are: animal Protection; arts; community development; education; environment; health and safety; human services; historical preservation; emergency recovery; infants & children; literacy; parks & recreation; public radio; public service; religion; technology; youth; women & girls. Grant, as well as scholarship requests are made by first contacting the President & CEO of the foundation.
Requirements: Grantees are required to provide ongoing status reports. CFCC Board members are assigned grant projects to mentor. Additionally, grantees are required to report on the actual outcomes of the funding versus the original anticipated outcomes.
Restrictions: Funding opportunities are generally restricted to the Crawford County.
Geographic Focus: Indiana
Contact: Bonita Embry Coots; (812) 633-2077; fax (812) 633-2429; BECoots@cf-cc.org
Internet: http://www.cf-cc.org/page31105.cfm
Sponsor: Community Foundation of Crawford County
9430 East State Road 64, P.O. Box 205
Milltown, IN 47145

Community Foundation of Dutchess County Grants 1421
The fund awards grants to eligible New York nonprofit organizations to improve the quality of life in Dutchess County. Areas of interest include health, education, and social services. Types of support include building construction/renovation, equipment acquisition, program development, conferences and seminars, and seed money grants. Application forms are available online.
Requirements: Dutchess County, NY, 501(c)3 tax-exempt organizations are eligible.
Geographic Focus: New York
Date(s) Application is Due: Mar 31; Sep 30
Amount of Grant: 100 - 100,000 U.S.D
Contact: Nevill Smythe; (845) 452-3077; fax (845) 452-3083; cfdc5@hvc.rr.com
Internet: http://www.communityfoundationdc.org
Sponsor: Community Foundation of Dutchess County
80 Washington Street, Suite 201
Poughkeepsie, NY 12601

Community Foundation of East Central Illinois Grants 1422
The community foundation awards grants from its unrestricted funds to Champaign County, IL, nonprofit organizations to use for charitable purposes. Representative categories include arts and humanities, environmental concerns, education, health, human services, research, urban affairs, church, and youth programs. Types of support include continuing support, annual campaigns, building/renovation, equipment, publication, consulting services, and scholarships.
Requirements: Champaign County, IL, nonprofit organizations are eligible.
Geographic Focus: Illinois
Date(s) Application is Due: Aug 31
Contact: Joan Dixon, President & CEO; (217) 359-0125; fax (217) 352-6494; joandixon@cfeci.org or info@cfeci.org
Internet: http://www.cfeci.org/non-profit/grant-application
Sponsor: Community Foundation of East Central Illinois
404 West Church Street
Champaign, IL 61820-3411

Community Foundation of Grant County Grants 1423
The Community Foundation of Grant County Grants address the needs of Grant County in the fields of community development, education, and health and human services. Nonprofit organizations, coalitions, community associations, and other civic groups may apply if they provide services within the county. The project must be located in or directly serve the people in Grant County, and meet all other criteria in the guidelines and application. All proposals must be received by the Foundation on the last Friday in April, July, October, and January, with committee reviews in May, August, November, and February.
Requirements: The Board will only accept written proposals for consideration after an applicant has first consulted with the Foundation's staff to find if the project is suitable for funding. The organization should keep the following guidelines in mind when preparing their written proposal: the project's purpose; whether the Foundation will be the sole funder; how many will be served or affected by the project, if there is a broad base of support for it, or if services are duplicated by another source; if an important need has been shown; whether the objectives are realistic and measurable; is there a viable plan for future support; who the key people are and if they are available for the long-term; and if all financial information is included and makes sense.
Restrictions: To be eligible for a grant, the project must be located in or directly serve the people and natural resources in Grant County, Indiana. Funding is not available for: profit making enterprises; political activity; operating budgets, except for limited experimental or demonstration periods; salaries; sectarian or religious purposes; endowment purposes unless for special promotions and/or matching challenges; multi-year funding requests; or capital improvements to church owned facilities or properties.
Geographic Focus: Indiana
Contact: Sherrie Stahl, Office and Program Director; (765) 662-0065; fax (765) 662-1438; sstahl@comfdn.org
Internet: http://www.comfdn.org/grants.htm
Sponsor: Community Foundation of Grant County
505 West Third Street
Marion, IN 46952

Community Foundation of Greater Birmingham Grants 1424
The community foundation serving the greater Birmingham, AL, area awards grants in arts and culture, education, environment, health, and welfare. Types of support include capital campaigns, operating support, building/renovation, equipment, program development, publication, seed money, curriculum development, and matching funds. Grant requests are considered twice a year at distribution committee meetings. The foundation requires that grant requests be submitted in writing. There is no application form. Applicants are asked to submit a cover letter and a complete statement of the purpose of the grant, including the need for the project and population and number of people to be served, project budget, other funding sources, length of time for which foundation aid is needed, a method to evaluate the project's success, description of applicant organization including annual budget, copy of IRS determination letter, and names of the board of directors or trustees.
Requirements: IRS 501(c)3 organizations that provide services in the greater Birmingham metropolitan area (Jefferson, Shelby, St. Clair, Blount and Walker counties) are eligible.
Restrictions: Grants are not made to or for individuals, operating expenses of organizations, religious organizations for religious purposes, national fund-raising drives, conference or seminar expenses, tickets for benefits, political organizations or candidates for public office, organizations with IRS 501(h) status, budget deficits, replacement of government funding cuts, or scholarships or endowment funds.
Geographic Focus: Alabama
Date(s) Application is Due: Mar 15; Sep 15
Samples: Exchange Club Family Skills Center, $2,200—program support; Catholic Family Services, $4,600— to expand outpatient psychiatric care for uninsured women and children in metro Birmingham; Birmingham Civil Rights Institute, $20,000—toward operating costs as the institute continues to serve the general public, tourists and students, particularly in kindergarten through second grade.
Contact: James McCrary, Senior Program Officer; (205) 327-3812 or (205) 327-3800; fax (205) 328-6576; jmccrary@foundationbirmingham.org or guidelines@foundationbirmingham.org
Internet: http://www.foundationbirmingham.org/page30508.cfm
Sponsor: Community Foundation of Greater Birmingham
2100 First Avenue N, Suite 700
Birmingham, AL 35203-4223

Community Foundation of Greater Flint Grants 1425

The community foundation is interested in funding charitable organizations that are able to demonstrate they have planned their projects with respect to the community's needs in the areas of arts and humanities, community services, education, environment and conservation, ethics, health, human and social services, and philanthropy. The foundation's current objectives for the allocation of discretionary funds give top priority to programs that: advance the health and well-being of children; and/or improve the capacity of public education. Types of support include general operating support, program development, seed money, scholarship funds, and technical assistance. Applications are available on the Web site.
Requirements: IRS 501(c)3 organizations with programs of direct relevance to the residents of Genesee County, Michigan are eligible. Types of support include general operating support.
Restrictions: Grants will not be made to individuals. In general, requests for sectarian religious purposes, budget deficits, routine operating expenses of existing organizations, litigation, endowments, and other capital fund drives or projects are not funded.
Geographic Focus: Michigan
Samples: Greater Flint Health Coalition, $40,000—towards the Flint Healthcare Employment Opportunities (FHEO) program, which helps unemployed or low-income individuals in Genesee County to obtain employment opportunities or to advance in the healthcare industry; Genesee Chamber Foundation,$30,000— towards the Summer Youth Initiative, a summer employment and leadership development program for Genesee County youth ages 14-18; United Way of Genesee County, $66,000- in support of the BEST Project (Building Excellence, Sustainability and Trust) BEST provides technical assistance and resources designed to build the capacity of area nonprofit organizations.
Contact: Lynn Larkin, Senior Program Officer; (810) 767-8270; fax (810) 767-0496; llarkin@cfgf.org
Internet: http://www.cfgf.org/page32610.cfm
Sponsor: Community Foundation of Greater Flint
500 South Saginaw Street
Flint, MI 48502-1206

Community Foundation of Greater Fort Wayne - Community Endowment and Clarke Endowment Grants 1426

The Foundation encourages projects or programs that are developed in consultation with other agencies and planning groups that increase coordination and cooperation among agencies and reduce unnecessary duplication of services. Preference is given to projects or programs that: address priority community concerns; encourage more effective use of community resources; test or demonstrate new approaches and techniques in the solution of community problems; are intended to strengthen the management capabilities of agencies; promote volunteer participation and citizen involvement in community affairs. Contact program staff for current guidelines and deadlines. The deadlines listed are for concept letters, with final invited proposals due by dates provided by the Foundation.
Requirements: Nonprofit organizations in Allen County, Indiana, are eligible to apply. Applicants should mail hardcopies of the application package including the following: concept letter fact sheet; the original concept letter; a detailed program budget or agency budget; current financial statements; a copy of the organization's 501(c)3 IRS determination letter; and a list of the board of directors and their principal affiliations.
Restrictions: Grants do not support: annual fund drives; operating deficits or after-the-fact support; endowment funds, except for endowment-building matching grants for funds held at the Community Foundation; direct or grassroots lobbying; religious purposes; hospitals, medical research, or academic research; public, private, or parochial educational institutions except in special situations when support is essential to projects/programs that meet critical community needs; governmental agencies, including public school systems, except in special situations when support is essential to projects/programs that meet critical community needs; limited, special interest organizations except when such support significantly benefits the disadvantaged; and funding for sponsorships, special events, commercial advertising, films or videos, television programs, conferences, group uniforms, or group trips.
Geographic Focus: Indiana
Date(s) Application is Due: Jan 9; Apr 9; Jul 9
Samples: Salvation Army of Fort Wayne (Fort Wayne, Indiana)—for remodeling of building, $50,000; United Way of Allen County (Fort Wayne, Indiana)—for diversity and special urban initiatives, $100,000; Local Education Fund (Fort Wayne, IN)—for training materials, $38,314.
Contact: Annette Smith, Grants Manager; (260) 426-4083; fax (260) 424-0114; asmith@cfgfw.org
Internet: http://www.cfgfw.org/
Sponsor: Community Foundation of Greater Fort Wayne
555 East Wayne Street
Fort Wayne, IN 46802

Community Foundation of Greater Fort Wayne - Lilly Endowment Scholarships 1427

The Community Foundation of Greater Fort Wayne administers the Lilly Endowment Scholarships for the Allen County area. High school seniors in the top 40% of their class are eligible to apply. The awards cover full tuition for a four-year degree, and includes an $800 per year book stipend. Any college major is eligible.
Requirements: Students should contact their Allen County high school guidance office for application and deadline information.
Restrictions: Grant recipients must enroll in an Indiana college or university.
Geographic Focus: Indiana
Contact: Lydia Von Uderitz, Scholarship Manager; (260) 426-4083; fax (260) 424-0114; lvon@cfgfw.org
Internet: http://www.cfgfw.org/scholarships/scholar-college.html
Sponsor: Community Foundation of Greater Fort Wayne
555 East Wayne Street
Fort Wayne, IN 46802

Community Foundation of Greater Fort Wayne Scholarships 1428

The Community Foundation of Greater Fort Wayne offers a wide variety of scholarship funding for Allen County students attending public and private schools. Scholarships are available for every school age ranging from elementary school through college, and are categorized by school, major, financial need, GPA, class rank, businesses, and local organizations. Application deadlines vary, but are generally from January to May.
Requirements: Students can contact their guidance office or the Foundation for an application if it is not posted online. Students should consult the individual listings for requirements, information needed, and deadlines.
Restrictions: Most of the scholarships require that students are residents of Allen County.
Geographic Focus: Indiana
Contact: Lydia Von Uderitz, Scholarship Manager; (260) 426-4083; fax (260) 424-0114; lvon@cfgfw.org
Internet: http://www.cfgfw.org/scholarships/scholar_apply.html
Sponsor: Community Foundation of Greater Fort Wayne
555 East Wayne Street
Fort Wayne, IN 46802

Community Foundation of Greater Greensboro Grants 1429

The foundation supports community projects that strengthen the community, embrace the diversity of the community, demonstrate community support and in-kind investment, outline well-established proposed plans, and focus on collaborative efforts. Grants are awarded in the areas of community building, cultural and civic programs, arts, community development, health, environment, and social services. Types of support include general support, program development, seed grants, and emergency community funding. Annual deadline dates may vary; contact program staff for exact dates.
Requirements: Nonprofit organizations in Greensboro, NC, are eligible.
Restrictions: The Foundation does not award grants from the Community Grants Program for individuals, debt retirement, political organizations or campaigns, for-profit entities, capital campaigns, programming that promotes religious instruction or doctrine, or endowments.
Geographic Focus: North Carolina
Date(s) Application is Due: Feb 15; Jun 15; Oct 15
Amount of Grant: 10,000 U.S.D
Samples: Eastern Music Festival (Greensboro, NC)—for the 2003 Piedmont Jazz Festival, $10,000; Guilford County Council of PTAs (Greensboro, NC)—to promote greater parental involvement in schools, $40,000; Piedmont Authority of Regional Transportation (Greensboro, NC)—to coordinate a regional medical-transportation program, $10,000 over two years; North Carolina Ctr for Nonprofits (Raleigh, NC)—for scholarships to enable local nonprofit groups to attend a statewide conference, $10,000.
Contact: Melissa Johnson, Program Officer; (336) 379-9100, ext. 124; fax (336) 378-0725; mjohnson@cfgg.org
Internet: http://www.cfgg.org/guidelines-new.htm
Sponsor: Community Foundation of Greater Greensboro
P.O. Box 20444
Greensboro, NC 27420

Community Foundation of Greater Lafayette - Robert and Dorothy Hughes Scholarships 1430

The Community Foundation of Greater Lafayette - Hughes Scholarships are available to help students earn a baccalaureate degree in nursing. Applicants must live or work in Tippecanoe County; plan to attend and obtain a baccalaureate degree from an accredited nursing school; and have financial need.
Requirements: The application is available online. Students must include a completed financial form; a signed certification; high school and/or college transcripts; a two page essay; and three letters of recommendation, then submit the application packet to the Foundation.
Restrictions: The scholarships are renewable, although students must re-apply to be considered for a renewal.
Geographic Focus: Indiana
Date(s) Application is Due: Jan 13
Amount of Grant: 1,000 U.S.D
Contact: Cheryl Ubelhor, Program Manager; (765) 742-9078; fax (765) 742-2428; cheryl@cfglaf.org
Internet: http://www.cfglaf.org/scholarship/index.htm#Hughes
Sponsor: Community Foundation of Greater Lafayette
1114 East State Street
Lafayette, IN 47905-1219

Community Foundation Of Greater Lafayette Grants 1431

The Community Foundation of Greater Lafayette Grants help meet the ever-changing needs of the community. Funding priorities include education, children/youth, health, diversity, physical environment, and arts and culture. Charitable organizations that serve Tippecanoe and the surrounding counties are eligible to apply. Most grants are awarded to nonprofit organizations that are located in and serve Tippecanoe County. Priority is given to projects that reach as many people as possible; improve the ability of organizations to serve the community over the long-term; serve the Greater Lafayette area; and are run by non-profit organizations. Grant seekers are encouraged to contact the

Foundation office to be certain their proposal is appropriate for consideration. Proposal deadlines are listed according to dollar amount request, so grant seekers should carefully review the website to judge when to submit their proposal.
Requirements: Organizations should submit the following information along with the online application: contact and background information for the organization; a concise narrative about the project; amount requested from the foundation; and a project budget. Also included is a board of directors list; project estimates/bids; and financial statements. Applicants may be asked for further information or a site visit to clarify the request.
Restrictions: The Foundation does not fund: programs that are sectarian or religious in nature; political organizations or candidates; endowments; ongoing operating expenses; government agencies or public institutions; programs that taxpayers would normally support; individuals; special events (i.e. parades, festivals, sporting activities, fundraisers); programs already completed; multi-year grants; debt or deficit reduction; and projects funded in a previous year (unless invited to resubmit).
Geographic Focus: Indiana
Date(s) Application is Due: Apr 1; Sep 1; Dec 1
Contact: Cheryl Ubelhor, Program Director; (765) 742-9078; fax (765) 742-2428; info@cfglaf.org or cheryl@cfglaf.org
Internet: http://www.glcfonline.org/grantseekers/index.htm
Sponsor: Community Foundation of Greater Lafayette
1114 East State Street
Lafayette, IN 47905-1219

Community Foundation of Greater New Britain Grants 1432
Grants from the Community Foundation of Greater New Britain support organizations and programs benefiting the residents of Berlin, New Britain, Plainville and Southington, Connecticut. The foundation supports organizations involved with arts, culture and heritage; community and economic development; early childhood development; education; health and human services.
Requirements: Grants are available to non-profit organizations that have tax-exempt status under Section 501(c)3 of the IRS Code or that are a qualified entity eligible to receive grants from community foundations under the IRS Code. The foundation will also consider funding a grassroots group if it consists of at least five people, has a governing body and a fiscal agent with the appropriate tax status. All grant applications must begin with a one or two page Letter of Intent summarizing the program or project for which funding is being sought. Submitted Letters of Intent will be reviewed by staff. Applicants selected by staff to submit complete, formal applications will be contacted directly to begin the proposal.
Restrictions: The Foundation does not make grants for: sectarian or religious activities; previously incurred expenses; annual or endowment campaigns; sponsorships or fundraisers; political activities; direct support of individuals; camperships or scholarships for academic and/or enrichment programs. In addition, the Foundation generally does not make grants for performances or one-time events, conferences or advertising.
Geographic Focus: Connecticut
Date(s) Application is Due: Feb 15; Jun 15; Oct 15
Amount of Grant: 5,000 - 25,000 U.S.D
Samples: New Britain Museum of American Art, New Britain, Connecticut, $60,000 - fund for support of the Museum's Education and Outreach program; Municipal Economic Development Agency, New Britain, Connecticut, $15,000 - to develop an economic development website for the City of New Britain featuring economic development tools and information that will promote business in New Britain; New Britain Public Library/Jefferson Family Resource Center, New Britain, Connecticut, $12,000 - for the Raise a Reader program, aimed at improving literacy skills among hard-to-reach New Britain families with children under age five.
Contact: Joeline Wruck; (860) 229-6018, ext. 307; jwruck@cfgnb.org
Internet: http://www.cfgnb.org/GrantsScholarships/tabid/68/Default.aspx
Sponsor: Community Foundation of Greater New Britain
74A Vine Street
New Britain, CT 06052

Community Foundation of Greater Tampa Grants 1433
The Community Foundation of Tampa Bay is a nonprofit, tax-exempt organization which administers funds established by individuals, corporations, private foundations, and nonprofit organizations to support the charitable needs of the Tampa Bay area. Program interests of the Foundation include: arts and culture, community development, education, environment, animals, health and, human services.
Requirements: Florida 501(c)3 organizations based in and serving communities in Hillsborough, Pasco, Pinellas and/or Hernando counties are eligible.
Restrictions: The Foundation is generally not interested in requests for: funding of ongoing operating costs; grants for capital campaigns or expenditures; tickets for any fundraising event, conference, or advertising space in programs or other publications; legislative lobbying or political campaigns; medical research; religious or sectarian purposes; loans or assistance to individuals; a multiple year funding commitment.
Geographic Focus: Florida
Date(s) Application is Due: Mar 1; Sep 1
Amount of Grant: Up to 7,500 U.S.D
Samples: Bay Springs—for the equipment, uniforms, shoes, and transportation of adolescent boys in foster care to participate in city and community sports leagues; Salesian Youth Center/Boys and Girls Club—for the Literacy and Technology Initiative at the Salesian Youth Center to offer reading comprehension and computer skills training to children 5-18.
Contact: Ann Berg; (813) 282-1975; fax (813) 282-3119; aberg@cftampabay.org
Internet: http://www.cftampabay.org/nonprofit_resources/grant_app_guidelines.html
Sponsor: Community Foundation of Greater Tampa
550 North Reo Street, Suite 301
Tampa, FL 33609

Community Foundation of Greenville-Greenville Women Giving Grants 1434
Greenville Women Giving (GWG) is a special initiative of Community Foundation of Greenville. GWG is a philanthropic organization, founded on the idea that women, informed about philanthropy and the needs in their community, collectively can make a real difference. GWG is committed to strengthening the community through the collective resources of it's members by awarding high-impact grants in five areas: arts and culture, education, health, human services, and environment. GWG will award grants ranging from a minimum of $40,000 to $100,000 to community organizations in Greenville County, South Carolina. Consideration will be given to applications requesting that funding be spent over a two year period. Additional guidelines and application are available on the GWG website: http://greenvillewomengiving.org/faq.aspx. The completed application should be mailed or delivered to Greenville Women Giving, c/o Community Foundation of Greenville on or before January 15, at 4 p.m.
Requirements: Applicants must qualify as tax-exempt under Section 501(c)3 of the Internal Revenue Code or be classified as a unit of government, located and providing services in Greenville County, South Carolina. Organizations must not discriminate on the basis of age, race, national origin, ethnicity, gender, physical ability, sexual orientation, political affiliation or religious belief.
Restrictions: Support is not available for the following: endowment campaigns; travel expenses and conferences; individuals; projects or programs that promote religious or political views; organizations for re-granting purpose.
Geographic Focus: South Carolina
Date(s) Application is Due: Jan 15
Amount of Grant: 40,000 - 100,000 U.S.D
Samples: Camp Opportunity, $25,000—to fund a week of summer camp for children in DSS protective custody; Greenville Literacy Association, $29,000—for their young adult job-readiness program; Graduate Greenville, $50,000—for meeting its aim to increase the high school graduation percentage of Greenville County students.
Contact: Debbie Cooper, Director of Donor Services; (864) 233-5925 or (864) 331-8414; fax (864) 242-9292; dcooper@cfgg.com or Cfgg@cfgg.com
Internet: http://greenvillewomengiving.org/guidelines.aspx
Sponsor: Greenville Women Giving
27 Cleveland Street, Suite 101
Greenville, SC 29601

Community Foundation of Greenville Community Enrichment Grants 1435
The Community Foundation of Greenville supports qualified nonprofit organizations in the areas of the arts and humanities, education, early childhood education, religion, and environmental programs. The Foundation is committed to the responsible allocation of the Foundation's unrestricted funds. As an organization, the Foundation seeks ways to make meaningful, sustained differences in the quality of life for the citizens of Greenville County. The goal is to seek out community-based projects where a modest grant can make a significant impact. The Foundation encourages proposals that: propose practical solutions to community problems; promote cooperation among agencies without duplicating services; use modest funds to make a significant impact; address prevention as well as remediation; make use of matching funds; strengthen an agency's effectiveness and stability; generate community support, both professional and volunteer; state project objectives in measurable terms. The application deadline is 4:00 pm on Friday, March 5.
Requirements: All 501(c)3 non-profit groups can apply for Community Foundation grants to fund innovative projects that have a direct relevance to the Greenville County, South Carolina area. An agency can receive one community enrichment grant per year.
Restrictions: The Foundation will: not make grants to individuals; not fund capacity building; will give low grant making priority to capital fund projects.
Geographic Focus: South Carolina
Date(s) Application is Due: Mar 5
Contact: Debbie Cooper, Director of Donor Services; (864) 233-5925 or (864) 331-8414; fax (864) 242-9292; dcooper@cfgg.com or Cfgg@cfgg.com
Internet: http://www.cfgreenville.org/page18740.cfm
Sponsor: Community Foundation of Greenville
27 Cleveland Street, Suite 101
Greenville, SC 29601

Community Foundation of Greenville Hollingsworth Funds Program/Project Grants 1436
The Hollingsworth Funds, Inc. has a competitive grant process that is administered by the Community Foundation. Consistent with the purposes established by its benefactor, John D. Hollingsworth, Jr., an astute textile executive and real estate investor. The Hollingsworth Funds will serve a broad range of charitable, educational, religious, literary and cultural purposes. The Hollingsworth Funds has adopted initial guidelines in order to assist its evaluation of charitable organizations for funding. The Hollingsworth Funds will make grants: to health and human service agencies dedicated to improving quality of life for residents of Greenville County and particularly for those which deliver services to the poor, homeless, and illiterate; to interdenominational faith-based programs that benefit a broad cross-section of the community; to a wide variety of arts and education related initiatives including high quality preschool and after school programs; for grants for capital projects that provide a public benefit on a nondiscriminatory basis; to increase the organizational capacity of a non-profit organization; to public/private partnerships. The Hollingsworth Funds will give priority to new and innovative projects that are outside of the scope of an organization's ongoing operating budget. To submit a request for funding, a charitable organization must deliver or mail two (2) complete application

packets to the Community Foundation of Greenville by 4:00 pm on September 3. All applicants will be notified of their status by December 31.
Requirements: A 501(c)3 charitable organization can submit one grant application per year for an amount not to exceed $50,000 to fund innovative projects for the benefit of charitable uses within Greenville County, South Carolina, or for the benefit of residents of Greenville County.
Restrictions: The Hollingsworth Funds will not make grants: to sponsor special fundraising events, celebration functions, dinners or annual meetings; for areas that are traditionally the primary responsibility of local, state or federal governments; for core operating expenses of public, private or parochial schools; that primarily benefit the religious activities of a church, synagogue, mosque or other house of worship; to pay off existing debts; scholarship awards or grants to individuals.
Geographic Focus: South Carolina
Date(s) Application is Due: Sep 3
Contact: Debbie Cooper, Director of Donor Services; (864) 233-5925 or (864) 331-8414; fax (864) 242-9292; dcooper@cfgg.com or Cfgg@cfgg.com
Internet: http://www.cfgreenville.org/page18741.cfm
Sponsor: Community Foundation of Greenville
27 Cleveland Street, Suite 101
Greenville, SC 29601

Community Foundation of Henderson County Grants 1437
Grantmaking interest areas of the community foundation in Henderson, NC, are arts and humanities, civic affairs, conservation, education, health, and social services. Projects are encouraged that promote cooperation among organizations without duplicating services; promote volunteer involvement; demonstrate practical approaches to current community issues; enhance or improve organizations that serve people whose needs are not met by existing services and encourage independence, self-sufficiency, and effectiveness; and emphasize prevention. Types of support include equipment acquisition, program development, seed grants, publication, curriculum development, scholarships, technical assistance, and matching funds. The foundation favors projects that affect a broad segment of the population, are relevant to overall community needs and available resources, are pilot programs that can clearly be used as a model for others, have reasonable prospects for future support, and move the community to a higher cultural awareness. Grants are ordinarily made for one year only. Grant recipients will be notified by mail.
Requirements: IRS 501(c)3 organizations in Henderson County, North Carolina are eligible.
Restrictions: Unrestricted funds normally will not be used for programs outside Henderson County unless directly benefiting the area; routine operating support for ongoing programs; annual fund campaigns; religious or political purposes; individuals (including scholarships); organizations whose primary function is to allocate funds to other charitable organizations or projects; research; lobbying; second- or multiyear funding; conferences, seminars, and other short-term events; travel; non-tax-exempt 509(a)91 and 170B(1)(a) organizations under IRS codes; augmenting endowments; underwriting for fund-raising events and performances; or loans.
Geographic Focus: North Carolina
Date(s) Application is Due: Mar 1; Jun 1; Sep 1; Dec 1
Amount of Grant: 500 - 2,000 U.S.D
Contact: Kathryn McConnell, Program Director; (828) 697-6224; fax (828) 696-4026; kmcconnell@cfhendersoncounty.org or info@cfhendersoncounty.org
Internet: http://www.cfhendersoncounty.org
Sponsor: Community Foundation of Henderson County
P.O. Box 1108
Hendersonville, NC 28793

Community Foundation of Howard County Grants 1438
The Community Foundation of Howard County Grants fund initiatives that improve the quality of life for citizens of Howard County. The Foundation addresses needs that generally fall into the following categories: health and medical; social services; education; cultural affairs; civic affairs; and community beautification. The Foundation uses the following evaluation criteria: does the project fit the purpose of the organization; is there an established need for the project; how well the project's purpose has been defined; does it fit with the Foundation's guidelines; and what kind of impact the project will have on the community.
Requirements: All applicants must receive pre-qualification prior to submitting an application. A letter of inquiry addressed to the Program Director should contain a brief statement of the applicant's need for assistance, estimate of total cost of project, and enough information so that the Foundation can determine whether the application falls within the guidelines of its grants program. The organization may then submit the online application with the following information: the grant application and cover page; project budget; copy of IRS determination 501(c)3 letter; current month and year-to-date financial statement; year-end financial statements; itemized list of board members; and evidence of board approval.
Restrictions: The Foundation does not award grants for: normal operating expenses and/or salaries; individuals; seminars or trips; sectarian religious purposes; endowment purposes of recipient organizations; projects which have been proposed by individuals or organizations responsible to advisory bodies or persons; and new projects and/or equipment which were purchased prior to the grant application being approved.
Geographic Focus: Indiana
Date(s) Application is Due: Mar 5; May 7; Sep 3; Nov 5
Contact: Kim Abney; (765) 454-7298 or (800) 964-0508; kim@cfhoward.org
Internet: http://www.cfhoward.org/grants.html
Sponsor: Community Foundation of Howard County
215 West Sycamore Street
Kokomo, IN 46901

Community Foundation of Madison and Jefferson County Grants and Scholarships 1439
The Community Foundation of Madison and Jefferson County was established in 1991 by a group of local volunteers. Since its birth, more than 100 permanent funds have been established in the Foundation. The main interest are: animals, community conservation and beautification, community service, education, the elderly, fine arts, religion, health and youth. The Community Foundation has over 30 scholarships available to graduating seniors of Jefferson County high schools. To see what scholarships are available at your school, visit the Foundations website. The deadline for most scholarships is in March. Because of the current economic climate, the Board of Directors of the Community Foundation of Madison and Jefferson County has started a special Help Your Neighbor campaign instead of conducting its traditional competitive Spring and Fall Grants programs. The Foundations Help Your Neighbor initiative will make grants to 501(c)3 charities that assist Jefferson County residents struggling to meet day-to-day expenses. Up to $30,000 is available. Awards of up to $2,000 each will be made. Contact the Foundation for more information about Help Your Neighbor grants.
Requirements: Grants given to 501(c)3 charities that assist Jefferson County residents only.
Geographic Focus: Indiana
Contact: Louise Markel, Executive Director; (812) 265-3327; fax (812) 273-0181: email: jeffcom; jeffcom@cfmjc.org
Internet: http://www.cfmjc.org/index.html
Sponsor: Community Foundation of Madison and Jefferson County
214 E Main Street, P.O. Box 306
Madison, IN 47250

Community Foundation of Morgan County Grants and Scholarships 1440
The foundation supports student education, builds community, improves health care services, develops programs for youth, preserves the environment, supports the arts, and strengthens the nonprofit sector. Several scholarship programs assist Morgan County students in achieving their educational goals. Scholarship amounts and deadlines vary. Information is available online.
Requirements: Morgan County nonprofit organizations are eligible for grants. Morgan County high school graduates are available to apply for scholarships.
Geographic Focus: Indiana
Contact: Edward Thiebe, Executive Director; (877) 822-6958 or (317) 831-1232; fax (317) 831-2854; ethiebe@cfmconline.org
Internet: http://cfmconline.org/GrantsScholarships/Grants/tabid/63/Default.aspx
Sponsor: Community Foundation of Morgan County
250 N Monroe Street
Mooresville, IN 46158

Community Foundation of Mount Vernon and Knox County Grants 1441
The Community Foundation of Mount Vernon/Knox County, awards grants and scholarships from a variety of funds. The Foundations mission is to improve the quality of life in Mount Vernon and, Knox County through charitable giving. To assess and respond to emerging and changing community needs in the fields of: education, youth services, recreation, arts and culture, social services, and civic and community development. All grant applications are considered by the Foundation board at regularly scheduled meetings in the months of February, April, June, August, October and December. Applicants should be aware grants are awarded on a competitive basis, and that deadlines for submission of requests must be strictly observed. See the Foundations website for additional guidelines, deadline dates and two different grant applications (small and major.)
Requirements: Most grants are made to tax-exempt, private agencies classified as 501(c)3 organizations, and to public charities as defined by the U.S. Internal Revenue Service. Public schools, governmental entities and Knox County colleges and universities are also eligible to apply for Foundation grants.
Restrictions: Certain projects and organizations are not eligible for funding, including but not limited to: grants to individuals; endowment campaigns; ongoing operating expenses; religious organizations (for religious purposes); purchase of computers or other equipment (that is not part of a broader project); existing obligations, debts or liabilities; police and fire protection; staff positions for government agencies travel (when it is the proposal's primary focus) organizations that do not operate programs in Knox County, or for the benefit of Knox County residents; political campaigns.
Geographic Focus: Ohio
Samples: Knox County Historical Society, $30,000— for a building addition project; Hospice of Knox County, $75,000— for its new administrative and training facility; City of Mount Vernon, $3,000— for its historic district street sign placement project.
Contact: Sam Barone, Executive Director; (740) 392-3270; fax (740) 399-5296; thefoundation@firstknox.com or sbarone@mvkcfoundation.org
Internet: http://www.mvkcfoundation.org/Grant/index.html
Sponsor: Community Foundation of Mount Vernon and Knox County
1 S Main Street, P.O. Box 1270
Mount Vernon, OH 43050

Community Foundation of Muncie and Delaware County Grants 1442
The Community Foundation of Muncie and Delaware County Grants provide funding in the following areas of interest: arts and culture, human services, economic development, education, and community betterment. The Foundation focus on: new and innovative projects and programs for which there is a demonstrable need or community benefit; capital needs of community institutions and organizations; emerging needs of Muncie and Delaware County; establishment of community priorities; monitoring of community services to avoid duplication and ineffective programs; acting as a catalyst for action and community participation. Applications are encouraged for types of projects that address

one or more of the following: yield substantial benefits to the community for the resources invested; promote cooperation among agencies without duplicating services; enhance or improve institutional or organizational self-sufficiency; provide "seed money" for innovative community programs; encourage matching gifts or additional funding from other donors; and reach a broad segment of the community with needed services which are presently not provided. The Board of Directors review applications on a quarterly basis in the months of February, May, August, and November.
Requirements: First-time applicants must contact the Foundation to discuss their proposal prior to submission. Organizations are required to submit 18 copies of the application packet, but must submit different items depending on the type of funding requested. See website for submission details.
Restrictions: Requests of $25,000 or more are preferred during the first cycle of each year. The Foundation does not make grants to individuals or grants for religious purposes, budget deficits, for travel, fundraising events, endowments or projects normally the responsibility of a government agency.
Geographic Focus: Indiana
Amount of Grant: Up to U.S.D
Samples: Animal Rescue Fund, to defray the cost of a van to transport animals to the vet and pick up strays in danger, $15,000 (2011); United Way of Delaware County, support to the Technology Wrap-Around Project, a pilot project to create access to computers and the Internet to ensure that those using these services can improve their lives in education, income, and health, $40,000 (2011); Gateway Health Clinic, to increase clinic hours to reduce the waiting list of patients, $25,000 (2011).
Contact: Suzanne Kadinger, Program Officer; (765) 747-7181; fax (765) 289-7770; skadinger@cfmdin.org or info@cfmdin.org
Internet: http://www.cfmdin.org/main/grant-seekers/
Sponsor: Community Foundation of Muncie and Delaware County
201 East Jackson Street
Muncie, IN 47305

Community Foundation of Muncie and Delaware County Maxon Grants **1443**
The Community Foundation of Muncie and Delaware County Maxon Grants give back to the community by supporting worthy causes in Delaware County. Organizations are notified within eight weeks whether they have been funded.
Requirements: In addition to the online application, organizations must submit the following: the grant application cover sheet; the organization's mission statement; a list of the board of directors with their affiliations; a copy of the organization's Federal IRS tax exemption letter; and a letter of endorsement from the Board President, Principal or Chief Executive Officer. They must also submit a brief proposal that provides; project goals and objectives; implementation plan; project budget including expected revenue, in-kind contributions and other grants; staff involved in project; community benefits; and the organization's method of evaluation.
Geographic Focus: Indiana
Date(s) Application is Due: Jul 30
Amount of Grant: 5,000 - 15,000 U.S.D
Contact: Suzanne Kadinger, Program Officer; (765) 747-7181; fax (765) 289-7770; skadinger@cfmdin.org or info@cfmdin.org
Internet: http://www.cfmdin.org/main/grant-seekers/
Sponsor: Community Foundation of Muncie and Delaware County
201 East Jackson Street
Muncie, IN 47305

Community Foundation of North Central Wisconsin, Inc Grants **1444**
The foundation awards grants to eligible Wisconsin nonprofit organizations to enrich the quality life in the areas of arts, education, health care, historic and cultural preservation, and social services. Types of support include challenge/matching grants, program grants, and seed money grants. Applications are due the first business day each month.
Requirements: Wisconsin 501(c)3 nonprofit organizations in the greater Wausau area are eligible.
Geographic Focus: Wisconsin
Amount of Grant: Up to 28,000,000 U.S.D
Contact: Jean Tehan; (715) 845-9555; fax (715) 845-5423; info@cfoncw.org
Internet: http://www.cfoncw.org/grantsScholarships/index.cfm
Sponsor: Community Foundation of North Central Wisconsin
500 Third Street, Suite 310
Wausau, WI 55403-4857

Community Foundation of Randolph County Grants **1445**
The Community Foundation of Randolph County seeks to bring people and resources together to enrich the lives of Randolph County, Indiana residents. The Foundation makes grants to increase the capacity of Randolph County's not-for-profit organizations to respond effectively to the needs of the community. In general, the Foundation prefers funding for: start-up costs for new programs; one-time projects or needs; capital needs beyond an applicant's capabilities and means. The Foundation also funds projects in the areas of arts and culture, civic and community development, education and libraries, environmental and historical preservation, health and human services, youth; and the elderly.
Requirements: The Foundation makes grants to tax-exempt 501(c)3 organizations operating or proposing to operate programs for the benefit of Randolph County residents. Proposals shall include: completed application cover; one original and 10 copies of completed application; one list of your organization's/agency's officers or governing body; one copy of your federal tax exemption 501(c)3 letter; one copy of your last financial statement showing income and expenses (annual report); one copy of the total project budget; one copy of all pertinent supporting information. Applications can be obtained by contacting the Foundation office or downloading the form from the Foundation's website. Application forms should be mailed or delivered to the Foundation's office, since applications cannot be submitted online.
Restrictions: The Foundation will usually not fund: individuals other than scholarships; organizations for religious or sectarian purposes; make-up of operating deficits, post-event or after-the-fact situations; endowment campaigns; for any propaganda, political or otherwise, attempting to influence legislation or intervene in any political affairs or campaigns; services such as fire, police, schools, parks, etc. that are the responsibility of government and tax supported. However, the Foundation occasionally supports special projects of these agencies.
Geographic Focus: Indiana
Date(s) Application is Due: Mar 31; Sep 30
Contact: Ruth Mills, Executive Director; (765) 584-9077; fax (765) 584-7710; rmills@cfrandolphcounty.org
Internet: http://cfrandolphcounty.org/cfrc/jsp/GrantCenter/ApplicationForms/main.jsp
Sponsor: Community Foundation of Randolph County
213 South Main Street
Winchester, IN 47394

Community Foundation of Santa Cruz County Grants **1446**
The foundation promotes philanthropy to make greater Santa Cruz County a better place to live, now and in the future. Grants are designed to develop and strengthen services delivered by the nonprofit sector. Fields of interest include arts and humanities; community development; education; environment; health; historic preservation; human services; youth-led projects (11 to 14 years olds); lesbian, gay, bisexual, transgender community; and Vecinos project, to support residents' capacity to improve their neighborhoods. Management assistance grants are available for projects that strengthen the ability of an organization to fulfill its mission through improved governance and management practices. The listed application deadlines are for program grants; management assistance grants have no deadline. Complete guidelines are available online.
Requirements: To be eligible, an organization must be a nonprofit 501(c)3 organization or public agency, not classified as a private foundation; have a board-approved policy regarding staff employment, electing a governing board, and providing services without discrimination on the basis of race, religion, gender, sexual orientation, age, disability, or national origin; provide direct benefits to residents of Santa Cruz County, the Summit, or the Pajaro Valley and include representation of Santa Cruz County residents on its board of directors; and be current on all reports due to the community foundation for any previous grants.
Geographic Focus: California
Date(s) Application is Due: Feb 18; Apr 29
Amount of Grant: 1,000 - 15,000 U.S.D
Samples: Actors Theatre (CA)—to support the continuation of marketing and audience development efforts, $10,000; The Beat Within (CA)—to support a music education and recording program for county youth, ages 7 through 18, $5000.
Contact: Christina Cuevas, (831) 477-0800, ext. 213; fax (831) 477-0991; christina@cfsscc.org
Internet: http://www.cfscc.org/page17117.cfm
Sponsor: Community Foundation of Santa Cruz County
2425 Porter Street, Suite 17
Soquel, CA 95073

Community Foundation of South Alabama Grants **1447**
The Community Foundation of South Alabama is the platform for building community in Baldwin, Clarke, Conecuh, Choctaw, Escambia, Mobile, Monroe and Washington counties. The Foundation provides grant support to South Alabama nonprofit organizations in four major program areas: community and civic affairs, education, arts and recreation, and social services (health and human services). Priority consideration is given to projects that clearly provide innovative responses to community needs, are collaborative in nature when appropriate, and potentially affect broad segments of the community. Types of support include general operating support, capital campaigns, endowment funds, program development, and scholarship funds. The board of directors meets annually to consider grant requests. Application forms may be obtained from the office.
Requirements: Nonprofit organizations that have a 501(c)3, government entities and religious organizations that are located in Mobile, Baldwin, Clarke, Conecuh, Washington, Choctaw, Escambia and Monroe counties are eligible.
Restrictions: Grants generally are not provided for or to individuals, recurring requests for the same purpose for which foundation grant funds have already been awarded, research that is noncommunity-related or that does not have short-range results, films, conferences and workshops, lobbying activities, and tickets to fund-raising events.
Geographic Focus: Alabama
Amount of Grant: 2,500 - 28,000 U.S.D
Contact: Janine Phillips, Director and Program Officer; (334) 438-5591; fax (334) 438-5592; info@communityendowment.com or program@communityendowment.com
Internet: http://communityendowment.com/grants/grants.htm
Sponsor: Community Foundation of South Alabama
212 St. Joseph Street, P.O. Box 990
Mobile, AL 36601-0990

Community Foundation of Southeastern Connecticut Grants **1448**
Within its geographic area, comprising the 11 towns of southeastern Connecticut, the community foundation awards grants to assist charitable, educational, and civic institutions; promote health and general welfare; support environmental programs; provide human care services for the needy; secure the care of children and families; encourage artistic and cultural endeavors; and initiate planning of appropriate projects

within these areas. High-priority programs include those that strengthen families; improve access to area resources, especially for underserved populations; encourage residents to participate in the cultural life of the community; demonstrate collaborative efforts and inclusive practices; reinforce best practices or show innovative approaches; and add to the general well being of the community. Types of support include building/ renovation, equipment, emergency funds, program development, conferences and seminars, publication, seed grants, scholarship funds, technical support, and scholarships to individuals. Applications requesting support of normal operating expenses will not be considered. Proposals may be submitted between April 1 and November 15 each year.
Requirements: IRS tax-exempt organizations serving the towns of East Lyme, Groton, Ledyard, Lyme, Montville, New London, North Stonington, Old Lyme, Salem, Stonington, and Waterford, CT, including inner-city, suburban, and rural areas, are eligible.
Restrictions: The foundation does not consider requests for direct financial assistance to individuals; religious or sectarian programs; political or lobbying purposes; fundraising events; or debt retirement.
Geographic Focus: Connecticut
Date(s) Application is Due: Nov 15
Amount of Grant: 2,000 - 10,000 U.S.D
Contact: Jennifer O'Brien, Program Officer; (860) 442-3572; fax (860) 442-0584; jennob@cfsect.org
Internet: http://www.cfsect.org/grant.html
Sponsor: Community Foundation of Southeastern Connecticut
P.O. Box 769
New London, CT 06320

Community Foundation of South Puget Sound Grants 1449
The foundation awards grants to a variety of charitable, cultural, education, health, and welfare organizations. Grants from the unrestricted funds are considered for general operational or program support. Typical grants range from $1,000 to $7,500. Emergency grants may be considered on a case-by-case basis. Application and guidelines for grants and scholarships are available online.
Requirements: The foundation awards grants to Washington tax-exempt organizations primarily for use in Thurston, Mason, and Lewis Counties, except on instructions of the donor at the time of the gift or bequest.
Restrictions: Grants do not support religious organizations for religious purposes; individuals; annual campaigns of organizations (direct mail or special events); political or lobbying activities; organizations that discriminate based on race, creed, or ethnic group; capital campaigns for bricks and mortar or endowment funds; or for multiple year commitments.
Geographic Focus: Washington
Date(s) Application is Due: Apr 3; Oct 2
Amount of Grant: 1,000 - 7,500 U.S.D
Samples: Rochester Organization of Families, $7,000—to help maintain ROOF's Kid's Place academic enrichment program for 50 low -income, at-risk children in the first through fifth grades; Harmony Hill, $4,500—to purchase a computer server to maintain its extensive database of clients and donors; which in turn supports their mission to improve the quality of life for those affected by cancer; Thurston Mason Project Access, $4,000—to make donated medical services available to uninsured, low-income Thurston County residents with acute, urgent conditions.
Contact: Norma Schuiteman, Executive Director; (360) 705-3340; fax (360) 705-2656; legacy@thecommunityfoundation.com
Internet: http://www.thecommunityfoundation.com/grants.php
Sponsor: Community Foundation of South Puget
111 Market Street NE, Suite 375
Olympia, WA 98501

Community Foundation of St. Joseph County Lilly Endowment Community Scholarship 1450
The Lilly Endowment Community Scholarships offer four-year full tuition plus book stipend to any Indiana college or university. Four awards are made annually to residents of St. Joseph County. All college majors are eligible to apply.
Requirements: Applicants must be residents of St. Joseph County, Indiana and graduates from an accredited Indiana high school. They must intend to pursue a full time baccalaureate course of study at an accredited public or private Indiana college or university, and possess at least a 3.8 GPA or SAT score of 1800 out of 2400. All applicants must complete and submit the Free Application for Federal Student Aid (FAFSA) to the U.S. Department of Education (see Office of Financial Aid or visit http://www.fafsa.ed.gov for more information). In order for the application to be considered complete, students must forward the Student Aid Report (SAR) listing the Expected Family Contribution (EFC) to the Community Foundation. Applicants should refer to the St. Joseph County Foundation website for current applications and deadlines.
Restrictions: Students eligible to receive a tuition benefit due to their parent(s)/guardian(s) place of employment will not be considered for this scholarship if the benefit meets or exceeds the cost of tuition. In addition, family members of those serving on the scholarship nomination committee or Community Foundation staff are deemed ineligible (defined as spouses, lineal descendants and their spouses, siblings and their spouses).
Geographic Focus: Indiana
Contact: Emily Addis (574) 232-0041; fax (574) 233-1906; emily@cfsjc.org
Internet: http://www.cfsjc.org/scholarships/lilly_scholarship.html
Sponsor: Community Foundation of St. Joseph County
205 W Jefferson Boulevard, P.O. Box 837
South Bend, IN 46624

Community Foundation of St. Joseph County Scholarships 1451
The Community Foundation of St. Joseph County has more than 50 scholarship funds which award over $450,000 to local students each year. The scholarships were established for a variety of reasons—to assist low-income students, encourage children of employees to pursue their dreams, train medical professionals, teachers, and engineers, or in memory of a lost loved one. The Community Foundation has distributed more than $6 million in scholarships since its inception. Applicants should view the website for a variety of scholarships categorized by the school they attended in St. Joseph County or the college they plan to attend.
Geographic Focus: Indiana
Contact: Emily Addis (574) 232-0041; fax (574) 233-1906; emily@cfsjc.org
Internet: http://www.cfsjc.org/scholarships/index.html
Sponsor: Community Foundation of St. Joseph County
205 W Jefferson Boulevard, P.O. Box 837
South Bend, IN 46624

Community Foundation of St. Joseph County Special Project Challenge Grants 1452
The Special Project Challenge Grants assist public and other 501(c)3 agencies in their efforts to serve community needs. For every $1 raised by the chosen agency, the Community Foundation will match $1. The foundation encourages projects in the following areas: community development and urban affairs; health and human services; parks, recreation, and environment; and youth and education.
Requirements: In additional to the online application, all applicants must submit the following materials online: up to a two page proposal narrative; a detailed project budget; current board roster with officers identified; fiscal year income statement; proof of nonprofit status. Application materials must be submitted via email to grants@cfsjc.org in word processing format (narrative or budget) or Microsoft Excel (budget). Hard copy applications are no longer accepted.
Restrictions: Grants are not made to fund: operational phases of established programs; endowment campaigns; religious organizations for religious purposes; individuals directly; development or public relations activities (e.g. literature, videos, etc.); retirement of debts; camperships; annual appeals or membership contributions; travel for bands, sports teams, classes, etc; j) computers (unless presented as a necessary component of larger program or objective); and post-event or after-the-fact situations.
Geographic Focus: Indiana
Date(s) Application is Due: Mar 1; Oct 1
Contact: Angela Butiste; (574) 232-0041; fax (574) 233-1906; angela@cfsjc.org
Internet: http://www.cfsjc.org/grants/sproj/special_project_grants.html
Sponsor: Community Foundation of St. Joseph County
205 W Jefferson Boulevard, P.O. Box 837
South Bend, IN 46624

Community Foundation of Switzerland County Grants 1453
The Community Foundation of Switzerland County is a nonprofit organization created by and for the people of the Switzerland County area with one goal, making our community a better place to live, now and for future generations. Organizations applying for a grant may request up to $1,000. In addition, the Community Foundation will give priority to grant applications that focus on the basic needs of the community. While the term basic needs can be interpreted differently, in general, the Community Foundation uses the term to describe food, housing, shelter, health care, clothing, personal care, transportation services and other essential needs that people in our community may not have due to the weakened economy. While grant applications addressing basic needs of the community will be given priority, the Community Foundation still welcomes applications for other programs and projects that benefit Switzerland County. To make grant dollars available on a timely basis, there are no deadlines. Grant letters will be reviewed and decided upon monthly. Any request in by the 1st of each month will be notified of a decision by the end of that month.
Requirements: Any organization with a 501(c)3 or any organization that provides a program with charitable intent and has a fiscal agent is eligible to apply. To apply for a grant, an organization must write a one page letter explaining: who will be doing the requested program/project; what will be done (program, purchase, training, etc); when it will be done; where it will be done; and how much it will cost (short budget). If the cost is more than being requested from the Community Foundation, indicate where the rest of the money will come from. There will be no match required for these grants, so a program or project can be totally funded by a request or it may be part of a larger project.
Geographic Focus: Indiana
Amount of Grant: 1,000 U.S.D
Contact: Pam Acton; (812) 427-9160; fax (812) 427-4033; pacton@cfsci.org
Internet: http://www.cfsci.org/
Sponsor: Community Foundation of Switzerland County
303 Ferry Street, P.O. Box 46
Vevay, IN 47043

Community Foundation of Tampa Bay Grants 1454
The Community Foundation of Tampa Bay awards creative grants from its community fund with the goal of fostering positive changes in the lives of their citizens. The areas of interest include: arts and culture, community development, education, environment and animals, and health and human services. See the Foundations website for the Grant Application Form and guidelines: http://www.cftampabay.org/nonprofit_resources/grant_app_guidelines.html
Requirements: Non-profit organizations based in and serving communities in Hillsborough, Pasco, Pinellas and Hernando counties, with a 501(c)3 status in good standing from the IRS are eligible to apply.

Restrictions: The Foundation is generally not interested in requests for: funding of ongoing operating costs; grants for capital campaigns or expenditures; tickets for any fundraising event or advertising space in programs or other publications; legislative lobbying or political campaigns; medical research; religious or sectarian purposes; loans or assistance to individuals; a multiple year funding commitment, although in exceptional cases, such funding will be considered not to exceed three years.
Geographic Focus: Florida
Date(s) Application is Due: Mar 1; Sep 1
Amount of Grant: Up to 7,500 U.S.D
Contact: Ann Berg, Vice President of Grants; (813) 282-1975; fax (813) 282-3119; aberg@cftampabay.org
Internet: http://www.cftampabay.org/nonprofit_resources/grant_app_guidelines.html
Sponsor: Community Foundation of Tampa Bay
550 North Reo Street, Suite 301
Tampa, FL 33609

Community Foundation of the Eastern Shore Community Needs Grants 1455
The Community Foundation of the Eastern Shore awards grants from its discretionary fund, known as the Lower Shore Fund for Community Needs. Monetary grants are awarded to a wide range of tax exempt organizations in Somerset, Wicomico and Worcester Counties of Maryland, whose programs benefit: health; human services; arts and culture; community affairs; environmental conservation; and historic preservation. Through this grant program, the Foundation identifies high priority needs and seeks opportunities where a relatively modest amount of grant money can make a significant difference in the community. Often the grants provide start-up or short-term funding for innovative, potentially replicable projects that meet newly identified needs or demonstrate new solutions for previously identified needs.
Requirements: To be eligible for a grant from this program, an organization must meet two fundamental criteria: it must be a governmental unit, a religious organization engaged in a non-sectarian activity, or a non-profit, tax exempt public charity, as defined in Section 501(c)3 of the Internal Revenue Code; and it must provide services to benefit the residents of the lower three counties of Maryland's Eastern Shore: Worcester, Wicomico, and Somerset. Programs are not required to serve all three counties, but regional projects are encouraged.
Restrictions: The Community Needs Grant Program does not fund: endowment funds; ongoing/operating expenses; fundraising campaigns; sectarian religious programs; playground equipment; building campaigns; operational deficits; debt retirement; capital requests; political/lobbying programs; office equipment/staff training; and school-based programs. Grants through this program are made to organizations, not individuals.
Geographic Focus: Maryland
Date(s) Application is Due: Feb 1; Aug 1
Amount of Grant: 500 - 7,500 U.S.D
Contact: Erica N. Joseph; (410) 742-9911; fax (410) 742-6638; joseph@cfes.org
Internet: http://www.cfes.org/grants_community_needs.php
Sponsor: Community Foundation of the Eastern Shore
1324 Belmont Avenue, Suite 401
Salisbury, MD 21804

Community Foundation of the Ozarks Grants 1456
Grants from unrestricted funds of the community foundation are made to qualified nonprofit organizations that serve the Greene County area of Missouri. Areas of interest include arts, community betterment, education, and social justice. Preference will be given to projects that are exclusive of the ongoing operation of the organization; provide seed money for new programs; open the possibility of receiving matching funds; and offer the community foundation the opportunity to meet needs in full in diverse problem areas. Agencies are encouraged to ask for funding amounts that reflect their true needs. Submittal of a one-page concept letter is the first step in applying. Concept letters must be received by the August deadline for grants awarded in November, by the December deadline for grants awarded in March, and by the March deadline for grants awarded in June. Concept letters may be submitted to dmyers@cfozarks.org.
Requirements: IRS 501(c)3 organizations that serve the Greene County area in Missouri are eligible.
Restrictions: Capital campaigns and grants to individuals are not allowed.
Geographic Focus: Missouri
Date(s) Application is Due: Mar 1; Aug 1; Dec 1
Samples: Skate Park (Springfield, MO)—to complete construction of skate park for youth, $25,000; Congretional Health Ministries (Springfield, MO)—start up money, $5000; Springfield Symphony (Springfield, MO)—to purchase tickets for the elderly to the symphony, $3000.
Contact: Gary Funk, President; (417) 864-6199; fax (417) 864-8344; gfunk@cfozarks.org
Internet: http://www.cfozarks.org/grantseekers.html
Sponsor: Community Foundation of the Ozarks
901 St. Louis Street, Suite 701
Springfield, MO 65806

Community Foundation of the Verdugos Educational Endowment Fund Grants 1457
The Educational Endowment Fund provides financial support to innovative as well as traditional educational programs and projects in public and private schools and community organizations in the Crescenta Valley, California region. The goal is to enrich educational opportunities by supporting building, equipment, instruction, guidance, coaching, practical application, classroom activities, in-service, training and/or practice programs. Grants will support: equipment purchase, replacement & modernization; improvement to facilities including athletic facilities; printed materials; fundraising events or capital campaigns; classroom materials; public/private schools and colleges; child day care/development centers; libraries; hospitals; community enrichment projects; salaries. All qualified organizations are invited to submit grant applications once a year for grants that will be awarded at the Crescenta Valley Chamber of Commerce Installation Luncheon in early January. Average grant sizes are from $500 to $3,000. The grant application is available at the Community Foundation of the Verdugos website. The application must be submitted by October 9th.
Requirements: 501(c)3 IRS nonprofit public charities in the Crescenta Valley region of California (includes La Crescenta, Montrose, Tujunga and La Canada) and, organizations located outside the Crescenta Valley but delivering educational programs in the Crescenta Valley are eligible to apply. Please indicate the number of Crescenta Valley residents who will be served by your program. An organization may apply for no more than two grants during a grant cycle. Note, the president, school principal or leader must indicate awareness of multiple requests and each application must be for different types of support.
Restrictions: Grants will not support: uniforms or clothing; travel expenses; fiscal agents; individuals and individual scholarships; endowment funds; feasibility studies or consulting fees; advertising; research; political lobbying, voter registration or political campaigns; insurance or maintenance contracts; faith-based projects.
Geographic Focus: California
Date(s) Application is Due: Oct 9
Amount of Grant: 500 - 3,000 U.S.D
Samples: Crescenta Valley Arts Council, $500; Verdugo Hills Hospital Foundation, $1,000; YMCA of the Foothills & Rosemont School, $750.
Contact: Edna Karinski, Executive Director; (818) 241-8040; fax (818) 241-8045; info@communityfoundationoftheverdugos.org
Internet: http://www.glendalecommunityfoundation.org/grants_endowment.php
Sponsor: Community Foundation of the Verdugos
330 Arden Avenue, Suite 130
Glendale, CA 91203

Community Foundation of the Verdugos Grants 1458
The Community Foundation of the Verdugos welcomes grant requests from public agencies and nonprofit organizations that serve the Verdugo region including Burbank, Glendale, La Canada Flintridge, La Crescenta, Montrose, and Verdugo City of California. Grant recipients must have appropriate fiscal and program accountability. Average grant amounts range from $2,500 to $10,000. Grants are provided in the following areas of interest: arts and culture; civic; health and human services (examples include projects and services for the disabled/handicapped, general community health, or homeless services); education programs; senior services; student aid (scholarships); youth services; environment and; community. Grant application and, application guidelines are available at the Foundation's website. Grants are made three times a year. Deadline for grant applications are February 1, June 1, and September 1. Scholarship deadline is March 5. Only one proposal from an organization is permitted per grant cycle. Approval or denial of your request will be provided to you in writing.
Requirements: The Foundation funds nonprofit organizations that: have IRS nonprofit status; predominantly serve the Verdugo region of California (Glendale, La Canada Flintridge, La Crescenta, Montrose, Verdugo City) and adjacent areas; capital equipment that helps to increase your organization's long-term sustainability or services to increase impact in the Verdugo region; programs (including related overhead, supplies and administrative expenses) responsive to changing community needs and which increase impact in the Verdugo region. Requests that provide significance and impact for the Verdugo region in the areas of arts and culture, children/youth, education and literacy, health and human services, civic activities, senior services, and environmental/animal related programs.
Restrictions: The Foundation does not make grants or loans to individuals unless they are students receiving scholarships. The Foundation also does not provide funds for religious or political purposes, for budget deficits, or projects that are usually the responsibility of a public agency.
Geographic Focus: California
Date(s) Application is Due: Feb 1; Mar 5; Jun 1; Sep 1
Amount of Grant: 25,000 - 10,000 U.S.D
Samples: San Gabriel Valley Habitat for Humanity, $10,000—toward handicapped lifts; Salvation Army Glendale Corporation, $10,000—ZONE after school program for low-income and at-risk children; College View School, $10,000—toward a new playground structure.
Contact: Edna Karinski, Executive Director; (818) 241-8040; fax (818) 241-8045; info@communityfoundationoftheverdugos.org
Internet: http://www.glendalecommunityfoundation.org/grants.php
Sponsor: Community Foundation of the Verdugos
330 Arden Avenue, Suite 130
Glendale, CA 91203

Community Foundation of Wabash County Grants 1459
The goal of the Community Foundation of Wabash County Grants is to enrich the quality of life in Wabash County, Indiana, by responding to emerging and changing needs of the community. It also seeks to support existing organizations and institutions through grants in support of the following categories: arts and culture; community and civic development; education; environment; health and human services; and recreation. Types of support include: building or renovation; continuing support; curriculum development; endowments; equipment; general operating support; matching/challenge support; program development; program evaluation; scholarship funds; scholarships to individuals; seed money; and technical assistance.
Requirements: Proposals are accepted from organizations serving Wabash County that are defined as tax exempt under Section 501(c)3 of the IRS code or have comparable status and charitable causes. Grant selection is judged on program focus; program

design; benefits; reach; and organizational profile. In addition to the online application, organizations must submit the completed cover sheet, proposal budget, a list of member of the organization's current staff and governing board; current year-end financial statement; and copy of the tax exempt IRS letter. Eight copies of the application and attached documentation are then submitted to the Foundation office for review.
Restrictions: The following are not eligible for funding: national organizations (except for local chapters serving Wabash County); annual fund campaigns; or programs or products produced for resale. Faith-based organizations may apply for program funding, provided there is not a requirement to participate in religious instruction and/or take part in religious activities.
Geographic Focus: Indiana
Date(s) Application is Due: Mar 15; Jul 15; Nov 15
Amount of Grant: Up to U.S.D
Samples: Lighthouse Missions, holiday food baskets: $5,000 (2011); North Manchester Parks & Recreation "Keeping Indoor Fitness Fun" After School Program: $2,000 (2011); Wabash County Hospital Foundation (85 Hope Free Medical Clinic) $9,500 (2011).
Contact: Cathy McCarty, Program Assistant; (260) 982-4824; fax (260) 982-8644; cathy@cfwabash.org
Internet: http://www.cfwabash.org/nonprofits-grant-information/guidelines.html
Sponsor: Community Foundation of Wabash County
218 East Main Street
North Manchester, IN 46962-0098

Community Foundation of Western Massachusetts Grants **1460**
The general objectives of the community foundation's grant making are to serve a wide range of western Massachusetts community development interests in the areas of the arts, education (including adult basic education and literacy programs), environment, health, housing, and human services; encourage creative and collaborative responses to existing and emerging problems or opportunities; and leverage additional support for programs from other private and public funding sources. The foundation supports capital campaigns, building/renovations, equipment, endowment funds, program development, conferences and seminars, publication, seed money, fellowships, scholarship funds and scholarships to individuals, technical assistance, and matching funds. Proposals are considered at the distribution committee meetings held four times each year and should be received by the first Monday of February, May, August, and/or November. Applicants are encouraged to call the foundation for guidelines.
Requirements: Proposals from western Massachusetts nonprofit organizations, including Hampden, Hampshire, and Franklin Counties, should include the name of the organization; an application; a one-page description of the need, project, and personnel for the project; 501(c)3 and 509(a) determination letters; names and affiliations of the governing board; and a copy of operating budget and/or recent organization audit.
Restrictions: Grants are not usually made to or for individual recipients, including scholarships or loans; operating budget items in general; influencing specific legislation or election campaigns; religious or sectarian purposes; private secondary/private higher education except for programs that serve community needs with a high priority; endowments; fund-raising events; academic or medical research; replacement of governmental or other funding except for grants bridging periods before anticipated funding; requests for multiyear funding; or applications for the same project less than one year apart.
Date(s) Application is Due: Mar 31; Jun 1; Oct 1
Amount of Grant: Up to 100,000,000 U.S.D
Samples: Amherst Writers and Artists Press—to support training for low-income women, $10,000; Holyoke Community Land Trust—to support Holyoke landscape improvement, $1000; Greenfield Area Community Land Trust—to support a co-op housing ownership program, $2300; Arise—to support a study of the criminal justice system, $10,000.
Contact: Mary Jenewin-Caplin, Program Officer; (413) 732-2858; fax (413) 733-8565; grants@communityfoundation.org
Internet: http://www.communityfoundation.org
Sponsor: Community Foundation of Western Massachusetts
1500 Main Street, P.O. Box 15769
Springfield, MA 01115

Community Foundation of Western North Carolina Grants **1461**
The foundation serves an 18-county area, including Cherokee, Graham, Clay, Swain, Macon, Jackson, Haywood, Transylvania, Madison, Buncombe, Henderson, Yancey, McDowell, Rutherford, Polk, Burke, Mitchell, and Avery Counties. The community foundation awards grants for projects in the areas of arts and culture, children and youth, community development, education, environment, health, human services, and social justice. The foundation awards seed money grants to new or existing programs and technical assistance grants to strengthen nonprofit organizations. A workshop for grant applicants is held one month prior to the deadline at which time the guidelines are reviewed and questions are addressed; contact the foundation for date and times.
Requirements: Nonprofit 501(c)3 organizations serving areas within the 18-county region are eligible. Applicants must obtain foundation approval in order to apply for signature grants.
Restrictions: Grants are not made for the following: operating support (except for start-up); activities taking place before grant decisions are made; capital requests (except for modest budget items, such as equipment); endowments; sectarian purposes; private schools and clubs; political purposes; veteran affairs; deficit funding or debt retirement; and individuals except through scholarships. Multiple-year grants are not made. The foundation is only occasionally able to fund the second phase of a project. Organizations may not submit more than one application at a time; it is suggested that no more than one request be submitted during a 12-month period.
Geographic Focus: North Carolina
Date(s) Application is Due: Feb 1; Sep 1
Amount of Grant: Up to 10,000 U.S.D
Samples: Asheville Area Chamber of Commerce Community Betterment Foundation—for an expanded, outdoor Mountain Dance and Folk Festival, $5000; Citizens for Media Literacy—to produce a video to show nonprofits and local governments how to utilize public access television, $8000.
Contact: Tara Scholtz; (828) 254-4960; fax (828) 251-2258; scholtz@cfwnc.org
Internet: http://www.cfwnc.org/grants.html
Sponsor: Community Foundation of Western North Carolina
P.O. Box 1888
Asheville, NC 28802

Community Foundation Serving Boulder County Grants **1462**
The foundation awards grants to eligible Colorado nonprofit organizations in its areas of interest, including arts, civic involvement, education, the environment, and health and human services. Types of support include capital campaigns, endowments, general operating support, and program support. The foundation also offers a variety of programs to build capacity and the organizational structure of nonprofits. Deadlines vary depending on program, check Web site or contact staff for exact dates.
Requirements: Colorado 501(c)3 nonprofits serving the Boulder area are eligible.
Geographic Focus: Colorado
Contact: Carly Hare, (303) 442-0436, ext. 111; fax (303) 442-1221; Carly@commfound.org
Internet: http://www.commfound.org/grants/index.html
Sponsor: Community Foundation Serving Boulder County
1123 Spruce Street
Boulder, CO 80302

Community Foundation Silicon Valley Grants **1463**
The foundation supports California nonprofits in its areas of interest, including arts and cultural participation, education and lifelong learning, self-reliant individuals and family, and neighborhood and civic projects. The foundation also supports the Youth in Philanthropy program and early literacy grants. Preference is given to proposals that focus on the prevention of problems rather than the cure, equip people to become self-sufficient, encourage programs that take a comprehensive approach to addressing interrelated problems, promote cooperation and collaboration among agencies, encourage leveraging and matching grant opportunities, and provide seed funding for innovative projects. Additional types of support include program development, technical assistance, consulting services, and general operating grants. Grants generally are made for one year of support; long-term commitments are contingent upon annual review of progress and available funds. Guidelines and application are available online.
Requirements: Nonprofit organizations benefiting Santa Clara County and southern San Mateo County, CA, are eligible.
Restrictions: In general, the foundation does not consider fundraising events, capital equipment and endowment campaigns, on-going operating expenses, existing debts/obligations, for-profit schools, or individuals.
Geographic Focus: California
Amount of Grant: Up to 25,000 U.S.D
Samples: Child Advocates of Santa Clara and San Mateo Counties (San Jose, CA)—to recruit and train new child-advocacy volunteers, $225,000 over three years; Project H.E.L.P (High Expectations Learning Program) (Sunnyvale, CA)—to expand this early-intervention program that provides high-quality supplemental academic instruction to poorly performing students at public elementary schools in Santa Clara County, CA, $225,0000 over three years.
Contact: Jeff Sunshine, Director of Programs; (408) 278-2200; fax (408) 278-0280; jsunshine@cfsv.org or info@cfsv.org
Internet: http://www.cfsv.org/grants_focusareas.html
Sponsor: Community Foundation Silicon Valley
60 S Market Street, Suite 1000
San Jose, CA 95113

Community Health Scholars Program Fellowships **1464**
The Community Health Scholars Program (CHSP) is a postdoctoral fellowship program designed to meet the growing needs of schools of public health and other health professions for faculty with community competency. The 12-month fellowship enables scholars to develop and enhance skills in working with communities and engaging in participatory community-based research at institutions where these skills are present. Applications are welcome from junior faculty as well as those recently completing doctoral-level training so long as the above eligibility requirements are met. Women and other underrepresented groups are encouraged to apply.
Requirements: Applicants must be U.S. citizens located at one of three training sites (Johns Hopkins, University of Michigan, or University of North Carolina) and have completed all formal requirements for professional and postprofessional training (e.g., internships or residencies) at the doctoral level (e.g., PhD, MD, JD, or DrPH) by the date of entry into the program.
Restrictions: Scholars may not hold other training fellowships or be enrolled full time in another degree program during their tenure in this program.
Amount of Grant: 60,000 U.S.D
Contact: Saundra Bailey; (734) 647-3065; fax (734) 936-0927; chsp@umich.edu
Internet: http://www.sph.umich.edu/chsp
Sponsor: University of Michigan
109 Observatory Street, M4142
Ann Arbor, MI 48109-2029

Community Impact Fund 1465

The foundation awards grants aimed at meeting the needs and enhancing the lives of individuals in California's Riverside and San Bernardino Counties. Funding categories include health and human services—promoting access to healthcare for all residents and helping individuals and families obtain basic services to promote an improved quality of life; youth and families—enhancing opportunities that promote academic achievement and positive youth development and developing family support services that foster learning and growth; arts and culture—encouraging creative expression and providing opportunities for enjoyment of cultural activities and art forms; civic and public benefit—building a sense of community and promoting civic participation. Preference is given to projects that are perceived as a high need in the community being served; fill a gap in service; benefit a large number of residents; enhance collaboration and/or make the delivery of services more effective and efficient; have clear objectives and can document successful outcomes; expand successful programs to serve additional residents or new geographic areas within the two counties; serve remote areas or areas that have received little funding from the foundation. Contact the office for application deadlines and forms.

Requirements: The foundation will only accept applications from 501(c)3 organizations that can document on-going operations and nonprofit status for at least three years. Organizations may only apply once very 12 months.

Restrictions: Grants are generally not made for on-going operating expenses; retroactive funding for cost already incurred; paying off deficits or existing obligations; endowment, capital fund, or annual fund appeals; capital projects, i.e. construction of new buildings; direct support of individuals; sectarian programs or fraternal organizations; event sponsorships; research or development activities; school or college-based extracurricular activities; partisan activities; or re-granting purposes.

Amount of Grant: Up to 10,000 U.S.D

Contact: Celia Cudiamat; (951) 684-4194; ccudiamat@thecommunityfoundation.net

Internet: http://www.thecommunityfoundation.net/grants/grant_commImpact_04.html

Sponsor: Community Foundation Serving Riverside and San Bernardino Counties
3880 Lemon Street, Suite 300
Riverside, CA 92501

Community Technology Foundation of California Building Communities Through Technology Grants 1466

The foundation awards grants to eligible California nonprofit organizations in support of programs that bring communications technologies to traditionally underserved populations, including low-income, inner-city, minority, disabled, limited-English speaking, and low-income senior communities. Areas of support include community building, community economic development, education, and health. Current grantmaking focuses on the development and maintenance of—as well as the removal of barriers to—digital media, communications, and outreach for underserved Californians. Successful grantees will be those who effectively utilize ICT in a manner that enables their target communities to become more active and engaged participants in civic action. Types of support include fellowships, matching/challenge grants, and leveraged gifts. Applications for proposals are accepted on a rolling basis, with funding decisions made at quarterly meetings (usually held in January, April, July, and October). Guidelines are available online.

Requirements: California nonprofit organizations are eligible.

Restrictions: CTFC will specifically not fund projects in the following areas: general operating expenses (unless an applicant organization's sole mission and purpose is to conduct work that falls within the grantmaking guidelines); grants to individuals; grants to non-501(c)3 organizations (unless such organizations have a commitment in writing from a 501(c)3 sponsoring fiscal agent); requests for only distribution or dissemination of completed media projects; media projects used solely for distribution at agency special events, memorials, annual fundraisers, anniversary celebrations, etc; requests that are solely for equipment or software purchases without a corresponding content-based component; ICT projects that are only for organizational internal infrastructure upgrades or administrative/operational support; or any project that does not have a media, information, or communications technology component.

Geographic Focus: California

Amount of Grant: 10,000 - 50,000 U.S.D

Samples: Asian Perinatal Advocates (San Francisco, CA)—to develop a client-database system and interactive Web site, $84,500 over two years; California Small Business Education Foundation (Los Angeles, CA)—for a seminar about the availability of technology for small businesses, $15,000; Ctr for Academic and Social Advancement (San Diego, CA)—for a technology program for children, $84,000 over two years.

Contact: Administrator; (415) 371-8808; fax (415) 371-8818; info@zerodivide.org

Internet: http://www.zerodivide.org

Sponsor: Community Technology Foundation of California
101 Spear Street, Suite 218
San Francisco, CA 94105

Community Toolbox for Children's Environmental Health Grants 1467

Grants are awarded to community-based groups working on children's environmental health and on lead poisoning prevention. Priority will be given to requests from organizations serving children at the highest risk of lead poisoning, including children in urban, low income, and minority communities living in areas with older housing stock. The majority of grants will be awarded for capacity building only. The request for proposal and grant application are available on the Web site, as well as from the office. Applications for round one are due by September 13; invited applications for round two are due by January 21. Annual deadline dates may vary; contact program staff for exact dates.

Requirements: Nonprofit, community-based organizations are eligible to apply.

Restrictions: Grants do not support organizations with annual budgets greater than $250,000 for the past year and the current year; federal, state, or local government agencies; international organizations; academic institutions; for-profit businesses; or programs that seek to influence legislation.

Date(s) Application is Due: Jan 21; Sep 13

Samples: Echo Park Community Coalition (Los Angeles, CA)—to develop leadership skills, to conduct workshops on preventing lead poisoning, and to create a strategic plan, $40,000 over two years; Friends of McKinley (Los Angeles, CA)—to organize parents and teachers to advocate lead abatement in schools, $10,000; Josiah Hill III Clinic (P.O.rtland, OR)—for community education and outreach, strategic planning, and the development of a fundraising strategy, $12,000; Justus WeCare (Louisville, KY)—to build awareness about lead poisoning and risk reduction through home visits, screenings, and resident surveys, $30,000 over two years.

Contact: Program Contact; (800) 220-7630 or (415) 614-9533; fax (415) 614-9537; info@communitytoolbox.org

Internet: http://www.communitytoolbox.org/grantmk.htm

Sponsor: Community Toolbox for Children's Environmental Health
999 Sutter Street, 4th Floor
San Francisco, CA 94109

Compass Bank Foundation Grants 1468

The Foundation, established in 1981, supports organizations involved with arts and culture, higher and law education, health, heart disease, multiple sclerosis, medical research, youth development, human services, and philanthropy. Giving is primarily centered in the states of Alabama, Arizona, Colorado, Florida, New Mexico, and Texas. There are no specific deadlines with which to adhere. Contact the Foundation for further application information and guidelines.

Restrictions: No support is provided for religious organizations, and there are no grants to individuals (except for employee-related scholarships).

Geographic Focus: Alabama, Arizona, Colorado, Florida, New Mexico, Texas

Samples: University of Alabama, Tuscaloosa, AL, $375,000; United Way, Mile High, Denver, CO, $29,167; Alabama Symphonic Association, Birmingham, AL, $55,000; John P. McGovern Museum of Health and Medical Science, Houston, TX, $20,000;

Contact: Jerry W. Powell, Treasurer; (205) 297-3960

Sponsor: Compass Bank Foundation
P.O. Box 10566, MC AL/BI/CH/ACT
Birmingham, AL 35296-0002

Comprehensive Health Education Foundation Grants 1469

The foundation awards grants to support programs that address health inequities. The initial grantmaking effort will focus on Clark, Pierce, and Spokane Counties in Washington State. One-year grants of up to $20,000 each will be awarded to culturally appropriate, community-led collaborations to test their best idea on how to make it easier for people who suffer from health inequities to move more and eat healthier. Health inequities are defined as differences in the incidence, prevalence, mortality, and burden of diseases that exist for specific populations in the United States. Low-income individuals and people of color within the United States generally have higher rates of poor health and injury than those who are in higher-income groups and are Caucasian.

Requirements: 501(c)3 tax-exempt organizations located in Clark, Pierce, and Spokane Counties in Washington State and units of government that are nondiscriminatory in policy and practice regarding disabilities, age, sex, sexual orientation, race, ethnic origin, or creed are eligible.

Restrictions: Support will not be provided for building or land acquisitions; equipment or furniture purchases; endowment funds; emergency funds; grants to individuals; fellowships/scholarships; research; debt retirement; fundraising activities; general fund drives; indirect overhead; or CHEF programs or products.

Geographic Focus: Washington

Amount of Grant: 500 - 20,000 U.S.D

Contact: Kari L. Lewis, Grant Coordinator; (800) 323-2433, ext. 1899; fax (206) 824-3072; KariL@chef.org

Internet: http://www.chef.org/about/grants.php

Sponsor: Comprehensive Health Education Foundation
159 S Jackson Street, Suite 510
Seattle, WA 98104

ConAgra Foods Foundation Community Impact Grants 1470

The Community Impact Grants (CIG) program will award grants between $10,000 and $100,000 to impactful, grassroots organizations that leverage innovation and creativity to address childhood hunger and nutrition needs in communities where ConAgra Foods' employees live and work or states where 20% or more of children are food insecure. Organizations that demonstrate a strong alignment with the Foundation's giving strategies (i.e., direct service, capacity building, and advocacy) and core funding priorities have the greatest chance of receiving a grant. The CIG program is a two-step, competitive process that first requires the submission of an Letter of Inquiry (LOI) and then the subsequent completion of a full application if invited to apply for a grant. For more detailed program information and guidelines see: http://www.nourishkidstoday.org/downloads/pdf/CIGGuidelines.pdf.

Requirements: The preference of ConAgra Foods Foundation is to award Community Impact Grants to organizations located in states with a child food insecurity rate of 20% or more according to Feeding America's Child Hunger Study as well as those communities where ConAgra Foods has a significant employee presence.

Restrictions: ConAgra Foods Foundation does not fund: professional or amateur sports organizations and teams, or athletic events and programs; political organizations; terrorist organizations or those not compliant with the U.S.A Patriot Act; fundraising events; emergency funding; loans, debt reduction or operating deficits; individuals;

endowments; capital campaigns (unless solicited at the founder's discretion); memorial campaigns; elementary and secondary education.
Geographic Focus: Arizona, Arkansas, California, Colorado, District of Columbia, Florida, Georgia, Idaho, Illinois, Indiana, Iowa, Louisiana, Massachusetts, Michigan, Minnesota, Mississippi, Missouri, New Mexico, North Carolina, Ohio, Oregon, Pennsylvania, South Carolina, Tennessee, Texas, Washington, Wisconsin
Date(s) Application is Due: Jan 29
Amount of Grant: 10,000 - 100,000 U.S.D
Contact: Program Contact; foundation@conagrafoods.com
Internet: http://www.nourishkidstoday.org/about-us/application-guidelines.jsp
Sponsor: ConAgra Foods Foundation
One ConAgra Drive, CC-304
Omaha, NE 68102-5001

ConAgra Foods Foundation Nourish Our Community Grants **1471**
The Foundation awards Nourish Our Community Grants to non-profit organizations based on recommendations from employees. While any organization that is working to address community needs is eligible for funding, preference will be given to those that seek to provide children and their families with access to food and nutrition education. Organizations must be located in the communities where ConAgra Foods employees live and work in order to be considered for a Nourish Our Community grant. Nourish Our Community grants typically range from $5,000 to $25,000, with an average grant of $10,000. The grant requests are reviewed by a committee representative of a cross-section of employees within the company. Organizations can receive funding for up to three consecutive years and then must postpone applying for support for one grant making cycle. Applications for Nourish Our Community grant requests are reviewed annually. The application process for the current fiscal year is June 1-May 31. Additional guidelines are available at: http://www.nourishkidstoday.org/about-us/application-guidelines.jsp
Requirements: Non-profit organizations based in AR, AZ, CA, CO, DC, FL, GA, IA, ID, IL, IN, LA, MA, MI, MN, MO, MS, NC, NE, NM, OH, OR, PA, SC, TN, TX, WA, and WI are eligible to apply for this grant.
Restrictions: ConAgra Foods Foundation does not fund: professional or amateur sports organizations and teams, or athletic events and programs; political organizations; terrorist organizations or those not compliant with the U.S.A Patriot Act; fundraising events; emergency funding; loans, debt reduction or operating deficits; individuals; endowments; capital campaigns (unless solicited at the founder's discretion); memorial campaigns; elementary and secondary education.
Geographic Focus: Arizona, Arkansas, California, Colorado, Connecticut, District of Columbia, Florida, Georgia, Idaho, Illinois, Indiana, Iowa, Louisiana, Maine, Massachusetts, Michigan, Minnesota, Mississippi, Missouri, New Hampshire, New Mexico, North Carolina, Ohio, Oregon, Pennsylvania, Rhode Island, South Carolina, Tennessee, Texas, Vermont, Washington, Wisconsin
Date(s) Application is Due: May 31
Amount of Grant: 5,000 - 25,000 U.S.D
Contact: Program Contact; foundation@conagrafoods.com
Internet: http://www.nourishkidstoday.org/about-us/application-guidelines.jsp
Sponsor: ConAgra Foods Foundation
One ConAgra Drive, CC-304
Omaha, NE 68102-5001

Cone Health Foundation Grants **1472**
The Cone Health Foundation invests in the development and support of activities, programs, and organizations that measurably improve the health of those in the greater Greensboro, North Carolina area. The Foundation awards grants to eligible not-for-profit organizations, government agencies, public schools and academic and/or research institutions, directing resources to four funding priorities: access to necessary health services with particular emphasis on eliminating the barriers often encountered by people in need; adolescent pregnancy prevention; HIV/AIDS and other sexually transmitted infections; and mental health and substance abuse. Current applications and guidelines are available online.
Restrictions: The Foundation does not support: activities that exclusively benefit the members of sectarian or religious organizations; annual fund drives; political campaigns or other partisan political activity; direct financial assistance to meet the immediate needs of individuals; endowments; or retirement of debt.
Geographic Focus: North Carolina
Contact: Antonia Monk Reaves, Senior Program Director; (336) 832-9555; fax (336) 832-9559; antonia.reaves@conehealth.com
Internet: http://www.conehealth.com/ForGrantseekers
Sponsor: Cone Health Foundation
721 Green Valley Road, Suite 102
Greensboro, NC 27408

Connecticut Community Foundation Grants **1473**
The Connecticut Community Foundation serves the people of Greater Waterbury and Northwest Connecticut by supporting public and nonprofit organizations providing programs and services including those for the arts, human services, health care, environment, youth development and education. The Foundation gives priority to efforts that prevent problems, encourage community solutions or improve the organizational capability and financial stability of nonprofit agencies.
Requirements: Nonprofit organizations in Beacon Falls, Bethlehem, Bridgewater, Cheshire, Goshen, Litchfield, Middlebury, Morris, Naugatuck, New Milford, Oxford, Prospect, Roxbury, Southbury, Thomaston, Warren, Watertown, Washington, Wolcott, and Woodbury, CT, may submit applications.
Restrictions: Grants are not awarded for religious purposes, political activities, deficit financing, continuing support, fund-raising events, annual campaigns, newly established arts organizations, commissioning of new works of art, general operating support, or endowments.
Geographic Focus: Connecticut
Date(s) Application is Due: Jan 9; Mar 27; Aug 28
Amount of Grant: 8,000 - 10,000 U.S.D
Contact: Carol O'Donnell, Associate CEO/Director of Grants; (203) 753-1315; fax (203) 756-3054; info@conncf.org
Internet: http://www.conncf.org/grants/grants.htm
Sponsor: Connecticut Community Foundation
43 Field Street
Waterbury, CT 06702

Connecticut Health Foundation Health Initiative Grants **1474**
The foundation awards grants to organizations and institutions that directly respond to its current priority areas and result in improving the health status of Connecticut's underserved and unserved populations. Program priorities include children???s mental health—projects related to children???s mental health, including research, community grants, creating resources for clinical effective practices, and parent advocacy groups; oral health—improving oral health care access, quality, and utilization; and racial and ethnic health disparities—improving the diversity of the health care workforce, and increasing cultural competency in the existing workforce. The foundation awards two major types of grants: strategic and responsive. Application and guidelines are available online.
Requirements: Connecticut state and local units of government, community health centers, health advocacy organizations, community-based organizations, community and cultural groups, schools, and faith-based organizations are eligible. Applicants must have IRS 501(c)3 tax-exempt status or be public entities. Unincorporated organizations may apply through 501(c)3 fiscal agents.
Restrictions: Foundation grants do not support awards to individuals; construction of buildings; capital projects, endowments, or chairs associated with universities, and medical schools; conferences (unless part of a greater project or program); projects that do not benefit Connecticut residents; lobbying or influencing the outcomes of a proposed piece of legislation or election; and indirect cost for discretionary grants.
Geographic Focus: Connecticut
Date(s) Application is Due: Mar 15; Jun 15; Sep 15; Dec 15
Amount of Grant: 50,000 - 200,000 U.S.D
Samples: 1000 Friends of Connecticut, $10,000 to improve public health through the Smart Growth Education and Communications Campaign; Asian Family Services, Inc., Hartford $25,000 to integrate health-related policy issues into the community dialogue during the political season in 2006; Asian Family Services, Inc., Hartford, $10,000 to merge their services with Community Renewal Team, an anti-poverty multi-service agency that serves families and people throughout the Connecticut River Valley; Association of Yale Alumni in Public Health, New Haven, $30,000 to conduct an assessment of the policy, practices and procedures for recruiting and retaining historically under-represented faculty members at the Yale University School of Public Health.
Contact: Onell Jesus Calderas, Grants Administrator; (860) 224-2200; fax (860) 224-2230; onell@cthealth.org
Internet: http://www.cthealth.org/matriarch
Sponsor: Connecticut Health Foundation
74A Vine Street
New Britain, CT 06052

Connecticut Health Foundation Leadership Fellows Program **1475**
This is a one-year (October to June) knowledge and skill-building program designed to create leaders, particularly racial and ethnic community leaders, who will pursue careers in public policy, public health practice, community advocacy, academia, or any other related field. Fellows are selected through a competitive process, which consists of submitting an online application and, if invited, a personal interview. Fellows make a one-year commitment to attend two weekend retreats and monthly seminars and participate in a project that either demonstrates, illustrates, or replicates a solution to a health issue that impacts vulnerable communities. Benefits include the following: fellows receive a stipend to aid in their personal or professional development; fellows who are employed by nonprofit organizations receive a gift for their employer; fellows will enhance their knowledge and skills of leadership and health issues through two weekend retreats and monthly seminars; and fellows will meet with national and local health leaders and policymakers. Application and guidelines are available online.
Requirements: Applicant must be a resident of Connecticut, or primary work must directly impact the people of Connecticut; have experience or interest in working with health needs of minority populations; have strong evidence of leadership experience or potential, especially as related to community projects or health policy; and have the intention to pursue a career in public health practice, policy, academia, or community advocacy.
Geographic Focus: Connecticut
Date(s) Application is Due: May 15
Amount of Grant: 1,500 - 2,000 U.S.D
Contact: William Crimi, (860) 224-2200; fax (860) 224-2230; Will@cthealth.org
Internet: http://www.cthealth.org/matriarch/MultiPiecePage.asp?PageID=171&Page Name=LeadershipFellows
Sponsor: Connecticut Health Foundation
74A Vine Street
New Britain, CT 06052

Connelly Foundation Grants 1476

To achieve its mission to foster learning and improve the quality of life, Connelly Foundation provides grants toward costs associated with programs, direct services, general operations and capital projects to non-profit organizations and institutions working in the following fields: education; health and human services; arts and culture and civic enterprise. The Foundation supports non-profits with strong leadership, sound ideas, future viability, and attainable and well defined goals. It directs its philanthropy toward 501(c)3 organizations and institutions based in and serving Philadelphia and the counties of Bucks, Chester, Delaware, Montgomery and the City of Camden. The Foundation values the proposal process. Given its preference to review a comprehensive package as a primer for discussion, letters of inquiry or requests for pre-proposal discussions are not deemed necessary. Written proposals from nonprofit organizations are accepted and reviewed by the Connelly Foundation throughout the year, there are no deadlines.

Requirements: 501(c)3 organizations and institutions based in and serving in Philadelphia and its surrounding counties of Bucks, Chester, Delaware, and Montgomery in Pennsylvania and in the City of Camden, New Jersey are eligible to apply. There are determined parameters to the Foundation's financial support. It provides non-profit organizations only one grant within a twelve month period. As a general practice, it does not fund advocacy, annual appeals, charter schools, conferences, environmental projects, feasability or planning studies, general solicitations, historic preservation projects, national organizations, organizations focused on a single disease, public schools or research.

Restrictions: The foundation does not award grants to individuals, or political or national organizations; Nor does it respond to annual appeals or general letters of solicitation.

Geographic Focus: Pennsylvania

Contact: Emily C. Riley, Executive Vice President; (610) 834-3222; fax (610) 834-0866; info@connellyfdn.org

Sponsor: Connelly Foundation
1 Tower Bridge, Suite 1450
West Conshohocken, PA 19428

ConocoPhillips Foundation Grants 1477

The foundation makes charitable grants (primarily in the communities where it has operations) in support of education, medical programs, human services, civic, cultural, youth, and other services. Contributions will be considered for organizations such as: federated organizations; educational institutions, both public and private, primarily at the college level; youth organizations; hospital and medical facilities and programs such as hospital buildings and equipment, improvement campaigns and other medical facilities; cultural organizations; civic services; and human service organizations. Applicants may download the application form from the Web site. There are no application deadlines.

Requirements: 501(c)3 tax-exempt organizations and, where appropriate 170(c) organizations, and international nonprofit organizations are eligible. Proof of the exemption must be submitted with grant applications.

Restrictions: Grants do not support religious organizations for religious purposes; war veterans and fraternal service organizations; endowment funds; national health organizations and programs; grants or loans to individuals; fund-raising events; corporate memberships or contributions to chambers of commerce, taxpayer associations and other bodies whose activities are expected to directly benefit the company; or political organizations, campaigns and candidates.

Geographic Focus: Alaska, California, Illinois, Louisiana, Montana, New Jersey, Oklahoma, Pennsylvania, Texas, Washington

Date(s) Application is Due: Aug 1

Amount of Grant: Up to 46,000,000 U.S.D

Contact: Community Relations Manager

Internet: http://www.conocophillips.com/about/Contribution+Guidelines/index.htm

Sponsor: ConocoPhillips Foundation Grants
600 N Dairy Ashford
Houston, TX 77079

Conservation, Food, and Health Foundation Grants for Developing Countries 1478

The Conservation, Food, and Health Foundation's geographic focus is the developing world. Through grants to support research and through targeted grants to help solve specific problems, the foundation helps build capacity within developing countries in three areas of interest: conservation, food, and health. The foundation concentrates its grant-making on research, technical assistance, and training projects of benefit to the Third World; favors grants for pilot projects and special programs that have a potential for replication; prefers to support projects that employ and/or train personnel from the developing world; and favors research concerning problems of importance to the developing world. The foundation has two 4-month funding cycles each year. Concept papers received by January 1 will be considered for eligibility for the March 1 full proposal deadline; concept papers received by July 1 will be considered for eligibility for the September 1 full proposal deadline. Full proposals are by invitation. Concept applications must be submitted through the foundation's online application system. Detailed guidelines, instructions, and faqs are provided at the website.

Requirements: The following organizations are eligible to apply: 501(c)3 organizations that are not private foundations under section 509(a) of the United State Internal Revenue Code; "501(c)3 equivalent" foreign or domestic government units; and nongovernmental foreign organizations which can provide secure evidence of their nongovernmental status and charitable purpose.

Restrictions: The foundation does not provide support for buildings, land purchases, or vehicles; quantity purchases of durable medical equipment; endowments or fundraising activities; famine or emergency relief; direct medical care or treatment; feeding or food distribution programs; films, videos, or web-site production; scholarships, fellowships, or travel grants; conferences; re-granting through intermediaries; general operating support; or individuals (however, the foundation may support an individual engaged in research on a problem of significance to the developing world where the research is sponsored by an established, nonprofit organization such as an educational institution and conducted in close partnership with a local nongovernmental organization). The foundation does not consider the states of the former Soviet Union or former Eastern Bloc countries as within its geographic focus.

Geographic Focus: All States, All Countries

Date(s) Application is Due: Jan 1; Jul 1

Amount of Grant: Up to 25,000 U.S.D

Samples: Fundacion Natura Bolivia, Santa Cruz, Bolivia, $29,242 - to develop a training program that helps municipal watershed authorities in Bolivia to develop payment schemes that help upstream landowners protect forests valuable to downstream communities (2012); One Acre Fund, Falcon Heights, Minnesota, $25,000 - to refine a training, financing, education, and support program that helps rural Kenyan and Rwandan farmers double their harvests and income (2012); Hohenheim University, Stuttgart, Germany, $23,813 - for a study in collaboration with Veld Products Research & Development in Botswana to assess how strip intercropping wild grass with sorghum-cowpea contributes to the field water balance as well as effects on field-water-use efficiency (2012).

Contact: Prentice Zinn, Administrator; (617) 391-3091 or skype: prentice.zinn; pzinn@gmafoundations.com

Internet: http://cfhfoundation.grantsmanagement08.com/?page_id=5

Sponsor: Conservation, Food, and Health Foundation
77 Summer Street, 8th Floor
Boston, MA 02110-1006

Constantin Foundation Grants 1479

The foundation awards grants to nonprofit organizations in Texas, with an emphasis on higher and other education, including secondary school education, vocational education, and adult continuing education. Grants also support human, family, and social service agencies; hospitals and health organizations, including substance abuse; community development; and arts and culture. Grants are awarded for continuing support, capital campaigns, building construction/renovation, equipment acquisition, challenge/matching funds, and to develop programs. Request guidelines prior to applying.

Requirements: Nonprofit organizations in Dallas County, TX, may submit applications.

Restrictions: No support for tax-supported institutions, theater groups, churches, debt retirement, political organizations or second party requesters. No grants to individuals, or for endowments, research, debt retirement, operations, research, special events, fundraisers, or second party requests; no loans.

Geographic Focus: Texas

Date(s) Application is Due: Sep 15

Amount of Grant: 1,000 - 100,000 U.S.D

Contact: Cathy Doyle, Executive Director; (214) 522-9300 or (214) 522-9305; fax (214) 521-7023; constantinfdn@sbcglobal.net

Sponsor: Constantin Foundation
4809 Cole Avenue, LB 127
Dallas, TX 75205-3578

Constellation Energy Corporate Grants 1480

Constellation Energy provides its philanthropic resources to non-profit organizations that make an impact in these key focus areas: energy Initiatives; environment; education; economic development. Applications are accepted at various times throughout the year. Grant, Sponsorship, Banner Hanging, and In-Kind requests under $10,000 are reviewed on a rolling basis. The Corporate Contributions Committee meets bi-annually, in May and October, to review significant financial grant requests of $10,000 or more. Grant requests should be submitted by April 1 and September 1 respectively. Additional guidelines and the online applications are available at the companies website.

Requirements: The company makes charitable donations to 501(c)3 tax-exempt, nonprofit organizations.

Restrictions: Constellation Energy does not make grants to: individuals; churches or religious causes; individual schools; athletic teams or events; programs located outside Constellation Energy communities.

Date(s) Application is Due: Apr 1; Sep 1

Contact: Larry McDonnell, Director; (401) 470-7433; media@constellation.com

Internet: http://www.constellation.com/portal/site/constellation/menuitem.94939662e40191875fb60610025166a0/

Sponsor: Constellation Energy Corporate
100 Constellation Way, Suite 1000C
Baltimore, MD 21202

Consumers Energy Foundation 1481

Since its creation in 1990, the Consumers Energy Foundation has touched countless lives and communities through it's grant programs, corporate giving and employee volunteers. The Foundation accepts grant applications from nonprofit organizations for innovative projects and activities creating measurable impact in five areas: Social Welfare, Michigan Growth and Environmental Enhancement, Education, Community and Civic, and Culture/Arts. The Foundation's areas of support include, operating budgets and capital funds.

Requirements: The Consumers Energy Foundation provides financial support primarily to Michigan organizations classified by the Internal Revenue Service as tax-exempt under section 501(c)3 of the Internal Revenue Code.

Restrictions: The following are ineligible for funding: individuals; individual scholarships; individual sponsorship related to fund-raising; organizations that do not qualify as charitable organizations as defined by the Internal Revenue Service; organizations that practice discrimination on the basis of sex, age, height, weight, marital status, race, religion, sexual orientation, creed, color, national origin, ancestry, disability, handicap, or veteran status; organizations whose operating activities are already supported by the United Way (except when the request is approved by the appropriate community United Way organization); political organizations and political campaigns; religious organizations when the contribution will be used for denominational or sectarian purposes; labor or veterans organizations; fraternal orders; social clubs; sports tournaments; talent or beauty contests; loans for small business; debt reduction campaigns.
Geographic Focus: Michigan
Amount of Grant: 500 - 10,000 U.S.D
Samples: Saginaw Basin Land Conservancy, $6,000—for educational outreach efforts supporting wetlands preservation and land conservation within the Saginaw Bay Watershed; American Red Cross Mid-Michigan Chapter, $10,000—for a disaster response vehicle; Kalamazoo Department of Public Safety, $15,000—for development of a Kalamazoo Regional Police and Fire Training Center.
Contact: Carolyn Bloodworth, Secretary/Treasurer; (517) 788-0432; fax (517) 788-2281; foundation@consumersenergy.com
Internet: http://www.consumersenergy.com/welcome.htm
Sponsor: Consumers Energy
1 Energy Place, Room EP8-210
Jackson, MI 49201-2276

Cooke Foundation Grants 1482
Grants are awarded primarily for culture and the arts, social services, education, programs for youth and the elderly, humanities, health, and the environment. Organizations receiving grants must be located in Hawaii or serve the people of Hawaii. Preference will be given to requests from Ohau. Types of support include general operating support, capital campaigns, building and renovations, program development, seed money, and matching funds.
Requirements: Grant making is limited to the state of Hawaii.
Restrictions: Grants are not made to individuals, churches, or religious organizations, or for endowment funds, scholarships, or fellowships.
Geographic Focus: Hawaii
Date(s) Application is Due: Mar 2; Sep 1
Amount of Grant: 5,000 - 25,000 U.S.D
Samples: Maui Community Arts and Cultural Ctr (HI)—to support the capital campaign to complete construction, $10,000; Honolulu Dance Theatre (HI)—for a ballet performance, $2,500; Kauai Historical Society (HI)—to support the lecture series, $2,500; Honolulu Academy of Arts (HI)—annual grant, $100,000.
Contact: Carrie Shoda-Sutherland, Senior Program Officer; (808) 566-5524 or (888) 731-3863, ext. 524; fax (808) 521-6286; csutherland@hcf-hawaii.org
Internet: http://www.cookefdn.org/
Sponsor: Cooke Foundation, Limited
1164 Bishop Street, Suite 800
Honolulu, HI 96813

Cooley's Anemia Research Fellowships 1483
The grants-in-aid program supports research that will add to current knowledge about Cooley's anemia and related disorders. The ultimate goal is to find the cause and cure or effective treatment of these diseases. This involves a great deal of basic research in hematology, pathology, physiology, biochemistry, physical chemistry, and related sciences. Fellowships are also available and are awarded for one year, with possible renewal for a second year upon reapplication. The deadline for both fellowships and grants in aid of research is mid-March each year; awards are announced by June 1.
Requirements: Grants are made to institutions and researchers of any country. Individual applicant must have a doctoral degree and no more than five years postdoctoral training.
Date(s) Application is Due: Mar 1
Amount of Grant: Up to 40,000 U.S.D
Contact: Chairperson, Medical Advisory Board; (718) 321-2873 or (800) 522-7222; fax (718) 321-3340; info@cooleysanemia.org
Internet: http://www.cooleysanemia.org/sections.php?sec=58
Sponsor: Cooley's Anemia Foundation
129-09 26th Avenue, Suite 203
Flushing, NY 11354

Cooper-Levy Trust Grants 1484
The trust provides financial support for the United Way and other nonprofit agencies providing food, shelter, basic services, and medical care to low-income families and children in Washington state. Types of support include general operating support and project support. The majority of grants will be made in the Puget Sound area. Application guidelines are available online.
Requirements: Washington 501(c)3 nonprofit organizations in Washington's King, Pierce, Spokane, Snohomish, and Yakima Counties are eligible.
Restrictions: Funds are not awarded for capital campaigns; computer or office equipment; book, video, or website production; event sponsorship; scholarships; medical research; or to individuals. Grants are not directed toward projects serving older adults or indigent adults without dependent children.
Geographic Focus: Washington
Date(s) Application is Due: May 15; Oct 15
Amount of Grant: 5,000 - 10,000 U.S.D
Contact: Therese Ogle, Grants Consultant, c/o The Private Bank, Union Bank of California; (206) 781-3472; fax (206) 784-5987; oglefounds@aol.com
Internet: http://fdncenter.org/grantmaker/cooperlevy
Sponsor: Cooper-Levy Trust
P.O. Box 3123
Seattle, WA 98114

Cooper Industries Foundation Grants 1485
Cooper and the Cooper Industries Foundation annually donate more then $3 Million to nonprofit organizations serving the communities where their employees live and work. The Cooper Industries Foundation accepts and reviews grant requests throughout the year. There is no deadline; however, budgets are compiled annually each fall for the following year. Applicants must submit a brief letter explaining the purpose of the request with: concise description of organization and its mission; purpose and amount of request; budget information and other funding sources; evidence of 501(c)3 tax-exempt status; current listing of board members.
Requirements: Only 501(c)3 tax-exempt-status organizations are eligible. Programs must: benefit a community where Cooper is a significant employer; be endorsed by local Cooper management when applicable; not duplicate the efforts of the four Cooper-created programs; fulfill an important community need. Program objectives must coincide with that of the company.
Restrictions: The following types of organizations are generally ineligible for funding: United Way-funded organizations; national and state organizations; religious organizations; veterans organizations; political candidates, labor and lobbying organizations; hospitals; primary and secondary schools; scholarship organizations (except National Merit).
Geographic Focus: Alabama, Georgia, Illinois, Missouri, New York, North Carolina, South Carolina, Texas, Wisconsin, United Kingdom
Contact: Victoria B. Guennewig, VP Public Affairs; (713) 209-8800; fax (713) 209-8982; info@cooperindustries.com
Internet: http://www.cooperindustries.com/common/aboutCooper/corporateGiving.cfm?CFID=160327&CFTOKEN=61243618
Sponsor: Cooper Industries Foundation
P.O. Box 4446
Houston, TX 77210

Coors Brewing Corporate Contributions Grants 1486
The company has a firm commitment to giving back to its home-market communities—including Denver, CO; Memphis, TN; and Elkton, VA—and supports grassroots, nonprofit organizations that address community, civic and industry issues. The primary focus is on programs that enhance the quality of life. Vehicles of support include cash grants, Coors products for events, Coors logo items for fund-raisers, volunteer hours by Coors employees or retirees, or used equipment or in-kind services. Preference will be given to groups that focus on issues of national scope. Corporate Contributions usually reviews requests the first Wednesday of every month. A minimum of two months lead time is required before the event or funding need.
Requirements: IRS 501(c)3 nonprofit organizations located in Denver, CO; Memphis, TN; and Elkton, VA; are eligible.
Restrictions: Requests will not be considered for individuals in personal programs; individual scholarships; teams, groups, or races; travel expenses; third-party fund-raisers or sales promotions; political activities; or requests by telephone.
Geographic Focus: Colorado, Tennessee, Virginia
Contact: Buck Boze, Corporate Contributions Manager; (800) 642-6116 or (303) 277-5953; fax (303) 277-6132
Internet: http://www.coors.com
Sponsor: Coors Brewing Company
P.O. Box 4030, Department NH420
Golden, CO 80401

Copeland Family Foundation Grants 1487
The foundation awards grants to nonprofit organizations in its areas of interest, including animal welfare, Christian organizations, community development, education, hospitals, health associations, and social services.
Requirements: Nonprofit organizations are eligible.
Geographic Focus: Massachusetts
Amount of Grant: 5,000 - 15,000 U.S.D
Samples: Fairfield U (CT)—to develop a nonlinear-editing suite for use by students in the video-production program, $30,000.
Contact: Grants Administrator; (617) 698-5980
Sponsor: Copeland Family Foundation
1183 Randolph Avenue
Milton, MA 02186

Cornerstone Foundation of Northeastern Wisconsin Grants 1488
The community foundation supports nonprofit organizations primarily in Brown County, Wisconsin. The foundations area of interest are: education, cultural programs, social service and, youth agencies. With an additional interest in supporting healthcare facilities. The foundation offers the following types of support: annual campaigns; building/renovation; capital campaigns; continuing support; debt reduction; emergency funds; endowments; equipment; general/operating support; matching/challenge support; program; development. Contact the foundation by telephone for guidelines before submitting a proposal, no application form is required.

Requirements: Nonprofit organizations in Green Bay and northeastern Wisconsin may submit applications for grant support. The primary focus is Brown County.
Restrictions: No grants are available to: individuals, religious, or political organizations.
Geographic Focus: Wisconsin
Date(s) Application is Due: Jan 15; Sep 15
Amount of Grant: 10,000 - 400,000 U.S.D
Samples: Green Bay Symphony Orchestra, Green Bay, WI, $10,000—for youth orchestra program; Boys and Girls Club of Green Bay, Green Bay, WI, $400,000—for capital campaign; Salvation Army of Green Bay, Green Bay, WI, $24,900—for transitional housing program; N.E.W. Community Clinic, Green Bay, WI, $50,000—for medical expense program.
Contact: Sheri Prosser, Executive Director; (920) 490-8290; fax (920) 490-8620; cornerstone@cfnew.org
Sponsor: Cornerstone Foundation of Northeastern Wisconsin
111 North Washington Street, Suite 450
Green Bay, WI 54301-4208

Corning Foundation Community Service Grants **1489**
The foundation supports a variety of organizations that serve a broad base of constituents. Included in this category are hospitals and hospices, community foundations, youth and women's centers, YMCAs, local chapters of Girl Scouts and Boy Scouts of America, and selected United Ways.
Requirements: All requests to the foundation for support must be made in writing. Grant seekers are advised to submit a two- to three-page letter of inquiry, signed by the senior administrative officer of the organization.
Restrictions: Grants do not support individuals; political parties, campaigns, or causes; labor or veterans' organizations; religious or fraternal groups; volunteer emergency squads; athletic activities; courtesy advertising; or fundraising events.
Samples: Glove House (Elmira, NY)—to support programs for children and families at risk, $4000; Corning Meals on Wheels (New York, NY)—general program support, $2000.
Contact: Karen Martin, Associate Director; (607) 974-8746; fax (607) 974-4756; martinkc@corning.com
Internet: http://www.corning.com/inside_corning/foundation.asp
Sponsor: Corning Foundation
1 Riverfront Plaza, MP-BH-07
Corning, NY 14831

Courage Center Jay and Rose Phillips Award **1490**
The awards are presented each year to people with disabilities in Minnesota, Iowa, Wisconsin, North Dakota, and South Dakota who have achieved outstanding success in their vocation. The purpose of the awards is to encourage individuals with disabilities to prepare for and find employment in their chosen field and to encourage employers to recognize the advantages of hiring people with disabilities.
Requirements: Nominees must have a physical disability and be vocationally and financially independent (medical assistance excluded) for a minimum of three years. Self-employed individuals are eligible for the award, and employers are encouraged to nominate employees who meet nomination criteria.
Geographic Focus: Iowa, Minnesota, North Dakota, South Dakota, Wisconsin
Date(s) Application is Due: Aug 1
Amount of Grant: 1,500 U.S.D
Contact: Tom Kelley, (763) 520-0675; fax (763) 520-0562; tom.kelley@courage.org
Internet: http://www.courage.org/newsevents/newsdetail.asp?id=256&type=3
Sponsor: Courage Center
3915 Golden Valley Road
Minneapolis, MN 55422

Covenant Foundation Grants (Iowa) **1491**
Covenant Foundation, Inc. was created in 1988 through the merger of the Schoitz and St. Francis Hospital Foundations. The Foundation seeks to bring together the gifts and resources of the community to help meet the health care needs of those who are served in all areas of the hospital. Since its creation, the Foundation has served patients and families through the purchase of technical and specialized equipment, the construction and improvement of physical facilities, and community education program that promote health and illness prevention.
Requirements: 501(c)3 organizations of northeast Iowa are eligible to apply.
Geographic Focus: Iowa
Contact: Heather Bremer-Miller, Executive Director; (319) 272-7678; Heather.BremerMiller@wfhc.org
Internet: http://www.covhealth.com/covenant_foundation.asp
Sponsor: Covenant Foundation (Waterloo)
3421 W 9th Street
Waterloo, IA 50702

Covenant Matters Foundation Grants **1492**
The Covenant Matters Foundation, established in 2003, is the non-profit arm of Strategic Health Policy International, a group that organizes chaos and solves complex problems in health care. The Foundation supports research, strategic planning, marketing, sales, public affairs and other business operations, and it helps to fund, facilitate, negotiate, analyze, strategize, and mentor health care policy-makers. Currently, there are no specific application forms or deadlines, and potential applicants should contact the foundation in writing.
Contact: Glenna M. Crooks, President; (215) 646-8182 or (215) 646-0542; fax (215) 646-7368; GMC@strategichealthpolicy.com
Internet: http://www.strategichealthpolicy.com/OPENSOURCE/Grantseeking/tabid/123/Default.aspx
Sponsor: Covenant Matters Foundation
1075 Fort Washington Avenue
Fort Washington, PA 19034-1617

Cowles Charitable Trust Grants **1493**
The foundation awards grants, primarily in New York, Florida, and on the East Coast, for the arts and culture, including museums and the performing arts; environment; education, including early childhood education, secondary and higher education, medical school education, adult basic education and literacy, and adult continuing education; hospitals and AIDS programs, including research; social services, including family planning, human services, and federated giving; and community funds, including leadership development, civil rights, and race relations. Types of support include general operating support, capital campaigns, annual campaigns, equipment acquisition, endowment funds, continuing support, seed money, building construction/renovation, matching funds, professorships, and program development. Application forms are required; initial approach should be by letter. The board meets in January, April, July, and October.
Requirements: Nonprofit organizations may apply for grant support. Grants are awarded primarily along the Eastern Seaboard.
Geographic Focus: Florida, New York
Date(s) Application is Due: Mar 1; Jun 1; Sep 1; Dec 1
Amount of Grant: 1,000 - 40,000 U.S.D
Samples: Church of the Holy Comforter, Burlington, NC, $53,200; Miami Art Museum of Dade County, Miami, FL, $30,000; Allen-Stevenson School, New York, NY, $27,500.
Contact: Mary Croft, Treasurer; (732) 936-9826
Sponsor: Cowles Charitable Trust
P.O. Box 219
Rumson, NJ 07760

CPhA Centennial Award **1494**
The objective of these awards is to give firsthand information to qualified pharmacy students to improve and develop their knowledge of their chosen profession. The awards provide travel, accommodation, and registration for the CPhA annual conference. One student per university is selected on the basis of achievement up to midterm period of the particular year, the achievement especially related to student activity involvement, and leadership coupled with academic qualifications. Nine awards are made each year.
Requirements: Honors students in their second-last year of study from each faculty or school of pharmacy in Canada who have actively promoted the profession and faculty of pharmacy, and who have a good academic standing.
Geographic Focus: All States, Canada
Date(s) Application is Due: Jan 31
Amount of Grant: 300 CAD
Contact: Nancy Shurtleff, Manager of Board Affairs; (613) 523-7877; fax (613) 523-0445; Nshurtleff@cdnpharm.ca
Internet: http://www.pharmacists.ca/content/about_cpha/who_we_are/awards/centennial.cfm
Sponsor: Canadian Pharmaceutical Association
1785 Alta Vista Drive
Ottawa, ON K1G 3Y6 Canada

Crail-Johnson Foundation Grants **1495**
The Crail-Johnson Foundation (CJF) has defined itself as a children's charity, and the vast majority of grant-making is directed toward programs benefiting children, youth and families in the greater Los Angeles area. Proposals, which are not relevant to the foundation's mission and funding priorities, will not be considered. CJF provides grants for program initiatives and enhancements, general operating support and capital projects as well as selected endowments. CJF provides technical assistance to selected community-based initiatives benefiting children and families. The majority of Crail-Johnson Foundation funding supports organizations located in the greater Los Angeles area and projects that directly benefit Los Angeles area residents. National organizations providing services in Los Angeles are also considered. Occasionally, grants are made to programs and projects that are regional or national in scope, where potential benefits to children and families in Los Angeles can be clearly demonstrated. Initial contact with the Crail-Johnson Foundation should be in the form of a letter of inquiry, letters are accepted October through December each year for the following year's grant cycle and are generally considered in the order in which they are received. Organizations selected to submit proposals will be asked to complete a Grant Application Form provided by the Foundation.
Requirements: The foundation provides financial support primarily through grants to public non-profit organizations that are exempt under Section 501(c)3of the Internal Revenue code and are not a private foundation under Section 509(a).
Restrictions: Support is not granted for programs and projects benefiting religious purposes, university level education, research, events recognizing individuals or organizations, political causes or programs attempting to influence legislation. No grants are made directly to individuals.
Geographic Focus: California
Amount of Grant: 5,000 - 50,000 U.S.D
Samples: Heart of Los Angeles Youth, Los Angeles, CA, $25,000—Smart Start Elementary Education Program; Mentor LA, Los Angeles, CA, $25,000—staffing and special program enrichment; Strive Foundation, Houston, TX, $20,000—expansion of academic and arts Program.
Contact: Pat Christopher, Program Officer; (310) 519-7413; fax (310) 519-7221; pat-christopher@crail-johnson.org

Internet: http://www.crail-johnson.org/grants-application.htm
Sponsor: Crail-Johnson Foundation
222 W Sixth Street, Suite 1010
San Pedro, CA 90731

Cralle Foundation Grants **1496**
The foundation awards grants in the areas of education and higher education, children and youth services, community development, human services, and museums to Kentucky nonprofits. Emphasis is given to nonprofits serving residents of Louisville. Types of support include: building and renovation; capital campaigns; continuing support; endowments; equipment; operating support; matching and challenge support; program development; scholarship funds; and seed money. Interested organizations should initially send a letter requesting an application form. Applicants should submit four copies of their application. Application deadlines are March 1 and September 1. The foundation's board meets in April and October.
Requirements: Nonprofit organizations in Kentucky are eligible.
Restrictions: No grants to individuals.
Geographic Focus: Kentucky
Date(s) Application is Due: Mar 1; Sep 1
Amount of Grant: 5,000 - 50,000 U.S.D
Samples: Maryhurst School, Louisville, KY, $40,250 - for Seven Challenges Program (2011); Blessings in a Backpack, Louisville, KY, $25,000 - for students at Portland Elementary (2011); Family and Children's Place, Louisville, KY, $25,000 - for operating support (2011).
Contact: James Crain, Jr., Executive Director; (502) 581-1148; fax (502) 581-1937; jcrain37@bellsouth.net
Sponsor: Cralle Foundation, Inc.
614 West Main Sreet, Suite 2500
Louisville, KY 40202-4252

Cranston Foundation Grants **1497**
The foundation directs most of its grants to the geographic areas where Cranston Print Works Company has operations and a large number of employees, including Rhode Island, Massachusetts, Louisiana, and New York. Targeted priority areas are education, health and welfare, culture and the arts, and civic and community programs. Types of support include general operating support, employee matching gifts, and employee-related scholarships. All requests should be made in writing and should contain details and funding requirements pertaining to the project. The February deadline is for scholarships; the August deadline is for grants.
Requirements: IRS 501(c)3 organizations serving Rhode Island, Massachusetts, Louisiana, and New York are eligible.
Geographic Focus: Massachusetts, New York, Rhode Island
Date(s) Application is Due: Feb 15; Aug 31
Amount of Grant: 200 - 5,000 U.S.D
Contact: Trustee; (401) 943-4800
Sponsor: Cranston Foundation
1381 Cranston Street
Cranston, RI 02920

Crestlea Foundation Grants **1498**
The foundation supports Delaware nonprofits in its areas of interest, including secondary education, higher education, natural resource conservation, health care, housing development, community development, social services, and public affairs. Types of support include annual campaigns, general operating support, capital campaigns, building construction/renovation, and equipment acquisition. There are no application forms.
Requirements: Delaware nonprofits are eligible.
Geographic Focus: Delaware
Date(s) Application is Due: Nov 1
Amount of Grant: 5,000 - 50,000 U.S.D
Samples: Philadelphia Museum of Art (Philadelphia, PA)—for operating support, $82,500; Archmere Academy (Claymont, DE)—for capital campaign, $20,000.
Contact: Stephen Martinenza, Secretary-Treasurer; (302) 654-2477
Sponsor: Crestlea Foundation
100 W 10th Street, Suite 1109
Wilmington, DE 19801

CRH Foundation Grants **1499**
The foundation awards general operating grants to U.S. nonprofit organizations in its areas of interest, including health and human services. Religious organizations also are eligible. Preference is given to requests from New Jersey and New York. Applications are accepted each year between May 1 and October 31. Inquiries must be in writing.
Requirements: U.S. 501(c)3 nonprofit organizations may apply.
Date(s) Application is Due: Oct 31
Amount of Grant: Up to 74,000 U.S.D
Samples: United Jewish Community (River Edge, NJ)—$4000; New York Times Neediest Cases (New York, NY)—$1000; American Cancer Society (Atlanta, GA)—$1000.
Contact: Grants Administrator; (201) 568-9300; fax (201) 568-6374
Sponsor: CRH Foundation
175 N Woodland Street
Englewood, NJ 07631

CRI Clinical Investigator Award in Cancer Immunology **1500**
Investigator awards are available to six qualified scientists at the assistant professor level who are working in the field of clinical cancer immunology. The four-year award may be used at the recipient's discretion for salary, technical assistance, supplies, or capital equipment. CRI has no citizenship restrictions, and research supported by the award may be conducted anywhere in the United States or abroad, excluding for-profit institutions. Applications may be downloaded from the Web site.
Requirements: Candidate must hold a doctoral degree and be a tenure-track assistant professor or equivalent rank at the time of the award.
Date(s) Application is Due: Mar 1
Amount of Grant: 50,000 - 200,000 U.S.D
Contact: Lynne Harmer, Grants Administrator; (212) 688-7515; fax (212) 832-9376; grants@cancerresearch.org
Internet: http://www.cancerresearch.org/clinical.html
Sponsor: Cancer Research Institute
681 Fifth Avenue
New York, NY 10022-4209

CRI Postdoctoral Fellowships **1501**
Each year the institute awards postdoctoral fellowships to qualified individuals in the formative stages of their career who wish to receive training in cancer immunology with special emphasis on tumor immunology. Work may be carried out in the United States or abroad. The fellowships are awarded for a period of two years, with the possibility of renewal for one additional year. CRI requires all fellowship applicants to complete an electronic application as well as a paper application.
Requirements: Applicant of any nationality must have a doctoral degree and must conduct research under a sponsor who holds a formal appointment at the sponsoring institution.
Date(s) Application is Due: Apr 1; Oct 1
Amount of Grant: 11,200 - 130,500 U.S.D
Contact: Brian Brewer, Grants Administrator; (212) 688-7515, ext. 242 or (800) 992-2623; fax (212) 832-9376; grants@cancerresearch.org
Internet: http://www.cancerresearch.org/postdoc.html
Sponsor: Cancer Research Institute
681 Fifth Avenue
New York, NY 10022-4209

CRI Prostate Cancer Initiative Grants **1502**
The program supports Phase I and Phase I/II clinical trials that test novel therapies for advanced prostate cancer, with particular emphasis on hormonal therapy and immunotherapy. CRI does not provide funds for indirect costs. The completed application consists of 15 sets of the following items collated in the order specified: application form; biographical sketches for all key personnel; abstract of proposed project explaining the importance of the proposed research and ramifications; five-page description of research program, including summary of past work and future plans; and IRB-approved protocol. Applications must be typed or printed (using both sides of the paper, if possible). In addition, a self-addressed, stamped postcard is required if applicant would like to receive notification of receipt of application. Annual deadline dates may vary; contact the program office for exact dates.
Date(s) Application is Due: Apr 15
Amount of Grant: 150,000 U.S.D
Contact: Grants Administrator; (212) 688-7515 or (800) 992-2623; fax (212) 832-9376; info@cancerresearch.org
Internet: http://www.cancerresearch.org/pciinit.html
Sponsor: Cancer Research Institute
681 Fifth Avenue
New York, NY 10022-4209

Crohn's Disease and Colitis Research Fellowships **1503**
Three-year fellowships are awarded to encourage the development of individuals with research potential to help them prepare for careers of independent research in the areas of Crohn's disease and ulcerative colitis. Yearly renewal is dependent on the receipt of satisfactory progress reports. Application forms must be obtained from the foundation. Letters of intent are due May 1 and November 1. Submission deadlines for full applications are January 14 and July 1 of each year. Annual deadline dates may vary; contact program staff for exact dates.
Requirements: The applicant must hold an MD, PhD, or equivalent degree and be currently employed by an institution engaged in health care and/or health-related research within the United States. Candidates who have obtained MD degrees must have at least two years of postdoctoral experience, one year of which must be documented research experience relevant to inflammatory bowel disease. Candidates holding PhD degrees must have one year of postdoctoral experience and at least one year of documented research experience relevant to IBD.
Date(s) Application is Due: Jan 14; May 1; Jul 1; Nov 1
Amount of Grant: Up to 58,250 U.S.D
Contact: Carol Cox, Director of Research and Scientific Programs; (212) 685-3440 or (800) 932-2423; fax (212) 779-4098; ccox@ccfa.org
Internet: http://www.ccfa.org/research/?LMI=4
Sponsor: Crohn's and Colitis Foundation of America
386 Park Avenue S, 17th Floor
New York, NY 10016-8804

Crohn's Disease and Ulcerative Colitis Senior Research Grants **1504**

Grants are awarded to individuals for up to two years for research in the basic biomedical and clinical sciences that will increase the understanding of the cause, mechanisms of symptoms, and treatment of inflammatory bowel disease including Crohn's disease and ulcerative colitis. Support for additional one-year periods will be by competitive renewal. Application forms must be secured from the foundation. Submission deadlines for letters of intent are November 1 and May 1. Submission deadlines for full applications are January 14 and July 1 of each year, unless noted otherwise.

Requirements: Applicants must be established researchers in the field of inflammatory bowel disease.

Date(s) Application is Due: Jan 14; May 1; Jul 1; Nov 1

Amount of Grant: Up to 115,000 U.S.D

Contact: Carol Cox, Director of Research and Scientific Programs; (212) 685-3440 or (800) 932-2423; fax (212) 779-4098; ccox@ccfa.org

Internet: http://www.ccfa.org/research/?LMI=4

Sponsor: Crohn's and Colitis Foundation of America
386 Park Avenue S, 17th Floor
New York, NY 10016-8804

Crown Point Community Foundation Grants **1505**

The Crown Point Community Foundation is interested in funding projects which improve the quality of life for citizens in the Crown Point area. The foundation utilizes matching funds programs and challenge grants to stimulate the fundraising efforts of local organizations. The Crown Point Community Foundation considers grant applications for projects in the following areas: education; health and human services; civic affairs; arts and culture; and preservation/conservation. Grants are awarded in February, June, and September of each year. Applications received after the deadline will be held for consideration during the next cycle.

Requirements: The Foundation considers the following guidelines when reviewing grant applications: projects are limited within the territorial boundaries of the Crown Point Community School Corporation and the city of Crown Point; only one grant application per organization is eligible within a 12 month period, January through December; the potential community impact of the grant and the number of people who will benefit; the extent of local volunteers involvement and support for the project; the composition and level of commitment of the organization's directors/trustees; the organization's fiscal responsibility and management qualifications; the ability of the organization to obtain additional funding to implement the project; and the organization's ability to provide funding after a long-term grant has expired. Grants are usually made to 501(c)3 non-profit organizations. Grants are primarily awarded to underwrite program expenses or to fund capital expenditures. All applicants must submit the application form; a detailed three page narrative of the proposal; a proposed budget including revenue and expenses for the project; a recent financial audit or statement; a copy of the 501(c)3 tax exempt form; and a list of the organization's officers and board of directors.

Restrictions: Grants are not made for endowment purposes. No grants may be used for any political campaign, or to support attempts to influence the legislature of any governmental body. The Foundation funds only grant seekers who do not unlawfully discriminate as to age, race, religion, sex, disability, or national origin.

Geographic Focus: Indiana

Date(s) Application is Due: Feb 1; Jun 1; Sep 1

Samples: St. Mary's Catholic Community School, Crown Point, Indiana; $7,140: Lake of the Four Seasons Volunteer Fire Department, Crown Point, Indiana; $4,110: Crown Point Adult Learning Center, Crown Point, Indiana; $5,000.

Contact: Patricia Huber, President/Executive Director; (219) 662-7252; fax (219) 662-9493; cpcf@sbcglobal.net

Internet: http://crownpointcommunityfoundation.org/grants/guidelines/

Sponsor: Crown Point Community Foundation
115 South Court Street
Crown Point, IN 46308-0522

Crown Point Community Foundation Scholarships **1506**

The Crown Point Community Foundation Scholarships are made available through donations and endowment fund distributions. A donor committee helps review and score the applications, interview the students and award the scholarships. The website lists specific scholarships available from area schools, churches, country clubs, and individuals.

Requirements: Applicants should contact the foundation for applications and more information.

Restrictions: Several scholarships have specific requirements, such as the applicant must have attended the school, studying a particular major, or the applicant must be the child of a local steelworker.

Geographic Focus: Indiana

Amount of Grant: 500 - 1,500 U.S.D

Contact: Patricia Huber, President/Executive Director; (219) 662-7252; fax (219) 662-9493

Internet: http://crownpointcommunityfoundation.org/scholarships/available/

Sponsor: Crown Point Community Foundation
115 South Court Street
Crown Point, IN 46308-0522

Cruise Industry Charitable Foundation Grants **1507**

The foundation awards grants to improve the quality of life in U.S. cities and towns where the cruise industry maintains vessel operations, employs a significant number of individuals, and purchases products and services. Areas of interest include civic and community development, educational assistance and training programs, public health programs, and environmental initiatives. There are no application deadlines, see Foundation's website for application guidelines.

Requirements: U.S. 501(c)3 organizations are eligible to apply.

Restrictions: Requests will not be considered from individuals, fraternal organizations, religious organizations, political organizations, or organizations that conduct lobbying activity.

Samples: American Heart Association—Public Health Program; Armed Forces Foundation—Civic and Community Development Program; Page Ahead—Educational Assistance and Training Program.

Contact: Cynthia Colenda; (703) 522-3160; fax (703) 522-3161; cicf@iccl.org

Internet: http://www.cruisefoundation.org/requirements-and-guidelines.php

Sponsor: Cruise Industry Charitable Foundation
2111 Wilson Boulevard, 8th Floor
Arlington, VA 22201

Crystelle Waggoner Charitable Trust Grants **1508**

Born to a ranching family, Crystelle Waggoner raised cattle and thoroughbred horses. A patron of the arts and supporter of medical charities during her lifetime, she established a 50-year trust to benefit charitable organizations in the arts and social services. Applicants must apply online at the grant website. Applicants are strongly encouraged to do the following before applying: review the downloadable state application procedures for additional helpful information and clarifications; review the downloadable online-application guidelines at the grant website; review the trust's funding history (link is available from the grant website); review the online application questions in advance; and review the list of required attachments. These will generally include: a list of board members, financial statements (audited, reviewed, or compiled by independent auditor); an organization summary; a list of other funding sources; an IRS Determination letter; and other required documents. All attachments must be uploaded in the online application as PDF, Word, or Excel files. The Crystelle Waggoner Charitable Trust has bi-annual application deadlines of March 31 and September 30. Applications must be submitted by 11:59 p.m. on the deadline dates.

Requirements: Applicants must have 501(c)3 tax-exempt status.

Restrictions: Preference is given to charitable organizations in existence before January 24, 1982. The trust does not support requests from individuals, organizations attempting to influence policy through direct lobbying, or any political campaigns.

Geographic Focus: Texas

Date(s) Application is Due: Mar 31; Sep 30

Amount of Grant: 5,000 - 35,000 U.S.D

Samples: Texas Ballet Theater, Texas, $100,000 (2010); Trinity Valley School, Fort Worth, Texas, $150,000 (2010); Botanical Research Institute of Texas, $100,000 (2010).

Contact: Mark J. Smith, Philanthropic Relationship Manager; (817) 390-6028; tx.philanthropic@baml.com

Internet: https://www.bankofamerica.com/philanthropic/fn_search.action

Sponsor: Crystelle Waggoner Charitable Trust
500 West 7th Street, 15th Floor, TX1-497-15-08
Fort Worth, TX 76102-4700

CSL Behring Local Empowerment for Advocacy Development Grants **1509**

CSL Behring Local Empowerment for Advocacy Development (LEAD) Grants are community-based grants of approximately $10,000. These grants are intended to help local patient organizations achieve their advocacy objectives. CSL Behring will award LEAD Grants to organizations that submit proposals demonstrating that financial assistance would help them address important local and state advocacy issues and initiatives. Local patient organizations have already demonstrated significant leadership in tackling complex legislative and regulatory public policy issues. Advocacy can, however, be costly. CSL Behring LEAD Grants will help defray these costs. Moreover, CSL Behring will work with the grant recipients to advance their objectives in order to obtain success.

Requirements: In order to qualify for a LEAD Grant, a local organization must be: a recognized patient advocacy organization representing individuals who use plasma/recombinant therapies to treat bleeding disorders, immune disorders, Alpha-1 deficiency or other conditions; a non-profit organization with 501(c)3 tax status; and an organization currently addressing a specific advocacy issue or intending to address such an issue.

Date(s) Application is Due: Apr 30; Oct 31

Amount of Grant: Up to 10,000 U.S.D

Contact: Patrick Collins, Public Affairs Manager; (610) 878-4311; Patrick.Collins@cslbehring.com

Internet: http://www.cslbehring.com/s1/cs/enco/1199979063088/content/1199979062780/content.htm

Sponsor: CSL Behring
1020 First Avenue, P.O. Box 61501
King of Prussia, PA 19406-0901

CSRA Community Foundation Grants **1510**

The foundation awards grants to organizations that provide programs and services to the Greater Augusta area, including Richmond, Columbia, McDuffie, Burke, Aiken and Edgefield counties. Areas of interest include arts and cultural; children, youth, and family services; civic affairs; community development; economic development, education; environmental; health; and human services. Types of support include annual campaigns, capital campaigns, seed grants, scholarship funds, and matching funds. The board meets quarterly to consider requests. Applications are due by 11:00 am on the listed application deadline.

Requirements: Georgia 501(c)3 organizations, providing programs and services to the Greater Augusta area, including one or more of the following counties : Richmond, Columbia, McDuffie, Burke, Aiken and Edgefield are eligible to apply.

Restrictions: The Community Foundation discourages applicants from submitting applications in support of the following: computer/word processing hardware not directly related to project or services; grants for individuals; building campaigns; deficit financing and debt retirement; endowments; fraternal organizations, societies or orders; professional/association conferences or seminars (support for or attendance at); political organization or campaigns; lobbying legislators or influencing elections; special fundraising events/celebration functions; surveys, feasibility studies, marketing endeavors and personal research; travel for individuals or groups such as bands, sports teams, and classes; core operating expenses for public and private elementary and secondary schools and public and private colleges and universities; projects that are typically the funding responsibility of federal, state or local governments; fifteen passenger vans.
Geographic Focus: Georgia
Date(s) Application is Due: Jul 31
Amount of Grant: Up to 15,000 U.S.D
Samples: Full Circle Refuge, Inc., $15,000—for the support of the Gang Prevention Youth Mentoring Program; Georgia Committee on Employment of People with Disabilities, $9,600—to purchase equipment for 45 students with disabilities who are enrolled in the High School High Tech programs; Gertrude Herbert Institute of Art, $4,100—to present a collaborative project, Hidden Thoughts: Visual Communication, a yearlong program for children under the age of 18 who are victims of domestic violence and child abuse.
Contact: Cindy Arrant, Manager of Finance & Administration; (706) 724-1314; fax (706) 724-1315; info@csracf.com
Internet: http://www.cfcsra.org/common/content.asp?PAGE=336
Sponsor: CSRA Community Foundation
P.O. Box 31358
Augusta, GA 30903

CSTE CDC/CSTE Applied Epidemiology Fellowships 1511
The Council of State and Territorial Epidemiologists (CSTE), in collaboration with the Centers for Disease Control and Prevention (CDC), the Association of Schools of Public Health (ASPH), and the Health Resources and Services Administration (HRSA) has established the Applied Epidemiology Fellowship Program to train recent graduates in the expanding field of applied epidemiology in an effort to meet the nation's ongoing need for applied epidemiology workforce capacity in state and local health departments. In achieving the fellowship's goal which is to provide a high quality training experience and to secure long-term career placement for fellows at the state or local level, the program focuses on balancing three key concepts as identified in the Applied Epidemiology and Training Program (AETP) Development Handbook (McDonnell 2002): providing service to the sponsoring agency; creating and training a core group of public health workers; and strengthening the capacity in applied epidemiology across public health institutions. Participating fellows will receive two years of on-the-job training at a state health agency (starting in June - August of the program year). Fellows are placed under the guidance of highly-trained and experienced primary and secondary mentors and work closely with epidemiologists at the state, local, and federal levels. A primary mentor must have a doctoral-level degree and the secondary mentor is preferred to have doctoral training as well. Mentors must devote four hours per week to spend with the fellow during the first month of the fellowship and two hours per week for the duration. The mentors oversee the training, research and field activities of the fellow, ensure that the fellow is familiar with relevant techniques in a given specialty, help provide day-to-day activities that provide "hands on" epidemiology experience, and encourage the overall professional development of the fellow. Mentors receive an initial orientation conducted by the CSTE National Office prior to the fellow's arrival, and they participate in conference calls at various intervals during the course of the fellowship. Host health agencies are CDC- and CSTE-approved, with a demonstrated capacity to provide an Applied Epidemiology Fellow with technical training, research opportunities, and practical experience in the application of epidemiologic methods. Host sites are selected based on: the career interests of the fellow; the scope, quality, and diversity of the experience offered to the fellow; the experience of supervisors in management; mentors' number of years of experience in epidemiology; academic training of the mentors; the availability of office space; and letters of support for the agency seeking a fellow. Fellows and host agencies are also carefully matched based on the program areas of interest of the fellow and available opportunities of the host agency. Program areas include infectious disease, injury, maternal and child health, environmental heath, chronic disease, substance abuse, and occupational health. Host agencies are expected to integrate fellows into their worksites and to treat them as entry-level permanent employees. If employee programs are offered to regular permanent employees, host agencies are expected to provide comparable programs and financial support for the fellow. The fellowship program will provide M.P.H.-degree fellows with a bimonthly stipend of up to $39,000 annually. Doctoral-level candidates will be paid up to $56,000. Stipends will be set according to location, cost of living, and pay structure at the host agency. The fellowship program will make available up to $3,200 per year to supplement the fellow's individual health insurance coverage. Up to $1,000 will be provided for moving-related expenses. $970 per year is provided each fellow for professional development. Host health agencies and mentors are strongly encouraged to provide financial support and opportunities as well for the fellow to participate in other public health activities that will expand the fellow's scope and depth of epidemiologic knowledge and/or expand his/her job-related capabilities. CDC/CSTE Applied Epidemiology fellows will develop a comprehensive set of core skills through competency-based training. Applicants may view a list of core competencies at the fellowship website by clicking on the competencies tab. Applicants may apply online beginning in November. The link is available from the Application page of the fellowship website (click on "Application" on the left hand side of the screen). Applicants are encouraged to review the fellowship website for complete information on the fellowship and the application process.
Requirements: Applicants must have: an M.P.H., M.S.P.H., M.S., or an equivalent degree or advanced degree in a public-health-related field (ie. Ph.D. in epidemiology, biostatistics, or other pertinent field; an M.D. degree; or a D.V.M. degree); have completed coursework in at least three advanced graduate-level epidemiology courses and one graduate level biostatistics course; and a desire to pursue a long-term career at the state or local level.
Restrictions: United States citizenship
Date(s) Application is Due: Feb 1
Amount of Grant: 39,970 - 56,970 U.S.D
Contact: Ashlyn Beavor, Workforce and Fellowship Administrator; (770) 458-3811; fax (770) 454-8516; abeavor@cste.org
Internet: http://www.cste.org/dnn/ProgramsandActivities/FellowshipProgram/tabid/259/Default.aspx
Sponsor: Council of State and Territorial Epidemiologists
2872 Woodcock Boulevard Suite 303
Atlanta, GA 30341

CSX Corporate Contributions Grants 1512
The program is focusing its support on personal safety education and community safety. In an effort to keep kids safe, CSX will support nonprofit organizations and school programs which help educate children and their parents on issues of personal safety, help keep children safe on the internet, teach railroad saety to children, teens and adults, and provide safe havens that keep children and adults safe from abuse. First responders are the key to a timely and effective manner of keeping a community safe. Safety, disaster, high alert and Hazmat training, equipment requirements of first responders, and projects and activities that keep communities safe, such as a fence around a neighborhood playground are supported. Programs that protect air, land and water resources in the communities in which CSX operates, preserve natural resources, and teach environmental stewardship to children and adults. A limited number of grants to local social service agencies and arts and culture organizations will be supported. Organizations in the targeted cities must submit an on-line application for sponsorship.
Requirements: 501(c)3 charities and government institutions, such as schools and libraries in their areas of operations are eligible.
Restrictions: Grants do not support individuals; galas, concerts, sports tournaments, auctions, lunches, banquets or dinners; film and video projects; churches or faith-based organizations whose projects benefit wholly their members or adherents; or organizations geographically falling outside of CSX's targeted giving areas. Operating costs or capital improvements are seldom supported.
Geographic Focus: Alabama, Connecticut, Delaware, Florida, Georgia, Illinois, Indiana, Kentucky, Louisiana, Maryland, Massachusetts, Michigan, Mississippi, New Jersey, New York, North Carolina, Ohio, Pennsylvania, South Carolina, Tennessee, Virginia, West Virginia
Contact: Ellen M. Fitzsimmons; (904) 359-3200; corporatecontributionsd@csx.com
Internet: http://www.csx.com/?fuseaction=general.csxog_giving
Sponsor: CSX Corporation
Office of the Corporate Secretary, 500 Water Street, C160
Jacksonville, FL 32202

CTCRI Idea Grants 1513
Idea Grants are designed to encourage unique or original research that has the potential to advance knowledge in tobacco control. Grants will allow investigators with innovative ideas and observations to conduct pilot studies, to perform secondary analysis of data sets or to gather new evidence necessary to determine the viability of research directions or hypotheses. Innovative projects are a priority for the Idea Grant program. Preference will be given to proposals that address research priorities identified by the Canadian Tobacco Control Research Summit. Research employing new or unconventional methodologies is encouraged. Proposed methods must be appropriate to the research question(s). Each successful proposal may receive a one-time grant of up to $50,000. Work is expected to be completed in one year's time.
Requirements: Research proposals must meet the following eligibility criteria: The Principal Applicant (PA) is a Canadian citizen or legal resident (Co-applicants may be citizens or residents of other countries); The research proposal demonstrates basic relevance to tobacco abuse / nicotine addiction / tobacco control and addresses research priorities identified by the Canadian Tobacco Control Research Summit; Applicant(s) has disclosed other sources of funding; Commercial interests have been disclosed or applicant(s) has indicated no commercial interests; Applicant(s) has affirmed lack of support from the tobacco industry; Applicant(s) can demonstrate that he/she works in an environment that adequately supports research through ethical review, administration of funds, provision of space and equipment, etc; Research plans include gender analysis, or it has been demonstrated that this is not appropriate; Proposal avoids duplication of previous research, unless it can be demonstrated that replication is of value; Research proposal involves pilot testing of research methods, tools, or hypotheses by a research team in order to strengthen proposals prepared for submission to traditional funding sources; Proposal articulates preliminary plans for development of a full research proposal -OR- Research proposal constitutes a novel idea which falls outside the normal scope of traditional research, either in topic or methodology; Proposal demonstrates that adequate funding is not available from other sources for this purpose.
Restrictions: Members of the staff or Board of Directors of the CTCRI, or a staff member of any CTCRI funding partner organization are not eligible to apply.
Geographic Focus: All States, Canada
Date(s) Application is Due: Apr 1; Oct 1
Amount of Grant: Up to 50,000 U.S.D
Samples: Tony George, Centre for Addiction and Mental Health, Toronto, Canada, $49,995 - a nicotinic partial agonist for tobacco dependence treatment in bipolar disorder; Robert Schwartz, University of Toronto, Ontario Tobacco Research Unit, Toronto, Canada, $50,000 - cigarette pack as advertisement: Beyond light and mild.

Contact: Carol Bishop; (416) 934-5640; cbishop@cancer.ca
Internet: http://www.ctcri.ca/en/index.php?option=com_content&task=view&id=25&Itemid=44
Sponsor: Canadian Tobacco Control Research Initiative
10 Alcorn Avenue, Suite 200
Toronto, ON M4V 3B1 Canada

Cudd Foundation Grants 1514
The foundation awards grants to eligible nonprofit organizations in its areas of interest, including arts, culture, performing arts, children and youth, education, environment, health care, historic preservation, and social services. Types of support include annual campaigns, building construction/renovation, capital campaigns, continuing support, curriculum development, emergency grants, endowments, program development, research, and scholarship funds. The listed application deadline is for letters of intent; full proposals are by invitation.
Requirements: Louisiana and New Mexico nonprofit organizations are eligible.
Geographic Focus: Louisiana, New Mexico
Amount of Grant: 250 - 222,088 U.S.D
Contact: Amanda Stuermer, (505) 986-8416; fax (505) 986-8427; cuddfdn@aol.com
Sponsor: Cudd Foundation
P.O. Box 2322
Santa Fe, NM 87504

Cullen Foundation Grants 1515
The foundation awards grants to eligible Texas nonprofit organizations in its areas of interest, including arts and culture, education, health, and public service programs. There are no application deadlines or forms. Proposals should detail the purpose and scope of the grant; the amount requested; other sources of anticipated funding; a list of trustees, directors, and staff; financial statements and tax returns of the last two years; and anticipated project budgets. Preference is given to requests from Houston.
Requirements: Texas 501(c)3 or 170(c) nonprofit organizations are eligible.
Restrictions: The board prefers not to consider galas, testimonials, and various other types of fundraising events; organizations that in turn make grants to others; activities whose sole purpose is the promotion or support of a specific religion, denomination, or church; purchase of uniforms, equipment, or trips for school-related organizations or amateur sports teams; applications from an organization more frequently than once every 12 months, whether the previous application was approved or denied; applications from an organization that has received a multi-year grant from the foundation, until all payments of that grant have been made; or oral presentations from potential applicants.
Geographic Focus: Texas
Amount of Grant: Up to 100,000,000 U.S.D
Samples: Reasoning Mind (Houston, TX)—Bringing the internet-based 5th grade math curriculum to Houston students, $50,000; Child Advocates, Inc. (TX)—Operating support, $20,000; The Heritage Society (Houston, TX)—for the relocation and restoration of the 1868 Pilot House, $25,000; Camp for All Foundation (Houston, TX)—Improvements to the camp's drainage system and main thoroughfare, $72,950;
Contact: Sue Alexander, Grant Administrator; (713) 651-8837; salexander@cullenfdn.org
Internet: http://www.cullenfdn.org
Sponsor: Cullen Foundation
601 Jefferson, 40th Floor
Houston, TX 77002

Cullen Trust for Health Care Grants 1516
The trust awards grants to Texas nonprofits providing health care. Proposals must detail the purpose and scope of the grant; amount requested; other sources of anticipated funding; a list of trustees, directors, and staff; financial statements and tax returns of the last two years; and anticipated project budget. Grants are awarded in the Houston area, primarily.
Requirements: Texas tax-exempt public charities are eligible.
Geographic Focus: Texas
Samples: U of Texas Health Science Ctr (Houston, TX)—for its capital campaign to build and equip a new home for the Institute of Molecular Medicine for the Prevention of Human Diseases, and to recruit and support world-class genetic and molecular scientists, $3 million.
Contact: Grant Administrator; (713) 651-8835
Sponsor: Cullen Trust for Health Care
601 Jefferson, Suite 4000
Houston, TX 77002

Cultural Society of Filipino Americans Grants 1517
The CSFA is a non-profit organization dedicated to the promotion and preservation of Filipino culture. Grants are made to local, regional, foreign or international organizations engaged in non-profit activities designed to improve the economic, social, emotional, physical, educational and mental well-being of members of stated organizations in fulfillment of the prescribed responsibility of CSFA.
Requirements: Grants will be made only to organizations that have evidence of 501(c)3 tax-exempt status or to organizations with the endorsement and support of a fiscal agent. In general, the Charitable Committee will look for projects or programs that: (1) fill a real need; (2) have clear, attainable, measurable objectives; (3) are likely to produce results which can be replicated or marketed; (4) are likely to reduce costs, increase revenues or otherwise pay for themselves over time; (5) propose activities which will continue after the grant period without additional funding; and, (6) contain a simple, workable, evaluation plan. The committee will give special consideration to project/programs which propose collaboration with other community organization.
Restrictions: Funds will not be given to individuals; organizations whose primary function is to influence legislation or the local, state or federal levels of government; political parties or candidates; churches or sectarian organizations; however, funds may be given to provide humanitarian aid/services on completely non-sectarian reasons.
Amount of Grant: 500 - 1,000 U.S.D
Contact: Jose Rivera, President; webmaster@csfamn.org
Internet: http://www.csfamn.org
Sponsor: Cultural Society of Filipino Americans
P.O. Box 2773
St. Paul, MN 55102

Cumberland Community Foundation Grants 1518
Cumberland Community Foundation is a nonprofit 501(c)3 charitable foundation established in 1980 by Dr. Lucile Hutaff. The Foundation has defined the following as areas of interest: arts, culture, & recreation; civic engagement; economic development & community advancement; education; environment; health & human services. Proposals accepted on a rolling basis and by request only. Contact the Grants Manager at (910) 483-4449 to discuss the proposed project. If project meets the foundation's criteria and is being considered for funding, a Data Form and Letter of Intent may then be submitted.
Requirements: The foundation will normally make grants only to 501(c)3 organizations in Cumberland County and surrounding counties in North Carolina. However, the Community Foundation remains open to proposals that identify emerging needs, even where needs assessment indicators have not yet been identified.
Restrictions: No grants are made from the Community Funds to or for: individuals; national fundraising drives; tickets for benefits or fundraising events; political organizations or candidates for public office; lobbying activities; endowment funds; scholarships; budget deficits / failure to raise adequate annual operating support; membership dues; religious organizations for religious purposes (community grants do not support programs/projects with religious content or purpose or buildings owned by religious organizations).
Geographic Focus: North Carolina
Samples: Partnership for Children of Cumberland County, $50,000—to improve the lives of children and families by convening service providers, developing a strategic planning process and data system, and building collaborative change in the system; Vander Civic Association, $10,000— to develop a neglected property into a new park and community resource; Boys & Girls Club of Cumberland County, $18,000— to support after-school programming that includes academic assistance, character and leadership development for 300 underserved youth.
Contact: Rachel Stack Anderson, Financial Director; (910) 483-4449; fax (910) 483-2905; rachel@cumberlandcf.org
Internet: http://www.cumberlandcf.org/grant_seekers.php
Sponsor: Cumberland Community Foundation
308 Green Street, P.O. Box 2345
Fayetteville, NC 28302-2171

CUNA Mutual Group Foundation 1519
The corporation makes grants to programs and organizations that are of benefit to its subsidiaries; communities in which associates reside; and programs that are of benefit to Waverly, Bremer/Butler Counties, and Iowa (in that order). Priority areas include education, urban/civic services, human services, and the arts. Some multiyear grants will be awarded. The committee gives preference to programs in which associates and their families are involved and those involving more than just monetary funding.
Requirements: Iowa 501(c)3 organizations are eligible.
Restrictions: The committee will not consider grants for individuals; political parties, candidates, and partisan political campaigns; professional associations; operating expenses for organizations receiving United Way funding; or religious groups for religious purposes.
Geographic Focus: Iowa
Amount of Grant: 2,500 U.S.D
Contact: Executive Director; (608) 231-7908
Internet: http://www.cunamutual.com/cmg/freeFormDetail/0,1248,888,00.html
Sponsor: CUNA Mutual Group Foundation
P.O. Box 391
Madison, WI 53701-0391

Curtis Foundation Grants 1520
The Curtis Foundation provides funding primarily in the Longview, Texas region of the United States. The Foundation's fields of interest include: arts; community/economic development; education; health care; health organizations; human services.
Requirements: Texas 501 (c)3 nonprofit organizations are eligible for funding. There are no deadline dates, nor application form required, when submitting a proposal to the Foundation.
Restrictions: No grants to individuals.
Geographic Focus: Texas
Amount of Grant: 200 - 5,000 U.S.D
Samples: Texas Christian University, Forth Worth, TX, $5,000; City of Longview, Longview, TX, $1,000; Crisman School, Longview, TX, $600.
Contact: Sue Curtis, President; (903) 757-2408
Sponsor: Curtis Foundation
P.O. Box 3188
Longview, TX 75606-3188

CVS All Kids Can Grants 1521

CVS Caremark and the CVS Caremark Charitable Trust will support nonprofit organizations that provide innovative programs and services that are focused on helping children with disabilities learn, play and succeed in life. The goals of CVS Caremark All Kids Can are to provide medical rehabilitation and related services to children with disabilities, to build barrier-free playgrounds so children of all abilities can play side-by-side; and to raise awareness in school and in local communities about the importance of inclusion.
Date(s) Application is Due: Nov 30
Amount of Grant: 1,000 U.S.D
Contact: Jennifer Veilleux, Director; (401) 770-4517 or (401) 770-4209; jhveilleux@cvs.com or CommunityMailbox@cvs.com
Internet: http://info.cvscaremark.com/community/our-impact/all-kids-can
Sponsor: CVS Caremark Charitable Trust
One CVS Drive
Woonsocket, RI 02895

CVS Caremark Charitable Trust Grants 1522

The trust has a goal of positively impacting the culturally diverse populations in the communities where CVS stores are located. Health programs that serve children (under the age of 18) with disabilities and address awareness, accessibility, early intervention, health and rehabilitative services are funded. Public schools (grades pre-K through 12) that are expanding programs promoting inclusion of children with disabilities in all aspects of school functions, including: student academic activities, extracurricular programs, and physical activity/play will be considered. Disabilities are impairments that limit one or more routine activities of daily living. Physical, developmental, and sensory disabilities are supported. The primary focus is on children with disabilities and health care for the uninsured. Application must be submitted online.
Requirements: 501(c)3 public charities as determined by the IRS are eligible. Organizations must benefit the customers and communities in the areas where the corporation operates, and demonstrate the specific ways that the organization benefits the community. Religious organizations are eligible for nonsectarian, nondenominational programs.
Restrictions: The trust does not provide funding for sectarian and religious organizations that do not serve the general public on a nondenominational basis; organizations that discriminate on the basis of race, color, gender, sexual orientation, age, religion, national or ethnic origin or physical disability; political or fraternal groups, social clubs, or any other organization that engages in any kind of political activity; sponsorship of sports teams; journal or program advertisements; organizations geographically located outside of CVS markets; sponsorships for music, film, and art festivals; or business expositions/conferences.
Date(s) Application is Due: Jun 15
Contact: Jennifer Veilleux, Director; (401) 770-4517 or (401) 770-4209; jhveilleux@cvs.com or CommunityMailbox@cvs.com
Internet: http://info.cvscaremark.com/community/our-impact/charitable-trust
Sponsor: CVS Caremark Pharmacy Corporation
One CVS Drive
Woonsocket, RI 02895

CVS Community Grants 1523

CVS Caremark Pharmacy Community Grants target effective and innovative programs. Grants support programs that promote independence among children with disabilities including physical and occupational therapies, speech and hearing therapies, technology that will assist and recreational therapies. They also support public schools that promote inclusive programs in student activities and extracurricular programs and enrich the lives of children with disabilities. The Community Grants Program works to ensure that students are not left behind in school. Proposed programs must be fully inclusive where children with disabilities are full participants in early childhood, adolescent, or teenage programs along side their typically developing peers. CVS is also devoted to the principle of free play. The unstructured, spontaneous, voluntary activity that is so engaging for children's long been recognized as the most beneficial form. Proposed programs may include either physical activities or play opportunities and should address the specific needs of the population served. The CVS/pharmacy Community Grants Program assures that more uninsured people receive needed care that the care received is of higher quality, and that the uninsured are served by providers who participate in accountable community health care programs. There is no age limit on proposed programs that create greater access to health care services.
Requirements: All applicants must answer a number of questions before gaining access to the application. These questions help to determine if the program falls within the guidelines.
Restrictions: An EIN number must be provided before beginning the eligibility quiz for all non-profit organizations applying for a Community Grant for children under age 18 with disabilities. All public schools applying for a Community Grant for this type of program are not required to provide an EIN number. An EIN number must be provided before beginning the eligibility quiz for all non-profit organizations applying for a Community Grant for health care for uninsured people.
Date(s) Application is Due: Oct 31
Amount of Grant: 5,000 U.S.D
Contact: Jennifer Veilleux, Director; (401) 770-4517 or (401) 770-4209; jhveilleux@cvs.com or CommunityMailbox@cvs.com
Internet: http://info.cvscaremark.com/community/our-impact/community-grants
Sponsor: CVS Caremark Pharmacy Corporation
One CVS Drive
Woonsocket, RI 02895

Cyrus Eaton Foundation Grants 1524

The Cyrus Eaton Foundation is committed to providing financial support to qualifying non-profit organizations in Cleveland and northeast Ohio, whose programs enhance the quality of life in this area, and whose aims are in accord with those of our founder, the late Cyrus Eaton. The Foundation limits its funding to these priority areas: the arts; education; science; public affairs; conservation and restoration; health and social welfare. The Foundation's Board meets twice yearly to review grant applications: in June and in November. Submissions may be accepted by email, fax, or postal mail.
Requirements: Cleveland and northeast Ohio nonprofit organizations are eligible.
Restrictions: The foundation does not consider more than one proposal from the same organization within a twelve-month period, unless that organization is applying for an Impact Grant. Tickets or tables for events, municipalities or individuals will not be funded.
Geographic Focus: Ohio
Date(s) Application is Due: May 1; Oct 1
Amount of Grant: 1,000 - 15,000 U.S.D
Samples: Blessing House, Lorain, OH, $2,000; Cleveland Foodbank, Cleveland, OH, $5,000; U.S. Pugwash, Pugwash, Nova Scotia, Canada, $15,000—conferences on science and world affairs.
Contact: Raymond Szabo, President; (216) 320-2285; fax (216) 320-2287; cyrus.eaton.foundation@deepcove.org
Internet: http://www.deepcove.org/Grants/guidelines.html
Sponsor: Cyrus Eaton Foundation
2475 Lee Boulevard, Suite 2B
Cleveland Heights, OH 44118

Cystic Fibrosis Research Grants 1525

The trust was founded to find a complete solution for cystic fibrosis and to improve upon current methods of treatment. Research grants are awarded in the United Kingdom and internationally in the fields of science and medicine.
Requirements: Only researchers from recognized academic institutions are eligible to apply.
Contact: Research Administrator, Research Administrator; 020 8290 7906; researchgrants@cftrust.org.uk
Internet: http://www.cftrust.org.uk/scope/page/view.go?layout=cftrust&pageid=54
Sponsor: Cystic Fibrosis Trust
11 London Road
Bromley, BR1 1BY England

D.F. Halton Foundation Grants 1526

The foundation awards grants to nonprofit organizations in Charlotte, NC, and San Miguel County, CO, primarily in the areas of youth, education, social services, and the performing arts. Additional areas of interest include historical preservation, education, vocational education, business school education, substance abuse, cancer research, heart/circulatory diseases and research, human and family services, and community development. Grants are awarded for general operating support, annual campaigns, capital campaigns, and scholarship funds. There are no application deadlines or forms.
Requirements: Nonprofit organizations in North Carolina counties, including Mecklenburg, Union, Cleveland, Cabarrus, Stanly, Lincoln, and Gaston, may submit proposals. Nonprofit organizations in San Miguel county, CO. may also submit proposals.
Restrictions: No grants will be awarded to individuals.
Geographic Focus: Colorado, North Carolina
Contact: Dale Halton, President
Sponsor: D.F. Halton Foundation
P.O. Box 834
Ophir, CO 81426

D.V. and Ida J. McEachern Charitable Trust Grants 1527

The Trust was established to give a better start in life to all children, both educationally and physically. Most grants are made to social service agencies addressing basic human needs for children and youth. The Trust is also interested in providing creative and positive opportunities to enhance the lives of the region's children through a variety of artistic and cultural programs. The Advisory Committee prefers proposals from groups that clearly describe how their program makes a positive difference in the lives of children, provides a unique service, or addresses problems that are not being met by other agencies in the area. Grants will be made solely for capital projects and items of immediate and direct service to children. There is no letter of inquiry process. If your project fits the guidelines, submit a proposal letter with a narrative no longer than four pages.
Requirements: Grant organizations should: request capital support only (McEachern grants will generally not exceed 5% of a capital campaign budget); provide services to a broad cross-section of the population up to age 18; be established for at least five years (exceptions may be made for groups with strong community leadership and reputation, offering unique enrichment programs for children); serve Puget Sound region (particularly King, Pierce, and Snohomish Counties); have a secure and diverse funding base, with a majority of budget income derived from non-governmental sources.
Restrictions: Grants will generally not be made to endowment campaigns, individuals, private schools, public schools, day-care programs, political candidates, or religious institutions.
Geographic Focus: Washington
Date(s) Application is Due: Mar 8; Sep 8; Dec 8
Amount of Grant: 15,000 - 25,000 U.S.D
Contact: Therese Ogle, Grants Consultant; (206) 781-3472; fax (206) 784-5987; oglefounds@aol.com
Internet: http://fdncenter.org/grantmaker/mceachern/guide.html
Sponsor: D.V. and Ida J. McEachern Charitable Trust
P.O. Box 3123
Seattle, WA 98114

D. W. McMillan Foundation Grants 1528

The D. W. McMillan Foundation, established in 1956, supports organizations involved with youth services, health care, health organizations, homelessness, hospitals, human services, mental health/crisis services, residential/custodial care, hospices, people with disabilities, and the economically disadvantaged population. Giving is primarily centered in the states of Alabama, and Florida. There are no specific deadlines with which to adhere. Contact the Foundation for further application information and guidelines.
Geographic Focus: Alabama, Florida
Samples: Manna Food Bank, Pensacola, FL, $20,000; Baptist Health Care Foundation, Pensacola, FL, $76,000; Boy Scouts of America, Pensacola, FL, $20,000;
Contact: Ed Leigh McMillan II, Treasurer; (251) 867-4881
Sponsor: D. W. McMillan Foundation
329 Belleville Avenue
Brewton, AL 36426-2039

DAAD Research Stays for University Academics and Scientists 1529

These grants and scholarships aim to provide foreign academics and scientists working in higher education or at research institutes with an opportunity to carry out a research project at a state (public) or state-recognized higher education institution or non-university research institute in Germany. Depending on the applicant's work schedule, the research stay will last between one and three months. Depending on the applicant's academic status, the monthly award will amount to 1,840 Euros for assistant lecturers, assistant professors and young lecturers, and 1,990 Euros for professors. In some rare exceptions, 2,240 Euros may be available. In addition to these payments, the DAAD generally will pay an appropriate flat-rate travel allowance, unless these costs are covered by the home country or by another funding source. Applications for research grants are decided twice a year. Applications must have been submitted by 1 August or 15 January. At the earliest, grants can begin 4 months after the date of application.
Requirements: Applications for DAAD research stays are open to excellently-qualified academics and scientists who should generally hold a doctorate/PhD. All applicants must be working in higher education or at a research institute in their home country. Besides their previous academic achievements (for example, recent publications), the most important selection criterion is a convincing and well-planned research project to be completed during the stay in Germany. The application must provide proof of a workplace being provided at the host institute.
Restrictions: DAAD support for a research stay can only be awarded once in any three-year period. No travel expenses can be paid from the grant.
Amount of Grant: 1,840 - 6,720 EUR
Contact: Scholarship Administrator; (212) 758-3223; fax (212) 755-5780; kim@daad.org or daadny@daad.org
Internet: http://www.daad.de/deutschland/foerderung/stipendiendatenbank/00462.en.html?detailid=39&fachrichtung=15&land=44&status=4&seite=1&daad=1
Sponsor: German Academic Exchange Service (DAAD)
871 United Nations Plaza
New York, NY 10017

Dade Community Foundation Community AIDS Partnership Grants 1530

The Community AIDS Partnership, is a special funding initiative established in 1990 through the National AIDS Fund and local funders to increase the availability of private funds to address gaps in the local HIV/AIDS service system, particularly in the area of prevention. The Community AIDS Partnership represents one of the major sources of private philanthropic funds for HIV prevention in Miami-Dade County. As the result of a public/private partnership with Miami-Dade County, the Foundation also awards $350,000 in county funds, which represents a significant increase in the local public sector's commitment to funding HIV prevention. Through the Community AIDS Partnership, the Foundation supports quality programs that respond to the local population risk profile and that incorporate strategies with the greatest potential to effectively address the priority populations, communities and service needs related to HIV prevention in Miami-Dade County. Grant Size: $30,000 Single Organization, $60,000 for Collaborations.
Requirements: Eligible applicants include nonprofit tax-exempt organizations, as defined by the Internal Revenue Code, which are serving the residents of Miami-Dade County. Preference will be given to or- ganizations based in Miami-Dade County or if located outside the county, are working in partnership with an organization based in Miami-Dade.
Geographic Focus: Florida
Samples: Camillus House, $30,000 - to support a comprehensive HIV/AIDS program that targets homeless members of racial/ethnic minority communities who are either HIV positive or are at very high risk for HIV infection due to injection drug use and/or heterosexual contact; MUJER, $30,000 - to continue the Nosotras Viviremos project which seeks to increase awareness about HIV/AIDS in migrant farmworker communities in South Miami-Dade County;
Contact: Gianne Ewing-Chow, Senior Program Officer; (305) 371-2711; fax (305) 371-5342; gianne.ewingchow@dadecommunityfoundation.org
Internet: http://www.dadecommunityfoundation.org/Site/wc/wc145.jsp
Sponsor: Dade Community Foundation
200 S Biscayne Boulevard, Suite 505
Miami, FL 33131-5330

Dade Community Foundation Grants 1531

The funding for this program is made available through the Foundation's unrestricted and field of interest funds. This program is designed to honor both the donors interests and address significant community issues such as: education; health; human services; arts and culture; environment; economic development; at-risk youth; abused and neglected children; living with HIV/AIDS; homelessness; social justice; care of animals; heart disease and more.
Requirements: Eligible applicants include nonprofit tax-exempt organizations, as defined by the Internal Revenue Code, which are serving the residents of Miami-Dade County. Preference will be given to organizations based in Miami-Dade County or if located outside the county, are working in partnership with an organization based in Miami-Dade.
Restrictions: The Foundation does not provide grants to individuals, for memberships, fundraising events or memorials. Grants to government agencies are made on a very restricted basis.
Geographic Focus: Florida
Date(s) Application is Due: Nov 15
Amount of Grant: 7,500 U.S.D
Samples: Alliance for Musical Arts Productions, $7,500 - to provide after school and summer camp services to youth from low-income households living in Opa-locka; Chai Lifeline, $6,500 - to support Smile S'more, a unique program designed to bring fun and love to children who are being treated for cancer and other life-threatening illnesses on pediatric specialty units of hospitals; Earth Learning, $10,000 - to create an EarthFest network that will form an umbrella to unite Earth Day celebrations across South Florida under one banner.
Contact: Charisse Grant, Vice President for Programs; (305) 371-2711; fax (305) 371-5342; charisse.grant@dadecommunityfoundation.org
Internet: http://www.dadecommunityfoundation.org/Site/programs/overview.jsp
Sponsor: Dade Community Foundation
200 S Biscayne Boulevard, Suite 505
Miami, FL 33131-5330

DaimlerChrysler Corporation Fund Grants 1532

The fund contributes to organizations grouped under the general categories of education, health and human services, civic and community, religion, and culture and the arts. Within these categories, grants are made available for public welfare or for charitable, scientific, educational, environmental, safety, building, and affirmative action purposes. Higher education grants largely support science and engineering education and business management. A major interest of the corporation is the establishment of national certification standards for elementary and secondary teachers. Another area of concern is the encouragement of early reading skills, and a pilot project has been funded for research in this area. The fund earmarks funds for its future workforce initiatives, which support business and engineering departments, community-based job-skill training, and entry-level work preparation. Types of support include matching gifts, program grants, scholarships, annual campaigns, building construction/renovation, general support, and employee matching gifts. In considering requests, the fund evaluates each applicant organization on its own merits; considered are the programs in which it is engaged, constituencies served, operation procedures, services offered, quality of management, its accountability, finances, and fund-raising practices. Applications are accepted at any time.
Requirements: Eligible for support are nonprofit, tax-exempt educational, health, civic, and cultural organizations primarily in locations where the greatest number of employees of Chrysler and its U.S.-based subsidiaries live and work (Alabama, Delaware, Illinois, Indiana, Michigan, Missouri, New York, Ohio, and Wisconsin). Some support is targeted for national organizations as well.
Restrictions: Grants are not awarded to support endowments, conferences, trips, direct health care delivery, multiyear pledges, capital campaigns, fund-raising activities related to sponsorships, advertising, or debt retirement.
Samples: Charles H. Wright Museum of African American History (Detroit, MI)—$1 million to complete an exhibit area; Second Harvest Food Bank of Central Florida, $25,000 and food via convoy of Dodge Rams, and $50,000 to help residents affected by recent tornadoes.
Contact: Brian Glowiak; (248) 512-2502; fax (248) 512-2503; mek@dcx.com
Internet: http://www.fund.daimlerchrysler.com
Sponsor: DaimlerChrysler Corporation Fund
1000 Chrysler Drive
Auburn Hills, MI 48326-2766

Dairy Queen Corporate Contributions Grants 1533

The Foundation's philosophy toward awarding contribution grants is to assist those organizations which have an impact on the company's home state of Minnesota and its communities and have programs geared toward children. Foundation focus is on: culture and the performing arts, education, health care and hospitals, and social welfare and community services. Application information is available online.
Requirements: U.S. nonprofit organizations are eligible. Priority is given to Minnesota organizations.
Restrictions: Funds are not available for: strictly sectarian or denominational religious organizations; direct or indirect use for political purposes; lobbying activities; benefits and fund-raisers; commemorative, courtesy, institutional or any other form of advertising; organizations that tend to be for the benefit of the individual members, rather than the general public, such as fraternities, sororities, social clubs, labor organizations, parties and banquets; conferences, seminars or meetings; individuals or scholarships to individuals.
Geographic Focus: Minnesota
Date(s) Application is Due: Sep 1
Contact: Janelle Ianfolla, Corporate Contributions Manager; (952) 830-0207; fax (952) 830-0480; janelle.ianfolla@idq.com
Internet: http://www.dairyqueen.com/us-en/community/
Sponsor: Dairy Queen Corporation
7505 Metro Boulevard
Minneapolis, MN 55439-0286

Daisy Marquis Jones Foundation Grants 1534

The mission of the Foundation is to improve the well being of residents in Yates and Monroe counties, New York in particular, the City of Rochester. A central concern is disadvantaged children and families and the neighborhoods in which they live. The Foundation believes it can best serve those in need by granting time-limited support with programs or projects that provide access to health care, attend to the needs of young children, assist senior citizens, or help families develop economic security. The Foundation also looks for programs that: give people the tools they need to help themselves; encourage collaboration among agencies and between individuals and agencies; have measurable outcomes; and make long-term commitments to specific neighborhoods. A letter of inquiry is the first step to application and is available on line. The board reviews letters of inquiry once a month except July and August.
Requirements: Grants are awarded to qualified 501(c)3 nonprofit organizations that are located in Monroe and Yates Counties, NY.
Restrictions: Grants are not awarded to fund the arts, endowments, local chapters of national health-related organizations, private schools, religious projects, research, scholarships, projects by individuals, or organizations, projects not in Yates or Monroe Counties, or non 501(c)3 organizations.
Geographic Focus: New York
Samples: Advertising Council of Rochester, $10,000—to help support the Ad Council Academy, Enhancing Professional Know How, an educational program for non-profits; Community Partners for Youth, Inc., $45,000—to help support the Big Brothers Big Sisters program; George Eastman House, $10,000—to support the Senior Citizens Matinee Series at the Dryden Theatre.
Contact: Roger Gardner; (585) 461-4950; fax (585) 461-9752; mail@dmjf.org
Internet: http://www.dmjf.org/funding.asp
Sponsor: Daisy Marquis Jones Foundation
1600 South Avenue, Suite 250
Rochester, NY 14620

Dale and Edna Walsh Foundation Grants 1535

The foundation joins hands with effective charitable organizations to meet human need and promote the common good worldwide. Grants support ministries, religious activities, health, relief efforts, education, community services, and arts organizations. Guidelines are available online.
Requirements: Tax-exempt, nonprofit charities are eligible.
Restrictions: The foundation does not contribute toward normal church operation; culturally liberal activist causes or organizations that primarily seek to influence legislation or government spending; or political parties or associated political organizations.
Contact: Grants Administrator; (847) 230-0056; fax (847) 901-9193; Info@dewfoundation.org
Internet: http://dewfoundation.org
Sponsor: Dale and Edna Walsh Foundation
6461 Valley Wood Drive
Reno, NV 89523

Dallas Mavericks Foundation Grants 1536

The foundation of this National Basketball Association franchise makes grants in the Dallas, Texas, area for projects that provide education, good health, and leadership skills for kids. The foundation supports the programs and organizations in the Dallas-Fort Worth Metroplex that address the community's most pressing problems involving youth, specifically education, good health, and community service activity. Application guidelines are available online.
Requirements: To apply for an award, complete a grant application form which can be obtained on line. Include the supporting data: copy of IRS 501(c)3 final determination letter; fact sheet; current board roster; project timeline; project line-item detail budget; general operating budget. There is one grant cycle annually.
Restrictions: Grants will not be made to individuals, churches, public/private schools, or national organizations that do not have locally, financially independent chapters. Grants also will not be made for medical research, travel, salaries, operational phases of established programs, political campaigns or fundraising events, including the purchase of tables, tickets or advertisements. The Foundation would prefer not to commit funds for continued support of long-term programs lasting more than one year, endowment campaigns, administrative costs, advertising/fund raising drives, salaries for staff/individuals, intermediary funding agencies and/or research. The Foundation does not fund multi-year-grants.
Geographic Focus: Texas
Date(s) Application is Due: Jun 30
Contact: Dawn Holgate, (214) 658-7170
Internet: http://www.nba.com/mavericks/news/00405573.html
Sponsor: Dallas Mavericks Foundation
2909 Taylor Street
Dallas, TX 75226

Dallas Women's Foundation Grants 1537

The Foundation promotes women's philanthropy and raises money to support community programs that help women and girls realize their full potential. Priority will be given to funding programs that include elements of the following: effect long-term, positive changes to help women succeed in reaching their full potential; expanded choices and opportunities for women and girls; evaluation tools that include measurement for effectiveness with clean definition of program success; and programs specifically designed to take into consideration the gender-specific needs and differences of women and girls. The Foundation also encourages projects developed in consultation and collaboration with other agencies, and which promote coordination, cooperation and sharing among organizations. Effective use of volunteers, as well as Board diversity and involvement, are equally encouraged.
Requirements: To be eligible to receive a grant from the Foundation, applicants must meet all of the following criteria: be in receipt of a current 501(c)3 tax-exempt designation from the Internal Revenue Service (dated within the last 10 years); at least 50% of the population served must be residents of Dallas, Denton, or Collin County, with priority given to organizations serving residents of Dallas County; 75% of the clients benefiting from the grant funding must be women and/or girls.
Geographic Focus: Texas
Samples: Avance (Dallas, TX)—for equipment and operating support of this organization that promotes academic achievement among Hispanics in Dallas, $10,000; U of North Texas, Division of Equity and Diversity (Dallas, TX)—for a seminar on critical issues affecting lesbians, $15,000; Interfaith Housing Coalition (Dallas, TX)—to provide transitional housing, training, and support services to Dallas-area homeless families, $20,000; English Language Ministry (Dallas, TX)—to provide free child care and English-language instruction to non-English speaking adults, particularly women, $17,500.
Contact: Pat Alexander, (214) 965-9977; palexander@dallaswomensfoundation.org
Internet: http://www.dallaswomensfoundation.org/grants/highlights.html
Sponsor: Dallas Women's Foundation
4300 MacArthur Avenue, Suite 255
Dallas, TX 75209

Dammann Fund Grants 1538

The fund is committed to using its resources for the support of qualified charitable organizations operating programs in New York City, Southwestern Connecticut and Charlottesville, VA with a focus in any one or more of two areas: mental health - programs that foster living or independent living for the mentally ill; teen parenthood - programs that enable development of parenting skills in young parents. Grant recipients are eligible to receive funding for a single program for up to three consecutive years.
Requirements: Nonprofits in the greater New York City, Southwestern Connecticut and Charlottesville, VA.
Restrictions: Grants are not made to individuals and they are not made for capital or annual campaigns, endowments, loans, conferences or travel.
Geographic Focus: Connecticut, New York, Virginia
Date(s) Application is Due: Jun 30
Samples: Institute for Community Living (New York, NY)—to support residential and rehabilitation services for people with mental and physical disabilities, $12,000; Boston Medical Ctr Corp (Boston, MA)—for the Domestic Violence Guardian and Litem Program, $30,000; Louise Wise Services (New York, NY)—for the Teen Dad Program, $30,000; Northside Ctr for Child Development (New York, NY)—for therapeutic and educational programs for the Harlem community, $25,000.
Contact: Administrator; (212) 956-4118; fax (212) 262-9321; df@engelanddavis.com
Internet: http://www.thedammannfund.com/
Sponsor: Dammann Fund
521 Fifth Avenue, 31st Floor
New York, NY 10175

Damon Runyon Cancer Foundation Fellowships 1539

The foundation encourages all theoretical and experimental research relevant to the study of cancer and the search for cancer causes, mechanisms, therapies, and prevention. Candidates must apply for the fellowships under the guidance of a sponsor—a senior member of the scientific research community. The sponsor should be actively engaged in the planning, execution, and supervision of the proposed research and should encourage the fellow to report the results of the research in scientific journals and meetings. Awards are made to institutions for the support of the fellow under direct supervision of the sponsor.
Requirements: Applicants must have completed one or more of the following degrees or its equivalent: MD, PhD, DDS, and DVM.
Date(s) Application is Due: Mar 15; Aug 15
Amount of Grant: 134,000 - 174,000 U.S.D
Contact: Fellowship Administrator; (212) 455-0520; awards@drcrf.org
Internet: http://www.drcrf.org/apFellowship.html
Sponsor: Damon Runyon Cancer Research Foundation
675 Third Avenue, 25th Floor
New York, NY 10017

Dana Brown Charitable Trust Grants 1540

The trust awards grants to eligible Missouri nonprofit organizations in its areas of interest, including children and youth, health care, the arts, and human services.
Requirements: Missouri nonprofit organizations are eligible. Preference is given to requests from the greater Saint Louis metropolitan area.
Geographic Focus: Missouri
Date(s) Application is Due: Feb 15; May 15; Aug 15; Nov 15
Amount of Grant: 5,000 - 100,000 U.S.D
Samples: KETC/Channel 9 (Saint Louis, MO)—to develop and implement a three-part educational program for children, $1.2 million; Saint Vincent Home for Children (Saint Louis, MO)—for the Day Treatment program, which provides education and counseling to help children overcome emotional and behavioral problems and return to community schools; and for a temporary-placement program for children in crisis situations, $100,000.
Contact: Carol Eaves, (314) 418-8317
Sponsor: Dana Brown Charitable Trust
12935 N Outer Forty Drive
Saint Louis, MO 63141-8695

Dana Clinical Hypotheses in Brain and Immuno-Imaging Research Grants 1541
The program focuses on improving human brain and immune system functioning in health and disease. The program consists of two tracks. Track A is for conventional systems imaging of brain tissues (imaging white or gray matter in the brain). Track B is for the evolving field of cellular and molecular imaging of brain cells, or immune cells, or their interactions (imaging at the level of cells rather than tissues). Applicants undertaking cellular and molecular imaging also can employ systems imaging in their proposals to view the actions of cells within their system. Institutions may submit only one application per track. Both tracks are designed to support pilot-testing of promising but high-risk innovative ideas that have direct clinical application and that, when successful, could be supported on a larger scale by other funders. Although one goal of the program is to make support available to researchers early in their careers, any researcher with a promising new hypothesis is eligible. Guidelines are available online.
Requirements: Each U.S. medical school dean, and the presidents of the few selected biomedical research institutions that have been invited by letter, may nominate a total of two applicants, one for Track A, conventional brain systems imaging research, and one application for Track B, using cellular and molecular imaging techniques alone or in combination with brain tissue (systems) imaging techniques. Each application must be countersigned by the medical school dean or invited biomedical institution's president.
Date(s) Application is Due: May 27
Amount of Grant: Up to 100,000 U.S.D
Contact: Rebecca Husman, danainfo@dana.org
Internet: http://www.dana.org/grants/health/proposals/brainimaging.cfm
Sponsor: Charles A. Dana Foundation
745 Fifth Avenue, Suite 900
New York, NY 10151-0002

Dana Corporation Foundation Grants 1542
The foundation gives to communities throughout the country. Social services and education are top priorities. Most education grants are awarded to colleges and universities through a matching gifts program. Other grants go to civic affairs, environmental, youth, and health organizations with the smallest percentage awarded in the areas of culture and arts. Types of support include annual campaigns, building construction/renovation, continuing support, equipment acquisition, land acquisition, operating budgets, seed grants, and capital campaigns.
Requirements: 501(c)3 organizations in communities where Dana Corporation facilities are located are eligible.
Restrictions: The foundation does not make grants to individuals or to organizations that practice discrimination, religious groups for denominational purposes, political activities, or United Way-supported organizations for operating expenses. The foundation does not purchase tickets to charitable or fund-raising events or support goodwill advertising.
Amount of Grant: 1,000 - 20,000 U.S.D
Samples: Big Brothers/Big Sisters of Iredell County (Statesville, NC)—$10,000; Toledo Museum of Art (Toledo, OH)—$50,000.
Contact: Ed McNeal, (419) 535-4500
Internet: http://www.dana.com
Sponsor: Dana Corporation Foundation
P.O. Box 1000
Toledo, OH 43697

Dana Foundation Science and Health Grants 1543
The Foundation supports research in neuroscience, immunology, and the effects of arts training on cognition. Areas of focus are: Brain and Immuno-imaging; Human Immunology; Neuroimmunology; Clinical Neuroscience Research; and Arts and Cognition. All other Science and Health Grants are made solely by invitation.
Requirements: The Foundation requires institutions, in many cases, to share the cost of a project or raise matching funds.
Restrictions: The foundation makes no grants directly to individuals; does not support annual operating budgets of organizations, deficit reduction, capital campaigns, or individual sabbaticals; and does not schedule meetings with applicants, other than by specific invitation initiated by the foundation.
Samples: Dartmouth College (Hanover, NH)—to conduct research on how arts education affects learning and the brain, to be carried out by the Dana Arts and Cognition Consortium, a collaboration of six universities, $1.85 million over three years; Cold Spring Harbor Laboratory (NY)—to support the work of Hollis Cline, for research on fragile X syndrome, a genetic condition that causes mental impairment, $100,000 over 18 months.
Contact: Administrator; (212) 223-4040; fax (212) 317-8721; danainfo@dana.org
Internet: http://www.dana.org/grants
Sponsor: Charles A. Dana Foundation
745 Fifth Avenue, Suite 900
New York, NY 10151-0002

Dana Science and Health Research Grants 1544
Science and health grants support brain research in neuroscience and immunology and their interrelationship in human health and disease. Areas of interest include brain and immuno-imaging—anatomical, physiological, or cellular and molecular imaging techniques to pilot-test novel clinical hypotheses of the brain, immune cells, or their interactions; and human immunology—clinical studies that measure human immune system functioning in health and disease, including the measurement of immune system responses to experimental therapeutic trials supported by other sources. Guidelines are available online.
Requirements: RFPs for the brain and immuno-imaging category are sent twice yearly to deans of U.S. medical schools and other invited biomedical research institutions. Individual investigators as well as collaborating investigators are eligible to apply in the human immunology category.
Amount of Grant: Up to 300,000 U.S.D
Contact: Administrator; (212) 223-4040; fax (212) 317-8721; danainfo@dana.org
Internet: http://www.dana.org/grants/health
Sponsor: Dana Foundation
745 Fifth Avenue, Suite 900
New York, NY 10151-0002

Danellie Foundation Grants 1545
The foundation's primary areas of interest include services for the financially disadvantaged, including housing and social services. The types of support offered are: building/renovation; capital campaigns; continuing support; general/operating support; program development; scholarship funds and; sponsorships. Contact foundation for application guidelines. New York/New Jersey Area Common Application Form and New York/New Jersey Common Report Form accepted. Application form required.
Requirements: Nonprofit organizations in Southern New Jersey, (including Mercer and portions of Monmouth counties), the Greater Philadelphia, Pennsylvania area and, the Balimore, Maryland region are eligible to apply. The foundation also has an international interest in Guatemala and Haiti.
Restrictions: No support for political organizations or professional sports, libraries or museums. No grants to individuals, or for endowments or radio and television.
Geographic Focus: Maryland, New Jersey
Amount of Grant: 4,000 - 180,000 U.S.D
Samples: University of Medicine and Dentistry, Stratford, NJ, $71,000 - for Latino health clinic; Doctors of the World U.S.A, New York, NY, $65,000 - for medical needs in Third World countries; Aid for Friends, Philadelphia, PA, $20,000 - for needy, disabled, and the elderly homebound.
Contact: Nancy Dinsmore, (856) 810-8320; danelliefoundation@verizon.net
Sponsor: Danellie Foundation
P.O. Box 376
Marlton, NJ 08053

Daniel Mendelsohn New Investigator Award 1546
The Daniel P. Mendelsohn award is intended to encourage new investigators whose work is compatible with this mission. New investigators in the field of research or education whose work relates to health and well being are eligible for the award. Candidates, who may or may not be members of the FRI organization, must have demonstrated scholarly research potential compatible with the mission of FRI. They must show future promise, evidenced by energy and enthusiasm in research that addresses prevention, treatment or education activities. Candidates for the biennial award are reviewed by the Awards Review Committee of FRI that presents their recommendation to the Board of Directors at their October Annual Meeting. Deadline for recommending candidates is May 1 of even numbered years. The Award is a check for $5000 payable to the recipient for her or his personal or professional use. The award will be presented in the fall of the awarding year.
Date(s) Application is Due: May 1
Amount of Grant: 5,000 U.S.D
Samples: Michael S.Gordon, D.P.A., University of Baltimore, Baltimore, MD, $5,000—drug abuse treatment for criminal justice populations.
Contact: Janet Klein Brown, Chair; (410) 823-5116; fax (410) 823-5131; fri@friendsresearch.org
Internet: http://www.friendsresearch.org/mendelsohn_award.htm
Sponsor: Friends Research Institute
505 Baltimore Avenue, P.O. Box 10676
Baltimore, MD 21285

Daniels Fund Grants 1547
The fund focuses its efforts on providing a better life and greater opportunities to thousands of individuals in Colorado, New Mexico, Utah, and Wyoming. The fund also supports national programs. The grants program addresses the following areas of need: aging, alcoholism and substance abuse, amateur sports, disabilities, education, homeless and disadvantaged, and youth development. Generally, national grants are reserved for projects with national impact, rather than projects in a single state. Grant applications are accepted at any time throughout the year, however, decisions are made quarterly.
Requirements: 501(c)3 nonprofit organizations in Colorado, New Mexico, Utah, and Wyoming—as well as nationwide—are eligible.
Restrictions: The fund generally will not support medical or scientific research; arts and cultural programs; candidates for public office; endowments; special events or fundraising events; or debt elimination.
Geographic Focus: Colorado, New Mexico, Utah, Wyoming
Samples: U of Wyoming College of Business (Laramie, WY)—to establish a professorship and related curriculum in business ethics, $900,000; Gonzaga U (Spokane, WA)—to establish a concentration and certificate program in ethics and integrity at its business school, $94,000; New Mexico Military Institute (Roswell, NM)—for expansion projects on campus, as part of the institute's capital campaign, $2.5 million challenge grant.
Contact: Grants Program Officer; (303) 393-7220; fax (303) 393-7339; grantsinfo@danielsfund.org
Internet: http://www.danielsfund.org/Grants/index.asp
Sponsor: Daniels Fund
101 Monroe Street
Denver, CO 80206

Daphne Seybolt Culpeper Memorial Foundation Grants 1548

The Foundation, established in 1983, supports organizations involved with education, health care and human services. Giving is limited to Fairfield County, Connecticut and Palm Beach County, Florida. There are no specific deadlines with which to adhere. Contact the Foundation for further application information and guidelines.
Requirements: Nonprofits in Connecticut and Florida are eligible.
Restrictions: No grants to individuals, or for endowments, forums, conferences, seminars, gratuities, honorariums, travel, meals or lodging.
Geographic Focus: Connecticut, Florida
Amount of Grant: 1,000 - 100,000 U.S.D
Samples: Bethesda Hospital Foundation, Boynton Beach, FL, $100,000 - funding for new therapy system; Advent Lutheran Church, Boca Raton, FL, $50,000 - funding for a new school bus; Norwalk Community College Foundation, Norwalk, CT, $50,000 - for general support;
Contact: Nicholas Nardi, Secretary; (203) 762-3984
Sponsor: Daphne Seybolt Culpeper Memorial Foundation
129 Musket Ridge Road, P.O. Box 206
Norwalk, CT 06852-0206

DAR Medical Scholarships 1549

Available medical scholarships include the Alice W. Rooke Scholarship—awarded to students who have been accepted into or are pursuing an approved course of study to become a medical doctor (not pre-med) at an approved, accredited medical school; and Irene and Daisy MacGregor Memorial Scholarship—awarded to students of high scholastic standing and character who have been accepted into or are pursuing an approved course of study to become a medical doctor(not pre-med) at an approved, accredited medical school. This scholarship is also available to students who have been accepted into or who are pursuing an approved course of study in the field of psychiatric nursing, graduate level, at an accredited medical school. Qualified female applicants will receive preference.
Date(s) Application is Due: Apr 15
Amount of Grant: Up to 5,000 U.S.D
Contact: Scholarship Administrator; (202) 628-1776
Internet: http://www.dar.org/natsociety/edout_scholar.cfm
Sponsor: Daughters of the American Revolution
1776 D Street NW
Washington, DC 20006

DAR Nursing Scholarships 1550

Available nursing scholarships include the Caroline E. Holt Nursing Scholarship—awarded to students who are in financial need and desire to attend or are attending an accredited school of nursing; Mildred Nutting Nursing Scholarship—awarded to students who have been accepted or are currently enrolled in an accredited school of nursing. Preference will be given to candidates from the greater Lowell, MA area; and Occupational/Physical Therapy Scholarship—awarded to students who are in financial need and have been accepted or are attending an accredited school of occupational or physical therapy (including art or music therapy). It is only necessary to submit one application for consideration for any award in this group or for the Madeline Picket Halbert Cogswell Nursing Scholarship.
Date(s) Application is Due: Feb 15; Aug 15
Amount of Grant: 500 U.S.D
Contact: Scholarship Administrator; (202) 628-1776
Internet: http://www.dar.org/natsociety/edout_scholar.cfm
Sponsor: Daughters of the American Revolution
1776 D Street NW
Washington, DC 20006

David and Barbara B. Hirschhorn Foundation Grants 1551

Seeking to improve the lives of families and children and cultivate a level playing field, the David and Barbara B. Hirschhorn Foundation supports Jewish and secular initiatives that expand educational opportunity, address human service needs, and promote intergroup tolerance and understanding. Support is primarily given within the state of Maryland, providing grants of annual campaigns, capital campaigns, endowments, general/operating support. Letters of inquiry and proposals are accepted on a rolling basis. Your application will be acknowledged by postcard within two weeks.
Requirements: 501(c)3 non-profit organizations are eligible to apply. To begin the application process submit a letter of inquiry to the Foundation, no longer then 3 pages long. The initial application should include the following: information about the programs(s) for which funding is requested; need, purpose, activities, and evaluation plan of the proposed program(s); program budget (including sources of anticipated income as well as expenditure) and timeline; dollar amount of funding requested. Information about your organization: history, mission, and key accomplishments; information on Board members and key staff; current institutional operating budget (including major sources of revenue as well as expenditures); copy of IRS tax status determination letter or information about your fiscal agent. The Foundation will also accept initial applications that conform to the Association of Baltimore Area Grantmakers [ABAG] Common Grant Application.
Restrictions: No support available for: scholarships to individuals; unsolicited proposals for academic, scientific, or medical research; direct mail, annual giving, membership campaigns, fundraising and commemorative events.
Amount of Grant: 2,500 - 350,000 U.S.D
Samples: American Committee for Shaare Zedek Medical Center in Jerusalem, Baltimore, MD, $20,000; AMIT Women, Inc. New York, NY, $10,000; Institute for Christian and Jewish Studies, Inc., Baltimore, MD, $100,000;
Contact: Betsy Ringel; (410) 347-7103; fax (410) 347-7210; info@blaufund.org
Internet: http://www.blaufund.org/foundations/davidandbarbara_f.html
Sponsor: David and Barbara B. Hirschhorn Foundation
10 E Baltimore Street, Suite 1111
Baltimore, MD 21202

David Geffen Foundation Grants 1552

The foundation supports nonprofits in Los Angeles, California, and New York, New York, with some giving in Israel as well. The Foundation offers support in its five principal funding areas: AIDS/HIV, civil liberties, the arts, issues of concern to the Jewish community, and health care. Support is provided for general operations and special projects. There are no application deadlines. Applicants are asked to submit a letter of request.
Requirements: 501(c)3 tax-exempt organizations in Los Angeles, CA, and New York, NY, may submit a proposal letter (without folder or binder) including description of the project, objectives, constituents served, evaluation criteria; background on the organization including key staff, volunteers, and board; copy of IRS letter confirming tax-exempt status; financial information including line-item budget for the project; and copy of non-discrimination policy from the organization's hiring guidelines.
Restrictions: The Foundation does not fund individuals, or organizations based outside of the United States. The foundation generally does not support documentaries or other types of audio-publication of books or magazines.
Geographic Focus: California, New York, Israel
Samples: Natural Resources Defense Council, New York, NY, $50,000; Childrens Diabetes Foundation at Denver, Denver, CO, $25,000; Morehouse College, Atlanta, GA, $50,000.
Contact: J. Dallas Dishman, Program Director; (310) 581-5955; fax (310) 581-5949; ddishman@geffenco.com
Sponsor: David Geffen Foundation
12011 San Vicente Boulevard, Suite 606
Los Angeles, CA 90049-4926

David N. Lane Trust Grants for Aged and Indigent Women 1553

The David N. Lane Trust for Aged & Indigent Women was established in 1964 to support and promote quality human services and health care programming for underserved older women living in Fairfield and New Haven Counties, Connecticut. Grants from the Lane Trust are one year in duration. Applicants must apply online at the grant website. Applicants are strongly encouraged to do the following before applying: review the downloadable state application procedures for additional helpful information and clarifications; review the downloadable online-application guidelines at the grant website; review the trust's funding history (link is available from the grant website); review the online application questions in advance; and review the list of required attachments. These will generally include: a list of board members, financial statements (audited, reviewed, or compiled by independent auditor); an organization summary; a list of other funding sources; an IRS Determination letter; and other required documents. All attachments must be uploaded in the online application as PDF, Word, or Excel files. The deadline for application to the David N. Lane Trust for Aged & Indigent Women is July 1. Applications must be submitted by 11:59 p.m. on the deadline date. Applicants will be notified of grant decisions by letter within two to three months after the proposal deadline.
Requirements: Applicant organizations must have 501(c)3 tax-exempt status and serve senior women in the counties of Fairfield or New Haven, Connecticut. A breakdown of number/percentage of individuals served by specific towns will be required in the online application.
Restrictions: Grant requests for capital projects will not be considered. The trust does not support requests from individuals, organizations attempting to influence policy through direct lobbying, or any political campaigns.
Geographic Focus: Connecticut
Date(s) Application is Due: Jul 1
Samples: TEAM, Derby, Connecticut, $10,000 (2010); Wesley Heights, Shelton, Connecticut, $9,000 (2010); VNA Health at Home, Watertown, Connecticut $3,000 (2010).
Contact: Kate Kerchaert; (860) 657-7016; kate.kerchaert@baml.com
Internet: https://www.bankofamerica.com/philanthropic/fn_search.action
Sponsor: David N. Lane Trust for Aged and Indigent Women
200 Glastonbury Boulevard, Suite # 200, CT2-545-02-05
Glastonbury, CT 06033-4056

Daviess County Community Foundation Health Grants 1554

The Foundation considers proposals for grants on a yearly cycle, which begins each June. At the start of each cycle, a notice is mailed to nonprofit organizations that have applied for grants in the past, have received grants in the past, or have otherwise requested notification of the start of each cycle. Grants in the area of Health include activities that: improve and promote health outcomes; provide general and rehabilitative health services; offer mental health services; provide crisis intervention programs; strengthen associations or services associated with specific diseases, disorders, and medical disciplines; and support medical research. Applications are due by August 4, and all organizations that have submitted grant proposals will be notified of the outcome of the grants committee's deliberation in writing no later than November 1.
Requirements: The Foundation welcomes proposals from nonprofit organizations that are deemed tax-exempt under sections 501(c)3 and 509(a) of the Internal Revenue Code and from governmental agencies serving the County of Daviess, Indiana. Proposals from nonprofit organizations not classified as a 501(c)3 public charity may be considered provided the project is charitable and supports a community need.
Geographic Focus: Indiana
Date(s) Application is Due: Aug 4

Amount of Grant: 5,000 - 10,000 U.S.D
Samples: Mental Health America of Daviess County—for the Bridges out of Poverty program, $1,780; Pregnancy Care Cente—for the Essentials for Newborns parenting project, $6,000.
Contact: Jeanne Fields, Director; (812) 254-9354; fax (812) 254-9355; jeanne@daviesscommunityfoundation.org
Internet: http://www.daviesscommunityfoundation.org/program-areas
Sponsor: Daviess County Community Foundation
320 E Main Street, P.O. Box 302
Washington, IN 47501

Davis Family Foundation Grants **1555**
The foundation provides grants primarily to Maine-based educational, medical, and cultural/arts charitable organizations in support of a wide variety of worthwhile projects. Eligible educational organizations include: colleges, universities, and other educational institutions. Eligible medical organizations include: hospitals, clinics and medical research organizations (grant requests will also be considered from other similar health organizations for programs designed to increase the effectiveness or decrease the cost of medical care). Eligible cultural and arts organizations include: those agencies whose customary and primary activity is to promote music, theater, drama, history, literature, the arts or other similar cultural activities. Further guidelines are available online.
Requirements: Eligible educational organizations include colleges, universities, and other educational institutions (grants are not made to public elementary and secondary schools, nor to schools whose financial support is derived primarily from a church or other religious organization. Trustees will consider grant requests from other educational organizations whose purpose is to promote systemic change in education or to provide innovative programs whose objectives are to improve education). Medical organizations eligible for support include hospitals, clinics, and medical research organizations. Grant requests will also be considered from other similar health organizations for programs designed to increase the effectiveness or decrease the cost of medical care. Eligible cultural/arts organizations include organizations whose customary and primary activity is to promote music, theater, drama, history, literature, the arts, or other similar cultural activities.
Restrictions: The Foundation does not make grants to individuals, religious programs, fellowships, or in the form of loans. The Foundation does not normally provide support for annual giving campaigns or general operating needs. Grants to endowment campaigns have a low priority.
Geographic Focus: Maine
Date(s) Application is Due: Feb 10; May 10; Aug 10; Nov 10
Samples: Arthur L. Mann Memorial Library, West Paris, ME, $15,000 - expansion and renovation project; Maine Archaeological Society, Inc., Augusta, ME, $8,300 - Maine Archaeology Classroom Initiative; South Bristol Historical Society, South Bristol, ME, $20,000 - schoolhouse restoration.
Contact: Anne Vaillancourt; (207) 781-5504; info@davisfoundations.org
Internet: http://www.davisfoundations.org/site/family.asp
Sponsor: Davis Family Foundation
4 Fundy Road
Falmouth, ME 04105

Dayton Power and Light Foundation Grants **1556**
The foundation contributes general operating grants to nonprofit organizations in the greater Dayton, OH, area in support of education and charitable activities and programs. Education grants are made to support adult basic education/literacy-skills development and engineering school education. Direct donations are made to civic, cultural, and health and welfare organizations that do not participate in community funds, such as United Way or community chests, but serve a real need. There is no official application form. Requests should be made in writing and must include a description of the history, structure, purpose, and program of the organization and a summary of the support needed and how it will be used. Contributions are planned a year in advance and based on a calendar-year budget. The distribution committee normally meets quarterly.
Requirements: 501(c)3 organizations in the greater Dayton, OH, area are eligible.
Restrictions: The foundation prefers not to support capital campaigns; college fund-raising associations; conduit organizations; endowment or development funds; fraternal, labor, or veterans organizations; hospital operating budgets; individual members of federated campaigns; individuals; national organizations outside the DP&L service territory; religious organizations; sports leagues; or telephone or mass-mail solicitations. Grants are rarely made to tax-supported institutions.
Geographic Focus: Ohio
Amount of Grant: 1,000 - 20,000 U.S.D
Samples: Dayton Art Institute (OH)—to support the capital campaign, $25,000; United Way of the Greater Dayton Area (OH)—to support the annual campaign, $200,000.
Contact: Ginny Strausburg, Executive Director; (937) 259-7925; fax (937) 259-7923
Internet: http://www.waytogo.com/cc/cc.phtml
Sponsor: Dayton Power and Light Company Foundation
1065 Woodman Drive
Dayton, OH 45432

Daywood Foundation Grants **1557**
The Foundation distributes available funds primarily to community service organizations focusing on the arts, and health and human services. Grants are available in the following types: annual campaigns; building/renovation; capital campaigns; continuing support; debt reduction; emergency funds; equipment; general/operating support; matching/challenge support; seed money.
Requirements: Nonprofit 501(c)3 organizations situated in, and benefiting Charleston, Lewisburg, Barbour, Greenbrier and Kanawha counties Of West Virginia are eligible to apply. There is no standardized grant application form. Send proposal to: William W. Booker, 1500 Chase Tower, Charleston, WV 25301.
Restrictions: Grants do not support endowment funds, research, individuals, or individual scholarships or fellowships.
Geographic Focus: West Virginia
Date(s) Application is Due: Sep 15
Samples: Clay Center for the Arts and Sciences of West Virginia, Charleston, WV, $150,000—for endowment; University of Charleston, Charleston, WV, $125,000— for pharmacy school; West Virginia Symphony Orchestra, Charleston, WV, $100,000- for endowment.
Contact: William W. Booker, Treasurer; (304) 343-4841
Sponsor: Daywood Foundation
707 Virginia Street E, Suite 1600
Charleston, WV 25301

Deaconess Community Foundation Grants **1558**
The foundation awards grants to eligible Ohio nonprofit organizations in its areas of interest, including health, education, welfare, community, and social service activities. Proposals that are of greatest interest to the Foundation are those that have the strongest fit to the mission statement and that have some or all of the following characteristics: projects that have specific measurable outcomes and a tangible ability to evaluate results and measure success; projects that are supported by other funding sources; and projects that have identified potential for ongoing support beyond the life of the grant. Application information is available online.
Requirements: Only qualified non-profit organizations located in Cuyahoga County which are classified by the Internal Revenue Code as tax-exempt 501(c)3 organizations are eligible for funding consideration.
Restrictions: Grant requests for the following will not be considered: individuals, governmental agencies or any other organization that is not a tax exempt 501(c)3 organization; internal operations and capital campaigns of churches; research projects; or endowments. Grant funds may not be used to carry on propaganda or otherwise attempt to influence legislation, participate in, or intervene in, any political campaign on behalf of or in opposition to any candidate for public office, or to conduct, directly or indirectly, any voter registration drive (within the meaning of Section 4945(d)(2) of the Internal Revenue Code).
Geographic Focus: Ohio
Date(s) Application is Due: Jan 15; May 15; Sep 15
Samples: Junior Achievement, $10,000—Workforce Readiness & Entrepreneurship programs; Magnolia Clubhouse, $35,000—part-time senior development consultant, $2,000 —general operating support; The Intergenerational School, $28,000—establishment of a development department.
Contact: Deborah Vesy; (216) 741-4077; fax (216) 741-6042; dvesy@deacomfdn.org
Internet: http://www.deacomfdn.org/guidelines.html
Sponsor: Deaconess Community Foundation
7575 Northcliff Avenue, Suite 203
Brooklyn, OH 44144

Deafness Research Foundation Research Grants **1559**
The grants are intended for research directed to any aspect of the ear, e.g., investigation of function, physiology, biochemistry, genetics, anatomy, or pathology. Grants cover a range of subjects from basic neuroscience research to studies on corrective hearing/speech testing and training. Grants may be renewed for one to two years.
Requirements: Current policy favors awarding grants in support of projects directed by new investigators, so-called seed money support for studies in generally unexplored areas of research. It does not exclude grant support for new research by established researchers. Application should state whether the same project is receiving support from another source and whether an application has been submitted to another source of funding. The principal investigator and grantee institution must notify the DRF when and if this same project receives support from another granting agency.
Date(s) Application is Due: Dec 1
Amount of Grant: 20,000 U.S.D
Contact: Grants Administrator; (800) 829-5934 or (703) 610-9025; grants@drf.org
Internet: http://www.drf.org/grants/grants.htm
Sponsor: Deafness Research Foundation
8201 Greensboro Drive, Suite 300
McLean, VA 22102

Dean Foods Community Involvement Grants **1560**
The company's community support efforts are focused in three main areas: health/nutrition (including hunger); education/arts; and environmental stewardship/conservation. It supports worthy organizations both at the corporate level and locally through its network of processing facilities nationwide. Preference is given to supporting and participating in a meaningful way with a limited number of organizations that support these three focus areas, rather than spreading limited resources more broadly. Specifically, funding goes to programs that provide direct service to individuals and communities in need. The corporation places special emphasis on supporting organizations that assist children, particularly at-risk children or children with disabilities, or that are dedicated to serving their needs. As a point of contact, the corporate headquarters requests that no phone calls be made regarding its giving programs.
Requirements: 501(3)3 tax-exempt organizations are eligible. Applications must provide background information on the organization as well as how it relates to the mission of Dean Foods. The corporation supports initiatives only in the communities in which they operate and where employees live and work.

Geographic Focus: Alabama, California, Colorado, Connecticut, Florida, Georgia, Hawaii, Idaho, Illinois, Indiana, Kentucky, Louisiana, Maine, Maryland, Massachusetts, Michigan, Minnesota, Montana, Nebraska, Nevada, New Jersey, New Mexico, New York, North Carolina, North Dakota, Ohio, Oklahoma, Pennsylvania, South Carolina, South Dakota, Tennessee, Texas, Utah, Virginia, Wisconsin
Contact: Gregg Engles, Chief Executive Officer; (214) 303-3400; fax (214) 303-3499
Internet: http://www.deanfoods.com/our-company/about-us/corporate-responsibility.aspx
Sponsor: Dean Foods
2515 McKinney Avenue, Suite 1200
Dallas, TX 75201

Dearborn Community Foundation City of Lawrenceburg Community Grants 1561
The City of Lawrenceburg has generously allocated grant funds specifically for not-for-profit organizations that provide a benefit, direct or indirect, to the Lawrenceburg community and Dearborn County. The Dearborn Community Foundation is pleased to administer this program. The program consists of two phases, Phase I: grants requesting $5,000 or less will be accepted. Applications will be considered as long as funds are available. Phase II: grants requesting $5,001 - $100,000 will be excepted to the grant application deadlines. Applications will be considered as long as funds are available.
Requirements: The maximum grant award is $100,000 per organization per calendar year; grant funds are not granted to sustain ongoing programs/projects; and grant applications will be considered as long as funds are available.
Restrictions: Funds specifically for not-for-profit organizations that provide a benefit, direct or indirect, to the Lawrenceburg community and Dearborn County.
Geographic Focus: Indiana
Date(s) Application is Due: Feb 3; May 5; Aug 4; Oct 6
Amount of Grant: 5,000 - 100,000 U.S.D
Contact: Jerri Worthington, DCF Program Director; (812) 539-4115; fax (812) 539-4119; dcfgrants@comcast.net
Internet: http://www.dearborncf.org/grants/G_Lawrence.aspx
Sponsor: Dearborn Community Foundation
322 Walnut Street
Lawrenceburg, IN 47025

Dearborn Community Foundation County Progress Grants 1562
The Dearborn Community Foundation, Inc. (DCF) has been working to build the community's philanthropic resources since it was founded in December of 1997. The DCF County Progress Grant Program supports the interests of the community from non-profit organizations for charitable purposes in the fields of community service, social service, education, health, environment, and the arts. The maximum grant award allotted is $10,000 per organization per calendar year. To be considered, grant applications must be received in the DCF office no later than noon on the deadline date. All questions relating to available scholarship/educational grant opportunities should be directed to the DCF Program Director.
Requirements: Grants are awarded only to organizations whose programs benefit the residents of Dearborn County Indiana and which provide for a responsible fiscal agent and adequate accounting procedures.
Restrictions: Submitted grant applications shall contain one original application and ten copies. This must be done for each grant program from which the organization is seeking grant funds. (Absolutely no binders or folders). Applicants should also retain a copy for their records.
Geographic Focus: Indiana
Date(s) Application is Due: Feb 3; May 5; Aug 4; Oct 6
Amount of Grant: 10,000 U.S.D
Samples: Dillsboro Elementary School, Outdoor Learning Center, IN, $5,073;2007; Hamline Chapel United Methodist Church, Restoration of Boy Scout cabin, IN $2,500;2007;
Contact: Jerri Worthington, DCF Program Director; (812) 539-4115; fax (812) 539-4119; dcfgrants@comcast.net
Internet: http://www.dearborncf.org/grants/G_Progress.aspx
Sponsor: Dearborn Community Foundation
322 Walnut Street
Lawrenceburg, IN 47025

DeBakey Medical Foundation Fellowships and Scholarships 1563
The foundation fosters and encourages medical education and graduate training and promotes medical research in any field of medicine, but it is especially interested in investigations that have direct clinical application, particularly those concerned with cardiovascular disease. Financial aid may be extended to persons in the form of scholarships or fellowships and to institutions in the form of grants or funds for hospital care. Applications are accepted at any time.
Contact: Program Director; (713) 798-8600; fax (713) 793-1192; pa@bcm.tmc.edu
Sponsor: DeBakey Medical Foundation
1 Baylor Plaza, Baylor College of Medicine
Houston, TX 77030

Deborah Munroe Noonan Memorial Fund Grants 1564
The Deborah Munroe Noonan Memorial Fund was established in 1949 by her son, Walter Noonan, to support and promote quality educational, human-services, and health-care programming for underserved populations. Grant requests for general operating support are strongly encouraged. Program support will also be considered. Small, program-related capital expenses may be included in general operating or program requests. To better support the capacity of nonprofit organizations, multi-year funding requests are strongly encouraged. Applicants must apply online at the grant website. Applicants are strongly encouraged to do the following before applying: review the downloadable state application procedures for additional helpful information and clarifications; review the downloadable online-application guidelines at the grant website; review the foundation's funding history (link is available from the grant website); review the online application questions in advance; and review the list of required attachments. These will generally include: a list of board members, financial statements (audited, reviewed, or compiled by independent auditor); an organization summary; a list of other funding sources; an IRS Determination letter; and other required documents. All attachments must be uploaded in the online application as PDF, Word, or Excel files. The application deadline for the Deborah Munroe Noonan Memorial Fund is 11:59 p.m. on July 1. Applicants will be notified of grant decisions before September 30.
Requirements: Applicants must have 501(c)3 tax-exempt status and serve the people of Greater Boston.
Restrictions: The fund does not support requests from individuals, organizations attempting to influence policy through direct lobbying, or any political campaigns.
Geographic Focus: Massachusetts
Date(s) Application is Due: Jul 1
Samples: Community Music Center of Boston, Boston, Massachusetts, $21,000, general operating support of CMCB's 100th anniversary and to support a continuum of music education programs and music therapy for children, teens, adults (2010); Family and Childrens Services of Greater Lynn, Lynn, Massachusetts, $25,000, capacity-building for program services that promote early literacy (2010); Brockton Neighborhood Health Center, Brockton, Massachusetts, $29,000, Certified Medical Interpreter Program (2010).
Contact: Miki C. Akimoto, Vice President; (866) 773-6859; miki.akimoto@baml.com
Internet: https://www.bankofamerica.com/philanthropic/fn_search.action
Sponsor: Deborah Munroe Noonan Memorial Fund
225 Franklin Street, 4th Floor, MA1-225-04-02
Boston, MA 02110

DeKalb County Community Foundation - Garrett Hospital Aid Foundation Grants 1565
The DeKalb County Community Foundation administers the Garrett Hospital Aid Foundation Grants, which award funding to nonprofit organizations that serve residents of Garrett, Indiana. Grant proposals are accepted in January of each year and grants are awarded at the recommendation of the Garrett Hospital Aid Foundation board of directors.
Requirements: Organization should submit the online application to the DeKalb Foundation for consideration. They should include the organization's name and contact information, a description of their project and why they need it, in addition to a first and second priority they would consider.
Restrictions: The Foundation only funds projects in Garrett, Indiana.
Geographic Focus: Indiana
Date(s) Application is Due: Jan 1
Amount of Grant: 200 - 800 U.S.D
Samples: DeKalb Association for the Developmentally Disabled, job training to prepare individuals for workforce and independent living, $250 (2010); Junior Achievement programs serving Garrett School District, $500 (2010); St. Martin's Healthcare Clinic, medical and dental services, $800.
Contact: Rosie Shinkel, Program Manager; (260) 925-0311; rshinkel@dekalbfoundation.org
Internet: http://www.dekalbfoundation.org/g_garrett.php
Sponsor: DeKalb County Community Foundation
650 West North Street
Auburn, IN 46706

DeKalb County Community Foundation Grants 1566
The DeKalb County Community Foundation Grants support programs for DeKalb County, Indiana citizens that address today's needs and prepare for tomorrow's challenges. Grant guidelines are intentionally broad in order to meet the community's ever-changing charitable needs. Grants are awarded for charitable programs and projects in the following areas of interest: art and culture; community development; education; environment; health and human services; and youth development. Grants are also available for the general operating expenses of organizations that address local charitable needs. Applicants are encouraged to contact the Foundation before submitting a proposal to be certain it follows the grant guidelines. They are also encouraged to attend a free one hour workshop to help them understand the Foundation's grant process and learn basic proposal writing tips. The Foundation gives priority to grant proposals for programs/projects that: will be completed within one year of receiving a grant; strengthen the grant seeking organization; directly relate to the grant seeker's mission; project a high degree of community impact; benefit many local people; and are proactive rather than reactive.
Requirements: After reviewing the grant guidelines, applicants will fill out the online proposal form. Applicants should include their contact information, financial information, a brief summary of the request, their organization's mission statement, and a detailed explanation of the benefits they'll receive from the grant. They should also include their operating expenses, total budget, and their source of funds. Applicants will then email the proposal form to the Foundation contact person or mail a printed copy to the Foundation address.
Restrictions: Grants are less likely to be awarded for: repeat funding for a program/project that has received a Foundation grant within the last two years; or a funding debt. The Foundation grants to religious organizations for charitable purposes but does not award grants for religious purposes.
Geographic Focus: Indiana
Date(s) Application is Due: Jul 1
Amount of Grant: 500 - 7,000 U.S.D
Contact: Rosie Shinkel; (260) 925-0311; rshinkel@dekalbfoundation.org

Internet: http://www.dekalbfoundation.org/g_grantmaking.php
Sponsor: DeKalb County Community Foundation
650 West North Street
Auburn, IN 46706

Deland Fellowships in Health Care and Society **1567**
This one-year fellowship provides a training opportunity for graduates interested in pursuing careers in health care management. Appointees will explore the most challenging issues of health care delivery today, including problems of access to quality health care for the poor, uninsured, and underinsured; the changing roles of physicians, nurses, social workers, administrators, and allied health professionals; ethical, legal, and human considerations arising from new biomedical technologies; the impact of for-profit ventures on the work of teaching and community hospitals; and development of integrated delivery systems in the health care marketplace. Stipends are commensurate with the seniority, prior educational experience, and living requirements of the appointees. Fellowship tenure begins in July. For current deadline information, please email Kevin Gahagan at kgahagan@partners.org.
Requirements: It is anticipated that candidates will come from a variety of careers and educational backgrounds including business, law, economics, public policy, and medicine. Candidates are required to have an advanced degree.
Contact: Kevin Gahagan, (617) 732-5500; kgahagan@partners.org
Internet: http://www.brighamandwomens.org/general/Deland_Fellowship.asp
Sponsor: Brigham and Women's Hospital
75 Francis Street
Boston, MA 02115

Delaware Community Foundation-Youth Philanthropy Board for Kent County **1568**
The Youth Philanthropy Board (YPB) for Kent County of the Delaware Community Foundation is composed of 19 students from public, independent and diocesan high schools in Kent County. The Board is studying youth issues in their neighborhoods and schools, learning about community service and grant making, and will award a total of $10,000 to schools and qualified 501(c)3 organizations to carry out youth programs for residents of Kent County. This year, the Youth Philanthropy Board for Kent County will consider programs that provide support for minors who are affected by abusive situations. Additional consideration will be given to programs that encompass reinforcement of academic achievement and/or extracurricular activities.
Requirements: 501(c)3 organizations in Kent County, Delaware are eligible to apply.
Restrictions: Programs completed prior to May 1, are ineligible for funding.
Geographic Focus: Delaware
Date(s) Application is Due: Jan 7
Contact: Ann Frazier, (302) 856-4393; fax (302) 856-4367; frazier@delcf.org
Internet: http://www.delcf.org/Apply_4_1.htm
Sponsor: Youth Philanthropy Board for Kent County c/o Delaware Community Foundation
Southern Delaware Office, 36 The Circle
Georgetown, DE 19947

Delaware Community Foundation Grants **1569**
The Foundation supports organizations and programs throughout the State of Delaware that address a wide range of community needs, including, but not limited to, health and human services, the arts, humanities and culture, the environment, housing and community development. The foundation awards capital grant and program grants. Capital grants are for construction, major renovation or repair of buildings and/or the purchase of land; and also may include requests for equipment purchases. Capital grants typically range from $5,000 to $20,000, with a maximum award of $25,000. Equipment grants generally range from $2,000 to $7,000, with a maximum award of $10,000. Grant proposals must be submitted on the appropriate DCF application form (capital projects or equipment). One application per organization will be accepted. Application forms may be obtained from the DCF office or at the DCF Web site.
Requirements: IRS 501(c)3 non-profit organizations in Delaware are eligible to apply.
Restrictions: Ineligible for support: endowment; debt reduction; religious organizations for sectarian purposes (However, projects that serve the entire community, regardless of religious affiliation, are eligible for support); annual fundraising campaigns or general operating expenses; projects completed before June 30; sports clubs or leagues; educational institutions; purchase of vehicles; individuals.
Geographic Focus: Delaware
Date(s) Application is Due: Jan 29
Amount of Grant: 2,000 - 25,000 U.S.D
Contact: Beth Bouchelle, Director of Grants; (302) 504-5239; bbouchelle@delcf.org
Internet: http://www.delcf.org/Apply_4_1.htm
Sponsor: Delaware Community Foundation
P.O. Box 1636
Wilmington, DE 19899

Del E. Webb Foundation Grants **1570**
The Foundation applies its resources only for the benefit of the residents of Arizona, California and Nevada for improved and expanded medical services, medical research and education. Within these broad areas of interest, the Foundation draws upon the talent and experience of leaders from many walks of life to select organizations of demonstrated competence which have sound programs that will be able to reach and sustain high levels of performance. In choosing particular projects for support, the Foundation's Board acts on the basis of how the public welfare most effectively may be served. Application information is available online.
Requirements: Grants are confined to the support of nonpartisan, non-profit organizations that are operated in the public interest and have tax-exempt status as charitable organizations granted by the Internal Revenue Service and the states in which the respective organizations are incorporated and/or engage in activities.
Restrictions: Grants are not made to the following: governmental agencies or subdivisions; sectarian or religious organizations whose principle activity is exclusively for the benefit of their own members; organizations soliciting funds in support of projects or programs operated by organizations other than the applicant; expenditures before the recipient incurs and makes such expenditures, and, in the absence of compelling circumstances, the Foundation does not make grants to recipients to liquidate or reduce previously incurred obligations or operating deficits. No grants or loans are awarded or made to individuals for any purpose. Grants are not made for scholarships, student aid or medical assistance. Grants may be made to organizations that provide scholarships, student aid or medical assistance when the organization selects recipients in conformity with accepted standards. The Foundation does not participate in the administration of program(s) that the Foundation funds through the award of the application. The Foundation does not fund or underwrite: galas or gala-like events, testimonial or fund-raising luncheons or dinners, advertising in programs or similar fund-raising activities; organizations that in turn make grants to others; purchase of uniforms, equipment, or trips for school related organizations or amateur sports teams; honoraria for guest speakers or panelists; charities operated by service clubs; or educational seminars.
Geographic Focus: Arizona, California, Nevada
Date(s) Application is Due: Feb 28; May 31; Aug 31; Nov 30
Amount of Grant: 5,000 - 100,000 U.S.D
Samples: Grand Canyon National Park Foundation (Flagstaff, AZ)—to renovate the park's education center, $2.5 million; U of Southern California (Los Angeles, CA)—to establish a laboratory suite in the New Institute for Genetic Medicine, $1 million.
Contact: Program Contact; (928) 684-7223; fax (928) 684-5665
Internet: http://www.dewf.com/DEFAULT.shtml
Sponsor: Del E. Webb Foundation
300 E Willis, Suite C
Prescott, AZ 86301-3110

Della Martin Foundation Grants **1571**
The foundation awards grants to organizations located in southern California for research seeking to find the causes of and cures for mental illness. Types of support include challenge/matching grants, endowments, fellowships, professorships, and research grants. There is no application deadline, but the foundation usually makes grants for mental health research at the year's end.
Requirements: Grants are made only to California 501(c)3 organizations.
Geographic Focus: California
Amount of Grant: 263,000 U.S.D
Contact: Laurence Gould Jr., Director; (213) 617-4143; fax (213) 620-1398
Sponsor: Della Martin Foundation
333 S Hope Street, 48th Floor
Los Angeles, CA 90071

Dell Foundation Healthy Communities, Literate Communities, and Connected Communities Grants **1572**
Healthy Communities Grants: Through two-year partnerships with organizations, Dell helps empower its communities to address basic needs for children, such as food, shelter, safety, and healthcare. (March 15 deadline). Literate Communities Grants: Through two-year partnerships with organizations, Dell helps empower its communities to address basic needs of quality education, particularly in math, science, and literacy (May 15 deadline). Connected Communities Grants: Through two-year partnerships with organizations, Dell helps empower its communities to address technology access needs through computer labs for areas of the community with limited access to technology (July 15 deadline).
Requirements: Middle Tennessee and Central Texas 501(c)3 tax-exempt organizations are eligible.
Geographic Focus: Tennessee, Texas
Date(s) Application is Due: Mar 15; May 15; Jul 15
Amount of Grant: 50,000 U.S.D
Contact: Grants Administrator; (512) 338-4400; the_dell_foundation@dell.com
Internet: http://www1.us.dell.com/content/topics/global.aspx/corp/foundation/en/index?c=us&l=en&s=corp
Sponsor: Dell Foundation
1 Dell Way
Round Rock, TX 78682

Dell Foundation Open Grants **1573**
The foundation, the giving arm of Dell Computer, seeks to fund collaborative and innovative solutions to community and children's issues addressing youth (ages newborn to high-school seniors) in Texas, Tennessee, Idaho, and Oregon. Through these efforts, Dell strives to effect change in its local community, while providing lessons and best practices for communities everywhere. Funded areas include arts, education, social services, and health. The foundation supports specific, preventative, and measurable programs with cash, in-kind, and volunteer contributions. Grants are awarded quarterly. An online application form is available on the Web site.
Requirements: Eligibility is open to 501(c)3 nonprofit organizations in Texas—Travis, Williamson County, and McLennan County; Tennessee—Wilson, Davidson County; Idaho—Twin Falls County; and Oregon—Roseburg, Douglas County.
Restrictions: Grants do not support individuals; academic or research projects; civic, religious, or political institutions; school fundraisers; marketing opportunities; or sports events and organizations.

Geographic Focus: Idaho, Oregon, Tennessee, Texas
Date(s) Application is Due: Jan 15; Apr 15; Jul 15; Oct 15
Amount of Grant: Up to 5,000 U.S.D
Contact: Grants Administrator; (512) 338-4400; the_dell_foundation@dell.com
Internet: http://www1.us.dell.com/content/topics/global.aspx/corp/foundation/en/open_grants?c=us&l=en&s=corp
Sponsor: Dell Foundation
1 Dell Way
Round Rock, TX 78682

Dell Scholars Program Scholarships 1574
Dell Scholars demonstrate their desire and ability to overcome barriers and to achieve their goals. Applicants will be evaluated on their individual determination to succeed; future goals and plans to achieve them; ability to communicate the hardships they have overcome or currently face; self motivation in completing challenging coursework; and demonstrated need for financial assistance. Students may apply directly through the Scholar website.
Requirements: To be eligible to apply for the Dell Scholarship, applicants must: participate in a Michael and Susan Dell Foundation approved college readiness program for a minimum of two of the last three years; graduate from an accredited high school this academic year; earn a minimum of a 2.4 GPA; demonstrate need for financial assistance; plan to enter a bachelor's degree program at an accredited higher education institution in the fall directly after their graduation from high school; and be a citizen or permanent resident of the U.S.
Restrictions: Applications faxed, mailed, or emailed to the Michael and Susan Dell Foundation will not be considered. The official Dell Scholars Application can only be found at this web site and must be submitted on-line during the specified application period. email attachments will not be opened.
Date(s) Application is Due: Jan 15
Contact: Dell Scholar Contact; 800-294-2039; apply@dellscholars.org
Internet: http://www.dellscholars.org/Criteria.aspx
Sponsor: Michael and Susan Dell Foundation
P.O. Box 163867
Austin, TX 78716-3867

Delta Air Lines Foundation Grants 1575
The foundation awards grants to nonprofits in the areas of: Community Enrichment - focusing on involvement and participation in volunteer, civic, and social activities; Health and Wellness - with support to organizations that dedicate their efforts to the health and well-being of communities with a focus on broad-reaching research for cures and education around diseases that affect all walks of life; Youth Development - offering support to organizations focusing on keeping young people interested in math and science and helping them develop leadership skills and positive self esteem; and Arts and Culture to include fine art, theatre, music, or other creative endeavors which enhance a community's quality of life.
Requirements: For proposals which meet the foundation's area of focus, priority will be given to: programs meeting compelling needs in communities where Delta has a presence; proposals that exhibit clear, reasonable goals, and measurable outcomes; distinctive projects where the foundation's involvement will leave a legacy; projects that include collaboration or cooperation with other nonprofit organizations; projects that offer opportunities for Delta employee involvement. The foundation Board of Trustees reviews and approves funding in March, June, September, and November. The deadline for receiving completed proposals is the first day of each of these months.
Restrictions: The foundation will generally not consider: individual applicant's request for support of personal needs; religious activities; political organizations or campaigns; specialized single-issue health organizations; annual or automatic renewal grants; general operating expenses; endowment campaigns; capital campaigns; multi-year commitments; fraternal organizations, professional associations, or membership groups; fundraising events such as benefits; charitable dinners, or sporting events.
Contact: Administrator; (404) 715-5487; fax (404) 715-3267
Internet: http://www.delta.com/about_delta/community_involvement/delta_foundation/
Sponsor: Delta Air Lines
P.O. Box 20706, Department 979
Atlanta, GA 30320-6001

DeMatteis Family Foundation Grants 1576
The foundation makes grants in the New York metropolitan area to eligible institutions whose mission involves education, health and human services, medical research, social services, and the arts. Types of support include facilities construction, expansion, renovation; acquisition of capital equipment; scientific/medical research; projects and programs that enable the applicant to expand its mission through new or expanded programs to reach a greater segment of the community served; project-oriented capital campaigns. In general, most grants are made to cover projects that can be accomplished within one year. For construction and longer duration projects, grants may be structured to conform to identified milestones. For major projects, the payment of the grant may be over a number of years.
Requirements: Metropolitan New York 501(c)3 tax-exempt agencies, institutions, and organizations are eligible.
Restrictions: In general, the foundation does not support grants for operating deficits; general operating support; endowments; loans, or financing of any kind; annual appeals, dinner functions, and other special fund raising events; or unrestricted funds.
Geographic Focus: New York
Contact: Grants Administrator; (516) 705-4974
Internet: http://fdncenter.org/grantmaker/dematteis/about.html
Sponsor: DeMatteis Family Foundation
P.O. Box 25
Glen Head, NY 11545

Dennis and Phyllis Washington Foundation Grants 1577
The foundation supports a broad spectrum of worthy causes benefiting people of all ages primarily serving the State of Montana and surrounding states where the Washington Companies are located. Priorities for funding are direct service, youth oriented programs and the advancement of educational opportunities through scholarships to units of higher education in Montana. The Foundation also focuses on the needs of economical and socially disadvantaged people, troubled or at-risk youth and individuals with special needs.
Requirements: Eligible applicants must be a charitable, nonprofit entity with tax exempt status under Section 501(c)3. Organizations applying for support must be categorized in one of the four giving areas of education, health and human services, community service and arts and culture. The Foundation places particular emphasis on those organizations and programs that provide a direct service to economically and socially disadvantaged youth and their families, at-risk or troubled youth, and individuals with special needs. Preference is given to applicants who are able to demonstrate that a majority, if not all, of Foundation funds will be used for direct services. The Foundation prefers giving to organizations with no or low administrative costs. Organizations may apply for funding only in the year in which funds will be used. The Foundation prefers that organizations show evidence of substantial financial support from their community, constituency groups or other funding sources prior to applying.
Restrictions: Applications will not be considered for the following organizations or purposes: organizations that, in policy or practice, unfairly discriminate against race, ethnic origin, sex, creed, or religion; to fund loans, debt retirement or operational deficits; to fund on-going, general operations; to individuals, unless under an approved educational scholarship program; to sectarian or religious organizations for religious purposes where the principal activity is for the benefit of their own members or adherents; to veterans or fraternal organizations, unless their programs are available to members of the community as a whole; for travel expenses or trips; to general endowment funds, private or public foundations and most capital campaigns; for operation expenses of tax-supported groups; for sponsorships including auctions, dinners, tickets, advertising, or annual fundraising events; for political action or legislative advocacy groups or influencing legislation or elections; for operational costs or curriculum development at educational institutions; the purchase of motor vehicles or other forms of transportation; to organizations acting on behalf of, but without the authority of, qualified tax exempt organizations.
Geographic Focus: Montana
Contact: Mike Halligan, Executive Director; (406) 523-1325
Internet: http://www.dpwfoundation.org/home.htm
Sponsor: Dennis and Phyllis Washington Foundation
P.O. Box 16630
Missoula, MT 59808-6630

Dental Teaching and Research Fellowships 1578
Fellowships are awarded to individuals who plan to become dental teachers and/or researchers in dental schools in Canada and who must guarantee to pursue careers in teaching and research at a university for a minimum of two years following fellowship tenure. A formal application form must be filed with official transcripts and letters of support. The award is given for one year with possible renewal for a second year based on satisfactory progress.
Requirements: Candidates must be Canadian citizens or permanent residents who have completed an undergraduate course in dentistry, a dental hygiene program, or a program in science and who are also eligible for admission to a graduate or other advanced education program.
Geographic Focus: All States, Canada
Date(s) Application is Due: Feb 1
Amount of Grant: 1,800 - 8,000 CAD
Contact: Richard Munro; (613) 236-4763; fax (613) 236-3935; Information@dcf-fdc.ca
Internet: http://www.dcf-fdc.ca/grants.html
Sponsor: Dentistry Canada Fund
427 Gilmour Street
Ottawa, ON K2P 0R5 Canada

Denton A. Cooley Foundation Grants 1579
The foundation awards grants in Texas in support of health care, health education, hospitals, and medical research. Types of support include endowments, general operating grants, program grants, and research grants. There are no application forms or deadlines. The board meets quarterly to consider requests.
Restrictions: Grants do not support conferences, loans, individuals, scholarships, fellowships, or publication.
Geographic Focus: Texas
Amount of Grant: 100 - 250,000 U.S.D
Samples: U of Texas Medical Branch at Galveston (TX)—to establish a professorship in the division of cardiothoracic surgery, $250,000.
Contact: Grants Administrator; (713) 799-2700
Sponsor: Denton A. Cooley Foundation
6624 Fannin, Suite 1640
Houston, TX 77030

Denver Broncos Charities Fund Grants 1580
The fund supports qualified non-profit organizations that work to impact the quality of life for youth, health and the hungry and homeless. The emphasis of the fund is on programs designed to assist young people in the areas of education and youth football with a particular emphasis on programs aimed at disadvantaged and at-risk youth. The fund will also consider programs devoted to health and hunger issues. Grants may be stand-alone grants or magnet grants that attract other corporate dollars.
Requirements: Nonprofit organizations in the state of Colorado are eligible.

Geographic Focus: Colorado
Date(s) Application is Due: Jun 1
Samples: Denver Public Schools Foundation, $50,000—Prep League Football Program; Inner City Health Center, $50,000—Denver Rescue Mission; Meeker School District, $50,000—Community Fields Project; Darrent Williams Memorial Teen Center, $50,000—Darrent Williams Memorial Grant.
Contact: Charities Fund Manager; (720) 258-3000
Internet: http://www.denverbroncos.com/page.php?id=1157
Sponsor: Denver Broncos Charities Fund
1701 Bryant Street, Suite 1400
Denver, CO 80204

Denver Foundation Community Grants **1581**
The Denver Foundation carries out its mission of improving life in Metro Denver by investing in the vision, passion and expertise of hundreds of nonprofit organizations. The Community grants program awards hundreds of grants annually to nonprofit organizations in the seven-county Metro Denver area, which includes: Adams, Arapahoe, Boulder, Broomfield, Denver, Douglas, and Jefferson counties. Grants are awarded in the four major focus areas of arts and culture, health, human services and civics and education.
Requirements: To qualify for a grant an organization must be a 501(c)3 tax-exempt nonprofit organization, serve residents in seven specific Denver counties, and provide a service that falls under one of the four funding areas.
Restrictions: The Foundation will not consider requests to fund the following: requests from organizations that have received funding from the program for the three previous consecutive calendar years; organizations with fund balance deficits in their most recently completed fiscal year; organizations that discriminate on the basis of race, color, religion, gender, age, national origin; disability, marital status, sexual orientation, or military status; debt retirement; endowments or other reserve funds; membership or affiliation campaigns, dinners, or special events; conferences and symposia and related travel; grants that further political doctrine; grants that further religious doctrine; grants to individuals; scholarships or sponsorships; individual medical procedures; medical, scientific, or academic research; grants to parochial or religious schools; grants to governmental agencies, except public schools; requests from individual public schools that have not coordinated the request with their central school district administration; requests from foundations/organizations that raise money for individual public schools; creation or installation of art objects; development, production, or distribution of books, newspapers, or video productions; grants for re-granting programs; requests for capital campaigns that have not met 75% of their goal; activities, projects, or programs that will have been completed before funding becomes available; and multi-year funding requests.
Geographic Focus: Colorado
Date(s) Application is Due: Feb 2; Jun 1; Oct 1
Contact: Jeff Hirota, Vice President; (303) 300-1790, ext. 129; fax (303) 300-6547; jhirota@denverfoundation.org
Internet: http://www.denverfoundation.org/page17823.cfm
Sponsor: Denver Foundation
55 Madison, 8th Floor
Denver, CO 80206

Dept of Ed College Assistance Migrant Program (CAMP) **1582**
The College Assistance Migrant Program (CAMP) assists students who are migratory or seasonal farmworkers (or children of such workers) enrolled in their first year of undergraduate studies at an IHE. The funding supports completion of the first year of studies. Competitive five-year grants for CAMP projects are made to IHEs or to nonprofit private agencies that cooperate with such institutions. The grants funded under CAMP grantee serve approximately 2,400 participants each year.
Requirements: Institutions of higher education or private nonprofit agencies in cooperation with institutions of higher education may apply.
Restrictions: Institutions of higher education or a nonprofit private agency in cooperation with an IHE may apply.
Contact: David De Soto, (202) 260-8103; fax (202) 205-0089; david.de.soto@ed.gov
Internet: http://web99.ed.gov/GTEP/Program2.nsf/b39cd123fd4a045b85256 44400514f2b/07783e5ad0636ec5852563bc00540495?OpenDocument
Sponsor: U.S. Department of Education
400 Maryland Avenue SW, Room 3E225, FB-6
Washington, DC 20202-6135

Dept of Ed Rehabilitation Training Grants **1583**
The purpose of this program is to ensure that skilled personnel are available to serve the rehabilitation needs of individuals with disabilities assisted through vocational rehabilitation (VR), supported employment and independent living programs. The program supports training and related activities designed to increase the number of qualified personnel trained in providing rehabilitation services. Grants and contracts are awarded to states and public and nonprofit agencies and organizations, including institutions of higher education, to pay all or part of the cost of conducting training programs.
Requirements: Institutions of Higher Education (IHEs), Nonprofit Organizations, Other Organizations and/or Agencies may apply. Applicants may include state and public or nonprofit agencies and organizations and Indian tribes.
Contact: Ruth Brannon, Program Contact; (202) 245-7278; fax (202) 245-7591; ruth.brannon@ed.gov
Internet: http://www.ed.gov/programs/rsatrain/index.html
Sponsor: U.S. Department of Education
400 Maryland Avenue SW, Room 5052, PCP
Washington, DC 20202

Dept of Ed Safe and Drug-Free Schools and Communities State Grants **1584**
This program provides support to SEAs for a variety of drug and violence prevention activities focused primarily on school-age youths. Activities may include: developing instructional materials; providing counseling services and professional development programs for school personnel; implementing community service projects and conflict resolution, peer mediation, mentoring and character education programs; establishing safe zones of passage for students to and from school; acquiring and installing metal detectors; and hiring security personnel.
Requirements: State Education Agencies may apply. Local Education Agencies or intermediate education agencies or consortia must apply to the State Education Agency.
Contact: Paul Kesner, Office of Safe and Drug-Free Schools; (202) 205-8134; fax (202) 260-7767; paul.kesner@ed.gov
Internet: http://www.ed.gov/programs/dvpformula/index.html
Sponsor: U.S. Department of Education
400 Maryland Avenue SW, Room 3E230, FB-6
Washington, DC 20202-6450

Dept of Ed Special Education—Personnel Development to Improve Services and Results for Children with Disabilities **1585**
The objectives of this program are to help address state-identified needs for highly qualified personnel in special education, related services, early intervention, and regular education, to work with children with disabilities; and ensure that those personnel have the skills and knowledge, derived from practices that have been determined, through research and experience, to be successful, that are needed to serve those children. Awards are made to train personnel in the areas of: leadership; early intervention and early childhood; low-incidence; high-incidence; related services, speech/language, and adapted physical education; and programs in minority institutions.
Requirements: Institutions of Higher Education, Local Education Agencies, Nonprofit Organizations, Other Organizations and/or Agencies and State Education Agencies are eligible to apply. In addition to these categories, public charter schools that are LEAs under state law, other public agencies, private nonprofit organizations, for-profit organizations, outlying areas (American Samoa, Guam, the Northern Mariana Islands, and the U.S. Virgin Islands), freely associated states, and Indian tribes or tribal organizations are eligible to apply.
Contact: Patricia Wright, (202) 245-7620; fax (202) 245-7619; patricia.wright@ed.gov
Internet: http://www.ed.gov/programs/osepprep/index.html
Sponsor: U.S. Department of Education
400 Maryland Avenue SW, Room 4113, PCP
Washington, DC 20202-2600

Dept of Ed Special Education—Studies and Evaluations **1586**
This program is designed to assess progress in implementing IDEA, including the effectiveness of state and local efforts to provide: free appropriate public education to children with disabilities; and early intervention services to infants and toddlers with disabilities. This program supports studies, evaluation, and assessments.
Requirements: Institutions of higher education, local education agencies, nonprofit organizations, other organizations and/or agencies, and state education agencies are eligible. Public agencies, outlying areas (American Samoa, Guam, the Northern Mariana Islands, and the U.S. Virgin Islands), freely associated states, Indian tribes, tribal organizations, and for-profit organizations, if appropriate, may apply.
Contact: Jeffrey Payne; (202) 245-7473; fax (202) 245-7617; jeffrey.payne@ed.gov
Internet: http://www.ed.gov/programs/osepsae/index.html
Sponsor: U.S. Department of Education
400 Maryland Avenue SW, Room 4064, PCP
Washington, DC 20202-2500

Dept of Ed Special Education—Technical Assistance and Dissemination to Improve Services and Results for Children with Disabilities **1587**
The purpose of this program is to promote academic achievement and improve results for children with disabilities by providing technical assistance, model demonstration projects, dissemination of useful information, and implementation activities that are supported by scientifically based research. The program supports technical assistance and dissemination activities. Annual deadline dates may vary; contact the office for exact dates.
Requirements: Institutions of Higher Education, Local Education Agencies, Nonprofit Organizations, Other Organizations and/or Agencies, and State Education Agencies are eligible to apply. Public charter schools that are LEAs under state law, for profit organizations, outlying areas (American Samoa, Guam, the Northern Mariana Islands, and the U.S. Virgin Islands), freely associated states, and Indian tribes or tribal organizations may also apply.
Amount of Grant: 200,000 - 2,000,000 U.S.D
Contact: Claudette Carey, Office of Special Education and Rehabilitative Services; (202) 245-7291; fax (202) 245-7617; claudette.carey@ed.gov
Internet: http://web99.ed.gov/GTEP/Program2.nsf/a5b8d6c38fdd4ca085 25644400514f2c/02df9c7d71171b1785256571007ce61c?OpenDocument
Sponsor: U.S. Department of Education
400 Maryland Avenue SW, Room 4098, PCP
Washington, DC 20202

Dept of Ed Special Education-National Activities-Technology and Media Services for Individuals with Disabilities **1588**
The purpose of this program is designed to: improve results for children with disabilities by promoting the development, demonstration, and use of technology; support educational media services activities designed to be of education value in the classroom setting to

children with disabilities; and provide support for captioning and video description that is appropriate for use in the classroom setting.
Requirements: State Education Agencies; Local Education Agencies; public charter schools that are LEAs under state law; Institutions of Higher Education; other public agencies; private nonprofit organizations; outlying areas (American Samoa, Guam, the Northern Mariana Islands, The U.S. Virgin Islands); freely associated states; Indian tribes or tribal organizations; and for-profit organizations may apply.
Contact: Jeffrey Payne, Program Officer; (202) 245-7473; fax (202) 245-7617; jeffrey.payne@ed.gov
Internet: http://www.ed.gov/programs/oseptms/index.html
Sponsor: U.S. Department of Education
400 Maryland Avenue SW, Room 4064, PCP
Washington, DC 20202-2500

Dermatology Clinical Career Development Awards **1589**
The purpose of this award is to enhance the academic careers of clinician-scientists in the early stages of their career development. The program is aimed at junior investigators with significant creativity in clinically relevant research. Research must be conducted in the United States. Such studies could include research on diagnostic and/or prognostic indicators, as well as outcome and/or epidemiologic studies. Creativity in developing new treatment modalities, including novel pharmacologic approaches and clinical trials, is encouraged. Studies need not involve direct laboratory investigation. The primary criteria for selection of a successful awardee are the quality of the research proposal, the environment of the candidate, and ability of the proposal to clarify the place of dermatologic practice in a changing health care environment. Award recipients are expected to spend at least 75 percent of their time in clinical research. Awardees are encouraged to seek simultaneous grant support from other agencies to provide for the nonsalary components of the research being performed under the auspices of this award.
Requirements: Applications are accepted from faculty members in a department or division of dermatology who have completed clinical training in a U.S. dermatology residency program. Applicants must demonstrate a strong commitment to skin research and have already had appropriate initial training (a two- to three-year research fellowship or postdoctoral training) in relevant research.
Restrictions: The foundation does not fund awards to be performed as part of the U.S. government research program (with the exception of the Veterans Administration) or awards to private foundations without an academic affiliation to dermatology.
Date(s) Application is Due: Oct 17
Amount of Grant: 55,000 U.S.D
Contact: Sandra Rahn Benz, Executive Director; (847) 328-2256; fax (847) 328-0509; dfgen@dermatologyfoundation.org
Internet: http://www.dermfnd.org
Sponsor: Dermatology Foundation
1560 Sherman Avenue, Suite 870
Evanston, IL 60201

Dermatology Foundation Grants **1590**
Limited funds are available for investigators at the early stages of their career development to initiate research projects in dermatology and cutaneous biology. Applications are reviewed by the Medical and Scientific Committee and awarded on a competitive basis for projects not funded from other sources. The research has to be performed under the sponsorship of a division or department of dermatology. The maximum period of funding is for one year. Only one grant application will be considered from a single academic center.
Restrictions: Awards cannot be used for payment of indirect costs.
Date(s) Application is Due: Oct 17
Amount of Grant: 10,000 U.S.D
Contact: Sandra Rahn Benz, Executive Director; (847) 328-2256; fax (847) 328-0509; dfgen@dermatologyfoundation.org
Internet: http://dermatologyfoundation.org
Sponsor: Dermatology Foundation
1560 Sherman Avenue, Suite 870
Evanston, IL 60201

Dermatology Postdoctoral Research Fellowships **1591**
Financial support is available to postdoctoral fellows with commitments to careers in academic dermatology who desire research training. Applications will be accepted in two categories. Dermatologist Investigator Research Fellowships are designed to support dermatologists who desire research training and have commitments to careers in academic dermatology. Research Fellowships support training in research skills relevant to dermatology and cutaneous biology. Highest priority will be given to new applicants. A limited number of second-year awards may be given. Although it is recognized that fellowships will encompass specific research projects, they are intended primarily for the support of research training. Fellowship recipients must spend at least 75 percent of their total effort in cutaneous research.
Requirements: Applications for Dermatologist Investigator Research Fellowships are accepted from individuals holding the MD, MD-PhD, or DO degree who have completed their clinical training in U.S. dermatology residency programs. Research Fellowship applications are accepted from individuals holding the MD or PhD or equivalent.
Restrictions: In general, investigators at more than four years past their terminal degrees or two years beyond their residencies at the time of the initial application are not eligible. (Consideration will be given to individuals with substantial training in other areas and who are entering into skin research.) Individuals with academic appointments at the level of assistant professor or above are not ordinarily eligible.
Date(s) Application is Due: Oct 17
Amount of Grant: Up to 25,000 U.S.D
Contact: Sandra Rahn Benz, Executive Director; (847) 328-2256; fax (847) 328-0509; dfgen@dermatologyfoundation.org
Internet: http://dermatologyfoundation.org/rap
Sponsor: Dermatology Foundation
1560 Sherman Avenue, Suite 870
Evanston, IL 60201

Dermatology Research Career Development Awards **1592**
Awards are available to assist in the transition from fellowship to established investigator in cancer and other diseases of the skin, hair, and nails. The intent of the program is to advance the research careers of young individuals in dermatology and cutaneous biology, with the emphasis on research benefiting the dermatology community at large. Award recipients are expected to spend at least 75 percent of their time in cutaneous research. Awardees are encouraged to seek simultaneous grant support from other agencies to provide for the nonsalary components of the research being performed under the auspices of this award.
Requirements: Applicants must be faculty members in a department or division of dermatology who demonstrate a strong commitment to skin research and have already had appropriate initial training (a two- to three-year research fellowship or postdoctoral training) in biomedical research.
Restrictions: The foundation does not fund awards to be performed as part of the U.S. government research program (with the exception of the Veterans Administration) or awards to private foundations without an academic affiliation to dermatology.
Date(s) Application is Due: Oct 17
Amount of Grant: 55,000 U.S.D
Contact: Sandra Rahn Benz, Executive Director; (847) 328-2256; fax (847) 328-0509; dfgen@dermatologyfoundation.org
Internet: http://dermatologyfoundation.org/rap
Sponsor: Dermatology Foundation
1560 Sherman Avenue, Suite 870
Evanston, IL 60201

DeRoy Testamentary Foundation Grants **1593**
Established in 1979, the Foundation gives primarily in the state of Michigan. Giving primarily for youth development and services, education, human services, health care, and the arts. The Foundation offers support in the form of: scholarship funds; annual campaigns; building/renovation; continuing support; general/operating support; program development grants. There are no application deadlines. The board meets monthly.
Requirements: Michigan nonprofit organizations are eligible.
Restrictions: Grants are not made to individuals.
Geographic Focus: Michigan
Amount of Grant: 10,000 - 300,000 U.S.D
Samples: National Bone Marrow Transplant Link, Southfield, MI, $10,000; Oakland Family Services, Pontiac, MI, $65,000; Interlochen Center for the Arts, Interlochen, MI, $300,000.
Contact: Julie Rodecker Holly, Vice President; (248) 827-0920; fax (248) 827-0922; deroyfdtn@aol.com
Sponsor: DeRoy Testamentary Foundation
26999 Central Park Boulevard, Suite 160N
Southfield, MI 48076

Detlef Schrempf Foundation Grants **1594**
The focus of the foundation is on children, youth and families in the Pacific Northwest. This foundation seeks supporting organizations that can help underwrite the costs of events, enabling the maximum amount of proceeds to go directly to the benefiting charity.
Requirements: Typically, a qualifying charity partner will have a $5-million, or smaller, operating budget.
Geographic Focus: Idaho, Oregon, Washington
Contact: Nicole Morrison, Executive Director; (206) 464-0826; fax (206) 464-8020; info@detlef.com
Internet: http://www.detlef.com
Sponsor: Detlef Schrempf Foundation
1904 Third Avenue, Suite 339
Seattle, WA 98101

Detroit Lions Charities Grants **1595**
The organization supports charitable and community causes in Michigan. Funding interests include child abuse and domestic violence prevention, youth recreation, and spinal cord injury research. Grants have been awarded to support the Never, Never Shake a Baby billboard campaign; Pigskin Geography, a learning initiative for youth; Big Brothers/Big Sisters; Grand Rapids Metropolitan YMCA—for swimming lessons for inner-city youth; and Saint Joseph Mercy Hospital—for a domestic violence education program. Requests are accepted between October 1 and December 31.
Geographic Focus: Michigan
Date(s) Application is Due: Dec 31
Samples: Dominion Family Services (MI)— to battle child abuse and domestic violence in Michigan, $210,000.
Contact: Detroit Lions Community Affairs Information Hotline; (313) 216-4056
Internet: http://detroitlions.com/community/index.cfm?cont_id=49715
Sponsor: Detroit Lions Charities
222 Republic Drive
Allen Park, MI 48101

DHHS Abandoned Infants Assistance Grants 1596

Funds are provided to prevent the abandonment of infants and young children including the provision of services; to identify and address their needs, especially those who have been infected with the human immunodeficiency virus or who have been prenatally exposed to the virus or a dangerous drug; to assist such children to reside with their natural families or in foster care, as appropriate; to conduct residential programs for abandoned infants and young children who are unable to reside with their families; to provide respite care for families and caregivers; to recruit, train, and obtain foster families; and to recruit and train health and social services personnel to work with families. Contact the office for deadline dates.
Requirements: Eligible applicants are state or local governments, federally recognized Native American tribal governments, U.S. territories and possessions, and nonprofit organizations.
Date(s) Application is Due: Aug 12
Amount of Grant: Up to 475,000 U.S.D
Contact: Patricia Campiglia, Program Officer; (202) 205-8060; pcampiglia@acf.hhs.gov
Internet: http://a257.g.akamaitech.net/7/257/2422/01jan20051800/edocket.access.gpo.gov/2005/05-11592.htm
Sponsor: U.S. Department of Health and Human Services
330 C Street SW, Room 2428
Washington, DC 20447

DHHS Adolescent Family Life Demonstration Projects 1597

This program has multiple components of prevention, care and research. The prevention component focuses on development and testing of abstinence based programs designed to delay the onset of sexual activity reducing the incidence of adolescent pregnancy, STD transmission and HIV/AIDS. The care component is focused on providing comprehensive health, education and social services to pregnant adolescents, adolescent parents, their infants, male partners and their families. The primary focus of research is to improve understanding of the issues surrounding sexuality, pregnancy and parenting by examining the factors that influence adolescent sexual, contraceptive and fertility behaviors, the nature and effectiveness of care services for pregnant and parenting adolescents and why adoption is a little-used alternative among pregnant adolescents. Deadlines are announced in the Federal Register and on the Office of Population Affairs Web site (http://opa.osophs.dhhs.gov).
Requirements: Public and private non-profit organizations may apply.
Samples: Father Flanagan High School (Omaha, NE)—for the Boys Town Adolescent Parenting program; Teenage Alternative School (Lincoln Park, MI)—for a school for pregnant teens and adolescent parents; Mount Vernon Public Schools (Mount Vernon, NY)—for its Family Life Education program, a prevention initiative aimed at fifth- through eighth-grade students.
Contact: Andrea Brandon, Chief Grants Management Officer; (301) 594-6554
Internet: http://aspe.hhs.gov/SelfGovernance/inventory/OPA/111.htm
Sponsor: U.S. Department of Health and Human Services
4350 East-West Highway
Bethesda, MD 20814

DHHS AIDS and HIV Epidemiologic Research Studies Grants 1598

The purpose of this program is to encourage studies using the new rapid HIV tests in different settings (Operational Research), specifically focused on African American, Latino, and other racial and ethnic minorities that are underserved and/or disproportionally affected by the HIV epidemic, and conducted by researchers who have experience working with these populations; to learn more about the effects of rapid HIV testing on motivators and barriers to HIV testing at the individual, provider, and system levels; and to foster collaborations between organizations serving minority communities and their respective state and local health departments in the design and implementation of innovative practical strategies using rapid HIV tests to increase knowledge of HIV serostatus and facilitate entry into prevention and care systems. The research should contribute to the health services knowledge base from which empirically based information can be derived by policy makers, both immediately and over the coming decades. Assistance is available for a 12-month budget period within project periods ranging from one to five years. Contact the office for deadline dates.
Requirements: Applications may be submitted by public and private nonprofit organizations; community-based, national, and regional organizations; state and local governments or their bona fide agents or instrumentalities; federally recognized Indian tribal governments; and Indian tribes or organizations.
Amount of Grant: 5,000 - 270,000 U.S.D
Contact: JAllyn Nakashima, Program Supervisor; (404) 639-0900
Internet: http://www.dhhs.gov
Sponsor: U.S. Department of Health and Human Services
1600 Clifton Road NE, MS-E-45
Atlanta, GA 30333

DHHS AIDS Education and Training Centers Grants 1599

Grants are awarded for the establishment of AIDS education and training centers. The purpose is to provide education and training to primary care providers and others on the treatment and prevention of AIDS, in collaboration with health professions schools, local hospitals, and health departments; to provide updates of new and timely information about HIV infection to primary and secondary health care providers; and to serve as the support system for area health professionals through the AIDS hotline, clearinghouse, and referral activities. Awards will also support community-based organizations and community health clinics affiliated with accredited public and private nonprofit entities. Preference will be given to qualified projects that will train or result in the training of health professionals who will provide treatment for minority individuals with HIV, other individuals who are at high risk of contracting HIV, or to minority health professionals and allied health professionals who will provide treatment for persons with HIV. Awards will be made for a three-year project period. Deadlines are available from the HRSA Application Center. Annual deadline dates vary; contact program staff for exact dates.
Requirements: Public and nonprofit entities, schools, and academic health science centers are eligible to apply.
Amount of Grant: 450,000 - 51,000,000 U.S.D
Contact: Deborah Willis-Fillinger, Coordinator; (301) 443-6364; fax (301) 443-9887
Internet: http://hab.hrsa.gov
Sponsor: U.S. Department of Health and Human Services
5600 Fishers Lane, Parklawn Building
Rockville, MD 20857

DHHS AIDS Project Grants 1600

The grants are intended to fund programs that develop and implement surveillance, epidemiological research, health education, school health, and risk reduction activities of AIDS in states and major cities. The program also gives support for cooperative agreements for AIDS activities. The funding period will be from one to five years, renewable. Applicants should contact the CDC for deadline information.
Requirements: Public and private organizations, both nonprofit and for profit, state and local governments, U.S. territories and possessions, small and minority businesses, and businesses owned by women are encouraged to apply.
Amount of Grant: 45,000 - 2,000,000 U.S.D
Contact: Cheryl Maddux, Grants Manager; (770) 498-1911
Internet: http://www.dhhs.gov
Sponsor: U.S. Department of Health and Human Services
2920 Brandywine Road
Atlanta, GA 30341

DHHS Chiropractic Demonstration Project Grants 1601

The objective of the program is to make grants and enter into contracts with schools, colleges, and universities of chiropractic for the purposes of carrying out demonstration projects in which chiropractors and physicians collaborate to identify and provide effective treatment for spinal and lower back conditions. Grant funds may be used for personnel, equipment, supplies, domestic travel, consultants and guest lecturers, rental of space, renovations, and other costs directly related to the project as described in the approved application. Application materials are available on the World Wide Web at address: http://www.hrsa.gov/bhpr/grants.html.
Requirements: Public or private nonprofit schools, colleges, and universities of chiropractic medicine are eligible to apply.
Restrictions: Grant funds may not be used for construction of facilities, acquisition of land, foreign travel, or support of students, including fellowships, stipends, tuition, fees, or travel allowances.
Amount of Grant: 381,048 - 389,214 U.S.D
Contact: Jennifer Hannah, Administration; (301) 443-0908
Internet: http://bhpr.hrsa.gov/interdisciplinary/chiro.html
Sponsor: U.S. Department of Health and Human Services
5600 Fishers Lane, Parklawn Building
Rockville, MD 20857

DHHS Coal Miners Respiratory Impairment Treatment Clinics and Services Project Grants 1602

The program's goal is to develop high-quality, patient-oriented, integrated systems of care that assure access to and continuity of care with maximum use of existing resources; emphasize patient and family member education to maximize the patient's ability for self-care; and minimize the effects of respiratory and pulmonary impairments in coal miners. Grants may be used for the cost of providing diagnostic treatment procedures required in management of black lung and other occupational-related respiratory impairments in coal miners, and costs of equipment and facilities when these costs are demonstrated to be necessary to enable the implementation of services. Applications are accepted at any time.
Requirements: Eligible to apply are state and local government agencies and private nonprofit agencies. Support is available only in areas where it can be demonstrated that there are significant numbers of active and/or inactive coal miners.
Date(s) Application is Due: Mar 31
Amount of Grant: 150,000 - 12,000,000 U.S.D
Contact: Tamara Lewis-Johnson, Bureau of Primary Health Care; (301) 594-4497; Tjohnson@hrsa.gov
Internet: http://www.hrsa.gov/grants/preview/guidanceprimary/pin200508.htm
Sponsor: U.S. Department of Health and Human Services
5600 Fishers Lane, Parklawn Building
Rockville, MD 20857

DHHS Community Health Centers Grants 1603

The grants support the development and operation of community health centers that provide primary health services, supplemental health services, and environmental health services to medically underserved populations. Priorities are focused on providing services in the most medically underserved areas and maintenance of existing centers that are serving high-priority populations and have demonstrated sound fiscal and management capabilities, monitoring and assessment of project performance, development and implementation of mechanisms for improving quality of care, and maximizing third-party reimbursement levels through improved project administration and management. Applications should be designed to improve the availability, accessibility, and organization of health care within medically underserved communities. Funds may be

used for acquiring and modernizing existing buildings including the costs of amortizing the principal of and paying the interest on loans and for the purchase of special purpose equipment in excess of $100,000. Contact the office for deadline dates.
Requirements: Public or nonprofit private agencies, institutions, or organizations, and a limited number of state and local governments may apply.
Restrictions: Profit-making organizations are not eligible.
Amount of Grant: 45,000 - 88,000,000 U.S.D
Contact: Grants Management Officer; (877) 477-2123; hrsagac@hrsa.gov
Internet: http://bphc.hrsa.gov/chc/default.htm
Sponsor: U.S. Department of Health and Human Services
4350 East-West Highway
Bethesda, MD 20814

DHHS Comprehensive Community Mental Health Services Grants for Children with Serious Emotional Disturbances 1604
This program provides community-based systems of care for children and adolescents with serious emotional disturbances and their families. The program will ensure that services are provided collaboratively across child-serving systems; that each child or adolescent served through the program receives an individualized service plan developed with the participation of the family; that each individualized plan designates a case manager to assist the child and family; and that funding is provided for mental health services required to meet the needs of youngsters in these systems.
Requirements: States; political subdivisions of a state, such as county or local governments; and federally recognized Native American tribal governments are eligible to apply.
Amount of Grant: 200,064 - 35,000,000 U.S.D
Contact: Gary Blau, Chief, Child Adolescent and Family Branch; (301) 443-1333
Internet: http://mentalhealth.samhsa.gov/publications/allpubs/CA-0013/default.asp
Sponsor: U.S. Department of Health and Human Services
5600 Fishers Lane, Parklawn Building
Rockville, MD 20857

DHHS Developmental Disabilities University Affiliated Project Grants 1605
Grants assist with the cost of administration and operation of facilities for providing interdisciplinary training for personnel concerned with developmental disabilities, demonstrations of the provision of exemplary services related to the developmentally disabled, demonstration of technical assistance for generic and specialized agencies, dissemination of findings related to the provision of services to researchers and government agencies, and generation of information on the need for further service-related research. Grants may cover salaries for administrators, coordinators, and others needed to operate a training facility, such as clerical and financial personnel, and maintenance and housekeeping personnel; overhead expenses and expenses required to start up new programs; and faculty for training programs that will meet critical personnel shortages and are not eligible for support from other sources. Deadline dates are available from the office.
Requirements: Public and private nonprofit organizations and agencies are eligible to apply.
Date(s) Application is Due: Sep 1
Amount of Grant: 480,000 U.S.D
Contact: Jennifer Johnson, Administration on Developmental Disabilities, Administration for Children and Families; (202) 690-5982
Internet: http://www.dhhs.gov
Sponsor: U.S. Department of Health and Human Services
370 L'Enfant Promenade SW
Washington, DC 20447

DHHS Emergency Medical Services for Children (EMSC) Program 1606
The purpose of this program is to support demonstration projects for the expansion and improvement of emergency medical services for children who need treatment for trauma or critical care. It is expected that maximum distribution of projects among the states will be made and that priority will be given to projects targeted toward populations with special needs, including Native Americans, minorities, and the disabled. Annual deadline dates may vary; contact program staff for exact dates.
Requirements: States and accredited schools of medicine are eligible to apply.
Amount of Grant: 200,000 U.S.D
Contact: Dan Kavanaugh, (301) 443-1321; dkavanaugh@hrsa.gov
Internet: http://mchb.hrsa.gov/programs/emsc
Sponsor: U.S. Department of Health and Human Services
5600 Fishers Lane, Parklawn Building
Rockville, MD 20857

DHHS Emerging Leaders Program Internships 1607
The Emerging Leaders Program (ELP) is a competitive, two-year paid, federal internship with the U.S. Department of Health and Human Services (HHS). The Program provides a unique opportunity to develop enhanced leadership skills in one of the largest federal agencies in the nation. Upon successful completion of the ELP, participants will be eligible for non-competitive conversion to a permanent appointment. The ELP offers participants the following: formal competency-based leadership training and professional development; challenging developmental rotational assignments; mentorship; fast-paced and diverse work environment; and promotion to a permanent career track that targets the following mission-critical occupation specialties: scientific, social science, human resources, administrative, information technology, public health, and law enforcement. To begin the application process, applicants apply to one or more career track(s) that match their education and/or experience specialty. Later applicants may apply to the specific positions under their chosen career track(s). At the date of this writing, the website states that the ELP is not currently accepting applications, but encourages those interested to check back for future updates.
Requirements: Applicants must qualify for a minimum of one career track to move on to the next phase of the selection process.
Contact: Grants Coordinator; (877) 696-6775; ELP@hhs.gov
Internet: http://hhsu.learning.hhs.gov/elp/howtoapply.asp
Sponsor: U.S. Department of Health and Human Services
200 Independence Avenue SW
Washington, DC 20201

DHHS Family Planning and Personnel Training Grants 1608
The purpose of this program is to provide job specific training for personnel to improve the delivery of family planning services. Grants are awarded to provide training to Title X project staffs; to improve utilization and career development of paraprofessional and paramedical manpower in family planning services, particularly in rural areas. Applications are available on the Web site.
Requirements: Any public or private nonprofit organization or agency is eligible.
Restrictions: Funds may not be used in programs where abortion is a method of family planning.
Amount of Grant: 238,000 - 472,000 U.S.D
Contact: Office of Grants Management; (240) 453-8822; fax (240) 453-8823
Internet: http://opa.osophs.dhhs.gov/titlex/ofp.html
Sponsor: U.S. Department of Health and Human Services
1101 Wootton Parkway, Suite 550
Rockville, MD 21029

DHHS Geriatric Training for Physicians and Dentists Grants 1609
These grants are intended to assist in the operation of postdoctoral training preparing current and future faculty for leadership roles in geriatric medicine and dentistry and to provide support, including traineeships and fellowships, for geriatric medicine training projects to train physicians and dentists who plan to teach geriatric medicine or geriatric dentistry. Geriatric training is to be provided through one or both of the following projects: one grant is offered for a one-year retraining program in geriatrics for physicians who are faculty members in departments of internal medicine, family medicine, gynecology, geriatrics, and psychiatry at schools of medicine and osteopathy; and dentists who are faculty members at schools of dentistry or at hospital departments of dentistry. A grant is offered for a two-year internal medicine or family medicine fellowship program with emphasis in geriatric research for physicians who have completed graduate medical education programs in internal medicine, family medicine, psychiatry, neurology, gynecology, geriatrics, or rehabilitation medicine; and dentists who have completed postdoctoral dental education programs. Each project for which a grant is made must be staffed by full-time teaching physicians who have experience or training in geriatric medicine, be staffed by full-time or part-time teaching dentists who have experience or training in geriatric dentistry, and be based in a graduate medical education program in a department of internal medicine or family medicine or a department of geriatrics that has been in existence over a year. Each project must provide participants with exposure to a diversified population of elderly individuals and provide training in geriatrics and exposure to the physical and mental disabilities of elderly individuals through a variety of service rotations, such as geriatric consultation services, acute care services, dental services, geriatric psychiatry units, day and home care programs, rehabilitation services, extended care facilities, geriatric ambulatory care and comprehensive evaluation units, and community care programs for elderly mentally retarded individuals. Applications and deadline date information can be obtained on the Web site or by contacting the office.
Requirements: Grants will be made to accredited public or private nonprofit schools of medicine, schools of osteopathic medicine, teaching hospitals, and graduate medical education programs.
Amount of Grant: 178,721 - 632,825 U.S.D
Contact: Kathleen Bond, Department of Health and Human Services; (301) 443-8681
Internet: http://www.dhhs.gov
Sponsor: U.S. Department of Health and Human Services
5600 Fishers Lane, Parklawn Building
Rockville, MD 20857

DHHS Grants for Dental Public Health 1610
The purpose of this program is to promote the postgraduate education of dentists in public health. Grants are intended to assist in meeting the costs of planning and developing new dental public health residency training programs and providing financial assistance to trainees enrolled in the programs. Grants may not be used for construction or for providing direct patient services. Annual deadline dates may vary; contact program staff for exact dates.
Requirements: Any accredited public or private school of public health or dentistry may apply. Applicants must demonstrate that it has or will have available full-time faculty members with training and experience in the field of preventive dental public health and support from other faculty members trained in public health and other relevant specialties and disciplines.
Amount of Grant: 46,962 - 153,323 U.S.D
Contact: Raymond Lala, Division of Medicine and Dentistry; (301) 443-1707; rlala@hrsa.gov
Internet: http://www.dhhs.gov
Sponsor: U.S. Department of Health and Human Services
5600 Fishers Lane, Parklawn Building
Rockville, MD 20857

DHHS Great Lakes Human Health Effects Research Program Grants 1611

The objectives of this program are to build upon and amplify the results from past and ongoing research; develop information, databases, and/or research methodology that will provide long-term benefit to the Great Lakes human health research effort; develop directions and methodology for future human health effects research; provide health information to the subjects of the research and their medical professions; and increase the public awareness of the health implications of the toxic pollution problem in the Great Lakes. Grant funds may be used to conduct research on the impact on human health of fish consumption in this region and to extend the knowledge of the effects of contaminants on human reproductive/developmental, behavioral, neurological, and endocrinological adverse health. Contact the office for deadline dates.

Requirements: Eligible applicants are the official public health agencies or their bona fide agents or instrumentalities and political subdivisions thereof, which may include state universities, state colleges, state research institutions, state and local health departments, and federally recognized Native American tribal governments located in the Great Lakes states (Illinois, Indiana, Michigan, Minnesota, Ohio, Pennsylvania, New York, and Wisconsin).

Geographic Focus: Illinois, Indiana, Michigan, Minnesota, New York, Ohio, Pennsylvania, Wisconsin

Amount of Grant: 200,000 U.S.D

Contact: Heraline Hicks, Program Supervisor; (404) 489-0717; fax (404) 498-0094; HEH2@cdc.gov

Internet: http://www.atsdr.cdc.gov/grtlakes.html

Sponsor: U.S. Department of Health and Human Services
1600 Clifton Road NE
Atlanta, GA 30333

DHHS Health Careers Opportunity Program Grants 1612

The grants award funds to identify, recruit, and select individuals from disadvantaged backgrounds for education/training in a health or allied health professions school; facilitate their entry into such a school; provide counseling or other services designed to assist selected individuals so that they may successfully complete their education/training; provide preliminary education for a period prior to entry into the regular course of health or allied health professions education, designed to assist students in completing successfully regular course of education or to refer the appropriate individuals to institutions providing preliminary education; and publicize existing sources of financial aid. Annual deadline dates may vary; contact program staff for exact dates.

Requirements: Eligible to apply are health professions schools and public or nonprofit private health or educational entities including schools of medicine, osteopathy, public health, dentistry, veterinary medicine, optometry, pharmacy, allied health, chiropractic and podiatry, public and nonprofit private schools that offer graduate programs in clinical psychology, and other public or private nonprofit health or educational entities.

Restrictions: Grant funds may not be used for tuition and fees, training of program staff or retraining of health professionals, sectarian instruction, religious purposes, construction, land acquisition, or foreign travel.

Amount of Grant: 116,237 - 1,000,000 U.S.D

Contact: Lawrence Poole, Director, Division of Grants Management; (301) 443-2385

Internet: http://www.dhhs.gov

Sponsor: U.S. Department of Health and Human Services
5600 Fishers Lane, Parklawn Building
Rockville, MD 20857

DHHS Health Care Systems Cost and Access Research and Development Grants 1613

Grants are made to support health services research to create new knowledge and better understanding of the process by which health services are made available and how they may be provided more efficiently and effectively. The AHCPR has a broad legislative mandate to support general health services research on problems related to health care cost, quality, and access to health services. Major categories of research issues include consumer decision making, managed care and the health care marketplace, primary care, rural health services, and AIDS. Annual deadline dates may vary; contact program staff for exact dates.

Requirements: Government agencies (federal, state, and local), federally recognized tribal governments, U.S. territories, sponsored organizations, nongovernment organizations, minority groups, specialized groups, public or private institutions of higher education, and other public or nonprofit private agencies, institutions, or organizations are eligible. Research project grants may also be awarded to individuals.

Restrictions: Profit-making organizations are not eligible for grants.

Date(s) Application is Due: Feb 1; Jun 1; Oct 1

Amount of Grant: 5,000 - 28,000,000 U.S.D

Contact: Mable Lam, Grants Management Officer; (301) 427-1447

Internet: http://www.dhhs.gov

Sponsor: U.S. Department of Health and Human Services
2101 E Jefferson Street, Executive Office Center
Rockville, MD 20852

DHHS Health Center Grants for Homeless Populations 1614

Grants are awarded for the purpose of enabling grantees, directly or through contracts, to provide for the delivery of primary health services and substance abuse (alcohol and/or illicit drugs) services to homeless individuals. Funds may be used to provide for primary health care and substance abuse services at locations accessible to homeless individuals; 24-hour emergency primary health and substance abuse services; referral, as appropriate to medical facilities, for necessary hospital services; referral of homeless individuals who are mentally ill to entities that provide mental health services unless the applicant will provide such services directly; outreach services to inform homeless individuals of the availability of primary health and substance abuse services; and aid in establishing eligibility for assistance, and in obtaining services under entitlement programs. Podiatry, dental (including dentures), and vision services are supplemental services and may be provided where medically necessary to the extent that the level of delivery of the required services is not diminished. The grants may be used to continue to provide the services listed above for up to 12 months to individuals who have obtained permanent housing if services were provided to these individuals when they were homeless. Deadline dates are specified in the application instructions.

Requirements: Nonprofit private organizations and public entities including state and local governmental agencies are eligible to apply.

Amount of Grant: 84,500 - 100,000,000 U.S.D

Contact: Dr. Regan Crump, Acting Director; (301) 594-4420

Internet: http://www.federalgrantswire.com/health_center_grants_for_homeless_populations.html

Sponsor: U.S. Department of Health and Human Services
4350 East-West Highway
Bethesda, MD 20814

DHHS Health Centers Grants for Migratory and Seasonal Farmworkers 1615

The grants support the development and operation of migrant health centers and projects that provide primary health care services, supplemental health services, and environmental health services that are accessible to migrant and seasonal agricultural farm workers and their families as they move and work. Funds may be used for the planning, development, and operation of clearly defined migrant health centers and programs; for the acquisition and modernization of existing buildings; and for training related to the management of programs assisted through grants and contracts.

Requirements: Any public or nonprofit private entity may apply. Preference will be given to applications submitted by community-based organizations that are representative of the populations to be served.

Restrictions: For-profit organizations are ineligible.

Amount of Grant: 50,000 - 2,500,000 U.S.D

Contact: Grants Management Officer, Bureau of Primary Health Care; (301) 594-4235

Internet: http://www.bphc.hrsa.gov

Sponsor: U.S. Department of Health and Human Services
4350 East-West Highway
Bethesda, MD 20814

DHHS Health Centers Grants for Residents of Public Housing 1616

Grants are awarded to improve minority access to primary care services and reduce infant mortality by enabling grantees, directly or through contracts, to provide to public housing residents, especially pregnant residents and their infants, primary health services, including health screening, education, and counseling. Applications should be designed to improve the availability, accessibility, and provision of primary health care services, including comprehensive perinatal care, to residents of public housing. Contact the office for deadline dates.

Requirements: Eligible applicants are public and nonprofit private entities that have the capacity to administer a grant and are located in a high infant mortality area as determined by the Healthy Start Initiative criteria. Preference is given to community health centers, health care for the homeless programs, and resident management corporations.

Amount of Grant: 140,000 - 400,000 U.S.D

Contact: Office of Grants Management, Bureau of Primary Health Care; (301) 594-4235

Internet: http://www.federalgrantswire.com/health_centers_grants_for_residents_of_public_housing.html

Sponsor: U.S. Department of Health and Human Services
4350 East-West Highway
Bethesda, MD 20814

DHHS Health Conference Support Program for Toxic Substances and Disease Registry 1617

Funds are available through this program for partial support for nonfederal conferences on disease prevention, health promotion, and projects related to hazardous substances. The purpose of this program is to work closely with state, local, and other federal agencies to reduce or eliminate illness, disability, and death resulting from exposure of the public and workers to toxic substances at spill and waste disposal sites.

Amount of Grant: 130,000 - 300,000 U.S.D

Contact: Mildred Garner, Grants Management Branch; (770) 488-2745; fax (770) 488-2777

Internet: http://www.dhhs.gov

Sponsor: U.S. Department of Health and Human Services
1600 Clifton Road NE
Atlanta, GA 30333

DHHS Health Education and Training Centers Project Grants 1618

The objective of this program is to provide assistance for planning, developing, establishing, maintaining, and operating health education and training centers (HETC). Grants are to assist in meeting costs of the program that cannot be met from other sources. Not less than 75 percent of the total funds provided to a school or schools of allopathic medicine or osteopathic medicine will be expended in the development and operation of the HETC in the service area of such program. To the maximum extent feasible, the schools of medicine are to obtain from nongovernmental sources the amount of the total operating funds for such programs that are not provided by DHHS. Grants or contracts will not provide funds solely for the planning or development of a HETC program for a period in excess of two years, not more than 10 percent of the annual budget of each program may be utilized for the renovation and equipping of clinical teaching sites, and grants or contracts shall not provide funds to be used outside the United States

except as the DHHS may prescribe for travel and communications purposes related to the conduct of a border HETC. The first project period will be up to three years to allow for the development of the HETC. Annual deadline dates may vary; contact program staff for exact dates.
Requirements: Accredited schools of allopathic or osteopathic medicine, or the parent institution on behalf of such schools or a consortium of such schools, may apply.
Amount of Grant: 92,000 - 525,000 U.S.D
Contact: Louis Coccodrilli, Division of Medicine; (301) 443-6950
Internet: http://www.dhhs.gov
Sponsor: U.S. Department of Health and Human Services
5600 Fishers Lane, Parklawn Building
Rockville, MD 20857

DHHS Health Professions Preparatory Scholarship Program for Indians 1619
The program makes scholarship grants for a maximum of two years to individuals of Native American or Alaskan Native descent for the purpose of completing compensatory preprofessional education to enable the recipient to qualify for enrollment or re-enrollment in a health professions school. Grants for stipends and books are made directly to the individual applicant; tuition payments are made to the college or university. New applications are usually available in February with an annual submission deadline in April. Annual deadline dates may vary; contact program staff for exact dates.
Requirements: Scholarship awards are made to individuals of Native American or Alaskan Native descent who have successfully completed high school education or high school equivalency and who have been accepted for enrollment in a compensatory, preprofessional general education course or curriculum.
Date(s) Application is Due: Apr 1
Amount of Grant: 17,500 - 26,019 U.S.D
Contact: Lois Hodge, Grants Management Officer; (301) 443-0243
Internet: http://www.ihs.gov
Sponsor: U.S. Department of Health and Human Services
12300 Twinbrook Parkway, Twinbrook Metro Plaza
Rockville, MD 20852

DHHS Health Professions Scholarship Program 1620
The program makes scholarship grants to Native Americans and other students for the purposes of completing health professional education—nursing; medicine including allopathic, osteopathic, and veterinary medicine; dentistry; x-ray technology; optometry; pharmacy; public health nutrition (graduate); medical social work (graduate); speech pathology/audiology (graduate); podiatry; health care administration; and other allied health professions—to obtain health professionals to serve Native Americans. Not all disciplines participate each year. Upon completion, grantees are required to fulfill an obligated service payback. Maximum length of funding is four years. Grants for stipends and books are made directly to the individual applicant; tuition payments are made to the college or university. New applications are available in February with a submission deadline in April. Annual deadline dates may vary; contact program staff for exact dates.
Requirements: Priority consideration for scholarship awards is granted to persons of Native American or Alaskan Native descent. Applicants for new awards must be accepted by an accredited U.S. educational institution for a full-time course of study leading to a degree in medicine, osteopathy, dentistry, or other participating health profession that is deemed necessary by the Indian Health Service; be eligible for or hold an appointment as a commissioned officer in the regular or reserve corps of the Public Health Service; or be eligible for civilian service in the Indian Health Service.
Date(s) Application is Due: Apr 1
Amount of Grant: 24,128 - 38,222 U.S.D
Contact: Lois Hodge, Grants Management Officer; (301) 443-0243
Internet: http://www.dhhs.gov
Sponsor: U.S. Department of Health and Human Services
12300 Twinbrook Parkway, Twinbrook Metro Plaza
Rockville, MD 20852

DHHS Health Professions Training in Geriatrics Grants 1621
Grants provide funds to develop regional resource centers focused on strengthening multidisciplinary training of health professionals in geriatric health care. Annual deadline dates may vary; contact program staff for exact dates.
Requirements: Any accredited health professions school or school of allied health or any program for the training of physician assistants may apply.
Restrictions: Grant funds may not be used for trainee costs, construction of buildings, initial equipment, additions, extensions, or the acquisition of land. Grants may not be spent for sectarian instruction or for any religious purpose.
Amount of Grant: 47,072 - 432,000 U.S.D
Contact: Jennifer Hannah, (301) 443-0908
Internet: http://www.dhhs.gov
Sponsor: U.S. Department of Health and Human Services
5600 Fishers Lane, Parklawn Building
Rockville, MD 20857

DHHS Health Promotion and Disease Prevention Research and Demonstration Centers Grants 1622
Grants are awarded to qualified schools to establish, maintain, and operate academic-based centers for high-quality research and demonstration with respect to health promotion and disease prevention; to establish linkages between ongoing basic research in a wide array of fields and applied research in disease prevention and health promotion; to bring the knowledge and expertise of academic health centers to bear on practical public health problems; to field test and rigorously evaluate more cost-effective methods and strategies for preventing unnecessary illness and promoting good health; and to shorten the time lag between the development of new and proven effective disease prevention and health promotion techniques and their widespread application. The length of assistance will be from one to five years, renewable based on competitive applications and availability of funds. Deadlines will be announced in the Federal Register.
Requirements: Eligible applicants are schools of medicine, schools of osteopathy, and schools of public health.
Restrictions: Grantees may not award subgrants but may enter into consortia agreements or contracts as necessary to achieve the aims of the program.
Amount of Grant: 740,000 - 770,000 U.S.D
Contact: Dr. Eduardo Simoes, Program Director; (770) 488-5919
Internet: http://www.cdc.gov/prc
Sponsor: U.S. Department of Health and Human Services
4770 Buford Highway, K-45
Atlanta, GA 30333

DHHS Healthy Tomorrows Partnership for Children 1623
This program encourages community-based partnerships to coordinate health resources for pregnant women, infants, and children. Grantees are encouraged to seek additional support from the private sector and foundations. Applicants must provide a dollar for dollar cash match after the end of the first year of the grant.
Requirements: Eligible applicants include public and private entities, community-based organizations, Indian tribes, and tribal organizations. Priority is given to projects in all U.S. territories and in Alabama, Delaware, Florida, Georgia, Hawaii, Idaho, Indiana, Iowa, Kansas, Louisiana, Maine, Maryland, Massachusetts, Mississippi, Missouri, Montana, Nevada, New Hampshire, New Jersey, North Carolina, North Dakota, South Carolina, Texas, West Virginia, Wisconsin, and Wyoming.
Geographic Focus: Alabama, Delaware, Florida, Georgia, Hawaii, Idaho, Indiana, Iowa, Kansas, Louisiana, Maine, Maryland, Massachusetts, Mississippi, Missouri, Montana, Nevada, New Hampshire, New Jersey, North Carolina, North Dakota, South Carolina, Texas, West Virginia, Wisconsin, Wyoming
Amount of Grant: Up to 50,000 U.S.D
Contact: Lawrence Poole, Director, Division of Grants Management; (301) 443-2385
Internet: http://mchb.hrsa.gov/programs/training/healthytomorrows.htm
Sponsor: U.S. Department of Health and Human Services
200 Independence Avenue SW
Washington, DC 20201

DHHS HIV Demonstration Program for Children, Adolescents, and Women 1624
Grants are awarded to support, improve, and expand the system of comprehensive care services for children, youth, women, and families who are infected with or affected by HIV and AIDS and to link comprehensive care systems with clinical research. Projects should promote collaboration between research institutions and primary care medical and social service programs. Funding priority will be given to projects that demonstrate established models of care that are comprehensive and coordinated, sensitive to differing cultural needs, and family and community based. Grants may be made for up to three-year project periods, renewable competitively after the first year. Contact the office for deadline dates.
Requirements: All public and private nonprofit organizations that provide or arrange for primary health care are eligible for grants. This includes state or local health departments, hospitals, community health centers, drug abuse treatment agencies, school-based clinics, tribal health programs, colleges, and hemophilia treatment centers.
Amount of Grant: 230,384 - 25,000,000 U.S.D
Contact: Dr. Jose Rafael Morales, Division of Community Based Programs/Title IV, HIV/AIDS Bureau; (301) 443-9051
Internet: http://www.dhhs.gov
Sponsor: U.S. Department of Health and Human Services
5600 Fishers Lane, Parklawn Building
Rockville, MD 20857

DHHS HIV Prevention Cooperative Agreements Grants 1625
The purpose of this program is to provide assistance to local, regional, and national nonprofit organizations to develop and implement effective community-based Human Immunodeficiency Virus (HIV) prevention programs related to achieving national goals; to promote coordination for primary and secondary HIV prevention efforts among community organizations, HIV education/prevention service agencies, and public organizations including local and state health departments and substance abuse agencies; and to evaluate the HIV prevention programs for which support is provided.
Requirements: Nongovernmental public and nonprofit private entities are eligible to apply.
Restrictions: Funds may not be used for direct medical care or to supplant existing funding.
Amount of Grant: 62,000 - 100,000,000 U.S.D
Contact: Allyn Nakashima; (404) 639-0900
Internet: http://www.cdc.gov/about/default.htm
Sponsor: U.S. Department of Health and Human Services
1600 Clifton Road NE, MS E-27
Atlanta, GA 30333

DHHS Indian Health Service Health Management Development Grants 1626
The program intends to raise to the highest possible level the health of the Native Americans and Native Alaskans by providing a full range of curative, preventive, and rehabilitative health services and to improve the management capability of Native Americans and Native Alaskans so they may assume operation of all or part of existing Indian Health Service

direct-operated health care programs. Grant applications must be submitted 90 days prior to the proposed project start date. Contact the office for deadline dates.
Requirements: Federally recognized tribes and tribal organizations may apply.
Restrictions: Tribal management grants may not be used to support operational programs, or to supplant existing public and private resources.
Amount of Grant: 80,000 U.S.D
Contact: Lois Hodge; (301) 443-2214; Lois.Hodge@ihs.gov
Internet: http://www.ihs.gov/NonMedicalPrograms/gogp
Sponsor: U.S. Department of Health and Human Services
801 Thompson Avenue, Suite 120
Rockville, MD 20852

DHHS Maternal and Child Health Projects Grants **1627**

The goal of the program is to improve the quality and use of genetic services in maternal and child health care, integrate services for children with special health care needs, support state fetal and infant mortality review centers, and enhance the use of data in problem solving for maternal and child health. The four grant programs currently accepting applications are genetic services, integrated services for children with special health care needs, state fetal/infant mortality review support centers, and data utilization and enhancement for state/community infrastructure building and managed care. Genetic services projects should improve the quality, availability, accessibility, and use of genetic services as an integral part of comprehensive maternal and child health care. Integrated services for children with special health care needs should demonstrate innovative and replicable models of community-based services. State fetal/infant mortality review support centers should provide training and technical assistance to state maternal and child health agencies on their particular needs. Data utilization and enhancement for state/community infrastructure building and managed care projects should enhance the use of qualitative and quantitative analytic methods in solving health problems for women and children. Contact the office for deadlines.
Requirements: Eligible applicants are public and private nonprofit agencies, including hospitals, Native American tribes, and social service agencies. The bureau will give extra attention to applications from historically black colleges and universities and Hispanic-serving institutions.
Amount of Grant: 14,000 - 15,000,000 U.S.D
Contact: Program Contact; (301) 443-2170; fax (301) 443-1797
Internet: http://mchb.hrsa.gov/programs/default.htm
Sponsor: U.S. Department of Health and Human Services
5600 Fishers Lane, Parklawn Building
Rockville, MD 20857

DHHS Maternal and Child Health Research and Training Programs Grants **1628**

Grants fund special projects that develop new data and improve the professional development of health practitioners who serve women and children. Research grants will fund projects that address minority and disadvantaged populations, health-promoting behaviors, quality outcome measures, and systems integration and reform. Long-term training grants will be awarded to train health professionals at the graduate and postgraduate levels. Training should focus on family-centered, community-based care. Continuing education grants support continuing education for workers in the maternal and child health field. Priority will be given to projects focusing on emergency medical services for children, violence prevention in schools, and core public health. Application kits may be obtained by calling toll-free (877) 477-2123.
Requirements: Training grants may be made to public or nonprofit private institutions of higher learning. Research grants may be made to public or nonprofit institutions of higher learning and public or nonprofit private agencies and organizations engaged in research, maternal and child health, or programs for children with special health care needs.
Amount of Grant: 14,000 - 15,000,000 U.S.D
Contact: Program Contact, Maternal and Child Health Bureau; (301) 443-2170
Internet: http://www.dhhs.gov
Sponsor: U.S. Department of Health and Human Services
5600 Fishers Lane, Parklawn Building
Rockville, MD 20857

DHHS Minority Community Health Coalition Demonstration Grants **1629**

These grants are offered to support minority community health coalitions to develop, implement, and conduct demonstration projects that coordinate integrated community-based screening and outreach services and include linkages for access and treatment to minorities in high-risk, low-income communities. Awards must focus on at least one of the following major minority groups: African Americans, Hispanics, Native Americans, Alaska Natives, and Asian/American Pacific Islanders or subgroups of these. These projects are to address sociocultural and linguistic barriers to care and should have the potential for replication in similar communities.
Requirements: Eligible applicants are public and private nonprofit organizations that will serve as the grantee organization for the coalition and have a history of Office of Minority Health coalition grant support.
Restrictions: Grant funds are not to be used for medical treatment; construction of facilities, including additions and extensions; acquisition of land; sectarian instruction or any religions purposes; or lobbying.
Amount of Grant: 132,041 - 150,000 U.S.D
Contact: Karen Campbell, Grants Management Officer; (301) 594-0758
Internet: http://www.omhrc.gov/omh/aids/2k2/0993_4C2.htm
Sponsor: U.S. Department of Health and Human Services
5515 Security Lane, Suite 1000
Rockville, MD 20852

DHHS National Health Promotion Project Grants **1630**

These grants support projects that are intended to educate the public about environmental, occupational, societal, and behavioral factors that affect health, so that individuals may make informed decisions about health-related behavior. The National Health Promotion program is the federal focal point to promote the development, implementation, and coordination of programs that promote good health habits and programs that are designed to prevent disease and disability. Projects supported include, but are not limited to, those identifying or developing materials for health promotion programs for use by schools, medical treatment sites, work sites, and community health promotion programs; adding to the scientific database; and identifying the needs of special population groups such as African Americans, Hispanics, Asian/Pacific Islanders, Native American/Alaskan Natives, the disabled, and elderly in the United States. Contact the office for deadline dates.
Requirements: Public or private nonprofit organizations are eligible to apply.
Amount of Grant: 50,000 - 75,000 U.S.D
Contact: Grants Management Specialist; (301) 594-0758
Internet: http://www.healthypeople.gov
Sponsor: U.S. Department of Health and Human Services
200 Independence Avenue SW
Washington, DC 20201

DHHS National Health Service Corps Loan Repayment Awards **1631**

To help ensure an adequate supply of trained health professionals for the National Health Service Corps, this program provides for the repayment of educational loans for participants who agree by written contract to serve an applicable period of time in a health-related workforce shortage area or in an Indian health program or facility. Awards provide payments toward participants' qualified government and commercial health professions education loans during each year of practice at a selected NHSC loan repayment service site with a two-year service minimum. Health professions given priority for selection are those determined by the workforce needs of the NHSC and the Indian Health Service. Beneficiaries include primary care physicians; dentists; certified nurse midwives; certified nurse practitioners; physicians assistants; clinical psychologists; clinical social workers; psychiatric nurse specialists; marriage and family therapists; licensed professional counselors; and dental hygienists. Deadline dates are published in the Federal Register.
Requirements: U.S. citizens who are enrolled as full-time students in the final year of study in an accredited health profession education institution or possess a health professions degree and are either enrolled in postgraduate health professions training or are in professional practice are eligible to apply. They must hold an unrestricted health professions license in a state and be eligible for or hold an appointment as a commissioned officer in the regular or reserve corps of the Public Health Service or be eligible for selection for a federal civil service appointment.
Date(s) Application is Due: Mar 26
Amount of Grant: 25,039 U.S.D
Contact: Chief, Loan Repayment Programs Branch, Division of Health Services Scholarships; (301) 594-4400 or (800) 435-6464
Internet: http://www.hrsa.gov
Sponsor: U.S. Department of Health and Human Services
5600 Fishers Lane, Parklawn Building
Rockville, MD 20857

DHHS National Health Service Corps Scholarships **1632**

To obtain adequate numbers of trained physicians, dentists, and other health-related specialists for the National Health Service Corps, scholarships are given in exchange for service in health personnel shortage areas within the United States. Disciplines have included allopathic and osteopathic medicine, dentistry, nursing (baccalaureate and graduate), public health nutrition (graduate), medical social work (graduate), speech pathology/audiology (graduate), veterinary medicine, optometry, podiatry, and pharmacy. Not all disciplines participate each year. Deadlines for receipt of applications for each academic year are the last Friday in March. A student may not receive more than a total of four years of support. Each award covers one academic year of support (including monthly living stipend, tuition, and fees), as specified by signed contracts, with annual continuation awards if funds are available.
Requirements: Applicants for new awards must be accepted by accredited U.S. educational institutions for full-time courses of study leading to degrees in medicine, osteopathy, or dentistry; be former recipients of the federal scholarship program for first-year students of exceptional financial need; be eligible for or hold appointments as commissioned officers in the regular or reserve corps of PHS or be eligible for civilian service in the NHSC; and submit applications and signed contracts to accept payment of scholarships and to serve for the applicable period of obligated service in health workforce shortage areas.
Restrictions: Each year of scholarship support incurs a year of federal service obligation with the minimum obligation being two years. Failure to fulfill the service obligation incurs financial damages, payable in one year, calculated at three times the scholarship benefits plus interest.
Amount of Grant: 1,128 U.S.D
Contact: Program Contact; (301) 594-4400 or (800) 221-9393
Internet: http://www.dhhs.gov
Sponsor: U.S. Department of Health and Human Services
4350 East-West Highway
Bethesda, MD 20814

DHHS Native American Health Professionals Recruitment Grants **1633**

The program makes awards to Native American tribes and tribal and Native American health organizations to enable them to recruit, place, and retain health professionals to fill critical vacancies and to meet the staffing needs of Native American health programs

and facilities including those administered by the Indian Health Service. Grants are awarded for one year; continuation applications must be submitted annually. Contact the office for deadline dates.
Requirements: Eligibility is open to any federally recognized Native American tribe or tribal or Native American health organization.
Amount of Grant: 60,000 - 100,000 U.S.D
Contact: Lois Hodge, Grants Management Officer; (301) 443-5204
Internet: http://www.dhhs.gov
Sponsor: U.S. Department of Health and Human Services
12300 Twinbrook Parkway, Twinbrook Metro Plaza
Rockville, MD 20852

DHHS Native American Health Service Loan Repayment Awards 1634
To help ensure an adequate supply of trained health professionals for IHS facilities, the program provides for the repayment of educational loans for participants who agree by written contract to serve an applicable period of time at a facility IHS has designated as a retention/recruitment priority site or in a designated specialty at a site with an appropriate position. Recipients must agree to serve an applicable period of time in such a site with a minimum period of participation of two years. Contact the program office for deadline dates.
Requirements: Eligible applicants must be individuals who are enrolled as full-time students in the final year of a course of study or program leading to a degree in allopathic or osteopathic medicine, dentistry, or other health profession in a state; are enrolled in an approved graduate training program in allopathic or osteopathic medicine, dentistry, or other health profession; or have a degree in allopathic or osteopathic medicine, dentistry, or other health profession and have completed an approved graduate training program and have a current and valid license to practice such health profession in a state. In addition, applicants must be eligible for appointment as commissioned officers in the regular or reserve corps of Public Health Service (PHS) or be eligible for selection for civilian service in IHS, submit an application to participate in the loan repayment program, and sign and submit it to DHHS at the time of agreeing to accept repayment of educational loans, and serve for the applicable period of service in a retention/recruitment site as determined by DHHS.
Amount of Grant: 3,000 - 48,000 U.S.D
Contact: Jackie Sanitago, Program Chief; (301) 443-3396
Internet: http://www.dhhs.gov
Sponsor: U.S. Department of Health and Human Services
12300 Twinbrook Parkway, Twinbrook Metro Plaza
Rockville, MD 20852

DHHS Neurodevelopmental Test Methods Research Grants 1635
The purpose of the program is to determine and validate a battery of neurodevelopmental tests for use in assessing the effects of prenatal or postnatal exposure to developmental toxicants. The battery of tests should be applicable to a wide range of potential neurodevelopmental toxicants found at waste sites and in the environment including metals and solvents; be applicable to a wide range of exposure levels found in the environment; and cover a broad range of developmental domains including cognitive function, sensory function, motor function, and complex multi-tasking performance. These research methods will address an agency goal to develop methods and tools for evaluating human health consequences from exposure to toxic substances in the environment.
Requirements: Applications may be submitted by official public health agencies of the states, or their bona fide agents. This includes the District of Columbia, American Samoa, the Commonwealth of Puerto Rico, the Virgin Islands, the Federated States of Micronesia, Guam, the Northern Mariana Islands, the Republic of the Marshall Island, the Republic of Palau, federally recognized Indian tribal governments, public and private nonprofit and for-profit universities, colleges, and research institutions.
Date(s) Application is Due: Jul 15
Amount of Grant: 130,000 - 300,000 U.S.D
Contact: Mildred Garner, Grants Management Officer; (770) 488-2745; fax (770) 488-2777; info@cdc.gov
Internet: http://www.cdc.gov
Sponsor: U.S. Department of Health and Human Services
2920 Brandywine Road, Suite 3000, MS E13
Atlanta, GA 30341-4146

DHHS Oral Health Promotion Research Across the Lifespan 1636
The National Institute of Dental and Craniofacial Research has invited proposals for improving the oral health of people of all ages. The research team must include someone with extensive experience in health promotion, behavioral and/or social science research. The health promotion intervention proposed for funding must be based on a previously conducted assessment of the epidemiology, social, behavioral and/or environmental factors related to the disease or condition under study. Research could focus on maternal and child health, adolescent and young adult health, or health of adults with complex diseases. For example, an applicant might propose to study approaches to involving families, social networks, communities, or neighborhoods in behaviors that promote and improve oral health; improving patient-provider communication related to oral preventive measures; or effective ways to train oral health professional students to communicate with diverse patient populations.
Requirements: Any person with the skills, knowledge, and resources necessary to carry out the proposed research as the project director/principal investigator (PD/PI) is invited to work with his/her organization to develop an application for support. Applications must be submitted electronically through Grants.gov (http://www.grants.gov) using the SF424 research and related forms and the SF424 application guide.
Date(s) Application is Due: Mar 5; Jul 5; Nov 5
Contact: Maria Teresa Canto, (301) 594-5497; fax (301) 480-8322; maria.canto@nih.gov
Internet: http://grants.nih.gov/grants/guide/pa-files/PA-07-225.html
Sponsor: National Institute of Dental and Craniofacial Research
45 Center Drive, Natcher Building
Bethesda, MD 20892

DHHS Podiatric Primary Care Residency Training Grants 1637
This program intends to promote the postgraduate education of podiatrists in primary care podiatric practice. Grants are to assist in meeting the costs of the program that cannot be met from other sources. Grants may include support for the program only, residents only, or support for both the program and residents but are not intended to absorb costs of existing or new positions that can be supported from other available funds. Project period is a maximum of 36 months; a noncompeting continuation application is required at the end of each budget period. Contact the office for deadline dates.
Requirements: Public or nonprofit private accredited schools of podiatric medicine and teaching hospitals may apply.
Amount of Grant: 205,291 - 286,362 U.S.D
Contact: Chris McLaughlin, Division of Medicine; (301) 443-1568; cmclaughlin@hrsa.gov
Internet: http://www.dhhs.gov
Sponsor: U.S. Department of Health and Human Services
5600 Fishers Lane, Parklawn Building
Rockville, MD 20857

DHHS Policy Research and Evaluation Grants 1638
Priorities for this program include issues of long-term care, disability, and personal assistance services, including informal care giving; health care delivery issues including health care financing; welfare service delivery issues and policies affecting children and youth; management coordination of DHHS-wide policy development and policy support activities; science policy development; and the reduction of poverty. Contact the office for deadlines.
Requirements: Assistance may be provided to nonprofit organizations and institutions, government agencies, and individuals. Occasionally awards are made to for-profit organizations when deemed to be consistent with the legislative intent and purpose of the program.
Amount of Grant: 300,000 - 12,000,000 U.S.D
Contact: Grants Officer; (202) 690-8794
Internet: http://www.acf.hhs.gov/programs/opre
Sponsor: U.S. Department of Health and Human Services
200 Independence Avenue SW
Washington, DC 20201

DHHS Preventive Medicine 1639
The grants promote the postgraduate education of physicians in preventive medicine. Grants are intended to assist in meeting the costs of planning and developing new preventive medicine programs; maintaining or improving existing residency training programs in preventive medicine; and providing financial assistance to trainees enrolled in such programs. Contact the office for deadlines and application materials.
Requirements: Any accredited public or private school of medicine, osteopathy, or public health may apply. To be eligible for a grant, the applicant must demonstrate that it has or will have available, full-time faculty members with training and experience in the fields of preventive medicine.
Restrictions: Grants may not be used for construction or patient services.
Amount of Grant: 101,897 - 430,188 U.S.D
Contact: Dr. Douglas Lloyd, Public Health and Dental Education Branch; (301) 443-0157
Internet: http://www.dhhs.gov
Sponsor: U.S. Department of Health and Human Services
5600 Fishers Lane, Parklawn Building
Rockville, MD 20857

DHHS Primary Medical Care National Research Service Awards (NRSA) 1640
This program is intended to promote postdoctoral research training programs in primary medical care. NRSAs are made directly to individuals for research training in primary medical care; in addition, grants may be made to institutions to enable them to award NRSAs to individuals selected by them. Each individual who receives an NRSA is obligated, upon termination of the award, to comply with certain service and payback provisions. Recipients agree to engage in primary medical care research and/or teaching for a period equal to the period of the NRSA support in excess of 12 months. Contact the office for deadline dates.
Requirements: Domestic public or private nonprofit organizations may apply for training grants. States or local governments and U.S. territories are eligible. Individual applicants for fellowships must have received doctoral degrees. All persons supported as fellows or trainees must be citizens or noncitizen nationals of the United States or have been lawfully admitted for permanent residence. Eligibility is limited to individuals affiliated with entities having received grants or contracts under Sections 780, 784, or 786 of the Public Health Service Act.
Amount of Grant: 258,100 - 538,167 U.S.D
Contact: Christopher McLaughlin and Anne Patterson, Division of Medicine; (301) 443-6785
Internet: http://www.dhhs.gov
Sponsor: U.S. Department of Health and Human Services
5600 Fishers Lane, Parklawn Building
Rockville, MD 20857

DHHS Priority Health Conditions Studies Initiative 1641

The primary purpose of this program is to solicit scientific proposals designed to study the occurrence of and risk factors for priority adverse health conditions. The priority health conditions are birth defects and reproductive disorders, cancers, immune function disorders, kidney dysfunction, liver dysfunction, lung and respiratory diseases, and neurotoxic disorders. Each fiscal year, one of the seven priority health conditions will be emphasized; however, applications may address any of the seven. This will improve recipients' ability to address potential public health problems related to exposure to hazardous substances. Contact the office for deadline dates.
Requirements: Eligible applicants are state health departments and the District of Columbia; the Commonwealth of Puerto Rico; the Virgin Islands; Guam; the Federated States of Micronesia; the Republic of the Marshall Islands; the Republic of Palau; the Northern Mariana Islands; American Samoa; the political subdivisions thereof, which may include state universities, state colleges, and state research institutions; and federally recognized Native American tribal governments.
Amount of Grant: 50,000 - 500,000 U.S.D
Contact: Caroline McDonald, Agency for Toxic Substances and Disease Registry; (404) 498-0270; fax (404) 498-0059; COS4@cdc.gov
Internet: http://www.dhhs.gov
Sponsor: U.S. Department of Health and Human Services
1600 Clifton Road NE
Atlanta, GA 30333

DHHS Professional Nurse Traineeships Program 1642

Grants are awarded to eligible institutions to provide financial support through traineeships for registered nurses enrolled in advanced education nursing programs to prepare nurse practitioners, clinical nurse specialists, nurse midwives, nurse anesthetists, nurse administrators, nurse educators, public health nurses and nurses in other specialties determined by the Secretary to require advanced education.
Requirements: Schools of nursing and public health, public or nonprofit private hospitals, and other public or nonprofit entities are eligible for funds. Trainees are selected by participating institutions. A candidate must be a U.S. citizen or have been lawfully admitted to the United States for permanent residence; a graduate of a state-approved school of nursing; licensed as a professional nurse in a state or territory; and able to enroll full-time in a course of study.
Amount of Grant: 1,000 - 235,000 U.S.D
Contact: Karen Breeden, Division of Nursing, Bureau of Health Professions; (301) 443-6333; fax (301) 443-8586; kbreeden@hrsa.gov
Internet: http://www.dhhs.gov
Sponsor: U.S. Department of Health and Human Services
5600 Fishers Lane, Parklawn Building
Rockville, MD 20857

DHHS Project Grants for Renovation or Construction of Nonacute Health Care Facilities 1643

Grants are intended to be used to renovate, expand, repair, equip, or modernize nonacute care intermediate and long-term care facilities for patients with AIDS. Contact the program office for deadline dates.
Requirements: Applicants must be public or nonprofit entities.
Amount of Grant: 100,000,000 U.S.D
Contact: Paul Murhy, Division of Facilities Compliance and Recovery Contact; (301) 443-5656
Internet: http://www.dhhs.gov
Sponsor: U.S. Department of Health and Human Services
5600 Fishers Lane, Parklawn Building
Rockville, MD 20857

DHHS Promoting Safe and Stable Families Grants 1644

The National Center on Child Abuse and Neglect is interested in research on the impact of community-based family support and family preservation programs on child abuse and neglect. Research should focus on expanding the current knowledge base, build on previous research, and provide insights into new approaches to preventing child maltreatment and preserving families through support and preservation services. The center is particularly interested in projects that address specific populations and outcomes. The populations are families who receive family support services but have had no previous contact with child protective services; families who have been referred to child protective services whose cases were unsubstantiated but were found to need services and were referred to family support programs; families who have been in the system whose child abuse or neglect cases were substantiated, who received family preservation or support services, and whose cases are now closed; and families who have open cases whose children have not been removed and who are receiving family preservation services. Outcomes of interest are case finding, which involves families who were not referred to child protective services, and the impact of family support and/or preservation services on prevention, recidivism, and removal of children. Applicants should plan and design the proposed research in collaboration with state and local CPS and Title IV-B agencies as well as community-based entities providing family support services, such as family resource centers. Contact the office for deadline dates.
Requirements: Agencies of state and local governments, public and private nonprofit agencies, and institutions engaged in child and family welfare activities and research are eligible for these grants.
Amount of Grant: 194,000 - 48,000,000 U.S.D
Contact: Joseph Bock, Program Supervisor; (202) 205-8618
Internet: http://www.acf.hhs.gov/programs/cb/programs/fpfs.htm
Sponsor: U.S. Department of Health and Human Services
330 C Street SW, Room 2428
Washington, DC 20447

DHHS Public Health Students Traineeships and Other Graduate Public Health Programs 1645

The programs support traineeships for students in graduate educational programs in schools of public health or other public or nonprofit educational entities that offer graduate programs for training in biostatistics or epidemiology; health administration, health planning, or health policy analysis and planning; environmental or occupational health; dietetics or nutrition; maternal and child health; or approved residency training in preventive medicine or dentistry. Deadlines and application kits are forwarded to all eligible applicants. Annual deadline dates may vary; contact program staff for exact dates.
Requirements: Accredited schools of public health and other public or nonprofit educational entities providing graduate or specialized training in public health may apply. Trainees must be U.S. citizens or noncitizen nationals possessing visas permitting permanent residence in the United States and must be pursuing a graduate degree in an eligible school of public health or be enrolled in an eligible program.
Amount of Grant: 8,109 - 148,850 U.S.D
Contact: Cecelia Maryland, Public Health Branch, Division of State, Community and Public Health; (301) 443-1973
Internet: http://www.dhhs.gov
Sponsor: U.S. Department of Health and Human Services
5600 Fishers Lane, Parklawn Building
Rockville, MD 20857

DHHS Quentin N. Burdick Rural Program for Interdisciplinary Training Grants 1646

The objective of this program is to promote recruitment to and retention of health professionals in rural states through interdisciplinary team training in rural health care settings. Grant funds are awarded to eligible applicants to carry out their interdisciplinary training projects using various methods including student stipends, postdoctoral fellowships, faculty training, and the purchase or rental of transportation and telecommunications equipment where appropriate to the rural area. Grants may be awarded for project periods of up to three years. If additional support is desired to continue a grant beyond the approved project period, an application for competing continuation must be submitted for review in the same manner as a new application. Contact the office for deadlines.
Requirements: Each application must be jointly submitted by at least two of the following: local health departments; nonprofit organizations; or public or nonprofit colleges, universities, or schools of medicine, nursing, or allied health.
Restrictions: Profit-making entities are not eligible to obtain funds, either directly or through a subcontract or subgrant.
Amount of Grant: 133,709 - 344,077 U.S.D
Contact: Marcia Starbecker, Program Officer, Division of State, Community, and Public Health; (301) 443-6867
Internet: http://bhpr.hrsa.gov/interdisciplinary/rural.html
Sponsor: U.S. Department of Health and Human Services
5600 Fishers Lane, Parklawn Building
Rockville, MD 20857

DHHS Refugee Assistance Voluntary Agency Program Grants 1647

The objective of the program is to assist refugees in becoming self-supporting and independent members of U.S. society, by providing grant funds to private nonprofit organizations to support case management, transitional assistance, and social services for new arrivals. Federal funds of up to $1000 per refugee are provided on a dollar-for-dollar matching basis. Required services include job development and maintenance support for up to four months. Assistance may also be used for English language training, job training, and medical support. Annual deadline dates vary; contact the office for specific dates.
Requirements: Grant awards are limited to private nonprofit organizations that have a Reception and Placement Grant with the Department of State and are to provide the prescribed services to eligible recipient refugees.
Contact: Ron Munia, Office of Refugee Resettlement; (202) 401-4559; fax (202) 401-5487; rmunia@acf.hhs.gov
Internet: http://www.acf.hhs.gov/programs/orr/funding/cfunding.htm
Sponsor: U.S. Department of Health and Human Services
370 L'Enfant Promenade SW
Washington, DC 20447

DHHS Residency Training in Family Practice, General Internal Medicine, or General Pediatrics Grants 1648

Grants are awarded to assist graduate training programs in family medicine, general internal medicine and/or general pediatrics to expand and improve the quality of residency training programs that prepare graduates to enter primary care practice. Residency training programs should emphasize national innovations aimed at primary care residency education across disciplines.
Requirements: An accredited public or private nonprofit school of medicine, a school of osteopathic medicine, and a public or private nonprofit hospital or other entity located in a state is eligible to apply. Each allopathic program must be fully or provisionally approved by the Accreditation Council for Graduate Medical Education. Each osteopathic program must be approved by the American Osteopathic Association.
Restrictions: Grants may not be used for construction, patient services, or student assistance.
Amount of Grant: 27,000 - 657,344 U.S.D
Contact: Dr. P. Preston Reynolds, (301) 443-1467

Internet: http://www.dhhs.gov
Sponsor: U.S. Department of Health and Human Services
5600 Fishers Lane, Parklawn Building
Rockville, MD 20857

DHHS Rural Telemedicine Grants 1649
The objective of this program is to demonstrate and collect information on the feasibility, costs, appropriateness, and acceptability of telemedicine for improving access to health services for rural residents and reducing the isolation of rural practitioners, and to demonstrate how telemedicine can be used as an effective tool for the development of integrated systems of health care for rural areas. Application kits may be obtained by calling toll-free (888)300-4772. Annual deadline dates may vary; contact program staff for exact dates.
Requirements: Eligible applicants are nonprofit public (nonfederal) or private health care providers or a consortium of providers, nonprofit or for profit, that are members of an existing or proposed telemedicine network.
Restrictions: Not more than 40 percent of grant funds may be expended for equipment. Not more than 20 percent of grant funds may be expended for direct costs. Grant funds may not be used for purchasing and installing telecommunications transmission equipment.
Amount of Grant: Up to 250,000 U.S.D
Contact: Monica Cowan, Administrative Assistant, Office for the Advancement of Telehealth; (301) 443-1730
Internet: http://www.telehealth.hrsa.gov
Sponsor: U.S. Department of Health and Human Services
4350 East-West Highway
Bethesda, MD 20814

DHHS Technical and Non-Financial Assistance to Health Centers 1650
The grants assist community health centers in the areas of collaborative activities on state or regional issues; promotion of support and involvement of state agencies in primary care; provision of, or arrangement for, training and technical assistance; and development of shared services and joint purchasing arrangements—for purposes including, but not limited to, primary care provider retention and recruitment, clinical development, assessment of community health needs, expertise in dealing with maternal and child health and other special populations, and management and maximization of nonfederal sources. Recipients will be expected to provide certain technical and nonfinancial assistance to community health centers. Support is recommended for a specified project period, not in excess of five years. Deadlines are specified in application instructions.
Requirements: Grants will be made to private nonprofit entities, including state and regional primary care associations.
Amount of Grant: 580,000 U.S.D
Contact: Cicely Nelson, (301) 594-4300; cnelson@hrsa.gov
Internet: http://www.hrsa.gov/grants/preview/primary.htm
Sponsor: U.S. Department of Health and Human Services
4350 East-West Highway
Bethesda, MD 20814

DHHS Thimerosal Pharmacokinetics: Assessment of the Distribution, Metabolism, and Excretion 1651
The purpose of this program is to define and characterize the appropriate pharmacokinetics of both methyl mercury and thimerosal distribution, metabolism, and elimination that are needed to accurately characterize the comparative neurotoxicity of these substances and develop recommendations for future extrapolation of these results to possible human exposure scenarios.
Requirements: Applications may be submitted by official public health agencies of the states, or their bonafide agents. This includes the District of Columbia, American Samoa, Guam, the Commonwealth of Puerto Rico, the Commonwealth of the Northern Mariana Islands, the Federated States of Micronesia, the Virgin Islands, the Federated States of Marshall Island, the Republic of Palau, federally-recognized Indian tribal governments, public and private nonprofit universities and colleges.
Date(s) Application is Due: Aug 15
Amount of Grant: 130,000 - 300,000 U.S.D
Contact: Mildred Garner, Grants Management Officer; (770) 488-2745; fax (770) 488-2777; info@cdc.gov
Internet: http://www.dhhs.gov
Sponsor: U.S. Department of Health and Human Services
200 Independence Avenue SW
Washington, DC 20201

DHHS Tuberculosis Demonstration, Research, Public, and Professional Education Cooperative Agreements 1652
The purpose of the program is to assist states, political subdivisions of states, and other public and nonprofit private entities to conduct research into the prevention and control of tuberculosis, especially research concerning strains of tuberculosis resistant to drugs and research concerning cases of tuberculosis that affect certain populations; conduct demonstration projects for the prevention and control of tuberculosis; provide public information and provide education programs for prevention and control of tuberculosis; and develop education, training, and clinical skills improvement activities in the prevention and control of tuberculosis for health professionals, including allied health personnel. Project periods are for one to five years with 12-month budget periods. Applications will be evaluated on the extent of the tuberculosis problem; the establishment of specific and measurable objectives; and the development of a sound operational plan that will ensure the implementation of each program element. Contact the office for deadline dates.
Requirements: States, political subdivisions of states, and other public and nonprofit private entities are eligible.
Amount of Grant: 150,000 - 200,000 U.S.D
Contact: William Ryan , Grants Management Officer; (770) 488-2717
Internet: http://www.dhhs.gov
Sponsor: U.S. Department of Health and Human Services
1600 Clifton Road NE
Atlanta, GA 30333

DHHS Universal Newborn Hearing Screening and Intervention 1653
This program supports the implementation of universal physiologic newborn hearing screening prior to hospital discharge with linkages to medical home, ongoing family-to-family support, diagnostic evaluation by three months of age, and enrollment in a program of early intervention by six months of age for those infants identified with hearing loss. Applicants are expected to notify the Maternal and Child Health Bureau's Division of Services for Children with Special Health Care Needs by November 10; the deadline for receipt of applications is December 8.
Requirements: This program is open to state agencies with the capacity to implement a statewide universal newborn hearing screening and intervention program for all newborn infants in the state.
Contact: Lawrence Poole, Grants Management Officer; (301) 443-0354
Internet: http://www.dhhs.gov
Sponsor: U.S. Department of Health and Human Services
5600 Fishers Lane, Parklawn Building
Rockville, MD 20857

DHHS Urban Indian Health Services Grants 1654
The objective of the program is to provide health-related services to urban Native Americans including alcohol and substance abuse prevention, treatment, rehabilitation, and education; mental health needs assessment and services; health promotion and disease prevention services; immunization services; and child abuse prevention and treatment. Contact the office for deadline dates.
Requirements: Urban Native American organizations with whom the Secretary has entered into a contract or grant under Title V of the Indian Health Care Improvement Act are eligible.
Contact: Lois Hodge, Grants Management Officer, Division of Acquisition and Grants Management; (301) 443-5204
Internet: http://www.ihs.gov
Sponsor: U.S. Department of Health and Human Services
801 Thompson Avenue, Suite 200
Rockville, MD 20852

DHL Charitable Shipment Support 1655
The corporation provides free shipment of materials supporting charitable programs. Applicants must submit a complete web questionnaire.
Requirements: Application by internet only.
Restrictions: Religious organizations are ineligible for religious sectarian activities, and political organizations are not eligible for political purposes.
Contact: Technical Support; (800) 527-7298
Internet: http://www.dhl-usa.com
Sponsor: DHL International, Ltd.
1220 South Pine Island Road, Suite 600
Plantation, FL 33324

Dickson Foundation Grants 1656
The foundation awards grants primarily to local nonprofit organizations in its areas of interest, including education (secondary schools, higher education),and health care. Types of support include building construction/renovation and general operating support. There are no application deadlines or forms. Submit a letter of interest for consideration.
Requirements: Giving is primarily in the Southeast.
Restrictions: Grants are not awarded to individuals, or for building or endowment funds.
Geographic Focus: North Carolina
Contact: Susan Patterson, (704) 372-5404; fax (704) 372-6409
Sponsor: Dickson Foundation
301 S Tryon Street, Suite 1800
Charlotte, NC 28202

Dietetic Outcomes Research Study Grants 1657
The program seeks proposals for a prospective, controlled trial study on the effectiveness of nutrition services. The objective is to investigate the cost and clinical effectiveness of nutrition services provided by registered dietitians compared to other health practitioners or health care teams with and without dietitians in an ambulatory setting. The study must focus on one of four disease states: diabetes, hyperlipidemia, congestive heart failure, or obesity with metabolic complications. Key personnel requirements include an experienced research dietitian and ADA members as a principal or co-principal investigator. Specifications are contained in the RFP posted on the Web site.
Amount of Grant: Up to 225,000 U.S.D
Contact: Elisabeth Puga, (800) 877-1600, ext. 4803; epuga@eatright.org
Internet: http://www.eatright.org/Public/7713_7808.cfm
Sponsor: American Dietetic Association Foundation
120 S Riverside Plaza, Suite 2000
Chicago, IL 60606

Disable American Veterans Charitable Grants 1658

The trust is dedicated to building better lives for U.S. disabled veterans and their families. Grants support physical and psychological rehabilitation programs, aid efforts to meet the special needs of veterans with specific disabilities such as amputation and blindness, and support programs and shelters for homeless veterans. Priority is given to long-term service projects providing direct assistance to disabled veterans and their families. The trust also awards grants to service programs that assist veterans suffering from substance abuse and post traumatic stress disorder, health promotion programs for aging disabled veterans, and nursing and health programs designed to ensure quality healthcare for veterans.

Requirements: Nonprofit organizations are eligible.

Restrictions: Grants generally do not support individuals; goodwill advertising, souvenir journals, or dinner programs; political causes, candidates, organizations, or campaigns; or endowments or capital campaigns.

Date(s) Application is Due: Jan 31; May 1; Aug 1; Nov 1

Contact: Grants Administrator; (877) 426-2838, ext. 2057; fax (859) 442-2088; cst@davmail.org

Internet: http://www.dav.org/cst/index.html

Sponsor: Disabled American Veterans Charitable Service Trust
3725 Alexandria Pike
Cold Spring, KY 41076

DOD HBCU/MI Partnership Training Award 1659

This program's goal challenges the scientific community to design innovative research that will foster new directions for, address neglected issues in, and bring new investigators to the field of breast cancer research. The program focuses its funding on innovative projects, particularly those involving multidisciplinary and/or multi-institutional collaborations and alliances that have the potential to make a significant impact on breast cancer. Proposals that address the needs of minority, low-income, rural, and other underrepresented and/or medically underserved populations are strongly encouraged. The listed application deadline is for preproposals. Preproposals must be submitted through the CDMRP eReceipt system. Guidelines are available online.

Requirements: Applicants must be HBCU/MI doctoral-level faculty members. All individuals, regardless of ethnicity, nationality, or citizenship status, may apply as long as they are employed by, or affiliated with, an eligible HBCU/MI institution. Eligible institutions are those approved as HBCU/MIs by the Department of Education.

Contact: Commander, U.S. Army Medical Research and Materiel Command; (301) 619-7079; fax (301) 619-7792; cdmrp.pa@det.amedd.army.mil

Internet: http://cdmrp.army.mil/bcrp/default.htm

Sponsor: U.S. Department of Defense
3701 North Fairfax Drive
Arlington, VA 21702-1714

DOE Environmental Remediation Grants 1660

The objectives of the program relate to environmental processes affected by energy production and use. The program develops information on the physical, chemical, and biological processes that cycle and transport energy-related material, particularly contaminates that arose during nuclear weapons production, through the earth's surface and subsurface. Emphasis is put on the development of a strong basis for understanding and implementing the appropriate and efficient use of bioremediation, particularly at the department's sites. Applications are accepted at any time and are available on the Web site.

Amount of Grant: 10,000 - 2,000,000 U.S.D

Contact: Martin Rubinstein, Office of Science; (301) 903-5212

Internet: http://www.science.doe.gov/production/grants/grants.html

Sponsor: U.S. Department of Energy
19901 Germantown Road, MS FE-3
Germantown, MD 20874-1290

DOE Health Effects and Life Sciences Research Grants 1661

The objectives of this program of basic and biological research are to create and apply new technologies and resources in mapping, sequencing, and information management for characterizing the molecular nature of the human genome; to develop and support DOE national user facilities for use in fundamental structural biology; to use model organisms to understand human genome organization, human gene function and control, and the functional relationships between human genes and proteins; to characterize and exploit the genomes and diversity of microbes with potential relevance for energy, bioremediation, or global climate; to understand and characterize the risks to human health from exposures to low levels of radiation and chemicals; to develop novel technologies for high throughput determination of protein structure; and to anticipate and address ethical, legal, and social implications arising from genome research. Applications are accepted at any time and are available on the Web site.

Requirements: Unsolicited research proposals should be preceded by an information preproposal of no more than two or three pages that briefly describes the research project and the proposed level of funding.

Amount of Grant: 10,000 - 2,000,000 U.S.D

Contact: Martin Rubinstein; (301) 903-5212

Internet: http://www.science.doe.gov/production/grants/grants.html

Sponsor: U.S. Department of Energy
19901 Germantown Road, MS FE-3
Germantown, MD 20874-1290

DOE Human Genome Program—Ethical, Legal, and Social Implications 1662

DOE announces its interest in receiving applications in support of the Ethical, Legal, and Social Implications (ELSI) subprogram of the Human Genome Program (HGP). Applications should focus on issues of genetics and the workplace, storage of genetic information and tissue samples, education, or complex or multigenic traits. The HGP is a coordinated, multidisciplinary, directed research effort aimed at obtaining a detailed understanding of the human genome at the molecular level.

Amount of Grant: 10,000 - 2,000,000 U.S.D

Contact: Martin Rubinstein, Office of Science; (301) 903-5212

Internet: http://www.sc.doe.gov/production/grants/grants.html

Sponsor: U.S. Department of Energy
19901 Germantown Road, MS FE-3
Germantown, MD 20874-1290

DOE Low Dose Radiation Research Program 1663

Research is specifically sought for pilot projects that involve innovative collaborations between experimentalists and modelers to model the mechanisms of key radiation-induced biological responses and to describe or identify strategies for developing biologically-based risk models that incorporate information on mechanisms of radiation-induced biological responses. Potential applicants should submit a one-page preapplication referencing Program Notice 01-17 by February 1; formal applications are due May 1. Annual deadline dates may change; contact program staff for exact dates.

Date(s) Application is Due: Feb 1; May 1

Amount of Grant: 10,000 - 2,000,000 U.S.D

Contact: Martin Rubinstein, Grants and Contracts Division, Office of Science; (301) 903-5212

Internet: http://www.science.doe.gov/grants

Sponsor: U.S. Department of Energy
19901 Germantown Road, MS FE-3
Germantown, MD 20874-1290

DOE Medical Applications and Measurement Science Grants 1664

The objectives of this program are to develop technologies for the beneficial applications of radiation and in-vivo radiotracer detection in the study, diagnosis, and treatment of human diseases and disorders; to develop new instrumentation for biological and medical research; and to develop new concepts and techniques for detecting and measuring the hazardous agents of biochemical, physical, and environmental consequences related to energy production. Applications are accepted at any time; application forms are available on the Web site.

Amount of Grant: 10,000 - 2,000,000 U.S.D

Contact: Martin R. Rubinstein; (301) 903-5212

Internet: http://www.er.doe.gov/production/grants/guide.html

Sponsor: U.S. Department of Energy
19901 Germantown Road, MS FE-3
Germantown, MD 20874-1290

DOI Biological Resource Division Brucellosis Program 1665

Applications are invited for a research project on the improvements in ballistic delivery systems for brucellosis vaccination of free-ranging elk and bison of the Greater Yellowstone area. The purpose of this project is to develop methods of ballistic delivery that improve the distance, reliability, ease, and/or rapidity of brucella vaccine parenteral delivery.

Requirements: Applications may be submitted by educational institutions, private firms, private foundations, individuals, and agencies of state and local governments.

Amount of Grant: 1,000 - 582,000 U.S.D

Contact: Branch of Acquisition and Federal Assistance Director; (703) 648-7478 or (703) 648-7361

Internet: http://www.usgs.gov/contracts

Sponsor: U.S. Department of the Interior
P.O. Box 25046, MS 204B
Denver, CO 80225

DOJ Gang-Free Schools and Communities Intervention Grants 1666

The objectives of the program include support to prevent and reduce the participation of juveniles in the activities of gangs that commit crimes. Such programs and activities may include individual, peer, family, and group counseling, including provision of life skills training and preparation for living independently, which shall include cooperation with social services, welfare, and health care programs; education and social services designed to address the social and developmental needs of juveniles; crisis intervention and counseling to juveniles, who are particularly at risk of gang involvement, and their families; the organization of the neighborhood and community groups to work closely with parents, schools, law enforcement, and other public and private agencies in the community; and training and assistance to adults who have significant relationships with juveniles who are or may become members of gangs, to assist such adults in providing constructive alternatives to participating in the activities of gangs. Deadlines are published in program announcements.

Requirements: Public or private nonprofit agencies, organizations, or individuals may apply.

Contact: Heidi Hsia, Office of Juvenile Justice and Delinquency Prevention; (202) 307-5924; hsiah@ojp.usdoj.gov

Internet: http://www.ojp.usdoj.gov/FinGuide/part4chap2.htm#ojjdp

Sponsor: U.S. Department of Justice
633 Indiana Avenue NW
Washington, DC 20531

DOJ Mental Health and Juvenile Justice 1667

The purpose of this initiative is to substantially improve knowledge about the nature and prevalence of mental health and co- occurring substance abuse disorders among youth

in the juvenile justice system and to develop and test a service delivery model based on theoretical and empirical information regarding best practices for meeting the needs of this population. The model will incorporate both current theory and best practices in the areas of screening, assessment, service provision, and quality assurance. The model should include youth at all stages of the juvenile justice system, including, but not limited to, arrest, intake, adjudication, detention, secure corrections, probation, and community-based treatment.
Requirements: Public or private agencies or organizations with a demonstrated capacity to carry out the requirements of this initiative may apply. Private, for-profit organizations must agree to waive any profit or fee. The organization must have demonstrated experience in conducting extensive literature reviews, conducting in-depth sample surveys of juvenile or criminal justice offenders and facilities, carrying out research and program development, and writing reports.
Amount of Grant: 1,000,000 U.S.D
Contact: Demonstration Division, Office of Juvenile Justice and Delinquency Prevention; (202) 307-5940
Internet: http://ojjdp.ncjrs.org/funding/funding.html
Sponsor: U.S. Department of Justice
2277 Research Boulevard, MS 2K
Rockville, MD 20850

Dolan Foundation Grants **1668**
The foundation awards community service grants with a focus on Long Island, NY. Areas of support include human services, disability services, mental health, schools, rehabilitation, hospitals, and health facilities. Grants provide program support. There are no application deadlines.
Requirements: Schools and other nonprofits in New York are eligible. Applicants outside New York should call or write prior to submitting proposals.
Geographic Focus: New York
Date(s) Application is Due: Mar 1; Jun 1; Sep 1; Nov 1
Amount of Grant: 25,000 - 375,000 U.S.D
Samples: John Carroll U (Cleveland, OH)—to construct a science and technology center, $20,000.
Contact: Robert Vizza, President; (516) 629-2103; fax (516) 629-2183
Sponsor: Dolan Foundations
1 Media Crossways
Woodbury, NY 11797

Dole Food Company Charitable Contributions **1669**
Dole's charitable program is dedicated to bringing about positive change in the area of nutrition education for children. The goal of the program is to help the next generation of adults prevent many diseases by teaching the value of good nutrition to children through improved and interactive teaching curriculum for schools nationwide. Please contact Corporate Contributions for application process.
Requirements: Dole's program makes grants based on the following guidelines: the program is restricted to funding 501(c)3 charitable organizations only; documentation from the I.R.S. confirming 501(c)3 status must be provided; and the organization should provide nutrition education programs that benefit schools on a nationwide basis as opposed to assisting an individual school.
Restrictions: Grants are restricted for the following: individuals; religious, fraternal, sports or political groups; legislative or lobbying efforts or groups who have a primary focus of changing laws; and sporting events and sponsorships.
Contact: Contribution Program Manager; (818) 879-6600; fax (818) 879-6615
Internet: http://dolecsr.com/Principles/CharitableGiving/tabid/410/Default.aspx
Sponsor: Dole Food Company
P.O. Box 5132
Westlake Village, CA 91361

DOL Occupational Safety and Health—Susan Harwood Training Grants **1670**
Grants provide funds for programs to train workers and employers to recognize, avoid, and prevent safety and health hazards in their workplaces. The program emphasizes three areas: educating workers and employers in small businesses; training workers and employers about new OSHA standards; and training workers and employers about high risk activities or hazards identified by OSHA through its Strategic Management Plan, or as part of an OSHA special emphasis program.
Requirements: Nonprofit organizations, including community-based and faith-based organizations, that are not an agency of a State or local government are eligible to apply. State or local government supported institutions of higher education are eligible to apply in accordance with 29 CFR part 95. Eligible organizations can apply independently for funding or in partnership with other eligible organizations, but in such a case, a lead organization must be identified. Sub-contracts must be awarded in accordance with 29 CFR 95.40-48, including OMB circulars requiring free and open competition for procurement transactions.
Restrictions: Grants may not be used for: any activity that is inconsistent with the goals and objectives of the Occupational Safety and Health Act of 1970; training individuals not covered by the Occupational Safety and Health Act; training workers or employers from workplaces not covered by the Occupational Safety and Health Act; training on topics that do not cover the recognition, avoidance, and prevention of unsafe or unhealthy working conditions; assisting workers in arbitration cases or other actions against employers, or assisting employers and workers in the prosecution of claims against Federal, State or local governments; duplicating services offered by OSHA, a State under an OSHA approved State Plan, or consultation programs provided by State designated agencies under section 21(d) of the Occupational Safety and Health Act; generating membership in the grantee's organization; administrative costs exceeding 25% of the total grant budget.
Contact: Cynthia Bencheck, (847) 297-4810; bencheck.cindy@dol.gov
Internet: http://www.osha.gov/dcsp/ote/sharwood.html
Sponsor: U.S. Department of Labor
2020 S Arlington Heights Road
Arlington Heights, IL 60005-4102

Dominion Foundation Human Needs Grants **1671**
Dominion Foundation grants are made in four focus areas, and they support a variety of programs: food banks, homeless shelters, land and habitat preservation, STEM (science, technology, engineering, math) education, and neighborhood revitalization, to name a few. Special consideration is given to programs with an energy conservation or energy efficiency component. In the area of Human Needs, the Foundation focus is on: providing warmth and cooling; alleviating hunger; ensuring energy-efficient shelter; providing access to medicine and basic health care; supporting communities through the united way; and disaster assistance. Since the Dominion Foundation supports a wide range of charitable programs, most grants are in the $1,000 to $15,000 range. Higher amounts may be awarded when a program is an exceptional fit with corporate business or giving priorities, or when there is significant employee involvement in the effort. Requests are considered quarterly by the Foundation's Community Investment Boards – statewide and regional committees comprised of Dominion employees representing key geographic, business and functional areas.
Requirements: Foundation grants are limited to organizations defined as tax-exempt under Section 501(c)3 of the IRS code. Additional grants occasionally may be made directly from the corporation to sponsor special events that benefit a non-profit organization.
Restrictions: Awards are throughout a 10-state area to include: Connecticut, Maryland, Massachusetts, North Carolina, Ohio, Pennsylvania, Rhode Island, Texas, Virginia and West Virginia. Information pertaining to the specific counties served in each state is on the website.
Geographic Focus: Connecticut, Illinois, Indiana, Maryland, Massachusetts, North Carolina, Ohio, Pennsylvania, Rhode Island, Texas, Virginia, West Virginia
Amount of Grant: 1,000 - 250,000 U.S.D
Contact: James C. Mesloh, Executive Director; (800) 730-7217 or (412) 237-2973; fax (412) 690-7608; Educational_Grants@dom.com
Internet: http://www.dom.com/about/education/grants/index.jsp
Sponsor: Dominion Foundation
501 Martindale Street, Suite 400
Pittsburgh, PA 15222-3199

Donaghue Investigator Program Grants **1672**
The program supports particularly promising medical researchers holding faculty appointments at Connecticut institutions. The funding emphasis is on the investigator and his or her program of research rather than on a specific study. Approximately four to six awards will be made each year.
Requirements: A prototypical recipient will be committed to pursuing his or her career in Connecticut for the foreseeable future; show potential for an outstanding independent research career, and for leadership in his or her field(s) of research; be accepting of the importance of earning support for research by actively explaining to the public his or her field of work and specific results and their value; be enthusiastic about and committed to disciplinary and interinstitutional collaboration in integrative research; be strongly supported by his or her institution; and possess evidence of external recognition of his or her work.
Geographic Focus: Connecticut
Amount of Grant: Up to 100,000 U.S.D
Contact: Lynne Garner, (860) 521-9011, ext. 11; fax (860) 521-9018; garner@donaghue.org
Internet: http://www.donaghue.org/granttype.htm
Sponsor: Patrick and Catherine Weldon Donaghue Medical Research Foundation
18 N Main Street
West Hartford, CT 06107-1919

Donaghue Practical Benefit Initiatives Grants **1673**
The foundation initiates research projects showing particular promise for producing practical benefit to human health and works in an interactive process with prospective investigators to develop projects for funding. Organizations with project ideas aimed at productive ways to deliver practical benefit to the public through health research should submit a short summary letter. There is no specific timeline for application or designated grant award amount.
Contact: Director of Operations; (860) 521-9011; fax (860) 521-9018; office@donaghue.org
Internet: http://www.donaghue.org
Sponsor: Patrick and Catherine Weldon Donaghue Medical Research Foundation
18 N Main Street
West Hartford, CT 06107-1919

Donaghue Research in Clinical and Community Health Issues Grants **1674**
The program funds biomedical, behavioral, and other health-related research projects that address the major medical conditions and social problems influencing the health of individuals, groups, and communities. The foundation is especially interested in clinical, epidemiological, and health-service research studies that focus on the development and evaluation of more effective methods for preventing, diagnosing, and treating illnesses and conditions that have a major impact on the people of Connecticut. Each year several one- to three-year grants are awarded.Deadline listed is for the letter of intent; full proposals are by request.
Requirements: 501(c)3 tax-exempt organizations are eligible.
Restrictions: Grants do not support religious organizations or individuals.
Date(s) Application is Due: Sep 2
Amount of Grant: 50,000 - 240,000 U.S.D

Contact: Jacque Daniel, Program Administrator; (860) 521-9011; fax (860) 521-9018; office@donaghue.org
Internet: http://www.donaghue.org/granttype.htm
Sponsor: Patrick and Catherine Weldon Donaghue Medical Research Foundation
18 N Main Street
West Hartford, CT 06107-1919

Donald and Sylvia Robinson Family Foundation Grants **1675**
The foundation awards grants to U.S. nonprofit organizations in its areas of interest, including animal and wildlife protection, arms control, arts (general, performing arts, and visual arts), environmental conservation and protection, eye diseases and eye research, family planning and human reproductive health, food distribution, international affairs, Israel, Jewish social services, and social service delivery programs. Types of support include annual campaigns, building construction/renovation, capital campaigns, and general operating support. There are no application deadlines or forms.
Requirements: National nonprofit organizations are eligible to apply.
Restrictions: Individuals are ineligible.
Geographic Focus: Pennsylvania
Amount of Grant: 500 - 5,000 U.S.D
Samples: Phipps Conservatory and Botanical Gardens, Pittsburgh, PA, $5,000; Jewish Residential Services, Pittsburgh, PA, $5,000; Jewish Family and Childrens Service of Pittsburgh, Pittsburgh, PA, $500.
Contact: Donald Robinson, Treasurer; (412) 661-1200; fax (412) 661-4645
Sponsor: Donald and Sylvia Robinson Family Foundation
6507 Wilkins Avenue
Pittsburgh, PA 15217

Donald W. Reynolds Foundation Aging & Quality of Life Grants **1676**
The Foundation launched its Aging and Quality of Life Program in 1996. Its goal remains improving the quality of life for America's elderly by preparing physicians to provide better care for frail older people. Most physicians today lack adequate training to meet the needs of the frail elderly patient. Such patients typically suffer from interacting physical, social and psychological conditions – both acute and chronic – that limit their independence and threaten their capacity to function in daily life. As part of its geriatrics training initiative, the Foundation has provided support to the Association of Directors of Geriatric Academic Programs (ADGAP) to carry out important facilitating activities including the hosting of annual meetings of the Foundation's geriatrics training grantees and the facilitation of information exchange among those in America's academic health centers who are engaged in strengthening geriatrics training of physicians.
Contact: Rani Snyder, Senior Program Officer; (702) 804-6000; fax (702) 804-6099; rani.snyder@dwrf.org or GeneralQuestions@dwrf.org
Internet: http://www.dwreynolds.org/Programs/National/Aging/Aging.htm
Sponsor: Donald W. Reynolds Foundation
1701 Village Center Circle
Las Vegas, NV 89134-6303

Donald W. Reynolds Foundation Cardiovascular Clinical Research Grants **1677**
In 1998, the trustees adopted a grant-making plan to improve Americans' health with an investment in clinical research to fight heart disease, the nation's number one killer. The plan called for supporting collaborative, multidisciplinary research and its application to improve the prevention, diagnosis and treatment of atherosclerotic heart disease through a network of interacting institutions and scientists. The centerpiece of the program is a network of Donald W. Reynolds Cardiovascular Clinical Research Centers, with the following goals: contribute new knowledge that will speed the progress toward a cure for atherosclerotic heart disease; effectively translate new knowledge into applications that will improve health and prevent cardiovascular diseases; demonstrate new intellectual and organizational strategies for facilitating translational research and clinical trials, health services research, epidemiology and biostatistics to combat atherosclerotic heart disease; and create innovative and productive approaches for collaborating among scientists at multiple institutions and across disciplines in the conduct of translational research and clinical trials, health services research, epidemiology and biostatistics in the fight against atherosclerotic heart disease. In October 2005 the trustees initiated a new phase of the Cardiovascular Clinical Research Program in their decision to limit future support to those Center projects offering the greatest promise for clinical impact within a reasonable period of time in the following three areas: Atherosclerosis and associated inflammation, heart failure, and sudden death.
Contact: Courtney Latta Knoblock, Program Director; (702) 804-6000; fax (702) 804-6099; capitalquestions@dwrf.org or GeneralQuestions@dwrf.org
Internet: http://www.dwreynolds.org/Programs/National/Cardio/Cardiovascular.htm
Sponsor: Donald W. Reynolds Foundation
1701 Village Center Circle
Las Vegas, NV 89134-6303

Donna K. Yundt Memorial Scholarship **1678**
The Donna K. Yundt Memorial Scholarship was established in memory of a teacher in the Lake Central Schools. This one time scholarship is awarded to a student who attended who attended Peifer Elementary. All college majors are eligible.
Requirements: Applicants must be high school graduating seniors who attended Peifer Elementary. The application is available at the Lake Central High School Guidance office, along with the Dollars for Scholars application. Applicants should check with the guidance office for current application deadline.
Geographic Focus: Indiana
Amount of Grant: 2,500 U.S.D
Contact: Yundt Scholarship Contact; (219) 736-1880; fax (219) 736-1940
Internet: http://www.legacyfoundationlakeco.org/scholarships/index.htm
Sponsor: Legacy Foundation
1000 East 80th Place, 302 South
Merrillville, IN 46410

Dora Roberts Foundation Grants **1679**
The foundation awards general operating grants to eligible Texas organizations in its areas of interest, Arts, Social Services, Health and Education. Most grants are awarded in Big Spring, TX. There are no application forms. Trustees meet annually in the Fall to review all requests received prior to the deadline.
Requirements: The Foundation can distribute grants only to qualified charitable organizations in the State of Texas. Persons who represent an organization which they believe might qualify for our support are welcome to submit an application in letter form containing: a brief narrative history of the organization's purpose and work; a specific description of the program or project for which support is asked; a statement of the amount of funds requested; proof of tax exempt status; a list of Trustees or Directors and principal staff; and budgetary information pertaining to the requested grant.
Restrictions: The Foundation does not give or lend money to individuals.
Geographic Focus: Rhode Island
Date(s) Application is Due: Sep 30
Contact: Konnie Darrow, c/o JP Morgan Chase Bank, N.A., (817) 884-4772
Sponsor: Dora Roberts Foundation
P.O. Box 2050
Fort Worth, TX 76113

Doreen Kronick Scholarship **1680**
This scholarship is awarded annually to a graduate student in a program that will lead to the recipient's being able to assist persons with learning disabilities. Application forms are available from the association or the faculty of graduate studies at Canadian universities.
Requirements: Any student, 18 or older, enrolled in any graduate program of any university in Canada is eligible.
Restrictions: Scholarship recipients will not be eligible to reapply for three years.
Geographic Focus: All States, Canada
Date(s) Application is Due: May 15
Amount of Grant: 500 CAD
Contact: Kronick Scholarships Coordinator; (613) 238-5721; fax (613) 235-5391; information@ldac-taac.ca
Internet: http://www.ldac-taac.ca/Scholarships/scholarships-e.asp
Sponsor: Learning Disabilities Association of Canada
323 Chapel Street, Suite 200
Ottawa, ON K1N 7Z2 Canada

Doree Taylor Charitable Foundation **1681**
The mission of the Doree Taylor Charitable Foundation is to support charitable organizations that: provide relief to people the form of basic needs (including the provision of food, housing, shelter); promote the humane care of animals; provide health-care services for the underserved; and conduct Public radio or television. Occasional support will also be provided to colleges and universities as well as to environmental charitable organizations in Maine. The foundation will make grants throughout Maine, but has a priority for charitable organizations or projects located in the areas of Boothbay Harbor, Southport, and Brunswick. From time to time, the foundation may support organizations with a national scope. Grant requests for general operating support or program support are strongly encouraged and preferred; however, small capital requests may also be considered. The majority of grants from the Taylor Foundation are one year in duration; on occasion, multi-year support is awarded. Applicants must apply online at the grant website. Applicants are strongly encouraged to do the following before applying: review the downloadable state application procedures for additional helpful information and clarifications; review the downloadable online-application guidelines at the grant website; review the foundation's funding history (link is available from the grant website); review the online application questions in advance; and review the list of required attachments. These will generally include: a list of board members, financial statements (audited, reviewed, or compiled by independent auditor); an organization summary; a list of other funding sources; an IRS Determination letter; and other required documents. All attachments must be uploaded in the online application as PDF, Word, or Excel files. The application deadline for the Doree Taylor Charitable Foundation is 11:59 p.m. on April 1. Applicants will be notified of grant decisions before August 31.
Requirements: 501(c)3 charitable organizations and municipalities which meet the mission criteria are eligible to apply.
Restrictions: The foundation will not contribute to endowments or consider grant requests from individuals, organizations attempting to influence policy through direct lobbying, or political campaigns.
Geographic Focus: All States, Maine
Date(s) Application is Due: Apr 1
Contact: Miki C. Akimoto, Vice President; (866) 778-6859; miki.akimoto@baml.com
Internet: https://www.bankofamerica.com/philanthropic/fn_search.action
Sponsor: Doree Taylor Charitable Foundation
225 Franklin Street, 4th Floor, MA1-225-04-02
Boston, MA 02110

Doris and Victor Day Foundation Grants **1682**
The foundation supports local community organizations providing food, shelter, clothing, medical care, and education in Illinois/Iowa Quad Cities region. Preference will be given to preventive programs and projects that foster pride in the local community. Types of

support include general operating support, building/renovation, equipment, emergency funds, seed grants, and scholarship funds.
Requirements: Illinois/Iowa Quad Cities region nonprofit organizations may apply.
Restrictions: The Foundation is committed to programs that are non-sectarian, and therefore, will not contribute toward programs and capital projects for religious purposes, except for modest contributions to the churches in which the Days held membership. However, clearly non-sectarian, community serving programs of religious organizations will be considered for funding.
Geographic Focus: Illinois, Iowa
Date(s) Application is Due: May 1
Samples: Transitions Mental Health Rehabilitation, $4,000—Client Home Renovation; Rock Island Economic Growth Corporation, $50,000—Neighborhood Stabilization Program; Martin Luther King Center, $175,000 — expansion and renovation.
Contact: Program Contact; (309) 788-2300; info@dayfoundation.org
Internet: http://www.dayfoundation.org/guide.htm
Sponsor: Doris and Victor Day Foundation Grants
1800 3rd Avenue, Suite 302
Rock Island, IL 61201-8019

Doris Duke Charitable Foundation Clinical Interfaces Award Program 1683
The program seeks to catalyze activity at the interface of clinical and other research disciplines by: supporting the formation of new collaborations and strengthening existing collaborations of outstanding scientists across disciplines; demonstrating successful models for clinical research at the interface of multiple disciplines; and supporting interdisciplinary and inter-institutional endeavors that go beyond the program project mindset. Full grants, awarded over five years, are made to established teams with key investigators from at least three disciplines. Planning grants are awarded to new teams for the development of full proposals over 18 months. The listed application deadline is for pre-proposals; full proposals are by invitation. Though new grants are not being offered continually, potential applicants can sign up to be notified of future competitions.
Requirements: Teams of at least three key investigators whose primary expertise lie in different disciplines are eligible to apply. Key investigators must have advanced degrees (MD, PhD, MD/PhD, or the equivalent), and one of the key investigators must be a clinical researcher. The team leader must work in a U.S. nonprofit institution, such as an academic medical center. The team may include investigators at other institutions in the United States and overseas.
Restrictions: Planning grants will not be awarded in the current competition.
Date(s) Application is Due: Nov 2
Amount of Grant: Up to 100,000,000 U.S.D
Contact: Elaine K. Gallin; (212) 974-7104; fax (212) 974-7590; egallin@ddcf.org
Internet: http://www.ddcf.org/page.asp?pageId=299
Sponsor: Doris Duke Charitable Foundation
650 Fifth Avenue, 19th Floor
New York, NY 10019

Doris Duke Charitable Foundation Clinical Research Fellowships for Medical Students 1684
The fellowship program is designed to encourage medical students to pursue careers in clinical research by giving exceptional students the opportunity to take a year to experience clinical research first hand. The CRF program is available at the following medical schools: Columbia University College of Physicians and Surgeons; Harvard Medical School; Mount Sinai School of Medicine; University of California at San Francisco School of Medicine; University of Iowa Roy J. and Lucille A. Carver College of Medicine; University of North Carolina at Chapel Hill School of Medicine; University of Pennsylvania School of Medicine; University of Texas Southwestern Medical Center at Dallas; Washington University Medical School; and Yale University School of Medicine. An additional 11-12 fellowships will be available to medical students interested in conducting clinical research in Africa. Each participating medical school provides medical students with a one-year fellowship experience in clinical research that includes both didactic and research components; matches students to outstanding clinical research mentors; and offers fellowships to at least five students per year.
Requirements: Medical students matriculated at any U.S. medical school who have completed two or more years of medical school prior to the start of the fellowship and who have completed some clinical experience are eligible to apply to any of the participating schools.
Date(s) Application is Due: Jan 18
Amount of Grant: 27,000 U.S.D
Contact: Elaine K. Gallin; (212) 974-7104; fax (212) 974-7590; egallin@ddcf.org
Internet: http://www.ddcf.org/page.asp?pageId=292
Sponsor: Doris Duke Charitable Foundation
650 Fifth Avenue, 19th Floor
New York, NY 10019

Doris Duke Charitable Foundation Clinical Scientist Development Award 1685
Grants are awarded to junior physician-scientists to facilitate their transition to independent clinical research careers. The program is designed to help prepare and support new investigators with an MD or MD/PhD as they begin their careers as independent clinical researchers. The program is aimed at conducting clinical research in any disease area. The listed application deadline is for preproposals, which must be submitted electronically.
Requirements: Applicants must: be a physician-scientist conducting clinical research in any disease area; have received an M.D. or a foreign equivalent from an accredited institution; be working in a U.S. degree-granting institution, but do not have to be a U.S. citizen; have a full-time faculty level position not higher than the Assistant Professor level; and have been appointed to their first full-time faculty level position between January 31, 2002 and January 31 of the grant year. (All full-time post-fellowship Instructor level positions will be considered full-time faculty level appointments).
Restrictions: Funds cannot be used on experiments that utilize animals or primary tissues derived from animals. An award will not be made if, prior to the commencement of the award, the applicant becomes the principal investigator on a federal government, peer-reviewed, research or career development award or any non-government award averaging more than $126,000 per year in direct costs and of a duration of three or more years.
Date(s) Application is Due: Jan 23
Amount of Grant: 135,000 U.S.D
Contact: Elaine K. Gallin, Program Director; (212) 974-7104; fax (212) 974-7590; egallin@ddcf.org
Internet: http://www.ddcf.org/page.asp?pageId=291
Sponsor: Doris Duke Charitable Foundation
650 Fifth Avenue, 19th Floor
New York, NY 10019

Doris Duke Charitable Foundation Clinical Scientist Development Award Bridge Grants 1686
The purpose of the bridge grant is to help Doris Duke Clinical Scientist Development Award (CSDA) recipients successfully transition to independent National Institutes of Health (NIH) R01 funding. The Medical Research Program created these grants to bridge the time between when a CSDA investigator learns of a very strong but unfunded score on an NIH application to when s/he is able to successfully obtain an R01 grant. The bridge grants provide protected time and support for CSDA recipients to continue their research projects, collect supportive data, keep patients enrolled if applicable, and strengthen their proposals to resubmit to the NIH. The foundation hopes that the bridge grants will help keep promising physician-scientists committed to clinical research careers by providing stable support during their critical transition to independent R01 funding. The foundation will award up to four bridge grants of $135,000 each to be spent only on direct costs over 12 to 18 months. Eligible CSDA recipients are invited to submit applications throughout the year up to the deadline of September 1.
Requirements: This competition is open only to former CSDA grantees or current CSDA grantees in the last year of their award (or in a no-cost extension period).
Date(s) Application is Due: Sep 1
Amount of Grant: Up to 135,000 U.S.D
Contact: Elaine Gallin, Program Director for Medical Research; (212) 974-7104; fax (212) 974-7590; egallin@ddcf.org
Internet: http://www.ddcf.org/page.asp?pageId=291
Sponsor: Doris Duke Charitable Foundation
650 Fifth Avenue, 19th Floor
New York, NY 10019

Doris Duke Charitable Foundation Distinguished Clinical Scientist Award 1687
The purpose of the program is to recognize outstanding physician-scientists who are engaged in applying the latest basic science advances to the prevention, diagnosis, treatment, and cure of disease, and to enable the physician-scientist to support and mentor the next generation of physician-scientists conducting translational clinical research. Awards will be granted to physician-scientists conducting translational clinical research in any disease area; this award cycle is not limited to specific disease areas.
Requirements: Grantees must hold an M.D. degree from an accredited institution in the United States (holders of M.D./Ph.D. degrees are also eligible, as are holders of M.D.-equivalent degrees from non-U.S. institutions); hold a full-time university faculty appointment at the level of Associate Professor or above as of the date of nomination; have been appointed to their first full-time, faculty-level position for no more than 15 years; and have an established translational clinical research program.
Restrictions: Experiments that utilize animals or primary tissues derived from animals are not eligible for support through this award program. An award will not be made if, prior to the commencement of the foundation's award, the applicant becomes the principal investigator of a federal government, peer-reviewed, research or career development award or of any nongovernment award averaging $100,000 or more per year and of three years or more duration.
Date(s) Application is Due: Feb 14
Amount of Grant: Up to 15,000,000 U.S.D
Contact: Elaine Gallin, Program Director for Medical Research; (212) 974-7104; fax (212) 974-7590; egallin@ddcf.org
Internet: http://www.ddcf.org/page.asp?pageId=297
Sponsor: Doris Duke Charitable Foundation
650 Fifth Avenue, 19th Floor
New York, NY 10019

Doris Duke Charitable Foundation Operations Research on AIDS Care and Treatment in Africa (ORACTA) Grants 1688
The Program seeks to improve the care and treatment of AIDS patients in resource-limited settings, inform antiretroviral therapy (ART) policy and practice, and improve outcomes of the roll-out and scale-up of ART in Africa. ORACTA grants provide two-year grants of up to $100,000 per year to teams of investigators conducting health operations research on AIDS care and treatment in Africa. Though new grants are offered only periodically, applicants should sign up for the Medical Research Program's mailing list.
Restrictions: Indirect costs are not supported by this award.
Amount of Grant: 100,000 U.S.D
Contact: Elaine Gallin, Program Director for Medical Research; (212) 974-7104; fax (212) 974-7590; egallin@ddcf.org

Internet: http://www.ddcf.org/page.asp?pageId=486
Sponsor: Doris Duke Charitable Foundation
650 Fifth Avenue, 19th Floor
New York, NY 10019

Dorothea Haus Ross Foundation Grants **1689**

The foundation awards grants to eligible nonprofit organizations that work to relieve suffering among children who are sick, handicapped, injured, disfigured, orphaned, or otherwise vulnerable. Types of support include direct services, medical research, equipment and supplies, and small renovation projects. There are no application deadlines. Grants have supported projects for children in Native American schools and schools that serve populations of children who are disadvantaged or underserved.
Requirements: U.S. Charities may apply if: they have 501(c)3 status; they are listed in the current edition of the Cumulative List of Charities published by the U.S. Department of the Treasury; they are a Catholic organization listed in the current edition of the Catholic Director; or they are listed in the Free Methodist Yearbook, or other Protestant Denomination Directory that has a group ruling for tax exemption from the IRS. Although grants are made internationally, Foundation by-laws prohibit sending money directly to foreign charities.
Restrictions: The Foundation does not fund day-to-day operations, individuals, conferences, day care, or public education. Although the Foundation makes international grants, there are restrictions in some countries for the following reasons: war, widespread violence, or breakdown of law and order; or countries where grants are restricted by the U.S. Government due to a boycott or other reason.
Contact: Wayne Cook, Executive Director; (716) 473-6006; fax (716) 473-6007; rossfoundation@frontiernet.net
Internet: http://www.dhrossfoundation.org/index.html
Sponsor: Dorothea Haus Ross Foundation
1036 Monroe Avenue
Rochester, NY 14620

Dorothy Hooper Beattie Foundation Grants **1690**

The Foundation primarily supports higher education, health care, and religious programs in and around the Greenville, South Carolina, region. Its program interests include: camps; cancer treatment; Catholicism; children and youth programs; higher education; health services; literacy; and the environment. Grants range from $620 to $6,500, and there are no specific deadlines or application forms. Initial approach should be by letter, with telephone contacts discouraged. Mail applications to: Bank of America, 7 North Laurens Street, Greenville, SC 29601.
Geographic Focus: South Carolina
Amount of Grant: 620 - 6,500 U.S.D
Samples: The Bernardine Fund St. Francis, Greenville, SC, $620; Upstate Forever, Greenville, SC, $1,345; Greenville Free Medical Clinic, Greenville, SC, $4,000; United Ministries, Greenville, SC, $6,500.
Contact: Mary W. Green, Bank of America Contact; (864) 271-5789
Sponsor: Dorothy Hooper Beattie Foundation
101 South Tryon Street
Charlotte, NC 28255

Dorothy Rider Pool Health Care Grants **1691**

The Foundation's intent is to serve as a means to improve the quality of life in the Lehigh Valley community, to build on community strengths and add to its vitality, and to increase the capacity of the community to serve the needs of all its citizens. Within this objective the Foundation's funding program is focused on education, health and welfare, culture and art and community development. Interested applicants should submit a letter of intent of five pages or less.
Requirements: Allentown, PA, nonprofit organizations are eligible.
Restrictions: The Foundation is restricted from providing funds to individuals, legislative or lobbying efforts, political or fraternal organizations or organizations outside the United States and its territories. The Foundation as a policy does not provide operating or capital funds to Sectarian institutions, organizations or programs in which funds will be used primarily for the propagation of religion, hospitals or United Way member agencies. Further, the Foundation does not underwrite charitable or testimonial dinners, fund-raising events or related advertising or the subsidization of books, mailings or articles in professional journals.
Geographic Focus: Pennsylvania
Date(s) Application is Due: Apr 1; Aug 15
Amount of Grant: 100,000 U.S.D
Samples: Lehigh Valley Hospital and Health (Allentown, PA)—for its department of obstetrics and gynecology, $225,000 over three years; Lehigh Valley Hospital and Health Network (Allentown, PA)—for the department of emergency medicine's sexual assault response team, $150,000 over three years; Lehigh Valley Hospital and Health Network (Allentown, PA)—for the Physician Leaders in the Lehigh Valley program, $120,000 over three years.
Contact: Ronald Dendas, Program Manager; (610) 770-9346; fax (610) 770-9361; drpool@ptd.net
Internet: http://www.pooltrust.com
Sponsor: Dorothy Rider Pool Health Care Trust
1050 S Cedar Crest Boulevard, Suite 202
Allentown, PA 18103

Dorrance Family Foundation Grants **1692**

The Dorrance Family Foundation was founded by Bennett Dorrance, co-owner of the Campbell Soup Company. The Foundation gives primarily in the state of Arizona, offering support in the areas of education and natural resource conservation.
Requirements: No formal application form is required, non-profits operating in Arizona are eligible to apply for these grants. Applicants should submit a proposal consisting of a detailed description of project and amount of funding requested.
Restrictions: No funding available to individuals.
Geographic Focus: Arizona
Samples: Arizona Community Foundation, Phoenix, AZ, $1,980,412; scholarship program,(2008). Nature Conservancy, Phoenix, AZ, $1,778,000; precious lands and water preservation.
Contact: Carolyn O'Malley, Executive Director; (480) 367-7000
Sponsor: Dorrance Family Foundation
7600 East Doubletree Ranch Road, Suite 300
Scottsdale, AZ 85258

Do Something BR!CK Awards **1693**

The awards honor dynamic young people for service in the areas of community building, health, and the environment. BR!CK winners are leaders who identify and realize solutions to problems facing local communities across America. Nine winners will receive a minimum of $10,000 in community grants and scholarships (if applicable). Of those nine winners, 1 will be selected by a national, online vote as a Golden BR!CK Award winner. That Golden BR!CK Award winner receives a total of $100,000 in community grants. Winners are announced and recognized at an annual gala in New York City. Application and guidelines are available online.
Requirements: Applicants must be 25 years old or younger. Only winners who are age 18 and under are eligible for a scholarship of $5,000 and a $5,000 community grant (total= $10,000 BR!CK Award). Winners age 19-25 receive their entire award in the form of a community grant.
Date(s) Application is Due: Dec 15
Amount of Grant: 10,000 - 100,000 U.S.D
Samples: Ashley Rhodes-Courter (21) - GOLDEN BR!CK - North American Council on Adoptable Children; William Hwang (22) - United InnoWorks Academy; Kelly Voigt (14) - Prevent the Bite; Kimmie Weeks (25) - GOLDEN BR!CK - Youth Action International; Cheryl Perera (21) - OneChild Network & Support Systems Inc; Ruth DeGolia (24) - Mercado Global; Jennifer Staple (25) - GOLDEN BR!CK - Unite For Sight; Jennifer Zwilling (16) - Tourette Syndrome Association Youth Ambassador Program; David Fajgenbaum (21) - National Students of AMF (Ailing Mothers & Fathers); Divine Bradley (24) - GOLDEN BR!CK - Team Revolution; Jacob Komar (14) - Computers for Communities; Hannah Taylor (11) - The Ladybug Foundation, Inc.
Contact: Jordyn Wells, (212) 254-2390, ext. 238; jwells@dosomething.org
Internet: http://www.dosomething.org/programs/awards
Sponsor: Do Something
32 Union Square E, Suite 4L
New York, NY 10003-3209

DPA Promoting Policy Change Advocacy Grants **1694**

The program seeks to broaden public and political support for drug policy reform and will fund strategic and innovative approaches to increase such support. Proposals should be designed to: educate the public and policymakers about the negative consequences of current local, state or national drug policies; promote better awareness and understanding of alternatives to current drug policies; and broaden awareness and understanding of the extent to which punitive prohibitionist policies are responsible for most drug-related problems around the country.
Requirements: 501(c)3 tax-exempt organizations and organizations with 501(c)3 fiscal sponsors are eligible. All grantmaking will be directed to organizations working within the United States, and possibly Canada, with particular emphasis on state-based activity.
Date(s) Application is Due: Jun 15
Amount of Grant: 50,000 U.S.D
Contact: Asha Bandele, (212) 613-8020; fax (212) 613-8021; abandele@drugpolicy.org or grants@drugpolicy.org
Internet: http://www.drugpolicy.org/about/jobsfunding/grants/index.cfm
Sponsor: Drug Policy Alliance
70 West 36th Street, 16th Floor
New York, NY 10018

Dr. and Mrs. Paul Pierce Memorial Foundation Grants **1695**

The Dr. and Mrs. Paul Pierce Memorial Foundation was established in 1963 to support and promote quality education, human-services, and health-care programming for underserved populations. Special consideration is given to charitable organizations that serve the people of Grayson County, Texas. Grants from the Pierce Memorial Foundation are one year in duration. Applicants must apply online at the grant website. Applicants are strongly encouraged to do the following before applying: review the downloadable state application procedures for additional helpful information and clarifications; review the downloadable online-application guidelines at the grant website; review the foundation's funding history (link is available from the grant website); review the online application questions in advance; and review the list of required attachments. These will generally include: a list of board members, financial statements (audited, reviewed, or compiled by independent auditor); an organization summary; a list of other funding sources; an IRS Determination letter; and other required documents. All attachments must be uploaded in the online application as PDF, Word, or Excel files. The Dr. and Mrs. Paul Pierce Memorial Foundation has four deadlines annually: March 1, June 1,

September 1, and December 1. Applications must be submitted by 11:59 p.m. on the deadline dates. Grant applicants are notified as follows: March deadline applicants will be notified of grant decisions by June 30; June applicants will be notified by September 30; September applicants will be notified by December 31; December applicants will be notified by March 31 of the following year.
Requirements: Applicants must have 501(c) tax-exempt status.
Restrictions: The foundation does not support requests from individuals, organizations attempting to influence policy through direct lobbying, or any political campaigns.
Geographic Focus: Texas
Date(s) Application is Due: Mar 1; Jun 1; Sep 1; Dec 1
Samples: Denison Helping Hands, Denison, Texas, $20,000, purchase of food for families living on restricted incomes (2011); Hyde Park Elementary School, Denison, Texas, $5,000, purchase of new books for school library (2011); Grayson County Womens Crisis Line, Sherman, Texas, $10,000, funding in support of the Shopping Card Program (2010).
Contact: David Ross, Senior Vice President; tx.philanthropic@baml.com
Internet: https://www.bankofamerica.com/philanthropic/fn_search.action
Sponsor: Dr. and Mrs. Paul Pierce Memorial Foundation
901 Main Street, 19th Floor, TX1-492-19-11
Dallas, TX 75202-3714

Dr. Bob and Jean Smith Foundation Grants **1696**
The foundation awards grants to Texas nonprofit organizations in its areas of interest, including arts and culture, higher education, medical education, medical research, and health care. Types of support include annual campaigns, building construction/renovation, capital campaigns, general operating support, matching/challenge grants, and scholarship funds. There are no application deadlines; the board meets quarterly.
Requirements: Texas nonprofit organizations are eligible. Preference is given to Dallas-based organizations.
Restrictions: Individuals are ineligible.
Geographic Focus: Texas
Amount of Grant: 100 - 500,000 U.S.D
Samples: Southwestern Medical Foundation (Dallas, TX)—to conduct neuromuscular research at the U of Texas Southwestern Medical Ctr at Dallas, $1 million.
Contact: Sally Smith, Grants Administrator; (214) 521-3461
Sponsor: Dr. Bob and Jean Smith Foundation
3811 Turtle Creek Center, #2150 LB 53
Dallas, TX 75219

Dr. Courtney W. Shropshire Scholarship Grant **1697**
All qualified applicants will be considered without regard to race, creed, sex, or national origin. The grant is to be used only for tuition and approved purposes that are invoiced from the school. Necessary forms are available annually after September 1 from the Civitan International Foundation; please include a self-addressed business envelope with postage to cover two-ounce U.S. mailing.
Requirements: Candidates must be a Civitan member (or a Civitan's immediate family member) and must have been a Civitan for at least two years and/or must be or have been a junior Civitan for no less than two years. Candidates must be enrolled in a degree or certificate program at an accredited community college, vocational school, four-year college, or graduate school.
Date(s) Application is Due: Jan 31
Amount of Grant: Up to 1,000 U.S.D
Contact: Shropshire Scholarship Coordinator; (205) 591-8910; fax (205) 592-6307; civitan@civitan.org
Internet: http://www.civitaninternational.com/templates/cuscivitan/details.asp?id=23844&PID=155457
Sponsor: Civitan International Foundation
P.O. Box 130744
Birmingham, AL 35213-0744

Dr. J. Roberto Villavicencio Foundation Grants **1698**
The foundation operates in Rosario and its surrounding area in the fields of education, medicine and health, and science and technology. Types of support include research grants to institutions and individuals, scholarships and fellowships, prizes, conferences, training courses, and publications.
Contact: Dra. Ana Maria Uriarte, Director; 54-341-449 0152; info@villavicencio.org.ar
Internet: http://www.villavicencio.org.ar
Sponsor: Dr. J. Roberto Villavicencio Foundation
Alvear 854
Rosario, Santa Fe, 2000 Argentina

Dr. John Maniotes Scholarship **1699**
The Dr. John Maniotes Scholarship was established to benefit the children or grandchildren of current employees of Lake County Government Center in Crown Point. The Scholarship awards two nonrenewable scholarships of $2,500 each year to fund tuition and books. Any college major is eligible.
Requirements: Applicants are eligible if they meet the following criteria if they are: a U.S. citizen and resident of Indiana; a graduating high school senior; a parent or grandparent is a current employee of Lake County Government Center in Crown Point; accepted as an incoming freshman to an accredited public or private college or university; and seeking a four year baccalaureate degree or two year associate degree at a college or university. Selection criteria is based on academic performance (SAT 1400; ACT 19; and GPA 2.5), character, essay, and extracurricular activities. Applicants must submit the completed application along with an original essay, two recommendation letters, the completed Scholastic Profile Form filled out by the guidance counselor; and current high school transcript.
Geographic Focus: Indiana
Date(s) Application is Due: Apr 1
Amount of Grant: 2,500 U.S.D
Contact: Maniotes Scholarship Contact; (219) 736-1880; fax (219) 736-1940; legacy@legacyfoundationlakeco.org
Internet: http://www.legacyfoundationlakeco.org/scholarships/index.htm
Sponsor: Legacy Foundation
1000 East 80th Place, 302 South
Merrillville, IN 46410

Dr. John T. Macdonald Foundation Grants **1700**
The foundation awards grants to eligible Florida nonprofit organizations in Miami-Dade County for medical and health-related programs with priority given to those serving children, youth, and economically disadvantaged individuals. The foundation supports medical rehabilitation, disease prevention, and health education. Types of support include capital grants, matching/challenge grants, program development grants, seed money grants, and training grants. Deadline listed is for letters of intent; full proposals are by request. Annual deadline dates may vary; contact program staff for exact dates.
Requirements: Florida nonprofit organizations serving the health care needs of people in Miami-Dade County are eligible. Priority will be given to projects in the Coral Gables community.
Restrictions: Grants do not support national projects, multiyear funding requests, for-profit organizations, political candidates or campaigns, religious projects, individuals, or other grantmaking foundations.
Geographic Focus: Florida
Date(s) Application is Due: Apr 15
Samples: Voices for Children Foundation (Hollywood, FL)—to hire a case coordinator for the Guardian Ad Litem Program, which provides advocacy for abused, neglected, and abandoned children, $30,000; U of Miami (Coral Gables, FL)—to create a new genetics-research center, and for a public school health program that will serve as a national model for school-based health centers, $12.5 million over five years.
Contact: Kim Greene, Executive Director; (305) 667-6017; fax (305) 667-9135; info@jtmacdonaldfdn.org
Internet: http://www.jtmacdonaldfdn.org
Sponsor: Dr. John T. Macdonald Foundation
1550 Madruga Avenue, Suite 215
Coral Gables, FL 33146

Dr. Leon Bromberg Charitable Trust Fund Grants **1701**
The trust awards grants to U.S. nonprofit organizations in its areas of interest, including arts, education, hospitals and health organizations, and social services. Types of support include building construction/renovation, capital campaigns, conferences and seminars, equipment acquisition, professorships, scholarship funds, and social services. Preference will be given to requests from Texas nonprofits in Bryan, Galveston, Crockett, and Houston. Applications must be received by the 10th of each month for consideration; the board meets monthly.
Requirements: U.S. nonprofit organizations are eligible.
Restrictions: Grants are not awarded to individuals or for operating grants, endowment funds, or debt reduction.
Samples: U of Texas Medical Branch (Galveston, TX)—to endow a mentor program in the sciences that pairs graduate students, postdoctoral students, and faculty members with high-school students, $250,000.
Contact: Program Contact; (409) 762-5890
Sponsor: Dr. Leon Bromberg Charitable Trust Fund
2200 Market Street, Suite 710
Galveston, TX 77550-1532

Dr. R.T. White Scholarship **1702**
The R.T. White Scholarship was established in memory of a beloved principal at Elliott Elementary School. It is awarded each year to graduating seniors from Munster High School who attended Elliott Elementary. The number of awards varies, but the awards are usually $750 each.
Requirements: Applicants must be graduating seniors from Munster High School who attended Elliott Elementary School. Any college major is eligible. The application is available in the Munster High School Guidance Office.
Geographic Focus: Indiana
Date(s) Application is Due: Apr 1
Amount of Grant: 750 U.S.D
Contact: R.T. White Scholarship Contact; (219) 736-1880; fax (219) 736-1940; legacy@legacyfoundationlakeco.org
Internet: http://www.legacyfoundationlakeco.org/scholarships/index.htm
Sponsor: Legacy Foundation
1000 East 80th Place, 302 South
Merrillville, IN 46410

Dr. Scholl Foundation Grants **1703**
Applications are considered in the following areas: private education, including elementary, secondary, college and university level; programs for children, developmentally disabled, senior citizens, civic and cultural institutions, social service agencies, hospitals and health care, environmental organizations and religious institutions. The general areas of

interest are not intended to limit the interest of the foundation from considering other worthwhile projects. An application form is required and may be obtained from the foundation office. All grant applications must be requested in writing. Telephone, email or fax requests will not be accepted.
Requirements: Grants are awarded on an annual basis to valid IRS 501(c)3 organizations. Non-U.S. applicants without a 501(c)3 must complete a notarized affidavit.
Restrictions: Funding is not available for: organizations that do not have a valid IRS 501(c)3 determination letter; political organizations, political action committees, or individual campaigns whose primary purpose is to influence legislation; foundations that are themselves grantmaking bodies; grants to individuals, endowments or capital campaigns; grants for loans, operating deficit reduction, the liquidation of a debt, or general support; event sponsorships including the purchase of tables, tickets or advertisements; or more than one request from the same organization in the same year.
Date(s) Application is Due: Mar 1
Contact: Pamela Scholl, President; (847) 559-7430
Internet: http://www.drschollfoundation.com/procedures.htm
Sponsor: Dr. Scholl Foundation
1033 Skokie Boulevard, Suite 230
Northbrook, IL 60062

Dr. Stanley Pearle Scholarships **1704**
OneSight Research Foundation (formerly known as the Pearle Vision Foundation) will award ten (10) scholarships in the amount of $2,000 each through the Dr. Stanley Pearle Scholarship Fund. Invitation to apply is extended to second, third, and fourth year full-time students pursuing graduate studies leading to a Doctor of Optometry degree. Scholarships will be awarded on a competitive basis, based on scholastic performance and potential and evidence of commitment to a career in the optometric profession and to community service. Applicants will be required to describe accomplishments they believe would qualify them for recognition in general, and specifically, accomplishments in the areas of leadership, community service, and extracurricular activities.
Requirements: Students accepted to, or enrolled in, a fully accredited college or university that is an Approved Educational Institution pursuing full-time graduate studies in an optometry program accredited by the Accreditation Council on Optometric Education of the American Optometric Association (ACOE) are eligible to apply. Scholarships are not renewable, but an individual may re-apply for a Scholarship. OneSight values diversity. Members of racial and ethnic minorities are especially encouraged to apply. All scholarships will be awarded on an objective, non-discriminatory basis.
Restrictions: Limited to students in graduating class 2010, 2011 or 2012.
Date(s) Application is Due: Apr 15
Amount of Grant: 2,000 U.S.D
Contact: Trina Parasiliti, Administrator; (972) 277-6191; fax (972) 277-6414; tparasil@onesight.org
Internet: http://www.onesight.org/northamerica/na/
Sponsor: OneSight Research Foundation
2465 Joe Field Road
Dallas, TX 75229

Dr. Sydney Segal Research Grants **1705**
Research grants are awarded to enable students to pursue full-time higher degree studies at a Canadian institution researching the possible causes, effects, and/or prevention of sudden infant death syndrome. The research grants are awarded for one year in any discipline—medical, psychological, biological, sociological, nursing, or other. Contact the foundation for applications.
Requirements: Applications are invited from suitably qualified graduate students who are undertaking full-time training in research in the health sciences leading to an MSC, PhD, or equivalent degree.
Date(s) Application is Due: Jun 1
Amount of Grant: Up to 35,000 CAD
Contact: Executive Director; (800) 363-7437 or (416) 488-3260; fax (416) 488-3864; sidsinfo@sidscanada.org
Internet: http://www.sidscanada.org/segal_research_grants.htm
Sponsor: Canadian Foundation for the Study of Infant Deaths
586 Eglinton Avenue E, Suite 308
Toronto, ON M4P 1P2 Canada

Dreitzer Foundation Grants **1706**
The foundation awards general operating grants to nonprofit organizations in its areas of interest, including health, homeless, social services, and youth programs. Grants are awarded nationally, with preference given to New York nonprofits. Applications are available by mail or fax. Phone calls are strongly discouraged.
Requirements: 501(c)3 tax-exempt organizations in New York are eligible.
Geographic Focus: New York
Date(s) Application is Due: Jul 1
Amount of Grant: 5,000 - 65,000 U.S.D
Samples: Partnership for the Homeless (New York, NY)—grant recipient, $50,000; Ethical Culture Society of Queens (Bayside, NY)—grant recipient, $25,000.
Contact: Alan Seget c/o Davidson, Dawson and Clark LLP, (212) 557-7700; fax (212) 286-8513
Sponsor: Dreitzer Foundation
60 E 42nd Street, 38th Floor
New York, NY 10165

Drs. Bruce and Lee Foundation Grants **1707**
The Foundation's goal is to advance the general welfare and the quality of all life in the Florence, S.C. area by providing economic support to qualified programs and non-profit organizations. The Foundation will support a broad range of charitable purposes including, but not limited to: medical; health; human services; education; arts; religion; civic affairs; and the conservation, preservation and promotion of cultural, historical and environmental resources. There are no application deadlines. Contact the office for application materials and guidelines.
Requirements: Florence, SC, nonprofit organizations are eligible.
Restrictions: The Foundation does not purchase tickets for fundraising events. Grants to individuals will not be considered.
Geographic Focus: South Carolina
Samples: Francis Marion U (Florence, SC)—to construct a facility for the new bachelor's-degree nursing program, $5 million.
Contact: L. Bradley Callicott, Executive Director; (843) 664-2870; blfound@bellsouth.net
Sponsor: Drs. Bruce and Lee Foundation
181 East Evans Street, BTC Box 022
Florence, SC 29506

DSO Controlling Pathogen Evolution Workshop Grants **1708**
The goal of the workshop is to explore: the potential of computational methods to precisely predict pathogen evolution; and vaccination and/or therapeutic strategies to control direction of pathogen evolution. The workshop will investigate tools for predicting protective epitopes, pathogen evolution and protein structure, and then review strategies to force pathogens into an evolutionary trap that prevents escape through mutation. From there, vaccines and therapies can be developed against the particular pathogen strain. Workshop participants are strongly encouraged to prepare posters describing previous or potential research in the area in order to facilitate discussions and/or formation of well-rounded teams.
Requirements: Attendees should have specific expertise in infectious disease, epitope prediction (computational and/or immunomic), vaccinology, mathematical modeling, protein structure, pathogen evolution, drug design, and/or animal modeling.
Date(s) Application is Due: Mar 16
Contact: Dr. Michael V. Callahan, Program Manager; (571) 218-4596; fax (571) 218-4553; michael.callahan@darpa.mil
Internet: http://www.darpa.mil/baa/SN07-21.html
Sponsor: Defense Sciences Office within the Defense Advanced Research Projects Agency
3701 N Fairfax Drive
Arlington, VA 22203-1714

DSO Radiation Biodosimetry (RaBiD) Grants **1709**
The DSO is seeking proposals for new technologies for rapid, high-throughput, portable and low-cost biodosimeters to determine radiation dose to individuals after acute radiation exposure. This technology would provide rapid identification of individuals who have been exposed to high-dose radiation in order to accurately assess radiation exposure levels. The Program is a single phase, 15-month program with a goal of revolutionizing radiation exposure detection. The first 12 months consist of scientific research and development during which the proposer will develop a non- or minimally-invasive radiological biodosimeter and demonstrate accurate radiological detection from biological samples into quartiles of doses for humans with a detection time less than 10 minutes. This will include radiological dose/detection curves, as well as multipoint data regarding decay of biological signal as a function of time after radiation exposure. A successful proposal will demonstrate a clear path to these deliverables, including a timeline for technology development, experimentation, and delivery of results.
Requirements: All proposers must provide evidence of a competent team capable of developing a new radiation biodosimetry technology meeting all program objectives and milestones. The Program requires that each proposal include a team with demonstrated (or established) capability. At a minimum, proposers are expected to possess expertise or demonstrate collaboration with professionals in the following areas: radiation biologist/radiation oncologist—with expertise in radiation exposure to animal models; and engineering—with developmental expertise needed to translate biological sampling to radiation readout.
Restrictions: Only unclassified proposals will be accepted in response.
Date(s) Application is Due: Jul 9
Contact: Dr. Mildred Donlon, Program Manager; (703) 696-2289; fax (703) 741-3896; mildred.donlon@darpa.mil
Internet: http://www.darpa.mil/dso/solicitations/baa07-29.htm
Sponsor: Defense Sciences Office within the Defense Advanced Research Projects Agency
3701 N Fairfax Drive
Arlington, VA 22203-1714

DTE Energy Foundation Health and Human Services Grants **1710**
The DTE Energy Foundation is at the core of DTE Energy's commitment to the communities and customers it is privileged to serve. The DTE Energy Foundation is dedicated to strengthening the health and human services sector of these communities. Priority will be given to supporting organizations that are in the forefront of addressing the critical, acute human needs brought on by the economic downturn. Grant amounts generally range from $500 to $100,000, and the application process is distinct for each of the following ranges: $500 to $2,000; $2,001 to $10,000; and any amount greater than $10,000. Applications must be submitted electronically by the stated deadlines.

Requirements: Eligible applicants must meet all of the following criteria: be located in or provide services to a community in which DTE Energy does business; and be a nonprofit (i.e. be exempt for federal income tax under section 501(c)3 of the Internal Revenue Code and not a private foundation, as defined in Section 509(a) of the Code).
Restrictions: The Foundation does not provide support to: individuals; political parties, organizations or activities; religious organizations for religious purposes; organizations that are not able to demonstrate commitment to equality and diversity; student group trips; national or international organizations, unless they are providing benefits directly to our service-area residents; projects that may result in undue personal benefit to a member of the DTE Energy Foundation board, or to any DTE Energy employee; conferences unless they are aligned with DTE Energy's business interests; single purpose health organizations; hospitals, for building or equipment needs.
Geographic Focus: Michigan
Date(s) Application is Due: Apr 13; Jul 13; Oct 12; Dec 28
Amount of Grant: 500 - 100,000 U.S.D
Samples: Center for Autism, Ann Arbor, Michigan, $100,000 - general operating support (2010); Girl Scouts of Southeastern Michigan, Detroit, Michigan, $75,000 - general operating costs (2010); Forgotten Harvest, Oak Park, Michigan, $60,000 - general operating costs (2010).
Contact: Karla D. Hall, Vice-President; (313) 235-9271 or (313) 235-9416; foundation@dteenergy.com
Sponsor: DTE Energy Foundation
One Energy Plaza, 1046 WBC
Detroit, MI 48226-1279

Dubois County Community Foundation Grants 1711
The central purpose of the community foundation is to serve the needs of Dubois County, Indiana and the philanthropic aims of donors who wish to better their community. The Foundations fields of interests are: arts, education, environment, beautification programs, health care, human services, recreation, youth development.
Requirements: Application: obtain grant application form from the Dubois County Community Foundation offices. Completed Grant Application Form, Grant Agreement, and Grant Application Certification. Attachments: Program Narrative - expansion upon the summary including program procedures, indication of whom the program will serve and an estimated number of individuals so served, personnel involved, anticipated results, and internal evaluation procedures; IRS 501(c)3 Determination Letter; Governing Body Information - names of the organization's officers, directors, and key project personnel; Detailed Project Budget - identification of all projected expenses AND anticipated income associated with the program in justification of the amount being requested; Most recent Audit Report and/or Financial Report; Visual materials may be submitted as appendices. Submit one original and 11 copies of your complete application (a total of 12)
Restrictions: Giving concentrated in Dubois County, Indiana. No support for the operational expenses of government units or agencies. No grants for operating expenses of nonprofits, funding after the fact, annual fund raising, sponsorship of events, or debt retirement, no loans.
Geographic Focus: Indiana
Date(s) Application is Due: Sep 15
Contact: Marilyn Rumbach Weisheit, Executive Director; (812) 482-5295; fax (812) 482-7461; dccf@fullnet.com
Internet: http://www.duboiscountycommunityfoundation.org/grant process.htm
Sponsor: Dubois County Community Foundation
323 East Sixth Street, P.O. Box 269
Jasper, IN 47547-0269

Duchossois Family Foundation Grants 1712
The foundation focuses its efforts in the Chicago metropolitan area and gives primary consideration to nonprofit organizations that contribute to the community in the area of health. The foundation supports organizations involved with mental health/crisis services, cancer, HIV/AIDS, cancer research, AIDS research and, human services. A one-page summary-request letter should include a description of the organization and its specific needs and purposes, the amount requested, and a list of members of the board of directors and their business/professional affiliations.
Requirements: 501(c)3 tax-exempt public charities serving the Chicago metropolitan area are eligible.
Geographic Focus: Illinois
Samples: American Cancer Society, Chicago, IL, $50,000; Metropolitan Family Services, Chicago, IL, $125,000; Barrington Area Council on Aging, Barrington, IL, $2,500.
Contact: Iris Krieg, Executive Director; (312) 641-5765; iriskriegl@aol.com
Sponsor: Duchossois Family Foundation (Chicago)
203 N Wabash Avenue, Suite 1800
Chicago, IL 60601

Duke Endowment Health Care Grants 1713
The purpose of the program is to: improve the quality and safety of health care in the areas of patient quality and safety programs/intervention, workforce development, and health information technology; improve access to health care in the areas of care for the medically indigent/uninsured, mental health care, and where racial disparities are evident; and expand prevention and early intervention in the areas of child health and early intervention programs, wellness and healthy lifestyle programs, and disease management.
Requirements: The endowment awards grants in North Carolina and South Carolina to: not-for-profit hospitals; academic health centers; health education centers; and select health organizations in counties without an eligible hospital.
Geographic Focus: North Carolina, South Carolina
Date(s) Application is Due: Jun 15; Dec 15
Contact: Mary L. Piepenbring; (704) 376-0291; mpiepenbring@tde.org
Internet: http://www.dukeendowment.org/program-areas/health-care
Sponsor: Duke Endowment
100 N Tryon Street, Suite 3500
Charlotte, NC 28202-4012

Duke Energy Foundation Grants 1714
The Duke Energy Foundation gives primarily in areas of company operations in Indiana, Kentucky, North Carolina, Ohio and South Carolina. At the core of The Duke Energy Foundation is its commitment to the community, with a focus in three areas: Environment and Energy Efficiency; Economic Development; and Community Vitality.
Requirements: Organizations with a 501(c)3 verification from the IRS or are a part of a governmental entity are eligible. All organizations applying for a grant must have: completed the Online Grant Application; a clear reason for making the contribution that relates to the areas of focus; regular reports on the measurable results of the project.
Restrictions: Foundation funds are not provided for: organizations that discriminate by race, creed, gender, age or national origin; political activities and organizations; grants to individual agencies of the United Way or the Charlotte Arts and Science Council; capital campaigns and endowments, except in extremely rare and specialized situations that relate directly to our areas of expertise in business; individuals; athletics, including individual sports teams and all-star teams; underwriting of films, video and television productions; reducing the cost of utility service; sectarian or religious activities; conferences, trips or tours; fraternal, veteran or labor groups serving only their members; advertising; membership fees or association fees, either personal or corporate; dinners or tables at fund-raisers are rarely considered; family foundations.
Geographic Focus: Indiana, Kentucky, North Carolina, Ohio, South Carolina
Amount of Grant: 1,000 - 1,250,000 U.S.D
Contact: Deanita McCall, Vice President; (704) 382-7200; fax (704) 382-7600
Internet: http://www.duke-energy.com/community/foundation.asp
Sponsor: Duke Energy
P.O. Box 35469
Charlotte, NC 28254

Duke University Adult Cardiothoracic Anesthesia and Critical Care Medicine Fellowships 1715
Nine one year fellowship positions are available each year in the ACGME Accredited Fellowship. Second year research and clinical fellowship opportunities are available; programs are individualized to meet the goals of successful applicants. Generally clinical experience is the focus of the first fellowship year, along with involvement in clinical research. Fellows opting for a two year fellowship continue with one clinical day per week in their second year, in addition to a structured plan involving laboratory and/or clinical research and teaching. Second year clinical specialization in pediatric cardiac anesthesiology or intensive care is also available. The Adult cardiothoracic Anesthesia Fellowship is highly recommended as formal training for specialization in cardiac and/or thoracic anesthesiology. The Fellowship is designed to develop clinical, consultant, and in-depth research expertise. Applications are usually considered starting 18 months prior to the start date
Requirements: Complete application (see Duke University website), in addition to the references required, include a letter from your Chairperson/Program Director and two letters from other faculty members who have worked with you closely.
Contact: Mark S. Smith; (919) 681-5046 or (919) 416-3853; fax (919) 681-8993
Internet: http://medschool.duke.edu/modules/som_rt/index.php?id=1
Sponsor: Duke University Medical Center
P.O. Box 3094
Durham, NC 27710

Duke University Ambulatory and Regional Anesthesia Fellowships 1716
The Fellowship in Ambulatory/Regional Anesthesia is designed to provide advanced training in anesthesia in the ambulatory surgery setting. This program will be an extension to the previous training with an emphases on the continued educational component to include research, clinical and teaching responsibilities.
Requirements: The selection process for fellows includes the completion of a Fellowship Application (see website) which is to be substantiated by three letters of recommendation from experts in the field. In addition to the letters of reference the application is to include a current CV and a letter of affirmation stating interest in the program.
Date(s) Application is Due: Aug 15
Contact: Stephen M. Klein, Director; (919) 668-2056; klein006@mc.duke.edu
Internet: http://anesthesiology.duke.edu/modules/anes_flwshp/index.php?id=1
Sponsor: Duke University Medical Center
P.O. Box 3094
Durham, NC 27710

Duke University Clinical Cardiac Electrophysiology Fellowships 1717
The Clinical Cardiac Electrophysiology fellowship training program is a comprehensive two-year program including training in catheter ablation of supraventricular tachycardias, ventricular tachycardias, and complex arrhythmias such as atrial fibrillation, ischemic VT, and arrhythmias associated with congenital heart disease. Comprehensive training is also provided in lead extraction and device therapy including permanent pacing, implantable defibrillators, and bi-ventricular devices. Fellows are provided with significant opportunity for research activity through protected block time and are expected to participate in and complete a research project with a faculty mentor. Applicants must have completed at least three years in an ACGME approved Cardiology fellowship program.

Requirements: Applications for training to begin July 1, will be accepted beginning January 1; the application deadline will be September 1. To apply, please include with the completed CCE (EP) Application form (available on the Duke University website), a current curriculum vitae, personal statement (not to exceed 250 words), and the $40 (check or money/order no cash please) processing fee.
Date(s) Application is Due: Sep 1
Contact: Dr. Patrick Hranitzky, Program Director; (919) 684-2304; fax (919) 684-4322; Patrick.hranitzky@duke.edu
Internet: http://fellowships.medicine.duke.edu/modules/fellows_cardio/index.php?id=8
Sponsor: Duke University School of Medicine
P.O. Box 3878
Durham, NC 27710

Duke University Hyperbaric Center Fellowships 1718
The clinical Hyperbaric fellow works in the Hyperbaric Center and F.G. Hall Environmental Laboratory, both of which are an integral part of Duke University Medical Center. Fellowships are available for 12 months although they may be extended for additional time. During the period of training the Hyperbaric fellow will be trained in the clinical use of hyperbaric oxygen. This includes the emergency treatment of clostridial myonecrosis, necrotizing fasciitis, arterial gas embolism, decompression sickness, carbon monoxide poisoning and certain other less common conditions such as acute vasculitis with ischemia. The fellow will receive training in the appropriate critical care interventions necessary for the treatment of these patients. This will include hemodynamic monitoring, ventilatory support, fluid resuscitation and the interpretation of arterial blood gases during hyperbaric therapy. The fellow will also receive training in the use of hyperbaric oxygen therapy for non-emergency conditions. The fellow will be responsible for all inpatient and outpatient consults, under the direction of the clinical Hyperbaric faculty. During the training period the fellow will also be required to complete a research project. This will be completed under the guidance of one of the faculty members in the Hyperbaric Center. Part of the didactic program includes regular hyperbaric conferences. The fellow will participate in these and normally be required to present at least one topic during his year of training. The fellow will also have the opportunity to attend the annual two-week course in Hyperbaric and Diving Medicine given by the National Oceanic and Atmospheric Administration in Seattle, Washington. In addition to these activities the applicant may participate in numerous other teaching activities within Duke Medical Center, for example those sponsored by the Departments of Anesthesiology, Medicine, Surgery and Emergency Medicine. Clinical participation under the aegis of those departments can be arranged in some cases, depending upon the training of the individual concerned.
Requirements: Applicants must be eligible for license as a trainee in the state of North Carolina. The Duke Hyperbaric Center Fellowship Application can be downloaded at: http://hyperbaric.mc.duke.edu/education.htm
Contact: Tonya Manning, Assistant; (919) 681-1685 or (919) 684-6726
Internet: http://hyperbaric.mc.duke.edu/education.htm
Sponsor: Duke University Medical Center for Hyperbaric Medicine and Environmental Physiology
P.O. Box 3823
Durham, NC 27710

Duke University Interventional Cardiology Fellowships 1719
The Interventional Cardiology Fellowship Training Program at Duke University Medical Center focuses on training leaders in Interventional Cardiology committed to a career in academic cardiology. The length of the fellowship is 12 months. During the fellowship, all fellows will develop both clinical and technical skills in interventional cardiology. All fellows will gain experience in diagnosing, selecting therapies, performing diagnostic and interventional procedures, and judging the effectiveness of treatments for inpatients and outpatients with chronic ischemic heart disease, acute ischemic syndromes, and valvular heart disease. All fellows are expected to participate in research efforts. During the fellowship, each fellow will work with a mentor on a research project which will help foster their long-term goals.
Requirements: Individuals applying for only a 1 year Interventional Cardiology Fellowship position will be required to provide his/her own funding resources for salary support during their training. This funding must be institutional funding (no personal or private funds). A processing fee of $40 (check or money/order made payable to Duke University- no cash please), letter documenting funding support, current curriculum vitae, and personal statement will be required to accompany the completed Interventional Cardiology Application (see Duke University website for application).
Date(s) Application is Due: Feb 1
Contact: Sherolyn Patterson, Fellowship Program Coordinator; (919) 684-2304; fax (919) 684-4322; sherolyn.patterson@duke.edu
Internet: http://fellowships.medicine.duke.edu/modules/fellows_cardio/index.php?id=8
Sponsor: Duke University School of Medicine
P.O. Box 3878
Durham, NC 27710

Duke University Obstetric Anesthesia Fellowships 1720
The Department of Anesthesiology at Duke University Medical Center offers 1 - 2 positions of a one year Fellowship in Obstetric Anesthesia. Emphasis is on clinical research, advanced clinical training and teaching. Duke University is a tertiary center with approximately 3,000 deliveries per year. The division is currently made up of 8 faculty staff with diverse international pedigree.
Contact: Yemi Olufolabi, Fellowship Director; (919) 681-6535 or (919) 668-6266; fax (919) 668-6265; olufo001@mc.duke.edu
Internet: http://anesthesiology.duke.edu/modules/anes_flwshp/index.php?id=1
Sponsor: Duke University Medical Center
P.O. Box 3094
Durham, NC 27710

Duke University Pain Management Fellowships 1721
The Fellowship Program is based in the Duke Health Center at Morreene Road (MRC). Fellows will: cover the Pain Clinic and Durham VAMC, and Acute Pain Service at Duke North; evaluate new patients; perform comprehensive or neurologic studies; formulate diagnosis and plan; become familiar with procedures; learn intervention of the lumbar spine and its application e.g. facet rhizotomy, discography, and intradiscal annuloplasty; implantation therapies; dictation of clinic and procedure notes. Fellows will be asked to participate in research, journal clubs, round table discussions, grand rounds and lectures. There are opportunities for the fellow to become involved in research projects already in progress or to develop an original project. The clinical objectives are to build knowledge, skills and professionalism.
Requirements: Applicants must meet eligibility specified by the ACGME. Acceptance into the program is based on demonstrated academic qualification, interpersonal skills, and evidence of ability to function in a multidisciplinary setting. Fellowship applications may be downloaded from http://anesthesia.duhs.duke.edu/. Submit application, CV and three letters of reference.
Date(s) Application is Due: Jun 30
Contact: Dianne L. Scott, Fellowship Director; (919) 684-6736
Internet: http://anesthesiology.duke.edu/modules/anes_flwshp/index.php?id=1
Sponsor: Duke University
932 Morreene Road, Room 232
Durham, NC 27705

Duke University Pediatric Anesthesiology Fellowship 1722
The pediatric anesthesia fellowship program provides an opportunity for one anesthesia to develop skills in pediatric anesthesia. The goals of the program are to foster the development of the following for the fellow: a broad clinical experience in pediatric anesthesia to allow proficiency in providing care for neonates, infants, children, and adolescents undergoing a wide variety of surgical, diagnostic, and therapeutic procedures; technical skills in pediatric airway management; technical expertise in pediatric invasive monitoring; a comprehensive understanding of the developmental, pharmacological, anatomic, physiologic, and psychological changes that occur with age and disease; an understanding of pediatric pain and its management; an understanding of critical perioperative care and advanced life support; and competence as a perioperative pediatric anesthesiologist and consultant. The goals of the program will be met by providing fellows with: graded clinical experiences; didactic lectures, seminars, and conferences; and clinical electives and research opportunities.
Requirements: The fellowship application can be downloaded from : http://anesthesia.duhs.duke.edu/education/fellows/fellowapp.pdf. Fellowship selection is based on the process used for application to the core residency program. All applicants are considered without regard to race, color, religion, gender, or national origin and must have completed an ACGME accredited program in Anesthesiology or it's equivalent. Documentation of completion will be required. In addition to the completed application, all candidates are required to submit: completed fellowship application; three letters of recommendation; curriculum vitae; one page personal statement; medical school transcript; letter of completion from the Anesthesiology residency chairman or program director or copy of Anesthesiology training certificate.
Contact: B. Craig Weldon M.D., Director; (919) 668-0976 or (919) 668-4208; fax (919) 668-4228; weldo@mc.duke.edu
Internet: http://anesthesiology.duke.edu/modules/anes_flwshp/index.php?id=1
Sponsor: Duke University Medical Center
Department of Anesthesiology, DUMC 3094
Durham, NC 27710

Duke University Postdoctoral Research Fellowships in Aging 1723
The goal of the postdoctoral research training program is to produce highly skilled research scientists who have the potential for leadership in gerontological research. In the Duke Research Training Program (RTP), much of the training for each fellow is provided by that person's faculty mentor(s) in a research apprenticeship program. A fellow carries out his/her own research as a junior colleague in the mentor's research program or laboratory. In addition to working in their mentor's programs, all fellows attend a weekly interdisciplinary didactic seminar. The stipend amount is determined by years of relevant postdoctoral experience. The program also provides health insurance for the fellow (using the University's insurance plan), and partial travel support when presenting research at a professional meeting.
Requirements: Applicants must be citizens of the United States, or have been lawfully admitted to the United States for permanent residence. Applicants must have a completed doctoral degree when joining the program. All course work must be completed, and if applying as a Ph.D., final orals must be passed, and the dissertation signed before an applicant can begin the program.
Date(s) Application is Due: May 1
Contact: James A. Blumenthal, (919) 660-7517; fax (919) 684-8569; blume003@mc.duke.edu
Internet: http://www.geri.duke.edu/post_doc/
Sponsor: Duke University School of Medicine
P.O. Box 3878
Durham, NC 27710

Duncan L. Gordon Fellowships 1724
Fellowships are available for research in pediatric medicine including nutrition, infectious diseases, mental retardation, environmental health, and clinical pharmacology, tenable in any agreed institution within Canada for one or two years. Candidates should be nominated by the head of the department in which they are employed or in which they will be employed upon completion of the fellowship. Nominating institutions must affirm their intention to employ the fellow upon completion of studies and to continue or establish a program in the field concerned. Award includes a full-time stipend, paid on a quarterly basis, plus a modest transportation allowance to take up the fellowship if required.
Requirements: Fellowships are available to Canadian citizens or landed immigrants at a post-doctoral level and who are of outstanding academic achievement in a pediatric related health care field and who can provide evidence of aptitude for pediatric patient care, teaching, and research.
Date(s) Application is Due: Oct 1
Amount of Grant: Up to 47,500 CAD
Contact: Pam Gilliland, Coordinator, National Grants Program; (416) 813-6166, ext. 2354; fax (416) 813-7311; national.grants@sickkids.ca
Internet: http://www.sickkidsfoundation.com
Sponsor: Sick Kids Foundation
555 University Avenue
Toronto, ON M5G 1X8 Canada

Duneland Health Council Incorporated Grants 1725
The Duneland Health Council, Inc. is a private foundation focused on improving the health and general welfare of the greater Michigan City, Indiana community.
Restrictions: Funding limited primarily in the metropolitan Michigan City, IN area. No support for religious organizations. No grants to individuals, or for fund-raising, endowments or advertising.
Geographic Focus: Indiana
Contact: Norm Steider, Director; (219) 874-4193; fax (219) 873-2416; normsteider@yahoo.com
Sponsor: Duneland Health Council Incorporated
P.O. Box 9327
Michigan City, IN 46361-9327

Dunspaugh-Dalton Foundation Grants 1726
The foundation awards grants to eligible nonprofit organizations in the areas of higher, secondary, and elementary education; social services; youth; health associations and hospitals; cultural programs; and civic affairs. Types of support include capital campaigns, continuing support, endowment funds, matching funds, operating support, professorships, and special projects. The board meets monthly to consider requests.
Requirements: U.S. nonprofit organizations are eligible. The foundation primarily supports programs in California, Florida, and North Carolina.
Restrictions: Individuals are not eligible.
Geographic Focus: California, Florida, North Carolina
Amount of Grant: 5,000 - 50,000 U.S.D
Samples: Barry U—for program support, $1.5 million over five years; Bertha Abess Children's Ctr (Miami, FL)—for operating support, $3000.
Contact: William Lane Jr, (305) 668-4192; fax (305) 668-4247
Sponsor: Dunspaugh-Dalton Foundation
1533 Sunset Drive, Suite 150
Coral Gables, FL 33143-5700

DuPage Community Foundation Grants 1727
The DuPage Community Foundation is a community-based philanthropic organization that identifies, promotes, and supports programs that raise the quality of life in DuPage County for this generation and future generations. To this end, the Foundation seeks to address the broad needs of DuPage County through grants in support of the following program categories: Arts and Culture, Education, Environmental, Health, and Human Services.
Requirements: Grant applications will be accepted from non-profit charitable organizations that: are classified as exempt from federal income taxes under section 501(c)3 of the Internal Revenue Code; and are located in DuPage County. Previous grantees must have complied with all reporting requirements and be in good standing before submitting another grant request.
Restrictions: The Foundation does not generally award grants to: organizations located outside of DuPage County (exceptions, if any, will be made by the director of grants); individuals; governmental agencies, including supporting foundations; programs for religious purposes; endowments; disease-specific organizations; private foundations and private operating foundations; or public, parochial, private or charter schools (pre-K through 12th grade).
Geographic Focus: Illinois
Date(s) Application is Due: Feb 2; Aug 3
Samples: To be divided among 44 organizations in DuPage County (IL)—to support nonprofit groups working in the fields of the arts and culture, civic affairs, education, health, and human services, and for two special projects, $288,000; Willowbrook Wildlife Foundation (Glen Ellyn, IL)—to help establish a permanent endowment at the foundation, $10,000 challenge grant.
Contact: Bonnie Heydorn, (630) 665-5556; fax (630) 665-9571; bheydorn@dcfdn.org
Internet: http://www.dcfdn.org
Sponsor: DuPage Community Foundation
2100 Manchester Road, Building A, Suite 303
Wheaton, IL 60187-4584

Dysautonomia Research Grants 1728
Grants are awarded to promote research into control and/or cure for familial dysautonomia with particular emphasis on genetic screening; prenatal and/or carrier tests are necessary. Funds awarded are variable and depend on research proposed.
Requirements: Postdoctoral candidates must be connected with a recognized medical and/or teaching institution. There are no geographic limitations.
Date(s) Application is Due: Nov 30
Amount of Grant: 100,000 U.S.D
Contact: Grants Administrator; (212) 949-6644; fax (212) 682-7625; info@familialdysautonomia.org
Internet: http://www.familialdysautonomia.org/researchgrantprogram.htm
Sponsor: Dysautonomia Foundation
633 Third Avenue, 12th Floor
New York, NY 10017

Dyson Foundation Mid-Hudson Valley General Operating Support Grants 1729
General operating support grants are sometimes referred to as core support or unrestricted grants. Many organizations use general operating support grants to cover day-to-day activities or ongoing expenses such as administrative salaries, utilities, office supplies, technology maintenance, etc. Other organizations use this type of funding to cover project costs, capital, technology purchases, and professional development. The use of these funds is totally at the discretion of the organization's board and/or executive staff, although the Foundation expects all organizational expenditures to be part of a board-approved annual budget. The first step in applying for a general operating support grant is to submit a letter of inquiry.
Requirements: 501(c)3 nonprofit organization or libraries based in the Mid-Hudson Valley (Columbia, Dutchess, Greene, Orange, Putnam, and Ulster counties) are eligible. To apply for a general operating support grant, an organization must: review the principles of best practice relative to nonprofit governance, finance, public disclosure, and programming; have been a recipient of a Dyson Foundation project grant within the past three years of their general operating support request; have a mission and programs that are consistent with core funding interests of the Dyson Foundation; and have demonstrated at least three years of stable executive leadership.
Restrictions: The Foundation places no restrictions on the use of these funds, unlike project or management technical assistance grants that are restricted by the Foundation to a particular project or activity. Colleges and universities, hospitals, faith-based institutions, and organizations with annual budgets in excess of $15 million dollars are not eligible to apply for general operating support grants.
Geographic Focus: New York
Amount of Grant: 1,000 - 1,000,000 U.S.D
Contact: Diana M. Gurieva, Executive Vice President; (845) 790-6312 or (845) 677-0644; fax (845) 677-0650; dgurieva@dyson.org or info@dyson.org
Jennifer Drake, Grants Program Coordinator; (845) 790-6318 or (845) 677-0644; fax (845) 677-0650; jdrake@dyson.org or info@dyson.org
Internet: http://www.dysonfoundation.org/grantmaking/general-operating-support-grants
Sponsor: Dyson Foundation
25 Halcyon Road
Millbrook, NY 12545-9611

Dyson Foundation Mid-Hudson Valley Project Support Grants 1730
In 2009 the Dyson Foundation changed its grantmaking priorities to best support the people and communities in its region most vulnerable to the economic downturn. The Foundation's funding now focuses on organizations and activities that address basic needs such as food, housing, health care, and other human services. In limited circumstances, it will also make grants to faith-based organizations, government entities and libraries. The Foundation will also consider limited funding to arts organizations or projects that provide management support or training to other arts organizations, and to arts organizations or projects that can demonstrate the potential to increase local tourism and employment and/or other local economic development as a result of their efforts. Note that there are separate guidelines for Dutchess County and for the other Mid-Hudson Valley counties.
Requirements: IRS 501(c)3 nonprofits that support Hudson Valley, New York, and are not classified as foundations under section 509(a)of the code are eligible. The region is defined as Columbia, Dutchess, Greene, Orange, Putnam, and Ulster counties. Occasionally the foundation awards grants to fiscal sponsors of non-qualifying organizations.
Restrictions: The Foundation is not currently funding certain areas including: the environment, historic preservation, and capital projects. Grants do not support: individuals for any purpose; dinners, fund raising events, tickets, or benefit advertising; direct mail campaigns; service clubs and similar organizations; debt or deficit reduction; governmental units; or international projects or to organizations outside of the United States.
Geographic Focus: New York
Amount of Grant: 1,000 - 1,000,000 U.S.D
Contact: Diana M. Gurieva, Executive Vice President; (845) 790-6312 or (845) 677-0644; fax (845) 677-0650; dgurieva@dyson.org or info@dyson.org
Jennifer Drake, Grants Program Coordinator; (845) 790-6318 or (845) 677-0644; fax (845) 677-0650; jdrake@dyson.org or info@dyson.org
Internet: http://www.dysonfoundation.org/grantmaking/project-grants
Sponsor: Dyson Foundation
25 Halcyon Road
Millbrook, NY 12545-9611

Dystonia Medical Research Grants and Fellowships 1731

The research grants provide assistance in supporting investigations in specified areas of biomedical research pertinent to dystonia and related disorders. Research grants are being sought to develop important tools for DYT1 dystonia research; these include cell culture models, proteomics, genetic animal models of DYT1 dystonia, including transgenic and knock-in/ knock-out mice, Drosophila, C. elegans, and Zebrafish models. Grants are also being sought to develop new assays suitable for high throughput drug screening. Funding is available up to $40,000 for one-year pilot studies. A limited number of two-year grants are also available for larger studies with funding up to $75,000 per year. Fellowships assist post-doctoral students in establishing careers in dystonia research.

Requirements: Eligible for research grants are all nonprofit organizations or institutions. Eligible for research fellowships are investigators with MD or PhD degrees. For all investigations involving humans, approval by the institution's human subject protection committee is necessary. The regulations pertaining to recombinant DNA research and animal welfare as established by PHS have been adopted by DMRF.

Restrictions: Grant support will not be given for medical care of dystonia patients or for constructions and alterations.

Date(s) Application is Due: Dec 30

Amount of Grant: 100,000 - 150,000 U.S.D

Contact: Dr. Robert McAlister, Executive Director; (312) 755-0198; fax (312) 803-0138; dystonia@dystonia-foundation.org

Internet: http://www.dystonia-foundation.org/research/other.asp?id=1

Sponsor: Dystonia Medical Research Foundation

One E Wacker Drive, Suite 2430

Chicago, IL 60601-2001

E.B. Hershberg Award for Important Discoveries in Medicinally Activev Substances 1732

This award, supported by Schering-Plough Corporation, is given biennially in odd-numbered years to recognize and encourage outstanding discoveries in the chemistry of medicinally active substances. The discovery for which the award is given should have been made during the past two decades. Nominations are accepted in even-numbered years, and the award consists of $3,000 and a certificate.

Requirements: Any individual, except a member of the award committee, may submit one nomination or seconding letter for the award in any given year. The nominating documents consist of a letter of not more than 1000 words containing an evaluation of the nominee's accomplishments and a specific identification of the work to be recognized, a biographical sketch including date of birth, and a list of publications and patents authored by the nominee.

Restrictions: Self-nominations are not accepted.

Date(s) Application is Due: Nov 1

Amount of Grant: 4,000 U.S.D

Contact: Felicia Dixon, Awards Administrator; (800) 227-5558 or (202) 872-4408; fax (202) 776-8008; f_dixon@acs.org or awards@acs.org

Internet: http://portal.acs.org/portal/acs/corg/content?_nfpb=true&_pageLabel=PP_ARTICLEMAIN&node_id=1319&content_id=CTP_004515&use_sec=true&sec_url_var=region1

Sponsor: American Chemical Society

1155 Sixteenth Street, NW

Washington, DC 20036-4801

E. Balzan Foundation International Prizes 1733

The foundation awards prizes to encourage the worthiest humanitarian and cultural activities, without regard to nationality, race, or religion. Prizes are awarded for literature, moral sciences, arts, physics, mathematics, natural sciences, medicine, and peace.

Amount of Grant: 1,000,000 CHF

Contact: Grants Administrator; 41-1-2014822; fax 41-1-2014829; info@balzan.com

Internet: http://www.balzan.com/en/preise/index.cfm

Sponsor: E. Balzan International Prize Foundation

Claridenstrasse 35

Zurich, CH-8002 Switzerland

E.D.Thomas Fellowship 1734

The foundation supports research into the diagnosis, prevention, and cure of leukemia and related hematological malignancies. The fellowship is awarded for one year, renewable yearly for two additional years upon satisfactory performance. The award supports costs not to exceed 8% and salary not to exceed $35,000 including fringe benefits, with the remainder for supplies and/or equipment. Applications must be requested by October 15.

Requirements: Candidates must hold an MD or PhD degree and have completed at least three years postdoctoral training but must be less than 10 years post their first doctoral degree when the award begins. There are no restrictions based on nationality, but only one application will be considered from each sponsoring institution. Candidates must be committed to the research goals of the foundation, must be able to devote at least 80% of their time to the project and must have a sponsoring institution with the academic environment to provide adequate support for the proposed project.

Date(s) Application is Due: Nov 2

Amount of Grant: 50,000 U.S.D

Contact: Grants Administrator; 34 93 414 55 66; fax 34 93 201 0588; fundacio@fcarreras.es

Internet: https://www.fcarreras.org/eng/seccio.php?opcion=5

Sponsor: Jose Carreras International Leukemia Foundation

Muntaner, 383 2n

Barcelona, 08021 Spain

E.J. Grassmann Trust Grants 1735

The trust awards grants in Georgia and Union County, NJ, in support of higher and secondary education, hospitals and health organizations, historical associations, environmental conservation, and social welfare. Types of support include capital campaigns, building construction/renovation, equipment acquisition, and endowment funds. Groups with low administrative costs, that have outside funding, and that encourage self-help are preferred. Grants are awarded in May and November. Written proposals should not be longer than four pages.

Restrictions: Grants are not awarded to individuals or for operating expenses, current scholarship funds, conferences, or workshops.

Geographic Focus: Georgia

Date(s) Application is Due: Apr 20; Oct 15

Amount of Grant: 5,000 - 20,000 U.S.D

Contact: William Engel, Executive Director; (908) 753-2440

Sponsor: E.J. Grassmann Trust

P.O. Box 4470

Warren, NJ 07059

E.L. Wiegand Foundation Grants 1736

The foundation provides grants to develop and strengthen programs and projects at Arizona, California, the District of Columbia, Idaho, Nevada, New York, Oregon, Utah, and Washington educational institutions in the academic areas of science, business, fine arts, and law; and medicine and health organizations in the areas of heart, eye, and cancer surgery, treatment, and research. The foundation also considers requests for projects that enrich children, communities, public policy, and art. Grants for education, including funds for computers and scientific equipment, are awarded. The board of trustees meets in February, June, and October to choose recipients, but applications may be submitted at any time. Application guidelines are available upon request.

Requirements: Nonprofit organizations in Arizona, California, District of Columbia, Idaho, Nevada, New York, Oregon, Utah, and Washington State are eligible.

Geographic Focus: Arizona, California, District of Columbia, Idaho, Nevada, New York, Oregon, Utah, Washington

Amount of Grant: 10,000 - 200,000 U.S.D

Samples: Roman Catholic Diocese of Reno (NV)—to help build a new dining hall at a planned complex several blocks east of the downtown casino district that will provide services to homeless people, $1 million; Saint Bonaventure U, School of Business (NY)—to develop internships at 10 local nonprofit organizations for business students, $43,000.

Contact: Kristen Avansino, Executive Director; (775) 333-0310; fax (775) 333-0314

Sponsor: E.L. Wiegand Foundation

165 W Liberty Street, Suite 200

Reno, NV 89501

E. Rhodes and Leona B. Carpenter Foundation Grants 1737

The Foundation considers grant requests from: public charities which had direct relationships with Leona or Rhodes Carpenter during their lifetime; requests in support of graduate theological education from U.S. and Canadian organizations which are public charities; public charities in communities where Carpenter Co. has had long-time manufacturing facilities; museums which have a permanent collection for the purchase, restoration and conservation of Asian art; charities involved in providing hospice care; and projects (e.g., programs, conferences, resources, etc.) offering support to lesbian, gay, bisexual and transgender persons of faith, or endeavoring to insure faith communities' understanding, affirmation, and inclusion of such persons.

Requirements: IRS 501(c)3 organizations are eligible.

Restrictions: Generally, the Foundation will not consider grant requests to support private secondary education, individuals or large public charities and will not transfer funds from its endowment to the endowment of another organization.

Date(s) Application is Due: Jan 31; Jul 31

Samples: Norton Museum of Art (West Palm Beach, FL)—for an exhibition entitled Masterworks of Chinese Painting: In Pursuit of Mists and Cloud, $25,000; Historic Deerfield (MA)—to conserve its collection of Chinese export art, $30,000.

Contact: Grants Administrator; (215) 979-3221

Sponsor: E. Rhodes and Leona B. Carpenter Foundation

1735 Marker Street, Suite 3420

Philadelphia, PA 19103

Eastern Bank Charitable Foundation Grants 1738

The goal of the foundation is to contribute to the health and vitality of northeastern Massachusetts communities served by Eastern Bank. The program provides financial support to nonprofit medical, cultural, educational, community service, or voluntary relief organizations operating within communities served by the bank. Types of support include capital grants, demonstration grants, development grants, endowments, general operating grants, and program grants. Requests for major gifts ($10,000-$25,000) must be submitted in the Associated Grantmakers of Massachusetts Application Form plus Eastern Bank donation request summary. Listed application deadlines are for major grants; there are no deadlines for community grants (less than $10,000). Guidelines are available online.

Requirements: Nonprofit organizations operating in northeastern Massachusetts communities served by Eastern Bank are eligible for funding.

Restrictions: Recipients of major gifts generally will be considered ineligible to reapply for a major gift for a period of three years.

Geographic Focus: Massachusetts

Date(s) Application is Due: May 1; Nov 1

Amount of Grant: 100 - 25,000 U.S.D

Samples: Special Children's Outing Committee (MA)—charitable support, $1000; Northeastern U (MA)—pledge payment, $2000; Visiting Nurse Assoc Foundation (MA)—program support, $1000.
Contact: Laura Kurzrok, Foundation Coordinator; (781) 598-7888; fax (978) 740-6329; Lkurzrok@easternbk.com
Internet: http://www.easternbank.com/a_charitable_foundation.html
Sponsor: Eastern Bank Charitable Foundation
1 Eastern Place, 195 Market Street
Lynn, MA 01901

Eastman Chemical Company Foundation Grants 1739
The Foundation supports organizations and programs that help to promote its efforts of enhancing the quality of life in Eastman communities. Contributions are concentrated on the following areas: education, health and human services, civic and community, and culture and the arts. Grants may be considered outside of these areas.
Requirements: Contributions are generally restricted to organizations which have been granted 501(c)3 tax exempt status and are usually directed toward the communities in which the company has operating units.
Restrictions: Contributions will not be made to the following: individuals, athletic teams or sports related events, choirs, bands, drill teams, labor, veteran, fraternal, social or political organizations; individual agencies supported by the area United Way organization, except capital fund drives which are normally supported by the community, not the United Way; organizations that discriminate on the basis of race, color, sex or national origin; national organizations whose programs do not directly serve the region's needs; places of worship and institutions devoted solely to religious instruction; travel related, including student trips or tours.
Geographic Focus: Pennsylvania, South Carolina, Tennessee, Texas
Contact: CeeGee McCord, Contributions Manager; (423) 229-2000; fax (423) 229-6974; CeeGeeMcCord@kingsporthousing.org
Internet: http://www.eastman.com/Company/Sustainability/communities/Philanthropy/Pages/Philanthropy.aspx
Sponsor: Eastman Chemical Company
P.O. Box 431
Kingsport, TN 37662

East Stroudsburg University Graduate Assistantships 1740
Graduate assistantships are available in a variety of academic disciplines including pedagogy, sciences (biology, computer science, and general), social sciences (history and political science), and health sciences and human performance (public health, community health, movement studies, athletic training, cardiac rehabilitation, and exercise sciences). Stipends are based on 10-hour and 20-hour assignments with tuition waiver for the academic year. Minorities and women are strongly encouraged to apply.
Requirements: A minimum 2.5 overall GPA and 3.0 GPA in applicant's major field are required.
Date(s) Application is Due: Jun 15
Amount of Grant: 2,500 - 5,000 U.S.D
Contact: Program Office; (866) 837-6130 or (570) 422-3536; fax (570) 422-3506; grad@po-box.esu.edu
Internet: http://www.esu.edu/grants/gradcat/newasst.shtml
Sponsor: East Stroudsburg University
200 Prospect Street
East Stroudsburg, PA 18301-2999

eBay Foundation Community Grants 1741
The mission of the Foundation is to make investments that improve the economic and social well-being of local communities. The Foundation works to fulfill its mission by collaborating with non-profit organizations and funding innovative programs primarily in microenterprise development. The Foundation also provides support to community organizations in areas where employees are located.
Requirements: 501(c)3 nonprofit organizations in communities where eBay has a major employment base, which includes San Jose, CA, and Salt Lake City, UT, are eligible.
Geographic Focus: California, Utah
Amount of Grant: 1,000 - 15,000 U.S.D
Samples: Friends of Farm Drive (San Jose, CA)—for operating support of this organization, which is dedicated to improving the quality of life for residents of a neighborhood rampant with drug dealing and gang activity; U Research Expedition Program (U of California)—for scholarships to send California elementary and high school teachers on UREP trips, where they learn by working with university scientists and local residents in countries around the world, to increase awareness of current events and to foster understanding of different cultures among both students and teachers.
Contact: Grants Administrator; ebayfdn@cfsv.org
Internet: http://pages.ebay.com/community/aboutebay/foundation/grantapp.html
Sponsor: eBay Foundation
60 South Market Street, Suite 1000
San Jose, CA 95113

Eberly Foundation Grants 1742
The foundation awards grants to eligible nonprofit organizations in the areas of: undergraduate educational opportunities; arts programming and institutions; health and human services projects; primary and secondary supplemental educational programming; and economic development activities and organizations that benefit the residents of Fayette County. The initial approach should be submittal of a letter of request.
Requirements: Nonprofit organizations in Pennsylvania are eligible.
Geographic Focus: Pennsylvania
Samples: Pennsylvania State U (University Park, PA)—to construct a community center on the Fayette campus that will house performance spaces, a cafeteria and dining hall, a fitness center, an auditorium, and other features needed for large gatherings, $6.9 million; Indiana U of Pennsylvania, College of Business (PA)—to support technology initiatives, $599,000 challenge grant; Western Pennsylvania Conservancy (Pittsburgh, PA)—for restoration and protection of Fallingwater, $1 million.
Contact: Carolyn Blaney, President; (724) 438-3789; fax (724) 438-3856
Sponsor: Eberly Foundation
2 West Main Street, Suite 101
Uniontown, PA 15401-3448

Echoing Green Fellowships 1743
The program awards full-time fellowships to emerging entrepreneurs to create innovative domestic or international public service projects that seek to catalyze positive social change. The proposed project may be in any public service area, including but not limited to, the environment, arts, education, health, youth service and development, civil and human rights, and community and economic development. The fellowship provides a two-year stipend, health care benefits, online connectivity, access to Echoing Green's network of social entrepreneurs, training, and technical assistance.
Requirements: Applicants must be at least 18 years old and commit to leading the project for at least two years. Partnerships of up to two individuals also are eligible.
Restrictions: Faith-based, research projects, and lobbying activities are not eligible.
Date(s) Application is Due: Dec 2
Amount of Grant: 60,000 - 90,000 U.S.D
Contact: Rich Leimsider, Director of Fellow and Alumni Programs; (212) 689-1165; fax (212) 689-9010; Rich@echoinggreen.org or apply@echoinggreen.org
Internet: http://www.echoinggreen.org/fellowship
Sponsor: Echoing Green Foundation
494 Eighth Avenue, Second Floor
New York, NY 10001

Eckerd Corporation Foundation Grants 1744
The Foundation's philanthropic mission is to monetarily support nonprofit organizations that directly benefit health care, education, women and children through programs and events. Funding proposals are submitted to the board for review on a quarterly basis.
Requirements: Nonprofit organizations in company operating areas are eligible.
Contact: Tami Alderman, Public Relations; (727) 395-7091; fax (727) 395-7934; service@eckerd.com
Internet: http://www.eckerd.com/content.asp?content=company/news/presscontact
Sponsor: Eckerd Corporation
8333 Bryan Dairy Road
Largo, FL 33777

Eddie C. and Sylvia Brown Family Foundation Grants 1745
The Brown Family Foundation will focus its grant making in three general areas: the arts, education, and health (HIV/AIDS and cancer) The foundation's grant making will support programs that: provide opportunities for youth involvement in meaningful art experiences—this can include support for arts organizations or youth serving organizations that have developed programs which expose young people to the arts (programs that infuse the arts as a tool for leadership development, academic enrichment, and are built on strong youth development principles are encouraged to apply); provide access to educational opportunities for disadvantaged children and youth—this can include local organizations that work to improve PreK-12 public education and to offer enhancement and enrichment opportunities to Baltimore City students (educational opportunities are defined in the broadest sense to include, but are not limited to afterschool programs, mentoring, alternative education, computer technology, scholarship support, as well as traditional in-school K-12 programs); address the impact of HIV/AIDS on the African American community—programs that provide direct services and intervention including testing and counseling, case management strategies, and access to treatment will be considered; and address the prevention and treatment of cancer—this can include local organizations that provide cancer screening, prevention education, and treatment modalities.
Requirements: Organizations (or their fiscal agents) serving the Baltimore area that qualify as public charities under section 501(c)3 of the Internal Revenue Code and do not discriminate on the basis of race, creed, national origin, color, physical handicap, gender or sexual orientation can apply.
Restrictions: The Foundation does not make grants for: start-up programs; capital campaigns; individuals; multiple years; institutions of higher education; and organizations outside the Baltimore region.
Geographic Focus: Maryland
Date(s) Application is Due: Sep 26
Contact: Maya Smith, Program Administrative Assistant; (410) 332-4172, ext. 142; fax (410) 837-4701; msmith@bcf.org
Internet: http://www.bcf.org/ourgrants/ourgrantsdetail.aspx?grid=2
Sponsor: Eddie C. and Sylvia Brown Family Foundation
2 East Read Street, 9th Floor
Baltimore, MD 21202

Eden Hall Foundation Grants 1746
Eden Hall Foundation is a private foundation established pursuant to the will of Sebastian Mueller, a Pittsburgh philanthropist and vice president and director of the H. J. Heinz Company. During his lifetime, Mr. Mueller gave substantial support to improve conditions of the poor and disadvantaged, the promotion of sound education, and the support of

health facilities and projects. Today, the trustees of Eden Hall Foundation continue his stewardship in the areas of social welfare, health, education and the arts. Eden Hall Foundation seeks to improve the quality of life in Pittsburgh and western Pennsylvania through support of organizations whose missions address the needs and concerns of the area. Proposals and projects of primary interest to the Foundation are: educational programs dedicated to the advancement and dissemination of useful knowledge. Support of schools is generally confined to four-year privately funded and controlled colleges, universities and other educational institutions; social welfare, women's issues, and the improvement of conditions of the poor and disadvantaged; and the advancement of better health through support of organizations dedicated to health issues. Eden Hall Foundation will entertain grant proposals for capital projects, scholarship funds, research projects, programming and, in limited cases, endowments directed toward these objectives. A specific application form is not required. Therefore, grant requests should be submitted by letter proposal to the Foundation. Acknowledgements will be made promptly, advising applicants whether or not their requests meet Eden Hall Foundation's criteria.

Requirements: Requests for grants should include: a brief statement of the background and purpose of the requesting organization; a copy of that organization's most recent financial statements, as prepared by its certified public accountants; the specific purpose for which the requested grant is to be used, including a detailed budget, timetable for implementation and proposed method of evaluation; IRA certification of tax-exempt and charitable status under sections 501(c)3 and 509(c) of the Internal Revenue Code; information concerning the status of grants solicited from other foundations in the past twelve-month period (indicating the name of each foundation and amount requested); and a listing of the organization's officers and directors, their primary occupations and the responsibilities to the organization.

Restrictions: Grants cannot be made to individuals or private foundations. Not eligible for funding are sectarian or denominational religious organizations, except those providing direct educational or health care services to the public, tax-supported organizations, political parties and fraternal organizations. Requests to cover operating expenses, endowments or accumulated deficits are discouraged.

Geographic Focus: Pennsylvania

Contact: Sylvia Fields, Executive Director; (412) 642-6697; fax (412) 642-6698

Internet: http://www.edenhallfdn.org

Sponsor: Eden Hall Foundation

600 Grant Street, Suite 3232

Pittsburgh, PA 15219

Edina Realty Foundation Grants **1747**

The Foundation extends financial support to organizations primarily in Minnesota, which provide housing and related services such as counseling and medical care to homeless children, families and individuals. Application forms may be submitted at anytime, see Foundations website for Grant Application forms: http://www.edinarealty.com/Content/Content.aspx?ContentID=148855

Requirements: Nonprofits 501(C)(03) operating in Minnesota are eligible.

Geographic Focus: Minnesota

Amount of Grant: 500 - 2,000 U.S.D

Samples: Home of the Good Shepherd, MN, $15,000 - operates two transitional housing programs in St. Paul for homeless and at-risk women; Spare Key of Minnesota, MN, $13,000 - provides mortgage assistance to Minnesota homeowners with critically ill or seriously injured children by making a mortgage payment directly to the lender on behalf of the family;

Contact: Susan Cowsert, Director; (952) 928-5900; susancowsert@edinarealty.com

Internet: http://www.edinarealty.com/Content/Content.aspx?ContentID=187015

Sponsor: Edina Realty Foundation

6800 France Avenue South, Suite 670

Edina, MN 55435-2017

Edna G. Kynett Memorial Foundation Grants **1748**

The purpose of the Foundation is to improve cardiovascular health. Grants support programs and services that educate primary care practitioners and community members and promote cardiovascular health. Grants are made to medical schools, hospitals and other nonprofit health organizations. Programs that have received support in recent years include innovative curricula, lectures and seminars, a tobacco prevention program, a summer Community Health Internship program and clinical research projects. Innovative projects are preferred.

Requirements: Applicants must be nonprofit organizations in the Delaware Valley, including Philadelphia, Bucks, Delaware, Chester, Montgomery, Berks, Lehigh and Camden counties and New Castle County, Delaware. Applicants must be either public entities or tax-exempt under Section 501(c)3 of the Internal Revenue Code and not private foundations as defined under Section 509(a).

Geographic Focus: Pennsylvania

Amount of Grant: Up to 50,000 U.S.D

Samples: Penn Presbyterian Medical Center, Philadelphia, PA, $5,000 - Cardiology update; Drexel University College of Medicine, Philadelphia, PA, $40,000 - Cardiovascular CME programs; FOCU.S. on Health and Leadership for Women (Trustees of the University of Pennsylvania) Philadelphia, PA, $39,000 - Junior Faculty Investigators and Medical Student Fellowship in cardiovascular health.

Contact: Judith L. Bardes, Manager; (610) 828-8145; fax (610) 834-8175; judyl@aol.com

Sponsor: Edna G. Kynett Memorial Foundation

P.O. Box 540

Plymouth Meeting, PA 19462-0540

EDS Foundation Grants **1749**

The Foundation wishes to ensure information technology champions cultural and civic change which will enrich the education of current and future generations, while enhancing the communities they serve. The Foundation is interested in supporting comprehensive technology solutions that increase performance and productivity in educational institutions and community organizations globally. Grant seekers must complete the EDS Foundation Application for Funding and have a current EDS employee volunteer partnership.

Requirements: Eligible applicants must be verified as a non-profit charitable organization (or equivalent) according to each country's rules and regulations. For the U.S., the organization must have a 501(c)3 IRS tax-exempt code.

Restrictions: Funding will not be provided for: individuals; operating deficits; foundations that are grant-making institutions; sponsorships; trips or tours; local athletic teams or events; fraternal, social or labor organizations; political or partisan organizations; journal or program advertising; private foundations; or organizations that do not comply with all applicable laws prohibiting discrimination and EDS Equal Opportunity policy.

Geographic Focus: California, Colorado, Georgia, Illinois, Michigan, Texas, Virginia

Contact: Diane Spradlin, Executive Director; (972) 605-8429; diane.spradlin-eds@eds.com

Internet: http://www.eds.com

Sponsor: EDS Foundation

5400 Legacy Drive, H3-6F-47

Plano, TX 75024

Educational Foundation of America Grants **1750**

The foundation makes grants only for specific projects in its areas of interest, including, but not limited to, the environment, reproductive freedom, theatre, education, medicine, drug policy reform, democracy, peace & national security issues and human services. Important characteristics considered by EFA are an organization's record of achievement, intended broad impact, sound financial practices, increasing independence, and correspondence with EFA objectives.

Requirements: U.S. nonprofits are eligible for grant support. Applicants are required to send the EFA Inquiry Form via email as the first step.

Restrictions: Grants are not made for endowment or endowed faculty chairs, building/capital programs, religious purposes, grants to individuals, annual fund-raising campaigns, indirect costs, overhead or general support. The Foundation prefers not to fund projects located outside the U.S..

Contact: Diane Allison, Executive Director; (203) 226-6498; fax dallison@efaw.org

Internet: http://www.efaw.org/Inquiry%20Guidelines.htm

Sponsor: Educational Foundation of America

35 Church Lane

Westport, CT 06880

Education Network to Advance Cancer Clinical Trials Grants **1751**

The goal of the program (PEP), which is made possible by the Lance Armstrong Foundation, is to demonstrate how a community-driven outreach and education program can increase awareness about cancer clinical trials, enhance their acceptability, and improve access to them. Grants will provide financial support, training, and technical assistance to three partnerships as they meeting the following objectives: change knowledge, attitudes, and beliefs about cancer clinical trials among healthcare providers and community leaders; enhance the acceptability of cancer clinical trials as a treatment option among community members; increase the number of inquiries about cancer treatment clinical trials; increase the number of referrals for cancer treatment clinical trials; and reduce local barriers to access cancer treatment clinical trials. Preliminary application form and guidelines are available online. The listed application deadline is for preliminary applications.

Requirements: Partnerships must have a track record of at least three years of collaborative efforts to address health issues, with at least one of those issues related to cancer; be led by a broad-based group that represents the community; have a membership that includes, in roughly equal proportion, people affected by a disease or health condition the partnership has addressed, healthcare professionals, and leaders of community-based organizations that work on health, economic, housing, labor, or other social justice issues; be located in an area(s) that already has a cancer treatment clinical trial infrastructure; be a 501(c)3 organization or have a fiscal sponsor; and be based in the United States or one of its territories.

Date(s) Application is Due: Jul 11

Amount of Grant: 150,000 U.S.D

Contact: Grants Administrator; (301) 562-2774; info@enacct.org

Internet: http://www.enacct.org/pep.htm

Sponsor: Education Network to Advance Cancer Clinical Trials

1010 Wayne Avenue, Suite 770

Silver Spring, MD 20910

Edward and Helen Bartlett Foundation Grants **1752**

Established by a single donor, Edward E, Bartlett, in Oklahoma in 1961, the Foundation primarily offers grant support for: education, particularly public schools; community programs and services; health care; children and youth; and social services. There are no specific application forms or deadlines with which to adhere, and applicants should begin by forwarding a letter of application to the contact listed. In the recent past, grant amounts have ranged between $5,000 to $150,000.

Requirements: Preference id given to non-profit 501(c)3 organizations located in, or serving the residents of, Oklahoma.

Restrictions: No grants are to individuals directly.

Geographic Focus: Oklahoma

Amount of Grant: 5,000 - 150,000 U.S.D

Samples: Clarehouse, Tulsa, Oklahoma, $150,000; Creek County Literacy Program, Sapulpa, Oklahoma, $30,000; Retired Senior Volunteer Program of Tulsa, Tulsa, Oklahoma, $20,000.
Contact: Bruce A. Currie, (918) 586-5273
Sponsor: Edward and Helen Bartlett Foundation
P.O. Box 3038
Milwaukee, WI 53201-3038

Edward Bangs Kelley and Elza Kelley Foundation Grants **1753**
The Foundation's interest is primarily in Barnstable County, Massachusetts. The Foundation has been the leader in improving the health and welfare of the community. Grants are made to a great variety of health, social and human service agencies, as well as to cultural and environmental organizations. Town libraries, theatre and art groups, and musical organizations have been supported. Grants are sometimes utilized as seed money by young organizations.
Requirements: Grant applicants must be tax-exempt organizations which are not private foundations and must be located in Barnstable County. The proposed project/program must have a direct benefit to the inhabitants of Barnstable County. If the tax-exempt organization is located outside of Barnstable, but the proposed project/program will have a direct and substantial benefit to the inhabitants of Barnstable, the grant application may be eligible for funding.
Restrictions: Applicants must be residents of Barnstable County, Massachusetts, or must demonstrate very significant ties to Barnstable County.
Geographic Focus: Massachusetts
Contact: Henry Murphy Jr., Administrative Manager; (508) 775-3117; contact@kelleyfoundation.org
Internet: http://www.kelleyfoundation.org
Sponsor: Edward Bangs Kelley and Elza Kelley Foundation
243 South Street, P.O. Drawer M
Hyannis, MA 02601

Edward F. Redies Foundation Grants **1754**
The foundation awards grants to eligible Michigan nonprofit organizations in its areas of interest, including community services, education, health, and social services. Grants are awarded for one year and are renewable. Submit a copy of the IRS tax-determination letter with application.
Requirements: Michigan 501(c)3 organizations in Washtenaw County are eligible.
Geographic Focus: Michigan
Date(s) Application is Due: Mar 30
Amount of Grant: 1,000 - 50,000 U.S.D
Contact: R&B Machine Tool Company; (734) 429-9421; fax (734) 429-4965
Sponsor: Edward F. Redies Foundation
P.O. Box 411
Saline, MI 48176

Edward N. and Della L. Thome Memorial Foundation Awards in AMD Research **1755**
The Edward N. and Della L. Thome Memorial Foundation was created in 2002 to advance the health of older adults through the support of direct-service projects and medical research on diseases and disorders affecting older adults. In keeping with the foundation's mission, the goal of the research program is to support translational research that will lead to improved therapies for individuals suffering from age-related macular degeneration (AMD). Successful research proposals will extend recent basic research findings regarding the underlying mechanisms of AMD. Examples of funding areas include, but are not limited to, genetically-engineered animal models, the discovery and testing of small molecule therapies directed at promising targets including immune or inflammation related pathways associated with AMD, local drug delivery systems, and neuroprotective strategies. Basic research and clinical trials are outside the scope of this program. However, clinical studies that involve the evaluation of new imaging modalities with no potential risks to human subjects are eligible. Three-year awards of up to $750,000 ($250,000 per year) will be made to independent investigators working in nonprofit academic, medical, or research institutions within the United States (all awards inclusive of 10% indirect costs). The application deadline for the initial proposal is June 15, 12:00 Noon, U.S. Eastern Time. Applications must be submitted through the online application system. Downloadable application guidelines, instructions, and other details are available at the grant website. Applicants invited to submit a full proposal will be notified in early August. Applicants will be notified of grant decisions by December 15. This grant has a biennial cycle and alternates with the Edward N. and Della L. Thome Memorial Foundation, Bank of America, N.A., Trustee, Awards in Alzheimer's Disease (AD) Drug Discovery Research. Prospective applicants are advised to check the Medical Foundation website to verify current deadline dates for these grants. Organizations interested in applying for funding through the Edward N. and Della L. Thome Memorial Foundation's direct services program should go to the Bank of America's philanthropic website (https://www.bankofamerica.com/philanthropic/index.action).
Requirements: All applicants must hold a faculty appointment at a nonprofit academic, medical, or research institution in the United States. United States citizenship is not required. Eligibility is not limited to those investigators currently working in age-related macular degeneration research; scientists who have conducted research exploring the biologic causes of related disorders and/or similar translational research programs are encouraged to apply. Applicants do not need to be nominated by their institutions. Preference will be given to originality of ideas, regardless of faculty seniority. The research proposal must clarify how the study may improve current therapeutic strategies or reveal new therapeutic approaches.
Date(s) Application is Due: Jun 15
Amount of Grant: Up to 750,000 U.S.D
Samples: Rajendra Apte, M.D., Ph.D., Washington University School of Medicine, St. Louis, Missouri, "The Role of Cholesterol in Regulating the Pro-Angiogenic Properties f Senescent Macrophages in Age-Related Macular Degeneration"; Peter Campochiaro, M.D., Johns Hopkins University School of Medicine, Baltimore, Maryland, "Sustained Delivery of Antiangiogenic Peptides for Neovascular Age-Related Macular Degeneration"; Constance Cepko, Ph.D., Howard Hughes Medical Institute, Chevy Chase, Maryland, "Promotion of Photoreceptor Survival using HDACs and Gluconeogenic Genes".
Contact: Linda Lam, Program Officer; (617) 279-2240, ext. 710; LLam@hria.org
George Thorn, Vice President; (312) 828-4154; ilgrantmaking@bankofamerica.com
Internet: http://hria.org/tmfservices/tmfgrants/thome.html
Sponsor: Medical Foundation, a Division of Health Resources in Action
95 Berkeley Street, Suite 208
Boston, MA 02116

Edward N. and Della L. Thome Memorial Foundation Direct Services Grants **1756**
The Edward N. and Della L. Thome Memorial Foundation was established in 2002 by Robert P. Thome to honor the memory of his parents, Edward and Della Thome. The mission of the foundation is twofold: to advance the health of older adults through (1) the support of direct service projects and (2) medical research on diseases and disorders affecting older adults. The goal of the foundation's direct services program is to support organizations in Maryland and Michigan that provide direct services addressing one or more of the following critical issues facing older adults: health care, housing, family services, neighborhood involvement, workforce opportunities, and aging with dignity at home. Grant requests for general operating support are strongly encouraged. Program support will also be considered. Program-related capital expenses may be included in general operating or program requests. To better support the capacity of nonprofit organizations, multi-year funding requests are encouraged. Applicants must apply online at the direct-services grant website above. Applicants are strongly encouraged to do the following before applying: review the downloadable state application procedures for additional helpful information and clarifications; review the downloadable online-application guidelines at the grant website; review the foundation's funding history (link is available from the grant website); review the online application questions in advance; and review the list of required attachments. These will generally include: a list of board members, financial statements (audited, reviewed, or compiled by independent auditor); an organization summary; a list of other funding sources; an IRS Determination letter; and other required documents. All attachments must be uploaded in the online application as PDF, Word, or Excel files. There is a rolling deadline for organizations applying for a direct services grant from the Thome Foundation. Applicants will generally be notified of grant decisions within six months of proposal submission. The research aspect of the foundation's mission is managed by the Medical Foundation, a Division of Health Resources in Action (HRIA). Applicants interested in applying for research funding through the Edward N. and Della L. Thome Memorial Foundation's research program will find application information at the Medical Foundation's website.
Requirements: The Thome Foundation direct services program is most interested in expanding services for older adults. Requests should clearly state how many previously-unserved older adults will now be served as a result of funding from the Thome Foundation. Applicants must have 501(c)3 tax-exempt status.
Restrictions: The foundation does not support requests from individuals, organizations attempting to influence policy through direct lobbying, or any political campaigns.
Geographic Focus: Maryland, Michigan
Contact: George Thorn; (312) 828-4154; ilgrantmaking@bankofamerica.com
Internet: https://www.bankofamerica.com/philanthropic/fn_search.action
Sponsor: Edward N. and Della L. Thome Memorial Foundation - BAML
231 South LaSalle Street, IL1-231-13-32
Chicago, IL 60604

Edward S. Moore Foundation Grants **1757**
The foundation awards grants to nonprofits in Connecticut and New York in its areas of interest, including youth, hospitals, education, cultural programs, museums, and Christian religion. Types of support include operating budgets, continuing support, annual campaigns, seed money, emergency funds, building funds, equipment, land acquisition, endowment funds, matching funds, internships, scholarship funds, special projects, and research. There are no application deadlines. The board meets in January, April, July, and October to consider proposals.
Requirements: Nonprofits in Connecticut and New York may submit proposals.
Restrictions: Grants are not awarded to individuals or for deficit financing, publications, or conferences.
Geographic Focus: Connecticut, New York
Amount of Grant: 10,000 - 50,000 U.S.D
Samples: Children's Aid Society (New York, NY)—$50,000; Metropolitan Opera (New York, NY)—$35,000; Stamford Hospital Foundation (Stamford, CT)—$430,000.
Contact: John W. Cross III, President; (203) 629-4591
Sponsor: Edward S. Moore Foundation
30 Lismore Lane
Greenwich, CT 06831

Edwards Memorial Trust Grants **1758**
The trust awards grants to eligible Minnesota nonprofit organizations in support of health care for people without health insurance or who are underinsured, preventive health care for children, and programs for the disabled. Areas of support include health care and hospitals, mental health crisis services, social services, children and youth, and the disabled. Types of support include building construction/renovation, equipment acquisition, general

operating grants, and program development. A copy of the tax determination letter and most recent audited financial statements must accompany applications.
Requirements: Minnesota 501(c)3 tax-exempt organizations in the greater Saint Paul area are eligible.
Geographic Focus: Minnesota
Date(s) Application is Due: May 1; Nov 1
Amount of Grant: 2,000 - 50,000 U.S.D
Contact: Cheryl Nelson, c/o U.S. Bank; (651) 466-8441; fax (651) 244-4267
Sponsor: Edwards Memorial Trust
P.O. Box 64713
Saint Paul, MN 55164-0713

Edward W. and Stella C. Van Houten Memorial Fund Grants **1759**
Stella C. Van Houten resided in Bergen County, New Jersey. This foundation, providing funding for health and human services, education, education of medical professionals, and the care of children, was established in 1978 in memory of her husband and herself. The Van Houten's had a particular fondness for the Valley Hospital of Ridgewood, New Jersey and for the Rollins College in Florida. The Foundation continues to honor their preferences with grants to these two organizations in addition to grants to other organizations. The Foundation's mission is to: supports agencies, institutions and services in Passaic and Bergen Counties, New Jersey, having to do with the care or cure of sick or disabled persons or for the care of orphaned children or aged persons; educates students in the medical profession; support for educational purposes; support for the care of children. A target of 10% of the grants each year is for medical scholarships.
Requirements: Passaic and Bergen Counties, New Jersey non-profits are eligible to apply. The application form & guidelines are available online at the Wachovia website. The applications must be submitted by January 31 for a March meeting & August 1 for an October meeting.
Geographic Focus: New Jersey
Date(s) Application is Due: Jan 31; Aug 1
Amount of Grant: 6,000 - 100,000 U.S.D
Contact: Trustee, c/o Wachovia Bank; grantinquiries2@wachovia.com
Internet: https://www.wachovia.com/foundation/v/index.jsp?vgnextoid=00c78689fb0aa110VgnVCM1000004b0d1872RCRD&vgnextfmt=default
Sponsor: Edward W. and Suitella C. Van Houten Memorial Fund
190 River Road, NJ3132
Summit, NJ 07901

Edwin S. Webster Foundation Grants **1760**
The policy of the foundation is to support charitable organizations that are well known to the trustees, with emphasis on special projects and capital programs, or operating income for hospitals, medical research, education, youth agencies, cultural activities, and programs addressing the needs of minorities. Types of support include operating budgets, continuing support, annual campaigns, building funds, equipment, land acquisition, endowment funds, matching funds, scholarship funds, professorships, internships, fellowships, special projects, and research.
Requirements: The Foundation confines its grants primarily to the New England area. Grantees must provide evidence of their tax-exempt status. The AGM common proposal format, available on the Internet at http://agmconnect.org is suitable for submission of proposals but not required. There are no set deadlines, but for consideration at the spring meeting, proposals should arrive by May 1 and by November 1 for consideration at the fall meeting. Proposals received after the trustees meet will be held for consideration at the next meeting.
Restrictions: Grants are not made to organizations outside the United States or to individuals.
Geographic Focus: Massachusetts
Date(s) Application is Due: May 1; Nov 1
Amount of Grant: 15,000 - 50,000 U.S.D
Samples: Massachusetts Institute of Technology, Office of Minority Education, Cambridge, MA, $40,000 - Mentor Advocate Partnership; Eastern Virginia Medical School, Norfolk, VA, $25,000 - general operations and interdisciplinary research; Museum of Science, Boston, MA, $50,000 - capital support for the Hayden Planetarium.
Contact: Michelle Jenney, Administrator; (617) 391-3087; fax (617) 426-7080; mjenney@gmafoundations.com
Sponsor: Edwin S. Webster Foundation
GMA Foundations, 77 Summer Street, 8th Floor
Boston, MA 02110-1006

Edwin W. and Catherine M. Davis Foundation Grants **1761**
The foundation awards grants to U.S. nonprofit organizations in its areas of interest, including arts, elderly, environment, higher education, housing, mental health, religion, social services,and youth. Types of support include annual campaigns, endowment funds, fellowships, operating grants, research grants, and scholarship funds. There are no application deadlines. The board meets in May or June; submit a proposal that is three pages or less in length.
Requirements: U.S. nonprofit organizations are eligible.
Geographic Focus: Washington
Amount of Grant: 1,000 - 10,000 U.S.D
Contact: Gayle Roth, Grants Administrator; (651) 215-4408; fax (651) 228-0776
Sponsor: Edwin W. and Catherine M. Davis Foundation
30 East 7th Street, Suite 2000
Saint Paul, MN 55101-1394

Edyth Bush Charitable Foundation Grants **1762**
The Foundation is operated exclusively for charitable, religious, literary, and other exempt purposes. Grants are available for challenge and development purposes, construction and renovation, equipment and expansion of functions, pilot projects and seed start-up of new programs, and study or planning grants. The Foundation has broad interests in human service, education, health care and a limited interest in the arts. Requests for grants or other funds should be submitted in writing. Contact the Program Officer before beginning any proposals or applications.
Requirements: The Foundation welcomes grant requests from otherwise eligible tax-exempt organizations under IRS Sections 501(c)3 and Section 509(a) headquartered within the Orlando MSA of Orange, Seminole, and Osceola Counties, Florida. Other grant requests should have special interest or support from one or more of our Directors.
Restrictions: The Foundation will ordinarily deny grant requests: from chiefly tax-supported institutions, or their support foundations; for individual scholarships or for individual research grants even if through an exempt or otherwise qualified educational organization; for alcoholism or drug abuse programs or facilities; for routine operating expenses; to pay off deficits or pre-existing debt; for foreign organizations or for foreign expenditure; for travel projects or fellowships; for chiefly church, sacramental, denominational or inter-denominational purposes, except outreach projects for elderly, indigents, needy, youth, or homeless regardless of belief, race, color, creed, or sex; for endowment funds or other purely revenue generating funds; advocacy organizations or advocacy component funding; for cultural or arts organizations unless their collections, exhibits, projects or performances are of demonstrated nationally recognized quality; from organizations having receipts of revenues from memberships and/or contributions of less than $25,000 in the previous year, or from any organization whose IRS Sec. 509(a) publicly supported status will need renewal in the next six (6) months.
Geographic Focus: Florida
Amount of Grant: 5,000 - 50,000 U.S.D
Samples: Jewish Community Ctr of Greater Orlando (Maitland, FL)—to renovate its playground for toddlers and to help raise additional funds for a project to renovate the elementary school playground, $75,867 partial challenge grant; Community Communications, Inc., WMFE-TV/FM (Orlando, FL)—for hurricane emergency relief effort, $12,000; Holocaust Memorial Resource and Education Center (Maitland, FL)—to support work in addressing intolerance, $2500; Winter Garden Heritage Foundation (Winter Garden , FL)—for leadership training, $350.
Contact: Deborah Hessler, Program Officer; (888) 647-4322 or (407) 647-4322; fax (407) 647-7716; dhessler@edythbush.org
Internet: http://www.edythbush.org
Sponsor: Edyth Bush Charitable Foundation
P.O. Box 1967
Winter Park, FL 32790-1967

EF Behavioral Sciences Student Fellowships **1763**
This fellowship stimulates individuals to pursue careers in epilepsy in either research or practice settings. Appropriate fields include sociology, social work, psychology, anthropology, nursing, economics, vocational rehabilitation, counseling, political science, and others relevant to epilepsy research or practice. Annual deadline dates may vary; contact program staff for exact dates.
Requirements: Applicant must be actively enrolled in a degree program at the time the fellowship is to be undertaken. The fellowship must be undertaken during a free period in the student's year, e.g., during the summer. Both graduate and undergraduate students are eligible provided all other eligibility criteria are met. The project must be carried out at an approved facility.
Date(s) Application is Due: Mar 1
Amount of Grant: 3,000 U.S.D
Samples: Cathleen Buchanan, U of Wisconsin, Madison (WI)—for an examination of attention deficit hyperactivity disorder in children with epilepsy; David Sbarra, U of Virginia (VA)—for an assessment of patterns of self-reliance, social and emotional adaptation, and parent-child interactions and how they affect children with epilepsy.
Contact: Fellowships Office; (301) 459-3700; fax (301) 577-2684; grants@efa.org
Internet: http://www.epilepsyfoundation.org/research/grants.cfm
Sponsor: Epilepsy Foundation
4351 Garden City Drive, Suite 406
Landover, MD 20785-7223

Effie and Wofford Cain Foundation Grants **1764**
The foundation gives primarily for higher and secondary education, medical research, and public service organizations. Grants also are awarded to religious organizations (Baptist, Christian, Episcopal, Presbyterian, Salvation Army, and United Methodist), and for aid for the handicapped. Additional fields of interest include elementary and secondary education, early childhood development and education, medical school education, nursing school education, hospitals and general health care and health organizations, religious federated giving programs, government and public administration, African Americans, Latinos, the disabled, the aging, and economically disadvantaged and homeless. Types of support include general operating support, continuing support, annual campaigns, capital campaigns, building/renovations, equipment acquisition, endowment funds, program development, seed money, curriculum development, fellowships, internships, scholarship funds, research, and matching funds. Organizations may reapply for funding every other fiscal year.
Requirements: The foundation only makes grants to 501(c)3 tax-exempt organizations in Texas.
Restrictions: Individuals are ineligible.
Geographic Focus: Texas
Amount of Grant: 1,000 - 100,000 U.S.D

Samples: U of Texas at Austin (TX)—to acquire works of art for its collection, $500,000; Texas College (Tyler, TX)—for its concert choir, $100,000 over four years.
Contact: Lynn Fowler, Executive Director; (512) 346-7490; fax (512) 346-7491
Sponsor: Effie and Wofford Cain Foundation
4131 Spicewood Springs Road, Suite A-1
Austin, TX 78759

EF Health Sciences Student Fellowships **1765**
Three-month fellowships are awarded to medical and health science students for work on epilepsy study projects. The project may be carried out at the U.S. institution of the student's choice where there are ongoing programs of research, training, or service in epilepsy. Application forms and guidelines are available upon request. Annual deadline dates may vary; contact program staff for exact dates.
Requirements: Predoctoral training students in the Health Sciences may be accepted at any point in their schooling-following acceptance but before beginning the first year, or in the period immediately following their final year. A supervisor or preceptor must accept responsibility for the supervision of the student program and for the submission of a full report to the institute upon completion of the program.
Date(s) Application is Due: Mar 1
Amount of Grant: 3,000 U.S.D
Samples: Rosalind Carter Brown, U of Virginia Medical School (VA)—for an examination of the mechanisms of action of diazepam (Valium) in the treatment of status epilepticus, prolonged seizures; Daniel Campbell, Pennsylvania State U College of Medicine (PA)—for an analysis of the role played by one type of calcium-related molecule in a genetic model of epilepsy.
Contact: Fellowships Office; (301) 459-3700; fax (301) 577-2684; grants@efa.org
Internet: http://www.epilepsyfoundation.org/research/grants.cfm
Sponsor: Epilepsy Foundation
4351 Garden City Drive, Suite 406
Landover, MD 20785-7223

EF Making a Difference Award **1766**
The award recognizes women with epilepsy who have contributed in a significant way at the local, regional, or national level to support and advance quality of life for women with epilepsy; or inspire people with epilepsy and those with other chronic disabilities; or enhance public understanding of epilepsy and its impact on women. The winner will receive a monetary award and expense-paid trip to the foundation's national conference, in October. Information and nomination form are available online.
Amount of Grant: 1,500 U.S.D
Contact: Awards Coordinator; (301) 577-4941
Internet: http://www.epilepsyfoundation.org/programs/wei/index.cfm
Sponsor: Epilepsy Foundation
4351 Garden City Drive, Suite 406
Landover, MD 20785-7223

EF Postdoctoral Research Training Fellowships **1767**
The foundation awards grants and fellowships for cutting-edge research into the causes of epilepsy. Program goals include understanding basic mechanisms, developing new therapeutic approaches, understanding the behavioral and psychosocial aspects of having epilepsy, and encouraging the professional development of scientists and healthcare professionals. The purpose of the fellowship is to support predoctoral students with dissertation research related to epilepsy. Applications from all fields of research pertinent to epilepsy will be considered.
Requirements: Physicians or PhD neuroscientists who desire postdoctoral research experience are eligible for these fellowships, which must be carried out at an approvable facility where there is an ongoing epilepsy research program. Applicants must be licensed to practice medicine in the United States or Canada.
Restrictions: Individuals holding faculty appointments at the level of assistant professor or higher are ineligible.
Date(s) Application is Due: Sep 1
Amount of Grant: Up to 40,000 U.S.D
Samples: Dr. Jon Backstrom, Vanderbilt U School of Medicine (TN)—for an investigation of the anatomical interaction between GABA and seratonin, and its effect on seizures; Dr. Sarah Caddick, Duke U Medical Ctr (NC)—for an investigation of the function of zinc in certain systems of the brain that exhibit high concentrations of ion zinc and are highly prone to seizures.
Contact: Research and Professional Education Office; (301) 459-3700; fax (301) 577-2684; grants@efa.org
Internet: http://www.epilepsyfoundation.org/research/grants.html
Sponsor: Epilepsy Foundation
4351 Garden City Drive, Suite 406
Landover, MD 20785-7223

EF Predoctoral Research Training Fellowship **1768**
The predoctoral research training fellowship supports predoctoral students with dissertation research related to epilepsy, thus strengthening their interest in establishing epilepsy research as a career direction. Annual deadline dates may vary; contact program staff for exact dates.
Requirements: Graduate students pursuing a PhD degree in neuroscience, physiology, pharmacology, psychology, biochemistry, genetics, nursing, or pharmacy may apply.
Date(s) Application is Due: Sep 1
Amount of Grant: 20,000 U.S.D
Contact: Research and Professional Education Office; (301) 459-3700; fax (301) 577-2684; grants@efa.org
Internet: http://www.epilepsyfoundation.org/research/grants.cfm
Sponsor: Epilepsy Foundation
4351 Garden City Drive, Suite 406
Landover, MD 20785-7223

EF Research and Training Fellowships for Clinicians **1769**
These fellowships offer qualified individuals the opportunity to develop expertise in epilepsy research through a one-year training experience and involvement in an epilepsy research project. The purpose of the fellowships is to develop academic clinicians to teach patient care of persons with epilepsy and advance knowledge about epilepsy through clinical investigation. Applications from individuals interested in acquiring experience in the conduct of human clinical studies and who plan to study epilepsy as a human disorder are encouraged. Emphasis is placed on individuals who will be trained in clinical research in epilepsy rather than use epilepsy as a tool in their research in other fields. Applications from women, members of minority groups, and people with disabilities are especially encouraged.
Requirements: Individuals with the MD or DO degree who will have completed residency training in neurology, neurosurgery, pediatrics, internal medicine, or psychiatry by the time the fellowship commences and desire additional postdoctoral clinical research experience are eligible. Applicants planning to work in the United States receive preference.
Restrictions: These are training fellowships and will not be granted to individuals who have received faculty appointments at the level of assistant professor or higher. Individuals may not receive concurrent support for more than one EFA or AES grant or fellowship.
Date(s) Application is Due: Oct 14
Amount of Grant: 50,000 U.S.D
Contact: Research and Professional Education Office; (301) 459-3700; fax (301) 577-2684; grants@efa.org
Internet: http://www.epilepsyfoundation.org/research/grants.cfm
Sponsor: Epilepsy Foundation
4351 Garden City Drive, Suite 406
Landover, MD 20785-7223

EIF Community Grants **1770**
Grants are made in the greater Los Angeles area for education and prevention programs, creative and performing arts, environmental programs, literacy and education, family care and welfare, substance abuse prevention, and youth welfare. Applicants must apply annually and an organization may submit only one application per year. Grants fund only specific programs or special projects of an organization. Applicants must call the foundation to have an application sent to their organization.
Requirements: Grants are made to nonprofit 501(c)3 organizations in the greater Los Angeles area. Organizations must be responsive to the changing needs of the constituents served and the local community.
Restrictions: The foundation does not give funds for political organizations, capital campaigns, public or private schools, fund-raising events, or religious or sectarian purposes.
Geographic Focus: California
Amount of Grant: 3,000 - 15,000 U.S.D
Contact: Grants Administrator; (818) 760-7722; fax (818) 760-7898; info@eifoundation.org
Internet: http://www.eifoundation.org/grants/wp_grants/how_to_apply.asp
Sponsor: Entertainment Industry Foundation
1201 West 5th Street, Suite T-700
Los Angeles, CA 90017

Eileen J. Garrett Scholarship **1771**
The foundation awards an annual scholarship for research and study in parapsychology. Any student attending an accredited college or university who has already demonstrated an academic interest in parapsychology and plans to pursue the subject as a career may apply. Applications and collateral material may be submitted via post, fax, or as MS Word or Adobe Acrobat attachments. Guidelines are available online.
Date(s) Application is Due: Jul 15
Amount of Grant: 3,000 U.S.D
Contact: Garrett Scholarship, (212) 628-1550; fax (212) 628-1559; office@parapsychology.org
Internet: http://www.parapsychology.org/dynamic/040200.html
Sponsor: Parapsychology Foundation
P.O. Box 1562
New York, NY 10021

Eisner Foundation Grants **1772**
The Eisner Foundation exists to provide access and opportunity for disadvantaged children and the aging of Los Angeles County. The Eisner Foundation has been funding innovative and effective non-profit organizations that improve and enrich the lives of underserved children in Southern California since 1996. In 2008, the Foundation recognized that many of the same attributes that the children unfortunately possess physical and emotional vulnerability, extreme poverty, lack of advocacy on their behalf, minimal access to the arts, and general powerlessness also applied to many members of our rapidly aging population in the community. The Foundation elected to broaden the funding focus to include those at both ends of the spectrum of life, the young and the old. The Foundation's now dedicated to bringing about lasting changes in the lives of disadvantaged and vulnerable people starting and ending their lives in Los Angeles County.
Requirements: California nonprofit organizations serving Los Angeles and Orange Counties are eligible to apply. Applicants can apply at any time, but should know that

these proposals will be reviewed and approved at the June or December board meetings. Applying to the Foundation is a two step process. The first step is to submit a Letter of Inquiry (see Eisner Foundation website for guidelines). If Letter of Inquiry meets the Foundation's criteria, a full application will be sent to the applicant for completion and submission, completing the second step.
Restrictions: Proposals for endowments are rarely excepted.
Geographic Focus: California
Amount of Grant: 5,000 - 100,000 U.S.D
Samples: LA Scores, Los Angeles, CA, $100,000—to provide project support for an after-school program that combines creative writing, soccer, and community service; California Institute of the Arts, Los Angeles, CA, $1,250,000—to support the Community Arts Partnership program which provides new media arts education programs to local youth; Friendship Circle, Los Angeles, CA, $25,000—to provide support for an organization that addresses the challenges of families with special-needs children in the South Bay; Eisner Pediatric & Family Medical Center, Los Angeles, CA, $1,000,000—to support the endowment campaign for a pediatric and family medical center serving low-income, uninsured and under-insured families; Aquarium of the Pacific, Los Angeles, CA, $100,000—to provide support for a project that increases science learning access to Title I school communities in the South Bay; St. Barnabas Senior Center, Los Angeles, CA, $75,000—to support a program that provides education on sustained nutrition, adult day care, and other health promotion services for 3,000 seniors in the greater downtown Los Angeles area.
Contact: Trent Stamp, Executive Director; (310) 228-6808; trent.stamp@eisnerfoundation.org
Internet: http://www.eisnerfoundation.org/what_we_do/
Sponsor: Eisner Foundation
233 South Beverly Drive
Beverly Hills, CA 90212

Eleanor Naylor Dana Charitable Trust Grants **1773**
The trust supports biomedical research and the performing arts. In the biomedical program, grants are awarded to support clinical investigation by established scientists to pursue innovative projects designed to improve medical practice or prevent disease. Grants do not support large-scale field studies of a therapeutic or epidemiological nature. Grants for the performing arts support U.S. programs benefiting the public with long-range potential. Giving primarily is in areas east of the Mississippi River. The foundation is interested in the organization's history and its public service. The board meets in March, June, September, and December.
Requirements: Established scientists working under a qualified U.S. institution's approval and U.S. arts organizations are eligible to apply.
Restrictions: The biomedical program does not make grants for instrumentation alone or solely for conference or individual travel. The arts program will not fund deficits; exhibits, publications, or conclaves; or individuals.
Date(s) Application is Due: Feb 1; May 1; Sep 1; Nov 1
Amount of Grant: 5,000 - 50,000 U.S.D
Samples: New York City Opera (NY)—program support, $100,000; Lincoln Ctr Theater (NY)—program support, $50,000.
Contact: Trust Administrator; (212) 754-2890; fax (212) 754-2892
Sponsor: Eleanor Naylor Dana Charitable Trust
P.O. Box. 1803, Murray Hill Station
New York, NY 10156

Elihu Foundation Charitable Trust Grants **1774**
The foundation awards grants to eligible Alaska community-based organizations in its areas of interest, including community-scale economic development; community concern for environmental quality; affordable housing; and health and human services, including health care, vocational training, drug and alcohol counseling, and assistance to the hungry and homeless. Applications are accepted annually between February 1 and June 1.
Requirements: Alaska nonprofit, community-based organizations are eligible.
Geographic Focus: Alaska
Date(s) Application is Due: Jun 1
Amount of Grant: Up to 11,000 U.S.D
Samples: Pratt Museum/Homer Society (Homer, AK)—for vocational training and efforts to preserve cultural heritage, $10,000; Native Village Brevig Mission (AK)—to assist village residents in their efforts to revive traditional native dances, $10,000.
Contact: Grants Administrator, c/o John Egan, McKenna, Storer, Rowe, White and Farrug; (312) 558-8346; fax (312) 558-8348; danab@berkeley.edu or ljhberkeley@comcast.net
Sponsor: Elihu Foundation Charitable Trust
33 N LaSalle Street, Suite 1400
Chicago, IL 60602

Eli Lilly & Company Foundation Grants **1775**
Discretionary grants are awarded in two categories: nonprofit groups aligned with company interests, and grants in Indianapolis and several other communities with significant employee populations. Company-aligned giving focuses on public policy research, health and human services aligned with major therapeutic interests, and academic relations. Community-aligned giving focuses on culture, K-12 education and youth development, locally aligned healthcare organizations, fencerow neighborhood groups, and diversity. The foundation supports organizations within these categories that have a well-defined sense of purpose, a demonstrated commitment to maximizing available resources, and a reputation for meeting objectives and delivery quality programs and services.
Restrictions: Grants do not support individuals; endowments; debt reduction; religious or sectarian programs for religious purposes; bands or fraternal, labor, athletic, or veterans organizations; political contributions; beauty or talent contests; fundraising activities related to individual sponsorship; conferences or media productions; nonaccredited education groups; or memorials.
Date(s) Application is Due: Jun 30; Dec 31
Samples: American National Red Cross (Washington, DC)—for relief efforts in South Asia and Africa, $1 million; Oklahoma Medical Research Foundation Foundation (Oklahoma City, OK)—to endow two professorships in biomedical research, $6 million.
Contact: Thomas King, President; (317) 276-3177; fax (317) 277-2025
Internet: http://www.lilly.com
Sponsor: Eli Lilly and Company Foundation
734 N LaSalle Street, #1167
Indianapolis, IN 46285

Eli Lilly & Company Foundation Schizophrenia Reintegration Awards **1776**
The program honors contributions made by the mental health community, and contributions of individuals battling mental illness as they progress toward the goal of reintegration. Three awards are given for people living with schizophrenia and bipolar disorder. The Artistic Contribution Award honors individual accomplishment in self-expression, in areas such as painting, video, dance performance, music, design, and fashion. The Mentorship Award recognizes individuals who have risen above their own personal challenges to lend a hand to a friend in need. The Inspiration Award honors individuals whose improvements over time have allowed for important personal achievements. Seven treatment team awards are made. The Clinical Medicine Award honors those who combine effective protocols and compassionate strategies to achieve success in clinical outcomes. The On the Job Award recognizes those who help individuals in their care develop the skills needed to find a meaningful job and the determination to keep it. The Home Sweet Home Award celebrates the accomplishment of helping someone make a home. The Advocacy Award honors those who demonstrate their dedication by speaking up for people living with schizophrenia and bipolar disorder. The Keep Learning Award pays tribute to those who help consumers achieve their educational goals. The Social Support Award recognizes the exceptional dedication of those who help individuals in their care adopt the basic social and coping skills required for community interaction. The Lifetime Achievement Award honors a healthcare professional who has made remarkable contributions to many lives over the course of time. In addition, three honorary categories will recognize individuals who have worked to raise awareness of schizophrenia and related disorders, and/or bipolar disorder, on a local and national level in the past year. There will be one recipient per category, which includes government, media and the public eye. Guidelines and applications are available online.
Requirements: The public is invited to nominate individuals, caregivers, health care professionals, treatment groups, treatment programs, and consumer advocates.
Date(s) Application is Due: Jul 8
Amount of Grant: 2,500 - 5,000 U.S.D
Contact: Lilly Secretariat; (800) 809-8202; lillyawards@reintegration.com
Internet: http://www.reintegration.com/resources/scholarships
Sponsor: Eli Lilly and Company Foundation
734 N LaSalle Street, #1167
Indianapolis, IN 46285

Eli Lilly & Company Welcome Back Awards **1777**
The awards annually recognize excellence and courage in the mental health community and are designed to help fight the stigma associated with mental illness and show others that depression is treatable. The awards acknowledges individuals whose accomplishments in the areas of psychiatry, primary care, community service, and destigmatization have helped give new hope to those suffering from clinical depression. The program also presents a Lifetime Achievement Award to honor those people who have overcome depression to lead fulfilling and productive lives. A donation is made to the nonprofit organization of each awardee's choice.
Requirements: Nominations may be made by anyone wishing to recognize an individual for outstanding contributions to the depression community. Anyone may be nominated, including mental health consumers, advocates, physicians, research scientists, and employees of corporate, civic or academic institutions.
Date(s) Application is Due: Nov 14
Amount of Grant: 10,000 - 15,000 U.S.D
Contact: Welcome Back Awards Coordinator; (800) 463-6440; fax (212) 884-0628
Internet: http://www.lilly.com/about/awards/wba/index.html
Sponsor: Eli Lilly and Company
P.O. Box 845
New York, NY 10268-0845

Elisabeth Severance Prentiss Foundation Grants **1778**
The foundation awards the majority of its support in the field of health and medicine. Scientific and medical research, hospitals, health projects, medical education, and care and support of the elderly are funded. Types of support include building programs, equipment needs, research, operating budgets, continuing support, seed grants, endowment funds, and projects/programs. Decision makers favor specific projects over general operating support. The maximum term of a commitment is five years, with preference for proposals for three years or less. Special consideration will be given to requests from applicants in the Cuyahoga County area. The board meets in June and December.
Requirements: Grants are awarded to promote and improve medical services in the greater Cleveland, OH, area.
Restrictions: Grants are not awarded to individuals for scholarships, fellowships, or grants in aid; or to organizations for fund-raising campaigns, surveys, assessments, studies, or planning activities.

Geographic Focus: Ohio
Date(s) Application is Due: May 15; Nov 15
Amount of Grant: 10,000 - 100,000 U.S.D
Samples: University Hospitals of Cleveland (OH)—to create an inpatient psychiatric unit for children and adolescents at Rainbow Babies and Children's Hospital, $5 million.
Contact: Michael Galland, National City Bank Contact; (216) 222-2736
Sponsor: Elisabeth Severance Prentiss Foundation
P.O. Box 94651
Cleveland, OH 44104

Elizabeth Glaser Pediatric AIDS Foundation International Leadership Awards 1779

The Award program is designed to invest in trained individuals in resource-poor countries who have the potential to develop programs which will have a direct impact on the pediatric HIV epidemic in their country, but that lack the resources to do so. This program will complement existing training programs by providing support after training is completed. The Award specifically seeks to identify individuals who are likely to have an enduring impact on control of the epidemic. Applicants may apply for support to create and or scale-up effective pediatric HIV/AIDS research and/or implementation programs including those addressing epidemiology, prevention of mother-to-child transmission, treatment of infected mothers and infants, continuum of care and scale up of prevention services to include care and treatment, development of national policies or strategies to combat pediatric HIV, assessment of the economic impact of disease and its prevention, or to help train additional in-country and regional scientists and program leaders. Award recipients will be asked to mentor a minimum of three additional people who would benefit from the experience and could help them achieve their goals. These are individuals who would be hired and trained to work on the project proposed by the applicant. The Award provides up to $100,000 in total costs per year for three years for a total of $300,000. Funding in years 2 and 3 is dependent on progress in year one and is not guaranteed. Letters of Intent are due by August 24th, with full applications solicited by August 31st and full applications due on September 28th. Awards are made to institutions on behalf of individuals.
Requirements: Applicants must be from either Africa or India, and performing work in either Africa or India in order to apply. Special interested will be given to applications from the African nations of South Africa and Tanzania, but the foundation will consider applicants from other African countries as well. Applicants must demonstrate commitment to continuing to work in a developing country on HIV/AIDS when the award is complete, which is a requirement of the award. Applicants are expected to be in the process of establishing successful careers and programs. Most applicants will have completed training within the past 5 years. Senior investigators such as Full Professors will usually not be eligible for the award. Applicants must hold a position at a public or private non-profit institution that will allow him or her adequate time and provide appropriate facilities to perform the work proposed.
Restrictions: U.S. citizens or European citizens living in these countries are not eligible to apply. Previous International Leadership Award recipients will not be eligible for this round of funding. Grants can not be made directly to individuals.
Date(s) Application is Due: Mar 20
Amount of Grant: 100,000 U.S.D
Contact: Chris Hudnall, Senior Program Coordinator; (310) 491-3154 or (310) 314-1459, ext. 154; fax (310) 314-1469; Chris@pedaids.org or research@pedaids.org
Internet: http://www.pedaids.org/GrantsandAwards/Awards/International-Leadership-Award.aspx
Sponsor: Elizabeth Glaser Pediatric AIDS Foundation
11150 Santa Monica Boulevard, Suite 1050
Los Angeles, CA 90025

Elizabeth Glaser Pediatric AIDS Foundation Operations Research Grants 1780

The Foundation, in its commitment to improve programs to prevent mother-to-child transmission of HIV (PMTCT), announces the availability of a second and final round of funding for operations research (OR) projects under the new Operations Research for Enhancing PMTCT Program. The RFP will further develop an OR program that is designed to utilize Foundation-supported sites and data to generate studies capable of ultimately improving PMTCT program activities. For the purposes of this EGPAF Request for Proposals (RFP), Operations Research is defined broadly and includes the use of analytical techniques to achieve better health outcomes, assessment of the relative contribution of components of service delivery to achieve desired outcomes, and the measurement of cost-effective systems to achieve desired goals. One year awards will be funded up to a maximum of $150,000 inclusive of 5% (maximum) indirect costs. Two year awards will provide up to $125,000 per year to a total of $250,000 in research support per 2 year award including 5% indirect costs. Letters of Intent (LOIs) are due April 14th, with full applications due at the Foundation office by June 6.
Requirements: Proposals will be accepted from investigators who are working with health care teams that are providing PMTCT services to EGPAF-supported sites around the world. Foundation staff, Foundation partners including host-country government and non-governmental organizations and/or combinations of the above are eligible to apply.
Amount of Grant: 150,000 - 250,000 U.S.D
Contact: Chris Hudnall, Senior Program Coordinator; (310) 491-3154 or (310) 314-1459, ext. 154; fax (310) 314-1469; Chris@pedaids.org or research@pedaids.org
Internet: http://www.pedaids.org/GrantsandAwards/Awards/Operations-Research-Award.aspx
Sponsor: Elizabeth Glaser Pediatric AIDS Foundation
11150 Santa Monica Boulevard, Suite 1050
Los Angeles, CA 90025

Elizabeth Glaser Pediatric AIDS Foundation Pediatric HIV Vaccine Research Grants 1781

The Foundation's Pediatric HIV Vaccine Research Program seeks to address basic and pre-clinical research relevant to pediatric HIV vaccine design and development, including, but not limited to, the following topics: the study of breast-milk transmission of HIV; pediatric immune responses to HIV; identification of potential vaccines for pediatric populations; and identification of obstacles to the conduct of clinical trials in infants.
Samples: Marylyn Addo, M.D., Ph.D., Massachusetts General Hospital, Charlestown, Massachusetts; Grace Aldrovandi, M.D. Children's Hospital, Los Angeles, U.S.C Los Angeles, California.
Contact: Chris Hudnall, Senior Program Coordinator; (310) 491-3154 or (310) 314-1459, ext. 154; fax (310) 314-1469; Chris@pedaids.org or research@pedaids.org
Internet: http://www.pedaids.org/GrantsandAwards/Awards/Pediatric-HIV-Vaccine-Program.aspx
Sponsor: Elizabeth Glaser Pediatric AIDS Foundation
11150 Santa Monica Boulevard, Suite 1050
Los Angeles, CA 90025

Elizabeth Glaser Pediatric AIDS Foundation Two-Year International Scholar Award 1782

The Foundation is offering a postdoctoral fellowship for clinicians/scientists from developing countries. The program is aimed at health care professionals who have specific training or experience with HIV/AIDS and hold an M.D. or Ph.D. Applicants must demonstrate commitment to continuing to work in a developing country on HIV/AIDS when the award is complete, as the emphasis of this program is on building long-term scientific capacity in developing countries. The subjects for these projects could include, but are not limited to, the following: epidemiology; issues related to implementation of either prevention or therapy programs; transmission of the HIV virus from mother to infant, including breast-milk transmission and prevention; pediatric HIV vaccine development; antiretroviral drug treatment including pharmacology, complications, and resistance; and microbicide research.
Requirements: Applicants must be from a developing country, and performing work in a developing country. The Foundation encourages those who are determined to improve care in the international setting, as well as those interested in basic science, to apply.
Restrictions: United States or European citizens living in developing countries are not eligible to apply.
Amount of Grant: 57,225 U.S.D
Contact: Chris Hudnall, Senior Program Coordinator; (310) 491-3154 or (310) 314-1459, ext. 154; fax (310) 314-1469; Chris@pedaids.org or research@pedaids.org
Internet: http://www.pedaids.org/GrantsandAwards/Awards/Two-Year-International-Award.aspx
Sponsor: Elizabeth Glaser Pediatric AIDS Foundation
11150 Santa Monica Boulevard, Suite 1050
Los Angeles, CA 90025

Elizabeth Glaser Pediatric AIDS Foundation Two-Year Scholar Awards 1783

The Awards are postdoctoral fellowships that provide two years of salary support, plus supplies and/or travel costs toward medical research on pediatric HIV/AIDS. Annual salaries - based on years of postdoctoral experience - range from $30,000 to $46,000. These awards provide a powerful incentive for a new generation of scientists to take an early professional interest in studying pediatric HIV/AIDS. Scientists are supported by a mentor experienced in HIV/AIDS research. Through this mentoring relationship, a growing number of talented investigators are focusing their efforts on helping to bring an end to HIV/AIDS in children.
Requirements: Scholars must select an experienced sponsor with appropriate qualifications to oversee the proposed research. The sponsor should be a faculty member with an advanced degree (M.D. or Ph.D.) who will oversee the research, provide guidance to the applicant, and can provide evidence of the institution's commitment to the career development of the applicant.
Amount of Grant: Up to 60,000 U.S.D
Samples: Melissa Farrow, Ph.D., College of the Holy Cross—Investigation of the Expression of Anti-HIV AP.O.BEC3 Family Members; Vandana Kalia, Ph.D., Emory University School of Medicine—Development of CD8 T Cell Immunological Memory; Surojit Sarkar, Ph.D., Emory University School of Medicine—Immune Reconstitution and Reversal of T Cell Exhaustion During Chronic Infection.
Contact: Chris Hudnall, Senior Program Coordinator; (310) 491-3154 or (310) 314-1459, ext. 154; fax (310) 314-1469; Chris@pedaids.org or research@pedaids.org
Internet: http://www.pedaids.org/GrantsandAwards/Awards/Two-YearScholar.aspx
Sponsor: Elizabeth Glaser Pediatric AIDS Foundation
11150 Santa Monica Boulevard, Suite 1050
Los Angeles, CA 90025

Elizabeth Glaser Scientist Award 1784

The Award is the Foundation's highest scientific honor, and is an investment in the most promising HIV/AIDS researchers at a critical stage in their careers. These scientists, who represent the best and brightest investigators from the international medical science community, are selected on the basis of their knowledge, innovation, and dedication. By providing up to $682,500 in research funding over a five-year period, this award enables recipients to focus their long-term efforts on issues specific to pediatric HIV/AIDS. The Award fosters an unprecedented spirit of collaboration among these scientists. Each year, the Elizabeth Glaser Scientists meet with our internationally renowned advisory board to stimulate ideas, report on current programs, and plan future research. Since the program's inception, the Foundation has built an invaluable network of scientists whose

work in vaccine development, immune response, breast milk transmission, and other critical areas impacts the entire field of HIV/AIDS research. Interested applicants must contact the foundation to request a letter of intent cover sheet. Letters of intent are due by the listed deadline date; full applications are by invitation.
Requirements: Candidates must have an MD, PhD, DDS, or DVM degree and be at the assistant professor level or above.
Restrictions: The foundation will not fund large therapeutic trials or the pharmocologic development of drugs.
Amount of Grant: Up to 682,500 U.S.D
Samples: David Evans, Ph.D., Department of Microbiology and Molecular Genetics, New England National Primate Research Center—Oral Transmission of Single-Cycle SIV in Neonates; Margaret Feeney, M.D., M.Sc., Massachusetts General Hospital—The Immune Response to Acute Perinatal HIV Infection; Alexandra Trkola, Ph.D., Division of Infectious Diseases and Hospital Epidemiology, University Hospital Zurich—The Humoral Immune Response to HIV.
Contact: Chris Hudnall, Senior Program Coordinator; (310) 491-3154 or (310) 314-1459, ext. 154; fax (310) 314-1469; Chris@pedaids.org or research@pedaids.org
Internet: http://www.pedaids.org/GrantsandAwards/Awards/ElizabethGlaserScientist.aspx
Sponsor: Elizabeth Glaser Pediatric AIDS Foundation
11150 Santa Monica Boulevard, Suite 1050
Los Angeles, CA 90025

Elizabeth McGraw Foundation Grants **1785**
The foundation awards grants nationally, with a focus on the U.S. Northeast, in its areas of interest, including cancer research, elementary and secondary education, art museums, performing arts (opera and theater), and culture. There are no application deadlines or specific guidelines. Send letter containing a detailed description of project and amount of funding requested to: John McGraw, Pres., 157 Eel River Road, Osterville, MA 02655
Geographic Focus: Florida, Massachusetts, New York
Contact: Paul J. Bisset, Vice President, Deutsche Bank Trust Co. of NY; (212) 454-8607
Sponsor: Elizabeth McGraw Foundation
Deutsche Bank Trust Company of New York, P.O. Box 1297
New York, NY 10008-1297

Elizabeth Morse Genius Charitable Trust Grants **1786**
Established in 1992, the Elizabeth Morse Genius Charitable Trust honors the memory of Elizabeth Morse Genius, the daughter of Charles Hosmer Morse, a nineteenth century Chicago financier, industrialist, and land developer. The trust supports and promotes charitable organizations that: encourage the principles of individual self-reliance, self-sacrifice, thrift, industry, and humility; relieve human suffering through scientific research and education regarding disease; provide assistance to youths with troubled childhoods and emotional disorders; attend to the care of the elderly; provide assistance to humankind during times of natural and man-made disasters; foster individual self-worth and dignity, with a broad emphasis on the classical fine arts; develop physical health and spiritual well-being through vigorous athletic activity; and promote world peace and understanding through the improvement of national and international means of travel by air, rail, and sea. The majority of grants from the Genius Charitable Trust are one year in duration; on occasion, multi-year support is awarded. Applicants must first submit a letter of inquiry. Downloadable application guidelines are available at the Sponsor's website by clicking on the "Grant Application Process Button", or prospective applicants may also call the second phone number given to obtain application information. A synopsis of the grant, contact information, and a link to the trust's giving history is available at the grant website. The Elizabeth Morse Genius Charitable Trust has a rolling application deadline. In general, applicants will be notified of grant decisions three to four months after proposal submission.
Requirements: Nonprofit organizations serving Chicago and Cook County are eligible.
Restrictions: The trust generally does not fund individuals; organizations outside the metropolitan Chicago city area; organizations attempting to influence policy through direct lobbying; capital campaigns; endowment campaigns; or political campaigns.
Geographic Focus: Illinois
Amount of Grant: 10,000 - 300,000 U.S.D
Samples: WTTW Channel 11, Chicago, Illinois, $145,400 (2010); Chicago High School for the Arts, Chicago, Illinois, $125,000 (2010); Chicago Community Trust, Chicago, Illinois, $50,000 (2010).
Contact: Kristin Carlson Vogen, Senior Vice President; (312) 828-5554 or (312) 828-1029; ilgrantmaking@bankofamerica.com
Internet: https://www.bankofamerica.com/philanthropic/fn_search.action
Sponsor: Elizabeth Morse Genius Charitable Trust
231 South LaSalle Street, IL1-231-13-32
Chicago, IL 60604

Elkhart County Community Foundation Fund for Elkhart County **1787**
The Fund for Elkhart County is the county's core grant making program and has the broadest guidelines designed to meet the Foundation's mission to produce a brighter future for all people in Elkhart County in Indiana. Grants are awarded from the fund's earnings on a quarterly basis. More than $550,000 has been awarded from this fund to assist non-profits in Elkhart County.
Requirements: The Fund for Elkhart County grants funding to after-school programs, temporary help for abused women and children, and services for the mentally and physically challenged. Other programs include health and dental care for the underprivileged, services for senior citizens, support of local arts and cultural events, environmental and wildlife services, historical preservations, and capacity building. Applicants should carefully review the website's grant guidelines before applying.
Geographic Focus: Indiana
Date(s) Application is Due: Nov 1
Contact: Fund for Elkhart County Contact; (574) 295-8761
Internet: http://www.elkhartccf.org/File/static/grant_seekers/FundForElkhartCounty.shtml
Sponsor: Elkhart County Community Foundation
101 S Main Street, P.O. Box 2932
Elkhart, IN 46615

Elkhart County Community Foundation Grants **1788**
The foundation is looking for innovative programs or projects that address community issues in Indiana's Elkhart County. Community collaboration is encouraged, where organizations work together toward a shared goal with shared responsibility, accountability, and resources. Grants are awarded in the areas of arts and culture, community development, education, youth development, board development and succession planning, and health and human services in addition to continuing support, technical assistance, and matching funds. Most grants are single-year awards, although multi-year grants also will be considered. The staff is available for advice in the preparation of a proposal. Applicants are encouraged to read the Foundation website's "Helpful Tips for Grant Writers." Letters of interest are due to the Foundation by January 1, April 1, June 1, and August 1.
Requirements: Elkhart County non-profit organizations, or those seeking such status, and some governmental agencies, such as the public library or public school system, are eligible.
Restrictions: Support generally will not be given for continuing operating costs of established programs, projects, and agencies. The foundation does not offer support for religious or sectarian purposes. No grants are awarded to individuals (other than scholarships), or for operating budgets or budget deficits, annual funds, conferences, scholarly research, endowments, personal travel, or films.
Geographic Focus: Indiana
Date(s) Application is Due: Mar 1; Jun 1; Sep 1; Nov 1
Contact: Jim Siegmann, Program Officer; (574) 295-8761; fax (574) 389-7497; jim@elkhartccf.org
Internet: http://www.elkhartccf.org/File/static/grant_seekers/WhatWeFund.shtml
Sponsor: Elkhart County Community Foundation
101 S Main Street, P.O. Box 2932
Elkhart, IN 46615

Elkhart County Foundation Lilly Endowment Community Scholarships **1789**
The Lilly Endowment Community Scholarship Program strives to enhance the quality of life for residents in Indiana by raising the level of educational achievement. The scholarship covers full time, four year tuition with an $800/per year book stipend. All college majors are eligible.
Requirements: Scholarship candidates must be U.S. citizens and high school graduates from Elkhart County. They are first generation college students (parents do not have a four year degree) with plans to enroll in a full time, four year field of study. Applicants should contact the Elkhart Community Foundation for an application and instructions for submitting the application packet to their high school guidance counselor.
Geographic Focus: Indiana
Date(s) Application is Due: Feb 1
Contact: Peter McCown, President; (574) 295-8761; fax (574) 389-7497
Internet: http://www.elkhartccf.org/static/students/lilly_scholarships.shtml
Sponsor: Elkhart County Community Foundation
101 S Main Street, P.O. Box 2932
Elkhart, IN 46615

Ellie Fund Grants **1790**
The foundation awards grants to organizations located in Cuyahoga County, OH, whose activities benefit the county's residents. Applications outside of the area must be preapproved by a trustee. Preference is given to services for children within these priorities: crisis intervention, literacy, mental health, physical health, safety, and shelter. The trustees favor requests to support program development, as well as operating needs (especially for start-up, emergency, or bridge funding). Types of nonprofit organizations that receive support include start-up organizations; grassroots and community-based organizations; smaller organizations that do not receive wide public support; religious organizations, but only for secular programs that provide social services to the general community; and established organizations, but only for innovative programs. Organizations that have received a foundation grant for four consecutive years must go on a one-year hiatus before they can reapply for a grant.
Requirements: Grants are awarded only to tax-exempt, nonprofit organizations, and never to individuals.
Restrictions: Requests for annual appeals, fundraisers, symposia, and seminars will not be considered.
Geographic Focus: Ohio
Date(s) Application is Due: Aug 15
Amount of Grant: 1,000 - 7,500 U.S.D
Contact: Foundation Management Services; (216) 621-2901; ellie@fmscleveland.com
Internet: http://www.fmscleveland.com/ellie/guidelines.cfm
Sponsor: Ellie Fund
1422 Euclid Avenue, Suite 627
Cleveland, OH 44115-1952

Elliot Foundation Inc Grants 1791

The Elliot Foundation Inc. formerly known as Elliot Foundation for Medical Research and Education, Inc. is a independent foundation, operating primarily in the state of Indiana. The foundation offers funding in the form of general/operating support. The fields of interest include: Christian agencies & churches; crime/violence prevention; environment; natural resources; child abuses; education.
Requirements: Contact Foundation for more information.
Geographic Focus: Indiana
Contact: Richard E. Bond, Secretary; (575) 293-1165
Sponsor: Elliot Foundation
2210 East Jackson Boulevard
Elkhart, IN 46516-1165

Ellison Medical Foundation/AFAR Julie Martin Mid-Career Award in Aging Research 1792

The Ellison Medical Foundation and AFAR developed this program for outstanding mid-career scientists who propose novel directions of high importance to biological gerontology. Proposals in areas where NIH awards or other traditional sources are unlikely because the research is high risk, are particularly encouraged if they have the potential for leading to major new advances in our understanding of basic mechanisms of aging. Projects investigating age-related diseases are also supported, but only if approached from the point of view of how basic aging processes may lead to these outcomes. Projects concerning mechanisms underlying common geriatric functional disorders are also encouraged, as long as these include connections to fundamental problems in the biology of aging. Two four-year awards of $500,000 will be given, at the level of $125,000 per year. In addition, up to 10% ($50,000) may be requested for administrative/indirect costs. Recipients of this award are expected to attend the AFAR Grantee Conference. The purpose of the meeting is to promote scientific and personal exchanges among recent AFAR grantees and experts in aging research.
Requirements: The applicant must be an Associate Professor who achieved tenured status after December 1st. Non-tenured Associate Professors at institutions with tenure (even if tenure is only offered at the Full Professor level) are not eligible. Applicants at institutions that do not offer tenure must demonstrate that their appointment is equivalent to that of an Associate Professor who received tenure status after December 1st. The proposed research must be conducted at any type of not-for-profit setting in the United States. The deadline of receipt of applications and all supporting materials is December 16th at 5:00 p.m. EST. Refer to the Julie Martin Mid-Career Award instruction sheet and application at: http://afar.org/grants.html for complete application procedures. Incomplete applications cannot be considered.
Restrictions: Projects that deal strictly with clinical problems such as the diagnosis and treatment of disease, health outcomes, or the social context of aging are not eligible. Applicants who are employees in the NIH Intramural program are not eligible.
Date(s) Application is Due: Dec 16
Contact: Grants Manager; (212) 703-9977 or (888) 582-2327; fax (212) 997-0330; grants@afar.org or info@afar.org
Internet: http://afar.org/Ellison%20Mid-Career.html
Sponsor: American Federation for Aging Research
55 West 39th Street, 16th Floor
New York, NY 10018

Ellison Medical Foundation/AFAR Postdoctoral Fellows in Aging Research 1793

The Ellison Medical Foundation, in partnership with the American Federation for Aging Research (AFAR), created the Ellison/AFAR Postdoctoral Fellows in Aging Research Program to encourage and further the careers of postdoctoral fellows with outstanding promise in the basic biological and biomedical sciences relevant to understanding aging processes and age related diseases and disabilities. The award is intended to provide significant support to permit these postdoctoral fellows to become established in the field of aging. Projects concerned with understanding the basic mechanisms of aging will be considered. Projects investigating age related diseases are also supported, if approached from the point of view of how basic aging processes may lead to these outcomes. Projects concerning mechanisms underlying common geriatric functional disorders are also considered. It is anticipated that up to 15 one year grants will be awarded, ranging from $44,850 for a first year fellow, up to $58,850 for a fellow with more than 7 years of training. Of the award, up to $7,850 may be requested for expenses such as research supplies, equipment, health insurance and travel to scientific meetings. Refer to the Ellison/AFAR instruction sheet and application on the AFAR website for complete application procedures. Recipients of this award are expected to attend the AFAR Grantee Conference. The purpose of the meeting is to promote scientific and personal exchanges among recent AFAR grantees and experts in aging research.
Requirements: The applicant must be a postdoctoral fellow (MD and/or PhD degree) at the start date of the award (July 1). Any former Ellison/AFAR postdoctoral award recipient may apply for this award. All candidates must submit applications endorsed by their institutions. The proposed research must be conducted at any type of not-for-profit setting in the United States.
Restrictions: Employees in the NIH Intramural program are not eligible. Fellows may not hold any concurrent foundation or not-for-profit funding.
Date(s) Application is Due: Dec 16
Amount of Grant: 44,000 - 58,000 U.S.D
Contact: Grants Manager; (212) 703-9977 or (888) 582-2327; fax (212) 997-0330; grants@afar.org or info@afar.org
Internet: http://www.afar.org/ellisonpostdoc.html
Sponsor: American Federation for Aging Research
55 West 39th Street, 16th Floor
New York, NY 10018

Ellison Medical Foundation/AFAR Senior Postdoctoral Fellows Research 1794

The program seeks to encourage and further the careers of senior postdoctoral fellows with outstanding promise in the basic biological and biomedical sciences relevant to understanding aging processes and age-related diseases and disabilities. The award is intended to provide significant support to permit these postdoctoral fellows to become established in the field of aging. Projects concerned with understanding the basic mechanisms of aging will be considered. Projects investigating age-related diseases are also supported, if approached from the point of view of how basic aging processes may lead to these outcomes. Projects concerning mechanisms underlying common geriatric functional disorders are also considered. It is anticipated that three two-year grants will be awarded each year.
Requirements: Postdoctoral fellows, both MDs and PhDs, are eligible. Fellows with at least three and not more than five years of prior postdoctoral training are eligible.
Date(s) Application is Due: Dec 15
Amount of Grant: 100,000 U.S.D
Contact: Grants Manager; (212) 703-9977; fax (212) 997-0330; grants@afar.org
Internet: http://www.afar.org/ellison.html
Sponsor: American Federation for Aging Research
55 West 39th Street, 16th Floor
New York, NY 10018

Ellison Medical Foundation Senior Scholar Awards 1795

The awards enable established investigators working at U.S. institutions to conduct up to four years of research in the basic biological and clinical sciences relevant to understanding aging processes and age-related diseases and disabilities. The award is intended to provide significant support to established investigators in order to allow the development of new, creative research programs by investigators who may not currently be conducting aging research or who may wish to develop new research programs in aging. Acceptable uses for award funds include salary, other personnel, equipment, supplies, resource acquisition, and travel. (Carry-overs in excess of $25,000 must be approved by the Ellison Medical Foundation Scholars Program Office.) Full indirect costs at the NIH negotiated rate will be provided. Ten awards are made annually. Application deadline is for letters of intent.
Requirements: Established investigators working at U.S. institutions are eligible.
Date(s) Application is Due: Mar 9
Amount of Grant: Up to 150,000 U.S.D
Samples: To be divided among 11 established investigators at U.S. institutions—to pursue innovative research on aging that might not otherwise be financed by traditional resources because of its perceived novelty or risk, $6.6 million maximum over four years.
Contact: Dr. Richard Sprott, Executive Director; (301) 657-1830; fax (301) 657-1828; rsprott@ellison-med-fn.org
Internet: http://www.ellison-med-fn.org/emf_applications.jsp
Sponsor: Ellison Medical Foundation
4710 Bethesda Avenue, Suite 204
Bethesda, MD 20814

Ellison Medical Foundation Senior Scholar Awards in Global Infectious Diseases 1796

The program funds up to four years of basic research on molecular and cellular mechanisms of parasitology and infectious diseases that result from microbial, protozoan, or viral pathogenesis—with special focus on tuberculosis, malaria, and parasitoses. Awards are intended to inspire new directions that may entail substantial risk. Ten awards are made annually. The listed application deadline is for letters of intent; full applications are by invitation.
Requirements: Investigators at least four years beyond their postdoctoral fellowship, employed by U.S. 501(c)3 institutions, or U.S. colleges or universities, are eligible to apply. Whereas the foundation only makes awards to U.S. nonprofit institutions, the program encourages formation of research consortia between U.S. institutions and those in other disease-endemic countries, as through a subcontract mechanism, when such collaborations will benefit the proposed research.
Restrictions: The awards are not intended to supplement ongoing, already funded programs.
Date(s) Application is Due: Mar 9
Amount of Grant: Up to 150,000 U.S.D
Samples: To be divided among 10 established investigators at U.S. institutions—to pursue innovative research on global infectious diseases that might not otherwise be financed by traditional resources because of its perceived novelty or risk, $6 million maximum over four years.
Contact: Dr. Richard Sprott, Executive Director; (301) 657-1830; fax (301) 657-1828
Internet: http://www.ellison-med-fn.org/emf_applications.jsp
Sponsor: Ellison Medical Foundation
4710 Bethesda Avenue, Suite 204
Bethesda, MD 20814

Elmer L. and Eleanor J. Andersen Foundation Grants 1797

The foundation exists to enhance the quality of the civic, cultural, educational, environmental, and social aspects of life in Minnesota, primarily in the metropolitan area of Saint Paul and Minneapolis. Types of support include general operating support, continuing support, annual campaigns, capital campaigns, building construction/renovation, endowment funds, program development, deficit reduction, publication, seed money, curriculum development, research, technical assistance, and matching funds. The board meets four times annually.
Requirements: Minnesota nonprofit organizations are eligible.
Geographic Focus: Minnesota
Date(s) Application is Due: Feb 1; May 1; Aug 1; Nov 1
Amount of Grant: 500 - 75,000 U.S.D

Contact: Mari Oyanagi Eggum, Foundation Administrator; (651) 642-0127; fax (651) 645-4684; eandefdn@mtn.org
Sponsor: Elmer L. and Eleanor J. Andersen Foundation
2424 Territorial Road
Saint Paul, MN 55114

El Paso Community Foundation Grants **1798**
The Foundation is a facilitator linking the generosity of the region's donors to local non-profit organizations to meet the charitable needs of the El Paso area. In this capacity, the Foundation helps to provide funding for programs and initiatives of these organizations for the broader good of the El Paso Community. Major areas of interest are: arts and humanities; education; public benefit; health and disabilities; environment;animals and; human services. Types of support offered include: equipment acquisition; general/operating support; management development/capacity building; matching/challenge support; program development; scholarship funds; seed money; technical assistance. Priority is given to: more effective ways of doing things; projects where a moderate amount of grant money can have an impact; and projects that show collaboration with other organizations. Application deadlines are February 1 and August 1.
Requirements: Grant requests will be considered only from agencies located within or offering services to the citizens of our community, which includes far west Texas, southern New Mexico and northern Chihuahua, Mexico. Applicants must be exempt from income taxes under Section 501(c)3 of the Internal Revenue Service Code. Application form is available on the El Paso Community Foundation website.
Restrictions: Funding is not available for/to: individuals; capital campaigns; fundraising events or projects; religious organizations for religious purposes; annual appeals and membership contributions; organizations that are political or partisan in purpose; travel for individuals or groups; organizations outside the El Paso geographic area; endowment funds; past operating deficits.
Geographic Focus: Texas
Date(s) Application is Due: Feb 1; Aug 1
Amount of Grant: 3,000 - 10,000 U.S.D
Samples: American Cancer Society, 42,000—provide local patients transportation to and from treatments; Angelo Catholic School, $11,000— teachers' salary; Animal Rescue League of El Paso, Inc., $8,000—animal rescue, rehabilitation and placement.
Contact: Bonita Johnson, Grants Manager; (915) 533-4020; fax (915) 532-0716; info@epcf.org
Internet: http://www.epcf.org/grant_guidelines
Sponsor: El Paso Community Foundation
P.O. Box 272
El Paso, TX 79943

El Paso Corporate Foundation Grants **1799**
El Paso Corporate Foundation funds initiatives that strengthen the communities where employees live, work, and volunteer. The primary focus areas of the foundation are education, health and human services and the community. The Foundation's secondary focus areas are the environment and arts and culture. Organizations must be invited by the Foundation to submit a letter of inquiry for funding consideration.
Requirements: Giving is limited to nonprofits in Alabama, Colorado and Texas.
Restrictions: The following are ineligible requests for funding: organizations without 501(c)3 status; organizations that already have an active El Paso Corporate Foundation grant; capital campaigns; multi-year commitments; religious organizations for religious purposes; individuals; hospitals and medical research; athletics or youth sports organizations; national or state-wide initiatives; war veterans and fraternal service organizations; endowment funds; pass-through grants; fundraising events, such as dinners, luncheons, and golf tournaments; political organizations, campaigns, and candidates; and technology grants, i.e. computers, software, or related hardware.
Geographic Focus: Alabama, Colorado, Texas
Date(s) Application is Due: Jan 31; May 31; Sep 30
Contact: Community Relations Manager; (713) 420-2878; fax (713) 420-5312; foundation@elpaso.com
Internet: http://www.elpaso.com/community/default.shtm
Sponsor: El Paso Corporate Foundation
P.O. Box 2511
Houston, TX 77252-2511

El Pomar Foundation Awards and Grants **1800**
The Foundation is one of the largest and oldest private foundations in the Rocky Mountain West and contributes annually through direct grants and Community Stewardship Programs to support Colorado nonprofit organizations. The Foundation's primary focus is in health, human services, education, arts and humanities, and civic and community initiatives.
Requirements: The foundation makes grants to 501(c)3 nonprofit Colorado organizations and for activities that take place within the state.
Restrictions: The Foundation does not accept grant applications for: other foundations or nonprofits that distribute money to recipients of its own selection; endowments; individuals; organizations that practice discrimination of any kind; organizations that do not have fiscal responsibility for the proposed project; organizations that do not have an active 501(c)3 nonprofit IRS determination letter; camps, camp programs, or other seasonal activities; religious organizations for support of religious programs; cover deficits or debt elimination; cover travel, conferences, conventions, group meetings, or seminars; influence legislation or support candidates for political office; produce videos or other media projects; fund research projects or studies; primary or secondary schools (K-12).
Geographic Focus: Colorado
Amount of Grant: 500 - 500,000 U.S.D
Samples: Four Miles Historical Park (Denver, CO)—to construct a visitors center on the history of Denver, $50,000; Saint Paul Catholic Church (Colorado Springs, CO)—to construct a community center, $333,333; Pikes Peak Hospice and Palliative Care (Colorado Springs, CO)—for a new system for managing patient information, $75,000; Colorado Assoc of Nonprofit Organizations (Denver, CO)—for general operating support for statewide nonprofit programs, $20,000.
Contact: Executive Office; (719) 633-7733 or (800) 554-7711; fax (719) 577-5702; grants@elpomar.org
Internet: http://www.elpomar.org/grant2.html
Sponsor: El Pomar Foundation
10 Lake Circle
Colorado Springs, CO 80906

Elsa U. Pardee Foundation Grants **1801**
The foundation has three main grantmaking focuses: ongoing research grants to established institutions for cancer research, the funding of worthwhile new cancer research projects, and grants to committees providing financial assistance to cancer victims in specified areas. The foundation particularly welcomes innovative, small-scale, short-term projects that may be difficult to fund elsewhere until some interesting results are obtained. Application forms are available on written request and must accompany all project proposals. Applications are accepted at any time.
Requirements: Application must include completed form, description of project, itemized budget, approval from the institution where the project will be done, amount requested, duration of project, and any other supportive material deemed helpful.
Restrictions: Grants are limited to research toward the cure and control of cancer and in general do not provide for building funds, equipment (except that used in a specific project), fellowships, or fund-raising campaigns.
Amount of Grant: 50,000 - 200,000 U.S.D
Samples: U of Maryland (MD)—for the study of synthetic, conformational, and enzymatic factors of asn-linked glycoproteins, $34,760; John Muir Foundation (CA)—for chemotherapy program for cancer patients in Contra Costa County, CA, $150,000.
Contact: James Kendall, Secretary; (989) 832-3691; info@pardeefoundation.org
Internet: http://www.pardeefoundation.org
Sponsor: Elsa U. Pardee Foundation
P.O. Box 2767
Midland, MI 48641-2767

Elsie H. Wilcox Foundation Grants **1802**
The Foundation provides partial support to programs and projects of tax-exempt, public charities in Hawaii to improve the quality of life in the state, particularly the island of Kauai. Areas of interest to the Foundation, include: education, health organizations, people with disabilities; human services, performing arts, theater, religion, YM/YMCAs & YM/YWHAs. Types of support include building/renovation; equipment; general/operating support. Grants average from $5,000 - $15,000.
Requirements: 501 (c)3 nonprofit organizations in Hawaii are eligible to apply. The Foundation places a special emphasis on the island of Kauai. Contact Paula Boyce to acquire the cover sheet/application forms and any additional guidelines required to begin the application process. Proposals must be submitted by October 1st.
Restrictions: No grants to individuals, or for endowments.
Geographic Focus: Hawaii
Date(s) Application is Due: Oct 1
Amount of Grant: 5,000 - 15,000 U.S.D
Samples: Girl Scouts of Hawaii, Honolulu, HI, $5,000—scholarships for Kauai girls in grades K-5; Bishop Museum, Honolulu, HI, $5,000—educational program support; American Cancer Society, Hawaii Pacific Division, Hawaii, Honolulu, HI, $5,000—quality of life services for cancer patients on Kauai.
Contact: Paula Boyce, c/o Bank of Hawaii; (808) 538-4944; fax (808) 538-4647; pboyce@boh.com
Internet: http://www.hawaiicommunityfoundation.org/index.php?id=290
Sponsor: Elsie H. Wilcox Foundation
Bank of Hawai'i, Foundation Administration Department 758
Honolulu, HI 96802-3170

Elsie Lee Garthwaite Memorial Foundation Grants **1803**
Established in 1943, the Foundation supports organizations primarily in Philadelphia, Chester, Montgomery and Delaware counties of Pennsylvania. Giving to organizations that: provide for the physical and emotional well-being of children and young people; seek to enable young people, particularly the needy, to reach their fullest potential through education, empowerment, and exposure to the arts; are smaller organizations, with budgets under $1 million per year.
Requirements: Non-profits are eligible in the Philadelphia, Chester, Montgomery and Delaware counties, PA. Contact the Foundation at least 30 days prior to deadlines with a Letter of Intent. An application form can be obtained from the Foundation after reviewing the Letter of Intent, unsolicited applications will not be excepted. Contact the Foundation for further guidelines.
Restrictions: No grants to individuals, public, private, or parochial schools, colleges and universities.
Geographic Focus: Pennsylvania
Date(s) Application is Due: Mar 31; Aug 31
Amount of Grant: 3,000 - 16,000 U.S.D
Samples: United States Fund for UNICEF, New York, NY, $15,000 - for Pakistan earthquake victim relief; Citizen Schools, Boston, MA, $10,000 - for program support;

Center for Grieving Children, Teens and Families, Philadelphia, PA, $3,000 - for general operating support;
Contact: Thomas Kaneda, Secretary; (610) 527-8101; fax (610) 527-7808
Sponsor: Elsie Lee Garthwaite Memorial Foundation
1234 Lancaster Avenue, P.O. Box 709
Rosemont, PA 19010-0709

EMBO Installation Grants **1804**
The aim of this scheme is to strengthen science in participating EMBC member states. At present applications are accepted from Croatia, the Czech Republic, Estonia, Hungary, Portugal, Poland and Turkey. Additional member states have expressed an interest to join in the near future. Grants will help scientists to relocate, set up their labs and rapidly establish a reputation in the European scientific community. The program is entirely funded by the participating EMBC member states and successful applicants receive an annual support of Euro $50,000 for three to five years. Each participating country sets an upper limit to the number of grants available per year. Application forms should be completed by the individual scientists and the receiving institute.
Requirements: Eligible scientists should: have an excellent publication record; have spent at least two consecutive years prior to the application deadline, outside the country in which they are applying to establish their lab; and be negotiating a full-time position at an institute/university in participating member states by the date of application. No age limit applies.
Geographic Focus: All States, Albania, Andorra, Armenia, Austria, Azerbaijan, Belarus, Belgium, Bosnia & Herzegovina, Bulgaria, Croatia, Cyprus, Czech Republic, Denmark, Estonia, Finland, France, Georgia, Germany, Greece, Hungary, Iceland, Ireland, Italy, Kosovo, Latvia, Liechtenstein, Lithuania, Luxembourg, Macedonia, Malta, Moldova, Monaco, Montenegro, Norway, Poland, Portugal, Romania, Russia, San Marino, Serbia, Slovakia, Slovenia, Spain, Sweden, Switzerland, The Netherlands, Turkey, Ukraine, United Kingdom, Vatican City
Date(s) Application is Due: Apr 15
Amount of Grant: 150,000 - 250,000 EUR
Contact: Hermann Bujard, Executive Director; 49 0 6221 8891, ext. 101; fax 49 0 6221 8891, ext. 200; embo@embo.org
Internet: http://www.embo.org/sdig/index.html
Sponsor: European Molecular Biology Organization
Meyerhofstrasse 1
Heidelberg, D-69117 Germany

EMBO Long-Term Fellowships **1805**
Long-Term Fellowships are awarded for prolonged visits (12 to 24 months) and are intended for advanced training through research. As the aim of the EMBO is to promote international research, it follows that mobility is a crucial element in deciding the eligibility of an application. All applications must involve a laboratory of origin or a receiving institute or applicant's nationality from one of the member states. Stipend rates depend on the country being visited, marital status, etc.
Requirements: Applicants must hold a doctorate degree or equivalent before the start of the fellowship but not necessarily when applying. For clarity, it is deemed that this requirement is fulfilled only when the thesis has been stamped as submitted for final examination by the university. As a minimum, applicants must have at least one first author publication in press or published in an international peer reviewed journal at the time of application.
Restrictions: Applications are only considered from candidates with a maximum of 3 years postdoctoral experience.
Geographic Focus: All States, Germany
Date(s) Application is Due: Feb 15; Aug 15
Contact: Liselott Maidment, 49 0 6221 8891, ext. 116 or 49 0 6221 8891 200; fellowships@embo.org
Internet: http://www.embo.org/fellowships/long_term.html
Sponsor: European Molecular Biology Organization
Meyerhofstrasse 1
Heidelberg, D-69117 Germany

EMBO Short-Term Fellowships **1806**
Short-term fellowships are established to advance molecular biology research by helping scientists to visit another laboratory with a view to applying a technique not available in the home laboratory. Such fellowships are not awarded for exchanges between two laboratories within the same country, and are intended for joint research work rather than consultations. The fellowships cover travel plus subsistence of the fellow only and not of any dependents. These fellowships are intended for visits of one week to three months duration (for non-European applicants the short-term fellowships are intended for a fixed duration of 90 days). Stipends vary widely from country to country,
Requirements: Applicants must be either: post-doctoral scientists with less than 10 years of professional experience since finishing their Ph.D. degree (or the equivalent); or pre-doctoral scientists who have not yet completed a Ph.D. degree (or the equivalent).
Restrictions: All applications must involve either a laboratory of origin or a receiving institute from one of the EMBC member states.
Geographic Focus: All States, Germany
Contact: Program Director; 49 0 6221-8891, ext. 115; fax 49 0 6221-8891, ext. 200; fellowships@embo.org
Internet: http://www.embo.org/fellowships/short_term.html
Sponsor: European Molecular Biology Organization
Meyerhofstrasse 1
Heidelberg, D-69117 Germany

Emerson Charitable Trust Grants **1807**
Established in 1944 in Missouri as the Emerson Electric Manufacturing Company Charitable Trust, the Foundation supports: arts and culture—fine arts and cultural institutions to enrich the diversity, creativity, and liveliness of the community; education—programs designed to promote educational systems at all levels; health and human services—programs designed to help individuals and families in times of need, including sickness, old age, family crisis, and natural disasters; civic affairs—programs designed to protect citizenry; further economic health of the community; and build and maintain assets such as parks and zoos; and youth—programs designed to give young people the opportunity to recognize their potential, confidence, and skills to achieve their dreams. The foundation awards college scholarships to children and step-children of employees of Emerson Electric. No specific application form is required, and initial approach should be the complete proposal,
Amount of Grant: 5,000 - 300,000 U.S.D
Samples: Alvin J. Siteman Ctr, Washington U School of Medicine and Barnes-Jewish Hospital (Saint Louis, MO)—for a new cancer-research facility and for cancer research, $6 million challenge grant.
Contact: Jo Ann Harmon, Senior Vice President; (314) 553-3722; fax (314) 553-1605
Internet: http://www.emerson.com/en-us/about_emerson/company_overview/pages/our_approach_to_corporate_philanthropy.aspx
Sponsor: Emerson Charitable Trust
8000 W Florissant Avenue, P.O. Box 4100
Saint Louis, MO 63136

Emerson Electric Company Contributions Grants **1808**
Emerson believes it is important to help support organizations and institutions that play significant roles in enhancing the quality of life in the communities where its facilities are located and where its employees and their families live and work. As a complement to its foundation, Emerson makes charitable contributions to nonprofit organizations directly. Support is given on a national basis in areas of company operations, with emphasis on St. Louis, Missouri. Its primary fields of interest include the arts, education, health care, and public affairs. An application form should be obtained from the nearest company facility.
Contact: Robert M. Cox, Senior Vice President; (314) 553-2000; fax (314) 553-1605
Internet: http://www.emerson.com/en-us/about_emerson/company_overview/pages/our_approach_to_corporate_philanthropy.aspx
Sponsor: Emerson Electric Company
8000 W Florissant Avenue, P.O. Box 4100
Saint Louis, MO 63136

Emily Davie and Joseph S. Kornfeld Foundation Grants **1809**
The Foundation's selection criteria for all programs are: evidence of strong and visionary leadership; development of new initiatives that are replicable and sustainable; creation of new ways to address existing critical issues; interest in building connections and creative partnerships; desire to improve and enrich lives of individuals (rather than capital improvements); and original, unusual, dynamic and focused proposals. Proposal letters should include a background on the organization, a detailed description of the proposed project and financial information.
Requirements: Eligible applicants must provide a copy of a 501(c)3 determination letter.
Date(s) Application is Due: Mar 1; Jul 15; Nov 15
Contact: Bobye List; (718) 624-7969; fax (718) 834-1204; office@kornfeldfdn.org
Internet: http://fdncenter.org/grantmaker/kornfeld
Sponsor: Emily Davie and Joseph S. Kornfeld Foundation
41 Schermerhorn Street, Suite 208
Brooklyn, NY 11021

Emma B. Howe Memorial Foundation Grants **1810**
The Emma B. Howe Memorial Foundation makes grants through the Minneapolis Foundation. The focus of all grants is to improve: the health and well-being of children, youth and families; opportunities for educational achievement; access to quality affordable housing; and economic vitality throughout the region. Information on the application process is available online.
Requirements: Eligible organizations include 501(c)3 nonprofits, public institutions; and emerging groups organized for nonprofit purposes.
Restrictions: Funds are not available for: individuals; organizations/activities outside of Minnesota; conference registration fees; memberships; direct religious activities; political organizations or candidates; direct fundraising efforts; telephone solicitations; courtesy advertising; financial deficits.
Geographic Focus: Minnesota
Samples: Kids in Distressed Situations (KIDS)—(New York, NY)—to provide donated products to low-income children and families in the Minneapolis area through Bridging Inc, $75,000 over three years.
Contact: Grants Manager; (612) 672-3836; grants@mplsfoundation.org
Internet: http://www.mplsfoundation.org/partners/emma.htm
Sponsor: Emma B. Howe Memorial Foundation
80 S Eighth Street
Minneapolis, MN 55402

EMMIS Communications/Radio One Community Fund Grants **1811**
The goal of the fund is to make charitable contributions to nonprofit organizations that enrich the quality of life for the minority community and make Central Indiana a better community for all citizens. The fund will consider grants that address the following issues: economic—projects that support minority business development, financial planning programs, and home ownership programs; and health—projects that address minority health care issues, with special consideration given to sickle cell anemia,

HIV/AIDS, and prenatal care. Between two and five grants will be awarded. Submit the following information: a brief summarary of the organization, a brief project description; the community need or problem being addressed; how effectiveness will be measured; constituent outreach efforts; how the program will be funded; and three funding levels requested, including the impact each level can make. Attach a list of board members; copy of 501(c)3 determination letter; most recent audited statement, or most recent IRS Form (990), if an audit statement is not available.
Requirements: Central Indiana nonprofit organizations that closely match the fund's guidelines are eligible.
Restrictions: Grants do not support organizations without 501(c)3 status; organizations that discriminate by race, color, religion, gender, sexual orientation, or national origin; conferences, workshops, or seminars; political, fraternal, and labor organizations; veteran's organizations; endowments or memorials; fundraising events including but not limited to tickets to dinners, golf outings, or sporting events; advertisements in programs, yearbooks, or souvenir publications; second-party giving, including service organizations raising money for charitable purposes; projects or groups benefiting an individual or few persons; or travel funds for groups or individuals. Low priority will be given to capital campaign requests.
Geographic Focus: Indiana
Date(s) Application is Due: Mar 3
Amount of Grant: 10,000 - 50,000 U.S.D
Contact: Fund Administrator; (317) 266-9700; fax (317) 684-2021; communityrelations@emmis.com
Internet: http://www.wibc.com/inside/emmiscommunityfund.aspx
Sponsor: EMMIS Communications/Radio One Community Fund
40 Monument Circle
Indianapolis, IN 46204

Emy-Lou Biedenharn Foundation Grants 1812

The foundation awards grants to eligible Louisiana nonprofit organizations in the areas of education and medicine. The foundation also supports the Biedenharn Museum and Gardens. Application are reviewed in October for the following year.
Requirements: Quachita Parrish, LA, nonprofit organizations are eligible.
Geographic Focus: Louisiana
Contact: Ralph Calhoun, Executive Director; (318) 387-5281 or (800) 362-0983; rcalhoun@bayou.com
Sponsor: Emy-Lou Biedenharn Foundation
2006 Riverside Drive
Monroe, LA 71201

Ensworth Charitable Foundation Grants 1813

The Ensworth Charitable Foundation was established in 1948 to support and promote educational, cultural, human services, religious, and health care programming for underserved populations. The foundation specifically serves the people of Hartford, Connecticut, and its surrounding communities. Grants from the Ensworth Charitable Foundation are primarily one year in duration. On occasion, multi-year support is awarded. Applicants must apply online at the grant website. Applicants are strongly encouraged to do the following before applying: review the downloadable state application procedures for additional helpful information and clarifications; review the downloadable online-application guidelines at the grant website; review the foundation's funding history (link is available from the grant website); review the online application questions in advance; and review the list of required attachments. These will generally include: a list of board members, financial statements (audited, reviewed, or compiled by independent auditor); an organization summary; a list of other funding sources; an IRS Determination letter; and other required documents. All attachments must be uploaded in the online application as PDF, Word, or Excel files. The Ensworth Charitable Foundation application deadline is January 15 at 11:59 p.m. Applicants will be notified of grant decisions by letter within three to four months after the deadline.
Requirements: Applicants must have 501(c)3 tax-exempt status and serve the people of Hartford, Connecticut, and its surrounding communities.
Restrictions: Applicants will not be awarded a grant for more than 3 consecutive years. The foundation does not support requests from individuals, organizations attempting to influence policy through direct lobbying, or any political campaigns.
Geographic Focus: Connecticut
Date(s) Application is Due: Jan 15
Amount of Grant: 1,000 - 25,000 U.S.D
Samples: Teach for America, New York, New York, $25,000 (2010); Malta House of Care, Hartford, Connecticut, $28,000 (2010); Hartford Stage Company, Hartford, Connecticut, $20,000 (2010).
Contact: Amy R. Lynch; (860) 657-7015; amy.r.lynch@baml.com
Internet: https://www.bankofamerica.com/philanthropic/fn_search.action
Sponsor: Ensworth Charitable Foundation
200 Glastonbury Boulevard, Suite # 200
Glastonbury, CT 06033-4056

Entergy Corporation Micro Grants 1814

The Foundation will bestow a monetary award for projects that effectively impact arts and culture, community improvement and enrichment, education and literacy, and healthy families. Organizations should be located within Entergy's service territory in Arkansas, Louisiana, Mississippi, Massachusetts, Michigan, New Hampshire, New York, Texas, or Vermont. Micro Grant applications are accepted on an ongoing basis. Applicants should allow at least 6 (six) to 8 (eight) weeks for review and notification of the result of a request.
Requirements: 501(c)3 tax-exempt organizations, schools, hospitals, governmental units, and religious institutions are eligible. Each organization may submit only one application per year.
Restrictions: Entergy will not fund: groups without 501(c)3 or similar non-profit status; administrative expenses (e.g., salaries, office equipment) or recurring expenses that exceed 15% of the requested amount; capital project funding (i.e., building campaigns); political candidates or groups; purchase of uniforms or trips for school-related organizations; amateur sports teams; activities whose sole purpose is promotion or support of a specific religion, denomination, or religious institution; grants to individuals or loans of any type; or any organization owned or operated by an employee of Entergy.
Geographic Focus: Arkansas, Louisiana, Massachusetts, Michigan, Mississippi, New Hampshire, New York, Texas, Vermont
Amount of Grant: Up to 1,000 U.S.D
Contact: Jennifer Quezergue, (504) 576-2674 or (504) 576-6980; jquezer@entergy.com Christine Jordan, (504) 576-7705 or (504) 576-6980; cminor@entergy.com
Internet: http://www.entergy.com/our_community/micro_grant_guidelines.aspx
Sponsor: Entergy Corporation
639 Loyola Avenue
New Orleans, LA 70161-1000

Entergy Corporation Open Grants for Healthy Families 1815

Entergy Corporation's Open Grants program focuses on improving communities as a whole. The Corporation believes that children need a good start to grow into healthy, well-adjusted adults. With that in mind, it gives to programs that have a direct impact on children educationally and emotionally. The Corporation is also interested in family programs, like those that better prepare parents to balance the demands of work and home. In considering requests for grants, priority is placed on programs in specific counties/parishes, including areas of: Arkansas, Louisiana, Massachusetts, Michigan, Mississippi, New Hampshire, New York, Texas, and Vermont. Applicants should contact the contributions coordinator in their region, which are listed on the web site. Open grant applications are accepted on an ongoing basis. Applications should be submitted at least three months prior to the time the funding is needed.
Requirements: Grants from the Entergy Corporation will only be made to the following types of organizations: non-profit organizations that are tax exempt under section 501(c)3 of the Internal Revenue Code; or schools, hospitals, governmental units and religious institutions that hold nonprofit status similar to that of 501(c)3 organizations.
Restrictions: Entergy will not fund: groups without 501(c)3 or similar non-profit status; administrative expenses (e.g., salaries, office equipment) or recurring expenses that exceed 15% of the requested amount; capital project funding (i.e., building campaigns); political candidates or groups; purchase of uniforms or trips for school-related organizations; amateur sports teams; activities whose sole purpose is promotion or support of a specific religion, denomination, or religious institution; grants to individuals or loans of any type; or any organization owned or operated by an employee of Entergy.
Geographic Focus: Arkansas, Louisiana, Massachusetts, Michigan, Mississippi, New Hampshire, New York, Texas, Vermont
Amount of Grant: Up to 1,000 U.S.D
Contact: Jennifer Quezergue, (504) 576-2674 or (504) 576-6980; jquezer@entergy.com Christine Jordan, (504) 576-7705 or (504) 576-6980; cminor@entergy.com
Internet: http://www.entergy.com/our_community/Grant_Guidelines.aspx
Sponsor: Entergy Corporation
639 Loyola Avenue
New Orleans, LA 70161-1000

EPA Baseline Assessment of Risks of Exposure to Lead Poisoning of Native American Children 1816

This program supports efforts to identify the risks of lead poisoning among children by conducting a baseline assessment of potential lead exposures. Tribes may then use this data to determine whether there is a need to implement an authorized lead-reduction program.
Requirements: Federally recognized Indian tribes, tribal consortia, and tribal organizations are eligible to apply.
Restrictions: Funds may not be used to purchase property, provide lead hazard reduction activities, conduct renovation or construction activities, or cover case management costs.
Amount of Grant: Up to 75,000 U.S.D
Contact: Darlene Watford, (202) 566-0516; fax (202) 566-0469; Watford.Darlene@epa.gov
Internet: http://www.epa.gov/lead
Sponsor: Environmental Protection Agency
1200 Pennsylvania Avenue NW
Washington, DC 20460

EPA Children's Health Protection Grants 1817

The objectives of this program are to catalyze community-based and regional projects and other actions that enhance public outreach and communication; assist families in evaluating risks to children and in making informed consumer choices; build partnerships that increase a community's long-term capacity to advance the protection of children's environmental health and safety; leverage private and public investments to enhance environmental quality by enabling community efforts to continue past EPA's ability to provide assistance to communities; and promote protection of children from environmental threats through lessons learned. There are no deadline dates.
Requirements: Eligible applicants include community groups, public nonprofit institutions/organizations, tribal governments, specialized groups, profit organizations, private nonprofit institutions/organizations, and municipal and local governments. Potential applicants are strongly encouraged to discuss proposed projects with or submit preapplications to program staff prior to the completion of a full proposal.
Amount of Grant: 5,000 - 250,000 U.S.D

Contact: Bettina Fletcher, Office of Children's Health Protection; (202) 564-2188; fax (202) 564-2733; fletcher.bettina@epa.gov
Internet: http://yosemite.epa.gov/ochp/ochpweb.nsf/content/grants.htm
Sponsor: Environmental Protection Agency
1200 Pennsylvania Avenue NW, Mail Code 1107A, Room 2512 Ariel Rios N
Washington, DC 20004

EPA Great Lakes Program Grants **1818**

The objectives of the program are to restore and maintain the chemical, physical, and biological integrity of the Great Lakes Basin Ecosystem. Specifically, the program's goals are to reduce the level of toxic substances in the Great Lakes; to protect and restore vital habitats; and to protect human health and restore and maintain stable, diverse, and self-sustaining populations. Funds must be matched. Annual deadline dates may vary; contact program staff for exact dates.
Requirements: State water pollution control agencies, interstate agencies, other public or nonprofit private agencies, institutions, organizations, and individuals for the purposes stated are eligible.
Amount of Grant: 5,000 - 660,000 U.S.D
Contact: Mike Russ, Great Lakes National Program Office; (312) 886-4013; russ.michael@epa.gov
Internet: http://www.epa.gov/glnpo
Sponsor: Environmental Protection Agency
77 W Jackson Boulevard, G-9J
Chicago, IL 60604-3590

EPA Lead Awareness Outreach for Native American Tribes **1819**

This program provides financial assistance to tribes to launch outreach efforts that would educate Native American families about the dangers of exposure to lead-based paint hazards among children, distribute educational information and encourage families to have their children screened and homes tested for lead hazards. Activities may include providing medical training, developing culturally specific lead outreach materials, distributing pamphlets and establishing an in-home education program.
Requirements: Federally recognized Indian tribes, tribal consortia, and tribal organizations are eligible to apply.
Restrictions: Funds may not be used to purchase property, provide lead hazard reduction, conduct renovation or construction activities, or cover case management costs.
Amount of Grant: Up to 50,000 U.S.D
Contact: Darlene Watford, Office of Pollution Prevention and Toxics; (202) 566-0516; fax (202) 566-0469; Watford.Darlene@epa.gov
Internet: http://www.epa.gov/lead
Sponsor: Environmental Protection Agency
1200 Pennsylvania Avenue NW
Washington, DC 20460

Epilepsy Postdoctoral and Clinical Research Fellowships **1820**

A few fellowships will be awarded to scientists (PhD or MD) wishing to carry out full-time research projects in the field of epilepsy. The period of support may begin at any time between May and October. The grant will be for one year but will be renewable once and exceptionally twice, upon request.
Requirements: Fellowships are available to Canadian citizens or for projects conducted in Canada.
Geographic Focus: All States, Canada
Date(s) Application is Due: Jan 15
Amount of Grant: 30,000 CAD
Contact: Caroline Savoy, (450) 358-9779; epilepsy@savoy-foundation.ca
Internet: http://www.savoy-foundation.ca/eng/mission/defaut.htm
Sponsor: Savoy Foundation
230 Foch Street
St Jean Sur Richelieu, QC J3B 6B2 Canada

Epilepsy Research Foundation New Therapy Grants **1821**

The foundation supports innovative research leading to new insights into a cure for epilepsy. The primary focus of this program is to bring new approaches and therapies to patients through translational research. Other areas of consideration include but are not limited to: meaningful grants to senior-level scientific and clinical investigators on the brink of new discoveries, working at the nation's leading academic and research institutions, and in private industry; seed funding for preliminary work necessary to explore novel approaches; building innovative platform technologies; research to bring new approaches and therapy to children; innovative cutting edge projects that could lead to breakthrough discoveries; support for applying work in other areas with potential for promise in epilepsy; research programs that might not otherwise be funded through traditional sources; projects that encourage collaboration among scientists and industry; and proposals dealing with commercializing academic research projects. Grant funds may be used to support the direct costs of research such as salary support for personnel directly engaged in the project, consumable supplies and services, and travel essential to the conduct of the research. Multi-year awards will be considered. Guidelines are available online.
Requirements: Investigators must hold a relevant advanced degree (MD and/or PhD) and have completed all research training. The program is open to investigators at corporations as well as the academic/university arena. All applications will be judged on a competitive basis, without regard to the academic rank or title of the investigator. Applicants whose research will involve patient care or direct involvement with patients must be licensed to practice their profession at their institution and must obtain Institutional Review Board (IRB) approval of their research. IRB approval must be included with the original application.
Restrictions: Grant funds may not be used for indirect costs or institutional overhead, and the purchase of permanent equipment that exceeds 10 percent of the budget without prior approval.
Date(s) Application is Due: Apr 1; Oct 1
Amount of Grant: 100,000 - 200,000 U.S.D
Contact: Research Department; (800) 332-1000 or (301) 459-3700; grants@efa.org or EpilepsyCure@aol.com
Internet: http://www.epilepsytdp.org/sec/support_translational
Sponsor: Epilepsy Foundation
4351 Garden City Drive, Suite 406
Landover, MD 20785-7223

Eppendorf & Science Prize for Neurobiology **1822**

The Eppendorf & Science Prize for Neurobiology acknowledges the increasingly active and important role of neurobiology in advancing our understanding of the functioning of the brain and the nervous system — a quest that seems destined for dramatic expansion in the coming decades. This international prize, established in 2002, encourages the work of promising young neurobiologists by providing support in the early stages of their careers. It is awarded annually for the most outstanding neurobiological research by a young scientist, as described in a 1,000-word essay based on research performed during the past three years. The winner is awarded money and publication of his or her essay in 'Science'. The essay and those of up to three finalists are also published on 'Science Online'. The award is announced and presented at a ceremony at the annual meeting of the Society for Neuroscience. Eppendorf provides financial support to help enable the grand prize winner to attend the meeting.
Requirements: Entrants must be a neurobiologist with an advanced degree and not older than 35 years. The research described in the entrant's essay must be based on the methods of molecular and cell biology. The entrant must have performed or directed the work described in the essay. The research must have been performed during the previous three years.
Restrictions: Employees of Eppendorf AG, its subsidiaries, Science and AAAS, and their relatives are not eligible for the prize.
Date(s) Application is Due: Jun 15
Amount of Grant: 25,000 U.S.D
Samples: Mauro Costa, $25,000- for essay Mattioli Switching Memories ON and OFF; Rachel Wilson, $25,000- for essay Neural Circuits Underlying Chemical Perception; Doris Tsao, $25,000- for essay A Dedicated System for Processing Faces.
Contact: Maryrose Madrid, mmadrid@aaas.org or eppendorfscienceprize@aaas.org
Internet: http://www.sciencemag.org/feature/data/prizes/eppendorf/eppenprize.dtl
Sponsor: American Association for the Advancement of Science
1200 New York Avenue NW
Washington, DC 20005-3920

Eppley Foundation Postdoctoral Grants **1823**

The foundation supports research in advanced scientific subjects, both in the biological and physical sciences, and is interested in supporting research where federal support is not available or to provide seed money toward larger undertakings. Work must be original and independent. Approximately 12 grants are awarded each year.
Requirements: Individuals who have several years of postdoctoral research experience in the biological and physical sciences are eligible. Grant proposals from foreign countries are considered only when a U.S. institution will administer the grant on their behalf.
Restrictions: Grants are not made to individuals and rarely for support of newly awarded PhDs or MDs. Social sciences, computer sciences, and education programs will not be supported. Under most circumstances, the foundation will not support heart, HIV, cancer, or DNA research.
Amount of Grant: 15,000 U.S.D
Contact: Huyler Held, Secretary-Treasurer; (212) 448-1100 or (212) 371-1660; fax (212) 448-6260; bblauner@Mclaughlinstern.com
Sponsor: Eppley Foundation for Research
260 Madison Avenue
New York, NY 10016

Equal Sweetener Foundation Grants **1824**

The foundation invests in programs that improve the health and well being of people throughout the world, and supports programs in which Merisant employees are actively involved as contributors or volunteers. The foundation prefers to support projects that fit within its focus area—diabetes research and education. Collaborative projects that achieve stated, measurable objectives receive preference. There are no deadlines for preliminary requests; full proposals are by invitation.
Requirements: 501(c)3 tax-exempt organizations are eligible. The grantseeking organization must be reputable, experienced, and show capable, responsible financial management.
Restrictions: Grants do not support individual aid or personal support; religious activities, politically partisan groups or activities, or fraternal orders; veterans or labor organizations; athletic teams; capital campaigns; endowments; scholarship programs; or projects from start-up organizations. The foundation will not contribute to any organization that discriminates on the basis of race, color, creed, sex, marital status, or physical ability.
Contact: Jo-Ann Digman, President
Internet: http://www.merisant.com/pages/about_merisant/career_center/community_involve.asp
Sponsor: Equal Sweetener Foundation
#1 N Brentwood Boulevard, Suite 510
Clayton, MO 63105

Eric Berne Fund for the Future Grants 1825

Grants support research or projects that demonstrate an interest in the evaluation of the effectiveness of various applications of transactional analysis (TA) theory. Preference is given to projects that are expected to generate particular benefits within a region or community, that extend TA to new areas of the world, support development of measurable professional standards of competency, and/or build bridges between TA and other disciplines. The deadline for making nominations for the EBMA is December 1 of the year preceding the award.
Requirements: Grants are open to individuals or nonprofit organizations. If profit-making organizations wish to submit proposals, special royalty or profit-sharing agreements may be required.
Contact: Fund Distribution Committee; (925) 600-8110; fax (925) 600-8112
Internet: http://www.itaa-net.org/itaa/grants.htm
Sponsor: International Transactional Analysis Association
2186 Rheem Drive, #B-1
Pleasanton, CA 94588

Erie Chapman Foundation Grants 1826

Established in 2007 by Erie D. Chapman III, the former CEO of Baptist Hospital System in Nashville, Tennessee. Giving primarily in Tennessee and Florida, the Foundation's mission is to support the arts and radical loving care in health care. The Foundation also provides funding for the arts in general, as well as to Christian agencies and churches. There are no specific application forms or deadlines with which to adhere, and applicants should begin the process by forwarding a letter in inquiry to Erie D. Chapman III.
Geographic Focus: Florida, Tennessee
Amount of Grant: Up to 20,000 U.S.D
Contact: Erie D. Chapman III, Chief Executive Officer and President; erie.chapman@live.com or erie_chapman@hotmail
Internet: www.eriechapmanfoundation.org
Sponsor: Erie Chapman Foundation
500 Madison Avenue
Nashville, TN 37208

Erie Community Foundation Grants 1827

The Erie Community Foundation is a collection of charitable endowments operating under the administrative umbrella of a single public charity. Grants will be awarded in four program areas: quality of life; human services; health; and education. The Foundation's goal is to help local charities more effectively and efficiently accomplish their missions. Application information is available online. Applicants are strongly encouraged to contact the Foundation's program officer prior to submitting an application.
Requirements: Erie, PA, 501(c)3 tax-exempt organizations are eligible.
Restrictions: The Foundation does not fund: program ads; fund raising events/sponsorships; more than 50 percent of the cost of a vehicle; deficit reduction; sectarian religious activities; government funding cuts; start-up organizations; school playgrounds; fire departments or nursing homes.
Geographic Focus: Pennsylvania
Date(s) Application is Due: Feb 2; Aug 3; Oct 16
Amount of Grant: 2,000 - 15,000 U.S.D
Contact: David Gonzalez, Program Officer; (814) 454-0843; fax (814) 456-4965; mbatchelor@eriecommunityfoundation.org
Internet: http://www.eriecommunityfoundation.org/for-grant-seekers/
Sponsor: Erie Community Foundation
459 West 6th Street
Erie, PA 16507

Essex County Community Foundation Discretionary Fund Grants 1828

The grant awards funds across the broad areas of nonprofit activity, including arts and culture, education, environment, health, social and community services and youth services. The Trustees will award Discretionary Fund grants to assist nonprofits with projects which strengthen the capacity of the organization to perform its work more effectively. ECCF is interested in helping organizations improve their infrastructure so that they can better serve their communities and clients. ECCF will consider projects which address capacity building such as: strengthening Board leadership; managing organizational change and growth; providing for strategic organizational planning; supporting leadership transition; strengthening fiscal management; and improving staff skills or other organizational functions which will improve the organization's long-term capacity.
Requirements: Eligible organizations must be Massachusetts 501(c)3 agencies that serve Essex County citizens with operating budgets of less than $500,000.
Restrictions: Funds are not available for: individuals; costs associated with programs or services provided to citizens outside of Essex County; sectarian or religious purposes; political purposes; debt or deficit reduction; capital campaigns for buildings, land acquisition or endowment; or to support academic research. Funding for equipment is limited to purchases that resolve a specific problem, are part of an overall capacity building project and will strengthen the operation of the organization. Equipment for programmatic purposes will not be granted. In general, staff salaries will not be eligible for funding unless the salary is directly tied to developing the capacity of the organization.
Geographic Focus: Massachusetts
Date(s) Application is Due: Feb 10
Contact: Julie Bishop, Vice President of Grants and Nonprofit Services; (978) 777-8876, ext. 28; fax (978) 777-9454; j.bishop@eccf.org or info@eccf.org
Internet: http://eccf.org/discretionary-fund-42.html
Sponsor: Essex County Community Foundation
175 Andover Street, Suite 101
Danvers, MA 01923

Essex County Community Foundation Merrimack Valley General Fund Grants 1829

The fund responds to new and important community needs as they arise. The Merrimack Valley General Fund (MVGF) considers requests from agencies serving Eastern Merrimack Valley communities in most fields of interest including: arts and culture, education, social and community services, and youth service. Particular interest is in programs which: benefit children, particularly those fostering youth development; provide health care services to those without health care coverage and address the need for affordable housing. Other areas of interest include: early childhood education; adult literacy; homelessness; elder care; job development; and food, arts and culture.
Requirements: Only non-profit organizations, recognized as tax exempt under section 501(c)3 are eligible for consideration. The MVGF provides grants to the communities of: Lawrence, Methuen, Haverhill, Andover, North Andover, Boxford, Georgetown, Groveland, W. Newbury, Merrimac, Amesbury, Newburyport, Newbury and Salisbury.
Restrictions: Funds are not available for: individuals; state or local government agencies; political purposes; and sectarian or religious purposes. Generally grants are not awarded for: debt or deficit reduction; replacing public funding, or for purposes which are generally a public sector responsibility; supporting academic research; and traveling outside the region.
Geographic Focus: Massachusetts
Date(s) Application is Due: Oct 1
Amount of Grant: 1,000 - 5,000 U.S.D
Contact: Julie Bishop, Vice President of Grants and Nonprofit Services; (978) 777-8876, ext. 28; fax (978) 777-9454; j.bishop@eccf.org or info@eccf.org
Internet: http://eccf.org/merrimack-valley-general-fund-44.html
Sponsor: Essex County Community Foundation
175 Andover Street, Suite 101
Danvers, MA 01923

Essex County Community Foundation Webster Family Fund Grants 1830

The Webster Family Fund provides grants to nonprofit organizations in which a substantial majority of clients reside in the following Greater Lawrence Communities: Lawrence, Methuen, Andover or North Andover. The fund's competitive grants are for capital projects. Preference is given to organizations that provide: youth development; health care; and arts and culture programming or services. The basis of the capital project can be an emergency situation or a capital improvement plan and associated capital campaign. Capital projects include large equipment whose useful life is over five years, as well as asset improvements (ie. bricks and mortar). For other capital grant requests, preference is given to projects that are part of an organization's strategic plan or have resulted from a feasibility study or long-term assessment of needs. For emergency capital projects, applicants should note that the grant review process for the Fund takes three and a half months, making this fund unsuitable for extremely time sensitive emergency requests. Only one-year grants are awarded. Grants typically range from $3,000 to $15,000.
Requirements: Only nonprofit organizations, recognized as tax exempt under section 501(c)3 of the Internal Revenue Code, are eligible for consideration.
Restrictions: Grants are not made to individuals, for political purposes, or for sectarian or religious purposes.
Geographic Focus: Massachusetts
Date(s) Application is Due: Sep 1
Amount of Grant: 3,000 - 15,000 U.S.D
Contact: Julie Bishop, Vice President of Grants and Nonprofit Services; (978) 777-8876, ext. 28; fax (978) 777-9454; j.bishop@eccf.org or info@eccf.org
Internet: http://eccf.org/webster-family-fund-48.html
Sponsor: Essex County Community Foundation
175 Andover Street, Suite 101
Danvers, MA 01923

Essex County Community Foundation Women's Fund Grants 1831

The Women's Fund seeks to identify existing programs or projects which offer the most significant impact for the grant dollar. The focus will be to identify programs or projects that serve women and girls in the following fields of funding interest: leadership and empowerment; economic self-sufficiency and security; and health and well being. Priority will be given to organizations that involve women and/or girls in program development and organizational leadership. The Women's Fund Advisory Board will select one field of funding interest per year. By April 1st of each year, the Women's Fund will make grants with a 1 to 3 year commitment to agencies within that year's Field of Funding Interest. Grants will be awarded up to $15,000. Agencies which receive a multiple year grant under one Field of Funding Interest may apply for other grants in subsequent years under other Field of Funding Interests.
Requirements: Non profit organizations located in Essex County which serve women and girls are eligible. Programs sponsored by religious organizations are eligible provided the enrollment is open to all qualified women and girls and the program is free of mandatory sectarian religious instruction.
Restrictions: Generally, funds will not be awarded: for debt or deficit reduction; to individuals; for political purposes; for sectarian or religious purposes; to state and local government agencies; for endowment or capital campaigns; for research; for feasibility studies; or to agencies with pending 501(c)3 status or using fiscal sponsors.
Geographic Focus: Massachusetts
Date(s) Application is Due: Dec 1
Contact: Julie Bishop, Vice President of Grants and Nonprofit Services; (978) 777-8876, ext. 28; fax (978) 777-9454; j.bishop@eccf.org or info@eccf.org
Internet: http://eccf.org/womens-fund-45.html
Sponsor: Essex County Community Foundation
175 Andover Street, Suite 101
Danvers, MA 01923

Estee Lauder Grants 1832
The company has long supported numerous institutions and programs that reflect the interests and concerns of their consumers and employees around the world. Over the years, corporate and brand philanthropy programs and community initiatives have focused on advancing the activities and growth of numerous organizations dedicated to health and human services, education, the environment and the arts. Contributions take many forms, both global and local. Submit a letter of request on company letterhead that describes the request, organization, and intended use of funds.
Amount of Grant: Up to 85,000,000 U.S.D
Contact: Contribution Program Manager; (212) 572-4200; fax (212) 527-7955
Internet: http://www.elcompanies.com/citizenship/corporate_responsibility/corporate_philanthropy.asp
Sponsor: Estee Lauder
767 Fifth Avenue, Box 100
New York, NY 10153

Ethel S. Abbott Charitable Foundation Grants 1833
The Foundation, established in 1972, offers support for art and culture, education, health, federated giving, and human service programs. The Board of Trustees will generally consider grant requests for capital projects and endowment projects more favorably than grant requests for operating funds. However, the Board will occasionally make an operating grant for the start up of a new 501(c)3 or a program which will become self-sustaining after the start up period. The application and its guidelines can be downloaded from the web site.
Requirements: The primary grant areas are: Lincoln, Nebraska and the surrounding 50 mile radius; Omaha, Nebraska and the surrounding 50 mile radius up to the Missouri River; and Western, Nebraska, which is defined as all parts of the state in the Mountain Time Zone.
Geographic Focus: Nebraska
Amount of Grant: Up to 250,000 U.S.D
Samples: University of Nebraska Foundation, Lincoln, Nebraska $1,000,000—endowment fund for program development for the Sheldon Art Gallery; Quality Living, Omaha, Nebraska, $500,000—for endowment fund for handicap accessible housing in Omaha, Nebraska.
Contact: Del Lienemann, President; (402) 435-4369; fax (402) 435-4371; info@abbottfoundation.org
Internet: http://www.abbottfoundation.org/request_a_grant/index.html
Sponsor: Ethel S. Abbott Charitable Foundation
P.O. Box 81407
Lincoln, NE 68501-1407

Ethel Sergeant Clark Smith Foundation Grants 1834
The activities of the Ethel Sergeant Clark Smith (ESCS) Memorial Fund focuses on grants to organizations located in Southeastern Pennsylvania, with primary emphasis on those serving community needs in Delaware County. Grants will be made for capital projects, operating expenses and special programs in amounts that are meaningful to the success of the individual endeavors of the organizations. However, operating expense grants are typically awarded for charities without capital requirement and under circumstances where continuing funding is not expected. Grants will be made in areas of medical, educational, cultural, arts, health and welfare, and such other areas as the trustee shall identify and determine from time to time, to be responsive to changes in community needs. Application forms are available online and must be submitted by March 1 or September 1 annually.
Requirements: Southeastern Pennsylvania 501(c)3 non-profit organizations with primary emphasis on those serving community needs in Delaware County, Pennsylvania are eligible to apply. Complete applications should include the following: one original copy of the Proposal which includes the purpose and general activities of the organization should be included as well as a description of the proposed project and its justification, a budget and timetable for the project are also required; one copy of audited financial statements for the last fiscal year (or if not audited, Internal Revenue Service form 990) plus an operating budget for the current period and budgets for future period if appropriate; copy of the Internal Revenue Service tax determination letter which shows the organization is tax-exempt under Section 501(c)3 and that it is not a private foundation under section 509 (a) of the Internal Revenue Code. Any organization that is awarded a grant will be required to sign a Grant Agreement Form prior to the distribution of funds. Approximately one year after a grant has been awarded, a Progress Report should be completed by the organization. This information must be submitted prior to the consideration of any new proposals.
Restrictions: Grants will not be considered for the following: deficit financing; construction or renovations to real estate not owned by the charitable entity; salaries; professional fund raiser fees; multi-year grants over three years; to any organization more than once in a given year; to any organization more than three years in succession; any organization receiving a grant over a three year period or in three successive years will not be eligible for a future grant until two years transpire after the three year period.
Geographic Focus: Pennsylvania
Date(s) Application is Due: Mar 1; Sep 1
Contact: Wachovia Bank, N.A., Trustee; grantinquiries4@wachovia.com
Internet: https://www.wachovia.com/foundation/v/index.jsp?vgnextoid=3b3852199c0aa110VgnVCM1000004b0d1872RCRD&vgnextfmt=default
Sponsor: Ethel Sergeant Clark Smith Foundation
620 Brandywine Parkway, Mail Code PA 5042
West Chester, PA 19380

Ethical Research Grants and Fellowships 1835
Graduate fellowships are available for investigations into the discovery, development, validation, and implementation of alternatives to the use of live animals in research, testing, and teaching. Areas of interest include tissue cultures, bacteria cultures, protozoan studies, gas chromatography, mass spectrometry, radioimmunoassay, mathematical and computer models, quantum pharmacology, mechanical models, and clinical and epidemiological surveys. Awards include stipend plus supplies. Students from all over the world are eligible to apply.
Requirements: Nonprofit educational and research institutions are eligible for grants. Students who are enrolled in master's and PhD programs in the sciences, humanities, psychology, and journalism are eligible for graudate fellowships.
Date(s) Application is Due: Mar 15
Amount of Grant: Up to 15,000 U.S.D
Contact: Peter O'Donovan, Executive Director; (312) 427-6025; fax (312) 427-6524; ifer@navs.org
Internet: http://www.ifer.org
Sponsor: International Foundation for Ethical Research
53 W Jackson Boulevard, Suite 1552
Chicago, IL 60604

Ethyl Corporation Grants 1836
The corporation supports nonprofit organizations internationally in company operating locations. Education grants are awarded to public and private institutions at the secondary, undergraduate, and graduate levels. The program matches employee grants of $25-$2500 and also supports but does not administer a scholarship fund. Economic education grants for all age groups are awarded. Health and welfare grants support programs in the areas of health care, social services, recreation, and physical fitness. Requests also are considered for culture/arts and civic affairs, including federated giving. Types of support include general operating support, building funds, community advertising, employee-matching gifts, endowment support, professorships, and scholarships. Applicants should submit proposals, including a statement of purpose of the organization, a brief description of the organization and list of current board members, proof of tax-exempt status, and a brief statement on why Ethyl should contribute to the organization.
Requirements: Organizations in company-operating locations worldwide are eligible, including the United States, Australia, Belgium, Brazil, Canada, England, France, Germany, Japan, Russia, Saudi Arabia, and Singapore.
Restrictions: Grants are not awarded to religious organizations for religious purposes, individuals for personal gain, or fraternal groups.
Date(s) Application is Due: Sep 1
Amount of Grant: 25 - 2,500 U.S.D
Contact: Human Resources and External Affairs; (804) 788-5720; fax (804) 788-5636; contributions@ethyl.com
Internet: http://www.ethyl.com/About+Ethyl/Our+History.htm
Sponsor: Ethyl Corporation
P.O. Box 2189
Richmond, VA 23218

Eugene and Agnes E. Meyer Foundation Grants 1837
The Foundation awards grants to nonprofit organizations that serve the people and communities in the Washington, D.C. region. Grants are awarded in ten program areas: arts; heritage, and culture; children, youth and families; civic engagement; education; employment and skills training; health and mental health; homelessness and hunger; housing and community development; immigrant communities; and law and justice. Candidates are required to first submit a letter of inquiry. Instructions for completion of LOI are available online.
Requirements: Eligible applicants must be a nonprofit organization with tax-exempt status and be located within and primarily serve the Washington, D.C. region which is defined as: the District of Columbia; the Maryland counties of Montgomery, Prince George's, Calvert, Charles, and St. Mary's; and the Virginia counties of Arlington, Fairfax, Loudoun, Prince William, and Stafford and cities of Alexandria, Falls Church, Manassas, and Manassas Park in Virginia.
Restrictions: The Foundation generally will not consider requests to fund: individuals, either through scholarships or other forms of financial assistance; scientific or medical research; sectarian purposes (programs that promote religious doctrine); special events or conferences; endowments.
Geographic Focus: District of Columbia, Maryland Virginia
Date(s) Application is Due: Feb 9; Jun 8; Oct 1
Contact: Julie Rogers; (202) 483-8294; fax (202) 328-6850; jrogers@meyerfdn.org
Internet: http://www.meyerfoundation.org
Sponsor: Eugene and Agnes E. Meyer Foundation
1400 16th Street, NW Suite 360
Washington, DC 20036

Eugene B. Casey Foundation Grants 1838
The foundation awards grants to nonprofit organizations in the District of Columbia and Maryland. Nonprofit organizations of the Christian and Roman Catholic faiths also are eligible. Organizations receiving grants include colleges and universities, community service organizations, government agencies, medical centers, parochial schools, and religious groups. Types of support include capital grants and general support grants. There are no application forms or deadlines. Applicants should submit written proposals that include an annual report, the purpose for which the funds are requested, and the amount requested compared with the total sought.
Requirements: Nonprofits in Maryland or the District of Columbia are eligible to apply.

Geographic Focus: District of Columbia, Maryland
Amount of Grant: 10,000 - 100,000 U.S.D
Samples: Redlands Christian Migrant Association, Immokalee, FL, $10,000—for general operating support; Cystic Fibrosis Foundation, Bethesda, MD, $20,000—for general operating support; United States Department of State, Washington, DC, $50,000—for general operating support.
Contact: Betty Brown Casey, Treasurer; (301) 948-4595
Sponsor: Eugene B. Casey Foundation
800 South Frederick Avenue, Suite 100
Gaithersburg, MD 20877-1701

Eugene G. & Margaret M. Blackford Memorial Fund Grants **1839**
The Eugene G. & Margaret M. Blackford Memorial Fund was established in 1981 to support and promote quality educational, human-services, and health-care programming for the blind or visually impaired. The deadline for application is October 1. Application materials are available for download at the grant website. Applicants will be notified of the grant decision by letter within 2 to 3 months after the proposal deadline. Applicants are strongly encouraged to review the downloadable state application guidelines before applying. Applicants are also encouraged to view the fund's funding history (link is available at the grant website).
Requirements: Applicant organizations must have 501(c)3 tax-exempt status and serve blind or visually impaired individuals living in Connecticut.
Restrictions: Applicants will not be awarded a grant for more than 3 consecutive years. The fund does not support requests from individuals, organizations attempting to influence policy through direct lobbying, or any political campaigns.
Geographic Focus: Connecticut
Date(s) Application is Due: Oct 1
Samples: National Federation of the Blind, East Hartford, Connecticut, $5,000 (2010); Fidelco Guide Dog Foundation, Bloomfield, Connecticut, $7,000 (2010); Connecticut Radio Information System, Windsor, Connecticut, $15,000 (2010).
Contact: Carmen Britt, Senior Officer; (860) 657-7019; carmen.britt@baml.com
Internet: https://www.bankofamerica.com/philanthropic/fn_search.action
Sponsor: Eugene G. & Margaret M. Blackford Memorial Fund
200 Glastonbury Boulevard, Suite # 200, CT2-545-02-05
Glastonbury, CT 06033-4056

Eugene M. Lang Foundation Grants **1840**
The foundation awards grants in New York and Pennsylvania in its areas of interest, including education (early childhood education and higher education), medical and health programs, arts, health organizations, medical research, minorities, and performing arts. Types of support include annual campaigns, conferences and seminars, continuing support, fellowships, general operating support, internship funds, professorships, program development, scholarship funds, and seed money. The foundation favors social services such as those helping homeless or single mothers. Locally based groups wanting support must involve a Lang family member. There are no application deadlines; initial approach should be by letter.
Requirements: Organizations in New York and Pennsylvania are eligible to apply.
Restrictions: Grants are not made to individuals, or for building funds, equipment and materials, capital or endowment funds, deficit financing, publications, or matching gifts.
Geographic Focus: New York, Pennsylvania
Amount of Grant: 500 - 50,000 U.S.D
Contact: Program Contact; (212) 949-4100
Sponsor: Eugene M. Lang Foundation
535 5th Avenue, Suite 906
New York, NY 10017

Eugene McDermott Foundation Grants **1841**
The foundation awards grants to Texas nonprofit organizations in its areas of interest, including children and youth, community development, education (early childhood through higher education), health care/organizations, international human rights, medical research, minorities, and social service delivery programs. Types of support include annual campaigns, building construction/renovation, capital campaigns, equipment/land acquisition, general operating grants, matching/challenge grants, professorships, programs/project support, research grants, scholarship funds, and seed grants. There are no application deadlines or forms. The board meets quarterly to consider requests.
Requirements: Texas nonprofit organizations are eligible.
Geographic Focus: Texas
Amount of Grant: 1,000 - 25,000 U.S.D
Samples: Nasher Sculpture Center (Dallas, TX)—for endowment, $500,000; U of Texas Southwestern Medical Center (Dallas, TX)—to refurbish Medical School, $400,000; U of Texas Southwestern Medical Ctr (Dallas, TX)—to establish a chair in cardiothoracic anesthesiology, $1 million.
Contact: Grants Administrator; (214) 521-2924
Sponsor: Eugene McDermott Foundation
3808 Euclid Avenue
Dallas, TX 75205

Eugene Straus Charitable Trust **1842**
The Eugene Straus Charitable Trust was established in 1974. Eugene Straus was born to a pioneer Dallas family who settled in Dallas County in the 1840's. Mr. Straus was a successful homebuilder and real estate developer. He created this trust under his will for Dallas County charitable institutions to erect and maintain permanent building improvements. The majority of grants from the Straus Charitable Trust are one year in duration; on occasion, multi-year support is awarded. Applicants must apply online at the grant website. Applicants are strongly encouraged to do the following before applying: review the downloadable state application procedures for additional helpful information and clarifications; review the downloadable online-application guidelines at the grant website; review the trust's funding history (link is available from the grant website); review the online application questions in advance; and review the list of required attachments. These will generally include: a list of board members, financial statements (audited, reviewed, or compiled by independent auditor); an organization summary; a list of other funding sources; an IRS Determination letter; and other required documents. All attachments must be uploaded in the online application as PDF, Word, or Excel files. The application deadline for the Eugene Straus Charitable Trust is 11:59 p.m. on July 31. Applicants will be notified of grant decisions before November 30.
Requirements: Applicants must have 501(c)3 tax-exempt status.
Restrictions: Grants are only awarded to organizations for the purpose of erecting and maintaining permanent building improvements in Dallas County. Requests for general operating or program-related grants will not be considered.
Geographic Focus: Texas
Date(s) Application is Due: Jul 31
Samples: Hope Cottage Pregnancy and Adoption Center, Dallas, Texas, $6,000, building improvements (2011); Dallas Children's Advocacy Center, Dallas, Texas, $6,000, building/capital campaign for a new facility (2011); Saint Philips School and Community Center, Dallas, Texas, $6,000, maintenance of campus facilities (2011).
Contact: David Ross, Senior Vice President; tx.philanthropic@baml.com
Internet: https://www.bankofamerica.com/philanthropic/fn_search.action
Sponsor: Eugene Straus Charitable Trust
901 Main Street, 19th Floor, TX1-492-19-11
Dallas, TX 75202-3714

Eva L. and Joseph M. Bruening Foundation Grants **1843**
The Foundation gives priority to grant requests that address one of the four program areas: services for vulnerable older adults; services for physically and mentally impaired individuals; secondary/higher education; and social services for the economically disadvantaged. There is no application or proposal form. Information regarding the application process is available online. Applicants are encouraged to contact the office for further clarification of the Foundation's grantmaking policies.
Requirements: Grants are awarded only to tax-exempt, nonprofit organizations located within Cuyahoga County, OH, usually for program, capital, and start-up operating support.
Restrictions: Grants are not awarded for endowment, general operating support, research, symposia or seminars. No grants are awarded to individuals, nor does the Foundation respond to mass mailings or annual campaign solicitations.
Geographic Focus: Ohio
Date(s) Application is Due: Mar 1; Jul 1; Oct 1
Samples: Jewish Family Service Assoc (OH)—to create a centralized, relational database system, $50,000; Youth Opportunities Unlimited (OH)—to help match Workforce Investment Act funds for the Jobs for Ohio's Graduates program in East Cleveland, $25,000; Towards Employment, Inc (OH)—to integrate Cleveland Works' legal assistance and ex-offender employment programs into Towards Employment's program structure, $50,000; Lutheran Chaplaincy Service (OH)—to support the full-time development director position, $25,000.
Contact: Janet Narten, Executive Director; (216) 621-2632; fax (216) 621-8198
Internet: http://www.fmscleveland.com/bruening
Sponsor: Eva L. and Joseph M. Bruening Foundation
1422 Euclid Avenue, Suite 627
Cleveland, OH 44115-1952

Evan and Susan Bayh Foundation Grants **1844**
Established in 2002 by Senator Evan Bayh and his wife Susan, the foundation gives to various education and human services programs mostly in Indiana.
Requirements: There are no deadlines and no specific forms to use. It is suggested that you contact the foundation prior to submitting an application.
Geographic Focus: Indiana
Amount of Grant: 2,500 - 5,000 U.S.D
Samples: Indiana University, Matching the Promise Campaign - $50,000; American Cancer Society, Atlanta, GA - $2,500; Deaconess Hospital, Evansville, IN - $5,000; Fisher House Foundation, Rockville, MD - $5,000; Munster Medical Research Foundation, Munster, IN - $5,000; Nature Conservancy, Arlington, VA - $2,500; St. Albans School of Public Service, Washington, DC - $5,000; Salvation Army, Indianapolis, IN - $2,500; United Way of Central Indiana, Indianapolis, IN - $2,500.
Contact: G. Frederick Glass, (317) 237-0300; fred.glass@bakerd.com
Sponsor: Evan and Susan Bayh Foundation
300 N Meridian Street, Suite 2700
Indianapolis, IN 46204-1750

Evanston Community Foundation Grants **1845**
Grants awarded by the Foundation: encourage and support new initiatives and innovative approaches to addressing community needs; build the capacity of local nonprofit organizations to fulfill their missions more effectively; encourage collaborative ventures that will strengthen the community; provide initial support of projects that will have impact beyond the scope and timeline of the proposed project; build community partnerships and resources; and strengthen the area's nonprofit community. Grants typically provide initial seed money to launch new projects, capstone dollars for a larger project that will grow over a longer period, and support a one-time activity or initial phase of a new program. The Foundation's area of interests include: arts and culture;

basic human needs; community development; education; environment; health; women and girls and youth and families. The Foundation's RFP and application instructions are available online.
Requirements: IRS 501(c)3 organizations serving the Evanston, Illinois, community are eligible.
Geographic Focus: Illinois
Date(s) Application is Due: Feb 26
Samples: Evanston Youth Bank (Evanston, IL)—for Northwestern University Undergraduate Leadership Program, $1000.
Contact: Sara Schastok, Executive Director; (847) 492-0990; fax (847) 492-0904; schastok@evcommfdn.org or info@evcommfdn.org
Internet: http://www.evcommfdn.org/grant_making.htm
Sponsor: Evanston Community Foundation
1007 Church Street, Suite 108
Evanston, IL 60201

Evert B. and Ruth Finley Person Foundation Grants 1846
The foundation supports nonprofits in Northern California, primarily Sonoma County, in the areas of music, musical arts, health associations, higher education, journalism, engineering, religion, and science education. Types of support include operating budgets, matching funds, and scholarship funds. There are no application deadlines.
Requirements: 501(c)3 nonprofit organizations in northern California may apply.
Geographic Focus: California
Amount of Grant: 5,000 - 500,000 U.S.D
Contact: Evert Person, Trustee; (707) 545-3136; fax (707) 575-5778
Sponsor: Evert B. and Ruth Finley Person Foundation
1400 N Dutton Avenue, Suite 12
Santa Rosa, CA 95401-4644

Everyone Breathe Asthma Education Grants 1847
The program was created by Sunovion Pharmaceuticals Inc., in partnership with the Asthma and Allergy Foundation of America (AAFA), to help improve the quality of asthma care and educate students about asthma. Parents or guardians of children with asthma are invited to nominate their child and child's school to receive an Everyone Breathe Asthma Education savings bond and grant. Ten applications will be selected as grand prize winners: the nominated child with asthma will be awarded a $2,500 savings bond; the school nominated in the winning application will receive a $5,000 grant to improve the quality of asthma care and asthma education in the school. Fifty additional applications will be selected as second-place winners and those nominated schools will receive asthma nebulizer kits. The grants and savings bonds will be awarded based on the level of commitment demonstrated by parents and school administrators to implement and sustain the proposed plan to improve the quality of asthma care and education at the schools. Prior to applying, parents are strongly encouraged to reach out to the administrators at their child's school to make them aware of the grant and engage them in the process of determining how the grant funds would be best utilized to improve the quality of asthma care at the school.
Requirements: Entries must be received from U.S. citizens aged 18 years or older who have a child or are a legal guardian of a child with asthma. All children nominated for the $2,500 savings bond must have asthma and be enrolled in the school (grades kindergarten through 12) that was nominated to receive the $5,000 grant. The Everyone Breathe Asthma Education Grants program is open to schools located in the 50 United States and the District of Columbia. Only public and private schools (excluding home schools) that plan to use the funds to improve the quality of asthma care and educate students about the condition are eligible. Winning schools must utilize the full amount of grant money received within 12 months from the date of the award.
Restrictions: Funding will not be considered for the following: groups that qualify as a government agency; religious schools that restrict entry to those of a certain faith or those who belong to a specific denomination or sect; schools that discriminate because of race, color, creed, gender, sexual orientation, gender identity or expression, ancestry, national origin, disability, veteran status, citizenship or any other legally protected category; capital campaigns or endowment funds; political activities or organizations; or Sunovion employees.
Date(s) Application is Due: Jul 29
Amount of Grant: 7,500 U.S.D
Contact: Grants Administrator; 888-394-7377; AsthmaAwards@Sunovion.com
Internet: http://www.everyonebreathe.com
Sponsor: Asthma and Allergy Foundation of America
8201 Corporate Drive, Suite 100
Landover, MD 20785

Evjue Foundation Grants 1848
The Foundation contributes each year to worthy educational, cultural and charitable organizations that contribute to the quality of life in Madison and Dane County. The Foundation's interests include: theater, hunger, troubled youth, and universities. The Foundation looks favorably upon projects responding to overall community needs and priorities that do not duplicate existing services. Application and guidelines are available online.
Requirements: Grants are only made to nonprofit organizations.
Restrictions: Grants are not made to individuals. All grants for scholarships are given to educational institutions, which have full responsibility for selecting the individual recipients. Grants are not usually made to establish or add to endowment funds. Grants are not ordinarily made to fund specific medical or scientific research inquiries, nor to support operating expenses and general administrative expenses of organizations.
Geographic Focus: Wisconsin
Date(s) Application is Due: Mar 5
Contact: Arlene Hornung, Director; (608) 252-6401; ahornung@madison.com
Internet: http://host.madison.com/ct/about/evjue/article_36a7a66e-a46a-11de-ac74-001cc4c002e0.html
Sponsor: Evjue Foundation
1901 Fish Hatchery Road
Madison, WI 53713

Ewing Halsell Foundation Grants 1849
The foundation awards grants to eligible Texas nonprofit organizations in its areas of interest, including education, environment, health care and health organizations, medical research, social services, and youth services. Types of support include annual campaigns, building construction/renovation, equipment acquisition, land acquisition, publication, research, seed grants, and technical assistance. There are no application deadlines or forms.
Requirements: Texas nonprofit organizations are eligible. Preference is given to requests from southwestern Texas, particularly San Antonio.
Restrictions: The foundation's grants do not support individuals or requests for deficit financing, emergency funds, general endowments, matching gifts, scholarships, fellowships, demonstration projects, general purposes, conferences, or loans.
Geographic Focus: Texas
Samples: San Antonio Botanical Society (TX)—to endow the Lucile Halsell Conservatory, $1 million matching grant.
Contact: Grants Administrator; (210) 223-2640
Sponsor: Ewing Halsell Foundation
711 Navarro Street, Suite 537
San Antonio, TX 78205

Expect Miracles Foundation Grants 1850
The MMLC Funding Award is given on an annual basis on behalf of the Miracle Maker Leadership Council to cancer fighting causes and patient care programs where such support is not always readily available. The proposed program should have the ability to materially impact an organization or a specific program of a qualified organization.
Requirements: Funding Award recipients must have 501(c)3 status as defined by the IRS. Applicants must provide service in at least one of the following areas: cancer patient care support programs and initiatives; cancer awareness and educational programs; or cancer prevention/health projects. Priority will be given to initiatives demonstrating sustainable benefits to the communities where the Foundation events are held and where Foundation supporters and the Miracle Maker Leadership Council (MMLC) members live.
Restrictions: Although there are no specific guidelines on the size, annual budget, or annual fundraising of a recipient organization, the program is not designed to assist in funding large institutional/national/society entities. The goal of the MMLC is to select programs and causes where funding is not always readily available. MMLC guidelines prohibit funding for the following: private pursuits; political parties associations and representatives of advocacy groups; organizations that discriminate by race, creed, gender, sexual orientation, age, religion, or national origin; religious causes; advertising, promotion, or sponsorship; and donation to an individual or team fundraising initiative or program.
Geographic Focus: California, Connecticut, District of Columbia, Illinois, Maine, Maryland, Massachusetts, New Hampshire, New Jersey, New York, Pennsylvania, Rhode Island, Vermont
Date(s) Application is Due: Mar 5
Amount of Grant: Up to 27,000 U.S.D
Contact: Alana Chin, (617) 391-9235 or (617) 827-7463 achin@expectmiraclesfoundation.org
Internet: http://www.expectmiraclesfoundation.org/
Sponsor: Expect Miracles Foundation
6 Quail Run
Hingham, MA 02043

Express Scripts Foundation Grants 1851
The Foundation provides financial and volunteer support for numerous causes. Requests are considered from organizations that advance medical and health-related causes, particularly those intended to support youth and strengthen families. Educational activities that enrich education and help create tomorrow's leaders are of particular interest. Programs designed to provide support that enables at-risk children to complete their educations are considered. The Foundation will consider requests from organizations that fall outside the established criteria if they enhance the communities in which the Foundation operates and make strong appeals for community support. Interested applicants should complete an expression of interest form provided online.
Requirements: All projects or programs requesting funding must be tax-exempt as described in both sections 501(c)3 and 509(a)1, 509(a)3 of the Internal Revenue Code. Organizations and programs must have broad community support and a record of fiscal and administrative stability. Preference is given to innovative projects that develop or expand services, or improve the quality of an organization's program. Typically, support will be awarded only for a term of one year; organizations must reapply for any subsequent year in which they are seeking funding.
Restrictions: The Foundation does not fund: organizations that discriminate on the basis of race, color, sex, or national origin; political causes, candidates, organizations or campaigns; fundraising activities such as benefits, charitable dinners, galas, social clubs; athletic teams or events; individual needs, including scholarships, sponsorships and other forms of financial aid; and endowment funds or capital-fund drives.
Contact: Grants Administrator
Internet: http://www.express-scripts.com/ourcompany/aboutus/foundation
Sponsor: Express Scripts Foundation
13900 Riverport Drive, Suite 20S
Maryland Heights, MO 63043

ExxonMobil Grants 1852

The corporation favors grants that relate to the conduct of U.S. business in general and the conduct of the petroleum and chemical industries in particular. The company is also concerned about social and economic conditions in corporate areas. Philanthropic giving is concentrated in the areas of the environment, public policy and public research; health; united appeals and civic and community-service organizations; minority- and women-oriented service organizations; arts, museums, and historical associations; and education. There are no application forms; submit a letter of inquiry that includes a brief history of the organization, the operating budget, a list of the board of directors and contributors, and a copy of the tax-exemption letter.

Requirements: Grants are made to tax-exempt organizations within principal company-operating areas.

Restrictions: Grants are not made to individuals for scholarships, fellowships, research, or travel. Organizations that are primarily religious in nature are ineligible.

Amount of Grant: Up to 100,000,000 U.S.D

Samples: U of Oklahoma (Norman, OK)—for a new facility that will enable engineering students to apply what they have learned in the classroom to real-world scenarios, $5 million; Path Malaria Vaccine Initiative (Bethesda, MD)—for its efforts to increase financial and political support for the development of a malaria vaccine, $500,000; To be divided among 13 organizations—for 24 community projects intended to prevent, control, and treat malaria in nine African countries; to promote the research in development of new medicines; and to advocate international support for such efforts, $5 million; American National Red Cross (Washington, DC)—for relief efforts in South Asia and Africa, $5 million.

Contact: Public Affairs; (972) 444-1103; contributions@exxonmobil.com

Internet: http://www.exxon.mobil.com

Sponsor: ExxonMobil
5959 Las Colinas Boulevard
Irving, TX 75039-2298

Eye-Bank for Sight Restoration and Fight for Sight Summer Student Research Fellowship 1853

Beginning in 2011, The Eye-Bank for Sight Restoration and Fight for Sight co-sponsored a Summer Student Research Fellowship. Annual grants of $2,100 will support full-time research conducted from June to August. The goal of this award is to advance the skills needed to initiate and carry out work in a scientific environment.

Requirements: Applicants must be currently enrolled undergraduates, medical students or graduate students studying in the United States who wish to explore ophthalmology or eye research as a career. Students are expected to complete a short, independent project during the summer months under the guidance of a senior scientist or clinician. Candidates are invited to apply by visiting www.fightforsight.org for an application.

Restrictions: Lab employees are not eligible for this fellowship.

Date(s) Application is Due: Feb 1

Amount of Grant: 2,100 - 2,100 U.S.D

Contact: Janice Benson; (212) 679-6060; fax (212) 679-4466; janice@fightforsight.org

Internet: http://www.eyedonation.org/scholarship_program.html#fight_sight

Sponsor: Eye-Bank for Sight Restoration
120 Wall Street
New York, NY 10005-3902

Ezra M. Cutting Trust Grants 1854

The Ezra M. Cutting Trust was established in 1965 under the will of Ezra Cutting, who was born in Marlborough, Massachusetts. The Trust supports charitable organizations serving residents of Marlborough. Areas of special interest include agencies serving youth as well as those with programs fostering economic growth and the general quality of life in the City of Marlborough. Applicants must apply online at the grant website. Applicants are strongly encouraged to do the following before applying: review the downloadable state application procedures for additional helpful information and clarifications; review the downloadable online-application guidelines at the grant website; review the trust's funding history (link is available from the grant website); review the online application questions in advance; and review the list of required attachments. These will generally include: a list of board members, financial statements (audited, reviewed, or compiled by independent auditor); an organization summary; a list of other funding sources; an IRS Determination letter; and other required documents. All attachments must be uploaded in the online application as PDF, Word, or Excel files. The Ezra M. Cutting Trust's deadline is 11:59 p.m. on January 15. Applicants will be notified of decisions by mid-March.

Requirements: Applicants must have 501(c)3 tax-exempt status.

Restrictions: Grants are restricted to organizations located in Marlborough, Massachusetts (and to other organizations which serve Marlborough residents) and are generally made for special projects and programs, with a limited number of capital gifts made each year. The trust does not support requests from individuals, organizations attempting to influence policy through direct lobbying, or any political campaigns.

Geographic Focus: Massachusetts

Date(s) Application is Due: Jan 15

Amount of Grant: 2,500 - 10,000 U.S.D

Samples: Friends of the Marlborough Public Library, Marlborough, Massachusetts, $5,000 (2010); Young Entrepreneurs Alliance, Roxbury, Massachusetts, $5,000 (2010); Employment Options, Marlborough, Massachusetts, $2,500 (2010).

Contact: Michealle Larkins; (866) 778-6859; michealle.larkins@baml.com

Internet: https://www.bankofamerica.com/philanthropic/fn_search.action

Sponsor: Ezra M. Cutting Trust
225 Franklin Street, 4th Floor, MA1-225-04-02
Boston, MA 02110

F.J. O'Neill Charitable Corporation Grants 1855

The corporation awards general operating grants to eligible Ohio nonprofit organizations in its areas of interest, including higher and secondary education, medical research, and Roman Catholic organizations and churches. There are no application deadlines or forms.

Requirements: Ohio nonprofit organizations serving the Cleveland area are eligible.

Geographic Focus: Ohio

Amount of Grant: 25,000 - 100,000 U.S.D

Contact: Grants Administrator; (216) 464-2121

Sponsor: F.J. O'Neill Charitable Corporation
3550 Lander Road
Cleveland, OH 44124

F.M. Kirby Foundation Grants 1856

Foundation grants are made to a wide range of nonprofit organizations in education, health and medicine, the arts and humanities, civic and public affairs, as well as religious, welfare and youth organizations. Grantees are largely in geographic areas of particular interest to the Kirby family. The Foundation has no required application format and applications will be accepted throughout the year. No solicitations by fax or email are accepted.

Requirements: North Carolina, New Jersey, and Pennsylvania tax-exempt organizations are eligible.

Restrictions: Grants are not made to individuals, public foundations or to underwrite fund-raising activities such as benefits, dinners, theater or sporting events.

Geographic Focus: New Jersey, North Carolina, Pennsylvania

Date(s) Application is Due: Oct 31

Samples: Wake Forest U (Winston-Salem, NC)—to construct a wing at the Wayne Calloway School of Business and Accountancy, $5 million; Children's Hospital of Boston, Division of Neuroscience (Boston, MA)—for research on cerebral palsy, epilepsy, Alzheimer's disease, and other diseases and degenerative disorders, $2 million; Carolina Theatre (Durham, NC)—for general operating support, $40,000.

Contact: S. Dillard Kirby, Executive Director; (973) 538-4800

Internet: http://www.fdncenter.org/grantmaker/kirby

Sponsor: F.M. Kirby Foundation
17 DeHart Street, P.O. Box 151
Morristown, NJ 07963-0151

FACES Assistance Grants 1857

The grants provide financial aid toward travel to and from comprehensive medical centers for families of patients with severe facial deformities resulting from birth defects, accidents, or disease. Applicants should submit a medical history, recent photograph, and financial statement with their request for assistance. Applications are accepted at any time.

Contact: Lynne Mayfield, Executive Director; (800) 332-2373; fax (423) 267-3124; faces@faces-cranio.org

Internet: http://www.faces-cranio.org

Sponsor: National Association for the Craniofacially Handicapped
P.O. Box 11082
Chattanooga, TN 37401

FACHE Albert W. Dent Scholarships 1858

These scholarships have been established to provide financial aid and to increase the enrollment of minority students in health care management graduate programs and to encourage students, through structured, formalized study, to obtain positions in middle and upper levels of health care management. Applications are accepted between January 1 and March 31, and scholarship awards are announced in July. Interested persons should request application forms from the director of their graduate program in health care administration or from the foundation.

Requirements: To be eligible, an applicant must be a U.S. or Canadian student associate in good standing in the ACHE, a minority student who has been enrolled for full-time study for the upcoming fall term, which is his/her final year in a health care management graduate program. Financial need also must be demonstrated.

Restrictions: Previous recipients are ineligible.

Date(s) Application is Due: Mar 31

Amount of Grant: 3,500 U.S.D

Contact: Membership Office; (312) 424-2800; fax (312) 424-0023; membership1@ache.org

Internet: http://www.ache.org/Faculty_Students/dent_scholarship.cfm

Sponsor: Foundation of the American College of Healthcare Executives
1 N Franklin Street, Suite 1700
Chicago, IL 60606-3491

Fairfield County Community Foundation Grants 1859

As a community foundation, FCCF makes discretionary grants to nonprofits in the broad program areas of economic opportunity (including affordable housing, neighborhood development, and workforce development); children, youth and families; health and human services; the environment; arts and culture; and nonprofit organizational effectiveness. The Foundation is particularly interested in proposals focused on: economic opportunity; education and youth development; advancing school readiness in Fairfield County; organizational effectiveness; and regionalism. Applicants must first submit a letter of inquiry. Application information is available online.

Requirements: Nonprofit organizations in Fairfield County, CT, are eligible.

Restrictions: Funds are not used to provide support for religious or political purposes, deficit financing, annual appeals, fundraising events, open space purchases, for-profit, parochial, charter or private schools, or nonprofit endowments. Grants are not given to individuals.

Geographic Focus: Connecticut

Samples: Cardinal Shehan Center, Bridgeport, CT, $20,000—to support the Made for Talent Program that offers youth arts education throughout the year; ACORN Housing Corporation, Bridgeport, CT, $30,000—to support ACORN's foreclosure counseling program in Bridgeport that prevents mortgage foreclosures as well as assists homeowners impacted by the subprime loan crisis; Danbury Youth Services, Danbury, CT, $20,000—to support the Target After School Program that offers after-school education to children who live in Danbury public housing.
Contact: Karen Brown, Vice President of Programs; (203) 750-3200; fax (203) 750-3232; kbrown@fccfoundation.org
Internet: http://www.fccfoundation.org/cm/grantseekers/what_we_fund.html
Sponsor: Fairfield County Community Foundation
383 Main Avenue
Norwalk, CT 06851-1543

Fairlawn Foundation Grants 1860

The Foundation is dedicated to supporting excellence and innovation in health care delivery, education and research in the Worcester area. The Foundation was established to: provide financial support to organizations, both public and private, that have specific, well-defined proposals to improve and/or expand the practice and delivery of medical and allied health care services in the Worcester area; and to support the education of Worcester area residents in health care related areas.
Requirements: Funding preference is given to institutions involved in higher education, medical or scientific research, hospitals and health care providers, and other organizations capable of making a long-term impact on health care through innovation or research. Applicants must be tax-exempt, nonprofit organizations as recognized by IRS code 501(c)3.
Restrictions: Except in unusual circumstances, the Foundation generally will not approve grants: to any program or individual in excess of $120,000; to programs that require more than five years to complete; for projects supported by other funds in which the Fairlawn Foundation contribution would comprise less than half of total funding; or for pre-existing projects as a minor participant.
Geographic Focus: Massachusetts
Date(s) Application is Due: Mar 15; Sep 15
Amount of Grant: Up to 120,000 U.S.D
Contact: Lois Smith; (508) 755-0980, ext. 107; lsmith@greaterworcester.org
Internet: http://www.greaterworcester.org/grants/Fairlawn.htm
Sponsor: Greater Worcester Community Foundation
370 Main Street, Suite 650
Worcester, MA 01608-1738

Fairview Lutheran Foundation Grants 1861

The foundation's mission is to advance the health and well-being of the Cleveland community by supporting Fairview and Lutheran Hospitals in their commitment to provide superior health-care services. The foundation also awards grants to Cleveland-area agencies and organizations that contribute to those efforts to meet the health-care needs of Cleveland's West Side. Grants are awarded usually for one year only. Application and guidelines are available online.
Requirements: Ohio 501(c)3tax-exempt organizations located in the greater western Cleveland area are eligible.
Restrictions: No funding is available for capital needs, endowments, or individuals.
Geographic Focus: Ohio
Date(s) Application is Due: Feb 16; May 18; Aug 24; Nov 16
Contact: Grants Administrator; (216) 476-7060; fax (216) 476-9730; linda@flfonline.org
Internet: http://www.flfonline.org/highlights.htm
Sponsor: Fairview Lutheran Foundation
20545 Center Ridge Road, Suite 448
Rocky River, OH 44116

Fallon OrNda Community Health Fund Grants 1862

The purpose of the health fund is to advance projects that increase access to health care or health promotion services that improve the health status of vulnerable populations. Of particular interest are projects that result in: the support or creation of primary care outreach services to vulnerable populations; the development of continuing managed care services rather than episodic or uncoordinated care; and the removal of barriers that prevent people from receiving services such as lack of transportation, cultural competency of providers, language differences, or others. Grants may be for operational expenditures such as personnel costs, or for construction, renovation, equipment purchase or other physical improvements. Past grants have ranged from $6,500 to $30,000.
Requirements: Funding will not be provided for long-term underwriting of operational costs for any one program.
Geographic Focus: Massachusetts
Date(s) Application is Due: Feb 15
Amount of Grant: 6,500 - 30,000 U.S.D
Contact: Lois Smith; (508) 755-0980, ext. 107; lsmith@greaterworcester.org
Internet: http://www.greaterworcester.org/grants/Fallon.htm
Sponsor: Greater Worcester Community Foundation
370 Main Street, Suite 650
Worcester, MA 01608-1738

Families U.S.A Wellstone Fellowship for Social Justice 1863

The annual, one-year fellowship is designed to foster the advancement of social justice through participation in health care advocacy work that focuses on the unique challenges facing many communities of color. The fellowship, from August through July, will be based in the Families U.S.A office in Washington, DC, and will afford the fellow the opportunity to learn about Medicare, Medicaid, efforts to achieve universal coverage, and other important health policy issues. At the same time, the fellow will learn about conducting health care campaigns through communication and collaboration with a network of state grassroots advocates and organizations. The goals of the program are to address disparities in access to health care; inspire fellows to continue to work for social justice throughout their lives; and increase the number and racial and ethnic diversity of up-and-coming social justice advocates and leaders. Guidelines are available online.
Requirements: Candidates must demonstrate an interest in both health care and grassroots organizing and a commitment to contributing to social justice work following their year of hands-on experience as a fellow. Preference will be given to applicants who have experience with, or demonstrate a keen interest in, working with communities of color. While there is no bias in favor of any specific academic discipline, a college degree is preferred.
Date(s) Application is Due: Jan 7
Amount of Grant: 35,000 U.S.D
Contact: Melissa Rosenblatt, (202) 628-3030; fax (202) 347-2417; wellstonefellowship@familiesusa.org
Internet: http://www.familiesusa.org/site/PageServer?pagename=Wellstone_Fellowship_About
Sponsor: Families U.S.A
1334 G Street NW
Washington, DC 20005

FAMRI Clinical Innovator Awards 1864

The purpose of the Clinical Innovator Awards (CIA) from the Flight Attendant Medical Research Institute (FAMRI) is to stimulate novel medical and clinical scientific research studies on the topic of second hand tobacco smoke. Of high priority to FAMRI are new technologies to detect second hand tobacco-caused illnesses and novel therapies to treat these diseases. FAMRI is interested in clinical and translational research, from basic molecular mechanisms of disease to occupational health. These areas include: studies on the etiology and pathogenesis of tobacco-caused respiratory diseases in humans; therapeutic interventions; clinical trials; and research to establish a protocol for inquiry regarding second hand tobacco smoke exposure by all health care professionals. FAMRI plans to provide awards of up to $100,000 per year for direct costs plus no more than 8.5% for indirect costs, subject to the availability of funds (FAMRI cannot guarantee annual and total levels of support). Awards usually will be made for three years with a possibility of a no-cost extension for one year upon request. Applicants must first submit an online letter of intent by the due date. Upon invitation, they may then submit online proposals. Instructions and links for submitting the letter of intent and proposal are available under "Current and Upcoming Awards" at the FAMRI website. Deadlines may vary from year to year. To be kept up to date of all current and future funding opportunities, interested applicants are encouraged to sign up for the FAMRI ListServ (click on "listserv signup" under the researchers tab at the FAMRI website).
Requirements: Applicants must be certain that their institution can agree to the terms of the FAMRI grant agreement. A copy of the agreement is available for review under "Current and Upcoming Awards" at the FAMRI website. Grantees must devote a minimum of 50% of professional time to research during the period of this award; at least 15% of the grantee's time (or a total of 15% of the combined times of the co-grantees) should be devoted to the research project described in the application. Researchers with an advanced degree (MD, PhD, or MD/PhD, or the equivalent) are eligible to apply. Individuals who apply to FAMRI for funding need not be U.S. citizens. They normally apply through and with an institution or organization that is qualified under United States tax laws to receive grants from FAMRI, a U.S.-based private foundation. The applying institution receives and administers FAMRI's funds and supervises the work of the Principal Investigator. If the applying institution is organized under the laws of a country other than the United States, it must provide FAMRI any and all documentation that FAMRI may require, in form and content acceptable to FAMRI, to ensure that any grant from FAMRI shall be a "qualified distribution" under Internal Revenue Service laws and regulations, including but not limited to affiliation with a qualified U.S.-based institution, a current affidavit from the applicant or an opinion of qualified legal counsel stating the facts and circumstances of the applicant's organization and activities and certifying that any grant from FAMRI shall be a "qualified distribution." Research projects that involve human subjects must follow Public Health Service guidelines (or its equivalent in non-U.S. institutions) and must have the approval of the appropriate Institutional Review Boards (IRB) or foreign equivalent prior to the start of the research. The IRB (or foreign equivalent) approval must be maintained and be in effect throughout the research project. All research must be performed in accordance with all relevant institutional and federal policies and guidelines (or foreign equivalent) relating to human subject protection, radiation and environmental health and safety. All FAMRI grantees are encouraged to collaborate with fellow grantees and may be required to participate periodically in FAMRI's collaborative efforts.
Restrictions: The Clinical Innovator Award must not exceed $100,000 per year plus 8.5 % indirect costs for a total of $108,500 per year. FAMRI will only support animal research that utilizes mice or rats. Research involving other animals will not be considered unless the animals will not be sacrificed as a result of the research. Pending Association for Assessment of Laboratory Animal Care (AALAC) approval must be submitted prior to funding of a grant involving animal subjects. To avoid all conflicts, any principal investigator or other participating applicant who is currently accepting tobacco industry funding, has recently applied for tobacco funding, or has any other association or consulting position with a tobacco company or affiliate of the tobacco industry is not eligible for FAMRI funding. All prior funding and/or association with tobacco companies must be disclosed and will be considered on a case-by-case basis. An award will not be made or will be withdrawn if the applicant becomes the Principal Investigator on a peer-reviewed grant with the same general aims as those proposed in the application to FAMRI. FAMRI reserves the right to announce the award publicly in a manner and at a time of its choosing.

Geographic Focus: All States, All Countries
Date(s) Application is Due: Sep 26
Amount of Grant: Up to 325,500 U.S.D
Samples: Diego A. Preciado, MD, PhD, Children's National Medical Center, Washington, D.C. - Tobacco Smoke Exposure Effects on Sinonasal Mucosa (2011); Noam A. Cohen, MD, PhD, Philadelphia Research & Education Foundation, Philadelphia, Pennsylvania - Induction of Bacterial Biofilms by Tobacco Smoke (2011); Jean Kim, MD, PhD, Johns Hopkins University, Baltimore, Maryland - Nasal Epithelial Growth Dysfunction by Second Hand Smoke in Chronic Sinusitis (2011)
Contact: Elizabeth Kress, Executive Director; (305) 379-7007; fax (305) 577-0005; ekress@famri.org
Internet: http://www.famri.org/current_awards/index.php
Sponsor: Flight Attendant Medical Research Institute
201 S Biscayne Boulevard, Suite 1310
Miami, FL 33131

FAMRI Young Clinical Scientist Awards **1865**
The Young Clinical Scientist Award (YSCA) is intended to prepare and support new clinical investigators with an MD or PhD as they begin their careers as independent researchers in the field of smoking-related disorders. Awards are offered to two groups of scientists: those individuals who are currently faculty members, or who expect to be before the start date of the award (YCSA faculty), and those individuals who do not expect to become faculty until after the start date of the award (YCSA fellow). Applicants must initially submit an online letter of intent (LOI) by 5 p.m. on the LOI due date, and upon subsequent invitation submit a full proposal by 5 p.m. on the proposal due date. Instructions and links for submitting the letter of intent and proposal are available under "Current and Upcoming Awards" at the FAMRI website. YCSA awards can be extended to a total of no more than five years. Renewal forms will be sent to current grantees in early February of the appropriate year and will be due late March. Grantees who are fellows should notify both the Flight Attendant Medical Research Institute (FAMRI) and the American Institute of Biological Sciences (AIBS) of any change in status. Grantees are encouraged to call or email the FAMRI Administrative Assistant to obtain current AIBS contact information. Deadlines may vary from year to year. To be kept up to date of all current and future funding opportunities, interested applicants are encouraged to sign up for the FAMRI ListServ (click on "listserv signup" under the researchers tab at the FAMRI website).
Requirements: Applicants must be certain that their institution can agree to the terms of the FAMRI grant agreement. A copy of the agreement is available for review under "Current and Upcoming Awards" at the FAMRI website. Each individual other than the institutional official must submit a tobacco disclosure form. A tobacco disclosure form is also required from the applicant's institution. Tobacco disclosure forms may be downloaded from the FAMRI website and after they are signed, they may be scanned and uploaded. Researchers with an advanced degree (M.D., Ph.D., or M.D./Ph.D., or the equivalent) are eligible to apply. A YCSA grantee must devote a minimum of 75% of professional time to research during the period of this award and at least 50% of the grantee's time should be devoted to the research project described in the application. Individuals who apply to FAMRI for funding need not be U.S. citizens. They normally apply through and with an institution or organization that is qualified under United States tax laws to receive grants from FAMRI, a U.S.-based private foundation. The applying institution receives and administers FAMRI's funds and supervises the work of the Principal Investigator. If the applying institution is organized under the laws of a country other than the United States, it must provide FAMRI any and all documentation that FAMRI may require, in form and content acceptable to FAMRI, to ensure that any grant from FAMRI shall be a "qualified distribution" under Internal Revenue Service laws and regulations, including but not limited to affiliation with a qualified U.S.-based institution, a current affidavit from the applicant or an opinion of qualified legal counsel stating the facts and circumstances of the applicant's organization and activities and certifying that any grant from FAMRI shall be a "qualified distribution." Research projects that involve human subjects must follow Public Health Service guidelines (or its equivalent in non-U.S. institutions) and must have the approval of the appropriate Institutional Review Boards (IRB) or foreign equivalent prior to the start of the research. The IRB (or foreign equivalent) approval must be maintained and be in effect throughout the research project. All research must be performed in accordance with all relevant institutional and federal policies and guidelines (or foreign equivalent) relating to human subject protection, radiation and environmental health and safety. All FAMRI grantees are encouraged to collaborate with fellow grantees and may be required to participate periodically in FAMRI's collaborative efforts.
Restrictions: The program is limited to the development of young researchers in smoking-related disorders. The YCSA/fellow award must not exceed $75,000 per year plus 8.5 % indirect costs for a total of $81,375 per year. The YCSA/faculty award must not exceed $100,000 per year plus 8.5 % indirect costs for a total of $108,500 per year. FAMRI will only support animal research that utilizes mice or rats. Research involving other animals will not be considered unless the animals will not be sacrificed as a result of the research. Pending Association for Assessment of Laboratory Animal Care (AALAC) approval must be submitted prior to funding of a grant involving animal subjects. To avoid all conflicts, any principal investigator or other participating applicant who is currently accepting tobacco industry funding, has recently applied for tobacco funding, or has any other association or consulting position with a tobacco company or affiliate of the tobacco industry is not eligible for FAMRI funding. All prior funding and/or association with tobacco companies must be disclosed and will be considered on a case-by-case basis. An award will not be made or will be withdrawn if the applicant becomes the Principal Investigator on a peer-reviewed grant with the same general aims as those proposed in the application to FAMRI. FAMRI reserves the right to announce the award publicly in a manner and at a time of its choosing.
Geographic Focus: All States, All Countries
Date(s) Application is Due: Sep 26
Amount of Grant: 162,750 - 542,500 U.S.D
Samples: Muragappa Ramanathan, Jr., M.D., John Hopkins University, Baltimore, Maryland - "Nrf2 and Tobacco Smoke Exacerbated Chronic Sinusitis" (2011); Monica Goldklang, M.D., Columbia University, New York, New York - "The Role of MMP—13 in Emphysema Development" (2011); Fabienne Gally, Ph.D., National Jewish Health, Denver, Colorado - "Cigarette Smoke Impairs FABP5—Mediated Bacterial Clearance" (2011).
Contact: Elizabeth Kress, Executive Director; (305) 379-7007; fax (305) 577-0005; ekress@famri.org
Internet: http://www.famri.org/current_awards/index.php
Sponsor: Flight Attendant Medical Research Institute
201 S Biscayne Boulevard, Suite 1310
Miami, FL 33131

Fan Fox and Leslie R. Samuels Foundation Grants **1866**
The foundation's areas of funding are the performing arts and healthcare. Healthcare funding supports patient-based and social service activities that directly help the elderly of New York City. The foundation supports performing arts organizations in the City of New York, principally, but not exclusively, in the borough of Manhattan. The foundation's primary mission is to support major performing arts institutions of national or international eminence. In addition to providing direct support, the foundation also assists presenting entities that have the requisite expertise, knowledge, and artistic judgment to present groups or individuals, new works, varied repertoire, and arts-in-education projects that will be contributions to the aesthetic and intellectual life of New York. Application guidelines are available on the Web site or upon request.
Requirements: The foundation funds organizations in the New York City area only. Only 501(c)3 tax-exempt organizations are invited to apply.
Restrictions: The foundation does not give grants to individuals or for scholarships, and does not support research, film, or video, nor does it fund education or social services. The foundation no longer actively solicits applications for support of arts-in-education programs at the primary and secondary level.
Geographic Focus: New York
Date(s) Application is Due: Mar 1; Jun 1; Dec 1
Amount of Grant: 25,000 - 250,000 U.S.D
Samples: Hebrew Home for the Aged at Riverdale (New York, NY)—for a comprehensive center dedicated to preventing and treating elder abuse, including a shelter for abused senior citizens, $200,000 over two years; Jamaica Hospital Medical Ctr (New York, NY)—to establish an interdisciplinary family-medicine program for predominantly poor, minority, and elderly patients who do not have access to palliative-care services, $150,000; Metropolitan Jewish Health System (New York, NY)—to establish a palliative-medicine program based on home visits, $201,000 over two years; Calvary Hospital (New York, NY)—for a program to increase access to comprehensive hospice and palliative care for nursing-home residents, $250,000.
Contact: Joseph Mitchell, President; (212) 239-3030; fax (212) 239-3039; info@samuels.org
Internet: http://www.samuels.org
Sponsor: Fan Fox and Leslie R. Samuels Foundation
350 Fifth Avenue, Suite 4301
New York, NY 10118

Fannie E. Rippel Foundation Grants **1867**
The foundation aids, assists, funds, equips, and provides maintenance for corporations, institutions, associations, organizations, or societies maintained for the relief and care of aged women; provides funds for the building, equipping, and maintenance of hospitals; and provides funds for corporations, institutions, and other organizations existing for treatment of and/or research on heart disease or cancer. The foundation gives emphasis to the equipment and programmatic needs of major teaching medical centers and local rural hospitals, particularly where opportunities exist for leveraging the expertise or capabilities of the medical centers/rural hospitals. Programs should reach underserved rural and urban groups, advocate preventive care, present strategies to change behaviors of the people served, and promote humanistic medicine and mind-body-spirit connections in the healing process. Preference also is given to proposed projects where the benefits can be leveraged through challenge grants.
Requirements: Organizations, associations, institutions, and hospitals in the Northeast are eligible.
Restrictions: Grants are not awarded to individuals.
Geographic Focus: Connecticut, Maine, Massachusetts, New Hampshire, Rhode Island, Vermont
Amount of Grant: 50,000 - 300,000 U.S.D
Contact: Leigh Scherrer, Foundation Associate; (973) 540-0101, ext. 305; fax (973) 540-0404; lscherrer@rippelfoundation.org
Internet: http://rippelfoundation.org/
Sponsor: Fannie E. Rippel Foundation
14 Maple Avenue, Suite 200
Morristown, NJ 07960

FAR Fund Grants **1868**
The FAR Fund Project is a New Orleans-based program exploring Hurricane Katrina's effects on New Orleans therapists and therapeutic practice. It was designed by and for clinicians. The project's mission is two fold: to offer support and concrete help to local clinicians, and to develop a model for better understanding how shared trauma affects therapists and therapy. Through this project, the FAR Fund wishes to unite and revitalize clinician communities following large-scale disasters wherever they occur, starting in New Orleans.
Requirements: The FAR Fund Project is open to New Orleans psychotherapists of all disciplines and theoretical orientations. Their mission is two-fold: to offer support and

concrete help to local mental health clinicians, and to develop a psychodynamic model to better understand how shared trauma affects therapists and therapy.
Geographic Focus: Louisiana
Contact: Shirlee Taylor, Director; (212) 982-8400; fax (212) 982-8477; FARFund@mac.com
Sponsor: Far Fund
928 Broadway, Suite 902
New York, NY 10010

Fargo-Moorhead Area Foundation Grants **1869**
The Foundation awards grants each year according to current and emerging community needs. The Board also prioritizes grants that: effect a broad segment of our community; leverage support from other sources; promote collaboration, without duplicating services; strengthen organization, self-sufficiency and long term stability; focus on problem solving; and show realistic planning and management. The Foundation's focus is on arts and culture, education, environment and animals, health, human services and public/society benefit. Applicants are required to use the FMAF grant application forms available online.
Requirements: The foundation welcomes grant requests from 501(c)3 organizations in Cass and Clay Counties in North Dakota and Minnesota or those serving residents of these counties.
Restrictions: Grants are not normally funded for: annual appeals or membership drives; capital debt reduction; capital campaigns; individuals or for-profit organizations; organizations with outstanding final reports from previous FMAF grants; organizations with outstanding due diligence reports from FMAF; political projects; religious groups for religious purposes; travel for groups and retroactive funding for any project expenses incurred before the decision date.
Geographic Focus: Minnesota, North Dakota
Date(s) Application is Due: Apr 20; Jun 1
Amount of Grant: 1,000 - 10,000 U.S.D
Contact: Cher Hersrud, Advancement Officer; (701) 234-0756; fax (701) 234-9724; cher@areafoundation.org
Internet: http://www.areafoundation.org/page5345.cfm
Sponsor: Fargo-Moorhead Area Foundation
502 First Avenue N, Suite 202
Fargo, ND 58102

Fargo-Moorhead Area Foundation Woman's Fund Grants **1870**
The Women's Fund, of the Fargo-Moorhead (FM) Area Foundation, is an endowment that will fund projects and programs that will improve the lives of women and girls in the FM area. The Fund's top three funding priorities are: children in need of care—i.e, significant percent of high risk behavior in high school girls, 13,000 children in need of after school care (Cass and Clay); 11,000 children between 0 and 5 needing child care (Cass and Clay); economic & physical well-being of women— i.e, the poverty rate and homelessness of women and households of lead by women, abuse stats; women in leadership to positively impact civic and business policy— i.e., exec roles in business, representation on boards, etc. lack of women in elected positions. The Women's Fund generally gives priority to projects that: seek to meet identified new or emerging women's or girl's issues identified by research; leverage additional support from other sources; promote cooperation among organizations without duplicating services; use a preventative approach to solving problems; show evidence of realistic organizational planning and management; will be completed within a year's time.
Requirements: The foundation welcomes grant requests from 501(c)3 tax-exempt organizations in Cass County, North Dakota or Clay County, Minnesota. Once an announcement has been made indicating that grant applications are being accepted, the Women's Fund staff will have grant applications packets available to be picked up at the Women's Fund office, accessed electronically at www.areafoundation.org, or can be emailed upon request.
Restrictions: The Women's Fund does not make grants to individuals & ordinarily doe's not support: ongoing operation expenses; annual appeals in membership drives and capital campaigns; religious groups for religious purposes; capital debt reduction; political projects; travel for groups; organizations with outstanding reports or requests; organizations outside Cass County in North Dakota and Clay County in Minnesota. Organizations are not encouraged to submit more than one request, but for those that choose to do so, they must prioritize. Grants in successive years are also discouraged. A narrative and fiscal report will be required upon completion of the project.
Geographic Focus: Minnesota, North Dakota
Amount of Grant: 300 - 2,500 U.S.D
Contact: Cher Hersrud, Advancement Officer; (701) 234-0756; fax (701) 234-9724; cher@areafoundation.org
Internet: http://www.areafoundation.org/page24016.cfm
Sponsor: Fargo-Moorhead Area Foundation
502 First Avenue N, Suite 202
Fargo, ND 58102

Farmers Insurance Group of Companies Grants **1871**
The corporate community relations program awards grants in the areas of education, public safety, arts and culture, civic improvement, and health and human services. Education giving focuses on literacy programs, mentoring programs, adopt-a-school programs, employee matching grants, and aid-to-education undergraduate scholarships. Public safety awards support tougher laws against drunk driving, drug/alcohol-free graduation night parties, neighborhood crime prevention, highway safety, and earthquake relief. Arts and culture funding supports children's programs and public television. Civic improvement focuses on recognizing exemplary youth, voter registration drives, adopt-a-highway programs, and community paint-a-thons. Health and human services giving supports March of Dimes, United Way, aid for families with cancer, aid to migrant farmworkers, and feeding the hungry. There are no application deadlines. Requests for contributions should be in the form of a letter outlining the purpose of the organization or program. The letter also should include the amount requested, its intended use, and a description of how Farmers' support will be recognized. Additional information should include a budget, annual report, proof of tax-exempt status, and a roster of the board of directors.
Requirements: 501(c)3 tax-exempt organizations are eligible.
Restrictions: Farmers does not make charitable contributions to individuals, political candidates, or religious groups or for sports events, advertising or raffle tickets, construction projects, or international programs.
Amount of Grant: Up to 43,000,000 U.S.D
Samples: John Wayne Institute for Cancer Treatment and Research (Santa Monica, CA)—for its cancer research and treatment programs, $50,000; Los Angeles Urban League (Los Angeles, CA)—to upgrade the Information Training Ctr, which teaches computer skills to low-income, minority residents, $10,000.
Contact: Angela Easton; (213) 932-3518
Internet: http://www.farmers.com/FarmComm/content/CC010153.jsp
Sponsor: Farmers Insurance Group of Companies
4680 Wilshire Boulevard
Los Angeles, CA 90010

Farm Foundation Extension Graduate Training Fellowships **1872**
The foundation provides a number of extension fellowships for graduate training in extension administration and the social sciences, with emphasis on agricultural economics, rural sociology, psychology, political science, and agricultural geography, to improve the capability of active extension workers to function in managerial and supervisory roles in the extension service. The period of study may be one quarter, one semester, or nine months. The amount of each award will be determined individually on the basis of each applicant's qualifications and needs, ranging from tuition only to tuition and additional expenses including family subsistence. Extension fellowships are granted effective June 1 for the ensuing year.
Requirements: Active federal-state agricultural extension workers in the United States are eligible to apply. Priority is given to administrators and supervisory personnel.
Date(s) Application is Due: Mar 1
Amount of Grant: Up to 5,500 U.S.D
Contact: Walter Armbruster, President; (630) 571-9393; fax (630) 571-9580; walt@farmfoundation.org
Internet: http://www.farmfoundation.org
Sponsor: Farm Foundation
1211 W 22nd Street, Suite 216
Oak Brook, IL 60523

Faye McBeath Foundation Grants **1873**
The Faye McBeath Foundation is a private, independent foundation providing grants to tax-exempt nonprofit 501 (c)3 organizations in the metropolitan Milwaukee, Wisconsin area, including Milwaukee, Waukesha, Ozaukee and Washington counties. The major areas of interest are: children; aging and elders; health; health education; civic and governmental affairs. The Foundation's Trustees are primarily interested in promising or established programs, operated by well-managed organizations. Benchmarks that will be applied to projects and organizations in all five interest areas are the following: strength of proposal - Foundation guidelines match; quality and creativity in project design, implementation and evaluation; quality of the applicant nonprofit's leadership and organizational capacity; potential to stretch or leverage McBeath funds; specific interests of Trustees and/or staff. The majority of grants support specific programs. On occasion, operating and capacity-building grants are awarded. Capital grants are limited to projects with community-wide impact that reflect the program interests of the Foundation. Any invited requests for operating or capacity-building support must include an organization's results-oriented annual plan, covering both programs and management. The trustees meet four times each year to consider grant proposals. Written notification of the Trustees' decision regarding a grant proposal will be sent within ten days of the meeting. Organizations awarded grants will receive written notification of grant conditions, payment dates and reporting requirements
Requirements: 501(c)3 nonprofit organizations in the metropolitan Milwaukee area are eligible. To begin the application process, submittal of a letter of intent, 1-2 pages long that includes: a brief description of the applicant organization, the program to be funded, the corresponding match between the Foundation's guidelines and the request, the amount requested and the estimated total project budget. Include the name and address of the organization, and the name, phone number and email address of a contact person. There are no deadlines for letters of intent. Upon receipt of a letter of intent, Foundation staff will formally acknowledge by email or letter. If the request is clearly outside the current program or geographical focus of the Foundation, a decline letter will be sent informing you of this determination. If you are formally invited by McBeath staff to submit a full proposal, you will be asked to use the Wisconsin Common Grant Application Form. The Foundation requires that a volunteer serving on the Board sign the application.
Restrictions: The Foundation does not award grants for annual fund drives, scholarships, support of individuals, sponsorship of fundraising events, or emergency funds. The Foundation does not consider or acknowledge general solicitation letters. Basic health sciences research is not funded by the Foundation. Single disease or condition organizations are typically not funded by the Foundation. An exception may be made for a creative project involving one or more of the following: service to multiple populations, collaboration and/or creation of a replicable model. There are no published deadlines

for letters of intent or grant proposals. Proposal deadlines are established based upon discussion with Foundation staff. Organizations serving the disabled will be considered only if the project targets children or older adults. Any discussion of the availability of funding in a subsequent year should be initiated by the grantee and discussed with staff at least 60 days in advance of a project anniversary.
Geographic Focus: Wisconsin
Amount of Grant: 1,000 - 50,000 U.S.D
Samples: Aurora Sinai Medical Center, Milwaukee, WI, 45,000—for improvement of access to Health Care Services; Greater Milwaukee Foundation, Milwaukee, WI, $25,000—for the Camps For Kids Forever Fund; NAMI Waukesha, Waukesha, WI, $10,000—for Housing Support and Advocacy Program.
Contact: Scott Gelzer, Executive Director; (414) 272-2626; fax (414) 272-6235; info@fayemcbeath.org
Internet: http://www.fayemcbeath.org/ProgramPriorities/index.html
Sponsor: Faye McBeath Foundation
101 West Pleasant Street, Suite 210
Milwaukee, WI 53212

Fayette County Foundation Grants **1874**
Founded in 1986, the Fayette County Foundation is the community's resource for charitable giving. The Foundation serves the entire Fayette County by assisting donors and meeting community needs. To be eligible to receive funding from the Fayette County Foundation a letter of intent must be submitted by March 1, July 1 or October 1 and followed by a completed grant application by the appropriate grant deadline. All grant seekers must have prior governing board approval for the project seeking funding. The approval of signed minutes and a signed letter from governing board must be available to the Fayette County Foundation. Faith-based organizations may apply as long as the project does not mandate participation in a religious activity as a condition for receiving services.
Requirements: Projects must have public access for all people. Completed grant application (original plus eight copies) must be received by the Foundation by the last Friday in March, the last Friday in July or the last Friday in October at Noon to be considered for funding.
Restrictions: Legal requirements forbid staff, trustees, directors, committee members and their families from profiting financially from any philanthropic grant. All persons actively connected with the Foundation will consistently strive to avoid self-interest in the processing and disposition of grant request. Only one successful application per 12 months will be considered. Repeat funding for the same project may be considered five years after the initial project on a case by case basis.
Geographic Focus: Indiana
Date(s) Application is Due: Mar 1; Jul 1; Oct 1
Samples: Crossroads Shelter, to replace roof and siding — $10,000.00; Fayette County Cancer Society, medicine, dietary needs, wigs — $6,000.00
Contact: Loree Crowe, Program Officer; (765) 827-9966; fax (765) 827-5836; info@fayettefoundation.com
Internet: http://www.fayettefoundation.com/default.asp?PageIndex=128
Sponsor: Fayette County Foundation
521 N Central Avenue, Suite A, P.O. Box 844
Connersville, IN 47331

FCD Child Development Grants **1875**
The foundation awards grants to support programs for children, particularly the disadvantaged, and promote their well-being through basic and policy-relevant research about the factors that promote optimal development of children and adolescents; policy analysis, advocacy, services, and public education to enhance the discussion and adoption of social policies that support families in their important child-raising responsibilities; and leadership development activities linked to the programmatic focus of the foundation. Grants focus on the integration of research, policy, and advocacy in two areas: the availability of and access to early childhood education programs and health care for children. Most grants support research, but a small number of direct service grants are made for New York City-based projects that advance the foundation's research and policy analysis efforts. There are no application deadlines; submit a brief letter of inquiry. Full proposals are by invitation.
Requirements: Nonprofit organizations are eligible.
Restrictions: The foundation does not consider requests for scholarships or support for individuals, capital campaigns, building purchase or renovation, or equipment purchase. The foundation does not make grants outside the United States.
Samples: Barnard College (New York, NY)—to create an internship program that will encourage and prepare undergraduate students to pursue public-service careers that benefit children and their families, $30,000; Harvard Education Letter (Cambridge, MA)—to produce and distribute nine articles, including a special edition, on the most effective ways to educate children age three to eight, $120,750; U of Wisconsin at Madison (WI)—to prepare a working paper that describes and assesses issues related to the structure, financing, and implementation of Chicago Parent-Child Centers, $30,000; National Education Assoc (Washington, DC)—to survey existing contract language, develop a policy brief on the survey results, and recommend strategies for bargaining units of the NEA to use contract language to support improved educational opportunities for children in pre-kindergarten through third grade, $104,479.
Contact: Grants Administrator; (212) 213-8337; fax (212) 213-5897; info@fcd-us.org
Internet: http://www.ffcd.org/ourwork/g-how.html
Sponsor: Foundation for Child Development
145 E 32nd Street, 14th Floor
New York, NY 10016

FCER Chiropractic Research Grants **1876**
The grants support basic and clinical research and research program development related to chiropractic. Types of support include awards/prizes, challenge/matching grants, conferences and seminars, demonstration grants, fellowships, research grants, residencies, and research contracts. Fellowships are for a maximum of five years.
Date(s) Application is Due: Mar 1; Oct 1
Amount of Grant: 10,000 - 130,000 U.S.D
Contact: Dr. Anthony Rosner, Director of Research; (617) 734-3397; fax (617) 734-0989; rosnerfcer@aol.com
Internet: http://www.fcer.org/html/Research/Grants/guidelines.htm#Instructions%20for%20Application
Sponsor: Foundation for Chiropractic Education and Research
1330 Beacon Street, Suite 315
Brookline, MA 02446-3202

FCER Research Grants and Fellowships **1877**
The fellowships support research training primarily for graduate chiropractors enrolled in academic programs in basic sciences and clinical health-related areas. Grants are for basic research and related expenses.
Requirements: Priority for funding will be based on merit and the relevance/importance of the proposed research of chiropractic theory and practice, as well as the ability of the investigator and sponsoring institutions to bring the project to successful completion.
Amount of Grant: 10,000 - 120,000 U.S.D
Samples: Robert Maust, Research Dimensions—for research on referral rates to chiropractors, $80,502; Deborah Callahan, Foundation for Chiropractic Education and Research—for research on the treatment of back pain, $9961.
Contact: Dr. Anthony Rosner, Director of Research & Education; (617) 734-3397; fax (617) 734-0989; rosnerfcer@aol.com
Internet: http://www.fcer.org/research.htm
Sponsor: Foundation for Chiropractic Education and Research
1330 Beacon Street, Suite 315
Brookline, MA 02446-3202

FDA Orphan Products Clinical Studies Grants **1878**
The objective of the grants is to support clinical trials on the safety and effectiveness of products to treat a rare disease or condition. These include conditions for which no current therapy exists or the therapy that does exist is in need of improvement. The FDA provides grants to conduct clinical studies intended to provide data that will either result in or substantially contribute to approval of these medical products. All studies of new drugs and biological products must be conducted under the FDA's investigational new drug procedure. Studies of medical devices must be conducted under the investigational device exemption procedures.
Requirements: Public or private for-profit or nonprofit organizations may apply. Profit organizations must commit to excluding fees or profit in their request for support.
Amount of Grant: Up to 100,000 U.S.D
Contact: Rosemary Springer, Chief Grants Management Officer; (301) 827-7182; fax (301) 827-7101; rspringe@oc.fda.gov
Internet: http://www.fda.gov
Sponsor: U.S. Department of Health & Human Services
10903 New Hampshire Avenue
Silver Spring, MD 20993

FDA Scientific Conferences Grants **1879**
The FDA accepts applications for grants for scientific conferences held in the United States and Canada on topics relating directly to the agency's mission and funding priorities. This initiative is intended to support and encourage state food regulatory agencies to establish (or provide support of existing) regularly scheduled Food Safety Task Force meetings. These meetings should foster communication and cooperation within the state among state and local food safety regulatory agencies and is part of the President's Food Safety Initiative (FSI).
Requirements: Awards can be made to any public or private nonprofit university, college, hospital, laboratory, or other institution, including state and local units of government. Commercial and nonprofit organizations are also eligible.
Amount of Grant: 5,000 - 3,000,000 U.S.D
Contact: Rosemary Springer, Chief Grants Management Officer; (301) 827-7182; fax (301) 827-7101; rspringe@oc.fda.gov
Internet: http://www.fda.gov
Sponsor: U.S. Department of Health & Human Services
10903 New Hampshire Avenue
Silver Spring, MD 20993

FDHN Bridging Grants **1880**
Awards of $50,000 each are available to assist investigators were not awarded their first R01 or the first renewal of their R01 to continue their gastroenterology related research. Two awards will be made following each of the three annual NIH review cycles and availability of Summary Statements. The primary objective of the award is to provide interim support to AGA Members who have submitted grants to NIH that were approved on the basis of scientific merit, but received priority scores out of the funding range.
Requirements: Applicants must: be AGA members who are applying for their first R01 or their second R01 either as a competitive renewal or a new R01 (this includes investigators going from a K award to their first R01); hold an MD, PhD, or equivalent degree (e.g., MB, ChB, MBBS, DO); hold full-time faculty positions at North American universities or professional institutes at the time of application; and have submitted a proposal to

NIH that has undergone the peer review process and was either approved and not funded or received high commendation but not funded.
Restrictions: Applicants cannot receive more than $75,000 of any additional extramural research support. Up to 25% of total amount the award may be used for salary support of the Principal Investigator. Indirect costs, including travel are not allowed.
Date(s) Application is Due: Feb 14; Jun 14; Oct 14
Amount of Grant: 50,000 U.S.D
Contact: Research Awards Manager; (301) 222-4012; fax (301) 652-3890; awards@fdhn.org
Internet: http://www.fdhn.org/wmspage.cfm?parm1=119
Sponsor: Foundation for Digestive Health and Nutrition
4930 Del Ray Avenue
Bethesda, MD 20814

FDHN Centocor International Research Fellowship in Gastrointestinal Inflammation & Immunology 1881
This award provides $50,000 to enable promising young investigators from outside the United States to spend a year at a U.S. institution engaged full-time in research related to the fundamental processes in gastrointestinal inflammation and immunology. The primary objectives of the award are: to provide the opportunity for young investigators from outside the U.S. to participate in basic research on inflammatory disease processes at prominent institutions in the U.S. (the award provides salary support for fulltime research); to initiate future international collaborative research efforts in inflammatory digestive diseases; and to address an important question in such a way as to provide a meaningful answer.
Requirements: Candidates must hold an MD or equivalent degree (eg, MB. ChB, MBBS). Applicants should currently be on the faculty of an academic institution outside the U.S. Applicants must be AGA members or be eligible for and submit an application for membership.
Restrictions: At the time of application, candidates are not allowed to hold any other similar research grant. These awards may not be renewed. No more than one application will be accepted from any institution in a given year.
Date(s) Application is Due: Jan 14
Amount of Grant: 50,000 U.S.D
Samples: Bo-In Lee, MD, PhD, Cleveland Clinic Foundation Cleveland, OH; Seamus J. Murphy, BSc, PhD, MB, BCh, Mount Sinai School of Medicine New York, NY.
Contact: Research Awards Manager; (301) 222-4012; fax (301) 652-3890; awards@fdhn.org
Internet: http://www.fdhn.org/wmspage.cfm?parm1=105
Sponsor: Foundation for Digestive Health and Nutrition
4930 Del Ray Avenue
Bethesda, MD 20814

FDHN Designated Outcomes Award in Geriatric Gastroenterology 1882
Two awards in the amount of $35,000 for one year are available to support investigator-initiated outcomes research in geriatric gastroenterology. In general, outcomes studies examine clinical outcomes, patient satisfaction, quality of life, economic evaluation, quality of care, functional status, appropriateness of care, conformance of recommended/desirable standards of performance, or change in practice patterns. The objective of this award is to promote research by young investigators in the area of outcomes, broadly defined above, as it relates to geriatric gastroenterology. Please review the AGA Future Trends Committee Report: Effects of Aging of the Population on Gastroenterology Practice, Education and Research to learn more about research topics encouraged for study. Funds may be used for salary support of personnel and technicians only, supplies and/or equipment/services. Women and minorities are strongly encouraged to apply.
Requirements: Investigators must possess an MD, PhD or equivalent and must hold faculty positions at accredited North American academic institutions by the time of the start date of the award. MD applicants: No more than five years should elapse following the completion of your clinical training (GI fellowship or equivalent) and the start date of this award. Applicants must be AGA Members.
Restrictions: The award is intended for junior faculty; therefore, established investigators are not eligible. Candidates may not hold awards on a similar topic from other agencies. Indirect costs are not allowed.
Date(s) Application is Due: Sep 5
Amount of Grant: 35,000 U.S.D
Contact: Awards Manager; (301) 222-4012; fax (301) 652-3890; awards@fdhn.org
Internet: http://www.fdhn.org/wmspage.cfm?parm1=234
Sponsor: Foundation for Digestive Health and Nutrition
4930 Del Ray Avenue
Bethesda, MD 20814

FDHN Designated Research Award in Geriatric Gastroenterology 1883
This award provides $75,000 per year for three years (total $225,000) for young investigators working toward independent careers in academic research related to geriatric gastroenterology. The overall objective is to enable young investigators to develop independent and productive research careers, with a focus on research related to geriatric gastroenterology, by ensuring that a major proportion of their time is protected for research. Non-recipient applicants for this award will be considered for the Research Scholar Awards. Applicants should review the AGA Future Trends Committee Report: Effects of Aging of the Population on Gastroenterology Practice, Education and Research to learn more about research topics encouraged for study. The award is not intended for fellows, but for young faculty who have demonstrated unusual promise and have some record of accomplishment in research.
Requirements: Candidates must hold an MD, PhD, or equivalent degree (e.g., MB, ChB, MBBS, DO). Applicants must hold full-time faculty positions at North American universities or professional institutes at the time award begins. Applicants must be Members of the AGA. # MD applicants: no more than five years shall have elapsed following the completion of your clinical training (GI fellowship or its equivalent) and the start date of this award. PhD applicants: no more than five years shall have elapsed from the completion of your postdoctoral training and the start date of this award. Candidates must devote at least 70 percent of their efforts to research related to geriatric gastroenterology.
Restrictions: Candidates should be in the beginning years of their careers; therefore, established investigators are not appropriate candidates.
Date(s) Application is Due: Sep 15
Amount of Grant: 75,000 U.S.D
Contact: Research Awards Manager; (301) 222-4012; fax (301) 652-3890; awards@fdhn.org
Internet: http://www.fdhn.org/wmspage.cfm?parm1=224
Sponsor: Foundation for Digestive Health and Nutrition
4930 Del Ray Avenue
Bethesda, MD 20814

FDHN Designated Research Award in Research Related to Pancreatitis 1884
The overall objective of this Award is to enable young investigators to develop independent and productive research careers, with a focus on pancreatic disease, by ensuring that a major proportion of their time is protected for research. The Award provides $75,000 per year for three years (total $225,000) for young investigators working toward independent careers in academic research related to understanding, improving treatments or curing Pancreatitis.
Requirements: Candidates must hold an MD, PhD, or equivalent degree (e.g., MB, ChB, MBBS, DO). Applicants must hold full-time faculty positions at North American universities or professional institutes at the time award begins. Applicants must also be members of the AGA. The award is not intended for fellows, but for young faculty who have demonstrated unusual promise and have some record of accomplishment in research. Candidates should be in the beginning years of their careers; therefore, established investigators are not appropriate candidates. MD applicants: No more than five years shall have elapsed following the completion of your clinical training (GI fellowship or its equivalent) and the start date of this award. PhD applicants: No more than five years shall have elapsed from the completion of your postdoctoral training and the start date of this award.
Date(s) Application is Due: Apr 17
Amount of Grant: 75,000 U.S.D
Contact: Awards Manager; (301) 222-4012; fax (301) 652-3890; awards@fdhn.org
Internet: http://www.fdhn.org/wmspage.cfm?parm1=100
Sponsor: Foundation for Digestive Health and Nutrition
4930 Del Ray Avenue
Bethesda, MD 20814

FDHN Fellow Abstract Prizes 1885
Two awards of $1000 each will be given to fellows who have submitted abstracts chosen to be presented during Digestive Disease Week. Awards will be presented at a ceremony during DDW. The primary objective of this award is to stimulate interest in GI research careers through competition and recognition
Requirements: Qualified candidates are MD or PhD postdoctoral fellows who are trainee members the AGA. Women and minority investigators are strongly encouraged to apply. Applicants must: be sponsored by an AGA member; and be the first author of an abstract accepted for presentation at DDW and provide evidence of abstract acceptance. Individuals with faculty appointments are not eligible. Applicants may only submit one abstract for consideration.
Restrictions: Fellows who have been awarded must present the abstract. No substitute presenters are allowed. Abstracts must have been selected for presentation at DDW. A letter of recommendation from the sponsor is required.
Date(s) Application is Due: Mar 21
Amount of Grant: 1,000 U.S.D
Contact: Awards Manager; (301) 222-4012; fax (301) 652-3890; awards@fdhn.org
Internet: http://www.fdhn.org/wmspage.cfm?parm1=124
Sponsor: Foundation for Digestive Health and Nutrition
4930 Del Ray Avenue
Bethesda, MD 20814

FDHN Fellowship to Faculty Transition Awards 1886
This award provides $40,000 per year for two years for current trainees in gastroenterology related fields so they may gain additional research training in gastrointestinal, liver function or related diseases. The objective is to prepare physicians for independent research careers in digestive diseases. The award provides salary support for additional full-time research training in basic science to acquire modern laboratory skills. The additional two years of research training provided by this award should broaden and expand the scope of investigative tools available to the recipient, generally in basic disciplines such as cell or molecular biology or immunology.
Requirements: Applicants must be MDs or MD/PhDs currently in a gastroenterology-related fellowship at an accredited North American institution, committed to academic careers. They will have completed two years of research training at the start of this award. Applicants must be AGA Trainee Members and be sponsored by an AGA Member.
Restrictions: Individuals who hold a PhD are ineligible. Although the institution may supplement the award, the applicant may not concurrently hold a similar training award or grant from another organization, such as the NIH, ALF, CCFA, or the Glaxo Institute of Digestive Health.
Date(s) Application is Due: Sep 5

Amount of Grant: 40,000 U.S.D
Samples: John Chang, MD, Fellow in Gastroenterology University of Pennsylvania, Philadelphia, PA; Lee Peng, MD, PhD, Massachusetts University, Boston, MA; Carl B. Rountree, Jr., MD, Children's Hospital, Los Angeles, CA.
Contact: Awards Manager; (301) 222-4012; fax (301) 652-3890; awards@fdhn.org
Internet: http://www.fdhn.org/wmspage.cfm?parm1=102
Sponsor: Foundation for Digestive Health and Nutrition
4930 Del Ray Avenue
Bethesda, MD 20814

FDHN Funderburg Research Scholar Award in Gastric Biology Related to Cancer 1887

This grant of $25,000 per year for two years (total $50,000) is awarded to an established investigator working on novel approaches in gastric cancer, including the fields of gastric mucosal regeneration and regulation of cell growth (not as they relate to peptic ulcer disease inflammation (including Helicobacter pylori) as precancerous lesions; genetics of gastric oncogenes in gastric epithelial malignancies; epidemiology of gastric cancer; etiology of malignancies; or clinical research in the diagnosis or treatment of gastric carcinoma. The primary objective of the award is to support an active, established investigator in the field of gastric biology who enhances the fundamental understanding of gastric cancer pathobiology in order to ultimately develop a cure for the disease.
Requirements: Applicants must hold faculty positions at accredited North American institutions and must have established themselves as independent investigators in the field of gastric biology. Women and minority investigators are strongly encouraged to apply. Applicants must be Members of the AGA.
Date(s) Application is Due: Sep 5
Amount of Grant: 25,000 U.S.D
Samples: Xiaolu Yang, PhD, Associate Professor, Department of Cancer Biology, Associate Investigator, Abramson Family Cancer Research Institute, University of Pennsylvania School of Medicine, PA; Steven Itzkowitz, MD, FACP, FACG The Dr. Burrill B. Crohn Professor of Medicine, Director, Gastroenterology Fellowship Program, Associate Director, Division of Gastroenterology, Mount Sinai School of Medicine, New York City, NY; JeanMarie Houghton, PhD, University of Massachusetts Medical School, Worchester, MA.
Contact: Awards Manager; (301) 222-4012; fax (301) 652-3890; awards@fdhn.org
Internet: http://www.fdhn.org/wmspage.cfm?parm1=90
Sponsor: Foundation for Digestive Health and Nutrition
4930 Del Ray Avenue
Bethesda, MD 20814

FDHN Graduate Student Awards 1888

Two awards of $20,000 a year for two years are offered to fund graduate students undertaking research in the biology and epidemiology of diseases of the gastrointestinal tract, liver or pancreas. The award includes $18,000 for stipend and $2,000 to be used towards fringe benefits such as medical insurance and travel to a national meeting. The primary objective of the award is to provide a salary stipend for graduate students performing doctoral research related to the gastrointestinal tract, liver or pancreas.
Requirements: Applicants should have completed at least one year and no more than three years of training towards the doctoral degree and have selected and confirmed the laboratory or department in which they will conduct doctoral research. Research is to be conducted at an accredited academic institution within North America and the research advisor must be a Member of the AGA. Applicants should be U.S. citizens, permanent residents or overseas students who have a current visa to pursue education within North America.
Restrictions: Indirect costs, excluding travel, are not allowed. Tuition costs are expected to be covered by the institution or host department and laboratory.
Date(s) Application is Due: Mar 14
Amount of Grant: 20,000 U.S.D
Contact: Awards Manager; (301) 222-4012; fax (301) 652-3890; awards@fdhn.org
Internet: http://www.fdhn.org/wmspage.cfm?parm1=154
Sponsor: Foundation for Digestive Health and Nutrition
4930 Del Ray Avenue
Bethesda, MD 20814

FDHN Jon I. Isenberg International Research Scholar Award 1889

This award provides a total of $50,000 for one year of study, $25,000 from the AGA and $25,000 in matching funds from the applicant's national GI society. The award provides for non-U.S. citizen young investigators to spend one year performing GI-related research at an American institution under the tutelage of an AGA member. Four awards will be given annually. The primary objective of this award is to promote international scholarship, increase AGA's involvement within the international GI community and foster international collaboration in training GI investigators.
Requirements: An eligible candidate must be nominated by his or her national GI Society. The national GI society must match the $25,000 contribution from the AGA. Before submitting an application, a candidate must identify a sponsoring American institution and a research preceptor who agrees in writing to supervise his or her training and research. Qualified candidates must possess a doctoral degree, either an MD, or equivalent degree, and/or a PhD degree. In keeping with the intent to support the career development of young investigators, the candidate must be within five (5) years of completing GI training or, if a PhD, within five (5) years of the receipt of the degree, at the time of initiation of the award. A documented parental leave of absence will not be counted towards the five (5) years of eligibility.
Date(s) Application is Due: Mar 1
Amount of Grant: 50,000 U.S.D
Samples: Maria Cristina, Almansa Menchero Hospital Clinico Universitario San Carlos, Madrid, Spain, to the Mayo Clinic School of Medicine, Jacksonville, FL; Koji Nozaki, MD, PhD, Department of Gastrointestinal Surgery, The University of Tokyo, Tokyo, Japan; Revital Kariv, MD, Case Western Reserve University, Beachwood, OH.
Contact: Awards Manager; (301) 222-4012; fax (301) 652-3890; awards@fdhn.org
Internet: http://www.fdhn.org/wmspage.cfm?parm1=101
Sponsor: Foundation for Digestive Health and Nutrition
4930 Del Ray Avenue
Bethesda, MD 20814

FDHN June & Donald O. Castell MD, Esophageal Clinical Research Award 1890

One award of $35,000 is made annually to provide research and/or salary support for junior faculty involved in clinical research in esophageal diseases. This award is funded by the June and Donald O. Castell, MD, Gastroenterology Research and Education Trust. The primary objective of the award is to support investigators who have demonstrated high potential to develop independent, productive research careers.
Requirements: Candidates must hold a MD or PhD or equivalent. Applicants must hold a full-time faculty position at a North American universities or professional institute. Applicants must be members of the AGA. The recipient must be at or below the level of assistant professor, and his/her initial appointment to the faculty position must have been within seven (7) years of the time of application. This award is not intended for fellows, but for junior faculty who have demonstrated unusual promise; have some record of accomplishment in research; and have established independent research programs at the time of the award. Candidates must devote at least 50 percent of their efforts to research related to esophageal function or diseases.
Restrictions: If an award recipient receives notification of another award with overlapping scientific objectives, prior to the start date of an AGA or FDHN award, the applicant must choose between the two awards.
Date(s) Application is Due: Jan 14
Amount of Grant: 35,000 U.S.D
Samples: John Pandolfino, MD, Northwestern University, Chicago, IL; Marcelo Vela, MD, Medical University of South Carolina, Charleston, SC; Braden Kuo, MD, Massachusetts General Hospital Boston, MA.
Contact: Awards Manager; (301) 222-4012; fax (301) 652-3890; awards@fdhn.org
Internet: http://www.fdhn.org/wmspage.cfm?parm1=104
Sponsor: Foundation for Digestive Health and Nutrition
4930 Del Ray Avenue
Bethesda, MD 20814

FDHN Moti L. & Kamla Rustgi International Travel Awards 1891

This program awards grants to young basic, translational and clinical investigators to support their travel and related expenses to attend Digestive Disease Week. Two awards of $500 each will be given to selected individuals residing outside North America. The primary objective of the award is to enable young investigators outside of North American (U.S. or Canada) institutions to attend Digestive Disease Week and encourage international trainees to become more involved in digestive tract, pancreatic and liver disease research.
Requirements: Candidates must: be MD or PhD or MD PhD postdoctoral fellows who are international trainee members of the AGA; be sponsored by an International member of AGA; be 35 years of age or younger at the time of the meeting; be fluent in English; and have a sufficient number of scientific papers (impact factor) published and/or poster presentations.
Date(s) Application is Due: Mar 21
Amount of Grant: 500 U.S.D
Contact: Awards Manager; (301) 222-4012; fax (301) 652-3890; awards@fdhn.org
Internet: http://www.fdhn.org/wmspage.cfm?parm1=126
Sponsor: Foundation for Digestive Health and Nutrition
4930 Del Ray Avenue
Bethesda, MD 20814

FDHN Non-Career Research Awards 1892

This award provides approximately $10,000 (per symposia) for travel support for young investigators and selected established investigators to participate in symposia on gastrointestinalrelated topics. The primary objective of the award is to foster interactions and enhance the exchange of information between clinical and basic science investigators, and established and junior investigators working in gastrointestinal research. Eligibility women and minority organizers are strongly encouraged to apply. Travel support may be provided for: junior investigators who are within 5 years of completing their clinical training in GI or from completion of their PhD doctoral thesis and are at or below the rank of assistant professor; and up to two established investigators who are invited speakers. Biosketches of these individuals and a letter addressing their contribution to the scientific content and atmosphere of the meeting should be submitted with the application.
Requirements: Within 60 days of the conclusion of the meeting, a one-page summary of the highlights of the meeting and an accounting of the use of funds must be submitted. The names of participants and speakers in the symposium, addresses, dates of birth and academic ranks of the attending scientists and their itemized expenses, must be sent to the Foundation. Any unexpended funds must be returned.
Restrictions: Indirect costs are not allowed.
Date(s) Application is Due: Feb 1; May 1; Oct 1
Contact: Awards Manager; (301) 222-4012; fax (301) 652-3890; awards@fdhn.org
Internet: http://www.fdhn.org/wmspage.cfm?parm1=123
Sponsor: Foundation for Digestive Health and Nutrition
4930 Del Ray Avenue
Bethesda, MD 20814

FDHN Non-Career Research Grants 1893

A research initiative grant of $25,000 for one year is offered to investigators to support pilot research projects in gastroenterology- or hepatology-related areas. The primary objective of the award is to provide non-salary funds for new investigators to help them establish their research careers or to support pilot projects that represent new research directions for established investigators. The intent is to stimulate research in gastroenterology- or hepatology-related areas by permitting investigators to obtain new data that can ultimately provide the basis for subsequent grant applications of more substantial funding and duration. Women and minorities are strongly encouraged to apply.
Requirements: Investigators must possess an MD or PhD degree or equivalent and must hold faculty positions at accredited North American institutions. Candidates may not hold awards on a similar topic from other agencies. Applicants must be AGA Members. Applicants for this award may not simultaneously apply for the AGA/Miles and Shirley Fiterman Foundation Basic Research Award or the AGA June and Donald O. Castell, MD, Esophageal Clinical Research Award.
Restrictions: If an award recipient receives notification of another award with overlapping scientific objectives, prior to the start date of an AGA or FDHN award, the applicant must choose between the two awards.
Date(s) Application is Due: Jan 14
Amount of Grant: 25,000 U.S.D
Contact: Awards Manager; (301) 222-4012; fax (301) 652-3890; awards@fdhn.org
Internet: http://www.fdhn.org/wmspage.cfm?parm1=121
Sponsor: Foundation for Digestive Health and Nutrition
4930 Del Ray Avenue
Bethesda, MD 20814

FDHN Research Scholar Awards 1894

These awards provide salary support for young investigators working in any area of gastrointestinal, liver function, or related diseases. The primary intent of the program is to support physician-investigators who have a high potential to develop independent, productive research careers in gastroenterology and hepatology. Candidates must devote at least 70 percent of their effort to research related to the gastrointestinal tract or liver. There must be a strong commitment from the candidate's division and department to support the candidate by protecting time for research and providing adequate laboratory space and facilities.
Requirements: Applicants must hold full-time faculty positions at North American universities or professional institutes. Nonphysician candidates with a PhD will also be considered. Candidates should be early in their research careers and commonly will have recently completed their fellowship training.
Restrictions: Indirect costs are not allowed. Candidates who have been at the assistant professor level or equivalent for more than five years are not eligible. Nor can applicants hold, or have held, an RO1, R29, K11, K08, or VA research award or any award with similar objectives from nonfederal sources (such as ALF, CCFA, or Glaxo Institute of Digestive Health). However, awards or grants obtained after receipt of this award need not be surrendered.
Date(s) Application is Due: Sep 5
Amount of Grant: 75,000 U.S.D
Samples: Michael Beyak, BSc, MD, Queen's University at Kingston—Ion channels underlying the actions of the satiety hormone CCK on gastrointestinal vagal afferents; Sean Koppe, MD, Northwestern University Feinberg School of Medicine—The Role of Fatty Acids in the Pathogenesis of Non-Alcoholic Steatohepatitis (NASH).
Contact: Awards Manager; (301) 222-4012; fax (301) 652-3890; awards@fdhn.org
Internet: http://www.fdhn.org/wmspage.cfm?parm1=103
Sponsor: Foundation for Digestive Health and Nutrition
4930 Del Ray Avenue
Bethesda, MD 20814

FDHN Student Abstract Prizes 1895

Ten travel awards of $500 each will be given to high school, college, graduate and medical students who have submitted abstracts chosen to be presented during Digestive Disease Week. The three best student abstracts submitted will receive a $1,000 prize. Awards will be presented at a ceremony during DDW. The primary objective of the award is to stimulate interest in gastroenterology research careers through competition and recognition.
Requirements: Any high school, undergraduate medical, premedical, pre-doctoral student or medical resident (up to and including postgraduate year three) who has performed original research related to gastroenterology or hepatology. Applicants must be sponsored by an AGA member.
Restrictions: Post doctoral fellows, technicians, visiting scientists and MD research fellows are not eligible for this award. Applicants may only submit one abstract for consideration and must be the designated presenter or first author of the abstract.
Date(s) Application is Due: Mar 21
Amount of Grant: 500 - 1,000 U.S.D
Contact: Awards Manager; (301) 222-4012; fax (301) 652-3890; awards@fdhn.org
Internet: http://www.fdhn.org/wmspage.cfm?parm1=152
Sponsor: Foundation for Digestive Health and Nutrition
4930 Del Ray Avenue
Bethesda, MD 20814

FDHN Student Research Fellowships 1896

This program offers financial support for students to spend a minimum of 10 weeks performing research in digestive diseases or nutrition and is intended to stimulate interest in research careers in these areas. The work may take place at any time during the year. Up to 20 fellowships will be available for full-time research with a preceptor, who must be a faculty member who directs a research project in a gastroenterology-related area at an accredited North American institution. A complete financial statement and scientific progress report are required upon completion of the program. The AGA has recognized the need to attract and encourage minority individuals to enter and pursue gastroenterology research careers. In response to this concern, seven of the student research fellowship awards will be reserved for underrepresented minority students. For the purpose of this award, minorities have been defined as African American, Hispanic, Native American/Alaskan Native, and Pacific Islander.
Requirements: Candidates may be high school, undergraduate, medical, or graduate students (not yet engaged in thesis research) in accredited North American institutions. Women and minority students are strongly encouraged to apply. The preceptor must be a faculty member who directs a research project in a gastroenterology-related area at an accredited North American institution.
Restrictions: Candidates may not hold similar salary support from other agencies: e.g., American Liver Foundation, Crohn's and Colitis Foundation.
Date(s) Application is Due: Mar 5
Amount of Grant: 2,000 - 3,000 U.S.D
Contact: Awards Manager; (301) 222-4012; fax (301) 652-3890; awards@fdhn.org
Internet: http://www.fdhn.org/wmspage.cfm?parm1=115
Sponsor: Foundation for Digestive Health and Nutrition
4930 Del Ray Avenue
Bethesda, MD 20814

FDHN TAP Endowed Designated Research Award in Acid-Related Diseases 1897

This award provides $75,000 per year for three years (total $225,000) for young investigators working toward independent careers in acid-related diseases. The overall objective is to enable young investigators to develop independent and productive careers in acid-related research by ensuring that a major proportion of their time is protected for research.
Requirements: Candidates must hold an MD, PhD, or equivalent degree (e.g., MB, ChB, MBBS, DO). Applicants must hold full-time faculty positions at North American universities or professional institutes at the time award begins. Applicants must be Members of the AGA. The award is not intended for fellows, but for young faculty who have demonstrated unusual promise and have some record of accomplishment in research. Candidates should be in the beginning years of their careers; therefore, established investigators are not appropriate candidates. Candidates must devote at least 70 percent of their efforts to research related to geriatric gastroenterolgy.
Restrictions: MD applicants: no more than five years shall have elapsed following the completion of your clinical training (GI fellowship or its equivalent) and the start date of this award. PhD applicants: no more than five years shall have elapsed from the completion of your postdoctoral training and the start date of this award.
Date(s) Application is Due: Sep 5
Amount of Grant: 75,000 U.S.D
Contact: Awards Manager; (301) 222-4012; fax (301) 652-3890; awards@fdhn.org
Internet: http://www.fdhn.org/wmspage.cfm?parm1=132
Sponsor: Foundation for Digestive Health and Nutrition
4930 Del Ray Avenue
Bethesda, MD 20814

FDHN Translational Research Awards 1898

One award of $100,000 per year for two years will be made annually to support translational research in gastroenterology and/or hepatology. Translational research will be defined as the process of applying ideas, insights and discovery generated through basic science research to the diagnosis, treatment or prevention of human disease. The primary objective of the award is to enhance interaction between researchers with basic science and clinical backgrounds with the goal of accelerating the pace of discovery that is directly applicable to patient care. The creation of teams including both a Ph.D. and an M.D. researcher is particularly encouraged.
Requirements: This award must be applied for jointly by a team of researchers (typically consisting of two members), including at least one researcher with significant training and experience in a basic science discipline (including areas outside the traditional biomedical sciences, such as physical sciences and engineering) and at least one researcher who is qualified to provide direct clinical care to patients. Junior investigators in either or both categories are particularly encouraged to apply. Applicants must be members of the AGA. Research must be conducted at an accredited North American institution.
Restrictions: Candidates may not hold support for the same project from another agency. If a proposed award recipient receives notification of another award with overlapping scientific objectives prior to the start date of an AGA or FDHN award, the applicant must choose only one of the awards to accept.
Date(s) Application is Due: Mar 14
Amount of Grant: 100,000 U.S.D
Contact: Awards Manager; (301) 222-4012; fax (301) 652-3890; awards@fdhn.org
Internet: http://www.fdhn.org/wmspage.cfm?parm1=148
Sponsor: Foundation for Digestive Health and Nutrition
4930 Del Ray Avenue
Bethesda, MD 20814

Federal Express Corporate Contributions Program 1899

The mission of the FedEx Social Responsibility department is to actively support the communities served and to strengthen global reputation through strategic investment of people, resources, and network. The foundation directs its philanthropic efforts to health and welfare programs, education, cultural/arts and civic assistance. Charitable shipping is limited to emergency, disaster or life-threatening situations coordinated through a nonprofit organization, disaster relief agency, or agency of the federal, state, or local government. FedEx is especially interested in supporting nonprofit organizations that

request 5 percent or less of a total project budget; contingency grants; or seed monies with the thought that other sources will contribute matching amounts. Organizations must show evidence of competent management, low administrative/fundraising expense ratios, and a nondiscriminatory program benefiting broad segments of the community. FedEx Community Relations responds to all requests in writing. Requests are accepted year-round and generally are reviewed within three weeks of receipt.
Requirements: Charities must be registered 501(c)3 organizations in good financial and public standing.
Contact: Grants Administrator, Community Relations; (901) 369-3600
Internet: http://www.fedex.com/us/about/responsibility/community/guidelines.html?link=4
Sponsor: Fedex Corporation
3610 Hacks Cross Road, Building A, 1st Floor
Memphis, TN 38125

Feldman Foundation Grants **1900**
Giving is concentrated in the areas of community development, social welfare, and Jewish causes. Types of support include annual campaigns, building construction/renovation, general operating support, program development, research, and scholarship funds. Initial contact should be by proposal letter. Foundation trustees meet in the spring and fall to review proposals under serious consideration. Applications are accepted at any time.
Requirements: Organizations based in New York and Texas are eligible to apply.
Restrictions: Grants will not be awarded to individuals.
Geographic Focus: New York, Texas
Amount of Grant: 10,000 - 50,000 U.S.D
Contact: Robert Feldman, Trustee; (214) 689-4337
Sponsor: Feldman Foundation
P.O. Box 1046
Dallas, TX 75221

Fellowships in Infectious Diseases **1901**
The program encourages and assists young qualified physicians to become specialists and investigators in the field of infectious diseases. The fellowship will be awarded to individuals who do not or will not have training or research grant support during the period of this grant and who are seeking support for one to three years of postdoctoral fellowship experience. Preference will be given to those applying for the third year of such support. The applicant must be sponsored by a university affiliated medical center. A letter from the chair of the infectious diseases department expressing a willingness to assume responsibility for training the applicant must accompany the application. In addition, each application must be accompanied by a letter from the host laboratory and signed by the laboratory director or appropriate department chair attesting willingness to accept the applicant and provide guidance, research space, and necessary research equipment. One thousand dollars of the award amount may be used for travel and supplies.
Requirements: Applicants must be U.S. physicians who have satisfactorily completed three or more years of postgraduate medical training (internal medicine, surgery, pediatrics, epidemiology, etc.).
Restrictions: The fellowship will not be awarded if the applicant has received or will receive a major fellowship, research grant, or traineeship in excess of the amount of this award from the federal government or another foundation.
Date(s) Application is Due: Feb 15
Amount of Grant: 25,000 U.S.D
Contact: Executive Director; (301) 656-0003; fax (301) 907-0878; info@nfid.org
Internet: http://www.nfid.org/fellow
Sponsor: National Foundation for Infectious Diseases
4733 Bethesda Avenue, Suite 750
Bethesda, MD 20814-5278

Ferree Foundation Grants **1902**
The foundation is a private family foundation dedicated to promoting and supporting excellence in the arts, culture and history, education, youth engagement, health, human services, and local economic and community development. Paul W. Ware is the Chairman of Ferree Foundation and continues the Ware family tradition of using philanthropy to enhance the lives of families and institutions primarily throughout Lancaster County and, occasionally, Southeastern Pennsylvania and beyond. Grant requests should be in writing and mailed to the executive director. Guidelines are available online.
Requirements: 501(c)3 nonprofit organizations are eligible. Preference is given to organizations serving Pennsylvania's Chester and Lancaster Counties.
Restrictions: The foundation does not make grants to individuals, nor does it make loans.
Geographic Focus: Pennsylvania
Date(s) Application is Due: Sep 15
Contact: Deb Arrive, Grants Administrator; (717) 735-8288; darrive@Ferree-Foundation.org
Internet: http://www.Ferree-Foundation.org/apply.html
Sponsor: Ferree Foundation
229 North Duke Street
Lancaster, PA 17602

FIC/NIH AIDS International Training and Research Program **1903**
The primary goal of this program is to build research capacity. This program is intended to complement ongoing HIV/AIDS and TB research efforts of the NIH and to the extent possible, of other government, nongovernment, and international organizations. Programs are encouraged to develop human resources in those developing and other participating countries that currently are or are likely to be sites of HIV/AIDS-related research on HIV and/or TB vaccines, anti-HIV and TB drugs, and other HIV and TB interventions. A letter of intent is requested by November 21; full application is due December 21. Annual deadline dates may change; contact program staff for exact dates. PAR-05-140)
Requirements: Countries in Central and Eastern Europe and the New Independent States are eligible to participate, as are countries in Africa, the Americas, Asia, and the Pacific Region. The grantee institution must be a U.S. nonprofit private or public institution capable of meeting the objectives in this RFA.
Restrictions: Institutions that currently hold an award are not eligible to submit a second application.
Date(s) Application is Due: Nov 21; Dec 21
Contact: Dr. Jeanne McDermott, Program Officer; (301) 496-1653; fax (301) 402-0779; mcdermoj@mail.nih.gov
Internet: http://www.fic.nih.gov/programs/aitrp/aitrp.html
Sponsor: Fogarty International Center
31 Center Drive, Building 31
Bethesda, MD 20892-2220

FIC/ORMH Minority International Research Training Grants **1904**
FIC and ORMH jointly support scientific training programs that offer international research training opportunities to qualified minority undergraduates and graduate/medical students underrepresented in biomedical and behavioral research careers. Applications are invited from innovative programs that provide international research and training to encourage underrepresented minority students to pursue biomedical research careers, broaden minority research efforts and scientific training to encompass international health problems, stimulate novel approaches to studying health problems that disproportionately affect underserved groups in the United States or in developing countries, and assist minority scientists to participate in international collaborative relationships and work effectively in the rapidly evolving global scientific environment. A letter of intent is requested by December 10; full application is due January 14. Annual deadline dates may vary; contact program staff for exact dates. RFA: TW-00-001
Requirements: Applicant institution and its associated consortia institutions must be colleges or universities that offer baccalaureate degrees in fields relevant to biomedical science. Only one application per institution will be accepted. Priority is given to consortia which include historically black colleges and universities, hispanic serving institutions, or tribal colleges and universities. Participant students and faculty must be U.S. citizens or permanent residents.
Date(s) Application is Due: Jan 14; Dec 10
Amount of Grant: Up to 18,000,000 U.S.D
Contact: Dr. Barbara Sina, International Training and Research Coordinator; (301) 496-1653; fax (301) 402-0779; barbara_sina@nih.gov
Internet: http://www.fic.nih.gov/programs/mirt.html
Sponsor: Fogarty International Center
31 Center Drive, Building 31
Bethesda, MD 20892-2220

FIC International Cooperative Biodiversity Groups Grants **1905**
Funding for this program has been provided by six components of the National Institutes of Health (NIH), the Biological Sciences Directorate of the National Science Foundation (NSF) and the Foreign Agriculture Service of the U.S.DA. The cooperating NIH components are the Fogarty International Center (FIC), National Cancer Institute (NCI), National Institute of Allergy and Infectious Diseases (NIAID), National Institute of Mental Health (NIMH), National Institute on Drug Abuse (NIDA) and the National Heart, Lung, and Blood Institute (NHLBI). The purpose of these groups will be to address the interdependent issues of biodiversity conservation, sustained economic growth, and human health in terms of drug discovery for cancer, infectious diseases including AIDS, cardiovascular diseases, mental disorders, and diseases of primary concern to developing countries. A group, under a single group leader (principal investigator), is expected to be a consortium of associate programs working together to form a multidisciplinary and/or multi-institutional team from academic, nonprofit, and/or commercial organizations. At least one of the group's associate programs must be located in a developing country. It is strongly advised that applicants contact program staff early in the planning process to discuss prospective applications and to obtain supplemental clarifying information and instructions. RFA-TW-04-004
Requirements: Public and private nonprofit institutions, governments and their agencies, and foreign institutions are eligible. Applicant institutions must be in the United States or in a participating developing country. For-profit institutions may participate as members of the group.
Amount of Grant: 500,000 - 600,000 U.S.D
Contact: Dr. Joshua Rosenthal, Director; (301) 496-1653; fax (301) 402-2056; Joshua_Rosenthal@nih.gov
Internet: http://www.fic.nih.gov/programs/grants.html#research
Sponsor: Fogarty International Center
31 Center Drive, Building 31
Bethesda, MD 20892-2220

FIC International Malaria Research Training Program Award **1906**
FIC invites applications to train or expand the capabilities of scientists and health professionals from malaria endemic developing countries to engage in malaria research. Proposals are requested that would create innovative, collaborative malaria research training programs and that would contribute to the long-term goal of harnessing scientific knowledge and skills to enhance efforts to prevent malaria-related morbidity and mortality and to control malaria transmission in endemic developing countries. To

develop sustainable research capacity in endemic developing countries, FIC will support malaria research training efforts as part of ongoing, productive malaria research projects in collaboration with internationally recognized research institutions in Africa and other highly endemic regions of the world.Annual deadlines may vary; contact program staff for exact dates. RFA: TW-00-006

Requirements: Applications may be submitted by U.S. nonprofit public and private institutions, such as universities, colleges, hospitals, laboratories, and eligible agencies of the federal government capable of meeting the objectives of the RFA.

Restrictions: Principal Investigators currently supported by other FIC malaria focused research training awards (D43) are not eligible to apply.

Amount of Grant: Up to 45,000 U.S.D

Contact: Grants Administrator; (301) 496-2075; fax (301) 594-1211; ficinfo@nih.gov

Internet: http://www.fic.nih.gov/programs/malaria.html

Sponsor: Fogarty International Center
31 Center Drive, Building 31
Bethesda, MD 20892-2220

FIC International Research and Training in Population and Health Grants 1907

FIC, NICHD, and NIA have developed this program to support international research and training in population-related sciences and to enable NIH grant recipients to extend the geographic base of research and training efforts to developing nations, in support of international population priorities. Types of training for foreign scientists or health professionals may include predoctoral training in research related to population; postdoctoral training in laboratory procedures and research projects and techniques conducted at the host U.S. institutions or in the trainee's home country; and participation in advanced research training conducted by U.S. faculty in the host country and also short-term in-country training for foreign scientists and health professionals in the host country. A letter of intent is requested by December 13; full application is due January 13. Annual deadline dates may vary; contact program staff for exact dates. RFA-TW-05-002)

Requirements: Applicants must be U.S. principal investigators on at least one NIH-sponsored research grant, cooperative agreement, or contract. The grantee institution must be a U.S. nonprofit private or public institution.

Date(s) Application is Due: Jan 13; Dec 13

Amount of Grant: 139,000 - 185,000 U.S.D

Contact: Dr. Jeanne McDermott, Division of International Training and Research; (301) 496-1492; fax (301) 402-0779; mcdermoj@mail.nih.gov

Internet: http://grants.nih.gov/grants/guide/rfa-files/RFA-TW-05-002.html

Sponsor: Fogarty International Center
31 Center Drive, Building 31
Bethesda, MD 20892-2220

FIC International Research Fellowship Program 1908

The program provides opportunities for non-U.S. postdoctoral biomedical or behavioral scientists in the formative stages of their research careers to gain further research experience by working in the laboratories of distinguished U.S. scientists on problems of mutual interest. Nominations are made by the national nominating committee in each participating country. These committees submit applications to the Fogarty International Center. More than 50 countries or regions in the Americas, Africa, Asia, the Far East, Australia, Europe, and New Zealand participate in the program. Fellowships are awarded for a minimum of 12 months and provide stipend, travel, and institutional allowance. Fellowships provide round-trip travel for the fellow only. In addition, the host institution receives a modest allowance to cover such costs as the fellow's health insurance, supplies, equipment, and travel to scientific meetings or laboratories in the United States or Canada.

Requirements: Candidate must hold a doctoral degree or equivalent in a health science field at the time of submission of application to a nominating committee, have demonstrated the ability to engage in independent basic or clinical research, have a sponsor in the United States at a nonprofit institution who has accepted the applicant for research on his/her proposed project, and have assurance from a nonprofit institution in the home country that there is a position for him/her upon return. Applicant must be proficient in spoken and written English.

Contact: Program Officer; (301) 496-1653; fax (301) 402-0779/2056; nugentr@mail.nih.gov

Internet: http://www.fic.nih.gov/programs/grants.html#fellowships

Sponsor: Fogarty International Center
31 Center Drive, Building 31
Bethesda, MD 20892-2220

FIC International Research Scientist Development Award 1909

The award supports basic research for behavioral and clinical scientists who are committed to a career in international health research and would benefit from an additional period of mentored research as part of a strong, established collaboration between a U.S. sponsor and leading scientists at a developing country center of scientific excellence. The specific research training may be new to the candidate or in an area that would demonstrably enhance the candidate's pursuit of a career focusing on international research pertinent to health in developing countries. Applications are encouraged to address one of the global health research priorities defined by the World Health Organization (http://www.who.ch), which include both infectious and chronic/degenerative conditions and mental health disorders. It is expected that following this experience, the candidate will be able to pursue an independent and productive international research career involving ongoing collaboration with developing country scientists to more effectively pursue research relevant to stemming a major global health problem. Annual deadlines may vary; contact program staff for exact dates. PAR-04-058

Requirements: An applicant must be a U.S. citizen or non-citizen national; hold doctoral or medical degree or the equivalent in a health science field; have earned the relevant degree within seven years of the application receipt date (not including clinical training); have demonstrated a commitment and competence in international health research as well as the potential to engage in independent and productive basic biomedical, behavioral, or epidemiological/clinical research in the period following the award; and have a sponsor in an internationally recognized developing country research institution addressing one or more of that country's major health problems. Eligible countries are in regions that include Africa, Asia (except Japan, Singapore, South Korea, and Taiwan), Central and Eastern Europe, Latin America and the Caribbean, the Middle East (except Israel) and the Pacific Ocean Islands (except Australia and New Zealand). Contact the program office or visit the Web site for full eligibility requirements.

Date(s) Application is Due: Feb 16

Amount of Grant: Up to 70,000 U.S.D

Contact: Dr. Rachel Nugent, Division of International Training and Research Contact; (301) 496-8733; fax (301) 402-0779; nugentra@mail.nih.gov

Internet: http://www.fic.nih.gov/programs/irsda.html

Sponsor: Fogarty International Center
31 Center Drive, Building 31
Bethesda, MD 20892-2220

FIC International Training and Research in Emerging Infectious Diseases Grants 1910

The Fogarty Center and NIAID have developed this program to train laboratory scientists and public health workers in developing countries and the United States in emerging and re-emerging infectious diseases research, control, and prevention strategies and their implementation and evaluation; facilitate international collaboration on emerging infectious diseases research, including the conduct of research in developing countries; enhance domestic infectious diseases research programs and improve the protection of the U.S. population from infectious diseases by early detection and response to epidemics internationally and nationally; and develop international leaders in science and public health. Training will include predoctoral and postdoctoral programs, participation in advanced and/or short-term in-country training conducted by U.S. faculty in the host country; and participation in training of health workers in the diagnosis, patient management, control, and prevention of disease. RFA: TW-96-001

Requirements: Applicant must be a U.S. principal investigator of at least one NIH-sponsored research grant related to infectious diseases.

Contact: Dr. Joel Breman, International Training and Research Contact; (301) 496-0815; Joel_Breman@.nih.gov

Internet: http://www.fic.nih.gov/programs/erid.html

Sponsor: Fogarty International Center
31 Center Drive, Building 31
Bethesda, MD 20892-2220

FIC International Training and Research in Environmental and Occupational Health 1911

This program, sponsored by the Fogarty International Center, NIEHS, NIOSH, and CDC, was developed to train foreign health scientists, clinicians, epidemiologists, toxicologists, engineers, industrial hygienists, chemists, and allied health workers from developing countries and emerging democracies in both general environmental health and occupational health. Types of training may include training in epidemiology concepts and methods, environmental monitoring, industrial hygiene, field studies, and other research related to environmental and occupational health that will lead to the MS or PhD degree for individuals with previous field research experience; short-term comprehensive courses in epidemiology, toxicology, chemistry, industrial hygiene, and environmental and safety engineering, with an emphasis on control of occupational injuries and illnesses, for health and safety professionals to be given in the United States; training in laboratory procedures and research techniques related to environmental and occupational health for individuals with the MS or PhD degree to be given in the United States; and postdoctoral research training for foreign scientists who want to expand their abilities in the epidemiology, diagnosis, prevention, and treatment of environmental and occupational disease and injury. Postdoctoral training can take place both in the United States and in foreign countries. RFA: TW-01-002.

Requirements: The grantee institution must be a U.S., nonprofit private or public institution. Although most applications will be from academic institutions, a nonacademic, nonprofit institution may apply.

Contact: Dr. Christopher Schonwalder, Senior Environmental Health Advisor to the Director; (919) 541-4794; fax (919) 541-2583; cs64c@nih.gov

Internet: http://www.fic.nih.gov/programs/environ.html

Sponsor: Fogarty International Center
31 Center Drive, Building 31
Bethesda, MD 20892-2220

FIC Israeli Ministry of Health Postdoctoral Research Fellowships 1912

The program provides postdoctoral fellowships to U.S. health scientists to conduct biomedical research in Israel. The purpose is to enhance the exchange of research experience and information in the biomedical and behavioral sciences with emphasis on heart diseases, aging, cancer, human reproduction, and child development. Short- and long-term fellowships are available for appointments at Hebrew University, Hadassah Medical School; Sackler Faculty of Medicine, Tel Aviv University; Weizmann Institute of Science; Technion-Israel Institute of Technology; and Ben Gurion University of the Negev. Information and applications are available from NIH Fogarty International Center; funding is from the Israeli Ministry of Health.

Requirements: Applicants must be citizens or permanent residents of the United States; have doctorates in one of the clinical, biomedical, or behavioral sciences; and have professional experience in the proposed area.

Restrictions: Support is not provided for brief observational visits, attendance at scientific meetings, or independent study.
Contact: Bruce Butrum, Grants Manager; (301) 496-1653
Internet: http://www.fic.nih.gov
Sponsor: Fogarty International Center
31 Center Drive, Building 31
Bethesda, MD 20892-2220

FIC Swedish Medical Research Council Fellowships 1913
The Swedish Medical Research Council (SMRC) provides a limited number of research fellowships to U.S. health scientists to conduct research in Sweden and to enhance the exchange of research experience and information in the biomedical and behavioral sciences. Activities supported include collaboration in basic or clinical research and the familiarization with or utilization of special techniques and equipment not otherwise available to the applicant. The program is administered by the Fogarty International Center and funded by the SMRC; applications may be obtained from the center.
Requirements: Candidate must be a citizen or permanent resident of the United States; have a doctorate in clinical, behavioral, or biomedical science; have 10 years or less of postdoctoral experience; and have professional experience in the health sciences for at least two of the last four years. A letter of invitation from the host sponsor is required.
Restrictions: Support is not provided for brief observational visits, attendance at scientific meetings, or independent study.
Amount of Grant: 30,000 - 45,000 U.S.D
Contact: Bruce Butrum, Grants Manager; (301) 496-1653
Internet: http://www.fic.nih.gov/programs/grants.html#fellowships
Sponsor: Fogarty International Center
31 Center Drive, Building 31
Bethesda, MD 20892-2220

Fidelity Foundation Grants 1914
The foundation awards grants to nonprofits in community-operating areas working in the fields of community development, cultural affairs, education, and health. Fields of interest include arts and cultural programs, education, health care and health organizations, medical research, and community development. Types of support include capital campaigns, building/renovations, endowment funds, program development, technical assistance, and employee matching gifts.
Requirements: Nonprofit organizations in community-operating areas, particularly in Covington, KY, Boston and Marlborough, MA; Merrimack, NH; New York, NY; Cincinnati, OH; Smithfield, RI; Dallas/Fort Worth, TX; Salt Lake City, UT; and Toronto, Canada, are eligible.
Restrictions: Grants are not awarded to support sectarian organizations, disease-specific associations, or public school systems. Support does not go to individuals or for scholarships, civic or start-up organizations, corporate memberships, operating support, or participation in benefit events, film, or video projects. Multiyear requests will be denied.
Geographic Focus: Kentucky, Massachusetts, New Hampshire, New York, Ohio, Rhode Island, Texas, Utah, Canada
Samples: Ctr for Information and Crisis Services (Lantana, FL)—for organizational development and to improve their volunteer programs, $50,000; YMCA of Metropolitan Dallas (Dallas, TX)—for capital support of the new Irving Family YMCA at North Lake, $40,000.
Contact: Margaret Morton, Senior Vice President, Program; (617) 563-6806
Internet: http://www.fidelityfoundation.org
Sponsor: Fidelity Foundation
82 Devonshire Street, S2
Boston, MA 02109

Field Foundation of Illinois Grants 1915
The Foundation awards grants only to institutions and agencies operating in the fields of urban and community affairs, culture, education, community welfare, health, and environment, primarily serving the people of Chicago with extremely limited grant making in the metropolitan area. Preference will be given to funding innovative approaches for addressing program areas.
Requirements: Applicants must reside in Illinois. Grant applications are not provided; however, a formal proposal is required. The Foundation does not accept grant requests via email or fax.
Restrictions: No grants will be made to support: endowments; individuals; medical research or national health agency appeals; propaganda organizations or committees whose efforts are aimed at influencing legislation; printed materials, video or computer equipment; fund-raising events or advertising; appeals for religious purposes; other granting agencies or foundations for ultimate distribution to agencies or programs of its own choosing; custodian afterschool programs or tutoring; most disease specific programs, research or activities; or operating support of established neighborhood health centers or clinics, day care centers for children, or small cultural groups. Requests for computer equipment will not be considered.
Geographic Focus: Illinois
Date(s) Application is Due: Jan 15; May 15; Sep 15
Amount of Grant: 50,000 U.S.D
Contact: Joann Ross, (312) 831-0910; fax (312) 831-0961; jross@fieldfoundation.org
Internet: http://www.fieldfoundation.org
Sponsor: Field Foundation of Illinois
200 S Wacker Drive, Suite 3860
Chicago, IL 60606

Field Psychology Research Grants 1916
Grants support promising graduate students exploring the history, contributions, and further development of perceptual (field) psychology in relationship to the research and writings of Arthur W. Combs, PhD. Applications of Dr. Combs's theoretical position to the improvement of educational, community, conflict-resolution, therapeutic efforts, etc. are also encouraged.
Date(s) Application is Due: Jan 31; Oct 5
Amount of Grant: 500 - 1,500 U.S.D
Contact: Anne Richards, Trustee; (770) 834-8143; arichard@westga.edu
Internet: http://www.fieldpsychtrust.org/application.asp
Sponsor: Field Psychology Trust
301 Dixie Street
Carrollton, GA 30117

Fifth Third Foundation Grants 1917
The foundation awards grants to eligible nonprofit organizations in its areas of interest, including community development, education, health and human services, and arts and culture. Proposals are favored that are likely to make a substantial difference in the quality of community life; strengthen families and communities; expand meaningful civic engagement and build social capital; use volunteers; help nonprofit organizations build capacity and become more effective; include financial and other strategic commitments from other funding organizations; and leverage change in the capacity of community-wide systems rather than individual organizations. Interested applicants should initially contact the Foundation via a brief letter detailing the organization, its mission, the project it seeks funding for, and an approximate grant amount to be requested. Guidelines are available online.
Requirements: 501(c)3 nonprofit, tax-exempt organizations operating in Fifth Third's geographic regions—Ohio, Kentucky, Indiana, Michigan, Illinois, Tennessee, West Virginia, and Florida— are eligible.
Restrictions: Capital campaigns for individual churches; publicly supported entities, such as public schools or government agencies; elementary schools; and individuals are ineligible.
Geographic Focus: Florida, Illinois, Indiana, Kentucky, Michigan, Ohio, Tennessee, West Virginia
Contact: Grants Administrator; (513) 534-7001
Internet: http://www.53.com/wcm/connect/FifthThirdSite/About+53/In+the+Community/Foundation+Office/
Sponsor: Fifth Third Foundation
38 Fountain Square Plaza, MD 1090CA
Cincinnati, OH 45263

Fight for Sight-Streilein Foundation for Ocular Immunology Research Award 1918
Beginning in 2011, Fight for Sight and the Streilein Foundation for Ocular Immunology co-sponsored a new annual Summer Student Research Fellowship, Fight for Sight-Streilein Foundation for Ocular Immunology Research Award, for the study of ocular immunology. Candidates for the award, an annual $2,100 grant for fulltime research conducted from June to August, are invited to apply by visiting www.fightforsight.org for an application.
Requirements: The award is open to undergraduates, graduate and medical students studying in the United States. Applications will be reviewed by the Fight for Sight Scientific Review Committee and Streilein Foundation for Ocular Immunology and a grantee will be selected based upon the committee's recommendations to the Boards of both organizations.
Date(s) Application is Due: Feb 1
Amount of Grant: 2,100 - 2,100 U.S.D
Contact: Janice Benson; (212) 679-6060; fax (212) 679-4466; janice@fightforsight.org
Internet: http://www.streilein-foundation.org/fellowships.html
Sponsor: Streilein Foundation for Ocular Immunology
P.O. Box 6104
Boston, MA 02114

Fight for Sight Grants-in-Aid 1919
Fight for Sight was founded in 1946 by Mildred Weisenfeld, a young woman with retinitis pigmentosa, to encourage and fund research in ophthalmology, vision and related sciences. The goal of Fight for Sight is to encourage and facilitate research in detection, understanding, prevention, treatment and cures of visual disorders especially those diseases leading to impaired sight or blindness. To that end, FFS funds Grants-in-Aid to academic researchers within the first three years of their faculty appointments. FFS primarily supports new investigators, promoting the development of scientific skills and enabling the development of preliminary scientific findings and pilot studies necessary to successfully apply for more substantial federal and private funding such as that provided by the National Eye Institute and other divisions of the NIH. Support may be used to defray costs of personnel (but not the applicant), equipment and consumable supplies needed for the specific research project. One year awards of $20,000 are provided and may start between July 1 and September 1.
Requirements: Applications will only be considered from researchers who have received their first faculty or research appointment in eye/vision within the previous three years. Fringe benefits are not included and institutional overhead charges are not covered. Travel costs are generally not supported.
Date(s) Application is Due: Jan 1
Contact: Janice Benson; (212) 679-6060; fax (212) 679-4466; janice@fightforsight.org
Internet: https://www.fightforsight.org/Grants/Research-Award-Types/
Sponsor: Fight for Sight (U.S.A)
381 Park Avenue S, Suite 809
New York, NY 10016

Fight for Sight Post-Doctoral Awards 1920
Post-Doctoral Awards support individuals with a doctorate (Ph.D., M.D., O.D., Dr.PH, or D.V.M.) who are interested in academic careers in basic or clinical research in ophthalmology, vision or related sciences. This funding is intended to offer those interested in an academic career the opportunity to spend a year engaged in vision and eye research under the supervision of a senior scientist/clinician mentor. Clinical post-doctoral researchers are required to spend sufficient time on the funded research project to carry out the proposed objectives while basic researchers are expected to work full-time. One year grants of $20,000 are awarded for start dates between July 1 and September 1.
Requirements: Applications are considered from individuals who are within three years of their doctoral degrees or clinical residency training and have not received a previous Fight for Sight fellowship award. If at the time of application a doctorate has not yet been obtained, a cover letter must be submitted by the conferring institution advising when such degree is expected to be awarded. Recipients may supplement their awards with institutional or other funds however any anticipated supplemental support must be disclosed at the time of application. Total combined salary support must not exceed the annual stipend level set by the NIH for National Research Service Award recipients.
Restrictions: Fringe benefits are not provided by Fight for Sight.
Date(s) Application is Due: Feb 1
Amount of Grant: 20,000 - 20,000 U.S.D
Contact: Janice Benson, Assistant Director; (212) 679-6060; fax (212) 679-4466; janice@fightforsight.org
Internet: https://www.fightforsight.org/Grants/Research-Award-Types/
Sponsor: Fight for Sight (U.S.A)
381 Park Avenue S, Suite 809
New York, NY 10016

Fight for Sight Research Grants and Fellowships 1921
The primary purpose of the program is to encourage and finance research in ophthalmology, vision, and related sciences. The goal is to restore and preserve sight through research in detection, prevention, treatment, and curing of visual disorders, as well as diseases leading to impaired sight and partial or total blindness. Funding priority is given to pilot projects for research relating to clinically important eye diseases. Fight for Sight will continue to support younger investigators, promoting the development of scientific skills that will lead to more substantial funding from other sources. Preference will be given to clinical and disease-oriented grants and training fellowships in the areas of age-related macular degeneration, diabetic retinopathy, retinitis pigmentosa, glaucoma, amblyopia, strabismus, AIDS-related eye disorders, cataract, and uveal and corneal inflammation. Award mechanisms include grants in aid, postdoctoral fellowships, and student fellowships.
Requirements: Applicant must have a doctoral degree in medicine, optometry, or basic science and have less than one year of postdoctoral training. Fellowships are available to U.S. and Canadian residents and are not offered for study abroad. Undergraduate, graduate, and medical students may apply for student fellowships.
Restrictions: An applicant may not currently be financed by other research funds.
Date(s) Application is Due: Mar 1
Amount of Grant: 25,000 - 50,000 U.S.D
Contact: Program Coordinator, Fight for Sight Research Division; (847) 843-2020; fax (847) 843-8458; info@preventblindness.org
Internet: http://www.preventblindness.org/research
Sponsor: Prevent Blindness America
500 E Remington Road
Schaumburg, IL 60173

Fight for Sight Summer Student Fellowships 1922
Fight for Sight was founded in 1946 by Mildred Weisenfeld, a young woman with retinitis pigmentosa, to encourage and fund research in ophthalmology, vision and related sciences. The goal of Fight for Sight is to encourage and facilitate research in detection, understanding, prevention, treatment and cures of visual disorders especially those diseases leading to impaired sight or blindness. Unrestricted awards of $2,100 are given for two to three months of full-time research, usually during June-August.
Requirements: Fight for Sight's primary mission is to support and encourage promising scientists early in their careers. FFS primarily supports new investigators, promoting the development of scientific skills and enabling the development of preliminary scientific findings and pilot studies necessary to successfully apply for more substantial federal and private funding such as that provided by the National Eye Institute and other divisions of the NIH. Summer Student Fellowships are available to undergraduates, graduate and medical students who are interested in pursuing eye-related clinical or basic research. For most students, this is their first exposure to eye or vision research and the experience has resulted in many students choosing academic ophthalmology or eye research as a full-time career.
Restrictions: Students receiving stipends from other sources are generally not eligible.
Date(s) Application is Due: Feb 1
Amount of Grant: 2,100 - 2,100 U.S.D
Contact: Janice Benson; (212) 679-6060; fax (212) 679-4466; janice@fightforsight.org
Internet: https://www.fightforsight.org/Grants/Research-Award-Types/
Sponsor: Fight for Sight (U.S.A)
381 Park Avenue S, Suite 809
New York, NY 10016

Fireman's Fund Foundation Grants 1923
Program interests are primarily in support of education, human needs, arts and culture, and civic community activities, especially programs for youth, the elderly, and minority groups. Support is provided primarily to organizations in Marin County in the San Francisco Bay area. Limited funding is provided to Sonoma County. There is no deadline for applications. Proposals are reviewed four times a year. Once an application is submitted, a decision is usually made within 90 days. Guidelines and application are available online.
Requirements: To apply, a nonprofit, tax-exempt organization should send a letter to the director setting forth the request for funding and the amount desired, enclosing a financial statement, IRS letter, and a list of board members. Support is provided primarily to organizations in Marin County in the San Francisco Bay area.
Restrictions: The foundation does not make grants for capital campaigns; endowment funds; general operating expenses; individuals; religious organizations; fraternal, veteran, or sectarian groups; medical research and health organizations; trips or tours; advertising; video, television, or film productions; fundraising events; sporting events; political organizations; subscription fees or admission tickets; or public sector services (with the exception of some school district programs).
Geographic Focus: California
Date(s) Application is Due: Apr 1; Oct 1
Amount of Grant: 5,000 - 15,000 U.S.D
Samples: North Bay Children's Ctr (CA)—for program support, $15,000; Youth in Arts (CA)—for program support, $5000.
Contact: Phyllis Secosky, Executive Director; (415) 899-2757 or (415) 899-2000; phyllis_secosky@ffic.com
Internet: http://www.firemansfund.com/servlet/dcms?c=about&rkey=131
Sponsor: Fireman's Fund Foundation
777 San Marin Drive
Novato, CA 94998-1406

FirstEnergy Foundation Community Grants 1924
The FirstEnergy Foundation's contributions to local nonprofit organizations help strengthen the social and economic fabric of our communities. Funded solely by FirstEnergy, the Foundation extends the corporate philosophy of providing community support. The Foundation traditionally funds these priorities: help improve the vitality of our communities and support key safety initiatives; promotion of local and regional economic development and revitalization efforts; and support of FirstEnergy employee community leadership and volunteer interests.
Requirements: 501(c)3 organizations within the FirstEnergy Corporation operating companies' service areas - Ohio Edison Company, the Cleveland Electric Illuminating Company, the Toledo Edison Company, Pennsylvania Power Company, Metropolitan Edison Company, Pennsylvania Electric Company, Jersey Central Power & Light Company, Monongahela Power Company, the Potomac Edison Company, West Penn Power Company, FirstEnergy Solutions Corporation, FirstEnergy Generation, and FirstEnergy Nuclear Operations - are eligible to apply.
Restrictions: Funding is not considered for: direct grants to individuals, political or legislative activities; organizations that receive sizable public tax funding; fraternal, religious, labor, athletic, social or veterans organizations - unless the contribution is earmarked for an eligible program or campaign open to all beneficiaries, including those not affiliated with the host organization; national or international organizations; organizations supported by federated campaigns, such as United Way; research; equipment purchases; loans or second party giving, such as endowments, debt retirement, or foundations; or public or private schools.
Geographic Focus: New Jersey, Ohio, Pennsylvania
Contact: Dee Lowery, President; (330) 761-4246; fax (330) 761-4302
Internet: http://www.firstenergycorp.com/community
Sponsor: FirstEnergy Foundation
76 South Main Street
Akron, OH 44308-1890

Fisa Foundation Grants 1925
The foundation's mission is to build a culture of respect and improve the quality of life for three populations in southwestern Pennsylvania: women, girls, and people with disabilities. Emphasis is given to addressing gaps and unmet needs in the community. The foundation will consider support for new and ongoing programs and projects, operating needs, capacity building, equipment, and capital expenditures. Applicants are encouraged to submit a letter of inquiry outlining the needs to be addressed, a brief description of the proposed project or program, and the amount of the request. Letters of inquiry are reviewed on a monthly basis; full proposals are by invitation. Application and guidelines are available online.
Requirements: 501(c)3 and 509(a) nonprofit organizations in the 10-county southwestern Pennsylvania area, including Allegheny, Armstrong, Beaver, Butler, Fayette, Greene, Indiana, Lawrence, Washington and Westmoreland Counties, are eligible.
Restrictions: The foundation does not make grants in support of individuals; political campaigns or lobbying; religious organizations for religious purposes; travel, study and scholarships; services for people with disabilities that result from the aging process; renovations to improve physical accessibility of buildings; programs with the primary mission of serving individuals with mental illness; clinical research; or parenting programs, unless they specifically relate to parents with disabilities.
Geographic Focus: Pennsylvania
Amount of Grant: 5,000 - 75,000 U.S.D
Samples: City Theatre (Pittsburgh, PA)—for disability-awareness training for employees and for the development of audio-description services for visually impaired audience members, $10,000; Women and Girls Foundation of Southwest Pennsylvania (Pittsburgh, PA)—to train high-school girls to be grantmakers, $19,000 over three years; Hosanna House (Pittsburgh, PA)—to provide permanent housing and coordinated support services for families with disabled members, $15,000; Gwen's Girls (Pittsburgh, PA)—for a long-term residential facility for teenage mothers who are in the foster-care or juvenile-justice systems and their babies, $40,000.

Contact: Mary Delaney, (412) 456-5550; fax (412) 456-5551; info@fisafoundation.org
Internet: http://www.fisafoundation.org/grtmk.ivnu
Sponsor: Fisa Foundation
1001 Liberty Avenue, Suite 650
Pittsburgh, PA 15222

Fischelis Grants for Research in the History of American Pharmacy 1926

The institute supports a wide range of scholarly activities as long as they promise to contribute significantly to historical understanding and are clearly related to the modern practice of pharmacy in the United States. Projects must be connected directly to American pharmacy practice. Preference will be given to 20th-century topics, although studies of earlier periods will be considered. Historical discussions of current practice issues are also eligible. An application received after the deadline date will be evaluated individually if any funds remain unawarded; otherwise it will be considered in the next year's program.
Requirements: Each application is referred for evaluation to a committee of the institute. Principal criteria used in evaluating an application are a candidate's qualifications and record relevant to the intended purpose, relevance to the history of modern American pharmacy, and the adequacy of resources to fulfill the stated purpose.
Date(s) Application is Due: Mar 1
Amount of Grant: Up to 5,000 U.S.D
Contact: Gregory Higby, Director; (608) 262-5378; grants@aihp.org
Internet: http://www.pharmacy.wisc.edu/aihp
Sponsor: American Institute of the History of Pharmacy
777 Highland Avenue
Madison, WI 53705-2222

Fisher Foundation Grants 1927

The foundation awards grants to Connecticut nonprofits that benefit education, health and human services, housing, community needs, and arts and culture. The majority of grants are single year awards. Before submitting an application, a letter of inquiry or conversation with staff to discuss the specific purpose for which the funds are being requested is strongly recommended. Application information and guidelines are available online. Do not submit applications by fax or email.
Requirements: Nonprofits located in, and/or serving the residents of the Greater Hartford area, including: Andover, Avon, Bloomfield, Bolton, Canton, East Hartford, East Granby, East Windsor, Ellington, Enfield, Farmington, Glastonbury, Granby, Hartford, Hebron, Manchester, Marlborough, Newington, Rocky Hill, Simsbury, Somers, South Windsor, Suffield, Tolland, Vernon, West Hartford, Wethersfield, Windsor, and Windsor Locks are eligible.
Restrictions: Foundation policy does not allow funding for: organizations which are not tax-exempt under IRS Code section 501(c)3; organizations which have the IRS private foundation designation; individuals; performances, conferences, retreats, one-time events, trips; annual campaigns.
Geographic Focus: Connecticut
Date(s) Application is Due: Jan 15; Apr 15; Sep 15
Contact: Beverly Boyle, Executive Director; (860) 570-0221; fax (860) 570-0225; bboyle@fisherfdn.org
Internet: http://www.fisherfdn.org/application/instructions.htm
Sponsor: Fisher Foundation
36 Brookside Boulevard
West Hartford, CT 06107

Fisher House Foundation Newman's Own Awards 1928

Grants are made to implement innovative programs to improve the quality of life for military families and their communities. This program is administered by Fisher House, a charitable organization that provides a home-like place to stay for United States military families while a family member is receiving treatment at a major military or Veterans Administration medical facility. One award will be given to an organization supporting the Army, Navy, Air Force, Marines, or Coast Guard The organization with the most innovative project submitted will receive up to $15,000. The other four organizations will receive lesser amounts, as determined by the judges. Annual deadline dates may vary; contact program staff for exact dates.
Requirements: Eligible applicants must support an Active Duty, National Guard, or Reserve unit or installation, and be tax-exempt, a private organization, or a volunteer organization. Private, nonprofit organizations as defined in Department of Defense (DoD) Instruction 1000.15 and approved for operation on a DoD installation by the installation commander are eligible.
Date(s) Application is Due: Apr 30
Amount of Grant: 1,000 - 15,000 U.S.D
Contact: James D. Weiskopf, Executive Vice President; (888) 294-8560 or (301) 294-8560; fax (301) 294-8562; info@fisherhouse.org
Internet: http://www.fisherhouse.org/programs/newmans.shtml
Sponsor: Fisher House Foundation
1401 Rockville Pike, Suite 600
Rockville, MD 20852-1402

Fishman Family Foundation Grants 1929

The Foundation considers grants for: research, education, and cultural development of and for the community; scholarships related to Jewish services, education, social, and community activities; medical and scientific research; providing resources to meet critical needs in Israel; and educational grants and scholarships. Proposals are reviewed in April and October. Application information is available online.
Requirements: 501(c)3 nonprofits are eligible.
Date(s) Application is Due: Mar 31; Sep 30
Contact: Betty Fishman, President; info@fishman.org
Internet: http://www.fishman.org/apply.html
Sponsor: Fishman Family Foundation
730 E Cypress Avenue
Monrovia, CA 91016

Flinn Foundation Grants 1930

The foundation's grantmaking programs, limited to Arizona, include enhancing community-based solutions to local health care needs, especially those for children and youth; strengthening medical education and biomedical research programs in Arizona; strengthening Arizona's universities through an undergraduate scholarship program for outstanding Arizona high school students; and enhancing the visibility and long-term artistic mission of Arizona's principal visual and performing arts organizations. Types of support include program development, seed money, scholarship funds, and research grants.
Requirements: Arizona-based institutions or organizations whose programs are operated for the benefit of Arizona institutions and individuals are eligible to apply. Applications are accepted at any time. There is no application form, but a preliminary letter of inquiry or phone call is requested to determine the appropriateness of a full submission.
Restrictions: The foundation rarely provides grants to individuals, building projects (capital campaigns), purchase of equipment, endowment projects, annual fund-raising campaigns, ongoing operating expenses, or deficit needs. Requests to support conferences and workshops, publications, or the production of films and video are considered only when these activities are an integral component of a larger foundation initiative.
Geographic Focus: Arizona
Amount of Grant: 2,500 - 150,000 U.S.D
Samples: Northern Arizona U (Flagstaff, AZ)—to expand teacher-preparation programs in the biological sciences, $517,100 over three years; Homeward Bound (Phoenix, AZ)—to coordinate access to health care services for disadvantaged children, $10,000; Ballet Arizona (Phoenix, AZ)—for 15 months of support, $200,000.
Contact: JoAnn Fazio; (602) 744-6800; fax (602) 744-6815; info@flinn.org
Internet: http://www.flinn.org/about/grants.cms
Sponsor: Flinn Foundation
1802 N Central Avenue
Phoenix, AZ 85004-1506

Flinn Foundation Scholarships 1931

The Flinn Foundation scholarships, in partnership with Arizona's three state universities, provide enriched educational offerings that expand a recipient's life and career options. Students receive a financial package for their entire undergraduate study that includes free tuition, room, and board, funding for study abroad, mentorship from faculty, exposure to world leaders, and fellowship in a community of current and alumni Scholars. Total dollar value exceeds $50,000, in addition to the cash value of tuition provided by the universities. Scholars begin as a group, with a three-week seminar in Central Europe. Each Scholar also receives a stipend for at least one international summer seminar, or a semester or year at a foreign university.
Requirements: Recipients must rank in the top 5 percent of their high school graduating class with at least a 3.5 grade point average; score a minimum of 29 on the ACT or 1300 on the SAT (critical reading and math sections only); demonstrate leadership in a variety of extracurricular activities; and hold U.S. citizenship and residency in Arizona for the two years prior to application. All majors may apply.
Restrictions: Applicants must apply to one of three Arizona universities: Arizona State University, Northern Arizona University, or University of Arizona.
Geographic Focus: Arizona
Date(s) Application is Due: Oct 21
Amount of Grant: 50,000 U.S.D
Contact: Flinn Scholarship Contact; (602) 744-6800; fscholars@flinn.org
Internet: http://www.flinnscholars.org/news/977
Sponsor: Flinn Foundation
1802 N Central Avenue
Phoenix, AZ 85004-1506

Florence Hunt Maxwell Foundation Grants 1932

The mission of the Florence Hunt Maxwell Foundation is to support charitable organizations that provide for the underserved and indigent community. Grants from the Florence Hunt Maxwell Foundation are primarily one year in duration; on occasion, multi-year support is awarded. Applicants must apply online at the grant website. Applicants are strongly encouraged to do the following before applying: review the downloadable state application procedures for additional helpful information and clarifications; review the downloadable online-application guidelines at the grant website; review the foundation's funding history (link is available from the grant website); review the online application questions in advance; and review the list of required attachments. These will generally include: a list of board members, financial statements (audited, reviewed, or compiled by independent auditor); an organization summary; a list of other funding sources; an IRS Determination letter; and other required documents. All attachments must be uploaded in the online application as PDF, Word, or Excel files. The Florence Hunt Maxwell Foundation application deadline is 11:59 p.m. on April 1. Applicants will be notified of grant decisions by letter within one to two months after the deadline.
Requirements: Applicants must have 501(c)3 tax-exempt status and serve residents of the Metro Atlanta area.
Restrictions: The foundation does not support requests from individuals, organizations attempting to influence policy through direct lobbying, or any political campaigns.
Geographic Focus: Georgia

Date(s) Application is Due: Apr 1
Samples: Center for the Visually Impaired, Atlanta, Georgia, $7,500 (2010); Project Open Hand Atlanta, Atlanta, Georgia, $3,000 (2010); Camp Kudzu, Smyrna, Georgia, $2,500 (2010).
Contact: Quanda Allen, Vice President; (404) 264-1377; quanda.allen@baml.com
Internet: https://www.bankofamerica.com/philanthropic/fn_search.action
Sponsor: Florence Hunt Maxwell Foundation
3414 Peachtree Road, N.E., Suite 1475, GA7-813-14-04
Atlanta, GA 30326-1113

Florian O. Bartlett Trust Grants 1933
The Florian O. Bartlett Trust was established in 1937 to support and promote quality educational, human services, and health care programming for underserved populations. The application deadline for the Florian O. Bartlett Trust is April 1. Applicants will be notified of grant decisions before June 30. Grant requests for general operating support are strongly encouraged. Program support will also be considered. Small, program-related capital expenses may be included in general operating or program requests. The majority of grants from the Bartlett Trust are 1 year in duration. On occasion, multi-year support is awarded.
Geographic Focus: Massachusetts
Date(s) Application is Due: Apr 1
Contact: Michealle Larkins; (866) 778-6859; michealle.larkins@baml.com
Internet: https://www.bankofamerica.com/philanthropic/fn_search.action
Sponsor: Florian O. Bartlett Trust
225 Franklin Street, 4th Floor, MA1-225-04-02
Boston, MA 02110

Florida BRAIVE Fund of Dade Community Foundation 1934
The Florida BrAIve Fund at Dade Community Foundation will award grants to non-profits to assist eligible military personnel and family members residing in Miami-Dade, Monroe, Broward, Palm Beach and Martin counties. BRAIVE Fund grants will address needs of military personnel and their families in the target population related to pre-deployment, during deployment, and their return from duty. To obtain more information and the application forms, go to http://www.dadecommunityfoundation.org/Site/wc/wc143.jsp.
Requirements: General criteria for evaluating applications to The Florida BrAIve Fund of Dade Community Foundation: applicant is a 501(c)3 organization; applicant organization has a proven record of effective service to its clientele; applicant organization demonstrates the management and financial capacity to efficiently achieve the grant purpose; grants will be made to nonprofit organizations, not individuals; programs or services can only benefit current and former military personnel serving in Iraq or Afghanistan and their families; program services must be provided in the area of Florida to be covered by Dade Community Foundation as indicated in The Florida BrAIve Fund map; successful organizations will demonstrate that the agency is currently serving military personnel and their families in Florida or has the experience to address an unmet need.
Geographic Focus: Florida
Date(s) Application is Due: Jun 8
Contact: Claudianna Williams, Administrative Assistant; (305) 371-2711; fax (305) 371-5342; claudianna.williams@dadecommunityfoundation.org
Internet: http://www.dadecommunityfoundation.org/Site/wc/wc143.jsp
Sponsor: Dade Community Foundation
200 S Biscayne Boulevard, Suite 505
Miami, FL 33131-5330

Florida BRAIVE Fund of The Community Foundation in Jacksonville 1935
The Florida BrAIve Fund of The Community Foundation in Jacksonville was created to provide assistance to North Florida military personnel and families affected by deployment to Afghanistan or Iraq. Grants will be made to non-profit agencies in North Florida to support services that assist military personnel and their families before, during, and after deployment. Specific priority target populations include: military personnel returning from deployment to Afghanistan or Iraq and their families; military personnel with physical or mental health needs as a result of deployment to Afghanistan or Iraq; families of military personnel currently deployed or preparing for imminent (within 120 days) deployment to Afghanistan or Iraq; families of those killed in action during deployment to Afghanistan or Iraq; and individual augmentees or others deployed without a typical support structure. Grant applications for The Florida BrAIve Fund of The Community Foundation in Jacksonville will be considered on a rolling basis as they are received. You may submit an application at any time. Once the Foundation have received your application and all supplemental materials, you may expect to receive a decision from The Community Foundation in Jacksonville within 60 days. See the Foundations website for a application form and additional guidelines.
Requirements: Nonprofit 501(c)3organization providing services to meet the unmet needs of North Florida military personnel and families affected by deployments to Iraq and Afghanistan as part of Operation Enduring Freedom or Operation Iraqi Freedom, may be eligible to apply for a grant.
Geographic Focus: Florida
Contact: Cheryl Riddick, Vice President of Grantmaking Services; (904) 356-4483; info@jaxcf.org
Internet: http://www.jaxcf.org/NetCommunity/Page.aspx?pid=430
Sponsor: Community Foundation in Jacksonville
121 W Forsyth Street, Suite 900
Jacksonville, FL 32202

Florida Division of Cultural Affairs Arts In Education Arts Partnership Grants 1936
Arts In Education project grants are designed to cultivate learning and artistic development for all students and teachers by promoting, encouraging, and supporting arts and culture as an integral part of education and lifelong learning for residents and visitors. This includes but is not limited to: the learning and artistic development of pre-kindergarten through grade 12 students and teachers; or initiatives and proposals that help applicants to work as individuals or partners to carry out community programs and school reform through the arts. The Arts Partnership funding category is designed to support projects that will advance arts education and the development of long-term partnerships through effective collaboration between community arts and cultural organizations, social service agencies, and educational entities. Applicants may request up to $25,000 for arts partnership projects that have completed planning and design work and are ready for implementation or expansion. If proposals include computer, video, and technology equipment, applicants must show how technology equipment, systems, and programs are integrated into their specific arts education partnership. Focus areas for the Arts Partnership project may include the following: school-based arts education; programming that integrates the arts into areas not usually associated with the arts (non-arts curriculum, school-to-work initiatives, the criminal justice system, the healthcare system, community care for the elderly, underserved populations, and adult-continuing education programs); programming that brings together different generations; arts and technology programming in music, visual arts, theatre, dance, media and/or literary arts; and community arts education.
Restrictions: Arts Partnership projects are not intended to fund the same project year after year. However panelists have the discretion to recommend funding for on-going projects.
Geographic Focus: Florida
Contact: Laura Lewis Blischke, Program Manager; (850) 245-6475; llblischke@dos.state.fl.us
Megan Burke, Program Manager; (850) 245-6458; mcburke@dos.state.fl.us
Sarah T. Stage, Program Manager; (850) 245-6459; sstage@dos.state.fl.us
Internet: http://www.florida-arts.org/documents/guidelines/2012-2013.scp.guidelines.cfm#aie
Sponsor: Florida Division of Cultural Affairs
500 South Bronough Street, 3rd Floor
Tallahassee, FL 32399-0250

Florida High School/High Tech Project Grants 1937
Florida High School/High Tech (HS/HT) is designed to provide high school students with all types of disabilities the opportunity to explore jobs or postsecondary education leading to technology- related careers. HS/HT links youth to a broad range of academic, career development and experiential resources and experiences that will enable them to meet the demands of the 21st century workforce. HS/HT is a community-based partnership made up of students, parents and caregivers, businesses, educators and rehabilitation professionals. It has been shown to reduce the high school dropout rate and increase the overall self-esteem of participating students. If you are interested in participating, there is a role for you. The program provides step-down funding while the program is gaining independent funding for the future of their program.
Requirements: Any established, Florida nonprofit corporation, agency, organization or association that has been granted exemption from federal income tax under Section 501(c)3 of the IRS Tax Code is eligible to submit a proposal for review.
Geographic Focus: Florida
Contact: Guenevere Crum, Senior Vice President; (850) 224-4493; fax (850) 224-4496; guenevere@abletrust.org or info@abletrust.org
Internet: http://www.abletrust.org/hsht/
Sponsor: Able Trust
3320 Thomasville Road, Suite 200
Tallahassee, FL 32308

Floyd A. and Kathleen C. Cailloux Foundation Grants 1938
The foundation awards grants in the areas of civic and cultural, education and youth, family and community service, and health and rehabilitation. Grants support specific projects or programs, technical assistance, and capital projects. Grants for general operations are less common and usually small. Endowment grants are rare. The foundation only reviews grant proposals from applicants whose letters of inquiry have been approved. Applications are taken online.
Requirements: Grants support 501(c)3 nonprofit organizations. Preference is given to requests from Texas Hill Country.
Restrictions: The Foundation does not fund church or seminary construction, or church related entities or activities other than ecumenically oriented projects/programs that otherwise meet Foundation guidelines. The general policy of the Foundation is not to make grants to an organization more than once within any 12-month period. Grants or loans to individuals are never made.
Contact: Grants Administrator; (830) 895-5222; info@caillouxfoundation.org
Internet: http://www.caillouxfoundation.org/grant_guidelines.htm
Sponsor: Floyd A. and Kathleen C. Cailloux Foundation
P.O. Box 291276
Kerrville, TX 78029-1276

Fluor Foundation Grants 1939
The Fluor Corporation achieves its contribution objectives through the Fluor Foundation and corporate giving. The Foundation's areas of interest are: education; human services; cultural outreach; and public/civic affairs. The Foundation considers requests for operating, program, capital or endowment support. Priority is given to funding organizations with employee volunteer participation. Application information is available online.

Requirements: Fluor's giving programs focus on community organizations in those locations around the world where the company has a presence. See website for local contact information: http://www.fluor.com/sustainability/community/fluor_giving/Pages/applying_for_fluor_foundation_grants.aspx
Restrictions: Funding is not available for: film production/publishing activities; individuals; sports organizations/programs; veterans, fraternal, labor or religious organizations; or lobbying/political organizations or campaigns.
Geographic Focus: Alaska, California, Louisiana, New Jersey, New York, North Carolina, Pennsylvania, South Carolina, Tennessee, Texas, Virginia, Washington, Albania, Andorra, Armenia, Australia, Austria, Azerbaijan, Belarus, Belgium, Bosnia & Herzegovina, Bulgaria, Canada, Caribbean, China, Croatia, Cyprus, Czech Republic, Denmark, Estonia, Finland, France, Georgia, Germany, Greece, Hungary, Iceland, Ireland, Italy, Japan, Kosovo, Latvia, Liechtenstein, Lithuania, Luxembourg, Macedonia, Malta, Mexico, Moldova, Monaco, Montenegro, New Zealand, Norway, Peru, Philippines, Poland, Poland, Portugal, Romania, Russia, Russia, San Marino, Serbia, Slovakia, Slovenia, Spain, Sweden, Switzerland, The Netherlands, Turkey, Ukraine, United Kingdom, Vatican City, Venezuela
Contact: Suzanne Esber, Executive Director of Community Relations; (949) 349-7847; fax (949) 349-7694; suzanne.esber@fluor.com
Internet: http://www.fluor.com/sustainability/community/fluor_giving/Pages/applying_for_fluor_foundation_grants.aspx
Sponsor: Fluor Foundation
3 Polaris Way
Aliso Viejo, CA 92698

FMC Foundation Grants **1940**
The foundation supports education, community improvement, urban affairs, health and human services, and public issues/economic education. Higher education is supported through scholarships and employee matching gifts. Education support tends to be in business, engineering, chemistry, and some minority education programs. Eligible applicants need to contact the FMC in their geograhic area. Each individual FMC location determines how their contributions will be given.
Requirements: Grants are awarded to 501(c)3 organizations in FMC-plant communities. U.S.-based organizations with an international focus are also eligible.
Restrictions: Grants are not made to individuals.
Geographic Focus: Arizona, California, Delaware, Florida, Illinois, Louisiana, Maine, Maryland, Missouri, New Jersey, New York, North Carolina, Pennsylvania, Tennessee, Texas, West Virginia, Wyoming, Canada
Contact: Program Contact; (215) 299-6000; fax (215) 299-6140
Internet: http://www.fmc.com
Sponsor: FMC Foundation
1735 Market Street
Philadelphia, PA 19103

Foellinger Foundation Grants **1941**
The foundation awards grants primarily in Fort Wayne and Allen County, IN, to improve the quality of life in the areas of early child development, youth development, strengthening family services, strengthening organizations, and community concerns. Special emphasis is given to projects/programs that help children and their families, particularly those with the greatest economic need and least opportunity. By doing so, the Foundation hopes to help children and their families move from dependence to independence. In addition to programs and projects, the foundation also awards grants for general operating, capital, seed money, renovation projects, conferences and seminars, consulting services, equipment, challenge/matching, and planning purposes.
Requirements: Indiana organizations are eligible to apply.
Restrictions: The foundation does not fund scholarships, travel assistance, conference fees, religious groups, public or private elementary or secondary schools, sponsorships, special events, advertising, or endowments.
Geographic Focus: Indiana
Date(s) Application is Due: Feb 1; May 1; Nov 1
Samples: Public Broadcasting of Northeast Indiana (IN)—for general operating support; Arthur J. Blaising Social Services (IN)—for its after-school program and its 10-week summer day camp, $75,585; Turnstone (IN)—for a transportation program that serves elderly and disabled people, $187,500 over 18 months; Taylor U—for the Allen County Postsecondary Education Consortium, $219,690 over two years.
Contact: Cheryl Taylor; (219) 422-2900; fax (219) 422-9436; info@foellinger.org
Internet: http://www.foellinger.org
Sponsor: Foellinger Foundation
520 E Berry Street
Fort Wayne, IN 46802

Fogarty International Center/Ellison Medical Foundation Overseas Fellowships in Global Health and Clinical Research **1942**
The fellowship is a one-year mentored clinical research training experience for graduate-level U.S. students in the health professions. The program enables individuals to experience mentored research training at top-ranked NIH-funded research centers in developing countries. Fellowship sites include Bangladesh; Botswana; Brazil; Haiti; Chennai, India; China; Vellore, India; Kenya; Mali, West Africa; Lima, Peru; Durban, South Africa; Pretoria, South Africa; Russia; Tanzania; Thailand; Uganda; and Zambia. Each fellowship is for a one-year period. The term will begin with an intensive orientation program on the NIH campus in Bethesda, MD, in July. This will be followed by approximately 10-plus months of intense research training at the foreign site. Application and guidelines are available online.
Requirements: This program is designed for students meeting all of the following qualifications: a strong interest in, and potential for, a career in international health activities and/or clinical research; advanced standing in a U.S. medical (M3) or osteopathic school, or enrollment in a doctoral-level program at a U.S. school of public health, nursing, or dentistry; and support of their home academic institution, including a committed mentor. Applicants must have strong academic records and must be U.S. citizens or permanent U.S. residents. Medical and osteopathic students must have completed their basic science courses and one year of clinical clerkship; public health doctoral students must have completed their coursework and passed their qualifying exams prior to the beginning of the fellowship.
Date(s) Application is Due: Dec 28
Amount of Grant: 24,000 U.S.D
Contact: Yolanda Thomas, fic-fellowship@aamc.org
Internet: http://www.aamc.org/students/medstudents/overseasfellowship/2006fogartyapp.doc
Sponsor: Association of American Medical Colleges
2450 N Street NW
Washington, DC 20037-1127

Fondren Foundation Grants **1943**
The foundation awards grants to Texas-based programs in the areas of education, youth services, health care, and human services. The board meets quarterly to consider requests.
Requirements: Texas nonprofit organizations, with an emphasis on Houston and the Southwest are eligible to apply.
Restrictions: The Foundation does not allow funding to individuals, or for operating or annual fund drives.
Geographic Focus: Texas
Amount of Grant: 10,000 - 100,000 U.S.D
Contact: Melanie Scioneaux, Assistant Secretary
Internet: http://www.tcc.state.tx.us/cancerfunding/regional/05.html
Sponsor: Fondren Foundation
7 TCT 37, P.O. Box 2558
Houston, TX 77252-8037

Food, Pharmaceutical, and Bioengineering Division Award in Chemical Engineering **1944**
This award is given to recognize outstanding chemical engineering contributions and achievements by an individual in those industries involved in food, pharmaceutical, and bioengineering activities. The recipient need not be a member of AIChE, and there is no age limit. The individual's contributions may have been made in industrial, governmental, or academic areas, or with other organizations.
Requirements: All members, as well as other interested persons, are urged to nominate deserving candidates.
Date(s) Application is Due: Mar 25
Amount of Grant: 4,000 U.S.D
Contact: AIChE Awards Programs Director; (212) 591-7107; fax (212) 591-8890; awards@aiche.org
Internet: http://www.aiche.org/awards/awarddtl.asp?AwardID=36
Sponsor: American Institute of Chemical Engineers
3 Park Avenue
New York, NY 10016

Ford Motor Company Fund Grants **1945**
The fund supports not-for-profit organizations in three major areas: innovation and education; community development and American legacy, and auto-related safety education. The fund seeks to build partnerships with organizations that have a well-defined sense of purpose, a demonstrated commitment to maximizing available resources, and a reputation for meeting objectives and delivering quality programs and services. Priority is placed on the support and development of organizations that promote diversity and inclusion. Requests for support are accepted and reviewed throughout the year. The fund now implements an online application system. Details are available online.
Restrictions: Ford does not fund: advocacy-directed programs; animal-rights organizations; beauty or talent contests; day-to-day business operations; debt reduction; donation of vehicles; efforts to influence legislation, or the outcome of any elections or any specific election of candidates to public office or to carry on any voter registration drive; endowments; fraternal organizations; general operating support to hospitals and health care institutions; individual sponsorship related to fundraising activities; individuals; labor groups; loans for small businesses; loans to program-related investments; organizations that do not have 501(c)3 status; organizations that unlawfully discriminate in their provision of goods and services based on race, color, religion, gender, gender identity or expression, ethnicity, sexual orientation, national origin, physical challenge, age, or status as a protected veteran; political contributions; private K-12 schools; profit-making enterprises; religious programs or sectarian programs for religious purposes; species-specific organizations; sports teams.
Samples: American Red Cross, Southeastern Michigan Chapter (Detroit, MI)—for disaster-relief efforts in Florida related to Hurricane Charley, $50,000.
Contact: Ford Fund Coordinator; (888) 313-0102; Fordfund@ford.com
Internet: http://www.ford.com/en/goodWorks/fundingAndGrants/fordMotorCompanyFund/default.htm
Sponsor: Ford Motor Company Fund
P.O. Box 1899
Dearborn, MI 48121-1899

Forrest C. Lattner Foundation Grants 1946

The Foundation's primary objectives are in six areas of interest: arts and humanities; education; environment; health and social services; historic preservation; and medical research. The Foundation also wishes to encourage the development of innovative model programs. Application guidelines are available online.

Restrictions: The foundation is unable to fulfill grant requests to individuals, or to support programs that are the primary responsibility of the public sector.

Geographic Focus: Florida, Georgia, Kansas, Rhode Island, Texas

Date(s) Application is Due: Mar 1; Sep 1

Samples: Woodmere Art Museum (Philadelphia, PA)—for programs and exhibitions at the Helen Millard Children's Gallery, $20,000.

Contact: Susan Lattner Lloyd, President; (561) 278-3781; fax (561) 278-3167; lattner@bellsouth.net

Internet: http://www.lattnerfoundation.org

Sponsor: Forrest C. Lattner Foundation
777 E Atlantic Avenue, Suite 317
Delray Beach, FL 33483

Foster Foundation Grants 1947

The Foundation funds effective, nonprofit organizations that can provide tangible benefits and ongoing support to improve the quality of life for individuals, families and communities within the Pacific Northwest. The Foundation's primary field of interests are: basic human welfare issues, education, medical research, treatment and care and cultural activities. The Foundation places special emphasis on meeting the needs of the underserved and disadvantaged segments of the population, especially children, women and seniors.

Requirements: Eligible applicants must: be a nonprofit located in or serving the populations of the Pacific Northwest; address as their mission or project intent one of the Foundation's priority issues for funding; not have any delinquent final reports due to the Foundation from previous grant cycles.

Restrictions: The Foundation will not consider requests for: direct grants, scholarships or loans for the benefit of specific individuals; projects of organizations whose policies or practices discriminate on the basis of race, ethnic origin, sex, creed or sexual orientation; contributions or program support for sectarian or religious organizations whose activities benefit only their members; or loans.

Geographic Focus: Alaska, Idaho, Montana, Oregon, Washington

Contact: Jill Goodsell, Executive Director; (206) 726-1815; fax (206) 903-0628; info@thefosterfoundation.org

Internet: http://thefosterfoundation.org/Grants_Guide.asp

Sponsor: Foster Foundation
601 Union Street, Suite 3707
Seattle, WA 98101

Foster G. McGaw Prize 1948

This prize, established by the Baxter Foundation and administered by the Hospital Research and Educational Trust, the research and development affiliate of the American Hospital Association, is awarded to a community hospital that has undertaken an identifiable program or project that improves the status or enhances the quality of life of community members, particularly the poor. The winner will receive a monetary prize; trophy; recognition at a special awards ceremony at the AHA annual meeting; and coverage in AHA News, Hospitals and Health Networks, and other health care publications. Up to three finalists will each receive a monetary prize and mention in AHA News, Hospitals and Health Networks, and other health care publications. Application and guidelines are available online.

Requirements: Nominations can be made by either the hospitals themselves or by other community organizations.

Date(s) Application is Due: Apr 8

Amount of Grant: Up to 10,000 U.S.D

Contact: AHA Member Relations; (312) 422-3932

Internet: http://www.aha.org/aha/awards-events/foster/application/application.html#o

Sponsor: Hospital Research and Educational Trust
1 N Franklin, 32nd Floor
Chicago, IL 60606

Foundation Fighting Blindness Marjorie Carr Adams Women's Career Development Awards 1949

The Program's primary purpose is to support female clinical research scientists of superior dedication and talent to pursue vigorous research programs to find the cures for retinal degenerative diseases research (RDD) i.e. inherited orphan retinal degenerative diseases and dry age-related macular degeneration. The goal of this program is to facilitate advances in the laboratory and clinical research, to elucidate the mechanisms for the etiology and pathogenesis of RDDs and to develop innovative strategies to prevent, treat and cure these diseases. This award provides support for mentored study and research for clinically trained biomedical research professionals who have the potential to assume leadership roles as clinician-scientists in RDD research.

Requirements: Female clinical scientists possessing an, M.D., D.O., or recognized equivalent foreign degrees and who are in their first, second or third year of a junior faculty appointment are eligible to apply for the Adams WCDA. Applicants do not have to be U.S. citizens.

Date(s) Application is Due: Nov 14

Amount of Grant: 65,000 - 325,000 U.S.D

Contact: Cindy Settar, (410) 568-0150 or (800) 683-5555; csettar@Fightblindness.org or grants@Fightblindness.org

Internet: http://www.blindness.org/index.php?option=com_content&view=section&id=9&Itemid=107#career-development

Sponsor: Foundation Fighting Blindness
11435 Cronhill Drive
Owings Mills, MD 21117-2220

Foundation for a Healthy Kentucky Access to Health Care Initiative 1950

The grants program is directed at all nonprofit organizations in Kentucky that have the ability to make a significant contribution to the development of public policy regarding access to health services and health care. The foundation invites letters of intent from organizations that are interested in enhancing and expanding access to health care and health services for low-income, uninsured, and underinsured Kentuckians. The listed application deadline is for letters of intent; full proposals are by invitations.

Requirements: Kentucky nonprofit organizations are eligible.

Geographic Focus: Kentucky

Date(s) Application is Due: Mar 18

Contact: Mary Jo Dike, Program Manager; (502) 326-2583; fax (502) 326-5748; mdike@healthyky.org or info@healthyky.org

Internet: http://www.healthyky.org

Sponsor: Foundation for a Healthy Kentucky
9300 Shelbyville Road, Suite 1305
Louisville, KY 40222

Foundation for a Healthy Kentucky Grants 1951

The Foundation seeks to address the unmet health care needs of the people of Kentucky through strategic grants designed to improve health status and access to care. Funding is concentrated in the following areas: fitness and nutrition for children and families; youth smoking prevention; and youth substance abuse prevention. The Foundation also hopes to enhance access to health care for low-income and uninsured populations; health care for rural populations; and integrated mental health and medical services.

Requirements: The Foundation accepts proposals for funding only in direct response to a specific Request for Proposals (RFP) or Request for Quotes (RFQ) or grant solicitation issued by the Foundation. Organizations should familiarize themselves with the Foundation's What We Fund page, and the Grant Guidelines to determine whether they are eligible to receive funding. They should sign up for the Foundation's mail/email list to receive grant opportunities and free educational forums. Once the organization receives an RFP announcement, they should review it carefully to see that the goals of the RFP fit with their particular project. If it is acceptable, applicants should submit their proposal according to the guidelines and RFP due dates.

Restrictions: The Foundation will not review unsolicited grant requests except for requests for Matching Grant and Conference Support funds. Grants do not support direct patient care, except as part of demonstration or replication projects; capital campaigns or requests for bricks and mortar (although project related equipment may be included in requests); overhead expenses except in limited amounts for specific projects; organizations that discriminate on the basis of race, gender, age, religion, national origin, sexual orientation, disability, military, or marital status in hiring; multi-year commitments; expenses related to registered legislative agents for the purpose of lobbying; endowment funds; individuals; private, for-profit entities; religious organizations for religious purposes; political causes; or retroactive expenses, deficit reduction, or forgiveness.

Geographic Focus: Kentucky

Contact: Susan Zepeda, President/Chief Executive Officer; (502) 326-2583 or (877) 326-2583; fax (502) 326-5748; szepeda@healthyky.org or info@healthy-ky.org

Internet: http://www.healthyky.org

Sponsor: Foundation for a Healthy Kentucky
9300 Shelbyville Road, Suite 1305
Louisville, KY 40222

Foundation for Appalachian Ohio Bachtel Scholarships 1952

Dr. Harry Keig grew up to become a successful physician and surgeon in the Omaha area but never forgot his hometown or the teachers who educated him. He also never forgot the life-shaping influence of his friend, Forrest Bachtel, longtime teacher and coach at Middleport High School in Meigs County. When Dr. Keig passed away in 2003, his bequest created a $500,000 endowed fund at the Foundation for Appalachian Ohio in honor of his friend, Coach Bachtel, to provide college scholarships for Meigs High School students. Each scholarship recipient receives $2,500 to assist with college education.

Requirements: Available to Meigs High School graduates, located in Pomeroy, Ohio.

Geographic Focus: Ohio

Amount of Grant: 2,500 U.S.D

Contact: Phyllis Moody, Scholarship Program Coordinator; (740) 753-1111; fax (740) 753-3333; pmoody@ffao.org

Internet: http://www.appalachianohio.org/scholarships/index.php?section=258&page=296

Sponsor: Foundation for Appalachian Ohio
36 Public Square, P.O. Box 456
Nelsonville, OH 45764

Foundation for Appalachian Ohio Susan K. Ipacs Nursing Legacy Scholarships 1953

The Susan K. Ipacs Nursing Legacy Scholarship Fund was established at the Foundation for Appalachian Ohio to honor the work and life of Susan Ipacs – nurse, instructor, mother, wife and friend. Sue never lost her love of nursing and never forgot the circumstances that led to her college education. As a nursing instructor and associate dean, Sue strove to ease the burdens and obstacles facing students, while nurturing the passion to help others in all who surrounded her, both students and fellow colleagues. She felt a personal responsibility to recognize and attend to the underprivileged and non-

traditional student. As so many depended on Sue for her compassion and dedication to the profession, it is the hope that students pursuing their nursing degree remember the reason behind the studies, and the importance of compassion for people.
Geographic Focus: Ohio
Contact: Phyllis Moody, Scholarship Program Coordinator; (740) 753-1111; fax (740) 753-3333; pmoody@ffao.org
Internet: http://www.appalachianohio.org/scholarships/index.php?section=258&page=459
Sponsor: Foundation for Appalachian Ohio
36 Public Square, P.O. Box 456
Nelsonville, OH 45764

Foundation for Appalachian Ohio Zelma Gray Medical School Scholarship **1954**
Before her death in 2007, Zelma Gray established her support for medical professionals and care in Guernsey County through a scholarship fund with WesBanco Trust and Investment Services. Designed to support Guernsey County residents' pursuit of medical school education as well as encourage that talent to remain in, or return to, Guernsey County to serve its citizens upon completion of medical training. Funding will cover the cost of tuition, lab fees and text books while pursuing a Doctor of Medicine (M.D.) or Doctor of Osteopathic Medicine (D.O.) at an accredited medical school within the United States, in hopes that he or she will practice medicine in Guernsey County upon completion of his or her degree and training.
Requirements: Current and recent Guernsey County residents are eligible for the Zelma Gray Medical School Scholarship. Awards will be determined through objective criteria, including MCAT scores and college transcripts.
Geographic Focus: Ohio
Contact: Phyllis Moody, Scholarship Program Coordinator; (740) 753-1111; fax (740) 753-3333; pmoody@ffao.org
Internet: http://www.appalachianohio.org/scholarships/index.php?section=258&page=442
Sponsor: Foundation for Appalachian Ohio
36 Public Square, P.O. Box 456
Nelsonville, OH 45764

Foundation for Health Enhancement Grants **1955**
The purpose of the Foundation for Health Enhancement is to improve care in the United States by the development of the science of delivery of all kinds of health care, including medical, surgical, dental, nursing, health education, and other curative and preventive health services. The majority of grants from the Foundation for Health Enhancement are one year in duration. Applicants must apply online at the grant website. Applicants are strongly encouraged to do the following before applying: review the downloadable state application procedures for additional helpful information and clarifications; review the downloadable online-application guidelines at the grant website; review the foundation's funding history (link is available from the grant website); review the online application questions in advance; and review the list of required attachments. These will generally include: a list of board members, financial statements (audited, reviewed, or compiled by independent auditor); an organization summary; a list of other funding sources; an IRS Determination letter; and other required documents. All attachments must be uploaded in the online application as PDF, Word, or Excel files. The Foundation for Health Enhancement has biannual deadlines of April 1 and September 1. Applications should be submitted by 11:59 p.m. on the deadline dates. In general, applicants will be notified of grant decisions 3 to 4 months after proposal submission.
Requirements: The Foundation places special emphasis on: preventive health services; smaller qualifying applicants in an area geographically proximate to Chicago and/or the Midwest; and service organizations as opposed to research organizations.
Geographic Focus: Illinois
Date(s) Application is Due: Apr 1; Sep 1
Samples: I-PLU.S.: Independent Positive Living Under Supervision, North Chicago, Illinois, $10,000 (2010); Saint Anthony Hospital Foundation, Chicago, Illinois, $10,000 (2010); SOS Children's Villages Illinois, Chicago, Illinois, $10,000 (2010).
Contact: George Thorn; (312) 828-4154; ilgrantmaking@bankofamerica.com
Internet: https://www.bankofamerica.com/philanthropic/fn_search.action
Sponsor: Foundation for Health Enhancement
231 South LaSalle Street, IL1-231-13-32
Chicago, IL 60604

Foundation for Medical Research Grants **1956**
The foundation promotes all forms of scientific medical research, in particular clinical research and that connected with the basic biological sciences directly or indirectly related to medicine. Grants are awarded to individuals and institutions in France and on an international basis. Types of support include subsidies for travel and study abroad and equipment acquisition. Major annual awards include the Rosen Prize for Cancer Research and the Delahautemaison Prize, which is awarded every two years, alternately for research in nephrology and cancer; the annual Grand Prix; and five prizes for molecular biology, endocrinology, immunology, clinical investigation, and neurobiology.
Amount of Grant: Up to 10,000,000 U.S.D
Contact: Program Contact; 33-1-44-39-75-75; fax 33-1-44-39-75-99; frm@frm.org
Internet: http://www.frm.org
Sponsor: Foundation for Medical Research
54 rue de Varenne
Paris, 75007 France

Foundation for Physical Therapy Mary McMillan Doctoral Scholarships **1957**
The purpose of this program is to assist physical therapists with outstanding potential for doctoral studies in the first year of graduate studies towards a doctorate.
Requirements: Applicants must be entering post professional full-time or part-time doctoral study during the current academic year. Applicants must possess a license to practice physical therapy in a U.S. jurisdiction or have met all the requirements for licensure in a U.S. jurisdiction, including having received a passing score on the licensure exam.
Date(s) Application is Due: Aug 15
Amount of Grant: 5,000 U.S.D
Samples: Mark Diehl, PT, MPT, Virginia Commonwealth, $5000; Shauna D Dudley-Javoroski, PT, MPT, U of Iowa (IA)—scholarship recipient, $5000; Jill Stewart, PT, MPT, MS, NCS, University of Southern California CA)—scholarship recipient, $5000.
Contact: Lucy Dickson; (800) 999-2782, ext. 8505; foundation@apta.org
Internet: http://www.apta.org/AM/Template.cfm?Section=Awards_and_Scholarships&TEMPLATE=/CM/ContentDisplay.cfm&CONTENTID=23362
Sponsor: Foundation for Physical Therapy
1055 N Fairfax Street, Suite 350
Alexandria, VA 22314-1488

Foundation for Physical Therapy Research Grants **1958**
The foundation provides sole support or matching funds to individuals and groups of researchers to pursue scientifically based and clinically related physical therapy research, focused on studies initiated by emerging investigators. Emerging investigators are individuals who have had some previous research experience, may or may not have completed graduate or postgraduate work, or are beginning a new field of research or making strategic changes in research direction within an existing line of research. Proposed studies should add to or refine the body of theoretical, scientific, and clinical knowledge on which physical therapy practice is based, using any of a variety of recognized investigative methods such as experimental, descriptive, or correlational. Annual deadline dates may vary; contact program staff for exact dates.
Requirements: Proposals may be submitted by individuals or groups of investigators independently or through a sponsoring institution or organization with which they are affiliated. Groups must designate one member as the principal investigator responsible for directing the project. This person must be eligible for licensure as a physical therapist. Other members of the group may be physical therapists, physical therapist assistants, or persons from other fields that are relevant to the proposed study.
Date(s) Application is Due: Aug 15
Amount of Grant: Up to 40,000 U.S.D
Contact: Lucy Dickson, Scientific Program Administrator; (800) 999-2782, ext. 8505; foundation@apta.org
Internet: http://www.apta.org/AM/Template.cfm?Section=Awards_and_Scholarships&TEMPLATE=/CM/ContentDisplay.cfm&CONTENTID=23363
Sponsor: Foundation for Physical Therapy
1055 N Fairfax Street, Suite 350
Alexandria, VA 22314-1488

Foundation for Psychocultural Research Doctoral Fellowships **1959**
The foundation will competitively award limited dissertation research fellowships aimed at advancing interdisciplinary research projects and scholarship at the intersection of psychology, culture, neuroscience, and psychiatry, with emphasis on psycho-cultural factors as central, not peripheral. The dissertation topic will involve substantial engagement in Trauma and the Interaction of Early Relational Experiences, Social Context, and Developmental Trajectories. The interdisciplinary approach will emphasize the cultural and neuroscientific approaches to the study of PTSD. Grants will be awarded on a one-time basis. An institutional allowance is allotted to the host institution on a monthly basis for one year, to be used at the sponsor's discretion to help pay for the fellow's research supplies, travel to scientific meetings, or health insurance. If a deadline falls on a weekend, applications will be accepted on the following Monday. Application guidelines are available online.
Requirements: Candidates must be enrolled at an accredited U.S. university or institution. Applicant must be a doctoral candidate in anthropology, psychology, neuroscience, or health sciences, and should have interest in completing a dissertation involving interdisciplinary research in culture, neuroscience, and psychiatry.
Date(s) Application is Due: Feb 15
Amount of Grant: 25,000 - 26,250 U.S.D
Contact: Director; (310) 454-1417; info@thefpr.org
Internet: http://www.thefpr.org/programs_funding/applications/doctoral_app.html
Sponsor: Foundation for Psychocultural Research
P.O. Box 826
Pacific Palisades, CA 90272

Foundation for Psychocultural Research Postdoctoral Fellowships **1960**
The foundation funds research aimed at advancing interdisciplinary research projects and scholarship at the intersection of psychology, culture, neuroscience, and psychiatry, with emphasis on psychocultural factors as central, not peripheral. The current topical focus is Interdisciplinary Approach to Post Traumatic Stress and Resilience. The research will integrate cultural and neuroscientific approaches to the study of PTSD, in its design and/or clinical application. Candidates must conduct their proposed research under a sponsor who holds an academic appointment at an accredited U.S. university or institution. The fellowship may be extended up to the maximum of three years, based on the evaluation of the previous year's performance. An institutional allowance is allotted monthly to the host institution, to be used at the sponsor's discretion to help pay for the fellow's

research supplies, travel to scientific meetings, or health insurance. If a deadline falls on a weekend, applications will be accepted on Monday. Guidelines are available online.
Requirements: Applicants must have a doctoral or MD degree and should have interest in pursuing a career involving interdisciplinary research in neuroscience, culture, and psychiatry.
Date(s) Application is Due: Feb 15
Amount of Grant: 41,500 - 124,500 U.S.D
Contact: Director; foundationpsycul@aol.com
Internet: http://www.thefpr.org/programs_funding/applications/post_doctoral_app.html
Sponsor: Foundation for Psychocultural Research
P.O. Box 826
Pacific Palisades, CA 90272

Foundation for Science and Disability Grant Program for Physically Disabled Students in the Sciences 1961
A grant may be awarded annually to a disabled student who is interested in obtaining a graduate degree in science, mathematics, medicine, engineering, or computer science. Selection is based on financial need, sincerity of purpose, and scholarship and/or research ability. Funds may be used for an assistive device or instrument, or as financial support to work with a professor on an individual research project, or for some other special need. The number of grants awarded each year varies depending on funds available. Application must include a 250-word essay on professional goals and objectives, as well as the specific purpose for which the grant monies will be used, and two letters of recommendation from faculty members, one of whom must be the student's academic research advisor.
Requirements: College seniors who have some physical or sensory disability and who have been accepted to graduate or professional school, as well as students already in graduate or professional schools, are eligible to apply.
Restrictions: Undergraduate students are not eligible.
Date(s) Application is Due: Dec 1
Amount of Grant: 1,000 U.S.D
Contact: Dr. Richard Mankin, Co-Chair, Science Student Grant Committee; (352) 374-5774; fax (352) 374-5804; rmankin@nervm.nerdc.ufl.edu
Internet: http://www.as.wvu.edu/~scidis/organize/fsdinfo.html
Sponsor: Foundation for Science and Disability
503 NW 89th Street
Gainesville, FL 32607

Foundation for Seacoast Health Grants 1962
The mission of the foundation is to invest its resources to improve the health and well being of Seacoast residents. The foundation considers a very limited number of new grant initiatives which address one or more of the following prioritized health needs; access to affordable mental health services; access to preventative and restorative dental services; access to affordable child care and after school care; access to affordable primary medical care; and coordination and dissemination of health information related to identified priority needs. The deadline for the submission of the letter of intent is March 1 for Infants, Children and Adolescence and June 1 for Promoting Health and Preventing Disease. Annual deadline dates may vary; contact program staff for exact dates.
Requirements: Nonprofit organizations in cities and towns in the New Hampshire/Maine Seacoast area, including Greenland, New Castle, Newington, North Hampton, Portsmouth, and Rye, NH; and Eliot, Kittery, and York, ME, may submit applications.
Restrictions: Grants do not support ongoing general operating expenses, deficit elimination, political activities, travel, conferences, or lodging.
Geographic Focus: Maine, New Hampshire
Date(s) Application is Due: Mar 1; Jun 1
Samples: School Administrative Unit #50—for a nutrition education program at Rye Elementary School, $631; Seacoast Mental Health Ctr—for the New Heights junior staff training, $10,000.
Contact: Susan Bunting, President; (603) 422-8200; ffsh@communitycampus.org
Internet: http://www.ffsh.org/grants.cfm
Sponsor: Foundation for Seacoast Health
100 Campus Drive, Suite 1
Portsmouth, NH 03801

Foundation for Spirituality and Medicine Grants 1963
The foundation awards grants to nonprofit organizations in its areas of interest, including health, health education, and religious leaders. Preference is given to requests for matching funds. Awards are made primarily in Maryland. As part of the application process, the foundation may request an oral presentation to the board.
Requirements: Nonprofit organizations whose mission is the training of professionals in the health care field, the delivery of health care, or the provision of spiritual services are eligible.
Date(s) Application is Due: Jan 15; Jul 15
Amount of Grant: 50,000 - 100,000 U.S.D
Contact: Grants Administrator; (410) 938-3309
Sponsor: Foundation for Spirituality and Medicine
19 Glenbrook Drive
Phoenix, MD 21131

Foundation for the Carolinas Grants 1964
The foundation supports nonprofit organizations in North and South Carolina, especially in the greater Charlotte, NC, area. The foundation awards grants in the following program areas: education, religion, human services, the environment, historic preservation, public and civic affairs, health, the arts, youths, and senior citizens. Building Youth grants help school-age children make the transition from youth to adulthood.
Requirements: The foundation awards grants to nonprofit organizations serving the following areas in North Carolina: Cebarrus, Cleveland, Iredell, Lincoln, Mecklenburg, Richmond, Stanley, and Union Counties and the cities of Lexiington and Salisbury; and Cherokee, Lancaster, and York Counties in South Carolina.
Restrictions: Grants are not awarded for capital campaigns, operating budgets, endowments, publications, equipment, videos, or conferences, nor to individuals.
Geographic Focus: North Carolina, South Carolina
Samples: Family Ctr (Charlotte, NC)—for a 10-week course designed to strengthen family relationships and to increase parents' involvement in their chldren's preeschool education, $62,940; Richmond County Hospice (Rockingham, NC)—to construct a new residential hospice facility and an administrative center, $50,000; Boys and Girls Clubs (Gaffney, SC)—to purchase a sound system and wireless microphones for the drama program, $2380.
Contact: Dr. Don Jonas, Senior Vice President of Community Philanthropy; (704) 973-4500 or (800) 973-7244; fax (704) 973-4599; djonas@fftc.org
Internet: http://www.fftc.org/grants
Sponsor: Foundation for the Carolinas
P.O. Box 34769
Charlotte, NC 28234-4769

Foundation for the Mid-South Grants 1965
The Foundation for the Mid South invests in people and strategies that build philanthropy and promote racial, social and economic equality in Arkansas, Louisiana, and Mississippi. The Foundation is focused on four priorities: community development; education; health and wellness; and wealth building. Types of funding include: general support; project support; and planning.
Requirements: To be eligible for grants, the applying organization or institution, as well as the purpose of the proposed project, must meet both of the following criteria: possess tax-exempt status under section 501(c)3 of the Internal Revenue Code; possess a certificate from the Mississippi Secretary of State that designates it as a public charity or exempt for the state of Mississippi. For those applying to the Foundation for the Mid South, regardless of the state where your organization is established or operates: All organizations MU.S.T register and receive a certificate that designates it as a public charity or exempt for the state of Mississippi when submitting a full proposal. The application Form URS to apply to the Mississippi Secretary of State is available on their website: http://www.sos.state.ms.us/forms/forms.asp?Unit=Charities. An eligible organization may submit an initial inquiry using the Grant Inquiry Form, which is available on The Foundation for the Mid South website. If the review of the initial inquiry is favorable, applicants will be contacted by the appropriate program officer to discuss the request in greater detail. Inquiries are reviewed throughout the year. Within six weeks of the Foundation's receipt of your unsolicited Grant Inquiry Form, you will receive one of the following: a request to submit a full proposal; a request for more information; a letter informing you that your project will not be supported at this time.
Restrictions: The Foundation do not award grant funds to individuals or make grants for personal needs or business assistance. Additionally, funds are not awarded for lobbying activities; ongoing general operating expenses or existing deficits; endowments,; capital costs including construction, renovation, or equipment; or international programs.
Geographic Focus: Arkansas, Louisiana, Mississippi
Contact: Denise Ellis, Grants & Technology Officer; (601) 355-8167; fax (601) 355-6499; bdellis@fndmidsouth.org
Internet: http://www.fndmidsouth.org/grants_funding.htm
Sponsor: Foundation for the Mid South
134 East Amite Street
Jackson, MS 39201

Foundation Northwest Grants 1966
Grants are awarded to organizations in the Inland Northwest—eastern Washington and northern Idaho—for charitable and educational purposes in the fields of music and drama; education; civic categories which include culture, education, civic and community development; and human services. The foundation has five distinct grant programs and types of support include matching funds, special projects, seed money, and occasional operating support and capital funding. Deadlines vary according to area and contributing sponsorship; contact program support for exact dates.
Requirements: Grant applications are accepted for projects that will enrich the quality of life in the communities of eastern Washington and northern Idaho. Eligible Washington counties include Adams, Asotin, Columbia, Ferry, Garfield, Lincoln, Pend Oreille, Spokane, Steven, and Whitman. Eligible Idaho counties include Benewah, Bonner, Boundary, Clearwater, Idaho, Kootenai, Latah, Lewis, Nez Perce, and Shoshone.
Restrictions: Grants do not support endowments, debt retirement, lobbying, sectarian religious purposes, individuals, travel, sports teams and classes, projects that taxpayers or commercial interests normally support, fundraising campaigns, publications or films (unless an integral part of a foundation-supported program), or nondonor-supported research.
Geographic Focus: Idaho, Washington
Amount of Grant: 2,500 - 7,500 U.S.D
Contact: Program Contact; (509) 624-2606; fax (509) 624-2608; admin@foundationnw.org
Internet: http://www.foundationnw.org
Sponsor: Foundation Northwest
221 Wall Street N, Suite 624
Spokane, WA 99201-0826

Foundation of CVPH April LaValley Fund Grants 1967
In December of 1997, Mrs. Caroline LaValley announced the creation of the April LaValley Endowment Fund. She said that it was April's wish that money which was donated to her family to assist them with expenses related to April's care should be used for other families battling cystic fibrosis. Income generated by the fund is used to help pay for transportation and lodging expenses incurred by immediate family members who travel to be with a family member receiving care for cystic fibrosis. Expenses associates with travel and lodging of family members are allowed.
Requirements: Residents of Clinton, Essex and Franklin counties in New York are eligible for consideration. The person for whom medical expenses are incurred must have a diagnosis of cystic fibrosis.
Restrictions: Funds are not intended to be used for direct medical expenditures.
Geographic Focus: New York
Contact: Roger G. Ahrens, Executive Director; (518) 562-7168 or (518) 561-2000; fax (518) 561-0881; rahrens@cvph.org
Internet: http://www.cvph.org/Foundation/NamedFunds/AprilLaValleyFund/
Sponsor: Foundation of Champlain Valley Medical Center
75 Beekman Street
Plattsburgh, NY 12901

Foundation of CVPH Chelsea's Rainbow Fund Grants 1968
In December of 1999, the Kinblom family announced the creation of the Chelsea's Rainbow Fund. Their seven year old daughter, Chelsea, fought a long courageous battle against brain cancer. Friends throughout the North Country contributed to a fund to assist with expenses not covered by health insurance. The Kinblom family wishes to help other families that face similar battles against childhood cancer. Income generated by this fund will be used to help pay for transportation, food and housing expenses incurred by immediate family members who travel with a child receiving care for cancer. Expenses associates with travel and lodging of family members are allowed.
Requirements: Residents of Clinton, Essex, and Franklin counties in New York are eligible for consideration. Families that have children 18 years old and younger who are battling cancer are eligible for assistance.
Restrictions: Funds are not intended to be used for direct medical expenditures.
Geographic Focus: New York
Contact: Roger G. Ahrens, Executive Director; (518) 562-7168 or (518) 561-2000; fax (518) 561-0881; rahrens@cvph.org
Internet: http://www.cvph.org/Foundation/NamedFunds/ChelseasRainbow/
Sponsor: Foundation of Champlain Valley Medical Center
75 Beekman Street
Plattsburgh, NY 12901

Foundation of CVPH Melissa Lahtinen-Penfield Organ Donor Fund Grants 1969
The Fund was created in memory of Melissa Lahtinen-Penfield, whose wish was to assist families with expenses related to organ donation and educate the community of its vital importance. In July 2000, William Santa and members of the Hometown WIRY team, established the annual Melissa Lahtinen-Penfield (MLP) Organ Donor Golf Tournament in her memory. Proceeds of the tournament, together with donations made directly to the fund, are used to promote organ donation awareness and defray those costs associated with organ transplants which are not covered by health insurance. Expenses considered for reimbursement include medical expenses not covered by insurance, travel and lodging expenses, and meals not including alcohol. Financial need will be the primary determinant upon which the application will be weighed. The size of grants is limited to the income generated by the fund.
Requirements: Residents of Clinton County, New York, are eligible to receive consideration for grants of up to $2,500/annually for expenses associated with organ transplants not covered by health insurance.
Restrictions: Funds are not intended to be used for direct medical expenditures.
Geographic Focus: New York
Amount of Grant: Up to 2,500 U.S.D
Contact: Roger G. Ahrens, Executive Director; (518) 562-7168 or (518) 561-2000; fax (518) 561-0881; rahrens@cvph.org
Internet: http://www.cvph.org/Foundation/NamedFunds/OrganDonor/
Sponsor: Foundation of Champlain Valley Medical Center
75 Beekman Street
Plattsburgh, NY 12901

Foundation of CVPH Rabin Fund Grants 1970
In December of 2000, Pearlie and Mark Rabin established a fund to assist patients with fundamentals of care at CVPH Medical Center. With a gift which endowed the fund, they outlined their wish to offer patients who experience financial hardship, services such as essential articles of clothing and other inexpensive incidentals, which might make their stay in the hospital more pleasant. Throughout the year, CVPH staff takes care of patients who are not able to afford these types of items. In the past, staff members pooled their personal resources and purchased the item or the patient managed without it. The Rabin's Fund provides about $3,500 annually, which assists friends and neighbors from throughout the North Country.
Requirements: Residents of Clinton, Essex, and Franklin counties in New York are eligible for consideration.
Restrictions: Funds are not intended to be used for direct medical expenditures.
Geographic Focus: New York
Contact: Roger G. Ahrens, Executive Director; (518) 562-7168 or (518) 561-2000; fax (518) 561-0881; rahrens@cvph.org
Internet: http://www.cvph.org/Foundation/NamedFunds/RabinFund/
Sponsor: Foundation of Champlain Valley Medical Center
75 Beekman Street
Plattsburgh, NY 12901

Foundation of CVPH Roger Senecal Endowment Fund Grants 1971
The Roger Senecal Endowment Fund exists because of the North Country's love for children, particularly when a child is in need. Income generated by the fund is used to help pay for transportation and housing expenses incurred by an ill or injured child's family in order to be near the child when vital care is being given. Applications will be evaluated and approved by a committee appointed by the Foundation of CVPH Medical Center. Financial need is the primary determinant upon which application will be weighed. The size of the grants is limited to the income generated by the endowment fund on a quarterly basis.
Requirements: Residents of Clinton, Essex, and Franklin counties in New York are eligible for consideration. The person for whom medical expenses are incurred must be 19 years of age or younger.
Restrictions: Funds are not intended to be used for direct medical expenditures.
Geographic Focus: New York
Contact: Roger G. Ahrens, Executive Director; (518) 562-7168 or (518) 561-2000; fax (518) 561-0881; rahrens@cvph.org
Internet: http://www.cvph.org/Foundation/NamedFunds/RogerSenecal/
Sponsor: Foundation of Champlain Valley Medical Center
75 Beekman Street
Plattsburgh, NY 12901

Foundation of CVPH Travel Fund Grants 1972
In 2001, the Board of Directors of the Foundation decided to provide funding for adults who need to travel outside the community to receive health care. The fund provides reimbursement for travel, food and lodging expenses. In many ways, it mimics the Roger Senecal Endowment Fund established in the mid 1980s to assist the families of children who need to leave the North Country to receive health care.
Requirements: Residents of Clinton, Essex, and Franklin counties in New York are eligible for consideration. The person for whom medical expenses are incurred must be 19 years of age or older.
Restrictions: Only one adult may apply for travel funds in addition to the patient. The fund will not provide funds for children to travel. No health care expenses, medications or medical equipment will be funded. Maximum income for the family is limited to $45,000.
Geographic Focus: New York
Amount of Grant: Up to 1,000 U.S.D
Contact: Roger G. Ahrens, Executive Director; (518) 562-7168 or (518) 561-2000; fax (518) 561-0881; rahrens@cvph.org
Internet: http://www.cvph.org/Foundation/NamedFunds/TravelFund/
Sponsor: Foundation of Champlain Valley Medical Center
75 Beekman Street
Plattsburgh, NY 12901

Foundation of the American Thoracic Society Research Grants 1973
The American Thoracic Society established the ATS Foundation Research program to foster the development and training of future leaders in science and biomedical research. Through collaborations with non-profit and corporate partners the research program is able to provide research and career development grants and mentored fellowships to encourage young investigators to devote their talents and time to generating novel scientific ideas leading to new life-saving therapies for patients and their families. Since its founding in 2002, the Foundation of the ATS has awarded $10 million in grants to 105 young researchers investigating a wide spectrum of lung diseases, ranging from asthma and COPD to pulmonary fibrosis and alpha-1 antitrypsin. The program has also funded the career development of 24 pulmonary and critical care fellows. The grants amounts vary and range from $5,000 a year for 1 year to $50,000 per year for two years.
Restrictions: Grants are not awarded to individuals.
Amount of Grant: 5,000 - 50,000 U.S.D
Contact: Alyssa Chase, Senior Manager; (212) 315-8600; fax (212) 315-6498; achase@thoracic.org
Internet: http://foundation.thoracic.org/programs/research-program.php
Sponsor: Foundation of the American Thoracic Society
25 Broadway, 18th Floor
New York, NY 10004

Foundations of East Chicago Health Grants 1974
The Foundations of East Chicago are committed to improving the lives of every resident of its city. Conceived by the citizens of East Chicago to be independent, citizen-run, private foundations, it derives funding from East Chicago's local casino, and uses this money to support local churches, schools, and nonprofit organizations who know the community best and put in the money in action where it can do the most good. Health grants are specifically focused on programs or projects which improve the quality of health education, practices and/or services for East Chicagoans.
Requirements: Applicants must be registered 501(c)3 organizations located in East Chicago, Indiana. If an applicant is not located in East Chicago, it may still qualify if it complies with at least one of the following *Requirements:* the program that an applicant is applying for operates in East Chicago; the funding that an applicant is applying for will go toward assisting East Chicago residents.
Restrictions: All applications must be submitted via Foundations of East Chicago website. The Foundations will not accept any applications in person.
Geographic Focus: Indiana

Amount of Grant: Up to 15,000 U.S.D
Contact: Russell G. Taylor, Executive Director; (219) 392-4225; fax (219) 392-4245; grantinfo@foundationsofeastchicago.org
Internet: http://foundationsofeastchicago.org/apply-now
Sponsor: Foundations of East Chicago
100 W Chicago Avenue
East Chicago, IN 46312

Fourjay Foundation Grants 1975
The Foundation supports only those organizations whose chief purpose is to improve health and/or promote education, within Philadelphia, Montgomery, and Bucks counties in southeastern Pennsylvania. The Fourjay Foundation requires no specific application form. It will consider proposals that address a well-defined need, offer a concrete plan of action, and request a specific amount, from organizations whose staff has the ingenuity, commitment, and motivation to carry out the proposal's objectives. Requests may be for operating support, project specific funds, or capital funds.
Requirements: Organizations serving Philadelphia, Montgomery, and Bucks Counties in southeastern Pennsylvania are eligible.
Restrictions: Funding is not available for: charities operating outside Montgomery, Bucks, or Philadelphia counties; individuals; elementary or secondary educational institutions; museums, musical groups, theaters, or cultural organizations; religious organizations in support of their sacramental or theological functions; political groups or related think tanks; athletic organizations or alumni associations; libraries; public radio or television; United Way or the YMCA; civic organizations; organizations that have applied or been funded within the last 12 month period.
Geographic Focus: Pennsylvania
Date(s) Application is Due: Mar 1; Jun 1; Sep 1; Dec 1
Amount of Grant: 1,000 - 10,000 U.S.D
Contact: Ann T. Bucci; (215) 830-1437; fax (215) 830-0157; abucci@fourjay.org
Internet: http://www.fourjay.org/grantGuidelines.asp
Sponsor: Fourjay Foundation
2300 Computer Avenue, Building G, Suite 1
Willow Grove, PA 19090-1753

Fox Family Foundation Grants 1976
The foundation awards grants to organizations in Saint Louis, MO, and in Israel for support of projects that meet basic human needs including food, shelter, job training, and adult basic education and literacy. Support also is given for cultural programs that provide experiences to individuals who otherwise would not have access. Additional areas of interest include health care, agriculture, housing, Jewish agencies and temples, and the economically disadvantaged and homeless. Consideration will be given to requests for seed money, program development, and matching funds. Contact foundation for annual deadlines.
Restrictions: Grants are not made to individuals or for capital or annual campaigns, endowment funds, deficit financing, or operating expenses.
Geographic Focus: Missouri, Israel
Amount of Grant: 50 - 300,000 U.S.D
Samples: Saint Louis Zoo Foundation (Saint Louis, MO)—for capital campaign, $40,000; Saint Louis Art Museum (Saint Louis, MO)—for gala, $25,000.
Contact: Cheri Fox, Executive Director; (314) 889-0890; fax (314) 727-7314; fff@harbourgroup.com
Sponsor: Fox Family Foundation
7701 Forsyth Boulevard, Suite 600
Saint Louis, MO 63105

Fragile X Syndrome Postdoctoral Research Fellowships Program 1977
The purpose of this program is to promote research aimed at finding a specific treatment for fragile X syndrome, the primary inherited cause of mental retardation. The foundation is particularly interested in preclinical studies of potential pharmacological and genetic treatments for fragile X and studies aimed at understanding the function of the FMR1 gene. The goal is to bring practical treatment into current medical practice as quickly as possible; therefore, preference will be given to research projects that have a clear practical application and the results of which will be shared with other qualified researchers in a timely fashion. Fellowships are awarded for one year and may be renewed for a second year.
Requirements: Tax-exempt 501(c)3 institutions are eligible for funding. Individuals are nominated by applicant institutions for the fellowships and should have training and experience at least equal to the PhD or MD level.
Date(s) Application is Due: May 1; Dec 1
Amount of Grant: Up to 40,000 U.S.D
Contact: Katie Clapp, (978) 462-1866; fax (978) 463-9985; kclappFRAXA@comcast.net
Internet: http://www.fraxa.org/research_howtoapply.aspx
Sponsor: FRAXA Research Foundation
45 Pleasant Street, 2nd Floor
Newburyport, MA 01950

Frances and John L. Loeb Family Fund Grants 1978
The fund is committed to improving the quality of American life. To that end, its grant-making program seeks to address significant issues of society. The fund has chosen as its initial priorities the fields of education, health and family planning, including the public policy questions relating to them. In making grants, the fund favors proposals that offer new knowledge of and innovative approaches to problems rather than palliative measures. It also prefers applications that call for challenge grants and provide evaluation of results. There is no formal application form. Interested applicants should submit a preliminary request in letter form.
Requirements: Funding is directed primarily to organizations and institutions operating in the New York City metropolitan area that are tax exempt under the provisions of Section 501(c)3 of the Internal Revenue Code. Exceptions to this geographical limitation will be made for proposals that have the promise of national implication or extensive replication.
Restrictions: The fund will not consider proposals for annual or capital campaigns, building or renovation projects or loan or emergency funds. Neither will it make grants to individuals or for sectarian or religious purposes or political activities such as lobbying or propaganda. It will not fund any organization that discriminates on the basis of race, sex, religion, national origin or sexual preference.
Contact: John L. Loeb, President; (212) 588-9052; fax (212) 838-6470; loebff@aol.com
Internet: http://www.geocities.com/CapitolHill/5601/family_fund_brochure.html
Sponsor: Frances and John L. Loeb Family Fund
375 Park Avenue, Suite 801
New York, NY 10152

Frances L. and Edwin L. Cummings Memorial Fund Grants 1979
The foundation supports nonprofits in New York City and northern New Jersey in the areas of education, especially programs that serve public school children from disadvantaged backgrounds; social welfare concerns; and campaigns to build endowments through establishment of challenge grants. Other areas of interest include elementary education, secondary education, vocational education, adult basic education and literacy, and higher education; hospitals, medical care, AIDS, and cancer; children, youth, and human services; and community development. Support is provided for endowment funds, seed money, consulting services, matching funds, technical assistance, professorships, and program development. The board meets in June and December. Application forms are not required.
Requirements: Grants are awarded to nonprofit organizations in the metropolitan New York, NY, area, with emphasis on New York City, southern Westchester County, and northern New Jersey.
Restrictions: Grants are not awarded to individuals or for capital building campaigns, general operating support, moving expenses, conferences, surveys, annual fund-raising campaigns, or research conducted by individuals or private institutions.
Geographic Focus: New Jersey, New York
Amount of Grant: 10,000 U.S.D
Samples: Little Sisters of the Assumption Family Health Service (New York, NY)—to establish an endowment fund, $100,000; Neighborhood Initiatives (Bronx, NY)—for staff support to expand a teen center program, $30,000; Korean Community Service of Metropolitan New York (Woodside, NY)—to support a health educator and outreach worker for an HIV/AIDS prevention and education program, $40,000.
Contact: Elizabeth Costas; (212) 286-1778; fax (212) 682-9458
Sponsor: Frances L. and Edwin L. Cummings Memorial Fund
501 Fifth Avenue, Suite 708
New York, NY 10017-6103

Frances W. Emerson Foundation Grants 1980
The Foundation, in the care of Goldman Tax Service of Bellingham, Massachusetts, offers grants throughout Massachusetts and Vinalhaven, Maine. Its primary fields of interest include community development, economic development, education, and health care. There is no application form required or deadlines to adhere to, and applicants should send a letter or the full proposal to Eaton Vance management in Boston.
Geographic Focus: Maine, Massachusetts
Contact: Thomas Huggins, Vice President, Eaton Vance Management; (617) 482-8260
Sponsor: Frances W. Emerson Foundation
255 State Street
Boston, MA 02109

Francis A. and Jacquelyn H. Harrington Foundation Grants 1981
The foundation supports requests from health, educational, and cultural organizations that are providing solutions to, or prevention of, problems within the immediate Worcester, MA, community area and that are receiving additional support from other funding sources. Types of support include capital campaigns, equipment acquisition, general operating support, and program development. A proposal should be in letter form and should include the goals and objectives of the request together with a plan for accomplishing the same. The trustees meet twice each year, once in late summer to review requests received to date and again in December to review requests and confirm grants. The foundation does not make human service grants. Instead, all human service requests should be sent to the Greater Worcester Community Foundation, where the foundation has established the Francis A. and Jacquelyn H. Harrington Fund.
Requirements: The foundation provides grants to 501(c)3 tax-exempt organizations that are based in the Worcester, MA, area.
Geographic Focus: Massachusetts
Date(s) Application is Due: Jun 1
Samples: Quinsigamond Community College (Worcester, MA)—to create a one-stop enrollment center in the college's administration building, $1 million.
Contact: Sumner Tilton Jr., Trustee; (508) 798-8621
Sponsor: Francis A. and Jacquelyn H. Harrington Foundation
370 Main Street, 12th Floor
Worcester, MA 01608

Francis C. Wood History of Medicine Fellowships 1982
The institute offers fellowships to scholars engaged in projects requiring personal use of the historical collections of the library of the College of Physicians of Philadelphia and/or the collections of the Mutter Museum. The historical collections constitute one of the

largest medical history repositories in the United States, documenting the evolution of medicine from the medieval period to the present. Its resources include rare medical books and journals; manuscript case records, papers, and lecture notes of many physicians; and prints, engravings, and photographs of medical subjects. The Mutter Museum is renowned for its collections of medical artifacts, instruments, and pathological and anatomical specimens. Recipients may be requested to present a seminar at the institute and will be required to submit a report on their research. Awards will be announced in March.
Requirements: Letters of application outlining the proposed project (proposal should not exceed five pages); length of residence; historical materials to be used; and a budget with specific information on travel, lodging, and research expenses should be submitted, along with a curriculum vita and two letters of recommendation.
Date(s) Application is Due: Apr 1
Amount of Grant: Up to 1,000 U.S.D
Contact: Margaret Patton, Program Contact; (215) 563-3737, ext. 305; mpatton@collphyphil.org
Internet: http://www.collphyphil.org/woodfell.htm
Sponsor: College of Physicians of Philadelphia
19 S 22nd Street
Philadelphia, PA 19103

Francis L. Abreu Charitable Trust Grants **1983**
The Francis L. Abreu Charitable Trust was established under the will of May Patterson Abreu in honor of her husband, Francis, who died in 1969. The trust supports Atlanta-area nonprofit organizations in its areas of interest, including arts and cultural programs, secondary education, higher education, health associations, human services, and children and youth services. Types of support include capital campaigns, seed money, program development, and matching funds. Requests are reviewed at April and October trustee meetings. Application forms are available online.
Requirements: Georgia nonprofit organizations serving the greater Atlanta area are eligible to apply.
Restrictions: The foundation does not approve requests for operating costs or grants to individuals.
Geographic Focus: Georgia
Date(s) Application is Due: Mar 31; Sep 30
Contact: Peter Abreu, Chairman; (404) 549-6743; fax (404) 549-6752
Internet: http://www.abreufoundation.org
Sponsor: Francis L. Abreu Charitable Trust
P.O. Box 502407
Atlanta, GA 31150

Francis T. & Louise T. Nichols Foundation Grants **1984**
The Foundation gives primarily in the areas of education, health care, and human services. Its major fields of interest include: children and youth services; distaster relief; fire prevention and control; elementary and secondary education programs; health care; hospital support; and human/community service programs. There are no deadlines.
Requirements: Applicants should approach the Foundation initially by letter.
Restrictions: Giving primarily in Hancock County, Maine. There are no grants made to individuals.
Geographic Focus: Maine
Amount of Grant: Up to 80,000 U.S.D
Samples: George Stevens Academy, Blue Hill, Maine, $80,000; Maine Children's Home for Little Wanderers, Waterville, Maine, $17,000; Hancock County Homecare, Blue Hills, Maine, $17,000.
Contact: Calvin E. True, Treasurer; (207) 947-0111
Sponsor: Francis T. and Louise T. Nichols Foundation
P.O. Box 1210
Bangor, ME 04402-1210

Frank B. Hazard General Charity Fund Grants **1985**
The Frank B. Hazard General Charity Fund was established in 1924 to support charitable organizations that work to improve the lives of "the poor, or the poor sick." Organizations receiving support from the fund must be managed and/or governed by individuals a majority of whom are of the Protestant religious faith. Grant requests for general operating support or program support are strongly encouraged. The majority of grants from the Hazard General Charity Fund are one year in duration. Applicants must apply online at the grant website. Applicants are strongly encouraged to do the following before applying: review the downloadable state application procedures for additional helpful information and clarifications; review the downloadable online-application guidelines at the grant website; review the trust's funding history (link is available from the grant website); review the online application questions in advance; and review the list of required attachments. These will generally include: a list of board members, financial statements (audited, reviewed, or compiled by independent auditor); an organization summary; a list of other funding sources; an IRS Determination letter; and other required documents. All attachments must be uploaded in the online application as PDF, Word, or Excel files. The application deadline for the Frank B. Hazard General Charity Fund is 11:59 p.m. on December 1. Applicants will be notified of grant decisions before January 31 of the following year.
Requirements: Applicants must have 501(c)3 tax-exempt status.
Restrictions: The Hazard General Charity Fund specifically supports charitable organizations that serve the people of Providence, Rhode Island. The fund does not support requests from individuals, organizations attempting to influence policy through direct lobbying, or any political campaigns.
Geographic Focus: Rhode Island
Date(s) Application is Due: Dec 1
Samples: Family Service, Providence, Rhode Island, $10,000 (2010); Community Preparatory School, Providence, Rhode Island, $17,000 (2010); Rhode Island Free Clinic, Providence, Rhode Island, $10,000 (2010).
Contact: Emma Greene, Director; (617) 434-0329; emma.m.greene@baml.com
Internet: https://www.bankofamerica.com/philanthropic/fn_search.action
Sponsor: Frank B. Hazard General Charity Fund
225 Franklin Street, 4th Floor, MA1-225-04-02
Boston, MA 02110

Frank E. and Seba B. Payne Foundation Grants **1986**
The foundation awards grants in the Chicago, IL, area and in Pennsylvania in its areas of interest, including AIDS prevention, children and youth, cultural activities, education, and hospitals. Types of support include building construction/renovation, equipment acquisition, and general operating support.
Requirements: Nonprofit organizations in the greater Chicago, IL, metropolitan area and in Pennsylvania are eligible.
Restrictions: Grants are not made to individuals.
Geographic Focus: Illinois, Pennsylvania
Contact: M. Catherine Ryan, c/o Bank of America, (312) 828-1785
Sponsor: Frank E. Payne and Seba B. Payne Foundation
231 S LaSalle Street
Chicago, IL 60697

Frank G. and Freida K. Brotz Family Foundation Grants **1987**
The Foundation supports Wisconsin nonprofits of the Christian, Lutheran, and Roman Catholic faiths, including Salvation Army. Types of support include capital support and general support. Organizations eligible to receive support include colleges and universities, community service groups, hospitals, ministries, parochial schools and religious education organizations, religious welfare organizations, and youth organizations. There are no application forms or deadlines. Applicants should submit a brief letter of inquiry that contains a description of the organization, its purpose, purpose of grant, and proof of tax-exempt status.
Requirements: Grants are awarded to publicly supported, Section 170(c) IRS qualified organizations, primarily in the State of Wisconsin.
Restrictions: No grants are awarded to individuals or organizations that require expenditure responsibility under Treasury Regulations.
Geographic Focus: Wisconsin
Contact: Stuart W. Brotz, Grants Contact; (920) 458-2121; fax (920) 458-1923
Sponsor: Frank G. and Freida K. Brotz Family Foundation
3518 Lakeshore Road, P.O. Box 551
Sheboygan, WI 53082-0551

Frank J. Antun Foundation Grants **1988**
The foundation awards grants to nonprofits in its areas of interest, including health associations, minorities, immigrants, and social services. Most awards will be made in the greater New York, NY, metropolitan area. There are no application deadlines, but application forms are required.
Requirements: Giving primarily in the greater New York, NY, area
Restrictions: Individuals are ineligible.
Geographic Focus: New York
Amount of Grant: 1,000 - 10,000 U.S.D
Contact: Josephine Alex, Manager
Sponsor: Frank J. Antun Foundation
100 Crossways Park W, Suite 205
Woodbury, NY 11797

Franklin County Community Foundation Grants **1989**
The foundation was established in 1998 in Indiana. The Franklin County Foundation seeks to build permanent partnerships in giving that effectively connect donors and community projects to promote an enriched future for all citizens of Franklin County, Indiana. The Foundation also administers separate scholarship funds established by individuals, families, corporations, and organizations to assist local students in pursuit of advanced education. Together, these scholarships make up our Community Scholarship Program. Eligibility, criteria, and supplemental requirements are different for each scholarship. To apply for one or more of these scholarships contact the Foundation's office.
Geographic Focus: Indiana
Contact: Shelly Lunsford, Director; (765) 647-6810; fax (765) 647-0238; fcfoundation@yahoo.com
Internet: http://www.franklincountyindiana.info/2005%20Grants.htm
Sponsor: Franklin County Community Foundation
527 Main Street
Brookville, IN 47012-1284

Franklin H. Wells and Ruth L. Wells Foundation Grants **1990**
The Foundation awards grants to support the arts, community and economic development, education, health care, and human services. Funding is also available for emergency funds, equipment, program development, and seed money. The Foundation board meets in April and October, with funding notification in May and November.
Requirements: Applications are not required. Organizations should initially submit a letter of inquiry, and if accepted, one copy of their proposal.
Restrictions: The Foundation gives primarily to Dauphin, Cumberland, and Perry counties in Pennsylvania. Funding is not available for religious activities, individuals, endowments, debts, and capital campaigns.

Geographic Focus: Pennsylvania
Date(s) Application is Due: Mar 15; Sep 15
Contact: Miles Gibbons Jr., Executive Director; (866) 398-9023; mgibbons989@earthlink.net
Sponsor: Franklin H. Wells and Ruth L. Wells Foundation
One M and T Plaza, 8th Floor
Buffalo, NY 14203-2309

Franklin I. and Irene List Saemann Foundation Grants **1991**
The foundation regularly makes grants that benefit libraries, volunteer programs, hospitals, colleges, and other related organizations. Special consideration is given to organizations based in Kosciusko County, IN. The board of trustees meets in June to consider grant requests.
Requirements: A brief proposal plus supporting documents should include a description of the organization, amount requested and how it will be used, a current budget and anticipated expenses and sources of income, a financial statement of the previous full year, a 501(c)3 determination letter, and a list of the current board of directors.
Geographic Focus: Indiana, Iowa, Wisconsin
Date(s) Application is Due: Apr 1
Samples: Carthage College (Kenosha, WI)—for a curriculum resource and development center in the new Hedberg Library, $1 million; Wartburg College (Waverly, IA)—to renovate and expand the college's library, $300,000.
Contact: Walter Kilgus, Foundation Manager; (815) 772-2121; fax (815) 772-2026; saemann@soltec.net
Sponsor: Franklin I. and Irene List Saemann Foundation
P.O. Box 105
Morrison, IL 61270

Frank Loomis Palmer Fund Grants **1992**
The Frank Loomis Palmer Fund was established in 1936 to support and promote quality educational, cultural, human-services, and health-care programming for underserved populations. The Palmer Fund specifically serves the people of New London, Connecticut. Grants from the Palmer Fund are primarily one year in duration; on occasion, multi-year support is awarded. Applicants must apply online at the grant website. Applicants are strongly encouraged to do the following before applying: review the downloadable state application procedures for additional helpful information and clarifications; review the downloadable online-application guidelines at the grant website; review the fund's funding history (link is available from the grant website); review the online application questions in advance; and review the list of required attachments. These will generally include: a list of board members, financial statements (audited, reviewed, or compiled by independent auditor); an organization summary; a list of other funding sources; an IRS Determination letter; and other required documents. All attachments must be uploaded in the online application as PDF, Word, or Excel files. The application deadline for the Frank Loomis Palmer Fund is 11:59 p.m. on November 15. Applicants will be notified of grant decisions by letter within three to four months after the proposal deadline.
Requirements: Applicants must have 501(c)3 tax-exempt status and serve the people of New London, Connecticut.
Restrictions: Applicants will not be awarded a grant for more than three consecutive years. The fund does not support requests from individuals, organizations attempting to influence policy through direct lobbying, or any political campaigns.
Geographic Focus: Connecticut
Date(s) Application is Due: Nov 15
Samples: Garde Arts Center, New London, Connecticut, $50,000, 8th payment of multi-year grant to support addition of theater space (2010); Center: A Drop-In Community Learning and Resource Center, New London, Connecticut, $50,000, Before Care and After School Enrichment Programs to augment childcare tuition of low-income working families and children (2010); New London Homeless Hospitality Center, New London, Connecticut, $60,000, to meet increased demand for overnight emergency shelter during economic downturn (2010).
Contact: Amy Lynch, Senior Manager; (860) 657-7015; amy.r.lynch@baml.com
Internet: https://www.bankofamerica.com/philanthropic/fn_search.action
Sponsor: Frank Loomis Palmer Fund
200 Glastonbury Boulevard, Suite # 200, CT2-545-02-05
Glastonbury, CT 06033-4056

Frank Reed and Margaret Jane Peters Memorial Fund I Grants **1993**
The Frank Reed & Margaret Jane Peters Memorial Fund I was established in 1935 to support and promote quality educational, human-services, and health-care programming for underserved populations. Special consideration is given to charitable organizations that serve youth and children. The Peters Memorial Fund I is a generous supporter of the Associated Grant Makers (AGM) Summer Fund. The AGM Summer Fund is a collaborative group of donors that provides operating support for summer camps serving low-income urban youth from Boston, Cambridge, Chelsea, and Somerville. Excluding the grant made to the AGM Summer Fund, the typical grant range is $10,000 to $40,000. Grant requests for general operating support are strongly encouraged. Program support will also be considered. Small, program-related capital expenses may be included in general operating or program requests. The majority of grants from the Peters Memorial Fund I are one year in duration; on occasion, multi-year support is awarded. Applicants must apply online at the grant website. Applicants are strongly encouraged to do the following before applying: review the downloadable state application procedures for additional helpful information and clarifications; review the downloadable online-application guidelines at the grant website; review the foundation's funding history (link is available from the grant website); review the online application questions in advance; and review the list of required attachments. These will generally include: a list of board members, financial statements (audited, reviewed, or compiled by independent auditor); an organization summary; a list of other funding sources; an IRS Determination letter; and other required documents. All attachments must be uploaded in the online application as PDF, Word, or Excel files. The application deadline for the Frank Reed & Margaret Jane Peters Memorial Fund I is 11:59 p.m. on September 1. Applicants will be notified of grant decisions before November 30.
Requirements: Applicants must have 501(c)3 tax-exempt status.
Restrictions: The fund does not support requests from individuals, organizations attempting to influence policy through direct lobbying, or any political campaigns.
Geographic Focus: Massachusetts
Date(s) Application is Due: Sep 1
Amount of Grant: 10,000 - 40,000 U.S.D
Samples: Generations, Inc., Boston, Massachusetts, $20,000, general operating support (2010); Boston Scholars Program, Boston, Massachusetts, $15,000, general operating support (2010); Family Service, Lawrence, Massachusetts, $15,000, Youth Mentoring Department program development (2010).
Contact: Miki C. Akimoto, Vice; (866) 778-6859; miki.akimoto@baml.com
Internet: https://www.bankofamerica.com/philanthropic/fn_search.action
Sponsor: Frank Reed and Margaret Jane Peters Memorial Fund I
225 Franklin Street, 4th Floor, MA1-225-04-02
Boston, MA 02110

Frank Reed and Margaret Jane Peters Memorial Fund II Grants **1994**
The Frank Reed & Margaret Jane Peters Memorial Fund II was established in 1935 to support and promote quality educational, human-services, and health-care programming for underserved populations. In the area of education, the fund supports academic access, enrichment, and remedial programming for children, youth, adults, and senior citizens that focuses on preparing individuals to achieve while in school and beyond. In the area of health care, the fund supports programming that improves access to primary care for traditionally underserved individuals, health education initiatives and programming that impact at-risk populations, and medical research. In the area of human services the fund tries to meet evolving needs of communities. Currently the fund's focus is on (but is not limited to) youth development, violence prevention, employment, life-skills attainment, and food programs. Grant requests for general operating support are strongly encouraged. Program support will also be considered. Small, program-related capital expenses may be included in general operating or program requests. The majority of grants from the Peters Memorial Fund II are one year in duration; on occasion, multi-year support is awarded. Applicants must apply online at the grant website. Applicants are strongly encouraged to do the following before applying: review the downloadable state application procedures for additional helpful information and clarifications, review the downloadable online-application guidelines at the grant website; review the foundation's funding history (link is available from the grant website); review the online application questions in advance; and review the list of required attachments. These will generally include: a list of board members, financial statements (audited, reviewed, or compiled by independent auditor); an organization summary; a list of other funding sources; an IRS Determination letter; and other required documents. All attachments must be uploaded in the online application as PDF, Word, or Excel files. The application deadline for the Frank Reed & Margaret Jane Peters Memorial Fund II is 11:59 p.m. on July 1. Applicants will be notified of grant decisions before September 30.
Requirements: Applicants must have 501(c)3 tax-exempt status.
Restrictions: In general, capital requests are not advised. The fund does not support endowment campaigns, events such as galas or award ceremonies, and costs of fundraising events. The fund does not support requests from individuals, organizations attempting to influence policy through direct lobbying, or any political campaigns.
Geographic Focus: Massachusetts
Date(s) Application is Due: Jul 1
Samples: Epiphany School, Dorchester, Massachusetts, $20,000, general operating support (2010); Esther R. Sanger Center for Compassion, Wollaston, Massachusetts, $15,000, for general operating support (2010); Boston Learning Center, Boston, Massachusetts, $20,000, general operating support (2010).
Contact: Michealle Larkins; (866) 778-6859; michealle.larkins@baml.com
Internet: https://www.bankofamerica.com/philanthropic/fn_search.action
Sponsor: Frank Reed and Margaret Jane Peters Memorial Fund II
225 Franklin Street, 4th Floor, MA1-225-04-02
Boston, MA 02110

Frank S. Flowers Foundation Grants **1995**
The Frank S. Flowers Foundation primarily serves the Gloucester County, New Jersey area. The Foundation's area of interest include: Education—supporting public high schools of Gloucester County, New Jersey, to provide scholarships for college or graduate study, vocational or technical training; Youth—supporting chapters or councils or branches of Y.M.C.A. and Boys Scouts of America located in Gloucester and/or Salem Counties, grants are also considered for organizations having branches or offices in Gloucester County, New Jersey which treat and educate children with special needs; Health-Related—support to non-profit Gloucester County hospitals; Religious organizations—support to churches in the boroughs of Paulsboro and Wenonah, New Jersey. The Foundation also has a specific interest in The Shriner's Hospital for Crippled Children in Philadelphia, Pennsylvania and the Masonic Home Charity Foundation of New Jersey. Grants range from $1,000 - $8,000. Application deadline date is February 15th, application available online. Requestors will receive a letter acknowledging the receipt of their request.
Requirements: Qualifying tax-exempt 501(c)3 organizations are eligible for grants if they meet the purpose of the foundation. Proposals should be submitted in the following format:

completed Common Grant Application Form; an original Proposal Statement*; an audited financial report and a current year operating budget; a copy of your official IRS Letter with your tax determination; a listing of your Board of Directors. *Proposal Statement should answer these questions: what are the objectives and expected outcomes of this program/project/request; what strategies will be used to accomplish your objective; what is the timeline for completion; if this is part of an on-going program, how long has it been in operation; what criteria will you use to determine your success; if the request is not fully funded, what other sources can you engage. A Proposal budget should be included if this is for a specific program within your annual budget. Please describe any collaborative ventures.
Restrictions: Grants are not made for political purpose, nor to organizations which discriminate on the basis of race, ethnic origin, sexual or religious preference, age or gender.
Geographic Focus: New Jersey, Pennsylvania
Date(s) Application is Due: Feb 15
Amount of Grant: 1,000 - 8,000 U.S.D
Contact: Gale Y. Sykes, Trustee c/o Wells Fargo Bank; (908) 598-3576; grantinquiries2@wachovia.com
Internet: https://www.wachovia.com/foundation/v/index.jsp?vgnextoid=68d78689fb0aa110VgnVCM1000004b0d1872RCRD&vgnextfmt=default
Sponsor: Frank S. Flowers Foundation
190 River Road
Summit, NJ 07901

Frank Stanley Beveridge Foundation Grants **1996**
The foundation welcomes proposals in the areas of: animal care; arts, culture and humanities; civil rights, social action, advocacy; education; employment/jobs; environmental quality, protection and beautification; food, nutrition and agriculture; health; housing; human services; medical research; mental health; philanthropy; safety; recreation; religion; science; social services; and youth development. The board meets in October and April to consider requests. Multiyear grants are rare. Contact the foundation via the Web site only. No phone or written inquiries will be accepted.
Requirements: Applicants must be 501(c)3 nonprofit organizations or foundations in Massachusetts's Hampden and Hampshire Counties.
Restrictions: The Foundation prefers not to support: awards or prizes; commissioning of new artistic work; conferences/seminars; curriculum development; debt reduction; employee matching gifts; employee-related scholarships; endowment funds; exhibitions; faculty/staff development; fellowship funds; fellowships to individuals; film/video/radio production; foundation administered programs; general operating support; grants to individuals; income development; internship funds; management development; performance/production costs; professorships; program-related investment/loans; publications; scholarships to individuals; student aid; technical assistance.
Geographic Focus: Massachusetts
Date(s) Application is Due: Feb 1; Aug 1
Amount of Grant: 50,000 U.S.D
Contact: Philip Caswell, President; (800) 229-9667; fax (561) 748-0644; administrator@beveridge.org or caswell@beveridge.org
Internet: http://www.beveridge.org/
Sponsor: Frank Stanley Beveridge Foundation
1340 U.S. Highway 1, Suite 102
Jupiter, FL 33469

Frank W. and Carl S. Adams Memorial Fund Grants **1997**
The Frank W. and Carl S. Adams Memorial Fund was established in 1925 to support and promote quality educational, human services, and health care programming for underserved populations. Annual gifts are also awarded to the Harvard University Medical School and the Massachusetts Institute of Technology for student scholarships. Grant requests for general operating support are strongly encouraged. Program support will also be considered. Small, program-related capital expenses may be included in general operating or program requests. To better support the capacity of nonprofit organizations, multi-year funding requests are strongly encouraged. The application deadline is February 1. Applicants will be notified of grant decisions before May 31.
Requirements: 501(c)3 organizations serving the residents of Massachusetts are eligible.
Geographic Focus: Massachusetts
Date(s) Application is Due: Feb 1
Contact: Miki C. Akimoto, Vice President; (866) 778-6859; miki.akimoto@baml.com
Internet: https://www.bankofamerica.com/philanthropic/fn_search.action
Sponsor: Frank W. and Carl S. Adams Memorial Fund
225 Franklin Street, 4th Floor, MA1-225-04-02
Boston, MA 02110

Fraser-Parker Foundation Grants **1998**
The Foundation awards funds in its areas of interest, including Christian religion organizations, education and higher education, and hospitals. There are no application forms and no deadlines.
Amount of Grant: 5,000 - 50,000 U.S.D
Samples: Visiting Nurse/Hospice Atlanta, Atlanta, Georgia, $25,000, chaplain residency training program (2010); Colonial Williamsburg Foundation, Williamsburg, Virginia, $5,000, operating support (2010); and University of Rochester, Rochester, New York, $65,000, therapeutic programs (2010).
Contact: John Stephenson, Executive Director; (404) 827-6529
Sponsor: Fraser-Parker Foundation
3050 Peachtree Road NW, Suie 270
Atlanta, GA 30305

Fred & Gretel Biel Charitable Trust Grants **1999**
The Fred & Gretel Biel Charitable Trust was established in 2004 to support and promote quality educational, human-services, and health-care programming for underserved populations. Special consideration is given to organizations that provide food and clothing to low-income individuals and families. Consideration is also given to organizations that serve the economically disadvantaged through the provision of housing, legal assistance, or day-care services. Grant requests for general operating and capital support are encouraged. Grants from the Biel Charitable Trust are one year in duration. Application materials are available for download at the grant website. Applicants are strongly encouraged to review the state application guidelines for additional helpful information and clarifications before applying. Applicants are also encouraged to review the trust's funding history (link is available from the grant website). The application deadline for the Biel Charitable Trust is May 1. Applicants will be notified of grant decisions by June 30.
Requirements: The Biel Charitable Trust typically supports organizations serving the people of King and Snohomish Counties in the Puget Sound region of Washington. Occasionally grants will be made outside of the Puget Sound area. Applicant organizations must have 501(c)3 tax-exempt status.
Restrictions: Requests to assist with debt retirement or to correct an operating deficit will not be considered. Applicants who have received a grant for three consecutive years must wait two years before reapplying to the trust. The trust does not support requests from individuals, organizations attempting to influence policy through direct lobbying, or any political campaigns.
Geographic Focus: Washington
Date(s) Application is Due: May 1
Samples: Edmonds School District, Lynnwood, Washington, $2,500 (2010); Haller Lake Christian Health Clinic, Seattle, Washington, $5,000 (2010); North Helpline, Seattle, Washington, $2,500 (2010).
Contact: Nancy Atkinson, Vice President; (800) 848-7177 or (206) 358-0912; nancy.l.atkinson@baml.com
Internet: https://www.bankofamerica.com/philanthropic/fn_search.action
Sponsor: Fred and Gretel Biel Charitable Trust
800 5th Avenue
Seattle, WA 98104

Fred and Louise Latshaw Scholarship **2000**
The Fred and Louise Latshaw Scholarship provides funding to employees of Lake Area United Way Agencies (see application for list of eligible agencies) or employees of a nonprofit agency that has an endowment fund with Legacy Foundation (see application for list of eligible agencies). A scholarship of $1,000 is awarded for tuition, fees, and/or books and is renewable by reapplication each year of undergraduate study.
Requirements: Applicants will be considered for the scholarship if they meet the following criteria: they are U.S. citizens and residents of Lake County, Indiana; they are a high school graduate and have deferred their education and wish to return as a part time student or are currently enrolled; they are currently employed by a Lake Area United Way Agency or by a not-for-profit Agency that has an endowment fund with the Legacy Foundation; they are currently enrolled or a newly admitted student at an accredited public or private college, university, or vocational school. All college majors are eligible.
Geographic Focus: Indiana
Date(s) Application is Due: Jun 1
Amount of Grant: 1,000 U.S.D
Contact: Latshaw Scholarship Contact; (219) 736-1880; fax (219) 736-1940; legacy@legacyfoundationlakeco.org
Internet: http://www.legacyfoundationlakeco.org/scholarships/index.htm
Sponsor: Legacy Foundation
1000 East 80th Place, 302 South
Merrillville, IN 46410

Fred Baldwin Memorial Foundation Grants **2001**
The Foundation supports programs and projects that benefit the people of Maui County. Projects in the arts, education, environment, health, and human services are of greatest interest. Funding is available for capital improvement projects. There are two annual deadlines, which fall on the first business day in February and August. Application information is available online.
Requirements: Eligible applicants must have 501(c)3 status, or must apply through a fiscal sponsor with 501(c)3 status.
Restrictions: The Foundation does not fund loans or debt service, endowments, funds for re-granting, scholarships, grants to individuals or units of government, or activities that have already occurred.
Geographic Focus: Hawaii
Date(s) Application is Due: Feb 1; Aug 1
Amount of Grant: 10,000 U.S.D
Contact: Carrie Shoda-Sutherland, Senior Program Officer; (808) 566-5524 or (888) 731-3863, ext. 524; fax (808) 521-6286; csutherland@hcf-hawaii.org
Internet: http://www.fredbaldwinfoundation.org/
Sponsor: Fred Baldwin Memorial Foundation
1164 Bishop Street, Suite 800
Honolulu, HI 96813

Fred C. and Katherine B. Andersen Foundation Grants **2002**
Fred C. and Katherine B. Andersen Foundation, formerly known as the Andersen Foundation gives on a national basis for higher education. The foundation provides funds locally in Minnesota and western Wisconsin for all areas of interest, which include: arts;

health care; higher education; hospitals; and youth development. Funding is available in the forms of: capital campaigns; general/operating support and; program development.
Requirements: 501(c)3 tax-exempt organizations are eligible. Giving is on a national basis although preference is given to requests from Minnesota.
Restrictions: The foundation does not make grants to institutions that receive federal funding.
Date(s) Application is Due: Mar 18; Jul 22; Oct 21
Amount of Grant: 5,000 - 12,000,000 U.S.D
Contact: Mary Gillstrom, Director; (651) 264-5150
Sponsor: Fred C. and Katherine B. Andersen Foundation
P.O. Box 8000
Bayport, MN 55003-0080

Freddie Mac Foundation Grants **2003**

The Foundation focus is on children whose families have limited resources and who are vulnerable to poor outcomes. The Foundation emphasizes the integration of services that focus on family strengthening and youth development in order to maximize the benefit to children and their families. Grants will be made for direct service projects, general operating support, capacity building, public awareness, planning and capital projects. The following funding priorities will be considered: the early years; elementary school years; junior high and high school years; children and families in crisis; and public awareness education. Proposals must be submitted online.
Requirements: Eligible organizations must be tax exempt under IRS code 501(c)3 and defined as a public charity. The Foundation's grantmaking program services the following metropolitan Washington, DC areas: District of Columbia; Virginia - the counties of Arlington, Fairfax, Loudoun and Prince William and the cities of Alexandria, Falls Church, Manassas Park, and Leesburg; Maryland - the counties of Charles, Frederick, Howard, Montgomery, and Prince George's.
Restrictions: The Foundation will not fund organizations that discriminate in the provision of services or in employment practices based on race, color, religion, ethnicity, sex, age, national origin, disability, sexual orientation, marital status, and any other characteristics protected by applicable law. Unless approved by the board, the Foundation does not fund: individuals; training in/promotion of religious doctrine; incurring a debt liability; endowment campaigns.
Geographic Focus: District of Columbia, Maryland, Virginia
Date(s) Application is Due: Sep 9
Amount of Grant: 5,000 - 50,000 U.S.D
Samples: Distributed among nonprofit organizations to aid support-relief efforts following Hurricane Katrina, $10 million; CrossChildren's Law Ctr (Washington, DC)—to support the caregivers of children in the Washington, DC, foster-care system who are seeking adoption and guardianship, $90,000; Ctr for Multicultural Human Services (Falls Church, VA)—to acquire five transitional-housing units that will provide 16 homeless families in northern Virginia, many of whom are recent immigrants, with safe housing and intensive support services, $550,000; American National Red Cross (Washington, DC)—to assist tsunami-relief efforts, $150,000.
Contact: Ralph F. Boyd, President and CEO; fax (703) 918-8888; freddie_mac_foundation@freddiemac.com
Internet: http://www.freddiemacfoundation.org
Sponsor: Freddie Mac Foundation
8250 Jones Branch Drive, MS A40
McLean, VA 22102-3110

Frederick Gardner Cottrell Foundation Grants **2004**

Giving primarily in Woods Hole, Massachusetts, Hunt Valley, Maryland, Brunswick, Maine, and Houston, Texas. The foundation awards grants to educational and scientific organizations to advance research and chair teaching positions. The foundation's areas of interest include eye research, higher education, marine science, and science. There are no application deadline. Submit a written request for application.
Geographic Focus: Maine, Maryland, Massachusetts, Texas
Amount of Grant: 100,000 - 300,000 U.S.D
Contact: Gary Munsinger, President; (520) 748-4400; fax (520) 748-0025; munsinger@rctech.com
Sponsor: Frederick Gardner Cottrell Foundation
5210 East Williams Circle, Suite 240
Tucson, AZ 85711-4410

Frederick McDonald Trust Grants **2005**

The Frederick McDonald Trust was established in 1950 to support and promote quality educational, human-services, and health-care programming for underserved populations. Grant requests for general operating support, program, project and capital support will be considered. The majority of grants from the McDonald Trust are one year in duration. On occasion, multi-year support is awarded. Applicants must apply online at the grant website. Applicants are strongly encouraged to do the following before applying: review the downloadable state application procedures for additional helpful information and clarifications; review the downloadable online-application guidelines at the grant website; review the trust's funding history (link is available from the grant website); review the online application questions in advance; and review the list of required attachments. These will generally include: a list of board members, financial statements (audited, reviewed, or compiled by independent auditor); an organization summary; a list of other funding sources; an IRS Determination letter; and other required documents. All attachments must be uploaded in the online application as PDF, Word, or Excel files. The application deadline for the Frederick McDonald Trust is 11:59 p.m. on May 1. Applicants will be notified of grant decisions before August 31.
Requirements: Applicants must have 501(c)3 tax-exempt status.
Restrictions: Grants are made only to those organizations located in, or serving the people of Albany City. The trust does not support requests from individuals, organizations attempting to influence policy through direct lobbying, or any political campaigns.
Geographic Focus: New York
Date(s) Application is Due: May 1
Samples: Equinox, Albany, New York, $15,000 (2010); Albany Center for Economic Success, Albany, New York, $10,000 (2010); Girls, Inc. of the Greater Capital Region, Schenectady, New York, $15,000 (2010).
Contact: Christine O'Donnell; (646) 855-1011; christine.l.o'donnell@baml.com
Internet: https://www.bankofamerica.com/philanthropic/fn_search.action
Sponsor: Frederick McDonald Trust
One Bryant Park, NY1-100-28-05
New York, NY 10036

Frederick W. Marzahl Memorial Fund Grants **2006**

The Frederick W. Marzahl Memorial Fund was established in 1974 to support and promote quality educational, human-services, and health-care programming for underserved populations in Woodbury, Connecticut. Grants from the Marzahl Memorial Fund are one year in duration. Application materials are available for download at the grant website. Applicants are strongly encouraged to review the state application guidelines for additional helpful information and clarifications before applying. Applicants are also encouraged to review the foundation's funding history (link is available from the grant website). The deadline for application to the Frederick W. Marzahl Memorial Fund is November 1. Applicants will be notified of grant decisions by letter within two to three months after the proposal deadline.
Requirements: Applicants must have 501(c)3 tax-exempt status.
Restrictions: Applicants will not be awarded a grant for more than three consecutive years. The fund does not support requests from individuals, organizations attempting to influence policy through direct lobbying, or any political campaigns.
Geographic Focus: Connecticut
Date(s) Application is Due: Nov 1
Samples: Connecticut Junior Republic Association, Litchfield, Connecticut, $7,000 (010); Flanders Nature Center, Woodbury, Connecticut, $5,400 (2010); North Congregational Church, Woodbury, Connecticut, $3,000 (2010).
Contact: Carmen Britt, Senior Officer; (860) 657-7019; carmen.britt@baml.com
Internet: https://www.bankofamerica.com/philanthropic/fn_search.action
Sponsor: Frederick W. Marzahl Memorial Fund
200 Glastonbury Boulevard, Suite # 200, CT2-545-02-05
Glastonbury, CT 06033-4056

Frederic Stanley Kipping Award in Silicon Chemistry **2007**

This award, supported by Dow Corning Corporation, is given biennially in even-numbered years to recognize distinguished achievement in research in silicon chemistry and to stimulate the creativity of others toward further advancement of this field of chemistry. A nominee must have accomplished distinguished achievement in research in silicon chemistry during the preceding 10 years. The measure of this achievement should focus primarily on the nominee's significant publications in the field of silicon chemistry but may include consideration of contributions to the related field of organometallic chemistry. There are no limits on age or nationality. Applications are accepted in odd-numbered years. The award consists of $5,000 and a certificate.
Requirements: Any individual, except a member of the award committee, may submit one nomination or seconding letter for the award in any given year. Nominating documents consist of a letter of not more than 1000 words containing an evaluation of the nominee's accomplishments and a specific identification of the work to be recognized, a biographical sketch including date of birth, and a list of publications and patents authored by the nominee.
Restrictions: Self-nominations are not accepted.
Date(s) Application is Due: Nov 1
Amount of Grant: 5,000 U.S.D
Contact: Felicia Dixon, Awards Administrator; (800) 227-5558 or (202) 872-4408; fax (202) 776-8008; f_dixon@acs.org or awards@acs.org
Internet: http://portal.acs.org/portal/acs/corg/content?_nfpb=true&_pageLabel=PP_ARTICLEMAIN&node_id=1319&content_id=CTP_004523&use_sec=true&sec_url_var=region1
Sponsor: American Chemical Society
1155 Sixteenth Street, NW
Washington, DC 20036-4801

Fred L. Emerson Foundation Grants **2008**

The foundation gives grants to nonprofits in Auburn, Cayuga County, and upstate New York to improve the quality of life in the areas of education (primarily private, higher education), hospital and health programs, community agencies, churches, cultural institutions, youth and community service programs and social welfare agencies. An application form is not required. Proposals may be submitted in letter form detailing the project for which support is being sought.
Requirements: New York nonprofit organizations in Auburn, Cayuga County, and the upstate area are eligible.
Restrictions: Grants do not support individuals, operating budgets, or loans.
Geographic Focus: New York
Amount of Grant: 200 - 250,000 U.S.D
Contact: Daniel J. Fessenden; (315) 253-9621; dan@emersonfoundation.com
Sponsor: Fred L. Emerson Foundation
5654 South Street Road, P.O. Box 276
Auburn, NY 13021-9602

Fremont Area Community Foundation Amazing X Grants **2009**

The Amazing X Charitable Trust, a supporting organizaton of the Fremont Area Community Foundation (FACF), was established in the late 1970s by members of the Gerber family to benefit people with disabilites and address general charitable needs in Newaygo County, Michigan. Grant requests are accepted for: projects or programs that serve people with disabilities; projects or programs that address general charitable needs. Preferred programs are innovative, collaborative, and have a significant impact on the residents of Newaygo County. Grants range from $1,000 - $60,000 range. Grant applications are due each year by July 15.

Requirements: Michigan 501(c)3 organizations located in or supporting Newaygo County are eligible for funding. When submitting your proposal, include your organizations: mission, history, description of current programs, activities, and accomplishments; purpose of the grant (describe in detail and include supporting evidence); grant proposal budget form/narrative (form available at the Foundation's website). The following list of attachments must also be included: a copy of the current IRS 501(c)3 determination letter; roster of current governing board, including addresses and affiliations; finances: organization's current annual operating budget, including all expenses and revenues, audited financial statement (most recently completed), IRS Form 990 (most recently filed), annual report, if available; resumes and job descriptions of the key project personnel; organizational chart; letters of support (up to five).

Restrictions: The foundation typically does not fund requests for projects that have previously been denied, grants to individuals, programs that advocate specific religions, or services which are government or school obligations.

Geographic Focus: Michigan

Date(s) Application is Due: Jul 15

Amount of Grant: 1,000 - 60,000 U.S.D

Samples: Newaygo County Commission on Aging, $32,000—Alzheimer Congregate Respite program; Grant Area District Library, $6,500—circulating materials for people with disabilities; Pioneer Resources, $2,500—Pioneer Trails Day Camp for Newaygo County youth.

Contact: Jeff Jahr; (231) 924-5350; fax (231) 924-5391; jjahr@tfacf.org

Internet: http://www.tfaf.org/grants/amazingx.html

Sponsor: Fremont Area Community Foundation
P.O. Box B
Fremont, MI 49412

Fremont Area Community Foundation Elderly Needs Grants **2010**

The Foundation is focusing its grantmaking resources primarily on expanding opportunities that enhance the well being of residents from the Newaygo County, Michigan area. The purpose of the Elderly Needs Fund is to make grants to support health and enrich aging for the elderly of Newaygo County. The strategies of the Elderly Needs Fund include: promote the physical health of the elderly; promote the mental/emotional well being of the elderly and their caregivers; promote the social enrichment and prevent the social isolation of the elderly; promote the provision of basic human services for the elderly. Application deadlines are February 15 and October 15 each year.

Requirements: Michigan 501(c)3 organizations located in or supporting Newaygo County are eligible for funding. When submitting your proposal, include your organizations: mission, history, description of current programs, activities, and accomplishments; purpose of the grant (describe in detail and include supporting evidence); grant proposal budget form/narrative (form available at the Foundation's website). The following list of attachments must also be included: a copy of the current IRS 501(c)3 determination letter; roster of current governing board, including addresses and affiliations; finances: organization's current annual operating budget, including all expenses and revenues, audited financial statement (most recently completed), IRS Form 990 (most recently filed), annual report, if available; resumes and job descriptions of the key project personnel; organizational chart; letters of support (up to five).

Geographic Focus: Michigan

Date(s) Application is Due: Feb 15; Oct 15

Amount of Grant: 1,000 - 200,000 U.S.D

Samples: Newaygo County MSU Extension, $10,830—Senior Project Fresh food vouchers for use at area farmer's markets; Newaygo County Commission on Aging, $281,000—senior programs; Baldwin Family Health Care, $6,200—Adult Screening and Prevention Program (ASAP).

Contact: Jeff Jahr; (231) 924-5350; fax (231) 924-5391; jjahr@tfacf.org

Internet: http://www.tfaf.org/grants/elderlyneeds.html

Sponsor: Fremont Area Community Foundation
P.O. Box B
Fremont, MI 49412

Fremont Area Community Foundation Grants **2011**

The Foundation is focusing its grantmaking resources primarily on expanding opportunities that enhance the well being of children, youth and families in Newaygo County. The Foundation's areas of interest include: arts and culture, community development, education, the environment, and human services. Type of support offered include: building/renovation; capital campaigns; conferences/seminars; consulting services; continuing support; curriculum development; emergency funds; employee matching gifts; endowments; equipment; general/operating support; management development/capacity building; matching/challenge support; program-related investments/loans; program development; program evaluation; scholarship funds; scholarships-to individuals; seed money; technical assistance. Grant applications have a deadline date of February 15, May 15 and, September 15. Scholarship applications are due on March 15th.

Requirements: Michigan 501(c)3 organizations located in or supporting Newaygo County are eligible for funding. When submitting your proposal, include your organizations: mission, history, description of current programs, activities, and accomplishments; purpose of the grant (describe in detail and include supporting evidence); grant proposal budget form/narrative (form available at the Foundation's website). The following list of attachments must also be included: a copy of the current IRS 501(c)3 determination letter; roster of current governing board, including addresses and affiliations; finances: organization's current annual operating budget, including all expenses and revenues, audited financial statement (most recently completed), IRS Form 990 (most recently filed), annual report, if available; resumes and job descriptions of the key project personnel; organizational chart; letters of support (up to five).

Restrictions: The foundation typically does not fund requests for projects that have previously been denied, grants to individuals, programs that advocate specific religions, or services which are government or school obligations.

Geographic Focus: Michigan

Date(s) Application is Due: Feb 15; Mar 15; May 15; Sep 15

Samples: Grant Public Schools, Grant, MI, $13,750—for fitness equipment for student and community use; Newaygo County Regional Educational Service Agency, Fremont, MI, $305,000—for Newaygo County Parents as Teachers Plus program; Newaygo County Community Services, Fremont, MI, $566,130—for general operating and program support.

Contact: Jeff Jahr, (231) 924-5350; fax (231) 924-5391; jjahr@tfacf.org

Internet: http://www.tfaf.org/grants.html

Sponsor: Fremont Area Community Foundation
P.O. Box B
Fremont, MI 49412

Frey Foundation Grants **2012**

The foundation is committed to excellence and accountability in its role as a catalyst for productive social change and makes grants to nonprofit organizations based in and serving residents of Kent, Emmet, and Charlevoix Counties in Michigan. Primary areas of interest include enhancing the lives of children and families, protecting natural resources, encouraging civic progress, nurturing community art, capital projects, and philanthropy. Community-based projects receive priority consideration. Types of support include program development, seed money, technical assistance, program evaluation, and employee matching gifts. Send a one- to two-page letter of inquiry. Requests for less than $30,000 are reviewed monthly; requests over $30,000 must be approved by the board. Guidelines are available online.

Requirements: Nonprofit organizations based in and serving residents of Kent, Emmet, and Charlevoix Counties in Michigan are eligible.

Restrictions: Individuals are not eligible to apply.

Geographic Focus: Michigan

Date(s) Application is Due: Feb 15; May 15; Aug 15; Nov 15

Samples: Nature Conservancy, Michigan Chapter (Lansing, MI)—for the Northern Great Lakes Project, which will protect 271,000 acres across eight counties in Michigan's Upper Peninsula and link more than 2.5 million acres of protected federal, state, and natural areas, $1 million.

Contact: Teresa Crawford, Grants Manager; (616) 451-4565; fax (616) 451-8481; freyfdn@freyfdn.org

Internet: http://www.freyfdn.org

Sponsor: Frey Foundation
40 Pearl Street NW, Suite 1100
Grand Rapids, MI 49503

Frist Foundation Grants **2013**

The foundation supports a variety of organizations in the fields of health, human services, civic affairs, education, and the arts. Grants support building the capacity of nonprofit organizations by strengthening their management structure and systems. Types of support include capital, management, operating, project, and technology grants. Deadline listed is for technology grants.

Requirements: 501(c)3 nonprofits in Davidson County, TN, area may apply.

Restrictions: The foundation does not support: individuals or their projects, private foundations, political activities, or advertising or sponsorships. The foundation does not ordinarily support: projects, programs, or organizations that serve a limited audience or a relatively small number of people; organizations during their first three years of operation; disease-specific organizations seeking support for national projects and programs; biomedical or clinical research; hospitals; organizations whose principal impact is outside of Middle Tennessee; endowments; social events or similar fund-raising activities; or religious organizations for religious purposes.

Geographic Focus: Tennessee

Date(s) Application is Due: Apr 1

Amount of Grant: 500 - 10,000 U.S.D

Samples: Domestic Violence Intervention Ctr (TN)—to translate domestic violence educational materials into Spanish, $3000; Mockingbird Theatre (TN)—for general support, $5000; Metropolitan Nashville Public Schools (TN)—to establish a pilot reading program at Haywood Elementary School, $65,000; Faith Family Medical Clinic of Nashville (TN)—to help establish a medical clinic for low-income working people, $105,000.

Contact: Peter Bird, (615) 292-3868; fax (615) 292-5843; askfrist@fristfoundation.org

Internet: http://www.fristfoundation.org/grants/general.asp

Sponsor: Frist Foundation
3100 W End Avenue, Suite 1200
Nashville, TN 37203

Fritz B. Burns Foundation Grants **2014**

The foundation supports nonprofit organizations primarily in southern California by awarding grants for education, with an emphasis on buildings, equipment, endowments (except for ordinary operating expenses), student scholarship and loan funds, and faculty fellowships; to/for hospitals, hospital equipment, and medical research; and religious

organizations (Christian, Jewish, Latter-day Saints, nondenominational, Presbyterian, Protestant, Roman Catholic, and Salvation Army). Proposals should be concise, containing a brief description of what is planned, with a clear statement of the objective sought; IRS letter certifying tax exemption; financial statements; and a list of officers and directors. No formalized application or proposal format is required. Proposals are considered on a quarterly basis.
Requirements: Nonprofit, tax-exempt organizations in southern California may apply.
Restrictions: Grant requests are not considered from individuals nor from tax-supported entities.
Geographic Focus: California
Date(s) Application is Due: Sep 30
Amount of Grant: 10,000 - 250,000 U.S.D
Contact: Joseph Rawlinson, President; (818) 840-8802
Sponsor: Fritz B. Burns Foundation
4001 West Alameda Avenue, Suite 203
Burbank, CA 91505-4338

Fuji Film Grants **2015**
Grants are awarded to proposals that have relevance to Fujifilm's strategic interests which are: environment; education; arts and culture and health and human services. Organizations are limited to one grant per calendar year. Applications are reviewed monthly and applicants should anticipate an award or decline within six to eight weeks after submitting a request. Applicants should submit a corporate giving application online.
Requirements: 501(c)3 tax-exempt organizations are eligible. Organizations must serve or be located within the communities where Fujifilm is a member. Preference is given to those proposals that have relevance to the company's strategic interests.
Restrictions: The program does not fund: individuals or personal use; political organizations, causes, candidates or campaigns; legislative advocacy groups; religious groups; military agencies; fraternal, veterans or social organizations.
Amount of Grant: 1,000 - 5,000 U.S.D
Samples: Sea World and Busch Gardens Conservation Fund (Clayton, MO)—for the Environmental Excellence Awards program, $125,000; Westchester Community College (Valhalla, NY) and Rhode Island School of Design (Providence, RI)—for scholarships, $25,000 and $50,000, respectively.
Contact: Contribution Program Manager; (800) 755-3854; contributions@fujifilm.com
Internet: http://www.fujifilm.com/JSP/fuji/epartners/givingGuidelines.jsp?nav=2
Sponsor: Fuji Photo Film U.S.A
200 Summit Lake Drive, Floor 2
Valhalla, NY 10595

Fulbright Alumni Initiatives Awards Program Grants **2016**
The objective of the AIA program is to help translate the individual Fulbright experience into long-term institutional impact. To this end, the program provides small institutional grants to Fulbright alumni to develop innovative projects that will foster institutionally supported linkages and sustainable, mutually beneficial relationships between the Fulbright scholar's home and host institutions. Just about any activity resulting in the creation or fostering of a sustainable institutional relationship that will have an impact on both the Fulbright alum's home institution and the Fulbright host institution abroad and which both institutions are prepared to support in both the long and short terms will be considered. The program has been temporarily suspended, and applicants should check back for updates.
Requirements: Applicants must be U.S. citizens. The program is open to eligible alumni whose grants occurred at any time from the 1998/1999 academic year through the present academic year.
Restrictions: This program is limited to application by a Fulbright alum (whether U.S. or Visiting Scholar) in partnership with his/her host institution colleague. However, other faculty members from both applicants' institutions may be part of the proposal and participate in the project.
Contact: Stacey Bustillos, Program Officer; (202) 686-6252; sbustillos@cies.iie.org
Internet: http://www.cies.org/aia/
Sponsor: Institute of International Education, Council for International Exchange of Scholars
3007 Tilden Street NW, Suite 5L
Washington, DC 20008-3009

Fulbright Distinguished Chairs Program Grants **2017**
The Fulbright Distinguished Chairs Program comprises approximately forty distinguished lecturing, distinguished research and distinguished lecturing/research awards ranging from three to 12 months. Awards in the Fulbright Distinguished Chairs Program are viewed as among the most prestigious appointments in the Fulbright Scholar Program. Candidates should be senior scholars and have a significant publication and teaching record. Applicants should submit hard copies of the Distinguished Chairs Application Form (one page), a letter of interest (about three pages), a curriculum vitae (maximum eight pages) and, if required, a sample syllabus (maximum four pages). Chairs are available in Australia, Austria, Brazil, Canada, Denmark, Finland, France, Germany, Hungary, Ireland, Israel, Italy, Netherlands, Poland, Portugal, Russia, and Sweden. Because an objective of the Fulbright Program is to provide an educational exchange experience for those not previously afforded such an opportunity, preference will usually be given to candidates who have not had substantial recent experience in the country to which they are applying.
Requirements: Applicants must meet all of the following eligibility *Requirements:* U.S. citizenship at the time of application (permanent resident status is not sufficient); possess a Ph.D. or equivalent professional terminal degree at the time of application (for professionals and artists outside academe, recognized professional standing and substantial professional accomplishments); have college or university teaching experience at the level and in the field of the proposed lecturing activity as specified in the award description (for Distinguished Chairs awards, candidates should be senior scholars with a significant publication and teaching record); have foreign language proficiency only if specified in the award description or required for the completion of the proposed project; be of sound physical and mental health; and disclose prior conviction or current indictment for commission of a felony.
Restrictions: Previous Fulbright scholar grantees are eligible to apply only if five years will have elapsed between the ending date of one scholar award and the beginning date of the new scholar award. This rule does not apply if the previous grant was for less than two months. Employees, spouses or dependent children of the United States Department of State or public and private organizations under contract to the United States Department of State are ineligible to apply for a Fulbright grant until one year after the employee's termination.
Date(s) Application is Due: Aug 3
Contact: Maria Bettua, Program Officer; (202) 686-6245; mbettua@cies.iie.org
Internet: http://www.cies.org/Chairs/
Sponsor: Institute of International Education, Council for International Exchange of Scholars
3007 Tilden Street NW, Suite 5L
Washington, DC 20008-3009

Fulbright German Studies Seminar Grants **2018**
The program allows participation in a group seminar on current German society and culture. The program will begin in Berlin and include visits to other cities in Germany. The focus will be on the formation of policies in current issues at the core of modern society such as climate change, food technology, gene technology, stem cell research and the broad scope of education. The seminar will explore how Germany and its European neighbors view the role of science in their societies, examining the factors that result in differing approaches to these issues and the challenges of harmonizing different standards under a uniform EU-policy umbrella. With an eye to the global marketplace and worldwide competition in science, the seminar will further examine the multiple interests which structure relations between national governments, economic corporations, political and supranational bodies as well as research and development institutions. Scholars from U.S. universities, colleges, and community colleges who hold full-time teaching appointments and meet other academic requirements (Ph.D., Ph.D. candidacy or other equivalent degree or qualifications) are eligible. The seminar lasts two weeks, typically during the month of June.
Requirements: Applicants must be U.S. citizens with permanent residence in the United States at the time of application.
Geographic Focus: All States, Germany
Date(s) Application is Due: Nov 1
Contact: Maria Bettua, Assistant Director; (202) 686-6245; mbettua@cies.iie.org
Internet: http://www.iie.org/cies/us_scholars/us_awards/GSS/index.html
Sponsor: Institute of International Education, Council for International Exchange of Scholars
3007 Tilden Street NW, Suite 5L
Washington, DC 20008-3009

Fulbright International Education Administrators (IEA) Seminar Grants **2019**
The IEA seminars are designed to introduce participants to the society, culture and higher education systems of these countries through campus visits, meetings with foreign colleagues and government officials, attendance at cultural events and briefings on education. Participants in the International Education Administrators Program gain a firsthand look into the host country's academic infrastructure and culture. They gain new perspective on the need to internationalize U.S. campuses and insight into how it can be done. The deadline for the German program is February 1, while the deadline for both Japan and Korea is November 1.
Requirements: To be eligible, applicants must: be U.S. citizens; be international education professionals and senior university administrators (e.g., deans, provosts, vice presidents) with significant responsibility for international programs and activities; have an affiliation with an accredited college or university or nonprofit international exchange organization administering postsecondary student or faculty exchange; and have a minimum of three years of work experience in international education. Applicants for the Japan program must be affiliated with a four-year college or university, while Germany and Korea will consider applicants from both two- and four-year institutions.
Restrictions: Employees, spouses or dependent children of the United States Department of State or public and private organizations under contract to the United States Department of State are ineligible to apply for a Fulbright grant until one year after the employee's termination. TEFL administrators are ineligible for these seminar programs, but they are encouraged to view a listing of other Fulbright opportunities.
Date(s) Application is Due: Feb 1; Nov 1
Contact: Maria Bettua; (202) 686-6245; mbettua@cies.iie.org
Internet: http://www.cies.org/IEA/
Sponsor: Institute of International Education, Council for International Exchange of Scholars
3007 Tilden Street NW, Suite 5L
Washington, DC 20008-3009

Fulbright New Century Scholars (NCS) Program Grants **2020**
The NCS Program brings 30 top academics and professionals from around the world together each year to collaborate on an issue of global importance. Of the thirty, approximately one-third will be U.S. citizens while the remaining two thirds will be visiting scholars from countries with an operational Fulbright Scholar Program. NCS will provide a platform for scholars from the U.S. and around the world to engage in

debate and dialogue based on multidisciplinary research and to develop new global models for understanding the social context within which nations and communities shape their responses to the many challenges of the 21st century. This particular aspect of the New Century Scholars program is a unique feature that distinguishes it from the core Fulbright Scholar Program.
Requirements: Applicants must be conducting current research relevant to the program's theme and objectives, be open to exploring and incorporating comparative, interdisciplinary approaches in their investigations, and interested in developing collaborative activities with other NCS Scholars. U.S. applicants must have U.S. citizenship and be residing permanently in the United States. For academic applicants, a Ph.D. or equivalent terminal degree in a relevant field is required. For applicants in the professional fields, the appropriate terminal degree in a relevant field is required. Non-U.S. applicants must be citizens or permanent residents of and residing in the country from which they are applying at the time of application. All applicants must have fluency in English.
Restrictions: Non-U.S. applicants holding permanent residency green cards, whether or not they reside in the U.S., are not eligible.
Contact: Jonathan Looper, Program Associate; (202) 686-6235; fax (202) 362-3442; jlooper@cies.iie.org
Internet: http://www.cies.org/NCS/
Sponsor: Institute of International Education, Council for International Exchange of Scholars
3007 Tilden Street NW, Suite 5L
Washington, DC 20008-3009

Fulbright Specialists Program Grants **2021**
The Program is designed to provide short-term academic opportunities (two to six weeks) for U.S. faculty and professionals. Shorter grant lengths give specialists greater flexibility to pursue a grant that works best with their current academic or professional commitments. Applications for the Program are accepted on a rolling basis, and peer review of applications is conducted eight times per year. Program goals include: to increase the participation of leading U.S. scholars and professionals in Fulbright academic exchanges; to encourage new activities that go beyond the traditional Fulbright activities of lecturing and research; and to promote increased connections between U.S. and non-U.S. post-secondary academic institutions.
Requirements: Applicants must: be U.S. citizen at the time of application (permanent resident status is not sufficient; if a naturalized citizen, applicant must provide actual date of naturalization); possess a Ph.D. or equivalent professional/terminal degree at the time of application; have a minimum of five years of post-doctoral teaching or professional experience in the field in which you are applying (for professionals and artists outside academe, recognized professional standing and substantial professional accomplishments plus a minimum of five years of professional experience in the field in which you are applying); disclose prior conviction or current indictment for commission of a felony; and be residing in the United States at the time they are approved for a grant and intend to return to their U.S. institution after the grant's completion.
Restrictions: Employees, spouses or dependent children of the United States Department of State or public and private organizations under contract to the United States Department of State exchange programs are ineligible to apply for a Fulbright grant until one year after the employee's termination.
Date(s) Application is Due: Jan 28; Feb 24; Mar 25; Apr 21; May 20; Jul 15; Sep 9; Nov 4; Dec 30
Contact: Ryan Hathaway, Senior Program Coordinator; (202) 686-4026; rhathaway@cies.iie.org
Internet: http://www.cies.org/specialists/#top
Sponsor: Institute of International Education, Council for International Exchange of Scholars
3007 Tilden Street NW, Suite 5L
Washington, DC 20008-3009

Fulbright Traditional Scholar Program in Europe and Eurasia **2022**
The traditional Fulbright Scholar Program sends 800 U.S. faculty and professionals abroad each year. Grantees lecture and conduct research in a wide variety of academic and professional fields. Distribution of awards to countries in the region will vary annually according to the caliber of the applicants. Grants are available to: Albania, Andorra, Armenia, Austria, Azerbaijan, Belarus, Belgium, Bosnia and Herzegovina, Bulgaria, Croatia, Cyprus, Czech Republic, Denmark, Estonia, European Union, Finland, France, Georgia, Germany, Greece, Hungary, Iceland, Ireland, Italy, Latvia, Lithuania, Luxembourg, Macedonia, Moldova, Netherlands, Norway, Poland, Portugal, Romania, Russia, Slovakia (Slovak Republic), Slovenia, Spain, Sweden, Switzerland, Turkey, Ukraine, and United Kingdom. Only countries Considered to be in the Eastern Europe or Eurasia regions may be part of a multi-country application (2 to 3 countries). Multi-country proposals are not permitted for Western Europe. Language requirements vary by country, and prior knowledge of the local language may not be required, particularly where language is not commonly taught in the U.S.
Requirements: Applicants must: be U.S. citizens at the time of application (permanent residents are not eligible); be in good health (grantees will be required to submit a satisfactory Medical Certificate of Health from a physician); have sufficient proficiency in the written and spoken language of the host country (where required) to communicate with the people and to carry out the proposed study; and hold a B.A. degree or the equivalent before the start of the grant (applicants who have not earned a B.A. degree or the equivalent, but who have extensive professional study and/or experience in fields in which they wish to pursue a project, may be considered). In the creative and performing arts area, four years of professional study and/or experience meets the basic eligibility requirement. All candidates for the Europe and Eurasia region are required to obtain their own affiliations, generally established with an educational and/or research institution in the host country or countries. Candidates are responsible for securing their own research clearance as required by the host country.
Restrictions: The following persons are ineligible: anyone who has already held a U.S. Department of State-funded Fulbright student grant of any type; anyone who has previously received a Department of Education-funded Doctoral Dissertation Research Abroad (Fulbright Hays grant); employees of the U.S. Department of State, and their immediate families, for a period ending one year following termination of such employment; employees of private and public agencies (excluding educational institutions) under contract to the U.S. Department of State to perform administrative or screening services on behalf of the U.S. Department of State's exchange program, for a period ending one year following the termination of their services for the U.S. Department of State provided such employees have been directly engaged in performing services related to the exchange programs; applicants holding a doctoral degree at the time of application; applicants seeking enrollment in a medical degree program abroad; or applicants currently residing in the countries of Australia, Belgium/Luxembourg, Canada, Chile, Finland, Hungary, Mexico, Netherlands, New Zealand, Sweden, or Switzerland.
Geographic Focus: All States, Albania, Andorra, Armenia, Austria, Azerbaijan, Belarus, Belgium, Bosnia & Herzegovina, Bulgaria, Croatia, Cyprus, Czech Republic, Denmark, Estonia, Finland, France, Georgia, Germany, Greece, Hungary, Iceland, Ireland, Italy, Kosovo, Latvia, Liechtenstein, Lithuania, Luxembourg, Macedonia, Malta, Moldova, Monaco, Montenegro, Norway, Poland, Portugal, Romania, Russia, San Marino, Serbia, Slovakia, Slovenia, Spain, Sweden, Switzerland, The Netherlands, Turkey, Ukraine, United Kingdom, Vatican City
Date(s) Application is Due: Aug 1
Contact: Rachel Holskin, IIE Program Manager; (212) 984-5326; rholskin@iie.org
Internet: http://us.fulbrightonline.org/program_regions_countries.php?id=3
Sponsor: Institute of International Education
809 United Nations Plaza
New York, NY 10017-3580

Fulbright Traditional Scholar Program in Sub-Saharan Africa **2023**
The traditional Fulbright Scholar Program sends 800 U.S. faculty and professionals abroad each year. Grantees lecture and conduct research in a wide variety of academic and professional fields. Selection for countries in sub-Saharan Africa, with the exception of South Africa, will be made based on the quality of the applications, rather than per-country quotas. Distribution of awards to countries in the region will vary annually according to the caliber of the applicants. Grants are available to: Benin, Botswana, Burkina Faso, Cameroon, Chad, Eritrea, Ethiopia, Ghana, Guinea, Ivory Coast, Kenya, Madagascar, Malawi, Mali, Mauritius, Mozambique, Namibia, Niger, Nigeria, Senegal, South Africa, Swaziland, Tanzania, Togo, Uganda, Zambia, and Zimbabwe. All countries within the sub-Saharan Africa Region, with the exception of South Africa, may be part of a multi-country application (2 to 3 countries). For unlisted countries, applications may be considered on a case-by-case basis, but for dissertation research only.
Requirements: Applicants must: be U.S. citizens at the time of application (permanent residents are not eligible); be in good health (grantees will be required to submit a satisfactory Medical Certificate of Health from a physician); have sufficient proficiency in the written and spoken language of the host country to communicate with the people and to carry out the proposed study; and hold a B.A. degree or the equivalent before the start of the grant (applicants who have not earned a B.A. degree or the equivalent, but who have extensive professional study and/or experience in fields in which they wish to pursue a project, may be considered). In the creative and performing arts area, four years of professional study and/or experience meets the basic eligibility requirement. All candidates for Africa are required to obtain their own affiliations, generally established with an educational and/or research institution in the host country or countries.
Restrictions: Grants are not available to: Burundi, Central African Republic, Guinea-Bissau, Liberia, Somalia, or Sudan. The following persons are ineligible: anyone who has already held a U.S. Department of State-funded Fulbright student grant of any type; anyone who has previously received a Department of Education-funded Doctoral Dissertation Research Abroad (Fulbright Hays grant); employees of the U.S. Department of State, and their immediate families, for a period ending one year following termination of such employment; employees of private and public agencies (excluding educational institutions) under contract to the U.S. Department of State to perform administrative or screening services on behalf of the U.S. Department of State's exchange program, for a period ending one year following the termination of their services for the U.S. Department of State provided such employees have been directly engaged in performing services related to the exchange programs; applicants holding a doctoral degree at the time of application; applicants seeking enrollment in a medical degree program abroad; or applicants currently residing in the countries of Australia, Belgium/Luxembourg, Canada, Chile, Finland, Hungary, Mexico, Netherlands, New Zealand, Sweden, or Switzerland.
Geographic Focus: All States, Algeria, Angola, Benin, Botswana, Burkina Faso, Burundi, Cameroon, Cape Verde, Central African Republic, Chad, Comoros, Congo, Congo, Democratic Republic of, Cote d' Ivoire (Ivory Coast), Djibouti, Egypt, Equatorial Guinea, Eritrea, Ethiopia, Gabon, Gambia, Ghana, Guinea, Guinea-Bissau, Kenya, Lesotho, Liberia, Libya, Madagascar, Malawi, Mali, Mauritania, Mauritius, Morocco, Mozambique, Namibia, Niger, Nigeria, Rwanda, Sao Tome & Principe, Senegal, Seychelles, Sierra Leone, Somalia, South Africa, Sudan, Swaziland
Date(s) Application is Due: Aug 1
Contact: Jermaine Jones, Program Manager; (212) 984-5341; jjones@iie.org
Internet: http://us.fulbrightonline.org/program_regions_countries.php?id=1
Sponsor: Institute of International Education
809 United Nations Plaza
New York, NY 10017-3580

Fulbright Traditional Scholar Program in the East Asia/Pacific Region 2024

The traditional Fulbright Scholar Program sends 800 U.S. faculty and professionals abroad each year. Grantees lecture and conduct research in a wide variety of academic and professional fields. Distribution of awards to countries in the region will vary annually according to the caliber of the applicants. Grants are available to: Australia, Cambodia, China, Hong Kong, Indonesia, Japan, Korea, Laos, Macau, Malaysia, Mongolia, New Zealand, Philippines, Singapore, Taiwan, Thailand, and Vietnam. All countries within the East Asia/Pacific region, with the exception of China, may be part of a multi-country application (2 to 3 countries). For unlisted countries, applications may be considered on a case-by-case basis, but for dissertation research only. Language requirements vary by country, and prior knowledge of the local language may not be required, particularly where language is not commonly taught in the U.S.

Requirements: Applicants must: be U.S. citizens at the time of application (permanent residents are not eligible); be in good health (grantees will be required to submit a satisfactory Medical Certificate of Health from a physician); have sufficient proficiency in the written and spoken language of the host country to communicate with the people and to carry out the proposed study; and hold a B.A. degree or the equivalent before the start of the grant (applicants who have not earned a B.A. degree or the equivalent, but who have extensive professional study and/or experience in fields in which they wish to pursue a project, may be considered). In the creative and performing arts area, four years of professional study and/or experience meets the basic eligibility requirement. All candidates for the East Asia/Pacific region are required to obtain their own affiliations, generally established with an educational and/or research institution in the host country or countries. Candidates are responsible for securing their own research clearance as required by the host country.

Restrictions: Grants are not available to: Brunei, the Cook Islands, East Timor, Fiji, Myanmar, the Pacific Island Nations, Papua New Guinea, or Western Samoa. The following persons are ineligible: anyone who has already held a U.S. Department of State-funded Fulbright student grant of any type; anyone who has previously received a Department of Education-funded Doctoral Dissertation Research Abroad (Fulbright Hays grant); employees of the U.S. Department of State, and their immediate families, for a period ending one year following termination of such employment; employees of private and public agencies (excluding educational institutions) under contract to the U.S. Department of State to perform administrative or screening services on behalf of the U.S. Department of State's exchange program, for a period ending one year following the termination of their services for the U.S. Department of State provided such employees have been directly engaged in performing services related to the exchange programs; applicants holding a doctoral degree at the time of application; applicants seeking enrollment in a medical degree program abroad; or applicants currently residing in the countries of Australia, Belgium/Luxembourg, Canada, Chile, Finland, Hungary, Mexico, Netherlands, New Zealand, Sweden, or Switzerland.

Geographic Focus: All States, Abkhazia, Afghanistan, Armenia, Australia, Azerbaijan, Bahrain, Bangladesh, Bhutan, British Indian Ocean Territory, Brunei, Burma (Myanmar), Cambodia, China, Christmas Island, Cocos, Cyprus, Georgia, Hong Kong, India, Indonesia, Iran, Iraq, Israel, Japan, Jordan, Kazakhstan, Kuwait, Kyrgyzstan, Laos, Lebanon, Macau, Malaysia, Maldives, Mongolia, Nagorno-Karabakh, Nepal, North Korea, Northern Cyprus, Oman, Pakistan, Palestinian Authority, Philippines, Qatar, Russia, Saudi Arabia, Singapore, South Korea, South Ossetia, Sri Lanka, Syria, Taiwan, Tajikistan, Thailand, Timor-Lester, Turkey, Turkmenistan, United Arab Emirates, Uzbekistan, Vietnam, Yemen

Date(s) Application is Due: Aug 1

Contact: Jonathan Akeley, Program Manager; (212) 984-5487; jakeley@iie.org

Internet: http://us.fulbrightonline.org/program_regions_countries.php?id=2

Sponsor: Institute of International Education
809 United Nations Plaza
New York, NY 10017-3580

Fulbright Traditional Scholar Program in the Near East and North Africa Region 2025

The traditional Fulbright Scholar Program sends 800 U.S. faculty and professionals abroad each year. Grantees lecture and conduct research in a wide variety of academic and professional fields. Selection for countries in the Near East and North Africa region will be made based on the quality of the applications, rather than per-country quotas. Distribution of awards to countries in the region will vary annually according to the caliber of the applicants. Grants are available to: Bahrain, Egypt, India, Jordan, Kuwait, Morocco, Oman, Qatar, Syria, Tunisia, United Arab Emirates, and Yemen. Multi-country applications (2 to 3 countries) are available, except in Egypt, India, Jordan, and Morocco. For unlisted countries, applications may be considered on a case-by-case basis.

Requirements: Applicants must: be U.S. citizens at the time of application (permanent residents are not eligible); be in good health (grantees will be required to submit a satisfactory Medical Certificate of Health from a physician); have sufficient proficiency in the written and spoken language of the host country to communicate with the people and to carry out the proposed study; and hold a B.A. degree or the equivalent before the start of the grant (applicants who have not earned a B.A. degree or the equivalent, but who have extensive professional study and/or experience in fields in which they wish to pursue a project, may be considered). In the creative and performing arts area, four years of professional study and/or experience meets the basic eligibility requirement. Candidates are responsible for securing research clearance, as required. In countries with Fulbright Commissions, assistance may be provided.

Restrictions: Grants are not available to: Algeria, Iran, Iraq, Lebanon, Saudi Arabia, the West Bank, or Gaza. The following persons are ineligible: anyone who has already held a U.S. Department of State-funded Fulbright student grant of any type; anyone who has previously received a Department of Education-funded Doctoral Dissertation Research Abroad (Fulbright Hays grant); employees of the U.S. Department of State, and their immediate families, for a period ending one year following termination of such employment; employees of private and public agencies (excluding educational institutions) under contract to the U.S. Department of State to perform administrative or screening services on behalf of the U.S. Department of State's exchange program, for a period ending one year following the termination of their services for the U.S. Department of State provided such employees have been directly engaged in performing services related to the exchange programs; applicants holding a doctoral degree at the time of application; applicants seeking enrollment in a medical degree program abroad; or applicants currently residing in the countries of Australia, Belgium/Luxembourg, Canada, Chile, Finland, Hungary, Mexico, Netherlands, New Zealand, Sweden, or Switzerland.

Geographic Focus: All States, Algeria, Angola, Benin, Botswana, Burkina Faso, Burundi, Cameroon, Cape Verde, Central African Republic, Chad, Comoros, Congo, Congo, Democratic Republic of, Cote d' Ivoire (Ivory Coast), Djibouti, Egypt, Equatorial Guinea, Eritrea, Ethiopia, Gabon, Gambia, Ghana, Guinea, Guinea-Bissau, Israel, Kenya, Lesotho, Liberia, Libya, Madagascar, Malawi, Mali, Mauritania, Mauritius, Morocco, Mozambique, Namibia, Niger, Nigeria, Rwanda, Sao Tome & Principe, Senegal, Seychelles, Sierra Leone, Somalia, South Africa, Sudan, Swaziland, Uzbekistan

Date(s) Application is Due: Aug 1

Contact: Jermaine Jones, Program Manager; (212) 984-5341; jjones@iie.org

Internet: http://us.fulbrightonline.org/program_regions_countries.php?id=4

Sponsor: Institute of International Education
809 United Nations Plaza
New York, NY 10017-3580

Fulbright Traditional Scholar Program in the South and Central Asia Region 2026

The traditional Fulbright Scholar Program sends 800 U.S. faculty and professionals abroad each year. Grantees lecture and conduct research in a wide variety of academic and professional fields. Selection for countries in the South and Central Asia region will be made based on the quality of the applications, rather than per-country quotas. Distribution of awards to countries in the region will vary annually according to the caliber of the applicants. Grants are available to: Bangladesh, India, Kazakhstan, Kyrgyz Republic, Nepal, Sri Lanka, Tajikistan, and Uzbekistan. Multi-country applications (2 to 3 countries) are not available. Language requirements vary by country. Applicants to India who are recommended for final review will need to submit research visa applications in January or February.

Requirements: Applicants must: be U.S. citizens at the time of application (permanent residents are not eligible); be in good health (grantees will be required to submit a satisfactory Medical Certificate of Health from a physician); have sufficient proficiency in the written and spoken language of the host country to communicate with the people and to carry out the proposed study; and hold a B.A. degree or the equivalent before the start of the grant (applicants who have not earned a B.A. degree or the equivalent, but who have extensive professional study and/or experience in fields in which they wish to pursue a project, may be considered). In the creative and performing arts area, four years of professional study and/or experience meets the basic eligibility requirement. Candidates are responsible for securing research clearance, as required. In countries with Fulbright Commissions, assistance may be provided.

Restrictions: Grants are not available to: Afghanistan, Bhutan, Pakistan, Republic of Maldives, or Turkmenistan. The following persons are ineligible: anyone who has already held a U.S. Department of State-funded Fulbright student grant of any type; anyone who has previously received a Department of Education-funded Doctoral Dissertation Research Abroad (Fulbright Hays grant); employees of the U.S. Department of State, and their immediate families, for a period ending one year following termination of such employment; employees of private and public agencies (excluding educational institutions) under contract to the U.S. Department of State to perform administrative or screening services on behalf of the U.S. Department of State's exchange program, for a period ending one year following the termination of their services for the U.S. Department of State provided such employees have been directly engaged in performing services related to the exchange programs; applicants holding a doctoral degree at the time of application; applicants seeking enrollment in a medical degree program abroad; or applicants currently residing in the countries of Australia, Belgium/Luxembourg, Canada, Chile, Finland, Hungary, Mexico, Netherlands, New Zealand, Sweden, or Switzerland.

Geographic Focus: All States, Kazakhstan, Uzbekistan

Date(s) Application is Due: Aug 1

Contact: Jonathan Akeley; (212) 984-5487; jakeley@iie.org

Internet: http://us.fulbrightonline.org/program_regions_countries.html?id=6

Sponsor: Institute of International Education
809 United Nations Plaza
New York, NY 10017-3580

Fulbright Traditional Scholar Program in the Western Hemisphere 2027

The traditional Fulbright Scholar Program sends 800 U.S. faculty and professionals abroad each year. Grantees lecture and conduct research in a wide variety of academic and professional fields. Selection for countries in Central America and the Caribbean will be made based on the quality of the applications, rather than per-country quotas. Distribution of awards to countries in the region will vary annually according to the caliber of the applicants. Grants are available to: Argentina, Barbados and the Eastern Caribbean, Bolivia, Brazil, Canada, Chile, Colombia, Costa Rica, Dominican Republic, Ecuador, El Salvador, Guatemala, Honduras, Jamaica, Mexico, Nicaragua, Panama, Paraguay, Peru, Trinidad and Tobago, Uruguay, and Venezuela. English Teaching Assistantships are available in Argentina, Brazil, Chile, Uruguay and Venezuela. Language proficiency may be preferred or required. Applicants for English Teaching Assistantships can apply to only one country. Multi-country applications (2 to 3 countries) are available in all other disciplines throughout the Western Hemisphere. For unlisted countries, applications may be considered on a case-by-case basis, but for dissertation research only.

Requirements: Applicants must: be U.S. citizens at the time of application (permanent residents are not eligible); be in good health (grantees will be required to submit a

satisfactory Medical Certificate of Health from a physician); have sufficient proficiency in the written and spoken language of the host country to communicate with the people and to carry out the proposed study; and hold a B.A. degree or the equivalent before the start of the grant (applicants who have not earned a B.A. degree or the equivalent, but who have extensive professional study and/or experience in fields in which they wish to pursue a project, may be considered). In the creative and performing arts area, four years of professional study and/or experience meets the basic eligibility requirement. Candidates are responsible for securing research clearance, as required. In countries with Fulbright Commissions, assistance may be provided.
Restrictions: Grants are not available to: Bahamas, Belize, Cuba, French Guiana, Guyana, Haiti, Martinique, or Suriname. The following persons are ineligible: anyone who has already held a U.S. Department of State-funded Fulbright student grant of any type; anyone who has previously received a Department of Education-funded Doctoral Dissertation Research Abroad (Fulbright Hays grant); employees of the U.S. Department of State, and their immediate families, for a period ending one year following termination of such employment; employees of private and public agencies (excluding educational institutions) under contract to the U.S. Department of State to perform administrative or screening services on behalf of the U.S. Department of State's exchange program, for a period ending one year following the termination of their services for the U.S. Department of State provided such employees have been directly engaged in performing services related to the exchange programs; applicants holding a doctoral degree at the time of application; applicants seeking enrollment in a medical degree program abroad; or applicants currently residing in the countries of Australia, Belgium/Luxembourg, Canada, Chile, Finland, Hungary, Mexico, Netherlands, New Zealand, Sweden, or Switzerland.
Geographic Focus: All States, Antigua & Barbuda, Argentina, Bahamas, Barbados, Belize, Bolivia, Brazil, Canada, Chile, Colombia, Costa Rica, Cuba, Dominica, Dominican Republic, Ecuador, El Salvador, Grenada, Guatemala, Guyana, Haiti, Honduras, Jamaica, Mexico, Nicaragua, Paraguay, Peru
Date(s) Application is Due: Aug 1
Contact: Jody Dudderar, Program Manager in South America, Mexico and Canada; (212) 984-5565; fax (212) 984-5325; jdudderar@iie.org
Internet: http://us.fulbrightonline.org/program_regions_countries.php?id=5
Sponsor: Institute of International Education
809 United Nations Plaza
New York, NY 10017-3580

Fuller E. Callaway Foundation Grants **2028**
The foundation awards grants to nonprofit organizations and individuals in LaGrange and Troup County, GA. Grants are awarded in the areas of religion, higher and other education, social services, youth, and health. Types of support include general operating budgets, annual campaigns, building funds, equipment, matching funds, and student aid. Letters of application from organizations are accepted and have deadlines at the end of December, March, June, and September. Application forms are required for the scholarship program. Scholarship deadlines are February 15 and June 30 for law school scholarships. The board meets in January, April, July, and October to consider scholarship requests.
Requirements: Nonprofit organizations and individuals in LaGrange and Troup County, GA, are eligible for support.
Geographic Focus: Georgia
Amount of Grant: 500 - 25,000 U.S.D
Contact: H. Speer Burdette, President; (706) 884-7348; fax (706) 884-0201; hsburdette@callaway-foundation.org
Sponsor: Fuller E. Callaway Foundation
P.O. Box 790
LaGrange, GA 30241

Fuller Foundation Grants **2029**
The Fuller Foundation is a family foundation, inspired by its forward-thinking founder, Alvan T. Fuller. It's purpose is to support non-profit agencies which improve the quality of life for people, animals and the environment. The Foundation also funds the Fuller Foundation of New Hampshire which supports horticultural and educational programs for the public at Fuller Gardens. The geographic focus area is predominately the Boston area and the immediate seacoast area of New Hampshire. Through these grants the foundation strives to effect change, make an impact on the community, and inspire good deeds. Application information is available online. Faxed or emailed grant requests will not be accepted.
Requirements: Nonprofits in the Boston, MA, area and the immediate seacoast region of New Hampshire are eligible.
Restrictions: Funding is not available for: capital projects (unless in the opinion of the Trustees, the Foundation gift will have significant impact); individuals; or multi-year grants. Incomplete grants are not considered.
Geographic Focus: Massachusetts, New Hampshire
Date(s) Application is Due: Jan 15; Jun 15
Amount of Grant: 3,500 - 7,500 U.S.D
Samples: Nature Connection, Concord, MA, $3,000—for education and outreach programs; Dolphin Research Center, Grassy Key, FL, $10,000—for Dolphin Child Pathway programs; Womens Crisis Center of Greater Newburyport, Newburyport, MA, $5,000—for Children Overcoming Violence Program.
Contact: John Bottomley, Executive Director; (603) 964-6998; fax (603) 964-8901; atfuller@aol.com
Internet: http://www.fullerfoundation.org/FullerFoundation/HomePage.cfm?page=appprocedures
Sponsor: Fuller Foundation
P.O. Box 479
Rye Beach, NH 03871

Fulton County Community Foundation 4Community Higher Education Scholarship **2030**
The primary purpose of the 4Community Higher Education Scholarship is to support the charitable intentions of the Fulton County 4Community Partners in promoting higher education for GED program participants. The Scholarship will be awarded to Fulton County residents who are GED recipients. Scholarship recipients must be accepted into an accredited college, university, trade, or vocational school in Indiana. All majors are eligible to apply. There is no deadline; applicants may apply at any time of the year.
Requirements: Applicants must be U.S. citizens who are Fulton County residents. They must have earned a minimum standard score of 520 on the GED exam. Along with the online application, they must also submit: proof of their score; evidence that they have participated in community and work activities; a letter of recommendation from their GED instructor; and an essay expressing their personal reasons for wanting a post-high school education.
Geographic Focus: Indiana
Contact: Corinne Becknell Lucas, Scholarship Coordinator; (574) 223-2227 or (877) 432-6423; corinne@nicf.org
Internet: http://www.nicf.org/fulton/scholarships.html
Sponsor: Fulton County Community Foundation
715 Main Street, P.O. Box 807
Rochester, IN 46975

Fulton County Community Foundation Grants **2031**
Fulton County Community Foundation is part of the Northern Indiana Community Foundation (NICF). Grant making areas of interest are: education; health; human services; arts and culture; environment; and civic and recreation. The Foundation favors activities that: reach a broad segment of the community, especially those citizens whose needs are not being met by existing services; request seed money to meet innovative opportunities in the community; stimulate and encourage additional funding; promote cooperation and avoid duplication of effort; help make charitable organizations more effective, efficient and self-sustaining; and one time projects or needs. Applications are available online. Applicants are encouraged to contact the Program Coordinator to discuss their project before applying.
Restrictions: The foundation will not consider grants for: religious organizations for the sole purpose of furthering that religion; political activities or those designated to influence legislation; national organizations (unless the monies are to be used solely to benefit citizens of Fulton County); grant that directly benefit the donor or the donor's family; fundraising projects; contributions to endowments
Geographic Focus: Indiana
Date(s) Application is Due: Sep 30
Contact: Corinne Becknell Lucas, Program Coordinator; (574) 223-2227 or (877) 432-6423; fax (574) 224-3709; corinne@nicf.org
Internet: http://www.nicf.org/fulton/grants.html
Sponsor: Fulton County Community Foundation
715 Main Street, P.O. Box 807
Rochester, IN 46975

Fulton County Community Foundation Lilly Endowment Community Scholarships **2032**
The Lilly scholarships provide full tuition, plus an $800 book stipend for four years of full time undergraduate study, leading to a baccalaureate degree at any accredited Indiana public or private college or university. Applicants must be U.S. citizens and residents of Fulton County. Applications are available at high school guidance offices. Any college major may apply.
Geographic Focus: Indiana
Contact: Corinne Becknell Lucas, Scholarship Coordinator; (574) 223-2227 or (877) 432-6423; fax (574) 224-3709; corinne@nicf.org
Internet: http://www.nicf.org/fulton/availablescholarships.html
Sponsor: Fulton County Community Foundation
715 Main Street, P.O. Box 807
Rochester, IN 46975

Fulton County Community Foundation Paul and Dorothy Arven Memorial Scholarship **2033**
The Paul and Dorothy Arven Memorial Scholarship is offered to students pursuing careers in nursing or agriculture. Students must be accepted into an accredited college or university, U.S. citizens and residents of Fulton County. They must possess good citizenship in school, work, and community; good academic accomplishments; and likely to start and complete their course of study with the scholarship's financial assistance.
Requirements: Along with the online application, students must submit their estimated expenses and resources, academic certification form, signature page, two letters of recommendation, the FAFSA's Student Aid Report, and an answer to the essay question in the application.
Restrictions: Public high school students must return application to the guidance department of their high school. All other student should mail, fax or drop off the application to the attention of Corinne Becknell Lucas at the NICF office.
Geographic Focus: Indiana
Date(s) Application is Due: Mar 14
Contact: Corinne Becknell Lucas, Scholarship Coordinator; (574) 223-2227 or (877) 432-6423; fax (574) 224-3709; corinne@nicf.org
Internet: http://www.nicf.org/fulton/scholarships.html
Sponsor: Fulton County Community Foundation
715 Main Street, P.O. Box 807
Rochester, IN 46975

Fulton County Community Foundation Women's Giving Circle Grants **2034**
The Women's Giving Circle Grants help fund non-profit organizations in Fulton County. Organizations receiving their funds include Camp-We-Can, the County Council on Aging, and the County Cancer Fund.
Requirements: Applicants may contact Brian Johnson, the director of development at the Fulton County Community Foundation, for information about whether their project qualifies for application.
Geographic Focus: Indiana
Contact: Brian Johnson, Development Director; (574) 224-3223; fulton@nicf.org
Internet: http://www.nicf.org/fulton/grants.html
Sponsor: Fulton County Community Foundation
715 Main Street, P.O. Box 807
Rochester, IN 46975

Fyssen Foundation Postdoctoral Study Grants **2035**
The foundation awards postdoctoral research grants for research on cognitive mechanisms, including thought and reasoning, that underlie animal and human behavior; their biological and cultural bases; and phylogenetic and ontogenetic development. Research should be in ethology and psychology, neurobiology, anthropology/ethnology, and human paleontology. Grants support French scientists for research abroad and foreign researchers conducting research in France. Annual deadline dates may vary; contact program staff for exact dates.
Date(s) Application is Due: Mar 31
Contact: Grants Manager; 331 42975316; fax 331 42601795; secretariat@fondation-fyssen.org
Internet: http://www.fondation-fyssen.org
Sponsor: Fyssen Foundation
194 rue de Rivoli
Paris, 75001 France

G.A. Ackermann Memorial Fund Grants **2036**
The G. A. Ackermann Memorial Fund was established under the will of Mrs. Mary A. Ackermann in 1937 to support quality health care and human services programming for underserved populations. According to Mrs. Ackermann's wishes, the Ackermann Memorial Fund supports charitable organizations operated or controlled by the Roman Catholic church and/or members of the Roman Catholic Church, and charitable organizations operated or controlled by a Protestant Church and/or its members. Special consideration is given to hospitals. The Memorial Fund has biannual application deadlines of January 15 and June 1. Applicants for the January 15 deadline will be notified of grant decisions by June 30, and applicants for the June 1 deadline will be notified by December 31. To better support the capacity of nonprofit organizations, multi-year funding requests are considered.
Requirements: Organizations must be geographically located within the city limits of New York City or Chicago.
Restrictions: The Foundation will consider requests for general operating support only if the organization's operating budget is less than $1 million. In general, grant requests for individuals, endowment campaigns, capital projects, or research will not be considered.
Geographic Focus: Illinois, New York
Date(s) Application is Due: Jan 15; Jun 1
Contact: George Thorn, Vice President; (312) 828-4154; ilgrantmaking@bankofamerica.com
Internet: https://www.bankofamerica.com/philanthropic/fn_search.action
Sponsor: G.A. Ackermann Memorial Fund
231 South LaSalle Street, IL1-231-13-32
Chicago, IL 60604

G.D. Searle Charitable Trust Grants **2037**
Searle awards grants through its charitable trust to tax-exempt organizations for education at all levels, health, social services, arts and humanities, and civic affairs. A major part of the funding is to United Way/Crusade of Mercy and employee matching gift programs, preventive medicine, and medical research. Most projects funded have a duration of two years or less. Applications outlining the project for which funds are needed should be submitted in the first quarter of each year. Address applications to Searle Charitable Trust, 5200 Old Orchard Road, Skokie, IL 60077.
Requirements: Grants are not made to individuals for personal education or to organizations funded by United Way.
Amount of Grant: 500 - 30,000 U.S.D
Contact: Judith Van der Vort, Administrator; (708) 982-7000
Sponsor: G.D. Searle Charitable Trust
50 S LaSalle Street
Chicago, IL 60675

G.N. Wilcox Trust Grants **2038**
The trust provides partial support to programs and projects of tax-exempt, public charities in Hawaii to improve the quality of life in the state, particularly the island of Kauai. Grants of one year's duration are awarded in categories of interest to the trust, including education, literacy programs and adult basic education, health, Protestant religion, delinquency and crime prevention, social services, youth services, and culture and the performing arts. Types of support include general operating grants, capital grants, equipment acquisition, seed grants, scholarship funds, and challenge/matching grants. Deadlines dates for general grants are: January 1; April 1; July 1; October 1. The deadline date for scholarships is February 15th.
Requirements: Giving is limited to Hawaii, with emphasis on the island of Kauai. To begin application process, contact Paula Boyce for additional guidelines.
Restrictions: Grants are not awarded to support government agencies (or organizations substantially supported by government funds), individuals, or for endowment funds, research, deficit financing, or student aid in scholarships or loans.
Geographic Focus: Hawaii
Date(s) Application is Due: Jan 1; Feb 15; Apr 1; Jul 1; Oct 1
Contact: Paula Boyce, c/o Bank of Hawaii; (808) 538-4944; fax (808) 538-4647; pboyce@boh.com or emoniz@boh.com
Sponsor: G.N. Wilcox Trust
Bank of Hawai'i, Foundation Administration Department 758
Honolulu, HI 96802-3170

Galveston Shriners Hospital Postdoctoral Fellowships **2039**
The program awards postdoctoral clinical and research fellowships to individuals interested in trauma, critical care, and burn-related practice or research. Fellowships are for a minimum of one year (clinical) and two years (research). The UTMB Trauma and Burn Services have NIH- supported active investigators with comprehensive laboratories working in the fields of wound healing, metabolism, cardiopulmonary pathophysiology, molecular biology, and immunology. Graduate programs, including master's and PhD degrees, are also available.
Requirements: Applicants who have attained an MD, DO, PhD, or DVM degree are eligible. Applicants for clinical fellowships should be board eligible or board certified in general surgery; however, senior residents from academically oriented university programs will be considered or those who have completed U.S.MLE part III.
Amount of Grant: 30,000 - 45,000 U.S.D
Contact: Dr. David Herndon, Shriners Burns Hospital; (409) 770-6731; fax (409) 770-6919; dherndon@utmb.edu
Internet: http://www.shrinershq.org/shc/galveston/postdoc.html
Sponsor: University of Texas Medical Branch-Galveston
815 Market Street
Galveston, TX 77550

Gamble Foundation Grants **2040**
Founded in 1968, The Gamble Foundation is primarily interested in supporting organizations that serve disadvantaged children and youth in San Francisco, Marin and Napa counties. Within the field of youth development, the Foundation focuses on literacy, educational and personal enrichment programs designed to open doors of opportunity for at-risk youth in order to help them succeed in school and become productive, self-sufficient members of society. The Foundation is particularly interested in agricultural/environmental education, financial & computer literacy, vocational training and programs that prevent substance abuse and teen violence. To a lesser degree, the Foundation supports environmental organizations that focus on land preservation and sustainability, animal welfare and management, and pollution control. The foundation is interested in promoting green concepts that increase awareness of science based solutions that help reduce consumption of finite resources. The Foundation prefers to fund specific projects rather than annual appeals. Grants range from $5,000 to $20,000.
Requirements: Northern California 501(c)3 nonprofit organizations, with an emphasis on San Francisco, Marin, and Napa Counties, are eligible to apply. The Board meets in the spring each year and makes grants in late summer. The Foundation will accept proposals for the April meeting from January 25 - February 10. The Foundation encourages submission of proposals and attachments by email. For those submitting by email, the proposal and required attachments should be emailed in PDF format only. Send your proposals to Fiona Barrett at fbarrett@pfs-llc.net. If you do not receive an email within 24 hours confirming that your proposal has been received, please contact Fiona at (415) 561-6540, ext. 221. The Foundation will also accept a proposal submitted by mail as long as it is postmarked on or before February 10th. Proposals must include the following in the order listed: cover letter, on organization letterhead with address and phone number, including a brief summary of the request and a list of attachments (not more than one page); proposal narrative (not to exceed 5 pages); concise description of the organization (not more than two pages) including: relevant history, mission, geography and populations served, overview of programs; description of the project (not more than three pages) including: need, purpose, goals, timeline, project budget, including secured and projected sources of funding; financial statement, including actual revenue and expenses for the organization's most recently completed fiscal year; organizational budget for the present year, detailing proposed expenditures and projected sources of funding (not more than two pages); list of major public and private funders, identifying both secured and planned for funds; list of the Board of Directors, with affiliations; copy of the agency's IRS 501(c)3 tax-exempt determination letter. When submitting proposal, clasp the proposal materials with a single binder clip; do not use staples. Do not send audio-visual materials, binders, folders, or pamphlets unless requested. Receipt of proposals will be acknowledged with a written response within a reasonable period of time. Should additional information be required, applicants will be contacted.
Restrictions: In general, the Foundation does not support medical research, individuals, endowments, or capital improvements.
Geographic Focus: California
Date(s) Application is Due: Feb 10
Amount of Grant: 5,000 - 20,000 U.S.D
Contact: Eric Sloan, Grants Manager; (415) 561-6540, ext. 205; fax (415) 561-6477; esloan@pfs-llc.net
Internet: http://www.pfs-llc.net/gamble/gamble.html
Sponsor: Gamble Foundation
1660 Bush Street, Suite 300
San Francisco, CA 94109

Gates Award for Global Health 2041

This annual award recognizes an organization that has made a major and lasting contribution to the field of global health. Nominations will be considered by a jury consisting of health professionals from developing countries as well as the Global Health Council's board of directors. A winner will be selected by the jury, and the award will be presented in Washington, DC, at a special awards ceremony during the Global Health Council's annual international conference.

Requirements: Any organization from any country in the world that has substantively improved the health and the lives of people in need may be nominated for the Gates Award; the organization may be a charitable institution, a private company, or a public entity.

Restrictions: Self-nominations are not accepted.

Date(s) Application is Due: Oct 31

Amount of Grant: 1,000,000 U.S.D

Samples: African Medical and Research Foundation (Nairobi, Kenya)—for its efforts to improve public health in Africa's poorest communities, $1 million; Bangladesh Rural Advancement Committee (Dhaka, Bangladesh—for its pioneering community-based health programs, which serve 31 million poor Bangladeshis, almost a quarter of the country's population, and have influenced other health initiatives around the world, $1 million award.

Contact: Terry Fisher, (802) 649-1340, ext. 2128; fax (802) 649-1396; tfisher@globalhealth.org

Internet: http://www.globalhealth.org/view_top.php3?id=237

Sponsor: Bill and Melinda Gates Foundation
P.O. Box 23350
Seattle, WA 98102

GATX Corporation Grants 2042

Local community investment programs support organizations and projects that improve the quality of life in GATX communities across the U.S.. Each year, GATX contributes to nonprofit organizations committed to the issues surrounding culture, education, the environment, families, healthcare and social services. GATX has active community investment committees in Buffalo, Chicago, and San Francisco, which serve their respective regions as well as several other locations. Types of support include challenge grants, employee matching gifts, general operating support, project grants, and seed grants. Applications are accepted year-round. Funding decisions are made by the contributions committee, which meets quarterly in February, May, August, and November. Proposals received after these dates will be held for a subsequent meeting.

Requirements: Eligible to apply are nonprofit 501(c)3 tax-exempt organizations located in company-operating areas and that serve the economically disadvantaged.

Restrictions: Grants are not awarded to individuals, political organizations, religious organizations, private foundations, capital campaigns, endowment funds, health research, national organizations, or for fund-raisers.

Geographic Focus: Illinois

Amount of Grant: 7,000 - 80,000 U.S.D

Samples: Suzuki-Orff School for Young Musicians (Chicago, IL)—for the Clap, Sing, and Read program, which uses music to promote reading skills among at-risk children, $15,000; Chicago Anti-Hunger Federation (Chicago, IL)—for the Oliver's Kitchen Employment Training Program, which provides culinary training to formerly homeless and low-income individuals, $20,000; National Lekotek Ctr (Evanston, IL)—to provide learning and recreational services for disabled children in Chicago's West Humboldt Park neighborhood, as well as supportive services for their parents, $19,534.

Contact: Allison Dean, Community Affairs Specialist; (312) 621-4274; fax (312) 621-8062; community@gatx.com

Internet: http://www.gatx.com/common/about/community/about.asp

Sponsor: GATX Corporation
500 W Monroe
Chicago, IL 60661

GE & Science Prize for Young Life Scientists 2043

The GE & Science Prize for Young Life Scientists has been established to provide support to scientists at the beginning of their careers, because Science/AAAS and GE Healthcare believe that such support is critical for continued scientific progress. The prize will recognize outstanding graduate students in molecular biology from all regions of the world. This international prize will be awarded for the outstanding thesis in the general area of molecular biology as described in a 1000-word essay. The winning essay will be published in Science; essays of the regional award winners will appear in the online version of Science. For judging purposes, the essays will be grouped according to the geographic location of the degree-granting institution: North America, Europe, Japan, and all other countries. Initial screening of the submissions will be done by regional judges. Essays will be judged on the quality of the research and the entrant's ability to articulate the contribution of the research to the field of molecular biology. The top five essays from each geographic region will be forwarded to a panel of judges. All regional winners will compete for the grand prize of U.S.$25,000. The regional winners who do not receive the grand prize will be awarded U.S.$5,000. Winners will be announced in Science and the prize will be awarded in a location to be announced. The grand prize essay will be published in 'Science', and essays of the regional winners will be published on the online version of 'Science'.

Requirements: Entrants must have been awarded their Ph.D. between January, 1 and, December 31, of the previous year. Candidates for M.D./Ph.D. degrees are eligible to compete for the prize in either the year the Ph.D. is awarded or the year the final degree is awarded. The research described in the entrant's thesis must be in the field of molecular biology as described above. The prize will recognize only work that was performed while the entrant was a graduate student. The prize will be awarded without regard to sex, race, or nationality.

Restrictions: Employees of GE, Science and AAAS, and their relatives are not eligible for the prize.

Date(s) Application is Due: Aug 1

Amount of Grant: 25,000 U.S.D

Samples: Ethan Garner, a North American regional winner, $25,000—for his essay Understanding a Minimal DNA-Segregating Machine; Xu Tan, regional winner, $5,000—for his essay Plant Hormone Auxin Functions as Novel Molecular Glue; Sabrina Buttner, regional winner, $5,000—for her essay Endonuclease G Regulates Cellular Fate; Kaori Yamada, regional winner, $5,000—for her essay Moving PIP3 Regulates Cell Polarity; Sarel Fleishman, regional winner, $5,000—for his essay Modeling at the Gates of the Cell.

Contact: Sylvia Kihara, Prize Coordinator; (202) 326-6507; fax (202) 289-7562; skihara@aaas.org or gescienceprize@aaas.org

Internet: http://www.sciencemag.org/feature/data/prizes/ge/index.dtl

Sponsor: American Association for the Advancement of Science
1200 New York Avenue NW
Washington, DC 20005-3920

Gebbie Foundation Grants 2044

The Foundation's mission is to support appropriate charitable and humanitarian programs to improve the quality of life, primarily in Chautauqua County, New York by focusing on: children, youth and education; arts; human services; and community development. The foundation's strategic focus is to rejuvenate downtown Jamestown, New York, through economic development. Types of funding include: annual campaigns; building/renovation; capital campaigns; continuing support; endowments; equipment; general/operating support; matching/challenge support; program-related investments/loans; scholarship funds and; seed money.

Requirements: Organizations requesting funding must be approved as (or sponsored by) a 501(c)3 organization. The Board meets quarterly and proposals are accepted throughout the year. Before preparing a complete proposal, applicants are urged to submit a letter of inquiry addressed to the Executive Director. The letter should contain a descriptive title for the project, the project's intent, objectives, outcome measures, short and long term funding needs, and other relevant factual information. After reviewing the letter of inquiry, a detailed proposal may be requested. It should not be assumed that such a request is an indication of funding. The full proposal should include a detailed narrative, budgets, a board list, an IRS determination letter, other funding sources and financial documentation. In addition, the proposal should describe expected future funding sources and should indicate whether and to what extent additional funding by the Foundation will be necessary for the proposed project to be successful. The Foundation will respond to all inquiries from eligible organizations.

Restrictions: Grants are not made to individuals or sectarian or religious organizations. Because the Foundation makes annual contributions to the United Way of Southern Chautauqua County, applications for assistance from United Way-funded agencies will not be considered unless there is a strong link to the strategic focus.

Geographic Focus: New York

Amount of Grant: 1,000 - 1,000,000 U.S.D

Contact: John C. Merino, Executive Director; (716) 487-1062; fax (716) 484-6401; jmerino@gebbie.org or info@gebbie.org

Internet: http://www.gebbie.org

Sponsor: Gebbie Foundation
111 West Second Street, Suite 1100
Jamestown, NY 14701

GEICO Public Service Awards 2045

This program observes the many accomplishments of federal employees. These employees are making tremendous differences in the quality and efficiency of services and are responsible for the success of many scientific, medical, and technical programs. The foundation recognizes four active and one retired federal employee who have made outstanding achievements in one of the four fields of endeavor: Substance Abuse Prevention and Treatment, Fire Prevention and Safety, Physical Rehabilitation, and Traffic Safety and Accident Prevention. Nomination information is available online.

Requirements: All career civil service employees are eligible, including employees of the Library of Congress, the General Accounting Office, the Office of the Architect of the Capital, the Government Printing Office, the Administrative Office of the U.S. Courts, the Smithsonian Institution, the Botanic Garden, and the Office of Homeland Security.

Amount of Grant: 2,500 U.S.D

Contact: Program Contact; (877) 206-0215; federal@geico.com

Internet: http://www.geico.com/insproducts/fedpsa.html

Sponsor: GEICO Philanthropic Foundation
One GEICO Plaza
Washington, DC 20076

GenCorp Foundation Grants 2046

The Foundation is dedicated to supporting the communities where our employees live, work and volunteer. While the Foundation's primary focus is education, it also supports human services, civic and arts organizations. Emphasis is on science education. Funding preference is given to projects that involve issues important to employees and requests that are recommended by Foundation coordinators at company facilities. Interested applicants should submit a letter of inquiry.

Requirements: Non-profit organizations are eligible to apply. Priority is given to organizations that are in the GenCorp communities: Huntsville, AL; Camden, AR; Sacramento, CA; Socorro, NM; Jonesborough, TN; Clearfield, UT; Gainesville and Orange, VA; or Redmond, WA.

Geographic Focus: Alabama, Arkansas, California, New Mexico, Tennessee, Utah, Virginia, Washington
Samples: Keep Akron Beautiful (Akron, OH)—for continued support, $1000; Challenger Ctr for Space Science Education (Sacramento, CA)—for funding of space science education at the Sacramento Science Ctr, $20,000.
Contact: Program Contact; (916) 355-3600; gencorp.foundation@gencorp.com
Internet: http://www.gencorp.com/pages/gcfound.html
Sponsor: GenCorp Foundation
P.O. Box 15619
Sacramento, CA 95852-0619

Genentech Corporate Charitable Contributions Grants **2047**

The Foundation's primary focus areas are health science education, patient education/advocacy, and community. The Foundation supports nonprofits through its contributions in two primary ways: project-specific or general support funding to organizations whose mission aligns with the Foundation's focus areas; and sponsorships of selected events that fall into the Foundation's focus areas. Interested applicant's are encouraged to submit applications online.
Requirements: Nonprofit organizations recognized by the IRS as tax exempt, public charities, located in the United States are eligible to apply. Eligible grantees may include public elementary and secondary schools, as well as public colleges and universities and public hospitals.
Restrictions: The Foundation does not provide funding to organizations that discriminate on the basis of age, political affiliation, race, national origin, ethnicity, gender, disability, sexual orientation or religious beliefs. Funding is not provided for: advertising journals or booklets; alumni drives; capital campaigns/building funds; continuing medical education; infrastructural requests (e.g. salaries, equipment); memorial funds; memberships; organizations based outside of the United States; political or sectarian organizations; professional sports events or athletes; religious organizations; scholarships; yearbooks.
Geographic Focus: California
Samples: To be distributed among 19 organizations (CA)—for science programs for students in the San Francisco Bay area, $1 million.
Contact: Program Manager; (650) 467-9494; give@gene.com
Internet: http://www.gene.com/gene/about/community/overview.jsp
Sponsor: Genentech
1 DNA Way
South San Francisco, CA 94080-4990

General Mills Champions for Healthy Kids Grants **2048**

The program aims to improve youth nutrition and fitness in the United States through grants to community-based groups. The initiative consists of grants to community-based groups as well as three additional components: sponsorship of the President's Active Lifestyle Awards, developing nutrition and fitness mentoring models, and sharing best practices. Grants are intended to encourage the improvement of eating and physical activity patterns of young people, ages two to 20. Priority is given to programs that demonstrate significant potential impact on youth groups with special needs or for having an impact on large populations of youth. Guidelines are available online.
Requirements: U.S. 501(c)3 and 509(a) tax-exempt organizations are eligible. Proposed projects should reflect the practices and concepts recommended by the Dietary Guidelines for American 2000 and the American Dietetic Association, and should include at least one nutritional behavioral objective and at least one physical activity behavioral objective.
Date(s) Application is Due: Jan 15
Amount of Grant: Up to 10,000 U.S.D
Samples: Adams Park Community Ctr (Saint Louis, MO)—for Healthy Futures, a diabetes- prevention curriculum that inspires eight- to 11-year olds, from a low-income African-American community to maintain eating habits and physical activity conducive to a healthy lifestyle, $10,000; Fitness Forward Foundation (Durham, NC)—a program in which students earn points by meeting four daily behaviors: limiting TV and computer time to one hour, being physically active for one hour, not drinking sugar-sweetened beverages and getting eight to 11 hours of sleep, $10,000.
Contact: Beth Labrador, Program Contact; (800) 877-1600, ext. 4821; fax (763) 764-4114; blabrador@eatright.org
Internet: http://www.generalmills.com/corporate/commitment/champions.aspx
Sponsor: General Mills Foundation
1 General Mills Boulevard, P.O. Box 1113
Minneapolis, MN 55440

General Mills Foundation Grants **2049**

The Foundation was created to focus its philanthropic resources on community needs. Strategic objectives are to: demonstrably improve the quality of life in communities with GM facilities and employees; initiate innovative solutions and approaches to improve youth nutrition and fitness; and to support GM employees and retirees giving to United Way, education, and arts and culture organizations through gift matching. Funding priorities include: social services; youth nutrition and fitness; education; and arts and culture. Priority is given to organizations meeting the following criteria: their mission is closely related to the Foundation's priorities; programs or activities are based in communities with GM operations and employees; programs or activities involve GM employees and retirees; and services create sustainable community improvement. Applications are accepted at any time.
Requirements: U.S. and Canadian charitable 501(c)3 and 509(a) nonprofits in communities where General Mills operates (California, Georgia, Illinois, Indiana, Iowa, Maryland, Massachusetts, Missouri, Minnesota, Missouri, Montana, New Jersey, New York, Ohio, Oklahoma, Pennsylvania, Tennessee and Wisconsin) are eligible.
Restrictions: The Foundation does not support: organizations without 501(c)3 and 509(a) status; organizations that do not comply with the Foundation's non-discrimination policy; individuals; social, labor, veterans, alumni or fraternal organizations serving a limited constituency; travel by groups; recreational, sporting events or athletic associations; religious organizations for religious purposes; basic research; organizations seeking underwriting for advertising; political causes, candidates or legislative lobbying efforts; conferences, seminars and workshops; campaigns to eliminate or control specific diseases; publications, films or television programs; underwriting for program sponsorship.
Geographic Focus: Arizona, Arkansas, California, Georgia, Illinois, Indiana, Iowa, Maryland, Massachusetts, Michigan, Minnesota, Missouri, Montana, New Jersey, New Mexico, New York, Ohio, Oklahoma, Pennsylvania, Tennessee, Wisconsin
Contact: Christina L. Shea, President; (763) 764-2211; fax (763) 764-4114
Internet: http://www.generalmills.com/corporate/commitment/foundation.aspx
Sponsor: General Mills Foundation
1 General Mills Boulevard, P.O. Box 1113
Minneapolis, MN 55440

General Motors Foundation Grants Support Program **2050**

With a strong commitment to diversity in all areas the targeted areas of focus for the Foundation are: education; health and human services; civic and community; public policy; arts and culture; and environment and energy. Primary consideration is given to requests that meet the following criteria: exhibit a clear purpose and defined need in one of the foundation's areas of focus; recognize innovative approaches in addressing the defined need; demonstrate an efficient organization and detail the organization's ability to follow through on the proposal; and, explain clearly the benefits to the foundation and the plant city communities. Paper applications are no longer accepted. Completion of an online eligibility quiz is the first step in the application process.
Requirements: Nonprofit, tax-exempt organizations and institutions are eligible to apply. Applications must be made online.
Restrictions: The Foundation not not support organizations that discriminate on the basis of race, religion, creed, gender, age, veteran status, physical challenge or national origin. Contributions are generally not provided for: individuals; religious organizations; political parties or candidates; U.S. hospitals and health care institutions (general operating support); capital campaigns; endowment funds; conferences, workshops or seminars not directly related to GM's business interests.
Samples: American National Red Cross (Washington, DC)—for relief efforts in South Asia and Africa, $1 million; Pierre Chambon, College de France (Paris, France) and Ronald Evans, Salk Institute for Biological Studies (La Jolla, CA)—to honor their contributions to the diagnosis, prevention, and treatment of cancer, $250,000 jointly.
Contact: Grant Coordinator; (313) 556-5000
Internet: http://www.gm.com/company/gmability/community/guidelines/index.html
Sponsor: General Motors Foundation
300 Renaissance Center, P.O. Box 300
Detroit, MI 48265-3000

General Service Foundation Human Rights and Economic Justice Grants **2051**

The goal of the General Service Foundation's Human Rights and Economic Justice program is to support efforts that protect, promote and create good jobs with living wages for workers, including low-wage workers in the United States and Mexico. The Foundation seeks to: strengthen worker voices; promote public policies that protect labor rights; and democratize corporate power and promote corporate accountability. In assessing potential grantees, the Foundation will apply the following criteria: supporting organizations that work to address needs identified by the underrepresented and low-income communities that area directly impacted, connecting at the local and national level; and concentrating on organizations that take risks and try new ideas to enhance their skills, and reflect the diversity of their constituency.
Requirements: All applicants, regardless of their prior grant history with the Foundation, start their application process by submitting a letter of inquiry in the spring and fall. Organizations should determine if their project is an appropriate fit for the Foundation, then send a letter of inquiry via the Foundation's online application. Letters of inquiry should include: a political and situational analysis, statement of the issues to be addressed under the proposed project, the history and goals of the organization, and an explanation of why the organization or coalition is the entity that is most likely to achieve success; a brief summary of the project, short and long term goals and anticipated outcomes; the approximate starting date and duration of the proposed activities, the total amount of funding needed, the amount requested from the Foundation, a budget, and information about other sources of support; and a copy of IRS tax exempt 501(c)3 letter. If the project is judged appropriate, the organization will be offered an application to submit to the Foundation.
Restrictions: The Foundation does not fund: organizations based outside the United States or Mexico; projects without significant promise of impact beyond a local or state level; research and publications not directly linked to policy outcomes; direct service delivery; development or relief projects.
Geographic Focus: All States, Mexico
Date(s) Application is Due: Feb 1; Sep 1
Amount of Grant: 2,500 - 35,000 U.S.D
Samples: Jobs with Justice Education Fund, Washington, D.C., to build a strong, progressive labor movement that works with community, faith, and student organizations to build a broader movement for economic and social justice, $60,000, (2011); As You Sow, San Francisco, California, to support labor rights and forced labor programs, $25,000 (2011); National Day Laborer Organizing Network, Los Angeles, California, to improve the lives of day laborers in the U.S., $30,000 (2011).
Contact: Holly Bartling, Program Contact-Mexico grants; (202) 232-1005 or (970) 920-6834; fax (970) 920-4578; holly@generalservice.org

Mary Estrin, Program Contact-U.S. grants; (970) 920-6834; fax (970) 920-4578; maryestrin@gmail.com
Internet: http://www.generalservice.org/International%20Peace.htm
Sponsor: General Service Foundation
557 North Mill Street, Suite 201
Aspen, CO 81611

General Service Foundation Reproductive Justice Grants **2052**

The General Service Foundation believes that in order to be effective, the reproductive health and rights movement must address the needs of all women, including women of color, low-income women, and young women. The Reproductive Justice Grants are interested in supporting organizations that work on a broad range of reproductive health and rights issues that includes, but is not limited to, abortion. The Foundation supports organizations and projects engaged in leadership development, organizing, education, policy research and advocacy. In assessing potential grantees, the Foundation places priority on organizations and projects that have a broad approach to reproductive health, try new ideas and strategies, reflect the diversity of its constituency, and address needs identified by people directly impacted by the issues and policies.
Requirements: All applicants, regardless of their prior grant history with the Foundation, start their application process by submitting a letter of inquiry in the spring and fall. Organizations should determine if their project is an appropriate fit for the Foundation, then send a letter of inquiry via the Foundation's online application. Letters of inquiry should include: a political and situational analysis, statement of the issues to be addressed under the proposed project, the history and goals of the organization, and an explanation of why the organization or coalition is the entity that is most likely to achieve success; a brief summary of the project, short and long term goals and anticipated outcomes; the approximate starting date and duration of the proposed activities; the total amount of funding needed, the amount requested from the Foundation, a budget, and information about other sources of support; and a copy of IRS tax exempt 501(c)3 letter. If the project is judged appropriate, the organization will be offered an application to submit to the Foundation.
Restrictions: The Foundation does not fund: direct service delivery, unless it is linked to policy and advocacy; projects without significant promise of impact beyond a local or state level; research and publications not directly linked to policy outcomes; projects outside the United States; and NARAL or Planned Parenthood affiliates.
Date(s) Application is Due: Feb 1; Oct 1
Amount of Grant: 25,000 - 70,000 U.S.D
Samples: Center for Reproductive Rights-U.S. Legal Program, New York, New York, to establish and strengthen enforceable legal norms that protect reproductive rights as fundamental rights and ensure access to reproductive healthcare for all women, $30,000 (2011); National Advocates for Pregnant Women, New York, New York, to secure the human and civil rights, health and welfare of all women, $70,000 (2011).
Contact: Holly Bartling, Program Contact; (970) 920-6834; fax (970) 920-4578; holly@generalservice.org
Internet: http://www.generalservice.org/Reproductive%20Health.htm
Sponsor: General Service Foundation
557 North Mill Street, Suite 201
Aspen, CO 81611

Genesis Foundation Grants **2053**

The Foundation funds projects and program in the areas of health and formal education, benefiting children 0 to 18 years old, as well as training programs in these two areas. Capital, operating, program and general grants are funded. Approximately eighty percent of grant-making is devoted to Colombian entities, with the remainder devoted to projects in the U.S., specifically in Southern Florida, Washington, DC and the New York Metropolitan area.
Requirements: Preliminary criteria to apply for a grant include: must be located in Colombia, Southern Florida, Washington, DC or the New York Metropolitan Area; must be incorporated as a non-profit, charitable institution in the appropriate jurisdiction; must have been in operation for a minimum of two (2) full years; must include at least three (3) people in decision-making process at all times; must benefit a minimum of 50 people per year; in Colombia, must have raised at a minimum the equivalent of U.S. $10,000 in the previous two years; in the U.S. must have raised a minimum of $20,000 in the previous two years; must have an independent financial auditor; must be able to provide audited financial statements for the previous three years; must be able to provide a budget for the organization and the project for which the grant is sought; must not be a political organization; and must not have a religious purpose.
Restrictions: Individuals and political or religious causes are ineligible.
Geographic Focus: District of Columbia, Florida, New York, Colombia
Date(s) Application is Due: Mar 15; Sep 15
Amount of Grant: 10,000 - 100,000 U.S.D
Samples: Asociacion Social Popular (Bello, Colombia)—to purchase an educational facility, $35,000; Fundacion Estructurar (Bucaramanga, Colombia)—to provide education to working children, with an emphasis on training in design and graphic arts, $15,000; Fundacion Neumologica Colombiana (Bogota, Colombia)—to create an asthma center for children, $240,000; Nutrir (Bogota, Colombia)—to construct an additional facility that will provide nutritional and social services to underserved children, $40,000.
Contact: Grants Administrator; (212) 763-3703; fax (917) 591-7858; genesis@genesid-foundation.org
Internet: http://www.genesis-foundation.org
Sponsor: Genesis Foundation
140 East 45th Street, 2 Grant Central Tower 18th Floor, Suite A
New York, NY 10017

Genuardi Family Foundation Grants **2054**

The Foundation supports direct providers of services in the areas of education, health, human services, and culture. The Foundation gives preference to projects which receive broad-based community support and provides for reasonable costs associated with conducting the proposed project or program. Grant proposals may include the allocation of a reasonable percentage of grant monies towards general support of the recipient organization. On a limited basis, the Foundation will also consider grants in support of capital campaigns for facilities or equipment, and organizational capacity building.
Requirements: The Foundation only considers grant requests from organizations designated as tax exempt under Section 501(c)3 of the Internal Revenue Code. Preference will be given to non-profit organizations based in the Greater Philadelphia Area.
Restrictions: The Foundation's funding will not process grant requests from: individuals; fraternal and/or civic organizations; political candidates or to influence legislation; other foundations; general fund raising and endowment drives; debt reduction; environmental issues or initiatives; annual appeals or letters of solicitation; or public, private or parochial schools that serve the general public.
Geographic Focus: Pennsylvania
Date(s) Application is Due: Aug 1
Amount of Grant: 5,000 - 50,000 U.S.D
Contact: Meredith A. Huffman, Executive Director; (610) 834-2030; info@genuardifamilyfoundation.org
Internet: http://www.genuardifamilyfoundation.org/genuardi_priorities.html
Sponsor: Genuardi Family Foundation
Blue Bell Executive Campus, 470 Norristown Road, Suite 300
Auburn, NY 13021

George A. and Grace L. Long Foundation Grants **2055**

The George A. and Grace L. Long Foundation was established in 1960 to support and promote quality educational, cultural, human-services, and health-care programming for underserved populations in Connecticut. Grants from the Long Foundation are one year in duration. Applicants must apply online at the grant website. Applicants are strongly encouraged to do the following before applying: review the downloadable state application procedures for additional helpful information and clarifications; review the downloadable online-application guidelines at the grant website; review the foundation's funding history (link is available from the grant website); review the online application questions in advance; and review the list of required attachments. These will generally include: a list of board members, financial statements (audited, reviewed, or compiled by independent auditor); an organization summary; a list of other funding sources; an IRS Determination letter; and other required documents. All attachments must be uploaded in the online application as PDF, Word, or Excel files. The George A. and Grace L. Long Foundation has biannual deadlines of March 15 and September 15. Applications must be submitted by 11:59 p.m. on the deadline dates. Applicants will be notified of grant decisions by letter within two to three months after each respective proposal deadline.
Requirements: Applicant organizations must have 501(c)3 tax-exempt status and serve the people of Connecticut.
Restrictions: Applicants will not be awarded a grant for more than three consecutive years. The foundation does not support requests from individuals, organizations attempting to influence policy through direct lobbying, or any political campaigns.
Geographic Focus: Connecticut
Date(s) Application is Due: Mar 15; Sep 15
Samples: Christian Activities Council, Hartford, Connecticut, $3000 (2010); Greater Hartford Arts Council, Hartford, Connecticut, $5,000 (2010); Camp Courant, Hartford, Connecticut, $10,000 (2010).
Contact: Carmen Britt, Senior Foundation Officer; (860) 657-7019; carmen.britt@baml.com
Internet: https://www.bankofamerica.com/philanthropic/fn_search.action
Sponsor: George A. And Grace L. Long Foundation
200 Glastonbury Boulevard, Suite # 200, CT2-545-02-05
Glastonbury, CT 06033-4056

George and Ruth Bradford Foundation Grants **2056**

The foundation awards grants to local nonprofit organizations in the San Francisco Bay Area and Mendocino, California region. Areas of interest include: children and youth; families; education/higher education; environmental/wildlife conservation; housing and; social services. Types of support include general operating support and scholarship funds.
Requirements: California nonprofit organizations in the San Francisco peninsula and Mendocino, California region are eligible to apply. There is no deadline date to adhere to. The Board meets monthly, letters of inquiry may be submitted throughout the year for review.
Restrictions: No grants to individuals.
Geographic Focus: California
Samples: Frank R. Howard Foundation, Willits, CA, $10,000-hospital building fund; Ukiah Valley Cultural & Recreation Center. Ukiah, CA, $5,000—youth recreation grant; MC Aids Volunteer Network, Ukiah, CA, $1,000—health services grant.
Contact: Myrna Oglesby, Director; (707) 462-0141; fax (707) 462-0160
Sponsor: George and Ruth Bradford Foundation
P.O. Box 720
Ukiah, CA 95482-0720

George A Ohl Jr. Foundation Grants **2057**

The purpose of the George A Ohl Jr. Foundation is to improve the well-being of the citizens of the State of New Jersey through science, health, recreation, education and increased good citizenship. Grants are made to organizations engaged in such work whether through research, publications, health, school or college activities. The Foundation's mission is the

relief of the poor; the improvement of living conditions; the care of the sick, the young, the aged, the homeless, the incompetent and the helpless. The foundation will target: 35% of its grants to Community Redevelopment; 35% to Health and Human Services organizations; 15% for Arts and culture; 15% for Educational requests. Application deadlines are: January 22 for a March meeting and, June 15 for an August meeting. Application forms are available online. Applicants will receive notice acknowledging receipt of the grant request, and subsequently be notified of the grant declination or approval.
Requirements: New Jersey 501(c)3 nonprofit organizations are eligible to apply. Proposals should be submitted in the following format: completed Common Grant Application Form; an original Proposal Statement; an audited financial report and a current year operating budget; a copy of your official IRS Letter with your tax determination; a listing of your Board of Directors. Proposal Statements (second item in the above Format) should answer these questions: what are the objectives and expected outcomes of this program/project/request; what strategies will be used to accomplish your objective; what is the timeline for completion; if this is part of an on-going program, how long has it been in operation; what criteria will you use to measure success; if the request is not fully funded, what other sources can you engage; an Itemized budget should be included; please describe any collaborative ventures. Prior to the distribution of funds, all approved grantees must sign and return a Grant Agreement Form, stating that the funds will be used for the purpose intended. Progress reports and Completion reports must also be filed as required for your specific grant. All current grantees must be in good standing with required documentation prior to submitting new proposals to any foundation.
Restrictions: Grants are not made for political purposes, nor to organizations which discriminate on the basis of race, ethnic origin, sexual or religious preference, age or gender
Geographic Focus: New Jersey
Date(s) Application is Due: Jan 22; Jun 15
Amount of Grant: 3,000 - 35,000 U.S.D
Samples: Mountainside Hospital Foundation, $35,000—Modernization of Internal Medicine Residents' Learning Center; Bloomfield College, $10,000—Improving Academic Success Through Technological Support Project; Medical Missions for Children, $5,000—Giggles Theater Programming Support (12 performances).
Contact: Wachovia Bank, N.A., Trustee; grantinquiries2@wachovia.com
Internet: https://www.wachovia.com/foundation/v/index.jsp?vgnextoid=e0f78689fb0a a110VgnVCM1000004b0d1872RCRD&vgnextfmt=default
Sponsor: George A Ohl Jr. Foundation
190 River Road, NJ3132
Summit, NJ 07901

George B. Storer Foundation Grants **2058**
The foundation awards grants, primarily in Florida, in its areas of interest, including higher education, social services—particularly for the blind, youth organizations, conservation, hospitals, and cultural programs. Types of support include building construction/renovation, capital campaigns, general operating support, matching/challenge grants, and research grants. There are no application forms. Applications should be submitted between October 15 and the listed deadline.
Date(s) Application is Due: Nov 15
Samples: U of Miami (FL)—for marine-science and business-education programs, $205,000.
Contact: Grants Administrator, c/o Thomas McDonald
Sponsor: George B. Storer Foundation
P.O. Box 1040
Tavernier, FL 33070

George E. Hatcher, Jr. and Ann Williams Hatcher Foundation Grants **2059**
The George E. Hatcher and Ann Williams Hatcher Foundation was created to support charitable organizations that provide for the relief of diseased people and the relief of human suffering which is due to disease, ill health, physical weakness, physical disability and/or physical injury. In addition, the foundation supports organizations that aid in the promotion and prolongation of life and that support the principle of dying with dignity. Grants from George E. Hatcher and Ann Williams Hatcher Foundation are primarily one year in duration; on occasion, multi-year support is awarded. Applicants must apply online at the grant website. Applicants are strongly encouraged to do the following before applying: review the downloadable state application procedures for additional helpful information and clarifications; review the downloadable online-application guidelines at the grant website; review the foundation's funding history (link is available from the grant website); review the online application questions in advance; and review the list of required attachments. These will generally include: a list of board members, financial statements (audited, reviewed, or compiled by independent auditor); an organization summary; a list of other funding sources; an IRS Determination letter; and other required documents. All attachments must be uploaded in the online application as PDF, Word, or Excel files. The George E. Hatcher and Ann Williams Hatcher Foundation application deadline is 11:59 p.m. on May 31. Applicants will be notified of grant decisions by letter within one to two months after the deadline.
Requirements: Disbursements are authorized to institutions or organizations located in the Middle Georgia area (Bibb County & surrounding communities) which provide health care and/or shelter for individuals (especially children) who cannot otherwise obtain such services due to circumstances beyond their control. A breakdown of number/percentage of people served by specific counties is required on the online application.
Restrictions: The foundation does not support requests from individuals, organizations attempting to influence policy through direct lobbying, or any political campaigns.
Geographic Focus: Georgia
Date(s) Application is Due: May 31
Samples: Macon Volunteer Clinic, Macon, Georgia, $25,000 (2010); Rebuilding Macon, Macon, Georgia, $25,000 (2010); Methodist Home of the South Georgia Conference, Macon, Georgia, $10,000 (2010).
Contact: Quanda Allen, Vice President; (404) 264-1377; quanda.allen@baml.com
Internet: https://www.bankofamerica.com/philanthropic/fn_search.action
Sponsor: George E. Hatcher, Jr. and Ann Williams Hatcher Foundation
3414 Peachtree Road, N.E., Suite 1475, GA7-813-14-04
Atlanta, GA 30326-1113

George F. Baker Trust Grants **2060**
Grants are awarded nationwide, with preference given to nonprofits in the eastern United States, primarily for K-12, higher, and secondary education; hospitals; social services; private foundations; and zoos/zoological societies. Types of support include general operating support and matching funds. There are no application forms or deadlines. The board meets in June and November to consider requests.
Requirements: An application form is not required. Along with a letter of inquiry and brief outline of a proposal, applicants should submit the following: signature and title of chief executive officer; a copy of the organization's IRS determination letter; a detailed description of the project and amount of funding requested; and a listing of additional sources and amount of support. As a result of the enormous number of applications received and limited number of grants, only those applicants who receive a grant will be notified.
Restrictions: Funding is given primarily in Connecticut, Florida, Massachusetts, and New York. Grants are not made to individuals for scholarships or loans.
Geographic Focus: Connecticut, Florida, Massachusetts, New York
Amount of Grant: 1,000 - 50,000 U.S.D
Contact: Rocio Suarez, Executive Director; (212) 755-1890; fax (212) 319-6316; rocio@bakernye.com
Sponsor: George F. Baker Trust
477 Madison Avenue, Suite 1650
New York, NY 10022

George Family Foundation Grants **2061**
The mission of the Foundation is to foster wholeness in mind, body, spirit, and community in order to enhance the development of human potential. The Foundation's areas of interest are: integrative medicine, leadership development, educational opportunities, social justice and spirituality. Non- solicited proposals will not be accepted. Invited applicants are encouraged to submit a letter of inquiry or to discuss their request with staff before submitting an application.
Requirements: 501(c)3 tax-exempt organizations are eligible. Primary giving is restricted to the Twin Cities metropolitan area.
Restrictions: Grants do not support endowment or capital campaigns, programs in which the family has no active interest or involvement; individuals; memberships, special events, or other fundraisers; or programs that address debt retirement or recovery of operating losses.
Geographic Focus: Minnesota
Date(s) Application is Due: May 30; Oct 30
Amount of Grant: 1,000 - 200,000 U.S.D
Samples: WATCH (Minneapolis, MN) for efforts to improve the criminal-justice system's handling of cases involving domestic violence and sexual assault, $15,000; Twin Cities RISE (Minneapolis, MN)—for scholarships for participants in its skills-training and job-acquisition programs, $50,000; Plymouth Congregation Church (Minneapolis, MN)—to add an outreach worker to the church's staff roster, $17,000; Sigma Chi Foundation (Evanston, IL)—for the Horizons Leadership project, $25,000.
Contact: Grants Administrator; (612) 377-8400; fax (612) 377-8407
Internet: http://www.georgefamilyfoundation.org/
Sponsor: George Family Foundation
1818 Oliver Avenue South
Minneapolis, MN 55405

George Foundation Grants **2062**
The George Foundation strive to support organizations and programs that assist in developing strong, stable families across Fort Bend County region of Texas. The Foundation's areas of interest include: family stability; scholarships; foundation initiated programs, current programs include, Youth in Philanthropy (YIP), Leadership Excellence for Non-Profits, Integrated Mental Health Care, and Transportation. The Foundation prefers to fund the following types of grants to support Fort Bend organizations in their delivery of services to the community (listed in order of priority): program/project support; foundation initiated; general operating; capital. Proposal deadlines for making grant applications to the Foundation are January 15, April 15, July 15 and October 15 of each year. All proposals for capital support will be grouped together for review. The deadline for capital proposals will be October 15 of each year.
Requirements: Non-profit organizations in Fort Bend County, Texas may submit grant proposals.
Restrictions: The Foundation does not fund: grants to organizations that do not have a current 501(c)3 determination letter; churches or other organized religious bodies; grants to another organization that distributes money to recipients of its own selection, i.e., a regranting organization; regional, national or international programs; grants for research or studies; grants for travel, conferences, conventions, group meetings, or seminars; the purchase of event tickets, tables, ads or sponsorships; support to fairs and festivals; religious or private schools; request for funds to develop films, videos, books or other media projects; direct mail campaigns; loans of any kind; grants to individuals; grants to fraternal organizations; political interests of any kind; and institutions that discriminate on the basis of race, creed, gender, national origin, age, disability or sexual orientation in policy or in practice.

Geographic Focus: Texas
Date(s) Application is Due: Jan 15; Apr 15; Jul 15; Oct 15
Amount of Grant: 1,000 - 4,000,000 U.S.D
Samples: Wharton County Junior College, $110,000—to provide scholarships for Fort Bend County residents; YMCA of the Greater Houston Area, $250,000—to assist in acquiring facility to offer outdoor education/recreation experiences to Fort Bend families and children; OakBend Medical Center, $4,000,000—four year grant to provide indigent healthcare to Fort Bend County residents.
Contact: Dee Koch, Grant Officer; (281) 342-6109; fax (281) 341-7635; dkoch@thegeorgefoundation.org
Internet: http://www.thegeorgefoundation.org
Sponsor: George Foundation
310 Morton Street, PMB Suite C
Richmond, TX 77469

George Frederick Jewett Foundation Grants **2063**

The foundation is concerned primarily with people and values. The grants program focuses on the future and on stimulating and supporting activities and projects of established, voluntary, nonprofit organizations that are of importance to human welfare. Grants are made in the fields of arts and humanities, conservation and preservation, education, health care and medical services, population, religion, and social welfare. The foundation may support research on and studies of important problems of public concern solely for the purpose of aiding in the gathering and presenting of facts that may assist the public to better understand such problems and to arrive at realistic and effective solutions to them. From time to time, support may be given to the scholarship, fellowship, and research programs of established institutions when sufficient evidence is available to establish clearly that the applicant organization is awarding such grants in accordance with the regulations established by the IRS. Grants are awarded to support activities in progress, research into potential projects, building and equipment, general operations, program development, seed funding, research, technical assistance, and matching funds. Inquiries for clarification of the foundation's policy and program emphasis are encouraged.
Requirements: Preference is given to public charities or nonprivate foundations. The foundation confines its grants largely to requests from eastern Washington and the San Francisco Bay area.
Restrictions: Grants do not support advertising; advocacy, athletic, international, religious, political, or veterans organizations; or individuals.
Geographic Focus: California, Washington
Amount of Grant: 5,000 - 50,000 U.S.D
Samples: American U of Beirut (New York, NY)—general support, $15,000; Committee to Restore the Opera House (San Francisco, CA)—for a three-year commitment for restoration of the San Francisco War Memorial Opera House, $120,000.
Contact: Ann Gralnek, (415) 421-1351; fax (415) 421-0721; ADGjewettf@aol.com or tfbjewettf@aol.com
Sponsor: George Frederick Jewett Foundation
235 Montgomery Street, Suite 612
San Francisco, CA 94104

George Gund Foundation Grants **2064**

The Foundation's long-standing interests include: arts; economic development; community revitalization; education; environment; and human services. The Foundation considers global climate change an urgent issue that cuts across all of the Foundation's programs. The Foundation takes seriously it's own responsibility and wants to hear from grant applicants what they are doing or considering to reduce or to eliminate their organizational impact on climate change. The Foundation also supports special projects grants, which currently include: Retinal Degenerative Diseases research grants, making an annual commitment for research on the causes, nature and prevention of inherited retinal degenerative diseases and; philanthropic services grants, offering support to organizations that strengthen the infrastructure of the nonprofit and philanthropic communities. The George Gund Foundation also supports capital requests but only for projects that are clearly aligned with their program priorities and that meet Green Building Council LEED (Leadership in Energy and Environmental Design) certification. The Foundation's green building policy covers both planning and construction grants. In addition, the Foundation supports opportunities that cross program boundaries and that integrate elements of the Foundation's interests. Although the Foundation's focus is centered in the Greater Cleveland, Ohio region, a portion of their grantmaking will continue to support state and national policy making that bolsters their work. Proposals should be mailed directly to the George Gund Foundation. All proposals are screened and evaluated by the staff before presentation at Trustee Meetings. Receipt of proposals will be acknowledged by mail.
Requirements: 501(c)3 nonprofit organizations are eligible to apply for funding, with a special interest in Greater Cleveland, Ohio region. Proposals are accepted three times: March 15, July 15 and November 15. Proposals are due the next business day if a deadline falls on a weekend. The grant application form is available on the George Gund Foundation website. All proposals must include a climate change statement, the Foundation's website includes resources to assist grantees with this task. Applicants also must include a completed Grant Application Cover Sheet, which is signed by the organization's board chair and executive director. Proposals should also include: organizational background; history; mission; types of programs offered; constituencies served; project description; justification of need; specific goals and objectives; activities planned to meet goals and objectives; project time line; qualifications of key personnel; methods of evaluation; project budget; anticipated expenses, including details about how Foundation funds would be used; anticipated income, including information about other sources approached for funding; organizational budget; previous and current year budget and proposed budget for project year(s) showing both income and expenses; the organization's most recent audited financial statement, do not include IRS 990 forms; supporting documents; list of current trustees; letters of support; readily available printed material about organization such as annual reports and brochures; IRS letter confirming Internal Revenue Code 501(c)3 status and classification as a public charity or information confirming status as a government unit or agency. The Foundation also will accept the Ohio Common Grant Form, available at www.ohiograntmakers.org, if organizations are using it to apply to multiple funders. Faxed or electronic proposals are not accepted.
Restrictions: Do not submit proposals in notebooks, binders or plastic folders and print proposals on both sides of each sheet of paper. The Foundation normally does not consider grants for endowments. Capital requests must meet the Foundation's program goals and also adhere to green building standards of environmental sustainability. Details on these requirements are available from the Foundation. Grants are not made for debt reduction or to fund benefit events. The Foundation does not make grants to individuals, nor does it administer programs it supports. Grants are limited to organizations located in the United States.
Date(s) Application is Due: Mar 15; Jul 15; Nov 15
Samples: Progressive Arts Alliance Incorporated, Cleveland, OH, $20,000—database capacity building project (2010); Cleveland Restoration Society, Incorporated, Cleveland, OH, $20,000—preservation programs (2010); Center for Community Solutions, Cleveland, OH, $360,000—two year grant for policy and fiscal analysis and GroundWork early childhood campaign (2010).
Contact: David Abbott; (216) 241-3114; fax (216) 241-6560; info@gundfdn.org
Internet: http://www.gundfdn.org/what.asp
Sponsor: George Gund Foundation
1845 Guildhall Building, 45 Prospect Avenue, West
Cleveland, OH 44115

George H.C. Ensworth Memorial Fund Grants **2065**

The George H.C. Ensworth Memorial Fund was established in 1949 to support charitable organizations that focus on health and human services, youth services, enjoyment of the natural environment, education, religion, and the arts. Applicants must apply online at the grant website. Applicants are strongly encouraged to do the following before applying: review the downloadable state application procedures for additional helpful information and clarifications; review the downloadable online-application guidelines at the grant website; review the fund's funding history (link is available from the grant website); review the online application questions in advance; and review the list of required attachments. These will generally include: a list of board members, financial statements (audited, reviewed, or compiled by independent auditor); an organization summary; a list of other funding sources; an IRS Determination letter; and other required documents. All attachments must be uploaded in the online application as PDF, Word, or Excel files. The George H. C. Ensworth Memorial Fund has an annual deadline of May 15 at 11:59 p.m. Applicants will be notified of the grant decisions by letter within two months after the proposal deadline.
Requirements: Applicants must have 501(c)3 status and serve the people of Glastonbury, Connecticut. A breakdown of number/percentage of people served will be required in the online application.
Restrictions: Grant requests for capital projects will not be considered. Applicants will not be awarded a grant for more than 3 consecutive years. The fund does not support requests from individuals, organizations attempting to influence policy through direct lobbying, or any political campaigns.
Geographic Focus: Connecticut
Date(s) Application is Due: May 15
Samples: Friends of Glastonbury Youth, Glastonbury, Connecticut, $5,000 (2010); Rushford Center, Glastonbury, Connecticut, $6,000 (2010); Connecticut Childrens Medical Center Foundation, Hartford, Connecticut, $5,000 (2010).
Contact: Kate Kerchaert; (860) 657-7016; kate.kerchaert@baml.com
Internet: https://www.bankofamerica.com/philanthropic/fn_search.action
Sponsor: George H. C. Ensworth Memorial Fund
200 Glastonbury Boulevard, Suite # 200, CT2-545-02-05
Glastonbury, CT 06033-4056

George H. Hitchings New Investigator Award in Health Research **2066**

Provides awards to provide flexible support for young researchers in their second year of graduate school. Students planning to follow a career path in academic teaching and research that is closely tied to the health of North Carolinians and the strength of North Carolina Science are particularly encouraged to apply.
Requirements: Applications/nominations must be submitted by the institution on behalf of the student(s). Nomination applications are accepted from Duke, NCCU, NCSU, and UNC-Chapel Hill.
Restrictions: Only one nomination/application per graduate program/department within the university will be accepted.
Geographic Focus: North Carolina
Date(s) Application is Due: Mar 5
Amount of Grant: 10,000 U.S.D
Contact: Libby Long, Program Coordinator; (919) 474-8370, ext. 134; fax (919) 941-9208; libby@trialglecf.org
Internet: http://www.trianglecf.org/DatasetRecord.cfm?recordID=10001417&returnURL=%2Fpage10000237%2Ecfm&returntoname=View%20all%20Scholarships&sidepageid=10000237&thetitle=George%20H.%20Hitchings%20New%20Investigator%20Award%20in%20Health%20Research&Ds_PagepropId=1
Sponsor: Triangle Community Foundation
324 Blackwell Street, Suite 1220
Durham, NC 27701

George K. Baum Foundation Grants **2067**

The foundation supports nonprofits in the greater Kansas City, MO, area in its areas of interest, including arts and cultural programs, education, health care, Jewish federated giving, and general charitable giving. There are no application deadlines or forms.
Requirements: Nonprofits in the metropolitan Kansas City, MO, area are eligible.
Geographic Focus: Missouri
Amount of Grant: Up to 392,000 U.S.D
Samples: Jewish Federation of Greater Kansas City (MO)—general support, $20,000; University of Missouri, Conservatory of Music—general support, $7500.
Contact: Jonathan Baum, Trustee; (816) 474-1100 or (800) 821-7195; fax (816) 283-5171
Sponsor: George K. Baum Foundation
120 W 12th Street, Suite 800
Kansas City, MO 64105

George Kress Foundation Grants **2068**

The foundation awards grants to eligible Wisconsin nonprofit organizations in its areas of interest, including arts and culture; boys and girls clubs; children and youth services; Christian agencies and churches; community/economic development; education; family services; health organizations; higher education; historic preservation/historical societies; hospitals; human services; libraries; recreation; United Ways and Federated Giving Programs; and YM/YWCAs and YM/YWHAs. Types of support include annual campaigns; building/renovation; capital campaigns; continuing support; professorships; program development; and research. Preference is given to nonprofit organizations that benefit Green Bay and Madison.
Requirements: There are no application forms or deadlines. Interested applicants should submit a letter of inquiry describing their proposed project.
Geographic Focus: Wisconsin
Contact: John Kress, President; (920) 433-3109
Sponsor: George Kress Foundation
c/o Green Bay Packaging Company
Green Bay, WI 54301

George P. Davenport Trust Fund Grants **2069**

Established in Maine in 1927, the Trust awards grants for education, religion, temperance and needy children. The focus of the Trust is on the economically disadvantaged. Types of support include: building and renovation, emergency funding, general operating funds, capacity building, matching/challenge funds, and seed money. It is recommended that applicants contact the Trust office to make an initial inquiry regarding projects before beginning the application process. Interested applicants should use the standard Maine Philanthropy Center common grant application which is available at mainephilanthropy.org. Grant applications are accepted all year, with no deadlines.
Requirements: Only nonprofit organizations serving the residents of Bath, Maine, and its surrounding area are eligible to apply.
Geographic Focus: Maine
Amount of Grant: 125 - 10,000 U.S.D
Samples: Bath Elementary PTA, Bath, Maine, $6,550 - camp scholarships for youth (2010); Brunswick Area Respite Care, Topsham, Maine, $10,000 - adult disabilities of aging support (2010); Good Shepherd Food Bank, Auburn, Maine, $5,000 - food for the needy (2010).
Contact: Barry M. Sturgeon, Trustee; (270) 443-3431; fax (800) 665-5510; davenporttrust@verizon.net
Sponsor: George P. Davenport Trust Fund
65 Front Street
Bath, ME 04530-2508

George S. and Dolores Dore Eccles Foundation Grants **2070**

The foundation awards grants to eligible Utah organizations in its areas of interest, including arts, children and youth, economics, higher education, hospitals, medical research, performing arts, visual arts, and social services. Types of support include building construction/renovation, capital campaigns, equipment acquisition, general operating grants, matching/challenge grants, professorships, program development, research grants, and scholarship funds. A request for application is available online.
Requirements: Giving primarily in Utah.
Restrictions: Funding requests will not be considered from the following types of organizations: those that have not received a tax exemption letter establishing 501(c)3 status from the Internal Revenue Service, unless they are a unit of government, in which case such a letter is not required; other private foundations; those of a political nature that attempt to influence legislation and/or candidacy of persons for elected public office; conduit organizations, unified funds, or those that use funds to make grants to support other organizations; those that do not have fiscal responsibility for the proposed project. Funds will also not be considered for: contingencies, deficits, or debt reduction; general endowment funds; direct aid to individuals; conferences, seminars, or medical research; requests which do not fall within the Foundation's specified areas of interest.
Geographic Focus: Utah
Samples: Westminster College of Salt Lake City (UT)—for undergraduate scholarships, $1.2 million over three years.
Contact: Director; (801) 246-5340; fax (801) 350-3510; gseg@gseccles.org
Internet: http://www.gsecclesfoundation.org
Sponsor: George S. and Dolores Dore Eccles Foundation
79 South Main Street, 12th Floor
Salt Lake City, UT 84111

George W. Brackenridge Foundation Grants **2071**

The foundation awards grants in Texas to nonprofit colleges and universities, religious education organizations, accredited K-12 schools, and youth organizations such as YMCA/YWCA. Types of support include endowment funds, project grants, research grants, and scholarships. The foundation's Areas of Interest include: arts; arts education; biomedicine research; children/youth, services; christian agencies & churches; education; elementary/secondary education; higher education; human services and; the performing arts.
Requirements: Texas nonprofit organizations in the San Antonio and the surrounding area are eligible to apply. When applying for funding, submit a letter of request on the organization's letterhead. Upon review of letter of request, you may be invited to submit a proposal. Proposal should include: signature and title of chief executive officer; copy of IRS Determination Letter; copy of most recent annual report/audited financial statement/990; detailed description of project and amount of funding requested; organization's charter and by-laws; four copies of the proposal. There are no application deadlines. The board meets in March, June, September, and December each year.
Restrictions: Grants are not made to or for: individuals; general purposes; continuing support; seed money; emergency funds; land acquisition; renovation projects; building funds; operating budgets; annual campaigns; deficit financing; matching gifts; loans.
Geographic Focus: Texas
Contact: Emily D. Thuss, Treasurer; (210) 224-1011
Sponsor: George W. Brackenridge Foundation
711 Navarro Street, Suite 535
San Antonio, TX 78205-1746

George W. Codrington Charitable Foundation Grants **2072**

The foundation's support is limited to public charitable or educational projects within Cuyahoga County, OH, and immediately adjacent areas. Areas of interest include museums, performing arts, arts and cultural programs, higher education, other education, hospitals (general), and children and youth services. Types of support include continuing support, general operating support, annual campaigns, capital campaigns, equipment, program development, and research. The application should state fully but briefly the amount requested, need for the grant, area served by the applicant, brief history of the applicant organization, description of applicant's contribution to the area, listing of the applicant's officers and trustees, and evidence of 501(c)3 status. Grants generally are made on an annual basis. Meetings of the supervisory board are usually held in March, June, September, November and December, at which time pending applications are considered.
Requirements: IRS 501(c)3 organizations serving residents of Cuyahoga County, OH, and immediately adjacent areas are eligible.
Restrictions: Grants are not made to individuals.
Geographic Focus: Ohio
Amount of Grant: 500 - 25,000 U.S.D
Samples: Cleveland Ctr for Economic Education (Cleveland, OH)—for project support, $5000; Educational Television Assoc of Metropolitan Cleveland (Cleveland, OH)—for project support, $25,000; Ohio Chamber Orchestra (Cleveland, OH)—for support of the arts, $2000.
Contact: Raymond Sawyer, Chair; (216) 566-5837; fax (216) 566-5800
Sponsor: George W. Codrington Charitable Foundation
127 Public Square, 39th Floor
Cleveland, OH 44114-1216

George W. Wells Foundation Grants **2073**

The George W. Wells Foundation was established in 1934 to support and promote quality educational, human-services, and health-care programming for underserved populations. In the area of education, the fund supports academic access, enrichment, and remedial programming for children, youth, adults, and senior citizens that focuses on preparing individuals to achieve while in school and beyond. In the area of health care, the fund supports programming that improves access to primary care for traditionally underserved individuals, health education initiatives and programming that impact at-risk populations, and medical research. In the area of human services the fund tries to meet evolving needs of communities. Currently the foundation's focus is on (but is not limited to) youth development, violence prevention, employment, life-skills attainment, and food programs. Special consideration is given to charitable organizations that serve the people of Southbridge, Massachusetts, and its surrounding communities. Grant requests for general operating support are strongly encouraged. Program support will also be considered. Small, program-related capital expenses may be included in general operating or program requests. The majority of grants from the Wells Foundation are one year in duration; on occasion, multi-year support is awarded. Applicants must apply online at the grant website. Applicants are strongly encouraged to do the following before applying: review the downloadable state application procedures for additional helpful information and clarifications; review the downloadable online-application guidelines at the grant website; review the foundation's funding history (link is available from the grant website); review the online application questions in advance; and review the list of required attachments. These will generally include a list of board members, financial statements (audited, reviewed, or compiled by independent auditor); an organization summary; a list of other funding sources; an IRS Determination letter; and other required documents. All attachments must be uploaded in the online application as PDF, Word, or Excel files. The application deadline for the George W. Wells Foundation is 11:59 p.m. on October 15. Applicants will be notified of grant decisions before December 31.
Requirements: Applicants must have 501(c)3 tax-exempt status.
Restrictions: The foundation does not support requests from individuals, organizations attempting to influence policy through direct lobbying, or any political campaigns.
Geographic Focus: Massachusetts
Date(s) Application is Due: Oct 15

Contact: Miki C. Akimoto, Vice President; (866) 778-6859; miki.akimoto@baml.com
Internet: https://www.bankofamerica.com/philanthropic/fn_search.action
Sponsor: George W. Wells Foundation
225 Franklin Street, 4th Floor, MA1-225-04-02
Boston, MA 02110

Georgiana Goddard Eaton Memorial Fund Grants **2074**

Established in 1917, the Trust supports organizations involved with improving the lives of low-income individuals of Boston, Massachusetts. The fields of interest are: education; employment; family services; homelessness; housing/shelter; human services; legal services; urban/community development. Contact the Foundation for further application information and guidelines.
Requirements: Non-profits in the Boston, MA area are eligible. Applications should be addressed to: c/o Grants Mgmt. Assocs., 77 Summer Street, 8th Floor, Boston, MA 02110-1006
Restrictions: No grants to individuals (except former employees of Community Workshops, Inc.), or for endowment funds, or matching gifts; no loans.
Geographic Focus: Massachusetts
Date(s) Application is Due: Mar 1; Sep 1
Contact: Philip Hall, Administrator; (617) 426-7080; fax (617) 426-7087; phall@grantsmanagement.com
Sponsor: Georgiana Goddard Eaton Memorial Fund
45 School Street
Boston, MA 0002108-3206

Georgia Power Foundation Grants **2075**

Giving is focused on issues that directly affect customers, employees, business and shareholders. These include: improving the quality of education by partnering with organizations to assist students with personal development, mentoring and career exploration; protecting the environment by promoting programs to improve air and water quality, preserve natural resources and protect endangered species; preventing cancer; and promoting diversity. The Foundation gives strong preference to Georgia-based organizations and programs that seek to improve the quality of life for the state's residents. Applicants may apply online or with a written proposal.
Requirements: The foundation makes grants to tax-exempt organizations that seek to improve the quality of life for Georgia's residents.
Restrictions: The Foundation does not provide grants to individuals, private elementary or secondary schools, and religious organizations, nor political campaigns or causes. The Foundation does not provide multi-year funding commitments.
Geographic Focus: Georgia
Date(s) Application is Due: Feb 15; May 15; Aug 15; Nov 15
Amount of Grant: 10,000 U.S.D
Contact: Grants Administrator; (404) 506-6784; gpfoundation@southernco.com
Internet: http://www.georgiapower.com/community/apply.asp
Sponsor: Georgia Power Foundation
241 Ralph McGill Boulevard NE, Bin 10131
Atlanta, GA 30308-3374

Geraldine R. Dodge Foundation Morris County Grants **2076**

To help foster a more livable Morristown area, the Foundation will support organizations that help their neighbors and build community by doing one or more of the following: serve those whose needs and potential are great, especially children and youth; foster meaningful connections among people of diverse ethnicities, cultures, ages, and socio-economic groups; build collaborations among government, the business sector and nonprofit organizations that generate entrepreneurial solutions to current challenges; and help create a sustainable Morristown for the future through efforts such as inclusive community visionary processes, the development and promotion of local assets, and/or the promotion of green design. The Foundation also invites proposals for mini-grants in the range of $3,000 to $5,000 that feature new or recent connections and collaborations. These can be stand-alone proposals from two or more groups, or part of a broader proposal from a single organization.
Requirements: The foundation awards grant to 501(c)3 nonprofits in Morristown New Jersey.
Restrictions: Grants are not made to individuals or for the support of religion, higher education, health, international programs, conduit organizations, capital projects, deficit financing, or endowment funds.
Geographic Focus: New Jersey
Date(s) Application is Due: May 1
Amount of Grant: Up to 150,000 U.S.D
Contact: David Grant; (973) 540-8442; fax (973) 540-1211; info@grdodge.org
Internet: http://www.grdodge.org/morriscounty/index.htm
Sponsor: Geraldine R. Dodge Foundation
14 Maple Avenue, P.O. Box 1239
Morristown, NJ 07962-1239

Gerber Foundation Grants **2077**

The foundation awards grants to support national programs that have a significant impact on issues facing infants and young children. Areas of interest include pediatric health—promoting health and preventing disease, including projects geared toward research or interventions that will reduce the incidence of serious chronic illnesses (e.g., diabetes, heart disease, obesity, or cancer) or improve cognitive, social, and emotional aspects of development; pediatric nutrition—assuring adequate nutrition for infants and young children through projects of research or interventions; and the effects of environmental hazards. The listed deadlines are for letters of inquiry. Guidelines and application are available online.
Requirements: 501(c)3 nonprofit organizations are eligible. Priority is given to projects that improve infant and young children nutrition, care, and development from the first year before birth to three years of age.
Restrictions: The foundation does not make grants or loans to individuals. Outside the West Michigan area, the foundation does not support capital campaigns, operating support, national child welfare programs, international based programs, or food/baby products giveaway programs.
Date(s) Application is Due: Jun 1; Dec 1
Amount of Grant: Up to 1,000,000 U.S.D
Samples: Children's Hospital of Michigan (Detroit, MI)—to evaluate a pain-assessment technique for use in infants, $59,798; Cincinnati Children's Hospital (OH)—to evaluate a potential treatment for necrotizing enterocolitis, $247,331; U of Minnesota-Twin Cities (Minneapolis, MN)—for a study evaluating oxygen therapy in neonates and oxidative stress, $223,288; Van Andel Research Institute (Grand Rapids, MI)—for a study involving protein profiling in congenital abnormalities of infants, $743,665.
Contact: Program Contact; (231) 924-3175; fax (231) 924-7906; cobits@ncresa.org
Internet: http://www.gerberfoundation.org
Sponsor: Gerber Foundation
4747 W 48th Street, Suite 153
Fremont, MI 49412-8119

Gertrude and William C. Wardlaw Fund Grants **2078**

The fund awards general operating grants to Georgia nonprofit organizations in its areas of interest, including cultural activities, community development, education and higher education, and health and hospitals.
Requirements: Georgia nonprofit organizations are eligible.
Geographic Focus: Georgia
Date(s) Application is Due: Feb 15
Amount of Grant: Up to 675,546 U.S.D
Contact: Mark Drake, Secretary
Sponsor: Gertrude and William C. Wardlaw Fund
P.O. Box 4655, MC 217
Atlanta, GA 30302

Gertrude B. Elion Mentored Medical Student Research Awards **2079**

An annual award granted to support women medical students interested in pursuing health-related research projects.
Requirements: Students must have the support of a faculty mentor and must conduct their research at one of the four medical schools in North Carolina. Applications are accepted from Duke University Medical Center, East Carolina University School of Medicine, UNC-Chapel Hill School of Medicine or Wake Forest University Baptist Medical Center only.
Restrictions: Candidates must be women, enrolled as full-time students, and will have completed at least one year of medical school prior to the start of the award. They must conduct their research at the applying institution. Candidates must be citizens or permanent residents of the United States or Canada at the time of application. Documentation of permanent residency status must be provided with the application. Persons who have applied for permanent residency but have not received their government documentation by the time of application are not eligible.
Geographic Focus: North Carolina
Date(s) Application is Due: Mar 5
Amount of Grant: 10,000 U.S.D
Contact: Libby Long, Program Coordinator; (919) 474-8370, ext. 134; fax (919) 941-9208; libby@trianglecf.org
Internet: http://www.trianglecf.org/page10000237.cfm
Sponsor: Triangle Community Foundation
324 Blackwell Street, Suite 1220
Durham, NC 27701

Gertrude E. Skelly Charitable Foundation Grants **2080**

The primary mission of the Foundation is to provide educational opportunities, primarily at colleges and universities, and needed medical care for those who cannot afford them. The advisory committee will meet in October to review applications. There is no formal application. Interested applicants will contact the Foundation to receive application instructions and guidelines.
Requirements: 501(c)3 tax-exempt organizations, including colleges and universities, are eligible.
Restrictions: Funding is not available for endowments or capital projects. Multi-year commitments are generally not favored. Grants will not be made to private foundations, non-charitable organizations or directly to individuals.
Date(s) Application is Due: Jul 31
Amount of Grant: Up to 50,000 U.S.D
Contact: Bettee M. Collister, Executive Director; (561) 276-1008; fax (561) 272-2793; skelly@erikjoh.com
Sponsor: Gertrude E. Skelly Charitable Foundation
4600 North Ocean Boulevard, Suite 206
Boynton Beach, FL 33435-7365

Getty Conservation Guest Scholar Grants **2081**

The program at the Getty Conservation Institute supports new ideas and perspectives in the field of conservation, with an emphasis on the visual arts—including sites, buildings, and objects—and the theoretical underpinnings of the field. The program provides an opportunity for professionals to pursue scholarly research in an interdisciplinary

manner across traditional boundaries in areas of wide general interest to the international conservation community. Guest scholars are in residence at the Getty Center for three to nine consecutive months. In addition to the stipend, the grant also includes a workstation at the Conservation Institute, research assistance, airfare to Los Angeles, an apartment in the Getty scholar housing complex, and health benefits.
Requirements: These grants are for established conservators, scientists, and professionals who have attained distinction in conservation and allied fields. Applications are welcome from researchers of all nationalities.
Restrictions: Grants are not intended to fund research for the completion of an academic degree. Nonresidential research and travel are not supported by this grant.
Date(s) Application is Due: Nov 1
Amount of Grant: 10,500 - 31,500 U.S.D
Contact: Program Coordinator; (310) 440-7374; fax (310) 440-7703; researchgrants@getty.edu
Internet: http://www.getty.edu/grants/research/scholars/conservation.html
Sponsor: J. Paul Getty Trust
1200 Getty Center Drive, Suite 800
Los Angeles, CA 90049-1685

Gheens Foundation Grants **2082**
The Foundation makes grants in Metro Louisville, Kentucky and LaFourche and Terrebonne Parishes in Louisiana. The Foundation supports a wide variety of endeavors at all levels, including education, economic development, medical, arts, social/health services, handicapped, and mental health programs. Foundation focus is on education and Christian works. There are no deadlines for applications. Guidelines and application information are available online.
Requirements: The Foundation contributes only to 501(c)3 organizations that are not private foundations.
Restrictions: The Foundation does not give to individuals.
Geographic Focus: Kentucky, Louisiana
Samples: Western Kentucky U (Bowling Green, KY)—to provide scholarships to minority teacher-education students who are committed to teaching in Jefferson County, KY, public schools, $80,000 over four years; Louisville Seminary (Louisville, KY)—to construct the seminary's new academic building, $625,000 (2000); U of Louisville (KY)—to create a center where scientists will work to develop new gene-based technologies for studying a wide variety of age-related diseases, $2.5 million over five years.
Contact: Carl Thomas, Executive Director; (502) 584-4650; fax (502) 584-4652; carl@gheensfoundation.org
Internet: http://gheensfoundation.org/grant_application
Sponsor: Gheens Foundation
705 One Riverfont Plaza, 401 West Main Street
Louisville, KY 40202

Giant Eagle Foundation Grants **2083**
The foundation awards grants in Pennsylvania in its areas of interest, including community development; health and human services; Jewish agencies, charitable organizations, and temples; performing arts; and philanthropy. Types of support include project grants and employee-related scholarship. There are no application deadlines. The board meets four times each year to review requests.
Requirements: Pennsylvania nonprofit organizations are eligible. Preference will be given to requests from Pittsburgh.
Geographic Focus: Ohio, Pennsylvania
Amount of Grant: 1,000 - 2,000 U.S.D
Samples: U of Pittsburgh (PA)—for educational and community-outreach programs of the Cancer Institute, the College of Arts and Sciences, the Graduate School of Business, the School of Pharmacy, and the School of Social Work, $1.4 million.
Contact: David Shapira, Coordinator; (412) 963-6200
Sponsor: Giant Eagle Foundation
101 Kappa Drive
Pittsburgh, PA 15238

Giant Food Charitable Grants **2084**
Giant Food Charitable Grants provide funding and in-kind assistance to hundreds of charitable events and causes in schools, churches, synagogues, and civic community groups. Giant's focus is hunger relief programs, education, and wellness initiatives. Grant amounts vary. There are no application deadlines. Submit requests by mail only.
Requirements: Organizations must send requests in writing on their letterhead, along with proof of their 501(c)3 status and a description of the project.
Restrictions: Funding is available only in areas where Giant Food operates: District of Columbia, Maryland, Virginia, and Delaware.
Geographic Focus: Delaware, District of Columbia, Maryland, Virginia
Contact: Jamie Miller; (301) 341-8776; jmiller@giantofmaryland.com
Internet: http://www.giantfood.com/about_us/community/index.htm
Sponsor: Giant Food Corporation
8301 Professional Place, Suite 115
Landover, MD 20785

Giant Food Corporation Scholarships **2085**
The corporation awards scholarships for pharmacy education leading to a bachelor of science or doctorate degree. Scholarships include the Izzy Cohen Memorial Scholarship and the Esther Peterson Memorial Education Scholarship. Information regarding these programs is posted online in February.
Amount of Grant: 1,000 - 4,000 U.S.D
Contact: Barry Scher, Vice President; (301) 341-4171; fax (301) 618-4972
Internet: http://www.giantfood.com/corporate/company_charitable.htm
Sponsor: Giant Food Corporation
8301 Professional Place, Suite 115
Landover, MD 20785

Gibson Foundation Grants **2086**
The Foundation is committed to making the world a better place for children by creating, developing and supporting programs and other non-profit organizations in their efforts to advance education, music and the arts, the environment and health & welfare causes. The Foundation actively seeks out programs that will be a direct mission-fit and further its goals.
Requirements: Applicants must be 501(c)3 organizations.
Restrictions: The Foundation does not support religious or political affiliations, and does not award individual scholarships.
Contact: Nina Miller, Executive Director; (615) 871-4500, ext. 2114; fax (615) 884-9597; nina.miller@gibson.com
Internet: http://www.gibson.com/en-us/Lifestyle/GibsonFoundation/
Sponsor: Gibson Foundation
309 Plus Park Boulevard
Nashville, TN 37217

Gill Foundation - Gay and Lesbian Fund Grants **2087**
In partnership with grantees, the Fund works to create a vibrant, healthy Colorado, where all families can thrive, while being a vehicle for cultural change toward equality for all. Programs of interest are: arts and culture; civic leadership; healthy families; and public broadcasting. Grantmaking is carried out in three ways: events sponsorship; invitation only; and opportunity grants. Requests may not exceed 15% of an organization's total budget, or for event sponsorships, may not exceed 50% of the event budget. Application information is available online.
Requirements: Eligible recipients must: be a 501(c)3 public charity organization; have a board-approved employment nondiscrimination policy in the organization's by-laws, employee manual or other official source stating explicitly that employees will not be discriminated against based on sexual orientation; provide services or programming within the state of Colorado.
Restrictions: Funding is not available for: organizations that are not a 501(c)3 public charity; lesbian, gay, bisexual, and transgender, or HIV/AIDS organizations; housing developments; emergency requests; endowments; individuals; major capital campaigns or projects; golf tournaments; film production or promotion.
Geographic Focus: Colorado
Samples: National Assoc of People With AIDS (Washington, DC)—for programs related to HIV/AIDS, $15,000; Human Rights Campaign Foundation (Washington, DC)—for programs supporting lesbian, gay, bisexual, and transgender people, $100,000; Gay and Lesbian Fund of Colorado (CO)—for leadership-development activities, $26,500 distributed among seven organizations; AIDS Resource Ctr of Wisconsin (Milwaukee, WI)—for HIV/AIDS prevention and human services programs, $60,000.
Contact: Grants Manager; (800) 964-5643 or (719) 473-4455; info@gayandlesbianfund.org
Internet: http://www.gillfoundation.org/glfc_grants/glfc_grants_show.htm?doc_id=408610&cat_id=1394
Sponsor: Gill Foundation
315 East Costilla Street
Colorado Springs, CO 80903

Gilroy and Lillian P. Roberts Charitable Foundation Grants **2088**
The foundation awards grants to eligible nonprofit organizations in its areas of interest, including arts and fine arts, health care and health organizations, higher education, Jewish temples and organizations, and social services. Types of support include annual campaigns, capital campaigns, continuing support, fellowships, general operating grants, professorships, program development, and scholarships. Most grants are awarded in Montgomery and Delaware Counties, PA. There are no application deadlines or forms.
Restrictions: Individuals are not eligible.
Geographic Focus: Pennsylvania
Amount of Grant: Up to 408,750 U.S.D
Contact: Stanley Merves, Treasurer; (610) 668-1998
Sponsor: Gilroy and Lillian P. Roberts Charitable Foundation
101 W Elm Street, Suite 500
Conshohocken, PA 19428

Gina Finzi Memorial Student Summer Fellowship **2089**
Fellowships are awarded to encourage clinical, basic or psychosocial research related to the causes, treatments, prevention, or cure of lupus. Awards are made in May to the top 10 applicants for support during the summer months. Application materials are available in December of each year. There are no renewals; applicant must submit a new proposal.
Requirements: Junior investigators, defined as assistant professor and below rank if in academic medicine, are eligible to apply.
Date(s) Application is Due: Mar 15
Amount of Grant: 2,000 U.S.D
Contact: Grants Coordinator; (202) 349-1155; fax (202) 349-1156; Finzifellowship@lupus.org
Internet: http://www.lupus.org/research/studentannounc.html
Sponsor: Lupus Foundation of America
2000 L Street NW, Suite 710
Washington, DC 20036-4916

Ginn Foundation Grants 2090

The foundation's mission is to address educational and community-based health care needs through supporting effective programs and services that bring about long-term solutions for individuals and the community, principally in Cuyahoga County, OH. The Foundation will consider not only grants to academic institutions, but also to organizations that meet non-academic educational needs, such as, programs that address issues of disease avoidance, child and family counseling, after-school training, arts, housing, and employment. Preference in all of these areas will be given to organizations and programs that serve low-income recipients. Faxed applications will not be considered. Application information is available online.

Requirements: Nonprofit organizations in Cuyahoga County, OH, may apply. Consideration will also be given to trustee-sponsored grants to similar types of organizations in the Chicago, Washington, DC, and Minneapolis-St. Paul metropolitan areas.

Restrictions: The foundation will not fund requests for support of advocacy activities. Nor will it make grants to endowment, capital, or annual fund campaigns. The foundation will not fund special events or attendance at conferences or symposia.

Geographic Focus: District of Columbia, Illinois, Minnesota, Ohio

Date(s) Application is Due: Mar 15; Sep 15

Amount of Grant: 5,000 - 30,000 U.S.D

Samples: Bellflower Center for the Prevention of Abuse, Cleveland, OH, $10,000—general operating support for support services directly addressing healthy child development through the prevention & treatment of child abuse; Cleveland Playhouse, Cleveland, OH, $15,000—collaboration with Cleveland School of the Arts; Cleveland Hearing & Speech, Cleveland, OH, $15,000—Speech-Language Pathology Program.

Contact: Walter Pope Ginn, Trustee; info@ginnfoundation.org

Internet: http://www.ginnfoundation.org/index.html

Sponsor: Ginn Foundation
13938 A Cedar Road, P.O. Box 239
Cleveland Heights, OH 44118

Giving in Action Society Children & Youth with Special Needs Grants 2091

Giving in Action Society is a charitable organization established by Vancouver Foundation in 2006. The Society provides grants directly to individuals and families in B.C. The Children and Youth with Special Needs Fund provides grants to families who have children or youth (newborn to 19 years) with special needs living at home. The Fund offers one-time capital grants to help enhance or improve the individual's health, development or ability to participate in daily activities at home, in school and in the community. Eligible expenses may include such things as home renovations and vehicle modifications. This Fund defines children and youth with special needs as those who have significant impairments in one or more of the following areas: health, cognition, communication, sensory motor, social/emotional/behavioural or self help. The Children and Youth with Special Needs Fund helps address family needs that are not currently met through government-funded programs. If a family receives support from other government-funded programs, they can apply to the Fund for grants but only if the funds will complement â€" versus duplicate â€" the government support. The Fund was established with financial assistance from the B.C. Ministry of Children and Family Development.

Requirements: Only residents of B.C. are eligible to apply. The Children and Youth with Special Needs Fund provides grants to families who have children or youth (newborn to 19 years) with special needs living at home. There is a two-stage application process. The first stage, submit a Letter of Inquiry, consisting of an informal proposal to determine basic suitability and eligibility. The second stage is a full application. If your Letter of Inquiry fits within the funding guidelines you will be sent a grant application form with specific instructions for completion. You have 6 months to submit your application.

Geographic Focus: All States, Canada

Contact: Faye Wightman, President; (866) 523-3157 or (604) 683-3157; fax (604) 683-3134; info@vancouverfoundation.ca

Internet: http://www.givinginaction.ca/ourprograms/specialneeds.htm

Sponsor: Giving in Action Society, Vancouver Foundation
Harbour Centre, 555 W Hastings Street, Suite 1200, Box 12132
Vancouver, BC V6B 4N6 Canada

Giving in Action Society Family Independence Grants 2092

Giving in Action Society is a charitable organization established by Vancouver Foundation in 2006. The Society provides grants directly to individuals and families in B.C. The Family Independence Fund helps families throughout the province who have children or adults with developmental disabilities living at home. Grants from the Family Independence Fund help with the ongoing care of the relative by providing support for projects such as home renovations - including lifts, elevators, ramps, flooring, door widening or vehicle modifications - that enable the individual with the developmental disability to live in the family home and access their community. Preference is given to families that indicate their request for support is part of a long-term plan for their relative. Note that families who receive support from other government-funded programs are still eligible for Family Independence Fund grants. However these additional funds must complement - versus duplicate - the government support.

Requirements: Only residents of B.C. are eligible to apply. The Family Independence Fund supports a child/relative who has a developmental disability, defined as: concurrent with impaired adaptive functioning; low IQ; manifests before the age of 18. There is a two-stage application process. The first stage, submit a Letter of Inquiry, consisting of an informal proposal to determine basic suitability and eligibility. The second stage is a full application. If your Letter of Inquiry fits within the funding guidelines you will be sent a grant application form with specific instructions for completion. You have 6 months to submit your application.

Geographic Focus: All States, Canada

Contact: Faye Wightman, President; (866) 523-3157 or (604) 683-3157; fax (604) 683-3134; info@vancouverfoundation.ca

Internet: http://www.givinginaction.ca/ourprograms/familyindependencefund.htm

Sponsor: Giving in Action Society, Vancouver Foundation
Harbour Centre, 555 W Hastings Street, Suite 1200, Box 12132
Vancouver, BC V6B 4N6 Canada

Giving Sum Annual Grant 2093

Giving Sum is an organization of next generation givers working together to identify challenges in the community and to address them through education, engagement, inspiration and grants. Giving Sum is a charitable giving vehicle pooling resources raised through memberships, sponsorships, and fundraisers to establish annual grants for recipients selected by the organization and its members. Each year Giving Sum will grant 100 percent of its membership contributions to a nonprofit organization in Indianapolis working towards change. Members will be proactive in searching out these organizations in the greater Indianapolis area and having them apply.

Requirements: To be eligible, organizations must be recognized by the Internal Revenue Service as a public charity under sections 501(c)3 and 509(a)(1), (2), or (3) of the Internal Revenue Code, and provide a copy of their determination letter from the Internal Revenue Service. Organizations or the proposed project must serve audiences within the greater Indianapolis area, including the counties of: Boone, Hamilton, Hancock, Hendricks, Johnson, Marion, Morgan and Shelby. Eligible nonprofit organizations may submit only one application to request a grant of $50,000 (Money), volunteer hours (Time) and advocacy efforts (Voice) from Giving Sum to support general operations or special projects in one of Giving Sum's five Focus Areas (Arts, Culture & Humanities, Civic & Community Development, Education, Environment, and Health & Human Services). Each organization must select one of the above focus areas in which to apply. Applicants should select the focus area which best describes the primary operations or mission of their organization. If your organization and/or request is multi-disciplinary, select the category which best describes the activities that will be carried out using the Giving Sum grant. Preliminary review of the applications occurs within each of the five focus areas. Preference will be given to applications that meaningfully and creatively engage Giving Sum members in innovative, high-impact initiatives.

Restrictions: Grant funds may not be used to lobby or otherwise attempt to influence legislation, to influence the outcome of any public election, or to carry on any voter registration drive.

Geographic Focus: Indiana

Amount of Grant: 50,000 U.S.D

Samples: 2009 - Starfish Initiative: $50,000 grant for a leadership camp for up to 300 academically promising, economically disadvantaged high school students, attending more than 30 different high schools in greater Indianapolis. 2008 - Coburn Place: $50,000 toward a playground for the children of Coburn Place.

Contact: Ryan Brady, Program Director; (317) 634-2423; grants@givingsum.org

Internet: http://givingsum.org/grants.asp

Sponsor: Giving Sum
615 N Alabama Street, Suite 119
Indianapolis, IN 46204-1498

Gladys Brooks Foundation Grants 2094

At the beginning of each year, the Foundation Board determines a limited scope of activities for which it will consider grant applications. Current giving will be in the fields of libraries, education, hospitals, and clinics. Applications will be considered where: outside funding (including governmental) is not available; the project will be largely funded by the grant unless the grant request covers a discrete component of a larger project; and the funds will be used for capital projects including equipment or endowments. The Foundation's application form must be used and is available online. Electronic submissions are not acceptable.

Requirements: The foundation makes grants to nonprofit, private, publicly supported, tax-exempt organizations located in Connecticut, Delaware, Florida, Illinois, Indiana, Louisiana, Maine, Maryland, Massachusetts, New Hampshire, New Jersey, New York, Ohio, Pennsylvania, Rhode Island, Tennessee, and Vermont.

Restrictions: Applications for direct salary support will not be accepted. No portion of grants shall be appropriated as an administrative or processing fee, for overseeing the project or for its general overhead.

Geographic Focus: Connecticut, Delaware, Florida, Illinois, Indiana, Louisiana, Maine, Maryland, Massachusetts, New Hampshire, New Jersey, New York, Ohio, Pennsylvania, Rhode Island, Tennessee, Vermont

Date(s) Application is Due: Jun 1

Amount of Grant: 50,000 - 100,000 U.S.D

Samples: Hobart and William Smith College (Geneva, NY)—for endowed arts scholarships, $100,000.

Contact: Jessica Rutledge, (516) 746-6103

Internet: http://www.gladysbrooksfoundation.org

Sponsor: Gladys Brooks Foundation
1055 Franklin Avenue, Suite 208
Garden City, NY 11530

Glaucoma Foundation Grants 2095

The foundation funds research initiatives to determine the causes of glaucoma, to improve methods of treatment, and to develop cures for the various kinds of glaucoma. Two areas of particular focus are: optic nerve rescue and restoration—research into new approaches designed to protect the optic nerve against glaucomatous damage, to restore vision lost to glaucoma and eventually reverse blindness by restoring or regenerating the function of

the optic nerve cells, and to explore the feasibility of achieving transplantation of optic nerve cells; and molecular genetics—research into the genetic causes of the various forms of glaucoma, particularly the identification of the responsible genes, with the long-term goal of finding ways to reverse these genetic defects. Grants are awarded for a one-year period and are renewable. Guidelines and application are available online.
Requirements: Applicants must have a full time faculty position or the equivalent.
Restrictions: The foundation does not provide funds for investigator salaries, travel, overhead, or other indirect costs.
Date(s) Application is Due: Mar 1; Sep 1
Amount of Grant: 40,000 - 50,000 U.S.D
Contact: Kira A. Zmuda, Director; (212) 651-2509; kzmuda@glaucomafoundation.org
Internet: http://www.glaucomafoundation.org/Grant_Application.htm
Sponsor: Glaucoma Foundation Grants
80 Maiden Lane, Suite 700
New York, NY 10038

Glaucoma Research Pilot Project Grants **2096**
These grants provide funds to encourage innovative and pilot research, as well as to aid ongoing studies that seek to protect and restore the optic nerve, accurately monitor glaucoma's progression, find the genes responsible for the glaucoma, understand the intraocular pressure system and develop better treatments, and determine the risk factors for glaucoma damage.
Requirements: Applicants must have a graduate degree.
Restrictions: Funding is not granted for equipment purchases, overhead commercial applications, or indirect costs.
Date(s) Application is Due: Aug 1
Amount of Grant: 40,000 U.S.D
Contact: Jennifer Rulon, (800) 826-6693 or (415) 986-3162; fax (415) 986-3763; research@glaucoma.org
Internet: http://www.glaucoma.org/research/index.php
Sponsor: Glaucoma Research Foundation
251 Post Street, Suite 600
San Francisco, CA 94108

GlaxoSmithKline Corporate Grants **2097**
In the United States, GlaxoSmithKline Corporate makes charitable contributions through its U.S. Contributions Committee. Requests for U.S.-based community partnerships with nonprofit organizations should address issues in one of four general areas: education—science education, K-12 literacy, teacher professional development; health and human services—child health or prevention and access to health care for women related to breast or gynecologic cancers, targeting the needs of underserved and diverse populations; arts and culture—local organizations in the Greater Philadelphia and the Research Triangle Park areas, based on local needs, focusing on public school educational outreach; and civic and community—local organizations in the Greater Philadelphia and the Research Triangle Park areas, based on local needs.
Requirements: Organizations based in the United States that have a 501(c)3 IRS designation and that meet GSK's corporate criteria for funding may complete an application. Funding for the arts and civic programs is concentrated exclusively in the Greater Philadelphia and Research Triangle Park areas where employees live and work.
Restrictions: As a matter of policy, grants are not provided for general operating expenses or capital building costs, and they are not made to individuals. Grants are not given to political, religious, fraternal, profit-making, discriminatory, hobby-oriented, or tax-subsidized organizations.
Geographic Focus: Pennsylvania
Amount of Grant: Up to 100,000,000 U.S.D
Contact: Mary Anne Rhyne, Corporate Press; (919) 483-2319; fax (919) 483-8765; community.partnership@gsk.com
Internet: http://us.gsk.com/html/community/community-grants-corporate.html
Sponsor: GlaxoSmithKline Corporation
1 Franklin Plaza, FP2130, P.O. Box 7929
Philadelphia, PA 19101-7929

GlaxoSmithKline Foundation IMPACT Awards **2098**
The awards reward excellence in the delivery of nonprofit community health care. Awards are made to small and mid-size community-based health care organizations in the greater Philadelphia area and, the Research Triangle Park, North Carolina area. Selection criteria include: commitment to serving people in need; facilitating access to health care delivery, education, creative partnerships, and policy development; and demonstration of a solid record of achievement, management, and leadership. The application process for the GlaxoSmithKline 1st Annual IMPACT Awards in the Research Triangle Park, North Carolina region closed on May 15. July 15 is the deadline date for the greater Philadelphia region.
Requirements: Nominees in the following Philadelphia counties are eligible: Berks, Bucks, Chester, Delaware, Lancaster, Montgomery, or Philadelphia or in the City of Camden, NJ. Nominees in the following Research Triangle Park, North Carolina counties are: Chatham, Durham, Orange and Wake Counties. Organizations must have an operating or program budget of under $2 million and have been in existence at least five years. Nonprofit healthcare organizations must be nominated by individuals knowledgeable of the complexities and challenges of healthcare delivery in their communities.
Geographic Focus: Pennsylvania
Date(s) Application is Due: May 15; Jul 15
Amount of Grant: 40,000 U.S.D
Contact: Mary Anne Rhyne, Corporate Press; (919) 483-2319; fax (919) 483-8765; community.partnership@gsk.com
Internet: http://www.gsk-us.com/html/community/community-healthcare-awards.html
Sponsor: GlaxoSmithKline Foundation
1 Franklin Plaza, FP2130, P.O. Box 7929
Philadelphia, PA 19101-7929

Glazer Family Foundation Grants **2099**
The foundation awards grants to eligible Florida nonprofit organizations for programs focused on youth and families in the areas of general health, safety, education, and recreation. Types of support include grants, contributions, tickets, and special programs. Proposals are accepted May 1 through July 1 and September 1 through November 1. Annual deadline dates may vary; contact program staff for exact dates.
Requirements: Florida nonprofit organizations serving the greater Tampa Bay and West Central areas are eligible.
Restrictions: Grants do not support general fundraising drives, celebrations, administrative/training costs, capital expenditures, sponsorships, individual medical needs, scholarships, research, political activities, conferences, and university capital campaigns.
Geographic Focus: Florida
Date(s) Application is Due: Jul 1; Nov 1
Amount of Grant: Up to 5,000 U.S.D
Contact: Grants Administrator; (813) 870-2700, ext. 275; fax (813) 554-1324
Internet: http://www.glazerfamilyfoundation.com/GuidelinesforGrants.aspx
Sponsor: Glazer Family Foundation
One Buccaneer Pl
Tampa, FL 33607

Glenn/AFAR Breakthroughs in Gerontology Awards **2100**
The goal of the program is to provide timely support to a small number of pilot research programs that may be of high risk but offer significant promise of yielding transforming discoveries in the fundamental biology of aging. The hope is that one or more of the funded research projects will lead to major new insights into the molecular factors that coordinate aging in multiple cells and tissues, and the ways in which the aging process is differentially timed in long-lived species. Projects that focus on genetic controls of aging and longevity, on delay of aging by pharmacological agents or dietary means, or that elucidate the mechanisms by which alterations in hormones, anti-oxidant defenses, or repair processes promote longevity are all well within the intended scope of this competition. Projects that focus instead on specific diseases or on assessment of health care strategies will receive much lower priority, unless the research plan makes clear and direct connections to fundamental issues in the biology of aging. Studies of invertebrates, mice, human clinical materials, or cell lines are all potentially eligible for funding. Although preliminary data are always helpful for evaluating the feasibility of the experiments proposed, the emphasis in review will be on creativity and the likelihood that the findings will open new vistas and approaches to aging research that might merit intensive follow-up studies. Applications from individuals not previously engaged in aging research are particularly encouraged, as long as the research proposals show high promise for leading to important new discoveries in biological gerontology. The proposed research must be conducted at any type of nonprofit setting in the United States. Guidelines are available online.
Requirements: Applicants must at the time they submit their proposal be full-time faculty members at the rank of assistant professor or higher. A strong record of independent publication beyond the postdoctoral level is a requirement.
Restrictions: Applicants who are employees in the NIH Intramural program are not eligible.
Date(s) Application is Due: Dec 16
Amount of Grant: 200,000 U.S.D
Contact: Grants Manager; (212) 703-9977 or (888) 582-2327; fax (212) 997-0330; grants@afar.org or info@afar.org
Internet: http://afar.org/grants.html
Sponsor: American Federation for Aging Research
55 West 39th Street, 16th Floor
New York, NY 10018

Glenn/AFAR Research Grants for Postdoctoral Fellows **2101**
AFAR funds research projects concerned with understanding the basic mechanisms of aging. Projects investigating age-related diseases are also supported, especially if approached from the point of view of how basic aging processes may lead to these outcomes. Projects concerning mechanisms underlying common geriatric functional disorders are also encouraged. Examples of promising areas of research include aging and immune function; genetic control of longevity; neurobiology and neuropathology of aging; invertebrate or vertebrate animal models; cardiovascular aging; aging and cellular stress resistance; metabolic and endocrine changes; age-related changes in cell proliferation; caloric restriction and aging; DNA repair and control of gene expression; biology of the menopause; aging and apoptosis; and biodemographic analysis of aging. Funding begins July 1. Guidelines and application are available online.
Requirements: Postdoctoral fellows (MDs and PhDs) who will have had at least two and not more than five years of prior postdoctoral training at the time of the award are eligible.
Restrictions: Funds may not be requested for overhead or indirect costs. Fellows may not hold any concurrent foundation or not-for-profit funding.
Date(s) Application is Due: Dec 15
Amount of Grant: 60,000 U.S.D
Contact: Grants Manager; (212) 703-9977; fax (212) 997-0330; amfedaging@aol.com
Internet: http://www.afar.org/grants.html
Sponsor: American Federation for Aging Research
55 West 39th Street, 16th Floor
New York, NY 10018

Global Fund for Children Grants 2102

The Global Fund for Children's (GFC) mission is to advance the dignity of children and youth around the world through small grants to innovative community-based organizations working with some of the world's most vulnerable children and youth. The Fund selects its grantee partners based on their potential to grow in effectiveness and to become valuable resources or models for others. Selection criteria include exceptional leadership, sound management, strong community participation, innovative and effective programs, and direct engagement with the most vulnerable children. Fund focus is on four specific issues: learning, enterprise, safety, and healthy minds and bodies. Organizations may submit a letter of inquiry at any time. If a letter of inquiry falls within GFC's priorities, GFC will follow up with the organization to solicit a full proposal.
Requirements: Nonprofit organization worldwide are eligible.
Amount of Grant: 5,000 - 20,000 U.S.D
Samples: Anandan, Kolkata, India, $8,000—education grant; Benishyaka Association, Kigali, Rwanda, $15,000—academic scholarships; Children in the Wilderness, Lilongwe, Malawi, $15,000—education grant.
Contact: Andrew Barnes, Grants Manager; (202) 331-9003; fax (202) 331-9004; info@globalfundforchildren.org
Internet: http://www.globalfundforchildren.org/applyforagrant/index.html
Sponsor: Global Fund for Children
1101 Fourteenth Street, North West, Suite 420
Washington, DC 20005

Global Fund for Women Grants 2103

The Global Fund for Women supports women's groups that advance the human rights of women and girls. Grants support women's groups based outside the United States by providing small, flexible, and timely grants for operating and project expenses. Grantees address issues that include but are not limited to building peace and ending gender-based violence; advancing health and sexual and reproductive rights; expanding civic and political participation; ensuring economic and environmental justice; increasing access to education; and fostering social change philanthropy. In addition urgent requests for support to organize or participate in local, regional, or international meetings and conferences will be considered outside of the normal grant cycle. These types of requests must come from organizations, not individuals, and must be received at least eight weeks before the event. Funds for these types of grants are limited. Applications and guidelines are available online. Organizations should refer to the staff website for specific contacts according to their applicant origin.
Requirements: The applicant group must be based in a country outside the United States; demonstrate a strong commitment to women's equality and human rights that is clearly reflected in its activities; be a group of women working together; and be governed, directed, and led by women.
Restrictions: Grants do not support individuals; scholarships; academic research; groups based and working primarily or only in the United States; international organizations proposing projects with local partners; groups without a strong women's rights focus; groups headed and managed by men, or without women in the majority of leadership positions; groups whose sole purpose is to generate income or to provide charity to individuals; or political parties or election campaigns.
Geographic Focus: All Countries
Amount of Grant: 5,000 - 50,000 U.S.D
Contact: Program Assistant; (415) 248-4800; fax (415) 202-4801
Internet: http://www.globalfundforwomen.org/apply-for-a-grant/types-of-grant
Sponsor: Global Fund for Women
222 Sutter Street, Suite 500
San Francisco, CA 94108

Global Fund to Fight AIDS, Tuberculosis, and Malaria Grants 2104

The fund provides grants to locally developed programs to prevent and treat AIDS, tuberculosis, and malaria. Countries and organizations may apply for funding by submitting proposals in ongoing funding rounds. Proposals should support the scale up of effective existing programs and innovative projects that meet the fund's criteria. Proposals submitted should also clearly demonstrate how the resources sought from the fund will achieve additional results in partnership with existing programs. Application and guidelines are available online.
Requirements: Proposals for funding must generally be made through a Country Coordinating Mechanism. These country-level partnerships develop and submit grant proposals to the fund based on priority needs at the national level.
Date(s) Application is Due: Jun 10
Contact: Geneva Secretariat; 41 22 791 17 00; fax 41 22 791 17 01; proposals@theglobalfund.org
Internet: http://www.theglobalfund.org/en/apply/call5
Sponsor: Global Fund to Fight AIDS, Tuberculosis and Malaria
53, Avenue Louis-Casai
Geneva-Cointrin, CH-1216 Switzerland

Goddess Scholars Grants 2105

The purpose of these grants is to develop young researchers at the entry faculty level (Instructor or Assistant Professor) who will focus a research career on studying the unique aspects of stroke in women. Such research may be basic or clinical in orientation, including areas such as mechanisms of ischemic brain injury, neuroprotective treatments, primary or secondary stroke prevention, epidemiological studies, educational programs for patients or health care providers, rehabilitation, and outcome research. The grant provides funding for two years, with potential non-competitive renewal for a third year should satisfactory progress be demonstrated. These grants are intended to provide bridging funds for individuals who are finishing or have completed residency or postdoctoral training and require additional support before their subsequent establishment as independent investigators.
Requirements: Goddess Scholar grants are targeted at individuals who: hold an M.D., Ph.D., or equivalent degree; are at the fellow, instructor, or assistant professor rank; and are within 5 years of completion of residency or post-doctoral fellowship at the time of grant activation. Individuals who are currently enrolled in a fellowship program may apply if there is assurance of a faculty position at the time of the grant's activation.
Amount of Grant: 65,000 U.S.D
Contact: Erin Tower, Program/Operations Assistant; (212) 713-6789; fax (212) 288-2160; erin@thegoddessfund.org
Internet: http://www.thegoddessfund.org/grants.html
Sponsor: Hazel K. Goddess Fund for Stroke Research in Women
785 Park Avenue
New York, NY 10021-3552

Godfrey Foundation Grants 2106

The Foundation, established in Wisconsin in 1945 and supported by a donation from the Fleming Companies, Inc., offers grants primarily in the Milwaukee and Waukesha, Wisconsin, region. Its major field of interest include: education; health care; human services; community development and services; and youth development programs. The major type of support comes in the form of general operations funding. An application form is required, and interested organizations should contact the Foundation for a copy. There are no particular deadlines with which to adhere, and grant amounts have ranged from $500 to $30,000.
Requirements: Applicants must be 501(c)3 organizations serving the residents of Wisconsin.
Geographic Focus: Wisconsin
Amount of Grant: 500 - 30,000 U.S.D
Contact: Terry L. Daniels, Director; (262) 275-0458
Sponsor: Godfrey Foundation
680 Kenosha Street, P.O. Box 810
Walworth, WI 53184-0810

Golden State Warriors Foundation Grants 2107

The foundation committed to positively impacting the communities of Oakland and the greater San Francisco Bay Area by providing financial assistance and opportunities to other non-profit civic and community organizations that benefit and enrich the lives of children, youth and those in need. By providing financial support and unique resources, the Foundation endeavors to meet the social, educational and cultural needs of the community. Although the Warriors Foundation will not be accepting grant applications until the beginning of the 2007-08 fiscal year, it remains committed to providing assistance and in-kind support to Bay Area nonprofit organizations through the donation of tickets, autographed memorabilia, promotional items, sponsorships, and player and talent appearances. Application information and forms are available on the Web site.
Requirements: 501(c)3 nonprofit organizations that provide programs and services in the greater San Francisco Bay area are eligible.
Restrictions: Grants do not support individuals; schools; political, labor, religious, or fraternal activities; endowments; fundraising events; government agencies; or organizations with discriminatory or unlawful practices.
Geographic Focus: California
Amount of Grant: Up to 375,000 U.S.D
Contact: Angela Cohan, President; (510) 986-5307; community_web@gs-warriors.com
Internet: http://www.nba.com/warriors/community/Warriors_Foundation-011024.html
Sponsor: Golden State Warriors Foundation
1011 Broadway
Oakland, CA 94607-4019

Goldhirsh Foundation Brain Tumor Research Grants 2108

The Goldhirsh Foundation is interested in providing strategic investment in both pediatric and adult brain tumor research to accelerate progress toward more effective treatment for malignant diffuse glioma tumors. Applications are encouraged from investigators working in the continuum between basic research and clinical application, integrating and translating knowledge in various disciplines into meaningful progress for patients. The Initial Proposal Application must be submitted by Thursday, January 7, 12:00 Noon, U.S. and Canada Eastern Time (GMT-5), to the online application system. Applicants who are invited to submit Full Proposals will be notified by email in early March. Up to three investigators will receive grants of $400,000 U.S.D (inclusive of 10% indirect costs) over two years at $200,000 U.S.D per year. The Foundation also makes awards of $100,000 U.S.D (inclusive of 10% indirect costs) for one-year pilot studies. Please note that The Goldhirsh Foundation is not offering the three-year $600,000 U.S.D award for the 2010 Grant Cycle.
Requirements: The Goldhirsh Foundation is accepting Initial Proposal Applications from applicants who meet the following eligibility *Requirements:* have an M.D. and/or Ph.D. degree(s) or equivalent degree; hold a faculty appointment at a nonprofit academic, medical or research institution in the United States, Canada or Israel. The applicant may collaborate with investigators from other institutions and these institutions may include for-profit companies; preference will be given to originality of ideas, regardless of applicant seniority; eligibility is not limited to those investigators currently working in brain tumor research. Investigators from other fields are encouraged to apply with proposals directly relevant to malignant diffuse glioma tumors; projects must be relevant to tumors within the category of malignant diffuse gliomas, i.e., glioblastomas, diffuse astrocytomas (WHO grade III), oligoastrocytomas (WHO grade III), and oligodendrogliomas (WHO grade III).

Restrictions: Projects focusing on WHO grade I-II gliomas, non-glial neuroepithelial tumors, meningeal tumors, nerve sheath tumors, CNS lymphomas, germ-cell tumors and tumors of the sella region are beyond the scope of the grant program. Projects that characterize genes and gene products of normal cellular development are not eligible, nor are epidemiological studies.
Date(s) Application is Due: Jan 7
Amount of Grant: Up to 600,000 U.S.D
Samples: Lara Collier, Ph.D., University of Wisconsin-Madison, $600,000— Genetic Approaches to Identify and Characterize Potential Drug Targets for High-grade Gliomas; Al Charest, Ph.D., Tufts Medical Center, $100,000—Targeting Cancer Stem Cells in a Pre-Clinical Mouse Model of Glioblastoma Multiforme Using Therapeutic Nanoparticles.
Contact: Sally E. McNagny, Director; (617) 279-2254; fax (617) 423-4619; smcnagny@goldhirshfoundation.org
Internet: http://www.goldhirshfoundation.org/application_information.htm
Sponsor: Goldhirsh Foundation
95 Berkeley Street, Suite 201
Boston, MA 02116

Goldman Philanthropic Partnerships Program **2109**
Since its founding, the Program has solicited, validated and helped donors co-fund innovative research with the potential to quickly find cures for life-altering diseases, and supported these donors using for-profit business tools to insure that the research and philanthropic goals are met through proper project and resource management. The Program provides funding research for cures for a range of catastrophic diseases by exploring groundbreaking research in all disciplines of medicine, including those areas of research outside of conventional medicine such as alternative medicine. The Program usually co-funds a venture with at least one organizational, institutional, or corporate partner. The Partnerships maintains a strong interest in diseases of children and young adults, although it will fund research into any patient population where it believes there is a high likelihood of return on disease prevention, treatment or cure.
Contact: Dr. Bruce E. Bloom, President; (312) 601-8856 or (312) 780-3440; fax (312) 780-3459; bruce@goldmanpartnerships.org
Internet: http://www.goldmanpartnerships.org/aboutus.html
Sponsor: Goldman Philanthropic Partnerships
70 West Madison Street, Suite 1500
Chicago, IL 60602

Goodrich Corporation Foundation Grants **2110**
The Foundation makes charitable grants in four categories: education; arts and culture; civic and community; and health and human services. Preference shall be given to requests for projects or programs in areas having a significant number of employees, employees serving on boards of charitable organizations or other noticeable corporate presence. The Foundation staff accepts and reviews grant requests quarterly. To request funding, applicants will need to complete the application form which is available online. Telephone and email requests or inquiries are not accepted.
Requirements: 501(c)3 tax-exempt organizations are eligible.
Restrictions: The foundation generally will not support: multiyear grants in excess of five years; individuals, private foundations, endowments, churches or religious programs, fraternal/social/ labor/veterans organizations; groups with unusually high fundraising or administrative expenses; political parties, candidates, or lobbying activities; travel funds for tours, exhibitions, or trips by individuals or special interest groups; organizations that discriminate because of race, color, religion, national origin, or areas covered by applicable federal, state, or local laws; local athletic/sports programs or equipment, courtesy advertising benefits, raffle tickets and other fundraising events; organizations that receive sizable portions of their support through municipal, county, state, or federal dollars; individual United Way agencies that already benefit from Goodrich contributions to the United Way; or international organizations.
Date(s) Application is Due: Feb 1; Aug 1
Contact: Foundation Contact, Foundation Contact; (704) 423-7011
Internet: http://www.goodrich.com/CDA/GeneralContent/0,1136,59,00.html
Sponsor: Goodrich Foundation
Four Coliseum Center, 2730 West Tyvola Road
Charlotte, NC 28217-4578

Good Samaritan Foundation Inc Grants **2111**
The foundation supports health care projects aimed at the low-income population in Kentucky. It generally supports primary and preventive care and professional training for medical personnel who will work in an underserved area, and limits its grants to charitable and educational activities related to health care, health education, and research in Kentucky. Preference is given to activities serving Lexington, and central and eastern Kentucky. Grants should be used to make significant contributions to a community or area by improving access, delivery, affordability, and the quality of health care in Kentucky. Guidelines are available on the Web site.
Requirements: An applicant must be 501(c)3 tax-exempt organization or an appropriate public charity in Kentucky.
Restrictions: The foundation does not support staff funding within an organization; individuals; endowment funds; indirect costs; major organizations (e.g. heart funds, cancer societies, etc.); hospitals, colleges, or universities; activities that serve a small number of people or have a high per person cost; activities that compete with or duplicate a publicly supported program such as Medicare, Medicaid, Health Department, or other; building/capital improvement projects; custodial care, adult or child day care, retirement facilities, or nursing homes; benefit fund-raisers, testimonials, recognition dinners, or events; program advertising, tickets, or raffles.
Geographic Focus: Kentucky
Date(s) Application is Due: Jan 10
Contact: Joyce Triantafilou, Administrative Assistant; (859) 255-1691; fax (859) 254-7337; joyce@gsfky.org
Internet: http://www.gsfky.org/grants.htm
Sponsor: Good Samaritan Foundation
270 S Limestone Street
Lexington, KY 40508-2566

Good Samaritan Inc Grants **2112**
Good Samaritan, Inc. is an independent foundation, which was incorporated in 1938, in Delaware. This foundation gives primarily in the northeastern region of the United States: Pennsylvania, New Jersey, Delaware and Maryland. The foundation field of interest include: higher education; cancer research; secondary school/education; human services; hospitals (general); and the economically disadvantaged. Support is offered in the form of: endowments; professorships; program development; seed money: and general/operating support grants.
Requirements: 501(c)3 nonprofit organizations are eligible.
Restrictions: No grants to individuals, or for building funds, capital assets or conferences.
Date(s) Application is Due: Apr 1; Oct 1
Amount of Grant: 10,000 - 200,000 U.S.D
Contact: Edmund Carpenter II, President; (302) 654-7558; fax (302) 654-2376
Sponsor: Good Samaritan
600 Center Mill Road
Wilmington, DE 19807-1502

Goodyear Tire Grants **2113**
Requests for charitable gifts are required to meet guidelines of strategic giving based on safety by focusing on safety programs plus at least one additional category (civic and community; culture and the arts, education, or health and human services) of giving and/or provide volunteer opportunities for company associates which are appropriately aligned with the company's strategy.
Requirements: 501(c)3 tax-exempt organizations in Goodyear communities are eligible. Organizations in the greater Akron, OH area should send requests to: Faith Stewart, Director, Goodyear Tire and Rubber Company, 1144 East Market Street, D/798, Akron, OH, 44316. Organizations in U.S. communities where Goodyear has plants and major facilities should contact the local plant manager's office for mailing instructions.
Restrictions: Grants cannot be made to: organizations outside of communities in which plants or principal offices are located; individuals or to organizations on behalf of an individual; national, political, labor, fraternal, social, veterans organizations; individual schools within public or private school systems; athletic programs/extracurricular activities, travel or exchange programs; endowments or for debt reduction; religious organizations or endeavors; major hospitals or medical center; second parties except specialized organizations.
Contact: Faith Stewart, Director
Internet: http://www.goodyear.com/corporate/about/about_community.html
Sponsor: Goodyear Tire and Rubber Company
1144 East Market Street D/798
Akron, OH 44316

Google Grants Beta **2114**
The Google Grants program supports organizations sharing their philosophy of community service to help the world in areas such as science and technology, education, global public health, the environment, youth advocacy, and the arts. Google Grants has awarded AdWords advertising to non-profit groups whose missions range from animal welfare to literacy, from supporting homeless children to promoting HIV education. Recipients use their award of free AdWords advertising on Google.com to raise awareness and increase traffic. Application available online only.
Requirements: In the United States - Organizations must have current 501(c)3 status, as assigned by the Internal Revenue Service to be considered for a Google Grant. Outside the U.S - currently accepting applications from eligible charitable organizations based in Australia, Brazil, Canada, Denmark, France, Germany, India, Ireland, Italy, Japan, the Netherlands, Spain, Sweden, Switzerland and the UK.
Restrictions: Organizations already participating in the Google AdSense program are not eligible for Google Grants consideration. In addition, organizations that are either religious or political in nature are not eligible, including those groups focused primarily on lobbying for political or policy change.
Samples: Three award recipients have achieved these results: - Room to Read, which educates children in Vietnam, Nepal, India and Cambodia, attracted a sponsor who clicked on its AdWords ad. He has donated funds to support the education of 25 girls for the next 10 years. - The U.S. Fund for UNICEF's e-commerce site, Shop UNICEF, has experienced a 43 percent increase in sales over the previous year. - CoachArt, supporting children with life-threatening illnesses through art and athletics programs, has seen a 60 to 70 percent increase in volunteers.
Contact: Google Grants Team Contact; googlegrants@google.com
Internet: http://www.google.com/grants/
Sponsor: Google
1600 Amphitheatre Parkway
Mountain View, CA 94043

GPP Charles E. Culpeper Foundation Biomedical Pilot Projects Grants **2115**

The initiative is designed to encourage the investigation of novel ideas in cures for disease, particularly in the areas of molecular genetics, bio-engineering, and molecular pharmacology. Research into complimentary and alternative medicine will also be considered. Grants will be made on a one-time basis with the possibility for renewal for a second year upon reapplication. No more than 8 percent of the grant may be allocated to cover indirect costs. The purpose of these grants is to explore new and even untested hypotheses, thus substantial preliminary information is not required. Proposals are accepted throughout the year and each successful
Requirements: Nonprofit health care organizations, accredited medical schools, and universities in the United States are eligible.
Date(s) Application is Due: Nov 16
Amount of Grant: Up to 25,000 U.S.D
Samples: Dr. Natalia Broude, Boston U (MA)—to support a study that will develop an innovative test for SARS, which would hook a molecule that glows in the dark to a protein that would attach a specific spot on the SARS gene, $25,000; Ji-Xin Cheng, PhD, Purdue University (West Lafayette, IN)—to support a project that will develop a new highly sensitive microscope for the imaging of molecules using vibrational imaging, $25,000; Dr. Gregory Crawford, Brown U (Providence, RI)—to create a device that will take a photo of the lining of the lower eyelid using a small camera, resulting in accuracy similar to the current lab test when the physician draws blood from the patient, $25,000. .
Contact: Grants Administrator; (847) 948-5512; fax (847) 948-5516
Internet: http://www.goldmanpartnerships.org/Culpeper/CulpeperBiomedicalPilotInitiativegrants.htm
Sponsor: Goldman Philanthropic Partnerships
155 N Pfingsten Road, Suite 109
Deerfield, IL 60015

Grace and Franklin Bernsen Foundation Grants **2116**

The Grace and Franklin Bernsen Foundation provide grants primarily within the metropolitan Tulsa, Oklahoma area. Areas of interest supported are: religious; charitable; scientific; literary and; educational purposes. Grant applications for programs and projects that will provide a defined benefit such as capital projects, building programs, specific program needs or ongoing operations from time to time are all considered. Grant applications from elementary or secondary education institutions will be considered if they involve programs for at-risk, handicapped or learning-disabled children; or if they are innovative and apply to all schools in the system.
Requirements: Applicant should submit a narrative summary no more than three pages in length. This letter should be addressed to the Trustees of the Foundation, see website for detailed description of narrative requirements. IRS 501(c)3 non-profit organizations in the metropolitan Tulsa, Oklahoma area are eligible to apply. Application may be made at any time for support of activities consistent with the Foundation's guidelines. The Foundation's Board meets monthly to review grant applications. Applications received on or before the 12th day of the month (unless the 12th falls on a Saturday, Sunday, or a holiday, in which event applications are due on the preceding business day) preceding a next regularly scheduled Board meeting are normally considered at such meeting.
Restrictions: Grant applications for individuals are not considered. The Foundation discourages applications for general support or reduction of debt, or for continuing or additional support for the same programs. The Foundation will only review one grant request per agency during our fiscal year
Geographic Focus: Oklahoma
Amount of Grant: 2,000 - 1,000,000 U.S.D
Samples: Parent Child Center of Tulsa, Tulsa, OK, $10,000—program funds; St. John Medical Center Foundation, Tulsa, OK, $106,000—funds for programs and capital improvements; Tulsa Air & Space Museum, Tulsa, OK, $10,000—provide funds for capital improvements.
Contact: Margaret Skyles, Administrator; (918) 584-4711; fax (918) 584-4713; mskyles@bernsen.org or info@bernsen.org
Internet: http://www.bernsen.org/grant.html
Sponsor: Grace and Franklin Bernsen Foundation
15 West Sixth Street, Suite 1308
Tulsa, OK 74119-5407

Grace Bersted Foundation Grants **2117**

The Grace Bersted Foundation was established in 1986 to support and promote quality educational, human services, and health care programming for underserved populations. Special consideration is given to charitable organizations that serve the needs of children or the disabled. Applicants must apply online at the grant website. Applicants are strongly encouraged to do the following before applying: review the downloadable state application procedures for additional helpful information and clarifications; review the downloadable online-application guidelines at the grant website; review the foundation's funding history (link is available from the grant website); review the online application questions in advance; and review the list of required attachments. These will generally include: a list of board members, financial statements (audited, reviewed, or compiled by independent auditor); an organization summary; a list of other funding sources; an IRS Determination letter; and other required documents. All attachments must be uploaded in the online application as PDF, Word, or Excel files. The deadline for application to the Foundation is 11:59 p.m. on September 1, and grant decisions will be made by December 31.
Requirements: Applicant organizations must have 501(c)3 tax-exempt status and an office located in one of the following counties: DuPage, Kane, Lake, or McHenry.
Restrictions: The foundation does not support requests from individuals, organizations attempting to influence policy through direct lobbying, or any political campaigns.
Geographic Focus: Illinois
Date(s) Application is Due: Sep 1
Contact: Debra Grand, Senior Vice President; (312) 828-4154; ilgrantmaking@bankofamerica.com
Internet: https://www.bankofamerica.com/philanthropic/fn_search.action
Sponsor: Grace Bersted Foundation
231 South LaSalle Street, IL1-231-13-32
Chicago, IL 60604

Graco Foundation Grants **2118**

The Foundation's goal is to help organizations grow their ability to serve community needs through grants specifically aimed at expanding or enhancing services to clients, with particular focus on capital projects and technology needs. The Foundation addresses the needs of the community in the following areas: self-sufficiency - emphasizing educational programs, human service programs that promote self-sufficiency, and sports/youth development programs; and civic projects - focusing on healthy communities. Projects that improve the community will be given special consideration. Application information is available online.
Requirements: IRS 501(c)3 organizations in company-operating areas may apply.
Restrictions: Grants are not awarded for: organizations or causes that do not directly impact Graco communities; political campaigns; individuals; religious organizations for religious purposes; fund-raising; travel; special events, dinners, courtesy advertising; fraternal organizations; national or local campaigns for disease research; first time general operating grants; or product donations.
Geographic Focus: Minnesota, Ohio, South Dakota
Date(s) Application is Due: Feb 1; May 1; Aug 1; Nov 1
Contact: Kristin R. Ridley; (612) 623-6684; kridley@graco.com
Internet: http://www.graco.com/Internet/T_Corp.nsf/SearchView/Foundation2006
Sponsor: Graco Foundation
P.O. Box 1441
Minneapolis, MN 55440-1441

Grammy Foundation Grants **2119**

The foundation awards grants in two program areas. Research Projects grants are made to organizations and individuals to support efforts that advance the research and/or broad-reaching implementations of original research projects related to the impact of music study on early childhood development, the human development, and the medical and occupational well-being of music professionals. Archiving and Preservation Projects grants are made to organizations and individuals to support efforts that advance the archiving and preservation of the music and recorded sound heritage of the Americas. The foundation funds preservation of original, preexisting media and source material; preservation projects that follow the recommended methodology; projects of historical, artistic, cultural, and or/national significance; and archiving projects including the rescue, organization of, and access to preexisting media and materials. Application and guidelines are available online.
Restrictions: The foundation does not fund Recording Academy chapters, trustees, governors, officers, or staff; organizations that discriminate on the basis of race, sex, religion, national origin, disability, or age; projects promoting advocacy issues; a single organization or individual for more than three consecutive years; organizations or individuals not based in the Americas; purchase of collections; recording projects, demo tapes, or performance events; proposals for commercial purposes (i.e., CD reissue or textbook/ A/V package); purchase of repairs of equipment; purchase or repairs of musical instruments; maintenance or upgrading of computer systems; competitions or any expense associated with competitions; work toward academic degrees; music education or in-residence programs; documentaries; endowments and fundraising; buildings and facilities; marketing, publicity, design costs; or projects where copyright status is unknown.
Date(s) Application is Due: Oct 1
Amount of Grant: 10,000 - 40,000 U.S.D
Samples: The Institute for Music & Brain Science, Boston, MA, $20,000 - this project will test whether music decreases behavioral, neurophysiological and endocrinological pain and stress caused by medically-necessary procedures. In addition, the hypothesis that humans innately prefer consonant over dissonant music will be tested; Elliott Leib, San Diego, CA, $5,000 - a plan will be developed to digitally preserve material from the Trade Roots Reggae Collection; Bob Moog Memorial Foundation for Electronic Music, Asheville, N.C., $15,000 - musical and historical content relative to the unique legacy of synthesizer pioneer Dr. Robert Moog will be cleaned, restored, re-housed and transferred to digital format for accessibility and long-term storage.
Contact: Kristin Murphy, (310) 392-3777, ext. 8662; fax (310) 392-2188; grants@grammy.com
Internet: http://www.grammy.com/GRAMMY_Foundation/Grants/
Sponsor: Grammy Foundation
3030 Olympic Boulevard
Santa Monica, CA 90405

Grand Haven Area Community Foundation Grants **2120**

The community Foundation awards grants to meet the changing and emerging needs of residents of the tri-cities area in such fields as arts, education, health, the environment, youth, social services and other human needs. The Foundation encourages programs which: are collaborative, comprehensive and have potential for continuity; encourage leveraging and matching grant opportunities involving multiple funding sources; support seed money requests to assist innovative projects; encourages programs and projects focused on problem prevention rather than cure. Applicants are encouraged to call and talk with a staff member before beginning a proposal. Application information is available online.
Requirements: IRS 501(c)3 organizations working to enhance the quality of life for residents in the Tri-Cities of Grand Haven, Spring Lake and Ferrysburg in West Michigan.

Restrictions: The Foundation does not generally fund: general operating expenses; private individuals; profit making organizations; elimination of existing financial obligations/debts/liabilities; religious programs that serve, or appear to serve, specific religious denominations; fund raising events.
Geographic Focus: Michigan
Date(s) Application is Due: Jan 7; Mar 13; Jun 26; Oct 10
Contact: Carol Bedient, Grants and Programs Director; (616) 842-6378; fax (616) 842-9518; cbedient@ghacf.org or bpost@ghacf.org
Internet: http://www.ghacf.org/grants.htm
Sponsor: Grand Haven Area Community Foundation
1 South Harbor Drive
Grand Haven, MI 49417

Grand Rapids Area Community Foundation Grants **2121**
The Foundation enables donors to make a difference in the community through grants, scholarships and programs. The deadline for applications is September 15th, and grants are distributed during the month of October. Specific funds include: Alzheimer's Family Support Fund; Fund for the Arts; Fund for the Community; Betty Kauppi Music Fund; Streufert Peace and Safety Fund; Warren Youngdahl Memorial Fund; and the Fund for Women. Amount of funding ranges from $500 to $3,000
Restrictions: No funding is available for individuals.
Geographic Focus: Minnesota
Date(s) Application is Due: Sep 15
Amount of Grant: Up to 3,000 U.S.D
Contact: Wendy Roy; (218) 999-9100; fax (218) 999-7430; wroy@gracf.org
Internet: http://www.gracf.org/index.php/grants-and-programs
Sponsor: Grand Rapids Area Community Foundation
350 NW First Avenue, Suite E
Grand Rapids, MN 55744

Grand Rapids Area Community Foundation Nashwauk Area Endowment Fund Grants **2122**
The Fund is interested in supporting innovative and effective ideas that support the greater Nashwauk, Minnesota area. This area includes the areas of: Nashwauk, Nashwauk Township, Buck Lake and Pengilly. The main cycle offers grants of up to $15,000, with a deadline for application November 15th. The Mini-Grant cycle offers grants of up to $5,000, with the deadline for applications being September 15th. Nonprofit organizations that serve or benefit these areas are encouraged to submit a Letter of Inquiry.
Restrictions: Grants can only be made to charitable organizations, and cannot be made to individuals.
Geographic Focus: Minnesota
Date(s) Application is Due: Sep 15
Amount of Grant: 5,000 - 15,000 U.S.D
Contact: Wendy Roy; (218) 999-9100; fax (218) 999-7430; wroy@gracf.org
Internet: http://www.gracf.org/index.php/grants-and-programs
Sponsor: Nashwauk Area Endowment Fund
350 NW First Avenue, Suite E
Grand Rapids, MN 55744

Grand Rapids Area Community Foundation Wyoming Grants **2123**
The Wyoming Community Foundation gives priority to projects that address the areas of art & culture, community development, education, environment, health, or social needs and that: represent an innovative, start-up effort or are capital in nature (e.g., construction, renovation, equipment); promote cooperation among agencies without duplicating services; obtain the necessary additional funding to implement and maintain the project; serve the greater city of the Wyoming. Michigan area; strengthen or improve agency self-sufficiency or efficiency; yield substantial community benefits for the resources invested; serve a broad segment of the community; encourage additional and permanent funding or matching gifts from other donors; have non-profit, 501(c)3 status. The Foundation will consider proposals for the following only if they are highly unique, collaborative, and community-oriented, have limited access to other resources, demonstrate substantial impact, or will address the needs of a substantial or underserved portion of the community: hospitals; nursing or retirement facilities; K-12 schools; computers; child care centers; motorized vehicles; commonly accepted community services already supported by tax dollars.
Requirements: Michigan IRS 501 (c)3 non-profit organizations in or serving the Wyoming, Michigan area are eligible to apply. General applications will be requested in Fall. For more information on how to request funding for projects in the City of Wyoming Michigan, please contact Lillian VanderVeen, Board Chair, at (616) 534-9625.
Restrictions: The Foundation generally does not support: annual fund-raising drives; films, videos or television projects; endowments or debt reduction; capital projects without site control; medical research; individuals; ongoing operating expenses of established institutions; organizations located outside greater Wyoming; political projects, or those that are primarily cause-related; religious organizations for religious purposes; sabbatical leaves; scholarly research; travel, tours or trips; underwriting of conferences; venture capital for competitive profit-making activities.
Contact: Lillian VanderVeen, Board Chair; (616) 534-9625; lillian@lengertravel.com
Internet: http://www.grfoundation.org/wyoming
Sponsor: Grand Rapids Area Community Foundation-Wyoming Community Foundation
185 Oakes Street SW
Grand Rapids, MI 49503

Grand Rapids Area Community Foundation Wyoming Youth Fund Grants **2124**
The Wyoming Community Foundation Youth Advisory Committee seeks to improve the community by teaching about the giving of time, talent, and treasure while funding programs that help youth gain advantages and build character. The Wyoming Community Foundation Youth Advisory Committee is offering student groups and local non-profit organizations (including schools, churches, and community groups with 501(c)3 non-profit status), an opportunity to apply for grants. These grants will be made with income derived from the Wyoming Community Foundation Youth Fund. Primary consideration will be given to projects that address drug and/or alcohol use, teen pregnancy, stress, smoking, and sexual abuse. Suggested projects or activities include: real life learning experiences, or experiential learning activities; competitive sports; arts, writing, or music-focused activities; programs that explore jobs, career options or job preparation; programs that teach self-defense strategies or martial arts; outdoor experiences, camps, or challenge courses; classes or groups to help quit smoking or using drugs. Typically, the Request for Proposals is available late December of the year and the proposals are due mid-February. Grant awards are announced the following May.
Requirements: Student groups and local non-profit organizations (including schools, churches, and community groups with 501(c)3 non-profit status), are eligible to apply for these funding opportunities.
Geographic Focus: Michigan
Date(s) Application is Due: Dec 31
Contact: Cris Kooyer, Education & Youth Program Officer; (616) 454-1751, ext. 118; ckooyer@grfoundation.org
Internet: http://www.grfoundation.org/wyoming-yac.php
Sponsor: Grand Rapids Area Community Foundation-Wyoming Community Foundation Youth Fund
185 Oakes Street SW
Grand Rapids, MI 49503

Grand Rapids Community Foundation Grants **2125**
The Foundation impacts a wide range of nonprofit organizations in the communities they serve. Funding is in six broad categories: academic achievement; high quality of life; healthy people; economic prosperity; vibrant neighborhoods; and environmental integrity.
Requirements: IRS 501(c)3 organizations supporting residents of Kent County are eligible to apply, applications are available at the Foundation's website.
Restrictions: The Foundation does not fund one-time, special or annual events, annual operating funds, political or religious causes, or endowments.
Geographic Focus: Michigan
Contact: Marcia Rapp, Vice President, Programs; (616) 454-1751, ext. 104; fax (616) 454-6455; mrapp@grfoundation.org or grfound@grfoundation.org
Internet: http://www.grfoundation.org/grants.php
Sponsor: Grand Rapids Community Foundation
185 Oakes Street SW
Grand Rapids, MI 49503

Grand Rapids Community Foundation Ionia County Grants **2126**
The Ionia County Community Foundation gives priority to projects that address the areas of art & culture, community development, education, environment, health, or social needs and that: represent an innovative, start-up effort or are capital in nature (e.g., construction, renovation, equipment); promote cooperation among agencies without duplicating services; obtain the necessary additional funding to implement and maintain the project; are located in Ionia County, Michigan; strengthen or improve agency self-sufficiency or efficiency; yield substantial community benefits for the resources invested; serve a broad segment of the community; encourage additional and permanent funding or matching gifts from other donors; have non-profit, 501(c)3 status. The Foundation will consider proposals for the following only if they are highly unique, collaborative, and community-oriented; have limited access to other resources; demonstrate substantial impact; or will address the needs of a substantial or underserved portion of the community: hospitals; nursing or retirement facilities; K-12 schools; computers; child care centers; motorized vehicles; commonly accepted community services already supported by tax dollars.
Requirements: Non-profit, IRS 501(c)3 status organizations, located in Ionia County, Michigan may apply.
Restrictions: The Foundation generally does not support: annual fundraising drives; film, video or television projects; endowments or debt reduction; capital projects without site control; medical research; individuals; ongoing operating expenses of established institutions; organizations located outside Ionia County; political projects, or those that are primarily cause-related; religious organizations for religious purposes; sabbatical leaves; scholarly research; travel, tours or trips; underwriting of conferences; venture capital for competitive profit-making activities.
Geographic Focus: Michigan
Date(s) Application is Due: Dec 31
Contact: Kate Luckert Schmid; (616) 454-1751, ext. 117; kluckert@grfoundation.org
Internet: http://www.grfoundation.org/ionia-general.php
Sponsor: Grand Rapids Community Foundation
185 Oakes Street SW
Grand Rapids, MI 49503

Grand Rapids Community Foundation Ionia County Youth Fund Grants **2127**
The Ionia County Community Foundation's Youth Advisory Committee is committed to addressing and improving problems concerning young people, by funding programs directed toward youth. The Youth Advisory Committee is affiliated with the Ionia County Community Foundation. Current funding priorities are: Drug, alcohol, tobacco use; Teen pregnancy; Depression/suicide; Parent education; Service-learning; Guest

speakers that address youth issues; Literacy; Obesity; Teen driver safety. The Youth Advisory Committee is especially interested in youth projects that assist students in bridging the achievement gap. Grant proposals should assist the applying organization to implement a project or program that benefits youth. Proposals developed by youth or with youth involvement in planning are encouraged. The proposed project or program should: have clear goals that are specific, measurable, attainable, realistic, and timely; impact a significant number of Ionia County youth from birth to age 18; avoid duplication with other projects and programs in the community; address an issue and make a significant difference for youth. In addition, the applying organization and its participants will not be discriminated against on the basis of, but not limited to, gender, religion, physical disability, sexual orientation, or ethnicity. Grants will be made with income derived from the Ionia County Community Foundation Youth Fund.
Requirements: Student groups and local IRS 501(c)3 nonprofis are eligible apply for grants.
Geographic Focus: Michigan
Date(s) Application is Due: Dec 31
Contact: Cris Kooyer, Education & Youth Program Officer; (616) 454-1751, ext. 118; ckooyer@grfoundation.org
Internet: http://www.grfoundation.org/ionia-yac.php
Sponsor: Grand Rapids Community Foundation
185 Oakes Street SW
Grand Rapids, MI 49503

Grand Rapids Community Foundation Lowell Area Fund Grants **2128**
The Lowell Area Community Fund gives grants to organizations that assist in fulfilling its mission: to assure community cooperation and participation that supports a healthy, dynamic community. The fund places an emphasis on broad educational initiatives, but also supports initiatives in the areas of: Arts & Culture, Community Development, Environment, Health, Human Services, and Recreation. The Lowell area is currently defined as the City of Lowell, Michigan, the Township of Lowell, and the Township of Vergennes. Grant applications are accepted three times a year on the third Friday of April, August, and December. Additional guidelines and grant application are available at the Foundation's website.
Requirements: 501(c)3 tax-exempt organizations are in the Lowell area are eligible.
Restrictions: The fund generally does not support: political projects, or those that are primarily cause-related; individuals; religious organizations for religious purposes; or profit-making activities.
Geographic Focus: Michigan
Amount of Grant: 2,000 - 75,000 U.S.D
Contact: Kate Luckert Schmid, Program Director; (616) 454-1751, ext. 117; kluckert@grfoundation.org
Internet: http://www.grfoundation.org/lowell.php
Sponsor: Grand Rapids Community Foundation
185 Oakes Street SW
Grand Rapids, MI 49503

Grand Rapids Community Foundation Southeast Ottawa Grants **2129**
The Southeast Ottawa Community Foundation gives priority to projects that address the areas of art & culture, community development, education, environment, health, or social needs and that: represent an innovative, start-up effort or are capital in nature (e.g., construction, renovation, equipment); promote cooperation among agencies without duplicating services; obtain the necessary additional funding to implement and maintain the project; are located in the greater Georgetown Township, Hudsonville and Jamestown Township area of Michigan; strengthen or improve agency self-sufficiency or efficiency; yield substantial community benefits for the resources invested; serve a broad segment of the community; encourage additional and permanent funding or matching gifts from other donors; have non-profit, 501(c)3 status. The Foundation will consider proposals for the following only if they are highly unique, collaborative, and community-oriented, have limited access to other resources, demonstrate substantial impact, or will address the needs of a substantial or underserved portion of the community: hospitals; nursing or retirement facilities; K-12 schools; computers; child care centers; motorized vehicles; commonly accepted community services already supported by tax dollars.
Requirements: Michigan IRS 501 (c)3 non-profit organizations that are located in the greater Georgetown Township, Hudsonville and Jamestown Township area are eligible to apply. General applications are due August 31st. For more information on how to request funding for projects, contact Larry Bergman at (616) 896-8769.
Restrictions: The Foundation generally does not support: annual fund-raising drives; films, videos or television projects; endowments or debt reduction; capital projects without site control; medical research; individuals; ongoing operating expenses of established institutions; organizations located outside greater Georgetown Township, Hudsonville and Jamestown Township area; political projects, or those that are primarily cause-related; religious organizations for religious purposes; sabbatical leaves; scholarly research; travel, tours or trips; underwriting of conferences; venture capital for competitive profit-making activities.
Geographic Focus: Michigan
Date(s) Application is Due: Aug 31
Contact: Larry Bergman, (616) 896-8769; lbergman@charter.net
Internet: http://www.grfoundation.org/seottawa
Sponsor: Grand Rapids Area Community Foundation-Southeast Ottawa Community Foundation
185 Oakes Street SW
Grand Rapids, MI 49503

Grand Rapids Community Foundation Southeast Ottawa Youth Fund Grants **2130**
The Southeast Ottawa Community Foundation's Youth Advisory Committee is seeking to fund projects or programs that positively impact youth within Georgetown, Hudsonville, and Jamestown Townships. Groups or organizations made up of (or serving) youth ages 12 - 20 are invited to apply. The committee is particularly interested in programs or activities that address youth issues, particularly: alcohol and drug use (including smoking); impaired driving; coping with external pressures from and conflicting expectations of family and peers; sexual activity, teen pregnancy, and harassment; time management related to academics, employment, and other activities; depression, self-esteem, and body image. Typically, the Request for Proposals is available late December of the year and the proposals are due mid-February. Grant awards are announced the following May.
Requirements: Any non-profit organization with 501(c)3 status located within Georgetown, Hudsonville, and Jamestown Townships of Michigan, that is in need of funding for programs that benefit youth in this area are eligible to apply.
Restrictions: Grants proposals from religious organizations for a religious purpose will not be considered.
Geographic Focus: Michigan
Date(s) Application is Due: Dec 31
Contact: Cris Kooyer, Education & Youth Program Officer; (616) 454-1751, ext. 118; ckooyer@grfoundation.org
Internet: http://www.grfoundation.org/seottawa-yac.php
Sponsor: Grand Rapids Community Foundation
185 Oakes Street SW
Grand Rapids, MI 49503

Grand Rapids Community Foundation Sparta Grants **2131**
The Sparta Community Foundation gives priority to projects that address the areas of art & culture, community development, education, environment, health, or social needs and that: represent an innovative, start-up effort or are capital in nature (e.g., construction, renovation, equipment); promote cooperation among agencies without duplicating services; obtain the necessary additional funding to implement and maintain the project; are located in Sparta and its surrounding areas of Michigan; strengthen or improve agency self-sufficiency or efficiency; yield substantial community benefits for the resources invested; serve a broad segment of the community; encourage additional and permanent funding or matching gifts from other donors; have non-profit, 501(c)3 status. The Foundation will consider proposals for the following only if they are highly unique, collaborative, and community-oriented, have limited access to other resources, demonstrate substantial impact, or will address the needs of a substantial or underserved portion of the community: hospitals; nursing or retirement facilities; K-12 schools; computers; child care centers; motorized vehicles; commonly accepted community services already supported by tax dollars.
Requirements: Michigan IRS 501 (c)3 non-profit organizations that are located in Sparta and its surrounding areas are eligible to apply. General applications are due September 4th. For more information on how to request funding for projects, contact Becky Cumings, Grant Committee Chair, at (616) 887-8428.
Restrictions: The Foundation generally does not support: annual fund-raising drives; films, videos or television projects; endowments or debt reduction; capital projects without site control; medical research; individuals; ongoing operating expenses of established institutions; organizations located outside greater Georgetown Township, Hudsonville and Jamestown Township area; political projects, or those that are primarily cause-related; religious organizations for religious purposes; sabbatical leaves; scholarly research; travel, tours or trips; underwriting of conferences; venture capital for competitive profit-making activities.
Geographic Focus: Michigan
Date(s) Application is Due: Sep 4
Contact: Becky Cumings, Chair; (616) 887-8428; cumingst@hotmail.com
Internet: http://www.grfoundation.org/sparta
Sponsor: Grand Rapids Area Community Foundation-Sparta Community Foundation
185 Oakes Street SW
Grand Rapids, MI 49503

Grand Rapids Community Foundation Sparta Youth Fund Grants **2132**
The Sparta Community Foundation's Youth Advisory Committee is seeking to fund projects or programs that positively impact youth in the Sparta community. Groups or organizations made up of (or serving) youth ages 12 - 20 are invited to apply. The committee is particularly interested in programs or activities that address the following issues: alcohol and drug use (including smoking); impaired driving; coping with external pressures from and conflicting expectations of family and peers; sexual activity, teen pregnancy, and harassment; time management related to academics, employment, and other activities; depression, self-esteem, and body image. The projects should be youth initiated and youth driven.
Requirements: Any non-profit organization with 501(c)3 status located in or serving the Sparta area that is in need of funding for programs that benefit youth in the Sparta community.
Restrictions: Grants proposals from religious organizations for a religious purpose will not be considered.
Geographic Focus: Michigan
Date(s) Application is Due: Dec 31
Contact: Cris Kooyer, Education & Youth Program Officer; (616) 454-1751, ext. 118; ckooyer@grfoundation.org
Internet: http://www.grfoundation.org/sparta-yac.php
Sponsor: Grand Rapids Community Foundation
185 Oakes Street SW
Grand Rapids, MI 49503

Grass Foundation Marine Biological Laboratory Fellowships 2133
The foundation awards fellowships to encourage independent research for scientific discovery by investigators early in their careers and to increase research opportunities for persons trained for careers in neurobiological investigation. Neurobiological approaches supported include neurophysiology, membrane biophysics, integrative neurobiology and neuroethology, neuroanatomy, neuropharmacology, systems neuroscience, cellular and developmental neurobiology, and computational approaches to neural systems. Fellowships provide funds to support an investigator, his/her spouse or domestic partner, and dependent children for one summer at the Marine Biological Laboratory in Woods Hole, MA. Laboratory research space, equipment, housing and board, and travel expenses to and from the MBL are covered. Guidelines and application are available online.
Requirements: Applicants may be researchers in the late stages of predoctoral training or postdoctoral researchers. Priority is given to applicants who have demonstrated a commitment to a research career, who have no more than five years of postdoctoral research and who have no prior research experience at the Marine Biological Laboratory. All U.S. citizens, permanent residents, and foreign applicants with or able to obtain J-1 Visas for the duration of the fellowship are eligible to participate in this program. The foundation must have a copy of the J-1 Visa prior to participation in the program.
Restrictions: Applicants should not attempt to combine a fellowship at the Marine Biological Laboratory with writing a dissertation (PhD thesis).
Date(s) Application is Due: Dec 15
Contact: Administrator; (781) 843-0219; fax (781) 843-0474; grassfdn@aol.com
Internet: http://www.grassfoundation.org/fellowships/grass_fellowship/index.html
Sponsor: Grass Foundation
400 Franklin Street, Suite 302
Braintree, MA 02184

Great-West Life Grants 2134
The program's objectives are to support organizations that meet the most vital needs in the greater Denver metro community, including access to affordable health care, shelter, nutrition and education, and to support our employees' involvement in their communities. Funding priorities are on preventive and ongoing health services, homeless shelters, food services, and keeping children in school. Support is limited to organizations providing direct services, free or charge or on a sliding scale basis, that demonstrate effectiveness in addressing the communities vital issues. There is no formal application process. Interested applicants should submit a brief concept paper.
Requirements: To be eligible for support an organization: must be a U.S. based, charity and tax-exempt under 501(c)3 of the Internal Revenue Code; must not discriminate on the basis of race, religion, gender, age, national origin, disability, or sexual orientation; must show demonstrated success in meeting the program's objectives and fall within the program's focus areas; must serve residents in areas with a Great-West presence.
Restrictions: Funding is not provided for: capital campaigns; debt financing; endowments; advertising; grant-making foundations; individuals; political organizations; exclusively religious organizations; veterans organizations; fraternal organizations.
Contact: Human Resources; (800) 537-2033; greatwestcomments@gwl.com
Internet: http://www.greatwest.com/about/community_supp.htm
Sponsor: Great-West Life and Annuity Insurance Company
8515 East Orchard Road
Greenwood Village, CO 80111

Greater Cincinnati Foundation Priority and Small Projects/Capacity-Building Grants 2135
Grants are awarded on a quarterly basis and support projects addressing community needs and priorities in six areas: arts and culture; community progress; education; environment; health; and human services. The Foundation looks for applicants that are established and fiscally sound with strong governance and a demonstrated organizational capacity/track record. Application information is available online. Nonprofits seeking grants of up to $20,000 can get a decision in eight weeks or less; all other grants above $20,000 will be awarded twice a year.
Requirements: The Foundation awards grants to qualified 501(c)3 nonprofit organizations in eight counties: Butler, Clermont, Hamilton, and Warren, Ohio; Boone, Kenton and Campbell, Kentucky; and Dearborn, Indiana.
Restrictions: Funds are not available for: individuals; units of government or government agencies; religious organizations for religious purposes; ongoing operating expenses of nonprofit organizations; annual fund raising drives, event sponsorship or underwriting; stand-alone books, films or videos unless part of a more comprehensive program activity; stand-alone travel expenses for conferences or educational purposes unless part of a more comprehensive program activity; regular operating costs or capital projects for individual public, private or parochial schools, universities, hospitals, nursing or retirement homes; endowments; scholarships; loans; scholarly or medical research; partisan political advocacy; debt retirement or funding of an activity after it is completed.
Geographic Focus: Indiana, Kentucky, Ohio
Amount of Grant: 10,000 - 100,000 U.S.D
Contact: Kay Pennington, Community Investment Coordinator; (513) 241-2880; fax (513) 852-6886; penningtonk@greatercincinnatifdn.org
Internet: http://www.greatercincinnatifdn.org/page10004265.cfm
Sponsor: Greater Cincinnati Foundation
200 West Fourth Street
Cincinnati, OH 45202

Greater Green Bay Community Foundation Grants 2136
The Foundation provides assistance to worthy charities serving the people of Brown, Kewaunee, and Oconto counties, Wisconsin. The Foundation's areas of interest are: arts and culture; community development; cultural and ethnic diversity; health and human services; historic preservation; education; and the environment. The Foundation also awards special funds to projects that address issues involved with Alzheimer's Disease, Diabetes, and Hospice. Applicants should share their ideas with the Foundation before completing a formal application. Application information is available online.
Requirements: Eligible recipients meet the following criteria: are tax exempt and have 501(c)3 status; primarily serve residents of Brown, Oconto, and Kewaunee counties; and operate, choose a governing board, and provide services without discrimination.
Restrictions: Funding is not available for: annual or capital campaigns; projects that promote any religious belief; individuals; debt retirement; lobbying; activities that occur prior to the awarding of the grant; efforts substantially serving people outside of Brown, Kewaunee, and Oconto counties.
Geographic Focus: Wisconsin
Date(s) Application is Due: Jan 15; Apr 15; Jul 15; Oct 15
Amount of Grant: 1,000 - 10,000 U.S.D
Samples: Children's Museum of Green Bay (Green Bay, WI)—for the Pets are People Too Exhibit, $7,000; Greater Green Bay YMCA (Green Bay, WI)—for Swim Smart & Family Nights, $3,367.
Contact: Lora Warner; (920) 432-0800; fax (920) 432-5577; lora@ggbcf.org
Internet: http://www.ggbcf.org
Sponsor: Greater Green Bay Community Foundation
301 W Walnut Street, Suite 350
Green Bay, WI 54303

Greater Milwaukee Foundation Grants 2137
The Foundation places its highest funding priority on supporting creative efforts to address issues of poverty in the community, particularly grants focused on education, employment, and strengthening children, youth and families. Lower priority is given to projects that do not meet the above criteria and/or do not address issues of persistent poverty. In addition, the Foundation places special emphasis on programs that accomplish the following: improving understanding among people of different backgrounds through support of efforts addressing issues of racial, cultural and economic diversity; and strengthening the voluntary sector by supporting efforts to enhance the management capacities of nonprofit organizations, promote philanthropy, and encourage civic involvement and community service. Additional application information is available online.
Requirements: Grants are made only to 501(c)3 nonprofit organizations and, on occasion, to governmental agencies. Geographically, funding for the discretionary grantmaking program is limited to projects that will significantly improve the lives of people living in Milwaukee, Waukesha, Ozaukee and Washington counties.
Restrictions: Grants for ongoing operational costs or to individuals are not eligible for support from the Foundation's discretionary funds. The Foundation does not provide support for debt reduction, sectarian religious purposes, medical or scientific research, fund drives for sustaining support or organizations that are discriminatory in their practices.
Geographic Focus: Wisconsin
Samples: Milwaukee Chamber Orchestra, Milwaukee, Wisconsin, $25,000 - to support staff and programming expenses as part of its rebuilding efforts (2011); Layton Boulevard West Neighbors, Milwaukee, Wisconsin, $25,000 - to partially fund three staff positions that help support the agency's work as part of the Foundation's Milwaukee Healthy Neighborhoods Initiative (2011).
Contact: Fran Kowalkiewicz, Grants Manager; (414) 272-5805; fax (414) 272-6235; fkowalkiewicz@greatermilwaukeefoundation.org or info@greatermilwaukeefoundation.org
Internet: http://www.greatermilwaukeefoundation.org/grant_seekers/
Sponsor: Greater Milwaukee Foundation
101 West Pleasant Street, Suite 210
Milwaukee, WI 53212

Greater New Orleans Foundation Grants 2138
The Greater New Orleans Foundation is the community foundation serving the 13-parish region of metropolitan New Orleans. Funding is Focused on: housing, workforce development, education, rural and regional initiatives, healthcare, and arts and culture
Requirements: Nonprofits serving the metropolitan New Orleans area are eligible to apply.
Restrictions: The Foundation does not fund the ongoing operational expenses of an established not-for-profit organization.
Geographic Focus: Louisiana
Samples: The Bayou District Foundation,$500,000—to purchase individual lots and vacant homes in St. Bernard; Greater New Orleans Afterschool Partnership, $2,500—to fund a government transparency project; The Urban Conservancy, $20,000—to implement continuity planning for New Orleans' local businesses.
Contact: Dr. Albert Ruesga, President; (504) 598-4663; fax (504) 598-4676; albert@gnof.org or info@gnof.org
Internet: http://www.gnof.org/nonprofits/
Sponsor: Greater New Orleans Foundation
1055 St. Charles Avenue, Suite 100
New Orleans, LA 70130

Greater Saint Louis Community Foundation Grants 2139
The Foundation makes grants annually in the broad areas of the arts, community development, education, environment, health, human services and religion. The Foundation administers its discretionary grant program through a periodic series of initiatives related to four areas of interest: a multi-year, issue-focused initiative; capacity building for nonprofits; knowledge development and dissemination related to current community issues; and knowledge development and dissemination related to

philanthropy and volunteerism. Organizations interested in learning more about the Foundation's discretionary grant program are encouraged to contact the Director of Community Investment.
Requirements: While the Foundation predominately supports St. Louis area nonprofits, grants are also made to national and international charities.
Geographic Focus: Illinois
Contact: Dwight D. Canning, Vice President; (314) 588-8200, ext. 139; fax (314) 588-8088; dcanning@gstlcf.org
Internet: http://www.stlcf.org/page11503.cfm
Sponsor: Greater Saint Louis Community Foundation
319 North Fourth Street, Suite 300
Saint Louis, MO 63102

Greater Tacoma Community Foundation Grants **2140**
The Foundation invests in the community through grants to vital nonprofit agencies in Pierce County. While grants are made across all fields, principal funding areas include: arts and culture; civics; education; the environment; health; and human services. The Foundation also makes capital, equipment, project and operating support grants. Application information is available online.
Requirements: Eligible applicants must meet the following criteria: attest to non-discrimination on the basis of race, sex, age, national origin, religion, physical or mental handicap, veteran status or sexual orientation; provide a service primarily for residents of Pierce County; qualify as tax-exempt under section 501(c)3 of the IRS Code; and apply no more than once a year for any one governing or umbrella organization.
Restrictions: Grants will not be made for: fundraising events or fundraising feasibility projects; individuals; religious organizations for sacramental/theological purposes; production of books, videos, films, or other publications; annual campaign appeals; travel; political or lobbying activities; endowments; debt reduction; events or programs that occur prior to the board of directors decision/notification dates.
Geographic Focus: Washington
Contact: Rose Lincoln Hamilton; (253) 383-5622; fax (253) 272-8099; rlincoln@gtcf.org
Internet: http://www.tacomafoundation.org
Sponsor: Greater Tacoma Community Foundation
950 Pacific Avenue, Suite 1220, P.O. Box 1995
Tacoma, WA 98402

Greater Tacoma Community Foundation Ryan Alan Hade Endowment Fund **2141**
The Fund is to provide support and care to children who have been through traumatic experiences of abuse and disfigurement. The Foundation is looking for nonprofit partners who serve abused or disfigured children in the Greater Tacoma area to continue or implement programming that will ease these children's lives. Interested applicants should submit a one page letter of intent. LOI information is available online.
Requirements: Nonprofit organizations in the Greater Tacoma area are eligible to apply.
Geographic Focus: Washington
Contact: Kristen Corning, Coordinator; (253) 383-5622; kcorning@gtcf.org
Internet: http://www.tacomafoundation.org/newsarticle.cfm?articleid=126285&ptsidebaroptid=0&returnto=page19554.cfm&returntoname=Apply%20for%20a%20Grant&siteid=1735&pageid=19538&sidepageid=19554&banner1img=banner%5F1%2EJPG&banner2img=banner%5F2%2EJPG&bannerbg=banner
Sponsor: Greater Tacoma Community Foundation
950 Pacific Avenue, Suite 1220, P.O. Box 1995
Tacoma, WA 98402

Greater Worcester Community Foundation Discretionary Grants **2142**
Grants are made to nonprofit organizations to build healthy and vibrant communities. Several large field-of-interest funds are also included in this process, providing support for culture and the arts, academic achievement for disadvantaged youth, progressive education, conservation, homelessness and affordable housing, and health and wellness for women and children. These grants enable us to be responsive to community needs and priorities.
Requirements: The Foundation considers applications only from nonprofit, tax-exempt organizations for activities serving the people of the Central Massachusetts region. Organizations not incorporated as tax-exempt may apply through an established organization that agrees to provide fiscal oversight.
Geographic Focus: Massachusetts
Date(s) Application is Due: Mar 15; Sep 15
Amount of Grant: 3,000 - 25,000 U.S.D
Contact: Pamela B. Kane, (508) 755-0980; pkane@greaterworcester.org
Internet: http://www.greaterworcester.org/grants/disc-grants.htm
Sponsor: Greater Worcester Community Foundation
370 Main Street, Suite 650
Worcester, MA 01608-1738

Greater Worcester Community Foundation Jeppson Memorial Fund for Brookfield Grants **2143**
The Fund provides money to civic and community projects that help improve the lives of residents and enrich the cultural environment. The Fund provides support for: cultural or artistic performances; public seminars; festivals or exhibitions; services that help frail or vulnerable citizens, or that contribute to public health and safety; opportunities for educational enrichment; youth involvement in recreation, sports and the arts; and projects that foster community awareness and connections among different groups. Application information is available online.
Requirements: Any nonprofit or civic organization that serves the residents of Brookfield may apply.
Restrictions: Grant funds may not be used for expenses already incurred by the applicant. Fund awards are not intended to replace municipal funds.
Geographic Focus: Massachusetts
Date(s) Application is Due: Jul 15
Contact: Pamela B. Kane, (508) 755-0980; pkane@greaterworcester.org
Internet: http://www.greaterworcester.org/grants/Jeppson.htm
Sponsor: Greater Worcester Community Foundation
370 Main Street, Suite 650
Worcester, MA 01608-1738

Green Bay Packers Foundation Grants **2144**
The Green Bay Packers organization has enjoyed tremendous fan support through its long and storied history. To give back to the community, the team created the Green Bay Packers Foundation in December 1986. The Foundation assists in a wide variety of activities and programs that benefit education, civic affairs, health services, human services and youth-related programs.
Requirements: 501(c)3 nonprofits in Wisconsin may apply for grant support.
Restrictions: No substantial part of the activities of the organization shall involve carrying on propaganda, or otherwise intervening in any political campaign on behalf of any candidate for public office.
Geographic Focus: Wisconsin
Samples: Packers Scholarship Program, $15,000—$7,500 to Scholarships, Inc., for distribution to students in four-year colleges and $7,500 to Northeast Wisconsin Technical College for distribution to students in two-year associate degree or apprenticeship trades programs; Cooks for Room at the Inn, Green Bay, WI, $2,000—New Community Shelter; Hometown Huddle Program, Green Bay, WI, $5,000—to provide a new playground for the city's Beaumont Elementary School Park.
Contact: Margaret Meyers, Grant Coordinator; (920) 569-7315; fax (920) 569-7309; meyersm@packers.com
Internet: http://www.packers.com/community/packers_foundation
Sponsor: Green Bay Packers Foundation
P.O. Box 10628
Green Bay, WI 54307-0628

Greenburg-May Foundation Grants **2145**
Grants are almost entirely awarded for medical research, primarily in the fields of cancer, heart, and neurological research. In addition, some support is given to hospitals, Jewish welfare funds, temple support, and the aging. Types of support include general operating support, continuing support, annual campaigns, endowment funds, emergency funds, program development, internships, scholarship funds, research, and consulting services. Organizations in Florida and New York City receive the largest part of the funds. Applications may be submitted at any time.
Requirements: Nonprofit organizations in southern Florida and New York are eligible.
Restrictions: Grants are not given to individuals or for endowment funds, special projects, publications, or conferences, and generally not for scholarships and fellowships.
Geographic Focus: Florida, New York
Amount of Grant: Up to 295,282 U.S.D
Samples: Miami Jewish Home and Hospital for the Aged (Miami, FL)—$75,200; Greater Miami Opera (Miami, FL)—$4,220.
Contact: Isabel May, President; (305) 864-8639
Sponsor: Greenburg-May Foundation
P.O. Box 54-5816
Miami Beach, FL 33154

Green Diamond Charitable Contributions **2146**
Green Diamond Resource Company is a fifth-generation, family-owned forest-products company that has operations in the states of California and Washington. The mission of Green Diamond's contributions program is to improve the quality of life in communities where the company has a significant number of employees living and working; and to serve as a catalyst for employees to become involved and to provide leadership in their communities. The company has supported a broad range of organizations in its operating communities that address key community needs such as education, health services, economic development, the arts, and more. The company also provides support to environmental and conservation projects, such as environmental education, habitat restoration, and research relevant to the timber industry, as well as other projects that impact the timber business. Contributions are generally made in locations where the company has operations. To the extent possible, contributions will support organizations of interest to, or recommended by, Green Diamond employees. Green Diamond prefers to make capital contributions that will benefit the operating communities for the long term as opposed to contributing operating funds. Generally, support is committed for one year at a time and in amounts less than $5,000. The company supports the United Way in all of its operating communities. The company accepts applications twice a year, generally in May and in late summer/early fall. Interested applicants should call or mail the contact person given for their state to request application materials. More detailed information about application deadlines will be communicated at that time.
Requirements: Organizations that serve Pierce, Mason, Thurston, and Grays Harbor counties in Washington or Del Norte and Humboldt counties in California are eligible to apply. Criteria taken into account in determining the amount of any contributions are as follows: degree of support from company employees; relative size and importance of company operations in the community (balance among Green Diamond communities); needs of organization or program for which funding is requested; amount of previous company contributions to the organization; amount committed by other companies, foundations, and/or governments (projects should demonstrate broad-based community

support); and proximity of the requesting organization to Green Diamond operations or administrative offices.
Geographic Focus: California, Washington
Amount of Grant: Up to 5,000 U.S.D
Contact: Jacki Deuschle, Public Affairs Manager; (707) 668-4488; fax (707) 668-4402; jdeuschle@greendiamond.com
Patti Case, Public Affairs Manager; (360) 426-3381; pcase@greendiamond.com
Internet: http://www.greendiamond.com/charitable-giving/
Sponsor: Green Diamond Resource Company
1301 Fifth Avenue, Suite 2700
Seattle, WA 98101-2613

Greene County Foundation Grants **2147**
The Greene County Foundations mission is to work with charitably minded individuals and organizations to strengthen Greene County, Indiana, now and for generations to come. The Foundation administers several funding opportunities. Eligibility, criteria, and supplemental requirements are different for each grant. Contact foundation directly to begin application process.
Geographic Focus: Indiana
Contact: Kerry Conway, Executive Director; (812) 659-3142 or (812) 659-3144; fax (812) 659-3142; gcfound@yahoo.com
Internet: http://www.greenecountyfoundation.org/funds_available.php
Sponsor: Greene County Foundation
RR #2, Box 38X
Bloomfield, IN 47424

Greenfield Foundation of Maine Grants **2148**
The Foundation, established in 1999, offers funding primarily in the form of ongoing general operating support. Focus areas include: federated giving programs; health care; the environment; wildlife; and animal welfare. The geographic focus is primarily in Maine, with an initial approach by Letter of Interest (LOI) and no specific application forms. There are no annual deadlines, and individual grants average $500.
Requirements: Grantees must be 501(c)3 organizations with a sound track record.
Restrictions: No grants to individuals.
Geographic Focus: Maine
Amount of Grant: 50 - 1,000 U.S.D
Samples: Maine Audubon Society, $50; Animal Refuge League, $1,000.
Contact: Margaret Cary Curran, (207) 781-5360
Sponsor: Greenfield Foundation
2 Lady Cove
Falmouth, ME 04105-1956

Green Foundation Human Services Grants **2149**
The Foundation's mission is to uncover new opportunities, encourage growth and ultimately effect positive change within those institutions that best reflect the core focus areas and the communities it serves. Within the human services area, the Foundation focuses on institutions that provide hope and support to those least able to help themselves as well as the general community; including children, adolescents, the elderly, the homeless, and families who struggle with domestic abuse.
Requirements: 501(c)3 nonprofits (as per the IRS Service Code of 1986) are eligible. Most grant making is limited to institutions that serve the Los Angeles community; however the Foundation will consider requests beyond this geographic boundary for those institutions with the potential to impact communities statewide or nationally.
Restrictions: The Foundation does not provide funds for: those with net assets or fund balances of less than $100,000; multi-year commitments; annual meetings, conferences, and/or seminars; religious programs; capital campaigns; direct mail campaigns; conduit institutions, unified funds, fiscal agents, or institutions using grant funds from donors to support other institutions or individuals; private foundations; or individuals.
Amount of Grant: 1,000 - 150,000 U.S.D
Samples: Esperanza Community Housing, Los Angeles, California, $25,000 (2011); Foothill Family Service, Pasadena, California, $25,000 (2011); Interfaith Community Services, Escondido, California, $25,000 (2011).
Contact: Kylie Wright; (626) 793-6200, ext. 1; fax (626) 793-6201; kylies@ligf.org
Internet: http://www.ligf.org/humanservices.php
Sponsor: Green Foundation
225 South Lake Avenue, Suite 1410
Pasadena, CA 91101

GreenPoint Foundation Grants **2150**
The foundation awards grants in operating communities to organizations dedicated to increasing the availability of affordable housing and that meet basic health and human service needs. Submit grant requests on organization letterhead, which is signed by an officer, and include specific details, including budget, about the project or event; the organization's mission statement; details regarding the organization's services to low- to moderate-income individuals and/or its activities to revitalize/stabilize low- to moderate-income areas (include statistics if possible); and the IRS determination letter with the organization's federal (not state) tax ID number.
Requirements: 501(c)3 nonprofits that serve the needs of the communities in which GreenPoint Bank does business are eligible.
Restrictions: Grants do not support capital campaigns, endowments, scholarship funds, special events, other short-term projects, religious organizations, or membership organizations.
Geographic Focus: New York
Amount of Grant: Up to 28,000,000 U.S.D
Samples: Families of employees of Keefe, Bruyette and Woodf (New York, NY)—for families of employees killed in the September 11 attacks on the World Trade Center, $250,000; American Red Cross, Disaster Relief Fund (Washington, DC)—to provide disaster-relief services to victims of the September 11 terrorist attacks, $25,000.
Contact: Gwen Perry, Foundation Manager; (212) 834-1215; fax (212) 834-1406; gperry@GreenPoint.com
Internet: http://www.greenpoint.com/index.cfm?spPathname=static/com-grants5less.htm
Sponsor: GreenPoint Foundation
90 Park Avenue, 4th Floor
New York, NY 10016

Greenspun Family Foundation Grants **2151**
The foundation awards grants to eligible Nevada organizations in its areas of interest, including cancer research, children's services, higher education, the physically and mentally disabled, and religious welfare programs that provide for the homeless. Nevada nonprofit organizations are eligible. Preference is given to requests from the Las Vegas area.There are no application forms or deadlines. Applicants should submit a letter of inquiry.
Geographic Focus: Nevada
Contact: Jane Greenspun Gale, Vice President; (702) 259-4023
Sponsor: Greenspun Family Foundation
901 N Green Valley Parkway, Suite 210
Henderson, NV 89074

Greenwall Foundation Bioethics Grants **2152**
The Foundation provides funding for physicians, lawyers, philosophers, theologians and other professionals to address micro and macro issues in bioethics, providing guidance for those engaged in decision-making at the bedside as well as those responsible for shaping institutional and public policy. The Foundation is especially interested in the work of junior investigators and pilot projects that may lead to NIH support, and it is prepared to address issues regarded by some as sensitive or potentially controversial.
Restrictions: The Foundation is not normally interested in proposals to support equipment purchase, facility construction or renovation, or general operating expenses, and will not normally consider grants to private foundations, endowment funds, or individual applicants.
Date(s) Application is Due: Feb 1; Aug 1
Amount of Grant: 5,000 - 50,000 U.S.D
Contact: Sam Teigen, Grants Administrator; (212) 679-7266; fax (212) 679-7269; steigen@greenwall.org
Internet: http://www.greenwall.org/guidebio.htm
Sponsor: Greenwall Foundation
420 Lexington Avenue, Suite 2500
New York, NY 10170

Gregory B. Davis Foundation Grants **2153**
The Foundation's mission is to enhance the lives of youth and their families through the development and support of health related and educational programs and initiatives. More specifically, GBDF will support healthy life choices and well-being for rural and medically challenged families, caregivers, and youth affected by HIV/AIDS, Alzheimer's Disease, and other debilitating illnesses and those with limited academic and life skills opportunities. There are no specific deadlines, and applicants should forward an entire proposal.
Amount of Grant: 500 - 2,000 U.S.D
Contact: Edna Davis-Brown, CEO; (301) 593-0644; GBDF@gbdf.org
Internet: http://www.gbdf.org/programs.html
Sponsor: Gregory B. Davis Foundation
11404 December Way, Suite 403
Silver Springs, MD 20904-3623

Gregory Family Foundation Grants (Florida) **2154**
The Foundation, established in 1999, offers support to cancer research and prevention, Christian churches and agencies, and hospitals. Funding most often comes in the form of a donation for general operations. The Foundation's geographic restrictions are New York and Florida. A formal application form is required, which should include: a copy of most recent annual report, audited financial statement, or 990; a detailed description of project; and amount of funding requested. There are no specific deadlines.
Requirements: Applicants should be a 501(c)3 located in the state of Florida or New York.
Geographic Focus: Florida, New York
Amount of Grant: Up to 2,000 U.S.D
Samples: St. Ambrose Church, Boca Raton, Florida, $1,150—for general operations; Christ Child Society of Boca Raton, Boca Raton, Florida, $800—for general operations.
Contact: Dale M. Gregory, President; (561) 362-9868; gregory1@bellsouth.net
Sponsor: Gregory Family Foundation
101 Plaza Reale S, Apt. 618
Boca Raton, FL 33432-4849

Gregory L. Gibson Charitable Foundation Grants **2155**
The Foundation, established in Terre Haute, Indiana, in 2005, is interested in programs that support children, youth, and families. Its primary areas of interest include: community service programs; education; health care and health care access; hospice care; technology; and youth sports. The specific type of support is for general operations. There are no specific deadlines with which to adhere, and applicants should forward a submission letter.
Requirements: 501(c)3 organizations situated in, or serving the community of, Terre Haute, Indiana, are eligible to apply.

Geographic Focus: Indiana
Amount of Grant: Up to 30,000 U.S.D
Contact: Gregory L. Gibson, President; (812) 466-1233
Sponsor: Gregory L. Gibson Charitable Foundation
3200 E Haythorne Avenue
Terre Haute, IN 47805-002108

Griffin Family Foundation Grants **2156**

The Foundation, established in 1999, is primarily interested in supporting non-profit organizations in the state of New York. Fields of interest include: Catholic agencies and churches; education; health care clinics and centers; and hospitals. Funding most often comes in the form of general operating support. There are no particular forms, and applicants should submit a copy of the current year's organizational budget and/or the specific project budget.
Requirements: Applicants should have 501(c)3 status with the IRS.
Geographic Focus: New York
Amount of Grant: Up to 20,000 U.S.D
Contact: William E. Griffin, President; (914) 961-1300; fax (914) 961-9385
Sponsor: Griffin Family Foundation
51 Pondfield Road
Bronxville, NY 10708-3703

Griffin Foundation Grants **2157**

The Foundation, established in 1991, serves non-profit organizations in Fort Collins, Colorado, and Laramie, Wyoming. Its major fields of interest include: the arts; health care; higher education; performing arts; orchestras; and substance abuse service programs. Funding most often comes in the forms of building construction, renovation, and scholarships. Applicants should request an application form, and submit a detailed description of the project and budget needed.
Geographic Focus: Colorado, Wyoming
Contact: David L. Wood, President; (970) 482-3030; fax (970) 484-6648; carol.wood@thegriffinfoundation.org
Internet: http://www.thegriffinfoundation.org/index.shtml
Sponsor: Griffin Foundation
303 W Prospect Road
Fort Collins, CO 80526-2003

Grotto Foundation Project Grants **2158**

The Grotto Foundation works to improve the education and economic, physical and social well-being of citizens, with a special focus on families and culturally diverse groups. The Foundation funds support agencies and institutions dedicated to improving the quality of parenting and well-being of infants and children from birth to six years of age. The Foundation is further interested in increasing public understanding of American cultural heritage, the cultures of nations and the individual's responsibility to fellow human beings.
Requirements: Detailed applications for Native American grants are available online. Applicants are encouraged to contact the Foundation and discuss their early childhood development to be certain it is appropriate for funding.
Restrictions: Policy precludes grants being awarded for capital fund projects, travel, publication of books or manuscripts, undergraduate research projects, or grants to individuals.
Geographic Focus: Minnesota
Date(s) Application is Due: Jan 15; Mar 15; Jul 15; Nov 15
Amount of Grant: Up to 10,000 U.S.D
Contact: Jennifer Kolde, Grants Manager; (651) 209-8010; fax (651) 209-8014; jkolde@grottofoundation.org
Internet: http://www.grottofoundation.org
Sponsor: Grotto Foundation
1315 Red Fox Road, Suite 100
Arden Hills, MN 55112

Grover Hermann Foundation Grants **2159**

The foundation awards one-year grants to nonprofit organizations to support programs that meet the needs of society and improve the quality of life. Requests are considered quarterly, in March, June, September, and December. Preference is given to requests from Chicago, IL.
Requirements: 501(c)3 tax-exempt organizations are eligible. Religious organizations sponsoring nonsectarian programs also are eligible.
Geographic Focus: California, Illinois
Amount of Grant: 2,000 - 50,000 U.S.D
Contact: Katheryn Rhoads, Executive Director; (708) 246-8331
Sponsor: Grover Hermann Foundation
1000 Hillgrove, Suite 200
Western Springs, IL 60558

Gruber Foundation International Prizes **2160**

The Peter and Patricia Gruber Foundation honors and encourages educational excellence, social justice and scientific achievements that better the human condition. The Peter and Patricia Gruber Foundation is a private, United States-based philanthropic organization established in 1993 under the 501(c)3 section of U.S. Corporate Law. It is funded entirely by Peter and Patricia Gruber, who serve as its Chairman and President, respectively. The Foundation is headquartered in St. Thomas, United States Virgin Islands, and maintains a small branch office in New York City. A major focus of the Foundation's philanthropy is its International Prize Program, created to recognize excellence in science and humanities by highlighting five fields that create a better world: Cosmology, Genetics, Neuroscience, Justice, and Women's Rights. Each Prize is given annually, and carries an unrestricted cash award of $500,000 and a gold medal. Independent advisory boards comprised of experts from each field review and select a prize recipient at annual selection meetings held for each discipline. Only fully completed nominations are considered, and the Chair of each Selection Advisory Board screens the nominations, presenting to the Board only those of sufficient caliber. The Selection Advisory Board alone is responsible for the choice of recipient, making its decision from among officially nominated candidates following discussion and deliberation. The Peter and Patricia Gruber Foundation awards its international prizes to recipients only after following a process designed to ensure complete independence and integrity in selecting, as worthy of international recognition, scientific and human rights achievements at the highest levels.
Requirements: Only candidates who have been officially nominated on a fully completed nomination form will be considered for a Peter and Patricia Gruber Foundation International Prize. Additional guidelines and nomination forms are available on line.
Date(s) Application is Due: Dec 15
Amount of Grant: 500,000 U.S.D
Contact: Bernetia Akin, Foundation Contact; (340) 775-8035 or (340) 775-8014; fax (340) 775-8040; info@gruberprizes.org
Internet: http://www.petergruberfoundation.org/GruberPrizes/GruberPrizes.php
Sponsor: Peter and Patricia Gruber Foundation
6000 Estate Charlotte Amalie, Suite 4
St. Thomas, VI 00802

Grundy Foundation Grants **2161**

The Grundy Foundation's endowment, which sustains the operations of both the museum and library, also provides grantmaking. Grant support is given to projects that benefit the people and institutions of Pennsylvania. Upon availability, the Board of Trustees generally restricts grantmaking to Bucks County public charities, with special consideration to those of Bristol Borough. Grant applications are awarded primarily for capital projects to serve a wide area rather than a single neighborhood. Grantmaking activities include community development, arts and culture, education, environment, health, and human services. The average grant is $2,500.
Requirements: There are no applications or specific deadlines. The Foundation accepts and reviews written requests for funding throughout the year. In general, organizations having other support for core operating expenses and long-term costs of new projects are given priority. Organizations are encouraged to contact the Foundation prior to submission of a grant application. Prospective grantees may use the forms developed by the Delaware Valley Grant makers, if preferred. At a minimum, each proposal must include: One-page summary with contact name, executive director's name, organization name, address, telephone, email, and fax; project summary; amount requested; and total project budget amount; detailed proposal with mission and history of the organization; complete description of the proposed project; project budget, including other sources of support, with indication of whether support is in hand, pledged, requested, or to be requested; expected sources of support for this project in the future; and expected arrangements for future maintenance and repairs; the organization's financial report of IRS 990 filing for the most recent fiscal year (audited reports are preferred); copy of IRS 501(c)3 letter of determination or proof that the organization is a government agency; a list of officers and directors; a copy of report of the organization's activities over the most recent fiscal year. Proposals can be mailed, faxed or sent electronically (faxes and email versions require prior Foundation approval). Videotaped proposals will not be accepted.
Restrictions: The Foundation does not make grants to nonpublic schools, individuals, religious organizations, or for endowments, loans, research, or political activities.
Geographic Focus: Pennsylvania
Contact: Eugene Williams, Executive Director; (215) 788-5460; fax (215) 788-0915; info@grundyfoundation.com
Internet: http://www.grundymuseum.org/
Sponsor: Grundy Foundation
680 Radcliffe Street, P.O. Box 701
Bristol, PA 19007

Guidant Foundation Grants **2162**

The foundation supports charitable and educational programs that lead to improved quality of life for patients who are at risk or suffer from cardiovascular disease. The foundation awards grants in the areas of education, healthcare, and community/social services in Guidant communities and to leading healthcare institutions that provide educational services. Preference will be given to programs that develop a patient-support program, using the Internet, to increase patient awareness of advances in the treatment of cardiovascular disease. There are no application deadlines. Application is available online.
Requirements: 501(c)3 organizations are eligible.
Amount of Grant: 500,000 U.S.D
Samples: Cardiovascular Learning Partnership (Atlanta, GA); Denver Health Medical Ctr's Cardiovascular Risk Reduction Program (CO).
Contact: Pat Kuhl, (317) 971-2272; fax (317) 971-2118; pkuhl@guidant.com
Internet: http://www.guidant.com/foundation/index.shtml
Sponsor: Guidant Foundation
111 Monument Circle, Suite 2900
Indianapolis, IN 46244-2159

Guido A. and Elizabeth H. Binda Foundation Grants **2163**

The Foundation supports nonprofit organizations in Southwest Michigan. Areas of interest are educational projects, culture and arts, and human services. Capital campaigns and endowments are considered on a very limited basis. Initial contact may be made by letter describing the project for which funds are sought.

Requirements: Nonprofit organizations in Battle Creek and southwestern Michigan may request grant support.
Geographic Focus: Michigan
Date(s) Application is Due: May 1; Dec 1
Amount of Grant: 500 - 50,000 U.S.D
Samples: Battle Creek Community Foundation, Battle Creek MI, $1,150—health and human services grant(2008); CIR Community Inclusive Recreation Battle Creek, MI, $25,000—arts and culture grant(2008); Grandville Public Schools, Grandville, MI, $4,000—education grant.
Contact: Nancy Taber, Executive Assistant; (269) 968-6171; fax (269) 968-5126; grants@bindafoundation.org
Internet: http://www.bindafoundation.org/grants.html
Sponsor: Guido A. and Elizabeth H. Binda Foundation
15 Capital Avenue NE, Suite 205
Battle Creek, MI 49017

Guitar Center Music Foundation Grants **2164**
The Guitar Center Music Foundation makes the gift of music available to people across the country by providing funding and resources for music programs. Grants are awarded to music academies, schools, local music programs and national music programs across America. Awards are made in the following categories: In-school music classes, in which the students make music; After-school music programs that are not run by the school; and, Music therapy programs, in which the participants make the music. The intent of the program must be music instruction, not music appreciation or entertainment, and the participants/students cannot be professional or career musicians. The grant committee reviews all applications three times yearly, and grant awards range from $500 to $5,000 in value.
Requirements: Applications are accepted online throughout the year from 501(c)3 or governmental organizations. Qualifying applicants are established, ongoing and sustainable music programs in the United States, which provide music instruction for people of any age who would not otherwise have the opportunity to make music. Almost all of the grants awarded are traditional instruments and the equipment necessary to play them. In specific instances, microphones and PA systems have also been awarded. In order to be eligible for a grant, the instruments and/or equipment must stay in the program for the life of the instrument/equipment and must be played/used by the participants.
Restrictions: Requests that are specifically ineligible are: (a) General operating expenses, including rent, utilities, supplies, travel expenses, etc; (b) Items or funds to benefit individuals, including scholarships, stipends or salaries; (c) Seed money; (d) Event sponsorship; (e) Music production equipment and/or expenses; (f) Recording studio equipment and/or expenses, including those used for educational purposes; (g) Instrument repairs; (h) Uniforms; (i) Sheet music.
Geographic Focus: Oklahoma, Oregon, Pennsylvania, Puerto Rico, Rhode Island, South Carolina, South Dakota, Tennessee, Texas, Utah, Vermont, Virginia, Washington, West Virginia, Wisconsin, Wyoming
Amount of Grant: 500 - 5,000 U.S.D
Samples: Music in Schools Today - provides therapeutic percussion, world music and movement classes to the most underserved and at-risk youth, in Bay Area Schools and community centers; Hijos de la Musica Latina - Non-profit music program that empowers talented Latino youth that are primarily financially, economically and emotionally at-risk, who have a certified talent and gift for Latin music; Conejo Valley Youth Orchestras - serves an ethnically diverse group of families, with no students being turned away due to socio-economic status or ability, and continues to grow as the demand for music programs continues. The Creative Arts Program at Dana-Farber Cancer Institute Program - offers adult oncology patients and their families the opportunity to receive free music lessons that are especially helpful during difficult times of their treatment. Guitars in the Classroom - provides affordable music instruction in guitar, singing and song leading to educators in 20 states reaching over 3,000 educators and 200,000 students annually.
Contact: Larry E. Thomas, Chairman; (818) 735-8800, ext. 2112; fax (818) 735-7518; info@guitarcentermusicfoundation.org
Internet: http://www.guitarcentermusicfoundation.org/grants/index.cfm?sec=overview
Sponsor: Guitar Center Music Foundation
5795 Lindero Canyon Road
Westlake Village, CA 91362

Gulf Coast Community Foundation Grants **2165**
The Foundation is a public charity dedicated to the progressive development of the Mississippi Gulf Coast. Its primary mission is to increase philanthropy to worthy causes by providing donor services, grants and leadership for problem solving. The Foundation carries out donor wishes by making grants in the areas of health and human services, education, arts and culture, historic preservation, and neighborhood enrichment. Grant application and additional guidelines are available online at: http://www.gulfcoastfoundation.org/guidelines.html.
Requirements: Applicants must be non-profit, tax-exempt organizations that are involved in enhancing the quality of life for citizens of South Mississippi. Organizations that have not been recognized as tax-exempt by the IRS may apply if they have a fiscal agent relationship with a 501(c)3 nonprofit organization.
Restrictions: Grants do not support: capital, operating endowment fund drives; political activities; individuals; or religious activities.
Geographic Focus: Mississippi
Contact: Rose Dellenger, (228) 897-4841; fax (228) 897-4843; rdellenger@mgccf.org
Internet: http://www.gulfcoastfoundation.org/grants.html
Sponsor: Gulf Coast Community Foundation
P.O. Box 984
Gulfport, MS 39503

Gulf Coast Foundation of Community Operating Grants **2166**
Through grants and strategic community initiatives, Gulf Coast invests in the work of effective nonprofit organizations that improve quality of life in our region. The Foundation will award operating grants over $10,000, with grant funds to be expended within one year of approval. Currently, the Foundation has four grant cycles for operating grants, which fund the core operating needs of nonprofits and help make them stronger. This includes staff and training, database and accounting systems, marketing and fundraising operations. Organizations that want to diversify income streams and generate new revenue sources may apply for an operating – earned revenue grant. Examples of successful proposals in this category include using existing facilities to generate rental income or marketing and selling a packaged program/service to other organizations. Groups that want to reduce operating costs may apply for an operating – efficiency grant. Examples of successful proposals in this category include consolidating services with other nonprofits, implementing new or improved technologies such as databases or financial systems, or "greening" offices.
Requirements: The Foundation will make grants to qualified organizations classified as 501(c)3 tax-exempt public charities by the Internal Revenue Service in the counties of Brevard, Charlotte, Citrus, Collier, DeSoto, Glades, Hardee, Hendry, Hernando, Highlands, Hillsborough, Indian River, Lake, Lee, Manatee, Okeechobee, Orange, Osceola, Pasco, Pinellas, Polk, Sarasota, Seminole, Sumter, and St. Lucie. All operating – earned revenue grant applications must provide realistic revenue estimates that result in a return on investment that exceeds the grant amount. Operating – efficiency grant applications must provide a cost?benefit analysis showing how proposed funding approaches will produce cost savings.
Geographic Focus: Florida
Samples: Special Operations Warrior Foundation $500,000 - immediate financial assistance for special operations personnel, who have been severely wounded, to have his/her family travel and be bedside; Charlotte Behavioral Health Care, $200,000 - addressing mental health, substance abuse, and related psychosocial needs of veterans and their families.
Contact: Kirstin Fulkerson, Philanthropic Advisor; (941) 486-4600; fax (941) 486-4699; info@gulfcoastcf.org
Internet: http://www.gulfcoastcf.org/resources.php
Sponsor: Gulf Coast Foundation of Community
601 Tamiami Trail South
Venice, FL 34285

Gulf Coast Foundation of Community Program Grants **2167**
Gulf Coast Foundation of Community Program Grants address regional priorities identified through its Environmental Scan. The environmental scanning process looked broadly at trends and conditions in its region through interviews with community leaders and analysis of available data. The findings help Gulf Coast identify the best ways to lift the regional economy and sustain community through our grantmaking. Successful program grant applicants will clearly target important regional challenges or opportunities, provide measurable data that will be the basis for assessing their impact, and share stories of transformation through a variety of social and traditional media. Organizations must have strong leadership and competent staff to execute their program and will significantly leverage their own program dollars so that they have "skin in the game" to maximize their program's impact.
Requirements: The Foundation will make grants to qualified organizations classified as 501(c)3 tax-exempt public charities by the Internal Revenue Service in the counties of Brevard, Charlotte, Citrus, Collier, DeSoto, Glades, Hardee, Hendry, Hernando, Highlands, Hillsborough, Indian River, Lake, Lee, Manatee, Okeechobee, Orange, Osceola, Pasco, Pinellas, Polk, Sarasota, Seminole, Sumter, and St. Lucie.
Geographic Focus: Florida
Contact: Kirstin Fulkerson, Philanthropic Advisor; (941) 486-4600; fax (941) 486-4699; info@gulfcoastcf.org
Internet: http://www.gulfcoastcf.org/resources.php
Sponsor: Gulf Coast Foundation of Community
601 Tamiami Trail South
Venice, FL 34285

Gustavus and Louise Pfeiffer Research Foundation Grants **2168**
The Foundation makes grants to U.S. tax-exempt institutions for projects or programs for the advancement of medicine and pharmacy, including scientific research; postgraduate scholarship and fellowship assistance; and studies in nutrition, blindness, deafness and other physical disabilities. The letter of inquiry should not exceed two pages and should state the nature and purpose of the project, the field of interest of the foundation to which it relates, the amount requested, and a general description of the items to be funded; full proposals are by invitation. Applications are considered by the foundation's directors at two meetings a year, one in spring and one in fall The listed application deadlines are for letters of inquiry; full applications are by invitation.
Requirements: U.S. tax-exempt organizations are eligible.
Restrictions: Grants will not be made for delivery of health care services; that benefit a limited geographical area; for endowment, general institutional support, building programs, chairs, or the like; where a major part of the funds are for equipment, exhibits, conferences, seminars, lectures, workshops, sabbatical leave, or the like; for indirect costs; for general programs of research rather than specific research projects of limited duration; to other foundations or fund raising organizations for the purpose of their making grants to others; that are more appropriate for support by other sources, such as pharmaceutical companies for commercial applications of existing products; that involve animal experimentation (other than insects); or to institutions for projects not in the United States.
Date(s) Application is Due: Jan 8; Jul 25

Amount of Grant: Up to 75,000 U.S.D
Contact: Matt Herold Jr., Secretary; (973) 983-0480; fax (973) 586-3456
Internet: http://fdncenter.org/grantmaker/pfeiffer
Sponsor: Gustavus and Louise Pfeiffer Research Foundation
89 Diamond Spring Road, CN 3050
Danville, NJ 07834

Guy's and St. Thomas' Charity Grants **2169**
The Charity's grants program supports both small (under 20,000 British pounds) and large (between 20,000 and 1 million British pounds) projects under three themes: staff benefits and development; clinical and service innovation; and buildings and the environment. Applications for grants under 20,000 British pounds are considered by the Small Grants Committee, and grants over 20,000 British pounds are assessed by the New Services and Innovations Committee. The small grants deadline dates are: August, 14; September, 25; October 16. Deadline dates for the large grants are: December, 1; March, 6; June, 5; September, 4; November, 20.
Requirements: The Charity only considers applications from anyone working in one of the four NHS Trusts in Lambeth or Southwark, although there are some restrictions for mental health staff.
Geographic Focus: All States, United Kingdom
Date(s) Application is Due: Mar 6; Jun 5; Aug 14; Sep 4; Sep 25; Oct 16; Nov 20; Dec 1
Amount of Grant: 20,000 - 1,000,000 GBP
Samples: Julie Brown, Midwifery Practice Leader for Parent Education, 5,000 pounds; Dr. Danielle Harari, Consultant Physician in Elderly Medicine, 20,000 pounds; Improving dermatology training for primary care professionals, NHS Foundation Trust, 254,000 pounds.
Contact: Anne Rigby, Grants Manager; 020 7188 1227; anne.rigby@gsttcharity.org.uk
Internet: http://www.gsttcharity.org.uk/grants/index.html
Sponsor: Guy's and St. Thomas' Charity
Freepost Lon 15724
London, SE1 9YA England

Guy I. Bromley Trust Grants **2170**
The Guy I. Bromley Trust was established in 1964 to support and promote quality educational, cultural, human-services, and health-care programming. In the area of education the trust supports programming that: promotes effective teaching; improves the academic achievement of, or expands educational opportunities for disadvantaged students; improves governance and management; strengthens nonprofit organizations, school leadership, and teaching; and bolsters strategic initiatives of area colleges and universities. In the area of culture the trust supports programming that: fosters the enjoyment and appreciation of the visual and performing arts; strengthens humanities and arts-related education programs; provides affordable access; enhances artistic elements in communities; and nurtures a new generation of artists. In the area of human services, the trust supports programming that: strengthens agencies that deliver critical human services and maintains the community's safety net; and helps agencies respond to federal, state, and local public policy changes. In the area of health the trust supports programming that: improves the delivery of health care to the indigent, uninsured, and other vulnerable populations; and addresses health and health-care problems that intersect with social factors. Grant requests for general operating support and program support will be considered. Grants from the trust are one year in duration. There are no application deadlines for the Bromley Trust. Proposals are reviewed on an ongoing basis. Downloadable application materials are available at the grant website. Applicants are encouraged to review the downloadable state guidelines at the grant website for further information and clarification before applying. Applicants are also encouraged to view the trust's funding history (link is available at the grant website).
Requirements: Applicants must have 501(c)3 tax-exempt status and serve the residents of Atchison, Kansas and the Greater Kansas City Metropolitan area. Applications must be mailed.
Restrictions: Grant requests for capital support will not be considered. The trust does not support requests from individuals, organizations attempting to influence policy through direct lobbying, or any political campaigns.
Samples: Humane Society of Atchison, Atchison, Kansas, $5,000 (2010); Benedictine College, Atchison, Kansas, $30,000 (2010); Community of Christ, Independence, Missouri, $30,000 (2010).
Contact: Spence Heddens; (816) 292-4301; Spence.heddens@baml.com
Internet: https://www.bankofamerica.com/philanthropic/fn_search.action
Sponsor: Guy I. Bromley Trust
1200 Main Street, 14th Floor, P.O. Box 219119
Kansas City, MO 64121-9119

H&R Block Foundation Grants **2171**
The foundation supports the arts and culture, community development, education, and health and human services. Grants for operational support; special projects; and capital improvements for construction, renovation, and equipment will be considered. The foundation usually makes one-year grants but may consider requests for up to five years for special project funding. Guidelines and application are available online.
Requirements: 501(c)3 tax-exempt organizations are eligible. Major emphasis is placed on support of activities in the Kansas City metropolitan area.
Restrictions: Except in most unusual circumstances, the foundation does not make grants to individuals or businesses; publications; projects for which the foundation must exercise expenditure responsibility; single-disease agencies; travel or conferences; historic preservation; or telethons, dinners, advertising, or other fundraising events.
Geographic Focus: Kansas, Missouri
Date(s) Application is Due: Jan 28; Apr 29; Jul 29; Oct 14
Amount of Grant: 500 U.S.D
Contact: Grants Administrator; (816) 932-8324; fax (816) 753-1585; foundation@hrblock.com
Internet: http://www.hrblockfoundation.org/grants/index.html
Sponsor: HandR Block Foundation
4400 Main Street
Kansas City, MO 64111

H & R Foundation Grants **2172**
The Foundation prefers to support projects and programs that strive for excellence, improve service for clients and strengthen the organization. A major emphasis is placed on support of activities that serve underserved, low-income persons living in Jackson, Clay, and Platte counties in Missouri, and Wyandotte and Johnson counties in Kansas.
Requirements: Grants are made only to organizations that are tax-exempt from federal income taxation pursuant to Section 501(c)3 and that are not classified as private foundations within the code.
Restrictions: Except in the most unusual circumstances, the Foundation does not make grants to: organizations that knowingly discriminate on the basis of race, color, religion, gender, national origin, sexual orientation, age, disability, marital status, or status as a veteran; organizations that are not in compliance with laws and regulations that govern them; individuals or businesses; publications; projects for which the Foundation must exercise expenditure responsibility; single-disease causes; travel or conferences; historic preservation; telethons, dinners, advertising, sponsorships, or other special events; animal-related causes; sports-related causes.
Geographic Focus: Kansas, Missouri
Date(s) Application is Due: Feb 26; Apr 30; Jul 30; Oct 15
Samples: Greater Kansas City Community Foundation, Kansas City, MO, $200,000; Scholarship America, Saint Peter, MN, $178,845; Genesis School, Kansas City, MO, $18,495;
Contact: David Miles, President; (816) 854-4372 or (816) 854-4361; fax (816) 854-8025; davmiles@hrblock.com
Internet: http://www.blockfoundation.org/nonprofit_organizations/faqs.html
Sponsor: H and R Foundation
One H&R Bloack Way
Kansas City, MO 64105

H.A. and Mary K. Chapman Charitable Trust Grants **2173**
H. Allen Chapman was born in Colorado in 1919. In 1976, he established the H.A. and Mary K. Chapman Charitable Trust, a perpetual charitable private foundation that maintains endowments to fund charitable grants to public charities. The trustees and staff that administer the foundation also provide public stewardship through service to charitable organizations and causes. A major charitable focus of H. A. Chapman during his life, and the lives of his philanthropic parents, James A. and Leta Chapman, was education and medical research. Though not limited geographically, most grants and public service are within Oklahoma. Grants to human services and civic and community programs and projects are primarily focused in the area of Tulsa. There are two steps in the process of applying for a grant. The first is a Letter of Inquiry from the applicant. This letter is used to determine if the applicant will be invited to take the second step of submitting a formal Grant Proposal.
Requirements: IRS 501(c)3 non-profits are eligible to apply.
Restrictions: Grant requests for the following purposes are not favored: endowments, except as a limited part of a capital project reserved for maintenance of the facility being constructed; deficit financing and debt retirement; projects or programs for which the Chapman Trusts would be the sole source of financial support; travel, conferences, conventions, group meetings, or seminars; camp programs and other seasonal activities; religious programs of religious organizations; project or program planning; start-up ventures are not excluded, but organizations with a proven strategy and results are preferred; purposes normally funded by taxation or governmental agencies; requests made less than nine months from the declination of a previous request by an applicant, or within nine months of the last payment made on a grant made to an applicant; requests for more than one project.
Geographic Focus: Oklahoma
Amount of Grant: Up to 300,000 U.S.D
Samples: Gilcrease Museum Management, Tulsa, Oklahoma, $250,000 - general operating funds (2010); Tulsa Air and Space Museum, Tulsa, Oklahoma, $35,000 - general operating support (2010); Friends of the Fairgrounds Foundation, Tulsa, Oklahoma, $15,000 - general operating funds (2010).
Contact: Andrea Doyle, Program Officer; (918) 496-7882; fax (918) 496-7887; andie@chapmantrusts.com
Internet: http://www.chapmantrusts.org/grants_programs.html
Sponsor: H.A. and Mary K. Chapman Charitable Trust
6100 South Yale, Suite 1816
Tulsa, OK 74136

H.B. Fuller Company Foundation Grants **2174**
The Foundation makes grants within the corporate headquarters community, and other key H.B. Fuller communities throughout the world. Within Minnesota, the Foundation recently served as a catalyst in drawing attention to and support for early family literacy programs, and now focuses on science, technology, engineering, and math education. In other communities and cultures, the foundation provides critical grants in education, arts/culture, and health/human services. As an international company, the foundation's commitments extend well beyond the Minnesota state borders. Making significant

community grants to schools and non-governmental agencies (NGOs) in locations across Latin America, as well as key community investments in six North American cities where the H.B. Fuller Company has manufacturing operations. Most recently, expanding philanthropic giving into Asia.
Requirements: 501(c)3 organizations serving the communities where the company has operations are eligible. Organizations incorporated in countries other than the United States must qualify for tax-exempt status according to U.S. tax regulations and comply with national and/or state charity laws.
Restrictions: Funding is not available for: individuals, including scholarships for individuals; fraternal or veterans' organizations except for programs which are of direct benefit to the broader community; religious groups for religious purposes; political/lobbying organizations; travel; basic or applied research; disease specific organizations; courtesy, goodwill or public service advertisements; fundraiser events or sponsorships; general support education institutions.
Geographic Focus: California, Florida, Georgia, Illinois, Indiana, Kentucky, Michigan, Minnesota, Ohio, Texas, Washington, Abkhazia, Afghanistan, Armenia, Azerbaijan, Bahrain, Bangladesh, Bhutan, British Indian Ocean Territory, Brunei, Burma (Myanmar), Cambodia, China, Christmas Island, Cocos, Cyprus, Georgia, Hong Kong, India, Indonesia, Iran, Iraq, Israel, Japan, Jordan, Kazakhstan, Kuwait, Kyrgyzstan, Laos, Lebanon, Macau, Malaysia, Maldives, Mongolia, Nagorno-Karabakh, Nepal, North Korea, Northern Cyprus, Oman, Pakistan, Palestinian Authority, Philippines, Qatar, Russia, Saudi Arabia, Singapore, South Korea, South Ossetia, Sri Lanka, Syria, Taiwan, Tajikistan, Thailand, Timor-Lester, Turkey, Turkmenistan, United Arab Emirates, Uzbekistan, Vietnam, Yemen
Date(s) Application is Due: Feb 2; Jun 4; Oct 5
Contact: Keralyn Groff, Director; (651) 236-5104; keralyn.groff@hbfuller.com
Internet: http://www.hbfuller.com/About_Us/Community/000110.shtml#P0_0
Sponsor: H.B. Fuller Co
1200 Willow Lake Boulevard, P.O. Box 64683
Saint Paul, MN 55164-0683

H.J. Heinz Company Foundation Grants **2175**
The Foundation is committed to promoting the health and nutritional needs of children and families. Priority is given to programs in communities where Heinz operates with a special focus given to southwestern Pennsylvania. The Foundation proactively donates funds to develop and strengthen organizations that are dedicated to nutrition and nutritional education, youth services and education, diversity, healthy children and families, and quality of life. Application information is available online.
Requirements: Only organizations that have 501(c)3 tax status under the U.S. Internal Revenue Code are eligible for support domestically. International organizations are encouraged to submit a letter of inquiry prior to preparing a full proposal to determine eligibility. International organizations will need to provide a 501(c)3 determination letter from the United States Internal Revenue Service or sufficient documentation to demonstrate that the non-U.S. grantee is the equivalent of a U.S. public charity. Documentation should be provided in English. The Foundation will also accept proposals in the Common Grant Application Format. All proposals must be submitted in writing. Electronic submissions are not accepted at this time.
Restrictions: The Foundation will not provide grants to individuals nor make multi-year pledges except for major capital or grant campaigns. Generally, the Foundation does not make loans and does not provide grants for individuals, equipment, conferences, travel, general scholarships, religious programs, political campaigns, and unsolicited research projects.
Samples: King Edwards Memorial Hospital, Subiacio, Perth, Australia, $28,000—Micronutrient long-term development study; The Children's Medical & Research Foundation, Dublin, Ireland, $25,000—new adolescent Hematology/Oncology unit; Pittsburgh Public Theater, Pittsburgh, PA, $30,000—program support.
Contact: Tammy B. Aupperle, Program Director; (412) 456-5773; fax (412) 442-3227; heinz.foundation@hjheinz.com
Internet: http://www.heinz.com/sustainability.aspx/social/heinz-foundation.aspx
Sponsor: H.J. Heinz Company Foundation
P.O. Box 57
Pittsburgh, PA 15230-0057

H. Leslie Hoffman and Elaine S. Hoffman Foundation Grants **2176**
The foundation awards general operating grants to California non-profits, primarily focusing on education. Additional funding is also available in the following areas of interest: arts; social services; hospitals; health organizations; children/youth services, including children's hospitals and, social services. There are no application forms or deadlines.
Requirements: California IRS 501 (c)3 non-profits are eligible to apply. The majority of grants are funded in the Los Angeles area with a special emphasis on Pasadena.
Restrictions: Individuals are ineligible.
Geographic Focus: California
Amount of Grant: 1,000 - 200,000 U.S.D
Samples: All Saints Church, Pasadena, CA, $10,000—general funding; Hillsides Home For Children, Pasadena, CA, $5,000—general funding; Descanso Gardens, Lacanada, CA, $1,000—annual benefit.
Contact: J. Kristoffer Popovich, Treasurer; (626) 793-0043
Sponsor: H. Leslie Hoffman and Elaine S. Hoffman Foundation
225 S Lake Avenue, Suite 1150
Pasadena, CA 91101-3005

H. Schaffer Foundation Grants **2177**
The foundation awards grants to nonprofit organizations in New York in the fields of higher education, hospitals and health care, human services, international relief, theater, aging centers and services, federated giving programs, and Jewish agencies and temples. Eligible activites include recreation, children and youth, services, and general charitable giving. There are no application forms or deadlines.
Requirements: Nonprofits in New York, with an emphasis on Schenectady, are eligible.
Geographic Focus: New York
Amount of Grant: 1,000 - 250,000 U.S.D
Contact: Sonya Stall, President; (518) 580-0188
Sponsor: H. Schaffer Foundation
2 Claire Pass
Saratoga Springs, NY 12307

Hackett Foundation Grants **2178**
The Foundation is limited by its giving to the funding of grants for supplies and equipment. Grants are primarily given, but not limited to, Catholic Missions. The primary focus of the Foundation is to provide assistance to those Catholic Organizations which promote the Health, Welfare, Education and Independence of individuals. Application information is available online.
Restrictions: Grants to individuals, scholarships, endowments and fellowships will not be considered. Grants will not be given to supporting organizations having a 509(a)(3) status. Grants will not be used as reimbursement for previously purchased items. Grants are not given for salaries, administrative expenses or matching funds. Grants will not be considered for demonstration projects and/or capital campaigns. Grants will not be provided for non-U.S. based organizations or for taxes or shipping and handling charges.
Geographic Focus: New Jersey, New York, Pennsylvania
Amount of Grant: 1,000 - 20,000 U.S.D
Samples: School of the Holy Childhood (Rochester, NY)—to purchase a copier, printers, and power generators for this organization that provides education, training, and employment to developmentally disabled children and adults, $14,000.
Contact: Maggie Hackett, Grants Manager; (908) 238-9444
Internet: http://fdncenter.org/grantmaker/hackett
Sponsor: Hackett Foundation
P.O. Box 693
Pittstown, NJ 08867

Hadassah Foundation Grants **2179**
The foundation's mission is to improve the status, health, and well-being of women and girls; bring their contributions, issues, and needs from the margins to the center of Jewish concern; and encourage and facilitate their active participation in decision-making and leadership in all spheres of life. The foundation funds projects that empower Israeli women by providing them with the tools they need to become economically self-sufficient. The foundation also supports training for girls so that they will develop into self-confident adults and leaders. Letters of intent must meet the listed application deadline.
Requirements: 501(c)3 nonprofit organizations and organizations who have amutah or mossad tziburi status in Israel are eligible.
Date(s) Application is Due: Jul 6
Amount of Grant: Up to 100,000 U.S.D
Samples: Jewish Arab Economic Development (Herzliya Pituach, Israel)—to provide business-skills training and to encourage Jewish and Palestinian Israeli women to work together on business activities, $30,000; Economic Empowerment for Women (Haifa, Israel)—to teach women marketing skills to help them promote their businesses and reach a broader customer base, $50,000; FEGS (Syosset, NY)—to implement a program for Jewish girls and young women designed to prevent eating disorders, $37,000; Tides Ctr (Oakland, CA)—for a youth-led program that seeks to prevent dating violence among Jewish teenagers, $40,000.
Contact: Grants Administrator; (212) 451-6245; fax (212) 303-8282; Hadassahfoundation@hadassah.org
Internet: http://www.hadassahfoundation.org
Sponsor: Hadassah Foundation
50 W 58th Street
New York, NY 10019-2505

Hagedorn Fund Grants **2180**
William Hagedorn directed that the remainder of his estate be dedicated to this fund, in memory of his late wife, Tillie Hagedorn, that would support religious or charitable organizations in the New York City region. Funding interests include: health (including cancer, HIV/AIDS, blindness), gardens, social services, youth, education, senior services and housing and community development. All applications to the Hagedorn Fund must be submitted online, additional information is available at the Hagedorn Fund website.
Requirements: New York nonprofit organizations are eligible to apply.
Restrictions: No grants are made to individuals or private foundations or for matching gifts or loans.
Geographic Focus: New York
Date(s) Application is Due: Sep 1
Amount of Grant: 5,000 - 45,000 U.S.D
Contact: Erin K. Hogan, Fund Manager; (212) 464-2476; fax (212) 464-2305; erin.k.hogan@jpmorgan.com
Internet: http://fdncenter.org/grantmaker/hagedorn
Sponsor: Hagedorn Fund
270 Park Avenue, 16th Floor
New York, NY 10017

Hall-Perrine Foundation Grants 2181

The Hall-Perrine Foundation Grants funds charitable projects for nonprofit tax-exempt organizations that benefit the Linn County, Iowa community. The Foundation screens grant applications, and those meeting the established criteria are presented to the Board of Directors. The Board reviews proposals approximately four times a year. After a decision has been reached, the organization is notified in writing of whether they have been approved for funding.

Requirements: Potential grantees should first make a preliminary inquiry to determine the Foundation's interest in their request. This communication should briefly describe the background and purposes of the organization and outline the proposed project and its goals. If the Foundation is interested, a grant application will be given to the applicant to be completed in order that the proposal may be properly evaluated. Information requested includes: a description of the organization, including its legal name, history, purposes, and activities; a list of members of the governing board; a clear and detailed description of the purpose for which the grant is requested and the goals to be achieved; the total cost of the project and amount requested with a detailed budget; a list of current and potential sources of financial support; a copy of the organization's most recent audited financial statement or the last IRS Form 990 (income tax return of organization exempt from income tax); and a copy of the IRS determination letter indicating 501(x)3 tax-exempt status.

Geographic Focus: Iowa

Contact: Kristin Novak, Program Officer; (319) 362-9079; fax (319) 362-7220; kristin@hallperrine.org

Internet: http://www.hallperrine.org/

Sponsor: Hall-Perrine Foundation
115 Third Street SE, Suite 803
Cedar Rapids, IA 52401-1222

Halliburton Foundation Grants 2182

The foundation supports education at all levels and charitable organizations in the following ways: matching U.S.- based employee donations on a two-for-one basis up to $20,000 annually per employee for accredited junior colleges, colleges, and universities; matching U.S.-based employee donations to accredited elementary and secondary schools on a two-for-one basis up to $500 annually per employee; making direct donations to U.S.-based elementary and secondary schools and colleges and universities; and recognizing and supporting active U.S.-based employee volunteerism with direct donations through the Halliburton Volunteer Incentive Program. The corporate giving program makes donations to tax-exempt nonprofit organizations dedicated to education, health/welfare, civic issues, and arts and culture. The board meets quarterly. There are no application deadlines or forms.

Geographic Focus: Arizona, Texas

Contact: Margaret Carriere, (713) 676-3717; fhoufoundation@halliburton.com

Internet: http://www.halliburton.com/about/community.jsp

Sponsor: Halliburton Foundation
4100 Clinton Drive, Building 1, 7th Floor
Houston, TX 77020

Hallmark Corporate Foundation Grants 2183

The mission of the Foundation is to help create communities where: all children have the chance to grow up as healthy, productive and caring persons; vibrant arts and cultural experiences enrich the lives of all citizens; there is a strong infrastructure of basic institutions and services, especially for persons in need; and all citizens feel a responsibility to serve their community. Proposals are accepted and reviewed throughout the year. There are no deadlines. The application process can be completed online using the Hallmark General Grant Application, which is available on the Hallmark website.

Requirements: IRS Non-profit 501(c)3 organizations are eligible to apply in the following areas: Kansas City metropolitan area; Center, Texas; Columbus, Georgia; Enfield, Connecticut; Lawrence, Kansas; Leavenworth, Kansas; Liberty, Missouri; Metamora, Illinois and; Topeka, Kansas.

Restrictions: Funding is not available for: individuals for any purpose, including travel, starting a business or paying back loans; religious organizations unless they can demonstrate that services are provided to the community-at-large and separated from religious purposes; fraternal, international and veterans organizations; sports teams and athletic organizations; individual youth clubs, troops, groups or school classrooms; social clubs; disease-specific organizations whose local chapters primarily raise funds for national research; past operating deficits; endowment or foundation funds; conferences; scholarly or health-related research; scholarship funds.

Geographic Focus: Connecticut, Georgia, Illinois, Kansas, Missouri, Texas

Contact: Community Development Manager; (816) 545-6906; contributions@hallmark.com.

Internet: http://corporate.hallmark.com/Community/Community-Involvement-Program-Guidelines

Sponsor: Hallmark Corporate Foundation
Mail Drop 323, P.O. Box 419580
Kansas City, MO 64141-6580

Hamlet Trust Grants 2184

The trust awards grants in support of self-help initiatives for people with long-term mental health problems in Central and Eastern Europe and the countries of the New Independent States.

Contact: Grants Administrator; +44 (0) 20 7247 8666; fax +44 (0) 20 7247 1984; office@hamlettrust.plus.com

Internet: http://www.hamlet-trust.org.uk

Sponsor: Hamlet Trust
Norvin House, 45-55 Commercial Street
London, E1 6BD England

Hammond Common Council Scholarships 2185

The Hammond Common Council Scholarships are awarded to graduating Hammond high school seniors. Fifteen scholarships of $1,000 each are awarded annually. Any college major is eligible.

Requirements: Applicants must reside in Hammond, Indiana. They must also enroll in a college or university of their choice as a full time student with at least a 2.0 in a two or four year program. Applications are available at Hammond high school guidance offices between February 15 and April 15.

Geographic Focus: Indiana

Amount of Grant: 1,000 U.S.D

Contact: Hammond Council Scholarship Contact; (219) 736-1880; fax (219) 736-1940; legacy@legacyfoundationlakeco.org

Internet: http://www.legacyfoundationlakeco.org/scholarships/index.htm

Sponsor: Legacy Foundation
1000 East 80th Place, 302 South
Merrillville, IN 46410

Hampton Roads Community Foundation Developmental Disabilities Grants 2186

The Foundation's mission is to inspire philanthropy and transform the quality of life in southeastern Virginia. Developmental Disabilities Grants are funded by the Laura Turner Fund and the Jennifer Lynn Gray Fund to support people with cerebral palsy and other developmental disabilities. Applications are available on the website. Applicants from the Eastern Shore of Virginia should speak with a Program Officer before submitting a proposal.

Requirements: Nonprofit organizations serving residents of south Hampton Roads (Chesapeake, Franklin, Norfolk, Portsmouth, Suffolk, Virginia Beach and Isle of Wight County) and the Eastern Shore of Virginia are eligible.

Geographic Focus: Virginia

Date(s) Application is Due: Oct 15

Contact: Program Officer; (757) 622-7951; grants@hamptonroadscf.org

Internet: http://www.hamptonroadscf.org/nonprofits/specialInterestGrants.html

Sponsor: Hampton Roads Community Foundation
One Commercial Plaza, Suite 1410
Norfolk, VA 23510

Hampton Roads Community Foundation Faith Community Nursing Grants 2187

The Foundation's mission is to inspire philanthropy and transform the quality of life in southeastern Virginia. The Nightingale Fund was established in 2004 to support established Faith Community Nursing programs in the Hampton Roads area of Virginia. Programs were formerly known as Parish Nursing programs. Eligible expenses include: medical and educational equipment; specific programs; educational materials; and supplies. Applications are available on the website. Applicants from the Eastern Shore of Virginia should speak with a Program Officer before submitting a proposal.

Requirements: Eligible churches, synagogues or other religious institutions located in the Hampton Roads area of Virginia must have operated a Faith Community Nursing program for at least one year and be classified as tax exempt by the Internal Revenue Service.

Restrictions: Support is not available for: salaries or stipends; financial assistance to individuals; and activities normally considered to be the responsibility of the requesting religious institution.

Geographic Focus: Virginia

Date(s) Application is Due: Oct 31

Contact: Program Officer; (757) 622-7951; SusanSaundersRN@cox.net or grants@hamptonroadscf.org

Internet: http://www.hamptonroadscf.org/nonprofits/nightingdaleGrants.html

Sponsor: Hampton Roads Community Foundation
One Commercial Plaza, Suite 1410
Norfolk, VA 23510

Hampton Roads Community Foundation Health and Human Service Grants 2188

The Foundation's mission is to inspire philanthropy and transform the quality of life in southeastern Virginia. The Foundation believes the region will thrive only when its most vulnerable residents have the opportunities and support needed to succeed. Health and Human Services Grants focus on providing opportunities for disadvantaged people to become self-sufficient and support innovative programs that: prevent and alleviate homelessness; improve delivery of basic human services to those in need; improve access to medical care, including dental care and mental health; develop job skills and employment opportunities; and develop sound financial education and savings programs to help low income people build financial assets for long-term economic well-being. Applications are available on the website. Applicants from the Eastern Shore of Virginia should speak with a Program Officer before submitting a proposal.

Requirements: Nonprofit organizations serving residents of south Hampton Roads (Chesapeake, Franklin, Norfolk, Portsmouth, Suffolk, Virginia Beach and Isle of Wight County) and the Eastern Shore of Virginia are eligible. Proposals must articulate the program's desired outcomes and the measurement of effectiveness in reaching those outcomes.

Restrictions: Funding is generally not available for: individuals; fundraising events (such as tickets, raffles, auctions or tournaments), annual fundraising appeals or agency celebrations; ongoing operating support; capital projects and facilities and equipment upgrades that can be considered routine maintenance or replacements; houses of worship unless applying for the Nightingale Fund for faith community nursing or the E.K. Sloane Piano Fund; religious activities (organizations and activities that require religious participation by those receiving services); political or fraternal organizations; endowment building; existing obligations, debts/liabilities or costs that the agency has already incurred; scholarly research; scholarships, camper fees, fellowships or travel; passenger

vans for transporting youth; national or international organizations or purposes; hospitals and similar health-care facilities unless applying for a Special Interest Grant; projects or services normally considered the responsibility of government; private primary or secondary schools, daycare facilities or academies other than those whose primary purpose is for students with special needs; and capital campaign requests exceeding 5% of campaigns valued at $1 million or more.
Geographic Focus: Virginia
Date(s) Application is Due: Jul 15
Samples: Chesapeake Service Systems, Chesapeake, Virginia, $35,544 - to make laundry equipment safer and healthier for disabled workers (2011); Habitat for Humanity of South Hampton Roads, Norfolk, Virginia, $32,000 - new computers and service (2011).
Contact: Program Officer; (757) 622-7951; grants@hamptonroadscf.org
Internet: http://www.hamptonroadscf.org/nonprofits/communityGrants-Health.html
Sponsor: Hampton Roads Community Foundation
One Commercial Plaza, Suite 1410
Norfolk, VA 23510

Hancock County Community Foundation Grants **2189**
The Hancock County Community Foundation has been working to enhance the quality of life in Hancock County, Indiana since 1992. The Community Foundation has established an annual grant program that is available on a competitive basis to incorporated, 501(c)3 organizations serving Hancock County. Each grant program has its own criteria used to evaluate applications. The foundation welcome and encourage contact from organizations that are considering applying for a grant and are happy to clarify and respond to questions concerning policies, guidelines, or application procedures.
Requirements: Annual grant programs begin with a community announcement in which nonprofits are asked to submit a Letter of Intent. Upon review, the invitation to apply for a grant will be made to qualifying organizations. Proposals will be reviewed and compared with other proposals received. Those proposals which best meet our criteria are selected for funding.
Restrictions: The Community Foundation has established an annual grant program that is available on a competitive basis to incorporated, 501(c)3 organizations serving Hancock County.
Geographic Focus: Indiana
Contact: Rachel Strong, Program Officer; (317) 462-8870, ext. 226; rstrong@hccf.cc.
Internet: http://www.hccf.cc/granttypes.asp
Sponsor: Hancock County Community Foundation
312 Main Street
Greenfield, IN 46140

Hannaford Charitable Foundation Grants **2190**
The Foundation awards grants to nonprofit organizations in its areas of interest, including health and welfare, educational institutions, civic and cultural organizations, and other local charitable organizations. Types of support include capital campaigns, scholarship funds, and exchange programs. There are no application forms or deadlines. Preference for funding is given to organizations or programs that involve Hannaford associates, are located in Hannaford's marketing area, and have the potential to provide ongoing services for their customers. Small to medium requests are reviewed monthly. Allow three to four months for a response to larger requests ($50,00 or more), as these are reviewed quarterly. Large grants are usually reserved for capital drives by organizations with strong community-impact potential.
Requirements: To apply, the organization or program must: have an active and responsible board of trustees; exhibit ethical publicity methods and solicitation of funds; provide for an appropriate audit to reveal income and disbursements in reasonable detail; demonstrate long-term financial viability; be tax-exempt as described in both sections 501(c)3 and 509(a)1, 509(a)3, or 509(a)3 of the Internal Revenue Code. In addition to a letter of inquiry sent to the Foundation, organizations should also send ten copies of the following information: name, address and telephone number of your organization; contact person and title; amount requested; population and geographic area served; a two- or three-sentence mission statement of your organization, with a brief description of its history; a two- or three-page description of the specific project or program for which you are seeking funding; a list of current and potential funding sources; and a recent statement of revenues and expenses. They should also send one set of the following information: a copy of the organization's tax exemption letter, indicating both sections 501(c)3 and 509(1) status; most recent Form 990 return; and a letter attesting that the organization's tax-exempt status is current.
Restrictions: The Foundation does not offer support for the following: individuals; tax-supported institutions; institutions that, by virtue of their charters, programs or policies, are open to a relatively small or restricted segment of the public; operations of veterans, fraternal or religious organizations, except those that make their services fully available to the community on a nonsectarian basis; program advertising; operating expenses; scholarship programs outside of the Foundation's own; and organizations or events outside of the Foundation's marketing area.
Geographic Focus: Maine, Massachusetts, New Hampshire, New York, Vermont
Contact: Grants Administrator; (507) 931-1682
Internet: http://www.hannaford.com/content.jsp?pageName=charitableFoundation&leftNavArea=AboutLeftNav
Sponsor: Hannaford Charitable Foundation
P.O. Box 1000
Portland, ME 04104

Harden Foundation Grants **2191**
The Harden Foundation considers grants in the following focus areas: children, youth, and families; senior citizens; general health; agricultural education; animal welfare; environment; and arts and culture. The goals of the Foundation are to improve the well-being of youth; strengthen families; develop individual self-reliance and health; prevent inappropriate institutionalization of individuals; improve the quality of life through the cultural activities; encourage more humane treatment of animals; and eliminate duplication and improve coordination of social and community services in Monterey County areas, with emphasis on the Salinas Valley. Organizations may refer to the Foundation website for examples of funded projects in the Monterey area.
Requirements: Organizations should download the Common Grant application from the Foundation's website. Applicants must also include the following attachments: the completed grant application checklist; a completed and signed grant application form, with signature of an authorized representative; a detailed list of Board of Directors and staff roster; organizational chart, if available; current annual budget; a recent financial statement; year-to-date organizational financial statements; a detailed project budget; a list of the ten largest financial gifts received in the most recent fiscal year; a letter from the board chairperson or board member indicating approval for this application; and, if applicable, a completed grant report for a previous grant. The organization must also submit a 3-5 page narrative briefly describing the organization's history, major accomplishments, and their current programs and activities. They should also include a description of their constituency, how they are involved in their work, and how they benefit from the program. The Common Impact Evaluation Plan is not required for Harden Foundation applicants, but applicants must also include the following: a problem statement describing what problems, needs or issues are addressed, how this was determined, and how the project addresses and/or changes the underlying or root cause of the problem. If the funding requested is not for operating support, describe the project, why it is being pursued, and whether it is new or an expansion of an existing program. Include a list of all other grant requests, pending and approved, for this project, showing funding source and amount requested. Please refer to the Common Grant Guidelines for specific instructions on submitting the application. Applications may be mailed, dropped off, or emailed to the Foundation's office. Faxed applications are not accepted.
Restrictions: The Harden Foundation does not make grants to organizations headquartered outside of Monterey County unless they demonstrate service to the Monterey area. The Foundation does not fund: organizations that do not have a current 501(c)3 status; organizations that support sectarian religious programs; creation or addition to endowments; annual events, conferences, or fundraising events; academic or medical research or scholarships to individuals; foundations or associations established for the benefit of an organization which receives substantial tax support; individuals; political parties, candidates or partisan political organizations or activities.
Geographic Focus: California
Date(s) Application is Due: Mar 1; Sep 1
Amount of Grant: 5,000 - 100,000 U.S.D
Contact: Joseph Grainger, Executive Director; (831) 442-3005; fax (831) 443-1429; joe@hardenfoundation.org
Internet: http://www.hardenfoundation.org/grant-information-introduction.html
Sponsor: Harden Foundation
P.O. Box 779
Salinas, CA 93902

Harley Davidson Foundation Grants **2192**
The mission of the Foundation is to support communities with funding and employee volunteerism. The Foundation reaches out to build healthy, thriving communities by placing an emphasis on education, community revitalization, arts and culture, health, and the environment. The Foundation targets areas of greatest need among under-served populations to enhance the quality of life in their communities. The Foundation also supports selected national causes, including veterans initiatives. In making granting decisions, the Foundation looks closely for the following detail: the proposal's relevance to the Foundation's areas of interest; the proposal's clarity in stating the expected project or program outcomes and the strategy for achieving them; the extent to which the strategy involves a collaborative approach to solving a problem or issue affecting the targeted groups or organizations. The program you are nominating should focus on one of the Foundation's funding priorities: education (core curriculum, academic enhancers), community revitalization, job enablers, neighborhood, social services, arts & culture, health, environment. The Common Application Form is located on the company website.
Requirements: 501(c)3 or 170(c) organizations serving metro Milwaukee, Wauwatosa, Franklin, Menomonee Falls or Tomahawk, WI; York, PA; Talladega, AL; or Kansas City, MO are eligible.
Restrictions: The Foundation does not make grants to individuals, political causes or candidates, operating or endowment funds; athletic teams, or religious organizations (unless engaged in a major project benefiting the greater community). The Foundation does not fund conferences or capital campaigns.
Geographic Focus: Alabama, Missouri, Pennsylvania, Wisconsin
Date(s) Application is Due: Jan 23; May 22; Oct 24
Contact: Mary Anne Martiny, Secretary; (414) 343-4001; fax (414) 343-4254; ma.martiny@harley-davidson.com
Internet: http://www.harley-davidson.com/CO/FOU/en/foundation.asp?locale=en_U.S.&bmLocale=en_U.S.
Sponsor: Harley Davidson Foundation
P.O. Box 653
Milwaukee, WI 53201

Harold Alfond Foundation Grants 2193

The foundation awards grants to eligible organizations in its areas of interest, including secondary education, higher education, medical research, health care, and general charities. Grants support: education—public and private colleges and universities, as well as private secondary schools, to fund athletically oriented capital projects and scholarship endowments; medical research—individual research projects are considered when sponsored by a recognized medical research center; health—community support for capital campaigns and endowment funds; and general charities—community organizations that support youth, the arts, persons with disabilities, underprivileged, substance abuse rehabilitation, and annual fund drives of national organizations focusing on the above areas. There are no application deadlines or forms.

Restrictions: Grants are not made to individuals.

Geographic Focus: Florida, Maine

Amount of Grant: Up to 2,000,000 U.S.D

Contact: Gregory Powell, (207) 828-7999

Sponsor: Harold Alfond Foundation
Two Monument Square
Portland, ME 04101-4093

Harold Amos Medical Faculty Development Program 2194

This program offers four-year, postdoctoral research fellowships to minority physicians who have demonstrated superior academic and clinical skills and who are committed to careers in academic medicine. The program seeks to increase the number of minority faculty physician scientists who can achieve senior rank in academic medicine and who will encourage and foster the development of succeeding classes of minority physicians. Applications are accepted from African American, Mexican American, Native American, and mainland Puerto Rican physicians. (Mainland Puerto Ricans are those who have completed at least their postsecondary education on the mainland.) Fellows will pursue biomedical research, clinical investigation, or health services research. Participants will conduct research in association with senior faculty members at academic centers.

Requirements: Eligible to apply are minority physicians who are U.S. citizens at the time of application, have excelled in their postsecondary education within the United States, are now completing or will have completed formal clinical training, and are committed to academic careers.

Date(s) Application is Due: Mar 23

Amount of Grant: 91,350 U.S.D

Samples: Baylor College of Medicine and U of Utah's School of Medicine (UT)—for awards through the Minority Medical Faculty Development Program, $365,400 each; U of Medicine and Dentistry of New Jersey (NJ)—for programs to help minorities and disadvantaged people pursue careers in healthcare, $200,953; U of North Carolina at Chapel Hill (NC)—for faculty development programs for minority medical faculty members, $365,400; New York U (NY)—for faculty development programs for minority medical faculty members, $365,400.

Contact: National Program Office; (301) 565-4080; amfdp@starpower.net

Internet: http://www.amfdp.org/applicat.htm

Sponsor: Robert Wood Johnson Foundation
College Road E and Route 1, P.O. Box 2316
Princeton, NJ 08543-2316

Harold and Arlene Schnitzer CARE Foundation Grants 2195

The Foundation's principal purpose is to assist with Jewish, cultural, youth, education, medical, social service, and community activities. Foundation focus is on proposals for projects that enhance the quality of life in Oregon and Southwest Washington. The Foundation funds grant requests for operating expenses, special projects, matching grants, multiple-year grants and capital campaigns as well as for purchase of specific items. There is no formal application form. Interested applicants may contact the Foundation for guidelines.

Requirements: Nonprofit organizations in Oregon and Washington are eligible. Organizations located in Portland are given first funding priority. Approximately 90% of grant funds stay in the Portland metropolitan area.

Restrictions: The Foundation does not provide funds for individuals, non tax-exempt organizations, other private foundations or political groups.

Geographic Focus: Oregon, Washington

Date(s) Application is Due: Feb 28; May 31; Aug 31; Nov 30

Amount of Grant: 1,000 - 5,000,000 U.S.D

Samples: Mittleman Jewish Community Center, Portland, OR, $400,000—for operating support; Oregon Health and Science University Foundation, Portland, OR, $5,500,000— for Diabetes Center; Give Them Wings, Hood River, OR, $25,000—for Wyeth Work Camp renovation.

Contact: Barbara Hall, Vice President; (503) 973-0286; fax (503) 450-0810

Sponsor: Harold and Arlene Schnitzer Foundation
P.O. Box 2708
Portland, OR 97208-2708

Harold and Rebecca H. Gross Foundation Grants 2196

The Harold and Rebecca H. Gross Foundation was established in 2006 to support and promote charitable organizations that assist persons with physical disabilities to become better adjusted to their environments. Preference is given to organizations that provide direct services to physically disabled people. Grants from the Gross Foundation are primarily one year in duration. Multi-year grants may be considered on a case-by-case basis. The Harold and Rebecca H. Gross Foundation utilizes a two-phase application process. Interested organizations must submit a one-page concept paper as the first step in the process. Concept papers to the Gross Foundation are due on May 1. Applicants will be notified by letter of the Trustee's decision to invite a full proposal within four to six weeks after the concept paper deadline. Applications to the Gross Foundation are by invitation only and are due on October 1. Notification of the grant decision will be made in writing within two to three months after the application deadline. Detailed concept paper guidelines are available at the grant website. A link to the foundation's funding history is also available at the grant website.

Requirements: Applicants must have 501(c)3 tax-exempt status.

Restrictions: Grant requests for research or capital projects will not be considered. The foundation does not support requests from individuals, organizations attempting to influence policy through direct lobbying, or any political campaigns.

Geographic Focus: All States, Connecticut

Date(s) Application is Due: Oct 1

Amount of Grant: 10,000 - 60,000 U.S.D

Samples: Saint Jude Children's Hospital, Tennessee, $55,000 (2010); Hospital for Special Care, New Britain, Connecticut, $29,615; General Israel Orphan Home for Girls Jerusalem, New York, $20,000 (2010)

Contact: Kate Kerchaert, Senior Foundation Officer; (860) 657-7016 or (860) 952-7405; kate.kerchaert@baml.com

Internet: https://www.bankofamerica.com/philanthropic/fn_search.action

Sponsor: Harold and Rebecca Gross Foundation
200 Glastonbury Boulevard, Suite # 200, CT2-545-02-05
Glastonbury, CT 06033-4056

Harold Brooks Foundation Grants 2197

Harold Brooks, of Braintree, Massachusetts, was a successful business executive and entrepreneur who manufactured and sold prefabricated structures and underground bomb shelters during the Cold War. Mr. Brooks died in 1963 and the Foundation that bears his name was established in 1984. The Harold Brooks Foundation provides assistance to causes/organizations that help the largest possible number of residents of Massachusetts' South Shore communities, especially those that support the basic human needs of South Shore residents. The foundation supports nonprofit organizations that have the greatest impact on improving the human condition and/or that provide the neediest South Shore residents with "tools" that will help them restore their lives. The foundation focuses on five key areas: Education; Food, Agriculture, & Nutrition; Health; Housing & Shelter; and Mental Health. Multi-year (two-three years maximum) requests are welcome. Applicants must apply online at the grant website. Applicants are strongly encouraged to do the following before applying: review the downloadable Massachusetts state application procedures for additional helpful information and clarifications; review the downloadable online-application guidelines at the grant website; review the foundation's funding history (link is available from the grant website); review the online application questions in advance; and review the list of required attachments. These will generally include: a list of board members, financial statements (audited, reviewed, or compiled by independent auditor); an organization summary; a list of other funding sources; an IRS Determination letter; and other required documents. All attachments must be uploaded in the online application as PDF, Word, or Excel files. The Harold Brooks Foundation has biannual deadlines of April 1 and October 1. Applications must be submitted by 11:59 p.m. on the deadline date. Applicants for the April deadline will be notified of grant decisions before June 30. Applicants for the October deadline will be notified before December 31.

Requirements: Grants are made to 501(c)3 tax-exempt organizations that serve the following South Shore residents: Abington, Braintree, Bridgewater, Brockton, Carver, Cohasset, Duxbury, Hanover, Hanson, Hingham, Holbrook, Hull, Marshfield, Norwell, Pembroke, Plymouth, Quincy, Randolph, Rockland, Scituate, Weymouth, and Whitman.

Restrictions: Grants are made to support program/project expenses. General operating support and support for endowment campaigns are not provided. Only one request will be accepted from an organization during a twelve-month period. The foundation does not support requests from individuals, organizations attempting to influence policy through direct lobbying, or any political campaigns.

Geographic Focus: Massachusetts

Date(s) Application is Due: Apr 1; Oct 1

Samples: Catholic Charities South, Brockton, Massachusetts, $20,000 (2010); South Shore Health and Educational Foundation, Weymouth, Massachusetts, $60,000 (2010); Homeowners Options for Massachusetts Elders, Braintree, Massachusetts, $25,000 (2010).

Contact: Miki C. Akimoto, Vice President; (866) 778-6859; miki.akimoto@baml.com

Internet: https://www.bankofamerica.com/philanthropic/fn_search.action

Sponsor: Harold Brooks Foundation
225 Franklin Street, 4th Floor, MA1-225-04-02
Boston, MA 02110

Harold R. Bechtel Charitable Remainder Uni-Trust Grants 2198

Established in Davenport, Iowa, in 1987, the Harold R. Bechtel Charitable Remainder Uni-Trust supports programs in Scott County, Iowa, with giving primarily for health associations, education, and for children, youth, and social services. The Foundation's primary interest areas include: children and youth services; community and economic development; education; family services; health organizations and associations; heart and circulatory research; higher education; human services; museums, music; and performing arts. Types of support building construction and renovation, as well as program-related funding. There are no specific deadlines, and applicants should contact the office in writing or via telephone to request application forms.

Requirements: 501(c)3 organizations serving the residents of Rock County, Iowa, are eligible to apply.

Restrictions: No grants are given to individuals, or for endowment funds, debt retirement, past operating deficit, general or continuing support, or for scholarly research in an established discipline.

Geographic Focus: Iowa
Amount of Grant: Up to 150,000 U.S.D
Samples: River Music Experience, Davenport, Iowa, $150,000; Saint Ambrose University, Davenport, Iowa, $150,000; Figge Art Museum, Davenport, Iowa, $125,000.
Contact: R. Richard Bittner, Trustee; (563) 328-3333; fax (563) 328-3352; loseband@blwlaw.com
Sponsor: Harold R. Bechtel Charitable Remainder Uni-Trust
201 West 2nd Street, Suite 1000
Davenport, IA 52801-1817

Harold R. Bechtel Testamentary Charitable Trust Grants 2199

Established in Davenport, Iowa, in 1989, the Harold R. Bechtel Testamentary Charitable Trust Grants gives primarily to higher education and youth services. Major fields of interest include: children and youth services; community development; economic development; education; human services; and medical research. Applicants should contact the office in writing or via telephone to request application forms.
Requirements: 501(c)3 organizations serving the residents of Davenport, Iowa, Rock Island, Illinois, and Moline, Illinois, are eligible to apply.
Restrictions: No grants are given to individuals, or for endowment funds, debt retirement, past operating deficit, general or continuing support, or for scholarly research in an established discipline.
Geographic Focus: Iowa
Amount of Grant: Up to 150,000 U.S.D
Contact: R. Richard Bittner, Trustee; (563) 328-3333; fax (563) 328-3352; loseband@blwlaw.com
Sponsor: Harold R. Bechtel Testamentary Charitable Trust
201 West 2nd Street, Suite 1000
Davenport, IA 52801-1817

Harold Simmons Foundation Grants 2200

The foundation awards grants to Texas nonprofit organizations, with emphasis on social services, religion, health, the arts, and youth. Grants also support community programs and projects, child development, and adult basic education/literacy programs. The foundation also supports international development and relief efforts in Third World countries. Grants are awarded for general operating support, annual campaigns, capital campaigns, building construction/renovation, continuing support, seed money, and program development. Application forms are not required, and there are no deadline dates.
Requirements: Dallas, TX, nonprofits are eligible.
Restrictions: Grants are not awarded to support individuals or for endowment funds or loans.
Geographic Focus: Texas
Samples: Children's Medical Ctr Dallas (TX)—to establish a hospital within the medical center devoted to the needs of child cardiac patients, $5 million; U of Texas Southwestern Medical Ctr (Dallas)—to enhance the Simmons Comprehensive Cancer Center, including supporting the work of its newly recruited director, $15.4 million.
Contact: Lisa Simmons Epstein, President; (972) 233-2134
Sponsor: Harold Simmons Foundation
5430 LBJ Freeway, Suite 1700
Dallas, TX 75240-2697

Harris and Eliza Kempner Fund Grants 2201

The foundation provides grants primarily in the Galveston, TX, area to qualifying organizations in the broad areas of the arts, historic preservation, community development, education, health, and human services. The foundation gives preference to requests for seed money, operating funds, small capital needs, and special projects partnering with other funding sources. Application information is available online.
Requirements: Grants are made primarily to Texas qualifying organizations in the greater Galveston area.
Restrictions: Funding is not available for: fund-raising benefits; direct mail solicitations; grants to individuals; and grants to non-U.S.A based organizations.
Geographic Focus: Texas
Date(s) Application is Due: Mar 15; Oct 15
Samples: Galveston Art League, Galveston, TX, $5,000—matching grant to purchase gallery space; Avenue L Missionary Baptist Church, Galveston, TX, $20,000—renovation and restoration repairs to Fellowship Hall; Galveston Island Swim Team, Galveston, TX, $$4,600—program expansion to include all Galveston Island youth.
Contact: Harrette N. Howard, Grants Administrator; (409) 762-1603; fax (409) 762-5435; information@kempnerfund.org
Internet: http://www.kempnerfund.org/app/programs.html
Sponsor: Harris and Eliza Kempner Fund
2201 Market Street, Suite 601
Galveston, TX 77550-1529

Harris Bank Foundation Grants 2202

Grants are made to a broad range of cultural, educational, community, and civic organizations in Chicago, IL, including organizations that help ethnic and/or minority groups. Education grants are made to private four-year institutions to help defray operating expenses, limited to colleges and universities within Cook County that provide comprehensive accredited programs. Education-related grants considered are unrestricted for operating expenses of recognized, nonprofit Chicago educational institutions other than schools, such as museums, public television, career development programs, and so forth. Culture and fine arts grants are given to organizations that foster an appreciation of the arts such as opera companies, art institutes, and symphony orchestras. Health care grants are available to local health organizations that increase minority community access to primary health care. The bank will fund programming for handicapped, drug rehabilitative, and other health care services on a selective basis. For civic affairs grants, the foundation considers organizations whose work and service enhance the business and community spirit of greater Chicago. Harris Bank also supports the revitalization of Chicago's neighborhoods throughout the city, makes grants for initial expenses of new organizations or new projects of established organizations that provide social services, especially to minority groups, and also provides funding for capital expenses of metropolitan organizations whose objectives are to respond to charitable, educational, scientific, medical, or cultural needs. Grants are approved at bimonthly meetings, usually on a yearly basis; however, the committee may approve multiple-year grants.
Requirements: Grants are given only to tax-exempt organizations located in the Chicago metropolitan area. A proposal should include a statement of the proposed project; project budget and potential funding sources; background information about the organization; evidence of tax-exempt status; and current, audited financial statements.
Restrictions: Grants are not made for or to political activities, individuals, individual sectarian or religious organizations, fraternal organizations, nor for testimonial drives, advertisements in souvenir or program books, or raffle tickets.
Geographic Focus: Illinois
Samples: Art Institute of Chicago (IL)—for general operating support, $14,000; YMCA of Metropolitan Chicago (IL)—for Duncan YMCA renovation and expansion, $10,000; Habitat for Humanity Uptown Chicago (IL)—for general operating support, $10,000.
Contact: Mary Houpt, Secretary Treasurer; (312) 461-5834; fax (312) 293-4702
Sponsor: Harris Bank Foundation
111 W Monroe Street, P.O. Box 755
Chicago, IL 60690

Harrison County Community Foundation Grants 2203

The foundation awards grants to eligible Indiana nonprofit organizations in its areas of interest, including arts and culture, community projects, education, environment, animal protection, health and safety, historical preservation, human services, and recreation. The foundation also awards mini grants to support small projects outside of the regular semiannual grant cycles. Mini grants support unexpected costs such as emergency equipment repairs or purchases, special event support, or other non-routine, nonrecurring expenses. Grants to religious organizations are limited to projects that benefit the community at large, preserve a historical site (typically 100 years or older), or respond to an immediate health or safety related problem. Each spring, the foundation provides funds to each of the county's four high schools for scholarships. Scholarship recipients are selected by the individual high school scholarship selection committee, who also determine the amounts awarded.
Requirements: Nonprofit agencies providing services to residents of Harrison County, IN, are eligible to apply. 501(c)3 tax-exempt organizations and schools, religious organizations, and local governmental units are eligible. All graduating high school seniors in Harrison County are eligible for scholarships.
Restrictions: The foundation does not award grants to purchase real estate that has not been identified and an offer accepted; political activities or those designed to influence legislation; individuals (except scholarships); travel associated with a school-sponsored event; or religious organizations for the purpose of furthering that religion. Traditional equipment, routine maintenance, or facility improvements will not be funded.
Geographic Focus: Indiana
Date(s) Application is Due: Jan 15; Jul 15
Samples: South Central Elementary School (Elizabeth, IN)—to purchase books, $19,986; Ctr for Women and Families (Louisville, IN)—to provide domestic-violence and rape-crisis services in Harrison County, $12,000; Bellarmine U (Louisville, KY)—to provide scholarships to students from Harrison County, $16,000; Harrison County Fire Chiefs Assoc (Corydon, IN)—for training and equipment testing, $45,000.
*Contact:*Administrator; (812) 738-6668; fax (812) 738-6864; staff@hccfindiana.org
Internet: http://www.hccfindiana.org
Sponsor: Harrison County Community Foundation
1523 Foundation Way, P.O. Box 279
Corydon, IN 47112

Harry A. and Margaret D. Towsley Foundation Grants 2204

The foundation awards grants to Michigan organizations in its areas of interest, including environment, medical and preschool education, social services, continuing education, and research in the health sciences. Types of support include annual campaigns, building construction and renovation, capital campaigns, continuing support, employee matching gifts, endowments, general operating support, matching/challenge support, professorships, program development, research, and seed grants. There are no application forms; submit a letter of inquiry between January and the listed application deadline.
Restrictions: Grants are not awarded to individuals or for travel, scholarships, fellowships, conferences, books, publications, films, tapes, audio-visual or communication media, or loans.
Geographic Focus: Michigan
Date(s) Application is Due: Mar 31
Samples: U of Michigan at Ann Arbor, School of Music ((MI)—to construct a theater and performing-arts center, $1.5 million; Nature Conservancy, Michigan Chapter (Lansing, MI)—for the Northern Great Lakes Forest Project, which will protect 271,000 acres across eight counties in Michigan's Upper Peninsula and link more than 2.5 million acres of protected federal, state, and natural areas, $2 million.
Contact: Lynn White, President; (989) 837-1100; fax (989) 837-3240
Sponsor: Harry A. and Margaret D. Towsley Foundation
140 Ashman Street, P.O. Box 349
Midland, MI 48640

Harry B. and Jane H. Brock Foundation Grants 2205

The Harry B. and Jane H. Brock Foundation honors a former Fort Payne, Alabama, native who changed the structure of the banking industry in Alabama. Today, the Foundation awards grants to eligible Alabama organizations to support community development and higher education. Primary areas of interest include community service, community funds, education, the environment, higher education, volunteerism, women's services, social services, and cancer treatment and research. Types of support include: annual campaigns, capital campaigns, endowments, general operating support, program development, and research. Written proposals should describe the project and organization and include a copy of the IRS tax-determination letter. The annual deadline is November 1.
Requirements: 501(c)3 organizations serving the residents of Alabama are eligible.
Restrictions: No grants to individuals are awarded. Giving primarily in Birmingham and Huntsville, Alabama.
Geographic Focus: Alabama
Date(s) Application is Due: Nov 1
Amount of Grant: 1,000 - 25,000 U.S.D
Samples: Alabama Business Hall of Fame, Tuscaloosa, Alabame, $1,700 - general operating support (2010); Alabama Wildlife Rehab, Pelham, Alabama, $2,500 - general operating support (2010); Alexis De Tocqueville Society, Birmingham, Alabama, $25,000 - general administration (2010).
Contact: Harry B. Brock, Jr.; (205) 939-0236 or (205) 918-0833; fax (205) 939-0806
Sponsor: Harry B. and Jane H. Brock Foundation
2101 Highland Avenue, Suite 250, P.O. Box 11643
Birmingham, AL 35202-1643

Harry Bramhall Gilbert Charitable Trust Grants 2206

The Harry Bramhall Gilbert Charitable Trust supports tax-exempt organizations that contribute to the health, education, and cultural life of the Tidewater, Virginia region. The Trust currently and substantially funds nonprofits based in the cities of Norfolk, Chesapeake, and Virginia Beach, Virginia. To apply, submit a letter to the Trust (no application form is required) containing the following information: copy of IRS Determination Letter; copy of most recent annual report/audited financial statement/990; listing of board of directors, trustees, officers and other key people and their affiliations; detailed description of project and amount of funding requested. Include two copies.
Requirements: Virginia nonprofit organizations based in the cities of Norfolk, Chesapeake, and Virginia Beach are eligible.
Restrictions: No support for religious or political organizations. No grants to individuals.
Geographic Focus: Virginia
Date(s) Application is Due: Sep 30
Amount of Grant: 2,500 - 100,000 U.S.D
Samples: Chesapeake Public Library Foundation, Chesapeake, VA, $5,000—for additions to collections of books, magazines, newspapers, computers, and software at Central Library; University of Virginia, Charlottesville, VA, $100,000—for merit scholarships for undergraduate students in their fourth-year at UVA meeting certain criteria, who have graduated from Public High School of City of Chesapeake, VA; Virginia Symphony, Norfolk, VA, $25,000—for general support and the Virginia Symphony Endowment Fund.
Contact: Stuart D. Glasser, Treasurer; (757) 204-4858; sdglasser@cox.net
Internet: http://fdncenter.org/grantmaker/gilbert
Sponsor: Harry Bramhall Gilbert Charitable Trust
316 Scone Castle Loop
Chesapeake, VA 23322

Harry Edison Foundation 2207

Harry Edison Foundation was incorporated in Missouri in 2003 by one of the founders of Edison Brothers Stores, among five sons of a Latvian immigrant who had peddled shoes off a pack mule in southern Georgia. Giving is primarily centered around the Saint Louis, Missouri, region in the areas of education, hospitals, human services, Jewish organizations, and medical research. Major types of support include: annual campaigns, building and renovation costs, capital campaigns, professorships, research, and scholarship endowments. There are no specific application forms or annual deadlines, and applicants should begin by contacting the Foundation directly. Most recently, support has ranged from $100 to $120,000.
Requirements: Applicants should be 501(c)3 organizations that support the residents of St. Louis, Missouri, and its surrounding region.
Geographic Focus: Missouri
Amount of Grant: 100 - 120,000 U.S.D
Samples: John Burrough's School, St. Louis, Missouri, $120,000 - general research support (2010); St. Louis Community Schools, St. Louis, Missouri, $50,000 - general operating support (2010).
Contact: Bernard A. Edison, President; (314) 331-6504 or (314) 331-6505
Sponsor: Harry Edison Foundation
220 N. 4th Street, Suite A
St. Louis, MO 63102-1905

Harry Frank Guggenheim Foundation Dissertation Fellowships 2208

The Foundation awards ten or more dissertation fellowships each year to individuals who will complete their dissertations within the award year. These fellowships are designed to help doctoral candidates finish writing their dissertations rather than to support dissertation research. Questions that interest the Foundation concern violence and aggression in relation to social change, intergroup conflict, war, terrorism, crime, and family relationships, among other subjects. Priority will also be given to areas and methodologies not receiving adequate attention and support from other funding sources. Applicants may be citizens of any country and studying at colleges or universities in any country. The application deadline is February 1, with final decisions made by the Board of Directors at its June meeting for funding that fall.
Requirements: Applicants must submit the following to apply: title page; abstract page and survey; advisor's letter and abbreviated CV; applicant's CV and graduate school transcript; project description; protection of subjects; and their other support for funding. Applicants should refer to the guidelines for additional information and specific instructions for their abstract page and survey (based on their individual discipline), and contact the Foundation if they have questions.
Restrictions: Dissertations with no relevance to understanding human violence and aggression will not be supported. Applicants should only apply if they are entering the dissertation stage of graduate school (fieldwork or other research is complete and writing has begun). Recipients of the dissertation fellowship must submit a copy of the dissertation, approved and accepted by their institution, within six months after the end of the award year. Any papers, books, articles, or other publications based on the research should also be sent to the foundation.
Geographic Focus: All States, All Countries
Date(s) Application is Due: Feb 1
Amount of Grant: 20,000 U.S.D
Contact: Program Officer; (646) 428-0971; info@hfg.org
Internet: http://www.hfg.org/df/guidelines.htm
Sponsor: Harry Frank Guggenheim Foundation
25 West 53rd Street
New York, NY 10019-5401

Harry Frank Guggenheim Foundation Research Grants 2209

The Foundation invites proposals in any of the natural sciences, social sciences or humanities that promise to increase understanding of the causes, manifestations, and control of human violence and aggression. Highest priority is given to research that can increase understanding and improve urgent problems of violence and aggression in the modern world. Questions that interest the foundation concern violence and aggression in relation to social change, intergroup conflict, war, terrorism, crime, and family relationships, among other subjects. Priority will also be given to areas and methodologies not receiving adequate attention and support from other funding sources. Applicants may be citizens of any country. Most awards fall within the range of $15,000 to $40,000 per year for periods of one or two years. Applications for larger amounts and longer durations must be very strongly justified. Applications must be received by August 1, with a decision given in December.
Requirements: Applicants must mail two copies of the typed application in English to the Foundation. Applications may not be faxed or emailed. Along with two copies of the application, applicants must include the following: title page; abstract and survey; budget and its justification; personnel; research plan; other support; protection of subjects; and referee comments. Applicants should refer to the Foundation website for specific instructions on how to submit each of the attachments, and contact the Foundation if they have any questions.
Restrictions: Research with no relevance to understanding human problems will not be supported, nor will proposals to investigate urgent social problems where the Foundation cannot be assured that useful, sound research can be done.While almost all recipients of the Foundation grants possess a Ph.D., M.D., or equivalent degree, there are no formal degree requirements for the grant. The grant, however, may not be used to support research undertaken as part of the requirements for a graduate degree. Applicants need not be affiliated with an institution of higher learning, although most are college or university professors. The foundation awards research grants to individuals (or a few principal investigators at most) for individual projects, but does not award grants to institutions for institutional programs. Individuals who receive research grants may be subject to taxation.
Geographic Focus: All States, All Countries
Date(s) Application is Due: Aug 1
Amount of Grant: 15,000 - 40,000 U.S.D
Contact: Program Officer; (646) 428-0971; fax (646) 428-0981; info@hfg.org
Internet: http://www.hfg.org/rg/guidelines.htm
Sponsor: Harry Frank Guggenheim Foundation
25 West 53rd Street
New York, NY 10019-5401

Harry Kramer Memorial Fund Grants 2210

The Harry Kramer Memorial Fund, established in 1982, supports Jewish organizations in the United States and Israel, involved with the care of the sick or aged, education, and religious organizations. Giving primarily in the southern Florida area. Areas of interest include: aging, centers/services; arts; disasters, 9/11/01; higher education; human services; international terrorism; Jewish federated giving programs; disabled individuals.
Requirements: There are no specific deadlines with which to adhere. Contact the Foundation for further application information and guidelines.
Restrictions: No grants to individuals, or for operating budgets or continuing support.
Geographic Focus: All States, Israel
Amount of Grant: 1,800 - 15,000 U.S.D
Samples: Jewish Museum of Florida, Miami Beach, FL, $15,000; PEF Israel Endowment Funds, New York, NY, $5,000; American Friends of Israel Elwyn, Philadelphia, PA, $10,000;
Contact: Leslie J. August, Wachovia Bank Trustee; (305) 789-4645
Sponsor: Harry Kramer Memorial Fund c/o Wachovia Bank, N.A.
100 North Main Street, 13th Floor
Winston-Salem, NC 27150-0001

Harry S. Black and Allon Fuller Fund Grants 2211

The Harry S. Black and Allon Fuller Fund was established in 1930 to support quality health-care and human-services programming for underserved populations. The grantmaking focus is in the areas of health care and physical disabilities. The fund supports access to health care; health education; health/wellness promotion and disease prevention; health policy analysis and advocacy; access programs for physically disabled individuals; disability policy analysis and advocacy; workforce development programs; and programs that improve quality of life for the disabled. Emphasis will be placed on programs serving low-income communities. Grant requests for general operating support or program/project support are strongly encouraged. Applicants must apply online at the grant website. Applicants are strongly encouraged to do the following before applying: review the downloadable state application procedures at the grant website; review the downloadable online-application guidelines at the grant website; review the foundation's funding history (link is available from the grant website); review the online application questions in advance; and review the list of required attachments. These will generally include: a list of board members, financial statements (audited, reviewed, or compiled by independent auditor); an organization summary; a list of other funding sources; an IRS Determination letter; and other required documents. All attachments must be uploaded in the online application as PDF, Word, or Excel files. The Harry S. Black and Allon Fuller Fund has a deadline of July 31. Grant decisions will be made by December 31.
Requirements: Nonprofit organizations must be geographically located within the city limits of New York City or Chicago to be eligible to apply.
Restrictions: The fund generally does not support the following: projects in the areas of health care specific to medical/academic research; organizations or programs that primarily provide mental health services; programs that primarily provide services to either the mentally or developmentally disabled; endowment campaigns; and capital projects. The fund does not support requests from individuals, organizations attempting to influence policy through direct lobbying, or any political campaigns.
Geographic Focus: Illinois, New York
Date(s) Application is Due: Jul 31
Samples: Saint Bernard Hospital and Health Care Center, Chicago, Illinois, $6,250 (2010); American Civil Liberties Union Foundation, New York, New York, $15,000 (2010); Theater Development Fund, New York, New York, $10,000 (2010).
Contact: Christine O'Donnell; (646) 855-1011; christine.l.o'donnell@baml.com
Internet: https://www.bankofamerica.com/philanthropic/fn_search.action
Sponsor: Harry S. Black and Allon Fuller Fund
One Bryant Park, NY1-100-28-05
New York, NY 10036

Harry Shwachman Clinical Investigator Award 2212

This three-year award provides the opportunity for clinically trained physicians to develop into independent biomedical research investigators who have active involvement in cystic fibrosis-related areas. It is also intended to facilitate the transition from postdoctoral training to a career in academic medicine. Applications are due the first Wednesday in September.
Requirements: U.S. citizenship or permanent resident status is required.
Date(s) Application is Due: Sep 1
Amount of Grant: 85,000 U.S.D
Contact: Grants Management; (301) 951-4422; fax (301) 951-6378; grants@cff.org
Internet: http://www.cff.org/research/cystic_fibrosis_foundation_grants/research_grants
Sponsor: Cystic Fibrosis Foundation
6931 Arlington Road, 2nd Floor
Bethesda, MD 20814

Harry Sudakoff Foundation Grants 2213

The Foundation, established in 1956, offers funding primarily in the forms of ongoing support and capital campaigns. Focus areas include: federated giving programs; higher education; and human/social services. Of particular interest are the performing arts, including dance, music, and theater. The geographic focus is Sarasota, Florida, with an initial approach by Letter of Interest (LOI). Though there are no specific application forms, the annual deadline for application is August 15, with final notification by November 1st.
Requirements: Grantees must be 501(c)3 organizations with a sound track record.
Restrictions: Generally the Foundation will not provide grants for: endowments, deficit financing, debt reduction, or ordinary operating expenses; conferences, seminars, workshops, travel, surveys, advertising, fund-raising costs or research; annual giving campaigns; individuals; or projects that have already been completed.
Geographic Focus: Florida
Date(s) Application is Due: Aug 15
Amount of Grant: 46,875 U.S.D
Samples: Mental Health Community Centers, Sarasota, Florida—to provide a 15-passenger van for support programs at Prospect House, $30,000; Suncoast Communities Blood Bank, Sarasota, Florida, $72,500.
Contact: Janet L Dickens, Asst. V. P., Bank of America; (941) 952-2826; fax (941) 952-2768; janet.dickens@ustrust.com
Sponsor: Harry Sudakoff Foundation
Bank of America, 1605 Main Street
Sarasota, FL 34236-5840

Harry W. Bass, Jr. Foundation Grants 2214

The Dallas-based Harry W. Bass, Jr. Foundation seeks to enrich the lives of the citizens of Texas by providing support to qualified organizations in the areas of education, health, human services, civic & community, science, research, or, arts and culture. Grant applications for specific programs or projects, capital projects or, less often, general operations are considered. Endowment gifts are rare. The Harry W. Bass, Jr. Foundation also considers program-related investments as part of its grant-making activities. The foundation strives to be responsive to the needs of all eligible organizations and considers requests of any amount. There is no formal application form. Grant requests are accepted at any time throughout the year. Each organization is limited to one application within a twelve-month period. Requests are usually processed within three to four months. Electronic grant applications should be submitted to dcalhoun@hbrf.org (no file attachments, please). For more information, contact the Harry W. Bass, Jr. Foundation.
Requirements: Texas 501(c)3 tax-exempt organizations are eligible.
Restrictions: In general, grants are not made for purposes of: church or seminary construction; annual fundraising events or general sustentation drives; professional conferences and symposia; out-of-state performances or competition expenses; to other private foundations. Unsolicited grant requests are restricted to organizations based in the Greater Dallas area. In the event a grant is approved, the recipient organization will be unable to submit another grant request for two years due to the foundation policy of not providing subsequent year grants.
Geographic Focus: Texas
Contact: F. David Calhoun, Executive Director; (214) 599-0300; fax (214) 599-0405; dcalhoun@hbrf.org
Internet: http://www.harrybassfoundation.org/about.asp
Sponsor: Harry Bass Foundation
4809 Cole Avenue, Suite 252
Dallas, TX 75205

Hartford Aging and Health Program Awards 2215

The foundation's two principal programs are Health Care Cost and Quality and Aging and Health. The Health Care Cost and Quality program supports the community health management initiative, health care quality measures, and reducing inappropriate health care services. The Aging and Health program supports strengthening physicians' knowledge of geriatrics, reducing medication problems of the elderly, and demonstrating integrated financing and service delivery for comprehensive geriatric services. Types of support include operating budgets, continuing support, projects/programs, research, publications, and conferences and seminars. The foundation also welcomes inquiries regarding projects that may not fit these specific interests but would further its broad goal of improving health care in America. Types of support include general operations, continuing support, projects/programs, research, publications, and conferences and seminars. Applications are accepted at any time and are reviewed four times each year when the board meets.
Requirements: U.S. health, education, and social service organizations may apply.
Restrictions: Requests will be denied for general research or for projects lasting more than three years.
Samples: New York Academy of Medicine (NY)—to help start rotational field-training components, with a focus on working with older adults, in programs that offer master's degrees in social work, $5.1 million over four years; U of California (San Diego, CA) and U of Pittsburgh (PA)—to create new centers of excellence in geriatric medicine in order to increase the number of physician faculty members in the field of geriatrics, $450,000 each over three years; American Federation for Aging Research (New York, NY)—for the Medical Student Summer Research Training in Aging Program, $1.9 million over five years.
Contact: Corinne Rieder, Executive Director; (212) 832-7788; fax (212) 593-4913; mail@jhartfound.com
Internet: http://www.jhartfound.org
Sponsor: John A. Hartford Foundation
55 E 59th Street, 16th Floor
New York, NY 10022-1178

Hartford Courant Foundation Grants 2216

The Foundation established in 1950, supports organizations involved with education, the arts, community development, health and social services, with an emphasis on programs benefiting children, youth and families. Grants in the area of the arts will be made only for outreach or education programs for children or families. Grants may be awarded in support of program, capital and operating expenses, for seed money and as challenge or matching grants. The Hartford Courant Foundation serves: Andover, Avon, Bloomfield, Bolton, Canton, East Granby, East Hartford, East Windsor, Ellington, Enfield, Farmington, Glastonbury, Granby, Hartford, Hebron, Manchester, Marlborough, Middletown, New Britain, Newington, Rocky Hill, Simsbury, Somers, South Windsor, Suffield, Tolland, Vernon, West Hartford, Wethersfield, Windsor, and Windsor Locks.
Requirements: An organization wishing to apply for a grant from The Hartford Courant Foundation must be: within the Foundations service areas; a non-profit organization; submit its request using a Hartford Courant Foundation application form, available on the Foundations website; include supporting documentation.
Restrictions: The Hartford Courant Foundation's policies do not allow funding of: organizations which are not tax exempt under IRS Code section 501(c)3; organizations which have the IRS Private Foundation designation; individuals; endowment funds; religious institutions, other than for their provision of non-sectarian community services; capital projects related to the arts; performances, conferences, trips, one-time events; annual campaigns.
Geographic Focus: Connecticut
Date(s) Application is Due: Mar 15; Jun 15; Sep 15; Dec 15
Contact: Kate Miller, Executive Director; (860) 241-6472; fax (860) 520-6988; kmiller@hcfdn.org
Internet: http://www.hcfdn.org/application/guidelines.htm
Sponsor: Hartford Courant Foundation
285 Broad Street
Hartford, CT 06115

Hartford Foundation Regular Grants 2217

The Foundation awards grants for a broad range of purposes to a wide variety of nonprofit organizations in social services, health, education, early childhood and youth services, arts and culture, housing and neighborhood development, and other charitable fields. These grants: enhance or strengthen existing activities; provide start-up for organizations and new programs; provide support for capital improvements. Interested applicants are encouraged to phone a program contact to discuss their project before beginning the application process.
Requirements: Nonprofit organizations in the following Connecticut towns are eligible to apply for funding: Andover, Avon, Bloomfield, Bolton, Canton, East Granby, East Hartford, East Windsor, Ellington, Enfield, Farmington, Glastonbury, Granby, Hartford, Hebron, Manchester, Marlborough, Newington, Rocky Hill, Simsbury, Somers, South Windsor, Suffield, West Hartford, Wethersfield, Windsor, Windsor Locks, Tolland, and Vernon. Proposals for statewide programs may be considered when there is a substantial benefit to residents of these communities.
Restrictions: The Foundation does not make grants from its unrestricted funds for: sectarian or religious activities; grants directly to individuals; grants to private foundations; endowments or memorials; direct or grass-roots lobbying efforts; conferences; research; or informational activities on topics that are primarily national or international in perspective. In addition, the Foundation generally does not make grants for: federal, state, or municipal agencies or departments supported by taxation; sponsorship of or support for one-time events; liquidation of obligations incurred at a previous date; or sustaining support for recurring operating expenses.
Geographic Focus: Connecticut
Amount of Grant: Up to 500,000 U.S.D
Samples: Town of Farmington, Farmington, CT, $19,749 - for construction of a tower and zip line at the youth center and for the Farmington High School's McDonough School Math and Literacy Tutoring Program; The ARC of the Farmington Valley, Inc., Canton, CT, $375,000 - for remodeling, code compliance, and upgrades to the Alleluia property.
Contact: Cheryl L. Gerrish, Grants Manager; (860) 548-1888; fax (860) 524-8346; gerrish@hfpg.org or hfpg@hfpg.org
Internet: http://www.hfpg.org/GrantmakingPrograms/Overview/tabid/163/Default.aspx
Sponsor: Hartford Foundation
10 Columbus Boulevard, 8th Floor
Hartford, CT 06106

Hartford Geriatrics Health Outcomes Research Scholars Program 2218

This program supports physician-scientists committed to improving the health care of older adults during the critical transition from junior faculty to independent researcher. Outcomes research is defined as the study of functional status, impairments, perceptions, social opportunities, and health services utilization that are influenced by disease, injury, treatment, or health policy, including etiology, diagnosis, and intervention. Research must be focused on older adults and may address clinical strategies and effectiveness, innovative outcomes measures, and quality of life. Scholars must have a research sponsor who should have a strong background in training and outcomes research, as well as expertise and interest in the outcomes research that is being proposed. Guidelines and an application request form are available online.
Requirements: The applicant must be a U.S. citizen or permanent resident; be sponsored at a U.S. allopathic or osteopathic institution by a full-time faculty member; have an MD or DO degree; hold a full-time faculty appointment at the level of assistant professor for no longer than two years at the time the grant becomes effective; demonstrate the motivation and ability to devote at least 75 percent of his or her time to conduct outcomes research related to improving the care of older adults with the goal of devoting his/her research career to this area; be a geriatrician or geriatric psychiatrist who has completed all of the requirements to be eligible to sit for a Certificate of Added Qualifications by the time the award commences (July 1), or a general internist (or family physician, neurologist, or subspecialist of internal medicine) who is pursuing a career devoted to aging research and who has completed all the requirements to be eligible to sit for board certification in his/her discipline; and have at least two years of training with one year of clinical experience at an accredited U.S. institution, and one year of research training at a U.S. academic institution.
Date(s) Application is Due: Jan 15
Amount of Grant: 75,000 U.S.D
Contact: Funding Administrator; (800) 247-8779 or (212) 308-1414; fax (212) 832-8646; sreinthaler@americangeriatrics.org
Internet: http://www.healthinaging.org/research
Sponsor: American Geriatrics Society
350 Fifth Avenue, Suite 801
New York, NY 10118

Harvard Pilgrim Health Care Foundation Community Service Grants 2219

Community Service Grants emphasize public health practices that utilize community-based strategies to identify and respond to the health needs of selected populations, geographic areas, or residents that comprise a community group. Eligible programs will be focused on community-based approaches to health promotion and disease prevention that reduce health disparities among populations disproportionately affected by conditions related to leading health indicator(s), including overweight and obesity, (e.g., diabetes, cardiovascular health, etc.), physical activity, tobacco use, substance abuse, responsible sexual behavior, injury and violence, and mental health. Successful proposals will indicate how a grant from the foundation will assist the grantee organization or coalition to improve one or more of the selected health indicators. In addition, the proposal will reflect the concerns, appropriate strategies, and participation of the population(s) involved and the capacity of the applicant organization to address health disparities with cultural competence. Guidelines and application are available online. At the Web site, click on Foundation.
Restrictions: The foundation does not support capital expenditures or campaigns; endowment contributions; lobbying efforts; or programs or activities that promote specific religious beliefs, doctrines, or practices.
Geographic Focus: Maine, Massachusetts, New Hampshire
Amount of Grant: 1,000 - 20,000 U.S.D
Contact: Mike Devlin, (617) 509-9421; Mike_Devlin@harvardpilgrim.org
Internet: http://www.harvardpilgrim.org/portal/page?_pageid=213,84006&_dad=portal&_schema=P.O.RTAL
Sponsor: Harvard Pilgrim Health Care Foundation
93 Worcester Street, Suite 100
Wellesley, MA 02841-9181

Harvest Foundation Grants 2220

Formed by the sale of Memorial Health Systems, the foundation is managing an endowment to invest in programs and initiatives that will address local challenges in the areas of health, education, and welfare in Martinsville/Henry County. The foundation is committed to honoring the legacy of Memorial Hospital by emphasizing prevention, safety and access to health care; by facilitating opportunities for local citizens to help their community reach its potential; and by improving the learning environment for citizenship, academic, and vocational preparedness. Application and guidelines are available online.
Requirements: To be eligible for consideration, an organization must: be located within, or have its program focused within Martinsville and/or Henry County; have a letter from the IRS stating its 501(c)3 status; and propose a project within one of the Foundation's three interest areas of health, education or welfare.
Restrictions: The foundation does not fund organizations that discriminate based upon race, creed, gender, or sexual orientation; scholarships, fellowships, or grants to individuals; sectarian religious activities, political lobbying, or legislative activities; profit-making businesses; emergency needs or extremely time sensitive requests; or direct replacement of discontinued government support.
Geographic Focus: Virginia
Contact: Allyson Rothrock, Executive Director; (276) 632-3329; fax (276) 632-1878; arothrock@theharvestfoundation.org
Internet: http://www.theharvestfoundation.org/page.cfm/topic/strategic-map
Sponsor: Harvest Foundation
1 Ellsworth Street, P.O. Box 5183
Martinsville, VA 24115

Harvey Prizes 2221

Two prizes are awarded annually to outstanding personalities whose achievements in one of the following fields have served as a source of inspiration to many others: science and technology, human health, and outstanding contribution to peace in the Mideast. Recipients are invited to Israel as guests of the Technion and to spend time at the institute teaching their subjects. Those interested in nominating candidates should contact the society for deadline date and guidelines. Inquiries may also be directed to Ruth Eckstein in Israel at 011-972-4-8292863, email: andree@tx.technionac.il.
Requirements: Nominations may be made by members of the board of governors of Technion and by request from the prize committee. Persons making nominations are requested not to inform the proposed candidates of the nomination.
Restrictions: Candidates may not nominate themselves.
Date(s) Application is Due: Oct 1
Amount of Grant: 35,000 U.S.D
Contact: Miriam Golzman, (212) 407-6300; fax (212) 753-2925; info@ats.org
Internet: http://www.admin.technion.ac.il/harvey/Philosophy.html
Sponsor: American Technion Society - Israel Institute of Technology
55 E 59th Street
New York, NY 10022

Harvey Randall Wickes Foundation Grants 2222

The foundation awards grants to eligible Michigan nonprofit organizations in its areas of interest, including arts, civic affairs, education, health care, and recreation. Types of support include matching gifts and program grants. There are no application deadline dates. Applications must be received three weeks prior to board meetings in January, April, June, and October.
Requirements: Michigan 501(c)3 organizations in Saginaw County are eligible.
Geographic Focus: Michigan
Amount of Grant: 5,000 - 100,000 U.S.D
Contact: Hugo Braun, (517) 799-1850; fax (517) 799-3327; hrwickes@concentric.net
Sponsor: Harvey Randall Wickes Foundation
4800 Fashion Square Boulevard, Plaza N, Suite 472
Saginaw, MI 48604

Hasbro Children's Fund 2223

The mission of the Fund is to assist children triumphing over critical life obstacles as well as bringing the joy of play into their lives. Through the Fund's initiatives, the mission is achieved by supporting programs which provide terminal and seriously ill children respite and access to play, educational programs for children at risk, and basics for children in need. Interested applicants are asked to submit a letter of inquiry through the online system.
Requirements: Applications are accepted from nonprofits around the world.
Restrictions: Funding is not available for: religious organizations; individuals; cash free grants; research; political organizations; scholarships; travel stipends; loans; endowments;

goodwill advertising; sponsorship of recreational activities; fundraisers; auctions; and schools.
Geographic Focus: Massachusetts, Rhode Island, Washington
Date(s) Application is Due: Jun 1
Amount of Grant: 1,000 - 10,000 U.S.D
Contact: Grants Administrator; HCFinfo@hasbro.com
Internet: http://www.hasbro.org/default.cfm?page=grantmaking
Sponsor: Hasbro Children's Fund
1027 Newport Avenue
Pawtucket, RI 02862-1059

Hastings Center International Biomedical Ethics Research Program for International Scholars 2224

This program was inaugurated in response to the growing interest in comparative bioethics and to the number of foreign scholars and medical professionals interested in the work of the center. The center carries out an active research program on timely and crucial subjects, working in a variety of fields—law, medicine, science, philosophy, and religion, among others. The center is presently studying ethical issues related to abortion, neonatal care, surrogate motherhood, organ transplantation, occupational health, health professional-patient relationships, allocation of resources, animal welfare, aging, AIDS, care of the dying and termination of treatment, chronic illness, artificial reproduction, genetic screening, rehabilitation medicine, and justice in health care delivery. International scholars stay at the center for four to six weeks, dividing their time between their own research and the ongoing activities at the center. There is a modest stipend for foreign citizens who require financial assistance. Applications should be submitted four to six months in advance of proposed stay. Applications are accepted at any time.
Requirements: The program is designed for advanced scholars and practitioners who have made or will make a significant contribution to bioethics in their own countries.
Amount of Grant: 500 - 2,000 U.S.D
Contact: Director of Education; (845) 424-4040, ext. 203; fax (845) 424-4545; visitors@thehastingscenter.org
Internet: http://www.thehastingscenter.org/visitors/visitorsint.asp
Sponsor: Hastings Center
Route 9D, 21 Malcolm Gordon Road
Garrison, NY 10524-5555

Hastings Center International Visiting Scholars Program 2225

This program responds to the growing interest in comparative bioethics and to the number of foreign scholars and medical professionals interested in the center's work—examining critical ethical issues in medicine, the life sciences, and the professions. The program is designed for advanced scholars and practitioners who have made or will make significant contributions to bioethics in their own countries. International scholars usually stay two weeks to six weeks, dividing their time between their own research and ongoing activities at the center. There is a modest stipend for foreign citizens who require financial assistance. Applications are due four to six months in advance of proposed stay.
Amount of Grant: 500 - 1,500 U.S.D
Contact: Daniel Callahan, Director, International Programs; (845) 424-4040, ext. 222; fax (845) 424-4545; callahand@thehastingscenter.org
Internet: http://www.thehastingscenter.org/visitors/visitorsint.asp
Sponsor: Hastings Center
Route 9D, 21 Malcolm Gordon Road
Garrison, NY 10524-5555

Hawaii Community Foundation Health Education and Research Grants 2226

The program provides support from three distinct funds: Tobacco Prevention & Control Trust Fund; the Leahi Fund Research Fund; and the Medical Research Program. Funding will be provided for medical education and research in the fields of: cancer; heart disease; lung disease and research. Priority is given to projects that: demonstrate a foreseeable benefit to the people of Hawaii; support new investigators in Hawaii; and support collaborative efforts. Proposal submission information is available online.
Requirements: To be eligible for consideration an organization must be a 501(c)3 organization or a unit of government. The Principal Investigator must be based in Hawaii and conducting the research in Hawaii.
Geographic Focus: Hawaii
Date(s) Application is Due: Feb 27; Jul 17; Aug 14; Sep 10
Amount of Grant: 25,000 - 50,000 U.S.D
Contact: Christel Wuerfel, Philanthropic Services Assistant; (808) 537-6333 or (888) 731-3863; cwuerfel@hcf-hawaii.org
Internet: http://www.hawaiicommunityfoundation.org/index.php?id=71&categoryID=23
Sponsor: Hawai'i Community Foundation
1164 Bishop Street, Suite 800
Honolulu, HI 96813

Hawaii Community Foundation Reverend Takie Okumura Family Grants 2227

The Fund was established by members of the Okumura family to continue the charitable work of Reverend Okumura focusing on the healthy development of Hawaii's young children and youth. Proposals will be accepted in the following areas of focus: youth (ages 6-20 years old) and young children (ages birth to 5 years old). Priorities will be given to programs which: develop the ability to think critically; understand and appreciate one's own culture and those of others; develop the ability to settle differences peacefully; strengthen the early care and education community system. Proposal submission information is available online.
Requirements: Hawai'i organizations that are tax exempt, including nonprofit organizations, 501(c)3 organizations, religious organizations that are exempt from taxation, or units of government are eligible to apply.
Restrictions: Funding is not available for: major capitol projects; endowments; or on-going or general operating costs.
Geographic Focus: Hawaii
Date(s) Application is Due: Aug 5
Amount of Grant: 5,000 - 15,000 U.S.D
Contact: Christel Wuerfel, Philanthropic Services Assistant; (808) 537-6333 or (888) 731-3863; cwuerfel@hcf-hawaii.org
Internet: http://www.hawaiicommunityfoundation.org/index.php?id=71&categoryID=24
Sponsor: Hawai'i Community Foundation
1164 Bishop Street, Suite 800
Honolulu, HI 96813

Hawaii Community Foundation West Hawaii Fund Grants 2228

The West Hawaii Fund was established at the Hawaii Community Foundation in 1992 by a group of concerned citizens for the purpose of accepting charitable gifts for the benefit of the people and communities of West Hawaii, from North Kohala to Hawaiian Ocean View Estates. Today, the Fund continues to grow through additional contributions and planned gifts from donors in the community to respond to current and emerging needs and improve the quality of life for the residents of West Hawaii. The strongest proposals will be those that meet the following criteria: proposal articulates how the proposed project will benefit the West Hawaii community; proposal clearly describes how the project will be implemented and how these particular activities will address critical issues in the West Hawaii community; there are clearly stated outcomes and a plan for measuring and reporting results; project demonstrates collaboration among different sectors of the community; project budget is realistic, relates to project narrative and is reasonable in cost.
Requirements: 501(c)3 organizations - Grant awards will range up to $10,000. Community organizations not designated as 501(c)3 organizations -Grant awards will range up to $2,000. To be eligible for grants of more than $2,000, organization must be designated 501(c)3 or have a 501(c)3 fiscal sponsor.
Restrictions: Projects not likely to be funded: projects that do not benefit the residents of West Hawaii; funds for endowments or for the benefit of specific individuals; out-of-state travel expenses; start-up costs of a new organization.
Geographic Focus: Hawaii
Date(s) Application is Due: Oct 1
Amount of Grant: 2,000 - 10,000 U.S.D
Contact: Diane Chadwick, Fund Director; (808) 885-2174; fax (808) 885-1857; dchadwick@hcf-hawaii.org
Internet: http://www.hawaiicommunityfoundation.org/index.php?id=71&categoryID=22
Sponsor: Hawai'i Community Foundation
1164 Bishop Street, Suite 800
Honolulu, HI 96813

Hawn Foundation Grants 2229

The foundation awards grants to Texas arts, education, health care, hospitals, medical research, and social services organizations. Preference will be given to requests from organizations in Dallas, TX.
Restrictions: The Foundation does not award scholarships or fellowships.
Geographic Focus: Texas
Date(s) Application is Due: Jun 1
Amount of Grant: 1,000 - 200,000 U.S.D
Samples: U of Texas Southwestern Medical Ctr (Dallas, TX)—for medical research and for studies on age-related macular degeneration, $100,000.
Contact: Joe Hawn Jr., (214) 696-6595; fax (214) 696-6596
Sponsor: Hawn Foundation
5949 Sherry Lane, Suite 775
Dallas, TX 75225-8043

Hazel Corbin/Maternity Center Association Grant for Evidence-Based Midwifery Care 2230

The fund provides stipends for students accredited by the American College of Nurse Midwives and accepted in post-RN education programs for nurse-midwifery ending in certification eligibility. Post-RN programs are eight to 12 months in duration, and master's programs usually are from 12 to 24 months. Applicants must write for application forms, enclosing evidence of program acceptance and complete details of the program. Applications are accepted at any time; awards are made as funds permit.
Requirements: Applicants must be registered nurses who are preparing for nurse-midwifery certification in the United States. Students must have a license to practice nursing in a state of the United States and must already have been accepted in an approved nurse-midwifery basic or refresher program at a U.S. school.
Date(s) Application is Due: Apr 30
Amount of Grant: 4,500 U.S.D
Contact: Hazel Corbin Grant, (240) 485-1800; fax (240) 485-1818
Internet: http://www.midwife.org/foundation/award-corbin.cfm
Sponsor: A.C.N.M. Foundation
8403 Colesville Road, Suite 1550
Silver Spring, MD 20910

HCA Foundation Grants 2231

The Foundation is committed to the care and improvement of human life. Grants are made in the areas of health and well being, childhood and youth development, and the arts. Preference will be given to requests from organizations where an HCA employee volunteers or serves on the board. New applicants are asked to send a one-or two-page letter of inquiry to the Foundation, describing the proposed project, its goals and objectives and the approximate level of funding required. Foundation staff will review each request and will notify the organization as to whether the project coincides with funding priorities.

Requirements: 501(c)3 nonprofit organizations are eligible. Because the foundation focuses its giving in Middle Tennessee, all requests outside of the Nashville area should be submitted to the closest HCA facility location or division office. Organizations must have a full updated GivingMatters.com profile to be considered for funding. For more information please go to www.givingmatters.com.

Restrictions: Funding is not available for: individuals or their projects; private foundations; political activities; advertising or sponsorships of events; social events or similar fund-raising activities. The Foundation does not ordinarily support: organizations in their first three years of operation and organizations involved in research, sports, environmental, wildlife, civic and international affairs. The Foundation does not accept proposals from individual churches or schools, but will support broad faith-based initiatives consistent with it's mission and guidelines.

Geographic Focus: Tennessee

Date(s) Application is Due: Mar 12; Jun 11; Sep 10; Dec 12

Amount of Grant: 20,000 - 100,000 U.S.D

Contact: Lois Abrams, Grants Manager; (615) 344-2390; fax (615) 344-5722; lois.abrams@hcahealthcare.com

Internet: http://www.hcacaring.org/CustomPage.asp?guidCustomContentID=BBB7D8F2-B906-4302-A164-07643DB2582E

Sponsor: HCA Foundation
1 Park Plaza, Building 1, 4th Floor East
Nashville, TN 37203

HCR Manor Care Foundation Grants 2232

Grants are awarded to communities nationwide where Manor Care Health Services has operations. The foundation has two grantmaking interests: organizations involved in research of diseases and disorders affecting the elderly and community programs that offer support and outreach services to seniors. The foundation concentrates on research organizations that treat and look for cures for illnesses such as Alzheimer's disease and stroke. Grants also are made to community organizations that offer support and information services to seniors. Initial contact should be made through a letter of inquiry (five pages or less). Letters are reviewed on an ongoing basis. If the foundation is interested, a formal grant proposal will be invited, with application deadlines as indicated.

Requirements: Nonprofits must have 501(c)3 status, be located in a state where Manor Care Health Services operates, and have at least one other source of support for the project.

Restrictions: The foundation does not support fundraising events, advertising, individuals, for-profit organizations, building or capital campaigns, political purposes or campaigns, endowment funds, multiple year grants, overhead fees, or general purpose/operating grants.

Geographic Focus: Arizona, California, Colorado, Connecticut, Delaware, Florida, Georgia, Illinois, Indiana, Iowa, Kansas, Kentucky, Maryland, Michigan, Minnesota, Missouri, Nevada, New Jersey, New Mexico, North Carolina, North Dakota, Ohio, Oklahoma, Pennsylvania, South Carolina, South Dakota, Tennessee, Texas, Utah, Virginia, Washington, West Virginia, Wisconsin

Date(s) Application is Due: Apr 1; Oct 1

Amount of Grant: 8,300 - 35,000 U.S.D

Samples: Family Service of Champaign County (IL)—for the Senior Resource Center, which helps senior citizens maintain their independence, $10,000; Myers Research Institute (Beachwood, OH)—to develop and adopt a computer-based activities system for long-term care residents suffering from dementia who reside in a skilled-nursing unit and two assisted-living facilities, $30,000; Coda Alliance (San Jose, CA)—to conduct six Respecting Choices workshops, which provide training and education about end-of-life care, $26,000; National Hospice Foundation (Alexandria, VA)—for an effort to increase access to hospice and palliative care among underserved populations, $56,192.

Contact: Executive Director; (419) 252-5989; fax (419) 252-5521; foundation@hcr-manorcare.com

Internet: http://www.hcr-manorcare.org

Sponsor: HCR Manor Care Foundation
333 N Summit Street, P.O. Box 10086
Toledo, OH 43699-0086

Health Canada National Seniors Independence Program Grants 2233

The program is intended to provide financial assistance to Canadian seniors for projects that contribute to improving their health, well-being, and independence and to enhancing their quality of life. Particular attention will be given to projects that address the needs of women seniors, seniors living in rural and remote areas, and seniors who are less advantaged due to life circumstances. The program will provide funds for projects that support action leading to a new or improved resource or organization; test the value or feasibility of new program models; improve the quality and availability of information on particular subjects; and/or develop and strengthen the knowledge, skills, and abilities of seniors or people working with seniors. Groups and organizations interested in submitting an application should contact the appropriate regional office where program representatives are available to meet with groups to discuss their ideas and assist in the preparation of an application. Lists of regional offices are available from the national office.

Requirements: Proposals may be submitted by voluntary, nongovernmental, nonprofit groups and organizations; these may include health or social service agencies, higher educational institutions, professional associations, and other parapublic organizations.

Restrictions: Individuals, for-profit groups, and government departments are not eligible for funding.

Geographic Focus: All States, Canada

Contact: Stephanie Wilson, (613) 954-8549; fax (613) 954-7363; Stephanie_Wilson@hc-sc.gc.ca

Internet: http://www.hc-sc.gc.ca

Sponsor: Health Canada
Postal Locator 1912
Ottawa, ON K1A 1B4 Canada

Health Care Financing Research, Demonstrations, and Evaluations Grants 2234

Project grants (including cooperative agreements) are awarded to support analyses, experiments, demonstrations, and pilot projects in efforts to resolve major health care financing issues or to develop innovative methods for the administration of Medicare and Medicaid. HCFA has identified priority areas for the current fiscal year, including access and quality of care, coordinated care systems, provider payment, health care systems reform and financing, program evaluation and analysis, service delivery systems, and subacute and long-term care. Awardees are required to share in the cost of projects; normally, the minimum cost-sharing requirement is 5 percent of total project costs. Projects are generally funded on a 12-month basis, with support beyond the first year contingent upon acceptable evidence of satisfactory progress, continuing program relevance, and availability of funds. Applications may be submitted at any time.

Requirements: Grants or cooperative agreements may be made to private or public and nonprofit or for-profit agencies/organizations.

Amount of Grant: 25,000 - 1,000,000 U.S.D

Contact: Judy Norris, (410) 786-5130; jnorris1@cms.hhs.gov

Internet: http://www.cms.hhs.gov/contracts

Sponsor: Health Care Financing Administration
7500 Security Boulevard
Baltimore, MD 21244-1850

Healthcare Foundation for Orange County Grants 2235

The foundation's mission is to improve the health of the neediest and most underserved residents of Orange County, with particular emphasis on central Orange County. The foundation will address ways of improving the health of residents by advancing access to health information, prevention, and basic health care. Grantmaking is divided among initiatives that support nonprofit hospital working with community-based organizations (Partners for Health); grants directly to community-based organizations (Healthy Orange County); and increased understanding of community needs, and collaborative granting strategies (coalition projects.)

Requirements: California 501(c)3 nonprofit organizations in Orange County that administer health services consistent with the foundation's goals of providing aid to uninsured, poor families are eligible. Eighty-two percent of support goes to or through qualified nonprofit hospitals. Government agencies also are eligible.

Restrictions: Grants generally are not awarded for annual fund drives, building campaigns, major equipment, or biomedical research. Activities that exclusively benefit the members of a religious or fraternal organization are not funded.

Geographic Focus: California

Amount of Grant: Up to 10,000 U.S.D

Contact: William B. Stannard, Chairman of the Board; (714) 245-1650; fax (714) 245-1653; dflander@hfoc.org

Internet: http://www.hfoc.org/giving/

Sponsor: Healthcare Foundation for Orange County
1450 N Tustin Avenue, Suite 103
Santa Ana, CA 92705-8641

Healthcare Foundation of New Jersey Grants 2236

The Foundation's funding priorities are the vulnerable populations of the greater Newark New Jersey community; the emergent health needs of serving at-risk individuals and families in the MetroWest Jewish community; and clinical research/medical education initiatives that significantly and directly impact these populations. The Foundation seeks grant proposals that promise innovation and change or a significant enhancement of services. Proposals are accepted on a rolling basis throughout the calendar year. Proposal submission instructions are available online.

Requirements: Nonprofit organizations in the Greater Newark area are eligible to apply. The Foundation strongly suggests that application documents be submitted electronically.

Restrictions: Grants are made only to private nonprofit organizations that have tax-exempt status under Section 501(c)3 of the Internal Revenue Code and that are not private foundations. The Foundation does not make grants to individuals or government agencies. The Foundation does not typically fund the following: organizations outside of Essex, Morris or Union County, New Jersey; programs not related to health care; direct support of an individual's healthcare needs; fundraising events or endowment campaigns; advertising campaigns; lobbying; or scholarships.

Geographic Focus: New Jersey

Contact: Program Contact; (973) 921-1210; fax (973) 921-1274; info@hfnj.org

Internet: http://www.hfnj.org

Sponsor: Healthcare Foundation of New Jersey
60 East Willow Street, 2nd Floor
Millburn, NJ 07041

Healthcare Georgia Foundation Grants 2237

The foundation's mission is to advance the health of all Georgians and to expand access to affordable, quality healthcare for underserved individuals and communities. Grantmaking priority areas include addressing health disparities, strengthening nonprofit health organizations, and expanding access to primary healthcare. The foundation will award grants that support: policy development and advocacy, research and evaluation, technical assistance/capacity building, organizational improvement, leadership development and recognition, direct service delivery, professional development and training, and public education and awareness. The grant application process begins with the submission of a two-page letter of inquiry that describes the funding request specifying who, what, where, when, for whom, and for what cost. Full proposals are by invitation.

Requirements: Georgia organizations, including community-based/health service, education, government, public policy, and other nonprofits, are eligible.

Geographic Focus: Georgia

Samples: Lowndes County Partnership for Health (Atlanta, GA)—to expand the Well Workplace health promotion program, to implement a strategic plan, and to help a church-based program become a self-sustaining, community-based nonprofit group, $145,000; National Council of La Raza-Atlantic Program Office (GA)—for the Georgia Latino/Hispanic Health Agenda and Leadership Project, $235,000; Planned Parenthood of Georgia (Atlanta, GA)—to strengthen its ability to provide health services to low-income women and to enhance the self-sustainability of its programs in Augusta, $50,000; Meridian Educational Resource Group (Atlanta, GA)—to establish a fitness, health-promotion, and weight-loss program for children served by the Whitefoord and Coan School Health Clinics, $150,000.

Contact: Grants Administrator; (404) 653-0990; fax (404) 577-8386; info@healthcaregeorgia.org

Internet: http://www.healthcaregeorgia.org

Sponsor: Healthcare Georgia Foundation
50 Hurt Plaza, Suite 550
Atlanta, GA 30303

Health Foundation of Greater Cincinnati Grants 2238

The foundation awards grants to programs and activities that improve health in Cincinnati and 20 surrounding counties in Ohio, Kentucky, and Indiana. Health is broadly defined to include social, behavioral, environmental, and other dimensions beyond the absence of illness. Grant making is focused on strengthening primary care providers to the poor; school-based child health interventions in K-8 school settings; substance abuse; and severe mental illness. Grants also are awarded in response to requests from organizations seeking funds to support a broad range of health-related needs in the community. The Health Foundation of Greater Cincinnati considers projects for funding in three ways: through Requests for Proposals (RFPs), through invitations to submit proposals, and through grantee-initiated requests. In any case, proposals are considered using the same eligibility requirements and proposal evaluation criteria. For more information about the types of proposals, see the Foundations website or call the Grant Manager.

Requirements: 501(c)3 nonprofits in Adams, Brown, Butler, Clermont, Clington, Hamilton, Highland, and Warren Counties in Ohio; Boone, Bracken, Campbell, Gallatin, Grant, Kenton, and Pendleton Counties in Kentucky; and Dearborn, Franklin, Ohio, Ripley, and Switzerland Counties in Indiana are eligible.

Restrictions: The foundation does not normally fund capital campaigns, annual fundraising campaigns, endowments, event sponsorships, clinical research, scholarships, routine operational costs, or direct financial subsidy of health services to individuals or groups.

Geographic Focus: Indiana, Kentucky, Ohio

Contact: Shelly Stolarczyk-George, Grants Manager; (513) 458-6619; fax (513) 458-6610; sstolarczyk@healthfoundation.org

Internet: http://www.healthfoundation.org/grants

Sponsor: Health Foundation of Greater Cincinnati
3805 Edwards Road, Suite 500
Cincinnati, OH 45209-1948

Health Foundation of Greater Indianapolis Grants 2239

The foundation promotes health care for children, youth, and families in Marion County, IN, and seven contiguous Indiana counties. Programs of interest include school-based health, adolescent health, and HIV/AIDS Grants also are awarded to support the development of health careers and to promote cooperative effort between health professionals. Types of support include general operations, building/renovations, equipment, program/project development, conferences and seminars, seed money, technical assistance, and matching funds. Prior to submitting a proposal, contact Stephen L. Everett, Vice President of Programs, to determine if your program and proposal matches The Health Foundation of Greater Indianapolis' funding priorities and application guidelines.

Requirements: Grants are awarded to neighborhood-based service centers in Indiana's Marion County and the seven contiguous counties of Boone, Hamilton, Hancock, Hendricks, Johnson, Morgan and Shelby. Applicants must be a 501(c)3 group, organization or agency that provides health-related programs or services.

Restrictions: Grants will not be provided for: individuals; sectarian religious organizations; research projects; purchase of advertising or tickets to events; production and design of educational materials already available; endowments; short or long term loans or payment of financial obligations.

Geographic Focus: Indiana

Amount of Grant: 10,000 - 100,000 U.S.D

Samples: Indiana AIDS Fund (Indianapolis, IN)—for a fundraising event for HIV/AIDS organizations, $25,000; Indianapolis Medical Society Foundation (IN)—for a project designed to improve access to health care for low-income, uninsured Indianapolis residents, $40,000; Little Red Door Cancer Agency (Indianapolis, IN)—to provide baskets filled with personal-care items to people with cancer, $15,000; Damien Ctr (Indianapolis, IN)—to implement its HIV Care Network, $50,000.

Contact: Stephen Everett, Program Officer; (317) 630-1805; fax (317) 630-1806; severett@thfgi.org

Internet: http://www.thfgi.org

Sponsor: Health Foundation of Greater Indianapolis
429 East Vermont Street, Suite 300
Indianapolis, IN 46202

Health Foundation of Southern Florida Responsive Grants 2240

The foundation awards Responsive Grants through two grant cycles per year. With exceptions, the foundation focuses on providing one to three year grants that do not exceed $300,000 annually. The majority of grants are funded in the $50,000 to $150,000 range over one or two years. Funding is provided in four categories: Project Planning; Health Services; Organizational Capacity Building and Health System/Health Policy Development.

Requirements: Though the foundation welcomes proposal applications anytime, the applications are reviewed on a semi-annual basis. Applicant organizations must be tax-exempt nonprofit under section 501(c)3 of the Internal Revenue Code or a local or state governmental agency. The project must serve exclusively the residents of Broward, Miami-Dade and/or Monroe counties. Initially, a preliminary proposal is required (see the sponsor's website for specific details). If approved, the sponsor will then invite a full proposal. Download the sponsor's grant guide from the website.

Restrictions: The foundation does not fund: Biomedical research or other research that will not impact local residents within the immediate future (1-3 years) or that does not have a direct application to implementing a community-driven health intervention; Capital campaigns of over $1 million (versus grants toward specific health-related equipment or the 'build out' of a specific health-focused space); Secondary and tertiary services (versus preventive and primary medical, oral and behavioral health care services); Health promotion and/or health care with a high per capita cost (this figure will vary depending upon the type of intervention, but over $1,000 per person/year cost may be a rule of thumb); Service expansion or new projects without viable sustainability (unable to reach sustainability without the foundation's resources within a four-year period).

Geographic Focus: Florida

Date(s) Application is Due: Mar 13; Apr 24

Amount of Grant: Up to 300,000 U.S.D

Contact: Eliane Morales, (305) 374-7200; fax (305) 374-7003; emorales@hfsf.org

Internet: http://www.hfsf.org/ORIGHTML/responsive.html

Sponsor: Health Foundation of South Florida
2 South Biscayne Boulevard
Miami, FL 33131

Health Management Scholarships and Grants for Minorities 2241

This financial assistance program provides scholarships and grants to ethnic minority students for graduate study in health care administration/management. Annual tuition scholarships and grants are made to outstanding students in health services management and related programs. Grants are provided to cover the costs of books, tutorial assistance, and other nontuition educational aids. Students are eligible to receive a combination of awards during the same time period.

Requirements: To be eligible, applicants must be members of an ethnic minority group; must be U.S. citizens or have permanent resident status (birth certificate, green card, or current passport); they must be college juniors, seniors or graduate students; they must be health care administration/management majors; they must have a strong academic background (2.5 GPA on a 4.0 scale); and they must have strong extracurricular and community service activities.

Contact: Program Contact; (800) 233-0996 or (312) 422-2680; fax (312) 422-4566

Internet: http://www.diversityconnection.org/diversityconnection_app/career-center/internships/Standard-Internship-page.jsp?fll=S4

Sponsor: Institute for Diversity in Health Management
1 N Franklin, 30th Floor
Chicago, IL 60606

Hearst Foundation and William Randolph Hearst Foundation Grants 2242

The Hearst Foundations support well-established nonprofit organizations that address important issues within our major areas of interests: education, health, culture, social services that primarily serve large demographic and/or geographic constituencies. Within these areas, the Foundations generally provide endowment, program, and capital grant support. Private nonprofits with significant support from the philanthropic community are favored over those financed through government sources. Proposals may be submitted throughout the year; board meetings on grant decisions are held in March, June, September, and December. Only fully documented appeals will be considered. As a general policy, meetings are arranged between foundation staff and applicant only if serious consideration of the appeal is anticipated. Programs headquartered east of the Mississippi River should mail appeals to the Hearst Foundations at the New York address. Those west of the Mississippi River should mail requests to Hearst Foundations, 90 New Montgomery Street, Suite 1212, San Francisco, CA 94105; phone: (415) 543-0400; fax: (415) 348-0887.

Requirements: Grants are made only to 501(c)3 tax-exempt organizations that are not private foundations.

Restrictions: Funding is not available for: individuals; loans and program-related investments; multi-year grants; start-up or seed funding; equipment of any kind, including computers and transportation; publishing projects; radio, film, television or other media-related projects; conferences, workshops or seminars; public policy research; special events, tickets, tables or advertising for fundraising events.

Samples: The Archer School for Girls, Los Angeles, CA, $50,000—scholarship support; YMCA of Greater New York, New York, NY, $80,000—Y Scholars Program to prepare middle and high school students for college; Cerebral Palsy International Research Foundation, Washington, DC, $200,000—to support one investigator-initiated research project and two career development awards; The Village of Arts and Humanities Inc., Philadelphia, PA, $50,000—general support;
Contact: Mason Granger, Program Director; (212) 586-5404; fax (212) 586-1917
Internet: http://www.hearstfdn.org/fp_home.html
Sponsor: Hearst Foundation
300 West 57th Street, 26th Floor
New York, NY 10019-3741

Heart Rhythm Society Fellowships **2243**
Fellowships and research grants are available for projects pertaining to cardiac pacing and/or cardiac electrophysiology. All awards are made for one year, renewable in competition. Also available are traveling fellowships, for a one- to three-month duration, to enable a researcher to observe research at other institutions or attend seminars and conferences. One traveling Canadian fellowship is offered; the applicant must be a Canadian citizen or must train at a site in Canada. Annual deadline dates may vary; contact program staff for exact dates.
Requirements: Fellowships are open to applicants who have a doctor of medicine, philosophy, osteopathy, science, or equivalent training or experience in the field of biomedical engineering. Applicants must be citizens or permanent residents of a North American country and include cardiac pacing or cardiac electrophysiology as a central part of his/her long-range career goals.
Date(s) Application is Due: Sep 15
Contact: Stephanie Mascetta, (508) 647-0100, ext. 3046; fax (508) 647-0124; fellowship@HRSonline.org
Internet: http://www.hrsonline.org/society_in_action/fellowships/default.asp
Sponsor: Heart Rhythm Society
Six Strathmore Road
Natick, MA 01760-2499

Heckscher Foundation for Children Grants **2244**
The Heckscher Foundation for Children was founded in 1921 to promote the welfare of children, primarily in New York City. Funding organizations serving youth in the fields of education, family services, job training, health, arts and recreation. The Foundation's giving takes the form of program support, capacity-building, capital projects and general operating support. The Foundation does not participate in annual appeals, endowments, fundraising events or political efforts. The Heckscher Foundation provides support in the following categories: Education & Academic Support; Arts; Social Services; Health; Recreation and; Workforce Development. The Foundation accepts applications online only. To begin the application process. fill out the application template provided online. Do not submit additional information or materials unless asked to do so. All inquiring organizations will be notified of the outcome of the Foundation's review, and any further necessary materials will be requested at that time.
Requirements: 501(c)3 nonprofit organizations are eligible.
Restrictions: Funding is not available for annual appeals, endowments, fundraising events or political efforts.
Geographic Focus: New York
Amount of Grant: 25,000 - 500,000 U.S.D
Contact: Virginia Sloane, President; (212) 744-0190; fax (212) 744-2761
Internet: http://fdncenter.org/grantmaker/heckscher
Sponsor: Heckscher Foundation for Children
123 East 70th Street
New York, NY 10021

Hedco Foundation Grants **2245**
Incorporated in 1972 in California, the Hedco Foundation gives predominantly to qualified educational and health institutions, support is also available for social services. Types of support including: building/renovation; equipment; land acquisition and; matching/challenge grants.
Requirements: 501(c)3 nonprofit organizations are eligible.
Restrictions: No grants to: individuals, or for general support, operating budgets, endowment funds, scholarships, fellowships, special projects, research, publications, or conferences; no loans.
Geographic Focus: California
Amount of Grant: 10,000 - 1,000,000 U.S.D
Contact: Mary Goriup, Manager; (925) 743-0257
Sponsor: Hedco Foundation
P.O. Box 339
Danville, CA 94526-0339

HEI Grants **2246**
The institute awards research grants for scientific information about the health effects of motor vehicle emissions on the environment and on human health.
Amount of Grant: 70,000 - 300,000 U.S.D
Contact: Daniel Greenbaum, President; (617) 886-9330, ext. 311; fax (617) 886-9335; dgreenbaum@healtheffects.org
Internet: http://www.healtheffects.org/funding.htm
Sponsor: Health Effects Institute
120 Second Avenue
Boston, MA 02129-4533

Heineman Foundation for Research, Education, Charitable and Scientific Purposes **2247**
The purpose of the Heineman Foundation is to provide seed money to start-up projects and new projects within existing organizations for a maximum of three to five years. Preference will be given to organizations that we have not previously funded. The average range of our donations is $20,000.00 to $50,000.00, per annum. An organization must have 501(c)3 status and upload copies of corresponding IRS documents to the online application form in order for the application to be considered. The Foundation's general areas of interest are the following (in no particular order): programs that enable economically challenged women to enter and remain in the workplace; on site day care centers for women in the workplace; job training programs for women; language and leadership skills for women; environmental research that will help prevent, reduce and/or eliminate water degradation; music as education and preserver of culture; research into prevention of and treatment for childhood illnesses; programs that enable youth to think, create and communicate effectively; and programs that support and promote high achievement in music, science, and literature. Applications/proposals must be submitted online no later than September 1st.
Date(s) Application is Due: Sep 1
Amount of Grant: 20,000 - 50,000 U.S.D
Contact: Simon Rose, President; (212) 493-8000; info@heinemanfoundation.org
Internet: http://www.heinemanfoundation.org/guidelines
Sponsor: Heineman Foundation
140 Broadway
New York, NY 10005-1108

Heiser Leprosy and Tuberculosis Research Postdoctoral Fellowships **2248**
The program fosters research on leprosy and tuberculosis by supporting the research training of a limited number of young biomedical scientists interested in the field. The program is designed to encourage national and international cooperation and research interchange in the study of the diseases. A maximum of two years of support will be provided to successful applicants at stipend levels adjusted according to U.S. government guidelines for geographical location at which the training is to take place. There are no application forms. Application may be made either by the individual seeking training or by the head of a laboratory engaged in research related to leprosy. In the latter case, the head of the laboratory must name the prospective fellow in the application.
Requirements: Biomedical scientists with either the PhD or MD degree are eligible for these fellowships. Applicants should be at the beginning or in an early phase of their postdoctoral training and interested in obtaining research training in a basic science field directly related to the study of leprosy. There are no citizenship requirements.
Date(s) Application is Due: Mar 1
Amount of Grant: 28,500 - 73,500 U.S.D
Contact: Len McNally, Program Director; (212) 686-0010, ext. 556; lm@nyct-cfi.org
Sponsor: Heiser Program for Research in Leprosy and Tuberculosis
909 3rd Avenue
New York, NY 10022

Heiser Leprosy Research Grants **2249**
The grants provide support to laboratories involved in leprosy research training. Proposals of high scientific caliber related to leprosy will be considered. Seed grants may be requested for new projects or facilities that show promise of support from other sources within one year. Leprosy postdoctoral research fellowships and tuberculosis postdoctoral research fellowships also are awarded. There are no application forms. Applicants should submit proposals and budgets along with curriculum vita and relevant bibliographies of participating scientists.
Requirements: Senior investigators who are experienced in leprosy and tuberculosis research and associated with laboratories providing training opportunities in these areas are eligible.
Restrictions: Grants do not support clinical trials, salaries of personnel, or more than 10 percent of institutional overhead.
Date(s) Application is Due: Mar 1
Amount of Grant: Up to 35,000 U.S.D
Contact: Len McNally, Program Director; (212) 686-0010, ext. 556; lm@nyct-cfi.org
Sponsor: Heiser Program for Research in Leprosy and Tuberculosis
909 3rd Avenue
New York, NY 10022

Helen and Merrill Bank Foundation Grants **2250**
The foundation awards general operating grants to nonprofit organizations nationwide, including religious organizations (Christian, Jewish, and Salvation Army), hospitals, museums, organizations serving the disabled, parochial schools, temples, and organizations addressing issues affecting women. The foundation requests applications be made in writing. Include a description of the organization and activities for which funds are sought. There are no application deadlines.
Requirements: Nonprofit organizations in Florida and Maryland are eligible.
Geographic Focus: Florida, Maryland
Amount of Grant: Up to 548,025 U.S.D
Samples: Raymond F. Kravis Ctr for the Performing Arts (West Palm Beach, FL)—for operating support, $79,150; Temple Israel (Boston, MA)—for operating support, $45,000.
Contact: Herbert Bank, (410) 363-6767
Sponsor: Helen and Merrill Bank Foundation
8 Roland Mews
Baltimore, MD 21210

Helena Rubinstein Foundation Grants 2251

The Foundation supports programs in education, community services, arts/arts in education, and health with emphasis on projects which benefit women and children. Grants are primarily targeted to organizations in New York City. Although general operating grants are made, the Foundation prefers to support specific programs. Grant proposals are accepted throughout the year. There is no formal application form; however, the New York Common Application Form may be used. Organizations seeking funds are asked not to make telephone inquiries, but to submit a brief letter outlining the project.

Requirements: U.S. nonprofit organizations are eligible.

Restrictions: Support is not offered to individuals, or for film or video projects. Grants are rarely made to endowment funds and capital campaigns. The foundation does not make loans and cannot provide emergency funds. Funding of new proposals is limited by ongoing commitments and fiscal constraints.

Geographic Focus: New York

Samples: New York City Outward Bound Ctr (NY)—for programs to help high school students develop self-esteem, self-reliance, social responsibility, and improved academic performance, $10,000; Goodwill Industries of Greater New York (NY)—to support the capital campaign, $10,000; Winston Preparatory School (NY)—salary support for a vocational guidance counselor, $5000.

Contact: Diane Moss, President; (212) 750-7310

Internet: http://www.helenarubinsteinfdn.org/guide.html

Sponsor: Helena Rubinstein Foundation
477 Madison Avenue, 7th Floor
New York, NY 10022-5802

Helen Bader Foundation Grants 2252

Throughout her life, Helen Bader sought to help others. She played many roles - student, mother, businesswoman, and social worker - believing that everyone should have the opportunity to reach their fullest potential. Growing up in the railroad town of Aberdeen, South Dakota, Helen learned the value of hard work and self-reliance. The Great Depression and the sacrifices of World War II also taught her the importance of reaching out to those in need. Helen attended Downer College in Milwaukee, earning a degree in botany. She married Alfred Bader, a chemist from Austria, and together they started a family and created a business, the Aldrich Chemical Company. From the 1950s to the 1970s, their hard work helped build one of Wisconsin's most successful start-up enterprises of the era. The Baders' eventual divorce led Helen to again become self-reliant. She subsequently finished her Master of Social Work at the University of Wisconsin-Milwaukee. While doing her field work with the Legal Aid Society of Milwaukee, Helen met and helped many people in need, including single mothers and adults with mental illness. In the process, she gained a deeper appreciation for their everyday struggles. After graduation, she worked at the Milwaukee Jewish Home, where working with older adults brought home the many issues of aging. At a time when Alzheimer's disease was almost a complete mystery, she helped open the resident' minds and hearts through dance and music. Helen felt that the residents' quality of life depended upon the small details, so she was happy to run errands or escort them to the symphony. She found herself touched by the arts and studied the violin and guitar at the Wisconsin Conservatory of Music. Helen eventually faced cancer. As the illness began to sap her physical strength, she shared a wish with her family: to continue to aid those in need. She died in 1989. After her death, patterns of Helen's quiet style of philanthropy became more apparent. When she had come across an organization that impressed her, she would just pull out her checkbook without a lot of fanfare. In her name, the Helen Bader Foundation (HBF) supports worthy organizations working in key areas affecting the quality of life in Milwaukee, the state of Wisconsin, and Israel. The foundation also seeks to inspire the generosity in others, as every individual can make a difference through gifts of time, talent, and resources. The foundation will consider multiple-year requests with 24 or 36 month terms. Multi-year grants are subject to annual review before funds for subsequent years are released. The application deadline for the online preliminary proposal is January 5. The application deadline for organizations invited to complete a full proposal is February 2. The link to the online application system is available at the grant website. Application deadlines may vary from year to year. Prospective applicants are encouraged to visit the grant website to verify current deadline dates.

Requirements: Grants are awarded for projects consistent with one or more of the Helen Bader Foundation's program areas: Alzheimer's and aging (national in scope, with priority given to Wisconsin); economic development (restricted to the city of Milwaukee); community partnerships for youth (restricted to the city of Milwaukee); community initiatives (restricted to greater Milwaukee); arts (restricted to the city of Milwaukee); and directed grants and initiatives such as aid and support to Israel (for which proposals must be staff-solicited). Grants are given only to U.S. organizations which are tax exempt under Section 501(c)3 of the Internal Revenue Code or to government entities; grants will only be approved for foreign entities which meet specific charitable status requirements.

Restrictions: The Foundation does not provide direct support for individuals, such as individual scholarships.

Geographic Focus: All States, All Countries

Date(s) Application is Due: Jan 5; Feb 25

Amount of Grant: 10,000 - 100,000 U.S.D

Contact: Tamara Hogans, Grants Manager; (414) 224-6464; fax (414) 224-1441; tammy@hbf.org or info@hbf.org

Internet: http://www.hbf.org/apply.htm

Sponsor: Helen Bader Foundation
233 North Water Street, Fourth Floor
Milwaukee, WI 53202

Helen Gertrude Sparks Charitable Trust Grants 2253

Helen Sparks was born in Fort Worth, TX but went to college in the East. After graduation she returned to Fort Worth where she worked with the Fort Worth Little Theater. A cultured woman, she was interested in local artists and supported them by commissioning works. She also enjoyed literature and needlepoint. The Helen Gertrude Sparks Charitable Trust made its first distributions in 1971, and was created to benefit charitable organizations focused on: the elderly; those who are disabled in any way; children who are disabled, orphaned or disadvantaged; the arts, including performing and nonperforming arts; and education. This trust makes approximately 2-3 awards each year and grants are typically between $5,000 and $15,000. Applicants must apply online at the grant website. Applicants are strongly encouraged to do the following before applying: review the downloadable state application procedures for additional helpful information and clarifications; review the downloadable online-application guidelines at the grant website; review the trust's funding history (link is available from the grant website); review the online application questions in advance; and review the list of required attachments. These will generally include: a list of board members, financial statements (audited, reviewed, or compiled by independent auditor); an organization summary; a list of other funding sources; an IRS Determination letter; and other required documents. All attachments must be uploaded in the online application as PDF, Word, or Excel files. The Helen Gertrude Sparks Charitable Trust application deadline is 11:59 p.m. on September 30.

Requirements: Applicants must have 501(c)3 tax-exempt status.

Restrictions: A preference will be shown for organizations which do not receive the predominate portion of their funds from government sources. The trust does not support requests from individuals, organizations attempting to influence policy through direct lobbying, or any political campaigns.

Geographic Focus: Texas

Date(s) Application is Due: Sep 30

Amount of Grant: 5,000 - 15,000 U.S.D

Samples: Carter BloodCare, Bedford, Texas, $15,000, high school population health initiative (2011); Meals on Wheels of Tarrant County, Fort Worth, Texas, $10,000 (2011); Performing Arts Fort Worth, Fort Worth, Texas, $10,000, Children's Education Program (2010).

Contact: Mark J. Smith; (817) 390-6028; tx.philanthropic@baml.com

Internet: https://www.bankofamerica.com/philanthropic/fn_search.action

Sponsor: Helen Gertrude Sparks Charitable Trust
500 West 7th Street, 15th Floor, TX1-497-15-08
Fort Worth, TX 76102-4700

Helen Hay Whitney Foundation Postdoctoral Research Fellowships 2254

The program provides three-year fellowships for beginning postdoctoral training in basic biomedical research. The program is intended for individuals planning careers in biological or medical research to receive financial support of sufficient duration to help further their professional careers. It is expected that fellowship training will be obtained in an academic setting. The foundation will not consider applicants who plan tenure of the fellowship in the laboratory in which they have already received extensive predoctoral or postdoctoral training; the aim of the fellowship is to broaden postdoctoral training and experience. Applications are available online.

Requirements: Residents of the United States, Canada, and Mexico who are in the final stages of obtaining the MD, PhD, or equivalent degree and are seeking beginning postdoctocral training in basic biomedical research are eligible.

Restrictions: Applications from established scientists or advanced fellows will not be considered. Applicants who have already had one year of postdoctoral laboratory training by the listed deadline date will not be considered.

Date(s) Application is Due: Jul 15

Amount of Grant: 130,000 U.S.D

Samples: Dr. Sean Burgess, Harvard U, Department of Biochemistry and Molecular Biology. Dr. David Wassarman, National Institutes of Health, Department of Cell Biology and Metabolism.

Contact: Administrator; (845) 639-6799; fax (845) 639-6798; hhwf@earthlink.net

Internet: http://www.hhwf.org/HTMLSrc/ResearchFellowships.html

Sponsor: Helen Hay Whitney Foundation
20 Squadron Boulevard, Suite 630
New York, NY 10956

Helen Irwin Littauer Educational Trust Grants 2255

Mrs. Littauer was born in Fort Worth, Texas and was a descendant of the Cetti family, a prominent family in Fort Worth, Texas. She earned a degree in journalism and worked in New York as an editor. In 1952, she moved to Connecticut and became an active community volunteer and was involved in city government. As a result of her passion for teaching, she also worked with troubled youth. She was honored with numerous awards for her civic endeavors. Mrs. Littauer established the Educational Trust in 1969 and remained involved in grant decisions until her death in 1989. The trust is particularly interested in, but not limited to charitable organizations that focus on: scholarships that enable needy, but worthy boys and girls and young adults to attend school, college, or university, with a particular emphasis on making scholarships available for attending schools of journalism; promotion of art, education, and good citizenship; alleviating human suffering; medical care and treatment for all needy persons, including hospitals and clinics; providing care, education, recreation and/or physical training for needy, orphaned or disabled children; providing care of needy persons who are sick, aged or disabled; and improvement of living and working conditions of all persons. Applicants must apply online at the grant website. Applicants are strongly encouraged to do the following before applying: review the downloadable state application procedures for

additional helpful information and clarifications; review the downloadable online-application guidelines at the grant website; review the trust's funding history (link is available from the grant website); review the online application questions in advance; and review the list of required attachments. These will generally include: a list of board members, financial statements (audited, reviewed, or compiled by independent auditor); an organization summary; a list of other funding sources; an IRS Determination letter; and other required documents. All attachments must be uploaded in the online application as PDF, Word, or Excel files. The Helen Irwin Littauer Educational Trust has bi-annual application deadlines of March 31 and September 30. Applications must be submitted by 11:59 p.m. on the deadline dates.
Requirements: Applicants must have 501(c)3 tax-exempt status.
Restrictions: The trust considers requests primarily from charitable organizations that provide services to Tarrant County. The trust does not support requests from individuals, organizations attempting to influence policy through direct lobbying, or any political campaigns.
Date(s) Application is Due: Mar 31; Sep 30
Amount of Grant: 1,000 - 25,000 U.S.D
Samples: Texas Christian University, Fort Worth, Texas, $25,000 (2011); Fort Worth Opera Association, Fort Worth, Texas, $25,000 (2011); Presbyterian Night Shelter of Tarrant County, Fort Worth, Texas, $30,000 (2011).
Contact: Mark J. Smith; (817) 390-6028; tx.philanthropic@baml.com
Internet: https://www.bankofamerica.com/philanthropic/fn_search.action
Sponsor: Helen Irwin Littauer Educational Trust
500 West 7th Street, 15th Floor, TX1-497-15-08
Fort Worth, TX 76012-4700

Helen K. and Arthur E. Johnson Foundation Grants **2256**

The foundation makes grants to a wide variety of nonprofit organizations in an attempt to solve community problems and enrich the quality of life in Colorado in the following areas: education, youth, health, community and social services, civic and cultural, and senior citizens. Requests are welcomed throughout the year. However, to be considered at the next board meeting, complete written proposals must be received by the listed application deadline dates. Proposal submission information is available online.
Requirements: IRS 501(c)3 organizations serving Colorado residents may apply.
Restrictions: Funding is not available for: loans or fund endowments; individuals; conferences; scholarships to individuals; multiple year grants; fundraising dinners or special events. The Foundation does not support organizations whose primary purpose is to influence (directly or indirectly) the legislative or judicial process in any manner or for any cause. In addition, the Foundation will not consider grant requests that pass through the nominal grant recipient to another organization.
Geographic Focus: Colorado
Date(s) Application is Due: Jan 1; Apr 1; Jul 1; Oct 1
Contact: John H. Alexander, President; (800) 232-9931 or (303) 861-4127; fax (303) 861-0607; info@johnsonfoundation.org
Internet: http://www.johnsonfoundation.org
Sponsor: Helen K. and Arthur E. Johnson Foundation
1700 Broadway, Suite 1100
Denver, CO 80290-1718

Helen McWilliam Memorial Scholarship **2257**

This scholarship is open to full-time graduate students at the University of Calgary interested in pursuing careers in school psychology. Recipients will be selected on the basis of academic excellence, skill in the handling of children referred for testing or guidance, and evidence of interest and ability in the area of school psychology. The award will be based on grades and observations by faculty members of academic and practical work. Annual deadline dates may vary; contact program staff for exact dates.
Requirements: Applicants must be full-time graduate students enrolled in the school psychology program of the Department of Educational Psychology at the master's or doctoral level. Applications should be made at the end of the first academic year.
Geographic Focus: All States, Canada
Amount of Grant: 2,000 CAD
Contact: Program Contact, Faculty of Graduate Studies; (403) 220-5690; cbusch@acs.ucalgary.ca or ucawards@ucalgary.ca
Internet: http://www.grad.ucalgary.ca/funding/internal_scholarships/lev_4/mcwilliam.htm
Sponsor: University of Calgary
2500 University Drive NW, Earth Sciences Building, Room 720
Calgary, AB T2N 1N4 Canada

Helen Pumphrey Denit Charitable Trust Grants **2258**

The Helen Pumphrey Denit Trust for Charitable and Educational Purposes was established in 1988 to support charitable organizations that promote quality education, culture, human service, health service and arts opportunities. The grants are primarily made to organizations in the Baltimore region. Special consideration is given to the following three organizations: Montgomery General Hospital in Onley, Maryland; The George Washington University in Washington, D.C.; and The Wesley Theological Seminary in Washington, D.C. Grants for capital and program support are encouraged. Requests for general operating support will be received, but they will be given lower priority than other grants that have more specific purposes. Grants are primarily one year in duration. On occasion, multi-year grants will be awarded. Applicants must apply online at the grant website. Applicants are strongly encouraged to do the following before applying: review the downloadable state application procedures for additional helpful information and clarifications; review the downloadable online-application guidelines at the grant website; review the trust's funding history (link is available from the grant website); review the online application questions in advance; and review the list of required attachments. These will generally include: a list of board members, financial statements (audited, reviewed, or compiled by independent auditor); an organization summary; a list of other funding sources; an IRS Determination letter; and other required documents. All attachments must be uploaded in the online application as PDF, Word, or Excel files. The Helen Pumphrey Denit Trust has an application deadline of 11:59 p.m. on February 1. The applicants will be notified of grant decisions by June 30.
Requirements: Applicants must have 501(c)3 tax-exempt status and serve residents of Baltimore and surrounding communities.
Restrictions: The trust does not support requests from individuals, organizations attempting to influence policy through direct lobbying, or any political campaigns.
Geographic Focus: Maryland
Date(s) Application is Due: Feb 1
Samples: George Washington University, Washington, D.C., $50,000 (2010); Walters Art Museum, Baltimore, Maryland, $20,000 (2010); Montgomery General Hospital, Olney, Maryland, $50,000 (2010).
Contact: Sarah Kay, Vice President; (804) 788-2673; sarah.kay@baml.com
Internet: https://www.bankofamerica.com/philanthropic/fn_search.action
Sponsor: Helen Pumphrey Denit Charitable Trust
1111 E. Main Street, VA2-300-12-92
Richmond, VA 23219

Helen S. Boylan Foundation Grants **2259**

The Helen S Boylan Foundation is a private family foundation established in 1982 to continue the family tradition of commitment to enhancing the quality of life of the community through grants to qualified charitable organizations. In carrying out its mission, the Foundation considers a wide range of proposals within the following areas: arts, education, health, human services, environment, and public interest. Generally, grants are limited to projects that benefit the citizens of Jasper County, Missouri, Smith County, Texas and, Kansas City, Metropolitan area. Occasionally, projects that benefit the state of Missouri as a whole may be considered as well. The Foundation prefers to support proposals for new initiatives, special projects, expansion of current programs, capital improvements or building renovations. Grants from the Foundation are usually awarded for one year only. For projects in those areas in which the Foundation has a special interest, requests for multi-year funding and general operating support may be considered. The Board of Directors meet four times a year to consider grant requests. Applications must be received by March 31, June 30, September 30 or December 31 to be acted upon at the following meeting.
Requirements: IRS 501 (c)3 nonprofit organizations operating in the Carthage, Kansas City Metro and Lindale Texas area are eligible to apply for funding. Application form is available online at the Foundation's website.
Restrictions: No support for political organizations or religious activities. No grants to individuals, or for annual campaigns or endowments.
Geographic Focus: Missouri, Texas
Date(s) Application is Due: Mar 31; Jun 30; Sep 30; Dec 31
Amount of Grant: 500 - 50,000 U.S.D
Samples: Lindale Independant Schools, Lindale, TX, $52,000—supplies, systems, books, programs; Heartland Hoops, Kansas City, MO, $10,000—uniforms & fees; Art Central, Carthage, MO, $7,500—children's art camp.
Contact: James R. Spradling, President; (417) 358-4033; fax (417) 358-5937; spradlinglaw@hotmail.com
Internet: http://www.boylanfoundation.org/
Sponsor: Helen S. Boylan Foundation
320 Grant Street, P.O. Box 731
Carthage, MO 64836

Helen Steiner Rice Foundation Grants **2260**

The Foundation awards grants to worthy charitable programs that assist the needy and the elderly. Essential objectives for grant consideration are: basic necessities and human needs for the poor and elderly; preference for meeting the immediate needs of the poor; and innovative approaches. Organizations in the Greater Cincinnati area should use the contact information for Cincinnati. Organizations in Lorain County should *Contact:* Linda Weaver, Community Foundation Center of Lorain County, (440) 277-0142, Ext. 23. Application forms and instructions are available online.
Requirements: Nonprofit organizations in the greater Cincinnati area and Lorain, OH, may submit grant applications.
Restrictions: Funding is not available for building or endowment programs, direct gifts to individuals, or capital fund drives.
Geographic Focus: Ohio
Date(s) Application is Due: Jul 1
Contact: James D. Huizenga, Director; (513) 241-2880; fax (513) 768-6122; huizengai@greatercincinnatifdn.org or hrice@cincymuseum.org
Internet: http://www.helensteinerrice.com/grants.html
Sponsor: Helen Suiteiner Rice Foundation
200 West Fourth Street
Cincinnati, OH 45202-2602

Hendrick Foundation for Children Grants **2261**

The foundation supports programs designed to provide health, medical, social welfare, and educational services to benefit children with illness, disease, injury, pain, disability, incapacity, or other disadvantages; and improve quality of life for children with life-threatening or chronic injuries, illness, and disabilities. It is also contemplated that the Foundation will honor grants to other charitable and educational organizations sponsoring programs which are community oriented and focus on improving the quality

of life of children with life-threatening or chronic injuries, illnesses and disabilities. All requests shall be given careful and thorough consideration by the Hendrick Foundation for Children Grant Committee. Grant awards will be determined on the merits of the organization, its objectives and mission. All grant requests should be submitted in writing on the organization's letterhead
Amount of Grant: 3,500 - 250,000 U.S.D
Contact: Charles V. Ricks, President; (704) 568-5550
Internet: http://www.thehendrickfoundation.org/grants.html
Sponsor: Hendrick Foundation for Children
P.O. Box 240070
Charlotte, NC 28224-0070

Hendricks County Community Foundation Grants **2262**
The Hendricks County Community Foundation Grants provide funding for organizations or charitable projects that serve in the following program areas: arts and culture; community development; education; environment; health and human services; and youth. These grants enable organizations to provide effective programs and respond to needs of people in the Hendricks County community.
Requirements: A letter of intent should be submitted to the organization between December 1 and January 11. The Foundation uses the following criteria when reviewing applications: sustainability; effective operations; proven success; strong leadership; innovation and creativity; accessibility; collaboration; and engagement.
Restrictions: The Foundation will not fund: bands, sports teams, or other groups without a philanthropic project; annual appeals, galas or membership contributions; fundraising events such as golf tournaments, walk-a-thons, and fashion shows; grants to individuals; projects aimed at promoting a particular religion or construction projects for religious institutions; operating, program and construction costs at schools, universities and private academies unless there is a significant opportunity for community use or collaboration; organizations or projects that discriminate based upon race, ethnicity, age, gender, sexual orientation; political campaigns or direct lobbying efforts by 501(c)3 organizations; post-event, after-the-fact situations or debt retirement; medical, scientific or academic research; publications, films, audiovisual and media materials, programs produced for artistic purposes or produced for resale.
Geographic Focus: Indiana
Amount of Grant: 500 - 17,500 U.S.D
Samples: Farmers and Hunters Feeding the Hungry, general operating expenses, $3,000 (2011); Hendricks County Senior Services, transportation for the elderly and disabled, $5,000 (2011); Kingsway Community Care Center, general operating expenses, $2,500 (2011).
Contact: Susan Rozzi, Associate Director; (317) 718-1200; fax (317) 718-1033; janet@hendrickscountycf.org
Internet: http://www.hendrickscountycf.org/grants/oppfund_grants/index.shtml
Sponsor: Hendricks County Community Foundation
5055 East Main Street, Suite A
Avon, IN 46123

Henrietta Lange Burk Fund Grants **2263**
The Henrietta Lange Burk Fund is a private foundation created by Mrs. Henrietta Lange Burk in 1994 to support and promote quality arts, cultural, educational, health-care, and human-services programming for underserved populations. Special consideration is given to charitable organizations that address the health concerns of older adults, through either direct programming or research. Mrs. Burk created the Foundation as a memorial to her parents, Mr. and Mrs. Henry G. Lange, as well as to her husband, William Burk, and herself. Mrs. Burk was particularly interested in the performing and cultural arts; age-related health problems, research and care; as well as human services organizations, especially those that were connected to her Protestant background. The foundation will consider requests for general operating support only if the organization's operating budget is less than $1 million. The majority of grants from the Lange Burk Fund are one year in duration. Applicants must apply online at the grant website. Applicants are strongly encouraged to do the following before applying: review the downloadable state application procedures for additional helpful information and clarifications; review the downloadable online-application guidelines at the grant website; review the foundation's funding history (link is available from the grant website); review the online application questions in advance; and review the list of required attachments. These will generally include: a list of board members, financial statements (audited, reviewed, or compiled by independent auditor); an organization summary; a list of other funding sources; an IRS Determination letter; and other required documents. All attachments must be uploaded in the online application as PDF, Word, or Excel files. The Henrietta Lange Burk Fund has biannual deadlines of June 1 and November 1. Applications must be submitted by 11:59 p.m. on the deadline dates. Applicants for the June deadline will be notified of grant decisions by September 30. Applicants for the November deadline will be notified by March 31.
Requirements: Applicants must have 501(c)3 tax-exempt status and serve residents of the Chicago Metropolitan area.
Restrictions: In general, grant requests for individuals, endowment campaigns or capital projects will not be considered. Because requests for support usually exceed available resources, organizations can only apply to either the Lang Burk Fund or the Colonel Stanley NcNeil Foundation in the same calendar year. Grant requests to both foundations during the same calendar year will no longer be accepted. The foundation does not support requests from individuals, organizations attempting to influence policy through direct lobbying, or any political campaigns.
Geographic Focus: Illinois
Date(s) Application is Due: Jun 1; Nov 1
Samples: Salvation Army, Chicago, Illinois, $50,000 (2010); The Harbour, Park Ridge, Illinois, $15,000 (2010); Chicago Opera Theater, Chicago, Illinois, $10,000 (2010).
Contact: George Thorn; (312) 828-4154; ilgrantmaking@bankofamerica.com
Internet: https://www.bankofamerica.com/philanthropic/fn_search.action
Sponsor: Henrietta Lange Burk Fund
231 South LaSalle Street, IL1-231-13-32
Chicago, IL 60604

Henrietta Tower Wurts Memorial Foundation Grants **2264**
The Foundation supports organizations which are engaged in helping or caring for people in need, or alleviating the conditions under which they live, primarily for the elderly, women, family and child welfare services. Giving is limited to Philadelphia, Pennsylvania.
Requirements: Non-profits in Philadelphia, Pennsylvania are eligible. The initial approach should be a letter requesting an application form from the Foundation.
Restrictions: No grants to individuals, or for endowment funds, scholarships, fellowships, or matching gifts; no loans. Organizations or programs serving disadvantaged youth and the elderly in Philadelphia, Pennsylvania must have an annual budget of less than 3 million.
Geographic Focus: Pennsylvania
Date(s) Application is Due: Feb 1; May 1; Sep 1
Amount of Grant: 1,000 - 7,000 U.S.D
Samples: Empowerment Group, Philadelphia, PA, $7,000; Juvenile Law Center, Philadelphia, PA, $5,000; Supportive Older Womens Network, Philadelphia, PA, $4,000;
Contact: Andrew Swinney, President; (215) 563-6417
Sponsor: Henrietta Tower Wurts Memorial Foundation
1234 Market Street, Suite 1800
Philadelphia, PA 19107-3704

Henry and Ruth Blaustein Rosenberg Foundation Grants **2265**
The Henry and Ruth Blaustein Rosenberg Foundation provides support primarily in the Baltimore, Maryland region. Areas of interest are: arts and culture; youth development; health; and adult self-sufficiency. The goal of the arts and culture program is to enable cultural institutions in the Baltimore metropolitan area to reach underserved youth and to diversify audiences. Youth Development funding is provided to promote recreation, learning, and leadership development through after-school programs; prevent teen pregnancy; and advocate for policies and practices that improve outcomes for youth. Health funding is available to improve the prevention and treatment of breast cancer, diabetes, multiple sclerosis, and kidney disease. Adult Self-Sufficiency funding is provided to promote adult literacy; provide job training; improve parenting skills, especially for parents of children under age three; and eliminate domestic violence. The foundation meets three or four times a year to consider proposals. Letters of intent and proposals are accepted on a rolling basis. Applications that conform to the Association of Baltimore Area Grantmakers (ABAG) Common Grant Application are accepted and this form is available at the Henry and Ruth Blaustein Rosenberg Foundation website.
Requirements: 501(c)3 tax-exempt charitable organizations located in and/or primarily serving the metropolitan Baltimore area are eligible to apply. The initial application should include the following: information about the program(s) for which funding is requested; need, purpose, activities, and evaluation plan of the proposed program(s); program budget (including sources of anticipated income as well as expenditures) and timeline; dollar amount of funding requested; history, mission, and key accomplishments of your organization; information on Board members and key staff; current institutional operating budget (including major sources of revenue as well as expenditures); copy of your IRS tax status determination letter or information about your fiscal agent.
Restrictions: The foundation does not: make grants or scholarships to individuals; accept unsolicited proposals for academic, scientific, or medical research; support direct mail, annual giving, membership campaigns, fundraising and commemorative events. The foundations rarely make capital grants unless there is a prior relationship with the applicant organization.
Geographic Focus: Maryland
Amount of Grant: 1,000 - 40,000 U.S.D
Samples: Heifetz International Music Institute, Ellicott City, MD, $30,000—support for a six-week summer program for aspiring young musicians; Open Society Institute, Baltimore, MD, $15,000—sponsorship support for the Community Fellows Program; St. Agnes Hospital Foundation, Baltimore, MD, $45,000—support for the Beacon of Hope capital campaign.
Contact: Michael J. Hirschhorn; (410) 347-7201; fax (410) 347-7210; info@blaufund.org
Internet: http://www.blaufund.org/foundations/henryandruth_f.html
Sponsor: Henry and Ruth Blaustein Rosenberg Foundation
10 E Baltimore Street, Suite 1111
Baltimore, MD 21202

Henry County Community Foundation Grants **2266**
As a community foundation, the Henry County Community Foundation addresses the broad needs in Henry County, Indiana which include, but are not limited to, the following five categories: health and medical; social services; education; cultural affairs and civic affairs. All requests for grants are reviewed by the Henry County Community Foundation's Grants Committee, which is made up of members of the Board of Directors and several outside advisers. Reviews and recommendations are then presented to the full Board of Directors at its regularly scheduled meetings in April and October, at which time all final grant decisions are made. However, the Board reserves the right to consider individual requests at any regularly scheduled meeting, if deemed necessary. The Henry County Community Foundation's grant program will emphasize change-oriented and focused types of grants to achieve certain objectives such as becoming more efficient, increasing fund-raising capabilities, delivering better products, etc.

Requirements: Applications from organizations whose programs benefit the residents of Henry County will be accepted only from organizations who attend the spring or fall grant workshops held at the Henry County Community Foundation office. The Spring Grant Workshop is March 11th. The Fall Grant Workshop is August 12th. Organizations must be a 501c (3) tax-exempt or be sponsored by a 501c (3) tax-exempt organization. Grants will also be accepted from school and government entities.
Restrictions: Restrictions for use of grant money: grant money will not be awarded to cover operating cost; no grants will be made solely to individuals; generally, grants will not be made to enable individuals or groups of individuals to attend seminars or take trips except where there are special circumstances which will benefit the community; no grants will be made specifically for sectarian religious purposes but can be made to religious organizations for general community programs. No grants will be made exclusively for endowment purposes of recipient organizations; generally, no grant proposals will be considered which have been proposed by individuals or organizations that are not charitable organizations; generally, grants will not be made for programs and/or equipment which were committed to prior to the submission of grant application.
Geographic Focus: Indiana
Date(s) Application is Due: Mar 31; Aug 28
Contact: Jerry Schaeffer, Executive Director; (765) 529-2235; fax (765) 529-2284; jerry@henrycountycf.org
Internet: http://www.henrycountycf.org/index.php?submenu=Grants&src=gendocs&ref=Grants&category=Grants
Sponsor: Henry County Community Foundation
700 S Memorial Drive, P.O. Box 6006
New Castle, IN 47362

Henry J. Kaiser Family Foundation Grants **2267**
The foundation concentrates giving in the following areas: U.S. government's role in health, health of low-income and minority groups with major emphasis on HIV/AIDS policy, reproductive health policy, and health system innovation and reform in California. The foundation also operates a major program to improve health and health care and promote social justice in South Africa. Grants are awarded for one to three years and support a range of activities, including policy analysis, applied research to define and measure public health problems, demonstration and pilot projects, and communications activities that help sharpen health care debates and improve quality of health information. Prospective applicants should submit a preliminary letter (two to three pages in length) that briefly describes the proposed project, along with an estimate of the total budget and the amount requested from the foundation. There are no deadlines and no application forms. Inquiries for projects in South Africa may be addressed to Dr. Michael Sinclair, Senior Vice President, The Henry J. Kaiser Family Foundation, 1450 G St NW, Suite 250, Washington, DC 20005, (202) 347-5270, fax: (202) 347-5274.
Requirements: Grants in response to unsolicited proposals are made only to governmental agencies and to private organizations with IRS 501(c)3 tax-exempt status.
Restrictions: The foundation does not award grants to individuals. Support is not given to ongoing general operating expenses, indirect costs, capital campaigns, annual appeals or other fundraising events, construction, purchase or renovation of facilities, or equipment purchases.
Amount of Grant: 1,000 - 500,000 U.S.D
Samples: Alliance for Justice (Washington, DC)—for general operating support, $600,000 over three years; U of Florida, Tropical Conservation and Development Program (Gainesville, FL)—for a community-based conservation and training program for organizations working in protected areas of Colombia and Ecuador, $240,000 over three years; U of California at Los Angeles (CA)—for the Network on Youth Mental Health Care, $5 million over four years; Ctr for Defense Information (Washington, DC)—for policy research and advocacy on nuclear-weapons systems and their vulnerabilities, $650,000 over three years.
Contact: Renee Wells; (415) 854-9400; fax (415) 854-4800; rwells@kff.org
Internet: http://www.kff.org
Sponsor: Henry J. Kaiser Family Foundation
2400 Sand Hill Road
Menlo Park, CA 94025

Henry J. Kaiser Family Foundation Media Fellowships **2268**
The purpose of the fellowship is to help journalists and commentators do the best possible job of keeping the public informed about health issues at this critical time in the evolution of the health care system. The goal of the fellowship is to provide a highly flexible range of opportunities tailored to each fellow's interests; help journalists improve the quality of the work they do; provide time to research specific topics; and deepen participants' commitment to becoming specialists in health reporting. Fellows may want to participate in the fellowship program on a part-time basis and continue writing or reporting during their fellowship. Fellows are encouraged to do reporting based on their fellowship research and to experiment with different media or forms of writing. Most fellows work from home, or base themselves at a local academic or research institution. There is no obligation to pursue academic coursework, although that is an option open to all fellows. There is no application form. Guidelines are available online.
Requirements: Fellowships will be awarded to U.S. journalists interested in U.S. health policy issues.
Date(s) Application is Due: Mar 16
Amount of Grant: Up to 50,000 U.S.D
Contact: Penny Duckham; (650) 234-9220; fax (650) 854-4800; pduckham@kff.org
Internet: http://www.kff.org/about/mediafellowships.cfm
Sponsor: Henry J. Kaiser Family Foundation
2400 Sand Hill Road
Menlo Park, CA 94025

Henry L. Guenther Foundation Grants **2269**
The foundation awards grants to California nonprofit organizations in its areas of interest, including medical services, creating opportunities for youth, and community services. Grants support hospitals, medical research, and social services. The board meets in January and July. Contact the office for application procedures.
Requirements: California nonprofit organizations are eligible. Grants are awarded primarily in Lost Angeles.
Restrictions: Grants do not support government agencies, religious organizations for religious purposes, individuals, or debt-reduction requests.
Geographic Focus: California
Date(s) Application is Due: May 31; Nov 30
Amount of Grant: 10,000 - 100,000 U.S.D
Samples: Downey Community Hospital Health Services Corp (CA)—to build a multipurpose medical campus for seniors, $1 million; White Memorial Medical Ctr Charitable Foundation (Los Angeles, CA)—to purchase imaging equipment, $500,000; Saint Jude Medical Ctr (Fullerton, CA)—to update the cardiac catheterization laboratory, $600,000.
Contact: W.D. Milliken, Secretary; (310) 785-0658
Sponsor: Henry L. Guenther Foundation
2029 Century Park D, Suite 4392
Los Angeles, CA 90067

Herbert A. and Adrian W. Woods Foundation Grants **2270**
The Herbert A. and Adrian W. Woods Foundation was established on June 9, 1999 upon the death of Mrs. Adrian W. Woods. Mrs. Woods had a long history of charitable giving in the St. Louis community and wanted to establish this foundation to continue that legacy of giving. The Foundation supports charitable organizations primarily in the greater St. Louis, Missouri area. The Trustees will consider the donor's past giving history and any special needs of any of the charities she has given to in the past. The Trustees will also consider requests from charitable organizations that fall into one or more of the following categories: abused, neglected, or troubled children; the poor; the Episcopal Church and affiliates, including outreach programs; arts and culture in the Metropolitan St. Louis area; animal welfare (in Missouri); and victims of illness or disability, including research in this area. The following types of requests may be submitted: special projects; capital campaign requests (capital grants are awarded only as a source of support among a broad community of funders); challenge or matching grants; and general operation funding. Applicants must apply online at the grant website. Applicants are strongly encouraged to do the following before applying: review the downloadable state application procedures for additional helpful information and clarifications; review the downloadable online-application guidelines at the grant website; review the foundation's funding history (link is available from the grant website); review the online application questions in advance; and review the list of required attachments. These will generally include: a list of board members, financial statements (audited, reviewed, or compiled by independent auditor); an organization summary; a list of other funding sources; an IRS Determination letter; and other required documents. All attachments must be uploaded in the online application as PDF, Word, or Excel files. The application deadline for this foundation is 11:59 p.m. on September 1. Final decisions about grant rewards will be made by November 30.
Requirements: Applicants must have 501(c)3 tax-exempt status.
Restrictions: The fund will not consider requests for multi-year grants (pledges) or endowment creation and funding. The fund does not support requests from individuals, organizations attempting to influence policy through direct lobbying, or any political campaigns.
Geographic Focus: Missouri
Date(s) Application is Due: Sep 1
Samples: Grace Hill Settlement House, Saint Louis, Missouri, $20,000; Humane Society of Missouri, Saint Louis, Missouri, $15,000; Central Institute for the Deaf, Saint Louis, Missouri, $5,000.
Contact: Shanise Evans, Vice President; (314) 466-8027; shanise.evans@baml.com
Internet: https://www.bankofamerica.com/philanthropic/fn_search.action
Sponsor: Herbert A. and Adrian W. Woods Foundation
100 North Broadway MO2-100-07-15
Saint Louis, MO 63102-2728

Hereditary Disease Foundation John J. Wasmuth Postdoctoral Fellowships **2271**
Support is offered for research projects that contribute to identifying and understanding the basic defect of Huntington's disease. Areas of interest include trinucleotide expansions, animal models, gene therapy, neurobiology and development of the basal ganglia, cell survival and death, and intercellular signaling in striatal neurons. Awards may be renewed for a third year. To receive an application, submit a letter of intent. Letter of intent forms may be submitted electronically via the Web site.
Date(s) Application is Due: Feb 15; Jun 15; Oct 15
Amount of Grant: 45,000 U.S.D
Contact: Dr. Carl D. Johnson, (212) 928-2121; fax (212) 928-2172; carljohnson@hdfoundation.org
Internet: http://www.hdfoundation.org/funding/postdoct.php
Sponsor: Hereditary Disease Foundation
3960 Broadway, 6th Floor
New York, NY 10032

Hereditary Disease Foundation Lieberman Award **2272**
The Hereditary Disease Foundation announces a special Lieberman Award, to catalyze innovative proposals leading to the treatment and cure of Huntington's disease. A Lieberman Award can be funded for two years for up to $75,000 per year. Areas of interest include trinucleotide expansions, animal models, gene therapy, neurobiology

and development of the basal ganglia, cell survival and death, and intercellular signaling in striatal neurons.
Amount of Grant: 75,000 - 150,000 U.S.D
Contact: Dr. Carl D. Johnson, (212) 928-2121; fax (212) 928-2172; carljohnson@hdfoundation.org
Internet: http://www.hdfoundation.org/funding/lieberman.php
Sponsor: Hereditary Disease Foundation
3960 Broadway, 6th Floor
New York, NY 10032

Hereditary Disease Foundation Research Grants **2273**
The focus of the Hereditary Disease Foundation is on Huntington's disease. Support will be for research projects that contribute to identifying and understanding the basic defect in Huntington's disease. Grants are usually for one year, with the possibility of renewal for up to three years. Areas of interest include trinucleotide expansions, animal models, gene therapy, neurobiology and development of the basal ganglia, cell survival and death, and intercellular signaling in striatal neurons. In addition to its regular grants, the foundation also offers the Lieberman Award for innovative proposals accelerating the discovery of a treatment and cure of Huntington's disease. Grants that receive funding are considered seed money. If the project shows promise, it is hoped that other institutions will fund it thereafter. To obtain an application, submit a one-page letter of intent. Letter of intent forms may be submitted electronically via the Web site.
Date(s) Application is Due: Feb 15; Jun 15; Oct 15
Amount of Grant: Up to 50,000 U.S.D
Samples: Dr. Andrew Dwork, Columbia U—for research entitled Ultrastructural Immunohistochemistry of Cortical Biopsies in HD; Dr. Nansheng Chen, U of British Columbia—for research entitled Regulation of NMDA Receptor Function by Mutant Huntington in Neostriatal Neurons; Dr. Chris Huang, Massachusetts General Hospital—for research entitled Anti-Huntington Aggregation Agents as Potential Therapeutics for Huntingtons Disease; Dr. Lucius Passani, Massachusetts General Hospital—for research entitled Evaluation of HYPA/FBP-11, HYPB and HYPC as Candidates for Involvement in HD Pathogenesis.
Contact: Dr. Carl D. Johnson, (212) 928-2121; fax (212) 928-2172; carljohnson@hdfoundation.org
Internet: http://www.hdfoundation.org/funding/grants.php
Sponsor: Hereditary Disease Foundation
3960 Broadway, 6th Floor
New York, NY 10032

Herman Goldman Foundation Grants **2274**
The Herman Goldman Foundation strives to enhance the quality of life through innovative grants in four main areas: 1) health — to achieve effective delivery of physical and mental health care services; 2) social justice — to develop organizational, social, and legal approaches to those who are aid deprived or handicapped; 3) education — for new or improved counseling for effective preschool, vocational and paraprofessional training; and 4) the arts — to increase opportunities for talented youth to receive training and for less affluent individuals to attend quality presentations. Grantmaking is primarily to 501(c)3 organizations in the metropolitan New York area. Types of support include: annual campaigns; building and renovation; capital campaigns; continuing support; endowments; general operating support; program development; internship funds; research; and seed money.
Requirements: Applicants should submit a proposal along with a copy of the organization's IRS determination letter. There are no deadlines. The board meets monthly, with grants considered in April, July, and November. Organizations are notified within two to three months of submission.
Restrictions: The Foundation does not fund religious organizations, individuals, or emergency funds.
Geographic Focus: New York
Contact: Richard Baron, Executive Director; (212) 797-9090; goldfound@aol.com
Sponsor: Herman Goldman Foundation
44 Wall Street, Suite 1212
New York, NY 10005-2401

Hershey Company Grants **2275**
The Hershey Company remains committed to supporting the communities in which it operates and to society in general. Cash and product contributions are made to support a variety of worthy causes and non-profit organizations which support Education, Health & Human Services, Civic & Community initiatives, Arts & Culture and the Environment. Particular emphasis is placed upon causes that support kids and kids at risk. The Hershey Company has plants in the following areas: Hilo, HI; Robinson, IL; Hazleton, PA; Hershey, PA; Lancaster, PA; Memphis, TN; Stuarts Draft, VA; Sao Roque, Brazil; Guadalajara, Mexico; Monterrey, Mexico; Mumbai, India; Shanghai, China. Organizations located in communities in which The Hershey Company has manufacturing operations should direct requests for funding to the management of The Hershey Company facility in their area, see the Hershey website for additional guidelines and contact information.
Requirements: Nonprofit organizations are eligible. Preference is given to nonprofits in corporate-operating locations. All requests must be submitted in writing.
Restrictions: The following are ineligible for funding: organizations without an Internal Revenue Code 501(c)3 non-profit, tax-exempt status; individuals; organizations outside the immediate areas of The Hershey Company's manufacturing facilities, with the exception of national and state-wide organizations whose programs complement Hershey's funding priorities; political campaigns, political or lobbying organizations, or those supporting the candidacy of a particular individual; churches or religious organizations, including seminaries, Bible colleges and theological institutions; fraternal organizations; labor organizations; member agencies of United Way. Exception: requests for capital campaign funding will be considered; affiliate organizations of the Cultural Enrichment Fund in Central Pennsylvania. Exception: requests for capital campaign funding will be considered.
Geographic Focus: Hawaii, Illinois, Pennsylvania, Tennessee, Virginia, Brazil, China, Mexico
Contact: Grants Administrator; (800) 468-1714; fax (717) 534-6550
Internet: http://www.thehersheycompany.com/about/responsibility.asp
Sponsor: Hershey Company
Community Relations, 100 Crystal A Drive, P.O. Box 810
Hershey, PA 17033-0810

Heublein Foundation Inc Grants **2276**
The foundation offers grant support in the areas of preschool through higher education, including a scholarship program for dependents of employees, health and social care, culture and art, problems of the inner cities, and education to combat alcohol abuse. A large percentage of grants within all categories is targeted at disadvantaged populations. Heublein also contributes directly to the alcoholism prevention programs of several organizations and to research by noted universities. Application deadline for scholarships is December 1.
Restrictions: Grants may not be made to individuals, religious groups, political organizations, or for endowments. The foundation principally supports programs in communities where the company has major facilities greater Hartford, CT, area; Allen Park, MI; and Madera, Menlo Park, and Napa Valley, CA.
Geographic Focus: Connecticut
Date(s) Application is Due: Dec 1
Contact: Treasurer; (860) 231-5000
Sponsor: Heublein Foundation
P.O. Box 778
Farmington, CT 06142-0778

HFSF Concern Awards **2277**
The foundation awards grants to individuals in South Florida who are making a healthy difference in the lives of others. The awards recognize those whose volunteerism, philanthropic support, and or leadership efforts improved the quality of life for others in the tri-county area comprising Broward, Miami-Dade, and Monroe Counties. Two people each receive a grant, and four runners-up receive a grant for a health or health-related nonprofit organization of their choice. Contact the office for a nomination form.
Requirements: Community organizations and individuals are encouraged to nominate people who are making a real impact on the health status of people in Florida's Broward, Miami-Dade, and Monroe Counties.
Geographic Focus: Florida
Date(s) Application is Due: Oct 1
Amount of Grant: 10,000 - 25,000 U.S.D
Contact: Shari Gantman, (305) 374-9199; sgantman@hfsf.org
Internet: http://www.hfsf.org/index.cfm?fa=news&viewid=360
Sponsor: Health Foundation of South Florida
601 Brickell Key Drive, Suite 901
Miami, FL 33131

HFSF Grants **2278**
The foundation conducts two separate grant programs: responsive grants and strategic initiative grants. With exceptions, the foundation will focus on developing one- to two-year grants. The following responsive grant categories have been established: project planning—to support the planning of projects requiring outside expertise, significant amounts of inter-agency collaboration, or other aspects needing additional resources to develop; health system/health policy development—to support a variety of activities specifically designed to improve public policies, infrastructure, and/or inter-agency procedures leading to increased system effectiveness, efficiency, sustainability, and overall capacity; organizational capacity building/core operating support—to organizations with a clear plan for increasing their agencies' effectiveness, efficiency, and sustainability; and health services—to support a range of preventive and primary health care services. The listed deadlines are for preliminary proposals; full proposals are by invitation. Application and guidelines are available online.
Requirements: 501(c)3 tax-exempt organizations serving southern Florida are eligible.
Geographic Focus: Florida
Date(s) Application is Due: Mar 31; Oct 22
Amount of Grant: 50,000 - 300,000 U.S.D
Samples: Broward House (Fort Lauderdale, FL)—to support operational efficiency and improve its technology, $86,380; Community Committee for Developmental Handicaps (Miami, FL)—to develop a plan designed to strengthen its board development, marketing, staffing, and technology, $23,925; Human Services Coalition (Miami, FL)—to increase organizational capacity by improving communications and increasing membership, revenue, and donations, $20,000; Florida International U, Ctr on Aging (FL)—for efforts to improve the health of older adults who are at high risk for poor health due to obesity and physical inactivity, $263,394.
Contact: Peter Wood, Chief Program Officer; (305) 374-2282; fax (305) 374-7003; pwood@hfsf.org
Internet: http://www.hfsf.org
Sponsor: Health Foundation of South Florida
601 Brickell Key Drive, Suite 901
Miami, FL 33131

HHK Employment Grants 2279

The majority of the foundation's grants are directed toward increasing employment opportunities for people with physical disabilities. Grants are awarded to organizations that work toward improving the employment and career advancement of people with physical disabilities living in New Jersey. Areas of interest include programs that focus on transition to work for youth and adults; transportation to and from work; workplace accommodations—overcoming barriers to employment, such as access to assistive technology, providing accessible facilities and services, employee education on disabilities; equipment and training programs focusing on workplace preparation and vocational training; supported employment programs; mentoring; school-based enterprises; work-based learning; alternative work solution, such as telecommuting; and employment strategies to support disability recruitment, hiring practices, placement, and retention. Types of support include project/program funds and general operating funds. Grants are awarded for one year. Preference is given to organizations seeking matching funds and that collaborate with other organizations or entities. Guidelines are available online.

Requirements: 501(c)3 tax-exempt organizations are eligible.

Restrictions: The foundation does not support grants to individuals; grants to underwrite fundraising events, such as benefits, dinners, or theater events, ticket purchases, or advertising; endowment campaigns; scholarship funds; deficit reduction or loans; or international causes.

Geographic Focus: New Jersey

Amount of Grant: 20,000 - 50,000 U.S.D

Samples: Brain Injury Assoc of New Jersey (Edison, NJ)—for training, technical assistance, and resources for people with brain injuries, $49,980; Cheshire Home (Florham Park, NJ)—for salary support of a teacher for GED and job preparation and computer-skills training, $12,000; Employment Horizon (Cedar Knolls, NJ)—for salary support of a driver to transport people with disabilities to work centers in their community, $10,000.

Contact: Elaine Katz; (973) 324-8367; fax (973) 324-8373; ekatz@hhkfdn.org

Internet: http://www.hhkfdn.org

Sponsor: Henry H. Kessler Foundation

300 Executive Drive, Suite 150

West Orange, NJ 07052

HHMI-NIBIB Interfaces Initiative Grants 2280

HHMI and the National Institute of Biomedical Imaging and Bioengineering have formed a partnership to support biomedical research institutions in developing graduate-level research training programs in emerging interdisciplinary fields. The primary goal of this initiative is to train a cadre of PhD scientists who possess the knowledge and skills to conduct interdisciplinary research at the interface between the biomedical sciences and the physical science, computational, engineering, or mathematical disciplines. These fields may include, but are not limited to, chemistry, imaging, science, materials science, nanotechnology, and physics. Another goal is to reduce barriers to interdisciplinary graduate science education. Proposals should reflect the unique educational and scientific capabilities and strengths of the applicant institution(s) as well as address the specific goals of the initiative. The initiative consists of two phases: Phase I supports the establishment of new interdisciplinary training programs; Phase II will sustain the training programs through their critical years. Registration letters of intent to apply are due by January 20, with the proposal submission deadline June 15.

Requirements: All U.S. institutions that grant PhD degrees in appropriate science or engineering disciplines are eligible to apply. Collaborative programs between two or more institutions are acceptable.

Date(s) Application is Due: Jan 20; Jun 15

Amount of Grant: Up to 1,000,000 U.S.D

Contact: Maryrose Franko, Senior Program Officer; (800) 448-4882, ext. 8880; fax (301) 215-8888; interdisc@hhmi.org

Internet: http://www.hhmi.org/grants/institutions/nibib.html

Sponsor: Howard Hughes Medical Institute

4000 Jones Bridge Road

Chevy Chase, MD 20815-6789

HHMI-NIH Cloister Research Scholars Program 2281

The goal of this program, a joint venture of HHMI and NIH, is to expand the pool of medically trained researchers by encouraging medical students to pursue research careers. The program provides support for an intensive nine months to one year of full-time research, with a possible one-year extension, at NIH in Bethesda, MD. Most students enter after their second year of medical school. The Cloister, a residential facility for scholars, is available on the NIH campus. The research project is selected upon arrival at NIH, after a round of laboratory visits. Application forms may be downloaded from the Web site or requested from the program officer.

Requirements: Applicant must be in good standing at a medical or dental school in the U.S. or Puerto Rico and must receive permission from the school to participate in the program. The school does not have to be a major academic medical center or research-oriented school.

Restrictions: Enrollees in an M.D./Ph.D. or D.D.S./Ph.D. program are not eligible.

Date(s) Application is Due: Jan 10

Amount of Grant: 17,800 U.S.D

Contact: Office of Grants and Special Programs; (800) 424-9924 or (301) 215-8873; fax (301) 215-8888; research_scholars@hhmi.org

Internet: http://www.hhmi.org/research/cloister/program.html

Sponsor: Howard Hughes Medical Institute

4000 Jones Bridge Road

Chevy Chase, MD 20815-6789

HHMI/EMBO Start-up Grants for Central Europe 2282

The program encourages young scientists to obtain independent faculty positions in select countries that have ongoing HHMI and EMBO science programs. Applications are encouraged from scientists to set up their first independent laboratories in a Central European EMBC member state (Croatia, Czech Republic, Estonia, Hungary, Poland, and Slovenia). Application forms should be completed by the individual scientist and the receiving institute. It is expected that the institute make the applicant an offer that goes beyond that of a position and laboratory space. HHMI will contribute $50,000, while the participating member state must provide the remaining $25,000. Application and guidelines are available online.

Requirements: Eligible scientists should have an excellent publication record; have a maximum of 10 years postdoctoral experience; be negotiating a position at an institute/university in Central Europe by the date of the application; and be located outside Central Europe at the time of application.

Geographic Focus: All States, Austria, Czech Republic, Germany, Hungary, Liechtenstein, Poland, Slovakia, Slovenia, Switzerland

Date(s) Application is Due: Aug 1

Amount of Grant: 75,000 U.S.D

Contact: Grants Administrator; (301) 215-8500

Internet: http://www.embo.org/projects/yip/embo_hhmi_startup_grants.html

Sponsor: Howard Hughes Medical Institute

4000 Jones Bridge Road

Chevy Chase, MD 20815-6789

HHMI Biomedical Research Grants for International Scientists: Infectious Diseases and Parasitology 2283

The Program provides five-year grants to support promising basic biomedical research scientists working in the fields of infectious diseases and parasitology. The Institute seeks to build the strongest foundation of scientific knowledge and capability by creating a global network of scientists who make important contributions outside the U.S. research community. It support these leaders in their home countries to facilitate the application and dissemination of their knowledge and to strengthen research environments and educational opportunities.

Requirements: Applicants must hold a full-time appointment or have a pending appointment at a nonprofit scientific research organization in any country other than the United States and the United Kingdom. Researchers may not be citizens or permanent residents of the United States.

Restrictions: Only scientists working in the research areas specified in the program announcement are eligible to apply for grant support. The Program does not consider unsolicited grant proposals or award grants for clinical trials and research on health education, health-care delivery, or health services.

Date(s) Application is Due: Sep 15

Amount of Grant: 50,000 - 100,000 U.S.D

Contact: Jill G. Conley; (301) 215-8873; fax (301) 215-8888; parasite@hhmi.org

Internet: http://www.hhmi.org/grants/individuals/idap.html

Sponsor: Howard Hughes Medical Institute

4000 Jones Bridge Road

Chevy Chase, MD 20815-6789

HHMI Biomedical Research Grants for International Scientists in Canada and Latin America 2284

HHMI's International Research Scholars Program provides five-year grants to support promising basic biomedical research scientists working in Canada and Latin America. Eligible epidemiology research includes that directed toward an understanding of disease distribution in populations or of associations that may suggest causality or preventive strategies.

Requirements: The supported research must be conducted in one of the following eight eligible countries: Argentina, Brazil, Canada, Chile, Mexico, Peru, Uruguay, or Venezuela. Applicants must hold a full-time appointment or have a pending full-time appointment at a nonprofit scientific research organization in any one of the eight eligible countries.

Restrictions: Clinical trials and research on health education, health care delivery, or health services are not eligible fields.

Date(s) Application is Due: Sep 14

Amount of Grant: 50,000 - 100,000 U.S.D

Contact: Jill G. Conley; (301) 215-8873; fax (301) 215-8888; canlatam@hhmi.org

Internet: http://www.hhmi.org/grants/individuals/canlatam.html

Sponsor: Howard Hughes Medical Institute

4000 Jones Bridge Road

Chevy Chase, MD 20815-6789

HHMI Biomedical Research Grants for International Scientists in the Baltics, Central and Eastern Europe, Russia, and Ukraine 2285

The Program will award five-year grants to support promising basic biomedical research scientists working in the Baltics, Central and Eastern Europe, Russia, and Ukraine. Eligible epidemiology research includes that directed toward an understanding of disease distribution in populations or of associations that may suggest causality or preventive strategies. Allowable expense categories include salaries and stipends for award recipients, graduate students, postdoctoral fellows, and technicians, equipment and supplies, travel, and publication costs. Up to 10 percent of the grant may be used to pay for indirect costs (including administrative costs) of the scientist's institution.

Requirements: The supported research must be conducted in one of the following eligible countries: Bulgaria, Croatia, Czech Republic, Estonia, Hungary, Latvia, Lithuania, Poland, Romania, Russia, Slovak Republic, Slovenia, or Ukraine.

Restrictions: Researchers may not be citizens or permanent residents of the United States. Grant funds may not be used for laboratory renovations.
Geographic Focus: All States, Albania, Andorra, Armenia, Austria, Azerbaijan, Belarus, Belgium, Bosnia & Herzegovina, Bulgaria, Croatia, Cyprus, Czech Republic, Denmark, Estonia, Finland, France, Georgia, Germany, Greece, Hungary, Hungary, Iceland, Ireland, Italy, Kosovo, Latvia, Liechtenstein, Lithuania, Luxembourg, Macedonia, Malta, Moldova, Monaco, Montenegro, Norway, Poland, Portugal, Romania, Russia, Russia, San Marino, Serbia, Slovakia, Slovenia, Spain, Sweden, Switzerland, The Netherlands, Turkey, Ukraine, Ukraine, United Kingdom, Vatican City
Date(s) Application is Due: Nov 17
Amount of Grant: 50,000 - 100,000 U.S.D
Contact: Jill G. Conley, Program Director; (301) 215-8873; fax (301) 215-8888; bceeru@hhmi.org
Internet: http://www.hhmi.org/grants/individuals/bceeru.html
Sponsor: Howard Hughes Medical Institute
4000 Jones Bridge Road
Chevy Chase, MD 20815-6789

HHMI Grants and Fellowships Programs **2286**
The institute administers a grants program that focuses on improving science education from preschool through postdoctoral training, enhancing the science literacy of the general public, and supporting the research of biomedical scientists in selected foreign countries. Currently, grants are administered through five programs: an international program that supports biomedical scientists outside the United States and provides funding for selected courses and workshops; a precollege science-education program that offers grants for educational activities to science museums, botanical and zoological gardens, etc; a biomedical-research program that supports medical schools and research organizations; a graduate science-education program that provides fellowships for graduate students, medical students, and physicians, and supports special courses; and a biological-sciences education program that awards grants to selected undergraduate institutions.
Contact: Jill G. Conley, Program Director; (301) 215-8873; fax (301) 215-8888; grantswww@hhmi.org
Internet: http://www.hhmi.org/grants
Sponsor: Howard Hughes Medical Institute
4000 Jones Bridge Road
Chevy Chase, MD 20815-6789

HHMI International Research Scholars Program **2287**
This five year grant program supports non-U.S. biomedical scientists, scientific meetings for grant-supported scientists, and other international educational activities. Grants are awarded to promising biomedical scientists who have made significant contributions to fundamental research. Application information is available online.
Requirements: To be considered, individuals must hold a full-time academic or research appointment at a university, medical school, or other nonprofit scientific institution. They are not permitted to have major administrative responsibilities.
Contact: Jill G. Conley, Program Director; (301) 215-8873; fax (301) 215-8888
Internet: http://www.hhmi.org/grants/for_grantees/
Sponsor: Howard Hughes Medical Institute
4000 Jones Bridge Road
Chevy Chase, MD 20815-6789

HHMI Med into Grad Initiative Grants **2288**
HHMI will award grants to institutions to improve the understanding of medicine and pathobiology by scientists conducting biomedical research. Grants will be used to modify existing graduate training or initiate new programs to develop a cadre of Ph.D. researchers who understand pathophysiology and are committed to working at the interface of the basic sciences and clinical medicine. Applications will be accepted from any university in the United States that offers Ph.D.-level training in an appropriate science or engineering discipline. It is expected that as much as $10 million will be awarded in this competition. Awards will range from $400,000 to $1 million, and the grant funds will be allocated over four consecutive years.
Requirements: U.S. institutions that grant PhD degrees in the appropriate science and engineering disciplines are eligible to apply.
Date(s) Application is Due: Apr 27
Amount of Grant: 400,000 - 1,000,000 U.S.D
Contact: Maryrose Franko, Senior Program Officer; (800) 448-4882, ext. 8880; fax (301) 215-8888; medintograd@hhmi.org
Internet: http://www.hhmi.org/grants/institutions/medintograd.html
Sponsor: Howard Hughes Medical Institute
4000 Jones Bridge Road
Chevy Chase, MD 20815-6789

HHMI Physician-Scientist Early Career Award **2289**
Through this competitive grant initiative, HHMI awards five-year grants to selected alumni of the HHMI-NIH Research Scholars Program and the HHMI Research Training Fellowships for Medical Students Program to support these individuals as they begin careers as independent physician-scientists. The award provides $375,000 over a five-year period for direct research costs.
Requirements: Only alumni of the HHMI-NIH Research Scholars Program and the HHMI Research Training Fellowship for Medical Students Program who have received an M.D., M.D./Ph.D., D.D.S, or equivalent degree are eligible to apply.
Restrictions: The funds may not be used for the salary of the awardee or institutional indirect costs.
Amount of Grant: 375,000 U.S.D
Contact: Office of Grants and Special Programs; (800) 448-4882, ext. 8889 or (800) 424-9924; fax (301) 215-8888; earlycareer@hhmi.org
Internet: http://www.hhmi.org/grants/individuals/earlycareer.html
Sponsor: Howard Hughes Medical Institute
4000 Jones Bridge Road
Chevy Chase, MD 20815-6789

HHMI Research Training Fellowships **2290**
Medical fellowships are intended to strengthen and expand the nation's pool of medically trained researchers. These fellowships provide funds to help meet fellows' research-related expenses and education costs during the research training period. Fellowships are awarded annually to provide support for one year of full-time training in fundamental biomedical, clinical, translational, or applied research. The fellowship includes a stipend, a research allowance to meet research-related expenses, and a fellow's allowance to be used on behalf of the fellow for health care, tuition and fees, and other research-related expenses. Application information is available online.
Requirements: Fellowships are awarded to students enrolled in medical and dental programs in the U.S.
Date(s) Application is Due: Jan 11
Contact: Program Officer; (301) 215-8500 or (800) 424-9924; fax (301) 215-8888; medfellows@hhmi.org
Internet: http://www.hhmi.org/grants/individuals/medfellows.html
Sponsor: Howard Hughes Medical Institute
4000 Jones Bridge Road
Chevy Chase, MD 20815-6789

HHS Children's Hospitals Graduate Medical Education Payment Program **2291**
The program provides funds to freestanding children's hospitals to support the training of pediatric and other residents in graduate medical education (GME) programs. This program compensates for the disparity in the level of federal funding for teaching hospitals for pediatrics versus other types of hospitals.
Requirements: To be eligible, a children's hospital must meet all of the following criteria: participates in an approved graduate medical education (GME) program; has a Medicare provider agreement; is excluded from the Medicare inpatient prospective payment system (PPS) under section 1886(d)(1)(B)(iii) of the Social Security Act, and its accompanying regulations; and ss a freestanding hospital.
Date(s) Application is Due: May 1; Aug 1
Amount of Grant: Up to 40,000,000 U.S.D
Contact: Kara Castellow, (301) 443-1058; kcastellow@hrsa.gov
Internet: http://bhpr.hrsa.gov/childrenshospitalgme/default.htm
Sponsor: U.S. Department of Health and Human Services
5600 Fishers Lane, Parklawn Building
Rockville, MD 20857

Highmark BCBS Challenge for Healthier Schools in Western Pennsylvania **2292**
The program addresses the problem of overweight and inactive children and the subsequent health consequences of being overweight. Schools in Western Pennsylvania will receive support for innovative, evidence-based model program ideas dealing with increased physical activity or nutrition education. The nine most promising plans will win cash grants to be used to implement their programs at the school. Four cash awards will be made to elementary schools, three cash award will be made to middles schools, and two cash awards will be made to high schools. The program encourages collaborative efforts among teachers, students, and parents to create innovative approaches to student wellness.
Requirements: To be eligible, schools must be located within the 29 counties of the Western Pennsylvania region, and become a Caring Team School through the Western Pennsylvania Caring Foundation (http://www.highmarkcaringfoundation.com/chooseRegion.shtml).
Geographic Focus: Pennsylvania
Date(s) Application is Due: Jun 1
Amount of Grant: 10,000 - 20,000 U.S.D
Contact: Aaron Billger, (412) 544-7826; aaron.billger@highmark.com
Internet: https://www.highmark.com/hmk2/community/index.shtml
Sponsor: Highmark
120 5th Avenue, Suite 2112
Pittsburgh, PA 15222-3099

Highmark Corporate Giving Grants **2293**
The Foundation awards grants in the following categories: health; human services; community; education; and arts and culture. Support is expressed in cash grants, sponsorships and in-kind gifts to non-profit organizations. There are no maximum or minimum limits; however, awards to programs and services are granted based on importance to the Foundation's business, social mission and corporate citizen objectives, as well as the initiative's proposed impact. Interested applicants should submit a brief proposal. For additional information or proposal submission, applicants should contact the Community Affairs department in their county of residence. Contact information by county is available online.
Requirements: The Foundation awards grants to nonprofit organizations that are defined as tax exempt under section 501(c)3 of the Internal Revenue Code and as public charities under section 509(a) of that code.
Restrictions: Proposals are not accepted for: multiple-year grants; capital campaigns; individual causes; seed money or start-up organizations; endowment funds; political causes or campaigns; fraternal or civic groups; religious programming; organizations that discriminate on the basis of race, religion, sex, disability or national origin.

Geographic Focus: Pennsylvania
Contact: Mary Ann Papale, (412) 544-4032; mary.papale@highmark.com
Internet: https://www.highmark.com/hmk2/community/corpgiving/index.shtml
Sponsor: Highmark
120 5th Avenue, Suite 2112
Pittsburgh, PA 15222-3099

Highmark Healthy High 5 2294
This program is designed to support schools in the effort to introduce programs that will ultimately create healthier, more successful students. Schools or school districts are invited to submit proposals for innovative programming to improve nutrition education, nutrition services, or increase opportunities for physical activity or education. The School Challenge grant application packet is available online.
Requirements: All schools within the Foundation's 49-county, Pennsylvania service region are eligible to apply. All schools that apply must participate in the PANA Keystone Healthy Zone Program and complete the KHZ Assessment.
Geographic Focus: Pennsylvania
Date(s) Application is Due: Apr 14
Contact: Yvonne Cook, President; (412) 544-7000; fax yvonne.cook@highmark.com
Internet: https://www.highmark.com/hmk2/community/hmfoundation/index.shtml
Sponsor: Highmark
120 5th Avenue, Suite 2112
Pittsburgh, PA 15222-3099

Highmark Physician eHealth Collaborative Grants 2295
The program was created to encourage the adoption of health information technology used at the point of care to improve patient safety, quality of care, and cost efficiency for people in western and central Pennsylvania. Physicians may apply to receive grants, which must be used to acquire and use electronic technology systems such as a personal computer, a PDA, or electronic tablet or digital pen to generate and transmit electronically prescriptions to pharmacies. The collaborative will pay up to 75 percent of the cost for a physician's office to acquire, install, and implement the electronic technology system, with the physician's practice to pay the remaining balance. Physicians must apply online.
Requirements: Physicians must be licensed to practice medicine in Pennsylvania and must be a licensed prescriber.
Geographic Focus: Pennsylvania
Amount of Grant: Up to 7,000 U.S.D
Contact: Grants Administrator; info@highmarkehealth.org
Internet: http://www.highmarkehealth.org
Sponsor: Pittsburgh Foundation
1 PPG Place, 30th Floor
Pittsburgh, PA 15222-5401

High Meadow Foundation Grants 2296
The foundation grants funds in support of the performing arts, especially theater and music, and other cultural organizations. Organizations supported are usually located in Berkshire County, MA. Types of support include continuing support, annual campaigns, capital campaigns, building construction/renovation, equipment acquisition, emergency funds, program development, employee-related scholarships, and matching funds. Applications are accepted at any time. Forms are not necessary; a letter outlining the proposed project should include budget and administrative information.
Requirements: Western Massachusetts 501(c)3 tax-exempt organizations in Berkshire County are eligible.
Geographic Focus: Massachusetts
Samples: Massachusetts Museum of Contemporary Art, North Adams, MA, $100,000; Berkshire Theater Festival, Stockbridge, MA, $60,000; Lenox Library Association, Lenox, MA, $10,000;
Contact: Jane Fitzpatrick; (413) 298-5565 or (413) 243-1474; fax (413) 298-4058
Sponsor: High Meadow Foundation
30 Main Street
Stockbridge, MA 01262

Hilda and Preston Davis Foundation Grants 2297
The Foundation provides funds to charitable organizations whose programs advance the development of all areas of the lives of children and young adults. The Foundation places special emphasis on, and channels most of its financial resources toward, those organizations whose attention is concentrated on eating disorders and education for the underprivileged. Grants may range from $10,000 up to $100,000. In unique circumstances, the Foundation does consider a more significant grant for a program having a major impact in one or more of its areas of interest.
Requirements: 501(c)3 tax-exempt organizations are eligible.
Restrictions: The foundation generally will not provide grants to: non-501(c)3 tax-exempt organizations; individuals; general fundraising drives; endowments; government agencies; or organizations that subsist mainly on third-party funding and have demonstrated no ability or expended little effort to attract private funding.
Amount of Grant: 10,000 - 100,000 U.S.D
Contact: Grants Administrator; (203) 629-8552; fax (203) 547-6112; davis@fsllc.net or info@fsllc.net
Internet: http://www.hpdavis.org/application.htm
Sponsor: Hilda and Preston Davis Foundation
640 West Putnam Avenue, 3rd Floor
Greenwich, CT 06830

Hillcrest Foundation Grants 2298
The Hillcrest Foundation was created by Mrs. W.W. Caruth, Sr. (Mrs. Earle Clark Caruth) in 1958 to provide financial support to qualified Texas charitable organizations for the advancement of education, the promotion of health, and the relief of poverty. Mrs. W.W. Caruth, Sr. was from a pioneer family who settled in the Dallas area in 1848. Several Caruth family generations owned and managed farms and ranches, which the family later developed into real estate properties as Dallas became a major metropolitan area. Each succeeding generation has been characterized by a pioneering spirit, vision, courage, hard work and generosity. The Hillcrest Foundation was created to carry the Caruth family's generosity to the people of Texas. Approximately 90% of grant funds are paid to organizations in North Texas, with emphasis on charitable services in the Dallas area. The trustees will give priority consideration to the following types of grant requests: construction/improvements of permanent buildings; programs and special projects for education, health and poverty relief; capital campaigns; and buildings, facilities and equipment. Grant requests for capital and program support are strongly encouraged. Grant amounts range from $10,000 to $300,000 (multi-year payment) with an average grant of $35,000. The Trustees favorably consider proposals which are unique, necessary, and of high priority for the charitable organizations, and which do not duplicate other services which are available; proposals for which funding may not be readily available from other sources; and essential projects which are sufficiently described as worthwhile, important and of a substantive nature. Grants to meet challenges or matching funds have a special appeal. The majority of grants from the Hillcrest Foundation are one year in duration. On occasion, multi-year support is awarded. Applicants must apply online at the grant website. Applicants are strongly encouraged to do the following before applying: review the downloadable state application procedures for additional helpful information and clarifications; review the downloadable online-application guidelines at the grant website; review the foundation's funding history (link is available from the grant website); review the online application questions in advance; and review the list of required attachments. These will generally include: a list of board members, financial statements (audited, reviewed, or compiled by independent auditor); an organization summary; a list of other funding sources; an IRS Determination letter; and other required documents. All attachments must be uploaded in the online application as PDF, Word, or Excel files. The Hillcrest Foundation has three deadlines annually: February 28, July 31, and November 30. Applications must be submitted by 11:59 p.m. on the deadline dates. Grant applicants are notified as follows: February deadline applicants will be notified of grant decisions by June 30; July applicants will be notified by November 30; and November applicants will be notified by March 31 of the following year.
Requirements: 501(c)3 organizations in Texas may apply.
Restrictions: Grant requests for general operating support will not be considered. Organizations that receive a one-year grant from the foundation must skip a year before submitting a subsequent application. Organizations that receive a multi-year grant are not eligible to apply until one year after the close of their grant cycle. Organizations whose last request was declined must wait one year before applying again. The foundation does not support requests from individuals, organizations attempting to influence policy through direct lobbying, or any political campaigns.
Geographic Focus: Texas
Date(s) Application is Due: Feb 28; Jul 31; Nov 30
Amount of Grant: 10,000 - 300,000 U.S.D
Samples: Southern Methodist University, Dallas, Texas, $400,000 (2010); Planned Living Assistance Network of North Texas, Dallas, Texas, $250,000 (2010); Dallas Center for the Performing Arts Foundation, Dallas, Texas, $300,000 (2010); North Texas Food Bank, Dallas, Texas, $10,000 (2010).
Contact: David Ross, Senior Vice President; tx.philanthropic@baml.com
Internet: https://www.bankofamerica.com/philanthropic/fn_search.action
Sponsor: Hillcrest Foundation
901 Main Street, 19th Floor, TX1-492-19-11
Dallas, TX 75202-3714

Hill Crest Foundation Grants 2299
The Hill Crest Foundation awards grants to Alabama nonprofits, primarily health associations and human services and education, with some funding for the arts. Types of support include: building and renovation; capital campaigns; endowments; equipment; matching and challenge support; professorships; program development; publication; research; scholarship funds; seed money; and technical assistance. There are no deadlines or applications. The Board of Directors meets quarterly.
Requirements: Application forms are not required. Before submitting a proposal, organizations should submit a letter, detailing the project and the amount requested, along with descriptive literature about their organization.
Restrictions: Grantmaking is limited to Alabama. The Foundation does not fund individuals.
Geographic Focus: Alabama
Amount of Grant: 20,000 - 150,000 U.S.D
Samples: Methodist Homes for the Aging, Birmingham, Alabama, $100,000 payable over one year (2010); Alabama Symphonic Association, Birmingham, Alabama, $30,000 payable over one year (2010); Bell Center for Early Intervention Programs, Birmingham, Alabama, $50,000 payable over one year (2010).
Contact: Charles Terry, Chairperson; (205) 425-5800
Sponsor: Hill Crest Foundation
P.O. Box 530507
Mountain Brook, AL 35253-0507

Hill Foundation Grants 2300

The foundation awards grants mainly for special health care projects, higher education, equipment funds, and services for the elderly, disabled children, and cultural programs, primarily in Colorado. Some support is provided to social service agencies. Types of support include program development, scholarship funds, and matching funds. There are no application deadlines.
Requirements: Giving primarily in CO and WY.
Restrictions: Grants are not made to individuals nor for capital improvements. The foundation does not make multiyear commitments.
Geographic Focus: Colorado, Wyoming
Amount of Grant: 1,000 - 45,000 U.S.D
Contact: Grants Administrator, c/o Wells Fargo Bank; (720) 947-6820
Sponsor: Hill Foundation
1740 Broadway, MC 7300-483
Denver, CO 80274

Hillman Foundation Grants 2301

The foundation provides grants to nonprofit organizations in the city of Pittsburgh and the southwestern Pennsylvania region for programs and projects designed to improve the quality of life in the area. The foundation's areas of interest include community affairs, social services, culture and the arts, education at all levels, and youth services including medical and health. Grants range widely in size and are allocated for large and small capital projects, endowment, new and expanding programs, and on a limited basis, operating support. Applications should include an annual budget of the organization, project information, and evidence of tax-exempt status. Application information is available online.
Requirements: The Foundation considers requests only from organizations classified as tax-exempt under Section 501(c)3 of the U.S. Internal Revenue Code and designated as public charities under Section 509(a).
Restrictions: Grants are not made to individuals, organizations located outside of the United States, for travel expenses for groups, or in support of events, sponsorships and meetings such as conferences, institutes and seminars.
Geographic Focus: Pennsylvania
Samples: Strong Women, Strong Girls, INc., $10,000—towards program for at-risk and low-income girls; Greater Pittsburgh Community Food Bank, $65,000—toward food pantry and outreach program; Carnegie Institute/Museum of Natural History, $50,000—towards purchase of bournonite, amazonite, smoky quarts and rhodochrosite mineral specimens.
Contact: David K. Roger, President; (412) 338-3466; fax (412) 338-3463; foundation@hillmanfo.com
Internet: http://www.hillmanfdn.org/grantprograms.html
Sponsor: Hillman Foundation
330 Grant Street, Suite 2000
Pittsburgh, PA 15219

Hillsdale County Community Foundation Healthy Senior/Healthy Youth Fund Grants 2302

The Healthy Senior/Healthy Youth Fund provides grants to smoking cessation and prevention programs for young people (under 18) and seniors (over 65) in Hillsdale County. Organizations interested in applying are strongly encouraged to discuss their project with the Executive Director prior to submitting an application. If the project meets the basic eligibility criteria and is consistent with the Foundation's program interests, a formal application will be sent to the applicant. The Grantmaking Committee meets twice a year to review funding requests.
Requirements: The Foundation welcomes applications from the Hillsdale County area or outside Hillsdale County, Michigan, if a significant number of the people to be served reside within Hillsdale County. Applicants shall be tax exempt according to Section 501(c)3 of the Internal Revenue Code.
Geographic Focus: Michigan
Date(s) Application is Due: May 1; Nov 1
Contact: Sharon Bisher; (517) 439-5101; fax (517) 439-5109; s.bisher@abouthccf.org
Internet: http://www.abouthccf.org/funds.asp
Sponsor: Hillsdale County Community Foundation
2 S Howell Street, P.O. Box 276
Hillsdale, MI 49242

Hillsdale County Community General Adult Foundation Grants 2303

General Adult Foundation grant making is made from the Foundation's named unrestricted endowment funds. These grants focus on improving the quality of life for the citizens of Hillsdale County. Eligible projects generally fall within these categories: education, fine arts, social services, community development, recreation, environmental issues, health and wellness, and improvement in the physical, mental, and moral conditions of Hillsdale County residents. The Foundation aims to support creative approaches to community needs and problems that benefit the widest possible range of people.
Requirements: The Foundation requires applicants to call and discuss their project with the Executive Director prior to submitting the application. Applications from the Hillsdale County area or outside Hillsdale County, Michigan are welcome, if a significant number of the people to be served reside within Hillsdale County. Applicants shall be tax exempt according to Section 501(c)3 of the Internal Revenue Code.
Restrictions: The current objectives of the Foundation do not allow grants for: religious or sectarian purposes; individuals; legislative or political purposes; loans; capital campaigns; routine maintenance, including office equipment; administrative costs for maintaining the present operation of an organization, including, but not limited to, staff salaries, wages, and benefits; basic education materials including state mandated/benchmark core curriculum supplies and resources.
Geographic Focus: Michigan
Date(s) Application is Due: May 1; Nov 1
Samples: Hillsdale Arts Chorale, Hillsdale, MI, $7,500—to underwrite the 2007 Messiah Christmas Concert; Montgomery Fire and Rescue Department, Montgomery, MI, $4,200—to purchase a Self-Contained Breathing Apparatus; Mrs. Stock's Park, Hillsdale, MI, $4,000—to assist in funding costs associated with the 2007 summer festival musical performances.
Contact: Sharon Bisher, Executive Director; (517) 439-5101; fax (517) 439-5109; s.bisher@abouthccf.org
Internet: http://www.abouthccf.org/grants.asp
Sponsor: Hillsdale County Community Foundation
2 S Howell Street, P.O. Box 276
Hillsdale, MI 49242

Hillsdale Fund Grants 2304

The fund awards grants to nonprofit organizations in its areas of interest, including arts and culture, education, health care, social services delivery, and religion. Call for meeting dates and application materials.
Requirements: Tax-exempt organizations are eligible
Restrictions: Grants do not support indirect costs or overhead; routine, recurring operating expenses; conferences and seminars; travel and study; or individuals.
Geographic Focus: North Carolina
Amount of Grant: 3,000 - 50,000 U.S.D
Samples: College of the Atlantic, Bar Harbor, ME, $50,000; Rocky Mountain Field Institute, Colorado Springs, CO, $10,000; Piedmont Environmental Council, Warrenton, VA, $30,000.
Contact: Edward Doolan, Grants Administrator; (336) 274-5471
Sponsor: Hillsdale Fund
P.O. Box 20124
Greensboro, NC 27420-0124

Hilton Head Island Foundation Grants 2305

The Foundation exists to enhance the quality of life for individuals living and/or working in Southern Beaufort County (Hilton Head Island, Daufuskie Island, Bluffton, Okatie). Grantmaking is in the broad areas of: arts, education, and community development. The Board meets three times per year. Application information is available online.
Requirements: To be eligible for funding, an applicant must be: a nonprofit agency with a tax-exempt status under section 501(c)3 of the Internal Revenue code, or eligible to be classified as such, but are not classified as a private foundation; serve the people who live and/or work in the Hilton Head Island area community.
Restrictions: Funding is not available for: sectarian or religious activities; political activities or organizations; grants directly to individuals; endowments; annual fundraising campaigns; special events or fundraisers; scholarships for students in grades K-12.
Geographic Focus: South Carolina
Date(s) Application is Due: Apr 1; Aug 1; Dec 1
Contact: Cynthia Smith, Vice President for Grantmaking; (843) 681-9100; fax (843) 681-9101; csmith@cf-lowcountry.org
Internet: http://www.cf-lowcountry.org/receive/grants
Sponsor: Community Foundation of the Lowcountry
4 Northridge Drive, Suite A, P.O. Box 23019
Hilton Head Island, SC 29925

Hilton Hotels Corporate Giving Program Grants 2306

Hilton makes charitable contributions to nonprofit organizations involved with K-12 education, youth development, public policy, homelessness, and civic affairs. Support is given primarily in areas of company operations, with emphasis on California, including Los Angeles and San Francisco, and Tennessee, including Memphis; giving also to national organizations. Contact Ellen Gonda for additional information at: corporate_communications@hilton.com.
Requirements: 501(c)3 tax-exempt organizations are eligible.
Restrictions: No support for sport teams, religious organizations not of direct benefit to the entire community, government-supported organizations (over 20 percent of budget), hospitals, private schools, pre-schools, or day care facilities, film and video production, or promotional materials. No grants to individuals (except for employee-related scholarships), or for fellowships, sports activities, debt reduction, capital campaigns or endowments, film or video projects, or promotional merchandise.
Geographic Focus: California, Tennessee
Contact: Ellen Gonda, Senior Vice President; (310) 278-4321; fax (310) 205-7678; corporate_communications@hilton.com
Internet: http://www.hiltonworldwide.com
Sponsor: Hilton Hotels Corporation
9336 Civic Center Drive
Beverly Hills, CA 90210-3604

Hirtzel Memorial Foundation Grants 2307

The foundation awards grants to eligible organizations in New York and Pennsylvania in its areas of interest, including medical research and health care; community and neighborhood improvement; higher education; and social and human services. Types of support include building construction and renovation, capital campaigns, scholarships, equipment acquisition, general operating support, research, and scholarships. There are no application deadlines; contact the office for appropriate forms.

Requirements: New York nonprofit organizations in Ripley, Chautauqua County, and Pennsylvania nonprofit organizations in North East, Erie County, are eligible.
Geographic Focus: New York, Pennsylvania
Amount of Grant: 750 - 250,000 U.S.D
Samples: Mercyhurst College (Erie, PA)—to establish the Institute on Aging and Geriatric Health, $1 million.
Contact: Laurie Moritz, Grants Administrator, c/o Mellon Financial Corporation; (412) 234-0023
Sponsor: Orris C. Hirtzel and Beatrice Dewey Hirtzel Memorial Foundation
P.O. Box 185
Pittsburgh, PA 15230

HIV/AIDS Research Training Fellowship **2308**

This program supports the training of researchers in HIV/AIDS research. The goals of the program are: to increase the number of ethnic minority research scientists studying HIV/AIDS issues among ethnic minority populations; to increase the availability and quality of ethnic minority research training; and to increase the knowledge on issues related to ethnic minority HIV/AIDS prevention and treatment. Fellowships are awarded for one academic year and are renewable for two additional years. Students specializing in such research areas as HIV prevention, AIDS treatment adherence, provider education, or psychoneuroimmunology are encouraged to apply. Students of any specialty in psychology will be considered if they plan careers in HIV/AIDS research.
Requirements: An applicant must be a U.S. citizen or resident with a permanent visa; a member of an ethnic minority group including, but not limited to, African American, Alaskan Native, American Indian, Asian American, Hispanic, Pacific Islander, and Puerto Rican, and/or demonstrate commitment to a career in psychology with a specialty in HIV/AIDS research related to ethnic minority populations; and enrolled full-time in a doctoral degree program in psychology at the time the fellowship is awarded.
Date(s) Application is Due: Jan 15
Amount of Grant: 9,600 U.S.D
Contact: Dr. James Jones, Director; (202) 336-6127; fax (202) 336-6012; mfp@apa.org
Internet: http://www.apa.org/mfp/hprogram.html
Sponsor: American Psychological Association
750 First Street NE
Washington, DC 20002-4242

HNHfoundation Grants **2309**

The foundation awards general operating and program support to New Hampshire healthcare organizations. Preference will be given to programs providing services to children and youth. There are no application deadlines; obtain application forms and procedures from the office.
Requirements: New Hampshire nonprofit organizations are eligible.
Geographic Focus: New Hampshire
Amount of Grant: 250 - 250,000 U.S.D
Samples: Home Health and Hospice Care (Nashua, NH)—to support an outreach program for uninsured children, $10,000; North Country Health Consortium (NH)—for health insurance for small businesses, $35,361; Community Health Access Network (Newmarket, NH)—to support chronic-disease education programs, $34,433.
Contact: Sandi Van Scoyoc, President; (603) 229-3260; fax (603) 229-3259; info@hnhfoundation.org
Internet: http://www.hnhfoundation.org
Sponsor: HNHfoundation
14 Dixon Avenue
Concord, NH 03301

Hoblitzelle Foundation Grants **2310**

The Hoblitzelle Foundation was established by Karl and Esther Hoblitzelle in 1942. Grants made by the Directors are usually focused on specific, non-recurring needs of the educational, social service, medical, cultural, and civic organizations in Texas, particularly in the Dallas area. The Board meets three times per year. It is preferred that the initial approach be through a brief narrative letter describing the project for which funds are asked. Guidelines are available online.
Requirements: Nonprofit organizations in the State of Texas, primarily within the Dallas Metroplex, are eligible to apply.
Restrictions: No grants are made for religious purposes or to individuals. No grants are made outside the State of Texas. Contributions are not made toward operating budgets, debt retirement, research, media productions or publications, scholarships or endowments. The Foundation makes no loans.
Geographic Focus: Texas
Date(s) Application is Due: Apr 15; Aug 15; Dec 15
Samples: Dallas Arboretum, Texas Discovery Gardens, Dallas, TX—natural resources/environment grant; Dallas Services, Dallas TX—handicapped/disabled grant; Museum of Biblical Art, Dallas, TX—cultural grant (2010); Phoenix House, Dallas, TX-substance abuse grant (2010); Dallas Public Library, Dallas, TX—civic project grant (2010); Baylor Health Care System Foundation, Dallas, TX—medicine & health care grant (2010).
Contact: Paul W. Harris, President & CEO; (214) 373-0462; pharris@hoblitzelle.org
Internet: http://www.hoblitzelle.org/
Sponsor: Hoblitzelle Foundation
5956 Sherry Lane, Suite 901
Dallas, TX 75225

Hoffmann-La Roche Foundation Grants **2311**

The corporate contributions program supports nonprofits in company operating areas. Funds may be requested in two categories: health programs that focus on health promotion and health education and educational projects that encourage math and science literacy and promote secondary school teacher enrichment. Seed grants are awarded for new activities for which Hoffman-La Roche may spark other funders to pitch-in operating support.
Requirements: Nonprofits in company-operating communities are eligible to apply, with some preference given to groups near the New Jersey headquarters.
Restrictions: Funds will not be awarded in support of endowments; scholarships; facility expansion or construction; purchase of equipment; capital expenditures; good will advertising; and political, veterans, or labor organizations. Telephone, facsimile, and email requests will not be honored.
Contact: Vivian Beetle, Administrative Director; (973) 562-2055; fax (973) 562-2999
Internet: http://www.rocheusa.com/about/responsibility.html
Sponsor: Hoffmann-La Roche Foundation
340 Kingsland Street
Nutley, NJ 07110-1199

Hogg Foundation for Mental Health Grants **2312**

The Foundation accomplishes its mission through the funding of external and internal projects in strategically selected areas. The Foundation operates major initiatives in three priority areas: integrated health care, cultural competence, and workforce development. Funds available through the Foundation's major initiatives and projects are distributed through a Request for Proposals (RFP) process. The Foundation does not accept unsolicited grant proposals on any topic. To be notified of future RFP opportunities, send an email message including your name, affiliation, and mailing address to: HoggCommunications@austin.utexas.edu.
Geographic Focus: Texas
Samples: People's Community Clinic, Austin, TX, $275,200—paid over three years, this grant was awarded to People's Community Clinic (PCC) in Austin to treat adults with various mental health problems in partnership with Austin Travis County Mental Health Mental Retardation Center; Texas Children's Pediatric Associates, Houston, TX, $306,500—paid over three years, this grant will benefit children with attention-deficit hyperactivity disorder in its Ripley clinic, located in Houston's East End.
Contact: Program Contact; (888) 404-4336 or (512) 471-5041; fax (512) 471-9608; Hogg-Grants@austin.utexas.edu
Internet: http://www.hogg.utexas.edu/funding_grantmaking.html
Sponsor: Hogg Foundation for Mental Health
University of Texas at Austin, P.O. Box 7998
Austin, TX 78713-7998

Hoglund Foundation Grants **2313**

The foundation supports nonprofits working in the areas of education, health science and services, social services, and children's health and development. Grants are made primarily in Dallas, and Houston, Texas, with a limited number of grants awarded outside of this area. Types of support include project support, capital support, and general operating support. Application guidelines are available on the foundation's Web site.
Requirements: To be eligible for consideration, an organization must provide proof that they have received a determination letter from the Internal Revenue Service indicating that it is a tax exempt organization as described in Section 501(c)3 of the Internal Revenue Code of 1986 and is treated as other than a private foundation as within the meaning of Section 509(a) of the Code. An organization may also qualify under Section 170(c)(1) if the grant requested is to be used exclusively for public purposes as described in the Code.
Restrictions: Grants are not made to individuals.
Geographic Focus: Texas
Date(s) Application is Due: Mar 15; Jul 15; Nov 15
Amount of Grant: 1,000 - 200,000 U.S.D
Samples: University of Virginia Fund, Charlottesville, VA, $33,000-Global Scholarship Fund; Vickery Meadow Learning Center, Dallas, TX, $10,000—Family Literacy program; Victims Outreach, Dallas, TX, $5,000—program services for victims of crime & violence; Reasoning Mind, Inc., Houston, TX, $250,000—program expansion and general operating support.
Contact: Kelly Compton, Executive Director; (214) 987-3605; fax (214) 363-6507; info@hoglundfoundation.org
Internet: http://www.hoglundfoundation.org/grants.html
Sponsor: Hoglund Foundation
5910 North Central Expressway, Suite 255
Dallas, TX 75206

Holland/Zeeland Community Foundation Grants **2314**

This Foundation funds programs and projects that focus on the arts and culture, education, environmental issues, health and human services, youth, seniors, community and economic development, affordable housing, and others that enhance the well-being of all citizens in the community. Proposals that help organizations deliver services more effectively, enhance cooperation and collaboration, address emerging needs or try new approaches, positively impact a large number of people for the resources invested, or focus on prevention receive priority consideration. The Foundation generally awards grants one-time only for a specific program or project. To apply, first call to discuss the project and confirm it meets the Foundation guidelines.
Requirements: Qualifying organizations include nonprofits that are tax-exempt under Section 501(c)3 of the Internal Revenue Code, schools, municipalities, and other governmental entities that serve a charitable purpose in the greater Holland and Zeeland area.

Restrictions: Funding is not available for the following: requests in excess of $15,000; annual fund-raising drives; services which are commonly recognized as government or school obligations (however the foundation does consider applications from schools for pilot projects and innovative programs); endowments, loans, taxes or debt reduction; conference speakers, fieldtrips, travel or tours, for individuals or groups; religious programs that advocate specific religious doctrines or do not serve the broader community.
Geographic Focus: Michigan
Date(s) Application is Due: Jan 18; Jun 7; Sep 20
Samples: Macatawa Greenway Partnership, $5,000—Trailhead Kiosk: to design, fabricate and install a prototype kiosk along Zuidema bike path; to explore, test and develop potential uses for an efficiently designed trailhead/obelisk; Holland Free Health Clinic, $15,000— pilot program providing free or low cost diabetic supplies and education on the proper use of supplies; Tulipanes Latino Art & Film Festival, Inc., $5,000— to fund programs including purchase and rental of films and productions and live performances by local and regional artists.
Contact: Janet DeYoung, Executive Director; (616) 396-6590; fax (616) 396-3573; janet@cfhz.org or nfo@cfhz.org
Internet: http://cfhz.org/files/2010Funding_Guidelines.pdf
Sponsor: Community Foundation of the Holland/Zeeland Area
70 West 8th Street, Suite 100
Holland, MI 49423

HomeBanc Foundation Grants **2315**
Giving primarily in Florida and Georgia, the HomeBanc Foundation also, gives to national organizations. The Foundation has three main goals: provide support to cancer related causes through funding education, advocacy, and research; support the American dream of homeownership for those who might not reach it on their own; and provide college funding to students who strive to achieve their goals through education beyond high school. Scholarship forms are available online. Send a Letter of Inquiry to the Foundation, unsolicited applications are not excepted. The Foundation will contact you if your proposal fits their criteria.
Requirements: Nonprofits providing services in Georgia and Florida are eligible.
Geographic Focus: Florida, Georgia
Samples: Susan B. Koman Foundation, 5,000—general purposes; American Cancer Society, 75,1000—advocacy, research, education; Junior Acheivement, $5,000—hispanic outreach scholarships.
Contact: Amanda Albertelli; (404) 497-1000; aalbertelli@homebanc.com
Sponsor: HomeBanc Foundation
2002 Summit Boulevard, Suite 100
Atlanta, GA 30319-1497

Home Building Industry Disaster Relief Fund **2316**
The Home Building Industry Disaster Relief Fund (HBIDRF) makes funds available for direct contribution to other recognized charities aiming to meet similar primary needs as the HBIDRF - shelter, health care, education. The board welcomes recommendations of charitable organizations that are actively working in your region to rebuild communities. Guidelines and application form are available at the website.
Requirements: Charities can include corporations, funds, or foundations organized and operated exclusively for charitable, scientific, or educational purposes. These charitable organizations may be involved in making direct contributions to adversely impacted individuals through established process, or with coordinated building projects such as one-day builds, blitzes, etc. Other entities that could be eligible to receive funds might include vocational training programs, temporary shelters or clinics and educational ventures intended to help the local home building industry recover from disastrous events. To apply for funding, contact your local home builders association to determine if any partnering opportunities are available at this time.
Restrictions: The fund does not make assistance available directly to individuals.
Contact: Jerry Howard, Executive Officer; (800) 368-5242; fax (202) 266-8400
Internet: http://www.nahb.org/generic.aspx?sectionID=842
Sponsor: Home Building Industry Disaster Relief Fund
1201 15th Street NW
Washington, DC 20005-2800

Homer Foundation Grants **2317**
The Foundation seeks to fund innovative, creative projects that have a high likelihood of success and will have a long-term, positive impact on the community. Funding is available for arts and culture, community development, youth programs, education, health care, scholarships and the environment. Application information is available online.
Requirements: 501(c)3 nonprofit organizations and other qualified nonprofit entities located in the southwestern portion of the Kenai Peninsula, from Ninilchik south and including the communities across Kachemak Bay, are eligible. An individual who applies must be a resident of the region as described above, who will spend the funds on projects completed within this region. Projects taking place outside of this region, but by a resident, will be considered on the merit of the project and how it could benefit the community.
Geographic Focus: Alaska
Amount of Grant: 4,000 U.S.D
Contact: Joy Steward, Director; (907) 235-0541; fax (907) 235-0542; jsteward@homerfund.org
Internet: http://www.homerfund.org/grantmaking.html
Sponsor: Homer Foundation
P.O. Box 2600
Homer, AK 99603

Honda of America Manufacturing Foundation Grants **2318**
The Foundation supports community outreach in Ohio. Each year, the Foundation funds various community projects throughout the state. This involvement helps build the community where it does business and where its associates live. The Foundation supports programs and organizations in the areas of education, arts and culture, civic and community, health and human services, and the environment. It places its priorities on the development of programs, and not supporting annual operating budgets.
Requirements: Priority consideration will be given for projects within the West Central Ohio area, including: Allen, Auglaize, Champaign, Clark, Darke, Delaware, Franklin, Hardin, Logan, Madison, Marion, Mercer, Miami, Shelby and Union counties. In addition, Honda considers projects on a case-by-case basis outside this area if the projects are Ohio-based and reflect its priorities of interest.
Restrictions: Honda of America does not make grants to: individuals; religious or political groups; organizations which are not tax-exempt under the U.S. Internal Revenue Code paragraph 501(c)3; or programs outside the state of Ohio. Generally, the Foundation does not provide contributions in support of: conferences or workshops; seminars or pageants; field trips, extra-curricular school activities or sports teams; fraternal or veterans organizations; national health organizations; courtesy advertisements, lobbying organizations, memberships or legal advocacy; staff salaries; or pilot programs.
Geographic Focus: Ohio
Date(s) Application is Due: Mar 31; Jun 30; Sep 30; Dec 31
Contact: Ms. Ginny Milburn, Coordinator; (937) 645-8792; fax (937) 645-8787; ginny_milburn@ham.honda.com
Internet: http://www.ohio.honda.com/community/giving.cfm
Sponsor: Honda of America Manufacturing Foundation
24000 Honda Parkway
Marysville, OH 43040

Horace A. Kimball and S. Ella Kimball Foundation Grants **2319**
The Kimball Foundation makes grants almost exclusively to Rhode Island operatives (charities) or those benefitting Rhode Island residents and causes. Although, the Foundation considers gifts to all areas in the state, greater emphasis is placed on South County. Areas of interest for the Foundation are: human services, the environment, and health care.
Requirements: The Foundation will consider any organization which has proper 501(c)3 and 509(a) I.R.S. tax classification status. Application form and guidelines available on foundation Web site. Application form required. Applicants should submit the following: copy of IRS Determination Letter; brief history of organization and description of its mission; copy of most recent annual report/audited financial statement/990.
Restrictions: No support for religious organizations. No grants to individuals, or for feasibility studies, capital projects or multi-year commitments.
Geographic Focus: Rhode Island
Date(s) Application is Due: Jul 15
Amount of Grant: 1,000 - 50,000 U.S.D
Samples: Rhode Island Philharmonic Orchestra, East Providence, RI, $25,000; Westerly School Department, Westerly, RI, $10,000; South County Hospital Healthcare System, Wakefield, RI, $25,000;
Contact: Thomas F. Black III, President; (401) 364-3565
Internet: http://www.hkimballfoundation.org/index2.htm
Sponsor: Horace A. Kimball and S. Ella Kimball Foundation
130 Woodville Road
Hope Valley, RI 02832

Horace Moses Charitable Foundation Grants **2320**
The Horace Moses Charitable Foundation was established in 1923 to support and promote quality educational, human-services, and health-care programming for underserved populations. Special consideration is given to charitable organizations that serve the community of Springfield and its surrounding communities. Grant requests for general operating support are strongly encouraged. Program support will also be considered. Small, program-related capital expenses may be included in general operating or program requests. The majority of grants from the Moses Charitable Foundation are one year in duration; on occasion, multi-year support is awarded. Applicants must apply online at the grant website. Applicants are strongly encouraged to do the following before applying: review the downloadable state application procedures for additional helpful information and clarifications; review the downloadable online-application guidelines at the grant website; review the foundation's funding history (link is available from the grant website); review the online application questions in advance; and review the list of required attachments. These will generally include: a list of board members, financial statements (audited, reviewed, or compiled by independent auditor); an organization summary; a list of other funding sources; an IRS Determination letter; and other required documents. All attachments must be uploaded in the online application as PDF, Word, or Excel files. The application deadline for the Horace Moses Charitable Foundation is 11:59 p.m. on April 1. Applicants will be notified of grant decisions before June 30.
Requirements: Applicants must have 501(c)3 tax-exempt status.
Restrictions: The foundation does not support requests from individuals, organizations attempting to influence policy through direct lobbying, or any political campaigns.
Geographic Focus: Massachusetts
Date(s) Application is Due: Apr 1
Samples: Big Brothers Big Sisters of Hampden County, Springfield, Massachusetts, $15,000, Continuing Support (2010); Open Pantry Community Services, Springfield, Massachusetts, $10,000, for Loaves and Fishes Community Kitchen (2010); Drama Studio, Springfield, Massachusetts, $2,500, Acting Training and Community Outreach Program (2010).
Contact: Michealle Larkins; (866) 778-6859; michealle.larkins@baml.com

Internet: https://www.bankofamerica.com/philanthropic/fn_search.action
Sponsor: Horace Moses Charitable Foundation
225 Franklin Street, 4th Floor, MA1-225-04-02
Boston, MA 02110

Horizon Foundation for New Jersey Grants 2321

The Foundation's purpose is to promote health, well-being, and quality of life in New Jersey communities. Foundation goals are to improve health by promoting quality health care programs and access and to enhance arts and cultural opportunities. Application information is available online. Only electronic grant applications will be accepted. Application are accepted between January 1 and September 31.
Requirements: 501(c)3 tax-exempt organizations in New Jersey communities located in and served by Horizon Blue Cross Blue Shield of New Jersey are eligible.
Restrictions: Funding is not available for: capital campaigns; endowments; hospitals or hospital foundations; individuals; political causes; political candidates; political organizations; political campaigns.
Geographic Focus: New Jersey
Amount of Grant: 10,000 - 50,000 U.S.D
Samples: ChoiceOne Pregnancy & Sexual Health Resource Centers, Lawrenceville, NJ, $25,000—to support the Sexually Transmitted Diseases Education and Prevention component of its Straight Talk Program; Family Guidance Center, Hamilton, NJ, $15.000—to support its Behavioral Health Services-Depression Recovery Program; American Repertory Ballet, New Brunswick, HJ, $7,500—to support the Dance Power Program.
Contact: Michele Berry, Grants Coordinator; Foundation_Info@horizonblue.com
Internet: http://www.horizon-bcbsnj.com/foundation/about/funding.html?WT.svl=leftnav
Sponsor: Horizon Foundation for New Jersey
Three Penn Plaza East, PP-15V
Newark, NJ 07105-2200

Horizons Community Issues Grants 2322

Grants support program activities or projects serving lesbian, gay, bisexual, and transgender (LGBT) people of all ages in designated California counties focusing on the following issue areas: arts and culture; advocacy, awareness, and civil rights; children, youth, and families; and community and social services. Application information is available online.
Requirements: To be eligible, an organization must: be a nonprofit, 501(c)3 organization, or provide documentation that the organization is sponsored under a fiscal agent umbrella that has 501(c)3 status. Organizations or programs must request support for one or more of the following counties: Alameda; Contra; Costa; Marin; Napa; San Francisco; San Mateo; Santa Clara; Solano; Sonoma.
Restrictions: The following are not eligible for support: costs incurred prior to the date of the grant award; government agencies; capital support, including construction and renovation; fundraising or event sponsorship; individuals; Non-LGBT organizations with budget over $1 million.
Geographic Focus: California
Amount of Grant: 10,000 - 20,000 U.S.D
Contact: Jewelle Gomez, Program Officer; (415) 398-2333, ext. 116; jgomez@horizonsfoundation.org
Internet: http://www.horizonsfoundation.org/page/organizations/ci
Sponsor: Horizons Foundation
870 Market Street, Suite 728
San Francisco, CA 94102

Hormel Foods Charitable Trust Grants 2323

The Charitable Trust was established to benefit various organizations and projects that emphasize education, hunger and quality-of-life initiatives in and around Hormel Foods plant communities. The Program operates as a compliment to its foundation, making charitable contributions to nonprofit organizations directly. Support is given primarily in areas of company operations. Fields of interest include: disaster and preparedness services; education; general charitable giving; and health care. Types of support include: scholarships; matching gifts; general operations; and sponsorships. The Public Relations Department handles giving. Annual donations to a single organization typically do not exceed $15,000. A contributions committee reviews all written requests. Applicants should submit a detailed description of the project and amount of funding requested. The board meets once each month.
Requirements: Only organizations classified as 501(c)3 are eligible for support. Applicants should be located in or serve the communities of: Aurora, Illinois; Austin, Minnesota; Algona, Iowa; Alma, Kansas; Atlanta, Georgia; Beloit, Wisconsin; Chino, California; Fort Dodge, Iowa; Fremont, Nebraska; Knoxville, Iowa; Lathrop, California; Mendota Heights, Minnesota; New Berlin, Wisconsin; Osceola, Iowa; Rochelle, Illinois; Stockton, California; Turlock, California; and Wichita, Kansas.
Restrictions: Giving is primarily limited to areas of company operations.
Geographic Focus: California, Georgia, Illinois, Iowa, Kansas, Minnesota, Nebraska, Wisconsin
Contact: Julie H. Craven; fax (507) 437-5345; media@hormel.com
Internet: http://www.hormelfoods.com/responsibility/philanthropy/default.aspx
Sponsor: Hormel Foods Charitable Trust
1 Hormel Place
Austin, MN 55912-3680

Hormel Foundation Grants 2324

The Foundation, established in 1941, is a non-profit organization dedicated to supporting higher education, health care and research, social services, and community programs located in or serving the city of Austin, Minnesota. Applications for grants must be presented to the Foundation in writing. They should be as brief as appropriate to present the necessary facts about the applying organization and the project for which the grant is being sought. An application form is required, and must be received by the Foundation office by September 1st of each year.
Requirements: Only 501(c)3 organizations in Minnesota should apply.
Geographic Focus: Minnesota
Date(s) Application is Due: Sep 1
Contact: R.L. Knowlton, Chairperson; (507) 437-9800; fax (507) 434-6731
Internet: http://www.thehormelfoundation.com/requirements.asp
Sponsor: Hormel Foundation
301 North Main Street
Austin, MN 55912-3498

Hospital for Sick Children Foundation External Grants 2325

Grants are awarded for research in child health (up to age 18). Application forms should be obtained from the foundation. Listed application deadlines are for receipt of letters of intent. Annual deadline dates may vary; contact program staff for exact dates.
Requirements: Individuals affiliated with Canadian registered charities (including hospitals and universities) are eligible to apply.
Restrictions: The foundation does not support projects related to support of children outside Canada; established, ongoing research projects; computer hardware; building construction; operating expenses; or general endowments.
Geographic Focus: All States, Canada
Date(s) Application is Due: Mar 1; Sep 1
Amount of Grant: Up to 65,000 CAD
Contact: Grants Officer; (416) 813-6166, ext. 2354; fax (416) 813-5024; national.grants@sickkids.ca
Internet: www.sickkids.ca/foundation/grants.asp
Sponsor: Hospital for Sick Children Foundation
555 University Avenue
Toronto, ON M5G 1X8 Canada

Houston Endowment Grants 2326

The Endowment supports nonprofit organizations and educational institutions that improve life for the people of the greater Houston area. The Foundation funds programs in the arts, community enhancement, education, health, human services, the environment and neighborhood development. Requests for funding are accepted throughout the year. Further application information is available online.
Requirements: Nonprofit organizations that are recognized as charitable organizations by the Internal Revenue Code are eligible to apply. Funds are provided primarily to Harris County and contiguous counties. Grants seldom are given outside of Texas and never are made outside of the United States.
Restrictions: The Endowment does not select scholarship recipients or make direct scholarship awards to individuals. Inquiries about scholarship assistance should be made to financial aid offices at colleges and universities. Funding is not available for other grant-making organizations or to charities operated by service clubs. Funds are not available for religious activities; the purchase of uniform, equipment or trips for school-related organizations or sports teams; honoraria for speakers and panelists; fund-raising activities and galas; and memorials for individuals. The Foundation does not make loans or grants to individuals.
Geographic Focus: Texas
Amount of Grant: 2,500 - 1,000,000 U.S.D
Contact: Harriet W. Garland, Grant Manager; (713) 238-8100; fax (713) 238-8101; info@houstonendowment.org
Internet: http://www.houstonendowment.org/grants/grants_apply.htm
Sponsor: Houston Endowment
600 Travis, Suite 6400
Houston, TX 77002-3000

Howard and Bush Foundation Grants 2327

The foundation's funding is limited to Rensselaer County, NY. with an emphasis on arts and culture, education, civic and urban affairs, legal services, including a women's bar association, and social service and health programs that benefit the areas resident. Requests for equipment acquisition, programs development, matching funds, building/renovation and seed money will receive consideration. There are no deadline dates, funding guidelines are sent upon request. Contact staff prior to submitting a proposal. The board meets twice a year, in April and October. Notification of proposal acceptance will be within 14 days, following the board meeting.
Requirements: Nonprofit organizations benefiting Rensselaer County, NY, and its residents may request grant support.
Restrictions: Grants are not awarded to colleges, schools, churches, or individuals or for endowment purposes, operating budgets, or deficit financing. Generally no support for government or largely tax-supported agencies, or churches not connected with the founders.
Geographic Focus: New York
Amount of Grant: 750 - 75,000 U.S.D
Samples: Friends of Barker Park, $5,000—for park improvements; East Side Neighborhood Association, $5,000—for security system/modify heating system; Whitney M. Young Jr. Foundation, $10,000—for Troy oral health expansion project.
Contact: Deborah Byers, Program Contact; (518) 271-1134; dogclover@aol.com
Sponsor: Howard and Bush Foundation
2 Belle Avenue
Troy, NY 12180

Howard Gilman Foundation Grants 2328
The foundation is dedicated to the preservation of natural and cultural resources, with a focus on environmental, especially animal, conservation; the preservation and advancement of artistic and cultural endeavors; and medical research, especially in the fields of HIV/AIDS, cardiology, and sports medicine. The arts and culture component of the program is focused primarily in the New York metropolitan area. Types of support include general operating grants, program grants, research grants, and seed money grants. Staff prefer to respond to a brief letter of inquiry outlining the project's aim and general budgetary requirements before asking potential grantees to prepare full proposals.
Requirements: 501(c)3 nonprofit organizations are eligible.
Restrictions: Religious and political agencies are ineligible.
Geographic Focus: New York
Amount of Grant: 1,000 - 100,000 U.S.D
Samples: Brooklyn Academy of Music (Brooklyn, NY)—for endowment and operating support, $5 million.
Contact: Harry Brown, Program Associate; (212) 307-1073; fax (212) 262-4108; hbrown@gilman.com
Sponsor: Howard Gilman Foundation
111 W 50th Street, 40th Floor
New York, NY 10020

Howe Foundation of North Carolina Grants 2329
The Howe Foundation, established in 1966, supports organizations involved with arts and culture, education, human services, Christianity, and mentally disabled individuals. Giving limited to the Belmont, North Carolina area.
Requirements: No application form required, submit your proposal to: P.O. Box 749, Belmont, NC 28012.
Restrictions: No grants to individuals.
Geographic Focus: North Carolina
Amount of Grant: 200 - 144,000 U.S.D
Contact: Henry Howe, Treasure; (704) 825-5372
Sponsor: Howe Foundation
P.O. Box 227
Belmont, NC 28012-0227

HRB Clinical Research Training Fellowships 2330
These fellowships are designed to enable medical and dental graduates at any stage in their career to gain specialized clinical research training in a biomedical field in Ireland. Awards are available for up to two years. Fellowships are tenable only in institutions approved of by the board, such as teaching hospitals, universities and research institutes in Ireland. Applications are available on the Web site.
Requirements: Applicants should be graduates in medicine or dentistry from post-registration up to and including senior registrar or equivalent academic level. Applicants should be EU citizens or be graduates from outside the EU with permanent Irish resident status.
Date(s) Application is Due: Sep 16
Contact: Patricia Cranley, Research Grants Officer; (353) 1 6761176, ext. 120; fax (353) 1 6611856; pcranley@hrb.ie
Internet: http://www.hrb.ie/display_content.php?page_id=95
Sponsor: Health Research Board
73 Lower Baggot Street
Dublin 2, Ireland

HRB Health Services Research Fellowships 2331
Fellowships provide an opportunity to graduates with appropriate experience to undertake health services research in Ireland. Research may be interdisciplinary in nature and may involve clinical, epidemiological, public health, statistics, health economics, social science, operational, and management disciplines. Awards are available for up to three years. Application forms are available on the Web site.
Requirements: To be eligible, candidates must hold a primary degree in a discipline relevant to health services research; have acquired appropriate postgraduate experience in the field of health services research; have support from an approved academic department or center; have obtained the prior approval of a head of department for the research study being proposed; be European Union citizens or be graduates from outside the European Union with permanent Irish resident status.
Date(s) Application is Due: Oct 14
Amount of Grant: 53,200 - 56,792 EUR
Contact: Patricia Cranley, Research Grants Officer; (353) 1 6761176, ext. 120; fax (353) 1 6611856; pcranley@hrb.ie
Internet: http://www.hrb.ie/display_content.php?page_id=96
Sponsor: Health Research Board
73 Lower Baggot Street
Dublin 2, Ireland

HRB Postdoctoral Research Fellowships 2332
Fellowships enable researchers with a PhD, MD, or equivalent research experience to develop research careers at an advanced level in medicine, dentistry, health-related biological sciences, or epidemiology in Ireland. Awards are available for up to three years. Applications are available on the Web site.
Requirements: Applicants must have a PhD or MD in a related biomedical science field or have submitted a thesis for such at the time of application, or have equivalent research experience. Applicants should not have more than five years postdoctoral experience at the time of application and be European Union citizens or be graduates from outside the European Union with permanent Irish resident status. Applications should be made by the prospective fellow with the support of the head of an appropriate sponsoring laboratory in the Republic of Ireland.
Date(s) Application is Due: Oct 7
Amount of Grant: 53,200 - 56,792 EUR
Contact: Patricia Cranley, Research Grants Officer; (353) 1 6761176, ext. 120; fax (353) 1 6611856; pcranley@hrb.ie
Internet: http://www.hrb.ie/display_content.php?page_id=97
Sponsor: Health Research Board
73 Lower Baggot Street
Dublin 2, Ireland

HRK Foundation Grants 2333
The foundation focuses its resources on single-year funding for special projects in the areas of health care, AIDS, and children and youth services. Within the Twin Cities metro area, the Saint Croix Valley, and Ashland and Bayfield Counties of Wisconsin, the board will consider programming related to the issue of AIDS and programming that addresses children's health issues. Within the Saint Croix Valley and Ashland and Bayfield Counties in Wisconsin, the board will consider local community-specific projects. Types of support include general operating support, seed funding, community development, matching/challenge grants, equipment acquisition, and annual campaigns. Requests are considered two times each year.
Requirements: The foundation makes grants only to qualified IRS 501(c)3 organizations that specifically benefit people in the area surrounding Bayport and Saint Paul, MN, and western Wisconsin.
Restrictions: The foundation does not make loans or provide grants to individuals.
Geographic Focus: Minnesota, Wisconsin
Date(s) Application is Due: Mar 15; Sep 15
Amount of Grant: 1,000 - 25,000 U.S.D
Contact: Kathleen Fluegel, Foundation Director; (866) 342-5475 or (612) 293-9001; fax (612) 298-0551; info@HRKFoundation.org
Internet: http://www.hrkfoundation.org/grants/index.htm
Sponsor: HRK Foundation
345 St. Peter Street, Suite 1200
Saint Paul, MN 55102-1639

HRSA Abstinence Education Implementation and Planning Grants 2334
HRSA is awarding two types of grants, both of which focus on the teaching of sexual abstinence to adolescents. One-year planning grants are offered to support program planning, training, and community assessment, and three-year implementation grants support the development of the program in full. These grants seek to reduce the rate of births to adolescents, the proportion of those who have had sexual intercourse, and the proportion of those who have engaged in activities such as drug and alcohol use.
Requirements: Any public or private organization, including Indian tribes, faith-based organizations, and nonprofit agencies, are eligible to apply.
Restrictions: Applicants may not teach or promote religion in their projects.
Amount of Grant: 250,000 - 1,000,000 U.S.D
Contact: Lawrence Poole, Director; (301) 443-0654
Internet: http://www.acf.hhs.gov/programs/fysb
Sponsor: Health Resources and Services Administration
5600 Fishers Lane, Parklawn Building
Rockville, MD 20857

HRSA Basic Nurse Education and Practice Grants 2335
This program is intended to help schools of nursing and other institutions improve the quality and availability of nursing education through projects for specified purposes such as expanding enrollment in professional nursing programs, improving access to primary health care in noninstitutional settings, providing continuing education for nurses in medically underserved communities, and providing long-term care fellowships for certain paraprofessionals. Grant funds may be used for salaries of personnel specifically employed for the project, consultant fees, supplies and equipment necessary to conduct the project, and essential travel expenses and other expenses related to the project. Indirect administrative costs incurred as a result of the project are allowed but limited to 8 percent of direct costs. Applications are reviewed two times a year by the Advisory Council on Nurses Education. Assistance is available up to five years. Contact the office or visit the Web site for deadline dates.
Requirements: Public and nonprofit private schools of nursing and other public or nonprofit private entities may apply.
Amount of Grant: 250,000 U.S.D
Contact: Grants Administrator; (301) 443-5688; fax (301) 443-8586
Internet: http://bhpr.hrsa.gov/nursing
Sponsor: Health Resources and Services Administration
5600 Fishers Lane, Room 9-35
Rockville, MD 20857

HRSA Community Access Program Grants 2336
The goal of this program is to assist communities and their safety net providers in developing integrated health care delivery systems that serve the uninsured and underinsured with greater efficiency and improved quality of care. The program provides funding for approximately 20 communities to further their development of integrated delivery systems for the uninsured and underinsured. Grants will vary in size, based on the scope of the project and the size of the service area. The coordination of services through this grant will allow the uninsured and underinsured to receive efficient and higher quality care

and gain entry into a comprehensive system of care. The system will be characterized by effective collaboration, information sharing, and clinical and financial coordination among all levels of care in the community network. The system will be committed to continuous performance improvement, implementation of best practices, staff development, and real-time feedback of outcomes of care. Care management (e.g., case, disease) will be applied across the continuum for those with chronic illnesses, high-risk individuals, and high utilizers. The system will also strive to provide universal access to the target population, and to improve the health status of the community population.
Requirements: Applications may be submitted by public, private, and nonprofit entities that demonstrate a commitment to and experience with providing a continuum of care to uninsured individuals. Each applicant must represent a communitywide coalition that is committed to the project and includes safety net providers (where they exist) who have traditionally provided care to the community's uninsured and underinsured regardless of ability to pay. The communitywide coalition must consist of partners from all levels of care (i.e., primary, secondary, tertiary) and partners who represent a range of services (e.g., mental health and substance abuse treatment, maternal and child health care, oral health, HIV/AIDS). Agencies of state governments, multistate health systems, or special interest groups may submit applications on behalf of multiple communities if they demonstrate the ability to coordinate community health care delivery systems and bring resources to the community.
Date(s) Application is Due: Jun 14
Amount of Grant: 900,000 U.S.D
Contact: Program Office; (301) 594-4488; fax (301) 480-7833; capcentraloffice@hrsa.gov
Internet: http://bphc.hrsa.gov/cap/Default.htm
Sponsor: Health Resources and Services Administration
4350 East-West Highway, 3rd Floor
Bethesda, MD 20814

HRSA Eliminating Disparities in Perinatal Health Grants **2337**
The purpose of this program is to enhance a community's capacity to address significant disparities in several perinatal health indicators, with an emphasis on infant mortality rates.
Requirements: Public and private nonprofit community-based organizations are eligible to apply.
Restrictions: Entities that are currently receiving a Healthy Start implementation grant are not eligible to apply.
Amount of Grant: 100,000 - 23,000,000 U.S.D
Contact: Director; (301) 443-0543
Internet: http://www.hrsa.gov
Sponsor: Health Resources and Services Administration
5600 Fishers Lane, Parklawn Building
Rockville, MD 20857

HRSA Emergency Medical Services for Children Grants **2338**
State and accredited medical schools may be eligible for grants to support projects that expand and improve emergency medical services for children who need treatment for trauma or critical care. It is expected that maximum distribution of projects among the states will be made and that priority will be given to projects targeted toward populations with special needs, including Native Americans, minorities, and the disabled. HRSA-05-057
Requirements: States and accredited schools of medicine are eligible to apply.
Date(s) Application is Due: May 7
Amount of Grant: 700,000 U.S.D
Contact: Dan Kavanaugh, (301) 443-1321; dkavanaugh@hrsa.gov
Internet: http://www.hrsa.gov/grants/preview/mch.htm#hrsa05057
Sponsor: Health Resources and Services Administration
5600 Fishers Lane, Parklawn Building
Rockville, MD 20857

HRSA Geriatric Education Centers **2339**
The purpose of this program is to develop geriatric education centers to support development of collaborative arrangements involving several health professions schools and healthcare facilities. Geriatric centers facilitate the training of health professional faculty, students, and practitioners in the diagnosis, treatment, and prevention of disease, disability, and other health problems of the aged. Projects must improve training of health professionals; develop and disseminate curricula relating to treatment; support training and retraining of faculty and continuing education of health professionals who provide geriatric care; and provide students with clinical training in geriatrics in nursing homes, hospitals, ambulatory care centers, and senior centers. HRSA-05-077
Requirements: Accredited health professions schools, physicians assistants training programs, allied health schools, and nursing schools are eligible to apply.
Date(s) Application is Due: Mar 1
Amount of Grant: 200,000 U.S.D
Contact: Donna Nash; (301) 443-6960; fax (301) 443-6343; dnash@hrsa.gov
Internet: http://www.hrsa.gov/grants/preview/guidanceprofessions/hrsa05077.htm
Sponsor: Health Resources and Services Administration
5600 Fishers Lane, Room 11A-05
Rockville, MD 20857

HRSA Hansen's Disease National Ambulatory Care Program Grants **2340**
Grants are provided for screening and testing of contacts of patients diagnosed with Hansen's disease and for ambulatory care and treatment services to any person diagnosed with Hansen's disease who needs and requests care and treatment for that disease. Grants may be used for the cost of providing diagnostic treatment services to patients and contacts of patients diagnosed with Hansen's disease. Contact the office for deadline dates.
Requirements: Existing health care facilities are eligible to apply.
Amount of Grant: 35,000 - 375,000 U.S.D
Contact: Office of Grants Management; (800) 642-2477; fax (225) 756-3760
Internet: http://bphc.hrsa.gov/nhdp/AMBULATORY_CARE_MAIN_PAGE.htm
Sponsor: Health Resources and Services Administration
4350 East-West Highway
Bethesda, MD 20814

HRSA Health Professions Scholarships for Disadvantaged Students **2341**
The program makes funds available for grants to schools of medicine; osteopathic medicine; dentistry; optometry; pharmacy; podiatric medicine; veterinary medicine; nursing (associate, diploma, baccalaureate, and graduate degree); public health; chiropractic; allied health (baccalaureate or graduate degree programs of dental hygiene, medical laboratory technology, occupational therapy, physical therapy, radiologic technology, speech pathology, audiology, and registered dietitians); graduate programs in behavioral and mental health practice (includes clinical psychology, clinical social work, professional counseling, or marriage and family therapy); and programs providing training of physician assistants for the purpose of assisting such schools in providing scholarships to individuals from disadvantaged backgrounds who are enrolled (or accepted for enrollment) as full-time students in the schools. Notice of awards is issued to the schools, which, in turn, select and issue the awards to the eligible students. Consultation and technical assistance is provided to institutions wishing to apply. Annual deadline dates may vary; contact the office to confirm exact dates.
Requirements: Accredited public or private nonprofit schools of medicine, nursing, osteopathic medicine, dentistry, pharmacy, podiatric medicine, optometry, veterinary medicine, public or allied health, or schools offering graduate programs in clinical psychology are eligible.
Amount of Grant: 727 - 650,000 U.S.D
Contact: Andrea Stampone, (301) 443-4776
Internet: http://bhpr.hrsa.gov/dsa/sds.htm
Sponsor: Health Resources and Services Administration
5600 Fishers Lane, Parklawn Building
Rockville, MD 20857

HRSA Nurse Anesthetist Traineeships Program **2342**
Grants are awarded to eligible institutions to provide financial support through traineeships for licensed registered nurses enrolled as full-time students beyond the twelfth month of study in a master's nurse anesthesia program. HRSA-05-062
Requirements: Eligible applicants are schools of nursing, academic health centers, and other public or private nonprofit entities which provide registered nurses with full-time anesthetist master's education and have evidence of earned accreditation status from the American Association of Nurse Anesthetists (AANA) Council on Accreditation of Nurse Anesthesia Education Programs.
Date(s) Application is Due: Nov 18
Amount of Grant: 7,100 U.S.D
Contact: Karen Breeden, Division of Nursing, Bureau of Health Professions; (301) 443-6333; fax (301) 443-8586; kbreeden@hrsa.gov
Internet: http://www.hrsa.gov/grants/preview/professions.htm#hrsa05062
Sponsor: Health Resources and Services Administration
5600 Fishers Lane, Parklawn Building
Rockville, MD 20857

HRSA Nursing Education Opportunities for the Disadvantaged Project Grants **2343**
This program is intended to provide financial assistance to eligible schools of nursing and other applicants to meet the costs of special projects to increase nursing education opportunities for individuals from disadvantaged backgrounds. Funds may be used for salaries of personnel specifically employed for the project; consultant fees; supplies and equipment necessary to conduct the project; essential travel expenses; and other expenses related to the project. Indirect administrative costs are allowed, limited to 8 percent of direct costs. Grants may be made for up to three-year project periods and may be renewed competitively for an additional two years.
Requirements: Public and nonprofit private schools of nursing and other public or nonprofit private entities are eligible.
Amount of Grant: 58,800 - 518,600 U.S.D
Contact: Dr. Janice Young, Division of Nursing, Bureau of Health Professions; (301) 443-6193
Internet: http://bhpr.hrsa.gov/dsa/sds.htm
Sponsor: Health Resources and Services Administration
5600 Fishers Lane, Parklawn Building
Rockville, MD 20857

HRSA Physician Assistant Training **2344**
Grants are awarded for projects for the training of physician assistants and for the training of individuals who will teach in programs to provide such training. A project supported must meet the following definition of a training program for physician assistants as defined under Section 799B of the Public Health Service Act: has as its objective the education of individuals who will, upon completion of their studies in the program, be qualified to provide primary care under the supervision of a physician; extends for at least one academic year and consists of supervised clinical practice and at least four months (in the aggregate) of classroom instruction directed toward preparing students to deliver health care; has an enrollment of not less than eight students; and trains students in primary care, disease prevention, health promotion, geriatric medicine, and home health care. The program assists schools to meet the costs of projects to plan,

develop, and operate or maintain programs for the training of physician assistants and for the training of individuals who will teach in programs to provide such training. Programs must develop and use methods designed to encourage graduates of the program to work in health professional shortage areas. Programs also must develop and use methods for placing graduates in positions for which they have been trained.
Requirements: Public or nonprofit private hospitals, schools of medicine or osteopathic medicine, or public or private nonprofit entities are eligible to apply. Eligible physician assistant programs are those that are either accredited by the American Medical Association's Committee on Allied Health Education and Accreditation (AMA-CAHEA) or its successor organization, the Commission on Accreditation of Allied Health Education Programs (CAAHEP).
Amount of Grant: 13,487 - 257,297 U.S.D
Contact: Ellie Grant, (301) 443-5404; egrant@hrsa.gov
Internet: http://www.hrsa.gov
Sponsor: Health Resources and Services Administration
5600 Fishers Lane, Parklawn Building
Rockville, MD 20857

HRSA Rural Health Outreach Grants **2345**
The purpose of this program is to expand access to, coordinate, restrain the cost of, and improve the quality of essential health care services, including preventive and emergency services, through the development of integrated health care delivery systems or networks in rural areas and regions. Funds are available for projects to support the direct delivery of health care and related services, to expand existing services, or to enhance health service delivery through education, promotion, and prevention programs. The program's emphasis is on the actual delivery of specific services rather than the development of organizational capabilities. Projects may be carried out by networks of the same providers (i.e., all hospitals) or more diversified networks. Annual deadline dates may vary; contact program staff for exact dates.
Requirements: Rural public or nonprofit private organizations that include three or more health care providers or other entities that provide or support the delivery of health care services are eligible to apply. The administrative headquarters of the organization must be located in a rural county or in a rural census tract of an urban county, or an organization constituted exclusively to provide services to migrant and seasonal farmworkers in rural areas and supported under Section 330(g) of the Public Health Service Act. These organizations are eligible regardless of the urban or rural location of the administrative headquarters.
Date(s) Application is Due: Sep 23
Amount of Grant: Up to 185,000 U.S.D
Contact: Eileen Holloran, (301) 443-7529; fax (301) 443-2803; eholloran@hrsa.gov
Internet: http://ruralhealth.hrsa.gov/funding/outreach.htm
Sponsor: Health Resources and Services Administration
5600 Fishers Lane, Parklawn Building
Rockville, MD 20857

HRSA Rural Health Policy Analytic Centers **2346**
The purpose of this program is to fund rural health services policy analytic centers to conduct policy-relevant research on rural health services issues of multistate and national significance, and disseminate the findings of their research. The centers study the critical issues facing rural communities in their quest to secure adequate, affordable, high-quality health services. Rural health research findings are published in appropriate refereed journals and disseminated to a national audience.
Requirements: All public and private entities, both nonprofit and for-profit, are eligible to apply.
Amount of Grant: Up to 1,000,000 U.S.D
Contact: Joan Van Nostrand, Office of Rural Health Policy; (301) 443-0835; fax (301) 443-2803; jvan_nostrand@hrsa.gov
Internet: http://ruralhealth.hrsa.gov/policy/rhrccoop.htm
Sponsor: Health Resources and Services Administration
5600 Fishers Lane, Room 9A-55
Rockville, MD 20857

HRSA Rural Network Development Program Grants **2347**
The purpose of this program is to support the planning and development of vertically integrated health care networks in rural areas to overcome the fragmentation of health care delivery service. The emphasis of the program is on projects to develop the organizational capabilities of these vertically integrated networks, which must be composed of three different types of providers. Funding preference may be given to applicant networks that include a majority of the health care providers serving in the area or region to be served by the network; any federally qualified health centers, rural health clinics, and local public health departments serving in the area or region; outpatient mental health providers serving in the area or region; or appropriate social service providers, such as agencies on aging, school systems, and providers under the women, infants, and children program (WIC) to improve access to and coordination of health care services. Grants are awarded annually. More information is listed in the program preview available on the Web site.
Requirements: A rural public or nonprofit private organization that is or represents a network that includes three or more health care providers or other entities that provide or support the delivery of health care services is eligible to apply. The administrative headquarters of the organization must be located in a rural county or in a rural census tract of an urban county or be an organization constituted exclusively to provide services to migrant and seasonal farm workers in rural areas and supported under Section 330(g) of the Public Health Service Act. These organizations are eligible regardless of the urban or rural location of the administrative headquarters.
Date(s) Application is Due: Sep 8
Amount of Grant: 25,000 - 85,000 U.S.D
Contact: Michele Pray-Gibson, Program Analyst, Rural Network Development; (301) 443-7320; fax (301) 443-2803; mpray@hrsa.gov
Internet: http://www.hrsa.gov/grants/preview/guidancerural/hrsa06007.htm
Sponsor: Health Resources and Services Administration
5600 Fishers Lane, Parklawn Building
Rockville, MD 20857

HRSA Ryan White HIV Early Intervention Services Grants **2348**
To increase the capacity of entities that provide comprehensive primary care services to persons with Human Immunodeficiency Virus (HIV) infection, or who are at-risk of infection, and to offer more early intervention services. The services will include increased counseling and testing, partner involvement in risk reduction, transmission prevention, appropriate primary care diagnostic and treatment services, and, as needed, case-management to ensure that individual service needs are met. A planning and capacity building program provides assistance to organizations wishing to develop a new or expanded HIV program, particularly in rural and underserved areas and communities of color. HRSA-05-020
Requirements: Eligible for grants are public and private nonprofit entities, including migrant and community health centers, health care for the homeless grantees, family planning grantees, comprehensive hemophilia diagnostic and treatment centers, federally qualified health centers, and other groups that currently provide primary care services to people at risk for HIV.
Restrictions: Grant funds may not support construction, in-patient care, residential care, or cash payments to service recipients.
Date(s) Application is Due: Mar 1
Amount of Grant: Up to 150,000 U.S.D
Contact: Dr. Jose Morales, (301) 443-0493; jmorales@hrsa.gov
Internet: http://www.hrsa.gov/grants/preview/hivaids.htm#hrsa05020
Sponsor: Health Resources and Services Administration
5600 Fishers Lane, Parklawn Building
Rockville, MD 20857

HRSA Ryan White Title IV Funding for Access to Medical Care for HIV Patients **2349**
The purpose of the program is to improve access to primary medical care, research, and support services for children, youth, women, and families infected with HIV. Funded projects will link clinical research and other research with comprehensive care systems, and improve and expand the coordination of a system of comprehensive care for women, infants, children, and youth who are infected/affected by HIV. Funds will be used to support programs that cross established systems of care to coordinate service delivery, HIV prevention efforts, and clinical research and other research activities; and address the intensity of service needs, high costs, and other complex barriers to comprehensive care and research experienced by underserved at-risk and limited populations. Activities under these grants should address the goals of enrolling and maintaining clients in HIV primary care; increasing client access to research by linking HIV/AIDS clinical research trials and activities with comprehensive care; fostering the development and support of comprehensive, community-based and family-centered care infrastructures, and emphasizing prevention within the care system including the prevention of perinatal HIV transmission.
Requirements: Public or private nonprofit entities that provide or arrange for primary care are eligible.
Amount of Grant: 766,000 U.S.D
Contact: Lawrence Poole, Director, Division of Grants Management; (301) 443-2385
Internet: http://hab.hrsa.gov/grant.htm
Sponsor: Health Resources and Services Administration
5600 Fishers Lane, Parklawn Building
Rockville, MD 20857

Huber Foundation Grants **2350**
The Foundation focuses its grants program on three specific aspects of reproductive rights: keeping abortion safe, legal, and preferably increasingly rare; contraceptive choice and availability; and relevant education. The Foundation's major interest lies in funding organizations that will impact these issues on a national level. There is no formal application procedure. A letter describing the project, a budget for the project, and proof of tax-exempt status is required. Though the Foundation board meets four times annually, there are no fixed deadline dates.
Requirements: U.S. nonprofits, including advocacy groups, hospitals, legal defense groups, family planning agencies, educational organizations, universities, and women's groups may submit grant applications.
Restrictions: The foundation does not encourage proposals for projects that are local or regional in scope. It will not consider grants to individuals, foreign organizations, capital campaigns, scholarships, endowment funds, research, international projects, or film productions.
Amount of Grant: 10,000 - 100,000 U.S.D
Samples: Planned Parenthood Federation of America (New York, NY)—for general support and Responsible Choices Campaign, $700,000; Feminist Majority Foundation (Arlington, VA)—for the National Clinic Access Project, $100,000; Planned Parenthood of Northern New England (Williston, VT)—for general support, $90,000.
Contact: Lorraine Barnhart, Executive Director; (732) 933-7700
Sponsor: Huber Foundation
P.O. Box 277
Rumson, NJ 07760-0277

HUD Mainstream Housing Opportunities for Persons with Disabilities 2351

The purpose of this program is to provide Section 8 rental vouchers to enable persons with disabilities (elderly and nonelderly) to rent affordable private housing. The rental vouchers that HUD provides under this NOFA must be made available to eligible disabled families regardless of their type of disability. Annual deadline dates may vary; contact the program office for exact dates.
Requirements: PHAs and nonprofit disability organizations that provide services to disabled families are eligible to apply.
Restrictions: American Indian Housing Authorities and American Indian tribes and their tribally designated housing entities are not eligible to apply.
Amount of Grant: Up to 10,000 U.S.D
Contact: George Hendrickson, Program Specialist; (202) 708-0477, ext. 4064
Internet: http://www.hud.gov/library/bookshelf18/supernofa/nofa04/grpsec8.cfm
Sponsor: U.S. Department of Housing and Urban Development
451 Seventh Street SW
Washington, DC 20410

Hudson Webber Foundation Grants 2352

The purpose of the Foundation is to improve the vitality and quality of life of the metropolitan Detroit community. The Foundation concentrates its giving primarily within the City of Detroit and has a particular interest in the revitalization of the urban core because this area is a focus for community activity and pride and is of critical importance to the vitality of the entire metropolitan community. The Foundation presently concentrates its efforts and resources in support of projects within five program missions: Detroit Physical Revitalization; Economic Development; The Arts; Safe Community; and the Detroit Medical Center. A brief letter signed by a senior officer of the requesting organization is the preferred form of application. Information for completing the letter of request are available online.
Requirements: The foundation concentrates its giving within Detroit.
Geographic Focus: Michigan
Samples: Habitat for Humanity - Metro Detroit, Inc., Detroit, MI, $150,000—Partnership to Build Capacity; Detroit Renaissance Foundation, Inc., Detroit, MI, $100,000—Business Attraction Program; Detroit Science Center, Inc., Detroit, MI, $10,000—General program support; Michigan Community Service Commission, Detroit, MI, $75,000—Winning Futures Program-Mentor Michigan; Harper Hospital, Detroit, MI, $200,000—Renal Transplant Unit.
Contact: Katy Locker, Program Director; (313) 963-7777; fax (313) 963-2818; HWF@hudson-webber.org
Internet: http://www.hudson-webber.org/HowToApply_Instructions.html
Sponsor: Hudson Webber Foundation
333 West Fort Street, Suite 1310
Detroit, MI 48226-3134

Huffy Foundation Grants 2353

The Huffy Foundation gives primarily to areas in which the company has operations in California, Ohio, Pennsylvania, and Wisconsin. These areas include: the visual arts; museums; performing arts; theater; arts/cultural programs; all levels of education; hospitals (general); health care and associations; recreation; children and youth services; and federated giving (United Way). Types of support include general and operating support; continuing support; annual campaigns; capital campaigns; building/renovation programs; emergency funds; program development; seed money; consulting services; employee matching gifts; and matching funds. Requests should include a description of the organization, its history and purpose, a description of the people it serves, and a summary of total budget and funding.
Requirements: The Foundation provides a brochure which delineates the application and grant guidelines. There is no application form. The Foundation recommends that the initial approach be in the form of a letter or a proposal. One copy of the letter/proposal should be submitted. There are no deadlines. The Board meets in February, May, August and November.
Restrictions: Grants are not made to individuals, in support of political activities or of religious organizations for religious purposes, or organizations that are not tax exempt. Grants are seldom made for medical research; to endowments; or for operating funds for organizations located outside the corporation communities.
Geographic Focus: California, Ohio, Pennsylvania, Wisconsin
Contact: Pam Booher, Secretary; (937) 866-6251
Sponsor: Huffy Foundation
225 Byers Road
Miamisburg, OH 45342

Hugh J. Andersen Foundation Grants 2354

The mission of the Hugh J. Andersen Foundation is to give back to our community through focused efforts that foster inclusivity, promote equality, and lead to increased human independence, self sufficiency and dignity. To fulfill this mission, the Foundation acts as a grantmaker, innovator, and convener. The Foundation's primary geographic area of focus is the St. Croix Valley: Washington County in Minnesota and Pierce, Polk and St. Croix Counties in Wisconsin. Secondarily, there is an interest in St. Paul, Minnesota. From time to time the Foundation may consider programs in other parts of the Metro Area and Greater Minnesota. Please contact the Program Director prior to submitting a proposal to determine if the program might be of interest to the Foundation. The Board generally considers requests in June, September, December and February. Faxed or emailed applications will not be considered.
Requirements: The Hugh J. Andersen Foundation awards grants only to qualified charitable organizations that are designated as tax exempt under Internal Revenue Service Code 501(c)3 and are not classified as private foundations.
Restrictions: The Hugh J. Andersen Foundation: does not make loans, and does not provide grants or scholarships to individuals; does not provide grants for lobbying activities, fundraising dinners and events, or travel; will generally not consider the following types of organizations and programs for funding: agencies/divisions/councils/programs that have counterparts in St. Paul or the St. Croix Valley; arts organizations exclusively focused on music, dance or visual arts; athletic teams; business/economics education; child care centers; civic action groups; debt or after the fact situations; immigration/refugee issues and programs; independent media productions; political/voter education; private or alternative schools; religious institutions. In addition, the Foundation: will generally not fund the entire project budget, but prefers to be part of an effort supported by a number of sources; considers major endowment and capital requests for funding to be a low priority; does not provide funding through fiscal agents; considers letters of inquiry. Letters of inquiry are reviewed at the Foundation's next board meeting. If the board determines that the request falls within its guidelines and interests, a full proposal will be requested from the applicant for review at the following board meeting.
Geographic Focus: Minnesota, Wisconsin
Date(s) Application is Due: Mar 15; Jun 15; Aug 15; Nov 15
Contact: Brad Kruse, Program Director; (651) 275-4489 or (888) 439-9508; fax (651) 439-9480; hjafdn@srinc.biz
Internet: https://www.srinc.biz/hja/documents/HJAGuidelines09-10.pdf
Sponsor: Hugh J. Andersen Foundation
White Pine Building, 342 Fifth Avenue North
Bayport, MN 55003

Hugoton Foundation Grants 2355

The foundation awards one-year, nonrenewable grants to eligible New York and Florida nonprofit organizations in its areas of interest, including social services and improved patient care. To apply, submit a letter of inquiry and a copy of the IRS tax-exemption letter.
Requirements: 501(c)3 tax-exempt organizations serving the borough of Manhattan, NY; and Miami, FL, are eligible.
Geographic Focus: New York
Amount of Grant: 1,000 - 50,000 U.S.D
Contact: Joan Stout, President; (212) 734-5447
Sponsor: Hugoton Foundation
900 Park Avenue, Suite 17E
New York, NY 10021

Huie-Dellmon Trust Grants 2356

The trust awards grants to Louisiana nonprofits in its areas of interest, including hospitals, higher and secondary education, libraries, and Protestant churches and organizations. Types of support include general operating support, capital campaigns, building construction/renovation, equipment acquisition, program development, scholarship funds, research, and matching funds. There are no application forms or deadlines.
Requirements: Central Louisiana nonprofit organizations are eligible.
Geographic Focus: Louisiana
Amount of Grant: 1,000 - 100,000 U.S.D
Contact: Richard Crowell Jr., Trustee; (318) 748-8141
Sponsor: Huie-Dellmon Trust
P.O. Box 330
Alexandria, LA 71309-0330

Huisking Foundation Grants 2357

The foundation awards grants in its areas of interest, including Catholic higher and secondary education, church support and social services, hospitals, and religion associations. Types of support include general operating support, building construction/renovation, research grants, endowment funds, scholarship funds, special projects, and continuing support. The board meets in April and November; letters of intent are due in February and August.
Amount of Grant: 200 - 60,000 U.S.D
Samples: KITSAP Humane Society (OR)—for operating support, $1000; Georgetown U, Alumni Fund, (Washington, DC)—for the scholarship fund, $6000.
Contact: Frank Huisking, Treasurer; (203) 426-8618
Sponsor: Huisking Foundation
291 Peddlers Road
Guilford, CT 06437

Humana Foundation Grants 2358

The foundation was established to promote worthwhile organizations that improve the health and welfare of communities in its headquarters region of Louisville, KY. It supports nonprofit institutions primarily in the areas of domestic and international health, education, and civic and cultural development. Religious organizations are eligible for project-specific support (e.g., social services outreach) or funds for an accredited, church-affiliated educational institution. Grants also are awarded in Humana market areas outside of the headquarters region. Application information is available online.
Requirements: Applicant organizations must be 501(c)3 tax-exempt located in communities where Humana has a meaningful presence.
Restrictions: Grants are not given to organizations for seed money. The foundation does not contribute to social, labor, political, veterans, or fraternal organizations. Funds cannot be used solely to support an organization's salary expenses or other administrative costs. The foundation does not support lobbying efforts or political action committees.
Geographic Focus: Arizona, Colorado, Florida, Georgia, Illinois, Indiana, Kansas, Kentucky, Louisiana, Michigan, Ohio, Tennessee, Texas, Utah, Wisconsin
Date(s) Application is Due: Jan 15; Jun 15

Samples: Odyssey House, Louisiana, KY, $100,000—offers services and support systems through a behavioral healthcare facility with an emphasis on addiction treatment; The Warrick Dunn Foundation, Tampa Bay, FL, $100,000—provides opportunities for economically disadvantaged single parents and children who have demonstrated a commitment to achieve financial independence and stability; Kids on the Move, Orem, UT, $100,000—provides support and services for low-income families, especially those with special needs children.
Contact: Foundation Office; (502) 580-4140; fax (502) 580-1256; HumanaFoundation@humana.com
Internet: http://www.humanafoundation.org/philanthropy/
Sponsor: Humana Foundation
500 West Main Street, Suite 208
Louisville, KY 40202

Human Growth Foundation Small Grants **2359**
The foundation awards grants for investigation of human growth and its disorders, with special consideration given for the chondrodystrophies. Special consideration will be given to new investigators and ideas new to the field. Postdoctoral research dealing with all aspects of normal and abnormal human growth and development including, but not limited to, the areas of biological, psychological, educational, and diet and nutrition will be considered. Up to three grants will be awarded. A synopsis of the research results is required to be submitted the following year. An NIH-type biographical sketch and two-page letter of intent should be sent by May 15; full applications are due by September 1.
Restrictions: No portion of the grant can be used for administrative overhead.
Date(s) Application is Due: May 15; Sep 1
Amount of Grant: 7,500 - 10,000 U.S.D
Contact: Program Contact; (800) 451-6434 or (516) 671-4041; fax (516) 671-4055; hgf1@hgfound.org
Internet: http://www.hgfound.org/smallgrants.html
Sponsor: Human Growth Foundation
997 Glen Cove Avenue
Glen Head, NY 11545

Human Source Foundation Grants **2360**
The Human Source Foundation is the charitable, non for profit arm of PALMEA. The Foundation is organized to support charitable and educational initiatives that improve the human condition, by providing assistance and financial support for programs involved with botany, higher education, and human services, with grants coming in the form of general operating funds. Giving is restricted to the State of Texas, primarily in Denton and Fort Worth counties. There are no specific deadlines or application forms with which to adhere, and applicants should begin by forwarding a letter of request.
Requirements: 501(c)3 tax-exempt organizations serving Texas, primarily Denton and Fort Worth counties, are eligible to apply.
Geographic Focus: Texas
Contact: Mary G. Palko, (817) 926-2799; fax (817) 926-5202; mary@ftw.com or mpalko@compuserve.com
Sponsor: Human Source Foundation
2409 Winton Terrace West, P.O. Box 100423
Fort Worth, TX 76185-0423

Humboldt Area Foundation Technical Assistance Program (TAP) Grants **2361**
TAP is a small award program which provides nonprofits with one-on-one technical assistance in a variety of forms: consultants, training workshops, self-pace manuals, software, or any other form identified by applicants. TAP awards typically range from $500 to $2,000. Funds must be used to improve the organization, its leadership or operations, as a whole; not on programs or projects of the agency. Areas of assistance may include: strategic planning, financial management, operational assessments, board development, program assessments, personnel issues, collaborations, and fundraising. TAP applications and information are confidential and will not affect other potential Humboldt Area Foundation grant funding. TAP grant applications are accepted on the first of each month and grants that are awarded are paid directly to a consultant to cover the costs of their consulting services.
Restrictions: TAP will not pay for grant writing, programmatic development or for duties that are typically handled by staff (e.g. data entry). Organizations less than a year old and government entities are not eligible for TAP funds.
Geographic Focus: California
Date(s) Application is Due: Jan 1; Feb 1; Mar 1; Apr 1; May 1; Jun 1; Jul 1; Aug 1; Sep 1; Oct 1; Nov 1; Dec 1
Amount of Grant: 500 - 2,000 U.S.D
Contact: Peter Pennekamp, Executive Director; (707) 442-2993, ext. 304; fax (707) 442-3811; peterp@hafoundation.org
Internet: http://www.hafoundation.org/index.php?option=content&task=view&id=74
Sponsor: Humboldt Area Foundation
373 Indianola Road
Bayside, CA 95524

Hunter's Hope Foundation Grants **2362**
Hunter's Hope funds projects directly related to Krabbe disease or other leukodystrophies. Eligible studies may investigate new treatment approaches, or study basic mechanisms related to these diseases. Areas of relevant research include, among others, gene therapy, myelin development, neurodegeneration and transplantation, developmental neurobiology, and molecular biology. Applications are accepted in three project categories: postdoctoral fellowships provide stipend support for individuals who are within five years of receiving their terminal degree (MD, DVM, PhD, or equivalent degree). Fellows will be working with senior investigators within their sponsoring institution. Fellowships are awarded for two years. Pilot studies are one-year projects designed to test new concepts or ideas and can serve as a source of preliminary data for future studies. Major research grants provide support to senior investigators conducting major research studies. Projects may be funded for one to three years. Applications are reviewed twice annually.
Date(s) Application is Due: Apr 15; Sep 15
Amount of Grant: 30,000 - 100,000 U.S.D
Contact: Executive Director; (877) 984-4673 or (716) 667-1200; fax (716) 667-1212; info@huntershope.org
Internet: http://www.huntershope.org/research/default.asp
Sponsor: Hunter's Hope Foundation
P.O. Box 643
Orchard Park, NY 14127

Huntington's Disease Society of America Research Fellowships **2363**
Fellowships are designed to assist promising young postdoctoral investigators in the early stages of their careers who are engaged in research at the basic and clinical levels relating to the cause and treatment of Huntington's disease. The amount of the award will depend upon the policy of the sponsoring accredited medical school or university and the training and experience of the applicant. Fellowships are awarded for one year with the possibility of renewal. Application forms and requested information on curriculum vita, publications, research proposal, sponsorship, and other sources of funding are provided by the society.
Requirements: All applicants must have an MD or PhD degree or the equivalent and work must in some way be related to Huntington's disease. There is no citizenship requirement.
Amount of Grant: Up to 80,000 U.S.D
Samples: Dr. Tsugn Peng, Emory U (Atlanta, GA)—for research on impaired mitochortinal function, $30,000; Dr. Nansheng Chen, U of British Columbia (Vancouver, BC)—for research on the role of glutamate receptor-HD protein interactions, $30,000.
Contact: Robert Graze, Grants Administrator; (212) 242-1968 or (800) 345-4372, ext. 227; rgraze@hdsa.org
Internet: http://www.hdsa.org/research/grant-applications.html
Sponsor: Huntington's Disease Society of America
505 8th Avenue, Suite 902
New York, NY 10018

Huntington's Disease Society of America Research Grants **2364**
The Society awards research grants for one year for support of basic or clinical research related to Huntington's disease. Grant awards are provided as seed monies for new or innovative research projects in the hope that they will develop sufficiently to attract funding from other sources. Application forms are provided by the society and generally follow the format of NIH grant applications.
Requirements: Applicants may be doctorate or graduate students or equivalent professionals. There is no citizenship requirement.
Amount of Grant: Up to 100,000 U.S.D
Contact: Robert Graze, Grants Administrator; (212) 242-1968 or (800) 345-4372, ext. 227; rgraze@hdsa.org
Internet: http://www.hdsa.org/research/grant-applications.html
Sponsor: Huntington's Disease Society of America
505 8th Avenue, Suite 902
New York, NY 10018

Huntington Beach Police Officers Foundation Grants **2365**
The Foundation provides funds to various schools and community youth activities, benefits to spouses or children of police officers killed in the line of duty, and medical assistance to children of police officers in need of help. Giving is limited to Huntington Beach, California. There are no specific grant application forms or deadlines with which to adhere. Applicants should contact the Foundation office directly.
Requirements: Support is given only to groups situated in the Huntington Beach, California, region.
Geographic Focus: California
Contact: Kreg Muller, President; (714) 842-8851; fax (714) 847-0064; hbpof@hbpoa.org
Internet: http://www.hbpoa.org/foundation/hbpof.htm
Sponsor: Huntington Beach Police Officers Foundation
20422 Beach Boulevard, Suite 450
Huntington Beach, CA 92648-8301

Huntington Clinical Foundation Grants **2366**
The Foundation, established in 1986, supports health care programs and access, higher education, and medical research. Grant amounts range from approximately $4,000 to $8,000, with some larger amounts available. There are no specific applications or deadlines with which to adhere. The Board meets quarterly, and applicants should forward a detailed description of the project and amount of funding requested.
Requirements: Giving is limited to the Huntington, West Virginia, region.
Geographic Focus: West Virginia
Amount of Grant: 4,000 - 8,000 U.S.D
Contact: Dr. J. David Daniels, President; (304) 529-4217; fax (304) 523-6051
Sponsor: Huntington Clinical Foundation
P.O. Box 117
Huntington, WV 25706-0117

Huntington County Community Foundation - Make a Difference Grants 2367

The Huntington County Community Foundation - Make a Difference Grants fund charitable projects that make a positive impact on the residents of Huntington County. Grant areas to be considered include: arts and culture; community development; health and human services; education; and other charitable services. These grants are awarded in spring and fall cycles. Grant decisions on based on the following criteria: the project's beneficial impact on Huntington County, immediate and ongoing; known and anticipated community needs; number of persons benefitted/affected by the project; and a clear, complete, and comprehensible statement of particulars.

Requirements: Applicants should submit one original and five additional paper clipped sets of the online application packet to include the following: grant proposal form; detailed project and organizational budget; current year-end financial statement; strategic or long-range plan; roster of board members; IRS tax exempt status; and articles of incorporation.

Restrictions: Grants will not be awarded to fund: operational or ongoing recurring (within 60 months) cost of the program; political projects or campaigns; or projects of applicants, or project owners, with taxing authority (e.g. school corporations; units of government).

Geographic Focus: Indiana

Date(s) Application is Due: Apr 15; Oct 15

Samples: Boys and Girls Club, complete gymnasium floors and walls, $5,000 (2010); Friends of Upper Wabash Interpretive Services, construction of a bird shelter/sanctuary for rehabilitated hawks, owls, and an eagle, $4,000 (2010); American Red Cross, expand emergency information system, $400 (2010);

Contact: Michael Howell, Executive Director; (260) 356-8878; fax (260) 356-0921; michael@huntingtonccf.org

Internet: http://www.huntingtonccf.org/hccf_grant_opportunities.html

Sponsor: Huntington County Community Foundation
356 West Park Drive
Huntington, IN 46750

Huntington National Bank Community Affairs Grants 2368

The corporation supports a variety of community focused initiatives that benefit children, education, economic development, health, social service housing, and the arts in each community of its banking offices. A letter of request (one to two pages) can serve as an initial application for both single- and multi-year contributions. The letter should include a brief statement about the organization, its history, objectives, and goals; purpose of the project; visibility for Huntington as a participant in the project; list of officers and directors; target population(s) to be served by contribution; and annual organization budget, total project budget, amount requested, list of other contributors, and amounts of funds anticipated or committed.

Requirements: Tax-exempt organizations in Ohio, Kentucky, Indiana, Michigan, and West Virginia may apply.

Restrictions: Community donations may not be used for political campaigns involving individuals or issues; religious, ethnic, military, fraternal, or labor groups; individuals; organizations outside of the corporate market area; or organizations without IRS 501(c)3 tax-exempt status.

Geographic Focus: Indiana, Kentucky, Michigan, Ohio, West Virginia

Amount of Grant: 1,000 - 10,000 U.S.D

Contact: Elfi Di Bella, Community Affairs Director; (614) 480-4483; fax (614) 480-4973; elfi.dibella@huntington.com

Internet: http://www.huntington.com/us/HNB3150.htm

Sponsor: Huntington Bancshares
41 South High Street, Huntington Center 3413
Columbus, OH 43215

Hutchinson Community Foundation Grants 2369

The Hutchinson Community Foundation provides funding in the counties of Hutchinson and, Reno, Kansas, inluding the immediate area surrounding area. Areas of interest include, but are not, limited to: community needs in areas such as early childhood, youth, education, arts and culutre, and neighborhood development.

Requirements: Nonprofit organizations in Kansas are eligible to apply. Special purpose units of government can apply for support of innovative projects located in Reno County.

Restrictions: Grant proposals from individuals or non-qualifying organizations will not be considered.

Geographic Focus: Kansas

Contact: Audrey Abbott Patterson, President; (620) 663-5293; fax (620) 663-9277; aubrey@hutchcf.org

Internet: http://www.hutchcf.org/default.asp?sm=2&si=1

Sponsor: Hutchinson Community Foundation
One North Main, Suite 501, P.O. Box 298
Hutchinson, KS 67504-0298

Hut Foundation Grants 2370

Since 1998, the Hut Foundation has been funding nonprofits primarily in the states of California and Virginia, in the following areas of interest: arts; children/youth services; community/economic development; education; health care; youth development. Unsolicited requests for funds are not accepted, therefore initial contact should be through a Letter of Inquiry. If the proposal follows the Foundations guidelines, the applicant will be invited to submit an application.

Requirements: Must be a 501(c)3 nonprofit.

Restrictions: No grants to individuals.

Geographic Focus: California, Virginia

Amount of Grant: 5,000 - 10,000 U.S.D

Contact: Marcus Guerrero; (415) 834-2464; hutfoundation@gmail.com

Sponsor: Hut Foundation
19 Sutter Street
San Francisco, CA 94104

Hutton Foundation Grants 2371

The foundation awards grants throughout Santa Barbara County, California to nonprofit organizations in its areas of interest, including education, arts and culture, health and human services, and civic and community development. Preference is given to efforts that promise to bring about major change, build leadership and institutional capacity, and achieve lasting results. Types of support include standard, program related investments, marketing, media, and endowment grants. An application form is available online.

Requirements: California nonprofit organizations headquartered in Santa Barbara county are eligible to apply.

Restrictions: Funding is not available to individuals or organizations that discriminate on the basis of age, gender, race, ethnicity, sexual orientation, disability, national origin, political affiliation or religious belief.

Geographic Focus: California

Contact: Arlene R. Craig, Vice President; (805) 957-4740; fax (805) 957-4743; info@huttonfoundation.org

Internet: Arlene R. Craig, Vice President;

Sponsor: Hutton Foundation
26 West Anapamu Street, 4th Floor
Santa Barbara, CA 93101

Hyde and Watson Foundation Grants 2372

The foundation's primary concern is supporting capital projects that increase the quality, capacity, or efficiency of a grantee's programs or services, such as purchase or relocation of facilities, facilities improvements, capital equipment, instructive materials development, or certain medical research areas. Broad fields include health, education, religion, social services, arts, and humanities.

Requirements: Primary geographic areas served include the New York City metropolitan region and primarily Essex, Union, and Morris Counties in New Jersey.

Restrictions: Requests for endowment or operating support or from fiscal agents generally will be denied. Grants are not made to individuals.

Geographic Focus: New Jersey, New York

Date(s) Application is Due: Feb 15; Sep 15

Amount of Grant: 5,000 - 25,000 U.S.D

Samples: Ctr for Molecular Medicine and Immunology (Newark, NJ)—for building renovation and cancer research equipment. $200,000; Kent Place School (Summit, NJ)—to alter and modernize the dining hall, $50,000.

Contact: Hunter Corbin, President; (973) 966-6024; fax (973) 966-6404; Hcorbin@HydeandWatson.org

Internet: http://fdncenter.org/grantmaker/hydeandwatson

Sponsor: Hyde and Watson Foundation
437 Southern Boulevard
Chatham Township, NJ 07928

I.A. O'Shaughnessy Foundation Grants 2373

The Foundation is concerned that too many schools lack sufficient resources; that students in high-poverty areas have lower achievement scores, higher drop-out rates, and lower rates of college graduation; that low-income families lack the resources to choose better schools; and that the gap between the rich and the poor is increasing. The Foundation has set its current funding interest to help address these critical matters of public concern. The Foundation is currently interested in making grants to support high quality education that prepares students in disadvantaged communities for educational and life success. Priority is given to organizations that provide support networks; remove impediments to student success; are broadly supported by the community, and have a record of demonstrated success. Additional guidelines and application forms are available at the foundation's website.

Requirements: Eligible applicant must be: nonprofit organizations; mission-consistent; fiscally sound; demonstrate a need; and capable and accountable.

Restrictions: The Foundation will not fund: grants that are not consistent with Foundation values; national or umbrella organizations that raise funds through broad-based solicitations to the general public; organizations that become overly dependent on the Foundation for on-going operational support; capital campaign gifts exceeding 20% of the campaign goal; political campaigns, events, or organizations whose purpose is to promote political candidates; lobbying; individuals.

Geographic Focus: Illinois, Kansas, Minnesota, Texas

Contact: Eileen A. O'Shaughnessy, Secretary; (952) 698-0959; fax (952) 698-0958; info@iaoshaughnessyfdn.org

Internet: http://www.iaoshaughnessyfdn.org/guidelines.htm

Sponsor: I.A. O'Shaughnessy Foundation
2001 Killebrew Drive, Suite 120
Bloomington, MN 55425

i2 Foundation Grants 2374

The foundation promotes advancements in education, technology, environmental practices, medicine, and economic opportunity through programs improving the quality of life and creating a better society. Although the foundation's main priority is the development and education of youth, proposals related to a wide array of issues are accepted, including illiteracy, youth violence, early childhood development, poverty, improved healthcare, and scientific research. All future grant cycles are currently postponed and new proposals are not being accepted.

Requirements: Nonprofit organizations closely aligned with the foundation's goals are eligible.
Restrictions: Individuals, religious institutions, political organizations, and government entities are ineligible.
Amount of Grant: 5,000 - 50,000 U.S.D
Contact: Administrator; (469) 357-4200; fax (469) 357-7777; Melis_Jones@i2.com
Internet: http://www.i2.com/company/i2foundation/index.cfm
Sponsor: i2 Foundation
One i2 Pl, 11701 Luna Road
Dallas, TX 75234

IAFF Burn Foundation Research Grants **2375**
The IAFF Burn Foundation Research Grant Program continues to lead to new knowledge and innovative approaches to the prevention and treatment of the physical and psychological problems that impair the quality of life for a burn patient. Research grant proposals are accepted annually and awards are distributed during the American Burn Association Annual Meeting. Topics of interest include by are not limited to pain management, physical and psychological rehabilitation, and wound healing and scarring, especially in children. Clinical projects are recommended. The foundation's Board of Medical Advisors oversees the selection process, reviewing each of the submissions. While all of candidates are deserving, only seven to ten grants are chosen and submitted to the IAFF Burn Foundation for consideration. Typically, five to seven are funded.
Requirements: The foundation welcomes research grant applications from members of the American Burn Association. The areas of funded research fall into three categories: (1) Quality of Life - This area of research is the most common, focusing on issues such as pain management, nutrition, and physical and psychological rehabilitation. The major focus in this area is the care of burned children. (2) Burn Prevention - Research in this area is especially important in shaping prevention programs. (3) Basic Science - Research grants funded in this category surround nutrition, wound healing and scar formation. No animal studies are funded. The required application and detailed guidelines can be found at the sponsor's website.
Restrictions: No animal studies are funded. The grant does not include indirect costs.
Geographic Focus: All States, Canada
Amount of Grant: 25,000 U.S.D
Contact: Patrick Morrison, Director; (202) 824-8620; fax (202) 637-0839; burnfoundation@iaff.org
Internet: http://burn.iaff.org/grants.shtml
Sponsor: International Association of Fire Fighters
1750 New York Avenue NW
Washington, DC 20006

IARC Expertise Transfer Fellowship **2376**
The Fellowship is intended to enable an established investigator to spend normally from six to twelve months in an appropriate host institute in a low- to medium-resource country in order to transfer knowledge and expertise in a research area relevant for the host country and related to epidemiology, biostatistics, environmental chemical carcinogenesis, cancer etiology and prevention, infection and cancer, molecular cell biology, molecular genetics, molecular pathology and mechanisms of carcinogenesis. Applications should include a proposed collaborative research project, specifying the link to IARC's on-going activities and a letter of support from the host lab giving details of feasibility and anticipated benefit to the receiving institute. Priority will be given to projects directly linked to IARC's on-going research program, involving at least one contact at IARC. There will be an annual remuneration of up to $70,000, which will take into account the on-going salary of the Fellow. This amount may include limited support for the project. The cost of travel will also be met.
Requirements: Applicants should be established cancer researchers actively engaged in the field with appropriate scientific or medical qualifications and an excellent publications' record. They must also belong to the staff of a university or a research institution.
Date(s) Application is Due: Nov 30
Amount of Grant: Up to 70,000 U.S.D
Contact: Dr. Paolo Boffetta, +33-(0)472-73-84-48; fax +33 (0)472-73-80-80; fel@iarc.fr or vsa@iarc.fr
Internet: http://www.iarc.fr/en/education-training/expertisetransfer.php
Sponsor: International Agency for Research on Cancer
150 cours Albert-Thomas
Lyon Cedex 08, 69372 France

IARC Postdoctoral Fellowships for Training in Cancer Research **2377**
Applications for training fellowships are invited from junior scientists from low- or medium-resource countries wishing to complete their training in those aspects of cancer research related to coordinate and conduct both epidemiological and laboratory research into the causes of cancer. Disciplines covered include: epidemiology, biostatistics, environmental chemical carcinogenesis, cancer etiology and prevention, infection and cancer, molecular cell biology, molecular genetics, molecular pathology and mechanisms of carcinogenesis, with emphasis on interdisciplinary projects. The fellowship is for a period of one year, with the possibility of an extension for a second year subject to satisfactory appraisal.
Requirements: Candidates should have spent less than five years abroad (including doctoral studies) and have finished their doctoral degree within five years of the closing date for application or be in the final phase of completing their doctoral degree (M.D. or Ph.D.). They must provide evidence of their ability to return to their home country and keep working in cancer research. The working languages at IARC are English and French. Candidates must be proficient in English at a level sufficient for scientific communication.
Restrictions: Candidates already working as a postdoctoral fellow at the Agency at the time of application or who have had any contractual relationship with IARC during the 6 months preceding the application deadline or who have already spent more than one year at IARC cannot be considered.
Date(s) Application is Due: Nov 30
Amount of Grant: 32,000 EUR
Contact: Dr. Paolo Boffetta, +33-(0)472-73-84-48; fax +33-(0)472-73-80-80; fel@iarc.fr or vsa@iarc.fr
Internet: http://www.iarc.fr/en/education-training/postdoc.php
Sponsor: International Agency for Research on Cancer
150 cours Albert-Thomas
Lyon Cedex 08, 69372 France

IARC Visiting Scientist Award for Senior Scientists **2378**
The IARC is offering this Award for a qualified and experienced investigator with recent publications in international peer-reviewed scientific journals who wishes to spend from six to twelve months at the IARC working on a collaborative project in a research area related to the Agency's programs: epidemiology, biostatistics, environmental chemical carcinogenesis, cancer etiology and prevention, infection and cancer, molecular cell biology, molecular genetics, molecular pathology and mechanisms of carcinogenesis. There will be an annual remuneration of up to $80,000, which will take into account the on-going salary of the visiting scientist plus the cost of travel.
Requirements: Applicants must belong to the staff of a university or a research institution and should provide written assurance of a post to return to at the end of the period of award.
Date(s) Application is Due: Nov 30
Amount of Grant: Up to 80,000 U.S.D
Contact: Dr. Paolo Boffetta, +33-(0)472-73-84-48; fax +33-(0)472-73-80-80; vsa@iarc.fr or vsa@iarc.fr
Internet: http://www.iarc.fr/en/education-training/vsa.php
Sponsor: International Agency for Research on Cancer
150 cours Albert-Thomas
Lyon Cedex 08, 69372 France

IATA Research Grants **2379**
In order to enhance the National Athletic Trainer's Association's (NATA) Research and Education Foundation, the Illinois Athletic Trainers' Association has provided limited funding for athletic training related research in the state of Illinois. Grant monies from the IATA are awarded for projects concerning research in the domains relevant to athletic training within the state of Illinois. Grant applications will be reviewed and monies will be awarded during the Spring and Fall IATA BOD meetings.
Requirements: Applicants must be Certified/Licensed, members in good standing not only with the NATA but also the IATA. Applications must be submitted 3 months prior to the beginning of the research study, and the research must be relevant to the profession of athletic training within the state of Illinois. All applications will be analyzed bi-annually and as stated above must be submitted prior to the beginning of the project. Please note that there is limited funding for these awards, and you may be turned down due to lack of funding.
Restrictions: There will be no awards granted for recovery of cost involved with research that has already been initiated.
Geographic Focus: Illinois
Contact: Mike Sullivan; (630) 853-0820; webmaster@illinoisathletictrainers.org
Internet: http://www.illinoisathletictrainers.org/Researchgrts.htm
Sponsor: Illinois Athletic Trainers' Association
701 Thomas Road
Wheaton, IL 60187

IATA Scholarships **2380**
Two scholarships of $1,000, three scholarships of $500, and two scholarships of $250 are available from the Illinois Athletic Trainers' Association (IATA). They are available on an annual basis but do not have to be awarded annually, depending on the applicant's qualifications. Scholarships will be awarded by the Vice-President at the IATA State Meeting Awards Luncheon. Certificates will be awarded to the scholarship recipient and their sponsoring Athletic Trainer, or in the case of the offspring scholarship, the parental member of the IATA.
Requirements: All applicants must have a grade point average of 'B,' or its equivalent, or above. All undergraduate and graduate applicants must be a member of the IATA. Applicants for the continuing education scholarship must be a licensed member of the IATA for one year prior to application. The offspring applicant does not need to be a member. Application forms are available from the IATA Vice President. Application requests must be written and will be included in the applicant's application packet. Recipients will be selected based on academic achievement, letters of recommendation and the applicant's autobiographical sketch.
Restrictions: If a committee member is a scholarship applicant, he/she will recuse from the evaluation process. The President-elect or Past-president will gather the committee evaluations.
Geographic Focus: Illinois
Date(s) Application is Due: Aug 1
Amount of Grant: 250 - 1,000 U.S.D
Contact: Mike Overturf; (630) 575-6212; fax (630) 575-7412; moverturf@athletico.com
Internet: http://www.illinoisathletictrainers.org/Scholarinfo.htm
Sponsor: Illinois Athletic Trainers' Association
701 Thomas Road
Wheaton, IL 60187

IBEW Local Union #697 Memorial Scholarships 2381

The International Brotherhood of Electrical Workers (IBEW) Local Union #697 Memorial Scholarships are awarded in memory of deceased members of IBEW Local #697 and benefits the children of Local #697. The number of awards vary, but the individual amount is $1,000. Any college major is eligible.

Requirements: Applicants must be dependent children, age 25 or under, of IBEW workers. Eligibility is determined with a copy of a current tax return indicating the child/stepchild is a dependent. Applicants must be registered for at least 12 hours in the school of his or her choice. The application is available at the Local Union office. Recipients are chosen at the Union meeting held on the third Monday of July.

Geographic Focus: Indiana

Date(s) Application is Due: Jun 1

Amount of Grant: 1,000 U.S.D

Contact: IBEW Memorial Scholarship Contact; (219) 736-1880; fax (219) 736-1940; legacy@legacyfoundationlakeco.org

Internet: http://www.legacyfoundationlakeco.org/scholarships/index.htm

Sponsor: Legacy Foundation
1000 East 80th Place, 302 South
Merrillville, IN 46410

ICCP Outstanding Young Investigator Award 2382

The annual award recognizes the recent contributions of a promising young investigator and, in so doing, encourages and facilitates his/her lifelong commitment to the field of spinal cord repair. The award is intended to encourage young scientists to devote their careers to spinal cord research; promote and enrich the international research field in repair of the damaged spinal cord; and stimulate collaborative research in the field. The award supports inter-laboratory travel and expenses to complete a collaborative new study or a clearly defined project to learn a specific technique. The goal of the award is to benefit the grantee and the field for a spinal cord study or learning experience. Application and guidelines are available online.

Requirements: To be eligible, a candidate must be a postdoctoral fellow who possesses a doctoral degree at the time of proposal submission and who is in his/her first to seventh year of postdoctoral work. The winning nominee must be a postdoctoral fellow during the 12-month award period, and the study must be completed within one year of the award.

Date(s) Application is Due: Sep 1

Amount of Grant: 10,000 U.S.D

Contact: Dr. John Steeves, (604) 822-2924; steeves@icord.org

Internet: http://www.campaignforcure.org/Young%20Investigator%20Award.htm

Sponsor: International Campaign for the Cure for Spinal Cord Paralysis
ICORD at UBC, 2469-6270 University Boulevard
Vancouver, BC V6T 1Z4 Canada

Ida Alice Ryan Charitable Trust Grants 2383

The trust awards single-year grants for capital projects to eligible Georgia nonprofit organizations. Grants support programs of the Catholic Church and a wide variety of community causes including health organizations, legal aid societies, youth groups, educational programs, museums, and social services. Application and guidelines are available online.

Requirements: 501(c)3 tax-exempt organizations in the Atlanta area are eligible.

Geographic Focus: Georgia

Date(s) Application is Due: Feb 1; Aug 1

Contact: NA Trustee, c/o Wachovia Bank; grantinquiries8@wachovia.com

Internet: https://www.wachovia.com/foundation/v/index.jsp?vgnextoid=cf578689fb0aa110VgnVCM1000004b0d1872RCRD&vgnextfmt=default

Sponsor: Ida Alice Ryan Charitable Trust
3280 Peachtree Road NE, Suite 400, MC G0141-041
Atlanta, GA 30305

Idaho Community Foundation Eastern Region Competitive Grants 2384

The mission of the Foundation is to enrich life's quality throughout Idaho. Grants are awarded through the regional grant cycle for a wide range of organizations and for a wide range of projects consistent with that mission. Grants are made to fund activities, services, and projects of established organizations, as well as to provide assistance for new organizations to fill unmet and/or emerging community needs. Grant areas include: arts and culture; education; emergency services; libraries; natural sciences; health; recreation; social services; and public projects. Application information is available online.

Requirements: Nonprofit entities in Idaho are eligible to apply.

Restrictions: Funding is not available for: projects which replace school district responsibilities to students or that fund state or federally mandated programs; projects which are considered operating expenses or salaries normally paid by a school district; computer hardware used solely for pre-K through 12th grade educational purposes; religious organizations for the sole purpose of furthering that religion; political activities or those designed to influence legislation; national organizations; or grants that directly benefit a donor to ICF or a donor's family.

Geographic Focus: Idaho

Date(s) Application is Due: Apr 1

Amount of Grant: 5,000 U.S.D

Contact: Administrator; (208) 342-3535; fax (208) 342-3577; grants@idcomfdn.org

Internet: http://www.idcomfdn.org/pages/grant_regional_guidelines.htm

Sponsor: Idaho Community Foundation
P.O. Box 8143
Boise, ID 83707

Idaho Power Company Corporate Contributions 2385

The company awards grants in company operating territories in its areas of interest, including arts and culture, civic and community, education, and health and human services. Types of support include building construction/renovation, capital campaigns, continuing support, equipment, general operating support, matching gifts, scholarships, and sponsorships. The corporate contribution request form is available online. There are no application deadlines.

Requirements: Nonprofits in southern Idaho and eastern Oregon are eligible.

Restrictions: Grants do not support: individuals; loans or investments; churches or religious organizations for purposes of religious advocacy; tickets for contests, raffles, or other prize-oriented activities; organizations that discriminate for any reason, including race, color, religion, creed, age, sex, or national origin; individual school programs or projects with limited participation; fraternal or labor organizations; unrestricted operating funds; special occasion good-will advertising.

Geographic Focus: Idaho, Oregon

Contact: Contribution Program Manager; (208) 388-2200

Internet: http://www.idahopower.com/aboutus/community/corporateContributions.htm

Sponsor: Idaho Power Company
P.O. Box 70
Boise, ID 83707

Ida S. Barter Trust Grants 2386

The Ida S. Barter Trust was established in 1953 to support and promote quality educational, human services, and health care programming for underserved populations. Grant requests for general operating support are strongly encouraged. Program support will also be considered. Small, program-related capital expenses may be included in general operating or program requests. The application deadline is April 1, and applicants will be notified of grant decisions before June 30. The majority of grants from the Barter Trust are 1 year in duration. On occasion, multi-year support is awarded.

Geographic Focus: Massachusetts

Date(s) Application is Due: Apr 1

Contact: Michealle Larkins, Vice President; (866) 778-6859; michealle.larkins@baml.com

Internet: https://www.bankofamerica.com/philanthropic/fn_search.action

Sponsor: Ida S. Barter Trust
225 Franklin Street, 4th Floor, MA1-225-04-02
Boston, MA 02110

Iddings Foundation 2387

Grants are given to Ohio nonprofits for operating budget support, capital improvements, upper-class collegiate scholarships to designated institutions, expansion of programs, and equipment in the areas of physical and mental health; education; culture and the environment; community welfare; and human services. Types of support include challenge/matching grants, equipment acquisition, demonstration grants, development grants, seed grants, annual campaigns, conferences and seminars, publication, and scholarship funds. and project grants. Proposals are reviewed quarterly; the distribution committee meets the month following the application deadline date.

Requirements: Applicants must be Ohio-based, tax-exempt organizations whose purpose is to improve the community environment and lives of the citizens, and whose primary focus is on the greater Dayton area.

Restrictions: Grants are not made to individuals or to organizations outside Ohio. Additionally, there will be no endowment support.

Geographic Focus: Ohio

Date(s) Application is Due: Mar 1; Jun 1; Sep 1; Nov 1

Amount of Grant: 500 - 50,000 U.S.D

Samples: AIDS Foundation of Miami Valley (OH)—for staff support over two years, $19,125;Miami Valley Literacy Council (OH)—for salary support of a curriculum director, $5000.

Contact: Maribeth Graham, Administrator; (937) 224-1773; fax (937) 224-1871

Sponsor: Iddings Benevolent Trust
Ketting Tower, Suite 1620
Dayton, OH 45423

IDPH Emergency Medical Services Assistance Fund Grants 2388

This project provides for distribution of moneys in the EMS Assistance Fund to each of the eleven Regions in the State in accordance with protocols established in each Region's EMS Region Plan. Objectives of this grant will be to purchase any equipment requested and complete any education as requested. Expected outcomes and goals of this grant are to help improve EMS services and increase education to EMS personal. Measurements and outcomes of this grant will be met by showing timely response to EMS calls, proper use of equipment, education of new equipment, educational objectives completed. Funds might not be equally divided among the eleven regions; consequently, award decisions will not be made based on financial parity among regions.

Requirements: Any Illinois licensed/designated EMS participant that provides EMS service within the State of Illinois may apply for funds through their Regional EMS Advisory Committee. Programs, services, and equipment funded by the EMS Assistance Fund must comply with the Emergency Medical Services (EMS) Systems Act and the Regional EMS Plan in which the applicant participates. The grant cycle runs from July 1-June 30 of each year. All funds remaining at the end of the period of time grant funds are available for expenditure (June 30 of the fiscal year the grant was awarded) shall be returned to the State within 45 days. All applications from providers must be submitted to their respective Regional EMS Advisory Committee by the deadline required by each Regional Committee. No applications will be accepted by the Department directly from an applicant.

Restrictions: Due to limited amount of grant funds available, the Department will not consider applications for new vehicles, vehicle re-chassis, building projects or grant requests over $5000.
Geographic Focus: Illinois
Date(s) Application is Due: Jan 29
Amount of Grant: Up to 5,000 U.S.D
Contact: Mark Vassmer, PHEP Grant Coordinator; 217-558-0560
Internet: http://www.idph.state.il.us/fundop.htm
Sponsor: Illinois Department of Public Health
525-535 W. Jefferson Street
Springfield, IL 62761-0001

IDPH Hosptial Capital Investment Grants **2389**
The program allows qualifying hospitals to apply for grants to fund projects to improve or renovate a hospital's physical plant, or to improve, replace or acquire equipment or technology. Projects can include activities to satisfy building code, safety standards, or life safety code; maintain, improve, renovate, expand, or construct buildings or structures; maintain, establish, or improve health information technology; or maintain or improve safety, quality of care, or access to care. Two types of awards are available: Safety Net Grants ($4,600,000 to $7,000,000) and Community Hospital Grants (approximately $350,000 to $1,000,000).
Requirements: Prior to submitting a grant application, hospitals need to submit a letter of intent to the Department which must be received at least 10 calendar days prior to the submission of a grant application. The letter will help determine if a hospital qualifies to apply and will notify the Department of an impending application. The letter must contain the following information: name of the applicant; name of the hospital where grant funds will be used; site of the proposed project, including the address of the hospital where grant funds will be used; county where the hospital is located; description of the project; hospital's Medicaid inpatient utilization rate for the rate year beginning October 1, 2008; signature and contact information of an authorized official from the hospital; and, information on whether the project requires a CON or COE from the Health Facilities and Services Review Board. (The CON/COE Assessment of Applicability Internet site can assist in this determination: http://www.hfsrb.illinois.gov/pdf/checklist-revised.doc.) A hospital that applies for this grant shall be licensed by the Illinois Department of Public Health in accordance with the Hospital Licensing Act. The license shall be valid and the hospital shall be in operation when the grant application is submitted, when the grant agreement is executed and when the project is complete. Applications must be received by the Department within 120 calendar days of the date of the notice's publication. Applications received after this deadline will not be accepted.
Geographic Focus: Illinois
Date(s) Application is Due: Nov 12
Amount of Grant: 350,000 - 7,000,000 U.S.D
Contact: Grant Coordinator; 217-782-1624; dph.crh@illinois.gov
Internet: http://www.idph.state.il.us/fundop.htm
Sponsor: Illinois Department of Public Health
535 West Jefferson Street, Ground Floor
Springfield, IL 62761-0001

IDPH Local Health Department Public Health Emergency Response Grants **2390**
The Illinois Department of Public Health, Office of Preparedness and Response, Division of Disaster Planning and Readiness, is making available approximately $5 million in unspent CDC Public Health Emergency Response (PHER) grant funding to Illinois local health departments. The purpose of the funding is to continue previously approved PHER-funded activities, or conduct new activities that address or retest identified gaps in local health department pandemic flu response. Many local health departments have indicated that they could not use the original PHER funding for CDC's intended purpose. Therefore, grants will be awarded on a first-come, first-served basis to applicants that can best justify activities that will advance pandemic planning and preparedness based on the Grant Application Review Criteria. IDPH anticipates making about 20 to 40 awards based on the number and type of applicants. The maximum award for either single or multi-jurisdictional will be $300,000, or $1.70 per capita, per participating local health department, whichever is less.
Requirements: Applicant must be a single Illinois Certified Local Health Department. Regional partnerships may select either one certified local health department or an agent to receive the funding from IDPH and issue subgrants to other participating certified local health departments. Application must clearly describe reasonable and significant programmatic and financial participation by all partners. Funds may be used for the continuation of previously approved PHER-funded activities, or new activities that address or retest identified gaps in pandemic flu response, or new activities that directly advance pandemic planning and preparedness.
Geographic Focus: Illinois
Date(s) Application is Due: Jan 31
Amount of Grant: Up to 300,000 U.S.D
Contact: Mark Vassmer, PHEP Grant Coordinator; 217-558-0560
Internet: http://www.idph.state.il.us/fundop.htm
Sponsor: Illinois Department of Public Health
525-535 W. Jefferson Street
Springfield, IL 62761-0001

IEEE Canadian Foundation Special Grants **2391**
The foundation occasionally awards special grants; meetings are held quarterly (March, June, September, December) to consider applications for these grants. Applications should be received two weeks prior to the meetings. Some examples of special grants that have been awarded are the Ontario Engineering Design Competition, BEAM Robotic Olympics, and Control of Artificial Limbs at Ontario Crippled Children's Centre. Applications are available online.
Requirements: Anyone or any group with a proposal that is consistent with the mission of the foundation is eligible to apply.
Samples: Carleton University (Ottawa, Canada)—to provide travel and living costs for six attendees to attend the 12th International Conference of Women Engineers and Scientists, 2002.
Contact: Grants Administrator; (705) 743-7712; fax (705) 743-7712; grants@ieeecanadianfoundation.org
Internet: http://ewh.ieee.org/r7/cafoundation/EN/sp_grnts.htm
Sponsor: Institute of Electrical and Electronics Engineers Canadian Foundation
456 Rogers Street
Peterborough, ON K9H 1W9 Canada

Ignacio Martin-Baro Fund for Mental Health and Human Rights Grants **2392**
The organization supports children's programs in health and mental health and is a resource for advocacy, education, prevention, service, and action research. The fund has a strong preference for projects that are developing innovative and progressive ways to deal with the mental health consequences of violence and political repression. Applicants should understand that the foundation does not view mental health and human rights as issues that can be addressed separately, but are seeking to fund activities that explore the links between them. Groups with budgets under $25,000 receive priority consideration. The listed application deadline applies to letters of interest; full proposals are by invitation.
Date(s) Application is Due: Sep 1
Amount of Grant: Up to 7,000 U.S.D
Contact: Grants Administrator; (617) 469-7454; fax (617) 469-3379; apply@martinbarofund.org
Internet: http://www.martinbarofund.org/contact/applying.htm
Sponsor: Ignacio Mart-Bar Fund for Mental Health and Human Rights
P.O. Box 2122
Jamaica Plain, MA 02130

IHS Native American and Alaskan Native Health Research Centers **2393**
The purpose of this program is to establish centers that would encourage research on diseases that affect Native American populations and other health issues of importance to tribes and tribal organizations. Funds may be used to reduce health disparities that affect Native Americans and Alaskan Natives, train scientists and health professionals, help encouraging partnerships among health organizations, and for health research. A letter of intent is requested by October 1; full application is due by December 12. Annual deadline dates may vary; contact program staff for exact dates. RFA: GM-00-007
Requirements: Federally recognized tribes, tribal consortia, and nonprofit Indian health organizations are eligible to apply.
Date(s) Application is Due: Oct 1; Dec 12
Amount of Grant: 27,000 - 300,000 U.S.D
Contact: Lois Hodge, Grants Management Officer; (301) 443-5204
Internet: http://www.ihs.gov
Sponsor: Indian Health Service
12300 Twinbrook Plaza
Rockville, MD 29852

IIE 911 Armed Forces Scholarships **2394**
The 911 Armed Forces Scholarships provide scholarships for the dependent children of active duty U.S. military personnel who were killed in the September 11, 2001 terrorist attacks. Scholarships are one-time awards for study leading to a certificate, associate's or bachelor's degree in the United States or its equivalent at an accredited post-secondary institution in any country.
Requirements: Dependent children pursuing a certificate or undergraduate degree programs in a college or university in the U.S. or abroad are eligible to apply.
Restrictions: Awards will not be provided for study leading to degrees higher than a bachelor's degree.
Contact: Jonah Kokodyniak, Deputy Vice President; (212) 883-8200; fax (212) 984-5401; development@iie.org
Internet: http://www.iie.org/Programs/911-Armed-Forces-Scholarship-Fund
Sponsor: Institute of International Education
809 United Nations Plaza
New York, NY 10017-3580

IIE Adell and Hancock Scholarships **2395**
The Institute of International Education (IIE) sponsors the Adell and Hancock Scholarship. The purpose of the Adell & Hancock Fund is to provide supplemental support to U.S. and international students who are in need of additional funds to carry out their international educational plans. The amount awarded will depend on individual need as determined by IIE/RMRC scholarship committee to not exceed $2,400 per candidate. Applicants should refer to the IIE website for further information and current application information.
Geographic Focus: All States, All Countries
Date(s) Application is Due: Oct 15
Amount of Grant: Up to 2,400 U.S.D
Contact: Lauren Granstrom, Administrative Assistant; (303) 837-0788, ext. 27 or (303) 837-1409; rockymountainscholarships@iie.org or lgranstrom@iie.org
Internet: http://www.rockymountainiie.org/scholarships
Sponsor: Institute of International Education
809 United Nations Plaza
New York, NY 10017-3580

IIE AmCham Charitable Foundation U.S. Studies Scholarship 2396
The Institute of International Education sponsors the AmCham Charitable Foundation U.S. Studies Scholarship. The scholarship funds undergraduate study at a U.S. university for high school students from Hong Kong.
Requirements: Applicants must be permanent residents of Hong Kong; they must have received at least five years of continuous education in a Hong Kong high school where they are currently enrolled; and they must have gained admission to an undergraduate degree program in a U.S. college or university for the coming year. Current applications and deadlines are found on the website.
Restrictions: Only current residents of Hong Kong may apply.
Geographic Focus: Hong Kong
Date(s) Application is Due: Apr 15
Amount of Grant: 16,000 HKD
Contact: Linda Pham, Scholarship Contact; (852) 2530-6917; fax (852) 2810-1289; lpham@amcham.org.hk or amcham@amcham.org.hk
Internet: http://www.amcham.org.hk/index.php/U.S.-Studies-Scholarship.html
Sponsor: Institute of International Education
809 United Nations Plaza
New York, NY 10017-3580

IIE Brazil Science Without Borders Undergraduate Scholarships 2397
The Brazilian government's new Science Without Borders Program will provide scholarships to undergraduate students from Brazil for one year of study at colleges and universities in the United States. This program, administered by IIE, is part of the Brazilian government's larger initiative to grant 100,000 scholarships for the best students from Brazil to study abroad at the world's best universities.
Requirements: Applicants should meet the following criteria for eligibility: they should be citizens of Brazil, and reside in the home country at the time of application; they must have completed four semesters of their bachelor's degree; they must be in the first quarter of their graduation class, have superior academic abilities, well-rounded personalities, and participate in extracurricular activities; their transcripts should show a consistently high level of performance; and applicants should submit their TOEFL score report, and present official and translated transcripts from an institution recognized by the Ministry of Higher Education.
Restrictions: Scholarships will be given primarily to students in the Science, Technology, Engineering and Mathematics (STEM) fields. Students in the program will return to Brazil to complete their degrees. Candidates with U.S. or dual citizenship of the United States and Brazil are not eligible for this program.
Geographic Focus: Brazil
Contact: Edward Monks, Enrichment and Professional Development Director; (212) 984-5335; fax (212) 984-5578; undergraduatprogram@iie.org
Shahreen Rahman, Program Officer; (212) 984-5335; fax (212) 984-5578; undergraduatprogram@iie.org
Internet: http://www.iie.org/Programs/Brazil-Science-Without-Borders
Sponsor: Institute of International Education
809 United Nations Plaza
New York, NY 10017-3580

IIE CESRI Summer Research Fellowship 2398
The Central Europe Summer Research Institute (CESRI) is a fellowship opportunity for U.S. graduate students in science and engineering. With support from the National Science Foundation (NSF) and the German Academic Exchange Service (DAAD), CESRI will provide eight U.S. graduate students with a high-quality international research experience in Austria, the Czech Republic, Germany, Hungary, Poland or Slovakia. Two CESRI grants are specifically for CESRI Fellowships to Germany. The remaining six CESRI grants are funded by NSF for the entire Central Europe region, with preference for grants to countries other than Germany. The program is intended for scientific research projects. The CESRI program will be eight weeks in length, with the first four days spent as a group in Budapest, Hungary in a specially designed cultural and academic orientation to the region. Participants will spend the remaining 7.5 weeks working in individually-arranged placements in university labs or other appropriate sites where they can participate in creative research activities under the supervision of European mentors. In addition to the scholarship, recipients receive round trip airfare to Central Europe, room and board for 8 weeks, health and accident insurance, and a four day academic and cultural orientation in Budapest.
Requirements: All applicants must be U.S. citizens or permanent residents in the U.S., and be current Master's or Ph.D. students at a U.S. university in one of the following fields: Biology, Chemistry, Computer Science, Engineering, Environmental Science, or Mathematics. Candidates should submit the online application, along with the online self-evaluation form and two completed professional reference forms.
Restrictions: Projects involving field work, policy, ethics, or clinical research on human objects are not eligible for consideration.
Date(s) Application is Due: Feb 1
Amount of Grant: 2,000 U.S.D
Contact: CESRI Fellowship Contact; (212) 883-8200; fax (212) 984-5452
Internet: http://www.iie.org/Programs/CESRI
Sponsor: Institute of International Education
809 United Nations Plaza
New York, NY 10017-3580

IIE Chevron International REACH Scholarships 2399
The Chevron International REACH (Recognizing Excellence and Achievement) Scholarship Program is sponsored by Chevron for the sons and daughters of their employees and retirees. The program was established to recognize and assist outstanding children who plan to pursue post-secondary education. Renewable scholarships are offered each year for full-time undergraduate study at colleges, universities, and vocational schools. Four-year scholarships, ranging in value from $500 to $2,500, and honorariums, ranging from $1,000 to $1,500, will be offered to children of Chevron employees. The scholarships for children of non-U.S.-payroll employees are managed by the Institute of International Education West Coast Center in San Francisco. Application forms are available through the applicant's local Chevron human resources office.
Requirements: Applicants for the scholarship must be children of non-U.S. payroll active employees of Chevron and its wholly owned subsidiaries; children of non-U.S. payroll retirees with a minimum of 75 points as calculated by Human Resources or 20 years of service prior to retirement; the retired parent must not be an employee of competitive energy company; and the applicant must be currently enrolled in the final year of high school and planning to attend a college, university, or vocational-technical school as a full time student.
Restrictions: Students planning pre-university study are not eligible. Military academies are not eligible. Scholarships are intended to assist students as they pursue undergraduate study and are not intended to cover all educational expenses. Recipients are responsible for the balance of funding, including tuition and fees, books and supplies, living expenses, and transportation.
Geographic Focus: Angola, Argentina, Australia, Azerbaijan, Bangladesh, Brazil, Cambodia, Canada, Chad, China, Colombia, Indonesia, Kazakhstan, Kuwait, Netherlands, New Zealand, Nigeria, Philippines, Russia, Saudi Arabia, Singapore, South Africa, South Korea, Thailand, Trinidad and Tobago, United Kingdom, Venezuela, Vietnam
Amount of Grant: 500 - 2,500 U.S.D
Contact: Naoko Dunnigan, Director, International Scholarship Program; (415) 362-6520, ext. 201; fax (415) 392-4667; Chevron@iie.org
Internet: http://www.iie.org/Programs/Chevron-International-REACH-Scholarship-Program
Sponsor: Institute of International Education
809 United Nations Plaza
New York, NY 10017-3580

IIE David L. Boren Graduate Fellowships 2400
Boren Fellowships provide funding to U.S. graduate students to add an international and language component to their graduate education through specialization in area study, language study, or increased language proficiency. Boren Fellowships support study and research in areas of the world that are critical to U.S. interests, including Africa, Asia, Central & Eastern Europe, Eurasia, Latin America, and the Middle East. Boren Fellowships are funded by the National Security Education Program (NSEP), which focuses on geographic areas, languages, and fields of study that are critical to U.S. national security. Applicants should identify how their projects, as well as their future academic and career goals, will contribute to U.S. national security.
Requirements: Boren Fellowships promote long term linguistic and cultural immersion. Therefore, applicant preference will be given to those proposing overseas programs of 6 months or longer. However, applicants proposing overseas programs of 3-6 months, especially those in the STEM (science, technology, engineering, and mathematics) fields are encouraged to apply. Boren Fellowships are awarded with preference for countries, languages, and fields of study critical to U.S. national security. Preference is also given to students who will study abroad for longer periods of time, and who are highly motivated by the opportunity to work in the federal government. Applicants should refer to the website for current deadlines and detailed criteria for application.
Restrictions: Applicants must commit to a length of study of at last 12 weeks.
Amount of Grant: Up to 30,000 U.S.D
Contact: Michael Saffle; (800) 618-6737; fax (202) 326-7672; boren@iie.org
Internet: http://www.borenawards.org/program
Sponsor: Institute of International Education
809 United Nations Plaza
New York, NY 10017-3580

IIE Eurobank EFG Scholarships 2401
The Eurobank EFG Scholarship Program aims to identify and honor the best undergraduate students of Serbian state universities. An independent selection committee of experts evaluate all applications. Semi-finalists are invited to a personal interview conducted in English. The winners are publicly recognized at an award ceremony, with their names published in daily newspapers.
Requirements: The ideal candidate has high academic standing, leadership potential, and an interest in community development. Participation in extracurricular activities, good communication skills, and proficiency in foreign languages are also essential qualities. All applicants must have: a grade point average higher than 9.5; extracurricular activities; a demonstrated interest in community development with leadership potential; good communication skills; and proficiency in foreign languages.
Geographic Focus: Serbia
Amount of Grant: 1,000 EUR
Contact: Eurobank EFG Scholarship Team; (+36-1) 472-2283; fax (+36-1) 472-2255; eurobankefg-scholarship@iie.eu
Internet: http://www.iie.org/en/Programs/Eurobank-EFG-Scholarships
Sponsor: Institute of International Education
809 United Nations Plaza
New York, NY 10017-3580

IIE Freeman Foundation Indonesia Internships 2402
The Freeman Indonesia Nonprofit Internship (FINIP) addresses the limited knowledge among Indonesian students about the nonprofit sector. It also addresses the limited understanding and exposure of U.S. students to Indonesia, despite the increasing

importance of the country globally. The Freeman Indonesia Nonprofit Internship Program is an opportunity to strengthen the leadership of Indonesia's future nonprofits and deepen ties between America and Indonesia. The shared experience of collaborating to assist a local Indonesian nonprofit is a learning experience and a vehicle for cross-cultural understanding, building ongoing friendships and shared interests in the nonprofit sector in the U.S. and Indonesia.
Requirements: Applicants must meet all of the following eligibility *Requirements:* they must be a U.S. or Indonesian citizen; currently enrolled as a full time sophomore or junior pursuing their first bachelor's degree at a U.S. university; and be in good academic standing at their university. Potential applicants should refer to the website for specific instructions about the application process.
Geographic Focus: All States, Indonesia
Date(s) Application is Due: Mar 1
Contact: FINIP Contact; (212) 984-5542; fax (212) 984-5325; finip@iie.org
Internet: http://www.iie.org/Programs/FINIP
Sponsor: Institute of International Education
809 United Nations Plaza
New York, NY 10017-3580

IIE Hewlett Foundation/IIE Dissertation Fellowship **2403**
The Hewlett/IIE Dissertation Fellowship in Population, Reproductive Health, and Economic Development provides financial and research development support for dissertations on topics that examine how population dynamics, family planning, and reproductive health influence economic development. This can include economic growth, poverty reduction, and equity. Dissertations that address population and development issues pertinent to the African continent are especially encouraged. Fellowship recipients are awarded up to $20,000 per year (depending on tuition, research expenses, and cost of living) for a total of two years to cover expenses incurred while working on their dissertation. These expenses must be clearly specified on the budget component of the fellowship application. In addition to financial support, Fellows will actively engage with a network of researchers supported by the Hewlett Foundation, Population Reference Bureau, the Institute of International Education, and other funders. Network activities during the two-year fellowship can include an annual research conference, workshops on advanced methods in population-economic analysis, and writing workshops.
Requirements: The fellowship is intended for doctoral students enrolled in economics, economic demography, geography, and epidemiology, whose dissertation addresses population and development issues. All applicants are required to submit a hard copy and electronic application, an application cover sheet, vitae, statement of intent, budget plan, dissertation schedule, personal essay, and two recommendation letters. Students should refer to the website for specific information regarding the requirements and current due dates.
Restrictions: Students must have completed their graduate coursework by the start of the fellowship, and must be studying at a university in sub-Saharan Africa, the U.S., or Canada.
Geographic Focus: All States, Angola, Benin, Botswana, Burkina Faso, Cameroon, Canada, Cape Verde, Central African Republic, Chad, Comoros, Congo, Congo, Democratic Republic of, Cote d' Ivoire (Ivory Coast), Djibouti, Equatorial Guinea, Eritrea, Ethiopia, Gabon, Gambia, Ghana, Guinea, Guinea-Bissau, Kenya, Lesotho, Liberia, Madagascar, Malawi, Mali, Mauritania, Mauritius, Mozambique, Namibia, Niger, Nigeria, Reunion, Rwanda, Sao Tome & Principe, Senegal, Seychelles, Sierra Leone, Somalia, South Africa, Sudan, Swaziland, Tanzania, Togo, Uganda, Western Sahara, Zambia, Zimbabwe
Amount of Grant: Up to 20,000 U.S.D
Contact: Nancy Scally, Fellowship Contact; (212) 984-5342; popecondissfellows@iie.org
Internet: http://www.iie.org/Programs/Hewlett-IIE-Fellowship
Sponsor: Institute of International Education
809 United Nations Plaza
New York, NY 10017-3580

IIE Iraq Scholars and Leaders Scholarships **2404**
The scholarship is available to Iraqi citizens residing in Iraq. The program supports Bachelor's, Master's, or Doctoral levels of study at a U.S. university or college in one of the following fields: business, economics, education administration, engineering, geology, geophysics, law, public health, or public policy. Preference will be given to candidates pursuing studies in business, engineering or geosciences. Selected scholars will receive the following benefits: full tuition and placement assistance to attend a U.S. college or university; visa and academic support; transportation to the U.S. at the beginning of the program and return to Iraq at the end of the program; an extensive pre-academic program and medical insurance while in the U.S. The application deadline is November 1, with study beginning in the U.S. the following June.
Requirements: Applicants should carefully review the requirements for the scholarship. Candidates should submit a completed nine page application, TOEFL scores, resume, transcript, statement of purpose, field of study essays, and letters of recommendation.
Restrictions: Scholarship policy does not allow for a change in field of study once it has been submitted.
Geographic Focus: Iraq
Date(s) Application is Due: Nov 1
Contact: Iraq Scholars and Leaders Scholarship; (713) 621-6300, ext. 32; fax (713) 621-0876; islp@iie.org
Internet: http://www.iie.org/Programs/ISLP
Sponsor: Institute of International Education
1800 West Loop South, Suite 250
Houston, TX 77027

IIE Klein Family Scholarship **2405**
The Klein Family Scholarship is available to high school students from Hungary or another central or eastern European countries to pursue their four-year Bachelor's degree at The University of the South, located in the United States in Sewanee, Tennessee. The purpose of the scholarship is to provide full financial support for students to accomplish their educational goals and develop their potential. The Klein Family Scholarship will provide a full, four-year scholarship to one of these students.
Requirements: In order to qualify for the scholarship, candidates must be citizens of Hungary or of another central or eastern European country; be in their last year of high school and be able to pursue a four-year degree at Sewanee the following year; and have high academic achievements and leadership potential. The Institute of International Education's European Office (IIE) is managing the prescreening process for the Klein Family Scholarship. IIE will preselect 4 to 5 promising candidates, who will then be invited to apply to Sewanee directly. The ultimate scholarship winner will be selected by Sewanee's Admissions Office, as part of the college's regular annual admissions process.
Restrictions: Candidates must have a high level of English language skills adequate for study at a U.S. college.
Geographic Focus: Hungary
Date(s) Application is Due: Dec 15
Contact: Nora Nemeth, Program Assistant; (+36) 1 472-2297; fax (+36) 1 472-2294; nnemeth@iie.eu
Internet: http://www.iie.org/Programs/Klein-Family-Scholarship
Sponsor: Institute of International Education
809 United Nations Plaza
New York, NY 10017-3580

IIE Leonora Lindsley Memorial Fellowships **2406**
The Lindsley Scholarship was established for Leonora Lindsley, an American member of an international all-female combat unit on the Western Front during World War II, who was killed the day before the war officially ended. The scholarship was created in her honor by French citizens who are descendants from Resistance fighters to fund those pursuing educational studies in the United States. An award is granted to one or two students for graduate study for a 12 month period, and is renewable for a second year. The winner(s) are chosen every other year. The amount of the award varies from year to year, but ranges from $10,000 to $12,000.
Requirements: The Fulbright Commission in Paris gathers applications and sends them to IIE where they are reviewed for completeness and eligibility. Once this is done, a selection panel is convened and finalists chosen. Any major is eligible.
Restrictions: Only descendants of French Resistance fighters are eligible. Grantees must be at an accredited U.S. university.
Geographic Focus: France
Contact: Leonara Lindsley Fellowship Contact; (212) 984-5552
Internet: http://www.iie.org/Programs/Leonora-Lindsley-Memorial-Fellowship
Sponsor: Institute of International Education
809 United Nations Plaza
New York, NY 10017-3580

IIE Lingnan Foundation W.T. Chan Fellowship **2407**
The W.T. Chan Fellowships commemorate Professor Wing-Tsit Chan, former Dean of Lingnan University and distinguished Professor of Chinese Philosophy and Religion. The Fellowships extend the Foundation's commitment to higher education, increased international understanding, and personal growth. In addition, the fellowships address and explore the Lingnan motto "Education for Service." It is a step in a movement towards strengthening linkages between academia and communities.
Requirements: The fellowships are available to upper-level undergraduates and postgraduates of Sun Yat-sen University in Guangzhou and Lingnan University in Hong Kong. Applicants are encouraged to view the Lingnan Foundation web link for detailed videos and blogs about requirements in the application process. After their selection and orientation, each Fellow is assigned to work at non-profit organization in the United States. With support and guidance from program organizers, the internships are designed to help the Fellows gain practical experience in service work and facilitate inter-cultural cooperation. Fellows live in American homes for the duration of the program. While in the U.S., they also attend seminars on non-profit organization management and community development, cultural events, and weekly reflection meetings. Fellows are assigned to one of two sites in California. In Los Angeles, the program is managed by the Dashew Center for International Students and Scholars at the University of California-Los Angeles (UCLA). In Berkeley, the program is hosted by UC Berkeley Cal Corps Public Service Center.
Geographic Focus: China, Hong Kong
Date(s) Application is Due: Jan 31
Contact: Maria Luk, W.T. Chan Fellowships; +85 2 2103 1502; fax +85 2 2603 5765; wtchan!iiehongkong.org or scholarships@iiehongkong.org
Internet: http://www.iie.org/Programs/Lingnan-WT-Chan-Fellowships
Sponsor: Institute of International Education
809 United Nations Plaza
New York, NY 10017-3580

IIE Lotus Scholarships **2408**
The LOTU.S. Scholarships offers four year scholarships to high school Egyptian students. These scholarships are available for students to attend private Egyptian universities in fields of study that are important to Egypt's current and future development. The LOTU.S. Scholarships aim to educate young men and women so that they can become future leaders and passionate about their national and local communities. Recipients will attend one of the

five private participating universities: Ahram Canadian University in Sixth of October City; British University in Shorouk City; Future University in New Cairo; Modern Sciences and Arts University in Sixth of October City; or Pharos University in Alexandria.
Requirements: The LOTU.S. Scholarship Program seeks Egyptian applicants in their last year of public or Azhari high school who have outstanding credentials and high financial need. All majors are eligible to apply. Candidates should complete an application form in Arabic according to the requirements and guidelines in the application packet.
Restrictions: Egyptian citizens who are also U.S. citizens or holders of a U.S. Green Card are not eligible.
Geographic Focus: Egypt
Date(s) Application is Due: Jun 5
Contact: Lotus Scholarship Contact; +20 (02) 2524-2172 or +20 (02) 2528-5779; fax +20 (02) 2524-2175; MENA@iie-egypt.org or lotus@iie-egypt.org
Internet: http://www.iie.org/Programs/U.S.AID-Lotus-Scholarship-Program-English/About
Sponsor: Institute of International Education
809 United Nations Plaza
New York, NY 10017-3580

IIE Mattel Global Scholarship 2409

The Mattel Children's Foundation sponsors a scholarship program designed to assist and encourage children of Mattel employees worldwide to pursue higher education (see Mattel locations posted at http://corporate.mattel.com/about-us/locations). Scholarships are offered for full-time study at a college, university or vocational school, in any country of the student's choice. The awards granted are based on the number of applications received and the funds available for scholarship distribution.
Requirements: Dependent children of regular full-time and part-time employees who have a minimum of six months employment with Mattel as of the application deadline date are eligible to apply. Students who are currently enrolled in post-secondary programs are eligible to apply. Official proof of admission to a recognized educational institution is required before scholarship funds become available. All fields of study are acceptable under these scholarships if offered at recognized universities or vocational schools.
Restrictions: Children of employees with a title of vice president or high are not eligible to apply.
Date(s) Application is Due: Apr 30
Contact: Mattel Scholarship Contact; (415) 362-6520, ext. 201; fax (415) 392-4667; Mattel@iie.org
Internet: http://www.iie.org/Programs/Mattel-Global-Scholarship-Program
Sponsor: Institute of International Education
809 United Nations Plaza
New York, NY 10017-3580

IIE Nancy Petry Scholarship 2410

The Institute of International Education sponsors the Nancy Petry Scholarship. The Nancy Petry Scholarship for Study Abroad provides financial support for students enrolled in graduate programs at universities in Colorado. Students should demonstrate some proficiency with the language of their chosen country if the program proposed is for a non-English speaking country.
Requirements: Applicants should refer to the website for current deadlines and application.
Restrictions: Students are not restricted to certain countries or fields of study, but the committee encourages applicants in the fields of environmental studies, international business, or economics.
Geographic Focus: Colorado
Date(s) Application is Due: Oct 21
Amount of Grant: 5,000 U.S.D
Contact: Lauren Granstrom, Administrative Assistant; (303) 837-0788, ext. 27 or (303) 837-1409; lgranstrom@iie.org
Internet: http://www.rockymountainiie.org/scholarships
Sponsor: Institute of International Education
809 United Nations Plaza
New York, NY 10017-3580

IIE New Leaders Group Award for Mutual Understanding 2411

The Institute of International Education (IIE) sponsors the IIE New Leaders Group Award for Mutual Understanding to recognize the outstanding work of one current Fulbright grantee who actively promotes mutual understanding between the U.S. and another country. The individual whose work is judged to have the most impact will receive the Award to continue an ongoing project. The Award can be for work in any field. The intent of this Award is to encourage and recognize bright young professionals for their innovative ideas, valuable knowledge, and dedication to the cause of mutual understanding. The recipient will be honored and asked to speak about the impact of this Award on their project at an IIE New Leaders Group event in the spring.
Requirements: There are no limits on the disciplines or fields supported. Individuals from any academic field or disciplines are eligible. Applicants should send a project description, resume, two references, a news article or brochure about their project (if available), along with the application available online. Current information and deadlines are found on the website.
Restrictions: The project being recognized must be an ongoing activity that is currently underway; it cannot be a proposal for a project to be done in the future. The Award is limited to current Foreign and U.S. Fulbright Scholar grantees.
Geographic Focus: All States, All Countries
Amount of Grant: 5,000 U.S.D
Contact: Becky Yi, Award Contact; (212) 984-5456; fax (212) 984-5566; byi@iie.org
Internet: http://www.iie.org/Who-We-Are/Awards/New-Leaders-Group-Award
Sponsor: Institute of International Education
809 United Nations Plaza
New York, NY 10017-3580

IIE Rockefeller Foundation Bellagio Center Residencies 2412

The Bellagio Center Residency program offers scholars, artists, thought leaders, policymakers and practitioners a serene setting conducive to focused, goal-oriented work, and the opportunity to establish new connections with fellow residents across an array of disciplines and geographies. The Center sponsors three kinds of residencies—for scholars, creative artists and practitioners. The Center offers one-month residency for each category and is able to accommodate a group working on a particular project. The Center is particularly interested in applicants whose work connects in some way to the Foundation's issues areas (basic survival safeguards, global health, climate and environment, urbanization, and social and economic security).
Requirements: Applicants must submit an online application, a detailed project proposal of no more than 1,250 words, an abbreviated curriculum vitae, three reference letters, and work samples.
Date(s) Application is Due: Dec 1
Contact: Bellagio Center Committee
Internet: http://www.rockefellerfoundation.org/bellagio-center
Sponsor: Institute of International Education
809 United Nations Plaza
New York, NY 10017-3580

IIE Toyota International Teacher Professional Development 2413

The Toyota International Teacher Professional Development provides international, professional-development opportunities to U.S. secondary school teachers to advance environmental stewardship and global connectedness in U.S. schools and communities. Each year, the program sends educators overseas for short-term (2-3 week) study tours to countries that are at the forefront of innovative solutions to environmental challenges. Participating teachers explore social and environmental issues through hands-on activities, and apply what they learn to create interdisciplinary and solution-focused lesson plans. The objectives of this professional development are: to increase global awareness and understanding; to enhance environmental knowledge, awareness, and sensitivity; to instill a sense of responsibility in teachers to empower their students and community leaders to do the same; to facilitate and creative hands-on approach to environmental and global issues; and to strengthen ties and facilitate collaboration among local and international teachers, community members, and organizations. Toyota covers all costs of program materials, transportation, meals, and lodging. Toyota will also issue each participant's school a $500.00 stipend to help defray the costs of participation during the school year.
Requirements: Applicants must be U.S. educators and teacher-librarians (grades 7-12) who work in the U.S. and District of Columbia. Applicants must hold a U.S. citizenship; hold a position as a full time classroom teacher or teacher-librarian for grades 7-12; and have at least three years of full time teaching experience. Teachers of diverse backgrounds and experience are strongly encouraged to apply. All programs focus on ways in which U.S. teachers can incorporate global and environmental education into their classrooms and communities.
Geographic Focus: All States, District of Columbia
Date(s) Application is Due: Nov 19
Amount of Grant: 500 U.S.D
Contact: Toyota International Teacher Contact; (877) 832-2457; toyotateach@iie.org
Internet: http://www.iie.org/Programs/Toyota-International-Teacher-Program
Sponsor: Institute of International Education
809 United Nations Plaza
New York, NY 10017-3580

IIE Western Union Family Scholarships 2414

The Family Scholarships are intended to help two members of the same family move up the economic development ladder through education. Scholarships may be used for tuition for college/university education, language acquisition classes, technical/skill training, and/or financial literacy. Families must have overcome barriers to pursue their educational goals and demonstrate financial need, with specific plans to utilize the scholarship. Recipients are eligible to receive scholarships from $1,000 to $5,000. For example, one recipient may receive $3,500 for college tuition, while the other may receive $1,500 for an ESL course.
Requirements: Requirements for the scholarship include the following criteria: applicants must be at least 18 years of age and living in the U.S. for 7 years or less; both applicants must have been born outside the U.S. and be currently living in the U.S.; application must include educational providers for primary and secondary award recipients (must be two family members); and applicants must reside in one of the following U.S. locations — Los Angeles, California; San Francisco, California; Denver, Colorado; Chicago, Illinois; New York, New York; Washington, D.C., or Miami, Florida.
Restrictions: Western Union employees, Western Union Agents and dependents are not eligible to apply.
Geographic Focus: California, Colorado, District of Columbia, Florida, Illinois, New York
Amount of Grant: Up to 5,000 U.S.D
Contact: WU Family Scholarship Contact; (303) 837-0788; fax (303) 837-1409; wufamily@iie.org
Internet: http://corporate.westernunion.com/scholarship.html
Sponsor: Institute of International Education
809 United Nations Plaza
New York, NY 10017-3580

Ildaura Murillo-Rohde Scholarships **2415**
The association awards these scholarships for tuition to outstanding Hispanic associate, diploma, baccalaureate, or graduate students. Recipients will be selected on the basis of potential to graduate, current academic standing in an accredited school of nursing (minimum 3.0 GPA), potential for contribution to the profession, and financial need. The amount awarded each year depends on the availability of funds.
Requirements: An applicant must be currently enrolled in an accredited school of nursing and be a U.S. citizen or legal resident of the United States, Puerto Rico, or the U.S. territories. Applicant must be a member of the association.
Date(s) Application is Due: Apr 30
Contact: Scholarships Coordinator; (202) 387-2477; fax (202) 483-7183; info@thehispanicnurses.org
Internet: http://www.thehispanicnurses.org
Sponsor: National Association of Hispanic Nurses
1501 16th Street NW
Washington, DC 20036

Illinois Children's Healthcare Foundation Grants **2416**
The Illinois Children's Healthcare Foundation works to ensure that all children in Illinois have access to affordable and quality health care. The Foundation focuses its giving on three specific areas: improving the oral health of underserved children; addressing the mental health needs of children; and increasing the incidence of developmental screening in young children. Types of support include: building/renovation; conferences and seminars; curriculum development; emergency funds; equipment; program development and evaluation; and research. There are no specific deadlines or applications. The Board meets six times per year.
Requirements: Applicants must first contact the Foundation by phone or submit a letter of inquiry to discuss their proposed project.
Restrictions: Giving is limited to Illinois. The following are not eligible for funding: intermediary funding agencies; partisan, lobbying, political or denominational organizations; organizations not determined to be public charities; grants to individuals, endowments general medical research.
Geographic Focus: Illinois
Contact: Tammy Lemke, President; (630) 571-2555; fax (630) 571-2566; tammylemke@ilchf.org or info@ilchf.org
Internet: http://www.ilchf.org/
Sponsor: Illinois Children's Healthcare Foundation
1200 Jorie Boulevard, Suite 301
Oak Brook, IL 60523-2269

Illinois Tool Works Foundation Grants **2417**
The corporate foundation awards grants in areas of company operations, with emphasis on Chicago, Ilinoise. Grants support organizations that focus on: education, the arts, health and human services, social welfare, housing, environmental and youth issues. The Foundation contributes financial support to not-for-profit organizations through two major giving programs: a direct-giving program and a three-for-one matching gift program for employees.
Requirements: Not-for-profit organizations are eligible to apply.
Amount of Grant: 5,000 - 1,000,000 U.S.D
Samples: United Way of Metropolitan Chicago, Chicago, IL, $1,110,732; Scholarship America, Saint Peter, MN, $478,350; Scott & White Memorial Hospital / Scott Sherwood & Brindley Foundation, Temple, TX, $60,000.
Contact: Mary Ann Mallahan, Secretary; (847) 724-7500; fax (847) 657-4261; mmallahan@itw.com
Internet: http://www.itwinc.com/itw/corporate_citizenship/itw_foundation
Sponsor: Illinois Tool Works Foundation
3600 West Lake Avenue
Glenview, IL 60026

Impact 100 Grants **2418**
Impact 100 empowers women to dramatically improve lives by collectively funding significant grants that make a lasting impact in the Greater Cincinnati and Northern Kentucky community. During the first few two months of the year, non-profit organizations in the ten-county Greater Cincinnati/Northern Kentucky area region are encouraged to submit grant applications to Impact 100. The Grant Review Committees of Impact 100 review the applications and conduct site visits to narrow their choices. At the end of the review process, a finalist is chosen by each of the five committees (Education, Health & Wellness, Environment/Preservation & Recreation, Culture and Family). These five finalists present their grant requests at the Annual Awards Celebration. After the finalist presentations, each member casts a vote (or has sent an absentee vote in advance). The votes are immediately tabulated, and the Impact 100 grants are awarded at the end of the evening.
Requirements: Applications that are programmatic, endowment, capital, start-up, or research-oriented are excepted. Applications from non-profit organizations headquartered in any of the following counties are available to apply: Ohio - Adams, Brown, Butler, Clermont, Hamilton, Warren; Kentucky - Boone, Campbell, Kenton; Indiana - Dearborn. Every applying agency must have 501(c)3 status with the IRS.
Restrictions: The following types of applications are not available for funding: operating; partisan; individual churches; indigent care subsidy; travel; loans; or for individuals.
Geographic Focus: Indiana, Kentucky, Ohio
Date(s) Application is Due: Mar 1
Amount of Grant: 100,000 U.S.D
Contact: Luann Scherer, Coordinaor; (513) 624-9509; Grants@Impact100.org
Internet: http://impact100.org/Grant/Default.aspx
Sponsor: Impact 100
PMB314 2692 Madison Road NI
Cincinnati, OH 45208-1320

Implant Dentistry Research and Education Foundation Grants **2419**
The objective of the grant is to encourage new investigators by providing seed or start-up funding for promising research projects. Grants support basic research and education relating to dental implants. Areas include, but are not limited to, biology, biochemistry, biomechanics, biomaterials, and molecular biology. Grants are awarded after thorough review by an independent scientific advisory board. Grant applications are available upon request.
Requirements: The principal investigator or coprincipal investigator must hold an appropriate terminal degree. A dental resident or PhD candidate may apply if the appropriate department or division chair provides a letter confirming he/she is coming on staff the year a project is funded.
Date(s) Application is Due: Jan 1
Amount of Grant: 20,000 - 25,000 U.S.D
Contact: Craig Johnson, Executive Director, IDREF; (800) 442-0525 or (973) 783-6300; fax (973) 783-1175; icoi@dentalimplants.com
Internet: http://www.dentalimplants.com
Sponsor: Implant Dentistry Research and Education Foundation
248 Lorraine Avenue, #3
Upper Montclair, NJ 07043-1454

Inasmuch Foundation Grants **2420**
The foundation was established for charitable, scientific, and educational projects primarily in Oklahoma; however, select projects in Colorado Springs, CO, also are supported. The foundation consistently provides funding and support to educational, health and human service, cultural, artistic, historical, and environmental concerns. This funding is not available to individuals, but is available to formal organizations seeking capital and support for existing programs that meet the emerging needs of the community. In order to initiate a proposal, submit a one-page letter of inquiry by the listed application deadlines. Application information is available online.
Restrictions: The foundation generally does not fund endowments or scholarships.
Geographic Focus: Colorado, Oklahoma
Date(s) Application is Due: Feb 15; Aug 15
Samples: Colorado College (Colorado Springs, CO)—to construct a facility for arts education and performances, $4 million.
Contact: Program Contact; (405) 604-5292
Internet: http://www.inasmuchfoundation.org/application.html
Sponsor: Inasmuch Foundation
210 Park Avenue, Suite 3150
Oklahoma City, OK 73102

Independence Blue Cross Charitable Medical Care Grants **2421**
The Independence Blue Cross (IBC) Charitable Medical Care Grant Program awards financial and programmatic grant support to nonprofit, privately funded health clinics that provide free or nominal-fee care to the uninsured and medically underserved communities in Bucks, Chester, Delaware, Montgomery, and Philadelphia counties. See website for grant application.
Requirements: To be eligible for a grant through this program, an applicant must meet the following criteria: the mission of the requesting organization aligns with IBC's Social Mission and its funding priorities; the requesting clinic is a privately funded entity. (Note: A clinic may receive federal funding as part of its revenues.); clinic grantee provides direct medical care for free or for a nominal fee to uninsured and/or underinsured residents of southeastern Pennsylvania; all grantees must be located in Bucks, Chester, Delaware, Montgomery, or Philadelphia counties; organizations must be classified as tax-exempt nonprofit under Section 501(c)3 of the IRS code.
Restrictions: Funding is unavailable to: individuals; private foundations; fund-raising events; conferences or seminars; political causes, candidates, organizations, or campaigns; sports teams; endowments; social organizations; religious groups for religious purposes; organizations that do not meet IBC's Social Mission.
Geographic Focus: Pennsylvania
Amount of Grant: 10,000 - 100,000 U.S.D
Contact: Courtney Smith, Social Mission Program Analyst; (215) 241-4862; fax (215) 241-3543; courtney.smith@ibx.com
Internet: http://www.ibx.com/social_mission/medical_grants/index.html
Sponsor: Independence Blue Cross
1901 Market Street, 28th Floor
Philadelphia, PA 19103-1480

Independence Blue Cross Nurse Scholars Program **2422**
This program was created to address the severe threat posed by the current nursing shortage to the quality and cost of health care in southeastern Pennsylvania. The program provides financial assistance to aspiring nurse educators and undergraduate nursing students. It also includes summer internship programs for nursing students and funding for faculty positions at area nursing schools. Application information can be located at the Pennsylvania Higher Education Foundation Web site.
Geographic Focus: Pennsylvania
Samples: Nursing programs in southeastern Pennsylvania (PA)—to establish a scholarship program for graduate students pursuing careers as nursing instructors, $1.74 million over three years; Pennsylvania Higher Education Fdn (Harrisburg, PA)—for a scholarship program for undergraduate nursing students, $750,000.

Contact: Program Supervisor; (800) 377-4502
Internet: http://www.ibx.com/social_mission/nurse_scholars/index.html
Sponsor: Independence Blue Cross
1901 Market Street, 28th Floor
Philadelphia, PA 19103-1480

Independence Blue Cross Nursing Internships **2423**
The Independence Blue Cross (IBC) Nursing Internship Program is an integral part of IBC's Nurse Scholars Program and supports the IBC Social Mission commitment to address the nursing shortage and overall health care crisis in the southeastern Pennsylvania region. The IBC Nursing Internship Program is open to undergraduate nursing students who have completed at least one clinical rotation and who attend one of the 31 accredited nursing programs (see website for a complete listing of the participating nursing programs) supported by the IBC Nurse Scholars Program. Two unique student nursing internship opportunities are offered through this program. One is for nursing students who would like to gain experience at IBC in a health care/insurance administration setting; the other is for nursing students interested in exploring the field of public health by interning on site at a community health center clinic.
Requirements: Each program is open to undergraduate nursing students who have completed at least one clinical rotation and who attend an accredited nursing program in southeastern Pennsylvania supported by the IBC Nurse Scholars program. Applications are available on the IBC website. If you are selected to participate, you must be available for the entire ten week internship program beginning June 8 through August 14.
Geographic Focus: Pennsylvania
Date(s) Application is Due: Mar 31
Contact: Courtney Smith, Social Mission Program Analyst; (215) 241-4862; fax (215) 241-3543; courtney.smith@ibx.com
Internet: http://www.ibx.com/social_mission/nurse_scholars/nursing_internship/index.html
Sponsor: Independence Blue Cross
1901 Market Street, 28th Floor
Philadelphia, PA 19103-1480

Independence Community Foundation Community Quality of Life Grant **2424**
The Foundation makes grants to organizations that engage in community-based efforts that support important institutions or causes. Generally, the grants support: disease awareness and prevention; food pantries; neighborhood greening and graffiti removal; and counseling support for vulnerable populations. Interested applicants should submit an initial letter of inquiry. Application information is available online.
Requirements: Program or project grants are made on a competitive basis to nonprofit organizations located within New York City, Nassau, Suffolk and Westchester counties in New York, and Essex, Bergen, Union, Hudson, Middlesex, Ocean, and Monmouth counties in New Jersey.
Restrictions: The Foundation does not fund: individuals; political contributions; funding for religious purposes; purchase tickets for dinners, golf outings, or similar fundraising events.
Geographic Focus: New Jersey, New York
Date(s) Application is Due: Mar 30; Sep 30
Contact: Program Contact; (718) 722-2300; fax (718) 722-5757; inquiries@icfny.org
Internet: http://www.icfny.org/comm_quality.html
Sponsor: Independence Community Foundation
182 Atlantic Avenue
Brooklyn, NY 11201

Independence Foundation Grants **2425**
The foundation awards grants to Philadelphia nonprofit organizations in its areas of interest, including community-based nurse managed health care, health promotion, family planning, and comprehensive health care services where issues of quality, access, and cost are considered; arts and culture; legal aid; and assistance to the disadvantaged. Types of support include development grants, endowments, general operating grants, professorships, program development grants, seed money grants, and scholarships. A brief letter outlining a proposed project is a prerequisite to a formal application.
Requirements: 501(c)3 tax-exempt organizations serving Philadelphia and the surrounding counties including Bucks, Chester, Delaware, and Montgomery are eligible.
Restrictions: Grants will not be made to individuals or for building and development funds or to support travel, research, or publications.
Geographic Focus: Pennsylvania
Amount of Grant: 1,000 - 100,000 U.S.D
Samples: La Salle U, Neighborhood Nursing Center (Philadelphia, PA)—to provide healthcare and educational services to uninsured residents of northwest Philadelphia, $125,000.
Contact: Grants Administrator; (215) 985-4009; fax (215) 985-3989
Internet: http://www.independencefoundation.org
Sponsor: Independence Foundation
200 S Broad Street, Suite 1101
Philadelphia, PA 19102

Indiana 21st Century Research and Technology Fund Awards **2426**
The Indiana 21st Century Research and Technology Fund of the Indiana Economic Development Corporation (IEDC) is open to proposals from all public and private entities for technology-based commercialization activities encompassing science/technology creation, innovation, and transfer intended to have commercial impacts. The fund intends to increase the numbers, and rates of development, of new and expanding technology-based companies by funding promising opportunities that, in some cases, the financial markets might find too risky. The Fund makes awards in two broad categories: Science and Technology Commercialization and Centers of Excellence. In addition, the Fund provides cost-share on behalf of Federal proposals submitted by Indiana-based entities. Generally awards are made in multiples of $50,000 up to $2,000,000. Support for awards in excess of $2,000,000 will be rare.
Requirements: The IEDC defines a technology-based company as one that is involved in transferring advanced technology into products, developing technologies with the near-term intention of creating products, or using new or advanced technologies in its design, development, and/or manufacturing of products. The Fund emphasizes the creation of academic-sector - commercial-sector partnerships. In making awards, the Fund expects significant leverage from the partners involved in the projects. Important: before applying, contact Fund staff (email preferred) to discuss your interest in submitting a proposal and to discuss your technology and commercialization goals. While not a review criterion, the fund encourages the inclusion of interns from any academic institution, or participating commercial sector partner, in order to increase project-related involvement of students at all levels.
Restrictions: Only direct costs will be supported. Institutions will not be provided indirect (overhead) cost support. Entities with previous Fund awards that are not current with regard to financial or technical reporting requirements will be disqualified from making new submissions to the Fund. Resubmissions of previously declined proposals will be considered only if substantive changes have been made to the proposal. Fund staff will determine whether to review resubmissions.
Geographic Focus: Indiana
Amount of Grant: 500,000 - 2,000,000 U.S.D
Samples: 2K Corporation: $400,000 - Commercialization of a Neutron Based Explosives Detection Device (Car Bomb Detection). Arxan Technologies, Inc.: $1,944,096 - Protecting Critical IP in the 21st Century: Advancing Anti-Tamper Technologies. BioVitesse Inc.: $1,300,000 - Rapid Detection and Identification of Live Bacteria. CIS LLC: $1,075,000 - Commercialization of Satellite Radio for Cell Phone Applications.
Contact: Carla Phelps, Financial Manager; (317) 233- 4336; cphelps@21fund.org
Internet: http://www.21fund.org/
Sponsor: Indiana Economic Development Corporation
1 North Capitol Avenue, Suite 900
Indianapolis, IN 46204

Indiana AIDS Fund Grants **2427**
Two grant programs are available to Indiana organizations that provide HIV/AIDS advocacy, prevention and direct care services. The Direct Emergency Financial Assistance Fund gives grants to agencies and organizations specifically to help HIV/AIDS patients access emergency financial help for housing, transportation, food, and/or medical care. The Indiana AIDS Fund gives prevention grants in November each year to agencies and organizations that provide persons living with HIV/AIDS direct care services, health education and risk-reduction programs and counseling, testing, and referral programs. Proposal information is available online.
Requirements: Any Indiana-based non-profit 501(c)3 group, organization or agency that provides HIV/AIDS-related programs or services to local constituencies is eligible to apply for grants from the Indiana AIDS Fund. The applicant must function without discrimination or segregation because of race, gender, age, religion, national origin, sexual orientation, disability, military or marital status, in hiring, termination, assignment and promotion of staff, selection of board members or provision of service.
Restrictions: The fund will not make the following types of grants: to individuals; for sectarian religious purposes; to purchase advertising or tickets to events; to support research; for the purpose of deficit reduction; to produce currently available educational materials; to fund endowments; for short- or long-term loans; for multiple-year program funding; for mass media campaigns; for capacity building; or for services not directly related to primary or secondary prevention.
Geographic Focus: Indiana
Samples: LifeCare/Clarian Health Partners (Indianapolis, IN)—to assist HIV/AIDS patients with emergency needs related to food, housing, medical care, and transportation, $11,000; Planned Parenthood of Indiana (Indianapolis, IN)—for counseling, testing, referral, and partner-notification services, $15,000; Step-Up (Indianapolis, IN)—for a program that offers sessions on HIV/AIDS prevention to people over the age of 50, $14,000; Jameson Camp (Indianapolis, IN)—for a camp that serves children affected by HIV/AIDS, $10,000.
Contact: Grants Administrator; (317) 630-1805; severett@thfgi.org
Internet: http://www.indianaaidsfund.org/index.cfm?navigationid=1844
Sponsor: Indiana AIDS Fund
429 East Vermont Street, Suite 300
Indianapolis, IN 46202

Indiana Minority Teacher/Special Services Scholarships **2428**
The Minority Teacher Scholarship was created by the 1988 Indiana General Assembly to address the critical shortage of Black and Hispanic teachers in Indiana. In 1990 the Indiana General Assembly amended the Minority program to include the field of Special Education, and in 1991 the fields of Occupational and Physical Therapy were added. The maximum annual scholarship is $1,000. However, if a minority student applicant demonstrates financial need he/she may be eligible to receive up to $4,000 annually. Colleges will determine the actual amount when reviewing a scholar's financial aid package.
Requirements: To be eligible you must be: 1. A minority student (defined as Black or Hispanic) seeking a teaching certification; or a student seeking a Special Education teaching certification; or a student seeking an Occupational or Physical Therapy certification. 2. An Indiana resident and an U.S. Citizen. 3. Admitted to an eligible institution as a full time student or already attending as a full time student. 4. Pursuing or intend to pursue a course of study that would enable the student upon graduation

to teach in an accredited elementary or secondary school in Indiana. 5. Complete and file the Free Application for Federal Student Aid (FAFSA). 6. Not be in default on a State or Federally-funded Student loan. 7. Meet all minimum criteria established by the Commission. In addition, students who are already enrolled in college, must have a Grade Point Average (GPA) of at least 2.0 on a scale of 4.0 or the equivalent, or meet the minimum GPA requirements established at the college for its school of education, if it is higher. Scholars must agree to: 1. Pursue a program which leads to either a teaching or Occupational or Physical Therapy Certification. 2. Complete the Teacher or Occupational or Physical Therapy certification program within six (6) years from the time the first scholarship is received. 3. Teach on a full time basis in an accredited Indiana elementary or secondary school; or, 4. Practice in the field of Occupational or Physical Therapy in an accredited school, vocational rehabilitation center, community mental retardation or other developmental disabilities center, for 3 years out of the first 5 years after certification. Colleges are responsible for making the actual awards. Financial need may be considered but it is not a requirement to apply. Preference will be given to minority students and students enrolling in college for the first time. Applications should be submitted to the Office of Financial Aid at the school where the applicant plans to attend.
Restrictions: Scholarships are non-transferable between universities. Those who fail to fulfill the terms of their scholarship agreements must reimburse the State of Indiana.
Geographic Focus: Indiana
Amount of Grant: Up to 4,000 U.S.D
Contact: Ada Sparkman; (317) 232-2350; asparkman@ssaci.in.gov
Internet: http://www.in.gov/ssaci/2342.htm
Sponsor: State Student Assistance Commission of Indiana
150 W Market Street, Suite 500
Indianapolis, IN 46204

Indiana Nursing Scholarships **2429**
The Nursing Scholarship Fund was created by the 1990 Indiana General Assembly to encourage and promote qualified individuals to pursue a nursing career in Indiana. The scholarship can only be applied towards tuition and fees. Colleges will determine the actual award amount when developing a scholar's financial aid package. The maximum annual scholarship is $5,000. However, the amount of the scholarship may be affected by the level of other tuition specific grants and scholarships aid received by an applicant.
Requirements: To be eligible students must: 1) Be admitted to an approved institution of higher learning as a full time (12 hours or more) or part time (6-11 hours) nursing student; 2) Be an Indiana resident and an U.S. citizen; 3) Agree in writing to work as a nurse in an Indiana health care setting for at least the first two (2) years following graduation; 4) Demonstrate a financial need for the scholarship; 5) Have a minimum Grade Point Average (GPA) of at least 2.0 on a scale of 4.0 or the equivalent, or meet the minimum GPA requirements established for the college's nursing program if it is higher; 6) Not be in default on a state or federally sponsored student loan; 7) Complete and file the Free Application for Federal Student Aid (FAFSA); 8) Meet all other minimum criteria established by the Commission. In order to renew the scholarship, Scholars must: 1) Maintain the cumulative GPA required by the college for admission to its nursing program, or; 2) Maintain a GPA of at least 2.0 on a scale of 4.0 or the equivalent if there is no minimum at the college; and, 3) Reapply each year. Applications should be submitted to the Office of Financial Aid where the applicant plans to attend. Applications should not be submitted to SSACI. Colleges will make the award decisions. For a list of colleges and universities participating in the program, go to http://www.in.gov/ssaci/2368.htm.
Geographic Focus: Indiana
Date(s) Application is Due: Mar 10
Amount of Grant: Up to 5,000 U.S.D
Contact: Ada Sparkman; (317) 232-2350; asparkman@ssaci.in.gov
Internet: http://www.in.gov/ssaci/2343.htm
Sponsor: State Student Assistance Commission of Indiana
150 W Market Street, Suite 500
Indianapolis, IN 46204

Infinity Foundation Grants **2430**
The foundation awards grants in a broad array of areas including charitable, scientific, religious, educational, and holistic healing activities by organizations and public agencies that work to better the lives of people. Grants also support individuals' research and development of educational materials to improve the authenticity of the portrayal of Indic traditions in the educational system. Proposed projects could result in one or more of the following: books, curriculum development, articles, conferences, CD-Roms, digital slide shows, Internet presentations, and audio/video materials. Topics covered may include philosophy, history, religion, science, art, and sociology, as they pertain to the educational curricula on Indic traditions. Proposals should be submitted by email.
Requirements: Grantee may be a scholar, teacher, visionary, or spiritual leader whose work in the designated topics would be enhanced by a foundation grant.
Contact: Rajiv Malhotra, Founder and President; (609) 683-0548; fax (609) 683-0478; rm.infinity@gmail.com
Internet: http://www.infinityfoundation.com/callforgrantproposals.htm
Sponsor: Infinity Foundation
66 Witherspoon Street, Suite 400
Princeton, NJ 08542

Inland Foundation Inc Grants **2431**
Nonprofit organizations located in communities where Inland Container has facilities and employees are invited to apply for grants in the areas of community services, education, art and culture, and health. Types of support include employee-related scholarships, annual campaigns, continuing support, general operating budgets and special projects. Proposals should consist of a brief letter that includes a description of the sponsoring organization and project; statement of the importance of the project and why it meets the foundation's guidelines; geographical area served by the project; total funds needed and the amount requested from the foundation; other sources of income and funding; and proof of the organization's tax-exempt status. Grant proposals are accepted throughout the year.
Requirements: Grants are awarded to nonprofits in plant locations, including Fort Smith, AR; Buena Park, El Centro, Los Angeles, Neward, Santa Fe Springs, and Tracy, CA; Denver, CO; Orlando, FL; Rome, GA; Chicago, IL; Crawfordsville, Evansville, Indianapolis, and Newport, IN; Garden City and Kansas City, KS; Louisville and Maysville, KY; Minden, LA; Minneapolis, MN; Hattiesburg, MS; Saint Louis, MO; Edison and Spotswood, NJ; Middletown, OH; Biglerville, and Hazleton, PA; Vega Alta, Puerto Rico; Lexington and Rock Hill, SC; Elizabethton and New Johnsonville, TN; Dallas, Edinburg, and Orange, TX; and Petersburg, VA.
Restrictions: The foundation does not make grants to individuals, religious organizations for sectarian purposes, or individual professors or departments within universities.
Amount of Grant: 1,000 - 30,000 U.S.D
Contact: Evonne Nerren, Secretary/Treasurer; (936) 829-1721
Internet: http://www.templeinland.com
Sponsor: Inland Foundation
303 S Temple Drive
Diboll, TX 75941

Innovative Technology Development Award Program **2432**
The center annually funds biotechnology research projects that promise development of commercializable technologies. In this way, the center aims to further establish and strengthen ties between academic scientists and industry. Grants are for one year, with a continuation seed grant offered for the second and third year if necessary. Proposals are evaluated based on the quality of the science, the commercial potential, and how the project might impact New York state economy. Applications are posted on the Web site or are available by fax or mail.
Requirements: Faculty at any New York state research institution may apply for an ITD award.
Geographic Focus: New York
Amount of Grant: 40,000 - 80,000 U.S.D
Samples: Roswell Park Cancer Institute—for DNA damage kits, in collaboration with Zeptogen Corp, $78,000; Dept of Pathology, SUNY (Brooklyn, NY)—for rational design and testing of new anticancer agents, $32,000.
Contact: Dr. Anil Dhundale, Director of Scientific Affairs; (516) 632-8521; fax (516) 632-8577; anil.dhundale@sunysb.edu
Internet: http://www.biotech.sunysb.edu/techDev/ITD/index.html
Sponsor: Center for Biotechnology
Psychology A Building, 3rd Floor, SUNY Stony Brook
Stony Brook, NY 11794-5208

Institute for Agriculture and Trade Policy Food and Society Fellowships **2433**
The IATP Food and Society Fellows Program provides fellowships to professionals in food and agriculture from across North America, enabling them to use mass media channels to inform and shape the public agenda. The goal of the program is to create sustainable food systems that promote good health, vibrant communities, environmental stewardship, worker justice and accessibility for all. Fellows come from many disciplines: chefs, farmers, nutritionists, activists, public health professionals, fishers, policy experts and academics. Together they form an interdisciplinary team that works to: use communication to influence the issues that reach the public agenda, thereby creating policy changes at the personal, organizational and public policy levels that advance sustainable food and farming systems; increase the mass media communications on issues around sustainable food and farming systems that produce healthy, green, fair and affordable foods; raise the profile of the fellows as food system experts among media and policymakers; and build capacity, leadership, and cohesiveness in a group of experts who collaborate and communicate using mass media channels to bring sustainable food system issues to the public agenda.
Requirements: Applicants should be U.S. residents actively involved in a professional career that involves two or more of the following: agriculture, health promotion, youth development, food production, and/or policy analysis.
Contact: Mark Muller; (612) 870-3420; fax (612) 870-4846; mmuller@iatp.org
Internet: http://www.foodandsocietyfellows.org/about/page/about-us
Sponsor: Institute for Agriculture and Trade Policy
2105 First Avenue South
Minneapolis, MN 55404

Institute of Human Values in Health Care Interdisciplinary Fellowships **2434**
The program is designed to provide non-clinical professionals who work in health related fields (e.g., law, government, economics, sociology, philosophy, journalism, religion, and history) with first-hand exposure to and involvement with intensive inpatient and ambulatory care of the critically ill within a major academic medical center. The fellowship runs from January through May.
Requirements: Professionals who are active and established in their fields and can devote full time to the program for five months may submit fellowship applications. Postdoctoral candidates also will be considered.
Date(s) Application is Due: Sep 15
Amount of Grant: 7,500 U.S.D
Contact: Dr. Robert Sade; (843) 792-5278; fax (843) 792-8286; sader@musc.edu
Internet: http://www.values.musc.edu/news/if_announcement.htm

Sponsor: Medical University of South Carolina
96 Jonathan Lucas Street, Suite 409, P.O. Box 250612
Charleston, SC 29425

Institute of Mental Hygiene Grants **2435**
The program invests in children through its grantmaking programs, active involvement with grantees, and leadership in improving mental health programs and policies. Grants support organizations that are tax-exempt and provide services in New Orleans. The foundation supports the following programs: early childhood mental health grants, children's mental health grants, early childhood mini-grants, and technical assistance grants. Applications and guidelines, including deadline dates, are available online.
Requirements: Louisiana nonprofit organizations are eligible.
Restrictions: Capital projects, general fund drives, or grants to individuals will not be considered.
Geographic Focus: Louisiana
Samples: Safehouse Youth Services, Lifestyle Counseling Program (New Orleans, LA)—to work with African American youth ages nine - 21 who are enticed by or at risk of becoming involved in the drug trade, $50,000; Juvenile Justice Project of Louisiana (New Orleans, LA)—to visit model programs and to create a new model for effective legal representation and treatment for troubled youths that includes mobilizing their families and communities,$14,250; Louisiana State U Health Sciences Ctr, Dept of Psychiatry (New Orleans, LA)—to expand on-site, comprehensive mental-health consultation and treatment services to several early childhood centers, $194,000 over two years.
Contact: Nancy Freeman, Executive Director; (504) 566-1852; fax (504) 566-1853; nfreeman@imhno.org or imh@imhno.org
Internet: http://www.imhno.org/imh_guidelines.html
Sponsor: Institute of Mental Hygiene
1055 St. Charles Avenue, Suite 350
New Orleans, LA 70130

Intergrys Corporation Grants **2436**
Integrys supports the giving initiatives of Wisconsin Public Service Foundation in Michigan, Minnesota and Wisconsin. Other significant giving by Integrys occurs through programs operated by Peoples Gas and North Shore Gas in Illinois, including the city of Chicago and its 54 suburban communities in northeastern Illinois. Areas if interest include: arts and culture; education (all levels); human services and health; community and neighborhood development; and the environment. Employee volunteers allow staff to give back to the neighborhoods they cherish. Matching funds energize charitable involvement. Categories of giving include health and welfare, civic and community, higher education, and cultural. Most corporate contributions are in the form of unrestricted grants. Types of support include capital grants, endowments, general operating grants, matching gifts, and program grants. Before eligibility is determined, consideration will be given to the background of the organization, the organization's legal status, how the program will benefit the community, whether the organization receives broad community support, the quality of the organization's leadership, and the organization's financial status.
Requirements: Organizations in Illinois, Michigan, Minnesota, and Wisconsin are eligible.
Restrictions: Contributions will not be made to individuals; organizations that discriminate by race, color, creed, or national origin; political organizations or campaigns; organizations whose prime purpose is to influence legislation; religious organizations for religious purposes; agencies owned and operated by local, state, or federal governments; or for trips or tours, or special-occasion or goodwill advertising.
Geographic Focus: Illinois, Michigan, Minnesota, Wisconsin
Amount of Grant: Up to 5,000 U.S.D
Contact: Community Contributions Officer; (312) 240-7516 or (800) 699-1269; fax (312) 240-4389
Internet: http://www.integrysgroup.com/corporate/corporate_giving.aspx
Sponsor: Intergrys Corporation
130 E Randolph Drive, 18th Floor
Chicago, IL 60601

International Cancer Technology Transfer Fellowships **2437**
The fellowships are funded by a group of cancer institutes, leagues, societies, associations, foundations, and governmental agencies in North America, Europe, and Australia. The aims of the awards are threefold: to facilitate rapid international transfer of cancer research and clinical technology; exchange knowledge and enhance skills in basic, clinical, behavioral, and epidemiological areas of cancer research and in cancer control and prevention; and acquire up-to-date clinical management, diagnostic, and therapeutic expertise. Applications are particularly encouraged for projects in the fields of cancer control and prevention, epidemiology and cancer registration, public education, and behavioral sciences. The short-term fellowships permit successful candidates to spend up to three months at a suitable host institute abroad. They are particularly aimed at investigators and clinicians working in places where such teaching is not yet available and where the necessary facilities exist to apply and disseminate the new skills upon return. Applications may be submitted at any time, and funding decisions are generally communicated to candidates within 60 days of receipt of a complete application.
Requirements: Qualified cancer investigators should be at the early stages in their careers; clinicians should be well established in their oncology practice. Experts from any country who have been invited to teach specialized skills abroad also may apply.
Amount of Grant: 3,000 U.S.D
Contact: International Cancer Research Program; (4122) 809 18 40; fax (41-22) 809 18 10; fellows@uicc.org
Internet: http://fellows.uicc.org/fell4icr.shtml
Sponsor: International Union against Cancer
3, rue de Conseil-General
Geneva, CH-1205 Switzerland

International College of Surgeons Grants and Scholarships **2438**
The ICS makes grants and scholarships available to surgeons who wish to enhance their surgical skills through postgraduate training. Applicants are responsible for making their own arrangements for the program of study; must clearly indicate the place of study, with whom they will be studying, and acceptance by the institution and surgeon for the course of study as defined; and must indicate what it is they expect to achieve upon completion of the study program (e.g. learn a particular procedure, etc.). Guidelines are available online. No formal application form is required.
Requirements: Applicants must have graduated from an accredited medical school, completed their residency, and be licensed to practice surgery in their home country (documentation of licensure must be provided). Applicants need not be ICS fellows.
Contact: Max Downham, Executive Director; (312) 787-6274; fax (312) 787-9289; max@icsglobal.org
Internet: http://www.icsglobal.org/RandS/index.asp
Sponsor: International College of Surgeons
1516 N Lake Shore Drive
Chicago, IL 60610-1694

International Essential Tremor Foundation Grants **2439**
The foundation awards grants for clinical or basic research specific to essential tremor. Data to be derived should lead to grant proposals to governmental or other sources of funds. Two grants are available in the current year. Under exceptional circumstances, funding may be allowed for a second year as a competing renewal. No more than one award can be given to an individual investigator per year. Funding begins in June or July.
Requirements: Grants are awarded to MDs and/or PhDs.
Amount of Grant: 20,000 U.S.D
Contact: Catherine Rice, (888) 387-3667 or (913) 341-3880; fax (913) 341-1296; crice@essentialtremor.org
Internet: http://www.essentialtremor.org/research/index.php
Sponsor: International Essential Tremor Foundation
P.O. Box 14005
Lenexa, KS 66285-4005

International Fellowships in Medical Education **2440**
The program provides opportunities for faculty from schools of medicine outside the United States to study aspects of medical education in the United States that have the potential to improve medical education in their home country institutions. Mentoring will be provided by preceptors in U.S. basic and clinical science departments, medical education departments, and health system institutions. Eligible study areas include educational methodology, curriculum design, evaluation systems, medical school governance, development of basic and clinical science departments, and the design and operation of health care and public health system programs linked to medical education. Fellows may concurrently pursue collaborative research interests with U.S. basic science faculty or observe some clinical activity with clinical faculty. Fellowship allowance generally includes a monthly stipend, round-trip economy-class air fare for the fellow, and travel to one scientific meeting in the United States. Programs may range from six months to one year. The interactive, web-based application is now available online; paper applications are no longer accepted.
Requirements: Candidates must reside and work in their home countries at the time of application; have a graduate or professional degree in medicine or in a basic medical science; have not less than three years of work experience in their chosen field in the home country; hold an academic appointment as a faculty member in a school of medicine; have the ability to communicate effectively in English; have the endorsement of a home country medical school or organization; and have a position to return to in the home country medical school or organization.
Restrictions: Fellowships are not provided for programs in basic or clinical research; degree-granting educational programs that require acceptance to an institution and tuition payments; tuition grants for short-term courses; specialty training in residency programs, or solely for training in clinical procedures.
Amount of Grant: 14,400 - 28,800 U.S.D
Contact: Grants Administrator; (215) 386-5900; fax (215) 386-9196; info@ecfmg.org
Internet: http://www.ecfmg.org/about.html
Sponsor: Educational Commission for Foreign Medical Graduates
3624 Market Street
Philadelphia, PA 19104-2685

International Oncology Nursing Fellowships **2441**
The fellowships are funded by the Oncology Nursing Society and provide an opportunity for qualified nurses to augment their professional knowledge and experience through short-term observerships at a renowned comprehensive cancer center in Australia, North America, or the United Kingdom. A project description must provide full details of the particular cancer nursing skills that the candidate wishes to observe. It should elaborate on the treatment and nursing facilities at the home institute and how the newly acquired knowledge and techniques would be used and disseminated to other nursing staff upon return. Projects need the approval of the candidate's home and host supervisors. The maximum fellowship duration is three months; stipend support is for one month only. The program also contributes to the least expensive international two-way airfare or other appropriate form of transport. Financial support is not provided for dependents.

Requirements: English-speaking registered nurses who are actively engaged in the management of cancer patients in their home institutes and who come from developing or Eastern European countries are eligible, as are established oncology nurses from any country who wish to disseminate their skills in these regions.
Date(s) Application is Due: Nov 1
Amount of Grant: 2,800 U.S.D
Contact: International Cancer Research Program; (4122) 809 18 40; fax (41-22) 809 18 10; fellows@uicc.org
Internet: http://fellows.uicc.org/fell5ion.shtml
Sponsor: International Union against Cancer
3, rue de Conseil-General
Geneva, CH-1205 Switzerland

International Positive Psychology Association Student Scholarships **2442**
The International Positive Psychology Association offers scholarships to help fund student participation in the World Congress on Positive Psychology conference. The number and amount of awards are determined by the number of applications received. The scholarships help fund travel, registration, food, and lodging for the conference.
Requirements: All applicants must meet the following *Requirements:* complete the online application; submit an abstract proposal of less than 250 words; attach a letter of recommendation from a professor who works with positive psychology themes; attach a curriculum vitae of no more than two pages; provide proof of university or college affiliation and a promise via email to travel to the U.S. to present a poster; currently enrolled in an accredited university or college or at a community college with intent to transfer to a university/college for the upcoming year; and pursuing an undergraduate, master's, or PhD in psychology or related field. Applicants should refer to the website to check for a current online application.
Geographic Focus: All States, All Countries
Date(s) Application is Due: Jan 31
Contact: Gene Terry, Administrative Director; (856) 423-2862; fax (856) 423-3420; ippamtg@talley.com
Internet: http://community.ippanetwork.org/worldcongress/studentscholarshipinformation/
Sponsor: International Positive Psychology Association
19 Mantua Road
Mt. Royal, NJ 08061

International Spinal Research Trust Grants **2443**
The trust awards grants worldwide for research into the treatment of paralysis caused by spinal cord injury. Grants also are available for the purchase of relevant equipment.
Contact: John Cavanagh, Head of Research; 01483 898786; fax 01483 898763; research@spinal-research.org or john@spinal-research.org
Internet: http://www.spinal-research.org/display_section.asp?section=researchers
Sponsor: International Spinal Research Trust
Station Road, Bramley
Guildford, GU5 0AZ England

IOM Gustav O. Lienhard Award **2444**
The award recognizes creative or pioneering efforts that have appreciably improved personal health services, rather than on contributions to the science base of healthcare. Award recipients are selected on the basis of achievement in the area of personal health services; and achievement of national scope. Additional selection criteria include innovative, creative, and pioneering achievement; unique contributions by the nominee to that achievement; positive change over a sustained period not simply for the potential for such change through the nominee's achievement; a qualitative and quantitative impact; and success in overcoming barriers, based on resources available. The recipient must attend the ceremony to receive the award. Complete nomination guidelines and procedures are available online.
Requirements: Any individual or group may submit a nomination. Nominees are not considered on the basis of education or profession.
Date(s) Application is Due: Apr 30
Amount of Grant: 25,000 U.S.D
Contact: Susanne Stoiber, (202) 334-2177; fax (202) 334-1412; lienhard@nas.edu
Internet: http://www.iom.edu/lienhard
Sponsor: Institute of Medicine
500 Fifth Street NW
Washington, DC 20001

Ireland Family Foundation Grants **2445**
Established in 2000, the Ireland Family Foundation givings primarily in North Carolina with some funding available in California. The Foundations supports non-profit organizations involved with autism research, human services and education.
Requirements: The initial approach should be in the form of a letter, followed up with a proposal containing : a copy of IRS Determination Letter; a detailed description of project and amount of funding requested.
Restrictions: No grants to individuals.
Geographic Focus: California, North Carolina
Amount of Grant: 10,000 - 60,000 U.S.D
Samples: ARC of Orange County, Chapel Hill, NC., $35,000;
Contact: Lori Ireland, President; (919) 932-3556
Sponsor: Ireland Family Foundation
1434 Arboretum Drive
Chapel Hill, NC 27517-9161

Irene and Daisy MacGregor Memorial Scholarship **2446**
The scholarship is awarded to a student who has been accepted into an accredited school of medicine to pursue an MD degree. This scholarship is also available to students who have been accepted into or who are pursuing an approved course of study in the field of psychiatric nursing, graduate level, at accredited medical schools, colleges, or universities. There is a preference to females, if equally qualified.
Requirements: Applicants must be U.S. citizens and must attend an accredited U.S. college or university. All applicants must obtain letters of sponsorship from local DAR chapters.
Date(s) Application is Due: Apr 15
Amount of Grant: 5,000 - 20,000 U.S.D
Contact: Office of the Committees/Scholarships; (202) 628-1776; fax (202) 879-3252
Internet: http://www.dar.org/natsociety/edout_scholar.cfm
Sponsor: National Society of the Daughters of the American Revolution
1776 D Street NW
Washington, DC 20006-5303

Irvine Health Foundation Grants **2447**
The foundation supports IRS nonprofit organizations in support of projects and programs designed to meet the health care needs of Orange County, Ca, residents. Preference is given to activities addressing current unmet health care needs. Programs for youth, the elderly, and other groups also are supported. Types of support include challenge/matching grants, program grants, research grants, and seed money grants. There are no application deadlines; the board reviews requests ongoing.
Requirements: Orange County, CA, nonprofit organizations are eligible.
Restrictions: Grants are not awarded to individuals or for religious purposes, sporting activities, events, or endowments.
Geographic Focus: California
Samples: Laguna Beach Community Clinic (Laguna Beach, CA)—for expanded staffing, including a full-time receptionist and a part-time nurse and a clinical social worker, $37,000; U of California at Irvine, Biomedical Research Ctr (Irvine, CA)—for construction costs associated with the new Cancer and Genetics facility, $300,000; Catholic Charities of Orange County (Santa Ana, CA)—for health and social services delivery to poor residents of southern Orange County, CA, $60,000.
Contact: Executive Director; (949) 253-2959; fax (949) 253-2962; info@ihf.org
Internet: http://www.ihf.org/grant/index.asp
Sponsor: Irvine Health Foundation
18301 Von Karman Avenue, Suite 440
Irvine, CA 92612

Irving S. Gilmore Foundation Grants **2448**
The Irving S. Gilmore Foundation endeavors to develop and to enrich the Greater Kalamazoo community of Michigan and, its residents by supporting the work of nonprofit organizations. The Foundation's funding priorities are: health and well-being; arts; human services; education; community development; culture and humanities. Organizations that are first time Foundation applicants or have not received Foundation funding since 2007 must contact the Foundation at least four weeks prior to an applicable submission deadline.
Requirements: The Foundation supports Kalamazoo County projects, programs, and purposes carried out by charitable institutions, primarily public charities and governmental entities.
Restrictions: Grants are not made to individuals.
Geographic Focus: Michigan
Date(s) Application is Due: Jan 4; Mar 1; May 3; Jul 1; Sep 1; Nov 1
Samples: Michigan Festival of Sacred Music—Operational Support; Hospice Care of Southwest Michigan—Capital Campaign; Volunteer Services of Greater Kalamazoo—BoardConnect Program.
Contact: Janice C. Elliott, Vice President; (269) 342-6411; fax (269) 342-6465
Internet: http://www.isgilmorefoundation.org/communityinvolvement.htm
Sponsor: Irving S. Gilmore Foundation
136 East Michigan Avenue, Suite 900
Kalamazoo, MI 49007-3912

Irvington Institute Immunology Postdoctoral Fellowships **2449**
The institute will award up to 12 postdoctoral fellowships in immunology for three consecutive years of work in a U.S. laboratory or hospital for the award period beginning January 1 or July 1. Fellowships support basic medical research in the prevention and treatment of AIDS, cancer, diabetes, allergies, lupus, rheumatoid arthritis, and other immune system diseases. These fellowships are available to students from foreign countries if they have the proper visas for their stay.
Requirements: Medical doctors, PhDs, and MD/PhDs are encouraged to apply. Candidates with not more than three years' laboratory experience will receive priority. The institute does not have a citizenship requirement; however, applicants should indicate their citizenship and/or current visa status.
Restrictions: A laboratory may qualify for only one fellow per year.
Date(s) Application is Due: Jul 12
Amount of Grant: 45,000 U.S.D
Samples: Dr. Michael G. Brown, Jewish Hospital of Saint Louis, Washington U School of Medicine (Saint Louis, MO)—for research on the natural killer cell gene complex; Dr. Shigemi Kinoshita, Department of Molecular Pharmacology, Stanford U School of Medicine (Stanford, CA)—for research on the molecular secrets of long-term survivors of HIV infection.
Contact: Coordinator; (212) 576-1005; fax (212) 576-1006; irving1@ix.netcom.com

Internet: http://www.irvingtoninstitute.org/fellowships.html
Sponsor: Irvington Institute for Immunological Research
245 Fifth Avenue, Room 2101
New York, NY 10016

Irvin Stern Foundation Grants **2450**

The Foundation awards grants to nonprofits primarily in the following areas: human services; civic affairs; and Jewish welfare. Other grant requests should be in the Foundation's other areas of interest which include: aiding the underserved, poor, and disadvantaged; improving the quality of life in urban communities; and enhancing Jewish community, education and spirituality. Interested applicants may submit an online letter of inquiry or applications may be mailed to the Foundation office.
Requirements: The Foundation makes grants to tax-exempt organizations.
Restrictions: The Foundation does not contribute to endowments, capital campaigns, capital construction projects, and academic or medical research programs of any kind.
Samples: Book Worm Angels, Chicago, IL—general operating funds; Hand in Hand/ Center for Jewish-Arab Education in Israel, Portland OR—Hand in Hand Expansion Project. Project will expand capacity of Hand in Hand's bilingual, multicultural schools for Jewish and Arab students in Israel; Metropolitan College of New York, New York, NY—to support the Office of Career Services.
Contact: Christine Flood, Grants Administrator; christine@irvinstern.org
Internet: http://www.irvinstern.org/guidelines.html
Sponsor: Irvin Suitern Foundation
25 West Hubbard Street, 4th Floor
Chicago, IL 60610

Isabel Allende Foundation Esperanza Grants **2451**

The Foundation is guided by a vision of a world in which women have achieved social and economic justice. The vision includes empowerment of women and girls and protection of women and children. The Foundation feels that the way to achieve empowerment is: reproductive self-determination; health care; and education. Application information is available online.
Requirements: 501(c)3 organizations (and equivalent international organizations) are eligible to apply. Priority is given to programs in the San Francisco Bay Area and Chile.
Restrictions: The foundation does not fund capital campaigns, individual trips or tours, conferences, or events; and projects that benefit political, religious, and/or military organizations. Individuals are not eligible to receive grants and should not apply.
Geographic Focus: California
Date(s) Application is Due: Jan 1; Apr 1; Jul 1; Oct 1
Amount of Grant: 1,000 - 5,000 U.S.D
Contact: Lori Barra, Executive Director; (415) 289-0992; fax (415) 289-1154; lori@isabelallendefoundation.org
Internet: http://www.isabelallendefoundation.org/iaf.php?l=en&p=application
Sponsor: Isabel Allende Foundation
116 Caledonia Street
Sausalito, CA 94965

ISOA/Elan Research Program **2452**

The objective of this program, which targets academia and the biotechnology industry, is to facilitate the discovery, development, and clinical evaluation of effective therapies for Alzheimer's disease. The goal is to catalyze and accelerate the development of innovative and effective treatments by funding the development of lead compounds through preclinical in-vitro and in-vivo evaluation, including pharmacology, toxicology, pharmacokinetics, formulation chemistry, and the conduct of preclinical proof-of-concept studies. One-year grants will be awarded. Proposals covering two years may be submitted, although the second year of funding will be contingent on progress of the completed work, the investigator's second-year work plan, and the numbers and quality of competing proposals. Application and guidelines are available online.
Requirements: The principal investigator must hold an MD or PhD degree.
Date(s) Application is Due: Oct 14
Amount of Grant: 130,000 U.S.D
Contact: Wendy Ramos, Grant and Database Manager; (212) 901-8005; fax (212) 935-2408; wramos@aging-institute.org
Internet: http://www.aging-institute.org
Sponsor: Institute for the Study of Aging
1414 Avenue of the Americas, Suite 1502
New York, NY 10153

Ittleson Foundation AIDS Grants **2453**

In regards to AIDS, the foundation is particularly interested in new model, pilot, and demonstration efforts: addressing the needs of underserved at-risk populations and especially those programs recognizing the overlap between such programs; responding to the challenges facing community-based AIDS service organizations and those organizations addressing systemic change; providing meaningful school-based sex education; making treatment information accessible, available and easily understandable to those in need of it; or, addressing the psycho-social needs of those infected and affected by AIDS, especially adolescents.
Requirements: Tax-exempt organizations may apply.
Restrictions: The foundation generally does not provide funds for capital building projects, endowments, grants to individuals, scholarships or internships (except as part of a program), direct service programs (especially outside New York City), projects that are local in focus and unlikely to be replicated, continuing or general support, projects and organizations that are international in scope or purpose, or biomedical research.
Date(s) Application is Due: Sep 1
Samples: AIDS Alliance for Children Youth and Family, Washington, DC, $5,000—to transform the National Consumer Leadership Corps Training Program into one that can be replicated by local AIDS organizations around the nation; Cesar E. Chavez Institute, New York, NY, $40,000—one-time grant to develop new family interventions and a new family-related model of care to reduce risk for HIV and mental health problems in lesbian, gay, bisexual and transgender (LGBT) youth.
Contact: Anthony C. Wood, Executive Director; (212) 794-2008; fax (212) 794-0351
Internet: http://www.ittlesonfoundation.org/aids.html
Sponsor: Ittleson Foundation
15 E 67th Street, 5th Floor
New York, NY 10021

Ittleson Foundation Mental Health Grants **2454**

For this program, the foundation is interested in innovative, pilot, model and demonstration projects that are: fighting the stigma associated with mental illness and working to change the public's negative perception of people who have mental illness; utilizing new knowledge and current technological advances to improve programs and services for people who have mental illness; bringing the full benefits of this new knowledge and technology to those who presently do not have access to them; or, advancing preventative mental health efforts, especially those targeted to youth and adolescents, with a special focus on strategies that involve parents, teachers, and others in close contact with these populations.
Requirements: Tax-exempt organizations may apply.
Restrictions: The foundation generally does not provide funds for capital building projects, endowments, grants to individuals, scholarships or internships (except as part of a program), direct service programs (especially outside New York City), projects that are local in focus and unlikely to be replicated, continuing or general support, projects and organizations that are international in scope or purpose, or biomedical research.
Date(s) Application is Due: Sep 1
Samples: Active Minds, Inc., Washington, DC, $10,000—to dramatically expand the network of chapters on college campuses; Partnership with Children, New York, NY, $50,000—one-time grant to develop the Center for Capacity Building to enable the dissemination of the successful counseling and prevention care program for inner-city boys and girls, Open Heart-Open Mind to over 100 schools by 2012; Horticultural Society of New York, New York, NY, $30,000—to launch their new Nonprofit Partnership for Horticultural Therapy to formalize and expand their ability to help a wide range of organizations working with the mentally and physically ill, formerly homeless, HIV+, victims of substance abuse, the elderly, at-risk juveniles and those re-entering society from incarceration, use horticultural therapy programs to address the needs of these marginalized populations.
Contact: Anthony C. Wood, Executive Director; (212) 794-2008; fax (212) 794-0351
Internet: http://www.ittlesonfoundation.org/mental.html
Sponsor: Ittleson Foundation
15 E 67th Street, 5th Floor
New York, NY 10021

IWMF Public Health Fellowship **2455**

The goals of the fellowship are to offer women newspaper editors and radio producers the opportunity to enhance their ability to cover public health issues, build their journalism skills, and learn first-hand about standards and practices in the U.S. media. The six-month fellowship includes four months in the United States and two months in the fellows' home countries, where the fellows will implement a public health reporting project. Each applicant must clearly identify and describe a public health reporting project that she will focus on when she returns to her home country for that segment of the fellowship. Those demonstrating original approaches to covering HIV/AIDS will be given top consideration. The fellowship includes application costs for a U.S. visa; round-trip economy airfare from the fellows' home countries to Washington, DC; round-trip economy airfare from Washington, DC, to the fellowship site and health insurance; and a fixed stipend to cover lodging, meals, and ground transportation during the U.S. segment of the fellowship. For the home country segment, a fixed stipend will be provided to support the fellow and cover all costs associated with implementing her project. Application and guidelines are available online.
Requirements: The fellowship is open to women journalists who are working in one of the following countries: Asia (China and Indonesia), West Africa (Benin, Burkina Faso, Cameroon, Cape Verde, Chad, Gambia, Ghana, Guinea, Guinea Bissau, Ivory Coast, Liberia, Mali, Mauritania, Niger, Nigeria, Senegal, Sierra Leone, and Togo), and Southern Africa (Angola, Botswana, Lesotho, Malawi, Mozambique, Namibia, Swaziland, Zambia, and Zimbabwe). A successful applicant will be dedicated to a career in journalism and have five or more years of full-time experience in the profession, with at least three of those years having been focused on public health journalism. She must be currently working as a newspaper editor or radio producer and show a strong commitment to sharing knowledge and skills with colleagues upon returning home. The fellow must also have excellent written and spoken English skills.
Date(s) Application is Due: Apr 15
Contact: Administrator; (202) 496-1992; fax (202) 496-1977; fellowship@iwmf.org
Internet: http://iwmf.org/programs/8702
Sponsor: International Women's Media Foundation
1625 K Street NW, Suite 1275
Washington, DC 20006

Izumi Foundation Grants 2456

The mission of the foundation is to alleviate human suffering through improved health care, especially the poorest and the most vulnerable members of society. The foundation's current geographic focus area is a select number of countries in Africa and Central and South America. Program interests include disease, disabilities, and inadequate health care related to poverty. Within its expressed goals and designated geographic areas, the foundation seeks to support projects and programs that distribute medicines and vaccines, and provide direct services and care; provide interventions that encourage the most effective use of limited, local health care resources; address the underlying causes of disease and persistent health care problems; build health care capacity on the community level; provide cost-effective strategies for prevention; and advance health care partnerships and collaborations. The foundation does not accept unsolicited proposals. Potential applicants are asked to submit a letter of inquiry that does not exceed three pages in length. Letters of inquiry may be submitted any time during the year.

Requirements: Organizations within the United States must have 501(c)3 tax-exempt status to be eligible to apply. Agencies and organizations outside of the United States will also be required to furnish documentation verifying their nonprofit status.

Restrictions: The foundation does not fund medical research or other research-related activities; income generating, endowments, or fundraising activities; ongoing general operating expenses, indirect costs, or existing deficits; capital costs—the acquisition of land, buildings, or vehicles; direct support to individuals, scholarships, school fees, or clothing; or religious activities, media campaigns, or lobbying of any kind.

Amount of Grant: Up to 100,000 U.S.D

Contact: Catherine Bryant, Program Officer; (617) 303-0345; fax (617) 303-0339; cbryant@izumi.org

Internet: http://www.izumi.org

Sponsor: Izumi Foundation

One Financial Center, 28th Floor

Boston, MA 02111

J.B. Reynolds Foundation Grants 2457

The foundation makes grants in the local Kansas City, MO, area, up to a 150-mile radius of the city. Grants are awarded for building and equipment, community development, medical research, social welfare, and the arts and humanities. Some support is given to colleges and universities. Additional types of support include general operating support, continuing support, annual campaigns, endowment funds, publications, and research. The board meets in April and December of each year to consider letters of requests, and only invited proposals are reviewed. All grants are awarded in December.

Requirements: 501(c)3 organizations in the Kansas City, MO, area may submit applications.

Geographic Focus: Missouri

Amount of Grant: 5,000 - 50,000 U.S.D

Contact: Richard L. Finn, Secretary-Treasurer; (816) 753-7000; fax (816) 753-1354

Sponsor: J.B. Reynolds Foundation

P.O. Box 219139

Kansas City, MO 64141-6139

J.C. Penney Company Grants 2458

The corporate contributions program awards grants to 501(c)3 nonprofit organizations in the company's areas of interest, including improvement of K- through 12th-grade education through curriculum-based after school care, with a priority on JCPenney Afterschool; support/promotion of associate (employee) volunteerism, primarily through the James Cash Penney Awards for Community Service; and United Way (most of JCPenney's support for health and welfare issues is contributed through local United Ways). Funding for other types of programs and projects is limited. Funding is concentrated on projects and organizations that serve a broad sector of the community; national projects that have a multiplier effect by benefiting local organizations across the country; nonprofits offering direct services; and projects with proven track records. Proposals are accepted year-round. There is no formal grant application. Interested applicants should submit a brief letter of inquiry. Submission information is available online.

Requirements: JCPenney only considers grants to tax-exempt organizations with 501(c)3 status or organizations that are a political subdivision of the state as described 170(c)1 of the IRS.

Restrictions: JCPenney refrains from or limits support of: individuals (including family reunions); individual student exchange/travel programs; membership organizations, unless the project benefits the entire community; religious organizations, unless the project benefits the entire community; journal or program advertising; fundraising dinners, luncheons or other types of benefits; merchandise donations; door prizes, gift certificates, or other giveaways; conferences and seminars; capital campaigns and multiyear pledges; international projects except in countries with corporate business locations; pilot projects; film and video projects; research projects; scholarships, except at colleges and universities with which the company has a recruiting relationship; higher education institutions, unless the company has a business or recruiting relationship; individual K-12 schools, unless the company has a business partnership with the school (including proms, graduations, PTOs and PTAs); or donations of returned, damaged or excess merchandise. JCPenney does not offer a matching gifts program at this time.

Contact: Jeannette M. Siegel, Community Relations and Contributions Manager; (972) 431-1349; fax (972) 431-1355; jsiegel@jcpenney.com

Internet: http://www.jcpenney.net/about/social_resp/community/default.aspx

Sponsor: J.C. Penney Company

P.O. Box 10001

Dallas, TX 75301-8101

J.E. and L.E. Mabee Foundation Grants 2459

The purpose of the Foundation is to aid Christian religious organizations, charitable organizations, institutions of higher learning, hospitals and other organizations of a general charitable nature. The Foundation favors organizations that combine sound character and stability with progressiveness and purpose. There is no formal application procedure. Proposal submission information is available online.

Requirements: Nonprofit, tax exempt, and not tax- supported organizations in Arkansas, Kansas, Missouri, New Mexico, Oklahoma, and Texas are eligible.

Restrictions: The foundation does not generally favor grants for deficit financing and debt retirement (except a construction contract that was executed prior to the time of application); operating or program funds or annual fundraising campaigns (except Junior Achievement and United Way in Tulsa, OK, and Midland, TX); reserve purposes or for projects likely to be long delayed; endowments; governmental owned or operated institutions and/or facilities (such as state universities and municipal parks and libraries); educational institutions below the college level; furnishing or equipment (except major medical equipment); or churches.

Geographic Focus: Arkansas, Kansas, Missouri, New Mexico, Oklahoma, Texas

Contact: Thomas R. Brett, Director; (918) 584-4286

Internet: http://www.mabeefoundation.com/policies.htm

Sponsor: J.E. and L.E. Mabee Foundation

Mid-Continent Tower, 401 South Boston Avenue, Suite 3001

Tulsa, OK 74103-4017

J. Homer Butler Foundation Grants 2460

The foundation awards grants to Roman Catholic nonprofits nationwide and internationally, with a focus on New York. Recipient organizations include child welfare organizations, churches, community service organizations, health care organizations and hospitals, international ministries and relief organizations, missions, parochial schools and religious education organizations, seminaries, and women affairs organizations. Types of support include emergency funds, general support grant, multiyear grants, project development grants, special project grants, and scholarships. Application must be made in writing.

Date(s) Application is Due: Jun 30; Dec 31

Amount of Grant: 1,000 - 3,000 U.S.D

Samples: Xavier Jesuit Community (New York, NY)—for general support, $5000; Gospel Outreach (Hillsboro, MO)—for leprosy programs in India, $2000.

Contact: Grant Administrator; (212) 242-7340; fax (718) 442-5088

Sponsor: J. Homer Butler Foundation

30 W 16th Street

New York, NY 10011

J.L. Bedsole Foundation Grants 2461

The Foundation considers requests that most closely match its overall mission: to improve the quality of life for the citizens of Southwest Alabama and to strengthen the communities in which they live. The Foundation will consider only those grant applications that meet several of the following guidelines. The project addresses needs in at least one of the following areas: education, arts and culture, health and human services and economic development; potential for permanent, enduring benefits that will provide value to the community and the residents of Southwest Alabama; diverse groups are collaborating on the project to achieve common goals; the organization and the project clearly demonstrate sound fiscal management and accountability; the organization attracts multiple sources of support for the project; the project addresses underserved segments of the population, the economically disadvantaged or citizens of rural communities. The Foundation offers a scholarship program as well, see: http://www.jlbedsolefoundation.org/default.asp?ID=7 for details and guidelines.

Requirements: Nonprofit organizations in Mobile, Baldwin, Clarke, Monroe and Washington County, Alabama are eligible to apply.

Restrictions: The Foundation will not support the following: grants that support political activities or attempts to influence action on specific legislation; multiple year pledges nor make grants beyond the current year; grants to endowment funds of other organizations; grants are not made directly to individuals; grants are not made to organizations, programs or projects outside of the State of Alabama.

Geographic Focus: Alabama

Contact: Christopher L. Lee, Executive Director; (251) 432-3369; fax (251) 432-1134; chrislee@jlbedsolefoundation.org

Internet: http://www.jlbedsolefoundation.org/default.asp?ID=6

Sponsor: J.L. Bedsole Foundation

P.O. Box 1137

Mobile, AL 36633

J.M. Long Foundation Grants 2462

The Foundation is focused on providing support to organizations involved with health care, education, and conservation in the communities of Northern California and Hawaii. Preference will be given for new, innovative projects, which can be completed with the Foundation contribution. Applications for new grant requests are given by invitation only. In order to receive an invitation, an organization will need to submit to the Foundation a single page Request for Invitation letter using organization letterhead. This Request for Invitation letter should describe the organization's objective and the specific project for which the grant would be used. The applicant should be sure to include an estimated amount for the grant, as well as contact information including email address if available.

Requirements: Hawaiin and Northern California nonprofits are eligible to apply.

Restrictions: Please note that grants will not be made: for the support of solely religious, sacramental or theological functions; in support of political bodies or campaigns; to individuals or foreign organizations; for loans; for purposes of memorializing an

individual, although donations may be made to memorial funds as a means of achieving other purposes; for covering operating deficits; to organizations not qualifying as either a federal tax-exempt non-profit private organization or a qualifying public organization; to supporting organizations with a 509(a)(3) designation.
Geographic Focus: California, Hawaii
Amount of Grant: 1,000 - 50,000 U.S.D
Samples: California State Parks Foundation, Kentfield, CA - Park Education Legacy Program; Children's Discovery Center, Honolulu, HI - Ready, Set, Grow Program; The Creek, A Middle School Youth Center, Walnut Creek, CA - Start up Funding.
Contact: Brenda Kauten, Grants Administrator; (925) 935-4138; fax (925) 935-2092
Internet: http://www.jmlongfoundation.org/Grants.html
Sponsor: J.M. Long Foundation
P.O. Box 3827
Walnut Creek, CA 94598-2827

J.M. McDonald Foundation Grants 2463
Grants are awarded to nonprofit organizations for educational purposes; for the aged, orphans, and children who are sick, infirm, blind, or crippled; for youth and child care in an effort to combat juvenile delinquency; and to aid underprivileged, mentally or physically handicapped children. Grants are also given for building funds, equipment, and general purposes to educational institutions and hospitals. Application guidelines will be sent upon written request.
Requirements: New York organizations classified as IRS 509(a) tax-exempt are eligible.
Restrictions: Grants are not made to individuals, or for seminars, workshops, endowment funds, fellowships, travel, exhibits, or conferences.
Geographic Focus: New York
Date(s) Application is Due: Apr 15; Sep 15
Amount of Grant: 5,000 - 100,000 U.S.D
Contact: Donald McJunkin, President; (303) 674-9300; fax (303) 674-9216
Sponsor: J.M. McDonald Foundation
P.O. Box 3219
Evergreen, CO 80437

J.W. and Ida M. Jameson Foundation Grants 2464
The foundation supports higher and theological education, hospitals, medical research, cultural programs, and Protestant and Catholic religion. Types of support include research and general operating budgets. Applicants should submit a proposal.
Requirements: California nonprofits may submit applications for grant support.
Geographic Focus: California
Date(s) Application is Due: Feb 1
Amount of Grant: 5,000 - 50,000 U.S.D
Contact: Les Hugn, President; (626) 355-6973
Sponsor: J.W. and Ida M. Jameson Foundation
P.O. Box 397
Sierra Madre, CA 91025

J.W. Kieckhefer Foundation Grants 2465
The Kieckhefer Foundation awards grants to nation-wide 501(c)3 organizations in support of medical research, hospices, and health agencies; family planning services; social services; higher education; youth and child welfare agencies; ecology and conservation; community funds; and cultural programs. Types of support include the following: annual campaigns; building renovation; conferences and seminars; continuing support; emergency funds; endowments; equipment; general operating support; land acquisition; matching and challenge support; program development; publication; and research.
Requirements: Applications are not accepted. Organizations should submit a letter of inquiry to the Foundation, with a description of their project and amount requested.
Restrictions: Grants are not awarded to individuals.
Contact: John I. Kieckhefer, Trustee; (928) 445-4010
Sponsor: J.W. Kieckhefer Foundation
116 East Gurley Street
Prescott, AZ 86301-3821

J. Walton Bissell Foundation Grants 2466
The J. Walton Bissell Foundation Grants support 501(c)3 organizations in Connecticut, with emphasis on Hartford. The Foundation gives primarily to the arts and social services, including child welfare and programs for the blind. Funding requests should be submitted four months before funding is needed. Types of support include general operating support, program development, and seed money.
Requirements: Applicants should submit a letter of inquiry, along with a copy of their IRS determination letter, and a copy of their annual report, audited financial statement, or 990.
Restrictions: The Foundation does not grant funding to individuals or endowments.
Geographic Focus: Connecticut
Contact: J. Danford Anthony, Jr., President; (860) 586-8201
Sponsor: J. Walton Bissell Foundation
P.O. Box 370067
West Hartford, CT 06137

Jackson County Community Foundation Unrestricted Grants 2467
The Foundation awards grants to nonprofit organizations covering a full range of charitable activity such as health and human services, education, economic/community development, arts and culture, and environmental quality and protection. All grants are evaluated on a competitive basis and priority is given to those projects that show potential for providing maximum benefit to community members. Application information is available online.
Requirements: The Foundation makes grants to Jackson County, Michigan nonprofit, tax-exempt organizations.
Restrictions: While the Foundation remains flexible in trying to address the needs of the community, it do not make grants to individuals (except scholarships), programs that advocate specific religions, or services which are standard government or school obligations. Applications from organizations not serving Jackson County are not accepted.
Geographic Focus: Michigan
Date(s) Application is Due: Jan 15; May 15; Sep 15
Amount of Grant: 3,000 - 50,000 U.S.D
Samples: Smiles on Wheels Preventative Dental Health Care Project, $5,000—to purchase dental supplies and equipment; Jackson School of the Arts, $10,000—Operational support; Center for Family Health, $30,000—Jackson High Teen Health Center.
Contact: Diane McDonald, Grant & Scholarship Coordinator; (517) 787-1321; fax (517) 787-4333; jcf@jacksoncf.org
Internet: http://www.jacksoncf.org/grants.html
Sponsor: Jackson County Community Foundation
1 Jackson Square, 100 E Michigan Avenue, Suite 308
Jackson, MI 49201-1406

Jackson Laboratory Postdoctoral Research Training Program 2468
Fellowships are intended for postdoctoral associateships at Jackson Laboratory allowing the fellows to participate in research into gene regulation, developmental genetics, hematology and immunology, and other fields of mammalian biology, with the purpose of the recipient receiving research training under the guidance of experienced researchers. The fellowships are awarded for one year, renewable for up to two years. Applications are accepted at any time.
Requirements: Recent recipients of doctoral degrees in medicine, veterinary medicine, and the biological sciences are eligible to apply. Applicants may be of any nationality.
Contact: Suzanne Serreze, Manager, Sponsored Training Programs; (207) 288-6420; fax (207) 288-6079; sbs@jax.org
Internet: http://www.jax.org/education/postdoc.html
Sponsor: Jackson Laboratory
600 Main Street
Bar Harbor, ME 04609-1500

Jacob and Charlotte Lehrman Foundation Grants 2469
The foundation makes grants to establish scholarships and fellowships at institutions of learning and to foster research in medicine and science. Grants are also given for Jewish welfare funds, care of the aged and sick, the establishment of trade schools, the fostering of religious observance, recreation, and aid to refugees. Additional types of support include general operating grants and matching funds. The Foundation Board generally makes grants once a year, typically in October. The deadline for RFPs as well as proposals submitted from current grant recipients is April 1.
Requirements: Grants are made primarily to organizations in the greater metropolitan Washington, DC, area.
Restrictions: Grants are not made to individuals.
Geographic Focus: District of Columbia
Date(s) Application is Due: Apr 1
Amount of Grant: 1,000 - 10,000 U.S.D
Contact: Robert Lehrman, Vice President; (202) 338-8400; fax (202) 338-8405; info@lehrmanfoundation.org
Internet: http://www.lehrmanfoundation.org
Sponsor: Jacob and Charlotte Lehrman Foundation
1027 33rd Street NW, 2nd Floor
Washington, DC 20007

Jacob and Hilda Blaustein Foundation Grants 2470
The Foundation promotes social justice and human rights through its five program areas: Jewish life; strengthening Israeli democracy; health and mental health; educational opportunity; and human rights. Support is provided to organizations in the United States and abroad, with a preference to the Baltimore, Maryland area. The Foundation supports organizations that promote systemic change; involve constituents in planning and decision-making; encourage volunteer and professional development; and engage in ongoing program evaluation. Additional guidelines and application information is available online.
Requirements: Nonprofit organizations are eligible to apply.
Restrictions: Support is unavailable for the following: individuals; scholarships to individuals; unsolicited proposals for academic, scientific, or medical research; direct mail; annual giving; membership campaigns; fundraising; commemorative events.
Geographic Focus: Maryland
Samples: The Foundation for Jewish Campus Life, Washington, DC, $150,000—support offered over a two year period, for Tzedek Hillel, a program that promotes social justice activities among Jewish college students across the country, and support for the Tzedek Leadership Mission to Israel; Fund for Educational Excellence, Baltimore, MD, $250,000—support for over five years, for the implementation of a reform plan for Baltimore's neighborhood high schools; Mental Disability Rights International, Washington, DC, $70,000—support paid over two years, to promote the human rights of people with mental disabilities worldwide.
Contact: Betsy Ringel, Executive Director; (410) 347-7103; fax (410) 347-7210; info@blaufund.org
Internet: http://www.blaufund.org/foundations/jacobandhilda_f.html
Sponsor: Jacob and Hilda Blaustein Foundation
10 E Baltimore Street, Suite 1111
Baltimore, MD 21202

Jacob and Valeria Langeloth Foundation Grants 2471

The Foundation's grantmaking program is centered on the concepts of health and well-being. The Foundation's purpose is to promote and support effective and creative programs, practices and policies related to healing from illness, accident, physical, social or emotional trauma and to extend the availability of programs that promote healing to underserved populations. The Foundation is particularly interested in funding programs that address the health of individuals who, because of barriers to difficulty accessing care. Preference is given in the areas of: caregivers and; correctional health care.

Requirements: 501(c)3 organizations that promote physical and emotional healing, especially to underserved populations, such as: community-based organizations, health care providers and research institutions in the state of New York are eligible to apply.

Restrictions: The Foundation will not consider proposals for annual or capital campaigns, for building or renovation projects, for budgetary relief, or preventive medicine. The Foundation does not support projects that focus on children or end-of-life issues. Neither will the Foundation make grants to individuals or for sectarian or religious purposes, or for political activities or lobbying. The Foundation will not fund any organization that discriminates on the basis of age, gender, national origin, race, or sexual preference.

Geographic Focus: New York

Date(s) Application is Due: Feb 1; Aug 1

Samples: AARP Foundation, Washington, DC, $261,000— for professional partners supporting family caregivers; Foundation for Quality Care, Albany, NY, $171,000— for geriatric nursing assistants career development program; New York Academy of Medicine, New York, NY, $200,000— for Prison Health Reentry initiative.

Contact: Scott Moyer; (212) 687-1133; fax (212) 681-2628; smoyer@langeloth.org

Internet: http://www.langeloth.org/apply.php

Sponsor: Jacob and Valeria Langeloth Foundation

521 Fifth Avenue, Suite 1612

New York, NY 10175-1699

Jacob G. Schmidlapp Trust Grants 2472

The trust supports tax-exempt organizations in the greater Cincinnati, OH, area. The trust supports charitable or educational purposes; for relief in sickness, suffering, and distress; for the care of young children, the aged, or the helpless; and for the promotion of education to improve living conditions. Support is offered in the form of: endowments; equipment; land acquisition; program development; technical assistance; and seed money grants. An organization interested in submitting a grant proposal should initially contact the Foundation office via a letter of inquiry, an application form is required for proposals. Proposals are reviewed quarterly, March, June, September and, December.

Requirements: Nonprofits in the greater Cincinnati, Ohio area receive the majority of funding, however giving is also available in the surrounding states of Indiana, Kentucky and Michigan.

Restrictions: No support for religious or political purposes. No grants to individuals; no loans.

Geographic Focus: Indiana, Kentucky, Michigan, Ohio

Date(s) Application is Due: Feb 1; May 1; Aug 1; Nov 1

Amount of Grant: 2,000 - 500,000 U.S.D

Samples: Jewish Community Center, Cincinnati, OH, $500,000— for capital support; Spectrum Health Foundation, Grand Rapids, MI, $100,000— for capital support; Children Inc., Covington, KY, $100,000— for project/program support; Xavier University, Cincinnati, OH, $250,000—for scholarship.

Contact: Heidi B. Jark, Charitable Services, c/o Fifth Third Bank; (513) 534-4397

Internet: https://www.53.com:443/wps/portal/personal

Sponsor: Jacob G. Schmidlapp Trust

38 Fountain Square Plaza, MD 1090CA

Cincinnati, OH 45263-0001

Jacobs Family Foundation Village Neighborhoods Grants 2473

The Foundation primarily serves the four neighborhoods immediately surrounding The Village at Market Creek: Chollas View, Emerald Hills, Lincoln Park, and Valencia Park; proposals are considered from throughout the Fourth City Council District of southeastern San Diego. The Neighborhood Grants Program funds programs or strategies that: strengthen and expand health, environmental, education and family resources; foster opportunities for ownership and building assets through economic and business development; create vibrant places and spaces which express and enhance the cultures and environment of the community; and/or strengthen the ability of the community to get things done and include the voice of residents in decision-making, advocacy, and planning. The Village Neighborhoods Fund makes grants to resident groups and neighborhood organizations that serve or support The Village at Market Creek. Funding is also available for a limited number of grants that support projects outside the neighborhoods that demonstrate innovative community and economic development strategies and a commitment to share what works, what doesn't, and why. Applications are excepted year round, however unsolicited applications are not. Call the Foundation to discuss your project. If the project meets the Foundations criteria, you may be asked to send a brief letter of interest and/or complete a full Village Neighborhoods Fund application.

Requirements: 501(c)3 non-profits may apply that: work with the Jacobs Center to develop The Village at Market Creek in southeastern San Diego; provide programs or services that benefit the neighborhoods surrounding The Village at Market Creek; or provides neighborhood strengthening strategies or models that can inform and advance The Village at Market Creek work.

Geographic Focus: California

Amount of Grant: 10,000 - 75,000 U.S.D

Contact: Program Contact; (619) 527-6161 or (800) 550-6856; fax (619) 527-6162; communications@jacobscenter.org

Internet: http://www.jacobsfamilyfoundation.org/how.htm

Sponsor: Jacobs Family Foundation

Joe & Vi Jacobs Center, 404 Euclid Avenue

San Diego, CA 92114

James & Abigail Campbell Family Foundation Grants 2474

The James & Abigail Campbell Family Foundation embraces the values and beliefs of James and Abigail Campbell by investing in Hawwaii's people and the communities that nuture them. The Foundation supports projects in the following areas: Youth—programs that address the challenges of young people; Education—support for public schools, early childhood education and environmental stewardship; Hawaiian—support for programs that promote values and the health and welfare of Hawaiians. Priority is given to programs located in or serving communities in the following areas of West Oahu: Ewa/Ewa Beach, Kapolei, Makakilo and the Wai'anae Coast. The following types of requests are eligible for consideration: support for special projects that are not part of an organization's ongoing operations; program support when unforeseen circumstances have affected the financial base of an organization; financial assistance to purchase items such as office equipment and to fund minor repairs and renovations. Grants range from $5,000 - $50,000. Your grant application must be postmarked by February 1 for the April/May meeting, August 1 for the October/November meeting.

Requirements: The Foundation will only consider requests from organizations which qualify as non-profit, tax-exempt public charities under Section 501(c)3 and 170(b) of the Internal Revenue Code. To apply for a grant, summarize the following information in a two - three page proposal letter: the nature and purpose of your organization; the objectives of your program, include the grant amount requested and the proposed use of funds; a brief outline on how you plan to accomplish your objectives; a statement of a community problem, need or opportunity that this project will address; the duration for which Foundation funds are needed; other sources of funding currently being sought and future funding sources; methods used to measure the program's effectiveness. In addition to the proposal letter, submit a copy of the following: Internal Revenue Service notification of tax-exempt status; most recent annual financial statement; list of the current Board of Directors; the project's proposed budget; one (1) copy of your complete grant proposal package.

Restrictions: The Foundation will not consider funding for: individuals, endowments, sectarian or religious programs, loans, political activities or highly technical research projects. Only one request per organization will ordinarily be considered in a calendar year. Funds are usually not committed for more than one year at a time.

Geographic Focus: Hawaii

Date(s) Application is Due: Feb 1; Aug 1

Amount of Grant: 5,000 - 50,000 U.S.D

Samples: Kapolei High School, $20,000—scholarships grant; Kahikolu Ohana Hale O Wai'anae, $50,000—Hawaiian culture grant; KAMP Hawai'i, $5,000—youth grant; International Dyslexia Association, $10,000—educational grant.

Contact: D. Keola, Grant Manager; (808) 674-3167; fax (808) 674-3349; keolal@jamescampbell.com

Internet: http://www.campbellfamilyfoundation.org/grant_procedures.cfm

Sponsor: James and Abigail Campbell Family Foundation

1001 Kamokila Boulevard

Kapolei, HI 96707

James A. and Faith Knight Foundation Grants 2475

Primarily serving Jackson and Washtenaw counties, Michigan, the Foundation is dedicated to improving communities by providing grant support to qualified nonprofit organizations including, but not limited to, those that address the needs of women and girls, animals and the natural world, and internal capacity. In general, the foundation supports organizations that believe in good nonprofit practices including sound financial management, developed governance practices, and clarity of mission. Attention to gender, diversity, and outcomes are important. Giving is primarily for human services, including a neighborhood center, women's organizations, and family services; support also for nonprofit management, the United Way housing, the arts, education, and environmental conservation. Fields of interest include: adult education and literacy; basic skills; GED; arts; environment and natural resources; family services; housing development; human services; nonprofit management; and women's services. There are two annual deadlines, and initial approach should be by letter.

Requirements: Giving is limited to Michigan, with emphasis on Jackson and Washenaw counties.

Restrictions: No support is offered for religious or political organizations. No grants to individuals, or for conferences or special events, or for annual campaigns.

Geographic Focus: Michigan

Date(s) Application is Due: Jan 28; Sep 16

Contact: Margaret A. Talburtt, Executive Director; (734) 769-5653; fax (734) 769-8383; peg@KnightFoundationMi.org or info@knightfoundationmi.org

Internet: http://www.knightfoundationmi.org/guidelines.htm

Sponsor: James A. and Faith Knight Foundation

180 Little Lake Drive, Suite 6B

Ann Arbor, MI 48103-6219

James and Elsie Nolan Charitable Trust Grants 2476

The trust awards grants, primarily in southeastern Alaska, in its areas of interest, including museums, hospitals, and public administration. General operating and project grants are awarded. There are no application deadlines. The board meets in early October.

Requirements: Alaska nonprofit organizations are eligible.

Restrictions: Individuals are ineligible.

Geographic Focus: Alaska

Amount of Grant: Up to 22,000,000 U.S.D
Samples: Catholic Community Services of Juneau (AK)—for wheelchair-accessible vehicles, $17,000; Junior Achievement in Anchorage (AK)—for computer equipment; Wrangell Medical Ctr (AK)—for mammogram machine and endoscopes, $105,000; Irene Ingle Public Library (Wrangell, AK)—for library book purchases and a learning project, $4164.
Contact: Gwendolyn Feltis, c/o Wells Fargo Bank Alaska, N.A., N.A.
Sponsor: James and Elsie Nolan Charitable Trust
P.O. Box 927
Wrangell, AK 99929

James Ford Bell Foundation Grants **2477**
The Foundation supports organizations primarily in Minnesota. Emphasis is on cultural programs, support is also available for wildlife preservation and conservation, youth agencies, the environment, education, health and human services. A high priority is given to projects with historical connections to the Bell Family. The Trustees meet in the Spring and Fall. Contact the Foundation prior to sending in a proposal. Unsolicited requests for funds are not accepted.
Requirements: Nonprofit organizations in Minnesota are eligible to apply.
Restrictions: No grants are made directly to individuals, nor for scholarships, fellowships, or political campaigns. No funding is available to units of local government. The Foundation does not respond to requests for memberships, annual appeals, or special events and fundraisers.
Geographic Focus: Minnesota
Contact: Ellen George; (612) 377-8400; fax (612) 377-8407; ellen@fpadvisors.com
Internet: http://www.fpadvisors.com/jamesfordbell/jamesfordbell.htm
Sponsor: James Ford Bell Foundation
1818 Oliver Avenue South
Minneapolis, MN 55405-2208

James G.K. McClure Educational and Development Fund Grants **2478**
The fund devotes most of its resources to an ongoing scholarship program for the students of western North Carolina. When funds are available, the trustees make grants to organizations in the region, primarily those that enhance the lives of the rural citizens of the area. The trustees look favorably on requests from institutions helping to improve the lives of the people within its geographic area. Hospitals, colleges, libraries, and various schools are common recipients. The trustees meet twice each year, usually in May and October.
Requirements: Applicants must be residents of the following counties: Allegheny, Ashe, Avery, Buncombe, Burke, Caldwell, Cherokee, Clay, Graham, Haywood, Henderson, Jackson, Macon, Madison, McDowell, Mitchell, Polk, Rutherford, Swain, Transylvania, Watauga, and Yancey.
Restrictions: No student may apply for a scholarship unless he or she resides in one of the specified counties, and no grant money will be made to organizations outside this district.
Geographic Focus: North Carolina
Amount of Grant: 1,000 - 5,000 U.S.D
Contact: John Curtis Ager, Executive Director; (704) 628-2114; jager@ioa.com
Sponsor: James G.K. McClure Educational and Development Fund
11 Sugar Hollow Lane
Fairview, NC 28730

James Graham Brown Foundation Quality of Life Grants **2479**
The James Graham Brown Foundation fosters the Quality of Life of Louisville by helping to make it a community of choice for families and businesses. The Foundation gives priority to projects that: strengthen the impact of core human services and cultural organizations and agencies in the community; support efforts to improve and sustain quality neighborhoods and a thriving downtown; and support initiatives that strengthen the relationship between business and education to increase growth in human capital and jobs in higher wage, higher knowledge, and high technical areas. It actively supports and funds projects in the fields of education, economic development, health and social services, culture and humanities (excluding performing arts) with an emphasis on community-wide capital campaigns. The Board meets six times annually and determinations are made at the meetings. Funds are dispersed semi-annually.
Requirements: Formal applications must be submitted by the following deadline dates: March 2, Social Services requests; and May 8, Culture and Civic requests. Only organizations that have obtained a tax-exempt designation under Section 501(c)3 of the I.R.S. code may apply. The Foundation grants are limited to the Louisville metropolitan area and the state of Kentucky. Campaigns outside the Jefferson County and Louisville metropolitan area must show evidence of significant local community support before being considered by the foundation for funding.
Restrictions: The Foundation does not support the following: organizations outside of the state of Kentucky; those related either directly or indirectly to the performing arts; requests from religious organizations for religious purposes (including theological seminaries); requests from individuals; requests from political entities; requests from national organizations, even if for local projects.
Geographic Focus: Kentucky
Date(s) Application is Due: Mar 2; May 8
Contact: Dodie L. McKenzie, Grants Director; (502) 896-2440 or (866) 896-5423; fax (502) 896-1774; grants@jgbf.org
Internet: http://www.jgbf.org/Home/Grant-Seeking-Tips/Funding-Areas.aspx
Sponsor: James Graham Brown Foundation
4350 Brownsboro Road, Suite 200
Louisville, KY 40207

James H. Cummings Foundation Grants **2480**
The Foundation, gives exclusively for charitable purposes in advancing medical science, research, and education in selected cities in the U.S. and Canada, and for charitable work among underprivileged boys and girls, and aged and infirm persons in designated areas. Priority is given to medical proposals. The funding is limited to: Toronto, Ontario; Canada; Hendersonville, North Carolina and; Buffalo, New York. Grants are available, in the following types of support: building/renovation; capital campaigns; equipment; land acquisition; matching/challenge support; research and seed money.
Requirements: The giving program is limited to the vicinity of the cities of Buffalo, NY; Hendersonville, NC; and Toronto, ON, Canada.
Restrictions: No support for national health organizations. Grants are not awarded to individuals, nor for: loans, annual campaigns, program support, endowment funds, operating budgets, emergency funds, deficit financing, scholarships, fellowships, publications, conferences, or continuing support.
Geographic Focus: New York, North Carolina, Canada
Amount of Grant: 2,000 - 250,000 U.S.D
Samples: Child and Family Services Conners Children's Ctr (Buffalo, NY)—to renovate this center that provides residential care for severely disturbed boys between the ages of six and 14, $100,000; YMCA of Henderson and Henderson County (NC)—program support, $62,500; Margaret R. Pardee Hospital (NC)—capital support, $100,000.
Contact: William McFarland, Executive Director; (716) 874-0040; fax (716) 874-0040; cummings.foundation@verizon.net
Sponsor: James H. Cummings Foundation
1807 Elmwood Avenue, Room 112
Buffalo, NY 14207

James Hervey Johnson Charitable Educational Trust Grants **2481**
The goal of the trust is to expose religion as against reason, publicize nontheistic views of religion, and publicize James Hervey Johnson's views on health, which are centered primarily in vegetarianism, natural hygiene, and alternative medicine. Interested applicants should contact the Foundation prior to submitting a proposal. The Foundation requires an application form and, will provide guidelines. There are no geographic restrictions, but preference is given to requests from San Diego County, CA.
Requirements: 501(c)3 nonprofits that have been in existence for at least two years are eligible.
Date(s) Application is Due: Jun 30
Amount of Grant: 5,000 - 100,000 U.S.D
Samples: Institute for the Study of the Religious Right—for a Web site that provides information on how the religious right threatens separation of church and state, $9300; Action on Smoking and Health—for a letter-writing contest on why the U.S. president should proceed with plans to toughen anti-smoking regulations, $7000; American Natural Hygiene Society—for a new desktop publishing system for producing publications advocating natural hygiene, $8622.
Contact: Kevin V. Munnelly, Treasurer; (619) 297-9036; promotions@home.com
Sponsor: James Hervey Johnson Charitable Educational Trust
P.O. Box 16160
San Diego, CA 92176-6160

James J. and Angelia M. Harris Foundation Grants **2482**
The foundation's primary areas of interested include: higher and other education, health services and hospitals, social services, youth, Christian organizations and, Presbyterian churches. Types of support include: annual campaigns, building construction/renovation, capital campaigns, challenge/matching grants, program development, seed money and, scholarship funds. There are no application deadlines or forms. The board meets in May and November to consider requests. Letters of inquiry are due in April and October.
Requirements: Nonprofit organizations in Clarke County, GA, and Mecklenburg County, NC, are eligible.
Geographic Focus: Georgia, North Carolina
Samples: Covenant Presbyterian Church, Charlotte, NC, $35,000; Billy Graham Evangelistic Association, Charlotte, NC, $100,000; YMCA of Greater Charlotte, Charlotte, NC, $70,000.
Contact: Sherri Harrell, Grants Administrator; fax (704) 364-6046
Sponsor: James J. and Angelia M. Harris Foundation
P.O. Box 220427
Charlotte, NC 28222-0427

James L. and Mary Jane Bowman Charitable Trust Grants **2483**
Established in 1997 in Virginia through a donation by James L. Bowman, the James L. and Mary Jane Bowman Charitable Trust Supports both education and libraries within Virginia. The Trust's primary fields of interest include: Christian agencies and churches, higher education, human services, public libraries, and recreation. There are no specific application forms or deadlines with which to adhere, and applicants should forward a letter of application that includes a needs statement, population to be served, and the project budget.
Requirements: 501(c)3 nonprofit organizations serving residents of Virginia are eligible.
Restrictions: The foundation does not award grants to individuals in the form of scholarships or other direct support; orpolitical causes or candidates.
Geographic Focus: Virginia
Date(s) Application is Due: Apr 1
Amount of Grant: 1,000 - 15,000 U.S.D
Contact: Beverley B. Shoemaker, Trustee; (540) 869-1800; fax (540) 869-4225
Sponsor: James L. and Mary Jane Bowman Charitable Trust
P.O. Box 480
Stephens City, VA 22655-0480

James M. Collins Foundation Grants 2484

The foundation awards grants to Texas nonprofit organizations in its areas of interest, including: arts, economic development, health organizations, association, higher education, human services, museums, Salvation Army, secondary school/education. Types of support include research grants, program support, and social services. There are no application deadlines or forms.

Requirements: Texas nonprofit organizations are eligible.

Restrictions: Individuals are not eligible.

Geographic Focus: Texas

Samples: Smithsonian Institution, Washington, DC, $50,475; Wellesley College, Wellesley, MA, $200,250; Dallas Museum of Art, Dallas, TX, $17,950.

Contact: Dorothy Dann Collins Torbert, President; (214) 691-2032

Sponsor: James M. Collins Foundation
8115 Preston Road, Suite 680
Dallas, TX 75225

James M. Cox Foundation of Georgia Grants 2485

The Foundation awards grants in Cox Enterprises communities in its areas of interest, including journalism, visual and performing arts, child development, education, healthcare, environment, wildlife, and community and family services. Types of support include building construction/renovation, capital campaigns, and program development. There are no application forms or deadlines. Interested applicants should send a request to the Foundation outlining project needs and goals.

Requirements: Nonprofits in Cox Enterprises communities are eligible.

Samples: Colorado State U (Fort Collins, CO)—to endow a chair within the Animal Cancer Center, $3 million.

Contact: Leigh Ann Launius, Assistant Secretary; (678) 645-0929; fax (678) 645-1708

Internet: http://media.corporate-ir.net/media_files/IROL/76/76341/reports/AR_2003/community.html

Sponsor: James M. Cox Foundation of Georgia
6205 Peachtree Dunwoody Road
Atlanta, GA 30328

James McKeen Cattell Fund Fellowships 2486

The nonrenewable fellowships advance the science of psychology and its useful applications by enabling faculty members to take a full year sabbatical leave for independent study and research. Annual deadline dates may vary; contact program staff for exact dates.

Requirements: Eligibility is limited to residents of North America. The fellowships are intended for tenured faculty of a psychology department who are eligible for sabbatical leave. Matching institutions' support required.

Date(s) Application is Due: Dec 1

Amount of Grant: Up to 32,000 U.S.D

Contact: Dr. Christina Williams, Program Contact; williams@psych.duke.edu

Internet: http://www.cattell.duke.edu

Sponsor: James McKeen Cattell Fund
9 Flowers Drive, Box 90086, Duke University
Durham, NC 27708-0086

James R. Dougherty Jr. Foundation Grants 2487

The James R. Dougherty, Jr. Foundation, established in 1950, gives to organizations primarily in Texas. Areas of interest include: family services, domestic violence, human services and women. Support is offered in the form of: annual campaigns, building/renovation, capital campaigns, continuing support, curriculum development, endowments, equipment, general/operating support, income development, management development/capacity building, matching/challenge support, program development, program evaluation, research, scholarship funds, seed money and, technical assistance grants. The board meets twice a year, in the Spring and Fall. No application form is required, contact the Foundation, before submitting a proposal.

Requirements: Nonprofit organizations in Texas are eligible to apply.

Geographic Focus: Texas

Date(s) Application is Due: Mar 1; Sep 1

Contact: Daren Wilder, Grants Administrator; (512) 358-3560

Sponsor: James R. Dougherty Jr. Foundation
P.O. Box 640
Beeville, TX 78104-0640

James R. Thorpe Foundation Grants 2488

The Foundation is interested in supporting organizations, programs or projects which address the needs of the Elderly and Youth. The Thorpe Foundation is most likely to make general operating or program support grants. In the area of Youth, the Foundation supports organizations and programs that: engage youth in the arts; encourage character development; support academic and social development; provide safety and support to youth. In the area of the Elderly, the Foundation fosters the vital aging of seniors through support of services which help them to live independently. The Foundation will give special consideration to services which address the needs of economically disadvantaged seniors, immigrant seniors and the frail elderly. The Foundation supports organizations and programs that: provide transportation services; assist seniors in remaining in their own homes; educate seniors about housing options; provide social and recreational opportunities for seniors. Additional guidelines are available at: http://www.jamesrthorpefoundation.org/2010procedures.html.

Requirements: The Thorpe Foundation supports 501(c)3 non-profit organizations located in and serving the Minneapolis and the western Minneapolis metro suburbs of the Twin Cities.

Restrictions: The Foundation does not: make multi-year grants or fund individuals; make grants in the east metro area, greater Minnesota, or outside the State of Minnesota; make grants through fiscal agents; support endowment drives, conferences, seminars, tours, or fundraising events; support organizations with operating budgets over $2 million; support organizations with no paid staff.

Geographic Focus: Minnesota

Date(s) Application is Due: Mar 1; Sep 1

Contact: Kerrie Blevins, Foundation Manager; (612) 822-3412; kerrieblevins@jamesrthorpefoundation.org

Internet: http://www.jamesrthorpefoundation.org/2010guidelines.html

Sponsor: James R. Thorpe Foundation
318 West 48th Street
Minneapolis, MN 55419

James S. Copley Foundation Grants 2489

Incorporated in California in 1953, the Foundation serves as the philanthropic arm of the Copley Press, Inc., publishers of the Copley newspapers. The foundation offers support in the form of: scholarship funds; capital campaigns; equipment; employee matching gifts; endowments; equipment; building and; renovation grants. Funding organizations involved with arts and culture, education, animals and wildlife, health, recreation, and human services should apply. Giving is restricted primarily in areas of company operations in California, Illinois and Ohio.

Requirements: 501(c)3 organizations in the circulation areas of company newspapers/operations are eligible to apply. These areas include: California, Illinois and Ohio.

Restrictions: Ineligible funding opportunities include: religious, fraternal, or athletic organizations, government agencies, local chapters of national organizations, public elementary, secondary schools, public broadcasting systems, individuals, research, publications, conferences, general operating support, large campaigns and loans.

Geographic Focus: California, Illinois, Ohio

Date(s) Application is Due: Jan 2

Contact: Kim Koch, Secretary; (858) 454-0411, ext. 7671

Sponsor: James S. Copley Foundation
P.O. Box 1530
La Jolla, CA 92038-1530

James S. McDonnell Foundation Brain Cancer Research Collaborative Activity Awards 2490

The Foundation offers Collaborative Activity Awards to initiate interdisciplinary discussions on problems or issues, to help launch interdisciplinary research networks, or to fund communities of researchers/practitioners dedicated to developing new methods, tools, and applications of basic research to applied problems. All proposals submitted to the foundation must clearly link the experimental models and questions to human disease. Proposals primarily intending to characterize basic mechanisms of growth and development that may plausibly but are not yet known to be contributory to human brain cancer are not encouraged. Proposals testing molecules as possible treatment interventions should consider including tests designed to uncover unintended biological effects of such molecules that would disqualify future clinical usefulness. The Foundation is particularly interested in supporting novel research that will generate new knowledge leading to increased rates of survival and improve functional recovery for individuals with brain cancer.

Restrictions: The Foundation does not fund: undergraduate tuition, stipends, scholarships, fellowships, research or travel expenses, or other educational expenses; graduate or postdoctoral stipends, scholarships, fellowships, research or travel expenses, or other educational expenses (exceptions include requests that qualify as allowable budget items as part of specific Research Award applications); expenses tied to projects whose explicit goal is the publication of a book or other bound volume (although publication of a book or special issue of a journal may be one goal or outcome of work funded by the Foundation); expenses tied to the establishment or day-to-day running of a journal or small press; scientific meetings or workshops other than those JSMF puts together or which are affiliated with Collaborative Activity Awards; ongoing operational support for university-based centers, programs, or institutes; professional society meetings or specific sessions at such meetings; charitable functions, museum exhibitions, or similar causes or events; or charitable donations to individuals and organizations.

Contact: Cheryl Washington, Grants Manager; (314) 721-1532; fax (314) 721-7421; washington@jmsf.org or info@jmsf.org

Internet: http://www.jsmf.org/programs/bc/index.htm

Sponsor: James S. McDonnell Foundation
1034 S Brentwood Boulevard, Suite 1850
Saint Louis, MO 63117

James S. McDonnell Foundation Complex Systems Collaborative Activity Awards 2491

The Foundation offers Collaborative Activity Awards to initiate interdisciplinary discussions on problems or issues, to help launch interdisciplinary research networks, or to fund communities of researchers/practitioners dedicated to developing new methods, tools, and applications of basic research to applied problems. The Program supports scholarship and research directed toward the development of theoretical and mathematical tools that can be applied to the study of complex, nonlinear systems. It is anticipated that research funded in this program will address issues in fields such as biology, biodiversity, climate, demography, epidemiology, technological change, economic Complex Systems development, governance, or computation. While the program's emphasis is on the development and application of theoretical models used in these research fields and not on particular fields per se, JSMF is particularly interested in projects attempting to apply complex systems approaches to real world problems.

Proposals attempting to apply tools and models to problems where such approaches are not yet considered usual or mainstream (for example, differentiating normal physiology from disease) are encouraged.
Restrictions: The Foundation does not fund: undergraduate tuition, stipends, scholarships, fellowships, research or travel expenses, or other educational expenses; graduate or postdoctoral stipends, scholarships, fellowships, research or travel expenses, or other educational expenses (exceptions include requests that qualify as allowable budget items as part of specific Research Award applications); expenses tied to projects whose explicit goal is the publication of a book or other bound volume (although publication of a book or special issue of a journal may be one goal or outcome of work funded by the Foundation); expenses tied to the establishment or day-to-day running of a journal or small press; scientific meetings or workshops other than those JSMF puts together or which are affiliated with Collaborative Activity Awards; ongoing operational support for university-based centers, programs, or institutes; professional society meetings or specific sessions at such meetings; charitable functions, museum exhibitions, or similar causes or events; or charitable donations to individuals and organizations.
Contact: Cheryl Washington, Grants Manager; (314) 721-1532; fax (314) 721-7421; washington@jmsf.org or info@jmsf.org
Internet: http://www.jsmf.org/programs/cs/index.htm
Sponsor: James S. McDonnell Foundation
1034 S Brentwood Boulevard, Suite 1850
Saint Louis, MO 63117

James S. McDonnell Foundation Research Grants **2492**
The Awards are designed to support research projects with a high probability of generating new knowledge and insights. Projects submitted for funding consideration should be at an early, even preliminary stage of development that intend to break new ground or to challenge commonly-held assumptions. Projects submitted should be sufficiently novel, cross-disciplinary, or heterodox so that they have a strong likelihood of influencing the development of new ways of thinking about important problems. Awards provide adequate, flexible funding over a sufficient time period to allow investigators to pursue and develop innovative directions to their research programs. Funds can be expended over a minimum of 3 years or a maximum of 6 years. Smaller amounts of money expended over shorter amounts of time may be requested to help investigators pursue pilot projects or test the feasibility of an experimental approach. The applicant can apply the grant funds towards any research-based expense, including travel, equipment, and supplies. Funds can be used to support collaborative projects. A percentage of the funds can also be used to support small workshops organized by the applicant where the goal of the workshop is to gather expertise in support of the research objective.
Requirements: Applications for grants are considered only from 501(c)3 organizations which are not private foundations. Applications from foreign organizations will be considered only if they provide written legal opinion that they qualify as tax-exempt under these sections of the U.S. IRS code or if a qualifying American organization is authorized to receive funds for them under this code. Individuals also are eligible.
Amount of Grant: Up to 450,000 U.S.D
Contact: Cheryl Washington, Grants Manager; (314) 721-1532; fax (314) 721-7421; washington@jmsf.org or info@jmsf.org
Internet: http://www.jsmf.org/apply/research/index.htm
Sponsor: James S. McDonnell Foundation
1034 S Brentwood Boulevard, Suite 1850
Saint Louis, MO 63117

James S. McDonnell Foundation Scholar Awards **2493**
The Awards program derives from and is consistent with the foundation's commitment to supporting high quality research and scholarship leading to the generation of new knowledge and its responsible application. Currently, the Awards are only available in the Brain, Mind & Behavior program area, and provide largely unrestricted funding over a sufficient time period to allow investigators to pursue and develop new directions to their research programs.
Requirements: Applications must be sponsored by a nonprofit institution as defined by Section 501(c)3 of the United States Internal Revenue Tax Code. Eligible Scholar Award nominees must have completed all doctoral, postdoctoral, or fellowship training and hold an independent research position. Nominees in the earlier career stages are encouraged.
Restrictions: Researchers with current grant support from JSMF are not eligible to apply for a Scholar Award until all prior JSMF grant funds have been fully expended and a final grant report has been received and approved by the foundation office.
Amount of Grant: Up to 600,000 U.S.D
Contact: Cheryl Washington, Grants Manager; (314) 721-1532; fax (314) 721-7421; washington@jmsf.org or info@jmsf.org
Internet: http://www.jsmf.org/apply/scholar/index.htm
Sponsor: James S. McDonnell Foundation
1034 S Brentwood Boulevard, Suite 1850
Saint Louis, MO 63117

James S. McDonnell Foundation Understanding Human Cognition Awards **2494**
The long-standing interest of the James S. McDonnell Foundation on human mind/brain is reflected in changes made to the 2008 RFA, including the explicit emphasis on understanding human cognition. The program is intended to help investigators pursue experiments designed to answer well-articulated questions. JSMF Scholar Awards support research studying how neural systems are linked to and support cognitive functions and how cognitive systems are related to an organism's (preferably human) observable behavior. Studies with model organisms should justify why such models were selected and how data obtained from models advances our understanding of human cognition. Proposals proposing to use functional imaging to identify the neural correlates of cognitive or behavioral tasks (for example, mapping the parts of the brain that light up when different groups of subjects play chess, solve physics problems, or choose apples over oranges) are not funded through this program. In general, JSMF and its expert advisors have taken an unfavorable view of projects attempting too wide a leap in a single bound. Functional imaging studies using poorly characterized tasks as proxies for complex behavioral issues involving empathy, moral judgments, or social decision-making are generally not appropriate responses to this call for proposals. In past competitions, proposals structured along such lines were eliminated from funding consideration early in the review process.
Requirements: Aspects of proposals appropriate to the JSMF UHC program would include, but are not limited to: characterizing the cognitive operations involved in performing a task; studying how the brain identifies, extracts and uses relevant information from complicated environments; examining how manipulations and/or perturbations at one spatial or temporal scale are meaningful at finer or coarser levels of organization (e.g. does a synaptic change account for a change in network function and vice versa); re-examining common wisdom assumptions (such as the existence of critical periods in human learning); evaluating the usefulness of methodologies or improving the usefulness of methodologies commonly used in mind/brain research; applying approaches and knowledge from cognitive psychology or cognitive science to important problems in education, training, or rehabilitation; taking a comparative, evolutionary approach to characterizing the uniqueness of the human brain and of human cognition.
Restrictions: The Foundation does not fund: undergraduate tuition, stipends, scholarships, fellowships, research or travel expenses, or other educational expenses; graduate or postdoctoral stipends, scholarships, fellowships, research or travel expenses, or other educational expenses (exceptions include requests that qualify as allowable budget items as part of specific Research Award applications); expenses tied to projects whose explicit goal is the publication of a book or other bound volume (although publication of a book or special issue of a journal may be one goal or outcome of work funded by the Foundation); expenses tied to the establishment or day-to-day running of a journal or small press; scientific meetings or workshops other than those JSMF puts together or which are affiliated with Collaborative Activity Awards; ongoing operational support for university-based centers, programs, or institutes; professional society meetings or specific sessions at such meetings; charitable functions, museum exhibitions, or similar causes or events; or charitable donations to individuals and organizations.
Contact: Cheryl Washington, Grants Manager; (314) 721-1532; fax (314) 721-7421; washington@jmsf.org or info@jmsf.org
Internet: http://www.jsmf.org/programs/uhc/index.htm
Sponsor: James S. McDonnell Foundation
1034 S Brentwood Boulevard, Suite 1850
Saint Louis, MO 63117

Jane's Trust Grants **2495**
The Jane's Trust has particular interest in organizations and projects which primarily benefit underserved populations and disadvantaged communities. Jane's Trust will support grants for general operating purposes, as well, as it's Fields of Interest, which are: arts and culture, education, the environment, health and welfare. The Trust will make grants in the states of Florida, with a preference for southwest and central Florida; Massachusetts, with a preference for greater Boston and eastern Massachusetts; and in the northern New England states of Maine, New Hampshire and Vermont. Preference will be given to organizations located in those states for projects which will primarily provide benefits within those states. The application process begins by submitting of a concept paper. Guidelines are available at the Trusts website.
Requirements: 501(c)3 tax-exempt organizations in Florida, Maine, Massachusetts, New Hampshire, and Vermont are eligible.
Restrictions: Jane's Trust will not support: loans to charitable organizations; attempts to influence legislation; requests from individuals. Please note: Jane's Trust will normally not support public entities, such as municipalities, municipal departments, or public schools directly, but will entertain applications from tax-exempt fiscal agents or partners for collaborative projects with municipalities or schools. This does not apply to public colleges and universities.
Geographic Focus: Florida, Maine, Massachusetts, New Hampshire, Vermont
Date(s) Application is Due: Jan 25; Jul 15
Amount of Grant: 50,000 - 1,000,000 U.S.D
Contact: Susan Fish, (617) 227-7940, ext. 775; fax (617) 227-0781; sfish@hembar.com
Internet: http://www.hembar.com/selectsrv/janes/index.html
Sponsor: Hemenway and Barnes LLP
60 State Street
Boston, MA 02109-1899

Jane Beattie Memorial Scholarship **2496**
The purpose of the fund is to provide scholarships to subsidize travel to the U.S. for purposes of scholarly activity by a foreign scholar in the area of judgment and decision research, broadly defined. Attendance at the annual SJDM meetings is one example of an activity that would be appropriate for support, but by no means the only one. In most years, the Fund awards one or two scholarships in amounts of $400 - $700 each.
Requirements: Applicants should be scholars living and working in a country other than the U.S. who will use the award to help pay for travel to the U.S. for scholarly activities associated with research in judgment and decision making. It is anticipated that most awards will be granted to faculty or graduate students at colleges and universities, especially recent and soon-to-be Ph.D.'s, but others will also be considered. The required application form can be downloaded from: http://gsbwww.uchicago.edu/fac/joshua.klayman/more/BeattieForm.rtf
Restrictions: U.S. Scholars are not eligible to apply.

Date(s) Application is Due: Jul 26
Amount of Grant: 400 - 700 U.S.D
Contact: Joshua Klayman, (773) 834-9134; joshk@uchicago.edu
Internet: http://www.sjdm.org/content/beattie-award
Sponsor: Society for Judgment and Decision Making
College of Business, Florida State University, P.O. Box 3061110
Tallahassee, FL 32306-1110

Jane Bradley Pettit Foundation Health Grants **2497**

The Jane Bradley Pettit Foundation awards grants to initiate and sustain projects in the Greater Milwaukee community, with a focus on programs and projects that serve low-income and disadvantaged individuals, women, children, and the elderly. In the area of Health, the Foundation gives priority to community-based health care and prevention programs which address the physical and mental health needs of families, children, persons at-risk and the elderly. Support and advocacy for victims of abuse and neglect are also a priority. An initial application should be in the form of a letter of intent.
Requirements: Milwaukee-area charitable organizations are eligible.
Geographic Focus: Wisconsin
Date(s) Application is Due: Jan 15; May 15; Sep 15
Amount of Grant: 5,000 - 100,000 U.S.D
Contact: Heidi Jones, Director of Administration; (414) 982-2880 or (414) 982-2874; fax (414) 982-2889; hjones@staffordlaw.com
Internet: http://www.jbpf.org/guidelines/index.html
Sponsor: Jane Bradley Pettit Foundation
1200 N. Mayfair Road, Suite 430
Wauwatosa, WI 53226-3282

Jane Coffin Childs Memorial Fund for Medical Research Fellowships **2498**

The fellowships are for furthering of research into the causes, origins, and treatment of cancer. The major effort of the fund, at this time, is directed toward the support of a fellowship program for the training of young men and women at the postdoctoral level for independent research in the same field. Stipend, child dependency allowance, travel for fellow and family, and an institutional allowance are provided. Fellowships are awarded for two to three years.
Requirements: Applicants should not have more than one year of postdoctoral experience. They must either hold the MD degree or PhD degree in the field in which they propose to study. Applicants may be citizens of any country, but awards for foreign nationals will be made only for study in the United States. American citizens may hold fellowships either in the United States or in a foreign country.
Date(s) Application is Due: Feb 1
Amount of Grant: 42,500 - 131,500 U.S.D
Contact: Kim Roberts; (203) 785-4612; fax (203) 785-3301; info@jccfund.org
Internet: http://www.jccfund.org
Sponsor: Jane Coffin Childs Memorial Fund for Medical Research
333 Cedar Street, P.O. Box 208000
New Haven, CT 06520

Janesville Foundation Grants **2499**

The foundation is dedicated to improving the quality of community living in Janesville, WI, and the surrounding area and makes grants to support nonprofit educational, charitable, and health organizations in the area. Capital grants and grants for specific projects or programs will be awarded. Proposals are reviewed quarterly, in February, May, August, and November. Preliminary inquiries are encouraged.
Requirements: Nonprofit, tax-exempt organizations serving Janesville, WI, and its surrounding area are eligible.
Geographic Focus: Wisconsin
Amount of Grant: 1,000 - 3,500 U.S.D
Contact: Bonnie Lynne Robinson; (608) 752-1032; fax (608) 752-1952
Sponsor: Janesville Foundation
P.O. Box 8123
Janesville, WI 53547-8123

Janirve Foundation Grants **2500**

The Janirve Foundation, established in 1954, givings primarily to colleges and universities in western North Carolina. However, the Foundation also supports organizations involved with: children and youth services; community/economic development; environment; environment, natural resources; environment, plant conservation; family services; health organizations, association; higher education; hospitals (general); housing/shelter, development; human services.
Requirements: Applicants should contact Asheville, North Carolina, office for application procedures. Proposals from colleges and universities are considered only in the front quarter. Application form required. Applicants should submit the following: copy of IRS Determination Letter; detailed description of project and amount of funding requested; listing of additional sources and amount of support; how project's results will be evaluated or measured; how project will be sustained once grant support is completed; copy of current year's organizational budget and/or project budget; copy of most recent annual report/audited financial statement/990; listing of board of directors, trustees, officers and other key people and their affiliations; signature and title of chief executive officer; five copies of the proposal.
Restrictions: No support for public and private elementary schools, or churches and religious programs. No grants to individuals (except for scholarships), or generally for operating budgets, endowments or for research programs, publication of books or printed material, theatrical productions, videos, radio or television programs; no loans.
Geographic Focus: North Carolina
Date(s) Application is Due: Mar 1; Jun 1; Sep 1; Dec 1
Amount of Grant: 10,000 - 750,000 U.S.D
Samples: Health Adventure, Asheville, NC, $500,000; YMCA of Catawba County, Hickory, NC, $100,000; South Florida Science Museum, West Palm Beach, FL, $375,000;
Contact: E. Charles Dyson, Chairman; (828) 258-1877; fax (828) 258-1837; janirve@charterinternet.com
Sponsor: Janirve Foundation
1 North Pack Square, Suite 416
Asheville, NC 28801-3409

Janson Foundation Grants **2501**

The Janson Foundation was established in 1983 to provide capital grants to deserving charitable organizations located in Skagit County, Washington. Edward W. Janson had a longtime career as a public servant in Skagit County and was committed to the citizens of the county. He established his foundation to assist the ongoing capital needs of deserving charitable organizations for the purchase of equipment, the construction of facilities, or the acquisition of land. In addition, the foundation funds the repair of existing facilities and the replacement of aging equipment. The majority of grants from the Janson Foundation are one year in duration; on occasion, multi-year support is awarded. Application materials are available for download at the grant website. Applicants are strongly encouraged to review the state application guidelines for additional helpful information and clarifications before applying. Applicants are also encouraged to review the foundation's funding history (link is available from the grant website). The application deadline for the Janson Foundation is February 28. Applicants will be notified of grant decisions by April 15.
Requirements: Applicants must have 501(c)3 status. Applications must be mailed.
Restrictions: Grant requests for general operating or program support will not be considered. The foundation does not support requests from individuals, organizations attempting to influence policy through direct lobbying, or any political campaigns.
Geographic Focus: Washington
Date(s) Application is Due: Feb 28
Samples: Mount Vernon Christian School, Mount Vernon, Washington, $9,570 (2010); Crisis Pregnancy Center of Skagit County, Mount Vernon, Washington, $6,388 (2010); Youth Dynamics, Burlington, Washington, $10,004 (2010).
Contact: Heidi Gordon, Vice President; (800) 848-7177; heidi.e.gordon@baml.com
Internet: https://www.bankofamerica.com/philanthropic/fn_search.action
Sponsor: Janson Foundation
800 5th Avenue, WA1-501-33-23
Seattle, WA 98104

Janus Foundation Grants **2502**

The Foundation strives to help communities reach greater levels of self-sufficiency, and impact the lives of many in each community. The Foundation awards grants to nonprofit organizations in its areas of interest, including: at-risk youth through education; community service and volunteerism; and cultural institutions in the Denver-metro area. Interested applicants should complete the application available online. There is no application deadline. Please note that the Foundation accepts grant applications from nonprofit organizations throughout the U.S. for the first two giving areas. The third giving area only applies to cultural institutions that operate in the Denver, CO metro area.
Requirements: 501(c)3 tax-exempt organizations are eligible.
Samples: Denver Public Library, Denver, CO, $100,000; Denver Center for the Performing Arts, Denver, CO, $18,000; Heart of Los Angeles Youth, Los Angeles, CA, $5,000.
Contact: Traci Papantones, Vice President; (303) 333-3863; fax (303) 394-7797; janusfoundation@janus.com
Internet: http://ww4.janus.com/Janus/Retail/StaticPage?jsp=Janushome/JanusFoundation.jsp
Sponsor: Janus Foundation
151 Detroit Street, 4th Floor
Denver, CO 80206-4805

Japan Heart Foundation Grants **2503**

The foundation awards grants in Japan to individuals, organizations, and research institutes for research projects in the area of heart and blood vessel diseases.
Requirements: Research institutes for research in the area of heart and blood vessel diseases, individuals, and organizations in Japan are eligible to apply.
Geographic Focus: All States, Japan
Contact: Grants Administrator; 81-3-3201-0810; fax 81-3-3213-3920; info@jhf.or.jp
Internet: http://www.jhf.or.jp
Sponsor: Japan Heart Foundation
835-A, New Kokusai Building, 3-4-1, Marunouchi
Chiyoda-ku, Tokyo, 100-0005 Japan

Jasper Foundation Grants **2504**

The purpose of the Foundation is to assist donors in creating assets to meet the ongoing and changing charitable interests of those living in Jasper County, Indiana. Special emphasis is placed on programs that enrich the quality of life of our community in five areas: arts and culture; preservation of historic and cultural resources; education; health; and social concerns. Favored grant requests will generally reflect the following characteristics: projects which propose practical solutions to community needs; projects which promote cooperation among existing agencies without duplicating services; projects that will become self-sustaining without requiring on-going Foundation funds. Applications and guidelines are available at the Foundation office or website. The Foundation has two grant cycles in the spring and fall.

Requirements: Applicants must submit 5 complete copies of the grant application to include the following: the grant application and supporting materials; an outline that addresses the project's goals and objectives, plan of implementation, project budget, staff, method of evaluation, and why the proposal is important to the community; a list of the officers and board of directors; a current financial report; and a copy of the organization's 501(c)3 tax exempt IRS letter.
Restrictions: Grants are not made for budget deficits or sectarian religious activities.
Geographic Focus: Indiana
Date(s) Application is Due: Apr 1; Oct 1
Contact: Linda Reiners, Executive Director; (219) 866-5899; fax (219) 866-0555; jasper@liljasper.com
Internet: http://www.jasperfdn.org/grants.htm
Sponsor: Jasper Foundation
301 N Van Rensselaer Street, P.O. Box 295
Rensselear, IN 47978

Jay and Rose Phillips Family Foundation Grants **2505**
The Jay and Rose Phillips Family Foundation has five funding priorities that guide it's grant making: strengthening families; supporting training and education for lifelong success; improving health and wellness; promoting independence and inclusion for people with disabilities and the elderly; and fostering good relations and civic participation. These funding priorities are the combined product of responding to community needs, the legacy of our founders and the current passions and interests of it's Trustees. Across all of these funding priorities, the Foundation place's a high value on supporting efforts that serve the most vulnerable in the community and those with the least access to resources. The Foundation has three program grant rounds each year with deadlines of March 15, July 15 and November 15. All proposals be submitted electronically using an online proposal submission process. The online submission deadline is 11:59 p.m. on the deadline date. The Foundation accepts requests for capital grants once per year through an online Letter of Inquiry (LOI) process. LOI's must be submitted by 11:59 p.m. on January 8.
Requirements: The Foundation awards grants only to organizations which are tax-exempt and publicly supported under Section 501(c)3 of the Internal Revenue Service Code. Grants are awarded primarily in the Twin Cities metropolitan area. Unsolicited proposals are not accepted from organizations located outside of Minnesota.
Restrictions: The foundation does not make grants in support of individuals, for political campaigns, or for lobbying efforts to influence legislation.
Geographic Focus: Minnesota
Date(s) Application is Due: Jan 8; Mar 15; Jul 15; Nov 15
Amount of Grant: 10,000 - 100,000 U.S.D
Contact: Dana Jensen, Grants Manager; (612) 623-1652 or (612) 623-1654; fax (612) 623-1653; djensen@phillipsfnd.org
Internet: http://www.phillipsfnd.org/index.asp?page_seq=15
Sponsor: Jay and Rose Phillips Family Foundation
East Bridge Building, 10 Second Street NE, Suite 200
Minneapolis, MN 55413

JDF Career Development Awards **2506**
The program awards grants to applicants who have demonstrated superior scholarship and show the greatest promise for future achievement in diabetes research, including either clinically relevant research or basic research. The research plan should be suitable for a five-year project, and awardees must spend at least 75 percent of time and effort on the research project during the period of the award. Research may be conducted at foreign and domestic, for-profit and nonprofit, and public and private organizations such as universities, colleges, hospitals, laboratories, units of state and local governments, and eligible agencies of the federal government. Letters of reference are due two weeks before the application deadline, which are July 15 and January 16.
Requirements: Candidates must have the doctoral degree or the equivalent from an accredited institution, with three to seven years of total professional postdoctoral clinical and/or research experience by the projected start of the award period. Candidate's institution must be a university, medical school, or comparable institution with strong, well-established research and training programs in the chosen area. Research may be conducted at foreign and domestic, for-profit and nonprofit, public and private laboratories; units of state and local governments; and eligible agencies of the federal government.
Restrictions: Individuals holding academic positions of associate professor or professor at the time of the award are not eligible.
Date(s) Application is Due: Jan 16; Jul 15
Amount of Grant: 150,000 U.S.D
Contact: Administrator; (212) 479-7565; fax (212) 785-9595; fellowships@jdrf.org
Internet: http://www.jdrf.org/index.cfm?page_id=103291
Sponsor: Juvenile Diabetes Foundation International
120 Wall Street, 19th Floor
New York, NY 10005-4001

Jean and Louis Dreyfus Foundation Grants **2507**
The Jean and Louis Dreyfus Foundation was established in 1979 from the estate of Louis Dreyfus, a music publisher, and that of his wife, Jean. The mission of the Foundation is to enhance the quality of life of New Yorkers, particularly the aging and disadvantaged. The Foundation disburses grants mainly within the five boroughs of New York City, and supports programs in the arts, health and social services (including youth agencies, women, and the elderly), and education (including literacy). Support is given for program development and matching funds. Application forms are not required. The board meets each year in the spring and in the fall. Initial inquiries should consist of a one or two page letter describing the organization and outlining the project in question. The January 15th and July 15th deadlines are for Letters of Intent.
Requirements: New York City nonprofits are eligible.
Restrictions: Grants are never made to individuals.
Geographic Focus: New York
Date(s) Application is Due: Jan 15; Jul 15
Contact: Jessica Keuskamp, Program Director; (212) 599-1931; fax (212) 599-2956; jldreyfusfdtn@hotmail.com
Internet: http://foundationcenter.org/grantmaker/dreyfus/guide.html
Sponsor: Jean and Louis Dreyfus Foundation
420 Lexington Avenue, Suite 626
New York, NY 10170

Jeanne Humphrey Block Dissertation Award **2508**
The grant assists dissertation research on the psychological development of women and girls for a doctoral student who is a woman. Proposals should focus on the sex and gender differences or some other developmental issue of particular concern to women or girls. Priority will be given to projects that draw on or contribute to the resources of the Murray Research Center. Research concerned with the life experiences of racially and ethnically diverse populations within the United States is encouraged.
Requirements: Any female student currently enrolled in a doctoral program in a relevant field is eligible to apply. Dissertation proposals must be approved by an advisor or committee before the grant application is submitted.
Date(s) Application is Due: Apr 1
Amount of Grant: Up to 5,000 U.S.D
Contact: Program Director; (617) 495-8140; fax (617) 496-3993; mrc@radcliffe.com
Internet: http://www.murray.harvard.edu/mra/index.jsp
Sponsor: Radcliffe College Henry A. Murray Research Center
10 Garden Street
Cambridge, MA 02138

Jeffris Wood Foundation Grants **2509**
The Jeffris Wood Foundation funds grants to community-based organizations working to provide opportunities for urban youth and economically disadvantaged. The Foundation supports programs that: help urban youth improve their futures, explore their creativity and avoid unwanted pregnancies; help Native American youth connect with their cultural traditions; connect low-income urban youth to nature; and provide services for domestic violence survivors. Current deadlines are posted on the website.
Requirements: Organizations must first submit a one-page letter of inquiry that includes a description of the organization and the project. They should also attach a brief budget describing income and expenses, plus complete contact information. If the request is screened for further consideration, the organization will be asked to submit a full proposal. All materials must be sent by U.S. mail.
Restrictions: Grants are not made to: individuals; scholarships; schools; capital expenses or renovation projects; food or shelter programs (except domestic violence); research or publications; video & web productions; religious programs that are not all inclusive; or athletic events and sponsorships.
Geographic Focus: Washington
Amount of Grant: 1,000 - 3,000 U.S.D
Contact: Therese Ogle, Grants Consultant; (206) 781-3472; OgleFounds@aol.com
Internet: http://foundationcenter.org/grantmaker/jeffriswood/guide.html
Sponsor: Jeffris Wood Foundation
6723 Sycamore Avenue NW
Seattle, WA 98117

JELD-WEN Foundation Grants **2510**
The Foundation seeks to improve the quality of life in company-operating areas and awards grants in the areas of capital campaigns, education, youth activities, community development, health and medical, and arts and humanities. Types of support include building construction/renovation, equipment, general operating support, land acquisition, matching/challenge support, program development, scholarship funds, and seed money. Prescreening and application information are available online.
Requirements: Nonprofit organizations in the Foundation's area of operations.
Amount of Grant: 1,000 - 1,000,000 U.S.D
Samples: Knox Community Hospital, Mt. Vernon, OH, $500,000—grant to help build an outpatient cancer treatment center; Crater Lake Science and Learning Center, Crater Lake, OR, $500,000—to help build the Science and Learning Center at Crater lake National Park; Yakima Valley Museum, Yakima, WA, $250,000—to upgrade and improve the museum facility.
Contact: JELD-WEN Foundation Headquarters; (503) 478-4478; fax (503) 478-4474
Internet: http://www.jeld-wenfoundation.org/
Sponsor: Jeld-Wen Foundation
200 SW Market Street, Suite 550
Portland, OR 97201

Jenkins Foundation: Improving the Health of Greater Richmond Grants **2511**
The foundation is committed to expanding access to community-based services through programs and organizations that have the potential to make a significant impact on the quality of health, especially for the youth in its local area.
Requirements: The current areas of focus are: Expanding access to health care services for the uninsured and underserved; Substance abuse prevention services that promote healthy lifestyles and increase availability of services; Violence prevention services that promote safe and healthy environments for children and their families and work toward the

elimination of violence in the local communities; In addition, the foundation is committed to the long-term viability of the organizations it supports, and will consider capacity building grants that strengthen an agency's ability to better serve its clients. The foundation will also consider a limited number of proposals outside the above stated focus areas.
Restrictions: Proposals will be accepted from charitable organizations, which serve the residents of the City of Richmond and the counties of Chesterfield, Hanover, Henrico, Goochland, and Powhatan.
Geographic Focus: Virginia
Date(s) Application is Due: May 5; Nov 5
Amount of Grant: Up to 50,000 U.S.D
Contact: Elaine Summerfield, Program Officer; (804) 330-7400; fax (804) 330-5992; esummerfield@tcfrichmond.org
Internet: http://www.tcfrichmond.org/Page2954.cfm#Jenkins
Sponsor: Community Foundation Serving Richmond and Central Virginia
7501 Boulders View Drive, Suite 110
Richmond, VA 23225

Jennings County Community Foundation Grants 2512
The foundation seeks to serve philanthropic and charitable needs in Jennings County, Indiana, by offering endowment services, grant making, scholarships, donor estate and planned gift services to individuals and qualified organizations serving the community of Jennings County, Indiana. The foundations field of interest include: arts; community/economic development; education; health care; human services.
Requirements: Submit 6 copies of the grant proposal; funding is given primarily in the in the Jennings County, Indiana area; contact foundation office for more details
Restrictions: No grants to individuals (except for scholarships)
Geographic Focus: Indiana
Date(s) Application is Due: Mar 1; Sep 1
Contact: Jean Hahn; (812) 346-5553; fax (812) 346-5553; jcff@comcast.net
Internet: http://www.jenningsfoundation.net/index.html
Sponsor: Jennings County Community Foundation
265 E Main Street
North Vernon, IN 47265-1510

Jerome and Mildred Paddock Foundation Grants 2513
The Foundation was established in 1967 to benefit disadvantaged children and elderly populations. Primarily, the foundation is interested in funding programming needs that will improve the quality of life of those less fortunate. Field of interest include: aging; children and youth services; programs for economically disadvantaged; health care support; and human/social services. Though giving is focused in and around Sarasota, the foundation does extend funding beyond Florida. The deadline is January 10th, with final notification by March 15th.
Restrictions: No grants to individuals, or for endowments, debt reduction, operating expenses, conferences or seminars, workshops, travel, surveys, advertising, research, fund raising or for annual campaigns or capital campaigns.
Date(s) Application is Due: Jan 10
Amount of Grant: 15,000 U.S.D
Samples: Florida Studio Theatre, Sarasota, Florida, $10,000—to partially fund the VIP Performing Arts Program for special needs youth; Shodair Childrens Hospital, Helena, Montana, $30,000—to purchase a Coulter automated cell counter; Samaritan Counseling Services, Ann Arbor, Michigan, $10,000—to assist low income, under-insured and uninsured clients.
Contact: Joan Greenwood, Charitable Associate, Wachovia Bank; (941) 361-5803; joan.greenwood@wachovia.com
Sponsor: Jerome and Mildred Paddock Foundation
Wachovia Bank, NA., P.O. Box 267
Sarasota, FL 34230

Jessie B. Cox Charitable Trust Grants 2514
The Trust is dedicated to improving the environment and the quality of life for people living in the six New England States. To achieve its goals, the Trust pursues initiatives in three key fields of interest: education: early learning, out-of-school time; environment: habitat conservation, concentrating on fresh and marine water protection; health: access to health care. In all of these funding areas, the Trust is mindful of its special status as one of the few funders with an interest in all six of the New England states. Applications for funding are reviewed in light of the important issues of the region. The Grants Committee meets twice per year to award grants, in April and October. The Trust has a two-step system of application, beginning with a concept paper and progressing to a full grant proposal. Concept papers are due on March 15 and September 15. Applicants who have submitted a concept paper and been asked to submit a complete proposal must do so by July 1 (March applicants) or January 5 (September applicants).
Requirements: All grant applicants are asked to demonstrate a plan for measuring their success. This plan should identify benchmarks against which progress towards identified goals can be measured. Grantees must report on their progress within a year of receiving support and on project completion, with reference to the initial benchmarks.
Restrictions: The Cox Trust does not normally support: general operating support and ongoing maintenance; buildings, equipment or land acquisition; endowments, scholarship funds, or fundraising activities; grant requests of under $50,000; projects for which the Trust is the sole or predominant source of support; programs for which the public sector normally assumes funding responsibility; core educational programs of public, private, parochial, or charter schools; requests from individuals; sectarian religious activity.
Geographic Focus: Connecticut, Maine, Massachusetts, New Hampshire, Rhode Island, Vermont
Date(s) Application is Due: Jan 5; Mar 15; Jul 1; Sep 15
Contact: Kirstie David; (617) 391-3081; kdavid@gmafoundations.com
Internet: http://www.jbcoxtrust.org/?page_id=28
Sponsor: Jessie B. Cox Charitable Trust
GMA Foundations, 77 Summer Street, 8th Floor
Boston, MA 02110-1006

Jessie Ball Dupont Fund Grants 2515
The fund is a national foundation having a special, though not exclusive, interest in issues affecting the South. Grants awarded include competitive grants—program grants, institutional development, and capacity building; and feasibility grants—smaller grants that enable institutions to carefully explore and develop new concepts and programs. Areas of interest include six focus areas: strengthening the independent sector; organizing and nurturing philanthropy; building assets of people, families, and communities; building the capacity of eligible organizations; stimulating community problem solving; and helping people hold their communities accountable. The fund supports four initiatives: the religion initiative, the nonprofit initiative, the small liberal arts colleges initiative, and the independent schools initiative. Types of support include general operating support, building construction/renovation, equipment acquisition, program development, seed money grants, publication, consulting services, professorships, curriculum development, matching funds, and technical assistance. Grants are awarded generally for one year. The trustees meet in January, March, May, July, September, and November to consider requests.
Requirements: Applying organizations must have received a contribution from Mrs. DuPont between the five-year period of January 1, 1960, and December 31, 1964.
Amount of Grant: 2,000 - 5,000 U.S.D
Samples: Jacksonville Symphony Assoc (FL)—to implement a strategic marketing program, $50,000; Historical Society of Delaware (Wilmington, DE)—to hire a consultant to lead a community-based strategic-planning process, $52,690; Episcopal Diocese of Southern Ohio (Cincinnati, OH)—to support a partnership with Scioto Christian Ministries to conduct a needs-assessment in Ohio's Scioto County, $28,000; U of Richmond (VA)—to support the creation of an Internet-based resource center that will serve community organizations in the Northern Neck region of Virginia, $219,975.
Contact: Geana Potter, Grants Manager; (904) 353-0890 or (800) 252-3452; fax (904) 353-3870; contactus@dupontfund.org
Internet: http://www.dupontfund.org
Sponsor: Jessie Ball Dupont Fund
One Independent Drive, Suite 1400
Jacksonville, FL 32202-0511

Jessie Smith Noyes Foundation Grants 2516
The foundation is committed to preventing damage to the natural systems upon which all life depends and to strengthening individuals and institutions committed to protecting natural systems and ensuring a sustainable society. The foundation makes grants primarily in the interrelated areas of environment and reproductive rights. The program components include toxics, sustainable agriculture (both with emphasis on southern and Rocky Mountain states), and U.S. reproductive rights. In addition, a few grants are made in four areas of special concern: sustainable communities, U.S. environmental justice, strengthening the U.S. nonprofit sector, and environmental issues in the metropolitan New York region. Letters of inquiry are received at any time; proposals will be requested from the foundation after review.
Requirements: 501(c)3 tax-exempt organizations are eligible.
Restrictions: Normally, the foundation will not consider requests for direct service, endowments, loans or scholarships to individuals, capital construction funds, conferences, media events, production of media and TV programming, or general fund-raising drives. General research projects are not funded per se.
Samples: Northwest Women's Law Ctr (Seattle, WA)—for its contributions to the field of reproductive-health care and rights in Alaska, Idaho, Montana, Oregon, and Washington State, $100,000.
Contact: Millie Buchanan, (212) 684-6577; fax (212) 689-6549; noyes@noyes.org
Internet: http://www.noyes.org
Sponsor: Jessie Smith Noyes Foundation
6 E 39th Street, 12th Floor
New York, NY 10016-0112

Jewish Foundation for Education of Women Scholarships 2517
Scholarships are awarded to women pursuing undergraduate or graduate education. Requests for applications must be made in writing and specify whether aid is being sought for undergraduate or graduate studies. Financial assistance is offered every year depending on progress shown.
Requirements: Eligible to apply are women who are permanent residents of the five boroughs of New York City, although schools attended may be located anywhere.
Restrictions: No awards will be made for the pursuit of JD or MBA degrees. Scholarships will not be available for the upcoming school year.
Geographic Focus: New York
Date(s) Application is Due: Jun 1
Amount of Grant: Up to 5,000 U.S.D
Samples: Teachers College, Columbia U (New York, NY)—for scholarships for master's-degree candidates in mathematics education and science education who will teach in the New York City public schools, $135,000 over three years; Juilliard School (New York, NY)—to provide scholarships to arts students, $250,000 over four years; Jewish Foundation for Education of Women (New York, NY)—to provide scholarships to emigres from the former Soviet Union who are studying dentistry, medicine, nursing, and pharmacy at New York-area universities, $720,000 over three years.

Contact: Marge Goldwater, Executive Director; (212) 288-3931; fax (212) 288-5798; fdnscholar@aol.com
Internet: http://www.jfew.org
Sponsor: Jewish Foundation for Education of Women
135 E 64th Street
New York, NY 10021

Jewish Foundation for Education of Women Scholarships for Emigres in the Health Professions 2518

Scholarships are awarded to women emigres from the former Soviet Union for graduate and undergraduate education in medicine, dentistry, nursing, or pharmacy. Approximately 48 scholarships are awarded. Applications are available in late October.
Requirements: Eligible to apply are women who are emigres from the former Soviet Union, live within a 50-mile radius of New York City, and demonstrate financial need.
Restrictions: Pre-medical students are not eligible. Applications submitted directly from students are not accepted.
Geographic Focus: New York
Date(s) Application is Due: Jun 1
Amount of Grant: 5,000 U.S.D
Samples: Jewish Foundation for Education of Women (New York, NY)—to administer scholarships for emigres from the New Independent States who are enrolled in occupational therapy, physical therapy, or physician's assistant programs, $300,000 over three years.
Contact: Marge Goldwater, (212) 288-3931; fax (212) 288-5798; fdnscholar@aol.com
Internet: http://www.jfew.org/programs.html#HP
Sponsor: Jewish Foundation for Education of Women
135 E 64th Street
New York, NY 10021

Jewish Fund Grants 2519

The Fund awards grants to sustain, enrich, and address the overall health care needs of both the Jewish community and general community in the metropolitan Detroit area. The Fund is particularly interested in supporting projects that: address health care and social welfare needs of vulnerable/at-risk populations within the Jewish community; respond to priority capital and equipment needs of the Detroit Medical Center/Sinai Hospital; improve the health and well-being of vulnerable/at risk populations in the general community; support inclusion of people with special needs into the general activities of the community; enhance positive relationships between the Jewish community and the Detroit community. Highest priority is given to requests for programs that: address a critical need; impact the lives of residents of the Wayne, Oakland and Macomb counties; have a defined plan for sustaining the program beyond the grant period; include a financial or in-kind contribution from the organization; involve collaboration with others; have an outcomes-based evaluation plan; and can be funded and replicated by others. Samples of previously funded grants are available on the Fund website.
Requirements: The Jewish Fund will make grants to 501(c)3 organizations and other non-profits qualified as tax exempt under the Internal Revenue Code. Applicant organizations must provide a current audited financial statement. Applicants are encouraged to contact the Fund and discuss their proposed project with the executive director before applying for funding.
Restrictions: The Fund will usually not support: grants made directly to individuals; loans; grants to support religious activities or sectarian education; overseas projects; capital projects or equipment purchases (except equipment at the DMC/Sinai); endowments, annual fund drives, and fundraising events; and past operating deficits.
Geographic Focus: Michigan
Contact: Margo Pernick, Executive Director; (248) 203-1487; fax (248) 645-7879
Internet: http://thejewishfund.org/grant-request-guidelines.html
Sponsor: Jewish Fund
6735 Telegraph Road
Bloomfield Hills, MI 48301-2030

Jewish Healthcare Foundation of Pittsburgh Grants 2520

The foundation supports the provision of health care services and education and responds to the medical, custodial, and other health-related needs of children and of elderly, underprivileged, indigent, and underserved persons in both the Jewish and general communities throughout western Pennsylvania. Types of support include awards/prizes, challenge/matching grants, conferences and seminars, demonstration grants, program grants, seed money grants, and technical assistance. To be considered for support, programs must be health related and should address the three foundation priorities: special needs of the elderly and chronically ill, the health of children at risk, and women's health. Programs will receive special attention if they provide opportunities for new information and fresh perspectives about health problems, improve health care systems or utilization of health care, build partnerships among community institutions, and/or contain a community education component.
Requirements: Organizations in Pennsylvania are eligible to apply. A preliminary letter of intent not in excess of six pages should include problem description, program objectives, proposed intervention, timetable, anticipated outcome, innovative aspects, community education and research and/or evaluation components, likelihood of success, partnerships, and long-term plans. Additional information to include are the name and address of a contact person, institutional and personnel qualifications, budget, board of directors list, 501(c)3 letter of determination, and most recent financial statements.
Restrictions: Generally, the foundation does not fund non-tax-exempt organizations or organizations outside western Pennsylvania; programs without a health care component; general operations; endowment programs; capital needs; operating deficits or retirement of debt; political campaigns; scholarships, fellowships, individual research grants; or individual travel.
Geographic Focus: Pennsylvania
Amount of Grant: 5,000 - 825,000 U.S.D
Samples: Patient Safety Authority (Pittsburgh, PA)—for a seminar series for hospital officials on how to use data on clinical quality and safety incidents to improve the quality of care and financial performance of their organizations, $50,000 maximum; Primary Health Care Services (Pittsburgh, PA)—to enable seven federally financed health centers in western Pennsylvania focused on reducing health disparities to meet chronic-care standards, $100,000 over two years; U of Pittsburgh Cancer Institute (PA)—to evaluate a program that teaches pain-management skills to cancer patients, $72,000; U of Pittsburgh Ctr for Emergency Medicine (PA)—to develop a classification system to identify and prioritize medical errors that occur during emergency care, in order to access risks to patients, $75,000.
Contact: Dr. Karen Feinstein, President; (412) 594-2550; fax (412) 232-6240; feinstein@jhf.org or info@jhf.org
Internet: http://jhf.org/grants/guidelines/guidelines
Sponsor: Jewish Healthcare Foundation of Pittsburgh
650 Smithfield Street, Centre City Tower, Suite 2400
Pittsburgh, PA 15222

Jewish Institute of Religion Graduate Fellowships and Scholarships 2521

The Jewish Institute of Religion at Hebrew Union College awards graduate fellowships and tuition scholarships for MA and PhD studies in Bible, Semitics, History, Rabbinics, and Modern Jewish studies. Also, joint programs are offered with the University of Cincinnati in modern Hebrew, literature, and sociology.
Contact: Graduate Study Admissions; (513) 221-1875; fax (513) 221-0321
Internet: http://www.huc.edu/admissions/aid.shtml
Sponsor: Hebrew Union College
3101 Clifton Avenue
Cincinnati, OH 45220-2488

Jewish Women's Foundation of New York Grants 2522

The foundation provides support for unmet social, economic, and health needs of Jewish females in the New York metropolitan area and beyond. Strategic grants support cutting-edge projects that address Jewish education, training, and culture—projects related to education, training of educators, leadership development, ethics, spirituality, arts, sports, etc; mental and physical health—projects involving genetics, self-esteem, violence prevention, issues concerning the end of life, and other areas that relate to women's health as a quality of life; and economic empowerment—projects that give women more independence, knowledge, competency, and responsibility for their economic situations. Grant requests may address innovative programs, services, and/or research. The projects primary target group must be Jewish females. The foundation expects to award three to four grants. The listed deadline is for concept letters; full proposals are by invitation.
Requirements: 501(c)3 tax-exempt organizations in the New York metropolitan area (the five boroughs of New York City, Long Island and/or Westchester) are eligible.
Restrictions: The foundation will not provide grants for scholarships, equipment, capital campaigns, or ongoing support for existing programs.
Geographic Focus: New York
Date(s) Application is Due: Sep 30
Amount of Grant: Up to 25,000 U.S.D
Contact: Sherri Greenbach, Executive Director; (212) 836-1478; fax (212) 836-1831
Internet: http://www.jewishwomenny.org
Sponsor: Jewish Women's Foundation of New York
130 E 59th Street, Room 563
New York, NY 10022

Jim Moran Foundation Grants 2523

The Jim Moran Foundation Grants award funding to 501(c)3 organizations in Florida. The Foundation seeks to improve the quality of life for youth and families through the support of innovative programs and opportunities that meet the ever-changing needs of the community. The Foundation's funding focuses include: education; elder care programs; family strengthening programs; meaningful after school programs; and youth transitional living programs. Proposals for programs that improve the quality of life for those who are at-risk and economically disadvantaged (without extenuating medical or developmental disabilities) will receive priority consideration. Grants are primarily awarded to the Florida counties of Broward, Palm Beach, and Duval. Applicants submit an online letter of inquiry, and will be notified within 90 days if they qualify to submit the online application.
Requirements: The Foundation will consider only organizations that have received 501(c)3 tax-exempt status under the IRS code. Additionally, the organization must be appropriately recognized by state statutes, laws and regulations that govern tax exempt organizations.
Restrictions: The Foundation will consider requests for operating dollars, but only if they do not exceed 50% of the grant request. The Foundation will not consider requests for capital campaigns, capacity building, healthcare or medical research, or event sponsorships.
Geographic Focus: Florida
Contact: Melanie Burgess, Executive Director; (954) 429-2122; fax (954) 429-2699; information@jimmoranfoundation.org
Internet: http://www.jimmoranfoundation.org/GrantApplication.aspx
Sponsor: Jim Moran Foundation
100 Jim Moran Boulevard
Deerfield Beach, FL 33442

Jimmy Buffet's Singing for Change (SFC) Grants 2524

The foundation offers annual competitive grants to progressive U.S. nonprofit organizations that address the root causes of social or environmental problems. SFC is interested in funding projects that improve the quality of life for people and that empower individuals to effect positive change in their communities. Most likely to be considered are organizations that keep their overhead low and collaborate with other groups in their community to find innovative ways of solving common problems. Areas of interest include children and youth—health, education, and protection of children and their families; disenfranchised groups—projects that help people overcome social or economic barriers to education or employment, promote the empowerment of individuals toward self-sufficiency and provide opportunities for personal growth, and demonstrate human equality and encourage people to cross boundary lines to help others and the environment; and the environment—programs that promote environmental awareness and teach people methods of conservation, protection and the responsible use of natural resources. Submit a one-page letter of interest describing the organization and project; full proposals are by invitation.
Requirements: U.S. nonprofit organizations are eligible.
Amount of Grant: 500 - 10,000 U.S.D
Samples: Appalachian Mountain Club/Businesses for the Northern Forest (Boston, MA)—to protect wildlands, improve forest stewardship, and foster sustainable economic development through advocacy, networking, research, educational programs, and joint projects, $10,000.
Contact: Judith Ranger Smith, Executive Director; (816) 363-8132; fax (816) 363-1290; info@singingforchange.com
Internet: http://margaritaville.com/sfc
Sponsor: Jimmy Buffet's Singing for Change Charitable Foundation
P.O. Box 7210
Kansas City, MO 64113

JM Foundation Grants 2525

The foundation awards grants to eligible nonprofit organizations in its areas of interest, including education and research that fosters market-based policy solutions; developing state and national organizations that promote free enterprise, entrepreneurship, and private initiative; and identifying and educating young leaders. Types of support include internships, matching/challenge grants, program grants, publication, research grants, seed grants, and technical assistance. The foundation's board of directors meets bi-annually, usually in May and October. There are no formal proposal deadlines. Inquiries and proposals are processed on an ongoing basis.
Requirements: Public charities, including 501(c)3, 509(a)1, and 170(b)1(a)(vi) nonprofit organizations, that shares the foundation's priority interests is invited to submit a proposal.
Restrictions: Grants do not support individuals, the arts, government agencies, public schools, and international agencies; or requests for operating expenses, annual fundraising campaigns, capital campaigns, equipment, endowment funds, and loans.
Amount of Grant: 5,000 - 100,000 U.S.D
Samples: Acton Institute, Grand Rapids, Michigan, $20,000—in support of the Effective Compassion Initiative; Center of the American Experiment, Minneapolis, Minnesota, $40,000—for Intellectual Takeout; National Taxpayers Union Foundation, Alexandria, Virginia, $25,000—toward the National Taxpayers Conference.
Contact: Carl Helstrom, Executive Director; (212) 687-7735; fax (212) 697-5495
Internet: http://foundationcenter.org/grantmaker/jm/guide_jm.html
Sponsor: JM Foundation
654 Madison Avenue, Suite 1605
New York, NY 10065

Joan's Legacy Lung Cancer Research Grants 2526

Joan's Legacy invites grant applications for institutional research that studies lung cancer. The foundation is particularly interested in the genetic basis and biology of bronchoalveolar carcinoma as well as novel therapeutic approaches for the treatment of this disease. This funding is intended primarily as seed money for promising new work. It is not for supporting research where funding has either lapsed or has been previously disapproved. Preference will be given to applications where indirect costs are minimal or nonexistent. Grants are awarded for one or two years. Funding for promising research may be extended further at the discretion of the medical committee. Application and guidelines are available online.
Date(s) Application is Due: Jul 1
Amount of Grant: 50,000 U.S.D
Samples: For lung cancer research, Dana-Farber Cancer Institute (Boston, MA)—to support the work of Kwok-Kim Wong; Indiana U (Bloomington, IN)—to support the work of Chong-xian Pan; Mayo Clinic Cancer Ctr (Rochester, MN)—to support the work of Julian Monlina, $300,000 divided.
Contact: Grants Administrator; (212) 627-5500; fax (917) 661-7811
Internet: http://www.joanslegacy.org/grant_app.html
Sponsor: Joan Scarangello Foundation to Conquer Lung Cancer
27 Union Square W, Suite 205
New York, NY 10003

Joe W. and Dorothy Dorsett Brown Foundation Grants 2527

The foundation awards grants to nonprofit organizations in Louisiana and the Gulf Coast of Mississippi. Areas of interest include medical research; housing for the homeless; support for organizations who care for the sick, hungry or helpless; religious and educational institutions; and organizations and groups concerned with improving the local community. Types of support include operating budgets, research, and student aid. The foundation also supports Service Learning, a learn-by-doing approach to the curriculum. Students receive practical, hands-on experience in the subject matter studied by meeting identified community needs through active participation. The listed deadline date is for Service Learning grants.
Requirements: Louisiana and Mississippi nonprofit organizations, with a focus on South Louisiana, the New Orleans area, and the Mississippi Gulf Coast, are eligible. Service Learning grant applications are available yearly to sixth through 12th grades in the following parishes: Orleans, Jefferson, Plaquemines, Saint Bernard, Saint Charles, Tangipahoa, Saint James, Saint John, Saint Tammany, and Washington.
Geographic Focus: Louisiana, Mississippi
Date(s) Application is Due: Sep 30
Amount of Grant: 5,000 - 25,000 U.S.D
Contact: Beth Buscher, (504) 834-3433; BethBuscher@thebrownfoundation.org
Internet: http://www.thebrownfoundation.org
Sponsor: Joe W. and Dorothy Dorsett Brown Foundation
320 Hammond Highway, Suite 500
Metairie, LA 70005

John Ben Snow Memorial Trust Grants 2528

The mission of the Foundation is to make grants within specific focus areas to enhance the quality of life in Central and Northern New York State. Historically, the Foundation has made grants in the following program areas: arts and culture, community development, education, environment, historic preservation, and journalism. The Foundation responds to the ever-changing needs of various segments of the population, especially to the needs of young people and people who are disadvantaged either physically or economically. It is the Foundation's general policy to give preference to proposals seeking funds for new or enhanced programs, one-time, short-term grants to sustain a program until funding is stabilized, matching grants used to encourage the participation of other donors, and last dollars towards a capital campaign. There are no minimums or maximum grant amounts; however, most grants range from $5,000 to $15,000.
Requirements: Giving is focused in Maryland, Nevada, and central New York.
Restrictions: The Foundation will not accept proposals from individuals or for-profit organizations. Additionally, the Foundation does not encourage proposals from religious organizations or proposals for endowments, contingency funding, or debt reduction.
Date(s) Application is Due: Apr 1
Amount of Grant: 5,000 - 25,000 U.S.D
Contact: Jonathan L. Snow, (315) 471-5256; fax (315) 471-5256
Internet: http://www.johnbensnow.com/jbsmt
Sponsor: John Ben Snow Memorial Trust
50 Presidential Plaza, Suite 106
Syracuse, NY 13202

John Clarke Trust Grants 2529

Dating from April 20, 1676, this historic trust was created under the will of Dr. John Clarke, a Baptist clergyman and physician and one of the co-founders of the first European settlement on Aquidneck Island in 1638. He was born in 1609 and arrived in Boston in 1637 and was one of some three hundred persons that founded the colony on Aquidneck Island. He was the author of the Royal Charter of 1663 that maintained Rhode Island as a colony. Dr. Clarke had no surviving children. He practiced medicine, was a minister, and held public office, although his chief interest was that of the Christian ministry. He was also very interested in education and is thought to have been involved in establishing a free school for Newport in 1640. Dr. Clarke is buried in a small cemetery on West Broadway in Newport. In his will, written on the date of his death, John Clarke established a perpetual charitable trust. He directed that the income from the trust be used "for the relief of the poor or bringing up of children unto learning from time to time forever." He further instructed the trustees "to have a special regard and care to provide for those that fear the Lord." Applicants must apply online at the grant website. Applicants are strongly encouraged to do the following before applying: review the downloadable state application procedures for additional helpful information and clarifications; review the downloadable online-application guidelines at the grant website; review the trust's funding history (link is available from the grant website); review the online application questions in advance; and review the list of required attachments. These will generally include: a list of board members, financial statements (audited, reviewed, or compiled by independent auditor); an organization summary; a list of other funding sources; an IRS Determination letter; and other required documents. All attachments must be uploaded in the online application as PDF, Word, or Excel files. The semi-annual application deadlines are April 1 and November 1. Applications should be submitted by 11:59 p.m. on the deadline dates. Applicants will be notified of grant decisions by letter within two to three months after the proposal deadline. All applications will be acknowledged. Applicants are encouraged to call the second phone number given if they do not receive an acknowledgment within two to three weeks of proposal submission.
Requirements: The trustees of the John Clarke Trust have established a policy of giving preference to organizations located on Aquidneck Island, Rhode Island, and within the East Bay area. However, applications from any Rhode Island 501(c)3 charitable organizations are acceptable.
Restrictions: Capital grant requests will be considered ONLY for Aquidneck Island. The trust does not support requests from individuals, organizations attempting to influence policy through direct lobbying, or any political campaigns.
Geographic Focus: Rhode Island
Date(s) Application is Due: Apr 1; Nov 1
Samples: Martin Luther King Community Center, Newport, Rhode Island, $15,000 (2010); Community Preparatory School, Providence, Rhode Island, $12,600 (2010); Newport Hospital Foundation, Newport, Rhode Island, $20,000 (2010).
Contact: Emma Greene, Director; (617) 434-0329 or (888) 703-2345; emma.m.greene@baml.com

Internet: https://www.bankofamerica.com/philanthropic/fn_search.action
Sponsor: John Clarke Trust
225 Franklin Street, 4th Floor, MA1-225-04-02
Boston, MA 02110

John D. and Katherine A. Johnston Foundation Grants **2530**
The John D. and Katherine A. Johnston Foundation was established in 1928 to support charitable organizations that work to improve the lives of physically disabled children and adults. Special consideration is given to organizations that serve low-income individuals. Preference is given to charitable organizations that serve the people of Newport, Rhode Island. Capital requests that fund handicapped assistive devices (wheelchairs, walkers, etc.) or adaptive equipment (lift installation, ramp installation, etc.) are strongly encouraged. Grant requests for general operating or program support will also be considered. The majority of grants from the Johnston Foundation are one year in duration. The Johnston Foundation shares a mission and grantmaking focus with the Vigneron Memorial Fund. Both foundations have the same proposal deadline date of 11:59 p.m. on April 1. Applicants will be notified of grant decisions before May 31. Applicants must apply online at the grant website. Applicants are strongly encouraged to do the following before applying: review the downloadable state application procedures for additional helpful information and clarifications; review the downloadable online-application guidelines at the grant website; review the foundation's funding history (link is available from the grant website); review the online application questions in advance; and review the list of required attachments. These will generally include: a list of board members, financial statements (audited, reviewed, or compiled by independent auditor); an organization summary; a list of other funding sources; an IRS Determination letter; and other required documents. All attachments must be uploaded in the online application as PDF, Word, or Excel files.
Requirements: Applicants must have 501(c)3 tax-exempt status.
Restrictions: The foundation does not support requests from individuals, organizations attempting to influence policy through direct lobbying, or any political campaigns.
Geographic Focus: Rhode Island
Date(s) Application is Due: Apr 1
Samples: Rhode Island Hospital Foundation, Providence, Rhode Island, $7,500 (2010); CranstonArc, Cranston, Rhode Island, $4,000 (2010); Meals on Wheels of Rhode Island, Providence, Rhode Island, $3,500 (2010).
Contact: Emma Greene, Director; (617) 434-0329; emma.m.greene@baml.com
Internet: https://www.bankofamerica.com/philanthropic/fn_search.action
Sponsor: John D. and Katherine A. Johnston Foundation
225 Franklin Street, 4th Floor, MA1-225-04-02
Boston, MA 02110

John Deere Foundation Grants **2531**
The foundation invests in programs in education, health/human services, community improvement, arts and culture. Types of support include annual campaigns, building construction/renovation, continuing support, fellowships, general operating support, scholarship funds, and seed money grants. Foundation interest also includes support for Third World development through U.S.-based nonprofits with international building funds, research grants, general operating purposes, and continuing support. There are no application deadlines.
Requirements: Nonprofit organizations in communities with major John Deere operating units, and employee presence are eligible. Eligible U.S. locations are: Augusta, GA; Quad City Region, IL; Des Moines, IA, Dubuque, IA, Iowa Quad Cities, IA, Ottumwa, IA, Waterloo, IA, Coffeyville, KS, Lenexa, KS, Thibodaux, LA, Springfield, MO, Cary, NC, Fuquay-Varina, NC, Fargo, ND, Greeneville, TN, Madison, WI, Horicon, WI. Exceptions include: accredited colleges and universities; organizations focused on international development initiatives related to John Deere Solutions for World Hunger initiative. Because John Deere dealerships are owned and operated independently, their communities are not included in this geographic scope. Also eligible to apply are organizations and institutions of national or international scope that reflect the foundation's concerns.
Restrictions: Funds are not available for the following organizations or purposes: individual initiatives, including scholarships; sports teams, racing teams, athletic endeavors or scholarships designated for athletes; faith-based organizations for sectarian purposes; political candidates, campaigns or organizations; private clubs, fraternities or sororities; other foundations for purposes of building endowment; tax-supported entities.
Geographic Focus: Georgia, Illinois, Iowa, Kansas, Louisiana, Missouri, North Carolina, Tennessee, Wisconsin, Belarus, Brazil, Canada, Estonia, Latvia, Lithuania, Moldova, Ukraine
Contact: Amy Nimmer, Director; (309) 765-8000
Internet: http://www.deere.com/en_U.S./globalcitizenship/socialinvestment/index.html
Sponsor: John Deere Foundation
1 John Deere Place
Moline, IL 61265-8098

John Edward Fowler Memorial Foundation Grants **2532**
The foundation makes grants to qualified charitable organizations providing grassroots programs for people in need in the Washington, DC, metropolitan area. Preference is given to programs that address the issues of homelessness, hunger, at-risk children and youth (pre-school through high school), adult literacy, free medical care (prenatal to seniors), seniors aging in place, job training and placement. Types of support include general operating support, building construction/renovation, equipment acquisition, program development, and matching funds. The foundation is interested in supporting smaller, well-managed nonprofit organizations that have innovative ideas about how to help people help themselves. Contact the office for application forms. There are no application deadlines.
Requirements: Organizations in Washington, DC, and its close Maryland and Virginia suburbs are eligible.
Restrictions: Grants are not made outside the metropolitan Washington, DC, area, or to/for national health organizations; government agencies; medical research; public school districts; individuals; or arts (except for intensive arts-in-education programs that directly benefit at-risk children and youth).
Geographic Focus: District of Columbia, Maryland, Virginia
Amount of Grant: 5,000 - 20,000 U.S.D
Samples: Carpenter's Shelter, Alexandria, VA, $20,000 - housing with intensive supportive services for homeless adults and children; Bucknell University, Lewisburg, PA , $10,000 - general support; All Souls' Episcopal Church, Mechanicsville VA, $15,000 - Orphans Program of the Maseno Mission in Kenya.
Contact: Suzanne Martin, Grant Consultant; (301) 654-2700
Internet: http://fdncenter.org/grantmaker/fowler/about.html
Sponsor: John Edward Fowler Memorial Foundation
4340 East-West Highway, Suite 206
Bethesda, MD 20814

John G. Duncan Trust Grants **2533**
The trust awards grants in the following areas of interest: arts; education; health care; human services; religion; and support: building/renovation; capital campaigns; emergency funds; equipment; program development; seed money; and research. Contact the foundation, for the application guidelines and application form.
Requirements: IRS 501(c)3 organizations in Colorado are eligible.
Restrictions: No support for other grantmaking organizations, individuals, endowments, or for general operating expenses.
Geographic Focus: Colorado
Date(s) Application is Due: Jan 31; Apr 30; Jul 31; Oct 31
Amount of Grant: 1,000 - 10,000 U.S.D
Samples: Gospel Shelters for Women, Colorado Springs, CO, $3,250; Boulder Valley Womens Health Center, Boulder, CO, $2,000; Commerce City Community Health Services, Commerce City, CO, $5,000.
Contact: Jason Craig, c/o Wells Fargo Bank, N.A., (720) 947-6820
Sponsor: John G. Duncan Trust
1740 Broadway, MAC C7300-483
Denver, CO 80274-0001

John G. Martin Foundation Grants **2534**
The foundation awards grants to eligible Connecticut nonprofit organizations in the areas of education at all levels. Grants may be considered in other areas. Preferential consideration will be given to proposals whose benefits include assistance to the aging and elderly or assistance to youth and adolescents. The foundation is interested in the long-term stabilization and advancement of effective organizations specializing in these areas of interest. Types of support include building construction/renovation, capital campaigns, and matching/challenge grants.
Requirements: Nonprofits serving the Hartford, CT, area may submit applications.
Geographic Focus: Connecticut
Amount of Grant: 10,000 - 65,000 U.S.D
Contact: Frank Loehmann, President; (860) 677-4574; fax (860) 674-1490; frank@resmgtcorp.com
Sponsor: John G. Martin Foundation
2 Batterson Park Road
Farmington, CT 06032-2553

John H. and Wilhelmina D. Harland Charitable Foundation Grants **2535**
The foundation, offers support for: youth services, community services, cultural programs, education, and health. The focus is local rather than regional or national, and priority is given to institutions in metropolitan Atlanta, Georgia. Grants support: building/renovation; capital campaigns; equipment; general/operating support; matching/challenge support; and scholarship funds. The Foundation prefers a telephone call as opposed to a letter of inquiry for the initial approach.
Requirements: Grant support is available to nonprofit organizations in Georgia, with emphasis on the metropolitan Atlanta area.
Restrictions: Grants are not awarded to individuals.
Geographic Focus: Georgia
Date(s) Application is Due: Mar 1; Sep 1
Amount of Grant: 4,000 - 250,000 U.S.D
Contact: Jane Hardesty, Executive Director; (404) 264-9912
Sponsor: John H. and Wilhelmina D. Harland Charitable Foundation
2 Piedmont Court, Suite 106
Atlanta, GA 30305-1567

John H. Wellons Foundation Grants **2536**
Established in 1950, the John H. Wellons Foundation, is a private foundation. The foundation's support is limited to the Dunn, North Carolina, area. The foundation awards student loans to local area residents, provides housing for the elderly and/or handicapped, and contributes to charitable organizations. There are no specific deadlines with which to adhere. Contact the Foundation for further application information and guidelines.
Geographic Focus: North Carolina
Contact: John H. Wellons, President; (910) 892-0436
Sponsor: John H. Wellons Foundation
P.O. Box 1254
Dunn, NC 28335-1254

John I. Smith Charities Grants **2537**

The foundation supports nonprofit organizations, primarily in South Carolina. With its areas of interest, including: the visual and performing arts, higher education, medical education, theological education, literacy education and basic skills, Christian agencies and churches, child welfare, and programs for the disabled. Grants support general operations, capital campaigns, endowment funds, scholarship funds, and emergency funds. Application forms are not required. The board meets on a quarterly basis. Mail applications to: P.O. Box 1687, Greer, SC 29652

Requirements: Nonprofit organizations in South Carolina are eligible to apply.

Geographic Focus: South Carolina

Amount of Grant: 2,500 - 200,000 U.S.D

Samples: Furman University, Greenville, SC, $100,000; Alliance for Quality Education, Greenville, SC, $27,000; Columbia Theological Seminary, Decatur, GA, $200,000.

Contact: Jefferson Smith, President; (864) 879-2455

Sponsor: John I. Smith Charities
P.O. Box 40200, FL9-100-10-19
Jacksonville, FL 32203-0200

John J. Leidy Foundation Grants **2538**

The John J. Leidy Foundation was established in 1957. The foundation gives primarily in the metropolitan Baltimore, Maryland, area. Funding is available to people with disabilities, education, health, social services, and Jewish organizations in the form of, scholarship funds, building/renovation, equipment, general/operating support, and program development grants.

Requirements: Maryland non-profits are eligible to apply.

Restrictions: No grants to individuals.

Geographic Focus: Maryland

Samples: Maryland Food Bank, Baltimore, MD, $20,000; Baltimore Symphony Orchestra, Baltimore, MD, $5,000; Baltimore Hebrew Congregation, Baltimore, MD, $50,000.

Contact: Robert L. Pierson, President; (410) 821-3006; Leidyfd@attglobal.net

Sponsor: John J. Leidy Foundation
305 W Chesapeake Avenue, Suite 308
Towson, MD 21204-4440

John Jewett and Helen Chandler Garland Foundation Grants **2539**

Established in 1959, the John Jewett & Helen Chandler Garland Foundation gives primarily in California, with emphasis on southern California. The foundations area of interest include: the arts, education, health care, children and social services. There are no application deadlines.

Requirements: California nonprofit organizations are eligible.

Restrictions: No telephone inquiries excepted, submit a letter of inquiry as your initial approach.

Geographic Focus: California

Amount of Grant: 2,500 - 610,000 U.S.D

Samples: Huntington Library, Friends of the, San Marino, CA, $800,000 - for general support and educational programs at Botanical Center; Marlborough School, Los Angeles, CA, $25,000 - for general support and minority scholarships; Boys and Girls Clubs of Pasadena, Pasadena, CA, $110,000 - for general support and to make swimming pool accessible year-round.

Contact: Lisa M. Hausler, Manager

Sponsor: John Jewett and Helen Chandler Garland Foundation
P.O. Box 550
Pasadena, CA 91102-0550

John Lord Knight Foundation Grants **2540**

The Foundation, established in 1957, provides funding for program in the Lucas County, Ohio, region and throughout Florida. Its areas of interest include: cancer research; Christian agencies and churches; education; elementary education; federated giving programs; hospitals; human services; public libraries; and youth services. There are no specific deadlines, and applicants should submit a detailed description of project and amount of funding requested to the listed contact person in care of KeyBank.

Geographic Focus: Florida, Ohio

Contact: Erwin Diener, Key Bank, Trust Officer; (419) 259-8372

Sponsor: John Lord Knight Foundation
Three Seagate, P.O. Box 10099
Toledo, OH 43699-0099

John M. Lloyd Foundation Grants **2541**

The foundation has the following funding objectives: to increase funding from public and private sectors to address the HIV/AIDS pandemic, both globally and domestically; to improve domestic and international policies (universal protection of human rights to issues concerning HIV/AIDS, access to HIV/AIDS healthcare and treatment, and access to accurate information about HIV/AIDS); to amplify global awareness of HIV/AIDS and to facilitate broad-based change in attitudes to reduce stigma and change behavior; and to develop the leadership of organizations that fight HIV/AIDS, as well as to foster collaborations among those organizations and leaders. Projects that have promise of making a significant impact and those that are new and innovative receive preference. The listed application deadlines are for concept letters; full proposals are by invitation. Organizations may submit only one concept letter per year.

Requirements: 501(c)3 tax-exempt organizations are eligible. Grants are made to locally focused projects in California, U.S.-based projects with a national or global scope, and international projects.

Restrictions: In general, the foundation does not make contributions more than once per calendar year to any single organization; more than three consecutive years to any single project; to annual campaigns; to operating budgets of established organizations; to capital expenditures (physical plant, equipment, endowment); to indirect costs; to individuals; to locally focused projects in the United States with the exception of locally focused projects in California; for health care or service provision; or for general support

Date(s) Application is Due: Aug 15; Dec 15

Amount of Grant: Up to 20,000 U.S.D

Contact: Melanie Havelin, Executive Director; (310) 622-1050; fax (310) 622-1070; Mhavelin@johnmlloyd.org

Internet: http://www.johnmlloyd.org/grant_programs.html

Sponsor: John M. Lloyd Foundation
11777 San Vicente Boulevard, Suite 745
Los Angeles, CA 90049

John Merck Fund Grants **2542**

The foundation fosters innovative advocacy and problem solving in the fields of Developmental Disabilities, Environment, Reproductive Health, Human Rights and Job Opportunities. Its objective is to act as a catalyst, supporting organizations that can effect constructive and measurable change in each of these areas.

Requirements: The foundation actively seeks out projects and programs that may merit support, then requests grant applications on behalf of those it finds most promising. It does not encourage the submission of unsolicited proposals. However, organizations interested in obtaining support for work they do in one of the foundation's areas of interest are welcome to send a brief letter of inquiry. The foundation favors: outstanding individuals working on promising projects in organizations that may have difficulty attracting funds; pilot projects with potential for widespread application; advocacy, including litigation, capable of setting or protecting important precedents; smaller organizations, start-ups included; one-year grant requests (though multi-year grants of up to three years occasionally are made); matching-grant opportunities, particularly to help broaden support for fledgling initiatives.

Restrictions: The foundation does not provide grants for: endowment or capital-fund projects; large organizations with well-established funding sources (except those that need help launching promising new projects for which funding is not readily available); general support (except in the case of small organizations whose entire mission coincides with one of foundation's areas of interest); individuals (except if his or her project is sponsored by a domestic or foreign educational, scientific or charitable organization).

Contact: Ruth G. Hennig, Executive Director; (617) 556-4120; fax (617) 556-4130; info@jmfund.org

Internet: http://www.jmfund.org/program.html

Sponsor: John Merck Fund
2 Oliver Street, 8th Floor
Boston, MA 02108

John Merck Scholars Awards **2543**

Scholars are chosen from the ranks of the most promising assistant professors currently working, or planning to work, in neurobiological and cognitive sciences relating to the biology of mental disability and developmental disabilities, including developmental studies of cognition, perception, language, reading, learning and motor performance. The foundation will accept one application for a neurobiologist and one for a cognitive scientist from major universities and other research centers. Scholars will receive $75,000 per year for a four-year period, subject to an annual review of research progress. The sponsoring institution is responsible for fiscal management.

Requirements: The foundation will fund the most promising young researchers whose work illuminates neurodevelopmental disorders from the perspectives of (i) synapse formation and synaptic plasticity; (ii) learning and memory, and synaptic plasticity; (iii) perception, cognition and behavior; (iv) neurogenesis and pattern formation; and (v) genetics and early development. Also encouraged are proposals that (i) investigate the possible role of environmental chemicals in the origins of developmental disabilities, or that (ii) aim to distinguish subgroups within accepted diagnostic categories through the use of sophisticated behavioral and neuroimaging tests of perception, cognition, and emotions based on concepts from modern cognitive neuroscience. In all cases, they seek proposals from young scientists conducting research that is of the highest quality and that has the greatest chance of increasing understanding of neurodevelopmental disorders. Applicants must have the following: Academic rank in a university or medical school, or equivalent standing in a research institute or medical center; A record of research in areas relating to the foundation's interest in the underlying causes of developmental disabilities; Not more than four years of experience in an independent faculty position; Evidence of a commitment to a career in neuroscience or cognitive science. Request application forms by mail, email or phone.

Restrictions: Applicants may not: hold tenured positions or their equivalents; apply in more than two rounds of competition; apply for continued postdoctoral support. Holding other fellowships concurrently with the John Merck Scholars Award is discouraged. Prior approval by the foundation for an overlapping fellowship is required and will be given only in unusual circumstances.

Date(s) Application is Due: Sep 15

Amount of Grant: 75,000 U.S.D

Contact: Jason Bentsman, Staff Assistant; (617) 556-4120; fax (617) 556-4130; j bentsman@jmfund.org

Internet: http://www.jmfund.org/jm_scholars_program.html

Sponsor: John Merck Fund
2 Oliver Street, 8th Floor
Boston, MA 02108

John P. McGovern Foundation Grants 2544

The John P. McGovern Foundation, established in 1961, supports the charitable interests of the donor to support the activities of established nonprofit organizations, which are of importance to human welfare with special focus on children and family health education and promotion, treatment and disease prevention. The Foundation gives primarily in Texas with an emphasis on the Houston area, but also provides funding in the Southwest region as well. The types of support offered are: building/renovation; conferences/seminars; continuing support; curriculum development; emergency funds; endowments; general/operating support; matching/challenge support; professorships; publication; research; and scholarship funds.
Requirements: Non-profits in Texas are eligible to apply.
Geographic Focus: Texas
Date(s) Application is Due: Aug 31
Amount of Grant: 10,000 - 100,000 U.S.D
Samples: University of Texas Health Science Center, Houston, TX. $1,500,000 - for John P. McGovern, MD, Center for Health, Humanities, and Human Spirit/Certificate Program Endowment Fund; Community Family Centers, Houston, TX. $200,000- for John P. McGovern Community Sports and Recreation Building; Salvation Army of Houston, Houston, TX. $150,000 - for general operating fund.
Contact: Kathrine G. McGovern, President; (713) 661-4808; fax (713) 661-3031
Sponsor: John P. McGovern Foundation
2211 Norfolk Street, Suite 900
Houston, TX 77098-4044

John P. Murphy Foundation Grants 2545

The Foundation awards grants primarily in Cuyahoga County, Ohio, and its immediately adjacent counties. Areas of interest include education—college level, and primary and secondary education to improve educational programs benefiting low income and minority students, most often in Cleveland public schools; arts and culture; social services; community—community activities, agencies, and events; health—institutions and organizations such as hospitals, clinic, and nursing homes; and religion. Grants to national or regional institutions with broad public support will be made only for start-up programs where time is a factor or where the proposal addresses a specific local community need. Guidelines are available online. Contact the office for an application form.
Requirements: Ohio organizations properly classified by the IRS as being eligible to receive grants may apply. Each applicant must file its most recent IRS letter of classification with each proposal. National or regional institutions with broad public support may submit proposals that address a specific local community need.
Restrictions: No grants will be made to endowment funds or for scholarships.
Geographic Focus: Ohio
Amount of Grant: 5,000 - 75,000 U.S.D
Contact: Allan J. Zambie, Executive Vice President; (216) 623-4770 623-4771 or (216) 623-4771; fax (216) 623-4773
Internet: http://fdncenter.org/grantmaker/jpmurphy/interest.html
Sponsor: John P. Murphy Foundation
50 Public Square, Suite 924
Cleveland, OH 44113-2203

John R. Oishei Foundation Grants 2546

The foundation's primary mission is to support medical research and care and education, as well as cultural and social needs existing in the Buffalo Niagara region. The foundation favors creative programs that attempt to advance from the status quo, are strategically sound, and are strongly focused on excellence. Programs that provide opportunities for foundation support to be leveraged into greater support from other sources will be especially favored. The foundation generally does not fund operating expenses, though occasional exceptions may be made for organizations that address basic human needs. Requests for capital funds will not be considered unless they are an integral part of an otherwise eligible proposal. Contact the foundation to discuss eligibility before submitting an application. Grant applications are accepted throughout the year.
Requirements: It is the general policy of the foundation to confine its support to activities located in the Buffalo, NY, metropolitan region.
Restrictions: Grants do not support endowments; capital requests (buildings or equipment); deficit funding or loans; individual scholarships or fellowships (except within specific foundation programs); travel, conferences, seminars, or workshops; fundraising events; or 509(a) private foundations.
Geographic Focus: New York
Amount of Grant: 100,000 - 200,000 U.S.D
Contact: Blythe T. Merrill, Senior Program Officer; (716) 856-9490, ext. 3; fax (716) 856-9493; btmerrill@oisheifdt.org
Internet: http://www.oisheifdt.org/Home/Fund/WhatWeFundOverview
Sponsor: John R. Oishei Foundation
1 HSBC Center, Suite 3650
Buffalo, NY 14203-2805

John S. Dunn Research Foundation Grants and Chairs 2547

Established in 1985, the foundation funding area is limited to the Texas, providing support to health and medical-related organizations, especially hospitals; support also for cancer, other medical research, and freestanding clinics.
Requirements: Grants are made to nonprofit organizations within Texas having 501(c)3 tax-exempt status.
Restrictions: No grants to individuals, or for multi-year or seed money grants.
Geographic Focus: Texas
Amount of Grant: 10,000 - 100,000 U.S.D
Samples: Gulf Coast Consortia, Houston, TX, $1,180,000 - to support drug discovery program; University of Texas M.D. Anderson Cancer Center, Houston, TX, $650,000 - to fund radiological sciences and metabolic lab; Memorial Hermann Healthcare System, Houston, TX, $250,000 - to support life flight services to patients without insurance.
Contact: Dr. Lloyd Gregory; (713) 626-0368; fax (713) 626-3866
Sponsor: John S. Dunn Research Foundation
3355 West Alabama street, Suite 720
Houston, TX 77098-1718

Johns Manville Fund Grants 2548

The Johns Manville Fund, Inc. gives primarily in areas of company operations in Denver, Colarado, Etowah, Tennessee, and in Canada. The Fund offers support in the areas of: the arts; education; health care; human services; youth services and the American Red Cross. The grants are offered in the form of employee-related scholarships, employee volunteer services, general/operating support, and program development. Contact the Johns Manville Fund, Inc. for a Informational brochure, including application guidelines.
Requirements: Non-profits operating in Denver, Colorado, Etowah, Tennessee, and in Canada are eligible.
Restrictions: No support for religious organizations not of direct benefit to the entire community, hospitals, or non-special needs private educational organizations. No grants for special events.
Geographic Focus: Colorado, Tennessee, Canada
Amount of Grant: 1,000 - 27,000 U.S.D
Contact: Community Relations Manager; (303) 978-3863; fax (303) 978-2108
Internet: http://www.jm.com/corporate/careers/1241.htm
Sponsor: Johns Manville Fund
717 17th Street
Denver, CO 80202-3330

Johnson & Johnson Community Health Care Program Grants 2549

The program provides support for community public health initiatives impacting access and delivery of quality health care services for medically underserved populations, with emphasis on women and children (including infants and adolescents). Additional consideration will be given to programs focusing on the education and prevention of diabetes, obesity, and cardiovascular disease. Grants also will assist organizations in developing a broad-based public and private support network by extending public recognition of their efforts. Second-year continuation grants also are available.
Requirements: Grants are awarded to community/public health care organizations/agencies in Alabama (all areas); Arkansas (all areas); Boston, Massachusetts; Florida (all areas); New Jersey (all areas); Ohio (all areas); San Diego, California; and San Francisco, California.
Restrictions: Proposals from foundations, universities, and political advocacy groups will not be considered.
Geographic Focus: Alabama, Arkansas, California, Florida, Massachusetts, New Jersey, Ohio
Date(s) Application is Due: Oct 17
Amount of Grant: 150,000 U.S.D
Samples: Bread for the City (Washington, DC)—for community healthcare awards, which support innovative, community-based programs that provide healthcare services to medically underserved people, $150,000 over two years.
Contact: Sierra Veale, (443) 287-5138; fax (410) 510-1974; jandj@jhsph.edu
Internet: http://www.jhsph.edu/johnsonandjohnson/index.html
Sponsor: Johnson and Johnson
615 N Wolfe Street, Room W1100
Baltimore, MD 21205

Johnson & Johnson Corporate Contributions Grants 2550

Johnson & Johnson and its many operating companies support community-based programs that improve health and well-being. The Company works with community-based partners that have the greatest insight into the needs of local populations and the strategies that stand the greatest chances of success. Giving focuses on: saving and improving the lives of women and children; building on the skills of people who serve community health needs, primarily through education; and preventing diseases and reducing stigma and disability in underserved communities where we have a high potential for impact.
Requirements: Grants are awarded to nonprofit and tax-exempt local, national, and international organizations and institutions.
Restrictions: Grants are not awarded to individuals, for deficit funding, capital or endowment funds, demonstration projects, or publications.
Contact: Shaun Mickus, (732) 524-2086; smickus@corus.jnj.com
Internet: http://www.jnj.com/connect/caring/corporate-giving/
Sponsor: Johnson and Johnson
1 Johnson & Johnson Plaza
New Brunswick, NJ 08933-0001

Johnson and Johnson / SAH Arts and Healing Grants 2551

Johnson & Johnson and the Society for the Arts in Healthcare have provided program funding to 117 organizations in the U.S. and Canada. In 2008, the grant structure was updated to a three-year grant period with a focus on existing programs. This new structure promotes the evaluation and replication of promising models in order to strengthen and expand the arts in healthcare field.
Requirements: To be eligible for consideration, grant applicants must be: a current member of Society for the Arts in Healthcare; located in the U.S. or Canada (partners

may be international); a non-profit organization with federal tax exempt status and organized and operated for charitable purposes or a governmental agency (state or local, including education or institutions). Note: U.S. based non-profit organization applicants must be tax exempt under section 501(c)3 of IRS Code; Canada based applicants must be a registered Canadian charity; and able to show a demonstrated ability to work effectively in partnership with other organizations. To be eligible for consideration, proposed programs must: utilize a model that has documented positive impact; and be designed with proposed outcomes, which show initial impact. Program goals should be specific, measurable, attainable, realistic and timely; and Be sustainable based on existing resources and organizational capacity (adequate staffing, funding, space, etc.).
Restrictions: Individuals are ineligible to apply for scholarships or other forms of assistance. Members of the Society for the Arts in Healthcare board of directors, advisory board, or staff are ineligible to apply or partner with any applicant for the purposes of this grant.
Geographic Focus: All States, Canada
Date(s) Application is Due: Jan 13; Oct 15
Contact: Anita Boles; (202) 299-9770; fax (202) 299-9887; anita@thesah.org
Internet: http://www.thesah.org/template/page.cfm?page_id=15
Sponsor: Society for the Arts in Healthcare
2437 15th Street NW, B South
Washington, DC 20009

Johnson Controls Foundation Health and Social Services Grants 2552
The Johnson Controls Foundation provides financial gifts to select U.S.-based organizations located in the communities in which the company has a presence. Operating support for organizations in the health and social services category largely occurs in communities where Johnson Controls has a local presence and is often directed through contributions to United Way.
Requirements: In evaluating requests for funds, the Advisory Board has developed policies and guidelines for giving in health and social services. Contributions will be given financial assistance to federated drives, hospitals, youth agencies and other health and human service agencies. Ordinarily, operating support of health and social service agencies is reserved for Johnson Controls communities, and generally directed through contributions to United Way.
Restrictions: In general, no grants will be made to any political campaign or organization; any municipal, state, federal agency, or department, or to any organization established to influence legislation; any private individual for support of personal needs; any sectarian institutions or programs whose services are limited to members of any one religious group or whose funds are used primarily for the propagation of a religion; for testimonial dinners, fund raising events, tickets to benefits, shows, or advertising; to provide monies for travel or tours, seminars and conferences or for publication of books and magazines or media productions; for specific medical or scientific research projects; foreign-based institutions nor to institutions or organizations for use outside of the United States; fraternal orders or veteran groups; private foundations or to endowment funds. The foundation does not donate equipment, products or labor.
Date(s) Application is Due: Apr 30
Amount of Grant: 1,000 - 150,000 U.S.D
Contact: Charles A. Harvey, President; (414) 524-1200 or (414) 524-2296
Internet: http://www.johnsoncontrols.com/publish/us/en/about/our_community_focus/johnson_controls_foundation.html
Sponsor: Johnson Controls Foundation
5757 North Green Bay Avenue, P.O. Box 591
Milwaukee, WI 53201

Johnson County Community Foundation Grants 2553
The Community Foundation seeks to meet the challenges of the Johnson County, Indiana community as a whole. Special attention is given to requests that increase the capacity of not-for-profit organizations to serve the community and requests that demonstrate community support and in-kind investment. The foundation generally supports three kinds of requests: projects of service to the general community and pilot projects; seed money to enable projects to demonstrate their potential or enhance services; emergency funding for community needs.
Requirements: The grants committee will be only accepting applications this year from area organizations that plan to meet a specific need in the area of Health and Human Services. Applicants should submit a letter of inquiry (2-3 pages), also include a copy of your 501(c)3 Letter of Determination. Letters of inquiry should include: a brief statement of the organization's purpose and goals; a brief description of the project, the need and the target population it addresses; short- and long-term outcomes anticipated and plans for assessing achievements; grant amount needed; a statement about the total agency budget and the project budget; a statement about other funding sources for the agency and/or project, specifying both committed and projected sources of support. The staff will review the letter of inquiry to determine if the proposed effort fits within the community foundation's grant program. If so, you will receive a call or letter from a staff member requesting additional information or a full proposal for our next Grant Cycle deadline. The community foundation supports organizations that are tax exempt under Section 501(c)3 of the Internal Revenue Service Code and are not classified as private foundations under Section 509(a) of the Code. In selected cases, it may consider support for projects sponsored by governmental entities.
Restrictions: The Community Foundation does not: take multi-year commitments; support services commonly regarded as the responsibility of government; provide support for political or partisan purposes or for programs in which religious teachings are an integral part; consider more than one proposal from the same organization within a 12-month period; provide discretionary grant support to an organization's ongoing operating budget, building funds, capital campaigns, or endowments; fund raising events and functions; make grants to individuals except for scholarship or special award funds; fund existing obligations or to replenish resources (deficit funding) for such purposes.
Geographic Focus: Indiana
Date(s) Application is Due: Mar 6; Sep 4
Contact: Kim Minton; (317) 738-2213; fax (317) 738-9113; kimm@jccf.org
Internet: http://www.jccf.org/index.asp?p=37
Sponsor: Johnson County Community Foundation
398 S Main, P.O. Box 217
Franklin, IN 46131-2311

Johnson F. Hammond Memorial Scholarship 2554
The annual scholarship will be awarded to a medical student of high moral character and outstanding academic achievement with a demonstrated interest and involvement in medical journalism. The scholarship recipient will be selected from among nominees proposed by the deans of AMA approved medical schools, with financial need not being a consideration. Each school may propose one student.
Date(s) Application is Due: Feb 13
Amount of Grant: 3,000 U.S.D
Contact: Dina Lindenberg, Program Officer; (312) 464-4193; fax (312) 464-5973; dina.lindenberg@ama-assn.org or amafoundation@ama-assn.org
Internet: http://www.ama-assn.org/ama/pub/category/14772.html
Sponsor: American Medical Association Foundation
515 N State Street
Chicago, IL 60654

Johnson Foundation Wingspread Conference Support Program 2555
The Johnson Foundation at Wingspread sponsors grants to partially fund conferences focusing on subjects in the public interest, primarily health issues and the environment. Meeting facilities include Wingspread, the home designed by Frank Lloyd Wright, and formerly owned by Herbert Fisk Johnson of the Johnson and Johnson family. Conferences are intensive, one- to four-day meetings of small groups convened in partnership with nonprofit organizations, public agencies, universities, and other foundations. Strategic interests of the Foundation are education, sustainable development and environment, democracy and community, and family. The Foundation's usual contribution to a conference sponsored by one or more other organizations consists of the provision of the full conference facilities of Wingspread, planning and logistical support by the staff, meals and other amenities for the period of the meeting.
Requirements: To be invited to submit a full proposal, applicants first must submit a brief concept letter, consisting of: a clear statement of purpose; a draft agenda; the identification of key participants; and an estimated budget and schedule. The letter should describe how the conference will enhance collaboration and community, include diverse opinions and perspectives, identify solutions, and result in action.
Contact: Coordinator; (262) 639-3211; fax (262) 681-3327; info@johnsonfdn.org
Internet: http://www.johnsonfdn.org/guidelines.html
Sponsor: Johnson Foundation
33 East Four Mile Road
Racine, WI 53402

John Stauffer Charitable Trust Grants 2556
The trust awards grants to California higher education institutions and hospitals. Types of support include building construction/renovation, challenge/matching funds, endowments, equipment acquisition, fellowships, professorships, and scholarship funds. There are no application forms or deadlines.
Requirements: California nonprofits are eligible.
Geographic Focus: California
Amount of Grant: 100,000 - 500,000 U.S.D
Samples: California Institute of Technology, Pasadena, CA, $500,000 - for construction of research lab in new chemistry building; Loma Linda University, Loma Linda, CA, $250,000 - for classroom for emergency personnel; Occidental College, Los Angeles, CA, $159,320 - for endowment for summer research.
Contact: H. Jess Senecal; (626) 793-9400; fax (626) 793-5900; jess@lagerlof.com
Sponsor: John Stauffer Charitable Trust
301 N Lake Avenue, 10th Floor
Pasadena, CA 91101-4108

John V. and George Primich Family Scholarship 2557
The John V. and George Primich Family Scholarship was established to benefit the children or grandchildren of current employees of G.W. Berkheimer Company and South Central Company. The number of scholarships depends on the number of applicants, but they are a minimum of $2,500 each. Any college major is eligible. Contact the Legacy Foundation for application and requirement details.
Geographic Focus: Indiana
Date(s) Application is Due: Jun 1
Amount of Grant: 2,500 U.S.D
Contact: Primich Family Scholarship Contact; (219) 736-1880; fax (219) 736-1940; legacy@legacyfoundationlakeco.org
Internet: http://www.legacyfoundationlakeco.org/scholarships/index.htm
Sponsor: Legacy Foundation
1000 East 80th Place, 302 South
Merrillville, IN 46410

John W. Alden Trust Grants 2558
The trust awards grants to Massachusetts nonprofit organizations in its areas of interest, including education and therapy for children who are blind, disabled, retarded, or mentally or physically ill. Grants will normally be made for specific projects rather

than general operating purposes. There are no application deadlines. Grant applications should be submitted on-line at www.cybergrants.com/alden. Detailed instructions on the web page will guide you through the process. All questions should be directed to the grants coordinator.
Requirements: Eastern Massachusetts 501(c)3 tax-exempt organizations are eligible.
Restrictions: Grants are not made to individuals.
Geographic Focus: Massachusetts
Date(s) Application is Due: Jan 15; Apr 15; Jul 15; Oct 15
Amount of Grant: Up to 15,000 U.S.D
Samples: Hospice of the North Shore, Inc., $6,000 - Music Therapy Program; - University of Massachusetts Foundation Inc., $12,000 - NERCVE Assistive Technology Center; Franciscan Hospital for Children, $13,000 - aquatic aerobic conditioning program.
Contact: Susan Monahan, Grants Coordinator; (617) 951-1108; fax (617) 542-7437; smonahan@rackemann.com
Internet: http://www.cybergrants.com/alden
Sponsor: John W. Alden Trust
160 Federal Street, 13th Floor
Boston, MA 02110-1700

John W. and Anna H. Hanes Foundation Grants **2559**
The trust awards grants, primarily for the arts, to nonprofit organizations in North Carolina. Other interests that are funded are children and youth services; education; environment; health care; historic preservation; and human services. Types of support include annual and capital campaigns; building and renovation; emergency funds; endowments and seed money; equipment and land acquisition; matching and challenge support; and program development. The board meets in January, April, July, and October; application forms are required.
Requirements: 501(c)3 North Carolina nonprofit organizations are eligible.
Restrictions: Giving is limited to North Carolina, with emphasis on Forsyth County. No grants are made to individuals, or for operating expenses.
Geographic Focus: North Carolina
Date(s) Application is Due: Mar 15; Jun 15; Sep 15; Dec 15
Amount of Grant: 1,000 - 200,000 U.S.D
Contact: Christopher Spaugh, Vice President, Wachovia Bank NA; (336) 732-5991
Sponsor: John W. and Anna H. Hanes Foundation c/o Wachovia Bank N.A.
1525 West WT Harris Boulevard
Charlotte, NC 28288-5709

John W. Anderson Foundation Grants **2560**
The John W. Anderson Foundation is an independent foundation established in 1967, in Indiana. The trust was established by John W. Anderson, a manufacturing executive and inventor. Mr. Anderson was president of the Anderson Company. The foundation givings primarily to organizations serving youth; higher educational institutions; community funds; scientific or medical research for the purpose of alleviating suffering; care of needy, crippled or orphaned children; care of needy persons who are sick, aged or helpless; improving the health, and quality of life of all persons; human services; and the arts and humanities. The Foundations gives primarily in Lake and Porter counties in northwest Indiana.
Requirements: Applicants should submit the following: signature and title of chief executive officer; copy of IRS Determination Letter; brief history of organization and description of its mission; copy of most recent annual report/audited financial statement/990; listing of board of directors, trustees, officers and other key people and their affiliations; detailed description of project and amount of funding requested; copy of current year's organizational budget and/or project budget.
Restrictions: Applications sent by fax not considered. Application form not required. An organization may submit a request once in a 12-month period. No support for elementary and secondary schools, or for business or any for-profit organization, or for supporting organizations classified 509(a)3. No grants to individuals, or for endowment funds, multi-year grants, fund raising events, advertising, seed money, deficit financing; no loans.
Geographic Focus: Indiana
Date(s) Application is Due: Jan 20; Mar 20; May 20; Jul 20; Sep 20; Nov 20
Amount of Grant: 5,000 - 50,000 U.S.D
Contact: William N. Vinovich, Vice-Chair; (219) 462-4611
Sponsor: John W. Anderson Foundation
402 Wall Street
Valparaiso, IN 46383-2562

John W. Boynton Fund Grants **2561**
The John W. Boynton Fund was established in 1952 by Dora Carter Boynton in memory of her husband. He had a thriving business in tinware in Templeton, MA, and later in his life was a resident of Athol, MA. In her will, Mrs. Boynton asked that "organizations which benefit poor, needy, and deserving persons and particularly those of advanced years and gentility" be considered. She also expressed a desire that special consideration be given to charitable organizations serving the Town of Athol. Grant requests for new or special programs and capital projects are preferred. General operating support also will be considered. In come cases, general operating grants are made to make up for a temporary loss of public or private funding. A typical grant is $5,000. Applicants must apply online at the grant website. Applicants are strongly encouraged to do the following before applying: review the downloadable state application procedures for additional helpful information and clarifications at the grant website; review the downloadable online-application guidelines at the grant website; review the foundation's funding history (link is available from the grant website); review the online application questions in advance; and review the list of required attachments. These will generally include: a list of board members, financial statements (audited, reviewed, or compiled by independent auditor); an organization summary; a list of other funding sources; an IRS Determination letter; and other required documents. All attachments must be uploaded in the online application as PDF, Word, or Excel files. The application deadline is 11:59 p.m. on July 15. Applicants will be notified of decisions by the end of September.
Requirements: Applicants must have 501(c)3 tax-exempt status and serve the residents of Greater Boston.
Restrictions: The fund does not support requests from individuals, organizations attempting to influence policy through direct lobbying, or any political campaigns.
Geographic Focus: Massachusetts
Date(s) Application is Due: Jul 15
Amount of Grant: 2,500 - 25,000 U.S.D
Samples: Friends of the Athol Council on Aging, Athol, Massachusetts, $60,000 (2010); Associated Grant Makers, Boston, Massachusetts, $5,000 (2010); Womens Lunch Place, Boston, Massachusetts, $3,000 (2010).
Contact: Michealle Larkins; (866) 778-6859; michealle.larkins@baml.com
Internet: https://www.bankofamerica.com/philanthropic/fn_search.action
Sponsor: John W. Boynton Fund
225 Franklin Street, 4th Floor, MA1-225-04-02
Boston, MA 02110

John W. Speas and Effie E. Speas Memorial Trust Grants **2562**
The John W. Speas and Effie E. Speas Memorial Trust was established in 1943 to support and promote quality educational, cultural, human-services, and health-care programming. In the area of arts, culture, and humanities, the trust supports programming that: fosters the enjoyment and appreciation of the visual and performing arts; strengthens humanities and arts-related education programs; provides affordable access; enhances artistic elements in communities; and nurtures a new generation of artists. In the area of education, the trust supports programming that: promotes effective teaching; improves the academic achievement of, or expands educational opportunities for, disadvantaged students; improves governance and management; strengthens nonprofit organizations, school leadership, and teaching; and bolsters strategic initiatives of area colleges and universities. In the area of health, the trust supports programming that improves the delivery of health care to the indigent, uninsured, and other vulnerable populations and addresses health and health-care problems that intersect with social factors. In the area of human services, the trust funds programming that: strengthens agencies that deliver critical human services and maintains the community's safety net and helps agencies respond to federal, state, and local public policy changes. In the area of community improvement, the trust funds capacity-building and infrastructure-development projects including: assessments, planning, and implementation of technology for management and programmatic functions within an organization; technical assistance on wide-ranging topics, including grant writing, strategic planning, financial management services, business development, board and volunteer management, and marketing; and mergers, affiliations, or other restructuring efforts. Grant requests for general operating support and program support will be considered. Grants from the foundation are one year in duration. Application materials are available for download at the grant website. Applicants are strongly encouraged to review the state application guidelines for additional helpful information and clarifications before applying. Applicants are also encouraged to review the foundation's funding history (link is available from the grant website). There are no application deadlines for the Speas Memorial Trust. Proposals are reviewed on an ongoing basis.
Requirements: Applicants must have 501(c)3 tax-exempt status and serve the residents of the Greater Kansas City Metropolitan area. Applications must be mailed.
Restrictions: Grant requests for capital support will not be considered. The trust does not support requests from individuals, organizations attempting to influence policy through direct lobbying, or any political campaigns.
Geographic Focus: Missouri
Samples: TLC for Children and Families, Olathe, Kansas, $25,000, Technology Capacity Building Project (2010); Truman Medical Center Charitable Foundation, Kansas City, Missouri, $200,000, joint venture between Truman and Children's Mercy in support of Don Chisholm Hospital Hill Center (2010); Mid-America Regional Council Community Services Corporation, Kansas City, Missouri, $100,000, Kansas City Bi-State Health Information Exchange (2010).
Contact: Spence Heddens; (816) 292-4301; Spence.heddens@baml.com
Internet: https://www.bankofamerica.com/philanthropic/fn_search.action
Sponsor: John W. Speas and Effie E. Speas Memorial Trust
1200 Main Street, 14th Floor, P.O. Box 219119
Kansas City, MO 64121-9119

Joseph Alexander Foundation Grants **2563**
Established in 1960 in New York, the Foundation supports primarily the New York, New York region. Offering support for: higher education, health organizations and medical research, particularly optic nerve research, social services, and Jewish organizations.
Requirements: Nonprofit organizations are eligible to apply.
Restrictions: Grants are not made to individuals.
Geographic Focus: New York
Amount of Grant: 1,000 - 100,000 U.S.D
Samples: Bar-Ilan University in Israel, New York, NY, $50,000; Center for Jewish History, New York, NY, $15,000; Autism Speaks, New York, NY, $25,000.
Contact: Robert Weintraub, President; (212) 355-3688
Sponsor: Joseph Alexander Foundation
110 East 59th Street
New York, NY 10022-1304

Joseph and Luella Abell Charitable Trust Scholarships 2564

Established in 2003 in Indiana, this Trust (in the care of the Jackson County Bank) offers scholarship awards to individuals that are nursing school students or graduate students of the Protestant ministry throughout Indiana. The funding, which ranges from $1,000 to $5,000 annually, can be used to defray costs for tuition expenses, books, housing, and other costs associated with attending college. An application form can be secured by contacting the Jackson County Bank, and the annual deadline is April 1st.

Requirements: Only students enrolled in a nursing program or ministry in Indiana may apply.

Geographic Focus: Indiana

Date(s) Application is Due: Apr 1

Amount of Grant: 1,000 - 5,000 U.S.D

Contact: George F. Spray, Trustee Officer; (812) 522-3607

Sponsor: Joseph and Luella Abell Charitable Trust
Jackson County Bank, P.O. Box 1001
Seymour, IN 47274-1001

Joseph Collins Foundation Grants 2565

These grants enable men and women whose own resources are inadequate to attend accredited medical schools of their own choice that offer the MD degree. Consideration will include scholastic record; interest in cultural pursuits; intention to specialize in neurology, psychiatry, or general medicine; moral character; age; and geographical proximity (applicants within 200 miles of the medical school preferred). These awards are tenable at any accredited medical school east of the Mississippi River for one year, and renewable at the discretion of the foundation. Applications should be obtained from appropriate medical school authorities; applications are sent to the school, not to students. The school nominates one student per school year; however, trustees act favorably or unfavorably in their sole discretion. Student must interact with the financial aid officer at his/her school. Request applications in January.

Requirements: Grants are open to anyone attending an accredited medical school in states east of or contiguous to the Mississippi River and who intends to specialize in neurology, psychiatry, or general practice. Applicants must have satisfactorily completed the first year of medical school. Awards are not made to premedical or postgraduate medical students.

Date(s) Application is Due: Mar 1

Amount of Grant: Up to 10,000 U.S.D

Contact: Jack H. Nusbaum; (212) 728-8060; fax (212) 728-9060; jnusbaum@willkie.com

Sponsor: Joseph Collins Foundation
787 Seventh Avenue, Room 3950
New York, NY 10019-7099

Joseph Drown Foundation Grants 2566

The foundation makes contributions in the areas of education; community, health and social services; and arts and humanities. It supports programs dealing with such issues as the high school drop-out rate, teen pregnancy, lack of sufficient health care, substance abuse, and violence. Types of support include general operating support, program development, seed money, scholarship funds, and matching funds. Most grant making is limited to programs or organizations in California. Requests are considered each year at March, June, September, and December meetings. The foundation makes grants for both operating support and program support but does not make multiyear commitments. No special application form is required. Proposal should include a letter with information about the organization and the project, a copy of 501(c)3 determination letter, a budget for the organization and the project, the most recent audited financial statements, a copy of the most recent IRS Form 990, and a list of the current board of directors. Any additional materials, such as an annual report, may be attached. Questions should be directed to the program director; proposals should be sent to the foundation president.

Requirements: California 501(c)3 organizations may apply.

Restrictions: The foundation does not provide funds to individuals, endowments, capital campaigns, or annual funds. The foundation does not underwrite annual meetings, conferences, or special events, nor does it fund religious programs or purchase tickets to fund-raising events.

Geographic Focus: California

Date(s) Application is Due: Jan 15; Apr 15; Jul 15; Oct 15

Amount of Grant: 10,000 - 100,000 U.S.D

Samples: The Henry Mancini Institute, Los Angeles, CA, $15,000 - for support of the Community Outreach Initiative; The Museum of Tolerance, Los Angeles, CA, $50,000 - for Tolerance Youth Education programs; Friends of the Semel Institute for Neuroscience, Los Angeles, CA, $40,000 - for a postdoctoral fellowship; Los Angeles Leadership Academy, Los Angeles, CA, $30,000 - for a literacy coordinator and development director in initial year of operation.

Contact: Alyssa Eichelberger, Program Administrator; (310) 277-4488, ext. 100; fax (310) 277-4573; alyssa@jdrown.org

Internet: http://www.jdrown.org

Sponsor: Joseph Drown Foundation
1999 Avenue of the Stars, Suite 1930
Los Angeles, CA 90067

Joseph H. and Florence A. Roblee Foundation Grants 2567

The foundation awards grants to enable organizations to promote change by addressing significant social issues in order to improve the quality of life and help fulfill the potential of individuals. The foundation arises out of a Christian framework, and values ecumenical endeavors. The foundation particularly supports programs which work to break down cultural, racial, and ethnic barriers. Organizations and churches are encouraged to collaborate in achieving positive change through advocacy, prevention, and systemic improvements.

Requirements: Giving limited to nonprofit organizations in the greater bi-state St. Louis region, and Miami/Dade, FL. Contact Foundation for additional guidelines.

Restrictions: Support is not given to individuals or for annual campaigns, research, or loans.

Geographic Focus: Florida, Illinois, Missouri

Date(s) Application is Due: Jan 15; Jun 15

Amount of Grant: 1,000 - 30,000 U.S.D

Samples: Planned Parenthood of the Saint Louis Region, Saint Louis, MO, $20,000 - for comprehensive sex education programs for area teens; Christian Activity Center, East Saint Louis, IL, $15,000 - for Bridging the Gap after-school homework room program; Switchboard of Miami, Miami, FL, $30,000 - for Kevin Kline Suicide Awareness Initiative;

Contact: Peggy Thomas c/o Bank of America, N.A., Bank of America; (314) 466-1304; kathydc@robleefoundation.org.

Sponsor: Joseph H. and Florence A. Roblee Foundation
P.O. Box 14737, MO2-100-07-19
Saint Louis, MO 63178-4737

Josephine G. Russell Trust Grants 2568

The purpose of the trust is for the care, healing, and nursing of the sick and injured, the relief and aid of the poor, the training and education of the young, and any other manner of social service in the city of Lawrence, Massachusetts.

Requirements: Massachusetts nonprofit organizations serving the greater Lawrence metropolitan area are eligible.

Geographic Focus: Massachusetts

Date(s) Application is Due: Jan 31

Amount of Grant: Up to 40,000 U.S.D

Contact: Clifford Elias, Treasurer; (978) 500-3171

Sponsor: Josephine G. Russell Trust
59 Lucerne Drive
Andover, MA 01810-1719

Josephine Goodyear Foundation Grants 2569

The foundation makes grants in the greater Buffalo, NY, area for programs and projects benefiting indigent women and children, particularly with their physical needs. Grants also support hospitals, child welfare organizations, youth agencies, and community funds. Types of support include capital campaigns, building construction/renovation, equipment acquisition, land acquisition, emergency funds, program development, seed grants, research, employee matching gifts, and matching funds. Applicants are requested to submit a two-page proposal including the purpose of the organization, a description of the project/program or need, budget, other sources of funding, and an IRS determination letter.

Requirements: IRS 501(c)3 tax-exempt organizations serving the greater Buffalo, NY, area are eligible.

Restrictions: Grants are not awarded in support of individuals, continuing support, annual campaigns, deficit financing, endowment funds, scholarships, fellowships, or loans.

Geographic Focus: New York

Date(s) Application is Due: Apr 15; Aug 15

Amount of Grant: 500 - 30,000 U.S.D

Contact: E.W. Dann Stevens, President; (716) 566-1465; fax (716) 846-1236; ewdstevens@hiscockbarclay.com

Internet: http://www.cfgb.org/index.php/affiliates-and-initiatives/josephine-goodyear-foundation/97-josephine-goodyear-grants

Sponsor: Josephine Goodyear Foundation
3 Fountain Plaza, Suite 1100, M & T Center
Buffalo, NY 14203

Josephine S. Gumbiner Foundation Grants 2570

The charitable foundation functions for the benefit of women and children in the Long Beach area. The funder supports a wide array of programs, such as the arts; day care; education health care intervention, prevention, and direct services; housing; and recreation. Organizations are not be eligible for funding more than once in any 12-month period. As a general rule, the foundation will not grant funding to any organization for more than three consecutive years. The two-step application process begins with requesting a letter of intent questionnaire, which should be requested through email. Full applications are by invitation. There are no application deadlines. The board meets three or four times each year.

Requirements: Southern California nonprofit organizations are eligible.

Restrictions: Grants do not support political campaigns, lobbying efforts, programs that supplant tradition schooling, pass-through organizations, or groups with endowments greater than $5 million.

Geographic Focus: California

Amount of Grant: 5,000 - 50,000 U.S.D

Samples: Children's Dental, CA, $20,000 - I.V. Sedation for L.B. Children Program; Children Today, Long Beach, CA, $15,000 - LCSW Support; Helpline Youth Counseling, Norwalk, CA, $15,000 - L.B. Gang Prevention Program.

Contact: Grants Administrator; (562) 437-2882; fax (562) 437-4212; julie@jsgf.org

Internet: http://www.jsgf.org/

Sponsor: Josephine S. Gumbiner Foundation
333 West Broadway, Suite 302
Long Beach, CA 90802

Joseph P. Kennedy Jr. Foundation Grants 2571

The foundation concentrates its funding in the area of mental retardation. Grants providing seed funding are offered to encourage new methods of service and support for people with mental retardation and their families.
Requirements: Nonprofit organizations may apply.
Restrictions: The foundation will not support capital costs or costs of equipment for projects, nor will it pay for ongoing support or operations of existing programs. Individuals are not eligible.
Samples: George Washington U (Washington, DC)—for the Ctr for the Study and Advancement of Disability Policy, $200,000.
Contact: Program Contact; (202) 393-1250; fax (202) 824-0351; eidelman@jpkf.org or info@jpkf.org
Internet: http://www.jpkf.org/JPKF_Info/GRANT.HTML
Sponsor: Joseph P. Kennedy Jr. Foundation
1133 19th Street NW, 12th Floor
Washington, DC 20036-3604

Joshua Benjamin Cohen Memorial Scholarship 2572

The Joshua Benjamin Cohen Memorial Scholarship was established by Max and Susan Cohen in memory of their son. Up to two renewable scholarships are awarded to graduating seniors from Valparaiso High School who demonstrate a high standard of leadership, academic performance, athletic performance, and financial need. Any college major is eligible.
Requirements: The scholarship is awarded to a student who demonstrates a high standard of leadership, with a GPA of 3.0 or better. They must also submit an original essay and demonstrate financial need. Renewal of the scholarship is dependent on the student maintaining a B average. Students are required to provide an official copy of their transcript to the Legacy Foundation after each term. Failure to do so may jeopardize their scholarship. Applications are available online and in the Guidance Office of Valparaiso High School each year.
Geographic Focus: Indiana
Date(s) Application is Due: Mar 5
Amount of Grant: 2,000 U.S.D
Contact: Cohen Memorial Scholarship Contact; (219) 736-1880; fax (219) 736-1940; legacy@legacyfoundationlakeco.org
Internet: http://www.legacyfoundationlakeco.org/scholarships/index.htm
Sponsor: Legacy Foundation
1000 East 80th Place, 302 South
Merrillville, IN 46410

Josiah Macy Jr. Foundation Grants 2573

The foundation awards grants to improve the education of physicians and other health care professionals. Grant making focuses on improving medical and health professionals' education in the context of the changing health care system, increasing the number of minority physicians, boosting teamwork between health care professionals, and using educational strategies to increase care for under-served populations. Current initiatives cover the fields of pediatric residency, inter-professional curriculum, improved clinical training, increasing diversity among health care professions, Macy Scholars program, women's health in medical curricula, women in medicine, and funding for a New York post-baccalaureate program. Applications may be submitted at any time. A preliminary letter of inquiry is useful in assisting the foundation staff in determining whether submission of a full proposal is appropriate.
Requirements: Grants are made to colleges, universities, and other 501(c)3 nonprofit organizations for activities within established program areas. Grant proposals, addressed to the president, should include the name of the sponsoring agency or institution; description of the project; names and qualifications of the person who will be responsible for the project; expected cost and duration of the project including an itemized budget; documents substantiating the tax-exempt status of the sponsoring institution; and a letter of endorsement from the sponsoring institution.
Restrictions: Grants are not awarded to individuals or for general undesignated support or for construction/renovation projects.
Amount of Grant: Up to 5,000,000 U.S.D
Contact: George E. Thibault; (212) 486-2424; fax (212) 644-0765; jmacyinfo@josiahmacy
Internet: http://www.josiahmacyfoundation.org/index.php?section=grant_guidelines
Sponsor: Josiah Macy Jr. Foundation
44 East 64th Street
New York, NY 10021

Josiah W. and Bessie H. Kline Foundation Grants 2574

The foundation was established for charitable, scientific, literary, and educational purposes. The objectives for its grantmaking activities are to aid blind or incapacitated persons or crippled children. Grants are made to Pennsylvania colleges and universities and to hospitals, to institutions for crippled children, or to any other benevolent or charitable institution. Generally, grants are made to such institutions that are located in south-central Pennsylvania. Grants also are made for scientific and medical research. Types of support include continuing support, annual campaigns, capital campaigns, equipment acquisition, building/renovation, curriculum development, scholarship funds, matching funds, research, and emergency funds. The foundation does not have an application form. Instead, a letter of request must be submitted with the following information: a description of the need and purpose of the applicant organization, a project budget, amount requested from the foundation and the dates of the need, and a copy of the 501(c)3 tax-exemption letter. Requests may be submitted at any time. The board generally meets in March and November to consider requests.
Requirements: 501(c)3 organizations serving south-central Pennsylvania are eligible.
Restrictions: The foundation does not make loans and does not make grants to individuals or to normal operational phases of established programs, endowments, campaigns or national organizations, or religious programs.
Geographic Focus: Pennsylvania
Amount of Grant: 500 - 150,000 U.S.D
Contact: John Obrock; (717) 561-0820 or (717) 561-4373; fax (717) 561-0826
Sponsor: Josiah W. and Bessie H. Kline Foundation
515 S 29th Street
Harrisburg, PA 17104

Journal Gazette Foundation Grants 2575

The foundation awards grants to northeastern Indiana nonprofit organizations in its areas of interest, including community funds, education, higher education, health organizations and hospitals, social services, Christian agencies and churches, and youth. Types of support include general operations and capital campaigns. There are no application deadlines; the board meets quarterly.
Requirements: Indiana nonprofit organizations are eligible. Preference is given to requests from northeastern Indiana.
Restrictions: Grants are not made to individuals.
Geographic Focus: Indiana
Amount of Grant: 25 - 112,000 U.S.D
Samples: Lincoln Museum (Fort Wayne, Indiana)—for operating support, $6000.
Contact: Jerry Fox, (260) 424-5257
Sponsor: Journal Gazette Foundation
701 S Clinton
Fort Wayne, IN 46802-1883

Joyce Foundation Grants 2576

The foundation's program areas are education, employment, environment, gun violence, money and politics, and culture. Grantmaking is focused on initiatives that promise to have an impact on the Great Lakes region, specifically the states of Illinois, Indiana, Michigan, Minnesota, Ohio, and Wisconsin. A limited number of environment grants are made to organizations in Canada. Education grantmaking focuses on Chicago, Cleveland, and Milwaukee. Culture grants are primarily focused on the Chicago metropolitan area. The foundation is committed to improving public policy through its grant program. Public policy grants provide general operating support or fund educational advocacy such as nonpartisan research, technical assistance, or examinations of broad social issues. Grant applicants are encouraged to describe the nature of advocacy activities in their grant applications and reports, so the foundation can ensure that it is in compliance with federal tax laws. Guidelines are available online.
Requirements: Applications should be made only by organizations in the Great Lakes region that can demonstrate their tax-exempt status and that can supply complete financial information and permit the results of a grant to be audited.
Restrictions: The foundation generally does not support capital projects, endowment campaigns, religious activities, direct services, or scholarships.
Geographic Focus: Illinois, Indiana, Iowa, Michigan, Minnesota, Ohio, Wisconsin, Canada
Date(s) Application is Due: Apr 15; Aug 16; Dec 12
Amount of Grant: 20,000 - 100,000 U.S.D
Samples: Environmental Defense (New York, NY)—to implement federal farm-conservation programs in Midwestern states in an effort to protect Great Lakes water resources, $500,000 over two years; Chicago Metropolis 2020 (Chicago, IL)—for efforts to revise the Illinois criminal code, $150,000 over two years; U of Michigan at Ann Arbor, Great Lakes Radio Consortium (MI)—to cover water-related stories in the region, $200,000 over two years; National-Louis U, Ctr for Early Childhood Leadership (IL)—to educate policy makers and others about the importance of highly qualified directors of early-childhood programs in achieving universal preschool, $200,000 over two years.
Contact: Veronica Salter; (312) 782-2464; fax (312) 782-4160; info@joycefdn.org
Internet: http://www.joycefdn.org/seekingagrant/seekingmain-fs.html
Sponsor: Joyce Foundation
70 W Madison Street, Suite 2750
Chicago, IL 60602

Joy Family Foundation Grants 2577

The foundation awards grants to nonprofit organizations and individuals in the city of Buffalo, NY, and the counties of Erie and Niagara, NY, to improve the quality of life. Grants are awarded in support of human and family services, education and literacy, AIDS research, substance abuse, alcoholism, the aging, the economically disadvantaged, hospitals, women and children, and Roman Catholic churches and organizations. Grants will be awarded for programs and projects, general operating support, capital campaigns, building/renovation, technical assistance, and consulting services. The board meets quarterly.
Requirements: New York nonprofit organizations are eligible.
Geographic Focus: New York
Amount of Grant: 100 - 100,000 U.S.D
Samples: Burchfield-Penny Art Center (Buffalo, NY)—for capital campaign, $25,000; Christ the King Seminary (East Aurora, NY)—grant recipient, $10,000; Computers for Children (Buffalo, NY)—for computer refurbishment, $10,000.
Contact: Marsha Sullivan, Executive Director; (716) 633-6600; fax (716) 633-0600; info@joyfamilyfoundation.org
Internet: http://www.joyfamilyfoundation.org
Sponsor: Joy Family Foundation
5436 Main Street, Suite 1
Williamsville, NY 14221

Judith and Jean Pape Adams Charitable Foundation ALS Grants 2578

Jean Pape Adams was interested in supporting charitable organizations actively engaged in discovering the causes of Amyotrophic Lateral Sclerosis (ALS) and in finding a cure. The Foundation encourages research that is collaborative, innovative and aggressive. These initiatives may fill gaps not currently being investigated and may involve novel approaches. The research must be of high scientific merit in finding a cure for ALS. Proposals may be longer than two pages. A lay summary is also required. Applications are reviewed by an Advisory Committee comprised of scientific researchers, neurologists and others who have had close relationships with friends and family members afflicted with ALS. The Committee makes funding recommendations to the Trustees.
Requirements: Public agencies and organizations that are classified as a charitable organization described in Section 501(c)3 of the Internal Revenue Code and as a public charity under Section 509(a) of the Internal Revenue Code may apply.
Restrictions: Support groups, education and medical equipment are not supported.
Date(s) Application is Due: Aug 15
Amount of Grant: 1,000 - 250,000 U.S.D
Samples: University of Texas Health Science Center, San Antonio, Texas, $65,000 - ALS research (2010); Oklahoma Medical Research Foundation, Oklahoma City, Oklahoma, $100,000 - medical research, (2010).
Contact: Marcia Y. Manhart, Executive Director; (830) 997-7347; fax (830) 997-9888; mmanhart@jjpafoundation.com
Sue Mayhue, (316) 383-1795
Internet: http://www.jjpafoundation.com/guidelines.html
Sponsor: Judith and Jean Pape Adams Charitable Foundation
105 S Cherry Street
Fredericksburg, TX 78624-3094

Judith and Jean Pape Adams Charitable Foundation Tulsa Area Grants 2579

The Foundation was established in 2004 as a private foundation and is involved in making distributions to charitable organizations on an annual basis. It encompasses two areas of support: organizations and agencies predominantly in Tulsa County, Oklahoma, and national Amyotrophic Lateral Sclerosis (ALS) research. For the former, primary areas of interest include arts and culture, human services and education. Support is given for operations, programs, capital projects, and maintenance reserve funding. The annual deadline is August 15.
Requirements: Public agencies serving the Tulsa, Oklahoma, region that are classified as a charitable organization described in Section 501(c)3 of the Internal Revenue Code and as a public charity under Section 509(a) of the Internal Revenue Code may apply.
Geographic Focus: Oklahoma
Date(s) Application is Due: Aug 15
Amount of Grant: 500 - 250,000 U.S.D
Samples: Oklahoma Humanities Council, Oklahoma City, Oklahoma, $1,250 - operating support (2010); Clarehouse, Tulsa, Oklahoma, $20,000 - operating support (2010); Safari Joe's Wildlife Ranch, Adair, Oklahoma - operating support (2010).
Contact: Marcia Y. Manhart, Executive Director; (830) 997-7347; fax (830) 997-9888; mmanhart@jjpafoundation.com
Sue Mayhue, (316) 383-1795
Internet: http://www.jjpafoundation.com/guidelines.html
Sponsor: Judith and Jean Pape Adams Charitable Foundation
105 S Cherry Street
Fredericksburg, TX 78624-3094

Julia Temple Davis Brown Foundation Grants 2580

The Foundation, established in 1969, is interested in supporting children and youth programs. Because of the limited size of the trust, and limited income available for distribution, the general policy has been to give fewer, but more meaningful grants, rather than spreading the income in smaller amounts to more organizations. The priority has been to the education of youth, from pre-school through high school, with emphasis on the lower age youngsters. Most recently, the Distribution Committee revised the grant emphasis to continue the focus on youth, but to stress youth programs relating to culture and arts, and the environment.
Requirements: Only 501(c)3 organization serving Hawaii are eligible to apply.
Geographic Focus: Hawaii
Date(s) Application is Due: Apr 1
Contact: Paula Boyce, AVP & Grants Administration Officer; (808) 538-4945 or (808) 538-4540; fax (808) 538-4006; pboyce@boh.com
Sponsor: Julia Temple Davis Brown Foundation
Foundation Administration #758, P.O. Box 3170, Dept. 715
Honolulu, HI 96802-3170

Julius N. Frankel Foundation Grants 2581

The foundation supports Chicago-area nonprofits in the areas of arts and performing arts; higher education; hospitals; human services; and medical school education. Recipients have included hospitals, universities, cultural organizations, and social service providers, with an emphasis on large, established organizations. There are no application deadlines. The initial approach should be by letter.
Requirements: Chicago-area nonprofits are eligible.
Restrictions: Individuals are not eligible.
Geographic Focus: Illinois
Amount of Grant: 10,000 - 330,000 U.S.D
Contact: Anne Blickensderfer, Trustee; (312) 461-3733
Sponsor: Julius N. Frankel Foundation
P.O. Box 755, Tax Division
Chicago, IL 60690

June Allyson Foundation Research Fellowships 2582

The purpose of this program is to identify potential future academic leaders and to stimulate their development through the support of research fellowships. The fellowship award is to be used to provide research training and experience for a candidate who exhibits significant evidence of talent and dedication to research. It is expected that the fellow will remain at the same institution for the entire funded period and present his/her research findings at the annual AUGS meeting the year after the fellowship has ended. A full written report will be due one year after the fellowship ends.
Requirements: Grant applicants must have an MD degree and demonstrate that grant proposals are part of ongoing programs representing a serious commitment to reducing or eliminating incontinence and/or its associated problems. The applicant's personal long-term career goals should include a significant emphasis on incontinence research, treatment, education, etc.
Date(s) Application is Due: Feb 28
Amount of Grant: 26,000 U.S.D
Contact: Grant Administrator; (202) 367-1167; fax (202) 367-2167; info@augs.org
Internet: http://www.augs.org/i4a/pages/index.cfm?pageid=229
Sponsor: June Allyson Foundation
2025 M Street NW, Suite 800
Washington, DC 20036

June Pangburn Memorial Scholarship 2583

This scholarship was established in memory of June Pangburn, who served as the Director of Nursing at St. Mary's Hospital (Hobart) and as a mentor to many nursing students. The current application and deadline is available in the Hobart High School Guidance Office.
Requirements: Applicants must be Hobart High School seniors who wish to pursue careers in Nursing or any field in Allied Health.
Geographic Focus: Indiana
Amount of Grant: 1,000 U.S.D
Contact: Pangburn Memorial Scholarship Contact; (219) 736-1880; fax (219) 736-1940; legacy@legacyfoundationlakeco.org
Internet: http://www.legacyfoundationlakeco.org/scholarships/index.htm
Sponsor: Legacy Foundation
1000 East 80th Place, 302 South
Merrillville, IN 46410

K. M. Hunter Charitable Foundation Social Welfare Grants 2584

The Foundation provides a number of major grants to social and health services programs in Ontario. In addition, a series of smaller grants are given to organizations, including housing shelters and hospices for the sick. The Foundation requests that applicants make all contacts through its web site. There are no specific applications or deadlines, and the Board meets two or three times annually to make its funding decisions.
Requirements: Grants are made to registered charitable organizations based in Ontario, with the exception of grants that support AIDS in Africa.
Geographic Focus: All States, Canada
Contact: Sarah Hunter, Vice President
Sponsor: K.M. Hunter Charitable Trusr
P.O. Box 38, Station E
Toronto, ON M6H 4E1, Canada

K21 Health Foundation Cancer Care Fund Grants 2585

The Kosciusko County Cancer Care Fund, administered by K21 Health Foundation, provides assistance to financially-eligible residents of Kosciusko County who are suffering from cancer. The purpose of the fund is to relieve some of the financial strain that often accompanies that dreaded diagnosis. The assistance provided includes but is not limited to items such as rent or mortgage payments, utilities, insurance, food, car payments, and prescription medications.
Requirements: Eligibility Requirements for Assistance: documented resident of Kosciusko County; verified cancer diagnosis within the last three months; can demonstrate a financial need; completion of the required application (contact the CCF Director for application).
Geographic Focus: Indiana
Contact: Clare Sessa, CCF Director; (574) 372-3500; csessa@hcfkc.org
Internet: http://www.k21foundation.org/cancer-care-fund/index.cfm
Sponsor: K21 Health Foundation
2170 North Pointe Drive, P.O. Box 1810
Warsaw, IN 46582

K21 Health Foundation Grants 2586

K21 Health Foundation exists for the benefit of Kosciusko County, Indiana citizens to ensure health care services are provided, and to advance prevention and healthy lifestyles. This will be accomplished by identifying health needs in our community, and maintaining an endowment so funding is available, through investments and grants, for those needs. Organizations are strongly encouraged to review K21's mission statement as that is the primary tool used to evaluate each application. When submitting an application, organizations must clearly explain how K21's mission will be advanced by supporting the project, program, or service, and the intended benefit(s) to residents of Kosciusko County.
Requirements: To qualify for a grant from K21, your organization must meet the following *Requirements:* must be a non-profit agency with verified IRS tax-exempt status, or a governmental agency; must be in good standing with the Indiana Secretary of State as indicated in a Business Entity Report; your Bylaws must require term limits for Directors and a rotating board is strongly encouraged.

Restrictions: To qualify for a grant from K21, your request must meet the following *Requirements:* the project, program, or service for which you are seeking funding must benefit residents of Kosciusko County; the project, program, or service for which you are seeking funding should provide direct health services or advance prevention and healthy life solutions.
Geographic Focus: Indiana
Date(s) Application is Due: Feb 1; May 1; Aug 1; Nov 1
Contact: Holly Swoverland, Grant Coordinator; (574) 269-5188, ext. 102; fax (574) 269-5193
Internet: http://www.k21foundation.org/apply-for-grant/index.cfm
Sponsor: K21 Health Foundation
2170 North Pointe Drive, P.O. Box 1810
Warsaw, IN 46582

Kahuku Community Fund 2587

The Kahuku Community Fund was established by the Estate of James Campbell in 2005 to be used for charitable and community purposes within the geographic district of Kahuku, bounded by Turtle Bay and Malaekahana. Preference is given to projects that address educational opportunities; recreational opportunities; economic sufficiency; social conditions; health care; strategic action plan around future development of the Kahuku community; housing opportunities; cultural arts, practices and values of the Ko'olauloa moku. Grant range is between $1,000 to $25,000 with an average of $10,000.
Requirements: Nonprofit, 501(c)3 organizations, schools, units of government, neighborhood groups or projects are eligible to apply. Community organizations without 501(c)3 status are eligible to apply for a grant up to $5,000, provided the activities to be supported are charitable. To be eligible for a grant of more than $5,000, a group must be a tax-exempt 501(c)3 organization or have a 501(c)3 fiscal sponsor. Program or project must benefit the Kahuku community.
Restrictions: Projects not likely to be funded: costs relating to establishing a new 501((c)(3) organization; general operating support, although up to 10% for indirect costs will be considered; funds for an endowment; funds for the benefit of specific individuals; except for emergency assistance through a 501(c)3 organization; full personnel costs (staffing costs are less likely to be funded) of a project or program; major capital improvements, although minor capital improvements required to implement a project may be considered; travel out of state.
Geographic Focus: Hawaii
Date(s) Application is Due: Jan 19
Amount of Grant: 1,000 - 25,000 U.S.D
Contact: Amy Luersen, (808) 566-5550 or (888) 731-3863; aluersen@hawaiicommunityfoundation.org or aluersen@hcf-hawaii.org
Internet: http://www.hawaiicommunityfoundation.org/index.php?id=71&categoryID=22
Sponsor: Hawai'i Community Foundation
1164 Bishop Street, Suite 800
Honolulu, HI 96813

Kaiser Permanente Cares for Communities Grants 2588

Kaiser Permanente partners with nonprofit organizations to improve the health of its members and the communities it serves. The company's four-pronged approach to improve community health includes: increasing access to coverage and care for vulnerable populations, including children and adults who are ineligible for existing state or private programs due to family income or immigration status, and community-based clinics; evidence-based medicine—successful practices in managing chronic diseases such as diabetes, asthma, depression, and heart failure; education of health care professionals and consumers—graduate medical education programs and nursing scholarships, partnerships with national consumer organizations, and educational theater programs; and national public health policy. Application and guidelines are available online.
Requirements: 501(c)3 nonprofit organizations that have national reach are eligible. Organizations must be nonpartisan and nondenominational.
Amount of Grant: Up to 100,000,000 U.S.D
Samples: Institute for Healthcare Improvement (Boston, MA)—for a scholarship endowment to enable health care professionals to participate in the institute's health-care-improvement programs, including conferences and training courses, $8 million; Doctors Without Borders (New York, NY)—to assist tsunami-relief efforts, $500,000; American National Red Cross (Washington, DC) and Operation U.S.A (Los Angeles, CA)—to assist tsunami-relief efforts, $250,000 each; Archdiocesan Health Care Network (Washington, DC)—to increase the number of volunteers who provide specialty care to low-income, uninsured people, $15,000.
Contact: Community Relations Manager; (510) 271-5685; fax (510) 271-6493
Internet: http://www.kaiserpermanentejobs.org/aboutus/community.asp
Sponsor: Kaiser Permanente
1 Kaiser Plaza, 21B
Oakland, CA 94612

Kalamazoo Foundation Grants 2589

The foundation awards unrestricted grants in four priority areas: individuals and families, economic development, education and learning, and community engagement and youth development. Types of support include general operating support, capital campaigns, building construction/renovation, equipment acquisition, emergency funds, program development, seed grants, scholarship funds, technical assistance, employee matching gifts, and matching funds. The period for grant support is normally one year. Guidelines and application are available online.
Requirements: Kalamazoo, MI, organizations recognized (or in the process of applying for recognition) under IRS code 501(c)3 are encouraged to contact foundation staff before submitting a request to determine eligibility.
Restrictions: Generally, the foundation does not provide funding for debt retirement, endowments, individuals, travel for individuals or groups, religious organizations for religious purposes, meetings, conferences, publications, films, or television and radio programming.
Geographic Focus: Michigan
Samples: Interfaith Strategy for Advocacy and Action in the Community (Kalamazoo, MI)—for a voter-registration and community-engagement project, $40,000; Catholic Family Services (Kalamazoo, MI)—for the Kalamazoo Caring Network's Read Me a Story program, which promotes reading skills and literacy among teenage parents and their children, $25,297; Michiana Addictions and Prevention Services (Kalamazoo, MI)—to support the merger of this substance-abuse-treatment center with the Guidance Clinic in Kalamazoo, $100,000; Kalamazoo College (MI)—for a postgraduate fellowship position at the Mary Jane Underwood Stryker Institute for Service Learning, $45,000.
Contact: Program Contact; (269) 381-4416; fax (269) 381-3146; info@kalfound.org
Internet: http://www.kalfound.org/page8490.cfm
Sponsor: Kalamazoo Foundation
151 S Rose Street
Kalamazoo, MI 49007

Kaman Corporate Giving Program Grants 2590

The corporate giving program focuses on the needs of the handicapped and disabled, but funding is provided to cover a broad range of human needs. Grants are made for capital building, renovation campaigns, and selected operating programs. The majority of grants are awarded to nonprofits in the Hartford, CT, area. Requests are accepted throughout the year. Requests for funding should include the organization's latest financial report; a description of (and budget for) the project to be funded; information concerning the status of funding toward the goal; a listing of the organization's board of directors and key staff; and the name, address, and phone number of the person to be contacted for further information. Other material the organization feels would be helpful is welcomed.
Requirements: Nonprofits in the greater Hartford region of Connecticut are eligible.
Restrictions: Grants do not support endowment funds, recipients of state grants, agencies receiving funds from the United Way/Combined Health Appeal, or galas and events.
Geographic Focus: Connecticut
Amount of Grant: 25,000 - 500 U.S.D
Contact: Russell Jones, Vice President; (860) 243-6308; fax (860) 243-6365; info.kamancorp@kaman.com
Internet: http://www.kaman.com
Sponsor: Kaman Corporation
1332 Blue Hills Avenue
Bloomfield, CT 06002

Kanawha Valley Foundation Grants 2591

The foundation makes grants to charitable organizations in the areas of the arts, education, health, human services, and public recreation primarily in the greater Kanawha Valley, WV. Innovative and special projects will receive priority consideration. Grant applications are reviewed quarterly. The foundation does not require an application form. Applicants should submit a letter describing the project, limited if possible to two pages, budget, and description of applicant organization. Deadline dates may be obtained by calling the office.
Requirements: IRS 501(c)3 organizations whose programs or projects benefit people living in the greater Kanawha Valley, WV, area are eligible. If the applicant is a public school, written approval of the school's faculty senate must be included.
Geographic Focus: West Virginia
Amount of Grant: 8,000 U.S.D
Contact: Kim Tieman, Senior Program Officer; (304) 346-3620; fax (304) 346-3640; ktieman@tgkvf.org
Internet: http://www.tgkvf.org/grant.htm
Sponsor: Kanawha Valley Foundation
One Huntington Square, 16th Floor, P.O. Box 3041
Charleston, WV 25331-3041

K and F Baxter Family Foundation Grants 2592

The foundation focuses on the special educational challenges encountered by multiracial children; policy recommendations in support of revised school curricula that will be more supportive of multiracial children without sacrificing either basic skills training or appropriate performance standards; and developing, by means of policy studies and analysis, a realistic template for education reform focused on the student rather than the school. In addition, the foundation will consider funding worthwhile projects in the arts, health care, and other areas related to education. Guidelines are available online.
Requirements: Biracial children grants are awarded to nonprofit organizations nationwide. Education grants are limited to schools in California Unified School Districts, including Berkeley, Los Angeles, Oakland, West Contra Costa Counties.
Date(s) Application is Due: Feb 1; Aug 1
Amount of Grant: 90 - 150,000 U.S.D
Contact: Stacey Bell, Executive Director; (510) 524-8145; fax (510) 524-4101; kfbaxterfound@aol.com
Internet: http://www.kfbaxterfoundation.com
Sponsor: K and F Baxter Family Foundation
P.O. Box 13053
Berkeley, CA 94712-4053

Kansas Health Foundation Major Initiatives Grants 2593
The Kansas Health Foundation is committed to supporting strategies that will make Kansas a healthier place. The Foundation's focus areas are: promoting the healthy behaviors of Kansans; strengthening the public health system; improving access to health care for Kansas children; growing community philanthropy; providing health data and information to policymakers; and building civic leadership.
Requirements: The Foundation accepts one-page letters of inquiry from organizations that believe they have innovative work that meets the Foundation's mission. The letter of inquiry form is located at the Foundation's website. Organizations may also review samples of previously funded major initiatives on the website.
Geographic Focus: Kansas
Samples: University of Kansas School of Medicine, Wichita, Kansas, supporting the Department of Preventative Medicine and Public Health in its efforts to enhance the public health components of its physicians' education curriculum, $400,000 (2011); Worksite Wellness Evaluation, to engage local leaders and organizations in communities throughout Kansas to complete the training necessary to develop worksite wellness plans and to fund a final report of the results, $800,000 (2011).
Contact: Nancy Claassen, Grants Manager; (316) 262-676; fax (316) 262-2044; nclaassen@khf.org
Internet: http://www.kansashealth.org/grant_type/major_initiatives
Sponsor: Kansas Health Foundation
309 East Douglas
Wichita, KS 67202-3405

Kansas Health Foundation Recognition Grants 2594
The Kansas Health Foundation Recognition Grants expand the Foundation's support to a broad range of health-related organizations throughout the state. The Foundation defines health broadly, and looks at all the aspects that affect health, including the social factors that contribute to a healthy population (a state of complete physical, mental, and social well-being and not merely the absence of disease or infirmity). While the majority of the Foundation's funding is through invited proposals, the Recognition Grants program is designed to fund unsolicited requests. It is targeted for organizations and agencies proposing meaningful and charitable projects that fit within the Foundation's mission of improving the health of all Kansans.
Requirements: Kansas 501(c)3 nonprofit health organizations are eligible. The application is at the Foundation website and must be submitted online.
Restrictions: The Foundation does not support: medical research; capital campaigns; operating deficits or retirement of debt; endowment programs not initiated by the Foundation; political advocacy of any kind; vehicles, such as vans or buses; medical equipment; construction projects or real estate acquisitions; direct mental health services; or direct medical services.
Geographic Focus: Kansas
Date(s) Application is Due: Mar 15; Sep 15
Amount of Grant: Up to 25,000 U.S.D
Contact: Nancy Claassen, Grants Manager; (800) 373-7681 or (316) 262-7676; fax (316) 262-2044; info@khf.org
Gina Hess, Grant Assistant; (316) 262-7676 or (800) 373-7681; rinfo@khf.org
Internet: http://www.kansashealth.org/grantmaking/recognitiongrants
Sponsor: Kansas Health Foundation
309 East Douglas
Wichita, KS 67202-3405

Karma Foundation Grants 2595
The foundation awards grants on an international basis to support the following broad interest areas: arts and culture, education and literacy, health and human services, and development and enrichment of Jewish life. Types of support include operating expenses, special projects, capital grants, seed grants, equipment and materials. and disaster relief. The foundation also provides technical assistance and assists in proposal development. There are no application deadlines.
Requirements: National and international nonprofit organizations are eligible.
Restrictions: Grants do not support travel expenses for bands or sports teams, political or lobbying activities, advertising for fundraising events, litigation, charter schools, or loans.
Geographic Focus: California, New Jersey, Wisconsin
Amount of Grant: 500 - 25,000 U.S.D
Samples: Wheaton Village—to purchase a computerized collection management system and to train personnel at the Museum of AmericanGlass, $15,900; Children's Hospital of Los Angeles (CA)—for medical equipment for the emergency transport team, $18,081; First Concern (Somerset, NJ)—to refurbish the kitchen and laundry room, $25,000; National Jewish Medical and Research Ctr—for asthma research, $10,000.
Contact: Dina Karmazin Elkins, Executive Director; (609) 924-5939; fax (609) 924-2714; info@karmafoundation.org
Internet: http://www.karmafoundation.org
Sponsor: Karma Foundation
140 Arreton Road
Princeton, NJ 08540

Kate B. Reynolds Charitable Trust Health Care Grants 2596
The Trust responds to health care and wellness needs and invests in solutions that improve the quality of health for financially needy residents throughout North Carolina. The Health Care Division seeks impact through two program areas: providing treatment and supporting prevention. The trust requires advanced consultation by phone or in writing. Application materials are available online, but applications are not accepted electronically.
Requirements: Nonprofit 501(c)3 organizations in North Carolina are eligible.
Restrictions: Grants are not awarded to individuals.
Geographic Focus: North Carolina
Date(s) Application is Due: Mar 15; Sep 15
Amount of Grant: 20,000 - 200,000 U.S.D
Contact: John H. Frank, Director; (336) 397-5502 or (866) 551-0690; fax (336) 723-7765; john@kbr.org
Internet: http://www.kbr.org/health-care-division-fund.cfm
Sponsor: Kate B. Reynolds Charitable Trust
128 Reynolda Village
Winston-Salem, NC 27106-5123

Katharine Matthies Foundation Grants 2597
The Katharine Matthies Foundation was established in 1987 to support and promote quality educational, human-services, and health-care programming for underserved populations. Special consideration is given to organizations that work to prevent cruelty to children and animals. The majority of grants from the Matthies Foundation are one year in duration; on occasion, multi-year support is awarded. Applicants must apply online at the grant website. Applicants are strongly encouraged to do the following before applying: review the downloadable state application procedures for additional helpful information and clarifications; review the downloadable online-application guidelines at the grant website; review the foundation's funding history (link is available from the grant website); review the online application questions in advance; and review the list of required attachments. These will generally include: a list of board members, financial statements (audited, reviewed, or compiled by independent auditor); an organization summary; a list of other funding sources; an IRS Determination letter; and other required documents. All attachments must be uploaded in the online application as PDF, Word, or Excel files. The deadline for application to the Katherine Matthies Foundation is 11:59 p.m. on May 1. Applicants will be notified of grant decisions by letter within three to four months after the proposal deadline.
Requirements: Applicant organizations must have 501(c)3 tax-exempt status and serve the people of the following Connecticut towns: Seymour, Ansonia, Derby, Oxford, Shelton, or Beacon Falls. A breakdown of number/percentage of people served by specific towns will be required in the online application. Special consideration will be given to organizations that serve the people of Seymour, Connecticut.
Restrictions: The Matthies Foundation specifically serves people of the Lower Naugatuck Valley. The foundation does not support requests from individuals, organizations attempting to influence policy through direct lobbying, or any political campaigns.
Geographic Focus: Connecticut
Date(s) Application is Due: May 1
Samples: Town of Seymour, Seymour, Connecticut, $22,000 (2010); Griffin Health Services Corporation, Derby, Connecticut, $66,000 (2010); Connecticut Hurricanes, Derby, Connecticut, $17,540 (2010).
Contact: Amy Lynch; (860) 657-7015; amy.r.lynch@baml.com
Internet: https://www.bankofamerica.com/philanthropic/fn_search.action
Sponsor: Katharine Matthies Foundation
200 Glastonbury Boulevard, Suite # 200
Glastonbury, CT 06033-4056

Katherine John Murphy Foundation Grants 2598
The foundation awards grants to Georgia tax-exempt organizations in its areas of interest, including arts and culture, child welfare, health care, higher education, hospitals, youth, and zoos. Types of support include annual campaigns, building construction/renovation, capital campaigns, continuing support, general operating support, seed grants, and project development. Grants are awarded primarily in Atlanta. Submit a letter of request.
Requirements: Organizations in Atlanta, GA are eligible to apply.
Restrictions: No grants to individuals, or for research, or matching gifts.
Geographic Focus: Georgia
Date(s) Application is Due: Mar 30; Jun 30; Sep 30; Dec 15
Amount of Grant: 250 - 50,000 U.S.D
Contact: Brenda Rambeau, (404) 589-8090; info@murphyfoundation
Internet: http://www.kjmurphyfoundation.org
Sponsor: Katherine John Murphy Foundation
50 Hurt Plaza, Suite 1210
Atlanta, GA 30303

Kathryne Beynon Foundation Grants 2599
Founded in California in 1967, the Kathryne Beynon Foundation provides support primarily for: hospitals (with a special interest in Asthma); youth agencies; child welfare; Roman Catholic church; higher education, Types of support include: general operating support; building construction/renovation; endowment funds; and scholarship funds. The Board meets quarterly to review grant requests. Applicants should contact the office in writing, outlining their proposal. Application is by invitation only, and there is no deadline date when submitting grant proposals. Contact the Foundation directly for additional guidelines before submitting a full proposal.
Requirements: 501(c)3 southern California tax-exempt organizations are eligible. Preference is given to requests from Pasadena. There are no: deadline dates; formal application form required to submit proposal.
Restrictions: No support to individuals.
Geographic Focus: California
Amount of Grant: 500 - 50,000 U.S.D
Samples: Artists of America, San Pedro, CA, $17,800—Laurel Elementary School, Prep art classes; Assistance League of Southern California, Hollywood, CA, $4,000—

children's club/ day nursery; Scripps College, Claremont, CA, $20,000—science scholarship program.
Contact: Robert D. Bannon, Trustee; (626) 584-8800
Sponsor: Kathryne Beynon Foundation
1111 South Arroyo Parkway, Suite 470
Pasadena, CA 91105-3239

Katrine Menzing Deakins Charitable Trust Grants **2600**
Katrine Deakins was executive secretary to Amon G. Carter and helped found the Amon G. Carter Foundation. She was the foundation's executive director and throughout her life was very active in numerous professional, social and charitable organizations. Katrine then established her own trust in 1987, the Katrine Menzing Deakins Charitable Trust to benefit several favored charities in addition to other charitable organization requests selected each year by the trustees. Grants are typically between $1,000 and $25,000.The Katrine Menzing Deakins Charitable Trust has bi-annual application deadlines of March 31 and September 30. Applications must be submitted by 11:29 p.m. on the deadline dates. Applicants must apply online at the grant website. Applicants are strongly encouraged to do the following before applying: review the downloadable state application procedures for additional helpful information and clarifications; review the downloadable online-application guidelines at the grant website; review the trust's funding history (link is available from the grant website); review the online application questions in advance; and review the list of required attachments. These will generally include: a list of board members, financial statements (audited, reviewed, or compiled by independent auditor); an organization summary; a list of other funding sources; an IRS Determination letter; and other required documents. All attachments must be uploaded in the online application as PDF, Word, or Excel files.
Requirements: Applicants must have 501(c)3 tax-exempt status.
Restrictions: The trust does not support requests from individuals, organizations attempting to influence policy through direct lobbying, or any political campaigns.
Geographic Focus: Texas
Date(s) Application is Due: Mar 31; Sep 30
Amount of Grant: 1,000 - 25,000 U.S.D
Contact: Mark J. Smith, Philanthropic Relationship Manager; (817) 390-6028; tx.philanthropic@baml.com
Internet: https://www.bankofamerica.com/philanthropic/fn_search.action
Sponsor: Katrine Menzing Deakins Charitable Trust
500 West 7th Street, 15th Floor, TX1-497-15-08
Fort Worth, TX 76102-4700

Kavli Foundation Research Grants **2601**
The Kavli Foundation supports research in the areas of astrophysics, nanoscience, neuroscience, and theoretical physics three scientific frontiers that promise great discoveries in the 21st century and beyond. This mission is implemented through an international program of research institutes, professorships, symposia and workshops, as well as prizes in the fields of astrophysics, nanoscience, and neuroscience. Dedicated to advancing science for the benefit of humanity and promoting increased public understanding and support for scientists and their work, the Foundation also makes available materials on these sciences at no cost to educational institutions.
Contact: Grants Administrator; (805) 983-6000; fax (805) 988-4800
Internet: http://www.kavlifoundation.org/areasofscience/
Sponsor: Kavli Foundation
1801 Solar Drive, Suite 250
Oxnard, CA 93031

Kellwood Foundation Grants **2602**
The foundation makes grants to 501(c)3 charitable scientific, educational, and literary organizations for programs that provide basic human needs and improve the quality of life. Primary focus areas include cultural and performing arts; education; health services; and social welfare, civic, and community programs. Types of support include general operating support, annual campaigns, capital campaigns, program development, and employee matching gifts. The board meets in March, June, September, and December to consider requests. Requests should be submitted on organization letterhead and contain the name, address, and phone number of the contact person; statement of the problem to be addressed; identification of geographic area and number of people served; statement of the purpose for which funds are requested; specific dollar request; explanation of how the program differs from others providing similar services in the same geographic area; annual organizational budget; identification of sources of annual operating funds; identification of board of directors and key staff members; and copy of IRS 501(c)3 determination letter.
Requirements: Giving is primarily made to the greater St. Louis, MO, area.
Geographic Focus: Missouri
Amount of Grant: 25 - 70,000 U.S.D
Samples: Central Institute for the Deaf (Saint Louis, MO)—for endowment, $10,000; Girl Scouts of the U.S.A (Saint Louis, MO)—for program support, $3000; American Red Cross (Saint Louis, MO)—for operating support, $10,000.
Contact: Terri Grandcolas, (314) 576-3431; fax (314) 576-3439
Sponsor: Kellwood Foundation
600 Kellwood Parkway
Chesterfield, MO 63017

Kelsey Trust Grants **2603**
The trust awards grants within its geographical territory to support organizations in its primary areas of interest, including child development education, adult basic education and literacy, vocational education, health care and health organizations, the environment and conservation, and services for children and families. Grants are awarded for general operating support, program development, seed money, scholarship funds, matching funds, and emergency funds.
Requirements: Nonprofit organizations in the Lake Champlain Valley drainage basin; eastern Adirondacks, in New York; and western Vermont, north of Rutland, may submit applications for grant support.
Restrictions: The trust does not typically make grants for most capital projects, debt reduction, normal operating budgets, to individuals, for sectarian religious purposes, for endowments, or where the trust may become a predominant source of support.
Geographic Focus: New York, Vermont
Date(s) Application is Due: Feb 1; Sep 1
Amount of Grant: 2,500 - 10,000 U.S.D
Samples: Planned Parenthood of Northern Adirondack (Plattsburgh, NY)—for program support, $10,000; Fletcher Allen Health Care (Burlington, VT)—for program support, $3000; Bristol Family Center (Bristol, VT)—for program support, $1000.
Contact: Paula Johnson, c/o Vermont Community Foundation; (802) 388-3355
Internet: http://www.vermontcf.org/guidelines-forms/kelsey-trust.html
Sponsor: Kelsey Trust
P.O. Box 30
Middlebury, VT 05753

Kelvin and Eleanor Smith Foundation Grants **2604**
The foundation awards grants to Ohio nonprofits in its areas of interest, including nonsectarian education, the performing and visual arts, health care, and environmental conservation and protection. Types of support include general operating support, continuing support, annual campaigns, capital campaigns, building construction/renovation, and equipment acquisition. There are no required application forms or deadlines.
Requirements: Nonprofit organizations in the greater Cleveland, OH, area are eligible.
Restrictions: Grants are not made in support of individuals or for endowment funds, scholarships, fellowships, matching gifts, or loans.
Geographic Focus: Ohio
Amount of Grant: 3,000 - 150,000 U.S.D
Samples: Case Western Reserve University, Kelvin Smith Library (Cleveland, OH)—for capital campaign, $200,000; Cleveland Museum of Art (Cleveland, OH)—for annual fund, $125,000.
Contact: Carol Zett, Grants Manager; (216) 591-9111; fax (216) 591-9557
Sponsor: Kelvin and Eleanor Smith Foundation
30195 Chagrin Boulevard, Suite 275
Cleveland, OH 44124

Kendrick Foundation Grants **2605**
Kendrick Foundation was created to support health-related programs in Morgan County, Indiana. Support includes making grants to health-related organizations and charities, which may include community health care programs, hospice programs, health care education and training, and tax-exempt medical and health programs. Prior to submitting a grant application, a Letter of Intent must be submitted. Letters of Intent are available for download for a limited amount of time during the opening of the grant cycle. Letters of Intent for the spring grant cycle must be physically received in the office of the Community Foundation of Morgan County by March 15, no later than 4 p.m., with invited applications due April 30. Letters of Intent for the fall grant cycle must be physically received in the office of the Community Foundation of Morgan County, Inc. by October 15, no later than 4 p.m., with invited applications due November 30.
Requirements: The Kendrick Foundation will only accept full grant applications from those organizations serving Morgan County, Indiana, which were approved through a Letter of Intent.
Geographic Focus: Indiana
Date(s) Application is Due: Mar 15; Apr 30; Oct 15; Nov 30
Contact: Tom Zoss, Executive Director; (317) 831-1232 or (877) 822-6958; fax (317) 831-2854; tzoss@cfmconline.org
Internet: http://www.cfmconline.org/kendrick/Grants/tabid/129/Default.aspx
Sponsor: Kendrick Foundation
250 N Monroe Street
Mooresville, IN 46158

Kenneth T. and Eileen L. Norris Foundation Grants **2606**
The foundation supports Los Angeles County nonprofits in the areas of medicine—to improve access to health care, increase knowledge through research, and provide facilities for those activities to take place; youth—to provide constructive activities, positive role models, and opportunities for disadvantaged, disabled, and misguided children; community—to support law enforcement agencies, good citizenship, and environmental conservation; arts and culture—to support museums, symphony orchestras, and dance and theater companies; and education and science—to focus on private education, especially secondary and college levels. Types of support include general operating support, continuing support, building construction/renovation, equipment acquisition, endowment funds, program development, professorships, scholarship funds, research, and matching funds. Education/science projects are due between May 1 and June 30; youth requests are due between February 15 and March 31; community and cultural (the arts) requests are due between December 1 and January 31; and medicine proposals are due between May 1 and June 30.
Requirements: Grants are awarded to organizations in southern California.
Geographic Focus: California
Date(s) Application is Due: Jan 31; Mar 31; Jun 30
Amount of Grant: 5,000 - 25,000 U.S.D

Samples: Los Angeles Regional Foodbank (Los Angeles, CA)—for general operating support, $10,000.
Contact: Ronald Barnes, Executive Director; (562) 435-8444; fax (562) 436-0584; grants@ktn.org
Internet: http://www.norrisfoundation.org/grant_funding.html
Sponsor: Kenneth T. and Eileen L. Norris Foundation
11 Golden Shore, Suite 450
Long Beach, CA 90802

Kenny's Kids Grants **2607**

The foundation awards grants to U.S. nonprofits in company-operating locations with a focus on programs that improve life for children and young adults. The primary target of grants is giving support to organizations which seek to give youth greater opportunities outside of the classroom to learn and grow, to teach them skills that will enable them to thrive in a technology-oriented society, and to offer them the guidance and attention necessary to develop such skills. Also consideration for youth-oriented organizations and other organizations that promote the welfare of youth, such as those that offer hope and encouragement to the sick and terminally ill, protect those who have suffered from abuse, and offer after-school programs for the underprivileged. There are no application deadlines. Allow eight to 12 weeks for proposal review.
Requirements: Nonprofits in company operating areas of Illinois are eligible.
Restrictions: No support for private foundations, schools (public or private), or corporations. Grants are not made to individuals, or for political campaigns, film, video, or audio productions.
Geographic Focus: Illinois
Amount of Grant: Up to 573,902 U.S.D
Contact: Nicholas Pontikes, President
Sponsor: Kenny's Kids
1212 W Lill Street
Chicago, IL 60614

Ken W. Davis Foundation Grants **2608**

The foundation awards grants in company-operating areas to support charities that directly provide for human welfare, basic needs, and quality of life—food, shelter, clothing, healthcare, disabilities, and childcare. Recipients may be asked to submit a post-grant evaluation of the project or program. Requests must be in writing and are reviewed and approved throughout the year. Submit requests as early as possible before October 31, the end of the foundation's fiscal year.
Requirements: Texas nonprofits in Midland-Odessa and Forth Worth are eligible.
Restrictions: The foundation does not support religious organizations for religious purposes; activities in support of social political causes; organizations that receive a substantial amount of federal and/or state funding; or organizations (other than the United Way) that receive wide public support.
Geographic Focus: Texas
Contact: Alan Tucker Davis, (817) 332-4081; fax (817) 332-4095; akd@kwdf.org
Internet: http://fdncenter.org/grantmaker/davis/proposal.html
Sponsor: Ken W. Davis Foundation
P.O. Box 3419
Fort Worth, TX 76113-3419

Kettering Family Foundation Grants **2609**

The Kettering Family Foundation (KFF) is a family grant-making organization that supports both Trustee-endorsed and unsolicited proposals that meet specific criteria. Unsolicited requests will be reviewed only one time per KFF fiscal year (ending December 31) from the same requesting organization and/or program. The following categories are of interest to KFF Trustees: cultural and arts; education; environment; medical health; and social/human services. A request summary (letter of interest) is required (by January 31 and July 31) and, if interested, trustees will ask for a full proposal.
Requirements: 501(c)3 tax-exempt organizations are eligible.
Restrictions: Requests for funding for the following areas are not generally approved unless endorsed by a trustee: capital construction; foreign-based or foreign- purpose organizations; religious organizations for religious purposes; individual public elementary or secondary schools or public school districts; local chapters of national organizations; grants, loans, scholarships, fellowships, or memberships; multiyear grants; travel expenses; community drives, event sponsorships; or conduit organizations.
Date(s) Application is Due: Jan 31; Mar 1; Jul 31; Sep 1
Contact: Charles F. Kettering; (303) 756-7664; info@ketteringfamilyfoundation.org
Internet: http://www.ketteringfamilyfoundation.org/main.html
Sponsor: Kettering Family Foundation
2833 S Colorado Boulevard
Denver, CO 80222

Kettering Fund Grants **2610**

The mission of the Kettering Fund is to support scientific, medical, social and educational studies and research conducted by nonprofit, charitable organizations that are located and/or provide service in Ohio. The Distribution Committee meets on a semi-annual basis to make funding decisions, generally in May and November. The committee will consider support for the following purposes: capital needs (such as the purchase of land, new construction, renovations and equipment); seed grants for new organizations or programs; project support; endowments; research; and scholarship funds.
Requirements: 501(c)3 nonprofit organizations that provide services in or are located in Ohio may apply.
Restrictions: The fund does not support: religious organizations for religious purposes; activities of specific individuals; public elementary or secondary schools; loans; travel; deficit reduction; benefit events; or efforts to carry on propaganda or otherwise attempt to influence legislation.
Geographic Focus: Ohio
Amount of Grant: 5,000 - 200,000 U.S.D
Samples: Kettering U (Flint, MI)—to support the Kettering Scholars Program, which awards scholarships to Ohio residents, $2.3 million.
Contact: Judith M. Thompson, Executive Director; (973) 228-1021; fax (973) 228-2399; info@ketteringfund.org
Internet: http://www.ketteringfund.org/main.html
Sponsor: Kettering Fund
1480 Kettering Tower
Dayton, OH 45423

Kevin P. and Sydney B. Knight Family Foundation Grants **2611**

The Foundation, established in 1997 by the founders of Knight Transportation of Phoenix, Arizona, is focused on the greater Phoenix area. Giving programs are aimed at: children and youth; Catholic agencies and churches; athletics/sports activities (particularly baseball); education; health care; and Mormon affiliated groups. Primary types of support include: community development, general research, and youth programs. There are no specific deadlines or application forms, and initial approach should be by letter.
Restrictions: Funding is restricted primarily to 501(c)3 organization located in Phoenix, Arizona.
Geographic Focus: Arizona
Amount of Grant: 1,000 - 2,000 U.S.D
Contact: Kevin P. Knight, President; (602) 269-2000; fax (602) 269-8409
Sponsor: Kevin P. and Sydney B. Knight Family Foundation
5601 W Buckeye Road
Phoenix, AZ 85043-4603

Key Foundation Grants **2612**

The foundation's objective is to improve the quality of life and economic vibrancy of the places where KeyCorp customers, employees, and shareholders live and work. The foundation supports programs that enhance economic independence in support of two strategic thrusts: financial education and workforce initiatives. To a lesser degree, the foundation will continue to support organizations that enhance quality-of-life in Key communities in the Eastern, Midwestern, and Western United States; and greater Cleveland, OH area. The foundation recognizes the importance of diversity and, accordingly, will make contributions to organizations that demonstrate a sincere commitment to individual diversity. Guidelines and application are available online.
Requirements: 501(c)3 tax-exempt organizations are eligible.
Restrictions: The foundation does not contribute to individuals, lobbying or political organizations, veteran or fraternal groups, or organizations that discriminate in any way with national equal opportunity policies. The foundation also does not buy journal advertisements or memberships.
Samples: Cleveland Museum of Art (OH)—to support programs that explore how diversity and inclusion influence the arts, to establish the KeyBank Summer Arts Internship Program, and to develop a bilingual distance-learning program that will focus on the art of the ancient Americans, $1.5 million; Cleveland Institute of Music (OH)—for a distance-learning program designed to boost academic and professional development among students and educators, $25,000; Cuyahoga Community College (Cleveland, OH)—for a new center that will work to facilitate business and economic growth across northeastern Ohio, $1 million over three years.
Contact: Administrator; (216) 689-5458; fax (216) 689-3865; key_foundation@keybank.com
Internet: http://www.key.com/community
Sponsor: Key Foundation
127 Public Square, 7th Floor
Cleveland, OH 33113-1306

KeySpan Foundation Grants **2613**

The foundation awards grants to eligible nonprofit organizations in its areas of interest, including health and human services; education—K-12 classroom education, scholarships and university partnerships for the underserved, skills training; environment—sustainability of natural resources, educate youth about their connection to the environment, and preserve open spaces for the future; community development; and the arts. Proposals are accepted and reviewed on a first come/first served basis through October. The foundation's board of directors meets on a quarterly basis to review requests and authorize grants.
Requirements: New Hampshire, Massachusetts, and New York nonprofits are eligible.
Restrictions: The foundation does not support religious or political groups, capital or endowment campaigns, advertisements, tables or tickets at dinners, and groups whose combined administrative, management, or fundraising expenses exceed 30 percent of the operating budget.
Geographic Focus: Connecticut, Maine, Massachusetts, New Hampshire, New York, Rhode Island, Vermont
Date(s) Application is Due: Oct 31
Amount of Grant: 5,000 - 25,000 U.S.D
Contact: Robert Keller, Executive Director; (516) 545-5147
Internet: http://www.keyspanenergy.com/corpinfo/community/foundation_all.jsp
Sponsor: KeySpan Foundation
175 E Old Country Road
Hicksville, NY 11801

KFC Allied Health Research Grants **2614**

The foundation supports research into the incidence and cure of kidney and urinary tract disease, patient services, and public education. Preference will be given to allied health personnel undertaking research related to the kidney and urinary tract and to a new research project not supported by any other agency. Applications from investigators in related specialty areas also are considered. Types of support include capital grants, fellowships, general operating grants, program development grants, scholarships, and travel assistance. One- and two-year renewable grants are awarded.
Requirements: Application is open to individuals affiliated with Canadian universities for research conducted within Canada.
Restrictions: Grants will not be awarded for equipment only.
Geographic Focus: All States, Canada
Date(s) Application is Due: Oct 15
Amount of Grant: Up to 50,000 CAD
Contact: National Research Programs Manager; (514) 369-4806, ext. 223 or (800) 361-7494, ext. 223; research@kidney.ca
Internet: http://www.kidney.ca/english/research/allied-funding-03.htm
Sponsor: Kidney Foundation of Canada
300-5165 Sherbrooke Street W
Montreal, QC H4A 1T6 Canada

KFC Allied Health Scholarship Awards **2615**

The fellowships support full-time academic and research preparation at the doctoral level in the field of nephrology/urology in Canada or abroad. Eligibility is open to nephrology/ urology nurses and technicians, social workers, dietitians, and transplant coordinators. Fellowships are awarded for one year and may be renewed upon successful reapplication a total of four years. If study is outside of Canada, the fellow must intend to return to Canada following the fellowship period. Obtain application forms.
Requirements: The applicant must be accepted into a full-time course of study, have a minimum of two years direct clinical practice, and agree to return to Canada if study is undertaken outside the country.
Geographic Focus: All States, Canada
Date(s) Application is Due: Mar 15
Amount of Grant: Up to 31,000 CAD
Contact: National Research Programs Manager; (800) 361-7494, ext. 232 or (514) 369-4806, ext. 232; fax (514) 369-2472; research@kidney.ca
Internet: http://www.kidney.ca/english/research/allied-funding-03.htm#allied-scholar
Sponsor: Kidney Foundation of Canada
300-5165 Sherbrooke Street W
Montreal, QC H4A 1T6 Canada

KFC Biomedical Fellowships **2616**

Fellowships support full-time, postdoctoral research training in the renal and urinary tract field for training in Canada or abroad. The award amount is determined by level of experience and the holding of MD, DDS, DVM, and/or PhD degrees. Fellows are expected to devote 12 consecutive months and 80 percent of their time toward fulfilling the objectives of the fellowship. Awards are applied for through the National Office of the Kidney Foundation. Obtain application forms.
Requirements: Applicants must be Canadian citizens or landed immigrants and apply through a university department or affiliated institution.
Geographic Focus: All States, Canada
Date(s) Application is Due: Oct 15
Amount of Grant: 45,000 CAD
Contact: National Research Programs Manager; (514) 369-4806, ext. 225 or (800) 361-7494, ext. 225; research@kidney.ca
Internet: http://www.kidney.ca/english/research/biomed-funding-03.htm
Sponsor: Kidney Foundation of Canada
300-5165 Sherbrooke Street W
Montreal, QC H4A 1T6 Canada

KFC Biomedical Research Grants **2617**

The program fosters and encourages research about the kidney and urinary tract and the understanding of kidney diseases. Types of support include general operating grants, research grants, materials and supplies, and travel assistance. One- and two-year renewable grants are offered.
Requirements: Canadian citizens or landed immigrants may apply. Research must be conducted in Canada by individuals holding staff appointments at Canadian universities or academic institutions.
Restrictions: Requests for equipment only are denied.
Geographic Focus: All States, Canada
Date(s) Application is Due: Oct 15
Amount of Grant: Up to 50,000 CAD
Contact: National Research Programs Manager; (514) 369-4806, ext. 225 or (800) 361-7494, ext. 225; fax (514) 369-2472; research@kidney.ca
Internet: http://www.kidney.ca/english/research/biomed-funding-03.htm
Sponsor: Kidney Foundation of Canada
300-5165 Sherbrooke Street W
Montreal, QC H4A 1T6 Canada

KFC Nephrology and Urology Scholarships **2618**

Nephrology and urology scholarships provide salary support for up to two years of an initial faculty appointment at an approved medical school in Canada. Successful applicants must use the scholarship within three years of the date of their first appointment as assistant professor and must commit to spending 75 percent or more time in research.
Geographic Focus: All States, Canada
Date(s) Application is Due: Oct 15
Amount of Grant: 45,000 CAD
Contact: National Research Programs Manager; (514) 369-4806, ext. 225 or (800) 361-7494, ext. 225; fax (514) 369-2472; research@kidney.ca
Internet: http://www.kidney.ca/english/research/biomed-funding-03.htm#bio-scholar
Sponsor: Kidney Foundation of Canada
300-5165 Sherbrooke Street W
Montreal, QC H4A 1T6 Canada

KFF Barbara Jordan Health Policy Scholars Program **2619**

The program brings talented African American, Latino, American Indian/Alaska Native, and Asian/Pacific Islander college seniors and recent graduates to Washington, DC, where they are placed in Congressional offices and learn about health policy. Through the nine-week program (May 23 through July 28), scholars gain knowledge about federal legislative procedure and health policy issues, while further developing their critical thinking and leadership skills. Scholars also participate in seminars and site visits to augment their knowledge of health care issues, and write and present a health policy research memo. Scholars receive a stipend upon completion of the program; a daily expense allowance for meals and local transportation; transportation/airfare to and from Washington, DC; and lodging at Howard University in Washington, DC. Application and guidelines are available online.
Requirements: Eligible candidates must be U.S. citizens who will be seniors or recent graduates of an accredited U.S. college or university in the fall. Successful applicants will show evidence of academic achievement and will demonstrate the following: be a member of a population that is adversely affected by racial or ethnic disparities in health; and have experience working in or with programs that address health disparities (e.g., local health departments, community health centers, inner city/rural hospitals, or organizations that conduct research on or develop policy related to health or health care disparities) or have done previous academic work (e.g., taken courses, written papers, or participate in research projects) related to health disparities.
Restrictions: Currently enrolled law, medical, and graduate students are not eligible to apply.
Date(s) Application is Due: Jan 9
Amount of Grant: 7,000 U.S.D
Contact: Grants Administrator; (650) 854-9400; fax (650) 854-4800
Internet: http://www.kff.org/about/jordanscholars.cfm
Sponsor: Henry J. Kaiser Family Foundation
2400 Sand Hill Road
Menlo Park, CA 94025

Kimberly-Clark Community Grants **2620**

The corporation contributes significant philanthropic resources to organizations and programs that provide vital information, resources, and services that strengthen today's families; involve parents and others in activities that help children grow physically, mentally, and socially; and develop creative, caring options to cope with the issues facing our older relatives. Kimberly-Clark limits its charitable contributions to organizations serving communities where employees live and work. The corporation also supports YMCA of the U.S.A, UNICEF, United Way, and disaster- relief efforts. Send requests in corporate-operating cities to the site manager's attention. All decisions are made locally, not at the corporate level. Guidelines are available online.
Requirements: Nonprofit organizations in Kimberly-Clark communities are eligible, including Alabama (Mobile), Arkansas (Conway, Maumelle), California (Escondido, Fullerton), Connecticut (New Milford), Georgia (LaGrange, Roswell), Idaho (Pocatello), Kentucky (Owensboro), Michigan (Munising), Mississippi (Corinth, Hattiesburg), North Carolina (Hendersonville, Lexington), Ohio (Piqua), Oklahoma (Jenks), Pennsylvania (Chester), South Carolina (Beech Island), Tennessee (Knoxville, Loudon), Texas (Dallas, Del Rio, Fort Worth, Paris, San Antonio), Utah (Draper, Ogden), Washington (Everett), and Wisconsin (Marinette, Neenah, Whiting).
Restrictions: As a general rule, no unsolicited requests for national or international programs will be considered.
Geographic Focus: Alabama, Arkansas, California, Connecticut, Georgia, Idaho, Kentucky, Michigan, Mississippi, North Carolina, Ohio, Oklahoma, Pennsylvania, South Carolina, Tennessee, Texas, Utah, Washington, Wisconsin
Samples: UNICEF (New York, NY)—to assist tsunami-relief efforts, $500,000; YMCA of the U.S.A (Chicago, IL)—to redistribute to 42 YMCAs nationwide for programs designed to strengthen families, $420,000.
Contact: Contribution Program Manager; (888) 525-8388
Internet: http://www.kimberly-clark.com/aboutus/contrib_program.asp
Sponsor: Kimberly-Clark Corporation
P.O. Box 2020, Department INT
Neenah, WI 54957-2020

Kimberly-Clark Foundation Grants **2621**

The foundation supports nonprofits within 50 miles of a Kimberly-Clark manufacturing plant or facility in a wide range of areas, including social welfare, medicine, and health; education; and civic and cultural activities. In the social welfare and health category, grants have supported groups such as boys' and girls' clubs, the Salvation Army, and visiting nurse associations. Educational funding includes NAACP and higher education institutions. The category of civic and cultural activities includes support for public broadcasting, museums, performing arts, and environmental groups. Types of support include annual campaigns, building construction funds, continuing support, equipment

acquisition, operating budgets, seed grants, scholarship funds, and capital campaigns. Requests for the company's application form must be in writing.
Requirements: Nonprofit organizations in Kimberly-Clark locations are eligible.
Amount of Grant: 1,000 - 75,000 U.S.D
Samples: Marquette U (Milwaukee, WI)—for the Raynor Library's Information Commons and the Thompson Center for Excellence in Education, $1.1 million over five years; U.S. Fund for Unicef (New York, NY)—to provide education, health, legal, and mental-health services to African children orphaned because of AIDS, $2.6 million over four years; UNICEF—to provide services to the more than 13 million children worldwide orphaned as a result of AIDS, including health and nutritional services, community care, group homes, access to education, grief counseling, and legal advice on how to keep siblings together, $2.6 million.
Contact: Carolyn Mentesana, Vice President; (972) 281-1200; fax (972) 281-1490
Internet: http://www.kimberly-clark.com/aboutus/kc_foundation.asp
Sponsor: Kimberly-Clark Foundation
P.O. Box 619100
Dallas, TX 75261-9100

Kimmel Scholar Award **2622**
The foundation seeks to improve the basic understanding of cancer biology and to develop new cancer treatment and prevention methods. Grants support programs of accomplished young cancer researchers that emphasize basic cancer research, rapid translation of basic science concepts into potential therapeutic applications, and clinical research with novel or innovative treatment strategies. Up to 10 awards are made each year. Additional grants are awarded to physicians engaged in translational science. Application and guidelines are available online.
Requirements: Applicants must hold an MD, PhD, or equivalent graduate degree. Researchers who have achieved the equivalent rank of assistant professor on or after July 2001. Physicians who have both clinical and research responsibility are eligible if appointed on or after July 2000. Research must be conducted at U.S. nonprofit organizations.
Date(s) Application is Due: Dec 6
Amount of Grant: 100,000 U.S.D
Samples: Sheil Steart, Washington U (Saint Louis, MO)—for research on cancer, $200,000 over two years; Brian Schaefer, Uniformed Services U of the Health Sciences (Bethesda, MD)—for research on cancer, $200,000 over two years.
Contact: Dr. Gary Cohen, (443) 849-3729; fax (443) 849-3719; gcohen@gbmc.org
Internet: http://www.kimmel.org/About41.html
Sponsor: Sidney Kimmel Foundation for Cancer Research
6569 N Charles Street, Suite 201
Baltimore, MD 21204

King Faisal Foundation Grants **2623**
The foundation operates in Saudi Arabia and internationally in the fields of science and medicine and arts and humanities. King Faisal International Prizes are awarded in the areas of Arabic literature, Islamic studies, service to Islam, and science and medicine. Nominations are due May 31. The foundation also sponsors a scholarship program intended to offer opportunities to outstanding Muslim graduates in the Islamic World to complete their higher education in the most prestigious universities of the world.
Requirements: Institutions and organizations around the world nominate candidates for the prizes. Scholarship applicants must be Muslim; have completed university studies with an average of 86 percent or above; have acceptance from an accredited European or North American university that is highly regarded in the particular field of specialization; not exceed 40 years of age; be fluent in the language of his/her studies; have specialization in one of the fields of medicine, engineering, and sciences (physics, chemistry, and geology); and have a graduation date of not more than three years.
Date(s) Application is Due: May 31
Amount of Grant: 200 - 200,000 U.S.D
Contact: Grants Administrator; 966-1-465-2255; fax 966-1-465-6524; info@kff.com
Internet: http://www.kff.com
Sponsor: King Faisal Foundation
P.O. Box 352
Riyadh, 11411 Saudi Arabia

King Foundation Grants **2624**
The foundation awards grants to eligible nonprofit organizations in its areas of interest, including scholarships, education, youth organizations, the indigent, cultural and historic community resources, medical and scientific research, and the aging population. Priority is given to agencies with innovative programs, a proven record of quality service delivery, and policies consistent with the foundation's mission. Single-year projects, rather than multiyear projects, are favored. After an application has been received and reviewed, the foundation may request further information, request a meeting, or request a site visit.
Requirements: Texas and Arkansas 501(c)3 tax-exempt organizations are eligible.
Restrictions: Grants are not made to individuals; organizations outside of Texas and Arkansas; church or seminary construction or religious programs other than social service-based initiatives; ongoing operating expenses or funds to offset operating losses; loan financing; endowments; professional conferences or symposia; balls, events, or galas benefiting charitable organizations; or organizations without tax-exempt status.
Geographic Focus: Arkansas, Texas
Samples: Communities in Schools-Dallas Region (TX)—to expand services available to at-risk children and their families at Maple Lawn Elementary School, $15,000; Saint Philip's School and Community Ctr (Dallas, TX)—for student scholarships, $10,000; Genesis Women's Shelter (Dallas, TX)—to renovate its shelter facilities, $35,000; Goodwill Industries of Dallas (TX)—to purchase a trailer truck for use in its Contributed Goods Program, $35,000.
Contact: Michelle Monse, Executive Director; (214) 750-1884; fax (214) 750-1651
Sponsor: Carl B. and Florence E. King Foundation
2929 Carlisle Street, Suite 222
Dallas, TX 75204

Kinsman Foundation Grants **2625**
The foundation awards grants to eligible nonprofit organizations in its areas of interest, including historic preservation, animal/wildlife welfare, and health care policy. Types of support include annual campaigns, building construction/renovation, capital campaigns, challenge/matching grants, conferences and seminars, consulting services, continuing support, curriculum development, endowments, equipment acquisition, general operating support, internships, program development, publication, research, seed grants, and technical assistance. Contact the office for application forms and guidelines.
Requirements: Oregon and southern Washington nonprofit organizations are eligible.
Geographic Focus: Oregon, Washington
Date(s) Application is Due: Aug 1
Samples: Kam Wah Chung and Company Museum (John Day, OR)—for restorations at this museum, a former trading post built in 1875, that documents the lives and culture of Chinese immigrants in eastern Oregon, $110,000.
Contact: Grants Administrator; (503) 654-1668; fax (503) 654-1759; grants@kinsmanfoundation.org
Sponsor: Kinsman Foundation
3727 SE Spaulding Avenue
Milwaukie, OR 97267-3938

Kiplinger Foundation Grants **2626**
The foundation awards grants to educational, health, welfare, civic, and cultural organizations, primarily in the Washington, DC, area. Building grants are also awarded to organizations in these areas of endeavor. The foundation also provides matching grants on a two-to-one basis to secondary schools and college-level institutions on behalf of Kiplinger Washington Editors employees and retirees of up to $5000 per year per employee or retiree. Areas of particular interest to the foundation include the arts and programs that provide better conditions for troubled and deprived youth. Grants also support annual campaigns, capital campaigns, continuing support, endowment funds, and special projects. Application should be a brief letter of proposal outlining the project for which funds are sought and objectives that the project should accomplish. Most grants are made for a period of one year.
Requirements: Organizations with IRS 501(c)3 status in the greater Washington, DC area are eligible.
Restrictions: The foundation does offer money for seed grants nor does it support scholarships.
Geographic Focus: District of Columbia
Amount of Grant: Up to 15,000,000 U.S.D
Contact: Andrea Wilkes, Secretary; (202) 887-6559; foundation@kiplinger.com
Sponsor: Kiplinger Foundation
1729 H Street NW
Washington, DC 20006

Kitsap Community Foundation Grants **2627**
The foundation strives to address basic human needs in the Kitsap community. The foundation is primarily interested in funding programs in its focus areas that build on existing strengths of an organization; focus areas include arts and culture, civic and community improvement, education, social services, health, recreation, environment, and youth. Grants will be considered for general operational support and for capital needs of qualified institutions or organizations. Decisions are based on a number of factors including, but not limited to the number of people served, the potential for funding from other sources, whether the grant falls within the funding priority areas, the originality of the project, whether any other services of its kind exist, and the demonstrated need of the organization. Application guidelines and procedures are available online.
Requirements: Washington 501(c)3 nonprofit organizations in Kitsap County and its neighboring communities are eligible.
Restrictions: Generally, the foundation does not make grants to annual campaign appeals; endowments; individuals; political or lobbying activities; religious organizations for sacramental or theological purposes; publications except those that grow out of research and experiments underwritten by the foundation; or deficit reduction.
Geographic Focus: Washington
Date(s) Application is Due: Feb 15
Amount of Grant: Up to 2,000 U.S.D
Contact: Executive Director; (360) 698-3622; kcf@kitsapfoundation.org
Internet: http://www.kitsapfoundation.org/grantgd.html
Sponsor: Kitsap Community Foundation
P.O. Box 3670
Silverdale, WA 98383

Kiwanis International Foundation Grants **2628**
The foundation supports nonprofits worldwide for programs and activities that meet the needs of young children. Past grants have supported a bloodmobile playground, Reading is Fundamental project, lifetime learning, and an effort to provide pediatric trauma kits. Applicants should contact the local chapter for additional information.
Requirements: U.S. and international nonprofits are eligible.
Restrictions: The foundation does not award grants for capital improvements or programs for more than three years.

Date(s) Application is Due: Apr 15; Jul 15; Nov 15
Amount of Grant: Up to 50,000 U.S.D
Samples: Unicef (New York, NY)—for its work to eliminate iodine-deficiency disorders in developing nations where iodine deficiency is a leading cause of preventable mental and physical retardation, $1.165 million.
Contact: Program Contact; (800) 549-2647; fax (317) 879-0204; foundation@kiwanis.org
Internet: http://kif.kiwanis.org/kif/grants/grants.asp
Sponsor: Kiwanis International Foundation
3636 Woodview Trace
Indianapolis, IN 46268

Klingenstein Fellowship Awards in the Neurosciences 2629
The purpose of these awards is to support, in the early stages of their careers, young investigators engaged in basic or clinical research that may lead to a better understanding of epilepsy. The fund recognizes that to accomplish this goal it is necessary to encourage a variety of new approaches. Several areas within the neurosciences are of particular interest to the fund: Cellular and molecular neuroscience—Studies of the mechanisms of neuronal excitability and development, and of the genetic basis of seizure disorders. Neural systems—Studies of the integrative function of the nervous system. Clinical research—Studies designed to improve the prevention, diagnosis, treatment and our understanding of the causes of epilepsy. Application may be made at any time.
Requirements: Applicants must hold the PhD and/or MD degrees, and have completed all research training, including post-doctoral training. U.S. citizenship is not a requirement, but it is expected that candidates will be permanent residents of the U.S. and that their research will be carried out in U.S. institutions.
Restrictions: The fund does not contribute to endowments and rarely contributes to buildings or other kinds of capital projects.
Date(s) Application is Due: Dec 10
Samples: Teachers College Columbia U (New York, NY)—for program support, $578,595; American Jewish Committee (New York, NY)—for work on issues of separation of church and state, $40,000; Foundation for Biomedical Research (Washington, DC)—for animal biomedical research, $25,000.
Contact: Kathleen Pomerantz; (212) 492-6181; kathleen.pomerantz@klingenstein.com
Internet: http://www.klingfund.org
Sponsor: Esther A. and Joseph Klingenstein Fund
787 7th Avenue, 6th Floor
New York, NY 10019

Klingenstein Neuroscience Fellowships 2630
The purpose of the fellowships is to support, in the early stages of their careers, young investigators engaged in basic or clinical research that may lead to a better understanding of epilepsy. Several areas within the neurosciences are of particular interest: cellular and molecular neuroscience, neural systems, and clinical research. Up to 10 fellows will be appointed. While U.S. citizenship is not a requirement, it is expected that candidates will be permanent residents of the United States and that their research will be carried out in U.S. institutions. The award may be used for salary support, research assistants, equipment, or for activities that promote the scientific advancement of the fellow. Payments will be made to the investigator's institution. Holding an award from another foundation concurrently with a Klingenstein award requires the approval of the advisory committee.
Requirements: Investigators must hold the PhD and/or MD degrees and have completed all research training, including postdoctoral training. The candidate should be an independent investigator, holding a tenure-track academic rank (but not yet tenured) in a university or medical school, or the equivalent standing in a research institute or medical center.
Date(s) Application is Due: Dec 9
Amount of Grant: 150,000 U.S.D
Contact: John Klingenstein, President; (212) 492-6181; fax (212) 492-7007
Internet: http://www.klingfund.org
Sponsor: Esther A. and Joseph Klingenstein Fund
787 7th Avenue, 6th Floor
New York, NY 10019

Klingenstein Third Generation Foundation Grants in Depression 2631
The foundation strives to improve the lives of families afflicted by clinical depression, including childhood and adolescent depression, and Attention Deficit/Hyperactive Disorder (ADHD). Grants are awarded in the areas of intervention and referral, prevention, public education/training, and infrastructure. The foundation will consider making capital grants, operating grants, challenge grants, seed grants, and grants given over more than a one-year period. The foundation also sponsors fellowships for postdoctoral research in depression. Funding proposals will be accepted from any geographic area, but the board expects to fund primarily in the New York, Chicago, and District of Columbia metropolitan areas. There is no formal application procedure or form. Applicants should submit a brief letter that succinctly describes the project and the amount sought and explains why the project is significant. The foundation will request a more detailed proposal.
Requirements: U.S. tax-exempt organizations are eligible.
Restrictions: Grants do not support direct services, unless they include a research or program evaluation component, or unless they represent the application of research findings to a clinical setting.
Amount of Grant: 5,000 - 35,000 U.S.D
Samples: American Academy of Child and Adolescent Psychiatry (Washington, DC)—for general support, $15,000; Columbia U, Carmel Hill Ctr for Early Diagnosis and Treatment (New York, NY)—to create brochures on mental-health disorders written in different languages and at a low reading level, $15,000; State U of New York (Buffalo, NY)—fellowship support for James Waxmonsky, for research on child and adolescent depression, $30,000 over two years.
Contact: Sally Klingenstein Martell, Executive Director; (212) 492-6179; fax (212) 492-7007; sally@ktgf.org
Internet: http://www.ktgf.org/depress.html
Sponsor: Klingenstein Third Generation Foundation
787 Seventh Avenue, 6th Floor
New York, NY 10019-6016

Kluge Center David B. Larson Fellowship in Health and Spirituality 2632
The Library of Congress invites qualified scholars to apply for a postdoctoral fellowship in the field of health and spirituality. It seeks to encourage the pursuit of scholarly excellence in the scientific study of the relation of religiousness and spirituality to physical, mental, and social health. The fellowship provides an opportunity for a period of six to 12 months of concentrated use of the collections of the Library of Congress, through full-time residency in the Library's John W. Kluge Center. If necessary, special arrangements may be made with the National Library of Medicine for access to its materials as well. Applicants must submit a formal application packet, including an application form, a two-page curriculum vita that should indicate major prior scholarship, a one-paragraph project summary, a bibliography of basic sources, a research proposal of no more than 1500 words, and three letters of reference (in English) from people who have read the research proposal.
Requirements: Applicants must by U.S. citizens or permanent residents and must possess a doctoral degree (PhD, MD, ScD, DrPH, DSW, PPsy, DST, ThD, and JD) awarded by the deadline date.
Date(s) Application is Due: Apr 17
Amount of Grant: 25,200 - 50,400 U.S.D
Contact: Larson Fellowships Coordinator; (202) 707-3302; fax (202) 707-3595; scholarly@loc.gov
Internet: http://www.loc.gov/loc/kluge/fellowships/larson.html
Sponsor: Library of Congress
101 Independence Avenue SE
Washington, DC 20540-4860

Knight Family Charitable and Educational Foundation Grants 2633
The Foundation operates in Michigan, and funding is reserved for elementary and secondary education, religious agencies, and community health and human service programs. Giving is primarily reserved for Michigan-based non-profit organizations. There are no specific deadlines or application forms to adhere to, and final approval is given approximately two months after submission.
Requirements: Applicants should be 501(c)3 organizations based in Michigan.
Amount of Grant: 500 - 10,000 U.S.D
Contact: N.J. Glasser, Secretary-Treasurer; (517) 547-6131
Sponsor: Knight Family Foundation
215 North Talbot Street
Addison, MI 49220-9698

Knox County Community Foundation Grants 2634
Founded in 1996, the Knox County Community Foundation is a nonprofit, public charity created by and for the people of Knox County, Indiana. The foundation helps, nonprofits fulfill their missions by, strengthening their ability to meet community needs through grants that assist charitable programs, address community issues, support community agencies, launch community initiatives, and support leadership development. Grant proposals are accepted once each year according to the grant cycle. Proposal requirements may change from year to year; therefore, grant seekers are advised to contact the foundation or see the foundations website, prior to beginning the grant application process. Grants are normally given as one-time support of a project but may be considered for additional support for expansions or outgrowths of an initial project. At the start of each cycle, a notice is mailed to nonprofit organizations that have applied for grants in the past, have received grants in the past, or have otherwise requested notification of the start of each cycle. The grant cycle runs from January through May each year. Proposals will be accepted from January through the March deadline. All organizations that have submitted grant proposals will be notified of the outcome of the grants committee's deliberation in writing no later than June 1.
Requirements: The Foundation welcomes proposals from nonprofit organizations that are deemed tax-exempt under sections 501(c)3 and 509(a) of the Internal Revenue Code and from governmental agencies serving the county. Proposals from nonprofit organizations not classified as a 501(c)3 public charity may be considered provided the project is charitable and supports a community need. Proposals submitted by an entity under the auspices of another agency must include a written statement signed by the agency's board president on behalf of the board of directors agreeing to act as the entity's fiscal sponsor, to receive grant monies if awarded, and to oversee the proposed project. Program areas considered for funding are: arts and culture; community development; education; health; human services; other civic endeavors, such as the environment, recreation, and youth development.
Restrictions: Project areas not considered for funding are: religious organizations for religious purposes; political parties or campaigns; endowment creation or debt reduction; operating costs; capital campaigns; annual appeals or membership contributions; travel requests for groups or individuals such as bands, sports teams, or classes. A six-month progress report and a final report at project completion are required by organizations whose proposals are approved for funding. Instructions and appropriate forms will be provided at the time the grant is awarded.
Geographic Focus: Indiana
Samples: Indiana Military Museum, $1,500 - for Website Development and Deployment Project; Walnut Grove United Methodist Church, $2,000 - to support their Back To School Mall Project; Freelandville Community Home, $5,000 - for Ambulance Bay Project;

Contact: Kathy Rinsch, Director; (812) 886-0093; kathy@knoxcountyfoundation.org
Internet: http://www.knoxcountyfoundation.org/disc-grants-program
Sponsor: Knox County Community Foundation
20 N Third Street, Suite 301, P.O. Box 273
Vincennes, IN 47591

Kodak Community Relations Grants **2635**
The company awards grants in Kodak communities to support community revitalization, education, and health and human services. Education grants support preschools, K-12, higher education, diversity initiatives, and umbrella education funds. University contributions are focused on recruitment, technology, and market development. Community revitalization grants support economic development, culture, and the arts. The company awards grants to international nonprofits for relief efforts. There are no application deadlines or forms.
Requirements: Nonprofits in company-operating areas nationwide are eligible.
Restrictions: Kodak does not support individuals; a commitment beyond three to five years; endowed chairs; university capital campaigns; event sponsorships; operating costs of organizations that receive funds from a Kodak supported United Way; legislators, political organizations, or campaigns; or sectarian organizations whose programs are limited to members of one religious group.
Date(s) Application is Due: Apr 30
Amount of Grant: Up to 11,000,000 U.S.D
Samples: American National Red Cross (Washington, DC)—for relief efforts in South Asia and Africa, $1 million.
Contact: Community Relations Manager; (716) 724-3041; fax (716) 724-1376
Internet: http://www.kodak.com/U.S./en/corp/community.shtml
Sponsor: Eastman Kodak Corporation
343 State Street
Rochester, NY 14650-0517

Koessler Family Foundation Grants **2636**
The family foundation awards grants to western New York nonprofits, with emphasis on Buffalo, NY. Areas of interest include higher education, hospitals, and hospices. Types of support include general operating support, building construction/renovation, capital campaigns, and matching grants. Submit a letter of request.
Requirements: Western New York nonprofits are eligible.
Geographic Focus: New York
Date(s) Application is Due: Nov 30
Amount of Grant: Up to 342,000 U.S.D
Samples: Canisius College (Buffalo, NY)—for its capital campaign, $1 million matching grant; Childrens Hospital of Buffalo (Buffalo, NY)—for general support, $60,000; Kenmore Mercy Foundation (Kenmore, NY)—to construct a new patient tower, $24,000.
Contact: Stephen Juhasz, Grant Manager
Sponsor: Kenneth L. and Katherine G. Koessler Family Foundation
124 Brantwood Road
Snyder, NY 14226

Komen Greater NYC Clinical Trials Enrollment Grants **2637**
The Greater New York City Affiliate of Susan G. Komen for the Cure (Komen Greater NYC) fulfills its mission of eradicating breast cancer as a life-threatening disease, in part, by offering grants to enhance the capacity of National Cancer Institute- and Department of Defense-approved breast cancer clinical research conducted in the Komen Greater NYC area to increase the number of women enrolled in breast cancer clinical research studies. Komen Greater NYC seeks to fund projects that implement evidence-based strategies to overcome barriers to enrollment in breast cancer clinical research. Collaborative projects between institutions and community-based organizations are encouraged.
Requirements: To apply, the organization must be: a not-for-profit organization (a charitable or educational tax-exempt organization), a government agency, or an Indian tribe; conducting National Cancer Institute (NCI) or Department of Defense (DOD) approved breast cancer clinical research (this includes studies through the Clinical Trials Cooperative Group Program including but not limited to ACOSOG, ECOG, SWOG, NSABP, ACRIN, and RTOG; and located in the Komen Greater NYC service area, which includes the five boroughs of New York City, Long Island, Westchester County, and Rockland County.
Geographic Focus: New York
Date(s) Application is Due: Oct 9
Amount of Grant: 200,000 U.S.D
Contact: Mary Haviland; (212) 560-9590, ext. 230; mhaviland@komennyc.org
Internet: http://www.komennyc.org/site/PageServer?pagename=grants_clinicaltrials
Sponsor: Susan G. Komen for the Cure Foundation
341 W 38th Street, 10th Floor
New York, NY 10018-2918

Komen Greater NYC Community Breast Health Grants **2638**
The mission of Komen Greater NYC is to eradicate breast cancer as a life-threatening disease by advancing research, education, screening, and treatment. The Greater New York City Affiliate is, therefore, soliciting applications from non-profit organizations that have current or planned breast health programs within the five boroughs of New York City, Long Island, Westchester County, and Rockland County. The purpose of the Community Breast Health Grants Program (formerly known as Screening, Treatment, and Education Program- STEP) is to support community-based programs that provide the medically under-served with access to breast health education, screening coordination, diagnostic, treatment, and support services.
Requirements: Applicant organizations must be: a not-for-profit organization (a charitable or educational tax-exempt organization), a government agency, or an Indian tribe; located in the Komen Greater NYC service area, which includes the five boroughs of New York City, Long Island, Westchester County, and Rockland County; and a current or planned provider of breast health services.
Geographic Focus: New York
Date(s) Application is Due: Oct 9
Amount of Grant: Up to 100,000 U.S.D
Contact: Mary Haviland; (212) 560-9590, ext. 230; mhaviland@komennyc.org
Internet: http://www.komennyc.org/site/PageServer?pagename=grants_breasthealthgrants
Sponsor: Susan G. Komen for the Cure Foundation
341 W 38th Street, 10th Floor
New York, NY 10018-2918

Komen Greater NYC Conference Grants **2639**
Komen Greater NYC provides grants to non-profit organizations to support educational conferences, symposiums or meetings. To be considered for funding, events must be specific to breast health or help further the body of knowledge related to breast cancer research, education, screening, diagnosis, treatment and/or support. Grants are considered three times per year, in May, August, and January. Awards will not exceed $5,000. Award notifications are generally made within 6 weeks after receipt of the application.
Requirements: To apply, the organization must be: a not-for-profit organization (a charitable or educational tax-exempt organization), a government agency, or an Indian tribe; conducting National Cancer Institute (NCI) or Department of Defense (DOD) approved breast cancer clinical research (this includes studies through the Clinical Trials Cooperative Group Program including but not limited to ACOSOG, ECOG, SWOG, NSABP, ACRIN, and RTOG; and located in the Komen Greater NYC service area, which includes the five boroughs of New York City, Long Island, Westchester County, and Rockland County.
Geographic Focus: New York
Date(s) Application is Due: May 14; Jul 25; Dec 1
Amount of Grant: Up to 5,000 U.S.D
Contact: Cara Thunder; (212) 461-6185; cthunder@komennyc.org
Internet: http://www.komennyc.org/site/PageServer?pagename=grants_conference
Sponsor: Susan G. Komen for the Cure Foundation
341 W 38th Street, 10th Floor
New York, NY 10018-2918

Komen Greater NYC Small Grants **2640**
Komen Greater NYC is currently offering small grants up to $5,000 to support small and pilot projects in breast health, organize conferences, and travel scholarships. Small Grant Applications are accepted three times a year. The deadlines are announced on the Komen Greater NYC website at: www.komennyc.org/grants. Requests for support of travel scholarships, conferences, or other events must be received at least 3 months prior to the event. Award notifications are generally made within 6 weeks after receipt of the application.
Requirements: Applicants and institutions must conform to the following eligibility criteria to be considered for funding: 1) Applicants must ensure that all past and current Komen-funded grants or awards are up-to-date and in compliance with Komen requirements; 2) Institutions must be located in or providing services to one or more of the following locations: the five boroughs of New York City; Long Island (Suffolk and Nassau Counties); Westchester County; Rockland County; 3) Project must be specific to breast health and/or breast cancer; 4) the applicant must be a non-profit organization with federal tax exemption. Individuals may not receive grants; 5) Small Grant applicants are restricted to a maximum combined award of $5,000 per Komen Greater NYC grant year (April through March), per program for any type of Small Grant- Travel, Conference and/or Small Grant; 6) current Grantees are eligible to apply for small grants.
Geographic Focus: New York
Amount of Grant: 5,000 U.S.D
Contact: Cara Thunder; (212) 461-6185; cthunder@komennyc.org
Internet: http://www.komennyc.org/site/PageServer?pagename=grants_small_grants_program
Sponsor: Susan G. Komen for the Cure Foundation
341 W 38th Street, 10th Floor
New York, NY 10018-2918

Kosair Charities Grants **2641**
In 1923, a committee of members from Louisville's Kosair Shrine Temple, doctors, lawyers, and community leaders formed Kosair Charities. The mission was to provide the highest quality health care possible for children who had nowhere else to turn. Since conception, Kosair Charities has became the largest charity in the history of the region that provides for the medical care of children. All of this is possible thanks to the generous support of individuals and companies. Kosair Charities receives no assistance from United Way, the Crusade for Children, or any governmental entity. Annual grants are awarded in the early summer of each year and applications are due no later than May 1.
Requirements: Grants are only made to those non-profit agencies and organizations whose program falls within the above stated mission. Two grant requests applications are available for download on the Kosair Charities website in amounts of $1,500 or less and $1,500 or more. With each completed application, attach the following: Copy of IRS determination letter; current year's budget; income and expense statement last two years; copy of last completed audit; copy of roster of Board of Directors; descriptive literature on agency and its services; if the proposal is for an existing or continuing program, submit latest self-evaluation of program; if the proposal is for one that Kosair Charities has funded in the past, demonstrate use of Kosaid Charities Supported Agency Logo and

public acknowledgements; detailed proposal budget; original grant application plus five copies for a total of six copies of all requested information.
Restrictions: Grants are available throughout Kentucky and the southern Indiana region. Grants are made for one year only and do not represent any future commitment on the part of Kosair Charities. Grant recipients are required to use the Kosair Charities Supported Agency logo on all printed and electronic materials and media used by their agency.
Geographic Focus: Indiana, Kentucky
Date(s) Application is Due: May 1
Contact: Jo Barrett; (502) 637-7696 or (888) 454-3752; jbarrett@kosair.org
Internet: http://www.kosair.org/grants.html
Sponsor: Kosair Charities
982 Eastern Parkway, P.O. Box 37370
Louisville, KY 40233-7370

Kosciusko County Community Foundation Grants 2642
The Kosciusko County Community Foundation serves Kosciusko County, Indiana. Nonprofit organizations serving Kosciusko County are eligible to apply in seven areas of interest: arts and culture, human services, civic projects, recreation, environment, health, and education. Grant applications and guidelines can be obtained at the Foundations office or website. Grant awards are announced nine weeks after each deadline.
Requirements: Grant seekers are strongly encouraged to call the Foundation's program staff to discuss a grant proposal before submitting a formal application. Once the proposal has been discussed, complete and submit a grant application with the required attachments: 6 copies of the original application; 1 copy of the IRS determination letter; 7 copies of the board of directors listing with names and addresses for all; 7 copies of staff listing with names and addresses for all; 7 copies of current internal financial statements; and 7 copies of program/project budget. Do not provide copies of news articles, brochures or other miscellaneous supporting information.
Restrictions: The Foundation will not consider grants for: individuals; political activities or those designated to influence legislation; national organizations (unless the monies are to be used solely to benefit citizens of Kosciusko County); fundraising projects;the direct benefit of the donor or the donor's family; religious organizations for the sole purpose of furthering that religion (this prohibition does not apply to funds created by donors who have specifically designated religious organizations as beneficiaries of the funds); contributions to endowments.
Geographic Focus: Indiana
Date(s) Application is Due: Jan 15; May 15; Sep 15
Contact: Stephanie Overbey, Communication & Program Director; (574) 267-1901; fax (574) 268-9780; stephanie@kcfoundation.org
Internet: http://www.kcfoundation.org/grants.html
Sponsor: Kosciusko County Community Foundation
102 E Market Street
Warsaw, IN 46580

Kosciusko County Community Foundation REMC Operation Round Up Grants 2643
The Kosciusko County Community Foundation serves the residents of Kosciusko County, Indiana. Kosciusko Rural Electric Membership Corporation (REMC) encourages its members to round up their electric bills to the nearest whole dollar. The extra funds are deposited into the Kosciusko REMC Operation Round Up Grants, which supports a variety of charitable causes in communities served by Kosciusko REMC. Applications are due the 15th of the following months: February, April, June, August, October, and December. Grant notifications take place within six weeks after the deadline.
Requirements: Applicants must submit the online application, along with a list of the organization's board of directors, officers, or trustees, and their phone numbers; a one page cover letter that specifies the amount requested and details about how the funds will be used locally; a copy of the organization's 501(c)3 letter; and a copy of the organization's most current financial statements.
Geographic Focus: Indiana
Date(s) Application is Due: Feb 15; Apr 15; Jun 15; Aug 15; Oct 15; Dec 15
Amount of Grant: 500 - 5,000 U.S.D
Samples: Boys and Girls Club of Kosciusko County; provide meals to children, $2,500: Fort Wayne Philharmonic, Inc; support of three concerts, $1,000: College Mentors for Kids; launch a College Mentors for Kids at Grace College, $5,000.
Contact: Stephanie Overbey, Communication and Program Director; (574) 946-0906; fax (574) 946-0971; Stephanie@kcfoundation.org
Internet: http://www.kcfoundation.org/seekingfunds/remc.php
Sponsor: Kosciusko County Community Foundation
102 East Market Street
Warsaw, IN 46580

Kosciuszko Foundation Dr. Marie E. Zakrzewski Medical Scholarship 2644
The Dr. Marie E. Zakrzewski Medical Scholarship of $3,500 is awarded each year to a young woman of Polish ancestry for first, second, or third year of medical studies at an accredited school of medicine in the United States. Funding is for the academic year September through May. Selection is based on academic excellence, the applicant's academic achievements, interest, motivation, interest in Polish subjects and involvement in the Polish American community. Financial need is taken into consideration. First preference is given to residents of the state of Massachusetts. Qualified residents of New England are considered if no first preference candidates apply. The scholarship is non-renewable.
Requirements: United States citizens of Polish descent and Polish citizens with permanent residency status in the United States who are entering first, second or third year of M.D. studies in the upcoming academic year and who have a minimum GPA of 3.0. This scholarship is open to female residents of the state of Massachusetts.
Geographic Focus: Connecticut, Maine, Maryland, Massachusetts, New Hampshire, Vermont
Date(s) Application is Due: Jan 5
Amount of Grant: 3,500 U.S.D
Contact: Addy Tymczyszyn , Program Officer, Scholarship & Grants for Americans; (212) 734-2130, ext. 210; fax (212) 628-4552; addy@thekf.org
Internet: http://www.thekf.org/scholarships/tuition/mzms/
Sponsor: Kosciuszko Foundation
15 East 65th Street
New York, NY 10065

Kosciuszko Foundation Grants for Polish Citizens 2645
The Kosciuszko Foundation annually awards a number of grants to Poles for study and research at universities and other institutions of higher learning in the United States. The Foundation provides a cost-of-living stipend which includes: transatlantic travel, housing allowance, health and accident insurance coverage, and, when warranted, domestic travel. Grants are awarded to those without doctoral degrees. In determining awards, in addition to overall excellence, consideration is given to number of other factors such as scholarly affiliation or geographic location in order to achieve fair and reasonable diversity among awardees. There are no restrictions as to fields of study or research supported, and each proposal is evaluated individually.
Requirements: Only Polish citizens residing permanently in Poland are eligible to apply. This program is not for Poles residing outside of Poland, nor those residing temporarily outside of Poland, whether or not engaged in research or study. Dual citizens (United States and Poland) as well as individuals applying or holding United States permanent residency status (green card) are not eligible. Applicants must have strong English language proficiency to communicate and to carry out proposed projects. The level of English proficiency will be determined during the personal interviews in Warsaw.
Restrictions: There is no allowance for dependents (spouse or child) support. Awardees are solely responsible for all expenses of accompanying dependents.
Geographic Focus: Poland
Date(s) Application is Due: Oct 15
Amount of Grant: 7,650 - 25,500 U.S.D
Contact: Addy Tymczyszyn, Program Officer, Scholarship & Grants for Americans; (212) 734-2130, ext. 210; fax (212) 628-4552; addy@thekf.org
Program Officer; (48) (22) 621-7067; fax (48) (22) 621-7067
Internet: http://www.thekf.org/scholarships/exchange-us/
Sponsor: Kosciuszko Foundation
15 East 65th Street
New York, NY 10065

Kovler Family Foundation Grants 2646
Established in 1967 and formerly known as the Harry and Maribel G. Blum Foundation, the Kovler Family Foundation awards grants in the arts, education, medical research (particularly diabetes research), and children and youth services. General operating or research grants are awarded primarily in the Chicago metropolitan area. There are no application forms. Applicants should submit a one to two page written proposal with a copy of their IRS determination letter by mid-November to considered for the current year. Typical grant awarded is between $1,000-$5,000.
Requirements: Illinois nonprofit organizations are eligible to apply.
Restrictions: The foundation does not award grants to individuals.
Geographic Focus: Illinois
Samples: Challenge Aspen, Aspen, Colorado, $1,000; Intonation Music Workshop, Chicago, Illinois, $4,000; PAWS Chicago, Chicago, Illinois, $26,500.
Contact: Hymen Bregar, Secretary; (312) 664-5050
Jonathan Kovler, President and Treasurer; (312) 664-5050
Internet: http://www.neh.gov/millenniumlibraries
Sponsor: Kovler Family Foundation
875 N. Michigan Avenue
Chicago, IL 60611-1958

Kresge Foundation Bricks and Mortar Grants 2647
The foundation awards challenge grants for capital projects, most often for construction or renovation of buildings, but also for the purchase of major equipment and real estate. Although Kresge's grant dollars support an organization's bricks and mortar project, their review goes beyond an evaluation of the particular capital need and incorporates a close scrutiny of program and organizational strengths. As challenge grants, they are intended to stimulate new, private gifts in the midst of an organized fund-raising effort. They offer special opportunities to build capacity, both in providing enhanced facilities in which to present programs and in strengthening the organization's ability to generate private support. Applications may be submitted at any time throughout the year. The board meets in March, June, September, and December.
Requirements: Tax-exempt, charitable organizations operating in the fields of higher education, health care and long-term care, human services, science and the environment, arts and humanities, and public affairs may apply.
Amount of Grant: 150,000 - 600,000 U.S.D
Samples: U of Maine at Farmington (ME)—to construct an education center, $150,000 challenge grant; Harding U (Searcy, AR)—to renovate and expand the building that houses the College of Education, $700,000 challenge grant; Campbellsville U (KY)—for a student-center facility, $1 million challenge grant; Swarthmore College (PA)—to help renovate Parrish Hall, the college's oldest building, $1 million challenge grant.

Contact: Sandra McAlister Ambrozy, Senior Program Officer; (248) 643-9630; fax (248) 643-0588; SMAmbrozy@Kresge.Org
Internet: http://www.kresge.org/programs/index.htm
Sponsor: Kresge Foundation
3215 W Big Beaver Road, P.O. Box 3151
Troy, MI 48007-3151

Kresge Foundation Science Initiative Grants **2648**
The program's current funding goal is to improve teaching and/or research laboratories by replacing and/or upgrading their equipment and, when necessary, the facilities where the equipment is housed. The foundation will provide funding in two equal installments, the first for one-third of the equipment cost, and the second for one-fifth of new endowment funds for future updating of such equipment and facilities. It is necessary to raise funds equal to two-thirds of the equipment prior to applying. Project plans should be specific at the time of application, including firm costs of both the equipment and space renovation. The foundation will waive the minimum unit cost restriction associated with its traditional equipment grants. Organizations may submit applications for traditional bricks-and-mortar challenge grants within the same 12-month period. For details, request a pamphlet—the Kresge Foundation Science Initiative. Send for the Policies and Application Procedures brochure before applying.
Requirements: Colleges and universities, teaching hospitals, medical schools, and research institutions are eligible.
Date(s) Application is Due: May 15
Amount of Grant: 200,000 - 800,000 U.S.D
Samples: U of Evansville (IN)—to purchase equipment for the Koch Center for Engineering and Science, $228,000 challenge grant; Monell Chemical Senses Ctr (Philadelphia, PA)—to begin the renovation of its facilities, $600,000 challenge grant; Methodist College (Fayetteville, NC)—to expand its science building, $550,000 challenge grant; Clark U (Worcester, MA)—to purchase a magnetometer and diffractometer, $205,920 challenge grant.
Contact: Grants Administrator; (248) 643-9630; fax (248) 643-0588
Internet: http://www.kresge.org
Sponsor: Kresge Foundation
3215 W Big Beaver Road, P.O. Box 3151
Troy, MI 48007-3151

L. A. Hollinger Respiratory Therapy and Nursing Scholarships **2649**
The intent of this scholarship is to promote his strong commitment to education with the emphasis on continually improving the quality of patient care and ensuring the continued preparation of future nurses and respiratory therapists to meet that objective. The scholarship is awarded to those who are currently pursuing a Nursing or Respiratory Therapy undergraduate or graduate degree. All applications must be submitted or postmarked by May 15th.
Requirements: All applicants must have a cumulative GPA of 2.5. All nursing applicants must be a recipient of a Trinity Lutheran Hospital School of Nursing Alumnae Scholarship for the current year. Respiratory Therapy applicants must be enrolled in the Respiratory Therapy program at either Johnson County Community College or the University of Kansas.
Date(s) Application is Due: May 15
Contact: Becky Schaid, Executive Director; (816) 276-7515 or (816) 276-7555; fax (816) 926-2261; becky@btllf.org
Internet: http://www.btllf.org/funding.html
Sponsor: Baptist-Trinity Lutheran Legacy Foundation
6601 Rockhill Road
Kansas City, MO 64131

L. W. Pierce Family Foundation Grants **2650**
Established in 1997, The Foundation supports organizations involved with health, social, and educational services in the areas of alcohol and drug abuse, hospice care and children's welfare. Giving is limited to the Vero Beach, FL and Philadelphia, PA areas.
Requirements: Non-profits in the Vero Beach, FL and Philadelphia, PA areas are eligible.
Restrictions: No grants to individuals.
Geographic Focus: Florida, Pennsylvania
Date(s) Application is Due: Mar 1
Amount of Grant: 1,000 - 50,000 U.S.D
Samples: Homeless Family Center, Vero Beach, FL, $50,000; Senior Resource Association, Vero Beach, FL, $20,000; Thomas Jefferson University, Philadelphia, PA, $20,000;
Contact: Constance Buckley, President; (610) 862-2105; fax (610) 862-2120
Sponsor: L. W. Pierce Family Foundation
8 Tower Bridge, Suite 1060, 161 Washington Street
Conshohocken, PA 19428-2060

La-Z-Boy Foundation Grants **2651**
The foundation makes grants to nonprofit organizations in communities where it maintains plants and within a 15-mile radius of its corporate headquarters in Monroe, MI. Grants are awarded to public charities, civic organizations, and hospitals and health organizations. Support is given for general operating costs, building construction/renovation, education programs, and preservation/restoration. Potential applicants are asked to write to the office and describe the purpose of their project, time span, estimated costs, and benefits. Requests are reviewed at quarterly staff meetings held in March, June, September, and December.
Requirements: 501(c)3 nonprofit charities are eligible. Grants are awarded primarily in communities of company operations, including Siloam Springs, AR; Redlands, CA; Monroe, MI; Neosho, MO; Newton and Leland, MS; Lincolnton, NC; Florence, SC; Dayton, TN; and Tremonton, UT.
Geographic Focus: Arkansas, California, Michigan, Mississippi, Missouri, North Carolina, South Carolina, Tennessee, Utah
Date(s) Application is Due: Mar 1; Jun 1; Sep 1; Dec 1
Amount of Grant: 250 - 50,000 U.S.D
Contact: Donald Blohm, Administrator; (734) 242-1444; fax (734) 457-2005
Internet: http://www.la-z-boy.com
Sponsor: La-Z-Boy Foundation
1284 N Telegraph Road
Monroe, MI 48162

LaGrange County Community Foundation Grants **2652**
The mission of the LaGrange County Community Foundation (LCCF) is to inspire and sustain leadership, generosity and service. Its purpose is to help community service organizations sponsor plans to meet critically important needs. The Foundation funds grants for innovative and creative projects and programs that are responsive to changing community needs in the areas of, but not restricted to: health and human services, environment, arts and culture, and recreation.
Requirements: Non-profit organizations, schools and qualifying government agencies serving the citizens of LaGrange County are invited to apply for a grant through the foundation's application process. In addition to the online application, applicants should submit a complete list of their organization's board of directors and their occupations; a copy of their specific line item budget with projected income and expenses; a copy of the organization's most recent operating budget; documentation to prove the organization's non-profit status.
Restrictions: The foundation does not make grants to individuals, except in the form of academic scholarships. Grants are generally given one-time only for specific purposes and will typically not be awarded to provide annual operating expenses or support. A grant will not be awarded to replenish funds previously expended. Grants are made with the understanding that the foundation has no obligation or commitment to provide additional support to the grantee. Grants may not be used for any political campaign, or to influence legislature of any government body other than through making available the results of nonpartisan analysis, study, and research.
Geographic Focus: Indiana
Date(s) Application is Due: Aug 1
Contact: Laura Lemings, (260) 463-4363; fax (260) 463-4856; llemings@lccf.net
Internet: http://www.lccf.net/grants.html
Sponsor: LaGrange County Community Foundation
109 E Central Avenue, Suite 3
LaGrange, IN 46761

LaGrange County Community Foundation Scholarships **2653**
The LaGrange County, Indiana Community Foundation has developed a scholarship program that offers significant financial support for a wide range of educational and career interests. The Foundation offers a variety of 30 separately funded new scholarships, with several offering multiple awards.
Requirements: Eligibility, criteria, and supplemental requirements vary for each scholarship. Scholarships are categorized by area schools and also include scholarships for non-traditional and home schooled students. See website for detailed information on each scholarship, as well as applications. Additional information can also be obtained by contacting the foundation office.
Geographic Focus: Indiana
Contact: Laura Lemings, Executive Director; (260) 463-4363; fax (260) 463-4856; llemings@lccf.net
Internet: http://www.lccf.net/scholarships.html
Sponsor: LaGrange County Community Foundation
109 E Central Avenue, Suite 3
LaGrange, IN 46761

LaGrange County Lilly Endowment Community Scholarship **2654**
The LaGrange County Lilly Endowment Community Scholarship may be used for any area of study and offers full tuition and books in return for a commitment to four years of full-time baccalaureate study at any accredited Indiana college or university. Contact the Foundation office for specific information about the scholarship, including application and deadline.
Requirements: Applicants must have been accepted at an Indiana four-year college or university; taken the SAT and have those scores recorded on their official high school transcript; participated in community service confirmed by an adult sponsor or supervisor.
Restrictions: The scholarship does not include room and board.
Geographic Focus: Indiana
Contact: Laura Lemings, Executive Director; (260) 463-4363; fax (260) 463-4856; lccf@lccf.net
Internet: http://www.lccf.net/scholarships.html
Sponsor: LaGrange County Community Foundation
109 E Central Avenue, Suite 3
LaGrange, IN 46761

Lake County Athletic Officials Association Scholarships **2655**
This scholarship was established to award funding to Lake County Athletic Official Association members and their children who are attending any accredited four year institution. The number of scholarships varies, but each scholarship is set at $500. All college majors are eligible. Contact Legacy Foundation about current applications.

Geographic Focus: Indiana
Date(s) Application is Due: Mar 12
Amount of Grant: 500 U.S.D
Contact: Athletic Officials Association Scholarship Contact; (219) 736-1880; fax (219) 736-1940; legacy@legacyfoundationlakeco.org
Internet: http://www.legacyfoundationlakeco.org/scholarships/index.htm
Sponsor: Legacy Foundation
1000 East 80th Place, 302 South
Merrillville, IN 46410

Lake County Community Fund Grants **2656**
The Lake County Community Grant was established to be responsive to community projects throughout Lake County in the areas of arts and culture, civic affairs, community development, education, the environment, health, human services, and youth services. The Foundation will make grants to non-profit organizations implementing projects with the most potential to improve the quality of life of a substantial number of residents of Lake County. Geographic distribution may be considered in awarding grants. Grants typically range from $1,000 to $25,000.
Requirements: Funding priorities include projects that: develop or test new solutions to community problems; address prevention as well as remediation; assist underserved community resources; provide a sustained effect for a substantial number of residents; improve the efficiency of non-profit groups; provide a favorable ratio between the amount of money requested and number of people served; facilitate collaboration among organizations without duplicating services; encourage volunteerism, civic engagement, and development. Applicants must submit the online detailed application, along with the grant narrative, project budget, a list of applicant's board of directors, summary of the organization's current fiscal year operating budget as well as financial audit or review; evidence of Board approval of this application, copy of tax exempt status, and organization's profile on GuideStar (www.guidestar).
Restrictions: The Legacy Foundation does not support: general operating expenses; endowment campaigns, annual campaigns, or fundraising events; travel grants; grants for individual schools or sponsorship of sports teams; previously incurred debt or retroactive funding for current projects; individuals and independent scholarly research projects; and religious or sectarian programs, political parties, or campaigns.
Geographic Focus: Indiana
Date(s) Application is Due: Mar 1; May 1; Sep 1; Nov 1
Amount of Grant: 1,000 - 25,000 U.S.D
Contact: Barry Tyler, Jr., Community Initiatives Officer; (219) 736-1880; fax (219) 736-1940; legacy@legacyfoundationlakeco.org or btyler@legacyfdn.org
Internet: http://www.legacyfoundationlakeco.org/grantsfundingopps.html
Sponsor: Legacy Foundation
1000 East 80th Place, 302 South
Merrillville, IN 46410

Lake County Lilly Endowment Community Scholarships **2657**
The Lilly Endowment Community Scholarships were designed to help raise the level of educational attainment in Indiana and to leverage further the ability of Indiana's community foundations to enhance the quality of life of the state's residents. Lilly Endowment Community Scholars will receive scholarships for full tuition, required fees, and a special allocation of up to $800 per year for required books and required equipment for four years of undergraduate study on a full-time basis, leading to a baccalaureate degree at any Indiana public or private college or university accredited by the North Central Association of Colleges and Schools. All college majors are eligible to apply.
Requirements: Applicants: must be U.S. citizens, high school seniors, and residents of Lake County, Indiana; must apply for a full time four-year course of study at an accredited public or private college or university in Indiana; must be available for an interview that will be scheduled for mid-February. Applicants should consult the website and their high school guidance office for the detailed procedures of the application process.
Restrictions: These scholarships do not cover room and board expenses.
Geographic Focus: Indiana
Date(s) Application is Due: Jan 5
Contact: Lilly Endowment Scholarships Contact; (219) 736-1880; fax (219) 736-1940; legacy@legacyfoundationlakeco.org
Internet: http://www.legacyfoundationlakeco.org/scholarshipslilly.html
Sponsor: Legacy Foundation
1000 East 80th Place, 302 South
Merrillville, IN 46410

Lalor Foundation Anna Lalor Burdick Program Grants **2658**
The program seeks to educate young women about human reproduction in order to broaden and enhance their options in life. Preference will be given to programs that focus on young women who have inadequate access to information regarding reproductive health, including the subjects of contraception and pregnancy termination. The program emphasizes support for projects and initiatives that demonstrate realistic plans to achieve greater financial self-sufficiency; new or smaller organizations, including grassroots efforts; collaborative efforts among nonprofit organizations; organizations that can demonstrate a proven ability to reach out to, include, and involve young women with inadequate access to information regarding reproductive health; and new ideas, initiatives, and demonstration projects, which, if proven effective, may be successfully replicated or provide multiple benefits. Normally grants are awarded for one year only. Under special circumstances, renewals are considered. Guidelines are available online.
Requirements: 501(c)3 tax-exempt organizations not classified as private foundations under section 509(a) of the IRS code are eligible. Applicants not located in the United States and that have not already been classified by the U.S. Internal Revenue Service should contact the office.
Restrictions: Grants are not normally made to full proposals submitted in advance of a concept paper that has been reviewed and approved by the trustees for further consideration; individuals, or for individual research projects and scholarships; requests for endowment or major capital support; prior grantees that have failed to provide grant reports; or organizations with no track record or no personnel known to the trustees or to the staff at Grants Management Associates.
Date(s) Application is Due: May 15; Nov 15
Amount of Grant: 10,000 - 50,000 U.S.D
Samples: Hesperian Foundation (Berkeley, CA)— for revision and update of a book for midwives and un libro para parteras, $20,000; Planned Parenthood League of Massachusetts, Inc (Boston, MA)—for the Emergency Contraception Awareness and Access Campaign, $25,000; Self Reliance Foundation (Washington, DC)—for the Hispanic Emergency Contraception Awareness Campaign, $25,000; Whittier Street Health Ctr (Roxbury, MA)—to add reproductive health to the Whittier Youth Service and Enrichment Decisions Program (WYSE), including counseling on contraceptives and options, $25,000.
Contact: Pamela Labonte Maksy, (617) 426-7080, ext. 314; fax (617) 426-5441; pmaksy@grantsmanagement.com
Internet: http://www.lalorfound.org/tlfalbp.html
Sponsor: Lalor Foundation
77 Summer Street, 8th Floor
Boston, MA 02110-1006

Lalor Foundation Postdoctoral Research Grants **2659**
Grants are awarded to institutions for basic postdoctoral research in mammalian reproductive physiology and biochemistry yielding improved methods of contraception, termination of pregnancy, and/or sterilization. Funds may be used to cover fellowship stipends, institution overhead, and laboratory and miscellaneous expenses. Less money is awarded for projects requiring less than one year.
Requirements: Applications must come from tax-exempt institutions; individuals nominated must have training and experience at least equal to the PhD, or MD level. People who have held the doctoral degree less than five years are preferred.
Date(s) Application is Due: Jan 13
Amount of Grant: Up to 35,000 U.S.D
Contact: Pamela Labonte Maksy, (617) 426-7080, ext. 314; fax (617) 426-5441; pmaksy@grantsmanagement.com
Internet: http://www.lalorfound.org/guidelines.html
Sponsor: Lalor Foundation
77 Summer Street, 8th Floor
Boston, MA 02110-1006

LAM Foundation Established Investigator Awards **2660**
The Established Investigator Awards support research directly related to lymphangioleiomyomatosis (LAM) and is targeted at established investigators who focus on the abnormal proliferation of smooth muscle that occurs in the disease. Awards provide a maximum of $50,000 per year, renewable for up to two additional years. The structure and terms of this award are identical to the LAM Fellowship Award except that with the LAM Established Award, faculty level investigators are eligible to receive funding for technician support and supplies.
Requirements: U.S. citizens and foreign nationals with appropriate immigrant visa status are eligible. At the time of application, an applicant must hold a doctoral degree and faculty appointment at the level of assistant or associate professor and be undertaking a project related to LAM.
Date(s) Application is Due: Sep 30
Contact: Francis X. McCormack, Director; (513) 558-4831; fax (513) 558-4858; frank.mccormack@uc.edu
Internet: http://www.thelamfoundation.org/research/apply-for-lam-funding
Sponsor: LAM Foundation
231 Albert Sabin Way, MSB Room 6053
Cincinnati, OH 45267-0564

LAM Foundation Pilot Project Grants **2661**
Pilot project awards are granted for one year to enable investigators to gather sufficient preliminary data for more substantial LAM funding. Funding is intended to support faculty-level investigators, or postdoctoral-level investigators who will be performing LAM research in the laboratory of an established scientist who is an expert in areas that are directly pertinent to LAM. Letters of intent (LOI) are required prior to submission of grant proposals to the LAM Foundation. Application materials and details are available online.
Requirements: Applications are accepted from investigators of all nationalities who have MDs or PhDs, but candidates must possess visas that allow for completion of the proposed project in the original laboratory.
Date(s) Application is Due: Sep 30
Amount of Grant: Up to 25,000 U.S.D
Contact: Francis X. McCormack, Director; (513) 558-4831; fax (513) 558-4858; frank.mccormack@uc.edu
Internet: http://www.thelamfoundation.org/research/apply-for-lam-funding
Sponsor: LAM Foundation
231 Albert Sabin Way, MSB Room 6053
Cincinnati, OH 45267-0564

LAM Foundation Postoctoral Fellowships 2662

The foundation offers postdoctoral fellowships for the study of the cellular and molecular basis of the abnormal smooth muscle proliferation that occurs in the disease, Lymphangioleiomyomatosis (LAM). Fellowships are renewable for up to two additional years. The fellowship is intended to support postdoctoral level investigators who will be performing LAM research in the laboratory of an established scientist who is an expert in areas that are directly pertinent to LAM. Letters of intent (LOI) are required prior to submission of grant proposals to the LAM Foundation. Application materials and details are available online.

Requirements: Applications are accepted from investigators of all nationalities who have MDs or PhDs, but candidates must possess visas that allow for completion of the proposed project in the original laboratory.

Restrictions: The award is generally not available to individuals who have attained faculty rank, or who are seeking transition funds for the establishment of a laboratory in a new setting

Date(s) Application is Due: Sep 30

Amount of Grant: Up to 50,000 U.S.D

Contact: Francis X. McCormack, MD, Director; (513) 558-4831; fax (513) 558-4858; frank.mccormack@uc.edu

Internet: http://www.thelamfoundation.org/research/apply-for-lam-funding

Sponsor: LAM Foundation
231 Albert Sabin Way, MSB Room 6053
Cincinnati, OH 45267-0564

LAM Foundation Research Grants 2663

The program provides seed money to investigators seeking to understand the abnormal smooth muscle proliferation that occurs in lymphangioleiomyomatosis (LAM). The grant supports clinical, laboratory, epidemiological, or any other kind of research that directly relates to LAM. Grants are subject to annual review and may be granted for up to two years. Scientists who which to apply for LAM Foundation funding must submit a Letter of Intent (LOI). Letters of Intent must reach The LAM Foundation by July 30th.

Requirements: U.S. citizens and foreign nationals with appropriate immigrant visa status are eligible. At the time of application, an applicant must hold a doctoral degree and faculty appointment with an academic institution and have completed two years of research training.

Restrictions: Residents, interns, postdoctoral fellows, students enrolled in degree-granting programs, and established investigators are ineligible.

Date(s) Application is Due: Sep 30

Amount of Grant: Up to 40,000 U.S.D

Contact: Francis X. McCormack, Director; (513) 558-4831; fax (513) 558-4858; frank.mccormack@uc.edu

Internet: http://www.thelamfoundation.org/research/apply-for-lam-funding

Sponsor: LAM Foundation
231 Albert Sabin Way, MSB Room 6053
Cincinnati, OH 45267-0564

Lance Armstrong Foundation Community Grants 2664

The foundation awards grants through its community program to nonprofit organizations in support of community-centered, cancer survivorship initiatives that serve the needs identified by the National Action Plan for Cancer Survivorship. Two types of grants will be awarded. One-year planning grants will be considered in two areas: practical issues of cancer survivorship, and outreach to underserved populations. This grant will support the development of a design for the proposed project and a complete plan for its implementation (and evaluation). One-year evaluation grants, awarded in the amount of up to 10 percent of the total project budget, will be considered in five areas: cancer pain, palliative and end-of-life care; physical activity and nutrition; survivorship education; survivorship support; and practical issues of cancer survivorship. The applicant should already have a well-developed design for the project and a complete plan for both its implementation and its impact and effectiveness evaluation. This grant will support the evaluation of the project. The listed deadline is for receipt of applications, accepted only through the online application system. Only the first 100 applications submitted after 9:00 am on August, 29th, will be accepted. Guidelines are available online.

Requirements: Applicants must be U.S. nonprofit organizations, classified as other than a private foundation under the IRS code. Eligible programs include any programmatic project (such as service delivery or education) that fits the guidelines described in the current RFP. For evaluation grants, eligibility includes projects that are finished with implementation, in the midst of implementation, or about to begin implementation—with funding for implementation already secured.

Restrictions: Ineligible programs include nonprogrammatic projects (such as an advertising campaign or a fundraising event or campaign), and other projects that do not fit the guidelines described in the RFP. For evaluation grants, ineligibility includes projects that have not yet been implemented and do not already have funding secured for implementation.

Date(s) Application is Due: Sep 9

Samples: U of Alabama, School of Public Health (Birmingham, AL)—to provide support services to people with cancer and to cancer survivors, $150,000; New York U School of Medicine, NYU Cancer Institute (NY)—for education programs to help people deal with the emotional, physical, and practical challenges of cancer, $20,000.

Contact: Grants Administrator; (512) 236-8820; fax (512) 236-8482

Internet: http://www.livestrong.org/site/c.jvKZLbMRIsG/b.739141/k.9A5A/Community_Program_RFP__Fall.htm

Sponsor: Lance Armstrong Foundation
P.O. Box 161150
Austin, TX 78716-1150

Lance Armstrong Foundation Research Grants 2665

The foundation funds and disseminates research that builds on the body of knowledge and services focused on improving the quality of life for cancer survivors across physical, emotional, and practical challenges. The research program also supports research focused on testicular cancer. The foundation funds research that is not readily fundable from traditional sources and that encourages and supports the efforts of both young and established investigators. Young investigators are defined as within eight years of completing a terminal degree or within five years of initiating independent research within a mentored laboratory, while general research awards support new research projects initiated by established investigators. Proposals are accepted from all countries. The listed application deadline is for letters of intent; full RFPs are available online.

Date(s) Application is Due: May 13

Amount of Grant: Up to 75,000 U.S.D

Contact: Amy Nunn, Grants Coordinator; (512) 236-8820, ext. 181; amy.nunn@laf.org or research.grants@laf.org

Internet: http://www.livestrong.org/site/c.jvKZLbMRIsG/b.739079/k.9EDA/Research_Grants.htm

Sponsor: Lance Armstrong Foundation
P.O. Box 161150
Austin, TX 78716-1150

Land O'Lakes Foundation Mid-Atlantic Grants 2666

The Land O'Lakes Foundation Mid-Atlantic Grants were developed specifically for the company's dairy communities in Maryland, New Jersey, New York, Pennsylvania, and Virginia. The program works to improve quality of life by supporting worthy projects and charitable endeavors initiated by our Mid-Atlantic dairy-member leaders. Community organizations applying for grants may be eligible for donations of $500 to $5,000 for local projects and programs. Funds could be used to support such worthwhile projects as: backing local food pantries or emergency feeding efforts; aiding 4-H or FFA programs; building a new park pavilion for the community; establishing a local wetland preserve; or purchasing books for the community library. Application procedures are available online.

Requirements: Applications are initiated by Land O'Lakes farmer-members. An Area Procurement Specialist (APS) review is also part of this process. To be considered, grant proposals must demonstrate how the donation will be used to help improve community quality of life. Mid-Atlantic grants are generally restricted to organizations that have been granted tax-exempt status under Section 501(c)3 of the Internal Revenue Code. The Foundation awards grants to projects that best address the following areas: hunger relief; youth and education; rural leadership; civic improvements; soil and water preservation; and art and culture.

Restrictions: Grants will not be awarded for the following purposes: scholarship funds, gifts or fund raisers for individuals, or non-public religious use.

Geographic Focus: Maryland, New Jersey, New York, Pennsylvania, Virginia

Amount of Grant: 500 - 5,000 U.S.D

Contact: Martha Atkins-Sakry, Executive Assistant; (651) 481-2470 or (651) 481-2212; MLAtkins-Sakry@landolakes.com

Internet: http://www.landolakesinc.com/company/corporateresponsibility/foundation/midatlanticgrants/default.aspx

Sponsor: Land O'Lakes Foundation
P.O. Box 64101
St. Paul, MN 55164-0150

Landon Foundation-AACR Innovator Award for Cancer Prevention Research 2667

This Award was established and first given in 2008 to recognize the outstanding achievement of an early career Assistant Professor, and to provide support for cancer prevention research of significant scientific merit in any discipline across the continuum of research. The goals of the program are to: encourage younger investigators to pursue cancer prevention research of significant scientific merit; provide the support necessary to sustain and enhance highly meritorious cancer prevention research; foster interactions between and among cancer scientists and disseminate the scientific knowledge about cancer prevention research; and contribute to a global impact against cancer. A two-year grant of $100,000 ($50,000 per year) will be provided to the recipient's institution to support direct research expenses; salary and benefits; and attendance at the AACR Annual Meeting, Cancer Prevention Conference, or other AACR meeting relevant to cancer prevention for the purpose of participating in scholarly exchange.

Requirements: Candidates must be scientists at the level of Assistant Professor who completed postdoctoral studies or clinical fellowships on or after July 1. Candidates cannot be tenured or under consideration for tenured academic positions at the time of application. A candidate whose title is not Assistant Professor but who believes that he or she meets the eligibility requirements for this Award must attach a letter from the institution's dean certifying that the position held is equivalent to Assistant Professor. Both AACR members and nonmembers are eligible to apply. However, nonmembers must submit a satisfactory application for AACR Active membership on or before October 31.

Restrictions: Employees or subcontractors of a national government or private industry are not eligible for this grant.

Date(s) Application is Due: Nov 20; Nov 25

Amount of Grant: 100,000 U.S.D

Contact: Julia Laurence, Program Assistant; (215) 440-9300, ext. 102 or (267) 646-0655; fax (215) 440-9372; julia.laurence@aacr.org or awards@aacr.org

Internet: http://www.aacr.org/home/scientists/research-funding—fellowships/landon-aacr-innovator-award-for-cancer-prevention-research.aspx

Sponsor: American Association for Cancer Research
615 Chestnut Street, 17th Floor
Philadelphia, PA 19106-4404

Landon Foundation-AACR Innovator Award for International Collaboration 2668
The Award will support highly meritorious research that is being conducted collaboratively by investigators in different countries around the world. The goals of the program are to: promote international cancer research collaboration as an effective means to accelerate progress against cancer; provide the support necessary to sustain and enhance highly meritorious international cancer research collaborations; foster interactions between and among cancer scientists and disseminate the scientific knowledge gained from international collaboration; and contribute to a global impact against cancer. The Award of $100,000 will be provided to the recipients' institutions in the form of a two-year grant to support the collaborative research project, and may be used for such purposes including, but not limited to: direct research expenses attributable to the proposed research, which may include the salary of research assistants or technicians; expenses for travel to the institutions of the collaborators relevant to the research project; expenses for training in new techniques that will benefit the research project, such as training received through a scholar exchange/visiting professor program; and expenses related to the presentation of research data at scientific meetings or through other means that will contribute to the dissemination of the scientific knowledge gained.
Requirements: Two or more independent researchers who working within an established international cancer research collaboration involving institutes in multiple countries may apply. Proposals to initiate a collaborative project will not be accepted. The co-investigators may be affiliated with any institution involved in cancer research, cancer medicine, or cancer-related biomedical science anywhere in the world. There is no limit to the number of co-investigators within each of the collaborating institutes. There are no national or residency status restrictions. Applicants need not be members of the AACR. Applications must be completed online using the proposalCENTRAL website, with one paper copy submitted to the AACR office. Online applications must be completed by 12:00 noon (United States Eastern Daylight Time) on November 20. The paper copy with original signatures and all supporting documents must be postmarked no later than November 25.
Restrictions: Scientific investigators or health professionals who are funded by the tobacco industry for any project, or whose named mentors in the case of mentored grants are funded by the tobacco industry for any project, may not apply and will not be eligible for AACR grants.
Date(s) Application is Due: Nov 20; Nov 25
Amount of Grant: 100,000 U.S.D
Contact: Hanna Hopfinger; (267) 646-0665; fax (215) 440-9372; hanna.hopfinger@aacr.org
Internet: http://www.aacr.org/home/scientists/research-funding—fellowships/landon-aacr-innovator-award-for-international-collaboration.aspx
Sponsor: American Association for Cancer Research
615 Chestnut Street, 17th Floor
Philadelphia, PA 19106-4404

Lands' End Corporate Giving Program 2669
The Corporation awards grants to nonprofits for youth and family services programs in their area of company operations in Wisconsin. Areas of interest include education, community development, environment, and health and human services.
Requirements: Wisconsin nonprofits are eligible. Organizations should submit the following: a timetable for implementation and evaluation of the project; statement of the problem that the project will address; population and geographic area to be served; name, address and phone number of organization; copy of IRS determination letter; copy of most recent annual report/audited financial statement/990; how the project's results will be evaluated or measured; list of company employees involved with the organization; detailed description of project and amount of funding requested; contact person; copy of current year's organizational budget and/or project budget; and listing of additional sources and amount of support. Applicants should also include a description of their past involvement with Lands' End, if any.
Restrictions: The Foundation does not consider grants for organizations without nonprofit status; individuals; political organizations, campaigns, or candidates for public office; lobbying groups; advertising in programs, bulletins, yearbooks, or brochures; testimonial/awards dinners; endowments; loans; religious groups for religious purposes; pageants; purchasing of land; salaries; administrative costs; international programs; research programs; or general operating expenses.
Geographic Focus: Wisconsin
Date(s) Application is Due: Mar 31; Jun 30; Sep 30; Dec 31
Contact: Jessica Winzenried, Corporate Giving Manager; (608) 935-6776 or (608) 935-6728; fax (608) 935-6432; donate@landsend.com
Sponsor: Lands' End
2 Lands' End Lane
Dodgeville, WI 53595

Larry L. Hillblom Foundation Grants 2670
The mission of the foundation is to provide philanthropic support exclusively for charitable, religious, scientific, literary, and educational purposes. Primarily, the foundation will support medical research, including basic scientific research, clinical research, and research related to patient self-care and management. Medical research priorities include the areas of diabetes mellitus, diseases associated with aging, and children's physical and neurological diseases. Specific grants include building/institute grants, start-up research grants, network grants, fellowships, endowed chairs, and clinical research grants.
Date(s) Application is Due: Jan 14
Amount of Grant: 250,000 - 10,000,000 U.S.D
Samples: U of California (Oakland, CA)—to endow a chair in geriatric medicine at the U of California, San Diego, in La Jolla, $2 million; U of California at San Francisco (CA)—to create a center for research on the basic biology of aging, and to support research on diabetes, eye disorders, and neurodegenerative diseases, $12 million.
Contact: Grants Administrator; (707) 762-6691; fax (707) 762-6694; som@llhf.org
Internet: http://www.llhf.org
Sponsor: Larry L. Hillblom Foundation
755 Baywood Drive, Suite 180
Petaluma, CA 94954

Laura B. Vogler Foundation Grants 2671
The Vogler Foundation supports innovative programs and projects in New York City and Long Island in the areas of education, health care, and social services. The Foundation is particularly interested in organizations that serve and support children, the elderly, and the disadvantaged. Types of support include general operating support, program development, research grants, and seed money grants. Grants provide one-time, nonrenewable support. The application is available on the website.
Requirements: Nonprofit organizations in New York City and Long Island, NY, may submit proposals.
Restrictions: Grants are not awarded to support building or endowment funds, annual fund-raising campaigns, or matching gifts. Requests for funds for conferences, seminars, or loans are not accepted.
Geographic Focus: New York
Date(s) Application is Due: Mar 1; Jul 1; Nov 1
Amount of Grant: 1,000 - 3,500 U.S.D
Samples: Brooklyn Children's Museum Corporation, Brooklyn, New York, an after-school program, $3,500 (2011); New Life Church, Staten Island, New York, computer laboratory for young people, $3,000 (2011).
Contact: Lawrence D'Amato, President; (718) 423-3000; fax (718) 631-4808
Internet: http://fdncenter.org/grantmaker/vogler
Sponsor: Laura B. Vogler Foundation
P.O. Box 610508
Bayside, NY 11361-0508

Laura J. Niles Foundation Grants 2672
The foundation's mission is to encourage and support efforts that offer learning and economic growth opportunities for the motivated poor. Of equal importance are charitable initiatives that foster life enrichment through canine and other types of animal companionship. The foundation has a particular interest in education, economic self-sufficiency, and programs that alleviate unhealthy dependencies. With regard to animals, most notably dogs, the foundation's areas of concentration center around canine health research, animal protection and adoption, search and rescue training, human assistance, and similar fields of interest. A most favorable outcome is one where both people and animals benefit simultaneously from an initiative. Applications are accepted throughout the year. Guidelines are available online.
Requirements: 501(c)3 nonprofit organizations are eligible. The majority of grants are awarded in the northeastern United States, although, occasionally, grants may be made in other regions of the country and/or abroad.
Restrictions: Grants do not support general fund-raising drives, endowments, government agencies, or organizations that subsist on third-party funding.
Amount of Grant: 2,000 - 50,000 U.S.D
Contact: Rose Anna Miller, (203) 629-7314; rmiller@fcsn.com
Internet: http://www.ljniles.org/index.htm
Sponsor: Laura J. Niles Foundation
88 Field Point Road
Greenwich, CT 06830

Laura Moore Cunningham Foundation Grants 2673
The Laura Moore Cunningham Foundation is dedicated to advancing the State of Idaho. Priorities include rural healthcare, educational programs for children, programs in underserved communities, and programs for underserved populations. Each year the Foundation accepts applications from throughout the State, allowing organizations of all types to express their need. The Foundation is interested in organizations that run in a cost-effective manner, serving large numbers of people who are truly in need.
Requirements: Eligible applicants must be 501(c)3 Idaho organizations. The Foundation does not limit giving to a certain type of program or need; however administrative costs are not preferred.
Restrictions: Individuals are ineligible.
Geographic Focus: Idaho
Date(s) Application is Due: May 15
Samples: The Peregrine Fund, Boise, Idaho, $100,000 - education programs (2010); Teton Valley Health Care Foundation, Driggs, Idaho, $30,000 - boiler upgrade (2010); and Jerome Public Library, Jerome, Idaho, $4,000 - new library books (2010).
Contact: Harry L. Bettis, President; (208) 472-4066; lmcf_idaho@msn.com
Internet: http://lauramoorecunningham.org/Applying_for_Grants.html
Sponsor: Laura Moore Cunningham Foundation
P.O. Box 1157
Boise, ID 83701

Lavelle Fund for the Blind Grants 2674
The fund is dedicated to supporting programs that promote the spiritual, moral, intellectual, and physical development of blind and low-vision people of all ages, together with programs that help people avoid vision loss. Priority is given to agencies that concentrate on serving the New York City metropolitan area. The fund gives preference to program creation, expansion, or improvement. (General operating support is rarely

provided.) Capital support is generally awarded only in connection with specific program support needs. Application guidelines are available online.
Requirements: 501(c)3 tax-exempt organizations are eligible.
Restrictions: The fund does not make grants to: individuals; medical research programs; conferences or media events (unless they are an integral part of a broader program of direct service); efforts to influence legislation or elections; or organizations seeking support for deficit reduction or emergency funding.
Samples: Cancer Care (New York, NY)—to create a financial-assistance and support program for people with ocular cancers, $122,750 over 18 months; Catholic Near East Welfare Assoc (New York, NY)—for projects at five Palestinian schools for the blind or at vision-rehabilitation groups in the Jerusalem area and the West Bank and Gaza, $209,316 over three years; International Trachoma Initiative (New York, NY)—for a program that will provide sight-saving surgeries in Mali and Niger, two countries with extremely high rates of trachoma, $300,000 Over two years; Recording for the Blind and Dyslexic (Princeton, NJ)—to create 300 digitally recorded textbooks on mathematics, science, and technology, $320,000 over two years.
Contact: Andrew Fisher, Executive Director; (212) 668-9801; fax (212) 668-9803; afisher@lavellefund.org
Internet: http://fdncenter.org/grantmaker/lavellefund
Sponsor: Lavelle Fund for the Blind
80 Maiden Lane, Suite 1207
New York, NY 10038

Lawrence Foundation Grants **2675**

The foundation awards grants to nonprofit organizations that address the following areas of interest: education (U.S. headquartered organizations for programs in the United States), environment (U.S. headquartered organizations operating programs in the United States or elsewhere in the world), health (U.S. headquartered organizations operating programs in the United States with national scope or specific programs in the Los Angeles area), disaster relief (U.S. headquartered organizations responding to disasters in the United States or elsewhere in the world on an occasional basis), and other topics (U.S. headquartered organizations operating programs in the United States or elsewhere in the world). Grant amounts vary. The duration of grants will be typically for a single year.
Requirements: Nonprofit organizations that qualify for 501(c)3 public charity status, public schools and libraries are eligible for contributions or grants.
Restrictions: For-profit businesses are not eligible. The foundation does not typically make grants for the following purposes: computers or software; audio or video equipment; designing and producing videos, kiosks, or promotional material; music programs or musical instruments; gardening programs or equipment; physical education programs or equipment; recreational programs; theater arts or performance arts programs or equipment; hospice or old age home programs; religious organizations; religious or charter schools; individuals for any purpose; international organizations that do not have a qualified domestic 501(c)3 representative; political lobbying activities or other political purposes; private foundations; dinners, balls, or other ticketed events; or purposes outside of the foundation's funding priorities.
Geographic Focus: All States, All Countries
Date(s) Application is Due: May 31; Nov 1
Amount of Grant: 500 - 100,000 U.S.D
Contact: Lori Mitchell, Executive Director; (310) 451-1567; fax (310) 451-7580; info@thelawrencefoundation.org
Internet: http://www.thelawrencefoundation.org/grant/index.html
Sponsor: Lawrence Foundation
530 Wilshire Boulevard, Suite 207
Santa Monica, CA 90401

Lawrence S. Huntington Fund Grants **2676**

The Foundation, established in 1997 by its namesake, Lawrence J. Huntington, is dedicated to supporting a historical seaport, as well as for environmental conservation, hospitals and human service programs. Mr. Huntington, who serves as the Chairman of St. Luke's Roosevelt Hospital, Chairman of the Woods Hole Research Center, and Vice Chairman of the Board of the South Street Seaport Museum, is primarily interested in preserving the environment and historical landmarks. There are no specific applications or deadlines with which to adhere, and applicants should approach the Foundation in writing.
Requirements: Giving is primarily restricted to the states of New York and Massachusetts, as well as the District of Columbia.
Geographic Focus: New York
Amount of Grant: Up to 80,000 U.S.D
Samples: Woods Hole Research Center, Falmouth, Massachusetts, $78,000; New York Law School, New York, New York, $15,000.
Contact: Lawrence S. Huntington, Director; (212) 717-8633 or (212) 486-2424
Sponsor: Lawrence S. Huntington Fund
46 E 70th Street, 4th Floor
New York, NY 10021-4928

Layne Beachley Aim For The Stars Foundation Grants **2677**

The Layne Beachley Aim for the Stars Foundation was created to inspire girls and women across Australia to dream and achieve. The Foundation hopes to prevent girls and women alike from having to go through adversity and to encourage, motivate and provide for all aspiring women. Encompassing academic, sport, cultural and community pursuits, Aim for the Stars offers ambitious and dedicated females an opportunity to receive financial and moral support to help them achieve their goals. Support will give them the opportunity to maintain a determined focus on their goal, to achieve their dreams earlier in life and allow them to further their ambitions and aim for the stars. Grants will be awarded, at the discretion of the Foundation, to deserving applicants who meet the criteria and best demonstrate their aspirations to further their educational and/or personal development. Females who aim to achieve their ultimate goal in their field of choice, 12 years of age and above, are eligible to apply.
Requirements: Applicants must be citizens of Australia, a minimum of 12 years old, and female.
Geographic Focus: All States, Australia
Date(s) Application is Due: Nov 15
Amount of Grant: 1,500 - 6,000 U.S.D
Samples: Gabrielle King, Maclean NSW—Sailing; Tayla Walsh—Waterpolo; Nicole Parks, Jindabyne NSW—Mogul Skiing; Anabelle Smith—Diving; Caitlin Gilliman—Soccer; Georgia Gibbs, Dardanup, WA—Rowing; Jessica Brown—Ballet; Samantha Balaton-Chrimes, Wantirna Sth, Victoria—African Studies; Emma—Speed Skating; Julie—Speed Skating; Tatiana—Speed Skating; Melissa Caldecutt—Speed Skating; Bianca Hertel—Rodeo Riding; Bernadette Branca, Meadow Hts, Victoria—Taekwondo.
Contact: Program Coordinator; (042) 260-0733; info@aimforthestars.com.au
Internet: http://www.aimforthestars.com.au/WhoCanApply/ApplicationChecklist/tabid/1854/language/en-U.S./Default.aspx
Sponsor: Layne Beachley Aim For the Stars Foundation
P.O. Box 666
Forestville, NSW 2087 Australia

Leadership in Education in Neurodevelopmental Disabilities Postdoctoral Fellowships **2678**

Munroe-Meyer Institute, an administrative unit of the Nebraska Medical Center, offers postdoctoral fellowships supported by the Maternal and Child Health Bureau of the Health Resources and Services Administration. Fellowships are available in administration, nursing, nutrition, occupational therapy, pediatrics, physical therapy, psychology, social work, and speech language pathology. The program provides interdisciplinary training for careers and leadership roles in developmental disabilities and children with special needs. The applicant is encouraged to contact the home department director prior to submitting materials to determine if applicant interest is in line with department mission, goals, and objectives. The recruitment/selection committee meets every two weeks, beginning in January, to review materials. Applicants offered an interview will be scheduled to provide a scholarly presentation to the personnel and students within the institute.
Amount of Grant: 18,600 U.S.D
Contact: Patricia Gromak, Munroe-Meyer Institute Contact; (800) 656-3937 or (402) 559-6402; fax (402) 559-5737
Internet: http://www.unmc.edu/dept/mmi/index.cfm?L2_ID=60&L1_ID=17&CONREF=14
Sponsor: University of Nebraska Medical Center
985450 Nebraska Medical Center
Omaha, NE 68198-5450

Leahi Fund **2679**

The purpose of the Fund is to support programs of research and education in, and the prevention of, pulmonary disease. Highest priority will be given to: research and education with foreseeable clinical application to pulmonary problems that exist in Hawai'i; scientific research in the prevention or treatment of pulmonary diseases and infirmities. Preference will be given to supporting new investigators and to new areas of investigation for established researchers. Application information is available online.
Requirements: Research grants can only be approved for individuals affiliated with tax-exempt organizations classified by the Internal Revenue Services as 501(c)3 type, or with units of government.
Restrictions: Funding is not available for: salary for the principal investigator; patient care; travel; major equipment purchases; printing or publication costs.
Geographic Focus: Hawaii
Date(s) Application is Due: Jun 21
Amount of Grant: 30,000 U.S.D
Contact: Christel Wuerfel, Philanthropic Services Assistant; (808) 537-6333 or (888) 731-3863; cwuerfel@hcf-hawaii.org
Internet: http://www.hawaiicommunityfoundation.org/index.php?id=71&categoryID=23
Sponsor: Hawai'i Community Foundation
1164 Bishop Street, Suite 800
Honolulu, HI 96813

Learning and Behavioral Disorders Research Grants **2680**

The foundation's purpose is to provide funds for research and dissemination of research information which targets the proving, diagnosing, or treating of biologically-based disordered behavior. Areas of special interest include nutrition, food and chemical intolerance, neurometrics, neurochemistry, and genetics. Studies ascertaining validity of innovative biomedical therapies are especially encouraged. Initial contact from parties interested in funding an appropriate research project should be by letter which briefly explains the proposed project and states approximate cost. A researcher need not have a doctoral degree to apply.
Amount of Grant: 1,000 - 50,000 U.S.D
Contact: John Wacker, President; (214) 373-3308; fax (214) 373-3308
Sponsor: Wacker Foundation
10848 Strait Lane
Dallas, TX 75229

LEGENDS Scholarship 2681

The LEGENDS scholarship celebrates the individual who overcomes challenges and obstacles in the pursuit of a college education and lifelong learning. The LEGENDS Scholar must demonstrate one or more of the virtues of Exploration, Courage, Creativity and/or Innovation. The recipient will receive a one-time amount of $1,000 toward higher education and his or her name will be displayed at the Lake County Convention and Visitor's Bureau as a LEGENDS Scholar. All college majors are eligible.
Requirements: The applicant must be a current student in good standing, have completed a minimum of 24 credit hours, and intends to be enrolled in an undergraduate or graduate degree program at one of the following institutions for the coming academic year: Calumet College of St. Joseph, Indiana University Northwest, Ivy Tech Community College of Indiana Northwest, Purdue University Calumet, Purdue University North Central, or Valparaiso University. Applicants must submit: an application form; two recommendation letters; and an essay that addresses how exploration, courage, creativity, and innovation have enabled them to overcome obstacles in pursuing a college education, then how they will use these values to improve the quality of life in northwest Indiana.
Geographic Focus: Indiana
Date(s) Application is Due: Mar 15
Amount of Grant: 1,000 U.S.D
Contact: LEGENDS Scholarship Contact; (219) 736-1880; fax (219) 736-1940; legacy@legacyfoundationlakeco.org
Internet: http://www.legacyfoundationlakeco.org/scholarships/index.htm
Sponsor: Legacy Foundation
1000 East 80th Place, 302 South
Merrillville, IN 46410

Legler Benbough Foundation Grants 2682

The mission of the Foundation is to improve the quality of life of the people in the City of San Diego. To accomplish that mission, the Foundation focuses on three target areas: providing economic opportunity; enhancing cultural opportunity; and providing a focus for health, education and welfare funding. Interested applicants may submit an initial letter requesting funds. When the Foundation wishes to pursue the request, it will provide an application form to the applicant. Initial letters preceding applications are due by February 15 and August 15. Applications (for invited applicants) are due by March 15 and September 15.
Requirements: Funding focuses on activities in support of San Diego city arts and cultural institutions, scientific or research organizations, and health, education and welfare programs.
Restrictions: Awards are made only in the Foundation's focus areas. The following are not funded: capital projects unless there is a special situation where capital expenditure is the best way to achieve a stated objective; awards to individuals; awards for special events, fundraising or recognition events; and projects in the area of the homeless, AIDS, alcohol or drug rehabilitation or treatment and seniors.
Geographic Focus: California
Contact: Peter Ellsworth, President; (619) 235-8099; fax (619) 235-8077; peter@benboughfoundation.org
Thomas E. Cisco; (619) 235-8099; fax (619) 235-8077; thomas@benboughfoundation.org
Internet: http://www.benboughfoundation.org/criteria.php
Sponsor: Legler Benbough Foundation
2550 5th Avenue, Suite 132
San Diego, CA 92103-6622

Leland Fikes Foundation Grants 2683

Grants are awarded for medical research, mental health programs, social welfare, education, reproductive health care, and arts and culture. Building programs and equipment acquisition are supported in these areas. Support is rarely given to programs located outside of the Dallas area. Additional types of support include general operating support, continuing support, annual campaigns, capital campaigns, endowment funds, emergency funds, program development, professorships, seed money grants, scholarship funds, research, and matching funds. Contact the foundation for guidelines.
Requirements: Giving is primarily to Texas organizations.
Restrictions: Grants are not awarded to individuals.
Geographic Focus: Texas
Amount of Grant: 5,000 - 75,000 U.S.D
Samples: Saint Marks School of Texas (Dallas, TX)—for capital campaign, $200,000; Texas Partnership for End-of-Life Care (Austin, TX)—for general support, $175,000; Planned Parenthood (TX)—for operating support.
Contact: Nancy Solana, (214) 754-0144
Sponsor: Leland Fikes Foundation
3050 Lincoln Plaza, 500 N Akard Street
Dallas, TX 75201

Leland McSpadden Memorial Fund for Medical and Scientific Research Grants 2684

The fund invites research proposals aimed at achieving a better understanding of the metabolic basis of Lowe syndrome and/or developing a better understanding and better treatments for the major complications of Lowe syndrome, a genetic disorder that affects boys. Grants are awarded for one year with possible renewal for a second year.
Requirements: The program invites applications from researchers of all types, including but not limited to universities, hospitals, and other nonprofit research organizations.
Restrictions: The grant may not be used as salary for the principal investigator, for honoraria for consultants, for the purchase of standard, nonconsumable laboratory equipment, to purchase services from other research laboratories, or for overhead or indirect costs.
Date(s) Application is Due: Oct 15
Amount of Grant: 25,000 U.S.D
Contact: Mary Tietz, President; (612) 770-8438; info@lowesyndrome.org
Internet: http://www.lowesyndrome.org/Research/index.html
Sponsor: Lowe Syndrome Association
18919 Voss Road
Dallas, TX 75287

Lemelson-MIT Lifetime Achievement Award 2685

The award recognizes outstanding individuals whose pioneering spirit and demonstrated inventiveness throughout their careers has improved our society and inspired others. Candidates are living role models in the fields of science, engineering, medicine, and technology. They are U.S. citizens or permanent residents. Candidates are suggested to the program by the inventor community and other national leaders in science and technology. The winner is selected by the Lemelson-MIT board and notified via telephone by the director. Annual deadline dates may vary; contact program staff for exact dates.
Amount of Grant: 100,000 U.S.D
Samples: Dr. Ruth Rogan Benerito—for her patented processes in textiles for easy-care cotton and her lifelong commitment to education.
Contact: Lemelson-MIT Program Office; (617) 253-3352 or (617) 253-3490; idudek@mit.edu
Internet: http://web.mit.edu/invent/a-lifetime.html
Sponsor: Massachusetts Institute of Technology
77 Massachusetts Avenue, Room E60-215
Cambridge, MA 02139-4307

Lena Benas Memorial Fund Grants 2686

The Lena Benas Memorial Fund was established in 1986 to provide for the health, human services, and housing needs of underserved people living in Litchfield, Connecticut, and its surrounding communities. The deadline for application to the Fund is November 1, and applicants will be notified of grant decisions by letter within 2 months after the proposal deadline. Special consideration is given to organizations that provide housing maintenance and human services programming to needy populations.
Requirements: Preference is given to organizations serving the people of Litchfield, Connecticut. Organizations serving the towns contiguous to Litchfield, including Cornwall, Goshen, Harwinton, Morris, Thomaston, Torrington, Warren, Washington, and Watertown will also be considered.
Restrictions: Grants from the Benas Memorial Fund are 1 year in duration.
Geographic Focus: Connecticut
Date(s) Application is Due: Nov 1
Contact: Kate Kerchaert; (860) 657-7016; kate.kerchaert@baml.com
Internet: https://www.bankofamerica.com/philanthropic/fn_search.action
Sponsor: Lena Benas Memorial Fund
200 Glastonbury Boulevard, Suite #200, CT2-545-02-05
Glastonbury, CT 06033-4056

Leo Goodwin Foundation Grants 2687

The Goodwin Foundation awards grants in the areas of children, youth, and health organizations. The Foundation gives some funding for the arts and education.
Requirements: Florida nonprofit organizations, primarily in Broward County, are eligible. Applicants should submit the following: population and geographic area to be served; copy of IRS determination letter; copy of most recent annual report/audited financial statement/990; listing of board of directors, trustees, officers and other key people and their affiliations; detailed description of project and amount of funding requested; copy of current year's organizational budget and/or project budget; and a listing of additional sources and amount of support.
Restrictions: Individuals are not eligible.
Geographic Focus: Florida
Date(s) Application is Due: Oct 31
Contact: Helen Furia, President; (954) 772-6863
Sponsor: Leo Goodwin Foundation
800 Corporate Drive, Suite 500
Fort Lauderdale, FL 33334-3621

Leonard and Helen R. Stulman Charitable Foundation Grants 2688

The Leonard and Helen R. Stulman Charitable Foundation supports work in four areas of interest. Ninety five percent (95%) of funding from The Foundation is reserved for programs in greater Baltimore and the State of Maryland. The foundation will make a limited number of new grants each year to projects that address one or more of the following: research as to causes and treatment of mental illness and services for people with mental illness and their families; initiatives for an older adult and aging population and programs for geriatric medicine; programs engaged in health care treatment and prevention as well as those involved in medical research; and higher educational institutions. To apply to the Foundation for funding, applicants should submit a two-page letter of inquiry, with basic background on your organization, the identified needs your project proposes to address, an overview of the proposed project, and the amount you intend to request.
Requirements: Organizations (or their fiscal agents) that qualify as public charities under section 501(c)3 of the Internal Revenue Code and do not discriminate on the basis of race, creed, national origin, color, physical handicap, gender or sexual orientation are eligible to apply.
Amount of Grant: 50,000 U.S.D
Contact: Cathy Brill; (410) 332-4171; fax (410) 837-4701; cathybrill@comcast.net

Internet: http://www.bcf.org/ourgrants/ourgrantsdetail.aspx?grid=4
Sponsor: Leonard and Helen R. Stulman Charitable Foundation
2 East Read Street, 9th Floor
Baltimore, MD 21202

Leo Niessen Jr., Charitable Trust Grants 2689

Leo Niessen lived in Abington Township, Montgomery County, Pennsylvania. In 1993, his Foundation was funded from a testamentary bequest. He was a charitable man, who also made substantial philanthropic gifts to Holy Redeemer Hospital during his lifetime. He had a special affinity for Red Cloud Indian School of Pine Ridge, South Dakota. To this day, the Co-trustees of his Foundation continue to support this school, as well as the Hospital and The Society for the Propagation of the Faith. All grants are made in the memory of Leo Niessen and his family. The Foundation also supports organizations: that provide health services for all ages; which educate the needy and educable at all academic levels, without regard to age; working for and on behalf of youth and the elderly, and which provide assistance to the homeless and economically disadvantaged; which provide spiritual and emotional guidance. Application Deadlines are, January 31 and July 31. Application forms are available online. Applicants will receive notice acknowledging receipt of the grant request, and subsequently be notified of the grant declination or approval.
Requirements: Pennsylvania 501(c)3 nonprofit organizations are eligible to apply. Proposals should be submitted in the following format: completed Common Grant Application Form; an original Proposal Statement; an audited financial report and a current year operating budget; a copy of your official IRS Letter with your tax determination; a listing of your Board of Directors. Proposal Statements (second item in the above Format) should answer these questions: what are the objectives and expected outcomes of this program/project/request; what strategies will be used to accomplish your objective; what is the timeline for completion; if this is part of an on-going program, how long has it been in operation; what criteria will you use to measure success; if the request is not fully funded, what other sources can you engage; an Itemized budget should be included; please describe any collaborative ventures. Prior to the distribution of funds, all approved grantees must sign and return a Grant Agreement Form, stating that the funds will be used for the purpose intended. Progress reports and Completion reports must also be filed as required for your specific grant. All current grantees must be in good standing with required documentation prior to submitting new proposals to any foundation.
Restrictions: Grants are not made for political purposes, nor to organizations which discriminate on the basis of race, ethnic origin, sexual or religious preference, age or gender. The Niessen Foundation normally does not consider grants for endowment.
Geographic Focus: Pennsylvania
Date(s) Application is Due: Jan 31; Jul 31
Amount of Grant: 10,000 - 60,000 U.S.D
Contact: Wachovia Bank, N.A., Trustee; grantinquiries3@wachovia.com
Internet: https://www.wachovia.com/foundation/v/index.jsp?vgnextoid=345852199c0aa110VgnVCM1000004b0d1872RCRD&vgnextfmt=default
Sponsor: Leo Niessen Jr., Charitable Trust
Wachovia Bank, N A. PA 1279, 1234 East Broad Street
Philadelphia, PA 19109-1199

Lesbian Health Fund Grants 2690

The fund supports research into lesbian health, including research with the following goals: determination or rates and risk factors for cancers and other diseases among lesbians; issues of access to medical care for lesbians; study of mental health concerns of lesbians and their families; definition and investigation of lesbian family issues; and methodologic exploration of diversity in the lesbian population. Recently funded projects have included the first-ever longitudinal study of lesbian mothers; a survey of lesbian health curricula in U.S. internal medicine residencies; and an effort to develop the phone-survey method as a way of identifying lesbians missed by other survey approaches. Applications are available online.
Restrictions: Grants are not awarded for completed research or for projects in progress.
Date(s) Application is Due: May 1; Oct 1
Amount of Grant: 500 - 10,000 U.S.D
Contact: Program Manager; (415) 255-4547; fax (415) 255-4784; info@glma.org
Internet: http://www.glma.org/programs/lhf/index.shtml
Sponsor: Lesbian Health Fund
459 Fulton Street, Suite 107
San Francisco, CA 94102

Lester E. Yeager Charitable Trust B Grants 2691

Established in 1989 in Kentucky, the Lester E. Yeager Charitable Trust B is an independent foundation. Funding is limited to areas of southern Indiana that are adjacent to Kentucky with emphasis on Daviess and Henderson counties in Indiana. The Trust gives primarily for health and human services. Contact Trust coordinator for additional information and to request an application form.
Requirements: Application form required. Applicants should submit the following: detailed description of project and amount of funding requested; additional materials/documentation; one page cover letter with application; four copies of proposal.
Restrictions: No grants are available to individuals. Grant recipients are expected to file a report when project is completed.
Geographic Focus: Indiana, Kentucky
Amount of Grant: 800 - 13,000 U.S.D
Contact: Trust Coordinator; fax (270) 686-8254
Sponsor: Lester E. Yeager Charitable Trust B
P.O. Box 964
Owensboro, KY 42302-0964

Lester Ray Fleming Scholarships 2692

The North Carolina Governor's Institute on Alcohol and Substance Abuse, Inc. has established scholarships for North Carolina community college students in memory of Lester Ray Fleming, son of Eugene and Mary Fleming and father of Tristan and Curtis. Mr. Fleming was an employee of the Institute at the time of his death in August 2005. It is the hope that these scholarships will benefit and prepare students to educate and serve others in their communities. Those selected will not only receive a $250.00 scholarship but may have the opportunity to attend other venues sponsored by the Institute free of charge where they will gain valuable insight meeting leading researchers, policymakers, and administrators in the field. Scholarship recipients may also have the opportunity to participate in a project at the discretion of the Governor's Institute.
Requirements: Applications can be downloaded from the website.
Geographic Focus: North Carolina
Date(s) Application is Due: Mar 19
Amount of Grant: 250 U.S.D
Contact: Larry Woodard, (919) 256-7412; fax (919) 990-9518; lwoodard1@mindspring.com
Internet: http://www.governorsinstitute.org/index.php?option=com_content&task=view&id=103&Itemid=66
Sponsor: North Carolina Governor's Institute on Alcohol and Substance Abuse
1730 Varsity Drive, Suite 105
Raleigh, NC 27606

Lettie Pate Whitehead Foundation Grants 2693

Annual grants are made to accredited educational institutions in the nine specified states in the southeast to fund scholarships for the education of women. The grantee institutions award scholarships on the basis of need to Christian women who reside in one of the named states. Thousands of women receive financial assistance each year in individual amounts determined by the recipient institutions. While most scholarships go toward undergraduate higher education, the foundation maintains a special interest in health education. A significant number of its scholarship grants support education in the medical, nursing, and allied health care fields.
Requirements: Grants are awarded to institutions in Alabama, Florida, Georgia, Louisiana, Mississippi, North Carolina, South Carolina, Tennessee, and Virginia.
Geographic Focus: Alabama, Florida, Georgia, Louisiana, Mississippi, North Carolina, South Carolina, Tennessee, Virginia
Samples: Roanoke College (Salem, VA)—for financial-aid scholarships for women, $138,000. Catawba College (Salisbury, NC)—for scholarships for female students from nine Southern states, $63,000; Wofford College (Spartanburg, SC)—for scholarships, $72,000.
Contact: Program Contact; (404) 522-6755; fax (404) 522-7026; fdns@woodruff.org
Internet: http://www.lpwhitehead.org
Sponsor: Lettie Pate Whitehead Foundation
50 Hurt Plaza, Suite 1200
Atlanta, GA 30303

Leukaemia Research Fund Grants 2694

The fund was founded to encourage, promote, and assist research into the leukemias, Hodgkin's disease, and other related conditions. The fund supports clinical fellowships and research projects at hospitals and university medical centers throughout the United Kingdom.
Geographic Focus: All States, United Kingdom
Contact: Grants Administrator; (020) 7405 0101; fax (020) 7405 3139; info@lrf.org.uk
Internet: http://www.lrf.org.uk/en/1/science.html
Sponsor: Leukaemia Research Fund
43 Great Ormond Street
London, WC1N 3JJ England

Leukemia and Lymphoma Society Career Development Awards 2695

The program grants help scientists in various stages of their careers and encourage them to investigate promising new directions and devote their careers to curing blood-related cancers. The four awards in the program—Scholar, Scholar in Clinical Research, Fellow, and Special Fellow—provide stipends that allow investigators to perform basic research into the cause and prevention of leukemia, lymphoma, Hodgkin's lymphoma and myeloma. A preliminary application is mandatory and must be completed online by September 15; full applications must be received at the office by October 1.
Requirements: Scientists pursuing the investigation of promising new directions to curing blood-related cancers are eligible.
Date(s) Application is Due: Sep 15; Oct 1
Amount of Grant: 50,000 - 110,000 U.S.D
Contact: Grants Administrator; (914) 821-8843 or (914) 821-8859; fax (914) 821-8946; researchprograms@leukemia-lymphoma.org
Internet: http://www.leukemia-lymphoma.org/all_page?item_id=11618
Sponsor: Leukemia and Lymphoma Society of America
1311 Mamaroneck Avenue
White Plains, NY 10605

Leukemia and Lymphoma Society Research Grants 2696

The society's research program is based on the belief that all scientifically sound approaches toward a cure or control of leukemia, lymphoma, and myeloma should be encouraged on a worldwide basis. Basic laboratory research and its application to a clinical setting are supported.
Samples: Baylor College of Medicine, Ctr for Cell and Gene Therapy (TX)—for blood-cancer research, through the Specialized Center of Research grant program, $5 million

over five years; To be divided among 32 research projects—to develop new treatments for leukemia, lymphoma, and myeloma, through the Translational Research Program, $11.3 million; Fred Hutchinson Cancer Research Ctr (Seattle, WA)—to support the work of Irwin Bernstein to develop less toxic and more effective immune-based therapies for treating advanced leukemia and lymphoma, $7 million.
Contact: Grants Administrator; (914) 949-5213; fax (914) 949-6691
Internet: http://www.leukemia-lymphoma.org/all_page?item_id=7427
Sponsor: Leukemia and Lymphoma Society of America
1311 Mamaroneck Avenue
White Plains, NY 10605

Levi Strauss Foundation Grants **2697**

The Levi Strauss Foundation Community Partnership program awards contributions to nonprofit community agencies in the following areas: reducing poverty; youth and women; economic development (job creation and community-based economic development); job training, placement, and access; leadership development aimed at strengthening the economic development capacity of community organizations; international education, health care, and peace; and microenterprise. The program also supports AIDS and disease prevention (assistance to people with AIDS and their caregivers); risk reduction education for those with high-risk behaviors; and services to populations severely affected by AIDS. Contributions are also made in the field of social justice (programs that seek to remove racial and other discriminatory barriers, ease tension between groups, promote diversity in community leadership, and prevent violent acts of racial and cultural prejudice). Types of support include matching funds, continuing support, seed money, equipment, exchange programs, employee-related scholarships, employee matching gifts, technical assistance, special projects, and operating budgets. Contact the foundation for more information and application guidelines.
Requirements: Giving is on a national basis with emphasis on areas of company operations. Agencies receiving funding must be accredited charitable organizations or public entities and have boards or advisory groups that reflect the population or communities being served.
Restrictions: The company does not fund projects by individuals or for political, sectarian, or religious purposes; tickets for dinners or other special events; requests for sponsorships; or courtesy advertising. Research and conferences are generally not considered for funding unless they are an integral part of a larger effort that the company is supporting. The foundation does not accept unsolicited proposals.
Samples: American Institute for Public Service (Wilmington, DE)—for support for 2003 Jefferson Awards, $20,000; The Asia Foundation (San Francisco, CA)— for the Labor Projects Workshop, $27,000; Citizens' Scholarship Foundation of America (Saint Peter, MN)—for the Families of Freedom Scholarship Fund, to provide undergraduate scholarships for the spouses and children of people injured or killed as a result of the September 11 terrorist attacks, $500,000.
Contact: Theresa Fay-Bustillos, Executive Director; (415) 501-3577; fax (415) 501-6575; lsf@levi.com
Internet: http://www.levistrauss.com/responsibility/foundation/guidelines.htm
Sponsor: Levi Strauss Foundation
1155 Battery Street, P.O. Box 7215
San Francisco, CA 94111

Lewis H. Humphreys Charitable Trust Grants **2698**

The Lewis H. Humphreys Charitable Trust was established in 2004 to support and promote quality educational, cultural, human-services, and health-care programming for underserved and disadvantaged populations. In the area of arts, culture, and humanities, the trust supports programming that: fosters the enjoyment and appreciation of the visual and performing arts; strengthens humanities and arts-related education programs; provides affordable access; enhances artistic elements in communities; and nurtures a new generation of artists. In the area of education, the trust supports programming that: promotes effective teaching; improves the academic achievement of, or expands educational opportunities for, disadvantaged students; improves governance and management; strengthens nonprofit organizations, school leadership, and teaching; and bolsters strategic initiatives of area colleges and universities. In the area of health, the trust supports programming that improves the delivery of health care to the indigent, uninsured, and other vulnerable populations and addresses health and health-care problems that intersect with social factors. In the area of human services, the trust funds programming that: strengthens agencies that deliver critical human services and maintains the community's safety net and helps agencies respond to federal, state, and local public policy changes. In the area of community improvement, the trust funds capacity-building and infrastructure-development projects including: assessments, planning, and implementation of technology for management and programmatic functions within an organization; technical assistance on wide-ranging topics, including grant writing, strategic planning, financial management services, business development, board and volunteer management, and marketing; and mergers, affiliations, or other restructuring efforts. Grant requests for general operating support, program support, and capital support will be considered. Grant requests for capital support, such as for buildings, land, and major equipment should meet a compelling community need and offer a broad social benefit. Grants from the trust are one year in duration. Application materials are available for download at the grant website. Applicants are strongly encouraged to review the state application guidelines for additional helpful information and clarifications before applying. Applicants are also encouraged to review the trust's funding history (link is available from the grant website). Grant applications can be submitted between August 1 and September 30. Applicants will be notified of grant decisions by November 30.
Requirements: Applicants must have 501(c)3 tax-exempt status and serve the residents of Kansas. Grant application materials must be mailed.
Restrictions: The trust does not support requests from individuals, organizations attempting to influence policy through direct lobbying, or any political campaigns.
Geographic Focus: Kansas
Date(s) Application is Due: Sep 30
Samples: Kansas Childrens Discovery Center, Topeka, Kansas, $50,000 (2010); Hetlinger Developmental Services, Emporia, Kansas, $50,000 (2010); Boy Scouts of America, Topeka, Kansas, $75,000 (2010).
Contact: James Mueth, Vice President; (816) 292-4342; james.mueth@baml.com
Internet: https://www.bankofamerica.com/philanthropic/fn_search.action
Sponsor: Lewis H. Humphreys Charitable Trust
1200 Main Street, 14th Floor, P.O. Box 219119
Kansas City, MO 64121-9119

Liberty Bank Foundation Grants **2699**

The foundation's charitable giving is focused primarily on organizations that provide meaningful programs and activities that benefit people within Liberty Bank's market area. Of particular interest are programs and activities that provide assistance and opportunities to improve the quality of life for people of low income, especially families in crisis or at-risk. Top priorities for funding include community and economic development—affordable housing for low/moderate-income individuals and families, community and neighborhood capacity-building, and community services targeted to low/moderate-income individuals; education—programs that address the needs of low/moderate-income individuals; health care and human services—outreach and educational programs on health issues, quality child care, homeless shelters and services, services for victims of domestic violence, and transitional housing assistance; and arts and culture—programs that increase access to arts and culture for people of low income who might not otherwise be able to participate in them. Grants generally support specific programs rather than capital projects, equipment, or general operating expenses. Organizations that have received funding in two consecutive calendar years should refrain from reapplying for one calendar year. Contact the office to discuss the project prior to applying. Guideline and application are available online.
Requirements: 501(c)3 tax-exempt organizations are eligible.
Restrictions: Individuals, fraternal groups, and organizations that are not open to the general public are ineligible. Grants do not support annual funds of colleges, universities or hospitals; trips, tours, or conferences; scientific or medical research; deficit spending or debt liquidation; lobbying or otherwise influencing the outcome of the legislative or electoral process; religious groups, except for nonsectarian programs; or endowments or other foundations.
Date(s) Application is Due: Mar 31; Jun 30; Sep 30; Dec 31
Amount of Grant: 250 U.S.D
Samples: Windham InterFaith Ministry—to support the Rooming House rent fund, $2500; Children's Home—to support the summer jobs program, $3000; Hospice of Southeastern Connecticut—to support the Live Alone program, $5000.
Contact: Grants Administrator; (860) 704-2181; smurphy@liberty-bank.com
Internet: http://www.liberty-bank.com/liberty_foundation.asp
Sponsor: Liberty Bank Foundation
P.O. Box 1212
Middletown, CT 06457

Libra Foundation Grants **2700**

The Libra Foundation awards grants to Maine nonprofits in its areas of interest, including art, culture, and humanities; education; health; human services; environment; justice; public/society benefit; and religion. The Foundation makes grants to organizations that it expects to develop innovative and sustainable Maine-based business initiatives and programs that provide for the welfare and betterment of children. The aforementioned activities comprise the majority of the Foundation's charitable giving. The application is available online.
Requirements: Organizations must be in Maine and 501(c)3 nonprofits to apply.
Restrictions: Individuals are ineligible. The Foundation does not provide funding to supplement annual campaigns, regular operating needs, multi-year projects, individuals, scholarships, or travel.
Geographic Focus: Maine
Date(s) Application is Due: Feb 15; May 15; Aug 15; Nov 15
Samples: Farnsworth Art Museum, Rockland, Maine, general operating support, $10,000 (2011); Portland Museum of Art, Portland, Maine, general operating support, $20,000 (2011); MaineHealth, Portland, Maine, start-up of a leading center for childhood health, $100,000 (2011); Pineland Farms, New Gloucester, Maine, develop, operate and staff Pineland Farms, Inc., for agricultural promotion, education, and research in Maine, $2,895,000 (2011).
Contact: Elizabeth Flaherty, Executive Assistant; (207) 879-6280
Internet: http://librafoundation.org/application-procedures
Sponsor: Libra Foundation
Three Canal Plaza, Suite 500
Portland, ME 04112-8516

Libraries for the Future Get Real, Get Fit Grants **2701**

Libraries for the Future, with support from MetLife Foundation, is sponsoring library-based activities and discussions focused on physical fitness and healthy eating for teenagers and families. The program is designed to use a health-related episode on In the Mix, the award-winning topical TV series for teens on PBS, to stimulate discussions and activities that educate teens and their parents about the benefits of choosing a healthy lifestyle. The program is designed as a one-time, replicable program, to be held at least four times during the year of the grant. A description of the In the Mix video and a list of possible topics and program formats is included in the RFP for the use of libraries in preparing proposals. However, applicants should not feel restricted by these suggestions. Libraries are encouraged to draw on their knowledge of the interests and needs of their

target population(s), and to design their Get Real, Get Fit! program to match the local context. Program details and guidelines are available online. Annual deadline dates may vary; contact program staff for exact dates.
Requirements: Applications are encouraged from libraries of all sizes and types, as well as libraries that have not previously engaged in systematic outreach to teens and their families. Applicant libraries must assign an individual to be responsible for program implementation and to participate in a national training, work with one or more community organizations or schools as partners, and agree to contribute to and benefit from the experiences of other libraries selected for the Get Real, Get Fit! Program.
Date(s) Application is Due: Dec 10
Amount of Grant: 1,500 - 2,500 U.S.D
Contact: Marilyn Ratner, (914) 723-7923; mratner@lff.org
Internet: http://lff.org/news/GRGF_rfp.html
Sponsor: Libraries for the Future
27 Union Square W, Suite 204
New York, NY 10003

Life Sciences Research Foundation Postdoctoral Fellowships 2702
Each year approximately 12-15 postdoctoral fellowships are awarded in the medical and life sciences to provide training and support of the highest quality young scientists in the best research environments. Research must be performed at nonprofit institutions. The fellowships are awarded for three-year periods. Send a self-addressed envelope for information and an application, or visit the Web site.
Requirements: Applicant must have an MD, PhD, or equivalent degree. Persons working on a second postdoctoral are eligible only if they are transferring to a different supervisor's laboratory and embarking on a new project not connected to their previous research. All U.S. citizens are eligible with no geographic restrictions. Foreign applicants will be eligible for study in U.S. laboratories only.
Restrictions: Candidate must not hold a faculty/permanent appointment. The LSRF does not award an institutional allowance for overhead.
Date(s) Application is Due: Oct 1
Amount of Grant: Up to 51,000 U.S.D
Contact: Susan DiRenzo; (609) 258-3551; sdirenzo@molbio.princeton.edu
Internet: http://www.lsrf.org/geninfo.htm
Sponsor: Life Sciences Research Foundation
Lewis Thomas Laboratory, Princeton University, Washington Road
Princeton, NJ 08544

Liliane Foundation Grants 2703
The foundation was founded to provide aid for disabled children in developing countries in Africa, Asia, and Central and South America. Grants are awarded to individuals in the field of medical and social rehabilitation.
Contact: G.B. Dollevoet; 31-73-511-90-29; fax 31-73-511-73-54; liliane@euronet.nl
Internet: http://www.lilianefonds.nl
Sponsor: Liliane Foundation
St. Catharinastraat 1, P.O.B 75
Vlijmen, 5250 AB Netherlands

Lillian S. Wells Foundation Grants 2704
The Lillian S. Wells Foundation established in 1976, primarily supports the Fort Lauderdale, Florida, and Chicago, Illinois regions with funding interests that include: medical research, with emphasis on brain cancer research, women's health, substance abuse, and at-risk youth. Additional funding for education and the arts. Contact the Foundation for an application form, the Board meets quarterly in January, April, July, and October.
Requirements: Nonprofit organizations in Chicago, Illinois and Fort Lauderdale, Florida area are eligible to apply.
Geographic Focus: Florida, Illinois
Date(s) Application is Due: Mar 15; Jun 15; Sep 15; Dec 15
Amount of Grant: 1,000 - 1,900,000 U.S.D
Contact: Patricia F. Mulvaney; patricia.mulvaney@thewellsfamily.net
Sponsor: Lillian S. Wells Foundation
600 Sagamore Road
Fort Lauderdale, FL 33301-2215

Lillian Sholtis Brunner Summer Fellowships for Historical Research in Nursing 2705
These summer fellowships for historical research in nursing are offered by the Center for the Study of the History of Nursing in the School of Nursing at the University of Pennsylvania. Selection of the scholars will be based on evidence of preparation and/or productivity in historical research related to nursing. It is expected that the research and new materials produced by Brunner scholars will help ensure the growth of scholarly work focused on the history of nursing. The fellowship will support up to eight weeks of residential study and use of the center's collections. The application should be sent via email to either Patricia Antonio, PhD, RN, FAAN (dantonio@nursing.upenn.edu) or Barbra Mann Wall, PhD, RN (wallbm@nursing.upenn.edu).
Date(s) Application is Due: Dec 31
Amount of Grant: 2,500 U.S.D
Contact: Dr. Karen Buhler-Wilkerson, Director; (215) 898-4725; fax (215) 573-2168; karenwil@nursing.upenn.edu
Internet: http://www.nursing.upenn.edu/history/Pages/Brunner_Fellowship.aspx
Sponsor: University of Pennsylvania
School of Nursing, 418 Curie Boulevard
Philadelphia, PA 19104-6020

Lincoln Community Foundation Grants 2706
The foundation offers program, capital, and quick response grants. Program grants are available to fund the creation of new programs or the enhancement of existing programs within nonprofit organizations. These grants are given in support of programs that focus on the areas of basic needs, children and youth, older adults, human diversity, family issues, environmental enhancement and cultural arts. Capital grants are available to help nonprofit organizations fund the construction of a new facility or facility addition, the renovation or purchase of an existing facility, and/or the purchase of equipment (including furnishings and computer equipment). Quick response grants are available to nonprofit organizations to provide quick solutions to unexpected needs or special opportunities. These grants give nonprofits the opportunity to meet requests that could not typically be responded to due to the traditional application process and timeline or lack of funding sources. Deadlines listed are for program grants.
Requirements: IRS 501(c)3 organizations serving the Lincoln, NE, area are eligible.
Restrictions: Grants are not awarded to individuals, and requests are generally not considered for religious purposes, political purposes, endowments, programs outside the Lincoln/Lancaster County area, routine operating expenses, large capital expenditures, budget deficits, and projects with future annual commitments.
Geographic Focus: Nebraska
Date(s) Application is Due: Jun 1; Dec 1
Amount of Grant: 50 - 50,000 U.S.D
Samples: 7th Street Loft Consortium (Lincoln, NE)—to develop a capital campaign, $11,500; Good Neighbor Community Ctr (Lincoln, NE)—for salary support of a part-time coordinator responsible for resource development and volunteer management, $14,000; Lincoln Literacy Council (Lincoln, NE)—for a program that promotes greater parental involvement in their children's education and activities, $14,736.
Contact: Debra Shoemaker, Program Director; (402) 474-2345; fax (402) 476-8532; debs@lcf.org or lcf@lcf.org
Internet: http://www.lcf.org
Sponsor: Lincoln Community Foundation
215 Centennial Mall S, Suite 200
Lincoln, NE 68508

Lincoln Financial Group Foundation Grants 2707
As one of the country's premier financial services organizations, Lincoln Financial Group is committed to providing its customers with the life insurance, retirement and investment products and services they need to maintain their quality of life. As a concerned corporate citizen, the company's foundation has sustained a strategic philanthropic program to improve the quality of life in the communities where employees live and work. These communities include: Chicago, Illinois; Concord, New Hampshire; Fort Wayne, Indiana; Greensboro, North Carolina; Hartford, Connecticut; Omaha, Nebraska; Philadelphia, Pennsylvania.
Requirements: The Lincoln Financial Foundation welcomes grant applications from organizations that meet our geographic, strategic focus, and eligibility guidelines. Local Charitable Contribution Committees meet and review grant applications. Only nonprofit organizations that have a 501(c)3 designation from the Internal Revenue Service are eligible to apply for funding. To apply for a grant, complete the Lincoln Financial Foundation application by using the online form and submitting it electronically. Grant submission deadlines are: Arts and Culture grants applications must be received by January 15; Education/Workforce Development grant applications must be received by April 21; Human Services grant applications must be received by July 21. Applications are reviewed by the local contribution committees within three to four months after the cycle deadline. Notification will be given soon after decisions are rendered.
Restrictions: Qualified organizations are eligible for one grant per calendar year. In general, the Lincoln Foundation will not award grants to: individuals; religious organizations; public or private elementary or secondary schools or school foundations; hospitals or hospital foundations; fraternal, political, or war veteran organizations; general operating support; capital funding; endowments.
Geographic Focus: Connecticut, Illinois, Indiana, Nebraska, New Hampshire, North Carolina, Pennsylvania, United Kingdom
Date(s) Application is Due: Jan 15; Apr 21; Jul 21
Amount of Grant: 5,000 - 75,000 U.S.D
Contact: Jean Vrabel, Program Officer; (260) 455-3868; fax (260) 455-4004; Jean.Vrabel@LFG.com
Internet: http://www.lfg.com/LincolnPageServer?LFGPage=/lfg/lfgclient/abt/fingrp/hta/index.html
Sponsor: Lincoln Financial Group Foundation
1300 S Clinton Street, P.O. Box 7863
Fort Wayne, IN 46802

Lindbergh Grants 2708
Grants are awarded to individuals whose proposed projects represent a significant contribution toward the achievement of a better balance between technology and our human and natural environment. Nine to 10 grants are awarded each year. The current funding program is interested in increasing representation in the following areas: aviation/aerospace; agriculture; arts and humanities; biomedical research; conservation of natural resources; exploration; health and population sciences; intercultural communication; oceanography; water resources management; waste disposal management; wildlife preservation; and adaptive technology. Grant application can be obtained by writing the fund; enclose a self-addressed, stamped envelope with request. The Web site contains additional information.
Requirements: Grants are international in scope and open to citizens of all countries. Letters of request, proposals, endorser reports, and required progress and final reports must be submitted in English. The fund welcomes candidates who may or may not be

affiliated with academic or nonprofit institutions. Grants are made to individuals, not to affiliated institutions for institutional programs. Note: When sending any materials by fax, seven hard copies via mail or delivery are requested. The deadline for grant applications is the second Thursday of June in the year preceding the awarding of funds.
Restrictions: The fund does not provide support for overhead costs of organizations, tuition, scholarships, fellowships, or travel related to such.
Date(s) Application is Due: Jun 9
Amount of Grant: Up to 10,580 U.S.D
Samples: Dr. Ernest Diedrich, Saint John's College (Collegeville, MN)—for Balancing Energy Consumption with Resource Conservation: A Wind Power Study for Central Minnesota, an Educational Model for the World;Dr. Ping Liu, Eastern Illinois U (Charleston, IL)—for Analysis of Recycled Plastic Composites to Aid in the Development and Design of Recycled Plastic Products;Dr. Vivian Charles McAlister, Dalhousie U (Halifax, Nova Scotia)—for Use of Human Response to Parasites to Improve Organ Transplant Survival;Dr. Carl Wulfman, U of the Pacific (Stockton, CA)—for Clean Energy into Clean Energy: Toward a New Method of Using Sunlight to Produce Electricity.
Contact: Grants Coordinator; (763) 576-1596; fax (763) 576-1664; info@lindberghfoundation.org
Internet: http://www.lindberghfoundation.org/grants
Sponsor: Charles A. and Anne Morrow Lindbergh Foundation
2150 Third Avenue N, Suite 310
Anoka, MN 55303-2200

Linford and Mildred White Charitable Fund Grants **2709**
The Linford and Mildred White Charitable Fund was established in 1956 to support and promote quality educational, human-services, and health-care programming for underserved populations within the city of Waterbury, Connecticut, and its surrounding communities. Grants from the White Charitable Fund are one year in duration. Applicants must apply online at the grant website. Applicants are strongly encouraged to do the following before applying: review the downloadable state application procedures for additional helpful information and clarifications; review the downloadable online-application guidelines at the grant website; review the foundation's funding history (link is available from the grant website); review the online application questions in advance; and review the list of required attachments. These will generally include: a list of board members, financial statements (audited, reviewed, or compiled by independent auditor); an organization summary; a list of other funding sources; an IRS Determination letter; and other required documents. All attachments must be uploaded in the online application as PDF, Word, or Excel files. The deadline for application to the Linford and Mildred White Charitable Fund is 11:59 p.m. on July 1. Applicants will receive notification of grant decisions by letter within two to three months after the proposal deadline.
Requirements: Applicant organizations must have 501(c)3 status and serve the people of Waterbury, Connecticut, and its vicinity.
Restrictions: Grant requests for capital projects will not be considered. Applicants will not be awarded a grant for more than 3 consecutive years. The fund does not support requests from individuals, organizations attempting to influence policy through direct lobbying, or any political campaigns.
Geographic Focus: Connecticut
Date(s) Application is Due: Jul 1
Samples: Boys and Girls Clubs of Greater Waterbury, Waterbury, Connecticut, $2,000, purchase of a bounce house for Bounce to Be Fit (2010); Covenant to Care for Children, Bloomfield, Connecticut, $1,500, Waterbury Adopt A Social Worker Program (2010); Palace Theater Group, Waterbury, Connecticut, $2,000, Ticket and Travel Subsidy Program for Title One schools (2010).
Contact: Carmen Britt; (860) 657-7019; carmen.britt@baml.com
Internet: https://www.bankofamerica.com/philanthropic/fn_search.action
Sponsor: Linford and Mildred White Charitable Fund
200 Glastonbury Boulevard, Suite # 200, CT2-545-02-05
Glastonbury, CT 06033-4056

Linn-Henley Charitable Trust Grants **2710**
The charitable trust awards grants to Alabama nonprofit organizations to improve the quality of life in its areas of interest, including arts and culture, higher education, libraries and library science, religion, and hospitals. Types of support include capital campaigns, research grants, and matching funds.
Requirements: Jefferson County, AL, nonprofit organizations are eligible.
Geographic Focus: Alabama
Amount of Grant: 2,000 - 120,000 U.S.D
Samples: Southern Museum of Flight (AL)—$10,000; U of Alabama (AL)—to support the Linn-Henley International Awards banquet, $7000.
Contact: Collins Compere, Trustee, c/o Regions Bank, Trust Dept; (205) 326-7224
Sponsor: Linn-Henley Charitable Trust
P.O. Box 10247
Birmingham, AL 35202

Lisa and Douglas Goldman Fund Grants **2711**
Established in 1992 the Lisa and Douglas Goldman Fund is a private foundation committed to providing support for charitable organizations that enhance society. As natives of San Francisco, the Goldmans place a high priority on projects that have a positive impact on San Francisco and the Bay Area. Interests and priorities include: children and youth; civic affairs; civil and human rights; education; environmental affairs; health; Jewish affairs; literacy; organizational development; population; social and human services; and sports and recreation.
Requirements: After reviewing the Fund's interests and priorities interested applicants may submit an initial letter of inquiry. Applicants who receive a favorable response will be invited to submit a formal proposal with supporting materials. Applicants are encouraged to contact the Fund directly with questions regarding the appropriateness of a project. There are no deadlines.
Restrictions: The following are ineligible: grants to individuals; documentaries and films; events/conferences; books and periodicals; research; and deficit budgets. Applications for annual support are not accepted. Organizations may submit only one request per year.
Samples: Little Kids Rock, Cedar Grove, New Jersey, $10,000 - Bay Area Expansion Initiative, to increase the number of low-income schools participating in the Little Kids Rock music program (2010); Ignite, San Francisco, California, $13,250 - to support the expansion of Ignite's programs at San Francisco schools and foster the development of young women's political knowledge and ambition (2010); and Basel Action Network, Seattle, Washington, $25,000 - for the e-Stewards Initiative, a program to promote responsible recycling of electronic waste through the certification of e-waste recyclers (2010).
Contact: Nancy D. Kami, Executive Director; (415) 771-1717; fax (415) 771-1797
Internet: http://fdncenter.org/grantmaker/goldman
Sponsor: Lisa and Douglas Goldman Fund
One Daniel Burnham Court, Suite 330C
San Francisco, CA 94109-5460

Littlefield-AACR Grants in Metastatic Colon Cancer Research **2712**
The grants will provide support for innovative cancer research projects designed to accelerate the discovery and development of new agents to treat metastatic colon cancer and/or for pre-clinical research with direct therapeutic intent. Special emphasis will be placed on research that holds promise for leading to individualized therapeutic options for treatment in the near future or for developing promising new cancer therapeutics for metastatic colon cancer, which will translate into clinical applications within a one- to two-year period.
Requirements: The prizes are open to all cancer researchers who are affiliated with any institution involved in cancer research, cancer medicine, or cancer-related biomedical science anywhere in the world. Such institutions include those in academia, industry, or government. Candidates must be active researchers and have a record of recent publications.
Date(s) Application is Due: Feb 1
Amount of Grant: 500,000 - 1,000,000 U.S.D
Contact: Julia Laurence, Program Assistant; (215) 440-9300, ext. 102 or (267) 646-0655; fax (215) 440-9372; julia.laurence@aacr.org or laurence@aacr.org
Internet: http://www.aacr.org/home/scientists/research-funding—training-grants/research-funding/littlefield-aacr-grants-for-metastatic-colon-cancer-research.aspx
Sponsor: Littlefield 2000 Trust and the American Association for Cancer Research
615 Chestnut Street, 17th Floor
Philadelphia, PA 19106-4404

Little Life Foundation Grants **2713**
The Little-Life Foundation is a 501(c)3 charitable organization founded in 2002 whose primary purpose is to support medical care and research for increasing the survival rates of prematurely born babies. Saint Barnabas Medical Center, where more than 7,500 babies are born each year, will be the primary beneficiary of the Foundation's efforts.
Geographic Focus: New Jersey
Contact: Richard P. Rizzuto, Director; (973) 277-2329; fax (973) 401-9799
Internet: http://www.littlelifefoundation.org/
Sponsor: Little Life Foundation
P.O. Box 765
New Vernon, NJ 07976-0765

Little River Foundation Grants **2714**
The foundation supports nonprofit organizations in the Mid-Atlantic states, primarily New Jersey, New York, and Pennsylvania. Grants are awarded for hospitals, AIDS research, substance abuse services, higher education, secondary education, early childhood education, adult basic education, building funds, literacy programs, animal welfare, legal education, international studies in Africa, environmental conservation, legal services, religious organizations, and community funds. Grants are awarded for general operating support, annual campaigns, endowment funds, conferences and seminars, research, and building construction/renovations. Application forms are not required. The board meets each year in the fall.
Requirements: Only residents of Maryland, New Jersey, New York, Pennsylvania, and Virginia may apply.
Geographic Focus: California, New York, Virginia
Amount of Grant: 100 - 175,000 U.S.D
Samples: Grace Episcopal Church (The Plains, VA)—for facilities improvement, $25,000; Piedmont Environmental Council (Warrenton, VA)—for the building program, $87,000; Johns Hopkins U (Baltimore, MD)—for brain tumor research, $25,000; Princeton U (Princeton, NJ)—for annual support, $25,000.
Contact: Dale Hogoboom, Assistant Treasurer
Sponsor: Little River Foundation
101 Park Avenue, Suite 3500
New York, NY 10178-0061

Livingston Memorial Foundation Grants **2715**
The foundation awards grants to promote medical and health-related services of benefit to the underserved and uninsured people of Ventura County, CA. General operating grants support well-established programs that continue to have positive results. Special projects grants support organizations that can demonstrate the need for assistance in funding a special project or with the cost of medical equipment that promises significant positive

results. Organizations should submit a brief letter of inquiry that includes a concise statement of the need for the funds, the amount requested, pertinent factual information, and state the desired type of grant. Include a copy of the tax-exemption letter.
Requirements: California nonprofit organizations serving Ventura County are eligible.
Restrictions: Grants are not awarded to individuals or for projects normally financed from government sources, conferences, seminars, workshops, exhibits, travel, or publication.
Geographic Focus: California
Amount of Grant: 2,500 - 156,000 U.S.D
Contact: Laura McAvoy, (805) 495-7489; Lmcavoy@jdplaw.com
Sponsor: Livingston Memorial Foundation
2815 Townsgate Road, Suite 200
Westlake Village, CA 91361

Lloyd A. Fry Foundation Grants **2716**
The foundation supports organizations with the strength and commitment to address persistent problems of urban Chicago resulting from poverty, violence, ignorance, and despair. The foundation is specifically interested in programs that improve conditions for low-income, underserved communities. Grants are made in four major fields: education, community services, health, and arts and culture. Generally, priority is given to proposals for the initiation of new programs or the strategic expansion or enhancement of existing programming. Programs that build the capacity of the individuals served also are encouraged. Former and current grantee organizations may submit proposals for projects that strengthen an organization's capacity, including requests for strategic planning, program evaluation, board and staff development, fundraising, and the application of technology. An organization that has received three years of funding for either general operating support or a specific program will generally be asked to take a year off in seeking support during the following year. Guidelines are available online. We are also interested in funding advocacy activities that focus on ensuring that low-income communities and individuals in Chicago are being treated fairly and are getting the services they deserve
Requirements: Grants are made only to tax-exempt organizations and are rarely made to organizations outside the Chicago metropolitan area.
Restrictions: The foundation does not fund government entities, political or religious activities, fundraising events, or medical research.
Geographic Focus: Illinois
Date(s) Application is Due: Mar 1; Jun 1; Sep 1; Dec 1
Samples: Changing Worlds (Chicago, IL)—for the Literacy and Cultural Connections Program and the Summer Institute for Teachers, $10,000; Hamdard Ctr for Health and Human Services (Addison, IL)—for the Employment Training Program serving Chicago residents, $20,000; Planned Parenthood of the Chicago Area (IL)—for the Community Partnership with Chicago Public Schools, $20,000; Chicago Children's Choir (IL)—for the in-school and neighborhood choir programs, $20,000.
Contact: Unmi Song, Executive Director; (312) 580-0310; fax (312) 580-0980; usong@fryfoundation.org
Internet: http://www.fryfoundation.org/guidelines.html
Sponsor: Lloyd A. Fry Foundation
120 S La Salle Street, Suite 1950
Chicago, IL 60603

Lockheed Martin Philanthropic Grants **2717**
Lockheed Martin funds grants that enhance the communities where Lockheed Martin employees work and live. Lockheed Martin will consider grant requests that best support the Corporation's strategic focus areas and reflect effective leadership, fiscal responsibility, and program success. Those focus areas include: education—K-16 science, technology, engineering and math (STEM) education; customer and constituent relations—causes of importance to customers and constituents, including the U.S. military and other government agencies; community relations—building partnerships between employee volunteers and the civic, cultural, environmental, and health and human services initiatives that strengthen the communities where employees work. Applications are accepted year-round. Evaluations are typically performed quarterly.
Requirements: To be considered for grant funding, organizations must meet all of the following criteria: apply through Lockheed Martin's online CyberGrants system; have a non-profit tax exempt classification under Section 501(c)3 of the Internal Revenue Service Code, or equivalent international non-profit classification, or be a public elementary/secondary school, or be a qualifying U.S.-based institute of higher education; align with one or more of Lockheed Martin's three strategic focus areas: delivering standards-based science, technology, engineering and math (STEM) education to students in K-16; investing in programs that support the long term success of the military and their families; and supporting the vitality of the communities where employees live and work; agree to act in accordance with Lockheed Martin's contribution acknowledgement *Requirements:* organization/grantee will comply with all applicable requirements of the Patriot Act and the Voluntary Anti-Terrorist Guidelines and will not use any portion of the grant funds for the support, direct or indirect, of acts of violence or terrorism or for any organization engaged in or supporting such acts; be located or operate in a community in which Lockheed Martin has employees or business interests; demonstrate fiscal and administrative responsibility and have an active, diverse board, effective leadership, continuity and efficiency of administration; be limited to one grant per year, except in unusual circumstances.
Restrictions: Some grant applications may not be able to be considered until the next year's budget cycle, particularly those received in the second half of the year. Grants are generally not made to: organizations that unlawfully discriminate on the basis of race, ethnicity, religion, national origin, age, military veteran's status, ancestry, sexual orientation, gender identity or expression, marital status, family structure, genetic information, or mental or physical disability; private K-12 schools, unless the contribution is in acknowledgement of employee volunteer service provided to the school; home-based child care/educational services; individuals; professional associations, labor organizations, fraternal organizations or social clubs; social events sponsored by social clubs; athletic groups, clubs and teams, unless the contribution is in acknowledgement of employee volunteer service provided to the school; religious organizations for religious purposes; or advertising in souvenir booklets, yearbooks or journals unrelated to Lockheed Martin's business interests.
Geographic Focus: California, Colorado, Florida, Georgia, Louisiana, Maryland, Minnesota, Mississippi, New Jersey, New Mexico, New York, Ohio, Pennsylvania, South Carolina, Texas, Virginia, Canada
Contact: David Phillips, Manager of Corporate Philanthropy; (301) 897-6292; fax (301) 897-6485; david.e.phillips@lmco.com
Internet: http://www.lockheedmartin.com/aboutus/community/philanthropy/funding.html
Sponsor: Lockheed Martin
6801 Rockledge Drive
Bethesda, MD 20817

Long Island Community Foundation Grants **2718**
The Foundation awards grants to eligible New York nonprofit organizations in the following program areas: arts; community development; education; environment; health; mental health; hunger; technical assistance; and youth violence prevention. Projects are preferred that accomplish specific tasks, solve problems, address needs of the disadvantaged, help a large number of people, and use community resources. Specific guidelines for each category and the application are available on the Foundation website. Applicants are encouraged to view recent grants on the website in order to judge if their project is appropriate for the Foundation.
Requirements: New York nonprofit organizations are eligible.
Restrictions: Grants are not made for the following: individuals; building or capital campaigns; medical or scientific research; equipment purchases; budget deficits; endowments; event sponsorships; re-granting purposes; or religious or political purposes.
Geographic Focus: New York
Date(s) Application is Due: Aug 24
Samples: Long Island Arts Alliance, Long Island, New York, strengthen the Long Island arts infrastructure through arts education advocacy forums, and a study about the economic significance of the nonprofit arts community, $30,000 (2011); Citizens Campaign Fund for the Environment, Long Island, New York, to analyze and grade 16 sewage treatment plants on Long Island, $25,000 (2011); Mental Health Association of Nassau County, Long Island, New York, to support a geriatric mental health training program that will increase access to quality mental health for the elderly, $15,000 (2011).
Contact: Nancy Arnold, Grants Administrator; (516) 348-0575; fax (516) 348-0570; narnold@licf.org
Internet: http://www.licf.org/grants
Sponsor: Long Island Community Foundation
1864 Muttontown Road
Syosset, NY 11791

Longwood Foundation Grants **2719**
Limited grants are made generally to educational institutions, to hospitals for construction purposes, and to social and youth agencies primarily in Delaware, with emphasis on Wilmington and with some giving also in Pennsylvania. Grants are given for annual campaigns, building funds, equipment (including computers), land acquisition, and research.
Requirements: Delaware nonprofit organizations and hospitals serving southern Chester County in Pennsylvania are eligible.
Restrictions: Grants are not given to individuals or for special projects. Grants generally will not be awarded to support K-12 schools or public high schools.
Geographic Focus: Delaware, Pennsylvania
Date(s) Application is Due: Mar 15; Sep 15
Amount of Grant: 25,000 - 500,000 U.S.D
Samples: Christiana Care Health System (Wilmington, DE)—$4 million; Connections Community Support Programs (Wilmington, DE)—$1 million; Exceptional Care (Wayne, PA)—$500,000.
Contact: Peter Morrow, Executive Director; (302) 654-2477; fax (302) 654-2323
Sponsor: Longwood Foundation
100 W 10th Street, Suite 1109
Wilmington, DE 19801

Lorain County Grants and Scholarships **2720**
The foundation awards medical scholarships and educational grants for postgraduate studies in the field of medical arts to residents of Lorain County, OH, who intend to return to practice in Lorain County.
Requirements: Applicant must be a Lorain County, OH, resident for at least two years and demonstrate financial need.
Geographic Focus: Ohio
Amount of Grant: Up to 2,000 U.S.D
Contact: Lynne Hammes, Bank One Ohio Trust Co. Administrator; (216) 781-2386 or (800) 945-2265; lynne.h.hammes@jpmorgan.com
Sponsor: James R. Nicholl Memorial Foundation
600 Superior Avenue
Lorain, OH 44114

Lotus 88 Foundation for Women and Children Grants 2721

The foundation's mission is to promote the empowerment of women and children through supporting their economic, emotional, and spiritual development. The foundation's focus area is American Indian Country. Grants are awarded in Indian Country for two strategic purposes: revitalizing the council tipis as spiritual, cultural, and service centers; and providing the basic needs through community building. Community building grants are intended to promote and support community building in Indian Country to improve basic living conditions and to encourage a more positive future. Tipi project grants are made to help tribal women living on reservation or off reservation in building community. Grants support cultural and social services, tribal gatherings, educational programs, healing and purification ceremonies, and retreats. Each proposal should identify the specific needs and uses for the tipi and should identify a nonprofit program partner working in the tribal area who will work with the foundation on the project. Contact the Foundation directly for application and additional guideline information.

Requirements: Projects in American Indian communities are eligible.

Contact: Patricia Stout, President; (510) 841-4123; fax (510) 841-4093; benita@lotus88.org

Internet: http://lotus88.net/

Sponsor: Lotus 88 Foundation for Women and Children
127 University Avenue, P.O. Box 10728
Berkeley, CA 94710

Louetta M. Cowden Foundation Grants 2722

The Louetta M. Cowden Foundation was established in 1964 to support and promote quality educational, cultural, human-services, and health-care programming. In the area of culture the foundation supports programming that: fosters the enjoyment and appreciation of the visual and performing arts; strengthens humanities and arts-related education programs; provides affordable access; enhances artistic elements in communities; and nurtures a new generation of artists. In the area of education the foundation supports programming that: promotes effective teaching; improves the academic achievement of, or expands educational opportunities for disadvantaged students; improves governance and management; strengthens nonprofit organizations, school leadership, and teaching; and bolsters strategic initiatives of area colleges and universities. In the area of health the foundation supports programming that: improves the delivery of health care to the indigent, uninsured, and other vulnerable populations; and addresses health and health-care problems that intersect with social factors. In the area of human services, the foundation supports programming that: strengthens agencies that deliver critical human services and maintains the community's safety net; and helps agencies respond to federal, state, and local public policy changes. There are no application deadlines for the Cowden Foundation. Proposals are reviewed on an ongoing basis. Grant requests for general operating support and program support will be considered. Grants from the Foundation are one year in duration. Application materials are available for download at the grant website. Applicants are strongly encouraged to review the downloadable state application guidelines before applying. Applicants are also encouraged to view the foundation's funding history (link is available at the grant website).

Requirements: Applicants must have 501(c)3 tax-exempt status and serve the residents of Kansas City, Missouri.

Restrictions: Grant requests for capital support will not be considered. The foundation does not support requests from individuals, organizations attempting to influence policy through direct lobbying, or any political campaigns.

Geographic Focus: Missouri

Samples: Nelson-Atkins Museum of Art, Kansas City, Missouri, $100,000 (2010); Truman Medical Center Charitable Foundation, Kansas City, Missouri, $50,000 (2010); YouthFriends, Kansas City, Missouri, $50,000 (2010).

Contact: Spence Heddens; (816) 292-4301; Spence.heddens@baml.com

Internet: https://www.bankofamerica.com/philanthropic/fn_search.action

Sponsor: Louetta M. Cowden Foundation
1200 Main Street, 14th Floor, P.O. Box 219119
Kansas City, MO 64121-9119

Louie M. and Betty M. Phillips Foundation Grants 2723

The foundation supports a variety of organizations, primarily in Nashville, TN, in the fields of health, human services, civic affairs, education, and the arts. Types of support include annual operating grants for selected organizations contributing significantly to the Nashville area; one-year project and program grants for specific projects or equipment; and capital support (five-years maximum) for major capital projects of organizations with strong records of community service. Highest priority is given to nonprofits that were supported by Mr. and Mrs. Phillips.

Requirements: Nonprofit organizations are eligible. With rare exceptions, grants are limited to organizations in the greater Nashville, TN, area.

Restrictions: The foundation does not support individuals or their projects, private foundations, political activities, advertising, or sponsorships. In general, the foundation does not support projects, programs, or organizations that serve a limited audience or a relatively small number of people; disease-specific organizations; biomedical or clinical research; organizations whose principal impact is outside the Nashville area; or tax-supported institutions.

Geographic Focus: Tennessee

Date(s) Application is Due: Jun 1; Oct 31

Amount of Grant: 1,000 - 20,000 U.S.D

Contact: Louie Buntin, (615) 385-5949; fax (615) 385-2507; louie@phillipsfoundation.org

Internet: http://www.phillipsfoundation.org

Sponsor: Louie M. and Betty M. Phillips Foundation
3334 Powell Avenue, P.O. Box 40788
Nashville, TN 37204

Louis and Elizabeth Nave Flarsheim Charitable Foundation Grants 2724

The Louis & Elizabeth Nave Flarsheim Charitable Foundation was established to support and promote quality educational, cultural, human-services, and health-care programming. In the area of arts, culture, and humanities, the foundation supports programming that: fosters the enjoyment and appreciation of the visual and performing arts; strengthens humanities and arts-related education programs; provides affordable access; enhances artistic elements in communities; and nurtures a new generation of artists. In the area of education, the foundation supports programming that: promotes effective teaching; improves the academic achievement of, or expands educational opportunities for, disadvantaged students; improves governance and management; strengthens nonprofit organizations, school leadership, and teaching; and bolsters strategic initiatives of area colleges and universities. In the area of health, the foundation supports programming that improves the delivery of health care to the indigent, uninsured, and other vulnerable populations and addresses health and health-care problems that intersect with social factors. In the area of human services, the foundation funds programming that: strengthens agencies that deliver critical human services and maintains the community's safety net and helps agencies respond to federal, state, and local public policy changes. In the area of community improvement, the foundation funds capacity-building and infrastructure-development projects including: assessments, planning, and implementation of technology for management and programmatic functions within an organization; technical assistance on wide-ranging topics, including grant writing, strategic planning, financial management services, business development, board and volunteer management, and marketing; and mergers, affiliations, or other restructuring efforts. Grant requests for general operating support and program support will be considered. Grants from the foundation are one year in duration. Application materials are available for download at the grant website. Applicants are strongly encouraged to review the state application guidelines for additional helpful information and clarifications before applying. Applicants are also encouraged to review the foundation's funding history (link is available from the grant website). There are no application deadlines for the Flarsheim Charitable Foundation. Proposals are reviewed on an ongoing basis.

Requirements: The Flarsheim Foundation supports organizations that serve the residents of Kansas City, Missouri. Applications must be mailed.

Restrictions: Grant requests for capital support will not be considered.

Geographic Focus: Missouri

Samples: Paul Mesner Puppets, Kansas City, Missouri, $10,000 (2010); Kansas City Chorale, Kansas City, Missouri, $5,000 (2010); Jewish Community Center Charitable Supporting Foundation, Overland Park, Kansas, $2,500.

Contact: Spence Heddens; (816) 292-4301; Spence.heddens@baml.com

Internet: https://www.bankofamerica.com/philanthropic/fn_search.action

Sponsor: Louis & Elizabeth Nave Flarsheim Charitable Foundation
1200 Main Street, 14th Floor, P.O. Box 219119
Kansas City, 64121-9119

Louis H. Aborn Foundation Grants 2725

The Louis H. Aborn Foundation was established in Connecticut, in 1975, by the longtime president of Tams-Witmark Music Library, the famed company that licenses Broadway musical scripts and scores to stock, amateur and professional theaters. Giving is primarily centered around public health projects, education and child welfare. Giving is restricted to Connecticut and New York. There are no particular applications or deadlines with which to adhere, and applicants should begin by contacting the Foundation directly.

Requirements: 501(c)3 organizations serving the residents of Connecticut and New York are eligible to apply.

Geographic Focus: Connecticut, New York

Contact: Hermine F. Aborn, Vice-President; (203) 661-4046

Sponsor: Louis H. Aborn Foundation
46 Wilshire Road
Greenwich, CT 06831-2723

Louis R. Cappelli Foundation Grants 2726

The foundation awards grants to eligible New York nonprofit organizations in its areas of interest, including arts—educational programs for at-risk children and youth in the visual and performing arts, art history, and public libraries; education—programs addressing at-risk children and youth; and health—programs that provide innovative, high-quality health education or organized sports programs serving at-risk children and youth. All requests must be in writing; phone and fax requests are not accepted.

Requirements: New York 501(c)3 tax-exempt organizations located in Sullivan or Westchester Counties are eligible.

Geographic Focus: New York

Date(s) Application is Due: Jun 18

Contact: Director; (914) 769-6500; fax (914) 747-9268; jfevola@cappelli-inc.com

Internet: http://www.cappelli-inc.com/lrc.html

Sponsor: Louis R. Cappelli Foundation
115 Suitevens Avenue
Valhalla, NY 10595

Louis R. Lurie Foundation Grants 2727

The foundation awards one-year renewable grants to nonprofits in the metropolitan Chicago area and in the San Francisco Bay area to support local arts, community services, education, and health care programs. Types of support include program development, seed money, equipment, and general operating support. Applications are accepted by invitation only.

Requirements: Nonprofits serving the metropolitan Chicago area and the San Francisco Bay area are eligible for grant support.

Geographic Focus: California, Illinois

Amount of Grant: 5,000 - 60,000 U.S.D
Samples: Chicago Children's Museum (IL)—for museum relocation, $25,000; Chicago Metro History Education Ctr (IL)—for general operating support, $1000.
Contact: Nancy Terry, Administrator; (415) 392-2470
Internet: http://fdncenter.org/grantmaker/lurie
Sponsor: Louis R. Lurie Foundation
555 California Street, Suite 5100
San Francisco, CA 94104

Love Thy Neighbor Course Competition for Secondary School Faculty 2728
The competition encourages academically rigorous secondary school courses that focus on unselfish love of neighbor as a spiritual and practical ideal. The program supports innovative courses that couple the study and practice of the spirituality of love with the sciences, such as the neurology of compassion, the evolution of altruism, cosmology, or the enhancement of psychological well-being and health. The winning courses must combine the study of unselfish love as understood within spiritual traditions and scientific frameworks, such as physics, cosmology, evolution, biology, political science, the social sciences, and health. The institute encourages a deeply integrative approach to the topic, which may also include experiential learning. Two hard copies of the application should be mailed to Dr. Post, and a copy must also be emailed to him as attachments. Course ideas, additional information, and guidelines are available online.
Requirements: Faculty in religious studies and/or theology, humanities, and the sciences are encouraged to apply. Courses co-taught by two faculty members representing religion and science might be especially compelling in some cases, but team teaching is by no means imperative. In addition to new individual course development, it is possible to apply on the basis of refocusing an already existing course on the theme of unselfish love with significant attention to religion and science. In some cases, several faculty teaching in an integrative core humanities course might work together to include the topic as a major and sustained theme.
Date(s) Application is Due: Jul 15
Amount of Grant: 5,000 U.S.D
Contact: Dr. Stephen Post, (216) 368-6205; Stephen.Post@Case.edu
Internet: http://www.unlimitedloveinstitute.org/competition.html
Sponsor: Institute for Research on Unlimited Love - Altruism, Compassion, Service
10900 Euclid Avenue
Cleveland, OH 44106-4976

Lowe Foundation Grants 2729
The foundation awards grants to organizations in Texas, primarily for programs that support the critical needs of women and children. The arts and higher education also receive support. Grants are awarded to support capital campaigns, building construction/renovation, and general operations. The board meets in April of each year. The December 1 deadline listed is for pre-proposals, with full proposals by request due on December 31.
Requirements: 501(c)3 organizations in Texas may apply for grant support.
Geographic Focus: Texas
Date(s) Application is Due: Dec 1; Dec 31
Amount of Grant: 5,000 - 225,000 U.S.D
Samples: U of Texas Southwestern Medical Ctr at Dallas (Dallas, TX)—to endow a professorship in pediatric critical care, $100,000.
Contact: Clayton Maebius, Trustee; (512) 322-0041; fax (512) 322-0061; info@thelowefoundation.org
Internet: http://www.thelowefoundation.org/guidelines.htm
Sponsor: Lowe Foundation
1005 Congress Avenue, Suite 895
Austin, TX 78701

Lowell Berry Foundation Grants 2730
Mr. Lowell W. Berry established The Lowell Berry Foundation in 1950 with the primary purpose of assisting in strengthening Christian ministry at the local church level. Mr. Berry's secondary purpose was that of assisting social service programs in the areas in California where he lived and operated his business. Guidelines are available at the Foundation's website.
Requirements: 501(c)3 tax-exempt organizations may apply, The Foundation provides funding primarily in Contra Costa and Alameda counties of California.
Restrictions: No grants to individuals, or for building or capital funds, equipment, seed money, or land acquisition.
Geographic Focus: California
Amount of Grant: 5,000 - 200,000 U.S.D
Contact: Katherine Sanders, Office Manager; (925) 284-4427; fax (925) 284-4332; info@lowellberryfoundation.org
Internet: http://www.lowellberryfoundation.org/process.html
Sponsor: Lowell Berry Foundation
3685 Mount Diablo Boulevard, Suite 269
Lafayette, CA 94549-3776

Lown Cardiovascular Center Developing World Fellowship 2731
The fellowship offers training in noninvasive and preventive cardiology and public health for physicians from developing countries. Participants receive training in a model of treatment that maximizes risk factor modification and medical management in order to minimize intervention. After completing training, fellowship participants return to their home countries prepared to address the mounting epidemic of cardiovascular disease. This two-year fellowship program program is conducted in the United States. Fellowship recipients travel to the United States on a J-1 visa. Guidelines are available online.
Requirements: The following criteria are essential and cannot be waived: valid ECFMG certificate; training in internal medicine and/or cardiology; formal affiliation with a supporting institution in the country of origin; and commitment to returning to medical practice in the home country after completing the fellowship program. Candidates from developing countries who meet the specified criteria are invited to apply.
Amount of Grant: 40,000 U.S.D
Contact: Dr. Shmuel Ravid, (617) 732-1318; fax (617) 734-5763
Internet: http://www.lowncenter.org/fellowship.htm
Sponsor: Lown Cardiovascular Research Foundation
21 Longwood Avenue
Brookline MA, MA 02446

LRF New Investigator Leukemia Research Grants 2732
The foundation supports leukemia-related research in a broad sense, including such areas as biochemistry, microbiology, virology, molecular biology, genetics, and immunology. Projects related to significant aspects of leukemia, lymphoma, or myelodysplasia also are considered. The goal of the program is to support new investigators; preference will be given to applicants proposing new lines of investigation. Types of support include research grants and postdoctoral fellowships. Grants and fellowships are for a one-year period and may be renewable for a second year. Annual deadline dates may vary; contact program staff for exact dates.
Requirements: Grants are restricted to researchers who are within five years of the end of their training or of their first independent faculty appointment at the time of the proposed starting date of the grant.
Restrictions: The funds may not be used for salary support of the principal investigator.
Date(s) Application is Due: Feb 15
Amount of Grant: 75,000 U.S.D
Contact: Kelli Fitzgerald, Grant Manager; (847) 424-0600; fax (847) 424-0606; Kelli@LRFMail.org
Internet: http://www.leukemia-research.org/phd/application.asp
Sponsor: Leukemia Research Foundation
2700 Patriot Boulevard, Suite 100
Glenview, IL 60026-8021

LRF Physician-Scientist Postdoctoral Fellowships 2733
The foundation awards physician-scientist postdoctoral fellowships to help physicians prepare for independent careers in research of leukemia and related disorders and obtain significant research results. Fellowships support two years of research. Annual deadline dates may vary; contact program staff for exact dates.
Requirements: Fellows (MD, DO, DVM, or equivalent) not yet established as independent investigators are eligible.
Date(s) Application is Due: Feb 15
Amount of Grant: 30,000 U.S.D
Contact: Kelli Fitzgerald, Medical Grants Administrator; (847) 424-0600; fax (847) 424-0606; Kelli@LRFMail.org
Internet: http://www.leukemia-research.org/phd/application.asp
Sponsor: Leukemia Research Foundation
2700 Patriot Boulevard, Suite 100
Glenview, IL 60026-8021

LRF Postdoctoral Fellowships 2734
The program awards postdoctoral fellowship support for trainees who wish to work in a research area related to leukemia and related disorders. Fellowships are awarded for two years. Annual deadlines dates may vary; contact program staff for exact dates.
Requirements: Fellows (MD or PhD) with up to three years of training (excluding medical residency) prior to receiving the fellowship are eligible.
Date(s) Application is Due: Feb 15
Amount of Grant: Up to 30,000 U.S.D
Contact: Kelli Fitzgerald, (847) 424-0600; fax (847) 424-0606; Kelli@LRFMail.org
Internet: http://www.leukemia-research.org/phd/application.asp
Sponsor: Leukemia Research Foundation
2700 Patriot Boulevard, Suite 100
Glenview, IL 60026-8021

LSA Fellowships 2735
Fellows of the Leukemia Society are promising investigators with no or minimal prior postdoctoral experience working in fields relevant to leukemia. They will have had less than two years of postdoctoral research training at the time of review. Fellows are encouraged to embark on academic careers involving clinical or fundamental research in leukemia, Hodgkin's disease, the lymphomas, and multiple myeloma. Fellowships are awarded for three years and are nonrenewable; however, an application may be made for a special fellowship in the third year of the fellowship award. The preliminary, two-page application must meet the September application deadline; complete applications must be received by the October date.
Requirements: Applicant must hold a PhD, MD, or equivalent degree at the time a grant is to become effective. There are no restrictions as to citizenship, age, race, sex, or creed.
Restrictions: An award will not be made to an individual with full institutional salary support; applicants must demonstrate lack of or minimal/limited salary support as per institutional policies. Awards are payable to the academic institution on behalf of the investigator.
Date(s) Application is Due: Sep 15; Oct 1
Amount of Grant: 50,000 U.S.D
Contact: Director of Research Administration; (914) 821-8843; fax (914) 821-8946; researchprograms@leukemia-lymphoma.org

Internet: http://www.leukemia-lymphoma.org/all_page?item_id=11618
Sponsor: Leukemia and Lymphoma Society of America
1311 Mamaroneck Avenue
White Plains, NY 10605

LSA Scholar Award for Clinical Research **2736**
Applicants for this award have demonstrated their ability to design and conduct original research on leukemia, lymphoma, and myeloma. They are expected to hold an independent faculty-level or equivalent position for a period of not less than three years and to have obtained substantial research support from a national agency. This program is designed as a career development award and is not intended for the support of well-established or senior investigators. The award runs for five years; however, annual renewal is required and is based on a yearly noncompetitive progress report review.
Date(s) Application is Due: Sep 15; Oct 1
Amount of Grant: 110,000 U.S.D
Contact: Director of Research Administration; (914) 821-8843; fax (914) 821-8946; researchprograms@lls.org
Internet: http://www.leukemia-lymphoma.org/all_page?item_id=11618
Sponsor: Leukemia and Lymphoma Society of America
1311 Mamaroneck Avenue
White Plains, NY 10605

LSA Scholar Awards **2737**
This is a career development award for highly qualified scholars who have demonstrated over a five-year period the ability to conduct original research on leukemia, Hodgkin's disease, the lymphomas, and multiple myeloma. Recipients are expected to hold independent faculty-level or equivalent positions and will usually have successfully competed for research grant support for their work. The program has been designed to aid the individual rather than the project and to encourage continued investigation while ensuring the investigator an income. The applicant should have shown the capacity for sustained, original investigation in abnormal cell growth as it relates to leukemia, lymphoma, and myeloma and must desire an investigative career in the field of leukemia, lymphoma, and myeloma. Awards are usually given to individuals in the clinical or basic science departments of universities or associated research institutions. Scholarships are awarded for terms of five years and are nonrenewable. Preliminary, two-page applications must meet the September deadline; complete applications must meet the October deadline.
Requirements: Applicant must hold a PhD, MD, or equivalent degree at the time the grant is to become effective. There are no restrictions as to citizenship, age, race, sex, or creed.
Restrictions: An award will not be made to an individual with full institutional salary support; applicants must demonstrate lack of or minimal/limited salary support as per institutional policies. All awards must be payable to an academic institution.
Date(s) Application is Due: Sep 15; Oct 1
Amount of Grant: 110,000 U.S.D
Contact: Director of Research Administration; (914) 821-8843; fax (914) 821-8946; researchprograms@lls.org
Internet: http://www.leukemia-lymphoma.org/all_page?item_id=11618
Sponsor: Leukemia and Lymphoma Society of America
1311 Mamaroneck Avenue
White Plains, NY 10605

LSA Special Fellowships **2738**
Special Fellows of the Leukemia Society are scientists who have completed more basic training and education in the field than are required for a fellow but less than is necessary for a scholar. A special fellow has had at least two years of postdoctoral training at the time of review but does not yet hold an independent position and relies on the sponsor for research support. Investigators applying must be willing to dedicate their careers to research bearing on a cure or control of leukemia, Hodgkin's disease, the lymphomas, or multiple myeloma. The special fellowships are awarded for a period of three years and are nonrenewable; however, an application may be made for a scholarship in the third year of the special fellowship award. Proposed research projects must be related directly to the field of leukemia, lymphomas, and myeloma. Preliminary, two-page applications must meet the September deadline; complete applications must meet the October deadline.
Requirements: Applicant should hold a PhD, MD, or equivalent degree at the time a grant is to become effective. There are no restrictions as to citizenship, age, race, sex, or creed.
Restrictions: An award will not be made to an individual with full institutional salary support; applicants must demonstrate lack of or minimal/limited salary support as per institutional policies. Award must be payable to an academic institution.
Date(s) Application is Due: Sep 15; Oct 1
Amount of Grant: 60,000 U.S.D
Contact: Director of Research Administration; (914) 821-8843; fax (914) 821-8946; researchprograms@lls.org
Internet: http://www.leukemia-lymphoma.org/all_page?item_id=11618
Sponsor: Leukemia and Lymphoma Society of America
1311 Mamaroneck Avenue
White Plains, NY 10605

LSA Specialized Center of Research Grants **2739**
The purpose of the program is to increase the cure rate of or prevent leukemia, Hodgkin's and non-Hodgkin's lymphoma, and myeloma in order to foster interdisciplinary and synergistic research. The proposed center should be interdisciplinary, cohesive, and sharply focused and should be composed of at least three relevant scientific projects capable of interacting. The research may be fundamental or applied or an integrated combination of the two approaches. Basic research tied to a related translational research project is encouraged but not mandatory. The most promising ideas and the likelihood of a significant impact on the morbidity and mortality of these diseases will be viewed favorably regardless of the particular approach. The center grant will also support scientific core laboratories required by the component research programs. Preliminary applications are due November 1; full applications are due March 15. Annual deadlines may vary; contact program staff for exact dates.
Requirements: Applications may be submitted by individuals holding an MD, PhD, or equivalent, working in domestic or foreign nonprofit organizations, such as universities, colleges, hospitals, or laboratories. Applications may be multi-institutional in nature. Applicants need not be U.S. citizens.
Date(s) Application is Due: Mar 15; Nov 1
Amount of Grant: Up to 100,000,000 U.S.D
Contact: Director of Research Administration; (914) 821-8843; fax (914) 821-8946; researchprograms@lls.org
Internet: http://www.leukemia.org/all_page?item_id=11627
Sponsor: Leukemia and Lymphoma Society of America
1311 Mamaroneck Avenue
White Plains, NY 10605

LSA Translational Research Program Grants **2740**
The society sponsors research toward a cure or control of leukemia, lymphoma, and myeloma. Grants fund research that shows high promise for translating basic biomedical knowledge to prevention or new treatments and, ultimately, to prolong and enhance life. The goal of the program is to accelerate transfer of findings from the laboratory to clinical application. A renewal of the grant may be funded for an additional two years if it includes an approved clinical trial. A principal intent of the program is to permit the acquisition of sufficient data to apply to an appropriate agency or sponsor for sustained support. Guidelines and sample application are available online.
Requirements: Applications may be submitted by individuals holding an MD or PhD, or equivalent degree, working in domestic or foreign nonprofit organizations, such as universities, colleges, hospitals, laboratories, units of state and local governments, or eligible agencies of the federal government.
Date(s) Application is Due: Mar 1
Amount of Grant: Up to 200,000 U.S.D
Contact: Director of Research Administration; (914) 821-8843; fax (914) 821-8946; researchprograms@lls.org
Internet: http://www.leukemia-lymphoma.org/all_page?item_id=11620
Sponsor: Leukemia and Lymphoma Society of America
1311 Mamaroneck Avenue
White Plains, NY 10605

Lubbock Area Foundation Grants **2741**
The Lubbock Area Foundation is a nonprofit community foundation that manages a pool of charitable endowment funds, the income from which is used to benefit the South Plains community through grants to nonprofit organizations, educational programs and scholarships. Funding priorities are: art and culture; social services; civic and community; education; social services; health and human services. Grants may be made for start-up funding, general operating support, program support and demonstration programs. Typical grant awards range from $500 - $2,500 with $5,000 as the maximum from the unrestricted funds. Grant application and additional guidelines are available on the Foundation's website.
Requirements: The Foundation restricts its support to organizations in Lubbock and the surrounding South Plains area which are 501(c)3 or the government equivalent.
Restrictions: The Foundation does not make grants to individuals, for political purposes, to retire indebtedness or for payment of interest or taxes.
Geographic Focus: Texas
Date(s) Application is Due: Jan 1; Mar 1; May 1; Jul 1; Sep 1; Nov 1
Amount of Grant: 500 - 5,000 U.S.D
Samples: Louise Hopkins Underwood Center for the Arts Clay Studio, Lubbock, TX, $8,500—Firehouse Theatre and Marketing Staff Salaries; The Haven Animal Care Shelter, Lubbock, TX—$8,000— funding for veterinary care for animals; Lubbock Children's Health Clinic, Lubbock, TX, $3,600—Pharmacy Assistance Program.
Contact: Kathleen Stocco, Executive Director; (806) 762-8061; fax (806) 762-8551
Internet: http://www.lubbockareafoundation.org/grant.shtml
Sponsor: Lubbock Area Foundation
1655 Main Street, Suite 202
Lubbock, TX 79401

Lubrizol Foundation Grants **2742**
The Lubrizol Foundation makes grants in support of education, health care, human services, civic, cultural, youth and environmental activities of a tax-exempt, charitable nature. Scholarships, fellowships, and awards are generally made in the fields of chemistry and chemical and mechanical engineering at colleges and universities. Types of support include the following: annual campaigns; building/renovation; capital campaigns; continuing support; employee matching gifts; employee volunteer services; equipment; fellowships; general/operating support; scholarship funds. Priority is given to the greater Cleveland, Ohio and Houston, Texas areas. The Lubrizol Foundation typically reviews and decides upon requests quarterly. There are no deadlines by which you need to submit your proposal. Proposals may be submitted via postal mail or email, but not by fax. Applicants will receive written notification of the decision on their proposal.
Requirements: Written applications of established Ohio and Texas nonprofit charitable organizations will be considered on a case by case basis. Grant proposals should include

the following: a cover letter that summarizes the purpose of the request, signed by the executive officer of the organization or development office; a narrative of specific information related to the subject of the request; current audited financial statements and a specific project budget, if applicable; documentation of the organization's Federal tax-exempt status, e.g., a copy of the 501(c)3 determination letter. Additional descriptive literature (e.g., an annual report, brochures, etc.) that accurately characterizes the overall activities of the organization is appreciated. Upon review, further information may be requested including an interview and site visit.
Restrictions: Grants are not made for religious or political purposes, to individuals nor, generally, to endowments.
Geographic Focus: Ohio, Texas
Amount of Grant: 2,000 - 250,000 U.S.D
Samples: Big Brothers Big Sisters of Greater Cleveland, Cleveland, OH, $2,500 - for operating support; American Red Cross, Houston, Texas, $25,000 - for Hurricane Ike Disaster Relief Fund; Hospice of the Western Reserve, Cleveland, OH, $250,000 - toward construction of new faculty.
Contact: Karen Lerchbacher, Administrator; (440) 347-1797; fax (440) 347-1858; karen.lerchbacher@lubrizol.com
Internet: http://www.lubrizol.com/CorporateResponsibility/Lubrizol-Foundation.html
Sponsor: Lubrizol Foundation
29400 Lakeland Boulevard, 053A
Wickliffe, OH 44092-2298

Lucile Horton Howe and Mitchell B. Howe Foundation Grants 2743

The foundation supports youth organizations, social services, medical research, hospitals, religion, child welfare, drug abuse, education, and family services. Types of support include continuing support, general operating funds, and research. Requests from qualified organizations must be received prior to the second Tuesday of March for the grant year. There is only one consideration meeting a year.
Requirements: Only nonprofit organizations in the Prox-Pasadena area and the San Gabriel Valley of California are eligible. A brief letter and a copy of the organizations 501(c)3 form are the requirements for application.
Restrictions: No restricted grants will be funded by the foundation.
Geographic Focus: California
Date(s) Application is Due: Feb 28
Amount of Grant: 1,000 - 300,000 U.S.D
Samples: Huntington Medical Research Institutes, Pasadena, CA., $60,000; Childrens Hospital Los Angeles, Los Angeles, CA., $12,000; Upward Bound Group Home, Pioneer, CA., $10,000;
Contact: Mitchell B. Howe, President; (626) 792-2771; lhmbhowefoun@earthlink.net
Sponsor: Lucile Horton Howe and Mitchell B. Howe Foundation
180 South Lake Avenue
Pasadena, CA 91101-4932

Lucile Packard Foundation for Children's Health Grants 2744

The vision of the Foundation is that all children in the communities the Foundation serves are able to reach their maximum health potential. The Foundation invests in programs and projects that have the potential to improve California's systems of care for children with special health care needs and their families. Funding is provided for programs that address improving care for children with special needs in the following areas: system reform; care coordination/enhanced medical home; pediatric education; quality measures focused on children with special health care needs; data collection and/or analysis; and advocacy.
Requirements: Generally the Foundation invites applications for grant proposals. However, organizations that have not been invited may submit a letter of inquiry if the organization meets the organizational criteria and if their proposed program falls within the Foundation's focus area. Applicants should review program and organizational criteria first. Eligible organizations: are classified as tax exempt under 501(c)3 or are a public or educational entity, or collaborations of nonprofit and public agencies with a designated fiscal sponsor, or entities that have a charitable purpose; have demonstrated capacity to implement effective, culturally competent programs, reach intended populations, and achieve clear, reasonable and measureable goals and objectives; and promote and maintain nondiscriminatory policies in programs and the workplace. Eligible programs or projects are those that hold promise of improving the systems of care for Children with Special Health Care Needs (CSHCN). Eligible programs should: be focused primarily on California; have potential for national applicability; have the potential to improve the system of care for children with special needs; demonstrate an understanding of the complex system of care for CSHCN and builds on existing knowledge and systems; have potential to affect a large number of CSHCN; have potential for impact within five years; have reasonable evidence of sustainability; have reasonable evidence of replicability; and present opportunities for collaboration.
Restrictions: Unsolicited full proposals should not be submitted. The following are not funded: disease-specific projects; individuals; scholarships; support for candidates for political office; private foundations; religious organizations for religious purposes; fundraising sponsorships; annual fund appeals; capital campaigns; and basic scientific research.
Geographic Focus: California
Contact: Amanda Frederickson, Annual Giving Program Manager; (650) 736-0676; amanda.frederickson@lpfch.org
Internet: http://www.lpfch.org/programs/cshcn/process.html
Sponsor: Lucile Packard Foundation for Children's Health
400 Hamilton Avenue, Suite 340
Palo Alto, CA 94301

Lucy Downing Nisbet Charitable Fund Grants 2745

The Lucy Downing Nisbet Charitable Fund was established in 2002 to support and promote educational, health-and-human-services, and arts programming for underserved populations. The Foundation specifically serves organizations located in and serving the people of Vermont. Special consideration is given to organizations in the area of healthcare/nursing, domestic violence awareness, heart disease, and endangered species, and organizations located in and serving the people of Morrisville, Vermont. Grants are one year in duration. Applicants must apply online at the grant website. Applicants are strongly encouraged to do the following before applying: review the downloadable state application procedures for additional helpful information and clarifications; review the downloadable online-application guidelines at the grant website; review the foundation's funding history (link is available from the grant website); review the online application questions in advance; and review the list of required attachments. These will generally include: a list of board members, financial statements (audited, reviewed, or compiled by independent auditor); an organization summary; a list of other funding sources; an IRS Determination letter; and other required documents. All attachments must be uploaded in the online application as PDF, Word, or Excel files. The Lucy Downing Nisbet Charitable Fund application deadline is 11:59 p.m. on January 15. Applicants will be notified of grant decisions by letter within two to three months of the deadline.
Requirements: The foundation specifically serves organizations located in and serving the people of Vermont. Applicants must have 501(c)3 tax-exempt status.
Restrictions: The trustees do not make grants for deficit financing, annual giving, endowments or capital projects. The fund does not support requests from individuals, organizations attempting to influence policy through direct lobbying, or any political campaigns.
Geographic Focus: Vermont
Date(s) Application is Due: Jan 15
Amount of Grant: 5,000 - 20,000 U.S.D
Samples: Copley Health Systems, Morrisville, Vermont, $58,400, to provide computerized Physician's Order Entry component of Electronic Medical Record Software and to fund Nursing Education (2010); Vermont Foodbank, South Barre Vermont, $20,000, for U.S.DA Summer Food Service Program in high-poverty and extremely rural area (2010); River Arts of Morrisville, Morrisville, Vermont, $8,000, Arts for Everyone archive (2010).
Contact: Amy Lynch, Foundation Manager; (860) 657-7015; amy.r.lynch@baml.com
Internet: https://www.bankofamerica.com/philanthropic/fn_search.action
Sponsor: Lucy Downing Nisbet Charitable Fund
200 Glastonbury Boulevard, Suite # 200, CT2-545-02-05
Glastonbury, CT 06033-4056

Lucy Gooding Charitable Foundation Trust 2746

The trust awards grants to eligible Florida nonprofit organizations in its areas of interest, including child welfare, disabled children, disadvantaged people, homelessness, and hospices. Types of support include building construction/renovation, capital campaigns, equipment and land acquisition, general operating support, seed grants, and United Way. Contact the office for required application form.
Requirements: Florida 501(c)3 nonprofit organizations serving the Jacksonville area are eligible.
Geographic Focus: Florida
Date(s) Application is Due: Sep 30
Amount of Grant: 2,500 - 25,000,000 U.S.D
Samples: Wolfson Children's Hospital (Jacksonville, FL)—to establish a pediatric neurosurgery center, $5 million; Community Foundation in Jacksonville (FL)—to endow a fund to support organizations that serve children or provide relief services, with an emphasis on Florida's Clay and Duval Counties, $25 million.
Contact: Bonnie Smith, Trustee; (904) 786-4796; fax (904) 786-4796; bhsmith@bellsouth.net
Sponsor: Lucy Gooding Charitable Foundation Trust
P.O. Box 37349
Jacksonville, FL 32236-7349

Lucy Gooding Charitable Foundation Trust Grants 2747

The Foundation, established in 1988, supports organizations involved with children and youth, services, education, human services, residential/custodial care, hospices, economically disadvantaged, people with disabilities, and homelessness. The Foundation gives primarily in the five county area surrounding Jacksonville, Florida.
Requirements: Funding preference is for organizations in the five county area surrounding Jacksonville, Florida, with projects helping children. Contact the Foundation for application form.
Restrictions: No support for private foundations, religious organizations, or for adults-only services, and individuals.
Geographic Focus: Florida
Date(s) Application is Due: Sep 30
Amount of Grant: 5,000 - 1,000,000 U.S.D
Samples: Baptist Medical Center, Jacksonville, FL, $1,000,000 - for Pediatric Neurosurgical Center; Bridge of Northeast Florida, Jacksonville, FL, $60,000 - for youth services for disadvantaged; United Way of Northeast Florida, Jacksonville, FL, $450,000 - for establishment of full service site in local school;
Contact: Bonnie H. Smith, Treasurer; (904) 786-4796; fax (904) 786-4796; bhsmith@bellsouth.net
Sponsor: Lucy Gooding Charitable Foundation Trust
P.O. Box 37349
Jacksonville, FL 32236-7349

Ludwick Family Foundation Grants 2748

The foundation is a California nonprofit public benefit corporation established exclusively for charitable, scientific, literary, and educational purposes. The foundation tends to fund tangible types of items that will remain with and can be used repeatedly by the organization.
Requirements: Eligible applicants include U.S. or U.S.-based international organizations, 501(c)3 nonprofit public charities, and government agencies (any level). All grants are to be used exclusively for charitable, public benefit purposes.
Restrictions: The foundation does not grant requests for salaries, general operating expenses, scholarships, endowment funds, fundraising events or capital campaigns, feasibility studies, consulting fees, or advertising. The foundation will no longer accept any unsolicited requests for research or from public/private schools (K-12), universities/colleges, child daycare/development centers, hospitals, or libraries.
Date(s) Application is Due: Mar 31; Aug 31
Amount of Grant: 5,000 - 50,000 U.S.D
Samples: AbilityFirst (Pasadena, CA)—to purchase an American Disabilities Act accessible van to enable the transportation of people who are disabled; Alaska Public Radio Network (AK)—to purchase recorder kits and professional stereo headphones to enable reports of news, current affairs, and cultural information for Alaska's Native People.
Contact: Deanna Monaghan, (626) 852-0092; fax (626) 852-0776; ludwickfndn@ludwick.org
Internet: http://www.ludwick.org
Sponsor: Ludwick Family Foundation
P.O. Box 1796
Glendora, CA 91740

Lumpkin Family Foundation Healthy People Grants 2749

Historically, the Lumpkin Family Foundation has supported entities that provide services that help keep people in its geographic area healthy. Requests will be evaluated on how well they accomplish one or more of the following: support the creativity of nonprofit organizations by seeding new projects and encouraging experimentation and innovation; support organizations demonstrating outstanding leadership in their field or community; promote the effectiveness of organizations and the nonprofit sector by supporting planning, learning and the professional development of staff and board leaders; facilitate collaboration across traditional organization or sector boundaries for community benefit; and develop public understanding of issues and promote philanthropic support necessary to address issues of community importance.
Requirements: Grants are awarded to nonprofit organizations that serve the community without discrimination on the basis of race, sex, or religion. Special consideration will be given to organizations and programs in East Central Illinois.
Restrictions: Proposals will not be considered from organizations that are not 501(c)3 tax-exempt; organizations whose primary purpose is to influence legislation; political causes, candidates, organizations or campaigns; individuals; or religious organizations, unless the particular program will benefit a large portion of the community and does not duplicate the work of other agencies in the community.
Amount of Grant: 1,000 - 50,000 U.S.D
Contact: Bruce Karmazin, (217) 234-5915 or (217) 235-3361; fax (217) 258-8444; Bruce@lumpkinfoundation.org
Internet: http://www.lumpkinfoundation.org/bWHATbwefund/HealthyPeople.aspx
Sponsor: Lumpkin Family Foundation
121 South 17th Street
Mattoon, IL 61938

Lupus Foundation of America Research Grants 2750

Grants are awarded annually to encourage research in lupus erythematosus. It is the foundation's hope that this investment in research will produce new information that may directly lead to much larger projects and substantially increased funding from other sources, particularly the National Institutes of Health, in the future. Grants are awarded for up to two years.
Requirements: Awards are for junior investigators only (assistant professors or below, if in academic medicine). Teachers and aides at teaching centers, hospitals, and colleges are eligible.
Restrictions: First-year fellows are not eligible.
Date(s) Application is Due: Feb 15
Amount of Grant: 125,000 U.S.D
Contact: Grants Coordinator; (202) 349-1155; fax (202) 349-1156; lupusinfo@aol.com
Internet: http://www.lupus.org/research/grantproginfo.html
Sponsor: Lupus Foundation of America
2000 L Street NW, Suite 710
Washington, DC 20036-4916

Lustgarten Foundation for Pancreatic Cancer Research Grants 2751

The foundation aims to make rapid advances in the battle against pancreatic cancer through a combination of cutting-edge technology and novel ideas. Grants can be submitted in one of three areas: novel technologies for pancreatic cancer genetics; screening for the early detection of pancreatic cancer; and novel therapies in pancreatic cancer. Proposals that exhibit leveraging of institutional resources and the attainment of major new programs are encouraged. This initiative encourages, but does not require, the development of integrative and collaborative teams of investigators from within a single institution or among several institutions. Letters of intent must be received on or before 5:00 p.m. Eastern Time on July 31, with full applications due no later than 5:00 p.m. Eastern Time on August 18. Grants will be awarded for a one-year period for a maximum amount of $100,000, of which no more than 10% can be used for indirect costs.
Requirements: Applications will be accepted from individual investigators as well as collaborating investigators. Applicants must have an MD or a PhD degree to become principal investigators. Eligible organizations include nonprofit organizations; public or private institutions, such as universities, colleges, hospitals, and laboratories; and domestic or foreign institutions/organizations.
Amount of Grant: Up to 100,000 U.S.D
Samples: Johns Hopkins University, Allison Klein, Ph.D. - Identification of Pancreatic Cancer Susceptibility Genes; University of Michigan Medical School, Chandan Kumar, Ph.D. - Discovery of Recurrent Gene Fusions in Pancreatic Cancer using High-Throughput Sequencing; Johns Hopkins University School of Medicine, Joshua Mendell, M.D., Ph.D. - Ras-regulated microRNAs in Pancreatic Cancer.
Contact: Kerri Kaplan, Executive Director; (516) 803-2304 or (866) 789-1000; fax (516) 803-2303; kkaplan@cablevision.com or lsasso@cablevision.com
Internet: http://www.lustgarten.org/Page.aspx?pid=666
Sponsor: Lustgarten Foundation for Pancreatic Cancer Research
1111 Stewart Avenue
Bethpage, NY 11714

Lutheran Foundation Grants 2752

The foundation awards Regular, Partnership, and Emergency Grants. Regular grants are available to any congregation or nonprofit organization in northeast Indiana. Regular grants are awarded four times a year. The foundation will consider funding of one-time grants, single-year grants, and multi-year grants that do not exceed three years. Partnership grants are available to each of the foundation's member congregations. They are designed to assist congregations with their outreach ministries and also to enrich the internal life of their own membership. Emergency grants are designed to address unforeseen circumstances that disrupt the ongoing ministry or service provided by churches and organizations. The foundation also encourages, promotes, and supports Lutheran and Christian educational opportunities. Guidelines and application are available online.
Requirements: The foundation awards grants to ministries located in 10 counties in northeast Indiana, including Adams, Allen, DeKalb, Huntington, LaGrange, Noble, Steuben, Wabash, Wells, and Whitley counties.
Restrictions: Proposals for projects that have been completed, deficit financing, fundraising, and promotional items are not considered. In addition, the foundation does not fund individuals, endowments, other funds, or the general operating expenses of congregations or medical procedures that are destructive of life.
Geographic Focus: Indiana
Amount of Grant: Up to 10,000 U.S.D
Samples: Allen County Non-Public School Assoc (Fort Wayne, IN)—to fund school nurse program expenses for parochial and independent schools $57,565 three-year grant; Bethlehem Evangelical Lutheran Church and School (Fort Wayne, IN)—for renovations to cafeteria, basement classrooms, and Fellowship Hall for handicap accessibility $9870 matching grant; Camp Lutherhave (Albion, IN)—to purchase multi-passenger van, $22,000; Cancer Services of Allen County (Fort Wayne, IN)—for operational support, $75,000 three-year grant.
Contact: Terri Kortokrax, (800) 677-2451 or (260) 458-2112, ext. 302; fax (260) 458-3069; tkortokrax@thelutheranfoundation.org
Internet: http://www.thelutheranfoundation.org/grants.htm
Sponsor: Lutheran Foundation
3024 Fairfield Avenue, Suite 200
Fort Wayne, IN 46807-1697

Lydia deForest Charitable Trust Grants 2753

The Lydia Collins deForest Charitable Trust was established in 2002. The deForest Charitable Trust specifically supports: organizations that provide services to those who are visually limited; churches and organizations affiliated with the Protestant Episcopal Church in the United States and other religious organizations in union with or recognized by the Episcopal Church; and organizations that provide services to those who are homeless, unemployed, or substance-dependent. Special consideration is given to the following 3 organizations: The Lighthouse, Inc., in New York, New York; the Calvary Episcopal Church of Summit, New Jersey; and the Salvation Army in Union, New Jersey. Grant requests for general operating support are strongly encouraged. Program support will also be considered. Small, program-related capital expenses may be included in general operating or program requests. To better support the capacity of nonprofit organizations, multi-year funding requests are encouraged. Applicants must apply online at the grant website. Applicants are strongly encouraged to do the following before applying: review the downloadable state application procedures for additional helpful information and clarifications; review the downloadable online-application guidelines at the grant website; review the trust's funding history (link is available from the grant website); review the online application questions in advance; and review the list of required attachments. These will generally include: a list of board members, financial statements (audited, reviewed, or compiled by independent auditor); an organization summary; a list of other funding sources; an IRS Determination letter; and other required documents. All attachments must be uploaded in the online application as PDF, Word, or Excel files. The application deadline for the deForest Charitable Trust is 11:59 p.m., November 30. Applicants will be notified of grant decisions before February 28 of the following year.
Requirements: Applicants must have 501(c)3 tax-exempt status and serve the people of New Jersey. Occasional support is given to organizations within the Metro New York City area.
Restrictions: The trust does not support requests from individuals, organizations attempting to influence policy through direct lobbying, or any political campaigns.
Geographic Focus: New Jersey
Date(s) Application is Due: Nov 30

Samples: Family Promise, Summit, New Jersey, $40,000; Saint Philips Academy, Newark, New Jersey, $40,000; Salvation Army New Jersey Divisional Headquarters, Union, New Jersey, $40,000 (2010).
Contact: Ken Goody, Vice President; (646) 855-0956; kenneth.l.goody@baml.com
Internet: https://www.bankofamerica.com/philanthropic/fn_search.action
Sponsor: Lydia Collins deForest Charitable Trust
One Bryant Park, NY1-100-28-05
New York, NY 10036

Lymphoma Research Foundation Fellowships **2754**
Two-year fellowships uniquely permit the uninterrupted development of promising leads and help attract the nation's best scientific talent to careers in lymphoma. Its emphasis upon translational as well as purely basic research assures that breakthroughs are quickly transformed into viable treatment options.
Date(s) Application is Due: Sep 9
Amount of Grant: 100,000 - 110,000 U.S.D
Contact: Director of Medical and Scientific Outreach; (212) 349-2910; fax (212) 349-2886; researchgrants@lymphoma.org
Internet: http://www.lymphoma.org/site/pp.asp?c=bfIKIVMIG&b=38480
Sponsor: Lymphoma Research Foundation
111 Broadway, 19th Floor
New York, NY 10016

Lymphoma Research Foundation Grants **2755**
The foundation supports high-quality research that results in safer, more effective treatments and a cure for lymphoma. Applications are accepted for two programs. Fellowship research grants support young investigators (third-year fellows) pursuing careers in lymphoma research. Lymphoma-specific projects submitted by individuals associated with accredited academic institutions, JCAHO-accredited research hospitals, and other research organizations that have national or international reputations for excellence will be considered. Junior faculty research grants support researchers at the assistant or associate professor level. Mantle cell research projects also are funded at both levels.
Date(s) Application is Due: Sep 15
Amount of Grant: 45,000 - 75,000 U.S.D
Samples: U Hospital Grosshadern (Germany)—for research on mantle-cell lymphoma, a rare form of blood cancer, $890,000; U of Texas MD Anderson Medical Ctr (Houston, TX)—for research on mantle-cell lymphoma, a rare form of blood cancer, $2.4 million.
Contact: Grants Administrator; (212) 349-2910 or (800) 235-6848; ResearchGrants@lymphoma.org
Internet: http://www.lymphoma.org/site/pp.asp?c=bfIKIVMIG&b=36878
Sponsor: Lymphoma Research Foundation
8800 Venice Boulevard, Suite 207
Los Angeles, CA 90034

Lynde and Harry Bradley Foundation Fellowships **2756**
The Bradley Fellowship program is a major part of the Foundation's program interest in strengthening America's intellectual infrastructure at the higher-education level. Begun in 1986, the program is a model for influencing the intellectual framework of national life. It helps fulfill the two basic prerequisites for liberal education — good teaching and good curriculum — by enabling colleges and universities to offer students in the advanced stages of graduate work the opportunity to pursue serious studies with excellent teachers. For the program, the Foundation chooses institutions of higher education each year to support graduate and post-graduate students, who are designated as Bradley Fellows. The Fellows are chosen by each institution upon the recommendation of one or more distinguished professors in the humanities, social sciences, and the law.
Contact: Daniel P. Schmidt, Vice President for Programs; (414) 291-9915; fax (414) 291-9991
Internet: http://www.bradleyfdn.org/bradley_fellows.asp
Sponsor: Lynde and Harry Bradley Foundation
1241 North Franklin Place
Milwaukee, WI 53202-2901

Lynde and Harry Bradley Foundation Grants **2757**
The foundation aims to encourage projects that focus on cultivating a renewed, healthier, and more vigorous sense of citizenship among the American people and among peoples of other nations. Projects likely to be supported will generally treat free people as self-governing, personally responsible citizens, not as victims or clients; aim to restore the intellectual and cultural legitimacy of common sense, the wisdom of experience, everyday morality, and personal character; seek to reinvigorate and re-empower the traditional, local institutions—families, schools, churches, and neighborhoods—that provide training in and room for the exercise of genuine citizenship, that pass on everyday morality to the next generation, and that cultivate personal character; and encourage decentralization of power and accountability away from centralized, bureaucratic, national institutions back to states, localities, and revitalized mediating structures. Eligible projects may address any arena of public life—economics, politics, culture, or civil society; the problem of citizenship at home or abroad; Milwaukee and Wisconsin community and state projects that aim to improve the life of the community through increasing cultural and educational opportunities, grassroots economic development, and social and health services; the resuscitation of citizenship in the economic, political, cultural, or social realms; policy research and writing about approaches encouraging that resuscitation; academic research and writing that explore the intellectual roots of citizenship; and popular writing and media projects that illustrate for a broader public audience the themes of citizenship. The foundation supports programs that research the needs of gifted children and techniques of providing education for students with superior skills and/or intelligence; research programs investigating how learning occurs in gifted children; and demonstration programs of instruction.
Requirements: As an initial step, tax-exempt and nonprofit organizations should prepare a brief letter of inquiry presenting a concise description of their project, its objectives and significance, and the qualifications of the organizations and individuals involved. If the project appears to fall within the foundation's mandate, the applicant will be invited to submit a formal proposal. If invited to submit a formal proposal, the applicant should submit another letter. It should include a more-thorough, yet still concise description of the project, its objectives and significance, and the qualifications of the groups and individuals involved in it. It should also include a project budget, the specific amount being sought from Bradley, and a list of its other sources of support, philanthropic or otherwise.
Restrictions: The foundation favors projects that are normally not financed by public funds and will consider requests from religious organizations that are not denominational in character. Grants without significant importance to the foundation's areas of interest will only under special conditions be considered for endowment or deficit financing proposals. Grants will not be made to individuals, for overhead costs, or for fund-raising counsel.
Date(s) Application is Due: Feb 1; May 1; Aug 1; Nov 1
Amount of Grant: 25,000 - 300,000 U.S.D
Samples: Bel Canto Chorus of Milwaukee, Milwaukee, Wisconsin, $25,000 - to support general operations (2010); Penfield Children's Center, Milwaukee, Wisconsin, $25,000 - to support general operations (2010); Rawhide, Inc., New London, Wisconsin, $20,000 - to support the Residential Care Assistance Program (2010).
Contact: Daniel P. Schmidt, Vice President for Programs; (414) 291-9915; fax (414) 291-9991
Internet: http://www.bradleyfdn.org/program_interests.asp
Sponsor: Lynde and Harry Bradley Foundation
1241 North Franklin Place
Milwaukee, WI 53202-2901

Lynde and Harry Bradley Foundation Prizes: Bradley Prizes **2758**
The Bradley Prizes formally recognize individuals of extraordinary talent and dedication who have made contributions of excellence in areas consistent with The Lynde and Harry Bradley Foundation's mission. Up to four Prizes of $250,000 each are awarded annually to innovative thinkers and practitioners whose achievements strengthen the legacy of the Bradley brothers and the ideas to which they were committed. Each year, Bradley Prize nominations are solicited from a national panel of more than 100 prominent individuals involved in academia, public-policy research, journalism, civic affairs, and the arts. All nominees are carefully evaluated by a distinguished selection committee that makes recommendations to the Foundation's Board of Directors, which selects them. The Prize winners are then honored at a celebratory awards ceremony.
Amount of Grant: 250,000 U.S.D
Contact: Daniel P. Schmidt, Vice President; (414) 291-9915; fax (414) 291-9991
Internet: http://www.bradleyfdn.org/bradley_prizes.asp
Sponsor: Lynde and Harry Bradley Foundation
1241 North Franklin Place
Milwaukee, WI 53202-2901

M-A-C AIDS Fund Grants **2759**
The fund makes grants to registered charitable, nonprofit organizations that provide basic needs, such as food, clothing, housing, or shelter (short-term or transitional); direct services related to health care; social services, transportation (for medical visits, outpatient visits, and other social services); health-related recreational activities; and programs that bring HIV/AIDS education, awareness, and prevention to public attention. Proposals are reviewed quarterly and must arrive at least one month prior to the quarterly review date. Application guidelines and required format are available online.
Requirements: Grants are awarded to tax exempt, non-profit organizations that are 501(c)3 and directly associated with HIV/AIDS. A minimum of 80% of the total organization's budget must be is directed to program.
Restrictions: Grants will not fund: individuals; lobbying activities; ongoing general operating expenses or existing deficits; endowments, unless they provide a direct service to PWA's; capital costs; conferences, summits, briefings, PSA's; or research. With the exception of North American based charities, the Fund does not accept unsolicited international grant proposals.
Amount of Grant: 5,000 - 25,000 U.S.D
Contact: Program Coordinator, (212) 965-6300; fax (212) 372-6171, macaidsf@maccosmetics.com
Internet: http://www.macaidsfund.org/support/givingguidelines.html
Sponsor: M-A-C AIDS Fund
130 Prince Street, 2nd Floor
New York, NY 10012

M.B. and Edna Zale Foundation Grants **2760**
The M.B. & Edna Zale Foundation honors the tradition of its founders through grants that stimulate change. To accomplish this mission, the Foundation acts as a catalyst for collaboration and makes grants in communities where the Directors live or have an interest. Grants are made primarily in the communities of Dallas (Dallas County) and Houston (Harris County), Texas; Boca Raton, Florida; Portland, Oregon; and New York, including Long Island. The Foundation has an interest in four areas of funding: community services; health; education; Jewish heritage.
Requirements: 501(c)3 nonprofits organizations in the communities of Dallas (Dallas County) and Houston (Harris County), Texas; Boca Raton, Florida; Portland, Oregon;

and New York, including Long Island are eligible. Contact the Foundation for further application information and guidelines.
Restrictions: The Foundation does not ordinarily provide: major support for the arts; grants to individuals; scholarships and fellowships to individuals (except through colleges and universities).
Geographic Focus: Florida, New York, Oregon, Texas
Amount of Grant: 2,000 - 100,000 U.S.D
Contact: Leonard Krasnow, President; (214) 855-0627; fax (972) 726-7252; mail@zalefoundation.org
Sponsor: M.B. and Edna Zale Foundation
6360 LBJ Highway, Suite 205
Dallas, TX 75240

M. Bastian Family Foundation Grants **2761**
The foundation supports nonprofit organizations in its areas of interest, including music, the arts, higher education, health care and health organizations, religion (Christian and Latter-day Saints), social services, and wildlife conservation. Types of support include general operating support and scholarship funds. Contact the office for application forms. There are no application deadlines.
Requirements: Funding focus is primarily in Utah.
Restrictions: Grants are not made to individuals.
Amount of Grant: 2,000 - 200,000 U.S.D
Samples: Brigham Young U (Provo, UT)—for scholarship endowment, $200,000; Pediatric AIDS Foundation (Los Angeles, CA)—for general support, $20,000; American Diabetes Assoc (Alexandra, VA)—for research, $100,000.
Contact: McKay Matthews, Program Contact; (801) 225-2455
Sponsor: M. Bastian Family Foundation
51 W Center Street, Suite 305
Orem, UT 84057

M.D. Anderson Foundation Grants **2762**
The foundation supports Texas nonprofit organizations, including religious organizations, higher education institutions, parochial education institutions, hospitals, and religious welfare organizations. Types of support include capital grants, general support, program support, scholarships, and research grants. Applicants should send a brief letter describing the proposed project, along with an original and four copies of the proposal. If applying for a matching grant, indicate other sources of funding. There are no application deadlines. The board meets on the third Tuesday of each month. Applications should be received a month prior to the the meeting to be considered.
Requirements: Texas nonprofit organizations are eligible.
Geographic Focus: Texas
Amount of Grant: 25,000 - 100,000 U.S.D
Samples: U of Texas (Austin, TX)—for the law school, $800,000; U of Saint Thomas (Houston, TX)—to construct a humanities and education building, $250,000.
Contact: Toloria Allen, Secretary-Treasurer; (713) 216-4513
Sponsor: M.D. Anderson Foundation
P.O. Box 2558
Houston, TX 77252-8037

M.E. Raker Foundation Grants **2763**
The foundation was established in 1984, serving the Allen County, Indiana area. Areas of interest include: children/youth, services; education; environment; natural resources; health care; historic preservation/historical societies; human services people with disabilities. Support is offered in the following areas: building/renovation; general/operating support; matching/challenge support; Program development. No support for the arts or to individuals.
Requirements: Indiana nonprofit organizations are eligible. Emphasis is given to requests from Fort Wayne. Send a letter requesting an application to the foundation office. Application form required for grant proposals.
Geographic Focus: Indiana
Amount of Grant: 5,000 - 40,000 U.S.D
Contact: Jennifer Pickard, Grants Coordinator; (260) 436-2182
Sponsor: M.E. Raker Foundation
6207 Constitution Drive
Fort Wayne, IN 46804-1517

M.G. and Lillie A. Johnson Foundation Grants **2764**
The foundation awards grants to Texas organizations in the areas of health and education. Education grants are awarded primarily to institutions of higher education. Recipients have been exclusively several junior colleges and universities within the immediate areas surrounding Victoria, TX. Types of support include building construction/renovation, capital campaigns, equipment acquisition, land acquisition, and scholarship funds. Grants are made three times each year. Applications must be received one month prior to the board meeting.
Requirements: Grants are almost exclusively restricted to the Texas Gulf Coast between Wharton and Aransas Counties.
Restrictions: Grants are not awarded for operating funds, endowments, research fellowships, or to individuals.
Geographic Focus: Texas
Amount of Grant: 5,000 - 228,000 U.S.D
Samples: Wharton Community Junior College (TX)—for the radiologic technology program, $60,000; City of Beeville (TX)—for Zoll heart monitors, $50,000; Golden Crescent CASA (TX)—for computers and related equipment, $6279.
Contact: Robert Halepeska, Executive Vice President; (361) 575-7970; fax (361) 575-2264; mgj@cox-internet.com
Sponsor: M.G. and Lillie A. Johnson Foundation
P.O. Box 2269
Victoria, TX 77902

M.J. Murdock Charitable Trust General Grants **2765**
The M.J. Murdock Charitable Trust General Grants provides support for education, arts and culture, and health and human services. The Trust considers educational projects in formal and informal settings, emphasizing enhancement or expansion, as well as new educational approaches. Of special interest to the Trust are the following: performance and visual arts projects which enrich the cultural environment of the region and educational outreach efforts; and programs that emphasize preventative efforts which address physical, spiritual, social, and psychological needs, with a focus on youth. Grants are awarded for capital projects, program initiation, expansion, or for increased organizational capacity. Organizations are encouraged to view the grants awarded link for examples of previously funded grants.
Requirements: Before proceeding, interested parties should review the General Grant Application Guidelines to see if their organization is eligible or their project is appropriate for application to the Trust. After determining eligibility and appropriateness, organizations must submit a letter of inquiry before proceeding with the application. Upon approval, the organization will use the Trust's General Grant Application form and procedures.
Restrictions: The Trust only funds programs in the Pacific Northwest region: Alaska, Idaho, Montana, Oregon, and Washington. The following requests for funding are not considered: specific individuals and/or their personal benefit; individuals unauthorized to act on behalf of a qualified tax-exempt organization; funds that will ultimately be passed through to other organizations; propagandizing or for influencing legislation and elections; institutions that in policy or practice unfairly discriminate against race, ethnic, origin, sex, creed, or religion; sectarian or religious organizations whose principal activity is for the primary benefit of their own members; or for long-term loans, debt retirement, or operational deficits. The following funding requests are rarely considered: normal ongoing operations or the continuation of existing projects; endowments or revolving funds that act as such; continuation of programs previously financed from other external sources; urgent needs, emergency, or gap funding; organizations organized or operating outside any state or territory of the United States.
Geographic Focus: Alaska, Idaho, Montana, Oregon, Washington
Samples: Blachet House of Hospitality, Portland, Oregon, $450,000, to serve Portland homeless (2011); Billings District Council of the Society of St. Vincent de Paul, Billings, Montana, to expand client services, $100,000 (2011); Whidbey Institute, Clinton, Washington, for program expansion to enhance civic engagement, $150,000 (2011).
Contact: Marybeth Stewart Goon, Senior Assistant; (360) 694-8415; fax (360) 694-1819
Internet: http://www.murdock-trust.org/grants/general-grants.php
Sponsor: M.J. Murdock Charitable Trust
703 Broadway, Suite 710
Vancouver, WA 98660

Mabel A. Horne Trust Grants **2766**
The Mabel A. Horne Trust was established in 1957 to support and promote quality educational, human-services, and health-care programming for underserved populations. Grant requests for general operating support are strongly encouraged. Program support will also be considered. Small, program-related capital expenses may be included in general operating or program requests. The majority of grants from the Horne Trust are one year in duration; on occasion, multi-year support is awarded. Applicants must apply online at the grant website. Applicants are strongly encouraged to do the following before applying: review the downloadable state application procedures for additional helpful information and clarifications; review the downloadable online-application guidelines at the grant website; review the trust's funding history (link is available from the grant website); review the online application questions in advance; and review the list of required attachments. These will generally include: a list of board members, financial statements (audited, reviewed, or compiled by independent auditor); an organization summary; a list of other funding sources; an IRS Determination letter; and other required documents. All attachments must be uploaded in the online application as PDF, Word, or Excel files. The application deadline for the Mabel A. Horne Trust is 11:59 p.m. on February 1. Applicants will be notified of grant decisions before May 31.
Requirements: Applicants must have 501(c)3 tax-exempt status.
Restrictions: The trust does not support requests from individuals, organizations attempting to influence policy through direct lobbying, or any political campaigns.
Geographic Focus: Massachusetts
Date(s) Application is Due: Feb 1
Contact: Miki C. Akimoto, Vice President; (866) 778-6859; miki.akimoto@baml.com
Internet: https://www.bankofamerica.com/philanthropic/fn_search.action
Sponsor: Mabel A. Horne Trust
225 Franklin Street, 4th Floor, MA1-225-04-02
Boston, MA 02110

Mabel F. Hoffman Charitable Trust Grants **2767**
The Mabel F. Hoffman Charitable Trust was established in 1969 to support and promote quality educational, human-services, and health-care programming for underserved populations. Special consideration is given to charitable organizations that serve the people of Hartford, Connecticut. The trust makes approximately twelve, modest size grants per year. Grants from the Hoffman Charitable Trust are one year in duration. Applicants must apply online at the grant website. Applicants are strongly encouraged to do the following before applying: review the downloadable state application procedures

for additional helpful information and clarifications; review the downloadable online-application guidelines at the grant website; review the trust's funding history (link is available from the grant website); review the online application questions in advance; and review the list of required attachments. These will generally include: a list of board members, financial statements (audited, reviewed, or compiled by independent auditor); an organization summary; a list of other funding sources; an IRS Determination letter; and other required documents. All attachments must be uploaded in the online application as PDF, Word, or Excel files. The deadline for application to the Mabel F. Hoffman Charitable Trust is 11:59 p.m. on June 15. Applicants will be notified of grant decisions by letter within two to three months after the proposal deadline.
Requirements: First time applicants should contact the Program Officer before applying (see Contact Info below). Applicants must have 501(c)3 tax-exempt status.
Restrictions: Grant requests for capital projects will not be considered. Applicants will not be awarded a grant for more than three consecutive years. The trust does not support requests from individuals, organizations attempting to influence policy through direct lobbying, or any political campaigns.
Geographic Focus: Connecticut
Date(s) Application is Due: Jun 15
Samples: ConnectiKids, Hartford, Connecticut, $1,000 (2010); Capital Classics, West Hartford, Connecticut, $1,000 (2010); Riverfront Recapture, Hartford, Connecticut, $1,000 (2010).
Contact: Kate Kerchaert, (860) 657-7016; kate.kerchaert@baml.com
Internet: https://www.bankofamerica.com/philanthropic/fn_search.action
Sponsor: Mabel F. Hoffman Charitable Trust
200 Glastonbury Boulevard, Suite # 200, CT2-545-02-05
Glastonbury, CT 06033-4056

Mabel H. Flory Charitable Trust Grants 2768
The Mabel H. Flory Charitable Trust was established in 1979 and funds organizations primarily in Washington, DC., and Baltimore, MD. Support is available for research designed to develop new cures and methods of treatment: for children with hearing and sight problems; arthritis in both children and adults; and aphasia related problems.
Requirements: Your initial approach should be in the form of a letter, prior to the deadline date. Mail letter to the following address: c/o BB&T Co. of Virginia, Attn.: Trustee, 6400 Arlington Boulevard, Falls Church, VA, 22042.
Restrictions: No grants to individuals.
Geographic Focus: District of Columbia, Maryland
Date(s) Application is Due: Aug 31
Contact: BB&T Co. Trustee, (703) 531-2053
Sponsor: Mabel H. Flory Charitable Trust c/o BBandT Co. of Virginia
P.O. Box 2907
Wilson, NC 27894-2907

Mabel Y. Hughes Charitable Trust Grants 2769
The Trust awards grants to Denver-area nonprofit organizations in its areas of interest, including children and youth services; education; family services; health care; higher education; human services; children's and art museums; performing arts centers; performing arts, opera; and reproductive health and family planning. Types of support include: annual campaigns; continuing support; emergency funds; endowments; equipment; general/operating support; program development; research; and seed money.
Requirements: Organizations should submit a letter of inquiry to the Trust, and if their project is appropriate for funding, they will be asked to submit a proposal.
Restrictions: The Trust does not support funding for individuals, deficit financing, scholarships, fellowships, or loans.
Geographic Focus: Colorado
Amount of Grant: 10,000 - 30,000 U.S.D
Contact: Peggy Toal, c/o Wells Fargo Bank, Private Client Services; (720) 947-6725
Sponsor: Mabel Y. Hughes Charitable Trust
1740 Broadway
Denver, CO 80274

Macquarie Bank Foundation Grants 2770
The Foundation is one of Australia's oldest and largest corporate foundations contributing more than $150 million to more than 1500 community organizations world-wide since 1985. The Foundation focuses its resources in five core areas - the arts, education, environment, health, and welfare. The Foundation is also committed to projects specifically aimed at supporting indigenous communities. The Foundation's funding criteria is flexible and open. It welcomes applications from a diverse range of community organizations that are working in innovative ways to provide long-term benefits. Funding levels are flexible and are dictated by the needs of the organization and funding availability. Each application is assessed on its individual merit, with priority given to programs which support a broad section of the community at a regional, state or national level; have the involvement or potential for involvement of Macquarie Bank staff through volunteering, fundraising, pro bono work and board and/or management committee involvement; are located in cities/countries where Macquarie Bank staff are located; and deliver long-term benefits and build community sustainability. Prospective applicants are encouraged to check the Foundation website or contact Foundation Staff for more information on how to apply.
Geographic Focus: All States, All Countries
Amount of Grant: 100 - 500,000 U.S.D
Samples: Po Leung Kuk, Hong Kong, $HK190,000 - a charitable organization supporting disadvantaged children (2010); Oxfam Trailwalker, Melbourne, Australia, $A325,000 (2009-2010); Princess Margaret Hospital Foundation, Toronto, Ontario, Canada, $C350,000 - Ride to Conquer Cancer (2009-2010).
Contact: Heather Matwejev, Macquarie Bank Foundation, Asia; +61 2 8232 6951; fax +61 2 8232 0019; heather.matwejev@macquarie.com or foundation@macquarie.com
Internet: http://www.macquarie.com/mgl/com/foundation/about/application-guidelines
Sponsor: Macquarie Bank Foundation
GP.O. Box 4294
Sydney, NSW 1164 Australia

Madison County Community Foundation - City of Anderson Quality of Life Grant 2771
The Madison County Community Foundation - Quality of Life Grant is used as economic development quality of life assistance to non-profits. Generated through the City of Anderson's food and beverage tax, the grant is dispersed to those non-profits who have no voting representation on the committee; are not solely supporting non-secular activities with these funds; carry proof of 501(c)3 status; and are requesting funding for projects with outcomes rather than operational expenses. The grants must be used to increase quality of life for all rather than limited access.
Requirements: Application forms and guidelines are available from the Foundation office or may be downloaded from the Foundation website at www.madisonccf.org. The application will also be available at the City of Anderson website at www.cityofanderson.com.
Restrictions: Funding is limited to the city of Anderson, Indiana. Funding is not available for: annual fund campaigns; individuals; capital debt reduction; sectarian religious purposes; gifts to endowments; political campaigns; medical, scientific, or health research; student loans, scholarship/fellowship programs, or travel grants; programs and/or equipment which were committed prior to the grant application being submitted; organizations without responsible fiscal agents and adequate accounting procedures; schools and government agencies; normal operational expenses of the organization. Funding is limited to grants for one year.
Geographic Focus: Indiana
Date(s) Application is Due: Sep 15
Amount of Grant: 1,000 - 10,000 U.S.D
Contact: Tammy Bowman, Program Director; (765) 644-0002; fax (765) 662-1438; tbowman@madisoncf.org
Internet: http://www.madisonccf.org/index.php?submenu=grantNO&src=gendocs&ref=GrantProcess&category=Non_Profits
Sponsor: Madison County Community Foundation
33 West 10th Street, Suite 600
Anderson, IN 46015-1056

Madison County Community Foundation General Grants 2772
The Madison County Community Foundation General Grants are made to support projects and programs of non-profit agencies located in or serving residents of Madison County. Grants are typically made in the spring and fall and range from $500 to $10,000 with rare exceptions. Proposals are reviewed by a Grants Committee and those chosen are then approved by the Foundation Board. The priorities of the foundation are arts and culture, education, economic development, civic affairs, and health and human services.
Requirements: Organizations should complete the online application and include the following detailed information: their project and cost; organization information; a project narrative that includes the objective, financial need, justification; constituency; evaluation plan and community impact; additional and/or future funding; professional references and collaborative value. They should also include a list of project expenses and project income. Eight copies of the complete application package should be sent to the Program Director, along with one copy each of the board of directors, IRS tax exempt letter, and financial statement.
Geographic Focus: Indiana
Amount of Grant: 500 - 10,000 U.S.D
Contact: Tammy Bowman, Program Director; (765) 644-0002; fax (765) 644-3392; tbowman@madisoncf.org
Internet: http://www.madisonccf.org/index.php?submenu=Grants&src=gendocs&ref=Grants&category=Non_Profits
Sponsor: Madison County Community Foundation
33 West 10th Street, Suite 600
Anderson, IN 46015-1056

Maine Community Foundation Charity Fund 2773
As Maine's first community foundation, the Maine Charity Foundation was established in 1921 to receive and hold individual charitable funds and distribute the income for charitable purposes in the State. In September of 1995, administration of the Maine Charity Foundation was transferred to the Maine Community Foundation. The following are funding priorities: start-up money for an organization or project; projects that involve the disabled or economically disadvantaged, as long as the project is not supported by a national campaign or public money; libraries; symphonies; hospice care; projects that are related to the Friendship, Maine area; discrete projects as opposed to general operating support; and small requests from social service organizations for "bricks and mortar" or purchase of necessary equipment. Application is made online from the Foundation's website.
Requirements: Applicants must be 501(c)3 organizations eligible to accept tax-deductible donations as outlined in Section 170(c) of the Internal Revenue Code.
Restrictions: Religious groups are eligible but funding will not be provided for religious purposes. Funding is not provided for the following: political campaigns, or to support attempts to influence legislation of any governmental body other than through making available the results of non-partisan analysis, study and research; ongoing operating support; endowments or capital campaigns; camperships; or capital equipment over $250.
Geographic Focus: Maine
Date(s) Application is Due: Sep 15

Amount of Grant: 5,000 U.S.D
Samples: Big Brothers Big Sisters of Bath/Brunswick, Brunswick, Maine, $1,500, copier and supplies (2010); Community Health and Counseling Services, Bangor, Maine, $4,750, complete a business plan and feasibility study for the development of a hospice house in the greater Bangor area (2010); and Island Community Center, Stonington, Maine $2,500, start-up costs associated with opening The Community Cafe in Stonington (2010).
Contact: Cathy Melio, Grants Administrator; (877) 700-6800; cmelio@mainecf.org
Internet: http://www.mainecf.org/mainecharityfound.aspx
Sponsor: Maine Community Foundation
245 E Main Street
Ellsworth, ME 04605

Maine Community Foundation Hospice Grants **2774**
The purpose of the Foundation's Hospice Grants is to support hospice services in southern Maine, in particular by funding programs that support volunteer and bereavement services in Cumberland and York Counties. Preference is given to innovative projects that focus on increasing the utilization of end of life services and/or increasing the quality of end-of-life services. Application is made online from the Foundation's website.
Requirements: Applicants must be 501(c)3 organizations eligible to accept tax-deductible donations as outlined in Section 170(c) of the Internal Revenue Code. High priority is given to proposals that: support collaborative efforts among hospice programs and other community groups that enhance training, provide recognition, and or increase the recruitment of diverse volunteers for hospice; raise community awareness of hospice services; and increase hospice volunteer coordinators' capacity to interact with volunteers.
Restrictions: Religious groups are eligible but funding will not be provided for religious purposes. Funding is not provided for the following: political campaigns, or to support attempts to influence legislation of any governmental body other than through making available the results of non-partisan analysis, study and research; ongoing operating support; endowments or capital campaigns; camperships; or capital equipment over $250.
Geographic Focus: Maine
Date(s) Application is Due: Feb 15
Amount of Grant: 7,000 U.S.D
Samples: Camp Sunshine at Sebago Lake, Casco, Maine, $5,000, to support five Cumberland and/or York County families participate in the organization's bereavement program (2010); Home Health Visiting Nurses of Southern Maine, Saco, Maine, to support the organization's palliative care program (2010); and Maine Humanities Council, Portland, Maine, to bring the Philoctetes Project to three Maine sites (2010).
Contact: Pam Cleghorn, Administrator; (877) 700-6800; pcleghorn@mainecf.org
Internet: http://www.mainecf.org/hospicefund.aspx
Sponsor: Maine Community Foundation
245 E Main Street
Ellsworth, ME 04605

Maine Community Foundation Penobscot Valley Health Association Grants **2775**
The Penobscot Valley Health Association was established in 1906 to help sufferers of tuberculosis in the greater Bangor area. After newer drugs and methods of treating tuberculosis were developed, the association turned its charitable interests to more general health and welfare projects in the greater Bangor area. In 1996 the association moved its endowment and became a component fund of the Foundation. The fund now supports projects that propose innovative ways to strengthen the health and welfare of the greater Bangor community. Priority is given to proposals that involve the targeted population in the design, implementation and evaluation of the project, draw upon the strengths of the community, and foster collaboration between community groups. Grants average $9,680. Applications are available at the Foundation's website.
Requirements: Applicants must be 501(c)3 organizations eligible to accept tax-deductible donations as outlined in Section 170(c) of the Internal Revenue Code.
Geographic Focus: Maine
Date(s) Application is Due: May 15
Samples: 32nd Degree Masonic Learning Centers for Children, Bangor, Maine, $5,000, for tutoring services for children with dyslexia in the greater Bangor area (2010); Evaluation Practice Group, Newburgh, Maine, $9,560, to support the Hooves of Hope equine therapy program for children with disabilities (2010); and Hammond Street Senior Center, Bangor, Maine, $18,828, to support programs for healthy aging in the Bangor region (2010).
Contact: Grants Administrator; (877) 700-6800; grants@mainecf.org
Internet: http://www.mainecf.org/PVHAFund.aspx
Sponsor: Maine Community Foundation
245 E Main Street
Ellsworth, ME 04605

Manuel D. and Rhoda Mayerson Foundation Grants **2776**
The Manuel and Rhoda Mayerson Foundation Grants supports organizations in its areas of interest, including children and youth, people with disabilities, health and well-being, arts and culture, housing and homelessness, and Jewish life and culture. Types of support include building construction/renovation, conferences and seminars, matching/challenge grants, seed grants, and technical assistance. The board meets quarterly.
Requirements: Grants are only awarded to nonprofit, tax-exempt organizations as defined by Section 501(c)3 of the Internal Revenue Code, to organizations that are public charities under Section 509(a)(1),(2), or (3) of the Internal Revenue Code, and to those that comply with the requirement of Section 4945(d)(3) or (4) of the Code. In making funding decisions, the Foundation is responsive to sound strategic planning, organizational stability, leadership, creativity, entrepreneurial visioning, leveraging of resources, collaboration and empowerment of people. Additionally, there is a focus on providing funding to worthy efforts that otherwise struggle to find vital support. The application is located on the Foundation website.
Restrictions: The Foundation funds projects in Israel, in addition to Cincinnati, Ohio; Berkeley, California; and Boca Raton, Florida. The Foundation does not make grants in support of any non-charitable purpose or to organizations that promote racial, ethnic or religious disharmony, hatred or violence.
Geographic Focus: California, Florida, Ohio, Israel
Amount of Grant: 1,000 - 30,000 U.S.D
Contact: Manuel Mayerson, Foundation President; (513) 621-7500; fax (513) 621-2864; info@mayersonfoundation.org
Internet: http://mayersonfoundation.org/GrantMaking/HowtoApply/tabid/609/Default.aspx
Sponsor: Manuel D. and Rhoda Mayerson Foundation
312 Walnut Street, Suite 3600
Cincinnati, OH 45202

MAPFRE Foundation for Medicine Grants **2777**
The foundation operates nationally and internationally in the fields of education, medicine and health, science and technology, and social welfare. Types of support include research grants, grants to individuals and organizations, scholarships, fellowships, and prizes.
Contact: Grants Administrator; 34-1-6265704; fax 34-1-6265825; infofmm@mapfremedicina.es
Internet: http://www.mapfre.com/pmapfre/paraguas_mapfre/en/06/0608.shtml
Sponsor: MAPFRE Foundation for Medicine
Carretera de Pozuelo a Majadahonda s/n
Madrid, 28220 Spain

MAP International Medical Fellowships **2778**
The MAP International Medical Fellowship encourages lifelong involvement in global health issues by providing selected medical students firsthand exposure in a Christian context to the health, social and cultural characteristics of a developing world community. MAP International, through a grant from the founders of Reader's Digest, DeWitt and Lila Wallace, sponsors fourth year medical students by funding travel arrangements for medical students to their chosen destination. The field experience must be designed to provide the student exposure to community health in a Christian context. Students select a mission agency or hospital that has an outreach among the poor in a rural or urban setting. Students must spend at least eight weeks in the field (six weeks for residents and interns). The fellowship provides 100 percent of the approved round trip airfare to one destination. In most instances, students pay room and board as well as any in-country travel expenses. To obtain an application, check with the dean of your school or the Student Affairs Office. Applications are sent to all medical schools annually. You may also download the application in PDF or Microsoft Word format on the Map International website.
Requirements: Fourth-year medical students, residents and interns are eligible. Applications should be submitted during the academic year prior to travel.
Date(s) Application is Due: Mar 1
Contact: Margaret Stevenson, Medical Fellowship Coordinator; (404) 880-0540; fax (912) 265-6170; mstevenson@map.org
Internet: http://www.map.org/site/PageServer?pagename=what_Medical_Fellowship
Sponsor: Map International
4700 Glynco Parkway
Brunswick, GA 31525-6800

Marcia and Otto Koehler Foundation Grants **2779**
Grants are made to support arts and culture, medical, education, and social organizations in San Antonio, Texas. Proposals must demonstrate leadership in effecting positive change; encourage collaborative effort; serve large and diverse sectors of the population; and demonstrate vision, effectiveness, and good fiscal management. Types of support include general operating support, building construction/renovation, and research. Grants are made for one time only. Applicants must send a letter of request for application by March 1 to receive an application.
Requirements: Grants are awarded only to organizations in Bexar County, Texas.
Restrictions: Grants will not be made to individuals or to support other foundations or endowments; salaries; operating deficits; political organizations; or churches, synagogues, or parishes.
Geographic Focus: Texas
Date(s) Application is Due: Jun 1
Amount of Grant: 2,500 - 50,000 U.S.D
Contact: Thomas K. Killion, Senior Vice President, Bank of America; (210) 270-5422; fax (210) 270-5520; thomas.k.killion@ustrust.com
Sponsor: Marcia and Otto Koehler Foundation
P.O. Box 121
San Antonio, TX 78291-0121

Mardag Foundation Grants **2780**
The Mardag Foundation is committed to making grants to qualified nonprofit organizations in Minnesota that help enhance and improve the quality of life, inspire learning, revitalize communities, and promote access to the arts. The Foundation focuses their grantmaking in these priority areas: improving the lives of at-risk families, children, youth, and young adults; supporting seniors to live independently; building the capacity of arts and humanities organizations to benefit their communities; and supporting community development throughout the St. Paul area. Grants normally support; capital projects, program expansion and special projects of a time-limited nature; start-up costs for promising new programs that demonstrate sound management and clear goals relevant to community needs; support for established agencies that have temporary or transitional needs; funds to match contributions

received from other sources or to provide a challenge to raise new contributions. Applicants are encouraged to submit a brief summary of their project prior to preparation of a full proposal to see if the project fits the guidelines and interests of the foundation. The Foundation's grantmaking meetings are in April, August, and November. Generally, full proposals must be received approximately three months prior to a meeting date.
Requirements: To see if a project fits the guidelines and interests of the Foundation, applicants may submit a letter of inquiry describing the proposed project prior to preparing a full proposal. This letter should be limited to two to three pages, with less detail than a full proposal, and address the information included in the Proposal Narrative Guidelines posted on the website. Prior to submitting a grant request, applicants should obtain the Foundation's application packet from the website, which includes the grant application cover sheet, proposal narrative guidelines, proposal checklist, and optional organization and project or program budget forms.
Restrictions: The geographic focus of the Foundation's grantmaking is the East Metro area of Dakota, Ramsey and Washington counties. The Foundation does not fund: programs exclusively serving Minneapolis and the surrounding West Metro area; scholarships and grants to individuals; ongoing annual operating expenses; sectarian religious programs; medical research; federated campaigns; conservation or environmental programs; events and conferences; programs serving the physically, developmentally or mentally disabled; capital campaigns of private secondary schools; and capital and endowment campaigns of private colleges and universities. The Foundation will review, on their own merits, grant applications received from private secondary schools and private colleges and universities for purposes not excluded in the information above.
Geographic Focus: Minnesota
Date(s) Application is Due: May 1; Aug 1; Dec 31
Amount of Grant: 5,000 - 50,000 U.S.D
Contact: Lisa Hansen, Grants Administration Manager; (651) 224-5463 or (800) 875-6167; info@mardag.org
Internet: http://www.mardag.org/apply_for_a_grant/
Sponsor: Mardag Foundation
55 Fifth Street East, Suite 600
St. Paul, MN 55101

Margaret E. Burnham Charitable Trust Grants **2781**
The trust awards grants to Maine nonprofit organizations and to nonprofits serving the state in five program areas of interest, listed in order of priority: community/social services, health care, education, arts and culture, and the environment.
Requirements: Maine nonprofit organizations and nonprofit organizations serving the state are eligible.
Geographic Focus: Maine
Date(s) Application is Due: Nov 15
Amount of Grant: 1,000 - 10,000 U.S.D
Samples: Bowdoin Summer Music Festival (ME)—$2500; Southern Maine Medical Ctr (ME)—$5000; National Alliance for the Mentally Ill in Maine (ME)—$5000; Portland's Downtown District (ME)—$3000.
Contact: Thomas Pierce, Trustee, c/o H.M. Payson and Company; (207) 772-3761
Internet: http://www.megrants.org/Burnham.htm
Sponsor: Margaret E. Burnham Charitable Trust
P.O. Box 31
Portland, ME 04112-0031

Margaret L. Wendt Foundation Grants **2782**
The foundation awards grants to eligible New York nonprofit organizations in its areas of interest, including education, the arts, social services, churches and religious organizations, health associations, public interest organizations, and youth agencies. There are no application deadlines or forms. The board meets quarterly
Requirements: Western New York nonprofit organizations are eligible.
Restrictions: Grants are not awarded to individuals, including scholarships.
Geographic Focus: New York
Samples: D'Youville College (Buffalo, NY)—for its chiropractic program, including equipping a computerized diagnostic-imaging classroom and an off-campus community-based clinic, $350,000.
Contact: Grants Administrator; (716) 855-2146
Sponsor: Margaret L. Wendt Foundation
40 Fountain Plaza, Suite 277
Buffalo, NY 14202

Margaret T. Morris Foundation Grants **2783**
The Margaret T. Morris Foundation awards grants, primarily in Arizona, in its areas of interest, including: animal welfare; arts; children and youth, services; education; environment; higher education; homeless service; human services; marine science; medical research and education; mental health and crisis services; museums; performing arts; reproductive health and family planning; and hospices. The Foundation's types of support include: building renovation; capital campaigns; debt reduction; endowments; general operating support; land acquisition; matching and challenge support; and program development. The Board of Directors meets in August, December, and as needed.
Requirements: Applications are not accepted. Applicants should submit a letter of inquiry with their request for funding and a description of the project.
Geographic Focus: Arizona
Contact: Thomas Polk, Trustee; (928) 445-4010
Sponsor: Margaret T. Morris Foundation
P.O. Box 592
Prescott, AZ 86302-0592

Margoes Foundation Grants **2784**
The foundation awards grants to California nonprofits in its areas of interest, including cardiac research, rehabilitation of mentally ill patients, and scholarships for minority students and for Asian and African nationals studying in the United States. Types of funding include general operating support, continuing support, program development, fellowships, scholarship funds, research, and matching funds. There are no application forms. Applicants should call or submit a preliminary letter prior to submitting a full proposal.
Requirements: California nonprofits are eligible, with preference given to the San Francisco Bay area.
Restrictions: Grants are not awarded to agencies supported by federated campaigns or to individuals.
Geographic Focus: California
Date(s) Application is Due: Apr 1; Oct 1
Amount of Grant: 10,000 - 20,000 U.S.D
Samples: Community Music Center (San Francisco, CA)—for financial aid, $10,000; U of the Pacific (Stockton, CA)—to increase minority enrollment in the predentistry undergraduate programs, $20,000 over two years; Fred Finch Youth Ctr (Oakland, CA)—for salary support of a part-time counselor for a transitional independent living program for older adolescents and young adults with severe emotional problems, $15,000.
Contact: John Blum, Principal Manager; (415) 981-2966; fax (415) 981-5218; margoesfdn@aol.com
Sponsor: Margoes Foundation
57 Post Street, Suite 510
San Francisco, CA 94104

Marie C. and Joseph C. Wilson Foundation Rochester Small Grants **2785**
The Marie and Joseph Wilson Foundation strives to improve the quality of life through initiating and supporting projects that measurably demonstrate a means of creating a sense of belonging within the family and community. The Foundation considers 501(c)3 organization requests ranging from $1,000 to $25,000. Grant applications are accepted on an ongoing basis. Foundation board members review applications as they are received. The review committee meets once a month except for July and August. Because the Foundation receives a large number of applications, responses may take up to four months. Prior to the receiving funding, grant recipients are required to sign a grant agreement contract. Written progress reports are required at six months and one year following the date of the grant. Samples of the Foundation's previously funded grants are available online.
Requirements: The Foundation review committee looks for one or more of the following conditions in a proposal: the proposal is a well-planned approach to delivering services; Foundation support would be catalytic to the project's success; the proposal is efficient in its use of funds and expenses are reduced by sharing resources with other agencies or groups; and a collaborative network exists that multiplies the impact of the grant. Applicants may contact the Foundation for a current application form.
Restrictions: Grants are limited to 501(c)3 organizations serving the Rochester, New York area. Grants will not be made to individuals, partisan political organizations, or to support lobbying efforts. Requests for capital projects also will not be considered.
Geographic Focus: New York
Amount of Grant: 1,000 - 25,000 U.S.D
Contact: Megan Bell, Executive Director; (585) 461-4696; fax (585) 473-5206
Internet: http://www.mcjcwilsonfoundation.org/funding.cfm
Sponsor: Marie C. and Joseph C. Wilson Foundation
160 Allens Creek Road
Rochester, NY 14618-3309

Marie H. Bechtel Charitable Remainder Uni-Trust Grants **2786**
Established in Scott County, Iowa, in 1987, the Marie H. Bechtel Charitable Remainder Uni-Trust (formerly known as the Marie H. Bechtel Charitable Trust) supports youth services and education. Support is also available for the advancement of health care, maintenance of community cultural activities, and enhancement of the community by restoring its vitality and creating meaningful employment. Contact the Foundation for application forms.
Requirements: 501(c)3 organizations serving the residents of Scott County, Iowa, are eligible to apply.
Restrictions: No grants are given for endowment funds, past operating deficit or debt retirement, general or continuing operating support, or basic scholarly research.
Geographic Focus: Iowa
Amount of Grant: Up to 350,000 U.S.D
Contact: R. Richard Bittner, Trustee; (563) 328-3333; fax (563) 328-3352; loseband@blwlaw.com
Sponsor: Marie H. Bechtel Charitable Remainder Uni-Trust
201 West 2nd Street, Suite 1000
Davenport, IA 52801-1817

Marietta Community Foundation Grants **2787**
The foundation awards grants to Ohio nonprofit organizations in its areas of interest, including civic affairs, education, environment, health, and community welfare. Types of support include challenge/matching grants, conferences and seminars, development grants, program grants, research grants, scholarships, seed money grants, training grants, and work-study programs.
Requirements: Tax-exempt organizations serving residents of Marietta, OH, and the surrounding community are eligible.
Geographic Focus: Ohio
Date(s) Application is Due: Feb 7; Jun 7; Oct 7
Contact: Carol Wharff, Executive Director; (740) 373-3286; fax (740) 376-4490; mcf@marietta.edu or info@mariettacommunityfoundation.org

Internet: http://www.mariettacommunityfoundation.org/grants/guidelines.htm
Sponsor: Marietta Community Foundation
P.O. Box 77
Marietta, OH 45750-0077

Marijuana Policy Project Grants **2788**

The program supports efforts that promote the legalization and regulation of marijuana. Projects may encompass any aspect of marijuana-law reform, such as: ballot initiatives, grassroots organizing, litigating for the right to use marijuana for religious purposes, lobbying the American Medical Association, marijuana research, and revising the definition of marijuana abuse and dependence in the American Psychiatric Association's Diagnostic and Statistical Manual. Submit letters of inquiry; full proposals are by invitation.
Requirements: Preference is given to 501(c)3 tax-exempt organizations, but such status is not an eligibility requirement.
Restrictions: The grants program does not fund political parties or candidates for office, state ballot initiatives, or hemp-related projects.
Date(s) Application is Due: Jan 2; May 1; Sep 1
Amount of Grant: Up to 60,000 U.S.D
Contact: Stephanie Vogel, (202) 462-5747, ext.104; fax (202) 232-0442; stephanie@mpp.org
Internet: http://www.mpp.org/grants/index.html
Sponsor: Marijuana Policy Project
P.O. Box 77492
Washington, DC 20013

Marilyn Augur Family Foundation Grants **2789**

The foundation gives priority to organizations serving basic human needs primarily in the geographical area of Dallas, TX. Basic needs include food, shelter, clothing, health, and education aimed at transforming lives of those living in poverty or in prison. Christian organizations that minister to the needy are of particular interest. The foundation gives emphasis to children's programs. The foundation prefers to support organizations and projects directly rather than supporting fundraising events.
Requirements: Texas nonprofit organizations are eligible.
Restrictions: Generally, the foundation does not give grants in the area of arts and culture.
Geographic Focus: Texas
Amount of Grant: 1,000 - 3,000 U.S.D
Contact: Nancy Roberts, (214) 522-5586; fax (214) 526-0253; maf@waymark.net
Internet: http://fdncenter.org/grantmaker/augur/app.html
Sponsor: Marilyn Augur Family Foundation
4209 McKinney Avenue, Suite 202A
Dallas, TX 75205

Marin Community Foundation Improving Community Health Grants **2790**

The Improving Community Health Grants is interested in funding two complimentary priorities: strengthening the delivery of health services with direct health services or community clinics; and addressing the social determinants of health so that everyone has the opportunity to live a long, healthy life regardless of income, education, or racial/ethnic background. Community, neighborhood, or geographically-focused research projects that assess the role of poverty, housing, education, transportation, healthy food, physical activity, and/or other community factors are strongly encouraged to apply. Proposed research projects should contribute to the development and/or implementation of public and/or school policies aimed at addressing social, economic, and/or political factors that contribute to the community's health. The proposed assessments should include leadership, input, and direction from the community. In addition to thorough community/neighborhood assessments, the Foundation is interested in supporting the development, advocacy, and implementation of policies and/or practices that impact the health of a low-income community or community of color. Projects that propose building infrastructure to improve the social conditions, economic opportunities, and/or physical environments in which people live, work, learn, and play will also be considered (e.g., infrastructure and technology for farmers to accept food stamps at the farmers market).
Requirements: To be eligible for funding, organizations must have a nonprofit tax-exempt status or a fiscal sponsor with a nonprofit tax-exempt status. Applicants must serve Marin County. The program director can be emailed directly through the Foundation website.
Geographic Focus: California
Contact: Wendy Todd, Program Director; (415) 464-2541
Internet: http://www.marincf.org/grants-and-loans/grants/community-grants/improving-community-health
Sponsor: Marin Community Foundation
5 Hamilton Landing, Suite 200
Novato, CA 94949

Marine Biological Laboratory Summer Research Fellowships **2791**

Each year a number of summer research support opportunities for young investigators are awarded for studies and research to be carried out at the Marine Biological Laboratory (MBL) at Woods Hole, MA. The Robert Dale Allen Fellowship supports research in cell motility and cytoarchitecture; and the Frederik B. Bang Fellowship funds research on marine invertebrate immunology, pathology, infectious diseases, and molecular biology. The M.G.F. Fuortes Fellowship sponsors research in the neurosciences, and the Stephen W. Kuffler Fellowship supports research in neurobiology. The H. Burr Steinbach Fellowship funds research in cell biology and physiology. In addition, the Porter Fellowships for Minority Students/Investigators are offered for research in the field of physiology with senior investigators at the MBL. The MBL Summer Fellowships are offered to investigators to conduct independent or collaborative research in any area. Awarded funds may be used to pay laboratory rental fees, housing costs, travel to and from Woods Hole, and appropriate personal expenses. Notification of selection of fellows is made by March 1.
Requirements: Postdoctoral students, instructors, and assistant professors may apply for these fellowships.
Date(s) Application is Due: Jan 15
Contact: Fellowship Coordinator, Office of Research Administration and Education; (508) 289-7441; fax (508) 289-1924; skaufman@mbl.edu
Internet: http://www.mbl.edu/research/summer/fellowships.html
Sponsor: Marine Biological Laboratory
7 MBL Street
Woods Hole, MA 02543

Marion Gardner Jackson Charitable Trust Grants **2792**

The Marion Gardner Jackson Charitable Trust was established by Marion Gardner Jackson, the granddaughter of local industrialist, Robert W. Gardner, founder of the Gardner-Denver Company. Before her death in 1976, Mrs. Jackson set up a perpetual trust to aid religious, charitable, scientific, literary, and educational organizations in the Quincy, Illinois area and surrounding communities in Adams County. The trust supports capital, program, and operating grants. Applicants may apply for one- to five-year grants to support capital projects. Projects must support the mission of the organization and/or the stated outcome of a program or project. Applicants with organizational budgets of $1 million or less may apply for grants of up to $25,000 per year, for up to three years, for program or general operating support. Applicants must apply online at the grant website. Applicants are strongly encouraged to do the following before applying: review the downloadable state application procedures for additional helpful information and clarifications; review the downloadable online-application guidelines at the grant website; review the trust's funding history (link is available from the grant website); review the online application questions in advance; and review the list of required attachments. These will generally include: a list of board members, financial statements (audited, reviewed, or compiled by independent auditor); an organization summary; a list of other funding sources; an IRS Determination letter; and other required documents. All attachments must be uploaded in the online application as PDF, Word, or Excel files. The application deadline for the Marion Gardner Jackson Charitable Trust is 11:59 p.m., July 31. Applicants will be notified of grant decisions by January 15.
Requirements: As a condition for accepting grant funds, grantees must provide annual programmatic and financial reports detailing the use of the funds and documenting the progress of the capital campaign/capital project or documenting the impact achieved as a result of the funding. Receipt of annual funds from multi-year grants will be dependent on submission of an annual financial and project/programmatic grant report, due no later than December 15.
Restrictions: For program support grants, the yearly request may not be more than 50% of the program's budget. Organizations can submit one application per year and will not receive more than one award from the trust in any given year. Organizations receiving a multi-year award from the trust that continues into the next grant year are not eligible to apply for an additional grant until the end of the grant cycle. The trust does not support requests from individuals, organizations attempting to influence policy through direct lobbying, or any political campaigns.
Geographic Focus: Illinois
Date(s) Application is Due: Jul 31
Samples: John Wood Community College Foundation, Quincy, Illinois, $37,000 (2010); Quincy Society of Fine Arts, Quincy, Illinois, $25,000 (2010); Access Health Adams County, Quincy, Illinois, $25,000 (2010).
Contact: Debra Grand, Senior Vice President; (312) 828-4154; ilgrantmaking@bankofamerica.com
Internet: https://www.bankofamerica.com/philanthropic/fn_search.action
Sponsor: Marion Gardner Jackson Charitable Trust
231 South LaSalle Street, IL1-231-13-32
Chicago, IL 60604

Marion I. and Henry J. Knott Foundation Discretionary Grants **2793**

Founded in 1977, the Marion I. and Henry J. Knott Foundation is a Catholic family foundation committed to honoring its founders' legacy of generosity to strengthen the community within the Archdiocese of Baltimore. Henry J. Knott, the eldest of six boys, grew up in a lively household in the Baltimore area. His father was a hard-working carpenter. Marion Isabel Burk, who was orphaned at the age of eleven, grew up cooking and looking after the children in a small boarding house and received little formal education as a result. Henry and Marion met on a blind date arranged by a good friend in 1926, while Henry was taking classes at Loyola College, and were married in 1928. They went on to build a large family (thirteen children, one lost to cancer) and a thriving construction business. Henry was the first developer in Baltimore to employ the practice of prefabricating wall panels in a factory and then sending them out to construction sites. Projects moved at a blistering pace and eventually led Henry to become the President of Arundel Corporation. Henry and Marion who knew firsthand the challenge of raising a large family always practiced philanthropy. The foundation makes awards in five Program categories: Arts and Humanities; Catholic Activities; Education (Catholic schools, nonsectarian private schools specifically catering to special needs, and private colleges and universities); Health Care; and Human Services. Within the five program categories, the foundation funds within five project categories, including capital expenses, development, new and/or ongoing programs, operating expenses, and technology. In addition to its standard granting program, the Knott Foundation provides a limited number of Discretionary Grants (20-30) throughout the year. These grants, ranging between $500 to $2,500, are designed to increase the Foundation's grant-making options as well as its responsiveness to community needs. Grants are awarded based on the proposed project, the availability of funds, and other

current requests for funding. To apply for a discretionary grant, applicants should submit a brief (one page) Letter of Inquiry (LOI) on their organization's letterhead. The LOI should describe the applicant's project or program, detail the applicant's needs, and provide a timeframe for use of the award if granted. In addition to the LOI, the applicant should also submit a 501(c)3 status letter, a project budget if applicable, and a list of the board of directors. Discretionary requests are accepted and awarded on a rolling basis throughout the year. Although not guaranteed, approved funds are usually disbursed within one to two weeks of the discretionary grant's approval date. Interested applicants should visit the website for further details and guidelines.
Requirements: Discretionary grant requests must be in alignment with the foundation's areas of geographic and programmatic giving. Funding is limited to 501(c)3 organizations serving Baltimore City and the following counties in Maryland: Allegheny, Anne Arundel, Baltimore, Carroll, Frederick, Garrett, Harford, Howard, and Washington. Applicants may apply through a fiscal sponsor. The fiscal sponsor must be a 501(c)3 nonprofit organization that has a formal relationship and Memorandum of Understanding (MOU) with the applicant. Selected applicants will need to submit a copy of their most recent IRS 990 and/or audited financials.
Restrictions: The following will not be funded: organizations that have not been in operation for at least one year, scholarships, public education/public sector agencies, pro-choice or reproductive health programs, individuals, annual giving, political activities, one-time only events/seminars/workshops, legal services, environmental activities, medical research, day care centers, endowment funds for arts/humanities, national/local chapters for specific diseases, agencies that redistribute grant funds to other nonprofits, reimbursables or any prior expenses, or government agencies that form 501(c)3 nonprofits to fund public sector projects.
Geographic Focus: Maryland
Amount of Grant: 500 - 2,500 U.S.D
Samples: Museum of Ceramic Art, Baltimore, Maryland, $2,500 - to the after-school Middle School Ceramics Art Program though out Baltimore City (2012); Filbert Street Garden, Baltimore, Maryland, $980 - to suppor their Winter Greens: Student-Supported Agriculture program; Maryland Association of Nonprofits (MANO), Baltimore, Maryland, $5,000 - to support the Leaders Circle's forum on building better relationships between nonprofit executive directors and their board leadership.
Contact: Kathleen McCarthy, Grants Manager; (410) 235-7068; fax (410) 889-2577; knott@knottfoundation.org or info@knottfoundation.org
Internet: http://www.knottfoundation.org/what_we_do/grant_application_process/discretionary_grant_application_process
Sponsor: Marion I. and Henry J. Knott Foundation
3904 Hickory Avenue
Baltimore, MD 21211-1834

Marion I. and Henry J. Knott Foundation Standard Grants **2794**

Founded in 1977, the Marion I. and Henry J. Knott Foundation is a Catholic family foundation committed to honoring its founders' legacy of generosity to strengthen the community within the Archdiocese of Baltimore. Henry J. Knott, the eldest of six boys, grew up in a lively household in the Baltimore area. His father was a hard-working carpenter. Marion Isabel Burk, who was orphaned at the age of eleven, grew up cooking and looking after the children in a small boarding house and received little formal education as a result. Henry and Marion met on a blind date arranged by a good friend in 1926, while Henry was taking classes at Loyola College, and were married in 1928. They went on to build a large family (thirteen children, one lost to cancer) and a thriving construction business. Henry was the first developer in Baltimore to employ the practice of prefabricating wall panels in a factory and then sending them out to construction sites. Projects moved at a blistering pace and eventually led Henry to become the President of Arundel Corporation. Henry and Marion who knew firsthand the challenge of raising a large family always practiced philanthropy. The foundation makes both standard and discretionary awards in five Program categories: Arts and Humanities; Catholic Activities; Education (Catholic Schools, Nonsectarian private schools specifically catering to special needs, and private colleges and universities); Health Care; and Human Services. Within the five program categories, the foundation funds within five project categories, including capital expenses, development, new and/or ongoing programs, operating expenses, and technology. The Knott Foundation uses a two-step online application process for its standard-grants program. Step one requires the submission of an online Letter of Inquiry (LOI) along with a Financial Analysis Form. Applicants whose LOIs are approved will move on to step two which requires online submission of a full proposal. Applicants are given the opportunity to submit a draft of their proposal for comments and feedback prior to their final submission. The review process for the foundation's standard grants program takes approximately four months from the date of the LOI submission until a final funding decision is made. The Knott Foundation accepts standard-grant applications three times per year - February, June and October. LOIs and proposals must be received by 5 p.m. on the applicable deadline date. Complete details, guidelines, and links to the online submission system are available at the grant website.
Requirements: Funding is limited to 501(c)3 organizations serving Baltimore City and the following counties in Maryland: Allegheny, Anne Arundel, Baltimore, Carroll, Frederick, Garrett, Harford, Howard, and Washington. Applicants may apply through a fiscal sponsor. The fiscal sponsor must be a 501(c)3 nonprofit organization that has a formal relationship and Memorandum of Understanding (MOU) with the applicant.
Restrictions: Organizations that are denied funding at the LOI stage of the grant process are eligible to apply again during the next grant cycle; organizations that are denied funding after submitting a full grant proposal must wait one year before reapplying; organizations that receive a grant award must wait two years before reapplying. The following will not be funded: organizations that have not been in operation for at least one year, scholarships, public education/public sector agencies, pro-choice or reproductive health programs, individuals, annual giving, political activities, one-time only events/seminars/workshops, legal services, environmental activities, medical research, day care centers, endowment funds for arts/humanities, national/local chapters for specific diseases, agencies that redistribute grant funds to other nonprofits, reimbursables or any prior expenses, or government agencies that form 501(c)3 nonprofits to fund public sector projects.
Geographic Focus: Maryland
Date(s) Application is Due: Mar 7; Jul 9; Nov 12
Amount of Grant: 35,000 - 45,000 U.S.D
Samples: Pathfinders for Autism, Hunt Valley, Maryland, $32,294 - to support half of the Resource Center and Outreach Coordinator's salary and operating expenses related to expanding the Parent and Professional Training Series (2012); St. Elizabeth School, Baltimore, Maryland, $44,748 - to support the purchase and installation of five new SMART Boards (2012); Big Brothers Big Sisters of the Greater Chesapeake, Baltimore, Maryland, $45,000 - to support salaries for professionals engaged in the mentoring of children facing adversity in Baltimore City (2012).
Contact: Kathleen McCarthy, Grants Manager; (410) 235-7068; fax (410) 889-2577; knott@knottfoundation.org or info@knottfoundation.org
Internet: http://www.knottfoundation.org/what_we_do/grant_application_process
Sponsor: Marion I. and Henry J. Knott Foundation
3904 Hickory Avenue
Baltimore, MD 21211-1834

Marion Park Deaver and Harry Gilbert Deaver Foundation Grants **2795**

The mission of the foundation is to promote education, with major emphasis on financial support to parochial and private schools for facilities, scholarships, computing and network capability, technical support for the use of computers in education, development and presentation of a course for teachers in the use of network for education, and the Deaver Foundation Award for Excellence in Teaching. Other areas designated for financial assistance are religion, science, health, environment, literacy arts, and the indigent. The geographic areas for grants are Florida, Wisconsin, and Minnesota. There are no application forms or deadlines.
Requirements: Florida, Minnesota, and Wisconsin nonprofit organizations are eligible.
Geographic Focus: Florida, Minnesota, Wisconsin
Samples: SpanishNet College (Boynton Beach, FL)—to offer Spanish-language classes to U.S. teachers and other school employees and to nurses and other health-care providers, beginning in Forsyth County, NC, $225,000.
Contact: Grants Administrator, c/o Wachovia Bank
Internet: http://www.deaverfoundation.org/AboutTheDeaverFoundation.htm
Sponsor: Marion Park Deaver and Harry Gilbert Deaver Foundation
975 S Federal Highway
Boca Raton, FL 33432

Marjorie C. Adams Charitable Trust Grants **2796**

The Foundation, established in New York in 1987, is administered by the JP Morgan Chase Bank of Newark and supports the care and treatment of deaf and blind people, research into the causes and the treatment of deafness and blindness, and K-12 education. Types of support for these programs include: building and renovation projects, capital campaigns, and endowments. There are no specific application forms required or deadlines with which to adhere.
Geographic Focus: Maryland, New York
Amount of Grant: Up to 150,000 U.S.D
Samples: League for the Hard of Hearing, New York New York, $95,000; Foundation for Fighting Blindness, Owings Mills, Maryland, $102,000—for an endowment.
Contact: Frank Lemma, Vice President of JP Morgan Chase Bank; (212) 464-2439
Sponsor: Marjorie C. Adams Charitable Trust
345 Park Avenue
New York, NY 10154

Marjorie Moore Charitable Foundation Grants **2797**

The Marjorie Moore Charitable Foundation was established in 1957 to support and promote quality educational, cultural, human-services, environmental, and health-care programming for underserved populations. Grants from the Moore Foundation made in support of operations or programming are one year in duration. Multi-year grants for long-term capital projects will be considered on a case-by-case basis. Applicants must apply online at the grant website. Applicants are strongly encouraged to do the following before applying: review the downloadable state application procedures for additional helpful information and clarifications; review the downloadable online-application guidelines at the grant website; review the foundation's funding history (link is available from the grant website); review the online application questions in advance; and review the list of required attachments. These will generally include: a list of board members, financial statements (audited, reviewed, or compiled by independent auditor); an organization summary; a list of other funding sources; an IRS Determination letter; and other required documents. All attachments must be uploaded in the online application as PDF, Word, or Excel files. The Marjorie Moore Charitable Foundation has biannual deadlines of June 1 and December 1 Applications must be submitted by 11:59 p.m. on the deadline dates. Applicants will be notified of grant decisions by letter within two to three months after each respective proposal deadline.
Requirements: Applicant organizations must have 501(c)3 tax-exempt status and serve the people of Kensington or Berlin, Connecticut. A breakdown of number/percentage of people served by specific towns is required on the online application. Preference is given to organizations that provide human services or health care programming.
Restrictions: The foundation does not support requests from individuals, organizations attempting to influence policy through direct lobbying, or any political campaigns.
Geographic Focus: Connecticut

Date(s) Application is Due: Jun 1; Dec 1
Samples: Town of Berlin, Berlin, Connecticut, $18,000 (2010); Young Audiences of Connecticut, Hamden, Connecticut, $9,000 (2010); Futures, Inc., Middletown, Connecticut, $8,000 (2010).
Contact: Kate Kerchaert, Senior Foundation Officer; (860) 657-7016; kate.kerchaert@baml.com
Internet: https://www.bankofamerica.com/philanthropic/fn_search.action
Sponsor: Marjorie Moore Charitable Foundation
200 Glastonbury Boulevard, Suite # 200, CT2-545-02-05
Glastonbury, CT 06033-4056

Mars Foundation Grants **2798**
The foundation awards grants to eligible nonprofit organizations primarily within the Washington, DC, metropolitan area, including Virginia, Maryland, and the District of Columbia. Areas of support include arts and culture, animal/wildlife, civic and community, education, environment, health and human services, and historic preservation. Types of support include capital grants, challenge grants, general operating grants, and project grants. Grants are awarded once a year. Requests must be submitted on the foundation's application form.
Requirements: U.S. 501(c)3 nonprofit organizations are eligible.
Date(s) Application is Due: Oct 15
Amount of Grant: 1,000 - 100,000 U.S.D
Samples: Colonial Williamsburg (Williamsburg, VA)—for new 17th- and 18th-century costumes for the casts of the Cry Witch! program, $25,000.
Contact: Grants Administrator; (703) 821-4900; fax (703) 448-9678
Internet: http://www.mars.com
Sponsor: Mars Foundation
6885 Elm Street
McLean, VA 22101

Mary Allen Lindsey Branan Foundation Grants **2799**
Income of the trust is to be distributed at least annually to 501(c)3 organizations for the relief of suffering and for charitable purposes—e.g., for the victims of floods, disasters, for the care of the sick, feeble, helpless, and orphans. However, grants are awarded exclusively at the discretion of the Board of Trustees. There are no geographic restrictions on distributions, but it has been the practice of the trustee to make grants within the state of Georgia.
Samples: Moving in the Spirit (GA)—$10,000; Georgia Health Decisions (GA)—$10,000; Stepping Stones Educational Therapy Ctr (GA)—$25,000; Cool Girls (GA)—$5,000.
Contact: Lydia Whitman, c/o Wachovia Bank, (404) 332-6677; grantinquiriesga@wachovia.com
Internet: http://www.wachovia.com/corp_inst/charitable_services/0,,3297,00.html
Sponsor: Mary Allen Lindsey Branan Foundation
3414 Peachtree Road, 5th Floor, MC GA8023
Atlanta, GA 30326

Mary Baker Eddy Library Summer Research Fellowships **2800**
The library awards fellowships to scholars and other researchers for research in its collections in the fields of women's history, spirituality and health, religious studies, 19th-century history, and journalism—themes for which the library's collections hold special relevance and value. Consideration will be given to applicants with outstanding writing skills, evidence of having previous publications, and a definite plan for disseminating the research through publication, lecturing, etc. Fellowship recipients are expected to be in regular and continuous use of the collections at the library in Boston during the period of the fellowship (between June 13 and September 2). Fellows are responsible for making their own arrangements for housing in the Boston area. Application and guidelines are available online.
Requirements: Fellowships are open to doctoral and postdoctoral scholars, faculty members, and independent researchers.
Date(s) Application is Due: Feb 13
Amount of Grant: Up to 8,400 U.S.D
Contact: Administrator; (617) 450-7206; fellowships@marybakereddylibrary.org
Internet: http://www.marybakereddylibrary.org/programs/fellowshipapplications.jhtml
Sponsor: Mary Baker Eddy Library
200 Massachusetts Avenue
Boston, MA 02115

Mary Black Foundation Grants **2801**
The foundation makes grants to nonprofit organizations in the Spartanburg County, South Carolina, region in its areas of interest, including early childhood development, and active living. Grants support projects in preventive health care, community health and wellness, and education. The community health fund is a small grantmaking resource that allows the foundation to respond to grant applications outside of the active living and early childhood development priority areas, but still within the foundation's mission of improving health and wellness in Spartanburg County. Guidelines and applications are available on the foundation's Web site.
Requirements: Nonprofit organizations in Spartanburg County, SC are eligible.
Restrictions: The foundation does not accept applications from individuals or general fundraising solicitations.
Geographic Focus: South Carolina
Date(s) Application is Due: Mar 15; Jun 15; Sep 15; Dec 15
Samples: Health Resource Ctr (Spartanburg, SC)—to publish and distribute Parent Talk, a newsletter on child rearing, $15,000; Senior Ctrs of Spartanburg County (SC)—for a home-repair program for elderly people, $28,000; Healthy Spartanburg (SC)—for general operating support, $25,000; Stop the Violence Collaboration of Spartanburg (SC)—to hold a series of violence-prevention institutes for seventh and eighth graders, $25,000.
Contact: Program Contact; (864) 573-9500; fax (864) 573-5805; info@maryblackfoundation.org
Internet: http://www.maryblackfoundation.org/application.html
Sponsor: Mary Black Foundation
945 E Main Street
Spartanburg, SC 29302

Mary Ingraham Bunting Institute Biomedical Research Fellowships **2802**
Funded by the Burroughs Wellcome Fund, the fellowships support women scientists in all fields of biomedical research. Two awards will be made in the current year. Fellows may compete for a second year of funding. Fellowships provide office or studio space, auditing privileges, and access to libraries and most other resources of Radcliffe College and Harvard University. Residence in the Boston area and participation in the institute community are required during the fellowship appointment.
Requirements: Women scientists who are U.S. citizens or permanent residents and have held a doctorate (MD or PhD) degree for at least two years prior to appointment are eligible.
Date(s) Application is Due: Oct 1
Amount of Grant: 55,000 U.S.D
Contact: Fellowship Coordinator; (617) 496-1324; fellowships@radcliffe.edu
Internet: http://www.radcliffe.edu/fellowships/index.php
Sponsor: Bunting Institute of Radcliffe College
34 Concord Avenue
Cambridge, MA 02138

Mary K. Chapman Foundation Grants **2803**
Mary K. Chapman was born in Oklahoma in 1920. She graduated from the University of Tulsa and worked as a nurse before her marriage to Allen Chapman in 1960. After the death of her husband in 1979, Mary Chapman maintained her own personal charitable giving program. Before her death in 2002, she established The Mary K. Chapman Foundation, a charitable trust founded to perpetuate her own charitable giving program. This foundation was fully funded with a bequest from her estate in 2005. Mary K. Chapman was very interested in supporting education, but as a former nurse and a very compassionate person, much of her charity was directed to health, medical research, and educating and caring for the less fortunate and disadvantaged. There are two steps in the process of applying for a grant. The first is a Letter of Inquiry from the applicant. This letter is used to determine if the applicant will be invited to take the second step of submitting a formal Grant Proposal.
Requirements: IRS 501(c)3 non-profits are eligible to apply.
Restrictions: Grant requests for the following purposes are not favored: endowments, except as a limited part of a capital project reserved for maintenance of the facility being constructed; deficit financing and debt retirement; projects or programs for which the Chapman Trusts would be the sole source of financial support; travel, conferences, conventions, group meetings, or seminars; camp programs and other seasonal activities; religious programs of religious organizations; project or program planning; start-up ventures are not excluded, but organizations with a proven strategy and results are preferred; purposes normally funded by taxation or governmental agencies; requests made less than nine months from the declination of a previous request by an applicant, or within nine months of the last payment made on a grant made to an applicant; requests for more than one project.
Geographic Focus: Oklahoma
Amount of Grant: Up to 300,000 U.S.D
Contact: Andie Doyle, Program Officer; (918) 496-7882; fax (918) 496-7887; andie@chapmantrusts.com
Internet: http://www.chapmantrusts.org/grants_programs.php
Sponsor: Mary K. Chapman Foundation
6100 South Yale, Suite 1816
Tulsa, OK 74136

Mary Kay Ash Charitable Foundation Grants **2804**
The foundation, whose mission is to enrich women's lives around the world, supports research grants to help find a cure for cancers affecting women (breast, cervical, ovarian, and uterine). It also funds U.S. women's shelters and offers financial support to organizations that raise awareness and address the problem of violence against women. Proposals may be submitted at any time.
Requirements: Cancer research grants are awarded to accredited U.S. medical schools and schools of public health. Shelters must be U.S. tax-exempt entities.
Samples: U of Texas Southwestern Medical Ctr (Dallas, TX)—to create a professorship in dermatology, $1 million.
Contact: Grants Administrator; (877) 652-2737
Internet: http://www.mkacf.org
Sponsor: Mary Kay Ash Charitable Foundation
P.O. Box 799044
Dallas, TX 75379-9044

Mary Kay Foundation Cancer Research Grants **2805**
In May 2010, the Foundation awarded $1.3 million in grants to select doctors and medical scientists focusing on curing cancers that affect women. These 13 recipients from across the United States received a $100,000 grant to conduct cutting-edge research. Since 1996, the Foundation has given more than $14 million to support this effort. Grants are awarded each year to researchers at medical schools recommended by the The Mary Kay Foundation Research Review Committee, which is composed of prominent

doctors who volunteer their time to help the Foundation select the best recipients across the United States. After reviewing these recommendations, the Board of Directors at the Foundation selects the grant recipients. The Mary Kay Foundation accepts grant applications from November to mid-February each year.
Amount of Grant: 100,000 U.S.D
Samples: Emily Bernstein, Ph.D., Mount Sinai School of Medicine, $100,000 - Identifying epigenetic drivers of breast cancer (2011); Jeanette Gowen Cook, Ph.D., University of North Carolina at Chapel Hill, $100,000 - Predicting genome instability in breast cancer (2011); Zeng-Quan Yang, Ph.D., Wayne State University, $100,000 - Targeting histone demethylases as a therapeutic strategy in basal breast cancer (2011).
Contact: Lana Rowe, (972) 687-4822 or (877) 652-2737; Lana.Rowe@mkcorp.com or MKCares@marykayfoundation.org
Jennifer Cook , (972) 687-5889 or (972) 687-4822; Jennifer.cook@mkcorp.com or MKCares@marykayfoundation.org
Internet: http://www.marykayfoundation.org/Pages/CancerGrantProgram.aspx
Sponsor: Mary Kay Foundation
P.O. Box 799044
Dallas, TX 75379-9044

Mary L. Peyton Foundation Grants **2806**
The private operating foundation awards grants to provide health, welfare, and vocational education benefits to legal residents of El Paso, TX, who are unable to attain assistance elsewhere. The program supports three areas: short-term necessities of life—food, clothing, and shelter; medical items, except doctor bills, hospital bills, or prescription costs; and vocational training. The board meets the third Thursday of each month.
Requirements: Individuals in El Paso County, TX, are eligible.
Restrictions: Support is not given to groups or organizations. Grants are not awarded for building or endowment funds, fellowships, matching gifts, or loans.
Geographic Focus: Texas
Contact: James Day, Executive Administrator; (915) 533-9698
Sponsor: Mary L. Peyton Foundation
303 Texas Avenue, Bassett Tower, Suite 908
El Paso, TX 79901-1456

Mary Owen Borden Foundation Grants **2807**
The Mary Owen Borden Foundation Grants support programs that address the needs of economically disadvantaged youth and their families. This includes needs such as health, family planning, education, counseling, childcare, substance abuse, and delinquency. Other areas of interest for the foundation include affordable housing, conservation, environment, and the arts. Grants average $10,000, and the maximum grant is $15,000. In unique circumstances, the Foundation considers a more significant grant for a program having a major impact in their areas of interests.
Requirements: New Jersey nonprofits in Monmouth and Mercer Counties are eligible. Most of the Foundation's grant go to nonprofit entities in Trenton, Asbury Park, and Long Branch.
Geographic Focus: New Jersey
Date(s) Application is Due: Mar 15; Sep 15
Amount of Grant: Up to 15,000 U.S.D
Contact: Quinn McKean, Executive Director; (732) 741-4645; fax (732) 741-2542; qmckean@aol.com
Internet: http://fdncenter.org/grantmaker/borden/guide.html
Sponsor: Mary Owen Borden Foundation
4 Blackpoint Horseshoe
Rumson, NJ 07760

Mary S. and David C. Corbin Foundation Grants **2808**
The Corbin Foundation gives primary consideration to charitable organizations and/or local chapters of national charities located in Akron and Summit County, Ohio, although extremely worthy causes outside of this preferred area may be considered. Areas of interest include arts and culture; civic and community; education; health care; housing; social services; medical research; and youth. The Foundation meets at least semiannually to consider requests. Grant requests must be received no later than March 1 for consideration in May and September 1 for consideration in November. The application and guidelines are available online.
Restrictions: Telephone and personal interviews are discouraged unless requested by the foundation. The Foundation does not fund individuals; annual fundraising campaigns; ongoing requests for general operating support (although some repeat grants are made); operating deficits; or organizations which in turn make grants to others.
Geographic Focus: Ohio
Date(s) Application is Due: Mar 1; Sep 1
Contact: Erika J. May, Grants Administrator; (330) 762-6427; fax (330) 762-6428; corbin@nls.net
Internet: http://foundationcenter.org/grantmaker/corbin/guide.html
Sponsor: Mary S. and David C. Corbin Foundation
159 South Main Street
Akron, OH 44308

Mary Wilmer Covey Charitable Trust Grants **2809**
The Mary Wilmer Covey Charitable Trust was created to support charitable organizations that promote education including instruction and training that help to build human capabilities. It also supports organizations that focus on relieving human suffering due to disease, ill health, physical weakness, disability, or injury; and supports organizations that work to prolong life and improve the quality of life, especially for children. Grants from the Mary Wilmer Covey Charitable Trust are primarily one year in duration; on occasion, multi-year support is awarded. Applicants must apply online at the grant website. The annual application deadline is September 1 at 11:59 p.m. Applicants will be notified of grant decisions by letter within one to two months after the deadline. Applicants are strongly encouraged to do the following before applying: review the downloadable state-specific application procedures at the grant website; review the downloadable online-application guidelines at the grant website; review the trust's funding history (link is available from the grant website); review the online application questions in advance; and review the list of required attachments. These will generally include: a list of board members, financial statements (audited, reviewed, or compiled by independent auditor); an organization summary; a list of other funding sources; an IRS Determination letter; and other required documents All attachments must be uploaded in the online application as PDF, Word, or Excel files.
Requirements: Applicants must have 501(c)3 tax-exempt status. Preference is given to charitable organizations located in Macon, Georgia; Richmond, Virginia; and Chatham Hall, Virginia.
Restrictions: The trust does not support requests from individuals, organizations attempting to influence policy through direct lobbying, or any political campaigns.
Geographic Focus: Georgia, Virginia
Date(s) Application is Due: Sep 1
Samples: Medical Center of Central Georgia, Macon, Georgia, $50,000 (2010); Chatham Hall, Chatham, Virginia, $15,000 (2010); Childrens Hospital, Richmond, Virginia, $15,000 (2010).
Contact: Quanda Allen, Vice President; (404) 264-1377; quanda.allen@baml.com
Internet: https://www.bankofamerica.com/philanthropic/fn_search.action
Sponsor: Mary Wilmer Covey Charitable Trust
3414 Peachtree Road, N.E., Suite 1475, GA7-813-14-04
Atlanta, GA 30326-1113

Massage Therapy Foundation Community Service Grants **2810**
The Foundation Community Service grants are awarded to organizations that seek to provide massage therapy to communities or groups who currently have little or no access to such services. This program is designed to promote working partnerships between the massage therapy profession and community-based organizations. It benefits the recipient, the massage therapist, and the sponsoring organization by building stronger relationships between these parties. The Community Service deadline is April 1, annually. The normal award for 12 months is $500-$5000 and must be used in the specific time period for which it has been awarded.
Requirements: These grants are available for organizations or affiliates of organizations that have been in existence for at least one year in the respective state or province; are tax-exempt under schedule 501(c)3 in the U.S., non-profit charitable organization in other countries; currently provide some therapeutic or other service programs to the community; and have designated a qualified staff member to oversee the program.
Date(s) Application is Due: Apr 1
Amount of Grant: 500 - 5,000 U.S.D
Contact: Alison Pittas, Program Manager of Research and Grants; (847) 869-5019, ext. 167 or (847) 905-1667; fax (847) 864-1178; apittas@massagetherapyfoundation.org
Internet: http://www.massagetherapyfoundation.org/grants_community.html
Sponsor: Massage Therapy Foundation
500 Davis Street, Suite 900
Evanston, IL 60201

Massage Therapy Foundation Practitioner Case Report Contest **2811**
The Foundation has chosen to encourage the writing of case reports to provide an opportunity for massage therapists and bodyworkers to develop research skills and enhance their ability to provide knowledge-based massage to the public. This contest is intended to enhance professional development skills of the practitioner: writing case reports help develop communication skills, critical thinking skills, and could contribute to future research and clinical practice. Cash and publication recognition will be awarded to practitioners submitting the top reports.
Requirements: The Foundation requires that HIPAA guidelines for confidentiality are followed.
Date(s) Application is Due: Oct 8
Samples: Erika Larson—for Massage Therapy Effects in a Long-Time Prosthetic User with Fibular Hemimelia; Glenda Keller—for The effects of massage therapy in treatment of chronic plantar fasciitis; Robin B. Anderson—for Reduction and Stabilization in Parkinson-related Peripheral Edema with Therapeutic Massage.
Contact: Alison Pittas, Program Manager of Research and Grants; (847) 869-5019, ext. 167 or (847) 905-1667; fax (847) 864-1178; apittas@massagetherapyfoundation.org
Internet: http://www.massagetherapyfoundation.org/practitionercontest.html
Sponsor: Massage Therapy Foundation
500 Davis Street, Suite 900
Evanston, IL 60201

Massage Therapy Foundation Research Grants **2812**
One of the primary purposes of the Foundation is to fund solid research studies investigating the many beneficial applications of massage therapy. Foundation research grants are awarded to individuals or teams conducting studies that promise to advance our understanding of specific therapeutic applications of massage, public perceptions of and attitudes toward massage therapy, and the role of massage therapy in health care delivery. The Research Grant deadline is normally March 1 annually (or the following Monday if this falls on a weekend). The normal award for 12 months is $1,000 to $30,000, and must be used in the specific time period for which it has been awarded. This

grant supports high quality, independent research that contributes to the basic science of massage therapy application, including applied research investigating massage therapy as a health/mental health treatment and/or prevention modality.
Requirements: Research Grants are available to investigators who: have experience in the relevant field of research; are presently associated with, or have secured the cooperation of a university, independent research organization, health center, or other institution qualified and willing to function as a Sponsoring Organization for the purpose of this project; or new investigators without prior research experience must document support from an experience investigator willing to act as a collaborator.
Restrictions: Research Grants will not be awarded to spouses, domestic partners, children, descendant, spouses of descendants or any other individual related to any officers or trustees of the Massage Therapy Foundation, or to members of the Research Proposal Review Committee.
Date(s) Application is Due: Mar 1
Amount of Grant: Up to 30,000 U.S.D
Contact: Alison Pittas, Program Manager of Research and Grants; (847) 869-5019, ext. 167 or (847) 905-1667; fax (847) 864-1178; apittas@massagetherapyfoundation.org
Internet: http://www.massagetherapyfoundation.org/grants_research.html
Sponsor: Massage Therapy Foundation
500 Davis Street, Suite 900
Evanston, IL 60201

Massage Therapy Research Fund (MTRF) Grants **2813**
The purpose of the MTRF is to fund high quality research that investigates the efficacy, and clinical effectiveness of massage therapy and that contributes to our understanding of how massage therapy achieves its effects. A secondary goal is to increase research knowledge and capacity among massage therapists through their engagement in research practices at all levels. The MTRF invites applications in areas indicated above. For both competition cycles, a total funding amount of $50,000 is offered. Grant applications up to $15,000 will be considered.
Requirements: All properly submitted applications will be considered. Studies may be qualitative, quantitative, or mixed method. Initially, funding preference will be given to smaller, pilot, or creative study designs. In this way, the MTRF aims to assist the scientific and massage communities in building a foundation for the eventual construction of more complicated research projects. The MTRF encourages applications that also seek funding from other sources. The principal investigator must be affiliated with a Canadian charitable institution (university or hospital) that is willing to administer the finances of the grant. Funds are awarded to the charitable institution, not to an individual. The research team may include members affiliated with a non-Canadian institution. One member of the research team should be a massage therapist. Students at the masters level and above are encouraged to apply.
Geographic Focus: All States, Canada
Date(s) Application is Due: Mar 1; Sep 30
Amount of Grant: 15,000 U.S.D
Samples: 2008 Awards - University of Saskatchewan ($8,929) Donelda Gowan-Moody, RMT and Anne Leis, PhD: Massage Therapists' Research Utilization and Perceptions Towards Research; Carleton University ($15,000) Diane Sliz, MSc (Cand) and Dr. Shawn Hayley: Psychological and Neuro-immune Effects of a Long-term Massage Therapy Treatement in Individuals Suffering from Major Depressive Disorder; Wilfred Laurier University ($14,981) Peter Tiidus, PhD, Kimberley Dawson, PhD and Lance Dawson, RMT: Effectiveness of regular proactive massage therapy for novice recreational runners.
Contact: Keren Brown, Executive Director; (416) 778-4443 or (866) 778-4443; fax (416) 778-5438; info@holistichealthresearch.ca
Internet: http://www.holistichealthresearch.ca/massagefund.gk
Sponsor: Holistic Health Research Foundation of Canada
80 Carlton Street
Toronto, ON M5B 1L6 Canada

MassMutual Charitable Contributions Grants **2814**
As good corporate citizens, the MassMutual Financial Group companies are committed to improving its corporate communities. Philanthropic activities—primarily targeted at education—range from direct grants to charitable organizations, to significant scholarships and academic programs, to the thousands of hours employees as representatives devote to volunteerism. The foundation also supports arts and culture, and community services.
Requirements: 501(c)3 organizations in Springfield, MA and Hartford, CT, are eligible to apply.
Geographic Focus: Connecticut, Massachusetts
Amount of Grant: 1,000 - 5,000 U.S.D
Samples: U of Connecticut School of Law (Storrs, CT)—to endow a chair in insurance law, $400,000.
Contact: Pam Mathison, Community Relations; (413) 744-2169; pmathison@massmutual.com
Internet: http://www.massmutual.com/mmfg/about/community.html
Sponsor: MassMutual Foundation
1295 State Street
Springfield, MA 01111

MassMutual Foundation for Hartford Grants **2815**
Programs are supported in the areas of education, housing, and health. Grants in education support collaborative relationships between the Connecticut Mutual/Hartford Public High School and Connecticut Mutual/Quirk Middle School Alliances; an employee matching gifts contribution program; and scholarship funding earmarked for Hartford Public High School graduates granted directly to the institutions of higher education and training. In the area of health, the foundation gives special consideration to proposals concerned with increasing the efficiency and effectiveness of health care in such areas as health education and disease prevention, alternative modes of health care delivery, and programs for cost containment. In the area of housing, the foundation will give special consideration to contributions relating to the city of Hartford and the Asylum Hill neighborhood in particular. Types of support include general operating support, continuing support, capital campaigns, program development, seed grants, internships, scholarship funds, matching funds, employee matching gifts, and technical assistance. Application address: 1295 State Street, H356, Springfield, MA 01111-0001.
Requirements: Grants are limited to Connecticut tax-exempt organizations.
Restrictions: No support for sectarian groups; political, fraternal, labor, or veterans' organizations; or federated drives outside the local area. No grants to individuals (including direct scholarships), or for endowment funds, deficit financing, emergency funds, publications, land acquisition, fellowships, capital fund drives outside the local area, or goodwill advertising.
Geographic Focus: Connecticut
Amount of Grant: 500 - 300,000 U.S.D
Contact: Ronald Copes, Executive Director; (860) 987-2085
Sponsor: MassMutual Foundation for Hartford
140 Garden Street, H356
Hartford, CT 06154

Mass Youth Against Tobacco Mini-Grants **2816**
The Mass Youth Against Tobacco grants program is intended to support youth-led tobacco prevention initiatives. Recently, mini-grant categories were developed through a close collaboration between youth and adults. Three categories of application are available: Youth Create Change—strengthening policies for healthier communities; Role Models, Take Action; and Taking Back Our Communities—taking on big tobacco. Between 20 and 25 projects will be funded, ranging from $8,000 to $15,000. See the web site for details.
Restrictions: Monies awarded through the grant cannot be spent on lobbying (calls for action and/or direct support of state legislation) for a specific bill or for direct attacks on the tobacco companies or their employees. Youth groups and their sponsors cannot receive funding nor have an affiliation or contractual relationship with any tobacco company, its affiliates, subsidiaries or its parent company. Funds cannot be used for cessation programs or starting new youth groups or programs.
Geographic Focus: Massachusetts
Date(s) Application is Due: Sep 15
Amount of Grant: 8,000 - 15,000 U.S.D
Contact: Brittany Chen, Project Associate; (617) 279-2240, ext. 324 or (617) 451-0049; fax (617) 451-0062; bchen@tmfnet.org
Internet: http://www.makesmokinghistory.org/myat/mini_grants.htm
Sponsor: Medical Foundation, a Division of Health Resources in Action
95 Berkeley Street, Suite 208
Boston, MA 02116

Matilda R. Wilson Fund Grants **2817**
The fund awards grants, primarily in southeast Michigan, in support of the arts, youth agencies, higher education, hospitals, and social services. Types of support include building construction/renovation, endowments, equipment acquisition, general operating support, matching/challenge grants, program development, research, and scholarship funds. There are no application forms; initial approach should be a letter of request. The board considers requests at quarterly board meetings, in January, June, September, and December.
Requirements: Michigan tax-exempt organizations are eligible.
Restrictions: Grants are not made to individuals.
Geographic Focus: Michigan
Amount of Grant: 10,000 - 100,000 U.S.D
Contact: David Larsen, Grants Administrator; (313) 259-7777; fax (313) 393-7579; roosterveen@bodmanllp.com
Sponsor: Matilda R. Wilson Fund
100 Renaissance Center, 34th Floor
Detroit, MI 48243

Matsumae International Foundation Grants **2818**
The foundation operates worldwide in the fields of medicine/health and science/technology. Fields of study such as natural science, engineering, and medicine are given first priority. Types of support include grants, scholarships, and fellowships, tenable for three to six months. Awards include a monthly stipend; lump sum on arrival to assist with the cost of lodging, travel expense, etc; round-trip travel; and insurance.
Requirements: Applicants must apply from outside Japan, be under the age of 40, and be postdoctoral researchers or have a minimum of two years of research experience after receipt of a master's degree. Study in the areas of natural science, engineering, and medicine receive priority.
Date(s) Application is Due: Jul 31
Amount of Grant: 600,000 - 1,200,000 JPY
Contact: Grants Administrator; 81-3-3214-7611; fax 81-3-3301-7601; contact@matsumae-if.org
Internet: http://www.mars.dti.ne.jp/~mif
Sponsor: Matsumae International Foundation
4-14-46, Kamiogi
Suginami-ku, Tokyo, 167-0043 Japan

Mattel Children's Foundation Grants 2819

Charitable organizations that demonstrate that they directly serve children in need may be eligible for grants of $5,000 up to $20,000. Such organizations must have a mission that focuses on the direct service of children ages zero to 12 years. Applicants or programs must show creative and/or innovative methods to address locally defined need directly impacting children, and must be aligned with Mattel's philanthropic priorities: Learnin—increasing access to education for underserved children and in particular, innovative strategies to promote and address literacy; health—supporting the physical health and well-being of children, with particular emphasis on promoting healthy, active lifestyles; and Girl Empowerment—promoting self-esteem in young girls, up to age 12. Organizations may submit under a fiscal agent as long as the program has been in existence for at least two years.

Requirements: Applicants must be a 501(c)3 Public Charity organizations. If an organization is a project under a fiscal sponsor, that sponsor must have a valid tax exemption status. Organizations must serve children in communities within the U.S. Organizations must not discriminate against a person or a group on the basis of age, political affiliation, race, national origin, ethnicity, gender, disability, sexual orientation or religious beliefs. Organizations must have an annual operating budget of less than $1,000,000.

Restrictions: Organizations may not be affiliated in any way with national organizations, regardless of whether funds are received from the national entity. Examples of organizations that are ineligible to apply include Boys & Girls Clubs, United Ways, YMCA's, and disease-affiliated organizations such as the American Heart Association.

Amount of Grant: 5,000 - 20,000 U.S.D

Contact: Grants Administrator; (310) 252-2000; fax (310) 252-2180; foundation@mattel.com

Internet: http://corporate.mattel.com/about-us/philanthropy/programs.aspx

Sponsor: Mattel Children's Foundation
333 Continental Boulevard, M1-1418
El Segundo, CA 90245-5012

Mattel International Grants 2820

The corporation and its foundation award grants to charitable organizations that directly serve children in need. Funding priorities include international organizations or programs that creatively address a defined need directly impacting children in need, and international organizations or programs that align with Mattel's philanthropic priorities, including health—supporting the physical and mental health and well-being of children, and increasing access to health care services for children in need; education—increasing access to education, promoting literacy to children in need, and resources that promote after-school educational achievement; and girls empowerment—promoting self-esteem of girls and increasing access to education, health, and community resources for girls. Types of support include program-specific grants—funding for new programs or expansion of existing programs; and core operating support—support of organizations to sustain their programs. Applications must be submitted online. Applications will not be accepted by fax or mail. Submit questions via email.

Requirements: Organizations must serve children in non-U.S. communities; and must not discriminate against a person or a group on the basis of age, political affiliation, race, national origin, ethnicity, gender, disability, sexual orientation, or religious belief. Pilot projects and new organizations may be considered as long as all eligibility criteria are met; however, preference will be given to organizations that have at least two years of experience. Preference is given to organizations that have an annual operating budget of less than $1 million and are not affiliated with a national organization.

Restrictions: The program does not not fund capital funding for physical property purchase, renovation, or developments; individuals; political parties, candidates, or partisan political organizations; labor organizations, fraternal organizations, athletic clubs, or social clubs; sectarian or denominational religious organizations, except for programs that are available to anyone, broadly promoted, and free from religious orientation; schools and school districts; fundraising events (e.g., dinners, tournaments); advertising or marketing sponsorships; or research overhead and/or indirect costs (including fiscal sponsor fees) that exceed 15 percent of the direct project costs.

Amount of Grant: 5,000 - 25,000 U.S.D

Contact: Administrator; (310) 252-2000; fax (310) 252-2180; foundation@mattel.com

Internet: http://corporate.mattel.com/about-us/philanthropy/internationalgrants.aspx

Sponsor: Mattel Children's Foundation
333 Continental Boulevard, M1-1418
El Segundo, CA 90245-5012

Maurice J. Masserini Charitable Trust Grants 2821

The trust awards one-year grants to eligible San Diego County California nonprofit organizations in its areas of interest, including children and youth, aging, music, higher education, and marine sciences. Types of support include building construction/renovation, equipment acquisition, program development, research grants, matching grants, development grants, internships, and scholarships. Submit a letter of application, and include a copy of the IRS tax-exempt letter. Contact foundation by email for current guidelines.

Requirements: San Diego County, CA, 501(c)3 tax-exempt organizations are eligible.

Geographic Focus: California

Date(s) Application is Due: Feb 15; May 15; Aug 15; Nov 15

Amount of Grant: 2,500 - 25,000 U.S.D

Contact: Brian Murphy, c/o Wells Fargo Bank, San Diego Private Banking Group; (858) 597-4307; fax (858) 597-4336; brian.j.murphy@wellsfargo.com

Sponsor: Maurice J. Masserini Charitable Trust
4475 Executive Drive, Suite 100
San Diego, CA 92121-2130

Max A. Adler Charitable Foundation Grants 2822

The foundation awards grants to eligible New York nonprofit organizations serving the greater Rochester area in its areas of interest, including arts, children and youth services, Jewish services, health care, higher education, and social services. Types of support include building construction/renovation, capital campaigns, general operating support, and program development. There are no application forms or deadlines.

Requirements: New York 501(c)3 nonprofit organizations serving the greater Rochester area are eligible.

Geographic Focus: New York

Amount of Grant: 1,000 - 25,000 U.S.D

Samples: Nazareth College (Rochester, NY)—for its capital campaign to expand the size of its campus, $50,000.

Contact: David Gray, President; (585) 232-7290; fax (585) 232-7260

Sponsor: Max A. Adler Charitable Foundation
1010 Times Square Building
Rochester, NY 14614

Max and Victoria Dreyfus Foundation Grants 2823

The Foundation awards grants for: museums; cultural, performing, and visual arts programs, schools; hospitals, educational and skills training projects and programs for youth, seniors, and the disabled. Areas of interest include arts, education, health care, hospitals, social services, and medical research. Submit an inquiry letter (three-page maximum) that describes the nonprofit organization, its annual operating budget, the project to be funded, and the project budget; include proof of tax-exempt status.

Requirements: U.S. nonprofit organizations are eligible.

Restrictions: Foreign charitable organizations and individuals are ineligible.

Date(s) Application is Due: Mar 10; Jul 10; Nov 10

Amount of Grant: 5,000 - 20,000 U.S.D

Contact: John W. Hager, (202) 337-3300; fax (202) 337-4925

Sponsor: Max and Victoria Dreyfus Foundation
2233 Wisconsin Avenue NW, Suite 414
Washington, DC 20007-4122

Max Bell Foundation Grants 2824

The foundation focuses its grant making on innovative projects having substantial impact within the areas of health, Canada and the Asian Pacific, and veterinary sciences and education. Under health care, the foundation directs its support to the areas of health education and preventive health care. It is also interested in assisting health care professionals to update their education on a continuing basis. The foundation also supports research intended to build a better understanding of the Asian Pacific area on the part of Canadian policymakers and businesspeople and assists Canadian scholars to build ties to their Asian counterparts. Also funded are major programs at centers of Asian research in Canadian universities. Within the field of veterinary sciences and education, the foundation seeks to provide broad-based support, although it has developed a particular interest in equine research. Applications are reviewed three times annually, eight weeks before the board meets (usually in May September, and November).

Requirements: Funding is restricted to registered Canadian charities. Applicants should submit a brief letter describing the project for which they are seeking funds, an estimated budget, and the organization's charitable gift registration number. If the letter proves of interest to the foundation, a more detailed proposal will be invited.

Restrictions: Support will not be provided for capital projects or building campaigns; continuing operating expenses, funds to cover budget deficits, general fund-raising drives, or annual charitable appeals; academic or sabbatical leaves, internships, bursaries or scholarships; conferences, seminars or workshops, equipment purchases, publication costs, or travel costs.

Geographic Focus: All States, Canada

Amount of Grant: 20,000 - 25,000 CAD

Samples: AboutFace (Toronto, Ontario)—to provide support to children with facial disfigurements and their families, $C100,000; U of Toronto (Toronto, Ontario)—to develop a summer study program on the campus of the U of Hong Kong, $C105,000; U of Guelph (Guelph, Ontario)—to explore the biological and physical attributes of the Great Lakes and consider the continuing impact of humans upon them, $C205,000.

Contact: Alida White, Administrator; (403) 215-7310; fax (403) 215-7319; apply@maxbell.org

Internet: http://www.maxbell.org

Sponsor: Max Bell Foundation
639 Fifth Avenue SW, Suite 1230
Calgary, AB T2P OM9 Canada

Max Goldenberg Foundation Grants 2825

The foundation awards grants to Illinois nonprofit organizations in its areas of interest, including healthcare, health organizations, and hospitals; higher education; Jewish services; medical research; and social services delivery.

Requirements: Illinois nonprofit organizations are eligible.

Geographic Focus: Illinois

Amount of Grant: 2,500 - 40,000 U.S.D

Contact: Christopher McClear, c/o Harris Trust and Savings Bank; (312) 461-5992

Sponsor: Max Goldenberg Foundation
P.O. Box 755
Chicago, IL 60690

Maxon Charitable Foundation Grants 2826

The Maxon Charitable Foundation is a company-sponsored foundation that was established in 1987 in Indiana. The foundation supports hospitals and organizations involved with higher education, natural resources, recreation, human services, community development, and Christianity and awards college scholarships to individuals in the Muncie, Indiana area.

Requirements: Contact the Foundation's office prior to the initial proposal, the office will provide you with the proper Application form.

Geographic Focus: Indiana

Amount of Grant: 500 - 5,000 U.S.D

Contact: Jeffrey R. Lang, Secretary; (765) 284-3304

Sponsor: Maxon Charitable Foundation
201 E 18th Street
Muncie, IN 47302-4124

Maxtor Corporate Contributions Grants 2827

Maxtor supports charitable, nonprofit organizations whose programs improve the quality of life through education, health and human services, community services, and cultural endeavors. This program is designed to help eligible organizations in northern California, Massachusetts, and Colorado. Types of support include cash grants, Maxtor products, community service projects, and used equipment or in-kind services.

Requirements: 501(c)3 nonprofits serving northern California, Massachusetts, and Colorado are eligible.

Restrictions: Maxtor does not provide support to organizations without 501(c)3 status; individuals in personal programs; religious activities; school activities such as sporting events, bands, etc; individual scholarships; team, groups, races, or -athons; travel expenses; or political activities.

Geographic Focus: California, Colorado, Massachusetts

Contact: Program Contact; (408) 894-5000; PR@Maxtor.com

Internet: http://www.maxtor.com/en/about/corporate/contributions.htm

Sponsor: Maxtor Corporation
500 McCarthy Boulevard
Milpitas, CA 95035

May and Stanley Smith Trust Grants 2828

The trust awards grants to nonprofit organizations in its areas of interest, including children and youth, people with disabilities, the economically disadvantaged, education, the elderly, social services, and the visually impaired. Types of support include general operating grants and scholarship funds. Grants are awarded primarily in North America, with emphasis on the San Francisco Bay area, and Canada. Applications are accepted from January 1 through the listed deadline date; grants are considered beginning in April.

Restrictions: Grants are not made to individuals.

Date(s) Application is Due: Sep 30

Amount of Grant: 500 - 6,000 U.S.D

Contact: Dale Matheny, (415) 391-0292

Sponsor: May and Stanley Smith Trust
720 Market Street, Suite 250
San Francisco, CA 94102

Mayday Fund Grants 2829

The mission of the fund involves one purpose: the reduction of the physical and psychological toll of pain and its consequences. The fund particularly invites projects that will improve public policy to make it more hospitable to the better treatment of pain and that involve multidisciplinary teams including physicians, nurses, pharmacists, and administrators working to introduce routine assessment, documentation, and treatment of pain in hospital settings. Mayday also continues to have special interest in projects that relieve children's pain, involve primary care, and improve the treatment of pain in home care or other settings where a person is in pain after being discharged from a medical institution. There are no deadlines.

Requirements: Grants are made only to public charity organizations, officially recognized as such by the IRS.

Restrictions: Grants generally will not be made for endowments, capital projects, equipment, general operating expenses, ongoing activities, annual fund-raising drives, or in support of international or non-U.S. activities and institutions. Grants are not made to individuals.

Amount of Grant: 30,000 - 50,000 U.S.D

Samples: U of Rochester Medical School (Rochester, NY)—to enhance pediatric anesthesiology services, $27,564; Triumph Over Pain Foundation (Chicago, IL)—for start-up and operating expenses, $10,000; Saint Mary's Hospital for Children (Bayside, NY)—to develop a pain-management program for children with special health care needs, $150,000.

Contact: Christina Spellman; (212) 649-5800; fax (212) 649-5801; inquiry@maydayfund.org

Internet: http://www.painandhealth.org/mayday/mayday-home.html

Sponsor: Mayday Fund
136 W 21st Street, 6th Floor
New York, NY 10011

May Department Stores Charitable Contribution Grants 2830

The company has a long-standing commitment to responsible citizenship and to supporting its operating communities. The foundation contributes to worthy organizations that strive to improve the quality of life for customers, associates, and neighbors. Charitable contributions are concentrated in four areas: health and welfare, including the United Way; education; culture and the arts; and civic causes.

Requirements: Nonprofit organizations in May operating areas are eligible.

Amount of Grant: Up to 100,000,000 U.S.D

Samples: Cathedral of San Fernando (San Antonio, TX)—$10,000; Jazz at Lincoln Ctr (New York, NY)—$5000; Legacy Health System (Portland, OR)—$13,700; Crossroads School (Saint Louis, MO)—$5300.

Contact: Deidre Burke, Corporate Communications Officer; (314) 342-6299; fax (314) 342-4461; mayfoundation@may-co.com

Internet: http://www2.mayco.com/common/com_index.jsp

Sponsor: May Department Stores Company Foundation
611 Olive Street
Saint Louis, MO 63101

Mayo Foundation Multidisciplinary Training in Digestive Diseases Research Grants 2831

Postdoctoral fellowships in gastroenterological research are available for two or more years. Faculty members employ a multidisciplinary approach using biochemical, cell/molecular biological, immunological, physiological, and electrophysiological techniques. Applicants may focus on clinical and/or basic studies. A curriculum vita and the names of three references should be submitted to the program director.

Requirements: Applicant should have an MD or PhD degree and must be a U.S. citizen or have permanent visa status.

Contact: Dr. N.F. LaRusso, Program Director; (507) 284-1000

Internet: http://mayoresearch.mayo.edu/mayo/research/traininggrants

Sponsor: Mayo Foundation
200 First Street SW
Rochester, MN 55905

MBNA Foundation Helen F. Graham Grants 2832

The program provides funding for teaching programs and other initiatives that create opportunities for the education and development of Delaware children and adults with cognitive disabilities. The grants program is intended to supplement funds currently available to schools and community organizations. It funds new education programs, initiatives that lack a funding source, and current initiatives that could be expanded or enhanced through additional funding. A significant focus is to promote and support programs that assist people with special needs to become independent, self-sustaining individuals with skills that will allow them to succeed in the workplace and to realize their full potential.

Requirements: Proposals must be submitted by a public or tax-exempt 501(c)3 private or parochial school in Delaware with a classroom of at least 50 percent of the children having cognitive disabilities; or a Delaware nonprofit 501(c)3 community organization.

Restrictions: Examples of initiatives that would not ordinarily be funded include salaries for additional permanent staff, benefits, or tuition reimbursements for required courses or professional accreditation programs; capital projects, major repairs to buildings or equipment, and routine operating expenses; housing costs, tuition assistance, or other financial aid for an individual; private-employer work programs; or research projects.

Geographic Focus: Delaware

Amount of Grant: 250 U.S.D

Contact: Grants Administrator; (800) 205-8877, option 1; fax (302) 432-5288; grahamgrants@mbna.com

Internet: http://www.mbnafoundation.org/hfggrants.html

Sponsor: MBNA Foundation
1100 N King Street
Wilmington, DE 19884-0723

McCarthy Family Foundation Grants 2833

The McCarthy Family Foundation is a small private foundation established in 1988. Organized as a California public benefit nonprofit corporation, it operates from San Diego, California. The foundation's current program interest areas have been funded throughout its almost two decades history. These funding areas are: K-12 science education; HIV/AIDS research, education and direct services; assistance to homeless persons; and child abuse prevention and services for victims and families. The board will occasionally fund a special project outside these categorical areas and/or for regional, national or international charitable purposes. A small portion of the grantmaking budget is provided for director matching grants to encourage and amplify personal philanthropy by the foundation's board of directors. Proposals will be considered in the above general areas or for special project areas established by the Board. Proposals will only be accepted for programs within San Diego County. Multiple year proposals may be considered but it is not the foundation's intention to provide annual support. It is not normally the foundation's desire to be the sole source of funds for a project. Most grants are expected to be small, $5,000-$15,000, reflecting the foundation's limited budget although there are no fixed minimum or maximum amounts. Grant proposals received by March 15 will be considered for decision/funding in June. Grant proposals received by September 15 will be considered for decision/funding in December.

Requirements: Proposals will only be accepted for programs within San Diego County, California. All applicants should first submit a letter of inquiry, which can be submitted directly from the Foundation's web site. The letter (1-2 pages) must contain a brief statement describing the applicant and the need for funds with enough information for the foundation to determine whether or not the application falls within its program areas. Proposals submitted without an initial letter of inquiry will not be reviewed by the foundation. Applicants should briefly and clearly provide a statement of their needs and the specific request to the foundation, taking into account other possible sources of funding. Letters of inquiry will be acknowledged upon their receipt, but because the foundation operates without a professional staff, a more detailed response may be delayed. Applicants who receive a favorable response to their initial inquiry will be invited to submit a grant application. The foundation accepts the Common Grant Application of San Diego Grantmakers, which can be downloaded from the Foundation's web site. Five copies of the application (proposal) should be submitted. Only one copy of required attachments need be submitted. The application should be signed by the organization's board chair or the executive director (or equivalent individuals).

Restrictions: The foundation does not consider grants for individuals, scholarship funds, or sectarian religious activities. Normally the foundation does not consider requests for general fundraising drives. It does not make grants intended directly or indirectly to support political candidates or to influence legislation.

Geographic Focus: California

Date(s) Application is Due: Mar 15; Sep 15

Amount of Grant: 500 - 250,000 U.S.D

Samples: United Way of San Diego County, $10,000—general operating support; UCLA AIDS Institute, $50,000—stem cell research for HIV prevention; Elementary Institute of Science, $10,000—-science and technology youth education.

Contact: Rachel McCarthy Bender; (858) 485-0129; mail@mccarthyfamilyfdn.org

Internet: http://www.mccarthyfamilyfdn.org/guide.html

Sponsor: McCarthy Family Foundation

P.O. Box 27389

San Diego, CA 92198-1389

McCombs Foundation Grants 2834

The foundation makes grants to Texas nonprofit charitable, philanthropic, educational, and benevolent organizations. Areas of interest include arts, education and higher education, historic preservation, medical research, philanthropy, recreation and sports, youth services, and voluntarism. There are no application deadlines or forms.

Requirements: Texas nonprofit organizations are eligible. Application form not required. Applicants should submit the following: name, address and phone number of organization; brief history of organization and description of its mission; descriptive literature about organization; detailed description of project and amount of funding requested.

Restrictions: No grants to individuals.

Geographic Focus: Texas

Amount of Grant: 1,000 - 6,000,000 U.S.D

Samples: University of Texas, Austin, TX, $6,000,000; Saint Marys Hall, San Antonio, TX, $50,000; San Antonio Academy of Texas, San Antonio, TX, $100,000.

Contact: Gary Woods, Trustee; (210) 821-6523

Sponsor: McCombs Foundation

755 East Mulberry, Suite 600

San Antonio, TX 78212

McConnell Foundation Grants 2835

The foundation awards grants to California nonprofit organizations in its areas of interest, including arts and culture, community enrichment, education, environment, health care, recreation, and social services. Types of support include program development, seed money, equipment acquisition, scholarship funds, employee matching gifts, and matching funds. Application forms are available.

Requirements: California nonprofit organizations serving Shasta and Siskiyou Counties are eligible.

Restrictions: Grants do not support individuals, religious organizations, research institutions, endowment funds, annual fund drives, budget deficits, or building purchase/construction.

Geographic Focus: California

Amount of Grant: 30,000 - 100,000 U.S.D

Samples: Alliance of Redding Museums (CA)—for multiyear start-up support, $11,488; Shasta and Siskiyou County High Schools (CA)—for scholarships, $79,429 and $32,880 respectively.

Contact: Ana Diaz-Bachman, Program Associate; (530) 226-6235; fax (530) 226-6230; adiaz@mcconnellfoundation.org

Internet: http://www.mcconnellfoundation.org

Sponsor: McConnell Foundation

P.O. Box 492050

Redding, CA 96049-2050

McCune Foundation Grants 2836

The foundation supports nonprofit organizations in southwestern Pennsylvania with an emphasizes on the following areas: children and youth, community development; civic affairs; humanities; culture; higher education; health care; and education. Funding requests may be made in support of capital campaigns; building construction/renovation; endowment; employee matching gifts; equipment acquisition; land acquisition; management development/capacity building; technical assistance; program development; and seed money. To apply for consideration, an organization should send a brief (2-3 page) initial inquiry, preferably using the Foundation's website. If the project meets current funding priorities and sufficient resources are available, the Foundation's staff will contact the applicant to arrange a meeting and to provide further guidance regarding submission of a proposal for formal consideration by the Distribution Committee. The Distribution Committee meets quarterly in March, June, September and, December to review proposals, there is no application deadline date for these meetings.

Requirements: The foundation supports 501(c)3 organizations in southwestern Pennsylvania, with emphasis on the Pittsburgh area. This area includes the following counties: Allegheny, Beaver, Butler, Armstrong, Westmoreland and Washington. To apply for consideration, an organization should send a brief (2-3 page) initial inquiry, preferably using the Foundation's website. The letter should contain: project overview—describe what the proposed efforts are intended to achieve for the region as well as for the organization; what activities/actions are planned to meet the stated goals; project timeline; resources required—total cost of the project; anticipated income, including private and public funders; amount of funding requested from the McCune Foundation; IRS 501(c)3 Determination Letter—attach a copy of the organization's IRS 501(c)3 tax-exempt status letter (either scanned via email or hard copy via regular mail); a copy of the organization's latest audit (either scanned via email or hard copy via regular mail).

Restrictions: Grants are not awarded to individuals or for general operating purposes or loans. Unsolicited proposals from outside the funding area are not accepted.

Geographic Focus: Pennsylvania

Amount of Grant: 1,000 U.S.D

Samples: Allegheny Trail Alliance, $500,000—for the Great Allegheny Passage trail completion; Carnegie Museum of Art, $150,000—toward the Ailsa Mellon Bruce Galleries project; Imani Christian Academy, $325,000—to help purchase and renovate the former East Hills Elementary School.

Contact: Henry Beukema, Executive Director; (412) 644-8779; fax (412) 644-8059; info@mccune.org

Internet: http://www.mccune.org

Sponsor: McCune Foundation

750 Sixth PPG Place

Pittsburgh, PA 15222

McGraw-Hill Companies Corporate Contributions 2837

The foundation supports innovative programs and initiatives that increase the abilities of people around the world to learn; to grow intellectually; to master new skills; and to maximize their individual talents for school, work, and community. Projects are considered that promote and support excellence in education and learning, focus on initiatives that advance knowledge and work and home skills, contribute to the education of youth in the communities and markets where McGraw-Hill companies operate, utilize unique applications of new and developing technologies, and extend their reach globally. Types of support include matching, program, and research grants; and scholarship funds and employee-related scholarships.

Requirements: Qualified applicants are invited to submit preliminary proposal letters that should include a brief background of the organization, concise program description, copy of the most recent audited financial statement and annual report, current year's budget, budget of the program, and evidence of 501(c)3 status.

Restrictions: The foundation does not contribute to political activities or organizations established to influence legislation; individuals; publication of books, magazines, films, or videos; organizations included in United Way drives; sectarian or religious organizations; endowment funds; conferences, trips, or tours; nor loans of any kind.

Amount of Grant: 500 - 10,000 U.S.D

Samples: Hofstra U (Hempstead, NY)—for a business-technology laboratory at the university's Business Development Ctr; Friends of the National Journalism Library, National Press Club (Washington, DC)—for its news information center, as a research center for journalists and journalism students.

Contact: Susan Wallman, Manager, Corporate Contributions; (212) 512-6480; fax (212) 512-3611; susan_wallman@mcgraw-hill.com

Internet: http://www.mcgraw-hill.com/community/guidelines.shtml

Sponsor: McGraw-Hill Companies

1221 Avenue of the Americas, 20th Floor

New York, NY 10020-1095

McGregor Foundation Grants 2838

The program favors grants that address the following priority areas for elders in need: recruitment, education, training, and retention of health care professionals in geriatrics; access to care for underserved populations; promotion of health and wellness; and total quality of life for seniors. The board of directors particularly encourages requests for programs that increase the pool of workers trained in geriatrics, expand access to care for elders in need, or demonstrate effective new methods and approaches. In some instances, the foundation will consider requests for emergency operating support. The foundation will also consider requests for capital projects, with certain restrictions.

Requirements: Tax-exempt nonprofit organizations located within Cuyahoga County, OH, with preference given to communities traditionally served by The A.M. McGregor Home, are eligible. Currently no unsolicited grant applications from organizations outside Cuyahoga County will be accepted.

Restrictions: The foundation discourages capital requests from long-term residential care facilities or requests for debt reduction, annual funds, research, symposia, or endowments. No grants are awarded to individuals.

Geographic Focus: Ohio

Date(s) Application is Due: Mar 1; Sep 1

Samples: Access to the Arts (Cleveland, OH)—for art performances for elderly people in hospitals, nursing homes, and retirement communities, $15,270; Case Western Reserve U (Cleveland, OH)—to support the Gerontological Scholars Program of its Mandel School of Applied Social Sciences during the 2005-06 academic year, $20,000; Ohio Presbyterian Retirement Services (Cleveland, OH)—to implement a program on preventing and managing diabetes, $21,120; Recovery Resources (Cleveland, OH)—to collaborate with Fairhill Center to provide training for staff members caring for elderly people with a history of substance abuse and mental illness, $60,111.

Contact: Susan Althans, Executive Director; (216) 696-7273; fax (216) 621-8198; salthans@fmscleveland.com

Internet: http://mcgregorfoundation.org/guidelines.html

Sponsor: McGregor Foundation

1422 Euclid Avenue, Suite 627

Cleveland, OH 44115-1952

McGregor Fund Grants 2839

The fund makes grants to organizations in the following areas: human services—emphasis on activities in southeastern Michigan addressing emergency needs in housing, food, clothing, and other direct aid; education—early child education and development, colleges and universities in the Detroit metropolitan area, liberal arts colleges and universities in Michigan and Ohio; arts and culture in southeastern Michigan; health care—improving access to primary care for underserved populations in Detroit; and public benefit—civic and community organizations that improve the quality of life in southeastern Michigan. Types of support include project support, operating support, special projects, and capital support. Grant decisions are made by the board of trustees four times annually, in February, May, September, and November. Grant requests may be submitted at any time, but requests may take up to three months for staff review. Application guidelines are available online.

Requirements: Organizations in the metropolitan Detroit, MI area, are eligible. Requests will be considered from organizations located elsewhere for programs or projects that significantly benefit the metropolitan Detroit area (city of Detroit and Wayne, Oakland, and Macomb Counties).

Restrictions: The fund discourages proposals for student scholarships, travel, seminars, conferences, workshops, film or video projects, as well as disease-specific organizations and their local affiliates.

Geographic Focus: Michigan

Amount of Grant: 25,000 - 50,000 U.S.D

Samples: Detroit Institute of Arts (MI)—for its capital campaign to complete a building project, increase its endowment, and support operations, $2 million,

Contact: C. David Campbell, President; (313) 963-3495; fax (313) 963-3512; info@mcgregfund.org

Internet: http://www.mcgregorfund.org/guideline.html

Sponsor: McGregor Fund
333 W Fort Street, Suite 2090
Detroit, MI 48226

McInerny Foundation Grants 2840

The foundation awards grants to Hawaii nonprofit organizations in its areas of interest, including arts and culture, education, environment, health care, youth services, and social services. Types of support include general operating support, continuing support, building construction/renovation, equipment acquisition, program development, seed money, scholarship funds, and matching funds. The July 1 deadline is for capital funds; the November 15 deadline date is for scholarship funds. All other requests are accepted at any time.

Requirements: Hawaii nonprofit organizations are eligible.

Restrictions: Grants are not awarded in support of religious institutions, individuals, endowment funds, deficit financing, or research.

Geographic Focus: Hawaii

Date(s) Application is Due: Jul 1; Nov 15

Amount of Grant: 3,000 - 250,000 U.S.D

Samples: National Tropical Botanical Garden, Kalaheo, HI, $75,000—capital campaign for construction of Botanical Garden Research Center; Boys and Girls Club of Hawaii, Honolulu, HI, $50,000—for RALLY Project at Central Middle School; Honolulu Symphony Society, Honolulu, HI, $40,000-for youth music education program season.

Contact: Paula Boyce; (808) 538-4944; fax (808) 538-4006; pboyce@boh.com

Sponsor: McInerny Foundation
P.O. Box 3170, Department 758
Honolulu, HI 96802-3170

McKesson Foundation Grants 2841

The foundation seeks to enhance the health and quality of life in communities where McKesson HBOC operates and its employees live. Emphasis is focused on youth, especially health services for underserved populations, educational enrichment, the environment, recreation, and youth development activities. The foundation also funds emergency services for children and families, and a variety of social, educational and cultural programs. For culture and the arts, support is given primarily to organizations that reach out to youth and other populations that do not have easy access to such programs. Grants are made for specific projects and programs, and general operating support will be considered in some circumstances. Grants are awarded for one to three years. The Foundation does not accept uninvited applications. Initial approach should be to contact the Foundation directly.

Requirements: 501(c)3 tax-exempt organizations are eligible.

Restrictions: Grants are not made to endowment campaigns, individuals or individual scholarships, religious organizations for religious purposes, political causes or campaigns, advertising in charitable publications, research studies or health organizations concentrating on one disease.

Amount of Grant: 5,000 - 25,000 U.S.D

Samples: LA Theater Work (CA)—for a project that creates hands-on performing arts and literary workshops aimed at finding a positive outlet for the aggression of 3000 at-risk and incarcerated youths, $7500; Massachusetts Coalition for School-Based Health Ctrs (MA)—to hire a coordinator and to continue providing school-based health care, $125,000 over three years.

Contact: Marcia Argyris, Foundation President; (415) 983-8673 or (415) 983-8300; fax (415) 983-7590; marcia.argyris@mckesson.com or community.relations@mckesson.com

Internet: http://www.mckesson.com/en_us/McKesson.com/About%2BUs/Corporate%2BCitizenship/McKesson%2BFoundation.html

Sponsor: McKesson Foundation
1 Post Street, 32nd Floor
San Francisco, CA 94104-5203

McLean Contributionship Grants 2842

Originally established in 1951 as The Bulletin Contributionship for charitable, educational and scientific purposes, the Contributionship became The McLean Contributionship on May 1, 1980 when the association of the McLean family with the Bulletin ended. Independent Publications, also owned by the McLean family, continues as the main financial supporter of the Contributionship. The Contributionship favors projects that stimulate a better understanding of the natural environment and encourage the preservation of its important features; encourage more compassionate and cost-effective care for the ill and aging in an atmosphere of dignity and self-respect; or promote education, medical, scientific, or on occasion, cultural developments that enhance quality of life. In addition, the Trustees from time to time support projects which motivate promising young people to assess and develop their talents despite social and economic obstacles or encourage those in newspaper and related fields to become more effective and responsible in helping people better understand how events in their communities and around the world affect them. The Trustees meet several times a year. The Contribution accepts and processes applications for grants throughout the year. Applications must be received at least six weeks before a meeting date. Interested organizations may view the annual-meeting schedule on the Application Procedure page at the grant website. The Contributionship accepts the common-grant-application form of Delaware Valley Grantmakers Association (link is provided on the Application Procedure web page). Application may also be made by letter (guidelines are included on the Application Procedure web page).

Requirements: Applicants must show evidence of tax-exempt status.

Restrictions: Geographic area is limited to the following locations: Greater Philadelphia area; Nashua, New Hampshire; Dubois Pennsylvania; and Central Florida. The Contributionship does not fund the costs or expenses of existing staff allocated to a project it is asked to support.

Geographic Focus: Florida, New Hampshire, Pennsylvania

Amount of Grant: 2,000 - 100,000 U.S.D

Samples: A Little Taste of Everything, Philadelphia, Pennsylvania, $1,400 - towards maintaining and enhancing their living roof (2010); Visiting Nurse Association of Greater Philadelphia, Philadelphia, Pennsylvania, $50,000 - toward the installation of the Horizon Enterprise Content Management System (2010); National Constitution Center, Philadelphia, Pennsylvania - towards the upgrade of eight poorly-functioning exhibit cases housing artifacts uncovered during the Center's groundbreaking (2010).

Contact: Sandra McLean, Executive Director; (610) 527-6330; fax (610) 527-9733

Internet: http://fdncenter.org/grantmaker/mclean

Sponsor: McLean Contributionship
945 Haverford Road, Suite A
Bryn Mawr, PA 19010-3814

McMillen Foundation Grants 2843

The McMillen Foundation was incorporated in 1947 and serves the Fort Wayne and Allen Counties of Indiana. Fields of interest include: children/youth services; community/economic development; education; health care; health organizations, association; recreation.

Requirements: The foundation requires no application form and proposals may be submitted at anytime. Applicants should submit the following: copy of IRS Determination Letter; brief history of organization and description of its mission; copy of most recent annual report/audited financial statement/990; listing of board of directors, trustees, officers and other key people and their affiliations; detailed description of project and amount of funding requested; copy of current year's organizational budget and/or project budget; listing of additional sources and amount of support; one copy of the proposal.

Restrictions: Giving limited to Fort Wayne and Allen County, Indiana. No support for churches or religious groups. No grants to individuals.

Geographic Focus: Indiana

Amount of Grant: 4,000 - 400,000 U.S.D

Samples: Girl Scouts of the U.S.A., Fort Wayne, IN, $300,000; Junior Achievement of Northern Indiana, Fort Wayne, IN, $50,000; Fort Wayne Park Foundation, Fort Wayne, IN, $100,000.

Contact: Dorothy J. Robinson, Secretary; (260) 484-8631; fax (260) 484-2141

Sponsor: McMillen Foundation
6610 Mutual Drive
Fort Wayne, IN 46825-4236

McMurry Foundation Grants 2844

The foundation was established to have a significant and beneficial impact on the communities within Wyoming with special emphasis in Natrona County. Areas of interest include education, religion, children and advocacy for children, health and human services, the arts and humanities, and a favorable business environment. The foundation invests in innovative ventures as well as established community programs that have the potential to make a lasting difference. Funds provide seed money to start new programs; general funds to expand or improve services offered by established agencies; and organizational support, such as strategic planning, increasing management capacity, and board development. Priority will be given to promote strengthening the capacity of nonprofit organizations to do their work, leveraging the dollars of the McMurry Foundation, collaboration and efficient use of nonprofit resources, and addressing the causes or prevention of community problems. Application process and guidelines are available online.

Requirements: Wyoming 501(c)3 nonprofit organizations and organizations who have applied and anticipate receiving such status are eligible.

Restrictions: Grants designated to influence legislation or elect public officials are rejected.

Date(s) Application is Due: Jan 15; Apr 15; Jul 15; Oct 15

Samples: U of Wyoming (Laramie, WY)—to help renovate the College of Education facility, $250,000.

Contact: Trudi McMurry, Director; (307) 234-4631; trudie@mcmurry.net
Internet: https://online.foundationsource.com/public/home/mcmurryfoundation
Sponsor: McMurry Foundation
P.O. Box 2016
Casper, WY 82602-2016

MDA Development Grant 2845
The association will consider an application for a research grant from a candidate who may be a member of a research team in the laboratory of a senior investigator under whose guidance the applicant will be given flexibility to conduct a neuromuscular disease research project. Awards are for one to three years. Applications must be requested by the listed application deadlines. Guidelines are available online.
Requirements: An applicant must hold a Doctor of Medicine, Doctor of Philosophy, Doctor of Science or equivalent degree; be a professional member at an appropriate institution; be qualified to conduct a program of original research; have an acceptable research plan for a specific disease in MDA's program; have access to institutional resources necessary to conduct the proposed research project; and have eighteen months of postdoctoral research laboratory training at the time of application.
Date(s) Application is Due: Jun 15; Dec 15
Amount of Grant: Up to 45,000 U.S.D
Contact: Program Manager; (520) 529-2000; fax (520) 529-5454; grants@mdausa.org
Internet: http://www.mdausa.org/research/guidelines.html#developmentGrant
Sponsor: Muscular Dystrophy Association
3300 E Sunrise Drive
Tucson, AZ 85718

MDA Neuromuscular Disease Research Grants 2846
The association supports research into 40 diseases of the neuromuscular system to identify the causes of and effective treatments for the muscular dystrophies and related diseases, which include spinal muscular atrophies and motor neuron diseases, peripheral neuropathies, inflammatory myopathies, metabolic myopathies, and diseases of the neuromuscular junction. Awards cover periods of up to three years. The listed deadlines are for requesting applications.
Requirements: Applicants must be professionals or faculty members at appropriate educational, medical, or research institutions and qualified to conduct and supervise programs of original research; must have access to institutional resources necessary to conduct the proposed research project; and must have the MD, PhD, DSc, or equivalent degree. Preliminary letter requesting an application is required and must include name and address of institution(s) where the research is to be conducted, the specific aim of the proposed research, statement of relevance to the purposes of MDA, and a statement on whether human subjects or tissues will be used.
Date(s) Application is Due: Jun 15; Dec 15
Contact: Program Manager; (520) 529-2000; fax (520) 529-5454; grants@mdausa.org
Internet: http://www.mdausa.org/research/guidelines.html
Sponsor: Muscular Dystrophy Association
3300 E Sunrise Drive
Tucson, AZ 85718

MDBD Basic Research Grants 2847
Grants are awarded to qualified U.S. institutions to support medical and scientific research in the field of birth defects. Research supported by the foundation on the etiology of birth defects takes three main directions. The first supports investigation of the hereditary influences of birth defects. Requests may be to investigate the structure and function of chromosomes, their subunits, genes, and their supporting substructures. The second investigates environmental effects ranging from obvious factors such as drugs, radiation, infections, and malnutrition to more subtle influences such as cytoplasmic influences. The third is the interaction between heredity and environment such as enzymatic feedback and an individual's resistance to disease. Applications will be sent to those whose abstracts are considered relevant to this program. Abstracts must be received with a postmark on or before the listed application deadline date.
Requirements: A request for a new grant should be initiated by a 300-word abstract to the director of grants administration from the principal investigator of the proposed research program. This letter of inquiry, containing a brief protocol of the study, must be submitted before an application form will be provided. Letters of intent may be submitted online at http://researchgrants.modimes.org/LOI.
Date(s) Application is Due: Apr 30
Amount of Grant: 25,000 U.S.D
Contact: Vice President for Research; (914) 997-4555; fax (914) 997-4560; researchgrants@marchofdimes.com
Internet: http://www.marchofdimes.com/professionals/685_2160.asp
Sponsor: March of Dimes Birth Defects Foundation
1275 Mamaroneck Avenue
White Plains, NY 10605

MDBDF Agnes Higgins Award 2848
The award recognizes distinguished achievement in research, education, or clinical services in the field of maternal-fetal nutrition. In addition to an honorarium, winners are honored at a presentation and reception at the American Public Health Association Annual Meeting in the year of their award. The award is made every two years. Nominations are invited in even-numbered years. Nomination form and guidelines are available online.
Requirements: Candidates have been widely involved in maternal-fetal nutrition through teaching, research, or clinical practice for at least five years; shown a demonstrable effect in raising the quality of maternal-fetal nutritional care through scholarly pursuits, research, education, or practice; and demonstrated the ability to apply maternal-fetal nutritional standards of practice or facilitate their implementation by others.
Date(s) Application is Due: Mar 1
Amount of Grant: 3,000 U.S.D
Contact: Grants Administrator; (914) 997-4564 or (888) 663-4637; fax (914) 997-4560; researchgrants@marchofdimes.com
Internet: http://www.marchofdimes.com/professionals/685_1370.asp
Sponsor: March of Dimes Birth Defects Foundation
1275 Mamaroneck Avenue
White Plains, NY 10605

MDBD Research Grants and Basil O'Connor Starter Scholar Award 2849
The program is designed for young scientists at the level of instructor or assistant professor who are embarking on their independent research careers. Both awards support research on birth defects. Grants do not include the recipients' salaries, but provide support for technical help and supplies. Programs may be approved for two years only. Each potential applicant must submit a letter of nomination written by the chair of the department, director of the institute, dean, or other senior person; brief abstract of the proposed research; and curriculum vita. Candidates meeting the criteria of the program will receive application forms. Annual deadline dates may vary; contact program staff for exact dates.
Restrictions: Applicants may not submit simultaneously any other applications for March of Dimes research programs.
Date(s) Application is Due: Mar 19
Amount of Grant: Up to 75,000 U.S.D
Samples: Dr. Anne Calof, U of Iowa (Iowa City, IA)—for the research study Strategies for Identifying Genes that Regulate Neurogenesis in Mammals, $30,000; Dr. Piero Rinaldo, Yale U (New Haven, CT)—for research on sudden infant death syndrome, $10,000.
Contact: Grants Administrator; (914) 997-4564 or (888) 663-4637; fax (914) 997-4560; researchgrants@marchofdimes.com
Internet: http://www.marchofdimes.com/professionals/691_1452.asp
Sponsor: March of Dimes Birth Defects Foundation
1275 Mamaroneck Avenue
White Plains, NY 10605

MDBD Research Prize in Developmental Biology 2850
The prize is awarded to investigators whose research has profoundly advanced the science that underlines understanding of birth defects. Nomination form and guidelines are available online.
Date(s) Application is Due: Sep 15
Amount of Grant: 250,000 U.S.D
Contact: Dr. Michael Katz, (914) 997-4555 or (888) 663-4637; fax (914) 997-4560; mkatz@marchofdimes.com
Internet: http://www.marchofdimes.com/professionals/691_1442.asp
Sponsor: March of Dimes Birth Defects Foundation
1275 Mamaroneck Avenue
White Plains, NY 10605

Mead Johnson Nutritionals Charitable Giving Grants 2851
Mead Johnson Nutritionals considers requests for charitable giving from a broad range of charitable organizations. Requests for charitable giving must be made in connection with a general fund raising effort by the charitable entity, rather than a request directed only to Mead Johnson Nutritionals. In addition, the request for a contribution must come from the organization itself and not from an individual physician, pharmacist, or health care professional who is not an employee of the charity. Applicants should allow at least three months for the processing of a funding request. Grant applications submitted less than three months prior to an event may be rejected due to insufficient time to process the application. Organizations outside the U.S. that wish to request a charitable gift should refer to Mead Johnson contact information within their country or call the charitable giving department at (812) 429-7800.
Requirements: Charitable giving is permitted if the recipient is a tax-exempt organization under Section 501(c)3 of the Internal Revenue Code.
Restrictions: As with grants, charitable giving may not be tied, in any way, to past, present, or future prescribing, purchasing, or recommending (including formulary recommendations) of any product.
Contact: Grants Administrator; (812) 429-7831; fax (812) 647-8674; meadjohnson.grants1@bms.com
Internet: http://www.mjn.com/app/iwp/MJN/Content2.do?dm=mj&id=/MJN_Home2/mjnBtnSocialResponsibility2/mjnBtnCharitableGivingApplication&iwpst=B2C&ls=0&csred=1&r=3435833953
Sponsor: Mead Johnson Nutritionals
2400 West Lloyd Expressway
Evansville, IN 47721-0001

Mead Johnson Nutritionals Evansville-Area Organizations Grants 2852
Evansville area non-profit organizations of all varieties may submit charitable giving applications throughout the year. However, if a request exceeds $1,000 it is highly recommended that the applicant submit an application between June 1 and August 31 in the year prior to when the funds are needed. Requests greater than $1,000 are generally reviewed as a group, one time a year. If approved, funding will be awarded in the following year. Requests greater than $1,000 received after August 31 might not be considered and the applicant might be asked to reapply the following year.
Requirements: Evansville area 501(c)3 designated organizations are eligible to apply.
Geographic Focus: Indiana

Date(s) Application is Due: Aug 31
Contact: Grants Administrator; (812) 429-7831; fax (812) 647-8674; meadjohnson.grants1@bms.com
Internet: http://www.mjn.com/app/iwp/MJN/Content2.do?dm=mj&id=/MJN_Home2/mjnBtnSocialResponsibility2/mjnBtnCharitableGivingApplication&iwpst=B2C&ls=0&csred=1&r=3435833953#app
Sponsor: Mead Johnson Nutritionals
2400 West Lloyd Expressway
Evansville, IN 47721-0001

Mead Johnson Nutritionals Medical Education Grants **2853**
Mead Johnson Nutritionals has established a Grant program to support medical education and philanthropic initiatives. The program will consider supporting Independent Medical Education (IME) programs and activities designed to enhance the professional skills and knowledge of health care professionals involved in pediatrics, when the program is independently developed and conducted by a qualified medical education provider. Supported programs must be objective, balanced, and scientifically rigorous. Health care provider educational programming can be given for: medical education programs; scientific conferences; creation of health care publications; and other educational and scientific activities. In addition to the U.S. Medical Education Grants, Mead Johnson also funds International Medical Education Grants. Organizations outside the U.S. that wish to request an Independent Medical Education (IME) grant should refer to Mead Johnson contact information within their country or contact Mead Johnson at meadjohnson.grants1@bms.com.
Requirements: Eligible recipients for medical education grants include: academic medical centers; accredited medical education companies; community health centers; hospitals; managed care organizations; medical/professional societies; medical universities; patient advocacy groups; pharmacies; and similar entities.
Restrictions: Grants may not be tied, in any way, to past, present, or future prescribing, purchasing, or recommending (including formulary recommendations) of any product.
Contact: Grants Administrator; (812) 429-7831; fax (812) 647-8674; meadjohnson.grants1@bms.com
Internet: http://www.mjn.com/app/iwp/MJN/Content2.do?dm=mj&id=/MJN_Home2/mjnBtnSocialResponsibility2/mjnBtnMedicalEducationApplications&iwpst=B2C&ls=0&csred=1&r=3435836001
Sponsor: Mead Johnson Nutritionals
2400 West Lloyd Expressway
Evansville, IN 47721-0001

Mead Johnson Nutritionals Scholarships **2854**
Mead Johnson Nutritionals provides scholarship grants to medical students, residents, fellows, and other healthcare professionals in training to attend classes and major educational or scientific meetings of a national medical or professional association. The selection of the individuals to receive such support and the grant application must be made by the institution at which they are being trained or the organization hosting the event. Allow at least three months for the processing of your funding application. Independent Medical Education (IME) grant applications submitted less than three months prior to an event may be rejected, due to insufficient time to process the application.
Requirements: Eligible recipients for such grants: academic medical centers; hospitals; medical/professional societies; and medical universities.
Contact: Grants Administrator; (812) 429-7831; fax (812) 647-8674; meadjohnson.grants1@bms.com
Internet: http://www.mjn.com/app/iwp/MJN/Content2.do?dm=mj&id=/MJN_Home2/mjnBtnSocialResponsibility2/mjnBtnMedicalEducationApplications/MedScholarships&iwpst=B2C&ls=0&csred=1&r=3435836794
Sponsor: Mead Johnson Nutritionals
2400 West Lloyd Expressway
Evansville, IN 47721-0001

MeadWestvaco Foundation Grants **2855**
The foundation's primary focus is to address important community needs and improve the quality of life in communities where MeadWestvaco operates. Specific areas of emphasis include health, human services, education, stewardship, and arts and culture.
Contact: Grants Administrator; (203) 461-7400
Internet: http://www.meadwestvaco.com/corporate.nsf/mwvfoundation/applicationsGuidelines
Sponsor: MeadWestvaco Foundation
1 High Ridge Park
Stamford, CT 06905

Mead Witter Foundation Grants **2856**
Incorporated in Wisconsin in 1951, the Foundation awards grants to nonprofits in company operating locations, primarily in central and northern Wisconsin. Local community programs are funded, focusing on education, including scholarships in higher education and direct contributions to colleges and universities. Grants also are made to support the arts, health care, human services, youth organizations, environmental programs, and Christian and Roman Catholic organizations for nonreligious purposes. Types of support include general operating support, continuing support, annual campaigns, capital campaigns, building construction and renovation, equipment acquisition, endowment funds, professorships, seed grants, scholarship funds, and employee-matching gifts. The board meets twice annually to consider requests. Full proposal is by invitation only.
Requirements: Wisconsin 501(c)3 organizations may apply.
Restrictions: The foundation does not support religious, athletic, or fraternal groups, except when these groups provide needed special services to the community at large; direct grants or scholarships to individuals; community foundations; or flow-through organizations that redispense funds to other charitable causes.
Geographic Focus: Wisconsin
Amount of Grant: 25,000 - 18,000,000 U.S.D
Samples: Carroll University, Waukesha, Wisconsin, $1,030,000; Medical College of Wisconsin, Milwaukee, $512,000; Beloit College, Beloit, $300,000.
Contact: Cynthia Henke, President; (715) 424-3004; fax (715) 424-1314
Sponsor: Mead Witter Foundation
P.O. Box 39
Wisconsin Rapids, WI 54495-0039

Measey Foundation Medical Education Grants **2857**
The purpose of this grant program is to support and advance the medical education of physicians. Grants are made to medical schools in the immediate Philadelphia, PA, area that will determine eligible recipients. Types of support include fellowships, professorships, and scholarship funds. Grantmaking decisions are made at the quarterly meetings held in March, June, September, and December each year; grant requests should be received at least two weeks prior to the meetings.
Restrictions: Individuals are ineligible.
Geographic Focus: Pennsylvania
Amount of Grant: 15,000 - 400,000 U.S.D
Samples: Thomas Jefferson U, Jefferson Medical College (Philadelphia, PA)—to endow a professorship in surgery, $1 million.
Contact: James Brennan, Manager; (610) 566-5800
Sponsor: Benjamin and Mary Siddons Measey Foundation
P.O. Box 258
Media, PA 19063

Medica Foundation Grants **2858**
The foundation supports community-based initiatives and programs that support the needs of Medica's customers and the greater community by improving their health and removing barriers to health care services. Funding is directed toward initiatives that are true innnovations in how health care is delivered; have a high likelihood of creating systemic change that in turn will increase sustainability; reduce health care disparities and address social issues that drive health care costs; and build partnerships and collaborations between Medica and other community organizations and/or that build collaborations between community organizations. The foundation considers only single-year grant requests.
Requirements: 501(c)3 tax-exempt organizations and government entities in Minnesota, Western Wisconsin, North Dakota, South Dakota, and northern Iowa are eligible.
Restrictions: The foundation does not fund capital campaigns/capital expenditures; general and/or ongoing operational support; long-term financial support; projects where other viable funding sources are available; religious groups for religious purposes; projects in which administrative expenses exceed 10 percent of the total grant; or sports events and athletic groups.
Geographic Focus: Iowa, Minnesota, North Dakota, South Dakota, Wisconsin
Amount of Grant: 50,000 - 150,000 U.S.D
Contact: JoAnn Birkholz, (952) 992-2060; foundation@medica.com
Internet: http://www.medica.com/medica/default.aspx?section=visitor&page=medica_foundation2
Sponsor: Medica Foundation
P.O. Box 9310, Mail Rte CP230
Minneapolis, MN 55440-9310

Medicaid/SCHIP Eligibility Pilots **2859**
HCFA is soliciting project proposals from state Medicaid agencies and agencies administering the State Children's Health Insurance Program (SCHIP). The purpose of this project is to identify new and effective ways to simplify the application and enrollment process by piloting and evaluating innovative efforts on a small scale.
Requirements: State Medicaid or SCHIP agencies are eligible to apply. These agencies can collaborate with each other, advocates, schools, community-based organizations, consumers, provider organizations, and other entities.
Amount of Grant: Up to 80,000 U.S.D
Contact: Candice Hall, (410) 786-4453; chall2@hcfa.gov
Internet: http://www.cms.hhs.gov/schip/sho-letters/ch62600.asp
Sponsor: Health Care Financing Administration
7500 Security Boulevard
Baltimore, MD 21244-1850

Medicaid Infrastructure Grant Program to Support Employment of People with Disabilities **2860**
This program is designed to help states support the competitive employment of persons with disabilities by facilitating improvements to states' Medicaid programs. Request applications from Judith Norris,(410)786-5130 or jnorris@cms.hhs.gov.
Requirements: Single state Medicaid agencies and other state agencies in partnership with the single agency are eligible to apply.
Date(s) Application is Due: Jul 31
Amount of Grant: 500,000 U.S.D
Contact: Jeremy Silanskis, Project Manager; (410) 786-6126; jsilanskis@cms.hhs.gov
Internet: http://cms.hhs.gov/twwiia/infrastr.asp
Sponsor: Health Care Financing Administration
7500 Security Boulevard
Baltimore, MD 21244-1850

Medical/Nursing Education, Research, and Health Care Exchanges with Taiwan 2861
This organization provides for the exchange of medical and scientific knowledge and fosters cooperation between health scientists in Taiwan and the United States in biomedical teaching, research, and health care. Three separate programs are sponsored. The fellowship program assists Republic of China graduates or postgraduates to study and do research in the United States. The visiting specialists program provides for U.S. health specialists to teach and lecture in medical schools in Taiwan. The nursing program recruits and sponsors U.S. nurses for positions in Taiwan and places and sponsors nurses from Taiwan in U.S. institutions. In Taiwan, contact Executive Director, Room 9001, 212 Chung Hsiao Tung Lu, Section IV, Taipei, Taiwan 10646, Republic of China, 02-721-6511 ext 367, fax: 02-751-6639.
Requirements: Program recruitment is primarily by invitation.
Contact: Annie Kang c/o New York Academy of Medicine, (212) 860-1990
Sponsor: American Bureau for Medical Advancement in China
1216 Fifth Avenue, Room 585
New York, NY 10029

Medical Foundation Charles A. King Trust Postdoctoral Research Fellowships 2862
The Charles A. King Trust was established in 1936 to support and promote the investigation of human disease and the alleviation of human suffering through improved treatment. The King Trust specifically supports postdoctoral fellows in basic-sciences and clinical/health-services research. The complete application process requires an online submission as well as a mailed package containing the face sheet with original signatures, FOUR printed copies of the document uploaded online, and three confidential letters of recommendation. The online submission deadline for the Charles A. King Trust is 12:00 noon on December 15. Links to downloadable instructions and forms as well as a link to the online submission system are provided at the grant website. The mail submission deadline is 5:00 p.m. on December 20. Applicants will be notified of grant decisions before June 1. Deadline dates may vary from year to year. Applicants are encouraged to visit the grant website to verify current application deadlines. Bank of America acts as co-trustee for the King Trust, in conjunction with Mr. Edward Dane and Ms. Lucy West.
Requirements: Eligible applicants must work in an academic or medical research institution in the state of Massachusetts and hold a fellowship position under the supervision of a faculty member (mentor). Fellows applying in the area of clinical/health services must either have clinical responsibilities and have completed residency or have completed at least three and no more than five years of full-time postdoctoral research experience. Fellows applying in the area of basic science must hold a Ph.D., M.D., D.M.D., M.D./Ph.D. or equivalent degree and have completed at least three and no more than five years of full-time postdoctoral research experience. These requirements must be met by July 1st of each funding cycle. Applicants are strongly encouraged to download and read the guidelines (available at the grant website) for further requirements in each area. United States citizenship is not required; visa documentation is not required.
Restrictions: Only one new applicant per mentor may apply per application cycle; there are no institutional limitations on the number of applicants who may submit applications.
Geographic Focus: Massachusetts
Date(s) Application is Due: Dec 15; Dec 20
Amount of Grant: 43,500 - 51,000 U.S.D
Contact: Linda Lam, Program Officer; (617) 451-0049, ext. 702 or (617) 279-2240, ext. 710; LLam@hria.org
Miki C. Akimoto, Vice President; (866) 778-6859; miki.akimoto@baml.com
Internet: http://hria.org/tmfservices/tmfgrants/king.html
Sponsor: Medical Foundation, a Division of Health Resources in Action
95 Berkeley Street, Suite 208
Boston, MA 02116

Medical Foundation Deborah Munroe Noonan Memorial Research Fund Grants 2863
The Deborah Munroe Noonan Memorial Research Fund was established in 1947 by her son, Frank Noonan, to support innovative projects aimed at improving the quality of life for children and adolescents with physical and developmental disabilities. The Noonan Research Fund supports innovative clinical and service system research, demonstration projects, and pilot studies in the Boston area designed to improve the quality of life for children and adolescents with disabilities. The fund provides one-year grants of up to $75,000 (inclusive of 10% indirect costs). The complete application process requires an online submission as well as a mailed package. Links to downloadable instructions and forms as well as a link to the online submission system are provided at the grant website. The online submission deadline is 12:00 noon on February 1. The mail submission deadline is 5:00 p.m. on February 4. Applicants will be notified of grant decisions by June 30. Deadline dates may vary from year to year. Applicants are encouraged to visit the grant website to verify current application deadlines.
Requirements: Applicants must hold a position within a nonprofit institution or organization. New investigators are encouraged to submit proposals. Project must address the target age range of infant through 23 years old. Research projects must be conducted within the fund's geographic area of interest. A complete listing of the geographic eligibility area is contained in the downloadable guidelines available at the grant website.
Restrictions: The Noonan Memorial Research Fund specifically serves the people of Greater Boston. Proposals for basic science research will not be considered nor will applications for capital costs such as buildings, renovations, or major equipment items. Drug trials are rarely supported by the Noonan Research Fund. The Noonan Fund does not fund primary prevention projects.
Geographic Focus: Massachusetts
Date(s) Application is Due: Nov 1; Nov 4
Amount of Grant: Up to 75,000 U.S.D
Samples: Carrie Barlow, B.S., Franciscan Hospital for Children, Boston, Massachusetts, "Aquatic Physical Therapy for Children with Cerebral Palsy: A Pilot Intervention Study" (2010); Eugenia Chan, M.D., M.P.H., Children's Hospital, Boston, Massachusetts, "Assessing Quality of Life and Unmet Service Needs among Families of Children with Developmental and Behavioral Disorders: Implementation of a Web-based Tool" (2010); Yvonne Domings, M.Ed., CAST, Inc., Wakefield, Massachusetts, "Building Comprehension through Social Understanding in Students with Autism Spectrum Disorders" (2010); Inyang Isong, M.D., M.P.H., Massachusetts General Hospital, Boston, Massachusetts, "A Pilot and Feasibility Study of Electronic Screen Media to Reduce Dental Fear among Children with Autism Spectrum Disorders" (2010).
Contact: Jeanne Brown, Program Officer; (617) 279-2240, ext. 709; jbrown@hria.org
Miki C. Akimoto, Vice President; (866) 778-6859; miki.akimoto@baml.com
Internet: http://hria.org/tmfservices/tmfgrants/noonan.html
Sponsor: Medical Foundation, a Division of Health Resources in Action
95 Berkeley Street, Suite 208
Boston, MA 02116

Medical Foundation Edward N. and Della L. Thome Memorial Foundation, Awards in AD Drug Discovery Research 2864
The Edward N. & Della L. Thome Memorial Foundation was created in 2002 to advance the health of older adults through the support of direct service projects and medical research on diseases and disorders affecting older adults. In keeping with the foundation's mission, the goal of the research program is to support to support innovative drug discovery research that will lead to improved therapies for individuals suffering from Alzheimer's disease. Examples of funding areas include the design, synthesis, and development of target compounds or the modification of existing compounds to improve drug effectiveness and safety as well as other approaches within the field of medicinal chemistry. Researchers dedicated to the validation and testing of target compounds, small molecule therapies, nanotechnologies, or similar techniques are also encouraged to apply. Three-year awards of up to $750,000 ($250,000 per year) will be made (all awards inclusive of 10% indirect costs). The application deadline for the initial proposal is June 15, 12:00 Noon, U.S. Eastern Time. Applications must be submitted through the online application system. Downloadable application guidelines, instructions, and other details are available at the grant website. Applicants invited to submit a full proposal will be notified in late July. Applicants will be notified of grant decisions by December 31. This grant has a biennial cycle and alternates with the Edward N. and Della L. Thome Memorial Foundation, Bank of America N.A., Trustee, Awards in Age-Related Macular Degeneration Research. Prospective applicants are advised to check the Medical Foundation website to verify current deadline dates for these awards. Organizations interested in applying for funding through the Edward N. and Della L. Thome Memorial Foundation's direct services program should go to the Bank of America's philanthropic website (https://www.bankofamerica.com/philanthropic/index.action).
Requirements: All applicants must hold a faculty appointment at a nonprofit academic, medical, or research institution in the United States or U.S. territories. United States citizenship is not required. Applicants do not need to be nominated by their institutions. Preference will be given to originality of ideas, regardless of faculty seniority. The research proposal must clarify how the proposed research project may lead to improvements in current therapeutic strategies or create new approaches to treating Alzheimer's disease.
Restrictions: Basic research or new target discovery, genetic studies, biomarker research, neuroimaging, and clinical studies are currently outside the scope of this program.
Date(s) Application is Due: Jun 15
Amount of Grant: Up to 750,000 U.S.D
Samples: Ottavio Arancio, M.D., Ph.D., Taub Institute for Research on Alzheimer's Disease and the Aging Brain, New York, New York, "Histone Acetyltransferase (HAT) Activators as Chromatic Remodelers for the Treatment of Memory Loss in Alzheimer's Disease" (2010); Luciano D'Adamio, M.D., Albert Einstein College of Medicine of Yeshiva University, Bronx, New York, "Translation of BR12 Derived Peptides into Therapeutic Candidates" (2010); Todd Golde, M.D., Ph.D., Center for Translational Research in Neurodegenerative Diseases, University of Florida College of Medicine, Gainesville, Florida, "Amyloid Vaccines and Human Anti-Amyloid Antibodies" (2010).
Contact: Linda Lam, Program Officer; (617) 279-2240, ext. 710; llam@hria.org
Tyler C. Brown, Ph.D., Scientific Program Officer; (617) 279-2240, ext. 704; tylerbrown@hria.org
Sally E. McNagny, M.D., M.P.H., Director; (617) 279-2240, ext. 704; smcnagny@hria.org
George Thorn, Vice President; (312) 828-4154; ilgrantmaking@bankofamerica.com
Internet: http://hria.org/tmfservices/tmfgrants/thomead.html
Sponsor: Medical Foundation, a Division of Health Resources in Action
95 Berkeley Street, Suite 208
Boston, MA 02116

Medical Foundation Grants 2865
The foundation supports nonprofit organizations in the greater Boston area seeking support for projects that permit aging individuals to maintain independent lifestyles and function in their communities. Preference will be given to projects that are innovative, involve collaborations between research and service groups, and do not duplicate existing studies.
Requirements: Nonprofit organizations in the greater Boston, MA, area may apply.
Restrictions: Requests will not be considered for ongoing service delivery, basic science research, or capital costs for buildings or renovations.
Geographic Focus: Massachusetts
Amount of Grant: 40,000 U.S.D
Contact: Grants Administrator; (617) 451-0049, ext. 702; fax (617) 423-4619; research@tmfnet.org

Internet: http://www.tmfnet.org/grantmake.html
Sponsor: Medical Foundation, a Division of Health Resources in Action
95 Berkeley Street, Suite 208
Boston, MA 02116

Medical Foundation Health Initiatives **2866**
The program supports two major health initiatives in the Northeast. The Charles H. Hood Child Health Research Grant program supports New England medical or research institutions that initiate studies to diminish health problems affecting large numbers of children. Postdoctoral research fellowships in the biomedical sciences offer two-year grants to Massachusetts nonprofits addressing issues of health in the community. Proposals are judged on medical relevance, innovative ideas, background, and need for support.
Requirements: New England nonprofit organizations may apply for Charles H. Hood Child Health Research Grants. Massachusetts nonprofits addressing issues of health in the community may apply for postdoctoral research fellowships in the biomedical sciences.
Geographic Focus: Connecticut, Maine, Massachusetts, New Hampshire, Rhode Island, Vermont
Amount of Grant: 75,000 - 39,000 U.S.D
Contact: Grants Administrator; (617) 451-0049, ext. 702; fax (617) 423-4619; research@tmfnet.org
Internet: http://www.tmfnet.org/grantmake.html
Sponsor: Medical Foundation, a Division of Health Resources in Action
95 Berkeley Street, Suite 208
Boston, MA 02116

MedImmune Charitable Grants **2867**
The program supports educational and charitable activities that ultimately improve patient healthcare and further understanding of disease states and treatment options within the community. While the sponsor will consider initiatives in many areas, priority is given to those in the following areas of interest: (a) health and science education for young people; (b) influenza; (c) immunology; (d) pediatric infectious disease; (e) pediatric respiratory disease.
Requirements: Eligible programs include: (a) Community outreach designed to increase public awareness (e.g, fundraising walk or run, nonprofit community hospital or hospital foundation, etc.); (b) Fellowships; (c) Independent educational/scientific programs, accredited and non-accredited; (d) Medical or Scientific education for Healthcare Professionals; (e) Scholarships; or, (f) Academic Chair Endowments. Applications must be completed online at https://web.medimmune.com/corporategrants/, and must be submitted at least 6 weeks prior to the program date or event.
Restrictions: Applications will not be considered from: (1) individuals, physicians, or physician practices; (2) retroactive support; (3) exhibit or display booths at conventions/conferences; (4) textbooks; (5) operating expenses; (6) fraternal, social, leisure, labor or political organizations; (7) corporate foundations; (8) public or private schools or scholarship funds, except where a grant is specific to a science education initiative.
Contact: Grants Office; (866) 396-6235; corporatefunding@medimmune.com
Internet: http://www.medimmune.com/culture_grants.aspx
Sponsor: MedImmune
One MedImmune Way
Gaithersburg, MD 20878

Medtronic Foundation CommunityLink Health Grants **2868**
At Medtronic, the Foundation makes it a priority to enhance the vitality of the communities where its employees live and work. The CommunityLink program helps the Foundation accomplish this by supporting health, education and community programs throughout the world. The CommunityLink Health Grants support programs that improve the health and welfare of people, with a focus on Medtronic's areas of expertise. The Foundation gives priority to programs that help people develop and maintain health lifestyles, with particular interest in programs that reduces differences in health care and that support healthy lifestyles, such as nutrition and fitness. The application and guidelines are located on the Foundation website. Guidelines and funding varies depending on location.
Requirements: Only U.S. 501(c)3 nonprofit organizations and equivalent international organizations may apply.
Restrictions: The Foundation does not fund the following: Continuing Medical Education (CME) grants; 501(c)3 type 509(a)3 supporting organizations; capital or capital projects; fiscal agents; fundraising events/activities, social events or goodwill advertising; general operating support; general support of educational institutions; greater Twin Cities United Way supported programs; individuals, including scholarships for individuals; lobbying, political or fraternal activities; long-term counseling or personal development; program endowments; purchases of automatic external defibrillators (AEDs); religious groups for religious purposes; private foundations; or research.
Geographic Focus: Arizona, California, Colorado, Florida, Indiana, Massachusetts, Minnesota, Puerto Rico, Tennessee, Texas, Washington, Canada, Ireland, Japan, Netherlands, Switzerland
Date(s) Application is Due: Nov 16
Contact: Deb Anderson, Grants Administrator; (763) 514-4000 or (800) 633-8766; fax (763) 505-2648; deb.anderson@medtronic.com
Internet: http://www.medtronic.com/foundation/programs_cl.html
Sponsor: Medtronic Foundation
304 Landmark Center, 75 West Fifth Street
St. Paul, MN 55102

Medtronic Foundation Community Link Human Services Grants **2869**
At Medtronic, the Foundation makes it a priority to enhance the vitality of the communities where its employees live and work. The CommunityLink program helps the Foundation accomplish this by supporting health, education and community programs throughout the world. The CommunityLink Human Services Grants support human services in the U.S. that help individuals become more self-sufficient. Programs reach out to a wide range of people, such as: economic literacy programs to help low-income women gain economic stability; peer counseling for teens in crisis situations; homeless shelters; and subsidized childcare programs for low-income parents. The application and current guidelines are available at the website.
Requirements: Only U.S. 501(c)3 nonprofit organizations and equivalent international organizations may apply. Applicants must be located in U.S. areas where Medtronic has employees.
Restrictions: The Foundation does not fund the following: Continuing Medical Education (CME) grants; 501(c)3 type 509(a)3 supporting organizations; capital or capital projects; fiscal agents; fundraising events/activities, social events or goodwill advertising; general operating support; general support of educational institutions; greater Twin Cities United Way supported programs; individuals, including scholarships for individuals; lobbying, political or fraternal activities; long-term counseling or personal development; program endowments; purchases of automatic external defibrillators (AEDs); religious groups for religious purposes; private foundations; or research.
Geographic Focus: Arizona, California, Colorado, Florida, Indiana, Massachusetts, Minnesota, Tennessee, Texas, Washington
Date(s) Application is Due: Nov 16
Contact: Deb Anderson, Grants Administrator; (763) 514-4000 or (800) 633-8766; fax (763) 505-2648; deb.anderson@medtronic.com
Internet: http://www.medtronic.com/foundation/programs_cl.html
Sponsor: Medtronic Foundation
710 Medtronic Parkway
Minneapolis, MN 55432-5604

Medtronic Foundation Fellowships **2870**
The U.S. based Medtronic Fellows Program works to prepare the next generation of scientific innovators. Through the Foundation's partnerships with preselected universities such as Johns Hopkins, Duke University, Stanford University, and Massachusetts Institute of Technology, it supports college and university scholars and fellows in science, engineering and health-related fields. Applicants should review the website for specific university participants and contact the individual universities for more information.
Geographic Focus: All States, United Kingdom
Contact: Kris Fortman, U.S. Programs Manager; (763) 514-4000 or (800) 633-8766; fax (763) 505-2648; kris.fortman@medtronic.com
Internet: http://www.medtronic.com/foundation/programs_msp.html
Sponsor: Medtronic Foundation
710 Medtronic Parkway
Minneapolis, MN 55432-5604

Medtronic Foundation HeartRescue Grants **2871**
The HeartRescue Grants fund organizations and government agencies that provide community-based education, training, and awareness programs related to sudden cardiac arrest (SCA) and the use of automatic external defibrillators. Working with select partners, efforts are focused on developing an integrated community response to SCA, coordinating education, training and the application of high-tech treatments among the general public, first responders, emergency medical services, and hospitals. Funds may be used for all aspects of a systems-based approach to SCA, including human resources, data collection and measurement tools, training costs at all levels of the system, and technology. If your organization is interested in participating in the program, please submit a letter of inquiry.
Requirements: Nonprofit organizations and government agencies that provide education, training, and awareness programs related to sudden cardiac death, early defibrillation, and early intervention are eligible. Organizations should send a 1-2 page letter of inquiry with detailed information about their organization, their project, their staff, and plans for implementing their program.
Restrictions: The Foundation does not fund the following: Continuing Medical Education (CME) grants; 501(c)3 type 509(a)3 supporting organizations; capital or capital projects; fiscal agents; fundraising events/activities, social events or goodwill advertising; general operating support; general support of educational institutions; greater Twin Cities United Way supported programs; individuals, including scholarships for individuals; lobbying, political or fraternal activities; long-term counseling or personal development; program endowments; purchases of automatic external defibrillators (AEDs); religious groups for religious purposes; private foundations; or research.
Amount of Grant: 300,000 - 500,000 U.S.D
Contact: Deb Anderson, Grants Administrator; (763) 514-4000 or (800) 328-2518; fax (763) 505-2648; deb.anderson@medtronic.com
Joan Mellor, HeratRescue Consultant; (763) 514-4000 or (800) 328-2518; fax (763) 505-2648; joan.mellor@medtronic.com
Internet: http://www.medtronic.com/foundation/programs_hr.html
Sponsor: Medtronic Foundation
710 Medtronic Parkway
Minneapolis, MN 55432-5604

Medtronic Foundation Patient Link Grants **2872**
Through Patient Link grants, Medtronic partners with national and international patient organizations that educate, support, and advocate on behalf of patients and their families to improve the lives of people with chronic diseases. Grants are offered in the U.S.,

Canada, and Europe. In partnership with its Patient Link organizations, the Foundation: empowers patients to become active partners in their health care; engages patients from cultural communities who have not had access to health information and health care; and provides support to patients through what can be difficult emotional times. Each location requires different guidelines and gives a different range of funding.
Requirements: Health grants only fund 501(3)c U.S. or nonprofit international organizations. Public, private, government institutions, and educational institutions are the only exceptions to this rule. U.S. based organizations must gather and complete the application and corresponding documents. Organizations outside the U.S. complete a letter of inquiry to the corresponding contact near their location.
Restrictions: Grants are not available for the following: Continuing Medical Education (CME) grants; 501(c)3 type 509(a)3 supporting organizations; capital or capital projects; fiscal agents; fundraising events/activities, social events or goodwill advertising; general operating support; general support of educational institutions; greater Twin Cities United Way supported programs; individuals, including scholarships for individuals; lobbying, political or fraternal activities; long-term counseling or personal development; program endowments; purchases of automatic external defibrillators; religious groups for religious purposes; private foundations; or research.
Geographic Focus: All States, Belgium, Canada, Denmark, Finland, France, Germany, Great Britain, Iceland, Ireland, Italy, Japan, Lithuania, Luxembourg, Netherlands, Norway, Poland, Portugal, Spain, Sweden, Switzerland, The Netherlands, United Kingdom, Vatican City
Date(s) Application is Due: May 14; Sep 10; Nov 12
Amount of Grant: 10,000 - 175,000 U.S.D
Contact: Deb Anderson, Grants Administrator; (763) 514-4000 or (800) 633-8766; fax (763) 505-2648; deb.anderson@medtronic.com
Paige Bingham, Patient Link Consultant; (763) 514-4000 or (800) 633-8766; fax (763) 505-2648; paige.bingham@medtronic.com
Internet: http://www.medtronic.com/foundation/programs_pl.html
Sponsor: Medtronic Foundation
710 Medtronic Parkway
Minneapolis, MN 55432-5604

Medtronic Foundation Strengthening Health Systems Grants 2873
Non-communicable diseases (NCDs), including cardiovascular disease, diabetes, cancer, and chronic respiratory disease, are the leading cause of death and disability in the world. But while prevention initiatives and lifesaving treatments have been developed for NCDs, they are not widely available in many low- and middle-income countries. The Strengthening Health Systems Grants expand access to quality healthcare by strengthening health systems and integrating NCDs into primary care in developing countries. The Medtronic Foundation will work with a limited number of chronic disease centers and experts on a global level, and in our priority countries, to address the prevention and management of NCDs. Our emphasis will be on integrating NCDs into primary healthcare, and on improving global understanding of best practices in diabetes and cardiovascular disease. Grants are awarded for specific projects and programs. Being preselected as a potential grantee does not guarantee that a grant will be awarded, as projects are evaluated on their individual merits. Organizations should also consult the website for a detailed list of proposal criteria. Applications are accepted throughout the year, but review of an application usually takes four to five months.
Requirements: Organizations should be located in the countries served by this grant. After reviewing the general guidelines, applicants should send a two- or three-page letter of inquiry. After reviewing the letter, a Foundation representative may recommend that a full proposal is submitted.
Restrictions: Grants for endowment, equipment, or for capital projects are not considered. The Foundation does not support individuals, religious groups for religious purposes, fundraising events or activities, social events or goodwill advertising, reimbursable medical treatment, scientific research, lobbying, or political or fraternal activities.
Geographic Focus: Austria, Belarus, Brazil, Bulgaria, China, Croatia, Czech Republic, Germany, Hungary, India, Liechtenstein, Moldova, Poland, Romania, Russia, Serbia, Slovakia, Slovenia, South Africa, Switzerland, Ukraine
Amount of Grant: 50,000 - 250,000 U.S.D
Contact: Deb Anderson, Grants Administrator; (763) 514-4000 or (800) 633-8766; fax (763) 505-2648; deb.anderson@medtronic.com
Internet: http://www.medtronic.com/foundation/programs_shs_guidelines.html
Sponsor: Medtronic Foundation
710 Medtronic Parkway
Minneapolis, MN 55432-5604

MeHAF Program and Policy/Data Grants 2874
These grants offer single or multi-year support for strategic projects that initiate or expand coverage or models of integrated care delivery for Maine's uninsured and medically underserved. Program grants will support projects that implement, replicate, or expand successful models of coverage or care that improve system integration and collaboration, and promote evidence-based and outcome-effective services. Policy and data projects should focus on analyzing key issues that will support Maine's progress toward universal access to care. Policy or data grants should provide data-driven analysis and recommendations that can guide expansions of coverage and/or evidence-based and outcome-effective care delivery. Proposed projects should predominantly focus on issues pertaining to Maine's uninsured and/or medically underserved. The foundation uses a two-stage application process. Applicants must first submit a two-page concept letter that briefly describes their organization and the proposed project. Full grant applications are by invitation. The listed application deadline is for concept letters.
Requirements: 501(c)3 tax-exempt charitable organizations, educational institutions, governmental entities, and other public, nonprofit entities are eligible.
Restrictions: Individuals, fiscal sponsorships, organizations with pending nonprofit status, and 509(a) organizations are ineligible.
Date(s) Application is Due: Nov 8
Amount of Grant: 10,000 - 100,000 U.S.D
Samples: Jason Program (Saco, ME)—to improve access to pediatric palliative care in northern Maine for children with life-threatening illnesses, $40,000; Healthways and the Regional Medical Ctr at Lubec (ME)—to improve the delivery of health-care services at nursing homes and residential-care facilities, $40,000; Maine Primary Care Assoc (Augusta, ME)—to develop a plan for treating obesity among medically underserved Maine residents, $38,708; Women in Need (Portland, ME)—to expand access to reproductive health care for women who are immigrants or refugees, $39,986.
Contact: Grants Administrator; (866) 848-9210 or (207) 620-8266; fax (207) 620-8269; info@mehaf.org
Internet: http://www.mehaf.org/default.asp?m=&p=106
Sponsor: Maine Health Access Foundation
150 Capitol Street, Suite 4
Augusta, ME 04330

Melanoma Research Foundation Grants 2875
The grant supports promising medical research that will further the goal of developing an effective treatment and possible cure of malignant melanoma, while encouraging a new generation of scientists and clinicians to join in this mission. The program is directed primarily at young independent investigators, e.g., untenured assistant professors and the equivalent who have not yet established strong federal funding, who are working at U.S. institutions. In addition, emphasis will be placed on projects that explore innovative approaches to understanding melanoma and its treatment. Both basic and clinical research projects will be considered. Awards will be announced by December, to commence by January 1 of the following year. Complete guidelines are available online.
Requirements: Researchers who are either U.S. permanent residents or citizens, and hold a PhD or MD degree, and a title equivalent to assistant professor, or have at least four years of postdoctoral experience are eligible to apply. Applicants should have evidence indicating strong departmental or institutional support and commitment, and be within eight years of receiving their advanced degree. The proposed research must be conducted in a nonprofit research organization, medical organization, or educational institution located in the United States.
Date(s) Application is Due: Jul 1
Amount of Grant: 50,000 U.S.D
Contact: Grants Program Director; (800) 673-1290; fax (732) 821-5955; research@melanoma.org
Internet: http://www.melanoma.org
Sponsor: Melanoma Research Foundation
24 Old Georgetown Road
Princeton, NJ 08540

Memorial Foundation Grants 2876
The foundation awards grants each year to support nonprofit organizations that provide services in Middle Tennessee. Special emphasis is given to organizations that focus on health and rehabilitation; youth and children; senior citizens; education; human and social services; and substance abuse. Types of support include capital projects; general operating support; and start-up projects for new initiatives that address important, unmet community needs and that demonstrate a potential for ongoing operational support from other sources. The directors prefer that an applicant organization be able to demonstrate a broad base of financial support for a proposed grant from its own community and constituency prior to asking for support from the foundation. Multiyear grants are rarely awarded. There are no application deadlines. Guidelines are available online.
Requirements: 501(c)3 organizations that serve the Middle Tennessee area are eligible.
Restrictions: The foundation does not make grants to individuals; for newsletters, magazines, or books; to churches and religious organizations for projects that primarily benefit their own members (exception: church-based programs funded by multiple sources with broad community support); for tickets to charitable events or dinners for disease-specific organizations seeking support for national research projects and programs; to sponsor special events, productions, telethons, performances, or similar fundraising and advertising activities; for legislative lobbying or other political purposes; or to retire accumulated debt.
Geographic Focus: Tennessee
Contact: Grants Administrator; (615) 822-9499; fax (615) 822-7797
Internet: http://www.memfoundation.org
Sponsor: Memorial Foundation
100 Bluegrass Commons Boulevard, Suite 320
Hendersonville, TN 37075

Mental Retardation Distinguished Research Award 2877
The award recognizes an individual or individuals whose research has had a significant impact on the prevention or amelioration of mental retardation. The recipient will receive, in addition to the monetary award, an expense-paid trip to speak at the luncheon of the national convention. The location of the convention varies; the date of the convention is usually in the fall. Nominations must be received by the listed application deadline. Annual deadline dates may vary; contact program staff for exact dates. Nomination forms are available on the Web site.
Date(s) Application is Due: May 15
Amount of Grant: 1,000 U.S.D

Contact: Dr. Sharon Davis, (301) 565-5456; fax (301) 565-3843; davis@the arc.org or info@thearc.org
Internet: http://thearc.org
Sponsor: Arc
1010 Wayne Avenue, Suite 650
Silver Spring, MD 20910

Mercedes-Benz U.S.A Corporate Contributions Grants **2878**
The company has committed to conducting programs with the concentration of funds geared toward educational causes and organizations, in particular those that help to empower the next generation and underserved groups; diversity programs; and women's initiatives. Specific programs in the categories of health and human services, and civic and community also are supported. There are no application forms or annual deadlines. Applicants should begin by submitting a proposal idea to the nearest company facility.
Requirements: IRS 501(c)3, 4, 6, or 9 organizations whose primary influences and business operations are in the United States and who enhance the quality of life in Mercedes-Benz communities are eligible. These company communities include: Tuscaloosa, Alabama; Carson, California; Irvine, California; Rancho Cucamonga, California; Jacksonville, Florida; Carol Stream, Illinois; Itasca, Illinois; Rosemont, Illinois; Belcamp, Maryland; Baltimore, Maryland; Montvale, New Jersey; Parsippany, New Jersey; Robinsville, New Jersey; and Fort Worth, Texas.
Geographic Focus: Alabama, California, Florida, Illinois, Maryland, New Jersey, Texas
Contact: Geoff Day, Director of Communications & Events; (201) 573-2270 or (201) 573-0600; fax (201) 573-4787; geoff.day@mbusa.com
Internet: http://www.mbusa.com/mercedes/about_us/mbcommunity
Sponsor: Mercedes-Benz U.S.A
One Mercedes Drive
Montvale, NJ 07645

Merck Company Foundation Grants **2879**
The foundation advances the company's philanthropic outreach in helping to meet important needs of local and international communities that are consistent with Merck's overall mission to enhance the health and well-being of people around the world. Such needs include improving health care, fostering biomedical and science education; and supporting the arts, social services, and civic, environmental and other charitable organizations.
Requirements: Operating support of health and social service agencies is reserved for Merck communities and generally directed through annual contributions to United Way.
Restrictions: Grants are not made to political, labor, fraternal, or veterans organizations; sectarian groups; or to individuals. Except within foundation programs, grants are not given for elementary/secondary education, scholarships, fellowships, research, publications, conferences/seminars/symposia, and travel.
Samples: Alliance of Community Health Plans (Washington, DC)—to help healthcare groups that use culturally and linguistically appropriate health services demonstrate that those services are financially sound options, $235,000; American National Red Cross (Washington, DC), UNICEF (NY), and several local relief organizations—for relief efforts in South Asia and Africa, $3 million divided; Task Force for Child Survival and Development (Atlanta, GA)—to expand programs to treat people in Africa who are suffering from onchocerciasis, more commonly known as river blindness, as part of the Mectizan Donation Program, $1 million over five years.
Contact: Dr. Shuang Ruy Huang, (908) 423-1000; fax (908) 423-1987
Internet: http://www.merck.com/about/cr/policies_performance/social/philanthropy.html
Sponsor: Merck Company Foundation
P.O. Box 100 (WSIAF-35)
Whitehouse Station, NJ 08889-0100

Merck Institute of Aging and Health Grants **2880**
The institute is dedicated to improving the health and independence of older people around the world by communicating health information, educating the public and health professionals, and encouraging research on health in the aging field. Proposals are encouraged for demonstration and pilot programs in the following priority areas: innovative education of primary care providers about the care of older adults in community practice; research and evaluation of innovative health services provided for older adults in the community; and development and evaluation of innovative interventions that encourage health promotion and disease prevention for older adults. The institute currently serves the United States, Europe, and Latin America and may eventually expand into other areas of the world. Application guidelines are available online.
Requirements: Eligibility criteria include 501(c)3 designation and incorporation as an accredited school of medicine, nursing, osteopathy, pharmacy, or public health; a teaching hospital or institution affiliated with an accredited school of medicine, nursing or osteopathy; or a national health or health-care related organization, health professional organization, nursing or medical association, or consumer health or health-care organization. All non-U.S. organizations are required to provide evidence of nonprofit or NGO status, equivalent to a U.S. public charity.
Restrictions: Ineligible organizations and programs include organizations without 501(c)3 or NGO status; political organizations, campaigns, and activities; fraternal/labor/veteran's organizations and activities; religious organizations or groups whose activities are primarily sectarian in purpose; organizations that discriminate on the basis of race, color, sexual orientation, religion, marital status, age, national origin, veteran's status, or disability; fundraising events; capital or building campaigns and endowments; media productions; basic or clinical research projects; meetings, conferences, or symposia that are not part of a larger health-related program efforts; unrestricted general support; fellowship/tuition support for training purposes; programs that directly support marketing and/or sales objectives of Merck and Company; and programs for pre-college education.
Amount of Grant: 50,000 - 100,000 U.S.D
Contact: Julie Kennedy, Grants Manager; (202) 842-0525; fax (202) 354-6599
Internet: http://www.miahonline.org/aboutMIAH/grants.html
Sponsor: Merck Institute of Aging and Health
1100 New York Avenue NW, Suite 350W
Washington, DC 20005

Mericos Foundation Grants **2881**
The foundation primarily awards grants to organizations based in California with emphasis on Santa Barbara. Fields of interest include aging, animals/wildlife, environment, arts and arts education, child development, children and youth, elementary and secondary education, higher education, hospitals, medical care, medical research, libraries, and museums (art and natural history). Types of support include building/renovation, equipment, fellowships, general/operating support, matching/challenge support, and program development. Grants are usually initiated by the foundation. Interested organizations should contact the Vice-President for more information.
Requirements: Organizations in California are eligible to apply.
Restrictions: Grants are not made to individuals.
Amount of Grant: 15,000 - 200,000 U.S.D
Samples: Music Academy of the West, Santa Barbara, California, $125,000; Friends of Independent Schools and Better Education, Tacoma, Washington, $25,000; Los Angeles Children's Chorus, Pasadena, California, $25,000.
Contact: Linda Blinkenberg, Vice President; (626) 441-5188; fax (626) 441-3672
Sponsor: Mericos Foundation
625 S Fair Oaks Avenue, Suite 360
South Pasadena, CA 91030-2630

Meriden Foundation Grants **2882**
The foundation awards grants to eligible Connecticut nonprofit organizations in its areas of interest, including arts, children and youth, civic affairs, Christian organizations and churches, health organizations and hospitals, higher education, public libraries, and social services. Types of support include annual campaigns, general operating support, scholarships, and social services delivery. There are no application deadlines. A formal application is required, and the initial approach should be a letter, on organizational letterhead, describing the project and requesting an application.
Requirements: Connecticut nonprofit organizations in the Meriden-Wallingford area are eligible.
Geographic Focus: Connecticut
Amount of Grant: 175 - 10,000 U.S.D
Samples: Best Friends Animal Sanctuary, Kanab, Utah, $3,500 - general operations; Church of the Holy Angels, Meriden, Connecticut, $3,296 - general operations; Franciscan Home Care and Hospice, Meriden, Connecticut, $5,000 - to purchase laptop computers.
Contact: Jeffrey F. Otis, Director; (203) 782-4531; fax (203) 782-4530
Sponsor: Meriden Foundation
123 Bank Street
Waterbury, CT 06702-2205

Merkel Foundation Grants **2883**
The Merkel Foundation, established by Daniel A. and Betty Merkel in Sheboygan, Wisconsin, is interested in supporting a variety of Wisconsin organizations. The Foundation's primary interests are research, religion, education, social services, and relief programs. There are no specific application formats or deadlines with which to adhere. Applicants should begin the process by contacting the Foundation directly. Generally, support is given throughout Wisconsin and, ocassionally, across the country. Amounts range from $100 to $10,000.
Geographic Focus: Wisconsin
Amount of Grant: Up to 10,000 U.S.D
Samples: Common Hope, St. Paul, Minnesota, $3000 - support for international relief programs (2010); Wisconsin Foundation for Independent Colleges, Milwaukee, Wisconsin, $1,000 - support for religious programs (2010).
Contact: Betty Merkel, President; (920) 457-5051; fax (920) 457-1485; info@americanortho.com
Sponsor: Merkel Foundation
3712 Bismarck Circle
Sheboygan, WI 53083

Merrill Lynch Philanthropic Program Grants **2884**
The foundation supports a variety of charitable causes throughout the world, with a focus on education. Through 2005, the foundation has adopted children and youth as its global cause. Programs should meet the educational needs and interests of uderserved children and youth. Programs should support an ethnically diverse population of children and youth; provide direct services from early childhood to postsecondary education, and incorporate one or more of the following elements: academic support, global exchange, community service, internships, cultural enrichment, life skills, digital divide, mentoring, entrepreneurship, scholarships, personal finance, technology skills, GED programs linked with four-year colleges, and workforce preparation. The foundation gives priority to specific programs and projects, rather than general operating support. Preference will be given to requests from the five boroughs of New York City and national organizations that reflect that focus. Requests outside of New York City should be submitted to managers of local branches or offices.
Requirements: Tax-exempt 501(c)3 organizations are eligible.
Restrictions: Grants will not be made to private foundations; individuals; fundraising activities related to individual sponsorships (e.g., walk-a-thons, marathons); seed money for new

organizations; political causes, candidates, and campaigns as well as organizations designed specifically for lobbying; religious, fraternal, social, or other membership organizations providing services mainly to their own constituencies; athletic events and sports tournaments (Special Olympics is eligible); and fundraising events (dinners, luncheons).
Samples: Sesame Workshop (New York, NY)—for the WorldwideKids program, which will use multimedia tools and resources to educate three- to six-year olds about diverse people and cultures, $5 million; United Nations Assoc of the United States of America (Washington, DC)—to expand to 14 countries the Global Classrooms program, which supports Model United Nations activities in schools, and to develop a curriculum on global development, finance, and trade, $7.5 million.
Contact: Westina Matthews, Vice President, Philanthropic Programs; (212) 236-4319; fax (212) 236-8007; SDimaggio@exchange.ml.com
Internet: http://philanthropy.ml.com
Sponsor: Merrill Lynch and Company Foundation
2 World Financial Center, 5th Floor
New York, NY 10281-6100

Mervin Bovaird Foundation Grants 2885
The foundation awards grants, with a focus on Tulsa, OK, to nonprofit organizations such as churches, homeless shelters, medical centers, nursing homes, parochial schools, religious welfare organizations, the Salvation Army, and youth organizations. Areas of interest include arts, community development, education, environment, health care and health organizations, and social services. Types of support include general operating support, matching, project, and research grants. The foundation also funds a scholarship program for Tulsa County High School graduating seniors and graduates of Tulsa Community College at the University of Tulsa (not for graduate or professional study). Applicants should submit a brief letter of inquiry, including program and organization descriptions.
Requirements: Nonprofits of the Christian, Methodist, Roman Catholic, and United Methodist faiths in Tulsa, OK, are eligible.
Geographic Focus: Oklahoma
Amount of Grant: 1,000 - 250,000 U.S.D
Contact: R. Casey Cooper, President and Trustee, c/o Boesche, McDermott, and Eskridge; (918) 592-3300
Sponsor: Mervin Bovaird Foundation
401 S Boston Avenue, Suite 3300
Tulsa, OK 74103-4070

Mesothelioma Applied Research Foundation Grants 2886
The program seeks to stimulate translational research for the treatment of pleural or peritoneal mesothelioma. Grants support projects that apply cutting-edge general cancer research concepts specifically to mesothelioma, and translate directly to improved mesothelioma treatment. Eligible projects may relate to either benchwork research or clinical research, must not be presently funded or pending review, and may be conducted through any not-for-profit academic, medical, or research institution, in the United States or abroad. Grants are awarded for 12 months, renewable for 12 more months.
Requirements: The investigator must be affiliated with a nonprofit academic, medical, or research institution.
Date(s) Application is Due: Aug 15
Amount of Grant: 50,000 U.S.D
Contact: Danele Mangone, Grants Director; (805) 563-8400, ext. 7273; fax (805) 563-8411; dmangone@curemeso.org
Internet: http://www.curemeso.org/site/c.kkLUJ7MPKtH/b.4097447/k.E6A3/Mesothelioma_Research_Grants__Call_for_Applications.htm
Sponsor: Mesothelioma Applied Research Foundation
3944 State Street, Suite 340, P.O. Box 91840
Santa Barbara, CA 93105

Metro Health Foundation Grants 2887
The foundation awards grants in southeastern Michigan in its areas of interest, including children and youth, disabled, elderly, health associations, homeless, and women. Types of support include challenge grants, equipment acquisition, general operating support, project development, scholarship funds, seed grants, and federated giving. Obtain application materials from the office; the board meets in April and October.
Requirements: Nonprofit organizations in the metropolitan tri-county Detroit, MI, area are eligible to apply.
Restrictions: Grants do not support advertising, advocacy organizations, athletic groups, individuals, international organizations, political organizations, special events, benefit dinners, or state or local government agencies.
Geographic Focus: Michigan
Date(s) Application is Due: Feb 1; Aug 1
Amount of Grant: 1,000 - 25,000 U.S.D
Contact: Theresa Sondys, Program Contact; (313) 965-4220; fax (313) 965-3626; metrohealthfdn@aol.com
Sponsor: Metro Health Foundation
333 W Fort Street, Suite 1370
Detroit, MI 48226-3134

Metrowest Community Health Care Foundation Grants 2888
The foundation seeks to improve the health status of the MetroWest community and awards grants to support of youth and elders initiatives. The youth initiative provides grants to support new, innovative, or expanded programs and services that deal with child and adolescent health issues that include injuries and violence; child abuse; oral health care; asthma; depression and suicide; alcohol and substance abuse; access to health services; and health promotion and disease prevention. The elders initiative invites proposals that offer new, innovative, or expanded models for improving access by the elderly to medical and ancillary services, home health, and social services. Proposals also are invited that promote health and wellness services and mental health, alcohol, and substance abuse services to elders. Proposals must show a clear community need and demonstrated collaboration and/or partnership among community organizations in response to this need. Types of grants include but are not limited to project or program support, capacity building, community education, research and evaluation, and demonstration or pilot projects. The foundation will consider proposals for up to three years of funding, with funding beyond the first year contingent upon careful review.
Requirements: 501(c)3 tax-exempt organizations and organizations recognized as an instrumentality of state/local government in the following Massachusetts communities are eligible: Ashland, Bellingham, Dover, Framingham, Franklin, Holliston, Hopedale, Hopkinton, Hudson, Marlborough, Medfield, Medway, Mendon, Milford, Millis, Natick, Needham, Norfolk, Northborough, Sherborn, Southborough, Sudbury, Wayland, Wellesley, and Westborough.
Restrictions: Grants do not support individuals (except for scholarships—see guidelines online), capital fund drives, fundraising drives, retirement of debt or operating deficits, projects that directly influence legislation, political activities or candidates for public office, programs that are customarily operated by hospitals in Massachusetts, and programs that are outside the service area.
Geographic Focus: Massachusetts
Amount of Grant: 50,000 U.S.D
Samples: Hemenway School PTO (Framingham, MA)—for fitness and exercise programs for disabled students, $15,000; Autism Alliance of MetroWest (Natick, MA)—for organizational development, $24,000; Natick Senior Ctr (Natick, MA)—to continue its taxi-voucher program, which assists elderly people in getting to and from medical appointments, $22,970; Screening for Mental Health (Wellesley, MA)—to develop a middle-school curriculum on suicide prevention, $186,311.
Contact: Cathy Glover, Grants Administrator; (508) 879-7625; fax (508) 879-7628; cglover@mchcf.org or info@mchcf.org
Internet: http://www.mchcf.org
Sponsor: Metrowest Community Health Care Foundation
161 Worcester Road, Suite 202
Framingham, MA 01701

Metzger-Price Fund Grants 2889
The Metzger-Price Fund, established in New York in 1970, is an independent foundation trust which offers support to the handicapped, health services, child welfare, social service agencies, recreation, and the elderly. Its primary fields of interest include: services and centers for the aged, child and youth services, community and economic development, education, family support services, health care services and health care access, and human services. The Fund's target groups include the elderly, disabled, economically disadvantaged, homeless, women, and other minorities. Funding is directed toward continuing support, general operations, and program development. There are no specific application forms or deadlines, though the board meets to discuss proposals four times each year (January, April, July, and October). An applicant's initial approach should be in the form of an application letter submitted two months prior to each board meeting.
Requirements: 501(c)3 organizations serving the residents of New York, New York, are encouraged to apply.
Restrictions: No grants to individuals, or for capital campaigns or building funds; no multiple grants in single calendar year to same organization.
Geographic Focus: New York
Amount of Grant: 1,000 - 5,000 U.S.D
Contact: Isaac A. Saufer, (212) 867-9501 or (212) 867-9500; fax (212) 599-1759
Sponsor: Metzger-Price Fund, Inc.
230 Park Avenue, Suite 2300
New York, NY 10169-0005

Meyer Foundation Grants 2890
Meyer awards grants to nonprofit organizations that serve the people and communities of Greater Washington. The foundation values community-based organizations that work in concert with members of the community. Programs conducted by national organizations that operate in and with cooperation of the local community also will be considered. Grants fall under the following broad program areas: arts, heritage, and culture; community service—i.e., homelessness, domestic violence, immigrant communities, community building, hunger, and more; education—children and youth, and more; health and mental health; law and justice; neighborhood development and housing—civic engagement, advocacy, and more; management assistance; and nonprofit sector leadership. Eligible candidates are required to first submit a letter of inquiry. Listed deadlines are for letters of inquiry. Guidelines are available online.
Requirements: 501(c)3 nonprofit organizations organized and operated for charitable purposes are eligible. Meyer grantees must be located within and primarily serving the Washington, DC region (which includes Washington, DC; Maryland; and Virginia).
Restrictions: The foundation generally does not fund individuals, either through scholarships or other forms of financial assistance; scientific or medical research; programs that promote religious doctrine; special events or conferences; or endowments.
Date(s) Application is Due: Feb 10; Jun 9; Oct 2
Amount of Grant: 1,500 - 50,000 U.S.D
Contact: Program Contact; (202) 483-8294; fax (202) 328-6850; meyer@meyerfdn.org
Internet: http://www.meyerfoundation.org
Sponsor: Meyer Foundation
1400 16th Street NW, Suite 360
Washington, DC 20036

Meyer Memorial Trust Emergency Grants 2891

Emergency Grants provide expedited consideration of true emergencies. MMT's emergency fund is intended for sudden, unanticipated and unavoidable challenges that, if not addressed immediately, could threaten an organization's stability and/or ability to achieve its mission. Examples of Emergency Grants might include assistance for recovery from fire, flood, and other acts of nature. It is not intended to address failure of an organization to comply with legal requirements or problems that can be attributed to organizational neglect or failure to plan for likely contingencies, such as the breakdown of aging equipment.
Requirements: Grants are awarded to 501(c)3 nonprofit organizations in Oregon and Clark County, Washington.
Restrictions: Meyer does not fund general administrative overhead or indirect rate allocations.
Geographic Focus: Oregon, Washington
Contact: Barbara Gibbs, Grants Director; (503) 228-5512; barbara@mmt.org
Internet: http://www.mmt.org/grants_programs/
Sponsor: Meyer Memorial Trust
425 NW 10th Avenue, Suite 400
Portland, OR 97209

Meyer Memorial Trust Grassroots Grants 2892

Grassroots Grants enable smaller organizations, often without development departments, to compete for MMT grants. While organizations with budgets up to $2 million may apply for Grassroots Grants, historically, 70% of small grants have gone to organizations with budgets of $250,000 or less. Grants of $1,000 to $25,000 are awarded for projects with budgets of up to $150,000. Grassroots Grants fund programs, outreach, equipment, facility expansions or repairs, strategic planning, website development, and more. Decisions on Grassroots Grants are made by trustees at the program meetings in February, June and October.
Requirements: Grants are awarded to 501(c)3 nonprofit organizations in Oregon and Clark County, Washington.
Restrictions: Meyer does not fund general administrative overhead or indirect rate allocations.
Geographic Focus: Oregon, Washington
Date(s) Application is Due: Mar 15; Jul 15; Oct 15
Amount of Grant: 1,000 - 25,000 U.S.D
Samples: Assistance League of Klamath Basin, Klamath Falls, Oregon, $14,547—for safety improvements to its resale and consignment shop; Central Oregon Environmental Center, Bend, Oregon, $24,750—to support a merger with reSource to strengthen environmental work in the region.
Contact: Barbara Gibbs, Grants Director; (503) 228-5512; barbara@mmt.org
Internet: http://www.mmt.org/grants_programs/
Sponsor: Meyer Memorial Trust
425 NW 10th Avenue, Suite 400
Portland, OR 97209

Meyer Memorial Trust Operating Funds Grants 2893

MMT has launched an Operating Funds grant program to help small- and mid-sized organizations that have been hit particularly hard by the current economic crisis. The purpose of this initiative is organizational stability. Organizations for which a cash infusion is not critical to stability, but wish to request funds to maintain or expand current programming, should continue to apply for support through MMT's Responsive and Grassroots Grant programs, as usual. Applications will be accepted beginning September 1. Application deadline is October 1. Awards announcement is anticipated in December. Applications will be accepted through MMT's Responsive Grant inquiry process.
Requirements: Any 501(c)3 nonprofit organizations the operates in Oregon and Clark County, Washington is eligible to apply.
Restrictions: Meyers does not support the following: direct grants, scholarships, or loans to individuals; endowments; general fund drives, annual appeals, special events, or sponsorships; elimination of operating deficits; projects of sectarian or religious organizations that principally benefit their own members or adherents; direct replacement funding for activities previously supported by federal, state, or local public sources; acquisition of land for conservation purposes (except through Program Related Investments); hospital capital construction projects (except through Program Related Investments); animal welfare organizations (both domestic and wild animals); projects that primarily benefit students of a single K-12 school, unless the school is an independent alternative school primarily serving low-income and/or special needs populations; MMT funds cannot be earmarked for purposes of influencing legislation, and the grantee cannot expend any part of the grant in a way that violates its tax-exempt status.
Geographic Focus: Oregon, Washington
Date(s) Application is Due: Oct 1
Amount of Grant: Up to 50,000 U.S.D
Contact: Barbara Gibbs, Grants Director; (503) 228-5512; barbara@mmt.org
Internet: http://www.mmt.org/grants_programs/
Sponsor: Meyer Memorial Trust
425 NW 10th Avenue, Suite 400
Portland, OR 97209

Meyer Memorial Trust Responsive Grants 2894

Responsive Grants are awarded in the areas of social services, health, affordable housing, community development, conservation and environment, public affairs and social benefit, arts and culture, and education. They generally range from $30,000 to $500,000; grant periods are from one to three years in length. General purpose grants are given for a wide variety of projects in many fields of activity including program expansions, capacity building, capital projects, and start-ups. Final decisions on Responsive Grants proposals are made by trustees at program meetings every month except January, April and August. Final decisions on proposals for capital projects requesting more than $200,000 are made in May and November. Applicants should time their Initial Inquiry submission at least six months before they hope to have a final decision from MMT.
Requirements: General purpose and small grants are awarded to 501(c)3 nonprofit organizations in Oregon and Clark County, Washington.
Restrictions: Meyer does not fund general administrative overhead or indirect rate allocations.
Geographic Focus: Oregon, Washington
Amount of Grant: 30,000 - 500,000 U.S.D
Samples: Confederated Tribes of Warm Springs, Warm Springs, Oregon, $150,000—to help construct a building to co-locate the newspaper office and KWSO radio station; Oregon Ballet Theatre, Portland, Oregon, $200,000—for two years of operating support; South Lane Mental Health Services Cottage Grove, Oregon, $150,000—to purchase and renovate a building as a mental health clinic.
Contact: Barbara Gibbs, Grants Director; (503) 228-5512; barbara@mmt.org
Internet: http://www.mmt.org/grants_programs/
Sponsor: Meyer Memorial Trust
425 NW 10th Avenue, Suite 400
Portland, OR 97209

MfA Newton Fellowships 2895

The fellowships are available to mathematically sophisticated individuals who are interested in using their talents to make a difference in the lives of young people. The program seeks college graduates and mid-career professionals who can demonstrate a strong interest in teaching, have a bachelor's degree with substantial coursework in mathematics, and are willing to commit to a five-year program that includes one year of full-time graduate study and four years of teaching mathematics in New York City high schools. Fellows receive a stipend, full tuition to master's in education program, mentoring and coaching, and professional development. Applicants are required to take the ETS Praxis II Math subject tests. Application and guidelines are available online.
Requirements: U.S. citizens and permanent residents are eligible.
Restrictions: Individuals who are currently teaching, are certified to teach, or have completed an education degree program are not eligible.
Geographic Focus: New York
Date(s) Application is Due: Feb 10
Amount of Grant: 90,000 U.S.D
Contact: Fellowship Administrator; (212) 206-0053; fax (212) 206-1622; info@mathforamerica.org
Internet: http://www.mathforamerica.org
Sponsor: Math for America
101 Fifth Avenue, 5th Floor
New York, NY 10003

MFH Tobacco Prevention and Cessation Grants 2896

The foundation supports programs that address the impact of tobacco use as a major health issue affecting the residents of its service region. The initiative will support direct program implementation, policy development activities, and evaluation. Funded programs will include those involving public and private collaborations, as well as individual agencies that have the ability to carry out model programs supporting tobacco prevention and cessation efforts. This RFP is intended to stimulate programs that implement best practice comprehensive tobacco prevention and cessation programs; develop capacity in communities to continue program implementation on the local level; and ensure coordination across regional and community-based efforts. Guidelines are available online.
Requirements: Applicants will have demonstrated tobacco prevention and/or cessation programs with the capacity to develop materials, carry out activities on a regional basis, conduct studies, and assist community agencies in adapting programs to the specific needs of their communities.
Geographic Focus: Missouri
Date(s) Application is Due: Jun 30
Contact: Grants Administrator; (800) 655-5560 or (314) 345-5500; fax (314) 345-5599; info@mffh.org
Internet: http://www.mffh.org/tobaccoinit.htm
Sponsor: Missouri Foundation for Health
1000 St. Louis Union Station, Grand Central Building, Suite 400
Saint Louis, MO 63103

MGFA Post-Doctoral Research Fellowships 2897

The Myasthenia Gravis Foundation of America, Inc. announces a competitive twelve-month post-doctoral research fellowship for clinical or basic research pertinent to myasthenia gravis or related neuromuscular disorders. Research may be concerned with neuromuscular transmission, immunology, molecular or cell biology of the neuromuscular synapse, or the etiology/pathology or diagnosis of myasthenia gravis. See the Foundations website for application and, additional guidelines.
Requirements: The applicant must be: a permanent resident of the United States or Canada who has been accepted to work in the laboratory of an established investigator at an institution in the United States, Canada or abroad deemed appropriate by the Medical/Scientific Advisory Board of the Myasthenia Gravis Foundation of America, Inc; or a foreign national who has been accepted to work in the laboratory of an established investigator at an institution in the United States or Canada deemed appropriate by the Medical/Scientific Advisory Board of the Myasthenia Gravis Foundation of America, Inc.
Date(s) Application is Due: Oct 1
Amount of Grant: Up to 50,000 U.S.D

Contact: Tor Holtan, Chief Executive; (800) 541-5454 or (212) 297-2156; fax (212) 370-9047; Tor.Holtan@myasthenia.org or mgfa@myasthenia.org
Internet: http://www.myasthenia.org/hp_fellowships.cfm#nursing
Sponsor: Myasthenia Gravis Foundation of America
355 Lexington Avenue, 15th Floor
New York, NY 10017

MGFA Student Fellowships **2898**
These fellowships are awarded annually to current medical students or graduate students interested in the scientific basis of myasthenia gravis or related neuromuscular conditions, serving both to further scientific inquiries into the nature of these disorders and to encourage more research. The stipend is up to $5,000. See the Foundation's website for application form and guidelines.
Requirements: The fellowships are awarded to current medical students or graduate students. Briefly describe, in abstract form, the question that you propose to study, its association to myasthenia gravis or related neuromuscular conditions, and how you will approach the project. Applicants and sponsoring institutions must comply with policies governing the protection of human subjects, the humane care of laboratory animals and the inclusion of minorities in study populations.
Date(s) Application is Due: Mar 31
Amount of Grant: 5,000 U.S.D
Contact: Tor Holtan, Chief Executive; (800) 541-5454 or (212) 297-2156; fax (212) 370-9047; Tor.Holtan@myasthenia.org or mgfa@myasthenia.org
Internet: http://www.myasthenia.org/hp_fellowships.cfm#nursing
Sponsor: Myasthenia Gravis Foundation of America
355 Lexington Avenue, 15th Floor
New York, NY 10017

MGM Mirage Community Affairs Grants **2899**
It is the mission of MGM Mirage to strengthen communities in which the corporation operates. MGM Mirage's key community investment areas are: childhood development—(birth to 18 years old), with emphasis placed on at-risk/low income youth and early childhood development; community development—specific community issues and deliverable outcomes within targeted populations; diversity—including efforts that encourage economic development as well as professional and personal growth of our employees and programs that enhance community resources; and education—including efforts to strengthen public education, from kindergarten through higher education.
Requirements: Requests must benefit residents in communities which the Company operates (Nevada, Michigan, Mississippi).
Restrictions: The program does not support the following types of organizations or activities: requests which fall outside of communities in which MGM MIRAGE operates; requests for any type of gaming-casino items (new or previously used) for children; organizations that discriminate by race, color, creed, gender or national origin; grants or sponsorship to individuals; or athletic sponsorships of individuals and/or teams.
Geographic Focus: Michigan, Mississippi, Nevada
Date(s) Application is Due: Oct 1
Contact: Director of Corporate Philanthropy; (702) 650-7409; fax (702) 650-7401
Internet: http://www.mgmmiragevoice.com/pages/cg_giving.asp
Sponsor: MGM Mirage Corporation
3260 Industrial Road
Las Vegas, NV 89109

MGN Family Foundation Grants **2900**
The MGN Family Foundation makes grants to qualified 501(c)3 organizations specializing in the following areas: education; health care and medical research; children in need; armed service personnel. Areas of particular interest include: colleges, universities and private schools, examples would be: to fund a chair to provide lecturers in literature, philosophy or the arts, and provide scholarships; hospitals and clinics that specialize in excellent patient care and continuing medical research such as Memorial Sloan-Kettering Cancer Center and the Mayo and Cleveland Clinics are other examples, also under consideration would be hospice organizations that provide palliative care to the dying; organizations that support children in need whether due to emotional, physical abuse, neglect or disadvantaged circumstances. Support can be for basic necessities such as food, shelter, education, medical as well as, spiritual and emotional counseling; in light of recent events, this Foundation would like to offer assistance to our servicemen/ servicewomen and their families through those charities which give support to their unique needs. The examples provided above reflect the true mission and goals of the MGN Foundation. There should be no exclusions due to race or creed, provided all applicants have a strong moral base and core values in the areas of education, health care, welfare of children and service personnel and their families. The Foundation Board meets semi-annually in May and November. Applications are due by April 1st and October 1st. Application form is available online.
Requirements: Qualified 501(c)3 organizations specializing in the following areas: education; health care and medical research; children in need; armed service personnel. To apply, submit seven (7) sets of the following items: Grant application form completed, dated, and signed by the Chief Executive Officer or Chairman of the Board of the organization; list of Board of Directors; Financial Statement (audited if available), for the most recent complete fiscal year; copy of IRS 501(c)3 Determination Letter. If you wish, you may submit your proposal in a narrative format of not more than two pages in addition to the completed application form. Optional materials may be submitted but are not required (such as brochures discussing or depicting the activities of the organization).
Restrictions: Do not staple materials or place them in a bound notebook.
Date(s) Application is Due: Apr 1; Oct 1
Amount of Grant: 1,000 - 10,000 U.S.D
Samples: Agnes Scott College, $3,000—scholarship; Vermont National Guard Charitable Foundation, Inc., $3,000—wounded veteran program; Center for Molecular Medicine and Immunology/Garden State, $5,000—foster research for new technologies in the diagnosis, detection and treatment of cancer.
Contact: Pamela Nothstein, Grants Manager; (843) 937-4614; grantinquiries7@wachovia.com
Internet: https://www.wachovia.com/foundation/v/index.jsp?vgnextoid=61078689fb0aa110VgnVCM1000004b0d1872RCRD&vgnextfmt=default
Sponsor: MGN Family Foundation
16 Broad Street (SC1000)
Charleston, SC 29401

MHRC Graduate Studentships **2901**
A limited number of studentships are offered to qualified graduate students who are undertaking full-time master's or doctoral programs. Studentships are awarded on a competitive basis. Each candidate must arrange to have three individuals send supporting letters directly to the council. Studentships are initially awarded for a 12-month period and may be renewed. The maximum period of support by the council is two years for individuals enrolled in a master's program, three years for individuals holding a master's degree enrolled in a PhD program, and five years for individuals proceeding directly to a PhD from an honor's BS, BA, or equivalent degree. Recipients should begin their award period by July 1 at the earliest and September 1 at the latest.
Restrictions: In each competition a maximum of one studentship will be awarded per supervisor. Individuals who hold a faculty position are not eligible for graduate studentship awards.
Contact: Program Contact; (204) 775-1096; fax (204) 786-5401; mhrc@cc.umanitoba.ca
Internet: http://www.umanitoba.ca/academic_support/MHRC/PROGRAMS.HTM#contents
Sponsor: Manitoba Health Research Council
770 Bannatyne Avenue
Winnipeg, MB R3E 0W3 Canada

MHRC Operating Grants **2902**
Grants in aid of research are designed to defray the normal direct costs of research including, among others, personnel costs, supplies and expendable materials, experimental animals, equipment costing less than $C5000, computer costs, and publication costs. Up to five percent of the operating grant may be used for travel related to the investigator's research program. Grants are normally awarded for one year. The council gives priority to new investigators and new projects. The starting date for successful grants is July 1.
Requirements: Normally only investigators holding academic appointments or equivalent in the province of Manitoba are eligible to apply. Investigators who do not hold academic appointments should consult with the council to establish their eligibility before applying.
Restrictions: Individuals holding continuing operating grants from the council are not eligible to apply for other operating grants that would become effective during the tenure of their present grants. Expenditures for student and postdoctoral salaries will not be considered on operating grants.
Geographic Focus: All States, Canada
Amount of Grant: Up to 100,000 CAD
Contact: Program Contact; (204) 775-1096; fax (204) 786-5401; mhrc@cc.umanitoba.ca
Internet: http://www.umanitoba.ca/academic_support/MHRC/PROGRAMS.HTM#contents
Sponsor: Manitoba Health Research Council
770 Bannatyne Avenue
Winnipeg, MB R3E 0W3 Canada

MHRC Postdoctoral Research Fellowships **2903**
A limited number of fellowships are available to qualified candidates with doctoral degrees whose prior postdoctoral research training shall normally not exceed five years. Fellowships are awarded on a competitive basis and will carry the same monetary value as postdoctoral salaries on the council's operating grants. Only individuals licensed to practice in the province will be eligible for the stipend scale for professional degree holders; all others will receive the PhD stipend. Fellowships are awarded for a 12-month period and may be renewed for one additional year by reapplication and competitive review. Starting dates are between July 1 and September 1.
Requirements: Holders of professional doctoral degrees (MD, DDS, DVM) who will be registered as full-time master's or doctoral students may apply for either a postdoctoral fellowship or a graduate studentship; however, they are eligible for consideration in only one of these two categories and must indicate clearly for which category they wish to be considered. Each candidate must arrange to have three individuals send supporting letters to the council.
Restrictions: In each competition a maximum of one postdoctoral fellowship will be awarded per supervisor. Individuals who hold a faculty position are not eligible for postdoctoral fellowship awards.
Contact: Program Contact; (204) 775-1096; fax (204) 786-5401; mhrc@cc.umanitoba.ca
Internet: http://www.umanitoba.ca/academic_support/MHRC/PROGRAMS.HTM#contents
Sponsor: Manitoba Health Research Council
770 Bannatyne Avenue
Winnipeg, MB R3E 0W3 Canada

Miami County Community Foundation - Operation Round Up Grants **2904**

Miami County Community Foundation - Operation Round Up is a national program that allows participating electric cooperatives to partner with their members to provide charitable giving funds in their local community. Each month the total amount of the participating member's bill is rounded up to the nearest dollar and those extra pennies go into the fund designated for distribution to those in need in the area. Areas of consideration include: cultural, education, recreation, human services, health and medical, community development, and environmental awareness. Grant proposals are reviewed by the following criteria: is there an established need; what is the project solving; is it appropriate for the county; are their adequate resources; its benefit and timing to the community; what are the expected results; and will the organization work with other organizations to achieve their goals.

Requirements: Along with the application, organizations must include a one page budget for the amount requested, with justification; proof of 501(s)3 status; their most recent audited financial statement or annual report; and current organizational budget. A total of nine complete sets of applications and documentation must be submitted. Applications are available at the Foundation website and should be mailed to Miami-Cass REMC, Operation Round Up, 3086 West 100 North, P.O. Box 168, Peru, IN 46970.

Restrictions: Operation Round Up funds will not be used for paying utility bills, or funding political or private interests.

Geographic Focus: Indiana

Date(s) Application is Due: Mar 31; Jun 30; Sep 30; Dec 31

Contact: Mary Alexander, Director of Development; (765) 475-2859 or (877) 432-6423; fax (765) 472-7378; miami@nicf.org

Internet: http://www.nicf.org/miami/grantapplications.html

Sponsor: Miami County Community Foundation
13 East Main Street
Peru, IN 46970

Michael J. Fox Foundation Clinical Discovery Grants **2905**

The goal of this program is to encourage translational research by stimulating well-designed clinical research projects focused on potentially high-impact approaches to the field of Parkinson's disease. Through the program, researchers will be invited to submit grant applications to conduct new or novel clinical research in humans relevant to the cure, cause, prevention, or improved treatment of Parkinson's disease and its complications. The program will support a variety of innovative research studies that are directly related to patients or patient populations affected by the disease. Appliciants may request a project period of up to three years. The foundation employs an expedited, abbreviated granting mechanism to accelerate discoveries and hopefully translate findings to Parkinson's disease patients more quickly. A mandatory letter of intent is due on the listed application deadline; full applications are by invitation.

Requirements: Both U.S. and non-U.S. organizations are eligible. Applications may be submitted by public and private nonprofit entities, units of state and local governments, eligible agencies of the federal government, and for- profit entities. Collaborative efforts are welcome.

Restrictions: Postdoc students or fellows are not eligible to apply as principal investigators.

Date(s) Application is Due: Nov 21

Amount of Grant: Up to 200,000 U.S.D

Contact: Grants Administrator; research@michaeljfox.org

Internet: http://www.michaeljfox.org/research/index.php

Sponsor: Michael J. Fox Foundation for Parkinson's Research
P.O. Box 4777, Grand Central Station
New York, NY 10163

Michael J. Fox Foundation for Parkinson's Research Community Fast Track Grants **2906**

This program invites researchers to submit investigator-initiated grant applications to conduct research relevant to the cure, cause, prevention or improved treatment of Parkinson's disease and its complications. The intent of the program is to stimulate novel, innovative, and/or high-impact approaches to the field of Parkinson's disease as well as to fill funding gaps missed by more conventional funding sources. Responsibility for the planning, direction, and execution of the proposed project will be solely that of the principal investigator. The intent of this RFA is to stimulate work on potentially high-impact approaches to the field of Parkinson's disease research, as well as to fill funding gaps missed by more conventional funding sources. This program is restricted to projects and ideas that are conducted on a one-year timeframe; however, successful projects meeting all of their original specific aims may be allowed to submit proposals for an additional one to two years of funding. The listed application deadline is for mandatory letters of intent, which must be submitted via email.

Requirements: Applications may be submitted by U.S. and non-U.S. entities; public and private nonprofit entities, such as universities, colleges, hospitals, laboratories, units of state and local governments, and eligible agencies of the federal government; and for-profit entities. Collaborative efforts are welcome.

Restrictions: Postdoctoral students or fellows are not eligible to apply as principal investigators.

Date(s) Application is Due: Jul 27

Amount of Grant: Up to 100,000 U.S.D

Contact: Grants Administrator; research@michaeljfox.org

Internet: http://www.michaeljfox.org/research/grantsunderreview.php

Sponsor: Michael J. Fox Foundation for Parkinson's Research
P.O. Box 4777, Grand Central Station
New York, NY 10163

Michael J. Fox Foundation for Parkinson's Research LEAPS Program **2907**

Investigators are invited to apply for funding through a novel mechanism called LEAPS (Linked Efforts to Accelerate Parkinson's Solutions). LEAPS are multi-year, potentially multi-million dollar grant awards for collaborative, milestone-driven efforts to address questions that will have significant practical impact on the understanding and treatment of Parkinson's disease. LEAPS grants are goal-oriented and focused on answering a broad research question, one whose answer would represent a major, practical step forward in Parkinson's disease—a solution to a Parkinson's problem. Examples of such questions could be categorized under broad aims of: repairing the PD brain; slowing or halting the progression of PD; preventing or controlling levodopa-induced dyskinesias; and identifying the cause of PD. LEAPS are not intended to be conventional research grants. LEAPS investigators will be asked to interact regularly with foundation staff and advisors to discuss elements of the project, research directions, progress against stated milestones, and possible ways (other than through additional financial resources) the foundation may offer support or assistance. The RFP and preproposal cover sheet are available online.

Requirements: Applications may be submitted by U.S. and non-U.S. entities; public and private nonprofit entities, such as universities, colleges, hospitals, laboratories, units of state and local governments, and eligible agencies of the federal government; and for-profit entities.

Restrictions: Postdoctoral students or fellows are not eligible to apply as coordinating or principal investigators.

Date(s) Application is Due: Apr 29

Contact: Grants Administrator; (800) 708-7644

Internet: http://www.michaeljfox.org

Sponsor: Michael J. Fox Foundation for Parkinson's Research
P.O. Box 4777, Grand Central Station
New York, NY 10163

Michael J. Fox Foundation Parkinson's Cell Line Research Grants **2908**

Researchers are invited to submit investigator-initiated proposals using embryonic stem cells targeted at achieving optimized transplantation materials and protocols promoting graft survival, stability of dopaminergic phenotype, and long-term functional recovery in animal models of Parkinson's disease. The intent is to use the acquired knowledge to develop cell replacement therapy for PD patients. Work in animal or human cell lines is acceptable; however, a primary focus on human cell lines is required, and research on human embryonic stem cells is considered appropriate. Mandatory letters of intent are due August 9. Guidelines are available online.

Requirements: Investigators with scientific interests and demonstrated success in the development, culture, and manipulation of individual cell lines are invited to apply.

Date(s) Application is Due: Aug 9; Sep 20

Samples: Ernest Arenas of the Karolinska Institute (Stockholm Sweden), Jun Takahashi of the Kyoto U Graduate School of Medicine (Japan), Xuejun Wen of Clemson U (SC), and Su-Chung Zhang of the U of Wisconsin (Madison, WI)—for stem-cell research related to finding a cure for Parkinson's disease, $1 million divided.

Contact: Grants Administrator; (800) 708-7644; research@michaeljfox.org

Internet: http://www.michaeljfox.org/research/fundingopps.php?id=9

Sponsor: Michael J. Fox Foundation for Parkinson's Research
P.O. Box 4777, Grand Central Station
New York, NY 10163

Michael Reese Health Trust Grants **2909**

The trust awards grants in the metropolitan Chicago area to nonprofits for community-based health services provided to vulnerable groups, including children and youth, people with mental and physical disabilities, the elderly, immigrants and refugees, uninsured, medically indigent, and other underserved populations. Grants provide general and project support of innovative approaches to delivering health services to the target populations, efforts to improve community health through health promotion and disease prevention, and education to improve healthcare providers' ability to treat the underserved. Some research is funded, but emphasis is on direct service delivery. To apply, submit an inquiry letter. Deadlines listed are for letters of inquiry.

Requirements: Nonprofits operating in the metropolitan Chicago area are eligible.

Restrictions: Grants do not support capital needs (i.e., building construction/renovation, vehicles, and equipment), endowments, fund-raising events, individuals, scholarships or debt reduction.

Geographic Focus: Illinois

Date(s) Application is Due: Jun 15; Dec 15

Contact: Dorothy Gardner; (312) 726-1008; fax (312) 726-2797; programs@healthtrust.net

Internet: http://www.healthtrust.net/index.php?option=com_content&task=view&id=13&Itemid=42

Sponsor: Michael Reese Health Trust
20 N Wacker Drive, Suite 760
Chicago, IL 60606

Michigan Gateway Community Foundation Grants **2910**

The foundation awards grants to eligible Michigan nonprofit organizations in its areas of interest, including cultural programs, community development, education, and health care. Applicants may use the common grant application of the Council of Michigan Foundations. Applications should contain a copy of the IRS tax-exemption letter.

Requirements: Michigan 501(c)3 nonprofit organizations serving South Berrien and Cass Counties are eligible.

Geographic Focus: Michigan

Date(s) Application is Due: Feb 1; May 1; Aug 1; Nov 1

Amount of Grant: 100 - 20,000 U.S.D

Contact: Robert Habicht, President & CEO; (616) 695-3521; fax (616) 695-4250; mgcf@mgcf.org
Internet: http://www.mgcf.org/receive_a_grant.htm
Sponsor: Michigan Gateway Community Foundation
111 Days Avenue, P.O. Box 351
Buchanan, MI 49107

Michigan Psychoanalytic Institute and Society SATA Grants **2911**
The purpose of this grant is to partly sponsor one or more trainees in the mental health field in order for them to attend a national meeting of the American Psychoanalytic Association. The winter meeting of the American is held in New York City and the spring meeting is held in different cities. To apply for a partial scholarship, applicants should send a two-paragraph statement telling us about their background, interest in psychoanalytic thinking, and why he/she would like to attend a meeting of the American Psychoanalytic Association.
Contact: Kathleen Kunkel, Administrative Director; (248) 851-3380, ext. 2 or (248) 865-1164; fax (248) 851-1806
Internet: http://www.mpi-mps.org/joomla/index.php?option=com_content&task=view&id=32&Itemid=109
Sponsor: Michigan Psychoanalytic Institute and Society
32841 Middlebelt Road, Suite 411
Farmington Hills, MI 48334

Michigan Psychoanalytic Institute and Society Scholarships **2912**
The Continuing Education Division of the Michigan Psychoanalytic Institute offers two Programs, one in Ann Arbor and one in Farmington Hills. The Program is a one-year, clinically-based program which meets once weekly. The program is designed to introduce and further develop an understanding of psychoanalytic principles as applied to psychotherapy with adults, children, and adolescents. Participants meet as a class on a weekly basis with instructors from the faculty of the Michigan Psychoanalytic Institute. Each instructor teaches a 5-6 week segment. Instructors present clinical material to illustrate a psychoanalytic approach to clinical work. Discussion of the case material is utilized to examine psychoanalytic concepts and clinical technique as applied to psychoanalytic psychotherapy. Half of the clinical material presented during the year is from work with adult patients. The other half of the year, child analysts present and discuss clinical material from the treatment of children, adolescents, and their parents. The Scholarship application must be submitted by June 30.
Requirements: The Fellowship Programs are ideally suited for mental health clinicians who are in practice, recently graduated, or in training in their respective fields (psychiatry, psychology, social work, counseling, or nursing). It is also useful as preparation for clinicians who are considering more extensive training in psychoanalytic psychotherapy or psychoanalysis in the future.
Date(s) Application is Due: Jun 30
Contact: Jean Lewis, (248) 851-3380; fax (248) 851-1806; jlewis@mpi-mpi.org
Internet: http://www.mpi-mps.org/joomla/index.php?option=com_content&task=view&id=62&Itemid=216
Sponsor: Michigan Psychoanalytic Institute and Society
32841 Middlebelt Road, Suite 411
Farmington Hills, MI 48334

Michigan Society of Fellows Postdoctoral Fellowships **2913**
The society offers three-year postdoctoral fellowships to individuals selected for outstanding achievement, professional promise, and interdisciplinary interests in the humanities and arts, sciences, and professions. The fields of study include all departments within schools and colleges at the University of Michigan. Fellows are appointed as assistant professors in appropriate departments and are expected to teach for the equivalent of one academic year, to participate in the informal intellectual life of the society, and to devote time to their independent productive scholarship.
Date(s) Application is Due: Sep 30
Amount of Grant: 47,271 U.S.D
Contact: Michigan Society of Fellows Office; (734) 763-1259; society.of.fellows@umich.edu
Internet: http://www.rackham.umich.edu/Faculty/society.html
Sponsor: Michigan Society of Fellows
University of Michigan, 3572 Rackham Building, 915 E Washington Street
Ann Arbor, MI 48109-1070

Microsoft Research Cell Phone as a Platform for Healthcare Grants **2914**
Microsoft Research will support selected academic research with the goal of advancing the state-of-the-art in smart cell phones for health care applications. Successful projects must seek to increase the capacity of cell phone-based health care solutions in under-served rural and urban communities. They must also take into consideration the social context of application deployment. Topics to consider include: medical applications that are relevant, worldwide, for smart mobile phones (application and Web-enabled) in rural, and urban, communities; and appropriate services and infrastructures needed to provide affordable and accessible health care services. The total amount available under this request for proposals (RFP) is $1,000,000. Microsoft Research anticipates making approximately 10-12 awards averaging $80,000, with a maximum of $100,000 for any single award. Awards are made for the purpose of seed-funding larger initiatives, proofs of concept, or demonstrations of feasibility. It is important to understand that funding will continue after the first year only in exceptional circumstances, and that the principal investigators should therefore make every effort to leverage Microsoft Research funds as one component of a diverse funding base in a larger or longer-running project.
Requirements: All qualifying institutions are eligible without regard for geographic location. The proposing institution must be either: an accredited degree-granting college or university (or international equivalent) with non-profit status and awarding degrees at the baccalaureate level or above; or a research institution with non-profit status.
Restrictions: For all awards, payment of indirect costs (overhead) is not permitted.
Date(s) Application is Due: Oct 29
Amount of Grant: Up to 100,000 U.S.D
Contact: Rick Rashid, Senior Vice President; (800) 642-7676; fax (425) 936-7329; erpinq@microsoft.com
Internet: http://research.microsoft.com/en-us/um/redmond/about/collaboration/awards/cellphone-healthcare_awards.aspx
Sponsor: Microsoft Research
One Microsoft Way
Redmond, WA 98052-6399

Mid-Iowa Health Foundation Grants **2915**
The foundation is interested in supporting community-based health services in its geographic area of focus, comprising Iowa's Polk County and contiguous counties. A strong commitment exists toward programs involving children, elderly, disabled, and indigent populations, such as disadvantaged youth, isolated elderly, young mothers and their children, people with substance abuse, and chronically or terminally ill people. Types of support include general operating grants, capital campaigns, building construction/renovation, and matching funds. Request application forms by phone or mail.
Requirements: Organizations with nonprofit, tax-exempt status under IRS codes in Polk, IA, and contiguous counties are eligible.
Restrictions: The foundation does not provide grants for direct aid to individuals, religious programs, political organizations or projects, or dinners or tickets.
Geographic Focus: Iowa
Date(s) Application is Due: Apr 1; Oct 1
Amount of Grant: 15,000 - 40,000 U.S.D
Contact: Program Contact; (641) 277-6411; fax (515) 271-7579; info@midiowahealth.org
Internet: http://www.midiowahealth.org/grants.html
Sponsor: Mid-Iowa Health Foundation
550 39th Street, Suite 305
Des Moines, IA 50312

Middlesex Savings Charitable Foundation Capacity Building Grants **2916**
Since June 2000 the Middlesex Savings Charitable Foundation (MSCF) has provided more than $2 million in grants to over 150 non-profit organizations providing critical community services throughout Eastern and Central Massachusetts. Established as a nonprofit, private charitable foundation to further carry out the philanthropic mission of Middlesex Savings Bank, the Foundation aims to fund eligible organizations in the Bank's CRA market area in any economic environment. Middlesex Savings Bank initially provided $2 million on an endowment basis, and continues to add to the endowment annually. As of this writing, the endowment now totals $6.7 million. MSCF's Capacity Building Grants strengthen and increase the impact of local non-profits by improving their organizational capacity. Desired outcomes for non-profits selected to receive grants through this program include one or more of the following characteristics: improved governance and leadership; improved staff skills; improved management systems and practices; completed strategic plans; improved, expanded, or additional services; and expanded strategic assets, including financial and human resources. Successful applicants will be able to describe how their enhanced capacity will ultimately benefit the communities that they serve. Grant requests of up to $15,000 will be considered. Guidelines are available at the grant website. Applicants must apply online.
Requirements: Eastern Massachusetts 501(c)3 tax-exempt organizations serving one or more communities served by Middlesex Savings Bank, including Acton, Ashland, Ayer, Bedford, Bellingham, Berlin, Bolton, Boxborough, Carlisle, Chelmsford, Concord, Dover, Dunstable, Framingham, Franklin, Groton, Harvard, Holliston, Hopedale, Hopkinton, Hudson, Lexington, Lincoln, Littleton, Marlborough, Maynard, Medfield, Medway, Mendon, Milford, Millis, Natick, Needham, Newton, Norfolk, Northborough, Pepperell, Sherborn, Shirley, Southborough, Stow, Sudbury, Townsend, Tyngsborough, Upton, Walpole, Waltham, Wayland, Wellesley, Westborough, Westford, and Weston, are eligible. Projects for which support is requested should benefit people who live or work in the region. Given the continuing economic challenges faced by MSCF communities, the Board will be focusing upon, and giving preference to, grant submissions from organizations providing basic human services such as food and shelter.
Restrictions: The foundation will not fund political or sectarian activities.
Geographic Focus: Massachusetts
Date(s) Application is Due: Mar 1; Aug 1
Amount of Grant: Up to 15,000 U.S.D
Samples: Foundation for MetroWest, Natick, Massachusetts, $10,000 - support for the foundation's capacity-building training program which includes seminars for non-profits and non-profit boards, as well as special sessions for non-profit board chairs (2011); South Middlesex Opportunity Council, Framingham, Massachusetts, $10,000 - support for increased capacity of adult education programs at the MetroWest Career Center, and specifically for its GED program (2011); The Discovery Museum, Acton, Massachusetts, $5,000 - support for the Museum to make needed investment in its fundraising infrastructure as it prepares to launch a capital campaign (2011).
Contact: Dana Neshe, (508) 315-5360
Internet: http://www.middlesexbank.com/community/charitablefoundation/default.asp
Sponsor: Middlesex Savings Charitable Foundation
6 Main Street, P.O. Box 358
Natick, MA 01760

Middlesex Savings Charitable Foundation Community Development Grants **2917**
Since June 2000 the Middlesex Savings Charitable Foundation (MSCF) has provided more than $2 million in grants to over 150 non-profit organizations providing critical community services throughout Eastern and Central Massachusetts. Established as a nonprofit, private charitable foundation to further carry out the philanthropic mission of Middlesex Savings Bank, the foundation aims to fund eligible organizations in the Bank's CRA market area in any economic environment. Middlesex Savings Bank initially provided $2 million on an endowment basis, and continues to add to the endowment annually. As of this writing, the endowment now totals $6.7 million. MSCF's Community Development Grants support projects and programs that have a primary goal of improving the lives of low- and moderate-income families and individuals throughout the foundation's communities. Given the continuing economic challenges faced by MSCF communities, the foundation will focus upon, and give preference to, grant submissions from organizations providing basic human services such as food and shelter. Grant requests of up to $15,000 will be considered. Guidelines are available at the grant website. Applicants must apply online.
Requirements: Eastern Massachusetts 501(c)3 tax-exempt organizations serving one or more communities served by Middlesex Savings Bank, including Acton, Ashland, Ayer, Bedford, Bellingham, Berlin, Bolton, Boxborough, Carlisle, Chelmsford, Concord, Dover, Dunstable, Framingham, Franklin, Groton, Harvard, Holliston, Hopedale, Hopkinton, Hudson, Lexington, Lincoln, Littleton, Marlborough, Maynard, Medfield, Medway, Mendon, Milford, Millis, Natick, Needham, Newton, Norfolk, Northborough, Pepperell, Sherborn, Shirley, Southborough, Stow, Sudbury, Townsend, Tyngsborough, Upton, Walpole, Waltham, Wayland, Wellesley, Westborough, Westford, and Weston, are eligible. Projects for which support is requested should benefit people who live or work in the region.
Restrictions: The foundation will not fund political or sectarian activities.
Geographic Focus: Massachusetts
Date(s) Application is Due: Mar 1; Aug 1
Amount of Grant: Up to 15,000 U.S.D
Samples: Community Farms Outreach, Waltham, Massachusetts, $5,000 - support for the Fresh Food Access Program which grows over forty varieties of vegetables on eleven acres and donates tens of thousands of pounds annually to local partner agencies, including local food banks and pantries, as well as to local farmers' markets which are accessible to low income residents (2011); MetroWest Legal Services, Framingham, Massachusetts, $15,000 - support for Domestic Violence Project (2011); Natick Service Council, Natick, Massachusetts, $10,000 - support for food pantry (2011).
Contact: Dana Neshe, President; (508) 315-5360
Internet: http://www.middlesexbank.com/community/charitablefoundation/default.asp
Sponsor: Middlesex Savings Charitable Foundation
6 Main Street, P.O. Box 358
Natick, MA 01760

Miguel Aleman Foundation Grants **2918**
The foundation operates in Mexico in the fields of conservation and the environment, the arts and humanities, economic development, education, international affairs, medicine and health, science and technology, and social welfare. Types of support include research, grants to organizations, and prizes.
Contact: Fernando Castro y Castro, Director; 52-5-250-65-76; fax 52-5-250-10-43; 103503.1144@compuserve.com
Internet: http://www.miguelaleman.org.mx
Sponsor: Miguel Aleman Foundation
Ruben Dario 187
Col Chapultepec Morales, 11570 Mexico

Milagro Foundation Grants **2919**
Started by musician Carlos Santana and his wife Deborah, the foundation supports educational efforts that help youths live healthy, literate, and culturally enriched lives. Priority is helping at-risk and vulnerable populations acquire the skills to succeed in life. The foundation is especially interested in arts education. Unsolicited completed applications are not accepted. Preliminary letters of inquiry are accepted at any time. The board meets in February, June, and October.
Requirements: Nonprofit organizations are eligible. In order of priority, the foundation has an interest in helping children in San Rafael, the remainder of California, the United States, and other nations in which Carlos Santana performs.
Amount of Grant: 2,500 - 5,000 U.S.D
Samples: Friends of School of the Arts Foundation (San Francisco, CA)—to repair and replace instruments for the school orchestra and jazz bands at this public high school that serves artistic and creatively talented youths, $15,000; Huckleberry Youth Programs (San Rafael, CA)—to provide teenagers with health-education services, including information n HIV prevention, drug and alcohol abuse, and pregnancy prevention, $10,000; Jeanette Prandi Children's Ctr (San Rafael, CA)—to provide medical services to child victims of sexual assault, $12,000.
Contact: Grants Administrator; (415) 460-9939; fax (415) 460-6802; info@milagrofoundation.org or apply@milagrofoundation.org
Internet: http://www.milagrofoundation.org/apply.asp
Sponsor: Milagro Foundation
P.O. Box 9125
San Rafael, CA 94912-9125

Miles of Hope Breast Cancer Foundation Grants **2920**
The Fund provides programs and support services for women and families in the Hudson Valley, New York, affected by breast cancer. All funds raised for the Foundation are used in the Hudson Valley. The Miles of Hope Breast Cancer Foundation is currently offering grants for projects in the areas of breast health and breast cancer education, outreach, screening, treatment and support projects.
Requirements: Services must be provided in the eight counties served by the Miles of Hope Breast Cancer Foundation, which include: Columbia, Dutchess, Greene, Orange, Putnam, Rockland, Westchester and Ulster.
Restrictions: Project must be specific to breast health and/or breast cancer. Applicants must be a U.S. nonprofit (federally tax-exempt) organization. Nonprofit organizations, known educational institutions (i.e. college or university), government agencies, and Indian tribes are eligible. Equipment costs, if applicable, may not exceed 30% of direct costs and should be used exclusively on this project. Salaries, if requested, are for personnel related to this project only and not the general work of the employee. Funds will not be awarded to capital campaigns
Geographic Focus: New York
Contact: Program Administrator; (845) 527-6884 or (845) 264-2005; info@milesofhopebcf.org
Internet: http://www.milesofhopebcf.org/funds.html#grants
Sponsor: Miles of Hope Breast Cancer Foundation
P.O. Box 405
La Grangeville, NY 12540

Milheim Foundation Cancer Research Grants **2921**
Project grants provide support for research work for the prevention, treatment, and cure of cancer. In addition to a written proposal and completed application form, at least one letter of recommendation and one curriculum vita must be submitted. The grants are awarded for one year; renewal is possible but not encouraged.
Requirements: Organizations that are incorporated as nonprofit and operate within the continental limits of the United States are eligible. PI must hold MD, PhD, or the equivalent.
Restrictions: Grants do not support travel or presentation of papers, overhead expenses or support and maintenance of hospitals, salaries of academic personnel, or individuals.
Date(s) Application is Due: Mar 15
Amount of Grant: 10,000 - 22,000 U.S.D
Contact: Barbara Cole, c/o U.S. Bank PFS/Trust; (303) 316-5945 or (303) 316-5948; fax (303) 388-9387; jnelson@foundation.sdsu.edu
Sponsor: Milheim Foundation for Cancer Research
200 University Boulevard
Denver, CO 80206

Military Ex-Prisoners of War Foundation Grants **2922**
The Military Ex-Prisoners of War Foundation was founded primarily to assist Military Ex-P.O.W Veterans, and to fund their National Educational Scholarship Program for qualified heirs. The mission of the Military Ex-Prisoners of War Foundation is to support educational programs designed to inform Americans about the P.O.W experience; offer scholarships to children and grandchildren of former P.O.Ws and other such activities as may be approved by the Foundation's board of directors.
Requirements: Children, grandchildren, and great-grandchildren of former prisoners-of-war (P.O.Ws) who served after December 7, 1941 are eligible to apply. Applicants must include the following: completed application (available at the Foundations website); short autobiographical statement (not to exceed one typed page); Transcript *Requirements:* copy of official transcripts of high school grades including SAT/ACT test scores, official transcripts of college grades; two letters of recommendation, one must be from a present/former teacher; short essay, limited to 1,000 words, on the impact of WWII on society today; statement about your grandparent's experience as a prisoner of war, 1,000 words or less. Completed applications should be mailed to: Dorris Livingstone, Recording Secretary, Military Ex-Prisoners of War Foundation, 1561 Glen Hollow Lane South, Dunedin, FL 34698. If you have any questions, email your inquiry to: Dorris2001@aol.com with Foundation Scholarship Application/Question typed into the Subject line.
Restrictions: No support for individuals not related to former prisoners of war, religious, political causes, or for organizations.
Date(s) Application is Due: Apr 1
Amount of Grant: 2,000 U.S.D
Contact: F. Paul Dallas; (910) 867-2775; fax (910) 867-0339; threatt273@aol.com
Internet: http://www.militarypowfoundation.org/072409-scholarships.html
Sponsor: Military Ex-Prisoners of War Foundation
916 Bingham Drive
Fayetteville, NC 27803

Milken Family Foundation Grants **2923**
The purpose of the Milken Family Foundation is to discover and advance inventive and effective ways of helping people help themselves and those around them lead productive and satisfying lives. The Foundation advances this mission primarily through its work in education and medical research. In education, the Foundation is committed to: strengthening the profession by recognizing and rewarding outstanding educators, and by expanding their professional leadership and policy influence attracting, developing, motivating and retaining the best talent to the teaching profession by means of comprehensive, whole school reform; stimulating creativity and productivity among young people and adults through programs that encourage learning as a lifelong process; and building vibrant communities by involving people of all ages in programs that contribute to the revitalization of their community and to the well-being of its residents. In medical research, the Foundation is committed to: advancing and supporting basic and applied medical research, especially in the areas of prostate cancer and epilepsy, and recognizing and rewarding outstanding scientists in these areas; and supporting basic health care programs to assure the well-being of community members of all ages. Applicants may request funding at any time.

Requirements: Grants are made to 501(c)3 tax-exempt organizations. Grant recipients must have the financial potential to sustain the program for which funding is sought following the period of Foundation support. Preventive programs with long-range goals receive the closest consideration. Applicants should submit a brief written statement that includes: description of project, goals, procedure and personnel; brief background of organization, including number of years in operation, other areas of activity, applicant's qualifications for support, annual operating budget, and previous and current sources of funding; and a letter of exemption from the Internal Revenue Service.
Restrictions: Grants are not made directly to individuals.
Samples: Brentwood School, Los Angeles, California, $12,500, educational programs; AIDS Project Los Angeles, Los Angeles, California, $250, medical research; and Center for Jewish Community Studies, Balitomore, Maryland, $70,000, educational programs.
Contact: Richard Sandler, Executive Vice President; (310) 570-4800; fax (310) 570-4801; admin@mff.org
Internet: http://www.mff.org/about/about.taf?page=funding
Sponsor: Milken Family Foundation
1250 Fourth Street, 6th Floor
Santa Monica, CA 90401-1353

Miller Foundation Grants **2924**
The Miller Foundation focuses on assisting local nonprofit, charitable organizations and governmental agencies with projects that provide the following for the Battle Creek, Michigan area: economic development; education; health service; human service; neighborhood improvement; arts and culture; recreation and tourism; and leadership. Nonprofit 501(c)3 organizations located in and working to improve the Battle Creek community are eligible.
Requirements: Organizations should submit a preliminary letter of request, briefly describing their project, its estimated cost, amount requested, and funding from other sources. After reviewing the initial letter, the Foundation staff, if appropriate, will send a formal grant application to the requesting organization.
Restrictions: The Foundation seldom funds an entire project but rather joins with others as they work to improve the quality of life in the Battle Creek community. The Foundation does not make grants to individuals or for continuing operating funds of nonprofit organizations.
Geographic Focus: Michigan
Date(s) Application is Due: Jan 1; Mar 1; May 1; Jul 1; Sep 1; Nov 1
Contact: Diane Thompson, President and CEO; (269) 964-3542; fax (269) 964-8455
Internet: http://themillerfoundation.com/grants.htm
Sponsor: Miller Foundation
310 WahWahTaySee Way
Battle Creek, MI 49015

Millipore/Charles P. Schaufus Parenteral-Processing Technology Research Grant **2925**
The grant is intended to stimulate research in parenteral processing through the support of a graduate student. Parenteral processing includes the investigation of aspects in bulk manufacture of parenteral drugs, such as fermentation and scale-up; intermediate steps, such as crystallization and purification; equipment surfaces, such as their contribution to particulate matter and their resistance to corrosion; and computer-assisted manufacture, such as software and its validation. Emphasis of this research is to be on parenteral processing. The grant is for up to three years and includes cash and equipment each year. Applications are available on the Web site.
Requirements: Applicants must be associated with U.S. 501(c)3 and 509(a)(1)(2) or (3) tax-exempt organizations.
Date(s) Application is Due: Sep 15
Amount of Grant: 22,500 U.S.D
Samples: Dr. Bhogi Sheth, U of Tennessee (Memphis, TN)—three-year parenteral research grant recipient, $10,000 in cash and $10,000 in Millipore equipment.
Contact: Deanna Morton, (516) 829-5501; fax (516) 829-1008; DeeMorton@aol.com
Internet: http://www.pdafoundation.org/grant_offerings.html
Sponsor: Parenteral Drug Association Foundation for Pharmaceutical Sciences
111 Great Neck Road, Suite 308
Great Neck, NY 11021

Millipore Foundation Grants **2926**
The foundation supports nonprofit organizations in company operating communities in five main areas: education and research, social services, culture, health care, and public policy. Types of support include general operating support, employee matching gifts and employee-related scholarships. Applications are accepted throughout the year.
Requirements: 501(c)3 tax-exempt organizations are eligible.
Amount of Grant: 2,500 - 5,000 U.S.D
Contact: Charleen Johnson, Executive Director; (978) 715-1268; fax (978) 715-1385; Charleen_Johnson@millipore.com
Internet: http://www.millipore.com/corporate/milliporefoundation.nsf/foundationhome
Sponsor: Millipore Foundation
290 Concord Road
Billerica, MA 01821-7037

Milton Wexler Postdoctoral Fellowships **2927**
Support will be for research projects that contribute to identifying and understanding the basic defect of Huntington's disease. Areas of interest include trinucleotide expansions, animal models, gene therapy, neurobiology and development of the basal ganglia, cell survival and death, and intercellular signaling in striatal neurons. Awards may be renewed for a third year. To receive an application, submit a letter of intent. Letter of intent forms may be submitted electronically at any time via the Web site.
Date(s) Application is Due: Feb 15; Jun 15; Oct 15
Amount of Grant: 57,000 U.S.D
Contact: Dr. Carl Johnson, (310) 458-4183; fax (310) 458-3937; carljohnson@hdfoundation.org
Internet: http://www.hdfoundation.org/fundinga.htm
Sponsor: Hereditary Disease Foundation
3960 Broadway, 6th Floor
New York, NY 10032

Milwaukee Foundation Nonprofit Management Fund Grants **2928**
The Milwaukee Foundation, Faye McBeath Foundation, and Helen Bader Foundation have teamed up to make grants to small- and medium-sized nonprofit groups strictly for management improvement. Examples of eligible activities include strengthening boards, developing cost-sharing approaches, making long-range organization plans, improving financial management, or strengthening MIS systems. Types of support include capital campaigns, building/renovation, equipment acquisition, program development, conferences and seminars, seed grants, scholarship funds, technical assistance, and matching funds. Organizations applying for funds must state the results expected from funding and how it will improve services in the long run. Deadlines listed are for Waukesha County.
Requirements: The fund is available to nonprofits in Wisconsin that have been in operation for at least two years.
Restrictions: Grants are not awarded for the general uses of churches or for sectarian religious purposes, specific medical or scientific projects, operating budgets, continuing support, annual campaigns, endowment funds, deficit financing, or to individuals.
Geographic Focus: Wisconsin
Date(s) Application is Due: Feb 1; Aug 1
Samples: All Saints Catholic Parish—for a board planning retreat to streamline administration of community programs, $1600; Counseling Ctr of Milwaukee—for strategic planning, $5000; Hmong American Women's Assoc—to support merger negotiations with Hmong Educational Advancement, $7000; Martin Luther King Economic Development Corp—for development of a three-year plan, $5000.
Contact: Patricia Wyzbinski, Fund Advisor; (414) 271-4869
Internet: http://www.nonprofitmanagementfund.org/enter.html
Sponsor: Milwaukee Foundation
1020 N Broadway, Suite 112
Milwaukee, WI 53202

Mimi and Peter Haas Fund Grants **2929**
The Mimi and Peter Haas Fund supports early childhood development. Their primary focus is for activities that provide San Francisco's young (ages 2-5), low-income children and their families with access to high-quality early childhood programs that are part of a comprehensive, coordinated system. The Fund recognizes the importance of connecting the work of its direct service grants to the ongoing discussions of public policy and seek specific opportunities to collaborate with organizations to improve early childhood settings. The Fund will also continue trustee-initiated grantmaking to arts, education, public affairs, and health and human services organizations. Applicants should contact the trustee office to begin the application process. There are no particular application forms or deadlines with which to adhere.
Geographic Focus: California
Amount of Grant: Up to U.S.D
Contact: Lynn Merz, Executive Director; (415) 296-9249; fax (415) 296-8842; mphf@mphf.org
Sponsor: Mimi and Peter Haas Fund
201 Filbert Street, 5th Floor
San Francisco, CA 94133-3238

Minneapolis Foundation Community Grants **2930**
The foundation awards grants throughout Minnesota. Eligible activities include policy and systems change work in the following areas: affordable housing; economic opportunity; educational achievement; and the health and well-being of children, youth, and families. The foundation also focuses on systems and policy work that addresses the intersection of issues, such as housing and health. Types of support include program/project support, operating support, capital support (limited to the seven-county metro area); and some multiyear grants. Proposals are accepted throughout the year. Guidelines are available online.
Requirements: 501(c)3 nonprofit organizations located in the seven-county metropolitan area of Minneapolis and Saint Paul may apply.
Restrictions: The foundation does not fund individuals, organizations/activities outside of Minnesota, conference registration fees, memberships, direct religious activities, political organizations or candidates, direct fundraising activities, telephone solicitations, courtesy advertising, or financial deficits.
Geographic Focus: Minnesota
Date(s) Application is Due: Mar 15; Sep 15
Contact: Paul Verrette, Senior Grants Administrator; (612) 672-3836; pverrette@mplsfoundation.org
Internet: http://www.mplsfoundation.org/grants/guidelines.htm
Sponsor: Minneapolis Foundation
800 IDS Center, 80 South Eighth Street
Minneapolis, MN 55402

Minnie Pearl Cancer Foundation Grants 2931
Nonprofit organizations in Tennessee are eligible for grants to support programs in cancer research, education, and patient and family services. The foundation funds programs that offer the greatest benefit to cancer patients and their families.
Requirements: Nonprofits in Tennessee are eligible.
Restrictions: Grants are not normally funded for annual appeals or membership drives, capital campaigns, endowment funds, operating expenses of established organizations or debt retirement, or grants to individuals.
Geographic Focus: Tennessee
Amount of Grant: Up to 25,000 U.S.D
Samples: Cool Springs YMCA (TN)—for a program to help women after breast cancer treatment, $5000; Vanderbilt Ctr for Health Services (TN)—to support the Student Health Coalition, $10,000; Wellness Community East Tennessee (TN)—to develop its Newly Diagnosed Breast Cancer Support Group, $7500.
Contact: Patrick Clemens, (615) 467-1936; fax (615) 342-1745; pat@minniepearl.org
Internet: http://www.minniepearl.org/pages/grants.html
Sponsor: Minnie Pearl Cancer Foundation
2410 Patterson Street, Suite 110
Nashville, TN 37203

Minority Doctoral Fellowships in Science and Engineering 2932
The university accepts applications from members of minority groups for doctoral fellowships in the life sciences, physical sciences, and engineering. Contact the program director for detailed information on application and amounts of fellowships offered. Annual deadline dates may vary; contact program staff for exact dates. Contact program staff for program availability.
Requirements: Citizens or permanent residents of the United States who are African American, Hispanic, or Native American may apply. The recipient must have an earned baccalaurate degree and be accepted for admission to a doctoral program in the fields of life sciences, physical sciences, engineering, or mathematics at a participating Illinois institution. Upon completion, recipients must agree to seek and accept a full-time faculty appointment at an Illinois institution of higher learning for a period equal to the number of years of the award.
Date(s) Application is Due: Jan 23
Amount of Grant: 1,100 - 13,200 U.S.D
Contact: Karen Jennings, (618) 453-4555; fax (618) 453-4562; kjenning@siu.edu or gaoffice@siu.edu
Internet: http://www.siu.edu/gradschl/doctoral_fellowship.htm
Sponsor: Southern Illinois University at Carbondale
Graduate School, Woody Hall B130
Carbondale, IL 62901-4716

Missouri Foundation for Health Grants 2933
The foundation makes grants within five priority areas: elimination of health disparities; promotion of improved access to care; support for disease prevention and health promotion; strengthening the health care safety net; and enhancing the health of children. Proposals also are accepted for activities that improve the health of the people in the foundation's service region, but are outside the scope of the five priority areas. All applicants are required to complete: a preliminary proposal application, a preliminary proposal summary, and a budget sheet (forms available online). Detailed instructions, including deadline dates, will be provided to all applicants invited to submit a full proposal.
Requirements: 501(c)3 nonprofit organizations and organizations applying through a 501(c)3 fiscal sponsor are eligible.
Restrictions: The foundation will not fund ongoing general operating expenses of established programs; unspecified capital campaigns; annual appeals or other fundraising events; existing deficits incurred outside any foundation-funded project; restoration of funding cuts by government or other organizations; endowment building or development campaigns; basic biomedical research; research on drug therapies or devices; direct support of an individual's medical care; direct support of an individual's education or training; lobbying of any kind directly supported by foundation grant funds; purchase of health insurance for individuals or groups (other than as a part of employee fringes on approved projects); religious purposes; use of foundation funds to expand existing public insurance programs; or acquisition of real estate.
Geographic Focus: Missouri
Samples: Area Resources for Community and Human Services (Saint Louis, MO)—to provide eye exams and glasses free of charge to adults, $134,883; U of Missouri at Columbia (MO)—to expand a program that recruits and trains medical students to work in rural settings, $954,598; Food Outreach (Saint Louis, MO)—for a meal program for adults and children with HIV/AIDS, $150,000; National Alliance for the Mentally Ill of Southwest Missouri (Springfield, MO)—to help low-income individuals obtain prescriptions following hospitalization, $35,992 over two years.
Contact: Grants Administrator; (800) 655-5560 or (314) 345-5500; fax (314) 345-5599; info@mffh.org
Internet: http://www.mffh.org/for_grantees.html
Sponsor: Missouri Foundation for Health
1000 St. Louis Union Station, Grand Central Building, Suite 400
Saint Louis, MO 63103

Mitchell Kapor Foundation Grants 2934
The foundation accepts applications for grants in its environmental health program and the level playing field program. The environmental health program comprises healthcare without harm, the international treaty process to eliminate 12 highly toxic chemicals, and the international environmental health network. The level playing field program focuses on understanding and changing the ways in which attitudes, subcultures, and processes within educational and business environments operate in tandem to create barriers to full participation for disenfranchised groups based on race, ethnicity, culture, gender, sexual orientation, and other factors. Direct inquiries and applications related to the level playing field program to Institute for Inclusive Work Environments; 543 Howard Street, 5th Fl; San Francisco, CA 94105; (415) 946-3030, fax: (415) 946-3001.
Restrictions: The foundation rarely funds organizations that do not have 501(c)3 status or a similarly qualified fiscal sponsor; projects outside the United States that do not include a known internationally credible evaluation component; academic research not related to or supportive of other ongoing foundation interests; or civic programs that do not address the foundation's interest areas.
Amount of Grant: 5,000 - 25,000 U.S.D
Contact: Grants Administrator; (415) 561-2182; info@jaf.org
Internet: http://www.mkf.org
Sponsor: Mitchell Kapor Foundation
P.O. Box 29209
San Francisco, CA 94129

MLA Cunningham Memorial International Fellowship 2935
The Medical Library Association (MLA) Cunningham Memorial International Fellowship was established in 1967 with a bequest from the estate of Eileen R. Cunningham. The fellowship supports the education and training of a health-science librarian from a country outside the United States or Canada and includes a stipend plus approved travel for four months in either country. The fellowship also pays for the cost of attending MLA's Annual Meeting. Guidelines and application materials are available at the website. All materials and letters of support must be received by MLA by December 1. Founded in 1898, the Medical Library Association (MLA) is a nonprofit, educational organization of more than 1,100 institutions and 3,600 individual members in the health sciences information field, committed to educating health-information professionals, supporting health-information research, promoting access to the world's health-sciences information, and working to ensure that the best health information is available to all.
Requirements: Candidates must have both an undergraduate degree and a master's-level library degree (latter requirement may be waived) and must either be working in or preparing to work in a medical library in their country. The fellowship recipient must agree to return to his or her country and work in a medical library for a period of two years. In addition, a satisfactory score must be achieved on the TOEFL English competency examination.
Restrictions: Fellowship is not available to U.S. or Canadian citizens. Travel arrangements to and from the United States or Canada must be paid by the fellow. Previous recipients of the Cunningham fellowship are ineligible.
Geographic Focus: All Countries
Date(s) Application is Due: Dec 1
Amount of Grant: 8,000 U.S.D
Contact: Carla Funk, Program Contact; (312) 419-9094, ext. 14; fax (312) 419-8950; mlapd2@mlahq.org or grants@mlahq.org
Internet: http://www.mlanet.org/awards/grants/cunningham.html
Sponsor: Medical Library Association
65 East Wacker Place, Suite 1900
Chicago, IL 60601-7246

MLA Estelle Brodman Academic Medical Librarian of the Year Award 2936
The Estelle Brodman Award was established in 1986 and is sponsored by a bequest of Irwin H. Pizer to honor Estelle Brodman's exemplary career as an educator, seminal thinker, able administrator, technological innovator, and skillful practitioner. The Brodman award recognizes an academic medical librarian at mid-career who demonstrates significant achievement, the potential for leadership, and continuing excellence. The recipient receives a certificate at the association's annual meeting and a cash award of $500 after the annual meeting. The recipient assumes all costs of attending the meeting and the ceremony at which the presentation is made. Guidelines and a nomination form are available at the website. All materials and letters of support must be received by MLA via email, fax, or mail (in order of preference) by November 1. The recipient will be notified in March before the annual meeting. Founded in 1898, the Medical Library Association (MLA) is a nonprofit, educational organization of more than 1,100 institutions and 3,600 individual members in the health sciences information field, committed to educating health-information professionals, supporting health-information research, promoting access to the world's health-sciences information, and working to ensure that the best health information is available to all.
Requirements: Nominees must satisfy the following criteria: be members of MLA; be an academic health-sciences librarian at the time of the award; be in "mid-career" (defined as working at least five but nor more than fifteen years in an academic health sciences library) at the time of the award; and have worked in an academic health-sciences library for each of the last five years immediately preceding the nomination. Nomination may be made for outstanding national or international contributions to academic health-sciences librarianship as demonstrated by excellence in performance (leadership), publications, research, service, or a combination of these four elements.
Restrictions: The nominee may not be a library director.
Geographic Focus: All States, All Countries
Date(s) Application is Due: Nov 1
Amount of Grant: 500 U.S.D
Contact: Carla Funk, Program Contact; (312) 419-9094, ext. 14; fax (312) 419-8950; mlapd2@mlahq.org or awards@mlahq.org
Internet: http://www.mlanet.org/awards/honors/brodman.html
Sponsor: Medical Library Association
65 East Wacker Place, Suite 1900
Chicago, IL 60601-7246

MLA Graduate Scholarship 2937

The Medical Library Association (MLA) annually awards a $5,000 scholarship to a student who shows excellence in scholarship and potential for accomplishment in health-sciences librarianship. The scholarship is announced at the annual meeting of the association, where the recipient will also receive a one-year student membership in the Medical Library Association and free inclusive student registration at the association's annual meeting. The recipient will assume the other costs of attending the meeting and the ceremony at which the presentation is made. Guidelines and application materials are available at the website. All materials and letters of support must be received by MLA via email, fax, or mail by December 1. email applications and documents will be accepted as PDF or MS Word files only. The recipient will be notified in March before the annual meeting. Founded in 1898, the Medical Library Association (MLA) is a nonprofit, educational organization of more than 1,100 institutions and 3,600 individual members in the health sciences information field, committed to educating health-information professionals, supporting health-information research, promoting access to the world's health-sciences information, and working to ensure that the best health information is available to all.

Requirements: To be eligible applicants must be either citizens or permanent residents of the United States or Canada. Additionally, applicants must either be entering or have completed no more than one-half of the academic requirements of a Masters program at an American Library Association (ALA)-accredited graduate school of library science at the time of the granting.

Restrictions: Past recipients of the MLA Scholarship or the MLA Scholarship for Minority Students are not eligible to apply.

Geographic Focus: All States, Canada

Date(s) Application is Due: Dec 1

Amount of Grant: Up to 5,000 U.S.D

Contact: Carla Funk, Program Contact; (312) 419-9094, ext. 14; fax (312) 419-8950; mlapd2@mlahq.org or grants@mlahq.org

Internet: http://www.mlanet.org/awards/grants/scholar.html

Sponsor: Medical Library Association

65 East Wacker Place, Suite 1900

Chicago, IL 60601-7246

MLA Graduate Scholarship for Minority Students 2938

The Medical Library Association (MLA) annually awards a $5,000 scholarship to a student who shows excellence in scholarship and potential for accomplishment in health-sciences librarianship. The scholarship is announced at the annual meeting of the association, where the recipient will also receive a one-year student membership in the Medical Library Association and free inclusive student registration at the association's annual meeting. The recipient will assume the other costs of attending the meeting and the ceremony at which the presentation is made. Guidelines and application materials are available at the website. All materials and letters of support must be received by MLA via email, fax, or mail (in order of preference) by December 1. email applications and documents will be accepted as PDF or MS Word files only. The recipient will be notified in March before the annual meeting. Founded in 1898, the Medical Library Association (MLA) is a nonprofit, educational organization of more than 1,100 institutions and 3,600 individual members in the health sciences information field, committed to educating health-information professionals, supporting health-information research, promoting access to the world's health-sciences information, and working to ensure that the best health information is available to all.

Requirements: To be eligible the applicant must be a member of a minority group (Black or African-American, Hispanic or Latino, Asian, Aboriginal, North-American Indian or Alaskan Native, or Native Hawaiian or other Pacific Islander) and be a citizen or permanent resident of the United States or Canada. Additionally, the applicant must either be entering or have completed no more than one-half of the academic requirements of a Masters program at an American Library Association (ALA)-accredited graduate school of library science at the time of the granting.

Restrictions: Past recipients of the MLA Scholarship or the MLA Scholarship for Minority Students are not eligible to apply.

Geographic Focus: All States, Canada

Date(s) Application is Due: Dec 1

Amount of Grant: Up to 5,000 U.S.D

Contact: Carla Funk, Program Contact; (312) 419-9094, ext. 14; fax (312) 419-8950; mlapd2@mlahq.org or grants@mlahq.org

Internet: http://www.mlanet.org/awards/grants/minstud.html

Sponsor: Medical Library Association

65 East Wacker Place, Suite 1900

Chicago, IL 60601-7246

MLA Ida and George Eliot Prize 2939

The Ida and George Eliot Prize was established by friends of the Medical Library Association (MLA) Ida and George Eliot, owners of Eliot Health Sciences Books, Inc., Long Island City, New York. The award is presented annually for a work published in the preceding calendar year and which has been judged most effective in furthering medical librarianship. The Eliots' company was acquired in the mid-1980's by Login Brothers Book Company which continued to support the award. Currently the award is supported by MLA. The award recipient receives a certificate at the association's annual meeting and a cash award of $200 after the annual meeting. The recipient assumes all costs of attending the meeting and the ceremony at which the presentation is made. Guidelines and a nomination form are available at the website. All materials and letters of support must be received by MLA via email, fax, or mail (in order of preference) by November 1. The recipient will be notified in March before the annual meeting. Founded in 1898, the Medical Library Association (MLA) is a nonprofit, educational organization of more than 1,100 institutions and 3,600 individual members in the health sciences information field, committed to educating health-information professionals, supporting health-information research, promoting access to the world's health-sciences information, and working to ensure that the best health information is available to all.

Requirements: Nominations are accepted from the membership-at-large and from members of the Ida and George Eliot Prize Jury; the jury reviews publications other than those nominated and actively searches for publications, beginning the review process well before nominations may be received. It is preferred that nominees be members of MLA.

Geographic Focus: All States, All Countries

Date(s) Application is Due: Nov 1

Contact: Carla Funk, Program Contact; (312) 419-9094, ext. 14; fax (312) 419-8950; mlapd2@mlahq.org or awards@mlahq.org

Internet: http://www.mlanet.org/awards/honors/eliot.html

Sponsor: Medical Library Association

65 East Wacker Place, Suite 1900

Chicago, IL 60601-7246

MLA Janet Doe Lectureship Award 2940

The Janet Doe Lectureship was established in 1966 by an anonymous donation to support an annual lecture in honor of Janet Doe (1895-1985), former librarian of the New York Academy of Medicine, historical scholar, past president of the Medical Library Association (MLA), and editor of the first two editions of the Medical Library Practice handbook. The Janet Doe Lecturer is chosen for his or her unique perspective on the history or philosophy of medical librarianship. The lecturer receives a certificate at the annual meeting, a $250 honorarium, reimbursement for travel expenses to the site of the meeting, hotel for one night, and per diem for one day. Guidelines and a nomination form are available at the website. All materials must be received by MLA via email, fax, or mail (in order of preference) by November 1. The recipient will be notified in March, fourteen months before the lecture is to be given. Founded in 1898, the Medical Library Association (MLA) is a nonprofit, educational organization of more than 1,100 institutions and 3,600 individual members in the health sciences information field, committed to educating health-information professionals, supporting health-information research, promoting access to the world's health-sciences information, and working to ensure that the best health information is available to all.

Requirements: Nominations are accepted from the MLA membership at large and from the members of the Janet Doe Lectureship Jury. Nominees should have been active in the profession a sufficient number of years to have acquired a broad perspective of medical librarianship and possess the qualities to communicate their experiences and ideas articulately. Nominees should also be regular members of the MLA and have made a substantial contribution to both the profession and to the MLA's work in the following areas: leadership; development of or sustained work within a particular area within the association; dedicated service to medicine and the allied health sciences; contributions to research or scholarship; effectiveness in administration/management; contributions to technology application and library practice; and/or effectiveness in teaching and training.

Geographic Focus: All States, All Countries

Date(s) Application is Due: Nov 1

Contact: Carla Funk, Program Contact; (312) 419-9094, ext. 14; fax (312) 419-8950; mlapd2@mlahq.org or awards@mlahq.org

Internet: http://www.mlanet.org/awards/honors/doe.html

Sponsor: Medical Library Association

65 East Wacker Place, Suite 1900

Chicago, IL 60601-7246

MLA Lois Ann Colaianni Award for Excellence and Achievement in Hospital Librarianship 2941

The Lois Ann Colaianni Award for Excellence and Achievement in Hospital Librarianship was established in 1989 to recognize a hospital librarian who has made significant contributions to the profession through overall distinction (as demonstrated by excellence and achievement) in service, advocacy, leadership, publications, presentations, teaching, research, technology, administration, special projects, or any combination of these areas. The award was renamed in 1999 to honor Ms. Colaianni, former Associate Director of the National Library of Medicine. The award recipient receives a certificate at the association's annual meeting and a cash award of $500 after the annual meeting. The recipient assumes all costs of attending the meeting and the ceremony at which the presentation is made. Guidelines and a nomination form are available at the website. In order for the nomination to be considered, all materials and letters of support must be received by MLA via email, fax, or mail (in order of preference) by November 1. The recipient will be notified in March before the annual meeting. Founded in 1898, the Medical Library Association (MLA) is a nonprofit, educational organization of more than 1,100 institutions and 3,600 individual members in the health sciences information field, committed to educating health-information professionals, supporting health-information research, promoting access to the world's health-sciences information, and working to ensure that the best health information is available to all.

Requirements: It is preferred that nominees be members of MLA. Nominees must be hospital librarians at the time of the award and must have worked in a hospital library for at least five (5) years immediately preceding the award.

Geographic Focus: All States, All Countries

Date(s) Application is Due: Nov 1

Contact: Carla Funk, Program Contact; (312) 419-9094, ext. 14; fax (312) 419-8950; mlapd2@mlahq.org or awards@mlahq.org

Internet: http://www.mlanet.org/awards/honors/colaianni.html

Sponsor: Medical Library Association

65 East Wacker Place, Suite 1900

Chicago, IL 60601-7246

MLA Murray Gottlieb Prize Essay Award 2942
The Murray Gottlieb Prize was established in 1956 by Ralph and Jo Grimes of the Old Hickory Bookshop (Brinklow, Maryland) in memory of Murray Gottlieb, a New York antiquarian book dealer. The purpose of the award is to stimulate the health-sciences librarian's interest in the history of medicine. Currently the award is sponsored by the Medical Library Association (MLA) History of the Health Sciences Section and is awarded annually for the best unpublished essay on the history of medicine and allied sciences written by a health-sciences librarian. The author of the winning essay receives complimentary registration to the annual meeting, a certificate at the association's annual meeting, and a cash award of $100 afterwards. The recipient assumes all costs of attending the meeting and the ceremony at which the presentation is made. Guidelines and a submission form are available at the website. All materials must be received by MLA via email, fax, or mail (in order of preference) by November 1. The recipient will be notified in March before the annual meeting. Founded in 1898, the Medical Library Association is a nonprofit, educational organization of more than 1,100 institutions and 3,600 individual members in the health sciences information field, committed to educating health-information professionals, supporting health-information research, promoting access to the world's health-sciences information, and working to ensure that the best health information is available to all.
Requirements: Authors of papers submitted must be health-sciences librarians. The submitted paper must treat some aspect of the history of medicine or allied sciences. Entries will be judged according to quality of bibliographic research, quality of the experimental design or arguments developed to support a particular hypothesis, contribution to the study of the history of the health sciences, originality, and style (clarity, appearance, conciseness, and structure).
Restrictions: Papers may be under consideration for publication at the time of submission but cannot have been published.
Geographic Focus: All States, All Countries
Date(s) Application is Due: Nov 1
Amount of Grant: 100 U.S.D
Samples: Lindsay E. Blake, "The Tuskegee Syphilis Study: Medical Research versus Human Rights" (2010); Trenton Boyd, "The Lost History of American Veterinary Medicine: the Need for Preservation"; Heidi Heilemann, "Envisioning the Unborn: Art, Anatomy and the Printing Press in the Modern Era".
Contact: Carla Funk, Program Contact; (312) 419-9094, ext. 14; fax (312) 419-8950; mlapd2@mlahq.org or awards@mlahq.org
Internet: http://www.mlanet.org/awards/honors/gottlieb.html
Sponsor: Medical Library Association
65 East Wacker Place, Suite 1900
Chicago, IL 60601-7246

MLA Rittenhouse Award 2943
The award is presented annually by MLA for the best unpublished paper on health sciences librarianship or medical informatics written by a student in an ALA-accredited program of library and information studies or a trainee in an internship in health sciences librarianship or medical informatics.
Restrictions: Papers that have been published or are under consideration for publication are not eligible.
Date(s) Application is Due: Nov 1
Amount of Grant: 500 U.S.D
Contact: Lisa Fried, Program Contact; (312) 419-9094, ext. 28; fax (312) 419-8950; mlapd2@mlahq.org
Internet: http://www.mlanet.org/awards/honors/ritten.html
Sponsor: Medical Library Association
65 East Wacker Place, Suite 1900
Chicago, IL 60601-7246

MLA Thomas Reuters / Frank Bradway Rogers Information Advancement Award 2944
The Thomson Reuters/Frank Bradway Rogers Information Advancement Award is presented annually by the Medical Library Association in recognition of outstanding contributions in the following areas: the use of technology to deliver health sciences information; the science of information; and facilitating the delivery of health-sciences information. Originally sponsored by the Institute for Scientific Information (ISI), the award was presented for the first time in 1983 to Frank Bradway Rogers, M.D., for whom it was subsequently renamed. The recipient receives a certificate at the association's annual meeting and a cash award of $500 after the annual meeting. Guidelines and a nomination form are available at the website. In order for the nomination to be considered, all materials and letters of support must be received by MLA via email, fax, or mail (in order of preference) by November 1. The recipient will be notified in March before the annual meeting. Founded in 1898, the Medical Library Association (MLA) is a nonprofit, educational organization of more than 1,100 institutions and 3,600 individual members in the health sciences information field, committed to educating health-information professionals, supporting health-information research, promoting access to the world's health-sciences information, and working to ensure that the best health information is available to all.
Requirements: It is preferred that an individual nominee be a member of MLA, or if a small group is nominated for a combined effort, that at least one member of the group be a member of MLA. Nominations should be for innovative contributions to improved information handling and use. Preference will be given to pioneering efforts (i.e., new theories or applications).
Geographic Focus: All States, All Countries
Date(s) Application is Due: Nov 1
Amount of Grant: 500 U.S.D
Contact: Carla Funk, Program Contact; (312) 419-9094, ext. 14; fax (312) 419-8950; mlapd2@mlahq.org or awards@mlahq.org
Internet: http://www.mlanet.org/awards/honors/rogers.html
Sponsor: Medical Library Association
65 East Wacker Place, Suite 1900
Chicago, IL 60601-7246

MMRF Research Grants 2945
The foundation's mission is to accelerate a cure for multiple myeloma. The program supports three categories of research grants. Fellows awards support researchers entering the field of multiple myeloma working under the direction of a research sponsor (March 1 deadline). Senior research awards support investigators with an interest in myeloma who have been working in cancer research for a minimum of five years (May 2 deadline). Collaborative programs grants support multidisciplinary research in multiple myeloma. Cohesive and synergistic grant proposals that involve several research projects across one or more scientific core laboratories are sought (preliminary deadline for submitting a Collaborative Program Grant application is June 15, final application deadline is September 15).
Amount of Grant: 50,000 - 1,500,000 U.S.D
Samples: Johns Hopkins U (Baltimore, MD)—for research on immunotherapy of myeloma at the Sidney Kimmel Comprehensive Cancer Ctr, $1.5 million.
Contact: Jill Shook, Grants Administrator; (203) 972-1250; fax (203) 972-1259; shookj@themmrf.org
Internet: http://www.multiplemyeloma.org/research
Sponsor: Multiple Myeloma Research Foundation
3 Forest Street
New Canaan, CT 06840

MMS and Alliance Charitable Foundation Grants 2946
The foundation gives priority to funding projects addressing the health, benefit, and welfare of the community. Such issues include violence prevention, access to care, adolescent health, and health awareness. The foundation currently administers three grant programs: community action— funding for community action projects that creatively address issues affecting the health and well-being of the community; care for the medically uninsured—support for physician-led volunteer initiatives to provide free care to uninsured patients and increase access to care for the medically underinsured; international health studies—support for medical students and house officers who are members of MMS to pursue international education particularly focusing on underserved populations; and homelessness prevention—to provide health and mental health related services to women and children at risk of becoming homeless. Application and guidelines are available online. Annual deadline dates may vary; contact program staff for exact dates.
Geographic Focus: Massachusetts
Samples: Women of Means (Wellesley, MA)—to provide health care to homeless, low-income, and uninsured people, $30,000; Cape Cod Free Clinic (Falmouth, MA)—to create culturally appropriate medical and case-management services for uninsured Brazilian residents of Cape Cod, $26,400; Duffy Health Ctr (Hyannis, MA)—to provide primary health-care services to low-income individuals, $20,000; Injury Free Coalition for Kids of Worcester, U of Massachusetts Medical School (Worcester, MA)—for salary support of a coordinator who will work to establish a strong, community-based coalition focusing on the prevention of childhood injuries, $20,000.
Contact: Jennifer Day, (781) 434-7044; jday@mms.org or foundation@mms.org
Internet: http://www.massmed.org/pages/foundation_apply.asp
Sponsor: Massachusetts Medical Society and Alliance Charitable Foundation
860 Winter Street
Waltham, MA 02451

MNI Jeanne Timmins Fellowships 2947
The institute offers annual postdoctoral fellowships to support research and study in neurology, neurosurgery, or neuroscience for a period of one year at the institute. One renewal is possible. An application form may be requested in writing. Successful applicants will be notified by letter.
Requirements: Candidates must have MD or PhD degrees; those with MDs will ordinarily have completed clinical studies in neurology or neurosurgery.
Date(s) Application is Due: Oct 31
Amount of Grant: Up to 25,000 CAD
Contact: Assistant Director; (514) 398-1903; fax (514) 398-8248; john@mni.lan.mcgill.ca
Internet: http://www.mni.mcgill.ca/MNIfellowships.html
Sponsor: Montreal Neurological Institute
3801 University Street, Room 636
Montreal, QC H3A 2B4 Canada

Mockingbird Foundation Grants 2948
The Mockingbird Foundation (Mockingbird) is a non-profit organization founded by Phish (the rock band) fans in 1996. Since then Mockingbird has distributed over $750,000 to support music-education programs for children. The foundation provides funding through its competitive grants, emergency-related grants, and tour-related grants. Emergency-related grants and tour-related grants are unsolicited grants. Emergency grants come from an Emergency Fund in which 3% of Mockingbird's gross revenues are designated for music-education programs affected by disasters (e.g., hurricanes and tornadoes). Tour-related grants support music-education programs in communities touched by Phish tours and are intended to inspire support for music and arts education and to generate positive press coverage. Competitive grants are awarded to schools and nonprofit organizations through a two-tiered grant-application process. Applicants must first complete an initial inquiry form at Mockingbird's website. If an applicant's project is selected for further consideration, the applicant will then be invited to submit a full and formal proposal. Initial inquiries may be submitted at any time.

Mockingbird typically reviews inquiries in August and September, invites full proposals by Halloween, and announces new grants sometime between Christmas and the end of January. Mockingbird encourages projects with diverse or unusual musical styles, genres, forms, and philosophies; projects with unconventional outlets and forms of instruction as well as instruction in unconventional forms; projects that foster creative expression in any musical form (including composition, instrumentation, vocalization, or improvisation); and projects in which skills, as outcomes, are less assessable or even irrelevant. Projects should be experiential and directly engage students with creating and expressing music. Preference is given to programs that benefit disenfranchised groups, including those with low skill levels, income, education, disabilities, or terminal illness, and/or those in foster-care homes, hospitals, and prisons. Prospective applicants should review the instructions and guidelines given at the website for complete details before making application.
Requirements: Schools and 501(c)3 nonprofit organizations in the United States are eligible. Applicants may apply through a sponsor who meets these qualifications. Mockingbird is particularly interested in organizations with low overhead, innovative approaches, and/or collaborative elements to their work. The foundation encourages geographic diversity and has funded forty-three states to date. Mockingbird is interested in targeting children 18 years or younger, but will consider projects that benefit college students, teachers/instructors, or adult students.
Restrictions: Grants are made on a one-time basis and are non-renewable and non-transferable. Mockingbird does not normally support individuals, fund-raising organizations or events, research, and programs that promote or engage in religious or political doctrine. It is hoped that applicants for Mockingbird grants hire staff and provide services without discriminating on the basis of race, religion, gender, sexual orientation, age, national origin, or disability. Mockingbird supports the provision of instruments, texts, and office materials, and the acquisition of learning space, practice space, performance space, and instructors/instruction. Mockingbird is particularly interested in projects that foster self-esteem and free expression, but does not fund music therapy which is neither education nor music appreciation which does not include participation.
Amount of Grant: 100 - 5,000 U.S.D
Samples: Connecticut Percussive Arts Society, Bridgeport, Connecticut, $5,000 - for scholarships to their Student to Student program (2012); Rural Music Corps at Ethos, Portland, Oregon, $5,000 - for instruments (2012); Jefferson Middle School, Albuquerque, New Mexico, $4,716 - for equipment for their Rock and Rhythm Band (2012).
Contact: Ellis Godard, Executive Director; ellis@mbird.org
Kristin Godard, grants@mbird.org
Internet: http://mbird.org/funding/guidelines/
Sponsor: Mockingbird Foundation
6948 Luther Circle
Moorpark, CA 93021-2569

Molly Bee Fund Grants **2949**
The fund awards grants to support nonprofits in its areas of interest, including arts and cultural programs, education, and hospitals. Types of support include general operating support, capital campaigns, building construction/renovation, equipment acquisition, endowment funds, and program-related investments. Giving is primarily in Palm Beach and West Palm Beach, FL, and Cleveland, OH. There are no application forms or deadlines.
Restrictions: Grants are not made to individuals.
Amount of Grant: 5,000 - 100,000 U.S.D
Contact: Thomas Allen, (440) 331-8220
Sponsor: Molly Bee Fund
20325 Center Ridge Road, Suite 629
Rocky River, OH 44116

Molson Companies Donations Fund Grants **2950**
The corporation supports Canadian charitable organizations in its efforts to meet the changing needs of Canadian society. Areas of interest include the visual and performing arts, historic preservation, literacy, universities and community colleges, education research, basic science research, hospitals, the environment, and community and social services. Funding decisions are made throughout the year. Applicants are requested to apply by letter and to include descriptive material outlining activities of the organization and reason for the request, copy of latest annual report and/or audited financial statements, and charitable registration number. Decisions are made throughout the year.
Requirements: Organizations must be registered and provide a valid Canadian charitable registration number to be eligible.
Geographic Focus: All States, Canada
Contact: Program Contact; (416) 975-1786; fax (416) 975-1981
Internet: http://www.molson.com/en/community/donation/index.php?sec=5&com=3&mdf=1
Sponsor: Molson Companies
33 Carlingview Drive
Toronto, ON M9W 5E4 Canada

Monfort Family Foundation Grants **2951**
The foundation awards general operating grants to Colorado nonprofit organizations in its areas of interest, including education, health and medical research, arts and culture, agriculture, and general charitable giving. Application forms must be obtained.
Requirements: Colorado nonprofits, primarily in Weld County, are eligible.
Restrictions: Grants are not made to individuals.
Geographic Focus: Colorado
Date(s) Application is Due: May 1; Oct 1
Amount of Grant: 5,000 - 100,000 U.S.D
Samples: U of Northern Colorado, College of Business Administration (Greeley, CO)—to bring nationally prominent business leaders to campus to teach; to create a technology center; and to establish an entrepreneurship institute, a lecture series, and an instruction-improvement fund, $10.5 million.
Contact: Dave Evans, Program Contact; (970) 454-1357
Sponsor: Monfort Family Foundation
Box 337300
Greeley, CO 80633

Montana Power Foundation Grants **2952**
The foundation focuses its grant making in Montana on education, health and human services, civic and community, culture and the arts, and resource conservation. Education support is awarded primarily through gifts to university system foundations, scholarship programs, and employee matching gifts. Grants also are made to local colleges and special primary and secondary education programs in the fields of math, science, and youth leadership. Health and human services funding supports the United Way, youth homes, and community health and safety. Support is awarded for civic improvements, special community events, and youth and senior citizen organizations. Art grants support local museums, libraries, cultural centers, and the performing arts. The foundation also awards funding for resource conservation, namely habitat preservation and fish and wildlife protection. Application guidelines and forms are available on the Web site.
Requirements: Montana nonprofits in the service area of the power company and its subsidiaries may submit grant requests.
Restrictions: Health and human services funding does not support national health organizations, medical equipment purchases, or research.
Geographic Focus: Montana
Amount of Grant: Up to 314,000 U.S.D
Samples: Butte Family YMCA (Butte, MT)—operating support, $27,500; Montana Race for the Cure (Helena, MT)—program support, $2000.
Contact: Kathy Vogelsang, (406) 497-2602; fax (406) 497-2451
Sponsor: Montana Power Foundation
130 N Main Street
Butte, MT 59701-9394

Montgomery County Community Foundation Grants **2953**
The Montgomery County Community Foundation Grants help fund non-profit organizations and agencies in Montgomery County, Indiana. Significant grants have been awarded to organizations such as the Crawfordsville District Public Library, Boys and Girls Club, and the Family Crisis Shelter. Organizations should contact the Foundation for a grant proposal application and other documentation required.
Requirements: A strong proposal will have several or all of the following characteristics: an estimate of who and how many will benefit; show long term potential; address a community problem of some significance for which funding is not covered by the regular budget; present an innovative and practical approach to solve a community problem or project; identify possible future funding, if needed; give evidence of the stability and qualifications of the organization applying; show cooperation within the organization and avoid duplication effort.
Restrictions: The Foundation will usually not fund any of the following: grants to individuals; programs which are religious or sectarian in nature, except when the program is open to the entire community; operating expenses such as salaries and utilities; parades, festivals and sporting events; endowment funds; any propaganda, political or otherwise, attempting to influence legislation or intervene in any political affairs or campaigns; an organization's past debts or existing obligations; or post-event or after-the-fact situations.
Geographic Focus: Indiana
Date(s) Application is Due: May 10
Amount of Grant: Up to 50,000 U.S.D
Samples: Crawfordsville Main Street Program, $48,000 to ensure ongoing growth and fiscal sustainability of Athens of Indiana Arts Studio and Gallery (2011); Animal Welfare League, $17,000 to purchase a pre-owned pickup truck to be used for animal control (2011); Educational Foundation, $20,000 for teacher grants of up to $1,000 each for innovative and imaginative projects that go beyond the normal curriculum (2011); Lew Wallace Preservation Society, $9,500 to produce a Historic Structures Report considered vital for the preservation of significant structures (2011); Boys and Girls Club, $17,000 for operational costs of the After School Program (2011).
Contact: Cheryl Keim; (765) 362-1267; fax (765) 361-0562; cheryl@mccf-in.org
Internet: http://www.mccf-in.org/Granthomenewpage.html
Sponsor: Montgomery County Community Foundation
119 East Main Street
Crawfordsville, IN 47933

MONY Foundation Grants **2954**
The foundation awards grants in its areas of interest, including after-school community service and volunteer programs for youth, New York, NY; meeting the essential after-school needs of children and teens at risk, Syracuse, NY; and meeting the essential needs of minority children nationwide (with local MONY sales office). Types of support include employee matching gifts, employee-related scholarships, general operating support, in-kind gifts, matching/challenge support, and program development. Contact headquarters or local office by mail to obtain guidelines.
Restrictions: Grants do not support private foundations; fully participating members of the United Way; religious, fraternal, athletic, social, or veterans' organizations; individuals; capital fund drives; endowments; or deficit financing.
Amount of Grant: 1,500 - 10,000 U.S.D
Samples: To be distributed among 19 nonprofit organizations—for awards through the Field Grant Program, which provides fund-raising and financial assistance to groups that offer services to minority children, $95,000.

Contact: Administrator; (212) 708-2468; fax (212) 708-2001; lynn_stekas@mony.com
Internet: http://www.mony.com
Sponsor: MONY Foundation
1740 Broadway, Mail Drop 10-36
New York, NY 10019

Moody Foundation Grants 2955

Funds are awarded in the areas of education; arts, humanities, and religion; health and sciences; and community and social services, with primary emphasis on foundation-initiated projects in Galveston, TX. Grants are awarded for seed money, building funds, equipment, consulting services, technical assistance, matching funds, special projects, research, publications, conferences, and seminars. Applications are accepted at any time but must be received six weeks prior to each of four meetings every year.
Requirements: Grants are limited to Texas nonprofit organizations.
Restrictions: Grants are not awarded to individuals (except for students covered by one scholarship program in Galveston County), for continuing support, annual campaigns, deficit financing, or for operating budgets except for start-up purposes.
Geographic Focus: Texas
Amount of Grant: Up to 11,000,000 U.S.D
Samples: American Cancer Society, Inc./Cattle Baron's Ball (Dallas, TX)—to assist funding cancer research projects in Dallas County, $30,000; Communities in Schools (Dallas, TX)—for Bright Kids Reading Improvement Program for lower-elementary reading and financial literacy, $15,000; Galveston Independent School District, Educational Foundation, Inc. (Galveston, TX)—for assistance with the Good Start Early Childhood Preschool Program, $50,000.
Contact: Peter Moore, Director; (409) 763-5333; fax (409) 763-5564
Internet: http://www.moodyf.org/HTMLversion/grantappset.htm
Sponsor: Moody Foundation
2302 Post Office Street, Suite 704
Galveston, TX 77550

Moran Family Foundation Grants 2956

The foundation supports innovative interventions that promote healthy mental, physical, and emotional development of at-risk children and at-risk families in Northern Virginia and the District of Columbia. The foundation's goal is to enhance the opportunity for each child to reach his or her fullest potential by funding programs that address issues adversely impacting children and families, such as poverty; physical, emotional, or sexual abuse; addiction; limitations of community supports; and limitations of caregivers. Letters of inquiry are welcome. Grantseekers are asked to use the Common Grant Letter of Intent and the Common Grant Application designed by the Washington Regional Association of Grantmakers. These forms can be found on their web site at www.washingtongrantmakers.org. Forms should be submitted via email.
Requirements: 501(c)3 tax-exempt organizations are eligible.
Restrictions: Grants are not awarded to individuals or for projects that are sectarian in nature.
Geographic Focus: District of Columbia, Virginia
Contact: Grants Administrator; moranfamfdn@aol.com
Internet: http://fdncenter.org/grantmaker/moran
Sponsor: Moran Family Foundation
5965 Ranleigh Manor Drive
McLean, VA 22101

Morehouse PHSI Project Imhotep Internships 2957

IMHOTEP is an eleven-week internship (May - August) managed by Public Health Sciences Institute (PHSI) and designed to increase the knowledge and skills of rising juniors and seniors and recent graduates of an undergraduate institution in biostatistics, epidemiology, and occupational safety and health. The internship begins with two weeks of intense educational training. The purpose of this training is to equip interns with the academic coursework and information necessary to complete the program. During the remaining nine weeks, interns conduct public health research with experts at the Centers for Disease Control and Prevention (CDC), the National Institute for Medical Research (NIMR), Mexico Negro, and various other public health agencies. Since the inception of Project IMHOTEP, interns have been placed in various centers, institutes, and offices (CIOS) within the CDC and various domestic and international agencies, including, but not limited to: the National Center for Chronic Disease Prevention and Health Promotion (NCCDPHP); the National Center for Health Statistics (NCHS); the National Center for HIV, STD, and TB Prevention (NCHHSTP); the National Center for Infectious Diseases (now NCPDCID, NCIRD, NCZVED, and NCHHSTP); the National Institute for Occupational Safety and Health (NIOSH); the NIMR; Mexico Negro; and Sister Love, Incorporated. Throughout the internship, students participate in a wide variety of seminars, workshops, and other educational initiatives and must complete a required number of community service hours. Interns receive research support and consultation in data analysis and other project requirements. At the conclusion of the internship, interns deliver an oral presentation and submit a written manuscript suitable for publication in a scientific journal. Interns receive a financial stipend, housing (at Morehouse and site location) and a travel allowance (travel allowance includes travel to and from Atlanta, site location, and home residence). Interns are required to pay for all expenses incurred during the two-week training (e.g., books) and other miscellaneous expenses (e.g., food). In FY 2005, this cost totaled approximately $300. A link is provided at the IMHOTEP website to apply online. Applicants are encouraged to check the internship website and/or contact the PHSI staff for complete details on deadlines and how to apply for the internship.
Requirements: Applicants must be Juniors, Seniors or Recent Graduates with a cumulative GPA of 2.7 or higher. The selection process is very rigorous and extremely competitive. Candidates should have a genuine interest in pursuing a career in public health with a particular emphasis in biostatistics, epidemiology, or occupational safety and health. Interns are required to sign a legal contract, adhering to all rules, regulations, and program requirements.
Geographic Focus: All States, All Countries
Date(s) Application is Due: Jan 31
Contact: Cynthia Trawick, Director; (404) 681-2800; ctrawick@morehouse.edu
Internet: http://www.morehouse.edu/centers/phsi/imhotep/index.htm
Sponsor: Morehouse College
830 Westview Drive, S.W.
Atlanta, GA 30314

Morgan Adams Foundation Grants 2958

The Morgan Adams Foundation supports laboratory and clinical research in the area of pediatric cancer, with an emphasis on cancers of the brain and spine. The Foundation's particular interest is to provide seed and bridge grants for viable investigations not yet ripe enough for total funding by larger organizations. Additionally, the Foundation is able to partially fund translational studies, experimental therapeutics studies and/or Phase I and II clinical trials on a case by case basis. The Foundation is seeking out projects that are highly collaborative and/or multi-institutional in nature, hoping to facilitate through the strategic placement of funding, more rapid trials development in order to get the best possible treatment options into circulation with the greatest expediency possible. Through careful consideration and placement of its funding, the Foundation encourages and supports research intended to improve treatment effectiveness, improve treatment outcomes and improve the quality of life for children battling cancer.
Contact: Joan Slaughter, Executive Director; (303) 758-2130; fax (303) 758-2134; joan@morganadamsfoundation.org or info@morganadamsfoundation.org
Internet: http://www.morganadamsfoundation.org/fundsBecomeHope.html
Sponsor: Morgan Adams Foundation
5303 E Evans Avenue, Suite 202
Denver, CO 80222

Morgan Stanley Foundation Grants 2959

The foundation awards grants to nonprofit organizations in its areas of interest, including education, civic and community development, culture and the arts, and health and welfare. In the area of education, programs are supported that work to strengthen the teaching profession, address adult literacy, and retrain seniors and the disadvantaged. Economic development organizations with innovative programs designed to enhance and broaden the economic stability of communities receive consideration. In the area of health and welfare, applied research, educational programs, and other innovative programs in preventive health care, particularly those targeted to youth and seniors, receive support. Under arts and culture, performing arts and cultural programs in major metropolitan serving areas with outreach potential receive consideration. Types of support include capital campaigns, general operating support, continuing support, employee matching gifts, and in-kind gifts. The foundation prefers to fund project-specific proposals rather than requests for grants to underwrite operating or capital budgets.
Requirements: New York 501(c)3 organizations in company-operating areas are eligible.
Restrictions: Grants do not support United Way member agencies, individuals, endowment or building funds, capital campaigns, deficit financing, equipment acquisition, scholarships, fellowships, special projects, journal advertisements, research, publications, conferences, fund-raising dinners, or benefit events.
Geographic Focus: New York
Amount of Grant: 500 - 25,000 U.S.D
Contact: Director of Community Affairs; (212) 259-1235 or (212) 761-4000; fax (212) 259-1253; whatadifference@msdw.com
Internet: http://www.morganstanley.com/about/inside/community.html?page=about
Sponsor: Morgan Stanley Foundation
1601 Broadway, 12th Floor
New York, NY 10019

Morris and Gwendolyn Cafritz Foundation Grants 2960

The Morris & Gwendolyn Cafritz Foundation supports IRS-registered, tax-exempt, 501(c)3 organizations with a public charity status of 509(a)(1) or 509(a)(2) only. These organizations must serve residents in the District of Columbia, Prince George's and Montgomery Counties in Maryland, Arlington and Fairfax Counties, and the cities of Alexandria and Falls Church in Virginia. Grants are made in four program areas: Arts and Humanities, Community Services, Education and Health.
Requirements: Nonprofits serving residents in the District of Columbia, Prince George's and Montgomery Counties in Maryland, Arlington and Fairfax Counties, and the cities of Alexandria and Falls Church in Virginia may apply.
Restrictions: The Foundation does not generally fund the following projects: capital campaigns; endowments; multi-year grants; special events or tables for special events. Please also note, the Foundation does not fund: organizations that do not have 501(c)3 tax-exempt status with the IRS; private foundations; public charities with a non-private foundation status of 509(a)(3); individuals; organizations whose missions fall outside the Foundations funding priorities; organizations serving residents outside the Washington, DC metropolitan area.
Geographic Focus: District of Columbia, Maryland, Virginia
Date(s) Application is Due: Mar 1; Jul 1; Nov 1
Samples: Theater Downtown, Washington, DC, $200,000 - for Staging Our Future Campaign; Prince Georges Child Resource Center, Largo, MD, $25,000 - for Family Literacy Program; Spanish Catholic Center, Washington, DC, $75,000 - for oral health treatment and education for low-income, uninsured immigrants;

Contact: Rose Ann Cleveland, Executive Director; (202) 223-3100; fax (202) 296-7567; info@cafritzfoundation.org
Internet: http://www.cafritzfoundation.org/Applicant/app_guidelines.asp
Sponsor: Morris and Gwendolyn Cafritz Foundation
1825 K Sreet NW, Suite 1400
Washington, DC 20006

Morris Goldseker Foundation of Maryland Grants **2961**
The foundation makes grants to nonprofit organizations primarily in the metropolitan Baltimore, MD, area in its areas of interest, including community development, children and families, economic development, and neighborhoods. Target areas are modified from time to time in response to changing needs and problems in the area. Types of support include general operating grants, matching gifts, program development, scholarships, seed money grants, and technical assistance. Grants are generally awarded for a period of one year, though requests for longer periods will be considered. Telephone inquiries are accepted throughout the year. Proposals are considered at one of three annual meetings.
Requirements: IRS 501(c)3 and 509(a) tax-exempt organizations serving the metropolitan Baltimore area are eligible.
Restrictions: The foundation does not support endowments, individuals, deficit financing, annual giving, publications, arts and culture, religious programs or purposes, ongoing operating support, political action groups, specific disabilities or diseases, or projects normally financed with public funds.
Geographic Focus: Maryland
Date(s) Application is Due: Jan 1; Apr 1; Aug 1
Amount of Grant: Up to 35,000,000 U.S.D
Samples: Ctr for Poverty Solutions (MD)—to review and revise its financial and management systems, $50,000; Greenmount School (MD)—to enable this private school run by teachers and parents to seek donations, $90,000 over three years; Village Learning Place (MD)—to support the Village Learning Place, a library and after-school program in the Charles Village neighborhood, for board-development activities, $10,000.
Contact: Carol Gilbert, Program Officer; (410) 837-5100; fax (410) 837-7927; cgilbert@goldsekerfoundation.org
Internet: http://www.goldsekerfoundation.org
Sponsor: Morris Goldseker Foundation
1040 Park Avenue, Suite 310
Baltimore, MD 21201

Morris Stulsaft Foundation Grants **2962**
The foundation awards grants to California nonprofits for programs that focus on the needs of homeless children and youth; services for children living in subsidized housing; the needs of children and youth in foster care, especially as they prepare for emancipation; early intervention development programs for children with disabilities; child care for low-income families; and training for child care providers and staff. Types of support include operating budgets, building construction/renovations matching funds, research, special projects, and seed money. There are no application deadlines. Applicant should request an application form. At the full proposal stage, review turn around time is six months.
Requirements: Nonprofits in California counties, including Alameda, Contra Costa, Marin, San Francisco, Santa Clara, and San Mateo, are eligible.
Restrictions: Grants do not support individuals, deficit funding, emergency funding, endowments, fundraising, or special events/benefit dinners.
Geographic Focus: California
Amount of Grant: 5,000 - 15,000 U.S.D
Samples: Dance Palace (CA)—toward dance scholarships for low-income Latino children, $2500; Fort Milky Adventures Rope Course (CA)—to purchase and upgrade equipment for the Ropes Course, $5000; National Junior Tennis League of San Francisco (CA)—to expand summer after-school program for low-income minority children and youth, $7500.
Contact: Program Contact; (415) 986-7117; fax (415) 986-2521; Stulsaft@aol.com
Internet: http://www.stulsaft.org/approvedgrants.html
Sponsor: Morris Stulsaft Foundation
100 Bush Street, Suite 825
San Francisco, CA 94104-3911

Morton K. and Jane Blaustein Foundation Grants **2963**
The foundation makes grants in the United States and abroad. Support is provided in the program areas of educational opportunity—to improve educational opportunities for underserved K-12 children, with preference given to initiatives in Baltimore and New York City; health and mental health—to improve health and mental health care for underserved people, with preference given to programs in Baltimore and New York City; and human rights—to advance fundamental human rights both in the United States and abroad, with preference given to grassroots efforts.
Requirements: Education and human services organizations in Maryland and New York are eligible to apply.
Restrictions: No support for fundraising events, or direct mail solicitations. Grants are not made to individuals.
Geographic Focus: Maryland, New York
Samples: Johns Hopkins U (Baltimore, MD)—for a fellowship endowment at the School of Nursing, $1 million.
Contact: Grants Administrator; (410) 347-7205; info@blaufund.org
Internet: http://www.blaufund.org/foundations/mortonandjane_f.html
Sponsor: Morton K. and Jane Blaustein Foundation
10 E Baltimore Street, Suite 1111
Baltimore, MD 21202

Mount Sinai Health Care Foundation Grants **2964**
The foundation seeks to assist greater Cleveland's organizations and leaders to improve the health and well-being of the Jewish and general communities now and for generations to come. The foundation concentrates its efforts on several well-defined program areas, including health policy, health of the Jewish community, health of the urban community, and academic medicine and bioscience. In addition to these program areas, the foundation seeks to strengthen Cleveland's prominence as a national leader in medical education and research. Whenever possible, the foundation will encourage the active participation of other funders to increase the overall amount of support available.
Requirements: 501(c)3 nonprofits serving the greater Cleveland, OH, area may submit proposals for grant support.
Restrictions: In general, the foundation does not support general operating expenses, direct provision of health services, building or equipment expenses, fund-raising events, projects outside of greater Cleveland, endowment funds, lobbying, or program advertising.
Geographic Focus: Ohio
Date(s) Application is Due: Jan 1; Apr 1; Jul 1; Oct 1
Contact: Shelly Galvin, (216) 421-5500; fax (216) 421-5633; sgg2@case.edu
Internet: http://www.mtsinaifoundation.org/grants.html
Sponsor: Mount Sinai Health Care Foundation
11000 Euclid Avenue, 3rd Floor
Cleveland, OH 44106-1714

Mount Zion Health Fund Grants **2965**
The fund provides grants predominantly in the San Francisco Bay area to nonprofit organizations in its areas of interest, including UCSF at Mount Zion and its affiliates, the Institute on Aging, and the Bay Area Jewish community. MZHF will consider proposals for grants for up to three years in duration. For multiyear grants, grant distributions will be contingent upon annual review of the grantee's progress toward defined expectations. There are no deadlines for submission. Applicants will generally receive a response within three months after submitting a complete proposal.
Requirements: California 501(c)3 tax-exempt organizations and organizations with 501(c)3 fiscal sponsors are eligible.
Restrictions: Grants are not awarded to support the personal benefit of individuals; endowments, annual appeals, membership drives, and fundraising events; or the reduction of operating deficits.
Geographic Focus: California
Date(s) Application is Due: Feb 15; Sep 15
Amount of Grant: 5,000 - 100,000 U.S.D
Samples: Bay Area Jewish Healing Ctr (San Francisco, CA)—for Jewish spiritual care, outreach, and education, $145,000; Homeless Prenatal Program (San Francisco, CA)—to provide care to homeless pregnant women and mothers, $15,000; Jewish Home (San Francisco, CA)—for general support of this geriatric center, $100,000; Life Tapes Project (San Francisco, CA)—to produce videotaped interviews designed to help relatives of people diagnosed with cancer, $40,000.
Contact: Administrator; (415) 750-4100; fax (415) 495-6635; information@mzhf.org
Internet: http://sandbox.sfjcf.org/mountzion/grantapp
Sponsor: Mount Zion Health Fund
P.O. Box 194065
San Francisco, CA 94119-4065

MPAAT Grants **2966**
The organization's mission is to reduce the harm that tobacco causes the people of Minnesota. Requests for applications have been issued for intervention funding and for research funding. Intervention funding proposals are invited in the areas of Minnesota Tobacco Treatment Centers, and Minnesota Statewide Worksite Cessation Programs. In the research category, applications are sought for Research Funding for Projects to Reduce the Harm Tobacco Causes Minnesotans. Guidelines and application for research grants are available online.
Requirements: Minnesota nonprofit organizations are eligible.
Geographic Focus: Minnesota
Contact: Chris Tholkes, (952) 767-1409 or (952) 767-1400; fax (952) 767-1422; ctholkes@mpaat.org
Internet: http://www.mpaat.org
Sponsor: Minnesota Partnership for Action Against Tobacco
8011 34th Avenue S, Suite 400
Minneapolis, MN 55425

MRC Career Establishment Grants **2967**
Career establishment grants provide research support for scientists and clinicians newly appointed to academic posts in medical schools, teaching hospitals, and universities who are at an early stage in the development of their research career and who can show that they have the potential to become the research team leaders of the future. A career establishment grant is intended to help award holders to establish themselves as independent scientists capable of winning support in open competition by the time the grant ends.
Requirements: An applicant must be in an established academic post in a university, medical school, teaching hospital, or similar institution; have a PhD or DPhil (for scientifically qualified applicants) or an MD (for medically qualified applicants); have at least three years postdoctoral research experience; be planning to make a long-term career in biomedical or health-related research; and make their initial application for a career establishment grant within the first three years of appointment to their first established academic post.

Restrictions: Scientists who do not hold an established academic appointment are not eligible to apply. Grants are not intended for researchers whose career in research is already well established, as demonstrated by their publication track record and their success in winning research funding.
Geographic Focus: All States, United Kingdom
Samples: Dr. C.R. Byrne, School of Biological Science, U of Manchester—for research on the relationship between fetal skin patterning and epidermal differentiation, @ L310,000; Dr. S.L. Dickson, Physiological Laboratory, U of Cambridge—for research on hypothalamic circuits involved in body weight homeostasis, @L276,000; Dr. S.J. Talbot, Department of Medical Microbiology, U of Edinburgh—for research on characterization of HHV-8 latency associated genes and the development of an animal model for studying tumorigenesis, @L476,000.
Contact: Michele Michalczuk, Research Management Group; 020 7636 5422; fax 020 7436 6179; michele.michalczuk@headoffice.mrc.ac.uk
Internet: http://www.mrc.ac.uk/index/funding/funding-specific_schemes/funding-revised_grant_schemes.htm
Sponsor: Medical Research Council
20 Park Crescent
London, W1B 1AL England

Ms. Foundation for Women Ending Violence Grants **2968**
The Ms. Foundation works in partnership with grassroots, state, and national organizations to end gender-based violence by transforming policies, beliefs, and behaviors that threaten the well-being of individuals, families and communities nationwide. The Foundation supports a range of community-based strategies to stop violence before it occurs—to prevent violence directed against women, girls, and LGBTQ individuals. The Foundation also supports a movement to advance a community-based, social justice approach to child sexual abuse.
Requirements: Applicants must be tax exempt nonprofits organizations. The application process, current deadlines, and previous grant recipients can be found on the Foundation website.
Contact: Administrator; (212) 742-2300; fax (212) 742-1653; info@ms.foundation.org
Internet: http://ms.foundation.org/our_work/broad-change-areas/ending-violence
Sponsor: Ms. Foundation for Women
12 MetroTech Center, 26th Floor
Brooklyn, NY 11201

Ms. Foundation for Women Health Grants **2969**
The Ms. Foundation for Women supports organizing at grassroots, state and national levels to promote equitable access to health care and education for women and youth. The Foundation delivers strategic funding, technical assistance, and networking support to organizations that are working in their communities and beyond to address the urgent priorities of those most affected by failed health policies, especially low-income women, women of color, immigrant women, women living with HIV/AIDS, and LGBTQ youth. The Foundation strives to build social movements and advance policy and culture change in key areas: reproductive health, rights and justice; sexuality education; and women and AIDS.
Requirements: Applicants must be tax exempt nonprofit organizations. The application process, current deadlines, and previous grant recipients can be found on the Foundation website.
Contact: Grants Contact; (212) 742-2300; fax (212) 742-1653; info@ms.foundation.org
Internet: http://ms.foundation.org/our_work/broad-change-areas/womens-health
Sponsor: Ms. Foundation for Women
12 MetroTech Center, 26th Floor
Brooklyn, NY 11201

MSSC Career Development Awards **2970**
The awards further research into the cause, prevention, treatment, and cure of multiple sclerosis. Awards are intended to support independent research by the young, beginning researcher who holds a doctoral degree and intends to work full time at a Canadian school of medicine on research relevant to multiple sclerosis. Awards are normally made for three years with possible renewal.
Geographic Focus: All States, Canada
Date(s) Application is Due: Oct 1
Contact: Jackie Munroe, Research Grants Manager; (416) 967-3001; fax (416) 967-3040; jackie.munroe@mssociety.ca or info@mssoc.ca
Internet: http://www.mssociety.ca/en/research/researchfunding.htm
Sponsor: Multiple Sclerosis Society of Canada
175 Bloor Street E, Suite 700, N Tower
Toronto, ON M4W 3R8 Canada

MSSC Postdoctoral Fellowships **2971**
The fellowships further research into the cause, prevention, treatment, and cure of multiple sclerosis. These fellowships are intended for qualified scientists who have MDs or PhDs and intend to pursue further training in research related to multiple sclerosis. Fellowships are awarded for one year, renewable for up to three years if progress is deemed satisfactory.
Date(s) Application is Due: Oct 1
Contact: Jackie Munroe, Research Grants Manager; (416) 967-3001; fax (416) 922-7538; jackie.munroe@mssociety.ca or info@mssoc.ca
Internet: http://www.mssociety.ca/en/research/researchfunding.htm
Sponsor: Multiple Sclerosis Society of Canada
175 Bloor Street E, Suite 700, N Tower
Toronto, ON M4W 3R8 Canada

MSSC Research Grants **2972**
The grants further independent research into the cause, prevention, treatment, and cure of multiple sclerosis. Research grants are awarded for projects covering a one- to three-year period. Research must be conducted in or under the authority of an institution approved by the society. Grants are to be used to cover research costs, including funding of equipment up to $C5000.
Geographic Focus: All States, Canada
Date(s) Application is Due: Oct 1
Samples: Drs. Albert Aguayo and Garth Bray, McGill U—for Neuron-Glia Interactions: Experimental Manipulation of Trophic Factors in Demyelination, $88,100; Dr. Luanne Metz, U of Calgary—for The Cost of Multiple Sclerosis, $5050.
Contact: Jackie Munroe, Research Grants Manager; (416) 967-3001; fax (416) 922-7538; jackie.munroe@mssociety.ca or info@mssoc.ca
Internet: http://www.mssociety.ca/en/research/researchfunding.htm
Sponsor: Multiple Sclerosis Society of Canada
175 Bloor Street E, Suite 700, N Tower
Toronto, ON M4W 3R8 Canada

MSSC Research Studentships **2973**
Studentships are for doctoral candidates who require further training. The awards offer candidates an opportunity to acquire this training in specialized areas related to research into the cause, prevention, treatment, and/or cure of multiple sclerosis. Awards are for a maximum period of four years, renewable annually.
Geographic Focus: All States, Canada
Date(s) Application is Due: Oct 1
Contact: Jackie Munroe, Research Grants Manager; (416) 967-3001; fax (416) 922-7538; jackie.munroe@mssociety.ca or info@mssoc.ca
Internet: http://www.mssociety.ca/en/research/researchfunding.htm
Sponsor: Multiple Sclerosis Society of Canada
175 Bloor Street E, Suite 700, N Tower
Toronto, ON M4W 3R8 Canada

Multiple Sclerosis Foundation Brighter Tomorrow Grants **2974**
The goal of the grant is provide individuals with MS with goods or services (valued at up to $1000.00 per recipient) to improve their quality of life by enhancing safety, self-sufficiency, comfort, or well-being. Recipients have used the grant to purchase wheelchairs, walkers, eye glasses, computers, appliances, furniture, therapeutic equipment, hobby supplies, and for care repairs and various home modifications.
Requirements: To qualify, a person must be 18 years of age or older and diagnosed with MS, or the parent of a minor child diagnosed with MS, and be a permanent U.S. resident. They must not have any other means of fulfilling the need they express.
Date(s) Application is Due: Oct 10
Amount of Grant: 1,000 U.S.D
Contact: Program Services Department; (888) 673-6287 or (954) 776-6805; fax (954) 351-0630; support@msfocus.org
Internet: http://www.msfocus.org/Brighter-Tomorrow-Grant.aspx
Sponsor: Multiple Sclerosis Foundation
6350 North Andrews Avenue
Fort Lauderdale, FL 33309-2130

Musculoskeletal Rehabilitation Fellowships **2975**
The fellowship provides a unique opportunity for specialized training in clinical, education, and research aspects of musculoskeletal rehabilitation. Resources at the clinic are vast and include a multispecialty spine center, sports medicine center, biomechanics department, hand center, arthritis clinic, pain clinic, and orthopedic and general musculoskeletal rehabilitation. Fellows will have the opportunity to assist in medical coverage of sporting events in southeastern Minnesota to gain musculoskeletal management skills. Multiple educational opportunities are available, including the teaching of medical students and residents. Dedicated research time can be allotted with faculty mentorship and integrated use of other departments. Appointments and application deadlines vary each year.
Requirements: Candidate must have successfully completed an accredited residency program in physical medicine and rehabilitation and be board eligible or board certified.
Date(s) Application is Due: Aug 30
Contact: Cindy Weber, (507) 284-2946; fax (507) 284-3431; pmreduc@mayo.edu
Internet: http://www.mayo.edu/msgme/physmed-musculo-rch.html
Sponsor: Mayo Clinic
200 First Street SW
Rochester, MN 55905

Mutual of Omaha Community Support Grants **2976**
The program awards grants to eligible nonprofit organizations in its areas of interest, including education—institutions, capital projects, and matching support for associate donations; community betterment—community revitalization, and programs that address the challenges facing children, families, minority groups, and the disadvantaged; health and wellness—prevention and health education activities, and expanding access to quality health care; and arts and culture—performances, exhibits and youth education programs of established organizations. Requests for funding must be submitted through the online form. Guidelines are available online.
Requirements: The recipient organization must have tax-exempt status under section 501(c)3. In order of emphasis, the foundation funds programs that serve the corporate headquarters city of Omaha and surrounding eastern Nebraska and western Iowa areas; programs in cities where Mutual of Omaha has a significant presence; and national

programs that further the foundation's objectives (and can be implemented or replicated in cities where Mutual of Omaha has a substantial presence).
Restrictions: In general, the following are not considered eligible for funding from the foundation: endowment funds; projects of religious denominations or sects, other than eligible educational, health, or human-service related projects; support of individuals including education, research, or product development; travel for individuals or groups; team sponsorships or athletic scholarships; political actions, candidates, or lobbying efforts; organizations or programs whose principal funding comes from federal, state, or local taxes, except for certain activities of the state university system; requests for civic or commemorative advertising social clubs, labor, or fraternal organizations; festival participation; the creation or maintenance of monuments or memorials; programs or organizations outside the United States or its territories, except where Mutual of Omaha has significant business locations and would otherwise be eligible under policy guidelines. As a general rule, Mutual of Omaha does not purchase corporate tickets or tables. Requests will not be granted for gifts of services, materials, or use of Mutual of Omaha's equipments or systems, except when part of a Mutual of Omaha-sanctioned community activity.
Date(s) Application is Due: Mar 1; Jun 1; Oct 1; Dec 1
Contact: Contributions Committee, 9-Community Affairs Department; (402) 342-7600
Internet: http://www.mutualofomaha.com/community/index.html
Sponsor: Mutual of Omaha Foundation
Mutual of Omaha Plaza
Omaha, NE 68175

Myers Family Foundation Grants **2977**
The foundation awards grants to nonprofit organizations in its areas of interest, including arts and culture, education, hospitals, family services, and churches. Types of support include capital grants, general operating grants, programs grants, and seed money grants. There are no application deadlines.
Requirements: 501(c)3 nonprofits are eligible. Giving is primarily made in California.
Amount of Grant: 25 - 150,000 U.S.D
Contact: William Myers, Trustee; (805) 963-2690
Sponsor: Myers Family Foundation
1114 State Street, Suite 232
Santa Barbara, CA 93101-9650

NAF Fellowships **2978**
The foundation supports new and innovative studies that are relevant to the cause, pathogenesis, or treatment of the hereditary or sporadic ataxias. Postdoctoral fellowship awards serve as a bridge from postdoctoral positions to junior faculty positions. The award will permit individuals to spend an additional third year in a postdoctoral position and increase chances to establish an independent ataxia research program. Letters of Intent should arrive by August 15, with full proposals due by September 15. Application and guidelines are available online.
Requirements: Applicants should have completed at least one year of postdoctoral training, but not more than two at the time of application, and should have shown a commitment to research in the field of ataxia.
Restrictions: Indirect costs are not allowed.
Date(s) Application is Due: Aug 15; Sep 15
Amount of Grant: 35,000 U.S.D
Contact: Grants Administrator; (763) 553-0020; fax (763) 553-0167; naf@ataxia.org
Internet: http://www.ataxia.org/research/ataxia-research-grants.aspx
Sponsor: National Ataxia Foundation
2600 Fernbrook Lane, Suite 119
Minneapolis, MN 55447-4752

NAF Kyle Bryant Translational Research Award **2979**
The Friedreich's Ataxia Research Alliance (FARA) and the National Ataxia Foundation (NAF) invite proposals, under a competitive Request for Applications (RFA) process, to award a grant focusing on pre-clinical investigations that will facilitate clinical trials for Friedreich's Ataxia. Funding will be for one year. The total award is limited to $120,000 (direct costs only). The Foundation is requesting that applicants submit Letters of Intent by July 15, and full applications by July 31.
Requirements: Each application must include a project objectives and lay summary that will be used as FARA and NAF wishes and should not contain confidential information. The other parts of the grant application are considered confidential and will only be released to review committee, FARA and NAF staff and Board of Directors. Applications will be accepted from for-profit organizations, non-profit organizations, public or private institutions, and foreign institutions.
Date(s) Application is Due: Jul 15; Jul 31
Amount of Grant: Up to 120,000 U.S.D
Contact: Susan Hagen, (763) 553-0020; susan@ataxia.org
Internet: http://www.ataxia.org/research/ataxia-research-grants.aspx
Sponsor: National Ataxia Foundation
2600 Fernbrook Lane, Suite 119
Minneapolis, MN 55447-4752

NAF Research Grants **2980**
NAF research grants are for new and innovative studies that are relevant to the cause, pathogenesis or treatment of the hereditary or sporadic ataxias. Due to the larger availability of funding for Ataxia-Telangiectasia (A-T), those research proposals will receive a lower priority. However, a higher ranking will be given to those Ataxia-Telangiectasia research studies which lend themselves to an overall better understanding of the ataxia process. Research grants are offered primarily as seed monies to assist investigators in the early or pilot phase of their studies and as additional support for ongoing investigations on demonstration of need. It is hoped that these studies will be further developed to attract future funding from other sources. Grants are awarded for one year only. If funding for a second year is desired then another grant application should be submitted for that funding period. The average funds granted will be up to $15,000, but funding may be considered for up to $30,000 for projects deserving special consideration.
Restrictions: Indirect costs will not be funded.
Date(s) Application is Due: Aug 15
Amount of Grant: 5,000 - 15,000 U.S.D
Contact: Grants Administrator; (763) 553-0020; fax (763) 553-0167; naf@ataxia.org
Internet: http://www.ataxia.org/research/ataxia-research-grants.aspx
Sponsor: National Ataxia Foundation
2600 Fernbrook Lane, Suite 119
Minneapolis, MN 55447-4752

NAF Young Investigator Awards **2981**
The purpose of this award is to encourage young clinical and scientific investigators to pursue a career in the field of ataxia research. It is the foundation's hope that ataxia research will be invigorated by the work of young, talented individuals supported by stable multi-year funding. This award will be made to the sponsoring institution and will be used as salary support or for direct research expenses. The sponsoring institution must agree in writing to the following provision: a clinical investigator (MD) must be free to allocate a minimum of half time for a project; a non-clinician investigator (PhD) must allocate a minimum of three-fourths time for the project. Periodic progress reports will be required. Applicants are asked to submit a brief email with the intentions to apply for NAF funding, including a full title of the research proposal, on or before August 1. Deadline for full application is September 1.
Requirements: Candidates must have attained the MD or PhD degree, and have an appointment as a junior faculty member. Clinicians must have finished their residency no more than five years prior to applying, and investigators must have completed their postdoctoral training no more than five years prior to applying.
Restrictions: Individuals at the associate professor level are not eligible. Awards are for direct costs only, not for indirect costs or institutional overhead.
Date(s) Application is Due: Aug 1; Sep 1
Amount of Grant: 50,000 U.S.D
Contact: Grants Administrator; (763) 553-0020; fax (763) 553-0167; naf@ataxia.org
Internet: http://www.ataxia.org/research/ataxia-research-grants.aspx
Sponsor: National Ataxia Foundation
2600 Fernbrook Lane, Suite 119
Minneapolis, MN 55447-4752

Nancy Dickerson Whitehead Awards for Excellence in Reporting on Drug and Alcohol Issues **2982**
These awards honor Nancy Dickerson Whitehead's efforts to improve media coverage of drug issues and are the first awards in the nation to recognize excellence in reporting on drug and alcohol problems. The awards are given annually to two journalists, one from print media and one from broadcast (radio or television), who have demonstrated the highest standards of reporting on drug issues. Each winner receives a monetary award at an annual luncheon in New York. An application is available online.
Requirements: Entries must consist of a single article or broadcast, or a series of related articles or broadcasts, published or aired between June 2 of the previous year and June 1 of the current year.
Date(s) Application is Due: Jun 15
Amount of Grant: 10,000 U.S.D
Samples: ABC News & PJ Productions for Peter Jennings Reporting—for Ecstasy Rising, a documentary on the creation and rising popularity of the drug called Ecstasy and the U.S. government's ill-conceived efforts to prevent Ecstasy use by using flawed research and overstating the risks associated with taking the drug, $10,000.
Contact: Gretchen Wright, (202) 371-1999; prsolutions@prsolutionsdc.com
Internet: http://www.drugstrategies.org/awards.html
Sponsor: Drug Strategies
1150 Connecticut Avenue NW, Suite 800
Washington, DC 20036

Nancy J. Pinnick Memorial Scholarship **2983**
The Nancy J. Pinnick Memorial Scholarship was established by the Pinnick family to continually recognize her dedication and work as an educator in a meaningful way. A $1,000 scholarship will be awarded to a graduating senior from River Forest High School in Hobart, Indiana. The scholarship is designated for tuition and books and will be renewable for $1,000 for years two through four of the student's college studies. All college majors are eligible.
Requirements: The scholarship is awarded to a graduating high school senior with a cumulative GPA of at least 2.5. The student must be enrolled full time in a program leading to an associate or bachelor degree. The completed application package should include the application form, a current high school transcript, and answers, in essay form, to two questions asked on the application. A student will be selected based on his or her GPA, high school, and community activities.
Geographic Focus: Indiana
Date(s) Application is Due: Apr 15
Contact: Pinnick Scholarship Contact; (219) 962-7551; fax (219) 962-8338
Internet: http://www.legacyfoundationlakeco.org/scholarships/index.htm
Sponsor: Legacy Foundation
1000 East 80th Place, 302 South
Merrillville, IN 46410

Nancy L. Lynn Charitable Foundation Grants **2984**

The foundation awards grants to eligible Southern California organizations providing HIV/AIDS direct services and education, cancer organizations providing direct services and education, and child-abuse organizations providing direct services and prevention education. The foundation requires that grantees be able to ensure continued support after foundation involvement is discontinued, and that the grant will benefit others beyond the individual organization recipient of the grant.
Requirements: Southern California organizations are eligible.
Restrictions: Grants do not support capital fundraising or building construction/ renovation. In addition, the grant should not duplicate an existing foundation grant; involve the purchase of land; involve propagation activities or attempt to influence legislation; or restrict the services/ facilities to or employment of individuals based on discriminatory practices.
Geographic Focus: California
Amount of Grant: 5,000 - 15,000 U.S.D
Samples: Orange County Child Abuse Prevention Ctr (Santa Ana, CA)—for Welcome Baby, an in-home program for first-time parents, $10,000.
Contact: Grants Administrator; (949) 715-1825; fax (949) 499-1235; info@nllcf.org
Internet: http://www.nllcf.org/apply_for_a_grant.htm
Sponsor: Nancy L. Lynn Charitable Foundation
24040 Camino Del Avion, #A324
Monarch Beach, CA 92629

Nancy R. Gelman Foundation Breast Cancer Seed Grants **2985**

The program awards seed grants to fund projects aimed at improving outcomes for women with breast cancer. Proposed projects must: demonstrate a need for start-up, interim, or supplemental funding for a one-year period beginning April 1; not already be funded by other sources (projects that represent a clear new initiative extending beyond the scope of a currently funded project are permissible); and prospectively state an objective outcome measure that will be used to gauge project success. Types of projects seed grants might fund include: a mentored summer fellowship for a medical student to work in a laboratory, gaining experience in basic research on breast cancer mechanisms or therapies; community-based efforts to promote breast cancer awareness and early detection among populations at increased risk; development of information materials to encourage appropriate screening and to help guide newly diagnosed patients, or those whose cancer has recurred, through the challenging maze of treatment options; fledgling research studies to pave the way for further advances in breast cancer therapy; quality initiatives in cancer or emergency department nursing units to improve care for patients with breast cancer; and a summer research project studying end-of-life care for breast cancer patients and making recommendations to hospitals and hospice organizations to improve care. The application and guidelines are available online.
Date(s) Application is Due: Aug 5
Amount of Grant: Up to 1,500 U.S.D
Contact: Michael Gelman; (206) 441-7225; michael@nrgf.org or grants@nrgf.org
Internet: http://www.nrgf.org/grants.html
Sponsor: Nancy R. Gelman Foundation
1133 Broadway, Suite 706
New York, NY 10010

Naomi and Nehemiah Cohen Foundation Grants **2986**

The foundation awards grants primarily for Jewish organizations, especially welfare funds, community services, and educational institutions. Grants also are awarded to Christian nonprofits and to organizations for peace activities. Types of support include annual campaigns, building construction/renovation, capital campaigns, equipment acquisition, general operating support, project support, publications, seed money, and technical assistance. Nonprofits in the District of Columbia receive preference. There are no formal grant procedures or application deadlines. Initial approach should be a letter of inquiry.
Restrictions: Grants are not made to individuals.
Geographic Focus: All States, Israel
Amount of Grant: 1,000 - 50,000 U.S.D
Samples: Jewish Federation of Greater Washington (Rockville, MD)—for the United Jewish Endowment Fund's annual campaign, $4 million; Shefa Fund (Wyndmoor, PA)—to leverage investments from Jewish institutions and individuals for use by community development financial institutions, $360,000; United Jewish Appeal Federation (Rockville, MD)—operating support, $200,000.
Contact: Alison McWilliams; (202) 234-5454; fax (202) 234-8797; NNCF@erols.com
Sponsor: Naomi and Nehemiah Cohen Foundation
P.O. Box 73708
Washington, DC 20056

NAPNAP Foundation Elaine Gelman Scholarship **2987**

The purpose of this annual scholarship is to support the education of a nurse practitioner student who demonstrates the ability to articulate and follow through an innovative solution through clinical competence, academic achievement and involvement in political activism relating to a pediatric health care issue. Preference is given to applicants in an accredited PNP program or those with interest or experience in health policy or advocacy. The recipient will receive an award up to $1000 if all criteria are met. The recipient will be recognized in The Pediatric Nurse Practitioner newsletter and again at the Annual NAPNAP Conference Awards Breakfast.
Requirements: Applicant must be a full or part time NP Student enrolled in an accredited NP program with an expected graduation date of two years or less. Applicants must submit two (2) letters of support from professional colleagues addressing the applicant's clinical competence, academic achievement and involvement in political activism relating to health care issues.
Date(s) Application is Due: Jun 30
Amount of Grant: Up to 1,000 U.S.D
Contact: Dolores Jones, Director; (856) 857-9700; fax (856) 857-1600; djones@napnap.org or info@napnap.org
Internet: http://www.napnap.org/PNPResources/research/AvailGrantsScholarships.aspx
Sponsor: National Association of Pediatric Nurse Associates and Practitioners
20 Brace Road, Suite 200
Cherry Hill, NJ 08034-2634

NAPNAP Foundation Graduate Student Research Grant **2988**

The program supports research by graduate students currently enrolled in a pediatric nurse practitioner program. The purposes of the award are to: encourage graduate students to participate in the research process; and stimulate interest in research regarding children and their families. The recipient of the award will be invited to participate at the research presentation session at the annual conference in the spring following the award. Application and guidelines are available online.
Requirements: The graduate student must be a current member of NAPNAP.
Date(s) Application is Due: Apr 1
Amount of Grant: 1,000 U.S.D
Samples: Cathy Woodward—for Routine Chest Radiographs Post Chest Tube Removal in Children, $1,000 (2006-07); Sheri Vlam—for ADHD Assessment Methods Used by APRNs, $500 (2003-04); Jacqueline Rychnovsky—for A Pediatric Perspective of Maternal Fatigue Across the First Ten Weeks Postpartum, $1,000 (2002-03).
Contact: Dolores Jones, Director; (856) 857-9700; fax (856) 857-1600; djones@napnap.org or info@napnap.org
Internet: http://www.napnap.org/PNPResources/research/AvailGrantsScholarships.aspx
Sponsor: National Association of Pediatric Nurse Associates and Practitioners
20 Brace Road, Suite 200
Cherry Hill, NJ 08034-2634

NAPNAP Foundation Innovative Health Care Small Grant **2989**

The Innovative Health Care Small Grant Program is designed to encourage NAPNAP members and chapters to develop and implement innovative, feasible, and cost-effective programs to improve children's health services. These projects are intended to improve the accessibility of health services to underserved children and those in need of health care; enhance family education regarding the importance of health care maintenance and preventive care; increase utilization rates for children without services or who are underserved; and encourage providers to pursue new mechanisms in delivery of health services to children in need. Guidelines are available online.
Requirements: Each application must be submitted by a national NAPNAP member (or group of members) or by a NAPNAP chapter; represent a local initiative that is community-based, family-centered, and culturally appropriate; describe the project's target population; offer innovative or improved access to health education and/or health care services for children; clearly state project goal; define measurable objectives of the project and identify those activities that lead to their achievement; develop a simple evaluation mechanism reflecting extent to which goal and objectives have been achieved; and include a budget for the project (monies may not be used for salary, consultation, or gifts).
Date(s) Application is Due: Sep 15
Amount of Grant: Up to 4,000 U.S.D
Contact: Dolores Jones, Director; (856) 857-9700; fax (856) 857-1600; djones@napnap.org or info@napnap.org
Internet: http://www.napnap.org/PNPResources/research/AvailGrantsScholarships.aspx
Sponsor: National Association of Pediatric Nurse Associates and Practitioners
20 Brace Road, Suite 200
Cherry Hill, NJ 08034-2634

NAPNAP Foundation McNeil Grant-in-Aid **2990**

The purpose of the NAPNAP Foundation is to support nursing research and clinical practice efforts that contribute to the improvement of the quality of life for children and their families. Through the generosity of grants from the McNeil Consumer Healthcare Division of McNeil-PPC, Inc., the NAPNAP Foundation offers up to $5,000 for members and one student to attend the annual conference.
Date(s) Application is Due: Oct 4
Amount of Grant: 5,000 U.S.D
Contact: Dolores Jones, Director; (856) 857-9700; fax (856) 857-1600; djones@napnap.org or info@napnap.org
Internet: http://www.napnap.org/PNPResources/research/AvailGrantsScholarships.aspx
Sponsor: National Association of Pediatric Nurse Associates and Practitioners
20 Brace Road, Suite 200
Cherry Hill, NJ 08034-2634

NAPNAP Foundation McNeil PNP Scholarships **2991**

NAPNAP will award two annual McNeil Scholarships to students enrolled in pediatric nurse practitioner programs. Scholarships for each academic year are awarded for both the fall and spring semesters/quarters. Individuals who are enrolled in a program within the current academic year will be reviewed. Students who begin a program in January may also apply the following May if clinical course work will continue into the next academic year. Scholarship recipients will be determined by NAPNAP and notified within two months of the application deadline. The recipients will be formally recognized at NAPNAP's annual conference. Airfare and conference registration fees will be paid by NAPNAP. Instructions and application are available online.

Requirements: Applicant must be a registered nurse with at least three years previous work experience in pediatrics; have documented acceptance at a recognized PNP program; have no previous formal pediatric nurse practitioner education; demonstrate financial need, based on submission of information regarding resources; and state rationale for seeking PNP education that is consistent with NAPNAP's mission statement. In addition, students must maintain the following: full-time status (nine or more credit hours/semester) while in PNP program; GPA of 3.0 or higher; NAPNAP membership; provide NAPNAP with ongoing accounts regarding student status, as requested.
Date(s) Application is Due: Apr 30
Amount of Grant: 1,500 U.S.D
Contact: Dolores Jones, Director; (856) 857-9700; fax (856) 857-1600; djones@napnap.org or info@napnap.org
Internet: http://www.napnap.org/index.cfm?page=13&sec=102
Sponsor: National Association of Pediatric Nurse Associates and Practitioners
20 Brace Road, Suite 200
Cherry Hill, NJ 08034-2634

NAPNAP Foundation McNeil Rural and Underserved Scholarships 2992
Through the support of McNeil Consumer Healthcare Division of McNeil-PPC, Inc., NAPNAP members can apply for the McNeil Rural & Underserved Scholarship. This scholarship of up to $10,000 will cover a full academic year for a PNP student who will provide primary care to children and families in rural areas after graduating. The deadline for applications is June 15.
Requirements: To be eligible for the scholarship, applicants must meet the following criteria: plan on practicing in a rural or under served geographical region for the first two years after PNP education completed; be a registered nurse who has completed at least 2 semesters or quarters as defined by the university; show documented acceptance at a recognized Master's Degree program; have no previous formal pediatric nurse practitioner education; demonstrate financial need, based on submission of information regarding resources; and state rational for seeking PNP education that is consistent with NAPNAP's mission statement. In addition, students must maintain the following: full-time status (9 or more credit hours/semester) while in PNP program; G.P.A. of 3.0 or higher; NAPNAP membership; and provide NAPNAP with ongoing accounts regarding student status.
Date(s) Application is Due: Jun 15
Amount of Grant: 10,000 U.S.D
Contact: Dolores Jones, Director; (856) 857-9700; fax (856) 857-1600; djones@napnap.org or info@napnap.org
Internet: http://www.napnap.org/PNPResources/research/AvailGrantsScholarships.aspx
Sponsor: National Association of Pediatric Nurse Associates and Practitioners
20 Brace Road, Suite 200
Cherry Hill, NJ 08034-2634

NAPNAP Foundation Nursing Research Grants 2993
The foundation supports nursing research efforts that contribute to the improvement of the quality of life for children and their families. The foundation is primarily interested in the following areas of research: impact of nurse practitioner role in healthcare delivery; development/implementation of innovative methods for delivery of health care services in the areas of school health, community health, behavior and development, adolescent health, and chronic disease management; health care problems of children and their families; advanced nursing practice problems and issues; exploratory/evaluative research of PNP education and credentialing; and educational programs and materials for health care providers and/or children and families. Up to two grants will be awarded annually.
Date(s) Application is Due: Apr 1
Amount of Grant: 2,500 U.S.D
Contact: Dolores Jones, Director; (856) 857-9700; fax (856) 857-1600; djones@napnap.org or info@napnap.org
Internet: http://www.napnap.org/index.cfm?page=13&sec=102
Sponsor: National Association of Pediatric Nurse Associates and Practitioners
20 Brace Road, Suite 200
Cherry Hill, NJ 08034-2634

NAPNAP Foundation Reckitt Benckiser Student Scholarship 2994
Through a grant from Reckitt Benckiser Inc., the NAPNAP Foundation will award two grants of $1,600 to a Pediatric Nurse Practitioner Student to allow for attendance at the NAPNAP Annual Meeting in April. The postmark deadline for applications is November 2. The grant recipient will be determined by the NAPNAP Foundation and the recipient will be formally recognized at NAPNAP's Annual Conference.
Requirements: To be eligible for the grant, applicants must meet the following criteria: be a registered nurse and a graduate student who has completed at least 1 semester of graduate studies in a Pediatric Nurse Practitioner Program; provide documentation from the PNP Program Coordinator/Director that the student is enrolled in a recognized program of study leading to completion of qualifications for education and practice as a PNP; and be a member of NAPNAP.
Date(s) Application is Due: Nov 2
Amount of Grant: 1,600 U.S.D
Contact: Dolores Jones, Director; (856) 857-9700; fax (856) 857-1600; djones@napnap.org or info@napnap.org
Internet: http://www.napnap.org/PNPResources/research/AvailGrantsScholarships.aspx
Sponsor: National Association of Pediatric Nurse Associates and Practitioners
20 Brace Road, Suite 200
Cherry Hill, NJ 08034-2634

NAPNAP Foundation Shourd Parks Immunization Project Small Grants 2995
The NAPNAP Foundation is pleased to offer the Shourd-Parks Immunization Project Small Grant, as part of a national effort to improve health care for infants through young adults. Biennially, a $500 grant will be awarded to a NAPNAP member or chapter. To keep grant activities consistent with the goals of NAPNAP and the NAPNAP Foundation, grant applications should focus on improving health care in the area of child, adolescent and/or young adult immunizations. Four copies of the proposal and the application should be mailed to the NAPNAP National Office and postmarked no later than September 15.
Requirements: Each NAPNAP Foundation application should meet at least one of the following: 1) Improve access to immunizations for children and/or adolescents in any health care setting. 2) Strengthen existing or new health education programs in the area of immunizations. 3) Strengthen existing or new methods in the delivery of immunizations. 4) Incorporates NAPNAP's and NAPNAP Foundation priorities. Each application must meet these specific *Requirements:* 1) Be submitted by a national NAPNAP member (or group of members) or by a NAPNAP Chapter. 2) Describe the project's target population. 3) Clearly state the project purpose, goal(s) or aim(s). 4) Define one measurable outcome for the project. 5) Develop an evaluation mechanism for the one outcome. 6) Include a brief budget.
Date(s) Application is Due: Sep 15
Amount of Grant: 500 U.S.D
Contact: Dolores Jones, Director; (856) 857-9700; fax (856) 857-1600; djones@napnap.org or info@napnap.org
Internet: http://www.napnap.org/PNPResources/research/AvailGrantsScholarships.aspx
Sponsor: National Association of Pediatric Nurse Associates and Practitioners
20 Brace Road, Suite 200
Cherry Hill, NJ 08034-2634

NAPNAP Foundation Wyeth Pediatric Immunization Grant 2996
These grants will give NAPNAP members the opportunity to develop innovative programs to enhance the immunization rates in their communities. Examples of the ways in which a grant could be used include: parent education/awareness activities or materials; provider education workshops on current vaccines topics and true contra-indications to vaccinating; development of clinic policies to prevent missed opportunities; strategies for immunization registries; community outreach endeavors to assess and eliminate barriers to immunizations translation services for diverse communities; and evaluation research on effectiveness of vaccine programs to increase immunization rates. Two grants of $2,500 each and one grant of $7,500 are available.
Requirements: NAPNAP members are eligible to apply.
Date(s) Application is Due: Dec 15
Amount of Grant: 2,500 - 7,500 U.S.D
Contact: Dolores Jones, Director; (856) 857-9700; fax (856) 857-1600; djones@napnap.org or info@napnap.org
Internet: http://www.napnap.org/PNPResources/research/AvailGrantsScholarships.aspx
Sponsor: National Association of Pediatric Nurse Associates and Practitioners
20 Brace Road, Suite 200
Cherry Hill, NJ 08034-2634

Narcolepsy Network Seed Grants 2997
The network supports medical/scientific research of narcolepsy and awards seed/pilot grants in the following categories: tissue banks—facilitation of acquisition, storage, or utilization of tissue; subjects—addition of subjects with narcolepsy to an ongoing or recently completed study of some other condition; food—as regards understanding of short-term adverse effects in cataplectic-narcoleptic persons of some aspect of food consumption; and comorbidities—the sesarch for coexsisting structural or physiological alterations in persons with narcolepsy, which along with alterations in sleep reflect narcolepsy susceptibility gene(s). The network anticipates awarding one grant in each category. Complete application information is available online. Annual deadline dates may vary; contact program staff for exact dates.
Requirements: 501(c)3 tax-exempt organizations are eligible. A principal investigator with the MD, PhD, or DO degree must originate or be responsible for the projectt.
Date(s) Application is Due: Sep 15
Amount of Grant: 8,000 U.S.D
Contact: Grants Administrator; (401) 667-2523; fax (401) 633-6567; narnet@narcolepsynetwork.org
Internet: http://www.narcolepsynetwork.org
Sponsor: Narcolepsy Network
P.O. Box 294
Pleasantville, NY 10570

NARSAD Distinguished Investigator Grants 2998
The Brain and Behavior Research Foundation (formerly National Alliance for Research on Schizophrenia and Depression, or NARSAD) distributes funds for psychiatric brain and behavior disorder research. NARSAD Distinguished Investigator Grants provide support for experienced investigators (full professor or equivalent) conducting neurobiological and behavioral research. Areas of particular interest include patient populations with unique or unusual characteristics and central nervous system developments. A one-year grant of $100,000 is provided for established scientists pursuing innovative projects in diverse areas of neurobiological research.
Requirements: Applicants must be a full professor (or equivalent), and maintain peer reviewed competitively funded scientific programs. Guidelines and application deadlines are at the Foundation's website.
Restrictions: Previous grantees may not receive the Grant for a second time until five years have elapsed since the beginning date of the prior award. Grant requirements from all

previous awards must be met. Only one applicant may apply per application. Applicants may submit only one application per cycle.
Contact: Doris Ip, Manager Research Grants; (800) 829-8289 or (516) 829-5576; fax (516) 487-6930; grants@narsad.org
Internet: http://bbrfoundation.org/narsad-distinguished-investigator-grant
Sponsor: Brain and Behavior Research Foundation
60 Cutter Mill Road, Suite 404
Great Neck, NY 11021

NARSAD Independent Investigator Grants **2999**
The Brain and Behavior Research Foundation (formerly National Alliance for Research on Schizophrenia and Depression, or NARSAD) distributes funds for psychiatric brain and behavior disorder research. The NARSAD Independent Investigator Grants provide support for investigators during the critical period between the initiation of research and the receipt of sustained funding. Basic and/or clinical investigators are supported, but research must be relevant to schizophrenia, major affective disorders, or other serious mental illnesses. A two-year grant up to $100,000, or $50,000 per year, is provided to scientists at the associate professor level or equivalent, who are clearly independent and have won national competitive support as a principal investigator.
Requirements: Applicants must have a doctoral-level degree, be an associate professor (or equivalent), and have received, as Principal Investigator, competitive research support at a national level. Independent Investigators are eligible for a maximum of two Grants. Previous awardees at the associate professor level are encouraged to apply for a second Grant. Applications must be submitted electronically. Grant guidelines can be found at the Foundation's website.
Restrictions: The total budget request may not exceed $100,000. Awards cannot overlap in time, all prior grant requirements (including NARSAD Young Investigator Grants) must be met, and an applicant cannot submit more than one application per cycle. Investigators that have been promoted to associate professor, but have currently active NARSAD Young Investigator Grants, are not eligible.
Date(s) Application is Due: Mar 5
Contact: Doris Ip, Manager Research Grants; (800) 829-8289 or (516) 829-5576; fax (516) 487-6930; grants@bbrfoundation.org
Internet: http://bbrfoundation.org/narsad-independent-investigator-grants
Sponsor: Brain and Behavior Research Foundation
60 Cutter Mill Road, Suite 404
Great Neck, NY 11021

NARSAD Young Investigator Grants **3000**
The Brain and Behavior Research Foundation (formerly National Alliance for Research on Schizophrenia and Depression, or NARSAD) distributes funds for psychiatric brain and behavior disorder research. The Foundation's NARSAD Young Investigator Grants offer up to $30,000 a year for up to two years to enable promising investigators to either extend their research fellowship training or to begin careers as independent research faculty. Grants are intended to facilitate innovative research opportunities and support basic, as well as translational and/or clinical investigators. Research must be relevant to the understanding, treatment and prevention of serious psychiatric disorders such as schizophrenia, bipolar, mood and anxiety disorders, or child and adolescent psychiatric disorders.
Requirements: Applicants must have a doctoral level degree and already be employed in research training, or be in a faculty or independent research position. Applicants must have an on-site mentor or senior collaborator who is an established investigator in areas relevant to psychiatric disorders. Applications must be submitted online from the Foundation's website.
Restrictions: Pre-doctoral students or investigators at the rank of associate professor or equivalent are not eligible. Institutional overhead is not allowable.
Date(s) Application is Due: Sep 15
Contact: Doris Ip, Manager Research Grants; (800) 829-8289 or (516) 829-5576; fax (516) 487-6930; grants@bbrfoundation.org
Internet: http://bbrfoundation.org/narsad-young-investigator-grant
Sponsor: Brain and Behavior Research Foundation
60 Cutter Mill Road, Suite 404
Great Neck, NY 11021

Natalie W. Furniss Charitable Trust Grants **3001**
The Mission of the Furniss Foundation is to promote the humane treatment of animals by providing funding to societies for the prevention of cruelty to animals.
Requirements: Qualifying tax-exempt 501 (c)3 organizations operating in the state of New Jersey are eligible to apply. Applications are accepted Ocotber 1 - December 1. Complete the Common Grant Application form available at the Wachovi website to apply for funding.
Restrictions: Grants are not made for political purposes, nor to organizations which discriminate on the basis of race, ethnic origin, sexual or religious preference, age or gender.
Geographic Focus: New Jersey
Date(s) Application is Due: Dec 1
Samples: Oasis Animal Sanctuary, Inc., $4,200—animal sterilization program; Woodford Cedar Run Wildlife Refuge, Inc., $4,5000—winter operating support for wildlife rehab; Society for the Prtevention of Cruelty to Animals, $5,000—humane education in Cumberland Co. School system; Greyhound Friends of New Jersey, Inc., $5,000—broken leg program.
Contact: Wachovia Bank, N.A., Trustee; grantinquiries6@wachovia.com
Internet: https://www.wachovia.com/foundation/v/index.jsp?vgnextoid=61078689fb0aa110VgnVCM1000004b0d1872RCRD&vgnextfmt=default
Sponsor: Natalie W. Furniss Foundation
100 North Main Street
Winston Salem, NC 27150

Nathan Cummings Foundation Grants **3002**
The Nathan Cummings Foundation is rooted in the Jewish tradition and committed to democratic values and social justice, including fairness, diversity, and community. It seeks to build a socially and economically just society that values nature and protects the ecological balance for future generations; promotes humane health care; and fosters arts and culture that enriches communities. The Foundation's core programs include arts and culture; the environment; health; interprogram initiatives for social and economic justice; and the Jewish life and values/contemplative practice programs. Basic themes informing the Foundation's approach to grantmaking are: concern for the poor, disadvantaged, and underserved; respect for diversity; promotion of understanding across cultures; and empowerment of communities in need. The Board meets twice a year. Applicants should apply by January 15 to be considered for the spring Board meeting and by August 15 to be considered for the fall Board meeting.
Requirements: Eligible applicants must be 501(c)3 organizations. A two or three page letter of inquiry may be submitted with the following information: basic organizational information; contact person; grant purpose; key personnel; project budget and total organizational budget; amount requested and the length of time for which funds are being requested; and other funding sources. Projects that most closely fit with the Foundation's goals will be invited to submit a complete application.
Restrictions: The following is not funded: individuals; scholarships; sponsorships; capital or endowment campaigns; foreign-based organizations; specific diseases; local synagogues or institutions with local projects; Holocaust related projects; projects with no plans for replication; and general support for Jewish education.
Samples: Alliance for Justice, Washington, D.C., $10,000 - general support; American Friends of the Heschel Center, Inc., Jersey City, New Jersey, $150,000 - envisioning a political agenda for 21st century Israel; and American Jewish World Service, Inc., New York, New York, $100,000 - group leadership training program.
Contact: Armanda Famiglietti , Director of Grants Management; (212) 787-7300; fax (212) 787-7377; info@nathancummings.org
Internet: http://www.nathancummings.org/programs/index.html
Sponsor: Nathan Cummings Foundation
475 10th Avenue, 14th Floor
New York, NY 10018

National Blood Foundation Research Grants **3003**
Research grants support investigator-initiated original research in blood banking and transfusion medicine. Projects dealing with all aspects of blood banking, including transfusion sciences, blood banking administration, and education will be considered. Priority will be given to the following areas: transfusion medicine and immunology, hematology, immunohematology, infectious diseases, and medicine administration. Preference will be given to innovative new projects; applications from junior and less well-established investigators are encouraged. Grants will be awarded for periods of one or two years. Applications are available on the Web site.
Requirements: Preference is given to new investigators in the field. There is a $150 application fee.
Date(s) Application is Due: Dec 13
Amount of Grant: Up to 65,000 U.S.D
Contact: Program Contact; (301) 215-6552; fax (301) 907-6895; nbf@aabb.org
Internet: http://www.aabb.org/about_the_aabb/nbf/nbf2.htm#3
Sponsor: National Blood Foundation
8101 Glenbrook Road
Bethesda, MD 20814-2749

National Dairy Promotion and Research Board Scholarships **3004**
Scholarships are awarded annually to up to 19 students enrolled in college/university programs that emphasize dairy food science, technology, or marketing. Scholarships are awarded based upon outstanding academic achievement; a commitment to a career in a dairy food-related discipline; and demonstrated leadership, initiative, and integrity. Application forms are available on the Web site.
Requirements: Applications must be accompanied by an official transcript of all completed university credits and a letter of recommendation from a faculty member familiar with the applicant's scholastic performance. Sophomore, junior, and/or senior undergraduate students enrolled in college/university programs that emphasize dairy and majoring in communications, public relations, journalism, marketing, business, economics, nutrition, food science or ag education are eligible to apply.
Date(s) Application is Due: May 31
Amount of Grant: 1,500 U.S.D
Contact: Scholarship Administrator; (847) 803-2000, ext. 255 or (800) 853-2479
Internet: http://www.dairycheckoff.com/DairyCheckoff/about/scholarship.htm
Sponsor: National Dairy Council
10255 W Higgins Road, Suite 900
Rosemont, IL 60018-5616

National Emergency Medicine Association Grants **3005**
The association supports projects that foster the development and delivery of quality emergency services, with special emphasis given to heart attack and stroke. Proposals should address education or direct services to victims. Also to receive consideration are childrens' emergency issues and projects teaching good parenting skills. Programs must be related to health and emergency issues and prevention or diminishing of trauma events. Types of support include capital, program development, research, and training grants. Applications are accepted on an ongoing basis.
Samples: Beebe Medical Ctr (Lewes, DE)—to improve its trauma care services for people in southern Delaware, $83,778; National Stroke Assoc (Englewood, CO)—to update a

publication that educates emergency room physicians about emergency evaluation and treatment of stroke victims, $17,000.
Contact: Administrator; (410) 494-0300; fax (410) 494-0725; info@nemahealth.org
Internet: http://www.nemahealth.org/organization/grants.htm
Sponsor: National Emergency Medicine Association
306 W Joppa Road
Baltimore, MD 21204

National Headache Foundation Research Grants 3006
The National Headache Foundation is interested in research protocols that are objectively sound and whose results can, when published in the medical literature, contribute to the better understanding and treatment of headache and pain. Grants may be used for the purchasing of supplies needed for the study (chemical or pharmacological reagents, laboratory animals, tissue culture materials. forms, etc.) and for costs related to the recruiting of patients, data analysis, interpreting or reading results, etc. Applications are available online, and are due by December 1.
Requirements: Funds for a research project approved by the National Headache Foundation are awarded to the individual who is the principal investigator of the study. The funds are sent to the institution that employs the PI but designated for the PI and the approved study.
Restrictions: Grants from the NHF will not cover costs including overhead, salaries (payment for technicians, others collecting data, administrative services, etc.), purchasing of equipment, rent or other indirect expenses.
Date(s) Application is Due: Dec 1
Contact: Carolyn Smith, Executive Assistant; (312) 274-2652 or (312) 274-2650; fax (312) 640-9049; info@headaches.org
Internet: http://www.headaches.org/For_Professionals/Research_Grants
Sponsor: National Headache Foundation
820 N Orleans, Suite 217
Chicago, IL 60610-3132

National Headache Foundation Seymour Diamond Clinical Fellowship in Headache Education 3007
The one-year fellowship is intended to encourage, provide for, and engage physicians in the clinical management of headache patients. The program is an individualized hands-on approach at a headache clinic mutually selected by the committee and the applicant. The objectives are to develop competence in the management of common and complex headache problems, to publish a research paper keyed to the treatment of headache, and to prepare the applicant for a career in treating headache patients. The participating institution will be a specialized facility dedicated primarily to headache management, with facilities for inpatient and outpatient treatment. Up to 80 percent of the fellow's time will be clinical, 20 percent study/research.
Requirements: Applicant will be a graduate of an approved U.S. medical school and be four years or more postgraduate.
Date(s) Application is Due: Feb 1
Amount of Grant: 40,000 U.S.D
Contact: Fellowship Selection Committee; (312) 274-2650; fax (312) 640-9049; info@headaches.org
Internet: http://www.headaches.org/For_Professionals/Fellowship_Grants
Sponsor: National Headache Foundation
820 N Orleans, Suite 217
Chicago, IL 60610-3132

National Heart Foundation of New Zealand Grants 3008
The foundation is dedicated to the reduction of premature death and disability from heart and circulatory disease, including stroke. Types of support include research project grants, training fellowships enabling New Zealand cardiology trainees to study abroad, senior fellowships, and a chair in cardiovascular studies. Contact program staff for availability.
Geographic Focus: All States, New Zealand
Contact: Program Contact; (09) 571 9191; fax (09) 571 9190; info@nhf.org.nz
Internet: http://www.nhf.org.nz/index.asp?PageID=2145820407
Sponsor: National Heart Foundation of New Zealand
P.O. Box 17160
Greenlane, Auckland 1130 New Zealand

National Lottery Community Fund Grants 3009
The organization operates in the United Kingdom and internationally and awards grants to charitable, benevolent, and philanthropic organizations. The organization particularly wants to support projects that help people who face disadvantage that is severe, long-term, difficult to tackle, or who face a combination of needs. Grants support projects that try to prevent or reduce, as far as possible, future disadvantage.
Contact: Gerald Oppenheim, Director; 44-020-7-747-5299; fax 44-020-7-747-5214; enquiries.res@biglotteryfund.org.uk
Internet: http://www.community-fund.org.uk/about-us/our-grants-programmes/index.html
Sponsor: National Lottery Community Fund
Saint Vincent House, 16 Suffolk Street
London, SW1Y 4NL England

National MPS Society Grants 3010
The society awards research grants and fellowships to qualified individuals to promote medical research in the fields of mucopolysaccharidosis (MPS) and mucolipidosis (ML). The program supports MPS (I, II, III IV, V, and VI) research, ML II/III research, and general MPS research. Grants and fellowships are awarded for a two-year period. Application forms and guidelines are available online.
Requirements: Qualified medical researchers are eligible.
Date(s) Application is Due: May 1
Amount of Grant: 30,000 U.S.D
Samples: Mount Sinai School of Medicine (New York, NY)—to support the work of Caleogera Simanaro, for medical research on bone and joint diseases in mucopolysaccharide diseases, $100,000; Iowa State U (IA)—to support the work of Matthew Ellinwood, for medical research on Sanfilippo syndrome, a genetic disease that prevents the normal degeneration of cells and progressively complicates cognitive and physical function, $60,000.
Contact: Barbara Wedehase, Executive Director; (207) 947-1445; fax (207) 990-3074; barbara@mpssociety.org
Internet: http://www.mpssociety.org/grants.html
Sponsor: National MPS Society
P.O. Box 736
Bangor, ME 04402-0736

National Parkinson Foundation Clinical Research Fund Grants 3011
The National Parkinson Foundation (NPF)'s Clinical Research Fund (CRF) is designed to support cutting-edge clinical studies of new therapeutic options for Parkinson's patients as well as research comparing the effectiveness of existing treatment options. Conducted by or in collaboration with NPF's Centers of Excellence, CRF-funded research aims to discover the best treatment and care options for people living with Parkinson's disease. Broad clinical-research areas funded by the CRF program are as follows: care quality; imaging; cognition; physical therapy; care models; exercise; and biomarkers. CRF Requests for Application are issued as funding becomes available, but no less than once per year. Applicants may request two-year funding (or less) for their projects. An example of the CRF application process is as follows: an RFP is released in August; NPF schedules pre-submission conference calls for September and October; applicants submit their proposals in December; peer-review is conducted January through March; awards are announced in April. Specific deadline dates will vary from year to year and will be detailed in the RFP. More detailed information is generally posted in a listserv to which all applicants are invited to subscribe (address given in the RFP). Applications must be submitted electronically through NPF's online grant submission system. Registration is required.
Requirements: Principal investigators must be clinical investigators affiliated with a currently-certified NPF Center of Excellence. Management of the grant budget and reporting must be done at the Center of Excellence. Further, each application must be cosigned by the institutional officer responsible for grants and contracts. Principal investigators must meet their institution's criteria for eligibility to receive grants and oversee expenditures. Individual investigators at non-certified institutions may apply as collaborators. Applicants must reasonably expect to have Institutional Review Board approval shortly after the award date.
Restrictions: There are no limits on the number of grants an institution may request but each grant is limited to no more than $250,000 over a two-year period. Postdoctoral fellows, fellows, and students cannot serve as a principal investigator. Grants cannot be used to purchase equipment or to support conference fees or travel (except for travel to a Center of Excellence collaborating on the research project). Indirect costs will not be funded. NPF no longer funds basic laboratory research.
Geographic Focus: All States, All Countries
Amount of Grant: Up to 250,000 U.S.D
Samples: Antonio Strafella, M.D., Ph.D., F.R.C.P.C., Toronto Western Research Institute, Toronto, Ontario, Canada - "Early Detection of Cognitive Changes in the Brain" (2010); Sherrilene Classen, Ph.D., M.P.H, O.T.R./L., University of Florida, Gainesville, Florida - "Visual Attention Deficits as an Early Sign of Cognitive Change" (2010); David Houghton, M.D., M.P.H., University of Louisville, Louisville, Kentucky - "Comparing Physical Therapy Outcomes: LSVT-BIG© vs. Aquatic Methods" (2010).
Contact: Jorge Zamudio, (305) 243-4830; fax (305) 243-8144; jzamudio@parkinson.org or iiresearchgrants@parkinson.org
Internet: http://www.parkinson.org/Professionals/Grants—-Research.aspx
Sponsor: National Parkinson's Foundation
1501 N.W. Ninth Avenue / Bob Hope Road
Miami, FL 33136-1494

National Patient Safety Foundation Grants 3012
The foundation awards grants in the United States and Canada to health, medical, and research institutions to promote studies that prevent human errors, system errors, and preventable injuries in health care. Top priority will be given to proposals that apply to a broad spectrum of problems, are relevant to patient safety, and promise to improve conditions with a direct path to implementation into medical care. Of particular interest are projects that address how the relationship between physicians and patients contributes to or inoculate against error and how enhanced patient involvement in care may reduce errors. The foundation also encourages innovation, creative approaches, and cross-disciplinary research teams. Requests may be made for up to two years of support. The listed deadline date is for letters of intent.
Requirements: U.S. and Canadian health, medical, and research institutions are eligible.
Date(s) Application is Due: Jul 1
Amount of Grant: 100,015 U.S.D
Samples: Bruce Lambert, U of Illinois (Chicago, IL)—for research on the mechanisms that lead to patient injuries sustained in hospitals and on ways of reducing preventable healthcare errors, $99,923 over two years.
Contact: Emily Wilson, Projects Manager; (703) 506-3280; info@npsf.org
Internet: http://www.npsf.org/html/research/rfp.html
Sponsor: National Patient Safety Foundation
8405 Greensboro Drive, Suite 800
McLean, VA 22102-5120

National Patient Safety Foundation Patient Safety Research Grants 3013

NPSF seeks to stimulate new, innovative projects directed toward enhancing patient safety in the United States. Its objective is to promote studies leading to prevention of human errors, system errors, and any forms of preventable injuries in health care and the adverse consequences to patients that may result. In the first stage of a two-stage application process, letters of intent (LOIs) are solicited for research and development that is broadly related to identifying the causes of preventable injuries and errors and/or developing prevention strategies and methods to implement them. From these LOIs, a subset will be selected for preparation of a more detailed proposal. Highest priority is given to investigations that apply to a broad spectrum of problems and patients and that are clearly relevant to patient safety. Priority is also given to projects with the promise of improved methods of patient safety with a defined and direct path to implementation into health care. Areas of research include, but are not limited to: identifying, reporting on, analyzing, or otherwise assisting in the understanding of the causes of preventable injuries in health care; enhancing understanding of the impact of the changes in health care on the system, on human error, and on safety; developing methods for avoiding preventable injuries; assessing the cost effectiveness of techniques designed to increase patient safety; and developing educational tools and/or training materials designed to alter behaviors to prevent errors and increase safety. Innovation and creativity are strongly encouraged, as are cross-disciplinary research teams. Priority will be given to topics that do not have other available sources for funding. LOIs are due July 1; full applications are due December 2.
Requirements: Awards are made to a sponsoring institution, not to individuals or to departments. Any official member of a nonprofit institution in the United States may apply. Only one person should be listed as the principal investigator. All coinvestigators, collaborators, and consultants should be listed on the LOI.
Date(s) Application is Due: Jul 1; Dec 2
Amount of Grant: Up to 100,000 U.S.D
Contact: NPSF Program Coordinator; (703) 506-3280; fax (703) 506-3266; info@npsf.org.
Internet: http://www.npsf.org/html/research
Sponsor: American Medical Association Foundation
8405 Greensboro Drive, Suite 800
McLean, VA 22102

National Psoriasis Foundation Research Grants 3014

The foundation will award up to five $50,000 one year pilot project grants to support advancement of both basic and clinical research into the cause and cure of psoriasis and psoriatic arthritis. The purpose of pilot project grants is to provide support for innovative early stage ideas. These grants are designed to allow the investigator to determine whether their idea is worth pursuing and then to secure longer-term funding.
Requirements: Applicants should use their NIH biographical sketches. Download the required application form from the website.
Date(s) Application is Due: Oct 20
Amount of Grant: Up to 50,000 U.S.D
Contact: Bruce F. Bebo, Jr., Director of Research and Medical Programs; (503) 244-7404 or (800) 723-9166; research@psoriasis.org
Internet: http://www.psoriasis.org/research/grants/index.php
Sponsor: National Psoriasis Foundation
6600 SW 92nd Avenue, Suite 300
Portland, OR 97223-7195

National Sleep Foundation Pickwick Postdoctoral Fellowship 3015

The fellowships enable young researchers to devote full-time professional effort to research related to the study of sleep or sleep disorders. Fellows receive funding for one or two years. The second year of support is dependent upon satisfactory progress in the first year. Applicants must conduct research in recognized U.S. or Canadian programs of study or laboratories with strong mentorship in the appropriate area. Fellowships are available for basic, applied, or clinical research. Minorities are encouraged to apply.
Requirements: Candidates must hold either an MD, DVM, PhD, or DO degree in which the degree or subsequent training has been completed within the past five years.
Restrictions: Individuals in a faculty position or who have received an NIH grant are ineligible.
Date(s) Application is Due: Dec 1
Amount of Grant: 50,548 - 95,596 U.S.D
Contact: Jessica Steinitz, (202) 347-3471, ext. 209; fax (202) 347-3472; jsteinitz@sleepfoundation.org
Internet: http://www.sleepfoundation.org/nsfalert/index.php?id=31&secid=159#item159
Sponsor: National Sleep Foundation
1522 K Street NW, Suite 500
Washington, DC 20005

National Starch and Chemical Foundation Grants 3016

The foundation supports nonprofits in company-operating locations in the areas of higher education, children's services, and hospitals. Types of support include general operating support, continuing support, employee-related scholarships, and employee matching gifts. There are no application guidelines. Applicants are asked to submit a written proposal.
Requirements: Nonprofits in company operating areas such as Georgia, Illinois, Indiana, Missouri, North Carolina, New Jersey, Pennsylvania, South Carolina, and Tennessee may apply.
Geographic Focus: Georgia, Illinois, Indiana, Missouri, New Jersey, North Carolina, Pennsylvania, South Carolina, Tennessee
Amount of Grant: 500 - 10,000 U.S.D
Contact: Carmen Ortiz, (908) 685-5201; carmen.ortiz@nstarch.com
Internet: http://news.nationalstarch.com/NewsStory.asp?newsItemId=214
Sponsor: National Starch and Chemical Foundation
10 Finderne Avenue
Bridgewater, NJ 08807

National Student Nurses' Association Foundation Scholarship Program 3017

Scholarships are based on academic achievement, financial need, and involvement in nursing student organizations and community activities related to health care. Regular scholarships are open to all nursing students; career mobility scholarships are open to licensed practical nurses enrolled in programs leading to licensure as a registered nurse and to registered nurses enrolled in baccalaureate programs in nursing. Specialty scholarships have additional criteria set by the sponsors. Applicants submit one application and are considered for all scholarships for which they are eligible. Guidelines and application are available online.
Requirements: Students currently enrolled in state-approved schools of nursing or pre-nursing programs in baccalaureate, diploma, generic doctorate, and generic master's programs are eligible. Applications must be accompanied by a $10 processing fee.
Restrictions: Graduating high school seniors are not eligible. Funds are not available for graduate study, unless it is for a first degree in nursing.
Date(s) Application is Due: Jan 20
Amount of Grant: 1,000 - 2,500 U.S.D
Contact: Diane Mancino; (718) 210-0705; fax (718) 210-0710; diane@nsna.org
Internet: http://www.nsna.org/foundation/index.asp
Sponsor: National Student Nurses' Association Foundation
45 Main Street, Suite 606
Brooklyn, NY 11201

Nationwide Foundation Grants 3018

Grants are given to tax-exempt organizations located in central Ohio or recommended for funding by a state office or subsidiary/affiliate, In order of emphasis, grants support: health and human services, higher education, culture and arts, and civic and community. Funds are provided for arts programs, including fine arts, dance, music, museum, and theater programs; business and economics; children and youth; health programs, mental health, and programs dealing with alcoholism and drug abuse; assistance for and care of the aged; educational programs for minorities, persons with disabilities, and the disadvantaged; social service and social welfare programs; and urban affairs. Types of support include operating budgets, continuing support, annual campaigns, seed grants, projects and programs, employee matching gifts, and capital campaigns. Preference is given to organizations located in central Ohio.
Requirements: Seventy percent of grants are awarded to nonprofit organizations in the greater Columbus, OH, area, with the rest going to organizations in which Nationwide has a large number of employees and agents.
Restrictions: Individuals, organizations that support political candidates or political philosophies, public elementary and secondary schools, lobbying groups, and fraternal or veterans organizations are not eligible for these grants. Research is not funded.
Geographic Focus: Ohio
Date(s) Application is Due: Sep 1
Amount of Grant: 5,000 - 50,000 U.S.D
Samples: To be divided among more than 1000 organizations nationwide—to benefit these nonprofit groups, which include more than 700 United Way chapters, $14 million.
Contact: Grants Administrator; (614) 249-4310 or (614) 249-0039; fax (614) 249-3147
Internet: http://www.nationwide.com/aboutus/community/fndatn.htm
Sponsor: Nationwide Foundation
1 Nationwide Plaza, MD 1-22-05
Columbus, OH 43215-2220

NBME Stemmler Medical Education Research Fund Grants 3019

The purpose of the fund is to provide support for research or development of innovative evaluation methodology or techniques with the potential to advance assessment in medical education or practice, at eligible medical schools. Collaborative investigations within or among institutions are eligible, particularly as they strengthen the likelihood of the project's contribution and success. Proposed projects may request up to two years of support. Applicants for both new proposals and competing renewals may request up to $150,000 of NBME funding support for a project period of up to two years.
Requirements: Medical schools accredited by the Liaison Committee on Medical Education or the American Osteopathic Association are eligible.
Date(s) Application is Due: Oct 3
Amount of Grant: Up to 150,000 U.S.D
Contact: Officer of Research Support; (215) 590-9657; fax (215) 590-9488; stemmlerfund@nbme.org
Internet: http://www.nbme.org/research/stemmler/index.html
Sponsor: National Board of Medical Examiners
3750 Market Street
Philadelphia, PA 19104

NBNA Lauranne Sams Scholarships 3020

NBNA awards undergraduate and graduate nursing scholarships primarily to African Americans enrolled at any level in a nursing program. Each chapter must submit completed NBNA scholarship application forms filled out by each candidate it recommends. Scholarship winners are notified by July 1, and the awards are presented at the annual August conference.
Requirements: Candidate must be enrolled in a nursing program (AD, BSN, or LPN/LVN) and be in good scholastic standing at the time of application. Applications must be accompanied by two letters of recommendation (one from the school of nursing and

one from the local chapter). Letters from the school of nursing and a nurse in the area will suffice for candidates who have no local chapter in the area.
Date(s) Application is Due: Apr 15
Amount of Grant: 500 - 2,000 U.S.D
Contact: Scholarships Administrator; (800) 575-6298 or (301) 589-3200; fax (301) 589-3223; NBNA@erols.com
Internet: http://www.nbna.org/scholarship.htm
Sponsor: National Black Nurses Association
8630 Fenton Street, Suite 330
Silver Spring, MD 20910-3803

NCAA CHOICES Grants 3021
The NCAA and Anheuser-Busch Companies award grants to NCAA member institutions and conferences for the development, implementation, and evaluation of effective alcohol education and prevention programs on college campuses. Funded programs use athletes or athletic events to deliver the educational programs to the entire campus. Collaboration between athletic departments and other campus organizations involved in alcohol education is encouraged. Grants can be used to develop, implement, or evaluate programs. The Speakers Grant program offers grants of partial or full speaker fees for departments of athletics offering drug education or wellness programs for student athletes and athletics personnel. Application and guidelines are available online.
Requirements: NCAA member institutions are eligible.
Date(s) Application is Due: Feb 14
Amount of Grant: 5,000 - 15,000 U.S.D
Contact: Mary Wilfert, (317) 917-6319 or (317) 917-6339; mwilfert@ncaa.org
Internet: http://www1.ncaa.org/membership/ed_outreach/health-safety/drug_ed_progs/index
Sponsor: National Collegiate Athletic Association
P.O. Box 6222
Indianapolis, IN 46206-6222

NCCAM Exploratory Developmental Grants for Complementary and Alternative Medicine (CAM) Studies of Humans 3022
The National Center for Complementary and Alternative Medicine (NCCAM), the National Cancer Institute (NCI), the National Institute on Aging (NIA), and the Office of Dietary Supplements (ODS) invites high quality exploratory/developmental research grant applications of humans in all domains of complementary and alternative medicine (CAM), in order to obtain preliminary data that can be used as a foundation for a larger clinical study. NCCAM groups CAM practices into four domains: biologically-based practices; energy medicine; manipulative and body-based therapies, and mind-body medicine. In addition, NCCAM funds studies of whole medical systems that employ practices drawn from the four domains. The total project period for an application submitted in response to this funding opportunity may not exceed three years. Direct costs are limited to $400,000 over a three-year period, with no more than $250,000 in direct costs allowed in any single year.
Requirements: The following organizations and institutions may apply: public/state controlled institutions of higher education; private institutions of higher education; Hispanic-serving institutions; historically Black colleges and universities (HBCUs); tribally controlled colleges and universities (TCCUs); Alaska native and native Hawaiian serving institutions; nonprofits with 501(c)3 IRS status; nonprofits without 501(c)3 IRS status; small businesses; for-profit organizations; State governments; regional organizations; U.S. territories or possessions; Indian/Native American tribal governments; Indian/Native American tribally designated organizations; county governments; city or township governments; special district governments; independent school districts; public housing authorities/Indian housing authorities; eligible agencies of the Federal government; and faith-based or community based organizations.
Date(s) Application is Due: Feb 16; Jun 16; Oct 16
Amount of Grant: Up to 400,000 U.S.D
Contact: April Bower, (301) 451-3560; fax (301) 480-3621; bowera@mail.nih.gov
Internet: http://grants.nih.gov/grants/guide/pa-files/PAR-08-135.html#SectionVII
Sponsor: National Center for Complementary and Alternative Medicine
9000 Rockville Pike
Bethesda, MD 20892-5475

NCCAM Predoctoral Research Training in Complimentary and Alternative Medicine 3023
NCCAM provides awards to individuals for research training in specified areas of biomedical and behavioral research related to complementary and alternative medicine (CAM) to help ensure that highly trained scientists will be available in adequate numbers and in appropriate research areas and fields to meet the nation's health research needs. As such, this program is intended to provide a mechanism to train future generations of outstanding scientists committed to pursuing a research career in CAM. PA: PAR-00-023
Requirements: Applicants must be citizens or noncitizen nationals of the United States, or have been lawfully admitted to the United States for permanent residence and have in their possession an Alien Registration Receipt Card at the time of award. Applicants must have received, as of the activation date of the award, a baccalaureate degree and must be enrolled in a program leading to a research doctorate such as the PhD or DSc, or a combined clinical and research degree such as DC, DO, MD, or ND/PhD by the proposed fellowship activation date.
Date(s) Application is Due: Apr 5; Aug 5; Dec 5
Contact: Dr. Neal West; (301) 402-5867; fax (301) 402-4741; westn@od.nih.gov
Internet: http://grants.nih.gov/grants/guide/pa-files/PAR-00-023.html
Sponsor: National Center for Complementary and Alternative Medicine
9000 Rockville Pike
Bethesda, MD 20892-5475

NCCAM Ruth L. Kirschstein National Research Service Awards for Postdoctoral Training in Complementary and Alternative Medicine 3024
The National Center for Complementary and Alternative Medicine (NCCAM) awards individual postdoctoral research training fellowships to promising applicants with the desire and potential to become productive, independent investigators in the field of complementary and alternative medicine (CAM) research. The primary objective of this funding opportunity is to help ensure that diverse pools of highly trained scientists will be available in adequate numbers and in appropriate research areas to carry out the Nation's biomedical, behavioral, and clinical CAM research agendas. Applicants may come from a range of career backgrounds, including biomedical research, conventional allopathic medicine, or CAM healing practices. The number of awards and the total amount of funding that NCCAM expects to award through this announcement will depend on the scientific merit of applications received, relevance to NCCAM's program priorities, and the availability of funds.
Requirements: Eligible organizations include for-profit and non-profit organizations, public or private institutions such as universities, colleges, hospitals and laboratories, units of State and local governments, eligible agencies and labs of the Federal government including NIH intramural labs, and domestic or foreign institutions/organizations. Applicant fellows must be citizens or non-citizen nationals of the United States, or have been lawfully admitted to the United States for permanent residence and have in their possession an Alien Registration Receipt Card.
Restrictions: Individuals on temporary or student visas are not eligible.
Date(s) Application is Due: Apr 8; Aug 8; Dec 8
Contact: Nancy J. Pearson, (301) 594-0519; fax (301) 480-3621; pearsonn@mail.nih.gov
Internet: http://grants.nih.gov/grants/guide/pa-files/PAR-07-319.html
Sponsor: National Center for Complementary and Alternative Medicine
9000 Rockville Pike
Bethesda, MD 20892-5475

NCCAM Translational Tools for Clinical Studies of CAM Interventions Grants 3025
This Funding Opportunity Announcement (FOA) issued by the National Center for Complementary and Alternative Medicine (NCCAM) of the National Institutes of Health (NIH) encourages investigator(s)-initiated applications that propose to develop, enhance, and validate translational tools to facilitate rigorous study of complementary and alternative medicine (CAM) approaches that are in wide use by the public. Multidisciplinary studies and collaboration among investigators with expertise in appropriate disciplines are encouraged. NCCAM intends to award up to $3 million (total costs) per year, to support approximately 4 to 6 new grants. A maximum of $300,000 (direct costs) per year may be requested and the total project period for an application submitted in response to this FOA will be five years.
Requirements: The following organizations and institutions are eligible to apply: individuals; public/state controlled institutions of higher education; private institutions of higher education; Hispanic-serving institutions; historically Black colleges and universities (HBCUs); tribally controlled colleges and universities (TCCUs); Alaska Native and Native Hawaiian serving institutions; nonprofits with 501(c)3 IRS status; nonprofits without 501(c)3 IRS status; for-profit organizations; state governments; Indian/Native American tribal governments (federally recognized); Indian/Native American tribally designated organizations; county governments; city or township governments; special district governments; independent school districts; public housing authorities/Indian housing authorities; U.S. territory or possession; Indian/Native American tribal governments (other than federally recognized); regional organizations; non-domestic (non-U.S.) entities (foreign organizations); other eligible agencies of the federal government; and faith-based or community-based organizations.
Date(s) Application is Due: Jul 17
Amount of Grant: 300,000 U.S.D
Contact: Dr. Partap S. Khalsa, Division of Extramural Research Contact; (301) 594-3462; fax (301) 480-1587; khalsap@mail.nih.gov
Internet: http://grants.nih.gov/grants/guide/rfa-files/RFA-AT-09-002.html
Sponsor: National Center for Complementary and Alternative Medicine
9000 Rockville Pike
Bethesda, MD 20892-5475

NCHS National Center for Health Statistics Postdoctoral Research Appointments 3026
The National Center for Health Statistics (NCHS) is the Nation's principal health statistics agency and provides the information needed to develop the programs and policies that will improve the health of the American people. The objective of the NCHS Postdoctoral Research Program is to provide opportunities for postdoctoral candidates of unusual promise and ability to conduct research in problems of their choosing that are compatible with the interests of NCHS. A postdoctoral researcher is a temporary salaried employee of NCHS analogous to fellows or similar temporary researchers at the postdoctoral level in universities and other organizations. Appointments are two years in length, but extensions are possible with special consideration. Benefits include sick leave, annual leave, thrift savings, and health and life insurance. Flex time and alternate work schedules are available if desired, contingent on the supervisor's approval. Cost of relocation will be determined on an individual basis. An evaluation is conducted after one year to ensure that the postdoctoral researcher is making suitable progress. General areas of interest for research at NCHS include statistical theory, survey methodology, statistical computing, economics, demography, and social and behavioral science. Specific areas of interest are: confidentiality protection and related topics; data linkage; quality control, quality improvement, and its measurement; adjustment for nonresponse in sample surveys; telephone survey methods; data collection in a changing health care system; small-area estimation; cognitive processes and question response; data editing and imputation; statistical computing; sampling and estimation; questionnaire design and measurement research; human factors and usability

research; and data mining. To apply, applicants must submit a statement of research interest, a curriculum vitae, official transcripts, and three reference letters. The statement of research interest should describe the purpose and context of the research project, the research approach and methods to be used, key activities to be undertaken during the fellowship period, and the expected contributions of the proposed research. For fullest consideration, applicants should submit the application by the deadline date to the address given. However, applications will be accepted throughout the year until positions are filled. Evaluation of proposals will be conducted annually in March or April by a special panel convened for this purpose. Notification of awards will be made beginning in mid-April. The date on which an appointment would begin is negotiable on an individual basis but is expected to occur between June 1 and December 31. Applicants can find complete information on the postdoctoral research program and on application submission requirements at the program website. Applicants are encouraged to contact NCHS to obtain specific information on current research and available technical facilities and to speak to NCHS researchers working in the specific area of interest.
Requirements: Awardees must hold a Ph.D. or other earned research degree recognized in the United States as equivalent to the Ph.D. or must present acceptable evidence of having completed all the formal academic requirements for the degree before appointment. Applicants must have demonstrated ability for creative research. An applicant's training and research experience may be in any appropriate discipline or combination of disciplines required for the proposed research. Postdoctoral researchers must devote their full-time effort to the research program and must be in residence at NCHS during the program. No period of tenure may be spent in residence at another agency or institution.
Restrictions: Applicants must be U.S. citizens or legal permanent residents with a work authorization. Postdoctoral research positions are awarded only to persons who have held doctorates for less than 3 years or are in the process of receiving doctorate degrees at the time of application. No additional monetary aid or other remuneration may be accepted from another appointment, fellowship, or similar grant during the period of the program.
Date(s) Application is Due: Feb 28
Amount of Grant: 149,744 - 231,484 U.S.D
Contact: Cynthia Link , (800) 232-4636 or (301) 458-4344; fax (301) 458-4018; cynthia.link@cdc.hhs.gov
Internet: http://www.cdc.gov/nchs/about/postdoc.htm
Sponsor: National Center for Health Statistics
3311 Toledo Road
Hyattsville, MD 20782

NCI/NCCAM Quick-Trials for Novel Cancer Therapies 3027

This PA sponsored by NCI and NCCAM expands the initiative to all cancer sites and provides investigators with rapid access to support for pilot, phase I, and phase II cancer clinical trials testing new agents and patient monitoring and laboratory studies to ensure the timely development of new therapeutic approaches. QUICK-TRIAL will provide a new approach designed to simplify the grant application process and provide a rapid turnaround from application to funding. Features include a modular grant application and award process, inclusion of the clinical protocol within the grant application, and accelerated peer review with the goal of issuing new awards within five months of application receipt. Inclusion of the complete clinical protocol within the PHS 398 grant application is intended to simplify the application process by eliminating the need to duplicate protocol details in the research plan section. Investigators may apply for a maximum of two years of funding support using the exploratory/developmental (R21) grant mechanism for up to $250,000 direct costs per year. Therapeutic trials in human subjects employing new agents and therapeutic approaches, including CAM treatments, whether used as a single agent/modality or in combination, are appropriate. Preference will be given to clinical trials that include laboratory studies to validate mechanistic hypotheses or clinical correlates that can meaningfully guide further clinical development. A letter of intent is due one month prior to application receipt date. PA: PAR-04-155
Requirements: Applications may be submitted by foreign and domestic, for-profit and nonprofit organizations, both public and private; units of state and local governments; and eligible agencies of the federal government.
Date(s) Application is Due: Apr 9; Aug 9; Dec 9
Amount of Grant: Up to 250,000 U.S.D
Contact: Amy Connolly; (301) 496-8786; fax (301) 496-8601; amyconnolly@nih.gov
Internet: http://grants.nih.gov/grants/guide/pa-files/par-04-155.html
Sponsor: National Institutes of Health
6701 Rockledge Drive, Room 1040, MSC 7710
Bethesda, MD 20892-7710

NCI/NIDA Transdisciplinary Tobacco Use Research Centers Grants 3028

Transdisciplinary tobacco use research centers (TTURCs) are expected to conduct research that will lead to major scientific advances in knowledge of tobacco use, nicotine addiction, prevention and treatment interventions, and the translation of this knowledge into practice. There are several overarching goals: increase the number of investigators from relevant disciplines who focus on the study of tobacco use as part of transdisciplinary teams; generate basic research evidence to improve understanding of the etiology and natural history of tobacco use; produce evidence-based tobacco use interventions that can translate to the community and specific understudied or underserved populations; increase the number of evidence-based interventions that are novel, including the development, testing and dissemination of innovative behavioral treatments and prevention strategies based upon findings from basic research; train transdisciplinary investigators capable of conducting cutting-edge tobacco use research; and to increase the number of peer-reviewed publications in the area of tobacco use, nicotine addiction, and treatment. RFA-CA-04-012
Requirements: Applications may be submitted by public and private, domestic, for-profit and nonprofit organizations, units of state and local governments, and eligible agencies of the federal government. Domestic applications with foreign components are permitted. To be considered, applicant organizations must have: (1) an established investigator as a strong principal investigator who can oversee and conduct planning activities, provide direction to the TTURC, ensure a transdisciplinary research emphasis, and build a career development program; full institutional commitment from the applicant institution; clearly delineated and formally confirmed, subcontracted, collaborative scientific arrangements with scientists from other institutions, including foreign institutions, may be included.
Contact: Crystal Wolfrey, Grants Administration Branch, NCI; (301) 496-8634; fax (301) 496-8601; crystal.wolfrey@nih.gov or wolfreyc@mail.nih.gov
Internet: http://grants1.nih.gov/grants/guide/rfa-files/RFA-CA-04-012.html
Sponsor: National Cancer Institute
6130 Executive Boulevard, Executive Plaza N
Bethesda, MD 20892

NCI Academic-Industrial Partnerships for Development and Validation of In Vivo Imaging Systems and Methods for Cancer Investigations 3029

The overall goal of this Funding Opportunity Announcement (FOA) is to accelerate the transition of in vivo spectroscopic and imaging systems and methods into cancer research, clinical trials, and/or clinical practice. A related goal is to advance new technologies and methods for animal imaging. To promote these goals, the Cancer Imaging Program (CIP) of the National Cancer Institute (NCI) solicits applications from multi-institutional research partnerships formed by academic, industrial, and other investigators. The partners will establish an inter-disciplinary, multi-institutional research team for the purpose of developing imaging technology that is applicable to cancer research involving human subjects and/or animal models. The academic-industrial partnerships must be dedicated to the complete research and development cycle, culminating in the product prototype stage. The partnership model is expected to more readily overcome various barriers to the development of new imaging devices and technologies faced by either academia or industry working alone.
Requirements: Eligible institutions include: public and state controlled institutions of higher education; private institutions of higher education; nonprofits with 501(c)3 IRS status (other than institutions of higher education); nonprofits without 501(c)3 IRS status (other than institutions of higher education); small businesses; for-profit organizations (other than small businesses); state governments; U.S. territories or possessions; non-domestic (non-U.S.) entities (foreign organizations); and eligible agencies of the federal government.
Date(s) Application is Due: Feb 5; Jun 5; Oct 5
Contact: Guoying Liu, (301) 496-9531; fax (301) 480-3507; guoyingl@mail.nih.gov
Internet: http://grants.nih.gov/grants/guide/pa-files/PAR-07-214.html
Sponsor: National Cancer Institute
6130 Executive Plaza, EPN Suite 6000, MSC 7412
Bethesda, MD 20892-7412

NCI Application and Use of Transformative Emerging Technologies in Cancer Research 3030

This Funding Opportunity Announcement (FOA) solicits grant applications proposing exploratory research projects on evaluating the performance of emerging molecular and cellular analysis technologies utilizing an appropriate cancer-relevant biological system. Technologies proposed for evaluation may have broad applicability but must ultimately be suitable for cancer-relevant analyses such as the detection (at various molecular and cellular levels) of cancer-related characteristics and/or structural and/or functional alterations. Letters of intent are requested by January 23, April 27, and August 30; full proposal deadlines are February 23, May 27, and September 30. Annual deadline dates may vary; contact program staff for exact dates.
Requirements: Eligible institutions include: public and state controlled institutions of higher education; private institutions of higher education; nonprofits with 501(c)3 IRS status (other than institutions of higher education); nonprofits without 501(c)3 IRS status (other than institutions of higher education); small businesses; for-profit organizations (other than small businesses); state governments; U.S. territories or possessions; non-domestic (non-U.S.) entities (foreign organizations); and eligible agencies of the federal government.
Date(s) Application is Due: Jan 23; Feb 23; Apr 27; May 27; Aug 30; Sep 30
Amount of Grant: Up to 275,000 U.S.D
Contact: Richard Aragon, (301) 496-1550; fax (301) 496-7807; raragon@mail.nih.gov
Internet: http://grants.nih.gov/grants/guide/rfa-files/RFA-CA-09-006.html
Sponsor: National Cancer Institute
31 Center Drive, Building 31, Room 10A52, MSC 2580
Bethesda, MD 20892-2580

NCI Application of Metabolomics for Translational and Biological Research Grants 3031

The purpose of this Funding Opportunity Announcement (FOA), sponsored by multiple NIH Institutes/Centers (ICs), is to promote the use of metabolomic technologies in translational research in human health and disease for the purposes of enabling and improving disease detection, diagnosis, risk assessment, prognosis, and prediction of therapeutic responses. In addition to this overall goal, specific scopes and objectives of the participating ICs are presented below. Using the NIH Research Project Grant (R01) funding mechanism, this FOA focuses on well developed projects supported by preliminary data.
Requirements: Application can be made by the following institutions and organizations: public and state controlled institutions of higher education; private institutions of higher education; small businesses; for-profit organizations (other than small businesses); and non-domestic (non-U.S.) entities (foreign organizations).
Date(s) Application is Due: Feb 5; Feb 16; Jun 5; Jun 16; Oct 5; Oct 16
Contact: Padma Maruvada, (301) 496-3893; fax (301) 402-8990; maruvadp@mail.nih.gov
Internet: http://grants.nih.gov/grants/guide/pa-files/PA-07-301.html

Sponsor: National Cancer Institute
6130 Executive Boulevard, Executive Plaza N, Suite 3144, MSC 7362
Bethesda, MD 20892-7362

NCI Basic Cancer Research in Cancer Health Disparities Grants **3032**
Through this Funding Opportunity Announcement (FOA), the Center to Reduce Cancer Health Disparities (CRCHD) and the Division of Cancer Biology (DCB), at the National Cancer Institute (NCI), invite cooperative agreement research (U01) grant applications from investigators interested in conducting basic research studies into the causes and mechanisms of cancer health disparities. These awards will support pilot and feasibility studies, development and testing of new methodologies, secondary data analyses, and innovative mechanistic studies that investigate biological/genetic bases of cancer health disparities. This FOA is also designed to aid and facilitate the growth of a nationwide cohort of scientists with a high level of basic research expertise in cancer health disparities research who can develop resources and tools, such as biospecimens, cell lines and methods that are necessary to conduct basic research in cancer health disparities. The total project period for an application submitted in response to this funding opportunity may not exceed five years. Direct costs are limited to $250,000/year and may not exceed 5 years.
Requirements: The following organizations and institutions are eligible to apply: public/state controlled institutions of higher education; private institutions of higher education; Hispanic-serving institutions; historically Black colleges and universities (HBCUs); tribally controlled colleges and universities (TCCUs); Alaska Native and Native Hawaiian serving institutions; nonprofits with 501(c)3 IRS status; nonprofits without 501(c)3 IRS status; small businesses; for-profit organizations; state governments; Indian/Native American tribal governments (federally recognized); Indian/Native American tribally designated organizations; U.S. territories or possessions; Indian/Native American tribal governments (other than federally recognized); regional organizations; eligible agencies of the federal government; and units of local governments.
Date(s) Application is Due: Jun 23; Nov 23
Contact: Phillip J. Daschner, (301) 496-1951; fax (301) 496-2025; PD93U@nih.gov
Internet: http://grants.nih.gov/grants/guide/pa-files/PAR-09-161.html
Sponsor: National Cancer Institute
6130 Executive Boulevard, Room 5014, MSC 7344
Bethesda, MD 20892-7344

NCI Cancer Centers Support Grants (CCSG) **3033**
The purpose of this program is to provide an organizational focus and stimulus for the highest quality cancer research that effectively promotes interdisciplinary cancer research aimed toward the ultimate goal of reducing cancer incidence, mortality and morbidity. The Cancer Center Support Grant (CCSG) provides the resources and infrastructure to facilitate the coordination of interdisciplinary programs across a broad spectrum of research from basic laboratory research to clinical investigation to population science. The CCSG supports salaries for scientific leadership of the Center, shared resources for funded center investigators, certain administrative costs, planning and evaluation, and developmental funds for new recruitments and feasibility studies.Annual deadline dates may vary; contact program staff for exact dates.
Requirements: Universities, colleges, hospitals, public agencies, or nonprofit research institutions are eligible to apply.
Date(s) Application is Due: Feb 1; Jun 1; Oct 1
Amount of Grant: 15,950 - 100,000,000 U.S.D
Contact: Leo Buscher Jr., Grants Management Officer; (301) 496-7753
Internet: http://12.46.245.173/pls/portal30/CATALOG.PROGRAM_TEXT_RPT.show
Sponsor: National Cancer Institute
6116 Executive Boulevard, Suite 700
Bethesda, MD 20892

NCI Cancer Communications System Research Grants **3034**
The Division of Cancer Prevention and Control of NCI invites applications for studies to initiate cancer communications research activities. This project proposes to enable qualified investigators to identify a cancer communications issue or problem, and develop, implement, and evaluate a research project to address the identified issue or problem. The project should be targeted at specific audiences and may or may not utilize the resources of the Cancer Communications System. Investigators may propose new data collection activities and/or design studies comparing public knowledge, attitudes, and/or practices by location, ethnicity, age, and so forth. A specific group such as women, smokers, African Americans, or people over age 50 may be chosen for study. Projects may be proposed for durations of 18 to 36 months. Shared resources on the part of the applicant, such as contribution of staff time, should be described in the application. Annual deadline dates may vary contact program staff for exact dates. PA: PAR-99-141
Requirements: The awardee will be a university, college, hospital, public agency, nonprofit research institution, or for-profit organization that submits an application and receives a grant or cooperative agreement for support of research by a named principal investigator.
Restrictions: Restrictions are imposed against the use of funds for entertainment, foreign travel (unless specifically authorized), office equipment, and other items not normally necessary for the effective prosecution of such research.
Contact: Connie Dresser, Division of Cancer Control and Population Sciences; (301) 496-8520; fax (301) 480-6637; cd34b@nih.gov
Internet: http://www.grants.nih.gov/grants/guide/pa-files/PAR-99-141.html
Sponsor: National Cancer Institute
6120 Executive Boulevard, Executive Plaza S
Bethesda, MD 20892

NCI Cancer Construction Grants **3035**
The objective of this program is to renovate existing cancer research facilities and build new cancer research facilities to meet basic and clinical space requirements, laboratory safety, biohazard containment, and animal care standards in order to support the facility demands of NCI research programs.
Requirements: Except where otherwise prohibited by law, any nonprofit organizations competent to carry out cancer research, such as universities, colleges, hospitals, public agencies, or research institutions, are eligible to apply for support.
Date(s) Application is Due: Feb 1
Contact: Leo Buscher Jr., Grants Management Officer; (301) 496-7753
Internet: http://12.46.245.173/pls/portal30/CATALCG.PROGRAM_TEXT_RPT.show
Sponsor: National Cancer Institute
6120 Executive Boulevard, EPS-638
Bethesda, MD 20892

NCI Cancer Control Research Grants **3036**
Grants are awarded to programs that will reduce cancer incidence, morbidity, and mortality through an orderly sequence from research on interventions and their impact in defined populations to the broad, systematic application of the research results. Primary emphasis is on the inclusion of a cancer control intervention in any proposed study. Cancer control research studies are classified into one of five phases that represent the orderly progression noted in the definition: hypothesis development; methods development and testing; controlled intervention trials to establish cause-and-effect relationships; research in defined populations; and demonstration and implementation studies. Primary interest is in research on cancer control interventions in phases two through five. Cancer control programs include those in the following areas: chemoprevention; nutrition and diet; screening and early detection; community oncology, rehabilitation, and pain management; special populations research; public health applications; biobehavioral mechanisms; and surveillance research. Annual deadline dates may vary; contact program staff for exact dates.
Requirements: University, college, hospital, public agency, nonprofit research institutions or for-profit organization may apply.
Date(s) Application is Due: Feb 1; Jun 1; Oct 1
Amount of Grant: 15,000 - 100,000,000 U.S.D
Contact: Leo Buscher Jr., Grants Management Officer; (301) 496-7753
Internet: http://12.46.245.173/pls/portal30/CATALOG.PROGRAM_TEXT_RPT.show
Sponsor: National Cancer Institute
6120 Executive Boulevard, Executive Plaza S
Bethesda, MD 20892

NCI Cancer Detection and Diagnosis Research Grants **3037**
The purpose of this program is to improve screening and early detection strategies and to develop accurate diagnostic techniques and methods for predicting the course of disease in cancer patients. Screening and Early Detection research includes development of strategies to decrease cancer mortality by finding tumors early when they are more amenable to treatment. Diagnosis research focuses on methods to determine the presence of a specific type of cancer; to predict its course and response to therapy; and to monitor the effect of the therapy and the appearance of disease recurrence. These methods include diagnostic imaging and direct analyses of specimens from tumor or other tissues. Support is also provided for establishing and maintaining resources of human tissue to facilitate research. Annual deadline dates may vary; contact program staff for exact dates.
Requirements: The awardee will be a university, college, hospital, public agency, nonprofit research institution or organization, or a for-profit organization that submits an application and receives a grant or cooperative agreement for support of research by a named principal investigator.
Date(s) Application is Due: Feb 1; Jun 1; Oct 1
Amount of Grant: 60,000 - 38,000,000 U.S.D
Contact: Leo Buscher Jr., Grants Management Officer; (301) 496-7753
Internet: http://12.46.245.173/pls/portal30/CATALOG.PROGRAM_TEXT_RPT.show
Sponsor: National Cancer Institute
6120 Executive Boulevard, Executive Plaza S
Bethesda, MD 20892

NCI Cancer Education and Career Development Program **3038**
The purpose of the program is to support the development and implementation of curriculum-dependent programs to train predoctoral and postdoctoral candidates in cancer research settings that are highly interdisciplinary and collaborative. This program requires sustained leadership, dedicated faculty time, specialized curriculum development and implementation, interdisciplinary research environments, and more than one mentor per trainee to achieve career development research and education objectives. Areas of research particularly applicable but not all inclusive to interdisciplinary training are cancer prevention and control, nutrition, population sciences, behavioral sciences, imaging and molecular diagnostics. PA: PAR-03-148
Requirements: Applications may be submitted on behalf of the principal investigator by domestic, nonfederal, public or private organizations. Applications may include more than one institution to create a program through consortium agreements. Predoctoral and postdoctoral candidates must be citizens or noncitizen nationals of the U.S., or must have been lawfully admitted to the U.S. for permanent residence.
Restrictions: No application may request more than $500,000 in direct costs per annum without written approval from NCI.
Amount of Grant: Up to 500,000 U.S.D
Contact: Catherine Blount, Grants Administration Branch; (301) 496-3179; fax (301) 496-8601; blountc@mail.nih.gov
Internet: http://grants2.nih.gov/grants/guide/pa-files/PAR-03-148.html

Sponsor: National Cancer Institute
6130 Executive Boulevard, Executive Plaza N, Suite 3144, MSC 7362
Bethesda, MD 20892-7362

NCI Cancer Intervention and Surveillance Modeling Network (CISNET) 3039
The purpose of the project is to support collaborative research using simulation and other modeling techniques to describe the impact of interventions (i.e., primary prevention, screening, and treatment) in population-based settings in the United States or in non-U.S. settings that will shed light on U.S. population-based trends. It is well known that great progress can be made in the war against cancer by the complete use and adequate delivery of existing modalities of cancer control. A letter of intent is due by September 14; full proposal is due October 14. Annual deadline dates may vary; contact program staff for exact dates. PA: CA-05-018
Requirements: Applications may be submitted by domestic and foreign for-profit and nonprofit organizations, public and private, such as universities, colleges, medical centers, units of state and local governments, and eligible agencies of the federal government.
Date(s) Application is Due: Sep 14; Oct 14
Amount of Grant: Up to 15,000,000 U.S.D
Contact: Crystal Wolfrey, Grants Administration Branch; (301) 496-8634; fax (301) 496-8601; crystal.wolfrey@nih.gov or wolfreyc@mail.nih.gov
Internet: http://grants.nih.gov/grants/guide/rfa-files/RFA-CA-05-018.html
Sponsor: National Cancer Institute
6701 Rockledge Drive, Room 1040, MSC 7710
Bethesda, MD 20892

NCI Cancer NRSAs—Individual 3040
The awards offer health scientists with a doctoral degree the opportunity to receive full-time research training in cancer research. These fellowships may not be used for study leading to the MD, DO, DDS, or similar professional degree, or for the support of residency training. Awardees are subject to the payback provisions of the NRSA by which one year of research or teaching, or acceptable combination thereof, is required for each year of support (except for the first 12 months, which are free of the payback obligation). Annual deadline dates may vary; contact program staff for exact dates.
Requirements: Universities, colleges, hospitals, public agencies, or research institutions may apply for institutional grants, and individuals are eligible for fellowships.
Date(s) Application is Due: Apr 5; Dec 5
Contact: Program Contact; (301) 435-0714; grantsinfo@nih.gov
Internet: http://grants.nih.gov/grants/policy/nihgps_2001/part_iib_3.htm
Sponsor: National Cancer Institute
6120 Executive Boulevard, Executive Plaza S
Bethesda, MD 20892

NCI Cancer NRSAs—Institutions 3041
These grants are made to eligible institutions to develop or enhance pre- and post-PhD research training opportunities for individuals who are interested in preparing for careers in cancer research. The program is to ensure that an adequate number of highly competent basic and clinical cancer research specialists will be trained to meet needs in the following areas of cancer research: etiology, prevention, detection, diagnosis, treatment, and restorative care. Applications in surgical, radiation, and preventive oncology are of special interest. Proposed institutional training programs may be multidisciplinary. The training program director at the grantee institution will be responsible for selection and appointment of trainees and for the overall direction of the program. Trainees are subject to payback provisions of the NRSA by which one year of research or teaching, or acceptable combination thereof, is required for each year of support, except for the first 12 months, which are free from payback. Annual deadline dates may vary; contact program staff for exact dates.
Requirements: Universities, colleges, hospitals, public agencies, or nonprofit research institutions may apply for institutional grants, and individuals are eligible for fellowships.
Date(s) Application is Due: Jan 10; May 10; Sep 10
Contact: Program Contact; (301) 435-0714; GrantsInfo@nih.gov
Internet: http://grants.nih.gov/grants/policy/nihgps_2001/part_iib_3.htm
Sponsor: National Cancer Institute
6120 Executive Boulevard, Executive Plaza S
Bethesda, MD 20892

NCI Cancer Prevention Research Small Grants 3042
NCI invites applications that address developmental research in chemoprevention agent development, biomarkers, early detection, and nutrition science. The program is designed to aid and facilitate the growth of a nationwide cohort of scientists with a high level of research expertise in cancer prevention research. Annual deadline dates may vary; contact program staff for exact dates. PA: PAR-04-147
Requirements: Applications may be submitted by domestic, for-profit and nonprofit organizations, public and private, such as universities, colleges, hospitals, laboratories, units of state and local governments, eligible agencies of federal government, and small businesses.
Date(s) Application is Due: Mar 21; Jul 21; Nov 21; Dec 20
Amount of Grant: 50,000 - 100,000 U.S.D
Contact: Shane Woodward, Grants Administration Branch; (301) 846-1017; fax (301) 846-5720; woodwars@mail.nih.gov
Internet: http://grants.nih.gov/grants/guide/pa-files/PAR-04-147.html
Sponsor: National Cancer Institute
6116 Executive Boulevard
Bethesda, MD 20892

NCI Cancer Treatment Research Grants 3043
The objective of this program is to develop the means to cure as many cancer patients as possible and to control the disease in those patients who are not cured. Research includes the development and evaluation of improved methods of cancer treatment through the support and performance of both fundamental and applied laboratory and clinical research. Research is supported in the discovery, development, and clinical testing of all modes of therapy including: surgery, radiotherapy, chemotherapy, and biological therapy, both individually and in combination. In addition, research is carried out in areas of nutritional support, stem cell and bone marrow transplantation, studies to reduce toxicity of cytotoxic therapies, and other methods of supportive care that may supplement and enhance primary treatment. Annual deadline dates may vary; contact program staff for exact dates.
Requirements: Awardee will be a university, college hospital, public agency, nonprofit research institution, or for-profit organization that submits an application and receives a grant or cooperative agreement for support of research by a named principal investigator.
Date(s) Application is Due: Feb 1; Jun 1; Oct 1
Amount of Grant: 59,416 - 100,000,000 U.S.D
Contact: Leo Buscher Jr., Grants Management Officer; (301) 496-7753
Internet: http://aspe.os.dhhs.gov/cfda/p93395.htm
Sponsor: National Cancer Institute
6120 Executive Boulevard, Executive Plaza S
Bethesda, MD 20892

NCIC Canadian Breast Cancer Research Initiative Grants 3044
The initiative is a broad alliance of partners, comprising the Canadian Cancer Society, Health Canada through the National Health Research and Development Program, the Medical Research Council, the National Cancer Institute of Canada, the Breast Cancer International Centre, the Canadian Breast Cancer Foundation, and the Canadian Breast Cancer Network. The grants provide research support for up to three years to individuals formally affiliated with Canadian universities or other recognized Canadian institutions. Proposals must clearly state their relevance to the problem of breast cancer. Grantees are required to submit annual progress reports. Obtain application form 801 from the institute.
Restrictions: Applications may not be submitted in duplicate to sponsoring agencies. Applications for support of infrastructure will not be accepted.
Geographic Focus: All States, Canada
Contact: Dr. Pascale Macgregor, Research Program Director; (416) 596-6598, ext. 360; fax (416) 596-1714; pfmacgregor@cbcf.org
Internet: http://www.breast.cancer.ca/cbcra_research_program
Sponsor: National Cancer Institute of Canada
10 Alcorn Avenue, Suite 200
Toronto, ON M4V 3B1 Canada

NCIC Canadian Cancer Society Research Grants 3045
The grants are intended to stimulate Canadian investigators in the design and execution of clinical and psychosocial/behavioral research in cancer and in the exploration of all the areas of the basic sciences having relevance to cancer. Grants for research will be awarded to projects deemed worthy of support, provided that the basic equipment and research facilities are available in the institution concerned and that it will provide the necessary administrative services. Such awards will be granted for the purchase and maintenance of animals, for expendable supplies, minor items of equipment, payment of graduate students and other assistants, and for research travel. Researchers focusing on breast cancer are strongly encouraged to submit applications to the Canadian Breast Cancer Research Initiative.
Requirements: A researcher who is designated as the principal investigator must be based in, or formally affiliated with (but not necessarily receive salary support from), a Canadian host institution such as a university, research institute, or healthcare agency.
Restrictions: Funds may not be used for personal salary support of the grantee.
Geographic Focus: All States, Canada
Date(s) Application is Due: Oct 17
Amount of Grant: Up to 300,000 CAD
Samples: Ralph Durand, British Columbia Cancer Research Centre—for a three-year project beginning in 1998 entitled Modifying Tumour Physiology to Improve Response to Radiation, $C149,544 per year; Christopher Overall, U of British Columbia—for a three-year project beginning in 1998 entitled Cellular Activation of Human Gelatinase A, $C121,500 per year; Heather Sutherland, U of British Columbia—for a three-year project beginning in 1998 entitled Purification and Characterization of Human Acute Myeloid Leukemia Stem Cells, $C97,600 per year; Kevin Kane, U of Alberta—for a three year project beginning in 1998 entitled Ly-49 Interactions with Class I Major Histocompatibility Complex Molecules, $C95,390 per year.
Contact: Director of Research; (416) 961-7223; fax (416) 961-4189; research@cancer.ca
Internet: http://www.ncic.cancer.ca/ncic/internet/standard/0,3621,84658243_85805597__langId-en,00.html
Sponsor: National Cancer Institute of Canada
10 Alcorn Avenue, Suite 200
Toronto, ON M4V 3B1 Canada

NCIC Canadian Cancer Society Travel Award for Senior Level PhD Students 3046
The purpose of this program is to provide financial assistance to senior-level PhD or MD/PhD students by helping to defray the travel costs associated with making a scientific presentation at a conference, symposium, or other appropriate professional gathering. The travel award will be made to defray actual costs of travel, registration, accommodation, etc. Three competitions are held annually. Conference dates for which funding is being requested must occur within six months of the competition deadline. Individuals may receive this award only once. A maximum of one award per supervisor will be given in the current calendar year. Applications must be completed using form 405/96. Letters

announcing the results of eligible applications will be issued approximately one month following the deadline date.
Requirements: Applicants must be students enrolled in a PhD or MD/PhD program at a Canadian institution and be in their final phase of study. Candidates must be attending a conference for the purpose of presenting data (paper or poster presentation) from a cancer-related project on a first-author basis. It is not a requirement that the student and/or the supervisor be supported by an NCIC grant.
Geographic Focus: All States, Canada
Date(s) Application is Due: Apr 1; Sep 1; Dec 1
Amount of Grant: Up to 1,500 CAD
Contact: Program Contact; (416) 961-7223; fax (416) 961-4189; research@cancer.ca
Internet: http://www.ncic.cancer.ca/ncic/internet/standard/0,3621,84658243_85813951_91005239_langId-en,00.html
Sponsor: National Cancer Institute of Canada
10 Alcorn Avenue, Suite 200
Toronto, ON M4V 3B1 Canada

NCI Centers of Excellence in Cancer Communications Research **3047**
NCI invites applications for Centers of Excellence in Cancer Communications Research (CECCRs). The purpose is to identify areas of discovery that build upon important recent developments in knowledge and technology and that hold promise for making significant progress against all cancers. The centers must include three or more individual hypothesis-driven research projects, pilot or developmental research projects, shared resources, and a plan for career development. To be effective, the centers' research should integrate cancer communications appropriately into one or more contexts of the cancer continuum—from prevention through treatment to survivorship and end-of-life research. Communications research also is needed about challenging topics such as cancer information seeking, decision making under uncertainty, and genetic testing. Centers' research also should provide insight into mechanisms underlying how people process information. Annual deadline dates may vary; contact program staff for exact dates. RFA: CA-03-007
Requirements: Applications may be submitted by domestic for-profit and nonprofit organizations, both public and private; units of state and local governments; and eligible agencies of the federal government.
Amount of Grant: Up to 15,000,000 U.S.D
Contact: Crystal Wolfrey, Grants Administration Branch; (301) 496-8634; fax (301) 496-8601; crystal.wolfrey@nih.gov or wolfreyc@mail.nih.gov
Internet: http://grants2.nih.gov/grants/guide/rfa-files/RFA-CA-03-007.html
Sponsor: National Cancer Institute
6130 Executive Boulevard, Executive Plaza N, Suite 3144, MSC 7362
Bethesda, MD 20892-7362

NCIC Equipment Grants **3048**
The grants provide for the purchase of items of permanent equipment necessary for research projects supported by the institute, if such equipment is not otherwise available. Application for minor items of apparatus and equipment with a total cost for each item of less than $C10,000 should be included as part of the application for the research grant. Application for funds to purchase each item with a total cost of less than $C100,000 but more than $C5000 may be made either as part of an application for a grant for research or as a separate request if the applicant is the recipient of a continuing award. For items of specialized major equipment having a value of $C100,000 or more, the institute may have to contemplate alternative approaches to funding; therefore, a prior consultation with the institute is required.
Date(s) Application is Due: Oct 17
Amount of Grant: Up to 75,000 CAD
Contact: Grants Administrator; (416) 961-7223; fax (416) 961-4189; research@cancer.ca
Internet: http://www.ncic.cancer.ca/ncic/internet/standard/0,3621,84658243_85808828__langId-en,00.html
Sponsor: National Cancer Institute of Canada
10 Alcorn Avenue, Suite 200
Toronto, ON M4V 3B1 Canada

NCI Community Networks to Reduce Cancer Health Grants **3049**
Cooperative agreement awards will be made to eligible institutions to develop and implement a variety of community-based cancer control and prevention activities. The overall goals of this program are to significantly improve access to and utilization of beneficial cancer interventions in communities with cancer health disparities, thereby reducing these disparities. The CNP will be implemented in three phases. The goal of Phase I is to develop and increase capacity building to support community-based participatory education, research, and training to reduce cancer health disparities. The goal of Phase II is to develop community-based participatory research and training programs to reduce cancer health disparities. The goal of Phase III is to establish credibility and sustainability of the CNP. Annual deadline dates may vary; contact program staff for exact dates. RFA: CA-05-012
Requirements: Applications may be submitted by domestic organizations, public and private entities such as colleges, universities, hospitals, medical/nursing and sociological associations, biomedical research centers, and eligible agencies or units of federal, state, and local government. Applications are encouraged from existing Leadership Initiatives projects, and from subgroups throughout the country that focus on America's minority and underserved populations. Applicants should have a history of involvement with the community. Collaborative applications are encouraged; among collaborators, one must be designated as the lead applicant.
Restrictions: Foreign organizations are not eligible, nor are organizations that must rely primarily on consultants to conduct the project.
Date(s) Application is Due: Jun 14; Jul 13
Amount of Grant: 350,000 - 1,400,000 U.S.D
Contact: Crystal Wolfrey, Grants Administration Branch; (301) 496-8634; fax (301) 496-8601; crystal.wolfrey@nih.gov or wolfreyc@mail.nih.gov
Internet: http://grants.nih.gov/grants/guide/rfa-files/RFA-CA-05-012.html
Sponsor: National Cancer Institute
6130 Executive Boulevard, Executive Plaza N, Suite 3144, MSC 7362
Bethesda, MD 20892-7362

NCI Correlative Studies Using Specimens from Multi-Institutional Prevention and Treatment Trials **3050**
The objectives of this funding opportunity are to foster collaborations and interactions between basic researchers, scientists working in private industry, and clinical investigators to perform clinical translational research on promising predictive and prognostic markers. These studies should focus on clinical correlative or mechanistic studies that will be useful for cancer risk assessment, early detection, and prognosis, and the predicting responses to therapy and to prevention interventions. These studies should focus on correlations between biologic features of tissue specimens (collected from the NCI Cooperative Groups or other large multi-institutional clinical trials) and patient outcomes. Annual deadline dates may vary; contact program staff for exact dates. PA: PA-05-062
Requirements: Applications may be submitted by foreign and domestic, for-profit and nonprofit organizations, public and private. Applications may be from a single institution or may include arrangements with one or more institutions (e.g., consortia, NCI groups) if appropriate.
Date(s) Application is Due: Feb 1; Jun 1; Oct 1
Contact: Jill Rogers, Grants Management Branch; (301) 496-8699; fax (301) 496-8601; jr261m@nih.gov
Internet: http://grants.nih.gov/grants/guide/pa-files/PA-05-062.html
Sponsor: National Cancer Institute
6130 Executive Boulevard, Executive Plaza N
Bethesda, MD 20892

NCIC Terry Fox Postdoctoral Research Fellowships **3051**
Fellowships are designed to provide training in cancer research for outstanding candidates who plan careers in Canada in this field of investigation. Post-MD fellowships are made to provide support for specialized training in the research aspects of the broad field of oncology (including epidemiology, medical statistics, clinical trials, and other forms of biomedical research). Initial tenure of a fellowship will normally be two years. One or more years of additional support may be available to individuals who demonstrate a serious commitment to cancer research, e.g., by registration in a degree program. Post-MD fellows may wish to maintain their clinical skills by means of service to the host institution during the period of the award. Such activities are acceptable provided the service is directly related to the cancer-oriented plan of the fellow and neither exceeds 10 working hours per week nor adversely affects the quantity or quality of the research training.
Requirements: Candidates must be graduates of universities recognized by NCIC and must be accepted for postdoctoral training in the field of cancer research in an academic environment also recognized by NCIC.
Restrictions: In general, applicants will not be eligible for funding beyond five years of postdoctoral training.
Geographic Focus: All States, Canada
Date(s) Application is Due: Feb 1
Amount of Grant: 36,200 - 48,700 CAD
Contact: Director of Research; (416) 961-7223; fax (416) 961-4189; research@cancer.ca
Internet: http://www.ncic.cancer.ca/ncic/internet/standard/0,3621,84658243_85811512_90904221_langId-en,00.html
Sponsor: National Cancer Institute of Canada
10 Alcorn Avenue, Suite 200
Toronto, ON M4V 3B1 Canada

NCIC Terry Fox Research Studentships **3052**
Research studentships provide training to PhD students in cancer research and are offered to outstanding individuals who plan a career in cancer research in Canada. Awards are made for up to a maximum of four years and are not renewable. Research studentship award recipients must devote full time to the research and training program during the tenure of the studentship. If the supervisor considers a certain amount of teaching to be part of the training, such teaching and duties must be limited to not more than 10 hours per week. At the completion of each award year, a letter summarizing the progress to date, signed by the research student and the research supervisor, must be submitted to the institute.
Requirements: Applicants must be residents of Canada and be enrolled in a PhD program at a Canadian institution at the time of application. Candidates must have completed at least two years of research training at the graduate level or have equivalent research experience.
Geographic Focus: All States, Canada
Amount of Grant: Up to 22,700 CAD
Contact: Michael Wosnick, Executive Director; (416) 934-5672; fax (416) 961-4189; mwosnick@cancer.ca
Internet: http://www.ncic.cancer.ca/ncic/internet/standard/0,3621,84658243_85813949_90911828_langId-en,00.html
Sponsor: National Cancer Institute of Canada
10 Alcorn Avenue, Suite 200
Toronto, ON M4V 3B1 Canada

NCI Development and Application of Imaging in Therapeutic Studies 3053
NCI invites applications to apply imaging technologies in the assessment of investigational cancer therapeutic agents. This request for applications is designed to support research projects that address the development and application of labeled therapeutic agents as compounds for imaging studies, and/or the development and application of imaging agents as metabolic markers of response to newly developed therapeutic agents. The total project period for an application submitted in response to this RFA may not exceed four years. RFA: CA-98-024
Requirements: Applications may be submitted by domestic and foreign, for-profit and nonprofit organizations, public and private.
Amount of Grant: Up to 28,000,000 U.S.D
Contact: Dr. Anne Menkens, Cancer Treatment and Diagnosis Coordinator; (301) 496-9531; fax (301) 480-5785; am187k@nih.gov
Internet: http://www.grants.nih.gov/grants/guide/rfa-files/RFA-CA-98-024.html
Sponsor: National Cancer Institute
6130 Executive Boulevard, Executive Plaza N
Bethesda, MD 20892

NCI Development of Digital Mammography Displays and Workstations Grants 3054
The purpose of the program is to alert the investigator community to the need for a concerted effort to overcome the problems of display for digital mammograms. Integration of advanced hardware, software, and psychophysics research data is needed to optimize the early diagnosis of breast cancer using digital mammography. This program will utilize the Small Business Innovation Research (SBIR) and Small Business Technology Transfer (STTR) mechanisms, but will run in parallel with a program of identical scientific scope utilizing the R01 research project grant mechanism (PA-99-082). This PA encourages the small business community to develop displays for digital mammography. Funding will be based on the quality of the proposed project as determined by peer review, availability of funds, and program priority. Contact program staff for deadline dates. PA: PA-99-083
Requirements: Any small business, independently owned by U.S. citizens and located in the U.S., may apply.
Contact: Dr. Barbara Croft, Diagnostic Imaging Program Coordinator; (301) 496-9531; fax (301) 480-5785; bc129b@nih.gov
Internet: http://www.grants.nih.gov/grants/guide/pa-files/PA-99-083.html
Sponsor: National Cancer Institute
6130 Executive Boulevard, Executive Plaza N
Bethesda, MD 20892

NCI Development of Novel Imaging Technologies Grants 3055
The NCI invites applications for the development and delivery of novel image acquisition or enhancement technologies and methods for biomedical imaging and image-guided interventions and therapy, and which may incorporate limited pilot or clinical feasibility evaluations using either pre-clinical models or clinical studies. This initiative is primarily intended to facilitate the proof-of-feasibility, development, and delivery of novel imaging technologies for early detection, screening, diagnosis, image-guided interventions, and treatment of various diseases, and, secondarily, to facilitate limited evaluation studies to show proof-of-concept and functionality. Annual deadline dates may vary; contact program staff for exact dates. PA: PA-04-095
Requirements: Applications may be submitted by foreign and domestic, for-profit and nonprofit organizations, both public and private; units of state and local governments; and eligible agencies of the federal government.
Contact: Houston Baker, Program Contact; (301) 496-9531; fax (301) 480-3507; bakerhou@mail.nih.gov
Internet: http://grants.nih.gov/grants/guide/pa-files/PA-04-095.html
Sponsor: National Cancer Institute
6130 Executive Boulevard, Executive Plaza N
Bethesda, MD 20892

NCI Diagnostic Imaging and Guided Therapy in Prostate Cancer Award 3056
NCI invites applications on the development, risk assessment, and application of improved imaging methods for the localization, biopsy, and image guided biopsy or therapy of prostate cancer. Relevant investigations could include technology development, in vitro laboratory work, preclinical animal studies, or early feasibility testing in humans depending on the maturity of the methods proposed, or evaluation of the effects of age-associated changes and co-morbid conditions as they affect imaging diagnosis and treatment techniques. The development of several methodologies and their optimization for this particular organ system is required. The specific goals include the development and application of one or more of the following interrelated components: means for measuring local extent of disease using anatomic, metabolic or alternative novel imaging methods, means for improved image guided biopsy, staging or identification of aggressive cancers by metabolic or alternative novel imaging methods, and means for navigation, control of image guided therapy or measurement of early biological effects of therapy. Research is also encouraged on how age-associated differences in tumor characteristics and age-related changes in the prostate and adjacent tissues may affect the sensitivity, specificity, prognostic value, or the efficacy of imaging techniques in guiding therapy. The development of methods to increase sensitivity, specificity, prognostic value, and therapeutic applicability of these techniques across the full range of ages in which prostate cancer most frequently occurs, and in the presence of age-related co-morbid conditions in the prostate, other organs, and systems, is of particular interest. Annual deadline dates may vary; contact program staff for exact dates. CA-99-015
Requirements: Applications may be submitted by foreign and domestic, for-profit and nonprofit organizations, public and private, such as universities, colleges, hospitals, laboratories, industries, units of state and local governments, and eligible agencies of the federal government.
Contact: Barbara Croft, Diagnostic Imaging Program Coordinator; (301) 496-9531; fax (301) 480-5785; bc129b@nih.gov
Internet: http://grants2.nih.gov/grants/guide/rfa-files/RFA-CA-99-015.html
Sponsor: National Cancer Institute
6701 Rockledge Drive
Bethesda, MD 20892

NCI Diet, Epigenetic Events, and Cancer Prevention Grants 3057
The purpose of this Funding Opportunity Announcement (FOA) is to promote innovative preclinical and clinical research to determine how diet and dietary factors, including dietary supplements, impact epigenetic processes involved in cancer prevention. Although much evidence exists that dietary components are linked to cancer prevention, the specific nutrients and sites of action remain elusive. Diet, in fact, has been implicated in many of the pathways of cancer, including apoptosis, cell cycle control, differentiation, inflammation, angiogenesis, DNA repair, and carcinogen metabolism. These are also processes that are likely regulated by epigenetic events to impact gene function and chromatin stability. Thus, research supported by this initiative could address one of the following issues: How bioactive food components regulate epigenetic events for cancer prevention, how bioactive food components might alter aberrant epigenetic patterns or events and restore gene function, and how these components might circumvent or compensate for genes and pathways that are altered by epigenetic events. Another important aim of this FOA is to encourage collaborations between nutrition scientists and experts in epigenetics to study bioactive food components with cancer-preventative properties and to examine key epigenetic events in cancer processes (e.g., carcinogen metabolism, cell division, differentiation, and apoptosis) in order to begin to establish linkages between epigenetic events and patterns, and tumor incidences/behaviors. NIDDK is particularly interested in receiving applications that address normal and neoplastic conditions focusing on the impact of epigenetics on cell division, differentiation, and/or apoptosis in the gastrointestinal (GI) tract and related organs. It is also expected that each application demonstrate experience in nutrition and cancer prevention as well as in epigenetic research. Because the nature and scope of the proposed research will vary from application to application, it is anticipated that the size and duration of each award will also vary.
Requirements: Application can be made by the following institutions and organizations: public/state controlled institutions of higher education; private institutions of higher education; nonprofits with 501(c)3 IRS status (other than institutions of higher education); nonprofits without 501(c)3 IRS status (other than institutions of higher education); small businesses; for-profit organizations (other than small businesses); state governments; U.S. territories or possessions; Indian/Native American tribal governments (federally recognized); Indian/Native American tribal governments (other than Federally recognized); Indian/Native American tribally designated organizations; non-domestic (non-U.S.) entities (foreign organizations); Hispanic-serving institutions; historically Black colleges and universities (HBCUs); tribally controlled colleges and universities (TCCUs); Alaska Native and Native Hawaiian serving institutions; regional organization; eligible agencies of the Federal government; and faith-based or community based organizations.
Date(s) Application is Due: Feb 5; Jun 5; Oct 5
Contact: Vishnudutt Purohit, (301) 443-2689; fax (301) 594-0673; vpurohit@mail.nih.gov
Internet: http://grants.nih.gov/grants/guide/pa-files/PA-07-175.html
Sponsor: National Cancer Institute
6130 Executive Boulevard, Executive Plaza N, Suite 3157, MSC 7328
Bethesda, MD 20892-7328

NCI Division of Cancer Treatment Clinical Trials Cooperative Groups Grants 3058
NCI is accepting applications from institutions interested in conducting multi-institutional clinical trials in a cooperative group setting. Awards will be made using the cooperative agreement mechanism (U10). Potential applicants are encouraged to contact program staff to discuss and/or clarify any issues or questions regarding this announcement. The program announcement, which describes the research objectives, application procedures, review considerations, and award criteria for this solicitation, may be obtained through the NIH grant line (301-402-2221), and the NIH Web site, and by mail and email. Annual deadline dates may vary; contact program staff for exact dates. PA: PA-99-058
Date(s) Application is Due: Feb 1; Jun 1; Oct 1
Contact: Dr. Richard Ungerleider, Division of Cancer Treatment and Diagnosis Contact; (301) 496-2522; fax (301) 402-0557; ru4m@nih.gov
Internet: http://www.grants.nih.gov/grants/guide/pa-files/PA-99-058.html
Sponsor: National Cancer Institute
6130 Executive Boulevard, Executive Plaza N
Bethesda, MD 20892

NCI Early Detection Research Network: Biomarkers Validation Laboratories Research Grants 3059
The Division of Cancer Prevention invites applications for cooperative agreements to establish a national network that will have responsibility for the development, evaluation, and validation of biomarkers for earlier cancer detection and risk assessment. Biomarkers are defined as cellular, biochemical, molecular, or genetic alterations by which a normal, abnormal, or simply biologic process can be recognized or monitored. Biomarkers are measurable in biological media, such as in tissues, cells, or fluids. The purpose of the network is to establish a scientific consortium of investigators, academic as well as industrial, with resources for basic, translational, and clinical research. The consortium will have three main components—biomarkers developmental laboratories, biomarkers validation laboratories, and clinical/epidemiologic centers. The biomarkers developmental laboratories will have responsibility for the development and

characterization of new, or refinement of existing, biomarkers; the biomarkers validation laboratories will serve as a network resource for clinical and laboratory validation of biomarkers, which include technological development and refinement; and the clinical/epidemiology centers will conduct clinical and epidemiological research regarding the wide clinical application of biomarkers. A steering committee composed of the principal investigators in the network and appropriate NCI staff will coordinate the work of the consortium. Logistic support and informatics will be provided through an auxiliary data management and coordinating center. Approximately two awards will be made. Letter of intent is requested by July 16; full application is due August 16. Annual deadline dates may vary; contact program staff for exact dates. RFA: CA-05-009
Requirements: Domestic public and private for-profit and nonprofit organizations are eligible. Domestic institutions may propose collaborations/consortia with foreign institutions.
Restrictions: Applications will not be accepted from foreign institutions.
Date(s) Application is Due: Jul 16; Aug 16
Amount of Grant: Up to 1,000,000 U.S.D
Contact: Karen Chuang, Grants Administration Branch; (301) 496-2784; fax (301) 496-8601; chuangk@mail.nih.gov
Internet: http://grants.nih.gov/grants/guide/rfa-files/RFA-CA-05-009.html
Sponsor: National Cancer Institute
6130 Executive Boulevard, Executive Plaza N
Bethesda, MD 20892

NCI Early Detection Research Network: Clinical and Epidemiologic Centers Grants 3060

The Division of Cancer Prevention invites applications for cooperative agreements to establish a national network that will have responsibility for the development, evaluation, and validation of biomarkers for earlier cancer detection and risk assessment. Biomarkers are defined as cellular, biochemical, molecular, or genetic alterations by which a normal, abnormal, or simply biologic process can be recognized or monitored. Biomarkers are measurable in biological media, such as in tissues, cells, or fluids. The purpose of the network is to establish a scientific consortium of investigators, academic as well as industrial, with resources for basic, translational, and clinical research. The consortium will have three main components—biomarkers developmental laboratories, biomarkers validation laboratories, and clinical/epidemiologic centers. The biomarkers developmental laboratories will have responsibility for the development and characterization of new, or refinement of existing, biomarkers; the biomarkers validation laboratories will serve as a network resource for clinical and laboratory validation of biomarkers, which include technological development and refinement; and the clinical/epidemiology centers will conduct clinical and epidemiologic research regarding the wide clinical application of biomarkers. A steering committee composed of the principal investigators in the network and appropriate NCI staff will coordinate the work of the consortium. Logistic support and informatics will be provided through an auxiliary data management and coordinating center. Letter of intent is requested by May 14; full application is due June 14. RFA: CA-05-005
Requirements: Domestic public and private for-profit and nonprofit organizations are eligible. Domestic institutions may propose collaborations/consortia with foreign institutions.
Restrictions: Applications will not be accepted from foreign institutions.
Date(s) Application is Due: May 14; Jun 14
Contact: Karen Chuang, Grants Administration Branch; (301) 496-2784; fax (301) 496-8601; chuangk@mail.nih.gov
Internet: http://grants.nih.gov/grants/guide/rfa-files/RFA-CA-05-005.html
Sponsor: National Cancer Institute
6130 Executive Boulevard, Executive Plaza N
Bethesda, MD 20892

NCI Engineered Isogenic Cell Lines with Relevant Cancer Targets Research Grants 3061

This PA encourages the small business community to develop isogenic cell lines in which functions of a molecular activity or pathway found in human tumors can be analyzed in pair-wise fashion. Such cell lines could be used for drug screening with the intent of identification of compounds that influence the action of the identified target. The end result of the PA-supported projects would be a series of genetically engineered screening models that the small businesses could market directly or utilize in a license agreement or other type of collaboration. PAR-98-068 may be obtained through the NIH grant line (301-402-2221), the NIH Web site, mail, and email.
Contact: Toby Friedberg, Extramural Activities Division Conact; (301) 496-3428; fax (301) 402-0275; friedbet&7dea.nci.nih.gov
Internet: http://www.grants.nih.gov/grants/guide/pa-files/PAR-98-068.html
Sponsor: National Cancer Institute
6130 Executive Boulevard, Executive Plaza N
Bethesda, MD 20892

NCI Epidemiology Small Grants 3062

This is a short-term award, not to exceed three years, intended to provide support for pilot projects, testing of new techniques, or innovative or high-risk projects that could provide a basis for more extended research. Examples of research proposals that would apply to this PAR include, but are not limited to validating measurements in body fluids and tissues of exogenous exposures associated with the development of cancer; applying epigenetic approaches in cancer epidemiology; analyzing existing data that otherwise may have gone unexplored, such as pooled analyses of data from multiple studies coordinated into consortia; explore relationship between reported tobacco use and actual delivered amounts of tobacco constituents in epidemiologic studies and other large datasets in order to develop standardized markers of exposure in cancer risk; planning an epidemiological study to apply validated relevant carcinogenesis markers to better understand the natural history of neoplasms and identify individuals at high risk. Annual deadline dates may vary; contact program staff for exact dates. PA: PAR-04-159
Requirements: Applications may be submitted by domestic and foreign, for-profit and nonprofit organizations, both public and private; units of state and local governments; eligible agencies of federal government; and small businesses.
Date(s) Application is Due: Mar 20; Jul 20; Nov 21
Amount of Grant: Up to 100,000 U.S.D
Contact: Crystal Wolfrey, Grants Administration Branch, NCI; (301) 496-8634; fax (301) 496-8601; crystal.wolfrey@nih.gov or wolfreyc@mail.nih.gov
Internet: http://grants2.nih.gov/grants/guide/pa-files/PAR-04-159.html
Sponsor: National Cancer Institute
6120 Executive Boulevard, Executive Plaza S
Bethesda, MD 20892

NCI Established Investigator Award in Cancer Prevention, Control, Behavioral, and Population Sciences 3063

The purpose of this program is to provide established investigators protected time to devote to research and to act as mentors for new investigators and junior faculty members. The target candidates are outstanding established scientists who have demonstrated a sustained, high level of research productivity and significant contributions to cancer prevention, control, behavioral and/or population cancer research. They must demonstrate the need to develop and enhance their own research and a commitment to serve as mentors to new scientists. The award provides salary support for up to five years and it is renewable for one additional five-year period. The NCI is especially interested in training and retaining scientists who conduct high-quality research in cancer prevention, control, behavioral, and population sciences. Examples of relevant disciplines for this PA include any aspect of human cancer prevention (modifiable risk factors, new animal models, and extrapolation of these models to human cancer; the development of preventive interventions to decrease cancer risk behaviors and/or increase health-promoting behaviors; chemoprevention from studies of the identification and early-phase characterization of candidate agents to preventive intervention trials; studies of nutritional supplements or complementary/alternative interventions in relation to cancer prevention and control; health services research, including patient outcome studies, practice research, and medical decision analyses related to cancer prevention or to cancer care and patient outcomes; Palliative care studies, including interventions to improve quality of life; survivorship research, including studies of outcomes and quality of life as related to cancer course, treatments, and treatment side effects; and studies of the effectiveness of cancer health communications in reducing high risk behaviors or in increasing participation in screening activities. Annual deadline dates may vary; contact program staff for exact dates. PA: PAR-05-145
Requirements: Eligible applicants must be established cancer scientists with a strong track record of publications and successful competition for cancer research support; must have peer-reviewed, independent research support at the time of, and during the entire duration of, the award; must have a record of supervising junior researchers; must demonstrate the need for protected time to advance their research careers and mentoring activities; must be U.S. citizens or non-citizen nationals, or must have been lawfully admitted for permanent residence and possess an Alien Registration Receipt Card (I-151 or I-551) or some other verification of legal admission as a permanent resident.
Contact: Bill Wells, Grants Administration; (301) 496-8796; wellsw@mail.nih.gov
Internet: http://grants2.nih.gov/grants/guide/pa-files/PAR-05-145.html
Sponsor: National Cancer Institute
6120 Executive Boulevard, Executive Plaza S
Bethesda, MD 20892

NCI Exploratory Grants for Behavioral Research in Cancer Control 3064

This FOA will support innovative pilot projects or feasibility studies, which will facilitate the growth of research science in the cancer control continuum from a behavioral perspective. This FOA includes and incorporates the research interests of the Behavioral Research program, the Office of Cancer Survivorship, and the Community Oncology and Prevention Trials Research Group. This FOA is appropriate for testing timely interventions in pilot studies for feasibility or using rigorous qualitative research methods to assess the potential efficacy of an intervention. It is also appropriate for the psychometric evaluation of new measures or culturally appropriate ones to be adapted for use in populations where measures have not yet been developed or validated. This FOA encourages applications that include small cross-disciplinary teams of investigators who bring perspectives from the behavioral and social sciences, as well as other fields of public health.
Requirements: Eligible institutions and agencies include: public/state controlled institutions of higher education; private institutions of higher education; nonprofits with 501(c)3 IRS status (other than institutions of higher education); nonprofits without 501(c)3 IRS status (other than institutions of higher education); small businesses; for-profit organizations (other than small businesses); state governments; U.S. territories or possessions; regional organizations; non-domestic (non-U.S.) entities (foreign organizations); eligible agencies of the federal government; and units of local governments.
Date(s) Application is Due: Feb 16; Jun 16; Oct 16
Amount of Grant: Up to 275,000 U.S.D
Contact: Sabra F. Woolley, (301) 435-4589; fax (301) 594-0673; woolleys@mail.nih.gov
Internet: http://grants.nih.gov/grants/guide/pa-files/PA-09-130.html
Sponsor: National Cancer Institute
6130 Executive Boulevard, Room 4084, MSC 7365
Bethesda, MD 20850-7365

NCI Flexible System to Advance Innovative Research for Cancer Drug Discovery by Small Businesses Grants 3065

The objective of this RFA is to support the research activities required to enable small businesses to bring their research efforts for drug discovery and development to clinical evaluation. Annual deadline dates may vary; contact program staff for exact dates. PA: PAR-03-074

Requirements: Any small business, independently owned and operated by United States citizens or permanent resident aliens may apply. An eligible business must be organized for-profit, cannot be dominant in its field of expertise, and must have its principal place of business in the United States. Including any affiliates, the company can be the employer of no more than 500 people.

Amount of Grant: 18,000,000 U.S.D

Contact: Shane Woodward, Grants Administration Branch; (301) 496-8649; fax (301) 496-8601; woodwars@mail.nih.gov

Internet: http://grants2.nih.gov/grants/guide/pa-files/PAR-03-074.html

Sponsor: National Cancer Institute
6120 Executive Boulevard, Executive Plaza S
Bethesda, MD 20892

NCI Howard Temin Award 3066

The goal of this award is to bridge the transition from a mentored research environment to an independent research career for scientists who have demonstrated unusually high potential during their initial stages of training and development. This special award is aimed at fostering the research careers of outstanding junior scientists in basic research who are committed to developing research programs highly relevant to the understanding of human biology and human disease as it relates to the etiology, pathogenesis, prevention, diagnosis, and treatment of cancer. Candidates are offered up to five years to develop knowledge in the basic sciences and research skills, with up to three of the initial years (at least one year required) in a mentored environment followed by a transition to an unmentored independent investigator phase for the remaining time. More information may be obtained by mail and email or through the NIH grant line (301-402-2221) and the NIH Web site. PA: PAR-03-104

Requirements: Candidates must be U.S. citizens or noncitizen nationals, or must have been lawfully admitted for permanent residence by the time of award.

Restrictions: Former principal investigators on any NIH research grants or non-PHS peer reviewed grants over $70,000 direct costs per year are not eligible.

Amount of Grant: 105,000 - 605,000 U.S.D

Contact: Catherine Blount, Grants Administration Branch; (301) 496-3179; fax (301) 496-8601; blountc@mail.nih.gov

Internet: http://grants2.nih.gov/grants/guide/pa-files/PAR-03-104.html

Sponsor: National Cancer Institute
6120 Executive Boulevard, Executive Plaza S
Bethesda, MD 20892

NCI Improving Diet and Physical Activity Assessment Grants 3067

This Funding Opportunity Announcement (FOA) is aimed at advancing the quality of measurements of dietary intake and physical activity pertinent to cancer and/or other pathologies through support of research on improved instruments, technologies, and/or statistical/analytical techniques. Studies proposed in the grant applications should be aimed at optimizing the combined use of objective and self-report measures of physical activity and/or dietary intake for testing in both general and diverse populations. Applicants responding to this FOA, which uses the R01 grant mechanism, should have already tested and validated new approaches and have obtained feasibility data in the context of cancer or other relevant diseases. Because the nature and scope of the proposed research will vary from application to application, it is anticipated that the size and duration of each award will also vary.

Requirements: Application can be made by the following institutions and organizations: public/state controlled institutions of higher education; private institutions of higher education; nonprofits with 501(c)3 IRS status (other than institutions of higher education); nonprofits without 501(c)3 IRS status (other than institutions of higher education); small businesses; for-profit organizations (other than small businesses); state governments; U.S. territories or possessions; Indian/Native American tribal governments (federally recognized); Indian/Native American tribal governments (other than Federally recognized); Indian/Native American tribally designated organizations; non-domestic (non-U.S.) entities (foreign organizations); Hispanic-serving institutions; historically Black colleges and universities (HBCUs); tribally controlled colleges and universities (TCCUs); Alaska Native and Native Hawaiian serving institutions; regional organization; eligible agencies of the Federal government; and faith-based or community based organizations.

Restrictions: Applicants without sufficiently extensive preliminary data or those who wish to explore the utility of new untested dietary or physical activity assessment methods in pilot studies are advised to consider submitting applications under the parallel FOA, PAR-06-103, which uses the Exploratory/Developmental Grant (R21) award mechanism.

Date(s) Application is Due: Feb 5; Jun 5; Oct 5

Contact: Richard Troiano; (301) 435-6822; fax (301) 435-3710; troianor@mail.nih.gov

Internet: http://grants.nih.gov/grants/guide/pa-files/PAR-09-225.html

Sponsor: National Cancer Institute
6130 Executive Boulevard, Executive Plaza N, Room 4005, MSC 7344
Bethesda, MD 20892-7344

NCI Innovative Technologies for the Molecular Analysis of Cancer: Phased Innovation Award (R21/R33) Grants 3068

NCI invites applications for research projects to develop novel technologies that will support the molecular analysis of cancers and their host environment in support of basic, clinical, and epidemiologic research. Technology encompasses methods and tools that enable research including, but not limited to, instrumentation, techniques, devices, and analysis tools (e.g., computer software). Technology is distinct from resources such as databases and tissue repositories. Applications for support of such resources will not be considered to be responsive to this announcement. Technologies solicited include those that are suitable for the detection of alterations and instabilities of genomic DNA; measurement of the expression of genes and gene products; analysis and detection of gene and or cellular products including post-translational modification, and function of proteins; identification and characterization of exogenous infectious agents in cancer; and assaying the function of major signal transduction networks involved in cancer. This PA is intended to support the development of all required components of fully integrated systems for analysis including front end preparation of sample materials from cells, bodily fluids, and tumor specimens; novel chemistries or contrast agents; molecular detection systems; data acquisition methods; and data analysis tools. Technologies under consideration include those that will support molecular analysis either in vitro, in situ, or in vivo (by imaging or other methods) in the discovery process, as well as in preclinical models and clinical research. This solicitation will utilize the Phased Innovation Award mechanism (R21/R33). Letters of intent are requested by February, June, and October; full applications are due in March, July, and November. Annual deadline dates may vary; contact program staff for exact dates. PA: PAR-01-104

Contact: Dr. Jay George, Office of Technology and Industrial Relations; (301) 496-1550; fax (301) 496-7807; jgeorge@mail.nih.gov

Internet: http://grants.nih.gov/grants/guide/pa-files/PAR-01-104.html

Sponsor: National Cancer Institute
31 Center Drive
Bethesda, MD 20892

NCI Mentored Career Development Award 3069

The Comprehensive Minority Biomedical Program, Division of Extramural Activities, invites underrepresented minority research scientists who have been recipients of NIH Research Supplements for Underrepresented Minority Individuals in Postdoctoral Training (MIPT) or Minority Investigator Supplements (MIS), funded by NCI, who need extended periods of sponsored research as a way to gain scientific expertise while bridging the transition from mentored research environments to independent research/academic careers to submit applications. This award offers opportunities for a mentored peer review experience in cancer research which will enhance the candidate's knowledge and understanding of the peer review process with the intended purpose of developing skills with the expectation that the candidate will submit a grant application for nontargeted mechanisms (R01, R03, R21). Applicants are encouraged to contact NCI program staff regarding their eligibility. PA: PAR-03-016 may be obtained from the Web site and by mail and email.

Amount of Grant: 30,000 - 75,000 U.S.D

Contact: Barbara Fisher, Grants Administration Branch; (301) 846-1015; fax (301) 846-5720; fisherb@gab.nci.nih.gov

Internet: http://grants2.nih.gov/grants/guide/pa-files/PAR-03-016.html

Sponsor: National Cancer Institute
6130 Executive Boulevard, Executive Plaza N
Bethesda, MD 20892

NCI Mentored Career Development Award to Promote Diversity 3070

This mechanism establishes a pathway of recruiting, training, and retaining underrepresented investigators in research fields that address problems pertinent to the biology, etiology, pathogenesis, prevention, diagnosis, control, and treatment of human cancer and who can conduct independent competitive cancer research programs. The Mentored Career Development Award to Promote Diversity (K01) provides support for a sustained period of protected time for intensive research career development under the guidance of an experienced mentor, or sponsor, in the biomedical, behavioral or clinical sciences leading to research independence. The expectation is that through this sustained period of research career development and training, awardees will launch independent research careers and become competitive for new research project grant funding.

Requirements: The following organizations and institutions may apply: for-profit and non-profit organizations; public or private institutions, such as universities, colleges, hospitals, and laboratories; and domestic institutions.

Restrictions: Foreign institutions and organizations are not eligible to apply.

Date(s) Application is Due: Feb 12; Mar 12; Jun 12; Jul 12; Oct 12; Nov 12

Contact: Belinda Locke, (301) 496-7344; fax (301) 402-4551; Lockeb@mail.nih.gov

Internet: http://grants.nih.gov/grants/guide/pa-files/PAR-06-220.html

Sponsor: National Cancer Institute
6116 Executive Boulevard, Suite 7031, MSC 8350
Bethesda, MD 20892-8350

NCI Minorities in Clinical Oncology Grants 3071

The Comprehensive Minority Biomedical Program of NCI announces the availability of minority medical oncology awards. The purpose of these grants is to encourage recently trained underrepresented minority clinicians to acquire clinical training and research experience in medical oncology and to increase representation of minorities in medical oncology. More information may be obtained by mail or email or through the NIH grant line (301-402-2221), or the NIH Web site. Annual deadline dates may vary; contact program staff for exact dates. PA: PAR-03-002

Requirements: The award is intended for underrepresented minorities who have a health professional degree or its equivalent, or are doctorally trained oncology nurses to acquire research experience in clinical oncology. Individuals with the PhD or other doctoral degrees in clinical disciplines such as clinical psychology, nursing, clinical genetics, speech-language pathology, audiology and rehabilitation are also eligible.

Restrictions: Current and former PI,s on NIH research project grants, FIRST Awards, comparable career development awards, sub-projects of program project or center grants, and the equivalent are not eligible to apply.

Contact: Barbara Fisher, Grants Administration Branch, National Cancer Institute; (301) 846-1015; fax (301) 846-5720; fisherb@gab.nci.nih.gov
Internet: http://grants.nih.gov/grants/guide/pa-files/PAR-03-002.html
Sponsor: National Cancer Institute
6120 Executive Boulevard, Room 243
Bethesda, MD 20892

NCI Molecular Target Drug Discovery for Cancer **3072**
The goal of this program is the discovery and validation of molecular targets for cancer prevention or treatment intervention which can be developed into drug discovery assays. PA: PAR-01-046
Requirements: Applications may be submitted by domestic for-profit and nonprofit organizations, both public and private; units of state and local governments; and eligible agencies of the federal government.
Amount of Grant: Up to 250,000 U.S.D
Contact: Toby Friedberg extramural Activities, Extramural Activities Director; (301) 496-3428; fax (301) 402-0275; tf12w@nih.gov
Internet: http://grants.nih.gov/grants/guide/pa-files/PAR-01-046.html
Sponsor: National Cancer Institute
6116 Executive Boulevard
Bethesda, MD 20892

NCI National Cooperative Drug Discovery Research on Opportunistic Infections **3073**
Investigator-initiated research project grant applications are invited to support projects with adequate preliminary data for serious development of new therapies to treat AIDS- and cancer-associated opportunistic infections. Studies may include research on identified targets for rational design of inhibitors or projects with chemically identified candidate compounds with suitable efficacy and selectivity for exploration as candidate drugs. Applications that include collaborations with the private sector (e.g., pharmaceutical, chemical or biotechnological companies) are strongly encouraged. The opportunistic pathogens emphasized in this program announcement are human cytomegalovirus, Mycobacterium tuberculosis, Mycobacterium avium, Cryptosporidium parvum, systemic candidiasis, Aspergillus, and the Microsporida. Annual deadline dates may vary; contact program staff for exact dates. PA: PA-98-100
Requirements: Applications may be submitted by public and private, domestic and foreign, for-profit and nonprofit organizations.
Contact: Dr. Mary Wolpert, Division of Cancer Treatment and Diagnosis; (301) 496-8783; fax (301) 402-5200; mw8u@nih.gov
Internet: http://www.grants.nih.gov/grants/guide/pa-files/PA-98-100.html
Sponsor: National Cancer Institute
6130 Executive Boulevard, Executive Plaza N
Bethesda, MD 20892

NCI Nonmammalian Organisms as Models for Anticancer Drug Discovery Research Grants **3074**
This PA encourages the use of nonmammalian organisms to discover new cancer therapies. The goal is to identify key genes, enzymatic activities, components of signaling pathways, or cellular processes which are altered in human cancer, as potential intervention points that could be used in the design of new cancer drugs. Projects could focus on validating inferences already drawn from comparative study of cancerous and normal cells, exploring promising leads from the nature of mutations known to be associated with cancer development, or testing hypotheses generated by the identification of new and unknown gene products such as those sequenced through the NCI's Cancer Genome Project (CGAP). For example, projects may include development of organisms in which genes in key pathways or processes are mutated or in which human transgenes are expressed to simulate changes known to occur in human cancers. These genetic manipulations could present a proof of principle or validation of the importance of the target genes to the development of cancer. Some examples of genes that are well studied in model organisms and known or strongly suspected of involvement in cancer include oncogenes, proto-oncogenes, some cell cycle and signal transduction genes, and DNA repair genes. Investigations in response to this PA should provide insight into which genes or gene products in cancer-relevant pathways are potential targets for human cancer treatment. Projects of a general nature not specifically related to new cancer target or drug discovery will not be considered responsive. A search for small molecule inhibitors through screening is appropriate. Although this PA is being issued by NCI, NIGMS has an interest in supporting projects related to this topic. Studies of growth regulation, cell cycle control, or other basic studies that are not explicitly focused on tumor target cell systems may be assigned to NIGMS or another NIH institute. Annual deadline dates may vary; contact program staff for exact dates. PA: PA-01-042
Requirements: Applications may be submitted by foreign and domestic, public and private, for-profit and nonprofit organizations, units of state and local governments, and eligible agencies of the federal government. Small businesses are encouraged to consider a parallel program announcement of identical scientific scope that utilizes the SBIR and STTR mechanisms (PAR-99-020). It should be noted that a search for small molecule inhibitors through screening is appropriate for projects submitted for the SBIR program announcement.
Contact: Dr. George Johnson, Cancer Treatment and Diagnosis Division Contact; (301) 496-8783; fax (301) 402-5200; johnsong@exchange.nih.gov
Internet: http://grants.nih.gov/grants/guide/pa-files/PA-01-042.html
Sponsor: National Cancer Institute
6130 Executive Boulevard, Executive Plaza N
Bethesda, MD 20892

NCI Paul Calabresi Award for Clinical Oncology **3075**
The purpose of this program is to increase the number of medical doctors (MD's, DO's), and doctorally-degreed nurses (PhD's or equivalents) and basic scientists (PhD's DVM's, or equivalents) who are highly-motivated and trained to: perform clinical oncology therapeutic research that develops and tests scientific hypotheses based on fundamental and clinical research findings; design and test hypothesis-based, clinical therapeutic protocols and adjunct biological analyses and for clinician candidates to administer all phases (i.e., pilot/Phase I, Phase II, and Phase III) of cancer therapeutic clinical trials, and conduct cancer therapeutic research in team research settings in which basic and clinical scientists collaborate and interact to expedite the translation of basic research discoveries into patient-oriented therapeutic cancer research. The PCACO is not intended to train laboratory-based scientists whose research will emphasize the use of animal or other model systems. Letters of intent are requested by May 2; full application is due by June 1. Annual deadline dates may vary; contact program staff for exact dates. PA: PAR-04-096
Requirements: Applications may be submitted by domestic, for-profit and nonprofit organizations, both public and private; units of state and local governments; and eligible agencies of the federal government. The principal investigator and all clinician candidates must currently be physicians holding the MD or DO degrees, or be doctorally prepared oncology registered nurses, and must have completed their clinical training.
Date(s) Application is Due: May 2; Jun 1
Contact: Kimery Griffin, Grants Administration Branch; (301) 496-3196; fax (301) 496-8601; griffink3@mail.nih.gov
Internet: http://grants.nih.gov/grants/guide/pa-files/PAR-04-096.html
Sponsor: National Cancer Institute
6116 Executive Boulevard, Suite 7025, MSC 8346
Bethesda, MD 20892

NCI Planning Grants for NCI Cancer Research Centers **3076**
The Cancer Centers Branch of NCI invites planning grant applications for the development of Cancer Research Centers in a variety of organizational settings. The purpose is to expand the scientific, geographic, and demographic diversity of the Cancer Centers Program by encouraging research-oriented organizations to develop the qualities of a strong cancer research center and become competitive for a Cancer Center Support grant. Cancer center planning strategies may focus on a specific research theme (e.g., diagnosis, therapy, epidemiology) or integrate a broad spectrum of research to include the basic, clinical, prevention and control, and population sciences (i.e., an NCI-designated comprehensive cancer center). Information may be obtained electronically through the NIH Web site, email, and by the NIH grant line (301-402-2221) and mail. Annual deadline dates may vary; contact program staff for exact dates. PA: PAR-03-022
Requirements: Applications may be submitted by domestic for-profit and nonprofit organizations, both public and private; units of state and local governments; and eligible agencies of the federal government.
Contact: Eileen Natoli, Grants Administration Branch; (301) 496-8791; fax (301) 496-8601; cm113g@nih.gov
Internet: http://grants2.nih.gov/grants/guide/pa-files/PAR-03-022.html
Sponsor: National Cancer Institute
6130 Executive Boulevard, Executive Plaza N
Bethesda, MD 20892

NCI Priorities in Behavioral Research in Cancer Prevention and Control Grants **3077**
NCI and NIDR invite researchers to submit research project grant (R01) applications that address behavioral research issues in cancer prevention and control. The PA addresses recommendations made by a special Behavioral Research in Cancer Prevention and Control Working Group in 1996 that consisted of leading national experts whose role was to identify behavioral research needs in cancer prevention and control during the coming years. PA-97-055 may be obtained electronically through the NIH grant line (301-402-2221), the NIH Web site, and by mail and email.
Contact: Veronica Chollette, (301) 435-2837; vc24a@nih.gov
Internet: http://www.grants.nih.gov/grants/guide/pa-files/PA-97-055.html
Sponsor: National Cancer Institute
6130 Executive Boulevard, Executive Plaza N
Bethesda, MD 20892

NCI Prostate Cancer Aging, Race, and Ethnicity Research Grants **3078**
NIA, NCI, and NIEHS invite applications to expand the understanding of biological and clinical factors leading to the development, progression, and treatment of prostate cancer in aging men. The increased risk of prostate cancer with advancing age and its prominence in older-aged men are well-known characteristics of this tumor. The unusually high incidence and mortality rates of prostate cancer for older white and black American men and, by contrast, the much lower rates in men of Hispanic and Asian descent, provide the need for research that emphasizes the role of race and ethnic factors, as well as age, in early diagnosis, management, and etiology of this tumor. This PA is intended to stimulate research that applies the expanding scientific knowledge gained on prostate cancer to older men and to extend the knowledge base on age-related aspects of the etiology of this malignancy. The PA, which describes the research objectives, application procedures, review considerations, and award criteria, may be obtained electronically through the NIH grant line (301-402-2221), the NIH Web site, and by mail and email. PA: PA-97-019
Contact: Dr. Andrew Chiarodo, Organ System Coordinating Branch; (301) 496-5278; fax (301) 402-0181; ac53a@nih.gov
Internet: http://www.grants.nih.gov/grants/guide/pa-files/PA-97-019.html
Sponsor: National Cancer Institute
6130 Executive Boulevard, Executive Plaza N
Bethesda, MD 20892

NCI Review and Analysis of Tobacco Industry Documents Grants 3079

This program announcement is intended to stimulate research on a wide variety of scientific, technical, marketing, and tactical undertakings by the tobacco industry that were documented in papers, memos, and other records. The systematic, comprehensive analysis and evaluation of these documents will greatly contribute to the understanding of what the tobacco industry knew and will help researchers and the public health community identify effective strategies to reduce tobacco use. Annual deadline dates may vary; contact the program office for exact dates. PA: PAR-01-063

Requirements: Applications may be submitted by foreign and domestic, for-profit and nonprofit organizations, public and private. Applicants requesting budgets greater than $500,000 in direct costs are required to contact program staff prior to submitting applications.

Contact: Dr. Michele Bloch, Division of Cancer Control and Population Sciences; (301) 496-8584; fax (301) 496-8675; blochm@mail.nih.gov

Internet: http://grants.nih.gov/grants/guide/pa-files/PAR-01-063.html

Sponsor: National Cancer Institute
6130 Executive Boulevard, Executive Plaza N
Bethesda, MD 20892

NCI Ruth L. Kirschstein National Research Service Award (NRSA) Institutional Research Training Grants 3080

The National Institutes of Health (NIH) will award Ruth L. Kirschstein National Research Service Award (NRSA) Institutional Research Training Grants (T32) to eligible institutions as the primary means of supporting predoctoral and postdoctoral research training to help ensure that a diverse and highly trained workforce is available to assume leadership roles related to the Nation's biomedical, behavioral and clinical research agenda. The primary objective of the T32 program is to prepare qualified individuals for careers that have a significant impact on the health-related research needs of the Nation. This program supports predoctoral, postdoctoral and short term research training programs at domestic institutions of higher education with the T32 funding mechanism. Stipends are provided as a subsistence allowance for trainees to help defray living expenses during the research training experience and are based on a 12-month appointment period.

Requirements: The following organizations are eligible to apply: public/state controlled institutions of higher education; private institutions of higher education; nonprofits with 501(c)3 IRS status (other than institutions of higher education); nonprofits without 501(c)3 IRS status (other than institutions of higher education); for-profit organizations (other than small businesses); state governments; U.S. territories or possessions; Indian/Native American tribal governments (Federally recognized); Hispanic-serving institutions; historically Black colleges and universities (HBCUs); tribally controlled colleges and universities (TCCUs); Alaska Natives and Native Hawaiians serving institutions; regional organizations; eligible agencies of the federal government; faith-based or community based organizations.

Date(s) Application is Due: Apr 8; Aug 8; Dec 8

Contact: Crystal Wolfrey, Grants Administration Branch, NCI; (301) 496-8634; fax (301) 496-8601; crystal.wolfrey@nih.gov or wolfreyc@mail.nih.gov

Internet: http://grants.nih.gov/grants/guide/pa-files/PA-08-226.html

Sponsor: National Cancer Institute
6120 Executive Boulevard, Room 243, MSC 7150
Bethesda, MD 20892-7150

NCI Short-Term Scientist Exchange Program 3081

The program promotes collaborative research between established U.S. and foreign scientists by supporting, in part, exchange visits of U.S. cancer researchers to foreign laboratories, or foreign cancer researchers to U.S. laboratories. The visits may be from one week to six months in duration. The Office of International Affairs (OIA) funds an APEX fare airline ticket through an OIA contractor. The foreign host institution covers the in-country expenses of the U.S. visitor. In the case of foreign visitors, a foreign source pays for the airfare, while OIA provides a subsistence allowance. The subsistence allowance is generally the monthly equivalent of $32,004 per year (split equally by OIA and the sponsor). Cost sharing of the subsistence allowance by the sponsoring American laboratory is required.

Requirements: Candidates must have a minimum of three years postdoctoral experience in cancer research and have an invitation from a qualified sponsor. Sponsors or exchangees from the United States must be NCI-supported principal investigators, in either the intramural or extramural research programs. Foreign sponsors must be established scientists recognized in the international peer reviewed literature. Foreign candidates must be proficient in spoken and written English. Preference is given to applicants from not yet fully developed countries.

Amount of Grant: 37,100 U.S.D

Contact: Office of International Affairs; (301) 496-4761; fax (301) 496-3945; NC6@cu.nih.gov or cancergovstaff@mail.nih.gov

Internet: http://www.cancer.gov/about_nci/doc_img.aspx?viewid=8fc72aff-ab84-40b5-8fe9-20da67662417&docid=775a930b-99ec-40ff-ae9d-59beb7175527

Sponsor: National Cancer Institute
6130 Executive Boulevard, Executive Plaza N
Bethesda, MD 20892

NCI Small Grants Program for Behavioral Research in Cancer Control 3082

The small grants program is designed to aid and facilitate the growth of a nationwide cohort of scientists with a high level of research expertise in behavioral cancer control research. Small grants are short-term awards to provide support for pilot projects, development and testing of new methodologies, secondary data analyses, or innovative projects that provide a basis for more extended research. New investigators in relevant fields and disciplines (e.g., medicine, public health, health promotion, health communications and informatics, epidemiology, anthropology, social work, nursing research, nutrition, health policy, health services research, and behavioral sciences, such as psychology, health education, and sociology) may apply for small grants to test ideas or do pilot studies. The small grant award is not renewable. PA: PAR-04-020

Requirements: Applications may be submitted by public and private, domestic, for-profit and nonprofit organizations, units of state and local governments, and eligible agencies of the federal government. Eligible applicants include new investigators who have not previously been a principal investigator on an NCI-funded cancer control research grant (R03, R01, P01), or established scientists refocusing their research interests to behavioral research in cancer. Predoctoral investigators currently enrolled in an accredited doctoral degree program also are eligible to apply. All applicants should identify a mentor or sponsor from whom they will receive guidance regarding the proposed research. Small research grants may not be used to supplement research projects currently supported by federal or nonfederal funds, or to provide interim support of research project applications under review by the Public Health Service.

Restrictions: Foreign organizations are ineligible to apply.

Date(s) Application is Due: Apr 20; Aug 22; Dec 22

Amount of Grant: Up to 50,000 U.S.D

Contact: Veronica Chollette, Division of Cancer Control and Population Sciences; (301) 435-2837; vc24a@nih.gov

Internet: http://www.grants.nih.gov/grants/guide/pa-files/PAR-04-020.html

Sponsor: National Cancer Institute
6130 Executive Boulevard, Executive Plaza N
Bethesda, MD 20892

NCI Specialized Program of Research Excellence (SP.O.RE) in Human Cancer 3083

A SP.O.RE must address the scientific information base and provide focal points for sustaining and maintaining state-of-the-art research that will contribute to improved detection, diagnosis, treatment, and prevention of an organ-specific cancer. SP.O.RES not only are expected to conduct a wide spectrum of research activities, but also to contribute significantly to the development of specialized research resources, the development of improved research model systems and the expansion of the research base through collaborative research with scientists and clinicians in other institutions locally and nationwide. A letter of intent is due at least 60 days prior to the specific due date. Application receipt dates are February 1 for breast cancer SP.O.RES, June 1 for lung cancer SP.O.RES, and October 1 for prostate and genitourinary cancer SP.O.RES. PA: PAR-99-167

Requirements: Applications may be submitted by domestic for-profit and nonprofit organizations, both public and private; units of state and local governments; and eligible agencies of the federal government.

Date(s) Application is Due: Feb 1; Jun 1; Oct 1

Contact: Dr. Jorge Gomez; (301) 496-8528; jg1w@nih.gov

Internet: http://grants.nih.gov/grants/guide/pa-files/PAR-99-167.html

Sponsor: National Cancer Institute
6130 Executive Boulevard, Executive Plaza N
Bethesda, MD 20892

NCI Stages of Breast Development: Normal to Metastatic Disease Grants 3084

The National Cancer Institute (NCI), National Institute of Child Health and Human Development (NICHD), National Institute of Diabetes and Digestive and Kidney Diseases (NIDDK), National Institute on Aging (NIA), and National Institute of Environmental Health Sciences (NIEHS) invite investigator-initiated research grant applications to study the molecular, cellular, endocrine, and other physiological influences on the development and maturation of the normal mammary gland and alterations involved in early malignant and metastatic breast cancer. Multi-disciplinary collaborations, for example between cell biologists, molecular endocrinologists, bioengineers, geneticists, and mammary pathologists, are encouraged. Appropriate studies include, but are not limited to, documenting the role of dynamic hormonal influences and determining the role of cell growth, apoptosis, and differentiation in mammary gland maturation; integrating knowledge of cell signaling in breast tissue with whole organ biology; developing models of breast differentiation; and studies focusing on the characteristics of breast tumor physiology particularly in relation to metastasis. It is expected that a complete understanding of mammary gland development will form critical underpinnings for continued advances in detecting, preventing and treating breast cancer.

Requirements: Applications may be submitted by domestic and foreign, for-profit and non-profit organizations, public and private, such as universities, colleges, hospitals, laboratories, units of State or local governments, and eligible agencies of the Federal government.

Date(s) Application is Due: Feb 5; Jun 5; Oct 5

Amount of Grant: Up to 250,000 U.S.D

Contact: Padma Maruvada, (301) 496-3893; fax (301) 402-8990; maruvadp@mail.nih.gov

Internet: http://grants.nih.gov/grants/guide/pa-files/PA-99-162.html

Sponsor: National Cancer Institute
6130 Executive Boulevard, Executive Plaza N, Suite 3144, MSC 7362
Bethesda, MD 20892-7362

NCI Technologies and Software to Support Integrative Cancer Biology Research (SBIR) Grants 3085

This Funding Opportunity Announcement (FOA), issued by the National Cancer Institute (NCI), invites Small Business Innovation Research (SBIR) grant applications from small business concerns (SBCs) that propose the development of commercial software tools, computational/mathematical methods, and technologies that will enable integrative cancer biology research. Integrative cancer biology focuses on understanding cancer as a complex biological system by utilizing both computational and experimental biology to integrate heterogeneous data sources and ultimately to generate predictive computational models of cancer processes. Phase I applications may request up to 2 years of support with total costs of up to $300,000 for the duration of the award (not to exceed

$200,000 total costs in any one year). Budgetary requests of Phase II applications will conform to standard SBIR guidelines (up to $750,000 total costs for up to two years).
Requirements: Only United States small business concerns (SBCs) are eligible to submit SBIR applications.
Date(s) Application is Due: Apr 5; Aug 5; Dec 5
Amount of Grant: 200,000 - 750,000 U.S.D
Contact: Greg Evans, (301) 594-8807; fax (301) 480-4082; evansgl@mail.nih.gov
Internet: http://grants.nih.gov/grants/guide/pa-files/PA-09-188.html
Sponsor: National Cancer Institute
31 Center Drive, Building 31, Room 10A19, MSC 2580
Bethesda, MD 20892-2580

NCI Transition Career Development Award **3086**

The purpose of the program is to facilitate the transition of investigators from the mentored to the independent stage of their careers in cancer research. It applies to clinicians who are pursuing basic science careers; clinicians who are pursuing careers in patient-oriented research; and to individuals pursuing careers in the prevention, control, and population sciences. To apply, a candidate must have completed two years or more of postdoctoral, mentored research or have been in an independent position for less than one year at the time of the application. The unique feature of this award is that individuals may apply without a sponsoring institution while they are still in a mentored position. Successful postdoctoral applicants will be given up to 12 months to identify an independent, preferably tenure-track, position at a sponsoring institution before an award can be activated. Awardees must apply for an R01 research grant or equivalent prior to the end of the second year of the award. Annual deadline dates may vary; contact program staff for exact dates. PA: PAR-01-134
Requirements: Eligible candidates are doctorally degreed individuals who have been educated as clinicians (e.g., MDs, oncology nurses) or as prevention, control, and population scientists (e.g., PhDs, DPHs, MDs) and are ready to pursue independent careers in cancer research.
Restrictions: The award is not for basic scientists following a traditional basic science research career track (e.g., PhDs in molecular biology).
Date(s) Application is Due: Feb 1; Jun 1; Oct 1
Contact: Kimery Griffin, Grants Administration Branch; (301) 496-3196; fax (301) 496-8601; griffink3@mail.nih.gov
Internet: http://grants.nih.gov/grants/guide/pa-files/PAR-04-040.html
Sponsor: National Cancer Institute
6130 Executive Boulevard, Executive Plaza N
Bethesda, MD 20892

NCLIS Health Information Awards **3087**

The program recognizes libraries that provide exemplary consumer-health information programs and promote improved health for the populations they serve. The initiative will, over the course of one year, identify libraries in each state. The top 10 libraries selected from the 52 states receive an award. Five of the 10 awards will be reserved for specific categories of libraries: small library, medium-size library, large library, library with the most outstanding program for children and health, and library that does the most to address substance abuse prevention or treatment. The remaining five prizes will be open to all libraries. In May of the next year, a top winning library will be announced and representatives from that library will travel to Washington, DC, to receive the grand prize.
Requirements: All American libraries—including public, academic, specialized, government, school, large and small, rural, and urban or suburban—are eligible to apply.
Date(s) Application is Due: Jan 31
Amount of Grant: 1,000 - 20,000 U.S.D
Contact: Grants Administrator; (202) 606-9200; fax (202) 606-9203; info@nclis.gov
Internet: http://www.nclis.gov/award/award.cfm
Sponsor: National Commission on Libraries and Information Science
1800 M Street NW, Suite 350 N Tower
Washington, DC 20036-5841

NCRG Centers of Excellence in Gambling Research Grants **3088**

The purpose of the Centers of Excellence in Gambling Research Grants program is to: advance the understanding of gambling disorders through innovative, multi-disciplinary research investigations; and enhance the dissemination of research findings to the public. The center grants are intended to provide funding for a stable, long-term institutional focus on a complex set of gambling-related problems requiring an innovative, multi-disciplinary approach. Recipients are expected to provide leadership in the field by: conducting cutting-edge investigations of gambling-related disorders; translating research findings for non-academic audiences; and cultivating the next generation of gambling researchers by mentoring young investigators. Investigative efforts should be broadly based, encompassing a variety of areas, including biological, biomedical, social, behavioral, and/or clinical sciences. The proposed research investigation may focus on a broad range of research that develops and tests psychosocial or pharmacological approaches for prevention, intervention, treatment or relapse prevention of gambling disorders. The NCRG is especially interested in brief interventions targeted at underrepresented populations, such as minorities, young adults and persons with subclinical gambling disorders. Other priorities include the following topics: impact of Indian gaming, gambling and minorities, secondary data analysis,and technology and gambling.
Requirements: Both public and private nonprofit organizations are eligible to apply. Profit-making organizations should contact Christine Reilly (creilly@ncrg.org). Foreign institutions are required to collaborate with a U.S. institution. Applications involving a non-U.S. institution must have a principal investigator and fiscal agent based at a U.S. institution.
Geographic Focus: All States, All Countries
Date(s) Application is Due: Jul 2
Amount of Grant: Up to 350,000 U.S.D
Contact: Christine Reilly, Executive Director; (978) 338-6610; fax (978) 522-8452; creilly@ncrg.org
Internet: http://www.ncrg.org/research-center/apply-ncrg-funding/ncrg-centers-excellence-gambling-research
Sponsor: National Center for Responsible Gaming
900 Cummings Center, Suite 418-U
Beverly, MA 01915

NCRG Early Stage Investigator Grants for Gambling Researchers **3089**

The NCRG Early Stage Investigator Grants for Gambling Researchers is a career development award that will provide scientists with the mentoring and research experiences necessary for research independence. Applications for this award must propose a research plan that has: intrinsic research importance in the area of prevention and treatment of gambling disorders; and will serve as a suitable vehicle for learning the methodology, theories and concepts needed for a well-trained, independent researcher in the area of gambling research. Applicants may request up to $30,000 in direct costs for a period not to exceed 24 months.
Requirements: The Early Stage Investigator Grant is intended for individuals who are within 10 years of completing their terminal research degree or within 10 years of completing medical residency (or the equivalent). The candidate for this award may not have been a recipient of an NIH Career Development Award or an NCRG grant. Candidates who are not U.S. citizens but have an appointment at a U.S. institution are eligible to apply. Candidates based at foreign institutions are not eligible to apply.
Geographic Focus: All States, All Countries
Date(s) Application is Due: May 1
Amount of Grant: Up to 30,000 U.S.D
Contact: Christine Reilly, Executive Director; (978) 338-6610; fax (978) 522-8452; creilly@ncrg.org
Internet: http://www.ncrg.org/research-center/apply-ncrg-funding/early-stage-investigator-grants
Sponsor: National Center for Responsible Gaming
900 Cummings Center, Suite 418-U
Beverly, MA 01915

NCRG Exploration Grants for Research on Gambling Disorders **3090**

Exploration Grants provide quick access to funding for researchers focused on gambling disorders. Funds can be used to support a broad range of activities including: pilot data to complete an initial but unfunded project, pilot data needed for preliminary results,new direction on a current project not otherwise supported, and funded projects that did not build costs into budgets for services previously free or subsidized which now carry a cost. The proposed research investigation may focus on a broad range of research that develops and tests psychosocial or pharmacological approaches for prevention, intervention, treatment or relapse prevention of gambling disorders. The NCRG is especially interested in brief interventions targeted at underrepresented populations, such as minorities, young adults and persons with subclinical gambling disorders. Other priorities include the following topics: impact of Indian gaming; gambling and minorities; secondary data analysis; and technology and gambling.
Requirements: Both public and private nonprofit organizations are eligible to receive grants from the NCRG. Profit-making organizations should contact Christine Reilly (creilly@ncrg.org). Foreign institutions are required to collaborate with a U.S. institution. Applications involving a non-U.S. institution must have a principal investigator and fiscal agent based at a U.S. institution.
Geographic Focus: All States, All Countries
Amount of Grant: Up to 10,000 U.S.D
Samples: Yijun Liu, Ph.D., University of Florida, Gainesville, Florida, $5,400 - to develop a virtual Internet gambling website to be used as a simulation in a laboratory experiment to test pathological gambling in an online environment (2010).
Contact: Christine Reilly; (978) 338-6610; fax (978) 522-8452; creilly@ncrg.org
Internet: http://www.ncrg.org/research-center/apply-ncrg-funding/exploration-grants
Sponsor: National Center for Responsible Gaming
900 Cummings Center, Suite 418-U
Beverly, MA 01915

NCRG Independent Scientist in Gambling Research Grants **3091**

The Independent Scientist Award provides support for newly independent scientists who can demonstrate the need for a period of intensive research focus as a means of enhancing their research careers. The grant is intended to foster the development of outstanding scientists and to enable them to expand their potential to make significant contributions to the field of gambling disorders. Eligible applicants have independent peer-reviewed research support. Applicants may request up to $40,000 in direct costs per year for a period not to exceed 24 months.
Requirements: Both public and private nonprofit organizations are eligible to apply. Profit-making organizations should contact Christine Reilly (creilly@ncrg.org). Applications involving a non-U.S. institution must have a principal investigator and fiscal agent based at a U.S. institution. Candidates must have a doctoral degree and independent, peer reviewed research support at the time the award is made and must be willing to commit a minimum of one-third of full-time professional effort conducting research and relevant career development activities during the period of the award.
Geographic Focus: All States, All Countries
Date(s) Application is Due: May 1
Amount of Grant: Up to 40,000 U.S.D

Contact: Christine Reilly; (978) 338-6610; fax (978) 522-8452; creilly@ncrg.org
Internet: http://www.ncrg.org/research-center/apply-ncrg-funding/independent-scientist-grant
Sponsor: National Center for Responsible Gaming
900 Cummings Center, Suite 418-U
Beverly, MA 01915

NCRG Large Grants for Research on Gambling Disorders **3092**
Large Grants provide up to two years of support for discrete, specified, circumscribed research projects related to gambling disorders. The proposed research investigation may focus on a broad range of research that develops and tests psychosocial or pharmacological approaches for prevention, intervention, treatment or relapse prevention of gambling disorders. The NCRG is especially interested in brief interventions targeted at underrepresented populations, such as minorities, young adults and persons with subclinical gambling disorders. Other priorities include the following topics: impact of Indian gaming; gambling and minorities; secondary data analysis; and technology and gambling. Applicants may request up to $75,000 in direct costs per year for a period not to exceed 24 months.
Requirements: Both public and private nonprofit organizations are eligible to apply. Profit-making organizations should contact Christine Reilly, senior research director (creilly@ncrg.org). Foreign institutions are required to collaborate with a U.S. institution. Applications involving a non-U.S. institution must have a principal investigator and fiscal agent based at a U.S. institution.
Geographic Focus: All States, All Countries
Date(s) Application is Due: Jul 2
Amount of Grant: Up to 75,000 U.S.D
Contact: Christine Reilly, Executive Director; (978) 338-6610; fax (978) 522-8452; creilly@ncrg.org
Internet: http://www.ncrg.org/research-center/apply-ncrg-funding/large-grants
Sponsor: National Center for Responsible Gaming
900 Cummings Center, Suite 418-U
Beverly, MA 01915

NCRG Postdoctoral Fellowships for Gambling Researchers **3093**
The NCRG Postdoctoral Fellowship Grant is intended to help ensure that a pool of highly trained scientists is available to address the research needs of the field of gambling disorders. Eligible applicants are within five years of completing their terminal research degree or within five years of completing medical residency. The proposed research investigation may focus on a broad range of research that develops and tests psychosocial or pharmacological approaches for prevention, intervention, treatment or relapse prevention of gambling disorders. The NCRG is especially interested in brief interventions targeted at underrepresented populations, such as minorities, young adults and persons with subclinical gambling disorders. Other priorities include the impact of Indian gaming; gambling and minorities; secondary data analysis; and technology and gambling. Applicants may request up to $43,000 in direct costs for a period not to exceed 12 months.
Requirements: The Postdoctoral Fellowship is intended for individuals who are within five years of completing their terminal research degree or within five years of completing medical residency (or the equivalent). The candidate for this award may not have been a recipient of an NIH Career Development Award or an NCRG grant. Candidates who are not U.S. citizens but have an appointment at a U.S. institution are eligible to apply. Candidates based at foreign institutions are not eligible to apply.
Geographic Focus: All States, All Countries
Date(s) Application is Due: May 1
Amount of Grant: Up to 43,000 U.S.D
Contact: Christine Reilly; (978) 338-6610; fax (978) 522-8452; creilly@ncrg.org
Internet: http://www.ncrg.org/research-center/apply-ncrg-funding/postdoctoral-fellowships
Sponsor: National Center for Responsible Gaming
900 Cummings Center, Suite 418-U
Beverly, MA 01915

NCRG Seed Grants for Research on Gambling Disorders **3094**
NCRG Seed Grants support a variety of research activities, including: pilot and feasibility studies; secondary analysis of existing data; small, self-contained research projects; development of research methodology; development of new research technology.The proposed research investigation may focus on a broad range of research that develops and tests psychosocial or pharmacological approaches for prevention, intervention, treatment or relapse prevention of gambling disorders. The NCRG is especially interested in brief interventions targeted at underrepresented populations, such as minorities, young adults and persons with subclinical gambling disorders. Other priorities include the following topics: impact of Indian gaming, gambling and minorities, secondary data analysis and technology and gambling.
Requirements: Both public and private nonprofit organizations are eligible to receive grants from the NCRG. Applicants from profit-making companies should contact Christine Reilly, senior research director (creilly@ncrg.org). Foreign institutions are required to collaborate with a U.S. institution. Applications involving a non-U.S. institution must have a principal investigator and fiscal agent based at a U.S. institution.
Geographic Focus: All States, All Countries
Amount of Grant: Up to 25,000 U.S.D
Samples: T. Celeste Napier, Ph.D., Rush University Medical Center, Chicago, Illinois, $28,750 - to identify the potential for repurposing the atypical antidepressant, mirtazapine, as a pharmacological intervention for reducing risk-behavior and/or relapse prevention of gambling disorders (2011); Mark R. Dixon, Ph.D., Southern Illinois University, Carbondale, Illinois, $34,500 - attempt to demonstrate that brain activity in 20 under-age pathological gamblers who are exposed to near-misses that resemble large jackpot wins (2011).
Contact: Christine Reilly, Executive Director; (978) 338-6610; fax (978) 522-8452; creilly@ncrg.org
Internet: http://www.ncrg.org/research-center/apply-ncrg-funding/seed-grants
Sponsor: National Center for Responsible Gaming
900 Cummings Center, Suite 418-U
Beverly, MA 01915

NCRR Development of Nonmammalian Models and Related Biological Materials Research Grants **3095**
The Comparative Medicine area, NCRR, encourages the submission of applications for the development of nonmammalian models and associated biological materials for biomedical research. Models to be considered must be applicable to the research interests of two or more categorical disease institutes of NIH. The overall goal of these studies is to develop new nonmammalian research models of broad interest and expand on the usefulness of existing nonmammalian model systems. Accordingly, the mechanisms for research projects (R01), exploratory/developmental (R21) and resource-related research projects (R24) grants apply. PA: PAR-99-115
Requirements: Applications may be submitted by foreign and domestic, for-profit and nonprofit organizations, public and private; units of state or local governments; and eligible agencies of the federal government.
Amount of Grant: Up to 100,000 U.S.D
Contact: Dr. Jill Carrington, Division of Comparative Medicine Contact; (301) 435-0744; fax (301) 480-3819; jillc@ncrr.nih.gov
Internet: http://www.grants.nih.gov/grants/guide/pa-files/PAR-99-115.html
Sponsor: National Center for Research Resources
6701 Democracy Boulevard, MSC 4874
Bethesda, MD 20892-4874

NCRR Innovative Approaches to Developing New Technologies **3096**
The purpose of this PA is to encourage submission of new exploratory/developmental grant applications to support innovative approaches to the development of technologies exploring new research paradigms in engineering, instrumentation, physical sciences, mathematics, or computer science applicable to all areas of biomedical research. The projects should provide the opportunity to develop new technologies, methods, devices, and materials that can be used in research aimed at gaining a greater understanding of fundamental elements of biological phenomena as well as potentially providing a basis for the development of new products useful in all aspects of biomedical research and health care. Annual deadline dates may vary; contact program staff for exact dates. PA: PAR-98-047
Requirements: Applications may be submitted by foreign and domestic, nonprofit and for-profit organizations, public and private, such as universities, colleges, hospitals, laboratories, units of state and local governments, and eligible agencies of the federal government.
Date(s) Application is Due: Jun 1; Oct 1
Amount of Grant: Up to 75,000 U.S.D
Contact: Dr. Richard Dubois; (301) 435-0755; fax (301) 480-3659
Internet: http://www.grants.nih.gov/grants/guide/pa-files/PAR-98-047.html
Sponsor: National Center for Research Resources
6701 Democracy Boulevard, MSC 4874
Bethesda, MD 20892-4874

NCRR Institutional Animal Resources Developing and Improving Grants **3097**
The grants assist biomedical and behavioral research institutions in upgrading and developing their animal facilities. Support is limited to alterations and renovations to improve laboratory animal facilities and to the purchase of major equipment items for animal resources, diagnostic laboratories, transgenic animal resources, or similar associated activities. Facilities and administrative costs are not provided. Matching funds from nonfederal sources are required, equal to or exceeding one-half of the total allowable costs of the requested project. These matching funds must be applied to the specific project described in the application and cannot be met by citing other expenditures. The initial budget period may be either two or three years in length; the total project period may not exceed five years. PA: PAR-03-077
Requirements: Any domestic public or private nonprofit institution, organization, or association with one or more projects supported by the Public Health Service involving the use of animals is eligible to apply.
Restrictions: The program does not support requests for equipment used for teaching purposes and for housing non-research animals.
Amount of Grant: Up to 700,000 U.S.D
Contact: Paul Karadbil; (301) 435-0844; paulk@ncrr.nih.gov
Internet: http://grants2.nih.gov/grants/guide/pa-files/PAR-03-077.html
Sponsor: National Center for Research Resources
6701 Democracy Boulevard, MSC 4874
Bethesda, MD 20892-4874

NCRR National Stem Cell Resource Grants **3098**
The purpose of this RFA is to solicit applications to establish a national stem cell resource. The resource will be for deposit, maintenance, preservation, and distribution of nonhuman-derived embryonic and other stem cells from a variety of species, as well as for deposit and distribution of related reagents and information in order to make these critical research tools available for biomedical research. Annual deadline dates may vary; contact program staff for exact dates. RFA: RR-00-001
Requirements: Applications may be submitted by domestic, public and private, for-profit and nonprofit organizations, units of state and local governments, and eligible agencies of the federal government.

Restrictions: Foreign institutions are not eligible to apply.
Amount of Grant: 800,000 U.S.D
Contact: Dr. Jill Carrington, Division of Comparative Medicine Contact; (301) 435-0744; fax (301) 480-3819; jillc@ncrr.nih.gov
Internet: http://grants.nih.gov/grants/guide/rfa-files/RFA-RR-00-001.html
Sponsor: National Center for Research Resources
6701 Democracy Boulevard, MSC 4874
Bethesda, MD 20892-4874

NCRR Networks and Pathways Collaborative Research Projects Grants **3099**
This announcement solicits applications for research project grants that will leverage and complement the ongoing technology development being pursued in the National Technology Centers for Networks and Pathways (TCNPs), a program of the NIH Roadmap for Medical Research. These collaborative projects should focus either on addressing a challenging biological problem using the technology developed in one or more of the TCNPs, or on the development of technology that will complement that which is being developed in the centers. Applicants may request support for their own work as well as supplemental support for components pursued in the participating TCNP. While the nature and scope of these applications will vary, it is anticipated that the annual direct costs for these projects will typically range from $150,000 to $250,000.
Requirements: Application can be made by the following institutions and organizations: public/state controlled institutions of higher education; private institutions of higher education; nonprofits with 501(c)3 IRS status (other than institutions of higher education); nonprofits without 501(c)3 IRS status (other than institutions of higher education); small businesses; for-profit organizations (other than small businesses); state governments; U.S. territories or possessions; non-domestic (non-U.S.) entities (foreign organizations); historically Black colleges and universities (HBCUs); Tribally controlled colleges and universities (TCCUs); and other eligible agencies of the Federal government.
Date(s) Application is Due: Feb 5; Jun 5; Oct 5
Contact: Douglas M. Sheeley, (301) 594-9762; sheeleyd@mail.nih.gov
Internet: http://grants.nih.gov/grants/guide/pa-files/PA-07-266.html
Sponsor: National Center for Research Resources
6701 Democracy Boulevard, MSC 4874
Bethesda, MD 20892-4874

NCRR Novel Approaches to Enhance Animal Stem Cell Research **3100**
The purpose of this funding opportunity announcement (FOA) is to encourage the submission of applications for research to enhance animal stem cells as model biological systems. Innovative approaches to isolate, characterize and identify toti-potent and multi-potent stem cells from nonhuman biomedical research animal models, as well as to generate reagents and techniques to characterize and separate those stem cells from other cell types is encouraged.
Requirements: The following organizations and institutions may apply: public/state controlled institutions of higher education; private institutions of higher education; Hispanic-serving institutions; historically Black colleges and universities (HBCUs); tribally controlled colleges and universities (TCCUs); Alaska Native and Native Hawaiian serving institutions; nonprofits with 501(c)3 IRS status (other than institution of higher education); nonprofits without 501(c)3 IRS status (other than institution of higher education); non-domestic (non-U.S.) entities (foreign organizations); small businesses; for-profit organizations (other than small businesses).
Restrictions: Studies involving human subjects are not allowed.
Date(s) Application is Due: Feb 5; Jun 5; Oct 5
Contact: John D. Harding, (301) 435-0744; fax (301) 480-3819; hardingj@mail.nih.gov
Internet: http://grants.nih.gov/grants/guide/pa-files/PA-07-303.html
Sponsor: National Center for Research Resources
6701 Democracy Boulevard, MSC 4874
Bethesda, MD 20892-4874

NCRR Shared Instrumentation Grants **3101**
The objective of the program is to make available to institutions expensive research instruments that can only be justified on a shared-use basis and for which meritorious research projects are described. This program is designed to provide for the acquisition or updating of expensive shared-use instrumentation not generally available through other NIH mechanisms, such as the regular research project, program project, or center grant programs. Proposals for research on advancing the design or for the development of new instrumentation will not be considered. Annual deadline dates may vary; contact the NCRR for exact dates. PA: PAR-05-028
Requirements: Nonprofit organizations, public or private institutions, and domestic institutions/organizations are eligible to apply.
Restrictions: Federal institutions, foreign institutions, and for-profit institutions are not eligible to apply.
Date(s) Application is Due: Mar 22
Amount of Grant: Up to 500,000 U.S.D
Contact: Program Contact; (301) 435-0714; grantsinfo@nih.gov
Internet: http://grants.nih.gov/grants/guide/pa-files/PAR-05-028.html
Sponsor: National Center for Research Resources
6701 Democracy Boulevard, MSC 4874
Bethesda, MD 20892-4874

NEDA/AED Charron Family Research Grant **3102**
Committed to promoting study into the causes, cures and prevention of eating disorders, NEDA first established the Research Grants Program in 2002 in partnership with the Academy for Eating Disorders (AED). The aim of these grants is to expand eating disorders research while drawing promising new scientists into the field. In 2004 Shelley Charron-Bahniuk died from complications of her eating disorder while she was in recovery. The Charron Family Research Grant was created as a direct result of this illness that impacted both her immediate and extended family.
Requirements: Individuals working in the field of eating disorders who have completed their training, and have the doctorate degree or beyond are eligible.
Amount of Grant: 10,000 U.S.D
Contact: Dee Christoff, Development Director; (206) 382-3587, ext. 15; fax (206) 829-8501; info@nationaleatingdisorders.org
Internet: http://www.nationaleatingdisorders.org/p.asp?WebPage_ID=712
Sponsor: National Eating Disorders Association / Academy for Eating Disorders
603 Stewart Street, Suite 803
Seattle, WA 98101

NEDA/AED Joan Wismer Research Grant **3103**
Committed to promoting study into the causes, cures and prevention of eating disorders, NEDA first established the Research Grants Program in 2002 in partnership with the Academy for Eating Disorders (AED). The aim of these grants is to expand eating disorders research while drawing promising new scientists into the field.
Requirements: Individuals working in the field of eating disorders who have completed their training, and have the doctorate degree or beyond are eligible.
Amount of Grant: 10,000 U.S.D
Contact: Dee Christoff, Development Director; (206) 382-3587, ext. 15; fax (206) 829-8501; info@nationaleatingdisorders.org
Internet: http://www.nationaleatingdisorders.org/p.asp?WebPage_ID=818
Sponsor: National Eating Disorders Association / Academy for Eating Disorders
603 Stewart Street, Suite 803
Seattle, WA 98101

NEDA/AED Tampa Bay Eating Disorders Task Force Award **3104**
The National Eating Disorders Association sponsors a small grants program to support research in the areas of eating disorders etiology, prevention, and treatment. Through the grants, the organization seeks to expand innovative eating disorders research while supporting investigators in the early stages of their careers in the eating disorders field. Support can be included for: salary for technicians, research assistants and other ancillary personnel; laboratory, radiological and pharmaceutical costs in clinical research projects; rating instruments and questionnaires; equipment and supplies; acquisition and maintenance of laboratory animals; subject payment; and advertising.
Requirements: Any junior investigator worldwide who works in the field of eating disorders and has completed their terminal graduate level degree and training is eligible.
Restrictions: Trainees (e.g. students, grad students, residents) are not eligible and investigators must be within seven years following the end of their training. Individuals who previously served or currently serve as a principal investigator on a substantial externally funded grant (e.g. RO1, RO3, or K award in the U.S.) are ineligible. Funds cannot be used for the following purposes: salary support for residents, fellows or other trainees; faculty or investigator salary; secretarial salary and benefits; tuition for residents, graduate students or post-doctoral fellows; publication costs or purchase of reprints or books; room and board for clinical subjects; studies that are also sponsored by pharmaceutical firms; computers; or travel.
Amount of Grant: 10,000 U.S.D
Contact: Dee Christoff, Development Director; (206) 382-3587, ext. 15; fax (206) 829-8501; info@nationaleatingdisorders.org
Internet: http://www.nationaleatingdisorders.org/p.asp?WebPage_ID=816
Sponsor: National Eating Disorders Association / Academy for Eating Disorders
603 Stewart Street, Suite 803
Seattle, WA 98101

NEETF Competitive Grants **3105**
NEETF supports environmental education grants that explicitly connect environment, economy, and equity. Because of the immediate need to make this people/community/environment connection, NEETF is most interested in projects that educate adults through workplace programs, community initiatives, and in formal education settings. Youth-focused programs will be considered only where need can be clearly shown. Competitive grants are awarded in three program areas: health and environment, safe water, and environmental education innovation. The foundation's one-year challenge grants require a cash match; the preferred match ratio is two or more nonfederal dollars for each NEETF dollar awarded. To screen potential grant projects, NEETF uses a two-page preproposal application. Applicants will be invited to submit full proposals. Preproposals must be received by the listed application deadline date.
Requirements: Proposals may be submitted by public and private agencies, including nonprofit organizations such as schools and community-based organizations. Application forms are required.
Restrictions: Funds will not be provided for general administrative expenses, capital projects, scholarships or fellowships, basic research, political advocacy, or litigation projects.
Amount of Grant: 5,000 - 15,000 U.S.D
Contact: Samantha Blodgett, Grants Coordinator; (202) 833-2933; fax (202) 261-6464; blodgett@neetf.org
Internet: http://www.neetf.org/Education/index.htm
Sponsor: National Environmental Education and Training Foundation
1707 H Street NW, Suite 900
Washington, DC 20006-3915

NEI Clinical Study Planning Grant 3106

This program provides support for the development of a comprehensive research protocol for a large-scale clinical trial or epidemiologic study. This grant provides early peer review of the proposed clinical study in terms of its rationale, design, organizational structure, and implementation plan. The planning grant is used to support the development of a detailed MOP and may include preliminary studies to refine study procedures or to document recruitment potential. PAR-05-115

Requirements: Applications may be submitted by domestic for-profit and nonprofit organizations, both public and private; units of state and local governments; and eligible agencies of the federal government; domestic and foreign institutions.

Restrictions: Foreign institutions are not eligible.

Amount of Grant: Up to 150,000 U.S.D

Contact: Donald F. Everett, (301) 451-2020; fax: (301) 402-0528; dfe@nei.nih.gov *Bethesda, MD 20892-9300*

Internet: http://grants.nih.gov/grants/guide/pa-files/PAR-05-115.html

Sponsor: National Eye Institute
2020 Vision Place
Bethesda, MD 20892-3655

NEI Clinical Vision Research Development Award 3107

This award is intended to strengthen interactions among clinicians, biostatisticians, epidemiologists, statistical geneticists, and other clinical trial specialists to facilitate the design and conduct of clinical research projects, such as the development of coordinating center capabilities. The intent of this projects is to advance the understanding, prevention, or clinical management of visual system disorders. A number of experienced investigators and research groups are needed to provide leadership in the design and implementation of such clinical vision research projects. Therefore, NEI wishes to help institutions with NEI-funded investigators augment staff expertise and acquire the resources necessary to enhance their clinical vision research programs. PA: PAR-00-050

Requirements: Applications may be submitted by domestic for-profit and non-profit organizations, public and private, such as universities, colleges, hospitals, laboratories, units of State and local governments, and eligible agencies of the Federal government.

Restrictions: Foreign institutions and institutions having current or past NEI Core Grant for Vision Research (P30) support for a biostatistics module are not eligible to apply.

Amount of Grant: Up to 100,000 U.S.D

Contact: Mary Frances Cotch, (301) 496-5983; fax (301) 402-0528; mfcotch@nei.nih.gov or deverett@nei.nih.gov

Internet: http://grants.nih.gov/grants/guide/pa-files/PAR-00-050.html

Sponsor: National Eye Institute
2020 Vision Place
Bethesda, MD 20892-3655

NEI Innovative Patient Outreach Programs And Ocular Screening Technologies To Improve Detection Of Diabetic Retinopathy Grants 3108

This Funding Opportunity Announcement (FOA) solicits Small Business Innovation Research (SBIR) grant applications from small business concerns (SBCs) that propose to 1) develop educational outreach programs to create a greater awareness of the blinding consequences of diabetes; and, 2) develop tools and systems to be used for increasing patient access to eye exams for detecting Diabetic Retinopathy (DR). This FOA will utilize the SBIR (R43/R44) grant mechanisms for Phase I, Phase II, and Fast-Track applications. The estimated amount of funds available for support of 2-4 projects awarded as a result of this announcement is $2 million for this fiscal year.

Requirements: Only United States small business concerns (SBCs) are eligible to submit SBIR applications. A small business concern is one that, at the time of award of Phase I and Phase II, meets all of the criteria listed. Any individual(s) with the skills, knowledge, and resources necessary to carry out the proposed research as the PD/PI is invited to work with his/her organization to develop an application for support. Individuals from underrepresented racial and ethnic groups as well as individuals with disabilities are always encouraged to apply for NIH support. See website for details.

Date(s) Application is Due: Mar 23; Dec 23

Amount of Grant: 2,000,000 U.S.D

Contact: Jerome R. Wujek, (301) 451-2020; fax (301) 402-0528; wujekjer@nei.nih.gov

Internet: http://grants.nih.gov/grants/guide/rfa-files/RFA-EY-09-001.html

Sponsor: National Eye Institute
2020 Vision Place
Bethesda, MD 20892-3655

NEI Mentored Clinical Scientist Development Program Award (K12) 3109

The NEI Mentored Clinician Scientist Development Award (K12) is an award to an educational institution or professional organization to facilitate and support career development experiences which lead clinician scientists to research independence. It is expected that scholars appointed to the program will subsequently apply for their own individual research support. The program should be designed to accommodate candidates with varying levels of research experience who require an individualized career development program ranging from two to five years. Applications for this award should propose a structured, phased development plan overseen by a Program Director and a Program Advisory Committee. A designated period of didactic training should be followed by a period of supervised research experience. Candidates should be exposed to both research that is typically undertaken with a few investigators and to research approaches that require a multidisciplinary team to address complex research questions as appropriate. It is expected that at the end of the career development period, appointees will transition successfully into positions as independent investigators. This award provides five years of potentially renewable support. The subsequent continuation will depend in part on the progress made by the Scholars in achieving independent research support. Renewal applications must contain a progress report documenting the accomplishments and current career activities of former Scholars. PA# PAR-09-083

Requirements: The following organizations and institutions are eligible to apply: public/state controlled institutions of higher education; private institutions of higher education; hispanic-serving Institutions; historically black colleges and universities; tribally controlled colleges and universities; Alaska Native and Native Hawaiian serving institutions; nonprofits with 501(c)3 IRS Status (other than institutions of higher education); nonprofits without 501(c)3 IRS Status (other than institutions of higher education). Any individual with the skills, knowledge, and resources necessary to carry out the proposed research is invited to work with his/her institution to develop an application for support. Individuals from underrepresented racial and ethnic groups as well as individuals with disabilities are always encouraged to apply for NIH support.

Restrictions: Foreign institutions are not eligible to apply.

Date(s) Application is Due: May 13

Amount of Grant: 1,125,000 U.S.D

Contact: Neeraj Agarwal, (301) 451-2020; fax (301) 402-0528; agarwalnee@nei.nih.gov

Internet: http://grants.nih.gov/grants/guide/pa-files/PAR-09-083.html

Sponsor: National Eye Institute
2020 Vision Place
Bethesda, MD 20892-3655

NEI Research Grant For Secondary Data Analysis Grants 3110

The NEI Research Grant for Secondary Data Analysis (R21) program is designed to provide investigators with the support necessary to conduct secondary data analyses utilizing existing database resources. Applications may be related to, but must be distinct from, the specific aims of the original data collection. The total project period for an application submitted in response to this funding opportunity may not exceed two years. Direct costs are limited to $275,000 over an R21 two-year period, with no more than $200,000 in direct costs allowed in any single year. PA # PAR-06-326

Requirements: Eligible organizations: for-profit and non-profit organizations; public and private institutions (such as universities, colleges, hospitals, and laboratories); units of state and local governments; eligible agencies of the federal government; units of state; local tribal governments; domestic and foreign institutions. Eligible Project Directors/Principal Investigators (PD/PIs): any individual with the skills, knowledge, and resources necessary to carry out the proposed research. Individuals from underrepresented racial and ethnic groups as well as individuals with disabilities are always encouraged to apply for NIH support.

Date(s) Application is Due: May 8

Amount of Grant: 200,000 - 275,000 U.S.D

Contact: Natalie Kurinij, (301) 451-2020; fax (301) 402-0528; kurinijn@mail.nih.gov

Internet: http://grants.nih.gov/grants/guide/pa-files/PAR-06-326.html

Sponsor: National Eye Institute
2020 Vision Place
Bethesda, MD 20892-3655

NEI Ruth L. Kirschstein National Research Service Award Short-Term Institutional Research Training Grants 3111

The National Institutes of Health (NIH) will award Ruth L. Kirschstein National Research Service Award (NRSA) Short-Term Institutional Research Training Grants (T35) to eligible institutions to develop or enhance research training opportunities for individuals interested in careers in biomedical, behavioral and clinical research. Many of the NIH Institutes and Centers (ICs) use this grant mechanism exclusively to support intensive, short-term research training experiences for students in health professional schools during the summer. In addition, the Short-Term Institutional Research Training Grant may be used to support other types of predoctoral and postdoctoral training in focused, often emerging scientific areas relevant to the mission of the funding IC. The proposed training must be in either basic, behavioral or clinical research aspects of the health-related sciences. This program is intended to encourage graduate and/or health professional students to pursue research careers by exposure to and short-term involvement in the health- related sciences. The training should be of sufficient depth to enable the trainees, upon completion of the program, to have a thorough exposure to the principles underlying the conduct of research. PA # PA-08-227

Requirements: The following organizations/institutions are eligible to apply: public/state controlled institutions of higher education; private institutions of higher education; hispanic-serving institutions; historically black colleges and universities (HBCUs); tribally controlled colleges and universities (TCCUs); Alaska Native and Native Hawaiian serving institutions; nonprofits with 501(c)3 IRS Status (Other than Institutions of Higher Education); nonprofits without 501(c)3 IRS Status (Other than institutions of higher education); indian/native american tribal governments (federally recognized); indian/native american tribally designated organizations; U.S. Territory or possession; indian/native american tribal governments (other than federally recognized); faith-based or community-based organizations. Only domestic, non-profit, private or public institutions may apply for grants to support National Research Service Award (NRSA) short-term research training programs. Foreign institutions are not eligible to apply. The applicant institution must have a strong and high quality research program in the area(s) proposed for the research training and must have the requisite staff and facilities on site to conduct the proposed research training program. Any individual with the skills, knowledge, and resources necessary to organize and implement a high-quality research training program is invited to work with their institution to develop an application for support. Individuals from underrepresented racial and ethnic groups, individuals with disabilities, and individuals from disadvantaged backgrounds are always encouraged to apply for NIH support.

Restrictions: Foreign institutions are not eligible to apply.

Date(s) Application is Due: Sep 8

Contact: Neeraj Agarwal, (301) 451-2020; fax (301) 402-0528; agarwalnee@mail.nih.gov
Internet: http://grants.nih.gov/grants/guide/pa-files/PA-08-227.html
Sponsor: National Eye Institute
2020 Vision Place
Bethesda, MD 20892-3655

NEI Scholars Program 3112

This program provides an opportunity for outstanding individuals to obtain laboratory or clinical research training within NEI intramural environment and to facilitate the successful transition to continue their research career at an extramural institution as independent vision researchers. This is accomplished by first providing individuals with the necessary resources to receive high quality research training for three to four years at NEI. This is followed by providing extramural funding to support the research program for two years at the extramural institution to which NEI scholar is recruited. It is anticipated that NEI scholars will subsequently compete for independent funding to continue their research. During the intramural phase, the scholar is expected to spend full time on research. During the extramural phase, the scholar must spend a minimum of 75 percent of a full-time professional effort conducting research and engaging in research career development activities for the two years of the award. Applicants are encouraged to contact NEI extramural staff regarding their eligibility for this award and NEI intramural laboratory scientist regarding sponsorship information, prior to submitting an application. PA: PAR-98-107
Requirements: Individuals must have a research or health professional doctoral level degree or its equivalent and must have demonstrated the potential for a highly productive research career during predoctoral training and the immediate postdoctoral period, if applicable. Individuals at NEI or other NIH intramural laboratories who meet the eligibility requirements are eligible to apply. Before submitting an application for the program, the candidate must identify an individual in the NEI Division of Intramural Research who will serve as the sponsor and will be committed to supervise the training and research project. The sponsor must be an active investigator in the area of the proposed research and have research training experience and resources needed to support the NEI scholar.
Restrictions: Individuals who have had more than five years of postdoctoral research training at the time of application are not eligible to apply; however, clinical training does not count against the five years. Former principal investigators on NIH research project grants R01; SBIR/STTR awards R43, R44/R41, R42; subprojects of program project grants P01; center grants P30, P50; K08, K23, K24, or the equivalent are not eligible.
Date(s) Application is Due: Feb 1; Jun 1; Oct 1
Amount of Grant: Up to 175,000 U.S.D
Contact: Dr. Maria Giovanni, Division of Extramural Research; (301) 496-0484; fax (301) 402-0528; myg@nei.nih.gov
Internet: http://www.grants.nih.gov/grants/guide/pa-files/PAR-98-107.html
Sponsor: National Eye Institute
2020 Vision Place
Bethesda, MD 20892-3655

NEI Translational Research Program On Therapy For Visual Disorders 3113

This program continues support of collaborative, multidisciplinary research programs focused on new therapeutic approaches to restore or prevent the loss of function due to visual system diseases and disorders. The rapid and efficient translation of laboratory research findings into clinical development requires a comprehensive and highly integrated approach involving collaborative teams of scientists and clinicians with expertise in multiple disciplines. Such a collaborative approach would be particularly appropriate for research focused on pathways that will likely be targeted by biological intervention, such as gene therapy, cell-based therapy, pharmacological approaches, or development of appropriate delivery systems. The intention of this program is to make resources available to scientists from several disciplines to form research teams to address scientific and technical questions that would be beyond the capabilities of any one research group. It is anticipated that applications funded under this program will lead to therapies that can be tested in a clinical trial.
Requirements: Eligible Institutions, you may submit an application (s) if your organization has any of the following characteristics: for-profit organizations; non-profit organizations; public or private institutions, such as universities, colleges, hospitals, and laboratories; units of State government; units of local government; eligible agencies of the Federal government; domestic Institutions Foreign components to domestic applications will be allowed if the foreign component provides significant scientific input that is critical to the program and for which there is no similar resource in the United States. Eligible Individuals: any individual with the skills, knowledge, and resources necessary to carry out the proposed research is invited to work with their institution to develop an application for support. Individuals from underrepresented racial and ethnic groups as well as individuals with disabilities are always encouraged to apply for NIH programs.
Amount of Grant: Up to 2,000,000 U.S.D
Contact: Dr. Peter Dudley, (301) 451-2020; fax (301) 402-0528; pad@nei.nih.gov
Internet: http://grants.nih.gov/grants/guide/pa-files/PAR-05-110.html
Sponsor: National Eye Institute
2020 Vision Place
Bethesda, MD 20892-3655

NEI Vision Research Core Grants 3114

The primary objective of Core Grants for Vision Research is to provide groups of investigators who have achieved independent National Eye Institute (NEI) funding with additional, shared support to enhance their own and their institution's capability for conducting vision research. Secondary objectives of this program include facilitating collaborative studies and attracting other scientists to research on the visual system.
Requirements: Institutions applying for a Core Grant must hold, on the receipt date, a minimum of eight NEI research project grants (R01), Bioengineering Research Partnership Grants (R24), Mentored Clinical Scientist Development Awards (K08), Mentored Patient-Oriented Research Career Development Award (K23), or Midcareer Investigator Award in Patient-Oriented Research (K24), including any noncompeting extensions of these awards made with or without additional funds. No other mechanisms or source of research support will be considered in determining eligibility. Only one Core Grant will be made to any single applicant organization. For multicampus institutions, no more than one Core Grant will be made to each campus. Joint applications may be submitted by investigators at neighboring, independent institutions.
Date(s) Application is Due: Sep 25
Amount of Grant: 2,000,000 - 2,500,000 U.S.D
Contact: Hemin R. Chin, (301) 451-2020; fax (301) 402-0528; hemin@nei.nih.gov
Internet: http://www.nei.nih.gov/funding/p30.asp
Sponsor: National Eye Institute
2020 Vision Place
Bethesda, MD 20892-3655

Nelda C. and H.J. Lutcher Stark Foundation Grants 3115

The foundation awards grants in its areas of interest, including education, health and social services, medical and dental, community enrichment, arts and culture. Giving is primarily in Texas, with preference to programs that directly impact Southeast Texas and limited giving in Southwest Louisiana. There are no application forms. Submit a letter of proposal containing the following information: description of the project signed by the president or CEO; brief history, description, programs, and mission of the organization; project time line; specific amount requested; list of other funding sources and amounts; list of board of directors and occupations; form 990 for most recent year-end; most recent audited financial statements or organization prepared statements; and current copy of IRS determination letter certifying tax-exempt status. Guidelines are available online. All grants from the third quarter of 2008 and for 2009 have been suspended due to recovery efforts in which the Foundation must engage for its own operations as a result of damages from Hurricane Ike, together with the negative impact of the downturn in financial markets on the Foundation's corpus, the Stark Foundation will continue the prior suspension of the grant-making aspect of its operations through the entirety of Year 2009. Therefore, the Stark Foundation will not accept any grant applications for Year 2009, nor will be able to consider any grant applications that may be submitted notwithstanding this suspension.
Requirements: To be eligible for a grant, an organization must: be exempt from taxation under Section 501(c)3 of the Internal Revenue Code; not be a private foundation within the meaning of Section 509(a) of the Code. No grants are made for endowment purposes, to individuals, or to supporting organizations as described in Section 509(a)(3) of the Internal Revenue Code.
Geographic Focus: All States, Canada
Date(s) Application is Due: Mar 1; Jun 1; Oct 1
Samples: Lamar State College, Orange, Texas $128,269 - Funding to establish a regional EMT training program in cooperation with Acadian Ambulance Service; Texas Energy Museum, Beaumont, Texas $2,500 - Matching funds for an interactive science exhibit; Orange Community Band, Orange, Texas $12,340 - equipment and program support; Orange County Historical Commission, Orange, Texas $2,800 - matching funds to capture in digital format and place on-line the large collection of historical photographs belonging to the Historical Commission; The Conservation Fund, Austin, Texas $375,000 - funding for purchase of strategic conservation lands in Orange County, Texas for donation to the Big Thicket National Preserve for permanent protection;
Contact: Grant Department
Internet: http://www.starkfoundation.org/Grants/Grant-Guidelines.aspx
Sponsor: Nelda C. and H.J. Lutcher Stark Foundation
P.O. Box 909
Orange, TX 77631-0909

Nell J. Redfield Foundation Grants 3116

The foundation awards grants to eligible Nevada organizations in its areas of interest, including biomedical research, health care and health organizations, elderly, children with disabilities, education, human services, and religion. Types of support include building construction/renovation, capital campaigns, equipment acquisition, program development, and scholarship funds. Applications are accepted from January 1 to the listed application deadline. Contact the office for application form and guidelines.
Requirements: Nevada nonprofit organizations are eligible.
Geographic Focus: Nevada
Date(s) Application is Due: Jun 1
Amount of Grant: 1,000 - 500,000 U.S.D
Samples: Truckee Meadows Community College (Reno, NV)—to renovate the performing-arts center, $250,000.
Contact: Grants Administrator; (775) 323-1373
Sponsor: Nell J. Redfield Foundation
P.O. Box 61
Reno, NV 89502

Nell Warren Elkin and William Simpson Elkin Foundation 3117

The foundation awards grants to charitable organizations in its areas of interest, including universities, cancer clinics, specific disease associations, religious groups, and Bible studies. Application form and guidelines are available online.
Requirements: Nonprofit, charitable organizations are eligible.
Date(s) Application is Due: Feb 28; Jul 31
Amount of Grant: 2,000 - 10,000 U.S.D

Contact: Dale Welch, c/o SunTrust Bank, Atlanta; (404) 724-3773; rs.foundationinfo@suntrust.com
Internet: http://www.elkinfoundation.org
Sponsor: Nell Warren Elkin and William Simpson Elkin Foundation
P.O. Box 4655, MC 221
Atlanta, GA 30302

Nesbitt Medical Student Foundation Scholarship **3118**
Needy medical students residing in Illinois are eligible to apply for this scholarship, given to encourage them to enter general practice in DeKalb County or in any county in Illinois having a population of less than 50,000 residents. The scholarship covers one academic year and is renewable; a recipient must reapply each year.
Requirements: Applicants must be U.S. citizens, residents of Illinois, and accepted for admission or already full-time students at approved medical schools. Preference is given to women and former residents of DeKalb County.
Geographic Focus: Illinois
Date(s) Application is Due: May 1
Amount of Grant: 500 - 2,000 U.S.D
Contact: James Kirby, Vice President, National Bank & Trust Company; (815) 895-2125, ext. 228; fax (815) 895-2575
Internet: http://www.nabatco.com
Sponsor: Nesbitt Medical Student Foundation
230 W State Street
Sycamore, IL 60178

Nestle Foundation Research Grants **3119**
The Nestlé Foundation supports research in human nutrition with public health relevance in low-income and lower middle-income countries. The results of the research projects should ideally provide a basis for implementation and action which will lead to sustainable effects in the studied populations as generally applicable to the population at large. They should also enable institutional strengthening and capacity building in a sustainable manner in the host country and further cooperation and collaboration between institutions in developed and developing countries. Categories include: training grants, pilot grants, research grants (small and large), and re-entry grants. Training grants support a small research project such as a master's degree in science, Ph.D. thesis project, or a training endeavor. Pilot grants provide support for research with a high potential to lead to a subsequent full research grant. Small research grants provide support of a small research study which may represent a continuation of a training grant or a pilot grant. Large research grants include a full grant application of a complete research proposal. Re-entry grants encourage the return and re-establishment of post-graduate students into careers in their own countries. Applications are accepted all year and evaluated twice a year.
Requirements: The Foundation's work is primarily concerned with human nutrition research issues dealing with: maternal and child nutrition, including breastfeeding and complementary feeding; macro- and micronutrient deficiencies and imbalances; interactions between infection and nutrition; and nutrition education and health promotion. Interested scientists should first submit a letter of intent briefly describing the project and the estimated budget. Instructions are available on the Foundation's website. If approved applicants will receive an invitation to submit a full grant proposal. Guidelines and forms are available on our website. The letter of intent and the grant application should include detailed, evidence-based information about the public health relevance of the project as well as its immediate impact and sustainability. Funding is primarily based on the scientific quality, public health relevance in the short and long term, sustainability, capacity-building component, and budget considerations.
Restrictions: The Foundation expects research proposals to be primarily the initiative of local researchers from developing countries but considers applications jointly made by scientists from developed countries with those from developing countries provided it is clear that the initiative will result in capacity building and human resource development in the latter and the bulk of the budget is spent in the developing country.
Geographic Focus: All Countries
Date(s) Application is Due: Jan 10; May 10
Contact: Program Contact; 41-21-3203351; fax 41-21-3203392; nf@nestlefoundation.orgh
Internet: http://www.nestlefoundation.org/e/research.html
Sponsor: Nestle Foundation
4 Place de la Gare
Lausanne, CH-1001 Switzerland

Network Associates Community Giving Grants **3120**
The corporation supports the quality of life in its operating communities in the United States and abroad. Areas of support include education, health, youth, and technology. The company seeks opportunities to develop and disseminate quality programs; adapt and pilot for local-use programs proven elsewhere; encourage the development of new ideas and constructive alternatives to traditional solutions; foster collaborative efforts among public and private institutions; and contribute to the development of sound public policy, either directly or using a highly visible model. The corporation awards short-term funding, as well as multiyear funding to assure continuity as a program progresses from planning through implementation and, in some instances, dissemination. Network Associates also encourages programs to demonstrate community support, document their program's effectiveness, aid in future replication efforts, and improve the function of projects in order to receive on-going grants.
Requirements: 501(c)3 nonprofit organizations are eligible. Organizations located outside of the United States should contact the managing director of a Network Associates office in their respective regions.
Restrictions: Network Associates does not consider proposals for grants to individuals, religious organizations for sectarian purposes, political organizations, or political causes. It also does not provide funds for capital projects, organizational deficits, or general operating support.
Contact: Network Associates Community Relations; (888) 847-8766; community@nai.com
Internet: http://www.networkassociates.com/us/about/corporate/corp_giving.htm
Sponsor: Network Associates
3965 Freedom Circle
Santa Clara, CA 95054

Newcombe Scholarships for Physically Disabled Students **3121**
Grants support the physically disabled as they seek higher education. The Newcombe Foundation offers special disability expense scholarships, internship scholarships, and partial tuition scholarships through the counseling offices of colleges that provide excellent support services for disabled students. Preference is given to four-year colleges that render exceptional service to large numbers of physically disabled students. There is also a nationwide program for Presbyterian colleges with scholarships for minority students. Applications are available from June to October.
Requirements: Giving on a national basis, except for scholarship programs for mature women and students with disabilities which are limited to colleges and universities located within Pennsylvania, New Jersey, New York City, Delaware, Maryland, and Washington, and that provide specialized counseling and support services for physically disabled students, are eligible to apply. Application materials are available from June through October. Decisions are announced in May for funding in July.
Restrictions: Grants are not made directly to individuals. Funds are not awarded to community colleges, nor to individuals.
Date(s) Application is Due: Apr 1; Nov 1; Dec 1
Amount of Grant: 15,000 - 50,000 U.S.D
Samples: Gallaudet U (Washington, DC)—for scholarships for disabled students, $50,000; New York U (New York, NY)—for scholarships for disabled students, $50,000; Western Maryland College (Westminster, MD)—for scholarships for disabled students, $25,000.
Contact: Janet Fearon; (609) 452-7007 or (609) 924-7022; fax (609) 452-7828
Internet: http://www.woodrow.org/newcombe/newcombe_about_newcombe_founda.html
Sponsor: Charlotte W. Newcombe Foundation
35 Park Place
Princeton, NJ 08542-6918

New England Financial Community Involvement Grants **3122**
The corporation awards grants in the New England states, with preference given to requests from Boston, MA, in its areas of interest, including civic affairs—low-income areas, affordable housing, families,and youth; education—workplace readiness, higher education, and literacy; healthcare—healthy life styles and access to care for underserved individuals; and culture—arts outreach, education, and diverse audiences. There are no application deadlines.
Requirements: Northeast U.S. nonprofit organizations are eligible.
Geographic Focus: Connecticut, Maine, Massachusetts, New Hampshire, Rhode Island, Vermont
Amount of Grant: Up to 14,000,000 U.S.D
Contact: Charitable Contributions; (617) 578-2119; bspence@nefn.com
Internet: http://nef.metlife.com/Applications/Corporate/NEF/CDA/PageGenerator/0,1674,P255,00.html?FILTERNAME=%40URL&FILTERVALUE=%2FNEF%2F&IMAGE2.X=14&IMAGE2.Y=6
Sponsor: New England Financial
501 Boylston Street
Boston, MA 02116-3700

New Jersey Osteopathic Education Foundation Scholarships **3123**
Approximately five scholarships are awarded each year to deserving students entering their first year in an osteopathic college. They must be residents of New Jersey and have completed four years of pre-medical education. Selection of awardees is based on undergraduate academic achievement, financial need, motivation, and professional promise as an osteopathic physician.
Requirements: Student must be a resident of the state of New Jersey, must be osteopathically oriented, and must be a student member of the American Osteopathic Association and the New Jersey Association of Osteopathic Physicians and Surgeons throughout his/her medical education. Applicants should have a B average or a 3.0 grade point average or better on a 4.0 scale and be in the upper 25 percent in class standing. The required forms can be downloaded from the sponsor's website.
Geographic Focus: New Jersey
Date(s) Application is Due: Apr 30
Amount of Grant: 5,000 U.S.D
Contact: Robert Bowen, NJAOPS Executive Director; (732) 940-9000; fax (732) 940-8899; njaops@njosteo.com
Internet: http://www.njosteo.com/displaycommon.cfm?an=1&subarticlenbr=68
Sponsor: New Jersey Osteopathic Education Foundation
One Distribution Way, Suite 201
Monmouth Junction, NJ 08852-3001

Newman's Own Inc Grants **3124**
The corporate giving program awards general support grants to a broad array of charitable and educational causes, including children and youth, health, education, elderly, environment, the arts, handicapped, literacy, substance abuse education, programs for the needy (i.e., housing and food). More than 600 grants are made annually. Funding

levels vary each year based on after-tax profits. Actor and company president, Paul L. Newman, reviews each proposal and selects the groups and organizations to receive contributions. He is interested in organizations outside of the charity mainstream that help those who often cannot help themselves. Social service agencies in Australia and Canada also receive support. Early submission of proposals is encouraged. Faxed or emailed proposals are not accepted.
Requirements: IRS 501(c)3 nonprofits are eligible for grant support.
Restrictions: Paul Newman and the Newman's Own Foundation do not support individual donations and scholarships.
Amount of Grant: 2,000 - 15,000 U.S.D
Contact: Paul L. Newman, c/o Newman's Own
Internet: http://www.newmansownfoundation.org/giving/
Sponsor: Newman's Own
246 Post Road E
Westport, CT 06880-3615

Newton County Community Foundation Grants **3125**
The purpose of the Foundations is to assist donors in creating a source of assets to meet the ongoing needs and interests of the people living in Jasper and Newton County, Indiana communities. The Foundation welcomes grant applications from non-profit organizations whose programs benefit the residents of each county. Grant applications are available in the Foundation office or online at the Foundation's website.
Requirements: Favored grant requests for each county will generally reflect the following characteristics: projects which propose practical solutions to community needs; projects which promote cooperation among existing agencies without duplicating services; and projects that will become self-sustaining without requiring on-going Foundation funds. No restrictions are placed on gifts to these funds, allowing the Board of Directors to address community needs.
Restrictions: Grant are not normally made for budget deficits or sectarian religious activities.
Geographic Focus: Indiana
Date(s) Application is Due: Apr 1; Oct 1
Samples: Morocco Lions Club, Morocco, Indiana, $1,500 - funding for improvement of the Morocco Lions Club den and purchase of kitchen equipment (2010); Newton/Jasper Community Band, $3,000 - funds to help purchase a tuba that will enable the tuba section to function on a higher level (2010)
Contact: Linda Harris; (219) 866-5899; fax (219) 866-0555; jasper@liljasper.com
Internet: http://www.jasperfdn.org/newtgrants.htm
Sponsor: Newton County Community Foundation
301 N. Van Rensselaer Street, P.O. Box 295
Rensselaer, IN 47978

Newton County Community Foundation Lilly Scholarships **3126**
The Lilly Endowment Community Scholarships will provide full tuition, required fees, and a special allocation of up to $800 per year for required books and equipment for four years of undergraduate study leading to a baccalaureate degree at any Indiana public or private college or university. Any college major may apply.
Requirements: Any graduating senior who is a U.S. citizen and resident of Jasper or Newton County is eligible. The student must be an applicant for enrollment to an accredited Indiana university or college and pursuing a four year full time field of study. The student must demonstrate good citizenship qualities, high moral character, and be a positive role model to others. Also, he/she must have strong leadership skills and demonstrate their ability to be a good ambassador for the community. Along with the application, candidates must submit three letters of reference, and a privacy release statement. Candidates must then submit the application to their high school guidance counselor for scholastic information and their transcript, then forward the complete information to the Jasper Foundation for review. Candidates can contact the Jasper Foundation office for further information.
Geographic Focus: Indiana
Date(s) Application is Due: Jan 11
Contact: Rhonda Churchill; (219) 866-5899; fax (219) 866-0555; jasper@liljasper.com
Internet: http://www.jasperfdn.org/
Sponsor: Newton County Community Foundation
301 N. Van Rensselaer Street, P.O. Box 295
Rensselaer, IN 47978

Newton County Community Foundation Scholarships **3127**
The purpose of the Jasper Foundation is to assist donors in creating a source of assets to meet the ongoing and changing charitable needs and interests of the people living in Jasper and Newton County, Indiana communities. The Jasper Foundation, Inc. and the Newton County Community Foundation support education as a lifelong pursuit. Through the generosity of their contributors, the Foundations have been able to provide financial assistance to many young people in Jasper and Newton counties seeking a post-secondary education. Scholarships are awarded by a separate advisory committee, either designated by the donor or Foundation committee. Selections are based upon established criteria and guidelines which vary with each scholarship. See website for further details and applications for each program offered.
Geographic Focus: Indiana
Date(s) Application is Due: Jan 8
Contact: Rhonda Churchill, Chairperson; (219) 866-5899; fax (219) 866-0555
Internet: http://www.jasperfdn.org/
Sponsor: Newton County Community Foundation
301 N. Van Rensselaer Street, P.O. Box 295
Rensselaer, IN 47978

New York Life Foundation Grants **3128**
The foundation awards grants to nonprofit organizations in its areas of interest. Periodically, the foundation selects an area of special priority. The current focus, Nurturing the Children, directs the majority of the foundation's resources toward organizations, programs, and services aimed at helping young people. This initiative specifically focuses on safe places to learn and grow; educational enhancement; and mentoring. The foundation has initiated a new grant award program for nonprofit organizations based in New York City. Youth In Action recognizes innovative Nurturing the Children programs that engage youth in planning, coordinating and/or helping to provide services.
Requirements: Nonprofit 501(c)3 organizations may submit grant requests. The foundation funds projects in New York City, where New York Life's Home Office is located. The foundation also considers multi-site projects implemented by national organizations. These projects must serve two or more of the following locations: Atlanta; Cleveland; Clinton/Hunterdon County, New Jersey; Dallas; Kansas City; Minneapolis; New York City; Parsippany/Morris County, New Jersey; Tampa; and Westchester County, New York.
Restrictions: The foundation does not support sectarian or religious organizations or activities; fraternal, social, professional, athletic, or veterans groups; seminars or conferences; preschool, primary, or secondary education; endowments; memorials; basic or applied research; capital campaigns; or fund-raising activities.
Date(s) Application is Due: Dec 1
Samples: Coalition for the Homeless (New York, NY)—for its after-school programs at five New York City shelters and its summer camp programs for children living in homeless shelters, $140,000; Harlem Education Activities Fund (New York, NY)—to expand the Summer Quest Program, a six-week summer academic program for middle-school students, $290,750 over three years; New York Historical Society (New York, NY)—for a new educational program that will serve at-risk elementary- and middle-school students in New York, $160,000; Foundation for the National Archives (Washington, DC)—to produce a film and companion book for children as part of the National Archives Experience, an interactive learning program for students and families, and to expand exhibition spaces at the National Archives building, $1.2 million over three years.
Contact: Program Contact; (212) 576-7341; fax (212) 576-6220; NYLFoundation@newyorklife.com
Internet: http://www.newyorklife.com/foundation/index.html
Sponsor: New York Life Foundation
51 Madison Avenue, Room 1600
New York, NY 10010-1655

New York University Steinhardt School of Education Fellowships **3129**
The program awards fellowships for doctoral study in the field of education. Named fellowships include Founders Fellowships, Phyllis and Gerald LeBoff Fellowship in Media Ecology, and the Steinhardt Fellowship in Education and Jewish Studies. Fellowships provide up to three years of full-tuition support and annual stipends. Faculty in relevant disciplines are also invited to nominate qualified students for these awards for doctoral study in education, applied psychology, health, nursing, art, music, and media studies.
Requirements: Successful candidates are expected to have a history of collaborative and scholarly work with faculty, strong academic records, and an ability to draw on the resources of their program's faculty at the university.
Date(s) Application is Due: Jan 15
Amount of Grant: 10,000 - 18,000 U.S.D
Contact: Grants Administrator; (800) 771-4698, ext. K28 or (212) 998-5030; ed.gradadmissions@nyu.edu
Internet: http://www.education.nyu.edu
Sponsor: New York University
82 Washington Square E, 2nd Floor
New York, NY 10003-6680

New York Women's Foundation Grants **3130**
The foundation's goal is to support women helping low-income girls and women in New York City to achieve sustained economic self-sufficiency and self-reliance. The foundation's specific program areas of interest include community organizing and advocacy, economic security, girls' positive development, violence against women and girls, and women's health and reproductive rights. For the current year, only organizations whose constituencies consist of or who serve a majority of women and/or girls are eligible to apply for general operating support. As part of its grantmaking, the foundation is committed to supporting the efforts of organizations to develop their administrative and programmatic infrastructure. Capacity-building initiatives may be included as part of an organization's proposal for general or program support. The foundation also encourages the collaboration of organizations to increase the efficiency and effectiveness of service delivery and advocacy efforts and welcomes proposals for collaborative activities within its five program areas. Guidelines are available online.
Requirements: Organization/program criteria include: are within the five boroughs of New York City; work in the following issue areas: community organizing and advocacy, economic security, girls' positive development, violence against women and girls, and women's health and reproductive rights; are community-based, led by women (board and staff), and include recognized leaders of their community; are designed specifically for low-income women and girls, particularly women of color, immigrants, lesbians, the elderly, or other underserved populations; strive toward the philosophy of women helping women, an active, consensus-driven form of management; encourage the involvement of program participants in the decision-making; have limited access to ongoing traditional funding sources, i.e., government, foundations, and individual donors; and have been in existence for at least six months.
Restrictions: The foundation does not provide funds to individuals; start a new project (seed money); capital fund projects; endowments; fundraising events; feasibility studies; real estate purchases; public or private schools; colleges or universities; campaigns to elect

candidates to public office; programs to promote religious activities; statewide, national, or international organizations; programs that are primarily artistic or cultural; programs located outside the five boroughs of New York City; or programs that do not comply with federal, state, or local equal employment statutes.
Geographic Focus: New York
Date(s) Application is Due: Oct 14
Amount of Grant: 10,000 - 40,000 U.S.D
Contact: Grants Administrator; (212) 414-4342; fax (212) 414-5708; info@nwf.org
Internet: http://www.nywf.org/grant.html
Sponsor: New York Women's Foundation
34 W 22nd Street
New York, NY 10010

NFID Advanced Vaccinology Course Travel Grant 3131
These grants are intended to assist recent postdoctoral graduates who have a demonstrated interest in a career in vaccinology. This program provides participants with a comprehensive overview of vaccinology, from immunology to the social, economic, political, and ethical issues of vaccination. Practical approaches to the implementation of effective vaccination programs suitable for particular population groups will be discussed. To be eligible for this award the applicant must have made separate application to the course. Further information about the course can be obtained from: Betty Dodet, Fondation Merieux, 17, rue Bourgelat, BP 2021, Lyon Cedex 02, France. Annual deadline dates vary; contact program staff for exact dates. Note: Please do not contact NFID for information about the course. Contact the listed contact person.
Requirements: Individuals who are U.S. citizens who reside in the United States at the time of the course and who are within three years of obtaining a doctoral degree, or in the case of physicians, within two years of completing specialty training or a subspecialty fellowship, are eligible to apply. Candidates must have a demonstrated commitment to vaccinology as a career, and must be conducting research or working in a recognized and accredited U.S. institution of higher learning or in a governmental agency.
Date(s) Application is Due: Nov 14
Amount of Grant: 4,000 U.S.D
Contact: Betty Dodet; (33-4) 72 40 79 50; betty.dodet@fond-merieux.org
Internet: http://www.nfid.org/fellow/grantsfellow.pdf
Sponsor: National Foundation for Infectious Diseases
4733 Bethesda Avenue, Suite 750
Bethesda, MD 20814-5278

NFID Colin L. Powell Minority Postdoctoral Fellowship in Tropical Disease Research 3132
The program, sponsored by NFID and GlaxoSmithKline, invites applications from qualified minority researchers to become specialists and investigators in the field of tropical diseases. The one-year fellowship includes a stipend, $3000 of which may be used for travel and supplies. Fellowships commence on or about July 1, with notification on or about April 15. Annual deadline dates may vary; contact program staff for exact dates.
Requirements: The applicant must hold a doctorate from a recognized U.S. university and be a U.S. citizen or permanent resident or spouse of a U.S. citizen or permanent resident.
Date(s) Application is Due: Jan 6
Amount of Grant: 30,000 U.S.D
Contact: Senior Executive Director; (301) 656-0003; fax (301) 907-0878; info@nfid.org
Internet: http://www.nfid.org/fellow/grantsfellow.pdf
Sponsor: National Foundation for Infectious Diseases
4733 Bethesda Avenue, Suite 750
Bethesda, MD 20814-5278

NFID John P. Utz Medical Mycology Postdoctoral Fellowship 3133
The fellowship is intended to encourage and assist a qualified physician to become a specialist and investigator in medical mycology. Selection criteria for the one-year fellowship will include, but not be limited to, the applicant's record of scholarship performance and professional qualifications; scientific merit of the clinical research project proposed; and validity of research rationale and adequacy of facilities available to the applicant, host department, and host institution. The recipient will be required to submit two single-spaced, one-page progress reports during his/her year of clinical research. One report will be due after six months and the second after 12 months or upon completion of the fellowship year. Notification of award will be made on or about April 15.
Requirements: The applicant must be a physician who is a U.S. citizen. He/she must demonstrate aptitude and accomplishment in research and must confirm arrangements for conduct of the proposed research in a recognized host laboratory. A letter from the chairperson of the department of infectious diseases expressing a willingness to assume responsibility for training the applicant must accompany the application.
Restrictions: The fellowship may not be awarded if the applicant has received or will receive a major fellowship, research grant, or traineeship from the federal government or another foundation in excess of the amount of this award.
Date(s) Application is Due: Jan 6
Amount of Grant: 41,000 U.S.D
Contact: Senior Executive Director; (301) 656-0003; fax (301) 907-0878; info@nfid.org
Internet: http://nfid.org/fellow/grantsfellow.pdf
Sponsor: National Foundation for Infectious Diseases
4733 Bethesda Avenue, Suite 750
Bethesda, MD 20814-5278

NFID New Investigator Matching Grants 3134
Matching grants are intended to assist new investigators who are beginning their research. The proposal must be based on research relevant to infectious diseases, microbiology, clinical medicine, epidemiology, nursing, etc. Priority will be given to applications proposing research on HIV and AIDS; tuberculosis, influenza, and other respiratory diseases; hospital-acquired and opportunistic infections; viral hepatitis; gastrointestinal, diarrheal, and food-borne diseases; sexually transmitted diseases; antibiotic resistance; tropical, travelers', and parasitic diseases; diseases transferred from animals to humans; and new and reemerging infectious diseases. Each application must be accompanied by an agreement from the applicant's sponsoring institution to match the award with equal funds.
Requirements: A new investigator is defined as an individual with full-time junior faculty (instructor or assistant professor) status at a recognized and accredited institution of higher learning. Priority will be given to those who do not have research grants and whose studies represent pilot work with intent to further develop their research.
Date(s) Application is Due: Feb 14
Amount of Grant: 2,000 U.S.D
Samples: Dr. Nancy Connell, UMDNJ/New Jersey Medical School (NJ)—to support experiments to study the cell wall and membranes of mycobacteria, $2000; Dr. Christine Wanke, Harvard U Medical School (MA)—to support research on E. coli in adult patients infected with HIV with chronic diarrhea, $2000.
Contact: Senior Executive Director; (301) 656-0003; fax (301) 907-0878; info@nfid.org
Internet: http://www.nfid.org/fellow/grantsfellow.pdf
Sponsor: National Foundation for Infectious Diseases
4733 Bethesda Avenue, Suite 750
Bethesda, MD 20814-5278

NFID Nosocomial Infection Research and Training Postdoctoral Fellowship 3135
The fellowship is intended to encourage and assist a qualified physician researcher to become a specialist and investigator in the field of nosocomial infections. Priority will be given to the study of gram positive nosocomial infections. Each application must be accompanied by a letter from the host laboratory and signed by the laboratory director or appropriate department chairperson indicating a willingness to accept the applicant and to provide guidance, research space, and necessary research equipment. Overhead to the sponsoring institution will not be paid. One thousand dollars of the stipend may be used for travel and supplies.
Requirements: The applicant must be a physician and a citizen of the United States. He/she must demonstrate aptitude and accomplishment in research and must confirm arrangements for conduct of the proposed research in a recognized host laboratory. An applicant with no grant support will receive priority consideration.
Restrictions: Applicants receiving or to be awarded grants in the same academic year in excess of the amount of this award are ineligible.
Date(s) Application is Due: Jan 6
Amount of Grant: Up to 40,000 U.S.D
Contact: Senior Executive Director; (301) 656-0003; fax (301) 907-0878; info@nfid.org
Internet: http://nfid.org/fellow/grantsfellow.pdf
Sponsor: National Foundation for Infectious Diseases
4733 Bethesda Avenue, Suite 750
Bethesda, MD 20814-5278

NFID Postdoctoral Fellowships in Emerging Infectious Diseases 3136
This one-year fellowship is intended to encourage and assist a qualified physician to become a recognized authority on emerging infectious diseases and epidemiology. The fellow will be assigned to the National Center for Infectious Diseases, CDC, in Atlanta, GA, and will participate in analysis and reporting of the national burden of infectious diseases and potential threats of emerging infectious diseases.
Requirements: The applicant must be a physician who is a U.S. citizen who preferably will have completed an infectious diseases fellowship.
Restrictions: The fellowship may not be awarded if the applicant has received or will receive a major fellowship, research grant, or traineeship from the federal government or another foundation in excess of $20,000.
Date(s) Application is Due: Jan 6
Amount of Grant: 50,000 U.S.D
Contact: Senior Executive Director; (301) 656-0003; fax (301) 907-0878; info@nfid.org
Internet: http://nfid.org/fellow/grantsfellow.pdf
Sponsor: National Foundation for Infectious Diseases
4733 Bethesda Avenue, Suite 750
Bethesda, MD 20814-5278

NFL Charities Medical Grants 3137
NFL Charities supports medical research and enhancing scientific knowledge in order to benefit all who are actively involved in competitive sports and recreational athletic activities. These grants support research that addresses some of the risk factors that exist for all athletes and citizens with active lifestyles. The NFL Charities Board considers Medical Grant proposals that place emphasis on scientific merit, clinical relevance and significance to the NFL, especially projects concerning concussion and traumatic brain injury, cardiovascular research and MRSA. Grants will be available in the end of June or early July and will be open for approximately 8 weeks.
Requirements: Funding is available for non-profit educational and research institutions within the United States. Applications must be submitted online via the NFL Charities Grant Application Management System (G.A.M.S.) website. Every field in the application must be populated with requested information. Indirect costs should not be included in the project's total budget, and only direct research costs are funded by the grant. Salaries for Principal Investigators (PI's) and Co-Principal Investigators (Co-

PI's) (Ph.D's and M.D.'s) must follow NIH guidelines. PI's and Co-PI's must disclose all previous and/or current grant support and any potential overlap must be submitted. Postdoctoral fellows serving as PI or Co-PI must include assurances that they will remain at the same institution for the duration of the grant period; however, PI's and co-PI's may work at different institutions, but the grant must be submitted as associated with the institution of the primary PI.
Restrictions: For-profit groups are not eligible for funding. Only submission per PI or Co-PI will be funded, and only one submission per institution can be considered for NFL Charities Medical Grant funding. Prospective applicants first should be vetted through their institutional administrative offices in order to ensure that an institution only endorses and submits its top two proposals. Administrative contacts at the associated institution must sign off on the Compliance Certificate that accompanies each application(s) from a PI's institution, thereby ensuring that such vetting has occurred.
Date(s) Application is Due: Jun 1
Amount of Grant: Up to 100,000 U.S.D
Contact: Clare Graff, (212) 450-2435 or (917) 816-2885; clare.graff@nfl.com
Internet: http://www.nflcharities.org/grants/medical
Sponsor: NFL Charities
280 Park Avenue, 17th Floor
New York, NY 10017

NFX Patient-Oriented Research Grants 3138
The foundation supports a broad range of research endeavors that will lead to better recognition, treatment, and an eventual cure of children and adults with fragile X syndrome. The following forms of investigation are potentially appropriate for the current round of funding: educational or therapy interventions, medical treatment interventions, studies of family support and counseling, and diagnostic and reproductive studies. Basic research and patient-oriented research grants are awarded on a six-month alternating cycle. The January deadline is for basic science and clinical research applications. Applications should be submitted both as email attachments and as hard-copy documents.
Requirements: The program is open to all investigators with demonstrated ability to carry out research on fragile X syndrome.
Date(s) Application is Due: Jan 31
Amount of Grant: Up to 35,000 U.S.D
Samples: Dr. Emily Osterweil—for research, Investigating FMRP in mGluR-mediated Protein Synthesis and Synaptic Plasticity, $35,000; To be distributed between two researchers—for research on fragile X syndrome, a hereditary condition that causes mental impairment, $60,000.
Contact: Grants Administrator; (800) 688-8765 or (925) 938-9300; fax (925) 938-9315; NATLFX@FragileX.org
Internet: http://www.fragilex.org/html/request_for_application.htm
Sponsor: National Fragile X Foundation
P.O. Box 190488
San Francisco, CA 94119

NFX Summer Research Fellowships 3139
The foundation awards summer research fellowships in the area of fragile X syndrome. Proposals should be no longer than four pages, and should describe the research project in depth and the specific roles and activities of the summer fellow and the mentor. The curriculum vitas of the mentor and the fellow should be included as an addendum. Guidelines are available online.
Requirements: The fellow can be either an undergraduate or graduate student, and the mentor should be experienced in research concerning fragile X syndrome.
Date(s) Application is Due: Apr 17
Amount of Grant: 2,500 U.S.D
Contact: Grants Administrator; (800) 688-8765 or (925) 938-9300; fax (925) 938-9315; NATLFX@FragileX.org
Internet: http://www.fragilex.org/html/request_for_application.htm
Sponsor: National Fragile X Foundation
P.O. Box 190488
San Francisco, CA 94119

NHGRI Courses in Genomic Analysis and Interpretation and Ethical, Legal, and Social Implications (ELSI) Research Grants 3140
NHGRI invites applications for support of short, advanced-level courses in genomic analysis and interpretation and in ethical, legal, and social implications (ELSI) research. Of particular interest are courses that emphasize new laboratory techniques in genetic and physical mapping; technology and theory related to studying sequence variation; DNA sequencing technology applicable to large-scale projects, informatics as it relates to the Human Genome Project (HGP); new technologies for interpretation of the genome; interdisciplinary training in principles of genomic analysis for nonbiologists; and principles and methods of studying the ethical, legal, and social issues relevant to HGP for biologists. These courses are meant to enhance the skills of U.S. scientists and scholars who are interested in pursuing laboratory research that utilizes the data and materials generated by the HGP or scholarly research relevant to the goals of HGP. Faculty should consist of established investigators or scholars actively working in the area of instruction. Plans for the inclusion of individuals who are currently underrepresented in the field of genomic research, such as women, minorities, and persons with disabilities should be addressed. Courses may range from a few days to a few weeks and may be offered annually, although other terms will be acceptable. PA: PAR-99-059
Requirements: Applications may be submitted by domestic public and private organizations, such as universities, colleges, and research institutions. Applicants may initially request support for two years. Course offerers are expected to be academic or research institutions experienced in training; however, applications from for-profit institutions will also be considered.
Restrictions: Foreign institutions are not eligible for support under this mechanism.
Contact: Dr. Bettie Graham, Program Contact; (301) 496-7531; fax (301) 480-2770; bettie_graham@nih.gov
Internet: http://www.grants.nih.gov/grants/guide/pa-files/PAR-99-059.html
Sponsor: National Human Genome Research Institute
9000 Rockville Pike
Bethesda, MD 20892

NHGRI Curriculum Development Award in Genomic Research and Analysis 3141
NHGRI, through this initiative, supports the development of courses and curricula designed to train interdisciplinary scientists who combine knowledge of genomics and genetics research with expertise in computer sciences, mathematics, chemistry, physics, engineering, or closely related sciences. It is anticipated that these courses or curricula would be useful to students and scientists who wish to develop new conceptual approaches to genome research and analysis or organize, analyze, or interpret large data sets resulting from genomic and genetics research. Development of courses at the graduate and undergraduate levels is encouraged. As part of this program, awardees will be expected to develop and implement the courses or curricula in their institution. The courses or curricula are expected to be models that could be transferable to other institutions in whole or in part. PAR-98-063 may be obtained through the NIH grant line (301-402-2221), the NIH Web site, mail, and email.
Requirements: The principal investigator must be engaged in genomics/genetics research or research in one of the following areas: computer sciences, mathematics, chemistry, physics, engineering, or a related field, such as bioinformatics, statistics, biomathematics, computational biology, bioengineering with genomics research or genetics research.
Date(s) Application is Due: Feb 1; Jun 1; Oct 1
Contact: Jean Cahill, (301) 496-7531; (301) 402-0733; Jean_Cahill@nih.gov
Internet: http://grants.nih.gov/grants/guide/pa-files/PAR-98-063.html
Sponsor: National Human Genome Research Institute
9000 Rockville Pike
Bethesda, MD 20892

NHGRI Genome Scholar Development and Faculty Transition Award 3142
The purpose of the program is to enable promising new genome researchers to establish an independent research program in genomic research and analysis and to secure a tenure-track appointment in an academic institution in the United States. This is accomplished by providing the necessary resources to initiate an independent research program as a Genome Scholar in an extramural or an NIH intramural laboratory for up to two years followed by funding of their extramural research program for up to four years at an extramural institution to which she/he has been recruited as tenure-track faculty. The total number of years for both phases of the award cannot exceed five. This award has two phases—a Genome Scholar Development Phase and a Faculty Transition Phase. Applicants may apply for both phases or for the Faculty Transition Phase only. It is anticipated that awardees will subsequently obtain a research project grant (R01) to support the continuation of their research. PAR: PA-99-022
Requirements: At the time of application, individuals who apply to this program must have no more than five years postdoctoral experience, if applying for a Genome Scholar Development and Faculty Transition award; have no more than six years postdoctoral experience, if applying for a Faculty Transition award only; be in a postdoctoral or equivalent position; be engaged in fundamental or applied genomics research or technology development research as it relates to genomics; be able to devote at least 75 percent of time to research-related activities; not be a principal investigator on R01, R21, R29 or subproject of P grant, or recipient of a K award; and be a U.S. citizen or permanent resident of the United States.
Date(s) Application is Due: Feb 1; Jun 1; Oct 1
Contact: Jean Cahill, Grants Management; (301) 402-0733; Jean_Cahill@nih.gov
Internet: http://www.grants.nih.gov/grants/guide/pa-files/PAR-99-022.html
Sponsor: National Human Genome Research Institute
9000 Rockville Pike
Bethesda, MD 20892

NHGRI Genomic Mapping, Sequencing, and Analysis Pilot Project Grants 3143
NHGRI invites applications for exploratory/developmental grants (R21) to develop new and/or significantly improve existing technologies that will accelerate the genome mapping, sequencing, and analysis goals for the Human Genome Project in the most expeditious and economical manner. The purpose of this program announcement is to encourage high-risk/potential high-payoff applications that are not yet developed fully enough to successfully compete for a standard R01 grant. PA-97-045 may be obtained electronically through the NIH grant line (301-402-2221), the NIH Web site, and by mail and email.
Requirements: Applications may be submitted by domestic, for-profit and nonprofit organizations, public and private, such as universities, colleges, hospitals, laboratories, units of state and local governments, and eligible agencies of the federal government.
Amount of Grant: Up to 100,000 U.S.D
Contact: Dr. Bettie Graham, Program Contact; (301) 496-7531; fax (301) 480-2770; bettie_graham@nih.gov
Internet: http://grants.nih.gov/grants/guide/pa-files/PA-97-045.html
Sponsor: National Human Genome Research Institute
9000 Rockville Pike
Bethesda, MD 20892

NHGRI Individual Mentored Research Scientist Awards in Genomic Research and Analysis 3144

These awards are intended to foster the career development of individuals with expertise in scientific disciplines that would further technological developments critical to the success of the Human Genome Project and the understanding of the genetic basis of diseases. Applications for this award should propose a development plan that has both intrinsic research importance and will also serve as a suitable vehicle for learning the methodology, theories, and conceptualizations necessary for well-trained independent, interdisciplinary researchers. PA: PAR-98-061

Requirements: Applicants must hold a PhD or equivalent professional degree in a scientific discipline other than biology, such as engineering, mathematics, computer sciences, physics, chemistry, or a closely related scientific discipline (such as bioinformatics, computational biology, statistics, biomathematics, or bioengineering), and show evidence of expertise in their discipline.

Restrictions: Applicants who are or have been former principal investigators on NIH research projects, FIRST Awards, sub-projects of program project or center grants, or the equivalent, are not eligible to apply for a K01 award. Applicants may not concurrently hold any other PHS award that duplicates the provisions of this award.

Date(s) Application is Due: Feb 1; Jun 1; Oct 1

Contact: Jean Cahill, Grants Management; (301) 402-0733; Jean_Cahill@nih.gov

Internet: http://grants.nih.gov/grants/guide/pa-files/PAR-98-061.html

Sponsor: National Human Genome Research Institute
9000 Rockville Pike
Bethesda, MD 20892

NHGRI Mentored Research Scientist Development Award 3145

These awards are intended to foster the career development of individuals with expertise in scientific disciplines that would further technological developments critical to the success of the Human Genome Project and the understanding of the genetic basis of diseases. Up to three scientists may be selected and appointed to this program by the grantee institutions. Up to the Federal salary cap, plus fringe benefits, is available. Up to $20,000 of research support per year is also available.

Requirements: Eligibility is limited to individuals with degrees in computer sciences, mathematics, chemistry, engineering, physics, and closely related scientific disciplines, such as bioinformatics, computational biology, statistics, biomathematics, bioengineering. Applications will be accepted from domestic, nonfederal organizations of higher education that have strong, well-established research and training programs. The following organizations and institutions may apply: public/state controlled institutions of higher education; private institutions of higher education; Hispanic-serving institutions; historically Black colleges and universities (HBCUs); tribally controlled colleges and universities (TCCUs); Alaska native and native Hawaiian serving institutions; nonprofits with 501(c)3 IRS status; nonprofits without 501(c)3 IRS status; small businesses; for-profit organizations; State governments; regional organizations; U.S. territories or possessions; Indian/Native American tribal governments; Indian/Native American tribally designated organizations; eligible agencies of the Federal government; and faith-based or community based organizations.

Date(s) Application is Due: Feb 12; Mar 12; Jun 12; Jul 12; Oct 12; Nov 12

Contact: Bettie J. Graham, (301) 496-7531; bettie_graham@nih.gov

Internet: http://grants.nih.gov/grants/guide/pa-files/PAR-98-062.html

Sponsor: National Human Genome Research Institute
5635 Fishers Lane, Suite 4076
Rockville, MD 20852

NHGRI Technologies for Genomic Mapping, Sequencing, and Analysis Research Grants 3146

The NHGRI solicits applications for projects to develop new technologies, and/or significantly improve existing technologies, that will facilitate and accelerate the genome mapping, sequencing, and analysis goals of the Human Genome Project in the most expeditious and economical manner. The resources produced will be used to further studies of diseases and other biological phenomena. PA-97-044 may be obtained electronically through the NIH grant line (301-402-2221), the NIH Web site, and by mail and email.

Requirements: Applications may be submitted by domestic and foreign, for-profit and nonprofit organizations, public and private, such as universities, colleges, hospitals, laboratories, units of state and local governments, and eligible agencies of the federal government.

Contact: Dr. Bettie Graham, Program Contact; (301) 496-7531; fax (301) 480-2770; bettie_graham@nih.gov

Internet: http://grants2.nih.gov/grants/guide/pa-files/PA-97-044.html

Sponsor: National Human Genome Research Institute
9000 Rockville Pike
Bethesda, MD 20892

NHLBI Airway Smooth Muscle Function and Targeted Therapeutics in Human Asthma Grants 3147

The purpose of this funding opportunity announcement, issued by the NHLBI, NIH, is to solicit Research Project Grant applications from institutions and organizations that propose to investigate the complex role of airway smooth muscle (ASM) functions in the development of human asthma and identify novel therapeutic targets. Budgets for direct costs of up to $350,000 per year (exclusive of indirect costs associated with consortia) and a project duration of up to four years may be requested for a maximum of $1,400,000 direct costs over a four-year project period. Letters of Intent should be received by December 8, and full proposals are due by January 6.

Requirements: The following organizations and institutions are eligible to apply: public/state controlled institutions of higher education; private institutions of higher education; Hispanic-serving institutions; historically Black colleges and universities (HBCUs); tribally controlled colleges and universities (TCCUs); Alaska Native and Native Hawaiian serving institutions; nonprofits with 501(c)3 IRS status; nonprofits without 501(c)3 IRS status; for-profit organizations; state governments; Indian/Native American tribal governments (federally recognized); Indian/Native American tribally designated organizations; county governments; city or township governments; special district governments; independent school districts; public housing authorities/Indian housing authorities; U.S. territory or possession; Indian/Native American tribal governments (other than federally recognized); regional organizations; non-domestic (non-U.S.) entities (foreign organizations); other eligible agencies of the federal government; and faith-based or community-based organizations.

Date(s) Application is Due: Jan 6; Dec 8

Amount of Grant: Up to 350,000 U.S.D

Contact: Susan Banks-Schlegel, (301) 435-0202; Schleges@nhlbi.nih.gov

Internet: http://grants1.nih.gov/grants/guide/rfa-files/RFA-HL-09-007.html

Sponsor: National Heart, Lung, and Blood Institute
6701 Rockledge Drive RKL2/600 MSC 7902
Bethesda, MD 20892-7902

NHLBI Ancillary Studies in Clinical Trials 3148

The purpose of this Funding Opportunity Announcement (FOA) is to solicit research grant applications to conduct time-sensitive ancillary studies related to heart, lung, and blood diseases and sleep disorders in conjunction with ongoing NIH- or non-NIH-supported clinical trials. An applicant may request a project period of up to four years and a budget for direct costs up to $250,000 (10 modules) per year. Dates for Letters of Intent are: December 30, April 29, and August 31. Respective due dates fot full applications are: January 30, May 29, and September 30.

Requirements: The following organizations and institutions are eligible to apply: individuals; public/state controlled institutions of higher education; private institutions of higher education; Hispanic-serving institutions; historically Black colleges and universities (HBCUs); tribally controlled colleges and universities (TCCUs); Alaska Native and Native Hawaiian serving institutions; nonprofits with 501(c)3 IRS status; nonprofits without 501(c)3 IRS status; for-profit organizations; state governments; Indian/Native American tribal governments (federally recognized); Indian/Native American tribally designated organizations; county governments; city or township governments; special district governments; independent school districts; public housing authorities/Indian housing authorities; U.S. territory or possession; Indian/Native American tribal governments (other than federally recognized); regional organizations; non-domestic (non-U.S.) entities (foreign organizations); other eligible agencies of the federal government; and faith-based or community-based organizations.

Date(s) Application is Due: Jan 30; May 29; Sep 30

Amount of Grant: 250,000 - 1,000,000 U.S.D

Contact: Peter Kaufmann, (301) 435-0422; fax (301) 480-1773; kaufmannp@nhlbi.nih.gov

Internet: http://grants.nih.gov/grants/guide/rfa-files/RFA-HL-09-001.html

Sponsor: National Heart, Lung, and Blood Institute
6701 Rockledge Drive RKL2/600 MSC 7902
Bethesda, MD 20892-7902

NHLBI Bioengineering and Obesity Grants 3149

The purpose of this FOA is to solicit applications to develop and validate new and innovative engineering approaches to address clinical problems related to energy balance, intake, and expenditure. Novel sensors, devices, imaging, and other technologies, including technologies to detect biochemical markers of energy balance are expected to be developed and evaluated by collaborating engineers, physical scientists, mathematicians, and scientists from other relevant disciplines with expertise in obesity and nutrition. Basic and applied scientific applications with rigorous hypothesis-testing designs or for design-directed research are encouraged. The goal is to increase the number of useful technologies and tools available to scientists to facilitate their research in energy balance and health. Eventually these research tools should facilitate therapeutic advances and behavioral changes to address such problems as weight control and obesity.

Requirements: The following organizations and institutions are eligible to apply: public/state controlled institutions of higher education; private institutions of higher education; Hispanic-serving institutions; historically Black colleges and universities (HBCUs); tribally controlled colleges and universities (TCCUs); Alaska Native and Native Hawaiian serving institutions; nonprofits with 501(c)3 IRS status; nonprofits without 501(c)3 IRS status; for-profit organizations; state governments; Indian/Native American tribal governments (federally recognized); Indian/Native American tribally designated organizations; county governments; city or township governments; special district governments; independent school districts; public housing authorities/Indian housing authorities; U.S. territory or possession; Indian/Native American tribal governments (other than federally recognized); regional organizations; non-domestic (non-U.S.) entities (foreign organizations); other eligible agencies of the federal government; and faith-based or community-based organizations.

Date(s) Application is Due: Feb 5; Jun 5; Oct 5

Contact: J. Timothy Baldwin, (301) 435-0513; fax (301) 480-1335; BaldwinT@mail.nih.gov

Internet: http://grants.nih.gov/grants/guide/pa-files/PA-07-354.html

Sponsor: National Heart, Lung, and Blood Institute
6701 Rockledge Drive RKL2/600 MSC 7902
Bethesda, MD 20892-7902

NHLBI Bioengineering Approaches to Energy Balance and Obesity Grants for SBIR 3150

This funding opportunity will develop and validate new and innovative bioengineering technology to address clinical problems related to energy balance, intake, and expenditure. Novel sensors, devices, imaging, and other approaches are expected to be developed and evaluated by collaborating engineers, physical scientists, and scientists from other relevant disciplines with expertise in obesity and nutrition. The goal is to increase the number of useful technologies and tools available to scientists to facilitate their research in energy balance and health. Eventually these research tools should facilitate therapeutic advances and behavioral changes to address such problems as weight control and obesity. Because the nature and scope of the proposed research will vary from application to application, it is anticipated that the size and duration of each award will also vary. The total amount awarded and the number of awards will depend upon the mechanism numbers, quality, duration, and costs of the applications received.
Requirements: Only United States small business concerns (SBCs) are eligible to submit SBIR applications.
Date(s) Application is Due: Apr 5; Aug 5; Dec 5
Contact: J. Timothy Baldwin, (301) 435-0513; BaldwinT@mail.nih.gov
Internet: http://grants.nih.gov/grants/guide/pa-files/PA-07-435.html
Sponsor: National Heart, Lung, and Blood Institute
6701 Rockledge Drive RKL2/600 MSC 7902
Bethesda, MD 20892-7902

NHLBI Bioengineering Approaches to Energy Balance and Obesity Grants for STTR 3151

This funding opportunity will develop and validate new and innovative bioengineering technology to address clinical problems related to energy balance, intake, and expenditure. Novel sensors, devices, imaging, and other approaches are expected to be developed and evaluated by collaborating engineers, physical scientists, and scientists from other relevant disciplines with expertise in obesity and nutrition. The goal is to increase the number of useful technologies and tools available to scientists to facilitate their research in energy balance and health. Eventually these research tools should facilitate therapeutic advances and behavioral changes to address such problems as weight control and obesity. Because the nature and scope of the proposed research will vary from application to application, it is anticipated that the size and duration of each award will also vary. The total amount awarded and the number of awards will depend upon the mechanism numbers, quality, duration, and costs of the applications received.
Requirements: Only United States small business concerns (SBCs) are eligible to submit STTR applications.
Date(s) Application is Due: Apr 5; Aug 5; Dec 5
Contact: J. Timothy Baldwin, (301) 435-0513; BaldwinT@mail.nih.gov
Internet: http://grants.nih.gov/grants/guide/pa-files/PA-07-436.html
Sponsor: National Heart, Lung, and Blood Institute
6701 Rockledge Drive RKL2/600 MSC 7902
Bethesda, MD 20892-7902

NHLBI Biomedical Research Training Program for Individuals from Underrepresented Groups 3152

The NHLBI established the Biomedical Research Training Program for Individuals from Underrepresented Groups (BRTPUG) to offer opportunities for underrepresented post-baccalaureate individuals, undergraduate, and graduate students to receive training in fundamental biomedical sciences and clinical research disciplines. The purpose of the program is to enhance career opportunities in biomedical and behavioral research, including clinical and laboratory medicine, epidemiology, and biostatistics as applied to the etiology and treatment of heart, blood vessel, lung, and blood diseases. BRTPUG offers each participant the opportunity to work closely with leading research scientists in the Division of Intramural Research and extramural scientists in the Division of Prevention and Population Sciences. The program provides students with hands-on training in a research environment, which will prepare them to continue their studies and advance their careers in clinical and basic research. The trainee's appointment is a one-time appointment of 6 (2 summers) to 24 months over a 2 year period beginning the summer of selection. Each trainee is assigned to a mentor who is responsible for designing a carefully planned training program. For placements beginning in June of the calendar year or summer opportunities, NHLBI strongly encourages applicants to submit your application by January 15. For Post-Baccalaureate 1-2 year research internships beginning in July-September of the calendar year, applications should be submitted by March 1.
Requirements: Underrepresented post-baccalaureate individuals, and undergraduate and graduate students planning to pursue advanced degrees in the biomedical and behavioral sciences and meeting all other program criteria are eligible. Applicants must have a cumulative grade point average (GPA) or science course GPA of 3.3 or better on a 4.0 scale, or 4.3 or better on a 5.0 scale, and be U.S. citizens or permanent residents.
Date(s) Application is Due: Jan 15; Mar 1
Contact: Helena O. Mishoe, Deputy Director, Office of Minority Health Affairs; (301) 451-5081; fax (301) 480-0862; mishoeh@nhlbi.nih.gov
Internet: http://www.nhlbi.nih.gov/funding/training/redbook/undbrtum.htm
Sponsor: National Heart, Lung, and Blood Institute
6701 Rockledge Drive RKL2/600 MSC 7902
Bethesda, MD 20892-7902

NHLBI Blood Diseases and Resources Research Grants 3153

To foster research for the prevention, improved diagnosis, and treatment of non-malignant blood diseases such as Coolie's anemia, sickle cell disease, hemophilia, hemochromatosis, and disorders of hemostasis and thrombosis; to support research on hematopoietic stem cell biology and transplantation and on the use of transplantation to treat blood diseases; and to promote research in transfusion medicine for improving the availability, safety, and use of blood and blood products; and to develop new scientists for blood disease and resources research. Grants may support salaries, equipment, supplies, travel, and patient hospitalization as required to perform the research effort. National Research Service Awards (NRSAs) are made directly to individuals for research training in specified biomedical shortage areas. Grants may be made to institutions to enable them to make NRSAs to individuals selected by them.
Requirements: Any nonprofit organization engaged in biomedical research and institutions (or companies) organized for profit may apply for grants, with the exception of NRSAs. An individual may apply for a NRSA or, in some cases, may qualify for a research grant if adequate facilities in which to perform the research are available. SBIR grants can be awarded only to domestic small businesses (entities that are independently owned and operated for profit, are not dominant in the field in which research is proposed, and have no more than 500 employees).
Restrictions: Restrictions or limitations are imposed against the use of funds for entertainment, foreign travel, general-purpose equipment, alterations and renovations, and other items not regularly required for the performance of research.
Date(s) Application is Due: Feb 1; Jun 1; Oct 1
Amount of Grant: 67,988 - 29,000,000 U.S.D
Contact: Director; (301) 435-0080
Internet: http://www.federalgrantswire.com/blood-diseases-and-resources-research.html
Sponsor: National Heart, Lung, and Blood Institute
6701 Rockledge Drive RKL2/600 MSC 7902
Bethesda, MD 20892-7902

NHLBI Cardiac Development Consortium Grants 3154

The National Heart, Lung, and Blood Institute (NHLBI) and the Canadian Institutes of Health Research (CIHR) invite applications to participate in a new NHLBI Cardiac Development Consortium, a cooperative investigative group that will drive an integrated approach to the investigation of cardiovascular development. The purpose of this initiative is to support basic collaborative research leading to a comprehensive understanding of the regulatory networks controlling cardiovascular development. The consortium will consist of up to four Research Centers, scientific Cores as necessary, and a Steering Committee. Research Center applicants will propose projects that will select key regulatory pathways, identify the components of the pathways and targets, and rapidly disseminate data to the scientific community. The Research Centers will collaborate to maximize interoperability of the data. This Consortium is part of a new NHLBI translational program in pediatric cardiovascular disease, which includes the companion NHLBI Pediatric Cardiac Genomics Consortium and Administrative Coordinating Center. These consortia will interact with each other and the NHLBI Pediatric Heart Network to encourage translation of results from basic science to clinical research, and to provide clinical input on pressing needs for basic research. Maximum allowable direct costs for the research centers are $160,000 in the Year 1 planning phase, and up to $1,070,000 per year in years two through six. Letters of Intent should be received by January 6, and full proposals are due by February 6.
Requirements: The following organizations and institutions are eligible to apply: public/state controlled institutions of higher education; private institutions of higher education; Hispanic-serving institutions; historically Black colleges and universities (HBCUs); tribally controlled colleges and universities (TCCUs); Alaska Native and Native Hawaiian serving institutions; nonprofits with 501(c)3 IRS status; nonprofits without 501(c)3 IRS status; small businesses; for-profit organizations; state governments; and non-domestic (non-U.S.) entities in North America.
Date(s) Application is Due: Jan 6; Feb 6
Amount of Grant: Up to 5,510,000 U.S.D
Contact: Charlene Schramm, (301) 435-0510; SchrammC@nhlbi.nih.gov
Internet: http://grants.nih.gov/grants/guide/rfa-files/RFA-HL-09-002.html
Sponsor: National Heart, Lung, and Blood Institute
6701 Rockledge Drive RKL2/600 MSC 7902
Bethesda, MD 20892-7902

NHLBI Cardiac Translational Research Implementation Program Grants 3155

The National Heart, Lung, and Blood Institute (NHLBI) solicits Stage 1 exploratory planning (P20) applications to the Cardiac Translational Research Implementation Program (C-TRIP). The overall objective of the entire two-staged C-TRIP is to accelerate translation of promising new therapeutic interventions derived from fundamental research discoveries for the treatment and prevention of heart failure or arrhythmias through the planning and execution of well-designed clinical trials to demonstrate safety and efficacy. Stage 1 exploratory planning (P20) applications will support the planning and development of Stage 2 specialized center (P50) applications that support the conduct of the safety and efficacy trials recommended for support following a second phase of peer review. Subsequent Stage 2 (P50) applications for the conduct of the clinical trials planned and developed during Stage 1 will be solicited at a later time. The to-be-released solicitation for Stage 2 specialized center (P50) applications will be limited to successful Stage 1 applicants. Stage 2 (P50) applications will be anticipated 18 months following the initial award of Stage 1 applications. Support during Stage 2 of C-TRIP is expected to provide up to $1,515,000 direct costs per year for up to five years. During this period, applicants may request no more than $500,000 per year in direct costs for a maximum of $1,000,000 direct costs.
Requirements: The following organizations and institutions are eligible to apply: individuals; public/state controlled institutions of higher education; private institutions of higher education; Hispanic-serving institutions; historically Black colleges and universities (HBCUs); tribally controlled colleges and universities (TCCUs); Alaska Native and Native Hawaiian serving institutions; nonprofits with 501(c)3 IRS status;

nonprofits without 501(c)3 IRS status; for-profit organizations; state governments; Indian/Native American tribal governments (federally recognized); Indian/Native American tribally designated organizations; county governments; city or township governments; special district governments; independent school districts; public housing authorities/Indian housing authorities; U.S. territory or possession; Indian/Native American tribal governments (other than federally recognized); regional organizations; non-domestic (non-U.S.) entities (foreign organizations); other eligible agencies of the federal government; and faith-based or community-based organizations.
Date(s) Application is Due: Apr 28; May 28
Amount of Grant: 500,000 U.S.D
Contact: David A. Lathrop, (301) 435-0507; fax (301) 451-5458; lathropd@nhlbi.nih.gov
Internet: http://grants.nih.gov/grants/guide/rfa-files/RFA-HL-10-001.html
Sponsor: National Heart, Lung, and Blood Institute
6701 Rockledge Drive RKL2/600 MSC 7902
Bethesda, MD 20892-7902

NHLBI Career Enhancement Awards for Stem Cell Research 3156

The Career Enhancement Award for Stem Cell Research enables investigators to acquire new research capabilities in the use of human or animal embryonic, adult, or cord blood stem cells. Candidates must have a clinical or research doctoral degree and be actively engaged in research of interest to the NHLBI. Candidates must meet the NIH career development requirements. Actual salary is based on candidate's full-time, 12-month institutional salary, and the level of effort requested. Most recently, for mid-career and senior investigators, the maximum allowable salary was $186,600 for full-time effort. For junior faculty, maximum allowable salary was $75,000. Salary adjustments are announced in the NIH Guide. Up to $50,000 per year is allowed for research expenses, tuition, technical support, personnel consultant fees for mentor, and travel.
Requirements: Investigators at U.S. institutions with current or previous independent research support are eligible.
Date(s) Application is Due: Feb 12; Mar 12; Jun 12; Jul 12; Oct 12; Nov 12
Amount of Grant: Up to 200,000 U.S.D
Contact: Traci Heath Mondoro, (301) 435-0052; mondorot@nhlbi.nih.gov
Internet: http://www.nhlbi.nih.gov/funding/training/redbook/k18est-junior.htm
Sponsor: National Heart, Lung, and Blood Institute
6701 Rockledge Drive RKL2/600 MSC 7902
Bethesda, MD 20892-7902

NHLBI Career Transition Award 3157

The goal of the program is to provide highly qualified new investigators with an opportunity to receive research and research training experience in the NHLBI Division of Intramural Research and then to provide them with bridge funding to facilitate the transition of their research programs to extramural institutions. Total direct costs for the extramural phase cannot exceed $150,000 including fringe benefits per year in direct costs or $300,000 for the two-year period.
Requirements: The candidate must have a research or a health-professional doctorate, or its equivalent, and must have completed at least three and no more than five years of postdoctoral training, during which the potential for highly productive basic or clinical research was demonstrated.
Restrictions: Individuals on temporary or student visas are not eligible.
Date(s) Application is Due: Feb 5; Jun 5; Oct 5
Amount of Grant: 150,000 - 300,000 U.S.D
Contact: Herbert M. Geller, (301) 451-9440; direducation@nhlbi.nih.gov
Internet: http://grants.nih.gov/grants/guide/pa-files/PAR-07-317.html#SectionVII
Sponsor: National Heart, Lung, and Blood Institute
10 Center Drive, Building 10, Room 2N242, MSC 1754
Bethesda, MD 20892-1754

NHLBI Characterizing the Blood Stem Cell Niche Grants 3158

This FOA issued by the National Heart, Lung, and Blood Institute, National Institutes of Health, solicits Research Project Grant applications in a specific area of stem cell research, the blood stem cell niche, an area critical to advancing stem cell biology and its applications to cellular therapeutics including hematopoietic stem cell transplantation. This FOA is being initiated to foster collaborative research projects and insightful approaches to dissect the cellular components and factors involved in the hematopoietic stem cell niche. Development of conditional genetic knock-out models to test the role of factors in specific cell lineages and imaging technology to facilitate following stem cell engraftment in the niche in vivo are integral to this initiative. An individual research grant application may not request more than $250,000 direct cost a year unless collaborative in vivo cell imaging studies are included, in which case up to $350,000 direct cost may be requested in a year. An applicant may request a project period of up to 4 years. Letters of Intent should be received by December 8, and full proposals are due by January 6.
Requirements: The following organizations and institutions are eligible to apply: public/state controlled institutions of higher education; private institutions of higher education; Hispanic-serving institutions; historically Black colleges and universities (HBCUs); tribally controlled colleges and universities (TCCUs); Alaska Native and Native Hawaiian serving institutions; nonprofits with 501(c)3 IRS status; nonprofits without 501(c)3 IRS status; for-profit organizations; state governments; Indian/Native American tribal governments (federally recognized); Indian/Native American tribally designated organizations; county governments; city or township governments; special district governments; independent school districts; public housing authorities/Indian housing authorities; U.S. territory or possession; Indian/Native American tribal governments (other than federally recognized); regional organizations; non-domestic (non-U.S.) entities (foreign organizations); other eligible agencies of the federal government; and faith-based or community-based organizations.
Date(s) Application is Due: Jan 6; Dec 8
Amount of Grant: 250,000 - 350,000 U.S.D
Contact: John Thomas, (301) 435-0065; fax (301) 451-5453; ThomasJ@NHLBI.NIH.Gov
Internet: http://grants.nih.gov/grants/guide/rfa-files/RFA-HL-09-010.html
Sponsor: National Heart, Lung, and Blood Institute
6701 Rockledge Drive RKL2/600 MSC 7902
Bethesda, MD 20892-7902

NHLBI Childhood Obesity Prevention and Treatment Research Consortium Grants 3159

The goal of this Funding Opportunity Announcement (FOA) is to fund randomized trials testing intervention approaches to prevent excess weight gain in non-overweight youth and in those already overweight, and/or to reduce weight in obese and severely obese youth. The whole program is designed to promote collaboration, coordination and exchange of scientific information among institutions, and partnerships with local, state and national agencies. The ultimate goal is to improve children's health by preventing obesity-related morbidity and mortality from cardiovascular and other chronic diseases. Letters of Intent are due by September 8, with full proposals received by October 6.
Requirements: The following organizations and institutions are eligible to apply: individuals; public/state controlled institutions of higher education; private institutions of higher education; Hispanic-serving institutions; historically Black colleges and universities (HBCUs); tribally controlled colleges and universities (TCCUs); Alaska Native and Native Hawaiian serving institutions; nonprofits with 501(c)3 IRS status; nonprofits without 501(c)3 IRS status; for-profit organizations; state governments; Indian/Native American tribal governments (federally recognized); Indian/Native American tribally designated organizations; county governments; city or township governments; special district governments; independent school districts; public housing authorities/Indian housing authorities; U.S. territory or possession; Indian/Native American tribal governments (other than federally recognized); regional organizations; non-domestic (non-U.S.) entities (foreign organizations); other eligible agencies of the federal government; and faith-based or community-based organizations.
Date(s) Application is Due: Sep 8; Oct 6
Amount of Grant: Up to 11,100,000 U.S.D
Contact: Terry T-K Huang, Director, Obesity Research Strategic Core; (301) 594-1846; fax (301) 480-9791; huangter@mail.nih.gov
Internet: http://grants.nih.gov/grants/guide/rfa-files/RFA-HL-10-004.html
Sponsor: National Heart, Lung, and Blood Institute
6701 Rockledge Drive RKL2/600 MSC 7902
Bethesda, MD 20892-7902

NHLBI Childhood Obesity Prevention and Treatment Research Coordinating Unit Grants 3160

This funding opportunity announcement (FOA) invites applications from eligible institutions to participate as the Research Coordinating Unit (RCU) for the childhood obesity prevention and treatment research consortium, which runs in parallel to this FOA. Applicants for the clinical trial will conduct randomized controlled trials that are innovative and that would test innovative interventions to prevent overweight and obesity in youth and to treat existing overweight, obesity and severe obesity in youth. The participating organizations intend to award up to $5.25 million over seven years under this FOA. It is anticipated that one Research Coordinating Unit (RCU) award will be made. Budgets for direct costs of up to $500,000 per year and a project duration of up to seven years may be requested for a maximum of $3.5 million direct costs over a seven year project period. Letters of Intent are due by September 8, with full proposals received by October 6.
Requirements: The following organizations and institutions are eligible to apply: individuals; public/state controlled institutions of higher education; private institutions of higher education; Hispanic-serving institutions; historically Black colleges and universities (HBCUs); tribally controlled colleges and universities (TCCUs); Alaska Native and Native Hawaiian serving institutions; nonprofits with 501(c)3 IRS status; nonprofits without 501(c)3 IRS status; for-profit organizations; state governments; Indian/Native American tribal governments (federally recognized); Indian/Native American tribally designated organizations; county governments; city or township governments; special district governments; independent school districts; public housing authorities/Indian housing authorities; U.S. territory or possession; Indian/Native American tribal governments (other than federally recognized); regional organizations; non-domestic (non-U.S.) entities (foreign organizations); other eligible agencies of the federal government; and faith-based or community-based organizations.
Date(s) Application is Due: Sep 8; Oct 6
Amount of Grant: Up to 3,500,000 U.S.D
Contact: Charlotte A. Pratt, Program Director; (301) 435-0382; fax (301) 480-5158; prattc@nhlbi.nih.gov
Internet: http://grants.nih.gov/grants/guide/rfa-files/RFA-HL-10-005.html
Sponsor: National Heart, Lung, and Blood Institute
6701 Rockledge Drive RKL2/600 MSC 7902
Bethesda, MD 20892-7902

NHLBI Circadian-Coupled Cellular Function in Heart, Lung, and Blood Tissue Grants 3161

The purpose of this FOA is to stimulate Phase I translation (T1) of fundamental genome-based circadian biology to improve our understanding of heart, lung, and blood disease pathogenesis in non-neural tissues. For the purpose of this initiative, T1 research starts after gene discovery and has as its goal the development of molecular

models with application to an improved understanding of human health and disease pathogenesis, predictive testing for disease susceptibility, prognostic testing, or the selection of therapeutic strategies. Budget for direct costs up to $325,000 per year for project duration of up to four years may be requested. Letters of Intent should be received by December 8, and full proposals are due by January 6.
Requirements: The following organizations and institutions are eligible to apply: public/state controlled institutions of higher education; private institutions of higher education; Hispanic-serving institutions; historically Black colleges and universities (HBCUs); tribally controlled colleges and universities (TCCUs); Alaska Native and Native Hawaiian serving institutions; nonprofits with 501(c)3 IRS status; nonprofits without 501(c)3 IRS status; for-profit organizations; state governments; Indian/Native American tribal governments (federally recognized); Indian/Native American tribally designated organizations; county governments; city or township governments; special district governments; independent school districts; public housing authorities/Indian housing authorities; U.S. territory or possession; Indian/Native American tribal governments (other than federally recognized); regional organizations; non-domestic (non-U.S.) entities (foreign organizations); other eligible agencies of the federal government; and faith-based or community-based organizations.
Date(s) Application is Due: Jan 6; Dec 8
Amount of Grant: Up to 325,000 U.S.D
Contact: Aaron Laposky, Ph.D.; (301) 435-0199; laposkya@nhlbi.nih.gov
Internet: http://grants.nih.gov/grants/guide/rfa-files/RFA-HL-09-012.html
Sponsor: National Heart, Lung, and Blood Institute
6701 Rockledge Drive RKL2/600 MSC 7902
Bethesda, MD 20892-7902

NHLBI Clinical Centers for the NHLBI Asthma Network (AsthmaNet) Grants 3162
The purpose of this funding opportunity, issued by NHLBI, NIH, is to invite applications to participate in the NHLBI Asthma Network (AsthmaNet), a clinical research network that will develop and conduct multiple clinical trials to address the most important asthma management questions and new treatment approaches in pediatric and adult populations. AsthmaNet is designed to promote cooperation and coordination, facilitate scientific exchange, provides training opportunities, and leverage resources. AsthmaNet will include multiple Clinical Centers and one Data Coordinating Center. The protocols will include clinical trials to evaluate and/or compare existing or new therapeutic approaches to asthma management as well as a limited number of proof-of-concept studies to advance the development of novel therapies and studies that investigate the mechanistic bases for interventions examined in AsthmaNet. Letters of Intent should be received by December 30, and full proposals are due by January 30.
Requirements: The following organizations and institutions are eligible to apply: public/state controlled institutions of higher education; private institutions of higher education; Hispanic-serving institutions; historically Black colleges and universities (HBCUs); tribally controlled colleges and universities (TCCUs); Alaska Native and Native Hawaiian serving institutions; nonprofits with 501(c)3 IRS status; nonprofits without 501(c)3 IRS status; for-profit organizations; state governments; Indian/Native American tribal governments (federally recognized); Indian/Native American tribally designated organizations; county governments; city or township governments; special district governments; independent school districts; public housing authorities/Indian housing authorities; U.S. territory or possession; Indian/Native American tribal governments (other than federally recognized); regional organizations; non-domestic (non-U.S.) entities (foreign organizations); other eligible agencies of the federal government; and faith-based or community-based organizations.
Date(s) Application is Due: Jan 30; Dec 30
Contact: Virginia S. Taggart, (301) 435-0202; taggartv@nhlbi.nih.gov
Internet: http://grants.nih.gov/grants/guide/rfa-files/RFA-HL-08-010.html
Sponsor: National Heart, Lung, and Blood Institute
6701 Rockledge Drive RKL2/600 MSC 7902
Bethesda, MD 20892-7902

NHLBI Developmental Origins of Altered Lung Physiology and Immune Function Grants 3163
This FOA issued by the NHLBI, National Institutes of Health, solicits Research Project Grant applications from institutions and organizations that propose to perform research that will enhance the understanding of how the pre- and postnatal environments affect the interplay of the lung and immune system during development resulting in sustained changes in lung physiology and immune function that compromise respiratory health and outcomes. Budgets for direct costs of up to $350,000 per year (exclusive of indirect costs associated with consortia) and a project duration of up to five years may be requested for a maximum of $1,750,000 direct costs over a five-year project period. Letters of Intent should be received by September 19, and full proposals are due by October 21.
Requirements: The following organizations and institutions are eligible to apply: public/state controlled institutions of higher education; private institutions of higher education; Hispanic-serving institutions; historically Black colleges and universities (HBCUs); tribally controlled colleges and universities (TCCUs); Alaska Native and Native Hawaiian serving institutions; nonprofits with 501(c)3 IRS status; nonprofits without 501(c)3 IRS status; for-profit organizations; state governments; Indian/Native American tribal governments (federally recognized); Indian/Native American tribally designated organizations; county governments; city or township governments; special district governments; independent school districts; public housing authorities/Indian housing authorities; U.S. territory or possession; Indian/Native American tribal governments (other than federally recognized); regional organizations; non-domestic (non-U.S.) entities (foreign organizations); other eligible agencies of the federal government; and faith-based or community-based organizations.
Date(s) Application is Due: Sep 19; Oct 21
Amount of Grant: Up to 350,000 U.S.D
Contact: Patricia Noel, (301) 435-0202; noelp@nhlbi.nih.gov
Internet: http://grants.nih.gov/grants/guide/rfa-files/RFA-HL-08-009.html
Sponsor: National Heart, Lung, and Blood Institute
6701 Rockledge Drive RKL2/600 MSC 7902
Bethesda, MD 20892-7902

NHLBI Elucidating Nature's Solutions to Heart, Lung, and Blood Diseases and Sleep Disorder Processes Research Grants 3164
This Funding Opportunity Announcement (FOA) issued by the National Heart, Lung, and Blood Institute (NHLBI), National Institutes of Health (NIH), solicits grant applications proposing studies that elucidate the natural molecular and cellular adaptations of mammalian species to extreme environmental conditions that would rapidly evoke life-threatening cardiovascular or respiratory responses in other species, including humans. Extreme environmental conditions may include food scarcity, limited access to oxygen, or temperature extremes. The overall objective of the funding opportunity is to identify new targets for therapeutic interventions to treat and prevent heart, blood, and lung diseases and sleep disorders. There are no set aside specific dollars for this funding opportunity, and the amount of each award will be limited to $250,000 in direct costs for up to four years. Check the website for various deadlines.
Requirements: The following organizations and institutions are eligible to apply: individuals; public/state controlled institutions of higher education; private institutions of higher education; Hispanic-serving institutions; historically Black colleges and universities (HBCUs); tribally controlled colleges and universities (TCCUs); Alaska Native and Native Hawaiian serving institutions; nonprofits with 501(c)3 IRS status; nonprofits without 501(c)3 IRS status; for-profit organizations; state governments; Indian/Native American tribal governments (federally recognized); Indian/Native American tribally designated organizations; county governments; city or township governments; special district governments; independent school districts; public housing authorities/Indian housing authorities; U.S. territory or possession; Indian/Native American tribal governments (other than federally recognized); regional organizations; non-domestic (non-U.S.) entities (foreign organizations); other eligible agencies of the federal government; and faith-based or community-based organizations.
Restrictions: This FOA will not support the examination of animal models that simply mimic the progression of human cardiovascular and respiratory diseases.
Date(s) Application is Due: Jan 16; May 15
Amount of Grant: Up to 250,000 U.S.D
Contact: David Reiter, (301) 480-3310; reiterd@nhlbi.nih.gov
Internet: http://grants.nih.gov/grants/guide/pa-files/PAR-07-102.html
Sponsor: National Heart, Lung, and Blood Institute
6701 Rockledge Drive RKL2/600 MSC 7902
Bethesda, MD 20892-7902

NHLBI Exploratory Studies in the Neurobiology of Pain in Sickle Cell Disease 3165
The goal of this initiative is to foster novel basic and translational research of the neurobiology of pain in sickle cell disease. Priority will be given to the application of investigational techniques that have been utilized in other pain syndromes in both human and nonhuman studies. A multidisciplinary collaboration, between neurobiologists, clinical specialists in pain, and hematologists will be necessary to maximize the research effort.
Requirements: The following organizations and institutions are eligible to apply: public/state controlled institutions of higher education; private institutions of higher education; Hispanic-serving institutions; historically Black colleges and universities (HBCUs); tribally controlled colleges and universities (TCCUs); Alaska Native and Native Hawaiian serving institutions; nonprofits with 501(c)3 IRS status; nonprofits without 501(c)3 IRS status; for-profit organizations; state governments; Indian/Native American tribal governments (federally recognized); Indian/Native American tribally designated organizations; county governments; city or township governments; special district governments; independent school districts; public housing authorities/Indian housing authorities; U.S. territory or possession; Indian/Native American tribal governments (other than federally recognized); regional organizations; non-domestic (non-U.S.) entities (foreign organizations); other eligible agencies of the federal government; and faith-based or community-based organizations.
Date(s) Application is Due: Jan 5; Feb 3
Amount of Grant: Up to 300,000 U.S.D
Contact: Harvey Luksenburg, (301) 435-0050; luksenburgh@mail.nih.gov
Internet: http://grants.nih.gov/grants/guide/rfa-files/RFA-HL-09-008.html
Sponsor: National Heart, Lung, and Blood Institute
6701 Rockledge Drive RKL2/600 MSC 7902
Bethesda, MD 20892-7902

NHLBI Hyperglycemic Exacerbation of Ischemia/Reperfusion Injury After Ischemic Insult Grants 3166
This funding opportunity announcement (FOA) will stimulate a focus on the causal relationship between hyperglycemia and impaired cardioprotection in a cohesive, integrated manner. The concept of this initiative evolved from a survey of the grant portfolio in ischemic heart disease and from the literature and consultation with investigators in the scientific community. Appropriate topics for investigation under this FOA include but are not limited to the following: to elucidate mechanisms of acute hyperglycemic exacerbation of myocardial ischemic injury following an ischemic insult to identify new therapeutic targets; to examine the relationship between severity of hyperglycemia and increases in myocardial infarct size mediated by hyperglycemia-induced impairment of cardioprotective signal transduction; to determine the role of

inflammatory cells and oxidative stress in the hyperglycemic exacerbation of myocardial injury in infarction; to examine the interaction of metabolic and sex specific hormonal effects on endothelial and coronary microvascular function with acute hyperglycemia; to elucidate the specific mechanisms by which hyperglycemia affects glucose and lipid metabolism in the myocardium; and to study the pathogenesis of cardiovascular disease during acute hyperglycemia in suitable animal models.
Requirements: The following organizations and institutions are eligible to apply: individuals; public/state controlled institutions of higher education; private institutions of higher education; Hispanic-serving institutions; historically Black colleges and universities (HBCUs); tribally controlled colleges and universities (TCCUs); Alaska Native and Native Hawaiian serving institutions; nonprofits with 501(c)3 IRS status; nonprofits without 501(c)3 IRS status; for-profit organizations; state governments; Indian/Native American tribal governments (federally recognized); Indian/Native American tribally designated organizations; county governments; city or township governments; special district governments; independent school districts; public housing authorities/Indian housing authorities; U.S. territory or possession; Indian/Native American tribal governments (other than federally recognized); regional organizations; non-domestic (non-U.S.) entities (foreign organizations); other eligible agencies of the federal government; and faith-based or community-based organizations.
Date(s) Application is Due: Feb 5; Jun 5; Oct 5
Contact: Isabella Liang, (301) 435-0504; fax (301) 451-5458; Liangi@mail.nih.gov
Internet: http://grants.nih.gov/grants/guide/pa-files/PA-07-323.html
Sponsor: National Heart, Lung, and Blood Institute
6701 Rockledge Drive RKL2/600 MSC 7902
Bethesda, MD 20892-7902

NHLBI Hypersensitivity Pneumonitis Grants for Early Identification and Mechanisms of Disease **3167**
The objectives of this funding opportunity announcement (FOA) are to investigate basic mechanisms and early clinical identification of hypersensitivity pneumonitis (HP) interstitial lung disease, in order to elucidate how chronic disease progresses and how it can evolve into either interstitial fibrosis or airway obstructive disease.
Requirements: The following organizations and institutions are eligible to apply: individuals; public/state controlled institutions of higher education; private institutions of higher education; Hispanic-serving institutions; historically Black colleges and universities (HBCUs); tribally controlled colleges and universities (TCCUs); Alaska Native and Native Hawaiian serving institutions; nonprofits with 501(c)3 IRS status; nonprofits without 501(c)3 IRS status; for-profit organizations; state governments; Indian/Native American tribal governments (federally recognized); Indian/Native American tribally designated organizations; county governments; city or township governments; special district governments; independent school districts; public housing authorities/Indian housing authorities; U.S. territory or possession; Indian/Native American tribal governments (other than federally recognized); regional organizations; non-domestic (non-U.S.) entities (foreign organizations); other eligible agencies of the federal government; and faith-based or community-based organizations.
Date(s) Application is Due: Feb 5; Jun 5; Oct 5
Contact: Herbert Y. Reynolds, (301) 435-0222; Reynoldh@nhlbi.nih.gov
Internet: http://grants.nih.gov/grants/guide/pa-files/PA-07-300.html
Sponsor: National Heart, Lung, and Blood Institute
6701 Rockledge Drive RKL2/600 MSC 7902
Bethesda, MD 20892-7902

NHLBI Immunomodulatory, Inflammatory, and Vasoregulatory Properties of Transfused Red Blood Cell Units as a Function of Preparation and Storage Grants **3168**
The goal of this Funding Opportunity Announcement (FOA) in blood banking and transfusion medicine is to solicit novel basic and translational research projects which will identify the molecular and cellular changes that occur during red blood cell unit preparation and storage; and evaluate the immunomodulatory, inflammatory, and vasoactive effects of storage lesion elements from red blood cell units on the blood vessel wall, host cells, and tissue oxygenation. Only research applications involving human red blood cell units (and their components) suitable for transfusion in the U.S. and that are collected and stored according to methods approved by the U.S. FDA will be considered responsive to this FOA. Applications are encouraged that propose collaborations between investigators from different disciplines such as hematology, blood banking, transfusion medicine, immunology, cell biology, rheology, nanotechnologies, molecular biology, and other relevant disciplines. This program will promote interactions and resource sharing among the award recipients to maximize their research efforts. Letters of Intent should be received by April 21 (1st cycle) or December 19 (2nd cycle), and full proposals are due by either May 21 (1st cycle) or January 21 (2nd cycle).
Requirements: The following organizations and institutions are eligible to apply: public/state controlled institutions of higher education; private institutions of higher education; Hispanic-serving institutions; historically Black colleges and universities (HBCUs); tribally controlled colleges and universities (TCCUs); Alaska Native and Native Hawaiian serving institutions; nonprofits with 501(c)3 IRS status; nonprofits without 501(c)3 IRS status; small businesses; for-profit organizations; state governments; Indian/Native American tribal governments; Indian/Native American tribally designated organizations; and U.S. territories or possessions.
Date(s) Application is Due: Jan 21; Apr 21; May 21; Dec 19
Contact: Simone Glynn, (301) 435-0065; glynnsa@nhlbi.nih.gov
Internet: http://grants.nih.gov/grants/guide/rfa-files/RFA-HL-08-005.html
Sponsor: National Heart, Lung, and Blood Institute
6701 Rockledge Drive RKL2/600 MSC 7902
Bethesda, MD 20892-7902

NHLBI Improving Heart Failure Disease Management Grants **3169**
The purpose of this initiative is to stimulate innovative research to address inadequately answered questions about disease management in clinical practice for chronic heart failure (HF). The broader goal is to identify and improve, better implement, and disseminate clinically effective disease management tools into clinical practice to reduce HF morbidity and mortality and improve patient outcomes. Because the nature and scope of the proposed research will vary from application to application, it is anticipated that the size and duration of each award will also vary. The total amount awarded and the number of awards will depend upon the mechanism numbers, quality, duration, and costs of the applications received.
Requirements: The following organizations and institutions are eligible to apply: individuals; public/state controlled institutions of higher education; private institutions of higher education; Hispanic-serving institutions; historically Black colleges and universities (HBCUs); tribally controlled colleges and universities (TCCUs); Alaska Native and Native Hawaiian serving institutions; nonprofits with 501(c)3 IRS status; nonprofits without 501(c)3 IRS status; for-profit organizations; state governments; Indian/Native American tribal governments (federally recognized); Indian/Native American tribally designated organizations; county governments; city or township governments; special district governments; independent school districts; public housing authorities/Indian housing authorities; U.S. territory or possession; Indian/Native American tribal governments (other than federally recognized); regional organizations; non-domestic (non-U.S.) entities (foreign organizations); other eligible agencies of the federal government; and faith-based or community-based organizations.
Date(s) Application is Due: Feb 5; Jun 5; Oct 5
Contact: Patrice Desvigne-Nickens, (301) 435-0504; DesvignP@mail.nih.gov
Internet: http://grants.nih.gov/grants/guide/pa-files/PA-07-355.html
Sponsor: National Heart, Lung, and Blood Institute
6701 Rockledge Drive RKL2/600 MSC 7902
Bethesda, MD 20892-7902

NHLBI Independent Scientist Award **3170**
The award is a special salary-only grant designed to provide protected time for newly independent scientists who currently have non-research obligations such as heavy teaching loads, clinical work, committee assignments, service, and administrative duties that prevent them from having a period of intensive research focus. The award is targeted to persons with doctoral degrees who have completed their research training, have independent peer-reviewed research support, and who need a period of protected research time in order to foster their research career development. New applications are due February 12, June 12, and October 12, while renewal applications are due March 12, July 12, and November 12. Up to $75,000 per year plus fringe benefits is available. No other research development support funds are provided.
Requirements: The following organizations and institutions are eligible to apply: public/state controlled institutions of higher education; private institutions of higher education; nonprofits with 501(c)3 IRS status; nonprofits without 501(c)3 IRS status; small businesses; for-profit organizations; State governments; U.S. territories or possessions; Indian/Native American tribal governments (Federally recognized and other than Federally recognized); Indian/Native American tribally designated organizations; Hispanic-serving institutions; historically Black colleges and universities (HBCUs); tribally controlled colleges and universities (TCCUs); Alaska Native and Native Hawaiian serving institutions; regional organizations; and faith-based or community based organizations.
Restrictions: The award is not intended for investigators who already have full time to perform research, or have substantial publication records or considerable research support indicating that they are well established in their fields. Foreign institutions are not eligible to apply.
Date(s) Application is Due: Feb 12; Mar 12; Jun 12; Jul 12; Oct 12; Nov 12
Amount of Grant: 75,000 U.S.D
Contact: Traci Heath Mondoro, (301) 435-0052; mondorot@nhlbi.nih.gov
Internet: http://grants.nih.gov/grants/guide/pa-files/PA-09-038.html
Sponsor: National Heart, Lung, and Blood Institute
6701 Rockledge Drive RKL2/600 MSC 7902
Bethesda, MD 20892-7902

NHLBI Individual Postdoctoral National Research Service Awards **3171**
This program offers health scientists the opportunity to receive full-time research training for up to 3 years in areas that reflect the national need for biomedical, clinical and behavioral research in cardiovascular, pulmonary, hematologic, and sleep disorders.
Requirements: When the fellowship begins, the applicant must have received a doctoral degree and have arranged to work with a sponsor affiliated with an institution that has the staff and facilities needed for the proposed training. Applicants must be U.S. citizens, noncitizen nationals, or legal permanent residents of the U.S. Training can be conducted abroad if the site provides opportunities that are not available in this country.
Restrictions: These grants are not intended for study leading to the M.D., D.O., D.D.S., or equivalent professional degrees, nor do they support residency training. One month of payback for each month of training, up to a maximum of 12 months, is required. This requirement can be fulfilled by teaching or research (a minimum of 20 hours per week) on a continuous basis beginning within 2 years after support ends.
Date(s) Application is Due: Apr 8; Aug 0; Dec 8
Contact: Sandra Colombini Hatch, (301) 435-0222; Hatchs@nhlbi.nih.gov
Internet: http://www.nhlbi.nih.gov/funding/training/redbook/phdf32.htm
Sponsor: National Heart, Lung, and Blood Institute
6701 Rockledge Drive RKL2/600 MSC 7902
Bethesda, MD 20892-7902

NHLBI Individual Predoctoral Fellowships to Promote Diversity in Health-Related Research 3172

This program encourages students from underrepresented racial and ethnic groups, individuals with disabilities, and individuals from disadvantaged backgrounds to seek graduate degrees in the biomedical and behavioral sciences to help increase the number of well-trained scientists from underrepresented groups. The fellowship provides up to 5 years of support for research training leading to the Ph.D. or equivalent research degree, the combined M.D./Ph.D. degree, or other combined degrees in the biomedical or behavioral sciences.
Requirements: At the time of appointment, students must be U.S. citizens, noncitizen nationals, or lawfully admitted to the U.S. for permanent residence. Individuals on temporary or student visas are not eligible. Individuals with disabilities, or from racial and ethnic groups, or individuals from disadvantaged backgrounds pursuing advanced degrees in the biomedical and behavioral sciences are eligible.
Date(s) Application is Due: Apr 13; Aug 13; Dec 13
Contact: Ann Rothgeb, (301) 435-0202; fax (301) 480-3557; rothgeba@mail.nih.gov
Internet: http://www.nhlbi.nih.gov/funding/training/redbook/gradf31.htm
Sponsor: National Heart, Lung, and Blood Institute
6701 Rockledge Drive RKL2/600 MSC 7902
Bethesda, MD 20892-7902

NHLBI Integrating Lung Genetics and Genomics in Human Populations Grants 3173

The purpose of this Funding Opportunity Announcement (FOA) is to stimulate innovative approaches to merge genetic and genomic techniques to find genes associated with human complex traits for lung disease. Because the nature and scope of the proposed research will vary from application to application, it is anticipated that the size and duration of each award will also vary. The total amount awarded and the number of awards will depend upon the mechanism numbers, quality, duration, and costs of the applications received.
Requirements: The following organizations and institutions are eligible to apply: individuals; public/state controlled institutions of higher education; private institutions of higher education; Hispanic-serving institutions; historically Black colleges and universities (HBCUs); tribally controlled colleges and universities (TCCUs); Alaska Native and Native Hawaiian serving institutions; nonprofits with 501(c)3 IRS status; nonprofits without 501(c)3 IRS status; for-profit organizations; state governments; Indian/Native American tribal governments (federally recognized); Indian/Native American tribally designated organizations; county governments; city or township governments; special district governments; independent school districts; public housing authorities/Indian housing authorities; U.S. territory or possession; Indian/Native American tribal governments (other than federally recognized); regional organizations; non-domestic (non-U.S.) entities (foreign organizations); other eligible agencies of the federal government; and faith-based or community-based organizations.
Date(s) Application is Due: Feb 5; Jun 5; Oct 5
Contact: Susan Banks-Schlegel, (301) 435-0202; Schleges@nhlbi.nih.gov
Internet: http://grants.nih.gov/grants/guide/pa-files/PA-07-135.html
Sponsor: National Heart, Lung, and Blood Institute
6701 Rockledge Drive RKL2/600 MSC 7902
Bethesda, MD 20892-7902

NHLBI Intramural Research Training Awards 3174

The Office of Education of the Division of Intramural Research coordinates and assists the training program and provides administrative support and resources for all research training activities within the Division. Programs are available within the Laboratories and Branches within the Division to provide specialized research training for high school students, college students, graduate and medical students, postdoctoral fellows and medical residents and fellows. The Office of Education provides assistance to applicants at each of these levels. NHLBI is committed to improving the representation of underrepresented groups, including minorities, women, and scientists with disabilities in the mainstream of basic and clinical research.
Contact: Herbert M. Geller, (301) 451-9440; direducation@nhlbi.nih.gov
Internet: http://www.training.nih.gov/student/pre-irta/previewpostbac.asp
Sponsor: National Heart, Lung, and Blood Institute
10 Center Drive, Building 10, Room 2N242, MSC 1754
Bethesda, MD 20892-1754

NHLBI Investigator Initiated Multi-Site Clinical Trials 3175

The purpose of this Funding Opportunity Announcement is to provide a vehicle for submitting grant applications for investigator-initiated multi-site Phase II or Phase III randomized, controlled clinical trials. The trials may address any research question related to the mission and goals of NHLBI and may test clinical or behavioral interventions. The trials may randomize at the individual (patient) level or at a group or cluster level (e.g., randomization of clinics, schools, worksites, etc.).
Requirements: The following organizations and institutions are eligible to apply: public/state controlled institutions of higher education; private institutions of higher education; Hispanic-serving institutions; historically Black colleges and universities (HBCUs); tribally controlled colleges and universities (TCCUs); Alaska Native and Native Hawaiian serving institutions; nonprofits with 501(c)3 IRS status; nonprofits without 501(c)3 IRS status; for-profit organizations; state governments; Indian/Native American tribal governments (federally recognized); Indian/Native American tribally designated organizations; county governments; city or township governments; special district governments; independent school districts; public housing authorities/Indian housing authorities; U.S. territory or possession; Indian/Native American tribal governments (other than federally recognized); regional organizations; non-domestic (non-U.S.) entities (foreign organizations); other eligible agencies of the federal government; and faith-based or community-based organizations.
Restrictions: This Funding Opportunity Announcement is not intended for support of single-center studies or multi-center observational studies that are not testing an intervention. Foreign institutions are not eligible to apply.
Date(s) Application is Due: Feb 5; Jun 5; Oct 5
Contact: David Gordon, (301) 435-0564; fax (301) 480-7971; gordond@nhlbi.nih.gov
Internet: http://grants.nih.gov/grants/guide/pa-files/PAR-07-373.html
Sponsor: National Heart, Lung, and Blood Institute
6701 Rockledge Drive RKL2/600 MSC 7902
Bethesda, MD 20892-7902

NHLBI Lymphatic Biology in Health and Disease Grants 3176

This Funding Opportunity Announcement (FOA) issued by the National Heart, Lung, and Blood Institute (NHLBI) solicits Research Project Grant (R01) applications from qualified researchers to investigate the biology of the lymphatic system and potential new therapeutic implications. The purpose of this program announcement is to stimulate research on the biology of the lymphatic system at all biologic levels: molecular, cellular, tissue, organ, and whole body levels. Further, it includes research on innovations for identifying and intervening in lymphatic diseases across all age groups and disease states. Individuals with the skills, knowledge, and resources necessary to carry out the proposed research are invited to work with their institution/organization to develop an application for support. Individuals from underrepresented racial and ethnic groups as well as individuals with disabilities are always encouraged to apply for NIH support. Letters of Intent are due on August 17, while full ptoposals must arrive by September 18.
Requirements: The following organizations and institutions are eligible to apply: public/state controlled institutions of higher education; private institutions of higher education; Hispanic-serving institutions; historically Black colleges and universities (HBCUs); tribally controlled colleges and universities (TCCUs); Alaska Native and Native Hawaiian serving institutions; nonprofits with 501(c)3 IRS status; nonprofits without 501(c)3 IRS status; for-profit organizations; state governments; Indian/Native American tribal governments (federally recognized); Indian/Native American tribally designated organizations; county governments; city or township governments; special district governments; independent school districts; public housing authorities/Indian housing authorities; U.S. territory or possession; Indian/Native American tribal governments (other than federally recognized); regional organizations; non-domestic (non-U.S.) entities (foreign organizations); other eligible agencies of the federal government; and faith-based or community-based organizations.
Restrictions: Foreign institutions are not eligible to apply.
Date(s) Application is Due: Aug 17; Sep 18
Contact: Stephen Goldman, (301) 435-0560; goldmans@nhlbi.nih.gov
Internet: http://grants.nih.gov/grants/guide/pa-files/PAR-07-420.html
Sponsor: National Heart, Lung, and Blood Institute
6701 Rockledge Drive RKL2/600 MSC 7902
Bethesda, MD 20892-7902

NHLBI Mentored Career Award For Faculty At Minority Institutions 3177

The NHLBI Mentored Career Award for Faculty at Minority Institutions is a three-to-five-year non-renewable career development award made to faculty at a minority institution. The candidates are faculty members at a minority institution, must have research experience (length of time may vary), and be committed to developing into independent biomedical investigators in research areas relevant to the mission of the NHLBI (i.e, cardiovascular, pulmonary, hematologic, or sleep disorders research). The award will enable suitable faculty members holding doctoral degrees, such as the Ph.D., M.D., D.O., D.V.M., or an equivalent degree, to undertake special study and supervised research under a mentor who is an accomplished investigator in the research area proposed and has experience in developing independent investigators. Letters of Intent are due on June 21, while full proposals are due by July 19.
Requirements: Minority scientists and physicians with limited research experience needing guided coursework and supervised laboratory experiences, as well as minority faculty needing only an intensive research experience under the guidance of an established scientist, are eligible to apply. At the time of award, it is required that at least two years have elapsed since the receipt of the doctoral degree and that the candidate have at least one year of prior documented research experience.
Restrictions: Current or past principal investigators of an NIH grant or its equivalent, including the Clinical Investigator Award, Physician Scientist Award, Clinical Investigator Development Award, or Mentored Clinical Scientist Development Award are not eligible for the Mentored Research Scientist Development Award for Minority Faculty. Similarly, individuals serving as responsible investigators or project leaders on large grants, such as a Program Project Grant, are not eligible for the Mentored Research Scientist Development Award for Minority Faculty.
Date(s) Application is Due: Jul 19
Amount of Grant: Up to 200,000 U.S.D
Contact: Traci Heath Mondoro, (301) 435-0052; mondorot@nhlbi.nih.gov
Internet: http://grants.nih.gov/grants/guide/rfa-files/RFA-HL-05-016.html#SectionVII
Sponsor: National Heart, Lung, and Blood Institute
6701 Rockledge Drive RKL2/600 MSC 7902
Bethesda, MD 20892-7902

NHLBI Mentored Career Development Award to Promote Faculty Diversity/Re-Entry in Biomedical Research 3178

This award is offered to encourage underrepresented faculty members at U.S. institutions to enhance their research skills in areas of interest to the NHLBI and to increase

the number of highly trained investigators from diverse backgrounds or who have experienced an interruption in their research careers. It is geared toward individuals whose basic and clinical research interests are grounded in the advanced methods and experimental approaches needed to solve problems related to cardiovascular, pulmonary, hematological, and sleep disorders research. Salary support up to $75,000 per year plus fringe benefits is available, with research support up to $30,000 per year.
Requirements: Faculty members who have been determined by the grantee institution to be underrepresented on faculty in biomedical and behavioral research on a national or institutional basis, such as individuals from racial and ethnic minority groups, individuals with disabilities, individuals from disadvantaged backgrounds or individuals who have experienced an interruption in their research careers are eligible.
Date(s) Application is Due: Aug 15
Amount of Grant: Up to 105,000 U.S.D
Contact: Lorraine M. Silsbee, (301) 435-0709; Lorraine_Silsbee@nih.gov
Internet: http://www.nhlbi.nih.gov/funding/training/redbook/newsck01.htm
Sponsor: National Heart, Lung, and Blood Institute
6701 Rockledge Drive RKL2/600 MSC 7902
Bethesda, MD 20892-7902

NHLBI Mentored Clinical Scientist Research Career Development Awards 3179
This award enables candidates holding professional degrees (e.g., M.D., D.O., D.V.M., or equivalent degrees) to undertake 3 to 5 years of special study and supervised research with the goal of becoming independent investigators. The award also allows awardees to pursue a research career development program suited to their experience and capabilities under a mentor who is competent to provide guidance in the chosen research area. Institutions may submit applications on behalf of candidates who hold professional degrees. At least 2 years must have elapsed since the health professional degree was granted. Candidates can have varying levels of clinical training and research experience. New applications are due February 12, June 12, and October 12, while resubmitted applications are due March 12, July 12, and November 12. Up to $75,000 per year plus fringe benefits is available, and up to $25,000 per year for research support for new awards.
Requirements: All candidates must be U.S. citizens, non-citizen nationals, or legal permanent residents of the U.S. Persons with temporary or student visas are not eligible. The grantee institution must have a strong, well-established research and research training program in the chosen area, accomplished faculty in the basic and clinical sciences, and a commitment to the candidate's research development. The proposed program should include an appropriate mentor.
Date(s) Application is Due: Feb 12; Mar 12; Jun 12; Jul 12; Oct 12; Nov 12
Amount of Grant: Up to 100,000 U.S.D
Contact: Lorraine M. Silsbee, (301) 435-0709; Lorraine_Silsbee@nih.gov
Internet: http://www.nhlbi.nih.gov/funding/training/redbook/newrek08.htm
Sponsor: National Heart, Lung, and Blood Institute
6701 Rockledge Drive RKL2/600 MSC 7902
Bethesda, MD 20892-7902

NHLBI Mentored Patient-Oriented Research Career Development Awards 3180
This award supports the career development of investigators who are committed to patient-oriented research. It provides support for supervised study and research for clinically trained professionals who have the potential to develop into productive, clinical investigators focusing on patient-oriented research. This program provides research development opportunities for clinicians with varying levels of research experience. Support is provided for a minimum of 3 years and a maximum of 5 years. Up to $75,000 per year in salary, plus fringe benefits, is available. In addition, $30,000 for research development support per year is available (with sufficient justification up to $50,000 per year can be requested). New applications are due February 12, June 12, and October 12, while resubmitted applications are due March 12, July 12, and November 12.
Requirements: All candidates must be U.S. citizens, noncitizen nationals, or legal permanent residents of the U.S. Candidates must have an M.D., D.D.S., D.M.D., D.O., D.C., Pharm.D. or the Ph.D. or other doctoral degrees in disciplines such as clinical psychology, nursing, clinical genetics, and rehabilitation, or equivalent degree from an accredited domestic or foreign institution at the time of application. Certification by an authorized individual at the degree-granting institution that all degree requirements have been met is acceptable. Candidates must have completed their clinical training, including specialty and, if applicable, subspecialty training prior to receiving an award.
Restrictions: Persons with temporary or student visas are not eligible.
Date(s) Application is Due: Feb 12; Mar 12; Jun 12; Jul 12; Oct 12; Nov 12
Amount of Grant: Up to 125,000 U.S.D
Contact: Ann Rothgeb, (301) 435-0202; fax (301) 480-3557; rothgeba@mail.nih.gov
Internet: http://www.nhlbi.nih.gov/funding/training/redbook/newk23.htm
Sponsor: National Heart, Lung, and Blood Institute
6701 Rockledge Drive RKL2/600 MSC 7902
Bethesda, MD 20892-7902

NHLBI Mentored Quantitative Research Career Development Awards 3181
The goal of this program is to foster interdisciplinary collaboration in biomedical and behavioral research by supporting supervised research experiences for scientists with quantitative and engineering backgrounds. This award provides research and career development opportunities for scientists and engineers with little or no biomedical or behavioral research experience who are committed to establishing themselves in careers as independent biomedical or behavioral investigators. Examples of quantitative scientific and technical backgrounds outside of biology and medicine considered appropriate for this award include, but are not limited to: mathematics, statistics, computer science, informatics, physics, chemistry, and engineering. This mechanism is intended for research-oriented investigators from the postdoctoral level to the level of senior faculty. Institutions may submit applications on behalf of candidates who hold an advanced degree in a quantitative area of science or engineering. Up to $75,000 per year plus fringe benefits and up to $40,000 per year for research development support is available. New applications are due February 12, June 12, and October 12, while resubmitted applications are due March 12, July 12, and November 12.
Requirements: All candidates must be U.S. citizens, noncitizen nationals, or legal permanent residents of the U.S. The grantee institution must have a strong, well-established research and research training program in the chosen area, accomplished faculty in the basic and clinical sciences, and a commitment to the candidate's research development. The proposed program should include an appropriate mentor.
Date(s) Application is Due: Feb 12; Mar 12; Jun 12; Jul 12; Oct 12; Nov 12
Amount of Grant: Up to 115,000 U.S.D
Contact: Michael Commarato, (301) 435-0535; CommaraM@nhlbi.nih.gov
Internet: http://www.nhlbi.nih.gov/funding/training/redbook/newk25.htm
Sponsor: National Heart, Lung, and Blood Institute
6701 Rockledge Drive RKL2/600 MSC 7902
Bethesda, MD 20892-7902

NHLBI Microbiome of the Lung and Respiratory Tract in HIV-Infected Individuals and HIV-Uninfected Controls 3182
The National Heart, Lung, and Blood Institute (NHLBI) solicits grant applications under this Funding Opportunity Announcement (FOA) to characterize the microbiome of the lung (the airways and airspaces below the glottis and the lung parenchyma) alone or in combination with the nasal and/or oropharyngeal cavities in HIV-infected individuals and matched HIV-uninfected controls, including normal healthy controls, using molecular techniques to identify bacteria and if possible other organisms, e.g., viruses, cell-wall deficient organisms, protozoa, and fungi. Investigators should use high-throughput technology platforms to create a data set of sufficient quality and depth to allow analysis of how changes of microbiota relate to HIV lung disease progression/complications. These data will be used to examine the impact of changes in the respiratory microbiome on the pathogenesis and progression of HIV disease, on HIV-related respiratory complications, and the effects of anti-HIV therapies. The NHLBI also invites applications for four Clinical/Sequencing research sites and one supporting Data Coordinating Center (DCC). Each application for a clinical/sequencing research site should propose approaches to characterize the lung microbiome and applicants are encouraged to include at least one hypothesis-driven mechanistic aim. Direct costs up to $525,000 per year for project duration of up to five years may be requested. Letters of Intent are due by February 25, with full proposals to be received by March 25.
Requirements: The following organizations and institutions are eligible to apply: individuals; public/state controlled institutions of higher education; private institutions of higher education; Hispanic-serving institutions; historically Black colleges and universities (HBCUs); tribally controlled colleges and universities (TCCUs); Alaska Native and Native Hawaiian serving institutions; nonprofits with 501(c)3 IRS status; nonprofits without 501(c)3 IRS status; for-profit organizations; state governments; Indian/Native American tribal governments (federally recognized); Indian/Native American tribally designated organizations; county governments; city or township governments; special district governments; independent school districts; public housing authorities/Indian housing authorities; U.S. territory or possession; Indian/Native American tribal governments (other than federally recognized); regional organizations; non-domestic (non-U.S.) entities (foreign organizations); other eligible agencies of the federal government; and faith-based or community-based organizations.
Date(s) Application is Due: Feb 25; Mar 25
Amount of Grant: Up to 525,000 U.S.D
Contact: Sandra Colombini Hatch, (301) 435-0222; Hatchs@nhlbi.nih.gov
Internet: http://grants.nih.gov/grants/guide/rfa-files/RFA-HL-09-006.html
Sponsor: National Heart, Lung, and Blood Institute
6701 Rockledge Drive RKL2/600 MSC 7902
Bethesda, MD 20892-7902

NHLBI Midcareer Investigator Award in Patient-Oriented Research 3183
This award provides support for established clinicians within 15 years of their specialty training to further their research and mentoring of outstanding patient-oriented investigators. It will enable them to expand their potential for significant contributions to their field, and to act as mentors for beginning clinical researchers. This award provides protected time to enable the clinician to carry out these activities. The salary that may be requested and awarded is 25% to 50% of the recipient's institutional salary, exclusive of fringe benefits, up to the maximum allowable annual salary rate in effect at the time of award. New applications are due February 12, June 12, and October 12, while resubmitted applications are due March 12, July 12, and November 12.
Requirements: Patient-oriented investigators at U.S. institutions with independent research support are eligible.
Date(s) Application is Due: Feb 12; Mar 12; Jun 12; Jul 12; Oct 12; Nov 12
Contact: Sandra Colombini Hatch, (301) 435-0222; Hatchs@nhlbi.nih.gov
Internet: http://www.nhlbi.nih.gov/funding/training/redbook/estk24.htm
Sponsor: National Heart, Lung, and Blood Institute
6701 Rockledge Drive RKL2/600 MSC 7902
Bethesda, MD 20892-7902

NHLBI Minority Institutional Research Training Program Grants 3184
The NHLBI's Minority Institutional Research Training Program is a Ruth L. Kirschstein National Research Service Award Program intended to support training of predoctoral and health professional students and individuals in postdoctoral training

at minority schools having the potential to develop meritorious training programs in cardiovascular, pulmonary, hematologic, and sleep disorders. Predoctoral and health professional students and individuals in postdoctoral training in minority schools need further opportunities to develop biomedical and behavioral research skills. The NHLBI Minority Institutional Research Training Program is designed to attract students in their developmental stages; to increase their awareness of cardiovascular, pulmonary, hematologic, and sleep disorders research; and to encourage them to pursue research career opportunities in these areas. Trainee appointments will be determined by the institution and carried out in a manner that will include the recruitment of racial and ethnic minorities, individuals from disadvantaged backgrounds, individuals with disabilities, and women into heart, lung, blood, and sleep disorders scientific research areas. Letters of Intent are due on July 15, with full proposals due by August 15.
Requirements: The institution must be a domestic medical or non-medical college, university, or equivalent school in which students from underrepresented minority groups including African Americans, Hispanics, American Indians, Alaska Natives, and Pacific Islanders compose a majority or a significant proportion of the school enrollment. The individual to be trained must be a citizen of the United States, a noncitizen national, or have been lawfully admitted to the United States for permanent residence at the time of appointment for training, and have a baccalaureate degree. Trainees must be training at the post-baccalaureate level (i.e., predoctoral level) in a relevant biomedical or behavioral science and have made a strong commitment to completing a doctoral degree, be enrolled in a minority health professional school, or have a doctoral degree or equivalent in a biomedical or behavioral science. The program may not support studies leading to a health professional degree.
Date(s) Application is Due: Jul 15; Aug 15
Amount of Grant: Up to 250,000 U.S.D
Contact: Sandra Colombini Hatch, (301) 435-0222; Hatchs@nhlbi.nih.gov
Internet: http://grants.nih.gov/grants/guide/rfa-files/RFA-HL-08-017.html
Sponsor: National Heart, Lung, and Blood Institute
6701 Rockledge Drive RKL2/600 MSC 7902
Bethesda, MD 20892-7902

NHLBI Minority Institution Research Scientist Development Award **3185**
This program provides research support to faculty members at minority institutions who have the interest and potential to conduct high quality research in the areas of cardiovascular, pulmonary, hematologic, or sleep disorders. Important goals are to enhance the institution's science programs and to assist in the acquisition of hands-on research opportunities for minority students at the applicant's institution. Annual deadline dates may vary; contact the program office for exact dates.
Requirements: The institution must be a domestic medical or non-medical college, university, or equivalent school in which students of underrepresented minority ethnic groups including African Americans, Hispanics, American Indians, Alaska Natives, and Pacific Islanders compose a majority (more than 50 percent) of the school's enrollment. It must have ongoing staff and facilities required for the proposed program.
Date(s) Application is Due: Jul 21; Sep 12
Amount of Grant: Up to 400,000 U.S.D
Contact: Joyce Creamer, Division of Blood Diseases and Resources; (301) 435-0064; fax (301) 480-0867; CreamerJ@gwgate.nhlbi.nih.gov
Internet: http://www.grants.nih.gov/grants/guide/rfa-files/RFA-HL-00-017.html
Sponsor: National Heart, Lung, and Blood Institute
6701 Rockledge Drive RKL2/600 MSC 7902
Bethesda, MD 20892-7902

NHLBI Multidisciplinary Translational Research in Critical Care Grants **3186**
The purpose of this funding opportunity is to promote multidisciplinary translational research that will improve the treatment, diagnosis, and outcome of patients with critical illness and increase the understanding of fundamental processes causing critical illness in humans of all ages. The research team must include scientists from at least two medical or surgical specialties with direct relevance to critical care. Individuals with the skills, knowledge, and resources necessary to carry out the proposed research are invited to work with their institution/organization to develop an application for support.
Requirements: The following organizations and institutions are eligible to apply: individuals; public/state controlled institutions of higher education; private institutions of higher education; Hispanic-serving institutions; historically Black colleges and universities (HBCUs); tribally controlled colleges and universities (TCCUs); Alaska Native and Native Hawaiian serving institutions; nonprofits with 501(c)3 IRS status; nonprofits without 501(c)3 IRS status; for-profit organizations; state governments; Indian/Native American tribal governments (federally recognized); Indian/Native American tribally designated organizations; county governments; city or township governments; special district governments; independent school districts; public housing authorities/Indian housing authorities; U.S. territory or possession; Indian/Native American tribal governments (other than federally recognized); regional organizations; non-domestic (non-U.S.) entities (foreign organizations); other eligible agencies of the federal government; and faith-based or community-based organizations.
Date(s) Application is Due: Feb 5; Jun 5; Oct 5
Contact: Basil Eldadah, M.D., Ph.D; (301) 496-4996; eldadahb@mail.nih.gov
Internet: http://grants.nih.gov/grants/guide/pa-files/PA-07-233.html
Sponsor: National Heart, Lung, and Blood Institute
6701 Rockledge Drive RKL2/600 MSC 7902
Bethesda, MD 20892-7902

NHLBI National Research Service Award Programs in Cardiovascular Epidemiology and Biostatistics **3187**
This program enables research institutions to support pre- and post-doctoral training in areas of particular interest to the Institute and to prepare individuals for careers in biomedical and behavioral research. The cardiovascular epidemiology and biostatistics research training programs offer opportunities in the fields of epidemiology of cardiovascular diseases and statistical analysis. Trainees are selected through review procedures established by the program director at the grantee institution. Attention is given to recruiting individuals from minority groups that are under-represented nationally in the biomedical sciences (African Americans, Hispanics, Native Americans, Alaskan Natives, and Pacific Islanders).
Requirements: Students must be U.S. citizens, noncitizen nationals, or legal permanent U.S. residents. Trainees must have received a baccalaureate degree and must be training at the postbaccalauareate level in a program leading to a PhD, ScD, or equivalent degree.
Restrictions: These awards do not support study leading to the MD, DDS, or other similar professional degrees, nor do they support residency training. Persons on temporary or student visas are not eligible.
Date(s) Application is Due: Apr 5; Aug 5; Dec 5
Contact: Lorraine M. Silsbee, (301) 435-0709; Lorraine_Silsbee@nih.gov
Internet: http://www.nhlbi.nih.gov/funding/training/epi-bio/nrsa.htm
Sponsor: National Heart, Lung, and Blood Institute
6701 Rockledge Drive RKL2/600 MSC 7902
Bethesda, MD 20892-7902

NHLBI National Research Service Awards for Senior Fellowships **3188**
This program enables experienced scientists to change the direction of their research careers, broaden their scientific background, acquire new research capabilities, or enlarge their command of an allied research field. Selection is by national competition. All experienced scientists interested in changing the direction of their careers are eligible to apply.
Requirements: When the fellowship begins, an applicant must have received the Ph.D., M.D., D.O., D.D.S., D.V.M., O.D., D.P.M., Sc.D., D.Eng., DNSc., or equivalent degree. The applicant also must have at least 7 subsequent years of relevant postdoctoral research or professional training, and must have a sponsor affiliated with a public or private institution with appropriate facilities for the proposed training.
Restrictions: One month of payback for each month of training, up to a maximum of 12 months, is required. This requirement can be fulfilled by teaching or research (a minimum of 20 hours per week) on a continuous basis beginning within 2 years after support ends.
Date(s) Application is Due: Apr 8; Aug 0; Dec 8
Contact: Charlotte A. Pratt, Program Director; (301) 435-0382; fax (301) 480-5158; prattc@nhlbi.nih.gov
Internet: http://www.nhlbi.nih.gov/funding/training/redbook/estf33.htm
Sponsor: National Heart, Lung, and Blood Institute
6701 Rockledge Drive RKL2/600 MSC 7902
Bethesda, MD 20892-7902

NHLBI New Approaches to Arrhythmia Detection and Treatment Grants for SBIR **3189**
The purpose of this initiative is to improve our ability to detect, treat and prevent cardiac arrhythmias using new or improved methods, tools, and technologies. Applications are invited for the development and significant improvements of innovative monitoring tools, diagnostic methods, and therapeutic approaches for arrhythmia detection, treatment and prevention.
Requirements: Only United States small business concerns (SBCs) are eligible to submit SBIR applications.
Date(s) Application is Due: Apr 5; Aug 5; Dec 5
Contact: David A. Lathrop, PhD; (301) 435-0507; fax (301) 451-5458; LathropD@NHLBI.NIH.GOV
Internet: http://grants.nih.gov/grants/guide/pa-files/PA-07-032.html
Sponsor: National Heart, Lung, and Blood Institute
6701 Rockledge Drive RKL2/600 MSC 7902
Bethesda, MD 20892-7902

NHLBI New Approaches to Arrhythmia Detection and Treatment Grants for STTR **3190**
The purpose of this initiative is to improve our ability to detect, treat and prevent cardiac arrhythmias using new or improved methods, tools, and technologies. Applications are invited for the development and significant improvements of innovative monitoring tools, diagnostic methods, and therapeutic approaches for arrhythmia detection, treatment and prevention.
Requirements: Only United States small business concerns (SBCs) are eligible to submit STTR applications.
Date(s) Application is Due: Apr 5; Aug 5; Dec 5
Contact: David A. Lathrop, PhD; (301) 435-0507; fax (301) 451-5458; LathropD@NHLBI.NIH.GOV
Internet: http://grants.nih.gov/grants/guide/pa-files/PA-07-031.html
Sponsor: National Heart, Lung, and Blood Institute
6701 Rockledge Drive RKL2/600 MSC 7902
Bethesda, MD 20892-7902

NHLBI Nutrition and Diet in the Causation, Prevention, and Management of Heart Failure Research Grants **3191**
The purpose of this Funding Opportunity Announcement (FOA) is to encourage submission of investigator-initiated research applications on the role of nutrition and diet in the causation, prevention, and treatment of cardiomyopathies and heart failure. Basic,

translational, and applied interdisciplinary research applications with rigorous hypothesis-testing designs for projects in animals or humans are of interest. The overall goal is to develop a satisfactory science base for preventive approaches in high-risk individuals and for rational nutritional management of patients in various stages of heart failure.
Requirements: The following groups are eligible to apply: for-profit organizations on-profit organizations; public or private institutions, such as universities, colleges, hospitals, and laboratories; units of State government; units of local government; eligible agencies of the Federal government; foreign institutions; domestic institutions; faith-based or community-based organizations; units of State tribal government; and units of local Tribal government.
Date(s) Application is Due: Feb 16; Jun 16; Oct 16
Amount of Grant: Up to 200,000 U.S.D
Contact: Abby G. Ershow, (301) 435-0550; fax (301) 480-1338; ErshowA@mail.nih.gov
Internet: http://grants.nih.gov/grants/guide/pa-files/PA-06-136.html
Sponsor: National Heart, Lung, and Blood Institute
6701 Rockledge Drive RKL2/600 MSC 7902
Bethesda, MD 20892-7902

NHLBI Pathogenesis And Treatment Of Lymphedema And Lymphatic Diseases Grants 3192
The purpose of this Funding Opportunity Announcement (FOA) is to stimulate research on the lymphatic system, to characterize its function on the molecular, cellular, tissue, organ, and intact organism levels and pathophysiologic mechanisms that cause disease, to develop new methods for imaging and/or quantitating lymph flow, and to discover new therapeutic interventions, including nursing, complementary and alternative treatments. The scope of this research includes developmental biology and genetics of the lymphatic system to identify and characterize processes important for its organization and regulation. Such knowledge will help to improve early diagnosis of affected individuals, the choice and timing of treatment, and genetic counseling. Research is also needed on disorders of skin and subcutaneous tissue due to chronic lymphedema, and lymphedema which results from cancers or cancer treatment, with a goal to develop better therapies.
Requirements: The following organizations and institutions are eligible to apply: public/state controlled institutions of higher education; private institutions of higher education; Hispanic-serving institutions; historically Black colleges and universities (HBCUs); tribally controlled colleges and universities (TCCUs); Alaska Native and Native Hawaiian serving institutions; nonprofits with 501(c)3 IRS status; nonprofits without 501(c)3 IRS status; for-profit organizations; state governments; Indian/Native American tribal governments (federally recognized); Indian/Native American tribally designated organizations; non-domestic (non-U.S.) entities (foreign organizations); Hispanic-serving institutions; and regional organization.
Date(s) Application is Due: Feb 5; Jun 5; Oct 5
Contact: Stephen S. Goldman, (301) 435-0565; fax (301) 480-2849; goldmans@nih.gov
Internet: http://grants.nih.gov/grants/guide/pa-files/PA-07-165.html
Sponsor: National Heart, Lung, and Blood Institute
6701 Rockledge Drive RKL2/600 MSC 7902
Bethesda, MD 20892-7902

NHLBI Pathway to Independence Award 3193
The Pathway to Independence (PI) Award is designed to facilitate a timely transition from a mentored postdoctoral research position to a stable independent research position with independent NIH or other independent research support at an earlier stage than is currently the norm. The PI award will provide up to 5 years of support consisting of two phases. The initial phase will provide 1-2 years of mentored support for postdoctoral research scientists. This phase will be followed by up to 3 years of independent support contingent on securing an independent tenure-track or equivalent research position. A typical successful candidate will have an impressive record of research productivity and publications, and will have identified a unique research direction to pursue after the mentored phase of the award. Domestic for-profit or non-profit institutions and organizations, or public or private institutions, such as universities, colleges, hospitals and laboratories, eligible agencies of the federal government, and NIH intramural laboratories are eligible (all institution and organization types listed above are eligible for both the mentored and independent phase, with one exception: ineligible agencies of the Federal government, such as the NIH intramural program, are eligible only for the mentored phase). For the Mentored Phase, salary up to $75,000 plus fringe benefits per year is available, with up to $25,000 additional per year for support (F&A costs will be provided at 8% of modified direct costs). During the Independent investigator phase, total cost may not exceed $249,000 per year. New applications are due February 12, June 12, and October 12, while resubmitted applications are due March 12, July 12, and November 12.
Requirements: At the time of application submission (or resubmission) candidates for this award must have earned a terminal clinical or research doctorate (including Ph.D., M.D., D. O., D.C., N.D., D.D.S., D.V.M., Sc.D., D.N.S., Pharm. D., or equivalent doctoral degree, or a combined clinical and research doctoral degree); and have no more than 5 years (60 months) of research experience since completing the requirements of the doctoral degree (resubmissions must also comply with this requirement).
Restrictions: Foreign institutions are not eligible.
Date(s) Application is Due: Feb 12; Mar 12; Jun 12; Jul 12; Oct 12; Nov 12
Amount of Grant: 100,000 - 249,000 U.S.D
Contact: Herbert M. Geller, (301) 451-9440; fax (301) 594-8133; direducation@nhlbi.nih.gov
Internet: http://www.nhlbi.nih.gov/funding/training/redbook/phdk99r00.htm
Sponsor: National Heart, Lung, and Blood Institute
10 Center Drive, Building 10, Room 2N242, MSC 1754
Bethesda, MD 20892-1754

NHLBI Pediatric Cardiac Genomics Consortium Grants 3194
The purpose of this initiative is to support collaborative genetic and genomic research leading to a comprehensive understanding of the causes and outcomes of congenital heart disease in the human population. The Consortium will consist of up to 6 Research Centers, Cores, and a Steering Committee. Research Center applicants will propose detailed investigation of the genetics and genomics of one or more human cardiac malformations, with the goal of identifying genetic influences on outcome, as well as causative genes. Individual Research Centers will collaborate in recruitment of individuals with congenital heart disease and their families for all Research Center projects. Maximum allowable direct costs for the Research Centers per year are $160,000 in Year 1, and $535,000 per year in years 2-6. Letters of Intent should be received by January 6, and full proposals are due by February 6.
Requirements: The following organizations and institutions are eligible to apply: public/state controlled institutions of higher education; private institutions of higher education; Hispanic-serving institutions; historically Black colleges and universities (HBCUs); tribally controlled colleges and universities (TCCUs); Alaska Native and Native Hawaiian serving institutions; nonprofits with 501(c)3 IRS status; nonprofits without 501(c)3 IRS status; for-profit organizations; small businesses; state governments; non-domestic (non-U.S.) entities in North America; and other eligible agencies of the Federal Government.
Date(s) Application is Due: Jan 6; Feb 6
Amount of Grant: Up to 2,835,000 U.S.D
Contact: Gail D. Pearson, (301) 435-0510; pearsong@mail.nih.gov
Internet: http://grants.nih.gov/grants/guide/rfa-files/RFA-HL-09-003.html
Sponsor: National Heart, Lung, and Blood Institute
6701 Rockledge Drive RKL2/600 MSC 7902
Bethesda, MD 20892-7902

NHLBI Phase II Clinical Trials of Novel Therapies for Lung Diseases 3195
The purpose of this FOA is to solicit research grant applications to conduct Phase II clinical therapeutic trials that have the potential to advance development of novel therapies for a lung disease or a cardiopulmonary disorder of sleep. Each application will propose one Phase II interventional trial that will most likely use physiological or biochemical rather than clinical endpoints along with at least one smaller basic ancillary research study that is tightly related to the clinical question. Although definitive Phase III trials will not be supported, the proposed studies must provide proof of concept for a novel intervention that has high potential for modifying current treatments and could be disease modifying.
Requirements: The following organizations and institutions are eligible to apply: public/state controlled institutions of higher education; private institutions of higher education; Hispanic-serving institutions; historically Black colleges and universities (HBCUs); tribally controlled colleges and universities (TCCUs); Alaska Native and Native Hawaiian serving institutions; nonprofits with 501(c)3 IRS status; nonprofits without 501(c)3 IRS status; for-profit organizations; state governments; Indian/Native American tribal governments (federally recognized); Indian/Native American tribally designated organizations; county governments; city or township governments; special district governments; independent school districts; public housing authorities/Indian housing authorities; U.S. territories or possessions; Indian/Native American tribal governments (other than federally recognized); regional organizations; non-domestic (non-U.S.) entities (foreign organizations); other eligible agencies of the federal government; and faith-based or community-based organizations.
Date(s) Application is Due: Jan 29; May 18; Jun 18; Aug 15; Sep 15; Dec 30
Amount of Grant: Up to 1,515,000 U.S.D
Contact: Andrea L. Harabin, Ph.D.; (301) 435-0222; harabina@nhlbi.nih.gov
Internet: http://grants.nih.gov/grants/guide/rfa-files/RFA-HL-10-003.html
Sponsor: National Heart, Lung, and Blood Institute
6701 Rockledge Drive RKL2/600 MSC 7902
Bethesda, MD 20892-7902

NHLBI Prematurity and Respiratory Outcomes Program (PROP) Grants 3196
This RFA intends to fill the need for understanding the underlying pathophysiology by supporting the investigation of the molecular mechanisms that contribute to acute and chronic respiratory morbidity of the premature infant. The well phenotyped populations developed in this program will significantly increase the chance of studying groups with common pathophysiology and lay the groundwork for novel biomarker discovery. The program will use the cooperative agreement mechanism and require multidisciplinary expertise including neonatologists/pulmonologists, and physiological, molecular, biological, or bioinformatics scientists. Participating sites will plan and coordinate their own research and will develop and implement shared protocols for respiratory phenotyping and respiratory outcomes of the premature infant NICU graduate. Letters of Intent should be received by May 4, and full proposals are due by June 2 each year.
Requirements: The following organizations and institutions are eligible to apply: public/state controlled institutions of higher education; private institutions of higher education; Hispanic-serving institutions; historically Black colleges and universities (HBCUs); tribally controlled colleges and universities (TCCUs); Alaska Native and Native Hawaiian serving institutions; nonprofits with 501(c)3 IRS status; nonprofits without 501(c)3 IRS status; for-profit organizations; state governments; Indian/Native American tribal governments (federally recognized); Indian/Native American tribally designated organizations; county governments; city or township governments; special district governments; independent school districts; public housing authorities/Indian housing authorities; U.S. territories or possessions; Indian/Native American tribal governments (other than federally recognized); regional organizations; non-domestic (non-U.S.) entities (foreign organizations); other eligible agencies of the federal government; and faith-based or community-based organizations.

Date(s) Application is Due: May 4; Jun 2
Amount of Grant: 150,000 - 350,000 U.S.D
Contact: Carol J. Blaisdell, (301) 435-0222; blaisdellcj@nhlbi.nih.gov
Internet: http://grants.nih.gov/grants/guide/rfa-files/RFA-HL-10-007.html
Sponsor: National Heart, Lung, and Blood Institute
6701 Rockledge Drive RKL2/600 MSC 7902
Bethesda, MD 20892-7902

NHLBI Progenitor Cell Biology Consortium Administrative Coordinating Center Grants 3197

The goal of the Consortium is to identify and characterize progenitor cell lineages, to direct the differentiation of stem and progenitor cells to desired cell fates, and to develop new strategies to address the unique challenges presented by the transplantation of these cells. The Consortium will assemble multiple independent research projects, each with a multi-disciplinary team of Principal Investigators, core research support facilities, and skills development components to establish synergistic virtual hubs focused on progenitor cell biology. The total project period for an application submitted in response to this FOA is 7 years. Maximum allowable direct costs are $600,000 per year for ACC operations. Letters of Intent are dur by April 1, while full applications should arrive by May 1.
Requirements: The following organizations and institutions are eligible to apply: public/state controlled institutions of higher education; private institutions of higher education; Hispanic-serving institutions; historically Black colleges and universities (HBCUs); tribally controlled colleges and universities (TCCUs); Alaska Native and Native Hawaiian serving institutions; nonprofits with 501(c)3 IRS status; nonprofits without 501(c)3 IRS status; small businesses; for-profit organizations; and state governments.
Date(s) Application is Due: Apr 1; May 1
Amount of Grant: Up to 4,200,000 U.S.D
Contact: Denis Buxton, (301) 435-0513; Buxtond@nhlbi.nih.gov
Internet: http://grants.nih.gov/grants/guide/rfa-files/RFA-HL-09-005.html
Sponsor: National Heart, Lung, and Blood Institute
6701 Rockledge Drive RKL2/600 MSC 7902
Bethesda, MD 20892-7902

NHLBI Progenitor Cell Biology Consortium Research Hubs 3198

The National Heart, Lung, and Blood Institute (NHLBI) invites applications to participate in the NHLBI Progenitor Cell Biology Consortium. The goal of the Consortium is to identify and characterize progenitor cell lineages, to direct the differentiation of stem and progenitor cells to desired cell fates, and to develop new strategies to address the unique challenges presented by the transplantation of these cells. The Consortium will assemble clusters of synergistic research projects, each with a multi-disciplinary team of Principal Investigators, to establish virtual Research Hubs focused on progenitor cell biology. The Consortium will also contain core research support facilities and skills development components. The expected direct cost amount for the Consortium awards will be up to $750,000 per year. Letters of Intent are due by April 1, with full proposals received by May 1.
Requirements: The following organizations and institutions are eligible to apply: public/state controlled institutions of higher education; private institutions of higher education; Hispanic-serving institutions; historically Black colleges and universities (HBCUs); tribally controlled colleges and universities (TCCUs); Alaska Native and Native Hawaiian serving institutions; nonprofits with 501(c)3 IRS status; nonprofits without 501(c)3 IRS status; small businesses; for-profit organizations; and state governments.
Date(s) Application is Due: Apr 1; May 1
Amount of Grant: 750,000 U.S.D
Contact: Denis Buxton, (301) 435-0513; Buxtond@nhlbi.nih.gov
Internet: http://grants.nih.gov/grants/guide/rfa-files/RFA-HL-09-004.html
Sponsor: National Heart, Lung, and Blood Institute
6701 Rockledge Drive RKL2/600 MSC 7902
Bethesda, MD 20892-7902

NHLBI Protein Interactions Governing Membrane Transport in Pulmonary Health and Disease Grants 3199

This Funding Opportunity Announcement (FOA) issued by the National Heart, Lung, and Blood Institute, National Institutes of Health (NIH), solicits research grant applications to delineate the protein interactions and pathways governing membrane trafficking pathways operative in pulmonary health and disease and develop novel therapeutic interventions. Because the nature and scope of the proposed research will vary from application to application, it is anticipated that the size and duration of each award will also vary. The total amount awarded and the number of awards will depend upon the mechanism numbers, quality, duration, and costs of the applications received.
Requirements: The following organizations and institutions are eligible to apply: public/state controlled institutions of higher education; private institutions of higher education; Hispanic-serving institutions; historically Black colleges and universities (HBCUs); tribally controlled colleges and universities (TCCUs); Alaska Native and Native Hawaiian serving institutions; nonprofits with 501(c)3 IRS status; nonprofits without 501(c)3 IRS status; for-profit organizations; state governments; Indian/Native American tribal governments (federally recognized); Indian/Native American tribally designated organizations; county governments; city or township governments; special district governments; independent school districts; public housing authorities/Indian housing authorities; U.S. territory or possession; Indian/Native American tribal governments (other than federally recognized); regional organizations; non-domestic (non-U.S.) entities (foreign organizations); other eligible agencies of the federal government; and faith-based or community-based organizations.
Date(s) Application is Due: Feb 5; Jun 5; Oct 5
Contact: Susan Banks-Schlegel, (301) 435-0202; Schleges@nhlbi.nih.gov
Internet: http://grants.nih.gov/grants/guide/pa-files/PA-07-137.html
Sponsor: National Heart, Lung, and Blood Institute
6701 Rockledge Drive RKL2/600 MSC 7902
Bethesda, MD 20892-7902

NHLBI Research Demonstration and Dissemination Grants 3200

The purpose of this Funding Opportunity Announcement (FOA) is to encourage the scientific community to conduct Demonstration and Dissemination (D&D) studies to test the effectiveness of interventions in children, adolescents, and/or adults to: promote healthful behaviors; reduce risk factors for heart, lung, and blood diseases, and sleep disorders; improve the prevention or management of heart, lung, and blood diseases, and sleep disorders, including the delivery of health care services; and enhance understanding of the processes of intervention implementation and diffusion, or sustainability in a defined population, or defined clinical or community setting. The total project period for an application submitted in response to this funding opportunity may not exceed 5 years. Direct costs are limited to less than $500,000 in any single year, unless pre-approval is obtained from the Institute.
Requirements: The following organizations and institutions are eligible to apply: public/state controlled institutions of higher education; private institutions of higher education; Hispanic-serving institutions; historically Black colleges and universities (HBCUs); tribally controlled colleges and universities (TCCUs); Alaska Native and Native Hawaiian serving institutions; nonprofits with 501(c)3 IRS status; nonprofits without 501(c)3 IRS status; for-profit organizations; state governments; Indian/Native American tribal governments (federally recognized); Indian/Native American tribally designated organizations; county governments; city or township governments; special district governments; independent school districts; public housing authorities/Indian housing authorities; U.S. territory or possession; Indian/Native American tribal governments (other than federally recognized); regional organizations; non-domestic (non-U.S.) entities (foreign organizations); other eligible agencies of the federal government; and faith-based or community-based organizations.
Date(s) Application is Due: Jan 25; May 25; Sep 25
Amount of Grant: Up to 500,000 U.S.D
Contact: Charlotte Pratt, (301) 435-0382; fax (301) 480-5158; PrattC@nhlbi.nih.gov
Internet: http://grants.nih.gov/grants/guide/pa-files/PA-07-017.html
Sponsor: National Heart, Lung, and Blood Institute
6701 Rockledge Drive RKL2/600 MSC 7902
Bethesda, MD 20892-7902

NHLBI Research Grants on the Relationship Between Hypertension and Inflammation 3201

This Funding Opportunity Announcement (FOA) solicits grant applications that propose the study of the sequence of events in which the vascular inflammatory state contributes to the development and maintenance of hypertension. Ample evidence suggests that inflammation may play a role in the pathogenesis of hypertension or that it may characterize a functional state of the vessel wall as a consequence of high blood pressure. Angiotensin II (Ang II), a widely recognized vasoconstrictor and anti-natriuretic involved in blood pressure regulation, also acts as a pro-inflammatory factor in the cardiovascular system. Ang II stimulates the expression of several inflammatory cytokines, which in turn affect blood pressure. A potential linkage between Ang II and immuno-cytokines is their shared ability to induce the level of reactive oxygen species (ROS), which serve as second messengers for many intracellular signaling pathways. The production of ROS not only decreases bioavailability of nitric oxide (NO), a vasodilator, but also initiates the functional and morphological alterations, such as remodeling, in the vascular wall that accompany the hypertensive state over time. This FOA would provide an opportunity to bring focus on the potential causal relationship between hypertension and inflammation in a cohesive, integrated manner. A new understanding of hypertension and inflammation would provide novel opportunities to prevent and treat the disease.
Requirements: The following organizations and institutions are eligible to apply: public/state controlled institutions of higher education; private institutions of higher education; Hispanic-serving institutions; historically Black colleges and universities (HBCUs); tribally controlled colleges and universities (TCCUs); Alaska Native and Native Hawaiian serving institutions; nonprofits with 501(c)3 IRS status; nonprofits without 501(c)3 IRS status; for-profit organizations; state governments; Indian/Native American tribal governments (federally recognized); Indian/Native American tribally designated organizations; county governments; city or township governments; special district governments; independent school districts; public housing authorities/Indian housing authorities; U.S. territory or possession; Indian/Native American tribal governments (other than federally recognized); regional organizations; non-domestic (non-U.S.) entities (foreign organizations); other eligible agencies of the federal government; and faith-based or community-based organizations.
Date(s) Application is Due: Feb 5; Jun 5; Oct 5
Contact: Eser Tolunay, (301) 435-0560; tolunaye@nhlbi.nih.gov
Internet: http://grants.nih.gov/grants/guide/pa-files/PA-07-138.html
Sponsor: National Heart, Lung, and Blood Institute
6701 Rockledge Drive RKL2/600 MSC 7902
Bethesda, MD 20892-7902

NHLBI Research on the Role of Cardiomyocyte Mitochondria in Heart Disease: An Integrated Approach 3202

The National Heart, Lung, and Blood Institute (NHLBI) invites applications for collaborative research projects to develop an integrated understanding of cardiomyocyte mitochondria and its contributions to myocardial adaptations and heart disease progression by combining functional data with information derived from powerful new technologies.

Budgets for direct costs of up to $500,000 per year and a project duration of up to four years may be requested for a maximum of $2,000,000 direct costs over a four-year project period. Because the nature and scope of the proposed research will vary from application to application, it is anticipated that the size and duration of each award will also vary.
Requirements: The following organizations and institutions are eligible to apply: individuals; public/state controlled institutions of higher education; private institutions of higher education; Hispanic-serving institutions; historically Black colleges and universities (HBCUs); tribally controlled colleges and universities (TCCUs); Alaska Native and Native Hawaiian serving institutions; nonprofits with 501(c)3 IRS status; nonprofits without 501(c)3 IRS status; for-profit organizations; state governments; Indian/Native American tribal governments (federally recognized); Indian/Native American tribally designated organizations; county governments; city or township governments; special district governments; independent school districts; public housing authorities/Indian housing authorities; U.S. territory or possession; Indian/Native American tribal governments (other than federally recognized); regional organizations; non-domestic (non-U.S.) entities (foreign organizations); other eligible agencies of the federal government; and faith-based or community-based organizations.
Date(s) Application is Due: Apr 20; May 19
Amount of Grant: 500,000 - 2,000,000 U.S.D
Contact: Isabella Liang, (301) 435-0504; fax (301) 451-5458; Liangi@mail.nih.gov
Internet: http://grants.nih.gov/grants/guide/rfa-files/RFA-HL-10-002.html
Sponsor: National Heart, Lung, and Blood Institute
6701 Rockledge Drive RKL2/600 MSC 7902
Bethesda, MD 20892-7902

NHLBI Research Supplements to Promote Diversity in Health-Related Research (Postdoctoral Individuals/New Independent Researchers) 3203

Supplement enables principal investigators with eligible NHLBI research grants to include recently appointed junior faculty from underrepresented racial and ethnic groups and individuals with disabilities in their projects. Nationally, underrepresented groups in biomedical research careers include but are not limited to, African Americans, Hispanic Americans, American Indians/Alaska Natives, Native Hawaiians and Pacific Islanders. The research proposed for the investigator must be a logical extension of the parent grant. The investigator must be given opportunities to interact with other individuals on the grant, to contribute intellectually to the study, and to enhance his or her research skills and knowledge. Funds may be requested to make changes or adjustments in the laboratory setting that will enable the investigator with disabilities to perform the essential functions associated with his or her research project. Accommodation requests must be reasonable in relationship to the direct costs of the parent grant. There is no application receipt date. However, awards are generally made four times per year (December, March, June, and September). Principal Investigators are encouraged to synchronize the supplement requested start date with the parent grant budget period start date. A maximum of $75,000 per year in salary is available (plus fringe benefits), and a maximum of $10,000 per year for supplies and travel.
Requirements: All principal investigators at U.S. institutions with eligible NHLBI research grants are eligible to apply.
Amount of Grant: Up to 85,000 U.S.D
Contact: Helena O. Mishoe, Deputy Director, Office of Minority Health Affairs; (301) 451-5081; fax (301) 480-0862; mishoeh@nhlbi.nih.gov
Internet: http://www.nhlbi.nih.gov/funding/training/redbook/newmirs.htm
Sponsor: National Heart, Lung, and Blood Institute
6701 Rockledge Drive RKL2/600 MSC 7902
Bethesda, MD 20892-7902

NHLBI Research Supplements to Promote Diversity in Health-Related Research for Established Investigators Who Become Disabled 3204

This supplement enables principal investigators with eligible NHLBI research grants who become disabled to obtain additional support to continue their research. Funds may be requested to make changes or adjustments in the laboratory setting that will enable the disabled investigator to perform the essential functions associated with his or her research project and/or to hire an assistant. Accommodation requests must be reasonable in relationship to the direct costs of the parent grant. Applications can be submitted at any time, but no later than 3 months before the last 2 years remaining on the NHLBI research grant. Applicants may request support for special equipment, an assistant, or other modifications to facilitate reasonable accommodation to a disabling injury or illness that has occurred during the current project period.
Requirements: All principal investigators at U.S. institutions with eligible NHLBI research grants who become disabled are eligible to apply.
Contact: Helena O. Mishoe, Deputy Director, Office of Minority Health Affairs; (301) 451-5081; fax (301) 480-0862; mishoeh@nhlbi.nih.gov
Internet: http://www.nhlbi.nih.gov/funding/training/redbook/estdissu.htm
Sponsor: National Heart, Lung, and Blood Institute
6701 Rockledge Drive RKL2/600 MSC 7902
Bethesda, MD 20892-7902

NHLBI Research Supplements to Promote Diversity in Health-Related Research for Graduate Research Assistants 3205

This supplement enables principal investigators with eligible NHLBI research grants to include graduate students in their projects. Priority is given to applications requesting support for students with disabilities and/or from underrepresented racial and ethnic groups. The research activity proposed for the student must be part of the approved research for the parent grant or a logical extension. The student must be given opportunities to interact with other individuals on the grant and to contribute intellectually to the study. Applications can be submitted at any time, but no later than 3 months before the last 2 years remaining on the NHLBI research grant. Stipends include: salary (including fringe benefits) up to a maximum of $26,000 per year; and a maximum of $4,000 a year for supplies, travel (up to $1,000 for travel to scientific meetings), and other.
Requirements: All principal investigators at U.S. institutions with eligible NHLBI research grants who wish to offer research experiences to students with disabilities and/or students from underrepresented racial and ethnic groups who are pursuing a masters or doctoral degree in one of the biomedical or behavioral sciences or enrolled in a master's degree program in nursing are eligible to apply.
Amount of Grant: Up to 30,000 U.S.D
Contact: Helena O. Mishoe, Deputy Director, Office of Minority Health Affairs; (301) 451-5081; fax (301) 480-0862; mishoeh@nhlbi.nih.gov
Internet: http://www.nhlbi.nih.gov/funding/training/redbook/gradsup.htm
Sponsor: National Heart, Lung, and Blood Institute
6701 Rockledge Drive RKL2/600 MSC 7902
Bethesda, MD 20892-7902

NHLBI Research Supplements to Promote Diversity in Health-Related Research for Individuals in Postdoctoral Training 3206

This supplement enables principal investigators with eligible NHLBI research grants to include postdoctoral individuals from underrepresented racial and ethnic groups and individuals with disabilities in their projects. Nationally, underrepresented groups in biomedical research careers include but are not limited to African Americans, Hispanic Americans, American Indians/Alaska Natives, Native Hawaiians and Pacific Islanders. The research proposed for the individual must be a logical extension of the parent grant. The individual must be given opportunities to interact with other researchers on the grant, to contribute intellectually to the study, and to enhance his or her research skills and knowledge. Funds may be requested to make changes or adjustments in the laboratory setting that will enable the investigator with disabilities to perform the essential functions associated with his or her research project. Accommodation requests must be reasonable in relationship to the direct costs of the parent grant. There is no application receipt date. However, awards are generally made four times per year (December, March, June, and September). Principal Investigators are encouraged to synchronize the supplement requested start date with the parent grant budget period start date. Up to $40,000 per year is provided for salary and fringe benefits, plus up to $10,000 per year for supplies, travel, and other.
Requirements: The individual may be affiliated with the principal investigator's institution or another institution. Supplements are available for long-term research support for individuals with a postdoctoral degree. A minimum of two years and a maximum of four years full-time research effort (at least 40 hours a week) is required.
Restrictions: Normally, funding will not be provided for postdoctoral individuals to work with their former predoctoral mentors. If the postdoctoral mentor is the individual's predoctoral mentor, a justification for continued mentorship must be provided. Cumulative Postdoctoral research experience must not exceed 6 years.
Amount of Grant: Up to 50,000 U.S.D
Contact: Helena O. Mishoe, Deputy Director, Office of Minority Health Affairs; (301) 451-5081; fax (301) 480-0862; mishoeh@nhlbi.nih.gov
Internet: http://www.nhlbi.nih.gov/funding/training/redbook/phdressu.htm
Sponsor: National Heart, Lung, and Blood Institute
6701 Rockledge Drive RKL2/600 MSC 7902
Bethesda, MD 20892-7902

NHLBI Research Supplements to Promote Diversity in Health-Related Research for Undergraduate Students 3207

This supplement enables principal investigators with eligible NHLBI research grants to include undergraduate students in their projects. Priority is given to applications requesting support for students with disabilities and/or from underrepresented racial and ethnic groups. The research proposed for the student must be part of the approved research for the parent grant or a logical extension. The student must be given opportunities to interact with other individuals on the study, to contribute intellectually, and to enhance his or her research skills and knowledge. The program is designed to support undergraduate students who are interested in biomedical or behavioral research and who will devote at least 2 years (two 3-month periods) to the program. Support should be requested for at least 3 months during any 1 year. Applications can be submitted any time, but no later than 3 months before the last 2 years remaining on the NHLBI research grant. Stipend include: $10 per hour for salary; and supplies, travel (up to $1,000 for travel to scientific meeting(s) or between home/school and worksite), and other up to $2,400.
Requirements: All principal investigators at U.S. institutions with eligible NHLBI research grants are eligible.
Contact: Helena O. Mishoe, Deputy Director, Office of Minority Health Affairs; (301) 451-5081; fax (301) 480-0862; mishoeh@nhlbi.nih.gov
Internet: http://www.nhlbi.nih.gov/funding/training/redbook/undersup.htm
Sponsor: National Heart, Lung, and Blood Institute
6701 Rockledge Drive RKL2/600 MSC 7902
Bethesda, MD 20892-7902

NHLBI Right Heart Function in Health and Chronic Lung Diseases Grants 3208

This program will promote research on cellular, molecular and physiological determinants of right ventricular function in health, and function/ dysfunction in association with chronic lung diseases. The goal is to encourage research which concentrates on building a foundation of basic knowledge of right ventricular physiology and the basic molecular and biomechanical factors contributing to right ventricular dysfunction associated with pulmonary diseases. This knowledge base will be used in determining the best methods

of preventing, diagnosing and treating right heart failure. This FOA seeks to fill the current gaps that exist in our understanding of the right heart and how right heart function affects lung function in both health and disease. The emphasis on interaction of the heart and the lung during chronic lung disease in this program is intended to promote collaboration between the heart and lung research communities.
Requirements: The following organizations and institutions are eligible to apply: public/state controlled institutions of higher education; private institutions of higher education; Hispanic-serving institutions; historically Black colleges and universities (HBCUs); tribally controlled colleges and universities (TCCUs); Alaska Native and Native Hawaiian serving institutions; nonprofits with 501(c)3 IRS status; nonprofits without 501(c)3 IRS status; for-profit organizations; state governments; Indian/Native American tribal governments (federally recognized); Indian/Native American tribally designated organizations; county governments; city or township governments; special district governments; independent school districts; public housing authorities/Indian housing authorities; U.S. territory or possession; Indian/Native American tribal governments (other than federally recognized); regional organizations; non-domestic (non-U.S.) entities (foreign organizations); other eligible agencies of the federal government; and faith-based or community-based organizations.
Date(s) Application is Due: Feb 5; Jun 5; Oct 5
Contact: Elizabeth M. Denholm, (301) 435-0222; denholme@mail.nih.gov
Internet: http://grants.nih.gov/grants/guide/pa-files/PA-07-043.html
Sponsor: National Heart, Lung, and Blood Institute
6701 Rockledge Drive RKL2/600 MSC 7902
Bethesda, MD 20892-7902

NHLBI Ruth L. Kirschstein National Research Service Award (NRSA) Institutional Research Training Grants 3209

This award enables research institutions to support predoctoral research training in specific areas and fields of shortage. Selection of institutions is by national competition; trainees are selected through local review procedures established by the program director at the grantee institution. Institutional grants are renewable, through competition, for up to 5 years. The NHLBI will accept all types of competing T32 applications (new, renewal/competing continuation, resubmission/amended) until January 25. Only resubmission/amended T32 applications will be accepted until September 25.
Requirements: Trainees are required to pursue their research training full-time, and trainees in clinical areas are expected to confine their clinical duties to those that are part of their research training. Trainees must be U.S. citizens, noncitizen nationals, or legal permanent residents of the U.S. Persons on temporary or student visas are not eligible.
Date(s) Application is Due: Jan 25; Sep 25
Contact: Michael Commarato, (301) 435-0535; CommaraM@nhlbi.nih.gov
Internet: http://www.nhlbi.nih.gov/funding/training/redbook/gradt32.htm
Sponsor: National Heart, Lung, and Blood Institute
6701 Rockledge Drive RKL2/600 MSC 7902
Bethesda, MD 20892-7902

NHLBI Ruth L. Kirschstein National Research Service Awards for Individual Predoctoral MD/PhD Fellows 3210

The National Heart, Lung, and Blood Institute (NHLBI) provides Ruth L. Kirschstein National Research Service Awards (Kirschstein-NRSA) to individuals for combined MD/PhD fellowship training. The NHLBI awards this Kirschstein-NRSA individual fellowship to promising applicants with the potential to become productive, independent, highly trained physician-scientists, including patient-oriented physician-scientists in its scientific mission area. The Institute plans, conducts, fosters, and supports an integrated and coordinated program of basic research, clinical investigations and trials, observational studies, and demonstration and education projects. Research is related to the causes, prevention, diagnosis, and treatment of heart, blood vessel, lung, and blood diseases; and sleep disorders. The NIH will reimburse 60 percent of the cost of tuition up to $21,000.
Requirements: Applications may be submitted by: for-profit organizations; non-profit organizations; public or private institutions, such as universities, colleges, hospitals, and laboratories; eligible agencies of the Federal government; and domestic Institutions
Date(s) Application is Due: Apr 8; Aug 8; Dec 8
Contact: Jane D. Scott, (301) 435-0535; fax (301) 480-7404; scottj2@nhlbi.nih.gov
Internet: http://grants.nih.gov/grants/guide/pa-files/PA-08-021.html
Sponsor: National Heart, Lung, and Blood Institute
6701 Rockledge Drive RKL2/600 MSC 7902
Bethesda, MD 20892-7902

NHLBI Ruth L. Kirschstein National Research Service Award Short-Term Institutional Research Training Grants 3211

This program provides funds to research institutions to make awards to individuals in health professional schools for research opportunities that would not be available through their regular course of study. Awards are made to training institutions by national competition. The applicant institution may request the proportion of the NIH standard annual NRSA Training Related Expenses to help defray other costs of the short-term training experience, such as research supplies, tuition, fees, certain types of travel, and other expenses. Amounts for TRE are $4,200 annually ($350/month) for each predoctoral trainee and $7,850 annually ($654/month) for each postdoctoral trainee. Health insurance is an allowable expense that may be charged to the Trainee Related Expenses budget category but only to the extent that the same health insurance fees are charged to non-Federally students and postdoctoral individuals.
Requirements: Trainees are selected by the grantee institution and must be U.S. citizens, noncitizen nationals, or legal permanent residents of the U.S. Trainees should have successfully completed at least 1 semester at an accredited school of medicine, optometry, osteopathy, dentistry, veterinary medicine, pharmacy, or public health before entering the program. The award cannot be used to support courses that are required for the M.D., D.O., D.D.S., D.V.M., or similar professional degrees.
Date(s) Application is Due: Jan 25
Contact: Charlotte Pratt, (301) 435-0382; fax (301) 480-5158; PrattC@nhlbi.nih.gov
Internet: http://www.nhlbi.nih.gov/funding/training/redbook/gradt35.htm
Sponsor: National Heart, Lung, and Blood Institute
6701 Rockledge Drive RKL2/600 MSC 7902
Bethesda, MD 20892-7902

NHLBI Sarcoidosis: Research Grants into the Cause of Multi-Organ Disease and Clinical Strategies for Therapy 3212

The purpose of this Funding Opportunity Announcement (FOA) is to stimulate research on the etiology and management of sarcoidosis, an immune-mediated granulomatous inflammatory disorder. Studies supported by this FOA would include investigations to find the etiology(ies) and related host factors that might enhance susceptibility to sarcoidosis, especially development of symptomatic multi-organ disease that has a propensity to involve critical organs (such as lungs, heart, eyes, central/peripheral nervous system, liver, kidneys and other abdominal viscera) that create serious illness. Study of the still problematic management and the biological, behavioral, or psychosocial burden on individuals, families, and community is encouraged. Moreover, there is interest for approaches to risk reduction, psychological coping, and management of complications and side effects of treatment. Because the nature and scope of the proposed research will vary from application to application, it is anticipated that the size and duration of each award will also vary. The total amount awarded and the number of awards will depend upon the mechanism numbers, quality, duration, and costs of the applications received.
Requirements: The following organizations and institutions are eligible to apply: public/state controlled institutions of higher education; private institutions of higher education; Hispanic-serving institutions; historically Black colleges and universities (HBCUs); tribally controlled colleges and universities (TCCUs); Alaska Native and Native Hawaiian serving institutions; nonprofits with 501(c)3 IRS status; nonprofits without 501(c)3 IRS status; for-profit organizations; state governments; Indian/Native American tribal governments (federally recognized); Indian/Native American tribally designated organizations; county governments; city or township governments; special district governments; independent school districts; public housing authorities/Indian housing authorities; U.S. territory or possession; Indian/Native American tribal governments (other than federally recognized); regional organizations; non-domestic (non-U.S.) entities (foreign organizations); other eligible agencies of the federal government; and faith-based or community-based organizations.
Date(s) Application is Due: Feb 5; Jun 5; Oct 5
Contact: Herbert Y. Reynolds, (301) 435-0222; Reynoldh@nhlbi.nih.gov
Internet: http://grants.nih.gov/grants/guide/pa-files/PA-07-136.html
Sponsor: National Heart, Lung, and Blood Institute
6701 Rockledge Drive RKL2/600 MSC 7902
Bethesda, MD 20892-7902

NHLBI Short-Term Research Education Program to Increase Diversity in Health-Related Research 3213

This FOA issued by the National Heart, Lung, and Blood Institute, National Institutes of Health, solicits Research Education (R25) applications from institutions/organizations to promote diversity in undergraduate and health professional student populations by providing short-term research education support to stimulate career development in cardiovascular, pulmonary, hematologic, and sleep disorders research. The overall goal of the program is to provide research opportunities for individuals from disadvantaged backgrounds, racial and ethnic minorities, and others who are underrepresented in biomedical and behavioral research (e.g., individuals with disabilities), that will significantly contribute to a diverse research workforce in the future. The estimated funds (total costs) available for the first year of support for the entire program are expected to be $850,000. Up to eight new awards are anticipated, depending on the number of student participants requested and the availability of funds.
Requirements: The following organizations and institutions are eligible to apply: domestic institutions; non-profit organizations; public/state controlled institutions of higher education; private institutions of higher education; nonprofits with 501(c)3 IRS status; for-profit organizations (other than small businesses); state governments; county governments; city or township governments; special district governments; and faith-based or community-based organizations.
Restrictions: Applications from foreign institutions will not be accepted. Individuals on temporary or student visas and individuals holding PhD, MD, DVM, or equivalent doctoral degrees in the health sciences are not eligible.
Date(s) Application is Due: Jul 15; Aug 15
Amount of Grant: Up to 850,000 U.S.D
Contact: Dr. Michael Commarato, (301) 435-0530; michael_commarato@nih.gov
Internet: http://grants.nih.gov/grants/guide/rfa-files/RFA-HL-08-016.html
Sponsor: National Heart, Lung, and Blood Institute
6701 Rockledge Drive RKL2/600 MSC 7902
Bethesda, MD 20892-7902

NHLBI Stress Management Interventions to Reduce Risk of Coronary Artery Disease 3214

The purpose of this program announcement is to encourage applications that evaluate the efficacy of stress management interventions on intermediate outcomes purported to be in the biological or behavioral mechanistic pathway through which mental stress (MS) is thought to influence major adverse cardiac events (MACE) in patients

with confirmed coronary artery disease (CAD). These include subclinical measures of atherosclerosis, measures of cardiac or endothelial function, or measures such as cardiovascular reactivity or neurohumoral response to a provocative mental stress test. The ultimate goal is to determine whether stress management interventions can slow progression of CAD or reduce MACE and, if so, to translate effective stress management interventions to clinical practice.
Requirements: The following organizations and institutions are eligible to apply: individuals; public/state controlled institutions of higher education; private institutions of higher education; Hispanic-serving institutions; historically Black colleges and universities (HBCUs); tribally controlled colleges and universities (TCCUs); Alaska Native and Native Hawaiian serving institutions; nonprofits with 501(c)3 IRS status; nonprofits without 501(c)3 IRS status; for-profit organizations; state governments; Indian/Native American tribal governments (federally recognized); Indian/Native American tribally designated organizations; county governments; city or township governments; special district governments; independent school districts; public housing authorities/Indian housing authorities; U.S. territory or possession; Indian/Native American tribal governments (other than federally recognized); regional organizations; non-domestic (non-U.S.) entities (foreign organizations); other eligible agencies of the federal government; and faith-based or community-based organizations.
Date(s) Application is Due: Feb 5; Jun 5; Oct 5
Contact: Peter G. Kaufmann, (301) 435-2467; KaufmannP@mail.nih.gov
Internet: http://grants.nih.gov/grants/guide/pa-files/PA-07-322.html
Sponsor: National Heart, Lung, and Blood Institute
6701 Rockledge Drive RKL2/600 MSC 7902
Bethesda, MD 20892-7902

NHLBI Summer Institute for Training in Biostatistics II 3215
The National Heart, Lung, and Blood Institute (NHLBI) and the National Center for Research Resources (NCRR) invite applications for grants to develop, conduct, and evaluate summer courses in the basic principles and methods of biostatistics as employed in biomedical research. The courses would introduce participants to the field of biostatistics for the purpose of attracting new students into the field. The course will attract students from the entire U.S.A and will cover the fundamental concepts of probability, statistical reasoning and inferential methods motivated, in part, by examples that include data collected in studies of heart, lung, blood, and sleep disorders. The course will be taught during the summer, with appropriate modifications or refinements following each of the first two summers' sessions. Programs may choose an area of emphasis, such as clinical trials, statistical genetics, bioinformatics or epidemiology or take a general approach. Recent college graduates as well as advanced undergraduates should be eligible to participate provided they have not participated in SIBS previously. Direct costs are limited to $248,000 per year for a three-year period. Letters of Intent should be received by December 5, and full proposals are due by January 6.
Requirements: The following organizations and institutions are eligible to apply: public/state controlled institutions of higher education; private institutions of higher education; Hispanic-serving institutions; historically Black colleges and universities (HBCUs); tribally controlled colleges and universities (TCCUs); Alaska Native and Native Hawaiian serving institutions; nonprofits with 501(c)3 IRS status; nonprofits without 501(c)3 IRS status; for-profit organizations; state governments; Indian/Native American tribal governments (federally recognized); Indian/Native American tribally designated organizations; county governments; city or township governments; special district governments; independent school districts; public housing authorities/Indian housing authorities; U.S. territory or possession; Indian/Native American tribal governments (other than federally recognized); regional organizations; non-domestic (non-U.S.) entities (foreign organizations); other eligible agencies of the federal government; and faith-based or community-based organizations.
Date(s) Application is Due: Jan 6; Dec 5
Amount of Grant: Up to 248,000 U.S.D
Contact: Song Yang, (301) 435-0431; fax (301) 480-1862; yangso@nhlbi.nih.gov
Internet: http://grants.nih.gov/grants/guide/rfa-files/RFA-HL-09-009.html
Sponsor: National Heart, Lung, and Blood Institute
6701 Rockledge Drive RKL2/600 MSC 7902
Bethesda, MD 20892-7902

NHLBI Summer Internship Program in Biomedical Research 3216
The goal of the Summer Internship Program in Biomedical Research at NHLBI is to expose students to research investigation in a highly enriched environment that is devoted exclusively to biomedical research and training. The Program is open to high school and college, graduate or Medical and Dental Students. Participants join a research laboratory for a minimum of ten weeks between June and August and conduct research in selected areas of investigation under the guidance of an NHLBI intramural research scientist. To augment the training experience, students participate in a Seminar Series at the NHLBI, as well as the National Institutes of Health Summer Seminar Series and Summer Research Program Poster Day. While applications may be filed between Nov. 15 and March 1, early applications are encouraged.
Requirements: Applicants must be either U.S. citizens or permanent residents and be currently enrolled full-time in high school, college, professional or graduate school with a minimum GPA of 3.0 on a 4.0 scale. College graduates are eligible for the summer between graduation and beginning either graduate or professional school. High school students should be at least 16 years of age prior to June 1. Applications must be submitted online using the link for provided for Summer Internship Applicants.
Date(s) Application is Due: Mar 1
Contact: Llewellyn Jessica, Program Coordinator; (301) 451-9440; fax (301) 594-8133; llewellj@mail.nih.gov or direducation@nhlbi.nih.gov
Internet: http://dir.nhlbi.nih.gov/oe/summerprogs.asp
Sponsor: National Heart, Lung, and Blood Institute
6701 Rockledge Drive RKL2/600 MSC 7902
Bethesda, MD 20892-7902

NHLBI Targeted Approaches to Weight Control for Young Adults Grants 3217
The National Heart, Lung, and Blood Institute (NHLBI) of the National Institutes of Health (NIH) invites applications for cooperative agreements (U01) to conduct two-phase clinical research studies to develop, refine, and test innovative behavioral and/or environmental approaches for weight control in young adults at high risk for weight gain. Weight control interventions can address weight loss, prevention of weight gain, or prevention of excessive weight gain during pregnancy. The first phase will consist of formative research to refine the proposed intervention, recruitment, retention, and adherence strategies targeted to young adults. The second phase will consist of a randomized controlled trial to test the efficacy of the intervention. For the purpose of this FOA, young adults are defined as 18-35 years of age. The maximum award for the research component will be approximately $3,000,000 in direct costs, including direct costs of up to $500,000 for the first year, $650,000 for the second year, $650,000 for the third year, $650,000 for the fourth year, and $550,000 for the fifth year. Letters of Intent should be received by September 10, and full proposals are due by October 10.
Requirements: The following organizations and institutions are eligible to apply: public/state controlled institutions of higher education; private institutions of higher education; Hispanic-serving institutions; historically Black colleges and universities (HBCUs); tribally controlled colleges and universities (TCCUs); Alaska Native and Native Hawaiian serving institutions; nonprofits with 501(c)3 IRS status; nonprofits without 501(c)3 IRS status; for-profit organizations; Indian/Native American tribal governments; Indian/Native American Tribally Designated Organizations; eligible agencies of the federal government; faith-based or community-based organizations; and regional organizations.
Date(s) Application is Due: Sep 10; Oct 10
Amount of Grant: 500,000 - 650,000 U.S.D
Contact: Catherine Loria, (301) 435-0702; fax (301) 480-5158; loriac@nhlbi.nih.gov
Internet: http://grants.nih.gov/grants/guide/rfa-files/RFA-HL-08-007.html
Sponsor: National Heart, Lung, and Blood Institute
6701 Rockledge Drive RKL2/600 MSC 7902
Bethesda, MD 20892-7902

NHLBI Targeting Calcium Regulatory Molecules for Arrhythmia Prevention Grants 3218
The purpose of this Funding Opportunity Announcement (FOA) is to stimulate the development of therapeutic strategies for cardiac arrhythmias that target Ca2+ handling and Ca2+ signaling molecules. The overall goal is to exploit present knowledge of Ca2+ regulatory molecules and the role they play in arrhythmogenesis to develop novel therapeutic targets. Critical components of successful responses to this program announcement are: preclinical or early clinical testing of promising new therapeutic strategies to terminate and prevent cardiac arrhythmias in established models of human arrhythmogenic disease; and the feasibility to move positive findings to Phase I or II clinical trials. Grantees must budget for and attend two Project Director/Principal Investigator (PD/PI) meetings in Bethesda, MD, usually scheduled during the second and fourth year of the award.
Requirements: The following organizations and institutions are eligible to apply: public/state controlled institutions of higher education; private institutions of higher education; Hispanic-serving institutions; historically Black colleges and universities (HBCUs); tribally controlled colleges and universities (TCCUs); Alaska Native and Native Hawaiian serving institutions; nonprofits with 501(c)3 IRS status; nonprofits without 501(c)3 IRS status; for-profit organizations; state governments; Indian/Native American tribal governments (federally recognized); Indian/Native American tribally designated organizations; county governments; city or township governments; special district governments; independent school districts; public housing authorities/Indian housing authorities; U.S. territory or possession; Indian/Native American tribal governments (other than federally recognized); regional organizations; non-domestic (non-U.S.) entities (foreign organizations); other eligible agencies of the federal government; and faith-based or community-based organizations.
Date(s) Application is Due: Feb 5; Jun 5; Oct 5
Contact: Dennis A. Przywara, Coordinator; (301) 435-0506; fax (301) 480-1454; przywarad@nhlbi.nih.gov
Internet: http://grants.nih.gov/grants/guide/pa-files/PA-07-101.html
Sponsor: National Heart, Lung, and Blood Institute
6701 Rockledge Drive RKL2/600 MSC 7902
Bethesda, MD 20892-7902

NHLBI Transfusion-Related Acute Lung Injury Basic Research Grants 3219
This Funding Opportunity Announcement (FOA) issued by the National Heart, Lung, and Blood Institute (NHLBI), National Institutes of Health, solicits Research Project Grant (R01) applications from institutions/organizations that propose to study the mechanisms responsible for acute lung injury after blood transfusion and stimulate the development of strategies to improve the prevention, diagnosis, and treatment of the disorder for the four million transfusion recipients per year in the U.S. who may be at risk for developing Transfusion-Related Acute Lung Injury (TRALI).
Requirements: The following organizations and institutions are eligible to apply: individuals; public/state controlled institutions of higher education; private institutions of higher education; nonprofits with 501(c)3 IRS status; nonprofits without 501(c)3 IRS status; small businesses; for-profit organizations; and non-domestic (non-U.S.) entities.
Date(s) Application is Due: Feb 5; Jun 5; Oct 5
Contact: Traci Heath Mondoro, (301) 435-0052; mondorot@nhlbi.nih.gov

Internet: http://grants.nih.gov/grants/guide/pa-files/PA-07-385.html
Sponsor: National Heart, Lung, and Blood Institute
6701 Rockledge Drive RKL2/600 MSC 7902
Bethesda, MD 20892-7902

NHLBI Translating Basic Behavioral and Social Science Discoveries into Interventions to Reduce Obesity: Centers for Behavioral Intervention Development Grants 3220

This FOA solicits cooperative agreement (U01) applications from institutions/organizations that propose to translate findings from basic research on human behavior into more effective clinical, community, and population interventions to reduce obesity and improve obesity-related behaviors. This FOA will support Centers for Behavioral Intervention Development (CBIDs) in which interdisciplinary teams of basic and applied behavioral and social science researchers develop and refine novel interventions to reduce obesity and alter obesity-related health behaviors (e.g., diet, physical activity). The interventions to be developed include any of a wide range of strategies aimed at promoting weight loss and/or preventing weight gain. Interventions can be targeted to the individual, family, social network, community, environmental, clinical or population level, or combinations of these, and to any age group. Investigators at each CBID will conduct several types of studies during the funding period, including laboratory experimental studies, formative research, early phase trials and pilot/feasibility studies, in order to create promising new avenues for reducing obesity and improving obesity-related behaviors. The maximum award for the research component will be approximately $3,750,000 in direct costs across five years, with a limit of $750,000 in direct cost for any single year. Letters of Intent should be received by December 16, and full proposals are due by January 13.
Requirements: The following organizations and institutions are eligible to apply: public/state controlled institutions of higher education; private institutions of higher education; Hispanic-serving institutions; historically Black colleges and universities (HBCUs); tribally controlled colleges and universities (TCCUs); Alaska Native and Native Hawaiian serving institutions; nonprofits with 501(c)3 IRS status; nonprofits without 501(c)3 IRS status; for-profit organizations; state governments; Indian/Native American tribal governments (federally recognized); Indian/Native American tribally designated organizations; county governments; city or township governments; special district governments; independent school districts; public housing authorities/Indian housing authorities; U.S. territory or possession; Indian/Native American tribal governments (other than federally recognized); regional organizations; non-domestic (non-U.S.) entities (foreign organizations); other eligible agencies of the federal government; and faith-based or community-based organizations.
Date(s) Application is Due: Jan 13; Dec 16
Amount of Grant: Up to 750,000 U.S.D
Contact: Susan M. Czajkowski, (301) 435-0406; fax (301) 480-1773; czajkows@mail.nih.gov
Internet: http://grants.nih.gov/grants/guide/rfa-files/RFA-HL-08-013.html
Sponsor: National Heart, Lung, and Blood Institute
6701 Rockledge Drive RKL2/600 MSC 7902
Bethesda, MD 20892-7902

NHLBI Translational Programs in Lung Diseases 3221

The purpose of this program is to encourage interdisciplinary approaches to lung translational research, allowing for potentially high-impact research, which will translate basic observations to applied clinical research. An application to NHLBI requires a minimum of three interrelated research projects that focus a number of scientific disciplines on investigations of a complex biomedical theme or research question. Applicants will be required to propose research that will become progressively more clinical and translational during the life of the program project, and they are expected to assemble interactive, multidisciplinary teams that have the combined expertise to formulate a plan for successful translation. Research teams are not required to have prior collaborative experience but must be able to demonstrate an integrated, practical approach that will result in the effective progression of mechanistic/basic concepts toward application in the clinic. This program allows for a maximum of two 5-year awards (a new application and one competing renewal). There are two application cycles. Letters of Intent should be received by April 25 and August 25 (respectively), and full proposals are due by May 25 and September 25 (respectively).
Requirements: The following organizations and institutions are eligible to apply: public/state controlled institutions of higher education; private institutions of higher education; Hispanic-serving institutions; historically Black colleges and universities (HBCUs); tribally controlled colleges and universities (TCCUs); Alaska Native and Native Hawaiian serving institutions; nonprofits with 501(c)3 IRS status; nonprofits without 501(c)3 IRS status; for-profit organizations; state governments; Indian/Native American tribal governments (federally recognized); Indian/Native American tribally designated organizations; county governments; city or township governments; special district governments; independent school districts; public housing authorities/Indian housing authorities; U.S. territories or possessions; Indian/Native American tribal governments (other than federally recognized); regional organizations; non-domestic (non-U.S.) entities (foreign organizations); other eligible agencies of the federal government; and faith-based or community-based organizations.
Date(s) Application is Due: Apr 25; May 25; Aug 25; Sep 25
Contact: Patricia Noel, (301) 435-0202; noelp@nhlbi.nih.gov
Internet: http://grants.nih.gov/grants/guide/pa-files/PAR-09-185.html
Sponsor: National Heart, Lung, and Blood Institute
6701 Rockledge Drive RKL2/600 MSC 7902
Bethesda, MD 20892-7902

NHSCA Arts in Health Care Project Grants 3222

Arts in Health Care grants support arts activities, presentations and artist residencies that occur in health care facilities, rehabilitation centers and in centers serving the needs of the elderly. The overall goal of this grant category is to utilize the arts to enhance the quality of life and promote an environment conducive to healing for patients, residents, and/or clients. This grant category is in response to the Arts Council's commitment to meeting the needs of underserved populations, which can include the elderly, people with disabilities and people with health challenges. Requests may be made for $1,000 - $4,500.
Requirements: At a minimum, grants must be matched on a one-to-one basis. Health care facilities, hospitals, rehabilitation centers and facilities serving the elderly including: assisted living facilities, county and nonprofit nursing homes, veterans' homes, hospice care programs or visiting nurse associations may apply. Applicants must have 501(c)3 tax-exempt status from the IRS and not-for-profit incorporation in the State of New Hampshire and: make their programs accessible to people with disabilities; have submitted all required reports on past State Arts Council grants; and be in good standing with the N.H. Secretary of State's Office and the N.H. Attorney General's Office.
Restrictions: This grant does not fund: projects/activities that are not open to the general public; commercially viable for-profit publications, recordings or films; general operating expenses not directly related to the project; fundraising costs; projects already receiving funds from another State Arts Council grant category; or any cost item listed in the glossary under ineligible expenses.
Geographic Focus: New Hampshire
Date(s) Application is Due: Jun 2
Amount of Grant: 1,000 - 4,500 U.S.D
Contact: Catherine O'Brian, Program Coordinator; (603) 271-0795 or (603) 271-2789; fax (603) 271-3584; Catherine.R.OBrian@dcr.nh.gov
Internet: http://www.nh.gov/nharts/grants/partners/artsinhealthcare.htm
Sponsor: New Hampshire State Council on the Arts
19 Pillsbury Street, 1st Floor
Concord, NH 03301

NIA/NIEHS Biomedical Epidemiologic and Intervention Studies Planning Grants 3223

This program provides grant support for planning and protocol development of biomedical epidemiologic and intervention studies in research areas supported by NIA's geriatrics program. PA: PA-99-145
Requirements: Applications may be submitted by foreign and domestic for-profit and nonprofit organizations, both public and private; units of state and local governments; and eligible agencies of the federal government.
Amount of Grant: Up to 150,000 U.S.D
Contact: Dr. Gwen Collman, Program Administrator; (301) 541-4980; fax (919) 541-4937; Collman@niehs.nih.gov
Internet: http://grants.nih.gov/grants/guide/pa-files/PA-99-145.html
Sponsor: National Institute on Aging
Building 31, Room 5C27, 31 Center Drive, MSC 2292
Bethesda, MD 20892-7710

NIAAA Alcohol Research Centers Grants 3224

The program's objective is to provide long-term support for a concerted and multidisciplinary research effort into the problems of alcohol use and alcoholism by coordinating the activities of investigators from biomedical, behavioral, and social science disciplines around a basic research theme. Funds may be used for support of core resources including, for example, salaries of center personnel, services and facilities shared by center personnel, and program enrichment activities such as symposia. Support for individual components may include staff salaries, supplies, publication, travel, and equipment. Support will be provided for up to five years (renewable for subsequent periods) to qualifying centers selected for funding.
Requirements: State and local governments and any domestic (nonfederal) public or private nonprofit and for-profit institution may apply for a center grant. The proposed center must be affiliated with an institution, such as a university, medical center, or research center, that has the resources to sustain a long-term coordinated research program around a central theme related to problems of alcohol use and alcoholism. An applicant institution must demonstrate the ability to attract high-quality scientists, from biomedical, behavioral, and social science disciplines, who are willing to make a long-term commitment to alcohol research. In addition, the applicant must assure that alcoholism related research and clinical training opportunities will be available.
Restrictions: Funds may not be used for training, although the applicant is required to demonstrate an institutional training capacity related to alcohol. Funds may not be used for purchase of any land or the purchase, construction, preservation, or repair of any building, and grantees must have sufficient laboratory facilities and reference services to carry out the grant.
Amount of Grant: Up to 2,000,000 U.S.D
Contact: Dr. Ernestine Vanderveen, (301) 443-2531; fax (301) 584-0673; tvander@willco.niaaa.nih.gov
Internet: http://grants.nih.gov/grants/guide/rfa-files/rfa-aa-00-003.html
Sponsor: National Institute on Alcohol Abuse and Alcoholism
5635 Fishers Lane, MSC 9304
Bethesda, MD 20892-9304

NIAAA Hepatitis C Infection and Alcoholic Liver Disease 3225

NIAAA invites basic and clinical research grant applications that focus on the synergistic interactions between alcohol consumption and hepatitis C virus (HCV) infection to produce chronic liver disease. This announcement solicits grant applications for research

that elucidate mechanism(s) in the development and progression of liver disease in a setting of combined alcohol use and HCV infection. Studies that focus on cellular and molecular mechanisms of viral pathogenesis, immune responses and liver disease, in the context of gender, race, and age are encouraged. Development of animal models and in vitro experimental systems to study the interactive effects of alcohol exposure and HCV pathogenesis are highly encouraged. Annual deadline dates may vary; contact program staff for exact dates. PA: PAS-99-155
Requirements: Applications may be submitted by domestic and foreign, for-profit and nonprofit organizations, both public and private; units of state and local governments; and eligible agencies of the federal government.
Amount of Grant: Up to 2,000,000 U.S.D
Contact: Thomas Kresina, Division of Basic Research; (301) 443-6537; fax (301) 594-0673; tk13v@nih.gov
Internet: http://grants.nih.gov/grants/guide/pa-files/PAS-99-155.html
Sponsor: National Institute on Alcohol Abuse and Alcoholism
5635 Fishers Lane, MSC 9304
Bethesda, MD 20892-9304

NIAAA Independent Scientist Award 3226
The award is a special salary-only grant designed to provide protected time for newly independent scientists who currently have non-research obligations such as heavy teaching loads, clinical work, committee assignments, service, and administrative duties that prevent them from having a period of intensive research focus. The award is targeted to persons with doctoral degrees who have completed their research training, have independent peer-reviewed research support, and who need a period of protected research time in order to foster their research career development. New applications are due February 12, June 12, and October 12, while renewal applications are due March 12, July 12, and November 12. Up to 75% per year of base salary plus fringe benefits is available. Research support is not provided in most cases. Individuals engaged in predominantly theoretical work, such as modeling or computer simulation, may request up to $25,000 per year for research expenses.
Requirements: The following organizations and institutions are eligible to apply: public/state controlled institutions of higher education; private institutions of higher education; nonprofits with 501(c)3 IRS status; nonprofits without 501(c)3 IRS status; small businesses; for-profit organizations; State governments; U.S. territories or possessions; Indian/Native American tribal governments (Federally recognized and other than Federally recognized); Indian/Native American tribally designated organizations; Hispanic-serving institutions; historically Black colleges and universities (HBCUs); tribally controlled colleges and universities (TCCUs); Alaska Native and Native Hawaiian serving institutions; regional organizations; and faith-based or community based organizations. The candidate must have a doctoral degree and peer-reviewed, independent research support at the time the award is made. The candidate must spend a minimum of 75 percent effort conducting research during the period of the award.
Restrictions: The award is not intended for investigators who already have full time to perform research, or have substantial publication records or considerable research support indicating that they are well established in their fields. Foreign institutions are not eligible to apply.
Date(s) Application is Due: Feb 12; Mar 12; Jun 12; Jul 12; Oct 12; Nov 12
Amount of Grant: Up to 100,000 U.S.D
Contact: Michael Hilton, (301) 402-9402; mhilton@mail.nih.gov
Internet: http://grants.nih.gov/grants/guide/pa-files/PA-09-038.html
Sponsor: National Institute on Alcohol Abuse and Alcoholism
5635 Fishers Lane, MSC 9304
Bethesda, MD 20892-9304

NIAAA Mechanisms of Alcohol-Induced Hepatic Fibrosis Research Grants 3227
The Division of Basic Research of NIAAA invites grant applications that focus on the developmental mechanisms of hepatic fibrosis that occur in alcohol-induced liver disease. PA: PA-99-110
Requirements: Applications may be submitted by domestic and foreign, for-profit and nonprofit organizations, public and private. Applicants requesting $500,000 or more in direct costs for any year must contact program staff before submitting the application.
Amount of Grant: Up to 50,000 U.S.D
Contact: Dr. Vishnudutt Purohit; (301) 443-2689; vpurohit@willco.niaaa.nih.gov
Internet: http://www.grants.nih.gov/grants/guide/pa-files/PA-99-110.html
Sponsor: National Institute on Alcohol Abuse and Alcoholism
5635 Fishers Lane, MSC 9304
Bethesda, MD 20892-9304

NIAAA Mechanisms of Alcohol and Drug-Induced Pancreatitis Grants 3228
This Funding Opportunity Announcement (FOA) issued by the National Institute on Alcohol Abuse and Alcoholism (NIAAA), the National Institute of Diabetes and Digestive and Kidney Diseases (NIDDK), and the National Institute on Drug Abuse (NIDA), National Institutes of Health (NIH), solicits grant applications from institutions/organizations that propose to investigate the underlying molecular, biochemical, and cellular mechanisms by which long-term alcohol ingestion and drugs lead to the development of pancreatitis. Research is also encouraged to understand the role of various predisposing factors, including substance abuse, that make the pancreas susceptible to alcoholic injury. Understanding the mechanisms as well as the role of predisposing factors may help in developing strategies for the prevention or treatment of the disease. Because the nature and scope of the proposed research will vary from application to application, it is anticipated that the size and duration of each award will also vary.
Requirements: Application can be made by the following institutions and organizations: public/state controlled institutions of higher education; private institutions of higher education; nonprofits with 501(c)3 IRS status (other than institutions of higher education); nonprofits without 501(c)3 IRS status (other than institutions of higher education); small businesses; for-profit organizations (other than small businesses); state governments; U.S. territories or possessions; Indian/Native American tribal governments (federally recognized); Indian/Native American tribal governments (other than Federally recognized); Indian/Native American tribally designated organizations; non-domestic (non-U.S.) entities (foreign organizations); Hispanic-serving institutions; historically Black colleges and universities (HBCUs); tribally controlled colleges and universities (TCCUs); Alaska Native and Native Hawaiian serving institutions; regional organization; eligible agencies of the Federal government; and faith-based or community based organizations.
Date(s) Application is Due: Feb 5; Jun 5; Oct 5
Contact: Vishnudutt Purohit, (301) 443-2689; fax (301) 443-0673; vpurohit@mail.nih.gov
Internet: http://grants.nih.gov/grants/guide/pa-files/PA-07-067.html
Sponsor: National Institute on Alcohol Abuse and Alcoholism
5635 Fishers Lane, MSC 9304
Bethesda, MD 20892-9304

NIAAA Mentored Clinical Scientist Research Career Development Awards 3229
This award enables candidates holding professional degrees (e.g., M.D., D.O., D.V.M., or equivalent degrees) to undertake 3 to 5 years of special study and supervised research with the goal of becoming independent investigators. The award also allows awardees to pursue a research career development program suited to their experience and capabilities under a mentor who is competent to provide guidance in the chosen research area. Institutions may submit applications on behalf of candidates who hold professional degrees. At least 2 years must have elapsed since the health professional degree was granted. Candidates can have varying levels of clinical training and research experience. New applications are due February 12, June 12, and October 12, while resubmitted applications are due March 12, July 12, and November 12. Up to Up to 75 percent of base salary plus fringe benefits is available, and up to $50,000 per year for research support for new awards.
Requirements: All candidates must be U.S. citizens, non-citizen nationals, or legal permanent residents of the U.S. Persons with temporary or student visas are not eligible. The grantee institution must have a strong, well-established research and research training program in the chosen area, accomplished faculty in the basic and clinical sciences, and a commitment to the candidate's research development. The proposed program should include an appropriate mentor.
Date(s) Application is Due: Feb 12; Mar 12; Jun 12; Jul 12; Oct 12; Nov 12
Amount of Grant: 100,000 U.S.D
Contact: Judith A. Arroyo, (301) 402-0717; fax (301) 443-6077; jarroyo@mail.nih.gov
Internet: http://grants.nih.gov/grants/guide/pa-files/pa-06-512.html
Sponsor: National Institute on Alcohol Abuse and Alcoholism
5635 Fishers Lane, MSC 9304
Bethesda, MD 20892-9304

NIAAA Peptide Regulation of Alcohol Intake 3230
The purpose of this program is to identify peptides and their receptors that regulate alcohol consumption, ultimately leading to an understanding of their mechanism(s) of action. Although not exclusive, several areas have potential for further investigation including the hypothalamic-pituitary-adrenal axis, opioid peptides, the renin-angiotensin system, plus the group of peptides regulating food intake and energy balance, exemplified by leptin and neuropeptide Y. The goal of this initiative is to elucidate the role of peptides in pathological alcohol consumption, suggesting novel pharmacotherapeutic approaches for alcoholism. Annual deadlines may vary; contact program staff for exact dates. PA: PAS-99-156
Requirements: Applications may be submitted by domestic and foreign, for-profit and nonprofit organizations, both public and private; units of state and local governments; and eligible agencies of the federal government.
Amount of Grant: Up to 2,000,000 U.S.D
Contact: Jules Selden, Division of Basic Research; (301) 443-2678; fax (301) 594-0673; jselden@willco.niaaa.nih.gov
Internet: http://grants.nih.gov/grants/guide/pa-files/PAS-99-156.html
Sponsor: National Institute on Alcohol Abuse and Alcoholism
5635 Fishers Lane, MSC 9304
Bethesda, MD 20892-9304

NIAAA Secondary Analysis of Existing Health Services Data Sets Grants 3231
The NIAAA is soliciting exploratory/developmental research grant applications (R21) to support the secondary analysis of existing data sets relevant to developing a knowledge base to improve the delivery of services for alcohol-related problems, including both treatment and preventive interventions. Research objectives of this PA include, but are not limited to, five major areas: determining the effects of financing and reimbursement mechanisms on alcohol-related healthcare program availability, accessibility, delivery, organization, content, quality, and outcomes; assessing sources of variation in access to and utilization of treatment and prevention interventions for alcohol-related problems; identifying organizational and managerial factors that influence the delivery of treatment and prevention services for alcohol-related problems, either within or across regions, populations, and settings; evaluating the cost, cost-effectiveness, cost-benefit, and cost-utility of alcohol-related treatment and prevention services; and identifying and assessing the effectiveness and outcomes of alcohol-related treatment and preventive services. PA: PA-00-100
Requirements: Applications may be submitted by foreign and domestic, for-profit and nonprofit organizations, both public and private; units of state and local governments; and eligible agencies of the federal government.
Amount of Grant: Up to 100,000 U.S.D

Contact: Dr. Harold Perl, Division of Clinical and Prevention Research; (301) 443-0788; fax (301) 443-8774; hperl@willco.niaaa.nih.gov
Internet: http://grants.nih.gov/grants/guide/pa-files/PA-00-100.html
Sponsor: National Institute on Alcohol Abuse and Alcoholism
5635 Fishers Lane, MSC 9304
Bethesda, MD 20892-9304

NIAAA Strategies for the Prevention of DUI Recidivism Research Grants 3232
The purpose of this program announcement is to stimulate research that addresses the problem of recidivism among people convicted of drinking and driving offenses. This includes individuals who are apprehended for driving under the influence (DUI), driving while intoxicated (DWI), or related offenses such as driving after suspension for a DUI and violation of zero tolerance laws, and are either convicted or otherwise sanctioned for such illegal behavior. Applicants who respond to this program announcement are encouraged to develop research applications that design, develop, implement, and/or test interventions that have the potential of preventing or reducing DUI recidivism. Such interventions may be investigator-initiated and implemented in collaboration with appropriate members of judicial, administrative, and policy-making bodies specifically to test or compare their efficacy in preventing or reducing DUI recidivism. Alternatively, the interventions to be studied may occur naturally in society as laws, regulations, policies, and practices implemented by legally constituted bodies and may be tested for their effectiveness by researchers who use the established methodologies of studying natural experiments. PA PAS-99-023
Requirements: Applications may be submitted by domestic and foreign, for-profit and nonprofit, both public and private; units of state and local governments; and eligible agencies of the federal government.
Amount of Grant: 56,000 - 100,000,000 U.S.D
Contact: Dr. Ernestine Vanderveen, (301) 443-2531
Internet: http://grants2.nih.gov/grants/guide/pa-files/PAS-99-023.html
Sponsor: National Institute on Alcohol Abuse and Alcoholism
5635 Fishers Lane, MSC 9304
Bethesda, MD 20892-9304

NIA Aging, Oxidative Stress and Cell Death Grants 3233
The purpose of this Program Announcement is to encourage the submission of applications to support research on the relationship between oxidative stress and apoptosis, and how these biological processes are involved in aging and/or change with age. Although the National Institute of General Medical Sciences is not participating in this program announcement, that Institute continues to be interested in receiving applications on basic mechanisms of programmed cell death and oxidative stress when these applications do not focus on aging mechanisms. Modular Grant applications will request direct costs in $25,000 modules, up to a total direct cost request of $250,000 per year.
Requirements: The following organizations and institutions may apply: public/state controlled institutions of higher education; private institutions of higher education; Hispanic-serving institutions; historically Black colleges and universities (HBCUs); tribally controlled colleges and universities (TCCUs); Alaska native and native Hawaiian serving institutions; nonprofits with 501(c)3 IRS status; nonprofits without 501(c)3 IRS status; small businesses; for-profit organizations; State governments; regional organizations; U.S. territories or possessions; Indian/Native American tribal governments; Indian/Native American tribally designated organizations; county governments; city or township governments; special district governments; independent school districts; public housing authorities/Indian housing authorities; eligible agencies of the Federal government; and faith-based or community based organizations.
Date(s) Application is Due: Feb 5; Jun 5; Oct 5
Amount of Grant: Up to 250,000 U.S.D
Contact: Chyren Hunter, (301) 496-9322; fax (301) 402-0528; Hunterc@nia.nih.gov
Internet: http://grants.nih.gov/grants/guide/pa-files/PA-00-081.html
Sponsor: National Institute on Aging
Gateway Building, Room 2C218, 7201 Wisconsin Avenue, MSC 9205
Bethesda, MD 20892-9205

NIA Aging Research Dissertation Awards to Increase Diversity 3234
This Funding Opportunity Announcement (FOA) issued by the National Institute on Aging (NIA) of the National Institutes of Health (NIH) publicizes the re-issuance of an FOA that publicizes the availability of dissertation awards in aging research to increase the diversity of the research-on-aging workforce. These awards are available to all qualified Predoctoral students in accredited research doctoral programs in the United States (including Puerto Rico and other U.S. territories or possessions).
Requirements: Public or private institutions, such as universities, colleges, hospitals, and laboratories, that offer accredited research doctoral programs may submit an application.
Restrictions: Domestic institutions only are eligible.
Date(s) Application is Due: Feb 16; Jun 16; Oct 16
Amount of Grant: 50,000 U.S.D
Contact: Michael-David A.R.R. Kerns, (301) 402-7713; fax (301) 402-2945; michael-david.kerns@nih.hhs.gov
Internet: http://grants.nih.gov/grants/guide/pa-files/PAR-08-250.html
Sponsor: National Institute on Aging
Gateway Building, Room 2C218, 7201 Wisconsin Avenue, MSC 9205
Bethesda, MD 20892-9205

NIA AIDS and Aging: Behavioral Sciences Prevention Research Grants 3235
NIA, NIMH, and NINR invite qualified researchers to submit applications to investigate prevention issues relevant to AIDS in middle-aged and older populations. This announcement solicits AIDS prevention research proposals to study primary prevention of disease transmission as well as secondary and tertiary prevention of negative behavioral and social consequences of HIV/AIDS for persons with AIDS, their families, and communities. Thus, the primary goals are to identify social and behavioral factors associated with HIV transmission and disease progression in later life; examine behavioral and social consequences of HIV infection/AIDS across the life course; develop and evaluate age-appropriate behavioral and social interventions for preventing AIDS in middle-aged and older adults and/or ameliorating problems associated with older adult caregiver responsibilities and burdens; explore health care issues surrounding AIDS care; and strengthen existing research and evaluation methods.
Requirements: Applications may be submitted by domestic and foreign for-profit and non-profit organizations, public and private, such as universities, colleges, hospitals, laboratories, units of State and local governments, and eligible agencies of the Federal government. Applications may be submitted by single institutions or by a consortia of institutions.
Date(s) Application is Due: Feb 5; Feb 16; Jun 5; Jun 16; Oct 5; Oct 16
Contact: Ying Tian, (301) 496-6761; fax (301) 402-1784; tiany@nia.nih.gov
Internet: http://www.grants.nih.gov/grants/guide/pa-files/PA-97-069.html
Sponsor: National Institute on Aging
Gateway Building, Suite 3C307, 7201 Wisconsin Avenue, MSC 9205
Bethesda, MD 20892-9205

NIA Alzheimer's Disease Core Centers Grants 3236
The National Institute on Aging (NIA) invites applications from qualified institutions for support of Alzheimer's Disease Core Centers (ADCCs). These centers are designed to support and conduct research on Alzheimer's disease (AD), to serve as shared research resources that will facilitate research in AD and related disorders, distinguish them from the processes of normal brain aging and mild cognitive impairment (MCI), provide a platform for training, develop novel techniques and methodologies, and translate these research findings into better diagnostic, prevention and treatment strategies. New applications should request a project period of five years and a budget for direct costs of up to $750,000 per year. Competing renewal applications may request direct costs for all cores (both required and optional), and the other listed functions, i.e., satellites and pilot grants at a level not exceeding the combined direct costs of all funded activities awarded during the final year of the present funding period plus a 3% increase, or $750,000 per year, whichever is larger. Letters of Intent should arrive by the April 6 deadline, with full proposals due by May 5
Requirements: The following organizations and institutions are eligible to apply: public/state controlled institutions of higher education; private institutions of higher education; Hispanic-serving institutions; historically Black colleges and universities (HBCUs); tribally controlled colleges and universities (TCCUs); Alaska Native and Native Hawaiian serving institutions; nonprofits with 501(c)3 IRS status (other than institutions of higher education); and nonprofits without 501(c)3 IRS status (other than institutions of higher education).
Restrictions: At the time of award, the applicant institution cannot have another Alzheimer's Disease Center funded by the NIA.
Date(s) Application is Due: Apr 6; May 5
Amount of Grant: 750,000 U.S.D
Contact: Creighton H. Phelps, (301) 496-9350; fax (301) 496-1494; phelpsc@nia.nih.gov
Internet: http://grants.nih.gov/grants/guide/rfa-files/RFA-AG-10-001.html
Sponsor: National Institute on Aging
Gateway Building, Suite 350, 7201 Wisconsin Avenue, MSC 9205
Bethesda, MD 20892-9205

NIA Alzheimer's Disease Drug Development Program Grants 3237
The NIA and other NIH Institutes currently support research projects for the study of the epidemiology, etiology, diagnosis, and treatment of Alzheimer's disease (AD). Investigators in all relevant scientific disciplines and organizations are invited to develop and pre-clinically test compounds which can slow the disease progression, reverse the disease process, or delay the onset of or prevent AD, mild cognitive impairment (MCI), or age-related cognitive decline.
Requirements: The following organizations and institutions may apply: public/state controlled institutions of higher education; private institutions of higher education; nonprofits with 501(c)3 IRS status; nonprofits without 501(c)3 IRS status; small businesses; for-profit organizations; State governments; and non-domestic (non-U.S.) entities (foreign organizations).
Date(s) Application is Due: Jan 2; Feb 2; Sep 1; Oct 0
Amount of Grant: 300,000 - 800,000 U.S.D
Contact: Suzana Petanceska, (301) 496-9350; fax (301) 496-1494; petanceskas@nia.nih.gov
Internet: http://grants.nih.gov/grants/guide/pa-files/PAR-08-266.html
Sponsor: National Institute on Aging
Gateway Building, Suite 350, 7201 Wisconsin Avenue, MSC 9205
Bethesda, MD 20892-9205

NIA Alzheimer's Disease Pilot Clinical Trials 3238
The objective of the Alzheimer's Disease Pilot Clinical Trials initiative is to improve the quality of clinical research designed to evaluate interventions for the prevention and treatment of Alzheimer's disease (AD), mild cognitive impairment (MCI), and age-associated cognitive decline by stimulating applications for pilot clinical trials to test interventions aimed at delaying the onset of or preventing AD, MCI, and age-associated cognitive decline; slowing, halting, or, if possible, reversing the progressive decline in cognitive function; and modifying the cognitive and behavioral symptoms in AD and MCI. The goal is not to

duplicate or compete with pharmaceutical companies but to encourage, complement, and accelerate the process of testing new, innovative, and effective treatments. The pilot research project application should directly address how it will inform and advance the design of a subsequent full-scale clinical trial. The application should also address the intrinsic scientific merit of the pilot trial itself. The application may include the following: studies to refine the intervention strategy (e.g. drug dosage, duration, delivery system; behavioral intensity and duration); studies to define and refine the target population and ensure adequate enrollment, protocol adherence and subject retention; collection of preliminary data for establishing measures of clinical efficacy including clinical/neuropsychological/behavioral measures, neuroimaging measures, and other biological measures in blood and cerebrospinal fluid; assessment of efficacy (including primary and secondary outcome and assessment measures for both patients and caregivers); and studies of safety.
Requirements: The following organizations and institutions may apply: public/state controlled institutions of higher education; private institutions of higher education; Hispanic-serving institutions; historically Black colleges and universities (HBCUs); tribally controlled colleges and universities (TCCUs); Alaska native and native Hawaiian serving institutions; nonprofits with 501(c)3 IRS status; nonprofits without 501(c)3 IRS status; small businesses; for-profit organizations; State governments; regional organizations; U.S. territories or possessions; Indian/Native American tribal governments; Indian/Native American tribally designated organizations; county governments; city or township governments; special district governments; independent school districts; public housing authorities/Indian housing authorities; eligible agencies of the Federal government; and faith-based or community based organizations.
Date(s) Application is Due: Feb 5; Jun 5; Oct 5
Amount of Grant: 450,000 - 1,350,000 U.S.D
Contact: Laurie M. Ryan, (301) 496-9350; fax (301) 496-1494; ryanl@nia.nih.gov
Internet: http://grants.nih.gov/grants/guide/pa-files/PAR-08-062.html
Sponsor: National Institute on Aging
Gateway Building, Suite 350, 7201 Wisconsin Avenue, MSC 9205
Bethesda, MD 20892-9205

NIA Alzheimer's Disease Research Centers Grants 3239
The National Institute on Aging (NIA) invites applications from qualified institutions for support of Alzheimer's Disease Research Centers (ADRCs). These centers are designed to support and conduct research on Alzheimer's disease (AD), to serve as shared research resources that will facilitate research in AD and related disorders, distinguish them from the processes of normal brain aging and mild cognitive impairment (MCI), provide a platform for training, develop novel techniques and methodologies, and translate these research findings into better diagnostic, prevention and treatment strategies. New applications should request a project period of five years and a budget for direct costs of up to $1 million per year. Letters of Intent are due by April 6 annually, with full applications received by May 5.
Requirements: The following organizations and institutions are eligible to apply: public/state controlled institutions of higher education; private institutions of higher education; Hispanic-serving institutions; historically Black colleges and universities (HBCUs); tribally controlled colleges and universities (TCCUs); Alaska native and native Hawaiian serving institutions; nonprofits with 501(c)3 IRS status (other than institutions of higher education); and nonprofits without 501(c)3 IRS status (other than institutions of higher education).
Date(s) Application is Due: Apr 5; May 6
Amount of Grant: Up to 1,000,000 U.S.D
Contact: Creighton H. Phelps, (301) 496-9350; fax (301) 496-1494; phelpsc@nia.nih.gov
Internet: http://grants.nih.gov/grants/guide/rfa-files/RFA-AG-10-002.html
Sponsor: National Institute on Aging
Gateway Building, Suite 350, 7201 Wisconsin Avenue, MSC 9205
Bethesda, MD 20892-9205

NIA Archiving and Development of Socialbehavioral Datasets in Aging Related Studies Grants 3240
This Funding Opportunity Announcement (FOA), issued by the National Institute on Aging (NIA) is seeking small grant applications to stimulate and facilitate data archiving and development related to cognitive psychology, behavioral interventions in the context of randomized controlled trials (RCTs), demography, economics, epidemiology, behavioral genetics and other behavioral research on aging for secondary analysis.
Requirements: The following organizations and institutions may apply: public/state controlled institutions of higher education; private institutions of higher education; Hispanic-serving institutions; historically Black colleges and universities (HBCUs); tribally controlled colleges and universities (TCCUs); Alaska native and native Hawaiian serving institutions; nonprofits with 501(c)3 IRS status; nonprofits without 501(c)3 IRS status; small businesses; for-profit organizations; State governments; regional organizations; U.S. territories or possessions; Indian/Native American tribal governments; Indian/Native American tribally designated organizations; county governments; city or township governments; special district governments; independent school districts; public housing authorities/Indian housing authorities; eligible agencies of the Federal government; and faith-based or community based organizations.
Date(s) Application is Due: Feb 16; Jun 16; Oct 16
Amount of Grant: Up to 100,000 U.S.D
Contact: Partha Bhattacharyya, (301) 496-3136; fax (301) 402-0051; bhattacharyyap@mail.nih.gov
Internet: http://grants.nih.gov/grants/guide/pa-files/PA-08-252.html
Sponsor: National Institute on Aging
Gateway Building, Suite 350, 7201 Wisconsin Avenue, MSC 9205
Bethesda, MD 20892-9205

NIA Awards to Support Research on the Biology of Aging in Invertebrates Grants 3241
This Funding Opportunity Announcement (FOA) issued by the National Institute on Aging (NIA), National Institutes of Health, invites applications that propose the identification and development of new invertebrate models for use in biomedical research that is highly relevant to the biology of aging. Applications should focus on the identification, development and characterization of new invertebrate models that have short lifespans. Invertebrate models with tractable genetics and genome sequence available or in progress, negligible or induced senescence, and tissue regeneration in adults are of particular interest to NIA. Examples of invertebrate models of interest include, but are not limited to planaria, hydra, rotifers, and tunicates. These invertebrate models are currently not well developed for aging research. Within the scope of this FOA, invertebrate models other than those mentioned above currently used by the biomedical research community, but not yet fully developed or realized for aging research, will be considered responsive to this FOA. In particular, applications that focus on the characterization of these novel models for aging research utilizing molecular, genetic, cellular or physiological tools and approaches to identify genes involved in longevity, senescence and cellular pathways in tissue homeostasis during aging are encouraged. Letters of Intent are due by September 30, with final applications due by October 30.
Requirements: The following organizations and institutions are eligible to apply for support: public/state controlled institutions of higher education; private institutions of higher education; Hispanic-serving institutions; historically Black colleges and universities (HBCUs); tribally controlled colleges and universities (TCCUs); Alaska Native and Native Hawaiian serving institutions; nonprofits with 501(c)3 IRS status (other than institutions of higher education); nonprofits without 501(c)3 IRS status (other than institutions of higher education); Indian/Native American tribal governments (federally recognized); Indian/Native American tribally designated organizations; U.S. territories or possessions; and eligible agencies of the federal government.
Date(s) Application is Due: Sep 30; Oct 30
Amount of Grant: Up to 250,000 U.S.D
Contact: Anna M. McCormick, Ph.D.; (301) 496-6402; mccormia@nia.nih.gov
Internet: http://grants.nih.gov/grants/guide/rfa-files/RFA-AG-10-004.html
Sponsor: National Institute on Aging
Gateway Building, Suite 2C231, 7201 Wisconsin Avenue, MSC 9205
Bethesda, MD 20892

NIA Behavioral and Social Research Grants on Disasters and Health 3242
The purpose of this FOA is to stimulate research in the behavioral and social sciences on the consequences of natural and man-made disasters for the health of children, the elderly, and vulnerable groups, with an ultimate goal of preventing and mitigating harmful consequences and health disparities. Disasters include severe weather-related events, earthquakes, large-scale attacks on civilian populations, technological catastrophes or perceived catastrophes, and influenza pandemics. For the elderly and for children and youth, the health outcomes of greatest interest include mortality, disability and resilience, severe distress and clinically significant morbidity (as opposed to mild or transient symptoms and dysphoria), and economic hardship sufficient to harm health. For children and youth, long-term effects on development are also of interest.
Requirements: The following organizations and institutions may apply: for-profit organizations; non-profit organizations; public or private institutions, such as universities, colleges, hospitals, and laboratories; units of State government; units of Local government; units of State Tribal government; units of Local Tribal government; eligible agencies of the Federal government; foreign institutions; domestic institutions; faith-based or community-based organizations; Indian/Native American tribal governments (federally recognized); Indian/Native American tribal governments (other than federally recognized); and Indian/Native American tribally designated organizations.
Date(s) Application is Due: Feb 16; Jun 16; Oct 16
Contact: Rebecca L. Clark, (301) 496-1175; rclark@mail.nih.gov
Internet: http://grants.nih.gov/grants/guide/pa-files/PA-06-452.html
Sponsor: National Institute on Aging
Gateway Building, Room 533, 7201 Wisconsin Avenue, MSC 9205
Bethesda, MD 20892

NIA Bioenergetics, Fatigability, and Activity Limitations in Aging 3243
This FOA encourages Research Project Grant applications proposing to study bioenergetic factors underlying increased fatigability and activity limitations in aging. Increased fatigability is a significant cause of restricted physical and cognitive activity in older adults. Alterations in bioenergetics - the production and utilization of energy, and the regulation of these processes - may contribute significantly to increased fatigability. This FOA encourages applications that propose to: elucidate specific alterations in bioenergetics related to increased fatigability and activity limitations; develop and evaluate improved measures of fatigability related to bioenergetics; and evaluate interventions for increased fatigability and activity limitations that target alterations in bioenergetics and lead to improved quality of life.
Requirements: The following organizations and institutions may apply: public/state controlled institutions of higher education; private institutions of higher education; Hispanic-serving institutions; historically Black colleges and universities (HBCUs); tribally controlled colleges and universities (TCCUs); Alaska native and native Hawaiian serving institutions; nonprofits with 501(c)3 IRS status; nonprofits without 501(c)3 IRS status; small businesses; for-profit organizations; State governments; regional organizations; U.S. territories or possessions; Indian/Native American tribal governments; Indian/Native American tribally designated organizations; county governments; city or township governments; special district governments; independent

school districts; public housing authorities/Indian housing authorities; eligible agencies of the Federal government; and faith-based or community based organizations.
Date(s) Application is Due: Feb 5; Feb 16; Jun 5; Jun 16; Oct 5; Oct 16
Contact: Cheryl L. McDonald, (301) 435-0560; mcdonalc@nhlbi.nih.gov
Internet: http://grants.nih.gov/grants/guide/pa-files/PA-09-190.html
Sponsor: National Institute on Aging
Building 31, Room 5C27, 31 Center Drive, MSC 2292
Bethesda, MD 20892-7710

NIA Diversity in Medication Use and Outcomes in Aging Populations Grants 3244
The National Institute on Aging (NIA) in collaboration with the National Center for Complementary and Alternative Medicine, the Office of Research on Minority Health and the Office of Research on Women's Health invites qualified researchers to submit applications to investigate issues relevant to medication use and outcomes among older people. This program announcement solicits both basic and applied research proposals to investigate the multiple factors that influence medication use by the elderly. Goals of this announcement are to: identify the epidemiological, psychosocial, health care and clinical factors associated with medication use by older people, including the use of unconventional medical products; assess social, behavioral, psychological and cognitive factors that play a role in older people's understanding of and adherence to medication regimens; determine the role of medical and pharmaceutical professionals in facilitating or hindering proper use of prescribed and over-the-counter medications and unconventional medications including herbs and dietary supplements; refine methodologies to link these factors to medical or psychosocial outcomes; investigate interventions to improve medication adherence; and increase our understanding of biological factors which contribute to therapeutic outcomes in the use of medications by the elderly.
Requirements: Applications may be submitted by foreign and domestic for-profit and non-profit organizations, public and private, such as universities, colleges, hospitals, laboratories, units of state and local governments, and eligible agencies of the Federal government.
Contact: Laurie M. Ryan, (301) 496-9350; fax (301) 496-1494; ryanl@nia.nih.gov
Internet: http://grants.nih.gov/grants/guide/pa-files/PA-99-097.html
Sponsor: National Institute on Aging
Gateway Building, Suite 350, 7201 Wisconsin Avenue, MSC 9205
Bethesda, MD 20892-9205

NIA Effects of Gene-Social Environment Interplay on Health and Behavior in Later Life Grants 3245
The National Institute on Aging (NIA) invites applications for the development of multidisciplinary collaborations among existing longitudinal twin and family studies, with a focus on social and behavioral factors associated with aging outcomes. This FOA is intended to lay the foundation for future studies of the role of gene-environment interplay in accounting for links between social experiences and physical health, functionality, and psychological well-being in midlife and older age. Of particular interest are applications that can embed this foundation within a lifespan perspective. Applicants are encouraged to use this FOA as a starting point for: developing collaborations among existing data sets; harmonizing social phenotypes across a number of twin and family studies; conducting analyses aimed at further refining social phenotypes for genetic analysis; using animal models of gene-social environment interplay to inform work using twin studies; and/or establishing preliminary data on the feasibility of genetic/genomic approaches that could be applied to questions of how social experiences are biologically embedded and influence middle and later life outcomes. Use of existing data is strongly encouraged; funds from this award are not intended for new data collection, but a small percentage of funds may be used for adding modules to ongoing data collection efforts to augment phenotypic or genetic data. NIA expects to make 1-2 awards totaling $750,000 total costs in the first year, and $3,750,000 in total costs over the 5 year project period. Direct costs are expected to range from $200,000 -$500,000 in the first year. Letters of Intent are due by October 9, with final applications due by November 9.
Requirements: The following organizations and institutions may apply: public/state controlled institutions of higher education; private institutions of higher education; Hispanic-serving institutions; historically Black colleges and universities (HBCUs); tribally controlled colleges and universities (TCCUs); Alaska native and native Hawaiian serving institutions; nonprofits with 501(c)3 IRS status; nonprofits without 501(c)3 IRS status; small businesses; for-profit organizations; State governments; regional organizations; U.S. territories or possessions; Indian/Native American tribal governments; Indian/Native American tribally designated organizations; county governments; city or township governments; special district governments; independent school districts; public housing authorities/Indian housing authorities; eligible agencies of the Federal government; and faith-based or community based organizations.
Date(s) Application is Due: Oct 9; Nov 9
Amount of Grant: 200,000 - 500,000 U.S.D
Contact: Erica L. Spotts, (301) 451-4503; fax (301) 402-0051; spottse@mail.nih.gov
Internet: http://grants.nih.gov/grants/guide/rfa-files/RFA-AG-10-006.html
Sponsor: National Institute on Aging
Gateway Building, Room 533, 7201 Wisconsin Avenue, MSC 9205
Bethesda, MD 20892

NIA Factors Affecting Cognitive Function in Adults with Down Syndrome Grants 3246
The purpose of this solicitation is to encourage investigator initiated research to focus on various factors that maximize and maintain cognitive function in adults (>21 years of age) with Down syndrome. This solicitation encourages applications from investigators from very diverse backgrounds and interdisciplinary and transdisciplinary proposals, as well as those that involve systematic longitudinal study rather than study of samples of convenience. These studies could serve to form an evidence base that will inform not only health care management and treatment, but also issues such as quality of life, access to services, education, and ultimately affect public policy. This solicitation encourages applications that would inform several areas of inquiry: the natural history of medical and mental health issues in adults with Down syndrome, their incidence and onset; health disparities that occur, and their prevention or effective treatment; the development of better diagnostic tools to assess functional and cognitive impairment in adults with Down syndrome; improved quality of life measures; community-based interventions that improve individual educational, occupational, and social outcomes; the impact of early entry to services, type of intervention, and/or family history on quality of life for adults with Down Syndrome; and the impact of family structure and ancestry on critical life transitions, co-occurring conditions, and developmental changes of adults with Down syndrome through longitudinal studies rather than studies of samples of convenience. The requested budget is limited to $500,000 total costs per year for a maximum of five years. Letters of Intent are due by May 29, and full applications should arrive by June 30.
Requirements: The following organizations and institutions are eligible to apply: public/state controlled institutions of higher education; private institutions of higher education; nonprofits with 501(c)3 IRS status (other than institutions of higher education); nonprofits without 501(c)3 IRS status (other than institutions of higher education); for-profit organizations (other than small businesses); State governments; county governments; eligible agencies of the Federal government; and faith-based or community-based organizations.
Restrictions: Foreign institutions may not submit applications, but collaborators at foreign institutions are allowed.
Date(s) Application is Due: May 29; Jun 30
Amount of Grant: Up to 500,000 U.S.D
Contact: Mary Lou Oster-Granite, (301) 435-6866; fax (301) 496-3791; mo96o@nih.gov
Internet: http://grants.nih.gov/grants/guide/rfa-files/RFA-HD-09-028.html
Sponsor: Eunice Kennedy Shriver National Institute of Child Health and Human Development
6100 Executive Boulevard, Room 4B05L, MSC 7510
Bethesda, MD 20892-7510

NIAF Anthony Campitelli Endowed Fund Grants 3247
Anthony Campitelli immigrated to the U.S. with his family in 1928. He is a Washington D.C. architect and builder, but has maintained strong ties to his Italian birthplace, providing financial support to many of its local institutions. The Fund will support various humanitarian and community projects in and around Mr. Campitelli's birthplace, Castel Frentano (CH) in Abruzzo, Italy. These will include schools, churches, and eldercare facilities. Applications for scholarships for residents of Castel Frentano and the Chieti province will also be accepted. Grant recipients will be selected based on need and merit.
Geographic Focus: All States, Italy
Date(s) Application is Due: Jun 30
Contact: Alexandra Dall, Scholarships and Grants Manager; (202) 939-3118; fax (202 387-0800; adall@niaf.org
Internet: http://www.niaf.org/campitellifund/
Sponsor: National Italian American Foundation
1860 19th Street NW
Washington, DC 20009-5501

NIA Grants for Alzheimer's Disease Drug Discovery 3248
The objective of this solicitation is to stimulate preclinical research in the discovery, design, development and testing of novel compounds aimed at slowing, halting, or, if possible, reversing the progressive decline in cognitive function and modifying the behavioral symptoms in Alzheimer's disease as well as delaying the onset of or preventing AD. This initiative is intended to stimulate basic research and development efforts. The goal is not to duplicate or compete with pharmaceutical companies but to encourage, complement, and accelerate the process of discovering new, innovative, and effective compounds for the prevention and treatment of the cognitive impairment and behavioral symptoms associated with Alzheimer's disease. The development of compounds for ameliorating, modifying, or improving potential aberrations in neuronal cellular communication mechanisms is encouraged. These compounds should be designed to affect fundamental processes such as the neuronal dysfunction, death, and loss of connectivity associated with the disease by targeting molecules and mechanisms such as neurotransmitters, neuromodulators, and neurotrophins; receptors and ion channels; second and third messenger systems; all facets of relevant amyloid precursor protein, amyloid beta protein, and tau neurobiology; protein synthesis, aggregation, and degradation; energy utilization; and oxidative, immunological, and inflammatory mechanisms.
Requirements: The following organizations are eligible to apply: for-profit organizations; non-profit organizations; public or private institutions, such as universities, colleges, hospitals, and laboratories; units of State government; units of local government; eligible agencies of the Federal government; foreign institutions; and domestic institutions.
Date(s) Application is Due: Feb 16; Jun 16; Oct 16
Amount of Grant: Up to 275,000 U.S.D
Contact: Neil S. Buckholtz, (301) 496-9350; fax (301) 496-1494; Buckholn@nia.nih.gov
Internet: http://grants.nih.gov/grants/guide/pa-files/PAS-06-261.html
Sponsor: National Institute on Aging
Gateway Building, Suite 350, 7201 Wisconsin Avenue, MSC 9205
Bethesda, MD 20892-9205

NIA Harmonization of Longitudinal Cross-National Surveys of Aging Grants 3249
The National Institute on Aging (NIA) will provide support through the R21 mechanism to facilitate the ex-ante and ex-post harmonization of international longitudinal surveys

of aging populations to the Health and Retirement Study to maximize cross-national comparability of data. Facilitation will include, but is not limited to, meetings, pilots, and the development of methods to improve cross-national comparability of international aging surveys to the Health and Retirement Study and existing international data files that are harmonized with the HRS: the English Longitudinal Study on Ageing (ELSA); and the Survey of Health, Ageing, and Retirement in Europe (SHARE).
Requirements: The following organizations and institutions may apply: public/state controlled institutions of higher education; private institutions of higher education; nonprofits with 501(c)3 IRS status; nonprofits without 501(c)3 IRS status; for-profit organizations; State governments; and non-domestic (non-U.S.) entities (foreign organizations).
Date(s) Application is Due: Feb 16; Jun 16; Oct 16
Amount of Grant: Up to 275,000 U.S.D
Contact: John W. R. Phillips, (301) 496-3138; fax (301) 402-0051; phillipj@mail.nih.gov
Internet: http://grants.nih.gov/grants/guide/pa-files/PAS-07-387.html
Sponsor: National Institute on Aging
Gateway Building, Room 533, 7201 Wisconsin Avenue, MSC 9205
Bethesda, MD 20892

NIA Health Behaviors and Aging Grants 3250
The National Institute on Aging (NIA) invites qualified researchers to submit applications for research and research training on those health-related behaviors and attitudes of older adults, their families, and significant others, that can affect health and functioning as people grow older. Studies are sought that extend scientific understanding of how older adults health behaviors and attitudes develop under varying social conditions; how they relate to health promotion and disease prevention, care and treatment of disease, rehabilitation or death; and how they can be modified as relevant new scientific knowledge is developed.
Requirements: Applications may be submitted by foreign and domestic, for-profit and non-profit organizations, public and private, such as universities, colleges, hospitals, laboratories, units of State and local governments, and eligible agencies of the Federal government. Applications from minority individuals and women are encouraged.
Date(s) Application is Due: Feb 5; Jun 5; Oct 5
Amount of Grant: Up to 150,000 U.S.D
Contact: Chyren Hunter, (301) 496-9322; fax (301) 402-0528; Hunterc@nia.nih.gov
Internet: http://grants.nih.gov/grants/guide/pa-files/PA-93-064.html
Sponsor: National Institute on Aging
Gateway Building, Room 2C218, 7201 Wisconsin Avenue, MSC 9205
Bethesda, MD 20892-9205

NIA Healthy Aging through Behavioral Economic Analyses of Situations Grants 3251
The National Institute on Aging (NIA), National Institutes of Health, solicits Research Project Grant applications that propose to translate basic findings from Behavioral Economics into behavior change interventions targeting health behaviors associated with chronic health conditions of mid-life and older age. Applications should propose small pilot clinical trials or demonstration projects, ideally based on collaborations between individuals with expertise in behavioral economics and psychologists, psychiatrists, clinicians, or others with expertise in aging or implementing behavioral interventions. Depending on the availability of funding, the NIA expects to make 4 or 5 awards totaling $1,250,000 total costs in the first year, and $3,750,000 over the 3 year project period. The direct cost amount for each individual award will be no greater than $525,000 in total for three years, and no more than $175,000 in any year. Letters of Intent are due by October 2, with final applications due by November 2.
Requirements: The following organizations and institutions may apply: public/state controlled institutions of higher education; private institutions of higher education; Hispanic-serving institutions; historically Black colleges and universities (HBCUs); tribally controlled colleges and universities (TCCUs); Alaska native and native Hawaiian serving institutions; nonprofits with 501(c)3 IRS status; nonprofits without 501(c)3 IRS status; small businesses; for-profit organizations; State governments; regional organizations; U.S. territories or possessions; Indian/Native American tribal governments; Indian/Native American tribally designated organizations; county governments; city or township governments; special district governments; independent school districts; public housing authorities/Indian housing authorities; eligible agencies of the Federal government; and faith-based or community based organizations.
Date(s) Application is Due: Oct 2; Nov 2
Amount of Grant: Up to 175,000 U.S.D
Contact: Dr. Jonathan W. King, (301) 402-4156; fax (301) 402-0051; kingjo@nia.nih.gov
Internet: http://grants.nih.gov/grants/guide/rfa-files/RFA-AG-10-008.html
Sponsor: National Institute on Aging
Gateway Building, Room 533, 7201 Wisconsin Avenue, MSC 9205
Bethesda, MD 20892

NIA Higher-Order Cognitive Functioning and Aging Grants 3252
The National Institute on Aging (NIA) invites qualified researchers to submit new applications for research projects that focus on adulthood and aging related changes in the higher-order processes and strategies required for judgment, decision-making, reasoning, problem-solving, and processing complex information. As in earlier periods of life older adults continue to make decisions related to everyday life, but with advanced age, new, and sometimes even more complex, decision-making is required of them. Recent research indicates that age-related limitations in cognitive processing resources (e.g., speed and working memory) may impact decision-making. Research also indicates that some older adults experience growth in specific areas of cognitive functioning (e.g., expertise, semantic knowledge, emotional regulation) and continue to use adaptive intelligence, demonstrating multi-directionality in adult cognitive change. It is generally recognized that research on higher-order processing is underdeveloped in the field of aging. Research proposals are needed that examine the actual processes that are engaged when older adults make important decisions, how these processes change with age and context, and what environmental supports, interventions, and training may be necessary for optimal functioning. Research may investigate either individual or collaborative, or social processes. Modular Grant applications will request direct costs in $25,000 modules, up to a total direct cost request of $250,000 per year.
Requirements: Applications may be submitted by foreign and domestic, for-profit and non-profit, public and private organizations such as universities, colleges, hospitals, laboratories, units of State and local governments, and eligible agencies of the Federal government. Racial/ethnic minority individuals, women, and persons with disabilities are encouraged to apply as principal investigators.
Date(s) Application is Due: Feb 16; Jun 16; Oct 16
Amount of Grant: Up to 250,000 U.S.D
Contact: Chyren Hunter, (301) 496-9322; fax (301) 402-0528; Hunterc@nia.nih.gov
Internet: http://grants.nih.gov/grants/guide/pa-files/PA-00-052.html
Sponsor: National Institute on Aging
Gateway Building, Room 2C218, 7201 Wisconsin Avenue, MSC 9205
Bethesda, MD 20892-9205

NIA Human Biospecimen Resources for Aging Research Grants 3253
The purpose of this announcement is to disseminate information about available human biospecimen resources for age-related studies, and promote independent use of existing human biospecimen resources through collaborative and other arrangements with the studies that collect biospecimens. Websites for the participating studies are accessed via the Virtual Repository. NIA is interested in aging-related research that integrates biomarker analyses with the study of normal aging, age-associated diseases, and behavioral and psychosocial data. Budgets for direct costs of up to $50,000 per year and a project duration of up to two years may be requested for a maximum of $100,000 direct costs over a two-year project period.
Requirements: The following organizations and institutions may apply: for-profit organizations; non-profit organizations; public or private institutions, such as universities, colleges, hospitals, and laboratories; units of State government; units of Local government; units of State Tribal government; units of Local Tribal government; eligible agencies of the Federal government; foreign institutions; domestic institutions; faith-based or community-based organizations; Indian/Native American tribal governments (federally recognized); Indian/Native American tribal governments (other than federally recognized); and Indian/Native American tribally designated organizations.
Date(s) Application is Due: Feb 16; Jun 16; Oct 16
Amount of Grant: Up to 100,000 U.S.D
Contact: Nancy L. Nadon, (301) 402-7744; fax (301) 402-0010; nadonn@nia.nih.gov
Internet: http://grants.nih.gov/grants/guide/pa-files/PA-06-443.html
Sponsor: National Institute on Aging
Gateway Building, Suite 2C231, 7201 Wisconsin Avenue, MSC 9205
Bethesda, MD 20892

NIAID Allergy, Immunology, and Transplantation Program Research Grants 3254
This program concerns the immune system as it functions in the maintenance of health and as it malfunctions in the production of disease. Because of this dual focus, the program encompasses both basic research (immunology) and clinical research (allergic and immunologic diseases). Basic research is supported under immunobiology and immunochemistry to provide the information necessary for understanding immune system structure and function and under genetics and transplantation biology, directed more specifically at the histocompatibility antigens and their role in graft rejection, autoimmunity, immune responsiveness, and a variety of diseases. Clinical research is supported under asthma and allergic diseases, directed at determining the cause and improving the diagnosis, prevention, and treatment of a variety of common allergic diseases, and under immunologic diseases covering disorders of the human immune system and their underlying pathophysiologic inflammation and immunopathologic mechanisms and consequences. Clinical aspects of transplantation and genetic predisposition to disease are supported under genetics and transplantation biology. The small instrumentation program supports the purchase of relatively low-cost pieces of research equipment that generally are not funded in research project grants and that do not qualify for support under the NIH larger shared instrumentation program.
Requirements: Universities, colleges, hospitals, laboratories, and other public or private nonprofit domestic institutions, including state and local units of government, and individuals are eligible for grant support of research by a named principal investigator or a research career development candidate. Also, for-profit organizations are eligible.
Date(s) Application is Due: Feb 1; Jun 1; Oct 1
Amount of Grant: 100,000,000 U.S.D
Contact: Mary Kirker, Grants Management Officer; (301) 496-7075
Internet: http://12.46.245.173/pls/portal30/CATALOG.PROGRAM_TEXT_RPT.show
Sponsor: National Institute of Allergy and Infectious Diseases
6610 Rockledge Drive
Bethesda, MD 20892

NIAID Allergy, Immunology, Transplantation, Infectious Diseases, and AIDS NRSAs—Individual 3255
NRSAs are offered to increase the number of biomedical investigators who are conducting high-quality research in the areas of allergy, immunology, transplantation, infectious diseases, and AIDS. Predoctoral fellowships for minority and disabled students provide up to five years of support. NRSA postdoctoral fellowships provide support for up to three years. NRSA senior fellowships provide support for two years. Deadlines for

postdoctoral and senior fellowships are April 5, August 5, and December 5. Annual deadline dates may vary; contact program, staff for exact dates.
Requirements: Awards are given only to U.S. citizens and noncitizen nationals. Minority predoctoral fellowships are available to underrepresented minority students (African Americans, Hispanic, Native Americans, and Pacific Islanders) or scientists with disabilities who will be entering a PhD program. Postdoctoral fellowship applicants must have a doctoral degree (PhD, MD, DVM, DDS, OD, etc.). Senior fellowship applicants should plan to do research that will provide training the candidate does not have and will become an asset to her or his academic career.
Date(s) Application is Due: Apr 5; Aug 5; Dec 5
Amount of Grant: 7,586 - 743,606 U.S.D
Contact: Mary Kirker, Grants Management Officer; (301) 496-7075
Internet: http://12.46.245.173/pls/portal30/CATALOG.PROGRAM_TEXT_RPT.show
Sponsor: National Institute of Allergy and Infectious Diseases
6610 Rockledge Drive
Bethesda, MD 20892

NIAID Challenge Grants: Joint Ventures in Biomedicine and Biotechnology **3256**
The purpose of this initiative is to support research and development efforts whose outcomes could significantly reduce the impact of infectious diseases nationally and worldwide. NIAID has identified areas of high importance where it believes successful product research and development combined with existing infrastructures and federal challenge grant funding could significantly impact a major health or medical problem. Applications in response to this RFA for challenge grants are limited to these areas: Medicines for Malaria; Medicines for Tuberculosis; Vaccines for Influenza; Vaccines for Emerging and Resistant Infections; and Therapeutics for Emerging and Resistant Infections. Awards will be made in two Phases: Phase I awards will be short-term (two month), small dollar ($25,000) research grants for development of detailed plans for implementation of research and development efforts (i.e., a Phase II grant application); and Phase II awards will be cooperative agreement awards to those Phase I awardees whose research plans present the best opportunities, based on peer review and potential public health significance, for products that will benefit the public health. Annual deadline dates may vary; contact program staff for exact dates. RFA: AI-00-010
Requirements: Applications may be submitted by domestic for-profit and nonprofit organizations such as biotechnology, bioengineering and pharmaceutical companies with collaborators as necessary from universities, colleges, hospitals, and laboratories.
Restrictions: Phase I applications must directly address one or more of the research topics listed. Only Phase I awardees will be eligible for Phase II awards.
Amount of Grant: Up to 20,000,000 U.S.D
Contact: Program Officer; niaidchallenge@niaid.nih.gov
Internet: http://grants.nih.gov/grants/guide/rfa-files/RFA-AI-00-010.html
Sponsor: National Institute of Allergy and Infectious Diseases
6610 Rockledge Drive
Bethesda, MD 20892

NIAID Enhancing Vaccine-Elicited Protective Immunity in Malaria Research Grants **3257**
NIAID invites applications for research project (R01) grants to conduct innovative research to advance fundamental understanding of eliciting optimal protective immune responses with candidate malaria vaccines. As part of its Malaria Vaccine Development Plan, NIAID seeks to identify and validate methods and strategies for improved antigen delivery, processing, and presentation or enhancement of protective immunity with existing candidate antigens. Identification, characterization, and evaluation of new candidate vaccines, especially combination vaccines based on a rational strategy for antigen inclusion, are also encouraged. Because many of the poorest nations of the world are afflicted with malaria, to be effective any intervention must be inexpensive, cost-effective, and relatively easy to administer and maintain. Annual deadline dates may vary; contact program staff for exact dates. RFA: AI-99-006
Requirements: Applications may be submitted by domestic and foreign for-profit and nonprofit organizations and public and private institutions.
Amount of Grant: Up to 250,000 U.S.D
Contact: Dr. B.F. Hall, Microbiology and Infectious Diseases Coordinator; (301) 496-2544; fax (301) 402-0659; bh24q@nih.gov
Internet: http://www.grants.nih.gov/grants/guide/rfa-files/RFA-AI-99-006.html
Sponsor: National Institute of Allergy and Infectious Diseases
6610 Rockledge Drive
Bethesda, MD 20892

NIAID HIV Prevention Trial Units Grants **3258**
NIAID invites applications for HIV Prevention Trial Units (HPTUs), which will be integral parts of an HIV Prevention Trials Network (HPTN) designed to carry out a comprehensive scientific agenda on HIV prevention strategies. The HPTN will conduct domestic and international research on promising biomedical and behavioral strategies for the prevention of HIV transmission among adult, pediatric, and adolescent populations. Modalities of interventions may include, but are not limited to, antiretroviral drugs, microbicides, behavioral approaches, barrier and other contraceptive/STD prevention methods, vaccines (in perinatal settings), chemoprophylaxis, treatment of sexually transmitted diseases, and combined approaches. The HPTUs will contribute to the HPTN research agenda; conduct/participate in Phase I, II, and/or III prevention intervention clinical trials domestically and/or internationally; and expand for multiple prevention efficacy trials. Potential applicants should refer to the following Web site for further information on this initiative: http://www.niaid.nih.gov/daids/vtn-ptn. Annual deadline dates may vary; contact program staff for exact dates. RFA: AI-99-010
Requirements: Applications may be submitted by domestic and foreign for-profit and nonprofit organizations, public and private institutions, and eligible agencies of the federal government.
Contact: Rodney Hoff; (301) 496-6177; fax (301) 402-3684; rh25v@nih.gov
Internet: http://www.grants.nih.gov/grants/guide/rfa-files/RFA-AI-99-010.html
Sponsor: National Institute of Allergy and Infectious Diseases
6610 Rockledge Drive
Bethesda, MD 20892

NIAID HIV Vaccine Clinical Trial Units Grants **3259**
The Division of AIDS (DAIDS) of NIAID solicits applications for HIV Vaccine Trial Clinical Units (HVTUs) that are integral parts of an HIV Vaccine Trials Network (HVTN) designed to carry out a comprehensive scientific agenda on HIV vaccines. The HVTUs will contribute to the HVTN research agenda, participate in domestic and international Phase I and II clinical trials of candidate HIV vaccines, and participate in domestic and international Phase III efficacy trials of HIV vaccines. Potential applicants should refer to the following Web site for further information on this initiative: http://www.niaid.nih.gov/daids/vtn-ptn. In addition to the HVTN, NIAID supports the creation of the HIV Prevention Trials Network (HPTN) to perform Phase I, II, and III trials of other prevention strategies to prevent HIV-1 transmission, domestically and internationally. Clinical trials of vaccines focused on questions around maternal-infant transmission will be conducted within the HPTN in collaboration with the HVTN and, when desirable, the Pediatric AIDS Clinical Trials Network. Sites funded under the HPTN may serve as effective venues for the eventual implementation of larger HIV vaccine trials; therefore, linkages between HPTN and HVTN, in addition to other scientific organizations, are strongly encouraged. Annual deadline dates may vary; contact program staff for exact dates. RFA: AI-99-009
Requirements: Applications may be submitted by domestic and foreign for-profit and nonprofit organizations, public and private institutions, and eligible agencies of the federal government.
Contact: Dr. Jorge Flores; (301) 435-3578; fax (301) 402-3684; jf30t@nih.gov
Internet: http://www.grants.nih.gov/grants/guide/rfa-files/RFA-AI-99-009.html
Sponsor: National Institute of Allergy and Infectious Diseases
6610 Rockledge Drive
Bethesda, MD 20892

NIAID HIV Vaccine Research and Design—Program Project Grants **3260**
NIAID invites applications for research aimed at obtaining an efficacious vaccine against HIV or AIDS. NIAID supports all stages of AIDS vaccine research and development—from basic research to iterative product development through clinical trials. Applications may target any area of AIDS vaccine research. These areas include, but are not limited to, HIV antigen processing, correlates of immunity, animal model development, DNA vaccination, studies of the structure of HIV immunogens, development of virus and bacterial vaccine vectors, studies of existing vaccine cohorts, studies targeting initial infection (dendritic cells), and extensive modeling of vaccine concepts in macaques or chimps, etc. Clinical studies involving humans would not be considered appropriate for this announcement. Annual deadline dates may vary; contact program staff for exact dates. PA: PAR-00-093
Requirements: Applications may be submitted by domestic and foreign, for-profit and nonprofit organizations, public and private.
Date(s) Application is Due: Jun 10; Nov 10
Contact: Dr. Michael Pensiero, Division of AIDS; (301) 435-3749; fax (301) 402-3684; mpensiero@niaid.nih.gov
Internet: http://grants.nih.gov/grants/guide/pa-files/PAR-00-093.html
Sponsor: National Institute of Allergy and Infectious Diseases
6610 Rockledge Drive
Bethesda, MD 20892

NIAID Independent Scientist Award **3261**
The award is a special salary-only grant designed to provide protected time for newly independent scientists who currently have non-research obligations such as heavy teaching loads, clinical work, committee assignments, service, and administrative duties that prevent them from having a period of intensive research focus. The award is targeted to persons with doctoral degrees who have completed their research training, have independent peer-reviewed research support, and who need a period of protected research time in order to foster their research career development. New applications are due February 12, June 12, and October 12, while renewal applications are due March 12, July 12, and November 12. Up to $75,000 plus fringe benefits per year is available. Additional research is not generally provided.
Requirements: The following organizations and institutions are eligible to apply: public/state controlled institutions of higher education; private institutions of higher education; nonprofits with 501(c)3 IRS status; nonprofits without 501(c)3 IRS status; small businesses; for-profit organizations; State governments; U.S. territories or possessions; Indian/Native American tribal governments (Federally recognized and other than Federally recognized); Indian/Native American tribally designated organizations; Hispanic-serving institutions; historically Black colleges and universities (HBCUs); tribally controlled colleges and universities (TCCUs); Alaska Native and Native Hawaiian serving institutions; regional organizations; and faith-based or community based organizations. The candidate must have a doctoral degree and peer-reviewed, independent research support at the time the award is made. The candidate must spend a minimum of 75 percent effort conducting research during the period of the award.
Restrictions: The award is not intended for investigators who already have full time to perform research, or have substantial publication records or considerable research

support indicating that they are well established in their fields. Foreign institutions are not eligible to apply.
Date(s) Application is Due: Feb 12; Mar 12; Jun 12; Jul 12; Oct 12; Nov 12
Amount of Grant: 75,000 U.S.D
Contact: Milton J. Hernandez, (301) 496-3775; mh35c@nih.gov
Internet: http://grants.nih.gov/grants/guide/pa-files/PA-09-038.html
Sponsor: National Institute of Allergy and Infectious Diseases
6610 Rockledge Drive
Bethesda, MD 20892

NIAID Integrated Preclinical/Clinical AIDS Vaccine Development Grants 3262
NIAID invites applications for research aimed at obtaining a safe and efficacious vaccine against HIV or AIDS. The overall goal of this program is to support multidisciplinary research that will lead to a safe and protective AIDS vaccine. As such, IPCAVD applications must include human clinical studies within the term of the award. Applicants must address the following issues in their application regarding human studies: design of the clinical trial, plans for recruitment, enrollment, education and follow-up of clinical subjects, including attention to the possibility of social harms and discrimination arising from vaccine- induced positive screening tests that may be interpreted incorrectly as HIV infection, plus documentation of compliance with all applicable federal (FDA), state and local laws and regulations governing clinical research. Annual deadlines may vary; contact program staff for exact dates. PA: PAR-00-072
Requirements: Applications may be submitted by domestic for-profit and nonprofit organizations, both public and private; units of state and local governments; and eligible agencies of the federal government.
Contact: Dr. Michael Pensiero; (301) 435-3749; mpensiero@niaid.nih.gov
Internet: http://grants.nih.gov/grants/guide/pa-files/PAR-00-072.html
Sponsor: National Institute of Allergy and Infectious Diseases
6610 Rockledge Drive
Bethesda, MD 20892

NIAID Laboratory Methods to Assess Responses to HIV Vaccine Candidates Research Grants 3263
NIAID invites applications to conduct innovative research to develop and improve technologies for evaluation of immune responses to HIV vaccines and to expand the range and scope of immune functions currently measured in vaccine trials. This PA intends to complement the scope of vaccine research carried out in human volunteers and animals receiving experimental HIV/SIV vaccines in prophylactic and/or therapeutic trials funded through other programs. Development of new techniques that detect and quantitate changes in innate (nonspecific) and HIV-specific responses will be fostered. Research may include comparative analysis of samples acquired from individuals infected with HIV or who have been exposed to, but not infected with, HIV. Studies in animals, particularly primates that have received SIV/SHIV vaccines, are also encouraged. Applications to develop reagents (including non-clade B based materials), to adapt assays to field conditions and to improve on cell cryopreservation methods are also appropriate. The PA is not intended to directly support the conduct of human or animal clinical trials. Contact the program office for deadline dates. PA: PA-99-044
Requirements: Applications may be submitted by domestic and foreign, for-profit and nonprofit organizations, public and private.
Contact: Dr. Jorge Flores; (301) 435-3758; fax (301) 402-3684; jf30t@nih.gov
Internet: http://www.grants.nih.gov/grants/guide/pa-files/PA-99-044.html
Sponsor: National Institute of Allergy and Infectious Diseases
6610 Rockledge Drive
Bethesda, MD 20892

NIAID Microbiology and Infectious Diseases Research Grants and Contracts 3264
This program supports research with the broad aim of improving health by controlling diseases caused by infectious or parasitic agents. Investigator-originated and grant-supported research focuses on all aspects of microbiology, with particular emphasis on human pathogens. Projects range from studies of microbial physiology and antigenic structure to collaborative trials of experimental drugs and vaccines. NIAID also supports studies on the mechanisms of resistance to antibiotics as well as research dealing with epidemiologic observations in hospitalized patients or community populations. Program project grants and research contracts are used to fund selected projects of high institute relevance in the branches covering molecular microbiology; bacteriology, mycobacteriology, and mycology; virology; and parasitology.
Requirements: Universities, colleges, hospitals, laboratories, and other public or private nonprofit domestic institutions, including state and local units of government, are eligible. Individuals are eligible to make application for grant support of research by a named principal investigator or a research career development candidate.
Date(s) Application is Due: Feb 1; Jun 1; Oct 1
Amount of Grant: 2,000 - 100,000,000 U.S.D
Contact: Mary Kirker, Grants Management Officer; (301) 496-7075
Internet: http://12.46.245.173/pls/portal30/CATALOG.PROGRAM_TEXT_RPT.show
Sponsor: National Institute of Allergy and Infectious Diseases
6610 Rockledge Drive
Bethesda, MD 20892

NIAID Pediatric Clinical Trials Program for AIDS 3265
The purpose of this RFA is to solicit applications from institutions interested in establishing cooperative clinical trials groups to plan and direct pediatric therapeutics and prevention research. The awardees will have the capacity and expertise to conduct all phases of clinical trials involving children, adolescents, and pregnant women with all stages of infection. Studies focus on high priority research questions on the treatment and pathogenesis of HIV disease and its sequelae. The scope of activities ranges from studies requiring in-depth laboratory support to research in long-term clinical outcomes. Annual deadline dates may vary: contact program staff for exact dates. RFA: AI-00-015
Requirements: Applications may be submitted by domestic for-profit and nonprofit organizations, both public and private; units of state and local governments; and eligible agencies of the federal government.
Restrictions: Foreign institutions are not eligible to apply as primary grantees; however, foreign institutions may participate as part of a domestic application.
Amount of Grant: Up to 36,000,000 U.S.D
Contact: Dr. Frederick Batzold, (301) 402-0143; fax (301) 480-4582; fb10c@nih.gov
Internet: http://grants.nih.gov/grants/guide/rfa-files/RFA-AI-00-015.html
Sponsor: National Institute of Allergy and Infectious Diseases
6610 Rockledge Drive
Bethesda, MD 20892

NIAID Preparedness against Illegitimate Use of Bacterial Pathogens 3266
The purpose of this program is to support basic research on molecular and genetic aspects of pathogenesis of, and host immune/defense responses to, the following bacterial pathogens: Bacillus anthracis, Yersinia pestis, Francisella tularensis, Coxiella burnetii, or Brucella species. It is hoped that research from funded projects will be used to devise appropriate defense strategies that can be implemented shortly before or soon after bioterrorist acts. Annual deadline dates may vary; contact program staff for exact dates.
Requirements: Applications may be submitted by domestic for-profit and nonprofit organizations, both public and private; units of state and local governments; and eligible agencies of the federal government.
Amount of Grant: Up to 15,000,000 U.S.D
Contact: Dr. Phillip Baker, Division of Microbiology and Infectious Diseases; (301) 435-2855; fax (301) 402-2508; pb26o@nih.gov
Internet: http://grants.nih.gov/grants/guide/rfa-files/RFA-AI-01-002.html
Sponsor: National Institute of Allergy and Infectious Diseases
6610 Rockledge Drive
Bethesda, MD 20892

NIAID Research Scholar Development Award 3267
NIAID invites applications from outstanding intramural and extramural postdoctoral fellows for the Research Scholar Development Award (RSDA). The RSDA provides support for postdoctoral fellows who are moving to assistant professor positions in academic institutions. The purpose of the RSDA is to ease the transition to an academic position by enabling the awardee to focus on the establishment of his or her research laboratory prior to submitting applications for grant support. PAR-98-072 may be obtained through the NIH grant line (301-402-2221), the NIH Web site, mail, and email.
Requirements: The candidate must have a health professional doctoral degree and have demonstrated the capacity for highly productive, independent postdoctoral research in her or his chosen field prior to applying for this award.
Contact: Dr. Milton Hernandez, Division of Extramural Activities Contact; (301) 496-3775; fax (301) 402-0369; mh35c@nih.gov
Internet: http://www.grants.nih.gov/grants/guide/pa-files/PAR-98-072.html
Sponsor: National Institute of Allergy and Infectious Diseases
6610 Rockledge Drive
Bethesda, MD 20892

NIAID Small Business Innovation Research Advanced Technology Grants 3268
This PA invites grant applications for Small Business Innovation Research (SBIR) projects with award duration and amounts greater than those routinely allowed under the SBIR program. This PA provides examples of advanced technology areas in which NIAID will entertain SBIR applications and provides NIAID's interpretation and clarification of instructions relating to total cost, dates of project period, and page limitations for these applications. Information may be obtained through the NIH grant line (301-402-2221), the NIH Web site, mail, and email. PA: PA-04-127
Date(s) Application is Due: Apr 1; Aug 1; Dec 1
Contact: Gregory Milman, Division of Extramural Activities; (301) 496-8666; fax (301) 402-0369; gm16s@nih.gov
Internet: http://grants.nih.gov/grants/guide/pa-files/PA-04-127.html
Sponsor: National Institute of Allergy and Infectious Diseases
6610 Rockledge Drive
Bethesda, MD 20892

NIAID Statistical Methods in HIV and AIDS Research Grants 3269
NIAID and NIMH invite applications for the development of original statistical methods to advance the understanding, treatment, and prevention of HIV and AIDS. Traditional research project grant (R01) applications may be submitted in response to this PA. PA-98-054 may be obtained through the NIH grant line (301-402-2221), the NIH Web site, mail, and email.
Requirements: Applications may be submitted by domestic and foreign, for-profit and nonprofit organizations, public or private.
Contact: Mary Ledford, Grants Management; (301) 402-6446; fax (301) 480-3780; m128g@nih.gov
Internet: http://www.grants.nih.gov/grants/guide/pa-files/PA-98-054.html
Sponsor: National Institute of Allergy and Infectious Diseases
6610 Rockledge Drive
Bethesda, MD 20892

NIAID Therapeutics Research on AIDS-Associated Opportunistic Infections 3270
The goal of this program is the discovery of agents for treatment of AIDS-associated opportunistic infections. Given present advances in molecular biology, the principal objectives of the program are to stimulate original and innovative preclinical research to identify and characterize new, pathogen-specific molecular targets and to develop promising therapeutic approaches based on new targets. Program announcement is available on the Web site along with an addendum (http://grants2.nih.gov/grants/guide/notice-files/NOT-AI-00-001.html). PA: PA-99-124
Requirements: Applications may be submitted by domestic and foreign, for-profit and nonprofit organizations, both public and private; units of state and local governments; and eligible agencies of the federal government. Racial/ethnic minority individuals, women, and persons with disabilities are encouraged to apply.
Amount of Grant: 25,000 - 250,000 U.S.D
Contact: Dr. Chris Lambros, (301) 435-3769; clambros@niaid.nih.gov
Internet: http://www.grants.nih.gov/grants/guide/pa-files/PA-99-124.html
Sponsor: National Institute of Allergy and Infectious Diseases
6610 Rockledge Drive
Bethesda, MD 20892

NIA Independent Scientist Award 3271
The award is a special salary-only grant designed to provide protected time for newly independent scientists who currently have non-research obligations such as heavy teaching loads, clinical work, committee assignments, service, and administrative duties that prevent them from having a period of intensive research focus. The award is targeted to persons with doctoral degrees who have completed their research training, have independent peer-reviewed research support, and who need a period of protected research time in order to foster their research career development. New applications are due February 12, June 12, and October 12, while renewal applications are due March 12, July 12, and November 12. Up to 75% per year of base salary plus fringe benefits is available. No other research development support funds are provided.
Requirements: The following organizations and institutions are eligible to apply: public/state controlled institutions of higher education; private institutions of higher education; nonprofits with 501(c)3 IRS status; nonprofits without 501(c)3 IRS status; small businesses; for-profit organizations; State governments; U.S. territories or possessions; Indian/Native American tribal governments (Federally recognized and other than Federally recognized); Indian/Native American tribally designated organizations; Hispanic-serving institutions; historically Black colleges and universities (HBCUs); tribally controlled colleges and universities (TCCUs); Alaska Native and Native Hawaiian serving institutions; regional organizations; and faith-based or community based organizations. The candidate must have a doctoral degree and peer-reviewed, independent research support at the time the award is made. The candidate must spend a minimum of 75 percent effort conducting research during the period of the award.
Restrictions: The award is not intended for investigators who already have full time to perform research, or have substantial publication records or considerable research support indicating that they are well established in their fields. Foreign institutions are not eligible to apply.
Date(s) Application is Due: Feb 12; Mar 12; Jun 12; Jul 12; Oct 12; Nov 12
Contact: Chyren Hunter, (301) 496-9322; fax (301) 402-0528; Hunterc@nia.nih.gov
Internet: http://grants.nih.gov/grants/guide/pa-files/PA-09-038.html
Sponsor: National Institute on Aging
Gateway Building, Room 2C218, 7201 Wisconsin Avenue, MSC 9205
Bethesda, MD 20892-9205

NIA Malnutrition in Older Persons Research Grants 3272
NIA continues to encourage research on the full spectrum of issues related to the causes, prevention, and treatment as well as sociobehavioral and economic aspects related to malnutrition in older persons. Support for this program will be through the research project grant and mentored research scientist development award. Applications will compete for available funds on the basis of scientific merit with all other applications assigned to the institute.
Requirements: Applications may be submitted by foreign and domestic, for-profit and non-profit organizations, public and private, such as universities, colleges, hospitals, laboratories, unit of State and local governments, and eligible agencies of the Federal government. Applications from minority individuals and women are encouraged.
Date(s) Application is Due: Feb 12; Mar 12; Jun 12; Jul 12; Oct 12; Nov 12
Contact: Chyren Hunter, (301) 496-9322; fax (301) 402-0528; Hunterc@nia.nih.gov
Internet: http://grants.nih.gov/grants/guide/pa-files/PA-94-088.html
Sponsor: National Institute on Aging
Building 31, Room 5C27, 31 Center Drive, MSC 2292
Bethesda, MD 20892-7710

NIA Mechanisms, Measurement, and Management of Pain in Aging: from Molecular to Clinical 3273
This FOA encourages Research Project Grant applications from institutions and organizations that propose to: study biological, neurobiological, psychosocial, and clinical mechanisms and processes by which aging and/or age-related diseases affect the experience of pain; examine biological, neurobiological, psychosocial, and clinical factors that impact pain experience and prevalence in older people; evaluate existing pain assessment and/or management approaches in older adults; or develop new assessment methods and/or management strategies for pain with particular attention to the needs of older adults. Studies involving animal models or human subjects are appropriate under this program announcement.
Requirements: The following organizations and institutions may apply: public/state controlled institutions of higher education; private institutions of higher education; Hispanic-serving institutions; historically Black colleges and universities (HBCUs); tribally controlled colleges and universities (TCCUs); Alaska native and native Hawaiian serving institutions; nonprofits with 501(c)3 IRS status; nonprofits without 501(c)3 IRS status; small businesses; for-profit organizations; State governments; regional organizations; U.S. territories or possessions; Indian/Native American tribal governments; Indian/Native American tribally designated organizations; county governments; city or township governments; special district governments; independent school districts; public housing authorities/Indian housing authorities; eligible agencies of the Federal government; and faith-based or community based organizations.
Date(s) Application is Due: Feb 5; Feb 16; Jun 5; Jun 16; Oct 5; Oct 16
Contact: Wen G. Chen, (301) 496-9350; fax (301) 496-1494; chenw@nia.nih.gov
Internet: http://grants.nih.gov/grants/guide/pa-files/PA-09-193.html
Sponsor: National Institute on Aging
Gateway Building, Suite 350, 7201 Wisconsin Avenue, MSC 9205
Bethesda, MD 20892-9205

NIA Mechanisms Underlying the Links between Psychosocial Stress, Aging, the Brain and the Body Grants 3274
This FOA encourages multidisciplinary and interdisciplinary research to elucidate the mechanistic links between psychosocial stress and health in aging, as well as how the aging process and age-related diseases affect the responses to psychosocial stressors. Generally, research should be focused on: aging and how neural mechanisms respond to psychosocial stress and affect other body systems; characterizing the behavioral, psychological and social mechanisms and pathways involved in transducing psychosocial stressors into health outcomes; how stressors modulate physiological process underlying life-span, immune mechanisms, and metabolism; and how psychosocial stress contributes to the development or progression of geriatric syndromes, chronic medical conditions, and disabilities in later life. Research is strongly encouraged that aims to identify appropriate targets for intervention, at any level of analysis, from societal to molecular. Research spanning multiple levels of analysis is particularly encouraged.
Requirements: The following organizations and institutions are eligible to apply: public/state controlled institutions of higher education; private institutions of higher education; Hispanic-serving institutions; historically Black colleges and universities (HBCUs); tribally controlled colleges and universities (TCCUs); Alaska native and native Hawaiian serving institutions; nonprofits with 501(c)3 IRS status (other than institutions of higher education); nonprofits without 501(c)3 IRS status (other than institutions of higher education); small businesses; for-profit organizations (other than small businesses); state governments; Indian/Native American tribal governments (federally recognized); Indian/Native American tribally designated organizations; county governments; city or township governments; special district governments; independent school districts; public housing authorities/Indian housing authorities; U.S. territories or possessions; Indian/native American tribal governments (other than federally recognized); regional organizations; and non-domestic (non-U.S.) entities (foreign organizations).
Restrictions: Research focused on oxidative stress or on environmental or physical stressors of a non-psychosocial nature is not appropriate for this FOA.
Date(s) Application is Due: Mar 5; Jul 5; Nov 5
Contact: Suzana S. Petanceska, (301) 496-9350; petanceskas@mail.nih.gov
Internet: http://grants.nih.gov/grants/guide/pa-files/PA-09-216.html
Sponsor: National Institute on Aging
Gateway Building, Suite 350, 7201 Wisconsin Avenue, MSC 9205
Bethesda, MD 20892-9205

NIA Medical Management of Older Patients with HIV/AIDS Grants 3275
The National Institute on Aging (NIA), the National Institute of Allergy and Infectious Diseases (NIAID), the Center for Mental Health Research on AIDS of the National Institute of Mental Health (NIMH), and the National Institute of Nursing Research (NINR), invite Research Project Grant applications addressing clinical and translational health issues in the diagnosis and/or management of HIV infection and its consequences in older persons. The goal of this funding opportunity is to improve medical outcomes, functional status and quality of life in older patients with HIV/AIDS through improved understanding of interactions among aging processes, HIV viral infection, treatment effects and toxicities, and multiple morbidities commonly occurring in older persons.
Requirements: The following organizations and institutions may apply: public/state controlled institutions of higher education; private institutions of higher education; Hispanic-serving institutions; historically Black colleges and universities (HBCUs); tribally controlled colleges and universities (TCCUs); Alaska native and native Hawaiian serving institutions; nonprofits with 501(c)3 IRS status; nonprofits without 501(c)3 IRS status; small businesses; for-profit organizations; State governments; regional organizations; U.S. territories or possessions; Indian/Native American tribal governments; Indian/Native American tribally designated organizations; county governments; city or township governments; special district governments; independent school districts; public housing authorities/Indian housing authorities; eligible agencies of the Federal government; and faith-based or community based organizations.
Date(s) Application is Due: Jan 7; May 7; Sep 7
Contact: Susan G. Nayfield, (301) 496-6761; fax (301) 402-1784; nayfiels@nia.nih.gov
Internet: http://grants.nih.gov/grants/guide/pa-files/PA-09-017.html
Sponsor: National Institute on Aging
Gateway Building, Suite 3C307, 7201 Wisconsin Avenue, MSC 9205
Bethesda, MD 20892-9205

NIA Mentored Clinical Scientist Research Career Development Awards 3276
The overall objective of the NIH Research Career Development Award program is to prepare qualified individuals for careers that have a significant impact on the health-related research needs of the Nation. The objective of the NIH Mentored Clinical

Scientist Research Career Development Award (K08) program is to support didactic study and mentored research for individuals with clinical doctoral degrees (e.g., M.D., D.D.S., D.M.D., D.O., D.C., O.D., N.D., D.V.M., Pharm.D., or Ph.D. in clinical disciplines). The purpose of this Funding Opportunity Announcement (FOA) is to continue the long-standing NIH support of the K08 program. This award provides support and protected time for an intensive, mentored research career development experience in biomedical or behavioral research, including translational research. For the purpose of this award, translational research is defined as application of basic research discoveries toward the diagnosis, management, and prevention of human disease.
Requirements: The following organizations and institutions may apply: public/state controlled institutions of higher education; private institutions of higher education; Hispanic-serving institutions; historically Black colleges and universities (HBCUs); tribally controlled colleges and universities (TCCUs); Alaska native and native Hawaiian serving institutions; nonprofits with 501(c)3 IRS status; nonprofits without 501(c)3 IRS status; small businesses; for-profit organizations; State governments; regional organizations; U.S. territories or possessions; Indian/Native American tribal governments; Indian/Native American tribally designated organizations; county governments; city or township governments; special district governments; independent school districts; public housing authorities/Indian housing authorities; eligible agencies of the Federal government; and faith-based or community based organizations.
Restrictions: Individuals on temporary or student visas and organizations described in section 501(c)4 of the Internal Revenue Code that engage in lobbying are not eligible.
Date(s) Application is Due: Feb 12; Mar 12; Jun 12; Jul 12; Oct 12; Nov 12
Amount of Grant: Up to 100,000 U.S.D
Contact: Chyren Hunter, (301) 496-9322; fax (301) 402-0528; Hunterc@nia.nih.gov
Internet: http://grants.nih.gov/grants/guide/pa-files/PA-09-042.html
Sponsor: National Institute on Aging
Gateway Building, Room 2C218, 7201 Wisconsin Avenue, MSC 9205
Bethesda, MD 20892-9205

NIA Mentored Research Scientist Development Award **3277**

The award is for research scientists who have established careers in biomedical, behavioral, or social research and wish to change career direction toward aging research; for more junior researchers with training in aging research who need additional periods of mentored research experience prior to becoming fully independent; or for researchers with training and experience in some aspect of aging research who wish to gain complementary training to expand their research interests in aging. The primary purpose of the program is to provide support and protected time (three, four, or five years) for an intensive, supervised career development experience in the biomedical, behavioral, or clinical sciences leading to research independence.
Requirements: The candidate must have a research or health-professional doctorate or its equivalent and must have demonstrated the capacity or potential for highly productive independent research in the period after the doctorate. The candidate must identify a mentor with extensive research experience and must spend a minimum of 75 percent of full-time professional effort conducting research for the period of the award. The following organizations and institutions may apply: public/state controlled institutions of higher education; private institutions of higher education; Hispanic-serving institutions; historically Black colleges and universities (HBCUs); tribally controlled colleges and universities (TCCUs); Alaska native and native Hawaiian serving institutions; nonprofits with 501(c)3 IRS status; nonprofits without 501(c)3 IRS status; small businesses; for-profit organizations; State governments; regional organizations; U.S. territories or possessions; Indian/Native American tribal governments; Indian/Native American tribally designated organizations; eligible agencies of the Federal government; and faith-based or community based organizations.
Date(s) Application is Due: Feb 12; Mar 12; Jun 12; Jul 12; Oct 12; Nov 12
Amount of Grant: 75,000 U.S.D
Contact: Chyren Hunter, (301) 496-9322; fax (301) 402-0528; Hunterc@nia.nih.gov
Internet: http://grants.nih.gov/grants/guide/pa-files/PAR-09-060.html
Sponsor: National Institute on Aging
Gateway Building, Room 2C218, 7201 Wisconsin Avenue, MSC 9205
Bethesda, MD 20892-9205

NIA Minority Dissertation Research Grants in Aging **3278**

Small grants to support doctoral dissertation research will be available for underrepresented minority doctoral candidates. Grant support is designed to aid the research of such investigators and to encourage underrepresented minority individuals from a variety of academic disciplines and programs to conduct research related to aging. It is anticipated that a total of five to six new applications will be funded in each year. The maximum budget request should be limited to $25,000 in direct costs for the initial budget period. No more than $30,000 in total direct costs will be provided across the two year period.
Requirements: This research initiative is to provide underrepresented minority students assistance to complete their dissertation research on an aging-related topic and thereby increase their representation in aging research.
Date(s) Application is Due: Mar 15; Nov 15
Amount of Grant: Up to 30,000 U.S.D
Contact: Laurie M. Ryan, (301) 496-9350; fax (301) 496-1494; ryanl@nia.nih.gov
Internet: http://grants.nih.gov/grants/guide/pa-files/PAR-98-110.html
Sponsor: National Institute on Aging
Gateway Building, Suite 350, 7201 Wisconsin Avenue, MSC 9205
Bethesda, MD 20892-9205

NIAMS/NICHD Deep Infections of Total Joint Replacements **3279**

This PA seeks to stimulate the receipt of a broad range of basic science and clinical studies to better understand the pathophysiology, diagnosis, and treatment of deep infections around total joint replacement implants. PA: PA-00-014
Requirements: Applications may be submitted by domestic and foreign, for-profit and nonprofit organizations, both public and private; units of state and local governments; and eligible agencies of the federal government.
Contact: Dr. Ralph Nitkin, (301) 402-2242; fax (301) 402-0832; rn21e@nih.gov
Internet: http://grants.nih.gov/grants/guide/pa-files/PA-00-014.html
Sponsor: National Institute of Arthritis and Musculoskeletal and Skin Diseases
1 AMS Circle
Bethesda, MD 20892-3675

NIAMS Clinical Trial Planning Grants **3280**

The goal of this program is to provide support for planning activities critical for the successful implementation of clinical trials in areas of interest to NIAMS. In addition, the grant is intended to allow for early peer review of the rationale and design for complex or large-scale clinical trials and studies in those areas that are in the NIAMS mission; provide support for the development of a detailed clinical trial research plan, including a manual of operations and procedures, as a means of decreasing the long start-up time often needed for initiating large trials after award; and provide support to refine critical components of a clinical trial protocol, such as outcome measures, recruitment strategies, inclusion and exclusion criteria, etc. Annual deadline dates may vary; contact program staff for exact dates. RFA: AR-00-010
Requirements: Applications may be submitted by domestic for-profit and nonprofit organizations, both public and private; units of state and local governments; and eligible agencies of the federal government.
Amount of Grant: Up to 500,000 U.S.D
Contact: Dr. Joan McGowan, Division of Bone Diseases Contact; (301) 594-5055; fax (301) 480-4543; Joan_Mcgowan@nih.gov
Internet: http://grants.nih.gov/grants/guide/rfa-files/RFA-AR-00-010.html
Sponsor: National Institute of Arthritis and Musculoskeletal and Skin Diseases
1 AMS Circle
Bethesda, MD 20892-3675

NIAMS Collaborative Arthritis and Musculoskeletal and Skin Diseases Science Award **3281**

The objective of the Collaborative Arthritis and Musculoskeletal and Skin Diseases Science Award (CAMSSA) program is to develop and promote competitive scientific research programs in areas within the mission of the National Institute of Arthritis and Musculoskeletal and Skin Diseases at institutions with substantial minority enrollment. The CAMSSA program is designed to encourage collaborative investigations among scientists at institutions with substantial minority enrollment with grantees from research-intensive institutions who have grant support to conduct research in arthritis, musculoskeletal diseases or skin diseases. The nature of the collaborations will include joint research efforts, specialized training in research techniques, and participation in research seminars. The CAMSSA program should develop and expand scientific opportunities among the participating institutions for research in arthritis, musculoskeletal diseases or skin diseases. PAR 04-003
Requirements: Applicant institution must be an institution with substantial minority enrollment that may be a non-Federal, public or private nonprofit institution located in the United States, its possessions or territories. The collaborating institution must also be a non-Federal, public or private nonprofit institution located in the United States, its possessions or territories. Individuals are eligible to become principal investigators. Applicant investigator must have a doctoral degree in a basic or clinical science area, and have completed one or more years of postdoctoral research experience and must be a full-time employee of the applicant institution.
Date(s) Application is Due: Feb 1; Jun 1; Oct 1
Amount of Grant: Up to 750,000 U.S.D
Contact: Melinda Nelson, Chief Grants Management Officer; (301) 594-3535; fax (301) 480-5450; nelsonm@mail.nih.gov
Internet: http://grants2.nih.gov/grants/guide/pa-files/PAR-04-003.html
Sponsor: National Institute of Arthritis and Musculoskeletal and Skin Diseases
1 AMS Circle
Bethesda, MD 20892-3675

NIAMS Core Centers for Musculoskeletal Disorders Grants **3282**

NIAMS invites applications for research core centers in musculoskeletal disorders. The centers will provide the resources for a number of established, currently funded investigators, often from different disciplines, to adopt a multidisciplinary approach to common research problems in musculoskeletal disorders and to ensure greater productivity than from each of the separate projects. Annual deadline dates may vary; contact program staff for exact dates. RFA: AR-98-004
Requirements: Applications may be submitted by domestic for-profit and nonprofit organizations, public or private.
Restrictions: Foreign institutions are not eligible for center grants.
Amount of Grant: Up to 400,000 U.S.D
Contact: Dr. Julia Freeman, Centers Program; (301) 594-5052; fax (301) 480-4543; Julia_B_Freeman@nih.gov
Internet: http://grants.nih.gov/grants/guide/rfa-files/RFA-AR-98-004.html
Sponsor: National Institute of Arthritis and Musculoskeletal and Skin Diseases
1 AMS Circle
Bethesda, MD 20892-3675

NIAMS Exploratory Developmental Grants for Multidisciplinary Clinical Studies of Sarcopenia 3283

NIAMS, with NIA, invites exploratory developmental research grant (R21) applications for multidisciplinary clinical studies on the mechanisms underlying the physical functional consequences of sarcopenia in older persons. This exploratory grants program is intended to facilitate new collaborative efforts in the development of novel scientific hypotheses on the causes of changes in skeletal muscle morphology and/or metabolism that lead to physical disabilities in old age, by providing research support for the appropriate pilot/feasibility studies to refine these hypotheses. Multidisciplinary approaches may include collaborations among various types of clinical research expertise or between basic and clinical researchers, with or without previous aging research experience. PA-97-060 may be obtained electronically through the NIH grant line (301-402-2221), the NIH Web site, and by mail and email.
Contact: Dr. Richard Lymn, Muscle Biology and Musculoskeletal Fitness Program Director; (301) 594-5128; fax (301) 480-4543; lymnr@ep.niams.nih.gov
Internet: http://www.grants.nih.gov/grants/guide/pa-files/PA-97-060.html
Sponsor: National Institute of Arthritis and Musculoskeletal and Skin Diseases
1 AMS Circle
Bethesda, MD 20892-3675

NIAMS High Risk Arthritis and Musculoskeletal and Skin Diseases Research 3284

This program seeks to broaden the base of inquiry in fundamental biomedical, biobehavioral, and biomedical technology research by encouraging research projects that involve a high degree of innovation and novelty. It supports research in rheumatic diseases; cartilage biology and diseases; bone biology and diseases (e.g., osteoporosis, Paget's disease); skin biology and skin diseases; autoimmune diseases (e.g., lupus); connective tissue diseases; musculoskeletal diseases, injuries, and disorders; muscle biology and diseases (e.g., muscular dystrophy); exercise physiology and musculoskeletal fitness; sports injuries; occupational diseases and injuries; and orthopedics and bioengineering. Annual deadline dates may vary; contact program staff for exact dates. RFA: AR-01-004
Requirements: Applications may be submitted by domestic for-profit and nonprofit organizations, both public and private; units of state and local governments; and eligible agencies of the federal government.
Amount of Grant: Up to 50,000 U.S.D
Contact: Dr. Elizabeth Gretz; (301) 594-5032; gretze@mail.nih.gov
Internet: http://grants.nih.gov/grants/guide/rfa-files/RFA-AR-01-004.html
Sponsor: National Institute of Arthritis and Musculoskeletal and Skin Diseases
1 AMS Circle
Bethesda, MD 20892-3675

NIAMS Low-Back Pain and Common Spinal Disorders Research Grants 3285

The program is sponsored by NIAMS, NICHD, NINDS, and NIOSH at the CDC. This initiative invites applications directed to the study of the pathogenesis, epidemiology, prevention, and treatment of three common causes of low-back pain: herniated nucleus pulposus, spinal stenosis, and idiopathic low-back pain. The support mechanism for grants in this area will be the individual investigator-initiated research grant (R01). PA-97-058 may be obtained electronically through the NIH grant line (301-402-2221), the NIH Web site, and by mail and email.
Contact: Dr. James Panagis, Orthopaedics Program; (301) 594-5055; fax (301) 480-4543; panagisj@ep.niams.nih.gov
Internet: http://www.grants.nih.gov/grants/guide/pa-files/PA-97-058.html
Sponsor: National Institute of Arthritis and Musculoskeletal and Skin Diseases
1 AMS Circle
Bethesda, MD 20892-3675

NIAMS Mechanisms of Chondroprotection Research Grants 3286

Approximately six to eight awards will be made for research on mechanisms of chondroprotection. Applications may be for individual research projects or for exploratory/development grants. The objective of the exploratory/developmental mechanism is to encourage applications from individuals who are interested in testing innovative or conceptually creative ideas that are scientifically sound and may advance our understanding of chondroprotective mechanisms. Research should be specifically targeted towards identification and evaluation of chondroprotective agents to prevent cartilage destruction and/or facilitate its repair in such conditions as rheumatoid arthritis, juvenile rheumatoid arthritis, and osteoarthritis, where there is a decline in the structural integrity of the articular cartilage. This request for applications is for basic and applied research projects, but not epidemiological treatment projects. Annual deadline dates may vary; contact program staff for exact dates. RFA: AR-99-004
Requirements: Applications may be submitted by domestic and foreign public and private for-profit and nonprofit organizations, including universities, colleges, hospitals, laboratories, units of state and local government, and eligible agencies of the federal government.
Amount of Grant: Up to 15,000,000 U.S.D
Contact: Dr. Susana Serrate-Sztein, Arthritis Branch Coordinator; (301) 594-5032; fax (301) 480-4543; szteins@ep.niams.nih.gov
Internet: http://www.grants.nih.gov/grants/guide/rfa-files/RFA-AR-99-004.html
Sponsor: National Institute of Arthritis and Musculoskeletal and Skin Diseases
1 AMS Circle
Bethesda, MD 20892-3675

NIAMS Molecular Pathogenesis and New Interventions in Scleroderma 3287

The goal of this RFA is to foster developmental and traditional research projects to advance understanding of the pathogenesis of scleroderma and to promote the design, development, and pilot testing of hypothesis-driven innovative therapeutic approaches. Annual deadline dates may vary; contact program staff for exact dates. RFA: AR-00-007
Requirements: Applications may be submitted by domestic for-profit and nonprofit organizations, both public and private; units of state and local governments; and eligible agencies of the federal government.
Restrictions: Foreign institutions are not eligible to apply for these grants, however, applicants may collaborate, through consultation or contractual agreements, with investigators at foreign institutions.
Amount of Grant: Up to 15,000,000 U.S.D
Contact: Dr. Susana Serrate-Sztein, Rheumatic Diseases Branch Contact; (301) 594-5032; fax (301) 480-4543; szteins@mail.nih.gov
Internet: http://grants.nih.gov/grants/guide/rfa-files/RFA-AR-00-007.html
Sponsor: National Institute of Arthritis and Musculoskeletal and Skin Diseases
1 AMS Circle
Bethesda, MD 20892-3675

NIAMS Pathogenesis and Therapy of the Muscular Dystrophies Research Grants 3288

NINDS and NIAMS encourage investigator-initiated research grant applications to study the pathogenesis and therapy of the various forms of muscular dystrophy in children and adults. Responses to this PA may include studies in appropriate animal models or preclinical or clinical studies in patients with facioscapulohumeral dystrophy (FSH), limb-girdle muscular dystrophy (LGMD), myotonic dystrophy, congenital muscular dystrophy (CMD), Emery-Dreifuss muscular dystrophy (EMD), Duchenne muscular dystrophy (DMD), Becker muscular dystrophy (BMD), or other forms of muscular dystrophy. Support mechanisms for grants are research project grant (R01) and program project grant (P01). PA-98-044 may be obtained through the NIH grant line (301-402-2221), the NIH Web site, mail, and email.
Requirements: Applications may be submitted by domestic and foreign, for-profit and nonprofit organizations, public or private.
Contact: Sally Nichols, Grants Management Officer; (301) 594-3535; fax (301) 480-5450; nicholss@ep.niams.nih.gov
Internet: http://www.grants.nih.gov/grants/guide/pa-files/PA-98-044.html
Sponsor: National Institute of Arthritis and Musculoskeletal and Skin Diseases
1 AMS Circle
Bethesda, MD 20892-3675

NIAMS Pilot and Feasibility Clinical Research Grants in Arthritis and Musculoskeletal and Skin Diseases 3289

The goal of this initiative is to encourage novel clinical research aimed at improving the understanding of the causes, treatment or prevention of arthritis and musculoskeletal and skin diseases. This Pilot and Feasibility Clinical Research Grants Program is designed to allow initiation of exploratory, short-term clinical studies, intended to facilitate the development of new ideas which may be investigated without stringent requirements for preliminary data. Research aims appropriate for an investigator-initiated R01 grant, such as understanding underlying biological mechanisms of disease, would not be expected in an application under this FOA. Proposed studies should focus on research questions that are likely to gather critical preliminary data in support of a future, planned clinical trial or to benefit clinical research and trials more broadly. Overall goals of these studies are intended to generate data critical to future clinical research or trials such as by demonstrating the feasibility of a recruitment target or approach, obtaining data in support of potential inclusion and exclusion criteria, or to gather preliminary evidence of efficacy, tolerability and/or toxicity of an available drug, biologic or device in a new population relevant to the NIAMS mission. Examples include testing of: new prevention strategies; a new application of existing pharmacologic or non-pharmacologic interventions or their unique combination; a new cost effective or adaptive trial design in a specific population or across populations; and studies intended to expand the science of clinical trials overall. Results generated from this program may lead to critical input into a new clinical trial; trials that inform decision-making in urgent or high priority clinical practice needs for arthritis and musculoskeletal and skin diseases; or preliminary data for trials in understudied and rare diseases. A high priority is the use of such studies to help stimulate the translation of promising research developments from the laboratory into clinical practice. The total project period for an application submitted in response to this funding opportunity may not exceed 2 years. Although the size of award may vary with the scope of research proposed, it is expected that applications will stay within the budgetary guidelines for an exploratory/developmental project; direct costs are limited to $275,000 over an R21 two-year period, with no more than $200,000 in direct costs allowed in any single year.
Requirements: The following organizations/institutions are eligible to apply: public/state controlled institutions of higher education; private institutions of higher education; Hispanic-serving institutions; historically Black colleges and universities; tribally controlled colleges and universities; Alaska native and native Hawaiian serving institutions; nonprofits with 501(c)3 IRS status (other than institutions of higher education); nonprofits without 501(c)3 IRS status (other than institutions of higher education); small businesses; for-profit organizations (other than small businesses); state governments; Indian/Native American tribal governments (federally recognized); Indian/Native American tribally designated organizations; county governments; city or township governments; special district governments; independent school districts; public housing authorities/Indian housing authorities; U.S. territory or possession; Indian/Native American tribal governments (other than federally recognized); regional organizations; non-domestic (non-U.S.) entities (foreign organizations); eligible agencies of the federal government; and faith-based or community-based organizations.
Date(s) Application is Due: Mar 1; Jul 1; Nov 1
Amount of Grant: Up to 275,000 U.S.D
Contact: James Witter, M.D., Ph.D., (301) 594-1963; witterj@mail.nih.gov
William Phil Tonkins Jr., M.S., Dr.PH., (301) 594-4979; fax (301) 480-1284; tonkinsw2@mail.nih.gov

Internet: http://grants.nih.gov/grants/guide/pa-files/PAR-10-282.html#SectionVII
Sponsor: National Institute of Arthritis and Musculoskeletal and Skin Diseases
1 AMS Circle
Bethesda, MD 20892-3675

NIAMS Repetitive Motion Disorders Research Grants 3290
The program is sponsored by NIAMS, NICHD, NINDS, and NIOSH at the CDC. This initiative invites applications directed to the study of the pathogenesis, epidemiology, prevention, and treatment of repetitive motion disorders. The support mechanism for grants in this area will be the individual investigator-initiated research grant (R01). PA-97-059 may be obtained electronically through the NIH grant line (301-402-2221), the NIH Web site, and by mail and email.
Contact: Dr. James Panagis, Orthopaedics Program; (301) 594-5055; fax (301) 480-4543; panagisj@ep.niams.nih.gov
Internet: http://www.grants.nih.gov/grants/guide/pa-files/PA-97-059.html
Sponsor: National Institute of Arthritis and Musculoskeletal and Skin Diseases
1 AMS Circle
Bethesda, MD 20892-3675

NIAMS Research on Musculoskeletal Fitness and Sports Medicine Grants 3291
NIAMS and NINR invite investigator-initiated research grant applications to study a broad range of basic and clinical topics related to musculoskeletal fitness, exercise physiology, and sports medicine. The National Center for Medical Rehabilitation Research of NICHD encourages applications for both basic and clinical studies of musculoskeletal fitness and exercise physiology of persons with physical disabilities. The principal investigator and participating investigators will plan, direct, and perform the research. Applicants for program project grants are requested to contact NIAMS representative as early as possible in the planning stages. NINR will support individual research project grants. PA-97-025 may be obtained electronically through the NIH grant line (301-402-2221), the NIH Web site, and by mail and email.
Contact: Dr. Richard Lymn, Muscle Biology and Musculoskeletal Fitness Program Director; (301) 594-5128; fax (301) 480-4543; lymnr@ep.niams.nih.gov
Internet: http://www.grants.nih.gov/grants/guide/pa-files/PA-97-025.html
Sponsor: National Institute of Arthritis and Musculoskeletal and Skin Diseases
1 AMS Circle
Bethesda, MD 20892-3675

NIAMS Significance of Sarcopenia in Old Age Grants 3292
NIA and NIAMS invite research project applications to elucidate the metabolic/physiological and functional consequences of sarcopenia in old age. Sarcopenia is defined as the loss of skeletal muscle mass, quality, and strength. More information may be obtained by mail and email, or through the NIH grant line (301-402-2221) or the NIH Web site. PA: PA-96-038
Date(s) Application is Due: Feb 1; Jun 1; Oct 1
Amount of Grant: 10,000 - 15,000,000 U.S.D
Contact: Sally Nichols, Grants Management Officer; (301) 594-3535
Internet: http://www.grants.nih.gov/grants/guide/pa-files/PA-96-038.html
Sponsor: National Institute of Arthritis and Musculoskeletal and Skin Diseases
1 AMS Circle
Bethesda, MD 20892-3675

NIAMS Skin Diseases Research Core Centers Grants 3293
NIAMS is accepting applications for research core centers in skin diseases. The centers will provide the resources for a number of established, currently funded investigators, often from different disciplines, to adopt a multidisciplinary approach to common research problems in skin diseases and to ensure greater productivity through interaction among the investigators of individually funded research projects. New research directions are encouraged through a pilot and feasibility component. Broad advances in skin disease research can be effectively fostered by research core centers. Examples of advances include, but are not limited to, stratum corneum—biochemistry, structure, function; epidermis—differentiation, keratinization, cellular constituents; dermal-epidermal junction—structure, functions, diseases; skin and an immunological organ; and autoimmune skin diseases. Annual deadline dates may vary; contact program staff for exact dates. RFA: AR-99-009
Requirements: Applications may be submitted by domestic for-profit and nonprofit organizations, both public and private; units of state and local governments; and eligible agencies of the federal government. An established clinical and basic research program in skin diseases must be present.
Restrictions: Foreign institutions are not eligible for center grants.
Amount of Grant: Up to 400,000 U.S.D
Contact: Dr. Julia Freeman, Centers Program Contact; (301) 594-5052; fax (301) 480-4543; Julia_B_Freeman@nih.gov
Internet: http://grants.nih.gov/grants/guide/rfa-files/RFA-AR-99-009.html
Sponsor: National Institute of Arthritis and Musculoskeletal and Skin Diseases
1 AMS Circle
Bethesda, MD 20892-3675

NIAMS Specialized Centers of Research in Osteoarthritis and Systemic Lupus Erythematosus 3294
NIAMS invites applications for specialized centers of research in osteoarthritis and systemic lupus erythematosus. These centers should foster a coordinated research effort that strongly emphasizes basic disciplines but also involves significant interaction between basic research and clinical investigations. Annual deadline dates may vary; contact program staff for exact dates. RFA: AR-01-005
Requirements: Applications may be submitted by domestic for-profit and nonprofit organizations, both public and private; units of state and local governments; and eligible agencies of the federal government.
Contact: Melinda Nelson, Grants Management Officer; (301) 594-3535; fax (301) 480-5450; nelsonm@exchange.nih.gov
Internet: http://grants.nih.gov/grants/guide/rfa-files/RFA-AR-01-005.html
Sponsor: National Institute of Arthritis and Musculoskeletal and Skin Diseases
1 AMS Circle
Bethesda, MD 20892-3675

NIAMS Specialized Centers of Research in Osteoporosis, Rheumatoid Arthritis, and Scleroderma 3295
NIAMS invites applications for specialized centers of research in the following disease areas: osteoporosis, rheumatoid arthritis, and scleroderma. A center should foster a coordinated research effort that strongly emphasizes basic disciplines but also involves significant interaction between basic research and clinical investigations. More information may be obtained by mail and email or through the NIH grant line (301-402-2221), and the NIH Web site. Annual deadline dates may vary; contact program staff for exact dates. RFA: AR-98-001
Requirements: Domestic for-profit and nonprofit organizations, both public and private; units of state and local governments; and eligible agencies of the federal government may apply.
Amount of Grant: Up to 750,000 U.S.D
Contact: Sally Nichols, Grants Management Officer; (301) 594-3535; fax (301) 480-5450; nicholss@exchange.nih.gov
Internet: http://grants.nih.gov/grants/guide/rfa-files/RFA-AR-98-001.html
Sponsor: National Institute of Arthritis and Musculoskeletal and Skin Diseases
1 AMS Circle
Bethesda, MD 20892-3675

NIA Multicenter Study on Exceptional Survival in Families: The Long Life Family Study (LLFS) Grants 3296
This limited competition FOA is to continue the Long Life Family Study (LLFS). The LLFS was designed to determine the degree and patterns of familial transmission and aggregation of exceptional longevity and healthy survival to advanced age as characterized by a variety of phenotypic measures, with a further goal of also understanding potential genetic factors that contribute to exceptional survival. Five awards, to four study centers and one data management and coordinating center, were made in response to RFA-AG-03-004.
Date(s) Application is Due: Aug 30; Sep 30
Contact: Winifred K. Rossi, (301) 496-3836; fax (301) 402-1784; winnie_rossi@nih.gov
Internet: http://grants.nih.gov/grants/guide/rfa-files/RFA-AG-10-005.html
Sponsor: National Institute on Aging
Gateway Building, Suite 3C307, 7201 Wisconsin Avenue, MSC 9205
Bethesda, MD 20892-9205

NIA Nathan Shock Centers of Excellence in Basic Biology of Aging Grants 3297
The NIA invites applications for support of centers, known as Nathan Shock Centers of Excellence in Basic Biology of Aging. These center grants will provide funding for research and training activities related to basic biology of aging. They are intended for institutions with a substantial investment in and commitment to aging research, but they are not intended to directly support clinical research or clinical trials. The goal of this program is to enhance the ability of institutions with well-developed research programs in basic research on aging to utilize state-of-the-art research resources to provide the strongest environment for the conduct of research on aging. Thus, this RFA is intended to enhance the quality of research in the basic biology of aging, facilitate the planning and coordination of aging research activities, provide support and a suitable environment for investigators new to aging research to acquire research skills and experience at institutions that have demonstrated commitment to, and expertise in, basic biology of aging research, and to develop potential regional and/or national resource centers. Annual deadline dates may vary.
Requirements: Applications may be submitted by domestic nonprofit organizations and institutions, state and local governments and their agencies, and authorized federal institutions. The applicant institution must currently support a minimum of 15 peer-reviewed, externally funded research projects on aging to be eligible. In the case of currently funded program projects (P01s), or similar multiproject research grants, each research project will be deemed to be a separate project. (Core components do not qualify.)
Date(s) Application is Due: Apr 16; May 20
Contact: Dr. Huber Warner, Associate Director; (301) 496-4996; fax (301) 402-0010; warnerh@nia.nih.gov
Internet: http://grants.nih.gov/grants/guide/rfa-files/RFA-AG-04-010.html
Sponsor: National Institute on Aging
Gateway Building, Suite 2C231, 7201 Wisconsin Avenue, MSC 9205
Bethesda, MD 20892

NIA Network Infrastructure Support for Emerging Behavioral and Social Research Areas in Aging 3298
The purpose of this FOA is to provide infrastructure support in specific emerging interdisciplinary areas of behavioral and social research in aging using the NIH Resource-Related Research Project mechanism. The infrastructure support will facilitate research networks through meetings, conferences, small scale pilots, training, and dissemination to encourage growth and development in specified emerging areas and resources. Budgets for direct costs of up to $150,000 per year and project duration of 3-5 years may be requested with a maximum of $750,000 direct costs over a five-year project period. Letters of Intent are due by September 30, December 28, and April 26, respectively. Final applications due by October 30, January 25, and May 25, respectively..

Requirements: The following organizations and institutions may apply: public/state controlled institutions of higher education; private institutions of higher education; Hispanic-serving institutions; historically Black colleges and universities (HBCUs); tribally controlled colleges and universities (TCCUs); Alaska native and native Hawaiian serving institutions; nonprofits with 501(c)3 IRS status; nonprofits without 501(c)3 IRS status; small businesses; for-profit organizations; State governments; regional organizations; U.S. territories or possessions; Indian/Native American tribal governments; Indian/Native American tribally designated organizations; county governments; city or township governments; special district governments; independent school districts; public housing authorities/Indian housing authorities; eligible agencies of the Federal government; and faith-based or community based organizations.
Date(s) Application is Due: Jan 25; Apr 26; May 25; Sep 30; Oct 30; Dec 28
Amount of Grant: 150,000 - 750,000 U.S.D
Contact: Lis Nielsen, (301) 402-4156; fax (301) 402-0051; nielsenli@nia.nih.gov
Internet: http://grants.nih.gov/grants/guide/pa-files/PAR-09-233.html
Sponsor: National Institute on Aging
Building 31, Room 5C27, 31 Center Drive, MSC 2292
Bethesda, MD 20892-7710

NIA NRSAs—Individual Postdoctoral Fellowships **3299**
The fellowships support training for basic and clinical research on aging and the problems of the aging and the aged. The training should be directed toward long-term involvement in research in aging and gerontology. The institute will accept proposals in the behavioral and social sciences related to aging and in the biological and biomedical sciences related to aging. Support may be requested for up to three years. PA: PA-03-067
Requirements: Applicants must be citizens of the United States or permanent resident aliens who have received doctoral degrees before the beginning date of the fellowship.
Date(s) Application is Due: Apr 5; Aug 5; Dec 5
Amount of Grant: 26,256 - 41,268 U.S.D
Contact: Dr. Robin Barr , Office of Extramural Affairs; (301) 496-9322; fax (301) 402-2945; rb42h@nih.gov
Internet: http://grants.nih.gov/grants/guide/pa-files/PA-03-067.html
Sponsor: National Institute on Aging
Building 31, Room 5C27, 31 Center Drive, MSC 2292
Bethesda, MD 20892-7710

NIA NRSAs—Institutional Training Awards **3300**
NIA will award new and competing renewal NRSA (NRSA) Institutional Training Grants (T32) to eligible institutions to develop or enhance research training opportunities for individuals, selected by the institution, who are training for careers in specified areas of aging research. The purpose of this announcement is to help ensure a broad cadre of researchers trained in the content and methods of aging research and in the major biomedical, behavioral, and social areas of research most closely related to their field of study within aging. NIA supports both predoctoral and postdoctoral training in aging. The institute also supports short-term research training for students in health-professional programs as part of an overall T32 program. The NIA has a single receipt date each year for applications. PA: PA-00-057
Requirements: Applicant must be a citizen or noncitizen national of the United States or must have been lawfully admitted for permanent residence. Predoctoral trainees must have received a baccalaureate degree by the beginning date of their NRSA appointment, and must be training at the postbaccalaureate level and enrolled in a program leading to a PhD in science or in an equivalent research doctoral degree program. Postdoctoral trainees must have received, as of the beginning date of the NRSA appointment, a PhD, MD, or comparable doctoral degree from an accredited domestic or foreign institution. To be eligible for short-term predoctoral research training positions, health-professional students must have completed at least one quarter in a program leading to a clinical doctorate prior to participating in the program.
Contact: Dr. Robin Barr , Office of Extramural Affairs; (301) 496-9322; fax (301) 402-2945; rb42h@nih.gov
Internet: http://grants.nih.gov/grants/guide/pa-files/PA-00-057.html
Sponsor: National Institute on Aging
Building 31, Room 5C27, 31 Center Drive, MSC 2292
Bethesda, MD 20892-7710

NIA Paul Beeson Career Development Awards in Aging **3301**
The program will foster the independent research careers of clinically trained investigators whose research will enhance the health and quality of life of Americans, particularly older people. The aims of the program are to encourage and assist the development of future leaders in the field of aging by supporting junior and midcareer faculty members to gain additional research training as needed and to establish independent programs in aging research; deepen the commitment of research institutions to academic research in aging and to translating research outcomes to geriatric medicine by involving mentor and recipient in establishing and advancing the recipient's career in aging research; expand medical research on aging broadly defined as including the biology of aging, maintenance of health and independence in old age, diseases and disabilities of old age and issues in their clinical management, and systems of care for the elderly. Project periods of three, four, or five years are supported. A letter of intent is requested but not required by October 24. Program information and guidelines are available online. RFA-AG-06-005
Requirements: Eligible institutions are for-profit or nonprofit, public or private, units of state and local governments, and eligible agencies of the federal government. Individuals eligible to become principal investigators have a clinical doctoral degree (e.g., MD, DO, DDS) or its equivalent. At the time of award, applicants must be U.S. citizens or noncitizen nationals, or have been lawfully admitted to the United States for permanent residence.
Restrictions: Foreign institutions are not eligible to apply. Ineligible individuals include current and former principal investigators on NIH research projects, FIRST Awards, sub-projects of program project, or center grants. Individuals on temporary or student visas are not eligible.
Date(s) Application is Due: Nov 23
Amount of Grant: Up to 200,000 U.S.D
Contact: Office of Extramural Affairs; (301) 496-9322; fax (301) 402-2945; BarrR@nia.nih.gov
Internet: http://grants1.nih.gov/grants/guide/rfa-files/RFA-AG-06-005.html
Sponsor: National Institutes of Health
7201 Wisconsin Avenue, Room 2C218, MSC 9205
Bethesda, MD 20892-9205

NIA Pilot Research Grants **3302**
NIA is seeking small grant applications in specific areas to stimulate and facilitate the entry of promising new investigators into aging research or encourage established investigators to enter new targeted, high-priority areas in this research field. This small grant program provides support for pilot research that is likely to lead to subsequent individual research project grants and/or a significant advancement of aging research. Annual deadline dates may vary; contact program staff for exact dates.
Requirements: Applications may be submitted by domestic for-profit and nonprofit organizations, both public or private; units of state and local governments; and eligible agencies of the federal government. The program supports new and established investigators. For a new investigator to be eligible, the individual should be in the first five years of his or her independent research career. If the applicant is in the final stages of training it is permissible to apply for an R03, but no award will be made to individuals who are still in training or fellowship status at the time of award. For an established investigator to be eligible, the individual must propose research that is unrelated to a currently funded research project in which the investigator participates.
Restrictions: Foreign organizations and institutions are not eligible.
Amount of Grant: Up to 50,000 U.S.D
Contact: Dr. David Finkelstein, Biology of Aging Program Contact; (301) 496-6402; fax (301) 402-0010; BAPquery@extramur.nia.nih.gov
Internet: http://grants.nih.gov/grants/guide/pa-files/PA-99-049.html
Sponsor: National Institute on Aging
Gateway Building, Suite 2C231, 7201 Wisconsin Avenue, MSC 9205
Bethesda, MD 20892

NIA Postdoctoral Training in Research on Aging in Canada **3303**
The objective of the National Research Service Award (NRSA) F32 program is to provide support to promising postdoctoral applicants who have the potential to become productive and successful independent research investigators in scientific health-related fields relevant to the missions of the participating NIH Institutes and Centers (ICs). A primary purpose of this initiative is to offer training in research on aging that exposes developing American postdoctoral investigators to diverse combinations of expertise and resources singularly available in Canada through the Canadian Institutes of Health-Institute on Aging (CIHR-IA), combinations that also take into account the unique cultural and ethnic composition of Canada and its aging population.
Requirements: Applicants must choose Canadian institutions and mentors selected by CIHR-IA. These institutions and mentors can be located through the CIHR Funding Database. Any individual with the skills, knowledge, and resources necessary to carry out the proposed research training is invited to work with his/her sponsor and institution to develop an application for support. Individuals from underrepresented racial and ethnic groups as well as individuals with disabilities are always encouraged to apply for NIH support.
Restrictions: Candidates for the postdoctoral fellowship award must be citizens or non-citizen nationals of the United States, or must have been lawfully admitted to the United States for Permanent Residence.
Date(s) Application is Due: Apr 8; Aug 8; Dec 8
Contact: Michael-David A.R.R. Kerns, (301) 402-7713; fax (301) 402-2945; michael-david.kerns@nih.hhs.gov
Internet: http://grants.nih.gov/grants/guide/pa-files/PA-06-469.html
Sponsor: National Institute on Aging
Gateway Building, Room 2C218, 7201 Wisconsin Avenue, MSC 9205
Bethesda, MD 20892-9205

NIA Promoting Careers in Aging and Health Disparities Research Grants **3304**
The Mentored Research Scientist Development Award provides support for a sustained period of protected time for intensive research career development under the guidance of an experienced mentor, or sponsor, in the biomedical, behavioral or clinical sciences leading to research independence. The expectation is that through this sustained period of research career development and training, awardees will launch independent research careers and become competitive for new research project grant funding.
Requirements: The following organizations and institutions may apply: public/state controlled institutions of higher education; private institutions of higher education; Hispanic-serving institutions; historically Black colleges and universities (HBCUs); tribally controlled colleges and universities (TCCUs); Alaska native and native Hawaiian serving institutions; nonprofits with 501(c)3 IRS status; nonprofits without 501(c)3 IRS status; small businesses; for-profit organizations; State governments; regional organizations; U.S. territories or possessions; Indian/Native American tribal governments; Indian/Native American tribally designated organizations; county governments; city or township governments; special district governments; independent school districts; public housing authorities/Indian housing authorities; eligible agencies of the Federal government; and faith-based or community based organizations.

Date(s) Application is Due: Feb 12; Jun 12; Oct 12
Amount of Grant: 450,000 - 750,000 U.S.D
Contact: Dr. J. Taylor Harden, (301) 496-0765; Taylor_Harden@nih.gov
Internet: http://grants.nih.gov/grants/guide/pa-files/PAR-09-136.html
Sponsor: National Institute on Aging
Building 31, Room 5C27, 31 Center Drive, MSC 2292
Bethesda, MD 20892-7710

NIA Prostate Cancer Aging, Race, and Ethnicity Research Grants 3305
NIA invites research project grant applications to expand the understanding of biological and clinical factors leading to the development, progression, and treatment of prostate cancer in aging men. The increased risk of prostate cancer with advancing age and its prominence in older-aged men are well-known characteristics of this tumor. The unusually high incidence and mortality rates of prostate cancer for older white and black American men and, by contrast, the much lower rates in men of Hispanic and Asian descent, provide the need for research that emphasizes the role of race and ethnic factors, as well as age, in early diagnosis, management, and etiology of this malignancy. The purpose of this RFA is to support research that will provide information leading to better prevention, diagnosis, prognosis, and treatment of prostate cancer in the age range in which prostate cancer most frequently occurs in the diverse population groups at risk (i.e., in men 65 years and older). Proposals to address the issues indicated above may be focused on genetic and environmental risk factors, pre-malignant changes, tumorigenesis, detection of localized and advanced disease, prognostic indicators, disease progression, and response to treatment. Studies to evaluate or refine the ability of existing diagnostic, screening, or follow-up methods of persons of different age, race, and ethnicity are also responsive to this RFA. The PA, which describes the research objectives, application procedures, review considerations, and award criteria may be obtained electronically through the NIH grant line (301-402-2221), the NIH Web site, and by mail and email. RFA: AG-02-003
Contact: Dr. Rosemary Yancik, Geriatrics Program Coordinator; (301) 496-5278; fax (301) 496-2793; YancikR@31.nia.nih.gov
Internet: http://grants2.nih.gov/grants/guide/rfa-files/RFA-AG-02-003.html
Sponsor: National Institute on Aging
Building 31, Room 5C27, 31 Center Drive, MSC 2292
Bethesda, MD 20892-7710

NIA Renal Function and Chronic Kidney Disease in Aging Grants 3306
This Funding Opportunity Announcement (FOA) issued by the National Institute on Aging (NIA) and the National Institute of Diabetes and Digestive and Kidney Diseases (NIDDK), National Institutes of Health, invites applications that propose basic, clinical, and translational research on chronic kidney disease (CKD) and its consequences in aging and in older persons. Applications should focus on: biology and pathophysiology of CKD in animal models; etiology and pathophysiology of CKD in the elderly; epidemiology and risk factors for the development of CKD with advancing age; and/or diagnosis, medical management and clinical outcomes of CKD in this population. Research supported by this initiative should enhance knowledge of CKD and its consequences in the elderly and provide evidence-based guidance in the diagnosis, prevention, and treatment of CKD in older persons.
Requirements: The following organizations and institutions may apply: public/state controlled institutions of higher education; private institutions of higher education; Hispanic-serving institutions; historically Black colleges and universities (HBCUs); tribally controlled colleges and universities (TCCUs); Alaska native and native Hawaiian serving institutions; nonprofits with 501(c)3 IRS status; nonprofits without 501(c)3 IRS status; small businesses; for-profit organizations; State governments; regional organizations; U.S. territories or possessions; Indian/Native American tribal governments; Indian/Native American tribally designated organizations; county governments; city or township governments; special district governments; independent school districts; public housing authorities/Indian housing authorities; eligible agencies of the Federal government; and faith-based or community based organizations.
Date(s) Application is Due: Feb 5; Feb 16; Jun 5; Jun 16; Oct 5; Oct 16
Contact: Susan G. Nayfield, (301) 496-6761; fax (301) 402-1784; nayfiels@nia.nih.gov
Internet: http://grants.nih.gov/grants/guide/pa-files/PA-09-165.html
Sponsor: National Institute on Aging
Gateway Building, Suite 3C307, 7201 Wisconsin Avenue, MSC 9205
Bethesda, MD 20892-9205

NIA Research Grants on Chemical Senses: Neurotransmitters and Neuromodulators 3307
NIDCD and NIA sponsor research grants that support research fundamental to understanding the neurochemistry of pathways in the olfactory and gustatory systems throughout the life span. This research includes the identification and characterization of the neurotransmitters, neuromodulators, receptors, and secondary messengers throughout the chemosensory systems. More information may be obtained by mail and email, or through the NIH grant line (301-402-2221), and the NIH Web site. PA: PA-96-035
Requirements: Applications may be submitted by domestic and foreign, for-profit and non-profit organizations, public and private, such as universities, colleges, hospitals, laboratories, units of state and local governments, and eligible agencies of the federal government.
Contact: Dr. Judy Finkelstein, (301) 496-9350; FinkelsJ@gw.nia.nih.gov
Internet: http://www.grants.nih.gov/grants/guide/pa-files/PA-96-035.html
Sponsor: National Institute on Aging
Building 31, Room 5C27, 31 Center Drive, MSC 2292
Bethesda, MD 20892-7710

NIA Role of Apolipoprotein E, Lipoprotein Receptors and CNS Lipid Homeostasis in Brain Aging and Alzheimer's Disease 3308
This FOA encouarges multidisciplinary and interdisciplinary research to elucidate how Apolipoprotein E, lipoprotein receptors and CNS lipid homeostasis influence brain aging and the transition to neurodegeneration in Alzheimer's disease (AD). The ultimate goal is to gain an in depth understanding of the mechanisms by which the Apolipoprotein E e4 allele confers increased AD risk for the purpose of advancing the overall search for efficacious AD treatments and Apolipoprotein E e4-directed therapeutics in particular. To this end we encourage research spanning multiple levels of analysis in multiple species (from mice to man). Studies aimed at identifying new therapeutic targets for the treatment of AD and other age-related neurodegenerative conditions associated with lipid neurobiology are strongly encouraged. Also of great interest are projects that aim to use lipidomics, various types of imaging and other cutting edge technologies to identify and develop early biomarkers of neurodegeneration associated lipid dyshomeostasis. The estimated amount of funds available to support projects under this FOA is $2,500,000.
Requirements: The following organizations and institutions are eligible to apply: public/state controlled institutions of higher education; private institutions of higher education; Hispanic-serving institutions; historically Black colleges and universities (HBCUs); tribally controlled colleges and universities (TCCUs); Alaska native and native Hawaiian serving institutions; nonprofits with 501(c)3 IRS status (other than institutions of higher education); nonprofits without 501(c)3 IRS status (other than institutions of higher education); small businesses; for-profit organizations (other than small businesses); state governments; Indian/Native American tribal governments (federally recognized); Indian/Native American tribally designated organizations; county governments; city or township governments; special district governments; independent school districts; public housing authorities/Indian housing authorities; U.S. territories or possessions; Indian/native American tribal governments (other than federally recognized); regional organizations; and non-domestic (non-U.S.) entities (foreign organizations).
Date(s) Application is Due: Mar 5; Jul 5; Nov 5
Contact: Suzana S. Petanceska, (301) 496-9350; petanceskas@mail.nih.gov
Internet: http://grants.nih.gov/grants/guide/pa-files/PA-09-217.html
Sponsor: National Institute on Aging
Gateway Building, Suite 350, 7201 Wisconsin Avenue, MSC 9205
Bethesda, MD 20892-9205

NIA Role of Nuclear Receptors in Tissue and Organismal Aging Grants 3309
This Funding Opportunity Announcement (FOA) solicits grant applications proposing to conduct research into underlying biologic mechanisms involving nuclear receptors, their co-regulators and intracellular signaling systems in the process of aging and the connections of the aging process with pathophysiology in middle- and old-age. The focus of the proposed research must be on processes of aging and/or age-related changes, i.e., research proposed in applications responding to this FOA must have clear relevance to aging. Potential topic areas of interest to the NIA include, but are not limited to, the following: age-changes in levels, sensitivity, and gene expression activation profiles of NRs and co-regulators, and their effects on processes of aging (e.g., age-changes in tissue function, inflammation, glucose and lipid metabolism, adipogenesis, muscle, and brain function) and connections with disease processes for diseases common in older individuals; modification of lifespan in model organisms (e.g., the NR homologue DAF-12 in the insulin-IGF-like pathway in the nematode, and mammalian homologues) and the health-protecting, life-extending effects of caloric restriction or other life-span altering genetic mutations or transgenic models in mammalian organisms; and age-changes in steroid metabolism or in tissue sensitivity and/or response to exogenous sex steroid (or tissue-selective sex steroid receptor modulators) therapy in middle- and old-age.
Requirements: The following organizations and institutions may apply: public/state controlled institutions of higher education; private institutions of higher education; Hispanic-serving institutions; historically Black colleges and universities (HBCUs); tribally controlled colleges and universities (TCCUs); Alaska native and native Hawaiian serving institutions; nonprofits with 501(c)3 IRS status; nonprofits without 501(c)3 IRS status; small businesses; for-profit organizations; State governments; regional organizations; U.S. territories or possessions; Indian/Native American tribal governments; Indian/Native American tribally designated organizations; eligible agencies of the Federal government; and faith-based or community based organizations.
Date(s) Application is Due: Feb 5; Jun 5; Oct 5
Contact: Ronald Margolis, (301) 594-8819; fax (301) 435-6047; rm76f@nih.gov
Internet: http://grants.nih.gov/grants/guide/pa-files/PAS-07-267.html
Sponsor: National Institute on Aging
Gateway Building, Suite 2C231, 7201 Wisconsin Avenue, MSC 9205
Bethesda, MD 20892

NIA Significance of Sarcopenia in Old Age Grants 3310
NIA and NIAMS invite research project applications to elucidate the metabolic/physiological and functional consequences of sarcopenia in old age. Sarcopenia is defined as the loss of skeletal muscle mass, quality, and strength. More information may be obtained by mail and email, or through the NIH grant line (301-402-2221) and the NIH Web site. PA: PA-96-038
Requirements: Applications may be submitted by foreign and domestic, for-profit and nonprofit organizations, both public and private; units of state and local governments; and eligible agencies of the federal government.
Contact: Dr. Chhanda Dutta; (301) 496-6761; DuttaC@gw.nia.nih.gov
Internet: http://www.grants.nih.gov/grants/guide/pa-files/PA-96-038.html
Sponsor: National Institute on Aging
Building 31, Room 5C27, 31 Center Drive, MSC 2292
Bethesda, MD 20892-7710

NIA Sociobehavorial Data Analysis and Archiving in Aging 3311

NIA) is seeking small grant (R03) applications to stimulate and facilitate data archiving and secondary analyses of data related to caregiving, cognition, demography, economics, epidemiology, behavioral genetics and other behavioral research on aging; provide support for preliminary projects using secondary analysis that could lead to subsequent applications for other research project grant award mechanisms; provide support for rapid analyses of new databases and experimental modules for purposes such as informing the design and content of future study waves; provide support for the development, enhancement and assembly of new databases from existing data; and provide support for pilot research on under-utilized databases. PA: PA-04-123

Requirements: Applications may be submitted by domestic for-profit and nonprofit organizations, public and private, such as universities, colleges, hospitals, laboratories, units of state and local governments, and eligible agencies of the federal government.

Restrictions: Foreign organizations and institutions are not eligible.

Amount of Grant: Up to 850,000 U.S.D

Contact: Linda Whipp, Grants and Contracts Management Office; (301) 496-1472; fax (301) 402-3672; lw17m@nih.gov

Internet: http://grants.nih.gov/grants/guide/pa-files/PA-04-123.html

Sponsor: National Institute on Aging
Building 31, Room 5C27, 31 Center Drive, MSC 2292
Bethesda, MD 20892-7710

NIA Support of Scientific Meetings as Cooperative Agreements 3312

The purpose of the program announcement is to inform the scientific community that the National Institute on Aging (NIA) will now support scientific meetings as cooperative agreements in addition to the current practice of supporting them through the traditional grant mechanism. This program provides guidelines for when it is appropriate to request support of a meeting as a cooperative agreement and explains procedures for preparing and submitting such applications. The Principal Investigator will have the primary authority and responsibility to define objectives and approaches; plan, publicize, and conduct the scientific meeting; and publish the results of the meeting.

Requirements: Applications may be submitted by U.S. institutions, including scientific or professional societies eligible to receive grants from Public Health Service (PHS) agencies. In the case of an international conference, the U.S. representative organization of an established international scientific or professional society is the eligible applicant.

Restrictions: Foreign institutions are not eligible.

Contact: Chyren Hunter, (301) 496-9322; fax (301) 402-0528; Hunterc@nia.nih.gov

Internet: http://grants.nih.gov/grants/guide/pa-files/PA-00-128.html

Sponsor: National Institute on Aging
Gateway Building, Suite 2C231, 7201 Wisconsin Avenue, MSC 9205
Bethesda, MD 20892

NIA Symptom Management for Chronic Neurological Conditions Research Grants 3313

NINR, NIA, NICHD, NIDCD, and NINDS are interested in facilitating investigator-initiated research designed to deal with symptoms associated with chronic neurological conditions including stroke, epilepsy, brain and spinal cord injury, multiple sclerosis, and Parkinson's disease. The overall goals are to manage symptoms effectively, to prevent secondary effects of the condition when possible, to maximize the health potential, and to improve the health-related quality of life for patients, care givers, and families. The mechanism of support is the research project grant (R01). PA-98-030 may be obtained through the NIH grant line (301-402-2221), the NIH Web site, mail, and email.

Requirements: Applications may be submitted by domestic and foreign for-profit and nonprofit organizations, public or private.

Contact: Joseph Ellis, Grants Management Officer; (301) 496-1472; fax (301) 402-3672; ellisj@exmur.nia.nih.gov

Internet: http://www.grants.nih.gov/grants/guide/pa-files/PA-98-030.html

Sponsor: National Institute on Aging
Building 31, Room 5C27, 31 Center Drive, MSC 2292
Bethesda, MD 20892-7710

NIA Thyroid in Aging Grants 3314

The purpose of this funding opportunity announcement (FOA) is to encourage submission of investigator-initiated research applications on the thyroid in aging. This FOA is intended to promote basic, translational, and clinical studies leading to increased understanding of the physiology of the aging thyroid and improved diagnosis and management of thyroid disease in the elderly.

Requirements: The following organizations and institutions may apply: public/state controlled institutions of higher education; private institutions of higher education; Hispanic-serving institutions; nonprofits with 501(c)3 IRS status; nonprofits without 501(c)3 IRS status; small businesses; for-profit organizations; State governments; regional organizations; U.S. territories or possessions; Indian/Native American tribal governments; Indian/Native American tribally designated organizations; non-domestic (non-U.S.) entities (foreign organizations); Hispanic-serving institutions; historically Black colleges and universities (HBCUs); tribally controlled colleges and universities (TCCUs); Alaska native and native Hawaiian serving institutions; and regional organizations.

Date(s) Application is Due: Feb 5; Feb 16; Jun 5; Jun 16; Oct 5; Oct 16

Contact: Nancy J. Emenaker, PhD; (301) 496-0116; emenaken@mail.nih.gov

Internet: http://grants.nih.gov/grants/guide/pa-files/PA-08-037.html

Sponsor: National Institute on Aging
Gateway Building, Suite 3C307, 7201 Wisconsin Avenue, MSC 9205
Bethesda, MD 20892-9205

NIA Transdisciplinary Research on Fatigue and Fatigability in Aging Grants 3315

The purpose of this Funding Opportunity Announcement (FOA) is to encourage submission of research applications on fatigue and fatigability in aging. This FOA is intended to promote research studies employing transdisciplinary approaches that could lead to increased understanding of mechanisms contributing to, assessment of, or potential interventions for, increased fatigue or fatigability in older persons. Both animal models and humans are appropriate for study under this FOA.

Requirements: The following organizations and institutions may apply: public/state controlled institutions of higher education; private institutions of higher education; Hispanic-serving institutions; historically Black colleges and universities (HBCUs); tribally controlled colleges and universities (TCCUs); Alaska native and native Hawaiian serving institutions; nonprofits with 501(c)3 IRS status; nonprofits without 501(c)3 IRS status; small businesses; for-profit organizations; State governments; regional organizations; U.S. territories or possessions; Indian/Native American tribal governments; Indian/Native American tribally designated organizations; county governments; city or township governments; special district governments; independent school districts; public housing authorities/Indian housing authorities; eligible agencies of the Federal government; and faith-based or community based organizations.

Date(s) Application is Due: Feb 5; Feb 16; Jun 5; Jun 16; Oct 5; Oct 16

Contact: Kathleen Jett, PhD; (301) 594-2154; fax (301) 480-8260; jettk@mail.nih.gov

Internet: http://grants.nih.gov/grants/guide/pa-files/PA-08-161.html

Sponsor: National Institute on Aging
Gateway Building, Suite 3C307, 7201 Wisconsin Avenue, MSC 9205
Bethesda, MD 20892-9205

NIA Translational Research at the Aging/Cancer Interface (TRACI) Grants 3316

The purpose of this Funding Opportunity Announcement (FOA) is to enhance translational research in the overlapping areas of human aging and cancer by: integrating knowledge of basic processes in cancer biology and aging into clinical care of older patients with cancer (bench to bedside); and exploring clinical observations from the patient care setting at more basic and molecular levels (bedside to bench). Ultimately, information from the research supported by this initiative should lead to further improvements in prevention, diagnosis and disease management, improving the health and well-being of elderly patients at risk for, or diagnosed with, cancer and decreasing the functional impairment and morbidity associated with cancer in this population.

Requirements: The following organizations and institutions may apply: public/state controlled institutions of higher education; private institutions of higher education; Hispanic-serving institutions; historically Black colleges and universities (HBCUs); tribally controlled colleges and universities (TCCUs); Alaska native and native Hawaiian serving institutions; nonprofits with 501(c)3 IRS status; nonprofits without 501(c)3 IRS status; small businesses; for-profit organizations; State governments; regional organizations; U.S. territories or possessions; Indian/Native American tribal governments; Indian/Native American tribally designated organizations; county governments; city or township governments; special district governments; independent school districts; public housing authorities/Indian housing authorities; eligible agencies of the Federal government; and faith-based or community based organizations.

Date(s) Application is Due: Feb 5; Feb 16; Jun 5; Jun 16; Oct 5; Oct 16

Contact: Susan G. Nayfield, (301) 496-6761; fax (301) 402-1784; nayfiels@nia.nih.gov

Internet: http://grants.nih.gov/grants/guide/pa-files/PA-08-230.html

Sponsor: National Institute on Aging
Gateway Building, Suite 3C307, 7201 Wisconsin Avenue, MSC 9205
Bethesda, MD 20892-9205

NIA Vulnerable Dendrites and Synapses in Aging and Alzheimer's Disease Grants 3317

The purpose of this initiative is to solicit research proposals that will investigate the factors regulating synaptic plasticity and dysfunction with a particular emphasis on the age-dependent changes in the functions of dendrites, spines and synapses of key cell types in regions of brain that are selectively vulnerable in certain neurodegenerative disorders such as AD, or with aging using, for example, in vitro and in vivo models.

Requirements: The following organizations and institutions may apply: public/state controlled institutions of higher education; private institutions of higher education; Hispanic-serving institutions; historically Black colleges and universities (HBCUs); tribally controlled colleges and universities (TCCUs); Alaska native and native Hawaiian serving institutions; nonprofits with 501(c)3 IRS status; nonprofits without 501(c)3 IRS status; small businesses; for-profit organizations; State governments; regional organizations; U.S. territories or possessions; Indian/Native American tribal governments; Indian/Native American tribally designated organizations; county governments; city or township governments; special district governments; independent school districts; public housing authorities/Indian housing authorities; eligible agencies of the Federal government; and faith-based or community based organizations.

Date(s) Application is Due: Feb 5; Jun 5; Oct 5

Contact: D. Stephen Snyder, (301) 496-9350; fax (301) 496-1494; ss82f@nih.gov

Internet: http://grants.nih.gov/grants/guide/pa-files/PA-09-061.html

Sponsor: National Institute on Aging
Gateway Building, Suite 350, 7201 Wisconsin Avenue, MSC 9205
Bethesda, MD 20892-9205

NICHD Central Nervous System Effects of HIV Infection: Neurobiological, Neurovirological, and Neurobehavioral Studies Grants 3318

NIMH, NINDS, NIAAA, and NIDA, invites applications for support of research focused on determining the pathogenic mechanisms involved in Human Immunodeficiency Virus (HIV)-1 associated neurobehavioral and neurological dysfunction in domestic and

international settings. The objective of this cooperative effort is to foster investigations that will provide the foundation for the rapid development of therapeutic interventions to prevent and treat the effects of HIV-1 on the central nervous system (CNS). Applications ranging from basic research to clinical diagnosis and treatment are of interest. Multidisciplinary research teams and collaborative alliances are encouraged but not required. Applications will be accepted under PHS receipt dates for AIDS applications. PA: PA-04-154
Requirements: Applications may be submitted by domestic and foreign, for-profit and nonprofit organizations, both public and private; units of state and local governments; and eligible agencies of the federal government.
Amount of Grant: Up to 375,000 U.S.D
Contact: Dr. Jeymohan Joseph, (301) 443-3012; jjeymoha@mail.nih.gov
Internet: http://grants.nih.gov/grants/guide/pa-files/PA-04-154.html
Sponsor: National Institute of Child Health and Human Development
31 Center Drive, Building 31 Room 2A32, MSC 2425
Bethesda, MD 20892-2425

NICHD Clinical Trial Planning Grants for Pediatric Rehabilitation 3319
The purpose of this grant is to provide support for the organization of an effective research group and development of elements essential for a successful clinical trial. Applications may address the development of activities relevant to such elements as plans for recruitment of patients, experimental design and protocols, testing of procedures, data management, analysis techniques, facilities, administrative procedures, and collaborative arrangements. Detailed information regarding the rationale for the clinical trial based on preclinical and preliminary clinical research must be included in the application. Annual deadline dates may change; contact program staff for exact dates. RFA: HD-01-006
Requirements: Applications may be submitted by foreign and domestic, for-profit and nonprofit organizations, both public and private; units of state and local governments; and eligible agencies of the federal government.
Restrictions: Clinical trial planning support is for new projects and may not be used to supplement research projects already being supported or to provide interim support of projects under review.
Amount of Grant: Up to 75,000 U.S.D
Contact: Dr. Beth Ansel; (301) 402-2242; fax (301) 402-0832; Beth_Ansel@nih.gov
Internet: http://grants.nih.gov/grants/guide/rfa-files/RFA-HD-01-006.html
Sponsor: National Institute of Child Health and Human Development
31 Center Drive, Building 31 Room 2A32, MSC 2425
Bethesda, MD 20892-2425

NICHD Cooperative Specialized Infertility Research Centers 3320
The objective of this program is to expedite the development of new approaches for the diagnosis and treatment of human male or female infertility. Annual deadline dates may vary; contact program staff for exact dates. RFA: HD-01-001
Requirements: Applications may be submitted by domestic for-profit and nonprofit organizations, both public and private; units of state and local governments; and eligible agencies of the federal government.
Amount of Grant: Up to 3,000,000 U.S.D
Contact: Dr. Louis DePaolo, Reproductive Sciences Branch; (301) 435-6970; fax (301) 496-0962; ld38p@nih.gov
Internet: http://grants.nih.gov/grants/guide/rfa-files/RFA-HD-01-001.html
Sponsor: National Institute of Child Health and Human Development
31 Center Drive, Building 31 Room 2A32, MSC 2425
Bethesda, MD 20892-2425

NICHD Innovative Therapies and Clinical Studies for Screenable Disorders Grants 3321
This Funding Opportunity Announcement (FOA) issued by the National Institute of Child Health and Human Development (NICHD), the National Institute on Deafness and Other Communication Disorders (NIDCD), and the National Institute of Diabetes and Digestive and Kidney Diseases (NIDDK), National Institutes of Health (NIH), solicit applications for research relevant to the basic understanding and development of therapeutic interventions for currently screened conditions and high priority genetic conditions for which screening could be possible in the near future. Because the nature and scope of the proposed research will vary from application to application, it is anticipated that the size and duration of each award will also vary. The total amount awarded and the number of awards will depend upon the numbers, quality, duration, and costs of the applications received.
Requirements: Application can be made by the following institutions and organizations: public/state controlled institutions of higher education; private institutions of higher education; nonprofits with 501(c)3 IRS status (other than institutions of higher education); nonprofits without 501(c)3 IRS status (other than institutions of higher education); small businesses; for-profit organizations (other than small businesses); state governments; U.S. territories or possessions; Indian/Native American tribal governments (federally recognized); Indian/Native American tribal governments (other than Federally recognized); Indian/Native American tribally designated organizations; non-domestic (non-U.S.) entities (foreign organizations); Hispanic-serving institutions; historically Black colleges and universities (HBCUs); tribally controlled colleges and universities (TCCUs); Alaska Native and Native Hawaiian serving institutions; regional organization; eligible agencies of the Federal government; and faith-based or community based organizations.
Date(s) Application is Due: Feb 5; Mar 5; Jun 5; Jul 5; Oct 5; Nov 5
Contact: Gilian Engelson, (301) 451-2137; fax (301) 451-5512; engelsong@mail.nih.gov
Internet: http://grants.nih.gov/grants/guide/pa-files/PAR-07-184.html
Sponsor: National Institute of Child Health and Human Development
31 Center Drive, Building 31 Room 2A32, MSC 2425
Bethesda, MD 20892-2425

NICHD Low-Back Pain and Common Spinal Disorders Research Grants 3322
The program is sponsored by NIAMS, NICHD, NINDS, and NIOSH at the CDC. This initiative invites applications directed to the study of the pathogenesis, epidemiology, prevention, and treatment of three common causes of low-back pain: herniated nucleus pulposus, spinal stenosis, and idiopathic low-back pain. The support mechanism for grants in this area will be the individual investigator-initiated research grant (R01). Information may be obtained electronically through the NIH grant line (301-402-2221), the NIH Web site, and by mail and email. PA-97-058
Requirements: Applications may be submitted by foreign and domestic, for-profit and nonprofit organizations, both public and private; units of state and local governments; and eligible agencies of the federal government.
Restrictions: Foreign institutions or organizations in foreign countries are not eligible.
Contact: Dr. Stephen Tuel, National Center for Medical and Rehabilitation Research; (301) 402-2242; fax (301) 402-0832; tuels@hd01.nichd.nih.gov
Internet: http://www.grants.nih.gov/grants/guide/pa-files/PA-97-058.html
Sponsor: National Institute of Child Health and Human Development
31 Center Drive, Building 31 Room 2A32, MSC 2425
Bethesda, MD 20892-2425

NICHD Medical Rehabilitation Research 3323
This program announcement invites qualified researchers to submit grant applications for research on restoring, replacing, or enhancing the function of adults and children with disabilities. Medical rehabilitation research is directed towards restoration and improvement of functional capability lost as a consequence of injury, disease, or congenital disorders. The mission of the NCMRR is to increase the effectiveness of medical rehabilitation practices through research on improving functional mobility; promoting behavioral adaptation to functional losses; assessing the efficacy and outcomes of medical rehabilitation therapies and practices; developing improved assistive technology; understanding whole body system responses to physical impairments and functional changes; developing more precise methods to measure impairments, disabilities, and societal and functional limitations; and training research scientists in the field of medical rehabilitation. PA: PA-99-041
Requirements: Applications for research project grants (R01) may be submitted by domestic and foreign, for-profit and nonprofit organizations, both public and private; units of state or local governments; and eligible agencies of the federal government.
Contact: Dr. Stephen Tuel; (301) 402-2242; st100v@nih.gov
Internet: http://www.grants.nih.gov/grants/guide/pa-files/PA-99-041.html
Sponsor: National Institute of Child Health and Human Development
31 Center Drive, Building 31 Room 2A32, MSC 2425
Bethesda, MD 20892-2425

NICHD NRSAs—Individual 3324
The program offers health scientists with doctoral degrees the opportunity to receive full-time research training in areas that reflect the national need for biomedical and behavioral research on specific programs of mothers and children and population and reproductive research. Areas of concentration include pregnancy and infancy, developmental biology and nutrition, learning and behavior, mental retardation, fertility-infertility, fertility regulation, nutrition and reproduction, social and behavioral aspects of reproduction, and population change. These fellowships may not be used for study leading to an MD, DO, DDS, or similar degree, or for the support of residency training. Awardees are subject to the payback provisions of the NRSA by which one year of research or teaching, or acceptable combination thereof, is required for each year of support, except for the first 12 months, which are free of the payback obligation.
Requirements: Universities, colleges, medical, dental, and nursing schools and schools of public health, laboratories, hospitals, state and local health departments, other for-profit public or private institutions, and individuals may apply for research grants. NRSA support is provided for academic and research training only, in health and health-related areas that are periodically specified by the NIH. SBIR grants can be awarded only to domestic small businesses. STTR grants can be awarded only to domestic small business concerns that partner with research institutions in cooperative research and development.
Date(s) Application is Due: Apr 5; Aug 5; Dec 5
Contact: NICHD Office of Program and Public Liaison; (301) 496-0536 or (301) 435-3447
Internet: http://www.nichd.nih.gov/funding/funding.htm
Sponsor: National Institute of Child Health and Human Development
31 Center Drive, Building 31 Room 2A32, MSC 2425
Bethesda, MD 20892-2425

NICHD NRSAs—Institutional 3325
Awards provide opportunities for research training in the biological and/or behavioral science aspects of child health and human development. Primary concern in awarding institution awards will be given to multidisciplinary or interdisciplinary programs that cannot be provided through individual fellowships. Although major concern is for postdoctoral training, predoctoral training will be considered where a special case for support can be justified. In the Center for Research for Mothers and Children, areas covered include pregnancy and infancy, developmental biology and nutrition, learning and behavior, and mental retardation. In the Center for Population Research, areas covered include fertility-infertility, fertility regulation, nutrition and reproduction, social and behavioral aspects of reproduction, and population change. The training program director at the grantee institution will be responsible for the selection and appointment of trainees and for the overall direction of the program. These grants are not made for study leading to the MD, DDS, DVM, or other similar health professional degrees. Trainees are subject to the payback provisions of the NRSA by which one year of research or teaching, or acceptable combination thereof, is required for each year of support, except for the first 12 months, which are free of payback obligation.

Requirements: Universities; colleges; medical, dental, and nursing schools and schools of public health; laboratories; hospitals; state and local health departments; other for-profit public or private institutions; and individuals may apply for research grants. NRSA support is provided for academic and research training only, in health and health-related areas that are periodically specified by the National Institutes of Health. SBIR grants can be awarded only to domestic small businesses. STTR grants can be awarded only to domestic small business concerns that partner with research institutions in cooperative research and development.
Date(s) Application is Due: May 10
Contact: NICHD Office of Program and Public Liaison; (301) 496-0536 or (301) 435-3447
Internet: http://www.nichd.nih.gov/funding/funding.htm
Sponsor: National Institute of Child Health and Human Development
31 Center Drive, Building 31 Room 2A32, MSC 2425
Bethesda, MD 20892-2425

NICHD Population Research Grants 3326
The objective of the program is to seek solutions to the fundamental problems of the reproductive processes; to strive to develop and evaluate safer and more effective contraceptives; and to understand how population structure and change affect the health and well-being of individuals and society. The Center for Population Research supports biomedical research in the reproductive sciences, clinical and epidemiological studies of contraceptives and the risk of sexually transmitted diseases (STDs) including HIV, and behavioral and demographic studies on population processes and structure. Mechanisms are projects grants, small business innovation research (SBIR) grants, small business technology transfer (STIR) grants, and individual fellowships (NRSAs).
Requirements: Project grants: Universities, colleges, medical, dental, and nursing schools, and schools of public health; laboratories, hospitals, state and local health departments, and other nonprofit or for-profit public or private institutions; and individuals. NRSAs: Individuals with a professional or scientific degree are eligible (MD, PhD, DDS, DO, DVM, ScD, DEng, or equivalent domestic or foreign degree). SBIR and STIR grants can be awarded only to domestic small businesses (entities that are independently owned and operated for profit, are not dominant in the field in which research is proposed, and have no more than 500 employees).
Date(s) Application is Due: Feb 1; Jun 1; Oct 1
Amount of Grant: 50,000 - 5,000,000 U.S.D
Contact: Dr. Florence Haseltine, Director, Center for Population Research; (301) 496-1101; fax (301) 496-0962; haseltif@mail.nih.gov
Internet: http://www.nichd.nih.gov/cpr
Sponsor: National Institute of Child Health and Human Development
31 Center Drive, Building 31 Room 2A32, MSC 2425
Bethesda, MD 20892-2425

NICHD Repetitive Motion Disorders Research Grants 3327
The program is sponsored by NIAMS, NICHD, NINDS, and NIOSH at the CDC. This initiative invites applications directed to the study of the pathogenesis, epidemiology, prevention, and treatment of repetitive motion disorders. The support mechanism for grants in this area will be the individual investigator-initiated research grant (R01). Information may be obtained electronically through the NIH grant line (301-402-2221), the NIH Web site, and by mail and email. PA-97-059
Requirements: Applications may be submitted by foreign and domestic, for-profit and nonprofit organizations, both public and private; units of state and local governments; and eligible agencies of the federal government.
Restrictions: Foreign institutions or organizations in foreign countries are not eligible.
Contact: Dr. Stephen Tuel, National Center for Medical and Rehabilitation Research; (301) 402-2242; fax (301) 402-0832; tuels@hd01.nichd.nih.gov
Internet: http://www.grants.nih.gov/grants/guide/pa-files/PA-97-059.html
Sponsor: National Institute of Child Health and Human Development
31 Center Drive, Building 31 Room 2A32, MSC 2425
Bethesda, MD 20892-2425

NICHD Research on Musculoskeletal Fitness and Sports Medicine Grants 3328
NIAMS and NINR invite investigator-initiated research grant applications to study a broad range of basic and clinical topics related to musculoskeletal fitness, exercise physiology, and sports medicine. The National Center for Medical Rehabilitation Research of NICHD encourages applications for both basic and clinical studies of musculoskeletal fitness and exercise physiology of persons with physical disabilities. PA: PA-97-025
Requirements: Applications may be submitted by domestic and foreign, for-profit organizations, both public and private; units of state or local governments; and eligible agencies of the federal government.
Contact: Dr. Danuta Krotoski, National Center for Medical Rehabilitation Research; (301) 402-2242; krotoskd@hd01.nichd.nih.gov
Internet: http://www.grants.nih.gov/grants/guide/pa-files/PA-97-025.html
Sponsor: National Institute of Child Health and Human Development
31 Center Drive, Building 31 Room 2A32, MSC 2425
Bethesda, MD 20892-2425

NICHD Ruth L. Kirschstein National Research Service Award (NRSA) Institutional Predoctoral Training Program in Systems Biology of Developmental Biology & Birth Defects 3329
The objective of this NRSA T32 program is to provide research training to predoctoral students interested in establishing research careers that use systems biology approaches to study developmental biology and the formation of structural birth defects. Stipends are provided as a subsistence allowance to help defray trainees' living expenses during the research training experience. Funds are provided to offset the cost of tuition and fees. The NIH provides 60% of the combined requested costs, up to $16,000 annually per Ph.D. student, and up to $21,000 annually for students in formal programs leading to combined professional and Ph.D. degrees. Letters of Intent are due by April 25, with full applications due May 25.
Requirements: The following organizations are eligible to apply: public/state controlled institutions of higher education; private institutions of higher education; nonprofits with 501(c)3 IRS status (other than institutions of higher education); nonprofits without 501(c)3 IRS status (other than institutions of higher education); for-profit organizations (other than small businesses); state governments; U.S. territories or possessions; Indian/Native American tribal governments (Federally recognized); Hispanic-serving institutions; historically Black colleges and universities (HBCUs); tribally controlled colleges and universities (TCCUs); Alaska Natives and Native Hawaiians serving institutions; regional organizations; eligible agencies of the federal government; faith-based or community based organizations.
Date(s) Application is Due: Apr 25; May 25
Contact: James Coulombe, (301) 496-5541; fax (301) 480-0303; coulombej@mail.nih.gov
Internet: http://grants.nih.gov/grants/guide/pa-files/PAR-08-054.html
Sponsor: National Institute of Child Health and Human Development
31 Center Drive, Building 31 Room 2A32, MSC 2425
Bethesda, MD 20892-2425

NICHD Small Grants 3330
Applications for these small grants may be submitted by new or more established investigators for support of research in any scientific area relevant to the institute's programs in the Center for Population Research, the Center for Research for Mothers and Children, or the National Center for Medical Rehabilitation Research. Although proposals may involve a wide variety of biomedical, biobehavioral, or clinical disciplines, relevance to the mission of NICHD must be clear. Deadline dates are February 1, June 1, or October 1 for new applications; March 1, or July 1, or November 1 for revised applications. Additional information may be obtained by mail and email, or through the NIH grant line (301-402-2221), and the NIH Web site. PA: PAR-99-126
Requirements: Applications may be submitted by domestic, for-profit and nonprofit organizations, both public and private; units of state and local governments; and eligible agencies of the federal government.
Restrictions: Foreign institutions and organizations are not eligible.
Date(s) Application is Due: Feb 1; Jun 1; Oct 1
Amount of Grant: 50,000 - 100,000 U.S.D
Contact: Program Contact; (301) 435-0714; grantsinfo@nih.gov
Internet: http://grants.nih.gov/grants/guide/pa-files/PAR-99-126.html
Sponsor: National Institute of Child Health and Human Development
31 Center Drive, Building 31 Room 2A32, MSC 2425
Bethesda, MD 20892-2425

NICHD Socioeconomic Status and Health Across the Life Course 3331
NHLBI, NIA, NICHD, NIEHS, and NIMH seek research grant applications on the cumulative and contemporaneous relationships between socioeconomic status (SES) and physical and mental health and functioning over the life course and across generations. Given that the relationships between SES and physical and mental health, morbidity, disability, and mortality have been long and extensively documented, additional studies aimed at merely describing or demonstrating these relationships are outside the scope of this program announcement. Encouraged are studies relating to appropriate conceptualization and measurement of SES over the life course, across generations, and in various population groups; specification of the processes through which SES influences cumulatively and contemporaneously physical and mental health, disability, morbidity, and mortality outcomes over the life course, and how these outcomes, in turn, impact on SES; the relationship between SES and physical and mental health, disability, morbidity, and mortality over the life course in various population groups. PA: PA-98-098
Requirements: Applications may be submitted by public and private, foreign and domestic for-profit and nonprofit organizations, and eligible agencies of the federal government.
Contact: Dr. Rose Maria Li, Center for Population Research; (301) 496-1174; fax (301) 496-0962; Rose_Li@nih.gov
Internet: http://www.grants.nih.gov/grants/guide/pa-files/PA-98-098.html
Sponsor: National Institute of Child Health and Human Development
31 Center Drive, Building 31 Room 2A32, MSC 2425
Bethesda, MD 20892-2425

NICHD Symptom Management for Chronic Neurological Conditions Research Grants 3332
NINR, NIA, NICHD, NIDCD, and NINDS are interested in facilitating investigator-initiated research designed to deal with symptoms associated with chronic neurological conditions including stroke, epilepsy, brain and spinal cord injury, multiple sclerosis, and Parkinson's disease. The overall goals are to manage symptoms effectively, to prevent secondary effects of the condition when possible, to maximize the health potential, and to improve the health-related quality of life for patients, caregivers, and families. The mechanism of support is the research project grant (R01). Information may be obtained through the NIH grant line (301-402-2221), the NIH Web site, mail, and email. PA-98-030
Requirements: Research applications may be submitted by domestic and foreign, for-profit and nonprofit organizations, both public and private; units of state and local governments; and eligible agencies of the federal government.
Contact: Mary Ellen Colvin, Grants Management Branch; (301) 496-1303; fax (301) 402-0915; MC113B@nih.gov
Internet: http://www.grants.nih.gov/grants/guide/pa-files/PA-98-030.html
Sponsor: National Institute of Child Health and Human Development
31 Center Drive, Building 31 Room 2A32, MSC 2425
Bethesda, MD 20892-2425

Nicholas H. Noyes Jr. Memorial Foundation Grants 3333

The foundation awards grants to eligible Indiana nonprofit organizations in its areas of interest, including arts and culture, education from early childhood through higher education, disadvantaged, museums, social services, health, hospitals, family services, performing arts, and youth. Types of support include general operating support, endowment funds, and scholarship funds.

Requirements: Proposals are welcomed, and encouraged, to be submitted at any time after January 1st for the first funding cycle or after June 1st for the final funding cycle. Submit the following with your proposal: thirteen copies of the complete grant application (available at the foundations website or office) including attachments (i.e. applicable budget(s) as required). You may recreate the application form for ease in completing, but please limit your response to the space provided and use a minimum font of 12. It is also require that you use both sides of the paper i.e. side one is page 1 with page 2 on the flip-side etc; one copy of your organization's 501(c)3 tax determination letter from the IRS; one copy of your organization's most recent audited financial statement for all organizations requesting $25,000 or more.

Restrictions: Organizations may apply only one time in any calendar year. The Foundation does not make grants to individuals. The principle geographic region served by the Noyes Foundation is the greater Indianapolis area. If you intend to request a grant in excess of $50,000, please contact the Foundation's Program Officer before submitting your grant request.

Geographic Focus: Indiana

Date(s) Application is Due: Feb 1; Aug 29

Amount of Grant: Up to 150,000 U.S.D

Contact: Kelly Mills, Executive Administrator; (317) 844-8009; fax (317) 844-8099; kmills@noyesfoundation.org

Internet: http://www.noyesfoundation.org

Sponsor: Nicholas H. Noyes Jr. Memorial Foundation
1950 E Greyhound Pass, #18
Carmel, IN 46033-7730

Nick Traina Foundation Grants 3334

The Foundation supports organizations involved in the diagnosis, research, treatment, and/or family support of manic-depression, and other forms of mental illness, suicide prevention, child abuse and children in jeopardy, and provides assistance to struggling musicians in the areas of mental illness. The Foundation may give special consideration to proposals that address manic-depression in children and young adults. There are no deadlines and there is no formal application form. The Board meets four times a year to review proposals.

Requirements: Applicants must be 501(c)3 organizations. Proposals should be kept to three pages in addition to attachments and should include: the organization's purpose, including a description of its history and mission and the population served; the reason for the grant request, including the amount requested and how the funds will be used; whether the request is for general organizational support or for a specific project (including the duration of a specific project); a copy of the IRS determination letter; and a copy of the most recent annual report and the last two years' financial statements.

Restrictions: Organizations may apply for one grant each calendar year. Requests from individuals are ineligible.

Samples: At the Crossroads, San Francisco, California, $11,000; Homeless Youth Alliance, San Francisco, California, $27,000; and The Caduceus Society, Abilene, Texas, $1,000.

Contact: Danielle Steel, President; (415) 771-4224; info@nicktrainafoundation.org

Internet: http://nicktrainafoundation.com/grants.htm

Sponsor: Nick Traina Foundation
P.O. Box 470427
San Francisco, CA 94147-0427

Nicole Brown Charitable Foundation Grants 3335

Grants will be made to organizations whose primary purpose is to teach the public about the extent, dangers, and causes of domestic violence, including spousal and child abuse; intervene in the cycle of violence by providing alternative housing for victims of abuse; treat victims of domestic violence through medical care and counseling; and seek to end the abusive actions of abusers by appropriate treatment and intervention programs. Application forms are available on the Web site. Contact program staff for availability and deadlines.

Requirements: 501(c)3 public charities designated as shelters are eligible. Shelters must have been in existence for one full year.

Restrictions: Individuals are ineligible.

Contact: Grants Administrator; (949) 283-5330; info@nbcf.org

Internet: http://www.nbcf.org

Sponsor: Nicole Brown Charitable Foundation
Box 3777
Dana Point, CA 92629

NIDA Behavioral Science Track Awards for Rapid Transition 3336

NIDA invites newly independent investigators to submit applications for small-scale, exploratory (i.e., pilot) research projects related to NIDA's behavioral sciences mission. The award (B/START-NIDA) provides rapid review and funding decisions of applications. Experimentally based research applications are encouraged across a wide variety of behavioral factors in drug abuse, including neurocognitive, cognitive, and perceptual processes; psychosocial; and more broadly motivational, social, and community factors in drug abuse. Given the role that drug abuse plays in HIV/AIDS transmission, studies applying basic behavioral science models and methods to address this issue are especially encouraged. More information may be obtained by mail and email or through the NIH grant line (301-402-2221), and the NIH Web site. PAR-03-146

Amount of Grant: Up to 50,000 U.S.D

Contact: Gary Fleming, Grants Management Branch; (301) 443-6710; fax (301) 594-6847; gf6s@nih.gov

Internet: http://grants2.nih.gov/grants/guide/pa-files/PAR-03-146.html

Sponsor: National Institute on Drug Abuse
6001 Executive Boulevard, Room 5213
Bethesda, MD 20892-9561

NIDA Behavioral Therapies Development Program Grants 3337

The purpose of the program is to reaffirm NIDA's continued and ongoing commitment to a major program of research on behavioral therapies for drug abuse and dependence. The term behavioral therapy is used here in a broad sense and includes various forms of psychotherapy, behavior therapy, cognitive therapy, family therapy, skills training, counseling, and other rehabilitative therapies. Behavioral therapy research has been conceptualized, for the purposes of this initiative, to consist of three stages. Stage I, or early therapy development, involves research on the development, refinement, and pilot testing of behavioral interventions. Stage I includes research aimed at translating basic behavioral and cognitive research into novel behavioral therapies. Stage II involves the efficacy testing of therapies that show promise and includes research on the hypothesized mechanisms through which these therapies may have their effect. Stage III is research aimed at determining if and how efficacious behavioral therapies may be transported to community settings, including primary care sites. Stage III includes research examining methods of training therapists and counselors to administer new behavioral therapies. PA: PA-03-066

Requirements: Applications may be submitted by domestic and foreign, for-profit and nonprofit organizations, both public and private; units of state and local government; and eligible agencies of the federal government.

Amount of Grant: Up to 100,000 U.S.D

Contact: Gary Fleming, NIDA, Grants Management Branch; (301) 443-6710; fax (301) 594-6849; gf6s@nih.gov

Internet: http://grants2.nih.gov/grants/guide/pa-files/PA-03-066.html

Sponsor: National Institute on Drug Abuse
6001 Executive Boulevard, Room 5213
Bethesda, MD 20892-9561

NIDA Clinical Research Education Programs in Drug Abuse and Addiction 3338

Grants support educational programs developing the research skills of investigators who wish to focus on clinical, prevention, health services, or treatment issues in drug abuse; or developing a cadre of clinical drug abuse experts proficient in the use of research findings (for example, in public health, behavioral science, or medicine) to develop or implement early detection, diagnosis, treatment, or prevention interventions. This latter category would also include applications of health services research findings. Programs that address both of these aims are particularly desired. NIDA will grant awards to eligible institutions to develop or enhance research training opportunities for individuals, selected by the institution, who are training for careers as researchers, as clinicians (broadly defined), or (ideally) as a combination of the two. Educational partnerships between research institutions and drug abuse treatment organizations are highly encouraged in order to broaden research training and educational opportunities and as a result, increase the number of well-trained clinical and health services researchers in the field. Contact program staff for deadline dates. PA: PAR-04-054

Requirements: Applications may be submitted by domestic, for-profit and nonprofit organizations, both public and private; units of state and local governments; and eligible agencies of the federal government; All students must be U.S. citizens, nationals, or lawfully admitted permanent residents of the United States. Predoctoral students must have an MS, MSW, or equivalent degree in biostatistics, nursing, psychology, or other social sciences as a prerequisite to pursuing a doctorate degree.

Restrictions: This award will not support a master's level degree for a predoctoral student.

Amount of Grant: Up to 350,000 U.S.D

Contact: Gary Fleming, Chief, Grants Management Officer; (301) 443-6710; fax (301) 594-6847; gf6s@nih.gov

Internet: http://grants.nih.gov/grants/guide/pa-files/PAR-04-054.html

Sponsor: National Institute on Drug Abuse
6001 Executive Boulevard, Room 5213
Bethesda, MD 20892-9561

NIDA Cutting-Edge Basic Research Awards (CEBRA) 3339

The goal of this program is to foster highly innovative or conceptually creative research that advances our understanding of drug abuse and addiction and how to prevent and treat them. It is specifically designed to support research that is high-risk and potentially high-impact and that is underrepresented or not included in NIDA's current portfolio. The program is aimed at experienced drug abuse research investigators who wish to develop or adapt new methods or techniques and at new investigators or scientists with expertise in fields other than drug abuse who wish to establish new programs in drug abuse research. PA: PAR-01-047

Requirements: Applications may be submitted by domestic and foreign, for-profit and nonprofit organizations, both public and private; units of state and local governments; and eligible agencies of the federal government.

Restrictions: Amended R21 applications for CEBRA will not be accepted.

Amount of Grant: Up to 100,000 U.S.D

Contact: Dr. Rebekah Rasooly, (301) 443-6300; fax (301) 594-6043; rr185i@nih.gov

Internet: http://grants.nih.gov/grants/guide/pa-files/PAR-01-047.html

Sponsor: National Institute on Drug Abuse
6001 Executive Boulevard, Room 5213
Bethesda, MD 20892-9561

NIDA Drug Abuse Dissertation Research: Epidemiology, Prevention, Treatment, Services, and Women and Gender Differences 3340

NIDA announces the availability of small grants (R03) to support drug abuse doctoral dissertation research in epidemiology, prevention, treatment, services, and women and sex/gender differences, areas in which there is a significant need for new investigators. Grant support is designed to encourage doctoral candidates from a variety of academic disciplines and programs to conduct research in these areas of interest to NIDA. It is hoped that this program will ultimately facilitate the entry of promising new investigators into the field of drug abuse research. PA: PAR-05-083

Requirements: The applicant investigator must be enrolled in an accredited doctoral degree program in the behavioral, biomedical, or social sciences and must be conducting or intending to conduct research in one of the areas specified. The applicant must be a registered doctoral candidate student. All requirements for the doctoral degree other than the dissertation and clinical internship, if applicable, must be completed by the time of the award. This information and the approval of the dissertation topic by a named committee must be verified in a letter of certification from the thesis chairperson and submitted with the grant application. The applicant institution administering the grant on behalf of the proposed applicant must be domestic. Applications may be submitted by any public or private nonprofit university, college, or professional school. The doctoral candidate must be a citizen or noncitizen national of the United States or hold a permanent residence visa.

Restrictions: An applicant who receives support for dissertation research under a grant from NIDA may not hold concurrently another federally sponsored fellowship or similar Federal award, such as a National Research Service Award (NRSA) that provides a stipend or otherwise duplicates provisions of this grant program.

Contact: Gary Fleming, Grants Management; (301) 443-6710; fax (301) 594-6849; gfleming@nida.nih.gov

Internet: http://grants.nih.gov/grants/guide/pa-files/PAR-05-083.html

Sponsor: National Institute on Drug Abuse
6001 Executive Boulevard, Room 5213
Bethesda, MD 20892-9561

NIDA Drug Addiction Neuroscience Research Grants 3341

The intent of this program announcement is to continue to encourage investigator interest in the wide range of neuroscience research relevant to drug abuse, drug dependence, and drug addiction supported by NIDA. The research relevant to this announcement does not involve human subjects; for information about the neuroscience research program involving human subjects, contact Dr. Joseph Frascella, (301) 443-4877, jf80t@nih.gov. The scientific understanding gained by this research is anticipated ultimately to be applicable to improved treatment and prevention of drug abuse and drug addiction, and their consequences. The specific research topics emphasized in this program announcement are animal models to study the neurobiology of addiction; vulnerability to drug addiction; neuroadaptation and neurotoxicity; developmental neurobiology; drugs and learning, memory, and cognition; pain and analgesia; drug effects on sensory and perceptual systems; drugs of abuse; sleep and circadian rhythms; neuropsychopharmacology of drugs of abuse; and central nervous system (CNS) interaction with other systems. Contact the program officer for deadline dates. PA: PA-02-085

Requirements: Applications may be submitted by public and private, domestic and foreign, for-profit and nonprofit organizations, units of state and local governments, and eligible agencies of the federal government. Racial/ethnic minority individuals, women, and persons with disabilities are encouraged to apply as principal investigators.

Contact: Dr. Roger Brown, Basic Research Coordinator; (301) 443-1887; fax (301) 594-6043; rb99w@nih.gov

Internet: http://grants2.nih.gov/grants/guide/pa-files/PA-02-085.html

Sponsor: National Institute on Drug Abuse
6001 Executive Boulevard, Room 5213
Bethesda, MD 20892-9561

NIDA HIV Drug Therapy for HIV Users: Access, Adherence, Effectiveness 3342

NIDA invites applications for research on access, adherence, and effectiveness relevant to the treatment of drug users with HIV. This initiative supports research on a broad range of HIV care-related issues, including, but not limited to, access issues related to receipt of HIV-related treatment and outreach strategies to improve access; factors influencing initiation of HIV treatment and long-term adherence to HIV treatment regimens; and the development of intervention strategies to improve use of and adherence to complex medical regimens in a variety of settings. Annual deadline dates may vary; contact program staff for exact dates. RFA: DA-00-007

Requirements: Applications may be submitted by domestic and foreign for-profit and nonprofit organizations, both public and private; units of state and local governments; and eligible agencies of the federal government.

Amount of Grant: Up to 3,000,000 U.S.D

Contact: Dr. Elizabeth Lambert, (301) 402-1933; fax (301) 443-4100; el46i@nih.gov

Internet: http://grants.nih.gov/grants/guide/rfa-files/RFA-DA-00-007.html

Sponsor: National Institute on Drug Abuse
6001 Executive Boulevard, Room 5213
Bethesda, MD 20892-9561

NIDA International Studies on Drug Abuse and HIV/AIDS 3343

This program invites applications for international, cross-disciplinary research on HIV/AIDS, other blood-borne infections [i.e., hepatitis B virus (HBV), hepatitis C virus (HCV)], tuberculosis (TB), and comorbid conditions and consequences in injecting and noninjecting drug users, their sexual partners, and their children. This RFA is intended to build on lessons learned in developed countries in response to the intertwined epidemics of drug abuse and the spread of HIV/AIDS and other infectious diseases. It seeks to foster cross-national and international research collaborations that, through both independent research and the recruitment, training, and mentoring of new, multi-disciplinary researchers, lead to the development, adaptation, replication, and evaluation of effective interventions and approaches to slow or reverse the spread of HIV and other infections in vulnerable drug-using populations. Annual deadline dates may vary; contact program staff for exact dates. RFA: DA-01-006

Requirements: Applications may be submitted by domestic and foreign for-profit and nonprofit organizations, both public and private; units of state and local governments; and eligible agencies of the federal government.

Amount of Grant: Up to 15,000,000 U.S.D

Contact: Helen Cesari, (301) 402-1918; fax (301) 443-4100; hc30x@nih.gov

Internet: http://grants.nih.gov/grants/guide/rfa-files/RFA-DA-01-006.html

Sponsor: National Institute on Drug Abuse
6001 Executive Boulevard, Room 5213
Bethesda, MD 20892-9561

NIDA Medications Development for Cannabis-Related Disorder 3344

This program seeks grant applications focusing on the identification, evaluation, and development of safe and effective pharmacological treatments for cannabis-related disorders (CRDs), such as cannabis abuse or dependence, and cannabis-induced disorders (e.g., intoxication, delirium, psychotic disorder, and anxiety disorder), and their comorbidity with other medical and psychiatric disorders (e.g., depression), with special interest in the treatment of children and adolescents. Cannabis use includes marijuana, hashish, and other tetrahydrocannabinol (THC) containing substances. This RFA will support clinical trials to test potential medications in humans, and the preclinical development of new chemical entities (NCEs) for advancement to the clinical development stage. Applications may focus on the pharmacotherapy of cannabis abuse and they may target one or various CRDs or clinical manifestations of the disorders. Annual deadline dates may vary; contact program staff for exact dates. RFA-DA-04-014

Requirements: Applications may be submitted by domestic and foreign, for-profit and nonprofit organizations, both public and private; units of state and local governments; and eligible agencies of the federal government.

Contact: Gary Fleming, Grants Management Branch; (301) 443-6710; fax (301) 594-6847; gf6s@nih.gov

Internet: http://www.grants.nih.gov/grants/guide/rfa-files/RFA-DA-04-014.html

Sponsor: National Institute on Drug Abuse
6001 Executive Boulevard, Room 5213
Bethesda, MD 20892-9561

NIDA Mentored Clinical Scientist Research Career Development Awards 3345

The overall objective of the NIH Research Career Development Award program is to prepare qualified individuals for careers that have a significant impact on the health-related research needs of the Nation. The objective of the NIH Mentored Clinical Scientist Research Career Development Award (K08) program is to support didactic study and mentored research for individuals with clinical doctoral degrees (e.g., M.D., D.D.S., D.M.D., D.O., D.C., O.D., N.D., D.V.M., Pharm.D., or Ph.D. in clinical disciplines). The purpose of this Funding Opportunity Announcement (FOA) is to continue the long-standing NIH support of the K08 program. This award provides support and protected time for an intensive, mentored research career development experience in biomedical or behavioral research, including translational research. For the purpose of this award, translational research is defined as application of basic research discoveries toward the diagnosis, management, and prevention of human disease.

Requirements: The following organizations and institutions may apply: public/state controlled institutions of higher education; private institutions of higher education; Hispanic-serving institutions; historically Black colleges and universities (HBCUs); tribally controlled colleges and universities (TCCUs); Alaska native and native Hawaiian serving institutions; nonprofits with 501(c)3 IRS status; nonprofits without 501(c)3 IRS status; small businesses; for-profit organizations; State governments; regional organizations; U.S. territories or possessions; Indian/Native American tribal governments; Indian/Native American tribally designated organizations; county governments; city or township governments; special district governments; independent school districts; public housing authorities/Indian housing authorities; eligible agencies of the Federal government; and faith-based or community based organizations.

Date(s) Application is Due: Feb 12; Mar 12; Jun 12; Jul 12; Oct 12; Nov 12

Amount of Grant: Up to 140,000 U.S.D

Contact: Mimi M. Ghim, (301) 443-6071; ghimm@mail.nih.gov

Internet: http://grants.nih.gov/grants/guide/pa-files/PA-09-042.html#SectionIII

Sponsor: National Institute on Drug Abuse
6001 Executive Boulevard, Room 5213
Bethesda, MD 20892-9561

NIDA Microarray-Based Research on Drug Abuse 3346

With this announcement, NIDA plans to develop a strong, multidisciplinary basic research program—drawing from neuroscience, genetics, behavioral science, bioengineering, and computational biology—aimed at adapting and applying microarray techniques to the general problem of drug abuse. Eventually, it is anticipated that the identification of genes and gene products that may underlie the process of addiction will provide new targets for the development of new medications and therapies for drug abuse. Annual deadline dates may vary; contact program staff for exact dates. RFA: DA-00-003

Requirements: Applications may be submitted by domestic and foreign, for-profit and nonprofit organizations, both public and private; units of state and local governments; and eligible agencies of the federal government.

Amount of Grant: Up to 2,000,000 U.S.D

Contact: Dr. Rebekah Rasooly, (301) 443-6300; fax (301) 594-6043; rr185i@nih.gov
Internet: http://grants.nih.gov/grants/guide/rfa-files/RFA-DA-00-003.html
Sponsor: National Institute on Drug Abuse
6001 Executive Boulevard, Room 5213
Bethesda, MD 20892-9561

NIDA Minority Institutions Drug Abuse Research Development Grants 3347
In its efforts to broaden and strengthen drug abuse research in minority institutions, NIDA is initiating MIDARP. Grants will be provided to develop the capacity of minority institutions and their faculty, staff, and students to conduct rigorous drug abuse research in all areas of research supported by NIDA including neuroscience, behavioral, clinical, social science, public health, biological, HIV/AIDS, and health services areas. In addition, this PA seeks to broaden the scientific knowledge base in drug abuse in those areas where minority institutions may have particular interest, knowledge, and commitment. Awards under this PA will use the R24 grant mechanism. Applicants are responsible for the planning, direction, and execution of the proposed projects. The total project period for applications submitted in response to this PA may not exceed five years. PAR-98-001 may be obtained electronically through the NIH Web site, email, and by mail.
Requirements: An applicant institution must be a public or private nonprofit university, four-year college, or other institution with a traditionally high (more than 50 percent) minority student enrollment; an institution with a substantial enrollment of minority students, which provides evidence of efforts to recruit into scientific careers members of ethnic or racial groups who have been found to be underrepresented in biomedical and behavioral research; or a Native American tribe that, in conjunction with one or more institutions of higher learning, offers undergraduate and graduate degrees in ADM disciplines.
Contact: Dr. Pushpa Thadani, Division of Basic Research; (301) 443-6300; fax (301) 443-6043; pt24e@nih.gov
Internet: http://www.grants.nih.gov/grants/guide/pa-files/PAR-98-001.html
Sponsor: National Institute on Drug Abuse
6001 Executive Boulevard, Room 5213
Bethesda, MD 20892-9561

NIDA National Drug Abuse Treatment Clinical Trials Network 3348
NIDA invites cooperative agreement applications from established clinical investigators to participate in the National Drug Abuse Treatment Clinical Trials Network (CTN). Awardees will conduct and participate in coordinated multisite clinical trials of behavioral, pharmacological, and combined behavioral/pharmacological therapies for drug abuse and addiction, and conduct research on practices, i.e., studies of factors that impact upon successful adoption of new treatments. The overall goal of the program is to improve the quality of drug abuse treatment throughout the nation, thereby reducing the prevalence, morbidity, and mortality of drug abuse and addiction. Some of the program objectives include supporting studies of behavioral, pharmacological, and combined behavioral and pharmacological treatment interventions of established efficacy in rigorous, multisite clinical trials to determine effectiveness across a broad range of treatment settings and patient populations; encouraging research on the translation of science-based, effective treatment interventions into clinical practice by establishing effectiveness in community-based treatment programs and by determining effective strategies for transporting effective treatments; furthering the development of effective treatments by integrating behavioral, pharmacological, and practice research; investigating the impact of advances in treatment research on community-level treatment practices; and ensuring that treatment research in drug abuse and addiction meets the needs of the wider community, including minorities, women, children, and underserved populations. The complete RFA is available on the Web site. Annual deadline dates may vary; contact a program officer for exact dates. RFA: DA-00-002
Requirements: Applications may be submitted by domestic for-profit and nonprofit organizations, both private and public. Principal investigators should be able to document a substantial history of managing complex research initiatives in clinical sites and have an extensive publication record in clinical research.
Restrictions: Foreign applicants are not eligible, and domestic applications may not include a foreign site.
Contact: Dr. Jack Blaine, Division of Clinical and Services Research; (301) 443-0107; fax (301) 443-8674; jb167n@nih.gov
Internet: http://grants.nih.gov/grants/guide/rfa-files/RFA-DA-00-002.html
Sponsor: National Institute on Drug Abuse
6001 Executive Boulevard, Room 5213
Bethesda, MD 20892-9561

NIDA Neurobiological and Behavioral Research on Nicotine and Tobacco Products 3349
The purpose of this PA is to expand the basic science knowledge base on the neurobiological and behavioral effects of nicotine and associated tobacco chemicals, as part of continuing efforts to explain and prevent their use and to develop effective treatments for nicotine addiction. This PA encourages research on any aspect of the effects of nicotine and other tobacco components, using neurobiological, behavioral, or other methods in humans, animals, or in vitro systems, that seeks to explain nicotine use, addiction, or other effects in humans. This PA solicits research in two general categories, neurobiology and behavioral sciences. Neurobiological, cognitive, behavioral, and social processes, methods, and models are all relevant to this announcement. For both neurobiological and behavioral studies, approaches are encouraged that consider differences in gender, race, ethnicity, and age or developmental stage. PA: PA-00-045
Requirements: Applications may be submitted by domestic and foreign, for-profit and nonprofit organizations, both public and private; units of state and local governments; and eligible agencies of the federal government.
Contact: Dr. Thomas Aigner, Neurobiological Studies Coordinator; (301) 443-6975; fax (301) 594-6043; ta17r@nih.gov
Internet: http://grants.nih.gov/grants/guide/pa-files/PA-00-045.html
Sponsor: National Institute on Drug Abuse
6001 Executive Boulevard, Room 5213
Bethesda, MD 20892-9561

NIDA Origins and Pathways to Drug Abuse Research Grants 3350
The main goal of this PA is to stimulate innovative investigations that enhance our understanding of: drug use patterns and trends within and across populations; interplay of social interactions, social environment, structural context with individual behavioral characteristics and genetic vulnerability; the phenotypic heterogeneity of drug abuse; causal mechanisms leading to onset, maintenance, and remittance of drug abuse, as well as protective mechanisms that reduce the risk of drug abuse; and drug abuse over the life course, including developmental processes that influence drug use trajectories and behavioral, health, and social consequences of drug abuse. In addition, research is encouraged to develop methodologies to improve the accuracy, efficiency, scope, timeliness, and analytic yield of drug abuse epidemiologic data. PA: PA-04-100
Requirements: Applications may be submitted by foreign and domestic, for-profit and nonprofit organizations, both public and private; units of state or local governments; and eligible agencies of the federal government.
Restrictions: Foreign institutions are not eligible for program project or centers awards.
Contact: Gary Fleming, Grants Management; (301) 443-6710; fax (301) 594-6849; gfleming@nida.nih.gov
Internet: http://grants.nih.gov/grants/guide/pa-files/PA-04-100.html
Sponsor: National Institute on Drug Abuse
6001 Executive Boulevard, Room 5213
Bethesda, MD 20892-9561

NIDA Pilot and Feasibility Studies in Preparation for Drug Abuse Prevention Trials 3351
This FOA for R34 applications seeks to support: pilot and/or feasibility testing of new, revised, or adapted preventive intervention approaches targeting the initiation of drug use, the progression to abuse or dependence, and the acquisition or transmission of HIV infection among diverse populations and settings; and pre-trial feasibility testing for prevention services and systems research.
Requirements: The following organizations and institutions may apply: public/state controlled institutions of higher education; private institutions of higher education; Hispanic-serving institutions; historically Black colleges and universities (HBCUs); tribally controlled colleges and universities (TCCUs); Alaska native and native Hawaiian serving institutions; nonprofits with 501(c)3 IRS status; nonprofits without 501(c)3 IRS status; small businesses; for-profit organizations; State governments; regional organizations; U.S. territories or possessions; Indian/Native American tribal governments; Indian/Native American tribally designated organizations; county governments; city or township governments; special district governments; independent school districts; public housing authorities/Indian housing authorities; eligible agencies of the Federal government; and faith-based or community based organizations.
Restrictions: The NIDA R34 mechanism does not support the development of intervention protocols, manuals, or the standardization of protocols.
Date(s) Application is Due: Feb 16; Jun 16; Oct 16
Amount of Grant: 45,000 - 225,000 U.S.D
Contact: Aleta Meyer, Ph.D.; (301) 402-1725; fax (301) 443-2636; meyera2@nida.nih.gov
Internet: http://grants.nih.gov/grants/guide/pa-files/PA-09-146.html
Sponsor: National Institute on Drug Abuse
6001 Executive Boulevard, Room 5213
Bethesda, MD 20892-9561

NIDA Role of Limbic System and Brain Ontogeny in Drug Abuse Research Grants 3352
This initiative is designed to support basic research into the fundamental mechanisms of development of the midbrain and basal forebrain structures that mediate the euphoric properties of drugs as well as understanding how drugs of abuse affect the cellular and molecular mechanism underlying nervous system development. In the past, vertebrate model systems (such as chick, frog, mouse, and zebrafish) and invertebrate systems (such as Drosophila and C. Elegans) have provided insights into mechanisms of development and are also likely to provide new information about the formation and specification of the limbic system. Approaches using these or other model systems both in vitro and in vivo are applicable to this PA. PA-02-015 may be obtained through the NIH grant line (301-402-2221), the NIH Web site, mail, and email.
Requirements: Applications may be submitted by foreign and domestic for-profit and nonprofit organizations, public or private.
Amount of Grant: Up to 90,000 U.S.D
Contact: Gary Fleming; (301) 443-6710; fax (301) 594-6847; gf6s@nih.gov
Internet: http://grants2.nih.gov/grants/guide/pa-files/PA-02-015.html
Sponsor: National Institute on Drug Abuse
6001 Executive Boulevard, Room 5213
Bethesda, MD 20892-9561

NIDA Senior Scientist Award 3353
The Senior Scientist Award provides support to outstanding scientists who have demonstrated sustained, high levels of productivity and whose expertise, research accomplishments, and contributions to the field have been and will continue to be critical to the mission of the particular NIH center or institute. The award provides

salary support for award periods of five years. Contact the program office for more information. PA: PA-95-051
Requirements: The candidate must be a senior scientist and a recognized leader in the field with a distinguished record of original contributions, must have long-term support from a funding institute or center, and must have peer-reviewed grant support at the time of the award. Applications may be submitted on behalf of candidates by domestic, nonfederal organizations, both public or private; or other institutions. Candidates must be U.S. citizens, noncitizen nationals, or permanent residents.
Date(s) Application is Due: Feb 1; Jun 1; Oct 1
Amount of Grant: 1,000 - 100,000,000 U.S.D
Contact: Dr. Timothy Condon, Office of Science Policy, Education, and Legislation; (301) 443-6072; fax (301) 443-6277; tc52x@nih.gov
Internet: http://www.grants.nih.gov/grants/guide/pa-files/PA-95-051.html
Sponsor: National Institute on Drug Abuse
6001 Executive Boulevard, Room 5213
Bethesda, MD 20892-9561

NIDA Transition from Drug Abuse to Drug Addiction 3354
The intent of this program is to stimulate research on the processes and mechanisms underlying the transition from drug use to drug addiction from a wide variety of academic disciplines and approaches. Research from epidemiologic, basic, and clinical disciplines, including research on other transitional states with relevance to addiction, will be supported under this program. Annual deadline dates may vary; contact program staff for exact dates. RFA: DA-01-004
Requirements: Applications may be submitted by domestic, for-profit and nonprofit organizations, both public and private; units of state and local governments; and eligible agencies of the federal government.
Amount of Grant: Up to 400,000 U.S.D
Contact: Dr. Steven Gust, (301) 443-6480; fax (301) 443-9127; sgust@mail.nih.gov
Internet: http://grants.nih.gov/grants/guide/rfa-files/RFA-DA-01-004.html
Sponsor: National Institute on Drug Abuse
6001 Executive Boulevard, Room 5213
Bethesda, MD 20892-9561

NIDCD Clinical Trial Planning Grant 3355
NIDCD provides funding to those who are seeking to improve health by supporting well-executed clinical trials in its mission areas of hearing, balance, smell, taste, voice, speech, and language. Investigators who submit applications for large-scale clinical trials for consideration are expected to provide detailed information regarding the study rationale, design, protocols and procedures, analytical techniques, facilities and environment, administrative procedures, and collaborative arrangements. PA: PAR-02-158
Requirements: Applications may be submitted by domestic for-profit and nonprofit organizations, both public and private; units of state and local governments; and eligible agencies of the federal government.
Restrictions: Foreign organizations are not eligible to apply.
Amount of Grant: Up to 75,000 U.S.D
Contact: A. Julianna Gulya, Chief, Clinical Trials Branch; (301) 435-4085; fax (301) 402-6251; julie_gulya@nih.gov
Internet: http://grants2.nih.gov/grants/guide/pa-files/PAR-02-158.html
Sponsor: National Institute on Deafness and Other Communication Disorders
6120 Executive Boulevard, Executive Plaza S
Bethesda, MD 20892

NIDCD Communicative Competence of Users of Augmentative and Alternative Communication Systems Research Grants 3356
Awards are given for the support of research studies on issues related to the communicative competence of users of augmentative and alternative communication (AAC) systems. Such studies promise to lead to a better understanding of the impact of AAC on the development of or restoration of communicative competence of individuals with severe communication difficulties commonly seen in association with developmental or acquired disorders such as brain injury and other diseases or disorders affecting the oral motor system. It will also define the ways in which augmentative and alternative strategies can influence all phases of communication development and activity. Findings from these studies will direct the establishment and refinement of tools and intervention strategies enabling the development or recovery of communicative function. Additional information may be obtained by mail and email or through the NIH grant line (301-402-2221), and the NIH Web site. PA: PA-96-032
Date(s) Application is Due: Feb 1; Jun 1; Oct 1
Amount of Grant: 72,486 - 458,342 U.S.D
Contact: Sara Stone, Grants Management; (301) 402-0909
Internet: http://www.grants.nih.gov/grants/guide/pa-files/PA-96-032.html
Sponsor: National Institute on Deafness and Other Communication Disorders
6120 Executive Boulevard, Executive Plaza S
Bethesda, MD 20892

NIDCD Exploratory/Developmental Research Grants 3357
NIDCD invites grant applications from basic and clinical investigators who are interested in pursuing high-risk/high-impact (HR/HI) research focused on hearing, balance, smell, taste, voice, speech, or language related to the specific mission of the institute. HR/HI research involves pilot/feasibility studies in which the technological, methodological, or theoretical approach to the problem lacks a traditional historical basis or pilot data, but which could have a major impact on a scientific area or field. Additional information may be obtained electronically through the NIH grant line (301-402-2221), and by mail and email from the program contact listed. Annual deadline dates may vary; contact program staff for exact dates. PA: PAR-96-073
Date(s) Application is Due: Feb 1; Jun 1; Oct 1
Amount of Grant: 72,486 - 458,342 U.S.D
Contact: Sharon Hunt, Grants Manager; (301) 402-0909
Internet: http://www.grants.nih.gov/grants/guide/pa-files/PAR-96-073.html
Sponsor: National Institute on Deafness and Other Communication Disorders
6120 Executive Boulevard, Executive Plaza S
Bethesda, MD 20892

NIDCD Exploratory Research: Feasibility Pilot Studies Grants 3358
The NIDCD invites applications for exploratory, initial feasibility pilot studies focused on hearing, balance, smell, taste, voice, speech, and language, the organization's scientific mission areas. Exploratory research involves initial feasibility pilot studies in which the technological, methodological, or theoretical approach to a problem lacks sufficient preliminary/baseline data and a body of peer-reviewed publications, but whose successful outcome would make a significant contribution to a scientific area or field. It is anticipated that most applications will be submitted by investigators with ongoing research programs who wish to change the focus of their current research effort or move into a new area of research within the communication sciences but need additional funds to complete initial pilot studies. The NIDCD encourages applications from investigators conducting research outside the communication sciences who possess expertise with methodological or technological approaches which have not been typically applied to but which would greatly promote scientific progress within the communication sciences. Annual deadline dates may vary; contact program staff for exact dates. PA: PAR-01-103
Requirements: Applications may be submitted by domestic for-profit and nonprofit organizations, both public and private; units of state or local governments; and eligible agencies of the federal government.
Contact: Dr. Lynn Luethke; (301) 402-3461; lynn_huerta@nih.gov
Internet: http://grants2.nih.gov/grants/guide/pa-files/PAR-01-103.html
Sponsor: National Institute on Deafness and Other Communication Disorders
6120 Executive Boulevard, Executive Plaza S
Bethesda, MD 20892

NIDCD Health Promotion and Disease Prevention: Health Communication, Development, and Dissemination 3359
The purpose of this RFA is to encourage the submission of research grant applications that create, develop, and evaluate health communication information. Projects should be based in current health communication theory and directed to address important problems of disease prevention, health promotion, and improved understanding of the basis for human communication disorders through public information. Annual deadline dates may vary; contact program staff for exact dates. RFA: DC-00-003
Requirements: Applications may be submitted by domestic and foreign, for-profit and nonprofit organizations, both public and private; units of state and local governments; and eligible agencies of the federal government.
Amount of Grant: Up to 600,000 U.S.D
Contact: Dr. Amy Donahue, (301) 402-3458; Amy_Donahue@nih.gov
Internet: http://grants.nih.gov/grants/guide/rfa-files/RFA-DC-00-003.html
Sponsor: National Institute on Deafness and Other Communication Disorders
6120 Executive Boulevard, Executive Plaza S
Bethesda, MD 20892

NIDCD Mentored Clinical Scientist Research Career Development Award 3360
This award enables candidates holding professional degrees (e.g., M.D., D.O., D.V.M., or equivalent degrees) to undertake 3 to 5 years of special study and supervised research with the goal of becoming independent investigators. The award also allows awardees to pursue a research career development program suited to their experience and capabilities under a mentor who is competent to provide guidance in the chosen research area. Institutions may submit applications on behalf of candidates who hold professional degrees. At least 2 years must have elapsed since the health professional degree was granted. Candidates can have varying levels of clinical training and research experience. New applications are due February 12, June 12, and October 12, while resubmitted applications are due March 12, July 12, and November 12. Up to $105,000 per year plus fringe benefits is available, and up to $85,000 per year for research support for new awards.
Requirements: A candidate must have a clinical degree or its equivalent, must have initiated postgraduate clinical training, must identify a mentor with extensive research experience, and must be willing to spend a minimum of 75 percent of full-time professional effort conducting research. Applications may be submitted on behalf of candidates by domestic, nonfederal organizations, public or private, such as medical, dental, or nursing schools or other institutions of higher education. Candidates must be U.S. citizens, noncitizen nationals, or permanent residents.
Date(s) Application is Due: Feb 12; Mar 12; Jun 12; Jul 12; Oct 12; Nov 12
Amount of Grant: Up to 185,000 U.S.D
Contact: Dr. Daniel Sklare, (301) 496-1804; fax (301) 402-6251; sklared@nidcd.nih.gov
Internet: http://grants.nih.gov/grants/guide/pa-files/pa-06-512.html
Sponsor: National Institute on Deafness and Other Communication Disorders
6120 Executive Boulevard, Executive Plaza S
Bethesda, MD 20892

NIDCD Mentored Research Scientist Development Award 3361
The NIDCD utilizes the K01 Program for the retooling and transition of junior- and midcareer-level scientists and clinicians (generally below the rank of full professor, or the equivalent in nonacademic settings) into two research domains: translational

research, i.e., the application of basic research discoveries toward advancing the diagnosis, management and prevention of deafness and other communication disorders; and clinical research. Junior-level clinically-trained individuals should apply for the Mentored Clinical Scientist Development Award or the Mentored Patient-Oriented Research Career Development Award, as appropriate.
Requirements: The candidate must have a research or health-professional doctorate or its equivalent, and must have demonstrated the capacity or potential for highly productive independent research in the period after the doctorate. The candidate must identify a mentor with extensive research experience, and must be willing to spend a minimum of 75 percent of full-time professional effort conducting research and research career development activities for the period of the award. Applications may be submitted on behalf of candidates by domestic, public or private, nonfederal organizations. Candidates must be U.S. citizens or noncitizen nationals.
Date(s) Application is Due: Feb 12; Mar 12; Jun 12; Jul 12; Oct 12; Nov 12
Contact: Daniel A. Sklare, (301) 496-1804; sklared@nidcd.nih.gov
Internet: http://www.grants.nih.gov/grants/guide/pa-files/PAR-99-027.html
Sponsor: National Institute on Deafness and Other Communication Disorders
6120 Executive Boulevard, Executive Plaza S
Bethesda, MD 20892

NIDCD Minority Dissertation Research Grants in Human Communication 3362
NIDCD invites applications for small grants to support doctoral dissertation research in human communication for doctoral candidates who are members of racial and ethnic minority groups that are under-represented in biomedical and behavioral research. The goals of this program are to aid the research of new minority investigators and to encourage minority individuals from a variety of academic disciplines and programs to conduct research in hearing, balance, smell, taste, voice, speech, and language. These grants are not eligible for competitive renewal. Annual deadline dates may vary; contact the program office for exact dates. More information may be obtained by mail and email, or through the NIH grant line (301-402-2221) and the NIH Web site. RFA: DC-00-001
Requirements: Candidates for this award must be graduate students in good standing and must be identified by the grantee institution as a member of an ethnic or racial group that is underrepresented in biomedical or behavioral research.
Amount of Grant: 30,000 - 72,000 U.S.D
Contact: Daniel Sklare, Division of Extramural Research; (301) 496-1804; fax (301) 402-6251; daniel_sklare@nih.gov
Internet: http://grants2.nih.gov/grants/guide/rfa-files/RFA-DC-00-001.html
Sponsor: National Institute on Deafness and Other Communication Disorders
6120 Executive Boulevard, Executive Plaza S
Bethesda, MD 20892

NIDCD Molecular and Cellular Pathology of the Human Auditory and Vestibular Systems Grants 3363
These grants support cellular biologic, molecular biologic, and molecular genetic studies of newly acquired and archival human temporal bones and associated brain tissues of individuals who had well-documented evidence of disorders of hearing and/or balance before death. It is believed that studies of this kind can lead to a better understanding of the pathogenesis and pathophysiology of disorders of the human auditory and vestibular systems. Further information may be obtained by mail and email or through the NIH grant line (301-402-2221), and the NIH Web site. Annual deadline dates may vary; contact program staff for exact dates. RFA: DC-96-005
Requirements: Applications may be submitted by domestic and foreign for-profit and non-profit organizations, public and private; and eligible agencies of the federal government.
Amount of Grant: 72,486 - 458,342 U.S.D
Contact: Sharon Hunt, Grants Manager; (301) 402-0909
Internet: http://www.grants.nih.gov/grants/guide/rfa-files/RFA-DC-96-005.html
Sponsor: National Institute on Deafness and Other Communication Disorders
6120 Executive Boulevard, Executive Plaza S
Bethesda, MD 20892

NIDCD Neurobiology and Genetics of Autism Grants 3364
NICHD and NIDCD fund grants for research that will elucidate the biological basis of autism and contribute to the discovery of a biological and/or diagnostic marker for autism. The purpose of these grants is to provide support for collaborative, multidisciplinary, and methodologically rigorous programs of research that will use advanced techniques in the biological and behavioral sciences to study the diagnosis, biological etiology, pathophysiology, and developmental course of autism. Annual deadline dates may vary; contact program staff for exact dates. More information may be obtained by mail and email, or through the NIH grant line (301-402-2221), and the NIH Web site. RFA: HD-96-004
Date(s) Application is Due: Feb 1; Jun 1; Oct 1
Amount of Grant: 36,000 - 23,000,000 U.S.D
Contact: Marie Bristol, Center for Research for Mothers and Children Contact; (301) 496-1383; fax (301) 496-3791; BRISTOLM@HD01.NICHD.NIH.GOV
Internet: http://www.grants.nih.gov/grants/guide/rfa-files/RFA-HD-96-004.html
Sponsor: National Institute on Deafness and Other Communication Disorders
6120 Executive Boulevard, Executive Plaza S
Bethesda, MD 20892

NIDCD Physiologic and Molecular Bases of Tinnitus Research Grants 3365
Three to five exploratory/developmental grants will be awarded to support research addressing the physiologic and molecular bases of tinnitus. The goal is to stimulate research on pathologic processes in the auditory system that produce tinnitus. More information may be obtained by mail and email, or through the NIH grant line (301-402-2221), and the NIH Web site. Annual deadline dates vary; contact program staff for exact dates. RFA: DC-96-003
Amount of Grant: 72,486 - 458,342 U.S.D
Contact: Sharon Hunt, Grants Manager; (301) 402-0909
Internet: http://www.grants.nih.gov/grants/guide/rfa-files/RFA-DC-96-003.html
Sponsor: National Institute on Deafness and Other Communication Disorders
6120 Executive Boulevard, Executive Plaza S
Bethesda, MD 20892

NIDCD Program for Postdoctoral Research Training in Clinical Trials 3366
Grants support programs of postdoctoral research training in the design and conduct of clinical trials for individuals trained in the clinical disciplines of human communications. The purpose of this initiative is to foster the development of trained professionals into independent investigators capable of conducting clinical trials of the efficacy of treatment for diseases and disorders affecting hearing, balance, smell, taste, voice, speech, or language. More information may be obtained by mail and email, or through the NIH grant line (301-402-2221), and the NIH Web site. Annual deadline dates may vary; contact program staff for exact dates. PA: PA-96-024
Date(s) Application is Due: Feb 1; Jun 1; Oct 1
Amount of Grant: 72,486 - 458,342 U.S.D
Contact: Sharon Hunt, Grants Management; (301) 402-0909
Internet: http://www.grants.nih.gov/grants/guide/pa-files/PA-96-024.html
Sponsor: National Institute on Deafness and Other Communication Disorders
6120 Executive Boulevard, Executive Plaza S
Bethesda, MD 20892

NIDCD Research Core Centers 3367
NIDCD invites grant applications for research core centers. The overall goal of a core center is to promote a cooperative interaction among basic science and/or clinical investigators in a manner that will enrich the effectiveness of ongoing research and promote new research directions. Grants support centralized resources and facilities shared by investigators with existing research projects. A letter of intent is requested by April 1, August 1, and December 1; full application is due by February 1, June 1, and October 1. Annual deadline dates may vary; contact program staff for exact dates. PA: PA-01-011
Requirements: Applications may be submitted by domestic for-profit and nonprofit organizations, both public and private; units of state or local governments; and eligible agencies of the federal government.
Restrictions: Foreign organizations or organizations in foreign countries are not eligible; subcontracts to foreign organizations are not allowed.
Date(s) Application is Due: Feb 1; Apr 1; Jun 1; Aug 1; Oct 1; Dec 1
Contact: Dr. Amy Donahue, (301) 402-3458; fax (301) 402-6251; amy_donahue@nih.gov
Internet: http://grants.nih.gov/grants/guide/pa-files/PA-01-011.html
Sponsor: National Institute on Deafness and Other Communication Disorders
6120 Executive Boulevard, Executive Plaza S
Bethesda, MD 20892

NIDCD Research Grants on Chemical Senses: Neurotransmitters and Neuromodulators 3368
NIDCD and NIA sponsor research grants that support research fundamental to understanding the neurochemistry of pathways in the olfactory and gustatory systems throughout the lifespan. This research includes the identification and characterization of the neurotransmitters, neuromodulators, receptors, and secondary messengers throughout the chemosensory systems. More information may be obtained by mail and email, or through the NIH grant line (301-402-2221), and the NIH Web site. Annual deadline dates may vary; contact program staff for exact dates. PA: PA-96-035
Date(s) Application is Due: Feb 1; Jun 1; Oct 1
Amount of Grant: 72,486 - 458,342 U.S.D
Contact: Program Contact, Division of Human Communication; (301) 496-5061
Internet: http://www.grants.nih.gov/grants/guide/pa-files/PA-96-035.html
Sponsor: National Institute on Deafness and Other Communication Disorders
6120 Executive Boulevard, Executive Plaza S
Bethesda, MD 20892

NIDCD Small Grants 3369
NIDCD provides support for pilot research that is likely to lead to subsequent application for individual research project grants or first independent research support and transition grants. The research must be focused on one or more of the areas within the biomedical and behavioral scientific mission of NIDCD: hearing, balance, smell, taste, voice, speech, or language. Applicants may request support for up to two years, nonrenewable. For additional information, investigators are encouraged to contact NIDCD staff responsible for grants in the investigator's particular area of interest. PA: PAR-99-018
Requirements: This program is designed solely to support basic and clinical scientists with limited research experience who are at the beginning stages of their research careers. Former recipients of federally funded Training or Career Development awards are eligible for this program.
Restrictions: Current and previous principal investigators of NIH research grants or other federal funding agencies are ineligible.
Amount of Grant: Up to 35,000 U.S.D
Contact: Dr. Daniel Sklare; (301) 496-1804; Daniel_Sklare@nih.gov
Internet: http://www.grants.nih.gov/grants/guide/pa-files/PAR-99-018.html
Sponsor: National Institute on Deafness and Other Communication Disorders
6120 Executive Boulevard, Executive Plaza S
Bethesda, MD 20892

NIDCD Symptom Management for Chronic Neurological Conditions Research Grants 3370
NINR, NIA, NICHD, NIDCD, and NINDS are interested in facilitating investigator-initiated research designed to deal with symptoms associated with chronic neurological conditions including stroke, epilepsy, brain and spinal cord injury, multiple sclerosis, and Parkinson's disease. The overall goals are to manage symptoms effectively, to prevent secondary effects of the condition when possible, to maximize the health potential, and to improve the health-related quality of life for patients, caregivers, and families. The mechanism of support is the research project grant (R01). PA-98-030 may be obtained through the NIH grant line (301-402-2221), the NIH Web site, mail, and email.
Requirements: Applications may be submitted by domestic and foreign for-profit and nonprofit organizations, public or private.
Contact: Dr. Beth Ansel; (301) 402-3461; fax (301) 402-6251; Beth_Ansel@nih.gov
Internet: http://www.grants.nih.gov/grants/guide/pa-files/PA-98-030.html
Sponsor: National Institute on Deafness and Other Communication Disorders
6120 Executive Boulevard, Executive Plaza S
Bethesda, MD 20892

NIDCR Clinical Trial Pilot Grant 3371
The goal of this program is to identify effective preventive, diagnostic, and treatment approaches to craniofacial, oral and dental diseases, and disorders that will improve health by supporting well-executed clinical trials and to support pilot studies to obtain preliminary data necessary to address preliminary design and methodological issues for clinical trials. PA: PAR-99-158
Requirements: Applications may be submitted by foreign and domestic, for-profit and nonprofit organizations, both public and private; units of state and local governments; and eligible agencies of the federal government.
Amount of Grant: Up to 250,000 U.S.D
Contact: Dr. Ann Sandberg, (301) 594-2419; Ann.Sandberg@nih.gov
Internet: http://grants.nih.gov/grants/guide/pa-files/PAR-99-158.html
Sponsor: National Institute of Dental and Craniofacial Research
45 Center Drive, Natcher Building
Bethesda, MD 20892

NIDCR Clinical Trial Planning Grant 3372
The goal of this program is to identify effective preventive, diagnostic, and treatment approaches to craniofacial, oral and dental diseases, and disorders that will improve health by supporting well-executed clinical trials. PA: PAR-03-042
Requirements: Applications may be submitted by domestic and foreign, for-profit and nonprofit organizations, both public and private; units of state and local governments; and eligible agencies of the federal government.
Amount of Grant: Up to 250,000 U.S.D
Contact: Dr. Bruce Pihlstrom, Acting Director, Division of Clinical Research and Health Promotion; (301) 594-4830; fax (301) 480-8322; bruce.Pihlstrom@nih.gov
Internet: http://grants.nih.gov/grants/guide/pa-files/PAR-03-042.html
Sponsor: National Institute of Dental and Craniofacial Research
45 Center Drive, Natcher Building
Bethesda, MD 20892

NIDCR HIV/AIDS Associated Oral Viral Infections 3373
NIDCR invites new and experienced scientists to submit research grant applications for basic, behavioral, epidemiological, or clinical investigations to study the pathogenesis, infectivity, transmission, reactivation, and pathotropic spread of viral infections in the oral cavity of the immunocompromised host. This RFA is intended to solicit a broad range of research proposals on oral opportunistic infections with viruses, including human herpes virus, Epstein Barr virus, cytomegalovirus varicella-zoster virus, and human papillomavirus in HIV-infected/AIDS individuals. The institute encourages collaborative projects that bring together investigators from all relevant scientific disciplines including microbiology, immunology, biochemistry, and pathology interested in studying viral infections of the oral cavity. Applications that include collaboration with foreign scientists conducting unique research on this topic are also encouraged. The goal of this RFA is to increase our understanding of how virulence, infection, and host response is modified in the immune compromised patient to allow infection, activation, or reactivation of opportunistic pathogens. This understanding can lead to improved diagnosis, prevention, and treatment strategies, for the various opportunistic pathogens. Annual deadline dates may vary; contact program staff for exact dates. RFA: DE-00-002
Requirements: Applications may be submitted by domestic and foreign, for-profit and nonprofit organizations, both public and private; units of state and local governments; and eligible agencies of the federal government. Although an application must be submitted from a single institution, collaborative arrangements with other institutions are encouraged.
Amount of Grant: Up to 225,000 U.S.D
Contact: Dr. Denise Anne Russo, Infectious Diseases and Immunity Divion Coordinator; (301) 594-2617; fax (301) 480-8318; Denise.Russo@nih.gov
Internet: http://grants.nih.gov/grants/guide/rfa-files/RFA-DE-00-002.html
Sponsor: National Institute of Dental and Craniofacial Research
45 Center Drive, Natcher Building
Bethesda, MD 20892

NIDCR International Collaborative Oral Health Research Planning Grant 3374
The objective of this scientific program is to promote research that will advance understanding of the underlying causes of craniofacial dysmorphologies such as ectodermal dysplasia, enamel dysplasia, cleft lip and palate, and osteogenesis imperfecta. International collaboration in relevant areas such as genetic research, epidemiology, genetic registries, and treatment modalities can help speed progress toward the goal of improved prevention, diagnosis, and treatment by combining scientific talents from different countries to focus on selected populations at high risk. An international approach that includes all disciplines relevant to specific craniofacial anomalies might dramatically enhance the possibility of resolving the many complex issues surrounding these malformations. PA: PAR-00-012
Requirements: Applications may be submitted by foreign and domestic, for-profit and nonprofit organizations, both public and private; units of state and local governments; and eligible agencies of the federal government.
Amount of Grant: Up to 75,000 U.S.D
Contact: Dr. Judy Small, Craniofacial Anomalies and Injuries Branch; (301) 594-2425
Internet: http://grants.nih.gov/grants/guide/pa-files/PAR-00-012.html
Sponsor: National Institute of Dental and Craniofacial Research
45 Center Drive, Natcher Building
Bethesda, MD 20892

NIDCR Oral Microbiology/Immunology of Type 1 Diabetes 3375
The purpose of this program is to encourage submission of applications for two-year exploratory grants that will broaden research on the oral microbiology and immunology of Type 1 diabetes. In particular, research is needed on the oral microflora, cytokines, and antibodies in the subgingival crevice. Annual deadline dates may vary; contact program staff for exact dates. RFA: DE-01-001
Requirements: Applications may be submitted by domestic and foreign, for-profit and nonprofit organizations, both public and private; units of state and local governments; and eligible agencies of the federal government.
Amount of Grant: Up to 100,000 U.S.D
Contact: Dr. Dennis Mangan, Infectious Diseases and Immunity Branch Contact; (301) 594-2421; fax (301) 480-8318; Dennis.Mangan@nih.gov
Internet: http://grants.nih.gov/grants/guide/rfa-files/RFA-DE-01-001.html
Sponsor: National Institute of Dental and Craniofacial Research
45 Center Drive, Natcher Building
Bethesda, MD 20892

NIDCR Priorities in Behavioral Research in Cancer Prevention and Control Grants 3376
NCI and NIDCR invite researchers to submit research project grant (R01) applications that address behavioral research issues in cancer prevention and control. The PA addresses recommendations made by a special Behavioral Research in Cancer Prevention and Control Working Group in 1996 that consisted of leading national experts whose role was to identify behavioral research needs in cancer prevention and control during the coming years. PA-97-055 may be obtained electronically through the NIH grant line (301-402-2221), the Web site, and by mail and email.
Contact: Dr. Patricia Bryant, Behavior, Health Promotion, and Environment Program Contact; (301) 594-2095; fax (301) 480-8318; bryantp@de45.nidr.nih.gov
Internet: http://grants.nih.gov/grants/guide/pa-files/PA-97-055.html
Sponsor: National Institute of Dental and Craniofacial Research
45 Center Drive, Natcher Building
Bethesda, MD 20892

NIDDK Advances in Polycystic Kidney Disease Grants 3377
The National Institute of Diabetes and Digestive and Kidney Diseases (NIDDK) through its Division of Kidney, Urologic and Hematologic Diseases (DKUHD) invites experienced and new investigators to submit research grant applications to pursue basic and applied investigations in order to better understand the etiology and pathogenesis of Polycystic Kidney Disease (PKD), in both its autosomal dominant and autosomal recessive forms. Such applications may examine the genetic determinants, and cellular and molecular mechanisms, which disrupt normal kidney function; mechanisms of cyst formation and growth; development of experimental model systems; development of markers of disease progression; and the identification of innovative therapeutic interventions and gene targeted strategies to prevent progressive renal insufficiency due to this disorder. The intent of this funding opportunity is to intensify investigator-initiated research, to attract new investigators to the field, and to increase interdisciplinary research. The ultimate aim is to facilitate PKD-related research studies, which will provide the basis for new therapeutic approaches.
Requirements: The following organizations may apply: for-profit organizations; non-profit organizations; public or private institutions, such as universities, colleges, hospitals, and laboratories; units of State government; units of local government; eligible agencies of the Federal government; foreign institutions; domestic institutions; and faith-based or community-based organizations.
Date(s) Application is Due: Feb 5; Jun 5; Oct 5
Contact: Catherine M. Meyers, (301) 594-7717; fax (301) 480-3510; cm420i@nih.gov
Internet: http://grants.nih.gov/grants/guide/pa-files/PA-06-158.html
Sponsor: National Institute of Diabetes and Digestive and Kidney Diseases
6707 Democracy Boulevard
Bethesda, MD 20892-5460

NIDDK Adverse Metabolic Side Effects of Second Generation Psychotropic Medications Leading to Obesity and Increased Diabetes Risk Grants 3378
This Funding Opportunity invites investigator-initiated research grant applications for studies examining the adverse metabolic effects (i.e , obesity and diabetes) of psychotropic medications in animal models and across the human lifespan (including pediatric, adult and geriatric populations). Applications responsive to this FOA should focus on: increasing the understanding of the nature, rates, and pathophysiology of adverse metabolic effects

of psychotropic medications; elucidating biomedical and psychosocial risk factors for the development of metabolic adverse effects of psychiatric therapeutics; and develop interventions to prevent and/or mitigate metabolic adverse effects across the lifespan.
Requirements: The following organizations and institutions may apply: public/state controlled institutions of higher education; private institutions of higher education; Hispanic-serving institutions; historically Black colleges and universities (HBCUs); tribally controlled colleges and universities (TCCUs); Alaska native and native Hawaiian serving institutions; nonprofits with 501(c)3 IRS status; nonprofits without 501(c)3 IRS status; small businesses; for-profit organizations; State governments; regional organizations; U.S. territories or possessions; Indian/Native American tribal governments; Indian/Native American tribally designated organizations; eligible agencies of the Federal government; and faith-based or community based organizations.
Date(s) Application is Due: Feb 22; Jun 22; Oct 22
Amount of Grant: Up to 500,000 U.S.D
Contact: Christine Hunter, (301) 594-4728; hunterchristine@niddk.nih.gov
Internet: http://grants.nih.gov/grants/guide/pa-files/PAR-08-160.html
Sponsor: National Institute of Diabetes and Digestive and Kidney Diseases
6707 Democracy Boulevard
Bethesda, MD 20892-5460

NIDDK Ancillary Studies of Kidney Disease Accessing Information from Clinical Trials, Epidemiological Studies, and Databases Grants 3379

The National Institute of Diabetes and Digestive and Kidney Diseases (NIDDK) invites investigator-initiated research project applications for ancillary studies to ongoing or completed clinical trials and epidemiological studies of kidney disease as well as clinical trials and epidemiological studies for other diseases or populations that lend themselves to the study of kidney disease. These studies may range from new analyses of existing datasets of completed studies to additional collection of data and biological specimens in ongoing investigations. The goal of these studies should be to extend our understanding of the risk factors for developing kidney disease and their associated co-morbid illnesses such as malnutrition and cardiovascular disease, factors associated with rapid decline in kidney function among persons with chronic kidney disease, and the impact of these diseases on quality of life and mental and physical functioning. Studies of biomarkers for patients with acute renal failure and other kidney diseases, for example, are also appropriate topics for further investigation. Studies ancillary to both government and non-government supported clinical trials and epidemiological studies are encouraged. Analysis of large public access databases and other databases is also encouraged.
Requirements: The following organizations and institutions may apply: public/state controlled institutions of higher education; private institutions of higher education; Hispanic-serving institutions; historically Black colleges and universities (HBCUs); tribally controlled colleges and universities (TCCUs); Alaska native and native Hawaiian serving institutions; nonprofits with 501(c)3 IRS status; nonprofits without 501(c)3 IRS status; small businesses; for-profit organizations; State governments; regional organizations; U.S. territories or possessions; Indian/Native American tribal governments; Indian/Native American tribally designated organizations; eligible agencies of the Federal government; and faith-based or community based organizations.
Date(s) Application is Due: Feb 5; Jun 5; Oct 5
Contact: John W. Kusek, (301) 594-7735; fax (301) 480-3510; jk61x@nih.gov
Internet: http://grants.nih.gov/grants/guide/pa-files/PA-07-050.html
Sponsor: National Institute of Diabetes and Digestive and Kidney Diseases
6707 Democracy Boulevard
Bethesda, MD 20892-5460

NIDDK Ancillary Studies to Major Ongoing NIDDK and NHLBI Clinical Research Studies 3380

This Funding Opportunity Announcement (FOA) solicits grant applications from qualified investigators to conduct ancillary studies of selected ongoing major clinical research studies, including clinical trials, epidemiological studies and disease databases, supported by the National Institute of Diabetes and Digestive and Kidney Diseases and the National Heart, Lung, and Blood Institute, National Institutes of Health (NIH). The objective is to introduce NIDDK-oriented ancillary studies into either NIDDK or NHLBI-funded parent studies, or NHLBI-oriented ancillary studies into NIDDK-funded parent studies.
Requirements: The following organizations and institutions may apply: public/state controlled institutions of higher education; private institutions of higher education; Hispanic-serving institutions; historically Black colleges and universities (HBCUs); tribally controlled colleges and universities (TCCUs); Alaska native and native Hawaiian serving institutions; nonprofits with 501(c)3 IRS status; nonprofits without 501(c)3 IRS status; small businesses; for-profit organizations; State governments; regional organizations; U.S. territories or possessions; Indian/Native American tribal governments; Indian/Native American tribally designated organizations; eligible agencies of the Federal government; and faith-based or community based organizations.
Date(s) Application is Due: Feb 5; Jun 5; Oct 5
Contact: John W. Kusek, (301) 594-7735; fax (301) 480-3510; jk61x@nih.gov
Internet: http://grants.nih.gov/grants/guide/pa-files/PAR-07-024.html
Sponsor: National Institute of Diabetes and Digestive and Kidney Diseases
6707 Democracy Boulevard
Bethesda, MD 20892-5460

NIDDK Basic and Clinical Studies of Congenital Urinary Tract Obstruction Grants 3381

The purpose of this funding opportunity is to address the numerous scientific and clinical uncertainties related to the development, treatment and prognosis of congenital obstructive uropathy, by encouraging and facilitating research in diverse areas. These areas include: the development of objective prognostic markers; the genetic determinants of this congenital disorder; the development of reliable animal models of the disorder; and, evaluation of the long-term effectiveness of various treatment strategies. Because the nature and scope of the proposed research will vary from application to application, it is anticipated that the size and duration of each award will also vary. The total amount awarded and the number of awards will depend upon the mechanism numbers, quality, duration, and costs of the applications received.
Requirements: The following organizations and institutions may apply: public/state controlled institutions of higher education; private institutions of higher education; Hispanic-serving institutions; historically Black colleges and universities (HBCUs); tribally controlled colleges and universities (TCCUs); Alaska native and native Hawaiian serving institutions; nonprofits with 501(c)3 IRS status; nonprofits without 501(c)3 IRS status; small businesses; for-profit organizations; State governments; regional organizations; U.S. territories or possessions; Indian/Native American tribal governments; Indian/Native American tribally designated organizations; eligible agencies of the Federal government; and faith-based or community based organizations.
Date(s) Application is Due: Feb 5; Jun 5; Oct 5
Contact: Marva Moxey-Mims, (301) 594-7717; fax (301) 480-3510; mm726k@nih.gov
Internet: http://grants.nih.gov/grants/guide/pa-files/PA-07-059.html
Sponsor: National Institute of Diabetes and Digestive and Kidney Diseases
6707 Democracy Boulevard
Bethesda, MD 20892-5460

NIDDK Basic Research Grants in the Bladder and Lower Urinary Tract 3382

This funding opportunity encourages basic cellular, molecular, developmental and genetic research relevant to the bladder and lower urinary tract. Basic research studies that address age and gender differences in bladder and lower urinary tract function are also encouraged. New and established investigators from related fields of study are encouraged to apply their expertise to these problem areas. Investigators with diverse basic science and clinical backgrounds are encouraged to develop collaborative research relationships. Discoveries resulting from these studies may serve as the basis for the development of new agents, techniques, and strategies for detecting, preventing, and treating diseases of the bladder and lower urinary tract, as well as dealing with bladder complications resulting from other diseases, such as diabetes.
Requirements: Application can be made by the following institutions and organizations: for-profit organizations; non-profit organizations; public or private institutions (such as universities, colleges, hospitals, and laboratories); units of State government; units of local government; eligible agencies of the Federal government; foreign institutions; and domestic institutions.
Date(s) Application is Due: Feb 5; Jun 5; Oct 5
Contact: Chris Mullins, (301) 594-7717; fax (301) 480-3510; cm419z@nih.gov
Internet: http://grants.nih.gov/grants/guide/pa-files/PA-06-254.html
Sponsor: National Institute of Diabetes and Digestive and Kidney Diseases
6707 Democracy Boulevard
Bethesda, MD 20892-5460

NIDDK Calcium Oxalate Stone Diseases Grants 3383

The purpose of this Funding Opportunity Announcement (FOA) is to increase investigator interest in research into the genetics and heritability of oxalate regulation and the oxalate stone diseases, and to develop new treatments for the disorder. This initiative solicits research as well as pilot and feasibility studies that utilize new and innovative approaches to study the diagnosis, treatment and prevention of these disorders. Because the nature and scope of the proposed research will vary from application to application, it is anticipated that the size and duration of each award will also vary. The total amount awarded and the number of awards will depend upon the mechanism numbers, quality, duration, and costs of the applications received.
Requirements: The following organizations and institutions may apply: public/state controlled institutions of higher education; private institutions of higher education; Hispanic-serving institutions; historically Black colleges and universities (HBCUs); tribally controlled colleges and universities (TCCUs); Alaska native and native Hawaiian serving institutions; nonprofits with 501(c)3 IRS status; nonprofits without 501(c)3 IRS status; small businesses; for-profit organizations; State governments; regional organizations; U.S. territories or possessions; Indian/Native American tribal governments; Indian/Native American tribally designated organizations; eligible agencies of the Federal government; and faith-based or community based organizations.
Date(s) Application is Due: Feb 5; Jun 5; Oct 5
Contact: Rebekah S. Rasooly, (301) 594-6007; fax (301) 480-3510; rr185i@nih.gov
Internet: http://grants.nih.gov/grants/guide/pa-files/PA-07-051.html
Sponsor: National Institute of Diabetes and Digestive and Kidney Diseases
6707 Democracy Boulevard
Bethesda, MD 20892-5460

NIDDK Cell-Specific Delineation of Prostate and Genitourinary Development 3384

This RFA is intended to encourage the development of new research tools and methods to study the development and biology of the prostate and genitourinary tract. Strategies to be supported include systematic assessment of gene expression in specific cell types in the developing prostate and genitourinary tract; methods to tag individual cell types for purification, analysis, and characterization; tools to manipulate gene expression in vivo in individual cell types; and development of clinically relevant cell lines. The major goal of this initiative is to ensure that research tools are available so that a broad range of innovative methods can be applied to studying prostate development. Annual deadline dates may vary; contact program staff for exact dates.
Requirements: Application can be made by the following institutions and organizations: public/state controlled institutions of higher education; private institutions of higher

education; nonprofits with 501(c)3 IRS status (other than institutions of higher education); nonprofits without 501(c)3 IRS status (other than institutions of higher education); small businesses; for-profit organizations (other than small businesses); state governments; U.S. territories or possessions; Indian/Native American tribal governments (federally recognized); Indian/Native American tribal governments (other than Federally recognized); Indian/Native American tribally designated organizations; non-domestic (non-U.S.) entities (foreign organizations); Hispanic-serving institutions; historically Black colleges and universities (HBCUs); tribally controlled colleges and universities (TCCUs); Alaska Native and Native Hawaiian serving institutions; regional organization; eligible agencies of the Federal government; and faith-based or community based organizations.
Date(s) Application is Due: Feb 5; Mar 5; Jun 5; Jul 5; Oct 5; Nov 5
Amount of Grant: Up to 350,000 U.S.D
Contact: Dr. Robert Star, (301) 594-7715; StarR@extra.niddk.nih.gov
Internet: http://grants.nih.gov/grants/guide/rfa-files/RFA-DK-00-015.html
Sponsor: National Institute of Diabetes and Digestive and Kidney Diseases
31 Center Drive, Room 9A-35, MSC 2560
Bethesda, MD 20892-2560

NIDDK Co-Activators and Co-Repressors in Gene Expression Grants 3385
This initiative stems from an NIDDK workshop in molecular endocrinology and is designed to stimulate research that addresses the fundamental underlying mechanisms by which nuclear accessory proteins mediate signaling through hormone receptors at the level of the regulation of gene expression. Staff contact prior to submission is highly recommended.
Requirements: Applications may be submitted by domestic and foreign, for-profit and nonprofit organizations, both public and private; units of state and local governments; and eligible agencies of the federal government.
Date(s) Application is Due: Feb 16; Jun 16; Oct 16
Amount of Grant: 100,000 - 250,000 U.S.D
Contact: Dr. Ronald Margolis, (301) 594-8819; fax (301) 435-6047; rm76f@nih.gov
Internet: http://grants.nih.gov/grants/guide/pa-files/PA-99-111.html
Sponsor: National Institutes of Health
6707 Democracy Boulevard, Building 2DEM, Room 693
Bethesda, MD 20892

NIDDK Collaborative Interdisciplinary Team Science in Diabetes, Endocrinology and Metabolic Diseases Grants 3386
The purpose of the Collaborative Interdisciplinary Team Science Program described in this announcement is to provide support to enable strong investigative teams to do inter- and/or trans-disciplinary research on a complex problem in biomedical science relevant to Diabetes, Endocrinology and Metabolic Diseases. It is anticipated that 1-3 projects will be funded per year, but because the nature and scope of the proposed research will vary from application to application, it is anticipated that the size and duration of each award will also vary. The total amount awarded and the number of awards will depend upon the quality, duration, and costs of the applications received, as well as on the availability of funds.
Requirements: The following organizations and institutions are eligible to apply: public/state controlled institutions of higher education; private institutions of higher education; Hispanic-serving institutions; historically Black colleges and universities (HBCUs); tribally controlled colleges and universities (TCCUs); Alaska Native and Native Hawaiian serving institutions; and non-domestic (non-U.S.) entities (foreign organizations).
Date(s) Application is Due: Feb 24; Mar 24
Contact: Karen Salomon, (301) 594-7733; fax (301) 435-6047; ks495c@nih.gov
Internet: http://grants.nih.gov/grants/guide/pa-files/PAR-08-182.html
Sponsor: National Institute of Diabetes and Digestive and Kidney Diseases
6707 Democracy Boulevard
Bethesda, MD 20892-5460

NIDDK Developmental Biology and Regeneration of the Liver Grants 3387
The purpose of this FOA is to invite qualified scientific investigators to submit applications on liver development and regeneration to fully define the molecular and cellular mechanisms underlying these processes in health and disease and to apply these findings to developing improved therapies for liver disease. Because the nature and scope of the proposed research will vary from application to application, it is anticipated that the size and duration of each award will also vary. The total amount awarded and the number of awards will depend upon the mechanism numbers, quality, duration, and costs of the applications received.
Requirements: The following organizations and institutions may apply: public/state controlled institutions of higher education; private institutions of higher education; Hispanic-serving institutions; historically Black colleges and universities (HBCUs); tribally controlled colleges and universities (TCCUs); Alaska native and native Hawaiian serving institutions; nonprofits with 501(c)3 IRS status; nonprofits without 501(c)3 IRS status; small businesses; for-profit organizations; State governments; regional organizations; U.S. territories or possessions; Indian/Native American tribal governments; Indian/Native American tribally designated organizations; eligible agencies of the Federal government; and faith-based or community based organizations.
Date(s) Application is Due: Feb 5; Jun 5; Oct 5
Contact: Samir Zakhari, (301) 443-0799; fax (301) 594-0673; szakhari@mail.nih.gov
Internet: http://grants.nih.gov/grants/guide/pa-files/PA-07-026.html
Sponsor: National Institute of Diabetes and Digestive and Kidney Diseases
6707 Democracy Boulevard
Bethesda, MD 20892-5460

NIDDK Development of Assays for High-Throughput Drug Screening Grants 3388
The overall goal of the Molecular Libraries Initiative, part of the NIH Roadmap for Medical Research, is to offer public sector researchers opportunities to screen small molecules with the high-throughput chemical screening (HTS) methods commonly used by the private sector to develop therapeutic agents. The purpose of this FOA is to support the development of innovative assays that may ultimately be adapted for automated screening. The assay should aim to identify new tools for basic research or promising new avenues for therapeutics development, especially in areas related to the missions of the participating institutes. Awards issued under this FOA are contingent upon the availability of funds and the submission of a sufficient number of meritorious applications. The total amount awarded and the number of awards will depend upon the quality and costs of the applications received.
Requirements: The following organizations and institutions may apply: public/state controlled institutions of higher education; private institutions of higher education; Hispanic-serving institutions; historically Black colleges and universities (HBCUs); tribally controlled colleges and universities (TCCUs); Alaska native and native Hawaiian serving institutions; nonprofits with 501(c)3 IRS status; nonprofits without 501(c)3 IRS status; small businesses; for-profit organizations; State governments; regional organizations; U.S. territories or possessions; Indian/Native American tribal governments; Indian/Native American tribally designated organizations; eligible agencies of the Federal government; and faith-based or community based organizations.
Date(s) Application is Due: Feb 5; Jun 5; Oct 5
Contact: Laura K. Moen, (301) 594-4748; fax (301) 480-3510; lm232f@nih.gov
Internet: http://grants.nih.gov/grants/guide/pa-files/PA-07-320.html
Sponsor: National Institute of Diabetes and Digestive and Kidney Diseases
6707 Democracy Boulevard
Bethesda, MD 20892-5460

NIDDK Development of Disease Biomarkers Grants 3389
The goal of this Funding Opportunity Announcement (FOA) is to validate biomarkers for well-defined human diseases of liver, kidney, urological tract, digestive and hematologic systems, and endocrine and metabolic disorders, diabetes and its complications, and obesity, for which there are no or very few biomarkers, or for which standard biomarkers are currently prohibitively invasive or expensive. New biomarkers will stimulate bench to bedside translation by providing measures of the biological effects of potential new treatments. The ideal biomarker can be measured in a minimally invasive way, can be measured repeatedly over time, identifies early stages of disease, is indicative of disease prognosis, and correlates well with progression and response to therapy. Especially of interest would be studies designed to test the validity of candidate biomarkers or new technologies to monitor candidate biomarkers in patient tissue samples or small groups of well-characterized patients. For biomarkers already validated in human subjects, the larger effort might be to establish a reliable assay prior to pursuing larger patient studies. Priority will be given to those projects with high promise for improving clinical care in the relatively near future. Studies of potential biomarkers conducted in animal models of disease, and extended prospective clinical trials for marker validation against hard clinical endpoints, are outside the scope of this initiative. Because the nature and scope of the proposed research will vary from application to application, it is anticipated that the size and duration of each award will also vary.
Requirements: Application can be made by the following institutions and organizations: public/state controlled institutions of higher education; private institutions of higher education; nonprofits with 501(c)3 IRS status (other than institutions of higher education); nonprofits without 501(c)3 IRS status (other than institutions of higher education); small businesses; for-profit organizations (other than small businesses); state governments; U.S. territories or possessions; Indian/Native American tribal governments (federally recognized); Indian/Native American tribal governments (other than Federally recognized); Indian/Native American tribally designated organizations; non-domestic (non-U.S.) entities (foreign organizations); Hispanic-serving institutions; historically Black colleges and universities (HBCUs); tribally controlled colleges and universities (TCCUs); Alaska Native and Native Hawaiian serving institutions; regional organization; eligible agencies of the Federal government; and faith-based or community based organizations.
Date(s) Application is Due: Feb 5; Jun 5; Oct 5
Contact: Maren R. Laughlin, (301) 594-8802; fax (301) 480-0475; ml33q@nih.gov
Internet: http://grants.nih.gov/grants/guide/pa-files/PA-07-052.html
Sponsor: National Institute of Diabetes and Digestive and Kidney Diseases
6707 Democracy Boulevard
Bethesda, MD 20892-5460

NIDDK Diabetes, Endocrinology, and Metabolic Diseases NRSAs—Individual 3390
Individual National Research Service Awards (NRSAs) are given only at the postdoctoral level for research training in diabetes, endocrine, or metabolic diseases. To receive one, an individual must submit an application that describes a specific research project, which is guided and sponsored by a specific preceptor at a particular institution. The award may be for up to 36 months of training and consists of a stipend and an allowance to help meet expenses incurred during the course of fellowship.
Requirements: All awardees must be citizens or have been admitted to the United States for permanent residence. Individuals and public and private institutions, nonprofit and for-profit, that propose to establish, expand, and improve research activities in health sciences and related fields may apply for research grants. Individuals must be nominated and sponsored by public or nonprofit private institutions having staff and facilities appropriate to the proposed research training programs. SBIR grants can be awarded only to domestic small businesses. STTR grants can be awarded only to domestic small business concerns that partner with research institutions in cooperative research and development.
Date(s) Application is Due: Apr 5; Aug 5; Dec 5

Amount of Grant: 16,600 - 100,000,000 U.S.D
Contact: Dr. Judith Fradkin, Director; (301) 496-7348
Internet: http://grants.nih.gov/grants/guide/pa-files/PA-07-107.html
Sponsor: National Institute of Diabetes and Digestive and Kidney Diseases
45 Center Drive, Natcher Building
Bethesda, MD 20892-2560

NIDDK Diabetes Research Centers 3391
The mission of the Diabetes Centers is to serve as a key component of the NIDDK-supported research effort to develop new therapies and improve the health of Americans with, or at risk for, diabetes and related endocrine and metabolic disorders. The Centers promote new discoveries and enhance scientific progress through support of cutting-edge basic and clinical research related to the etiology and complications of diabetes, with the goal of rapidly translating research findings into novel strategies for the prevention, treatment and cure of diabetes and related conditions. This Funding Opportunity Announcement (FOA) solicits new and competing continuation applications for Diabetes Endocrinology Research Centers (DERCs) and Diabetes Research and Training Centers (DRTCs). Both types of Centers are designed to support and enhance the national research effort in diabetes and related endocrine and metabolic diseases. DERCs support three primary research-related activities: biomedical research cores, a Pilot and Feasibility (P&F) program, and an Enrichment program. DRTCs possess all elements of a DERC, with additional dedicated core services and P&F awards to support research in diabetes prevention and control. All activities pursued by Diabetes Centers are designed to enhance the efficiency, productivity, effectiveness and multidisciplinary nature of research in Diabetes Center topic areas.
Requirements: The following organizations and institutions are eligible to apply: public/state controlled institutions of higher education; private institutions of higher education; Hispanic-serving institutions; historically Black colleges and universities (HBCUs); tribally controlled colleges and universities (TCCUs); Alaska Native and Native Hawaiian serving institutions; nonprofits with 501(c)3 IRS status (other than institutions of higher education); nonprofits without 501(c)3 IRS status (other than institutions of higher education); for-profit organizations (other than small businesses); and state governments.
Date(s) Application is Due: Jun 22; Jul 15
Contact: James F. Hyde, (301) 594-7692; fax (301) 480-3503; jh486z@nih.gov
Internet: http://grants.nih.gov/grants/guide/rfa-files/RFA-DK-08-008.html
Sponsor: National Institute of Diabetes and Digestive and Kidney Diseases
6707 Democracy Boulevard
Bethesda, MD 20892-5460

NIDDK Diet Composition and Energy Balance Grants 3392
The goal of this funding opportunity announcement (FOA) is to invite Research Project Grant applications investigating the role of diet composition in energy balance, including studies in both animals and humans. Both short and longer-term studies are encouraged, ranging from basic studies investigating the impact of micro-or macronutrient composition on appetite, metabolism, and energy expenditure through clinical studies evaluating the efficacy of diets differing in micro- or macronutrient composition, absorption, dietary variety, or energy density for weight loss or weight maintenance. Because the nature and scope of the proposed research will vary from application to application, it is anticipated that the size and duration of each award will also vary. The total amount awarded and the number of awards will depend upon the mechanism numbers, quality, duration, and costs of the applications received.
Requirements: The following organizations and institutions may apply: public/state controlled institutions of higher education; private institutions of higher education; Hispanic-serving institutions; historically Black colleges and universities (HBCUs); tribally controlled colleges and universities (TCCUs); Alaska native and native Hawaiian serving institutions; nonprofits with 501(c)3 IRS status; nonprofits without 501(c)3 IRS status; small businesses; for-profit organizations; State governments; regional organizations; U.S. territories or possessions; Indian/Native American tribal governments; Indian/Native American tribally designated organizations; eligible agencies of the Federal government; and faith-based or community based organizations.
Date(s) Application is Due: Feb 5; Jun 5; Oct 5
Contact: Susan Z. Yanovski, (301) 594-8882; fax (301) 480-8300; sy29f@nih.gov
Internet: http://grants.nih.gov/grants/guide/pa-files/PA-07-218.html
Sponsor: National Institute of Diabetes and Digestive and Kidney Diseases
6707 Democracy Boulevard
Bethesda, MD 20892-5460

NIDDK Education Program Grants 3393
This funding opportunity announcement (FOA) solicits Research Education grant applications from applicant organizations that propose to create educational opportunities to attract undergraduate students, graduate students, and postdoctoral fellows to careers in areas of biomedical or behavioral research of particular interest to the NIDDK while fostering the career development of these students and fellows. The NIDDK is especially interested in attracting students and postdoctoral fellows from scientific disciplines underrepresented in disease-oriented biomedical research such as engineering, informatics, computer science, and computational sciences, to encourage them to apply their expertise to research relevant to diabetes and other endocrine and metabolic diseases, digestive and liver diseases, nutrition, obesity research and prevention, and kidney, urologic and hematologic diseases. The total project period for an application submitted in response to this funding opportunity may not exceed five years. Direct costs are limited to $100,000 per year.
Requirements: The following organizations and institutions may apply: public/state controlled institutions of higher education; private institutions of higher education; Hispanic-serving institutions; historically Black colleges and universities (HBCUs); tribally controlled colleges and universities (TCCUs); Alaska native and native Hawaiian serving institutions; nonprofits with 501(c)3 IRS status; nonprofits without 501(c)3 IRS status; and Indian/Native American tribally designated organizations.
Date(s) Application is Due: Jan 25; May 25; Sep 25
Amount of Grant: Up to 100,000 U.S.D
Contact: James F. Hyde, (301) 594-7692; fax (301) 480-3503; jh486z@nih.gov
Internet: http://grants.nih.gov/grants/guide/pa-files/PAR-06-554.html
Sponsor: National Institute of Diabetes and Digestive and Kidney Diseases
6707 Democracy Boulevard
Bethesda, MD 20892-5460

NIDDK Endoscopic Clinical Research Grants In Pancreatic And Biliary Diseases 3394
These small grants (R03) may be used as planning grants for full-scale multi-center clinical trials or for pilot studies that could lead to full-scale multi-center clinical trials designed to provide evidence for or against changes in the current standard of care. Pilot epidemiological studies are encouraged that could lead to more extended research that would provide evidence for or against changes in health policy, especially as related to disease and cancer prevention. It is expected that these R03 grants will serve as a basis for planning future multi-center research project grant applications (R01) or cooperative agreement (U01) awards. New and experienced investigators in relevant fields and disciplines may apply for these small grants. Investigators are encouraged to take advantage of recent endoscopic and laboratory developments. In addition, the small grant is a good mechanism for new and experienced investigators to become better equipped to perform clinical and epidemiological research. A project period of up to two years and a budget for direct costs of up to two $25,000 modules or $50,000 per year may be requested.
Requirements: The following organizations and institutions may apply: for-profit organizations; non-profit organizations; public or private institutions, such as universities, colleges, hospitals, and laboratories; units of State government; units of local government; eligible agencies of the Federal government; domestic institutions; and foreign institutions.
Date(s) Application is Due: Feb 16; Jun 16; Oct 16
Contact: Jose Serrano, (301) 594-8871; fax (301) 480-8300; js362q@nih.gov
Internet: http://grants.nih.gov/grants/guide/pa-files/PAR-06-171.html
Sponsor: National Institute of Diabetes and Digestive and Kidney Diseases
6707 Democracy Boulevard
Bethesda, MD 20892-5460

NIDDK Enhancing Zebrafish Research with Research Tools and Techniques 3395
This FOA encourages investigator-initiated applications designed to exploit the power of the zebrafish as a vertebrate model for biomedical and behavioral research. Applications proposing to develop new research tools or techniques that are of high priority to the zebrafish community and that will advance the detection and characterization of genes, pathways, and phenotypes of interest in development and aging, organ formation, neural processes, behavior, sensory processing, physiological processes, and disease processes are welcome. Budgets for direct costs under $500,000 (direct costs) per year and a project duration of up to five years may be requested. Letters of Intent should arrive by August 17, with full proposals due by September 17.
Requirements: The following organizations and institutions may apply: public/state controlled institutions of higher education; private institutions of higher education; Hispanic-serving institutions; historically Black colleges and universities (HBCUs); tribally controlled colleges and universities (TCCUs); Alaska native and native Hawaiian serving institutions; nonprofits with 501(c)3 IRS status; nonprofits without 501(c)3 IRS status; small businesses; for-profit organizations; State governments; regional organizations; U.S. territories or possessions; Indian/Native American tribal governments; Indian/Native American tribally designated organizations; eligible agencies of the Federal government; and faith-based or community based organizations.
Date(s) Application is Due: Aug 17; Sep 17
Contact: Dr. Rebekah S. Rasooly, (301) 594-6007; fax (301) 480-3510; rr185i@nih.gov
Internet: http://grants.nih.gov/grants/guide/pa-files/PAR-08-139.html
Sponsor: National Institute of Diabetes and Digestive and Kidney Diseases
6707 Democracy Boulevard
Bethesda, MD 20892-5460

NIDDK Erythroid Lineage Molecular Toolbox Grants 3396
The ultimate goals of this Funding Opportunity Announcement (FOA) will be to use erythroid cells at different stages of development and differentiation to assemble the complete collection of genes expressed, describe how they are expressed and elucidate physical parameters, including subcellular localization of the proteins translated during erythropoiesis. These reagents may then be used to discover the structure-function relationships that exist in erythroid cells with possible application to other cell types. Because the nature and scope of the proposed research will vary from application to application, it is anticipated that the size and duration of each award will also vary.
Requirements: Application can be made by the following institutions and organizations: public/state controlled institutions of higher education; private institutions of higher education; nonprofits with 501(c)3 IRS status (other than institutions of higher education); nonprofits without 501(c)3 IRS status (other than institutions of higher education); small businesses; for-profit organizations (other than small businesses); state governments; U.S. territories or possessions; Indian/Native American tribal governments (federally recognized); Indian/Native American tribal governments (other than Federally recognized); Indian/Native American tribally designated organizations; non-domestic (non-U.S.) entities (foreign organizations); Hispanic-serving institutions;

historically Black colleges and universities (HBCUs); tribally controlled colleges and universities (TCCUs); Alaska Native and Native Hawaiian serving institutions; regional organization; eligible agencies of the Federal government; and faith-based or community based organizations.
Date(s) Application is Due: Feb 5; Jun 5; Oct 5
Contact: Terry Rogers Bishop, (301) 594-7726; tb232j@nih.gov
Internet: http://grants.nih.gov/grants/guide/pa-files/PA-07-011.html
Sponsor: National Institute of Diabetes and Digestive and Kidney Diseases
6707 Democracy Boulevard
Bethesda, MD 20892-5460

NIDDK Erythropoiesis: Components and Mechanisms Grants **3397**
This FOA issued by the National Institute of Diabetes and Digestive and Kidney Diseases (NIDDK) the National Institute of Aging (NIA), and the National Heart, Lung, and Blood Institute (NHLBI), National Institutes of Health, encourages investigator-initiated R01 applications that propose hypothesis-driven research using erythroid cells. The aim of this program is to support research efforts towards a complete description of the molecular and cellular components of erythropoiesis and how these components contribute to erythropoiesis. Components include genes that are expressed (transcriptome) in erythroid cells, either during development or during differentiation, and the proteins (proteome) that are translated in erythroid cells, especially with post-translational modifications or subcellular localizations that are unique to erythroid cells. A long range goal of this program is to generate a concise description of erythropoiesis that unifies genetics, molecular processes and cytokine determinants in the erythroid lineages so that new therapeutics may be developed to measure and combat anemia. Budgets for direct costs of up to $500,000 per year and a project duration of up to five years may be requested.
Requirements: The following organizations and institutions may apply: public/state controlled institutions of higher education; private institutions of higher education; Hispanic-serving institutions; historically Black colleges and universities (HBCUs); tribally controlled colleges and universities (TCCUs); Alaska native and native Hawaiian serving institutions; nonprofits with 501(c)3 IRS status; nonprofits without 501(c)3 IRS status; small businesses; for-profit organizations; State governments; regional organizations; U.S. territories or possessions; Indian/Native American tribal governments; Indian/Native American tribally designated organizations; county governments; city or township governments; special district governments; independent school districts; public housing authorities/Indian housing authorities; eligible agencies of the Federal government; and faith-based or community based organizations.
Date(s) Application is Due: Feb 5; Jun 5; Oct 5
Amount of Grant: Up to 500,000 U.S.D
Contact: Terry Rogers Bishop, (301) 594-7726; tb232j@nih.gov
Internet: http://grants.nih.gov/grants/guide/pa-files/PA-09-255.html
Sponsor: National Institute of Diabetes and Digestive and Kidney Diseases
6707 Democracy Boulevard
Bethesda, MD 20892-5460

NIDDK Exploratory/Developmental Clinical Research Grants in Obesity **3398**
The goal of this initiative is to encourage exploratory/developmental clinical research that will accelerate the development of effective interventions for prevention or treatment of overweight or obesity in adults and/or children. The goal of this mechanism is to provide flexibility for initiating exploratory, short-term studies, thus allowing new ideas to be investigated in a more expeditious manner without stringent requirements for preliminary data. Such support is needed to encourage investigators to pursue new approaches, underdeveloped topics, or more creative avenues for research including new partnerships. Epidemiological research with a goal of informing translational/clinical research on prevention or treatment of obesity or overweight in adults and/or children is encouraged. The emphasis is thus on the development of exploratory clinical studies, pilot and feasibility studies, or small randomized clinical trials that will provide preliminary data for intervention and epidemiological studies that will inform translational/clinical research.
Requirements: The following organizations and institutions may apply: public/state controlled institutions of higher education; private institutions of higher education; Hispanic-serving institutions; historically Black colleges and universities (HBCUs); tribally controlled colleges and universities (TCCUs); Alaska native and native Hawaiian serving institutions; nonprofits with 501(c)3 IRS status; nonprofits without 501(c)3 IRS status; small businesses; for-profit organizations; State governments; regional organizations; U.S. territories or possessions; Indian/Native American tribal governments; Indian/Native American tribally designated organizations; eligible agencies of the Federal government; and faith-based or community based organizations.
Date(s) Application is Due: Feb 16; Jun 16; Oct 16
Amount of Grant: 200,000 - 275,000 U.S.D
Contact: Carolyn W. Miles, (301)451-3759; fax (301) 480-8300; cm294e@nih.gov
Internet: http://grants.nih.gov/grants/guide/pa-files/PA-09-124.html
Sponsor: National Institute of Diabetes and Digestive and Kidney Diseases
6707 Democracy Boulevard
Bethesda, MD 20892-5460

NIDDK Grants for Basic Research in Glomerular Diseases **3399**
This FOA invites applications from investigators with diverse scientific interests to apply their expertise in basic research to enhance the understanding of the pathogenesis of the various primary or secondary forms of glomerular disease. Recent observations regarding intrinsic glomerular cell biology, particularly in the podocyte, have provided exciting new insights into potential pathogenic mechanisms of human glomerular disease. Although both immune and nonimmune mechanisms of glomerular injury have been studied previously, experimental models of disease and recent techniques that provide tools for molecular and proteomic profiling show great promise for identifying glomerular disease biomarkers. Despite these recent advances, additional basic studies are needed to facilitate a better understanding of glomerular disease pathogenesis and ultimately impact on the treatment of human glomerulopathies. It is anticipated that applications submitted in response to this FOA could address a number of different aspects concerning the pathogenesis, natural history, therapy pre-emption or prevention of the various morphologic forms of experimental glomerular disease.
Requirements: The following organizations and institutions may apply: public/state controlled institutions of higher education; private institutions of higher education; Hispanic-serving institutions; historically Black colleges and universities (HBCUs); tribally controlled colleges and universities (TCCUs); Alaska native and native Hawaiian serving institutions; nonprofits with 501(c)3 IRS status; nonprofits without 501(c)3 IRS status; small businesses; for-profit organizations; State governments; regional organizations; U.S. territories or possessions; Indian/Native American tribal governments; Indian/Native American tribally designated organizations; eligible agencies of the Federal government; and faith-based or community based organizations.
Date(s) Application is Due: Feb 5; Jun 5; Oct 5
Contact: Marva Moxey-Mims, (301) 594-7717; fax (301) 480-3510; mm726k@nih.gov
Internet: http://grants.nih.gov/grants/guide/pa-files/PA-07-367.html
Sponsor: National Institute of Diabetes and Digestive and Kidney Diseases
6707 Democracy Boulevard
Bethesda, MD 20892-5460

NIDDK Health Disparities in NIDDK Diseases **3400**
It is recognized that there are many diseases and disorders that disproportionately affect the health of racial and ethnic minority populations in the United States. It is evident that African-Americans, Hispanic Americans, American Indians, Alaska Natives, some Asian-Americans, and Native Hawaiians and other Pacific Islanders experience much higher risks and poorer health status than the general population. Several of the diseases that disproportionately afflict minorities are high priority research areas for NIDDK, including diabetes, obesity, nutrition-related disorders, hepatitis C, gallbladder disease, H. Pylori infection, sickle cell disease, kidney diseases, and metabolic, gastrointestinal, hepatic, and renal complications from infection with HIV. NIDDK encourages efforts health disparity research to reduce the human and economic costs resulting from inequities in health care and health outcomes. Because the nature and scope of the proposed research will vary from application to application, it is anticipated that the size and duration of each award will also vary. The total amount awarded and the number of awards will depend upon the mechanism numbers, quality, duration, and costs of the applications received.
Requirements: The following organizations and institutions may apply: public/state controlled institutions of higher education; private institutions of higher education; Hispanic-serving institutions; historically Black colleges and universities (HBCUs); tribally controlled colleges and universities (TCCUs); Alaska native and native Hawaiian serving institutions; nonprofits with 501(c)3 IRS status; nonprofits without 501(c)3 IRS status; small businesses; for-profit organizations; State governments; regional organizations; U.S. territories or possessions; Indian/Native American tribal governments; Indian/Native American tribally designated organizations; eligible agencies of the Federal government; and faith-based or community based organizations.
Date(s) Application is Due: Feb 5; Jun 5; Oct 5
Contact: Myrlene Staten, (301) 402-7886; fax (301) 480-3503; ms808k@nih.gov
Internet: http://grants.nih.gov/grants/guide/pa-files/PA-07-027.html
Sponsor: National Institute of Diabetes and Digestive and Kidney Diseases
6707 Democracy Boulevard
Bethesda, MD 20892-5460

NIDDK Identifying and Reducing Diabetes and Obesity Related Health Disparities within Healthcare Systems **3401**
This FOA requests applications designed to identify or address factors or barriers that result in disparate outcomes within a healthcare system. All applications should measure the impact of identified factors or interventions on health outcomes. Research is sought that examines at least one of the following factors in a healthcare system and/or the interaction between these factors; healthcare professionals, healthcare organizations, and the patients and communities they serve.
Requirements: The following organizations and institutions may apply: public/state controlled institutions of higher education; private institutions of higher education; Hispanic-serving institutions; historically Black colleges and universities (HBCUs); tribally controlled colleges and universities (TCCUs); Alaska native and native Hawaiian serving institutions; nonprofits with 501(c)3 IRS status; nonprofits without 501(c)3 IRS status; small businesses; for-profit organizations; State governments; regional organizations; U.S. territories or possessions; Indian/Native American tribal governments; Indian/Native American tribally designated organizations; eligible agencies of the Federal government; and faith-based or community based organizations.
Date(s) Application is Due: Feb 5; Jun 5; Oct 5
Contact: Christine Hunter, (301) 594-4728; fax (301) 480-3503; ch514c@.nih.gov
Internet: http://grants.nih.gov/grants/guide/pa-files/PA-07-388.html
Sponsor: National Institute of Diabetes and Digestive and Kidney Diseases
6707 Democracy Boulevard
Bethesda, MD 20892-5460

NIDDK Insulin Signaling and Receptor Cross Talk Grants **3402**
This Funding Opportunity Announcement (FOA) solicits investigator-initiated research projects that will define the mechanistic basis of crosstalk between the insulin receptor signaling pathway and other signaling pathways, with the goal of defining how insulin action is influenced by the complex microenvironment found in insulin responsive tissues.

Requirements: The following organizations and institutions may apply: public/state controlled institutions of higher education; private institutions of higher education; Hispanic-serving institutions; historically Black colleges and universities (HBCUs); tribally controlled colleges and universities (TCCUs); Alaska native and native Hawaiian serving institutions; nonprofits with 501(c)3 IRS status; nonprofits without 501(c)3 IRS status; small businesses; for-profit organizations; State governments; regional organizations; U.S. territories or possessions; Indian/Native American tribal governments; Indian/Native American tribally designated organizations; eligible agencies of the Federal government; and faith-based or community based organizations.
Date(s) Application is Due: Feb 5; Jun 5; Oct 5
Contact: Kristin Abraham, (301) 451-8048; fax (301) 480-3503; ka136s@nih.gov
Internet: http://grants.nih.gov/grants/guide/pa-files/PA-07-058.html
Sponsor: National Institute of Diabetes and Digestive and Kidney Diseases
6707 Democracy Boulevard
Bethesda, MD 20892-5460

NIDDK Intestinal Failure, Short Gut Syndrome and Small Bowel Transplantation Grants **3403**
The National Institute of Diabetes and Digestive and Kidney Diseases (NIDDK) seeks grant applications to study the pathogenesis, natural history, treatment and complications of intestinal failure and its therapies, including parenteral nutrition and small bowel transplantation. The overall objective of this funding opportunity is to encourage basic and clinical research into intestinal failure, short gut syndrome and intestinal transplantation. Because the nature and scope of the proposed research will vary from application to application, it is anticipated that the size and duration of each award will also vary.
Requirements: Application can be made by the following institutions and organizations: public/state controlled institutions of higher education; private institutions of higher education; nonprofits with 501(c)3 IRS status (other than institutions of higher education); nonprofits without 501(c)3 IRS status (other than institutions of higher education); small businesses; for-profit organizations (other than small businesses); state governments; U.S. territories or possessions; Indian/Native American tribal governments (federally recognized); Indian/Native American tribal governments (other than Federally recognized); Indian/Native American tribally designated organizations; non-domestic (non-U.S.) entities (foreign organizations); Hispanic-serving institutions; historically Black colleges and universities (HBCUs); tribally controlled colleges and universities (TCCUs); Alaska Native and Native Hawaiian serving institutions; regional organization; eligible agencies of the Federal government; and faith-based or community based organizations.
Date(s) Application is Due: Feb 5; Jun 5; Oct 5
Contact: Dr. Michael K. May, (301) 594-8884; fax (301) 480-8300; mm102i@nih.gov
Internet: http://grants.nih.gov/grants/guide/pa-files/PA-07-010.html
Sponsor: National Institute of Diabetes and Digestive and Kidney Diseases
6707 Democracy Boulevard
Bethesda, MD 20892-5460

NIDDK Intestinal Stem Cell Consortium Grants **3404**
This FOA invites new applications to participate in the Intestinal Stem Cell Consortium (ISCC). The goals of this initiative are: to establish a network of individual research projects and a Coordinating Center focused on research on stem cells of the small intestine; to isolate, culture, characterize, functionally validate, and compare stem cell populations from the small intestine epithelium in vivo and in vitro; to share data, biomaterials, models, reagents, resources and methods between projects in the ISCC; and make these publicly available through a web site to be developed by the Coordinating Center. The objective of support for this consortium is to accelerate research on stem cells of the intestinal epithelium in order to facilitate understanding of the biology and function of the intestine and aid development of therapies for intestinal diseases and conditions where damage and replacement of this epithelium are important components. Direct costs are limited to $250,000 per year for a maximum five-year period for research projects and $350,000 per year for a maximum five-year period for the Coordinating Center.
Requirements: Application can be made by the following institutions and organizations: public/state controlled institutions of higher education; private institutions of higher education; nonprofits with 501(c)3 IRS status (other than institutions of higher education); nonprofits without 501(c)3 IRS status (other than institutions of higher education); small businesses; for-profit organizations (other than small businesses); state governments; U.S. territories or possessions; Indian/Native American tribal governments (federally recognized); Indian/Native American tribal governments (other than Federally recognized); Indian/Native American tribally designated organizations; non-domestic (non-U.S.) entities (foreign organizations); Hispanic-serving institutions; historically Black colleges and universities (HBCUs); tribally controlled colleges and universities (TCCUs); Alaska Native and Native Hawaiian serving institutions; regional organization; eligible agencies of the Federal government; and faith-based or community based organizations.
Date(s) Application is Due: Mar 18
Amount of Grant: Up to 250,000 U.S.D
Contact: Ronald Margolis, (301) 594-8819; fax (301) 435-6047; rm76f@nih.gov
Internet: http://grants.nih.gov/grants/guide/rfa-files/RFA-DK-08-010.html
Sponsor: National Institute of Diabetes and Digestive and Kidney Diseases
6707 Democracy Boulevard
Bethesda, MD 20892-5460

NIDDK Mentored Research Scientist Development Award **3405**
The overall objective of the NIH Research Career Development Award program is to prepare qualified individuals for careers that have a significant impact on the health-related research needs of the Nation. The objective of the NIDDK Mentored Research Scientist Development Award (K01) is to provide support for a sustained period of protected time for intensive research career development under the guidance of an experienced mentor, or sponsor, in the biomedical, behavioral or clinical sciences leading to research independence, thereby ensuring a future cadre of well-trained Ph.D. scientists working in research areas supported by NIDDK. The expectation is that through this sustained period of research career development and training, awardees will launch independent research careers and become competitive for new research project grant (R01) funding. By providing support for the critical transition period between postdoctoral training and independent R01 funding for non-clinical investigators, the NIDDK hopes to foster the careers of these investigators who are vital for the future excellence of the NIDDK research endeavor. Applicants must justify the need for a period of mentored research experience and provide a convincing case that the proposed period of support will substantially enhance their careers as independent investigators.
Requirements: The following organizations and institutions may apply: public/state controlled institutions of higher education; private institutions of higher education; Hispanic-serving institutions; historically Black colleges and universities (HBCUs); tribally controlled colleges and universities (TCCUs); Alaska native and native Hawaiian serving institutions; nonprofits with 501(c)3 IRS status; nonprofits without 501(c)3 IRS status; small businesses; for-profit organizations; State governments; regional organizations; U.S. territories or possessions; Indian/Native American tribal governments; Indian/Native American tribally designated organizations; eligible agencies of the Federal government; and faith-based or community based organizations.
Date(s) Application is Due: Mar 12; Jul 12; Nov 12
Contact: James F. Hyde, (301) 594-7692; fax (301) 480-3503; jh486z@nih.gov
Internet: http://grants.nih.gov/grants/guide/pa-files/PAR-09-060.html
Sponsor: National Institute of Diabetes and Digestive and Kidney Diseases
6707 Democracy Boulevard
Bethesda, MD 20892-5460

NIDDK Multi-Center Clinical Study Cooperative Agreements **3406**
This PAR provides for grant applications for investigator-initiated, multi-center clinical studies. NIDDK supports investigator-initiated, multi-center clinical studies through a two-part process that includes an implementation planning (U34) grant. Beginning with the receipt date, NIDDK will accept, peer review, and consider for funding applications for investigator-initiated, multi-center clinical studies only from NIDDK Multi-Center Clinical Study Implementation Planning (U34) grant awardees. Because the nature and scope of the proposed research will vary from application to application, it is anticipated that the size and duration of each award will also vary. The total amount awarded and the number of awards will depend upon the mechanism numbers, quality, duration, and costs of the applications received.
Requirements: The following organizations and institutions may apply: public/state controlled institutions of higher education; private institutions of higher education; Hispanic-serving institutions; historically Black colleges and universities (HBCUs); tribally controlled colleges and universities (TCCUs); Alaska native and native Hawaiian serving institutions; nonprofits with 501(c)3 IRS status; nonprofits without 501(c)3 IRS status; small businesses; for-profit organizations; State governments; regional organizations; U.S. territories or possessions; Indian/Native American tribal governments; Indian/Native American tribally designated organizations; eligible agencies of the Federal government; and faith-based or community based organizations.
Date(s) Application is Due: Feb 5; Jun 5; Oct 5
Contact: Catherine Meyers, (301) 451-4901; fax (301) 480-3510; meyersc@mail.nih.gov
Internet: http://grants.nih.gov/grants/guide/pa-files/PAR-08-058.html
Sponsor: National Institute of Diabetes and Digestive and Kidney Diseases
6707 Democracy Boulevard
Bethesda, MD 20892-5460

NIDDK Multi-Center Clinical Study Implementation Planning Grants **3407**
NIDDK supports investigator-initiated, multi-center clinical studies through a two-part process that includes an implementation planning grant. The planning grant is designed to: permit early peer review of the rationale for the proposed clinical study; permit assessment of the design/protocol of the proposed study; provide support for the development of a complete study protocol and associated documents including a manual of operations; and support the development of other essential elements required for the conduct of a clinical study. Completion of the required products of a U34 grant is a prerequisite for submission of a multi-center clinical study cooperative agreement application, which will support the actual conduct of the study. Because the nature and scope of the proposed research will vary from application to application, it is anticipated that the size and duration of each award will also vary. The total project period for an application submitted in response to this funding opportunity may not exceed two years. Applicants may request up to $250,000 in direct costs per year.
Requirements: The following organizations and institutions may apply: public/state controlled institutions of higher education; private institutions of higher education; Hispanic-serving institutions; historically Black colleges and universities (HBCUs); tribally controlled colleges and universities (TCCUs); Alaska native and native Hawaiian serving institutions; nonprofits with 501(c)3 IRS status; nonprofits without 501(c)3 IRS status; small businesses; for-profit organizations; State governments; regional organizations; U.S. territories or possessions; Indian/Native American tribal governments; Indian/Native American tribally designated organizations; eligible agencies of the Federal government; and faith-based or community based organizations.
Date(s) Application is Due: Mar 18; Jun 23
Contact: Catherine Meyers, (301) 451-4901; fax (301) 480-3510; meyersc@mail.nih.gov
Internet: http://grants.nih.gov/grants/guide/pa-files/PAR-08-057.html
Sponsor: National Institute of Diabetes and Digestive and Kidney Diseases
6707 Democracy Boulevard
Bethesda, MD 20892-5460

NIDDK Neurobiology of Diabetic Complications Grants 3408

NIDDK invites investigator-initiated research grant applications to study the mechanisms by which diabetes results in painful and disabling neuropathies and other neurological complications and to apply this information to the development of interventions to prevent, limit, or reverse these complications. The intent of this RFA is to attract basic neuroscientists to the study of diabetic neuropathy and neurobiology relevant to diabetes, and enhance interdisciplinary approaches to research in this area. Annual deadline dates may vary; contact the appropriate NIH institute for exact dates. Letters of intent are due by March 24, with full applications due by April 25 each year.
Requirements: Applications may be submitted by domestic and foreign, for-profit and nonprofit organizations, both public and private; units of state and local governments; and eligible agencies of the federal government.
Date(s) Application is Due: Mar 24; Apr 25
Amount of Grant: Up to 3,000,000 U.S.D
Contact: Dr. Paul Nichols, (301) 496-9964; fax (301) 402-2060; pn13w@nih.gov
Internet: http://grants.nih.gov/grants/guide/rfa-files/RFA-NS-00-002.html
Sponsor: National Institute of Diabetes and Digestive and Kidney Diseases
45 Center Drive, MSC 6600
Bethesda, MD 20892-6600

NIDDK Neuroimaging in Obesity Research Grants 3409

This FOA issued by the National Institute of Diabetes and Digestive and Kidney Diseases, the National Institute on Drug Abuse, and the National Institute of Biomedical Imaging and Bioengineering, of the National Institutes of Health, solicits Research Project Grant (R01) applications from institutions/ organizations that propose to use neuroimaging approaches in obesity research in human subjects and animal models. Many areas of the brain interact or communicate with other organs to control eating behavior, physical activity and energy metabolism, and functional neuroimaging holds enormous promise for expanding our understanding of how food intake and energy expenditure are mismatched in a setting of abundantly available nutrients, leading to excessive fat storage. The total amount of funding that the ICs expect to award through this announcement is up to $4.5 Million, to fund approximately 9-15 awards. The duration and requested direct costs will vary depending on the scope of the proposed research, but cannot exceed 5 years and $500,000 per year. Letters of intent are due by February 18, with full applications due by 5:00 p.m. on March 18.
Requirements: Application can be made by the following institutions and organizations: public/state controlled institutions of higher education; private institutions of higher education; nonprofits with 501(c)3 IRS status (other than institutions of higher education); nonprofits without 501(c)3 IRS status (other than institutions of higher education); small businesses; for-profit organizations (other than small businesses); state governments; U.S. territories or possessions; Indian/Native American tribal governments (federally recognized); Indian/Native American tribal governments (other than Federally recognized); Indian/Native American tribally designated organizations; non-domestic (non-U.S.) entities (foreign organizations); Hispanic-serving institutions; historically Black colleges and universities (HBCUs); tribally controlled colleges and universities (TCCUs); Alaska Native and Native Hawaiian serving institutions; regional organization; eligible agencies of the Federal government; and faith-based or community based organizations.
Date(s) Application is Due: Feb 18; Mar 18
Amount of Grant: Up to 500,000 U.S.D
Contact: Maren R. Laughlin, (301) 594-8802; fax (301) 480-0475; ml33q@nih.gov
Internet: http://grants.nih.gov/grants/guide/rfa-files/RFA-DK-08-009.html
Sponsor: National Institute of Diabetes and Digestive and Kidney Diseases
6707 Democracy Boulevard
Bethesda, MD 20892-5460

NIDDK New Technologies for Liver Disease SBIR 3410

The purpose of this Funding Opportunity Announcement (FOA) is to solicit Small Business Innovation Research (SBIR) grant applications from small business concerns (SBCs) that propose to develop resources, research tools, instrumentations, biomarkers, devices, drugs or new and innovative approaches to diagnosis, monitoring, management, treatment and prevention of liver diseases. Areas of interest include development of reliable and practical means of diagnosis of liver diseases; biomarkers for disease activity and stage; noninvasive tests for inflammation, fibrosis and fat in the liver; and drugs, complementary and alternative modalities, biologics or molecular reagents for the therapy or prevention of liver diseases. The goal of this announcement is to enlist members of the small business research community in advancing means of diagnosis, treatment and prevention of liver disease and facilitate the goals outlined in the trans-NIH Action Plan for Liver Disease Research.
Requirements: Only United States small business concerns (SBCs) are eligible to submit SBIR applications.
Date(s) Application is Due: Apr 5; Aug 5; Dec 5
Amount of Grant: 100,000 - 750,000 U.S.D
Contact: Christine L. Densmore, (301) 402-8714; fax (301) 480-8300; cd121z@nih.gov
Internet: http://grants.nih.gov/grants/guide/pa-files/PA-09-095.html
Sponsor: National Institute of Diabetes and Digestive and Kidney Diseases
6707 Democracy Boulevard
Bethesda, MD 20892-5460

NIDDK New Technologies for Liver Disease STTR 3411

The objective of this FOA is to encourage and enable scientists at small businesses to develop and evaluate new technologies, drugs, devices, and approaches to diagnosis, management, and prevention of liver disease. Development of new technologies as well as application of existing technologies may be proposed. Studies may include use of animal models or human participants or both. If appropriate, plans for manufacturing and clinical evaluation of developed technologies, drugs, devices and innovative approaches should be included in the application. However, clinical trials beyond Phase I studies will not be considered appropriate to this announcement.
Requirements: Only United States small business concerns (SBCs) are eligible to submit STTR applications.
Date(s) Application is Due: Apr 5; Aug 5; Dec 5
Amount of Grant: 100,000 - 750,000 U.S.D
Contact: Christine L. Densmore, (301) 402-8714; fax (301) 480-8300; cd121z@nih.gov
Internet: http://grants.nih.gov/grants/guide/pa-files/PA-09-094.html
Sponsor: National Institute of Diabetes and Digestive and Kidney Diseases
6707 Democracy Boulevard
Bethesda, MD 20892-5460

NIDDK Non-Invasive Methods for Diagnosis and Progression of Diabetes, Kidney, Urological, Hematological and Digestive Diseases 3412

This funding opportunity is a call for the application of imaging and other non- or minimally-invasive technologies to detect, characterize, diagnose, identify persons with predisposition to, or monitor treatment of diseases of interest to the National Institute of Diabetes and Digestive and Kidney Diseases. Also needed are new, robust surrogate markers for clinical trial endpoints, and new ways to characterize normal and pathological tissues in vivo. Diseases of interest include type 1 and 2 diabetes, obesity, and kidney, liver, urologic, hematologic, digestive, endocrine and metabolic diseases and their complications. Applicable techniques include molecular imaging and functional imaging approaches, imaging methods with high spatial, chemical or time resolution, and new spectroscopic or sensor array technologies for monitoring metabolic or physiological events.
Requirements: The following organizations and institutions may apply: for-profit organizations; non-profit organizations; public or private institutions, such as universities, colleges, hospitals, and laboratories; units of State government; units of local government; eligible agencies of the Federal government; foreign institutions; domestic institutions; and faith-based or community-based organizations.
Date(s) Application is Due: Feb 5; Jun 5; Oct 5
Contact: Maren R. Laughlin, (301) 594-8802; fax (301) 480-0475; ml33q@nih.gov
Internet: http://grants.nih.gov/grants/guide/pa-files/PA-06-143.html
Sponsor: National Institute of Diabetes and Digestive and Kidney Diseases
6707 Democracy Boulevard
Bethesda, MD 20892-5460

NIDDK Pancreatic Development and Regeneration Grants: Toward Cellular Therapies for Diabetes 3413

This Funding Opportunity Announcement (FOA) is intended to promote basic research in pancreatic development, stem cell biology and regeneration that will provide the foundation for the rationale design of future cell replacement therapies for diabetes. Appropriate topics for investigation include, but are not limited to: prospective isolation, purification and characterization of pancreatic stem/progenitor cells; identification of the signals, signaling pathway components and transcriptional factors that could direct the formation of pancreatic endoderm and progenitors, islet cell differentiation, morphogenesis, growth, function and survival in vitro; use of existing NIH-approved human ES cell lines or mouse embryonic cells, pancreatic stem cells, non-pancreatic adult stem cells such as those from liver or gut, human amniotic cells or bone marrow-derived cells to efficiently generate endoderm, pancreatic progenitor cells, or differentiated beta cells in culture; development of methods and reagents to assess quantitatively, the survival, migration, fate and function of transplanted animal or human pancreatic progenitors and their progeny in the living animal; determination of the factors important for beta cell regeneration and designing strategies to foster regeneration; development of new animal models that can used to identify exogenous or endogenous factors that might promote beta cell regeneration or neogenesis; characterization the lineage of cells that may contribute or participate in beta cell regeneration by genetic marking; and development of assays to screen chemical libraries for modifiers of beta cell regeneration, differentiation, and or expansion.
Requirements: The following organizations and institutions may apply: public/state controlled institutions of higher education; private institutions of higher education; Hispanic-serving institutions; historically Black colleges and universities (HBCUs); tribally controlled colleges and universities (TCCUs); Alaska native and native Hawaiian serving institutions; nonprofits with 501(c)3 IRS status; nonprofits without 501(c)3 IRS status; small businesses; for-profit organizations; State governments; regional organizations; U.S. territories or possessions; Indian/Native American tribal governments; Indian/Native American tribally designated organizations; eligible agencies of the Federal government; and faith-based or community based organizations.
Date(s) Application is Due: Feb 5; Jun 5; Oct 5
Contact: Sheryl M. Sato, (301) 594-8811; fax (301) 480-0475; ss68z@nih.gov
Internet: http://grants.nih.gov/grants/guide/pa-files/PA-07-056.html
Sponsor: National Institute of Diabetes and Digestive and Kidney Diseases
6707 Democracy Boulevard
Bethesda, MD 20892-5460

NIDDK Pilot and Feasibility Clinical Research Grants in Diabetes, Endocrine and Metabolic Diseases 3414

The goal of this initiative is to encourage exploratory/developmental clinical research related to the prevention or treatment of diabetes, obesity and endocrine and genetic metabolic diseases. The Pilot and Feasibility Clinical Research Grants Program is designed to allow initiation of exploratory, short-term clinical studies, so that new ideas may be investigated without stringent requirements for preliminary data. Such support can be used by experienced investigators, as well as new investigators, to pursue new approaches and investigate underdeveloped research areas. The short-term studies

should focus on research questions that are likely to have high clinical impact. They can include testing a new prevention strategy, a new intervention, or unique combinations of therapies. A high priority is the use of such studies to help stimulate the translation of promising research developments from the laboratory into clinical practice in diabetes, endocrine diseases and genetic metabolic diseases, including cystic fibrosis.
Requirements: The following organizations and institutions may apply: public/state controlled institutions of higher education; private institutions of higher education; Hispanic-serving institutions; historically Black colleges and universities (HBCUs); tribally controlled colleges and universities (TCCUs); Alaska native and native Hawaiian serving institutions; nonprofits with 501(c)3 IRS status; nonprofits without 501(c)3 IRS status; small businesses; for-profit organizations; State governments; regional organizations; U.S. territories or possessions; Indian/Native American tribal governments; Indian/Native American tribally designated organizations; eligible agencies of the Federal government; and faith-based or community based organizations.
Date(s) Application is Due: Feb 16; Jun 16; Oct 16
Amount of Grant: 200,000 - 275,000 U.S.D
Contact: Karen Salomon, (301) 594-7733; fax (301) 435-6047; ks495c@nih.gov
Internet: http://grants.nih.gov/grants/guide/pa-files/PA-09-133.html
Sponsor: National Institute of Diabetes and Digestive and Kidney Diseases
6707 Democracy Boulevard
Bethesda, MD 20892-5460

NIDDK Pilot and Feasibility Clinical Research Grants in Kidney or Urologic Diseases 3415
This funding opportunity announcement (FOA) is a re-issuance of PAR-06-113 and specifically encourages the submission of applications for small scale or pilot and feasibility clinical and translational research studies, including epidemiological studies or clinical trials related to kidney or urologic disease research that address important clinical and translational questions and are potentially of high impact. It is anticipated that some applications for pilot and feasibility studies may lead to full-scale clinical studies including evaluation of diagnostic strategies, epidemiologic studies, or trials in the diagnosis, prevention, or treatment of kidney or urologic diseases. These grants may be used to plan, evaluate the feasibility, or implement clinical trials that assess pharmacological, dietary, surgical, or behavioral interventions for the prevention or treatment of kidney or urologic disease. Small scale or pilot epidemiological studies are also encouraged.
Requirements: The following organizations and institutions may apply: public/state controlled institutions of higher education; private institutions of higher education; Hispanic-serving institutions; historically Black colleges and universities (HBCUs); tribally controlled colleges and universities (TCCUs); Alaska native and native Hawaiian serving institutions; nonprofits with 501(c)3 IRS status; nonprofits without 501(c)3 IRS status; small businesses; for-profit organizations; State governments; regional organizations; U.S. territories or possessions; Indian/Native American tribal governments; Indian/Native American tribally designated organizations; eligible agencies of the Federal government; and faith-based or community based organizations.
Date(s) Application is Due: Feb 16; Jun 16; Oct 16
Amount of Grant: 200,000 - 275,000 U.S.D
Contact: Catherine M. Meyers, (301) 594-7717; fax (301) 480-3510; cm420i@nih.gov
Internet: http://grants.nih.gov/grants/guide/pa-files/PAR-09-077.html
Sponsor: National Institute of Diabetes and Digestive and Kidney Diseases
6707 Democracy Boulevard
Bethesda, MD 20892-5460

NIDDK Pilot and Feasibility Clinical Research Studies in Digestive Diseases and Nutrition 3416
The goal of this pilot and feasibility mechanism is to provide flexibility for initiating preliminary, short-term clinical studies, thus allowing new ideas to be investigated in a more expeditious manner without stringent requirements for preliminary data. Such support is needed to encourage experienced investigators as well as new investigators to pursue new approaches, underdeveloped topics, or more creative avenues of research. If successful, these awards should lead to significant scientific advances in digestive disease research and should facilitate translation into clinical practice to improve patient outcomes.
Requirements: The following organizations and institutions are eligible to apply: public/state controlled institutions of higher education; private institutions of higher education; Hispanic-serving institutions; historically Black colleges and universities (HBCUs); tribally controlled colleges and universities (TCCUs); Alaska Native and Native Hawaiian serving institutions; nonprofits with 501(c)3 IRS status (other than institutions of higher education); nonprofits without 501(c)3 IRS status (other than institutions of higher education); small businesses; for-profit organizations (other than small businesses); state governments; Indian/Native American tribal governments (federally recognized); Indian/Native American tribally designated organizations; county governments; city or township governments; special district governments; independent school districts; public housing authorities; Indian housing authorities; U.S. territory or possession; Indian/Native American tribal governments (other than federally recognized); regional organizations; non-domestic (non-U.S.) entities (foreign organizations); other eligible agencies of the federal government; and faith-based or community-based organizations.
Date(s) Application is Due: Feb 16; Jun 16; Oct 16
Contact: Mary Evans, (301) 594-4578; fax (301) 480-8300; me189d@nih.gov
Internet: http://grants.nih.gov/grants/guide/pa-files/PA-09-151.html
Sponsor: National Institute of Diabetes and Digestive and Kidney Diseases
6707 Democracy Boulevard
Bethesda, MD 20892-5460

NIDDK Planning Grants For Translational Research For The Prevention and Control Of Diabetes And Obesity 3417
The National Institute of Diabetes and Digestive and Kidney Diseases (NIDDK) seeks to develop cost effective and sustainable interventions, that can be adopted in real world settings, for the prevention and control of diabetes and obesity. Research should be based on interventions already proven efficacious in clinical trials to prevent and reverse obesity and type 2 diabetes, to improve care of type 1 and type 2 diabetes and to prevent or delay its complications. The interventions proposed under this solicitation should have the potential to be widely disseminated to clinical practice, individuals and communities at risk. Relevant topics include: methods to improve health care delivery to patients with or at risk of diabetes; to enhance diabetes self management; methods to develop strategies to promote healthy lifestyles to reduce the risk of diabetes; and methods to promote lifestyle change that will prevent or reverse overweight and obesity, and 5) new cost effective ways to identify people with pre-diabetes and undiagnosed diabetes. This FOA will use the National Institutes of Health (NIH) clinical trial planning grant (R34) award mechanism, which will provide up to $150,000 in direct costs per year.
Requirements: Applications may be submitted by: for-profit organizations; non-profit organizations; public or private institutions, such as universities, colleges, hospitals, and laboratories; units of State government; units of local government; eligible agencies of the Federal government; domestic institutions; faith-based or community-based organizations; Indian/Native American tribal governments; and Indian/Native American tribally designated organizations.
Date(s) Application is Due: Feb 16; Jun 16; Oct 16
Amount of Grant: Up to 150,000 U.S.D
Contact: Sanford Garfield, (301) 594-8803; fax (301) 480-3503; sg50o@nih.gov
Internet: http://grants.nih.gov/grants/guide/pa-files/PAR-06-358.html
Sponsor: National Institute of Diabetes and Digestive and Kidney Diseases
6707 Democracy Boulevard
Bethesda, MD 20892-5460

NIDDK Proteomics: Diabetes, Obesity, And Endocrine, Digestive, Kidney, Urologic, And Hematologic Diseases 3418
This Funding Opportunity Announcement (FOA) issued by the National Institute of Diabetes and Digestive and Kidney Diseases (NIDDK), National Institutes of Health (NIH), solicits projects that advance research to identify and quantitate protein expression patterns, post-translational modification of proteins, and protein-protein interactions on cells, tissues, organ systems relevant to diabetes, obesity, endocrine and metabolic diseases, nutritional function and diseases of the alimentary tract, exocrine pancreas, liver, kidney, bladder and prostate and normal biological processes related to the function of these systems. The development and improvement of innovative proteomic technologies is also encouraged through their application to relevant biological questions related to the pathophysiology of endocrine glands, gastrointestinal tract, liver and kidney, bladder and prostate.
Requirements: The following organizations and institutions may apply: public/state controlled institutions of higher education; private institutions of higher education; Hispanic-serving institutions; historically Black colleges and universities (HBCUs); tribally controlled colleges and universities (TCCUs); Alaska native and native Hawaiian serving institutions; nonprofits with 501(c)3 IRS status; nonprofits without 501(c)3 IRS status; small businesses; for-profit organizations; State governments; regional organizations; U.S. territories or possessions; Indian/Native American tribal governments; Indian/Native American tribally designated organizations; eligible agencies of the Federal government; and faith-based or community based organizations.
Date(s) Application is Due: Feb 5; Jun 5; Oct 5
Contact: Salvatore Sechi, (301) 594-8814; fax (301) 480-2688; ss24q@nih.gov
Internet: http://grants.nih.gov/grants/guide/pa-files/PA-07-016.html
Sponsor: National Institute of Diabetes and Digestive and Kidney Diseases
6707 Democracy Boulevard
Bethesda, MD 20892-5460

NIDDK Race/Ethnic Disparities in the Incidence of Diabetes Complications 3419
This PA solicits research to investigate differences among contemporary populations in the United States, categorized by race-ethnicity and other factors, in risk factors for complications of diabetes and in rates of these complications; and the extent to which factors, including inherent metabolic and genetic variations, medical care, socioeconomic status, and behavioral factors, account for these differences. PA: PA-02-165
Requirements: Applications may be submitted by domestic for-profit and nonprofit organizations, both public and private; units of state and local governments; and eligible agencies of the federal government.
Contact: Dr. Kristin Abraham, (301) 451-8048; fax (301) 480-3503; ka136s@nih.gov
Internet: http://grants2.nih.gov/grants/guide/pa-files/PA-02-165.html
Sponsor: National Institute of Diabetes and Digestive and Kidney Diseases
45 Center Drive, Natcher Building
Bethesda, MD 20892-2560

NIDDK Research Grants for Studies of Hepatitis C in the Setting of Renal Disease 3420
It is the intent of this Funding Opportunity Announcement (FOA) to invite applications from investigators with diverse scientific interests, who wish to apply their expertise into basic and applied research to enhance the understanding of the pathogenesis, natural history, therapy,and prevention of hepatitis C in the setting of renal disease or renal transplantation. Because the nature and scope of the proposed research will vary from application to application, it is anticipated that the size and duration of each award will also vary. The total amount awarded and the number of awards will depend upon the mechanism numbers, quality, duration, and costs of the applications received.

Requirements: The following organizations and institutions may apply: public/state controlled institutions of higher education; private institutions of higher education; Hispanic-serving institutions; historically Black colleges and universities (HBCUs); tribally controlled colleges and universities (TCCUs); Alaska native and native Hawaiian serving institutions; nonprofits with 501(c)3 IRS status; nonprofits without 501(c)3 IRS status; small businesses; for-profit organizations; State governments; regional organizations; U.S. territories or possessions; Indian/Native American tribal governments; Indian/Native American tribally designated organizations; eligible agencies of the Federal government; and faith-based or community based organizations.
Date(s) Application is Due: Feb 5; Jun 5; Oct 5
Contact: Catherine M. Meyers, (301) 594-7717; fax (301) 480-3510; cm420i@nih.gov
Internet: http://grants.nih.gov/grants/guide/pa-files/PA-07-015.html
Sponsor: National Institute of Diabetes and Digestive and Kidney Diseases
6707 Democracy Boulevard
Bethesda, MD 20892-5460

NIDDK Research Grants on Improving Health Care for Obese Patients 3421
This funding opportunity announcement (FOA) solicits Research Project Grant applications from institutions and organizations that propose to conduct research to determine the barriers to optimal health care for obese patients, and to test innovations or modifications in care delivery to improve health outcomes for obese patients independent of weight loss. Awards issued under this FOA are contingent upon the availability of funds and the submission of a sufficient number of meritorious applications. Applications can be renewed by competing for additional project periods.
Requirements: The following organizations and institutions may apply: public/state controlled institutions of higher education; private institutions of higher education; Hispanic-serving institutions; historically Black colleges and universities (HBCUs); tribally controlled colleges and universities (TCCUs); Alaska native and native Hawaiian serving institutions; nonprofits with 501(c)3 IRS status; nonprofits without 501(c)3 IRS status; small businesses; for-profit organizations; State governments; regional organizations; U.S. territories or possessions; Indian/Native American tribal governments; Indian/Native American tribally designated organizations; eligible agencies of the Federal government; and faith-based or community based organizations.
Date(s) Application is Due: Feb 5; Jun 5; Oct 5
Contact: Susan Z. Yanovski, (301) 594-8882; fax (301) 480-8300; sy29f@nih.gov
Internet: http://grants.nih.gov/grants/guide/pa-files/PA-07-013.html
Sponsor: National Institute of Diabetes and Digestive and Kidney Diseases
6707 Democracy Boulevard
Bethesda, MD 20892-5460

NIDDK Role of Gastrointestinal Surgical Procedures in Amelioration of Obesity-Related Insulin Resistance and Diabetes Independent of Weight Loss 3422
This Funding Opportunity Announcement (FOA) issued by the National Institute of Diabetes and Digestive and Kidney Diseases, National Institutes of Health encourages Research Project Grant (R01) applications from institutions and organizations that propose to explain the underlying mechanism(s) by which various gastrointestinal surgical procedures ameliorate obesity-related insulin resistance and diabetes independent of the resultant weight loss. Awards issued under this FOA are contingent upon the availability of funds and the submission of a sufficient number of meritorious applications.
Requirements: The following organizations and institutions may apply: public/state controlled institutions of higher education; private institutions of higher education; Hispanic-serving institutions; historically Black colleges and universities (HBCUs); tribally controlled colleges and universities (TCCUs); Alaska native and native Hawaiian serving institutions; nonprofits with 501(c)3 IRS status; nonprofits without 501(c)3 IRS status; small businesses; for-profit organizations; State governments; regional organizations; U.S. territories or possessions; Indian/Native American tribal governments; Indian/Native American tribally designated organizations; eligible agencies of the Federal government; and faith-based or community based organizations.
Date(s) Application is Due: Feb 5; Jun 5; Oct 5
Contact: Carolyn W. Miles, (301) 451-3759; fax (301) 480-8300; cm294e@nih.gov
Internet: http://grants.nih.gov/grants/guide/pa-files/PA-08-014.html
Sponsor: National Institute of Diabetes and Digestive and Kidney Diseases
6707 Democracy Boulevard
Bethesda, MD 20892-5460

NIDDK Secondary Analyses in Obesity, Diabetes and Digestive and Kidney Diseases Grants 3423
The specific objectives of this announcement on Secondary Analyses In Obesity, Diabetes, Digestive and Kidney Diseases are to support the following: secondary analyses of data related to the epidemiology of disease areas of NIDDK; important and/or innovative hypotheses explored through analysis of existing data sets; secondary analyses designed to inform and support subsequent applications for individual research awards; rapid analyses of new databases and experimental modules to inform the design and content of future studies; the archiving of datasets to be made publicly available for research purposes related to disease areas of NIDDK, including both epidemiological studies and multi-center clinical trials. Research that employs analytic techniques that demonstrate or promote methodological advances in patient oriented and epidemiologic research is also of interest. International comparative analyses are encouraged. Applications that are innovative and high risk with the likelihood for high impact are especially encouraged.
Requirements: The following organizations and institutions may apply: public/state controlled institutions of higher education; private institutions of higher education; Hispanic-serving institutions; historically Black colleges and universities (HBCUs); tribally controlled colleges and universities (TCCUs); Alaska native and native Hawaiian serving institutions; nonprofits with 501(c)3 IRS status; nonprofits without 501(c)3 IRS status; small businesses; for-profit organizations; State governments; regional organizations; U.S. territories or possessions; Indian/Native American tribal governments; Indian/Native American tribally designated organizations; eligible agencies of the Federal government; and faith-based or community based organizations.
Date(s) Application is Due: Feb 16; Jun 16; Oct 16
Amount of Grant: 200,000 - 275,000 U.S.D
Contact: James E. Everhart, (301) 594-8878; fax (301) 480-8300; je17g@nih.gov
Internet: http://grants.nih.gov/grants/guide/pa-files/PA-09-131.html
Sponsor: National Institute of Diabetes and Digestive and Kidney Diseases
6707 Democracy Boulevard
Bethesda, MD 20892-5460

NIDDK Seeding Collaborative Interdisciplinary Team Science in Diabetes, Endocrinology and Metabolic Diseases Grants 3424
The purpose of the Seeding Collaborative Interdisciplinary Team Science Program described in this announcement is to provide initial support to enable strong new investigative teams to form, and to foster preliminary research activities. It is anticipated that research teams receiving seeding support under this program will be well positioned to compete for funds through the parent NIDDK Collaborative Interdisciplinary Research Program in Diabetes, Endocrinology and Metabolic Diseases solicitation. It is anticipated that 5-10 projects will be funded, but because the nature and scope of the proposed research will vary from application to application, it is anticipated that the size and duration of each award will also vary. Budgets for direct costs of up to $300,000 per year with a project duration of up to one year may be requested.
Requirements: The following organizations and institutions are eligible to apply: public/state controlled institutions of higher education; private institutions of higher education; Hispanic-serving institutions; historically Black colleges and universities (HBCUs); tribally controlled colleges and universities (TCCUs); Alaska Native and Native Hawaiian serving institutions; and non-domestic (non-U.S.) entities (foreign organizations).
Date(s) Application is Due: Jan 25; May 25; Sep 25
Contact: Karen Salomon, (301) 594-7733; fax (301) 435-6047; ks495c@nih.gov
Internet: http://grants.nih.gov/grants/guide/pa-files/PAR-08-181.html
Sponsor: National Institute of Diabetes and Digestive and Kidney Diseases
6707 Democracy Boulevard
Bethesda, MD 20892-5460

NIDDK Silvio O. Conte Digestive Diseases Research Core Centers Grants 3425
The objective of the Silvio O. Conte Digestive Diseases Research Core Centers is to bring together, on a cooperative basis, basic science and clinical investigators to enhance the effectiveness of their research related to digestive and/or liver diseases and their complications. A core center must be an identifiable unit within a single university medical center or a consortium of cooperating institutions, including an affiliated university. An existing program of excellence in biomedical research in the area of digestive and/or liver diseases is a prerequisite for applying. This research must be in the form of NIH research projects, program projects, or other peer-reviewed research that is already funded at the time of submission of a center grant application. Close cooperation, communication, and collaboration among all involved personnel of various professional disciplines are the ultimate objectives.
Requirements: The following organizations and institutions are eligible to apply: public/state controlled institutions of higher education; private institutions of higher education; nonprofits with 501(c)3 IRS status (other than institutions of higher education); nonprofits without 501(c)3 IRS status (other than institutions of higher education); and for-profit organizations (other than small businesses).
Date(s) Application is Due: Oct 12; Nov 9
Amount of Grant: 750,000 U.S.D
Contact: Judith Podskalny, (301) 594-8876; fax (301) 480-8300; jp53s@nih.gov
Internet: http://grants.nih.gov/grants/guide/rfa-files/RFA-DK-09-005.html
Sponsor: National Institute of Diabetes and Digestive and Kidney Diseases
6707 Democracy Boulevard
Bethesda, MD 20892-5460

NIDDK Small Business Innovation Research to Develop New Therapeutics and Monitoring Technologies for Type 1 Diabetes (T1D)—Towards an Artificial Pancreas (SBIR) 3426
A promising therapeutic option for the treatment of diabetes is a system (termed an artificial pancreas or closed-loop) that can mimic normal pancreatic beta cell function thereby restoring normal metabolic homeostasis without causing hypoglycemia. The design of any system capable of achieving this goal is complex and raises scientific, clinical and regulatory challenges. This FOA solicits Small Business Innovation Research (SBIR) grant applications from small business concerns (SBCs) for funding to perform research leading to the development of innovative technologies that may advance progress toward integrated long term glucose regulated insulin delivery system (artificial pancreas). Letters of Intent are due by March 16, with full proposals to be received by April 15.
Requirements: Only U.S. small business concerns (SBCs) are eligible.
Date(s) Application is Due: Mar 16; Apr 15
Amount of Grant: 500,000 - 1,000,000 U.S.D
Contact: Dr. Guillermo Arreaza, (301) 594-4724; fax (301) 480-3503; ga96b@nih.gov
Internet: http://grants.nih.gov/grants/guide/rfa-files/RFA-DK-09-001.html
Sponsor: National Institute of Diabetes and Digestive and Kidney Diseases
6707 Democracy Boulevard
Bethesda, MD 20892-5460

NIDDK Small Grant Program for K08/K23 Recipients 3427

The NIDDK invites recipients of its K08/K23 awards to apply for grant support during the final two years of their awards through the NIDDK Small Grant Program. This additional support will enable K awardees to either expand their current research objectives or to branch out to a closely related pilot study, thus demonstrating their growth as investigators and their independence. This should facilitate the transition to fully independent investigator status. The added grant support in the latter years of an award is expected to have the following benefits: increased fiscal independence for the K08/K23 award recipient as a precursor to complete independence; an opportunity for the recipient to generate additional publications and data to form the basis for an R01 application; an opportunity for a Scientific Review Group (SRG) to evaluate accomplishments made during the first two years of the K08/K23 award; an opportunity for the applicant to provide more detailed research plans for the last two years of the K award; and an opportunity for the applicant to demonstrate additional success in the peer review process during the course of their K08/K23 award. A project period of up to two years and a budget for direct costs of up to two $25,000 modules or $50,000 per year may be requested.

Requirements: This award is available only to those currently holding the K08 award from NIDDK. Details of the K08 award are given in NIH Program Announcement PA-00-003. Applications may be submitted on behalf of candidates by domestic non-federal nonprofit organizations, public or private.

Date(s) Application is Due: Feb 16; Jun 16; Oct 16

Amount of Grant: 50,000 U.S.D

Contact: Dr. Judith Podskalny, (301) 594-8876; fax (301) 480-8300; jp53s@nih.gov

Internet: http://grants.nih.gov/grants/guide/pa-files/PAR-06-172.html

Sponsor: National Institute of Diabetes and Digestive and Kidney Diseases
6707 Democracy Boulevard
Bethesda, MD 20892-5460

NIDDK Small Grants for Underrepresented Minority Scientists in Diabetes and Digestive and Kidney Diseases 3428

The goal of this program is to enable the applicant to accept a tenure-earning position, gain additional research experience and obtain preliminary data on which to base a subsequent research grant application in an area of diabetes, endocrinology, metabolism, digestive diseases, and nutrition, kidney, urology, or hematology. Annual deadline dates may vary; contact program staff for exact dates.

Requirements: Awards will be limited to minority citizens or non-citizen nationals of the United States or to individuals who have been lawfully admitted for permanent residence at the time of application. Applicants must also have a doctoral degree.

Restrictions: Applicants may not hold, nor apply concurrently for any other PHS research project grant at the time of this application.

Date(s) Application is Due: Mar 22

Amount of Grant: Up to 250,000 U.S.D

Contact: Dr. Charles H. Rodgers, (301) 594-7717; rodgersc@extra.niddk.nih.gov

Internet: http://grants.nih.gov/grants/guide/rfa-files/RFA-DK-00-007.html

Sponsor: National Institute of Diabetes and Digestive and Kidney Diseases
45 Center Drive, MSC 6600
Bethesda, MD 20892-6600

NIDDK Traditional Conference Grants 3429

Conference Grants are awarded to institutions and organizations (not individuals) to provide partial support for international or domestic meetings, conferences, and workshops to coordinate, exchange, and disseminate research information. NIDDK is especially interested in conference grants used to cover travel expenses for young and minority investigators, although meeting publications, salaries, consultant services, equipment rental, and supplies may be requested. Indirect costs normally are not allowed.

Requirements: Applicants must submit a letter of intent to apply at least 8 weeks before the application receipt date. This letter should include the names of the organizing committee, a preliminary agenda, a list of accepted or invited speakers, a proposed budget and other planned sources of funding, and a description of how the conference will address each of the above criteria.

Contact: Dr. M. Ken May, Nutrient Metabolism Program Director; (301) 594-8884; fax (301) 480-8300; maym@mail.nih.gov

Internet: http://www2.niddk.nih.gov/Funding/Grants/ApplicantGuidelines/ConferenceSupport.htm

Sponsor: National Institute of Diabetes and Digestive and Kidney Diseases
6707 Democracy Boulevard
Bethesda, MD 20892-5460

NIDDK Training in Clinical Investigation in Kidney and Urology Grants 3430

The Division of Kidney, Urology, and Hematology (DKUH) has long been interested in stimulating more training in clinical research methods in programs related to nephrology and urology. At the same time, there is a dearth of clinicians that have advanced training in clinical trial methodology. The purpose of this program announcement is to encourage applications for new training programs and to encourage program directors of existing Institutional Training Grants supported by DKUH to collaborate with schools of public health and departments of epidemiology, statistics, and biostatistics to offer didactic and practical training in clinical research methodologies. The goal of this program announcement is to increase the number of well-trained investigators who are capable of conceiving, designing, conducting, analyzing, and presenting the results of clinical research in areas pertaining to nephrology or urology.

Requirements: For new training programs, only domestic, nonprofit, private or public institutions may apply. The applicant institution must have the staff and facilities required for the proposed program. The research training program director at the institution will be responsible for the selection and appointment of trainees to receive National Research Service Award (NRSA) support and for the overall direction of the program.

Restrictions: Positions on NRSA institutional grants may not be used for study leading to the MD, DDS or other clinical health professional degrees except when those studies are a part of a formal combined research degree program such as the MD/PhD. Similarly, trainees may not accept NRSA support for studies which are a part of residency training leading to a medical specialty or subspecialty except when the residency program credits a period of full-time, postdoctoral research training toward board certification and the trainee intends to pursue a research career.

Date(s) Application is Due: Mar 10

Amount of Grant: 2,060 U.S.D

Contact: Dr. Charles Rodgers, (301) 594-7717; rodgersc@extra.niddk.nih.gov

Internet: http://grants.nih.gov/grants/guide/pa-files/PAR-99-012.html

Sponsor: National Institute of Diabetes and Digestive and Kidney Diseases
45 Center Drive, MSC 6600
Bethesda, MD 20892-6600

NIDDK Translational Research for the Prevention and Control of Diabetes and Obesity 3431

The National Institute of Diabetes and Digestive and Kidney Diseases (NIDDK), the National Institute of Nursing Research and the Office of Behavior and Social Sciences Research seek to develop cost effective and sustainable interventions that can be adopted in real world settings, for the prevention and control of diabetes and obesity. Research should be based on interventions already proven efficacious in clinical trials to prevent and reverse obesity and type 2 diabetes, to improve care of type 1 and type 2 diabetes and to prevent or delay its complications. The interventions proposed under this solicitation should have the potential to be widely disseminated to clinical practice, individuals and communities at risk. Relevant topics include: methods to improve health care delivery to patients with or at risk of diabetes; strategies to enhance diabetes self management; methods to develop strategies to promote healthy lifestyles to reduce the risk of diabetes; methods to promote lifestyle change that will prevent or reverse overweight and obesity; and new cost effective ways to identify people with pre-diabetes and undiagnosed diabetes.

Requirements: The following organizations and institutions may apply: public/state controlled institutions of higher education; private institutions of higher education; Hispanic-serving institutions; historically Black colleges and universities (HBCUs); tribally controlled colleges and universities (TCCUs); Alaska native and native Hawaiian serving institutions; nonprofits with 501(c)3 IRS status; nonprofits without 501(c)3 IRS status; small businesses; for-profit organizations; State governments; regional organizations; U.S. territories or possessions; Indian/Native American tribal governments; Indian/Native American tribally designated organizations; eligible agencies of the Federal government; and faith-based or community based organizations.

Date(s) Application is Due: Jan 25; May 25; Sep 25

Contact: Sanford Garfield, (301) 594-8803; fax (301) 480-3503; sg50o@nih.gov

Internet: http://grants.nih.gov/grants/guide/pa-files/PAR-06-532.html

Sponsor: National Institute of Diabetes and Digestive and Kidney Diseases
6707 Democracy Boulevard
Bethesda, MD 20892-5460

NIDDK Transmission of Human Immunodeficiency Virus (HIV) In Semen 3432

The National Institute of Diabetes and Digestive and Kidney Diseases (NIDDK) through its Division of Kidney, Urologic and Hematologic Diseases (DKUHD) and the National Institute of Child Health and Human Development, Center for Population Research invite investigators to submit research grant applications that will increase the basic and clinical knowledge of the biology of HIV in semen. Direct exposure to semen of HIV seropositive men is a major route for transmission of HIV type I. This Funding Opportunity Announcement (FOA) focuses on studies that will elucidate the factors that determine HIV shedding in the male genital tract. This includes studies that elucidate the infectivity of HIV in semen fractions, the effect of antiretroviral therapy on HIV infectivity in semen fractions, the relationship between the immunobiology of the male genital tract and HIV replication and infectivity, and factors such as genital tract inflammation which influence HIV transmission through semen. Of particular interest are studies of the anatomic sites of virus infected cells and virus production in the male genital tract. The intent of this Funding opportunity is to encourage investigator-initiated research, to attract new investigators to the field, and to increase interdisciplinary research to enhance the scope and effectiveness of research in this area. Because the nature and scope of the proposed research will vary from application to application, it is anticipated that the size and duration of each award will also vary. The total amount awarded and the number of awards will depend upon the mechanism numbers, quality, duration, and costs of the applications received.

Requirements: The following organizations and institutions are eligible to apply: public/state controlled institutions of higher education; private institutions of higher education; Hispanic-serving institutions; historically Black colleges and universities (HBCUs); tribally controlled colleges and universities (TCCUs); Alaska Native and Native Hawaiian serving institutions; nonprofits with 501(c)3 IRS status (other than institutions of higher education); nonprofits without 501(c)3 IRS status (other than institutions of higher education); small businesses; for-profit organizations (other than small businesses); state governments; Indian/Native American tribal governments (federally recognized); Indian/Native American tribally designated organizations; county governments; city or township governments; special district governments; independent school districts; public housing authorities; Indian housing authorities; U.S. territory or possession; Indian/Native American tribal governments (other than federally recognized); regional organizations; non-domestic (non-U.S.) entities (foreign organizations); other eligible agencies of the federal government; and faith-based or community-based organizations.

Date(s) Application is Due: Feb 5; Jun 5; Oct 5
Contact: Leroy M. Nyberg, Jr., (301) 594-7717; ln10f@nih.gov
Internet: http://grants.nih.gov/grants/guide/pa-files/PA-07-044.html
Sponsor: National Institute of Diabetes and Digestive and Kidney Diseases
6707 Democracy Boulevard
Bethesda, MD 20892-5460

NIDDK Type 2 Diabetes in the Pediatric Population Research Grants 3433
NIDDK and NICHD invite investigator-initiated research grant applications to study the epidemiology, natural history, pathophysiology, prevention, and treatment of type 2 diabetes in children in the United States. Type 2 diabetes has traditionally been viewed as a disease of adults. More recently, however, it has become apparent that an increasing number of cases of type 2 diabetes are being reported in children. New hypothesis-driven studies are needed to describe the epidemiology, refine the diagnosis, define the metabolic abnormalities, and formulate treatment options for type 2 diabetes in children. Annual deadline dates may vary; contact program staff for exact dates.
Requirements: Applications may be submitted by domestic for-profit and nonprofit organizations, both public and private; units of state and local governments; and eligible agencies of the federal government.
Contact: Dr. Gilman Grave, (301) 496-5593; fax (301) 480-9791
Internet: http://www2.niddk.nih.gov/Research/ScientificAreas/Diabetes/Type2Diabetes/DM2K.htm
Sponsor: National Institute of Diabetes and Digestive and Kidney Diseases
6707 Democracy Boulevard
Bethesda, MD 20892-5460

NIDRR Field-Initiated Projects 3434
The goals of this project include: developing methods and technologies that maximize the integration of individuals with disabilities into society; employment and family environments; and improving the effectiveness of services authorized under the Rehabilitation Act. Applicants who receive awards under this program may carry out either research or development activities, and applications for each type of project will be reviewed separately. The applicant is responsible for indicating which type of project is being proposed. Projects may last up to three years. Annual deadline dates may vary; contact program staff for exact dates.
Requirements: Proposals will be accepted from any public or private agency or organization.
Contact: Donna Nangle, (202) 245-7462; fax (202) 245-7323; Donna.Nangle@ed.gov
Internet: http://www.ed.gov/fund/grant/apply/nidrr/index.html
Sponsor: National Institute on Disability and Rehabilitation Research
550 12th Street SW, 6th Floor, Switzer Building
Washington, DC 20202-6510

NIEHS/NINDS Role of the Environment in Parkinson's Disease 3435
The objective of this program is to stimulate research on the relative roles of environmental, endogenous neurochemical and modifying genetic factors in the cause of Parkinson's disease. Annual deadline dates may vary; contact program staff for exact dates.
Requirements: Applications may be submitted by domestic and foreign, for-profit and nonprofit organizations, public and private, such as universities, colleges, hospitals, laboratories, units of state and local governments, and eligible agencies of the federal government.
Amount of Grant: Up to 4,000,000 U.S.D
Contact: Dr. Eugene Oliver, Stroke, Trauma and Neurodegenerative Disorders Coordinator; (301) 496-5680 or (301) 480-1080
Internet: http://grants.nih.gov/grants/guide/rfa-files/RFA-ES-00-002.html
Sponsor: National Institute of Environmental Health Sciences
Keystone Building, Room 3039, P.O. Box 12233, MD K3-14
Research Triangle Park, NC 27709

NIEHS Applied Toxicological Research and Testing Grants 3436
A small number of grants are supported to further the primary aims of the program to determine the toxicity and health hazards of chemicals by the best means available and to develop and validate better test methods. Toward the latter objective, the conventional two-year bioassays previously used to detect and assess carcinogenesis alone have been modified to assess other toxic responses in the same animals. Chemicals of interest are selected on the basis of recommendations from public and private sources including state and local government agencies, colleges and universities, trade unions, and industry. Testing priorities are established by an executive committee made up of personnel from key federal regulatory and research agencies. Testing of selected chemicals has been broadened and improved with new quality assessment standards, expanded screening for effects in various organs, on fertility and fetal development, and on distribution, metabolism, and excretion of the chemicals. Program emphases include development of reliable test methods, improvement of existing bioassay systems, and development of model test systems for xenobiotics.
Requirements: A university, college, hospital, state or local government, nonprofit research institution, or for-profit organization may submit an application and receive a grant for support or research by a named principal investigator.
Date(s) Application is Due: Feb 1; Jun 1; Oct 1
Amount of Grant: 141,500 - 16,000,000 U.S.D
Contact: Dr. William Suk; (919) 541-0797; suk@niehs.nih.gov
Internet: http://12.46.245.173/pls/portal30/CATALOG.PROGRAM_TEXT_RPT.show
Sponsor: National Institute of Environmental Health Sciences
Keystone Building, Room 3039, P.O. Box 12233, MD K3-14
Research Triangle Park, NC 27709

NIEHS Biological Response to Environmental Health Hazards Research Grants 3437
The aim of this program is to develop a better understanding of the mechanisms of injury and disease caused by environmental factors. Research focuses on perturbations by chemical and physical agents at the molecular, cellular, tissue, and organ levels. This is a broad area of inquiry that integrates the traditional disciplines of anatomy, physiology, cell biology, endocrinology, genetics, biochemistry, and reproduction. Problem areas include, but are not limited to, neurological and behavioral disorders, respiratory diseases, cardiovascular disease, biochemical and endocrine disturbances, kidney dysfunction, carcinogenesis, mutagenesis, reproductive disorders, and birth defects.
Requirements: A university, college, hospital, state or local government, nonprofit research institution, or for-profit organization may submit an application and receive a grant for support of research by a named principal investigator.
Date(s) Application is Due: Feb 1; Jun 1; Oct 1
Amount of Grant: 38,500 - 100,000,000 U.S.D
Contact: Dr. William Suk; (919) 541-0797; suk@niehs.nih.gov
Internet: http://12.46.245.173/pls/portal30/CATALOG.PROGRAM_TEXT_RPT.show
Sponsor: National Institute of Environmental Health Sciences
Keystone Building, Room 3039, P.O. Box 12233, MD K3-14
Research Triangle Park, NC 27709

NIEHS Carcinogenicity of Drinking Water Disinfection By-Products 3438
NIEHS, in collaboration with EPA, is conducting short-term and long-term toxicity studies of the disinfection by-products found in drinking water, on traditional rodent models with additional studies in genetically modified (transgenic) mice. Emphasis should be on whether lower concentrations, which may not cause cancer in rodents, can be either expected to pose a human health hazard or may be considered of minimal risk for the human population, including sensitive subpopulations. Annual deadline dates may vary; contact program staff for exact dates. RFA: ES-99-007
Requirements: Applications may be submitted by domestic, for-profit and nonprofit organizations, public and private, such as universities, colleges, hospitals, laboratories, units of state or local governments and eligible agencies of the federal government. Applications from minority individuals and women are encouraged.
Restrictions: Foreign institutions and organizations are not eligible.
Amount of Grant: Up to 450,000 U.S.D
Contact: Dr. Jerrold Heindel, (919) 541-0781; fax (919) 541-5064; heindelj@niehs.nih.gov
Internet: http://grants.nih.gov/grants/guide/rfa-files/RFA-ES-99-007.html
Sponsor: National Institute of Environmental Health Sciences
Keystone Building, Room 3039, P.O. Box 12233, MD K3-14
Research Triangle Park, NC 27709

NIEHS Community-Based Prevention Intervention Research in Environmental Health Sciences Grants 3439
NIEHS, NINR, and the Office of Behavioral and Social Sciences invite research grant applications addressing development of community-based strategies aimed at prevention and intervention activities in economically disadvantaged and/or underserved populations adversely impacted by an environmental contaminant. These funds support research activities that develop and implement improved prevention and intervention strategies related to underserved populations. In community-based research, active cooperation and participation of organizations within the community that is the focus of the study are essential components of the research. Hence, applicants must describe an existing or proposed involvement with one or more community-based organizations. These grants provide support to develop, test, and evaluate health service activities and to enhance the application or transfer of existing knowledge to efforts designed to prevent or control dysfunction or disease. Annual deadline dates may vary; contact program staff for exact dates. More information may be obtained by mail and email, or through the NIH grant line (301-402-2221), and the NIH Web site. RFA: ES-99-012
Requirements: Applications may be submitted by domestic and foreign, for-profit and nonprofit organizations, both public and private; units of state and local governments; and eligible agencies of the federal government.
Amount of Grant: Up to 300,000 U.S.D
Contact: Frederick Tyson, Chemical Exposures and Molecular Biology Branch; (919) 541-0176; fax (919) 316-4606; tyson2@niehs.nih.gov
Internet: http://grants.nih.gov/grants/guide/rfa-files/RFA-ES-99-012.html
Sponsor: National Institute of Environmental Health Sciences
Keystone Building, Room 3039, P.O. Box 12233, MD K3-14
Research Triangle Park, NC 27709

NIEHS Environmental Justice: Partnerships for Communication 3440
The purpose of this program is to strengthen the NIEHS effort that supports research aimed at achieving environmental equity for socioeconomically disadvantaged and medically underserved populations in the United States. One goal of the institute is to stimulate investigative efforts that attempt to address questions related to the influence of economic and social factors on the health status of individuals exposed to environmental toxicants. The main objective of this announcement is to establish a new model for linking members of a community who are directly affected by adverse environmental conditions with researchers and health care providers. The mechanism of support for this program will be the NIH Education Grant. The total project period for applications submitted in response to the announcement may not exceed four years, and projects are not renewable. Annual deadline dates may vary; contact the program office for exact dates. RFA: ES-99-005
Requirements: Applications may be submitted by domestic for-profit, nonprofit, public, and private institutions.
Amount of Grant: Up to 200,000 U.S.D

Contact: Dr. Frederick Tyson, Chemical Exposures and Molecular Biology Branch Contact; (919) 541-0176; fax (919) 541-4937; tyson2@niehs.nih.gov
Internet: http://www.grants.nih.gov/grants/guide/rfa-files/RFA-ES-99-005.html
Sponsor: National Institute of Environmental Health Sciences
Keystone Building, Room 3039, P.O. Box 12233, MD K3-14
Research Triangle Park, NC 27709

NIEHS Hazardous Materials Worker Health and Safety Training Grants 3441
This FOA issued by the National Institute of Environmental Health Sciences (NIEHS) invites applications for cooperative agreements to support the development of model programs for the training and education of workers engaged in activities related to hazardous materials and waste generation, removal, containment, transportation and emergency response. The major objective of this solicitation is to prevent work-related harm by assisting in the training of workers in how best to protect themselves and their communities from exposure to hazardous materials encountered during hazardous waste operations, hazardous materials transportation, environmental restoration of contaminated facilities or chemical emergency response. A variety of sites, such as those involved with chemical waste clean up and remedial action and transportation-related chemical emergency response may pose severe health and safety concerns to workers and the surrounding communities. These sites contain a multiplicity of hazardous substances, sometimes unknown substances, and often the site is uncontrolled. A major goal of the Worker Education and Training Program (WETP) is to provide assistance to organizations in developing their institutional competency to provide appropriate model training and education programs. The total project period for an application submitted in response to this funding opportunity must be 5 years. A new applicant may request a budget for direct costs of up to $700,000 for the first year. Letters of Intent (LOIs) are due by October 23, with full proposals due by November 23.
Requirements: The following organizations and institutions are eligible to apply: public/state controlled institutions of higher education; private institutions of higher education; Hispanic-serving institutions; historically Black colleges and universities (HBCUs); tribally controlled colleges and universities (TCCUs); Alaska Native and Native Hawaiian serving institutions; and nonprofits with 501(c)3 IRS status (other than institutions of higher education).
Date(s) Application is Due: Oct 23; Nov 23
Amount of Grant: Up to 700,000 U.S.D
Contact: Joseph Hughes; (919) 541-0217; fax (919) 541-0462; hughes3@niehs.nih.gov
Internet: http://grants.nih.gov/grants/guide/rfa-files/RFA-ES-09-004.html
Sponsor: National Institute of Environmental Health Sciences
Keystone Building, Room 3039, P.O. Box 12233, MD K3-14
Research Triangle Park, NC 27709

NIEHS Hazmat Training at Doe Nuclear Weapons Complex Grants 3442
The National Institute of Environmental Health Sciences (NIEHS) invites applications for cooperative agreements to support the development of model programs for the training and education of workers engaged in activities related to hazardous materials and waste generation, removal, containment, transportation and emergency response within the Department of Energy (DOE) Nuclear Weapons Complex. The major objective of this solicitation is to prevent work related harm by assisting in the training and education of workers in the DOE nuclear weapons complex. Safety and health training will transmit skills and knowledge to workers in how best to protect themselves and their communities from exposure to hazardous materials encountered during hazardous waste operations, facility decommissioning and decontamination, hazardous materials transportation, environmental restoration of contaminated facilities or chemical emergency response. The total project period for an application submitted in response to this funding opportunity must be 5 years. A new applicant may request a budget for direct costs of up to $700,000 for the first year. Letters of Intent (LOIs) are due by October 23, with full proposals due by November 23.
Requirements: The following organizations and institutions are eligible to apply: public/state controlled institutions of higher education; private institutions of higher education; Hispanic-serving institutions; historically Black colleges and universities (HBCUs); tribally controlled colleges and universities (TCCUs); Alaska Native and Native Hawaiian serving institutions; and nonprofits with 501(c)3 IRS status (other than institutions of higher education).
Date(s) Application is Due: Oct 23; Nov 23
Amount of Grant: Up to 700,000 U.S.D
Contact: Joseph Hughes; (919) 541-0217; fax (919) 541-0462; hughes3@niehs.nih.gov
Internet: http://grants.nih.gov/grants/guide/rfa-files/RFA-ES-09-003.html
Sponsor: National Institute of Environmental Health Sciences
Keystone Building, Room 3039, P.O. Box 12233, MD K3-14
Research Triangle Park, NC 27709

NIEHS Independent Scientist Award 3443
The award is a special salary-only grant designed to provide protected time for newly independent scientists who currently have non-research obligations such as heavy teaching loads, clinical work, committee assignments, service, and administrative duties that prevent them from having a period of intensive research focus. The award is targeted to persons with doctoral degrees who have completed their research training, have independent peer-reviewed research support, and who need a period of protected research time in order to foster their research career development. Strong preference is given to candidates with active research project grant support from the NIEHS. New applications are due February 12, June 12, and October 12, while renewal applications are due March 12, July 12, and November 12. Up to $75,000 per year of base salary plus fringe benefits is available. Research support of up to $25,000 is provided in some cases. Individuals engaged in predominantly theoretical work, such as modeling or computer simulation, may request up to $25,000 per year for research expenses.
Requirements: The following organizations and institutions are eligible to apply: public/state controlled institutions of higher education; private institutions of higher education; nonprofits with 501(c)3 IRS status; nonprofits without 501(c)3 IRS status; small businesses; for-profit organizations; State governments; U.S. territories or possessions; Indian/Native American tribal governments (Federally recognized and other than Federally recognized); Indian/Native American tribally designated organizations; Hispanic-serving institutions; historically Black colleges and universities (HBCUs); tribally controlled colleges and universities (TCCUs); Alaska Native and Native Hawaiian serving institutions; regional organizations; and faith-based or community based organizations. The candidate must have a doctoral degree and peer-reviewed, independent research support at the time the award is made. The candidate must spend a minimum of 75 percent effort conducting research during the period of the award.
Restrictions: The award is not intended for investigators who already have full time to perform research, or have substantial publication records or considerable research support indicating that they are well established in their fields. Foreign institutions are not eligible to apply.
Contact: Carol Shreffler, (919) 541-1445; fax (919) 541-5064; shreffl1@niehs.nih.gov
Internet: http://grants.nih.gov/grants/guide/pa-files/PA-09-038.html
Sponsor: National Institute of Environmental Health Sciences
Keystone Building, Room 3039, P.O. Box 12233, MD K3-14
Research Triangle Park, NC 27709

NIEHS Mentored Clinical Scientist Research Career Development Awards 3444
This award enables candidates holding professional degrees (e.g., M.D., D.O., D.V.M., or equivalent degrees) to undertake 3 to 5 years of special study and supervised research with the goal of becoming independent investigators. The award also allows awardees to pursue a research career development program suited to their experience and capabilities under a mentor who is competent to provide guidance in the chosen research area. Institutions may submit applications on behalf of candidates who hold professional degrees. At least 2 years must have elapsed since the health professional degree was granted. Candidates can have varying levels of clinical training and research experience. New applications are due February 12, June 12, and October 12, while resubmitted applications are due March 12, July 12, and November 12. Up to $75,000 per year plus fringe benefits is available, and up to $25,000 per year for research support for new awards.
Requirements: All candidates must be U.S. citizens, non-citizen nationals, or legal permanent residents of the U.S. Persons with temporary or student visas are not eligible. The grantee institution must have a strong, well-established research and research training program in the chosen area, accomplished faculty in the basic and clinical sciences, and a commitment to the candidate's research development. The proposed program should include an appropriate mentor.
Date(s) Application is Due: Feb 12; Mar 12; Jun 12; Jul 12; Oct 12; Nov 12
Amount of Grant: Up to 100,000 U.S.D
Contact: Dr. Carol Shreffler, (919) 541-1445; fax (919) 541-5064; shreffl1@niehs.nih.gov
Internet: http://grants.nih.gov/grants/guide/contacts/pa-06-512_contacts.htm
Sponsor: National Institute of Environmental Health Sciences
Keystone Building, Room 3039, P.O. Box 12233, MD K3-14
Research Triangle Park, NC 27709

NIEHS Mentored Research Scientist Development Award 3445
The MRSDA is for a research scientist who needs an additional period of sponsored research experience as a way to gain expertise in a research area new to the candidate or in an area that would demonstrably enhance the candidate's scientific career. Research projects proposed in response to this Announcement will be expected to have a defined impact on the environmental health sciences and be directly responsive to the mission of the NIEHS, which is distinguished from that of other Institutes by its focus on research programs seeking to link the effects of environmental exposures to the cause, mechanisms, moderation, or prevention of a human disease or disorder or relevant pathophysiologic process. Applications must include as part of the justification the specific human disease, dysfunction, pathophysiologic condition, or relevant human biologic process altered, and the specific environmentally relevant toxicant. Examples of environmentally relevant toxicants include industrial chemicals or manufacturing byproducts, metals, pesticides, herbicides, air pollutants and other inhaled toxicants, particulates or fibers, fungal, bacterial, or biologically derived toxins. Agents considered non-responsive to this announcement include, but are not limited to: alcohol, chemotherapeutic agents, radiation which is not the result of an ambient environmental exposure, drugs of abuse, and infectious or parasitic agents. Ecologic, biomonitoring, biotransformation or biodegradation studies are also not responsive. Applicants must demonstrate a significant career shift or career enhancement activity which will achieve research independence upon completion of the program. Preference will be given to candidates whose mentor has NEIHS research grant support.
Requirements: The candidate must have a research or health-professional doctorate or its equivalent and must have demonstrated the capacity or potential for highly productive independent research in the period after the doctorate. The candidate must identify a mentor with extensive research experience and must spend a minimum of 75 percent of full-time professional effort conducting research for the period of the award. Applications may be submitted on behalf of candidates by domestic, nonfederal organizations, public or private, such as medical, dental, or nursing schools or other institutions of higher education. Candidates must be U.S. citizens, noncitizen nationals, or permanent residents.
Date(s) Application is Due: Feb 12; Mar 12; Jun 12; Jul 12; Oct 12; Nov 12
Amount of Grant: 75,000 U.S.D
Contact: Carol Shreffler, (919) 541-1445; fax (919) 541-5064; shreffl@niehs.nih.gov
Internet: http://grants.nih.gov/grants/guide/pa-files/PA-09-040.html
Sponsor: National Institute of Environmental Health Sciences
Keystone Building, Room 3039, P.O. Box 12233, MD K3-14
Research Triangle Park, NC 27709

NIEHS SBIR E-Learning for HAZMAT and Emergency Response Grants 3446
This funding opportunity announcement (FOA) solicits Small Business Innovation Research (SBIR) grant applications from small business concerns (SBCs) that propose to further the development of Advanced Technology Training (ATT) Products for the health and safety training of hazardous materials (HAZMAT) workers, emergency responders, and skilled support personnel. These products would complement the goals and objectives of the Worker Education and Training Program (WETP). The major objective of the NIEHS/WETP is to prevent work related harm by assisting in the training of workers in how best to protect themselves and their communities from exposure to hazardous materials. There is a need to ensure that learning and training technologies are further developed, field tested and applied to real world situations. It is the intent of this solicitation to support the development of products to support e-collaboration, e-teaching, e-certification, and e-learning in safety and health training for workers engaged in hazardous materials response. A new applicant may request a budget for direct costs of up to $300,000 for the first year. Letters of Intent (LOIs) are not required, and full proposals are due by August 3.
Requirements: Only United States small business concerns (SBCs) are eligible to submit SBIR applications. A small business concern is one that, at the time of award of SBIR Phase I and Phase II, meets all of the following criteria: is organized for profit, with a place of business located in the United States, which operates primarily within the United States or which makes a significant contribution to the United States economy through payment of taxes or use of American products, materials or labor; is in the legal form of an individual proprietorship, partnership, limited liability company, corporation, joint venture, association, trust or cooperative, except that where the form is a joint venture, there can be no more than 49 percent participation by foreign business entities in the joint venture; is at least 51 percent owned and controlled by one or more individuals who are citizens of, or permanent resident aliens in, the United States, or it must be a for-profit business concern that is at least 51% owned and controlled by another for-profit business concern that is at least 51% owned and controlled by one or more individuals who are citizens of, or permanent resident aliens in, the United States, except in the case of a joint venture, where each entity to the venture must be 51 percent owned and controlled by one or more individuals who are citizens of, or permanent resident aliens in, the United States; and; has, including its affiliates, not more than 500 employees.
Date(s) Application is Due: Aug 3
Amount of Grant: Up to 300,000 U.S.D
Contact: Ted Outwater, (919) 541-2972; fax (919) 541-0462; outwater@niehs.nih.gov
Internet: http://grants.nih.gov/grants/guide/rfa-files/RFA-ES-09-006.html
Sponsor: National Institute of Environmental Health Sciences
Keystone Building, Room 3039, P.O. Box 12233, MD K3-14
Research Triangle Park, NC 27709

NIEHS Small Business Innovation Research (SBIR) Program Grants 3447
The focus of these grants is on understanding how chemical and physical agents cause pathological changes in molecules, cells, tissues, and organs and become manifested as respiratory disease; neurological, behavioral, and developmental abnormalities; cancer; and other disorders. By understanding the relationship between environmental exposures and the subsequent development of disease or biological injury, human health may be protected. SBIR program objectives are to stimulate technological innovation, use small businesses to meet federal research and development needs, increase private sector commercialization of innovations derived from federal research and development, and foster and encourage participation by minority and disadvantaged persons in technological innovation. Phase I awards, generally for six months, establish the technical merit and feasibility of proposed research efforts that may lead to commercial products or processes. Phase II awards are for the continuation of research initiated in Phase I.
Requirements: SBIR grants can be awarded only to domestic small businesses (entities that are independently owned and operated for profit, are not dominant in the field in which research is proposed, and have no more than 500 employees). Primary employment (more than one-half) of the principal investigator must be with the small business at the time of award and during the conducting of the proposed research. In both Phase I and II, the research must be performed in the United States.
Date(s) Application is Due: Apr 15; Aug 15; Dec 15
Amount of Grant: 100,000 - 750,000 U.S.D
Contact: Jo Anne Goodnight, SBIR/STTR; (301) 435-2688; fax (301) 480-0146; jg128w@nih.gov or sbir@od.nih.gov
Internet: http://grants.nih.gov/grants/funding/sbir.htm#sbir
Sponsor: National Institute of Environmental Health Sciences
Keystone Building, Room 3039, P.O. Box 12233, MD K3-14
Research Triangle Park, NC 27709

NIEHS Structure-Function Relationships of Environmentally Relevant Genetic Variants Research Grants 3448
The objective of this RFA is to encourage multidisciplinary projects to identify and characterize the gene products coded by allelic variants of environmental responsive genes. NIEHS is the principal funding agency that supports this RFA and NIA is the cosponsor. Data generated from the studies will allow for rapid developments of innovative approaches to elucidate the mechanisms of the xenobiotic-induced molecular alterations in specific target cells and organs. Collaborative research efforts between investigators in environmental health sciences and in molecular and structural biology, and/or molecular modeling disciplines are especially encouraged. The major research objectives for this RFA include: biochemical and structural biological studies of the proteins coded by the allelic variants of candidate genes; evaluation of the functional significance of single nucleotide polymorphisms or other allelic variations in candidate genes and their regulatory regions; characterization of the molecular mechanisms controlling temporal and cellular regulation of expression of the allelic variants of a specific class of genes; functional genomic studies to assess the expression patterns and functions of these classes of environmental responsive genes at the organ/tissue/cell levels; and studies on protein-protein interactions. RFA: ES-98-007
Requirements: Applications may be submitted by domestic and foreign, for-profit and nonprofit organizations, both public and private; units of state or local governments; and eligible agencies of the federal government. Applications from minority individuals and women are encouraged.
Amount of Grant: 35,000,000 U.S.D
Contact: Dr. Jose Velazquez, Division of Extramural Research and Training; (919) 541-4998; fax (919) 541-4937; velazqu1@niehs.nih.gov
Internet: http://www.grants.nih.gov/grants/guide/rfa-files/RFA-ES-98-007.html
Sponsor: National Institute of Environmental Health Sciences
Keystone Building, Room 3039, P.O. Box 12233, MD K3-14
Research Triangle Park, NC 27709

NIEHS Superfund Hazardous Substances Basic Research and Education Grants 3449
Grants support a wide range of research to address the broad public health concerns arising from the release of hazardous chemical wastes and other hazardous substances into the environment from waste-disposal sites. The primary objective is to expand the base of scientific knowledge needed for adequate assessment of exposure and health risks from environmental contamination, reduction in the amount and toxicity of hazardous substances, and ultimately to prevent adverse human health effects. Grants fund university-based research programs to develop advanced techniques for the detection, assessment, and evaluation of the effects of hazardous substances on human health; methods to assess the risks to human health presented by hazardous substances; methods and technologies to detect hazardous substances in the environment; and basic biological, chemical, and physical methods to reduce the amount and toxicity of hazardous substances. Advanced or graduate training should be integrated into the second year of the multidisciplinary research program to provide for training in environmental and occupational health and safety and the engineering aspects of hazardous wastes control, and graduate training in the geosciences including hydrogeology, geological engineering, geophysics, geochemistry, and related fields. Contact the program office for deadline dates. RFA: ES-99-001
Requirements: An accredited institution of higher education may submit an application and receive a grant for support of research by a named principal investigator. Subcontracts may be made with public and private organizations including generators of hazardous wastes; persons involved in the detection, assessment, evaluation, and treatment of hazardous substances; owners and operators of facilities at which hazardous substances are located; and state and local governments. Strong biomedical research is a requisite of this program. A minimum of three approved biomedical projects is required for funding.
Restrictions: Single-project applications will not be accepted. This program is not intended to support individual research project grants.
Amount of Grant: 182,000 - 3,800,000 U.S.D
Contact: Dr. William Suk, Chief, Office of Program Development; (919) 541-0797; fax (919) 541-2843; suk@niehs.nih.gov
Internet: http://www.grants.nih.gov/grants/guide/rfa-files/RFA-ES-99-001.html
Sponsor: National Institute of Environmental Health Sciences
Keystone Building, Room 3039, P.O. Box 12233, MD K3-14
Research Triangle Park, NC 27709

NIGMS/NIDCR Evolution of Infectious Diseases Research Grants 3450
The goal of this joint NIGMS/NIDCR request for applications is to encourage development of a predictive science of infectious diseases by applying the perspectives, theories, and methods of population and evolutionary biology to important issues of disease emergence, prevention, and treatment. To achieve this goal, this RFA seeks collaborations between two major groups of scientists: those with expertise in population and evolutionary biology and molecular phylogenetics, including mathematical modeling and complexity theory; and infectious disease experts such as clinicians, epidemiologists, immunologists, microbiologists, veterinarians, or plant pathologists. PA: PA-99-147
Requirements: Applications may be submitted by domestic and foreign, for-profit and nonprofit organizations, units of state and local governments, and eligible agencies of the federal government.
Restrictions: Foreign institutions are not eligible for program project grants.
Amount of Grant: Up to 250,000 U.S.D
Contact: Dr. Irene Anne Eckstrand, Division of Genetics and Developmental Biology; (301) 594-0943; fax (301) 480-2228; Irene_Eckstrand@nih.gov
Internet: http://grants2.nih.gov/grants/guide/pa-files/PA-99-147.html
Sponsor: National Institute of General Medical Sciences
45 Center Drive, Natcher Building
Bethesda, MD 20892

NIGMS Exploratory Studies for High Risk/High Impact Research Grants 3451
This program attempts to broaden the base of inquiry in fundamental biomedical research by encouraging applications for research projects that involve an especially high degree of innovation and novelty, such that their potential for highly significant outcomes may be difficult to judge by the standard criteria used in evaluating R01 proposals. Research projects proposed under this program may lack preliminary data establishing feasibility, but should present the opportunity for conceptual or technological breakthroughs. PA-03-100 may be obtained electronically through the NIH grant line (301-402-2221), the NIH Web site, and by mail and email.
Amount of Grant: Up to 275,000 U.S.D
Contact: Joe Ellis, Grants Management Officer; (301) 594-5135; fax (301) 480-1969; ellisj@nigms.nih.gov

Internet: http://grants2.nih.gov/grants/guide/pa-files/PA-03-100.html
Sponsor: National Institute of General Medical Sciences
45 Center Drive, Natcher Building
Bethesda, MD 20892

NIGMS Fellowships in Quantitative Biology **3452**
The NIGMS encourages applications for individual postdoctoral and senior fellowships from highly qualified individuals with doctoral training in both the traditional quantitative disciplines (such as mathematics, physics, engineering, and computer science) and biology. NIGMS supports individual postdoctoral and senior fellowships in those areas covered by its research grant portfolio. These areas include biomedical engineering, cell biology, molecular biophysics, genetics and developmental biology, pharmacology, physiology, especially burn and trauma research, and biological chemistry. PA 98-082
Requirements: Individuals must be citizens or noncitizen nationals of the United States, or have been lawfully admitted to the United States for permanent residence (i.e., possess a currently valid Alien Registration Receipt Card I-551, or other legal verification of such status) at the time of application. Applicants for individual postdoctoral fellowships must have completed all of the requirements for the doctoral degree by the time of award. Applicants for senior fellowships must have had at least seven years of relevant research or professional experience beyond the doctoral degree. In addition, a significant amount of retraining is required for senior fellowship applicants. Applications from women, underrepresented minorities, and individuals with disabilities are especially encouraged.
Restrictions: Individuals on temporary or student visas are not eligible.
Date(s) Application is Due: Apr 5; Aug 5; Dec 5
Amount of Grant: 21,000 - 33,000 U.S.D
Contact: Dr. James Cassatt, Division of Cell Biology and Biophysics; (301) 594-0828; fax (301) 402-2004; cassattj@nigms.nih.gov
Internet: http://www.grants.nih.gov/grants/guide/pa-files/PA-98-082.html
Sponsor: National Institute of General Medical Sciences
45 Center Drive, Natcher Building
Bethesda, MD 20892

NIGMS Individual Postdoctoral NRSA Fellowships in Clinical Pharmacology **3453**
Fellowships support high-quality physicians seeking research training in the field of clinical pharmacology. These fellowships are intended to support up to three years of advanced training to individuals with the MD degree to qualify them to perform basic laboratory and/or clinical investigations of the effects and mechanisms of drug actions in humans, in the scientific areas consistent with the mission of NIGMS. One goal of this initiative is to promote high-quality clinical pharmacology research training at academic health centers. Therefore, a special feature of this award is to encourage applicants at institutions with a General Clinical Research Center (GCRC) supported by the NCRR to submit a letter of endorsement from the director of the training institution's GCRC. The intent is to promote interaction between the applicant and the staff and director of the GCRC, expose the applicant to the extensive resources and rich research environment of the GCRC, and raise the visibility of this important area of translational research within the institution. PA: PA-99-078
Requirements: The candidate must have an MD degree or its equivalent and must meet the criteria for the NIH National Research Service Awards for Individual Postdoctoral Fellows (F32) described in program announcement PA-99-025, available on the Web site at http://www.nih.gov/grants/guide/pa-files/PA-99-025.htm l.
Contact: Dr. Alison Cole, Program Contact; (301) 594-1826; fax (301) 480-2802; colea@nigms.nih.gov
Internet: http://www.grants.nih.gov/grants/guide/pa-files/PA-99-078.html
Sponsor: National Institute of General Medical Sciences
45 Center Drive, Natcher Building
Bethesda, MD 20892

NIGMS Initiative for Minority Students: Bridges to the Baccalaureate Degree **3454**
NIGMS and the Office of Research on Minority Health (ORMH) reannounce two research initiatives directed at increasing the number of underrepresented minorities entering careers in biomedical research. The programs target two different underrepresented minority student populations: those in colleges and universities offering only master of science (MS) degree programs in biomedically-related sciences and those in two-year junior or community colleges. These have been identified as two key transition points for students considering careers in biomedical research. This RFA solicits new applications for a partnership program involving two-year colleges awarding the associate's degree and institutions awarding the baccalaureate degree. Annual deadline dates may change; contact program staff for exact dates. RFA: GM-01-002
Requirements: Applications may be submitted by domestic, private or public, educational institutions and state or local systems of higher education.
Amount of Grant: Up to 600,000 U.S.D
Contact: Dr. Irene Eckstrand, (301) 594-5402; fax (301) 480-2228; eckstrai@nigms.nih.gov
Internet: http://grants.nih.gov/grants/guide/rfa-files/RFA-GM-01-002.html
Sponsor: National Institute of General Medical Sciences
45 Center Drive, Natcher Building
Bethesda, MD 20892

NIGMS Initiative for Minority Students: Bridges to the Doctoral Degree **3455**
The program develops partnerships between educational institutions to support underrepresented minorities in the transition to doctoral programs. Besides reflecting the plans and priorities of the individual institutions, the partnership must address the needs of the minority graduate students. Support activities the institutions may implement include, but are not limited to, enhancing the curriculum of the master's institution; establishing a mentor program for master's students with faculty at the doctoral institution; enabling and encouraging students from either institution to take classes at the other institution; and providing research opportunities for master's students at the doctoral institution or in private industrial laboratories. Annual deadline dates may vary; contact program staff for exact dates. RFA: GM-01-003
Requirements: Partnerships involve a minimum of two educational institutions. One institution must have a significant enrollment of underrepresented minorities, and the highest postgraduate degree offered in the related sciences must be the master's degree. The other participating institution must be a research university and offer doctoral degrees in biomedical relevant areas. Applications also may represent several institutions within a single state, with one institution designated as the applicant institution.
Amount of Grant: Up to 20,000,000 U.S.D
Contact: Dr. Irene Eckstrand, (301) 594-5402; eckstrai@nigms.nih.gov
Internet: http://grants.nih.gov/grants/guide/rfa-files/RFA-GM-01-003.html
Sponsor: National Institute of General Medical Sciences
45 Center Drive, Natcher Building
Bethesda, MD 20892

NIGMS Institutional Research and Academic Career Development Award **3456**
The goal of this program is to prepare qualified individuals for careers that significantly impact the nation's research agenda. The objectives of this initiative are: to enhance research-oriented teaching at minority serving institutions (MSIs); to increase the research and other skills needed by developing scientists to conduct high quality research in an academic environment; and to promote linkages between research-intensive institutions and MSIs that can lead to further collaborations in research and teaching; and to increase the number of well-qualified underrepresented minority students entering competitive careers in biomedical research. Grant funds may be used to support the following expenses: salary/wages and benefits for a full-time, 12-month staff appointment; tuition, fees, and books; research expenses; supplies; travel to one training or scientific meeting per year; and statistical services. PAR-98-085.
Requirements: An application must be submitted on behalf of a consortium of domestic private or public educational institutions; the applicant institution should be the primary site of the postdoctoral research experience. An institution may not submit more than one application. A consortium must include one or more partners that are MSIs. Candidates appointed under this program award must be U.S. citizens or noncitizen nationals, or have been lawfully admitted for permanent residence and possess Alien Registration Receipt Cards (I-151 or I-551) or some other verification of legal status as permanent residents. Postdoctoral candidates must have received, as of the beginning date of the appointment to the program, the PhD, MD, or comparable doctoral degree from an accredited domestic or foreign institution.
Restrictions: Individuals on temporary or student visas are not eligible.
Contact: Dr. Clifton Poodry, Division of Minority Opportunities in Research; (301) 594-3900; fax (301) 480-2753; poodryc@nigms.nih.gov
Internet: http://www.grants.nih.gov/grants/guide/pa-files/PAR-98-085.html
Sponsor: National Institute of General Medical Sciences
45 Center Drive, Natcher Building
Bethesda, MD 20892

NIGMS Integrative and Collaborative Approaches to Research Grants **3457**
The purpose of this PA is to encourage collaborative and integrative approaches to research on multifaceted biological problems. While the traditional R01 research project grant remains the backbone of NIH and NIGMS support for biomedical science, individual R01 grants may be inadequate to support larger efforts that require collaborative approaches. In some cases, individual investigators with existing R01 or R37 support, or other sources of independent funding, will need to attract and coordinate expertise in different disciplines and approaches and may also require access to specialized resources, such as computational facilities, high-throughput technologies, and equipment. In order to provide a new mechanism to support such collaborative and interactive activities, this initiative will use the R24 consortium grant mechanism, designed for groups of currently funded investigators. This PA also strongly encourages consideration of existing and previously announced mechanisms whenever appropriate, to enhance collaborative and integrative approaches to research problems. These include administrative supplements to NIGMS-funded research grants, competitive supplements to NIGMS-funded research grants, individual R01 grants containing a consortium arrangement for a collaborative component, program project (P01) and center (P50) grants, and Interactive Research Project Grants (IRPG). PA: PA-00-099
Requirements: Applications may be submitted by domestic and foreign, for-profit and nonprofit organizations, both public and private; units of state and local governments; and eligible agencies of the federal government.
Amount of Grant: Up to 300,000 U.S.D
Contact: Dr. Marion Zatz, Division of Genetics and Developmental Biology; (301) 594-0943; fax (301) 480-2228; zatzm@nigms.nih.gov
Internet: http://grants.nih.gov/grants/guide/pa-files/PA-00-099.html
Sponsor: National Institute of General Medical Sciences
45 Center Drive, Natcher Building
Bethesda, MD 20892

NIGMS Large-Scale Collaborative Project Awards **3458**
The purpose of this program is to enable the solution of major problems in biomedical research and to facilitate the next evolutionary stage of integrative biomedical science. The intention is to make resources available for independently funded scientists to form research teams to solve a complex biological problem that would be beyond the means of any one research group, and that is of central importance to biomedical science and to the mission of NIGMS. It is expected that the participating investigators will already

hold externally peer reviewed and funded research grants in the area of the proposal, and, in general, support of new individual research projects will not be a part of these large-scale project awards. A high level of resources may be requested to allow participating investigators to extend their research efforts to form a consortium to approach a research problem of overarching importance in a comprehensive and highly integrated fashion. Grants for this initiative will be awarded in two phases. The purpose of the Phase I award is to provide resources for detailed planning to applicants who have demonstrated the selection of an appropriate complex biological problem, an innovative plan, and appropriate commitments to its solution from participating investigators and institutions. Phase I applicants will submit an overview of the proposed large-scale project for peer review. Successful Phase I applicants will receive a $25,000 (direct costs) planning grant, and those applicants who receive awards will be eligible to submit a more extensively planned and detailed application for a Phase II award to support the large-scale project itself. Annual deadline dates may vary; contact program staff for exact dates. RFA: GM-02-007
Requirements: Applications may be submitted by domestic, nonprofit organizations, both public and private; units of state and local governments; and eligible agencies of the federal government. Multi-institutional applications are encouraged. A collaborative project may include participating investigators from foreign and/or for-profit organizations. Participation by industry is also encouraged.
Contact: Dr. Michael Rogers, Division of Pharmacology, Physiology, and Biological Chemistry; (301) 594-3827; fax (301) 480-2802; rogersm@nigms.nih.gov
Internet: http://grants2.nih.gov/grants/guide/rfa-files/RFA-GM-02-007.html
Sponsor: National Institute of General Medical Sciences
45 Center Drive, Natcher Building
Bethesda, MD 20892

NIGMS MARC Undergraduate Student Training in Academic Research 3459
The Minority Access to Research Careers (MARC) Program Branch of the Division of Minority Opportunity in Research of NIGMS provides research training opportunities that target students and faculty from minority groups who are underrepresented in the biomedical sciences, including mathematics. As part of its ongoing commitment to the development of prospective underrepresented minority researchers and the enhancement of their training environment, the MARC Program provides support directed at underrepresented minority students to improve their preparation for graduate training in the biomedical sciences. PA: PAR-02-033
Requirements: Only domestic, nonprofit, private or public institutions which offer the baccalaureate degree and in which student enrollments are drawn substantially from ethnic minority groups that are underrepresented in the biomedical sciences, including mathematics, may apply. Only one grant per eligible institution will be awarded. The program is open to qualified undergraduate honors students majoring in the sciences with expressed interest in careers in biomedical research and intentions to pursue graduate education leading to the PhD, MD/PhD or other combined professional degree/PhD.
Contact: Dr. Adolphus Toliver, Division of Minority Opportunities in Research; (301) 594-3900; fax (301) 480-2753; tolivera@nigms.nih.gov
Internet: http://grants2.nih.gov/grants/guide/pa-files/PAR-02-033.html
Sponsor: National Institute of General Medical Sciences
45 Center Drive, Natcher Building
Bethesda, MD 20892

NIGMS Mentored Clinical Scientist Research Career Development Awards 3460
This award enables candidates holding professional degrees (e.g., M.D., D.O., D.V.M., or equivalent degrees) to undertake 3 to 5 years of special study and supervised research with the goal of becoming independent investigators. The award also allows awardees to pursue a research career development program suited to their experience and capabilities under a mentor who is competent to provide guidance in the chosen research area. Institutions may submit applications on behalf of candidates who hold professional degrees. At least 2 years must have elapsed since the health professional degree was granted. Candidates can have varying levels of clinical training and research experience. New applications are due February 12, June 12, and October 12, while resubmitted applications are due March 12, July 12, and November 12. Up to $75,000 per year plus fringe benefits is available, and up to $20,000 per year for research support for new awards.
Requirements: The candidate must have an MD degree or its equivalent, must have completed postgraduate clinical training and have secured a faculty appointment in an appropriate research-intensive environment, must identify a mentor with extensive research experience, and must be willing to spend a minimum of 75 percent of full-time professional effort conducting research and research career development. Applications may be submitted on behalf of candidates by domestic, nonfederal organizations, public or private, such as medical schools. Minorities and women are encouraged to apply. Candidates must be U.S. citizens or noncitizen nationals, or must have been lawfully admitted for permanent residence and possess Alien Registration Receipt Card (I-551) or some other verification of legal admission as permanent residents.
Restrictions: Former principal investigators on NIH research project (R01), sub-projects of program project (P01), or center grants (P50), or the equivalent, are not eligible. A candidate for the award may not concurrently apply for any other PHS award that duplicates the provisions of this award nor have another application pending award. K08 recipients are encouraged to apply for independent research grant support during the period of this award. The K08 recipient would be allowed to maintain the award if other PHS support is procured, as long as the new support does not interfere with the ability to meet the MCSDA requirements.
Date(s) Application is Due: Feb 12; Mar 12; Jun 12; Jul 12; Oct 12; Nov 12
Amount of Grant: Up to 95,000 U.S.D
Contact: Dr. Alison Cole, Program Contact; (301) 594-1826; fax (301) 480-2802; colea@nigms.nih.gov
Internet: http://www.grants.nih.gov/grants/guide/pa-files/PAR-98-084.html
Sponsor: National Institute of General Medical Sciences
45 Center Drive, Natcher Building, Room 2AS-49, MSC 6200
Bethesda, MD 20892-6200

NIGMS Minority Access to Research Careers Ancillary Training Activities Grants 3461
The MARC program provides research training opportunities for students and faculty from minority groups underrepresented in the biomedical and behavioral sciences relevant to biomedicine, including mathematics and engineering. Under the ancillary training initiative, the program provides support for ancillary research training activities that are harmonious with its overall objective of increasing the number of underrepresented minorities who participate fully in the biomedical enterprise of this country. Program-related scientific meetings, conferences, technical assistance workshops, and other ancillary training activities will be funded. Funds may be used for salaries, equipment, supplies, travel expenses, per diem costs, conference services, publication costs, registration fees, and speaker fees. Grant amounts are variable, based on the nature and scope of proposed activities. A letter of intent is requested by January 1, May 1, and September 1; full application is due February 1, June 1, and October 1. Annual deadline dates may vary; contact program officer for exact dates. PA: PAR-99-091
Requirements: Applications may be submitted by U.S. institutions, including scientific or professional societies, eligible to receive grants from the Department of Health and Human Services.
Restrictions: Foreign organizations are not eligible.
Date(s) Application is Due: Jan 1; Feb 1; May 1; Jun 1; Sep 1; Oct 1
Contact: Dr. Hinda Zlotnick, MARC Program Director; (301) 594-3900; fax (301) 480-2573; hindaz@nigms.nih.gov
Internet: http://www.grants.nih.gov/grants/guide/pa-files/PAR-99-091.html
Sponsor: National Institute of General Medical Sciences
45 Center Drive, Natcher Building
Bethesda, MD 20892

NIGMS Minority Access to Research Careers (MARC) Program 3462
The program's goals are to increase the number and capabilities of scientists from underrepresented minority groups who are engaged in biomedical research and to strengthen science curricula and student research opportunities at institutions with substantial minority enrollments in order to prepare minority students for research careers. The program offers postdoctoral and predoctoral fellowships, and undergraduate student training. PA: PAR-03-140
Requirements: Applications are accepted from domestic private and public universities, and research institutions or centers with graduate programs that offer a solid research environment as evidenced by a high level of faculty involvement in biomedical and/or behavioral research. To be eligible for participation in the program as a PREP scholar, individuals from underrepresented minority groups must have graduated with a baccalaureate degree in a biomedically relevant science from an accredited U.S. college or university no more than 36 months prior to the date of submission of the application. PREP scholars must be U.S. citizens or permanent residents.
Restrictions: An applicant institution may submit only one application for this program announcement.
Contact: Dr. Adolphus Toliver, Chief, MARC Branch; (301) 594-3900; fax (301) 480-8506; rm63f@nih.gov
Internet: http://grants.nih.gov/grants/guide/pa-files/PAR-03-140.html
Sponsor: National Institute of General Medical Sciences
45 Center Drive, Natcher Building
Bethesda, MD 20892

NIGMS Molecular Pharmacology of Anesthetic Action Research Grants 3463
NIGMS invites applications for the study of the molecular mechanisms of action of general and local anesthetics. The objective of this program is to encourage the submission of applications from individuals who have been trained either in basic or clinical research and who possess experience in research disciplines that are pertinent to answering questions germane to the understanding of anesthetic action such as the neurosciences, the pharmacological sciences, biophysics, biochemistry, and cell biology. Presumably, a wide variety of approaches and model systems could be employed to study anesthetic action at the molecular level. These model systems would include, but are not limited to, artificial membranes; ion channels, receptors, and other macromolecules in reconstituted membranes; electrophysiological measurements and biochemical determinations in isolated whole membrane systems; studies of whole cell responses to anesthetics; and actions of anesthetics on integrated systems such as brain slices and the spinal cord. Additionally, experimental designs utilizing recombinant DNA and molecular biological techniques, immunological approaches, and biophysical procedures whose aims are to elucidate structural features of anesthetic interactions with components of biological systems would also be of interest. There are no limits to the types of approaches that may be proposed to answer questions relevant to anesthetic action, so long as those approaches emphasize clarification of molecular aspects of anesthetic pharmacodynamics. Ideally, the ultimate goal of this research would be to develop hypotheses that might explain the relationship between molecular actions and physiological effects of general and local anesthetics. Additional information may be obtained by mail and email or through the NIH grant line (301-402-2221) or the NIH Web site. PA: PA-96-026
Requirements: Research project awards can be made to any public or private, for-profit or nonprofit university, college, hospital, laboratory, or other institution, including state and local units of government, qualifying small businesses, and individuals. SBIR grants can be awarded only to domestic small businesses. STTR grants can be awarded only to

domestic small business concerns that partner with research institutions in cooperative research and development.
Date(s) Application is Due: Feb 1; Jun 1; Oct 1
Contact: Dr. Alison Cole, (301) 594-1826; fax (301) 480-2802; colea@gm1.nigms.nih.gov
Internet: http://www.grants.nih.gov/grants/guide/pa-files/PA-96-026.html
Sponsor: National Institute of General Medical Sciences
45 Center Drive, Natcher Building
Bethesda, MD 20892

NIGMS MORE Faculty Development Awards 3464
The MORE program provides support for eligible faculty to spend a summer or one academic term every year for two to five years in full-time research in a research-intensive laboratory. Contact the program office for deadline dates. PA: PAR-95-048
Date(s) Application is Due: Feb 1; Jun 1; Oct 1
Contact: Dr. Adolphus Toliver, Predoctoral NRSA Awards Contact; (301) 594-3900
Internet: http://grants.nih.gov/grants/guide/pa-files/PAR-95-048.html
Sponsor: National Institute of General Medical Sciences
45 Center Drive, Natcher Building
Bethesda, MD 20892

NIGMS Pharmacogenomics Research Network Grants 3465
The purpose of this funding opportunity announcement is to allow new and renewal applications for research groups to compete to become part of the Pharmacogenomics Research Network (PGRN), and to create network resources that serve the entire PGRN. The PGRN is a consortium of interdisciplinary groups with a core mission to conduct research into understanding the genetic basis of variable drug responses, both therapeutic and adverse. The scientific objectives of the research network include definition of the pressing problems for safe and effective use of medications, establishment of truly cross-disciplinary research groups, application of the most appropriate genetic and genomic methods to important sample collections, discovery and replication studies that identify variation in critical genes and biological pathways, elucidation of the functional consequences of genetic variation for the drugs being studied, and understanding the clinical implications of genotyping to predict drug responses. The expected direct costs amount for each individual research group award is $1 to $1.5 million per year, and for each add-on network resource is $0.1 to $0.5 million per year. Letters of Intent are due by May 2, with full proposals received by June 2.
Requirements: The following organizations and institutions are eligible to apply: public/state controlled institutions of higher education; private institutions of higher education; Hispanic-serving institutions; historically Black colleges and universities (HBCUs); nonprofits with 501(c)3 IRS status (other than institutions of higher education); nonprofits without 501(c)3 IRS status (other than institutions of higher education); small businesses; and for-profit organizations (other than small businesses).
Date(s) Application is Due: May 2; Jun 2
Contact: Rochelle M. Long, (301) 594-3827; rochelle.long@nih.gov
Internet: http://grants.nih.gov/grants/guide/rfa-files/RFA-GM-10-001.html
Sponsor: National Institute of General Medical Sciences
45 Center Drive, Building 45
Bethesda, MD 20892

NIGMS Pilot Projects for the Protein Structure Initiative (Structural Genomics) 3466
This program funds research centers that serve as pilots to examine the best approach for developing subsequent integrated, large-scale research networks in structural genomics. This is part of the institute's Protein Structure Initiative (PSI), whose goal is the understanding of protein structural families, structural folds, and the relationship of structure and function. These research centers should include all the constituent tasks of structural genomics and should test strategies for large-scale high throughput structure determination by X-ray crystallography and/or NMR. A letter of intent is requested by November 3; full application is due February 12. Annual deadline dates may vary; contact program staff for exact dates. RFA: GM-00-006
Requirements: Applications may be submitted by domestic for-profit and nonprofit organizations, both public and private; units of state and local governments; and eligible agencies of the federal government.
Restrictions: Foreign institutions are not eligible for research center grants.
Date(s) Application is Due: Feb 12; Nov 3
Amount of Grant: Up to 3,000,000 U.S.D
Contact: Dr. John Norvell, Division of Cell Biology and Biophysics; (301) 594-0533; fax (301) 480-2004; norvellj@nigms.nih.gov
Internet: http://grants.nih.gov/grants/guide/rfa-files/RFA-GM-00-006.html
Sponsor: National Institute of General Medical Sciences
45 Center Drive, Natcher Building
Bethesda, MD 20892

NIGMS Predoctoral Training in Bioinformatics and Computation Biology Grants 3467
The purpose of this program is to develop a mechanism to train a cadre of scientists whose primary professional identification and disciplinary affiliation is bioinformatics or computational biology. For the purpose of this announcement, bioinformatics and computational biology are defined broadly to include the use of theory, computer implementation, and application to the full spectrum of basic research in the biomedical sciences. The terms thus include analysis of molecular sequence and structure, molecular function, cellular function, physiology, genomics, and genetics, as well as computational modeling of complex phenomena such as neural circuits and equilibrium phenomena, population biology, theoretical and mathematical biology, and the analysis of complex systems. The goal of this program is to train PhD students in the background theory, computational implementation, and biological application of information sciences (including computer science, statistics, mathematics and others), and to use this training to study problems relevant to biomedical research. PA: PAR-99-146
Requirements: Applications may be submitted by domestic public and private institutions with established programs leading to a PhD degree.
Amount of Grant: Up to 16,488 U.S.D
Contact: Dr. Marion Zatz, Division of Genetics and Developmental Biology; (301) 594-0943; fax (301) 480-2228
Internet: http://grants.nih.gov/grants/guide/pa-files/PAR-99-146.html
Sponsor: National Institute of General Medical Sciences
45 Center Drive, Natcher Building
Bethesda, MD 20892

NIGMS Quantitative Approaches to the Analysis of Complex Biological Systems 3468
The purpose of this program announcement is to advertise NIGMS and NIMH interest in supporting research projects that develop quantitative approaches to describe, analyze, and predict the behavior of complex biological systems, especially those requiring the integration of potentially large amounts of molecular, biochemical, cell biological, and physiological data. Such studies, adapted to the analysis of complex systems in humans, will ultimately have an impact on the treatment of human disorders and disease. These projects are expected to require the participation of individuals with diverse expertise and therefore to be of a collaborative and cross disciplinary nature. Additional information for this program may be obtained by phone, fax, email, or the Internet. PA: PA-98-077
Requirements: Applications may be submitted by foreign and domestic, for-profit and nonprofit organizations, both public and private. Racial/ethnic minority individuals, women, and persons with disabilities are encouraged to apply.
Restrictions: Foreign institutions are not eligible for program project (P01) grants.
Contact: Dr. Judith Greenberg, Division of Genetics and Developmental Biology; (301) 594-0943; fax (301) 480-2228; greenbej@nigms.nih.gov
Internet: http://www.grants.nih.gov/grants/guide/pa-files/PA-98-077.html
Sponsor: National Institute of General Medical Sciences
45 Center Drive, Natcher Building
Bethesda, MD 20892

NIGMS Ruth L. Kirschstein National Research Service Awards for Individual Predoctoral Fellows in PharmD/PhD Grants 3469
NIGMS is committed to increasing the number of pharmacy students who wish to pursue doctoral research in the pharmaceutical sciences. It recognizes that current career paths for Doctor of Pharmacy graduates have reduced the number of individuals who pursue academic research training. NIGMS's intent in issuing this Funding Opportunity Announcement (FOA) is to provide individual research training fellowships to promising predoctoral applicants who are enrolled in PharmD/PhD programs in Schools of Pharmacy who desire to become productive and successful independent investigators whose research meets the mission of NIGMS. This fellowship program will provide support for the dissertation stage of predoctoral students enrolled in joint PharmD/PhD programs. It is expected that the length of time for an award will be 3 years or less.
Requirements: The following organizations and institutions are eligible to apply: for-profit organizations; non-profit organizations; domestic institutions; public or private institutions, such as universities, colleges, and hospitals; units of State and local governments; eligible agencies of the Federal government.
Restrictions: Individuals on temporary or student visas are not eligible.
Date(s) Application is Due: Apr 8; Aug 8; Dec 8
Contact: Richard Okita, (301) 594-3827; fax (301) 480-2802; okitar@nigms.nih.gov
Internet: http://grants.nih.gov/grants/guide/pa-files/PA-09-029.html
Sponsor: National Institute of General Medical Sciences
45 Center Drive, Natcher Building, Room 2AS-49, MSC 6200
Bethesda, MD 20892-6200

NIGMS Short Courses on Mathematical, Statistical, and Computational Tools for Studying Biological Systems 3470
The goal of this FOA is to prepare researchers at all professional levels for multidisciplinary research on complex biological systems and phenotypes by improving integration of mathematical, statistical, and computational approaches into biological and/or behavioral research. Progress in contemporary biological and behavioral disciplines depends heavily on investigators who are skilled in the use of mathematics, computation, and statistics (collectively referred to as quantitative methods). This FOA issued by the National Institute of General Medical Sciences, the National Center for Complementary and Alternative Medicine, the National Institute on Aging and the National Institute of Mental Health encourages applications for Research Education Grants (R25) from institutions and organizations to conduct workshops and short courses to improve integration of mathematical, statistical, and computational approaches into biological and/or behavioral research. Support will be limited to activities that reach a wide audience of researchers.
Requirements: Application can be made by the following institutions and organizations: public/state controlled institutions of higher education; private institutions of higher education; nonprofits with 501(c)3 IRS status (other than institutions of higher education); nonprofits without 501(c)3 IRS status (other than institutions of higher education); small businesses; for-profit organizations (other than small businesses); state governments; U.S. territories or possessions; Indian/Native American tribal governments (federally recognized); Indian/Native American tribal governments (other than Federally recognized); Indian/Native American tribally designated organizations; non-domestic (non-U.S.) entities (foreign organizations); Hispanic-serving institutions;

historically Black colleges and universities (HBCUs); tribally controlled colleges and universities (TCCUs); Alaska Native and Native Hawaiian serving institutions; regional organization; eligible agencies of the Federal government; and faith-based or community based organizations.
Restrictions: The program announcement is not intended for university course or curriculum development.
Date(s) Application is Due: Jan 25; May 25; Sep 25
Amount of Grant: Up to 200,000 U.S.D
Contact: Matthew E. Portnoy, (301) 594-0943; mportnoy@nigms.nih.gov
Internet: http://grants.nih.gov/grants/guide/pa-files/PA-09-002.html
Sponsor: National Institute of General Medical Sciences
45 Center Drive, Building 45, 2AS.25P
Bethesda, MD 20892-6200

NIGMS Short Courses on Mathematical and Statistical Tools for the Study of Complex Phenotypes and Complex Systems **3471**
The purpose of this program is to provide support for short courses or workshops to assist scientists in preparing for research on complex phenotypes and complex systems. This program announcement solicits applications for education grants (R25), which provide for one to five years of support for short courses. Support is limited to courses that are highly focused and that reach a wide audience of scientists. The program announcement is not intended for university course or curriculum development. A variety of approaches will be considered. The nature of the activity, the needs of the target community, and the course goals should determine the instructional approach. These might include new approaches, such as Web-based instruction, or more traditional approaches, such as summer workshops or on-site instruction. PA: PA-98-083
Requirements: Applications may be submitted by domestic nonprofit and for-profit organizations, public and private.
Restrictions: Foreign organizations are not eligible.
Contact: Dr. Irene Anne Eckstrand, (301) 594-0943; fax (301) 480-2228; Irene_Eckstrand@nih.gov
Internet: http://www.grants.nih.gov/grants/guide/pa-files/PA-98-083.html
Sponsor: National Institute of General Medical Sciences
45 Center Drive, Natcher Building
Bethesda, MD 20892

NIGMS Supplements for the Study of Complex Biological Systems **3472**
The purpose of this initiative is to support new quantitative approaches to the study of complex, fundamental biological processes by encouraging nontraditional collaborations across disciplinary lines. NIGMS will provide supplements to existing NIGMS grants to support the salary and expenses of collaborating investigators such as physicists, engineers, mathematicians, and other experts with quantitative skills relevant to the analysis of complex systems. It is expected that the collaboration will result in new directions for the parent project, or new research projects that will compete for independent funding. PA-98-024 may be obtained through the NIH grant line (301-402-2221), the NIH Web site, mail, and email.
Requirements: Applications may be submitted by foreign, domestic, for-profit and nonprofit organizations, public or private.
Contact: Dr. Judith Greenberg, Division of Genetics and Developmental Biology; (301) 594-0943; fax (301) 480-2228; greenbej@nigms.nih.gov
Internet: http://www.grants.nih.gov/grants/guide/pa-files/PA-98-024.html
Sponsor: National Institute of General Medical Sciences
45 Center Drive, Natcher Building
Bethesda, MD 20892

NIGMS Support of Competitive Research (SCORE) Pilot Project Awards **3473**
The objective of the SCORE program is to foster the development of faculty at minority serving institutions (MSIs) in order to increase their research competitiveness and promote their transition to non-SCORE external sources of funding. This objective is expected to translate into an increase in the number of individuals from groups underrepresented in biomedical and behavioral research professionally engaged in these areas of research, and an enhancement of an institution's research base. SCORE SC grants are offered to eligible MSIs. For the purposes of this program, eligible MSIs are those with more than 50% student enrollment of individuals from groups underrepresented in biomedical and behavioral research. The SC2 awards may not exceed $300,000 (direct costs) for the entire length of the award which may be one to three years maximum.
Requirements: Eligible applicants include: public or private post secondary educational institution, such as a university or college awarding associate, undergraduate or graduate degrees located in the United States of America or its territories including Puerto Rico, Guam and Virgin Islands with more than 50 percent student enrollment from groups underrepresented in the biomedical and behavioral sciences (such as African American, Hispanic American, Native American, Alaska Natives or natives of the U.S. Pacific Islands); an Indian tribe that has a recognized governing body and that performs substantial governmental functions, or an Alaska Regional Corporation (ARC), as defined in the Alaska Native Claims Settlement Act (43 U.S.C. 1601 et. Seq.); or an institution that has a significantly high student enrollment from groups underrepresented in the biomedical or behavioral sciences but less than 50% (if the Secretary of the Department of Health and Human Services, through the MBRS Chief, determines that the institution has demonstrated special commitment to the retention and graduation of students from groups underrepresented in the biomedical and behavioral sciences and to the hiring and retention of science faculty from underrepresented groups).
Restrictions: Foreign institutions are not eligible to apply.
Date(s) Application is Due: Jan 25; May 25; Sep 25
Amount of Grant: 50,000 - 100,000 U.S.D
Contact: Hinda Zlotnik, (301) 594-3900; fax (301) 480-2753; zlotnikh@nigms.nih.gov
Internet: http://grants.nih.gov/grants/guide/pa-files/PAR-08-027.html
Sponsor: National Institute of General Medical Sciences
45 Center Drive, Room 2AS.37, MSC 6200
Bethesda, MD 20892-6200

NIGMS Support of Competitive Research (SCORE) Research Advancement Awards **3474**
The objective of the SCORE program is to foster the development of faculty at minority serving institutions (MSIs) in order to increase their research competitiveness and promote their transition to non-SCORE external sources of funding. This objective is expected to translate into an increase in the number of individuals from groups underrepresented in biomedical and behavioral research professionally engaged in these areas of research, and an enhancement of an institution's research base. SCORE SC grants are offered to eligible MSIs. For the purposes of this program, eligible MSIs are those with more than 50% student enrollment of individuals from groups underrepresented in biomedical and behavioral research. The anticipated amount of individual SC1 awards is expected to range between $125,000 to $250,000 maximum direct costs per year.
Requirements: Eligible applicants include: public or private post secondary educational institution, such as a university or college awarding associate, undergraduate or graduate degrees located in the United States of America or its territories including Puerto Rico, Guam and Virgin Islands with more than 50 percent student enrollment from groups underrepresented in the biomedical and behavioral sciences (such as African American, Hispanic American, Native American, Alaska Natives or natives of the U.S. Pacific Islands); an Indian tribe that has a recognized governing body and that performs substantial governmental functions, or an Alaska Regional Corporation (ARC), as defined in the Alaska Native Claims Settlement Act (43 U.S.C. 1601 et. Seq.); or an institution that has a significantly high student enrollment from groups underrepresented in the biomedical or behavioral sciences but less than 50% (if the Secretary of the Department of Health and Human Services, through the MBRS Chief, determines that the institution has demonstrated special commitment to the retention and graduation of students from groups underrepresented in the biomedical and behavioral sciences and to the hiring and retention of science faculty from underrepresented groups).
Restrictions: Foreign institutions are not eligible to apply.
Date(s) Application is Due: Jan 25; May 25; Sep 25
Amount of Grant: 125,000 - 250,000 U.S.D
Contact: Hinda Zlotnik, (301) 594-3900; fax (301) 480-2753; zlotnikh@nigms.nih.gov
Internet: http://grants1.nih.gov/grants/guide/pa-files/PAR-08-026.html
Sponsor: National Institute of General Medical Sciences
45 Center Drive, Room 2AS.37, MSC 6200
Bethesda, MD 20892-6200

NIH/EPA Mechanistic-Based Cancer Risk Assessment Methods Research Grants **3475**
This RFA solicits studies that focus on the development of cancer risk assessment methods and practices. Emphasis should be on the development of new methods and practices that reduce the uncertainties associated with cancer risk assessment, especially uncertainties that are encountered in extrapolating from animal species to humans. Studies that propose refinement, validation, or invalidation of existing methods and practices are also acceptable if they will result in a significant reduction in uncertainty. Methods of development can be based on in vitro and/or in vivo animal model systems, in vitro human cell model systems, and exploratory epidemiologic approaches. Short-term mechanistic-based methods that can be used to augment long-term bioassays are also encouraged. A clear and unambiguous demonstration of mechanistic commonality with humans is a requirement of any method based on animal model systems. Inquiries may be directed to the staff at the appropriate organization/institute: Dr. Roy Fleming, Director, (404) 639-3343, Research Grants Program, National Institute for Occupational Safety and Health, CDC, 1600 Clifton Road NE, Building 1, Atlanta, GA 30333; Dr. David Reese, (202) 564-6919, National Center for Environmental Research and Quality Assurance, EPA, 401 M St SW, Washington, DC 20460; and Dr. Kumiko Iwamoto, Epidemiology and Genetics Research Program, NCI, 6130 Executive Boulevard, Bethesda, MD 20892. Annual deadline dates may vary; contact program staff for exact dates. RFA: OH-99-003
Requirements: Applications may be submitted by domestic and foreign, public and private, nonprofit and for-profit organizations; governments and their agencies; and federally recognized Indian tribal governments, Indian tribes, or Indian tribal organizations.
Amount of Grant: Up to 250,000 U.S.D
Contact: Program Officer
Internet: http://www.grants.nih.gov/grants/guide/rfa-files/RFA-OH-99-003.html
Sponsor: National Institutes of Health
6701 Rockledge Drive
Bethesda, MD 20892

NIH Academic Career Awards **3476**
The awards support individuals interested in introducing or improving curriculum in a particular scientific field as a means of enhancing the educational or research capacity at the grantee institution. The award supports two types of activities. Development—support for more junior candidates who are interested in developing academic and research expertise in particular fields as a way to increase the overall pool of individuals capable of research or teaching in the identified area. Leadership—supports more senior individuals with acknowledged scientific expertise and leadership skills who are interested in improving the curricula and enhancing the research capacity within an academic institution. Inquiries may be directed to program staff at the appropriate

NIH institute: Dr. Robin Barr, Office of Extramural Affairs, (301) 496-9322, NIA, 7201 Wisconsin Avenue, Bethesda, MD 20892; Dr. Ernestine Vanderveen, Division of Basic Research, (301) 443-1273, NIAAA, 6000 Executive Boulevard, Bethesda, MD 20892; Dr. Richard Lymn, Extramural Programs, (301) 594-5128, NIAMS, Natcher Building, Bethesda, MD 20892; Dr. John Schneider, (301) 496-8580, NCI, Executive Plaza North, Bethesda, MD 20892; Dr. Annette Kirshner, (919) 541-0488, NIEHS, P.O. Box 12233, Research Triangle Park, NC 27709; and Dr. George Niederehe, (301) 443-3264, NIMH, 5600 Fishers Lane, Rockville, MD 20857. Annual deadline dates may vary; contact program staff for exact dates. PA: PA-95-052
Requirements: Candidate for this award must have a clinical or doctoral degree. Candidate for the development award must demonstrate the potential to develop into an excellent academician and must be able to identify a mentor who is an expert in the research field of interest. Candidate must also be able to devote 75 percent of full-time professional effort to the research program. For the leadership award, candidate must have sufficient clinical training, research, or teaching experience in the academic area of interest to implement a program of curriculum development within the applicant institution; must have an academic appointment at a level sufficient to exert an influence on the coordination of research, teaching, and clinical practice in an emerging field; and must be able to devote at least 25 percent effort to the program.
Date(s) Application is Due: Feb 1; Jun 1; Oct 1
Contact: Program Officer
Internet: http://www.grants.nih.gov/grants/guide/pa-files/PA-95-052.html
Sponsor: National Institutes of Health
6701 Rockledge Drive
Bethesda, MD 20892

NIH Academic Research Enhancement Awards (AREA) **3477**
NIH makes a special effort to stimulate research in educational institutions that provide baccalaureate training for a significant number of the nation's research scientists but that historically have not been major recipients of NIH support. The AREA funds are intended to support new research projects or expand ongoing research activities proposed by faculty members of these institutions in areas related to the health sciences. Funding decisions will be based on the proposed research project's scientific merit and relevance to NIH programs and the institution's contribution to the undergraduate preparation of doctoral-level health professionals. Annual deadline dates vary; contact program staff for exact dates. PA: PA-99-062
Requirements: All health professional schools/colleges and other academic components of domestic institutions offering baccalaureate or advanced degrees in the sciences related to health are eligible to apply.
Restrictions: Applicant principal investigators must not have active research grant support from either NIH or ADAMHA at the applicant institution at the time of award of an AREA grant; may not submit a regular NIH or ADAMHA research grant application for essentially the same project as a pending AREA application; are expected to conduct the majority of their research at their own institutions, although limited access to special facilities or equipment other institutions is permitted; and may not be awarded more than one AREA grant at a time nor be awarded a second AREA grant to continue research initiated under the first AREA grant.
Date(s) Application is Due: May 7; Jun 25; Sep 7
Amount of Grant: Up to 150,000 U.S.D
Contact: Denise Russo, Office of Extramural Research; (301) 451-7972; fax (301) 480-0146; drusso1@mail.nih.gov
Internet: http://grants.nih.gov/grants/guide/pa-files/PA-06-042.html
Sponsor: National Institutes of Health
6705 Rockledge Drive, Room 3524
Bethesda, MD 20892

NIH Alzheimer's Disease Clinical Trial Planning Grant **3478**
NIA, NIMH, and NINDS invite qualified investigators to submit research grant applications for clinical trial planning grants directed toward the drug treatment of the cognitive and behavioral symptoms of Alzheimer's disease (AD). The purpose of the grant is to allow for early peer review of the rationale and design for Alzheimer's disease clinical drug trials and to provide support for the development of a detailed clinical trial research plan, including a manual of operations and procedures. Note: There is a companion program announcement available for Alzheimer's Disease Pilot Clinical Trials (PA-99-038). Inquiries may be directed to program staff at the appropriate NIH institute: Dr. Neil Buckholtz, (301) 496-9350, Neuroscience and Neuropsychology of Aging Program, NIA, 7201 Wisconsin Avenue, Bethesda, MD 20892; Dr. Eugene Oliver, (301) 496-5680, Division of Stroke, Trauma, and Neurodegenerative Disorders, NINDS, 7550 Wisconsin Avenue, Bethesda, MD 20892; Dr. George Niederehe, (301) 443-1185, NIMH, 5600 Fishers Lane, Rockville, MD 20857. PA: PA-99-039
Requirements: Applications may be submitted by foreign and domestic, for-profit and nonprofit organizations, both public and private.
Restrictions: The award cannot be renewed.
Amount of Grant: Up to 150,000 U.S.D
Contact: Program Officer
Internet: http://www.grants.nih.gov/grants/guide/pa-files/PA-99-039.html
Sponsor: National Institutes of Health
6701 Rockledge Drive
Bethesda, MD 20892

NIH Autism and Autism Spectrum Disorders Research **3479**
The purpose of this program announcement is to encourage grant applications for the support of research designed to elucidate the diagnosis, epidemiology, etiology, genetics, treatment, and optimal means of service delivery in relation to autistic disorder (autism) and autism spectrum disorders. Inquiries may be directed to program staff at the appropriate NIH institute: Dr. Judith Rumsey, (301) 443-9264, NIMH, 5600 Fishers Lane, Rockville, MD 20857; Dr. Marie Bristol, (301) 496-1383, NICHD, 6100 Executive Boulevard, Bethesda, MD 20892; Dr. Judith Cooper, (301) 496-5061, NIDCD, Executive Plaza South, Bethesda, MD 20892; and Dr. Giovanna Spinella, (301) 496-5821, NINDS, 7550 Wisconsin Avenue, Bethesda, MD 20892. PA: PA-98-108
Requirements: Applications may be submitted by foreign and domestic, for-profit organizations, both public and private; units of state and local governments; and eligible agencies of the federal government.
Restrictions: Foreign institutions are not eligible to apply.
Contact: Program Officer
Internet: http://www.grants.nih.gov/grants/guide/pa-files/PA-98-108.html
Sponsor: National Institutes of Health
6701 Rockledge Drive
Bethesda, MD 20892

NIH Biobehavioral Pain Research Grants **3480**
The purpose of this PA is to inform the scientific community of the interests of the various institutes at the NIH and to stimulate and foster a wide range of basic and clinical studies on pain as they relate to the missions of these institutes. Applicants are encouraged to study individual differences in pain responses which may be due to factors such as genetic differences, endocrine activity, neural activity, immune function, psychological state, disability state, age, gender, and cultural background. Research is also needed in areas such as understanding the neuroanatomical pathways and the neurophysiological mechanisms in pain. The pain experience needs to be examined at all levels of research including the gene, molecule, cell, organ, and individual with the goal of developing biobehavioral interventions to manage or prevent pain. Inquiries may be directed to the appropriate NIH institute: Dr. Martha Hare, (301) 451-3874, Office of Extramural Programs, NINR, 6701 Democracy Boulevard, Room 710, Bethesda, MD 20892; Dr. Patricia Bryant, (301) 594-2095, Division of Extramural Research, NIDCR, Building 45, Bethesda, MD 20892; Dr. Judith Finkelstein, (301) 496-9350, Neuroscience and Neuropsychology of Aging Program, NIA, 7201 Wisconsin Avenue, Bethesda, MD 20892; Dr. Deborah Ader, (301) 594-5032, Behavior and Prevention Research Program Director, NIAMS, Building 45, Bethesda, MD 20892; Dr. Louis Quatrano, (301) 402-4221, Behavioral Sciences and Rehabilitation Engineering, NICHD, 6100 Executive Boulevard, Bethesda MD 20892; Dr. David Thomas, (301) 443-6975, Behavioral Neurobiology Research Branch, NIDA, 5600 Fishers Lane, Rockville, MD 20857; Dr. Peter Muehrer, (301) 443-4708, Prevention, Early Intervention, and Epidemiology Research Branch, NIMH, 5600 Fishers Lane, Rockville, MD 20857; Dr. Linda Porter, (301) 496-9964, Systems and Cognitive Neuroscience, NINDS, 6001 Executive Boulevard, Bethesda, MD 20892; and Dr. Nancy Pearson, (301) 594-0519, NCCAM, 6707 Democracy Boulevard, Bethesda, MD 20892. PA: PA-03-152
Requirements: Applications may be submitted by foreign and domestic, for-profit and nonprofit, both public and private organizations; units of state and local governments; and eligible agencies of the federal government.
Contact: Program Officer
Internet: http://grants2.nih.gov/grants/guide/pa-files/PA-03-152.html
Sponsor: National Institutes of Health
6701 Rockledge Drive
Bethesda, MD 20892

NIH Biobehavioral Research for Effective Sleep **3481**
The goal of this program announcement is to stimulate clinical and applied research on behavioral, psychosocial, and physiological consequences of acute and chronic partial sleep deprivation in either chronically ill or healthy individuals and to develop environmental, clinical management, and other interventions with the potential to reduce sleep disturbances and significantly improve the health of large numbers of people. Although sleep disorders are a cause of sleep loss in affected individuals, the questions to be addressed under this solicitation should focus on causes and consequences of sleep deprivation, apart from any sleep pathology. Inquiries may be directed to program staff at the appropriate NIH institute.
Requirements: Applications may be submitted by domestic and foreign, for-profit and non-profit organizations, public and private, such as universities, colleges, hospitals, laboratories, units of State and local governments, and eligible agencies of the Federal government. Racial/ethnic minority individuals, women, and persons with disabilities are encouraged to apply as Principal Investigators.
Date(s) Application is Due: Feb 16; Jun 16; Oct 16
Amount of Grant: Up to 250,000 U.S.D
Contact: Susan Grove; (301) 435-0714; sg16d@nih.gov or grantsinfo@od.nih.gov
Internet: http://grants.nih.gov/grants/guide/pa-files/PA-00-046.html
Sponsor: National Institutes of Health
6701 Rockledge Drive
Bethesda, MD 20892

NIH Bioengineering Nanotechnology Initiative **3482**
NIH invites grant applications for projects on nanotechnologies useful to biomedicine. Nanotechnology is defined as the creation of functional materials, devices, and systems through control of matter at the scale of one to 100 nanometers, and the exploitation of novel properties and phenomena at the same scale. Inquiries may be directed to program staff at the appropriate NIH institute. See web site for listing of program staff. PA: PA-02-125
Date(s) Application is Due: Apr 1; Aug 1; Dec 1
Amount of Grant: Up to 400,000 U.S.D

Contact: Program Officer
Internet: http://grants2.nih.gov/grants/guide/pa-files/PA-02-125.html
Sponsor: National Institutes of Health
6701 Rockledge Drive
Bethesda, MD 20892

NIH Biology, Development, and Progression of Malignant Prostate Disease Research Grants 3483

Investigator-initiated research grants will be made to examine a range of fundamental biological issues considered critical for progress in defeating prostate cancer. The purpose of funding is to encourage new projects focusing on the biology that underlies the development and progression of malignant prostatic disease. Multidisciplinary collaborations between basic and clinical scientists in projects ranging from exploratory basic biology studies to disease progression in appropriate biological systems or models are encouraged.
Requirements: Applications may be submitted by foreign or domestic for-profit and nonprofit organizations, both public and private; units of state and local government; and eligible agencies of the federal government.
Date(s) Application is Due: Feb 5; Feb 16; Jun 5; Jun 16; Oct 5; Oct 16
Amount of Grant: Up to 250,000 U.S.D
Contact: Frank Bellino, (301) 496-6402; fax (301) 402-0010; fb12a@nih.gov
Internet: http://grants.nih.gov/grants/guide/pa-files/PA-99-081.html
Sponsor: National Institutes of Health
Gateway Building, 7201 Wisconsin Avenue, Suite 2C231
Bethesda, MD 20892

NIH Biomedical Research on the International Space Station 3484

The National Institutes of Health (NIH) and the National Aeronautics and Space Administration (NASA) are cooperating to facilitate biomedical research in space for better understanding of human physiology and human health on Earth. Applications to this FOA should propose innovative biomedical research on the molecular or cellular level that is directly relevant to the NIH mission and can be carried out on the ISS. Awards made through this FOA will initially support milestone-driven, ground based preparatory studies (UH2 ground feasibility phase), with possible rapid transition to the second, ISS-based research phase (UH3 ISS experimental phase). The ground feasibility phase (UH2) will allow investigators to focus on ground-based preparatory work to meet scientific milestones and technical requirements leading to the ISS experimental phase (UH3). The UH3 phase will include preparing the experiments for launch, conducting them on the ISS, and the subsequent data analyses on Earth. UH3s will be awarded after administrative review of the eligible UH2s that have met the scientific milestones and feasibility requirements necessary to conduct research on the ISS. Applicants should contact individuals at the specific departments listed on the web site. Letters of Intent are due by August 31, and full applications should arrive by September 30.
Requirements: The UH2/UH3 application must be submitted as a single application, and applicants should note specific instructions for each phase in this FOA. The following organizations and institutions may apply: public/state controlled institutions of higher education; private institutions of higher education; Hispanic-serving institutions; historically Black colleges and universities (HBCUs); tribally controlled colleges and universities (TCCUs); Alaska native and native Hawaiian serving institutions; nonprofits with 501(c)3 IRS status; nonprofits without 501(c)3 IRS status; small businesses; for-profit organizations; State governments; regional organizations; U.S. territories or possessions; Indian/Native American tribal governments; Indian/Native American tribally designated organizations; county governments; city or township governments; special district governments; independent school districts; public housing authorities/Indian housing authorities; eligible agencies of the Federal government; and faith-based or community based organizations.
Date(s) Application is Due: Aug 31; Sep 30
Amount of Grant: 300,000 U.S.D
Contact: Susan Grove, Program Officer; (301) 435-0714; sg16d@nih.gov or grantsinfo@od.nih.gov
Internet: http://grants.nih.gov/grants/guide/pa-files/PAR-09-120.html
Sponsor: National Institutes of Health
6701 Rockledge Drive
Bethesda, MD 20892

NIH Bone and the Hematopoietic and Immune Systems Research Grants 3485

The participating institutes seek investigator-initiated projects that have the potential to illuminate functional interactions between bone and the hematopoietic and immune systems. Recent observations underscore the linkage between endochondral bone formation and the establishment of hematopoietic marrow, and suggest that interactions between bone, marrow, and the immune system persist in the mature skeleton. Marrow stromal cells include the precursors of the osteochondrogenic lineage, exert important influences on osteoclastogenesis and lymphopoiesis, and mediate the effects of some systemic factors on bone turnover. Recent evidence indicates that hematopoietic cells can influence the differentiation of osteogenic cells, and suggests that mature lymphocytes can influence osteoclastic and osteoblastic functions. In order to explore the mechanisms that underlie these interactions, the participating institutes support outstanding projects that have the potential to either clarify the importance of specific cell types and effector molecules or identify previously unrecognized cellular and molecular agents that influence bone physiology. Collaborations among bone biologists, hematologists, and immunologists, and between basic scientists and clinical investigators, are particularly encouraged. Inquiries may be directed to the staff at the appropriate NIH institute: Dr. William Sharrock, Musculoskeletal Diseases Branch, NIAMS, 45 Center Drive, Bethesda, MD 20892; Dr. Frank Bellino, (301) 496-6402, Biology of Aging Program, NIA, Gateway Building, Bethesda, MD 20892; Dr. Kenneth Gruber, (301) 584-4836, Chronic and Disabling Diseases Branch, NIDCR, 45 Center Drive, Bethesda, MD 20892; and Dr. Charles Peterson, (301) 435-0050, Division of Blood Diseases and Resources, NHLBI, 6701 Rockledge Drive, Bethesda, MD 20892. PA: PA-99-085
Requirements: Applications may be submitted by domestic and foreign, for-profit and nonprofit, public and private organizations, units of state and local governments, and eligible agencies of the federal government.
Restrictions: Foreign institutions are not eligible for program project grant support.
Amount of Grant: Up to 250,000 U.S.D
Contact: Frank Bellino, (301) 496-6402; fax (301) 402-0010; fb12a@nih.gov
Internet: http://grants.nih.gov/grants/guide/pa-files/PA-99-085.html
Sponsor: National Institutes of Health
Gateway Building, 7201 Wisconsin Avenue, Suite 2C231
Bethesda, MD 20892

NIH Building Interdisciplinary Research Careers in Women's Health Grants 3486

The NIH Office of Research on Women's Health (CRWH) and its cosponsors invite institutional career development award applications for Building Interdisciplinary Research Careers in Women's Health (BIRCWH) Career Development Programs, hereafter termed Programs. Programs will support mentored research career development of junior faculty members, known as BIRCWH Scholars, who have recently completed clinical training or postdoctoral fellowships, and who will be engaged in interdisciplinary basic, translational, behavioral, clinical, and/or health services research relevant to women's health or sex/gender factors. Direct costs are limited to $500,000 total costs for a five-year period. Letters of Intent are due by September 23, with final applications due by October 22.
Requirements: The following organizations and institutions are eligible to apply for support: public/state controlled institutions of higher education; private institutions of higher education; Hispanic-serving institutions; historically Black colleges and universities (HBCUs); tribally controlled colleges and universities (TCCUs); Alaska Native and Native Hawaiian serving institutions; nonprofits with 501(c)3 IRS status (other than institutions of higher education); nonprofits without 501(c)3 IRS status (other than institutions of higher education); Indian/Native American tribal governments (federally recognized); Indian/Native American tribally designated organizations; U.S. territories or possessions; and eligible agencies of the federal government.
Date(s) Application is Due: Sep 23; Oct 22
Amount of Grant: Up to 500,000 U.S.D
Contact: Joan Davis Nagel, (301) 496-9186; fax (301) 402-1798; joandav@mail.nih.gov
Internet: http://grants.nih.gov/grants/guide/rfa-files/RFA-OD-09-006.html
Sponsor: National Institutes of Health
6707 Democracy Boulevard, Suite 400
Bethesda, MD 20892

NIH Cancer Surveillance Using Health Claims-Based Data System Grants 3487

Three institutes of the NIH invite investigator-initiated grant applications for research to investigate the utility of health claims information as a reporting source for assessing the national cancer burden. Health claims include secondary data sources, for example fee-for-service insurance bills, managed care encounter data, and discharge summaries. Responses to this program announcement would initiate mechanisms to expand understanding of the capability of and methods needed to use claims data for cancer surveillance as well as applying these findings to the claims data in order to assess all aspects of cancer surveillance, such as cancer screening, incidence, treatment, and outcomes. The NIDCR's specific interests focus on health claims-based data as related to oral and pharyngeal cancers or the oral complications of cancer therapies. Inquiries may be directed to the program contact at the appropriate NIH institute: Dr. Joan Warren, (301) 496-5184, Applied Research Branch, NCI, Executive Plaza N, Bethesda, MD 20892; Dr. Patricia Bryant, (301) 594-2095, Division of Extramural Research, NIDCR, 45 Center Drive, Bethesda, MD 20892; and Dr. Bernard Friedman, (301) 594-6819, AHCPR, 2101 E Jefferson Street, Rockville, MD 20852. Annual deadline dates vary; contact program staff for exact dates. PA: PA-04-012
Requirements: Applications may be submitted by public and private, foreign and domestic, for-profit and nonprofit organizations.
Contact: Crystal Wolfrey, Grants Administration Branch; (301) 496-8634; fax (301) 496-8601; crystal.wolfrey@nih.gov or wolfreyc@mail.nih.gov
Internet: http://grants2.nih.gov/grants/guide/pa-files/PA-04-012.html
Sponsor: National Institutes of Health
6701 Rockledge Drive
Bethesda, MD 20892

NIH Career Development Awards: Child Abuse and Neglect Research 3488

The five sponsoring institutes invite applications in response to this announcement for the career development of investigators who have made a commitment to focus their research endeavors on child abuse and neglect through research career enhancement to conduct high-quality, multidisciplinary, clinically relevant research on basic biological, behavioral, and social aspects of child and adolescent abuse and neglect. While the focus of the career development program is on child abuse and neglect in human populations, the award may include complementary, appropriate laboratory and animal research related to the child abuse and neglect research proposed in the application. Inquiries may be directed to program staff at the appropriate NIH institute: Dr. Malcolm Gordon, (301) 443-4709, NIMH, 6001 Executive Boulevard, Bethesda, MD 20892; Dr. Cheryl A. Boyce, (301) 443-0848, NIMH, 6001 Executive Boulevard, Bethesda, MD 20892; Dr. Joseph S. Drage, (301) 496-4188, NINDS, 6001 Executive Boulevard, Suite 3309, Bethesda, MD 20892; Dr. G. Reid Lyon, (301) 496-9849, NICHD, Building 6100, Room 4B05, MSC 7510, Bethesda, MD 20892; Dr. Coryl Jones, (301) 443-6637, NIDA, 6001 Executive Boulevard, Room 5169, MSC 9589, Bethesda, MD 20892; and

Dr. Susan Martin, (301) 443-8767, NIAAA, 6000 Executive Boulevard, Suite 505, Rockville, MD 20892. PA: PA-99-133
Requirements: Prospective candidates should contact the program staff most relevant to the applicant's research interests to ascertain the conditions of the award supported by each institute.
Date(s) Application is Due: Feb 1; Jun 1; Oct 1
Contact: Program Officer
Internet: http://grants.nih.gov/grants/guide/pa-files/PA-99-133.html
Sponsor: National Institutes of Health
6701 Rockledge Drive, Room 1040
Bethesda, MD 20892

NIH Centers for AIDS Research (CFARs) 3489
Participating institutes of NIH invite applications for center core grants (P30) to support Centers for AIDS Research (CFARs). CFAR cores provide infrastructure and promote basic, clinical, behavioral and translational AIDS research activities at institutions that receive significant AIDS funding from multiple NIH institutes or centers. CFARs foster synergy and improve coordination of research, support emerging research opportunities, and promote economy of scale through resources shared by multiple independent laboratories. Inquiries may be directed to program staff at the appropriate NIH institute; see web site for staff listings. PA: PAR-03-089
Requirements: Applications may be submitted by domestic for-profit and nonprofit organizations, both public and private; units of state or local government; and eligible agencies of the federal government.
Restrictions: Foreign institutions are not eligible to apply.
Date(s) Application is Due: Jun 16
Contact: Lori Zimand, (301) 451-4580; fax (301) 493-0597; lz36k@nih.gov
Internet: http://grants2.nih.gov/grants/guide/pa-files/PAR-03-089.html
Sponsor: National Institutes of Health
6701 Rockledge Drive
Bethesda, MD 20892

NIH Centers for Dietary Supplements Research: Botanicals 3490
Applications are invited to establish specialized research centers to investigate the biological effects of botanicals including, but not limited to, botanicals available as dietary supplements. The creation of such centers is needed to advance the quality and quantity of scientific information on botanicals and to promote further research in this area. The major goal of this RFA is to foster interdisciplinary research to develop systematic evaluation of the safety and efficacy of botanicals. General inquiries may be directed to Dr. Christine Swanson, (301) 435-2920, Office of Dietary Supplements, NIH, Building 31, Bethesda, MD 20892. Programmatic inquiries may be directed to the staff at the appropriate NIH institute: Dr. Joshua Rosenthal, (301) 496-2516, Biodiversity Program, FIC, 31 Center Drive, Bethesda, MD 20892; Dr. Carolyn Clifford, (301) 496-8573, Division of Cancer Prevention, NCI, Executive Plaza N, Bethesda, MD 20892; Dr. Neal West, (301) 402-5867, NCCAM, Building 31, Bethesda, MD 20892; Dr. Momtaz Wassef, (301) 435-0550, Vascular Research Program, NHLBI, 6701 Rockledge Drive, Bethesda, MD 20892; Dr. Peter Preusch, (301) 594-5938, Pharmacology, Physiology, and Biological Chemistry Division, NIGMS, 45 Center Drive, Bethesda, MD 20892; Dr. Michael Ken May, (301) 594-8884, Division of Digestive Diseases and Nutrition, NIDDK, 45 Center Drive, Bethesda, MD 20892; Dr. Pamela Starke-Reed, (301) 496-6402, Office of Nutrition, NIA, 7201 Wisconsin Avenue, Bethesda, MD 20892; Dr. Jag Khalsa, (301) 443-1801, Center on AIDS and Other Medical Consequences of Drug Abuse, NIDA, 6001 Executive Building, Bethesda, MD 20892; Dr. Michael McClure, (919) 541-5327, Division of Extramural Research and Training, NIEHS, 111 TW Alexander Drive, Building 401, Research Triangle Park, NC 27709; and Dr. Kay Anderson, Research Program Director, (301) 402-1770, ORWH, 1 Center Drive, Bethesda, MD 20892. Annual deadline dates may vary; contact program staff for exact dates. RFA: OD-00-004
Requirements: Applications may be submitted by domestic for-profit and nonprofit organizations, both public and private; units of state and local governments; and eligible agencies of the federal government. Foreign organizations may participate if they are components of domestic U.S. organizations or via contractual or consortium agreements with domestic U.S. organizations.
Amount of Grant: Up to 15,000,000 U.S.D
Contact: Program Officer
Internet: http://grants.nih.gov/grants/guide/rfa-files/RFA-OD-00-004.html
Sponsor: National Institutes of Health
6701 Rockledge Drive
Bethesda, MD 20892

NIH Clinical Bioethics Postdoctoral Fellowships 3491
The Department of Clinical Bioethics of the Warren Grant Magnuson Clinical Center, NIH, offers a limited number of postdoctoral fellowship positions in bioethics. Fellows will participate in the activities and the intellectual life of the department and study ethical issues related to conduct of research, clinical practice, and health policy. Fellows will conduct their research under the guidance of senior faculty, participate in weekly bioethics seminars, case conferences, ethics consultations, and IRB deliberations, and have access to multiple educational opportunities at the NIH. Fellowships are two years in length, and stipends are based on the applicant's previous experience and the current U.S. government schedule. Guidelines are available online.
Requirements: Applicants should have a PhD, MD, JD, or other advanced degree in a relevant field. Fellows will be selected on the basis of their previous academic or professional achievements, commitment to scholarship, and the contribution they are likely to make in the field of bioethics. No prior experience in bioethics is necessary.
Date(s) Application is Due: Dec 30
Contact: Becky Chen, Department of Clinical Bioethics; (301) 496-2429; fax (301) 496-0760; bioethics-inquiries@mail.nih.gov
Internet: http://www.bioethics.nih.gov/opportunities/post-doc.html
Sponsor: National Institutes of Health
10 Center Drive, Building 10, Room 1C118
Bethesda, MD 20892-1156

NIH Clinical Bioethics Predoctoral Fellowships 3492
The Department of Clinical Bioethics of the Warren Grant Magnuson Clinical Center, NIH, supports a limited number of predoctoral fellowship positions in bioethics. Fellows will participate in the activities and intellectual life of the department, and study ethical issues related to the conduct of research, clinical practice, and health policy. Fellows will conduct their research under the guidance of the senior faculty, participate in weekly bioethics seminars, case conferences, ethics consultations, and IRB deliberations, and have access to multiple educational opportunities at NIH. Fellowships are two years in length. Stipends are based on the current U.S. government schedule. Application guidelines are available online.
Requirements: Applicants should have an undergraduate degree and should be planning postgraduate work. Fellows will be selected on the basis of their previous academic achievements, commitment to scholarship, and the contribution they are likely to make in the field of bioethics. No prior experience in bioethics is necessary.
Date(s) Application is Due: Jan 15
Contact: Program Contact, Department of Clinical Bioethics; (301) 496-2429; fax (301) 496-0760; bioethics-inquiries@mail.nih.gov
Internet: http://www.bioethics.nih.gov/opportunities/pre-doc.html
Sponsor: National Institutes of Health
10 Center Drive, Building 10, Room 1C118
Bethesda, MD 20892-1156

NIHCM Foundation Health Care Print Journalism Awards 3493
To encourage outstanding work from journalists furthering innovation in health care policy and management, The National Institute for Health Care Management Research and Educational Foundation (NIHCM Foundation) recognizes excellence in health care reporting and writing on the financing and delivery of health care and the impact of health care policy with two $10,000 awards. NIHCM Foundation awards prizes to the winning entry in the following two categories, Health Care Article or Series from General Circulation Publications and Health Care Article or Series from Trade Publications. The award prize will be presented to the winner at a dinner in Washington, DC in May.
Requirements: Articles must originally have been published during calendar year 2009. Entries will be accepted from all staff reporters and editors, as well as freelance writers for articles that appeared in a General Circulation or Trade Publication. Online articles will be accepted if produced and disseminated originally on the web in affiliation with a General Circulation or Trade Publication. Articles should be timely and produce new insights into health care financing and delivery or health care policy.
Date(s) Application is Due: Feb 26
Amount of Grant: 10,000 U.S.D
Contact: Nancy Chockley; (202) 296-4426; fax (202) 296-4319; nihcm@nihem.org
Internet: http://nihcm.org/awards/journalism
Sponsor: National Institute for Health Care Management Foundation
1225 19th Street NW, Suite 710
Washington, DC 20036

NIHCM Foundation Health Care Research Awards 3494
To encourage outstanding work from researchers furthering innovation in health care policy and management, the National Institute for Health Care Management (NIHCM) Research and Educational Foundation recognizes excellence in original and creative research with a $10,000 award. Selection criteria for the award includes the originality and creativity of the research. Impact that the research may have in advancing the public policy debate, improving health care delivery, management and the relevance to one or more of the subject areas. The award will be presented to the winner at a dinner in Washington, DC in May.
Requirements: Articles must have been published in a peer-reviewed journal or similar quality publication during calendar year 2009. Applications are encouraged from authors in diverse fields, including economics, health policy, political science, public health, etc. and from individuals in non-academic settings, such as research firms and policy organizations. All research articles must address some aspect of health care financing, delivery, organization or the implementation of health care policy.
Date(s) Application is Due: Feb 26
Amount of Grant: 10,000 U.S.D
Contact: Nancy Chockley; (202) 296-4426; fax (202) 296-4319; nihcm@nihem.org
Internet: http://nihcm.org/awards/research
Sponsor: National Institute for Health Care Management Foundation
1225 19th Street NW, Suite 710
Washington, DC 20036

NIHCM Foundation Health Care Television and Radio Journalism Awards 3495
The Third Annual NIHCM Foundation Health Care Television and Radio Journalism Award recognizes excellence in television and radio reporting on health care issues and policy. Selection criteria will include, new insights generated by the reporting, impact on health care policy and quality of writing and production. A $10,000 prize will be presented to the winner at a dinner in Washington, DC in May.
Requirements: Entries must originally have aired during calendar year 2009. Entries will be accepted from staff employees of national or local news organizations, nationally-distributed cable programs, or from freelance journalists who have produced reports

disseminated by same. Online stories will be accepted if produced for and disseminated originally on the web in affiliation with a television or radio news organization. Material repurposed from other media will not be accepted.
Date(s) Application is Due: Feb 26
Contact: Nancy Chockley; (202) 296-4426; fax (202) 296-4319; nihcm@nihcm.org
Internet: http://nihcm.org/awards/televison_and_radio_journalism
Sponsor: National Institute for Health Care Management Foundation
1225 19th Street NW, Suite 710
Washington, DC 20036

NIH Collaborations for Advanced Strategies in Complication of HIV Infection 3496
NIDA, NIAAA, NIDDK, and NIMH invite applicants to expand the clinical knowledge of the complications of HIV by expediting the translation of promising innovative preclinical findings into clinical research applications. The focus of the program is to support focused collaborative research between preclinical and clinical scientists to test, refine, and improve diagnostic, pathogenic, or therapeutic concepts. It anticipates such research will lead to innovative approaches for the prevention and management of complications causing significant morbidity or mortality in HIV infected persons, including the recently recognized metabolic complications, coincident pathogens such as hepatitis C virus, and the opportunistic pathogens. Inquiries may be directed to program staff at the appropriate NIH institute: Dr. Barbara Laughon, (301) 402-2304, NIAID, 6700 B Rockledge Drive, Bethesda, MD 20892; Dr. Jag Khalsa, Health Scientist Administrator, (301) 443-1801, NIDA, 6001 Executive Boulevard, Bethesda, MD 20892; Dr. Thomas Kresina, Chief, Biomedical Research Branch, (301) 443-6537, NIAAA, 6000 Executive Boulevard, Bethesda, MD 20892; Dr. Philip Smith, Senior Advisor, (301) 594-8816, NIDDK, 45 Center Drive, Bethesda, MD 20892; and Dr. Dianne Rausch, (301) 443-7281, NIMH, 6001 Executive Boulevard, Bethesda, MD 20892. Annual deadline dates vary; contact program staff for exact dates. PA: PA-00-048
Requirements: Applications may be submitted by domestic and foreign, for-profit and nonprofit organizations, both public and private; units of state and local governments; and eligible agencies of the federal government.
Contact: Program Officer
Internet: http://grants.nih.gov/grants/guide/pa-files/PA-00-048.html
Sponsor: National Institutes of Health
6701 Rockledge Drive
Bethesda, MD 20892

NIH Complementary and Alternative Medicine Postdoctoral Training Grants 3497
The Office of Alternative Medicine funds individual postdoctoral training awards using the National Research Service Award (NRSA) mechanism. The purpose is to provide a cadre of investigators capable of conducting systematic studies on safety, efficacy, cost effectiveness, or mechanisms of action of unconventional methods for treating major diseases and promoting well-being. This training is expected to attract postdoctoral candidates who are in the early stages of their careers. They will have obtained expertise in conventional research methodology and some familiarity with/or interest in alternative medical procedures. Prospective trainees will be expected to form alliances with established researchers to provide mutual learning experiences. Individuals may request up to three years of aggregate NRSA support at the postdoctoral level. Written and telephone inquiries are invited. PA: PA-01-088
Requirements: U.S. citizens or noncitizen nationals may apply if, prior to beginning the award, they have received the PhD, MD, DO, DDS, DVM, OD, DPM, ScD, D.Eng, DrPH, DNS, DPharm, or equivalent doctoral degree from an accredited domestic or foreign institution. Before submitting an application, the applicant must arrange for appointment to an appropriate accredited university, hospital, or other institution with facilities including staff for postdoctoral training. The candidate should be accepted by a sponsor who will actively supervise the training. The sponsor should have research experience in clinical medicine and/or basic preclinical research along with an involvement in the evaluation of alternative medicine. The sponsor should be qualified to supervise in the application of rigorous study design to the assessment of individual alternative therapies.
Date(s) Application is Due: Apr 5; Aug 5; Dec 5
Amount of Grant: 28,260 - 44,412 U.S.D
Contact: Dr. Nancy Pearson, (301) 594-0519; fax (301) 480-3621; pearsonn@mail.nih.gov
Internet: http://grants2.nih.gov/grants/guide/pa-files/PA-01-088.html
Sponsor: National Institutes of Health
6120 Executive Boulevard, Executive Plaza S
Rockville, MD 20892-9904

NIH Demographic Research on Sexual Behaviors Related to HIV 3498
NICHD, NIDR and NIMH invite qualified researchers to submit applications to study the social and behavioral aspects of the transmission of HIV through sexual intercourse, including oral sexual practices. Inquiries may be directed to program staff at the appropriate NIH institute: Dr. Susan Newcomer, (301) 496-1174, NICHD, 6100 Executive Boulevard, Bethesda, MD 20892; Dr. Willo Pequegnat, (301) 443-6100, NIMH, Parklawn Building, 5600 Fishers Lane, Rockville, MD 20857. Annual deadline dates may vary; contact program staff for exact dates. PAS-00-136
Requirements: Applications may be submitted by domestic for-profit and nonprofit organizations, both public and private; units of state and local governments; and eligible agencies of the federal government.
Contact: Program Officer
Internet: http://grants2.nih.gov/grants/guide/pa-files/PAS-00-136.html
Sponsor: National Institutes of Health
6701 Rockledge Drive
Bethesda, MD 20892

NIH Dietary Supplement Research Centers: Botanicals Grants 3499
ODS, NCCAM and NCI invite new and renewal applications to support research centers which will: promote collaborative integrated interdisciplinary study of botanicals, particularly those found as ingredients in dietary supplements; and conduct research of high potential for being translated into practical benefits for human health. This initiative is intended to advance the spectrum of botanical research activities, ranging from plant identification and characterization to early phase clinical studies. Preclinical research that will inform future clinical studies is encouraged as the primary research focus of this program. ODS has a particular interest in botanicals as part of health care for conditions relating to health maintenance or primary prevention. NCCAM is interested in the study of botanicals broadly used by the American public. The NCI Division of Cancer Prevention is interested in supporting research focused on mechanisms by which botanically derived bioactive food components might influence cancer risk and tumor behavior. An applicant may request a project period of up to five years; an annual budget in direct costs up to $1.0 million.
Requirements: The following organizations and institutions are eligible to apply: individuals; public/state controlled institutions of higher education; private institutions of higher education; Hispanic-serving institutions; historically Black colleges and universities (HBCUs); tribally controlled colleges and universities (TCCUs); Alaska Native and Native Hawaiian serving institutions; nonprofits with 501(c)3 IRS status; nonprofits without 501(c)3 IRS status; for-profit organizations; state governments; Indian/Native American tribal governments (federally recognized); Indian/Native American tribally designated organizations; county governments; city or township governments; special district governments; independent school districts; public housing authorities/Indian housing authorities; U.S. territory or possession; Indian/Native American tribal governments (other than federally recognized); regional organizations; non-domestic (non-U.S.) entities (foreign organizations); other eligible agencies of the federal government; and faith-based or community-based organizations.
Date(s) Application is Due: Dec 3
Amount of Grant: 1,000,000 U.S.D
Contact: Christine A. Swanson, Director, Botanical Research Centers Program; (301) 435-2920; fax (301) 480-1845; Swansonc@od.nih.gov
Internet: http://grants.nih.gov/grants/guide/pa-files/PAR-09-091.html
Sponsor: National Institutes of Health
6100 Executive Boulevard, Room 3B01, MSC 7517
Bethesda, MD 20892

NIH Earth-Based Research Relevant to the Space Environment 3500
The purpose of this Program Announcement (PA) is to stimulate ground-based research on basic, applied, and clinical biomedical and behavioral problems that are relevant to human space flight or that could use the space environment as a laboratory. Although none of the research supported under this initiative would be conducted in space, it is anticipated that it would form a basis for future competitively reviewed studies which could be conducted on the International Space Station, or other space flight opportunities, by skilled on-board specialists. Modular Grant applications will request direct costs in $25,000 modules, up to a total direct cost request of $250,000 per year.
Requirements: The following organizations and institutions may apply: public/state controlled institutions of higher education; private institutions of higher education; Hispanic-serving institutions; historically Black colleges and universities (HBCUs); tribally controlled colleges and universities (TCCUs); Alaska native and native Hawaiian serving institutions; nonprofits with 501(c)3 IRS status; nonprofits without 501(c)3 IRS status; small businesses; for-profit organizations; State governments; regional organizations; U.S. territories or possessions; Indian/Native American tribal governments; Indian/Native American tribally designated organizations; county governments; city or township governments; special district governments; independent school districts; public housing authorities/Indian housing authorities; eligible agencies of the Federal government; and faith-based or community based organizations.
Date(s) Application is Due: Feb 5; Jun 5; Oct 5
Amount of Grant: Up to 250,000 U.S.D
Contact: Susan Grove; (301) 435-0714; sg16d@nih.gov or grantsinfo@od.nih.gov
Internet: http://grants.nih.gov/grants/guide/pa-files/PA-00-088.html
Sponsor: National Institutes of Health
6701 Rockledge Drive
Bethesda, MD 20892

NIH Ecology of Infectious Diseases 3501
NIH, NSF, NASA, U.S.GS, and the Agricultural Research Service (ARS) invite applications for the establishment of research programs to elucidate the underlying mechanisms that govern the relationships between anthropogenic environmental changes and the transmission dynamics of infectious diseases. This RFA calls for the development of interdisciplinary research programs on the ecology of infectious diseases in the context of anthropogenic environmental changes such as biodiversity loss, habitat transformation, environmental contamination, climate change, and other influences. The focus of this RFA is on discovery of basic ecological and biological mechanisms and development of predictive models for the emergence and transmission of diseases in humans and other animals, and ultimately the development of strategies to prevent or control them. Annual deadline dates may vary; contact program staff for exact dates. RFA: TW-00-002
Requirements: Applications may be submitted by domestic and foreign, for-profit and nonprofit organizations, both public and private (including domestic institutions with foreign collaborators); units of state and local governments; and eligible agencies of the federal government.
Amount of Grant: Up to 4,000,000 U.S.D
Contact: Program Officer
Internet: http://grants.nih.gov/grants/guide/rfa-files/RFA-TW-00-002.html

Sponsor: National Institutes of Health
6701 Rockledge Drive
Bethesda, MD 20892

NIH Economic Studies in Cancer Prevention, Screening, and Care 3502

NCI and AHCPR invite investigator-initiated grant applications for research directed at increasing the knowledge base in the area of the economic aspects of cancer prevention, screening, and care. The goal of this program announcement is to generate new economic knowledge that will promote the optimal design of cancer prevention and control trial studies and interventions and will facilitate the formulation of effective health care policy related to cancer prevention and control. NIDCR has an interest in economic and health services studies as these relate specifically to oral and pharyngeal cancers and health care policies related to prevention and control of oral cancers. This initiative requests research applications on new methods development, the synthesis and extension of existing methods, and innovative data gathering strategies. Applications that propose to implement actual data collection on a pilot or full-scale basis as well as analytical studies that use existing data and methodology will be entertained. PA: PA-04-017
Requirements: Foreign and domestic, for-profit and nonprofit organizations, both public and private; units of state and local governments; and eligible agencies of the federal government may apply.
Contact: Crystal Wolfrey, Grants Administration Branch, NCI; (301) 496-8634; fax (301) 496-4801; crystal.wolfrey@nih.gov or wolfreyc@mail.nih.gov
Internet: http://grants1.nih.gov/grants/guide/pa-files/PA-04-017.html
Sponsor: National Institutes of Health
6701 Rockledge Drive
Bethesda, MD 20892

NIH Endocrine Disruptors: Epidemiologic Approaches Grants 3503

Through this program NIOSH, EPA, NIEHS, and NCI support research on the relationship between exposure to endocrine disruptors and adverse health effects in humans, particularly reproductive and developmental, with a focus on epidemiologic approaches. Synthetic chemicals released in the workplace and environment collectively called endocrine disruptors (EDs) or endocrine disrupting chemicals (EDCs) are thought to interfere with one or more aspects of the normal endocrine functions that are responsible for the maintenance of homeostasis and the regulation of developmental processes. The cooperating agencies support research that examines human health consequences of exposure to physical and chemical toxicants in the environment and in occupational settings as well as the underlying mechanisms involved in environmentally related disease and dysfunction. Research applications are sought that will investigate the relationships between exposure to endocrine disruptors and adverse health effects, particularly reproductive/developmental effects in humans. Annual deadline dates may vary; contact program staff for exact dates. Inquiries may be directed to program staff at the appropriate institute: Michael Galvin, Research Grants Program, (404) 639-3343, NIOSH, 1600 Clifton Road NE, Atlanta, GA 30333; David Reese, (202) 564-6919, EPA, 1200 Pennsylvania Ave NW, Washington, DC 20460; Kumiko Iwamoto, (301) 435-4911, NCI, 6130 Executive Boulevard, Bethesda, MD 20892; and Gwen Collman, (919) 541-4980, NIEHS, P.O. Box 12233, MD EC-21, Research Triangle Park, NC 27709. RFA: OH-01-001
Requirements: Applications may be submitted by domestic and foreign, public and private nonprofit and for-profit organizations, and by governments and their agencies; state and local governments or their bona fide agents; and federally recognized Indian tribal governments, Indian tribes, or Indian tribal organizations.
Amount of Grant: Up to 6,000,000 U.S.D
Contact: Program Officer
Internet: http://www.grants.nih.gov/grants/guide/rfa-files/RFA-OH-01-001.html
Sponsor: National Institutes of Health
6701 Rockledge Drive
Bethesda, MD 20892

NIH Enhancing Adherence to Diabetes Self-Management Behaviors 3504

This PA solicits applications for investigator-initiated research related to sociocultural, environmental, and behavioral mechanisms and biological/technological factors that contribute to successful and ongoing self-management in diabetes. Self-management is defined as client strategies and behaviors that contribute to blood glucose normalization, improved health, and prevention or reduction of complications. It is broader than adherence to specific regimen components and incorporates deliberate problem solving and decision-making processes. Applications are encouraged for both type 1 and type 2 diabetes; representative and minority populations; and all age groups. Inquiries may be directed to program staff at the appropriate NIH institute. Modular Grant applications will request direct costs in $25,000 modules, up to a total direct cost request of $250,000 per year.
Requirements: Applications may be submitted by domestic and foreign, for-profit and non-profit organizations, public and private, such as universities, colleges, hospitals, laboratories, units of State and local governments, and eligible agencies of the Federal government. Racial/ethnic minority individuals, women, and persons with disabilities are encouraged to apply as principal investigators.
Date(s) Application is Due: Feb 16; Jun 16; Oct 16
Amount of Grant: Up to 250,000 U.S.D
Contact: Susan Grove, (301) 435-0714; sg16d@nih.gov or grantsinfo@od.nih.gov
Internet: http://grants.nih.gov/grants/guide/pa-files/PA-00-049.html
Sponsor: National Institutes of Health
6701 Rockledge Drive
Bethesda, MD 20892

NIH Epidemiological and Behavioral Research in Oral Health 3505

This program seeks to stimulate epidemiological and behavioral research to improve oral health or increase the understanding of the determinants of oral health. It supports studies that expand and enhance the integration of population-based, genetic, social, and behavioral research. Specific research activities highlighted in the strategic plan include, but are not limited to, furthering the understanding of how genetic, behavioral, and environmental factors interact to contribute to disease susceptibility or resistance; assessing and bridging gaps in the knowledge, opinions, and practices of the public, educators, and health care professionals about oral diseases and their prevention; identifying risk factors and markers for the progression of disease to oral, dental, and craniofacial diseases and disorders; improving oral health literacy; and developing more effective strategies to encourage the appropriate and timely adoption of research findings into clinical practice. PA Number: PA-04-022
Requirements: Applications may be submitted by domestic and foreign, for-profit and nonprofit organizations, both public and private; units of state and local governments; and eligible agencies of the federal government.
Contact: Mary Daley; (301) 594-4808; fax (301) 480-3562; daleym@mail.nih.gov
Internet: http://www.grants.nih.gov/grants/guide/pa-files/PA-04-022.html
Sponsor: National Institutes of Health
6701 Rockledge Drive
Bethesda, MD 20892

NIH Ethical Issues in Human Studies Research Grants 3506

NIH and CDC invite research grant applications (R01) for support to conduct research on ethical issues that arise with research involving human participants. Grants may be requested in $25,000 direct cost modules. Annual deadline dates may vary; contact program staff for exact dates. PA: PA-02-103
Requirements: Applications may be submitted by any domestic and foreign, for-profit and nonprofit, public and private organizations; units of state and local governments; and eligible agencies of the federal government. Applicants are encouraged to visit the Web sites of the participating institutes and centers to learn more about their specific interests.
Restrictions: The PA does not solicit essays, editorials, literature or historical reviews of science, conferences, or speculative, theoretical, or policy analyses.
Date(s) Application is Due: Feb 1; Jun 1; Oct 1
Amount of Grant: Up to 250,000 U.S.D
Contact: Dr. Della Hann, Program Contact; (301) 402-2725; hannd@od.nih.gov
Internet: http://grants2.nih.gov/grants/guide/pa-files/PA-02-103.html
Sponsor: National Institutes of Health
6701 Rockledge Drive
Bethesda, MD 20892

NIH Exceptional, Unconventional Research Enabling Knowledge Acceleration (EUREKA) Grants 3507

The purpose of the EUREKA (Exceptional Unconventional Research Enabling Knowledge Acceleration) initiative is to foster exceptionally innovative research that, if successful, will have an unusually high impact on the areas of science that are germane to the mission of one or more of the participating NIH Institutes. Before submitting an application, it is extremely important to verify that the proposed research is of interest to at least one of the NIH Institutes that is participating in the EUREKA FOA, since applications that are not germane to the mission of one or more of the participating Institutes will be withdrawn without review.
Requirements: The following organizations and institutions are eligible to apply: public/state controlled institutions of higher education; private institutions of higher education; Hispanic-serving institutions; historically Black colleges and universities (HBCUs); tribally controlled colleges and universities (TCCUs); Alaska Native and Native Hawaiian serving institutions; nonprofits with 501(c)3 IRS status (other than institutions of higher education); nonprofits without 501(c)3 IRS status (other than institutions of higher education); small businesses; for-profit organizations (other than small businesses); non-domestic (non-U.S.) entities (foreign organizations).
Restrictions: EUREKA is for new projects, not for continuation of existing projects. Nor is EUREKA for support of pilot projects, i.e., projects of limited scope that are designed primarily to generate data that will enable the PI to seek other funding opportunities.
Date(s) Application is Due: Nov 24
Amount of Grant: Up to 800,000 U.S.D
Contact: Laurie Tompkins, (301) 594-0943; tompkinL@nigms.nih.gov
Internet: http://grants.nih.gov/grants/guide/rfa-files/RFA-GM-10-009.html
Sponsor: National Institutes of Health
45 Center Drive, MSC 6200
Bethesda, MD 20892-6200

NIH Exploratory/Developmental Research Grant 3508

The Exploratory/Developmental Grant (R21) mechanism is intended to encourage exploratory and developmental research projects by providing support for the early and conceptual stages of these projects. These studies may involve considerable risk but may lead to a breakthrough in a particular area, or to the development of novel techniques, agents, methodologies, models, or applications that could have a major impact on a field of biomedical, behavioral, or clinical research.
Requirements: Eligible organizations: for profit organizations; non-profit organizations; public or private institutions, such as universities, colleges, hospitals and laboratories; units of state government; units of local government; eligible institutions of the federal government; domestic institutions; foreign institutions; faith-based or community-based organizations; units of state tribal government; and units of local tribal government. Eligible Project Directors/Principal Investigators: any individual with the skills,

knowledge and resources necessary to carry out the proposed research, is invited to work with his/her institution to develop an application for support; individuals from underrepresented racial and ethnic groups; individuals with disabilities
Date(s) Application is Due: Feb 16; Jun 16; Oct 16
Amount of Grant: Up to 275,000 U.S.D
Contact: Ellen S. Liberman, (301) 451-2020; esl@nei.nih.gov
Internet: http://grants.nih.gov/grants/guide/pa-files/PA-06-181.html
Sponsor: National Eye Institute
2020 Vision Place
Bethesda, MD 20892-3655

NIH Exploratory/Development Grants for Mental Health Intervention Research 3509
NIMH is requesting research grant applications to develop somatic, pharmacologic, psychosocial, or combination interventions for the prevention, rehabilitation, treatment, and/or management of mental disorders and emotional or behavioral problems; research the co-occurrence of mental, physical, and substance abuse problems and the functional consequences of these problems across the life span; and to perform preliminary tests of feasibility, acceptability, and efficacy/effectiveness of interventions. PA-03-078
Requirements: Applications may be submitted by domestic and foreign, for-profit and nonprofit organizations, both public and private; units of state and local governments; and eligible agencies of the federal government.
Amount of Grant: 125,000 U.S.D
Contact: Joy Knipple, Grants Management Branch; (301) 443-8811; fax (301) 443-6885; jk173@nih.gov
Internet: http://grants.nih.gov/grants/guide/pa-files/PAR-03-078.html
Sponsor: National Institute of Mental Health
6701 Rockledge Drive, Room 1040, MSC 7710
Bethesda, MD 20892

NIH Exploratory Centers for Alternative Medicine Research Grants 3510
The objective of the program is to develop exploratory centers for the purpose of investigating and validating alternative therapies. These centers will support new interdisciplinary programs involving experienced investigators from conventional medicine with clinicians and investigators from alternative medicine and should provide clinical/scientific/technical assistance to alternative medicine clinicians who may have potentially useful accumulations of patient data that could be evaluated for treatment outcome. It will be important that the research faculty at these centers have expertise in broad areas such as biostatistics, computer processing, data management, and protocol design. Additionally, through a peer review process that will be established by the centers, small prospective pilot studies will be started. Program areas to be developed and evaluated can include nutrition, diet, and lifestyle/behavioral health changes; mind/body control therapies; traditional and ethnomedicine therapies; structural manipulations and energetic therapies; pharmacological and biological therapies; and bioelectromagnetic therapies. Annual deadline dates vary; contact program staff for exact dates. RFA: OD-94-004
Requirements: Applications may be submitted by domestic for-profit, nonprofit, public, and private organizations.
Amount of Grant: 20,000 - 350,000 U.S.D
Contact: Dr. John Spencer; (301) 402-2466; fax (301) 402-4741
Internet: http://www.grants.nih.gov/grants/guide/rfa-files/RFA-OD-94-004.html
Sponsor: National Institutes of Health
6120 Executive Boulevard, Executive Plaza S
Rockville, MD 20892-9904

NIH Exploratory Innovations in Biomedical Computational Science and Technology (BISTI) Grants 3511
The NIH is interested in promoting research and developments in computational science and technology that will support rapid progress in areas of scientific opportunity in biomedical research. As defined here, biomedical computing or biomedical information science and technology includes database design, graphical interfaces, querying approaches, data retrieval, data visualization and manipulation, data integration through the development of integrated analytical tools, and tools for electronic collaboration, as well as computational and mathematical research including the development of structural, functional, integrative, and analytical models and simulations. This particular FOA is intended to support exploratory biomedical informatics and computational biology research - applications should be innovative, with high risk/high impact in new areas that are lacking preliminary data or development. Because the nature and scope of the proposed research will vary from application to application, it is anticipated that the size and duration of each award will also vary.
Requirements: Application can be made by the following institutions and organizations: for-profit organizations; non-profit organizations; public or private institutions (such as universities, colleges, hospitals, and laboratories); units of state government; units of local government; eligible agencies of the Federal government; foreign institutions; domestic institutions; faith-based or community-based organizations; Indian/Native American tribal governments (federally recognized); Indian/Native American tribal governments (other than federally recognized); and Indian/Native American tribally designated organizations.
Date(s) Application is Due: Feb 16; Jun 16; Oct 16
Contact: Peter Lyster, (301) 451-6446; lysterp@mail.nih.gov
Internet: http://grants.nih.gov/grants/guide/pa-files/PAR-06-411.html
Sponsor: National Institutes of Health
45 Center Drive, Room 2AS.55k, MSC 6200
Bethesda, MD 20892-6200

NIH Fine Mapping and Function of Genes for Type 1 Diabetes Grants 3512
The purpose of this initiative is to bring together investigators with experience in genetics, immunology, and biochemistry to perform fine mapping of the loci discovered and study the function of the genes being identified and to identify the mechanisms by which newly discovered genes predispose to T1D This would lead to elucidation of the mechanisms whereby changes in the function or regulation of these genes are likely to provide crucial new insights into disease pathogenesis. The discovery of the genes would be relevant to developing a predictive strategy for individuals who may develop diabetes. In the future, genetic testing may lead to personalized treatment regimens by identifying the most appropriate class of drugs for particular patients. Total funding available for this five year program is $20 million to fund 4-10 applications, contingent upon the availability of funds and the submission of a sufficient number of meritorious applications. The maximum direct costs are $5 million that can be spent on a project period of up to 5 years. Letters of Intent are due by February 27, with final applications due by March 30. The office strongly urges communication from applicants via email.
Requirements: The following organizations and institutions are eligible to apply: public/state controlled institutions of higher education; private institutions of higher education; Hispanic-serving institutions; historically Black colleges and universities (HBCUs); tribally controlled colleges and universities (TCCUs); Alaska Native and Native Hawaiian serving institutions; nonprofits with 501(c)3 IRS status (other than institutions of higher education); nonprofits without 501(c)3 IRS status (other than institutions of higher education); small businesses; for-profit organizations (other than small businesses); non-domestic (non-U.S.) entities (foreign organizations); and other eligible agencies of the federal government.
Date(s) Application is Due: Feb 27; Mar 30
Amount of Grant: 1,500,000 - 5,000,000 U.S.D
Contact: Beena Akolkar, PhD.; (301) 594-8812; ba92i@nih.gov
Internet: http://grants.nih.gov/grants/guide/rfa-files/rfa-dk-08-006.html
Sponsor: National Institutes of Health
6707 Democracy Boulevard, Room 6105
Bethesda, MD 20892-5460

NIH Genetic Architecture of Complex Phenotypes 3513
The purpose of this initiative is to support new studies on the architecture of complex phenotypes, including research using human and model systems as well as research using theoretical approaches.The studies targeted by this program announcement are expected to expand our understanding of the roles of genetic and environmental variation and their interactions in causing phenotypic variation in populations; increase the quantity and quality of population-based data; lead to the development of mathematical and statistical tools for analyzing measured genotype data; lead to improvements in study design; and create biologically relevant models for understanding the origins, roles, and implications of genetic variation in causing variation in phenotypes. This program announcement encourages efforts including organization of DNA sequence variation in populations and subpopulations with the goal of understanding the causal forces that shape that organization; studies of the relationship of genotypic and phenotypic variation; data analysis; research on study design, including comparisons of different strategies for studying complex phenotypes; development of statistical and analytical tools; and construction of meaningful tests of relationships between genotypic and phenotypic data. Further information for this program is available on the Web site. Inquiries may be directed to program staff at the appropriate NIH institute; Dr. Irene Anne Eckstrand, (301) 594-0943, NIGMS, 45 Center Drive, Bethesda, MD 20892; Dr. Jose Velazquez, (919) 541-4998, NIEHS, P.O. Box 12233, Research Triangle Park, NC 27709; Dr. William Sharrock, (301) 594-5055, NIAMS, 45 Center Drive, Bethesda, MD 20892; Dr. Evan Hadley, (301) 435-3048, NIA, 7201 Wisconsin Avenue, Bethesda, MD 20892; Dr. Steven Moldin, (301) 443-2037, NIMH, 5600 Fishers Lane, Rockville, MD 20857; Dr. Harold Gordon, (301) 443-4877, NIDA, 5600 Fishers Lane, Rockville, MD 20857; Dr. Robert Karp, (301) 443-4223, NIAAA, 6000 Executive Boulevard, Bethesda, MD 20892; Dr. Ralph Nitkin, (301) 496-1383, NICHD, 6100 Executive Boulevard, Bethesda, MD 20892; Dr. Lisa Brooks, (301) 496-7531, NHGRI, 38 Library Drive, Bethesda, MD 20892; and Dr. Cashell Jaquish, (301) 435-0447, NHLBI, 6701 Rockledge Drive, Bethesda, MD 20892. PA: PA-02-110
Requirements: Applications may be submitted by domestic and foreign, for-profit and nonprofit organizations, both public and private. Racial/ethnic minority individuals, women, and persons with disabilities are encouraged to apply.
Restrictions: Foreign institutions are not eligible for program projects (P01) grants.
Contact: Program Officer
Internet: http://grants2.nih.gov/grants/guide/pa-files/PA-02-110.html
Sponsor: National Institutes of Health
6707 Democracy Boulevard, Room 6105
Bethesda, MD 20892-5460

NIH Genetics of Neurobehavioral Disorders in Existing Samples Grants 3514
The purpose of the program is to encourage applications for multidisciplinary, collaborative genetic research projects, employing state-of-the-art diagnostic procedures and analytic methods, to integrate different pedigree data sets that have already been collected for the study of a given neurobehavioral disorder. Applications will implement the assembling of data from multiple research groups for combined analysis, and will focus on one of the following: attention-deficit hyperactivity disorder, bipolar disorder or other related mood disorders, recurrent early-onset depression, eating disorders, obsessive-compulsive disorder or other anxiety disorders, panic disorder, schizophrenia or other psychotic disorders, personality disorders, or Tourette's syndrome. The creation and analysis of clinically well-characterized samples of sufficient size for linkage analyses and for linkage disequilibrium mapping studies in genetically isolated populations studied previously may be facilitated

by the establishment of international consortia. Full collaborations between American scientists and scientists at international institutions are encouraged, when scientifically appropriate. In these cases, awards may be made to international institutions or to domestic applications that include international components. Inquiries may be directed to program staff at the appropriate NIH institute: Dr. Steven Moldin, Division of Neuroscience and Basic Behavioral Science, NIMH, (301) 443-2037, 6001 Executive Boulevard, Bethesda, MD 20892; Dr. Lisa Freund, Child Development and Behavior Branch, NICHD, (301) 435-6879, Building 6100, 9000 Rockville Pike, Bethesda, MD 20892. Annual deadline dates may vary; contact program staff for exact dates. PA: PAR-99-120
Requirements: Applications may be submitted by domestic, for-profit and nonprofit organizations, both public and private; units of state and local governments; and eligible agencies of the federal government.
Date(s) Application is Due: Feb 1; Jun 1; Oct 1
Amount of Grant: Up to 250,000 U.S.D
Contact: Program Officer
Internet: http://www.grants.nih.gov/grants/guide/pa-files/PA-99-120.html
Sponsor: National Institutes of Health
6701 Rockledge Drive
Bethesda, MD 20892

NIH Genetic Susceptibility and Variability of Human Malformations Research Grants **3515**

NICHD, NIDCR, NIEHS, and EPA invite research grant applications to study the genetic susceptibility and mechanisms of human congenital malformations including research on the contribution of potential genetic and environmental factors, identified at the molecular level, to the etiology, distribution, and prevention of disease within families and across populations. The research conducted as a result of this RFA will provide the direction and focus for future initiatives to dissect the molecular and genetic mechanisms underlying specific human malformations. The current initiative is designed to encourage collaborative, interdisciplinary, and innovative studies that integrate the latest advances in developmental genetics, functional genomics, epidemiology, and high throughput biotechnology with the new mathematical, methodological, and statistical tools for analyzing and evaluating data related to the etiology and distribution of birth defects in populations. The primary objective of this initiative is to develop and use integrated multidisciplinary approaches to obtain information that will enhance our knowledge of and provide a basis for future initiatives focusing on the basic mechanisms and etiologies of human malformations. This RFA is part of a larger birth defects program which will include an upcoming initiative related to fundamental developmental mechanisms underlying the formation of structural anomalies. Inquiries may be directed to the staff at the appropriate institute/organization: Dr. Deborah Henken, (301) 496-5541, Center for Research for Mothers and Children, NICHD, Executive Building, 6100 Executive Boulevard, Bethesda, MD 20892; Dr. Judy Small, (301) 594-2425, Craniofacial Anomalies and Injury Branch, NIDCR, Building 45, Bethesda, MD 20892; Dr. Gwen Collman, (919) 541-4980, Chemical Exposures and Molecular Biology Branch, Division of Extramural Research and Training, NIEHS, P.O. Box 12233 Research Triangle Park, NC 27709; and Dr. Chris Saint, (202) 564-6909, National Center for Environmental Research & Quality Assurance, EPA, 1300 Pennsylvania Avenue, Washington, DC 20004. Annual deadline dates may vary; contact program staff for exact dates. RFA: HD-99-002
Requirements: Applications may be submitted by domestic and foreign, for-profit and nonprofit, public and private organizations; units of state and local governments; and eligible agencies of the federal government.
Restrictions: Applications limited to descriptive or traditional epidemiological approaches will be considered nonresponsive. The EPA will not be able to consider applications from for-profit organizations or federal agencies.
Amount of Grant: Up to 700,000 U.S.D
Contact: Program Officer
Internet: http://www.grants.nih.gov/grants/guide/rfa-files/RFA-HD-99-002.html
Sponsor: National Institutes of Health
6701 Rockledge Drive
Bethesda, MD 20892

NIH Health Promotion Among Racial and Ethnic Minority Males **3516**

This Funding Opportunity Announcement solicits Exploratory/Developmental grant applications from applicants that propose to stimulate and expand research in the health of minority men. Specifically, this initiative is intended to: enhance our understanding of the numerous factors (e.g., sociodemographic, community, societal, personal) influencing the health promoting behaviors of racial and ethnic minority males and their sub-populations across the life cycle; and solicit applications focusing on the development and testing of culturally and linguistically appropriate health-promoting interventions designed to reduce health disparities among racially and ethnically diverse males and their sub-populations age 21 and older. The total project period for an application submitted in response to this funding opportunity may not exceed two years. Direct costs are limited to $275,000 over an R21 two-year period, with no more than $200,000 in direct costs allowed in any single year.
Requirements: Application can be made by the following institutions and organizations: public/state controlled institutions of higher education; private institutions of higher education; nonprofits with 501(c)3 IRS status (other than institutions of higher education); nonprofits without 501(c)3 IRS status (other than institutions of higher education); small businesses; for-profit organizations (other than small businesses); state governments; U.S. territories or possessions; Indian/Native American tribal governments (federally recognized); Indian/Native American tribal governments (other than Federally recognized); Indian/Native American tribally designated organizations; non-domestic (non-U.S.) entities (foreign organizations); Hispanic-serving institutions; historically Black colleges and universities (HBCUs); tribally controlled colleges and universities (TCCUs); Alaska Native and Native Hawaiian serving institutions; regional organization; eligible agencies of the Federal government; and faith-based or community based organizations.
Date(s) Application is Due: Feb 16; Jun 16; Oct 16
Amount of Grant: 200,000 - 275,000 U.S.D
Contact: Dr. Charlotte Pratt, Program Director; (301) 435-0382; fax (301) 480-5158; prattc@nhlbi.nih.gov
Internet: http://grants.nih.gov/grants/guide/pa-files/PA-07-421.html
Sponsor: National Institutes of Health
6701 Rockledge Drive, Suite 10118, MSC 7936
Bethesda, MD 20892-7936

NIH Human Brain Project (Neuroinformatics) Phase I & II **3517**

The purpose of this initiative is to encourage and support investigator- initiated research on neuroscience informatics (neuroinformatics). This research will lead to the development of new web based databases, analytical tools, and knowledge management systems to foster sharing of data for all domains of neuroscience research. This program combines neuroscience and informatics (neuroinformatics) research to develop and apply advanced tools and approaches essential for efficient understanding of the structure, function and development in health and disorders of the nervous system, from the genetic to whole systems level. Annual deadline dates may vary; contact program staff for exact dates. PA: PAR-03-035
Requirements: Applications may be submitted by domestic and foreign, for-profit and nonprofit organizations, both public and private; units of state and local governments; and eligible agencies of the federal government.
Restrictions: Foreign institutions are not eligible to apply.
Date(s) Application is Due: Jan 21; May 20; Sep 22
Contact: Stephen Koslow, Office on Neuroinformatics, NIMH; (301) 443-1815; fax (301) 443-1867; koz@helix.nih.gov
Internet: http://grants2.nih.gov/grants/guide/pa-files/PAR-03-035.html
Sponsor: National Institutes of Health
6701 Rockledge Drive, Room 1040
Bethesda, MD 20892

NIH Human Connectome Project Grants **3518**

This Funding Opportunity Announcement is issued as an initiative of the NIH Blueprint for Neuroscience Research. The Neuroscience Blueprint is a collaborative framework through which 16 NIH Institutes, Centers and Offices jointly support neuroscience-related research, with the aim of accelerating discoveries and reducing the burden of nervous system disorders. The overall purpose of this five year Human Connectome Project (HCP) is to develop and share knowledge about the structural and functional connectivity of the human brain. This purpose will be pursued through the following specific efforts: existing, but cutting-edge, non-invasive imaging technologies will be optimized and combined to acquire structural and functional in vivo data about axonal projections and neural connections from brains of hundreds of healthy adults. Demographic data and data regarding sensory, motor, cognitive, emotional, and social function will also be collected for each subject, as will DNA samples and blood (to establish cell lines). Models to better understand and use these data will be developed. Connectivity patterns will be linked to existing architectonic data. Data and models will be made available to the research community immediately via a user-friendly system to include tools to query, organize, visualize and analyze data. Outreach activities will be conducted to engage and educate the research community about the imaging tools, data, models, and informatics tools. After five years, these specific efforts are expected to deliver: a set of integrated, non-invasive imaging tools to obtain connectivity data from humans in vivo; a high quality and well characterized, quantitative set of human connectivity data linked to behavioral and genetic data as well as to general, existing architectonic data, and associated models, from up to hundreds of healthy adult female and male subjects; and rapid, user-friendly dissemination of connectivity data, models, and tools to the research community via outreach activities and an informatics platform. Letters of Intent are due by October 24, with final applications due by November 24.
Requirements: The following organizations and institutions are eligible to apply for support: public/state controlled institutions of higher education; private institutions of higher education; Hispanic-serving institutions; historically Black colleges and universities (HBCUs); tribally controlled colleges and universities (TCCUs); Alaska Native and Native Hawaiian serving institutions; nonprofits with 501(c)3 IRS status (other than institutions of higher education); nonprofits without 501(c)3 IRS status (other than institutions of higher education); Indian/Native American tribal governments (federally recognized); Indian/Native American tribally designated organizations; U.S. territories or possessions; and eligible agencies of the federal government.
Date(s) Application is Due: Oct 24; Nov 24
Amount of Grant: Up to 6,000,000 U.S.D
Contact: Michael F. Huerta, (301) 443-1815; fax (301) 443-1731; mhuertl@mail.nih.gov
Internet: http://grants.nih.gov/grants/guide/rfa-files/RFA-MH-10-020.html
Sponsor: National Institutes of Health
6001 Executive Boulevard, Room 7202, MSC 9645
Bethesda, MD 20892-9645

NIH Human Models and Markers of Skeletal Aging Grants **3519**

NIA, NIAMS, NIDR, and NIDDK invite research grant applications to develop or refine models and markers using human bone, bone marrow, or bone cell constituents that more accurately reflect those age-related and/or pathophysiological processes occurring within the mature human skeleton that lead to osteoporosis and other age-related skeletal diseases. It is expected that improved human models and markers of age-related changes in skeletal tissue structure and function will enhance the accuracy and

reliability of diagnostic and prognostic capabilities and serve to expedite the development of more effective, targeted preventative, and therapeutic strategies. Annual deadlines vary; contact program staff for exact dates. Additional information may be obtained by mail and email, or through the NIH grant line (301-402-2221), and the NIH Web site. PA: PA-96-022
Requirements: Applications may be submitted by domestic and foreign, for-profit and nonprofit organizations, both public and private; units of state and local governments; and eligible agencies of the federal government.
Date(s) Application is Due: Feb 1; Jun 1; Oct 1
Contact: Dr. Richard Sprott, Biology of Aging Contact; (301) 496-4996
Internet: http://www.grants.nih.gov/grants/guide/pa-files/PA-96-022.html
Sponsor: National Institutes of Health
6701 Rockledge Drive
Bethesda, MD 20892

NIH Hyperaccelerated Award/Mechanisms in Immunomodulation Trials 3520
NIAID, NIA, NIAMS, NIDCR, NIDDK, and NINDS invite investigator-initiated research applications for mechanistic studies in clinical trials of: immunomodulatory interventions for immune system mediated diseases, including, but not limited to: asthma and allergic diseases; graft failure in solid organ, cell, tissue and stem cell transplantation; and chronic inflammatory, autoimmune, and immunodeficiency diseases; and preventative and therapeutic, vaccines for non-HIV/AIDS infectious diseases, including NIAID Category A, B, and C agents of bioterrorism and emerging/re-emerging infectious diseases. Applications will be accepted monthly on the ninth of each month; a letter of intent is due one month prior to application receipt date. RFA: AI-01-001
Requirements: Applications may be submitted by domestic and foreign, for-profit and nonprofit organizations, both public and private; units of state and local governments; and eligible agencies of the federal government.
Amount of Grant: Up to 2,000,000 U.S.D
Contact: Ann White-Devine; (301) 402-5601; fax (301) 480-3780; ad22x@nih.gov
Internet: http://grants.nih.gov/grants/guide/rfa-files/RFA-AI-01-001.html
Sponsor: National Institutes of Health
6701 Rockledge Drive
Bethesda, MD 20892

NIH Impact of HAART on HIV/CNS Disease 3521
This RFA solicits applications to study the effects of HAART on HIV/CNS disease with respect to the clinical, neurological, neuropsychological, virological, immunological, and neuropathological response to the drugs. This solicitation also includes studies to develop drug formulations with improved ability to cross the blood-brain barrier, studies to develop improved methods of delivering drugs to the CNS, and the interaction of anti-retrovirals with psychotropic drugs. Annual deadline dates vary; contact program staff for exact dates. Inquiries may be directed to program staff at the appropriate NIH institute: Dr. Dianne Rausch, (301) 443-7281, NIMH, 6001 Executive Boulevard, Bethesda, MD 20892; Dr. A.P. Kerza-Kwiatecki, (301) 496-1431, NINDS, 6001 Executive Boulevard, Bethesda, MD 20892; and Dr. Thomas Johnson, (301) 402-3461, NIDCD, 6120 Executive Boulevard, Bethesda, MD 20892. RFA: MH-00-003
Requirements: Applications may be submitted by domestic and foreign, for-profit and nonprofit organizations, both public and private; units of state and local governments; and eligible agencies of the federal government.
Amount of Grant: Up to 34,000,000 U.S.D
Contact: Program Officer
Internet: http://grants.nih.gov/grants/guide/rfa-files/RFA-MH-00-003.html
Sponsor: National Institutes of Health
6701 Rockledge Drive
Bethesda, MD 20892

NIH Independent Scientist Award 3522
The award is a special salary-only grant designed to provide protected time for newly independent scientists who currently have non-research obligations such as heavy teaching loads, clinical work, committee assignments, service, and administrative duties that prevent them from having a period of intensive research focus. The award is targeted to persons with doctoral degrees who have completed their research training, have independent peer-reviewed research support, and who need a period of protected research time in order to foster their research career development. New applications are due February 12, June 12, and October 12, while renewal applications are due March 12, July 12, and November 12.
Requirements: The following organizations and institutions are eligible to apply: public/state controlled institutions of higher education; private institutions of higher education; nonprofits with 501(c)3 IRS status; nonprofits without 501(c)3 IRS status; small businesses; for-profit organizations; State governments; U.S. territories or possessions; Indian/Native American tribal governments (Federally recognized and other than Federally recognized); Indian/Native American tribally designated organizations; Hispanic-serving institutions; historically Black colleges and universities (HBCUs); tribally controlled colleges and universities (TCCUs); Alaska Native and Native Hawaiian serving institutions; regional organizations; and faith-based or community based organizations. The candidate must have a doctoral degree and peer-reviewed, independent research support at the time the award is made. The candidate must spend a minimum of 75 percent effort conducting research during the period of the award.
Restrictions: The award is not intended for investigators who already have full time to perform research, or have substantial publication records or considerable research support indicating that they are well established in their fields. Foreign institutions are not eligible to apply.
Date(s) Application is Due: Feb 12; Mar 12; Jun 12; Jul 12; Oct 12; Nov 12
Amount of Grant: 75,000 - 100,000 U.S.D
Contact: Susan Grove; (301) 435-0714; sg16d@nih.gov or grantsinfo@od.nih.gov
Internet: http://grants.nih.gov/grants/guide/contacts/PA-09-038_contacts.html
Sponsor: National Institutes of Health
6701 Rockledge Drive
Bethesda, MD 20892

NIH Individual Postdoctoral Fellowships in Genomics and Related ELSI Topics 3523
The purpose of these fellowships is to train scientists who will have the multidisciplinary skills that will enable them to engage in research to accomplish the short- and long-term objectives of the HGP and similar genomic projects and to take full advantage of the resulting genomic data and resources to solve biomedical and bioethical problems. Broad areas of research that are relevant include genomic analysis (including technology development) and the ethical, legal, and social implications of human genetics research. The goal of the fellowship program in the area of genomic analysis and technology is to train highly skilled scientists who will use the expertise gained to develop research programs in the mapping and sequencing of the human genome and the genomes of other organisms, in sequence variation, in the analysis and interpretation of the resulting data, and in the development of biological, medical, or biotechnological applications based on the data, or in the development of technologies to support any of this research. Inquiries may be directed to program staff at the appropriate NIH institute: Dr. Bettie Graham, (301) 496-7531, Division of Extramural Research, NHGRI, Building 38A, Bethesda, MD 20892; Dr. Lisa Begg, (301) 496-8580, Office of Centers, Training and Resources, NCI, 6130 Executive Boulevard, Bethesda, MD 20892; Dr. Carol Shreffler, (919) 541-1445, Division of Extramural Research and Training, NIEHS, P.O. Box 12233, Research Triangle Park, NC 27709; Dr. Laurie Tompkins, (301) 594-0943, NIGMS, Building 45, Bethesda, MD 20892; and Dr. Hilary Sigmon, (301) 594-5970, Nursing Fellowships in Genetics, NINR, Building 45, Bethesda, MD 20815. Annual deadline dates vary; contact program staff for exact dates. PA: PA-99-122
Requirements: Individuals must meet the criteria for the NIH National Research Service Awards for Individual Postdoctoral Fellows (F32) described in program announcement PA-99-025, available in the NIH Guide for Grants and Contracts. By the time of award, individuals must be citizens or noncitizen nationals of the United States, or have been lawfully admitted to the United States for permanent residence. Before an NRSA award can be activated, the individual must have received a doctoral degree or its equivalent to be eligible for an individual postdoctoral fellowship (F32). To be eligible for a senior fellowship (F33), individuals must have received a doctoral degree and must have had at least seven subsequent years of relevant research or professional experience.
Restrictions: Individuals on temporary or student visas are not eligible.
Date(s) Application is Due: Apr 5; Aug 5; Dec 5
Amount of Grant: 26,256 - 41,268 U.S.D
Contact: Program Officer
Internet: http://www.grants.nih.gov/grants/guide/pa-files/PA-99-122.html
Sponsor: National Institutes of Health
6701 Rockledge Drive
Bethesda, MD 20892

NIH Innovations in Biomedical Computational Science and Technology Grants 3524
The NIH is interested in promoting research and developments in computational science and technology that will support rapid progress in areas of scientific opportunity in biomedical research. As defined here, biomedical computing or biomedical information science and technology includes database design, graphical interfaces, querying approaches, data retrieval, data visualization and manipulation, data integration through the development of integrated analytical tools, and tools for electronic collaboration, as well as computational and mathematical research including the development of structural, functional, integrative, and analytical models and simulations. Because the nature and scope of the proposed research will vary from application to application, it is anticipated that the size and duration of each award will also vary.
Requirements: Application can be made by the following institutions and organizations: for-profit organizations; non-profit organizations; public or private institutions (such as universities, colleges, hospitals, and laboratories); units of state government; units of local government; eligible agencies of the Federal government; foreign institutions; domestic institutions; faith-based or community-based organizations; Indian/Native American tribal governments (federally recognized); Indian/Native American tribal governments (other than federally recognized); and Indian/Native American tribally designated organizations.
Date(s) Application is Due: Feb 5; Jun 5; Oct 5
Amount of Grant: Up to 500,000 U.S.D
Contact: Peter Lyster, (301) 451-6446; lysterp@mail.nih.gov
Internet: http://grants.nih.gov/grants/guide/pa-files/PAR-09-218.html
Sponsor: National Institutes of Health
45 Center Drive, Room 2AS.55k, MSC 6200
Bethesda, MD 20892-6200

NIH Innovations in Biomedical Computational Science and Technology Initiative Grants for SBIR 3525
This funding opportunity announcement (FOA) solicits Small Business Innovation Research (SBIR) grant applications from small business concerns (SBCs) that propose innovative research in biomedical computational science and technology to promote the progress of biomedical research. There exists an expanding need to speed the progress of biomedical research through the power of computing to manage and analyze data and to model biological processes. The NIH is interested in promoting research and developments in biomedical computational science and technology that will support rapid progress in areas

of scientific opportunity in biomedical research. As defined here biomedical computing or biomedical information science and technology includes database design, graphical interfaces, querying approaches, data retrieval, data visualization and manipulation, data integration through the development of integrated analytical tools, and tools for electronic collaboration, as well as computational research including the development of structural, functional, integrative, and analytical models and simulations.
Requirements: Only United States small business concerns (SBCs) are eligible to submit SBIR applications.
Date(s) Application is Due: Apr 5; Aug 5; Dec 5
Contact: Peter Lyster, (301) 451-6446; lysterp@mail.nih.gov
Internet: http://grants.nih.gov/grants/guide/pa-files/PAR-07-160.html
Sponsor: National Institutes of Health
45 Center Drive, Room 2AS.55k, MSC 6200
Bethesda, MD 20892-6200

NIH Innovations in Biomedical Computational Science and Technology Initiative Grants for STTR 3526

This funding opportunity announcement (FOA) solicits Small Business Technology Transfer (STTR) grant applications from small business concerns (SBCs) that propose innovative research in biomedical computational science and technology to promote the progress of biomedical research. There exists an expanding need to speed the progress of biomedical research through the power of computing to manage and analyze data and to model biological processes. The NIH is interested in promoting research and developments in biomedical computational science and technology that will support rapid progress in areas of scientific opportunity in biomedical research. As defined here biomedical computing or biomedical information science and technology includes database design, graphical interfaces, querying approaches, data retrieval, data visualization and manipulation, data integration through the development of integrated analytical tools, and tools for electronic collaboration, as well as computational research including the development of structural, functional, integrative, and analytical models and simulations.
Requirements: Only United States small business concerns (SBCs) are eligible to submit STTR applications.
Date(s) Application is Due: Apr 5; Aug 5; Dec 5
Amount of Grant: 150,000 - 750,000 U.S.D
Contact: Peter Lyster, (301) 451-6446; lysterp@mail.nih.gov
Internet: http://grants.nih.gov/grants/guide/pa-files/PAR-07-161.html
Sponsor: National Institutes of Health
45 Center Drive, Room 2AS.55k, MSC 6200
Bethesda, MD 20892-6200

NIH Investigator-Initiated Interactive Research Project Grants 3527

This program provides support for formal, investigator-initiated, collaborative relationships. An IRPG group could be used constructively to support collaborative efforts to accelerate the development of fundamental knowledge and/or enhance the clinical application of that knowledge. The IRPG program may provide an effective means to fund applications for focused clinical research and related, correlative laboratory studies. Annual deadline dates may vary; contact program staff for exact dates. PA: PA-96-001
Requirements: Applications may be submitted by foreign and domestic, for-profit and nonprofit organizations, both public and private; state and local governments; and eligible agencies of the federal government.
Date(s) Application is Due: Feb 15; Jun 15; Oct 15
Contact: Program Contact; (301) 435-0715
Internet: http://www.grants.nih.gov/grants/guide/pa-files/PA-96-001.html
Sponsor: National Institutes of Health
6701 Rockledge Drive
Bethesda, MD 20892

NIH Low Birthweight in Minority Populations Research Grants 3528

This PA solicits research to expand our understanding of the underlying mechanisms that contribute to the ethnic variations in low birth weight (LBW) and strategies for prevention. The goals of this PA are the development of innovative strategies to prevent LBW in minority populations and to expand our understanding of how psychosocial and environmental factors affect or interact with the biologic mechanisms that influence pregnancy outcomes. Inquiries may be directed to the program staff at the appropriate NIH institute: Dr. Cara Krulewitch, (301) 594-2542, Division of Extramural Activities, NINR, Building 45, Bethesda, MD 20892; Dr. Dennis Mangan, (301) 594-2421, Division of Extramural Research, NIDCR, 45 Center Drive, Bethesda, MD 20892; Dr. Gwen Collman, (919) 541-4980, Division of Extramural Research and Training, NIEHS, Box 12233 Research Triangle Park, NC 27709; and Dr. Jonelle Rowe, (301) 496-5575, Center for Research for Mothers and Children, NICHD, 6100 Executive Boulevard, Rockville, MD 20852. PA: PA-99-045
Requirements: Applications may be submitted by foreign and domestic for-profit and nonprofit organizations, public and private.
Contact: Program Officer
Internet: http://www.grants.nih.gov/grants/guide/pa-files/PA-99-045.html
Sponsor: National Institutes of Health
6701 Rockledge Drive
Bethesda, MD 20892

NIH Mechanisms of Orofacial Pain: Anatomy, Genomics, and Proteomics 3529

NINDS and NIDCR invite applications to stimulate and support innovative, interdisciplinary research studies to elucidate the molecular mechanisms underlying orofacial pain, particularly the discovery of proteins and protein networks critical to processing nociceptive information. The purpose of this RFA is to encourage the use of genomic and proteomic approaches and imaging techniques to clarify the molecular events involved in acute orofacial pain, the transition from unrelieved acute pain to chronic pain (i.e. neuroplasticity), neuronal hyperexcitability as manifested by hyperalgesia and allodynia, and chronic orofacial pain disorders of an inflammatory and neuropathic origin. This improved understanding could lead to new therapeutic interventions to effectively treat chronic pain conditions. A letter of intent is requested by April 19; full proposal is due May 14. Annual deadline dates may vary; contact program staff for exact dates. RFA: RFA-DE-05-004
Requirements: Applications may be submitted by domestic and foreign, for-profit and nonprofit organizations, both public and private; units of state and local governments; and eligible agencies of the federal government.
Date(s) Application is Due: Apr 19; May 14
Amount of Grant: Up to 250,000 U.S.D
Contact: Mary Daley; (301) 594-4808; fax (301) 480-3562; daleym@mail.nih.gov
Internet: http://www.grants.nih.gov/grants/guide/rfa-files/RFA-DE-05-004.html
Sponsor: National Institutes of Health
45 Center Drive, Building 45, Room 4AN-44F
Bethesda, MD 20892-6402

NIH Mechanisms Underlying Individual Variations in Drug Responses 3530

The purpose of this program announcement is to stimulate research into identifying the critical candidate proteins and/or genes that play essential roles in determining individual variations in drug responses. The study of pharmacogenetic/pharmacogenomic variation presents opportunities to a wide range of researchers, working at levels ranging from the most molecular to the most clinical, in the fields of pharmacology, genetics, genomics, medicine, epidemiology, statistics, and computer science. Although NIGMS is leading the initiative, a number of other institutes of NIH are participating, and they are particularly interested in studies in the following areas: NCI—drug response proteins and/or genes involved in cancer, genetic/molecular epidemiologic studies of polymorphisms associated with cancer, and studies of nicotine susceptibility; NHLBI—drug response proteins and/or genes involved in cardiovascular, pulmonary, and hematologic systems; NICHD—development of drug response proteins and/or genes in children; NIEHS—drug response proteins and/or genes affected by environmental exposure; NIMH—drug response proteins and/or genes involved in the treatment of mental disorders and cognitive disorders; NIA—drug response proteins and/or genes involved in susceptibility to adverse drug reactions and drug-drug interactions in older persons; NIAAA—drug response proteins and/or genes that mediate the effects of alcohol and/or drugs that modify alcohol's effects. Applicants may contact the appropriate institute for information: Dr. Kumiko Iwamoto, (301) 496-9600, Division of Cancer Control and Population Sciences, NCI, Executive Plaza N, Rockville, MD 20852; Dr. Mary Wolpert, (301) 496-8783, Division of Cancer Treatment and Diagnosis, NCI, Executive Plaza N, Rockville, MD 20852; Dr. Peter Savage, (301) 435-0421, Division of Epidemiology and Clinical Applications, NHLBI, 2 Rockledge Center, Bethesda, MD 20817; Dr. George Giacoia, (301) 496-5589, Endocrinology, Nutrition and Growth Branch, NICHD, Building 62, Bethesda, MD 20892; Dr. Jose Velazquez, (919) 542-4998, Division of Extramural Research and Training, NIEHS, P.O. Box 12233, Research Triangle Park, NC 27709; Dr. Linda Brady, (301) 443-9875, Division of Basic and Clinical Science Research, NIMH, Parklawn Building, Rockville, MD 20857; Dr. Stanley Slater, (301) 496-6761, Geriatrics Program, NIA, Gateway Building, Bethesda, MD 20892; or Dr. Samir Zakhari, (301) 443-0799, Division of Basic Research, Willco Building, NIAAA, Bethesda, MD 20892. PA: PA-99-016
Requirements: Applications may be submitted by domestic and foreign, for-profit and nonprofit organizations, units of state and local governments, and eligible agencies of the federal government
Contact: Program Officer
Internet: http://www.grants.nih.gov/grants/guide/pa-files/PA-99-016.html
Sponsor: National Institutes of Health
45 Center Drive, Natcher Building
Bethesda, MD 20892

NIH Mechanisms Underlying Secondary Conditions in Mobility Disorders 3531

Primary disabling conditions are often associated with secondary conditions that further inhibit full participation, cause additional negative health consequences, and dramatically increase the burden of the disorder. Treatment or prevention of such secondary conditions (e.g., spasticity, pain, ulceration, infection, weakness and limited motion) will help to create a more enabling environment for persons with limited mobility, so that functioning of the whole person is enhanced. To accomplish these goals, additional research is required to investigate the underlying mechanisms that exacerbate secondary conditions in mobility disorders. Inquiries may be directed to program staff at the appropriate NIH institute: Dr. Mary Ellen Cheung, (301) 402-2242, NICHD, 6100 Executive Boulevard, Bethesda, MD 20892; Dr. Chhanda Dutta, (301) 435-3048, NIA, 7201 Wisconsin Avenue, Bethesda, MD 20892; Dr. James Panagis, (301) 594-5055, NIAMS, 45 Center Drive, Bethesda, MD 20892; Dr. Sanford Garfield, (301) 594-8803, NIDDK, 45 Center Drive, Bethesda, MD 20892; Dr. Scott Somers, (301) 594-5560, NIGMS, 45 Center Drive, Bethesda, MD 20892; and Dr. Karin Helmers, Program Director, (301) 594-2177, NINR, Building 45, Bethesda, MD 20892. Annual deadline dates vary; contact program staff for exact dates. PA: PA-99-096
Requirements: Applications may be submitted by domestic and foreign, for-profit and nonprofit organizations, both public and private; units of state or local governments; and eligible agencies of the federal government.
Contact: Program Officer
Internet: http://www.grants.nih.gov/grants/guide/pa-files/PA-99-096.html
Sponsor: National Institutes of Health
6701 Rockledge Drive
Bethesda, MD 20892

NIH Mentored Clinical Scientist Development Award 3532

This award enables candidates holding professional degrees (e.g., M.D., D.O., D.V.M., or equivalent degrees) to undertake 3 to 5 years of special study and supervised research with the goal of becoming independent investigators. The award also allows awardees to pursue a research career development program suited to their experience and capabilities under a mentor who is competent to provide guidance in the chosen research area. Institutions may submit applications on behalf of candidates who hold professional degrees. At least 2 years must have elapsed since the health professional degree was granted. Candidates can have varying levels of clinical training and research experience. New applications are due February 12, June 12, and October 12, while resubmitted applications are due March 12, July 12, and November 12.
Requirements: All candidates must be U.S. citizens, non-citizen nationals, or legal permanent residents of the U.S. Persons with temporary or student visas are not eligible. The grantee institution must have a strong, well-established research and research training program in the chosen area, accomplished faculty in the basic and clinical sciences, and a commitment to the candidate's research development. The proposed program should include an appropriate mentor.
Date(s) Application is Due: Feb 12; Mar 12; Jun 12; Jul 12; Oct 12; Nov 12
Contact: Susan Grove, Program Officer; (301) 435-0714; sg16d@nih.gov or grantsinfo@od.nih.gov
Internet: http://grants.nih.gov/grants/guide/pa-files/pa-06-512.html
Sponsor: National Institutes of Health
6701 Rockledge Drive
Bethesda, MD 20892

NIH Mentored Patient-Oriented Research Career Development Award 3533

The purpose of this program is to support the career development of investigators who have made a commitment to focus their research endeavors on patient-oriented research. This mechanism provides support for three to five years of supervised study and research for clinically trained professionals who have the potential to develop into productive, clinical investigators focusing on patient-oriented research. Inquiries may be directed to program staff at the appropriate NIH institute. See the following web page for specific contact information: http://grants.nih.gov/grants/guide/contacts/PA-09-043_contacts.html
Requirements: Applicants must have a clinical doctoral degree or its equivalent and must have completed their clinical training, specialty and, if applicable, sub-specialty training prior to receiving an award. However, candidates may submit an application prior to the completion of clinical training. Candidates must identify a mentor with extensive research experience, and must be willing to spend a minimum of 75 percent of full-time professional effort conducting research career development and clinical research.
Restrictions: Current and former principal investigators on NIH research project (R01), FIRST Awards (R29), comparable career development awards (K01, K07, or K08), sub-projects of program project (P01) or center grants (P50), and the equivalent are not eligible for funding under this grant.
Contact: Susan Grove; (301) 435-0714; sg16d@nih.gov or grantsinfo@od.nih.gov
Internet: http://grants.nih.gov/grants/guide/pa-files/PA-09-043.html#SectionVII
Sponsor: National Institutes of Health
6701 Rockledge Drive
Bethesda, MD 20892

NIH Mentored Research Scientist Development Awards 3534

The overall objective of the NIH Research Career Development Award program is to prepare qualified individuals for careers that have a significant impact on the health-related research needs of the Nation. The objective of the NIH Mentored Research Scientist Development Award is to provide support for a sustained period of protected time for intensive research career development under the guidance of an experienced mentor, or sponsor, in the biomedical, behavioral or clinical sciences leading to research independence. The expectation is that through this sustained period of research career development and training, awardees will launch independent research careers and become competitive for new research project grant funding. Inquiries should be directed to program staff at the appropriate NIH institute. See the website for specific contact information.
Requirements: Applications may be submitted, on behalf of candidates, by domestic, non-federal organizations, public or private, such as medical, dental, or nursing schools or other institutions of higher education. Candidates must be U.S. citizens or non-citizen nationals, or must have been lawfully admitted for permanent residence by the time of award. Candidates must have a research or a health-professional doctorate or its equivalent; have demonstrated the capacity or potential for highly productive independent research in the period after the doctorate; must identify a mentor with extensive research experience; and must be willing to spend a minimum of 75 percent of full-time professional effort conducting research and research career development during the entire award period.
Restrictions: Individuals on temporary or student visas are not eligible for this award.
Date(s) Application is Due: Feb 12; Mar 12; Jun 12; Jul 12; Oct 12; Nov 12
Amount of Grant: 75,000 - 100,000 U.S.D
Contact: Susan Grove, (301) 435-0714; sg16d@nih.gov or grantsinfo@od.nih.gov
Internet: http://grants.nih.gov/grants/guide/pa-files/PA-09-040.html
Sponsor: National Institutes of Health
6701 Rockledge Drive
Bethesda, MD 20892

NIH Midcareer Investigator Award in Mouse Pathobiology Research Grants 3535

The purpose of the program is to provide support for established pathobiologists to allow them protected time to devote to mouse pathobiology research and to act as mentors for beginning investigators. The target candidates are outstanding scientists engaged in pathobiology research who are within 15 years of their specialty training, who can demonstrate the need for a period of intensive research focus as a means of enhancing their research careers, and who are committed to mentoring the next generation of mouse pathobiologists. The award is intended to relieve mouse pathobiologists from time consuming service obligations and administrative responsibilities, thereby increasing the opportunities for their own research and for mentoring. Direct inquiries regarding programmatic issues to: Dr. Franziska Grieder, Comparative Medicine Area, (301) 435-0744, NCRR, 6705 Rockledge Drive, Bethesda, MD 20892. Inquiries regarding research and training related to age-related pathology may be directed to: Dr. Huber Warner, (301) 496-6402, Biology of Aging Program, NIA, Gateway Building, Bethesda, MD 20892. PA: PAR-01-064
Requirements: Most candidates for this award will have a Doctor of Veterinary Medicine degree (or equivalent) from an institution recognized by the American Veterinary Medical Association (AVMA). In addition, individuals holding other clinical (such as the MD) or research (such as the PhD) degree(s) may apply for the award if they have been certified or have demonstrated the necessary expertise to perform high quality, funded research in mouse pathobiology. Candidates must have completed their specialty or research training within 15 years of submitting the application, and there is no age limit for candidates. In exceptional circumstances, the period of eligibility may be extended if it can be demonstrated that candidates had an interruption in their career progression due to family or personal circumstances. Applications may be submitted on behalf of candidates by domestic, nonfederal organizations, public or private. At the time of award, candidates must be U.S. citizens or noncitizen nationals.
Restrictions: Individuals on temporary visas are not eligible.
Contact: Program Officer
Internet: http://grants.nih.gov/grants/guide/pa-files/PAR-01-064.html
Sponsor: National Institutes of Health
6701 Rockledge Drive
Bethesda, MD 20892

NIH Molecular Epidemiology of Prostate Carcinogenesis Grants 3536

NCI, NIDDK, and NIEHS invite investigator-initiated research grant applications of molecular epidemiologic studies for understanding prostate cancer development and progression. The purpose of this initiative is to stimulate development and application of biological markers of prostate cancer risk and tumor aggressiveness in developing strategies for prostate cancer prevention and control. Of special interest are studies of markers to elucidate multiethnic differences in prostate cancer susceptibility. Interdisciplinary collaborations, which may include multicenter study sites as well as joint effort between NIH/NCI intramural and extramural scientists, are encouraged. These collaborations may include utilization of shared laboratory and specimen resources whenever possible. Investigations in understudied populations and in populations of contrasting risk, including those who are residing outside of the United States, are strongly encouraged. This PA invites a range of interdisciplinary investigations, including molecular epidemiologic, family, or population-based studies of prostate cancer, utilizing available or new/novel biomarkers (e.g., biochemical, molecular, cellular, genetic) and sources of specimens (e.g., prostate tissue, prostatic fluid, blood components) whenever possible. Annual deadline dates vary; contact program staff for exact dates. PA PA-00-080
Requirements: Applications may be submitted by foreign and domestic, for-profit and nonprofit organizations, public and private.
Contact: Dr. Kumiko Iwamoto, Division of Cancer Control and Population Sciences; (301) 435-4911; fax (301) 402-4279; ki6n@nih.gov
Internet: http://grants2.nih.gov/grants/guide/pa-files/PA-00-080.html
Sponsor: National Institutes of Health
6701 Rockledge Drive
Bethesda, MD 20892

NIH Mouse Mutagenesis and Phenotyping: Developmental Defects Research Grants 3537

Funds will be awarded to establish a facility for large-scale mutagenesis and phenotyping of developmental defects in the laboratory mouse. The immediate objective of the facility is to produce and characterize mouse strains harboring mutations that affect normal developmental processes. Ultimately, the mutant mice produced are expected to help elucidate the basic cellular, molecular, and genetic mechanisms that direct embryonic and post-embryonic growth and function, as well as yield insights into the mechanisms of human disease. It is anticipated that this facility will devise and perform efficient genome-wide mutagenesis, devise and perform high-throughput phenotyping to screen for mutations that disrupt normal developmental processes, and devise and perform detailed characterization of mutants that display defects in development. Mutant mice, protocols, assays, assessment criteria, and other materials and information generated in projects will be made available to the wider biomedical community. The activities of this facility will be coordinated with the facility(s) established in response to RFA MH-99-007, Mouse Mutagenesis and Phenotyping: Nervous System and Behavior. Inquiries may be directed to program staff at the appropriate NIH institute: Dr. Steven Klein, (301) 496-5541, Developmental Biology, Genetics and Teratology Branch, NICHD, 6100 Executive Boulevard, Rockville, MD 20852; Dr. Judith Greenberg, (301) 594-0943, Division of Genetics and Developmental Biology, NIGMS, Building 45, Bethesda, MD 20892; Dr. Huber Warner, (301) 496-6402, Biology of Aging Program, NIA, Gateway Building, Bethesda, MD 20892; Dr. William Sharrock, (301) 594-5055, Musculoskeletal Diseases Branch, NIAMS, Building 45, Bethesda, MD 20892; Dr. Judy Small, (301) 594-2425, Inherited Craniofacial Diseases and Injury Program, NIDCR, Building 45, Bethesda, MD 20892; Dr. Josie Briggs, (301) 496-5325, Division of Kidney, Urology and Hemotology, NIDDK, Building 31, 31 Center Drive, Bethesda, MD 20892; Dr. John Fakunding, (301) 435-0544, Heart Research Program, Division of Heart and Vascular Disease, NHLBI, 6701 Rockledge Drive, Bethesda, MD 20892. A letter of intent is

requested by August 2; full application is due October 14. Annual deadline dates may vary; contact program staff for exact dates. RFA: HD-99-007
Requirements: Applications may be submitted by domestic public and private for-profit and nonprofit organizations such as universities, colleges, hospitals, laboratories, units of state and local government, and eligible agencies of the federal government.
Restrictions: Foreign institutions are not eligible to apply.
Date(s) Application is Due: Aug 2; Oct 14
Amount of Grant: Up to 2,800,000 U.S.D
Contact: Diane Watson, (301) 496-5001; fax (301) 402-0915; watsond@exchange.nih.gov
Internet: http://www.grants.nih.gov/grants/guide/rfa-files/RFA-HD-99-007.html
Sponsor: National Institutes of Health
6701 Rockledge Drive
Bethesda, MD 20892

NIH Mouse Mutagenesis and Phenotyping: Nervous System and Behavior Research Grants 3538
Funds will be awarded to establish facilities for large-scale mutagenesis and phenotyping of nervous system functions and complex behaviors in the laboratory mouse. Neuroscience-focused mutagenesis and phenotyping facilities established are expected to serve as a national resource by producing a bank of mouse strains that harbor a wide range of mutations affecting murine nervous system function and behavior. Data and biomaterials produced in projects supported will be made widely available to the scientific community. Projects may be funded for up to five years. Inquiries may be directed to program staff at the appropriate NIH institute: Dr. Hemin Chin, (301) 443-1706, Division of Neuroscience and Basic Behavioral Science, NIMH, 6001 Executive Boulevard, Bethesda, MD 20892; Dr. Jonathan Pollock, (301) 435-1309, Division of Basic Research, NIDA, 6001 Executive Boulevard, Bethesda, MD 20892; Dr. Robert Karp, (301) 443-4223, Division of Basic Research, NIAAA, 6000 Executive Boulevard, Bethesda, MD 20892; Dr.Bradley Wise, (301) 496-9350, Neuroscience and Neuropsychology of Aging Program, NIA, 7201 Wisconsin Avenue, Bethesda, MD 20892; Dr. Maria Giovanni, (301) 496-0484, Fundamental Retinal Processes, NEI, 6120 Executive Boulevard, Bethesda, MD 20892; Dr. Gabrielle Leblanc, (301) 496-5745, Division of Fundamental Neuroscience and Developmental Disorders, NINDS, 6001 Executive Boulevard, Bethesda, MD 20892; Dr. Rochelle Small, (301) 402-3464, Division of Human Communication, NIDCD, 6120 Executive Boulevard, Bethesda, MD 20892. A letter of intent is requested by August 2; full application is due October 14. Annual deadline dates vary; contact program staff for exact dates. RFA: MH-99-007
Requirements: Applications may be submitted by domestic public and private for-profit and nonprofit organizations such as universities, colleges, hospitals, laboratories, units of state and local government, and eligible agencies of the federal government.
Restrictions: Foreign institutions are not eligible.
Date(s) Application is Due: Aug 2; Oct 14
Amount of Grant: Up to 55,000,000 U.S.D
Contact: Robert Barnie, (301) 443-4116; fax (301) 443-0501; rb245s@nih.gov
Internet: http://www.grants.nih.gov/grants/guide/rfa-files/RFA-MH-99-007.html
Sponsor: National Institutes of Health
6701 Rockledge Drive
Bethesda, MD 20892

NIH Multidisciplinary Approach for Research on Oral Complications of HIV Infection 3539
This initiative is designed to encourage multidisciplinary research on the oral complications of HIV infection and their prevention that crosses several academic disciplines (e.g., virology, immunology, cell biology, pathology, epidemiology, biochemistry, pharmacology, medical imaging, etc.) It is anticipated that this research will increase the knowledge of the basic mechanisms involved in the pathogenesis of the oral disorders associated with AIDS and that it will identify novel strategies for prevention, treatment, and diagnosis of the oral manifestations of AIDS. A letter of intent is due by August 17; full applications are due September 14. Annual deadline dates may vary; contact program staff for exact dates. RFA Number: RFA-DE-05-003
Requirements: Applications may be submitted by domestic and foreign, for-profit and nonprofit organizations, both public and private; units of state and local governments; and eligible agencies of the federal government.
Date(s) Application is Due: Aug 17; Sep 14
Amount of Grant: Up to 11,000,000 U.S.D
Contact: Mary Daley; (301) 594-4808; fax (301) 480-3562; md74u@nih.gov
Internet: http://www.grants.nih.gov/grants/guide/rfa-files/RFA-DE-05-003.html
Sponsor: National Institutes of Health
Building 45, Room 4AN-44F
Bethesda, MD 20892-6402

NIH National Centers for Neurofibromatosis Research 3540
NINDS and NIDCD invite qualified investigators to submit grant applications for the establishment of National Centers for Neurofibromatosis Research. Recent discoveries have created important opportunities for basic, translational, and clinical research on the neurofibromatoses. The purpose of this Program Announcement (PA) is to encourage the formation and development of research centers that can capitalize on these opportunities, and ultimately develop therapeutic interventions for neurofibromatosis patients. These centers are intended to provide focused expertise and resources, and establish a multi-disciplinary environment that will accelerate research progress. PA Number: PAR-04-018
Requirements: Applications may be submitted by domestic and foreign, for-profit and nonprofit organizations, both public and private; units of state and local governments; and eligible agencies of the federal government.
Contact: Sara Stone, Division of Extramural Activities, NIDCD; (301) 402-0909; fax (301) 402-1758; stones@nidcd.nih.gov
Internet: http://grants1.nih.gov/grants/guide/pa-files/PAR-04-018.html
Sponsor: National Institutes of Health
6701 Rockledge Drive, Room 1040, MSC 7710
Bethesda, MD 20892-7710

NIH National Occupational Research Agenda Grants 3541
The purpose of the grants program is to develop knowledge that can be used in preventing occupational diseases and injuries and to better understand their underlying pathophysiology. The following types of research projects will be supported: causal research to identify and investigate the relationships between hazardous working conditions and associated occupational diseases and injury; the nature and magnitude of special risk factors experienced by older and/or minority workers; methods research to develop more sensitive means of evaluating hazards at work sites; and evaluations of the effectiveness of prevention and intervention programs, including new approaches or combinations of techniques such as control technologies, personal protective equipment, and changes in work organization factors which have been developed and implemented in workplaces. Inquiries may be directed to program staff at the appropriate NIH institute: Dr. Roy Fleming, (404) 639-3343, Research Grants, NIOSH, 1600 Clifton Road NE, Atlanta, GA 30333; Dr. Kumiko Iwamoto, (301) 435-4911, Epidemiology and Genetics Research, NCI, 6130 Executive Boulevard, Bethesda, MD 20892; Dr. Gail Weinmann, (301) 435-0202, Division of Lung Diseases, NHLBI, 6701 Rockledge Drive, Bethesda, MD 20892; Dr. Sidney Stahl, (301) 402-4156, Behavioral and Social Research, NIA, 7201 Wisconsin Avenue, Bethesda, MD 20892; Dr. Amy Donhaue, (301) 402-3458, Hearing and Balance/Vestibular Sciences, NIDCD, 6120 Executive Boulevard, Bethesda, MD 20892; Dr. Gwen Collman, (919) 541-4980, Division of Extramural Research and Training, NIEHS, P.O. Box 12233, Research Triangle Park, NC 27709. Awards will be made for projects lasting one to three years. Annual deadline dates may vary; contact program staff for exact dates. RFA: OH-99-002
Requirements: Applications may be submitted by domestic and foreign public and private nonprofit and for-profit organizations; state and local governments or their bona fide agents; and federally recognized Indian tribal governments, Indian tribes, or Indian tribal organizations.
Restrictions: Any organization involved in lobbying activities is not eligible.
Amount of Grant: Up to 250,000 U.S.D
Contact: Program Officer
Internet: http://www.grants.nih.gov/grants/guide/rfa-files/RFA-OH-99-002.html
Sponsor: National Institutes of Health
6701 Rockledge Drive
Bethesda, MD 20892

NIH National Research Service Awards—Senior Fellows 3542
This program awards senior fellowships to experienced scientists who wish to make major changes in the direction of their research careers or who wish to broaden their scientific background by acquiring new research capabilities. These awards will enable individuals with at least seven years of research experience beyond the doctorate, and who have progressed to the stage of independent investigator, to take time from regular professional responsibilities for the purpose of receiving training to increase their scientific capabilities. In most cases, this award is used to support sabbatical experiences for established independent scientists. This program is not designed for postdoctoral- level investigators seeking to prove their research potential prior to independence. Inquiries may be directed to program staff at the appropriate NIH institute: Dr. Robin Barr, (301) 496-9322, NIA; Dr. Ernestine Vanderveen, (301) 443-2530, NIAAA; Dr. Milton Hernandez, (301) 496-3775, NIAID; Dr. Richard Lymn, (301) 594-5128, NIAMS; Dr. Lisa Begg, (301) 496-8580, NCI; Dr. Steven Klein, (301) 496-5541, NICHD; Dr. Daniel Sklare, (301) 496-1804, NIDCD; Dr. James Lipton, (301) 594-2618, NIDR; Dr. Judith Podskalny, (301) 594-8876, NIDDK; Dr. Carol Shreffler, (919) 541-1445, NIEHS; Dr. Chyren Hunter, (301) 496-5301, NEI; Dr. John Norvell, (301) 594-0533, NIGMS; Dr. Sarah Knox, (301) 435-0409, NHLBI; Dr. Henry Khachaturian, (301) 496-4188, NINDS; Dr. Nell Armstrong, (301) 594-5973, NINR; Dr. Bettie Graham, (301) 496-7531, NCHGR; Dr. David Wilde, (301) 435-0799, NCRR; and Dr. Neal West, (301) 402-5867, NCCAM. Annual deadline dates may vary; contact program staff for exact dates. PA: PA-00-131
Requirements: Applicants must have at least seven years of postdoctoral research or professional experience in the field. By the time of award, individuals must be citizens or noncitizen nationals of the United States or have been lawfully admitted to the United States for permanent residence. The sponsoring institution may be private (profit or nonprofit) or public.
Date(s) Application is Due: Apr 5; Aug 5; Dec 5
Contact: Program Officer; grantsinfo@nih.gov
Internet: http://grants.nih.gov/grants/guide/pa-files/PA-00-131.html
Sponsor: National Institutes of Health
6701 Rockledge Drive
Bethesda, MD 20892

NIH New Directions in Pain Research Grants 3543
This program encourages investigator-initiated research project grant applications to study mechanisms underlying analgesic response and pain to advance the development of novel pain interventions, treatments, and management strategies. The purpose of this PA is to inform the scientific community of broad, shared interests in pain research encompassing the various components of the NIH, and to stimulate and foster a wide range of basic, translational, and patient-oriented clinical studies on pain. Applications are particularly encouraged to study pain throughout the lifespan from the perspectives

of molecular genetics, transcriptional controls, signal transduction, including cellular/ molecular mechanisms, innovative imaging technologies, plasticity, and from hormonal or gender influences. The pain experience needs to be examined at all levels of analysis from the gene, molecule, cell, tissue, and organ, to the individual, family, and community, with the ultimate goal of developing new insights into pain intervention, treatment, and management. Inquiries may be directed to program staff at the appropriate NIH institute: Dr. Cheryl Kitt, (301) 496-1431, Division of Convulsive, Infectious, and Immune Disorders, NINDS, Federal Building, Bethesda, MD 20892; Dr. Patricia Bryant, (301) 594-2095, Temporomandibular Joint Disorders and Neuropathies and Neurodegenerative Diseases, NIDRC, 45 Center Drive, Bethesda, MD 20892; Claudette Varricchio, (301) 496-8541, Division of Cancer Prevention, NCI, 6130 Executive Boulevard, Bethesda, MD, 20892; Dr. Judith Finkelstein, (301) 496-9350, Neuroscience and Neuropsychology of Aging Programs, NIA, 7201 Wisconsin Avenue, Bethesda, MD, 20892; Dr. John Killen, (301) 496-0545, Division of Acquired Immunodeficiency Syndrome, NIAID, 6003 Executive Boulevard, Rockville, MD 20892; Dr. James Panagis, (301) 594-5055, Orthopedics Program, NIAMS, 45 Center Drive, Bethesda, MD 20892; Dr. Rochelle Small, Division of Human Communication, (301) 402-3464, NIDCD, 6120 Executive Boulevard, Bethesda, MD 20892; Dr. David Thomas, (301) 443-4975, Division of Basic Research, NIDA, 5600 Fishers Lane, Rockville, MD 20857; Dr. Alison Cole, (301) 594-1826, Division of Pharmacology, Physiology, and Biological Chemistry, NIGMS, 45 Center Drive, Bethesda, MD 20892; Dr. Mary Leveck, (301) 594-5963, Program Director, NINR, Building 45, Bethesda, MD 20892; and Joyce Rudick, (301) 402-1770, Office of Research on Women's Health, NIH, Building 1, Bethesda, MD 20892. PA: PA-98-102
Requirements: Applications may be submitted by foreign and domestic, for-profit and nonprofit organizations, public and private.
Contact: Program Officer
Internet: http://www.grants.nih.gov/grants/guide/pa-files/PA-98-102.html
Sponsor: National Institutes of Health
9000 Rockville Pike
Bethesda, MD 20892

NIH Nonhuman Primate Immune Tolerance Cooperative Study Group Grants 3544
NIAID and NIDDK invite applications from single institutions or consortia of institutions to participate in a multi-center, cooperative research program to evaluate existing and new tolerance induction treatment regimens and to elucidate the underlying mechanisms of the induction, maintenance and/or loss of tolerance in nonhuman primate models of kidney and islet transplantation. The goals of this research program are to evaluate further the safety, toxicity, and efficacy of existing tolerance induction regimens; define the underlying mechanisms of action of the therapeutic approaches under investigation; and develop and validate immune and/or surrogate markers of the induction, maintenance, and loss of tolerance, graft function, graft acceptance, and graft survival. The deadline for Letters of Intent is August 22, with full applications due September 22.
Requirements: Applications may be submitted by domestic for-profit and nonprofit organizations and public and private institutions.
Restrictions: Foreign institutions are not eligible to apply. This RFA will not support: studies in animal models other than non-human primates; human studies or trials; xenotransplantation; preliminary development of a non-human primate transplantation model; studies to improve the isolation, preservation, or supply of organs, tissues or cells; hematopoietic stem cell transplantation, unless in the context of kidney or islet transplantation (i.e. bone marrow chimerism experiments); heart or lung transplantation models; or studies of non-immunological side effects of an immuno-modulatory agent.
Date(s) Application is Due: Aug 22; Sep 22
Contact: Kristy Kraemer, Division of Allergy, Immunology and Transplantation; (301) 496-5598; fax (301) 480-0693; kkraemer@niaid.nih.gov
Internet: http://grants.nih.gov/grants/guide/rfa-files/RFA-AI-06-018.html
Sponsor: National Institutes of Health
6610 Rockledge Drive, Room 3043, MSC-6601
Bethesda, MD 20892

NIH NRSAs for Individual Predoctoral Fellowships 3545
The awards are provided to individuals for research training in specified areas of biomedical and behavioral research to help ensure that highly trained scientists will be available in adequate numbers and in appropriate research areas and fields to meet the nation's mental health, drug abuse, alcohol abuse, and alcoholism research needs. As such, this program is intended to provide a mechanism to train future generations of outstanding scientists committed to pursuing a career in mental health, drug abuse, and/or alcohol abuse and alcoholism research. Inquiries may be directed to program staff at the appropriate NIH institute: Dr. Henry Khachaturian, (301) 443-4335, NIMH, 5600 Fishers Lane, Rockville, MD 20857; Dr. Cindy Miner, (301) 443-6071, NIDA, 5600 Fishers Lane, Rockville, MD 20850; and Dr. Tina Vanderveen, (301) 443-2531, NIAAA, 6000 Executive Boulevard, Bethesda, MD 20892. Annual deadline dates vary; contact program staff for exact dates. PA: PA-00-125
Requirements: Applicants must be citizens or non-citizen nationals of the United States, or have been lawfully admitted to the United States for permanent residence and have in their possession an Alien Registration Receipt Card at the time of award. Applicants must have received, as of the activation date of the award, a baccalaureate degree and must be enrolled in a program leading to a research doctorate such as the PhD or DSc, or a combined clinical and research degree such as MD/PhD, by the proposed fellowship activation date.
Contact: Program Officer
Internet: http://grants.nih.gov/grants/guide/pa-files/PA-00-125.html
Sponsor: National Institutes of Health
6701 Rockledge Drive
Bethesda, MD 20892

NIH NRSAs Predoctoral Fellowship Awards for Students with Disabilities 3546
The intent of this program is to encourage students with disabilities to seek graduate degrees and thus further the goal of increasing the number of scientists with disabilities who are prepared to pursue careers in biomedical and behavioral research. Inquiries may be directed to program staff at the appropriate NIH institute: Dr. Robin Barr, (301) 496-9322, NIA; Dr. Ernestine Vanderveen, (301) 443-1273, NIAAA; Dr. Milton Hernandez, (301) 496-7291, NIAID; Dr. Richard Lymn, (301) 594-5128, NIAMS; Dr. John Schneider, (301) 496-8580, NCI; Hildegard Topper, (301) 496-0104, NICHD; Dr. Daniel Sklare, (301) 496-1804, NIDCD; Dr. James Lipton, (301) 594-2618, NIDR; Dr. Ronald Margolis, (301) 594-8819, NIDDK; Dr. Charles Sharp, (301) 443-1887, NIDA; Dr. Michael Galvin Jr., (919) 541-7825, NIEHS; and Dr. Maria Giovanni, (301) 496-0484, NEI. Annual deadline dates may vary; contact program staff for exact dates. PA: PA-95-028
Requirements: Applicants must be citizens or non-citizen nationals of the United States, or have been lawfully admitted to the United States for permanent residence.
Restrictions: Individuals on temporary or student visas are not eligible.
Amount of Grant: 12,008 U.S.D
Contact: Program Officer
Internet: http://www.grants.nih.gov/grants/guide/pa-files/PA-95-028.html
Sponsor: National Institutes of Health
6701 Rockledge Drive
Bethesda, MD 20892

NIH Pathological Gambling: Basic, Clinical, and Services Research Grants 3547
NIMH, NIAAA, and NIDA invite investigator-initiated research relevant to their programs that will advance basic, clinical, and services research concerning pathological gambling, as well as enable more effective clinical assessment, prevention, and treatment. This program announcement encourages research that builds on extant research findings concerning pathological gambling and research that expands the breadth and depth of scientific knowledge through increased involvement of various disciplines, e.g., epidemiology; genetics; neuroscience; developmental psychopathology; and behavioral, cognitive, and social science. Multidisciplinary research is especially encouraged. Major questions include factors that predispose for pathological gambling and that distinguish types of pathological gamblers; models for explaining pathological gambling, e.g., social learning, addiction, impulse dyscontrol; developmental pathways and course; gender, age, and sociocultural factors; mechanisms involved in onset, continuation, and cessation; efficacious clinical interventions for the prevention of onset, for treatment, and for prevention and management of relapse; effectiveness of interventions in general use; and effects of the organization and financing of services. Inquiries may be directed to the appropriate NIH institute: Dr. James Breiling, (301) 443-3527, Division of Mental Disorders, Behavior, and AIDS Research, NIMH, 5600 Fishers Lane, Rockville, MD 20857; Darryl Bertolucci, (301) 443-4898, Epidemiology Branch, NIAAA, 6000 Executive Boulevard, Bethesda, MD 20892; and Dr. Meyer Glantz, Division of Epidemiology and Prevention Research, NIDA, 5600 Fishers Lane, Rockville, MD 20857. Annual deadline dates may vary; contact program staff for exact dates. PA: PA-98-106
Requirements: Applications may be submitted by public and private, foreign and domestic, for-profit and nonprofit organizations.
Date(s) Application is Due: Oct 19; Nov 16
Contact: Program Officer
Internet: http://grants.nih.gov/grants/guide/pa-files/PA-98-106.html
Sponsor: National Institutes of Health
6003 Executive Boulevard
Bethesda, MD 20892

NIH Predoctoral Fellowship Awards for Minority Students 3548
These fellowships are for well-qualified students from minority groups found to be underrepresented in the biomedical and behavioral sciences in the United States. The intent of this program is to encourage students from minority groups that are underrepresented in the biomedical and behavioral sciences to seek graduate degrees and thus further the goal of increasing the number of minority scientists who are prepared to pursue careers in biomedical and behavioral research. Inquiries may be directed to program staff at the appropriate NIH institute: Dr. Robin Barr, (301) 496-9322, NIA; Dr. Ernestine Vanderveen, (301) 443-1273, NIAAA; Dr. Milton Hernandez, (301) 496-7291, NIAID; Dr. Richard Lymn, (301) 594-5128, NIAMS; Dr. John Schneider, (301) 496-8580, NCI; Hildegard Topper, (301) 496-0104, NICHD; Dr. Daniel Sklare, (301) 496-1804, NIDCD; Dr. James Lipton, (301) 594-2618, NIDR; Dr. Ronald Margolis, (301) 594-8819, NIDDK; Dr. Charles Sharp, (301) 443-1887, NIDA; Dr. Michael Galvin Jr., (919) 541-7825, NIEHS; and Dr. Maria Giovanni, (301) 496-0484, NEI. Annual deadline dates may vary; contact program staff for exact dates. PA: PA-95-029
Requirements: Applicants must be citizens or non-citizen nationals of the United States, or have been lawfully admitted to the United States for permanent residence
Restrictions: Individuals on temporary or student visas are not eligible.
Amount of Grant: 12,008 U.S.D
Contact: Program Officer
Internet: http://www.grants.nih.gov/grants/guide/pa-files/PA-95-029.html
Sponsor: National Institutes of Health
6701 Rockledge Drive
Bethesda, MD 20892

NIH Protein Structure Initiative (Structural Genomics)—SBIR/STTR Grants 3549
The purpose of this SBIR/STTR program announcement is to encourage research on the development of methodology and technology underpinning the emerging field of structural genomics, whose goal is the understanding of protein structural families, structural folds, and the relation of structure and function. Projects related to high-

throughput structure determination by x-ray crystallography and/or NMR, as well as those addressing other constituent tasks of structural genomics, are relevant to this PA. PA-99-117
Requirements: Any small business concern, independently owned and operated by U.S. citizens or lawfully admitted permanent resident aliens, which is located in the United States and is organized for profit, may apply.
Date(s) Application is Due: Apr 1; Aug 1; Dec 1
Contact: Dr. John Norvell, Division of Cell Biology and Biophysics; (301) 594-0533; fax (301) 480-2004; norvellj@nigms.nih.gov
Internet: http://grants2.nih.gov/grants/guide/pa-files/PA-99-117.html
Sponsor: National Institutes of Health
6701 Rockledge Drive
Bethesda, MD 20892

NIH Protein Structure Initiative (Structural Genomics) Research Grants 3550
The purpose of this PA is to encourage research on the development of methodology and technology underpinning the emerging field of structural genomics, whose goal is the understanding of protein structural families, structural folds, and the relation of structure and function. Projects related to high throughput structure determination by x-ray crystallography and/or NMR, as well as those addressing other constituent tasks of structural genomics, are relevant to this PA. Inquiries may be directed to the staff at the appropriate NIH institute: Dr. John Norvell, (301) 594-0533, Division of Cell Biology and Biophysics, NIGMS, 45 Center Drive, Bethesda, MD 20892; Dr. Abraham Levy, (301) 435-0755, Biomedical Technology, NCRR, 6705 Rockledge Drive, Bethesda, MD 20892; and Dr. Jose Velazquez, (919) 541-4998, Division of Extramural Research and Training, NIEHS, P.O. Box 12233, Research Triangle Park, NC 27709. Annual deadlne dates vary; contact program staff for exact dates. PA: PA-99-116
Requirements: Applications may be submitted by domestic and foreign, for-profit and nonprofit organizations, public and private.
Restrictions: Foreign institutions are not eligible for program project (P01) grants.
Date(s) Application is Due: Feb 1; Jun 1; Oct 1
Contact: Program Officer
Internet: http://www.grants.nih.gov/grants/guide/pa-files/PA-99-116.html
Sponsor: National Institutes of Health
6701 Rockledge Drive
Bethesda, MD 20892

NIH Receptors and Signaling in Bone in Health and Disease 3551
The objective of this initiative is to elicit grant submissions that focus on systemic hormones, local growth factors, and bone-active cytokines, their receptors and mechanisms of signaling in bone. While the primary focus is on basic research, the long-term emphasis is on identifying mechanisms or processes related to hormone action with potential applicability as targets for therapeutic agents that may have efficacy in the treatment of diseases that adversely affect bone, such as osteoporosis and primary hyperparathyroidism. Modular Grant applications will request direct costs in $25,000 modules, up to a total direct cost request of $250,000 per year.
Requirements: Applications may be submitted by domestic and foreign for-profit and nonprofit organizations, public and private, such as universities, colleges, hospitals, laboratories, units of State and local governments, and eligible agencies of the Federal Government. Racial/ethnic minority individuals, women, and persons with disabilities are encouraged to apply as principal investigators.
Date(s) Application is Due: Feb 5; Jun 5; Oct 5
Amount of Grant: Up to 250,000 U.S.D
Contact: Susan Grove, Program Officer; (301) 435-0714; sg16d@nih.gov or grantsinfo@od.nih.gov
Internet: http://grants.nih.gov/grants/guide/pa-files/PA-00-017.html
Sponsor: National Institutes of Health
6701 Rockledge Drive
Bethesda, MD 20892

NIH Recovery Act Limited Competition: Academic Research Enhancement Awards 3552
The purpose of the Academic Research Enhancement Award (AREA) program is to stimulate research in educational institutions that provide baccalaureate or advanced degrees for a significant number of the Nation's research scientists, but that have not been major recipients of NIH support. These AREA grants create opportunities for scientists and institutions otherwise unlikely to participate extensively in NIH programs, to contribute to the Nation's biomedical and behavioral research effort. AREA grants are intended to support small-scale health-related research projects proposed by faculty members of eligible, domestic institutions. The requested budget is limited to $300,000 total costs for the entire three-year project (maximum time period). Applicants should contact individuals at the specific departments listed on the web site. No Letters of Intent are necessary, though full applications should arrive by September 24.
Requirements: The following institutions and organizations may apply: public/state controlled institutions of higher education; private institutions of higher education; Hispanic-serving institutions; historically Black colleges and universities (HBCUs); tribally controlled colleges and universities (TCCUs); Alaska Native and Native Hawaiian serving institutions; public or private institutions and components of institutions defined below as health professional schools or colleges; and other academic components.
Restrictions: Foreign organizations/institutions are not permitted as the applicant organization.
Date(s) Application is Due: Sep 24
Amount of Grant: Up to 300,000 U.S.D
Contact: Susan Grove; (301) 435-0714; sg16d@nih.gov or grantsinfo@od.nih.gov
Internet: http://grants.nih.gov/grants/guide/rfa-files/RFA-OD-09-007.html
Sponsor: National Institutes of Health
6701 Rockledge Drive
Bethesda, MD 20892

NIH Recovery Act Limited Competition: Biomedical Research, Development, and Growth to Spur the Acceleration of New Technologies (BRDG-SPAN) Pilot Program Grants 3553
This Funding Opportunity Announcement (FOA), supported by funds provided to the NIH under the American Recovery & Reinvestment Act of 2009 (Recovery Act or ARRA), Public Law 111-5, solicits grant applications for a new initiative called Biomedical Research, Development, and Growth to Spur the Acceleration of New Technologies (BRDG-SPAN) Pilot Program. The purpose of this pilot program is to address the funding gap between promising research and development (R&D) and transitioning to the market — often called the Valley of Death — by contributing to the critical funding needed by applicants to pursue the next appropriate milestone(s) toward ultimate commercialization; i.e., to carry out later stage research activities necessary to that end. This program aims to accelerate the transition of research innovations and technologies toward the development of products or services that will improve human health, help advance the mission of NIH and its Institutes and Centers (ICs), and create significant value and economic stimulus. This program also aims to foster partnerships among a variety of research and development (R&D) collaborators working toward these aims. The requested budget is limited to $1 million total costs per year for a maximum of three years. Applicants should contact individuals at the specific departments listed on the web site. Letters of Intent are due by August 3, and full applications should arrive by September 1.
Requirements: The applicant United States institution and organization must be located in the 50 states, territories and possessions of the U.S., Commonwealth of Puerto Rico, Trust Territory of the Pacific Islands, or District of Columbia.
Restrictions: Foreign organizations/institutions are not permitted as the applicant organization.
Date(s) Application is Due: Aug 3; Sep 1
Amount of Grant: Up to 1,000,000 U.S.D
Contact: Susan Grove; (301) 435-0714; sg16d@nih.gov or grantsinfo@od.nih.gov
Internet: http://grants.nih.gov/grants/guide/rfa-files/RFA-OD-09-008.html
Sponsor: National Institutes of Health
6701 Rockledge Drive
Bethesda, MD 20892

NIH Recovery Act Limited Competition: Building Sustainable Community-Linked Infrastructure to Enable Health Science Research 3554
Applications are invited from domestic (United States) institutions/organizations proposing to develop or expand needed infrastructures that will fundamentally transform collaboration and communication between academic health centers and local communities. Such collaborative infrastructures are essential to advance the health science research enterprise while ensuring that important research findings are effectively disseminated and implemented to improve public health and advance health care delivery. The primary intent is to fund projects to develop infrastructure for productive and sustainable academic-community research partnerships that can be leveraged in the future for efficiently conducting research that includes and is relevant to affected communities, and through which research findings can be disseminated in a manner that maximizes impact on public health. It is expected that each grantee will be able to leverage this Recovery Act funding into future research grants using the infrastructure partnership between itself and the participating community. Letters of Intent are due by November 12, with final applications due by December 11.
Requirements: The following organizations and institutions may apply: public/state controlled institutions of higher education; private institutions of higher education; Hispanic-serving institutions; historically Black colleges and universities (HBCUs); tribally controlled colleges and universities (TCCUs); Alaska native and native Hawaiian serving institutions; nonprofits with 501(c)3 IRS status; nonprofits without 501(c)3 IRS status; small businesses; for-profit organizations; State governments; regional organizations; U.S. territories or possessions; Indian/Native American tribal governments; Indian/Native American tribally designated organizations; county governments; city or township governments; special district governments; independent school districts; public housing authorities/Indian housing authorities; eligible agencies of the Federal government; and faith-based or community based organizations.
Restrictions: This RFA is not intended to fund research or evaluation projects, clinical trials, or public health campaigns. Foreign organizations/institutions are not permitted as the applicant organization.
Date(s) Application is Due: Nov 12; Dec 11
Contact: Dr. Suzanne Heurtin-Roberts, (301) 402-2277; sheurtin@mail.nih.gov
Internet: http://grants.nih.gov/grants/guide/rfa-files/RFA-OD-09-010.html
Sponsor: National Institutes of Health
1 Center Drive, Room B3-11
Bethesda, MD 20892

NIH Recovery Act Limited Competition: Small Business Catalyst Awards for Accelerating Innovative Research Grants 3555
This NIH Funding Opportunity Announcement, supported by funds provided to the NIH under the American Recovery and Reinvestment Act of 2009 (Recovery Act or ARRA), Public Law 111-5, invites grant applications from small business concerns that propose to accelerate innovation through high risk, high reward research and development (R&D) that has commercial potential and is relevant to the mission of the

NIH. The Small Business Catalyst Award is further expected to support entrepreneurs of exceptional creativity, drawn from scientific and technological environments beyond NIH, who propose pioneering and possibly transformative approaches to addressing major biomedical or behavioral challenges with the potential for downstream commercial development. The Small Business Catalyst Award for Accelerating Innovative Research funding opportunity seeks to encourage fresh research perspectives and approaches to serve the mission of NIH. In particular, applications from small business concerns without a history of NIH Small Business Innovation Research (SBIR) or Small Business Technology Transfer (STTR) support may receive funding priority. Solicited are applications for support for projects that have the potential to generate high impact results (e.g., products, processes or services) and/or innovative research applications, research tools, techniques, devices, inventions, or methodologies. The outcomes of the research supported should have potential to lead to products that will improve public health and create significant value and economic stimulus. This FOA solicits early-stage ideas that promise to lead to major leaps forward in capabilities important to serving the mission of NIH rather than incremental improvements of existing technologies. In accord with the funding priority of this initiative to attract applicants without a history of SBIR/STTR support from NIH, the focus of the projects solicited by this FOA is on early stage technology development. Budget requests are limited to $200,000 total costs for a maximum project period of one year. Applicants should contact individuals at the specific departments listed on the web site. Letters of Intent are due by August 3, and full applications should arrive by September 1.
Requirements: Only United States small business concerns (SBCs) are eligible to submit SBIR applications. A small business concern is one that, at the time of award of SBIR Phase I and Phase II, meets all of the following criteria: is organized for profit, with a place of business located in the United States, which operates primarily within the United States or which makes a significant contribution to the United States economy through payment of taxes or use of American products, materials or labor; is in the legal form of an individual proprietorship, partnership, limited liability company, corporation, joint venture, association, trust or cooperative, except that where the form is a joint venture, there can be no more than 49 percent participation by foreign business entities in the joint venture; is at least 51 percent owned and controlled by one or more individuals who are citizens of, or permanent resident aliens in, the United States, or it must be a for-profit business concern that is at least 51% owned and controlled by another for-profit business concern that is at least 51% owned and controlled by one or more individuals who are citizens of, or permanent resident aliens in, the United States, except in the case of a joint venture, where each entity to the venture must be 51 percent owned and controlled by one or more individuals who are citizens of, or permanent resident aliens in, the United States; and has, including its affiliates, not more than 500 employees.
Date(s) Application is Due: Aug 3; Sep 1
Amount of Grant: 200,000 U.S.D
Contact: Susan Grove; (301) 435-0714; sg16d@nih.gov or grantsinfo@od.nih.gov
Internet: http://grants.nih.gov/grants/guide/rfa-files/RFA-OD-09-009.html
Sponsor: National Institutes of Health
6701 Rockledge Drive
Bethesda, MD 20892

NIH Recovery Act Limited Competition: Supporting New Faculty Recruitment to Enhance Research Resources through Biomedical Research Core Centers Grants 3556

This initiative, which is supported by funds provided to the NIH under the Recovery Act, is designed to provide the necessary resources for U.S. academic institutions/organizations to enhance their biomedical research efforts through the development of Biomedical Research Core Centers. For this announcement, Biomedical Core Centers are defined as a community of multidisciplinary researchers focusing on areas of biomedical research relevant to NIH, such as centers, departments, programs, and/or trans-departmental collaborations or consortia. These Core Centers are designed to provide scientific and programmatic support for promising research faculty and their areas of research. The NIH invites applications that include plans to recruit and hire investigators to conduct biomedical research in all scientific disciplines, including the field of bioethics. Applicants should contact individuals at the specific departments listed on the web site. Letters of Intent are due by April 29, and full applications should arrive by May 29.
Requirements: The following institutions and organizations may apply: public/state controlled institutions of higher education; private institutions of higher education; Hispanic-serving institutions; historically Black colleges and universities (HBCUs); tribally controlled colleges and universities (TCCUs); Alaska Native and Native Hawaiian serving institutions; nonprofits with 501(c)3 IRS status (other than institutions of higher education); nonprofits without 501(c)3 IRS status (other than institutions of higher education); and Indian/Native American tribally designated organizations .
Restrictions: Foreign organizations/institutions are not permitted as the applicant organization.
Date(s) Application is Due: Apr 29; May 29
Contact: Susan Grove, (301) 435-0714; sg16d@nih.gov or grantsinfo@od.nih.gov
Internet: http://grants.nih.gov/grants/guide/rfa-files/RFA-OD-09-005.html
Sponsor: National Institutes of Health
6701 Rockledge Drive
Bethesda, MD 20892

NIH Research On Ethical Issues In Human Subjects Research 3557

The purpose of this funding opportunity announcement is to solicit research addressing the ethical challenges of human subjects research in order to optimize the protection of human subjects and enhance the ethical conduct of human subjects research. The research design for studies on ethical issues in human subjects research should be appropriate to the nature of the project(s) proposed and the disciplines involved. Given the conceptual and methodological complexity of many of these research questions, interdisciplinary and collaborative projects are encouraged, particularly those involving clinical researchers, ethicists, and behavioral/social scientists.
Requirements: The following organizations and institutions may apply: public/state controlled institutions of higher education; private institutions of higher education; nonprofits with 501(c)3 IRS status (other than institution of higher education); nonprofits without 501(c)3 IRS status (other than institution of higher education); small businesses; for-profit organizations (other than small business); State governments; U.S. territories or possessions; Indian/Native American tribal governments (Federally recognized); Indian/Native American tribal governments (other than Federally recognized); Indian/Native American tribally designated organizations; non-domestic (non-U.S.) entities (foreign organization); eligible agencies of the Federal government; and faith-based or community based organizations.
Date(s) Application is Due: Feb 5; Jun 5; Oct 5
Contact: Kim Witherspoon, (301) 496-8866; witherskk@ctep.nci.nih.gov
Internet: http://grants.nih.gov/grants/guide/pa-files/PA-07-277.html
Sponsor: National Institutes of Health
6130 Executive Boulevard, Executive Plaza N, Room 7009, MSC 7432
Bethesda, MD 20892-7432

NIH Research on Sleep and Sleep Disorders 3558

This Funding Opportunity Announcement (FOA) solicits grant applications proposing research to advance biomedical knowledge related to sleep or sleep disorders, improve understanding of the neurobiology or functions of sleep over the life-span, enhance timely diagnosis and effective treatment for individuals affected by sleep-related disorders, or implement and evaluate innovative community-based public health education and intervention programs. Because the nature and scope of the proposed research will vary from application to application, it is anticipated that the size and duration of each award will also vary. The total amount awarded and the number of awards will depend upon the mechanism numbers, quality, duration, and costs of the applications received.
Requirements: The following organizations and institutions may apply: public/state controlled institutions of higher education; private institutions of higher education; nonprofits with 501(c)3 IRS status; nonprofits without 501(c)3 IRS status; small businesses; State governments; non-domestic (non-U.S.) entities (foreign organizations); eligible agencies of the Federal government; and faith-based or community based organizations.
Date(s) Application is Due: Feb 5; Jun 5; Oct 5
Contact: Susan Grove; (301) 435-0714; sg16d@nih.gov or grantsinfo@od.nih.gov
Internet: http://grants.nih.gov/grants/guide/pa-files/pa-07-140.html
Sponsor: National Institutes of Health
6701 Rockledge Drive
Bethesda, MD 20892

NIH Research Project Grants 3559

The Research Project Grant (R01) is an award to support a discrete, specified, circumscribed project to be performed by named Project Directors/Principal Investigators (PDs/PIs) in areas representing the investigators' specific interests and competencies, based on the mission of the NIH. The R01 is the original and historically the oldest grant mechanism used by the NIH to support health-related research and development. The NIH awards R01 grants to institutions/organizations of all types. This mechanism allows the PDs/PIs to define the scientific focus or objective of the research based on particular areas of interest and competence. Although the PDs/PIs write the grant application and are responsible for conducting and supervising the research, the actual applicant is the research institution/organization. Research grant applications are assigned to an NIH IC based on receipt and referral guidelines and many applications are assigned to multiple ICs with related research interests. Each IC maintains a Web site with funding opportunities and areas of interest. Contact with an IC representative may help focus the research plan based on an understanding of the mission of the IC. For specific information about the mission of each NIH IC, see http://www.nih.gov/icd, which provides a brief summary of the research interests in each IC and access to individual IC home pages.
Requirements: The following organizations and institutions may apply: public/state controlled institutions of higher education; private institutions of higher education; Hispanic-serving institutions; historically Black colleges and universities (HBCUs); tribally controlled colleges and universities (TCCUs); Alaska native and native Hawaiian serving institutions; nonprofits with 501(c)3 IRS status; nonprofits without 501(c)3 IRS status; small businesses; for-profit organizations; State governments; regional organizations; U.S. territories or possessions; Indian/Native American tribal governments; Indian/Native American tribally designated organizations; eligible agencies of the Federal government; and faith-based or community based organizations.
Date(s) Application is Due: Feb 5; Jun 5; Oct 5
Contact: Grants Coordinator; (301) 435-0714; grantsinfo@od.nih.gov
Internet: http://grants.nih.gov/grants/guide/pa-files/PA-07-070.html
Sponsor: National Institutes of Health
6100 Executive Boulevard
Bethesda, MD 20892

NIH Rett Syndrome: Genetics, Pathophysiology, and Biomarkers Research Grants 3560

The Mental Retardation and Developmental Disabilities (MRDD) Branch and the Center for Research for Mothers and Children (CRMC) of NICHD and the Division of Fundamental Neuroscience and Developmental Disorders (DFNDD) of NINDS invite research grant applications that address the genetics, pathophysiological mechanisms, and the development of biomarkers for Rett syndrome. Areas of interest include developmental neurobiology, pathophysiology, genetics, epidemiology, structure-

function correlations, and clinical studies that have a direct link to Rett syndrome. Specific funds will be set aside to support investigator-initiated applications responsive to this grant solicitation. PA: PAS-99-037
Requirements: Applications may be submitted by domestic, for-profit and nonprofit organizations, both public and private; units of state or local governments; and eligible agencies of the federal government.
Amount of Grant: Up to 15,000,000 U.S.D
Contact: Dr. Mary Lou Oster-Granite, (301) 496-1383; fax (301) 496-3791; mo96o@nih.gov
Internet: http://www.grants.nih.gov/grants/guide/pa-files/PAS-99-037.html
Sponsor: National Institutes of Health
6701 Rockledge Drive
Bethesda, MD 20892

NIH Role of Microglia in Normal and Abnormal Immune Responses of the Nervous System 3561

NINDS, NIMH, NCRR, NIDCD, and NIDA invite applications to promote research into the role of microglia in the initiation and expansion of autoimmune processes of the central nervous system (CNS) and the resulting injury to CNS components. The intent of this PA is to intensify interest and investigator-initiated research to attract new investigators to this field and to mobilize interdisciplinary approaches. Inquiries may be directed to program staff at the appropriate NIH institute: Dr. A.P. Kerza-Kwiatecki, Neural Environment Team, (301) 496-1431, NINDS, 6001 Executive Boulevard, Bethesda, MD 20892; Dr. John Strandberg, Comparative Medicine, (301) 435-0744, NCRR, 1 Rockledge Ctr, Bethesda, MD 20892; Dr. Dianne Rausch, (301) 443-7281, NIMH, 6001 Executive Boulevard, Bethesda, MD 20892; Dr. Thomas Johnson, Program Officer, (301) 402-3461, NIDCD, 6120 Executive Boulevard, Bethesda, MD 20892; and Dr. Charles William Sharp, Division of Basic Research, (301) 443-1887, NIDA, 6001 Executive Boulevard, Bethesda, MD 20892. PA: PA-00-029
Requirements: Applications may be submitted by foreign and domestic, for-profit and nonprofit organizations, both public and private; units of state and local governments; and eligible agencies of the federal government.
Contact: Program Officer
Internet: http://grants.nih.gov/grants/guide/pa-files/PA-00-029.html
Sponsor: National Institutes of Health
6701 Rockledge Drive
Bethesda, MD 20892

NIH Roybal Centers for Translational Research on Aging 3562

The long-range objectives of the Roybal Centers are to improve the health, quality of life, and productivity of middle-aged and older people, through: facilitating the translation from the basic behavioral and social sciences (including human factors) to practical outcomes, including new technologies, for the benefit of the aged, and if possible also stimulating new use-inspired basic research in the behavioral and social sciences. A letter of intent is due by December 22; full application is due January 22. Annual deadline dates mayb vary; contact program staff for exact dates. RFA Number: RFA-AG-04-007
Requirements: An Edward R. Roybal Center Grant (P30) requires relevant pre-existing research activity at the institution. Any individual with the skills, knowledge, and resources necessary to carry out the proposed research is invited to work with their institution to develop an application for support.
Restrictions: Foreign organizations are not eligible to apply for P30 grants.
Date(s) Application is Due: Jan 22; Dec 22
Amount of Grant: Up to 360,000 U.S.D
Contact: Jean Richelsen; (301) 496-1472; richelsj@nia.nih.gov
Internet: http://grants1.nih.gov/grants/guide/rfa-files/RFA-AG-04-007.html
Sponsor: National Institutes of Health
6701 Rockledge Drive, Room 1040, MSC 7701
Bethesda, MD 20892-7710

NIH Ruth L. Kirschstein National Research Service Awards (NRSA) for Individual Senior Fellows 3563

The objective of the National Research Service Award (NRSA) F33 program is to provide senior fellowship support to experienced scientists who wish to make major changes in the direction of their research careers or who wish to broaden their scientific background by acquiring new research capabilities as independent research investigators in scientific health-related fields relevant to the missions of the participating NIH Institutes and Centers (ICs). Senior fellowship applicants for the F33 award must include a research training proposal that offers an opportunity for individuals to broaden their scientific background or to extend their potential for research in health-related areas as independent researchers. These awards will enable individuals with at least seven years of research experience beyond the doctorate, and who have progressed to the stage of independent investigator, to take time from regular professional responsibilities for the purpose of receiving training to increase their scientific capabilities. In most cases, this award is used to support sabbatical experiences for established independent scientists seeking support for retraining or additional career development.
Requirements: The following organizations are eligible to apply: for-profit organizations; non-profit organizations; public or private institutions, such as universities, colleges, hospitals, and laboratories; eligible agencies and labs of the Federal government, including NIH intramural labs; domestic institutions; and foreign institutions.
Restrictions: This program is not designed for postdoctoral level investigators seeking to enhance their research experience prior to independence.
Date(s) Application is Due: Apr 8; Aug 8; Dec 8
Contact: Nancy C. Lohrey, (301) 496-8580; fax (301) 402-4472; lohreyn@mail.nih.gov
Internet: http://grants.nih.gov/grants/guide/pa-files/PA-07-172.html
Sponsor: National Institutes of Health
6116 Executive Boulevard, Suite 7019, MSC 8346
Bethesda, MD 20892-8346

NIH Ruth L. Kirschstein National Research Service Awards for Individual Postdoctoral Fellows 3564

The purpose of the National Research Service Award Act (NRSA) is to help ensure that diverse pools of highly trained scientists will be available in adequate numbers and in appropriate research areas to carry out the Nation's biomedical, behavioral and clinical research agendas. The National Institutes of Health (NIH) awards individual postdoctoral fellowships (F32) to promising applicants who have the potential to become productive, independent investigators in fields related to the mission of participating NIH Institutes and Centers. The proposed postdoctoral training must be within the broad scope of biomedical, behavioral, or clinical research or other specific disciplines relevant to the research mission of the participating NIH Institutes and Centers. The proposed training must offer an opportunity to enhance the fellow's understanding of the health-related sciences and extend his/her potential for a productive research career.
Requirements: Applicant organizations must have any of the following characteristics: for-profit organizations; non-profit organizations; public or private institutions, such as universities, colleges, hospitals, and laboratories; units of State and local governments; eligible agencies and labs of the Federal government including NIH intramural labs; domestic institutions; and foreign institutions.
Date(s) Application is Due: Apr 8; Aug 8; Dec 8
Contact: Dr. Walter Schaffer, (301) 435-2687; fax (301) 480-0146; ws11q@nih.gov
Internet: http://grants.nih.gov/grants/guide/pa-files/PA-06-373.html
Sponsor: National Institutes of Health
45 Center Drive, Natcher Building
Bethesda, MD 20892

NIH Ruth L. Kirschstein National Research Service Awards for Individual Predoctoral Fellows 3565

The objective of this funding opportunity announcement is to help ensure that highly trained scientists will be available in adequate numbers and in appropriate research areas to carry out the Nation's biomedical, behavioral, and clinical research agenda. The participating Institutes of the National Institutes of Health (NIH) provide individual predoctoral research training fellowship awards to promising doctoral candidates who have the potential to become productive, independent investigators in research fields relevant to the missions of these participating NIH Institutes and Centers. Each participating NIH Institute has a unique scientific purview and different program goals and initiatives that evolve over time. Prior to preparing an application, it is critical that all applicants consult the appropriate Institute website for details of research areas supported by that Institute. Applicants should also contact the appropriate Institute scientific/research contact to obtain current information about specific program priorities and policies.
Requirements: The following organizations are eligible to apply: for-profit organizations; non-profit organizations; public or private institutions, such as universities, colleges, hospitals, and laboratories; eligible agencies and labs of the Federal government, including NIH intramural labs; domestic institutions; and foreign institutions.
Date(s) Application is Due: Apr 8; Aug 8; Dec 8
Contact: Michael-David Kerns, (301) 496-9322; kernsmd@mail.nih.gov
Internet: http://grants.nih.gov/grants/guide/pa-files/PA-07-002.html
Sponsor: National Institutes of Health
6120 Executive Boulevard
Bethesda, MD 20892-7180

NIH Ruth L. Kirschstein National Research Service Awards for Individual Predoctoral Fellowships to Promote Diversity in Health-Related Research 3566

The purpose of the F31 predoctoral fellowship to promote diversity in health-related research is to provide up to five years of support for research training leading to the PhD or equivalent research degree, the combined MD/PhD degree; or another formally combined professional degree and research doctoral degree in biomedical, behavioral, health services, or clinical sciences. These fellowships will enhance the diversity of the biomedical, behavioral, health services, and clinical research labor force in the United States by providing opportunities for academic institutions to identify and recruit students from diverse population groups to seek graduate degrees in health-related research and apply for this fellowship. The goal of this program is to increase the number of scientists from diverse population groups who are prepared to pursue careers in biomedical, behavioral, social, clinical, or health services research.
Requirements: The following organizations and institutions are eligible to apply: public/state controlled institutions of higher education; private institutions of higher education; Hispanic-serving institutions; historically Black colleges and universities (HBCUs); tribally controlled colleges and universities (TCCUs); Alaska Native and Native Hawaiian serving institutions; nonprofits with 501(c)3 IRS status (other than institutions of higher education); nonprofits without 501(c)3 IRS status (other than institutions of higher education); for-profit organizations (other than small businesses); non-domestic (non-U.S.) entities (foreign organizations); and other eligible agencies of the Federal Government.
Date(s) Application is Due: Apr 8; Aug 8; Dec 8
Contact: Susan Grove, Program Officer; (301) 435-0714; sg16d@nih.gov or grantsinfo@od.nih.gov
Internet: http://grants.nih.gov/grants/guide/pa-files/PA-09-209.html
Sponsor: National Institutes of Health
6701 Rockledge Drive
Bethesda, MD 20892

NIH Ruth L. Kirschstein National Research Service Awards for Individual Predoctoral MD/PhD and Other Dual Doctoral Degree Fellows 3567

The purpose of the Ruth L. Kirschstein National Research Service Awards (Kirschstein-NRSA) is to provide support to individuals for combined MD/PhD and other dual doctoral degree training (e.g. DO/PhD, DDS/PhD, AuD/PhD). The participating Institutes award this Kirschstein-NRSA individual fellowship to qualified applicants with the potential to become productive, independent, highly trained physician-scientists and other clinician-scientists, including patient-oriented researchers in their scientific mission areas. This funding opportunity supports individual predoctoral F30 fellowships with the expectation that these training opportunities will increase the number of future investigators with both clinical knowledge and skills in basic, translational or clinical research.
Requirements: The following organizations and institutions are eligible to apply: public/state controlled institutions of higher education; private institutions of higher education; Hispanic-serving institutions; historically Black colleges and universities (HBCUs); tribally controlled colleges and universities (TCCUs); Alaska Native and Native Hawaiian serving institutions; nonprofits with 501(c)3 IRS status (other than institutions of higher education); nonprofits without 501(c)3 IRS status (other than institutions of higher education); for-profit organizations (other than small businesses); and other eligible agencies of the Federal Government.
Restrictions: Non-U.S. entities are not eligible to apply.
Date(s) Application is Due: Apr 8; Aug 8; Dec 8
Contact: Nancy C. Lohrey, (301) 496-8580; fax (301) 402-4472; lohreyn@mail.nih.gov
Internet: http://grants.nih.gov/grants/guide/pa-files/PA-09-207.html
Sponsor: National Institutes of Health
6116 Executive Boulevard, Suite 7019, MSC 8346
Bethesda, MD 20892-8346

NIH Sarcoidosis: Research into the Cause of Multi-Organ Disease and Clinical Strategies for Therapy Grants 3568

The purpose of this Funding Opportunity Announcement (FOA) is to stimulate research on the etiology and management of sarcoidosis, an immune-mediated granulomatous inflammatory disorder. Studies supported by this FOA would include investigations to find the etiology(ies) and related host factors that might enhance susceptibility to sarcoidosis, especially development of symptomatic multi-organ disease that has a propensity to involve critical organs (such as lungs, heart, eyes, central/peripheral nervous system, liver, kidneys and other abdominal viscera) that create serious illness. Study of the still problematic management and the biological, behavioral, or psychosocial burden on individuals, families, and community is encouraged. Moreover, there is interest for approaches to risk reduction, psychological coping, and management of complications and side effects of treatment.
Requirements: The following organizations and institutions may apply: public/state controlled institutions of higher education; private institutions of higher education; Hispanic-serving institutions; historically Black colleges and universities (HBCUs); tribally controlled colleges and universities (TCCUs); Alaska native and native Hawaiian serving institutions; nonprofits with 501(c)3 IRS status; nonprofits without 501(c)3 IRS status; small businesses; for-profit organizations; State governments; regional organizations; U.S. territories or possessions; Indian/Native American tribal governments; Indian/Native American tribally designated organizations; eligible agencies of the Federal government; and faith-based or community based organizations.
Date(s) Application is Due: Feb 5; Jun 5; Oct 5
Contact: Herbert Y. Reynolds, (301) 435-0222; Reynoldh@nhlbi.nih.gov
Internet: http://grants.nih.gov/grants/guide/pa-files/PA-07-136.html
Sponsor: National Institutes of Health
6701 Rockledge Drive, Room 10018, MSC 7936
Bethesda, MD 20892-7952

NIH SBIR/STTR Study and Control of Microbial Biofilms Research Grants 3569

Several NIH institutes invite research grant applications to conduct studies on microbial biofilms leading to improved strategies and technologies to diagnose, prevent, and treat biofilm-associated infectious diseases. Inquiries may be directed to the staff at the appropriate NIH institute: Dr. Joyce Reese, (301) 594-2088, Division of Extramural Research, NIDCR, Building 45, Bethesda, MD 20892; Dr. Stephen Heyse, (301) 496-7728, Division of Microbiology and Infectious Diseases, NIAID, Solar Building, Bethesda, MD 20892; Dr. Kenneth Gruber, (301) 402-3458, Division of Human Communication, NIDCD, 6120 Executive Boulevard, Bethesda, MD 20892; Dr. James Panagis, (301) 594-5055, Orthopedics Program, NIAMS, 45 Center Drive, 4500 Bethesda, MD 20892; Dr. Scott Somers, (301) 594-5560, Division of Pharmacology, Physiology, and Biological Chemistry, NIGMS, 45 Center Drive, Bethesda, MD 20892; Dr. Susan Banks-Schlegel, (301) 435-0202, Airway Biology and Disease Program, NHLBI, 6701 Rockledge Drive, 7952 Bethesda, MD 20892; Dr. Lawrence Agodoa, (301) 594-7717, Division of Kidney, Urologic, and Hematologic Diseases, NIDDK, 45 Center Drive, Bethesda, MD 20892; Dr. Eugene Hayunga, (301) 402-1770, Office of Research on Women's Health, Building 1, Bethesda, MD 20892; Dr. Louis Quatrano, (301) 402-2242, National Center for Medical Rehabilitation Research, NICHD, 6100 Executive Boulevard, Rockville, MD 20852; Dr. Hilary Sigmon, (301) 594-5970, Division of Extramural Activities, NINR, 45 Center Drive, Bethesda, MD 20892; and Dr. William Heetderks, (301) 496-1447, Division of Stroke, Trauma, and Neurodegenerative Disorders, NINDS, Federal Building, Bethesda, MD 20892. Contact program staff for availability. PA: PA 99-084
Date(s) Application is Due: Apr 1; Aug 1; Dec 1
Internet: http://www.grants.nih.gov/grants/guide/pa-files/PA-99-084.html
Sponsor: National Institutes of Health
6701 Rockledge Drive
Bethesda, MD 20892

NIH School-based Interventions to Prevent Obesity 3570

This Funding Opportunity Announcement (FOA) encourages the formation of partnerships between academic institutions and school systems in order to develop and implement controlled, school-based intervention strategies designed to reduce the prevalence of obesity in childhood. This FOA also encourages evaluative comparisons of different intervention strategies, as well as the use of methods to detect synergistic interactions between different types of interventions. Because the nature and scope of the proposed research will vary from application to application, it is anticipated that the size and duration of each award will also vary. The total amount awarded and the number of awards will depend upon the mechanism numbers, quality, duration, and costs of the applications received.
Requirements: The following organizations and institutions may apply: public/state controlled institutions of higher education; private institutions of higher education; Hispanic-serving institutions; historically Black colleges and universities (HBCUs); tribally controlled colleges and universities (TCCUs); Alaska native and native Hawaiian serving institutions; nonprofits with 501(c)3 IRS status; nonprofits without 501(c)3 IRS status; small businesses; for-profit organizations; State governments; regional organizations; U.S. territories or possessions; Indian/Native American tribal governments; Indian/Native American tribally designated organizations; eligible agencies of the Federal government; and faith-based or community based organizations.
Date(s) Application is Due: Feb 5; Jun 5; Oct 5
Contact: Gilman Grave, (301) 496-5593; fax (301) 480-9791; graveg@mail.nih.gov
Internet: http://grants.nih.gov/grants/guide/pa-files/PA-07-180.html
Sponsor: National Institutes of Health
6100 Executive Boulevard, 4B-11, MSC 7510
Bethesda, MD 20892-7510

NIH Self-Management Strategies Across Chronic Diseases Grants 3571

The purpose of this Program Announcement (PA) is to solicit applications to expand research on established self-management interventions to multiple chronic diseases across the life-course. Interventions aimed at chronic disease self-management are numerous and many are well described in the literature. They are often presented as specific to a particular chronic disease. This PA encourages applicants to investigate the applicability of effective self-management interventions to a broader spectrum of chronic diseases. Chronic disease, for this announcement, is defined as illnesses that are prolonged, are rarely cured completely, and require self-management behaviors by affected individuals and/or their caretakers. Modular Grant applications will request direct costs in $25,000 modules, up to a total direct cost request of $250,000 per year.
Requirements: The following organizations and institutions may apply: public/state controlled institutions of higher education; private institutions of higher education; Hispanic-serving institutions; historically Black colleges and universities (HBCUs); tribally controlled colleges and universities (TCCUs); Alaska native and native Hawaiian serving institutions; nonprofits with 501(c)3 IRS status; nonprofits without 501(c)3 IRS status; small businesses; for-profit organizations; State governments; regional organizations; U.S. territories or possessions; Indian/Native American tribal governments; Indian/Native American tribally designated organizations; county governments; city or township governments; special district governments; independent school districts; public housing authorities/Indian housing authorities; eligible agencies of the Federal government; and faith-based or community based organizations.
Date(s) Application is Due: Feb 5; Jun 5; Oct 5
Amount of Grant: Up to 250,000 U.S.D
Contact: Susan Grove; (301) 435-0714; sg16d@nih.gov or grantsinfo@od.nih.gov
Internet: http://grants.nih.gov/grants/guide/pa-files/PA-00-109.html
Sponsor: National Institutes of Health
6701 Rockledge Drive
Bethesda, MD 20892

NIH Senior Scientist Award 3572

This award provides stability of support to outstanding scientists who have demonstrated a sustained, high level of productivity and whose expertise, research accomplishments, and contributions to the field have been and will continue to be critical to the mission of the particular NIH center or institute. The award provides salary support for award periods of up to five years as a means of enhancing the individual recipient's skills and dedication to his/her area of research. It permits NIH institutes and centers to identify and support exceptionally talented investigators who are well established in their field of research. Inquiries may be directed to program staff at the appropriate NIH institute: Dr. Ernestine Vanderveen, (301) 443-2531, NIAAA; Dr. Lester Gorelic, (301) 496-8580, NCI; Dr. Andrea Baruchin, (301) 443-6071, NIDA; Dr. Henry Khachaturian, (301) 443-4335, NIMH; and Dr. Neal West, (301) 402-5867, NCCAM. PA: PA-00-021
Requirements: Applications may be submitted on behalf of candidates by domestic, nonfederal organizations, public or private, such as medical, dental, or nursing schools or other institutions of higher education. At the time of award, candidates must be citizens or noncitizen nationals of the United States or must have been lawfully admitted for permanent residence. The candidate must be a senior scientist and a recognized leader in the field with a distinguished record of original contributions; must have a record of support from a funding institute or center; and must have peer-reviewed grant support at the time of the award to be eligible.
Restrictions: Individuals on temporary or student visas are not eligible.
Contact: Program Officer
Internet: http://grants.nih.gov/grants/guide/pa-files/PA-00-021.html
Sponsor: National Institutes of Health
6701 Rockledge Drive
Bethesda, MD 20892

NIH Small Business Innovation Research (SBIR): Animal Models of HCV Infection Research Grants 3573

This is a multi-institute solicitation targeting the development/identification of one or more small animal models of hepatitis C infection and disease progression including acute and chronic states, fibrosis/cirrhosis, and liver tumor development—not necessarily all in the same model. Identification of a small, inexpensive, reproducible, well-defined, and nonendangered mammalian animal model for hepatitis C research is critical to the progress of exploration of this pathogen. Such a model is needed to study the pathogenesis, immunology, natural history, and viral replication of hepatitis C, as well as to act as a model for the testing of new therapies and vaccine candidates. This RFA invites grant applications for SBIR projects with award duration and amounts greater than those routinely allowed under the SBIR program. Inquiries may be directed to the appropriate NIH institute: Dr. Leslye Johnson, (301) 496-7051, Division of Microbiology and Infectious Diseases, NIAID, 6003 Executive Boulevard, Bethesda, MD 20892; Dr. John Cole III, (301) 496-1718, Division of Cancer Biology, NCI, Executive Plaza N, Bethesda, MD 20892; Dr. Raymond O'Neil, (301) 435-0744, Comparative Medicine, NCRR, 6701 Rockledge Drive, Bethesda, MD 20892; and Dr. Thomas Kresina, (301) 443-6537, Biomedical Research Branch, NIAAA, 6000 Executive Boulevard, Bethesda, MD 20892. Annual deadline dates may vary; contact program staff for exact dates. RFA: AI-00-008
Requirements: Any small business, independently owned and operated by U.S. citizens or permanent resident aliens may apply. Such a business must be organized for profit, cannot be dominant in its field of expertise, and must have its principal place of business in the United States. Including any affiliates, the company can be the employer of no more than 500 people. Full eligibility details may be found in the NIH Omnibus Solicitation.
Amount of Grant: 32,000,000 U.S.D
Contact: Program Officer
Internet: http://grants.nih.gov/grants/guide/rfa-files/RFA-AI-00-008.html
Sponsor: National Institutes of Health
6701 Rockledge Drive
Bethesda, MD 20892

NIH Solicitation of Assays for High Throughput Screening (HTS) in the Molecular Libraries Probe Production Centers Network (MLPCN) 3574

he NIH Molecular Libraries Roadmap Initiative wishes to encourage HTS assay applications from investigators who have the interest and capability to work with the Molecular Libraries Probe Production Centers Network (MLPCN) for chemical probe development. This Funding Opportunity Announcement (FOA) promotes discovery and development of new chemical probes as research tools for use by scientists in both the public and private sectors to advance the understanding of biological functions and disease mechanisms. This initiative is one of the integrated components of the NIH Molecular Libraries Roadmap initiative that offers biomedical researchers access to large-scale automated high throughput screening (HTS) centers in the MLPCN, diverse compound libraries in the Small Molecule Repository (MLSMR) and information on biological activities of small molecules in the PubChem BioAssay public database. Letters of Intent (LOIs) are due on December 4, April 4, and August 3 respectively. Full application must be received by January 4, May 4, and September 3 respectively.
Requirements: The following organizations and institutions are eligible to apply: public/state controlled institutions of higher education; private institutions of higher education; Hispanic-serving institutions; historically Black colleges and universities (HBCUs); tribally controlled colleges and universities (TCCUs); Alaska native and native Hawaiian serving institutions; nonprofits with 501(c)3 IRS status (other than institutions of higher education); nonprofits without 501(c)3 IRS status (other than institutions of higher education); small businesses; for-profit organizations (other than small businesses); state governments; Indian/Native American tribal governments (federally recognized); Indian/Native American tribally designated organizations; county governments; city or township governments; special district governments; independent school districts; public housing authorities/Indian housing authorities; U.S. territories or possessions; Indian/native American tribal governments (other than federally recognized); regional organizations; non-domestic (non-U.S.) entities (foreign organizations); eligible agencies of the federal government; and faith-based or community-based organizations.
Date(s) Application is Due: Jan 4; May 4; Sep 3
Amount of Grant: Up to 50,000 U.S.D
Contact: Yong Yao, (301) 443-6102; yyao@mail.nih.gov
Internet: http://grants.nih.gov/grants/guide/pa-files/PAR-09-129.html#SectionIII
Sponsor: National Institutes of Health
6001 Executive Boulevard, Room 7175, MSC 9641
Bethesda, MD 20892-9641

NIH Structural Biology of Membrane Proteins Grants 3575

This PA solicits applications to develop research and methods to enhance the rate of membrane protein structure determination and to determine specific membrane protein structures. Innovative methods for expression, oligomerization, solubilization, stabilization, purification, characterization, crystallization, isotopic labeling, and structure determination of unique and biologically significant membrane proteins by x-ray diffraction, nuclear magnetic resonance (NMR), electron microscopic, mass spectrometry, and other biophysical techniques are encouraged. Projects that will lead in the near term to determining the structures of biologically important membrane proteins are also encouraged.
Requirements: Applicant organizations must be: for-profit organizations; non-profit organizations; public or private institutions, such as universities, colleges, hospitals, and laboratories; units of State government; units of local government; eligible agencies of the Federal government; foreign institutions; domestic institutions; faith-based agencies; or community-based organizations.
Date(s) Application is Due: Feb 5; Jun 5; Oct 5
Contact: Peter C. Preusch, (301) 594-3827; preuschp@nigms.nih.gov
Internet: http://grants.nih.gov/grants/guide/pa-files/PA-06-119.html
Sponsor: National Institutes of Health
45 Center Drive, Building 45, 2AS.55E, MSC 6200
Bethesda, MD 20892

NIH Studies of Ethical, Legal, and Social Implications of Research into Human Genetic Variation Grants 3576

The program solicits projects that examine the ethical, legal, and social implications of the study of human DNA sequence variation. Of particular interest are studies that explore how research on human genetic variation can be conducted in an ethically and culturally sensitive way; how the information that results from this research will interact with current concepts of race and ethnicity; how cultural and socioeconomic factors influence the interpretation and use of this information; and how this information may influence access to and use of genetic health services by various groups. The study of sequence variation and the exploration of related ethical, legal, and social issues have been identified as two of the new five-year goals of the U.S. Human Genome Project and will be of continuing interest to the NHGRI program over the next several years. Inquiries may be directed to program staff at the appropriate NIH institute: Jean Cahill, Division of Extramural Research, NHGRI, (301) 402-1733, 38 Library Drive, Bethesda, MD 20892; Sherry Dennison, Grants Management Branch, NIDCD, (301) 402-0909, 6120 Executive Boulevard, Bethesda, MD 20892; David Mineo, Division of Extramural Research and Training, NIEHS, (919) 541-1373, P.O. Box 12233, T.W. Alexander Drive, E Campus, Research Triangle Park, NC 27709; Antoinette Holland, Grants Administration Branch, NIGMS, (301) 594-5132, Building 45, Bethesda, MD 20892. Annual deadline dates vary; contact program staff for exact dates. RFA: HG-99-002
Requirements: Applications may be submitted by domestic and foreign, for-profit and nonprofit, public and private organizations; units of state and local governments; and eligible agencies of the federal government.
Amount of Grant: Up to 2,000,000 U.S.D
Contact: Program Officer
Internet: http://www.grants.nih.gov/grants/guide/rfa-files/RFA-HG-99-002.html
Sponsor: National Institutes of Health
6701 Rockledge Drive
Bethesda, MD 20892

NIH Summer Internship Program in Biomedical Research 3577

Summer programs at the National Institutes of Health (NIH) provide an opportunity to spend a summer working side-by-side with some of the leading scientists in the world, in an environment devoted exclusively to biomedical research. The NIH consists of the 240-bed Mark O. Hatfield Clinical Research Center and more than 1,200 laboratories located on the main campus in Bethesda, Maryland, as well as in Baltimore and Frederick, Maryland; Research Triangle Park, North Carolina; Phoenix, Arizona; Hamilton, Momtana; and Detroit, Michigan. Awards cover a minimum of eight weeks, with students generally arriving at the NIH in May or June. The stipends for trainees are adjusted yearly, with supplements for prior experience. Prospective candidates must apply online. The application is available from mid-November to March 1.
Requirements: The Summer Internship Program is for students who will be sixteen years of age or older at the time they begin the program and who are currently enrolled at least half-time in high school or an accredited U.S. college or university. Students who have been accepted into a college or university program may also apply.
Restrictions: To be eligible, candidates must be U.S. citizens or permanent residents.
Date(s) Application is Due: Mar 1
Contact: Arnetta Courtney, (301) 827-3724; arnetta.courtney@fda.hhs.gov
Internet: http://www.training.nih.gov/student/sip/info.asp
Sponsor: National Institutes of Health
31 Center Drive, Building 31, Room 9A06, MSC 2560
Bethesda, MD 20892-2560

NIH Supplements to Promote Reentry into Biomedical and Behavioral Research Careers 3578

NIH provides administrative supplements to research grants to support individuals with high potential to reenter an active research career after taking time off to care for children or parents or to attend to other family responsibilities. This program will provide administrative supplements to existing NIH research grants for the purpose of supporting full-time or part-time research by these individuals in a program geared to bring their existing research skills and knowledge up to date. For more information, contact the program officer at the appropriate institute. PA: PA-99-106
Requirements: Candidates must have a doctoral degree, such as MD, DDS, PhD, OD, DVM, or equivalent; at least two years of postdoctoral research experience; and must have had sufficient prior research experience to have qualified for a faculty appointment at the assistant professor or equivalent level at the time of their leaving active research.
Restrictions: Candidates who have begun the reentry process through a fellowship, traineeship, or similar mechanism are not eligible.
Amount of Grant: Up to 40,000 U.S.D
Contact: NIEHS Office; (919) 541-7723
Internet: http://www.grants.nih.gov/grants/guide/pa-files/PA-99-106.html
Sponsor: National Institutes of Health
6701 Rockledge Drive
Bethesda, MD 20892

NIH Targeting Research Gaps in HIV Therapeutics Research Grants 3579
NIAID, NIDDK, and NIMH invite applications to conduct applied studies in specific gap areas identified in current HIV therapeutics. Study areas include new delivery or formulation methods to enhance the clinical potential of anti-HIV drugs (FDA-approved or those undergoing clinical testing) in infected adults and children; validating new viral and cellular targets for HIV inhibition and for developing new drugs (or developing new agents against existing targets); and defining and validating the role of innate (nonacquired) immune responses in susceptibility to HIV infection, modulation of disease progression, and their potential use in therapeutic application in infected adults and children. Studies targeted in this PA should generate preclinical proof-of-concept on the therapeutic feasibility of the specific target or approach, and engage in applied preclinical studies. When relevant, applications with clinical studies (e.g., evaluating a drug delivery method) will be considered. Inquiries regarding eligibility and research scope may contact Dr. Nava Sarver, (301) 496-2970, Division of AIDS, NIAID, 6003 Executive Boulevard, Bethesda, MD 20892, email: ns18p@nih.gov. Inquiries regarding other programmatic issues may contact Dr. Dianne Rausch, (301) 443-7281, Office of AIDS Research, NIMH, 6001 Executive Boulevard, Bethesda, MD 20892; or Dr. Philip Smith, (301) 594-8816, Division of Diabetes, Endocrinology, and Metabolic Diseases, NIDDK, 45 Center Drive, Bethesda, MD 20892. PA-99-067
Requirements: Applications may be submitted by domestic and foreign for-profit and nonprofit organizations, both public and private; units of state and local governments; and eligible agencies of the federal government.
Contact: Program Officer
Internet: http://www.grants.nih.gov/grants/guide/pa-files/PA-99-067.html
Sponsor: National Institutes of Health
6701 Rockledge Drive
Bethesda, MD 20892

NIH Technologies to Improve the Utility of Animal Models Research Grants 3580
This PA encourages the small business community to develop technologies, reagents, and equipment to improve the utility of animal models for biomedical research. The results of projects supported under this PA are expected to be improvement of laboratory animal welfare, improved long-term maintenance and preservation of laboratory animal models, and improved technologies for generation of animal models of human diseases. Cost-effective methods, reagents, and equipment to improve animal health and to characterize, culture, and preserve biomedically important organisms are critical needs for biomedical research. Reagents to diagnose and control laboratory animal diseases, including those of aquatic animals, are needed. Equipment and techniques for cost-effective husbandry of animal models in controlled environments are needed for a variety of aquatic and nonaquatic laboratory animals. Improved technologies for the cryopreservation of embryos and gametes are needed for several scientific and economic reasons. Projects supported under the PA are to be directed along one or more of these objectives. This PA must be read in conjunction with the Omnibus Solicitations for SBIR and STTR applications (available on the Web site). Inquiries may be directed to the staff at the appropriate NIH institute: Director, Comparative Medicine, (301) 435-0744, NCRR, 6705 Rockledge Drive, Bethesda, MD 20892; Jo Anne Goodnight, (301) 496-5307, Division of Cancer Biology, NCI, Executive Plaza N, Bethesda, MD 20892; Dr. Christine Kelley, (301) 435-0565, Division of Heart and Vascular Diseases, NHLBI, 2 Rockledge Center, Bethesda, MD 20892; Dr. Joyce Reese, (301) 594-2088, Division of Extramural Research, NIDCR, Building 45, Bethesda, MD 20892; and Dr. Michael Galvin Jr., (910) 541-7825, Division of Extramural Research and Training, NIEHS, P.O. Box 12233, Research Triangle Park, NC 27709. PA: PA-99-048
Requirements: Any small business, independently owned by U.S. citizens and located in the United States, may apply.
Contact: Program Officer
Internet: http://www.grants.nih.gov/grants/guide/pa-files/PA-99-048.html
Sponsor: National Institutes of Health
6701 Rockledge Drive
Bethesda, MD 20892

NIH Transitional Career Development Award in Women's Health Research 3581
This award is designed to support career development experiences leading to independence for clinical investigators interested in patient-oriented or population-based research related to women's health. The career development program will provide an opportunity for investigators to develop solid clinical research skills during two years of study and research within the environment of the NIH Intramural Research Programs (IRP). The award will also include a follow-on two-year period of salary and research support at an academic institution of the candidate's choice. This period of extramural support will facilitate the transition to independence as a clinical researcher specializing in research issues related to women's health. Inquiries may be directed to program staff at the appropriate NIH institute: Dr. Robin Barr, (301) 496-9322, NIA, 7201 Wisconsin Avenue, Bethesda, MD 20892; Dr. Milton Hernandez, (301) 496-7291, NIAID, Solar Building, Bethesda, MD 20892; Dr. Richard Lymn, (301) 594-5128, NIAMS, Building 45, Bethesda, MD 20892; Dr. Lester Gorelic, (301) 496-8580, NCI, Executive Plaza N, Bethesda, MD 20892; Dr. Donna Vogel, (301) 496-6515, NICHD, Building 6100, Bethesda, MD 20892; Dr. James Lipton, (301) 594-2618, Natcher Building, Bethesda, MD 20892; Dr. Cora Lee Wetherington, (301) 443-1263, NIDA, 6001 Executive Boulevard, Bethesda, MD 20892; Dr. Carol Shreffler, (919) 541-5064, NIEHS, P.O. Box 12233, Research Triangle Park, NC 27709; and Dr. Beth Schucker, (301) 435-0535, NHLBI, 6701 Rockledge Drive, Bethesda, MD 20892. PA: OD-00-003
Requirements: Individuals must have a health-professional doctoral degree or its equivalent and must be able to demonstrate the potential for highly productive patient-oriented research. Candidates should possess aptitudes for independent, patient-oriented, or population-based research (e.g. epidemiology, behavior, or disease prevention) and should be able to demonstrate a commitment to the study of diseases that occur in women. Candidates must be U.S. citizens or noncitizen nationals, or must have been lawfully admitted for permanent residence and possess an Alien Registration Card.
Restrictions: Individuals on temporary or student visas are not eligible to apply.
Contact: Program Officer
Internet: http://grants.nih.gov/grants/guide/rfa-files/RFA-OD-00-003.html
Sponsor: National Institutes of Health
6701 Rockledge Drive
Bethesda, MD 20892

NIH Undergraduate Scholarships 3582
The program offers competitive one-year scholarships to students from disadvantaged backgrounds who are committed to careers in biomedical, behavioral, and social science health-related research. Scholarships support tuition, educational expenses, and reasonable living expenses. Recipients may reapply for additional years (not to exceed four scholarship years).The program is designed to improve access to education leading to research careers for those who have had fewer opportunities than others, and to provide an incentive for exceptional scholars to pursue research careers at the NIH. For each full or partial scholarship year, the recipient will incur two types of service obligation: 10-week summer laboratory experience; and employment at the NIH after graduation. Guidelines are available online.
Requirements: The applicant must be a U.S. citizen, national, or qualified noncitizen; enrolled or accepted for enrollment as a full-time student for the academic year at an accredited undergraduate institution; from a disadvantaged background; and must have a grade point average of 3.5 (on a 4.0 scale) or be within the top 5 percent of his or her class.
Date(s) Application is Due: Feb 28
Amount of Grant: Up to 20,000 U.S.D
Contact: Office of Loan Repayment and Scholarship; (800) 528-7689; fax (301) 480-3123; ugsp@nih.gov
Internet: http://ugsp.info.nih.gov/exesumfaq.htm
Sponsor: National Institutes of Health
2 Center Drive, MSC 0230
Bethesda, MD 20892-0230

NIH Urinary Incontinence Treatment Network 3583
NIDDK and NICHD invite cooperative agreement applications for investigators to design and implement a multicenter, prospective cohort study to clarify the benefits of the different surgical interventions for treatment of patients with urinary incontinence. This study will assess the long-term outcomes of the most commonly utilized surgical interventions to correct urinary incontinence in adult women and the utilization of concomitant medical and behavioral therapy. These cooperative agreements will support a collaborative network of continence treatment centers and a biostatistical coordinating center that will prospectively assess the outcomes of different surgical interventions for the treatment of urinary incontinence in women over a three- to four-year period. This group of investigators will develop standardized diagnostic and outcome measures that can be applied uniformly to all of the different surgical cohorts to be evaluated by this initiative. This includes the establishment of diagnostic entry criteria, identification of clinically and patient relevant outcomes, and the development of questionnaires to measure quality of life and other patient-reported factors. Inquiries may be directed to program staff at the appropriate NIH institute: Dr. John Kusek, Program Director, (301) 594-7717, NIDDK, 45 Center Drive, Bethesda, MD 20892; and Dr. Trent MacKay, (301) 435-6972, NICHD, 6100 Executive Boulevard, Bethesda, MD 20892. Annual deadline dates may vary; contact program staff for exact dates. RFA: DK-99-001
Requirements: Applications may be submitted by domestic, for-profit and nonprofit institutions, both public and private organizations; units of state and local government; and eligible agencies of the federal government. Racial/ethnic minorities, women, and persons with disabilities are encouraged to apply.
Restrictions: Foreign institutions are not eligible to apply.
Amount of Grant: 250,000 - 750,000 U.S.D
Contact: Program Officer
Internet: http://www.grants.nih.gov/grants/guide/rfa-files/RFA-DK-99-001.html
Sponsor: National Institutes of Health
6701 Rockledge Drive
Bethesda, MD 20892

NIH Xenobiotics and Cell Death/Injury in Neurodegenerative Disease Research Grants 3584
The objective of this program announcement is to stimulate research on the relative roles of environmental, endogenous, neurochemical, and genetic factors in the cause of neurodegenerative diseases (e.g., Parkinson's disease, Alzheimer's disease, and amyotrophic lateral sclerosis). Attempts are needed to understand the contribution of environmental exposure, endogenous susceptibility factors, and increasing age in the etiopathology of neurodegenerative disease and will require a concurrent advancement and refinement of methodologies and sciences. Of particular significance may be those approaches that can be used across species from lower animals to humans. Such approaches permit a precise characterization in animal models of alterations arising from defined environmental exposures which can serve as a cogent guide to underlying cellular and molecular mechanisms. The results of these investigations should help clarify the part xenobiotics play in the etiology of these diseases. Applications that are collaborations between basic and clinical neuroscientists and/or are interdisciplinary in nature are encouraged. Inquiries may be directed to the staff of the appropriate NIH institute: Dr. Annette Kirshner, (919) 541-0488, Division of Extramural Research and Training,

NIEHS, Research Triangle Park, NC 27709; Dr. Eugene Oliver, (301) 496-5680, Division of Stroke, Trauma, and Neurodegenerative Disorders, NINDS, 7550 Wisconsin Avenue, Bethesda, MD 20892; Dr. Bradley Wise, (301) 496-9350, Neuroscience and Neuropsychology of Aging Program, NIA, 7201 Wisconsin Avenue, Bethesda, MD 20892; and Dr. Linda Brady, Preclinical and Clinical, Therapeutics Research Branch, NIMH, 5600 Fishers Lane, Rockville, MD 20857. PA: PAS-99-054
Requirements: Applications may be submitted by domestic and foreign, for-profit and nonprofit organizations, public and private.
Amount of Grant: Up to 2,000,000 U.S.D
Contact: Program Officer
Internet: http://www.grants.nih.gov/grants/guide/pa-files/PAS-99-054.html
Sponsor: National Institutes of Health
6701 Rockledge Drive
Bethesda, MD 20892

NIH Zebrafish as an Animal Model for Development and Disease Research Grants 3585
The purpose of this PA is to solicit applications as part of an NIH initiative to increase support of the zebrafish as an animal model for research. The goals of this PA are to encourage new and innovative research and approaches using the zebrafish to identify the genes and elucidate the molecular and genetic mechanisms responsible for normal and defective development and disease. The mechanism of support for this PA will be the investigator-initiated research project grant (R01) award. Applications from new investigators are particularly encouraged. Inquiries may be directed to program staff at the appropriate NIH institute: Dr. Deborah Henken, (301) 496-5541, NICHD; Dr. David Badman, (301) 594-7717, NIDDK; Dr. Grace Shen, (301) 435-5226, NCI; Dr. Jill Carrington, (301) 435-0776, NCRR; Dr. Maria Giovanni, (301) 496-0484, NEI; Dr. Carol Letendre, (301) 435-0080, NHLBI; Dr. Adam Felsenfeld, (301) 496-7531, NHGRI; Dr. Bradley Wise, (301) 496-9350, NIA; Dr. Robert Karp, (301) 443-2239, NIAAA; Dr. Vicki Seyfert, (301) 496-7551, NIAID; Dr. Richard Lymn, (301) 594-5130, NIAMS; and Dr. Chyren Hunter, (301) 402-3461, NIDCD. PA: PA-01-095
Requirements: Applications may be submitted by domestic and foreign, for-profit and nonprofit organizations, public or private. The total project period for an application submitted in response to this PA may not exceed five years.
Contact: Program Officer
Internet: http://grants.nih.gov/grants/guide/pa-files/PA-01-095.html
Sponsor: National Institutes of Health
6705 Rockledge Drive
Bethesda, MD 20892

NIMH/NIDA Children's Services Program Grants 3586
The purpose of this program is to encourage research grant applications on services delivered to children, adolescents, and their families. Conducting the proposed research at one of the Children's Services Program sites (either existing sites or graduated sites) is a prerequisite for funding under this PA. This PA encourages studies of the effectiveness of treatments or services delivered at these sites, the nature and impact of routine clinical practice, and factors related to successful implementation of treatments or services. PA: PA-04-019
Requirements: Applications may be submitted by domestic and foreign, for-profit and nonprofit organizations, both public and private; units of state and local governments; and eligible agencies of the federal government.
Restrictions: Research grant funds may not be used to operate treatment, rehabilitation, or service programs.
Amount of Grant: 25,000 - 250,000 U.S.D
Contact: Dr. Heather Ringeisen, Chief, Child and Adolescent Services Research Program Contact; (301) 496-7227; fax (301) 443-4045; hringeis@mail.nih.gov
Internet: http://grants.nih.gov/grants/guide/pa-files/PA-04-019.html
Sponsor: National Institute of Mental Health
6001 Executive Boulevard
Bethesda, MD 20892

NIMH AIDS and Aging: Behavioral Sciences Prevention Research Grants 3587
NIA, NIMH, and NINR invite qualified researchers to submit applications to investigate prevention issues relevant to AIDS in middle-aged and older populations. This announcement solicits AIDS prevention research proposals to study primary prevention of disease transmission as well as secondary and tertiary prevention of negative behavioral and social consequences of HIV/AIDS for persons with AIDS, their families, and communities. Thus, the primary goals are to identify social and behavioral factors associated with HIV transmission and disease progression in later life; examine behavioral and social consequences of HIV infection/AIDS across the life course; develop and evaluate age-appropriate behavioral and social interventions for preventing AIDS in middle-aged and older adults and/or ameliorating problems associated with older adult caregiver responsibilities and burdens; explore health care issues surrounding AIDS care; and strengthen existing research and evaluation methods.
Requirements: Applications may be submitted by domestic and foreign for-profit and non-profit organizations, public and private, such as universities, colleges, hospitals, laboratories, units of State and local governments, and eligible agencies of the Federal government. Applications may be submitted by single institutions or by a consortia of institutions.
Date(s) Application is Due: Feb 5; Feb 16; Jun 5; Jun 16; Oct 5; Oct 16
Contact: Dr. Willo Pequegnat; (301) 443-6100; fax (301) 443-9719; wpequegn@nih.gov
Internet: http://www.grants.nih.gov/grants/guide/pa-files/PA-97-069.html
Sponsor: National Institute of Mental Health
5600 Fishers Lane, Parklawn Building
Rockville, MD 20857

NIMH Basic and Translational Research in Emotion Grants 3588
This Funding Opportunity Announcement encourages Research Project Grant applications to expand basic and translational research on the processes and mechanisms involved in the experience, expression, and regulation of emotion. The study of emotion encompasses a wide range of physiological, psychological, social, cognitive, and developmental phenomena. Emotional processes can be studied in human or animal subjects. Important objects of study include, but are not limited to: central and peripheral nervous system activity in the origins, expression, regulation and modulation of emotion; the contribution of emotional and motivational systems to cognitive faculties such as perception, attention, learning, and memory; investigations of overt behaviors, interpersonal relationships, communication and decision making; and the environmental circumstances and experiences that evoke and modulate different emotions.
Requirements: The following organizations and institutions may apply: public/state controlled institutions of higher education; private institutions of higher education; Hispanic-serving institutions; historically Black colleges and universities (HBCUs); tribally controlled colleges and universities (TCCUs); Alaska native and native Hawaiian serving institutions; nonprofits with 501(c)3 IRS status; nonprofits without 501(c)3 IRS status; small businesses; for-profit organizations; State governments; regional organizations; U.S. territories or possessions; Indian/Native American tribal governments; Indian/Native American tribally designated organizations; county governments; city or township governments; special district governments; independent school districts; public housing authorities/Indian housing authorities; eligible agencies of the Federal government; and faith-based or community based organizations.
Date(s) Application is Due: Feb 5; Jun 5; Oct 5
Contact: Wendy Nelson, (301) 435 4590; fax (301) 435 7547; nelsonw@mail.nih.gov
Internet: http://grants.nih.gov/grants/guide/pa-files/PA-09-137.html
Sponsor: National Institute of Mental Health
6001 Executive Boulevard, Room 7179, MSC 9637
Bethesda, MD 20892-9637

NIMH Behavior Genetics in Adulthood and Old Age Research Grants 3589
NIA and NIMH invite qualified researchers to submit research grant applications on behavior genetics in adulthood and old age. The goal of this PA is to further scientific knowledge about the relative contributions of hereditary and environmental factors in the health, well-being, and quality of life of adults (age 21 and over) and older persons. This PA encourages research that will address one or more of the following: behavior genetics and cognitive functioning/intellectual ability and behavior genetics and personality. PA: PA-03-128
Requirements: Applications may be submitted by foreign and domestic for-profit and nonprofit organizations, public or private.
Contact: Dr. Marilyn Miller, National Institute on Aging; (301) 496-9350; fax (301) 496-1494; Millerm@nia.nih.gov
Internet: http://www.grants.nih.gov/grants/guide/pa-files/PAS-03-128.html
Sponsor: National Institute of Mental Health
5600 Fishers Lane, Parklawn Building
Rockville, MD 20857

NIMH Career Opportunities in Research (COR) Education and Training Honors High School Research Education Grant 3590
This program encourages applications for training activities designed to stimulate interest among minority youth in research careers in biomedical and behavioral sciences relevant to mental health. This program is specifically designed to increase the participation of groups that are underrepresented in biomedical and behavioral sciences, including African Americans, Hispanics, Native Indians and Alaska Natives, and Asians/Pacific Islanders. The program is intended to support activities that allow hands-on experiences in research or science education and encourage high school students to make the choice of a research career in a mental health-related field. PA: PAR-01-009
Requirements: Applicant institutions must have a currently funded NIMH Career Opportunities in Research Education and Training Program Honors Undergraduate Research Training Grant. Such institutions are baccalaureate granting, four-year domestic, public or private, nonprofit colleges, universities, or health professional schools with at least 50 percent minority students.
Restrictions: Foreign organizations are not eligible to apply.
Date(s) Application is Due: May 10
Amount of Grant: 25,000 - 35,000 U.S.D
Contact: Dr. Carolyn Strete; (301) 443-2847; fax (301) 443-8552; cstrete@nih.gov
Internet: http://grants.nih.gov/grants/guide/pa-files/PAR-01-009.html
Sponsor: National Institute of Mental Health
6001 Executive Boulevard
Bethesda, MD 20892

NIMH Career Opportunities in Research (COR) Education and Training Honors Undergraduate Research Training Grant 3591
This program is intended to strengthen research and research training experiences of undergraduate minority students in scientific disciplines related to mental health. It is directed specifically to four-year colleges and universities that serve substantial numbers of students of one or more racial/ethnic minority groups, including African Americans, Hispanics, Native Americans and Alaska Natives, and Asians/Pacific Islanders. An applicant institution must propose a two-year COR Honors Undergraduate program for which six to 10 highly talented third-and fourth-year undergraduate students are selected. Proposed training programs should provide students with special research training experiences designed to improve their qualifications for entry into advanced research career training programs leading to doctoral-level or MD research career degrees. PA: PAR-01-008

Requirements: Applicant institutions must be four-year public or private, nonprofit colleges, universities, or heath professional schools with at least 50 percent racial/ethnic minority students; and be able to develop the faculty and institutional facilities in order to provide a suitable research environment and high-quality research training in areas relevant to mental health.
Restrictions: Foreign organizations are not eligible to apply.
Amount of Grant: Up to 300,000 U.S.D
Contact: Robert Mays, Office for Special Populations; (301) 443-3017; fax (301) 443-8022; rmays@mail.nih.gov
Internet: http://grants.nih.gov/grants/guide/pa-files/PAR-01-008.html
Sponsor: National Institute of Mental Health
6001 Executive Boulevard
Bethesda, MD 20892

NIMH Centers for AIDS Research (CFARs) 3592
NIMH is interested in establishing AIDS research centers to provide support for coordinated, multidisciplinary research programs on the mental health and drug abuse aspects of AIDS, ARC, and HTLV-III infection. AIDS research centers support may be requested for up to five years. The overall aims of the center must be clearly defined for the requested period of support. Areas of future development should be indicated and justified. Annual deadline dates may vary; contact program staff for exact dates. PA: PAR-03-089
Contact: Dr. Dianne Rausch, Office of AIDS Research; (301) 443-7281; fax (301) 443-9719; dr89b@nih.gov
Internet: http://grants2.nih.gov/grants/guide/pa-files/PAR-03-089.html
Sponsor: National Institute of Mental Health
5600 Fishers Lane, Parklawn Building
Rockville, MD 20857

NIMH Central Nervous System Effects of HIV Infection: Neurobiological, Neurovirological, and Neurobehavioral Studies Grants 3593
NIMH, with NINDS, NICHD, and NIDA, invites applications for support of research on the role of HIV in the central nervous system and its consequent effects on behavior. Applicants may request up to five years of support. Applications will be accepted under PHS receipt dates for AIDS applications. Potential applicants should contact the AIDS coordinator. PA: PA-04-154
Contact: Brian Albertini; (301) 443-0004; fax (301) 443-0219; albertib2@mail.nih.gov
Internet: http://grants.nih.gov/grants/guide/pa-files/PA-04-154.html
Sponsor: National Institute of Mental Health
5600 Fishers Lane, Parklawn Building
Rockville, MD 20857

NIMH Collaborative R01s for Clinical Studies of Mental Disorders 3594
NIMH invites applications for collaborative research efforts on clinical studies of mental disorders (CSMD), including schizophrenia, mood disorders, anxiety disorders, and other severe mental illnesses. This PA provides guidelines for the conceptualization, preparation, review, and administration of collaborative CSMD applications submitted to the NIMH. PA-01-123 may be obtained through the NIH grant line (301-402-2221), the NIH Web site, mail, and email.
Requirements: This mechanism requires more than one applicant organization, although multiple campuses of a single institution are also eligible. Applications may be submitted by domestic and foreign, for-profit and nonprofit organizations, public and private.
Contact: Dr. Barry Lebowitz, Services and Intervention Research Coordinator; (301) 443-1185; fax (301) 443-4045; blebowit@nih.gov
Internet: http://grants2.nih.gov/grants/guide/pa-files/PA-01-123.html
Sponsor: National Institute of Mental Health
5600 Fishers Lane, Parklawn Building
Rockville, MD 20857

NIMH Core Support Program for Mental Health/AIDS Research Grants 3595
The purpose of this program is to provide a mechanism to facilitate currently funded investigators in improving and expanding research infrastructure, including administrative coordination, subject recruitment, equipment, laboratories, statistical analysis, and database management. A letter of intent is requested by December 1; full application is due by January 2. Annual deadline dates may vary; contact program staff for exact dates. PA: PAR-03-142
Requirements: Applications may be submitted by domestic, for-profit and nonprofit organizations, both public and private; units of state and local governments; and eligible agencies of the federal government.
Date(s) Application is Due: Jan 2; Dec 1
Amount of Grant: Up to 2,000,000 U.S.D
Contact: Dr. Dianne Rausch; (301) 443-7281; fax (301) 443-9719; dr89b@nih.gov
Internet: http://grants.nih.gov/grants/guide/pa-files/PAR-03-142.html
Sponsor: National Institute of Mental Health
6001 Executive Boulevard
Bethesda, MD 20892

NIMH Curriculum Development Award in Neuroinformatics Research and Analysis 3596
The purpose of this program is to encourage and support applications from individuals, with the requisite scientific expertise and leadership, for the development of courses and curricula designed to train interdisciplinary neuroinformatics scientists at U.S. educational institutions. The field of neuroinformatics combines neuroscience research with informatics research developed from the computer sciences, mathematics, physics, engineering or closely related sciences. It is anticipated that these courses or curricula would be useful to students and scientists who wish: to develop new conceptual approaches to basic and/or clinical neuroscientific research and analysis; or to acquire, store, retrieve, organize, manage, analyze, visualize, manipulate, integrate, synthesize, disseminate, and share data about the brain and behavior. Development of courses at the graduate and undergraduate level is encouraged. As part of this program, awardees will be expected to develop and implement the courses or curricula in their institution. It is expected that such courses and curricula will be models that could be transferable to other institutions in whole or in part.
Requirements: The principal investigator must be engaged in neuroscience research or research in computer science, mathematics, physics, engineering, or a related informatics field. Collaborator(s) who will contribute to the interdisciplinary nature of the courses or curricula must be identified. The principal investigator must be willing to spend at least 20 percent of full-time professional effort on course(s) and curricula development during the period of the award. The principal investigator must also identify appropriate researcher(s) who will agree to collaborate on the development of course(s) and curricula.
Date(s) Application is Due: Feb 16; Jun 16; Oct 16
Contact: Stephen H. Koslow, (301) 443-1815; fax (301) 443-1867; koz@helix.nih.gov
Internet: http://grants.nih.gov/grants/guide/pa-files/PAR-99-135.html
Sponsor: National Institute of Mental Health
6001 Executive Boulevard, Room 6167, MSC 9613
Bethesda, MD 20892

NIMH Early Detection of HIV Grants 3597
NIMH offers funds to conduct research on ways to promote early detection of HIV, and the impact of these findings on prevention. Research should look at new strategies to increase HIV testing and communication of test results to at-risk individuals, as well as develop innovative prevention strategies associated with emerging technologies in the early detection of HIV. A letter of intent is requested by February 10; full proposal is due March 10. Annual deadline dates may vary; contact program staff for exact dates. RFA: MH-00-008
Requirements: Public and private nonprofits may apply.
Date(s) Application is Due: Feb 10; Mar 10
Amount of Grant: Up to 1,000,000 U.S.D
Contact: Louis Steinberg, (301) 443-6100; fax (301) 443-9719; lsteinbe@nih.gov
Internet: http://grants.nih.gov/grants/guide/rfa-files/RFA-MH-00-008.html
Sponsor: National Institute of Mental Health
5600 Fishers Lane, Parklawn Building
Rockville, MD 20857

NIMH Early Identification and Treatment of Mental Disorders in Children and Adolescents Grants 3598
This program supports research on early identification and treatment of mental disorders in children and adolescents. In particular, this announcement intends to encourage research on disorders such as schizophrenia, schizoaffective disorder, bipolar disorder, major depression, obsessive-compulsive disorder, and anorexia nervosa, alone or comorbid with other common mental or substance abuse disorders. PA: PA-00-094
Requirements: Applications may be submitted by domestic for-profit and nonprofit organizations, both public and private; units of state or local governments; and eligible agencies of the federal government.
Amount of Grant: 125,000 - 250,000 U.S.D
Contact: Rebecca Claycamp, Grants Management Officer; (301) 443-2811
Internet: http://www.nimh.nih.gov/researchfunding/grants.cfm
Sponsor: National Institute of Mental Health
6001 Executive Boulevard
Bethesda, MD 20892

NIMH HIV/AIDS, Severe Mental Illness, and Homelessness Research Grants 3599
This PA is sponsored by the NIMH to stimulate investigator-initiated research that targets persons with severe mental illness (SMI) either before or after HIV infection and to expand HIV-related research to homeless persons. It solicits studies on the SMI population and/or homeless persons with special attention to the development, implementation, and evaluation of effective HIV-prevention interventions and their dissemination and translation to the community and public health service organizations. also solicit studies on the SMI population and/or homeless persons that propose improved methodologies to study descriptive/analytic epidemiology of HIV infection and epidemiology of the behavior associated with risk and the spread of HIV infection An important objective of this PA is to encourage integration both across and within the different research areas by establishing multidisciplinary research teams and collaborative alliances. PA: PA-04-024
Requirements: Applications may be submitted by domestic and foreign, for-profit and nonprofit organizations, both public and private; units of state and local governments; and eligible agencies of the federal government.
Contact: Dr. David Stoff, (301) 443-4625; fax (301) 443-9719; dstoff@nih.gov
Internet: http://www.grants.nih.gov/grants/guide/pa-files/PA-04-024.html
Sponsor: National Institute of Mental Health
6701 Rockledge Drive
Bethesda, MD 20817

NIMH Informed Consent in Research Grants 3600
This program invites applications for a three-year research grant to stimulate investigations into the informed consent process in scientific research involving human participants. The grant is being offered because voluntary informed consent is the defining aspect of interactions between researchers and participants and is integral to the conduct of

scientific research. One of the goals of this project is to bring together perspectives of different agencies, because various areas of research reflect a diversity of issues relating to informed consent. The focus of the program will be to document the degree of understanding achieved by research participants regarding the identity of the sponsoring federal agency or agencies; the purposes for which the research is being conducted; the comprehension of a study's methods and procedures; the relative risks and benefits of deciding to consent or refuse participation; the confidentiality and any exceptions to confidentiality; the implications of withdrawal from a study; and the planned and other possible use of the data. The goal of the present initiative is to develop and test alternative strategies for obtaining informed consent in diverse populations and to determine optimal ways to obtain informed consent for research participation. Additional information may be obtained electronically through the NIH grant line (301-402-2221), the NIH Web site, and by mail and email from the program contact listed. Annual deadline dates may vary; contact program, staff for exact dates. RFA: OD-97-001
Requirements: Applications may be submitted by any domestic or foreign, for-profit or nonprofit organizations, both public or private; units of state and local governments; or eligible agencies of the federal government.
Contact: Dr. John Hsiao; (301) 443-1185; fax (301) 594-6784; jhsiao@heilix.nih.gov
Internet: http://www.grants.nih.gov/grants/guide/rfa-files/RFA-OD-97-001.html
Sponsor: National Institute of Mental Health
5600 Fishers Lane, Parklawn Building
Rockville, MD 20857

NIMH Interdisciplinary Behavioral Science Centers for Mental Health 3601
The purpose of these centers is to support collaborative, hypothesis-driven basic research activities that will extend the most cutting-edge theories and approaches in basic behavioral science to incorporate current approaches in neuroscience. A letter of intent is requested by January 18; full application is due February 18. Annual deadline dates may vary; contact program staff for exact dates. PA: PAR-04-004
Requirements: Applications may be submitted by domestic, for-profit and nonprofit organizations, both public and private; units of state and local governments; and eligible agencies of the federal government.
Date(s) Application is Due: Jan 18; Feb 18
Amount of Grant: Up to 15,000,000 U.S.D
Contact: Dr. Mary Ellen Oliveri, (301) 443-3942; fax (301) 443-9876; moliveri@nih.gov
Internet: http://grants.nih.gov/grants/guide/pa-files/PAR-04-004.html
Sponsor: National Institute of Mental Health
6001 Executive Boulevard, Room 7220, MSC 9651
Bethesda, MD 20892-9651

NIMH Interventions Infrastructure Program Research Grants 3602
NIMH is seeking to expand the capacity of the field to support state-of-the-art public health-oriented studies in the treatment, rehabilitation, and prevention of mental disorders in children, adolescents, and in mid- and late-life adults. The Interventions Infrastructure Program (IIP) encourages the expansion of intervention research collaborations between academic health centers and health care provider organizations that are oriented to community care, and have access to large, stable, and diverse patient populations. The purpose of the IIP is to provide the infrastructure for the support of research that has sufficient power to define standards of appropriate and cost-effective care for the diverse populations of patients with mental disorders seen in all health care settings. Applications are accepted annually. Annual deadline date may vary; contact program staff for exact dates. PA: PAR-04-015
Requirements: Applications may be submitted by domestic, for-profit and nonprofit organizations, public and private.
Restrictions: This program is nonrenewable.
Date(s) Application is Due: Jun 1
Contact: Dr. Junius Gonzales, (301) 443-3364; fax (301) 443-4045; jgonzale@mail.nih.gov
Internet: http://grants2.nih.gov/grants/guide/pa-files/PAR-04-015.html
Sponsor: National Institute of Mental Health
6001 Executive Boulevard
Bethesda, MD 20892

NIMH Jointly Sponsored Ruth L. Kirschstein National Research Service Award Institutional Predoctoral Training Program in the Neurosciences 3603
The Jointly Sponsored NIH Predoctoral Training Program in the Neurosciences supports broad and fundamental, early-stage graduate research training in the neurosciences via institutional NRSA research training grants (T32) at domestic institutions of higher education. Trainees are supported during years 1 and 2 of their graduate training when they are typically not committed to a dissertation laboratory. The primary objective is to prepare qualified individuals for careers in neuroscience that have a significant impact on the health-related research needs of the Nation. Because the nature and scope of the proposed research training will vary from application to application, it is anticipated that the size and duration of each award will also vary. The total amount awarded and the number of awards will depend upon the number, quality, duration, and costs of the applications received.
Requirements: The following organizations are eligible to apply: non-profit organizations; public or private institutions, such as universities, colleges, hospitals, and laboratories; and domestic institutions.
Date(s) Application is Due: Apr 25; May 25
Contact: Lindsey Grandison, (301) 443-0606; fax (301) 443-1650; lgrandis@mail.nih.gov
Internet: http://grants.nih.gov/grants/guide/pa-files/PAR-08-101.html
Sponsor: National Institute of Mental Health
5635 Fishers Lane, MSC 9304
Bethesda, MD 20892-9304

NIMH Mental Health Dissemination Research Grants 3604
NIMH invites applications for studies addressing factors influencing the dissemination of mental health research knowledge across mental health service settings. The purpose of this announcement is to stimulate research on the array of influences, and their interaction, that beneficially or adversely affect the adoption of valid mental health research findings into clinical practice; and to promote the development of a sound knowledge base on the effectiveness of knowledge dissemination interventions. PA: PA-02-131
Requirements: Applications may be submitted by domestic and foreign, for-profit and nonprofit organizations, public and private.
Restrictions: Foreign institutions are not eligible for small (R03) awards.
Contact: Joy Knipple; (301) 443-8811; fax (301) 443-6885; jk173r@nih.gov
Internet: http://grants2.nih.gov/grants/guide/pa-files/PA-02-131.html
Sponsor: National Institute of Mental Health
6001 Executive Boulevard
Bethesda, MD 20892

NIMH Minority Research Infrastructure Support Program Grants 3605
The purpose of this program is to increase the capacity of minority institutions and their faculty to conduct rigorous mental health-relevant research. The intent of the research infrastructure program is to strengthen the research environments of minority institutions through grant support to develop and/or expand existing capacities for conducting behavioral and neuroscience research in all fields related to mental health. The program provides for two types of activities: institutional development support to strengthen the institutional infrastructure and enhance the capability of individual faculty members to undertake mental health-related research; and individual investigator research support for developing research scientists to conduct small grant research activities that can lead to successful applications for funding under regular research grant mechanisms. PA: PAR-01-029
Requirements: Applications may be submitted by domestic for-profit and nonprofit organizations, public and private; units of state and local governments; and eligible agencies of state and local government.
Date(s) Application is Due: Jun 1
Amount of Grant: Up to 400,000 U.S.D
Contact: Dr. Robert Mays Jr., Office for Special Populations; (301) 443-3017; fax (301) 443-8022; rmays@mail.nih.gov
Internet: http://grants.nih.gov/grants/guide/pa-files/PAR-01-029.html
Sponsor: National Institute of Mental Health
5600 Fishers Lane, Parklawn Building
Rockville, MD 20857

NIMH National Research Service Awards (NRSA)—Individual 3606
A National Research Service Award to an individual supports research training experiences in mental health areas of biomedical and behavioral research. The NIMH research objective is to increase the knowledge of, and improve research methods on, mental and behavioral disorders; to generate information regarding basic biological and behavioral processes underlying these disorders and the maintenance of mental health; and to improve mental health services. PA: PA-99-025
Requirements: Applicants must be citizens or noncitizen nationals of the United States. Predoctoral applicants must have completed two or more years of graduate work and be enrolled in a doctoral degree program by the proposed activation date of the fellowship. Postdoctoral applicants must have received a doctoral or equivalent degree as of the proposed activation date, or submit certification by an authorized official of the degree-granting institution that all requirements have been met.
Date(s) Application is Due: Apr 5; Aug 5; Dec 5
Amount of Grant: 14,688 - 26,256 U.S.D
Contact: Dr. Robin Barr, (301) 443-2530; rb42h@nih.gov
Internet: http://www.grants.nih.gov/grants/guide/pa-files/PA-99-025.html
Sponsor: National Institute of Mental Health
5600 Fishers Lane, Parklawn Building
Rockville, MD 20857

NIMH New Minority Faculty Scientist Development Award 3607
The purpose of this program is to enable new minority faculty members to have the necessary time and assistance early in their academic careers to initiate a program of research and to help them to become outstanding independent investigators in mental health research. PA: PAR-99-169
Requirements: Applications may be submitted on behalf of the proposed candidate by domestic, nonprofit, public and private universities, colleges, and professional schools that are engaged in mental health research. The candidate must be a citizen or non-citizen national of the United States or hold a permanent residence visa. Candidates for this award must be new, non-tenured minority faculty in their first or second faculty position who have earned a doctoral degree (PhD, MD, DSc, etc.) by the time the award is made.
Contact: Henry Khachaturian, Office of Science Policy and Program Planning; (301) 443-4335; fax (301) 443-3225; hkhachat@mail.nih.gov
Internet: http://grants.nih.gov/grants/guide/pa-files/PAR-99-169.html
Sponsor: National Institute for Mental Health
6001 Executive Boulevard
Bethesda, MD 20892

NIMH Pilot Effectiveness Trials for Mental Disorders Grants 3608
The purpose of the program is to obtain needed preliminary data and to conduct studies to support subsequent effectiveness trials of interventions for mental disorders of children and adults. This is part of a series of four program announcements making up a broad-scale public health-oriented intervention research program. Projects may be funded for

up to three years. Annual deadline dates may vary; contact program staff for exact dates. PA: PAR-03-078
Requirements: Applications may be submitted by domestic, for-profit and nonprofit organizations, both public and private; units of state and local governments; and eligible agencies of the federal government.
Date(s) Application is Due: Feb 1; Jun 1; Oct 1
Amount of Grant: Up to 125,000 U.S.D
Contact: Joy Knipple, Grants Management Branch; (301) 443-8811; fax (301) 443-6885; jk173@nih.gov
Internet: http://www.grants.nih.gov/grants/guide/pa-files/PAR-03-078.html
Sponsor: National Institute of Mental Health
6001 Executive Boulevard
Bethesda, MD 20892

NIMH Quantitative Approaches to the Analysis of Complex Biological Systems 3609
The purpose of this program announcement is to advertise the interest of NIGMS and NIMH in supporting research projects that develop quantitative approaches to describe, analyze, and predict the behavior of complex biological systems, especially those requiring the integration of potentially large amounts of molecular, biochemical, cell biological, and physiological data. These projects are expected to require the participation of individuals with diverse expertise and therefore to be of a collaborative and cross-disciplinary nature. NIMH expresses particular interest in studies using mathematical, computational, or theoretical approaches to understanding the fundamental biological mechanisms underlying behavior. PA: PA-98-077
Requirements: Applications may be submitted by foreign and domestic, for-profit and nonprofit organizations, both public and private. Racial/ethnic minority individuals, women, and persons with disabilities are encouraged to apply.
Restrictions: Foreign institutions are not eligible for program project (P01) grants.
Contact: Dr. Dennis Glanzman, Basic and Clinical Neuroscience Research Division Contact; (301) 443-1576; fax (301) 443-4822; glanzman@helix.nih.gov
Internet: http://www.grants.nih.gov/grants/guide/pa-files/PA-98-077.html
Sponsor: National Institute of Mental Health
5600 Fishers Lane, Parklawn Building
Rockville, MD 20857

NIMH Risk Factors for Psychopathology Using Existing Data Sets Grants 3610
The program encourages extensive and innovative use of existing data sets to study the development of psychopathology or resilience in order to guide the development of prevention and early intervention strategies. The aims of the program are to mine the full potential of public use and other extant data sets to increase our knowledge of risk and protective factors for the development of psychopathology or resilience in community-based and clinical populations, and to encourage applications from new investigators to examine these research areas using state-of-the-art data analytic procedures. NIMH encourages proposals to conduct analyses using mental health research data that is in public-use format or that is privately held by a principal investigator. Annual deadline dates may vary; contact program staff for exact dates. PA: PA-03-044
Requirements: Applications may be submitted by domestic and foreign, for-profit and nonprofit organizations, both public and private; units of state and local governments; and eligible agencies of the federal government.
Restrictions: Foreign institutions are not eligible for NIMH R03 (small grant) awards.
Date(s) Application is Due: Feb 1; Jun 1; Oct 1
Amount of Grant: Up to 250,000 U.S.D
Contact: Karen Bourdon, (301) 443-5944; fax (301) 480-4415; kbourdon@nih.gov
Internet: http://www.grants.nih.gov/grants/guide/pa-files/PA-03-044.html
Sponsor: National Institute of Mental Health
6001 Executive Boulevard
Bethesda, MD 20892

NIMH Short Courses in Neuroinformatics 3611
The purpose of this program is to encourage and support short-term research for the development of short courses, seminars, and workshops on interdisciplinary neuroinformatics education . This short-term training will be provided to scientists seeking to combine knowledge about the various subdisciplines of neuroscience and behavioral science research with expertise in informatics research. It is anticipated that these short courses will allow the participants to acquire new conceptual approaches to basic neuroscience research and analyses; and learn to develop unique strategies for acquiring, storing, retrieving, organizing, managing, analyzing, visualizing, manipulating, integrating, synthesizing, disseminating, and sharing data about the brain and behavior.
Requirements: Any nonprofit or for-profit organization engaged in health-related education or research and located in the United States, its possessions, or territories is eligible to apply.
Date(s) Application is Due: Feb 16; Jun 16; Oct 16
Amount of Grant: Up to 150,000 U.S.D
Contact: Stephen H. Koslow, (301) 443-1815; fax (301) 443-1867; koz@helix.nih.gov
Internet: http://grants.nih.gov/grants/guide/pa-files/PAR-99-137.html
Sponsor: National Institute of Mental Health
6001 Executive Boulevard, Room 6167, MSC 9613
Bethesda, MD 20892

NIMH Short Courses on Mathematical and Statistical Tools for the Study of Complex Phenotypes and Complex Systems 3612
The purpose of this program announcement is to provide support for short courses or workshops to assist scientists in preparing for research on complex phenotypes and complex systems. This program announcement solicits applications for education grants (R25), which provide for one to five years of support for short courses. Support is limited to courses that are highly focused and that reach a wide audience of scientists. The program announcement is not intended for university course or curriculum development. A variety of approaches will be considered. The nature of the activity, the needs of the target community, and the course goals should determine the instructional approach. These might include new approaches, such as Web-based instruction, or more traditional approaches, such as summer workshops or on-site instruction. PA: PA-98-083
Requirements: Applications may be submitted by domestic nonprofit and for-profit organizations, both public and private.
Restrictions: Foreign organizations are not eligible.
Contact: Dr. Walter Goldschmidts, Division of Basic and Clinical Neurosciences Research; (301) 443-3563; fax (301) 443-1731; wg8u@nih.gov
Internet: http://www.grants.nih.gov/grants/guide/pa-files/PA-98-083.html
Sponsor: National Institute of Mental Health
5600 Fishers Lane, Parklawn Building
Rockville, MD 20857

NIMH Silvio O. Conte Centers for Neuroscience Research Grants 3613
NIMH invites research grant applications for centers. The primary goal of this initiative is to encourage investigators representing a variety of disciplines and/or approaches to bring to the field of mental health research the full range of expertise and advanced technologies available in the basic sciences. Centers for the feasibility of neuroscience research, centers for neuroscience research, and centers for the neuroscience of mental disorders will be represented. The letter of intent receipt date is July 1, and the application receipt date is October 20. PAR-02-122 may be obtained through the NIH grant line (301-402-2221), the NIH Web site, mail, and email.
Requirements: Applications may be submitted by domestic, for-profit and nonprofit organizations, public or private.
Contact: Carol Robinson; (301) 443-3858; fax (301) 443-6885; crobinso@mail.nih.gov
Internet: http://grants2.nih.gov/grants/guide/pa-files/PAR-02-122.html
Sponsor: National Institute of Mental Health
5600 Fishers Lane, Parklawn Building
Rockville, MD 20857

NIMH Socioeconomic Status and Health Across the Life Course 3614
NHLBI, NIA, NICHD, NIEHS, and NIMH seek research grant applications on the cumulative and contemporaneous relationships between socioeconomic status (SES) and physical and mental health and functioning over the life course and across generations. Given that the relationships between SES and physical and mental health, morbidity, disability, and mortality have been long and extensively documented, additional studies aimed at merely describing or demonstrating these relationships are outside the scope of this program announcement. Encouraged are studies relating to appropriate conceptualization and measurement of SES over the life course, across generations, and in various population groups; specification of the processes through which SES influences cumulatively and contemporaneously physical and mental health, disability, morbidity, and mortality outcomes over the life course, and how these outcomes, in turn, impact on SES; and the relationship between SES and physical and mental health, disability, morbidity, and mortality over the life course in various population groups. PA: PA-98-098
Requirements: Applications may be submitted by public and private, foreign and domestic for-profit and nonprofit organizations, and eligible agencies of the federal government.
Contact: Dr. Emeline Otey, Division of Mental Disorders, Behavioral Research, and AIDS; (301) 443-9400; fax (301) 443-9876; eotey@nih.gov
Internet: http://www.grants.nih.gov/grants/guide/pa-files/PA-98-098.html
Sponsor: National Institute of Mental Health
5600 Fishers Lane, Parklawn Building
Rockville, MD 20857

NIMH Supplements for Underrepresented Minorities 3615
NIMH recognizes the need to increase the number of underrepresented minority scientists participating in mental health research. Therefore, funds are made available for administrative supplements designed to provide a continuum of support for underrepresented minority scientists and students. These supplements may support research experience for minority undergraduate students, graduate research assistants, and minority investigators. The aim of these programs is to attract and encourage these minority individuals to pursue biomedical and behavioral research careers in areas within the mission of NIMH by providing supplemental funding to ongoing research grants. The minority individual must be given the opportunity to interact with individuals on the parent grant, to contribute intellectually to the research, and to enhance her/his research skills and knowledge regarding the particular area of biomedical or behavioral science. Applications are accepted at any time. Copies of the specific guidelines may be obtained upon request. PA: PA-99-104
Requirements: Any principal investigator at a domestic institution holding an active ADAMHA grant that has a minimum of two years of research support remaining at the time of a requested supplemental award is eligible. In making requests, the grantee institution, on behalf of the principal investigator of the parent grant and in cooperation with the minority individual, should submit the request for supplemental funds directly to the project officer for the parent grant.
Contact: Deputy Associate Director for Special Populations; (301) 443-2847; fax (301) 443-8552; sr14j@nih.gov
Internet: http://grants.nih.gov/grants/guide/pa-files/PA-99-104.html
Sponsor: National Institute of Mental Health
5600 Fishers Lane, Parklawn Building
Rockville, MD 20857

NIMH Therapeutic Effectiveness Protocol Development Program Grants 3616
The program is directed toward the development of infrastructure and resources for investigation of effectiveness of therapeutic interventions through support of protocol design. This program has been developed in response to a recommendation of the Clinical Treatment and Services Research Workgroup of the National Advisory Mental Health Council. The program will support effectiveness trial development that will lead to large-scale studies focusing on diseases and populations of major public health significance that are consistent with the mission of NIMH. This is part of a series of four program announcements. Projects may receive funding for up to three years. Annual deadline dates may vary; contact program staff for exact dates. PA: PAR-03-078
Requirements: Applications may be submitted by domestic, for-profit and nonprofit organizations, both public and private; units of state and local governments; and eligible agencies of the federal government.
Date(s) Application is Due: Feb 1; Jun 1; Oct 1
Amount of Grant: Up to 125,000 U.S.D
Contact: Dr. Linda Street, Division of Services and Intervention Research; (301) 443-0651; fax (301) 594-6784; Lstreet@nih.gov
Internet: http://www.grants.nih.gov/grants/guide/pa-files/PAR-03-078.html
Sponsor: National Institute of Mental Health
6001 Executive Boulevard
Bethesda, MD 20892

Nina Mason Pulliam Charitable Trust Grants 3617
The Nina Mason Pulliam Charitable Trust provides grants that focus on the areas of service Nina Pulliam supported during her lifetime: helping people in need; protecting animals and nature; enriching community life. The Trust awards grants for program projects and capital needs, and provides application opportunities three times during the calendar year. The Trust prefers to disperse funds as a one-year grant, but will consider projects of up to three years.
Requirements: Primary consideration is given to 501(c)3 charitable organizations that serve metropolitan Indianapolis and Phoenix. Secondary consideration is given to the states of Indiana and Arizona. National organizations whose programs benefit these priority areas and/or benefit society as a whole are occasionally considered.
Restrictions: Grants are not awarded for international purposes or academic research, Grants are not awarded to individuals, sectarian organizations for religious purposes, or to non-operating private foundations except in extraordinary circumstances.
Geographic Focus: Arizona, Indiana
Date(s) Application is Due: Jan 4; May 4; Sep 4
Contact: David Hillman, (317) 231-6075; fax (317) 231-9208
Internet: http://www.ninapulliamtrust.org
Sponsor: Nina Mason Pulliam Charitable Trust
135 N Pennsylvania Street, Suite 1200
Indianapolis, IN 46204

NINDS Central Nervous System Effects of HIV Infection: Neurobiological, Neurovirological, and Neurobehavioral Studies Grants 3618
NINDS, with NIMH, NICHD, and NIDA, invites applications for support of research on the role of the HIV virus in the central nervous system and its consequent effects on behavior. Applicants may request up to five years of support. Applications will be accepted under PHS receipt dates for AIDS applications. PA: PA-04-154
Requirements: Any public, private, nonprofit, or for-profit institution is eligible to apply for research grants. For-profit institutions are not eligible for institutional NRSAs but are eligible for individual NRSAs. SBIR grants can be awarded only to domestic small businesses. STTR grants can be awarded only to domestic small business concerns that partner with a research institution in cooperative research and development.
Date(s) Application is Due: Feb 1; Jun 1; Oct 1
Contact: Dr. Jeymohan Joseph, Chief, HIV Neurovirology; (301) 443-3012; fax (301) 443-9719; jjeymoha@mail.nih.gov
Internet: http://grants.nih.gov/grants/guide/pa-files/PA-04-154.html
Sponsor: National Institute of Neurological Disorders and Stroke
6701 Rockledge Drive, Room 1040, MSC 7710
Bethesda, MD 20892-7710

NINDS Cerebral Radiobiology and Neuroimaging of Brain Tumors Research Grants 3619
This program supports research that will increase our knowledge of the genetic, molecular, cellular, and physiological mechanisms of radiation-induced cell injury and recovery, and the sensitizing and protective mechanisms in the central nervous system under radiation treatment conditions for brain tumors. The intent of this program announcement is to encourage investigator-initiated applications to study tumor and normal brain cell injury and repair mechanisms induced by brain tumor radiotherapy including stereotactic radiosurgery procedures such as the Gamma Knife, altered fractionation and/or radioenhancing agents, using state-of-the-art neurobiological and neuroimaging approaches. An applicant planning to submit a new (Type 1) investigator-initiated grant application requesting $500,000 or more in direct costs for any year is advised that he or she must contact the program staff before submitting the application.
Requirements: Applications may be submitted by domestic and foreign, for-profit and nonprofit organizations, both public and private; units of state and local governments; and eligible agencies of the federal government.
Restrictions: Foreign institutions are not eligible for P01 grants.
Contact: Dr. Thomas Jacobs, Division of Stroke, Trauma, and Neurodegenerative Disorders Coordinator; (301) 496-4226; fax (301) 480-1080; TJ12G@nih.gov
Internet: http://www.grants.nih.gov/grants/guide/pa-files/PA-98-094.html
Sponsor: National Institute of Neurological Disorders and Stroke
6001 Executive Boulevard, Suite 2209
Bethesda, MD 20892

NINDS Low-Back Pain and Common Spinal Disorders Research Grants 3620
The program is sponsored by NIAMS, NICHD, NINDS, and NIOSH at CDC. This initiative invites applications directed to the study of the pathogenesis, epidemiology, prevention, and treatment of three common causes of low-back pain: herniated nucleus pulposus, spinal stenosis, and idiopathic low-back pain. The support mechanism for grants in this area will be the individual investigator-initiated research grant (R01). PA-97-058 may be obtained electronically through the NIH grant line (301-402-2221), the NIH Web site, and by mail and email.
Requirements: Applications may be submitted by foreign and domestic, for-profit and nonprofit organizations, both public and private; units of state and local governments; and eligible agencies of the federal government.
Restrictions: Foreign institutions or organizations in foreign countries are not eligible for First Independent Research Support and Transition (R29) awards.
Contact: Dr. Mary Ellen Cheung, Division of Stroke, Trauma, and Neurodegenerative Disorders; (301) 496-4226; fax (301) 480-1080; mm108w@nih.gov
Internet: http://www.grants.nih.gov/grants/guide/pa-files/PA-97-058.html
Sponsor: National Institute of Neurological Disorders and Stroke
6001 Executive Boulevard, Suite 2209
Bethesda, MD 20892

NINDS Optimization of Small Molecule Probes for the Nervous System Grants 3621
This Funding Opportunity Announcement (FOA) encourages Small Business Innovation Research (SBIR) grant applications from small business concerns (SBCs) that propose to develop new small molecule probes for investigating biological function in the nervous system via the application of advanced medicinal chemistry and the biological testing of compounds. Eligible SBCs will have identified probe candidates via screening of small molecule collections, using in vitro assays of biological activity developed to interrogate these collections, and be able to show that the structural features of these small molecules are related to their biological activity. Applications should nominate small molecule probe candidates from distinct structural series for the further, iterative design and testing of analogues in structure-activity relationship studies, using in vitro assays of biological function adapted to the medium throughput screening requirements of this work. These studies should have the goal of developing a small molecule probe possessing the attributes (eg: affinity, selectivity, activity) required for its use in future pharmacological studies proposed by the SBC. Applicants are strongly encouraged to utilize publicly available cheminformatic capabilities for the acquisition of compounds, and semi-custom synthesis of analogues.
Requirements: Only United States small business concerns (SBCs) are eligible to submit SBIR applications.
Date(s) Application is Due: Apr 5; Aug 5; Dec 5
Amount of Grant: Up to 150,000 U.S.D
Contact: Mark Scheideler, (301) 496-1779; scheidelerm@ninds.nih.gov
Internet: http://grants.nih.gov/grants/guide/pa-files/PAR-09-260.html
Sponsor: National Institute of Neurological Disorders and Stroke
6001 Executive Boulevard, NSC 2107
Bethesda, MD 20892-9527

NINDS Optimization of Small Molecule Probes for the Nervous System Grants 3622
This Funding Opportunity Announcement (FOA) encourages Small Business Technology Transfer (STTR) grant applications from small business concerns (SBCs) that propose to develop new small molecule probes for investigating biological function in the nervous system via the application of advanced medicinal chemistry and the biological testing of compounds. Eligible SBCs will have identified probe candidates via screening of small molecule collections, using in vitro assays of biological activity developed to interrogate these collections, and be able to show that the structural features of these small molecules are related to their biological activity. Applications should nominate small molecule probe candidates from distinct structural series for the further, iterative design and testing of analogues in structure-activity relationship studies, using in vitro assays of biological function adapted to the medium throughput screening requirements of this work. These studies should have the goal of developing a small molecule probe possessing the attributes (eg: affinity, selectivity, activity) required for its use in future pharmacological studies proposed by the SBC. Applicants are strongly encouraged to utilize publicly available cheminformatic capabilities for the acquisition of compounds, and semi-custom synthesis of analogues, which is required of these studies.
Requirements: Only United States small business concerns (SBCs) are eligible to submit STTR applications.
Date(s) Application is Due: Apr 5; Aug 5; Dec 5
Amount of Grant: Up to 150,000 U.S.D
Contact: Mark Scheideler, (301) 496-1779; scheidelerm@ninds.nih.gov
Internet: http://grants.nih.gov/grants/guide/pa-files/PAR-09-260.html
Sponsor: National Institute of Neurological Disorders and Stroke
6001 Executive Boulevard, NSC 2107
Bethesda, MD 20892-9527

NINDS Optimization of Small Molecule Probes for the Nervous System Grants 3623
This FOA issued by participating institutes of the National Institutes of Health, encourages research grant applications from institutions/ organizations that propose to develop new small molecule probes for investigating biological function in the nervous system via the application of advanced medicinal chemistry and the biological testing of compounds. Eligible investigators will have identified probe candidates via screening of small molecule

collections, using in vitro assays of biological activity developed to interrogate these collections, and be able to show that the structural features of these small molecules are related to their biological activity. Proposals should nominate small molecule probe candidates from distinct structural series for the further, iterative design and testing of analogues in structure-activity relationship studies, using in vitro assays of biological function adapted to the medium throughput screening requirements of this work. These studies should have the goal of developing a small molecule probe possessing the attributes (eg: affinity, selectivity, activity) required for its use in future pharmacological studies proposed by the investigator. Applicants are strongly encouraged to utilize publicly available cheminformatic capabilities for the acquisition of compounds, and semi-custom synthesis of analogues.
Requirements: The following organizations and institutions are eligible to apply: public/state controlled institutions of higher education; private institutions of higher education; Hispanic-serving institutions; historically Black colleges and universities (HBCUs); tribally controlled colleges and universities (TCCUs); Alaska native and native Hawaiian serving institutions; nonprofits with 501(c)3 IRS status (other than institutions of higher education); nonprofits without 501(c)3 IRS status (other than institutions of higher education); small businesses; for-profit organizations (other than small businesses); state governments; Indian/Native American tribal governments (federally recognized); Indian/Native American tribally designated organizations; county governments; city or township governments; special district governments; independent school districts; public housing authorities/Indian housing authorities; U.S. territories or possessions; Indian/native American tribal governments (other than federally recognized); regional organizations; non-domestic (non-U.S.) entities (foreign organizations); eligible agencies of the federal government; and faith-based or community-based organizations.
Date(s) Application is Due: Feb 16; Jun 16; Oct 16
Amount of Grant: Up to 150,000 U.S.D
Contact: Mark Scheideler, (301) 496-1779; scheidelerm@ninds.nih.gov
Internet: http://grants.nih.gov/grants/guide/pa-files/PAR-09-251.html
Sponsor: National Institute of Neurological Disorders and Stroke
6001 Executive Boulevard, NSC 2107
Bethesda, MD 20892-9527

NINDS Pathogenesis and Therapy of the Muscular Dystrophies Research Grants 3624
NINDS and NIAMS encourage investigator-initiated research grant applications to study the pathogenesis and therapy of the various forms of muscular dystrophy in children and adults. Responses to this PA may include studies in appropriate animal models or preclinical or clinical studies in patients with facioscapulohumeral dystrophy (FSH), limb-girdle muscular dystrophy (LGMD), myotonic dystrophy, congenital muscular dystrophy (CMD), Emery-Dreifuss muscular dystrophy (EMD), Duchenne muscular dystrophy (DMD), Becker muscular dystrophy (BMD), or other forms of muscular dystrophy. Support mechanisms for grants are research project grant (R01) and program project grant (P01). Information may be obtained through the NIH grant line (301-402-2221), the NIH Web site, mail, and email. PA: PA-98-044
Requirements: Applications may be submitted by domestic and foreign, for-profit and nonprofit organizations, both public and private; units of state and local governments; and eligible agencies of the federal government.
Contact: Dr. Paul Nichols, Division of Convulsive, Infectious, and Immune Disorders; (301) 496-1431; fax (301) 402-2060; pn13w@nih.gov
Internet: http://www.grants.nih.gov/grants/guide/pa-files/PA-98-044.html
Sponsor: National Institute of Neurological Disorders and Stroke
6001 Executive Boulevard, Suite 2209
Bethesda, MD 20892

NINDS Repetitive Motion Disorders Research Grants 3625
The program is sponsored by NIAMS, NICHD, NINDS, and NIOSH at CDC. This initiative invites applications directed to the study of the pathogenesis, epidemiology, prevention, and treatment of repetitive motion disorders. The support mechanism for grants in this area will be the individual investigator-initiated research grant (R01). PA-97-059 may be obtained electronically through the NIH grant line (301-402-2221), the NIH Web site, and by mail and email.
Requirements: Applications may be submitted by foreign and domestic, for-profit and nonprofit organizations, both public and private; units of state and local governments; and eligible agencies of the federal government.
Restrictions: Foreign institutions or organizations in foreign countries are not eligible for First Independent Research Support and Transition (FIRST) (R29) awards.
Contact: Dr. Paul Nichols, (301) 496-1431; fax (301) 402-2060; pn13w@nih.gov
Internet: http://www.grants.nih.gov/grants/guide/pa-files/PA-97-059.html
Sponsor: National Institute of Neurological Disorders and Stroke
6001 Executive Boulevard, Suite 2209
Bethesda, MD 20892

NINDS Specialized Neuroscience Research Programs (SNRP) on Health Disparity: HIV and the Nervous System 3626
The purpose of this program is to augment and strengthen the research capabilities of faculty, students, and fellows at minority institutions by supporting the development of new, and/or the enhancement of ongoing basic and clinical neuro-AIDS related research projects and programs. Applicants planning to submit an investigator-initiated new, competing continuation, competing supplement, or any amended/revised version of the preceding grant application types requesting $500,000 or more in direct costs for any year are advised that he or she must contact institute program staff before submitting the application. Annual deadline dates may vary; contact program staff for exact dates. RFA: NS-01-009
Requirements: Applications will only be accepted from domestic, public or private, nonprofit, academic institutions that offer PhD, MD, and/or equivalent health professional degrees, at which more than 50 percent of the students enrolled are from ethnic or racial minority groups. Each applicant institution must identify a collaborating institution that can be a domestic federal or nonfederal, public or private, nonprofit organization. The principal investigator, who serves as SNRP director, must be a U.S. citizen, permanent resident, or noncitizen national.
Amount of Grant: Up to 1,000,000 U.S.D
Contact: Dr. Joana Rosario, Office of Special Programs in Neuroscience; (301) 496-5489; fax (301) 594-5929; jr69z@nih.gov
Internet: http://grants.nih.gov/grants/guide/rfa-files/RFA-NS-01-009.html
Sponsor: National Institute of Neurological Disorders and Stroke
6001 Executive Boulevard, Suite 2209
Bethesda, MD 20892

NINDS Support of Scientific Meetings as Cooperative Agreements 3627
The purpose of the program announcement is to inform the scientific community that the National Institute of Neurological Disorders and Stroke (NINDS) will now support scientific meetings as cooperative agreements in addition to the current practice of supporting them through the traditional grant mechanism. This program provides guidelines for when it is appropriate to request support of a meeting as a cooperative agreement and explains procedures for preparing and submitting such applications. The Principal Investigator will have the primary authority and responsibility to define objectives and approaches; plan, publicize, and conduct the scientific meeting; and publish the results of the meeting.
Requirements: Applications may be submitted by U.S. institutions, including scientific or professional societies eligible to receive grants from Public Health Service (PHS) agencies. In the case of an international conference, the U.S. representative organization of an established international scientific or professional society is the eligible applicant.
Restrictions: Foreign institutions are not eligible.
Contact: Danilo A. Tagle, (301) 496-5745; fax (301) 402-2060; tagled@ninds.nih.gov
Internet: http://grants.nih.gov/grants/guide/notice-files/NOT-NS-02-003.html
Sponsor: National Institute of Neurological Disorders and Stroke
6001 Executive Boulevard, Neuroscience Center, Room 2114
Bethesda, MD 20892-9525

NINDS Symptom Management for Chronic Neurological Conditions Research Grants 3628
NINR, NIA, NICHD, NIDCD, and NINDS are interested in facilitating investigator-initiated research designed to deal with symptoms associated with chronic neurological conditions including stroke, epilepsy, brain and spinal cord injury, multiple sclerosis, and Parkinson's disease. The overall goals are to manage symptoms effectively, to prevent secondary effects of the condition when possible, to maximize the health potential, and to improve the health-related quality of life for patients, care givers, and families. The mechanism of support is the research project grant (R01). More information may be obtained through the NIH grant line (301-402-2221), the NIH Web site, mail, and email. PA: PA-98-030
Requirements: Research applications may be submitted by domestic and foreign, for-profit and nonprofit organizations, both public and private; units of state and local governments; and eligible agencies of the federal government.
Contact: Gladys Melendez-Bohler, Grants Management Branch; (301) 496-9231; fax (301) 402-0219; GB13Y@nih.gov
Internet: http://www.grants.nih.gov/grants/guide/pa-files/PA-98-030.html
Sponsor: National Institute of Neurological Disorders and Stroke
6001 Executive Boulevard, Suite 2209
Bethesda, MD 20892

NINR/NHGRI Nursing and Genetics Research Individual Postdoctoral and Senior Fellowships 3629
NINR and NHGRI announce the availability of long- and short-term support for individual postdoctoral and senior fellowships to scientists who are seeking training that will enable them to investigate the clinical implications of human genetics research. The three-fold purpose of this initiative is to incorporate genetic research into programs of research and research training for nurse scientists; integrate nursing research with research training programs for genetic scientists; and provide opportunities for addressing the ethical, legal, and social implications of genetic research training. PA-97-039 may be obtained electronically through the NIH grant line (301-402-2221), the NIH Web site, and by mail and email.
Contact: Dr. Hilary Sigmon, Scientific Program Administrator; (301) 594-5970; fax (301) 480-8260; hsigmon@ep.ninr.nih.gov
Internet: http://grants.nih.gov/grants/guide/pa-files/PA-97-039.html
Sponsor: National Institute of Nursing Research
31 Center Drive, Room 5B10
Bethesda, MD 20892-2178

NINR Acute and Chronic Care During Mechanical Ventilation Research Grants 3630
NINR seeks research applications for investigator-initiated research to improve the clinical management and care delivery of patients who require mechanical ventilation. For the purpose of this program announcement, the time course of mechanical ventilation is defined as beginning with endotracheal or tracheal tube intubation, maintenance of full and/or partial ventilatory support, discontinuation of mechanical ventilation via weaning techniques, and concluding with care of patients who require chronic mechanical ventilation. Two different, but related, types of basic science and/or clinical research on acute and/or chronic mechanical ventilation are encouraged: studies examining the effects of physiological, environmental, and/or psychological approaches to facilitate the

care of patients during any phase of mechanical ventilation, and studies directed toward examining the physical and psychological ramifications of technology dependency on patients and their families/significant others. Priority research areas include optimum ways to deliver patient care during mechanical ventilatory support; ethical issues related to initiation, maintenance, and withdrawal of life support technology; emotional support and communication during mechanical ventilation; methods of optimally managing the artificial airway; impact of care on family dynamics; and decision making regarding withdrawal of ventilatory support. Research is also needed to describe and intervene with care-giving resources, family functioning, and quality of life for patients on mechanical ventilation particularly chronically ill children and adults being maintained on home ventilators; optimum methods and techniques to deliver nursing care while critically ill patients receive mechanical ventilation; and determine optimal measures for promoting recovery as patients move through levels of care in acute, intermediate, and long-term care. This PA covers a wide range of topics relevant to clinical management and clinical problem solving issues for patients who require mechanical ventilation.
Requirements: Applications may be submitted by public and private, domestic and foreign, for-profit and nonprofit organizations; units of state and local governments; and eligible agencies of the federal government.
Restrictions: The total project period for an application may not exceed five years.
Date(s) Application is Due: Feb 5; Feb 16; Jun 5; Jun 16; Oct 5; Oct 16
Contact: Josephine Boyington, (301) 594-2542; boyingtonje@mail.nih.gov
Internet: http://grants.nih.gov/grants/guide/pa-files/PA-99-003.html
Sponsor: National Institute of Nursing Research
31 Center Drive, Room 5B10
Bethesda, MD 20892-2178

NINR Aging Women and Breast Cancer Grants 3631
NINR, NCI, and NIA are supporting grants that focus on the unique problems of older women with breast cancer. Their purpose is to inform the scientific community of the interests of NIA and NCI and to expand the knowledge base on breast cancer in older women through studies in the fields of biology, clinical medicine, epidemiology, and the behavioral and social sciences. More information may be obtained by mail and email or through the NIH grant line (301-402-2221), and the NIH Web site. PA: PA-00-001
Requirements: Foreign and domestic for-profit and nonprofit organizations, both public and private; units of state and local governments; and eligible agencies of the federal government are eligible to apply.
Date(s) Application is Due: Feb 1; Jun 1; Oct 1
Amount of Grant: 29,175 - 627,861 U.S.D
Contact: Dr. Janice Phillips, Program Director; (301) 594-6152; fax (301) 480-8260; Janice_Phillips@nih.gov
Internet: http://grants.nih.gov/grants/guide/pa-files/PA-00-001.html
Sponsor: National Institute of Nursing Research
31 Center Drive, Room 5B10
Bethesda, MD 20892-2178

NINR AIDS and Aging: Behavioral Sciences Prevention Research Grants 3632
NIA, NIMH, and NINR invite qualified researchers to submit applications to investigate prevention issues relevant to AIDS in middle-aged and older populations. This announcement solicits AIDS prevention research proposals to study primary prevention of disease transmission as well as secondary and tertiary prevention of negative behavioral and social consequences of HIV/AIDS for persons with AIDS, their families, and communities. Thus, the primary goals are to identify social and behavioral factors associated with HIV transmission and disease progression in later life; examine behavioral and social consequences of HIV infection/AIDS across the life course; develop and evaluate age-appropriate behavioral and social interventions for preventing AIDS in middle-aged and older adults and/or ameliorating problems associated with older adult caregiver responsibilities and burdens; explore health care issues surrounding AIDS care; and strengthen existing research and evaluation methods.
Requirements: Applications may be submitted by domestic and foreign for-profit and non-profit organizations, public and private, such as universities, colleges, hospitals, laboratories, units of State and local governments, and eligible agencies of the Federal government. Applications may be submitted by single institutions or by a consortia of institutions.
Date(s) Application is Due: Feb 5; Feb 16; Jun 5; Jun 16; Oct 5; Oct 16
Contact: Josephine Boyington, (301) 594-2542; boyingtonje@mail.nih.gov
Internet: http://www.grants.nih.gov/grants/guide/pa-files/PA-97-069.html
Sponsor: National Institute of Nursing Research
31 Center Drive, Room 5B10
Bethesda, MD 20892-2178

NINR Career Transition Award 3633
The overall goal of this program is to support outstanding individuals to obtain a research training experience at the NIH in the intramural research laboratories and to facilitate their successful transition to an extramural environment as independent researchers. Annual deadline dates may vary; contact program staff for exact dates. RFA: NR-01-003
Requirements: Registered Nurses (RN) who are recent recipients of a doctoral degree or its equivalent and who have demonstrated the potential for a highly productive research career are eligible to apply.
Amount of Grant: Up to 125,000 U.S.D
Contact: Hilary Sigmon, (301) 594-5970; fax (301) 480-8260; hilary_sigmon@nih.gov
Internet: http://www.grants.nih.gov/grants/guide/rfa-files/RFA-NR-01-003.html
Sponsor: National Institute of Nursing Research
31 Center Drive, Room 5B10
Bethesda, MD 20892-2178

NINR Chronic Illness Self-Management in Children and Adolescents Grants 3634
The purpose of this Funding Opportunity Announcement (FOA) issued by the NINR, NICHD, NIDDK, NCI, and the ODS is to solicit research to improve self-management and quality of life in children and adolescents with chronic illnesses. Biobehavioral studies of children within the context of family and family-community dynamics are encouraged. Children with a chronic disease and their families have a long-term responsibility for self-management. The child with the illness will have a life-long responsibility to maintain and promote health and prevent complications. Research related to sociocultural, environmental, and behavioral mechanisms as well as biological/technological factors that contribute to successful and ongoing self-management of particular chronic illnesses in children is encouraged. Because the nature and scope of the proposed research will vary from application to application, it is anticipated that the size and duration of each award will also vary.
Requirements: Application can be made by the following institutions and organizations: public/state controlled institutions of higher education; private institutions of higher education; nonprofits with 501(c)3 IRS status (other than institutions of higher education); nonprofits without 501(c)3 IRS status (other than institutions of higher education); small businesses; for-profit organizations (other than small businesses); state governments; U.S. territories or possessions; Indian/Native American tribal governments (federally recognized); Indian/Native American tribal governments (other than Federally recognized); Indian/Native American tribally designated organizations; non-domestic (non-U.S.) entities (foreign organizations); Hispanic-serving institutions; historically Black colleges and universities (HBCUs); tribally controlled colleges and universities (TCCUs); Alaska Native and Native Hawaiian serving institutions; regional organization; eligible agencies of the Federal government; and faith-based or community based organizations.
Restrictions: This FOA is restricted to studies of chronic illnesses in children and adolescents ages 8 to 21 grouped by developmental stages according to the discretion of the investigator. Studies of chronic mental illness or serious cognitive disability are beyond the scope of this FOA.
Date(s) Application is Due: Feb 5; Jun 5; Oct 5
Contact: Dr. Linda S. Weglicki, (301) 594-6908; weglickils@mail.nih.gov
Internet: http://grants.nih.gov/grants/guide/pa-files/PA-07-097.html
Sponsor: National Institute of Nursing Research
31 Center Drive, Room 5B10
Bethesda, MD 20892-2178

NINR Clinical Interventions for Managing the Symptoms of Stroke Research Grants 3635
NINR, NINDS, NIA, NIDCD, and NICHD seek research applications that address effective nonpharmacological approaches to managing the initial events and subsequent symptoms of a stroke. This research would focus on reducing the impairments, preventing secondary complications, and improving the functional independence and the quality of life for the individual following a stroke. Research areas that need to be explored include, but are not limited to: techniques to improve assessment during the initial stroke event and at each stage of recovery to guide nonpharmacologic treatment decisions and monitor patient progress; interventions to manage common symptoms, such as dysphagia, aphasia, bowel and bladder problems, sensorimotor deficits, depression, cognitive and perceptual deficits as well as difficulties in nutrition, hydration, skin integrity, and sleep; strategies to determine optimal timing, intensity, and duration of nonpharmacologic treatments; methods to prevent secondary complications, improve function, and address the sequelae of common comorbid conditions; strategies to promote self-care in both activities of daily living and instrumental activities of daily living and to enhance or maintain health promotion activities; measures to enhance family caregiving during all stages of recovery; methods to improve care during transition from one setting to another and care in specialized nurse managed units; and studies on the extent to which age and pre-existing conditions, such as specific diseases, sensory, motor, and cognitive declines, influence treatment outcome and management after stroke.
Requirements: Applications may be submitted by domestic and foreign, for-profit and nonprofit organizations, public and private.
Amount of Grant: Up to 250,000 U.S.D
Contact: Karin F. Helmers, (301) 594-2177; fax (301) 480-8260; karin_helmers@nih.gov
Internet: http://www.grants.nih.gov/grants/guide/pa-files/PA-99-088.html
Sponsor: National Institute of Nursing Research
31 Center Drive, Room 5B10
Bethesda, MD 20892-2178

NINR Clinical Trials: Collaborations for Nursing Research 3636
NINR invites nurse investigators to participate in an initiative that capitalizes on the substantial NIH resource of available study populations in clinical trials from the unique perspective of nursing research and special expertise of nurse researchers. The purpose of this RFA is to link substudies posed by nurse researchers to currently funded clinical research studies, or to link substudies by non-nurse researchers with ongoing clinical studies where nurse researchers are the principal investigators. Collaborations and consortia promoting the cross-fertilization of ideas are encouraged. The rationale for this RFA is that substudies are cost effective ways to stimulate ongoing research while capitalizing on resources expended on a parent clinical trial. Nurse researchers are to collaborate with non-nurse PIs and/or the steering committees on currently funded clinical trials where resources can be accessed to answer a clinical question or solve a nursing problem. Likewise, non-nurse researchers in various fields are encouraged to collaborate with an NIH-funded nurse researcher to access a study population in a scientific field of their choice. Annual deadline dates may vary; contact program staff for exact dates. RFA: NR-01-004
Requirements: Registered nurses with doctoral preparation who pose a research question from her/his ongoing program of research that can be answered by accessing a currently

funded clinical trial, or, a non-nurse doctorally prepared scientist who seeks to answer a clinical question or solve a nursing problem using the clinical study resources where the PI is a nurse researcher are eligible to apply. Applications may be submitted by domestic and foreign for-profit and nonprofit organizations, both public and private; units of state or local governments; and eligible agencies of the federal government.
Amount of Grant: Up to 15,000,000 U.S.D
Contact: Dr. Hilary Sigmon, Program Director; (301) 594-5970; fax (301) 480-8260; Hilary_Sigmon@nih.gov
Internet: http://www.grants.nih.gov/grants/guide/rfa-files/RFA-NR-01-004.html
Sponsor: National Institute of Nursing Research
31 Center Drive, Room 5B10
Bethesda, MD 20892-2178

NINR Community-Based Prevention Intervention Research in Environmental Health Sciences Grants 3637
NINR, NIEHS, and the Office of Behavioral and Social Sciences invite research grant applications addressing development of community-based strategies aimed at prevention and intervention activities in economically disadvantaged and/or underserved populations adversely impacted by an environmental contaminant. These funds support research activities that develop and implement improved prevention and intervention strategies related to underserved populations. In community-based research, active cooperation and participation of organizations within the community that is the focus of the study are essential components of the research. Hence, applicants must describe an existing or proposed involvement with one or more community-based organizations. These grants provide support to develop, test, and evaluate health service activities and to enhance the application or transfer of existing knowledge to efforts designed to prevent or control dysfunction or disease. Annual deadline dates may vary; contact program staff for exact dates. More information may be obtained by mail and email, or through the NIH grant line (301-402-2221), and the NIH Web site. RFA: ES-96-005
Date(s) Application is Due: Feb 1; Jun 1; Oct 1
Amount of Grant: 64,658 - 100,000,000 U.S.D
Contact: J. Taylor Harden; (301) 594-5976; THARDEN@EP.NINR.NIH.GOV
Internet: http://www.grants.nih.gov/grants/guide/rfa-files/RFA-ES-96-005.html
Sponsor: National Institute of Nursing Research
31 Center Drive, Room 5B10
Bethesda, MD 20892-2178

NINR Diabetes Self-Management in Minority Populations Grants 3638
This Program Announcement (PA) solicits applications for investigator-initiated research related to sociocultural, environmental, and behavioral mechanisms and biological/technological factors that contribute to successful and ongoing self-management of diabetes in minority populations. Applications that expand accepted intervention strategies in majority populations to minority populations are encouraged. Testing new interventions designed to promote self-management in minority diabetes populations will also be responsive to the PA. Self-management is defined as client strategies and behaviors that contribute to blood glucose normalization, improved health, and prevention or reduction of complications. The concept is broader than adherence to specific regimen components and incorporates deliberate problem solving and decision making processes. Applications are encouraged for both type 1 and type 2 diabetes and all age groups. Modular Grant applications will request direct costs in $25,000 modules, up to a total direct cost request of $250,000 per year.
Requirements: Applications may be submitted by domestic and foreign, for-profit and non-profit organizations, public and private, such as universities, colleges, hospitals, laboratories, units of State and local governments, and eligible agencies of the Federal government. Racial/ethnic minority individuals, women, and persons with disabilities are encouraged to apply as principal investigators.
Date(s) Application is Due: Feb 5; Jun 5; Oct 5
Amount of Grant: Up to 250,000 U.S.D
Contact: Dr. Karen Huss, (301) 594-5970; fax (301) 451-5649; husk@mail.nih.gov
Internet: http://grants.nih.gov/grants/guide/pa-files/PA-00-113.html
Sponsor: National Institute of Nursing Research
31 Center Drive, Room 5B10
Bethesda, MD 20892-2178

NINR Health Risk Reduction: Community-Based Strategies 3639
NINR is interested in facilitating investigator-initiated research into the clinical application of intervention strategies designed to reduce health risks at the community level. Community-based strategies targeting health problems of rural residents and of underserved minority groups are of particular interest. More information may be obtained by mail and email, or through the NIH grant line (301-402-2221), and the NIH Web site. PA: PA-96-037
Date(s) Application is Due: Feb 1; Jun 1; Oct 1
Contact: Grants Administrator; (301) 594-6906; fax (301) 480-8260; info@ninr.nih.gov
Internet: http://grants2.nih.gov/grants/guide/pa-files/PA-96-037.html
Sponsor: National Institute of Nursing Research
31 Center Drive, Room 5B10
Bethesda, MD 20892-2178

NINR Management of Symptoms at the End of Life Grants 3640
NINR, NCI, NIAID, NIMH, and the Office of Alternative Medicine, NIH, seek research grant applications concerning the clinical management of symptoms and syndromes that are associated with life-limiting illness, such as pain, dyspnea, delirium, cachexia, nausea, fatigue, and depression. The purpose of this initiative is to stimulate research that will lead to improved quality of life for those at the end of life and decreased distress for their caregivers. PA-98-019 may be obtained through the NIH grant line (301-402-2221), the NIH Web site, mail, and email.
Requirements: Applications may be submitted by domestic and foreign, for-profit and nonprofit organizations, public and private.
Contact: Jeff Carow, Grants Management Branch; (301) 496-7800, ext. 250; fax (301) 496-8601; Jeff_Carow@nih.gov
Internet: http://grants.nih.gov/grants/guide/pa-files/PA-98-019.html
Sponsor: National Institute of Nursing Research
31 Center Drive, Room 5B10
Bethesda, MD 20892-2178

NINR Mechanisms, Models, Measurement, & Management in Pain Research Grants 3641
The purpose of this Funding Opportunity Announcement (FOA), Mechanisms, Models, Measurement, and Management in Pain Research, is to inform the scientific community of the pain research interests of the various Institutes and Centers (ICs) at the National Institutes of Health (NIH) and to stimulate and foster a wide range of basic, clinical, and translational studies on pain as they relate to the missions of these ICs. New advances are needed in every area of pain research, from the micro perspective of molecular sciences to the macro perspective of behavioral and social sciences. Although great strides have been made in some areas, such as the identification of neural pathways of pain, the experience of pain and the challenge of treatment have remained uniquely individual and unsolved. Furthermore, our understanding of how and why individuals transition to a chronic pain state after an acute insult is limited. Research to address these issues conducted by interdisciplinary and multidisciplinary research teams is strongly encouraged, as is research from underrepresented, minority, disabled, or female investigators. Direct costs are limited to $275,000 over a two-year period, with no more than $200,000 in direct costs allowed in any single year.
Requirements: The following organizations and institutions are eligible to apply: public/state controlled institutions of higher education; private institutions of higher education; Hispanic-serving institutions; historically Black colleges and universities (HBCUs); tribally controlled colleges and universities (TCCUs); Alaska native and native Hawaiian serving institutions; nonprofits with 501(c)3 IRS status (other than institutions of higher education); nonprofits without 501(c)3 IRS status (other than institutions of higher education); small businesses; for-profit organizations (other than small businesses); state governments; Indian/Native American tribal governments (federally recognized); Indian/Native American tribally designated organizations; county governments; city or township governments; special district governments; independent school districts; public housing authorities/Indian housing authorities; U.S. territories or possessions; Indian/native American tribal governments (other than federally recognized); regional organizations; non-domestic (non-U.S.) entities (foreign organizations); eligible agencies of the federal government; and faith-based or community-based organizations.
Date(s) Application is Due: Feb 16; Jun 16; Oct 16
Amount of Grant: Up to 275,000 U.S.D
Contact: Dr. Susan Marden, (301) 496-9623; fax (301) 480-8260; mardens@mail.nih.gov
Internet: http://grants.nih.gov/grants/guide/pa-files/PA-10-007.html
Sponsor: National Institute of Nursing Research
31 Center Drive, Room 5B10
Bethesda, MD 20892-2178

NINR Mentored Research Scientist Development Award 3642
The objective of the NIH Mentored Research Scientist Development Award is to provide support for a sustained period of protected time for intensive research career development under the guidance of an experienced mentor, or sponsor, in the biomedical, behavioral or clinical sciences leading to research independence. The expectation is that through this sustained period of research career development and training, awardees will launch independent research careers and become competitive for new research project grant funding. NINR encourages mentored research scientist development in areas of symptom management, pulmonary, critical care, trauma, reproductive health, end-of-life and palliative care. NINR limits the length of the K01 award to a 3-year period and generally considers career development applications only from doctorally prepared applicants who will advance the science of nursing.
Requirements: The candidate must have a research or health-professional doctorate or its equivalent and must have demonstrated the capacity or potential for highly productive independent research in the period after the doctorate. The candidate must identify a mentor with extensive research experience and must spend a minimum of 75 percent of full-time professional effort conducting research for the period of the award. Applications may be submitted on behalf of candidates by domestic, nonfederal organizations, public or private, such as medical, dental, or nursing schools or other institutions of higher education. Candidates must be U.S. citizens, noncitizen nationals, or permanent residents.
Date(s) Application is Due: Feb 12; Mar 12; Jun 12; Jul 12; Oct 12; Nov 12
Amount of Grant: 50,000 U.S.D
Contact: David Banks, (301) 496-9558; fax (301) 480-8260; banksd@mail.nih.gov
Internet: http://grants.nih.gov/grants/guide/pa-files/PA-09-040.html
Sponsor: National Institute of Nursing Research
31 Center Drive, Room 5B10
Bethesda, MD 20892-2178

NINR Nursing National Research Service Awards (NRSA)—Individuals 3643
NRSAs are given to increase the opportunities for qualified nurses to engage in full-time graduate study and research training; to prepare professional nurses to conduct independent research, collaborate in interdisciplinary research, and stimulate and guide others in nursing research; to promote the availability and utilization of nurses with research training in nursing and/or the basic sciences to function as faculty in schools of nursing at

undergraduate and graduate levels; and to prepare nurses to conduct scientific inquiry in disciplines that have significance for nursing theory and practice. PA: PA-03-067
Requirements: Applicant must be a registered professional nurse with either a baccalaureate and/or a master's degree in nursing, must be a U.S. citizen or lawfully admitted for permanent residence. Applicants for postdoctoral study must have received PhD, DN, ScD, or equivalent prior to beginning date of proposed fellowship.
Restrictions: Those on temporary or student visas are ineligible.
Date(s) Application is Due: Apr 5; Aug 5; Dec 5
Contact: Dr. Nell Armstrong, (301) 594-5973; armstrongn@nih.gov
Internet: http://grants.nih.gov/grants/guide/pa-files/PA-03-067.html
Sponsor: National Institute of Nursing Research
31 Center Drive, Room 5B10
Bethesda, MD 20892-2178

NINR Nursing Research Exploratory Center Grants 3644
NINR invites applications for exploratory center grants in key clinical and basic areas of nursing research that establish a scientific basis for the care of individuals across the life span. NINR seeks to increase the numbers of research-intensive schools of nursing through this program. These grants augment the research resources in schools of nursing by facilitating the growth of infrastructure to centralize resources and facilities in a specific area of inquiry that has a beginning base of research funding. Annual deadline dates may vary; contact program staff for exact dates. RFA: NR-01-001
Requirements: Applications may be submitted by domestic, for-profit and nonprofit organizations both public and private; units of state and local governments; and eligible agencies of the federal government.
Restrictions: Foreign institutions are not eligible for the center program grants.
Contact: Carole Hudgings, (301) 594-5976; carole_hudgings@nih.gov
Internet: http://grants.nih.gov/grants/guide/rfa-files/RFA-NR-01-001.html
Sponsor: National Institute of Nursing Research
31 Center Drive, Room 5B10
Bethesda, MD 20892-2178

NINR Opportunities in Genetics and Nursing Research Grants 3645
The purpose of the PA is to integrate genetics and nursing research. Currently PHS-funded investigators with at least two years remaining on an NIH-funded research project grant (R01) are eligible. Nurse researchers are encouraged to collaborate with an ethicist, geneticist, molecular biologist, or an equivalent professional trained in genetic research. Likewise, PHS-funded genetic researchers in fields of ethics, genetics, molecular biology, or equivalent fields are encouraged to collaborate with a nurse researcher. Collaborations and consortia promoting the cross-fertilization of ideas are welcome. In such cases, each participant's contribution should be identified and well-integrated into the overall design. PA-97-047 may be obtained electronically through the NIH grant line (301-402-2221), the NIH Web site, and by mail and email.
Contact: Dr. Hilary Sigmon, Division of Extramural Activities; (301) 594-5970; fax (301) 480-8260; hsigmon@ep.ninr.nih.gov
Internet: http://www.grants.nih.gov/grants/guide/pa-files/PA-97-047.html
Sponsor: National Institute of Nursing Research
31 Center Drive, Room 5B10
Bethesda, MD 20892-2178

NINR Quality of Life for Individuals at the End of Life Grants 3646
The National Institute of Nursing Research (NINR) and 6 other ICs seek research grant applications that will generate scientific knowledge to improve the quality of life for individuals who are facing end-of-life issues and for their families. Research applications may include basic, clinical or care delivery studies focused on management of physical and psychological symptoms, patient-provider and patient-family communication, ethics and clinical decision-making, caregiver support, or the context of care delivery for those facing life-limiting illnesses. In a broad sense the purpose of this program announcement is to enhance the quality of life remaining for individuals who are nearing the end of their lives. Modular Grant applications will request direct costs in $25,000 modules, up to a total direct cost request of $250,000 per year.
Requirements: Applications may be submitted by domestic and foreign, for-profit and non-profit organizations, public and private, such as universities, colleges, hospitals, laboratories, units of State and local governments, and eligible agencies of the Federal government. Racial/ethnic minority individuals, women, and persons with disabilities are encouraged to apply as principal investigators.
Date(s) Application is Due: Feb 5; Jun 5; Oct 5
Amount of Grant: Up to 250,000 U.S.D
Contact: Nancy C. Lohrey, (301) 496-8580; fax (301) 402-4472; lohreyn@mail.nih.gov
Internet: http://grants.nih.gov/grants/guide/pa-files/PA-00-127.html
Sponsor: National Institute of Nursing Research
31 Center Drive, Room 5B10
Bethesda, MD 20892-2178

NINR Research on Musculoskeletal Fitness and Sports Medicine Grants 3647
NIAMS and NINR invite investigator-initiated research grant applications to study a broad range of basic and clinical topics related to musculoskeletal fitness, exercise physiology, and sports medicine. The National Center for Medical Rehabilitation Research of NICHD encourages applications for both basic and clinical studies of musculoskeletal fitness and exercise physiology of persons with physical disabilities. The principal investigator and participating investigators will plan, direct, and perform the research. Applicants for program project grants are requested to contact the NIAMS representative as early as possible in the planning stages. PA-97-025 may be obtained electronically through the NIH grant line (301-402-2221), the NIH Web site, and by mail and email.
Contact: Dr. Hilary Sigmon, Division of Extramural Activities; (301) 594-5970; fax (301) 480-8260; hsigmon@ep.ninr.nih.gov
Internet: http://www.grants.nih.gov/grants/guide/pa-files/PA-97-025.html
Sponsor: National Institute of Nursing Research
31 Center Drive, Room 5B10
Bethesda, MD 20892-2178

NINR Ruth L. Kirschstein National Research Service Award (NRSA) for Individual Predoctoral Fellows in Nursing Research 3648
This grant program will provide predoctoral training support for doctoral students. The applicant must propose a research training program and dissertation research that is consistent with the scientific mission of the NINR. Research topics and skills that will serve as a foundation for an ongoing program of research are of particular interest. The research training experience must enhance the applicant's conceptualization of research problems and research skills, under the guidance and supervision of a committed mentor who is an active and established investigator in the area of the applicant's proposed research. The research training program should be carried out in a research environment that includes appropriate human and technical resources and is demonstrably committed to the research training of the applicant in the program he/she proposes in the application. Up to five years of aggregate NRSA support may be provided. Fellowship awardees are required to pursue their research training on a full-time basis, devoting at least 40 hours per week to the training program.
Requirements: The following organizations are eligible to apply: for-profit organizations; non-profit organizations; public or private institutions, such as universities, colleges, hospitals, and laboratories; and domestic institutions.
Date(s) Application is Due: Apr 8; Aug 8; Dec 8
Contact: David Banks, (301) 496-9558; fax (301) 480-8260; banksd@mail.nih.gov
Internet: http://grants.nih.gov/grants/guide/pa-files/PAR-09-227.html
Sponsor: National Institute of Nursing Research
31 Center Drive, Room 5B10
Bethesda, MD 20892-2178

NINR Symptom Management for Chronic Neurological Conditions Research Grants 3649
NINR, NIA, NICHD, NIDCD, and NINDS are interested in facilitating investigator-initiated research designed to deal with symptoms associated with chronic neurological conditions, including stroke, epilepsy, brain and spinal cord injury, multiple sclerosis, and Parkinson's disease. The overall goals are to manage symptoms effectively, to prevent secondary effects of the condition when possible, to maximize the health potential, and to improve the health-related quality of life for patients, care givers, and families. The mechanism of support is the research project grant (R01). PA-98-030 may be obtained through the NIH grant line (301-402-2221), the NIH Web site, mail, and email.
Requirements: Applications may be submitted by domestic and foreign for-profit and nonprofit organizations, public or private.
Contact: Jeff Carow; (301) 594-6869; fax (301) 480-8260; jcarow@ep.ninr.nih.gov
Internet: http://grants.nih.gov/grants/guide/pa-files/PA-98-030.html
Sponsor: National Institute of Nursing Research
31 Center Drive, Room 5B10
Bethesda, MD 20892-2178

NIOSH Occupational Safety and Health Research and Demonstration Grants 3650
This program aims to increase knowledge about the underlying characteristics of occupational safety and health problems to prevent occupational diseases and injuries. Research projects supported include causal research to identify and investigate relationships between hazardous working conditions and associated occupational diseases and injuries; methods research to develop more sensitive means of evaluating hazards at work sites, as well as methods for measuring early markers of adverse health effects and injuries; control research to develop new protective equipment, engineering control technology, and work practices to reduce the risks of occupational hazards; and demonstrations to evaluate the technical feasibility or application of a new or improved occupational safety and health procedure, method, technique, or system. Funds support career development, small grants, and competing renewal and revised applications. PA: PA-04-038
Requirements: Eligible applicants include for-profit and nonprofit organizations, universities, colleges, research institutions, and other public and private organizations, including state and local governments, and small minority- and/or women-owned businesses.
Date(s) Application is Due: Mar 1; Jul 1; Nov 1
Amount of Grant: 30,000 - 400,000 U.S.D
Contact: Susan Board, Office of Extramural Programs; (404) 498-2512; fax (404) 498-2517; sboard@cdc.gov
Internet: http://grants.nih.gov/grants/guide/pa-files/PA-04-038.html
Sponsor: National Institute for Occupational Safety and Health
1600 Clifton Road NE
Atlanta, GA 30333

NIOSH Repetitive Motion Disorders Research Grants 3651
The program is sponsored by NIAMS, NICHD, NINDS, and NIOSH at CDC. This initiative invites applications directed to the study of the pathogenesis, epidemiology, prevention, and treatment of repetitive motion disorders. PA-97-059 may be obtained electronically through the NIH grant line (301-402-2221), the NIH Web site, and by mail and email.
Contact: Roy Fleming; (404) 639-2810; fax (404) 639-2196; rmf2@cdc.gov
Internet: http://www.grants.nih.gov/grants/guide/pa-files/PA-97-059.html

Sponsor: National Institute for Occupational Safety and Health
1600 Clifton Road NE
Atlanta, GA 30333

NKF Clinical Scientist Awards **3652**
The award supports investigators who have demonstrated outstanding clinical or basic research potential to promote their continued success as independent investigators. Scientists are expected to devote a minimum of 75 percent of their time to research. This undertaking includes the establishment of a new Study Section which will review all proposals involving clinical research for all funding mechanisms, based on the following definition from the Nathan Report, which describes three categories of clinical research: Patient-Oriented Research, defined as research conducted with human subjects (or on material of human origin such as tissues, specimens and cognitive phenomena) for which an investigator (or colleague) directly interacts with human subjects in either an outpatient or inpatient setting; Epidemiologic and Biobehavioral Studies; and Outcomes Research and Health Services Research. Applications must be submitted electronically no later than November 30.
Requirements: Applicants must be citizens or permanent residents of the United States at the time of application. Applicants must hold an M.D. or equivalent domestic or foreign degree and must be full-time staff members of a department within an institution in the United States at the time of the award. Applicants will generally have at least five, but not more than 10 years of postdoctoral research experience in nephrology by the beginning of the award date.
Date(s) Application is Due: Nov 30
Amount of Grant: 50,000 U.S.D
Contact: Emily B. Newell, Scientific Communications Director; (212) 889-2210, ext. 288 or (800) 622-9010; fax (212) 689-9261; emilyn@kidney.org or research@kidney.org
Internet: http://www.kidney.org/professionals/research/index.cfm
Sponsor: National Kidney Foundation
30 East 33rd Street
New York, NY 10016

NKF Franklin McDonald/Fresenius Medical Care Clinical Research Grants **3653**
The National Kidney Foundation is soliciting applications for the annual Franklin McDonald/Fresenius Medical Care Clinical Research Grant. Applicants should be able to spend a minimum of 50% of their time conducting research and 25% of effort in the direct care of maintenance dialysis patients and/or in teaching medical students, residents or nephrology fellows the pathophysiology and treatment of CKD, ESRD, or maintenance dialysis. Proposals are due on December 1.
Requirements: Applicants must hold a full-time faculty appointment at a university or an equivalent position as a scientist on the staff of a research organization.
Date(s) Application is Due: Dec 1
Contact: Emily B. Newell, Scientific Communications Director; (212) 889-2210, ext. 288 or (800) 622-9010; fax (212) 689-9261; emilyn@kidney.org or research@kidney.org
Internet: http://www.kidney.org/professionals/research/index.cfm
Sponsor: National Kidney Foundation
30 East 33rd Street
New York, NY 10016

NKF Professional Councils Research Grants **3654**
The purpose of the CNNT Research Grants Program is to further knowledge of nursing and technician issues in the management of kidney failure. The grants may cover basic or applied research on nursing and technical issues in the area of kidney failure; early intervention and treatment of chronic kidney disease; development and evaluation of education programs to enhance patient/family understanding of kidney failure treatment; demonstration projects related to kidney failure and rehabilitation. Letter of intent due on October 15, and full proposals must be received by December 1.
Requirements: Applicants must hold a regular membership in CNNT, a minimum of two years work experience as a nephrology nurse or technician, residence in the U.S. or its territories, written approval from the department head or facility director at the institution/facility where the research is to be conducted, and have prior research training or evidence of support from an individual with research experience.
Restrictions: Project support and stipend not to exceed 20% of requested amount; consultant fees not to exceed 15% of total requested funds.
Date(s) Application is Due: Oct 15; Dec 1
Amount of Grant: Up to 20,000 U.S.D
Contact: Beverly Sneed, Grant Contact; (214) 514-3846 or (800) 622-9010; fax (212) 689-9261; bevyann1@sbcglobal.net or research@kidney.org
Internet: http://www.kidney.org/professionals/research/profcouncil.cfm
Sponsor: National Kidney Foundation
30 East 33rd Street
New York, NY 10016

NKF Research Fellowships **3655**
The purpose of National Kidney Foundation (NKF) Research Fellowships is: to foster training of young and new investigators with the potential of making contributions to the understanding and cure of kidney diseases; to foster, through training fellows, kidney-related research that is of high merit; and to encourage high quality applicants who want to make a career change into academic Nephrology. The fellowships are open to qualified investigators of any nationality who are interested in a career in kidney research and who will not have had more than four and one-half years of research training prior to the commencement of the award. The one-year awards may be renewed.
Requirements: Applicants must have completed no more than 4.5 years of post-doctoral research training (after receipt of M.D., Ph.D, D.O., or equivalent degree) at time of the activation of the award. For those applying for a second year (competitive renewal), applicants must have completed no more than 5.5 years of post-doctoral training.
Restrictions: Anyone who has ever held a faculty appointment at the level of Assistant Professor is ineligible.
Date(s) Application is Due: Nov 30
Amount of Grant: 50,000 U.S.D
Contact: Emily B. Newell, Scientific Communications Director; (212) 889-2210, ext. 288 or (800) 622-9010; fax (212) 689-9261; emilyn@kidney.org or research@kidney.org
Internet: http://www.kidney.org/professionals/research/awards.cfm
Sponsor: National Kidney Foundation
30 East 33rd Street
New York, NY 10016

NKF Young Investigator Grants **3656**
The purpose of this program is to support research in the fields of nephrology, urology, and related disciplines by junior faculty now starting their investigative work, who are often hampered in their progress by difficulty in obtaining suitable financial support for their research during the formative periods of their careers. The applicant's research and career goals must be directed to the study of normal or abnormal kidney function or of diseases of the kidney or urinary tract.
Requirements: Applicants must have completed research fellowship training in nephrology, urology or closely related field prior to the start of the grant award, intend to pursue directly related research, and hold a full-time appointment to a faculty position at a university or an equivalent position as a scientist on the staff of a research oriented institution for no more than four years at the start of the award. Recipients typically hold an appointment as Assistant Professor.
Date(s) Application is Due: Nov 30
Amount of Grant: Up to 50,000 U.S.D
Contact: Emily B. Newell, Scientific Communications Director; (212) 889-2210, ext. 288 or (800) 622-9010; fax (212) 689-9261; emilyn@kidney.org or research@kidney.org
Internet: http://www.kidney.org/professionals/research/awards.cfm
Sponsor: National Kidney Foundation
30 East 33rd Street
New York, NY 10016

NLM Academic Research Enhancement Awards (AREA) **3657**
The purpose of the Academic Research Enhancement Award (AREA) program is to stimulate research in educational institutions that provide baccalaureate or advanced degrees for a significant number of the Nation's research scientists, but that have not been major recipients of support. NLM's participation is confined to applications that address research in biomedical informatics and bioinformatics. Only electronic applications are accepted. Deadlines for new applications are February 25, June 25 and October 25 each year.
Date(s) Application is Due: Feb 25; Jun 25; Oct 25
Contact: Dr. Hua-Chuan Sim, Clinical and Public Health Informatics; (301) 594-5983; fax (301) 402-1384; simh@mail.nih.gov
Internet: http://www.nlm.nih.gov/ep/area.html
Sponsor: National Library of Medicine
6705 Rockledge Drive, Suite 301
Bethesda, MD 20892

NLM Exceptional, Unconventional Research Grants Enabling Knowledge Acceleration **3658**
This RFA is for exceptionally innovative research on novel hypotheses or difficult problems, solutions to which would have an extremely high impact on biomedical or biobehavioral research that is germane to the mission of the National Library of Medicine. The RFA supports new projects, not continuation of projects that have already been initiated. For the EUREKA program, NLM seeks innovative applications in the following two discovery science areas: integrated discovery mining for biology and medicine; and integrated hypothesis testing for biology and medicine. Only electronic applications are accepted. The deadline for Letter of Intent is September 29, and the deadline for new applications is October 28.
Restrictions: This grant does not support pilot projects, i.e., projects of limited scope that are designed primarily to generate data that will enable the PI to seek other funding opportunities.
Date(s) Application is Due: Sep 29; Oct 28
Contact: Dr. Valerie Florance, (301) 594-5983; florancev@mail.nih.gov
Internet: http://www.nlm.nih.gov/ep/eureka.html
Sponsor: National Library of Medicine
6705 Rockledge Drive, Suite 301
Bethesda, MD 20892

NLM Exploratory/Developmental Grants **3659**
The purpose of this program is to support for early stage conceptual work and feasibility tests in biomedical informatics and bioinformatics. These studies may involve considerable risk but may lead to a breakthrough in a particular area, or to the development of novel techniques or approaches that could have major impact on a field of biomedical, behavioral, or clinical research. Preliminary data are not required, but applicants are urged to provide milestones by which progress on the project can be measured. Only electronic applications are accepted. Deadlines for new applications are February 16, June 16 and October 16 each year. Deadlines for revised applications are March 16, July 16 and November 16 each year.
Date(s) Application is Due: Feb 16; Mar 16; Jun 16; Jul 16; Oct 16; Nov 16

Contact: Dr. Jane Ye, Bioinformatics and Computational Biology; (301) 594-5983; fax (301) 402-1384; yej@mail.nih.gov
Internet: http://www.nlm.nih.gov/ep/GrantExDev.html
Sponsor: National Library of Medicine
6705 Rockledge Drive, Suite 301
Bethesda, MD 20892

NLM Express Research Grants in Biomedical Informatics and Bioinformatics 3660
The purpose of the program is to provide support for rigorous scientific research in biomedical informatics and bioinformatics. NLM Express Research Grant applications are limited to a 15-page research plan and to a modular budget. The scope of interests is outlined in the funding opportunity announcement. Investigators whose needs are not met by this announcement can also apply through the NIH Parent R01 (PA 07-070), with a cover letter requesting assignment to NLM. Those who do so should use PAR 08-080 as a source of guidance about NLM's research grant priorities and scope of interest. Deadlines for new applications are February 5, June 5 and October 5 each year. Deadlines for revised applications are March 5, July 5 and November 5 each year.
Date(s) Application is Due: Feb 5; Mar 5; Jun 5; Jul 5; Oct 5; Nov 5
Contact: Dr. Jane Ye, Bioinformatics and Computational Biology; (301) 594-5983; fax (301) 402-1384; yej@mail.nih.gov
Internet: http://www.nlm.nih.gov/ep/GrantResearch.html
Sponsor: National Library of Medicine
6705 Rockledge Drive, Suite 301
Bethesda, MD 20892

NLM Grants for Scholarly Works in Biomedicine and Health 3661
Small grants are awarded for up to three years to provide short-term assistance for the preparation of book-length manuscripts and, in some cases, the publication of important scientific information needed by U.S. health professionals. Grants are awarded for major critical reviews and analyses of current developments in important areas of the health sciences, historical studies, works about health sciences informatics, librarianship, and certain kinds of secondary information and literature tools in the health sciences. Publication may be in formats other than print-on-paper (e.g., electronic, film, etc.), and may involve new and innovative ways of organizing and presenting information. Only electronic applications are accepted.
Requirements: Grants may be made to public or private, nonprofit institutions on behalf of a principal investigator; or to unaffiliated individuals to support salaries, consultant fees, equipment and supplies, travel, and other justified costs.
Restrictions: The program does not support textbooks, the production of curriculum materials, the initial reporting of original biomedical scientific research findings, proceedings of annual meetings, projects highly likely to have substantial commercial sales (best-seller type publications), or projects of local interest only. The program does not provide support for the free distribution of publications resulting from grants.
Date(s) Application is Due: Feb 2
Amount of Grant: 50,000 - 150,000 U.S.D
Contact: Dr. Hua-Chuan Sim, Clinical and Public Health Informatics; (301) 594-5983; fax (301) 402-1384; simh@mail.nih.gov
Internet: http://www.nlm.nih.gov/ep/GrantPubs.html
Sponsor: National Library of Medicine
6705 Rockledge Drive, Suite 301
Bethesda, MD 20892

NLM Informatics Conference Grants 3662
The purpose of the program is to provide support for scientific meetings, conferences, and workshops that are relevant to scientific development in biomedical informatics, bioinformatics, computational biology, and information sciences. The typical award is $10,000 per year in direct costs. The maximum amount that can be requested for a single year is $20,000 in direct costs. Only electronic applications are accepted.
Restrictions: NLM will not support the annual meeting for an organization.
Date(s) Application is Due: Apr 12; Aug 12; Dec 12
Amount of Grant: Up to 20,000 U.S.D
Contact: Dr. Jane Ye, Bioinformatics and Computational Biology; (301) 594-5983; fax (301) 402-1384; yej@mail.nih.gov
Internet: http://www.nlm.nih.gov/ep/GrantConf.html
Sponsor: National Library of Medicine
6705 Rockledge Drive, Suite 301
Bethesda, MD 20892

NLM Internet Connection for Medical Institutions Grants 3663
NLM encourages the development of a communications infrastructure to promote the rapid interchange of medical information nationally and throughout the world. Internet access provides health professionals engaged in education, research, clinical care, and administration with a means of accessing remote databases and libraries, transferring files and images, and interacting with colleagues throughout the world. To accelerate the pace with which health-related institutions become part of the electronic information web, NLM offers grants to support Internet connections. Funds may be used for the purchase and installation of a gateway system and associated connection hardware. Funds may also be used to defray the cost of installation and leasing of communication circuits to connect the network. Annual deadline dates may vary; contact the program office for exact dates. RFA: LM-00-001
Requirements: Domestic public and private nonprofit institutions working in health sciences administration, education research, and/or clinical care services can apply for these grants. Groups that are health-related institutions are also encouraged to apply.
Amount of Grant: 30,000 - 50,000 U.S.D
Contact: Dr. Jane Ye, Bioinformatics and Computational Biology; (301) 594-5983; fax (301) 402-1384; yej@mail.nih.gov
Internet: http://grants.nih.gov/grants/guide/rfa-files/RFA-LM-00-001.html
Sponsor: National Library of Medicine
8600 Rockville Pike, Building 38, Room B1E08
Bethesda, MD 20894

NLM Limited Competition for Continuation of Biomedical Informatics Bioinformatics Resource Grants 3664
This limited competition funding opportunity announcement (FOA) solicits competitive renewal applications from investigators currently supported under PAR-04-142, NLM Resource Grant in Biomedical Informatics/Bioinformatics (P41) or its predecessors. NLM's Biomedical Informatics/Bioinformatics Resource Grant program provides support for five (5) resource centers that support activities, databases or software tools that advance research or practice in the biomedical sciences, clinical medicine or health services research. Direct costs for the first year may not exceed 5% above the current award of the active year of NLM funding at the time of application, or a maximum of $500,000, whichever is smaller. Applicants may include cost-of-living increases of up to 3% per year for direct costs in future years. The deadline for Letters of Intent is July 28, while the deadline for Renewal Applications is August 27.
Date(s) Application is Due: Jul 28; Aug 27
Contact: Dr. Valerie Florance, (301) 594-5983; florancev@mail.nih.gov
Internet: http://www.nlm.nih.gov/ep/grantp41.html
Sponsor: National Library of Medicine
6705 Rockledge Drive, Suite 301
Bethesda, MD 20892

NLM Manufacturing Processes of Medical, Dental, and Biological Technologies (SBIR) Grants 3665
NLM is encouraging eligible United States small business concerns to submit SBIR and STTR Phase I, Phase II, and Fast-Track grant applications whose biomedical research is related to advanced processing, manufacturing processes, equipment and systems, and manufacturing workforce skills and protection. Only electronic applications are accepted.
Date(s) Application is Due: Apr 5; Aug 5; Dec 5
Contact: Dr. Jane Ye, Bioinformatics and Computational Biology; (301) 594-5983; fax (301) 402-1384; yej@mail.nih.gov
Internet: http://www.nlm.nih.gov/ep/grantsbmanu.html
Sponsor: National Library of Medicine
6705 Rockledge Drive, Suite 301
Bethesda, MD 20892

NLM New Technologies for Liver Disease Grants 3666
The goal of this announcement is to enlist members of the small business research community in advancing means of diagnosis, treatment and prevention of liver disease and facilitate the goals outlined in the trans-NIH Action Plan for Liver Disease Research. Annual deadline dates are April 5, August 5 and December 5.
Date(s) Application is Due: Apr 5; Aug 5; Dec 5
Contact: Dr. Jane Ye, Bioinformatics and Computational Biology; (301) 594-5983; fax (301) 402-1384; yej@mail.nih.gov
Internet: http://www.nlm.nih.gov/ep/grantsbliver.html
Sponsor: National Library of Medicine
6705 Rockledge Drive, Suite 301
Bethesda, MD 20892

NLM Predictive Multiscale Models of the Physiome in Health and Disease Grants 3667
The goal of this funding opportunity announcement (FOA) is to move the field of biomedical computational modeling forward through the development of more realistic and predictive models of health and disease. NIH recognizes the need for sophisticated, predictive, computational models of development and disease that encompass multiple biological scales. These models may be designed to uncover biological mechanisms or to make predictions about clinical outcome and may draw on a variety of data sources including relevant clinical data. Ultimately the models and the information derived from their use will enable researchers and clinicians to better understand, prevent, diagnose and treat the diseases or aberrations in normal development. Specifically this FOA solicits the development of predictive multiscale models of health and disease states that must include higher scales of the physiome. The specific objectives are to develop multiscale models that are physiologically mechanistic and biomedically relevant, to bring together modeling and biomedical expertise to collaborate on building models, to validate and test models with standard datasets, and to develop models that can be explicitly shared with other modelers.
Date(s) Application is Due: Jan 14; May 14; Sep 15
Contact: Dr. Jane Ye, Bioinformatics and Computational Biology; (301) 594-5983; fax (301) 402-1384; yej@mail.nih.gov
Internet: http://www.nlm.nih.gov/ep/predictive.html
Sponsor: National Library of Medicine
6705 Rockledge Drive, Suite 301
Bethesda, MD 20892

NLM Research Project Grants 3668
The primary purpose of this program is to provide support for rigorous scientific research in biomedical informatics and bioinformatics. Medical informatics is an interdisciplinary field that combines health sciences specialty areas with the information and computer sciences, computational linguistics, and decision theory, among others. NLM's program

emphasizes knowledge issues without restriction as to specific categorical diseases. Proposed research must be clearly health relevant. Areas of interest include knowledge representation, more effective human-machine interfaces, and more efficient information storage and retrieval. The cognitive aspects of information-related behaviors are of particular interest. This area includes problems of medical decision analysis, medical decision support systems, and retrieval from image databases. Deadlines for new applications are February 5, June 5 and October 5 each year. Deadlines for revised applications are March 5, July 5 and November 5 each year.
Requirements: Institutions or organizations with research capabilities in the health information fields or in medical informatics are eligible to apply.
Date(s) Application is Due: Feb 5; Mar 5; Jun 5; Jul 5; Oct 5; Nov 5
Amount of Grant: 25,000 - 2,000,000 U.S.D
Contact: Dr. Jane Ye, Bioinformatics and Computational Biology; (301) 594-5983; fax (301) 402-1384; yej@mail.nih.gov
Internet: http://www.nlm.nih.gov/ep/GrantResearchParent.html
Sponsor: National Library of Medicine
6705 Rockledge Drive, Suite 301
Bethesda, MD 20892

NLM Research Supplements to Promote Reentry in Health-Related Research 3669
Principal Investigators holding NIH research grants of the specified types (see announcement) can apply for administrative supplements to support individuals with high potential to reenter an active research career after taking time off to care for children or attend to other family responsibilities. It is anticipated that at the completion of the supplement, the reentry scientist will be in a position to apply for a career development award, a research award, or some other form of independent research support. The grantee institution must submit the application for supplemental funds directly to the awarding component that supports the parent grant. The application must not be submitted to the NIH Center for Scientific Review.
Contact: Dr. Hua-Chuan Sim, Clinical and Public Health Informatics; (301) 594-5983; fax (301) 402-1384; simh@mail.nih.gov
Internet: http://www.nlm.nih.gov/ep/grantsupreentry.html
Sponsor: National Library of Medicine
6705 Rockledge Drive, Suite 301
Bethesda, MD 20892

NLM Understanding and Promoting Health Literacy Research Grants 3670
The goal of these Program Announcements (Understanding and Promoting Health Literacy Research Grants and Understanding and Promoting Health Literacy Small Project Grants) is to increase scientific understanding of the nature of health literacy and its relationship to healthy behaviors, illness prevention and treatment, chronic disease management, health disparities, risk assessment of environmental factors, and health outcomes including mental and oral health.
Date(s) Application is Due: Jan 25; May 25
Contact: Dr. Hua-Chuan Sim, Clinical and Public Health Informatics; (301) 594-5983; fax (301) 402-1384; simh@mail.nih.gov
Internet: http://www.nlm.nih.gov/ep/healthlit.html
Sponsor: National Library of Medicine
8600 Rockville Pike, Building 38, Room B1E08
Bethesda, MD 20894

NMF Aura E. Severinghaus Award 3671
One award is presented annually to an underrepresented minority (African American, Mexican American, mainland Puerto Rican, or Native American) student attending Columbia University's College of Physicians and Surgeons in recognition of outstanding academic achievement, leadership, and community service. A committee of faculty and administrators at the college selects the student most deserving of this award and presents the candidate's dossier to NMF for final approval.
Requirements: This award is open only to underrepresented minority students enrolled in their senior year at the College of Physicians and Surgeons at Columbia University.
Geographic Focus: New York
Amount of Grant: 2,000 U.S.D
Contact: Programs Department; (212) 483-8880; fax (212) 483-8897; nmf1@nmfonline.org
Internet: http://www.nmfonline.org/index.html
Sponsor: National Medical Fellowships
5 Hanover Square
New York, NY 10004

NMF Franklin C. McLean Award 3672
One award is presented annually to a senior underrepresented minority (African American, Mexican American, mainland Puerto Rican, and Native American) student enrolled in an accredited U.S. medical school in recognition of outstanding academic achievement, leadership, and community service. Annual deadline dates may vary; contact program staff for exact dates.
Requirements: Candidates must be nominated by medical school deans. Nominations are requested in June. Schools are required to submit letters of recommendation and official academic transcripts for each candidate.
Amount of Grant: 3,000 U.S.D
Contact: Programs Department; (212) 483-8880; fax (212) 483-8897; nmf1@nmfonline.org
Internet: http://www.nmf-online.org/Programs/MeritAwards/McLean/Overview.htm
Sponsor: National Medical Fellowships
5 Hanover Square
New York, NY 10004

NMF General Electric Medical Scholars Fellowships 3673
The sponsor has established a partnership with General Electric (GE) to provide need-based scholarship support to historically underrepresented minority medical students, and to allow them to engage in clinical work in underserved communities. The program also offers eligible students an opportunity to complete a two-month elective in Ghana, West Africa focusing on areas of critical need such as HIV/AIDS, diabetes, tuberculosis, heart disease, cancer and infant mortality. Five fellowships are awarded annually. Students will be matched with a mentor for the period of the program. Travel, hosing and a stipend will be provided for students.
Requirements: Students must be nominated by their medical school deans. Eligible candidates must be underrepresented minority students enrolled in their fourth year in an accredited U.S. medical school. Candidates must demonstrate outstanding academic achievement, leadership, and potential for community-based work or initiation of innovative projects in school or community. Additionally, candidates must also demonstrate potential for distinguished contributions to medicine. Complete application requirements can be found at the sponsor's website.
Contact: Dr. Stephen Keith, President; (240) 243-8000; fax (240) 465-0450
Internet: http://www.nmfonline.org/index.html
Sponsor: National Medical Fellowships
5 Hanover Square
New York, NY 10004

NMF Henry G. Halladay Awards 3674
Five supplemental scholarships are presented annually to African American men who have been accepted into the first-year classes of accredited U.S. medical schools and who have overcome significant obstacles to obtain a medical education. Staff members review NMF first-year general scholarship applicants and select five students who, on the basis of recommendations, personal statements, awards and financial need, are most deserving of these awards. Annual deadline dates may vary; contact program staff for exact dates.
Requirements: Applicants must have been accepted to AAMC or AOA-accredited U.S. medical schools for study leading to MD or DO degrees.
Amount of Grant: 760 U.S.D
Contact: Programs Department; (212) 483-8880; fax (212) 239-9718; nmf2@nmfonline.org
Internet: http://www.nmfonline.org/index.html
Sponsor: National Medical Fellowships
5 Hanover Square
New York, NY 10004

NMF Hugh J. Andersen Memorial Scholarship 3675
This scholarship program was established by the family of the late Hugh J. Andersen. Need-based merit scholarships are annually presented to Minnesota residents enrolled in any accredited U.S. medical school or students attending medical schools in Minnesota. Up to 5 scholarships of $2,500 are presented annually.
Requirements: Eligible candidates must be underrepresented minority students enrolled beyond their first year in an accredited U.S. medical school. Candidates must demonstrate outstanding academic achievement as indicated in academic transcripts, faculty evaluations, by receipt of special academic honors, fellowships, awards or induction into national medical honor societies, leadership (as indicated by active participation in community-based work or initiation of innovative projects in school or community), and potential for distinguished contributions to medicine as indicated by participation in research, publications or unique clerkships. Students must also provide documented proof of financial need. Students must be nominated by medical school deans. Schools are required to submit letters of recommendations and official academic transcripts. Applicants must submit scholarship applications and include documented proof of financial need. See the sponsor's website for the required application form and further instructions.
Geographic Focus: Minnesota
Amount of Grant: 2,500 U.S.D
Contact: Programs Department; (212) 714-0933; fax (212) 239-9718; nmf1@nmfonline.org
Internet: http://www.nmfonline.org/index.html
Sponsor: National Medical Fellowships
5 Hanover Square
New York, NY 10004

NMF Irving Graef Memorial Scholarship 3676
This two-year scholarship is presented annually to a third-year student and recognizes outstanding academic achievement, leadership, and community service. The scholarship is renewable in the fourth year if the award recipient continues in good academic standing. The scholarship is open only to rising third-year minority medical students who received NMF financial assistance during their second year. One new scholarship is awarded each year. Annual deadline date may vary; contact program staff for exact dates.
Requirements: Students must be nominated by medical school deans. Schools are required to submit letters of recommendation as well as official academic transcripts. Candidates must demonstrate outstanding academic achievement, submit the scholarship application (excluding proof of financial need), personal essay and current cirriculum vitae.
Amount of Grant: 2,000 U.S.D
Contact: Programs Department; (212) 483-8880; fax (212) 483-8897; nmf1@nmfonline.org
Internet: http://www.nmfonline.org/index.html
Sponsor: National Medical Fellowships
5 Hanover Square
New York, NY 10004

NMF Mary Ball Carrera Scholarship 3677

This need-based merit scholarship is annually presented to a Native American woman enrolled in any accredited U.S. medical school and recognizes outstanding academic achievement, leadership and community services. Recipients receive a one-time award of $2,500. Scholarships are granted on the basis of financial need as determined by the student's total resources (including parental and spousal support), education costs, and receipt of other scholarships and grants.

Requirements: Eligible candidates must be first- or second-year medical students. Students must provide proof of admission by an accredited medical school. Applicants must demonstrate and document financial need and submit a current financial aid award letter.

Restrictions: All applicants must reside within the 50 U.S.A. states and provide proof of citizenship.

Amount of Grant: 2,500 U.S.D

Contact: Programs Department; (212) 483-8880; fax (212) 483-8897; nmf1@nmfonline.org

Internet: http://www.nmfonline.org/index.html

Sponsor: National Medical Fellowships

5 Hanover Square

New York, NY 10004

NMF Metropolitan Life Foundation Awards Program for Academic Excellence in Medicine 3678

Need-based scholarships are awarded annually to third- and fourth-year underrepresented minority medical students who have demonstrated academic achievement and leadership. Applicants must attend medical schools or have legal residence in cities, designated each year; contact office for current eligible area. Nominations are requested in August. Annual deadline dates may vary; contact program staff for exact dates.

Requirements: Students must be nominated by medical school deans. Schools are required to submit letters of recommendation as well as official academic transcripts. Candidates are required to submit scholarship applications including personal statements and verification of financial need.

Amount of Grant: 4,000 U.S.D

Contact: Programs Department; (212) 483-8880; fax (212) 483-8897; nmf1@nmfonline.org

Internet: http://www.nmfonline.org/index.html

Sponsor: National Medical Fellowships

5 Hanover Square

New York, NY 10004

NMF National Medical Association Awards for Medical Journalism 3679

In general, these awards recognize academic achievement, leadership, and potential for distinguished contributions to medicine. Specifically, these awards recognize demonstrated skills in journalism. Deadlines vary - contact the sponsor for dates.

Requirements: Eligibility is limited to African-American medical students. Each medical school may submit up to two nominations. Candidates must be in the third or fourth year of medical school and must have published articles and photographs in, or been writers, editors, photographers on the staffs of: medical school newspapers; medical students' journals; recognized professional journals; or, other accredited scientific journal. Students who have written, produced or directed health-related films, commercials or videos are also encouraged to submit their applications and samples of their work. Applicants must also submit samples of their journalistic work (articles, photographs, audio tapes, video tapes, etc.) for review by the selection committee. Entries must have been published or produced within the last three years. The application form and more detailed guidelines can be found at the sponsor's website.

Amount of Grant: 2,500 U.S.D

Contact: Programs Department; (212) 483-8880; fax (212) 239-9718; nmf2@nmfonline.org

Internet: http://www.nmfonline.org/index.html

Sponsor: National Medical Fellowships

5 Hanover Square

New York, NY 10004

NMF National Medical Association Emerging Scholar Awards 3680

The National Medical Association and NMF sponsor need-based award programs that recognize and reward African American medical students for special achievements, academic excellence, leadership, and potential for outstanding contributions to medicine. The NMA Emerging Scholar Awards are presented to current first, second, and third-year medical students.

Requirements: Eligibility for these awards is limited to African-American medical students. Each school may submit two nominations. Applicants must demonstrate and document financial need. Deadlines vary - contact the sponsor for specific dates.

Amount of Grant: 2,250 U.S.D

Contact: Programs Department; (212) 483-8880; fax (212) 483-8897; nmf1@nmfonline.org

Internet: http://www.nmfonline.org/index.html

Sponsor: National Medical Fellowships

5 Hanover Square

New York, NY 10004

NMF National Medical Association Patti LaBelle Award 3681

This award is one of the highest honors presented to students by the National Medical Association. It recognizes academic achievement, leadership and potential for distinguished contributions to medicine. This award is presented to current first-, second-, and third-year medical students with demonstrated financial need.

Requirements: Eligibility is limited to African-American medical students. Application materials must include a letter of recommendation that fully discusses the candidate's academic performance, extracurricular activities and other qualifications; a complete scholarship application form; an essay of at least 500 words discussing the applicant's motivation for a career in medicine and plans over the next ten years; and, an official academic transcript.

Amount of Grant: 5,000 U.S.D

Contact: Programs Department; (212) 483-8880; fax (212) 483-8897; nmf1@nmfonline.org

Internet: http://www.nmfonline.org/index.html

Sponsor: National Medical Fellowships

5 Hanover Square

New York, NY 10004

NMF Need-Based Scholarships 3682

The program provides financial assistance to United States citizens from groups currently underrepresetned in the medical profession; specifically, African-Americans, Mexican-Americans, Native Americans, Alaska Natives, Native Hawaiians, and mainland Puerto Ricans who permanently reside within the 50 U.S. states. These awards are offered to first- and second-year medical students. Scholarships are granted on the basis of financial need as determined by the student's total resources (including parental and spousal support), cost of education, and receipt of other scholarships and grants. Awards have ranged from $500 to $10,000.

Requirements: Applicants must be U.S. citizens by the application deadline and must have been accepted to AAMC or AOA-accredited U.S. medical schools for study leading to M.D. or D.O. degrees. The application form and FAQs can be found at the sponsor's website. Applicants must demonstrate and document financial need by submitted complete copies of their parents', spouse's and their own most recent 1040, 1040A or 1040EZ tax forms. All nontaxable income (e.g. AFDC, ADC, Social Security benefits, etc.) must also be documented by the appropriate agency. Applicants with neither taxable nor nontaxable income must provide verification of means of support. All parental financial data must be submitted in order to evaluate dependent status.

Date(s) Application is Due: Aug 31

Contact: Programs Department; (212) 483-8880, ext. 304; fax (212) 483-8897; NMF1@NMFonline.org or info@NMFonline.org

Internet: http://www.nmfonline.org/programs.html

Sponsor: National Medical Fellowships

5 Hanover Square

New York, NY 10004

NMF Ralph W. Ellison Prize 3683

The prize is presented to a graduating underrepresented medical student who has demonstrated outstanding academic achievement, leadership, and potential to make significant contributions to medicine. One prize is presented each year (nonrenewable) and includes a certificate of merit and a stipend. Annual deadline date may vary; contact program staff for exact dates.

Requirements: African-Americans, mainland Puerto Ricans, Mexican-Americans, Native Hawaiians, Alaska Natives, and American Indians who are U.S. citizens attending M.D. degree-granting institutions accredited by the Liaison Committee on Medical Education of the Association of American Medical Colleges, or in D.O. degree-granting colleges of osteopathic medicine accredited by the Bureau of Professional Education of the American Osteopathic Association are eligible.

Amount of Grant: 500 U.S.D

Contact: Programs Department; (212) 483-8880; fax (212) 483-8897; nmf1@nmfonline.org

Internet: http://www.nmfonline.org/index.html

Sponsor: National Medical Fellowships

5 Hanover Square

New York, NY 10004

NMF William and Charlotte Cadbury Award 3684

This award is presented annually to a senior, underrepresented student enrolled in an accredited U.S. medical school. Candidates must demonstrate outstanding academic achievement, leadership, and community service. A certificate of merit is included in the award. Annual deadline dates may vary; contact program staff for exact dates.

Requirements: Students must be nominated by medical school deans. Schools are required to submit letters of recommendation and official academic transcripts for each candidate. Competition is open to senior, underrepresented minority students enrolled in accredited U.S. medical schools. Candidates must demonstrate outstanding academic achievement and leadership.

Amount of Grant: 2,000 U.S.D

Contact: Programs Department; (212) 483-8880; fax (212) 483-8897; nmf1@nmfonline.org

Internet: http://www.nmfonline.org/index.html

Sponsor: National Medical Fellowships

5 Hanover Square

New York, NY 10004

NMSS Daniel Haughton Senior Faculty Awards 3685

The award is granted to established investigators seeking support for specialized training in a field which they are not currently expert, inasmuch as such training will enhance their capacity to conduct research related to multiple sclerosis. Awards may be used for sabbatical support.

Date(s) Application is Due: Feb 13

Contact: Dr. Timothy Coetzee, Director; (212) 476-0478; timothy.coetzee@nmss.org

Internet: http://www.nationalmssociety.org/Research-srfaculty.asp

Sponsor: National Multiple Sclerosis Society

733 Third Avenue

New York, NY 10017-3288

NMSS Harry Weaver Neuroscience Scholar Awards 3686
These junior faculty awards provide higher support for qualified candidates who have concluded their research training and begun academic careers as independent investigators in areas of the neurosciences related to multiple sclerosis. The awards are designed to provide salary and grant support for a five-year period, thus permitting the awardee to establish competence in his/her chosen research area. Candidates may request up to 75 percent annual salary support, not to exceed $50,000 in the first year of the award, with a 5 percent annual increase in each of the remaining four years. Research support of $25,000 may be requested in the first year and up to $35,000 in the last year.
Requirements: Candidate must hold the MD, PhD, or equivalent with no more than five years of independent research experience and must have been offered a position equivalent to assistant professor at an approved institution.
Date(s) Application is Due: Feb 1
Contact: Timothy Coetzee, Director, Research Training Programs; (212) 476-0478; timothy.coetzee@nmss.org
Internet: http://www.nationalmssociety.org/Research-jrfaculty.asp
Sponsor: National Multiple Sclerosis Society
733 Third Avenue
New York, NY 10017-3288

NMSS Research Grants 3687
Grants are awarded for fundamental or applied research in scientific areas pertinent to multiple sclerosis. Grants are available for clinical or nonclinical studies, providing they show a reasonable relevance to the society's interests. Annual deadline dates may vary; contact program staff for exact dates.
Requirements: Qualified investigators affiliated with appropriate institutions are eligible to apply. Institutions are the official recipients of the grants.
Date(s) Application is Due: Feb 1; Aug 3
Amount of Grant: 35,568 - 51,036 U.S.D
Contact: Dr. Patricia O'Looney, Research Programs Department Contact; (212) 476-0413; fax (212) 986-7981; patricia.olooney@nmss.org
Internet: http://www.nationalmssociety.org/Research-Grants.asp
Sponsor: National Multiple Sclerosis Society
733 Third Avenue
New York, NY 10017-3288

NMSS Scholarships 3688
The society awards unrestricted scholarships to defray expenses related to post-high school education. Decisions are based on financial need, an academic record that shows the applicant is able to succeed in his or her chosen area, and on a personal essay discussing the impact of multiple sclerosis (MS) on the applicant's life.
Requirements: Application is open to high school seniors who have MS; high school seniors who have a parent or guardian with MS; and adults with MS who have never enrolled in a postsecondary institution.
Date(s) Application is Due: Mar 1
Amount of Grant: 2,000 - 3,000 U.S.D
Contact: Scholarship America; (800) 344-4467 or (510) 268-0572; info@msconnection.org
Internet: http://www.nationalmssociety.org/can/event/event_detail.asp?e=8108#scholarship
Sponsor: National Multiple Sclerosis Society
733 Third Avenue
New York, NY 10017-3288

NMSS Sylvia Lawry Physician Fellowships 3689
The society welcomes applications for support of physicians for training in key elements associated with conducting multiple sclerosis clinical trials. The purpose of this award is to help maintain and foster high quality clinical trials in multiple sclerosis and to ensure that there is a continuing pool of physicians who are trained in this area. In this program the individual with an MD or equivalent medical degree will acquire formal training, under the tutelage of an established investigator, in key elements associated with conducting clinical trials in MS. Annual deadline dates may vary; contact program staff for exact dates.
Requirements: Individuals must have received the MD or equivalent medical degree from an accredited institution and must be licensed to practice medicine in the United States.
Date(s) Application is Due: Feb 1
Amount of Grant: Up to 40,000 U.S.D
Contact: Dr. Timothy Coetzee, Director; (212) 476-0478; timothy.coetzee@nmss.org
Internet: http://www.nationalmssociety.org/Research-physfellow.asp
Sponsor: National Multiple Sclerosis Society
733 Third Avenue
New York, NY 10017-3288

NNEDVF/Altria Doors of Hope Program 3690
The goal of this initiative is to strengthen the safety net available to survivors of domestic violence through the provision of grant awards that support organizational excellence, innovation, and leadership in the provision of shelter and legal advocacy services. Direct service grants are awarded to domestic violence service providers for shelter services and advocacy/legal services. There is a two stage application process; stage one consists of registration online, which includes key questions that will be used to determine which applicants will be invited to apply for stage two. For the current funding year, grantmaking will be by invitation only. Applicants will be contacted if they are eligible to apply. Additional assistance and information is provided at www.nnedv.org.
Requirements: Eligible organizations must have a primary mission which includes the provision of direct services to survivors of domestic violence; provide shelter and/or legal advocacy services to survivors of domestic violence; be a U.S. based non-profit, non-governmental, tax-exempt organization; have been in operation for a minimum of three years with an organization budget greater than or equal to $250,000; and have nondiscrimination policies that include ethnicity, race, religious creed, national origin, disability, sexual orientation, marital status, age, and gender.
Restrictions: Applications will only be accepted through the online RFP.
Amount of Grant: 10,000 - 50,000 U.S.D
Samples: Hmong American Friendship Association (Milwaukee, WI)—to provide group facilitators, education groups in the workplace, and domestic violence training for Hmong elders, $24,000; Susquehanna Valley Women in Transition (Lewisburg, PA)—to implement and teach a curriculum of safe living and violence prevention to group home residents with mental disabilities, $19,000.
Contact: Administrator; (202) 543-5566; fax (202) 543-5626; altriadoorsofhope@nnedv.org
Internet: http://www.altria.com/responsibility/4_9_1_1_2_1_domviolprograms.asp
Sponsor: National Network to End Domestic Violence Fund
660 Pennsylvania Avenue SE, Suite 303
Washington, DC 20003

Noble County Community Foundation - Arthur A. and Hazel S. Auer Scholarship 3691
The Noble County Community Foundation offers 40 scholarships with a variety of requirements and specific criteria for eligible Noble County residents. Applicants can review the scholarship list and choose to apply for several scholarships through one application. The Arthur A. and Hazel S. Auer Scholarship helps fund two East Noble high school seniors preparing to enter a four year bachelor's program at any accredited college or university. The scholarship pays $500 per semester for the first two years for a total of $2,000. Any college major is eligible.
Requirements: Applicants must have a GPA of 3.7 or above. Candidates must apply through the Foundation's online application and follow the instructions for submission listing detailed information about their school performance, community service and volunteer activities, work experience, expected family contribution from the FAFSA, and three references.
Geographic Focus: Indiana
Date(s) Application is Due: Jan 28
Amount of Grant: 2,000 U.S.D
Contact: Jennifer Shultz, Scholarship Manager; (260) 894-3335; fax (260) 894-9020; jennifer@noblecountycf.org or info@noblecountrycf.org
Internet: http://www.noblecountycf.org/body.cfm?lvl1=servic&lvl2=schola&lvl3=availa
Sponsor: Noble County Community Foundation
1599 Lincolnway South
Ligonier, IN 46767

Noble County Community Foundation - Art Hutsell Scholarship 3692
The Noble County Community Foundation offers 40 scholarships with a variety of requirements and specific criteria for eligible Noble County residents. Applicants can review the scholarship list and choose to apply for several scholarships through one application. The Art Hutsell Scholarship helps fund a West Noble High School senior pursing advanced education with a one-time payment of $300.
Requirements: Applicants must be pursuing post-secondary education in emergency services involving fire, police, and/or EMS. Candidates must apply through the Foundation's online application and follow the instructions for submission by listing detailed information about their school performance, community service and volunteer activities, work experience, expected family contribution from the FAFSA, and three references.
Geographic Focus: Indiana
Date(s) Application is Due: Jan 28
Amount of Grant: 300 U.S.D
Contact: Jennifer Shultz, Scholarship Manager; (260) 894-3335; fax (260) 894-9020; jennifer@noblecountycf.org or info@noblecountrycf.org
Internet: http://www.noblecountycf.org/body.cfm?lvl1=servic&lvl2=schola&lvl3=availa
Sponsor: Noble County Community Foundation
1599 Lincolnway South
Ligonier, IN 46767

Noble County Community Foundation - Delta Theta Tau Sorority IOTA IOTA Chapter Riecke Scholarship 3693
The Noble County Community Foundation offers 40 scholarships with a variety of requirements and specific criteria for eligibility for Noble County residents. Applicants can review the scholarship list and choose to apply for several scholarships through one application. The Delta Theta Tau Sorority Tammy Riecke Scholarship offers funding for an East Noble high school senior who has worked to promote philanthropy and charity for the community. The scholarship amount is to be determined but is paid during the fall and spring semester of the first year.
Requirements: Candidates must apply through the Foundation's online application and follow the instructions for submission listing detailed information about their school performance, community service and volunteer activities, work experience, expected family contribution from the FAFSA, and three references. The application section for this scholarship includes an essay question, "In your own words, what is a humanitarian? List anything you have done that you feel is humanitarian related."
Geographic Focus: Indiana
Date(s) Application is Due: Jan 28
Contact: Jennifer Shultz, Scholarship Manager; (260) 894-3335; fax (260) 894-9020; jennifer@noblecountycf.org or info@noblecountrycf.org

Internet: http://www.noblecountycf.org/body.cfm?lvl1=servic&lvl2=schola&lvl3=availa
Sponsor: Noble County Community Foundation
1599 Lincolnway South
Ligonier, IN 46767

Noble County Community Foundation - Democrat Central Committee Scholarships 3694

The Noble County Community Foundation offers 40 scholarships with a variety of requirements and specific criteria for eligible Noble County residents. Applicants can review the scholarship list and choose to apply for several scholarships through one application. The Democrat Central Committee Scholarships are offered to several Noble County residents ad/or high school seniors pursuing post-high school education.
Requirements: Preference is given to current graduating high school seniors. Candidates must apply through the Foundation's online application and follow the instructions for submission by listing detailed information about their school performance, community service and volunteer activities, work experience, expected family contribution from the FAFSA, and three references.
Geographic Focus: Indiana
Date(s) Application is Due: Jan 28
Amount of Grant: 400 U.S.D
Contact: Jennifer Shultz, Scholarship Manager; (260) 894-3335; fax (260) 894-9020; jennifer@noblecountycf.org or info@noblecountrycf.org
Internet: http://www.noblecountycf.org/body.cfm?lvl1=servic&lvl2=schola&lvl3=availa
Sponsor: Noble County Community Foundation
1599 Lincolnway South
Ligonier, IN 46767

Noble County Community Foundation - Kathleen June Earley Memorial Scholarship 3695

The Noble County Community Foundation offers 40 scholarships with a variety of requirements and specific criteria for eligible Noble County residents pursuing higher education. Applicants can review the scholarship list and choose to apply for several scholarships through one application. The Kathleen June Earley Scholarship funds an East Noble high school senior planning to study at Ball State University with a one-time payment of $250. Any college major is eligible.
Requirements: Candidates must apply through the Foundation's online application and follow the instructions for submission listing detailed information about their school performance, community service and volunteer activities, work experience, expected family contribution from the FAFSA, and three references.
Geographic Focus: Indiana
Date(s) Application is Due: Jan 28
Amount of Grant: 250 U.S.D
Contact: Jennifer Shultz, Scholarship Manager; (260) 894-3335; fax (260) 894-9020; jennifer@noblecountycf.org or info@noblecountrycf.org
Internet: http://www.noblecountycf.org/body.cfm?lvl1=servic&lvl2=schola&lvl3=availa
Sponsor: Noble County Community Foundation
1599 Lincolnway South
Ligonier, IN 46767

Noble County Community Foundation - Lolita J. Hornett Memorial Nursing Scholarship 3696

The Noble County Community Foundation offers 40 scholarships with a variety of requirements and specific criteria for eligible Noble County residents. Applicants can review the scholarship list and choose to apply for several scholarships through one application. The Lolita Hornett Nursing Scholarship helps fund Noble County residents entering their first year in post-secondary education. A one-time payment of $500 is given at the completion of the first semester.
Requirements: Applicants must be East Noble High School seniors or graduates entering a nursing degree program as a freshman. They must have also maintained a 3.0 GPA in high school. Candidates must apply through the Foundation's online application and follow the instructions for submission by listing detailed information about their school performance, community service and volunteer activities, work experience, expected family contribution from the FAFSA, and three references.
Geographic Focus: Indiana
Date(s) Application is Due: Jan 28
Contact: Jennifer Shultz, Scholarship Manager; (260) 894-3335; fax (260) 894-9020; jennifer@noblecountycf.org or info@noblecountrycf.org
Internet: http://www.noblecountycf.org/body.cfm?lvl1=servic&lvl2=schola&lvl3=availa
Sponsor: Noble County Community Foundation
1599 Lincolnway South
Ligonier, IN 46767

Noble County Community Foundation Grants 3697

The Noble County Community Foundation makes grants for innovative, creative projects and programs responsive to changing community needs in the areas of health and human services, education, arts and culture, and civic affairs. The funding is not limited to these areas and grant seekers are encouraged to respond to emerging community needs. The Foundation considers only not-for-profit projects that: promote cooperation among organizations without duplicating services; promote volunteer involvement; demonstrate practical approaches to current community issues; enhance or improve an organization's self-sufficiency and effectiveness; emphasize prevention. In addition, the Foundation considers projects that: affect a broad segment of the population; are pilot programs clearly replicable in their design and have reasonable prospects for future support; serve people whose needs are not being met by existing services and which encourage independence; move the community to a higher cultural awareness. The Foundation offers grants in April, June, August, and December, with decisions made 60 days after each deadline. Applicants are encouraged to contact the Foundation before submitting a proposal to find out if the project is appropriate for funding. Applicants should refer to the website for further information.
Requirements: In applying for grants, the following points must be addressed (with explanation if information is not available): the specific purpose and the need for the funds requested; the need for the program/project in the community; the amount requested; a detailed description of how the money would be spent; a statement of how the project will improve life in Noble County and how the outcomes will be measured; recent grants received and applications pending for this program/project; a listing of the current board and/or project organizers and their contact information; a copy of the organization's IRS tax-exempt letter stating 501(c)3 status; and for schools applying, a letter of support from the organization's superintendent/principal. After the organization's request has been received, it may be contacted by a Foundation representative requesting a site visit or additional information.
Restrictions: Discretionary funds will not be used under any circumstances to support: deficit spending; political purposes; annual fund campaigns; lobbying; organizations whose primary function is to allocate funds to other charitable organizations or projects; projects that do not serve residents of Noble County; travel; augmenting endowments; underwriting for fundraising events; loans.
Geographic Focus: Indiana
Date(s) Application is Due: Mar 2; May 2; Jul 2; Nov 2
Amount of Grant: 500 - 5,000 U.S.D
Contact: Linda Yerick, Executive Director; (260) 894-3335; fax (260) 894-9020; linda@noblecountycf.org or info@noblecountrycf.org
Margarita White, Program Officer; (260) 894-3335; fax (260) 894-9020; margarita@noblecountycf.org or info@noblecountrycf.org
Internet: http://www.noblecountycf.org/body.cfm?lvl1=servic&lvl2=grants&lvl3=how2ap
Sponsor: Noble County Community Foundation
1599 Lincolnway South
Ligonier, IN 46767

Noble County Community Foundation Lilly Endowment Scholarship 3698

The Noble County Community Foundation Lilly Endowment Scholarship awards four years of full tuition and required fees and books to a Noble County high school student pursuing a baccalaureate degree at an Indiana public or private college or university. All college majors are eligible.
Requirements: Application criteria includes community service and at least a 3.5 high school GPA. Candidates may apply for the scholarship through the online joint scholarship application on the Foundation's website.
Restrictions: Applicants must be U.S. citizens and residents of Noble County for at least one year.
Geographic Focus: Indiana
Contact: Jennifer Shultz, Scholarship Coordinator; (260) 894-3335; fax (260) 894-9020; jennifer@noblecountycf.org or info@noblecountrycf.org
Internet: http://www.noblecountycf.org/body.cfm?lvl1=servic&lvl2=schola&lvl3=availa
Sponsor: Noble County Community Foundation
1599 Lincolnway South
Ligonier, IN 46767

Noble County Community Foundation Scholarships 3699

The Noble County Community Foundation helps build an enduring source of charitable assets to meet the changing needs of the people in Noble County, Indiana. The Foundation administers separate scholarship funds established by individuals, families, corporations, and organizations to assist local students in pursuit of advanced education. Eligibility, criteria, and supplemental requirements are different for each scholarship. See the Noble County Community Foundation website for a description of each scholarship offered. A joint scholarship application is available online so that applicants may apply for several scholarships at one time. Candidates are encouraged to apply as soon as possible, but may apply from mid-November through the following January.
Requirements: Applicants must submit 7 copies of the application to the Foundation office. Applicants are also required to agree to several criteria listed on the application introduction. Applicants are required to list detailed information about their school performance, community service activities, volunteering, work experience, expected family financial contribution to their college studies, and three references.
Geographic Focus: Indiana
Date(s) Application is Due: Jan 28
Contact: Jennifer Shultz, Scholarship Manager; (260) 894-3335; fax (260) 894-3335; jennifer@noblecountycf.org or info@noblecountrycf.org
Internet: http://www.noblecountycf.org/body.cfm?lvl1=servic&lvl2=schola&lvl3=availa
Sponsor: Noble County Community Foundation
1599 Lincolnway South
Ligonier, IN 46767

Norcliffe Foundation Grants 3700

The Norcliffe Foundation is a private nonprofit family foundation established in 1952 by Paul Pigott for the purpose of improving the quality of life for all people in the Puget Sound region. The foundation provides grants in the areas of health, education, social services, civic improvement, religion, culture and the arts, the environment, historic preservation and youth programs. Foundation funding types include capital campaigns for building and equipment, certain operating budgets, endowments, challenge/matching grants, land acquisition, new projects, start-up funds, renovation, and research. Requests for publication, videos/films, and website production may be

considered as may scholarships, fellowships / chairs, conferences / seminars, social enterprise development, and technical assistance. Multi-year and renewable funding may be considered. Applications are accepted year-round. One copy of a letter proposal and/or common grant application form should be directed to the President in care of the Foundation Manager at the contact information given. An initial phone call is optional. Guidelines, instructions, and a copy of the common grant proposal form are provided at the website. Funding decisions and notification generally occur three to six months after receipt of request.
Requirements: 501(c)3 organizations in the Puget Sound region in and around Seattle, Washington are eligible to apply.
Restrictions: The Norcliffe Foundation does not provide the following types of assistance: deficit financing, emergency funds, grants to individuals, matches to employee-giving, Program-Related Investments / Loans (PRIs), in-kind services, volunteer / loaned executive, and internships. Applicants may only submit one request per year from date of funding or denial. Applicants who have previously received grants of $50,000 or more from the Foundation must wait two years from final payment before submitting a new application.
Geographic Focus: Washington
Amount of Grant: 1,000 - 25,000 U.S.D
Contact: Arline Hefferline, Foundation Manager; (206) 682-4820; fax (206) 682-4821; arline@thenorcliffefoundation.com
Internet: http://www.thenorcliffefoundation.com/
Sponsor: Norcliffe Foundation
999 Third Avenue, Suite 1006
Seattle, WA 98104-4001

Nord Family Foundation Grants **3701**
The Nord Family Foundation, in the tradition of its founders, Walter and Virginia Nord, endeavors to build community through support of projects that bring opportunity to the disadvantaged, strengthen the bond of families, and improve the quality of people's lives. Grants are awarded in the fields of arts and culture, civic affairs, education and health and social services. Most grants are made for program related activities, with some grants to support capital improvements and capital campaigns when special criteria are met. Projects that are directed at the root causes of social problems are of special interest. Grants range from $2,000 to $50,000.
Requirements: Applicants must be 501(c)3 non-profit organizations or selected public sector activities. Awards are made primarily in the environs of Lorain County, Ohio, with additional support in Cuyahoga County, Ohio. Particular attention is given to Cuyahoga County non-profits that provide services to Lorain County residents. Occasionally awards are made outside of this area, to organizations located in areas where voting members of the Foundation reside or to organizations with a national or international focus. Application is made online from the Foundation's website.
Restrictions: The Foundation does not support debt reduction, research projects, and tickets or advertising for fundraising activities.
Geographic Focus: Ohio
Date(s) Application is Due: Apr 1; Aug 1; Dec 1
Samples: Augustana Arts, Denver, Colorado, $10,000, support of the City Strings youth outreach (2010); Amherst Downtown Betterment Association, Amherst, Ohio, commercial revitalization efforts (2010); and Boys and Girls Clubs of Loraine County, Elyria, Ohio, $50,000, capital improvement (2010).
Contact: John Mullaney, Executive Director; (800) 745-8946 or (440) 984-3939; fax (440) 984-3934; execdir@nordff.org or info@nordff.org
Internet: http://www.nordff.org
Sponsor: Nord Family Foundation
747 Milan Avenue
Amherst, OH 44001

NORD Orphan Diseases Research Grants **3702**
The program provides seed money in small grants to academic scientists studying new treatments or diagnostics for rare diseases. One-year clinical-research projects should focus on the early detection, diagnosis, or treatment of one of the following disorders: adenoid cystic carcinoma, alveolar capillary dysplasia, Cat Eye Syndrome, Ehlers-Danlos Syndrome, erythromelalgia, giant hypertrophic gastritis, and olivopontocerebellar atrophy or closely related diseases. The clinical researchers supported by the research grants provide preliminary data indicating that a treatment (drug, device, or medical food) may be safe and effective when used for a larger number of patients. Researchers can then use the preliminary data to apply for larger multiyear government grants or to attract a commercial sponsor who will manufacture an orphan product and get it approved for marketing by the U.S. Food and Drug Administration.
Requirements: Individuals who have earned an MD or MD/PhD in the past 10 years are eligible.
Samples: Dr. Cynthia Jackson and Dr. Douglas Gnepp, Rhode Island Hospital (RI)—to focus on the identification of genes as potentially new prognostic markers, as well as to study their contribution to the development of adenoid cystic carcinoma; Dr. Eli Chang, Doheny Eye Institute, Keck School of Medicine of the U of Southern California (CA)—to research the establishment of an animal model for adenoid cystic carcinoma of the lacrimal gland, which may lead to the development of new and improved treatment modalities.
Contact: Administrator; (203) 744-0100; fax (203) 798-2291; orphan@rarediseases.org
Internet: http://www.rarediseases.org/research/requests
Sponsor: National Organization for Rare Disorders
55 Kenosia Avenue
Danbury, CT 06813-1968

NORD Roscoe Brady Lysosomal Storage Disease Fellowships **3703**
The fellowships assist physicians who desire to establish careers in lysosomal storage diseases (LSDs) clinical medicine. Clinical research funded by the fellowships will be collaborative, involve multidisciplinary teams, and focus on genetics, new treatments and diagnostics, and/or epidemiology of LSDs in general, or for a specific LSD. The program's primary goal is to develop and support clinicians beginning their medical career that will specialize and focus on research, treatment, and coordination of multidisciplinary care for people suffering with LSDs. The research will be conducted in the United States, countries of the European Union, Canada, Australia, New Zealand, Brazil, Argentina, Chile, South Africa, Japan, or Israel, and where supervision of grant administration is possible. Complete guidelines and application are available online.
Requirements: Applicants should have earned the degree(s) of MD or MD/PhD within the past 10 years.
Date(s) Application is Due: Apr 13
Amount of Grant: 50,000 - 70,000 U.S.D
Contact: Fellowships Administrator; (203) 744-0100 or (800) 999-6673; fax (203)798-2291; orphan@rarediseases.org
Internet: http://www.rarediseases.org/research/roscoe
Sponsor: National Organization for Rare Disorders
55 Kenosia Avenue
Danbury, CT 06813-1968

Nordson Corporation Foundation Grants **3704**
The corporate foundation awards grants to nonprofit organizations in geographic areas where Nordson facilities and employees are located: Cuyahoga and Lorain counties in Ohio; the Greater Atlanta, Georgia area; California, Rhode Island and Southeastern Massachusetts. The foundation provides a source of stable funding for community programs and projects in the areas of education, human welfare, civic, and arts and culture. Educational support is generally limited to improving elementary and secondary schools and certain programs for public and private higher education. Human welfare grants focus on children and youth. The foundation considers other funding areas, including urban affairs, volunteerism, public policy, health and health organizations, and literacy. Grants are awarded for general operating support, continuing support, annual campaigns, capital campaigns, building construction/renovation, equipment acquisition, emergency funds, seed money, scholarship funds, employee matching gifts, and technical assistance. The board meets four times each year. Deadlines vary by state; contact program staff for appropriate deadlines. Application forms are available on the website and may only be submitted online.
Requirements: Private, nonprofit organizations in California, Georgia, Ohio, Rhode Island, and southeastern Massachusetts may apply.
Restrictions: Funding is not provided for organizations whose services are not provided within the foundation's geographic areas of interest; direct grants or scholarships to individuals; organizations not eligible for tax-deductible support; organizations not exempt under Section 501(c)3 of the Internal Revenue Code; political causes, candidates, organizations or campaigns; organizations that discriminate on the basis of race, sex, or religion; or special occasion, goodwill advertising, i.e., journals or dinner programs.
Geographic Focus: California, Georgia, Massachusetts, Ohio, Rhode Island
Date(s) Application is Due: Feb 15; May 15; Aug 15; Nov 15
Amount of Grant: Up to 15,000,000 U.S.D
Samples: Ctr for Leadership in Education (Lorain, OH)—for general operating support, $250,000; Salvation Army of Lorain (Lorain, OH)—for the capital campaign, $50,000; Neighborhood House Assoc of Lorain County (Lorain, OH)—for rejuvenation of the Cityview Ctr, $39,625.
Contact: Cecilia H. Render, Director; (440) 892-1580, ext. 5172; crender@nordson.com
Internet: http://www.nordson.com/Corporate/Community/Foundation
Sponsor: Nordson Corporation Foundation
28601 Clemens Road
Westlake, OH 44145-1119

Norfolk Southern Foundation Grants **3705**
The Norfolk Southern Foundation awards grants to nonprofits in the territory served by Norfolk Southern Railway Company. The foundation offers grants in four principal areas: educational programs, primarily at the post-secondary level; community enrichment focusing on cultural and artistic organizations; environmental programs; and health and human services (primarily food banks, homeless programs, and free clinics). Decision are made by December 31 for the following year's discretionary funding program, and applicants are notified early in the following calendar year.
Requirements: Applicants should include the following in their information packet: a valid 501(c)3 or 170(c)1 letter from the IRS; a short request outlining objectives of the organization, project, and intended use of funds; a listing of the applicant's board of directors; current and potential sources of funding; audited financial statements from the last three years; and the applicant's contact name, phone number, and email address.
Restrictions: Grant requests are only accepted between July 15 and September 30 for funding in the following calendar year. Applications will not be accepted from: organizations not in the NS territory (must be served by Norfolk Southern to be considered); organizations that do not have a 501(c)3 or 170(c)(1) IRS letter; individuals or organizations established to help individuals; religious, fraternal, social or veterans organizations; political or lobbying organizations; public or private elementary and secondary schools; fundraising events, telethons, races or benefits; sports or athletic organizations or activities; community or private foundations, or other organizations that merely redistribute to other eligible organizations aggregated contributions; disease-related organizations and hospitals; mentoring programs; Boys and Girls Scout programs or similar organizations; animal organizations; non-U.S. based charities; and organizations whose programs have national scope

Geographic Focus: Alabama, Connecticut, Delaware, District of Columbia, Florida, Georgia, Illinois, Indiana, Iowa, Kentucky, Louisiana, Maine, Maryland, Massachusetts, Michigan, Mississippi, Missouri, New Hampshire, New Jersey, New York, North Carolina, Ohio, Pennsylvania, Rhode Island, South Carolina, Tennessee, Texas, Vermont, Virginia, West Virginia, Wisconsin
Date(s) Application is Due: Sep 30
Amount of Grant: Up to U.S.D
Contact: Katie Fletcher; (757) 629-2881; fax (757) 629-2361; katie.fletcher@nscorp.com
Internet: http://www.nscorp.com/nscportal/nscorp/Community/NS%20Foundation/
Sponsor: Norfolk Southern Foundation
P.O. Box 3040
Norfolk, VA 23514-3040

Norman Foundation Grants **3706**
The Norman Foundation supports efforts that strengthen the ability of communities to determine their own economic, environmental and social well-being, and that help people control those forces that affect their lives. These efforts may: promote economic justice and development through community organizing, coalition building and policy reform efforts; work to prevent the disposal of toxics in communities, and to link environmental issues with economic and social justice; link community-based economic and environmental justice organizing to national and international reform efforts. The following is considered when evaluating grant proposals: does the project arise from the hopes and efforts of those whose survival, well-being and liberation are directly at stake; does it further ethnic, gender and other forms of equity; is it rooted in organized, practical undertakings; and is it likely to achieve systemic change. In pursuing systemic change, the Foundation would hope that: the proposed action may serve as a model; the spread of the model may create institutions that can survive on their own; their establishment and success may generate beneficial adaptations by other political, social and economic institutions and structures. The Foundation provides grants for general support, projects, and collaborative efforts. We also welcome innovative proposals designed to build the capacity of social change organizations working in our areas of interest. Priority is given to organizations with annual budgets of under $1 million.
Requirements: Programs must be 501(c)3 tax-exempt organizations that focused on domestic United States issues. Prospective grantees should initiate the application process by sending a short two or three page letter of inquiry to the Program Director (fax, email or regular mail). There are no set deadlines, and letters of inquiry are reviewed throughout the year. The Foundation only accepts full proposals upon positive response to the letter of inquiry. The letter of inquiry should briefly explain: the scope and significance of the problem to be addressed; the organization's proposed response and (if appropriate) how this strategy builds upon the organization's past work; the specific demonstrable effects the project would have if successful, especially its potential to effect systemic (fundamental, institutional and significant) change; how the project promotes change on a national level and is otherwise related to the foundation's guidelines; the size of the organization's budget. All inquiries will be acknowledged and, if deemed promising, the Foundation will request a full proposal.
Restrictions: The Foundation does not make grants to individuals or universities; or to support conferences, scholarships, research, films, media and arts projects; or to capital funding projects, fundraising drives or direct social service programs, such as shelters or community health programs. The Foundation's grant making is restricted to U.S.-based organizations.
Amount of Grant: 5,000 - 30,000 U.S.D
Samples: Direct Action Welfare Group, Charleston, West Virginia, $20,000 - renewed support of statewide organizing and empowerment of people living in poverty in West Virginia (2011); Interfaith Action of Southwest Florida, Immokalee, Florida, $20,000 - renewed support for joint work with Coalition of Immokalee Workers in support of low-wage farmworkers in Florida (2011); Alaska Community Action on Toxics, Anchorage, Alaska, $30,000 - renewed support for work with tribal villages and other communities to eliminate environmental contaminants (2011).
Contact: June Makela, Program Director; (212) 230-9830; fax (212) 230-9849; norman@normanfdn.org
Internet: http://www.normanfdn.org/
Sponsor: Norman Foundation
147 East 48th Street
New York, NY 10017

North Carolina Biotechnology Center Event Sponsorship Grants **3707**
This popular Biotechnology Center grant promotes and supports events advancing the understanding or application of biotechnology to benefit North Carolina. Events must promote information sharing and personal interaction focused on biotechnology research, education, or business. The highest amount awarded is $3,000. Awards typically cover costs such as speaker fees, travel expenses, event-site rental and publicity.
Requirements: All award money is disbursed directly to the applicant's organization. The money must be used solely to support the event outlined in this application.
Restrictions: This sponsorship does not provide money either to promote a specific product to benefit a few companies or individuals. This sponsorship does not cover food or refreshments for event attendees.
Geographic Focus: North Carolina
Amount of Grant: Up to 3,000 U.S.D
Contact: Ginny DeLuca, (919) 549-8842; virginia_deluca@ncbiotech.org
Internet: http://www.ncbiotech.org/services_and_programs/grants_and_loans/biotechnology_event_sponsorship/index.html
Sponsor: North Carolina Biotechnology Center
15 T.W. Alexander Drive, P.O. Box 13547
Research Triangle Park, NC 27709-3547

North Carolina Biotechnology Center Grantsmanship Training Grants **3708**
The Grantsmanship Training Grant (GTG) program is a one-time funding opportunity that provides financial support for grant writing training activities for faculty and staff of North Carolina colleges and universities and other non-profits who are involved in research or programs related to the life sciences. Funds can be requested to support a variety of grantsmanship training activities, including workshops and other training events. The following individuals and groups would be eligible for training support under the GTG program: academic faculty members who seek funding related to life science research or life science-related economic development; research administration staff affiliated with departments or institutions that regularly submit proposals in life science research or life science-related economic development; and staff of non-profit organizations involved in activities related to biotechnology or the life sciences. A maximum of $5,000 may be requested based on the type of training activity proposed.
Requirements: Proposed programs and activities must clearly relate to biotechnology research, development, commercialization, education, or training. Priority will be giving to proposals requesting support for innovative programs or activities in biotechnology. Applicants are required to contribute a cash match equaling the requested amount (dollar-for-dollar).
Restrictions: Awards are not made to individuals.
Geographic Focus: North Carolina
Amount of Grant: Up to 5,000 U.S.D
Contact: Deborah De, Program Manager; (919) 549-8845; fax (919) 549-9710; deborah_de@ncbiotech.org
Internet: http://www.ncbiotech.org/gtg/
Sponsor: North Carolina Biotechnology Center
15 T.W. Alexander Drive, P.O. Box 13547
Research Triangle Park, NC 27709-3547

North Carolina Biotechnology Center Meeting Grants **3709**
This grant promotes and supports national and international meetings which advance the understanding or application of biotechnology and focus national and international attention on the North Carolina scientific community. Meetings must promote information sharing and personal interaction focused on biotechnology research, education, or business. The highest amount awarded is $10,000. Awards typically cover costs such as speaker fees, travel expenses, event-site rental and publicity.
Requirements: All award money is given to the applicant's organization. The money must be used solely to support the event outlined in this application.
Restrictions: This sponsorship does not provide money either to promote a specific product to benefit a few companies or individuals. This sponsorship does not cover food or refreshments for event attendees.
Geographic Focus: North Carolina
Amount of Grant: Up to 10,000 U.S.D
Contact: Ginny DeLuca, (919) 549-8842; virginia_deluca@ncbiotech.org
Internet: http://www.ncbiotech.org/services_and_programs/grants_and_loans/BiotechnologyMeetingGrant.html
Sponsor: North Carolina Biotechnology Center
15 T.W. Alexander Drive, P.O. Box 13547
Research Triangle Park, NC 27709-3547

North Carolina Biotechnology Center Regional Development Grants **3710**
The goal of the Regional Development Grant Program (PDF) is to build capacity through collaborative projects, providing a foundational resource for biotechnology development in the community that was not there previously and that the community could not have achieved without Biotechnology Center funding. Preproposals are required: cycle 1 preproposals are due no later than January 28, while cycle 2 preproposals should be received by August 20. The deadlines for final proposals are April 1 (cycle 1) and October 15 (cycle 2).
Requirements: Proposed programs and activities must clearly relate to biotechnology research, development, commercialization, education, or training. Priority will be giving to proposals requesting support for innovative programs or activities in biotechnology. Only North Carolina academic institutions, non profit research organizations, or businesses may apply.
Restrictions: Awards are not made to individuals.
Geographic Focus: North Carolina
Date(s) Application is Due: Jan 28; Apr 1; Aug 20; Oct 15
Amount of Grant: Up to 75,000 U.S.D
Contact: Deborah De, Program Manager; (919) 549-8845; fax (919) 549-9710; deborah_de@ncbiotech.org
Internet: http://www.ncbiotech.org/services_and_programs/grants_and_loans/regional_development/index.html
Sponsor: North Carolina Biotechnology Center
15 T.W. Alexander Drive, P.O. Box 13547
Research Triangle Park, NC 27709-3547

North Carolina Biotechnology Centers of Innovation Grants **3711**
The Centers of Innovation Program is bringing together North Carolina's best scientific and technical minds in the life sciences. This program is designed to focus the state's efforts in biotechnology research, development and commercialization in targeted industrial sectors important to economic development and job creation. The North Carolina Biotechnology Center (NCBC) will work with university researchers, technology transfer offices, industrial partners, non-profit stake holders as well as regional and state-wide community leaders to establish nine Centers of Innovation (COI). Initial Centers of Innovation will complement efforts already under way in the state to align academic and industrial resources. The Phase I award is a 12-month $100,000 planning

grant that will allow the development of a business plan and organization plan. The Phase II award is funding for the implementation of the business plan. Only invited applicants may apply. Potential applicants will need to demonstrate that they have established a cohesive academic-industry consortium in order to receive an invitation to submit an application.
Geographic Focus: North Carolina
Amount of Grant: Up to 2,400,000 U.S.D
Contact: Mary Beth Thomas, (919) 541-9366; fax (919) 549-9710; marybeth_thomas@ncbiotech.org
Internet: http://www.ncbiotech.org/services_and_programs/COI/index.html
Sponsor: North Carolina Biotechnology Center
15 T.W. Alexander Drive, P.O. Box 13547
Research Triangle Park, NC 27709-3547

North Carolina Biotechnology Center Technology Enhancement Grants 3712
The Technology Enhancement Grant (TEG) Program provides funding to North Carolina Universities or other research institutions in the state through their respective technology transfer offices. Under this program, awards of up to $50,000 are available to fund research studies determined to be critical to enhancing the university's licensing position for specified technologies with commercial opportunity. Eligible technologies are those for which intellectual property rights are assigned to the university or institution and have not been committeed through license, option or letter-of-intent to license to any third party commercial interest at the time of award. Funds are intended to support an approved research project designed to yield a significant technical milestone in the evolution of a technology from basic science to licensable commercial product or service. Examples of such projects include proof-of-concept, validation, lead identification and optimization using a proprietary platform technology, etc. Due to legal constraints, funds provided under the TEG program can only be used to offset direct costs associated with the project.
Requirements: Technology transfer offices associated with any North Carolina university or not-for-profit research institution are eligible and encouraged to apply.
Restrictions: Prohibited charges include (but are not limited to): purchase or lease of real estate, building construction or renovations; salaries for administrative personnel; overhead or indirect costs; and atenting costs.
Geographic Focus: North Carolina
Amount of Grant: Up to 50,000 U.S.D
Contact: Dr. Rob Lindberg, (919) 541-9366; fax (919) 549-9710; rob_lindberg@ncbiotech.org
Internet: http://www.ncbiotech.org/services_and_programs/grants_and_loans/technology_enhancement/index.html
Sponsor: North Carolina Biotechnology Center
15 T.W. Alexander Drive, P.O. Box 13547
Research Triangle Park, NC 27709-3547

North Carolina Community Development Grants 3713
The foundation supports nonprofit organizations in North Carolina to enrich the quality of life in the state, with special concern for rural regions and Cabarrus County. Grants are awarded primarily for healthcare, higher education, and community service. Other more limited areas of interest are arts, culture, historic preservation, religion, and the environment. Grants generally will be awarded for capital improvements, challenge opportunities, or special purposes rather than for operating purposes. Grant requests are considered on a quarterly basis.
Requirements: IRS 501(c)3 organizations serving North Carolina communities are eligible. Churches and governmental agencies also may apply.
Restrictions: Normally, the foundation does not make multiyear grants or frequent grants to the same organization (except for select, historic relationships); or provide operating funds, start-up funding, or lead gifts. The foundation does not make grants to individuals, to endowment funds, for loans, or for scholarships or fellowships.
Geographic Focus: North Carolina
Date(s) Application is Due: Jan 5; Apr 5; Jul 5; Oct 5
Amount of Grant: 10,000 - 30,000 U.S.D
Samples: Lenoir-Rhyne College (Hickory, NC)—for equipment and technology upgrades, $156,000; Mars Hill College (Mars Hill, NC)—to purchase two boilers and a fire-alarm system, $150,000.
Contact: Frank Davis, Executive Director; (704) 786-8216; fax (704) 785-2051
Internet: http://www.thecannonfoundationinc.org/guidelines.asp
Sponsor: Cannon Foundation
P.O. Box 548
Concord, NC 28026-0548

North Carolina Community Foundation Grants 3714
The community foundation exists to meet the needs of nonprofit organizations and improve the lives of citizens of North Carolina counties, including Allegheny, Caldwell, Cherokee, Clay, Craven, Eastern Bank of Cherokee Indians, Franklin Community, Franklin, Graham, Harnett, Haywood, Jackson, Johnston, Madison, Montgomery, Moore, Pender, Pitt, Randolph, Rockingham, Swain, Wake, Watauga, and Wilkes. Support is available for programs that enrich the quality of life in the areas of civic affairs, education, health, religion, social services, arts, and the conservation and preservation of environmental, historical preservation, and cultural resources. Grants will be awarded to support programs and projects, endowments, conferences and seminars, and consulting services. Grants are awarded in local communities; use the Affiliate Webpage link to contact the office serving your area.
Requirements: Nonprofit organizations throughout North Carolina are eligible to apply.
Geographic Focus: North Carolina
Contact: Jennifer Tolle Whiteside, President; (800) 201-9533 or (919) 828-4387; fax (919) 828-5495; jtwhiteside@nccommunityfoundation.org
Internet: http://nccommunityfoundation.org/05_grants_available.php
Sponsor: North Carolina Community Foundation
4601 Six Forks Road, Suite 524
Raleigh, NC 27609

North Carolina GlaxoSmithKline Foundation Grants 3715
The foundation supports activities primarily in North Carolina that help meet the needs of today's society and future generations by funding programs that emphasize the understanding and application of health, science, and mathematics at all educational and professional levels. Although providing seed funds for new and worthwhile educational programs is the foundation's primary focus, requests for funding of ongoing projects will also receive consideration. Types of support include operating budgets, professorships, and conferences and seminars. Proposals may be submitted for either one-year or multi-year funding with a maximum of five years' duration. The foundation's board of directors meets four times a year to consider and award grants. Completed applications must be received by the listed deadline dates. If the stated deadline falls on a weekend or holiday, the next business day serves as the official deadline. All applicants will be notified of the board's decisions within 15 days of the board meeting.
Requirements: The foundation makes grants only to North Carolina 501(c)3 tax-exempt organizations and institutions or to governmental agencies.
Restrictions: Grants are not made to individuals for construction or restoration projects, or for international programs unless specifically exempted by the board. Funds are not ordinarily provided to programs that benefit a limited geographic region.
Geographic Focus: North Carolina
Date(s) Application is Due: Jan 1; Apr 1; Jul 1; Oct 1
Samples: North Carolina Center for Public Policy Research, $446,000; The Carolina Ballet, Healthier Living Through the Arts, $650,000; Prevent Blindness North Carolina, $100,000.
Contact: Marilyn Foote-Hudson, Executive Director; (919) 483-2140; fax (919) 315-3015; community.partnership@gsk.com
Internet: http://us.gsk.com/html/community/community-grants-foundation.html
Sponsor: North Carolina GlaxoSmithKline Foundation
5 Moore Drive, P.O. Box 13398
Research Triangle Park, NC 27709

North Central Health Services Grants 3716
North Central Health Services (NCHS) was created in 1984 to serve as the parent company of a family of corporations which included Lafayette Home Hospital, Home Hospital Foundation, and Service Frontiers Incorporated. Today, NCHS is a medical services company whose primary purpose is to operate an ambulatory surgery center, where sterilization services are made available to the Greater Lafayette community. NCHS also is a grant making organization providing primarily capital grants to 501(c)3 organizations serving the citizens of Benton, Carroll, Clinton, Fountain, Montgomery, Tippecanoe, Warren, and White counties in Indiana. Grants are awarded to organizations for projects that relate to Health, and Healthy Communities. NCHS prefers to fund projects that have a significant potential for positive impact on the community. NCHS encourages collaboration among agencies and organizations that efficiently utilize resources, and provide services that are not duplicative of existing programs. NCHS provides grants that will: advance technology and scientific development in health care; develop and promote community health education; provide direct, individual social welfare and human service; promote a Healthy Community; improve the quality of life for all the communities people.
Requirements: The first step for an organization to apply for a grant is to submit a written letter of inquiry. This letter should include a concise description of the organization, a statement of the problem or need being addressed, and an explanation of the project, estimated total costs, and the amount being requested from NCHS. The letter should be addressed to: North Central Health Services, Inc., Attn: Rita Smith, Program Director, P.O. Box 528, Lafayette, In. 47902. Upon receipt of the letter of inquiry, initial review will determine whether or not the request meets the funding focus and priorities of NCHS. You will be promptly notified in writing of the initial determination, and informed of application opportunities.
Restrictions: Beginning in 2008, NCHS implemented the following grant categories and cycles. Grant requests and cycles: letter of inquiry requesting funds in excess of $250,000, accepted May 1 to August 15 of each calendar year. If approved, funding would be available for distribution after January 1 of the following year; letter of inquiry requesting funds of $100,000 to $250,000 will be accepted throughout the calendar year. If approved, funding would generally be available for distribution within 90 days of board approval; letter of inquiry requesting funds of $25,000 to $100,000 will be accepted throughout the calendar year. If approved, funding would generally be available within 60 days of board approval; letter of inquiry requesting funds of less than $25,000 will be accepted throughout the calendar year. If approved, funding would generally be available within 60 days of approval.
Geographic Focus: Indiana
Samples: Lauramie Township EMS - Purchase Zoll E series AED to update equipment for EMS basic life support ambulance service, $8,693.00; St. Elizabeth Regional Health - Gifts for hospitalized children, $4,155.00; Indiana Regular Baptist Youth Camp, Inc. - Build a new Health Center for Twin Lakes Camp & Confererence Center, $150,000.00.
Contact: Rita Smith, Program Director; (765) 423-1604
Internet: http://www.nchsi.com/grantmaking.cfm
Sponsor: North Central Health Services
201 Main Street, Suite 606, P.O. Box 528
Lafayette, IN 47902

North Dakota Community Foundation Grants 3717
The Foundation serves North Dakota communities statewide with the goal of improving the quality of life for the state's citizens. The Foundation does not have a narrow area of focus. Each project is reviewed on its own merits with an emphasis on helping applicants who have limited access to other sources of funding. Applicants may submit a concise letter of request not to exceed two pages describing the organization, the project, the approximate project cost, and the amount requested from the Foundation. If the Board is interested in additional information, formal application materials will be sent.
Requirements: Eligible applicants must be 501(c)3 organizations in North Dakota.
Restrictions: The Foundation accepts applications by mail but not by fax or email. Only one request per agency per year may be submitted. The following is not funded: grants to organizations and projects that exist to influence legislation, carry on propaganda, participate in political campaigns, or which threaten to cause significant controversy or divisiveness; grants to individuals; and multi-year grant commitments. Low priority is given to the following: projects already substantially supported by government, or which in the opinion of the Board, can and should be provided for by taxes; grants for sectarian projects; grants to national organizations; and organizations which field substantial fund-raising each year with paid and volunteer staff.
Geographic Focus: North Dakota
Date(s) Application is Due: Aug 15
Samples: Save the Hens Foundation, Inc., Sheldon, North Dakota, $8,500 - community programming (2010); Park River Volunteer Fire Department, Park River, North Dakota, $7,500 - health promotion (2010); and Kiddie Korner Preschool and Day Care, Park River, North Dakota, $5,500 - children's programming (2010).
Contact: Kevin Dvorak; (701) 222-8349; fax (701) 222-8349; kdvorak@ndcf.net
Internet: http://www.ndcf.net/Information/GrantGuidelines.asp
Sponsor: North Dakota Community Foundation
309 North Mandan Street, Suite 2, P.O. Box 387
Bismarck, ND 58502-0387

Northeast Utilities Foundation Grants 3718
The foundation supports community efforts to improve education, social and economic conditions, and protect the quality of the environment. Areas of focus include education, civic and community, human services, environment, and culture and arts. Grant requests are evaluated using criteria such as organizational and financial management, program objectives, diverse number of people who will benefit, existing service, funding sources and stability, organizational history, and measures of success. The board meets in February, April, August, and November.
Restrictions: Grants do not support individuals; corporations that do not qualify as charitable organizations according to IRS codes; private foundations; projects benefiting limited groups—religious, fraternal, political; athletic outings; reducing or eliminating a preexisting debt; or health care issues or organizations.
Geographic Focus: Connecticut, Massachusetts, New Hampshire
Date(s) Application is Due: Jan 10; Apr 11; Jul 11; Oct 10
Samples: Saint Joseph College, West Hartford, CT, $500,000 — to reconfigure two science laboratories, and to establish the Fresh Water Institute.
Contact: Patricia McCulloughPresident, Executive Director; (888) 682-4639 or (860) 665-5822; fax (860) 721-4331; NUFoundation@NU.com
Internet: http://northeastutilitiesfoundation.org/what/index.html
Sponsor: Northeast Utilities Foundation
P.O. Box 270
Hartford, CT 06102-0270

Northern Chautauqua Community Foundation Community Grants 3719
The foundation awards grants to nonprofit, tax-exempt organizations located in Northern Chautauqua County, New York. Priority will be given to programs representing innovative and efficient approaches to serving community needs and opportunities, projects that assist citizens whose needs are not met by existing services, projects that expect to test or demonstrate new approaches and techniques in the solutions of community problems, and projects that promote volunteer participation and citizen involvement in the community. Consideration will be given to the potential impact of the request and the number of people who will benefit. Seed grants will be awarded to initiate promising new programs in the foundation's field of interest as well as challenge grants to encourage matching gifts. Except in unusual circumstances, grants are approved for one year at a time. Contact the foundation office for an application.
Requirements: IRS 501(c)3 organizations in Northern Chautauqua County, New York, are eligible.
Restrictions: Areas generally not funded are capital campaigns to establish or add to endowment funds, general operating budgets for existing organizations, publication of books, conferences, or annual fund-raising campaigns.
Geographic Focus: New York
Date(s) Application is Due: Mar 23; Sep 21
Amount of Grant: 12,500 U.S.D
Contact: Diane E. Hannum, Executive Director; (716) 366-4892; fax (716) 366-3905; dhannum@nncfoundation.org or info@nccfoundation.org
Internet: http://www.nccfoundation.org/Grantseekers/ApplyforaGrant/tabid/256/Default.aspx
Sponsor: Northern Chautauqua Community Foundation
212 Lake Shore Drive West
Dunkirk, NY 14048

Northern New York Community Foundation Grants 3720
The foundation is looking for innovative programs that address problems to be solved, or opportunities to be seized, in the northern New York area. Grants are made only for capital items and for seed money for new agencies, or new projects by established agencies.
Requirements: Grants are made to nonprofit, tax-exempt organizations in or serving Jefferson and/or Lewis counties in New York State. Governmental units are also eligible provided that the purpose of the grant goes beyond the expected limits of government service. The foundation does not make grants to churches and religious organizations except in cases where projects clearly benefit the entire community. Applicants should contact the office to provide a brief summary of the project and needs. The applicant will be notified of grant award or denial by letter after the board of directors meets, which is generally one month following application deadline dates. Grant requests under $5,000 may be approved at the committee level. Larger grant requests are referred to the board of directors with a committee recommendation.
Restrictions: Individuals are ineligible. The foundation does not fund operations or pay off deficits.
Geographic Focus: New York
Date(s) Application is Due: Jan 30; Apr 24; Aug 28; Oct 23
Samples: 4th Qtr 2008: The Handweaving Museum and Arts Center - $35,000 to help fund a new Director of Development position; Northern Area Health Education Center (NAHEC) - $3,575 for student internships in technical fields; Ogdensburg Command Performances - $7,000 to match a Land Trust Alliance grant to develop a Point Peninsula conservation plan; Pendragon Theatre - $2,784 for three Shakespeare performances at Jefferson Community College; Trinity Episcopal Church, Watertown - $2,500 for a Trinity Concert Series performance of Handel's Messiah; Watertown Housing Authority - $5,000 to help fund a new playground for East Hills and Meadowbrook family housing projects; American Red Cross of Northern New York - $5,306 for access to donor management software.
Contact: Alex Velto; (315) 782-7110; fax (315) 782-0047; info@nnycf.org
Internet: http://www.nnycf.org/grantguidelines.asp?mm=5
Sponsor: Northern New York Community Foundation
120 Washington Street, Suite 400
Watertown, NY 13601

Northern Trust Company Charitable Trust and Corporate Giving Program 3721
Contributions from the Charitable Trust support program, general operating, capital, or endowment priorities. Grants usually are confined to programs and agencies that focus on advancing the well-being of disadvantaged women and children, and people with disabilities. Priority is given to programs or agencies that serve the following Chicago neighborhoods: Chatham, Englewood, Humboldt Park, Logan Square, Loop, Washington Park, and West Town. Applications, including all material specified in the checklist must be received in the Community Affairs office before 5:00 pm on the deadline date. Deadlines for requests for funding are determined by area of service, as follows: social welfare—January 13 and August 12, education-—May 13; and arts and culture—May 13.
Requirements: Nonprofit organizations in Illinois are eligible. A proposal should include a short cover letter, an operating budget for the current year, a copy of the agency's most recent audited financial statement, a list of members of the agency's governing board, a list of Chicago-based corporate and foundation contributors and amounts each has contributed in the last calendar year, and a letter from the IRS indicating tax-exempt status.
Geographic Focus: Illinois
Date(s) Application is Due: Jan 13; May 13; Aug 12
Amount of Grant: 2,500 - 5,000 U.S.D
Samples: Parents United for Responsible Education (Chicago, IL)—for operating support, $5000; Chicago Reporter (Chicago, IL)—for operating support, $3000.
Contact: Dawn McGovern, Program Contact; (312) 444-4059
Internet: http://www.northerntrust.com/aboutus/community/charitable/index.html
Sponsor: Northern Trust Company
50 S LaSalle Street, Suite L-7
Chicago, IL 60675

North Georgia Community Foundation Grants 3722
The foundation serves Banks, Dawson, Fannin, Forsyth, Franklin, Habersham, Hall, Hart, Jackson, Lumpkin, Rabun, Stephens, Towns, Union, and White counties. The foundation looks favorably at efforts which promise to affect a broad constituency in its service area. The foundation is interested in those organizations which can demonstrate that they have planned their projects in light of overall community need, and that similar projects by other organizations do not duplicate the proposed services. A request which asks the foundation to provide a portion of the support for a project generally will receive greater priority than one which looks to it as the single funding source. The foundation is especially interested in leveraging additional income from other sources.
Requirements: Grants are made only to Georgia nonprofit tax-exempt organizations in the counties listed for charitable purposes, unless donor-advised funds are designated to qualifying entities outside the foundation's area of emphasis.
Restrictions: Funding is not generally provided for annual fund campaigns, lobbying, operating support other than start-up, or to individuals.
Geographic Focus: Georgia
Amount of Grant: Up to 3,000 U.S.D
Contact: Program Officer; (770) 535-7880; fax (770) 503-0439; info@ngcf.org
Internet: http://www.ngcf.org/main.asp
Sponsor: North Georgia Community Foundation
615F Oak Street, Suite 1300
Gainesville, GA 30501

Northrop Grumman Corporation Grants 3723
Northrop Grumman is committed to supporting communities throughout the U.S., especially those where its employees live and work. The Contributions Program seeks to address critical issues and needs by providing financial assistance to accredited schools and 501(c)3 nonprofit organizations. The majority of these contributions address education, services for veterans and the military, health and human services, and the environment.
Requirements: The following information must be included in the grant request: Tax Exempt Number; non-profit contact name, title, address, phone number, fax number; brief history of the organization, mission statement, goals and objectives; type and scope of services offered, and the geographical area served; specific details as to how requested funds would be used; project budget and amount requested; demographic impact; current operating budget including latest financial statement; list of other corporate funders; list of directors and/or officers, and their affiliations; Northrop Grumman employee sponsor, if applicable (name and phone number); and contact information.
Restrictions: Grants are not awarded to: religious organizations; political groups; fraternal organizations; individuals; athletic groups or activities, including charity-benefit sporting events; charter schools, unless they have open enrollment and hold the same standards as public schools; bands or choirs; capital campaigns; organizations providing services primarily to animals; communities outside of the United States; or organizations whose programs discriminate based on race, color, age, sex, religion, national origin, sexual orientation, disability, veteran status or any other characteristic protected by law.
Date(s) Application is Due: Apr 30; Sep 30
Contact: Cheryl Horn, cheryl.horn@ngc.com
Internet: http://www.northropgrumman.com/corporate-responsibility/corporate-citizenship/contribution-guidelines.html
Sponsor: Northrop Grumman Corporation
1840 Century Park East
Los Angeles, CA 90067

Northwest Airlines KidCares Medical Travel Assistance 3724
The program provides air travel to children age 18 and younger who are unable to receive treatment in their home area. The KidCares program is fueled by generous mileage donations of Northwest's WorldPerks members. Availability of the program is based on that donated WorldPerks mileage, and consideration of each request is based on the guidelines below. Each application received must be evaluated on a case-by-case basis.
Requirements: Guidelines are as follows: (1) Only one round-trip per family is permitted. (2) A letter from the child's physician must accompany the application, stating: (a) patient's diagnosis; (b) confirmation of the need for travel; (c) the patient has medical clearance for air travel; (d) dates of appointments/surgery; and (e) length of stay. (3) The patient must be 18 years old or younger at the time the travel request is made. (4) The medical treatment for which the child is traveling must be unavailable in the child's home location. (5) KidCares travel can only be provided by Northwest Airlines or one of its airline partners. (6) Families demonstrating financial need are given priority consideration for KidCares travel. (7) One adult companion may accompany the child on KidCares travel at no additional cost. Applications may be found on Northwest's Web site, nwa.com at the following location http://www.nwa.com/corpinfo/aircares/about/KidCares_Application.pdf. You may also call (612) 726-4206 and leave detailed information regarding where the application should be sent and to whom.
Restrictions: KidCares travel is not available to families with insurance coverage that covers the cost of travel for medical treatment. KidCares travel is not eligible for Northwest Airlines WorldPerks bonus mileage credit. KidCares travel is not provided to St. Jude Children's Research Hospital, which is covered by a separate Northwest Airlines travel program.
Contact: Community Relations Manager; (612) 726-4206; aircares@nwa.com
Internet: http://www.nwa.com/corpinfo/aircares/feature/
Sponsor: Northwest Airlines
2700 Lone Oak Parkway, Department A1300
St. Paul, MN 55121

Northwestern Mutual Foundation Grants 3725
The people of Northwestern Mutual have a long history of community involvement. To carry out its corporate contributions program, Northwestern Mutual created a private foundation, The Northwestern Mutual Foundation. The Northwestern Mutual Foundation focuses its funding in Milwaukee, where the company is headquartered. Therefore, applying for a grant from the Northwestern Mutual Foundation is limited to Milwaukee-area nonprofit organizations with programs centered on Education, Health & Human Services or Arts & Culture.
Requirements: Northwestern Mutual Foundation does not accept unsolicited grant requests. Groups new to the Foundation are asked to submit a written letter of intent, describing the organization, funding request, and provide information about the program for which you are seeking support. The Foundations team will review your letter of intent and, if they are interested in learning more about your project, you will receive an application form. Groups that have an active relationship with the Foundation and currently receive funding are allowed to apply annually for support. Once an organization submits an application, the review process takes about 90 days. The Foundation review's Milwaukee-area grants on a monthly basis.
Restrictions: Milwaukee-area nonprofits are eligible to apply. The Foundation defines the Milwaukee-area as Milwaukee County, with limited programmatic grants in Waukesha, Racine and Ozaukee Counties.
Geographic Focus: Wisconsin
Amount of Grant: 10,000 - 800,000 U.S.D
Samples: Boys and Girls Clubs of Greater Milwaukee, Milwaukee, WI, $800,000; United Performing Arts Fund, Milwaukee, WI, $605,000; Scholarship America, Saint Peter, MN, $205,425;
Contact: Edward J. Zore , President; (414) 271-1444
Internet: http://www.nmfn.com/tn/aboutus—fd_intro
Sponsor: Northwestern Mutual Foundation
720 East Wisconsin Avenue
Milwaukee, WI 53202-4797

Northwest Minnesota Foundation Women's Fund Grants 3726
Formed in 1997 as a tax exempt charitable component of the Northwest Minnesota Foundation (NMF), the Women's Fund serves as a catalyst for improving the quality of life for women and girls by promoting ideas and supporting programs for the continued strengthening and empowerment of all women. The Women's Fund will consider grant requests of up to $2,500 addressing one or more of the following areas meeting the needs of females: cultural interactions; economic development; education; healthcare; and housing.
Requirements: Minnesota counties served by NMF are Beltrami, Clearwater, Hubbard, Kittson, Lake of the Woods, Mahnomen, Marshall, Norman, Pennington, Polk, Red Lake, and Roseau. Priority is given to applications supporting the following goals: developing entrepreneurial and economic opportunities for women; supporting programs that lead to a safer environment for women and girls; building networks of women in leadership positions throughout the region; and encouraging and equipping women and girls to achieve their full potential.
Geographic Focus: Minnesota
Date(s) Application is Due: May 1
Amount of Grant: 2,500 U.S.D
Contact: Lisa Peterson, Communications Director; (800) 659-7859; lisap@nwmf.org
Internet: http://www.nwmf.org/component-funds/community-fund-websites/nw-mn-womens-fund/womens_fund_grants_and_programs.html
Sponsor: Northwest Minnesota Foundation
4225 Technology Drive NW
Bemidji, MN 56601

Northwood University/Alden B. Dow Creativity Center Summer Residency 3727
The program provides a 10-week residency at the university from mid-June through mid-August to allow each successful applicant to pursue innovative ideas having the potential for impact in his or her field. Project ideas must be specific with accomplishable goals for the 10-week period. Selection is based on newness and uniqueness of idea. Fellows are expected to be capable of independent research on their proposed projects. From education and literature to health care and chemistry, physics, engineering, and architecture and the arts, no creative area will be overlooked when a project idea may result in a significant advancement. Foreign citizens are also invited, provided they can communicate in written and spoken English. Project budgets are approved on an individual basis; however, ideas involving heavy developmental costs are discouraged. The grant includes travel within the United States (foreign applicants must provide transportation to the United States), room and board, and a stipend for personal expenses and/or project materials. Application fee is $10; forms and instructional brochures are available upon request.
Date(s) Application is Due: Dec 31
Amount of Grant: 750 U.S.D
Contact: Dr. Grover Proctor, Executive Director; (989) 837-4478; fax (989) 837-4468; creativity@northwood.edu
Internet: http://www.northwood.edu/abd
Sponsor: Northwood University
4000 Whiting Drive
Midland, MI 48640-2398

Norwin S. and Elizabeth N. Bean Foundation Grants 3728
The Norwin S. and Elizabeth N. Bean Foundation was established in 1967 as a general purpose foundation to serve the communities of Manchester and Amherst and, consistent with the wishes of the founders, grants are made in the fields of arts and humanities, education, environment, health, human services, and public/society benefit.
Requirements: Applications are accepted from nonprofit 501(c)3 organizations and municipal and public agencies serving the communities of Manchester and Amherst, New Hampshire. Priority consideration is given to organizations operating primarily in those two communities. However, the Foundation will consider applications from statewide or regional organizations which provide a substantial and documented level of service to Manchester and Amherst.
Restrictions: The foundation does not make grants to individuals or provide scholarship aid.
Geographic Focus: New Hampshire
Date(s) Application is Due: Apr 1; Sep 1; Dec 1
Contact: Kathleen Cook, Grant Manager; (603) 493-7257; KCook@BeanFoundation.org
Internet: http://www.beanfoundation.org/grants/bean-foundation-grant/application-criteria.aspx
Sponsor: Norwin S. and Elizabeth N. Bean Foundation
40 Stark Street
Manchester, NH 03301

Notsew Orm Sands Foundation Grants 3729
The Notsew Orm Sands Foundation was established in 1995 via an initial donation from Charles Burnett, III, a British-born race car driver who set the steam-powered land speed record in 2009. Giving is on a national and international basis, with some emphasis on Houston, Texas, and the United Kingdom. Primary fields of interest are: higher education, hospitals, human services, medical research, and Protestant agencies

and churches. There are no specific deadlines, and applicants should submit a detailed description of project and amount of funding requested, along with a copy of current year's organizational budget and/or project budget.
Restrictions: No grants are given to individuals.
Amount of Grant: 250 - 75,000 U.S.D
Samples: Baylor College of Medicine, Houston, Texas, $75,000; University of Saint Thomas, Houston, Texas, $61,000; Yale University, New Haven, Connecticut, $55,636; Centurion Ministries, Princeton, New Jersey, $20,000.
Contact: Charles Burnett, III, President and Secretary; (281) 497-0744
Sponsor: Notsew Orm Sands Foundation
2470 S. Dairy Ashford Street, Suite 802
Houston, TX 77077-5716

NSERC Brockhouse Canada Prize for Interdisciplinary Research in Science and Engineering Grant 3730

The Brockhouse Canada Prize for Interdisciplinary Research in Science and Engineering recognizes outstanding Canadian teams of researchers from different disciplines who have combined their expertise to produce achievements of outstanding international significance in the natural sciences and engineering. The prize, accompanied by a team research grant of $250,000, reflects NSERC's commitment to supporting Canadian research through strategic investments in people, discovery and innovation. The grant supports the direct costs of university-based research and/or the enhancement of research facilities. The grant may be distributed in one lump sum or up to five instalments, depending on the needs of the recipients.
Requirements: Research teams nominated for the Brockhouse Canada Prize must have at least two members who are independent researchers, one of whom must hold an NSERC grant. The team can be part of an international effort, but the majority of the nominated team members must be employed at a Canadian university or public or private organization. NSERC recognizes that teams may change between the time of the specific research achievements and the time of nomination. Nominations will be accepted when changes have occurred but only as long as the core of the team remains intact. Contributions must be primarily in the natural sciences and engineering, and of an interdisciplinary and collaborative nature.
Restrictions: Any Canadian citizen may nominate a research team for the Brockhouse Canada Prize. Self nominations will not be accepted. Current NSERC Council members cannot be nominated. Since the prize is for interdisciplinary research, nominators should consult NSERC's Guidelines for the Preparation and Review of Applications in Interdisciplinary Research. Nominators should also consult NSERC's Guidelines for the Preparation and Review of Applications in Engineering and the Applied Sciences. See website for deadline & hyper links to above mentioned guidelines.
Geographic Focus: All States, Canada
Date(s) Application is Due: Jun 1
Amount of Grant: 250,000 U.S.D
Contact: Administrator; (613) 995-5829; fax (613) 992-5337; brockhouse@nserc-crsng.gc.ca
Internet: http://www.nserc-crsng.gc.ca/Prizes-Prix/Brockhouse-Brockhouse/Index-Index_eng.asp
Sponsor: Natural Sciences and Engineering Research Council of Canada
350 Albert Street
Ottawa, ON K1A 1H5 Canada

NSF-NIST Interaction in Chemistry, Materials Research, Molecular Biosciences, Bioengineering, and Chemical Engineering 3731

This program is intended to facilitate interactions between faculty and students supported by the National Science Foundation (NSF) and scientists at the National Institute of Standards and Technology's (NIST) Chemical Science and Technology Laboratory (CSTL) and Materials Science and Engineering Laboratory (MSEL), including the NIST Center for Neutron Research (NCNR). Chemistry, materials research, molecular biology, bioengineering, and chemical engineering are centralized at NIST in these laboratories. Support may be requested for supplements to existing NSF awards to provide the opportunity for faculty and students to participate in research at NIST facilities. Support may be requested as supplements to existing NSF awards for travel expenses and per diem associated with work at these NIST laboratories for faculty, students and other personnel associated with the NSF/NIST activity. It is estimated that 10-20 supplements will be made annually, up to $20,000 each.
Requirements: Support may be requested only by organizations that have currently active awards in any of the participating NSF Divisions, viz. the Division of Bioengineering and Environmental Systems (BES), the Division of Chemical and Transport Systems (CTS), the Division of Chemistry (CHE), the Division of Materials Research (DMR), and the Division of Molecular and Cellular Biosciences (MCB). Only faculty who are principal investigators on current NSF awards from BES, CHE, CTS, DMR, or MCB are eligible to apply for supplements. Standard NSF Grant Proposal Guidelines apply (see http://www.nsf.gov/publications/pub_summ.jsp?ods_key=gpg).
Restrictions: Indirect costs (F&A) are not allowed.
Amount of Grant: Up to 20,000 U.S.D
Contact: Dr. William F. Koch, Deputy Director; (301) 975-8301; fax (301) 975-3845; william.koch@nist.gov
Internet: http://www.nsf.gov/publications/pub_summ.jsp?ods_key=nsf03568
Sponsor: National Science Foundation
4201 Wilson Boulevard
Arlington, VA 22230

NSF Accelerating Innovation Research 3732

To accelerate the process of innovation, NSF is undertaking two related, new activities. The first will encourage the translation of the numerous, technologically-promising, fundamental discoveries made by NSF researchers, while drawing upon and building the entrepreneurial spirit of the researchers and students. The second activity will foster connections between an existing NSF innovation research alliance (including consortia such as Engineering Research Centers (ERC), Industry University Cooperative Research Centers (I/UCRC), Partnerships for Innovation (PFI), Science and Technology Centers (STC), Nanoscale Science and Engineering Centers (NSEC), Materials Research Science and Engineering Centers (MRSEC) grantees) and other institutions, whose complementary focus will spur the development of discoveries into innovative technologies through collaboration. Both of these activities are designed to strengthen the U.S. innovation ecosystem.
Requirements: For the Technology Translation Competition, the Principal Investigator or a co-PI must be a current or prior NSF awardee and a faculty member at a U.S. college or university at the time of award in the current competition. Subject technology must be derived from a discovery research project already conducted or initiated by the National Science Foundation. For the Research Alliance Competition, one of the partners must be an NSF funded innovation research alliance (including Centers for Analysis and Synthesis, Centers for Chemical Innovation, Engineering Research Centers, Industry/University Collaborative Research Centers, Materials Research Science and Engineering Centers, Nanoscale Science and Engineering Centers, Science and Technology Centers, and Science of Learning Centers). The research alliance must be active at the time of award in the current competition. The other partner(s) may be another research entity (either NSF-funded, other government agency funded, or privately funded), a small business consortium, or a local or regional innovation entity.
Date(s) Application is Due: Feb 1
Amount of Grant: 350,000 - 1,000,000 U.S.D
Contact: Rathindra DasGupta, Program Director; (703) 292-8353; fax (703) 292-9057; rdasgupt@nsf.gov
Internet: http://www.nsf.gov/funding/pgm_summ.jsp?pims_id=503553
Sponsor: National Science Foundation
4201 Wilson Boulevard
Arlington, VA 22230

NSF Alan T. Waterman Award 3733

The annual award recognizes an outstanding young researcher in any field of science or engineering supported by the National Science Foundation. In addition to a medal, the awardee receives a grant of $500,000 over a three year period for scientific research or advanced study in the mathematical, physical, biological, engineering, social, or other sciences at the institution of the recipient's choice.
Requirements: Candidates must be U.S. citizens, normally 35 years of age or younger, or not more than seven years beyond receipt of the PhD degree by December 31 of the year nominated. Candidates should have demonstrated exceptional individual achievements in scientific or engineering research of sufficient quality to place them at the forefront of their peers. Criteria include originality, innovation, and significant impact on the field.
Date(s) Application is Due: Dec 31
Amount of Grant: 500,000 U.S.D
Samples: David Charbonneau, Professor of Astronomy, Harvard University; Terence Tao, Professor of Mathematics, University of California, Los Angeles; Peidong Yang, Associate Professor of Chemistry, University of California, Berkeley.
Contact: Mayra N. Montrose, Program Contact; (703) 292-8040; fax (703) 292-9040; mmontros@nsf
Internet: http://www.nsf.gov/od/waterman/waterman.jsp
Sponsor: National Science Foundation
4201 Wilson Boulevard
Arlington, VA 22230

NSF Animal Developmental Mechanisms Grants 3734

Animal Developmental Systems programmatic area supports research that seeks to understand the processes that result in the complex phenotype of animals. Because different organisms may be more amenable to certain approaches than others, analyses of development in a wide range of different species are encouraged. The Developmental Systems Cluster is also particularly interested in understanding how emergent properties result in the development of complex phenotypes and lead to the evolution of developmental mechanisms. Proposals may be submitted at any time; however, receipt by target dates will ensure review at the regular panel meeting. Full proposals should be submitted via the FastLane system at NSF or Grants.gov.
Requirements: The most frequent recipients of support are academic institutions and nonprofit research groups. In special circumstances, grants also are awarded to other types of institutions and to individuals.
Date(s) Application is Due: Jan 12; Jul 12
Contact: James Deshler, Program Director; (703) 292-8417; jdeshler@nsf.gov
Internet: http://www.nsf.gov/funding/pgm_summ.jsp?pims_id=501087
Sponsor: National Science Foundation
4201 Wilson Boulevard
Arlington, VA 22230

NSF Behavioral Neurosciences Research Grants 3735

This program focuses on the neural regulation of behavioral events ranging from simple movements to complex adaptive and interactive responses. Major activities are sensorimotor integration, biological rhythms, and cognitive functions such as attention, spatial representation, and learning and memory. Approaches that use novel techniques

to study behavior within an evolutionary and ecological context are encouraged. Proposals may be submitted at any time. For review within six months, proposals should be submitted in line with the listed target dates for applications. Refer to NSF brochure PD 98-1191
Requirements: The most frequent recipients of support are academic institutions and nonprofit research groups. In special circumstances, grants also are awarded to other types of institutions and to individuals.
Date(s) Application is Due: Jan 12; Jul 12
Contact: Judith Verbeke, Acting Division Director; (703) 292-8420; fax (703) 292-9153; jverbeke@nsf.gov
Internet: http://www.nsf.gov/pubs/progdesc/1998/bio/ibn/1191.htm
Sponsor: National Science Foundation
4201 Wilson Boulevard
Arlington, VA 22230

NSF Biological Physics (BP) **3736**
The program supports projects in which the analytical and experimental tools of physics are applied to the study of problems originating in the living world. Both experimental and theoretical projects will be considered, although the main focus of the program is in the experimental area. Of particular interest are projects in which new experimental approaches are brought to bear on a well-identified problem. These approaches should at the same time have the potential for broad applicability to a set of similar problems, thereby adding to the set of tools the scientist has for addressing biological problems in general. While the problems under study must be important to advancing understanding of the living world in a meaningful way, particular emphasis will be placed on those projects in which the lessons learned from the application serve to foster new concepts and ideas that expand the intellectual basis of physics. The program funds individual investigators, although collaborative proposals between physicists and biologists are welcome.
Requirements: Unrestricted (i.e., open to any type of entity)
Date(s) Application is Due: Jul 31
Amount of Grant: 135,000 - 500,000 U.S.D
Samples: Trustees of Boston University - ($195,000) Electronic Recognition of Gene Regulatory Proteins Bound to DNA; University of California-Berkeley - ($300,000) In vitro and in vivo Single Molecular Experiments of Biological Systems; University of Illinois at Chicago - ($180,000) Study of the Dynamics of Protein-DNA Interactions to Probe Site-Specific Recognition; University of California-Davis - ($135,000) Computational study of DNA repair enzyme photolyase.
Contact: Ramona Winkelbauer, Program Contact; (703) 292-7390; winkelb@nsf.gov
Internet: http://nsf.gov/funding/pgm_summ.jsp?pims_id=6673&org=NSF&sel_org=NSF&from=fund
Sponsor: National Science Foundation
4201 Wilson Boulevard
Arlington, VA 22230

NSF Biomedical Engineering and Engineering Healthcare Grants **3737**
The mission of the Biomedical Engineering and Engineering Healthcare cluster is to provide opportunities to develop novel ideas into projects that integrate engineering and life science principles in solving biomedical problems that serve humanity. The cluster focuses on high impact transforming technologies for deriving information from cells, tissues, organs, and organ systems, extraction of useful information from complex biomedical signals, new approaches to the design of structures and materials for eventual medical use, biophotonics, and new methods of controlling living systems. This cluster is also directed toward the characterization, restoration, and/or substitution of normal functions in humans.
Contact: Marcia Rawlings, Information Technology Specialist; (703) 292-7956; fax (703) 292-9013; mrawling@nsf.gov
Internet: http://www.nsf.gov/funding/pgm_summ.jsp?pims_id=501032
Sponsor: National Science Foundation
4201 Wilson Boulevard
Arlington, VA 22230

NSF BIO Minority Postdoctoral Research Fellowships **3738**
These fellowships seek to provide opportunities for recent minority doctoral scientists to obtain additional training, to gain research experience under the sponsorship of established scientists, and to broaden their scientific horizons beyond their research experiences during their undergraduate or graduate training. They are further designed to assist new scientists to direct their research efforts across traditional disciplinary lines and to avail themselves of unique research resources, sites, and facilities, including foreign locations. Fellows must affiliate with an appropriate research institution and are expected to devote themselves full time to the fellowship activities for the duration of the fellowship. At the conclusion of the fellowship, a fellow who accepts a tenure-track appointment at a U.S. institution eligible to receive NSF funds may apply for a research starter grant. Application materials are available on request. The deadline for FastLane submission of new postdoctoral fellowships is 5 p.m. local time on the first Monday in December each year. NSF 00-139
Requirements: Applicants must be U.S. nationals from underrepresented minority groups. Graduate students who will complete the PhD within one year or have completed it within the previous four years and who have not completed more than two years of postdoctoral support are eligible to apply.
Date(s) Application is Due: Dec 5
Amount of Grant: Up to 18,000,000 U.S.D
Contact: Carter Kimsey, Program Manager; (703) 292-8470; ckimsey@nsf.gov
Internet: http://www.nsf.gov/pubs/2001/nsf00139/nsf00139.txt
Sponsor: National Science Foundation
4201 Wilson Boulevard
Arlington, VA 22230

NSF Biomolecular Systems Cluster Grants **3739**
The Biomolecular Systems Cluster emphasizes the structure, function, dynamics, interactions, and interconversions of biological molecules. The context for such studies can range from investigations of individual macromolecules to the large-scale integration of metabolic and energetic processes. Research supported by this cluster includes development of cutting-edge technologies integrating theoretical, computational, and experimental approaches to the study of biological molecules and their functional complexes; mechanistic studies of the regulation and catalysis of enzymes and RNA, and higher-order characterization of the biochemical processes by which all organisms acquire, transform, and utilize energy from substrates. This cluster emphasizes the importance of multi-disciplinary research carried out at the interfaces of biology, physics, chemistry, mathematics and computer science, and engineering.
Requirements: Standard NSF Grant Proposal Guidelines apply (see http://www.nsf.gov/publications/pub_summ.jsp?ods_key=gpg).
Date(s) Application is Due: Jan 12; Jul 12
Contact: Kamal Shukla, Program Director; (703) 292-7131; fax (703) 292-9061; kshukla@nsf.gov
Internet: http://www.nsf.gov/funding/pgm_summ.jsp?pims_id=12771
Sponsor: National Science Foundation
4201 Wilson Boulevard
Arlington, VA 22230

NSF Cell Systems Cluster Grants **3740**
The Cellular Systems Cluster, one of three thematic areas within the Division of Molecular and Cellular Biosciences, supports research, across all taxa, into the structure and organization of cells and the dynamics of cellular processes. The Cellular Systems cluster is interested not only in traditional areas of cell biology (such as the organization, function, and dynamics of membranes, organelles and other subcellular compartments, and intracellular and transmembrane signal transduction mechanisms and cell-cell signaling processes) but also in the development of quantitative, theory-driven approaches to cell biology that integrate experimental studies at the molecular genetic, biochemical, biophysical, transcriptomic and proteomic levels. Network theory (e.g., as applied to signal transduction) and molecular dynamic modeling (e.g., as applied to the structure/function relationships of cellular structures) are also of particular interest.
Requirements: While proposals using approaches and model systems traditional in the field of cell biology are welcome, studies focused on novel, unique approaches and on non-traditional model organisms are encouraged.
Date(s) Application is Due: Jan 12; Jul 12
Contact: LaJoyce Debro, Program Director; (703) 292-7135; ldebro@nsf.gov
Internet: http://www.nsf.gov/funding/pgm_summ.jsp?pims_id=12772
Sponsor: National Science Foundation
4201 Wilson Boulevard
Arlington, VA 22230

NSF Chemical, Biochemical, and Biotechnology Systems Research Grants **3741**
The Chemical, Biochemical, and Biotechnology cluster deals with problems involved in the processing and manufacture of products of economic importance by effectively utilizing chemical resources and renewable resources of biological origin, often with the aid of bioinformatics originating from genomic and proteomic information. This cluster supports fundamental and applied research on catalytic processes and understanding the kinetics and mechanisms of pertinent reactions. Research in control, optimization, and process modeling are also supported. Chemical and Biological Separations, Biochemical Engineering, Biotechnology, Catalysis and Biocatalysis, and Process and Reaction Engineering are funded as part of this activity.
Amount of Grant: Up to 400,000 U.S.D
Contact: Marcia Rawlings, Information Technology Specialist; (703) 292-7956; fax (703) 292-9013; mrawling@nsf.gov
Internet: http://www.nsf.gov/funding/pgm_summ.jsp?pims_id=13481
Sponsor: National Science Foundation
4201 Wilson Boulevard, Room 565
Arlington, VA 22230

NSF Chemistry Research Experiences for Undergraduates (REU) **3742**
The program supports active research participation by undergraduate students in any of the areas of research funded by the National Science Foundation. The NSF Division of Chemistry funds about 20-25 REU Sites a year (depending on the availability of funds) as part of the NSF-wide REU activity. Projects involve students in meaningful ways in ongoing research programs or in research projects specifically designed for the program. Two mechanisms are featured for support of student research: (1) REU Sites are based on independent proposals to initiate and conduct projects that engage a number of students in research. REU Sites may be based in a single discipline or academic department, or on interdisciplinary or multi-department research opportunities with a coherent intellectual theme. Proposals with an international dimension are welcome. A partnership with the Department of Defense supports REU Sites in DoD-relevant research areas. (2) REU Supplements may be requested for ongoing NSF-funded research projects or may be included as a component of proposals for new or renewal NSF grants or cooperative agreements. NSF 07-569.
Requirements: For REU Site proposals, a single individual should be designated as the Principal Investigator. This individual will be responsible for overseeing all aspects of the award. However, one additional person may be designated as Co-Principal Investigator, if developing and operating the REU Site would involve such shared responsibility. Other anticipated research supervisors should be listed as Non-Co-PI Senior Personnel. Undergraduate student participants in either Sites or Supplements must be citizens or permanent residents of the United States or its possessions. An

undergraduate student is a student who is enrolled in a degree program (part-time or full-time) leading to a baccalaureate or associate degree. Students who are transferring from one college or university to another and are enrolled at neither institution during the intervening summer may participate. High school graduates who have been accepted at an undergraduate institution but who have not yet started their undergraduate study are also eligible to participate. Students who have received their bachelor's degrees and are no longer enrolled as undergraduates are generally not eligible to participate. For REU Sites, a significant fraction of the student participants should come from outside the host institution or organization. Some NSF directorates encourage inclusion in the REU program of K-12 teachers of science, technology, engineering, and mathematics. Please contact the appropriate disciplinary program officer for guidance.
Restrictions: Students may not apply to NSF to participate in REU activities. Students apply directly to REU Sites and should consult the directory of active REU Sites.
Contact: Wilfredo Colon, Program Director; (703) 292-8171; fax (703) 292-9061; wcolon@nsf.gov
Internet: http://www.nsf.gov/funding/pgm_summ.jsp?pims_id=503210
Sponsor: National Science Foundation
4201 Wilson Boulevard
Arlington, VA 22230

NSF Doctoral Dissertation Improvement Grants in the Directorate for Biological Sciences (DDIG) 3743
This program awards grants in selected areas of the biological sciences. These grants provide partial support of doctoral dissertation research to improve the overall quality of research. Allowed are costs for doctoral candidates to participate in scientific meetings, to conduct research in specialized facilities or field settings, and to expand an existing body of dissertation research. NSF 08-564.
Requirements: U.S. institutions and organizations that are eligible for awards from the National Science Foundation, including colleges, universities, and other nonprofit research organizations such as botanical gardens, marine and freshwater institutes, and natural history museums may submit proposals. See Chapter I, Section E of the NSF Grant Proposal Guide (GPG) for specific definitions of these categories of proposers. The NSF encourages collaborations with scientists at foreign organizations; however, primary support for any foreign participants' activities must be secured through their own national sources. A student must have advanced to candidacy for a Ph.D. degree before the submission deadline to be eligible to submit a proposal. A statement that the student has advanced to candidacy for a Ph.D., signed and dated by the department chairperson, graduate dean, or similar administrative official is required. The proposal must be submitted through regular organizational channels by the dissertation advisor(s) on behalf of a graduate student who is at the point of initiating or is already conducting dissertation research. The student must be enrolled at a U.S. institution, but need not be a U.S. citizen. Organizations should limit applications to outstanding dissertation proposals with unusual financial requirements that cannot be met otherwise. Preference may be given to projects that are underway and for which feasibility is demonstrated.
Restrictions: Indirect costs not allowed.
Date(s) Application is Due: Nov 20
Amount of Grant: Up to 15,000 U.S.D
Contact: IOS Program Officer; (703) 292-8423; ddig-ios@nsf.gov
Internet: http://www.nsf.gov/funding/pgm_summ.jsp?pims_id=5234&org=BIO&sel_org=BIO&from=fund
Sponsor: National Science Foundation
4201 Wilson Boulevard
Arlington, VA 22230

NSF Emerging Frontiers in Research and Innovation (EFRI) Grants 3744
The Directorate for Engineering at the National Science Foundation has established the Office of Emerging Frontiers in Research and Innovation (EFRI) to serve a critical role in focusing on important emerging areas in a timely manner. The EFRI Office is launching a new funding opportunity for interdisciplinary teams of researchers to embark on rapidly advancing frontiers of fundamental engineering research. For this program, they will consider proposals that aim to investigate emerging frontiers in the following two specific research areas: (1) BioSensing & BioActuation: Interface of Living and Engineered Systems (BSBA), and (2) Hydrocarbons from Biomass (HyBi). EFRI seeks proposals with transformative ideas that represent an opportunity for a significant shift in fundamental engineering knowledge with a strong potential for long term impact on national needs or a grand challenge. It is anticipated that 11 or more standard grants will be made (4-year awards).
Requirements: Proposals may be submitted by a single organization or a group of organizations consisting of a lead organization in partnership with one or more partner organizations. Only U.S. academic institutions with significant research and degree-granting education programs in disciplines normally supported by NSF are eligible to be the lead organization. Principal investigators are encouraged to form synergistic collaborations among researchers and with private and public sector organizations, government laboratories, and scientists and engineers at foreign organizations where appropriate, though no NSF funds will be provided to those organizations. Submission of Letters of Intent is required, due by October 14. Submission of Preliminary Proposals is required, due by December 2. Full proposals will be due by April 30. Each project team may receive support of up to a total of $500,000 per year for up to four years, pending the availability of funds. It is not expected that all awards will receive the maximum amount; the size of awards will depend upon the type of research program proposed. See the website for the detailed solicitation, which includes guidelines, requirements, and application instructions.
Restrictions: The principal investigator and co-principal investigators may participate in one proposal submitted to this program.
Date(s) Application is Due: Apr 30; Oct 14; Dec 2
Amount of Grant: Up to 500,000 U.S.D
Contact: Sohi Rastegar, Director, Office of Emerging Frontiers in Research and Innovation; (703) 292-8305; srastega@nsf.gov
Internet: http://www.nsf.gov/funding/pgm_summ.jsp?pims_id=13708
Sponsor: National Science Foundation
4201 Wilson Boulevard
Arlington, VA 22230

NSF Genes and Genome Systems Cluster Grants 3745
Grants support studies on genomes and genetic mechanisms in all organisms, whether prokaryote, eukaryote, phage, or virus. Proposals on the structure, maintenance, expression, transfer, and stability of genetic information in DNA, RNA, and proteins and how those processes are regulated are appropriate. Areas of interest include genome organization, molecular and cellular evolution, replication, recombination, repair, and vertical and lateral transmission of heritable information. Of equal interest are the processes that mediate and regulate gene expression, such as chromatin structure, epigenetic phenomena, transcription, RNA processing, editing and degradation, and translation. The use of innovative in vivo and/or in vitro approaches, including biochemical, physiological, genetic, genomic, and/or computational methods, is encouraged, as is research at the interfaces of biology, physics, chemistry, mathematics and computer science, and engineering.
Requirements: Standard NSF Grant Proposal Guidelines apply (see http://www.nsf.gov/publications/pub_summ.jsp?ods_key=gpg).
Date(s) Application is Due: Jan 12; Jul 12
Contact: Patrick P. Dennis; (703) 292-7145; fax (703) 292-9061; pdennis@nsf.gov
Internet: http://www.nsf.gov/funding/pgm_summ.jsp?pims_id=12780
Sponsor: National Science Foundation
4201 Wilson Boulevard
Arlington, VA 22230

NSF Grant Opportunities for Academic Liaison with Industry (GOALI) 3746
Grant Opportunities for Academic Liaison with Industry (GOALI) promotes university-industry partnerships by making project funds or fellowships/traineeships available to support an eclectic mix of industry-university linkages. Special interest is focused on affording the opportunity for: faculty, postdoctoral fellows, and students to conduct research and gain experience in an industrial setting; industrial scientists and engineers to bring industry's perspective and integrative skills to academe; and interdisciplinary university-industry teams to conduct research projects. This solicitation targets high-risk/high-gain research with a focus on fundamental research, new approaches to solving generic problems, development of innovative collaborative industry-university educational programs, and direct transfer of new knowledge between academe and industry. GOALI seeks to fund transformative research that lies beyond that which industry would normally fund.
Contact: Donald Senich; (703) 292-7082; fax (703) 292-9056; dsenich@nsf.gov
Internet: http://www.nsf.gov/funding/pgm_summ.jsp?pims_id=13706
Sponsor: National Science Foundation
4201 Wilson Boulevard
Arlington, VA 22230

NSF Instrument Development for Biological Research (IDBR) 3747
The IDBR program provides support for development of the following: (1) New instrumentation that addresses emerging biological research needs with the capacity to transform biological research; (2) Novel instrument configurations (hybrids, etc.) that demonstrably address a fundamental biological research challenge; (3) Instrumentation that leverages advances in micro- and nano-fabrication and device design to provide new capabilities in measurement and observation of biological phenomena; (4) Concept and proof-of-concept of novel instruments for biological research; (5) New instruments that provide new capabilities for detection, measurement, and/or observation of biological phenomena, or that greatly extend currently achievable sensitivity, accuracy or resolution; (6) Sensors that meet emerging biological research needs and that have the potential to transform biological research at any level of organization (7) Data acquisition and analysis tools that, through the development of novel devices/instruments, meet biological research needs. The program is not limited to the above mentioned activities. Any proposal that is designed to meet the goals of the Program will be considered. Proposals must, in all cases, demonstrate a strong connection to biological research needs through, for example, collaboration with biological researchers as well as demonstrate that the need for the instrumentation is firmly based on biological research drivers. NSF Program Solicitation 08-566.
Requirements: The program encourages proposals which conduct collaborative and planning activities such as workshops and the development of virtual organization frameworks. Those activities which promote interactions between the instrument development community and biological researchers as well as innovative networking strategies that foster research collaborations or enable new instrument development directions are especially encouraged. Activities which increase participation of colleagues at smaller institutions, minority-serving institutions, community colleges, and K-12 students and teachers are also recommended. Recognizing that the development and use of novel instrumentation have become increasingly integral to activities supported by all BIO programs, the IDBR program will place a higher priority on proposals that create novel instrumentation that enables biological research in a new and potentially transformative way and that serves a broad user community. Check the NSF Grant Proposal Guide (GPG) available electronically on the NSF website at: http://www.nsf.gov/publications/pub_summ.jsp?ods_key=gpg for other general eligibility guidelines.
Restrictions: The program does not support research or technique development activities, except to the extent these are required as part of the development of the new or improved

instrument, or for the testing of its utility. Projects emphasizing the development of new research techniques should be addressed to an appropriate research program. The anticipated uses of the instrumentation to be developed or improved should include areas of research that fall within the scope of the Directorate for Biological Sciences (see BIO Home Page at http://www.nsf.gov/bio). Projects aimed at instrumentation whose primary use will be in studies of the etiology, diagnosis or treatment of physical or mental disease, abnormality, or malfunction in human beings or animals, is not supported by IDBR. Similarly, the development or testing of drugs or of instruments whose primary application is in pharmaceutical chemistry are not eligible for support. Such projects should be addressed to an appropriate program in another NSF Directorate or to another agency. Projects in which the main portion of the instrument development activity will be subcontracted to a Federally Funded Research and Development Center (FFRDC) or a commercial (for profit) organization are also not eligible for support by IDBR, and should be addressed to an appropriate NSF program or to another agency.
Date(s) Application is Due: Aug 28
Contact: Robyn Hannigan, (703) 292-8470; fax (703) 292-9063; dbiiid@nsf.gov
Internet: http://www.nsf.gov/publications/pub_summ.jsp?ods_key=nsf08566&org=NSF
Sponsor: National Science Foundation
4201 Wilson Boulevard
Arlington, VA 22230

NSF Major Research Instrumentation Program (MRI) Grants **3748**
The MRI program assists in the acquisition or development of major research instrumentation that is, in general, too costly for support through other NSF programs. This program seeks to improve the quality and expand the scope of research and research training in science and engineering, and to foster the integration of research and education by providing instrumentation for research-intensive learning environments. For proposals over $2 million, requests must be for the acquisition of a single instrument. For proposals requesting $2 million or less, investigators may seek support for instrument development or for acquisition of a single instrument, a large system of instruments, or multiple instruments that share a common or specific research focus. Submission of Letters of Intent is required (only for acquisition requests between $2 million and $4 million). NSF 08-503
Requirements: Proposals may only be submitted by the following: (1) U.S. colleges, universities and organizations of higher education located in the U.S., its territories and possessions; (2) U.S. independent research museums located in the U.S., its territories and possessions; (3) U.S. independent nonprofit research organizations located in the U.S., its territories and possessions, including consortia whose members consist only of organizations described in items (1) and (2). (Requests for instrumentation that will be located at a Federally Funded Research and Development Center (FFRDC) must be submitted as consortium proposals. This is the only mechanism by which instrumentation, whether purchased or developed, can be placed at an FFRDC.); (4) U.S. small businesses located in the U.S., its territories and possessions are eligible for instrument development support as private sector partners with submitting organizations; they may not submit proposals as a lead organization. The deadline is the fourth Thursday in January. Proposers are required to prepare and submit all proposals for this program solicitation through use of the NSF FastLane system.
Restrictions: The MRI program will NOT support proposal requests for: Computer networks used for general-purposes; General laboratory equipment or assorted instruments that do not share a common or specific research or research training focus; Instrumentation used primarily for standard science and engineering courses; Renovation or modernization of research facilities, fixed equipment, or facilities such as research vessels, airplanes, large telescopes, and supercomputing centers. The term 'research facilities' refers to the bricks-and-mortar physical plant in which sponsored or unsponsored research activities (including research training) take place, including related infrastructure, systems (e.g., HVAC and power systems, toxic waste removal systems), and fixed equipment. The term 'fixed equipment' refers to the permanent components of a research facility that are integral (i.e., built in, rather than affixed) to the facility (e.g., clean rooms, fume hoods, elevators, laboratory casework); their removal would affect the integrity or basic operation of the facility.
Date(s) Application is Due: Jan 22
Amount of Grant: Up to 4,000,000 U.S.D
Contact: Randy Phelps, (703) 292-8040; rphelps@nsf.gov
Internet: http://nsf.gov/funding/pgm_summ.jsp?pims_id=5260
Sponsor: National Science Foundation
4201 Wilson Boulevard
Arlington, VA 22230

NSF Perception, Action & Cognition Research Grants **3749**
This program supports research on perception, action and cognition including the development of these capacities. Emphasis is on research strongly grounded in theory. Research topics include vision, audition, haptics, attention, memory, reasoning, written and spoken discourse, motor control, and developmental issues in all topic areas. The program encompasses a wide range of theoretical perspectives, such as symbolic computation, connectionism, ecological, nonlinear dynamics, and complex systems, and a variety of methodologies including both experimental studies and modeling. Research involving acquired or developmental deficits is appropriate if the results speak to basic issues of perception, action, and cognition.
Date(s) Application is Due: Feb 1; Aug 1
Amount of Grant: 1,100 - 49,000,000 U.S.D
Contact: Vincent R. Brown, Program Director; (703) 292-7305; vrbrown@nsf.gov
Internet: http://www.nsf.gov/funding/pgm_summ.jsp?pims_id=5686
Sponsor: National Science Foundation
4201 Wilson Boulevard
Arlington, VA 22230

NSF Postdoctoral Research Fellowships in Biology (PRFB) **3750**
The Directorate for Biological Sciences (BIO) awards Postdoctoral Research Fellowships in Biology to recent recipients of the doctoral degree for research and training in selected areas of biology supported by BIO to encourage independence early in their research careers and to permit them to pursue their research and training goals in the most appropriate research locations regardless of the availability of funding for the Fellows at that site. The fellowships are also designed to provide active mentoring of the Fellows by the sponsoring scientists who will benefit from having additional members in their research groups. The research and training plan of each fellowship must address important scientific questions in contemporary biology within the scope of the BIO Directorate and the specific guidelines in this fellowship program solicitation. Because the fellowships are offered only to postdoctoral scientists early in their careers, doctoral advisors are encouraged to discuss the availability of BIO fellowships with their graduate students early in their doctoral programs. Fellowships are awards to individuals, not institutions, and are administered by the Fellows. Fellowship stipends are being adjusted on a yearly basis. NSF 07-580.
Requirements: Only individuals may apply. NSF postdoctoral fellowships are awards to individuals, and applications are submitted directly by applicants to NSF. However, applications must include sponsoring scientists' statements and the applicants must affiliate with institutions (e.g., colleges and universities, and privately-sponsored nonprofit institutes and museums, government agencies and laboratories, and, under special conditions, for-profit organizations) anywhere in the world. Applicants must: (1) be U.S. citizens (or nationals) or legally admitted permanent residents of the United States (i.e., have a 'green card') at the time of application; (2) earn or plan to earn the doctoral degree in a scientific or engineering field prior to the requested start date of the fellowship; (3) either currently be a graduate student or, at the deadline date, have served in a position requiring the doctoral degree for no more than 12 full time months since earning the degree. There is a one time exception to this criterion for Biological Informatics in November, 2007 only, as described below; (4) must present a research and training plan that falls within the purview of BIO and includes the required information for the specific competition as described below; (5) select a host institution and sponsoring scientist different from the doctoral degree and current position or provide compelling justification for why such a change is not being proposed; (6) not have received Federal funding of more than $20,000 as PI or co-PI (except graduate fellowships and doctoral dissertation improvement grants); (7) not have submitted concurrently the same project to another NSF program; (8) not be a named participant on any other proposal submitted to NSF, including regular research proposals, concurrent with the fellowship application, regardless of who is the named principal investigator; AND (9) meet other limitations on eligibility imposed within the specific competitive areas, if any. There is no limit on the number of applicants that an institution may host. For now, submission can only be made using NSF's FastLane system, given that Grants.gov does not support all required forms. Applications are due the first Monday in November by 5 p.m. proposer's local time.
Date(s) Application is Due: Nov 0
Amount of Grant: 60,000 - 66,000 U.S.D
Contact: Peter McCartney, Program Director; (703) 292-8470; pmccartn@nsf.gov
Internet: http://www.nsf.gov/publications/pub_summ.jsp?ods_key=nsf07580&org=BIO
Sponsor: National Science Foundation
4201 Wilson Boulevard
Arlington, VA 22230

NSF Presidential Early Career Awards for Scientists and Engineers Grants **3751**
Each year NSF selects nominees for the Presidential Early Career Awards for Scientists and Engineers (PECASE) from among the most meritorious new CAREER awardees. Selection for this award is based on two important criteria: innovative research at the frontiers of science and technology that is relevant to the mission of the sponsoring organization or agency, and community service demonstrated through scientific leadership, education or community outreach. These awards foster innovative developments in science and technology, increase awareness of careers in science and engineering, give recognition to the scientific missions of the participating agencies, enhance connections between fundamental research and national goals, and highlight the importance of science and technology for the Nation's future. Individuals cannot apply for PECASE. These awards are initiated by the participating federal agencies. At NSF, up to twenty nominees for this award are selected each year from among the PECASE-eligible CAREER awardees who are most likely to become the leaders of academic research and education in the twenty-first century. The White House Office of Science and Technology Policy makes the final selection and announcement of the awardees.
Requirements: Applicant must be employed at an institution in the United States, its territories or possessions, or Puerto Rico, that awards a baccalaureate or advanced degree in a field supported by NSF; be in their first or second full-time tenure-track academic appointment and have begun the first tenure-track or appointment (at any institution) on or after October 1, 1996, and before October 1, 2001; not be tenured or have held tenure on or before October 1, 2001; and not be a previous recipient of a CAREER or PECASE award. Contact program officer for exact requirements.
Restrictions: No salary support for other Senior Personnel is permitted, in either the primary budget or in any sub awards.
Date(s) Application is Due: Jul 21; Jul 22; Jul 23
Amount of Grant: 400,000 - 500,000 U.S.D
Contact: Elizabeth L. Rom; (703) 292-7709; fax (703) 292-9085; elrom@nsf.gov
Internet: http://www.nsf.gov/funding/pgm_summ.jsp?pims_id=5262
Sponsor: National Science Foundation
4201 Wilson Boulevard
Arlington, VA 22230

NSF Small Business Innovation Research (SBIR) Grants 3752

The Small Business Innovation Research (SBIR) Program stimulates technological innovation in the private sector by strengthening the role of small business concerns in meeting Federal research and development needs, increasing the commercial application of federally supported research results, and fostering and encouraging participation by socially and economically disadvantaged and women-owned small businesses.
Requirements: Eligible to apply are small business firms—those organized for profit, individually owned or operated, not dominant in the field in which they are bidding, and with an average of no more than 500 employees in all affiliated firms. In addition, the primary employment of the principal investigator must be with the small business firm at the time of the award.
Amount of Grant: 5,000 - 4,000,000 U.S.D
Samples: Inrad Inc (Northvale, NJ)—Growth of potassium iodate single crystals for nonlinear applications (Phase I SBIR), $65,000; Metcal Inc (Menlo Park, CA)—Computer design of a new class of magnetic alloys (Phase I SBIR), $65,000.
Contact: Gregory T. Baxter, Program Director; (703) 292-7795; fax (703) 292-9057; gbaxter@nsf.gov
Internet: http://www.nsf.gov/funding/pgm_summ.jsp?pims_id=503361
Sponsor: National Science Foundation
4201 Wilson Boulevard
Arlington, VA 22230

NSF Social Psychology Research Grants 3753

Grants support laboratory and field research in all areas of human social behavior, including attitude formation and change, social cognition, personality processes, interpersonal relations and group processes, the self, emotion, social comparison and social influence, the social psychology of health, and the psychophysiological correlates of social behavior. This program also supports research on social development over the human life span. Proposals are accepted at any time; target dates indicate the approximate time of review panel meetings. PD 98-1332
Date(s) Application is Due: Jan 15; Jul 15
Contact: Amber Story, Program Director; (703) 292-7249; astory@nsf.gov
Internet: http://www.nsf.gov/funding/pgm_summ.jsp?pims_id=5712
Sponsor: National Science Foundation
4201 Wilson Boulevard
Arlington, VA 22230

NSF Systematic Biology and Biodiversity Inventories Grants 3754

The Systematic Biology and Biodiversity Inventories Cluster supports research in taxonomy and systematics that contributes to: 1) using phylogenetic methods to understand the evolution of life in time and space, 2) discovery, description, and cataloguing global species diversity, and 3) organizing information from the above in efficiently retrievable forms that best meet the needs of science and society. The Systematic Biology and Biodiversity Inventories Cluster funds projects within the two Programs, Systematic Biology and Biodiversity Inventories, in addition to the PEET and PBI solicitations listed below. In addition, the cluster participates in AToL and other related funding opportunities. The SBBI Cluster continues to encourage and support studies that seek to synthesize available and new species-level taxonomic information in the context of providing revisionary treatments and predictive classifications for particular groups of organisms. NSF PD 047374.
Requirements: The most frequent recipients are academic institutions and nonprofit research groups. In special circumstances, grants also are awarded to other types of institutions and to individuals. Standard NSF Grant Proposal Guidelines apply (see http://www.nsf.gov/publications/pub_summ.jsp?ods_key=gpg).
Date(s) Application is Due: Jan 9; Jul 9
Contact: Maureen Kearney, Program Director; (703) 292-7187; mkearney@nsf.gov
Internet: http://www.nsf.gov/publications/pub_summ.jsp?ods_key=pd047374&org=NSF
Sponsor: National Science Foundation
4201 Wilson Boulevard
Arlington, VA 22230

NSF Undergraduate Research and Mentoring in the Biological Sciences 3755

This program will fund projects that have strong research and mentoring activities designed to prepare students for successful entry into graduate programs. URM will support projects involving the recruitment, retention and development of undergraduate students, especially those from underrepresented groups, for the purpose of preparing them for graduate study in the biological sciences. Proposed projects are expected to create a URM program that will actively engage students in interesting and exciting research ideas, provide hands-on research experience, and develop their academic skills. URM will enable institutions to create innovative programs that will increase the number and diversity of students who enter graduate research programs in the biological sciences. URM will support well-defined research and mentoring activities that will enable students, especially underrepresented minorities, to become independent thinkers and effective communicators, as well as professional development activities that will expose students to exciting careers in biology. Students are expected to be recruited early in their academic career. The project should frame all student activities around an effective mentoring strategy. Factors that are known to increase the likelihood of success for students to enter, and complete, graduate research programs should be emphasized. Approximately $4,000,000 will be available for this program, subject to the availability of funds. Under the solicitation, requests may be submitted for funding amounts up to a total of $1,000,000 for up to 5 years. However, funding beyond the first year is contingent upon satisfactory progress in the program. The program expects to make at least eight awards, depending upon the quality of submissions and the availability of funds. NSF 06-591
Requirements: Preliminary Proposals (deadline: Sep. 18) are required prior to submitting a Full Proposal (deadline: Mar. 4). Proposals may only be submitted by academic institutions located in the United States. Collaborative Proposals from two or more institutions are allowed. The proposing institution is expected to foster an enriched and intellectually stimulating research and educational environment. URM proposals should involve year-round mentoring and include a major emphasis on direct student participation in research leading to publishable data. URM projects may be based in a single discipline or academic department, or on interdisciplinary or multi-department research opportunities. A proposal should reflect the unique combination of the proposing organization's interests and capabilities and those of any partnering organizations. Partnerships between minority-serving institutions, especially Historically Black Colleges and Universities (HBCU), Tribal Colleges and Universities (TCU), and Hispanic-Serving Institutions (HSI), community colleges, and research intensive institutions are encouraged, if such partnerships will result in a diverse pool of qualified participants as well as an enriched research environment for the students. Undergraduate student participants supported with NSF funds must be citizens or permanent residents of the United States or its possessions. An undergraduate student is one who is enrolled in a degree program (part-time or full-time) leading to a baccalaureate or associates degree. High school graduates who have not yet enrolled, and students who have received their bachelor's degrees and are no longer enrolled as undergraduates, are not eligible.
Restrictions: All collaborative proposals submitted as separate submissions from multiple organizations must be submitted via the NSF FastLane system.
Date(s) Application is Due: Mar 4; Sep 18
Amount of Grant: Up to 1,000,000 U.S.D
Samples: University of Arkansas (AR) $149,233 - RIG: Biophysical Characterization of Cdc42Hs Forms Associated with Cell Transformation; University of Montana (MT) $170,382 - Collaborative UMEB: Training American Indians in Environmental Biology; Salish Kootenai College (MT) $132,715 - Collaborative UMEB: Training American Indians in Environmental Biology; University of Maryland Eastern Shore (MD) $150,000 - Discovery of Novel Transcription Factor III Proteins in Trypanosomes
Contact: URM Program Director; (703) 292-8470; biourm@nsf.gov
Internet: http://www.nsf.gov/funding/pgm_summ.jsp?pims_id=500036
Sponsor: National Science Foundation
4201 Wilson Boulevard
Arlington, VA 22230

Nuffield Foundation Africa Program Grants 3756

The Commission for Africa highlighted the importance of strengthening Africa's science capacity, recommending that donors develop incentives for research and development in health that meet Africa's needs, and increase funding to African-led research. With that in mind, the Foundation has recently replaced its Commonwealth Program with Africa Program grants, which aims to improve services in health, education and civil justice in Southern and Eastern Africa through the development of the expertise and experience of practitioners and policy makers.
Geographic Focus: All States, Algeria, Angola, Benin, Botswana, Burkina Faso, Burundi, Cameroon, Cape Verde, Central African Republic, Chad, Comoros, Congo, Congo, Democratic Republic of, Cote d' Ivoire (Ivory Coast), Djibouti, Egypt, Equatorial Guinea, Eritrea, Ethiopia, Gabon, Gambia, Ghana, Guinea, Guinea-Bissau, Kenya, Lesotho, Liberia, Libya, Madagascar, Malawi, Mali, Mauritania, Mauritius, Morocco, Mozambique, Namibia, Niger, Nigeria, Rwanda, Sao Tome & Principe, Senegal, Seychelles, Sierra Leone, Somalia, South Africa, Sudan, Swaziland, United Kingdom
Contact: Sarah Lock, 44-171-631-0566; fax 44-171-323-4877; africaprogram@nuffieldfoundation.org
Internet: http://www.nuffieldfoundation.org/go/grants/commonwealth/page_100.html
Sponsor: Nuffield Foundation
28 Bedford Square
London, WC1B 3JS England

Nuffield Foundation Children and Families Grants 3757

The foundation was founded for the advancement of health through medical research and teaching and for the care of the elderly through scientific research and education. The Children and Families program supports work to help ensure that the legal and institutional framework is best adapted to meet the needs of children and families. At present, particular interests include (but are not limited to): work that links education and child development, either in the case of adolescent mental health or younger children; work that considers policies relevant to child welfare in a broader institutional context: parents' paid working patterns; childcare and early years provision; work that considers especially the well-being of children growing up in adverse conditions, and what institutional responses may be appropriate; work in family law, including cohabitation, child contact, child support; and work in child protection and placement (adoption and fostering) but only when it raises significant issues. Where a proposal is for a research study, Trustees are interested in the dispassionate examination of evidence. It notes that evidence is likely to be different in different cases, for different types of children and families, and are more likely to support work that takes this approach.
Requirements: Usually grants are made only to United Kingdom organizations, and support work that will be mainly based in the UK, although the Trustees welcome proposals for collaborative projects involving partners in European or Commonwealth countries.
Restrictions: The Foundation does not make grants for the running costs of voluntary bodies but will consider making a contribution to voluntary sector overheads on funded projects.
Geographic Focus: All States, United Kingdom
Date(s) Application is Due: Jan 8; May 7; Sep 3
Amount of Grant: 5,000 - 150,000 GBP

Samples: Gillian Douglas, Cardiff Law School—for Attitudes towards the law of inheritance in the context of changing family structures, 193,542 GBP; Ingrid Schoon, Department of Quantitative Social Sciences, Institute of Education, 152,771 GBP.
Contact: Sharon Witherspoon, 44-171-631-0566; fax 44-171-323-4877; sfwpa@nuffieldfoundation.org or info@nuffieldfoundation.org
Internet: http://www.nuffieldfoundation.org/go/grants/cplfj/page_30.html
Sponsor: Nuffield Foundation
28 Bedford Square
London, WC1B 3JS England

Nuffield Foundation Law and Society Grants **3758**
The foundation was founded for the advancement of health through medical research and teaching and for the care of the elderly through scientific research and education. The Foundation has long had an interest in the area of law. Its current objectives in this area are: to promote developments in the legal system that will improve its accessibility to all people; to promote wider access to legal services and advice and a better understanding of the obstacles to access to justice; to fund research and promote developments in alternative dispute resolution; to help promote a greater knowledge of the rights and duties of the individual, including those of the European citizen; to examine the implications of new human rights' obligations on civil (not criminal) justice; and to help promote a greater public understanding of the role of law in society and of the legal system.
Requirements: Usually grants are made only to United Kingdom organizations, and support work that will be mainly based in the UK, although the Trustees welcome proposals for collaborative projects involving partners in European or Commonwealth countries.
Geographic Focus: All States, United Kingdom
Date(s) Application is Due: Jan 8; May 7; Sep 3
Amount of Grant: 5,000 - 150,000 GBP
Contact: Sharon Witherspoon, 44-171-631-0566; fax 44-171-323-4877; sfwpa@nuffieldfoundation.org or info@nuffieldfoundation.org
Internet: http://www.nuffieldfoundation.org/go/grants/accesstojustice/page_57.html
Sponsor: Nuffield Foundation
28 Bedford Square
London, WC1B 3JS England

Nuffield Foundation Oliver Bird Rheumatism Program Grants **3759**
The Fund will offer grants to up to five United Kingdom Higher Education Institutions where students of outstanding potential will be recruited into doctoral training. The institutions will provide multidisciplinary environments for researching problems in rheumatic disease, and be able to develop both the students' awareness of the broader context of these diseases and the professional skills required of future scientists. For each studentship the grant will provide: a stipend of not less than 13,500GBP per annum plus United Kingdom fees; research expenses at a rate of 8,000GBP per annum; and a meeting and travel allowance of 1,000GBP per annum. In addition the Fund will provide a grant to each institution of 25,000GBP over the five years of the program to support such activities as: meetings within the Collaborative Center to promote the field and assist the recruitment and training of doctoral students; the hosting of a program meeting to be attended by all program participants from the other Collaborative Centers; and occasional bursaries for interested undergraduates to work alongside the doctoral students and potential supervisors or to attend meetings in the Collaborative Centers.
Geographic Focus: All States, United Kingdom
Date(s) Application is Due: Feb 28
Amount of Grant: 27,500 GBP
Contact: Vicki Hughes, 44-171-631-0566; fax 44-171-323-4877; oliverbird@nuffieldfoundation.org
Internet: http://www.nuffieldfoundation.org/go/grants/oliverbird/page_86.html
Sponsor: Nuffield Foundation
28 Bedford Square
London, WC1B 3JS England

Nuffield Foundation Open Door Grants **3760**
The Foundation keeps an open door to proposals of exceptional merit for research projects or practical innovations that lie outside our main program areas, but that meet Trustees' wider interests. These must have some bearing on the Foundation's widest charitable object. Particular interest is given to projects which identify change or interventions which will have a practical implications for policy or practice, or that will improve the quality of research evidence in areas of public debate. Through the Open Door, the Foundation may also identify emerging areas that justify more sustained attention.
Requirements: Usually grants are made only to United Kingdom organizations, and support work that will be mainly based in the UK, although the Trustees welcome proposals for collaborative projects involving partners in European or Commonwealth countries.
Geographic Focus: All States, United Kingdom
Date(s) Application is Due: Jan 8; May 7; Sep 3
Amount of Grant: 5,000 - 150,000 GBP
Samples: Dr Alex Bryson, National Institute of Economic & Social Research—for British Workplaces in the shadow of recession, 219,167 GBP; Colin Aitken, School of Mathematics, University of Edinburgh, 75,869 GBP.
Contact: Sharon Witherspoon, 44-171-631-0566; fax 44-171-323-4877; sfwpa@nuffieldfoundation.org or info@nuffieldfoundation.org
Internet: http://www.nuffieldfoundation.org/go/grants/opendoor/page_115.html
Sponsor: Nuffield Foundation
28 Bedford Square
London, WC1B 3JS England

Nuffield Foundation Small Grants **3761**
Most recently, the Foundation has decided that it will no longer try to fund across the full range of all social sciences, but will focus on work that more closely matches the wider interests of the Foundation. As a consequence, the budgetary ceiling for individual grants will be raised from the current maximum of 12,000GBP.
Geographic Focus: All States, United Kingdom
Date(s) Application is Due: Jul 17
Amount of Grant: 12,000 GBP
Contact: Sarah Lock, 44-171-631-0566; fax 44-171-323-4877; smallgrants@nuffieldfoundation.org
Internet: http://www.nuffieldfoundation.org/go/grants/smallgrants/page_123.html
Sponsor: Nuffield Foundation
28 Bedford Square
London, WC1B 3JS England

NWHF Community-Based Participatory Research Grants **3762**
The Northwest Health Foundation supports community-based participatory research as a tool to generate meaningful information about community health and build the capacity of groups united by common challenges to organize and advocate for change. The RFP will seek to support community-based participatory research (CBPR) projects that: build community capacity for policy advocacy; and apply research findings to support policy change that improves community health.
Requirements: 501(c)3 educational institutions and governmental entities are eligible. Generally speaking, foundation programs serve the entire state of Oregon, and the following Washington counties: Clark, Cowlitz, Pacific, Skamania, and Wahkiakum. Specific programs and funds may have additional geographic restrictions. Applying organizations must sign a statement of nondiscrimination in leadership, staffing, and services.
Restrictions: Support will not be considered for basic operations, or for salaries or other expenses associated with ongoing programs.
Geographic Focus: Oregon, Washington
Date(s) Application is Due: Sep 8
Contact: Judith Woodruff, Program Director; (503) 220-1955; fax (503) 220-1355; judith@nwhf.org
Internet: http://www.nwhf.org/index.php?/apply/cbpr
Sponsor: Northwest Health Foundation
221 NW Second Avenue, Suite 300
Portland, OR 97209

NWHF Health Advocacy Small Grants **3763**
Awards of up to $10,000 will be available for 15-20 projects dedicated to one of the following areas: grassroots organizing; coalition building; communications; and policy analysis. For example, organizations may request funds to develop communications materials or enhance their Web site, dedicate staff time to keep abreast of health-reform related activity in Salem during the legislative session, or build their organizational capacity to develop a coalition focused on health reform. In addition to grant dollars, organizations can apply for approximately two days of technical assistance from the Consumer Voices for Coverage team in one or more of the areas listed above.
Requirements: 501(c)3 educational institutions and governmental entities are eligible. Generally speaking, foundation programs serve the entire state of Oregon, and the following Washington counties: Clark, Cowlitz, Pacific, Skamania, and Wahkiakum. Specific programs and funds may have additional geographic restrictions. Applying organizations must sign a statement of nondiscrimination in leadership, staffing, and services.
Restrictions: Support will not be considered for basic operations, or for salaries or other expenses associated with ongoing programs.
Geographic Focus: Idaho, Oregon, Washington
Date(s) Application is Due: Feb 4
Amount of Grant: Up to 10,000 U.S.D
Contact: Chris DeMars; (971) 230-1292 or (503) 220-1955; cdemars@nwhf.org
Internet: http://www.nwhf.org/index.php?/apply/access
Sponsor: Northwest Health Foundation
221 NW Second Avenue, Suite 300
Portland, OR 97209

NWHF Kaiser Permanente Community Fund Grants **3764**
The Kaiser Permanente Community Fund (KPCF) at Northwest Health Foundation was established in late 2004 to advance the health of the communities served by Kaiser Permanente Northwest. The Fund intends to achieve this goal by addressing those factors in the social, policy, and physical environment that impact community health. Often referred to as the social determinants of health, these factors have been shown to play a major role in the development of health disparities based on race, ethnicity, and socio-economic status.
Requirements: To be eligible to apply for a grant from the Kaiser Permanente Community Fund, the service area of your project must fall within the geographic region roughly spanning from Longview, Washington to Salem, Oregon and portions of the Willamette Valley.
Geographic Focus: Oregon, Washington
Date(s) Application is Due: Jul 2
Contact: Judith Woodruff, Program Director; (503) 220-1955; fax (503) 220-1355; judith@nwhf.org
Internet: http://www.nwhf.org/index.php?/apply/kaiser
Sponsor: Northwest Health Foundation
221 NW Second Avenue, Suite 300
Portland, OR 97209

NWHF Mark O. Hatfield Research Fellowship **3765**

The Mark O. Hatfield Research Fellowship Program was established in 2005 to honor the legacy of Senator Mark O. Hatfield, founding Board Chairman of the Northwest Health Foundation (NWHF). The purpose of the Fellowship program is to advance the research career of health practitioners who serve communities in Oregon and southwest Washington. The fellowship is open to health professionals from a variety of disciplines who wish to launch, renew, or expand their research activities in biomedical, applied, or translational research. A secondary purpose is to increase the intellectual resources of research-based institutions. One two-year fellowship of up to $150,000 will be awarded. Funds will be distributed to the fellow's sponsoring institution in two annual disbursements of half of the amount awarded.

Requirements: Eligible candidates must: be a clinical, allied health, public health, or social service provider (early-career researchers specializing in translational research are also encouraged to apply); possess (or be in the process of completing) a doctoral degree or equivalent in an appropriate field related to medicine, nursing, health care, public health, or biomedical sciences; obtain sponsorship from a research-based institution that can act as fiscal manager of the award; and demonstrate a commitment to communities in Oregon and southwest Washington.

Date(s) Application is Due: Mar 16

Amount of Grant: Up to 150,000 U.S.D

Contact: R. David Rebanal, Program Officer; (503) 220-1955; drebanal@nwhf.org

Internet: http://www.nwhf.org/index.php?/apply/hatfield

Sponsor: Northwest Health Foundation
221 NW Second Avenue, Suite 300
Portland, OR 97209

NWHF Partners Investing in Nursing's Future Grants **3766**

Partners Investing in Nursing's Future is a collaborative initiative of the RWJ Foundation and Northwest Health Foundation that will address nursing issues at the local level through funding partnerships with community and regional foundations. The goal of this initiative is to enable local foundations to act as catalysts in developing the comprehensive strategies that are vital to establishing a stable, adequate nursing workforce. Grants support the capacity, involvement, and leadership of local foundations to advance nursing workforce solutions in their communities. Local foundations must match funds with at least $1 for every $1 provided by the program. Proposals must be submitted online. Guidelines are available online. Letters of Intent are due by October 2, and invited proposals are due on December 19.

Requirements: U.S. local or regional private, family, or community foundations are eligible. Eligible foundations are those classified under IRS codes as 501(c)3 tax exempt, a nonexempt charitable trust treated as a private foundation under 4947(a)1, or organizations claiming status as a private operating foundation under 4942(j)3 or 5. Government entities, corporations, or corporate grantmakers may participate in funding collaboratives but may not serve as the applicant organization.

Date(s) Application is Due: Oct 2; Dec 19

Amount of Grant: Up to 250,000 U.S.D

Contact: Judith Woodruff, Program Director; (503) 220-1955; fax (503) 220-1355; judith@nwhf.org

Internet: http://www.partnersinnursing.org/grants.html

Sponsor: Northwest Health Foundation
221 NW Second Avenue, Suite 300
Portland, OR 97209

NWHF Physical Activity and Nutrition Grants **3767**

The Foundation is soliciting applications from organizations engaged in creating community environments and policies that support healthy lifestyles, specifically with regards to healthy food choices and opportunities for regular physical activity. It is looking to fund initiatives that: identify the social, policy or environmental factors to be addressed, and how they will support healthy physical activity and nutrition behaviors; articulate goals, strategies and priorities that have been developed in full partnership with the community to be served; build and mobilize constituencies to advocate for improved physical activity and nutrition policies, practices and environments; reduce health disparities related to unequal access to healthy food and opportunities for daily physical activity; involve collaboration among different organizations or agencies, especially across different sectors, to achieve common goals; and implement recommendations proposed in peer-reviewed national and state reports on physical activity and nutrition. The Foundation's intent is to award a portfolio of five to eight grants of various sizes in the $25,000 to $100,000 range (total award amount - not per year). Projects will be funded for up to three years.

Requirements: The Northwest Health Foundation invites proposals from non-profit organizations, local government agencies and/or tribal organizations

Restrictions: For this RFP, the Foundation is not interested in funding curriculum development, research studies, clinical interventions, exercise or cooking classes, educational programs, capital construction costs or general operating support.

Geographic Focus: Oregon, Washington

Date(s) Application is Due: Mar 16

Amount of Grant: 25,000 - 100,000 U.S.D

Contact: Judith Woodruff, Program Director; (503) 220-1955; fax (503) 220-1355; judith@nwhf.org

Internet: http://www.nwhf.org/index.php?apply/obesity

Sponsor: Northwest Health Foundation
221 NW Second Avenue, Suite 300
Portland, OR 97209

NYAM Brock Lecture, Award and Visiting Professorship in Pediatrics **3768**

The New York Academy of Medicine invites pediatric program directors from New York area medical schools and medical centers to nominate candidates for the annual fall Millie and Richard Brock Lecture, Award and Visiting Professorship in Pediatrics. The Brock Lecture, Award and Visiting Professorship in Pediatrics was established by Millie and Richard Brock in 1995 on the 100th anniversary of The Academy's Section on Pediatrics to sponsor a nationally recognized leader in pediatrics to engage in a one to two day visiting professorship at a New York-area pediatrics training program, to deliver the annual Brock Lecture, and to receive the Brock Award for distinguished contributions to pediatrics. The one to two day visiting professorship to a New York area pediatric training program should be part of a program which include a grand rounds lecture as well as teaching interactions with medical students, pediatric residents, fellows, and members of the attending staff. The Brock Lecture held at the Academy must address issues concerned with providing care for underserved children. Suggested topics for discussion include, but are not restricted to: childhood nutrition, HIV and AIDS, substance abuse, child abuse, asthma, violence, and environmental issues such as lead poisoning. The Brock program includes an honorarium of $1,000 and all expenses for travel, lodging, and meals. Letters of nomination, including an outline of the proposed plan for the visit and the topic for the Brock Lecture, as well as the curriculum vitae of the nominee, should be forwarded via email to: brock@nyam.org.

Date(s) Application is Due: Mar 1

Amount of Grant: 1,000 U.S.D

Samples: Mark W. Kline, MD, Professor of Pediatrics and Chief of Retrovirology at the Baylor College of Medicine and Texas Children's Hospital, Academy Lecture: Pediatric AIDS: Worlds Apart, Visiting Professorship: Weill Cornell Medical Center; Glenn Flores, MD, Professor of Pediatrics and Public Health, Director of General Pediatrics, the Judith and Charles Ginsburg Chair in Pediatrics, and the Director of the Academic General Pediatrics Fellowship at University of Texas Southwestern and Children's Medical Center Dallas, Academy Lecture: Racial and Ethnic Disparities in U.S. Children's Health and Healthcare: How We Can Level the Playing Field, Visiting Professorship: New York University School of Medicine.

Contact: Program Coordinator; (212) 822-7204; fax (212) 822-7338; brock@nyam.org

Internet: https://www.nyam.org/grants/brock.shtml

Sponsor: New York Academy of Medicine
1216 Fifth Avenue
New York, NY 10029-5202

NYAM Charles A. Elsberg Fellowship **3769**

The Fellowship in Neurological Surgery was established to support research training in the specialty of neurological surgery for individuals who have completed, or will shortly complete, accredited residency training in neurological surgery, and who intend to use research training for continued development of academic careers in this field. The Elsberg Fellowship program seeks to support individuals in supervised programs that will develop the candidate's capacity to perform independent clinical or laboratory research.

Requirements: Candidates who have, or will have completed an accredited neurological surgery residency program by June 30, and whose career direction suggests eligibility for subsequent research support are invited to apply. In general, applicants are expected to be at the start of their academic careers, having completed their residency training within the past year or two. Candidates must be board certified or eligible for certification by the beginning of the grant period. The proposed research must be conducted at an institution located in the United States under the mentorship of a senior researcher.

Amount of Grant: 50,000 U.S.D

Samples: Aaron S. Dumont, MD, University of Virginia School of Medicine (2008-09); Ziv Williams, MD, Massachusetts General HospitalJason (2007-08); H. Huang, MD, University of Rochester School of Medicine and Dentistry (2006-07).

Contact: Program Administrator; (212) 822-7204; fax (212) 822-7338; elsberg@nyam.org

Internet: http://www.nyam.org/grants/elsberg.shtml

Sponsor: New York Academy of Medicine
1216 Fifth Avenue
New York, NY 10029-5202

NYAM David E. Rogers Fellowships **3770**

The Fellowship is meant to enrich the educational experiences of medical and dental students through projects that bear on medicine and dentistry as social enterprises, such as enterprises devoted to the capacity of these professions in any and all of their expressions to serve human needs. Particularly, the focus is on the needs of underserved or disadvantaged patients or populations. The content of the Fellowship might include clinical investigation, health policy analysis, activities linking biomedicine, the social infrastructure and human need, or community activities.

Requirements: Available to first-year medical and dental students for support of projects to be executed during the summer between the first and second years of medical or dental school.

Amount of Grant: 3,500 U.S.D

Samples: Abdulrahman Mohamed El-Sayed, University of Michigan Medical School; Patricia Kuan-Pei Foo, Stanford University School of Medicine; Christina Marie Hupman, Medical College of Georgia.

Contact: Office of Medical and Academic Affairs; (212) 822-7204; fax (212) 996-7338; rogers@nyam.org

Internet: http://www.nyam.org/grants/rogers.shtml

Sponsor: New York Academy of Medicine
1216 Fifth Avenue
New York, NY 10029-5202

NYAM Edward N. Gibbs Memorial Lecture and Award in Nephrology 3771

The New York Academy of Medicine seeks nominations of physician scientists who have dedicated their careers to advances in nephrology or are presently making cutting-edge discoveries in the field. The distinguished recipient of this honor will present his or her research at a lecture before a broad audience of scientists and clinicians, to be held at The New York Academy of Medicine in the fall. The award recipient will receive the Edward N. Gibbs Memorial Medal and an honorarium of $7,500; all travel expenses associated with the lecture will be paid by the Edward N. Gibbs Memorial Endowment. It is expected that the manuscript from this lecture will be submitted to a scholarly journal for publication.
Requirements: Candidates must hold an MD degree and be citizens of the United States to be considered for this award.
Date(s) Application is Due: Mar 1
Amount of Grant: 7,500 U.S.D
Contact: Program Coordinator; (212) 822-7204; fax (212) 822-7338; gibbs@nyam.org
Internet: https://www.nyam.org/grants/gibbs.shtml
Sponsor: New York Academy of Medicine
1216 Fifth Avenue
New York, NY 10029-5202

NYAM Edwin Beer Research Fellowship 3772

The Edwin Beer Research Fellowship seeks to increase the number of investigators in urology and urology-related fields by providing transitional funding in support of research conducted by qualified candidates. The program provides two-year grants of up to $80,000 for research projects commencing July of the award year. The grants are made to the awardees' institutions for the direct support of their salary and research activities. Fellows are required to devote at least fifty percent of their time to the research funded by the program.
Requirements: Candidates must hold an MD, PhD, or equivalent degree and are expected to conduct research in a supervised program in the United States. Applicants should have completed residency and fellowship or postdoctoral training, having demonstrated their ability to perform and direct research, and should currently be under the mentorship of a senior researcher. Fellowships are awarded to qualified candidates who are at a stage of transition prior to receiving major independent funding. Because the program seeks to increase the number of investigators in urology and urology-related fields, applicants must have a direct association with the department of urology at their sponsoring institution. Candidates are not required to be United States citizens; however, non-citizen applicants are required to provide assurances of employment eligibility to perform the proposed research.
Date(s) Application is Due: Feb 2
Amount of Grant: Up to 80,000 U.S.D
Samples: David C. Miller, MD, University of Michigan Medical School; Aria F. Olumi, MD, Massachusetts General Hospital; Jayoung Kim, PhD, Harvard Medical School.
Contact: Office of Medical and Academic Affairs; (212) 822-7204; fax (212) 822-7338; beer@nyam.org
Internet: http://www.nyam.org/grants/beer.shtml
Sponsor: New York Academy of Medicine
1216 Fifth Avenue
New York, NY 10029-5202

NYAM Ferdinand C. Valentine Fellowship 3773

The Fellowship offers one-year, $50,000 fellowships in support of research by individuals who have completed, or will shortly complete, residency training acceptable to the American Board of Urology, and who intend to pursue academic research careers in urology.
Requirements: Fellows must commit no less than fifty percent of their time to the supported research and preference is given to investigators conducting their research at institutions in the greater New York area.
Restrictions: Indirect costs and fringe benefits will not be paid.
Date(s) Application is Due: Feb 2
Amount of Grant: 50,000 U.S.D
Samples: Gerald J. Wang, MD, New York Presbyterian Hospital, Weill Cornell Medical Center (2009-10); Maurico Davalos, MD, New York Medical College (2008-09); Gerald Y. Tan, MD, Weill Medical College of Cornell University (2009-09).
Contact: Program Administrator; (212) 822-7204; fax (212) 822-7338; valentine@nyam.org
Internet: http://www.nyam.org/grants/valentine.shtml
Sponsor: New York Academy of Medicine
1216 Fifth Avenue
New York, NY 10029-5202

NYAM Lewis Rudin Glaucoma Grants 3774

This $50,000 prize is awarded for the most outstanding scholarly article on glaucoma published in a peer-reviewed journal during the previous calendar year. Nominations are solicited from leaders of the ophthalmologic community and reviewed in the spring of the award year. One prize winner is recommended for ratification by the Academy's Board of Trustees by the Lewis Rudin Glaucoma Prize Selection Committee, a group of nationally recognized experts in glaucoma research chaired by David H. Abramson, MD of Memorial Sloan-Kettering Cancer Center and Weill Medical College of Cornell University. The winner is announced in July of the award year.
Requirements: Nominees for this award must be the first or last author of the published work and must hold primary responsibility for the research. Authors may nominate themselves but may not submit multiple articles and must choose only one article to submit for consideration by the committee. All authors of the article receive recognition, however the monetary prize is granted solely to the primary researcher specified in the nomination. Copies of the published article must accompany the completed nomination form.
Date(s) Application is Due: May 1
Amount of Grant: 50,000 U.S.D
Samples: Oscar A. Candia, MD, Mount Sinai School of Medicine; Claude Burgoyne, MD, Devers Eye Institute; John Danias, MD, PhD, Mount Sinai School of Medicine; Cynthia Lee Grosskreutz, MD, PhD, Harvard Medical School and Massachusetts Eye and Ear Infirmary.
Contact: Program Coordinator; (212) 822-7204; fax (212) 822-7338; rudinglaucoma@nyam.org
Internet: https://www.nyam.org/grants/rudin.shtml
Sponsor: New York Academy of Medicine
1216 Fifth Avenue
New York, NY 10029-5202

NYAM Mary and David Hoar Fellowship 3775

The Mary and David Hoar Fund was established to promote research in the prevention and treatment of hip fractures. The Fund provides a two-year grant of $100,000 to support pilot programs in clinical, epidemiologic and health services research in the following areas related to hip fractures: clinical care and treatment; prevention; rehabilitation; and model program development.
Requirements: Candidates must hold an MD, PhD, or equivalent degree and are expected to conduct research in a supervised program in the greater New York area. Preferential consideration will be given to new investigators for whom this award is likely to provide interim support pending future funding, or to more senior investigators who are making a change in career or research direction for whom this award will facilitate the transition. Candidates are not required to be United States citizens; however, non-citizen applicants are required to provide assurances, such as permanent resident status, of legal eligibility for employment to perform the proposed research throughout the award period, regardless of the disposition of this or any other pending grants.
Amount of Grant: 100,000 U.S.D
Samples: Amanda S. Carmel, Weill Medical College of Cornell University (2008-10); Marcella D. Walker, Columbia University College of Physicians and Surgeons (2007-09); Knox H. Todd, Beth Israel Medical Center (2006-08).
Contact: Program Administrator; (212) 822-7204; fax (212) 822-7338; hoar@nyam.org
Internet: http://www.nyam.org/grants/hoar.shtml
Sponsor: New York Academy of Medicine
1216 Fifth Avenue
New York, NY 10029-5202

NYAM Student Essay Grants 3776

The New York Academy of Medicine invites entries for the New York Academy of Medicine Student Essay Prize, awarded to the best unpublished essay by a graduate student in a medical, public health, pharmacy, or nursing program in the United States. Essays should address topics in the history of public health or medicine as they relate to urban health issues; social or environmental factors in the health of urban populations, institutional histories, or specific diseases may be considered. The winner will receive $500, and the winning essay will receive expedited review for possible publication in the Journal of Urban Health. Honorable Mention prizes may also be awarded at the discretion of the Prize Committee.
Requirements: The contest is open to students in accredited professional degree programs in medicine, nursing, pharmacy, and public health. The writer must have been a student at the time the essay was written.
Amount of Grant: 500 U.S.D
Contact: Program Coordinator; (212) 822-7314; fax (212) 822-7338; historyessay@nyam.org
Internet: https://www.nyam.org/grants/studentessay.shtml
Sponsor: New York Academy of Medicine
1216 Fifth Avenue
New York, NY 10029-5202

NYC Managed Care Consumer Assistance Workshop Re-Design Grants 3777

The New York City Managed Care Consumer Assistance Program (NYC MCCAP), a program of the Community Service Society of NY invites you to submit a proposal for the workshop re-design project. We are seeking individuals or agencies that have interest and knowledge around training, organizational structuring and curriculum development expertise. The goal of the workshop project is to create a more flexible workshop structure that can be easily adapted to different trainers, audiences and settings ranging from busy hospitals and clinics to classrooms and health fairs.
Requirements: 501(c)3 social service agencies with a history of providing education and counseling for low-income communities, immigrants, and populations with special health care needs are encouraged to apply.
Date(s) Application is Due: May 21
Contact: Priya Mendon, (212) 614-5400; fax (212) 614-5305; pmendon@cssny.org or mccap@cssny.org
Internet: http://www.nycmccap.org/newsitems.html#rfp
Sponsor: Community Service Society of New York
105 East 22nd Street, 8th Floor
New York, NY 10010

NYCT AIDS/HIV Grants 3778

The Trust supports projects that strengthen preventive health care, improve access to services, promote the efficient use of health resources, and develop the skills and independence of people with special needs. The goal of this program is to address the complex social, medical, and legal problems of people with HIV. Since this issue

cuts across a number of program areas, projects may fall within a number of funding categories. A limited amount of support is available specifically for policy research and advocacy efforts that: increase public understanding of AIDS and HIV infection; improve the funding and delivery of services; and improve coordination among service organizations. The grant review process takes from two to six months. Grants range from $5,000 to $100,000; an average grant is around $60,000.
Requirements: Grants are made primarily to nonprofit organizations located in the five boroughs of New York City. Grants for programs outside the area are generally from funds designated for specific charities or have been made at the suggestion of donors.
Restrictions: The trust does not make grants to individuals, offer general or capital funding, endowments, building construction/renovation, deficit financing, films, or religion.
Geographic Focus: New York
Date(s) Application is Due: Sep 14
Amount of Grant: 5,000 - 100,000 U.S.D
Contact: Joyce M. Bove, Senior Vice President; (212) 686-0010, ext. 552; fax (212) 534-8528; info@nycommunitytrust.org or grants@nycommunitytrust.org
Internet: http://www.nycommunitytrust.org/HowtoApply/RequestsforProposals/NewYorkCityAIDSFund/tabid/408/Default.aspx
Sponsor: New York Community Trust
909 Third Avenue, 22nd Floor
New York, NY 10022

NYCT Biomedical Research Grants **3779**
The Trust supports projects that strengthen preventive health care, improve access to services, promote the efficient use of health resources, and develop the skills and independence of people with special needs. The Trust provides funding for biomedical research projects in four areas: blood diseases, cancer, leprosy and tuberculosis. Grants for these projects are awarded in response to requests for proposals.
Requirements: Grants are made primarily to colleges, universities, and research organizations located in the five boroughs of New York City. Grants for programs outside the area are generally from funds designated for specific charities or have been made at the suggestion of donors.
Restrictions: The trust does not make grants to individuals, offer general or capital funding, endowments, building construction/renovation, deficit financing, films, or religion.
Geographic Focus: New York
Amount of Grant: 5,000 - 100,000 U.S.D
Contact: Joyce M. Bove, Senior Vice President; (212) 686-0010, ext. 552; fax (212) 534-8528; info@nycommunitytrust.org or grants@nycommunitytrust.org
Internet: http://www.nycommunitytrust.org/ForGrantSeekers/GrantmakingGuidelines/HealthandPeoplewithSpecialNeeds/tabid/207/Default.aspx
Sponsor: New York Community Trust
909 Third Avenue, 22nd Floor
New York, NY 10022

NYCT Blindness and Visual Disabilities Grants **3780**
The Trust supports projects that strengthen preventive health care, improve access to services, promote the efficient use of health resources, and develop the skills and independence of people with special needs. The goals of this program are to support program innovation and reform and eliminate service gaps. Specifically, grants are made to projects that: improve services to those people with visual disabilities who are presently underserved, such as the elderly, minorities, people with multiple disabilities, people in institutions, and youth in transition from school to work (projects need not be limited to direct service, and may include advocacy and organizational improvement); involve people with visual disabilities more fully in community activities; expand programs that identify people with vision problems at an early stage and link them with appropriate resources; empower persons with visual disabilities by enabling them to participate in planning programs that affect them; and support research in the prevention and treatment of blindness (this may include clinical, epidemiological, and applied studies). The grant review process takes from two to six months. Grants range from $5,000 to $100,000; an average grant is around $60,000.
Requirements: Grants are made primarily to nonprofit organizations located in the five boroughs of New York City. Grants for programs outside the area are generally from funds designated for specific charities or have been made at the suggestion of donors.
Restrictions: The trust does not make grants to individuals, offer general or capital funding, endowments, building construction/renovation, deficit financing, films, or religion.
Geographic Focus: New York
Amount of Grant: 5,000 - 100,000 U.S.D
Contact: Joyce M. Bove, Senior Vice President; (212) 686-0010, ext. 552; fax (212) 534-8528; info@nycommunitytrust.org or grants@nycommunitytrust.org
Internet: http://www.nycommunitytrust.org/ForGrantSeekers/GrantmakingGuidelines/HealthandPeoplewithSpecialNeeds/tabid/207/Default.aspx
Sponsor: New York Community Trust
909 Third Avenue, 22nd Floor
New York, NY 10022

NYCT Blood Disease Research Grants **3781**
The New York Community Trust is a charitable foundation concerned with the needs of the Greater New York area. The NYCT has a special fund to be used in support of research in the field of blood diseases. A panel of recognized hematologists will convene to review proposals solicited from the major medical research institutions on a competitive basis. It is estimated that about $700,000 will be available this year to support such research projects. A total of no more than two projects may be submitted from any one organization. Basic laboratory research. Candidates may apply for support totaling $100,000 for up to 2 years, but funds for any one project year may not exceed $50,000. Second-year funding is contingent upon submission of a satisfactory progress report at the end of the first year. Research that may have application to New York City residents is of particular interest to The Trust. Young investigators will be given special consideration.
Requirements: Proposals are due no later than 5:00 p.m. on September 25th. Additional guidelines and forms are available online at the NYCT website.
Geographic Focus: New York
Date(s) Application is Due: Sep 25
Amount of Grant: 100,000 U.S.D
Contact: Joyce M. Bove, Senior Vice President; (212) 686-0010, ext. 552; fax (212) 534-8528; info@nycommunitytrust.org or grants@nycommunitytrust.org
Internet: http://www.nycommunitytrust.org/BloodDiseaseResearch/tabid/504/Default.aspx
Sponsor: New York Community Trust
909 Third Avenue, 22nd Floor
New York, NY 10022

NYCT Children and Youth with Disabilities Grants **3782**
The Trust supports projects that strengthen preventive health care, improve access to services, promote the efficient use of health resources, and develop the skills and independence of people with special needs. The goals of this program are to stimulate policymakers and service providers to improve existing services for children with disabilities and to encourage a service approach that emphasizes independence and the development of full potential. Specifically, grants are made to projects that: foster integration into community life, independent living, and improved self-image; improve early identification of disability, encourage early intervention, and increase access to early childhood education; provide comprehensive treatment, planning, and referral programs for children and their families; and assess needs, develop policy, and advocate to improve the delivery and coordination of services. The grant review process takes from two to six months. Grants range from $5,000 to $100,000; an average grant is around $60,000.
Requirements: Grants are made primarily to nonprofit organizations located in the five boroughs of New York City. Grants for programs outside the area are generally from funds designated for specific charities or have been made at the suggestion of donors.
Restrictions: The trust does not make grants to individuals, offer general or capital funding, endowments, building construction/renovation, deficit financing, films, or religion.
Geographic Focus: New York
Amount of Grant: 5,000 - 100,000 U.S.D
Contact: Joyce M. Bove, Senior Vice President; (212) 686-0010, ext. 552; fax (212) 534-8528; info@nycommunitytrust.org or grants@nycommunitytrust.org
Internet: http://www.nycommunitytrust.org/ForGrantSeekers/GrantmakingGuidelines/HealthandPeoplewithSpecialNeeds/tabid/207/Default.aspx
Sponsor: New York Community Trust
909 Third Avenue, 22nd Floor
New York, NY 10022

NYCT Girls and Young Women Grants **3783**
The Trust works in partnership with government and private agencies to develop the strengths of families and young people; to improve their living and working conditions; to improve family and child welfare services; and to advance social work practice. The goal of this Program is to improve conditions and opportunities for those who are poor and disadvantaged, particularly minorities and young, single mothers. Specifically, the program funds projects that: prepare girls and young women for economic self-sufficiency by providing the economic and personal support necessary to gain access to jobs and promoting innovative ways to include indigent and vulnerable young women in the workplace; and reduce the effects of health and social problems that threaten the lives and well-being of girls and young women through direct services and policy and advocacy efforts. Particular emphasis is placed on projects addressing the issues of teen pregnancy, homelessness, and AIDS. The grant review process takes from two to six months. Grants range from $5,000 to $100,000; an average grant is around $60,000.
Requirements: Grants are made primarily to nonprofit organizations located in the five boroughs of New York City. Grants for programs outside the area are generally from funds designated for specific charities or have been made at the suggestion of donors.
Restrictions: The trust does not make grants to individuals, offer general or capital funding, endowments, building construction/renovation, deficit financing, films, or religion.
Geographic Focus: New York
Amount of Grant: 5,000 - 100,000 U.S.D
Contact: Joyce M. Bove, Senior Vice President; (212) 686-0010, ext. 552; fax (212) 534-8528; info@nycommunitytrust.org or grants@nycommunitytrust.org
Internet: http://www.nycommunitytrust.org/ForGrantSeekers/GrantmakingGuidelines/ChildrenYouthandFamilies/tabid/205/Default.aspx
Sponsor: New York Community Trust
909 Third Avenue, 22nd Floor
New York, NY 10022

NYCT Grants for the Elderly **3784**
The Trust supports projects that strengthen preventive health care, improve access to services, promote the efficient use of health resources, and develop the skills and independence of people with special needs. The goals of this program are to enable elderly people to remain active in their communities and to meet the basic needs of those who are vulnerable and dependent. Resources are primarily targeted toward the most underserved elderly and those whose needs are most acute, including members of racial and ethnic minorities, the poor, and those with chronic illnesses or mental or functional disabilities. Specifically, grants are made to projects that: increase the number of elderly who are able to participate in community activities, particularly those involving contact

with young people; enable the elderly to assume a leadership role in planning, influencing, or changing the programs that serve them; improve the management, capacity, and resources of government and voluntary agencies serving the elderly, and encourage collaboration among agencies; and assess need, develop policy, and advocate to improve the delivery and coordination of services. The grant review process takes from two to six months. Grants range from $5,000 to $100,000; an average grant is around $60,000.
Requirements: Grants are made primarily to nonprofit organizations located in the five boroughs of New York City. Grants for programs outside the area are generally from funds designated for specific charities or have been made at the suggestion of donors.
Restrictions: The trust does not make grants to individuals, offer general or capital funding, endowments, building construction/renovation, deficit financing, films, or religion.
Geographic Focus: New York
Amount of Grant: 5,000 - 100,000 U.S.D
Contact: Joyce M. Bove, Senior Vice President; (212) 686-0010, ext. 552; fax (212) 534-8528; info@nycommunitytrust.org or grants@nycommunitytrust.org
Internet: http://www.nycommunitytrust.org/ForGrantSeekers/GrantmakingGuidelines/HealthandPeoplewithSpecialNeeds/tabid/207/Default.aspx
Sponsor: New York Community Trust
909 Third Avenue, 22nd Floor
New York, NY 10022

NYCT Health Services, Systems, and Policies Grants **3785**
The Trust supports projects that strengthen preventive health care, improve access to services, promote the efficient use of health resources, and develop the skills and independence of people with special needs. The primary goal of this program is to improve the effectiveness, responsiveness, and equity of health care in New York City. Specifically, the program makes grants: for services, policy research, advocacy, and technical assistance that promote the accessibility of basic health services, especially in minority and immigrant communities; to strengthen health service providers, especially those serving the city's poorest residents; and to promote healthy lifestyles. The grant review process takes from two to six months. Grants range from $5,000 to $100,000; an average grant is around $60,000.
Requirements: Grants are made primarily to nonprofit organizations located in the five boroughs of New York City. Grants for programs outside the area are generally from funds designated for specific charities or have been made at the suggestion of donors.
Restrictions: The trust does not make grants to individuals, offer general or capital funding, endowments, building construction/renovation, deficit financing, films, or religion.
Geographic Focus: New York
Amount of Grant: 5,000 - 100,000 U.S.D
Contact: Joyce M. Bove, Senior Vice President; (212) 686-0010, ext. 552; fax (212) 534-8528; info@nycommunitytrust.org or grants@nycommunitytrust.org
Internet: http://www.nycommunitytrust.org/ForGrantSeekers/GrantmakingGuidelines/HealthandPeoplewithSpecialNeeds/tabid/207/Default.aspx
Sponsor: New York Community Trust
909 Third Avenue, 22nd Floor
New York, NY 10022

NYCT Mental Health and Mental Retardation Grants **3786**
The Trust supports projects that strengthen preventive health care, improve access to services, promote the efficient use of health resources, and develop the skills and independence of people with special needs. The goals of this program are to foster the independence of people with mental illness and mental retardation, and to encourage a community-based system of care. Specifically, grants are made to projects that: improve the quality and availability of community housing and services for people with chronic mental illnesses; improve and expand the delivery of mental health services to children and adolescents, especially those that stress the early identification and remediation of problems; strengthen advocacy groups that promote reimbursement practices and allocation of mental health resources that correspond to community needs; and direct the attention of service providers, policy makers, and the general public to the mental health concerns of racial and ethnic minorities. The grant review process takes from two to six months. Grants range from $5,000 to $100,000; an average grant is around $60,000.
Requirements: Grants are made primarily to nonprofit organizations located in the five boroughs of New York City. Grants for programs outside the area are generally from funds designated for specific charities or have been made at the suggestion of donors.
Restrictions: The trust does not make grants to individuals, offer general or capital funding, endowments, building construction/renovation, deficit financing, films, or religion.
Geographic Focus: New York
Amount of Grant: 5,000 - 100,000 U.S.D
Contact: Joyce M. Bove, Senior Vice President; (212) 686-0010, ext. 552; fax (212) 534-8528; info@nycommunitytrust.org or grants@nycommunitytrust.org
Internet: http://www.nycommunitytrust.org/ForGrantSeekers/GrantmakingGuidelines/HealthandPeoplewithSpecialNeeds/tabid/207/Default.aspx
Sponsor: New York Community Trust
909 Third Avenue, 22nd Floor
New York, NY 10022

NYCT Substance Abuse Grants **3787**
The Trust works in partnership with government and private agencies to develop the strengths of families and young people; to improve their living and working conditions; to improve family and child welfare services; and to advance social work practice. The goal of this program is to address the treatment needs of the most under-served, with a focus on women, ex-offenders, and adolescents. Specifically, grants are made to projects that: promote coordinated drug treatment, particularly between treatment providers and other systems such as mental health and child welfare; expand treatment by building the capacity of mid-sized drug treatment programs. We will help groups strengthen and increase existing services and develop new capacity to accommodate people not being treated; and promote strategies that reduce or prevent substance abuse among elders. The grant review process takes from two to six months. Grants range from $5,000 to $100,000; an average grant is around $60,000.
Requirements: Grants are made primarily to nonprofit organizations located in the five boroughs of New York City. Grants for programs outside the area are generally from funds designated for specific charities or have been made at the suggestion of donors.
Restrictions: The trust does not make grants to individuals, offer general or capital funding, endowments, building construction/renovation, deficit financing, films, or religion.
Geographic Focus: New York
Amount of Grant: 5,000 - 100,000 U.S.D
Contact: Joyce M. Bove, Senior Vice President; (212) 686-0010, ext. 552; fax (212) 534-8528; info@nycommunitytrust.org or grants@nycommunitytrust.org
Internet: http://www.nycommunitytrust.org/ForGrantSeekers/GrantmakingGuidelines/ChildrenYouthandFamilies/tabid/205/Default.aspx
Sponsor: New York Community Trust
909 Third Avenue, 22nd Floor
New York, NY 10022

NYCT Technical Assistance Grants **3788**
The Trust focuses on relieving New York's chronic shortage of affordable housing, strengthening the local economy, and protecting the environment. It supports community-based agencies working on these issues at the neighborhood level, and government and nonprofit institutions developing strategies for the City as a whole. The goal of this program is to improve the management capacity of nonprofits and strengthen the nonprofit sector. Specifically, The Trust provides grants: up to $10,000 to current and prospective grantees to hire consultants to help with specific management needs; to advance public and nonprofit service by developing skills and expertise of professionals in the field; and to support service and umbrella organizations providing technical assistance to groups of nonprofits. The grant review process takes from two to six months. Grants range from $5,000 to $100,000; an average grant is around $60,000.
Requirements: Grants are made primarily to nonprofit organizations located in the five boroughs of New York City. Grants for programs outside the area are generally from funds designated for specific charities or have been made at the suggestion of donors.
Restrictions: The trust does not make grants to individuals, offer general or capital funding, endowments, building construction/renovation, deficit financing, films, or religion.
Geographic Focus: New York
Amount of Grant: 5,000 - 100,000 U.S.D
Contact: Joyce M. Bove, Senior Vice President; (212) 686-0010, ext. 552; fax (212) 534-8528; info@nycommunitytrust.org or grants@nycommunitytrust.org
Internet: http://www.nycommunitytrust.org/ForGrantSeekers/GrantmakingGuidelines/CommunityDevelopmentandtheEnvironment/tabid/204/Default.aspx
Sponsor: New York Community Trust
909 Third Avenue, 22nd Floor
New York, NY 10022

Oak Foundation Grants **3789**
The foundation awards grants worldwide to address international social and environmental issues, particularly those that have a major impact on the lives of the disadvantaged. Grants support the environment, especially conservation of the marine environment and reducing the threat of global warming by promoting implementation of energy-efficient and renewable energy technologies; international human rights; issues affecting women, particularly domestic violence and the challenges faced by single mothers; child abuse, including sexual abuse, child pornography, and related trafficking; housing and homelessness; and learning disabilities, such as dyslexia. Types of support include special projects, core costs, capital needs, technical assistance, and collaborative activities. Priority is given to leadership projects that target root causes; demonstrate solutions that can be adopted by permanent providers and government; mobilize financial commitment from a range of sources; promote collaboration among organizations and funders; involve the target population in the planning and implementation of the project; and demonstrate good financial organizational management. Oak Foundation's main office is in Geneva. It has several other offices which are located to serve its various programs (see "Contact Oak Page" on the Oak Foundation website for contact information relating to each program). All email inquiries should be directed to the appropriate Oak Foundation programme and letters addressed to the appropriate office. If in doubt please write to the Geneva office or email info@oakfnd.ch To begin the application process submit a letter of inquiry to the appropriate office.
Requirements: The Foundation will consider requests for funding of special projects, core costs, capital needs, technical assistance, or collaborative activities. The Foundation may also seek to stimulate initiatives in its areas of interest by: issuing calls for proposals on specific topics; convening policymakers and practitioners to explore solutions to critical issues; initiating its own programmes.
Restrictions: Grants will not be awarded for: religious organisations for religious purposes; supporting candidates for office; general fund-raising drives or events; or Amounts under U.S.$25,000 (except in special circumstances).
Amount of Grant: 25,000 - 2,000,000 U.S.D
Contact: Grants Manager; naep@oakfnd.org or info@oakfnd.org
Internet: http://www.oakfnd.org
Sponsor: Oak Foundation U.S.A
511 Congress Street, Suite 800
Portland, ME 04101

Oakland Fund for Children and Youth Grants 3790

Grants support Oakland, CA, nonprofit organizations in two categories. The general grants program awards grants to organizations serving children and youth ages 0-20 in four priority areas: support for children's success in school, child health and wellness, healthy transitions to adulthood, and youth empowerment. The Community Promotions Program for Service Organizations and the Youth Violence/Drug Prevention Program award grants to organizations providing parent/caregiver support services, youth drug prevention services, or work to prevent youth violence. Annual deadline dates may vary; contact program staff for exact date.
Requirements: Oakland, California, nonprofits serving children and youth are eligible.
Geographic Focus: California
Date(s) Application is Due: Jan 6
Amount of Grant: 20,000 - 250,000 U.S.D
Contact: Ayako Miyashita, Grant Coordinator; (510) 238-4913; fax (510) 238-4971; amiyashita@oaklandnet.com
Internet: http://www.ofcy.org
Sponsor: East Bay Community Foundation, Oakland Fund for Children and Youth
De Domenico Building, 150 Frank H. Ogawa Plaza, Suite 4216
Oakland, CA 94612

Obsessive-Compulsive Foundation Research Awards 3791

The foundation invites proposals for research projects on any aspect of obsessive-compulsive disorders. Applicants should submit a proposal limited to five pages and a budget. The foundation will request more detailed proposals following initial review.
Amount of Grant: 39,000 - 49,000 U.S.D
Contact: Jeannette Cole, (888) 302-6662 or (203) 401-2069; fax (203) 315-2196; cole@ocfoundation.org
Internet: http://www.ocfoundation.org/ocf1140a.htm
Sponsor: Obsessive-Compulsive Foundation
337 Notch Hill Road
North Branford, CT 06471

OBSSR Behavioral and Social Science Research on Understanding and Reducing Health Disparities Grants 3792

The purpose of this program is to encourage behavioral and social science research on the causes and solutions to health and disabilities disparities in the U. S. population. Health disparities between, on the one hand, racial/ethnic populations, lower socioeconomic classes, and rural residents and, on the other hand, the overall U.S. population are major public health concerns. Emphasis is placed on research in and among three broad areas of action: public policy; health care; and disease/disability prevention. Particular attention is given to reducing health gaps among groups. Proposals that utilize an interdisciplinary approach, investigate multiple levels of analysis, incorporate a life-course perspective, and/or employ innovative methods such as system science or community-based participatory research are particularly encouraged. The NIH anticipates supporting 20 to 30 awards. Because the nature and scope of the proposed research will vary from application to application, it is anticipated that the size and duration of each award will also vary.
Requirements: Application can be made by the following institutions and organizations: public/state controlled institutions of higher education; private institutions of higher education; nonprofits with 501(c)3 IRS status (other than institutions of higher education); nonprofits without 501(c)3 IRS status (other than institutions of higher education); small businesses; for-profit organizations (other than small businesses); state governments; U.S. territories or possessions; Indian/Native American tribal governments (federally recognized); Indian/Native American tribal governments (other than Federally recognized); Indian/Native American tribally designated organizations; non-domestic (non-U.S.) entities (foreign organizations); Hispanic-serving institutions; historically Black colleges and universities (HBCUs); tribally controlled colleges and universities (TCCUs); Alaska Native and Native Hawaiian serving institutions; regional organization; eligible agencies of the Federal government; and faith-based or community based organizations.
Date(s) Application is Due: Feb 5; Jun 5; Oct 5
Contact: Ronald P. Abeles, (301) 496-7859; fax (301) 435-8779; abeles@nih.gov
Internet: http://grants.nih.gov/grants/guide/pa-files/PAR-07-379.html
Sponsor: Office of Behavioral and Social Sciences Research
31 Center Drive, Building 31, Room B1C19
Bethesda, MD 20892-2027

OceanFirst Foundation Grants 3793

The foundation awards two types of grants (small and major) to nonprofit organizations in OceanFirst Bank's service area and its neighboring communities for programs that expand home ownership opportunities and provide access to affordable rental housing; support youth development programs to improve life options through education and work skills; support community organizations that contribute to the quality of life; and support nonprofit medical facilities. There are no application deadlines; requests are considered in the first, second, and fourth quarters of the year.
Requirements: New Jersey nonprofit organizations serving Middlesex, Monmouth, and Ocean Counties are eligible.
Geographic Focus: New Jersey
Amount of Grant: 3,500 - 40,000 U.S.D
Contact: Katherine Durante, Executive Director; (732) 341-4676; fax (732) 473-9641; kdurante@oceanfirstfdn.org
Internet: http://www.oceanfirstfdn.org
Sponsor: OceanFirst Foundation
1415 Hooper Avenue, Suite 304
Toms River, NJ 08753

Oceanside Charitable Foundation Grants 3794

The Oceanside Charitable Foundation will focus on supporting projects that inspire and strengthen Civil Society in its communities; in particular, projects that foster community participation, consensus-building and collective problem solving in the field of children and youth programs. The Oceanside Charitable Foundation will fund Civil Society projects for school-aged (K-12) children and youth that are managed by non-profit organizations or government agencies.
Requirements: Projects for which funding is requested must serve children and/or youth in the geographical limits of the City of Oceanside.
Restrictions: The Oceanside Charitable Foundation does not make grants for: annual campaigns and fund raising events for non-specific purposes; capital campaigns for buildings or facilities; stipends for attendance at conferences; endowments or chairs; for-profit organizations or enterprises; individuals unaffiliated with a qualified fiscal sponsor; projects that promote religious or political doctrine; research projects (medical or otherwise); scholarships; or existing obligations or debt.
Geographic Focus: California
Date(s) Application is Due: Dec 18
Contact: Trudy Amstrong, Regional Manager; (619) 814-1384; trudy@sdfoundation.org
Internet: http://www.endowoceanside.org/grants.html
Sponsor: Oceanside Charitable Foundation / San Diego Foundation
2508 Historic Decatur Road, Suite 200
San Diego, CA 92106

Office Depot Corporation Community Relations Grants 3795

The corporate giving program awards grants to nonprofit organizations aligned with Office Depot's mission to directly impact the health, education, and welfare of children. Office Depot supports nonprofit organizations at the local level with donations of products, contributions of funds, and efforts to encourage employees and customers to become involved. To request a monetary donation, provide a brief description of the organization, the federal tax-ID number, an explanation of what is being requested, and the rationale based on Office Depot's charitable giving guidelines. The request should be submitted on organization letterhead and sent to the office.
Requirements: 501(c)3 nonprofit organizations are eligible.
Restrictions: Grants do not support individuals, advertising, athletic teams or events, fashion shows, project graduation, capital campaigns, individual or group travel, political causes, film/video projects, or nonprofit organizations that spend more than 25 percent of their revenue on management overhead and fundraising expenses.
Contact: Melissa Perlman, Public Relations; (561) 438-0704; Melissa.Perlman@officedepot.com
Internet: http://www.community.officedepot.com/local.asp
Sponsor: Office Depot Corporation
2200 Old Germantown Road
Delray Beach, FL 33445

Ogden Codman Trust Grants 3796

The trust awards grants to eligible Massachusetts nonprofit organizations in its areas of interest, including environmental conservation, historic preservation, cultural activities, health care, and social services. Grants support special programs intended to improve the quality of life. Matching funds are sometimes requested
Requirements: Lincoln, MA, nonprofit organizations are eligible to apply.
Restrictions: General operating grants are rarely awarded.
Samples: Campaign for Drumlin Farm, Lincoln, MA, $25,000; Lincoln Public Schools, Lincoln, MA, $25,000.
Contact: Susan Monahan, Grants Coordinator; (617) 951-1108; fax (617) 542-7437; smonahan@rackemann.com
Sponsor: Ogden Codman Trust
160 Federal Street, 13th Floor
Boston, MA 02110-1700

Ohio County Community Foundation Grants 3797

Established in 1998, the Ohio County Community Foundation is charged with assisting donors in building, managing and distributing a lasting source of charitable funds for the good of Ohio County. The Community Foundation currently offers services that include endowment creation, scholarship administration and charitable gift funds. Currently, there are two annual grant cycles, with deadlines on April 15 and October 15 respectively.
Requirements: Grants will be made only to organizations whose programs benefit the residents of Ohio County.
Restrictions: No grants will be made solely to individuals but can be made for the benefit of certain individuals for such purposes as scholarships and special programs through educational institutions and other sponsoring recipient organizations. Generally grants will not be made to enable individuals or groups of individuals to take trips except where there are special circumstances which will benefit the larger community.
Geographic Focus: Indiana
Date(s) Application is Due: Apr 15; Oct 15
Amount of Grant: Up to 3,000 U.S.D
Contact: Stephanie Scott, Program Coordinator; (812) 438-9401; fax (812) 438-9488; sscott@occfrisingsun.com
Internet: http://www.occfrisingsun.com/grants.html
Sponsor: Ohio County Community Foundation
591 Smart Drive, P.O. Box 170
Rising Sun, IN 47040

Ohio County Community Foundation Mini-Grants **3798**
Established in 1998, the Ohio County Community Foundation is charged with assisting donors in building, managing and distributing a lasting source of charitable funds for the good of Ohio County. The Ohio County Community Foundation will seek and accept mini-grant applications from area non-profit organizations whose programs benefit the residents of Ohio County. Each mini-application will be reviewed after submission by the Program Coordinator to ensure the application is complete and the organization is eligible to make application.
Requirements: Grants will be made only to organizations whose programs benefit the residents of Ohio County.
Restrictions: No grants will be made solely to individuals but can be made for the benefit of certain individuals for such purposes as scholarships and special programs through educational institutions and other sponsoring recipient organizations. Generally grants will not be made to enable individuals or groups of individuals to take trips except where there are special circumstances which will benefit the larger community.
Geographic Focus: Indiana
Amount of Grant: Up to 100 U.S.D
Contact: Stephanie Scott, Program Coordinator; (812) 438-9401; fax (812) 438-9488; sscott@occfrisingsun.com
Internet: http://www.occfrisingsun.com/grants.html
Sponsor: Ohio County Community Foundation
591 Smart Drive, P.O. Box 170
Rising Sun, IN 47040

Ohio Learning Network Grants **3799**
The network funds programs to help Ohioans increase work-related knowledge and skills through college credits; assist faculty in teaching using emerging technologies; and better understand the role of technology in learning. Grants support emerging needs, learning communities, distance learning research, partnership grants, and technology initiatives/innovations.
Geographic Focus: Ohio
Amount of Grant: Up to 100,000 U.S.D
Contact: Judith M. Leach; (614) 485-6738 or (614) 995-3240; jleach@oln.org
Internet: http://www.oln.org/about_oln/grants.php
Sponsor: Ohio Learning Network
35 E Chestnut Street, 8th Floor
Columbus, OH 43215-2541

Oleonda Jameson Trust Grants **3800**
The trust awards grants to New Hampshire nonprofit organizations in its areas of interest, including arts/culture, community foundations, children and youth, health care, housing, social sciences, human services, federated giving programs and, social services. Types of support include capital campaigns, building construction/renovation, equipment acquisition, emergency funds, and scholarship funds. There are no application forms or deadlines. The board meets in March, June, September, and December.
Requirements: New Hampshire nonprofit organizations are eligible. Requests from Concord nonprofits receive preference.
Restrictions: Grants do not support endowments or general operating expenses.
Geographic Focus: New Hampshire
Amount of Grant: 1,000 - 60,000 U.S.D
Samples: New Hampshire Charitable Foundation, Concord, NH, $25,000; Riverbend Community Mental Health, Concord, NH, $15,000; Arts in Reach, Dover, NH, $7,500.
Contact: Grants Management Officer; (603) 226-0400
Sponsor: Oleonda Jameson Trust
11 South Main Street, Suite 500
Concord, NH 03301-4945

Olfactory Research Fund Grants **3801**
The fund sponsors innovative scientific research in the olfactory arts and sciences. Grants promote the study of olfaction and also support educational and outreach programs about the sense of smell. The fund is especially interested in supporting research that integrates the study of olfaction with current issues in developmental, perceptual, social and cognitive psychology, and related disciplines. The fund is currently seeking proposals for research that focuses on the influence of olfaction in the mental and physical aspects of performance. Contact organization for submission deadlines.
Requirements: Applicants must have a doctoral degree.
Restrictions: Generally, the following types of research are ineligible for support: studies not directly related to olfaction and/or the human reaction to odors; studies exclusively on animal olfaction without direct relevance to humans; or invasive human studies.
Contact: Theresa Molnar, Executive Administrator; (212) 725-2755; fax (212) 779-9072; tmolnar@senseofsmell.org or info@senseofsmell.org
Internet: http://www.senseofsmell.org/resources/research.php
Sponsor: Olfactory Research Fund
145 E 32nd Street
New York, NY 10016

Olin Corporation Charitable Trust Grants **3802**
The charitable trust supports nonprofits in company-operating areas. Grants focus on three priority categories: education, with a science and technology emphasis; programs that include Olin employees as volunteers and that impact communities in which Olin employees work and live; and conservation, environmental education, and environmental research. Types of support include general operating budgets, continuing support, annual campaigns, seed money, emergency funds, building funds, equipment, land acquisition, special projects, research, publications, conferences and seminars, internships, scholarship funds, employee-related scholarships, fellowships, and employee matching gifts. U.S.-based nonprofits with an international focus are supported primarily through matching gifts. There are no application deadlines. Applicants should submit a letter of request. The board meets in January to consider requests.
Requirements: Support is directed primarily to communities where Olin has a major employee presence.
Restrictions: Grants are not awarded to individuals or for endowments.
Amount of Grant: 50 - 50,000 U.S.D
Samples: Thirteen communities in Olin operating areas—for education programs, with a focus on computer technology and science education, $242,137 divided.
Contact: Program Contact; (314) 480-1400
Internet: http://www.olin.com/about/charitable.asp
Sponsor: Olin Corporation Charitable Trust
190 Carondelet Plaza, Suite 1530
Clayton, MO 63105-3443

Omaha Community Foundation Grants **3803**
The goal of the community foundation is to improve the quality of life in the Omaha, NE, area by supporting needs that are not being met in the areas of civic, cultural, health, educational, and social service. Types of support include continuing support, building construction/renovation, equipment, emergency funds, program development, conferences and seminars, publication, seed grants, scholarship funds, technical assistance, and matching funds. Proposals from organizations demonstrating broad community support for their proposed programs are given priority consideration. The foundation, as a general policy, gives less consideration to applications from tax-supported institutions, individual churches and similar religious groups, veterans and labor organizations, social clubs, and fraternal organizations.
Requirements: IRS 501(c)3 nonprofits serving the Omaha, Nebraska, area are eligible.
Restrictions: The foundation does not make grants to individuals or for endowments, deficit financing, annual fund drives, or fund-raising activities.
Geographic Focus: Nebraska
Date(s) Application is Due: Mar 1; Sep 1
Samples: Heritage Joslyn Foundation (Omaha, NE)—continuing support, $245,000; Opera/Omaha (Omaha, NE)—program support, $91,550; Salvation Army of Omaha (Omaha, NE)—continuing support, $110,575.
Contact: Anne Camp; (402) 342-3458; fax (402) 342-3582; anne@omahacf.org
Internet: http://www.omahacf.org
Sponsor: Omaha Community Foundation
1623 Farnham Street, Suite 600
Omaha, NE 68102

OMNOVA Solutions Foundation Grants **3804**
The foundation strives to be a strengthening thread in the fabric of society through donations to OMNOVA Solutions communities in the areas of education, civic enhancement, health and welfare, and arts and culture. Preference is given to community projects that involve OMNOVA employees and requests that are recommended by the foundation coordinators at OMNOVA facilities.
Requirements: Grants support nonprofit organizations in OMNOVA communities: Calhoun, GA; Fitchburg, MA; Columbus, MS; Salem, NH; Pine Brook, NJ; New York, NY; Charlotte, Greensboro, and Monroe, NC; Akron, Fairlawn, Maumee, and Mogadore, OH; Auburn and Jeannette, PA; Chester, SC; and Green Bay, WI.
Geographic Focus: Georgia, Massachusetts, Mississippi, New Hampshire, New Jersey, New York, North Carolina, Ohio, Pennsylvania, South Carolina, Wisconsin
Contact: Grants Administrator; (330) 869-4289
Internet: http://www.omnova.com/comm_content.htm
Sponsor: OMNOVA Solutions Foundation
175 Ghent Road
Fairlawn, OH 44333

OneFamily Foundation Grants **3805**
The foundation awards grants to eligible Washington nonprofit organizations working to improve the lives of women living in poverty and at-risk youth, for support services for abused women, and for efforts to end sexual abuse against women and children. Consideration is given to programs that provide training and skills development to low-income women and services providing basic needs such as shelter, counseling, food, and childcare; educational and mentoring projects to help prevent teen pregnancy; job-training programs for youth and school/community-based programs to help low-income youth complete their education; parenting, training, and education programs to help break the cycle of family violence; shelters and services to support abused and neglected children and women; and hands-on programs to encourage philanthropy among children and youth. Grants will support operating expenses, special projects, and minor capital costs necessary to assure the success of a funded project. For general grants, two-page preapplication letters are due on the third Friday in March, July, and November and final proposals are due the second Friday in January, May, and September. Annual deadlines may vary; contact program staff for exact dates.
Requirements: Nonprofit 501(c)3 organizations based in King County, Snohomish County, or the Olympic Peninsula of Washington State are eligible to apply for funding.
Restrictions: Grants are not made to: individuals, scholarships, schools, research, summer camps, athletic events, video or film projects, website development, book publications. No multi-year requests are considered. Groups who have been declined three times are ineligible to reapply.
Geographic Focus: Washington

Amount of Grant: 5,000 - 12,000 U.S.D
Samples: Amara Parenting and Adoption, Seattle, Washington, $12,000 - to foster-adoption the program, training for foster parents, and support for hard-to-place children (2010); First Step Family Support Center, Port Angeles, Washington, $11,154 - Parent-Child Interaction Therapy (2010); Indian Law Resource Center, Helena, Washington. $8,000 - Safe Women/Strong Nations, breaking the cycle of violence against Native women (2010).
Contact: Therese Ogle; (206) 781-3472; fax (206) 784-5987; Oglefounds@aol.com
Internet: http://fdncenter.org/grantmaker/onefamily
Sponsor: OneFamily Foundation
6723 Sycamore Avenue NW
Seattle, WA 98117

OneSight Research Foundation Block Grants **3806**
Formerly known as the Pearle Vision Foundation, the OneSight Research Foundation is dedicated to sight preservation through research and education. A significant portion of their grant dollars will be awarded to fund research and treatment for diabetic eye diseases, hoping to find a cure for diabetic retinopathy, the number one cause of adult blindness in the United States.
Requirements: U.S. 501(c)3 nonprofits are eligible for grant support. Requests for funding should cover a period of time not to exceed one year. Exceptions to this policy will be considered on an individual basis.
Restrictions: Grants will not be considered for endowments or operating expenses.
Date(s) Application is Due: Jun 30; Dec 31
Amount of Grant: 5,000 - 50,000 U.S.D
Samples: Retina Foundation of the Southwest (Dallas, TX)—to equip and furnish a new visual-testing laboratory at the Morchower Pediatric Eye Research Laboratory, $16,342; Lighthouse International (New York, NY)—to study a new reading method for people with macular degeneration and other diseases that result in central-vision loss, $10,000; New Eyes for the Needy (Short Hills, NV)—for eyeglass vouchers to serve individuals nationwide, $25,000.
Contact: Trina Parasiliti; (972) 277-6191; fax (972) 277-6414; tparasil@onesight.org
Internet: http://www.onesight.org/northamerica/na/
Sponsor: OneSight Research Foundation
2465 Joe Field Road
Dallas, TX 75229

OneStar Foundation AmeriCorps Grants **3807**
OneStar NSC seeks to fund programs that engage members to address health, public safety, homeland security, education, or human service needs in Texas communities. Funds support the state priority grant, professional corps grant, and education award grant. AmeriCorps education awards for members who successfully complete their full term of service will also be supported. Guidelines are available online.
Requirements: Local or statewide nonprofit organizations, state and local units of government, and institutions of higher education are eligible.
Contact: Grants Administrator; (512) 287-2000; fax (512) 473-8228; onestar@onestarfoundation.org
Internet: http://www.onestarfoundation.org/page/americorpstexas
Sponsor: OneStar Foundation
816 Congress, Stuite 900
Austin, TX 78701

ONF Career Development Awards **3808**
The awards support the career development of professional nurses in the field of oncology nursing. The awards allow attendance in an educational program focusing on clinical ethics that will further the nurse's professional goals and potential to contribute to oncology nursing. This award is not intended to supplement tuition in a graduate degree program. Preference will be given to applicants who are seeking beginning level knowledge in clinical ethics.
Requirements: All applicants must be registered nurses or nurse administrators actively involved in some aspect of cancer patient care, education, or research.
Amount of Grant: 1,000 U.S.D
Contact: ONS Foundation Office; (412) 859-6100, ext. 8503; customer.service@ons.org or member@ons.org
Internet: http://www.ons.org/awards/foundawards/ethics.shtml
Sponsor: Oncology Nursing Foundation
125 Enterprise Drive
Pittsburgh, PA 15275-1214

ONF Small Grants **3809**
These grant awards are designed to support oncology nursing research conducted by beginning and experienced researchers. The majority of the small grants are for general investigator-initiated oncology topics. Funding preference is given to research that addresses the ONS Research Priorities and Research Agenda. A letter of intent is requested by October 1; full proposal is due November 1. Annual deadline dates may vary; contact program staff for exact dates.
Requirements: The principal investigator must be actively involved in some aspect of cancer patient care, education, or research. Funding preference is given to projects that involve nurses in the design and conduct of the research activity and that promote oncology nursing theory and a scientific basis for practice.
Date(s) Application is Due: Oct 1; Nov 1
Amount of Grant: 10,000 U.S.D
Samples: Kimberly Kahn, U of Virginia Health Science Ctr (Charlottesville, VA)—nursing education research grant, The Informed Consent Process in Oncology Clinical Trials, $7500; Dr. June Eilers, U of Nebraska Medical Ctr (Omaha, NE)—research grant, Longitudinal Study of the Impact of BMT on Family Members, $5000.
Contact: Grants Administrator; research@ons.org
Internet: http://www.ons.org/awards/foundawards/smallGrants.shtml
Sponsor: Oncology Nursing Foundation
125 Enterprise Drive, RIDC Park W
Pittsburgh, PA 15275-1214

ONR Human Performance Division Research Grants **3810**
The mission of the Human Performance S&T Division is to direct, plan, foster and encourage Science and Technology in cognitive science, computational neuroscience, bio-inspired systems, social/organizational science, manpower & personnel, training, human factors, decision making and human systems integration as related to Naval needs. Major goals include: to sustain warfighter performance and enhanced decision making in all environments through training technologies, manpower management, and human systems integration; to create options for future (perhaps unanticipated) naval decisions, based upon fundamental understanding of the cognitive and neurosciences; to support integrated interdisciplinary research programs; and to cultivate transition of findings to government and commercial funding of applied research, advanced technology development, small business and acquisition projects. Submission of the Research Proposal should always be preceded by submission of a Preproposal and a discussion of that Preproposal with the assigned Program Officer.
Requirements: Grants are awarded to academic, industrial, and nonprofit organizations, with strong emphasis on university research.
Restrictions: New grants or contracts are typically started near the beginning of the fiscal year (October 1). New proposals for the next fiscal year should be submitted no later than June 1, unless otherwise directed by the BAA. Competitive renewals should be submitted no later than six months prior to the award termination date after contacting the Program Officer. Conference proposals should be submitted at least four months prior to the conference.
Amount of Grant: 70,000 - 150,000 U.S.D
Contact: Dr. John F. Tangney, (703) 696-4501 or (703) 588-2427; fax (703) 696-5940; john.f.tangney@navy.mil
Internet: http://www.onr.navy.mil/sci_tech/34/341/
Sponsor: U.S. Department of Defense
One Liberty Center, 875 N. Randolph Street, Suite 1425
Arlington, VA 22203-1995

ONS Major Research Grants **3811**
Applications are invited for research proposals which focus on increasing understanding of and management of understudied symptoms and/or symptom clusters in diverse populations and across settings. Priority will be given to proposals that address management of symptoms and/or symptom clusters that include dermatologic problems in cancer patients. Examples of symptom clusters include dermatitis, pain, and sleep/wake disturbances or fatigue, depression, and sleep/wake disturbances. Other symptom clusters and symptoms will be considered. Justification must be provided for selecting the cluster and/or the rationale for the research needed for the symptom. Membership in ONS is not required.
Requirements: The principal investigator must be actively involved in some aspect of cancer patient care, education, or research.
Amount of Grant: 50,000 U.S.D
Contact: Research Team; (412) 859-6298; fax (877) 369-5497; research@ons.org
Internet: http://www.ons.org/awards/foundawards
Sponsor: Oncology Nursing Society
125 Enterprise Drive
Pittsburgh, PA 15275-1214

ONS Scholarships **3812**
Nursing scholarships are awarded each year to enable registered nurses to continue their education in the area of oncology nursing. Each year, undergraduate scholarships, graduate scholarships, and doctoral scholarships are awarded. In addition Congress Scholarships (December 1 deadline) may be applied toward registration, travel to and from Congress, and per diem Congress expenses. Guidelines are available online.
Requirements: An applicant must have a current license to practice as a registered nurse and be enrolled in (or accepted in) a nursing-degree program in an NLN-accredited school of nursing.
Date(s) Application is Due: Feb 1
Amount of Grant: 1,000 - 5,000 U.S.D
Contact: Research Team; (412) 859-6100; foundation@ons.org
Internet: http://www.ons.org
Sponsor: Oncology Nursing Society
125 Enterprise Drive
Pittsburgh, PA 15275-1214

Oppenstein Brothers Foundation Grants **3813**
Grants are awarded in the metropolitan area of Kansas, City, Missouri, primarily for social services and early childhood, elementary, secondary, adult-basic, vocational, and higher education; family planning and services; social services and welfare agencies; Jewish welfare organizations; and programs for youth, the handicapped, disadvantaged, mentally ill, homeless, minorities, and elderly. Additional areas of interest include arts/cultural programs, the performing arts, museums, health care and health organizations, and AIDS research. The foundation considers requests for building and renovation, capital campaigns, curriculum development, emergency funds, equipment, general operating

support, program development, seed money, technical support, conferences and seminars, consulting services, and matching funds. Application guidelines are available on request. The board meets every other month. Deadlines are generally three weeks prior to board meetings. Notification of award will take place within two to four months.
Requirements: Nonprofit organizations serving the metropolitan area of Kansas City, Missouri are eligible to apply.
Restrictions: The foundation primarily supports 501(c)3 nonprofit organizations serving the metropolitan area of Kansas City, Missouri. Grants are not awarded to support individuals or for annual campaigns.
Geographic Focus: Missouri
Amount of Grant: 5,000 - 15,000 U.S.D
Samples: Associated Youth Services, Kansas City, Kansas, $10,000 (2011); Child Advocacy Services Center, Kansas City, Missouri, $25,000 (2011); Spay and Neuter Kansas City, Kansas City, Missouri, $5,000 (2011).
Contact: Beth Radtke, Program Officer; (816) 234-2577
Sponsor: Oppenstein Brothers Foundation
922 Walnut Street, Suite 200
Kansas City, MO 64106-1809

Oprah Winfrey Foundation Grants **3814**
The foundation focuses on education and on empowering children, women, and families. Grants are made in a range of interests, including art and community, children and youth, education, health, and social services. Types of support include pilot, program, and operating grants. There are no application deadlines.
Requirements: Nonprofit organizations are eligible, with preference given to Chicago.
Amount of Grant: 10,000 - 100,000 U.S.D
Contact: Grants Administrator; (312) 633-1000
Sponsor: Oprah Winfrey Foundation
110 N Carpenter Street
Chicago, IL 60607

Oracle Corporate Contributions Grants **3815**
The corporation awards grants for medical research in the areas of cancer, AIDS, and neuroscience; endangered animal protection; environmental protection, specifically for education and preservation of important open space; and K-12 education, with emphasis on mathematics, science, and technology programs. Grants are awarded in February and August.
Requirements: Nonprofits nationwide and in operating communities are eligible.
Restrictions: Grants generally do not support fundraising benefits, charitable dinners, sporting events, and marketing brochures. The corporation does not award multiyear grants and rarely approves funds for more than three consecutive years.
Samples: U of Arkansas, Sam M. Walton College of Business (Fayetteville, AR)—for the Information Technology Research Ctr, software valued at $7.078 million; Boys and Girls Club of the Peninsula (Menlo Park, CA)—for the Academic Program, which offers homework assistance, tutoring, and technology training, $1.4 million; United Negro College Fund (Fairfax, VA)—to strengthen the technical capacities of its 39 member institutions, $10 million, including cash donation, scholarships, internships, and curricular training, certification, and technology resources.
Contact: Corporate Giving Coordinator; (650) 506-7000
Internet: http://www.oracle.com/corporate/giving/community/index.html?giveform.html
Sponsor: Oracle Corporation
500 Oracle Parkway, MS 50P11
Redwood City, CA 94065

Orange County Community Foundation Grants **3816**
The Foundation's mission is to encourage, support and facilitate philanthropy in Orange County, California. One of our most important strategic priorities is strengthening the capacity of Orange County's nonprofit sector. Foundation Grants are made possible by the income earned from unrestricted, field-of-interest endowment funds and legacy funds. Grants funding is determined through research of community needs and approved by the Board. In addition, Grants are periodically administered in partnership with statewide and national foundations as well as Foundation donors to advance special projects and initiatives in Orange County. Fields of interest include arts and culture, environment, education, health, an human services.
Requirements: Nonprofits agencies in Orange County, California, are eligible. Unsolicited grant proposals are not accepted. Nonprofit agencies wishing to apply should pay special attention to the website for requests for proposal and deadlines. Every grant has a formal application and its own annual cycle.
Geographic Focus: California
Contact: Patricia Benevenia, Program Officer; (949) 553-4202, ext. 37; fax (949) 553-4211; penevenia@oc-cf.org
Internet: http://www.oc-cf.org/Page.aspx?pid=496
Sponsor: Orange County Community Foundation
4041 MacArthur Boulevard, Suite 510
Newport Beach, CA 92660

Ordean Foundation Grants **3817**
The Foundation, established in 1933, supports organizations involved with alcoholism, children/youth, services, crime/violence prevention, youth, education, family services, food services, health care, homeless, human services, housing/shelter, human services, medical care, rehabilitation, mental health/crisis services, substance abuse, services, YM/YWCAs & YM/YWHAs, aging, disabilities, economically disadvantaged. Grants will be awarded for up to three years. Applications are reviewed on an ongoing basis; the deadline is the 15th of each month.
Requirements: Non-profits in the Duluth and contiguous cities and townships in St. Louis County, Minnesota are eligible. The Foundation accepts the Minnesota Common Grant Application Form.
Restrictions: No support for direct religious purposes, or for political campaigns or lobbying activities. No grants to individuals (directly), or for endowment funds, travel, conferences, seminars or workshops, telephone solicitations, benefits, dinners, research, including biomedical research, deficit financing, national fund raising campaigns, or to supplant government funding.
Geographic Focus: Minnesota
Amount of Grant: 1,000 - 290,000 U.S.D
Samples: Lake Superior Community Health Center, Duluth, MN, $350,000; Safe Haven Shelter for Battered Women, Duluth, MN, $80,000; YMCA, Duluth Area Family, Duluth, MN, $150,000;
Contact: Stephen A. Mangan, Executive Director; (218) 726-4785; fax (218) 726-4848; ordean@computerpro.com
Sponsor: Ordean Foundation
501 Ordean Building
Duluth, MN 55802-4725

OREF/AAOS Clinical Research Training Fellowship **3818**
The goal of this program is to encourage the development of a cadre of orthopaedic surgeons who have the research skills necessary to undertake and manage various types of health services and outcomes research. The candidate is encouraged to earn a public health degree. Candidates who have completed an accredited North American orthopedic residency and are recommended by their department chairs may apply.
Requirements: Applicants must have completed an orthopedic residency at an accredited North American institution and be recommended by their department chairs.
Date(s) Application is Due: Oct 3
Amount of Grant: 70,000 U.S.D
Contact: Jean McGuire, (847) 698-9980; fax (847) 698-7806; mcguire@oref.org
Internet: http://www.oref.org/rga/rga.html#aaos
Sponsor: Orthopaedic Research and Education Foundation
6300 N River Road, Suite 700
Rosemont, IL 60018

OREF Educational Awards **3819**
The awards support continuing medical education programs and other projects of relevance to orthopedic surgeons of all ages developed in conjunction with a national, recognized organization. Types of support include conferences and seminars, visiting scholars, and program grants. Honorariums support visiting scholars for lectureships; program grants vary in amount.
Requirements: Departments and state orthopedic societies may apply for honorariums. Orthopedic departments and regional and national specialty societies may apply for special symposia, conferences, and workshops.
Date(s) Application is Due: Oct 3
Amount of Grant: Up to 25,000 U.S.D
Contact: Jean Maguire, VP Grants; (847) 698-9980; fax (847) 698-7806
Internet: http://www.oref.org/rga/rga.html#educational
Sponsor: Orthopaedic Research and Education Foundation
6300 N River Road, Suite 700
Rosemont, IL 60018

OREF Orthopedic Career Development Awards **3820**
Career development awards are granted to candidates who have demonstrated a commitment to scientific research as well as excellence in clinical training in the field of orthopedic surgery. Program requirements specify that recipients spend a minimum of 20 percent of a five-day week on the research project. A minimum of 50 percent of grant funds may be used for the candidate's salary. Applications must note budget requirements. Awards for up to two years, conditional upon annual review, will be considered. The grant period each year runs from July 1 through June 30 following the year of application. Recipients will be notified by letter after the annual board meeting held in the first quarter of each year.
Requirements: Candidates must have completed a residency in orthopaedic surgery and demonstrated a sustained interest in research and excellence in clinical training. Letters of nomination and support must offer convincing evidence of the candidate's potential to develop as an investigator.
Restrictions: Recipients of current or past NIH RO1 grants are ineligible.
Date(s) Application is Due: Oct 3
Amount of Grant: Up to 75,000 U.S.D
Contact: Jean McGuire, (847) 698-9980; fax (847) 698-7806; mcguire@oref.org
Internet: http://www.oref.org/rga/rga.html#career
Sponsor: Orthopaedic Research and Education Foundation
6300 N River Road, Suite 700
Rosemont, IL 60018

OREF Orthopedic Resident Research Awards **3821**
Applications will be considered for one year of laboratory or clinical research by orthopedic surgery residents in approved orthopedic programs at U.S. medical centers. The objective of these awards is to encourage young investigators by providing funding for research projects that would not ordinarily be funded. Grants provide funds for supplies and expenses but not for a resident's salary. The foundation will notify each

applicant by letter after the annual meeting of the board, held in the first quarter of each year.
Requirements: The principal investigator must be an orthopedic surgery resident in an approved U.S. or Canadian orthopedic program.
Date(s) Application is Due: Oct 3
Amount of Grant: Up to 15,000 U.S.D
Contact: Jean McGuire, (847) 698-9980; fax (847) 698-7806; mcguire@oref.org
Internet: http://www.oref.org/rga/rga.html#resident
Sponsor: Orthopaedic Research and Education Foundation
6300 N River Road, Suite 700
Rosemont, IL 60018

OREF Prospective Clinical Research Proposal Grants **3822**
The objective of these grants is to provide funding for promising prospective clinical research projects in areas of high clinical importance in orthopedic surgery. Subject to available funds, at least four applications will be considered each year for funding up to three years, conditional on annual review. Annual deadline dates may vary; contact program staff for exact dates.
Requirements: The principal or coprincipal investigator must be a suitably qualified orthopedic surgeon working at a U.S. institution. Investigators with an MD, PhD, or DVM degree may serve with the orthopedic surgeon.
Date(s) Application is Due: Oct 3
Amount of Grant: Up to 50,000 U.S.D
Contact: Jean McGuire, (847) 698-9980; fax (847) 698-7806; mcguire@oref.org
Internet: http://www.oref.org/rga/rga.html#prospective
Sponsor: Orthopaedic Research and Education Foundation
6300 N River Road, Suite 700
Rosemont, IL 60018

OREF Research Grants **3823**
The objective of the program is to encourage research into the causes, cure, and prevention of orthopedic-related injuries, illness, and conditions, and to encourage research by new investigators in the field of orthopedic musculoskeletal surgery. Grants provide seed and start-up funding for promising research projects for no more than two years, conditional upon annual review. Grant period runs from July 1st through June 30th for two years. Successful applicants will be notified by letter after the annual board meeting, which is held in the first quarter of each year.
Requirements: The principal or coprincipal investigator must be an orthopedic surgeon at a U.S. or Canadian institution. In addition, the candidate must have adequate protected time to accomplish the proposed grant within a 24-month time period.
Restrictions: Recipients of current or past NIH RO1 grants are ineligible.
Date(s) Application is Due: Oct 3
Amount of Grant: Up to 50,000 U.S.D
Contact: Jean McGuire, (847) 698-9980; fax (847) 698-7806; mcguire@oref.org
Internet: http://www.oref.org/rga/rga.html#research
Sponsor: Orthopaedic Research and Education Foundation
6300 N River Road, Suite 700
Rosemont, IL 60018

Oregon Community Foundation Grants **3824**
The community foundation supports nonprofit organizations in Oregon to meet educational, environmental, cultural, medical, social, and civic needs in all areas and at all levels of society throughout the state. Grants are awarded to support arts and cultural programs, adult basic education and literacy, reading education, health care and health organizations, human services, youth development, community development, and government/public administration. Types of funds include general operating support, building construction/renovation, equipment acquisition, seed money, land acquisition, program development, scholarship funds, matching funds, and technical assistance. Contact the office for application forms.
Requirements: Oregon nonprofit organizations may apply.
Restrictions: Grants are not awarded to individuals or religious organizations for religious purposes or for endowments, annual campaigns, deficit financing, research, publications, films, or conferences.
Geographic Focus: Oregon
Date(s) Application is Due: Feb 1; Aug 1
Amount of Grant: 5,000 - 35,000 U.S.D
Contact: Program Contact; (503) 227-6846, ext. 414; fax (503) 274-7771; info@ocfl.org
Internet: http://www.ocf1.org/grant_programs/grant_programs_fr.htm
Sponsor: Oregon Community Foundation
1221 SW Yamhill, #100
Portland, OR 97205

Orentreich Foundation for the Advancement of Science Grants **3825**
Occasional and limited grants are made for joint or collaborative biomedical research, usually initiated by OFAS, with emphasis in dermatology, aging, endocrinology, and serum markers for human diseases. The amount of the grants varies with joint interest and need. Unsolicited applications may be submitted at any time.
Requirements: Applicants must be at or above postgraduate level in science or medicine at accredited universities and research institutions in the United States. There is no restriction on citizenship. Submit an outline of proposed program, curriculum vita of principal investigator, and estimated budget.
Amount of Grant: Up to 50,000 U.S.D
Contact: Dr. Rozlyn Krajcik, (914) 265-4210; ofas@orentreich.org
Internet: http://www.orentreich.org/research.htm
Sponsor: Orentreich Foundation for the Advancement of Science
Cold Spring-on-Hudson
New York, NY 10516

ORISE Postgraduate Research Participation at the Agency for Toxic Substances and Disease Registry **3826**
An interagency agreement between DOE and ATSDR enables ORISE to provide opportunities and support for research in exposure and disease registries, health investigations, public health assessments, toxicological profiles, emergency response, and health education. The one-year, renewable grants are tenable at the ATSDR in Atlanta, GA. Stipends are based on research area and degree, and include reimbursement for inbound travel and moving. Applications are accepted at any time.
Requirements: Applicants should have completed graduate degree requirements within the last three years; others are considered on a case-by-case basis.
Date(s) Application is Due: Feb 1
Contact: Tasha Powell, Program Specialist; (865) 574-3911; powellt@orau.gov
Internet: http://www.orau.gov/orise/edu/ATSDR/gi-gugsrp.htm
Sponsor: Oak Ridge Institute for Science and Education
P.O. Box 117
Oak Ridge, TN 37831-0117

ORISE Postgraduate Research Participation at the National Center for Toxicological Research **3827**
An interagency agreement between DOE and FDA enables ORISE to administer this program, which provides opportunities and support for research in biological effects of potentially toxic chemicals and deriving solutions to problems in toxicology that have a major impact on human health and the environment. The program is tenable at the NCTR in Jefferson, AR. Stipends are based on research area and degree. Applications are accepted at any time for the one-year, renewable grants. Applications for summer appointments due March 1.
Requirements: Candidates should have completed graduate degree requirements within the last three years; others are considered on a case-by-case basis.
Contact: Tasha Powell, Program Specialist; (865) 574-3911; powellt@orau.gov
Internet: http://www.orau.gov/orise/edu/FDA/gi-gugNCTR.htm
Sponsor: Oak Ridge Institute for Science and Education
P.O. Box 117
Oak Ridge, TN 37831-0117

ORISE Postgraduate Research Program at the Center for Devices and Radiological Health **3828**
These one-year renewable grants provide opportunities and support for research concerning medical devices and radiation emissions from electronic products. This is an interagency agreement between the DOE and FDA, with the positions located at the Center for Devices and Radiological Health in Rockville, MD. Stipends are based on research area and degree and include limited reimbursement for inbound travel and moving.
Requirements: Applicants must have completed bachelor's or graduate degree within last three years; others are considered on a case-by-case basis.
Contact: Sherry Edwards, Program Specialist; (865) 576-7385; warrens@orau.gov
Internet: http://www.orau.gov/orise/edu/FDA/gi-rgpdCDRH.htm
Sponsor: Oak Ridge Institute for Science and Education
P.O. Box 117
Oak Ridge, TN 37831-0117

Oscar Rennebohm Foundation Grants **3829**
The foundation awards grants to nonprofit organizations in the Madison, WI area. Areas of interest, include: higher education, the arts, environmental conservation, health and social service agencies. Types of support include building construction/renovation, equipment acquisition, and research. There are no application forms or deadlines.
Geographic Focus: Wisconsin
Samples: University of Wisconsin Foundation, Madison, WI, $2,275,000; Second Harvest Foodbank of Southern Wisconsin, Madison, WI, $140,000; Bethel Community Services Corporation, Madison, WI, $40,000.
Contact: Grants Administrator; (608) 274-5991
Sponsor: Oscar Rennebohm Foundation
P.O. Box 5187
Madison, WI 53705

OSI-Baltimore Special Opportunities Program Grants **3830**
The Special Opportunities Program allows OSI-Baltimore to make occasional grants outside its usual target areas that further its mission. For example, a grant to Your Public Radio Corporation helped develop a local news department and underwrite Just Words, a new weekly show that highlights justice issues relevant to the region. In addition, the Association of Baltimore Area Grantmakers received support for its Neighborhood Indicators Alliance, an information system that helps community groups access data about their neighborhoods and use it to improve local policy making. There are no specific deadlines or guidelines.
Geographic Focus: Maryland
Contact: Diana Morris; (410) 234-1091, ext. 205; fax (410) 234-2816; dmorris@sorosny.org
Internet: http://www.soros.org/initiatives/baltimore/focus_areas/special_opportunities
Sponsor: Open Society Institute
201 N Charles Street, Suite 1300
Baltimore, MD 21201

OSI-Baltimore Tackling Drug Addiction Grants 3831

The initiative aims to increase the access of uninsured drug-dependent citizens to comprehensive services that respond to their individual needs. To achieve this objective, the Program is helping Baltimore City to develop a sustainable, high-quality treatment system that uses research-based clinical practices and that benefits from inter-agency coordination. The initiative also seeks to demonstrate to policymakers and the general public that sound treatment practices and policies will save lives, reduce crime, rebuild families and communities, and use public funds wisely. Grants seek to fund three major priorities: to make available drug treatment that integrates services targeting poverty-related barriers to self-sufficiency; to engage more drug-dependent citizens in the treatment system by supporting aggressive outreach efforts and by removing barriers to treatment entry; and to promote a comprehensive, high-quality public treatment system by supporting infrastructure development and advocacy and public education efforts.
Requirements: Baltimore area non-profit and for-profit entities may apply.
Geographic Focus: Maryland
Contact: Robert Schwartz, (410) 234-1091, ext. 207; rschwartz@sorosny.org
Internet: http://www.soros.org/initiatives/baltimore/focus_areas/drug_addiction/guidelines
Sponsor: Open Society Institute
201 N Charles Street, Suite 1300
Baltimore, MD 21201

OSI Access to Essential Medicines Grants 3832

The project supports efforts to increase access to essential medicines to developing countries and marginalized populations. Work has focused on three key areas: transparency and monitoring of the pharmaceutical industry; procurement mechanisms; and intellectual property. The project promotes civil society engagement with these issues and actively supports a civil society platform around the WHO Working Group on innovation for neglected diseases. The project works closely with OSI's palliative care and harm reduction initiatives. In an effort to strengthen needed capacity, the OSI programs are seeking applications from qualified organizations interested in receiving institutional support for their work on access to essential medicines. Grants for up to $75,000, renewable for up to three years, will be awarded to local NGOs with a mandate and demonstrated capacity to advance policy and civil society engagement related to access to medicines in Brazil, India, Kenya, Malaysia, South Africa, and Thailand. If requested, technical assistance will be offered throughout the length of the grant period. OSI may further sponsor seminars and meetings for grantees to share lessons learned. Applicants will be notified of final selection results no later than September 14th.
Requirements: Legally registered NGOs based in one of the following countries should apply: Brazil, India, Kenya, Malaysia, South Africa, and Thailand.
Date(s) Application is Due: Jul 16
Amount of Grant: 75,000 - 225,000 U.S.D
Contact: Jane Li, (212) 548-0600; fax (646) 557-2550; jli@sorosny.org
Internet: http://www.soros.org/initiatives/health/news/access_20070608
Sponsor: Open Society Institute
400 W 59th Street, 3rd Floor
New York, NY 10019

OSI European Commission Internships for Young Roma Graduates 3833

The European Commission, in cooperation with the OSI, offers internships for young Roma university graduates from all new EU member states (excluding Cyprus and Malta), Croatia, Macedonia, Serbia (including Kosovo), and Montenegro. The training periods involve work experience in one of the Commission's departments. The primary aims of the program are: to give Roma interns a general idea of the objectives and problems of European integration; to provide them with practical knowledge of the working of Commission departments; to enable them to acquire personal experience by means of contacts made in the course of their everyday work; and to enable them to further and put into practice the knowledge they have acquired during their studies or professional careers. This is a five-month internship, with a starting date of October 1. The grant covers: travel expenses to and from Brussels/Luxembourg (using the cheapest available means of transportation, actual cost being counted by OSI up to a maximum of 600 euro); and accommodation, subsistence, and health and accident insurance: 6,000 euro (1,200 euro per month).
Requirements: Nationals from all new EU member states (excluding Cyprus and Malta), Croatia, Macedonia, Serbia, including Kosovo, and Montenegro, who are of Roma origin may apply. Applications from women are encouraged. Preference will be given to candidates with proven experience in civil society work. Working knowledge of at least English or French is required, as is ability to draft reports in one of these languages. Minimum computer literacy is also required. The age limit for interns is minimum 25.
Restrictions: Candidates must have completed a course of university education and obtained a full degree or its equivalent by the closing date for applications.
Geographic Focus: All States, Albania, Andorra, Armenia, Austria, Azerbaijan, Belarus, Belgium, Bosnia & Herzegovina, Bulgaria, Croatia, Cyprus, Czech Republic, Denmark, Estonia, Finland, France, Georgia, Germany, Greece, Hungary, Iceland, Ireland, Italy, Kosovo, Latvia, Liechtenstein, Lithuania, Luxembourg, Macedonia, Malta, Moldova, Monaco, Montenegro, Norway, Poland, Portugal, Romania, Russia, San Marino, Serbia, Slovakia, Slovenia, Spain, Sweden, Switzerland, The Netherlands, Turkey, Ukraine, United Kingdom, Vatican City
Date(s) Application is Due: Jul 15
Amount of Grant: Up to 6,600 EUR
Contact: Bernard Rorke; +36 1 327 3100; fax +36 1 327 3101; brorke@osi.hu or rpp@osi.hu
Internet: http://www.soros.org/initiatives/roma/focus/internship/grants/internship
Sponsor: Open Society Institute
Oktober 6 u 12
Budapest, H-1051 Hungary

OSI Harm Reduction and Drug Use Grants 3834

Founded in 1995, the International Harm Reduction Development Program (IHRD) works to reduce HIV and other harms related to injecting drug use, and to press for policies that reduce stigmatization of illicit drug users and protect their human rights. IHRD, which has supported more than 200 programs in Central and Eastern Europe, the former Soviet Union, and Asia, bases its activities on the philosophy that people unable or unwilling to abstain from drug use can make positive changes to protect their health and that of their families and communities. Since 2001, IHRD has prioritized advocacy to expand availability of needle exchange, opiate substitution treatment, and treatment for HIV; to reform discriminatory policies and practices; and to increase the political participation of people who use drugs and those living with HIV.
Contact: Janna Ataiants, Program Coordinator; (212) 548-0600; fax (212) 548-4679; jataiants@sorosny.org
Internet: http://www.soros.org/initiatives/health/focus/ihrd/about
Sponsor: Open Society Institute
400 W 59th Street, 3rd Floor
New York, NY 10019

OSI Indonesia Project Grants 3835

In concert with the Tifa Foundation, OSI facilitates grants and operational programs through the Indonesia Project, which focuses on access to justice, media, local government capacity building, and public health. The Project accepts proposals for programs taking place in Indonesia in the following areas: access to justice; media; local government capacity building; and public health. Concept papers are accepted on a rolling basis. Papers should be 2-3 pages in length and give a detailed summary of project objectives and activities. Organizations whose concept papers fall within the Project funding priorities will be invited to submit complete proposals in accordance with project guidelines.
Restrictions: The program cannot fund the follow types of projects: unsolicited research projects, conference proposals, and organizational newsletters; social welfare projects, including health care, schools, orphanages, and provision of social services; cultural projects, including those related to cultural heritage and preservation, museums, literature, and fine and performing arts unless directly connected to the protection and promotion of civil and political rights; environmental projects unless directly connected to the protection and promotion of civil and political rights; development projects (i.e., construction of buildings, well digging); micro-finance and loans; or grants to individuals.
Contact: Sumaya Regal, +27 21 683 3489; fax +27 21 683 3550; sumaya@ct.org.za
Internet: http://www.soros.org/initiatives/bpsai/focus_areas/indonesia
Sponsor: Open Society Institute
400 W 59th Street, 3rd Floor
New York, NY 10019

OSI Integrating Health and Legal Services Grants 3836

The Law and Health Initiative supports the integration of legal and paralegal advocacy into community-based health services, both to increase access to justice and ensure comprehensive care for vulnerable groups. The aim is to ensure access to timely, effective, and affordable legal assistance for underserved populations such as women and girls, people who use drugs, sex workers, LGBT persons, and people living with HIV and other terminal illnesses. LAHI welcomes letters of intent from organizations seeking support for projects in one of five focus areas: Integrating Health and Legal Services; Human Rights in Patient Care; HIV/AIDS and Human Rights; Law and Health Capacity Development; and Legal Strategies in Health Monitoring.
Requirements: Geographically, LAHI limits its support to Central and Eastern Europe and the former Soviet Union, sub-Saharan Africa, and China and Southeast Asia.
Contact: Francoise Girard, (212) 548-0600; fax (212) 548-4679; fgirard@sorosny.org
Internet: http://www.soros.org/initiatives/health/focus/law/focus_areas/integrating/guidelines
Sponsor: Open Society Institute
400 W 59th Street, 3rd Floor
New York, NY 10019

OSI Mental Health Grants 3837

The Public Health Program's Mental Health Initiative provides grants to projects that stimulate the reform of national health, social welfare, education, and employment policies. The initiative also provides technical assistance and training in substantive areas to its grantees. Many grantees provide high-quality, community-based services which demonstrate that people with mental disabilities can live in their communities when they receive appropriate support.
Contact: Francoise Girard, (212) 548-0600; fax (212) 548-4679; fgirard@sorosny.org
Internet: http://www.soros.org/initiatives/health/focus/mhi/focus_areas
Sponsor: Open Society Institute
400 W 59th Street, 3rd Floor
New York, NY 10019

OSI Palliative Cancer Care Grants 3838

OSI's International Palliative Care Initiative supports organizations in resource-limited countries to integrate palliative care into national health care plans and policies, budgets, delivery systems, and professional and public education. The initiative also supports efforts to ensure that essential medications for pain relief are available to all patients who need them. This International Palliative Care Initiative funding program provides a two-year grant to a cancer center or general hospital with a focused oncology program, to support the salary of a clinical palliative care physician to develop a palliative care program.
Contact: Mary Callaway, Project Director; (212) 548-0600; fax (212) 548-4679; mcallaway@sorosny.org

Internet: http://www.soros.org/initiatives/health/focus/ipci/grants/cancer
Sponsor: Open Society Institute
400 W 59th Street, 3rd Floor
New York, NY 10019

OSI Supplementary Grant Program: Burma 3839
The Supplementary Grant Program - Burma (SGPB) was established in 1994 to address the needs of Burmese students, both inside and outside of the country, whose college education was disrupted due to political instability in Burma. The program provides supplemental financial assistance to Burmese students enrolled in academic programs at accredited colleges and universities leading to a degree. For individuals who are studying or intend to study in the United States, Canada, Europe and Australia, only those studying in the humanities and social sciences are eligible to apply with preference given to graduate study in the following fields: Public Health, Social Policy, Social Work, Education, Environmental Policy, Public Policy, Public Administration, Economic Policy, and Law. The March 15 deadline is for students studying in India, Asia and Australia. The May 15 deadline is for students studying in Europe and North America.
Requirements: Applicants must be between the ages of 18-40, must be accepted into a full-time study program, must already have partial support, and must intend to return to Burma.
Restrictions: Requests for support in the following fields are ineligible: Business Administration/Management Training, Computer Science, Finance/Banking/Marketing, Engineering, Hard & Natural Sciences (Physics, Biology, Chemistry, etc.), Mathematics, Medical Science, Fine and Performing Arts.
Date(s) Application is Due: Mar 15; May 15
Contact: Maureen Aung-Thwin, Director; (212) 548-0632; fax (212) 548-4655; maungthwin@sorosny.org
Internet: http://www.soros.org/initiatives/scholarship/focus_areas/supplementary_burma
Sponsor: Open Society Institute
400 W 59th Street, 3rd Floor
New York, NY 10019

Osram Sylvania Corporate Contributions Grants 3840
The corporation awards grants to nonprofit organizations in operating communities in its areas of interest, including education, health and human services, civic, arts and culture, and environment. Specific projects are funded, rather than operating costs or capital campaigns. Contact the community relations coordinator at the local operating facility. Guidelines are available online.
Requirements: 501(c)3 nonprofit organizations in operating communities are eligible, including Bangor and Waldoboro, ME; Central Falls, RI; Danvers, MA; Exeter, Hillsboro, and Manchester, NH; Maybrook, NY; Saint Marys, Towanda, Wellsboro, and Warren, PA; Seymour, IN; Winchester and Versailles, KY; and Watertown, CT.
Restrictions: Grants do not support religious or political groups, organizations with a limited constituency, travel, national conferences, sports events, or advertising.
Geographic Focus: Connecticut, Indiana, Kentucky, Maine, Massachusetts, New Hampshire, New York, Pennsylvania, Rhode Island
Contact: Corporate Communications Coordinator; (978) 777-1900; communications@sylvania.com
Internet: http://www.sylvania.com/aboutus/corpgiving/guidelines.htm#application
Sponsor: Osram Sylvania Corporation
100 Endicott Street
Danvers, MA 01923

Osserman-Sosin Postdoctoral Myasthenia Gravis Research Fellowships 3841
These postdoctoral fellowships are intended to provide some support to physicians, PhD scientists, and allied health professionals for conducting research into myasthenia gravis (MG) or related conditions. Clinical investigations as well as laboratory studies concerned with research on MG and with underlying mechanisms of neuromuscular transmission and the pathogenesis of myasthenia gravis are supported. A qualified candidate desiring training in MG-related research first finds a scientist with an established research program who will agree to supervise her/his training and the basic or clinical research to be carried out. The number of fellowships awarded per year is dependent on the amount of funds available.
Requirements: applicant must be a permanent resident of the United States or Canada who has been accepted to work in the laboratory of an established investigator at an institution in the United States, Canada, or abroad; or a foreign national who has been accepted to work in the laboratory of an established investigator a an institution in the United States or Canada.
Date(s) Application is Due: Oct 1
Amount of Grant: Up to 50,000 U.S.D
Samples: Dr. Marika Falcone, U of Arizona—to study the role of B-lymphocytes as antigen-presenting cells in the induction of the pathogenic T-cell responses against nicotinic acetylcholine receptor, $30,000; Dr. Brian Hare, Harvard Medical School—to study the solution structure determination of a single chain T-cell receptor and design of T-cell ligands with therapeutic potential, $30,000; Drs. Anthony Auerbach and Claudio Grosman, State U of New York at Buffalo—to study binding, gating, and desensitization of the acetylcholine receptor channel.
Contact: Fellowships Administrator; (651) 917-6256 or (800) 541-5454; fax (651) 917-1835; mgfa@myasthenia.org
Internet: http://www.myasthenia.org/research/MGFA_PostDoctoralFellowship.pdf
Sponsor: Myasthenia Gravis Foundation of America
1821 University Avenue W, Suite S256
St. Paul, MN 55104

Osteogenesis Imperfecta Foundation Research Grants 3842
Funding is available for fellowships for young investigators working to develop expertise in osteogenesis imperfecta (OI) research or seed grants for either basic or clinical studies with relevance to OI. Michael Geisman Research Fellowships provide support for investigator's salary and for supplies. (Second year funding may be available.) Seed grants may be funded for one year. (Grants do not support indirect costs or principal investigator salaries.) Clinical Seed Grants may be funded for two years. (Grants do not support indirect costs or principal investigator salaries.)
Date(s) Application is Due: Nov 1
Amount of Grant: 50,000 - 120,000 U.S.D
Contact: Research Department; (301) 947-0083; fax (301) 947-0456; bonelink@oif.org
Internet: http://www.oif.org/site/PageServer?pagename=OIFGrantApp
Sponsor: Osteogenesis Imperfecta Foundation
804 W Diamond Avenue, Suite 210
Gaithersburg, MD 20878

Oticon Focus on People Awards 3843
Oticon Focus on People Awards, a national awards program now past its 14th year, honors hearing-impaired students, adults, and advocacy volunteers who have demonstrated through their accomplishments that hearing loss does not limit a person's ability to make a difference in their families, communities and the world. By spotlighting people with hearing loss and their contributions, Oticon aims to change outdated stereotypes that discourage people from seeking professional help for their hearing loss. Focus on People Awards are offered in four categories: Student (for young people with hearing loss, ages 6-21, who are full-time students); Adults (for people with hearing loss, ages 21 and above); Advocacy (for adults with hearing loss, ages 21 and above who actively volunteer their time in advocacy or support efforts for the hearing-impaired and deaf community); and Practitioner (a special award for hearing-care professionals who go "above and beyond"). First-place category winners will receive a $1,000 award and a $1,000 donation by Oticon to a not-for-profit cause of their choice. First-place winners in the Student, Adult, and Advocacy categories will also receive a set of advanced-technology Oticon hearing solutions. Second place winners in each category will receive a $500 award and third-place winners will receive a $250 award. Guidelines and a nomination form are available online during the annual nomination period. The deadline for nomination may vary from year to year. Applicants are encouraged to visit the grant website to verify the current deadline date.
Requirements: Anyone may nominate themselves or another individual with a hearing loss. Nominees in the Practitioner category are not required to have a hearing loss to qualify. Full-time students who volunteer their time in advocacy should apply for the Student category.
Date(s) Application is Due: Jul 27
Amount of Grant: 250 - 2,000 U.S.D
Contact: Peer Lauritsen, President; (732) 560-1220; fax (732) 560-0029; peoplefirst@oticonusa.com
Internet: http://www.oticonusa.com/Oticon/Professionals/FocusOnPeople.html
Sponsor: Oticon Corporation
29 Schoolhouse Road
Somerset, NJ 08873

Otto Bremer Foundation Grants 3844
The foundation concentrates its grantmaking activity in communities served by Bremer-affiliated banks and provides financial assistance to nonprofit organizations whose work contributes to the well-being of these towns. The foundation looks to support programs that promote civil and political rights, including freedom of assembly, speech, and religion; economic and social rights, including the right to education, food, health care, and shelter; and cultural and environmental rights, including the right to live in a clean environment and participate in the cultural and political events of one's community. Types of support include program development, operating support, capital (including building and equipment), matching or challenge grants, and internships. Applications are accepted throughout the year. Most grants are given for a one-year period, although some multiyear grants are awarded. The foundation encourages initial telephone or written inquiries concerning its interest in a particular project. Applicants are also encouraged to contact foundation staff for assistance in the development of a proposal.
Requirements: Private nonprofit or public 501(c)3 tax-exempt organizations whose beneficiaries are residents of Minnesota, North Dakota, or Wisconsin with priority given to those communities or regions served by Bremer affiliates.
Restrictions: Requests for the following types of projects are discouraged: annual fund drives; benefit events; camps; commercial and business development; K-12 education; medical research; sporting activities; building endowments other than for the development of community foundations; capital requests for hospitals and nursing homes; theatrical productions, including motion pictures, books, and other artistic or media projects; municipal and government services; or historical preservation, museums, and interpretive centers.
Geographic Focus: Minnesota, North Dakota, Wisconsin
Amount of Grant: 552 - 259,235 U.S.D
Contact: Danielle Cheslog, Grants Manager; (651) 312-3717 or (651) 227-8036; fax (651) 312-3665; danielle@ottobremer.org or obf@ottobremer.org
Sponsor: Otto Bremer Foundation
445 Minnesota Stret, Suite 2250
Saint Paul, MN 55101-2107

OUT Fund for Lesbian and Gay Liberation Grants 3845
The organization is dedicated to a range of social justice, race, class, gender, and sexuality issues and supports radical organization projects to build community among lesbians, gay men, bisexuals, and transgender people. Proposals are sought that originate in

marginalized communities like the poor, young, old, or disabled; serve rural areas; develop long-term coalitions in and out of diverse communities; organize antiracism efforts; foster organization through cultural projects; and focus on HIV health policy, prevention, and education efforts. Proposals that identify and target multi-issues, such as race, class, gender, and poverty, are preferred.
Requirements: 501(c)3 tax-exempt organizations and organizations with a 501(c)3 fiscal sponsor are eligible. 501(c)4 lobbying organizations may be eligible under very specific circumstances.
Contact: Grants Administrator; (212) 529-5300; fax (212) 982-9272; info@fex.org
Internet: http://www.fex.org/grantmaking.shtml
Sponsor: Funding Exchange
666 Broadway, Suite 500
New York, NY 10012

Owen County Community Foundation Grants **3846**
The Owen County Community Foundation was established in 1996 and serves the Owen County, Indiana area. The Foundation connects donors with the causes they care about. The foundation provides support for organizations involved with arts and culture, education, animal welfare, health care, human services, and community development.
Requirements: A letter of intent is required prior to submitting a grant request. The letter of intent must be submitted before June 1, 4:00 pm and must be: one page in length; include name, address, and phone number of organization; include name and title of contact person; include mission statement of agency/organization; list project description; list anticipated results. A copy of tax-exempt verification must accompany the letter of intent as well. You will be contacted approximately one month after the deadline as to whether or not to submit a full proposal. A full proposal should include the following: grant application and agreement page (available at the foundation office or the foundations website); organizations background, including history, mission statement, number of staff, and number of persons served by agency or program; proposed program, please indicate desired outcomes; program methods; evaluation; funding the program; timetable; project budget; agency budget; governing organization; financial statement and balance sheet; tax exempt status; audit, with the original proposal only, provide one copy of your organization's most recently completed audit. Mail or hand deliver an original signed and completed grant proposal, along with eight copies with no permanent binding to the Owen Community Foundation office no later then September 15, 4:00 pm. The Board will contact you in late October with it's funding decision and grant checks will be presented in November.
Restrictions: Giving limited to Owen County, Indiana. No grants for endowments, deficit funding, conferences, publications, films, television, or radio programs, travel, annual appeals, religious purposes or membership contributions.
Geographic Focus: Indiana
Date(s) Application is Due: Jun 1; Sep 15
Contact: Maggie Burton, Program Manager; (812) 829-1725; fax (812) 829-9958; maggie@owencountycf.org
Internet: http://www.owencountycf.org/Grants.asp
Sponsor: Owen County Community Foundation
201 W Morgan Street, Suite 202, P.O. Box 503
Spencer, IN 47460

Oxalosis and Hyperoxaluria Foundation Grants **3847**
Grants are awarded for research on hyperoxaluria, the presence of excess oxalic acid or oxalates in the urine, and the related conditions oxalosis and calcium-oxalate kidney stones. The foundation seeks to fund grants that will increase the understanding of hyperoxaluria and oxalosis and improve the clinical management and treatment of these two orphan diseases. Investigation into alternative methods of treatment that will alleviate the need for transplantation therapy are encouraged. Applications for up to two years of support will be considered. These funds may be used for salaries for technical assistance, special equipment, animals, supplies, and travel. Other organizations that have been approached about supporting the same project should be identified.
Requirements: Applicants must hold an MD, PhD, or equivalent degree by the time the funding period begins.
Date(s) Application is Due: Oct 15
Amount of Grant: Up to 160,000 U.S.D
Contact: Program Contact; (212) 777-0470; fax (212) 777-0471; execdirector@ohf.org
Internet: http://www.ohf.org/prof_application.html
Sponsor: Oxalosis and Hyperoxaluria Foundation
201 E 19th Street, 12E
New York, NY 10003

PacifiCare Foundation Grants **3848**
The foundation's mission is to improve the quality of life for residents of areas where PacifiCare Health Systems does business. The foundation's focus areas are: child/youth, including child care, youth activity programs, at-risk youth, and counseling programs; education, including school programs that promote self-esteem, encourage academic achievement and the development of specific skills, literacy programs, training programs, and programs that improve the effectiveness of the educational system; health, including prevention, health education, access to health care, and improved quality of health care of targeted populations; human/social services, including housing, shelters, education, protection, community development, crime prevention, food, transportation, and other social services for targeted populations; and senior, including social services, nutrition, education, volunteer, and and adult day care. Preference will be given to proposals for specific projects. Requests for operating costs will be considered if the request is very specific and clearly defined. Seed grant requests also receive consideration. Organizations funded by the foundation are welcome to reapply annually. Application forms are available on the foundation's website.
Requirements: IRS 501(c)3 nonprofit organizations serving residents of PacifiCare regions in Arizona, California, Colorado, Nevada, Oklahoma, Oregon, Texas, and Washington are eligible. The proposal must include two copies of the following: application form; checklist form; cover letter accompanying the proposal, signed by either the CEO or appointee of the organization, summarizing the proposed project, the problem addressed, the amount requested, and the name and phone number of the contact person; the written proposal, which should not exceed 2-5 pages in length and should include background information, description of the problem, need or issue being addressed, and a complete description of the proposed project; most recent audited financial statement and 990; current operating budget and line item budget for the specific project; list of major funders and amounts; list of board of directors; and one paragraph summary of previous support from the PacifiCare Foundation.
Restrictions: The foundation will not consider grants for arts/cultural programs, associations, annual campaigns, associations (professional/technical), capital campaigns, challenge/matching grants, hosting/supporting conferences, individual support, private foundations, programs that promote religious doctrine, research, scholarships, or sponsorship of special events.
Geographic Focus: Arizona, California, Colorado, Nevada, Oklahoma, Oregon, Texas, Washington
Date(s) Application is Due: Jan 1; Jul 1
Amount of Grant: 2,000 - 10,000 U.S.D
Contact: Riva Gebel, Director; (714) 825-5233
Internet: http://www.pacificare.com/vgn/images/portal/cit_60701/127503Guidelines_for_Charitable_Giving.pdf
Sponsor: PacifiCare Foundation
P.O. Box 25186, MS LC03-159
Santa Ana, CA 92799

Pacific Life Foundation Grants **3849**
Pacific Life has long recognized the importance of helping communities where their employees reside and work and has a record of community involvement that spans the history of the company. The Foundation accepts proposals from agencies seeking funds for programs and projects in the areas of health and human services; education; arts and culture; and civic, community, and environment. Grant proposals are generally accepted from July 15 through August 15. See the Foundation's website to verify deadlines.
Requirements: Funding is made primarily in areas with large concentrations of Pacific Life employees: generally, the greater Orange County, California, area and other areas, such as Omaha, Nebraska. Ideally, agencies should serve a large area, usually including more than one city or community. Some California statewide and national organizations also receive support. General grants range from $5,000 to $10,000 for a one-year period of funding and are given to support programs, operating expenses, or collaborative programs with other agencies. Capital grants range from $10,000 to $100,000 and are paid over multiple years. Capital grants are generally given to an agency with an organized campaign already under way to raise substantial funds. More than fifty percent of the campaign goal (excluding in-kind donations, anonymous gifts and loans) must be pledged prior to consideration by the Foundation.
Restrictions: The following is not funded: individuals; political parties, candidates, or partisan political organizations; labor organizations, fraternal organizations, athletic clubs, or social clubs; K-12 schools, school districts, or school foundations; sectarian or denominational religious organizations, except for programs that are broadly promoted, available to anyone, and free from religious orientation; fundraising events; sports leagues or teams; and advertising sponsorship or conference underwriting/sponsorship.
Samples: Aquarium of the Pacific, Long Beach, California, $258,503 - civic, community and environment support (2010); Greater Santa Ana Vitality Foundation, Santa Ana, California, $10,000 - education (2010); and Southern California Public Radio, Pasadena, California, $40,000 - arts and culture (2010).
Contact: Brenda Hardwig, (949) 219-3787; PLFoundation@PacificLife.com
Internet: http://www.pacificlife.com/About+Pacific+Life/Foundation+or+Community
Sponsor: Pacific Life Foundation
700 Newport Center Drive
Newport Beach, CA 92660

PacifiCorp Foundation for Learning Grants **3850**
The foundation awards grants to eligible nonprofit organizations in four broad areas, including educational and research institutions, both public and private, from early childhood through university level; youth organizations; cultural enrichment organizations dedicated to the performing arts, the visual arts, historic preservation, cross-cultural education, and other such activities; and health and human services organizations, including chemical dependency treatment and prevention programs, senior citizen centers, runaway youth services, and domestic violence treatment and prevention programs. Deadline for education organizations is March 15; for civic, community, and organizations not covered in other categories, June 15; for culture and arts organizations, September 15; for health, welfare, and social service organizations, December 15. Guidelines and application are available online.
Requirements: 501(c)3 community organizations in Idaho, Oregon, Utah, California, Washington, and Wyoming are eligible.
Restrictions: Organizations and activities not eligible for foundation support include religious groups for religious purposes; political organizations, ballot measure campaigns, or candidates for public office; organizations that discriminate against individuals on the basis of creed, color, sex, age, national origin, or veteran status; individuals; contributions to or memberships in chambers of commerce, service clubs, taxpayer associations, and

other similar bodies; or any non-charitable purpose. As a general rule, the foundation also will not support establishment or support of endowments; coverage of operating deficits; capital campaigns; conferences, conventions, and events; projects that do not have a bearing on communities within the six-state region served by PacifiCorp subsidiaries or communities served by PacifiCorp Power Marketing.
Geographic Focus: California, Idaho, Oregon, Utah, Washington, Wyoming
Date(s) Application is Due: Mar 15; Jun 15; Sep 15; Dec 15
Amount of Grant: 1,000 - 10,000 U.S.D
Contact: Grants Manager; (503) 813-7257; pacificorpfoundation@pacificorp.com
Internet: http://www.pacificorpfoundation.org/Article/Article25110.html
Sponsor: PacifiCorp Foundation for Learning
825 NE Multnomah, Suite 2000
Portland, OR 97232-4116

Packard Foundation Children, Families, and Communities Grants **3851**
The Foundation's Children, Families, and Communities Grants strive to ensure that all children have the opportunity to reach their full potential. Strategies address two interrelated and fundamental needs that must be met for children to thrive: health and education. Grants have three goals. The central goal is to create publicly supported, high-quality preschool opportunities for all three- and four-year-olds. This is funded through Preschool for California's Children Grants. A second goal is to ensure that all children receive appropriate health care by creating nationwide systems that provide access to health insurance for all children. Work is done in over a dozen states and at the federal level through Children's Health Insurance Grants. A third goal is to strengthen California's public commitment to school-based, after-school programs, while also spurring the expansion of these programs into summer. Such expanded learning opportunities that are aligned with the school day promote positive youth development for elementary and middle school children. The Foundation emphasizes literacy, good nutrition, and out-of-door experiences in these expanded learning settings. Work is supported at the federal level through After School and Summer Enrichment Grants.
Requirements: The Foundation provides Grants for charitable, educational, or scientific purposes, primarily from tax-exempt, charitable organizations. See the Foundation's website for current opportunities.
Contact: Meera Mani, Director; (650) 948-7658; cfc@packard.org
Internet: http://www.packard.org/what-we-fund/children-families-and-communities/
Sponsor: David and Lucile Packard Foundation
300 Second Street, Suite 200
Los Altos, CA 94022

Packard Foundation Local Grants **3852**
The Foundation supports an array of nonprofit organizations in geographic areas that are significant to the Packard family. These include the five California counties that surround the Foundation's headquarters in Los Altos, California (San Mateo, Santa Clara, Santa Cruz, Monterey, and San Benito) as well as Pueblo, Colorado, the birthplace of David Packard. The goal in supporting these communities is to help make them stronger and more vibrant places where all families can thrive and reach their potential. To achieve this goal, Local Grants focus resources on addressing five fundamental issue areas: arts; children and youth; conservation and science; food and shelter; and population and reproductive health. There are no deadlines.
Requirements: The Foundation accepts grant proposals only for charitable, educational, or scientific purposes, primarily from tax-exempt, charitable organizations. An online letter of inquiry form is available at the Foundation's website. Applicants typically receive a response within three to six weeks. If accepted, further details about completing a full proposal will be provided.
Restrictions: The following is not funded: public policy work, capital campaigns, specific performances or productions, one-time events, event sponsorships, religious or business organizations, and individuals. Request should generally not exceed 25 percent of the organization's operating budget.
Geographic Focus: California, Colorado
Amount of Grant: 15,000 - 150,000 U.S.D
Contact: Irene Wong; (650) 917-7238; fax (650) 948-1361; local@packard.org
Internet: http://www.packard.org/what-we-fund/local-grantmaking/
Sponsor: David and Lucile Packard Foundation
300 Second Street, Suite 200
Los Altos, CA 94022

Packard Foundation Population and Reproductive Health Grants **3853**
The Foundation's Population and Reproductive Health Grants fund innovative work that addresses population growth and promotes positive reproductive health. The goals are to slow population growth rates in high-fertility areas, and to ensure individual reproductive health and rights in order to improve the quality of life for more people. Currently work is focused at the global level, regionally on South Asia and Sub-Saharan Africa, and on selected initiatives within the United States.
Requirements: Interested applicants may submit a letter of inquiry not to exceed three pages to population@packard.org. Only letters that clearly support a particular subprogram strategy will be considered. Letters should include a brief overview of the applicant organization and its mission; a description of the project for which funding is being requested, including goals, major strategies, and rationale; and monitoring and evaluation plans. Every attempt is made to review and respond within four to six weeks. If accepted, applicants will be invited to submit a full grant proposal for further review.
Samples: Walta Information Center, Addis Ababa, Ethiopia, $250,000 - population; Women's Health Care Foundation, Quezon City, Philippines, $100,000 - population; and Women's Health and Action Research Centre, Benin City, Nigeria, $120,912 - population.
Contact: Chris DeCardy, Vice President and Director of Programs; (650) 948-7658; fax (650) 948-1361; population@packard.org
Internet: http://www.packard.org/what-we-fund/population-reproductive-health/
Sponsor: David and Lucile Packard Foundation
300 Second Street, Suite 200
Los Altos, CA 94022

PAHO-ASM International Professorship for Latin America **3854**
The program provides funding support to an ASM member from the U.S. who is scientifically recognized in his/her area to teach a hands-on, highly interactive short course in the area of antimicrobial resistance at an institution of higher learning in Bolivia, Colombia, Costa Rica, Dominican Republic, Ecuador, El Salvador, Guatemala, Honduras, Nicaragua, Panama, Paraguay, or Peru. One Teaching Professorship offered per program year.
Requirements: The application must be made jointly between a hosting institution and the visiting professor. Eligibility for Host Institution: institution with graduate students enrolled in a masters, doctoral or equivalent program, post-doctoral fellows or residents, and teaching faculty; institution where at least 12 students will be enrolled full-time in the short course; commitment to maximizing use of the course, as demonstrated by the applicability of the course's contents to existing programs at the institution or to the field of research of the Research Professor. demonstrated commitment to international collaborations and partnerships; demonstrated commitment to advancing the microbiological sciences through interactions with professional organizations; institutions in developed nations, as defined by the UN high HDI group, are ineligible. Eligibility for Visiting Teaching Professor: scientifically recognized their area of microbiological expertise; actively engaged in teaching at the post-secondary level; commitment to international collaborations and partnerships; must be an ASM member during the term of the Professorship. Preference will be given to: applicants who will be able to make maximal use of the course, as demonstrated either by the applicability of the course's contents to existing programs at the host institution or to the field of research of the visiting Professor; applications that demonstrate a commitment to building an ongoing institutional relationship. It is strongly recommended that the course include a hands-on component, such as a wet lab or other practical activity.
Date(s) Application is Due: Apr 15; Oct 15
Amount of Grant: 4,000 U.S.D
Contact: Supervisor; (202) 942-9368; fax (202) 942-9328; international@asmusa.org
Internet: http://www.asm.org/International/index.asp?bid=57785
Sponsor: American Society for Microbiology
1752 N Street, N.W.
Washington, DC 20036-2904

PAHO/WHO Fellowships **3855**
Fellowship is a specially tailored training activity for an individual or a group with the purpose of fulfilling specific learning objectives. Fellowships may be of short or long duration and take place in a training institution or in the field inside or outside of the fellow's country. Each of the six regional WHO offices share responsibility for management and administration of the fellowship program within its region: regional office for Europe in Copenhagen, Denmark; regional office for Eastern Mediterranean in Alexandria, Egypt; regional office for the Western Pacific in Manila, the Philippines; regional office for Africa in Brazzaville, Congo; regional office for South-East Asia in New Delhi, India; and regional office for the Americas in Washington, DC. The Pan American Health Organization (PAHO) serves as the Regional WHO Office for all of the Americas.
Requirements: Fellowships are open to nationals of member states of the PAHO who are public health workers in all fields, especially deans or faculty members of schools of public health, medicine, dentistry, engineering, veterinary medicine, and nursing. Individual must be recommended by his/her own government and return to his/her home country for at least three years service on completion of studies. Contact program staff for specific eligibility.
Restrictions: Candidates normally should not be more than 55 years of age.
Contact: Regional Office for the Americas; (202) 974-3000; fax (202) 974-3663; postmaster@paho.org
Internet: http://www.paho.org/English/HSP/HSR/fellowships.htm
Sponsor: Pan American Health Organization
525 23rd Street NW
Washington, DC 20037

Pajaro Valley Community Health Health Trust Insurance/Coverage & Education on Using the System Grants **3856**
The Trust provides grants for projects that advance our mission to improve the health and quality of life for all people of the greater Pajaro Valley. The primary goals of the Health Insurance/Coverage & Education on Using the System Initiative are to increase the number of Pajaro Valley residents with health insurance, increase residents understanding and appropriate use of a medical home (family practitioner), and decrease inappropriate use of the emergency department. The Trust will support programs that increase the number of Pajaro Valley residents that have health insurance, as well as programs that improve access to health care for our community's more vulnerable populations. Additionally, the Trust will look at community-wide solutions to these issues.
Requirements: Applicant organizations must provide or plan to provide programs/services benefiting the health of residents in the Trust's primary geographic service area. Communities within this service area include Watsonville, Pajaro, Freedom, and Aromas. The home office of the applicant organization need not be located in the Pajaro Valley, but the applicant organization must demonstrate that it provides or plans to provide services that directly benefit residents of the Pajaro Valley. The applicant

organization must be a nonprofit, 501(c)3 tax-exempt organization; a school-based health program; or have a 501(c)3, tax-exempt organization as a fiscal sponsor.
Restrictions: In general, the Trust's Board of Directors prefers not to fund programs or projects administered by a city, county, state, or federal government with the exception of school-based health programs. In general, the Trust does not give grants to: projects that do not substantially benefit residents of the Pajaro Valley; projects and proposals unrelated to the Trust's mission, eligibility requirements, and current strategic plan funding priorities and objectives; individuals, with the exception of the Trust's scholarship programs; religious organizations for secular purposes; endowments, building campaigns, annual fund appeals, fundraising events, or celebrations; or commercial ventures.
Geographic Focus: California
Samples: La Manzana Community Resources, La Manzana, California, $12,000 - to assist low-income families in accessing public health benefits (2010); Healthy Kids of Santa Cruz County, Santa Cruz, California, $15,000 - to provide access to health insurance to children in the Pajaro Valley who have incomes at or below 300% of the federal poverty level (2010); Santa Cruz Community Counseling Center, Santa Cruz, $5,000 - to launch a health literacy campaign focusing on low-income families with children birth to five years old (2010).
Contact: Raquel Ramirez Ruiz, Director of Programs; (831) 763-6456 or (831) 761-5639; fax (831) 763-6084; info@pvhealthtrust.org or raquel_dhc@pvhealthtrust.org
Internet: http://www.pvhealthtrust.org/grants_core.html
Sponsor: Pajaro Valley Community Health Trust
85 Nielson Street
Watsonville, CA 95076

Pajaro Valley Community Health Trust Diabetes and Contributing Factors Grants 3857

The Trust provides grants for projects that advance our mission to improve the health and quality of life for all people of the greater Pajaro Valley. The primary goals of the Diabetes and Contributing Factors Initiative are to reduce the risk factors associated with diabetes, reduce complications related to diabetes, and decrease the prevalence of childhood and adult obesity in the Pajaro Valley. The Trust will mobilize communities in the tri-county area to prevent the increase of type-2 diabetes in youth and young adult populations; teach diabetes self-management, and provide medical nutrition therapy to people living with diabetes thereby preventing the life-threatening complications associated with diabetes. Further, the Trust will promote "best practices" in clinical management of diabetes throughout the region. The Trust will seek to minimize factors that contribute to diabetes, including obesity, poor nutrition, and lack of physical activity.
Requirements: Applicant organizations must provide or plan to provide programs/services benefiting the health of residents in the Trust's primary geographic service area. Communities within this service area include Watsonville, Pajaro, Freedom, and Aromas. The home office of the applicant organization need not be located in the Pajaro Valley, but the applicant organization must demonstrate that it provides or plans to provide services that directly benefit residents of the Pajaro Valley. The applicant organization must be a nonprofit, 501(c)3 tax-exempt organization; a school-based health program; or have a 501(c)3, tax-exempt organization as a fiscal sponsor.
Restrictions: In general, the Trust's Board of Directors prefers not to fund programs or projects administered by a city, county, state, or federal government with the exception of school-based health programs. In general, the Trust does not give grants to: projects that do not substantially benefit residents of the Pajaro Valley; projects and proposals unrelated to the Trust's mission, eligibility requirements, and current strategic plan funding priorities and objectives; individuals, with the exception of the Trust's scholarship programs; religious organizations for secular purposes; endowments, building campaigns, annual fund appeals, fundraising events, or celebrations; or commercial ventures.
Geographic Focus: California
Amount of Grant: 5,000 - 30,000 U.S.D
Samples: Community Action Board of Santa Cruz County, Santa Cruz, California, $10,000 - to support the REAL for Diabetes Prevention project (2010); Ecology Action, Willits, California, $9,000 - to support the Boltage program, a daily student biking and walking tracking program (2010); Mesa Verde Gardens, Mesa Verde, California, $15,000 - to support the start-up of a second Community Garden (2010).
Contact: Raquel Ramirez Ruiz, Director of Programs; (831) 763-6456 or (831) 761-5639; fax (831) 763-6084; info@pvhealthtrust.org or raquel_dhc@pvhealthtrust.org
Internet: http://www.pvhealthtrust.org/grants_core.html
Sponsor: Pajaro Valley Community Health Trust
85 Nielson Street
Watsonville, CA 95076

Pajaro Valley Community Health Trust Oral Health: Prevention & Access Grants 3858

The Trust provides grants for projects that advance our mission to improve the health and quality of life for all people of the greater Pajaro Valley. The primary goals of the Oral Health Initiative are to reduce the risk factors associated with oral health disease, increase the number of Pajaro Valley residents with ready access to comprehensive dental care, and decrease the prevalence of dental disease among residents of the Pajaro Valley. The Trust's goals include improving access to dental treatment and preventing dental disease. Through this initiative, the Trust will look at systematic issues facing oral health care, particularly in the areas of prevention and access to care, and work with others in the community to remove these barriers.
Requirements: Applicant organizations must provide or plan to provide programs/services benefiting the health of residents in the Trust's primary geographic service area. Communities within this service area include Watsonville, Pajaro, Freedom, and Aromas. The home office of the applicant organization need not be located in the Pajaro Valley, but the applicant organization must demonstrate that it provides or plans to provide services that directly benefit residents of the Pajaro Valley. The applicant organization must be a nonprofit, 501(c)3 tax-exempt organization; a school-based health program; or have a 501(c)3, tax-exempt organization as a fiscal sponsor.
Restrictions: In general, the Trust's Board of Directors prefers not to fund programs or projects administered by a city, county, state, or federal government with the exception of school-based health programs. In general, the Trust does not give grants to: projects that do not substantially benefit residents of the Pajaro Valley; projects and proposals unrelated to the Trust's mission, eligibility requirements, and current strategic plan funding priorities and objectives; individuals, with the exception of the Trust's scholarship programs; religious organizations for secular purposes; endowments, building campaigns, annual fund appeals, fundraising events, or celebrations; or commercial ventures.
Geographic Focus: California
Amount of Grant: 5,000 - 30,000 U.S.D
Samples: Dientes Community Dental Care, Santa Cruz, California, $10,000 - to support the delivery of dental services to low-income, uninsured adults from the Pajaro Valley (2010); Salud Para La Gente, Watsonville, California $10,000 - to support the delivery of dental services to low-income, uninsured adults from the Pajaro Valley (2010).
Contact: Raquel Ramirez Ruiz, Director of Programs; (831) 763-6456 or (831) 761-5639; fax (831) 763-6084; info@pvhealthtrust.org or raquel_dhc@pvhealthtrust.org
Internet: http://www.pvhealthtrust.org/grants_core.html
Sponsor: Pajaro Valley Community Health Trust
85 Nielson Street
Watsonville, CA 95076

Pajaro Valley Community Health Trust Promoting Entry & Advancement in the Health Professions Grants 3859

The Trust provides grants for projects that advance our mission to improve the health and quality of life for all people of the greater Pajaro Valley. The primary goals of the Promoting Entry & Advancement in the Health Professions Initiative are to increase the number of culturally competent healthcare workers in the Pajaro Valley, increase the proportion of bilingual/bicultural healthcare workers in the Pajaro Valley, and reduce the number of unfilled healthcare position in the Pajaro Valley. The Trust will support and promote programs that encourage individuals to choose the health professions as a career/professional path. It is a priority for the Trust to encourage all individuals to enter the health professions, particularly bilingual/bicultural individuals that will return to the Pajaro Valley to serve this community's healthcare needs.
Requirements: Applicant organizations must provide or plan to provide programs/services benefiting the health of residents in the Trust's primary geographic service area. Communities within this service area include Watsonville, Pajaro, Freedom, and Aromas. The home office of the applicant organization need not be located in the Pajaro Valley, but the applicant organization must demonstrate that it provides or plans to provide services that directly benefit residents of the Pajaro Valley. The applicant organization must be a nonprofit, 501(c)3 tax-exempt organization; a school-based health program; or have a 501(c)3, tax-exempt organization as a fiscal sponsor.
Restrictions: In general, the Trust's Board of Directors prefers not to fund programs or projects administered by a city, county, state, or federal government with the exception of school-based health programs. In general, the Trust does not give grants to: projects that do not substantially benefit residents of the Pajaro Valley; projects and proposals unrelated to the Trust's mission, eligibility requirements, and current strategic plan funding priorities and objectives; individuals, with the exception of the Trust's scholarship programs; religious organizations for secular purposes; endowments, building campaigns, annual fund appeals, fundraising events, or celebrations; or commercial ventures.
Geographic Focus: California
Amount of Grant: 2,000 - 10,000 U.S.D
Samples: Aptos Adult Education, Watsonville, California, $5,000 - to support scholarship funding to train individuals as Certified Nurse Assistants (2010); Watsonville High School Health Careers Academy, Watsonville, California, $6,000 - for enhancement of its program that provides a specialized course of study for high school students interested in entering the health professions (2010).
Contact: Raquel Ramirez Ruiz, Director of Programs; (831) 763-6456 or (831) 761-5639; fax (831) 763-6084; info@pvhealthtrust.org or raquel_dhc@pvhealthtrust.org
Internet: http://www.pvhealthtrust.org/grants_core.html
Sponsor: Pajaro Valley Community Health Trust
85 Nielson Street
Watsonville, CA 95076

Palm Beach and Martin Counties Grants 3860

The community foundation supports nonprofit organizations serving Florida's Palm Beach and Martin Counties in eight program areas: arts and culture, education, environment, health, human services, community development, the conservation and preservation of historical and cultural resources, and human and race relations. Guidelines and forms are available online.
Requirements: Florida organizations based in and serving Palm Beach and Martin Counties are eligible to apply.
Restrictions: Proposals will not be funded for routine operating expenses, endowments, building campaigns, computers, fund-raising events, celebration functions, feasibility studies, religious organizations for religious purposes, or recurring requests for the same purpose.
Geographic Focus: Florida
Amount of Grant: Up to 25,000 U.S.D
Samples: Pleasant City Family Reunion Committee (West Palm Beach, FL)—to support artists and instructors who work with disadvantaged youths, $15,000; TAB Development Corp (West Palm Beach, FL)—for classes and training courses for first-time homebuyers,

$15,000; FAI/FIU Joint Ctr for Environmental and Urban Problems (Jupiter, FL)—to integrate economic, environmental, and social concerns in community design, $15,000; Florida Resource Ctr for Women and Children (Riviera Beach, FL)—for salary and program expenses of an antiviolence program for teenagers, $25,000.
Contact: Linda Raybin, (561) 659-6800; fax (561) 832-6542; lraybin@cfpdmc.org
Internet: http://www.cfpbmc.org/grant/Guidelines.asp
Sponsor: Community Foundation for Palm Beach and Martin Counties
700 S Dixie Highway, Suite 200
West Palm Beach, FL 33401

Palmer Foundation Grants **3861**
The foundation considers proposals that empower young people ages 0 to 25. General categories include arts and culture, education, health, and human services. In making awards, the foundation places the highest priority on applications that demonstrate a level of cooperation with other organizations, including leveraging financial and in-kind support from other groups; clearly avoids duplication of existing services; where applicable, demonstrates potential for continued funding from internal or other sources after the grant period; and offer challenge grants in order to stimulate support of the project by other organizations or individuals. The board will review Letters of Intent each month through the listed deadlines; full proposals are by invitation. Letters of intent may be submitted online. Guidelines are available online.
Requirements: 501(c)3 tax-exempt organizations are eligible. The foundation's geographical region is usually limited to the Mid West states, the Mid Atlantic states. Guatemala, and El Salvador.
Restrictions: The foundation does not usually support the following types of activities: grants to individuals or scholarships; endowment drives or capital campaigns; general operating expenses for an existing project, except to support an innovative program during the initial years of operation; grants for political activities, sectarian religious purposes, or individual medical or scientific research projects; or repeated funding to the same program or organization for the same purposes on an annual or ongoing basis.
Date(s) Application is Due: Jan 31; Jul 31
Contact: Diane Pavela, (262) 842-0079; fax (262) 842-0069; thepalmerfoundation@mcleodusa.net
Internet: http://www.thepalmerfoundation.org
Sponsor: Palmer Foundation
1025 56th Street
Kenosha, WI 53140

Pancreatic Cancer Action Network-AACR Fellowship **3862**
The Pancreatic Cancer Action Network-AACR Fellowship represents a joint effort to attract young scientist to careers in pancreatic cancer research. The Fellowship is open to Postdoctoral Fellows and Clinical Research Fellows working at an academic, medical, or research institution who will be in the 1st, 2nd, or 3rd year of their postdoctoral training at the start of the grant term on July 1. The Fellowship provides a one-year grant of $45,000 to support the salary and benefits of the Fellow while working on a mentored pancreatic cancer research project. A partial amount of funds, up to 25% of the total grant, may be designated for direct research support. The research proposed for funding may be basic, translational, clinical or epidemiological in nature and must have direct application and relevance to pancreatic cancer. It is anticipated that one Fellowship will be funded. AACR requires applicants submit both an online and a paper application. The online application must be submitted by 12:00 noon (United Sates Eastern Time) on October 28, using the proposalCENTRAL website at http://proposalcentral.altum.com. The paper copy with original signatures and all required documents must be postmarked and sent no later then October 30.
Requirements: Applicants must have acquired a doctoral degree (including Ph.D., M.D., D.O., D.C., N.D., D.D.S., D.V.M., Sc.D., D.N.S., Pharm.D., or equivalent doctoral degree, or a combined clinical and research doctoral degree) in a related field and may not currently be candidate for another degree. Applicants must be in the 1st, 2nd, or 3rd year of Postdoctoral or Clinical Research Fellowship at an academic, medical, or research institution within the United States at the start of the Fellowship term on July 1.
Restrictions: Scientific investigators or health professionals who are funded by the tobacco industry for any project, or whose named mentors in the case of mentored grants are funded by the tobacco industry for any project, may not apply and will not be eligible for AACR grants.
Date(s) Application is Due: Oct 28; Oct 30
Amount of Grant: 45,000 U.S.D
Contact: Hanna Hopfinger, Grants Program Assistant; (267) 646-0665; fax (215) 440-9372; hanna.hopfinger@aacr.org
Internet: http://www.aacr.org/home/scientists/research-funding—fellowships/postdoctoral-fellowships.aspx
Sponsor: American Association for Cancer Research
615 Chestnut Street, 17th Floor
Philadelphia, PA 19106-4404

Pancreatic Cancer Action Network-AACR Pathway to Leadership Grants **3863**
The Pancreatic Cancer Action Network-AACR Pathway to Leadership Grant represents a joint effort to ensure the future leadership of pancreatic cancer research by supporting outstanding early career investigators beginning in their postdoctoral research positions and continuing through their successful transition to independence. Applicants must be in the first five years of their postdoctoral or clinical research fellowships at the time of application and not already have a full-time tenure track assistant professor position (or equivalent). The Pathway to Leadership Grant will provide up to five years of support, for a total of $600,000, consisting of two phases. The initial Mentored Phase lasts up up two years, during which time recipients will receive $75,000 per year and are expected to work closely with mentors to develop their research. During the subsequent three years, recipients are expected to be independent research positions and will be funded at $150,000 per year. AACR requires applicants to submit both an online and a paper application. Online applications must be completed by 12:00 noon (United States Eastern Daylight Time) on October 28, using the proposalCENTRAL website. The paper copy with original signatures and all supporting documents must be postmarked and sent no later than October 30.
Requirements: Applicants must be in mentored training positions at an academic, medical, or research institution within the United States. There are no citizenship requirements. The applicant institution will be the mentored phase institution. Employees or subcontractors of a government entity or for-profit private industry are not eligible. Exceptions may apply if an applicant holds a full-time position at a Veterans' Hospital or national laboratory in the U.S. Contact AACR before submitting an application to determine your eligibility. Applicants must be Postdoctoral or Clinical Research Fellows who completed all doctoral degree requirements no more then five years prior to the start of the grant term.
Date(s) Application is Due: Oct 28; Oct 30
Amount of Grant: 600,000 U.S.D
Contact: Elizabeth Martin, Program Assistant; (267) 646-0664; fax (215) 440-9372; grants@aacr.org
Internet: http://www.aacr.org/home/scientists/research-funding—fellowships/postdoctoral-fellowships.aspx
Sponsor: American Association for Cancer Research
615 Chestnut Street, 17th Floor
Philadelphia, PA 19106-4404

Parke County Community Foundation Grants **3864**
The Parke County Community Foundation Grants fund fields of interest which are community-enhancing such as agricultural interests, family support, fine arts/culture, handicapped persons, historic preservation, individual township interests, religion, scholarship/education, and youth/recreation. The Foundation accepts grant applications year-round.
Requirements: Applications for grants are accepted from any new or existing charitable organization or community agency with a charitable purpose in Parke County. Organizations should fill out the online application with the following information: a detailed description of their organization, project, and budget required; other organizations who might partner with them with a similar project. For requests of $1,000 or less, a one-page letter or email is acceptable. This letter should describe the project, sharing the organization's perception of need. It should also describe who and how many are likely to benefit and the factors the organization will use to evaluate its success. Applicants are encouraged to contact the Foundation to discuss if their project is appropriate for funding before submitting the application.
Restrictions: The Foundation usually funds only to nonprofit organizations, but may fund other organizations if the project is designed to assist the needy and promote well-being in Parke County.
Geographic Focus: Indiana
Contact: Brad Bumgardner, Executive Director; (765) 569-7223; fax (765) 569-5383; bradbum@yahoo.com or parkeccf@yahoo.com
Internet: http://www.parkeccf.org/Grants.html
Sponsor: Parke County Community Foundation
115 North Market Street
Rockville, IN 47872

Parker B. Francis Fellowships in Pulmonary Research **3865**
Grants are awarded to institutions for the purpose of providing stipends, fringe benefits, and modest travel expenses in support of qualified post-doctoral fellows or newly appointed assistant professors. Award recipients will thereby be enabled to devote the major part of their professional effort to research related to pulmonary disease and lung biology. Sponsorship of the fellow by an established investigator is required. There can be only two fellows in a department at any one time. Only one application will be accepted from a department annually. Fellowships are awarded for three years.
Requirements: Fellows must be U.S. or Canadian citizens, or intend to make permanent residence in the United States or Canada, must hold doctoral degrees, and must be sponsored by established investigators. Foreign nationals must document intent to become permanent residents or citizens of North America.
Date(s) Application is Due: Oct 17
Amount of Grant: 45,500 - 139,500 U.S.D
Samples: Dr. Zissis Chroneos, Pediatrics, Children's Hospital Medical Ctr—for research entitled Cloning of the SP-A Receptor Gene; Dr. Emily DiMango, Medicine, Columbia U—for research entitled Signal Transduction of Inflammation in Cystic Fibrosis; Dr. Michael DuVall, Anesthesiology, U of Alabama, Birmingham—for research entitled Effects of Reactive Nitrogen Species on C1-secretion in Airway Epithelium.
Contact: Dr. Joseph Brain, Director of the Fellowship Program; (617) 432-4099; fax (617) 277-2382; brain@hsph.harvard.edu
Internet: http://www.francisfellowships.org
Sponsor: Francis Families Foundation
Harvard School of Public Health, Room 1411, Building 1
Boston, MA 02115

Parkersburg Area Community Foundation Grants **3866**
The foundation supports programs directed toward improving charitable, educational, cultural, health, and welfare activities in the mid-Ohio Valley, WV. Areas of interest include museums, libraries, arts and cultural programs, and historic preservation; higher education, adult basic education, scholarships to individuals, and literacy; health care and

health associations; human and family services; and community development. Grants are awarded to support capital campaigns, building construction/renovation, equipment acquisition, scholarship funds and scholarships to individuals, seed money, program development, and matching funds. There are no application forms.
Requirements: Grants are awarded to 501(c)3 tax-exempt nonprofit organizations in the mid-Ohio Valley, WV.
Restrictions: Grants do not support religious purposes, travel, meetings, seminars, student exchange programs, annual campaigns, endowment funds, operating budgets, or debt reduction.
Geographic Focus: West Virginia
Date(s) Application is Due: Mar 1; Sep 1
Samples: United Way of the Mid-Ohio Valley (Parkersburg, WV)—program support, $10,598; Vienna Public Library (Vienna, WV)—program support, $3500; Parkersburg High School Foundation (Parkersburg, WV)—program support, $5545.
Contact: Judy Sjostedt, Executive Director; (866) 428-4438 or (304) 428-4438; fax (304) 428-1200; judys@pacfwv.com or info@pacfwv.com
Internet: http://www.pacfwv.com
Sponsor: Parkersburg Area Community Foundation
501 Avenuery Street, P.O. Box 1762
Parkersburg, WV 26102-1762

Partnership for Cures Charles E. Culpeper Scholarships in Medical Science 3867
The program awards scholarships in medical science to support the career development of academic physicians. In direct costs, $100,000 per year to the sponsoring institution will be committed for up to three consecutive years. It is expected that a significant portion of the award will be used as a contribution to the Scholar's salary. Up to $50,000 (plus fringe benefits) of this sum may be used for this purpose; the rest for support of research including personnel and some travel costs. Eight percent will be added to each award for indirect costs. All scientific research relevant to human health is eligible for consideration.
Requirements: Candidates must: be U.S. citizens or aliens who have been granted permanent U.S. residence (proof required); hold the M.D. degree from a U.S. medical school or the equivalent of an M.D. degree from an educational institution equivalent to a United States medical school; and be judged worthy of support by virtue of the quality of their research proposals and their potential for successful careers in academic medicine. Applicants must have at least one year of post-doctoral clinical training.
Restrictions: Individuals who have achieved a rank above assistant professor are ineligible for this program, and an awardee who is promoted above the rank of assistant professor before the initiation of the scholarship will be ineligible to receive the award.
Date(s) Application is Due: Aug 16
Amount of Grant: 108,000 U.S.D
Contact: Dr. Bruce E. Bloom; (312) 696-1366; fax (312) 263-0939; Bruce@4cures.org
Internet: http://www.4cures.org/home/culpeper_scholar_program_research
Sponsor: Partnership for Cures
70 West Madison Street, Suite 1500
Chicago, IL 60602

Partnership for Cures Two Years To Cures Grants 3868
The Initiative is intended to bring researchers, funders and institutions together to produce better treatments and cures for patients with disease right now. The Program works to quickly bring together: breakthrough scientific discoveries about diseases and treatments; FDA approved drugs and other medical therapies (Eastern and Western) with new potential uses; research teams at top institutions ready to proceed with urgency and focus; and independent validation, compressed time lines, and financial leverage. Projects must be able to create a direct patient impact in 24 months from the date of initiation of the project. Budgets can range from $35,000 to $500,000. The Program is accepting research projects in any catastrophic disease, but has special interest in the following diseases: lupus, Crohns disease, MS, ALS, protein intolerances in children, ischemic colitis, autism, type I diabetes, lung cancer, osteosarcoma, Fragile X syndrome, Jewish genetic diseases, and Parkinsons disease. There is no deadline for submissions, but funders are looking for projects right now.
Requirements: Each project must come with a minimum of 30% matching funds. Those funds can come from the institution itself, from government or other public funders, or from individuals, foundations or other private funders.
Restrictions: No more than 10% of the budget can be allocated to overhead, travel and publications.
Amount of Grant: 35,000 - 500,000 U.S.D
Contact: Dr. Bruce E. Bloom, President; (312) 696-1366; fax (312) 263-0939; Bruce@4cures.org
Internet: http://www.4cures.org/home/funding_opportunities
Sponsor: Partnership for Cures
70 West Madison Street, Suite 1500
Chicago, IL 60602

Partnership for Pediatric Epilepsy Research Grants 3869
This program stimulates epilepsy research by providing funding for investigators in the early stages of their careers. Seed grants are awarded to clinical investigators or basic scientists for support of biological or behavioral research which will advance the understanding, treatment, and prevention of epilepsy. All research must be conducted in the United States. The partnership is a consortium of professional organizations and individuals with the common objective of researching treatments and cures to pediatric epilepsy.
Requirements: Proposals are invited from investigators who hold an advanced degree (MD or PhD) and have completed all research training. Investigators may be based at corporations as well as academic/university settings.
Date(s) Application is Due: Sep 1
Amount of Grant: Up to 100,000 U.S.D
Contact: Grants Administrator; (301) 459-3700; fax (301) 577-2684; grants@efa.org
Internet: http://www.epilepsyfoundation.org/research/grants.cfm
Sponsor: Partnership for Pediatric Epilepsy
4351 Garden City Drive
Landover, MD 20785

Pasadena Foundation Average Grants 3870
The community foundation serves the greater Pasadena, CA, area and awards grants for capital improvements to agencies serving children, the elderly, and the disabled. Fields of interest include child development, the elderly, disabled, youth, drug abuse, family planning, education, arts and the humanities, mental health, community development, and the environment. Challenge/matching grants also are awarded. The deadline is the first business day in March.
Restrictions: Schools, religious institutions, and political organizations are ineligible.
Geographic Focus: California
Amount of Grant: Up to 8,000 U.S.D
Samples: Women at Work—for the acquisition of three computer systems, $6000; Pasadena Boys Choir—to purchase concert uniforms, $5725; All Saints Children's Ctr—to purchase cots and floor coverings, $2610.
Contact: Program Contact; (626) 796-2097; fax (626) 583-4738; pcfstaff@pasadenacf.org
Internet: http://www.pasadenacf.org/grant_reg.htm
Sponsor: Pasadena Foundation
260 S Los Robles Avenue, #119
Pasadena, CA 91101

Paso del Norte Health Foundation Grants 3871
The Paso del Norte Health Foundation (PdNHF) is one of the largest private foundations on the U.S. - Mexico border. It was established in 1995 from the sale of Providence Memorial Hospital to Tenet Healthcare Corporation for $130 million. The purpose of the Foundation is to improve the health and promote the wellness of the people living in West Texas, Southern New Mexico, and Ciudad Juárez, Mexico through education and prevention. The foundation currently issues RFPs for the following initiatives: physical activity and balanced nutrition; tobacco, alcohol, and illicit drug use; health care and mental health services; healthy families and social environments; and leadership. Interested organizations can sign up for the foundation's RFP mailing list at the grant website. The foundation does not accept unsolicited proposals but offers workshops when new RFPs are issued and also encourages organizations to contact the Program Officer with suggestions and ideas for new programs, especially as they relate to the foundation's initiatives. At the time of this writing, the foundation has an active RFP entitled "A Smoke-free Paso del Norte" which may be downloaded from the grant website. The application must be received by 2 p.m. on the due date. It may be hand-delivered or emailed in PDF format directly to the Program Director, but not faxed.
Requirements: Proposals must be solicited, and must be linked to an established initiative. Applicants must be 501(c)3 or equivalent organizations operated for charitable, educational, or religious purposes. Funding is principally for in-field intervention projects. Research, studies, or planning activities may be considered only if they directly assist in the implementation of a project. Ineligible expenses include general operating and overhead expenses, capital acquisition or construction (such as property or buildings), purchase of motorized vehicles, and the direct provision of medical services and medicines for acute care.
Restrictions: Funding is not provided to individuals. Proposals are only considered from the Paso del Norte region which encompasses El Paso and Hudspeth Counties in West Texas; Doña Ana, Luna, and Otero Counties in Southern New Mexico; and Ciudad Juárez and Chihuahua in Northern Mexico.
Geographic Focus: Texas, Mexico
Date(s) Application is Due: Jan 10
Samples: El Paso Independent School District, El Paso, Texas, $2.5 million over four years - to create an innovative health literacy program "Get HIP (Health Initiative Program) Now"; Texas Tech University Health Sciences Center, El Paso, Texas, $1.25 million over six years - for the Medical Student Loan Program; YWCA, El Paso, $803,128 over three years - for child and youth programming to decrease sedentary lifestyles in children.
Contact: Jon Law, Program Officer; (915) 544-7636; fax (915) 544-7713; jlaw@pdnhf.org or health@pdnhf.org
Internet: http://www.pdnhf.org/grantcenter.asp
Sponsor: Paso del Norte Health Foundation
221 N. Kansas, Suite 1900
El Paso, TX 79901

Patrick and Anna M. Cudahy Fund Grants 3872
The Patrick and Anna Cudahy Fund is a general purpose foundation which primarily supports organizations in Wisconsin and the metropolitan Chicago area. The principal areas of interest are social service, youth, and education with some giving for the arts, and other areas. Some support is also given for local and national programs concerned with public interest and environmental issues. A few grants are given for international programs but only to those which are represented by a United States based organization.
Requirements: Nonprofits organizations in Wisconsin and the metropolitan Chicago area are eligible. See Foundations website for additional guidelines and application form: http://www.cudahyfund.org/Guideliines.htm
Restrictions: Organizations may submit only one proposal during any calendar year. Requests are not considered for: organizations and projects primarily serving a local

constituency outside of Wisconsin and the Chicago metropolitan area; organizations outside the United States who are not represented by a United States based 501 [c]3 organization; grants to individuals; loans or endowments.
Geographic Focus: Illinois, Wisconsin
Date(s) Application is Due: Jan 5; Apr 5; Jul 5; Oct 5
Amount of Grant: 5,000 - 25,000 U.S.D
Samples: Angel of Hope Clinic, Milwaukee, WI, $10,000 - Angel of Hope Clinic; DePaul University College of Law Chicago,IL, $22,000 - Summer Programs; Helen Bader Institute, Milwaukee, WI, $10,000 - Scholarships;
Contact: Janet S. Cudahy, President; (847) 866-0760; fax (847) 475-0679
Internet: http://www.cudahyfund.org/Guideliines.htm
Sponsor: 1609 Sherman Avenue, #207
Evanston, IL 60201

Patron Saints Foundation Grants **3873**
The Patron Saints Foundation is a private foundation that provides grants to public charities that improve the health of individuals residing in the West San Gabriel Valley through health care programs which are consistent with the moral and religious teachings of the Roman Catholic Church. Grants will be made to qualified public charities to sponsor charitable, scientific or educational health care programs which are not inconsistent with the moral and religious teachings of the Roman Catholic Church, in the following health care categories: community health services (direct health care services); community health care education; capital expenditures; equipment and supplies; medical research. Grants typically range from $5,000 to $15,000. Larger or smaller grants are at the discretion of The Patron Saints Foundation Board of Directors based on the merits of the project.
Requirements: California 501(c)3 tax-exempt organizations serving the West San Gabriel Valley are eligible, including: the cities of Alhambra, Arcadia, Duarte, El Monte, La Canada Flintridge, Monrovia, Monterey Park, Pasadena, Rosemead, San Gabriel, San Marino, Sierra Madre, South El Monte, South Pasadena, Temple City and the unincorporated areas of Los Angeles County known as Altadena and South San Gabriel along with the unincorporated portions of Los Angeles County within the West San Gabriel Valley.
Restrictions: Grants are not awarded for endowment funds, political activities, travel, surveys or fund raising activities.
Geographic Focus: California
Date(s) Application is Due: Mar 1; May 1
Amount of Grant: 5,000 - 15,000 U.S.D
Contact: Kathleen T. Shannon, Executive Director; (626) 564-0444; fax (626) 564-0444; patronsaintsfdn@sbcglobal.net
Internet: http://www.patronsaintsfoundation.org/index.php?nav=Grant%20Guidelines
Sponsor: Patron Saints Foundation
260 S Los Robles Avenue, Suite 201
Pasadena, CA 91101-3614

Paul and Mary Haas Foundation Contributions and Student Scholarships **3874**
Grants are given to tax-exempt organizations in six areas: civic and charitable, education (including adult basic education, adult/family literacy training, curriculum development, and vocational programs), health, religion, the arts, and college scholarships. Funding is generally seed money for innovative services in the Corpus Christi, TX, area. Creative, innovative, and exploratory projects are encouraged, with less interest in building projects and areas covered by normal budgets. Additional types of support include general operating support, equipment acquisition, program development, conferences and seminars, and scholarship funds and scholarships to individuals. Requests for support are normally reviewed on a year-round basis in the order in which they are received. All grants are made for one year and are not automatically renewed. Applications for scholarships are due in August at the beginning of senior high school year; application for other grants are accepted at any time.
Requirements: Grants are made to Texas organizations with 501(c)3 status. Initial inquiries should be made in the form of a one- to two-page written proposal.
Geographic Focus: Texas
Amount of Grant: 35 - 250,000 U.S.D
Contact: Karen Wesson, Administrative Director; (361) 887-6955; fax (361) 883-5992; haasfdn@aol.com
Sponsor: Paul and Mary Haas Foundation
P.O. Box 2928
Corpus Christi, TX 78403

Paul Balint Charitable Trust Grants **3875**
The trust operates in the United Kingdom and internationally and awards grants to nonprofit charities in the fields of Jewish charities, medicine, disability, with a specific interest in aging and elderly and improving the daily lives of those with disabilities.
Contact: Andrew Balint, 020 7624 2098
Sponsor: Paul Balint Charitable Trust
26 Church Crescent
London, N20 0JP England

Paul E. and Klare N. Reinhold Foundation Grants **3876**
The foundation makes grants in Clay County, Florida, for projects in health, religion, children and youth, music appreciation and education, art appreciation and education, and public improvement. Types of support include building construction/renovation, capital campaigns, emergency funds, equipment, land acquisition, matching/challenge grants, operating grants, and seed money grants. Details are available on the Web site.
Requirements: 501(c)3 tax-exempt organizations in Florida's Clay County are eligible.
Restrictions: Grants do not support advertising for fund-raising campaigns, tickets, debt reduction, endowments, attempts to influence legislation, or basic operating costs.
Geographic Focus: Florida
Amount of Grant: Up to 10,000 U.S.D
Samples: Clay County Habitat For Humanity, Inc (FL)—to supplement a Capacity Building Grant to help pay the Executive Director's salary, $20,000; Alzheimer's Assoc (FL)—to support Caregiver Education By enhancing the resource library, $3000; Christ Episcopal Church Building Fund (FL)—dedication of family/gathering room in honor of Thomas E. Camp, III, $50,000; Keystone Heights Elementary School (FL)—for additional supplies for Project Playground, $500.
Contact: Program Contact; (904) 269-5857; fax (904) 269-8382; lhoke@reinhold.net
Internet: http://www.reinhold.org/html/guidelines.htm
Sponsor: Paul E. and Klare N. Reinhold Foundation
1845 Town Center Boulevard, Suite 105
Orange Park, FL 32003

Paul G. Allen Family Foundation Grants **3877**
The single foundation, created through the consolidation of Allen's six previous foundations (The Allen Foundation for the Arts, The Paul G. Allen Charitable Foundation, The Paul G. Allen Foundation for Medical Research, The Paul G. Allen Forest Protection Foundation, The Allen Foundation for Music, and The Paul G. Allen Virtual Education Foundation), will continue to focus on the Allen family's philanthropic interests in the areas of arts and culture, youth engagement, community development and social change, and scientific and technological innovation. The Arts and Culture Program fosters creativity and promotes critical thinking by helping strong arts organizations become sustainable and supporting projects that feature innovative and diverse artistic forms. The Youth Engagement Program improves the way young people learn by supporting organizations that use innovative teaching strategies and provide opportunities for children to address issues relevant to their lives. The Community Development and Social Change Program promotes individual and community development by supporting initiatives and organizations that provide access to resources and opportunities. The Scientific and Technological Innovation Program advances promising scientific and technology research that has the potential to enhance understanding and stewardship of the world in which we live. Organizations may only receive one grant per year. Organizations must not have any delinquent final reports due to any of the Paul G. Allen Foundations for previous grants. Grantseekers are encouraged to apply through the online application process, where basic organizational and project information will be requested. Guidelines are available online.
Requirements: 501(c)3 tax-exempt organizations, status from the Internal Revenue government entities, and IRS-recognized tribes are eligible. Eligible organizations must be located in, or serving populations of, the Pacific Northwest, which includes Alaska, Idaho, Montana, Oregon, and Washington.
Restrictions: In general, the foundation will not consider requests for general fund drives, annual appeals, or federated campaigns; special events or sponsorships; direct grants, scholarships, or loans for the benefit of specific individuals; projects of organizations whose policies or practices discriminate on the basis of race, ethnic origin, sex, creed, or sexual orientation; contributions to sectarian or religious organizations whose principle activity is for the benefit of their own members or adherents; loans or debt retirement; projects that will benefit the students of a single school; general operating support for ongoing activities; or projects not aligned with the foundation's specified program areas. 509(a) private foundations are ineligible.
Geographic Focus: Alaska, Idaho, Montana, Oregon, Washington
Date(s) Application is Due: Mar 31; Sep 30
Samples: Koahnic Broadcast Corp (Anchorage, AK)—to produce a weekly public-radio program of contemporary American Indian music, $225,000; Boys and Girls Club of Yellowstone County (Billings, MT)—for a new club at Castle Rock Middle School, $150,000; Housing Hope (Everett, WA)—for its campaign to build 270 low-cost housing units in Snohomish County, WA, $350,000; Mercy Corps (Portland, OR), Northwest Medical Teams (Tigard, OR), and World Vision (Federal Way, WA)—to assist tsunami-relief efforts, $500,000 each.
Contact: Grants Administrator; (206) 342-2030; fax (206) 342-3030; info@pgafamilyfoundation.org
Internet: http://www.pgafamilyfoundation.org
Sponsor: Paul G. Allen Family Foundation
505 Fifth Avenue S, Suite 900
Seattle, WA 98104

Paul Marks Prizes for Cancer Research **3878**
The prizes recognize outstanding young investigators who have made significant contributions to increase the understanding of cancer or improve the treatment of the disease through basic or clinical research. The winners will present their work at the center, be honored at a dinner, and share a cash award of $150,000. Nomination packets must include a letter from the nominator outlining the significance of the accomplishments for which the candidate should be recognized; a one-page scientific biography of the candidate; a list of up to eight of the candidate's significant published papers with a brief (fewer than 100 words) explanation of the importance of each one; and the candidate's curriculum vita. Up to three supporting letters may also be submitted. Annual deadline date may vary; contact program staff for exact dates.
Requirements: Nominees are required to be age 45 or younger at the time of the submission deadline.
Date(s) Application is Due: Apr 29
Amount of Grant: 150,000 U.S.D
Contact: Paizhe Pressley, Program Contact; (212) 639-6561; presslep@mskcc.org

Internet: http://www.mskcc.org/mskcc/html/53983.cfm
Sponsor: Memorial Sloan-Kettering Cancer Center
1275 York Avenue
New York, NY 10021

Paul Ogle Foundation Grants 3879

The Paul Ogle Foundation, Inc., was incorporated as an Indiana non-profit corporation in 1979 and is qualified under the IRS 501(c)3 as a private foundation. The Foundation's founder and original benefactor was Paul W. Ogle, who lived in Clark County, Indiana. The Foundation is headquartered in Jeffersonville, Indiana, and provides grants to deserving IRS 501(c)3 organizations in Clark, Floyd, Harrison, Switzerland, Scott, and Washington Counties in Indiana and Jefferson County, Kentucky. The Board of Directors has established guidelines for programs which they will consider for grant funding of capital projects and endowments. The guidelines are published on the Foundations website and make up the current basis for recognized charities to apply for grant funding.
Requirements: First read guidelines prior to contacting the foundations office, then send a letter of intent to the office. The letter should be typed on your institutions letterhead and should not exceed for typewritten pages in length. It should contain the following information about the proposed project: a brief description of the need that will be addressed by the proposal; a statement of the project's principal objectives; how the need will be met; the total estimated project budget and other sources of support that may be forthcoming, including a list of donors and amounts; the staff that will carry out the project and their qualifications; a timetable for the project; brief background on the organization; list of board members; the name of the primary contact person for follow-up; copy of IRS 501(c)3 exemption letter. Based on review of your letter of intent, the foundation may request a full proposal. If a proposal is requested, you will be provided with a grant form along with specific instructions. All projects are funded only on the authorization of the Foundation's Board of Director, this may take up to six months. Formal written communications from the foundation is the ONLY way you will be advised of a favorable funding decision.
Restrictions: Grants are not made to: private foundations; religious organizations for religious purposes; individuals; primary or secondary schools; political entities; national organizations, even if for local projects; annual operating support or debt reduction.
Geographic Focus: Indiana, Kentucky
Contact: Robert W. Lanum, Chair; (812) 284-5519; klanum@ogle-fdn.org
Internet: http://www.ogle-fdn.org/Guidelines.htm
Sponsor: Paul Ogle Foundation
321 East Court Avenue, P.O. Box 845
Jeffersonville, IN 47131

Paul Rapoport Foundation Grants 3880

The Paul Rapoport Foundation was established in 1987 with funds from the estate of Paul Rapoport, a founder of both New York City's LGBT Community Services Center and GMHC. For its final years of grantmaking the Foundation's focus will be on three populations of low or no income: transgender communities of color; LGTBQ youth of color, ages 24 and under; and LGTB seniors of color aged 60 and over. The Foundation will also consider funding programs of organizations not focused exclusively on the LGTB community if the number of LGTBQ clients of color served by the program is at least 50%. The foundation awards grants to nonprofits in the metropolitan area of New York, New York. Types of support include general operating support, continuing support, building construction and renovation, program development, conferences and seminars, publications, seed money, technical assistance, and matching funds. There are no application forms. The board meets in February, June, and October to consider requests. Grants of $50,000 and higher, per year are awarded.
Requirements: The Foundation funds only non-profit, charitable organizations as defined by the Internal Revenue Service Code Section 509(a). The Foundation funds primarily within the five boroughs of New York City, as well as on Long Island, in Westchester and nearby New Jersey. It will only fund national organizations when they request funding for programs specific to the New York metropolitan area.
Restrictions: The Foundation will no longer support start-up organizations. The Foundation does not support medical research, cultural and artistic activities, major building campaigns, endowments, grants or scholar-ships to individuals or to other foundations. The Foundation does not make grants for purposes of influencing elections or legislation, or for any other activity that may jeopardize the Foundation's tax-exempt status.
Geographic Focus: New York
Date(s) Application is Due: Feb 1; Jun 1; Oct 1
Samples: Ali Forney Center (AFC), New York, New York, $150,000 (1st year of 3-year grant totaling $450,000) - creation of an In-house Development Department (2011); Friends of Green Chimneys, New York, New York, $230,000 (1st year of 4-year grant totaling $665,000) - (a) Career Development Program for LGTB Foster Care and Runaway and Homeless Youth, (b) Infrastructure Development for NYC Division of Green Chimneys, (c) Engineering Study of Basement of Gramercy Building (2011).
Contact: Jane D. Schwartz, Executive Director; (212) 888-6578; fax (212) 980-0867
Ona M. Winet, Program Director; (212) 888-6578; fax (212) 980-0867
Internet: http://www.paulrapoportfoundation.org/guide.html
Sponsor: Paul Rapoport Foundation
220 E 60th Street, Suite 3H
New York, NY 10022

PC Reproductive Biomedicine Postdoctoral Fellowships 3881

The council awards fellowships to persons who wish to pursue advanced study in the physiology and biochemistry of reproduction. Fellows train in the laboratories of the council's Center for Biomedical Research. Training in any aspect of reproductive biology may be considered. Priority research topics include, but are not limited to, the following areas of reproductive biomedicine: physiology of hypothalamo-pituitary-gonadal axis; leydig cell development; cell and molecular biology of Sertoli cells; cell-cell communication in the seminiferous tubule; molecular biology of steroid-hormone action; biosynthesis and secretion of reproductive hormones; transmission of immunodeficiency viruses; genetics of male infertility; and sperm maturation. Fellowship awards are for full-time research training and study, and are made initially for one year. The aim of the program is to support two years of training, provided the preceptor and fellow agree on the second year. Fellowships can begin any time during the year. There are no deadlines for submission of applications. Guidelines are available online.
Requirements: Candidates must have successfully completed an advanced degree—MD, PhD, or equivalent—and be proficient in English. Applications may be made in anticipation of receiving the doctoral degree, but the fellowship cannot begin until after the degree has been awarded. Fellowships are open to individuals of all countries. If the applicant is other than a United States citizen and enters this country on a Population Council-sponsored J-1 visa, she/he must have a strong commitment to return to her/his own country upon completion of the fellowship.
Amount of Grant: 36,000 - 72,000 U.S.D
Contact: Coordinator; (212) 339-0500; fax (212) 755-6052; fellowships@popcouncil.org
Internet: http://www.popcouncil.org/slr/biofellowships.html
Sponsor: Population Council
1230 York Avenue, Box 273
New York, NY 10021

PDF-AANF Clinician-Scientist Development Awards 3882

The Parkinson's Disease Foundation (PDF) and the American Academy of Neurology Foundation (AANF) collaborated in 2007 to create the Clinician Scientist Development Award to support our mutual interest in training outstanding residency-level clinicians in the field of Parkinson's disease. Successful applicants will receive a three-year award of $85,000. This award supports three years of research training in an environment where talented young clinicians address problems in Parkinson's disease with the most current scientific tools. It is expected that upon completion of the program, participants will be committed to a research or combined clinical/research career in PD and will be in line to direct robust research programs relevant to PD. This award is given every three years. It will be offered again until 2011.
Requirements: Eligibility & Guidelines: applicants must hold an M.D., D.O., or equivalent clinical degree from an accredited institution; applicants must have completed residency training but be less than seven years from completion of residency when funding begins; there is no citizenship requirement; however, the individual applying for the award must be licensed to practice medicine in the United States at the time of application; may only submit one application per department. The department is considered to be the one at which the applicant's training and research program will occur; each applicant must agree to a phone interview with a member of the PDF-AANF Advisory Committee, if necessary, during the application review period.
Amount of Grant: 85,000 U.S.D
Contact: Terry Heinz, Grants Administrator; (651) 695-2746; theinz@aan.com
Internet: http://www.pdf.org/en/grant_funding_fellow#pdflabres
Sponsor: Parkinson's Disease Foundation
1359 Broadway, Suite 1509
New York, NY 10018

PDF International Research Grants 3883

The International Research Grants Program (IRGP) is designed to promote innovative research that has a high potential to significantly advance the knowledge of Parkinson's disease (PD) but little likelihood of securing funding through more traditional sources. By supporting novel, high risk/high reward research, the goal of the IRGP is to enable investigators to demonstrate the feasibility of their ideas while generating preliminary data necessary for the support of future funding. At its core, research supported by the IRGP must be directly relevant to the treatment and/or understanding of Parkinson's disease. Successful IRGP projects are: novel research hypotheses of Parkinson's disease; inventive in terms of methodology or approach; and clinical, preclinical, or basic research proposals that will directly impact Parkinson's disease or its treatment. IRGP awards will now reach a maximum of $75,000 per year for two years (compared to the previous maximum of $50,000 per year for one year of research), subject to review of first year progress. Research grants are now also eligible for an additional ten percent in indirect costs. All applications must be submitted online through the PDF website; no paper copies will be accepted. Visit grants.pdf.org, after Monday, November 2, to create an account, see more detailed instructions and submit your application. The deadline for submission of grant applications will be 11:59 PM EST on Tuesday, February 2.
Requirements: The IRGP is open to applicants both nationally and internationally and basic, translational and clinical research proposals are eligible for support. All applicants must possess a Ph.D. or a M.D. (or equivalent). Preference will be given to scientists who are at an early stage in their professional careers.
Restrictions: Investigators who receive funding from PDF may not receive funding for other projects supported by PDF during the same year, nor receive funding from other foundations or institutions for the same project during that year.
Date(s) Application is Due: Feb 2
Amount of Grant: 75,000 U.S.D
Contact: Robin Anthony Elliott, Executive Director; (212) 923-4700; fax (212) 923-4778; info@pdf.org
Internet: http://www.pdf.org/en/grant_funding_irg#focus
Sponsor: Parkinson's Disease Foundation
1359 Broadway, Suite 1509
New York, NY 10018

PDF Postdoctoral Fellowships for Basic Scientists 3884

The Postdoctoral Fellowships for Basic Scientists are one-year fellowships for young scientists, fresh from their Ph.D. training, to study at major research institutions. The programs grants funding in the amount of $40,000 for one year as well as a research allowance of $5,000 to be used at the discretion of the Fellow, with approval of the sponsor/mentor, to pay for such items as books, training courses, travel costs (up to $2,000) or a computer. There is no provision for the deduction of postdoctoral taxes, institutional overhead or fees. However, if the Fellow submits a written request, $3,000 of the research allowance may be used to defray the cost of premiums for health insurance. Fellows may competitively reapply for continued funding after their first year for a total of three years of support. In these instances, a progress report of the current research would be due along with the new application. All applications must be submitted online through the PDF website; no paper copies will be accepted. The deadline for submission of grant applications will be 11:59 PM ET on Tuesday, February 2. Visit grants.pdf.org after Monday, November 2, to create an account, see more detailed instructions and submit your application. Final grant decisions are expected by mid-May.

Requirements: The applicant must have completed a Ph.D. or M.D. and must identify a research mentor who will serve as mentor for the fellow and supervisor of the research.

Date(s) Application is Due: Feb 2

Amount of Grant: 45,000 U.S.D

Contact: Robin Anthony Elliott, Executive Director; (212) 923-4700; fax (212) 923-4778; info@pdf.org

Internet: http://www.pdf.org/en/grant_funding_fellow#pdflabres

Sponsor: Parkinson's Disease Foundation
1359 Broadway, Suite 1509
New York, NY 10018

PDF Postdoctoral Fellowships for Neurologists 3885

For young clinicians who have completed their neurology residency and are seeking clinical research experience, PDF offers the Postdoctoral Fellowships for Neurologists, one-year awards in the amount of $55,000. A research allowance of $5,000 may be used at the discretion of the Fellow, with approval of the mentor, to pay for such items as books, training courses, travel costs (up to $2,000) or a computer. There is no provision for the deduction of postdoctoral taxes, institutional overhead or fees. However, if the Fellow submits a written request, $3,000 of the research allowance may be used to defray the cost of premiums for health insurance. Fellows may competitively reapply for continued funding after their first year for a total of three years of support. In these instances, a progress report of the current research would be due along with the new application. All applications must be submitted online through the PDF website; no paper copies will be accepted. The deadline for submission of grant applications will be 11:59 PM EST on Tuesday, February 2. Visit grants.pdf.org, after Monday, November 2, to create an account, see more detailed instructions and submit your application. Final decisions are expected by mid-May.

Requirements: Applicants seeking a Postdoctoral Fellowship for Neurologists must have an M.D. and be completing or have completed a neurology residency.

Restrictions: Applicants may not have their own lab.

Date(s) Application is Due: Feb 2

Amount of Grant: 55,000 U.S.D

Contact: Robin Anthony Elliott, (212) 923-4700; fax (212) 923-4778; info@pdf.org

Internet: http://www.pdf.org/en/grant_funding_fellow#pdflabres

Sponsor: Parkinson's Disease Foundation
1359 Broadway, Suite 1509
New York, NY 10018

PDF Summer Fellowships 3886

PDF's Summer Fellowship Program is used to support students from advanced undergraduates to graduate and medical students in their pursuit of Parkinson's-related summer research projects. The goal of the Summer Fellowship is to cultivate an early interest in Fellows into the cause(s) and possible treatments for Parkinson's disease. Fellows work under the close supervision of a sponsor who is an expert in the Parkinson's community and oversees the project. Typically, summer fellowships are offered for 10 weeks of laboratory work with an award of $3,000. The application period for Summer Student Fellowships will open on Monday, November 2, apply by 11:59 PM ET on Friday, January 15.

Requirements: Undergraduate students, graduate students and medical students are all eligible for summer fellowships. Each applicant must identify a mentor, with whom he or she will conduct the proposed project.

Date(s) Application is Due: Jan 15

Amount of Grant: 3,000 U.S.D

Contact: Robin Anthony Elliott, Executive Director; (212) 923-4700; fax (212) 923-4778; info@pdf.org

Internet: http://www.pdf.org/en/grant_funding_fellow#pdflabres

Sponsor: Parkinson's Disease Foundation
1359 Broadway, Suite 1509
New York, NY 10018

Peabody Foundation Grants 3887

The Peabody Foundation was established in 1894 in Massachusetts. Grants provide care, treatment, rehabilitation, education, and assistance to children with physical disabilities as well as encouraging and supporting medical research in the causes of crippling disease, particularly in children. The majority of recipient organizations will be located in the Boston, Massachusetts area. Proposals may be submitted at any time and are reviewed in the spring. Funding ranges from $20,000 to $200,000. Interested organizations should contact the Administrative Director for application procedures and eligibility information.

Restrictions: Grants are not made to individuals. Giving is limited to Massachusetts, with emphasis on the Boston area.

Geographic Focus: Massachusetts

Date(s) Application is Due: Feb 1

Amount of Grant: 20,000 - 200,000 U.S.D

Samples: Agassiz Village, Lexington, Massachusetts, $35,000 - to benefit disabled children and/or related research; Brigham and Women's Hospital, Boston, Massachusetts, $75,000 - to benefit disabled children and/or related research; Cotting School for Handicapped Children, Lexington, Massachusetts, $41,000.

Contact: Judi Mullen, Administrative Director; (508) 728-8780; jemullen12@comcast.net
Jonathan Bashein, Office Administrator; (617) 345-1000 or (617) 310-4100; fax (617) 345-1300; jbashein@nixonpeabody.com

Sponsor: Peabody Foundation
100 Summer Street
Boston, MA 02110-2131

Peacock Foundation Grants 3888

Established by Henry B. Peacock, Jr. in 1947, the mission of Peacock Foundation, Inc. is to enhance and promote the good health and well being of children, families, and underprivileged persons in Southeast Florida, through contributions, gifts, and grants to eligible nonprofit organizations. The priorities of Peacock Foundation, Inc. include: making grants to human services providers that promote youth development, assist abused or neglected children, women, and the elderly, and seek to reduce abuse, prevent homelessness, and end hunger in our community; supporting educational programs in the arts and the environment, as well as special education for disabled persons; contributing to medical research, health care organizations, and hospitals.

Requirements: All applicants must be IRS recognized 501(c)3 public charities classified as not a private foundation, registered with the Department of Agriculture to solicit funds in Florida, when applicable, and located in and/or of significant benefit to residents to the Southeast Florida counties of Miami-Dade, Broward, or Monroe. In order for a proposal to be considered for funding, the applicant first must send a brief letter of inquiry. See Foundations website for letter of inquiry guidelines: http://www.peacockfoundationinc.org/review_progress.html.

Restrictions: Peacock Foundation, Inc. does not fund: capital campaigns, construction, or renovation projects; deficit financing or debt reduction; conferences or festivals; fundraising events or advertising; special events or athletic events; individuals; lobbying to influence legislation; religious organizations, unless engaged in a significant project benefiting the entire community.

Geographic Focus: Florida

Contact: Joelle Allen, Executive Director; (305) 373-1386

Internet: http://www.peacockfoundationinc.org/eligibility.html

Sponsor: Peacock Foundation
100 SE Second Street, Suite 2370
Miami, FL 33131

Pediatric Brain Tumor Foundation Research Grants 3889

The Pediatric Brain Tumor Foundation (PBTF) is a 501(c)3 nonprofit charitable organization whose mission is as follows: to find the cause of and cure for childhood brain tumors by supporting medical research; to increase public awareness about the severity and prevalence of childhood brain tumors; to aid in the early detection and treatment of childhood brain tumors; to support a national database on all primary brain tumors; and to provide educational and emotional support for children and families affected by this life-threatening disease. PBTF grants are meant to expand scientists' knowledge of the causes of pediatric brain tumors and should demonstrate the potential to advance discoveries into new treatment methodologies. Each year PBTF issues a Request for Application for Basic or Translational Research support. The foundation accepts requests for seed grants (start-up research programs) and for fellowships. Guidelines and application are available at the foundation website.

Requirements: Research projects must be hypothesis-driven basic or translational scientific research on pediatric brain tumors. Applicants with M.D.s must have no more than five years of post-residency training as of September of the current year of application. Ph.D. applicants must have no more than five years postdoctoral laboratory experience as of current year of application. Applicant organizations must be 501(c)3 institutions located in the United States, Canada, and Australia.

Restrictions: Grant requests must not exceed $50,000 over the period of one year.

Geographic Focus: All States, Australia, Canada

Date(s) Application is Due: Sep 30

Amount of Grant: 50,000 U.S.D

Contact: Dianne Traynor, President and Chairman of the Board; (800) 253-6530 or (828) 665-6891; fax (828) 665-6894; dtraynor@pbtfus.org or pbtfus@pbtfus.org

Internet: http://www.pbtfus.org/medcomm/research/requests-for-applications.html

Sponsor: Pediatric Brain Tumor Foundation of the United States
302 Ridgefield Court
Asheville, NC 28806

Pediatric Cancer Research Foundation Grants 3890

Since its establishment in 1982 as a grass-roots organization, the Pediatric Cancer Research Foundation (PCRF) has focused its efforts upon improving the care, quality of life, and survival rate of children with malignant diseases. With these ends in mind, PCRF concentrates on laboratory research that will translate into immediate treatment for children with cancer. PCRF primarily funds research that brings innovative new drugs and treatment regimens to children. Proposals should be based on molecular, cellular, or integrated systems; be conceptually innovative; and have a clear plan for study

and future clinical use of the factual data obtained. Current areas of foundation funding include stem-cell transplantation, stem-cell biology, molecular oncology, and molecular and cellular genetics. Awards will generally be made for a one- to three-year period at the Board's discretion. Renewals for additional years may be made for promising projects as recognized by the Scientific Review Committee. Prospective applicants can download the application form at the PCRF website. Completed applications must be received at the PCRF office by 5:00 PM on May 1st. If a deadline falls on a Saturday or Sunday, the applications are due the next business day.
Requirements: Applications may be submitted by individuals holding a M.D., Ph.D., or equivalent degree and working in domestic non-profit organizations such as universities, colleges, hospitals, or laboratories (the "sponsoring institution"). Applications may be multi-institutional in nature. Applicants currently holding a grant may re-apply for extension of similar research by submitting a Renewal Application.
Restrictions: No overhead or indirect costs to institutions will be funded.
Date(s) Application is Due: May 1
Samples: Columbia University, New York, New York, $740,548 (2010); Children's Hospital of Los Angeles, Los Angeles, California, $64,250 (2010); MD Anderson, Houston, Texas, $37,500 (2010).
Contact: Joseph M. Galosic, Director of Scientific Affairs; (949) 859-6312; fax (949) 859-6323; admin@pcrf-kids.org
Internet: http://www.pcrf-kids.org/
Sponsor: Pediatric Cancer Research Foundation
9272 Jeronimo Road, Suite 122
Irvine, CA 92618

Pedro Barrie de la Maza, Count of Fenosa Foundation Grants 3891
The foundation operates mainly in the Galician region and awards grants in its areas of interest, including education, social welfare and studies, science, medicine, and arts and humanities. Types of support include research grants to individuals and institutions, scholarships, and prizes.
Contact: Vicente Arias Mosquera, Secretary; 34-81-221525; fax 34-81-224448; info@fbarrie.org
Internet: http://www.fbarrie.org/fundacion/ingles/home.htm
Sponsor: Pedro Barrie de la Maza, Count of Fenosa Foundation
Canton Grande 9
La Coruna, Spain

Penny Severns Breast, Cervical and Ovarian Cancer Research Grants 3892
The Penny Severns Breast, Cervical and Ovarian Cancer Research Fund is a special fund within the state treasury that is used for breast, cervical and ovarian cancer research grants. Revenue sources include general revenue funds, income tax contributions and gifts, as well as grants and awards from private foundations, nonprofit organizations and other governmental entities or persons. Grants support research in areas related to breast, cervical and ovarian cancer prevention, etiology, pathogenesis, early detection, treatment and behavioral sciences. Research also may include clinical trials. One-year grants are available with the possibility of two subsequent 12-month renewals. Standard research grants are intended to develop and advance the understanding, techniques and modalities effective in early detection, prevention, cure, screening and treatment of breast, cervical and ovarian cancers and may include clinical trials. The maximum award for standard research grants will not exceed $75,000. Fellowship research grants support supervised post-doctoral research training. These grants are intended to further develop the skills necessary for a career in breast, cervical and ovarian cancer research. Fellowship research grants will receive maximum annual funding of $35,000. Special consideration is given to single-year applications and to applications for pilot projects that have the potential for subsequent funding from other sources. Continuation grants, with a maximum of two additional years, are available for those grantees that upon peer review demonstrate progress toward stated goals.
Requirements: Although a majority of the applications submitted are bio-medical in nature, researchers in the fields of behavioral and social sciences also are encouraged to apply. Funding is granted to institutions, not individuals. Research is conducted by an individual(s) under the authority of an institution.
Geographic Focus: Illinois
Amount of Grant: 20,000 - 75,000 U.S.D
Contact: Grant Coordinator; 217-524-6088; fax 217-557-3326; dph.owhline@illinois.gov
Internet: http://www.idph.state.il.us/about/womenshealth/grants/penny.htm
Sponsor: Illinois Department of Public Health
535 W. Jefferson Street, First Floor
Springfield, IL 62761-0001

Percy B. Ferebee Endowment Grants 3893
Grants from the Percy B. Ferebee Endowment are awarded to support charitable, scientific and literary projects, and in particular, governmental and civic projects designed to further the cultural, social, economic and physical well-being of residents of Cherokee, Clay, Graham, Jackson, Macon and Swain Counties of North Carolina and the Cherokee Indian Reservation. Grants are also awarded in the form of scholarships to assist worthy and talented young men and women who reside in said counties in pursuing their college/university degree education within the state of North Carolina. The deadline for the submission of a grant application is September 30. The deadline for scholarship applications is January 31. Application forms are available online. Applicants will receive notice acknowledging receipt of the grant request, and subsequently be notified of the grant declination or approval.
Requirements: 501 (c)3 non-profits in the Cherokee, Clay, Graham, Jackson, Macon and Swain Counties of North Carolina are eligible to apply for grants. Proposals should be submitted in the following format: completed Common Grant Application Form; an original Proposal Statement; an audited financial report and a current year operating budget; a copy of your official IRS Letter with your tax determination; a listing of your Board of Directors. Proposal Statements (second item in the above Format) should answer these questions: what are the objectives and expected outcomes of this program/project/request; what strategies will be used to accomplish your objective; what is the timeline for completion; if this is part of an on-going program, how long has it been in operation; what criteria will you use to measure success; if the request is not fully funded, what other sources can you engage; an Itemized budget should be included; please describe any collaborative ventures. Prior to the distribution of funds, all approved grantees must sign and return a Grant Agreement Form, stating that the funds will be used for the purpose intended. Progress reports and Completion reports must also be filed as required for your specific grant. All current grantees must be in good standing with required documentation prior to submitting new proposals to any foundation. Scholarship recipients must be a resident of these areas and must attend a college or university in the state of North Carolina. Contact the Foundation for additional application requirements.
Restrictions: Grants are not made for political purposes, nor to organizations which discriminate on the basis of race, ethnic origin, sexual or religious preference, age or gender.
Geographic Focus: North Carolina
Date(s) Application is Due: Jan 31; Sep 30
Amount of Grant: 3,000 - 15,000 U.S.D
Samples: Nantahala Regional Library, $15,000—general operating support; Macon County Historical Society, $7,500—general operating expenses; Reach, Inc. Resources Education Assistance Counseling & Housing, $4,000—general operating support.
Contact: Wachovia Bank, N.A., Trustee; grantinquiries6@wachovia.com
Internet: https://www.wachovia.com/foundation/v/index.jsp?vgnextoid=5d6852199c0a a110VgnVCM1000004b0d1872RCRD&vgnextfmt=default
Sponsor: Percy B. Ferebee Endowment
Wachovia Bank, NC6732, 100 North Main Street
Winston Salem, NC 27150

Perkins Charitable Foundation Grants 3894
The Perkins Charitable Foundation Trust was established in 1950, by members of the Perkins family. The Foundation gives primarily for education, the arts, environmental conservation, animals, wildlife, health and medical care, and children, youth and social services. They offer funding on a national basis, with some emphasis on Ohio and Vermont.
Requirements: Contact the Foundation's office by letter or phone prior to your proposal. No Application form is required.
Restrictions: No grants to individuals.
Contact: Marilyn Best, Secretary; (216) 621-0465
Sponsor: Perkins Charitable Foundation
1030 Hanna Building, 1422 Euclid Avenue
Cleveland, OH 44115-2001

Perpetual Trust for Charitable Giving Grants 3895
The Perpetual Trust for Charitable Giving was established in 1957 to support and promote quality educational, human-services, and health-care programming for underserved populations. Special consideration is given to medical aid and medical research organizations, as well as to institutions of higher learning. Grant requests for general operating support are strongly encouraged. Program support will also be considered. Small, program-related capital expenses may be included in general operating or program requests. To better support the capacity of nonprofit organizations, multi-year funding requests are strongly encouraged. Applicants must apply online at the grant website. Applicants are strongly encouraged to do the following before applying: review the downloadable state application procedures for additional helpful information and clarifications; review the downloadable online-application guidelines at the grant website; review the trust's funding history (link is available from the grant website); review the online application questions in advance; and review the list of required attachments. These will generally include: a list of board members, financial statements (audited, reviewed, or compiled by independent auditor); an organization summary; a list of other funding sources; an IRS Determination letter; and other required documents. All attachments must be uploaded in the online application as PDF, Word, or Excel files. The application deadline for the Perpetual Trust is 11:59 p.m. on September 1. Applicants will be notified of grant decisions before November 30.
Restrictions: The trust does not support requests from individuals, organizations attempting to influence policy through direct lobbying, or any political campaigns.
Geographic Focus: Massachusetts
Date(s) Application is Due: Sep 1
Contact: Miki C. Akimoto, Vice President; (866) 778-6859; miki.akimoto@baml.com
Internet: https://www.bankofamerica.com/philanthropic/fn_search.action
Sponsor: Perpetual Trust for Charitable Giving
225 Franklin Street, 4th Floor, MA1-225-04-02
Boston, MA 02110

Perry County Community Foundation Grants 3896
The Perry County Community Foundation is a nonprofit, public charity created by and for the people of Perry County, Indiana. We connect donors with the causes they care about. The Foundation will accept grant proposals from all charitable organizations seeking funding; however, special attention will be given to projects that promote healthy living and that offer training and/or support to potential small business entrepreneurship. The foundation considers proposals for grants on a yearly cycle which begins each July. At the start of each cycle, a notice is mailed to nonprofit organizations that have applied for grants in the past, have received grants in the past, or have otherwise requested notification of the start of each cycle. The foundations grant cycle runs from July through November each year.

Proposals will be accepted from July through the September deadline. Proposals submitted after the September deadline (date varies, contact foundation office for current deadline) will not be considered. The grants committee will make its recommendations on funding to the Foundation's Board of Trustees, which will make final funding recommendations to the board of directors of the Community Foundation Alliance (our governing body). All organizations that have submitted grant proposals will be notified of the outcome of the grants committee's deliberation in writing no later than December 1.
Requirements: The foundation welcomes proposals from nonprofit organizations that are deemed tax-exempt under sections 501(c)3 and 509(a) of the Internal Revenue Code and from governmental agencies serving the county. Proposals from nonprofit organizations not classified as a 501(c)3 public charity may be considered provided the project is charitable and supports a community need. Proposals submitted by an entity under the auspices of another agency must include a written statement signed by the agency's board president on behalf of the board of directors agreeing to act as the entity's fiscal sponsor, to receive grant monies if awarded, and to oversee the proposed project.
Restrictions: Project areas NOT considered for funding: religious organizations for religious purposes; political parties or campaigns; endowment creation or debt reduction; operating costs; capital campaigns; annual appeals or membership contributions; travel requests for groups or individuals such as bands, sports teams, or classes. A six-month progress report and a final report at project completion are required by organizations whose proposals are approved for funding. Instructions and appropriate forms will be provided at the time the grant is awarded.
Geographic Focus: Indiana
Contact: Sally Houzanme; (812) 547-3176; sallyh@perrycommunityfoundation.org
Internet: http://www.perrycommunityfoundation.org/disc-grants-program
Sponsor: Perry County Community Foundation
601 Main Street, Suite B, P.O. Box 13
Tell City, IN 47586

Pet Care Trust Sue Busch Memorial Award **3897**
Incorporated in Washington, D.C. in 1990, The Pet Care Trust is a non-profit, charitable, public foundation. The purpose of The Pet Care Trust is to help promote public understanding regarding the value of and right to enjoy companion animals, to enhance knowledge about companion animals through research and education, and to promote professionalism among members of the companion animal community. The Pet Care Trust invites you submit a letter of recommendation for a graduating veterinary technician student who has been selected by your faculty as having a keen interest in companion animals (dogs, cats, small pet mammals, birds, reptiles, amphibians or pond and aquarium fish) as well as dedicating time to community service related to animals while maintaining a high academic standard. Please include instances relating to the above mentioned criteria. If selected, the student you have recommended will receive one of three $500 Sue Busch Memorial Awards for the year. Our goal is not only to bring recognition to a fine student but recognition as well to the institution instrumental in educating that student. Please submit your recommendations on or before Feb 1 You will be notified on or before April 30 in order for the Check and Certificate to be presented at your graduation ceremony.
Date(s) Application is Due: Feb 1
Amount of Grant: 500 U.S.D
Contact: Steve Hellem, Executive Director; (202) 530-5910; shellem@navista.net
Internet: http://www.petcaretrust.org/i4a/pages/index.cfm?pageid=3297
Sponsor: Pet Care Trust
1155 Fifteenth Street NW, Suite 500
Washington, DC 20005

Peter and Elizabeth C. Tower Foundation Annual Intellectual Disabilities Grants **3898**
Elizabeth Nelson Clarke was born in Mt. Vernon, New York, in 1920, and Peter Tower in Niagara Falls, New York, in 1921. They met while attending Cornell University and married in the summer of 1942. Two daughters, Mollie and Cynthia, were born in 1944 and 1947. Peter entered the Army Air Force in January 1943. After service in Texas and Europe, he hoped to find work in the fledgling air transport business which he expected to thrive. Taking a "temporary" clerk job for the family's customhouse broker business C. J. Tower and Sons, he stayed on to see the partnership evolve into a corporation. In 1986 it processed a total of $45 billion worth of merchandise and Peter then sold the business to McGraw-Hill Inc. who later sold it to the Federal Express Company, which utilizes Tower skills and systems worldwide. Meanwhile, Liz, who had studied art at Cornell, had become a notable artist, working mostly in oils. True to other personal philosophies, Peter and Liz knew that with prosperity came responsibility. Their desire was to assure that the resources they had acquired over the years were put to good use and to see the benefits spread among many. Formed December 31, 1990, the Peter and Elizabeth C. Tower Foundation seeks to support community programming that will help children, adolescents, and young adults affected by substance abuse, learning disabilities, mental illness, and intellectual disabilities achieve their full potential. The foundation's funding objective in the intellectual disabilities category is to improve service delivery for children, adolescents, and young adults to age 26 with intellectual disabilities. The foundation defines an intellectual disability as a disability characterized by significant limitations both in intellectual functioning and adaptive behavior, which covers many everyday social and practical skills. This disability originates before the age of 18. The Foundation will give preference to projects addressing one or more of the following priority areas: reducing obstacles to seeking services/treatment; stigma reduction; transitional services; early identification and linkage to services; effective treatment/programming; co-occurring disorders; and provider workforce shortage and readiness (capacity building). The foundation will also consider other project ideas that have the potential to advance its objective in the category. Typical funding range is $25,000 - $75,000 per year. Multi-year grants are encouraged; projects should be sustainable after grant funding ends. Preference is given to projects where the majority of costs are both new to the organization and directly related to the proposed initiative. Interested organizations must submit a letter of inquiry by 5:00 p.m. June 8 to the Chief Program Officer. Guidelines for the letter of inquiry are provided at the grant website. Applicants will be notified by July 27 of the result of their letters of inquiry. Selected applicants must submit a full grant application by June 22. Final grantees will be informed by September 14. Deadline dates may vary from year to year. Interested organizations should visit the grant website to verify current deadline dates.
Requirements: Letters of inquiry must be mailed or hand-delivered. emailed and faxed copies will not be accepted. Signature of organization's Executive Director, Superintendent or Headmaster is required; designee signatures are not acceptable.
Restrictions: The Foundation accepts only one letter of inquiry per applicant. The Foundation makes grants only to: tax-exempt organizations with 501(c)3 tax-exempt status from the Internal Revenue Service that are neither private foundations nor described as 509(a)3 organizations; diocesan and public school districts; charter schools; and to nonprofit public benefit corporations. Organizations must be located in and primarily serve residents of one of the following geographic areas: Barnstable County, Massachusetts; Dukes County, Massachusetts; Essex County, Massachusetts; Nantucket County, Massachusetts; Erie County, New York; or Niagara County, New York. The Foundation does not provide funds that: may be used for the private benefit of any grant recipient or affiliated person; attempt to influence legislation; attempt to influence or intervene in any political campaign; support capital campaigns or improvements; support individual scholarships; provide general operating support; or that subsidize individuals for the cost of care.
Geographic Focus: Massachusetts, New York
Date(s) Application is Due: Sep 14
Amount of Grant: 25,000 - 75,000 U.S.D
Samples: Community Services for the Developmentally Disabled, Buffalo, New York, $189,071, "Children's Admission Services"; League for the Handicapped, Inc., Springville, New York, $31,424, to provide staff training in Brain Gym and Ooey Gooey (2010); Sandwich Public School District, Sandwich, Massachusetts, $143,060, to provide early intervention staff with training and consultation in the Center for Autism and Related Disorders (CARD) model (2010).
Contact: Tracy A. Sawicki, Executive Director; (716) 689-0370; fax (716) 689-3716; info@thetowerfoundation.org or tas@thetowerfoundation.org
Donald W. Matteson, Chief Program Officer; (716) 689-0370; fax (716) 689-3716; dwm@thetowerfoundation.org
Internet: http://www.thetowerfoundation.com/WhatWeFund/ID/IDGrantGuidelines
Sponsor: Peter and Elizabeth C. Tower Foundation
2351 North Forest Road
Getzville, NY 14068-1225

Peter and Elizabeth C. Tower Foundation Annual Mental Health Grants **3899**
Elizabeth Nelson Clarke was born in Mt. Vernon, New York, in 1920, and Peter Tower in Niagara Falls, New York, in 1921. They met while attending Cornell University and married in the summer of 1942. Two daughters, Mollie and Cynthia, were born in 1944 and 1947. Peter entered the Army Air Force in January 1943. After service in Texas and Europe, he hoped to find work in the fledgling air transport business which he expected to thrive. Taking a "temporary" clerk job for the family's customhouse broker business C. J. Tower and Sons, he stayed on to see the partnership evolve into a corporation. In 1986 it processed a total of $45 billion worth of merchandise and Peter then sold the business to McGraw-Hill Inc. who later sold it to the Federal Express Company, which utilizes Tower skills and systems worldwide. Meanwhile, Liz, who had studied art at Cornell, had become a notable artist, working mostly in oils. True to other personal philosophies, Peter and Liz knew that with prosperity came responsibility. Their desire was to assure that the resources they had acquired over the years were put to good use and to see the benefits spread among many. Formed December 31, 1990, the Peter and Elizabeth C. Tower Foundation seeks to support community programming that will help children, adolescents, and young adults affected by substance abuse, learning disabilities, mental illness, and intellectual disabilities achieve their full potential. The foundation has a particular interest in serious mental illnesses, including major depression, schizophrenia, bipolar disorder, obsessive-compulsive disorder (OCD), panic disorder, post-traumatic stress disorder (PTSD) and borderline personality disorder. The foundation's annual mental health grants seek to prevent or alleviate psychological disorders in children, adolescents, and young adults to age 26. The foundation will give preference to projects involving one or more of the following priorities: providing direct benefit to individuals with serious mental illness; reducing barriers to seeking treatment or services; reducing stigma and prejudice often associated with mental illness; fostering social and emotional development; providing early identification and linkage to services; treating co-occurring disorders; offering life-skills programming for persons with serious mental illness; building provider workforce readiness (capacity building); providing indicated or selected prevention programming; and providing effective treatment/programming. The foundation will also consider other project ideas that have the potential to advance its objective in the category. Typical funding range is $25,000 - $75,000 per year. Multi-year grants are encouraged; projects should be sustainable after grant funding ends. Preference is given to projects where the majority of costs are both new to the organization and directly related to the proposed initiative. Interested organizations must submit a letter of inquiry by 5:00 p.m. March 30 to the Chief Program Officer. Guidelines for the letter of inquiry are provided at the grant website. Applicants will be notified by May 11 of the result of their letters of inquiry. Selected applicants must submit a full grant application by June 22. Final grantees will be informed by September 7. Deadline dates may vary from year to year. Interested organizations should visit the grant website to verify current deadline dates.

Requirements: To be eligible to apply, organizations must operate programs that provide services for children, adolescents, or young adults to age 26 with psychological disorders; and/or offer programming to prevent the onset of psychological disorders in children, adolescents, and young adults to age 26 who, based on a range of socio-economic factors, have a higher-than-average likelihood of developing these conditions. Letters of inquiry must be mailed or hand-delivered. emailed and faxed copies will not be accepted. Signature of organization's Executive Director, Superintendent or Headmaster is required; designee signatures are not acceptable.
Restrictions: The Foundation accepts only one letter of inquiry per applicant. Universal programs (programs applied to general population groups without reference to or identification of those at particular risk) will not be considered for funding. Certain general restrictions apply to all grant-seeking organizations. The Foundation makes grants only to: tax-exempt organizations with 501(c)3 tax-exempt status from the Internal Revenue Service that are neither private foundations nor described as 509(a)3 organizations; diocesan and public school districts; charter schools; and to nonprofit public benefit corporations. Organizations must be located in and primarily serve residents of one of the following geographic areas: Barnstable County, Massachusetts; Dukes County, Massachusetts; Essex County, Massachusetts; Nantucket County, Massachusetts; Erie County, New York; or Niagara County, New York. The Foundation does not provide funds that: may be used for the private benefit of any grant recipient or affiliated person; attempt to influence legislation; attempt to influence or intervene in any political campaign; support capital campaigns or improvements; support individual scholarships; provide general operating support; or that subsidize individuals for the cost of care.
Geographic Focus: Massachusetts, New York
Amount of Grant: 25,000 - 75,000 U.S.D
Samples: Lynn Public Schools, Lynn, Massachusetts, $276,230, "Paving the Way to Independence, One Step at a Time" (2011); Child and Family Services of Erie County, Buffalo, New York, $191,930, a two-year training program for clinicians and counselors who work with children and adolescents that have experienced trauma in their lives; Catholic Charities of Buffalo, Buffalo, New York, $200,033, a comprehensive training program for clinicians and therapists who work for managing behavioral and emotional problems stemming from the experience of personal trauma (2011).
Contact: Tracy A. Sawicki, Executive Director; (716) 689-0370; fax (716) 689-3716; info@thetowerfoundation.org or tas@thetowerfoundation.org
Donald W. Matteson, Chief Program Officer; (716) 689-0370; fax (716) 689-3716; dwm@thetowerfoundation.org
Internet: http://www.thetowerfoundation.com/WhatWeFund/MentalHealth/AnnualMentalHealthGrantGuidelines
Sponsor: Peter and Elizabeth C. Tower Foundation
2351 North Forest Road
Getzville, NY 14068-1225

Peter and Elizabeth C. Tower Foundation Learning Disability Grants **3900**
Elizabeth Nelson Clarke was born in Mt. Vernon, New York, in 1920, and Peter Tower in Niagara Falls, New York, in 1921. They met while attending Cornell University and married in the summer of 1942. Peter entered the Army Air Force in January 1943. After serving in Texas and Europe, he took a "temporary" clerk job for the family's customhouse broker business C. J. Tower and Sons. Two daughters, Mollie and Cynthia, were born in 1944 and 1947. Although he had hoped to find work in the fledgling air transport business which he expected to thrive, he stayed with Tower and Sons to see the partnership evolve into a corporation. In 1986 it processed a total of $45 billion worth of merchandise and Peter then sold the business to McGraw-Hill Inc. who later sold it to the Federal Express Company, which utilizes Tower skills and systems worldwide. Meanwhile, Liz, who had studied art at Cornell, had become a notable artist, working mostly in oils. True to other personal philosophies, Peter and Liz knew that with prosperity came responsibility. Their desire was to assure that the resources they had acquired over the years were put to good use and that the benefits were spread among many. Formed December 31, 1990, the Peter and Elizabeth C. Tower Foundation supports community programming that helps children, adolescents, and young adults affected by substance abuse, learning disabilities, mental illness, and intellectual disabilities achieve their full potential. Learning disabilities are defined as neurological disorders affecting the brain's ability to receive, process, store, and respond to information. These constitute disorders in one or more of the basic psychological processes involved in understanding or using language, spoken or written, and may manifest themselves in the imperfect ability to listen, think, speak, read, write, spell or do mathematical calculations. These disorders do not include learning problems that are primarily the result of visual, hearing, or motor abilities, of mental retardation, of emotional disturbance, of traumatic brain injury, or of environmental, cultural, or economic disadvantage. The Tower Foundation is conducting research on the current state of the field. The foundation plans to offer grant opportunities to organizations that offer programming to help the learning disabled after it completes the study. These organizations are encouraged to check the grant website in 2012.
Requirements: Eligible organizations must focus on providing services and programming that will help children, adolescents, and young adults affected by substance abuse, learning disabilities, mental illness, and intellectual disabilities achieve their full potential.
Restrictions: Certain general restrictions apply to all grant-seeking organizations. The Foundation makes grants only to the following: tax-exempt organizations with 501(c)3 tax-exempt status that are neither private foundations nor described as 509(a)3 organizations; diocesan and public school districts; charter schools; and nonprofit public benefit corporations. Organizations must be located in and primarily serve residents of one of the following geographic areas: Barnstable County, Massachusetts; Dukes County, Massachusetts; Essex County, Massachusetts; Nantucket County, Massachusetts; Erie County, New York; or Niagara County, New York. The Foundation does not provide funds for the following purposes: the private benefit of any grant recipient or affiliated person; attempts to influence legislation; attempts to influence or intervene in any political campaign; capital campaigns or improvements; individual scholarships; general operating support; or subsidy of individuals for the cost of care.
Geographic Focus: Massachusetts, New York
Contact: Tracy A. Sawicki, Executive Director; (716) 689-0370; fax (716) 689-3716; info@thetowerfoundation.org or tas@thetowerfoundation.org
Donald W. Matteson, Chief Program Officer; (716) 689-0370; fax (716) 689-3716; dwm@thetowerfoundation.org
Internet: http://www.thetowerfoundation.com/WhatWeFund/LearningDisabilities
Sponsor: Peter and Elizabeth C. Tower Foundation
2351 North Forest Road
Getzville, NY 14068-1225

Peter and Elizabeth C. Tower Foundation Mental Health Reference and Resource Materials Mini-Grants **3901**
Elizabeth Nelson Clarke was born in Mt. Vernon, New York, in 1920, and Peter Tower in Niagara Falls, New York, in 1921. They met while attending Cornell University and married in the summer of 1942. Two daughters, Mollie and Cynthia, were born in 1944 and 1947. Peter entered the Army Air Force in January 1943. After service in Texas and Europe, he hoped to find work in the fledgling air transport business which he expected to thrive. Taking a "temporary" clerk job for the family's customhouse broker business C. J. Tower and Sons, he stayed on to see the partnership evolve into a corporation. In 1986 it processed a total of $45 billion worth of merchandise and Peter then sold the business to McGraw-Hill Inc. who later sold it to the Federal Express Company, which utilizes Tower skills and systems worldwide. Meanwhile, Liz, who had studied art at Cornell, had become a notable artist, working mostly in oils. True to other personal philosophies, Peter and Liz knew that with prosperity came responsibility. Their desire was to assure that the resources they had acquired over the years were put to good use and to see the benefits spread among many. Formed December 31, 1990, the Peter and Elizabeth C. Tower Foundation seeks to support community programming that will help children, adolescents, and young adults affected by substance abuse, learning disabilities, mental illness, and intellectual disabilities achieve their full potential. The Mental Health Mini-Grant initiative seeks to prevent or alleviate psychological disorders in children, adolescents, and young adults up to age 26, as well as to build mental health providers' capacity. The mini-grants will provide funds for the purchase of clinical reference and resource materials for these purposes. Grant requests of up to $7,500 will be considered for mental health providers requesting reference materials for multiple sites within an agency. Grant requests of up to $3,000 will be considered for mental health providers requesting reference materials for one site. The Foundation anticipates making 30-45 awards. Applicants must mail or hand-deliver a letter of application (along with required supporting materials) by October 31. Guidelines are available at the grant website. Questions may be directed to the Program Officer. A FAQ is also available at the grant website. Award checks are distributed November 30.
Requirements: Letters of application from community-based organizations must have the Executive Director/CEO's original signature. Requests from school districts, private schools, or charter schools must have the Superintendent/School Headmaster's original signature.
Restrictions: Only one request per community-based organization, charter school, private school or school district will be accepted. Mini-grant monies may not be used for: the purchase of office equipment, furniture or supplies; computers or smart boards; non-mental health reference materials or consumable items; screening or assessment tools; subscriptions to journals/newsletters; professional development or staff training; staff salaries or other general operating expense; or for capital improvements. Certain general restrictions apply to all grant-seeking organizations. The Foundation makes grants only to: tax-exempt organizations with 501(c)3 tax-exempt status from the Internal Revenue Service that are neither private foundations nor described as 509(a)3 organizations; diocesan and public school districts; charter schools; and to nonprofit public benefit corporations. Organizations must be located in and primarily serve residents of one of the following geographic areas: Barnstable County, Massachusetts; Dukes County, Massachusetts; Essex County, Massachusetts; Nantucket County, Massachusetts; Erie County, New York; or Niagara County, New York. The Foundation does not provide funds that: may be used for the private benefit of any grant recipient or affiliated person; attempt to influence legislation; attempt to influence or intervene in any political campaign; support capital campaigns or improvements; support individual scholarships; provide general operating support; or that subsidize individuals for the cost of care.
Geographic Focus: Massachusetts, New York
Date(s) Application is Due: Oct 31
Amount of Grant: 3,000 - 7,500 U.S.D
Contact: Tracy A. Sawicki, Executive Director; (716) 689-0370; fax (716) 689-3716; info@thetowerfoundation.org or tas@thetowerfoundation.org
Internet: http://www.thetowerfoundation.com/WhatWeFund/MentalHealth/MH-RRMaterialsMiniGrant
Sponsor: Peter and Elizabeth C. Tower Foundation
2351 North Forest Road
Getzville, NY 14068-1225

Peter and Elizabeth C. Tower Foundation Organizational Scholarships **3902**
Elizabeth Nelson Clarke was born in Mt. Vernon, New York, in 1920, and Peter Tower in Niagara Falls, New York, in 1921. They met while attending Cornell University and married in the summer of 1942. Peter entered the Army Air Force in January 1943. After serving in Texas and Europe, he took a "temporary" clerk job for the family's customhouse broker business C. J. Tower and Sons. Two daughters, Mollie and Cynthia, were born in 1944 and 1947. Although he had hoped to find work in the fledgling air transport business which he expected to thrive, he stayed with Tower and Sons to see the partnership evolve

into a corporation. In 1986 it processed a total of $45 billion worth of merchandise and Peter then sold the business to McGraw-Hill Inc. who later sold it to the Federal Express Company, which utilizes Tower skills and systems worldwide. Meanwhile, Liz, who had studied art at Cornell, had become a notable artist, working mostly in oils. True to other personal philosophies, Peter and Liz knew that with prosperity came responsibility. Their desire was to assure that the resources they had acquired over the years were put to good use and that the benefits were spread among many. Formed December 31, 1990, the Peter and Elizabeth C. Tower Foundation seeks to support community programming that will help children, adolescents, and young adults affected by substance abuse, learning disabilities, mental illness, and intellectual disabilities achieve their full potential. Recognizing the benefits of using existing community programming aimed at building organizational infrastructure, the Tower Foundation partners with local providers to offer organizational scholarships for selected programs at the Canisius College Center for Professional Development, the University at Buffalo Institute for Nonprofit Agencies, the Boston University Institute for Nonprofit Management and Leadership, the Merrimack College's Non-Profit Certificate Program, and the Martha's Vineyard Donors Collaborative. More information is available at the grant website.

Restrictions: Funding is limited to organizations whose service recipients are children, adolescents, and young adults to age 26 affected by mental illness, intellectual disabilities, learning disabilities, and/or substance abuse. The Foundation makes grants only to 501(c)3 organizations that are neither private foundations nor described as 509(a)3 organizations, diocesan and public school districts, charter schools, and nonprofit public benefit corporations. Organizations must be located in and primarily serve residents of one of the following geographic areas: Barnstable County, Massachusetts; Dukes County, Massachusetts; Essex County, Massachusetts; Nantucket County, Massachusetts; Erie County, New York; or Niagara County, New York.

Geographic Focus: Massachusetts, New York

Contact: Tracy A. Sawicki, Executive Director; (716) 689-0370; fax (716) 689-3716; info@thetowerfoundation.org or tas@thetowerfoundation.org

Donald W. Matteson, Chairperson; (716) 689-0370; dwm@thetowerfoundation.org

Internet: http://www.thetowerfoundation.com/WhatWeFund/OrganizationalCapacity-Building/OrganizationalScholarships

Sponsor: Peter and Elizabeth C. Tower Foundation
2351 North Forest Road
Getzville, NY 14068-1225

Peter and Elizabeth C. Tower Foundation Phase II Technology Initiative Grants 3903

The Peter and Elizabeth C. Tower Technology Initiative program connects not-for-profit organizations to technological expertise in order to develop and execute a strategic technology plan. A strategic technology plan aligns an organization's administrative and business needs (as outlined in the organization's strategic plan) with its technology needs thus allowing it to function more efficiently and effectively. The application process for the technology initiative grants occurs in two phases: in Phase I, the foundation provides funds to hire a technology consultant to conduct a technology inventory and needs assessment and to assist an organization in developing a three-to-five-year technology plan; in Phase II, the foundation provides a dollar-for-dollar match (up to $125,000) to organizations wishing to implement their technology plans. The Tower Foundation anticipates making multiple Phase II awards in each of the geographic areas it serves. Each matching grant will be for a time period of up to three years; funds will be disbursed each year, not as a single lump sum. Grant awards will vary. The foundation does not accept unsolicited proposals. Applicants must first call the foundation to clarify the intent, scope, and details of the technology initiative; establish the organization's eligibility for the initiative; and to discuss their capacity and readiness to implement their strategic technology plan. Based on this telephone call, the foundation will determine whether or not it wishes to request a full application from the organization. Arrangements for this pre-screening telephone call must be made prior to January 6 for the summer cycle, or July 6 for the winter cycle. The deadline for full proposals is March 14 for the summer cycle and September 12 for the winter cycle. Successful applicants will be notified by June 15 or December 15. More thorough information on the application process is available at the grant website. Details on Phase I of the application process are also available at the website.

Requirements: Organizations must be located in and primarily serve residents of one of the following geographic areas: Barnstable County, Massachusetts; Dukes County, Massachusetts; Essex County, Massachusetts; Nantucket County, Massachusetts; Erie County, New York; or Niagara County, New York. Organizations must focus on the following: providing services for children, adolescents and young adults to age 26 with intellectual disabilities; treating mental illness among children, adolescents and young adults to age 26; or preventing or treating substance abuse among children, adolescents and young adults to age 26. To be eligible for Phase II technology initiative grants, organizations must have already developed a strategic technology plan (based on an up-to-date strategic plan) that recommends specific technologies/policies/practices, identifies their relationship to administrative needs, and provides an estimated implementation budget for each recommendation. Preference will be given to technology plans that also do the following: identify the current state of the organization's technology (including network infrastructure, desktop/server hardware and software, staff skills and training needs, end-user support, periodic systems maintenance, and current practices, policies, procedures, and documentation); place a priority on each recommendation, as well as identify benefits associated with each recommendation and consequences of failing to implement recommendations; and propose an implementation timeline and replacement schedule. Organizations funded through this initiative are expected to provide dollar for dollar matching funds for the Tower Foundation's award. These funds may be obtained through any source of unrestricted funds or awards designated for the technology implementation project specifically.

Restrictions: The foundation makes technology initiative grants only to private and charter schools, nonprofit public benefit corporations, and organizations with 501(c)3 tax-exempt status that are neither private foundations nor described as 509(a)3 organizations. Diocesan and public school districts are not eligible to apply for technology initiative grants. Phase II grant money may not used for the following: the private benefit of any grant recipient or affiliated person; endowments; non-technology capital projects or campaigns; the development of an organization's strategic plan; the development of a strategic technology plan; the development of custom software applications or websites; advanced information technology training or certifications; general operating support; or individual scholarships. Please note that the RFP does not provide funds for hiring new staff nor does it provide funds for existing staff to conduct Phase II technology-initiative-grant activities.

Geographic Focus: Massachusetts, New York

Date(s) Application is Due: Mar 14; Sep 12

Amount of Grant: Up to 125,000 U.S.D

Contact: Tracy A. Sawicki, Executive Director; (716) 689-0370; fax (716) 689-3716; info@thetowerfoundation.org or tas@thetowerfoundation.org

Donald W. Matteson, Chief Program Officer; (716) 689-0370; fax (716) 689-3716; dwm@thetowerfoundation.org

Internet: http://www.thetowerfoundation.com/WhatWeFund/OrganizationalCapacity-Building/TechnologyInitiative/PhaseII2012

Sponsor: Peter and Elizabeth C. Tower Foundation
2351 North Forest Road
Getzville, NY 14068-1225

Peter and Elizabeth C. Tower Foundation Phase I Technology Initiative Grants 3904

The Peter and Elizabeth C. Tower Technology Initiative program connects not-for-profit organizations to technological expertise in order to develop and execute a strategic technology plan. A strategic technology plan aligns an organization's administrative and business needs (as outlined in the organization's strategic plan) with its technology needs thus allowing it to function more efficiently and effectively. The application process for the technology initiative grants occurs in two phases: in Phase I, the foundation provides funds to hire a technology consultant to conduct a technology inventory and needs assessment and to assist an organization in developing a three-to-five-year technology plan; in Phase II, the foundation provides a dollar-for-dollar match (up to $125,000) to organizations wishing to implement their technology plans. The Tower Foundation anticipates making multiple Phase I awards in each of the geographic areas it serves. Each grant will be for a one-year time period. Grant awards will range from $10,000 to $50,000 depending on the size of the organization and the complexity of its business needs. The foundation does not accept unsolicited proposals. Applicants must first call the foundation to clarify the intent, scope, and details of the technology initiative; establish the organization's eligibility for the initiative; and to discuss their capacity and readiness to undertake the technology planning process. Based on this telephone call, the foundation will determine whether or not it wishes to request a full application from the organization. Arrangements for this pre-screening telephone call must be made prior to January 6 for the summer funding cycle, or July 6 for the winter funding cycle. The deadline for full proposals is March 14 for the summer cycle and September 12 for the winter cycle. Successful applicants will be notified by June 15 or December 15. More information on the application process is available at the grant website. Details on Phase II of the application process are also available at the website.

Requirements: Organizations must be located in and primarily serve residents of one of the following geographic areas: Barnstable County, Massachusetts; Dukes County, Massachusetts; Essex County, Massachusetts; Nantucket County, Massachusetts; Erie County, New York; or Niagara County, New York. Organizations must focus on the following: providing services for children, adolescents and young adults to age 26 with intellectual disabilities; preventing or treating mental illness among children, adolescents and young adults to age 26; or preventing or treating substance abuse among children, adolescents and young adults to age 26. Additionally, organizations must have completed a strategic planning document within the past 36 months explicitly identifying technology as a focal area.

Restrictions: The foundation makes technology initiative grants only to private and charter schools, nonprofit public benefit corporations, and organizations with 501(c)3 tax-exempt status that are neither private foundations nor described as 509(a)3 organizations. Diocesan and public school districts are not eligible to apply for technology initiative grants. Phase I grant money may not used for the following: the private benefit of any grant recipient or affiliated person; endowments; capital projects or campaigns; staffing costs associated with technology planning or implementation; the development of an agency's strategic plan; general operating support; individual scholarships; or the purchase of computer hardware or software.

Geographic Focus: Massachusetts, New York

Date(s) Application is Due: Mar 14; Sep 12

Amount of Grant: 10,000 - 50,000 U.S.D

Contact: Tracy A. Sawicki, Executive Director; (716) 689-0370 x206; fax (716) 689-3716; info@thetowerfoundation.org or tas@thetowerfoundation.org

Donald W. Matteson, Chief Program Officer; (716) 689-0370; fax (716) 689-3716; dwm@thetowerfoundation.org

Internet: http://www.thetowerfoundation.com/WhatWeFund/OrganizationalCapacity-Building/TechnologyInitiative/PhaseI-2012

Sponsor: Peter and Elizabeth C. Tower Foundation
2351 North Forest Road
Getzville, NY 14068-1225

Peter and Elizabeth C. Tower Foundation Social and Emotional Preschool Curriculum Grants 3905

Elizabeth Nelson Clarke was born in Mt. Vernon, New York, in 1920, and Peter Tower in Niagara Falls, New York, in 1921. They met while attending Cornell University and married in the summer of 1942. Two daughters, Mollie and Cynthia, were born in 1944

and 1947. Peter entered the Army Air Force in January 1943. After service in Texas and Europe, he hoped to find work in the fledgling air transport business which he expected to thrive. Taking a "temporary" clerk job for the family's customhouse broker business C. J. Tower and Sons, he stayed on to see the partnership evolve into a corporation. In 1986 it processed a total of $45 billion worth of merchandise and Peter then sold the business to McGraw-Hill Inc. who later sold it to the Federal Express Company, which utilizes Tower skills and systems worldwide. Meanwhile, Liz, who had studied art at Cornell, had become a notable artist, working mostly in oils. True to other personal philosophies, Peter and Liz knew that with prosperity came responsibility. Their desire was to assure that the resources they had acquired over the years were put to good use and to see the benefits spread among many. Formed December 31, 1990, the Peter and Elizabeth C. Tower Foundation seeks to support community programming that will help children, adolescents, and young adults affected by substance abuse, learning disabilities, mental illness, and intellectual disabilities achieve their full potential. The foundation's Social and Emotional Preschool Curriculum Grants provide funds to implement selected preschool curricula to enhance the social-emotional development of preschool children. Eligible social-emotional curricula include: "Al's Pals: Kids Making Healthy Choices"; "The Incredible Years: Dina Dinosaur Classroom Curriculum Preschool"; "PATHS Preschool (Promoting Alternative-Thinking Strategies)"; "Second Step"; and "Tools of the Mind." Grants awards will be for two or three years with grant payments made annually. Funding levels will depend on the specific curriculum selected. In general funds may be used for the cost of training and technical assistance from program developers or publishers, including travel expense; direct program expense including materials and equipment supporting curriculum implementation; and substitute teacher or stipends for out-of-school time training expenses, if applicable. The Social and Emotional Preschool Curriculum Grants have a two-step application process. Applicants must submit a pre-application by 5:00 p.m. on December 1; selected applicants must submit a full proposal by February 9; grantees will be notified by May 6. Further information on the grant application process, including downloadable forms, are available at the grant website. Deadline dates may vary from cycle to cycle. Interested organizations are encouraged to visit the website to verify current deadline dates.
Requirements: The pre-application must be mailed or hand-delivered to the Chief Program Officer. The foundation requires organizations to commit to the highest level of staff training available. However, only those costs associated with the specific training and technical assistance recommended and provided by program developers or publishers will be permitted.
Restrictions: emailed or faxed proposals are not accepted. Social and Emotional Preschool Curriculum Grants do not provide funds for existing personnel expense or to hire new staff. The foundation will accept only one request for a single curriculum from each eligible organization. Certain general restrictions apply to all grant-seeking organizations. The Foundation makes grants only to: tax-exempt organizations with 501(c)3 tax-exempt status from the Internal Revenue Service that are neither private foundations nor described as 509(a)3 organizations; diocesan and public school districts; charter schools; and to nonprofit public benefit corporations. Organizations must be located in and primarily serve residents of one of the following geographic areas: Barnstable County, Massachusetts; Dukes County, Massachusetts; Essex County, Massachusetts; Nantucket County, Massachusetts; Erie County, New York; or Niagara County, New York. The Foundation does not provide funds that: may be used for the private benefit of any grant recipient or affiliated person; attempt to influence legislation; attempt to influence or intervene in any political campaign; support capital campaigns or improvements; support individual scholarships; provide general operating support; or that subsidize individuals for the cost of care.
Geographic Focus: Massachusetts, New York
Date(s) Application is Due: Feb 9
Samples: Cape Cod Child Development, Cape Cod, Massachusetts, $36,034, "Promoting Alternative Thinking Strategies (PATHS) Preschool Curriculum" (2011); Niagara Wheatfield Central School District, Niagara Falls, New York, $28,514, "PATHS (Promoting Alternative Thinking Strategies)" (2011); YMCA of the North Shore, Boston, Massachusetts, $25,829, implementation of the Second Step curriculum in the six preschool programs in five Essex County communities (2011).
Contact: Tracy A. Sawicki, Executive Director; (716) 689-0370; fax (716) 689-3716; info@thetowerfoundation.org or tas@thetowerfoundation.org
Internet: http://www.thetowerfoundation.com/WhatWeFund/MentalHealth/2011SocialandEmotionalPreschoolRFPs
Sponsor: Peter and Elizabeth C. Tower Foundation
2351 North Forest Road
Getzville, NY 14068-1225

Peter and Elizabeth C. Tower Foundation Substance Abuse Grants **3906**

Elizabeth Nelson Clarke was born in Mt. Vernon, New York, in 1920, and Peter Tower in Niagara Falls, New York, in 1921. They met while attending Cornell University and married in the summer of 1942. Peter entered the Army Air Force in January 1943. After serving in Texas and Europe, he took a "temporary" clerk job for the family's customhouse broker business C. J. Tower and Sons. Two daughters, Mollie and Cynthia, were born in 1944 and 1947. Although he had hoped to find work in the fledgling air transport business which he expected to thrive, he stayed with Tower and Sons to see the partnership evolve into a corporation. In 1986 it processed a total of $45 billion worth of merchandise and Peter then sold the business to McGraw-Hill Inc. who later sold it to the Federal Express Company, which utilizes Tower skills and systems worldwide. Meanwhile, Liz, who had studied art at Cornell, had become a notable artist, working mostly in oils. True to other personal philosophies, Peter and Liz knew that with prosperity came responsibility. Their desire was to assure that the resources they had acquired over the years were put to good use and that the benefits were spread among many. Formed December 31, 1990, the Peter and Elizabeth C. Tower Foundation supports community programming that helps children, adolescents, and young adults affected by substance abuse, learning disabilities, mental illness, and intellectual disabilities achieve their full potential. Substance Abuse is defined as the use of illegal drugs or the use of prescription or over-the-counter drugs or alcohol for purposes other than those prescribed, or in excessive amounts. Substance abuse may lead to social, physical, emotional, and job-related problems. The Tower Foundation is conducting research on the current state of the field. The foundation plans to offer grant opportunities to organizations that offer treatment/programming for substance abusers after it completes the study. These organizations are encouraged to check the grant website in 2012.
Requirements: Eligible organizations must focus on providing services and programming that help children, adolescents, and young adults affected by substance abuse, learning disabilities, mental illness, and intellectual disabilities to achieve their full potential.
Restrictions: Certain general restrictions apply to all grant-seeking organizations. The Foundation makes grants only to: tax-exempt organizations with 501(c)3 tax-exempt status that are neither private foundations nor described as 509(a)3 organizations; diocesan and public school districts; charter schools; and to nonprofit public benefit corporations. Organizations must be located in and primarily serve residents of one of the following geographic areas: Barnstable County, Massachusetts; Dukes County, Massachusetts; Essex County, Massachusetts; Nantucket County, Massachusetts; Erie County, New York; or Niagara County, New York. The Foundation does not provide funds for the following purposes: the private benefit of any grant recipient or affiliated person; attempts to influence legislation; attempts to influence or intervene in any political campaign; capital campaigns or improvements; individual scholarships; general operating support; or subsidy of individuals for the cost of care.
Geographic Focus: Massachusetts, New York
Contact: Tracy A. Sawicki, Executive Director; (716) 689-0370; fax (716) 689-3716; info@thetowerfoundation.org or tas@thetowerfoundation.org
Donald W. Matteson, Chief Program Officer; (716) 689-0370; fax (716) 689-3716; dwm@thetowerfoundation.org
Internet: http://www.thetowerfoundation.com/WhatWeFund/SubstanceAbuse
Sponsor: Peter and Elizabeth C. Tower Foundation
2351 North Forest Road
Getzville, NY 14068-1225

Peter F. McManus Charitable Trust Grants **3907**

The Trust supports alcohol and mental health research on the causes of alcoholism and other substance abuses. Proposed projects may examine basic, clinical, social, and environmental causes of substance abuse.
Requirements: The Trust gives primarily to 501(c)3 organizations in Illinois, New York, Pennsylvania, and Texas. Applicants should submit no more than a three page proposal that includes the following: the qualifications of the key personnel, including the principal investigator's bio; a copy of the IRS determination letter; a detailed description of the project and the amount requested; and a copy of the most current organizational budget and/or project budget. Applicants are notified within three months of submission.
Restrictions: No funding is available for treatment of alcohol or substance abuse. No more than 10% of the grant amount may be used for indirect costs.
Geographic Focus: Illinois, New York, Pennsylvania, Texas
Date(s) Application is Due: Aug 31
Amount of Grant: Up to 50,000 U.S.D
Contact: Katharine Lidz, Trust Administrator; (610) 647-4974
Sponsor: Peter F. McManus Charitable Trust
31 Independence Court
Wayne, PA 19087-5824

Peter Kiewit Foundation General Grants **3908**

The foundation supports nonprofits and individuals in designated geographic areas for arts and cultural programs, higher and other education, health care, human services, youth services, rural development, community development, and government/public administration. In the general purpose grants program, there are no limitations on the size or duration of the grants that may be requested. Any applicant may submit a total of up to two applications, for two separate projects in any 12 month period. All Peter Kiewit Foundation grants are awarded on a matching funds basis.
Requirements: Grant application guidelines and application forms are required to submit a funding request and these materials are available through the Foundation office only. Potential applicants should contact the Foundation to establish an organization's eligibility to apply and to discuss the proposed project. Nonprofit 501(c)3 organizations in Rancho Mirage, California; western Iowa; Nebraska; and Sheridan, Wyoming are eligible for a maximum of 50% of the total project cost. Units of government (tax supported) in Rancho Mirage, California; western Iowa; Nebraska; and Sheridan, Wyoming may apply for a maximum of 25% of the total project cost.
Restrictions: Grants are not awarded to support elementary or secondary schools, churches, or religious groups. Grants are not awarded to individuals (except for scholarships), or for endowment funds or annual campaigns.
Geographic Focus: California, Iowa, Nebraska, Wyoming
Date(s) Application is Due: Jan 15; Apr 15; Jul 15; Oct 15
Amount of Grant: 10,000 - 500,000 U.S.D
Contact: Lynn Wallin Ziegenbein, (402) 344-7890; fax (402) 344-8099
Internet: http://www.peterkiewitfoundation.org/page.aspx?id=13&pid=3
Sponsor: Peter Kiewit Foundation
8805 Indian Hills Drive, Suite 225
Omaha, NE 68114-4096

Peter Kiewit Foundation Small Grants 3909

The foundation supports nonprofits and individuals in designated geographic areas for arts and cultural programs, higher and other education, health care, human services, youth services, rural development, community development, and government/public administration. The small grants program allows the Trustees to assist a large number of worthy organizations with a broad array of small projects which are limited in scope but significant for the organization. Small grants are rarely awarded to large organizations. Small grants range in size from $500 to $10,000. The Trustees created this category of grants to support small, defined projects; not to contribute small amounts to much larger budgets.
Requirements: Grant application guidelines and application forms are required to submit a funding request and these materials are available through the Foundation office only. Potential applicants should contact the Foundation to establish an organization's eligibility to apply and to discuss the proposed project. Nonprofit 501(c)3 organizations in Rancho Mirage, California; western Iowa; Nebraska; and Sheridan, Wyoming are eligible for a maximum of 50% of the total project cost. Units of government (tax supported) in Rancho Mirage, California; western Iowa; Nebraska; and Sheridan, Wyoming may apply for a maximum of 25% of the total project cost.
Restrictions: Grants are not awarded to support elementary or secondary schools, churches, or religious groups. Grants are not awarded to individuals (except for scholarships), or for endowment funds or annual campaigns.
Geographic Focus: California, Iowa, Nebraska, Wyoming
Date(s) Application is Due: Jan 15; Apr 15; Jul 15; Oct 15
Contact: Lynn Wallin Ziegenbein, Executive Director; (402) 344-7890; fax (402) 344-8099
Internet: http://www.peterkiewitfoundation.org/page.aspx?id=14&pid=3
Sponsor: Peter Kiewit Foundation
8805 Indian Hills Drive, Suite 225
Omaha, NE 68114-4096

Petoskey-Harbor Springs Area Community Foundation Grants 3910

The foundation awards grants to eligible Michigan nonprofit organizations in its areas of interest, including arts, community development, education, environment, health, and social services. Types of support include scholarships and grants for charitable purposes. Grants are awarded for one year and are renewable. Request application forms from the office. Grant applications are reviewed twice a year, in April and October.
Requirements: Michigan nonprofit organizations serving Emmet County are eligible.
Geographic Focus: Michigan
Amount of Grant: 500 - 5,000 U.S.D
Contact: Maureen Nicholson, Executive Director; (616) 348-5820; fax (616) 348-5883; phsacf@freeway.net
Internet: http://www.petoskey-harborspringsfoundation.org/grantinfo.htm
Sponsor: Petoskey-Harbor Springs Area Community Foundation
616 Petoskey Street, Suite 100
Petoskey, MI 49770

Pew Charitable Trusts Biomedical Research Grants 3911

Pew invests in an array of programs related to science that are aimed at improving the quality of scientific research as well as making data widely available. These projects are working toward solutions to environmental, health and safety dilemmas. Pew also has a decades-long commitment to support groundbreaking research by biomedical researchers early in their careers. All of these initiatives are part of our broader effort to study and promote nonpartisan policy solutions for pressing and emerging problems affecting the American public and the global community. Pew's experts partner with leading authorities in science, the environment and associated fields to conduct research and advance fact-based solutions to compelling problems. Pew targets the frontiers of scientific knowledge where it can make a difference.
Requirements: Grants are made only to colleges and universities, as well as 501(c)3 tax-exempt organizations that are not private foundations.
Restrictions: Grants are not made to individuals or for endowments, capital campaigns, unsolicited construction requests, debt reduction, or scholarships or fellowships that are not part of a program initiated by the Trusts.
Contact: Susan A. Magill, Managing Director; (202) 552-2129 or (202) 552-2000; fax (202) 552-2299; smagill@pewtrusts.org or info@pewtrusts.com
Internet: http://www.pewtrusts.org/our_work_category.aspx?id=336
Sponsor: Pew Charitable Trusts
2005 Market Street, Suite 1700
Philadelphia, PA 19103-7077

Pew Charitable Trusts Children and Youth Grants 3912

The Pew Charitable Trusts seeks to study and promote nonpartisan policy solutions for pressing and emerging problems that affect the next generation. Pew supports initiatives, grounded in research and evidence, that aim to help children and youth become active, contributing members of society both in Philadelphia and around the country. At the national level, Pew supports efforts to prevent children from languishing in foster care without safe, permanent families. Pew works to provide access to high-quality preschool for all three- and four-year-olds, and supports a public health initiative that seeks to reduce young people's exposure to alcohol advertising. Locally, the Pew Fund for Health and Human Services in Philadelphia offers operating and project-specific support to a number of nonprofits that aid youth in the city and four nearby counties.
Requirements: Grants are made only to 501(c)3 tax-exempt organizations that are not private foundations.
Restrictions: Grants are not made to individuals or for endowments, capital campaigns, unsolicited construction requests, debt reduction, or scholarships or fellowships that are not part of a program initiated by the Trusts.
Contact: Susan A. Magill, Managing Director; (202) 552-2129 or (202) 552-2000; fax (202) 552-2299; smagill@pewtrusts.org or info@pewtrusts.com
Internet: http://www.pewtrusts.org/our_work_category.aspx?id=4
Sponsor: Pew Charitable Trusts
2005 Market Street, Suite 1700
Philadelphia, PA 19103-7077

Pew Scholars in the Biomedical Sciences 3913

This program is designed to support young investigators of outstanding promise in the basic and clinical sciences relevant to the advancement of human health. The funding of the awards is provided by The Pew Charitable Trusts. The award is intended to provide assured support, during their earlier years, for junior members of the faculty as they establish their laboratories. It is hoped that the assurance provided through the program will encourage successful applicants to be more venturesome in their research and future applications for support than would otherwise be likely. ALL applicants must be nominated by their institution and complete the online application.
Requirements: The program is open to individuals with a doctorate in medicine, other health related professions, or biomedical sciences. Candidates must hold full- time appointments at the rank of assistant professor or independent researcher on the faculty of the sponsoring institution.
Restrictions: A biomedical scholar cannot apply on his/her own for this scholarship but must apply through a selected institution.
Date(s) Application is Due: Nov 1
Amount of Grant: 60,000 U.S.D
Samples: U of California at San Francisco (CA)—for the Pew Scholars Program in the Biomedical Sciences, $3.6 million over four years.
Contact: Leslie Tucker, (215) 575-4850; shans@pewtrusts.com or info@pewtrust.com
Internet: http://futurehealth.ucsf.edu/biomed/scholdes.html
Sponsor: Pew Charitable Trusts
2005 Market Street, Suite 1700
Philadelphia, PA 19103-7077

PeyBack Foundation Grants 3914

The foundation was established by NFL quarterback Peyton Manning with the purpose of promoting the future success of disadvantaged youth by assisting programs that provide leadership and growth opportunities for children at risk (ages 6-18). The nature of the programs and their immediate long-term benefit shall be guiding considerations in funding grants. Although the foundation does not have a dollar limit on grant requests, most grant amounts range between $1,500 and $10,000.
Requirements: Due to the close association of Peyton Manning with the Indiana, Tennessee and the New Orleans Metropolitan area, programs and projects related to the youth in these areas are of primary concern to the foundation. Proposals will be only considered from organizations that have tax-exempt status under Section 501(c)3 of the Internal Revenue Code. A proposal asking to consider providing a portion of the support for a project will generally receive greater preference than one seeking exclusive funding. Download the required application form at the website.
Restrictions: The following are not areas that the PeyBack Foundation supports: organizations without 501(c)3 tax-exempt status will immediately be eliminated; Fundraising and sponsorship events (e.g., golf tournaments, telethons, banquets); Groups outside of Indiana, Tennessee and New Orleans, LA; Projects/groups benefiting an individual or just a few persons; Building/renovating expenses of any kind; To defray meeting, conferences, workshops or seminars expenses; Payment of travel of individuals or groups; Re-granting organizations; Post-event fundraising; Multi-year gifts.
Geographic Focus: Indiana, Louisiana, Tennessee
Date(s) Application is Due: Feb 1
Amount of Grant: 1,500 - 10,000 U.S.D
Samples: Boy Scouts of America, Indianapolis, IN, $10,000; Start the Adventure in Reading, New Orleans, LA, $10,000; Eighteenth Avenue Family Enrichment Center, Nashville, TN, $8,000.
Contact: Elizabeth Ellis; (877) 873-9225; PeyBack@PeytonManning.com
Internet: http://www.peytonmanning.com/Peyback/GrantGuidelines.html
Sponsor: PeyBack Foundation
6325 North Guilford, Suite 201
Indianapolis, IN 46220

Peyton Anderson Foundation Grants 3915

The Peyton Anderson Foundation initiates projects to meet needs in the community and reacts to requests from charitable organizations in Macon, Bibb County, and Middle Georgia.
Requirements: Applicants must serve Macon and/or Bibb County Georgia in order to be considered for funding. Preference is given to organizations that have a substantial presence in Bibb County. Applications can be downloaded from the Foundation's website and must contain the original application plus five (5) copies, including copies of all required application items. In addition, organizations must submit one (1) copy of the organization's current annual operating budget, IRS determination letter, and most recent annual financial statement.
Restrictions: Grants may not be made to private foundations, individuals, private schools, endowments, churches, or for festivals and trips. Grants are also only awarded to organizations with current 501(c)3 status that benefit Macon and Bibb County, Georgia. DO NOT include any forms or information other than what is required above.
Geographic Focus: Georgia
Date(s) Application is Due: Apr 1; Aug 1
Amount of Grant: 1,000 - 500,000 U.S.D

Samples: Meals On Wheels of Macon and Bibb, Macon, Georgia, $1,420 - support for the annual holiday feast for all; Mid-State Children's Challenge Projects, Inc., Macon, Georgia, $60,000 - funding to retire debt and complete renovation; Medcen Community Health Foundation, Inc., Macon, Georgia, $500,000 - for renovation and expansion of all critical care units, and medical/surgical, neuro and surgical trauma.
Contact: Juanita T. Jordan, President; (478) 743-5359; fax (478) 742-5201; jtjordan@pafdn.org or grants@pafdn.org
Internet: http://www.peytonanderson.org
Sponsor: Peyton Anderson Foundation
577 Mulberry Street, Suite 830
Macon, GA 31201

Pezcoller Foundation-AACR International Award for Cancer Research **3916**
The Award for Cancer Research was established in 1997 to annually recognize a scientist: who has made a major scientific discovery in basic cancer research OR who has made significant contributions to translational cancer research; who continues to be active in cancer research and has a record of recent, noteworthy publications; and whose ongoing work holds promise for continued substantive contributions to progress in the field of cancer. More than one scientist may be co-nominated and selected to share the Award when their investigations are closely related in subject matter and have resulted in work that is worthy of the Award. The Award consists of an unrestricted grant of Euro 75,000 and a commemorative plaque.
Requirements: Eligible candidates are cancer researchers affiliated with institutions in academia, industry, or government that are involved in cancer research, cancer medicine, or cancer-related biomedical science anywhere in the world.
Restrictions: Institutions or organizations are not eligible for the Award.
Date(s) Application is Due: Sep 15
Amount of Grant: 75,000 EUR
Contact: Monique P. Eversley, Staff Associate; (267) 646-0576 or (215) 440-9300, ext. 1400; fax (215) 440-9372; monique.eversley@aacr.org or awards@aacr.org
Internet: http://www.aacr.org/home/scientists/scientific-achievement-awards/pezcoller-aacr-international-award.aspx
Sponsor: Pezcoller Foundation / American Association for Cancer Research
615 Chestnut Street, 17th Floor
Philadelphia, PA 19106-4404

Pfizer/AAFP Foundation Visiting Professorship Program in Family Medicine **3917**
This program is designed to bring educational benefits to U.S. medical schools, teaching hospitals, and community hospitals. Recipient institutions may select and invite prominent physician scientists for three days of teaching and interaction with the faculty, students, residents, research fellows, and physicians involved in family medicine at the medical center. During the visit, the guest faculty may give lectures and participate in rounds, seminars, and conferences. Awards are intended to cover the Visiting Professor's honorarium, travel expenses, and other direct expenses incurred by the host institution in conducting program activities. Amounts allotted for each item are at the discretion of the host institution.
Requirements: This program is open to any family medicine department within a U.S. medical school or any accredited family medicine residency program. Applications from community-based programs are encouraged. Each hospital may submit only one application. Submissions must originate from the Chair of the Family Medicine Department or the Residency Director.
Restrictions: Only one visit per year per visiting professor will be funded.
Date(s) Application is Due: Mar 28
Amount of Grant: 7,500 U.S.D
Contact: Susie Morantz, Senior Program Manager; (800) 274-2237, ext. 4470; fax (913) 906-6095; smorantz@aafp.org
Internet: http://www.aafpfoundation.org/x500.xml
Sponsor: American Academy of Family Physicians Foundation
11400 Tomahawk Creek Parkway, Suite 440
Leawood, KS 66211-2672

Pfizer Foundation Epilepsy Scholarship Award **3918**
The scholarship honors 16 outstanding students who have overcome the challenges of epilepsy, been successful in school, done well in activities out of the classroom, and shown the ability to make the most out of college or graduate school. Experts in education and medicine will choose the winners. Information and application are available online.
Requirements: An applicant must be under a doctor's care for epilepsy; a high school senior who has applied to college; a freshman, sophomore, or junior in college; or a college senior who has applied to graduate school.
Date(s) Application is Due: Mar 1
Amount of Grant: 3,000 U.S.D
Contact: Scholarship Coordinator; (800) 292-7373; Scholarship@edencomgroup.com
Internet: http://www.epilepsy-scholarship.com
Sponsor: Eden Communications Group
515 Valley Street, Suite 200
Maplewood, NJ 07040

Pfizer Foundation Grants **3919**
The foundation's charitable giving programs provide grants through targeted initiatives focused primarily on health care and science education. Areas of interest include: global health—health literacy programs to improve the ability of low-income, low-literate populations to effectively comprehend and act on health information; HIV/AIDS—programs to provide Pfizer's antifungal medicine (Diflucan) to low-income HIV/AIDS patients in the world's least developed countries where the disease is most prevalent, the training of health-care workers in opportunistic infections related to HIV, and health-promotion programs that target those at high-risk of HIV infection living in countries where Pfizer is donating Diflucan; international trachoma initiative—programs to provide Pfizer's key antibiotic (Zithromax) to countries implementing public-health programs to combat trachoma, including community-based initiatives to provide surgery, antibiotics, promotion of clean faces, and education about the disease; disaster relief—requests for product donations following floods, earthquakes, droughts, wars, and other disasters domestically and abroad; K-12 science education—partnerships between Pfizer sites and local schools, with an emphasis on training teachers, developing curriculum, constructing science labs, and support of exhibits and educational outreach programs at science-rich institutions across the country; and community and cultural—programs in Pfizer communities, with a particular focus in the New York City for a range of organizations that address urgent community needs, seek long-term solutions to community issues, and strengthen the cultural vitality of diverse communities.
Requirements: Nonprofit 501(c)3 organizations and schools are eligible.
Restrictions: The following types of organizations or activities are excluded from support: fundraising events or activities such as telethons, walkathons, and races; specific performances or concerts; sporting events; endowment campaigns; building fund drives and capital campaigns (except science lab renovations in partner schools and universities); film, video, television, or radio projects (except as part of a health literacy program); requests for loans or debt retirement; trips, tours, or cultural exchange programs; conferences, seminars, briefing programs, and similar activities (except in instances where the event evolves from Pfizer programs in health care and science education); organizations already supported through United Way contributions; operating expenses of United Way local agencies (except through annual United Way campaigns); grants to individuals; political causes or candidates; anti-business organizations; or organizations that practice discrimination or limit membership on the basis of race, creed, gender, age, sexual preference, or national origin.
Samples: International and local relief organizations—to support relief efforts in South Asia and Africa, $25 million value in drugs and $10 million cash distributed; Western Michigan U (MI)—for unrestricted use, as part of its capital campaign, $1 million; Western Michigan U (MI)—for its life-sciences programs, $2.9 million value in scientific equipment.
Contact: Grants Administrator; (800) 733-4717 or (212) 573-2323
Internet: http://www.pfizer.com/subsites/philanthropy/caring/index.html
Sponsor: Pfizer
235 E 42nd Street
New York, NY 10017-5755

Pfizer Foundation Health Literacy Grants **3920**
The program invites applications for research to advance the understanding of health literacy and for an intervention that overcomes problems associated with low levels of health literacy. Examples of proposed projects include developing new measures or new ways of looking at the causes or impact of health literacy; developing programs that demonstrate improved health literacy; or creating policy initiatives that promote improved health outcomes. Proposals for community-based research and interventions are encouraged.
Requirements: Applications are encouraged from investigators in diverse fields such as health literacy, adult education, health economics, medicine, nursing, pharmacy, public health, anthropology, health policy, social work, management sciences, and ethics.
Amount of Grant: 65,000 U.S.D
Contact: Programs Administrator; (212) 733-4250
Internet: http://www.pfizerhealthliteracy.com/partnerships_awards.html
Sponsor: Pfizer Foundation
235 E 42nd Street
New York, NY 10017-5755

Pharmacia-ASPET Award for Experimental Therapeutics **3921**
The Pharmacia-ASPET Award in Experimental Therapeutics is given annually to recognize and stimulate outstanding research in pharmacology and experimental therapeutics - basic laboratory or clinical research that has had, or potentially will have, a major impact on the pharmacological treatment of disease. The winner will receive a $2,500 honorarium, a plaque, hotel and economy airfare for the winner and spouse to the award ceremony at the ASPET annual meeting.
Requirements: Nominations must be made by ASPET members; however, the nominee need not be a member. No restrictions exist as to age or institutional affiliation of nominee.
Date(s) Application is Due: Sep 15
Amount of Grant: 2,500 U.S.D
Contact: Christine Carrico, Executive Officer; (301) 634-7060; fax (301) 634-7061; ccarrico@aspet.org or info@aspet.org
Internet: http://www.aspet.org/public/awards/exp_ther_award.html
Sponsor: Pfizer / American Society for Pharmacology and Experimental Therapeutics
9650 Rockville Pike
Bethesda, MD 20814-3995

Phelps County Community Foundation Grants **3922**
The Phelps County Community Foundation's grants program provides a means by which not-for-profit charitable organizations may secure financial assistance for projects and programs which will enhance the quality of life for residents of Phelps County, Nebraska. The foundation awards grants in the areas of education, culture, human services, health and recreation, and community. Priority is given to seed grants to initiate promising new projects or programs, programs representing innovative and efficient approaches to serving community needs and opportunities, challenge grants, organizations that work

cooperatively with other community agencies, projects or programs where a moderate amount of grant money can effect a significant result, and projects or programs that enlist volunteer participation and citizen involvement. Types of support include general operating grants, continuing support grants, building construction/renovation, equipment acquisition, program development, publication, seed money, scholarship funds, and matching grants. Applicants must have a plan for future funding and support from other sources.
Requirements: Grants are made to 501(c)3 nonprofit organizations in Phelps County, Nebraska, and sometimes to governmental agencies for capital expenditures and/or capital improvements within Phelps County.
Restrictions: Grants are not made to individuals, to support political activities, to support operating expenses of well-established organizations or public service agencies, to establish new endowment funds, for travel or related expenses for individuals or groups, for operating support of governmental agencies, to religious groups for religious purposes, to profit-making enterprises, or to agencies serving a populace outside of Phelps County. In addition, grants are not made to support annual fund drives or to eliminate previously incurred deficits.
Geographic Focus: Nebraska
Date(s) Application is Due: Apr 1; Oct 1
Amount of Grant: Up to 25,000 U.S.D
Samples: American Legion Baseball, Holdrege, Nebraska, $22,000 - new concession stand; Bertrand Nursing Home, Bertrand, Nebraska, $7,632 - operating expenses.
Contact: Vickie Klein, Executive Director; (308) 995-6847; fax (308) 995-2146; vlpccf@phelpsfoundation.org
Internet: http://www.phelpsfoundation.org/grants.html
Sponsor: Phelps County Community Foundation
504 4th Avenue
Holdrege, NE 68949

Philadelphia Foundation Organizational Effectiveness Grants **3923**
Organizational Effectiveness Grants are designed to strengthen and improve the business and operational practices of nonprofit organizations in Bucks, Chester, Delaware, Montgomery and Philadelphia counties. As in business, these practices are critical to the success of nonprofits' ability to lead, adapt, manage and effectively operate their organizations. These core competencies are critical to nonprofits at all stages of development and size for the effective delivery of services, implementation of best practices, utilization of data to drive decision-making and the ability to adapt to new demands, risks and opportunities. new applications for General Operating Support or Organizational Effectiveness cannot be accepted until January 1st.
Requirements: IRS 501(c)3 tax-exempt organizations located in Bucks, Chester, Delaware, Montgomery, and Philadelphia Counties of Pennsylvania are eligible.
Restrictions: Grants are rarely made to affiliates of national or international organizations, government agencies, organizations not located in Southeastern Pennsylvania, organizations with budgets of more than $1.5 million, private schools, or umbrella-funding organizations. Requests usually are denied for capital campaigns, conferences, deficit financing, endowments, publications, research projects, tours, and trips. Individuals are ineligible.
Geographic Focus: Pennsylvania
Amount of Grant: 3,000 - 30,000 U.S.D
Contact: Libby Walsh, Program Associate; (215) 563-6417; fax (215) 563-6882; lwalsh@philafound.org or oeapplications@philafound.org
Internet: https://www.philafound.org/ForNonprofits/DiscretionaryGrantmaking/OrganizationalEffectiveness/tabid/238/Default.aspx
Sponsor: Philadelphia Foundation
1234 Market Street, Suite 1800
Philadelphia, PA 19107

Philanthropic Collaborative for Integrative Medicine Bravewell Leadership Award **3924**
The purpose of the award is to recognize, empower, and support champions of holistic health and integrative medicine in their efforts to transform the culture of healthcare. Any individual may submit a nomination (of self or other) by completing the online form and sending it with a cover letter and the nominee's curriculum vita to the office. Candidates will be invited to submit additional information.
Requirements: Candidates for the award are physicians and other doctoral level professionals from North America who are catalysts in advancing the field of integrative medicine; have made significant contributions to the field of medicine and have demonstrated positive influence among their colleagues and those they serve; embody and advance the principles of the Declaration for a New Medicine; have a history of collaboration across disciplines and healing philosophies; have a compelling vision for the future of medicine that inspires and encourages others; and are resilient change agents and role models in their communities.
Date(s) Application is Due: Mar 31
Amount of Grant: 100,000 U.S.D
Samples: Dr. Ralph Snyderman, president and chief executive officer and chancellor for health affairs, Duke U (Durham, NC)—award recipient, $100,000.
Contact: Awards Coordinator; (612) 377-8400; info@bravewell.org
Internet: http://www.pcintegrativemedicine.org/braveaward/braveaward.asp
Sponsor: Philanthropic Collaborative for Integrative Medicine
1818 Oliver Avenue South
Minneapolis, MN 55405

Phoenix Suns Charities Grants **3925**
Ranging in size from $1,000 to $10,000, Phoenix Suns Charities Program Grants are intended for Arizona non-profit organization whose programs and activities focus on helping children and families maximize their potential. The foundation's largest annual gift, the Playmaker Award, is a one-time $100,000 grant which can be used for capital or programs, or a combination of both. For this grant, Suns Charities looks favorably on collaborative ideas and naming or branding opportunities.
Requirements: Applications and supporting documents must be submitted electronically through ZoomGrants at the Suns website. Organizations must consider the prerequisites, answer all questions and carefully follow directions. Applications are available in November, evaluated by Suns board members in April and May, with funding in June.
Geographic Focus: Arizona
Date(s) Application is Due: Apr 1
Amount of Grant: 1,000 - 10,000 U.S.D
Contact: Kathryn Pidgeon, Executive Director; (602) 379-7948; fax (602) 379-7990; kpidgeon@suns.com
Internet: http://www.nba.com/suns/charities.html
Sponsor: Phoenix Suns
201 East Jefferson Street
Phoenix, AZ 85004

Phoenixville Community Health Foundation Grants **3926**
The foundation operates in Pennsylvania and seeks to become the champion of community health. Categories of funding include personal health, including physical/behavioral health of children, youth, adults, senior citizens, and those at risk for abuse, neglect, or violence; and community health, including environmental and civic health of the community-at-large and the promotion of educational opportunities for individuals aspiring to be in the health field. Requests must demonstrate that the program/organization is addressing a community need and not duplicating existing efforts. Types of support include program support, general operations, capital projects, and equipment acquisition. Both single- and multiyear initiatives are considered. Applicants should discuss their funding interest with foundation staff; a grant request packet will be given if the proposed initiative fits within foundation guidelines. The board meets bimonthly.
Requirements: Pennsylvania 501(c)3 tax-exempt organizations serving Chester and Montgomery Counties are eligible.
Restrictions: Grants are not awarded for direct scholarship support to individuals; purchases of tickets or advertising for benefit purposes; coverage of continuing operating deficits; document publication; pass-through grant support through a third party (except United Way); and fraternal organizations, political parties or candidates, veterans, labor or local civic groups, or groups engaged in influencing legislation.
Geographic Focus: Pennsylvania
Samples: Montgomery County Dept of Economic and Workforce Development (Norristown, PA)—to expand its CareerLink Employment Program to the Phoenixville Library, $30,800; Chester County Futures (West Chester, PA)—to establish a mentor program to help at-risk youths with the potential to attend college, $15,000; Phoenixville Area Police Athletic League (PA)—for program support, $10,000; Phoenixville Area Economic Development Corp (PA)—for general operating support, $35,000.
Contact: Carol Poinier, Grants, Operations Manager; (610) 917-9890; fax (610) 917-9861; pchf1@juno.com
Internet: http://www.pchf1.org/gmakeover.html
Sponsor: Phoenixville Community Health Foundation
1260 Valley Forge Road, Suite 102
Phoenixville, PA 19460

PhRMA Foundation Health Outcomes Post Doctoral Fellowships **3927**
The program provides stipend support for individuals engaged in a research training program that will create or extend their credentials in health outcomes. The intent of this program is to support post doctoral career development activities of individuals prepared (or preparing) to engage in research that will strengthen representation of health outcomes in schools of pharmacy, medicine, nursing and public health. To accomplish these goals, support will be provided for a two-year period to selected individuals who are beginning careers in health outcomes research and who give promise of outstanding development as researchers. The award, consisting of a $40,000 annual stipend, is made to the institution on behalf of the fellow. The award is intended solely as a stipend and may not be used otherwise.
Requirements: Eligible applicants include well-trained graduates from Pharm. D., M.D., and Ph.D. programs who seek to further develop and refine their research skills through formal postdoctoral training. They must also have a firm commitment from an accredited U.S. university and be a U.S. citizen or permanent resident. The application must include a research plan written by the applicant, the mentor's research record, and a description of how the mentored experience will enhance the applicant's career development in health outcomes research. The sponsor (mentor) of the post doctoral program must describe how the goals of the research training program will be accomplished and provide assurance that key collaborating mentors endorse and are willing to support the training plan.
Restrictions: Applications, however good, which do not meet the aims of the program will be disapproved.
Amount of Grant: 40,000 U.S.D
Contact: Eileen M. Cannon, Executive Director; (202) 572-7756; fax (202) 572-7799; info@phrmafoundation.org
Internet: http://www.phrmafoundation.org/awards/outcomes/postdoc.php
Sponsor: Pharmaceutical Research and Manufacturers of America Foundation
950 F Street NW, Suite 300
Washington, DC 20004

PhRMA Foundation Health Outcomes Pre-Doctoral Fellowships 3928
To provide some assistance in this training sequence, the PhRMA Foundation program aims at supporting promising students during their thesis research. The program provides a stipend and funds to cover costs incidental to the training. Applications will be accepted for a minimum of one year and a maximum of two years. The program is designed for candidates who expect to complete the requirements for the Ph.D. in the field of health outcomes in two years or less from the time the fellowship begins. The program assumes that the fellows will devote full-time (including summers) to their research. The award is made to the university on behalf of the fellow. The fellowship provides a stipend of $20,000 a year payable monthly for a minimum of one year and a maximum of two years. Of the $20,000 awarded annually, $500 a year may be used for incidentals directly associated with the thesis research preparation (e.g., secretarial help, artwork, books, travel, etc.). The second year of a fellowship is contingent upon certification by the thesis advisor that satisfactory progress is being achieved.
Requirements: The fellowship program of pre-doctoral support is designed to assist full-time, in-residence Ph.D. candidates in the fields of health outcomes who are enrolled in U.S. schools of medicine, pharmacy, dentistry or schools of public health. The program seeks to support advanced students who will have completed the bulk of their pre-thesis requirements (two years of study) and are starting their thesis research by the time the award is activated. Students just starting in graduate school should not apply. A candidate enrolled in an M.D./Ph.D. program should not be taking required clinical course work or clerkships during the tenure of the fellowship. Before an individual is eligible to apply for a PhRMA Foundation award, the applicant must first have a firm commitment from a U.S. university. Applications must be submitted by an accredited U.S. school, and all applicants must be U.S. citizens or permanent residents.
Restrictions: The fellowship may not be used as a supplement to funds from other fellowships, traineeships or assistantships. The focus of the program is to assist in the pre doctoral training of the candidate. It should not be viewed as a program to fund a research project. An individual may not simultaneously hold or interrupt any other fellowship providing stipend support during the PhRMA Foundation fellowship.
Amount of Grant: 20,000 U.S.D
Contact: Eileen M. Cannon, Executive Director; (202) 572-7756; fax (202) 572-7799; info@phrmafoundation.org
Internet: http://www.phrmafoundation.org/awards/outcomes/predoc.php
Sponsor: Pharmaceutical Research and Manufacturers of America Foundation
950 F Street NW, Suite 300
Washington, DC 20004

PhRMA Foundation Health Outcomes Research Starter Grants 3929
The purpose of the program is to offer financial support to individuals beginning their independent research careers at the faculty level. The program provides a research grant of $30,000 per year for up to two years. This program supports individuals beginning independent research careers in academia who do not have other substantial sources of research.
Requirements: Those holding the academic rank of instructor or assistant professor and investigators at the doctoral level with equivalent positions are eligible to apply, providing their proposed research is neither directly nor indirectly subsidized to any significant degree by an extramural support mechanism. The program is not intended for those in postdoctoral training programs. However, individuals in postdoctoral training scheduled to conclude and who will hold an academic appointment by January 1 may apply. Applicants must be sponsored by the department or unit within which the proposed research is to be undertaken. The grant is made to the university on behalf of the applicant and with the understanding that the university will administer the funds. Schools of medicine, pharmacy, public health, nursing, dentistry and schools of other areas where appropriate are eligible for this award. Applications must be submitted by an accredited school in the U.S., and all applicants must be U.S. citizens or permanent residents. Before an individual is eligible to apply for a PhRMA Foundation award, the applicant must first have a firm commitment from a U.S. university.
Restrictions: This program is not offered as a means to augment substantially funded research efforts. It is intended to offer support for researchers who are starting their independent research efforts. This award is granted in part based on need. If an individual currently has or is guaranteed substantial funding, they should not apply.
Amount of Grant: 30,000 U.S.D
Contact: Eileen M. Cannon, Executive Director; (202) 572-7756; fax (202) 572-7799; info@phrmafoundation.org
Internet: http://www.phrmafoundation.org/awards/outcomes/starter.php
Sponsor: Pharmaceutical Research and Manufacturers of America Foundation
950 F Street NW, Suite 300
Washington, DC 20004

PhRMA Foundation Health Outcomes Sabbatical Fellowships 3930
The fellowship provides stipend support for individuals engaged in a multidisciplinary research training program that will create or extend their credentials in health outcomes. The program enables faculty with active research programs to work outside of their home institution for periods of six months to one year to learn new skills or develop new collaborations that will enhance their research and research training capabilities in health outcomes. The applicant and mentor of the program must describe how the multidisciplinary goals of the research experiential program will be accomplished and provide assurance that key collaborating mentors endorse and are willing to support the plan. Matching funds must be provided by the home institution. Awards may be activated beginning January 1 or on the first day of any month thereafter, up to and including December 1. Guidelines and application are available online.
Requirements: Eligible applicants must (1) hold a PhD, Pharm.D., M.D. or Sc.D. degree in a field of study logically or functionally related to the proposed post doctoral activities, (2) hold a faculty appointment that imparts eligibility for a sabbatical leave from their home institution, (3) have institutional approval of a sabbatical plan that includes partial salary that matches the PhRMA stipend, (4) hold an endorsement from a mentor who agrees to sponsor the applicant's visiting scientist activity, and (5) be a U.S. citizen or permanent resident. Matching funds must be provided by the home institution.
Restrictions: The program provides no other subsidies (travel, tuition, fringe benefit costs, etc.), and indirect costs to the institution are not provided.
Amount of Grant: 40,000 U.S.D
Contact: Eileen M. Cannon, Executive Director; (202) 572-7756; fax (202) 572-7799; info@phrmafoundation.org
Internet: http://www.phrmafoundation.org/awards/outcomes/sabbatical.php
Sponsor: Pharmaceutical Research and Manufacturers of America Foundation
950 F Street NW, Suite 300
Washington, DC 20004

PhRMA Foundation Informatics Post Doctoral Fellowships 3931
The postdoctoral program provides stipend support for individuals engaged in a multidisciplinary research training program that will create or extend their credentials in informatics. The intent of this program is to support postdoctoral career development activities of individuals preparing to engage in research that will bridge the gap between experimental and computational approaches in genomic and biomedical studies. It is anticipated that this research training will be accomplished in academic and/or industrial laboratory settings where multidisciplinary teams are organized to address problems that span the range of biological complexity rather than focus on the application of single technologies. The award, consisting of a $40,000 annual stipend, is made to the institution on behalf of the fellow. The award is intended solely as a stipend and may not be used otherwise.
Requirements: Eligible applicants must either hold a PhD degree in a field of study logically or functionally related to the proposed postdoctoral activities, or expect to receive the PhD before activating the award. They must also have a firm commitment from an accredited U.S. university and be a U.S. citizen or permanent resident. The sponsor (mentor) of the postdoctoral program must describe how the multidisciplinary goals of the research training program will be accomplished and provide assurance that key collaborating mentors endorse and are willing to support the training plan.
Restrictions: The program provides no other subsidies (travel, tuition, fringe benefit costs, etc.) and indirect costs to the institution are not provided. The second year of this award is contingent upon a progress report approved by the Foundation and submission of a financial report.
Amount of Grant: 40,000 U.S.D
Contact: Eileen M. Cannon, Executive Director; (202) 572-7756; fax (202) 572-7799; info@phrmafoundation.org
Internet: http://www.phrmafoundation.org/awards/informatics/postdoc.php
Sponsor: Pharmaceutical Research and Manufacturers of America Foundation
950 F Street NW, Suite 300
Washington, DC 20004

PhRMA Foundation Informatics Pre Doctoral Fellowships 3932
The program is designed for candidates who expect to complete the requirements for the Ph.D. in the fields of informatics in two years or less from the time the fellowship begins. It aims at supporting promising students during their thesis research. The award is made to the university on behalf of the fellow. The fellowship provides a stipend of $20,000 a year payable monthly for a minimum of one year and a maximum of two years. Of the $20,000 awarded annually, up to $500 a year may be used for incidentals directly associated with the thesis research preparation (e.g., secretarial help, artwork, books, travel, etc.). Applications will be accepted for a minimum of one year and a maximum of two years. The program assumes that the fellows will devote full-time (including summers) to their research.
Requirements: The fellowship program of pre doctoral support is designed to assist full-time, in-residence Ph.D. candidates in the fields of informatics who are enrolled in U.S.A schools of medicine, pharmacy, dentistry or veterinary medicine. The program supports full-time advanced students who will have completed the bulk of their pre-thesis requirements (at least two years of study) and are engaged in thesis research as Ph.D. Candidates by the time the award is activated. Students just starting in graduate school should not apply. Highest priority will be given to applications who support the goal of the Informatics Programs. The goal of the program is to promote the use of informatics in an integrative approach to the understanding of biological and disease processes. Informatics awards will support career development of scientists engaged in computational and experimental research to integrate cutting-edge information technology with advanced biological, chemical, and pharmacological sciences. Applications must be submitted by an accredited U.S. school, and all applicants must be U.S. citizens or permanent residents.
Restrictions: The fellowship may not be used to supplement funds from other fellowships, traineeships or assistantships, unless necessary to make stipend levels compliant with institutional policy. The focus of the program is to assist in the predoctoral training of the candidate. It should not be viewed as a program to fund a research project. Indirect costs are not provided to the institution and PhRMA Foundation grant funds may not be used for this purpose.
Amount of Grant: 20,000 U.S.D
Contact: Eileen M. Cannon, Executive Director; (202) 572-7756; fax (202) 572-7799; info@phrmafoundation.org
Internet: http://www.phrmafoundation.org/awards/informatics/predoc.php
Sponsor: Pharmaceutical Research and Manufacturers of America Foundation
950 F Street NW, Suite 300
Washington, DC 20004

PhRMA Foundation Informatics Research Starter Grants 3933

This program supports individuals beginning independent research careers in academia who do not have other substantial sources of research. The areas of interest within this program consists of research that supports career development of scientists engaged in computational and experimental research to integrate cutting edge-edge information technology with advanced biological, chemical, and pharmacological sciences in: Genetics (Molecular, Medical-human, Pharmaco, Population); Genomics (Function, Structural, Toxico, Pharmaco, Comparative); Proteomics; Biological pathways. The program provides a research grant of $30,000 per year for up to two years. The 'starter' aspect of the program strives to assist those individuals who are establishing careers as independent investigators in the field of informatics. The program is not offered as a means to augment an ongoing research effort or is the grant intended to be used for any direct effort to obtain further extramural funding. The funds are to be used to conduct the proposed research.

Requirements: Those holding the academic rank of instructor or assistant professor and investigators at the doctoral level with equivalent positions are eligible to apply for these research starter grants, providing their proposed research is neither directly nor indirectly subsidized to any significant degree by an extramural support mechanism. The program is not intended for those in postdoctoral training programs. However, individuals in postdoctoral training scheduled to conclude and who will hold an academic appointment by January 1 may apply. Applicants must be sponsored by the department or unit in which the proposed research is to be undertaken. The grant is made to the university on behalf of the applicant and with the understanding that the university will administer the funds. Applications must be submitted by an accredited U.S. school, and all applicants must be U.S. citizens or permanent residents.

Restrictions: Research projects that extend or develop the proprietary value of specific drug products are not acceptable in this program. This exclusion does not preclude research in which specific drug products are used to test hypotheses that have a general applicability. The funds may not be used for salary support of the grantee, but may be used to support technical assistance. No more than $500 a year may be used for travel to professional meetings by the grantee. Indirect costs are not provided to the institution, and grant funds may not be used for this purpose.

Amount of Grant: 30,000 U.S.D

Contact: Eileen M. Cannon, Executive Director; (202) 572-7756; fax (202) 572-7799; info@phrmafoundation.org

Internet: http://www.phrmafoundation.org/awards/informatics/starter.php

Sponsor: Pharmaceutical Research and Manufacturers of America Foundation
950 F Street NW, Suite 300
Washington, DC 20004

PhRMA Foundation Informatics Sabbatical Fellowships 3934

The fellowship supports individuals engaged in a multidisciplinary research training program that will create or extend their credentials in informatics. The program enables faculty with active research programs to work outside of their home institution for periods of six months to one year to learn new skills or develop new collaborations that will enhance their research and research training capabilities in informatics. Applicants are expected to engage in research that supports career development of scientists engaged in computational and experimental research to integrate cutting edge information technology with advanced biological, chemical, and pharmacological sciences in the fields of genetics, genomics, and proteomics biological pathways. Emphasis will be placed on the development of new informatics technologies that demonstrate the translation of genomic data into an elucidation and understanding of biological and disease processes. The applicant and mentor of the program must describe how the multidisciplinary goals of the research experiential program will be accomplished and provide assurance that key collaborating mentors endorse and are willing to support the plan. The award provides up to $40,000 stipend support for mid-career scientists to engage in an academic year or calendar year experiential program intended to redirect their core research focus to an area of emerging importance to pharmaceutical research and development.

Requirements: Eligible applicants must hold a PhD or MD degree in a field of study logically or functionally related to the proposed postdoctoral activities; hold a faculty appointment that imparts eligibility for a sabbatical leave from their home institution; have institutional approval of a sabbatical plan that includes partial salary that matches the PhRMA stipend; and be a U.S. citizen or permanent resident. Preference will be given to those individuals whose research combines novel computational methods with experimental validation. Emphasis will be placed on the development of new informatics technologies that demonstrate the translation of genomic data into an elucidation and understanding of biological and disease processes. Matching funds must be provided by the home institution.

Restrictions: The program provides no other subsidies (travel, tuition, fringe benefit costs, etc.), and indirect costs to the institution are not provided.

Amount of Grant: 40,000 U.S.D

Contact: Eileen M. Cannon, Executive Director; (202) 572-7756; fax (202) 572-7799; info@phrmafoundation.org

Internet: http://www.phrmafoundation.org/awards/informatics/sabbatical.php

Sponsor: Pharmaceutical Research and Manufacturers of America Foundation
950 F Street NW, Suite 300
Washington, DC 20004

PhRMA Foundation Paul Calabresi Medical Student Research Fellowship 3935

The fellowship is offered to medical or dental students who have substantial interests in research and teaching careers in pharmacology - clinical pharmacology and who are willing to spend full-time in a specific research effort within a pharmacology or clinical pharmacology unit. Fellowships are available for a minimum period of three months or any period of time up to 24 months. The commitment must be full-time. The student may undertake this investigative effort at his/her own school or at another institution. Requests for a fellowship are to be submitted online by the appropriate representative of the school or university. The fellowship is made to the institution on behalf of the student and the institution must agree to administer the funds. The program offers funding of up to $1,500 per month. It is expected that the funds will be distributed to the fellow in regular payments over the period of the fellowship by the recipient school. The funds may be used towards the stipend, tuition or affiliated fees for the fellow. Fellowships are available for a minimum period of six months or any period of time up to 24 months. However, if the student's award period is 24 months, the maximum stipend is still $18,000.

Requirements: A candidate must be enrolled in a U.S. medical/dental school and have finished at least one year of the school curriculum. Priority consideration will be given to those candidates who project strong commitments to careers in the field of clinical pharmacology. The applicant must be sponsored by the pharmacology or clinical pharmacology program in which the investigative project is to be undertaken. The sponsoring unit must have a demonstrated commitment to training in the fields of pharmacology or clinical pharmacology. Applications must be submitted by an accredited U.S. school, and all applicants must be U.S. citizens or permanent residents.

Restrictions: There are no indirect costs provided to the institution nor may any of the funds be used for this purpose. Once an individual has received a Medical Student Fellowship, that individual may not reapply.

Amount of Grant: Up to 18,000 U.S.D

Contact: Eileen M. Cannon, Executive Director; (202) 572-7756; fax (202) 572-7799; info@phrmafoundation.org

Internet: http://www.phrmafoundation.org/awards/clinical/medstudent.php

Sponsor: Pharmaceutical Research and Manufacturers of America Foundation
950 F Street NW, Suite 300
Washington, DC 20004

PhRMA Foundation Pharmaceutics Postdoctoral Fellowships 3936

The purpose of the fellowships is to encourage graduates from PhD programs in pharmaceutics to continue to refine their research skills through formal postdoctoral training. Applicants should indicate a strong determination to continue their research careers in pharmaceutics following completion of the fellowship. Applications will be accepted for postdoctoral research training extending over a minimum of one year and a maximum of two years. Each fellowship carries an annual stipend of $40,000 which can be supplemented by the sponsor and/or his/her institution.

Requirements: To be eligible, applicants must either hold a PhD degree in pharmaceutics from a school of pharmacy accredited by the American Council on Pharmaceutical Education or expect to receive such a degree before activating the fellowship. The mentor of the fellow must be in an accredited school of pharmacy. Alternatively, a student with a Ph.D. from a school of pharmacy may use the fellowship to study with a mentor in any science or engineering department. An equally important aspect of eligibility is that suitable facilities for the necessary training and research must be available to the applicant. The program with which the applicant is, or will be, associated must provide an environment where the applicant's potential can be developed to the fullest extent possible.

Restrictions: The award is intended solely as a stipend and may not be used otherwise. If necessary, the institution may supplement the award to a level that is consistent with other postdoctoral fellowships it currently offers. Research projects that extend or develop the proprietary value of specific drug products are not acceptable in this program. The program provides no other subsidies (travel, tuition, fringe benefit costs, etc.) and indirect costs to the institution are not provided.

Amount of Grant: 40,000 U.S.D

Contact: Eileen M. Cannon, Executive Director; (202) 572-7756; fax (202) 572-7799; info@phrmafoundation.org

Internet: http://www.phrmafoundation.org/awards/pharmaceutics/postdoc.php

Sponsor: Pharmaceutical Research and Manufacturers of America Foundation
950 F Street NW, Suite 300
Washington, DC 20004

PhRMA Foundation Pharmaceutics Pre Doctoral Fellowships 3937

To encourage the entry of more qualified individuals into this area, the program aims at supporting promising students during their thesis research by providing assistance in the form of stipend and funds to cover costs incidental to the training. Applications will be accepted for a minimum of one year and a maximum of two years. The program is designed for applicants who expect to complete the requirements for the Ph.D. in pharmaceutics in two years or less from the time the fellowship begins. For the purposes of this program, pharmaceutics includes basic pharmaceutics, biopharmaceutics and pharmaceutical technology. The fundamental aspects of pharmacokinetics are not included since these are covered by PhRMA Foundation awards in pharmacology and toxicology.

Requirements: To be eligible, applicants must hold a B.S., M.S., or Pharm.D. degree in pharmacy or a related area such as chemistry or biology from an accredited school in the U.S. A candidate enrolled in a Pharm.D./Ph.D. program should not be taking required clinical course work or clinical clerkships during the tenure of the fellowship. Also, before an individual is eligible to apply for a PhRMA Foundation award, the applicant must first have a firm commitment from a U.S. university, and must be a U.S. citizen or permanent resident. The program assumes that the fellows will devote full-time (including summers) to their research. The fellowship may not be used as a supplement to funds from other fellowships or assistantships. However, this does not preclude institutions supplementing a fellowship with their own funds in order to bring the stipend within the normal range of the institution.

Restrictions: Research projects which extend or develop the proprietary value of specific drug products are not acceptable in this program. This exclusion does not preclude research in which specific drug products are used to test hypotheses which have a general

applicability. Applications, however good, which do not meet the aims of the program will be disapproved.
Contact: Eileen M. Cannon, Executive Director; (202) 572-7756; fax (202) 572-7799; info@phrmafoundation.org
Internet: http://www.phrmafoundation.org/awards/pharmaceutics/predoc.php
Sponsor: Pharmaceutical Research and Manufacturers of America Foundation
950 F Street NW, Suite 300
Washington, DC 20004

PhRMA Foundation Pharmaceutics Research Starter Grants **3938**
The purpose of the starter grants is to offer financial support to individuals beginning their independent research careers at the faculty level. The areas of interest within this program are research efforts in the fields of pharmaceutics. The program provides a research grant of $30,000 per year for up to two years. For the purposes of this program, pharmaceutics includes basic pharmaceutics, biopharmaceutics, pharmaceutical technology and pharmaceutical biotechnology. The fundamental aspects of pharmacokinetics are not included since they are covered by PhRMA Foundation programs in pharmacology and toxicology. The 'starter' aspect of the program strives to assist those individuals who are establishing careers as independent investigators in the fields listed above.
Requirements: Those holding the academic, tenure track, rank of instructor or assistant professor and investigators at the doctoral level with equivalent positions are eligible to apply for these research starter grants, providing their proposed research is neither directly nor indirectly subsidized to any significant degree by an extramural support mechanism. It is anticipated that this research experience will occur in an academic and/or industrial laboratory setting. Applications must be submitted by a school of pharmacy accredited by the American Council on Pharmaceutical Education, and all applicants must be U.S. citizens or permanent residents. The grant is made to the university on behalf of the applicant and with the understanding that the university will administer the funds. Only schools of pharmacy are eligible for this award.
Restrictions: The program is not offered as a means to augment an ongoing research effort. Nor is the grant intended to be used for any direct effort to obtain further extramural funding. The funds are to be used to conduct the proposed research. The funds may not be used for salary support of the grantee, but may be used to support technical assistance. No more than $500 a year may be used for travel to professional meetings by the grantee. Indirect costs are not provided to the institution, and grant funds may not be used for this purpose.
Amount of Grant: 30,000 U.S.D
Contact: Eileen M. Cannon, Executive Director; (202) 572-7756; fax (202) 572-7799; info@phrmafoundation.org
Internet: http://www.phrmafoundation.org/awards/pharmaceutics/starter.php
Sponsor: Pharmaceutical Research and Manufacturers of America Foundation
950 F Street NW, Suite 300
Washington, DC 20004

PhRMA Foundation Pharmaceutics Sabbatical Fellowships **3939**
The purpose of this program is to enable faculty with active research programs to work outside of their home institution for periods of 6 months to one year to learn new skills or develop new collaborations that will enhance their research and research training capabilities in pharmaceutics. For the purposes of this program, pharmaceutics includes basic pharmaceutics, biopharmaceutics, pharmaceutical technology and pharmaceutical biotechnology. The fundamental aspects of pharmacokinetics are not included since they are covered by PhRMA Foundation programs in pharmacology and toxicology. It is anticipated that this research experience will occur in an academic and/or industrial laboratory setting. The award provides up to $40,000 stipend support for mid-career scientists to engage in an academic year or calendar year experiential program intended to redirect their core research focus to an area of emerging importance to pharmaceutical research and development.
Requirements: Eligible applicants must hold a PhD degree in pharmaceutics from a School of Pharmacy accredited by the American Council on Pharmaceutical Education; hold a faculty appointment that imparts eligibility for a sabbatical leave from their home institution; have institutional approval of a sabbatical plan that includes partial salary that matches the PhRMA stipend; hold an endorsement from a mentor who agrees to sponsor the applicants visiting scientist activity; and be a U.S. citizen or permanent resident. The applicant and mentor of the program must describe how the multidisciplinary goals of the research experiential program will be accomplished and provide assurance that key collaborating mentors endorse and are willing to support the plan. Matching funds must be provided by the home institution.
Restrictions: The program provides no other subsidies (travel, tuition, fringe benefit costs, etc.) and indirect costs to the institution are not provided.
Amount of Grant: 40,000 U.S.D
Contact: Eileen M. Cannon, Executive Director; (202) 572-7756; fax (202) 572-7799; info@phrmafoundation.org
Internet: http://www.phrmafoundation.org/awards/pharmaceutics/sabbatical.php
Sponsor: Pharmaceutical Research and Manufacturers of America Foundation
950 F Street NW, Suite 300
Washington, DC 20004

PhRMA Foundation Pharmacology/Toxicology Post Doctoral Fellowships **3940**
The postdoctoral program provides stipend support for individuals engaged in a multidisciplinary research training program that will create or extend their credentials in pharmacology or toxicology. The intent of this program is to support postdoctoral career development activities of individuals prepared (or preparing) to engage in research that integrates information on molecular or cellular mechanisms of action with information on the effect of an agent in the intact organism. It is anticipated that this research training will be accomplished in academic and/or industrial laboratory settings where multidisciplinary teams are organized to integrate informatics, molecular, cell, and systems biology with pharmacology/toxicology research. Funding is for up to 2 years but not beyond the 3rd year of the existing Post Doc.
Requirements: Eligible applicants must either hold a PhD degree or appropriate terminal research doctorate in a field of study logically or functionally related to the proposed post doctoral activities; or expect to receive the PhD before activating the award. They must also have a firm commitment from an accredited U.S. university and be a U.S. citizen or permanent resident. The sponsor (mentor) must describe how the multidisciplinary goals of the research training program will be accomplished and provide assurance that key collaborating mentors endorse and are willing to support the training plan. The award, consisting of a $40,000 annual stipend, is made to the institution on behalf of the fellow. The award is intended solely as a stipend and may not be used otherwise. If necessary, the institution may supplement the award to a level that is consistent with other postdoctoral fellowships it currently offers.
Restrictions: The program provides no other subsidies (travel, tuition, fringe benefit costs, etc.) and indirect costs to the institution are not provided. The second year of this award is contingent upon a progress report approved by the foundation and submission of a financial report.
Amount of Grant: 40,000 U.S.D
Contact: Eileen M. Cannon, Executive Director; (202) 572-7756; fax (202) 572-7799; info@phrmafoundation.org
Internet: http://www.phrmafoundation.org/awards/pharmacology/postdoc.php
Sponsor: Pharmaceutical Research and Manufacturers of America Foundation
950 F Street NW, Suite 300
Washington, DC 20004

PhRMA Foundation Pharmacology/Toxicology Pre Doctoral Fellowships **3941**
The program is designed for candidates who expect to complete the requirements for the Ph.D. in the fields of pharmacology or toxicology in two years or less from the time the fellowship begins. The award is made to the university on behalf of the fellow. The fellowship provides a stipend of $20,000 a year payable monthly for a minimum of one year and a maximum of two years. Of the $20,000 awarded annually, up to $500 a year may be used for incidentals directly associated with the thesis research preparation (e.g., secretarial help, artwork, books, travel, etc.). Awards may be activated beginning January 1 or on the first day of any month thereafter, up to and including August 1. The second year of a fellowship is contingent upon certification by the thesis advisor that satisfactory progress is being achieved.
Requirements: The fellowship program of pre doctoral support is designed to assist full-time, in-residence Ph.D. candidates in the fields of pharmacology or toxicology who are enrolled in U.S.A schools of medicine, pharmacy, dentistry or veterinary medicine. The program supports full-time advanced students who will have completed the bulk of their pre-thesis requirements (at least two years of study) and are engaged in thesis research as Ph.D. Candidates by the time the award is activated. Students just starting in graduate school should not apply. In the field of pharmacology or toxicology, highest priority will be given to applications for research that attempts to integrate information on the mechanism of action of a drug or chemical at a molecular or cellular level with the drug's effects in the intact laboratory animal or human, encompassing the potential influences of biochemical, physiological or behavioral systems. The pharmacology/toxicology research and teaching mission of the unit awarding the degree must be apparent as a core mission component on their web site. Applications must be submitted by an accredited U.S. school, and all applicants must be U.S. citizens or permanent residents. The program assumes that the fellows will devote full-time (including summers) to their research. The PhRMA Foundation fellowship may not be used to supplement funds from other fellowships, traineeships or assistantships, unless necessary to make stipend levels compliant with institutional policy. The focus of the program is to assist in the predoctoral training of the candidate. It should not be viewed as a program to fund a research project.
Restrictions: A candidate enrolled in an M.D./Ph D. program should not be taking required clinical course work or clerkships during the tenure of the fellowship. The foundation will accept only one application per academic institution. Exceptions may be considered for institutions with multiple campuses. Indirect costs are not provided to the institution and PhRMA Foundation grant funds may not be used for this purpose.
Amount of Grant: 20,000 U.S.D
Contact: Eileen M. Cannon, Executive Director; (202) 572-7756; fax (202) 572-7799; info@phrmafoundation.org
Internet: http://www.phrmafoundation.org/awards/pharmacology/predoc.php
Sponsor: Pharmaceutical Research and Manufacturers of America Foundation
950 F Street NW, Suite 300
Washington, DC 20004

PhRMA Foundation Pharmacology/Toxicology Research Starter Grants **3942**
The purpose of the program is to offer financial support to individuals beginning their independent research careers at the faculty level. The areas of interest within this program are research efforts in the fields of pharmacology and drug toxicology. The program provides a research grant of $30,000 per year for up to two years. The 'starter' aspect of the program strives to assist those individuals who are establishing careers as independent investigators in the fields listed above. The program is not offered as a means to augment an ongoing research effort. The funds are to be used to conduct the proposed research. This program supports individuals beginning independent research careers in academia who do not have other substantial sources of research. Applicants will be judged on the scientific merit of the proposed research, and on the degree of financial need.
Requirements: Those holding the academic rank of instructor or assistant professor and investigators at the doctoral level with equivalent positions are eligible to apply for these research starter grants, providing their proposed research is neither directly nor indirectly

subsidized to any significant degree by a competitive extramural grant. The program is not intended for those in postdoctoral training programs. However, individuals in postdoctoral training scheduled to conclude and who will hold an academic appointment by January 1 may apply. Applicants must be sponsored by the department or unit in which the proposed research is to be undertaken. The sponsoring unit must have responsibility for pharmacology/toxicology teaching and research as part of its core mission. If the pharmacology/toxicology mission is not apparent on the sponsoring unit's web site, the sponsor letter must describe how the academic appointment will support the applicant's career development in pharmacology/toxicology. The grant is made to the university on behalf of the applicant, with the understanding that the university will administer the funds. Schools of medicine (human or veterinary), pharmacy and dentistry are eligible for this award. In the field of pharmacology or toxicology, highest priority will be given to applications for research that attempts to integrate information on the mechanism of action of a drug or chemical at a molecular or cellular level with the drug's effects in the intact laboratory animal or human, encompassing the potential influences of biochemical, physiological or behavioral systems. Applications must be submitted by an accredited U.S. school, and all applicants must be U.S. citizens or permanent residents.

Restrictions: Research projects that extend or develop the proprietary value of specific drug products are not acceptable in this program. This exclusion does not preclude research in which specific drug products are used to test hypotheses that have a general applicability. The funds may not be used for salary support of the grantee, but may be used to support technical assistance. No more than $500 a year may be used for travel to professional meetings by the grantee. Indirect costs are not provided to the institution, and grant funds may not be used for this purpose. Applicants holding multi-year national competitive awards or grants supporting their independent research will not be considered. If a national competitive grant is obtained after the application has been submitted, the Foundation must be so informed.

Amount of Grant: 30,000 U.S.D

Contact: Eileen M. Cannon, Executive Director; (202) 572-7756; fax (202) 572-7799; info@phrmafoundation.org

Internet: http://www.phrmafoundation.org/awards/pharmacology/starter.php

Sponsor: Pharmaceutical Research and Manufacturers of America Foundation
950 F Street NW, Suite 300
Washington, DC 20004

PhRMA Foundation Pharmacology/Toxicology Sabbatical Fellowships **3943**

This program provides stipend funding to enable faculty members at all levels with active research programs an opportunity to work at other institutions for periods of six months to one year to learn new skills or develop new collaborations that will enhance their research and research training activities in pharmacology/toxicology. Applicants are expected to engage in multidisciplinary research that integrates information on molecular or cellular mechanisms of action with information on the effect of an agent in the intact organism. It is anticipated that this research experience will occur in academic and/or industrial laboratory settings as part of a multidisciplinary team organized to integrate informatics, molecular, cell, and systems biology with pharmacology/toxicology research. The award provides up to $40,000 stipend support for mid-career scientists to engage in an academic year or calendar year experiential program intended to redirect their core research focus to an area of emerging importance to pharmaceutical research and development.

Requirements: Eligible applicants must hold a PhD degree or appropriate terminal doctorate and record of research accomplishment in a field of study logically or functionally related to the proposed post doctoral activities; hold a faculty appointment that imparts eligibility for a sabbatical leave from their home institution; have institutional approval of a sabbatical plan that includes partial salary that matches the PhRMA stipend; hold an endorsement from a mentor who agrees to sponsor the applicants visiting scientist activity; and be a U.S. citizen or permanent resident. The applicant and mentor of the program must describe how the multidisciplinary goals of the research experiential program will be accomplished and provide assurance that key collaborating mentors endorse and are willing to support the plan. Matching funds must be provided through the university.

Restrictions: The program provides no other subsidies (travel, tuition, fringe benefit costs, etc.) and indirect costs to the institution are not provided.

Amount of Grant: 40,000 U.S.D

Contact: Eileen M. Cannon, Executive Director; (202) 572-7756; fax (202) 572-7799; info@phrmafoundation.org

Internet: http://www.phrmafoundation.org/awards/pharmacology/sabbatical.php

Sponsor: Pharmaceutical Research and Manufacturers of America Foundation
950 F Street NW, Suite 300
Washington, DC 20004

Physicians Foundation for Health Systems Excellence Grants **3944**

PFHSE seeks to fund projects that match the foundation's funding priorities: practice management—to create and increase efficiency within the practice for delivery of high-quality health care; physician education—to enhance practice-based delivery of health care through the application of technological and clinical developments; and patient safety/disease management—to develop and implement innovative practice strategies aimed at improving the health status and well being of patients. Concept papers, which may be submitted electronically, may be submitted at any time; full proposals are by invitation. Application and guidelines are available online.

Requirements: 501(c)3 tax-exempt organizations are eligible.

Restrictions: Grants are not awarded for unrestricted general operating expenses; payment for services of a fiscal agent; endowment funds; religious purposes; fundraising activities or events; capital expenditures by the grantee; lobbying or political activities; activities related to litigation, arbitration, or other dispute resolution; medical research or animal research; research and/or development of drugs or medical devices; or any activity inconsistent with the foundation's status as a 501(c)3 charitable organization.

Amount of Grant: 100,000 - 1,000,000 U.S.D

Contact: Tracy Little, Grants Management Associates Contact; (617) 399-0417; physicians@grantsmanagement.com

Internet: http://www.physiciansfoundation.org/e_intro.htm#funding

Sponsor: Physicians Foundation for Health Systems Excellence
77 Summer Street, Suite 800
Boston, MA 02110-1006

Piedmont Health Foundation Grants **3945**

The Piedmont Health Care Foundation was established in 1985 through the sale of the first HMO in Greenville County, South Carolina. During the past 25 years, the foundation has invested more than $3.4 million in dozens of nonprofit organizations in the Greenville, South Carolina area. The Piedmont Health Care Foundation has played an important role in catalyzing and providing seed funding to critical projects, and it has provided operating and programmatic funds needed by local health-service organizations. In looking ahead, the foundation recognizes that much of what makes a healthy community takes place outside of the health-care system. So for its 25th anniversary, the foundation decided to change its name to the Piedmont Health Foundation. The foundation currently focuses on the area of policy, system, and environmental change to reduce childhood obesity rates. Applications are accepted on a quarterly basis. Downloadable guidelines and editable forms are provided at the foundation website. Applicants should email their completed forms to the address given by midnight of the deadline date.

Requirements: Nonprofits that serve Greenville County are eligible to apply.

Geographic Focus: South Carolina

Date(s) Application is Due: Jan 31; Apr 10; Jul 10; Oct 10

Amount of Grant: 1,000 - 15,000 U.S.D

Samples: Communities in Schools, Greenville, South Carolina, $5,500 - health initiative program in Title I Greenville County schools (2011); Greenville Forward, Greenville, South Carolina, $7,500 - Gardening for Good community gardens and urban farms (2011); Taylors Free Medical Clinic, Taylors, South Carolina, $15,000 - to support clinic's pharmacy operation (2011).

Contact: Katy Smith, Executive Director; (864) 370-0212; fax (864) 370-0212; katypughsmith@bellsouth.net or katysmith@piedmonthealthfoundation.org

Internet: http://www.phcfdn.org/grantmaking.php

Sponsor: Piedmont Health Foundation
P.O. Box 9303
Greenville, SC 29604

Piedmont Natural Gas Corporate and Charitable Contributions **3946**

Piedmont Natural Gas supports local non-profit organizations sponsored by its employees through matching gifts, financial assistance and volunteer support.

Requirements: Piedmont supports 501(c)3 organizations that are sponsored or assisted by its employees. Eligible organizations with employee sponsors are also able to apply for Employee Matching Gifts. Applications are available at Piedmont's website.

Restrictions: As a general rule, grants do not support: individuals or non-501(c)3 organizations; pre-college level private schools except through employee matching gifts; travel and conferences; third-party professional fund-raising organizations; controversial social causes; fraternal and veterans organizations or private clubs; religious organizations with programs limited to or expressly for their membership only; agencies already receiving corporate support through United Way or a united arts drive, with the exception of approved capital campaigns; or athletic events and programs.

Geographic Focus: North Carolina, South Carolina, Tennessee

Contact: George Baldwin, Managing Director; (704) 731-4063; fax (704) 731-4086; george.baldwin@piedmontng.com

Internet: http://www.piedmontng.com/ourcommunity/communityoutreach.aspx

Sponsor: Piedmont Natural Gas Corporation
4720 Piedmont Row Drive
Charlotte, NC 28210

Piedmont Natural Gas Foundation Health and Human Services Grants **3947**

Piedmont Natural Gas Foundation's Health and Human Services grant focuses funding toward: organizations providing outreach services to community members with basic needs, including shelter, food and clothing; substance abuse or mental illness; organizations providing services or programs for a range of human service needs including youth engagement and mentoring, special needs and disability assistance, substance abuse or mental illnesses, transitional housing and situational homelessness support, and gang violence prevention; organizations providing emergency/disaster relief; increased access to critical healthcare services and comprehensive medical treatment including preventative care, prescription medication, medical exams, screenings, immunizations and dental care; and increased access to mental health services.

Requirements: Organizations must be a 501(c)3 non-profit organization or a qualified government entity. All grant requests must be submitted through Piedmont's online grant application form located at Piedmont's website.

Restrictions: Piedmont Natural Gas Foundation will not fund: religious, fraternal, political or athletic groups; four-year colleges and universities; private foundations; or social or veterans' organizations. In addition to these restrictions, contributions will generally not be made to or for: individuals; pre-college level private schools, except through the Employee Matching Gifts program; travel and conferences; third-party professional fundraising organizations; controversial social causes; religious organizations with programs limited to or expressly for their membership only; athletic events and

programs; agencies already receiving Piedmont support through United Way or a united arts drive, with the exception of an approved capital campaign; or any proposal outside of the geographic area where Piedmont Natural Gas does business.
Geographic Focus: North Carolina, South Carolina, Tennessee
Amount of Grant: 500 - 30,000 U.S.D
Contact: George Baldwin, President; (704) 731-4063; fax (704) 731-4086; george.baldwin@piedmontng.com
Internet: http://www.piedmontng.com/ourcommunity/ourfoundation.aspx#guidelines
Sponsor: Piedmont Natural Gas Foundation
4720 Piedmont Row Drive
Charlotte, NC 28210

Pike County Community Foundation Grants 3948
Founded in 1998, the Pike County Community Foundation is a charitable organization formed to strengthen the Pike County, Indiana community by awarding grants to local nonprofits, by bringing individuals together to address community needs, and by offering personalized charitable gift planning services to our donors. The foundation considers proposals for grants on a yearly cycle which begins each July. At the start of each cycle, a notice is mailed to nonprofit organizations that have applied for grants in the past, have received grants in the past, or have otherwise requested notification of the start of each cycle. The foundations grant cycle runs from July through November each year. Proposals will be accepted from July through the September deadline. Proposals submitted after the September deadline (date varies, contact foundation office for current deadline) will not be considered. The grants committee will make its recommendations on funding to the Foundation's Board of Trustees, which will make final funding recommendations to the board of directors of the Community Foundation Alliance (our governing body). All organizations that have submitted grant proposals will be notified of the outcome of the grants committee's deliberation in writing no later than December 1.
Requirements: The Foundation welcomes proposals from nonprofit organizations that are deemed tax-exempt under sections 501(c)3 and 509(a) of the Internal Revenue Code and from governmental agencies serving Pike County. Proposals from nonprofit organizations not classified as a 501(c)3 public charity may be considered provided the project is charitable and supports a community need. Proposals submitted by an entity under the auspices of another agency must include a written statement signed by the agency's board president on behalf of the board of directors agreeing to act as the entity's fiscal sponsor, to receive grant monies if awarded, and to oversee the proposed project.
Restrictions: Project areas NOT considered for funding: religious organizations for religious purposes; political parties or campaigns; endowment creation or debt reduction; operating costs; capital campaigns; annual appeals or membership contributions; travel requests for groups or individuals such as bands, sports teams, or classes. A six-month progress report and a final report at project completion are required by organizations whose proposals are approved for funding. Instructions and appropriate forms will be provided at the time the grant is awarded.
Geographic Focus: Indiana
Samples: Evansville Philharmonic Orchestra - for Pike County third through fifth grade students to participate in the youth concert program, $300.00; Petersburg Ministerial Fellowship - to purchase clothing for the Clothe-a-Child program, $2,000.00; Tri-Cap - to assist with emergency financial assistance for families in need, $918.00.
Contact: Tom Dooley, Director; (812) 354-6797; fax (812) 354-6797; tom@pikecommunityfoundation.org
Internet: http://www.pikecommunityfoundation.org/grantmaking
Sponsor: Pike County Community Foundation
714 Main Street, P.O. Box 587
Petersburg, IN 47567

Pinellas County Grants 3949
The Pinellas Community Foundation was established in 1969, it distributes grants twice annually to a wide variety of non-profit agencies, organizations and programs that enhance and support the quality of life in Pinellas County, Florida. Areas of interest are: art, culture, health care, environment, community development, employment opportunities, the underserved, and social services.
Requirements: Eligibility: a non-profit, 501(c)3; be headquartered in Pinellas County; provide social services to people in Pinellas County; not have a large endowment or fund raising staff; provide recent audited financial statements. For Grant Applications call Pinellas Community Foundation: (727)531-0058.
Geographic Focus: Florida
Date(s) Application is Due: Jun 15; Oct 1
Contact: Julie Scales, Executive Director; (727) 531-0058; fax (727) 531-0053; info@pinellasccf.org
Internet: http://www.pinellasccf.org/pinellas-community-foundation-pcf-clearwater-fl-our-grants-programs.htm
Sponsor: Pinellas County Community Foundation
5200 East Bay Drive, Suite 202
Clearwater, FL 33764

Pinkerton Foundation Grants 3950
The Pinkerton Foundation is an independent grantmaking foundation established in 1966 by Robert Allan Pinkerton with the broad directive to reduce the incidence of crime and to prevent juvenille delinquency. The Foundation's principal program interests are focused on economically disadvantaged children, youth and families, and severely learning disabled children and adults of borderline intelligence. The Foundation supports efforts to strengthen and expand community-based programs for children, youth and families in New York City. The Foundation also occasionally funds research, demonstration and evaluation projects in its principal program areas. While grants for direct service projects are usually limited to New York City, those with potential for national impact or replication may go beyond this geographic limitation. The Foundation's Board of Directors have two grantmaking meetings per year, in May and in December. Letters of inquiry are welcome throughout the year. Additional guidelines are available at the Foundation's website.
Requirements: Grants are awarded primarily to New York City 501(c)3 nonprofit public charitable organizations.
Restrictions: The foundation does not grant requests for emergencies, medical research, direct provision of health care, religious education, conferences, publications, or capital projects.
Date(s) Application is Due: Feb 1; Sep 1
Amount of Grant: 25,000 - 250,000 U.S.D
Contact: Joan Colello, Executive Director; (212) 332-3385; fax (212) 332-3399; pinkfdn@pinkertonfdn.org
Internet: http://fdncenter.org/grantmaker/pinkerton
Sponsor: Pinkerton Foundation
610 Fifth Avenue, Suite 316
New York, NY 10020

Pioneer Hi-Bred Community Grants 3951
The international corporation supports community-based projects in the areas of agriculture, education, farm safety, and the environment. Priority consideration is given to projects located in Pioneer facility communities or rural agricultural regions and to organizations with active Pioneer management/employee participation and company-related expertise and interest. Types of support include capital grants, general operating grants, program development grants, and seed money grants. The company accepts proposals from nonprofit organizations nationwide but favors programs in its operating communities. Pioneer prefers to make direct contributions to organizations, rather than sponsorships, ticket or table purchases. This allows more funding to go directly toward the non-profit organization; however, we are willing to consider sponsorships when Pioneer employees are actively involved with the organization. The employee must present the request to Community Investment at least one month in advance. Due to the number of non-profit events held annually, there is a $1,000 maximum contribution level per event per year. The company favors proposals that demonstrate cooperation with other community-based programs, broad-based funding, community need, and positive results. Grant proposals are reviewed on a quarterly basis.
Requirements: Nonprofit organizations are eligible. All requests should be directed to the Pioneer Hi-Bred office within the local area. Otherwise, send to the contact listed. Pioneer does not respond favorably to verbal requests.
Restrictions: Grants are not made to individuals, religious or political organizations that promote a particular doctrine, elected officials, company marketing or advertising, or organizations where there is a conflict of interest with Pioneer Hi-Bred.
Amount of Grant: Up to 5,000 U.S.D
Contact: Grants Administrator; (800) 247-6803, ext. 3915; fax (515) 334-4415; community.investment@pioneer.com
Internet: http://www.pioneer.com/web/site/portal/menuitem.bb020a6d93d9d318bc0c0a03d10093a0/
Sponsor: Pioneer Hi-Bred International
6900 NW 62nd Avenue, P.O. Box 246
Johnston, IA 50131

Pioneer Hi-Bred Society Fellowships 3952
The Pioneer Hi-Bred Society Fellowship is designed to contribute to agricultural research and education by providing financial support for outstanding students studying for graduate level degrees in plant breeding. Leading societies that support plant improvement graduate programs in crops of interest to Pioneer have been selected by the Fellowship Sub-committee. The institutions are responsible for student selection and administering the program. Funded societies include: Agronomic Science Foundation; American Phytopathological Society; American Society of Plant Biologists; the Consultative Group on International Agricultural Research (CGIAR); and the Entomological Foundation.
Requirements: Pioneer prefers this grant to be used solely to fund a student(s) progressing towards a graduate degree in plant breeding from the funded institution. No university overhead will be paid from these funds.
Contact: David Harwood, (515) 270-3200 or (800) 247-6803, ext. 3915; fax (515) 334-4415; David.Harwood@pioneer.com
Internet: http://www.pioneer.com/web/site/portal/menuitem.07196fe7964ae318bc0c0a03d10093a0/
Sponsor: Pioneer Hi-Bred International
6900 NW 62nd Avenue, P.O. Box 246
Johnston, IA 50131

Piper Jaffray Foundation Communities Giving Grants 3953
The foundation supports organizations and programs that enhance the lives of people living and working in communities in which the company has offices. Of primary interest is support for organizations that increase opportunities for individuals to improve their lives and help themselves. Highest priority is given to family stability programs (including housing, family violence, responsible parenting), early childhood development, job training/career development, youth development, and adult education services. The foundation will also consider requests from organizations that work to increase citizen understanding or involvement in civic affairs or that enhance the artistic and cultural life of the community. Requests for general operating support from proven nonprofit organizations will be considered. Requests for project and capital support will be considered on a very selective

basis. Support for higher education and K-12 public and private schools is provided primarily through the company gift-matching program. Contact the foundation for deadline dates.
Requirements: IRS 501(c)3 organizations located in the Minneapolis/Saint Paul metropolitan area should submit requests directly to the foundation. Organizations located outside the Minneapolis/Saint Paul metropolitan area should submit requests to the nearest Piper Jaffray office for forwarding to the foundation. Offices are located in communities in Arizona, California, Colorado, Idaho, Illinois, Iowa, Kansas, Kentucky, Minnesota, Missouri, Montana, Nebraska, Nevada, North Dakota, Ohio, Oregon, South Dakota, Tennessee, Utah, Washington, Wisconsin, and Wyoming.
Restrictions: Requests will not be considered from newly formed nonprofit organizations; individuals; teams; religious, political, veterans, or fraternal organizations; or organizations working to treat or eliminate specific diseases. Support is not available for basic or applied research, travel, event sponsorship, benefits or tickets, or to eliminate an organization's operating deficit.
Geographic Focus: Arizona, Arkansas, California, Colorado, Idaho, Illinois, Iowa, Kansas, Kentucky, Minnesota, Missouri, Montana, Nebraska, Nevada, North Dakota, Ohio, Oregon, South Dakota, Tennessee, Utah, Washington, Wisconsin, Wyoming
Date(s) Application is Due: Mar 18
Amount of Grant: 1,000 - 5,000 U.S.D
Contact: Connie McCuskey, Vice President; (612) 303-1309; fax (612) 342-6085; communityrelations@pjc.com
Internet: http://www.piperjaffray.com/2col_largeright.aspx?id=127
Sponsor: Piper Jaffray Foundation
800 Nicollet Mall, Suite 800
Minneapolis, MN 55402

Pittsburgh Foundation Community Fund Grants **3954**
The Pittsburgh Foundation, established in 1945, is the 14th largest community foundation in the county. The foundation awards grants to organizations operating primarily in the Pittsburgh area and Allegheny County, PA. The Foundation supports five targeted areas of impact: education; economic development; families, children and youth; health care; arts.
Requirements: Grants in the five targeted areas for impact are awarded to nonprofit organizations that are defined as tax-exempt under Section 501(c)3 of the Internal Revenue Code and are located within Allegheny County, or can demonstrate that a significant majority of their population served is from Allegheny County.
Geographic Focus: Pennsylvania
Contact: Jane Downing, Senior Program Officer; (412) 391-5122; fax (412) 391-7259; downingj@pghfdn.org
Internet: http://www.pittsburghfoundation.org/page8894.cfm
Sponsor: Pittsburgh Foundation
5 PPG Place, Suite 250
Pittsburgh, PA 15222-5414

Pittsburgh Foundation Medical Research Grants **3955**
Through the generosity of individuals and families who have established medical research funds with The Pittsburgh Foundation, the Foundation provides support for this cutting edge research. Based on the donors' fields of interest, the Foundation provides grants for the following: Richard S. Caliguiri Fund - for research in the field of amyloidosis; The Walter L. Copeland Fund - for cranial research (disease of the brain/neurological disease), to be conducted within the Department of Neurological Surgery at the University of Pittsburgh; John F. and Nancy A. Emmerling Fund - for research in the field of neuropsychobiology (with emphasis on biochemistry) of mental disorders, with the goal of discovering the causes and cure, of amelioration, of mental disorders. The principal investigator must have full time status at the University of Pittsburgh; Albert B. Ferguson, M.D. Orthopaedic Fund - for orthopaedic research; GLBT Health and Wellness Fund - to assess the incidence, cause or cure of certain diseases (i.e., cancer, heart disease, diabetes, arthritis, AIDS and Alzheimer's disease) in the Gay, Lesbian, Bisexual and Transgender (GLBT) community; Medical Research Funding - for research in the field of cancer, heart disease, arthritis, diabetes, multiple sclerosis and geriatric diseases (with focuses particularly on Alzheimer's disease and other related diseases of the elderly).
Requirements: Pennsylvania not-for-profit institutions with 501(c)3 tax-exempt status are invited to apply for funding. Support is given to investigators in Allegheny County with a proven record of performance in the area under investigation. Guidelines and applications vary, see: http://www.pittsburghfoundation.org/page32428.cfm for a complete listing of each grant.
Geographic Focus: Pennsylvania
Contact: Judy Powell, Administrative Support; (412) 394-4294; powellj@pghfdn.org
Internet: http://www.pittsburghfoundation.org/page32428.cfm
Sponsor: Pittsburgh Foundation
5 PPG Place, Suite 250
Pittsburgh, PA 15222-5414

PKD Foundation Lillian Kaplan International Prize for the Advancement in the Understanding of Polycystic Kidney Disease **3956**
The prize is a joint venture between the foundation and the International Society of Nephrology. The winning nominee of the initial award received the prize in Berlin; it has been awarded every other year since. Each award consists of a monetary award, a framed certificate with an appropriate citation, and a designated sculpture to clinicians and scientists who have made outstanding discoveries in basic science or who have made outstanding clinical contributions to the care of individuals. This is the largest cash prize for a clinical medical specialty in the world signaling the prestige and importance it carries. PKD experts from both organizations judge nominations.
Requirements: Awards are made to residents of any country without restriction to gender, race, religion, creed, or nationality. Nomination letters should describe the key role of the nominee(s) in a biomedical advance relevant to PKD and should be accompanied by a full curriculum vita with a complete mailing address and at least two supporting letters that detail a major discovery or sustained advance. If two individuals are nominated to share the prize, the reasons for why the prize would be shared must be clearly delineated.
Date(s) Application is Due: Feb 1
Amount of Grant: 50,000 U.S.D
Samples: Dr. Corinne Antignac, Necker Hospital, Paris, France, $50,000; Dr. Lisa Guay-Woodford, University of Alabama at Birmingham, $50,000; Dr. Friedhelm Hildebrandt, University of Michigan, $50,000.
Contact: Dan Lara, Government Relations Manager; (816) 268-8480; fax (816) 931-8655; larad@pkdcure.org or pkdcure@pkdcure.org
Internet: http://www.pkdcure.org
Sponsor: PKD Foundation
9221 Ward Parkway, Suite 400
Kansas City, MO 64114-3367

PKD Foundation Research Grants **3957**
The PKD Foundation, established in 1982 by Joseph H. Bruening and Jared J. Grantham, leads the fight against polycistic kidney disease through vital research funding and patient education. Annually, the PKD Foundation funds more than $4.2 million in vital PKD research. Most recently, clinical trials in cardiovascular health, Rapamycin, Somatostatin, Tolvaptan, and more have offered help and hope to the 12.5 million worldwide with the disease.
Amount of Grant: 25,000 - 1,000,000 U.S.D
Contact: Dan Lara, Government Relations Manager; (816) 268-8480; fax (816) 931-8655; larad@pkdcure.org or pkdcure@pkdcure.org
Internet: http://www.pkdcure.org/Research/tabid/82/Default.aspx
Sponsor: PKD Foundation
9221 Ward Parkway, Suite 400
Kansas City, MO 64114-3367

Plastic Surgery Educational Foundation/AAO-HNSF Combined Grant **3958**
The PSEF/AAO-HNSF Combined Grant is designed to promote interdisciplinary research by encouraging a collaborative research effort between the specialties of Plastic Surgery and Otolaryngology. Recipients of this grant must demonstrate a potential for excellence in research and training and a commitment to an academic research career in either Plastic Surgery or Otolaryngology.
Requirements: Applicants must have demonstrated a potential for excellence in research and teaching and serious commitment to an academic research career in otolaryngology or plastic surgery. Priority will be given to senior residents, fellows or faculty who have completed residencies or fellowships within four years of the application receipt date.
Date(s) Application is Due: Dec 15
Amount of Grant: Up to 20,000 U.S.D
Contact: Carol V. Wargo, Director of Corporate Relations; (847) 228-3358 or (847) 228-9900; fax (847) 228-9131; cvw@plasticsurgery.org or cschmieden@plasticsurgery.org
Internet: http://www.plasticsurgery.org/medical_professionals/research/Grant-Applications.cfm
Sponsor: Plastic Surgery Educational Foundation
444 E Algonquin Road
Arlington Heights, IL 60005

Plastic Surgery Educational Foundation Basic Research Grants **3959**
The Plastic Surgery Educational Foundation and the American Society for Aesthetic Plastic Surgery offer annual research grants to provide seed money to U.S. and Canadian surgeons for research projects in the field of plastic and reconstructive surgery. Nonmembers of PSEF must submit a letter of sponsorship from a foundation member.
Requirements: Applicant must be a plastic surgeon, MD, or PhD working in plastic surgery.
Date(s) Application is Due: Jan 17
Amount of Grant: 5,000 U.S.D
Samples: Dr. James Anthony, U of California (San Francisco, CA)—for research on wound cytokine levels, $4600. Dr. Steven Buchman, U of Michigan (MI)—for research on cranial suture closure, $4998. Dr. James Doyle, U of Massachusetts Medical Ctr (MA)—for research questioning IL-10 as an immunosuppressive factor in chronic wound fluid, $5000.
Contact: Carol V. Wargo, Director of Corporate Relations; (847) 228-3358 or (847) 228-9900; fax (847) 228-9131; cvw@plasticsurgery.org or cschmieden@plasticsurgery.org
Internet: http://www.plasticsurgery.org/medical_professionals/research/Grant-Applications.cfm
Sponsor: Plastic Surgery Educational Foundation
444 E Algonquin Road
Arlington Heights, IL 60005

Plastic Surgery Educational Foundation Research Fellowship Grants **3960**
The Research Fellowship Grants are awarded for the purpose of encouraging research and academic career development in Plastic Surgery. The PSEF Research Fellowship is for training in any area of Plastic Surgery and is designed for those who wish to supplement one year of clinical training for a research experience. The PSEF/KCI Wound Care Research Fellowship is open to plastic surgery residents and recent plastic surgery resident graduates focusing on a project in wound care management.

Samples: Dr. Emmanuella Joseph (Boston, MA)—senior award winner for clinical research, $3000. Dr. Jayesh Panchal (Oklahoma City, OK)—junior award winner for clinical research, $3000. Dr. W.P. Andrew Leww (Boston, MA)—senior award winner for basic science, $3000. Dr. Vaishali Doolabh (Saint Louis, MO)—junior award winner for basic science, $3000.
Contact: Carol V. Wargo, Director of Corporate Relations; (847) 228-3358 or (847) 228-9900; fax (847) 228-9131; cvw@plasticsurgery.org or cschmieden@plasticsurgery.org
Internet: http://www.plasticsurgery.org/medical_professionals/research/Grant-Applications.cfm
Sponsor: Plastic Surgery Educational Foundation
444 E Algonquin Road
Arlington Heights, IL 60005

Playboy Foundation Grants **3961**
The Playboy Foundation seeks to foster social change by confining its grants and other support to projects of national impact and scope involved in fostering open communication about, and research into, human sexuality, reproductive health and rights; protecting and fostering civil rights and civil liberties in the United States for all people, including women, people affected and impacted by HIV/AIDS, gays and lesbians, racial minorities, the poor and the disadvantaged; and eliminating censorship and protecting freedom of expression and First Amendment rights. The Foundation does not accept unsolicited proposals, but welcomes letters of inquiry via post for the areas of interest noted above. Grants awarded by the Foundation are typically up to $10,000.
Restrictions: The foundation will not consider religious programs, individual needs, capital campaigns, endowments, scholarships, or fellowships; social services, including residential care, clinics, treatment, or recreation programs; national health, welfare, educational, or cultural organizations, or their state affiliates; or government agencies or projects.
Amount of Grant: 5,000 - 10,000 U.S.D
Contact: Executive Director; (312) 373-2437 or (312) 751-8000; fax (312) 751-2818; giving@playboy.com
Internet: http://www.playboyenterprises.com/foundation
Sponsor: Playboy Foundation
680 North Lake Shore Drive
Chicago, IL 60611

Plough Foundation Grants **3962**
The foundation awards grants to nonprofit organizations in Shelby County, Tennessee, (emphasis on Memphis) for charitable purposes that will benefit the greatest number of people in the state. Areas of interest include early childhood and elementary education, crime, health care, economic development, social service agencies, housing and homelessness, and the arts. Types of support include building/renovation, capital campaigns, endowments, equipment, land acquisition, management development/capacity building, matching/challenge support, professorships, program-related investments/loans, program development, program evaluation, and seed money. Applicants should submit a brief letter of three pages or less explaining the specific project for which funds are needed and the results anticipated if the project is funded. Organizations whose projects are within the focus area will be invited to submit full proposals. The board meets in February, May, August, and November to review requests.
Requirements: Nonprofit organizations in Shelby County, TN, with particular emphasis on Memphis, may submit letters of request.
Restrictions: Funding requests are denied for annual operating expenses, individuals, and projects outside the Memphis, TN, area.
Geographic Focus: Tennessee
Date(s) Application is Due: Jan 10; Apr 10; Jul 10; Oct 10
Amount of Grant: 1,000 - 1,250,000 U.S.D
Samples: Inmotion Musculoskeletal Institute, Memphis, TN, $1,033,333 - for general operating support; ArtsMemphis, Memphis, TN, $1,250,000 - for capital campaign; Le Bonheur Childrens Medical Center Foundation, Memphis, TN, $500,000 - for capital campaign;
Contact: Scott McCormick, Executive Director; (901) 529-4063; mail@plough.org
Sponsor: Plough Foundation
62 North Main Street, Suite 201
Memphis, TN 38103

Plum Creek Foundation Grants **3963**
The corporate foundation awards grants to improve the quality of life in company operating areas. The foundation supports human services for children, at-risk youth, seniors, low-income, and abused individuals; education, particularly public and private colleges; youth organizations; hospitals, equipment, and medical programs; arts and cultural organizations; environmental education and conservation efforts; and civic affairs such as crime prevention, parks, and community development. Applications are reviewed quarterly. Application materials are available on the Web site.
Requirements: Nonprofit organizations in Alabama, Arkansas, Florida, Georgia, Idaho, Louisiana, Maine, Mississippi, Montana, North Carolina, Oklahoma, Oregon, Pennsylvania, South Carolina, Texas, Virginia, Washington, West Virginia, and Wisconsin are eligible.
Geographic Focus: Alabama, Arkansas, Florida, Georgia, Idaho, Louisiana, Maine, Mississippi, Montana, North Carolina, Oklahoma, Oregon, Pennsylvania, South Carolina, Texas, Virginia, Washington, West Virginia, Wisconsin
Amount of Grant: 1,000 - 10,000 U.S.D
Contact: Foundation Administrator; (206) 467-3600; fax (206) 467-3799; foundation@plumcreek.com
Internet: http://www.plumcreek.com/community/default.php
Sponsor: Plum Creek Foundation
999 Third Avenue, Suite 4300
Seattle, WA 98104-4096

PMI Foundation Grants **3964**
The PMI Foundation awards grants nationally, with emphasis on California, to a wide range of organizations with the goal of expanding homeownership. The foundation also contributes generously to deserving causes and charities in the areas of arts and culture, health and human services, education, civic organizations, and community development. Application guidelines are available for download from the PMI website. There are no application deadlines. Check with foundation staff to verify whether they are currently accepting applications.
Requirements: 501(c)3 organizations are eligible. Requests must target the disadvantaged, the poor, and distressed populations. Requests must either focus on increasing affordable housing opportunities or directly contribute to the quality of life in under-served communities.
Restrictions: The PMI Foundation does not accept requests for the following purposes: individuals; fraternal, veteran, labor, athletic or religious organizations serving a limited constituency; political or lobbying organizations, or those supporting the candidacy of a particular individual; travel funds; and films, videotapes or audio productions.
Amount of Grant: 200 - 200,000 U.S.D
Samples: Habitat for Humanity (East Bay), Oakland, California, $25,000; Animal Rescue Foundation, Walnut Creek, California, $2,500; Consumer Credit Counseling Services, San Francisco, California, $35,000.
Contact: Laura Kinney, Foundation Grant Administrator; (925) 658-6562
Internet: http://www.pmifoundation.org/index.html
Sponsor: PMI Foundation
3003 Oak Road
Walnut Creek, CA 94597-2098

PNC Charitable Trust and Foundation Grants **3965**
The PNC Charitable Trust Grant Review Committee serves over thirty perpetual charitable trusts and private foundations for which PNC is trustee. PNC's mission is to implement the legacies set forth by the individual donors who established these trust relationships with PNC. The Committee distributes grants from these charitable trusts to worthy 501(c)3 non-profit organizations. In accordance with the provisions of the trust agreements, the Committee funds grant requests that demonstrate a capacity and commitment to charitable service and enhance the quality of life for the people and their communities. Inquiries and proposals should be directly addressed to the local PNC representative. Names and contact information are listed on an interactive map on the website.
Requirements: Priority is given to the following: organizations that are direct providers of services, with projects and programs that meet donors' intent; organizations that demonstrate a capacity to meet the stated objectives, with objectives that are clearly articulated and measurable; the critical nature of the need being addressed; and organizations with strong leadership, diversified income streams, and long-term sustainability that have demonstrated efficiencies in delivering services. PNC seeks to support projects and programs that: reflect the core social mission of the applicant organization; address a community need; have broad impact, or benefit under-served populations; have measurable outcomes and a plan for documenting results; have detailed project budgets and demonstrate effective financial accountability; involve collaboration with other organizations; are of longer duration than events, conferences, or seminars; and complements rather than duplicate the work of other non-profit organizations.
Restrictions: Funding is provided in communities where donors lived and/or engaged. See the website to verify specific locations. The following are ineligible: general operating costs; event, conference or seminar sponsorships or entry fees; pilot or new programming; multi-year commitments; for-profit organizations; individuals or private foundations; fraternal, political, advocacy or labor organizations; and organizations that discriminate based on race, color, creed, gender, or national origin.
Geographic Focus: District of Columbia, Indiana, Kentucky, Maryland, New Jersey, Ohio, Pennsylvania
Date(s) Application is Due: Feb 1; May 1; Aug 1; Oct 1
Contact: Eva T. Blum, Chairman; (412) 705-3584; eva.blum@pnc.com
Internet: http://www.pncsites.com/pncfoundation/charitable_trusts.html
Sponsor: PNC Foundation
630 Liberty Avenue, Tenth Floor
Pittsburg, PA 15222-2705

Pohlad Family Foundation **3966**
The Pohlad Foundation concentrates their funding on the arts, health and human services, capital grants, and youth programs. The Foundation strives to support programs that: support all types and sizes of art programs of interest to diverse audiences; provide essential and effective human services to disadvantaged children and families, fund a range of services from emergency needs to workforce development, and have a positive multi-year impact;
Requirements: Applications are considered only in response to a request for proposal or an invitation to apply. Organizations must be 501(c)3 non-profit organizations. Priority is given to organizations that demonstrate a commitment to their communities by ensuring that their governing bodies include representatives from within the community and projects that directly benefit communities in need. Organizations may view sample grant recipients and specific instructions on how to apply on the website.
Restrictions: Grants are focused in Minnesota, primarily St. Paul and Minneapolis. Funding is not available for: direct funding to individuals; health or housing-related emergency assistance to individuals; benefits, fundraisers, walk-a-thons, telethons,

galas, or other revenue generating events; advertising; organizations that discriminate on the basis of race, gender, religion, culture, age, physical ability or disability, sexual orientation, gender identity, status as a military veteran or genetic information; veterans' and fraternal organizations; political or lobbying organizations; replacement of government funding.
Geographic Focus: Minnesota
Amount of Grant: 5,000 - 20,000 U.S.D
Contact: Rose Peterson, Program Manager; (612) 661-3903; fax (612) 661-3715; rpeterson@pohladfamilygiving.org
Internet: http://pohladfoundation.org/giving/grant-guidelines.html
Sponsor: Pohlad Family Foundation
60 South Sixth Street, Suite 3900
Minneapolis, MN 55402

Polk Bros. Foundation Grants **3967**
The primary focus of the Foundation is programs that work with populations of need, particularly children, youth, and families in underserved Chicago communities. Very few awards are made to organizations located outside the city of Chicago. Grants are made for both new and ongoing initiatives in four program areas: social service, education, culture and health care. In all areas, proposals should address increased access to services and improvement of the quality of life for area residents. Grants are seldom made for capital support.
Requirements: Illinois 501(c)3 nonprofit organizations are eligible. Preference is given to requests from Chicago. An organization that has not previously received a grant from the Foundation should first call the Foundation office or complete the pre-application form available on the Foundation website. The Foundation will then mail an application form or contact the organization with further questions.
Restrictions: The Polk Bros. Foundation will not support: organizations that devote a substantial portion of their activities to attempting to influence legislation or to participating in campaigns on behalf of candidates for public office; religious institutions seeking support for programs whose participants are restricted by religious affiliation or whose services promote a particular creed; purchase of dinner or raffle tickets or advertising in dinner programs; medical, scientific, or academic research; grants to individuals; tax-generating entities (municipalities, school districts, etc.) for services within their normal responsibilities. The Foundation will not consider more than one request from an organization or its affiliates in a 12-month period, nor will it generally fund more than eight percent of an organization's operating budget.
Geographic Focus: Illinois
Amount of Grant: Up to 15,000 U.S.D
Samples: Chicago High School for the Arts, $750,000 - general operating; Advocate Illinois Masonic Medical Center, $90,000 - school-based health centers at Amundsen and Lake View High Schools; Albany Park Theater Project, $80,000 - youth development through theater program; America Scores Chicago, $25,000 - salary support for the education and soccer directors;
Contact: Suzanne Doombos Kerbow, Assistant Director; (312) 527-4684; fax (312) 527-4681; questions@polkbrosfdn.org
Internet: http://www.polkbrosfdn.org/guidelines.htm
Sponsor: Polk Brothers Foundation
20 West Kinzie Street, Suite 1110
Chicago, IL 60611

Polk County Community Foundation Grants **3968**
The community foundation is a public charitable corporation dedicated to improving the quality of life in Polk County, NC, and the surrounding areas. General funds of the foundation support nonprofit organizations serving community residents in their educational, cultural, and charitable efforts. Types of support include building construction/renovation, capital campaigns, conferences and seminars, equipment, internship funds, matching/challenge grants, program development, publication, scholarship funds and scholarships to individuals, and seed grants. Grants are made three times each year. Contact the office for an application.
Requirements: Charitable organizations serving residents of Polk County, NC, and surrounding areas are eligible.
Restrictions: The foundation does not award grants to individuals, except for educational awards. Grants are not allocated anywhere but where they directly benefit local citizens.
Geographic Focus: North Carolina
Date(s) Application is Due: Mar 30; Aug 17
Amount of Grant: 2,000 - 50,000 U.S.D
Contact: Elizabeth Nager, Executive Director; (828) 859-5314; fax (828) 859-6122; foundation@polkccf.org
Internet: http://www.polkccf.org
Sponsor: Polk County Community Foundation
255 S Trade Street
Tryon, NC 28782-3707

Pollock Foundation Grants **3969**
The foundation awards grants to Texas nonprofit organizations in its areas of interest, including cultural programs, dental education and schools, health care and health organizations, Jewish organizations and temples, libraries and library science, nursing, public health education and schools, social services, and youth development. Grants support program develoment and general operating expenses. There are no application deadlines. Contact the office for application materials.
Requirements: Texas nonprofit organizations are eligible. Preference is given to requests from Dallas.
Geographic Focus: Texas
Amount of Grant: 1,000 - 300,000 U.S.D
Contact: Robert Pollock, Trustee; (214) 871-7155; fax (214) 871-8158
Sponsor: Pollock Foundation
2626 Howell Street, Suite 895
Dallas, TX 75204

Porter County Foundation Lilly Endowment Community Scholarships **3970**
The Lilly Endowment Community Scholarship Program is designed to raise the level of educational attainment and improve the quality of life for Indiana residents. To help further this goal, Lilly Endowment is providing funding to enable Indiana community foundations to offer the scholarship program. Three full tuition scholarships will be awarded to residents of Porter County who have graduated from an accredited Indiana high school. Students must pursue a baccalaureate degree at any accredited public or private Indiana college or university. The scholarship covers a four-year degree program including full tuition, required fees, and a special allocation of up to $800 per year for required books and equipment. Students will be evaluated on financial need, academic accomplishments, ability to communicate, motivation and desire to succeed, along with activities and leadership capabilities. The scholarship will be awarded to those who demonstrate that the scholarship will make a difference in their lives. Scholarship recipients may transfer from one Indiana college or university to another during the four-year period.
Requirements: Applicants must be U.S. citizens residing in Porter County, and graduates of an accredited Indiana high school. They must also be accepted to begin a full time four year study at an accredited public of private Indiana college or university. All college majors are eligible. In addition to the application, students must submit two letters of recommendation; an official high school transcript; copy of notification of SAT/ACT score; their completed financial information sheet; and a one page essay addressing the importance of the scholarships in helping them achieve their future goals. Applicants should refer to the Porter County website for current applications and deadlines. The application and supporting materials should be submitted to the candidate's individual high school counselor.
Restrictions: The scholarship will not extend beyond the four-year period.
Geographic Focus: Indiana
Contact: submit application materials to Individual Guidance Counselor
Internet: http://www.portercountyfoundation.org/scholarships.html
Sponsor: Porter County Community Foundation
57 South Franklin Street, Suite 207, P.O. Box 302
Valparaiso, IN 46384

Porter County Health and Wellness Grant **3971**
The Porter County Health and Wellness Fund awards grants to nonprofit organizations that promote, support, and/or advance health care in Porter County. Funding priorities include: increasing health care access for the underserved; improving and promoting healthy lifestyles for youth; and improving the nonprofit's operational capabilities to provide health care services. The maximum grant amount is $25,000.
Requirements: In addition to the application, all grant application packets must include the following information: a grant request cover page; a grant narrative; a project budget; a current operating budget and financial statement; the names and principal occupations of the organization's Board of Directors; the organization's grant application approval by their Board of Directors; and a copy of the organization's 501(c)3 tax exemption ruling.
Restrictions: Grants will not be made to: individuals; membership contributions; event sponsorships; programs that are sectarian or religious in nature; political organizations or candidates; contributions to endowment campaigns; campaigns to reduce previously incurred debt; and programs already completed.
Geographic Focus: Indiana
Date(s) Application is Due: Jun 15
Amount of Grant: 25,000 U.S.D
Contact: Brenda Sheetz, Health/Wellness Fund Contact; (219) 465-0294; fax (219) 464-2733; bsheetz@portercountyfoundation.org
Internet: http://www.portercountyfoundation.org/grantprograms.html
Sponsor: Porter County Community Foundation
57 South Franklin Street, Suite 207, P.O. Box 302
Valparaiso, IN 46384

Porter County Health Occupations Scholarship **3972**
The Porter County Health Occupations Scholarship provides funding for those wishing to continue their studies in a health-related field at a 2 or 4 year college or university in Indiana. The scholarship is evaluated on the student's financial need, academics, and capacity to succeed.
Requirements: Applicants must be U.S. citizens and residents of Porter County, and must have already completed one year of study. They must be in good standing at an Indiana accredited university or college, and enrolled in a health-related field of study. Along with the application, candidates should include their college transcript, a copy of their current Student Aid Report from FAFSA, a letter of recommendation from a professor, employer, or other professional affiliation, a one page statement describing why they chose a health occupation, and where they see themselves in ten years.
Geographic Focus: Indiana
Date(s) Application is Due: May 31
Contact: Health Occupations Scholarship; (219) 465-0294; fax (219) 464-2733; info@portercountyfoundation.org
Internet: http://www.portercountyfoundation.org/scholarships.html
Sponsor: Porter County Community Foundation
57 South Franklin Street, Suite 207, P.O. Box 302
Valparaiso, IN 46384

Porter County Women's Grant 3973
The Women's Grant of Porter County seeks to improve the quality of life for women and children in Porter County by collectively funding high impact grants for charitable initiatives with the same purpose. The group will award a $45,000 grant to a nonprofit organization in support of a project or program that addresses the issues of women and children and is sustainable. Priority will be given to innovative programs that demonstrate positive outcomes in one of the following areas: education and training to promote economic security and self-sufficiency; leadership development and programs designed to build self-esteem; access to women's health services and healthy lifestyles; safe environments and freedom from violence; and access to affordable daycare services that will expand hours of service, increase the number served on a sustainable basis and/or improve quality.
Requirements: In addition to the online application form, all application packets must include the following information: a grant request cover page; a grant narrative (see application for narrative details); a project budget; a copy of the organization's current year operating budget; the organization's most recent financial statement; names and principal occupations of the Board of Directors; evidence of Board approval for this application; a copy of the organization's 501(c)3 tax exemption.
Restrictions: The grant will not fund the following: projects or programs that do not address issues facing women and/or children; scholarship programs including daycare and program participation fees; annual appeals or membership contributions; event sponsorships; programs that are sectarian or religious in nature; political organizations or candidates; contributions to endowment campaigns; campaigns to reduce previously incurred debt; individuals; programs already completed and/or equipment already contracted for; and travel for bands, sports teams and similar groups.
Geographic Focus: Indiana
Date(s) Application is Due: Apr 15
Amount of Grant: 45,000 U.S.D
Contact: Brenda Sheetz, Women's Fund Contact; (219) 465-0294; fax (219) 464-2733; bsheetz@portercountyfoundation.org
Internet: http://www.portercountyfoundation.org/grantprograms.html
Sponsor: Porter County Community Foundation
57 South Franklin Street, Suite 207, P.O. Box 302
Valparaiso, IN 46384

Portland Foundation - Women's Giving Circle Grant 3974
The mission of The Portland Foundation Women's Giving Circle is to build a community of women philanthropists through the pooling of knowledge and resources for the purpose of providing grants to Jay County organizations and initiatives that address mutually-agreeable issues. The Circle awards grants which focus on enhancing the capacity of Jay County's not-for-profit organizations to support programming that addresses needs for youth and families. The Circle strives to fund innovative endeavors to benefit Jay County residents from toddlers to senior citizens.
Requirements: Organizations are encouraged to contact the Foundation to discuss whether their project is appropriate for funding. The application is available at the Foundation site. In addition to the application, proposals should include 10 copies of the following information: a copy of the IRS determination letter confirming tax-exempt status; organization's most recent financial statement, including budget and year-to-date income and expenses; and a detailed list of the Board of Directors.
Restrictions: The following are excluded from funding: organizations for religious or sectarian purposes; make-up of operating deficits, post-event or after-the-fact situations; endowment or capital projects and campaigns; for any propaganda, political or otherwise, attempting to influence legislation or intervene in any political affairs or campaigns; and dinner galas, advertising or other special fundraising events.
Geographic Focus: Indiana
Amount of Grant: 200 - 1,000 U.S.D
Samples: Arts Place, Inc., Arts in the Parks Clay Camp and Family Pottery Nights, $525,; Jay County Public Library, 1000 Books Before Kindergarten, $525; State of the Heart Hospice, Camp BEARable, $200.
Contact: Douglas Inman, Executive Director; (260) 726-4260; fax (260) 726-4273; tpf@portlandfoundation.org
Internet: http://www.portlandfoundation.org/womens-giving-circle-grant
Sponsor: Portland Foundation
112 East Main Street
Portland, IN 47371

Portland General Electric Foundation Grants 3975
Portland General Electric is committed to improving the quality of life for all Oregonians. The Foundation focuses giving in three areas: education, arts and culture, and healthy families. Education funding dedicates awards to scholarships, innovation in classroom instruction, transitional bridges between grade levels, career readiness and at-risk youth programs. Arts and Culture awards support youth programs and adult cultural programs that enhances understanding in communities. Healthy Families funding promotes access and services in areas of health, domestic violence, parenting, foster care, and other services that benefit families.
Requirements: 501(c)3 nonprofits in Oregon that address issues in the three focus areas of the Foundation are eligible for funding. Applicants must first submit a Letter of Inquiry to the Foundation; applicants may then be invited to submit a full application. Both of these steps must be completed online.
Restrictions: The foundation does not fund: bridge grants, debt retirement or operational deficits; endowment funds; general fund drives or annual appeals; political entities, ballot measure campaigns or candidates for political office; organizations that discriminate against individuals on the basis of creed, color, gender, sexual orientation, age, religion or national origin; fraternal, sectarian and religious organizations; individuals; travel expenses; or conferences, symposiums, festivals, events, team sponsorships or user fees. The Foundation also does not directly fund public K-12 education.
Geographic Focus: Oregon
Date(s) Application is Due: Jan 11; Apr 5; Jul 5; Nov 1
Amount of Grant: 2,500 - 10,000 U.S.D
Samples: Artists Repertory Theatre, Oregon, $7,500 - for educational programming (2011); Chess for Success, Oregon, $5,000 - after-school chess programs and tournaments for elementary schools (2011); Lifeworks NW, Oregon, $10,000 - vocational assistance to adults living with severe mental illness (2011).
Contact: Paige Haxton, Grant Administrator; (503) 464-8818; fax (503) 464-2929; pgefoundation@pgn.com
Internet: http://www.pgefoundation.org/how_we_fund.html
Sponsor: Portland General Electric Foundation
121 SW Salmon Street, One World Trade Center, 3rd Floor
Portland, OR 97204

Posey Community Foundation Grants 3976
The Posey County Community Foundation is a nonprofit, public charity created by and for the people of Posey County, Indiana. The Foundation connects donors with the causes they care about. The Foundations discretionary grants program accepts grant proposals once a year according to the grant cycle. Proposal requirements may change from year to year; therefore, grant seekers are advised to visit The Foundations website prior to beginning the grant application process. Grants are normally given as one time support of a project but may be considered for additional support for expansions or outgrowths of an initial project. The Foundation considers proposals for grants on a yearly cycle which begins each January. At the start of each cycle, a notice is mailed or emailed to nonprofit organizations that have applied for grants in the past, have received grants in the past, or have otherwise requested notification of the start of each cycle. The requirements and instructions provided here are to be considered current and accurate only for the duration of the Foundation's grant cycle, which runs from January through May each year. Proposals will be accepted from January through the March deadline. Proposals submitted after the March deadline will not be considered. Contact the Foundations office for current deadlines.
Requirements: The Foundation welcomes proposals from nonprofit organizations that are deemed tax-exempt under sections 501(c)3 and 509(a) of the Internal Revenue Code and from governmental agencies serving the county. Proposals from nonprofit organizations not classified as a 501(c)3 public charity may be considered provided the project is charitable and supports a community need. Proposals submitted by an entity under the auspices of another agency must include a written statement signed by the agency's board president on behalf of the board of directors agreeing to act as the entity's fiscal sponsor, to receive grant monies if awarded, and to oversee the proposed project. Required Documents: Cover Sheet (see Online Forms at the Foundations website); Narrative addressing the following (two to four pages): project summary; funding; implementation; evaluation. Supporting Documents: The following documents are to be attached to the Cover Sheet and Narrative described above. It is important to provide ALL requested attachments and that they be labeled in the following manner: Attachment 1: Project Budget Worksheet (see website Online Forms page); Attachment 2: Mission statement and nondiscrimination policy; Attachment 3: Statement of qualifications of project personnel; Attachment 4: Current board roster with professional affiliations; Attachment 5: Copy of IRS determination letter establishing 501(c)3 and 509(a) status (and fiscal sponsor agreement letter, if applicable); Attachment 6: Organization's current annual operating budget; Attachment 7: Most recent financial statement (monthly, quarterly, etc.); Attachment 8: Most recent audited financial statement or completed IRS Form 990 (if available). The original and thirteen copies of the proposal must be postmarked or delivered to the Foundation office no later than 4:30 p.m. on the first Tuesday of March. Use 8 1/2 by 11 inch paper. Attach all required documents using paperclips or binder clips only (no binders or folders).
Restrictions: Project areas NOT considered for funding: religious organizations for religious purposes; political parties or campaigns; endowment creation or debt reduction; operating costs (costs that keep the organization running and the doors open, i.e., general computer software or hardware, staff salary, general office supplies or equipment, utilities, etc.); capital campaigns (Large capital projects that may take several years to complete where there is no defined area for use of the grant); annual appeals or membership contributions; travel requests for groups or individuals such as bands, sports teams, or classes. A six month progress report and a final report at project completion are required by organizations whose proposals are approved for funding. Instructions and appropriate forms will be provided at the time the grant is awarded.
Geographic Focus: Indiana
Contact: Julie Eickhoff, Director; (812) 838-0288; fax (812) 838-8009; julie@poseycommunityfoundation.org
Internet: http://www.poseycommunityfoundation.org/disc-grants
Sponsor: Posey Community Foundation
402 Main Street, P.O. Box 746
Mt. Vernon, IN 47620

Posey Community Foundation Womens Fund Grants 3977
The Posey County Community Foundation is a nonprofit, public charity created by and for the people of Posey County, Indiana. The Foundation connects donors with the causes they care about. The Women's Fund of Posey County grant program is used to make yearly grants to support a variety of projects or programs serving women and girls in Posey County, Indiana such as programs that prevent domestic violence, secure family supporting jobs, promote health and education, and develop confidence. The programs or projects for which funding is being requested must address the needs and opportunities which support the Fund's mission of providing opportunities, encouragement, knowledge and hope to our community's women and girls. Grant proposals are accepted once each

year according to the grant cycle. Proposal requirements may change from year to year; therefore, grant seekers are advised to contact the Foundations office or visit the Foundations website prior to beginning the grant application process. Grants are normally given as one time support of a project but may be considered for additional support for expansions or outgrowths of an initial project. The Foundation considers proposals for grants on a yearly cycle which begins each June. At the start of each cycle, a notice is mailed to nonprofit organizations that have applied for grants in the past, have received grants in the past, or have otherwise requested notification of the start of each cycle, which runs from June through October each year. Proposals will be accepted from June through the July deadline. Proposals submitted after the July deadline will not be considered, contact the Foundation office for current deadline dates.
Requirements: The Women's Fund of Posey County welcomes proposals from non-profit organizations that are tax-exempt under sections 501(c)3 and 509(a) of the Internal Revenue Code and from governmental agencies serving Posey County women and girls. Proposals from other non-profit organizations to carry out charitable projects or activities that address issues facing women and girls in Posey County may be accepted. Proposals submitted by an entity under the auspices of another agency must include a written statement signed by the agency's board president on behalf of the board of directors agreeing to act as the entity's fiscal sponsor, to receive grant monies if awarded, and to oversee the proposed project. When submitting a grant proposal, include the following: cover sheet; narrative; project summary; funding; implementation; evaluation. The following documents are to be attached to the cover sheet and narrative. It is important to provide ALL requested attachments and that they be labeled in the following manner: attachment 1: project budget worksheet; attachment 2: mission statement and nondiscrimination policy; attachment 3: statement of qualifications of project personnel; attachment 4: current board roster with professional affiliations; attachment 5: copy of IRS determination letter establishing 501(c)3 and 509(a) status (and fiscal sponsor agreement letter, if applicable); attachment 6: organization's current annual operating budget; attachment 7: most recent financial statement (monthly, quarterly, etc.); attachment 8: most recent audited financial statement or completed IRS Form 990 (if available). The cover sheet form and the project budget worksheet form are provided on the Foundations website. When submitting the proposal send the original and fifteen copies. The proposal must be postmarked or delivered to the Foundation office no later than 4:30 p.m. on the fourth Tuesday in July. Use 8 1/2 by 11 inch paper. Attach all required documents using paperclips or binder clips only (no binders or folders).
Restrictions: If the grant proposal is approved, the organization is required to submit a six-month progress report and a final report at project completion. Appropriate forms will be provided.
Geographic Focus: Indiana
Contact: Julie Eickhoff, Director; (812) 838-0288; fax (812) 838-8009; julie@poseycommunityfoundation.org
Internet: http://www.poseycommunityfoundation.org/wf-grantmaking
Sponsor: Posey Community Foundation
402 Main Street, P.O. Box 746
Mt. Vernon, IN 47620

Postdoctoral Research Fellowships in Gerontological Public Health **3978**
The Center for Research on Health and Aging (CRHA) at the university offers postdoctoral fellowships in gerontological public health. These are two- to three-year appointments sponsored by the National Institute on Aging. This program focuses on developing research skills through mentoring, a comprehensive curriculum, hands-on research experience, presentations and publications, and opportunities for grantsmanship. The goal of the program is to foster and hone the research skills of future faculty and academic researchers in health and aging. This is an integrated multidisciplinary program with academic faculty and researchers in public health, gerontology, geriatrics, epidemiology, biostatistics, psychology, sociology, disability and human development, medicine and occupational therapy. Each postdoc fellow will be matched with a mentor from the Training Program Core Faculty. This primary mentor will work with the fellow to guide and collaborate on research projects. The fellowship includes a salary, with reimbursement for health insurance and yearly travel support. Additional financial support is also available. All fellows will have office space, a computer, phone, and access to university resources through the CRHA. Applications are accepted at any time for open positions. Guidelines are available online.
Requirements: Candidates should have received (or be very close to receiving) a doctoral degree in public health, gerontology, sociology, medicine, disability studies, or a related discipline.
Amount of Grant: 35,568 U.S.D
Contact: Dr. Thomas Prohaska, (312) 413-9830; fax (312) 413-9835; prohaska@uic.edu
Internet: http://www.uic.edu/depts/ovcr/hrpc/centers/fellow_gphpost.html
Sponsor: University of Illinois at Chicago
1747 W Roosevelt Road
Chicago, IL 60608

Pott Foundation Grants **3979**
Established in 1963 by Herman T. Pott, The Pott Foundation is a non-profit organization supporting children, education and health and human services. Most of the grant recipients are located in Herman Pott's adopted city of St. Louis, Missouri. The Foundation distributes approximately $1.5 million annually to over 100 charities.
Requirements: Non-profit 501(c)3 organization are eligible to apply in Missouri. Mail grant applications to: The Pott Foundation, C/O U.S. Bank, N.A., The Private Client Reserve, Attn: Carol Eaves, Mail Loc: SL-MO-CTCS, 10 North Hanley, Clayton, MO 63105.
Geographic Focus: Missouri
Date(s) Application is Due: Apr 1
Contact: Carol Eaves, (314) 418-8317
Sponsor: Pott Foundation
10 North Hanley
Clayton, MO 63105

Potts Memorial Foundation Grants **3980**
The Potts Memorial Foundation is a private foundation, incorporated in 1922 in New York. The Foundation gives on a national and international basis. The foundation was established to provide for the care, treatment, and rehabilitation of persons afflicted with tuberculosis; support for tuberculosis eradication, including fellowship programs for physicians. Grants given for tuberculosis and related disorders.
Requirements: Application form not required, Applicants should submit a detailed description of project and amount of funding requested. Proposal must be submitted via mail.
Restrictions: No grants to individuals, or for endowment funds or matching gifts; no loans.
Date(s) Application is Due: Apr 15; Sep 15
Amount of Grant: 15,000 - 45,000 U.S.D
Samples: Johns Hopkins University, Baltimore, Maryland, $41,800; Columbia Memorial Hospital, Hudson, New York, $20,000;
Contact: Charles Inman, Secretary; (518) 828-3365
Sponsor: Potts Memorial Foundation
444 Warren Street
Hudson, NY 12534-2415

Powell Foundation Grants **3981**
The purpose of the Powell Foundation is to distribute funds for public charitable purposes, principally for the support, encouragement and assistance to education, health, conservation, and the arts with a direct impact within the Foundation's geographic zone of interest The Foundation places priority on organizations and programs that serve residents in Harris, Travis and Walker counties, Texas, principally in the fields of education, the arts, health and conservation. The Foundation's current emphasis is in the field of public education in the broadest sense. Other areas of interest continue to be community service projects focused on the needs of children, the disadvantaged, the urban environment, and the visual and performing arts, especially in the Greater Houston, Texas area. The Foundation operates on a calendar year and its Board meets twice a year in the spring and in the fall. Submission of proposals is required at least two months prior to a meeting for consideration at that meeting. To allow for optimum consideration and due diligence, those seeking grants are encouraged to apply to the foundation on an ongoing basis. Grants that do not make the deadline for one meeting will be carried forward to the next meeting. Each request must be in writing and should be accompanied by the proposal summary and the required list of attachments. See the foundation's website for additional guidelines.
Requirements: Texas IRS 501(c)3 tax-exempt organizations serving Harris, Walker, and Travis counties are eligible to apply.
Restrictions: Normally, the Foundation will not support: requests for building funds or grant commitments extending into successive calendar years; grants to religious organizations for religious purposes; fund raising events or advertising; grants to other private foundations; grants to cover past operating deficits or debt retirement; grants for support to individuals; grants that impose the exercise of responsibility upon the Foundation. For example: private operating foundations or certain supporting organizations.
Geographic Focus: Texas
Amount of Grant: 1,000 - 20,000 U.S.D
Samples: Alley Theatre, Houston, Texas - educational outreach support (2011); Travis Audubon Society, Houston, Texas - operational support (2011); Great Expectations Foundation, Tahlequah, Oklahoma - professional development and mentoring for teachers (2011).
Contact: Caroline J. Sabin, Executive Director; (713) 523-7557; fax (713) 523-7553; info@powellfoundation.org
Internet: http://www.powellfoundation.org/powellguide.htm
Sponsor: Powell Foundation of Houston
2121 San Felipe, Suite 110
Houston, TX 77019-5600

PPG Industries Foundation Grants **3982**
Funding requests for a variety of project proposals that advance the foundation's interests are eligible for consideration. These may include capital projects, operating grants and special projects. In general, the foundation gives priority to applications from organizations dedicated to enhancing the welfare of communities in which PPG is a resident. Each grant application is reviewed with regard to the: compatibility of the applicant's goals with the foundation's priorities and available resources; financial needs of the organization; past practices of the foundation with respect to that organization; capability and reputation of the applicant; funds available to the applicant from other sources; extent to which the work of the applicant duplicates that of other organizations; public scope and impact of the applicant's proposal; and the interest of other corporate foundations with respect to the applicant. Historically the foundation has supported nonprofits in the areas of human services, health and safety, civic and community affairs, education, and cultural and arts. Requests for funding are accepted year-round. Determinations are made by the foundation's screening committee and board of directors.
Requirements: Applicants must use PPG's online grant making system to apply. The link is on the website. PPG Industries Foundation will review applications on a regular basis and will contact all grantseekers with proposals of interest. Organizations located in the Pittsburgh area and organizations of national scope should direct any questions to the executive director of the foundation. Organizations serving communities where PPG

facilities are located should direct any questions to the local PPG Industries Foundation agent in their area. A list of these may be found at the PPG Foundation website under the Foundation Governance link.
Restrictions: The foundation will not award grants for: advertising or sponsorships; endowments; political or religious purposes; projects which would directly benefit PPG Industries, Inc.; or special events and telephone solicitations. Operating grants are not made to United Way agencies.
Geographic Focus: All States, Pennsylvania
Samples: YMCA of Metropolitan Milwaukee, South Shore Center, Milwaukee, Wisconsin, $3,000 - funding supports SPLASH swimming program for second-graders (2011); Carlisle Regional Performing Arts Center, location unspecified, $3,000 - funding supports maintenance, technology upgrade costs (2011); Robert Morris University, Moon Township, Pennsylvania, $350,000 - funding supports renovation of the Career and Leadership Development Center.
Contact: Sue Sloan; (412) 434-2453; fax (412) 434-4666; foundation@ppg.com
Internet: http://www.ppg.com/en/ppgfoundation/Pages/Grant_Policies.aspx
Sponsor: PPG Industries Foundation
One PPG Place
Pittsburgh, PA 15272

Prader Willi Syndrome Association Research Grants **3983**
The program awards grants to scientific researchers for projects aimed at discovering and developing treatments, cures, and technologies benefiting individuals with Prader-Willi syndrome (PWS). The foundation is open to insightful grants regarding PWS of any sort, but has identified health issues of meritorious research benefit, including obesity, respiratory, and gastrointestinal health issues. Funding may be used for technical support, supplies, equipment, and appropriate travel. Only minimal salary support for faculty level investigators will be provided. The listed application deadline is for letters of intent. Guidelines are available online.
Restrictions: Projects with limited direct connection to therapies that could make a difference for the PWS community will not be considered.
Date(s) Application is Due: Jun 1
Amount of Grant: 50,000 U.S.D
Contact: Grants Administrator; (800) 926-4797 or (941) 312-0400; fax (941) 312-0142; national@pwsausa.org
Internet: http://www.pwsausa.org/research/research_funding_opportunities.htm
Sponsor: Prader Willi Syndrome Association U.S.A
5700 Midnight Pass Road, Suite 6
Sarasota, FL 34242

Premera Blue Cross CARES Grants **3984**
Premera CARES, a corporate giving program of Premera Blue Cross, focuses on health promotion and disease prevention in the communities it serves in Washington and Alaska. The program supports action-oriented events and programs which promote wellness and prevent major health conditions (e.g., diabetes, heart disease and cancer) as well as supporting initiatives which address the emotional and social well-being of community members and organizations which provide services to people with low and no income. Eligible organizations can submit funding requests through Premera's online grants-application tool on an ongoing basis. Guidelines and a link to the online-application tool are available at the grant website. Organizations should allow 90 days for processing of each request.
Requirements: Applicants must assist residents in areas served by Premera in Alaska and Washington, and must be a 501(c)3 organization, a government agency (or program), or a public school or school district. Funds must be used exclusively for public purposes. In addition, preference will be given to proposals that do the following: focus on health or wellness, including exercise and stress management, as a means of addressing common risk factors and preventing such health conditions as diabetes, heart disease, and cancer; address issues of access to healthcare for uninsured, low-income and underserved; and/or meet a special regional community-health need. Premera is a corporate contributor to United Way. In general, proposals from individual agencies already funded by the United Way will be given lower priority than proposals from other eligible organizations that are not receiving United Way support.
Restrictions: Grant awards are limited to areas of company operations in Alaska and Washington. Premera funds do not support the following entities: individuals; organizations that are for profit; arts organizations; religious organizations (unless the award is designated to an ongoing secular community-service program such as a food bank, shelter, homelessness project, or low-income health clinic and is not used to propagate a belief in a specific faith); or organizations that discriminate against individuals on any impermissible basis. Premera funds are not used for capital campaigns.
Geographic Focus: Alaska, Washington
Contact: Johanna Raisch, Grants Coordinator; (917) 677-2426
Internet: https://www.premera.com/stellent/groups/public/documents/xcpproject/social_responsibility.asp
Sponsor: Premera Blue Cross
7001 220th Street, S.W., Building 1
Mountlake Terrace, WA 98043-2160

Presbyterian Health Foundation Bridge, Seed and Equipment Grants **3985**
The primary objective of bridge grants is to provide funding of limited duration to enhance faculty competitiveness for national extramural funding. The seed grant program provides start-up funding for young investigators or for investigators launching into a new research direction. The equipment grants program provides the funding for laboratory equipment needed to advance research.
Requirements: Oklahoma nonprofit organizations are eligible. Submit a letter of application. Applications are due the last week in March, June, September, and December.
Restrictions: Grants are not made to individuals, private foundations, for-profit organizations or for operating funds of organizations.
Geographic Focus: Oklahoma
Amount of Grant: 25,000 - 250,000 U.S.D
Samples: U of Oklahoma Health Sciences Ctr (OK)—for research on genetics, $1.2 million over two years,; U of Oklahoma Health Sciences Ctr (OK)—to build the Stanton L. Young Research Center, $5 million.
Contact: Michael D. Anderson, President; (405) 319-8150; fax (405) 319-8168; manderson@phfokc.com
Internet: http://www.phfokc.com/bridge.htm
Sponsor: Presbyterian Health Foundation
655 Research Park Way, Suite 500
Oklahoma City, OK 73104-3603

Presbyterian Patient Assistance Program **3986**
The goal program is to be able to provide temporary assistance in times of urgent and emergency need to patients and clinical clients and their immediate families. Funding can be used to assist with temporary medication, food, utility and housing assistance, and/or clinically-related travel. Costs, expenses and other related expenses are sometimes covered by these funds.
Requirements: Requests for funding come to the foundation via their Chaplains, social workers, case workers, and other related individuals. Each case is evaluated on an individual basis and applicants must be able to provide documentation that verifies the needs for which assistance is requested. Those having the most devastating direct impact on an patient will take precedence.
Contact: Program Coordinator; (505) 724-6580
Internet: http://www.phs.org/foundation/assistance.shtml
Sponsor: Presbyterian Healthcare Foundation
P.O. Box 26666
Albuquerque, NM 87125-6666

Price Chopper's Golub Foundation Grants **3987**
Price Chopper's Golub Foundation provides financial support to eligible charitable organizations with a current 501(c)3 tax exempt status. Contributions are made through planned, continued giving programs in the areas of health and human services, arts, culture, education, and youth activities, within Price Chopper marketing areas. To be considered for funding, mail a written request, on letterhead for the organization seeking the donation, six to eight weeks prior to needed support or response deadlines. The Foundation reviews capital campaign requests quarterly, so please allow three to four months for a response.
Requirements: The Foundation's six state marketing area includes a specific mile radius around its stores in New York (Albany, Broome, Cayuga, Chenango, Clinton, Columbia, Cortland, Delaware, Dutchess, Essex, Franklin, Fulton, Greene, Hamilton, Herkimer, Jefferson, Lewis, Madison, Montgomery, Oneida, Onondaga, Orange, Oswego, Otsego, Rensselaer, St. Lawrence, Saratoga, Schenectady, Schoharie, Sullivan, Tioga, Tompkins, Ulster, Warren, and Washington counties), Massachusetts (Berkshire, Hampden, Hampshire, Middlesex, and Worcester counties), Vermont (Addison, Bennington, Caledonia, Chittenden, Essex, Franklin, Grand Isle, Lamoille, Orange, Orleans, Rutland, Washington, Windham, and Windsor counties), Pennsylvania (Lackawanna, Luzerne, Pike, Susquehanna, Wayne, and Wyoming counties), Connecticut (Hartford, Litchfield, New Haven, Tolland, and Windham counties) and New Hampshire (Cheshire, Grafton, and Sullivan counties).
Restrictions: The Foundation does not support: individuals; annual meetings; endowments; film and video projects; program advertising; funding for travel; organizations or events outside of its marketing area; events to raise funds for groups outside of its local community; conferences, conventions, or symposiums; publishing; operating expenses; scholarship programs outside of its own; or capital campaigns of national, religious or political organizations.
Geographic Focus: Connecticut, Massachusetts, New Hampshire, New York, Pennsylvania, Vermont
Contact: Deborah Tanski; (518) 356-9450 or (518) 379-1270; fax (518) 374-4259
Internet: http://www.pricechopper.com/GolubFoundation/GolubFoundation_S.las
Sponsor: Price Chopper's Golub Foundation
P.O. Box 1074
Schenectady, NY 12301

Price Family Charitable Fund Grants **3988**
Established in 1983, the Price Family Charitable Fund serves the San Diego, Carlsbad and, San Marcos, California region. The foundation is interested in supporting the economically disadvantaged, giving primarily for education and philanthropy purposes. The types of support include: annual campaigns; fellowships; program evaluation; scholarship funds; scholarships to individuals and; general operating support. There's no formal application to submit. Potential grantees much first submit a letter of Inquiry including: project goals, objectives and expected results (maximum of one page); project narrative (maximum of five pages); project budget (maximum of one page); a list of grants, if any, received by organization in the last 12 months for this program/project (sources and amounts); a list of the organization's current board of directors, including each member's name, profession, and office help on the board, if any.
Requirements: Non-profit organizations with 501(c)3 status or governmental units such as public schools or city departments in are available for funding. The Foundation gives primarily in the following region of: San Diego; Carlsbad; San Marcos, California.

Restrictions: Grants are not made: to organizations whose primary purpose is religious, or for propagandizing, influencing legislation and/or elections, promoting voting registration, for political candidates, political campaigns or organizations engaged in political activities; or to federal appeals or to organizations the collect funds for redistribution to other nonprofit groups. Unsolicited requests for funds are not accepted.
Geographic Focus: California
Samples: George G. Glenner Alzheimer's Family Centers, Inc., San Diego, CA, $33,000—Dementia Care Training & Public Scholarships; Episcopal Community Services, San Diego, CA, $60,000—general support; A Reason to Survive, San Diego, CA, $100,000—capacity building.
Contact: Terry Malavenda, (858) 551-2330
Sponsor: Price Family Charitable Fund
7979 Ivanhoe Avenue, Suite 520
La Jolla, CA 92037-4513

Priddy Foundation Grants 3989
The foundation makes grants in three geographic areas: in Texas within a 120 mile radius of Wichita Falls, in Oklahoma within a 75 mile radius of Wichita Falls, and, to a limited degree, in the Dallas/Fort Worth area. Requests for grants outside the above geographic area should provide statewide benefit. Primary grant categories include arts and culture, community enhancement, education, health, human services, and youth programs. The application process consists of a brief preliminary grant application for the purpose of policy screening. Full proposals are by request. Deadlines listed are for preliminary applications.
Requirements: 501(c)3 Texas and Oklahoma nonprofit organizations are eligible.
Restrictions: The foundation does not normally provide funding for the following: annual operating budgets on an ongoing basis, research, endowments, lobbying activities, churches or church denominations with the limited exception of certain Presbyterian-sponsored organizations, individuals, or individual scholarships.
Geographic Focus: Oklahoma, Texas
Date(s) Application is Due: Feb 1; Aug 1
Amount of Grant: Up to 6,000,000 U.S.D
Contact: Debbie White, Grants Director; (940) 723-8720; fax (940) 723-8656; debbiecw@priddyfdn.org
Internet: http://www.priddyfdn.org
Sponsor: Priddy Foundation
807 Eighth Street, Suite 1010
Wichita Falls, TX 76301-3310

Pride Foundation Fellowships 3990
The Pride Foundation Fellowship Program seeks to cultivate leaders and strengthen the Pacific Northwest LGBTQ community. By matching exceptional Pride Foundation scholarship recipients and other LGBTQ and ally students with Pride Foundation grantees for a substantial project-based fellowship, this fellowship experience will provide an opportunity for professional development as well as an introduction to the work of community leadership organizations. There are six internship opportunities, with four in Seattle, one in Sedro Wolley, Washington, and one in Portland, Oregon. Applicants will be asked to rank their top choices to help the selection committee determine eligible matches. Fellows are paid a $3,500 stipend at the end of the fellowship for their work. Fellows are expected to commit to 200 hours of work at their internship site (20 hours per week for 10 weeks), and must be available from mid-June through mid-August for their assignments.
Requirements: This program is open to current college students only (this includes students enrolled in community college, four year public or private colleges/universities, graduate school, professional programs, vocational or trade programs) pursuing any degree or major. Along with the online application, candidates are asked to provide basic contact information, summer availability, a list of applicable skills, and areas of potential development. Two references are also required.
Date(s) Application is Due: Jan 31
Amount of Grant: 3,500 U.S.D
Contact: Fellowships Coordinator; (800) 735-7287 or (206) 323-3318; fellowship@pridefoundation.org
Internet: http://www.pridefoundation.org/scholarships/fellowship/
Sponsor: Pride Foundation
1122 East Pike, PMB 1001
Seattle, WA 98122

Pride Foundation Grants 3991
The Pride Foundation works to strengthen the lesbian, gay, transgender, and bisexual community primarily in Washington state and extending to the four neighboring states of Alaska, Idaho, Montana, and Oregon. The Foundation awards grants to projects in arts and recreation; education, advocacy, and outreach; health and community service; HIV/AIDS service delivery and prevention; lesbian health; and youth and family services. Applicants will first submit the letter of inquiry online applications by August 19. If organizations are invited to submit the full application, they will be notified by September 23. Funds will be available in December.
Requirements: IRS 501(c)3 tax-exempt organizations or organizations affiliated with tax-exempt organizations are eligible. Projects or programs must directly benefit the lesbian, gay, bisexual, and transgender community; people affected by HIV/AIDS; and/or their friends and families.
Restrictions: Grants to individuals cannot be considered.
Geographic Focus: Alaska, Idaho, Montana, Oregon, Washington
Date(s) Application is Due: Sep 20
Amount of Grant: Up to 5,000 U.S.D
Contact: Jeff Hedgepeth, Grants Program Director; (800) 735-7287 or (206) 323-3318; jeff@pridefoundation.org
Internet: http://www.pridefoundation.org/grants/overview/
Sponsor: Pride Foundation
1122 East Pike, PMB 1001
Seattle, WA 98122

Pride Foundation Scholarships 3992
Pride Foundation provides post-secondary educational scholarships to lesbian, gay, bisexual, transgender, queer, and straight-ally leaders and role models from Alaska, Oregon, Idaho, Montana, and Washington. The Foundation offers one of the largest programs of its kind in the U.S. with nearly 50 specific scholarships for: students from various geographic areas in the Northwest; students of color; and students raised by LGBT parents. Scholarships cover nearly any accredited post-secondary school or program, including: community colleges; 4-year public or private colleges and universities; trade or vocational training; creative studies programs; certificate programs; medical or law schools; graduate studies; or other accredited degree programs. Candidates may apply scholarships to any of the following: tuition; course, program, or student fees; on campus room and board charges; school organized study abroad programs; books or course supplies; or graduate research related expenses. The application is available online.
Requirements: Requirements and scholarship amounts vary for each scholarship, but only one application is necessary for all scholarships. Applicants may contact the Foundation for more information.
Restrictions: Candidates outside the five state area are not eligible. Scholarship funds cannot be applied to off-campus housing, personal expenses, transportation costs, dependent care, computer or software purchases (unless required by the program); or other expenses not posted on the Foundation website.
Geographic Focus: Alaska, Idaho, Montana, Oregon, Washington
Date(s) Application is Due: Jan 31
Contact: Anthony Papini, Director of Program Strategies; (800) 735-7287 or (206) 323-3318; anthony@pridefoundation.org
Internet: http://www.pridefoundation.org/scholarships/overview/
Sponsor: Pride Foundation
1122 East Pike, PMB 1001
Seattle, WA 98122

Prince Charitable Trusts Chicago Grants 3993
The trusts awards grants to eligible Chicago nonprofit organizations in its areas of interest: arts and culture, education, environment, health, and social services.
Requirements: The Trusts Chicago program only funds organizations within the city limits of Chicago (with the exception of grants made through the MacArthur Fund for Arts and Culture at Prince). The Trusts make grants only to charitable organizations that are exempt from federal income tax under Section 501(c)3 of the Internal Revenue Code and are classified as public charities under Sections 509(a)(1) or 509(a)(2). All grant applications must include a Prince Charitable Trusts cover sheet, see the Trusts website, http://foundationcenter.org/grantmaker/prince/chi_app.html for proper form and additional guidelines.
Restrictions: The Trusts do not fund projects that promote or proselytize any religion. While the Trusts do fund the projects of faith-based organizations, those projects must be secular in nature. The Trusts do not fund organizations that discriminate on the basis of ethnicity, race, color, creed, religion, gender, national origin, age, disability, marital status, sexual orientation, gender identity, or any veteran's status.
Geographic Focus: Illinois
Date(s) Application is Due: Jan 13; May 1; Jun 1
Contact: Sharon Robison, Grants Manager; (312) 419-8700; fax (312) 419-8558; srobison@prince-trusts.org
Internet: http://www.fdncenter.org/grantmaker/prince/chicago.html
Sponsor: Prince Charitable Trusts
303 West Madison Street, Suite 1900
Chicago, IL 60606

Prince Charitable Trusts District of Columbia Grants 3994
The trusts awards grants to eligible Washington D.C. nonprofit organizations in its areas of interest: arts and culture, community, environment, health, emergency services, youth and provide a limited number of capital grants each year.
Requirements: The Trusts make grants only to charitable organizations that are exempt from federal income tax under Section 501(c)3 of the Internal Revenue Code and are classified as public charities under Sections 509(a)(1) or 509(a)(2). Electronic proposals are preferred. Attachments may be mailed separately. Proposals should include the Prince Charitable Trust Grant Application Cover Sheet and the Common Grant Application Format of Washington Grantmakers. These forms and additional guidelines may be obtained at the Trusts website, http://foundationcenter.org/grantmaker/prince/dc_app.html.
Restrictions: The Trusts do not make grants to individuals, nor does it fund projects that promote or proselytize any religion. While the Trusts do fund the projects of faith-based organizations, those projects must be secular in nature.
Geographic Focus: District of Columbia
Date(s) Application is Due: Feb 1; Aug 10; Sep 1
Amount of Grant: 10,000 - 30,000 U.S.D
Contact: Kristin Pauly, Managing Director; (202) 728-0646; fax (202) 466-4726; kpauly@princetrusts.org
Internet: http://www.fdncenter.org/grantmaker/prince/dc_interest.html
Sponsor: Prince Charitable Trusts
816 Connecticut Avenue NW
Washington, DC 20006

Princeton Area Community Foundation Greater Mercer Grants 3995

The Greater Mercer Grants Program is a competitive program for projects benefiting residents of Mercer County, New Jersey, and the immediately adjoining areas of surrounding counties. There are three distinct categories. In addressing the needs of low-income individuals and families (Category 1), grants up to $15,000 are made to nonprofit organizations only. In support of community organizing in low-income neighborhoods (Category 2), grants of $500 to $20,000 are made to 501(c)3 nonprofit organizations, neighborhood associations, and groups of residents forming a neighborhood association to work on a project. And in support of building community within and among municipalities (Category 3), grants generally range from $10,000 to $35,000, but may be as much as $50,000, made to nonprofit organizations only.
Requirements: Nonprofit organizations that have tax-exempt status under Section 501(c)3 of the Internal Revenue Code are eligible. They must be registered with the State of New Jersey as a charity, unless they are religious organizations, or schools that file their curricula with the Department of Education and are exempt from the provisions of the New Jersey Charitable Registration and Investigation Act.
Restrictions: Nonprofits may submit one application per calendar year in each category, for a maximum of three. Neighborhood associations or groups of residents who will form a neighborhood association may apply only once annually for a Community Organizing in Low-Income Neighborhoods grant.
Geographic Focus: New Jersey
Amount of Grant: 500 - 35,000 U.S.D
Samples: Trenton Film Society, $15,000— expansion of Year-Round Trenton Film Society Programming; Rescue Mission of Trenton, $10,000—job training program for Rescue Mission residents; Womanspace, $10,000— Womanspace Non-Denominational Chaplaincy Program to provide faith-based resources to victims of domestic violence and sexual assault to meet the victim where they are at if that meeting point is their faith.
Contact: Deborah Thomas, Director of Grants & Programs; (609) 219-1800; fax (609) 219-1850; daubert-thomas@pacf.org
Internet: http://www.pacf.org/grants/polCalendarEvent.cfm?Program_Code=12
Sponsor: Princeton Area Community Foundation
15 Princess Road
Lawrenceville, NJ 08648

Principal Financial Group Foundation Grants 3996

The foundation addresses concerns in the areas of health and human services, education, arts and culture, environment, and recreation and tourism. The primary objective is to support, through charitable contributions, selected nonprofit organizations primarily located in the greater Des Moines, IA, area. The foundation also will consider requests from organizations located in areas where the corporation has offices, including Des Moines, Mason City and Cedar Falls, IA; Grand Island, NE; Spokane, WA; Wilmington, DE; Appleton, WI; and Phoenix, AZ. The objectives, priorities, and programs seek to reflect the needs and concerns of communities in which the corporation operates. Support is given for annual campaigns, building funds, capital campaigns, continuing support, employee matching gifts, in-kind gifts, performances and exhibitions, conferences and workshops, adult basic education, vocational programs, operating budgets, internships, demonstration grants, matching grants, and seed grants. Contribution requests are considered on a quarterly basis, in accordance with the following schedule: health and human services, March 1; education, June 1; arts and culture, September 1; and environment, recreation, and tourism, December 1.
Requirements: 501(c)3 tax-exempt organizations in company operating locations may apply.
Restrictions: Proposals for athletic groups, conferences, endowments, fellowships, festivals, fraternal organizations, health care facility fund drives, libraries, or religious groups are denied.
Geographic Focus: Delaware, Iowa, Nebraska, Washington
Date(s) Application is Due: Mar 1; Jun 1; Sep 1; Dec 1
Amount of Grant: 1,000 - 50,000 U.S.D
Contact: Laura Sauser, Community Investment Consultant; (515) 247-7227; fax (515) 246-5475
Internet: http://www.principal.com/about/giving/grant.htm
Sponsor: Principal Financial Group Foundation
711 High Street
Des Moines, IA 50392-0150

Procter and Gamble Fund Grants 3997

The fund supports nonprofit organizations in company-operating locations in the areas of education, health and social services, civic projects, cultural organizations, disaster relief, and environmental efforts. Grants are awarded to education initiatives in local communities, such as teacher training efforts, with a focus on economic teaching; and other efforts by public policy, research, and economic education organizations. Employee voluntarism is prevalent in many K-12 initiatives. Most health and human services funding supports the United Way. The Salvation Army, Red Cross, hospitals, food banks, and other social service organizations receive support. Community support is awarded through grants that bolster economic growth and enrichment, including support for a youth jobs program, libraries, zoos, and local chambers of commerce. Support is awarded to a variety of arts organizations, including theater, dance, music, and visual arts. Major environmental groups also receive support. Grant Application Cycles are July 1 through September 30 and December 1 through February 28, grant requests are only accepted during those times.
Requirements: 501(c)3 organizations in communities where Procter and Gamble Company manufacturing plants are located are eligible.
Amount of Grant: Up to 25,000,000 U.S.D
Contact: Brenda Ratliff, Administrator, Corporate Philanthropy; (513) 945-8454; fax (513) 945-5211; pgfund.im@pg.com
Internet: http://www.pg.com/company/our_commitment/grant_application_guidelines.shtml
Sponsor: Procter and Gamble Fund
P.O. Box 599
Cincinnati, OH 45201

Prospect Burma Scholarships 3998

By funding education for young Burmese, Prospect Burma seeks to help create a cadre of Burmese who will be competent to run the country once democracy returns. Its aim is to educate those who will bring their knowledge and skills back to Burma for its future development. Preference is given to applicants studying one of the following subjects: Agriculture; Civil Society; Development; Ecology/Conservation; Economics; Education/Teacher Training; Engineering; Human Rights; Nursing; Public Administration; Public Health; or Science subjects. Preference is also given to students at educational institutions in South-East Asia or South Asia.
Requirements: You may qualify for an award if you are of Burmese origin and have been accepted for, or enrolled in a degree course at an accredited college or university, and are: accepted for and/or enrolled in a first degree (undergraduate) course at a college or university; already studying, or accepted for, a Master's degree course; or a postgraduate student who has already started on a Doctorate, or has a confirmed offer of a place to read for a Doctorate.
Restrictions: The Program does not contribute to international travel, computer equipment or debts. Furthermore, it does not make a contribution to the costs of dependants (spouses, partners, children).
Date(s) Application is Due: Mar 31
Contact: Shona Kirkwood, Education Project Coordinator; bpeoshona@gmail.com
Internet: http://www.prospectburma.org/index.php?option=com_content&task=category§ionid=5&id=18&Itemid=39
Sponsor: Prospect Burma
Rivermead Court, Ranelagh Gardens
London, SW6 3SF England

Prospect Hill Foundation Grants 3999

The foundation's national giving focuses on environmental conservation, especially in the Northeast and in Latin America; nuclear weapons control; and population growth/control in relation to natural resources, food supply, and opportunity to live a quality life. Types of support include matching funds, operating budgets, seed grants, and projects/programs. Preference goes to proposals for projects rather than for general support. Grant requests may be submitted year-round. Information and guidelines are available online.
Requirements: Giving is primarily in the northeastern United States, including New York and Rhode Island. Social service grants are limited to New York groups.
Restrictions: The foundation does not consider grants to individuals nor for basic research, sectarian religious activities, or organizations that lack tax exemption under United States law.
Amount of Grant: 10,000 - 25,000 U.S.D
Samples: Bulletin of the Atomic Scientists (Chicago, IL)—to enhance coverage of international and security issues in the Bulletin of the Atomic Scientists, $15,000; San Miguel-CASA (Guanajuato, Mexico)—toward the Professional Midwifery School, $35,000; Cunningham Dance Foundation (New York, NY)—for general support, $25,000; Fifth Avenue Presbyterian Church (New York, NY)—toward the capital campaign, $50,000.
Contact: Laura Callanan, Executive Director; (212) 370-1165; fax (212) 599-6282
Internet: http://fdncenter.org/grantmaker/prospecthill
Sponsor: Prospect Hill Foundation
99 Park Avenue, Suite 2220
New York, NY 10016-1601

Prostate Cancer Foundation Research Grants 4000

The heart of the Prostate Cancer Foundation's effort to to find better treatments and a cure for recurrent prostate cancer is its research awards programs. Each year, the PCF invests millions of dollars in research projects that have the greatest potential to advance the fight against prostate cancer. The PCF's research awards program has three components: Competitive Awards, Special Awards and Donor-Directed Awards. The foundation favors research applications focused on translational and clinical programs with potential for near term benefit for patients with recurrent prostate cancer. The listed application deadline is for competitive awards. Application guidelines and form are available online.
Requirements: Applications from individuals with or without institutional affiliations will be considered.
Restrictions: For-profit organizations or government agencies are not eligible for consideration.
Date(s) Application is Due: Dec 2
Amount of Grant: 25,000 - 100,000 U.S.D
Contact: Grants Administrator; (310) 570-4700; fax (310) 570-4701
Internet: http://www.prostatecancerfoundation.org/site/pp.asp?c=itIWK2OSG&b=68790
Sponsor: Prostate Cancer Foundation
1250 Fourth Street, Suite 360
Santa Monica, CA 90401

Prudential Foundation Education Grants 4001

The foundation's areas of interest are ready-to-learn programs, ready-to-work programs, and ready-to-live programs. In order to promote sustainable communities and improve social outcomes for community residents, the Prudential Foundation focuses its strategy in the following educational areas: education leadership to support reform in public education by increasing the capacity of educators, parents, and community residents to

implement public school reform; and youth development to build skills and competencies needed for young people to be productive citizens (this includes expanding arts education opportunities and supporting effective out-of-school-time programs for young people). Finally, the Prudential Foundation also funds organizations whose efforts influence policy that adapts promising practices and evidence-based approaches to instruction and learning in schools. Types of support include operating support, continuing support, annual campaigns, seed money, matching funds and employee matching gifts, consulting services, technical assistance, employee-related scholarships, research, capital campaigns, conferences and seminars, and projects/programs. The foundation is especially interested in proposals that anticipate and address potential major problems. Funds are targeted to areas where Prudential has a strong presence. Applicant should make initial contact with a brief letter to determine whether a more detailed proposal would be acceptable.
Requirements: The Prudential Foundation supports nonprofit, charitable organizations, and programs whose mission and operations are broad and non-discriminatory. The Foundation focuses its resources to support organizations whose activities address social needs or benefit underserved groups and communities. Priority in order of preference goes to programs in Newark, New Jersey, and surrounding communities; Los Angeles, California; Jacksonville, Florida; Atlanta, Georgia; Minneapolis, Minnesota; Philadelphia, Pennsylvania; and Houston, Texas. Third priority are national programs that can be implemented or replicated in the above cities.
Restrictions: The Foundation does not fund: organizations that are not tax-exempt under paragraph 501(c)3 of the U.S. Internal Revenue Code; labor, religious, political, lobbying, or fraternal groups—except when these groups provide needed services to the community at large; direct grants or scholarships to individuals; support for single-disease health groups; or good will advertising.
Geographic Focus: Arizona, California, Florida, Georgia, Minnesota, New Jersey, Pennsylvania, Texas
Amount of Grant: Up to 1,000,000 U.S.D
Contact: Lata Reddy, Director of Programs and Operations; (973) 802-4791; community.resources@prudential.com
Internet: http://www.prudential.com/view/page/public/12373
Sponsor: Prudential Foundation
751 Broad Street, 15th Floor
Newark, NJ 07102-3777

PSG Mentored Clinical Research Awards **4002**

The Mentored Clinical Research Award (MCRA) for new investigators is funded by a grant from the Parkinson's Disease Foundation (PDF) to the Parkinson Study Group (PSG). The PDF has partnered with the PSG to encourage the professional and scientific development of young investigators on their path to independence. To this end, this grant will support a new investigator for a one year project in patient oriented research in Parkinson's disease or other parkinsonian disorders under the mentorship of an experienced investigator. The training should lead a junior investigator to gain skills in clinical research. The research plan should address unmet needs of people living with PD, have the potential for broad application among the PD community, and lead to advances in clinically relevant treatment options. An award of $75,000 will be available this year. An electronic copy of the proposal in Microsoft Word or pdf-format must be received by the PSG on or before Friday, March 26. Applications should be sent by email to: Roseanna.Battista@ctcc.rochester.edu with a cover note that includes the candidate's name and the title of proposal. The candidate's proposal should be formatted according to the guidelines indicated in the toolkit available on the PSG website http://www.parkinson-study-group.org/PSGToolkit.asp. Applicants will be notified of the results of these reviews by May 21.
Requirements: Appropriate applicants for the MCRA are clinicians and scientists who are within 5 years of having completed formal training (this includes PhD professionals working in the field of Parkinson's disease). Fellows may apply. See website for additional guidelines.
Date(s) Application is Due: Mar 26
Amount of Grant: 75,000 U.S.D
Contact: Roseanna Battista, PSG Mentoring Committee Administrator; roseanna.battista@ctcc.rochester.edu
Internet: http://www.parkinson-study-group.org/Resources/PSG%20SRC%20Mentored%20Clinical%20Research%20Award%20RFP%202010-2011%20FINAL.pdf
Sponsor: Parkinson Study Group
University of Rochester, 1351 Mt. Hope Avenue, Suite 223
New York, NY 14620

Public Welfare Foundation Grants **4003**

Grants are awarded primarily to grassroots organizations in the United States and abroad, with emphasis on the environment, disadvantaged elderly and youth, population and reproductive health, economic development, welfare reform, health, human rights and global security, criminal justice, and community development. Programs must serve low-income populations, with preference to short-term needs. Types of support include matching funds, operating budgets, seed money, continuing support, and special projects. Grant guidelines are available upon request. Proposal with cover letter should be addressed to the Steering Committee at the address listed.
Requirements: Nonprofit organizations and, in certain cases, organizations without 501(c)3 status, may apply for grant support. Eligible exceptions are listed in the guidelines (available upon request).
Restrictions: Grants will not be made to individuals or for religious purposes, building funds, capital improvements, endowments, scholarships, graduate work, foreign study, conferences, seminars, publications, research, workshops, consulting services, annual campaigns, or deficit financing.
Amount of Grant: 25,000 - 50,000 U.S.D
Samples: Fair Housing Agency of Alabama (Mobile, AL)—to provide temporary and long-term housing and other support to people in Mobile, AL, left homeless by Hurricane Katrina, $10,000; Louisiana Bucket Brigade (Baton Rouge, LA)—to train people to monitor environmental pollution in their neighborhoods, $25,000; Share Foundation (San Francisco, CA)—to provide relief and support for reconstruction efforts in El Salvador following Hurricane Stan, $25,000; Environmental Support Ctr (Washington, DC)—to provide small grants to environmental justice groups in the Gulf Coast region that are struggling to reestablish their programs in the wake of recent hurricanes, $75,000.
Contact: Grants Administrator; (202) 965-1800; fax (202) 265-8851; reviewcommittee@publicwelfare.org
Internet: http://www.publicwelfare.org/about/about.asp
Sponsor: Public Welfare Foundation
1200 U Street NW
Washington, DC 20009

Puerto Rico Community Foundation Grants **4004**

The Foundation wishes to develop the capacities of communities in Puerto Rico so that they may achieve social transformation and economic self-sufficiency, by stimulating investment in communities and maximizing the impact and yield of each contribution. Grants are awarded in the areas of: education; community development; financial development; development of social interest housing; and philanthropy. Types of support include: general operating support; emergency funds; conferences and seminars; professorships; publications; research; technical assistance; consulting services; and matching funds. There are no application deadlines; the board meets in March, June, September, and December to consider requests.
Requirements: Organizations applying for grants must comply with the conditions below in order to demonstrate eligibility: be duly incorporated and registered as a nonprofit organization, according to the laws of the Commonwealth of Puerto Rico; be located and offer services in Puerto Rico. Present a copy of the following documents: certificate of good standing from the State Department; certificate from the Treasury Department; statement of organization's total budget for the year for which funds are solicited; financial statements; list of current members of Board of Directors. Should include each member's address and phone number; resume or curriculum vitae of the Project Director.
Restrictions: The foundation does not make grants to support individuals, annual campaigns, seed money, endowments, deficit financing, scholarships, or building funds.
Geographic Focus: Puerto Rico
Amount of Grant: 1,000 - 40,000 U.S.D
Contact: Grants Administrator; (787) 721-1037; fax (787) 721-1673; fcpr@fcpr.org
Internet: http://www.fcpr.org/
Sponsor: Puerto Rico Community Foundation
P.O. Box 70362
San Juan, PR 00936-8362
Puerto Rico

Pulaski County Community Foundation Grants **4005**

The Pulaski County Community Foundation was established to serve the citizens of Pulaski County, Indiana. The Foundation welcomes grant requests from any nonprofit organization in Pulaski County. The Foundation also invites applications to help fund new organizations who meet demonstrated needs or benefit the community through creative and innovative projects and programs. Grant seekers may access the online application but are also encouraged to discuss their project with the Foundation office before submitting a grant proposal.
Requirements: The Foundation favors grant requests which: impact a substantial number of people in the Pulaski community; propose practical solutions to current problems or address a current community interest; examine and address underlying causes of local needs; encourage cooperation and elimination of duplicate services; build the capacity of the applying organization; are from established non-profit organizations.
Restrictions: Grants are made only to organizations that serve the Pulaski county area. As a general rule, the Foundation does not make grants from its discretionary funds for the following: ongoing operating expenses or annual fund raising drives; existing obligations, debt reduction or building campaigns; individuals or travel expenses; loans or endowments; political purposes. Grantees are required to complete a Final Report (program and financial) detailing how the grant funds were spent.
Geographic Focus: Indiana
Contact: Wendy Rose, Executive Director; (574) 946-0906; fax (574) 946-0971; wrose@pulaskionline.org
Internet: http://www.pulaskionline.org/content/view/97/432/
Sponsor: Pulaski County Community Foundation
127 E. Pearl Street, P.O.Box 407
Winamac, IN 46996

Pulaski County Community Foundation Lilly Endowment Community Scholarships **4006**

The Lilly Endowment Community Scholarships help to raise the level of educational access in Indiana and enhance the quality of life for the state's residents. It provides full tuition and an $800 book stipend for four years of undergraduate study on a full-time basis at any accredited Indiana public or private college or university. All majors are eligible to apply.
Requirements: Applicants must be graduating high school seniors from Pulaski county. Candidates should contact the Foundation office or their guidance office for an application and further information about application requirements.
Geographic Focus: Indiana

Contact: Kim Krause, Scholarship Coordinator; (574) 946-0906; fax (574) 946-0971
Internet: http://www.pulaskionline.org/content/view/109/436/
Sponsor: Pulaski County Community Foundation
127 E. Pearl Street, P.O.Box 407
Winamac, IN 46996

Pulaski County Community Foundation Scholarships **4007**
The Pulaski County Community Foundation offers nearly thirty scholarships to provide financial support to local students each year. Each scholarship has specific criteria and deadlines for application. All college majors may apply. Students should contact the Foundation or their guidance office for further information.
Geographic Focus: Indiana
Contact: Kim Krause, Scholarship Coordinator; (574) 946-0906; fax (574) 946-0971
Internet: http://www.pulaskionline.org/content/view/151/435/
Sponsor: Pulaski County Community Foundation
127 E. Pearl Street, P.O.Box 407
Winamac, IN 46996

Pulte Homes Corporate Contributions **4008**
The charitable giving program is committed to addressing the life challenges facing individuals and families by partnering with organizations that have a significant positive impact on the future of society. Pulte's focus is in four primary categories: housing—funds are allocated for this category through 2006; education—major capital improvements or professorship endowments to improve programs targeted toward the residential homebuilding industry at U.S. universities; environmental—sustainable conservation projects or major initiatives to improve the environment; and health and human services. In the health and human services category, requests for funds benefiting the metropolitan Detroit area will be considered at the listed office location. Requests for funds for other geographic areas should be forwarded to the local Pulte market office. Guidelines and application are available online.
Requirements: 501(c)3 tax-exempt organizations are eligible.
Restrictions: Grants do not support individuals (except for residential homebuilding-industry scholarships); organizations that discriminate on the basis of race, religion, ethnicity, sexual preference, or other unlawful classification; or religious, fraternal, or veterans' organizations unless they benefit the community at large. Appeals for unrestricted funds or for organizations that distribute funds to other charities will not be considered.
Samples: Mercy Housing (Denver, CO)—to further develop high-quality affordable housing naitonwide, $1 million; Virginia Tech, College of Architecture and Urban Studies (Blacksburg, VA)—to endow a professorship in the department of building construction, $1.5 million.
Contact: Charitable Giving Administrator; (248) 433-4534
Internet: http://www.pulte.com/about_us/contributions_missionstatement.asp?ss={4F1CC3DE-CDDA-4CA2-BBE7-800009464021}
Sponsor: Pulte Homes Corporation
100 Bloomfield Hills Parkway, Suite 300
Bloomfield Hills, MI 48304

Putnam County Community Foundation Grants **4009**
The Putnam County Community Foundation is a nonprofit public charity established in 1985 to administer funds, award grants and provide leadership, enriching the quality of life and strengthening community in Putnam County. The Community Foundation administers more than 174 funds for donors supporting a variety of charitable activities. Three types of grants offered include: community grants, non-profit endowment building grants, and youth philanthropy grants. Types of support include: building renovation; consulting services; income development; matching/challenge support; program-related investments/loans; program development; scholarship funds; seed money; and technical assistance. The first funding cycle has a preliminary grant application due date of February 1, with full grants due by March 9; the second funding cycle has a preliminary grant application due date of August 1, with full grants due by September 9.
Requirements: A final report is due at the time of project completion. Unless other arrangements are made, it is expected that projects will be completed within one year of the grant award.
Restrictions: In general, grants from the Community Grants Program will not be awarded for the following: individuals; ongoing operational expenses (salaries, rent, and utilities); projects that do not serve Putnam County citizens; projects normally fully funded by units of government; projects to build or fund an endowment; religious activities or programs that appear to serve one denomination and not the community at large; political organizations or campaigns; national and state-wide fund raising projects; projects that could be considered discriminatory; projects operated by for-profit companies; or projects requesting retroactive funding.
Geographic Focus: Indiana
Date(s) Application is Due: Feb 1; Mar 9; Aug 1; Sep 9
Contact: M. Elaine Peck, Executive Director; (765) 653-4978; fax (765) 653-6385; epeck@pcfoundation.org or info@pcfountation.org
Internet: http://www.pcfoundation.org/grant_what_we_fund.html
Sponsor: Putnam County Community Foundation
2 South Jackson Street, P.O. Box 514
Greencastle, IN 46135

QLT Corporate Sponsorships **4010**
Each year, corporate sponsorships support requests that are aligned with corporate objectives and fall into the following priority areas: local, national, and global health organizations or service agencies— ophthalmology, oncology, and dermatology research; science education—programs that encourage British Columbian K-12 students to explore science as a career, and university science education; community—in Vancouver, improving quality of life for the underprivileged or disadvantaged through social, athletic, artistic or cultural endeavors; and United Way—employee fundraising campaign and matching funds raised dollar for dollar. Guidelines and application are available online.
Samples: Johns Hopkins U, School of Medicine (Baltimore, MD)—to endow a professorship in ophthalmology at the Wilmer Eye Institute, $2 million.
Contact: Corporate Communications Coordinator; (800) 663-5486 or (604) 707-7000; fax (604) 707-7179; corpcomm@qltinc.com
Internet: http://www.qltinc.com/Qltinc/main/mainpages.cfm?InternetPageID=4
Sponsor: QLT
887 Great Northern Way
Vancouver, BC V5T 4T5 Canada

Quaker Chemical Foundation Grants **4011**
The foundation supports nonprofit organizations in company-operating areas whose activities may be described as cultural, educational, health and welfare, and civic and community. Types of support include program grants, matching gifts, and scholarships. As much as possible, a grant application should contain a project description with a pro-forma budget; a current agency annual operating budget and audited financial statements; a list of funding sources for the organization, including past major contributors with amounts, recent applications with results, and anticipated future funding sources; and a list of board of directors and officers. The foundation chooses to make smaller grants to a large number of organizations rather than larger grants to a small number of organizations.
Requirements: Applicant organizations must be tax-exempt.
Restrictions: The foundation generally does not support national organizations or bricks and mortar projects.
Date(s) Application is Due: Apr 30
Amount of Grant: 1,500 U.S.D
Contact: Cathy Lasota, Secretary; (610) 832-4127
Internet: http://www.quakerchem.com/about_us/foundation_guidelines.pdf
Sponsor: Quaker Chemical Foundation
One Quaker Pike, 901 Hector Street
Conshohocken, PA 19428-0809

Qualcomm Grants **4012**
The philanthropic endeavors of Qualcomm develop and strengthen communities worldwide. Qualcomm invests human and financial resources in inspirational, innovative programs that serve diverse populations. Specifically their goal is to create educated, healthy, sustainable, culturally vibrant communities and to support employees' commitment to global communities through various programs. The company focuses primarily in geographic regions where they have a business presence. There are three focus areas. First is educated communities. The company is committed to improving science, technology, engineering and math education for students during their primary, secondary, and higher education years, and to expanding educational opportunities for under-represented students. Second is healthy, sustainable communities. The company strives to better the lives of underserved populations by providing basic human needs, with a focus on enhancing the welfare of children. They are also committed to protecting and enhancing our global environment. Third is culturally vibrant communities. Through their support of arts education and outreach programs, they help young people develop innovative minds and expand cultural enrichment opportunities to in-need populations. Applicants may submit a letter of inquiry form online and some organizations will be invited to submit a proposal for funding consideration. The submission deadlines are based on the grantmaking focus areas. The schedule can be found on the website.
Requirements: Eligible applicants must be 501(c)3 organizations.
Restrictions: The following are not eligible: individuals; sporting events without a charitable beneficiary; sectarian or denominational religious groups; faith-based schools unless the school accepts students from all religious and non-religious backgrounds and the students are not required to adhere to or convert to any religious doctrine; faith-based organizations unless the programs are broadly promoted and the program's beneficiaries are not encouraged or required to learn about, adhere, or convert to any religious doctrine; organizations that advocate, support, or practice activities inconsistent with Qualcomm's non-discrimination policies, whether based on race, religion, color, national origin, ancestry, mental or physical disability, age, gender, gender identity and/or expression, sexual orientation, veteran status, pregnancy, medical condition, marital status, or other basis protected by law; primary and secondary schools (note, however, these entities may be eligible for employee matching grants as long as they are not deemed ineligible by other exclusions and restrictions); and political contributions.
Geographic Focus: California, Colorado, Georgia, New Jersey, North Carolina, Texas
Contact: Corporate Giving Administrator; (858) 651-3200; fax (858) 651-3255; giving@qualcomm.com
Internet: http://www.qualcomm.com/citizenship/global-social-responsibility/philanthropy/guidelines
Sponsor: Qualcomm
5775 Morehouse Drive
San Diego, CA 92121

Quality Health Foundation Grants **4013**
The foundation's focus is on programs that demonstrate improved health care delivery, quality, and outcomes from a consumer's perspective. Priority areas include improved treatment through the use of best practices; improved access to health care services; and improved communication of health knowledge. Types of projects might include

promotion of healthy lifestyles and quality of life; promotion of culturally and linguistically appropriate care; promotion of consumer-centered care and decision-making; improvements in dissemination of health-related information; promotion of patient safety; promotion of best practices; and programs that reach underserved populations (rural, homeless, uninsured, etc.). The foundation supports projects of many types, including service demonstrations and rapid cycle change projects that have the potential to produce high impacts on health outcomes. Grants are awarded annually, for a one-year timeframe. Projects may apply each year for continued funding.
Requirements: Applicants must be from Maryland or the District of Columbia, where Delmarva is the Quality Improvement Organization (QIO) for the area. All organizations must be committed to improving the health of individuals at the community level. Examples of organizations that may apply for funding include faith-based, nonprofit, government, hospital/provider, and educational.
Restrictions: The foundation will not fund reimbursable direct patient-care services; individuals; ongoing operating expenses; capital expenses; lobbying; fundraising activities; or conferences (unless associated with an approved grant).
Geographic Focus: District of Columbia, Maryland
Date(s) Application is Due: Jan 15
Amount of Grant: Up to 50,000 U.S.D
Contact: Mariana Lesher, (410) 763-6227; lesherm@dfmc.org
Internet: http://www.qualityhealthfoundation.org/html/How_to.html
Sponsor: Delmarva Foundation
9240 Centreville Road
Easton, MD 21601

Quantum Corporation Snap Server Grants **4014**
The Snap Server Donation program focuses its charitable efforts on increasing access to storage technology by supporting organizations in the areas of human services and civic development. Quantum makes grants of storage appliances to nonprofit organizations worldwide. Its goal is to help bring the benefits of storage technology to people and their communities, to provide support to the communities in which its employees live, and to support its employees by taking an active role in their community. Snap Server donations are focused on youth organizations, local community programs, science and technology organizations, and humanitarian causes.
Geographic Focus: California
Contact: Sean Lamb, Public Relations Manager; (408) 879-8776; fax (408) 371-1783; sean.lamb@quantum.com
Internet: http://www.quantum.com
Sponsor: Quantum Corporation
2001 Logic Drive
San Jose, CA 95035

Quantum Foundation Grants **4015**
The foundation makes grants to approved charitable organizations, as well as state and local governmental entities, serving Palm Beach County, Florida, in the areas of health—improving access to insurance, health care, and prevention programs and reforming the health care delivery system; education—early childhood development, along with academic and career achievement for K-12 students; and community betterment (by invitation only)—promoting diversity, helping special needs populations, reducing family violence, and reducing child abuse and neglect. Types of support include capital grants, challenge/matching grants, program development grants, and seed money grants. Submit concept papers before full proposals.
Requirements: Nonprofits serving West Palm Beach, Florida, are eligible.
Geographic Focus: Florida
Samples: Area Agency on Aging (West Palm Beach, FL)—for its Leadership Institute for Active Aging program, $120,000; Joe DiMaggio Children's Hospital (Hollywood, FL)—to provide services for children and families in Palm Beach County, FL, with sickle cell disease and sickle cell trait, $390,000; YMCA of South Palm Beach County (Boca Raton, FL)—for its Safe Water Instructional Module Project, which engages children and teenagers in a lifeguard and swimming instructor training program, $207,000; YMCA of the Palm Beach (Boynton Beach, FL)—to expand a program that provides children with information about water safety, $110,000.
Contact: Christine Koehn, Vice President for Programs; (561) 832-7497; fax (561) 832-5794; chrisk@quantumfnd.org
Internet: http://www.quantumfnd.org
Sponsor: Quantum Foundation
2701 North Australian Avenue, Suite 200
West Palm Bach, FL 33407

Questar Corporate Contributions Grants **4016**
The corporation funds health and social services in the West, including some parts of Colorado, Oklahoma, Utah, and Wyoming. Grants have been awarded to hospitals, especially to build emergency rooms; substance abuse prevention programs, especially in the schools; child abuse prevention programs; and HIV/AIDS prevention and care. Questar determines the size of the grants based on the number of its employees in its facilities in the area, as well as how many other organizations of the same type it has funded nearby. Requests should be submitted in writing and describe the problem, outline the project, and state the amount sought. Applications are accepted at any time.
Requirements: Nonprofits in the West are eligible.
Geographic Focus: Colorado, Oklahoma, Utah, Wyoming
Amount of Grant: 5,000 - 20,000 U.S.D
Contact: Jan Bates; (801) 324-5132 or (801) 324-5202; Jan.Bates@Questar.com
Internet: http://www.questar.com/about_us/community/contributions.html
Sponsor: Questar Corporation
180 East 100 South, P.O. Box 45433, Mailstop QB 811
Salt Lake City, UT 84145-0433

R.C. Baker Foundation Grants **4017**
The foundation makes grants to U.S. nonprofit organizations for projects and programs that support social services for youth and the elderly, crime prevention, education, religion (Christian, Episcopal, Friends, Jewish, Methodist, and Presbyterian), scientific research, culture, and health. Support will be provided for fellowships and scholarships, general operating grants, challenge/matching grants, emergency funds, building funds, equipment, continuing support, annual campaigns, capital campaigns, renovation projects, and special projects. Submit cover letter with proposal. The board meets in June and November to consider requests.
Requirements: Nonprofit organizations are eligible. $25,000 to
Restrictions: Anaheim Memorial Medical Center, Anaheim, CA - $40,000; Harvey Mudd College, Claremont, CA - $25,000;
Date(s) Application is Due: May 1; Oct 1
Amount of Grant: 1,000 - 280,000 U.S.D
Samples: Anaheim Memorial Medical Center, Anaheim, CA - $40,000; Harvey Mudd College, Claremont, CA - $25,000;
Contact: Frank Scott, Chairman; (714) 750-8987
Sponsor: R.C. Baker Foundation
P.O. Box 6150
Orange, CA 92863-6150

R.S. Gernon Trust Grants **4018**
The R.S. Gernon Trust was established in 1975 to support and promote quality educational, human-services, and health-care programming for underserved populations. Grants from the R.S. Gernon Trust are primarily one year in duration. On occasion, multi-year support is awarded. Applicants must apply online at the grant website. Applicants are strongly encouraged to do the following before applying: review the downloadable state application procedures for additional helpful information and clarifications; review the downloadable online-application guidelines at the grant website; review the trust's funding history (link is available from the grant website); review the online application questions in advance; and review the list of required attachments. These will generally include: a list of board members, financial statements (audited, reviewed, or compiled by independent auditor); an organization summary; a list of other funding sources; an IRS Determination letter; and other required documents. All attachments must be uploaded in the online application as PDF, Word, or Excel files. The R. S. Gernon Trust has biannual deadlines of February 15 and August 15. Applications must be submitted by 11:59 p.m. on the deadline dates. Applicants will be notified of grant decisions by letter within two to three months after each respective proposal deadline.
Requirements: Applicants must have 501(c)3 tax-exempt status and serve the people of Norwich, Connecticut.
Restrictions: Grant requests for capital projects are generally not considered. Applicants will not be awarded a grant for more than 3 consecutive years. The trust does not support requests from individuals, organizations attempting to influence policy through direct lobbying, or any political campaigns.
Geographic Focus: Connecticut
Date(s) Application is Due: Feb 15; Aug 15
Contact: Kate Kerchaert, (860) 657-7016; kate.kerchaert@baml.com
Internet: https://www.bankofamerica.com/philanthropic/fn_search.action
Sponsor: R.S. Gernon Trust
200 Glastonbury Boulevard, Suite # 200, CT2-545-02-05
Glastonbury, CT 06033-4056

R.T. Vanderbilt Trust Grants **4019**
The trust awards grants in its areas of interest, including the arts, education, environmental programs, health care, historic preservation and societies, hospitals, and human services. Types of support include building construction/renovation, endowments, general operating support, and program development. The board meets in April, June, September, and December.
Requirements: There are no applications to submit. Proposals should be submitted in writing.
Restrictions: Applications are not accepted. Gives primarily to Connecticut, with some giving in Maine and New York. The trust does not give grants to individuals.
Geographic Focus: Connecticut, Maine, New York
Amount of Grant: Up to U.S.D
Contact: Gloria Kallas, Trustee; (908) 598-3582; fax (336) 732-2024/(336) 747-8722
Sponsor: R.T. Vanderbilt Trust
1525 West WT Harris Boulevard
Charlotte, NC 28288-5709

RACGP Cardiovascular Research Grants in General Practice **4020**
Individual grants of up to $25,000 will be offered for a period of one year to fund projects investigating various aspects of cardiovascular disease including hypertension, stroke, lipid disorder, health promotion and lifestyle changes. The Cardiovascular Research Grants (CVRGs) in General Practice are a joint initiative of the RACGP and the Australian Association of Academic General Practice (AAAGP).
Requirements: To be eligible for the Cardiovascular Research Grants, the chief investigator must be a GP or GP registrar who is a member of the RACGP or the AAAGP. New and emerging researchers are particularly encouraged to apply.
Restrictions: Unless otherwise indicated, Grants must be used in the year for which they are awarded. Grants will be paid only to incorporated bodies that have an ABN and

ACN and not to individuals. By applying for a Grant each applicant will be taken to have consented to permit the RACGP to publicise their name, institution, and title of their research project.
Geographic Focus: All States, Australia
Date(s) Application is Due: May 8
Amount of Grant: Up to 25,000 U.S.D
Contact: Samantha Fernandes, Research and Grants Program Administrator; 03-8699-0496; fax 03-8699-0400; research@racgp.org.au
Internet: http://www.racgp.org.au/researchfoundation/awards/cvrg
Sponsor: Royal Australian College of General Practitioners
College House, 1 Palmerston Crescent
South Melbourne, VIC 3205 Australia

RACGP National Asthma Council Research Award **4021**
The Award is awarded annually by the National Asthma Council to the best presenter of an asthma related abstract or poster at the RACGP Annual Scientific Convention. The award is to the value of $500 and is presented at the closing ceremony.
Requirements: To be eligible for the award, an individual must be: a GP or GP registrar; a member of the RACGP; in the early stages of their research career; presenting an asthma-related paper or poster at the RACGP Annual Scientific Conference, and; a major contributor to the research study being presented.
Restrictions: By applying for the award each applicant will be taken to have consented to permit RACGP to publicise their name, their institution, and the title of their research project.
Geographic Focus: All States, Australia
Date(s) Application is Due: Sep 1
Amount of Grant: 500 U.S.D
Contact: Samantha Fernandes, Research and Grants Program Administrator; 03-8699-0496; fax 03-8699-0400; research@racgp.org.au
Internet: http://www.racgp.org.au/researchfoundation/awards/nac
Sponsor: Royal Australian College of General Practitioners
College House, 1 Palmerston Crescent
South Melbourne, VIC 3205 Australia

RACGP Vicki Kotsirilos Integrative Medicine Grants **4022**
The Vicki Kotsirilos Integrative Medicine Grant is to encourage research that assists the integration of safe and ethical forms of natural and complementary medicines or therapies into mainstream general practice. The Grant is designed to encourage research that assists the integration of safe and ethical forms of natural and complementary medicines or therapies into mainstream general practice. A Grant of up $5,000 is offered to applicants conducting general practice research in the following areas: clinical research that contributes to the knowledge base concerning the efficacy of natural and complementary medicines or therapies; any project that supports research in behavioural and lifestyle factors impacting on health eg. diet, exercise, stress management, and; research that contributes to the efficient dissemination of the evidence regarding integrative medicine.
Requirements: Applicants must be: a General Practitioner or General Practice Registrar, and; a member of the RACGP.
Restrictions: Unless otherwise indicated, the Grant must be used in the year for which it is awarded. The Grant will be paid only to incorporated bodies that have an ABN and ACN and not to individuals. By applying for the Grant each applicant will be taken to have consented to permit the RACGP to publicise their name, institution, and title of their research project.
Geographic Focus: All States, Australia
Date(s) Application is Due: May 8
Amount of Grant: 5,000 U.S.D
Contact: Samantha Fernandes, Research and Grants Program Administrator; 03-8699-0496; fax 03-8699-0400; research@racgp.org.au
Internet: http://www.racgp.org.au/researchfoundation/awards/kotsirilos
Sponsor: Royal Australian College of General Practitioners
College House, 1 Palmerston Crescent
South Melbourne, VIC 3205 Australia

Radcliffe Institute Carol K. Pforzheimer Student Fellowships **4023**
The Arthur and Elizabeth Schlesinger Library on the History of Women in America invites Harvard undergraduates to make use of the library's collections with competitive awards of amounts up to $2,500 for relevant research projects. Preference will be given to applicants pursuing research in the history of work and the family, community service and volunteerism, the culinary arts, or women's health. The research may be, but is not required to be, in connection with a project for academic credit. Applications will be evaluated on the significance of the research and the project's potential contribution to the advancement of knowledge as well as its creativity in drawing on the library's holdings. The awards may be used to cover photocopying, other incidental research expenses, or living expenses in lieu of term time or summer employment. See website for current deadlines and application.
Requirements: Each application must include: the Schlesinger Library Grants and Fellowships cover page; a project description (no longer than four double-spaced pages in twelve-point font) indicating the purpose of the research, the Schlesinger Library holdings to be consulted, and the significance of these holdings to the project overall; a one-page bibliography of principal secondary sources consulted in designing the research; a résumé no longer than two pages; the name of one reference who has agreed to send a supporting letter; and a proposed budget indicating how the funds requested will be spent.
Restrictions: Only Harvard undergraduates are eligible. Funding does not cover the purchase of durable equipment.
Amount of Grant: Up to 2,500 U.S.D
Contact: (617) 496-3048; science@radcliffe.edu
Internet: http://www.radcliffe.edu/schles/pforzheimer_grant.aspx
Sponsor: Radcliffe Institute for Advanced Study
10 Garden Street
Cambridge, MA 02138

Radcliffe Institute Individual Residential Fellowships **4024**
The Radcliffe Institute Fellowship Program is a scholarly community where individuals pursue advanced work across a wide range of academic disciplines, professions, and creative arts. Radcliffe Institute fellowships are designed to support scholars, scientists, artists, and writers of exceptional promise and demonstrated accomplishment who wish to pursue work in academic and professional fields and in the creative arts. In recognition of Radcliffe's historic contributions to the education of women and to the study of issues related to women, the Radcliffe Institute sustains a continuing commitment to the study of women, gender, and society. Applicants need not focus on gender, however. Women and men from across the United States and throughout the world are encouraged to apply. email fellowships@radcliffe.edu for questions regarding humanities, social sciences, or creative arts fellowships. email sciencefellowships@radcliffe.edu for questions regarding natural science and mathematics fellowships.
Requirements: Scholars in any field with a doctorate or appropriate terminal degree at least two years prior to appointment (by December of the prior year) in the area of the proposed project are eligible to apply. Only scholars who have published at least two articles or monographs are eligible to apply. Artists and writers need not have a PHD or an MFA to apply; however, they must meet other specific eligibility requirements. Fellows are expected to be free of their regular commitments so they may devote themselves full time to the work. Applicants must reside in the Boston area, with their primary office at the Institute. Applications are available online, but applicants should refer to website for current deadlines and application requirements for each academic discipline.
Restrictions: Former fellows of the Radcliffe Institute Fellowship Program (1999 to present) are not eligible to apply.
Geographic Focus: All States, All Countries
Amount of Grant: 35,000 - 70,000 U.S.D
Contact: (617) 496-1324
Internet: http://www.radcliffe.edu/fellowships/apply.asp
Sponsor: Radcliffe Institute for Advanced Study
10 Garden Street
Cambridge, MA 02138

Rainbow Endowment Grants **4025**
Once a year, the Rainbow Endowment invites non-profit organizations having a national impact on the LGBT community to submit a request for funding. The proposals are reviewed by the Endowment's Grant Committee, which makes funding recommendations to the Board of Directors. The endowment supports national efforts to encourage positive physical and mental health; promote increased visibility; and advance full participation and access to social, cultural, and civic life for the gay and lesbian community. The Endowment seeks to support LGBT organizations whose efforts are national in scope in the areas of health and community. Priority is given to efforts that have practical implications or applications. In the area of health, grants are made to promote awareness of gay and lesbian mental and physical health issues among consumers and health care providers; to advocate for and develop public policy for equal health care treatment or equal access to health care treatment; and to advocate for and develop public policy to prevent further HIV/AIDS infection and to improve the lives of people with HIV/AIDS. In the area of community, grants are made to protect lesbian and gay rights; promote coalitions that work to strengthen advocacy efforts to improve conditions for lesbians and gays; and to develop policies that benefit lesbian and gay youth and the children of lesbian and gay families. Proposals can be submitted at anytime; however, they are reviewed only once per year. The deadline is the third Friday in May.
Requirements: U.S. nonprofit 501(c)3 organizations are eligible to apply.
Restrictions: The endowment does not support direct services, state or local projects, government agencies, individuals, regularly held meetings, fund-raising events, K-12 schools, religious purposes, endowments, or capital projects. Organizations may not apply more than once each calendar year.
Contact: Jean E. Bochnowski, Executive Director; (215) 241-7280; fax (215) 241-7278; jeb35@aol.com or endowment@rainbowcard.com
Internet: http://www.rainbowendowment.org/what.html
Sponsor: Rainbow Endowment
1501 Cherry Street
Philadelphia, PA 19102-1403

Rainbow Fund Grants **4026**
The Trust was established in 1954 in Georgia. Giving is primarily focused on theological education and other Christian endeavors. Support is also offered for substance abuse treatment and music education through mentoring high school students and conservatory musicians. There are no specific guidelines or application forms, and potential applicants are advised to contact the office directly.
Requirements: Eligible applicants from Florida, Georgia, Kentucky, Mississippi, and Texas can apply.
Geographic Focus: Florida, Georgia, Kentucky, Mississippi, Texas
Amount of Grant: 25,000 - 2,000,000 U.S.D
Contact: Burton S. Luce; (954) 764-7724; fax (954) 764-3603; lluce@drmail.com
Sponsor: Rainbow Fund
2408 Sunrise Key Boulevard
Fort Lauderdale, FL 33304

Rajiv Gandhi Foundation Grants 4027

The foundation conducts activities in India in the areas of the arts and humanities, conservation and the environment, economic affairs, education, international affairs, law and human rights, medicine and health, science and technology, and social welfare. Main projects are concerned with literacy and primary education, empowering women and children, aid for the disabled, health care services in rural and poor urban areas, and decentralized government. Types of support include research grants to institutions and scholarships, fellowships, and prizes to individuals.

Contact: Jawahar Bhawan, Director; (091-11) 23755117 or (091-11) 23312456; fax (091-11) 23755119; info@rgfindia.com

Internet: http://www.rgfindia.com

Sponsor: Rajiv Gandhi Foundation
Dr. Rajendra Prasad Road
New Delhi, 110 001 India

Ralph F. Hirschmann Award in Peptide Chemistry 4028

This award, supported by Merck Research Laboratories, is given annually to recognize and encourage outstanding achievements in the chemistry, biochemistry, and biophysics of peptides. The award is granted without regard to age or nationality of the recipient. The award consists of $5,000 and a certificate. Up to $2,500 for travel expenses to the meeting at which the award will be presented will be reimbursed.

Requirements: Any individual, except a member of the award committee, may submit one nomination or seconding letter for each award in any given year. Nominating documents consist of a letter of not more than 1000 words containing an evaluation of the nominee's accomplishments and a specific identification of the work to be recognized, a biographical sketch including date of birth, and a list of publications and patents authored by the nominee.

Restrictions: Self-nominations are not accepted.

Date(s) Application is Due: Nov 1

Amount of Grant: 5,000 U.S.D

Samples: Morten P. Meldal—award winner, $5,000; William F. DeGrado—award winner, $5,000; Samuel H. Gellman—award winner, $5,000.

Contact: Felicia Dixon, Awards Administrator; (800) 227-5558 or (202) 872-4408; fax (202) 776-8008; f_dixon@acs.org or awards@acs.org

Internet: http://portal.acs.org/portal/acs/corg/content?_nfpb=true&_pageLabel=PP_ARTICLEMAIN&node_id=1319&content_id=CTP_004547&use_sec=true&sec_url_var=region1

Sponsor: American Chemical Society
1155 Sixteenth Street, NW
Washington, DC 20036-4801

Ralph M. Parsons Foundation Grants 4029

The Ralph M. Parsons Foundation strives to support and facilitate the work of the region's best nonprofit organizations, recognizing that many of those in need today will go on to shape the future of Southern California, to define it, redefine it, and help it set and achieve new goals. The Foundation focuses on four areas: social impact, civic and cultural programs, health, and higher education. Applicants may submit a letter of inquiry, and there are no deadlines. If the Foundation decides to explore specifics of the request in more detail, applicants will be asked to submit a full proposal. Guidance will be provided in writing. Approximately 200 awards are made per year ranging from $25,000 to $50,000.

Requirements: Eligible applicants are 501(c)3 organizations located in Los Angeles County. Some occasional exceptions are made in the area of higher education. Excellence, access for disadvantaged populations, and the active participation of volunteers, board and staff are key characteristics the Foundation seeks in its applicants. Funding of direct services is a priority.

Restrictions: The following is ineligible: fundraising events, dinners and mass mailings; direct aid to individuals; conferences, seminars, workshops, etc; sectarian, religious or fraternal purposes; federated fundraising appeals; support of candidates for political office or to influence legislation; for-profit organizations or businesses; organizations outside of Los Angeles County (with occasional exceptions in the area of higher education); animal welfare; environment; documentary filmmaking; and scientific and/or medical research. Scholarship support is provided only to nonprofit institutions; individuals are not eligible to apply.

Geographic Focus: California

Samples: 24th Street Theatre, Los Angeles, California, $50,000, general support; Archdiocese of Los Angeles Education Foundation, Los Angeles, California, $75,000, tuition assistance to disadvantaged, low income children; and Arts in Education Aid Council, Los Angeles, California, $10,000, capacity-building support for this agency which provides arts eduction at public schools in the San Fernando Valley.

Contact: Wendy Garen, President and Chief Executive Officer; (213) 362-7600

Internet: http://www.rmpf.org

Sponsor: Ralph M. Parsons Foundation
1888 West Sixth Street, Suite 700
Los Angeles, CA 90017

Ralphs Food 4 Less Foundation Grants 4030

The foundation supports nonprofits primarily in areas of company operations in southern California, from Santa Barbara to San Diego, for programs and activities to improve the well-being of youth through education, recreation, and health-related programs; expand cultural awareness and appreciation of the arts; strengthen neighborhoods; and assist communities in the aftermath of local disasters. Types of support include special projects and general operating expenses. There are no application deadlines. Applicants should submit a letter of application.

Requirements: To be eligible, your organization must be: a 501(3) tax exempt nonprofit; located in Southern California; working in one of the focus areas. Requests for funding must arrive eight (8) weeks prior to the event date or date of need.

Restrictions: Funding is not available to: individuals; capital campaigns; travel expenses; projects of sectarian or religious organizations whose principal benefit is for their own members or adherents; organizations that discriminate on the basis of sex, race, religion, sexual orientation or national origin; third party giving.

Geographic Focus: California

Samples: Los Angeles Rescue Mission, Los Angeles, CA - $10,000; Community Environmental Council, Santa Barbara, CA - to support its annual Earth Day Festival $1,250; The Museum of Tolerance (MOT), Los Angeles, CA - to fund the MOT's student programs $10,000;

Contact: Michelle Williams, Executive Director; (310) 884-6205 or (310) 900-3522

Internet: http://www.ralphs.com/corpnewsinfo_charitablegiving_art5.htm

Sponsor: Ralph's-Food 4 Less Foundation
P.O. Box 54143
Los Angeles, CA 90054

Rap-it-Up/BASS Film Competition 4031

The competition is looking for screenplays (fiction) or treatments and story outlines (non-fiction) for short films up to a maximum length of 25 minutes that highlight the issue of HIV prevention, testing and treatment, and the effect of AIDS on individuals and families in the African American, Afro-Caribbean, and/or Afro-Latin communities in the United States. The winning films will be shot on Digital Video. Ten finalists and one winner will be selected from the pool of submissions and the competition will contribute financially toward the production of the winning short. The winning short films will air on or around December 1— World AIDS Day. The winning films will also be submitted to film festivals around the world. A completed application package (one original and five copies) with the required $75.00 entry fee must be received on or before May 27. No exceptions will be made. A discounted entry fee of $50.00 will be accepted for application packages received on or before May 21.

Requirements: The competition is open to members of all races and nationalities. However, all submissions should focus on the issue of AIDS in the African American, Afro-Caribbean and/or Afro-Latin communities in the United States.

Date(s) Application is Due: May 1; May 27

Amount of Grant: Up to 25,000 U.S.D

Contact: Khalid Kamau, (877) 757-2437, ext. 103

Internet: http://www.blackaids.org/bafilms

Sponsor: Black AIDS Institute
1833 W Eighth Street, Suite 200
Los Angeles, CA 90057-4920

Raskob Foundation for Catholic Activities Grants 4032

The Raskob Foundation is an independent private Catholic family foundation that makes grants worldwide for projects and programs associated with the Catholic Church. Grants support elementary and secondary education, community action and development, missionary activities, ministries (including youth and parish), health care, social concerns, AIDS victims, finance and development, and relief services. Types of support include operating budgets, seed money, emergency funds, equipment, land acquisition, conferences and seminars, program-related investments, renovation projects, special projects, and matching funds. Deadlines are June 8 and August 8 for the fall meeting; and December 8 and February 8 for the spring meeting.

Requirements: Roman Catholic organizations listed in the Kenedy Directory of Official Catholic Organizations may apply. Organizations should refer to the application guidelines for specific instruction on how to apply and information to submit.

Restrictions: The Foundation does not accept applications for the following purposes: tuition, scholarships or fellowships; reduction of debt; endowment funds; grants made by other grantmaking organizations; individual scholarly research; lobbying or legislation; or projects completed prior to our board meetings (mid-May and late November).

Geographic Focus: All States, All Countries

Amount of Grant: 5,000 - 15,000 U.S.D

Contact: Maureen Horner; (302) 655-4440; fax (302) 655-3223; info@rfca.org

Internet: http://www.rfca.org/en/Grantmaking/tabid/63/Default.aspx

Sponsor: Raskob Foundation for Catholic Activities
P.O. Box 4019
Wilmington, DE 19807

Rasmuson Foundation Capital Grants 4033

The Rasmuson Foundation supports non-profit organizations which strive to improve the quality of life for people throughout the state of Alaska. Capital Grants provide funding for assets such as furnishings, buildings, audio and video equipment, books, medical equipment, technology, art supplies, sports equipment, musical instruments, and vehicles. The Foundation encourages applicants to discuss proposals prior to submission. Applications are available on the website and are accepted at any time.

Requirements: Alaskan organizations that have received 501c(3) status and are classified as not a private foundation under section 509(a) of the Internal Revenue Service Code, units of government, and federally-recognized tribes are eligible.

Restrictions: For religious organizations, only projects with a broad community impact are considered. For units of government and tribes, only projects with a broad community impact beyond traditional government functions are considered. The following is not eligible: general operations, administrative, indirect, or overhead costs; deficits or debt reduction; endowments; scholarships; fundraising events or sponsorships; in general, K-

12 education; reimbursement for items already purchased; and electronic health records and other emerging technologies.
Geographic Focus: Alaska
Amount of Grant: 25,000 U.S.D
Samples: Christian Health Associates, Anchorage, Alaska, $25,000 - Anchorage Project Access moving and equipment (2012); Inupiat Community of he Arctic Slope, $22,500 - purchase of vehicle (2012); and Aiding Women in Abuse and Rape Emergencies, Juneau, Alaska, $11,457 - purchase of office furnishings and equipment and maintenance of emergency safe shelter (2012).
Contact: Barbara Bach, Director of Grant Management; (877) 366-2700 or (907) 297-2700; fax (907) 297-2770; bbach@rasmuson.org or rasmusonfdn@rasmuson.org
Internet: http://www.rasmuson.org/index.php?switch=viewpage&pageid=32
Sponsor: Rasmuson Foundation
301 W Northern Lights Boulevard, Suite 400
Anchorage, AK 99503

Rathmann Family Foundation Grants **4034**
The foundation's main funding areas are education, with priority given to science and math; the arts; children and youth health organizations; and preservation of the environment. Types of support include general operating support, continuing support, capital campaigns, equipment acquisition, endowment funds, program development, conferences and seminars, seed grants, curriculum development, fellowships, internships, scholarship funds, research, and matching funds. There are no specific deadlines or application forms, and interested parties should begin by contacting the Foundation directly.
Requirements: Grants are awarded to organizations in the San Francisco Bay, California area; the Annapolis, Maryland area; the Seattle, Washington area; the Philadelphia, Pennsylvania area: and metropolitan Minneapolis/Saint Paul, Minnesota.
Restrictions: Grants are not awarded to/for private foundations; religious organizations for religious activities; civil rights; social action; advocacy organizations; fraternal groups; political purposes; mental health counseling; individuals; or fundraisers, media events, public relations, annual appeals, or propaganda.
Geographic Focus: California, Maryland, Minnesota, Pennsylvania, Washington
Amount of Grant: 1,000 - 100,000 U.S.D
Contact: Rick Rathmann, (410) 349-2376; fax (410) 349-2377
Sponsor: Rathmann Family Foundation
1290 Bay Dale Drive, P.O. Box 352
Arnold, MD 21012

Raymond John Wean Foundation Grants **4035**
The foundation awards grants to children's nonprofits in northeast Ohio and southwest Pennsylvania in the areas of education and human services. In education, the foundation supports programs helping teachers and schools prepare students for life and work, especially projects improving educational opportunities, standards, performance, and achievement. In the human services category, the foundation prefers programs strengthening families, providing opportunities for at-risk youth, and helping disadvantaged people. The foundation also is interested in improving healthcare access.
Requirements: Nonprofit organizations serving northeast Ohio and southwest Pennsylvania are eligible.
Geographic Focus: Ohio, Pennsylvania
Date(s) Application is Due: Mar 1; Jun 1; Sep 1; Dec 1
Amount of Grant: 1,000 - 5,000 U.S.D
Contact: Grants Administrator; (330) 394-5600; fax (330) 394-5601; info@rjweanfoundation.org
Internet: http://www.rjweanfoundation.org
Sponsor: Raymond John Wean Foundation
108 Main Avenue SW, Suite 1005
Warren, OH 44481-1058

Rayne Foundation Grants **4036**
The foundation operates in the United Kingdom and internationally and awards grants for charitable purposes in its areas of interest, including the arts, education, medicine, and social welfare (especially the welfare of young people and the elderly). Applications are accepted anytime.
Restrictions: Applications from individuals and unregistered organisations will not be accepted. Organisations working outside the UK are not funded.
Contact: Tim Joss, Director; 020 7487 9650 or 020 7487 9637; fax 44 171 935 3737; info@raynefoundation.org.uk
Internet: http://www.raynefoundation.org.uk/grants.htm
Sponsor: Rayne Foundation
33 Robert Adam Street
London, W1U 3HR England

Rayonier Foundation Grants **4037**
The foundation awards grants to eligible nonprofit organizations in its areas of interest, including children and youth, community development, disadvantaged (economically), education, engineering, environmental programs, families, health care and hospitals, libraries, minorities, performing arts, recreation science, social services, technology, and volunteerism. Types of support include annual campaigns, building construction/renovation, continuing support, employee matching gifts, employee-related scholarships, endowments, equipment acquisition, general operating support, matching/challenge grants, performing arts, program development, recreation, science, research grants, scholarship funds, and seed grants.
Requirements: Nonprofit organizations in Florida, Georgia, and Washington are eligible.
Geographic Focus: Florida, Georgia, Washington
Amount of Grant: 250 - 25,000 U.S.D
Samples: United Way of Clallam County, Port Angeles, WA - $4,243; Florida Community College at Jacksonville Foundation, Jacksonville, FL - $25,000; University of Virginia, Charlottesville, VA - $7,500;
Contact: Charles H. Hood, President; (904) 357-9100; fax (904) 357-9101; info@rayonier.com
Sponsor: Rayonier Foundation
50 North Laura Street, Stuite 1900
Jacksonville, FL 32202

RCF General Community Grants **4038**
Six times a year, the Richland County Foundation awards General Community Grants to nonprofit organizations through a competitive process. In doing so, the Foundation looks to partner with nonprofit organizations to respond to current community needs in the following areas: education; health services; arts and culture; community services; children, youth and families; human services; environment; and employment and economic development. Application deadlines are 5:00 p.m. on the first Friday of January, March, May, July, September, and November. Applications must be at the Foundation Office by 5:00 p.m. on the due date to meet the deadline. Final decisions on all grant applications are made by the Board of Trustees approximately 6 – 8 weeks following the grant deadline.
Requirements: 501(c)3 public charities, government entities, schools, and nonprofit medical facilities serving residents of Richland County are eligible to apply. The application procedure should begin with a telephone call to the Program Officer to schedule an initial meeting to discuss the project. The Foundation typically looks for several of the following key elements in submitted applications: a one-time grant, especially for a pilot project which can serve as a model or be replicated; a project in which the Foundation is a funding partner, rather than the sole funder; a project or program which promotes volunteer involvement; an organization which can demonstrate the ability to sustain the project in the future when Foundation grant dollars end; projects or programs which are a collaborative effort(s) among nonprofit organizations in the community which eliminate duplication of services; a project which is likely to make a clear difference in the quality of life of a substantial number of people; an organization which is proposing a practical approach to a solution of a current community problem; a project or program which is focusing on prevention; and a worthy community project for which a grant from the Foundation will most likely leverage additional financial support. Applicants who have a program or project that meets these criteria are encouraged to contact the Foundation to discuss submitting an application.
Restrictions: Community grants are awarded from endowed, unrestricted and field of interest funds. Richland County Foundation typically does not provide funding from unrestricted funds for the following: sectarian activities of religious organizations; operating expenses for annual drives or to eliminate debt; medical, scientific or academic research; individuals other than for college scholarships; travel to or in support of conferences, or travel for groups such as bands, sports teams and classes (unless through special grant programs such as Summertime Kids or the Teacher Assistance Program); capital improvements to building and property not owned by the organization or covered by a long term lease; computer systems; projects that taxpayers support or expected to support; and political issues.
Geographic Focus: Ohio
Samples: Ashland University, Ashland, Ohio, $250 000 - this grant is part of a $15.5 million campaign to build a College of Nursing facility in Richland County (2011); Richland Community Development Group, Mansfield, Ohio, $50,000 - this supports phase two of a pilot project to coordinate a county-wide collaborative effort for both economic and community development (2011).
Contact: Bradford Groves, President; (419) 525-3020; fax (419) 525-1590; bgroves@rcfoundation.org
Rebecca Smith, Program Officer; (419) 525-3020; fax (419) 525-1590; bsmith@rcfoundation.org
Internet: http://www.richlandcountyfoundation.org/grant-information/types-of-grants/community
Sponsor: Richland County Foundation
24 West Third Street, Suite 100
Mansfield, OH 44902-1209

RCF Individual Assistance Grants **4039**
Twice a year, the Richland County Foundation awards Individual Assistance Grants to nonprofit organizations which operate programs that assist the needy. Grant dollars may be used for emergency assistance, basic human needs and sustaining basic health. Application deadlines are 5:00 p.m. on the first Friday of March and September. Applications must be at the Foundation Office by 5:00 p.m. on the due date to meet the deadline. Final decisions on all grant applications are made by the Board of Trustees approximately 6 – 8 weeks following the grant deadline. Community grants are awarded from endowed, unrestricted and field of interest funds.
Requirements: 501(c)3 public charities, government entities, schools, and nonprofit medical facilities providing emergency assistance, basic human needs, and basic health programs for residents of Richland County are eligible to apply. The application procedure should begin with a telephone call to the Program Officer to schedule an initial meeting to discuss the project.
Geographic Focus: Ohio
Samples: Salvation Army, Mansfield, Ohio, $25,000 - to provide emergency rent and mortgage assistance for their clients (2011); Mansfield Memorial Homes, Mansfield, Ohio, two grants totaling $35,000 - for programs that assist the elderly (2011); The Homemaker Home Healthcare Program, Mansfield, Ohio, $20,000 - to pay for home health services that include personal care, light housekeeping, and limited errand-runs to assist the elderly(2011)

Contact: Bradford Groves, President; (419) 525-3020; fax (419) 525-1590; bsmith@rcfoundation.org
Rebecca Smith, Program Officer; (419) 525-3020; fax (419) 525-1590; bsmith@rcfoundation.org
Internet: http://www.richlandcountyfoundation.org/grant-information/types-of-grants/community
Sponsor: Richland County Foundation
24 West Third Street, Suite 100
Mansfield, OH 44902-1209

RCF Summertime Kids Grants **4040**
Richland County Foundation's Summertime Kids Grants provide funding to nonprofit organizations that develop creative, educational and fun-filled activities for Richland County children throughout the summer months. Proposals are due the second Friday in February and must be at the Foundation Office by 5:00 p.m. on the due date to meet the deadline. The application with original signatures plus 18 copies must be submitted (the Foundation requests that applicants refrain from attachments). Awards will be announced in early April.
Requirements: 501(c)3 organizations, schools, churches, government entities and health service organizations serving residents of Richland County are eligible to apply. Collaborating applicants should submit only one application and note on it the lead agency.
Restrictions: Grant dollars should be used for direct programming expenses rather than operational expenses.
Geographic Focus: Ohio
Amount of Grant: Up to 2,500 U.S.D
Samples: City of Mansfield, Mansfield, Ohio - "Hooked on Fishing" (2011); Ohio Bird Sanctuary, Mansfield, Ohio - Nature Camp (2011); Raemelton Therapeutic Equestrian Center, Mansfield, Ohio - Summer Horse Camp (2011).
Contact: Becky Smith, Program Officer; (419) 525-3020; fax (419) 525-1590; bsmith@rcfoundation.org
Internet: http://richlandcountyfoundation.spirecms.com/grant-information/types-of-grants/summertime-kids
Sponsor: Richland County Foundation
24 West Third Street, Suite 100
Mansfield, OH 44902-1209

RCF The Women's Fund Grants **4041**
Richland County Foundation makes grants annually from "The Women's Fund" a permanent endowment established in 1996 to promote the physical, intellectual, emotional, social, economic, and cultural growth of women of all ages. While the current grant cycle is open to any nonprofit organization for programs which benefit women and girls in Richland County, a preference will be given to programs addressing childhood obesity, which exclusively target girls up to the age of (18) eighteen. The Foundation offers a workshop in August for interested applicants. Details are available at the grant website. Grant applications will be available at the workshop and will be made available for download from the website afterwards. Awards will be announced at the annual Women's Fund Luncheon in November.
Requirements: 501(c)3 public charities, government entities, schools, and nonprofit medical facilities serving residents of Richland County are eligible to apply. Deadline dates may vary from year to year. Interested applicants are encouraged to check the website, call the Program Officer with any questions and attend the Foundation's annual workshop for this grant.
Geographic Focus: Ohio
Date(s) Application is Due: Sep 16
Amount of Grant: 250 - 10,000 U.S.D
Samples: Center for Individual and Family Services, Mansfield, Ohio - "Ladies in Recovery" (2010); Emergency Pregnancy Contact, Mansfield, Ohio - "Helping Mom Succeed III" (2010); Eastern Elementary School, Lexington, Ohio - "Women in Science Day" (2010)
Contact: Bradford Groves, President; (419) 525-3020; fax (419) 525-1590; bgroves@rcfoundation.org
Internet: http://richlandcountyfoundation.spirecms.com/womens-fund/womens-fund-application-process
Sponsor: Richland County Foundation
24 West Third Street, Suite 100
Mansfield, OH 44902-1209

RCPSC/AMS CanMEDs Research and Development Grants **4042**
The primary objective of the grant is to support research, development, and/or implementation of projects to enhance specialty education in the promotion of the key roles for specialists as defined by CanMEDS. Applications may be individual or joint, involving one or more departments, disciplines, or faculties of medicine. It is not necessary to be a Fellow of the RCPSC or a physician; however, a Fellow of the Royal College must contribute to the project. Eligible projects may include faculty development, course development and/or medical education research that both apply and promote the CanMEDS roles. Budgets may include requests to defray the costs for travel, accommodation and meals for investigators to disseminate the project findings. One or more grants of up to $25,000 will be awarded each year.
Requirements: Applicants must be involved in teaching and/or medical education research in Canada. Applications that include research must meet contemporary standards of educational research, which include the relevance of the research question, the quality of the research design, ethical review of the submission, accountability for funds received, and the ability to complete the project during the funding period. Applications that include project development must include an evaluative component assessing the impact of the project.
Restrictions: Budgets should not include salary or honoraria for investigators. Ineligible costs include tuition fees, capital equipment costs (e.g. computer hardware, office equipment), overhead costs, university or other organization administrative fees.
Geographic Focus: All States, Canada
Date(s) Application is Due: Mar 2
Amount of Grant: Up to 25,000 U.S.D
Contact: Allison Shaw, Awards and Grants Section; (800) 668-3740, ext. 355 or (613) 730-8177; fax (613) 730-2410; awards@rcpsc.edu
Internet: http://rcpsc.medical.org/awards/ams_e.php
Sponsor: Royal College of Physicians and Surgeons of Canada
774 Echo Drive
Ottawa, ON K1S 5N8 Canada

RCPSC/AMS Donald Richards Wilson Award **4043**
The Royal College of Physicians and Surgeons of Canada, in collaboration with Associated Medical Services, Inc. has established an award to honour and acknowledge the contribution of Dr. Donald R. Wilson (President, RCPSC, 1988-1990) to medical education. This award will be given annually to a medical educator (not necessarily a Fellow) or an identified leader of a team, program, or department who has demonstrated excellence in integrating the CanMEDS roles into a Royal College or other health related training programs. Nominations will be received from Deans, Department or program heads, peers, or trainees. The successful applicant(s) will receive an award of $2,000 and a suitably engraved recognition.
Restrictions: Self-applications are ineligible.
Geographic Focus: All States, Canada
Date(s) Application is Due: Apr 30
Amount of Grant: 2,000 U.S.D
Contact: Ines Martin, (800) 668-3740, ext. 212 or (613) 730-6212; fax (613) 730-2410; imartin@rcpsc.edu
Internet: http://rcpsc.medical.org/awards/wilson_e.php
Sponsor: Royal College of Physicians and Surgeons of Canada
774 Echo Drive
Ottawa, ON K1S 5N8 Canada

RCPSC Balfour M. Mount Visiting Professorship in Palliative Medicine **4044**
The Royal College of Physicians and Surgeons of Canada, in collaboration with the Robert Pope Foundation of Nova Scotia, has established this professorship in order to bring to the health science faculties in Canada the most distinguished international lecturers in the palliative care community to engage in discussion about, and to lecture on, the physical, psycho-social and spiritual needs of patients requiring palliative care. The professorship is meant to facilitate the work of the multidisciplinary team that contributes to palliative care expertise in the community and university being visited. The professor will be appointed in alternate years and will visit the health science faculties of at least two universities in Canada. Using a variety of teaching tools, the professor will bring to the host faculties knowledge and research related to the physical, psycho-social, and spiritual needs of patients requiring palliative care.
Geographic Focus: All States, Canada
Contact: Ines Martin, (800) 668-3740, ext. 212 or (613) 730-6212; fax (613) 730-2410; imartin@rcpsc.edu
Internet: http://rcpsc.medical.org/awards/bmm-palliative_e.php
Sponsor: Royal College of Physicians and Surgeons of Canada
774 Echo Drive
Ottawa, ON K1S 5N8 Canada

RCPSC Canadian Research Awards for Specialty Residents (Medicine, Surgery) **4045**
The purpose of the awards is to provide national recognition for original work by postgraduate trainees. Any original work in the field of clinical investigation, in the basic sciences relating to medicine or surgery, in medical education, or in medical ethics, is acceptable. Work described in manuscripts published or submitted for publication during the 12 months preceding January 31 is acceptable, providing the other conditions are fulfilled. The manuscripts must be received no later than January 29, at 5 p.m. The recipients will receive a certificate that bears the logo or crest of the three participating organizations together with a check for $2,000. The prizes are presented at the Royal College Annual Conference.
Requirements: The candidate must be a resident (postgraduate trainee registered with a university office of postgraduate education) and enrolled in a residency program accredited by the Royal College. Those certified in a primary specialty but continuing their residency towards a subspecialty in an approved residency program are also eligible. The work must have been done mainly in Canada. Residents who have not previously won this award in either category are eligible.
Geographic Focus: All States, Canada
Date(s) Application is Due: Jan 29
Amount of Grant: 2,000 U.S.D
Contact: Ines Martin, (800) 668-3740, ext. 212 or (613) 730-6212; fax (613) 730-2410; imartin@rcpsc.edu
Internet: http://rcpsc.medical.org/awards/residents_e.php
Sponsor: Royal College of Physicians and Surgeons of Canada
774 Echo Drive
Ottawa, ON K1S 5N8 Canada

RCPSC Detweiler Traveling Fellowships 4046

Fellowships are made to improve the quality of medical and surgical practice in Canada. The fellowships will enable the recipients to visit medical centers in Canada or abroad to study or gain experience in the use or application of new knowledge or techniques in their fields or to further the pursuit of a fundamental or clinical research problem. Five or more fellowships offered annually to a maximum of $21,000 will be reserved for junior applicants. Two additional fellowships offered annually will be reserved for senior applicants: one will be granted to a Fellow who has held Royal College certification for five years or more and has been in private practice for at least three years, and one will be granted to a Fellow who has held Royal College certification for five years or more and has been in academic practice for at least three years. Each award will be to a maximum of $10,500.
Requirements: Applicants must be Fellows of the Royal College in good standing, or residents in the final year of their training who will be admitted to Fellowship in the Royal College prior to December 31st.
Restrictions: Travel fellowships will not be awarded retroactively.
Geographic Focus: All States, Canada
Date(s) Application is Due: Sep 30
Amount of Grant: 10,500 - 21,000 U.S.D
Contact: Allison Shaw, Awards and Grants Section; (800) 668-3740, ext. 355 or (613) 730-8177; fax (613) 730-2410; awards@rcpsc.edu
Internet: http://rcpsc.medical.org/awards/detweiler_e.php
Sponsor: Royal College of Physicians and Surgeons of Canada
774 Echo Drive
Ottawa, ON K1S 5N8 Canada

RCPSC Duncan Graham Award 4047

One of the notable and outstanding awards that The Royal College of Physicians and Surgeons of Canada may bestow upon an individual is the Duncan Graham Award. The Duncan Graham Award was introduced in 1969 in honor of the late Dr. Duncan Graham, for many years a professor of medicine at the University of Toronto. The Award is conferred upon any individual, whether physician or not, in recognition of outstanding lifelong contribution to medical education. The award is in the amount of $1,000 accompanied by a suitably engraved memento. It is usually conferred each year at the Convocation of the Royal College Annual Meeting. Nominations are solicited from the Deans of the Faculties of medicine. The Awards Committee of the Royal College selects the recipient of the Duncan Graham Award. At its discretion, the Awards Committee may add to the list of nominations.
Geographic Focus: All States, Canada
Date(s) Application is Due: Sep 30
Amount of Grant: 1,000 U.S.D
Contact: Ines Martin, (800) 668-3740, ext. 212 or (613) 730-6212; fax (613) 730-2410; imartin@rcpsc.edu
Internet: http://rcpsc.medical.org/awards/duncan-graham_e.php
Sponsor: Royal College of Physicians and Surgeons of Canada
774 Echo Drive
Ottawa, ON K1S 5N8 Canada

RCPSC Events, Speakers and Visiting Professors Grants 4048

The purpose of the grant is to help defray the cost of special events, speakers, and visiting professors. Events must be primarily organized for Fellows and residents of the RCPSC. The RACs represent the following regions of Canada: Region 1—British Columbia, Alberta, Yukon Territory, and Northwest Territories; Region 2—Saskatchewan and Manitoba; Region 3—Ontario and Nunavut; Region 4—Quebec; and Region 5—New Brunswick, Nova Scotia, Prince Edward Island, and Newfoundland and Labrador. Up to $1,200 is available in regions 1, 2, 3 and 5, and up to $1,000 is available in region 4.
Restrictions: No consideration given to applications submitted or received after the event. The application must be submitted by a Fellow of the RCPSC in good standing.
Geographic Focus: All States, Canada
Amount of Grant: Up to 1,200 U.S.D
Contact: Allison Shaw, Awards and Grants Section; (800) 668-3740, ext. 355 or (613) 730-8177; fax (613) 730-2410; awards@rcpsc.edu
Internet: http://rcpsc.medical.org/awards/cme_e.php
Sponsor: Royal College of Physicians and Surgeons of Canada
774 Echo Drive
Ottawa, ON K1S 5N8 Canada

RCPSC Faculty Development Grants 4049

The objective of the grant is to support the creation, promotion, and evaluation of, and/or research on, new and innovative methods of faculty development to foster the training of quality teachers. Proposals to create new and innovative faculty development programs must, at a minimum, contain an evaluative component in order to be considered. Innovative pilot research projects focused on faculty development will also be considered. Although not a requirement, applicants are encouraged to seek matching funds from the participating university. Preference will be given to projects that are generic, practical, applicable to postgraduate medical education, and relevant to the defined objective of the grant.
Requirements: Applicants must be members of a Canadian faculty of medicine involved in teaching and/or educational research. It is not necessary to be a Fellow of the RCPSC or a physician.
Restrictions: Budgets should not include salary, honoraria, travel, accommodation or meals for investigators; however, a maximum of $500 may be requested to defray the costs for travel, accommodation, meals or conference registration for investigators to disseminate the project findings. Ineligible costs include tuition fees, capital equipment costs (e.g. computer hardware, office equipment), overhead costs, university or other organization administrative fees.
Geographic Focus: All States, Canada
Date(s) Application is Due: Mar 2
Amount of Grant: Up to 5,000 U.S.D
Contact: Allison Shaw, Awards and Grants Section; (800) 668-3740, ext. 355 or (613) 730-8177; fax (613) 730-2410; awards@rcpsc.edu
Internet: http://rcpsc.medical.org/awards/faculty_e.php
Sponsor: Royal College of Physicians and Surgeons of Canada
774 Echo Drive
Ottawa, ON K1S 5N8 Canada

RCPSC International Traveling Fellowship 4050

The objective of the fellowship is to enable fellows residing outside of Canada to study in a Canadian medical center in order to acquire a new knowledge and skills applicable to clinical practice, undertake clinical research, or to enable fellows residing in Canada to practice and teach in a less developed country. The fellowship is applicable for visits of six to 12 months duration. Visits may be broken into separate time blocks but must be completed within two years. An amount equivalent to 75% of the award will be given at the beginning of the fellowship and the remainder upon receipt of the final report at the end of the fellowship. The fellowship may be used to supplement financial assistance granted from other sources.
Requirements: Applicants must be Fellows of the Royal College in good standing, or residents in the final year of their training who will be admitted to Fellowship in the Royal College prior to December 31.
Geographic Focus: All States, Canada
Date(s) Application is Due: Sep 30
Amount of Grant: 10,500 - 2,100 U.S.D
Contact: Allison Shaw, Awards and Grants Section; (800) 668-3740, ext. 355 or (613) 730-8177; fax (613) 730-2410; awards@rcpsc.edu
Internet: http://rcpsc.medical.org/awards/international_e.php
Sponsor: Royal College of Physicians and Surgeons of Canada
774 Echo Drive
Ottawa, ON K1S 5N8 Canada

RCPSC James H. Graham Award of Merit 4051

In 1987, the Council of The Royal College of Physicians and Surgeons of Canada recommended that an award of merit be given to a person whose outstanding achievements reflect the aims and objectives of the Royal College. This person need not be a physician. The Awards Committee will be the selection committee for the award of merit. The selection committee will receive nominations (with curriculum vitae) from the RCPSC Council (limit of two), and the Regional Advisory Committees (one per RAC). The presentation for the award of merit will be made at the Annual Meeting of the Royal College. The winner will receive a suitably engraved memento. The travel and maintenance expenses of the award winner and spouse will be paid in accordance with Royal College guidelines.
Geographic Focus: All States, Canada
Date(s) Application is Due: Sep 30
Samples: Dr. Robert Y. McMurty, London, Canada; Dr. Paul M. O'Byrne, Ancaster, Canada; Dr. Eldon R. Smith, Calgary, Canada.
Contact: Ines Martin, (800) 668-3740, ext. 212 or (613) 730-6212; fax (613) 730-2410; imartin@rcpsc.edu
Internet: http://rcpsc.medical.org/awards/james-graham_e.php
Sponsor: Royal College of Physicians and Surgeons of Canada
774 Echo Drive
Ottawa, ON K1S 5N8 Canada

RCPSC Janes Visiting Professorship in Surgery 4052

The purpose of this visiting professorship is to stimulate and increase an interest in the latest developments in surgery and surgical techniques. The visiting professor, preferably a Canadian surgeon of national stature, spends up to one week visiting one to three medical schools according to a mutually agreed schedule. The professor participates in undergraduate and postgraduate teaching and in the exchange of ideas with practicing surgeons, residents and students. The professorship is supported by a grant of $7,000 and is usually selected every second year.
Geographic Focus: All States, Canada
Date(s) Application is Due: Sep 30
Amount of Grant: 7,000 U.S.D
Contact: Ines Martin, (800) 668-3740, ext. 212 or (613) 730-6212; fax (613) 730-2410; imartin@rcpsc.edu
Internet: http://rcpsc.medical.org/awards/janes-surgery_e.php
Sponsor: Royal College of Physicians and Surgeons of Canada
774 Echo Drive
Ottawa, ON K1S 5N8 Canada

RCPSC Kilborn Memorial Visiting Professorship 4053

The Royal College of Physicians and Surgeons of Canada, in collaboration with the Kilborn Memorial Fund, has established a visiting professorship to West China Centre of Medical Sciences (WCCMS), formerly West China University of Medical Sciences (WCUMS), to honour the contributions made by members of the Kilborn family to the founding and development of the faculty of medicine and health sciences at WCCMS in Chengdu, Sichuan, China. The purpose of the visiting professorship is to enable Canadian specialists to visit WCCMS for a predetermined time (e.g., 2-4 weeks), and to contribute to the education of students and the professional development of the faculty of the school of medical sciences and/or related health sciences in the discipline identified by the host center.
Geographic Focus: All States, Canada

Contact: Ines Martin, (800) 668-3740, ext. 212 or (613) 730-6212; fax (613) 730-2410; imartin@rcpsc.edu
Internet: http://rcpsc.medical.org/awards/kilborn_e.php
Sponsor: Royal College of Physicians and Surgeons of Canada
774 Echo Drive
Ottawa, ON K1S 5N8 Canada

RCPSC KJR Wightman Award for Scholarship in Ethics **4054**
The Royal College of Physicians and Surgeons of Canada awards a prize for the best scholarly paper presented by a resident registered in a postgraduate program accredited by the Royal College. Papers arising from a clinical case, addressing ethical issues pertinent to the CanMEDS roles (medical expert, communicator, collaborator, manager, health advocate, scholar, professional), or on the topic of equity in any aspect relevant to the work of the College are welcome. The recipient will receive $1,000 as well as a certificate. The award will be presented during the Convocation Ceremony at the Annual Conference. The travel expenses of the recipient and guest to the Annual Conference will be paid in accordance with Royal College guidelines.
Requirements: Submissions for the award must be accompanied by a curriculum vitae and submitted electronically. The manuscript must be between 2500 and 5000 words, excluding the bibliography.
Geographic Focus: All States, Canada
Date(s) Application is Due: Apr 30
Amount of Grant: 1,000 U.S.D
Samples: Dr. Michel Shamy, University of Toronto, Toronto, Ontario—Narrative, Argument and Evidence: An Ethical Justification for the Use of Suggestion in the Treatment of Psychogenic Illness; Dr. Trevor Langhan, Calgary, Alberta—Simulation training for emergency medicine procedural skills: Should we move away from practicing on people?.
Contact: Ines Martin, (800) 668-3740, ext. 212 or (613) 730-6212; fax (613) 730-2410; imartin@rcpsc.edu
Internet: http://rcpsc.medical.org/awards/wightman_e.php
Sponsor: Royal College of Physicians and Surgeons of Canada
774 Echo Drive
Ottawa, ON K1S 5N8 Canada

RCPSC McLaughlin-Gallie Visiting Professorship **4055**
The Royal College of Physicians and Surgeons of Canada established the McLaughlin-Gallie Visiting Professorship in 1960. The professorship commemorates the outstanding contributions to surgical science of Dr. William Edward Gallie of Toronto. McLaughlin-Gallie Visiting Professors may be selected from any country in the world and may represent any branch of medicine. They spend up to one week at each of two Canadian faculties of medicine. They participate in undergraduate and postgraduate teaching, and in the exchange of ideas with the staff, the researchers and the undergraduate and postgraduate students.
Amount of Grant: 10,000 U.S.D
Contact: Ines Martin, (800) 668-3740, ext. 212 or (613) 730-6212; fax (613) 730-2410; imartin@rcpsc.edu
Internet: http://rcpsc.medical.org/awards/mclaughlin-gallie_e.php
Sponsor: Royal College of Physicians and Surgeons of Canada
774 Echo Drive
Ottawa, ON K1S 5N8 Canada

RCPSC Medical Education Research Grants **4056**
The primary objective of the grant is to support quality research in Canada relevant to the field of postgraduate medical education or continuing professional development. Applications may be individual or joint, involving one or more departments, disciplines, or faculties of medicine. It is not necessary to be a Fellow of the RCPSC or a physician; however, a Fellow of the Royal College must contribute to the research project. The following areas of research are encouraged: curriculum design and implementation; instructional methods; assessment; program evaluation; continuing education; professional development; and faculty development. S
Requirements: Applicants must be involved in teaching and/or medical education research in Canada. Applications must meet contemporary standards of educational research, which include the relevance of the research question, the quality of the research design, ethical review of the submission, accountability for funds received, and the ability to complete the project during the funding period. One or more grants, of up to $25,000 per year of study, will be awarded annually.
Restrictions: Ineligible costs include tuition fees, capital equipment costs (e.g. computer hardware, office equipment), overhead costs, university or other organization administrative fees.
Geographic Focus: All States, Canada
Date(s) Application is Due: Mar 2
Contact: Allison Shaw, Awards and Grants Section; (800) 668-3740, ext. 355 or (613) 730-8177; fax (613) 730-2410; awards@rcpsc.edu
Internet: http://rcpsc.medical.org/awards/MERG_e.php
Sponsor: Royal College of Physicians and Surgeons of Canada
774 Echo Drive
Ottawa, ON K1S 5N8 Canada

RCPSC Medical Education Traveling Fellowships **4057**
The objective of the fellowship is to enable the recipient to acquire knowledge and expertise in the field of medical education. Upon completion, the recipient will be able to use education principles to develop educational programs and proceed to the evaluation of learning experiences at the undergraduate, postgraduate, or continuing education level. Fellowships are applicable for visits of six to 12 months duration. Up to 75 percent of the fellowship stipend will be given at the beginning of the fellowship and the remainder upon receipt of the final report at the end of the fellowship. The fellowship may be used to supplement financial assistance granted from other sources.
Requirements: Applicants must be Fellows of the Royal College in good standing, or residents in the final year of their training who will be admitted to Fellowship in the Royal College prior to December 31st.
Geographic Focus: All States, Canada
Date(s) Application is Due: Sep 30
Amount of Grant: 15,000 - 30,000 U.S.D
Contact: Allison Shaw, Awards and Grants Section; (800) 668-3740, ext. 355 or (613) 730-8177; fax (613) 730-2410; awards@rcpsc.edu
Internet: http://rcpsc.medical.org/awards/medical_e.php
Sponsor: Royal College of Physicians and Surgeons of Canada
774 Echo Drive
Ottawa, ON K1S 5N8 Canada

RCPSC National Specialty Society Annual Meeting Grants **4058**
The National Specialty Society Annual Meeting Grants program was established to assist Royal College Fellows in their professional development activities by providing some financial support to national forums on specialty medicine. Funding can support: the speaker of the Royal College of Physicians and Surgeons of Canada; and postgraduate coursework at the time of the national specialty society's annual meeting. Up to $1,200 including expenses and honorarium.
Requirements: The national specialty society must be duly constituted and must represent a specialty or a sub-specialty recognized by the Royal College. Consideration may also given to multidisciplinary societies if a large percentage of its members are Fellows of the Royal College. The application form must be submitted by a Fellow of the Royal College at least eight (8) weeks before the society's annual meeting.
Geographic Focus: All States, Canada
Contact: Allison Shaw, Awards and Grants Section; (800) 668-3740, ext. 355 or (613) 730-8177; fax (613) 730-2410; awards@rcpsc.edu
Internet: http://rcpsc.medical.org/awards/nssgrant_e.php
Sponsor: Royal College of Physicians and Surgeons of Canada
774 Echo Drive
Ottawa, ON K1S 5N8 Canada

RCPSC Royal College Fellowship for Studies in Medical Education **4059**
The goal of this fellowship is to increase the number and quality of professionally trained medical educators in Canada by providing training in the science of medical education to selected promising candidates. The program is to help specialists acquire knowledge and skills in the field of medical education in order to develop educational programs and evaluation methods that apply to any phase of the continuum of medical education. This fellowship funding is for one year. Applicants may reapply for one additional year of funding; however, repeat or renewal applications will be considered on a competitive basis with all other applications received in the same year, and will require an outline of the progress of the previous year's work.
Requirements: Applicants must register in a Masters or PhD level university program leading to a degree in Education, or a subject field/program of study closely related to medical education. The candidate must be registered and attending university as a full time or part time student, either within Canada or abroad. Applicants must be Canadian citizens or permanent residents who, are Fellows of the Royal College in good standing or residents who, at the time of the application, are registered in a program accredited by the Royal College.
Geographic Focus: All States, Canada
Date(s) Application is Due: Sep 30
Amount of Grant: 45,000 U.S.D
Contact: Allison Shaw, Awards and Grants Section; (800) 668-3740, ext. 355 or (613) 730-8177; fax (613) 730-2410; awards@rcpsc.edu
Internet: http://rcpsc.medical.org/awards/fellow_e.php
Sponsor: Royal College of Physicians and Surgeons of Canada
774 Echo Drive
Ottawa, ON K1S 5N8 Canada

RCPSC Royal College Medal Award in Medicine **4060**
The Royal College of Physicians and Surgeons of Canada offers annually a competition for the RCPSC Medal in Medicine. This award provides national recognition for original work by a clinical investigator who has completed his/her training within the past 10 years. Any original work in clinical investigation or in the basic sciences relating to medicine is acceptable. The deadline for receipt of the manuscript is November 28, at 5 p.m. The recipient will receive a gold medal and $5,000.
Requirements: The candidate must be a Fellow of the Royal College and the work must have been done mainly in Canada. Only applicants who have not previously won the Medal are eligible.
Restrictions: Work which is directed or overseen by a supervisor such as in a graduate or post-doctoral training program is not eligible.
Geographic Focus: All States, Canada
Date(s) Application is Due: Nov 30
Amount of Grant: 5,000 U.S.D
Contact: Ines Martin, (800) 668-3740, ext. 212 or (613) 730-6212; fax (613) 730-2410; imartin@rcpsc.edu
Internet: http://rcpsc.medical.org/awards/medals_e.php
Sponsor: Royal College of Physicians and Surgeons of Canada
774 Echo Drive
Ottawa, ON K1S 5N8 Canada

RCPSC Royal College Medal Award in Surgery **4061**
The Royal College of Physicians and Surgeons of Canada offers annually a competition for the RCPSC Medal in Surgery. This award provides national recognition for original work by a clinical investigator who has completed his/her training within the past 10 years. Any original work in clinical investigation or in the basic sciences relating to surgery is acceptable. The deadline for receipt of the manuscript is November 30, at 5 p.m. The recipient will receive a gold medal and $5,000.
Requirements: The candidate must be a Fellow of the Royal College and the work must have been done mainly in Canada. Only applicants who have not previously won the Medal are eligible.
Restrictions: Work which is directed or overseen by a supervisor such as in a graduate or post-doctoral training program is not eligible.
Geographic Focus: All States, Canada
Date(s) Application is Due: Nov 30
Amount of Grant: 5,000 U.S.D
Samples: Dr. Mohit Bhandari, FRCSC, Burlington, Ontario, $5,000 - Trial of Reamed and Unreamed Intramedullary Nailing of Tibial Shaft Fractures; D Robert Nam, Toronto, Ontario, $5,000 - Expression of the TMPRSS2:ERG fusion gene predicts cancer recurrence after surgery for localized prostate cancer; Dr. Marc Ruel, Ottawa, Canada, $5,000 - Comparative effects of mesenchymal progenitor cells, endothelial progenitor cells or their combination on myocardial infarct regeneration and cardiac function.
Contact: Ines Martin, (800) 668-3740, ext. 212 or (613) 730-6212; fax (613) 730-2410; imartin@rcpsc.edu
Internet: http://rcpsc.medical.org/awards/medals_e.php
Sponsor: Royal College of Physicians and Surgeons of Canada
774 Echo Drive
Ottawa, ON K1S 5N8 Canada

RCPSC Royal College Visiting Professorships in Medical Research **4062**
The intent of this grant is to provide medical students and residents with the opportunity to meet established researchers from other centers in order to promote medical research as a career choice. It is intended that the visiting professor have the opportunity to interact with both undergraduate medical students and residents through rounds and other teaching sessions, and that the visit not be department-specific. Such interaction will provide the visiting professor with opportunities to expound on the challenges and rewards of a career in medical research. A $3,000 grant is available for each medical faculty.
Geographic Focus: All States, Canada
Date(s) Application is Due: Sep 30
Amount of Grant: 3,000 U.S.D
Contact: Ines Martin, (800) 668-3740, ext. 212 or (613) 730-6212; fax (613) 730-2410; imartin@rcpsc.edu
Internet: http://rcpsc.medical.org/awards/visiting-medresearch_e.php
Sponsor: Royal College of Physicians and Surgeons of Canada
774 Echo Drive
Ottawa, ON K1S 5N8 Canada

RCPSC Teasdale-Corti Humanitarian Award **4063**
The purpose of the Teasdale-Corti Humanitarian Award is to acknowledge and celebrate Canadian physicians who, while providing health care or emergency medical services, go beyond the accepted norms of routine practice, which may include exposure to personal risk. The recipient's action will exemplify altruism and integrity, courage and perseverance in the alleviation of human suffering. The award may be bestowed upon Canadian physicians whose current practice reflects the above description or upon those who have made such a contribution in the past. The Award will be open to Canadian physicians worldwide. Nominations will be widely solicited and may be submitted by any Fellow of the Royal College. The recipient will be presented with a suitably engraved memento that will be presented at the Royal College Annual Conference or other suitable venue. The travel and maintenance expenses of the award winner and companion will be paid in accordance with Royal College guidelines. The Royal College may make a donation to the relief organization with which the recipient is affiliated.
Geographic Focus: All States, Canada
Date(s) Application is Due: Sep 30
Contact: Ines Martin, (800) 668-3740, ext. 212 or (613) 730-6212; fax (613) 730-2410; imartin@rcpsc.edu
Internet: http://rcpsc.medical.org/awards/index.php
Sponsor: Royal College of Physicians and Surgeons of Canada
774 Echo Drive
Ottawa, ON K1S 5N8 Canada

RCPSC Traineeships **4064**
These Traineeships are provided for periods of personalized professional development (to learn a skill, surgical procedure or expand an area of competence). Applications should be received at least eight (8) weeks before the start of the traineeship. Up to $3,000 is available per applicant.
Restrictions: Applications are restricted to Fellows of the RCPSC in good standing.
Geographic Focus: All States, Canada
Amount of Grant: Up to 3,000 U.S.D
Contact: Allison Shaw, Awards and Grants Section; (800) 668-3740, ext. 355 or (613) 730-8177; fax (613) 730-2410; awards@rcpsc.edu
Internet: http://rcpsc.medical.org/awards/traineeships_e.php
Sponsor: Royal College of Physicians and Surgeons of Canada
774 Echo Drive
Ottawa, ON K1S 5N8 Canada

RCPSC Walter C. MacKenzie Fellowship in Surgery **4065**
The fellowship will enable the recipient to visit one or more centers to acquire surgical research or clinical surgical training. Knowledge gained should have a direct application to the recipient's clinical surgical practice. One or more fellowship are offered annually, depending on budget. Tenable for visits of two weeks to 12 months duration; visits may be broken into separate time blocks but must be completed within two years. An amount equivalent to 75% of the award will be given at the beginning of the fellowship, and the remainder upon receipt of a final report at the end of the fellowship. The fellowship may be used to supplement financial assistance granted from other sources.
Requirements: Applications are accepted from practicing surgeons who are fellows of the Royal College in good standing at the time of application and who reside in Canada. Practicing surgeons are those who have spent at least 50 percent of their professional time in active clinical surgery over the last five years.
Geographic Focus: All States, Canada
Date(s) Application is Due: Sep 30
Amount of Grant: 7,500 - 180,000 U.S.D
Contact: Allison Shaw, Awards and Grants Section; (800) 668-3740, ext. 355 or (613) 730-8177; fax (613) 730-2410; awards@rcpsc.edu
Internet: http://rcpsc.medical.org/awards/mackenzie_e.php
Sponsor: Royal College of Physicians and Surgeons of Canada
774 Echo Drive
Ottawa, ON K1S 5N8 Canada

RCPSC Wightman Visiting Professorship in Medicine **4066**
The Wightman Visiting Professor participates in undergraduate and postgraduate teaching and in discussions with peers, both in internal medicine and in the professor's particular area of academic interest. The visiting professor spends one or two weeks visiting two or three Canadian faculties of medicine. The Professorship is supported by a grant of $7,000.
Geographic Focus: All States, Canada
Date(s) Application is Due: Sep 30
Amount of Grant: 7,000 U.S.D
Contact: Ines Martin, (800) 668-3740, ext. 212 or (613) 730-6212; fax (613) 730-2410; imartin@rcpsc.edu
Internet: http://rcpsc.medical.org/awards/visiting-wightman_e.php
Sponsor: Royal College of Physicians and Surgeons of Canada
774 Echo Drive
Ottawa, ON K1S 5N8 Canada

Reader's Digest Partners for Sight Foundation Grants **4067**
The foundation awards grants to nonprofit organizations that serve the blind and visually impaired. Multiyear grants and smaller one-year grants are made to organizations that work on the local level. The listed deadline is for preliminary applications for multiyear grants ($1 million maximum); preliminary applications for one-year grants may be submitted at anytime. Grants are awarded every three years.
Requirements: 501(c)3 nonprofit organizations are eligible.
Date(s) Application is Due: Sep 15
Amount of Grant: Up to 1,000,000 U.S.D
Contact: Dianna Kelly-Naghizadeh, (914) 244-5830 or (800) 877-5293; fax (914) 244-7481; PartnersForSight@rd.com
Internet: http://www.rd.com/corporate/rd_fundblind.html
Sponsor: Reader's Digest Partners for Sight Foundation
Reader's Digest Road
Pleasantville, NY 10570-7000

Regence Foundation Access to Health Care Grants **4068**
Grants in this category support the health care safety-net and insurance access programs, including efforts to: increase access to health and medical care for low-income, uninsured or underinsured individuals; provide a medical home to those who would otherwise go without; improve medical outcomes; and help eligible individuals and families navigate the process of accessing available health insurance coverage.
Requirements: The Foundation funds: organizations that are nonprofit and tax-exempt under 501(c)3 of the Internal Revenue Service Code (IRC) and defined as a public charity under 509(a) 1, 2, or 3 (types I, II, or a functionally integrated type III); organizations that serve residents of Idaho, Oregon, Washington and/or Utah; accredited schools or universities; units of government and public agencies; and programs and services that align with our mission and program areas.
Restrictions: The Foundation does not fund: capital construction; award dinners, athletic events, competitions, special events or tournaments; conferences or seminars; religious organizations for religious purposes; political causes, candidates, organizations or campaigns; grants to individuals; grants to 509(a)3, type III supporting organizations that are not functionally integrated; or grants to organizations that are not doing significant work in Idaho, Oregon, Washington and/or Utah.
Geographic Focus: Idaho, Oregon, Utah, Washington
Amount of Grant: 50,000 - 115,000 U.S.D
Contact: Monique Barton, Executive Director; (503) 276-1965; mxbarto@regence.com or RegenceFoundation@regence.com
Internet: http://www.regencefoundation.org/programs/transformation.html
Sponsor: Regence Foundation
100 SW Market Street, MS E-8T
Portland, OR 97201

Regence Foundation Health Care Community Awareness and Engagement Grants 4069

At the core of this program area is a laser-focus on increasing the accessibility and usability of information for consumers to make informed health care decisions. Grants in this category support efforts that encourage consumers to seek information about the value and quality of their health care, and develop tools to access that information, such as: community collaborative pilot projects; campaigns promoting consumer education and engagement; and research on consumer use of health care information.

Requirements: The Foundation funds: organizations that are nonprofit and tax-exempt under 501(c)3 of the Internal Revenue Service Code (IRC) and defined as a public charity under 509(a) 1, 2, or 3 (types I, II, or a functionally integrated type III); organizations that serve residents of Idaho, Oregon, Washington and/or Utah; accredited schools or universities; units of government and public agencies; and programs and services that align with our mission and program areas.

Restrictions: The Foundation does not fund: capital construction; award dinners, athletic events, competitions, special events or tournaments; conferences or seminars; religious organizations for religious purposes; political causes, candidates, organizations or campaigns; grants to individuals; grants to 509(a)3, type III supporting organizations that are not functionally integrated; or grants to organizations that are not doing significant work in Idaho, Oregon, Washington and/or Utah.

Geographic Focus: Idaho, Oregon, Utah, Washington

Amount of Grant: 45,000 - 150,000 U.S.D

Contact: Monique Barton, Executive Director; (503) 276-1965; mxbarto@regence.com or RegenceFoundation@regence.com

Internet: http://www.regencefoundation.org/programs/transparency.html

Sponsor: Regence Foundation
100 SW Market Street, MS E-8T
Portland, OR 97201

Regence Foundation Health Care Connections Grants 4070

Grants in this category support general health programs that are innovative and solve community-identified problems, including efforts to: manage chronic conditions; make healthy lifestyle choices; and serve a target population whose health care needs may be unmet and/or who may be experiencing health disparities.

Requirements: The Foundation funds: organizations that are nonprofit and tax-exempt under 501(c)3 of the Internal Revenue Service Code (IRC) and defined as a public charity under 509(a) 1, 2, or 3 (types I, II, or a functionally integrated type III); organizations that serve residents of Idaho, Oregon, Washington and/or Utah; accredited schools or universities; units of government and public agencies; and programs and services that align with our mission and program areas.

Restrictions: The Foundation does not fund: capital construction; award dinners, athletic events, competitions, special events or tournaments; conferences or seminars; religious organizations for religious purposes; political causes, candidates, organizations or campaigns; grants to individuals; grants to 509(a)3, type III supporting organizations that are not functionally integrated; or grants to organizations that are not doing significant work in Idaho, Oregon, Washington and/or Utah.

Geographic Focus: Idaho, Oregon, Utah, Washington

Amount of Grant: 15,000 - 50,000 U.S.D

Contact: Monique Barton, Executive Director; (503) 276-1965; mxbarto@regence.com or RegenceFoundation@regence.com

Internet: http://www.regencefoundation.org/programs/transformation.html

Sponsor: Regence Foundation
100 SW Market Street, MS E-8T
Portland, OR 97201

Regence Foundation Improving End-of-Life Grants 4071

The Regence Foundation works to transform health care and address core problems in our health care system with innovative solutions. The Foundation supports a transformational health care system where consumers partner with their physicians and other providers to make health care decisions based on what's valuable to them - such as end-of-life decisions - just as they do when making other economic decisions.

Requirements: The Foundation funds: organizations that are nonprofit and tax-exempt under 501(c)3 of the Internal Revenue Service Code (IRC) and defined as a public charity under 509(a) 1, 2, or 3 (types I, II, or a functionally integrated type III); organizations that serve residents of Idaho, Oregon, Washington and/or Utah; accredited schools or universities; units of government and public agencies; and programs and services that align with our mission and program areas.

Restrictions: The Foundation does not fund: capital construction; award dinners, athletic events, competitions, special events or tournaments; conferences or seminars; religious organizations for religious purposes; political causes, candidates, organizations or campaigns; grants to individuals; grants to 509(a)3, type III supporting organizations that are not functionally integrated; or grants to organizations that are not doing significant work in Idaho, Oregon, Washington and/or Utah.

Geographic Focus: Idaho, Oregon, Utah, Washington

Contact: Monique Barton, Executive Director; (503) 276-1965; mxbarto@regence.com or RegenceFoundation@regence.com

Internet: http://www.regencefoundation.org/programs/endOfLife.html

Sponsor: Regence Foundation
100 SW Market Street, MS E-8T
Portland, OR 97201

Regence Foundation Tools and Technology Grants 4072

At the core of this program area is a laser-focus on increasing the accessibility and usability of information for consumers to make informed health care decisions. Grants in this category support innovative technology solutions that work to overcome practical barriers and improve the quality and safety of health care. Projects might include: multi-stakeholder technology collaborations; not-for-profit technology initiatives that include providers in development and testing; research on the usage and effects of technology collaborations in health care; and approaches that address patients who receive services from multiple providers across the spectrum.

Requirements: The Foundation funds: organizations that are nonprofit and tax-exempt under 501(c)3 of the Internal Revenue Service Code (IRC) and defined as a public charity under 509(a) 1, 2, or 3 (types I, II, or a functionally integrated type III); organizations that serve residents of Idaho, Oregon, Washington and/or Utah; accredited schools or universities; units of government and public agencies; and programs and services that align with our mission and program areas.

Restrictions: The Foundation does not fund: capital construction; award dinners, athletic events, competitions, special events or tournaments; conferences or seminars; religious organizations for religious purposes; political causes, candidates, organizations or campaigns; grants to individuals; grants to 509(a)3, type III supporting organizations that are not functionally integrated; or grants to organizations that are not doing significant work in Idaho, Oregon, Washington and/or Utah.

Geographic Focus: Idaho, Oregon, Utah, Washington

Contact: Monique Barton, Executive Director; (503) 276-1965; mxbarto@regence.com or RegenceFoundation@regence.com

Internet: http://www.regencefoundation.org/programs/transparency.html

Sponsor: Regence Foundation
100 SW Market Street, MS E-8T
Portland, OR 97201

Regenstrief General Internal Medicine - General Pediatrics Research Fellowship 4073

The Regenstrief Institute and the Division of General Internal Medicine offers a 2-year research oriented fellowship (with an optional third year) to prepare physicians for academic careers in general internal medicine and general pediatrics. Fellows receive formal didactic training through classes and seminars taught by faculty from both the Institute as well as Indiana University.

Requirements: Candidates for the Fellowship must hold the M.D. or equivalent degree, and must be U.S. citizens or permanent residents. Minority applicants are encouraged to apply.

Contact: Kurt Kroenke, Program Director; (317) 630-7447; fax (317) 630-6611; KKroenke@regenstrief.org

Internet: http://www.regenstrief.org/training/general-internal-medicine-pediatric-research-fellowship-program

Sponsor: Regenstrief Institute
410 West 10th Street, Suite 2000
Indianapolis, IN 46202-3012

Regenstrief Geriatrics Fellowship 4074

The Department of Medicine and the Division of General Internal Medicine and Geriatrics offer a 2-year fellowship in Geriatric Medicine (with an optional third year). Residents who have completed three years of training in internal medicine or family practice and who are board eligible or board certified may apply for the fellowship. Fellows will be trained in clinical geriatrics, research, teaching, and/or administration in preparation for a successful career in academic geriatric medicine. Formal instruction will be provided by a structured curriculum with goals and objectives that will be met by clinical experiences in a variety of settings, didactic seminars to augment the clinical experiences, and seminars in research, teaching and administration. For further information on the Geriatric Fellowship contact the Program Director.

Contact: Glenda Westmorland, Program Director; (317) 630-6906; fax (317) 630-2667; GWestmorland@regenstrief.org

Internet: http://www.regenstrief.org/training/geriatrics-fellowship-program

Sponsor: Regenstrief Institute
410 West 10th Street, Suite 2000
Indianapolis, IN 46202-3012

Regenstrief Master Of Science In Clinical Research/CITE Program 4075

Indiana University has received funding from the National Institute of Health through a K-30 grant to develop and implement the Clinical Investigation and Translational Education (CITE) program. The purpose of this program is to prepare health care professionals for a career in clinical research. Following completion of the program, graduates can embark on a career in clinical research with the skills necessary to successfully compete for grant funding, conduct and analyze research findings, and publish their work in scientific journals. By participating in the program, CITE trainees will complete a two-year formal clinical research curriculum, at the end of which they will receive a Master of Science in Clinical Research degree. The program is designed for participants that already have or are in training for a doctoral degree (e.g., MD, PhD, DNS, DDS, DPT).

Requirements: Most participants will already have a doctoral degree (e.g., MD, PhD, DNS, DDS, or DPT) or be in training for a doctoral degree. Potential applicants include, but are not limited to, the following: fellows or other health care professional trainees who have substantial protected time for clinical research; junior faculty who have career awards or support from their division or department head for participation in the CITE program; other faculty who previously have focused on basic research or other academic activities but now are committed to pursuing a career in clinical research CITE constitutes the formal didactic requirements for certain types of federal training

grants (such as K-23's) and other career awards. Individuals in these programs would be expected to participate in CITE unless they were enrolled in some other type of formal degree program.
Contact: Kurt Kroenke, Program Director; (317) 630-7447; fax (317) 630-6611; KKroenke@regenstrief.org
Internet: https://www.indianactsi.org/cite/citeapplication
Sponsor: Regenstrief Institute
1050 Wishard Boulevard, RG-6
Indianapolis, IN 46202-2872

Rehabilitation Nursing Foundation Research Grants 4076
The Foundation makes available research grants totaling $44,500. RNF offers funding of up to $30,000 for rehabilitation nursing research which is awarded in the form of multiple grants. A New Investigator Award, offered to encourage nurses who are novice researchers, will be awarded for up to $10,000. Up to two grants from the remaining amount will be awarded to recipients, who will be designated RNF Research Fellows. Research proposals that address clinical practice or educational or administrative dimensions of rehabilitation nursing are accepted. Quantitative and qualitative research projects will be reviewed. Funding may be awarded in the form of multiple grants. The Research Grant Committee reviews proposals once a year. Proposals must be postmarked by February 1. Grant recipients will be notified in September, and funding begins January of the following year for the selected proposals.
Requirements: The principal investigator for the research project must be a registered nurse who is active in rehabilitation or who demonstrates interest in and significant contributions to rehabilitation nursing. Membership in the Association of Rehabilitation Nurses is not required. Graduate student researchers are eligible for funding.
Date(s) Application is Due: Feb 1
Amount of Grant: Up to 44,500 U.S.D
Samples: Maryann Abendroth, Research Title: Understanding Caregivers' Decision Process When Choosing to Institutionalize a Relative with Parkinson's Disease; Cheryl Gies, Research Title: Web-based Psychosocial Assessment for Caregivers of Persons with Dementia: A Feasibility Study; Susan Newman, Research Title: Community Integration after Spinal Cord Injury: Identifying Barriers and Facilitators through Photovoice.
Contact: Program Director; (847) 375-4710 or (800) 229-7530; info@rehabnurse.org
Internet: http://www.rehabnurse.org/research/researchgrants.html
Sponsor: Rehabilitation Nursing Foundation
4700 W Lake Avenue
Glenview, IL 60025-1485

Rehabilitation Nursing Foundation Scholarships 4077
ARN offers financial assistance in the form of two distinct types of scholarships: a $1,000 award to nurses working toward a Bachelor of Science in Nursing (BSN) degree; and the Mary Ann Mikulic Scholarship, which covers one full tuition to the Professional Rehabilitation Nursing Course. Each award has a deadline of June 1.
Requirements: The BSN Scholarship eligibility requirements include: membership and involvement in ARN; enrollment in a Bachelor of Science in Nursing (BSN) program in good standing and successful completion of at least one course; current practice in rehabilitation nursing; and a minimum of 2 years experience in rehabilitation nursing. The Mary Ann Mikulic Scholarship eligibility requirements include: being a registered nurse with a current license; currently practicing in the specialty of rehabilitation nursing; and the ability to meet all other financial responsibilities incurred by participating in the course.
Date(s) Application is Due: Jun 1
Amount of Grant: 1,000 U.S.D
Contact: Gayle Elliott, (847) 375-4710 or (800) 229-7530; gelliott@connect2amc.com
Internet: http://www.rehabnurse.org/awards/allscholar.html
Sponsor: Rehabilitation Nursing Foundation
4700 W Lake Avenue
Glenview, IL 60025-1485

Rehab Therapy Foundation Grants 4078
The foundation awards grants to North Carolina nonprofit organizations. Areas of priority include organizations that serve children with developmental disabilities and their families; and impact inadequately insured or underserved children with developmental disabilities and their families. Grant applications are available year round at the Foundations website however, applications will only be accepted once a year beginning in January and must be received by February 28th. Award recipients will be notified by mail on or about May 15th.
Requirements: Applicant must be a nonprofit 501(c)3 organization or public school in North Carolina.
Restrictions: Proposals benefiting religious organizations, university level education, research, events, individuals, trusts, or political causes will not be funded.
Geographic Focus: North Carolina
Date(s) Application is Due: Feb 28
Amount of Grant: Up to 50,000 U.S.D
Contact: Kimberly D. Reilly, Wachovia Philanthropic Services Supervisor; info@rehabtherapyfoundation.org
Internet: http://www.rehabtherapyfoundation.org/grant.htm
Sponsor: Rehab Therapy Foundation
150 Fayetteville Street, Seventh Floor
Raleigh, NC 27602

Reinberger Foundation Grants 4079
Clarence T. Reinberger was born in 1894 on Cleveland's west side, and began his business career in the 1920's as a pioneer in the automobile replacement parts field. Starting as a clerk in the Cleveland retail store of the National Automotive Parts Association's (NAPA) Automotive Parts Company, he became president of that company in 1948. In the 1960's the Automotive Parts Company merged with the Genuine Parts Company of Atlanta. Mr. Reinberger held the position of Chairman of the Board of Genuine Parts Company until his death in 1968. Louise Fischer Reinberger was born in Germany. After graduation from high school in the United States, she was employed by the Halle Brothers Company, a large Cleveland department store. The Reinberger Foundation was established by Mr. and Mrs. Reinberger in 1966. Mr. Reinberger left a substantial bequest to the Foundation at his death in 1968. Upon Mrs. Reinberger's death in 1984, the major portion of her estate was also bequeathed to the Foundation. Although the Reinbergers had no children, the foundation continues to be managed by several generations of Mr. Reinberger's family. Since its inception, the foundation has distributed over $91,000,000 to the non-profit community. The foundation divides its support among the following program areas: Arts, Culture, and Humanities; Education; Human Service - Health; and Human Service - Other. Categories supported under Arts, Culture and Humanities include museums, visual arts, performing arts, media and communication, arts education, zoos, and public recreation. Categories supported under Education include K-12 schools, early childhood education, adult education/literacy, libraries, and workforce development. Categories supported under Human Service - Health include hospitals and clinics, substance abuse prevention/treatment, medical and disease research, disease prevention, and speech and hearing. Categories supported under Human Service - Other include children and youth services, residential and home care, emergency food programs, youth development, domestic violence, and temporary housing/homeless shelters. Letters of Inquiry for grants of any type may be submitted according to the following program-area schedule: Education - March 1; Human Service (Health) - June 1; Human Service (Other) - September 1; and Arts, Culture, and Humanities - December 1. Letters of Inquiry may either be sent by U.S. Mail or emailed and are due in the foundation office by the deadline date. Full proposals are accepted only at the request of the foundation.
Requirements: Applicants must be 501(c)3 organizations. Preferential consideration is given to organizations serving Northeast Ohio, or the greater-Columbus area.
Restrictions: No loans are made, nor are grants given to individuals. The Reinberger Foundation does not make more than one grant to a particular organization during a given calendar year, nor will new proposals be considered until existing multi-year commitments have been paid.
Geographic Focus: Ohio
Date(s) Application is Due: Mar 1; Jun 1; Sep 1; Dec 1
Samples: Cleveland Zoological Society, Cleveland, Ohio, $100,000 (2010); Cuyahoga Community College Foundation, Cleveland, Ohio, $37,500 (2010); The Free Medical Clinic of Greater Cleveland, Cleveland, Ohio, $50,000 (2010); Community Kitchen, Inc., Columbus, Ohio, $5,000 (2010).
Contact: Karen R. Hooser, President; (216) 292-2790; fax (216) 292-4466; info@reinbergerfoundation.org
Internet: http://reinbergerfoundation.org/apply.html
Sponsor: Reinberger Foundation
30000 Chagrin Boulevard #300
Cleveland, OH 44122

Religion and Health: Effects, Mechanisms, and Interpretation RFP 4080
The Center seeks proposals to elucidate how religious involvement (religious attendance, worship, altruistic and caring activities) influences individual and community health. The goals of this research program are to: quantitatively document effects on health (where health is broadly defined); clarify the biological, social, psychological mechanisms involved; and interpret what the findings mean for individual, congregational, and community health through trans-disciplinary collaboration and deliberation. This RFP focuses on research that examines the contribution of context and caring as part of the mechanism that accounts for the effects of religion on health, in particular the effects of membership and practices within a faith community. Letters of Intent are due July 15. Thirty finalists will be invited to submit full proposals; seven $200,000 grants will be awarded to outstanding research proposals. Finalists will be selected by Aug 5, with full proposals due by November 1, and awards announced on January 1.
Date(s) Application is Due: Jul 15
Amount of Grant: 200,000 U.S.D
Contact: Catherine Craver, Assistant Director; (919) 660-7578; fax (919) 684-8569; catherine.craver@duke.edu
Internet: http://www.dukespiritualityandhealth.org/proposals/
Sponsor: Center for Spirituality, Theology and Health at Duke University Medical Center
Box 3825, Duke University Medical Center
Durham, NC 27710

Retirement Research Foundation General Program Grants 4081
The Retirement Research Foundation, based in Chicago, was established in the 1950s and endowed in 1978 by the late John D. MacArthur. The Foundation is devoted solely to serving the needs of older persons in the U.S. and enhancing their quality of life. The Foundation supports a range of programs and special initiatives designed to: improve access to and quality of community-based and residential health and long-term care; promote economic security for all older adults by strengthening social insurance, pension, and personal savings programs; and support adequate training of, and compensation for, those already working directly with older persons and their families to bring higher quality of care to larger numbers of older adults. The Foundation's historic interest in innovative projects continues. RRF also has a strong interest in projects that implement or adapt proven models

that address clearly identified needs and gaps. Requests for support of projects focusing on advocacy, research, or education and training will be considered from anywhere in the United States. Direct service requests will be considered only from organizations in these seven states: Illinois, Indiana, Iowa, Kentucky, Missouri, Wisconsin, and Florida.
Requirements: The Retirement Research Foundation does not use a standard application form, proposals for funds should include the following elements: cover page; Summary; project significance; statement of objectives; description of methodology; dissemination; budget and timetable; plans for continued support; list of personnel; applicant organization; tax exempt status. Application submission must include: the signature of the chief executive officer of the applicant organization; the proposal and attachments must be submitted in triplicate; only one copy of the annual report and financial/audit are needed.
Restrictions: Funding for vans is limited to applicants from Illinois and Florida. Funding is not generally available for: biomedical research; computer equipment; conferences, publications or travel unless they are components of other larger Foundation-funded projects; construction of facilities; dissertation research; endowment or developmental campaigns; general operating expenses of established organizations; grants or scholarships to individuals; production of films and videos; projects of governmental organizations except for state universities, area agencies on aging, and programs of the Veterans Administration; projects outside the United States
Date(s) Application is Due: Feb 1; May 1; Aug 1
Samples: Alivio Medical Center, Chicago, Illinois $255,000 - three-year grant to design and implement a chronic disease self-management program for seniors; Family Caregiver Alliance, San Francisco, California $148,650 - An 18-month grant to facilitate advocacy around state and federal policies designed to assist caregivers through organization and dissemination of policy and resources; Organization of New City, Chicago, Illinois $150,000 - Three-year grant to help low-income elderly families cope with safety, utility, and housing issues;
Contact: Irene Frye; (773) 714-8080; fax (773) 714-8089; info@rrf.org
Internet: http://www.rrf.org/generalProgram.htm
Sponsor: Retirement Research Foundation
8765 W Higgins Road, Suite 430
Chicago, IL 60631-4170

Rett Syndrome Research Foundation Grants **4082**
The foundation (RSRF) awards research grants and postdoctoral fellowships for international biomedical research for the treatment and cure of Rett Syndrome. Relevant disciplines include but are not limited to developmental biology, molecular biology, genetics, biochemistry, cellular physiology, neurology, immunology, pharmacology, gastroenterology, orthopedics, and experimental neurotherapeutics. Treatment-related research is especially encouraged. RSRF seeks to encourage research that encompasses innovative approaches and techniques that will lead to NIH and other funding sources. The foundation is committed to funding senior investigators as well as junior, provided they have a supportive environment. Teams working together from different disciplines, institutions as well as industry are encouraged to apply. Grants are made for technical support, supplies, equipment, and relevant travel. No salary support is provided for the Principal Investigator. Indirect costs of up to 10 percent of direct costs may be included. Letters of intent must be received by the listed application deadline; full proposals are by invitation. Guidelines are available online.
Date(s) Application is Due: Jan 31
Amount of Grant: Up to 50,000 U.S.D
Samples: Dr. Bruria Ben Zeev and Dr. Eva Gak, Sheba Medical Ctr (Israel)— for research, Molecular Diagnosis of Rett Syndrome: An Alternative Scheme, $50,000 one-year award; Dr. Reiko Maki Fitzsimonds, Yale U School of Medicine (New Haven, CT)—for research, Relationship Between Synaptic Structure and Plasticity in Dendritic Neuropathologies Associated with Rett Syndrome, $100,000 two-year award; Dr. Michael Greenberg, Children's Hospital Boston (MA)—for research Neuronal Activity-Dependent Regulation of MeCP2, $100,000 two=year award.
Contact: Monica Coenraads, Director of Research; (203) 445-9233; fax (203) 445-9234; monica@rsrf.org
Internet: http://www.rsrf.org/researchers/4.7.1.html
Sponsor: Rett Syndrome Research Foundation
4600 Devitt Drive
Cincinnati, OH 45246

Reynolds and Reynolds Associate Foundation Grants **4083**
Reynolds has a long-standing reputation as a leading corporate citizen within the communities where it has a large presence. The company encourages its associates to roll up their sleeves and get involved in organizations they believe in, and in which they often are involved personally. The mission of The Foundation is to provide a vehicle for Dayton associates to improve health and human services in the local communities. Since 1956, the Foundation has provided financial support to area health and human service agencies throughout an eight-county area. Focus areas include: health and human services with emphasis on domestic violence and child abuse; youth-at-risk; hunger; issues of the elderly; homelessness; literacy; and life-threatening illnesses.
Requirements: Giving is limited to the eight-county region in and around Dayton, Ohio.
Restrictions: There is no support available for non 501(c)3 organizations, sectarian organizations with a predominately religious purpose, fraternal or veterans' organizations, individual primary or secondary schools (except for occasional special projects), or tax-supported universities and colleges (except for occasional special projects) No grants are offered to individuals, or for endowments, courtesy advertising, fund raising events, deficit or debt retirement, or capital campaigns.
Geographic Focus: Ohio
Date(s) Application is Due: Jan 7; Mar 11; Jun 10; Sep 9
Amount of Grant: Up to 40,000 U.S.D
Samples: Dayton Fund for Home Rehab, $2,500 - care for the Elderly, Homelessness, Youth at Risk; Project READ, $4,000 - Literacy, Youth at Risk; A Special Wish Foundation, $3,500 - Life Threatening Illness.
Contact: Alice Davisson, Treasurer; (937) 485-4409 or (937) 485-8138; alice_davisson@reyrey.com
Internet: http://www.reyrey.com/company/community/associate_foundation.asp
Sponsor: Reynolds and Reynolds Associate Foundation
P.O. Box 2608
Dayton, OH 45401-2608

RGK Foundation Grants **4084**
RGK Foundation awards grants in the broad areas of Education, Community, and Medicine/Health. The foundation's primary interests within Education include programs that focus on formal K-12 education (particularly mathematics, science and reading), teacher development, literacy, and higher education. Within Community, the foundation supports a broad range of human services, community improvement, abuse prevention, and youth development programs. Human service programs of particular interest to the foundation include children and family services, early childhood development, and parenting education. The foundation supports a variety of Community Improvement programs including those that enhance non-profit management and promote philanthropy and voluntarism. Youth development programs supported by the foundation typically include after-school educational enrichment programs that supplement and enhance formal education systems to increase the chances for successful outcomes in school and life. The foundation is also interested in programs that attract female and minority students into the fields of mathematics, science, and technology. The foundation's current interests in the area of Medicine/Health include programs that promote the health and well-being of children, programs that promote access to health services, and Foundation-initiated programs focusing on ALS.
Requirements: Although there are no geographic restrictions to the foundation's grantmaking program, the foundation no longer accepts unsolicited requests for international agencies or programs. All applicants must complete an electronic Letter of Inquiry from the Web site as the first step. RGK Foundation will entertain one electronic Letter of Inquiry (LOI) per organization in a twelve-month period. While the foundation occasionally awards grants for operating expenses, capital campaigns, endowments, and international projects, such grants are infrequent and usually initiated by the foundation. Grants are made only to nonprofit organizations certified as tax exempt under Sections 501(c)3 or 170(c) of the IRS code. Hospitals, educational institutions, and governmental institutions meeting these requirements are eligible to apply.
Restrictions: The foundation refrains from funding annual funds, galas or other special-event fundraising activities; debt reduction; emergency or disaster relief efforts; dissertations or student research projects; indirect/administrative costs; sectarian religious activities, political lobbying, or legislative activities; institutions that discriminate on the basis of race, creed, gender, or sexual orientation in policy or in practice; or loans, scholarships, fellowships, or grants to individuals; unsolicited requests for international organizations or programs or for ALS research projects.
Amount of Grant: 1,000 - 25,000 U.S.D
Samples: 2008: COMMUNITY - Ark House Foundation (Richardson, TX) to provide low cost, temporary housing for patients and their families while undergoing long-term medical treatment in the Dallas area; Rider University (Lawrenceville, NJ) to expand the Minding Our Business after-school and summer entrepreneurship program serving 6th-8th grade students in Trenton public schools; YWCA of Greater Atlanta (Atlanta, GA) expansion of Teen Girls in Technology Program to middle school girls at Title One schools. EDUCATION - BookSpring (Austin, TX) to build regional alliance's capacity following merger with Reading is Fundamental of Austin; Generations Incorporated (Boston, MA) to expand the Classroom Literacy program using older adults as literacy aids in Boston Public elementary schools; The Field Museum (Chicago, IL) to improve curriculum in Chicago Public Schools as part of the High School Transformation Project; MEDICINE/HEALTH - Ambulatory Surgery Access Coalition (San Francisco, CA) to provide essential outpatient surgery to uninsured through a volunteer network; Marathon Kids (Austin, TX) to enhance national program capacity through technology upgrades for the website; Travis County Medical Society Foundation (Austin, TX) support program and ongoing enhancements for Project Access that will add dental services and Bilingual Patient Service Coordinator.
Contact: Suzanne Haffey, Grants Associate; (512) 474-9298; fax (512) 474-7281; shaffey@rgkfoundation.org
Internet: http://www.rgkfoundation.org/public/guidelines
Sponsor: RGK Foundation
1301 West 25th Street, Suite 300
Austin, TX 78705-4236

Rhode Island Foundation Grants **4085**
The foundation seeks to promote philanthropic activities that will improve the living conditions and well-being of the inhabitants of Rhode Island. Grants for capital and operating purposes principally to agencies working in the fields of education, health care, the arts and cultural affairs, youth, the aged, social services, urban affairs, historic preservation, and the environment.
Requirements: Rhode Island nonprofits organizations may apply. See Foundations website for additional guidelines.
Restrictions: The foundation does not make grants for endowments, research, religious groups for religious purposes, hospital equipment, capital needs of health organizations, or to educational institutions for general operating expenses.
Geographic Focus: Rhode Island

Samples: Ocean State Action Fund, Cranston, RI, $50,000 - for continued funding for grassroots organizing, coalition work, and public communications in support of affordable, quality health care for all Rhode Islanders; Community Housing Land Trust of Rhode Island, Providence, RI, $50,000 - for new program targeting foreclosed properties or those close to being foreclosed upon; Sakonnet Preservation Association, Little Compton, RI, $25,000 - for three-community partnership to preserve drinking water quality and open space in Newport County;
Contact: Owen Heleen; (401) 427-4009 or (401) 274-4564; oheleen@rifoundation.org
Internet: http://www.rifoundation.org/Nonprofits/GrantOpportunities/tabid/175/Default.aspx
Sponsor: Rhode Island Foundation
1 Union Station
Providence, RI 02903

Rice Foundation Grants **4086**
The foundation awards grants to Illinois nonprofit organizations in its areas of interest, including civic affairs, higher education, hospitals, libraries, medical education, and youth programs. Project and operating support are available. There are no application forms or deadlines.
Requirements: Applicants should submit a proposal with directly to the Foundation including the following: statement of problem project will address; copy of IRS Determination Letter; brief history of organization and description of its mission; detailed description of project and amount of funding requested.
Geographic Focus: Illinois
Amount of Grant: 25,000 - 1,000,000 U.S.D
Samples: John G. Shedd Aquarium, Chicago, IL $1,000,000 - for general support; Northwestern University, Evanston, IL $386,700 - for general support; Big Shoulders Fund, Chicago, IL $30,000 - for general support;
Contact: Peter Nolan, President; (847) 581-9999
Sponsor: Daniel F. and Ada L. Rice Foundation
8600 Gross Point Road
Skokie, IL 60077-2151

Richard and Helen DeVos Foundation Grants **4087**
The foundation supports nonprofit organizations primarily in western Michigan and central Florida in its areas of interest, including religious agencies and churches and education and outreach, social services, the arts, public policy, and health care. Types of support include general operating support, continuing support, annual campaigns, capital campaigns, building construction/renovation, program development, seed grants, and matching funds.
Requirements: Application forms are not required. Applicants should submit the following: copy of IRS Determination Letter; copy of current year's organizational budget and/or project budget. The Board meets every 3 months, submit your proposal two weeks prior to the review (contact Foundation for deadline dates). Mail applications to: 126 Ottawa Avenue, N.W., Suite 500, Grand Rapids, MI 49503
Restrictions: No grants to individuals.
Geographic Focus: Florida, Michigan
Amount of Grant: 300 - 5,000,000 U.S.D
Samples: Michigan State University, East Lansing, Michigan, $5,000,000 - For general support, payable over 1 year); Salvation Army of West Palm Beach, West Palm Beach, Florida, $50,000 - For general support, payable over 1 year;
Contact: Ginny Vander Hart; (616) 643-4700; fax (616) 774-0116; virginiav@rdvcorp.com
Sponsor: Richard and Helen DeVos Foundation Grants
P.O. Box 230257
Grand Rapids, MI 49523-0257

Richard and Rhoda Goldman Fund Grants **4088**
The fund is interested in supporting programs that will have a significant positive impact in an array of fields, including culture, the environment, population, Jewish affairs, violence prevention, children and youth, the elderly, and social and human services. Types of support include general operating support, continuing support, capital campaigns, program development, and seed money. Funds will be allocated primarily in the San Francisco Bay area of California and to organizations benefiting this area. There are no application deadlines.
Requirements: Nonprofit organizations serving primarily the San Francisco Bay area of California are eligible.
Restrictions: The fund does not accept applications for research or award grants or scholarships for individuals, conferences, documentary films, or fund-raisers. Unsolicited proposals for support of arts organizations or institutions of primary, secondary, or higher education will not be accepted.
Geographic Focus: California
Amount of Grant: 1,000 - 500,000 U.S.D
Samples: San Francisco Beautification (CA)—for efforts to increase green schoolyards, $10,000; American Jewish Joint Distribution Committee (New York, NY)—for relief efforts following the earthquake and tsunamis that struck Southeast Asia in 2004, $30,000; United States Public Interest Research Group Education Fund (Washington, DC)—for efforts to increase attention to the role of hard money in federal political campaigns, $20,000; Community Awareness and Treatment Services (San Francisco, CA)—for support of this 24-hour shelter and transitional-housing program for women, $25,000.
Contact: Sam Salkin; (415) 345-6300; fax (415) 345-9686; info@goldmanfund.org
Internet: http://www.goldmanfund.org
Sponsor: Richard and Rhoda Goldman Fund
211 Lincoln Boulevard, P.O. Box 29924
San Francisco, CA 94129

Richard and Susan Smith Family Foundation Grants **4089**
The foundation awards grants to Massachusetts nonprofit organizations in its areas of interest, including arts and culture, children and youth, disadvantaged (economically), education (early childhood, elementary, and higher), minorities, biomedical research, health care and social services delivery. Types of support include annual campaigns, building construction/renovation, capital campaigns, curriculum development, general operating grants, program/project support, seed grants, and service delivery programs. There are no application forms.
Requirements: Massachusetts nonprofit organizations serving the greater Boston area are eligible.
Restrictions: Grants do not support political activities; religious activities; individuals; or requests for deficit financing or endowment funds.
Geographic Focus: Massachusetts
Date(s) Application is Due: Mar 15; Aug 15
Amount of Grant: 5,000 - 238,000 U.S.D
Samples: Compassionate Care ALS, West Falmouth MA $25,000 - for the purchase of a cargo van to transport adaptive equipment to and from patient homes; United Teen Equality Center, Lowell, MA $944,780 - funding provides violence prevention, youth development, education and community organizing programs for the youth center, the grant is payable over 5 years;
Contact: David Ford, Executive Director; (617) 278-5200; fax (617) 278-5250; dford@smithfamilyfoundation.net
Internet: http://www.smithfamilyfoundation.net
Sponsor: Richard and Susan Smith Family Foundation
1280 Boylston Street, Suite 100
Chestnut Hill, MA 02467

Richard D. Bass Foundation Grants **4090**
The Richard D. Bass Foundation's area of interest include: arts/cultural-programs, including music and dance companies; Catholic/Protestant agencies & churches; community/economic development; education; health organizations, association. The type of support available include: annual campaigns; building/renovation; capital campaigns; general/operating support. Giving primarily in the metropolitan Dallas, Texas area. Grants range from $500 - $15,000. Application deadlines are March 31 and September 30 annually.
Requirements: IRS 501 (c) 3 nonprofit organizations are available to apply. The Foundations gives primarily to the Dallas, Texas region of the United States but funding is not limited to Texas. Grant proposals are available through out the U.S., contact the Foundation directly before submitting a proposal to access the likely hood of funding for your project. There is no application form required when applying for funding. Submit the proposal with one copy of the organizations IRS Determination Letter.
Restrictions: Funding is not available for: private foundations; individuals.
Date(s) Application is Due: Mar 31; Sep 30
Amount of Grant: 500 - 10,000 U.S.D
Samples: Dallas Opera, Dallas, TX, $5,000—arts grant; Metro Dallas Homeless Alliance, Dallas, TX, $2,500—community grant; Yale University, New haven, CT, $5,000—education grant; First Baptist Church, Valley Mills, TX, $500—religious grant; American Foundation for the Blind, Dallas, TX, $2,500—health grant.
Contact: Barbara B. Moroney, Treasurer; (214) 351-6994
Sponsor: Richard D. Bass Foundation
4516 Wildwood Road
Dallas, TX 75209-1926

Richard E. Griffin Family Foundation Grants **4091**
The Foundation, established in 1997, is dedicated to supporting higher education, human and social services, and federated giving programs. Funding most often comes in the form of general operating support. Though there are no specific application forms or deadlines with which to adhere, applicants should make an initial approach by letter outlining the project and budgetary need.
Amount of Grant: 1,000 - 100,000 U.S.D
Samples: Nantucket Historical Association, Nantucket, Massachusetts, $36,000—for general operations; Massachusetts College of Pharmacy and Health Sciences, Boston, Massachusetts, $55,000—for general operations.
Contact: Richard E. Griffin, (603) 472-4652
Sponsor: Richard E. Griffin Family Foundation
75 Federal Street, Suite 413
Boston, MA 02110-1904

Richard King Mellon Foundation Grants **4092**
On a national basis, the foundation makes grants to acquire and preserve key tracts of land in danger of being lost to urban growth and environmentally insensitive development. Mellon gives to nonprofits in Pittsburgh and throughout southwestern Pennsylvania to improve human services, education, medical care, civic affairs, and cultural activities. Types of support include capital grants, challenge/matching grants, general operating grants, project grants, and seed money grants. Application and guidelines are available online.
Requirements: Projects originating in Pittsburgh and southwestern Pennsylvania are given special priority.
Restrictions: The Foundation gives priority to projects and programs that have clearly defined outcomes and an evaluation component. It does not consider requests on behalf of individuals, or from outside the Untied States. The Foundation does not encourage requests from outside Pennsylvania.
Geographic Focus: Pennsylvania

Samples: Saint Vincent College (LaTrobe, PA)—for a business and conference center, $5 million; Westminster College (New Wilmington, PA)—to renovate an academic and administration building, $300,000; Carnegie Museums of Pittsburgh (PA)—to renovate and expand dinosaur exhibits at the Carnegie Museum of Natural History, $3 million; Direct Relief International (Santa Barbara, CA)—to assist tsunami-relief efforts, $500,000.
Contact: Michael Watson, Senior Vice President; (412) 392-2800; fax (412) 392-2837
Internet: http://foundationcenter.org/grantmaker/rkmellon/
Sponsor: Richard King Mellon Foundation
500 Grant Street, Suite 4106
Pittsburgh, PA 15219-2502

Richard Lounsbery Foundation Grants **4093**
The foundation aims to enhance national strengths in science and technology through support of programs in the following areas: science and technology components of key U.S. policy issues; elementary and secondary science and math education; historical studies and contemporary assessments of key trends in the physical and biomedical sciences; and start-up assistance for establishing the infrastructure of research projects. Among international initiatives, the foundation has a long-standing priority in Franco-American scientific cooperation. The foundation generally provides seed money or partial support, rarely renews grants for continuing activities, does not normally fund endowments or laboratory research, and aims to achieve high impact by funding novel projects and forward-looking leaders.
Restrictions: Requests are denied for support of capital or building funds, loans, or individuals.
Amount of Grant: 10,000 - 50,000 U.S.D
Samples: State U of New York-Empire State College (Saratoga Springs, NY)—to support the work of Isaac Rabino, to study the current views of working scientists and clinicians about the ethics of research involving human embryonic stem cells, $170,000 over two years.
Contact: Grants Administrator; (202) 872-8080; fax (202) 872-9292; proposals@rlounsbery.org
Internet: http://www.rlounsbery.org
Sponsor: Richard Lounsbery Foundation
1020 19th Street NW, Suite LL60
Washington, DC 20036

Richard M. Fairbanks Foundation Grants **4094**
The Richard M. Fairbanks Foundation, Inc. was established in 1986 by Richard M. Fairbanks, founder and owner of Fairbanks Communications. An independent private foundation granting funds to qualifying nonprofit organizations, programs, and projects in the greater Indianapolis, Indiana area. Exceptions to this geographic limitation are normally made only for organizations historically supported by the Fairbanks family or for national disasters. If you are interested in learning if your organization and project/program matches the Richard M. Fairbanks Foundation's areas of interest and funding guidelines, you may contact the foundation by telephone, email or brief written letter of inquiry. Please note that the Fairbanks Foundation does not accept unsolicited proposals.
Requirements: The Foundation: makes grants only to tax exempt public charities as defined in Sections 501(c)3 & 509(a)(1)(2)(3) of the Internal Revenue Code, except as prohibited for 509(a)(3) Type III organizations; will not consider requests for loans and grants to individuals; does not make grants or give support to conferences, seminars, media events, or workshops unless they are an integral part of a broader program; ordinarily considers grant requests only from organizations located in Greater Indianapolis, Indiana.
Geographic Focus: Indiana
Contact: Claire Fiddian-Green, Grants Officer; (317) 663-4189 or (317) 846-7111; fax (317) 844-0167; Fiddiangreen@rmfairbanksfoundation.org
Internet: http://www.rmfairbanksfoundation.org/default.asp?p=3
Sponsor: Richard M. Fairbanks Foundation
9292 North Meridian Street, Suite 304
Indianapolis, IN 46260

Richard P. Covert Scholarship **4095**
Two scholarships are awarded annually, one each at the undergraduate and graduate levels, in the field of health care information and management systems. Funding is provided by HIMSS. Completed application form, available from the society, must be accompanied by relevant transcripts, a technical paper and/or an essay, and a faculty recommendation. Annual deadline dates may vary; contact the program office for exact dates.
Requirements: Applicant must be at least a junior when the scholarship is awarded and must be a student member of HIMSS. Previous winners are not eligible, although an undergraduate winner is eligible if pursuing a graduate degree.
Date(s) Application is Due: Oct 31
Amount of Grant: 5,000 U.S.D
Contact: John Page; (312) 644-4467; fax (312) 664-6143; jpage@himss.org
Internet: http://www.himss.org/ASP/scholarship_ms.asp
Sponsor: Healthcare Information and Management Systems Society
230 E Ohio Street, Suite 500
Chicago, IL 60611-3269

Rich Foundation Grants **4096**
The foundation awards grants to nonprofit organizations in Atlanta, GA, to support community funds, performing arts, cultural programs, higher education, social services, homeless, youth, hospitals, heart disease, theater, and AIDS. Types of support include annual campaigns, building funds, continuing support, equipment acquisition, operating budgets, technical assistance, and research. Contact the office for application forms. The board meets quarterly to review requests.
Requirements: Nonprofits in the Atlanta, Georgia, area are eligible for grant support.
Geographic Focus: Georgia
Date(s) Application is Due: Mar 15; Jun 15; Sep 15; Dec 15
Amount of Grant: 5,000 - 100,000 U.S.D
Contact: Anne Poland Berg, Grant Consultant; (404) 262-2266
Sponsor: Rich Foundation
11 Piedmont Center, Suite 204
Atlanta, GA 30305

Richland County Bank Grants **4097**
The Richland County Bank has had a long tradition of community involvement, dedication and volunteerism. The Bank is committed to supporting the community in which it serves. In addition to monetary contributions, the bank encourages its employees to take an active role in our community. As a corporate sponsor, we support various organizations in Richland and surrounding counties, including: Vernon, Crawford, Grant, Iowa, and Sauk counties. Areas of interest include: health care; higher education; agricultural agencies; community development; children and youth programs; and scholarship funds.
Requirements: Applicants must be 501(c)3 organizations serving the Wisconsin counties of Richland, Vernon, Crawford, Grant, Iowa, or Sauk.
Geographic Focus: Wisconsin
Contact: Gail Surrem, Vice President; (608) 647-6306
Internet: http://www.richlandcountybank.com/aboutInvolvement.cfm
Sponsor: Richland County Bank
195 West Court Street
Richland Center, WI 53581

Richmond Eye and Ear Fund Grants **4098**
The fund is committed to serving the city of Richmond, Colonial Heights, Hopewell and Petersburg, and the counties of Chesterfield, Hanover, Henrico, Goochland, New Kent, Charles City, Lancaster, Northumberland, Richmond, Westmoreland and Powhatan in the fields of ophthalmology, otolaryngology, and oral maxillofacial surgery for qualified organizations. Grantmaking focuses on underserved, uninsured, or medically indigent children and adults in need of ophthalmological (eye), otolaryngological (ear, nose, and throat), and/or oral maxillofacial services (conditions, defects, injuries, and aesthetic aspects of the mouth, teeth, and jaws); improving access and outreach as it relates to the stated services mentioned; and healthcare education/preventive care consistent with purposes of this fund. Guidelines are available online. The Fund???s grantmaking focuses on: Underserved, uninsured or medically indigent children and adults in need of ophthalmological (eye), otolaryngological (ear, nose and throat), and/or oral maxillofacial services (conditions, defects, injuries, and aesthetic aspects of the mouth, teeth, and jaws). Improving Access and Outreach as it relates to the services mentioned above. Healthcare Education/preventive care consistent with purposes of this fund.
Requirements: Proposals will be accepted from charitable organizations, which serve the residents of the focus area.
Geographic Focus: Virginia
Date(s) Application is Due: May 5; Nov 5
Amount of Grant: Up to 50,000 U.S.D
Contact: Elaine Summerfield, Program Officer; (804) 330-7400; fax (804) 330-5992; esummerfield@tcfrichmond.org
Internet: http://www.tcfrichmond.org/Page2954.cfm#ear
Sponsor: Community Foundation
7501 Boulders View Drive, Suite 110
Richmond, VA 23225

Riley Foundation Grants **4099**
The foundation's purpose is to make grants, distributions, and/or loans to charitable and governmental organizations for charitable purposes and to provide financial resources and assistance for community wide projects and programs in health care; education; and the betterment of cultural, environmental, and economic conditions for the people of Meridian and Lauderdale County, Mississippi. Applications are reviewed at quarterly meetings in January, April, July, and October and must be in by noon on the listed deadline days.
Requirements: Mississippi governmental agencies and 501(c)3 organizations that are not private foundations as defined under Section 509(a) are eligible.
Restrictions: Best of Both Worlds Addiction Center, Inc. $75,000 - Operating Support; Meridian Community College $5,000 - Lucile Reisman Rosenbaum Memorial Scholarship Endowmen;
Geographic Focus: Mississippi
Date(s) Application is Due: Feb 15; May 15; Aug 15; Nov 15
Amount of Grant: 2,000 - 300,000 U.S.D
Contact: Becky Lewis, (601) 481-1430; fax (601) 481-1434; info@rileyfoundation.org
Internet: http://www.rileyfoundation.org/guide.htm?46,10
Sponsor: Riley Foundation
4518 Poplar Springs Drive
Meridian, MS 39305

Ringing Rocks Foundation Discretionary Fund Grants **4100**
The foundation's mission is to explore, document, and preserve indigenous cultures and healing practices. Grants are awarded to organizations throughout the world that promote indigenous healing, work with indigenous cultures, and educate the public at large about these topics. Grants can be used for start-up costs, program development, and general operating expenses. Interested organizations should be able to show substantial leadership from the community they serve at both the board and staff levels. Submit a completed eligibility statement along with a one-page letter of intent in English briefly

describing the organization along with type and amount of funding sought. Application materials are available online. The listed application deadline is for letters of request; full proposals are by invitation.
Requirements: Applicant organizations must be recognized as a charity by their governments. The project must have support from the indigenous community it serves.
Restrictions: Faxed letters of Intent, eligibility forms, and applications cannot be accepted.
Date(s) Application is Due: Feb 2
Amount of Grant: 500 - 5,000 U.S.D
Samples: To be distributed among 13 organizations—for efforts to conserve indigenous healing practices and to educate young people and inform the public about these traditions, $50,000.
Contact: Grants Administrator; (928) 282-1298; fax (928) 282-1327; grants@ringingrocks.org
Internet: http://www.ringingrocks.org/www/index.php?grants_fundProgram
Sponsor: Ringing Rocks Foundation
1890 West Highway 89A, Suite B-1
Sedona, AZ 86336-5571

Ripley County Community Foundation Grants **4101**
The Ripley County Community Foundation, Inc. was formed in February 1997. The effort to create the Foundation was a result of the vision of a group of philanthropic minded Ripley County residents who wanted to improve the quality of life for Ripley County, Indiana residents. Organizations seeking grants from the Ripley County Community Foundation must file an Intent to Apply form before preparing a complete grant application. This process will allow the Foundation to review and determine if the grant request meets Ripley County Community Foundation guidelines and/or whether the grant is likely to be funded. This Intent to Apply form must be received at the Foundation's office no later then 4:00 PM on August 14th. A grant application will be mailed within two weeks if you are eligible to apply. Grant applications must be received at the Foundation's office no later then 4:00 PM on September 11th. If you have any questions, call the Foundations office.
Requirements: Grant applicants must qualify as exempt organizations under Internal Revenue Code Section 501(c), or be sponsored by such organizations, or qualify as governmental or educational entities or possess similar attributes per IRS Code Section 509(a). Grants will be made only to organizations that have a responsible fiscal agent and adequate accounting procedures.
Restrictions: No grants will be made: solely to individuals; to defray travel or lodging expenses to enable individuals or groups to attend seminars or take trips; exclusively for endowment purposes of recipient organizations; for programs which will have been completed and funded prior to the Ripley County Community Foundation publicized date for grant decisions; to repay acquisition costs for equipment already purchased or paid for; for acquisition of weapons, firearms or destructive devices as defines in the Indiana Code under IC 35-41-1-8 or IC 35-47.5-2-4; specifically for sectarian religious purposes, but can be made to religious organizations for general community programs, or to preserve an historical site; organizations who have been awarded prior grants must have complied with all reporting requirements and be in good standing before submitting another grant application. Non-compliance will result in exclusion from future granting; to attempt to influence legislation or to intervene in any political campaign on behalf of any candidate for public office. Ripley County Community Foundation reserves the right to refuse any and all grant applications. Grant recipients and fiscal agents will sign and abide by the respective conditions as set forth in the foundation's " Grant Award Agreements" and "Grant Report Forms".
Geographic Focus: Indiana
Samples: Morris Fireman's Assoc. - self-contained breathing apparatus, $4,500; SEICHN, Inc. - Breast Health Screening, $4,600; American Legion Post 337 - chairs and cart, $1,115.
Contact: Jane Deiwert, Program Officer; (877) 234-5220 or (812) 933-1098; fax (812) 933-0096; jdeiwert@rccfonline.org
Internet: http://www.rccfonline.org/grants.asp
Sponsor: Ripley County Community Foundation
132 South Main Street
Batesville, IN 47006

Robert and Joan Dircks Foundation Grants **4102**
The foundation supports non-profit organizations that enrich and improve the quality of life for individuals primarily located in the New Jersey area. The focus is to encourage innovative programs and projects that benefit and improve the lives of children and individuals who are physically mentally or economically disadvantaged.
Requirements: Submit your inquiry online, using the Grant Request Information form provided on the Robert and Joan Dircks Foundation website: http://www.dircksfoundation.org/application2.asp
Restrictions: The Robert and Joan Dircks Foundation does not award grants for political or lobbying activities, environmental or cultural projects, capital or annual campaigns, endowments, operating budgets, deficit or debt reduction, loans or housing projects. Also, grants are not made to programs of national or international scope.
Geographic Focus: Connecticut, Maine, Massachusetts, New Hampshire, Rhode Island, Vermont
Amount of Grant: 1,000 - 15,000 U.S.D
Contact: Grants Administrator; grants@dircksfoundation.org
Internet: http://www.dircksfoundation.org/guidelines.html
Sponsor: Robert and Joan Dircks Foundation
P.O. Box 6
Mountain Lakes, NJ 07046

Robert and Polly Dunn Foundation Grants **4103**
The foundation awards grants in the U.S. Southeast, with emphasis on the Atlanta, GA, area, for projects in the area of youth, child development, and higher education. Types of support include capital campaigns, general operating support, and scholarship funds. There are no application forms. The board meets in June and December to consider requests.
Restrictions: Grants do not support individuals and not awarded to support endowment funds, loans, or program-related investments
Geographic Focus: Georgia
Date(s) Application is Due: Apr 30; Sep 30
Amount of Grant: Up to 587,925 U.S.D
Contact: Karen Wilbanks, Executive Director; (404) 816-2883; fax (404) 816-2883
Sponsor: Robert and Polly Dunn Foundation
P.O. Box 723194
Atlanta, GA 31139

Robert B McMillen Foundation Grants **4104**
The Robert B McMillen Foundation is a non-profit charitable foundation. The Foundation offers funding in the states of Washington and Alaska. The two areas of intent are: medical and social enhancement. Fifty percent of the annual funding is earmarked for medical research. The Foundation will consider making grants to non-profit organizations involved in researching cardiology, lipid and organ transplants. Twenty-five percent of annual giving is earmarked for social areas including, but not limited to, Goodwill, Salvation Army & United Way. Preference is given to organizations and/or programs that use art as the vehicle to impact communities and change individual lives.
Requirements: The Foundation does not accept uninvited medical or social proposals. Contact the Foundations office for information on, how to be invited to make a medical or social proposal to the McMillen Foundation.
Restrictions: The Foundation do not make grants to: religious, fraternal or other organizations primarily benefiting their own members; other private foundations nor political organizations; for-profit entities; other than scholarships, the foundation will not make multi-year grants.
Geographic Focus: Alaska, Washington
Samples: Seattle Children's Hospital - Pediatric Heart Failure, Organ Transplantation and Neovascularization; ArtsEd Washington - PAL Program and Arts Partners Project; Seattle Goodwill - General Support.
Contact: Cassandra Town, President; (425) 313-5711; fax (425) 313-8955; cassandra@mcmillenfoundation.org
Internet: http://www.mcmillenfoundation.org/mission.htm
Sponsor: Robert B McMillen Foundation
55 1st Place NW, Suite 2, P.O. Box 1523
Issaquah, WA 98027

Robert E. and Evelyn McKee Foundation Grants **4105**
The foundation awards grants in the following categories: civic, cultural, and religion; education and scholarships; hospitals; medical and medical research; welfare, rehabilitation, and mental health; United Way; and youth activities. Grants awarded for scholarships are made to a few selected high schools and universities, who select scholarship recipients. Religious contributions are basically limited to the local Episcopal churches. The majority of funding is made to local charitable organizations or through local affiliates of national or statewide organizations.
Requirements: 501(c)3 tax-exempt organizations are eligible.
Restrictions: Funding is not available to organizations outside the United States or those limited by race or ethnic origin. The foundation refrains from awarding grants for endowments or deficit financing. Grants are not made to other foundations, except for a local community foundation.
Geographic Focus: Texas
Date(s) Application is Due: Dec 15
Amount of Grant: 1,000 - 10,000 U.S.D
Samples: Saint Francis on the Hill Episcopal Church (El Paso, TX)—$1000; U of Texas at El Paso (TX)—for Department of Engineering scholarships, $8000; Texas Tech U Health Sciences Ctr (El Paso, TX)—for the University Breast Care Center, $3000; Special Olympics, Area 19 (El Paso, TX)— $1000.
Contact: Grants Administrator; (915) 581-4025; fax (915) 833-3714; mckeefoundation@directway.com
Internet: http://www.mckeefoundation.org
Sponsor: Robert E. and Evelyn McKee Foundation
5835 Cromo Drive, Suite 1
El Paso, TX 79912

Robert F. Stoico / FIRSTFED Charitable Foundation Grants **4106**
The foundation was established to support the local communities of southeastern Massahusetts and Rhode Island. The foundation's main areas of interest are affordable housing, economic/job development, educational programs, arts and cultural programs that benefit the public or the traditionally underserved, and accessible health care for all. Small grants (under $2501) are reviewed throughout the year on a rolling basis. Large grant requests are reviewed four times a year (in April, July, October and January). Application and guidelines are available online.
Requirements: Massachusetts and Rhode Island 501(c)3 nonprofit organizations in the counties where FIRSTFED America Bancorp had business involvement are eligible.
Restrictions: The foundation does not support sports programs, political organizations, individual schools, individuals, or organizations that are not classified 501(c)3.
Geographic Focus: Massachusetts, Rhode Island

Samples: Thomas Chew Memorial Boys and Girls Club (Fall River, MA)—for its capital campaign, $1 million.
Contact: Stacie Charbonneau, Manager; (508) 235-1368; fax (508) 235-1854; info@stoicofirstfed.org
Internet: http://www.stoicofirstfed.org/apply.htm
Sponsor: Robert F. Stoico / FIRSTFED Charitable Foundation
P.O. Box 438
Swansea, MA 02777

Robert J. Kleberg Jr. and Helen C. Kleberg Foundation Grants **4107**
The foundation awards grants on a national basis for medical research, community services, education, health services, arts/humanities, wildlife, veterinary, and animal sciences. Local community organizations also are supported. Types of support include building construction/renovation, equipment acquisition, conferences and seminars, research, and matching funds. There are no application deadlines or forms. Initial approach should be by letter on organization?s letterhead. Send one copy of proposal. The board usually meets in June and December, with final notification in six months.
Requirements: 501(c)3 tax-exempt organizations are eligible. Requests from universities must be transmitted by the appropriate offices, such as the chancellor, president, or director of development.
Restrictions: Grants do not support community organizations outside Texas; organizations limited by race or religion; individuals; general purposes; endowments; deficit financing; ongoing operating expenses; overhead or indirect costs; scholarships or fellowships; or loans.
Samples: U of Texas MD Anderson Cancer Ctr (Houston, TX)—to establish a center that will focus on investigating molecular markers, the genes and proteins that may help identify cancer in its early stages or help predict how cancer will respond to treatment, $5 million.
Contact: Robert Washington, Grants Coordinator; (210) 271-3691
Internet: http://www.texascancercouncil.org/cancerfunding/state/04.html
Sponsor: Robert J. Kleberg Jr. and Helen C. Kleberg Foundation
700 N Saint Mary?s Street, Suite 1200
San Antonio, TX 78205

Robert Leet and Clara Guthrie Patterson Trust Grants **4108**
The Robert E. Leet and Clara Guthrie Patterson Trust was established in 1980 to support and promote the investigation of human disease and the alleviation of human suffering through improved treatment. While the funding history shows that the trust has recently awarded grants, the grant website states that the trust is not currently accepting applications. Individuals desiring more information about funding opportunities from the Robert E. Leet and Clara Guthrie Patterson Trust are advised to contact the Senior Foundation Officer for further information.
Geographic Focus: Connecticut, New Jersey, New York
Samples: Health Resources in Action, Boston, Massachusetts, $440,000; Health Resources in Action, Boston, Massachusetts, $300,000, postdoctoral fellowships in brain circuitry - various organizations (2010); Multiple Myeloma Research Foundation, Norwalk, Connecticut, $100,000, to fund a senior researcher (2011).
Contact: Carmen Britt; (860) 657-7019; carmen.britt@baml.com
Internet: https://www.bankofamerica.com/philanthropic/fn_search.action
Sponsor: Robert Leet and Clara Guthrie Patterson Trust
200 Glastonbury Boulevard, Suite # 200, CT2-545-02-05
Glastonbury, CT 06033-4056

Robert Mapplethorpe Foundation Grants **4109**
The foundation provides grants in two unrelated major areas: medical research to advance the cure and treatment of AIDS and HIV infection and the creation or expansion of photography departments in museums and other public institutions. Types of support include equipment acquisition, program development, publication, and research. A formal application is not required. The foundation board meets regularly to consider requests.
Restrictions: Individual artists will not receive support. The foundation does not award grants for social services.
Contact: Grants Administrator; (212) 755-3025; fax (212) 941-4764
Internet: http://www.mapplethorpe.org/foundation-guidelines.html
Sponsor: Robert Mapplethorpe Foundation
120 Wooster Street, 4th Floor
New York, NY 10012-0000

Robert P. & Clara I. Milton Fund for Senior Housing **4110**
The purpose of this fund is to support innovative programs focusing on issues related to senior housing. Projects will be considered that make a meaningful impact while promoting dignity, independence and quality of life. This fund strives to make St. Joseph County a regional model for senior housing by acting as a catalyst through the funding of innovative approaches. Senior housing may range from institutional care to in-home care, as well as the various services related to this issue. Special consideration will be given to projects that serve the most vulnerable of the aging population. Priorities for funding include education/training; home and community based services; programs and project serving vulnerable seniors; and public policy issues. Grants are judged by community impact, concept/idea, improved solutions to current and projected needs, the ability to implement the project, the quality of need, and its sustainability.
Requirements: In addition to the online application, the following attachments are required: up to a two page proposal narrative; a detailed project budget; current board roster with officers identified; and proof of nonprofit status. Detailed guidelines and application are available at the website.
Restrictions: Annual appeals or membership contributions, grants to individuals directly or special event underwriting will not be funded.
Geographic Focus: Indiana
Date(s) Application is Due: May 1; Nov 1
Samples: Housing Authority, South Bend, Indiana: salary of the Senior Assessment Coordinator, $30,000: Logan Center, South Bend, Indiana: renovation of their senior day program facility, $22,000.
Contact: Christopher Nanni; (574) 232-0041; fax (574) 233-1906; chris@cfsjc.org
Internet: http://www.cfsjc.org/initiatives/senior/milton_grants.html
Sponsor: Community Foundation of St. Joseph County
205 W Jefferson Boulevard, P.O. Box 837
South Bend, IN 46624

Robert R. Meyer Foundation Grants **4111**
The foundation supports 501(c)3 charitable organizations in the Birmingham, AL, area for programs and activities in the areas of the arts, humanities, civic affairs, health, and social services delivery. Requests must be in writing and describe the organization and grant request.
Requirements: Giving is limited to organizations in the metropolitan Birmingham, AL, area.
Geographic Focus: Alabama
Date(s) Application is Due: Mar 15; Sep 15
Amount of Grant: 5,000 - 100,000 U.S.D
Contact: Carla Gale, Trust Officer, AmSouth Bank; (205) 326-5382
Sponsor: Robert R. Meyer Foundation
P.O. Box 11426
Birmingham, AL 35202

Robert Sterling Clark Foundation Reproductive Rights and Health Grants **4112**
The objective of the Foundation's work in this field is to protect and expand women's access to reproductive health services in the belief that the ability to control one's fertility and prevent unintended pregnancy is fundamental to the advancement of women's opportunities. The strategy employed by the Foundation to achieve this objective is to support policy analysis, advocacy, litigation, research, message development, and/or organizing aimed at: promoting the implementation of laws, policies and practices that will enable all women to have access to comprehensive reproductive health information and services, including emergency contraception and abortion; challenging laws and legal decisions that undermine reproductive rights and developing legal theory in support of these rights; mobilizing an informed grassroots constituency; developing new pro-choice messages, based on current opinion research, that resonate with a broad public; informing the development of judicial selection criteria that will produce non-ideological courts; and promoting the use of comprehensive sexuality education curricula in the schools.
Requirements: The Foundation makes support available to national and regional organizations working on these issues.
Amount of Grant: 1,000 - 200,000 U.S.D
Contact: Margaret C. Ayers; (212) 288-8900; fax (212) 288-1033; rcsf@rsclark.org
Internet: http://www.rsclark.org/index.php?page=reproductive-right-and-health
Sponsor: Robert Suiterling Clark Foundation
135 East 64th Street
New York, NY 10065-7045

Robert W. Woodruff Foundation Grants **4113**
The foundation is interested in supporting programs and projects in the areas of elementary, secondary, and higher education; health care and education; Human services, particularly for children and youth; economic development and civic affairs; art and cultural activities; and conservation of natural resources and environmental education. Preference is given to one-time capital projects, matching/challenge grants, specific projects, construction, land acquisition, and equipment purchases.
Requirements: Most grants are awarded to tax-exempt organizations in Georgia. Grants to qualified public charities headquartered outside Georgia are considered when it is demonstrated that the proposed project or program will have particular impact in Georgia. Application forms are not required. Proposals should be made in letter form and describe the organization, including its purposes, staffing, and governing board; latest financial statement and most recent audit report; description of the proposed project and justification for funding; itemized project budget, including other sources of support; and evidence of tax-exempt status. The board meets in April and November to consider requests.
Restrictions: Grants are not awarded to support individuals, annual operating support, festivals and performances, conferences, films and documentaries, start-up funding or seed money, churches or their denominational programs, or youth services outside Atlanta.
Geographic Focus: Georgia
Date(s) Application is Due: Feb 1; Sep 1
Amount of Grant: 50,000 - 750,000 U.S.D
Samples: The Atlanta Lyric Theater, $75,000 for costs associated with the Lyric's move to the Strand Theater in Cobb County; Georgia Tech Foundation, Inc., $2,500,000 for construction of the Innovative Learning Resource Center; Arch Foundation, $250,000 for construction of a new horticulture complex at the State Botanical Garden of Georgia; Bobby Dodd Institute, $500,000 for establishment of an Enterprise Fund to provide working capital for Bobby Dodd's social enterprises.
Contact: P. Russell Hardin; (404) 522-6755; fax (404) 522-7026; fdns@woodruff.org
Internet: http://www.woodruff.org/appGuidelines_rww.aspx
Sponsor: Robert W. Woodruff Foundation
50 Hurt Plaza, Suite 1200
Atlanta, GA 30303

Robin Hood Foundation Grants 4114
The foundation's mission is to end poverty in New York City. Grants are awarded in the areas of early childhood—child development, comprehensive healthcare, parenting skills, and school preparation; education—public education reform; after-school programs—academic tutoring linked with healthcare, sex education, counseling, and job training; job training—job-traning and -placement for people with histories of substance abuse, crime, and homelessness; and basic survival initiatives—issues such as hunger, AIDS, domestic violence, and inadequate housing.
Geographic Focus: New York
Amount of Grant: Up to 50,000,000 U.S.D
Contact: Karen Moody, Grants Manager; (212) 227-6601; fax (212) 227-6698; grants@robinhood.org
Internet: http:/www.robinhood.org
Sponsor: Robin Hood Foundation
826 Broadway, 9th Floor
New York, NY 10003

Robins Foundation Grants 4115
The foundation awards grants to nonprofit organizations in the Richmond, VA, area. Areas of interest include general grants—including cultural, charitable, scientific, environmental, and educational programs, and at-risk youth; and early childhood/quality improvement grants— for organizations that devote a major portion of their resources to young children and their families. The goal is to help improve the quality of services by providing funding for accreditation, staff training, facilities improvements, and similar initiatives. The foundation seeks projects that meet well-defined community needs, use effective approaches that build on proven programs, develop models with potential for wider application, foster self-reliance and/or end dependency, and focus on prevention as well as treatment. Types of support include capital grants and endowments.
Requirements: 501(c)3 organizations based in Virginia are eligible. Organizations based in Virginia that have or support programs outside Virginia or the United States also are eligible.
Restrictions: In general, the foundation does not make grants to support annual operating funds or budgets, special events or fundraising benefits, or religious purposes unless they are otherwise compatible with the objectives of the foundation.
Geographic Focus: Virginia
Amount of Grant: 10,000 - 100,000 U.S.D
Contact: William Roberts Jr., Executive Director; (804) 697-6917; fax (804) 697-6797; wlrjr@robins-foundation.org
Internet: http://www.robins-foundation.org
Sponsor: Robins Foundation
Capitol Station, P.O. Box 1124
Richmond, VA 23218-1124

Rochester Area Community Foundation Grants 4116
The foundation seeks to improve the quality of life in the community and awards grants in the areas of child development, education, the environment, arts and cultural programs, health services, community development, and social services and general charitable giving. Types of support include general operating support, building construction/renovation, equipment acquisition, program development, conferences and seminars, publication, seed grants, scholarship funds, technical assistance, consulting services, and scholarships to individuals. There are no application deadlines. The board meets in January, February, March, May, June, July, October, and November.
Requirements: Nonprofits in Monroe, Livingston, Ontario, Orleans, Genessee, and Wayne Counties, NY, are eligible.
Restrictions: Grants do not support partisan political organizations or religious projects, individuals, annual campaigns, deficit financing, land acquisition, endowments, or emergency funds.
Geographic Focus: New York
Samples: Heritage Christian Services, $5,000—to support the annual Sheep Shearing Festival at Springdale Farm; Step by Step of Rochester, $12,000—to support the merger of Step by Step and Samaritan Women. Both organizations provide support to incarcerated women and their children; Arts & Cultural Council for Greater Rochester, $42,500—for Community Arts Grants, which are administered by the Arts & Cultural Council, have a capacity-building focus, and are aimed at smaller arts organizations; Lifespan of Greater Rochester, $10,000—for a financial management program which trains volunteers to provide in-home assistance with bill paying, budgeting, government benefits (such as Medicare and Medicaid), and other financial tasks for physically and/or mentally frail older adults.
Contact: Marlene Cole, Administrative Assistant; (585) 341-4333; fax (585) 271-4292; mcole@racf.org
Internet: http://www.racf.org/page10000903.cfm
Sponsor: Rochester Area Community Foundation
500 East Avenue
Rochester, NY 14607-1912

Rochester Area Foundation Grants 4117
The foundation makes grants in the fields of arts and culture, community development, education, human services and recreation in the greater Rochester, Minnesota area.
Requirements: Organizations must be one of the following to be eligible to receive grant funding: a tax-exempt 501(c)3 organization, a government unit (city, township, county), or a government-created organization such as a public agency. Download the required pre-application form from the website. If your pre-application is approved, you will be notified and asked to submit a full application to the foundation. (The required full application form is also available at the website.)
Geographic Focus: Minnesota
Date(s) Application is Due: Jan 2; May 1; Sep 1
Amount of Grant: 1,000 - 25,000 U.S.D
Samples: Rochester Public Library, $7,500—matching grant for an online tutoring program available to all students with a library card; Children's Dental Health Services, $8,000—for dental equipment required to meet growing demand for services; Child Care Resource & Referral, $5,300—for a new van for transporting participants of income-eligible services.
Contact: Steve Thornton, Executive Director; (507) 424-2400 or (507) 424-3755, ext. 102; fax (507) 282-4938; Steve@RochesterArea.org
Internet: http://www.rochesterarea.org/grant-resources/index.html
Sponsor: Rochester Area Foundation
400 South Broadway, Suite 300
Rochester, MN 55904

Rockefeller Archive Center Research Grants 4118
The foundation awards grants for travel and research at the Rockefeller Archive Center, which contains records of the Rockefeller Foundation, Rockefeller University, and other affiliated philanthropic organizations and individuals. The center also will award grants to support research on the history of basic medical research, the life of the Nobel laureate Paul Ehrlich, and on postcolonial studies.
Requirements: Scholarly reseearchers, particularly students preparing dissertations, are eligible.
Restrictions: The Archive Center's programs do not support research at other institutions, and they do not provide general tuition support.
Contact: Camilla Harris, Grants Administrator; (914) 366-6311 or (914) 631-4505; fax (914) 631-6017; charris@rockarch.org or archive@rockarch.org
Internet: http://www.rockarch.org/grants/
Sponsor: Rockefeller Foundation
15 Dayton Avenue
Sleepy Hollow, NY 10591-1598

Rockefeller Brothers Fund Peace and Security Grants 4119
The Fund's Peace and Security Program focuses on two factors that may be key to advancing or undermining global problem solving: the content and style of U.S. global engagement in the face of new perils and opportunities; and the strength and quality of relationships between Muslim and Western societies. In addition, peace and security is a theme that may be identified for attention in one or more of the Fund's pivotal places. Two specific goals have emerged: to advance U.S. policies and behaviors that reflect a broadly shared vision of constructive, cooperative, principled, farsighted, and effective global engagement; and to reduce the divisive and destabilizing tensions that exist between much of the Islamic world and the West, particularly the United States, and to increase the potential for collaboration among Muslim and Western societies on behalf of a better, safer world.
Requirements: A prospective grantee in the United States or foreign counterpart must be either a tax-exempt organization or an organization seeking support for a project that would qualify as educational or charitable.
Restrictions: The fund does not make grants to individuals, nor does it as a general rule support research, graduate study, or the writing of books or dissertations by individuals.
Amount of Grant: 25,000 - 300,000 U.S.D
Contact: Benjamin R. Shute; (212) 812-4200; fax (212) 812-4299; info@rbf.org
Internet: http://www.rbf.org/programs/
Sponsor: Rockefeller Brothers Fund
437 Madison Avenue, 37th Floor
New York, NY 10022-7001

Rockefeller Brothers Fund Pivotal Places Grants: South Africa 4120
The Fund will support human advancement through its work to improve basic education and to meet the developmental needs of orphans and vulnerable children. In addition, the RBF is exploring linkages between HIV/AIDS and the Fund's other substantive concerns: democratic practice, sustainable development, and peace and security. While the Fund recognizes the critical importance of HIV/AIDS education, prevention, and treatment, many other donors are active in these fields. The RBF seeks to complement these efforts by contributing to the development of integrated responses to the broad societal impacts of the pandemic. Specifically, the Fund seeks to: improve the quality and accessibility of basic education for children in the areas of early childhood development and primary learning; assist orphans and vulnerable children in achieving their full potentials individuals and as members of society; and strengthen understanding of the linkages between HIV/AIDS and democratic practice, sustainable development, and peace and security issues in South Africa, and to promote integrated responses to the broad societal impacts of the HIV/AIDS pandemic.
Requirements: A prospective grantee in the United States or foreign counterpart must be either a tax-exempt organization or an organization seeking support for a project that would qualify as educational or charitable.
Restrictions: The fund does not make grants to individuals, nor does it as a general rule support research, graduate study, or the writing of books or dissertations by individuals.
Geographic Focus: All States, Algeria, Angola, Benin, Botswana, Burkina Faso, Burundi, Cameroon, Cape Verde, Central African Republic, Chad, Comoros, Congo, Congo, Democratic Republic of, Cote d' Ivoire (Ivory Coast), Djibouti, Egypt, Equatorial Guinea, Eritrea, Ethiopia, Gabon, Gambia, Ghana, Guinea, Guinea-Bissau, Kenya, Lesotho, Liberia, Libya, Madagascar, Malawi, Mali, Mauritania, Mauritius, Morocco, Mozambique, Namibia, Niger, Nigeria, Rwanda, Sao Tome & Principe, Senegal, Seychelles, Sierra Leone, Somalia, South Africa, Sudan, Swaziland

Contact: Nancy L. Muirhead, Program Director; (212) 812-4200; fax (212) 812-4299; nmuirhead@rbf.org or info@rbf.org
Internet: http://www.rbf.org/programs/
Sponsor: Rockefeller Brothers Fund
437 Madison Avenue, 37th Floor
New York, NY 10022-7001

Rockefeller Foundation Grants **4121**
The foundation provides grants to institutions and individuals seeking to improve the lives of poor people with a focus on the issues and region where the foundation works. The foundation works globally but provides the majority of its grants to organizations whose work is focused in Southern and Eastern Africa, Southeast Asia, and North America through programs that address agriculture, health, employment, housing, education, arts and culture, and global policy. The foundation also operates special programs, including a conference and study program at its Bellagio Center, the Program Venture Experiment, and the Philanthropy Workshop.
Requirements: Nonprofit organizations are eligible.
Restrictions: The foundation does not give or lend money for personal aid to individuals, support attempts to influence legislation, or, except in rare cases, provide general institutional support, fund endowments, or contribute to building and operating funds.
Samples: Criterion Ventures, Haddam, CT—toward the start-up costs of a new mission-driven venture that will provide healthcare financial services to groups of American workers who are struggling with the negative economic and health impacts of paying cash for health services, $500,000; Water and Sanitation for the Urban Poor, London, England—toward the costs of building the capacity of community and local service groups and of testing innovative water and sanitation delivery systems in Gatwekera, Nairobi and the surrounding municipality to provide sustainable, equitable and affordable water and sanitation services, $249,900
Contact: Grants Administrator; (212) 869-8500; fax (212) 764-3468
Internet: http://www.rockfound.org/grants/grants.shtml
Sponsor: Rockefeller Foundation
420 Fifth Avenue
New York, NY 10018-2702

Rockwell Collins Charitable Contributions **4122**
The corporation has identified two distinct funding priorities, and special consideration will be given to proposals that integrate these issues: education and youth development, with emphasis in math, science and engineering; and culture and the arts, with emphasis on youth educational programs. Rockwell Collins also contributes to health, human services, and civic organizations. Special consideration is given to qualifying organizations where employees volunteer; however, the majority of gifts to these organizations are made through Rockwell Collins United Way corporate contributions or employee campaigns.
Requirements: Applying organizations must be tax exempt under the federal code; able to provide current full, certified, audited financial statements; and not be private foundations. Non-U.S. organizations must provide nongovernmental organization (NGO) documentation.
Restrictions: Funding will not be considered for general endowments, deficit reduction, grants to individuals, federated campaigns, religious organizations for religious purposes, or fraternal or social organizations.
Amount of Grant: Up to 5,000 U.S.D
Contact: Cindy Dietz, Manager Community Relations; (319) 295-7444; fax (319) 295-9374; communityrelations@rockwellcollins.com
Internet: http://www.rockwellcollins.com/about/community/index.html
Sponsor: Rockwell Collins Corporation
400 Collins Road NE
Cedar Rapids, IA 52498-0001

Rockwell Fund, Inc. Grants **4123**
Rockwell Fund, Inc. was founded in 1931 from the estate of James M. Rockwell. Rockwell's grant making is concentrated on the following issue areas: community health, concentrating on health issues that affect the broader community, including mental and behavioral health; education, concentrating on dropout prevention strategies that target the intermediate and middle school years; employment, concentrating on training/placement and jobs creation/enterprise development opportunities; and supportive housing, concentrating on affordable housing coupled with onsite, sustained services to support individuals and families in achieving self-sufficiency.
Requirements: Eligible applicants must be nonprofit tax exempt organizations. To be eligible, an organization must: have a determination letter from the Internal Revenue Service indicating that it is an organization described in Section 501(c)3 of the Internal Revenue Code or be a church or political subdivision that is not required to obtain a Section 501(c)3 designation in order to be a permitted donee of a private foundation. Funding is provided for the Houston area or in some cases for a purpose that will benefit a Houston cause. Interested applicants must first complete an on-line inquiry form. Applicants will then either be invited to submit a full grant application for further consideration or declined. Inquiries are accepted throughout the year.
Restrictions: The following projects or programs are ineligible: feasibility studies; annual fund drives; direct mail or other mass solicitations; grants that impose the exercise of "expenditure responsibility" upon Rockwell Fund, for example, private operating foundations or certain support organizations; houses of worship; individuals; medical or scientific research projects or organizations that target a specific disease; parochial or private primary and secondary schools; and underwriting for benefits, dinners, galas, golf tournaments or other fundraising or special events.
Geographic Focus: Texas
Contact: Judy Ahlgrim, Grants Administrator; (713) 341-5338; fax (713) 629-7702; jahlgrim@rockfund.org
Internet: http://www.rockfund.org/giving/howToApply.shtml
Sponsor: Rockwell Fund, Inc.
770 S. Post Oak Lane, Suite 525
Houston, TX 77056

Rockwell International Corporate Trust Grants **4124**
Grants are awarded in the areas of education and youth development, with emphasis in math, science, and engineering; and culture and the arts, with emphasis on youth educational programs. Rockwell also contributes to health, human services, and civic organizations and also gives special consideration to qualifying organizations in which employees are involved as volunteers. Types of support include capital grants, challenge/matching grants, conferences and seminars, development grants, endowments, research grants, scholarship funds, and general operating grants. Awards include single-year, multiple-year, and provisional continuing support of specific programs. Applicants should submit full, detailed proposals. Proposals are accepted at any time. At the Web site, click on About Us, then Corporate Citizenship, and then Rockwell Automation Corporation Trust.
Restrictions: Organizations are not eligible if they have not received a permanent, tax-exempt ruling determination from the federal government; if they cannot provide current full, certified, audited financial statements; or if they are private foundations. Funding will not be considered for the following purposes: general endowments, deficit reduction, grants to individuals, federated campaigns, organizations or projects outside the United States, religious organizations for religious purposes, or fraternal or social organizations.
Amount of Grant: 1,000 - 50,000 U.S.D
Contact: Rockwell International Corporation General Information; (562) 797-3311
Internet: http://www.rockwell.com
Sponsor: Rockwell International Corporate Trust
1201 S Second Street
Milwaukee, WI 53204

Roeher Institute Research Grants **4125**
The institute, through the support of the Scottish Rite Charitable Foundation of Canada, offers research grants to associates, associations, and agencies in a broad range of fields relating to human services and intellectual disability. The renewable grants are offered for one year. Research projects must be under the direction of the Roeher Institute academic associates or other university faculty or under researchers approved by the Roeher Institute. Areas constituting research priorities for the Roeher Institute are issues affecting people with intellectual disabilities; integration of people who have an intellectual disability into society; prevention, early identification, and minimization of disabling conditions; and strategies for social change that improve the quality of life of persons with intellectual disabilities. Applicants are asked to submit a summary of their research project, budget, and financial requirements.
Requirements: Applications are invited from university faculty, department heads, supervisors, Roeher Institute associates, and/or consultants.
Amount of Grant: Up to 10,000 CAD
Contact: Cameron Crawford, President; (416) 661-9611; fax (416) 661-5701; cameronc@roeher.ca or cameronc@worldchat.com
Internet: http://www.roeher.ca/english/about/about.htm
Sponsor: Roeher Institute
4700 Keele Street, York University, Kinsmen Building.
North York, ON M3J 1P3 Canada

Rogers Family Foundation Grants **4126**
The Rogers Family Foundation funds nonprofit organizations that provide educational, medical, artistic, and religious services within Massachusetts' Merrimack Valley and North Shore and southeastern New Hampshire. See website for specific areas served.
Requirements: Applicants must be tax exempt under section 501(c)3 of the Internal Revenue Code and classified as "not a private foundation" under Section 509(a) of the Code. All applications are submitted on-line. The online application follows the Common Proposal designed and published by Associated Grant Makers, Inc. See website for detailed online application and instructions.
Restrictions: Traditionally the Foundation awards grants to organizations located in Massachusetts' Merrimack Valley and North Shore and southeastern New Hampshire. From time to time grants may be made outside of this normal geographic area.
Geographic Focus: Massachusetts, New Hampshire
Date(s) Application is Due: Mar 1; Sep 1
Amount of Grant: 25,000 U.S.D
Contact: Susan Haff, (617) 426-7080; fax (617) 426-7087
Internet: http://www.rogersfamilyfoundation.com/app/
Sponsor: Rogers Family Foundation
c/o GMA Foundations
Boston, MA 02110

Roget Begnoche Scholarship **4127**
The Roger Begnoche Scholarship was established to encourage young men to pursue nursing or education careers. Scholarships of $4,000 per year for a maximum of 4 years will be awarded. A student is eligible for the renewable scholarship if they remain in good academic standing, continue toward a nursing or education degree, and submit the appropriate paperwork to the Legacy Foundation in a timely manner.
Requirements: Applicants are eligible to apply if they meet the following criteria: they are a U.S. citizen and resident of Indiana; they are a current high school graduate; they are male and have been accepted as a full time student at Purdue University Calumet,

Indiana University Northwest, Calumet College of St. Joseph, or Sisters of St. Francis nursing program; they have been accepted into the School of Nursing or are pursuing a degree in education; and they have an SAT score above 900 for a nursing student or 850 for an education student. Selection of scholarship recipients will be based on: academics, school activities, community or volunteer involvement; financial need; work experience; a recommendation letter from an adult; and an original essay.
Restrictions: Only specific colleges in Indiana will meet the selection criteria.
Geographic Focus: Indiana
Date(s) Application is Due: May 15
Amount of Grant: 4,000 U.S.D
Contact: Begnoche Scholarship Contact; (219) 736-1880; fax (219) 736-1940; legacy@legacyfoundationlakeco.org
Internet: http://www.legacyfoundationlakeco.org/scholarships/index.htm
Sponsor: Legacy Foundation
1000 East 80th Place, 302 South
Merrillville, IN 46410

Rohm and Haas Company Grants **4128**
The company awards grants in five key philanthropic categories: education, environment, civic and community, health and human services, and arts and culture. Education grants support programs linking education in science, technology, and math to workplace and career opportunities; and educational enrichment for students, particularly in the areas of math, science, technology, and the environment either during or after school hours (including weekends, summer, and other times when school is not in session). Environment grants support after-school programs that build environmental awareness and understanding; sustainability programs that educate and promote development that meets the needs of the present society without compromising the ability of future generations to meet their own needs; pollution prevention and waste management, environmental conservation and biodiversity, and energy and water conservation; and community beautification. Civic and community grants support regional efforts that build local competitiveness and strengthen the economic and social base in Rohm and Haas host communities; programs that cut across multiple giving categories and/or that impact key stakeholders in Rohm and Haas communities; and programs that focus on volunteerism. Health and human services grants support public safety awareness; disaster preparedness; structured youth development such as leadership training; and structured youth activities and recreation. Arts and culture grants support organizational capacity building in the form of plan implementation—technology implementation and/or enhancement, communications and marketing, outreach and audience development, financial management and fundraising, board development, strategic development, professional development, and educational program development. Application and guidelines are available online.
Requirements: Eligible national and international organizations include performing and visual arts, humanities and historical societies, and museums. Preference is given to requests from Chicago, IL, and the greater Philadelphia area.
Restrictions: Grants are not awarded to individuals or for fundraising, advertising, or testimonials.
Amount of Grant: 250 - 50,000 U.S.D
Contact: Alexandra Samuels, (215) 592-3644; fax (215) 592-6808; alexandra_samuels@rohmhaas.com
Internet: http://www.rohmhaas.com/community/giving/guidelines/guidelines.html
Sponsor: Rohm and Haas Company
100 Independence Mall W
Philadelphia, PA 19106-2399

Rolex Awards for Enterprise Program **4129**
The program rewards individuals undertaking groundbreaking work in any of five fields: science and medicine, technology and innovation, exploration and discovery, the environment, and cultural heritage. The awards support individuals in carrying out their work rather than acknowledging success after the project's completion. Projects must expand knowledge of the world, improve the quality of life, or contribute in some way to the betterment of humankind. Deadline dates are May 31, for projects from Asia, the Pacific, and the Middle East; July 31, for projects from Africa and North, Central, and South America; and September 30, for projects from Western and Eastern Europe.
Requirements: The program is open to anyone of any age or nationality.
Date(s) Application is Due: May 31; Jul 31; Sep 30
Amount of Grant: 100,000 U.S.D
Contact: Contribution Program Manager; (212) 758-7700; fax (212) 759-6521; secretariat@rolexawards.com
Internet: http://www.rolexawards.com/awards/how-to-apply/index.html
Sponsor: Rolex Watch U.S.A
665 Fifth Avenue
New York, NY 10022-5385

Rollin M. Gerstacker Foundation Grants **4130**
The primary purpose of the foundation is to furnish financial aid to charities of all types supported by Mr. and Mrs. R.M. Gerstacker during their lifetimes. These charities are concentrated in Michigan and Ohio. The foundation's areas of interest include churches, community development, education, environment, health care, music and arts, research, social services, and youth, and grants are also awarded for services to the elderly and the emotionally disturbed. Some funds are designated for cultural programs and for colleges. Applications are accepted at any time.
Restrictions: Grants are not awarded to individuals, or for scholarships or fellowships
Geographic Focus: Michigan, Ohio
Date(s) Application is Due: Apr 15; Aug 15; Nov 15
Samples: Alma College (MI)—to construct and maintain a residence hall, $1 million; Nature Conservancy, Michigan Chapter (Lansing, MI)—for the Northern Great Lakes Forest Project, which will protect 271,000 acres across eight counties in Michigan's Upper Peninsula and link more than 2.5 million acres of protected federal, state, and natural areas, $800,000.
Contact: E.N. Brandt, Vice President; (989) 631-6097; fax (989) 832-8842
Sponsor: Rollin M. Gerstacker Foundation
P.O. Box 1945
Midland, MI 48641-1945

Rollins-Luetkemeyer Foundation Grants **4131**
The foundation awards grants to eligible Maryland nonprofit organizations in its areas of interest, including education (early childhood, elementary school, and higher education), health care and health organizations, historic preservation, Protestant churches and organizations, and social services. Types of support include annual campaigns, building construction/renovation, general operating support, and project support.
Requirements: The Foundation gives funding mainly for building funds for early childhood, elementary and secondary schools. They also support certain health organizations and historical preservation. There is no deadline or application form, so the applicant should submit a letter of intent.
Restrictions: Maryland nonprofit organizations, with preference given to the Baltimore area, are eligible. No grants to individuals.
Geographic Focus: Maryland
Contact: John A. Luetkemeyer, Jr., President; (443) 921-4358
Sponsor: Rollins-Luetkemeyer Foundation
1427 Clarkview Road, Suite 500
Baltimore, MD 21209

Ronald McDonald House Charities Grants **4132**
Ronald McDonald House Charities provides grants to nonprofit, tax-exempt organizations whose national or global programs help improve the health and well being of children under 21. Types of support include capital grants, challenge/matching grants, formula grants, professorships, program grants, research grants, scholarships, seed money grants, and visiting scholars. The board meets quarterly to review, select, and award grants to organizations that have demonstrated an ability to respond to the needs of specific groups of children in a definitive, hands-on manner that yields measurable results.
Requirements: To be considered for funding, an applicant must be designated a not-for-profit, tax-exempt charitable organization. U.S. charities must have a current 501(c)3 tax-exempt status letter on file with the Internal Revenue Service. Organizations seeking funding should have: a broad base of funding support; management and staff capacity to effectively execute the project; a longer-term organizational strategic plan (evidence to ensure program and organization sustainability); and a clear, concise plan for project evaluation with outcome measurement that shows potential for meaningful change. Organizations seeking funding should have a specific program that: directly improves the health and well being of children; addresses a significant funding gap or critical opportunity; has long-term impact in terms of replication or reach; produces measurable results; and is sustainable without relying on RMHC funding. Prior to submitting a full proposal, organizations must submit a letter of inquiry which will be reviewed by the Board. If interested, the Board will invite organizations to submit a formal grant proposal.
Restrictions: The charity does not fund to projects/programs that are local in scope and that do not benefit children under 21; to reduce debt; for annual fund appeals; for medical research; for for-profit organizations; for fundraising sponsorships; for programs/projects administered by activist groups; for projects that have already been completed; for sectarian or religious purposes; for scholarships and fellowships to individuals; to propagandize or influence elections or legislation; for advertising or fundraising drives; to intermediary funding agencies; for endowment campaigns; for individuals.
Date(s) Application is Due: Mar 5; Sep 7
Samples: 2008 Recipients: Andean Health & Development, Big Shoulders Fund, Childrens Hospital Los Angeles, Foundation for the Children of the Californias, Hospice St. Joseph for the Diocese of Norwich, ORBIS International, RMHC China.
Contact: Michael Singer, Executive Director; (630) 623-7048; fax (630) 623-7488; grants@rmhc.org
Internet: http://rmhc.org/what-we-do/grants/how-to-apply/
Sponsor: Ronald McDonald House Charities
1 Kroc Drive
Oak Brook, IL 60523

Rosalinde and Arthur Gilbert Foundation/AFAR New Investigator Awards in Alzheimer's Disease **4133**
The major goal of this program is to support important research in areas in which more scientific investigation is needed to improve the prevention, diagnosis, and treatment of Alzheimer's Disease. The program will also serve to encourage junior investigators in the United States and Israel to pursue research and academic careers in the neurosciences, and Alzheimer's disease in particular. Projects in basic and translational research related to Alzheimer's disease (AD) that are clinically relevant, will be considered. Examples of promising areas of research include, but are not limited to: basic mechanisms of aging in the central nervous system; genetics of AD; neuroimaging and precursors of AD; cellular and Molecular pathways of AD; biological Markers of AD; exercise and dietary factors; neurogenesis and AD; impact of environmental agents in CNS aging and early AD; inflammation; cardiovascular and cerebrovascular factors; drug discovery. It is anticipated that 5 grants of up to $75,000 each will be awarded. Applicants may propose to use the award over the course of one or two years as justified by the proposed research. Funds may not be requested for overhead or indirect costs. Funding will begin July 1.

Requirements: The applicant must be an independent investigator with assigned independent space and must have received a junior faculty appointment (instructor, assistant professor or equivalent) no sooner then July 1st. The proposed research must be conducted at any type of not-for-profit setting in the United States or Israel. Five criteria are used to determine the merit of an application: qualifications of the applicant; quality of the proposed research; relevance of the proposal to how mechanisms of aging may lead to AD; excellence of the research environment; likelihood that the project will advance the applicant's career in basic research on the mechanisms of aging and AD. All candidates must submit applications endorsed by their institution. The deadline for receipt of applications and supporting materials from candidates based in the U.S. is December 16th, at 5:00 p.m. EST. The deadline for receipt of applications and supporting materials from candidates based in Israel is December 30th. See website for instruction sheet and application for complete application procedures. Incomplete applications cannot be considered.
Restrictions: The Gilbert/AFAR Research Grant Program does not provide support for: Postdoctoral fellows in the laboratory of a senior investigator; investigators who have already received major extramural funding for research on Alzheimer's disease and/or aging (such as an R01 grant); Senior faculty, i.e. at the rank of Associate Professor level or higher; NIH Intramural program employees. Applicants cannot apply for the AFAR Research Grant Program or Glenn/AFAR BIG Program.
Geographic Focus: All States, Israel
Date(s) Application is Due: Dec 16; Dec 30
Amount of Grant: 75,000 U.S.D
Contact: Grants Manager; (212) 703-9977 or (888) 582-2327; fax (212) 997-0330; grants@afar.org or info@afar.org
Internet: http://afar.org/Gilbert.html
Sponsor: American Federation for Aging Research
55 West 39th Street, 16th Floor
New York, NY 10018

Rosalynn Carter Institute Georgia Caregiver of the Year Awards **4134**
To honor their work and to focus public attention on the contributions of caregivers throughout the State of Georgia, the Rosalynn Carter Institute on Caregiving in cooperation with the Georgia CARE-NETS initiated the Georgia Caregiver of the Year Awards in 2007. Awardess are presented with a gilded rose and a check for $1,000.
Requirements: Nominees must be caregivers from the State of Georgia.
Geographic Focus: Georgia
Amount of Grant: 1,000 U.S.D
Contact: Laura Bauer Granberry; (229) 931-2034; fax (229) 931-2663; lbgran@gsw.edu
Internet: http://www.rosalynncarter.org/caregiver2/
Sponsor: Rosalynn Carter Institute for Caregiving
800 GSW Drive
Americus, GA 31709-4379

Rosalynn Carter Institute John and Betty Pope Fellowships **4135**
The John and Betty Pope Fellowship in Caregiving is a unique program of the Rosalynn Carter Institute for Caregiving at Georgia Southwestern State University. The fellowship provides financial support for outstanding individuals pursuing training and careers in fields related to caregiving, and provides a structure and course of study intended to prepare them for leadership in the field. The certificate program of study includes courses in psychology, nursing, sociology and special education taken over up to four years of study at GSW. The amount of the award is determined annually and is currently set at $3750 per academic year.
Requirements: Award of Pope Fellowships is currently limited to students pursuing a degree program at Georgia Southwestern State University and who are or will be enrolled in the Caregiver Specialist Certificate at GSW. Applicants must have a grade point average of 3.0 as well as an established interest in academic study related to caregiving or a record of outstanding professional or informal caregiving. In addition, Pope Fellows must: enroll for minimum of 12 credit hours each semester (undergraduate) or 9 credit hours (graduate) for a full award; attend Pope Fellowship orientation and other required meetings; and make steady progress in fulfilling requirements of the Caregiver Specialist Certificate.
Date(s) Application is Due: Feb 15
Amount of Grant: 6,000 - 24,000 U.S.D
Contact: Leisa Easom, Pope Fellowship Director; (229) 928-1234; fax (229) 931-2663; leasom@gsw.edu or rci@canes.gsw.edu
Internet: http://www.rosalynncarter.org/PopeFellowship/
Sponsor: Georgia Southwestern State University
800 Wheatley Street
Americus, GA 31709-4379

Rosalynn Carter Institute Mattie J. T. Stepanek Caregiving Scholarships **4136**
The Mattie Stepanek Caregiving Scholarship was created in memory of renowned poet and peacemaker Mattie J.T. Stepanek. The purpose of the Scholarship is to provide financial assistance to family, professional, or paraprofessional caregivers who are seeking training or education in specific skills, procedures and strategies that lead to more effective care, while also protecting the health and well-being of the caregiver.
Contact: Laura Bauer Granberry; (229) 931-2034; fax (229) 931-2663; lbgran@gsw.edu
Internet: http://www.rosalynncarter.org/08mattie/
Sponsor: Rosalynn Carter Institute for Caregiving
800 GSW Drive
Americus, GA 31709-4379

Rose Community Foundation Aging Grants **4137**
In its Aging program area, Rose Community Foundation promotes change in how communities organize care and support for both seniors and caregivers, with particular attention to the needs of low- and moderate-income seniors. In addition to funding community-based programs, the Foundation plays a leadership role in strengthening the existing network of aging resources, including initiatives that bring together community and government partners to address key issues in aging. The Foundation is especially interested in projects that address the following priorities: direct in-home and community-based services; transportation; and end-of-life care.
Requirements: Colorado 501(c)3 tax-exempt organizations serving the residents of Adams, Arapahoe, Boulder, Denver, Douglas, and Jefferson Counties are eligible.
Geographic Focus: Colorado
Contact: Therese Ellery, Senior Program Officer; (303) 398-7413 or (303) 398-7400; fax (303) 398-7430; tellery@rcfdenver.org
Internet: http://www.rcfdenver.org/programs_aging.htm
Sponsor: Rose Community Foundation
600 South Cherry Street, Suite 1200
Denver, CO 80246-1712

Rose Community Foundation Health Grants **4138**
The Rose Community Foundation recognizes that improving access to quality care requires well-informed, visionary leaders. For this reason, the Foundation promotes initiatives that develop health-policy and public-health leadership. While the Foundation invests in health leadership that can improve access to quality care over the long term, it also works to effect more immediate change. Committed to improving access to care for low-income children, youth and families, the Foundation continues to support efforts to enroll them in programs such as Child Health Plan Plus (CHP+), a publicly supported health insurance program. Because at least half of all premature deaths are the result of lifestyle choices that individuals may be able to change, the Foundation also supports health promotion and disease prevention programs that encourage healthy choices and discourage behaviors that lead to illness and injury. In addition, the Foundation funds community organizations that provide information about the steps individuals can take to prevent such conditions as diabetes, teen pregnancy, HIV, heart disease and cancer.
Requirements: Colorado 501(c)3 tax-exempt organizations serving the residents of Adams, Arapahoe, Boulder, Denver, Douglas, and Jefferson Counties are eligible.
Geographic Focus: Colorado
Contact: Whitney G. Connor, Program Officer; (303) 398-7410 or (303) 398-7400; fax (303) 398-7430; wconnor@rcfdenver.org
Internet: http://www.rcfdenver.org/programs_health.htm
Sponsor: Rose Community Foundation
600 South Cherry Street, Suite 1200
Denver, CO 80246-1712

Rose Hills Foundation Grants **4139**
The foundation supports Southern California nonprofit organizations that promote the welfare of humankind. This includes but is not limited to arts and culture, civic and community services, education, community-based health programs, youth activities, and the advancement of knowledge that the foundation believes will benefit humankind. Preference is given to well-managed organizations that are self-sustaining, show promise of measurable impact, provide matching funds, and reach the greatest number of people at the most reasonable cost. Proposals are accepted and processed throughout the year with the expectation of awarding grants quarterly.
Requirements: Qualified Southern California tax-exempt charitable organizations are eligible. Emphasis is given to the San Gabriel Valley and East Los Angeles area.
Restrictions: Grants will not be given for propagandizing, influencing legislations and/or elections, or promoting voter registration; political candidates, political campaigns, or organizations engaged in political activities; programs that promote religious doctrine; individuals (except permitted by the IRS); government agencies; or endowment.
Geographic Focus: California
Contact: Victoria Rogers, President; (213) 439-9690, ext. 3; fax (213) 439-9694; vbrogers@rosehillsfoundation.org
Internet: http://www.rosehillsfoundation.org
Sponsor: Rose Hills Foundation
444 S Flower Street, Suite 1450
Los Angeles, CA 90071

Rotary Foundation Grants **4140**
The program provides grants to Rotary clubs and districts to implement humanitarian projects. The following grants are available: district simplified grants—to support short-term, humanitarian projects that benefit the community; individual grants—subsidize travel of individual or small groups for qualified international humanitarian service in Rotary countries for periods up to 60 days; matching grants—support international service projects involving Rotary clubs or districts in two or more countries; health, hunger, and humanity grants (3-H)—to fund long-term, self-help, and grassroots development projects that are too large for one club or district to implement on its own; and Blane community immunization grants—help clubs and districts in the United States address disease immunizations in their local community. Guidelines and deadlines are available online.
Contact: Grants Administrator; (847) 866-3000; fax (847) 866-9759; Grants@rotaryintl.org
Internet: http://www.rotary.org/foundation
Sponsor: Rotary Foundation
One Rotary Center, 1560 Sherman Avenue
Evanston, IL 60201

Roy A. Hunt Foundation Grants 4141

The foundation awards general grants in the areas of arts and culture; education; environmental conservation; health; human services; international affairs, development, and peace; public/safety benefit; and religion. Most grants are awarded for general operating support. The foundation also funds special initiatives in the areas of community development, environment, and youth violence prevention. Special initiative grants are administered according to specific guidelines and have earlier application deadlines. Guidelines for both categories are available online. First-time applicants are advised to submit a preliminary letter of inquiry. Proposals for general grants from previously funded organizations are considered only in November and do not require a preliminary letter of inquiry. Deadlines are April 15 and September 15 for general grants; and March 1 and August 1 for special initiatives.

Requirements: Tax-exempt organizations are eligible.
Geographic Focus: Massachusetts, Pennsylvania
Amount of Grant: 100 - 25,000 U.S.D
Samples: Peace Development Fund (Amherst, MA)—general operating support, $4000; Shadyside Hospital Foundation (Pittsburgh, PA)—for the Hillman Cancer Ctr capital campaign, $10,000; Buddhist Ray (New York, NY)—for the Tricycle magazine, $5000; Ligonier Valley Library (PA)—for general operating support, $4000.
Contact: Grants Administrator; (412) 281-8734; fax (412) 255-0522; info@rahuntfdn.org
Internet: http://www.rahuntfdn.org/programs.shtml
Sponsor: Roy A. Hunt Foundation
1 Bigelow Square, Suite 630
Pittsburgh, PA 15219-3030

Royal Norwegian Embassy Kavli Prizes 4142

The Kavli Prize is awarded every two years and includes three separate prizes awarded for outstanding scientific research in the fields of astrophysics, neuroscience and nanoscience. The Prize, which consists of a cash award, a scroll, and a medal, is intended not only as an accolade for the individual scientists concerned, but also as a recognition of important research efforts. It aims to promote international cooperation between researchers and to increase awareness among the general public of the importance of research. The Norwegian Academy of Science and Letters has full responsibility for selecting the prize winners to ensure that this process is independent of the funding organization. The Academy has set up three selection committees made up of top international scientists within the fields for which the prize is awarded. Four of the committee members are Nobel Laureates. The committee members have been chosen in close cooperation with the corresponding institutions in the U.S.A, China, France, the UK and Germany.

Requirements: The nomination process is open to all who wish to nominate candidates. The prize can be awarded to a single person for profound scientific achievements or shared for closely related fundamental contributions. Specific guidelines and a current nomination form, which must be completed online in one session, are available at the website.
Restrictions: Self nominations are not accepted, and deceased persons may not be nominated.
Geographic Focus: All States, All Countries
Date(s) Application is Due: Dec 1
Amount of Grant: 1,000,000 U.S.D
Contact: Eirik Lislerud; +47 221236640; eirik.lislerud@dnva.no
Internet: http://www.norway.org/studywork/research/prizes/Kavli/
Sponsor: Royal Norwegian Embassy in Washington
2720 34th Street NW
Washington, DC 20008

Roy and Christine Sturgis Charitable Trust Grants 4143

The Roy and Christine Sturgis Charitable Trust was established in 1981 to support and promote quality educational, cultural, human-services, and health-care programming for all people. Roy Sturgis was one of ten children of an Arkansas farmer and homemaker. He dropped out of school after the tenth grade to join the Navy during World War I. Mr. Sturgis returned to his family home in southern Arkansas after the war and went to work in the local sawmills. In 1933, he married Texas native Christine Johns. They became very successful in the timber, lumber, and sawmill industries in Arkansas, owned other prosperous business enterprises, and had notable success managing their investments. Mr. and Mrs. Sturgis spent most of their lives in Arkansas and Dallas, Texas. They did not have children, but were particularly interested in educational opportunities for young people. In addition, Mr. and Mrs. Sturgis supported organizations working in the areas of health, social services, and the arts. The Sturgis Charitable Trust encourages requests for the following types of grants: capital, project-related, medical research, and endowment campaign. Funding for start-up programs and limited general-operating requests will also be considered. Approximately 65% of the trust's annual distributions are made within the state of Arkansas. The remaining 35% of grants are distributed within the state of Texas with strong preference given to organizations located in the Dallas area. The majority of grants from the Sturgis Charitable Trust are one year in duration; on occasion, multi-year support is awarded. Applicants must apply online at the grant website. Applicants are strongly encouraged to do the following before applying: review the downloadable state application procedures for additional helpful information and clarifications; review the downloadable online-application guidelines at the grant website; review the trust's funding history (link is available from the grant website); review the online application questions in advance; and review the list of required attachments. These will generally include: a list of board members, financial statements (audited, reviewed, or compiled by independent auditor); an organization summary; a list of other funding sources; an IRS Determination letter; and other required documents. All attachments must be uploaded in the online application as PDF, Word, or Excel files. The application deadline for the Roy and Christine Sturgis Charitable Trust is 11:59 p.m. on December 31. Applicants will be notified of grant decisions before July 31.

Requirements: 501(c)3 nonprofit organizations in Texas and Arkansas are eligible.
Restrictions: Former grantees must skip a year before submitting a subsequent application. The trust does not support requests from individuals, organizations attempting to influence policy through direct lobbying, or any political campaigns.
Geographic Focus: Arkansas, Texas
Date(s) Application is Due: Dec 31
Samples: University of Arkansas, Little Rock, Arkansas, $150,000, for Endowed Professorship at Nanotechnology and Educational Center (2010); Bishop Lynch High School, Dallas, Texas, $50,000, to construct new science labs (2010); Access Group, Little Rock, Arkansas, $25,000, ACCESS Evaluation and Resource Center (program support) (2010).
Contact: David Ross, Senior Vice President; (214) 209-1965; tx.philanthropic@baml.com
Internet: https://www.bankofamerica.com/philanthropic/fn_search.action
Sponsor: Roy and Christine Sturgis Charitable Trust
901 Main Street, 19th Floor, TX1-492-19-11
Dallas, TX 75202-3714

Roy J. Carver Charitable Trust Medical and Science Research Grants 4144

The trust supports Iowa and western Illinois nonprofit organizations and projects in the area of medical and scientific research. The goal of the program in medical and scientific research is to provide support for innovative investigation that may hold great promise for advancing scientific knowledge and improving human health. In this context, the Trust has recognized a variety of university-based scientific endeavors, most notably in the fields of medicine, engineering and the natural sciences. Although grants are occasionally awarded to address institutional capital needs, the focus of this program area is typically on the work of specific laboratories, often emphasizing multidisciplinary approaches and the development of untried, inherently risky research directions. Applicants are encouraged to contact staff prior to the development of a proposal. The board considers requests at quarterly meetings in January, April, July, and October. Approved grant payments are made semiannually in April and October.

Requirements: Grants are made to 501(c)3 tax-exempt scientific organizations in Iowa and western Illinois.
Restrictions: The trust does not support annual campaigns or ongoing operations, direct grants to individuals, religious activities, organizations without 501(c)3 status, fund-raising benefits or program advertising, or political parties or candidates.
Geographic Focus: Illinois, Iowa
Date(s) Application is Due: Feb 15; May 15; Aug 15; Nov 15
Amount of Grant: 10,000 - 150,000 U.S.D
Contact: Dr. Troy Ross, Executive Administrator; (563) 263-4010; fax (563) 263-1547; info@carvertrust.org
Internet: https://www.carvertrust.org/index.php?page=56
Sponsor: Roy J. Carver Charitable Trust
202 Iowa Avenue
Muscatine, IA 52761-3733

RRF General Program Grants 4145

The program funds service, education, research, and advocacy projects. The foundation is particularly interested in innovative projects that have the potential to change practice, policy, or delivery systems. The foundation's programs seek to improve the availability and quality of community-based and institutional long-term care programs; expand opportunities for older persons to play meaningful roles in society; support selected basic, applied, and policy research into the causes and solutions of significant problems of the aged; and increase the number of professionals and paraprofessionals adequately prepared to serve the elderly. Proposals fall into three general categories: research, model projects and service, and education and training. Guidelines are available online.

Requirements: Direct service projects that seek to improve the availability and quality of community-based and institutional long-term care programs; expand opportunities for older persons to play meaningful roles in society; support selected applied and policy research into the causes and solutions of significant problems of the aged; increase the number of professionals and paraprofessionals adequately prepared to serve the elderly.
Restrictions: The foundation will only consider proposals from applicants in Illinois, Indiana, Iowa, Kentucky, Missouri, Wisconsin, and Florida. Priority is given to nonprofit organizations serving the Chicago metropolitan area.
Geographic Focus: Florida, Illinois, Indiana, Iowa, Kentucky, Missouri, Wisconsin
Date(s) Application is Due: Feb 1; May 1; Aug 1
Amount of Grant: Up to 15,000,000 U.S.D
Contact: Marilyn Hennessy, President; (773) 714-8080; fax (773) 714-8089; info@rrf.org
Internet: http://www.rrf.org/generalProgram.htm
Sponsor: Retirement Research Foundation
8765 W Higgins Road, Suite 430
Chicago, IL 60631-4170

RRF Organizational Capacity Building Grants 4146

The foundation provides grants for capacity-building activities that will help a nonprofit organization improve its management or governance to better serve the elderly. The following are examples of activities that will be supported but are not meant to exclude other ideas: planning (strategic, business, and financial plans) to steer an organization through changes; organizational assessments to identify problem areas that need to be strengthened; evaluation to develop outcome measures and determine progress in order to strengthen accountability to constituents and funding sources; public relations, communications, and marketing to increase awareness and utilization of services; financial management and resource development to improve financial systems, explore

revenue options, and develop a stronger funding base; human resources management to enhance professional development, do team building, and improve the working environment; information systems management to identify and incorporate technology to improve efficiency; restructuring and building relations among organizations to strengthen service delivery, reduce costs, share resources, develop joint evaluations, or achieve other benefits; and board assessment, recruitment, training and structuring to strengthen the governance of the organization. All organizations that receive a grant are eligible for additional funding over the course of the grant, to be used flexibly for technical assistance opportunities that arise during the course of the grant. These may include seminars, workshops, short-term courses, publications, or other training opportunities related to organizational capacity building.
Requirements: For an organization to be eligible, it must meet the following criteria: be a nonprofit organization, but not a unit of government; be located in Cook, DuPage Kane, Kendall, Lake, or McHenry County, IL; provide services to the elderly; and intentionally give a high priority to the elderly.
Restrictions: The foundation will not fund building, renovation or capital improvements; emergency needs; deficits; or one-time events that do not build long-term capacity, i.e. fundraising events or one-time conferences.
Geographic Focus: Illinois
Date(s) Application is Due: Feb 1; May 1; Aug 1
Amount of Grant: 30,000 - 75,000 U.S.D
Contact: Mary O'Donnell, Coordinator; (773) 714-8080; info@rrf.org
Internet: http://www.rrf.org/organizationalCapacityBuildingProgram.htm
Sponsor: Retirement Research Foundation
8765 W Higgins Road, Suite 430
Chicago, IL 60631-4170

RSC Abbyann D. Lynch Medal in Bioethics **4147**
The award was created to honour bioethicist and Past-President of AMS, Professor Abbyann D. Lynch, C.M., Ph.D., for her substantial achievements and support of biomedical ethics in Canada. Its purpose is to raise the profile of ethics in medicine and science by honouring leading Canadian exponents in these fields and bring recognition to their work. The bronze medal and a cash amount of $2,000 are offered every year if there is a suitable candidate.
Requirements: The medal is awarded for a major contribution in bioethics by a Canadian. The contribution may be a book, a report, a scholarly article, a monograph or series of articles, which will normally be published in the two years preceding the nomination. Candidates shall be Canadian citizens or persons having permanent resident status, and who have lived in Canada during the three years preceding the date of nomination. Nominations shall be made by three persons with the initiator being a Fellow of the Society. The form for nomination can be downloaded from the sponsor's website.
Restrictions: Self-nomination for an award is not permitted.
Geographic Focus: All States, Canada
Date(s) Application is Due: Dec 1
Amount of Grant: 2,000 CAD
Contact: Erin Hearty; (613) 991-6990; fax (613) 991-6996; fellowship@rsc-src.ca
Internet: http://www.rsc.ca/index.php?page_id=61&lang_id=1&award_id=16
Sponsor: Royal Society of Canada
170 Waller Street
Ottawa, ON K1N 9B9 Canada

RSC Jason A. Hannah Medal **4148**
The Jason A. Hannah Medal was established in 1976 by RSC with the assistance of Associated Medical Services Incorporated, through its Hannah Institute for the History of Medicine. The medal was established to honor the late President and Managing Director of AMS Inc., Dr. Jason A. Hannah, and to bring recognition to the work of Canadian research in the history of medicine. The award is normally made for a book published in the two years preceding its nomination. The bronze medal and a cash amount are offered every year if there is a suitable nomination.
Requirements: The medal is awarded for an important Canadian publication in the history of medicine. In this context, the word 'Canadian' refers to the citizenship or residence of the author, or to a content clearly relevant to Canadian medicine and health care. Nominations shall be made by three persons with the initiator being a Fellow of the Society.
Geographic Focus: All States, Canada
Date(s) Application is Due: Mar 1
Amount of Grant: 1,500 CAD
Samples: Michael Bliss, 'Harvey Cushing: A Life in Surgery'. James Opp, 'Home, Work, and Play: Situating Canadian Social History, 1840-1980'. David and Rosemary R. Gagan, 'or Patients of Moderate Means: A Social History of the Voluntary Public General Hospital in Canada, 1890-1950'. F. J. Paul Hackett, 'A Very Remarkable Sickness: Disease and Epidemics in the Petit Nord, 1670-1846'.
Contact: Erin Hearty; (613) 991-6990; fax (613) 991-6996; fellowship@rsc-src.ca
Internet: http://www.rsc.ca/index.php?page_id=61&lang_id=1&award_id=14
Sponsor: Royal Society of Canada
170 Waller Street
Ottawa, ON K1N 9B9 Canada

RSC McLaughlin Medal **4149**
This medal and cash award are given annually for important research of sustained excellence in any branch of medical sciences. Nominations must be received by the listed application deadline.
Requirements: Candidates are to be Canadian citizens or persons who have been Canadian residents for the five years prior to award date. Nominations must be made by three persons, with the initiator being a fellow of the society, and consist of citation of not more than 250 words stating concisely why the nominee should be considered, curriculum vita, list of publications, and detailed appraisal of nominee's work. The nomination dossier will remain valid for three years from the year of first nomination.
Restrictions: Self-nomination for an award is not permitted.
Geographic Focus: All States, Canada
Date(s) Application is Due: Dec 1
Amount of Grant: 2,500 CAD
Contact: Erin Hearty; (613) 991-6990; fax (613) 991-6996; fellowship@rsc-src.ca
Internet: http://www.rsc.ca/index.php?page_id=61&lang_id=1&award_id=17
Sponsor: Royal Society of Canada
170 Waller Street
Ottawa, ON K1N 9B9 Canada

RSC McNeil Medal for the Public Awareness of Science **4150**
The McNeil Medal for the Public Awareness of Science was established in 1991 through McNeil Consumer Healthcare. It is intended to highlight the important role that science plays within our society and to encourage the communication of science to students and the public. The bronze medal and a cash amount are awarded every year if there is a suitable candidate.
Requirements: The medal is awarded to a candidate who has demonstrated outstanding ability to promote and communicate science to students and the public within Canada (the term public is defined in its broadest sense). Candidates shall be Canadian citizens or persons having permanent resident status, and who have lived in Canada during the three years preceding the date of nomination. Nominations shall be made by three persons with the initiator being a Fellow of the Society.
Restrictions: Self-nomination for an award is not permitted.
Geographic Focus: All States, Canada
Date(s) Application is Due: Dec 1
Amount of Grant: 1,500 CAD
Contact: Erin Hearty; (613) 991-6990; fax (613) 991-6996; fellowship@rsc-src.ca
Internet: http://www.rsc.ca/index.php?page_id=61&lang_id=1&award_id=18
Sponsor: Royal Society of Canada
170 Waller Street
Ottawa, ON K1N 9B9 Canada

RSC Thomas W. Eadie Medal **4151**
The medal is awarded annually in recognition of major contributions to any field through engineering or applied science, with preference given to those having an impact on communications, in particular the development of the Internet. Nominations must be received by the listed application deadline. The award includes a bronze medal and a cash amount.
Requirements: Candidates are to be Canadian citizens or persons who have been Canadian residents for the five years prior to award date. Nominations must be made by three persons, with the initiator being a fellow of the society, and consist of citation of not more than 250 words stating concisely why the nominee should be considered, curriculum vita, list of publications, and detailed appraisal of nominee's work. The nomination dossier will remain valid for five years from the year of first nomination.
Restrictions: Self-nomination for an award is not permitted.
Geographic Focus: All States, Canada
Date(s) Application is Due: Dec 1
Amount of Grant: 3,000 CAD
Samples: 2007 - Hussein T. Mouftah; 2006 - Alberto Leon-Garcia; 2005 - Norman C. Beaulieu; 2004 - Vijay K. Bhargava; 2003 - Morrel P. Bachynski.
Contact: Erin Hearty; (613) 991-6990; fax (613) 991-6996; fellowship@rsc-src.ca
Internet: http://www.rsc.ca/index.php?page_id=61&lang_id=1&award_id=5
Sponsor: Royal Society of Canada
170 Waller Street
Ottawa, ON K1N 9B9 Canada

Rudolf Steiner Foundation Grants **4152**
The fund makes grants to initiatives and new programs that fall within its mission statement and that demonstrate the capacity to further social and cultural development. The fund supports initiatives that encourage growth and eventual financial independence in the following areas: education and the arts, science and care of the earth, social responsibility, medical and religious renewal, and associative economic relationships. Projects that break new social or cultural ground on the basis of spiritual insight within the framework of the foundtation's mission receive preference.
Requirements: 501(c)3 tax-exempt organizations are eligible.
Date(s) Application is Due: Mar 15; Sep 15
Amount of Grant: 1,000 - 5,000 U.S.D
Contact: Kerri Kleven; (415) 561-6161; kerri.kleven@rsfoundation.org
Internet: http://www.rsfoundation.org/index.asp?nav=STRE&page=strengthen_grants.htm
Sponsor: Rudolf Suiteiner Foundation
P.O. Box 29915
San Francisco, CA 94129-0915

Rural Community Foundation Grants **4153**
The foundation awards grants for social, educational, or other charitable purposes in a 12-county region in northern Missouri and southern Iowa. General grants are awarded in four target areas: education, health, rural development, and agriculture/environment. These four target areas are not static and may change as the needs of the communities served change. Types of support include program development, equipment, building construction/renovation, social services (abuse shelters, food pantires, hospices, and family/human services), business development, and grassroots leadership development. The board meets quarterly.

Requirements: Giving is limited to nonprofit organizations in a 12-county area, including Harrison, Mercer, Grundy, Linn, Livingston, Daviess, Gentry, Putnam and Sullivan Counties in Missouri and Ringgold, Wayne and Decatur Counties in Iowa.
Geographic Focus: Iowa, Missouri
Date(s) Application is Due: Mar 1; May 1; Aug 1; Dec 1
Contact: Scott Long, Executive Director; (660) 359-1404; fax (660) 359-6030; scott@ruralcommunity.org
Internet: http://www.ruralcommunity.org/index.html
Sponsor: Rural Community Foundation
4100 Oklahoma Avenue, P.O. Box 462
Trenton, MO 64683

Rush County Community Foundation Grants **4154**
The Rush County Community Foundation, is a nonprofit public charity established in 1991 to serve donors, award grants and scholarships, and provide leadership to enrich and enhance the quality of life in Rush County, Indiana. The Rush County Community Foundation now holds over 130 funds, and have permanent endowment assets of over $6.5 million. As a public foundation, it helps donors provide grant making dollars for not-for-profit organizations that serve Rush County citizens. For additional information contact the Foundations office.
Requirements: Funding opportunities must serve the Rush County citzens.
Geographic Focus: Indiana
Date(s) Application is Due: Feb 15; Jun 15; Oct 15
Contact: Garry Cooley, Executive Director; (765) 938-1177; fax (765) 938-1719; rccf@verizon.net
Internet: http://www.rushcountyfoundation.org/funds.php
Sponsor: Rush County Community Foundation
117 North Main Street
Rushville, IN 46173

Russell Berrie Foundation Grants **4155**
The foundation supports nonprofit organizations primarily in Israel and the New York and New Jersey metropolitan areas. Grants are awarded for education (colleges and universities, Jewish education, divinity schools, parochial elementary and secondary schools, and religious higher education), health care (hospitals and respite homes), international programs (development programs, ministries, and missions), religious organizations (religious welfare, synagogues, and temples), social issues, and youth welfare. Types of support include capital support and general operating support. Applicants should submit a brief letter of inquiry and include a description of the program and the organization, including a financial statement. There are no application deadlines.
Requirements: Christian, Jewish, and Roman Catholic organizations in New York, New Jersey, and Israel are eligible.
Geographic Focus: New Jersey, New York, Israel
Amount of Grant: 5,000 - 20,000 U.S.D
Samples: Pave the Way Foundation (New York, NY)—for unrestricted support of this organization that promotes good will and reconciliation between different religions, $100,000; American Technion Society (New York, NY)—to establish a nanotechnology institute at the Technion-Israel Institute of Technology, in Haifa, $26 million.
Contact: Susan Strunk, Administrative Director; (201) 928-1880
Internet: http://www.russberrie.com/foundation.html
Sponsor: Russell Berrie Foundation
300 Frank Burr Boulevard, Building E, 7th Floor
Teaneck, NJ 07666-6704

Ruth Anderson Foundation Grants **4156**
The Ruth Anderson Foundation awards grants to Florida nonprofit organizations in Dade County. The Foundation's funding interests include AIDS research; children and youth services; homeless/housing shelters; human services; and substance abuse services. There are no application forms or deadlines. The initial approach should consist of a brief exploratory letter; full proposals will be by invitation only. The board meets throughout the year to consider requests for funding.
Requirements: Miami Dade County 501(c)3 tax-exempt organizations are eligible.
Restrictions: Grants are not awarded to individuals. Funding is limited to the Miami area.
Geographic Focus: Florida
Amount of Grant: 1,000 - 10,000 U.S.D
Contact: Funding Contact; (305) 789-4929
Sponsor: Ruth Anderson Foundation
1525 W. W.T. Harris Boulevard, D1114-044
Charlotte, NC 28288-1161

Ruth and Vernon Taylor Foundation Grants **4157**
The foundation awards grants to nonprofit organizations in the areas of arts and humanities, civic and public affairs, secondary schools, higher education, environment, hospitals, human services, health, youth services, and social services. Types of support include general operating support, building construction and renovation, endowment funds, and research. The foundation suggests that initial contact be made in writing, since unsolicited requests for funds are not accepted. The Board meets in May and September.
Requirements: Organizations located in Colorado, Illinois, Montana, New Jersey, New York, Pennsylvania, Texas, or Wyoming are eligible.
Restrictions: Grants are not awarded to individuals.
Geographic Focus: Colorado, Illinois, Montana, New Jersey, New York, Pennsylvania, Texas, Wyoming
Amount of Grant: 1,000 - 20,000 U.S.D
Contact: Douglas Taylor, Trustee; (303) 893-5284; fax (303)893-8263
Sponsor: Ruth and Vernon Taylor Foundation
518 17th Street, Suite 1670
Denver, CO 80202

Ruth Eleanor Bamberger and John Ernest Bamberger Memorial Foundation Grants **4158**
The Foundation is dedicated to fulfilling the Founders' desires to help people reach their individual potential. The Foundation assists people of all ages, but especially children and young people through educational opportunities and scholarships, supporting crisis care and protective services, dental aid, after school programs, etc.
Requirements: Only residents of Utah may apply.
Restrictions: Grants and scholarships are given only to organizations, not to individuals personally.
Geographic Focus: Utah
Date(s) Application is Due: Mar 16; Sep 28
Amount of Grant: 1,000 - 20,000 U.S.D
Contact: Eleanor Roser; (801) 364-2045; bambergermemfdn@qwestoffice.net
Internet: http://www.ruthandjohnbambergermemorialfdn.org/
Sponsor: Ruth Eleanor Bamberger and John Ernest Bamberger Memorial Foundation
136 S Main, Suite 418
Salt Lake City, UT 84101-1690

Ruth H. and Warren A. Ellsworth Foundation Grants **4159**
The foundation awards grants to nonprofits in Massachusetts in support of the arts, children and youth, community development, education and higher education, and health care and hospitals. Types of support include general operating support, continuing support, annual campaigns, building construction/renovation, equipment acquisition, emergency funds, and seed grants. There are no application forms.
Requirements: Nonprofits in the Worcester, MA, area are eligible.
Restrictions: Grants are not awarded to individuals or for endowment funds, scholarships, fellowships, research, publications, conferences, matching gifts, or loans.
Geographic Focus: Massachusetts
Amount of Grant: 500 - 100,000 U.S.D
Samples: Bancroft School, Worcester, MA, $125,000; UMass Memorial Foundation, Worcester, MA, $50,000;
Contact: Sumner Tilton Jr., Trustee; (508) 798-8621
Sponsor: Ruth H. and Warren A. Ellsworth Foundation
370 Main Street, 12th Floor, Suite 1250
Worcester, MA 01608-1723

Ruth Mott Foundation Grants **4160**
The community is invited to submit grant requests for the program areas of arts—culture, music, theater, dance, media, graphics, photography, storytelling, and crafts; beautification—enhance neighborhoods and quality of life in the greater Flint area; and health promotion—healthy and active lifestyles and environments such as physical activity, proper nutrition, responsible sexual behavior, and overall well-being. Types of support include special projects, capacity building (planning, technical assistance, and resource development), evaluation, and general purposes. Priority is given to requests that bring diverse segments of the community together to share ideas and common interests and to respectfully address and resolve differences; enable the engagement of citizens in civic life, community service and leadership; ensure that all stakeholders are informed and involved in decisions affecting them; promote life-long learning and involvement; share and celebrate community members' unique talents, histories, cultures, and contributions; and benefit low-income segments of the community and reduce social, economic, and racial disparities. Concept papers may be submitted at any time throughout the year; full proposals are by invitation.
Requirements: Michigan nonprofit organizations are eligible. Preference is given to requests from Flint, MI.
Amount of Grant: Up to 65,000,000 U.S.D
Samples: Flint Institute of Music (MI)—for the Bradley's Tap Festival for Youth, $72,000; Genesis Pentecostal Church of God (MI)—for the Peace Garden, $57,800; F.A.C.E.D. (MI)—to build organizational capacity, $25,000; American Arab Heritage Council (MI)—to support the Family Health and Art Fair, $28,000.
Contact: Grants Administrator; (810) 233-0170
Internet: http://www.ruthmottfoundation.org
Sponsor: Ruth Mott Foundation
111 E Court Street, Suite 3C
Flint, MI 48502-1649

RWJF Bridging the Gap: Research Informing Practice and Policy for Healthy Youth Behavior **4161**
RWJF has developed three integrated strategies to reverse the childhood obesity epidemic: evidence, action and advocacy. The Bridging the Gap: Research Informing Practice and Policy for Healthy Youth Behavior program is intended to improve understanding of school, community, state and national policies and environmental factors affecting youth diet, physical activity, obesity, and tobacco, alcohol and drug use, and to evaluate the effectiveness of interventions to prevent youth obesity and tobacco use.
Requirements: Eligible applicants for research grants are scholars associated with educational institutions, research organizations, health care providers or other public or nonprofit organizations. Preference will be given to public entities or nonprofit organizations tax exempt under Section 501(c)3 of the Internal Revenue Code. Applicants

may be independent of or associated with organizations currently receiving grants. All proposals must be submitted only through the RWJF Grantmaking Online system.
Contact: C. Tracy Orleans, (609) 627-5962 or (877) 843-7953; cto@rwjf.org
Internet: http://www.rwjf.org/grants/npo.jsp?FUND_ID=55100
Sponsor: Robert Wood Johnson Foundation
College Road E and Route 1, P.O. Box 2316
Princeton, NJ 08543-2316

RWJF Changes in Health Care Financing and Organization Grants **4162**
The program supports policy analysis, research, evaluation and demonstration projects that provide policy leaders timely information on health care policy and financing issues. Supported projects include: examining significant issues and interventions related to health care financing and organization and their effects on health care costs, quality and access; and exploring or testing major new ways to finance and organize health care that have the potential to improve access to more affordable and higher quality health services. The grant period is 12 or fewer months for small grants and 12 or more months for large grants. There are no application deadlines. Program information and the online application process are available online.
Requirements: Researchers, as well as practitioners and public and private policy-makers working with researchers, are eligible to apply. Projects may be initiated from within many disciplines, including health services research, economics, sociology, political science, public policy, public health, public administration, law and business administration. Multidisciplinary teams and researchers who are just beginning their careers - perhaps teaming with a more senior researcher to develop the analytic approach - are especially encouraged to apply. Preference will be given to applicants that are public agencies or are tax-exempt under Section 501(c)3 of the Internal Revenue Code and are not private foundations as defined under Section 509(a). Proposals for this program must be submitted through the RWJF Grantmaking Online system.
Restrictions: Grant funds may not be used to subsidize individuals for the cost of health care, to construct or renovate facilities, for lobbying, or as a substitute for funds currently being used to support similar activities.
Samples: U of Southern Maine (ME)—to support the research of Dr. J. William Thomas, Using Physician Profiling Software to Evaluate the Practice Efficiency of Physician Specialists, $102,829; Georgetown U (WA)—to support the research of Mila Kofman, JD, Private Insurance Markets: The Missing Link-Association Health Plans and Other Pooled Purchasing Arrangements, $230,547; U of Kansas Medical Center Research Institute (KS)—to support the research of Dr. Mary Zimmerman, Local Funding for Health Services in Rural Counties, $100,000.
Contact: Program Officer; (202) 292-6700; fax (202) 292-6800; hcfo@academyhealth.org
Internet: http://www.rwjf.org/applications/solicited/cfp.jsp?ID=19274
Sponsor: Robert Wood Johnson Foundation
College Road E and Route 1, P.O. Box 2316
Princeton, NJ 08543-2316

RWJF Clinical Scholars Program Grants **4163**
The program fosters the development of leaders who will transform U.S. health and health care. These future leaders will be equipped to work with communities and to conduct innovative research as a result of the didactic and experiential learning they receive at each of four chosen training sites (University of California, Los Angeles; University of Michigan; University of Pennsylvania; and Yale University). The program's focus emphasizes community-based research and leadership training. The goal of the program is the integration of scholars' clinical training with skills in program development and research methodology to help them find solutions for issues in public policy, community health, and health services research. The program offers graduate-level study and research as part of a university-based postresidency training program. Up to 29 scholars will be selected for appointments beginning the following year. Under the program, the universities provide Scholars with their stipends and health insurance. Stipends currently range from $49,000 to $51,500 per year, depending on the training site. Limited financial support is also available at each university for expenses associated with Scholars' research, including travel.
Requirements: To be eligible, physicians must be U.S. citizens; be committed to a career consonant with the program's purposes and priorities; be highly regarded by those responsible for their clinical training; and plan to complete the clinical requirements of their residency/fellowship training by the date of entry into the program (except for surgeons). Scholars may not hold appointments as sub-specialty fellows during their tenure in the program.
Date(s) Application is Due: Feb 15
Amount of Grant: 49,000 - 51,500 U.S.D
Contact: Desmond Kimo Runyan, Program Director; (919) 843-8261; fax (650) 566-2340; drunyan@med.unc.edu or rwjcsp_admin@med.unc.edu
Internet: http://rwjcsp.unc.edu/
Sponsor: Robert Wood Johnson Foundation
333 MacNider Hall
Chapel Hill, NC 27599-7105

RWJF Community Health Leaders Awards **4164**
Each year, the program honors ten outstanding individuals who overcome daunting odds to expand access to health care and social services to underserved and isolated populations in communities across the United States. Awards include a personal stipend and a grant for program enhancement. Potential nominators must submit a letter of intent; if a letter of intent is approved, a packet is sent to the nominator. For more information, contact the program office.
Requirements: Candidates must be residents of the United States, be affiliated with a 501(c)3 tax-exempt organization, have a five- to 15-year record of work in the community-health field, have experience as a direct provider of health-care services or be involved in the improvement of health-care access and delivery, demonstrate a full-time commitment to working with needy people, and have received no significant national recognition.
Restrictions: Nominations are not accepted from development and public relations departments or professional grant writers. Self-nominations are not accepted.
Date(s) Application is Due: Oct 15
Amount of Grant: 125,000 U.S.D
Samples: Elisabeth Arenales, J.D. - Healthcare Program Director, Colorado Center on Law and Policy; Sharon Baskerville - Executive Director, DC Primary Care Association; Gregg Croteau, M.S.W., L.C.S.W. - Executive Director, United Teen Equality Center, Inc.; Monty Fakhouri, M.S.C.H.S., C.H.E.S. - Director of Public Health and Youth Programs/Services, Arab American Chaldean Council.
Contact: Helen Dundas, (609) 627-5947 or (609) 627-5809; fax (609) 419-8347; hdundas@rwjf.org
Internet: http://www.rwjf.org/applications/solicited/cfp.jsp?ID=20902
Sponsor: Robert Wood Johnson Foundation
College Road E and Route 1, P.O. Box 2316
Princeton, NJ 08543-2316

RWJF Evaluating Innovations in Nursing Education Grants **4165**
Evaluating Innovations in Nursing Education (EIN) will support evaluations of interventions that expand teaching capacity or promote faculty recruitment and retention in nursing schools. The program aims to increase the number of nursing school graduates by evaluating strategies that address the nurse faculty shortage. An important activity of the program will be disseminating successful strategies so they can be replicated in other nursing education settings.
Requirements: Nursing schools or health care organizations who will act as the applicant institution and subcontract to the evaluation team are eligible. Preference will be given to applicant organizations that are tax-exempt under Section 501(c)3 of the Internal Revenue Code and are not private foundations as defined under Section 509(a). The Robert Wood Johnson Foundation makes grants to organizations in the United States and its territories only.
Date(s) Application is Due: Nov 24
Amount of Grant: Up to 300,000 U.S.D
Contact: Program Director; (732) 932-4670; info@evaluatinginnovationsinnursing.org
Internet: http://www.evaluatinginnovationsinnursing.org/
Sponsor: Robert Wood Johnson Foundation
College Road E and Route 1, P.O. Box 2316
Princeton, NJ 08543-2316

RWJF Executive Nurse Fellowships **4166**
This program is an advanced leadership program for nurses in senior executive roles who are aspiring to lead and shape the U.S. health care system of the future. The three-year fellowships are intended to offer participating nurses the experiences, insights, competencies, and skills necessary to achieve or advance in executive leadership positions in a health care system undergoing unprecedented change. Each Fellow's individual learning activities are supported by a leadership development account of $15,000 ($7,000 in year one, $4,000 in year two, and $4,000 in year three). These funds support self-directed learning activities, independent study, and access to an electronic communications network. The program office also will provide matching support for the Fellow's leadership project of $20,000 over the first two years of the program ($10,000 in year one and $10,000 in year two). The Fellow's employing organization must match the project resources of $20,000 dollar for dollar, either in cash or in-kind. Applications must be submitted online.
Requirements: Nurses in executive roles in the fields of health services, public and community health, or nursing education who possess a strong professional record reflecting positions of increasing leadership responsibility are eligible.
Date(s) Application is Due: Feb 1
Amount of Grant: 15,000 U.S.D
Samples: Jeannette Andrews, PhD, APRN, BC-Chair and Associate Professor, Biobehavioral Nursing Medical College of Georgia (Augusta, GA); Kim Hoover, PhD, RN-Associate Dean for Research and Evaluation, School of Nursing, University of Mississippi Medical Center (Jackson, MS); Susan Bakewell-Sachs, PhD, RN, APRN, BC-Dean and Professor, School of Nursing, Health & Exercise Science, The College of New Jersey (Ewing, NJ); Jill Briggs, RN, MS, MPH-Manager, Community and Family Health Division, Minnesota Department of Health (St. Paul, MN)
Contact: Maryjoan D. Ladden, Program Officer; (877) 843-7953
Linda Cronenwett, Ph.D., R.N., F.A.A.N., Nursing Program Director; (919) 966-6840; lcronenwett@unc.edu
Internet: http://www.executivenursefellows.org./
Sponsor: Robert Wood Johnson Foundation
College Road E and Route 1, P.O. Box 2316
Princeton, NJ 08543-2316

RWJF Finding Answers: Disparities Research for Change Grants **4167**
The program was created to support rigorous evaluations of practical and replicable solutions designed to reduce and eliminate disease specific racial and ethnic health care disparities. It seeks to improve the quality of health care provided to patients affected by racial and ethnic disparities by: Granting funds to discover and evaluate practical and replicable solutions designed to reduce and eliminate disease specific racial and ethnic health care disparities; Focusing on interventions aimed at health care delivery for one or more of the following health concerns: cardiovascular disease, depression, or diabetes; Conducting systematic reviews of the literature regarding racial and ethnic health care disparities interventions; and Disseminating results from these research efforts and systematic reviews to encourage health care systems to address racial and ethnic gaps in care.

Requirements: For-profit organizations are eligible to apply, however preference will be given to applicants that are public agencies or are tax-exempt under Section 501(c)3 of the Internal Revenue Code and are not private foundations as defined under Section 509(a). Proposals must focus on innovative interventions that reduce racial and ethnic disparities in the treatment of cardiovascular disease, depression, and diabetes. There are no age restrictions for the target populations of the interventions proposed for evaluation using program funds. Although there are no restrictions on the age of target populations, proposals targeting pediatric and geriatric populations are encouraged. Applications are accepted in two phases. The first phase is a brief proposal submission open to all interested research teams. The second phase is a full proposal submission and will be accepted by invitation only. All applications, both the brief and full proposals, must be submitted through the Robert Wood Johnson Foundation's Grantmaking Online system.
Restrictions: Individuals are not eligible to apply. Stand-alone prevention interventions such as tobacco cessation or exercise programs are not eligible unless they are one part of a larger, more comprehensive, intervention. The program will only directly fund health care delivery institutions. However, eligible institutions may partner/sub-contract with academic or research groups.
Amount of Grant: 100,000 - 275,000 U.S.D
Contact: Marshall H. Chin, Director; (866) 344-9800 or (773) 702-4769; mchin@medicine.bsd.uchicago.edu
Internet: http://www.solvingdisparities.org/
Sponsor: Robert Wood Johnson Foundation
College Road E and Route 1, P.O. Box 2316
Princeton, NJ 08543-2316

RWJF Generalist Physician Faculty Scholars **4168**
The national career development program is designed to develop leaders in primary care academics by providing funding for junior faculty research. Each year the program provides $300,000 for 15 junior faculty members in family medicine, internal medicine and general pediatrics, to conduct four-year research projects nationwide. Under the program, the scholars, sponsored by their medical schools, receive support to enhance their research productivity, develop innovative mentoring programs for medical students, and participate in the design and implementation of primary education initiatives. The program coordinates several activities including an annual meeting for the scholars to present their research findings and conference calls for the scholars to discuss professional issues.
Requirements: Accredited colleges and universities within the United States are eligible to apply.
Amount of Grant: Up to 300,000 U.S.D
Samples: Dartmouth Medical School Center for the Evaluative Clinical Sciences, $240,906; University of California, San Francisco, School of Medicine, $300,000; University of Colorado at Denver and Health Sciences Center, $299,980; Mount Sinai School of Medicine of New York University, $300,000; University of Pennsylvania School of Medicine, $299,993.
Contact: Cynthia D. Mulrow, (210) 877-2275; mulrowc@uthscsa.edu
Internet: http://www.rwjf.org/applications/solicited/npo.jsp?FUND_ID=55121
Sponsor: Robert Wood Johnson Foundation
College Road E and Route 1, P.O. Box 2316
Princeton, NJ 08543-2316

RWJF Harold Amos Medical Faculty Development Scholars **4169**
The Harold Amos Medical Faculty Development Program supports basic/biomedical, clinical, and health services/epidemiology research. The purpose of the AMFDP award is to facilitate the transition of the newly-trained clinician who wishes to develop into an independent investigator. A completed application consists of two parts: the completion of the online proposal process, and submission of other (paper) supporting documents.
Requirements: To be eligible, applicants must be physicians from historically disadvantaged backgrounds (ethnic, financial, or educational) who are U.S. citizens or permanent residents at the time of application deadline; have excelled in their education; are now completing or have completed their formal clinical training; are prepared to devote four consecutive years to research; are committed to pursuing academic careers; and are committed to improving the health status of the underserved, decreasing health disparities, or serving as role models for students and faculty from historically disadvantaged backgrounds Preference will be given to physicians who have recently completed their formal clinical training. In order to pursue the advanced research training required by this program, applicants must identify faculty mentors with whom a research plan will be developed. The mentor's experience in the supervision of trainees and the adequacy of the mentor's research environment and support will be strongly considered. Deadline for receipt of completed online applications - March 23. Deadline for postmark of all supporting documents and hard copies - March 24. Complete guidelines and application instructions are available at the website.
Date(s) Application is Due: Mar 23; Mar 24
Amount of Grant: 104,140 U.S.D
Contact: Nina Ardery, Program Deputy Director; (301) 565-4080 or (317) 278-0500; fax (301) 565-4088; nardery@erols.com or amfdp@starpower.net
Internet: http://www.amfdp.org/applicat.htm
Sponsor: Robert Wood Johnson Foundation
College Road E and Route 1, P.O. Box 2316
Princeton, NJ 08543-2316

RWJF Health & Society Scholars **4170**
The program is a two-year fellowship designed to build the nation's capacity for research, leadership, and action to address the broad range of factors affecting health. The goal of the program is to improve health by training scholars to rigorously investigate the connections between biological, behavioral, environmental, economic, and social determinants of health; and develop, evaluate, and disseminate knowledge and interventions based on integration of these determinants. Each year the program will enable up to 18 outstanding individuals who have completed doctoral training to engage in an intensive two-year program at one of six nationally prominent universities. Applications and guidelines are available online.
Requirements: Individuals must apply only through the foundation's online system (http://www.hsscholars.org). To be eligible, scholars must have completed doctoral training by the time of entry into the program (August or September) in one of a variety of fields including, but not limited, to the behavioral and social sciences, the biological and natural sciences, health professions, public policy, public health, history and ethics; have significant research experience; clearly connect their research interests to substantive population health concerns; and be citizens or permanent residents of the United States or its territories. Applications from candidates with diverse backgrounds are encouraged.
Restrictions: Individuals placed in the program must be willing to relocate to the institution designated during the match process for the duration of the two-year program.
Date(s) Application is Due: Oct 2
Amount of Grant: 89,000 U.S.D
Contact: Pamela G. Russo, Program Officer; (212) 419-3566; fax (212) 419-3569; hss@nyam.org
Internet: http://www.healthandsocietyscholars.org
Sponsor: Robert Wood Johnson Foundation
College Road E and Route 1, P.O. Box 2316
Princeton, NJ 08543-2316

RWJF Health Care Costs: Research to Inform Policy Grants **4171**
The Robert Wood Johnson Foundation, through its Changes in Health Care Financing and Organization (HCFO) initiative, is issuing a special topic solicitation for proposals to conduct research and analyses on the complex causes and potential solutions to rising health care costs. The solicitation seeks research contributing to an evidence base that defines and addresses the problem of rising health care costs, and how it relates to health insurance coverage. Research should focus on identifying sources of continued cost growth, examining promising cost control efforts, and exploring the relationship between health care costs and health insurance. The research questions are broken down into four major categories (Cost Controls, Affordability, Pricing, and Cost Estimates). The questions are not an exhaustive list, but have been identified among the priorities for the foundation. However, these should not be viewed as prescriptive.
Requirements: Researchers should have a demonstrated track record in health services research, economics, financing, or public policy. Junior researchers may apply as the principal investigator. However, to be competitive, the proposal must provide evidence that there is sufficient senior research oversight and support. All applicants are required to submit a brief proposal.
Restrictions: Grants are not available for ongoing general operating expenses or existing deficits; endowment or capital costs, including construction, renovation, or equipment purchases; basic biomedical research; conferences, symposia, publications, or media projects, unless they are clearly related to the foundation's goals; research on unapproved drug therapies or devices; international programs and institutions; or direct support to individuals.
Contact: Bonnie J. Austin, Deputy Director; (202) 292-6700; fax (202) 292-6800; Bonnie.Austin@academyhealth.org or hcfo@academyhealth.org
Internet: http://www.hcfo.net/grants.htm
Sponsor: Robert Wood Johnson Foundation
College Road E and Route 1, P.O. Box 2316
Princeton, NJ 08543-2316

RWJF Health Impact Project Grants **4172**
The Health Impact Project, a collaboration of the Robert Wood Johnson Foundation and The Pew Charitable Trusts, encourages the use of health impact assessment (HIA) to help decision-makers better assess proposed policies, projects and programs with respect to their impact on health, in order that they may avoid adverse health consequences and costs and improve health. The foundations are issuing this call for proposals to demonstrate the effectiveness of HIAs and promote their incorporation into local, state, tribal and federal decision-making. Grants will range from $25,000 to $150,000 and must be completed within 24 months. Proposals for more than $150,000 may be considered under rare and exceptional circumstances.
Requirements: Eligible applicant organizations include: state, tribal or local agencies; tax-exempt educational institutions; or publicly supported charitable organizations that are exempt from federal income tax as an organization described by Section 501(c)3 of the Code. Applicant organizations must be located in the United States or its territories at the time of application.
Amount of Grant: 25,000 - 150,000 U.S.D
Contact: Program Director; (202) 540-6012; healthimpactproject@pewtrusts.org
Internet: http://www.rwjf.org/applications/solicited/cfp.jsp?ID=20921
Sponsor: Robert Wood Johnson Foundation
College Road E and Route 1, P.O. Box 2316
Princeton, NJ 08543-2316

RWJF Health Policy Fellowships **4173**
The purpose of this fellowship is to give outstanding mid-career health professionals in academic and community-based settings an opportunity to learn more about the legislative process through direct participation. Fellows actively contribute to the formulation of national health policies and accelerate their careers as leaders in health policy. Up to six grants of up to $165,000 each will be made. Each fellow will receive up to $94,000 for the Washington stay in salary plus fringe benefits or fellowship stipend. Fellows will receive

an additional allowance for relocation subject to limitations provided in detail on the program's Web site. November 13 is the deadline for receipt of applications submitted online, with January 29 the deadline for receipt of full proposals.
Requirements: Applicants may have backgrounds in the following disciplines: allied health professions; biomedical sciences; dentistry; economics or other social sciences; health services organization and administration; medicine; nursing; public health; or social and behavioral health. Applicants must be U.S. citizens or have permanent residency status at the time of application. Potential applicants must obtain documented sponsorship from an eligible institution to support their application. Nonprofit health care organizations and academic centers are eligible sponsors. Eligible institutions may sponsor more than one candidate, but no more than one individual will be selected from any one sponsoring organization.
Restrictions: No additional direct costs (such as laptops or PDAs) or indirect costs are paid for during the first year of this program.
Date(s) Application is Due: Jan 29; Nov 13
Amount of Grant: Up to 165,000 U.S.D
Contact: Marie Michnich; (202) 334-1506; fax (202) 334-3862; mmichnich@nas.edu
Internet: http://www.rwjf.org/applications/solicited/cfp.jsp?ID=20904
Sponsor: Robert Wood Johnson Foundation
College Road E and Route 1, P.O. Box 2316
Princeton, NJ 08543-2316

RWJF Healthy Eating Research Grants 4174
Healthy Eating Research is a national program that will support research to identify, analyze, and evaluate environmental and policy strategies that can promote healthy eating among children and prevent childhood obesity. Special emphasis will be given to research projects that benefit children in the low-income and racial/ethnic populations at highest risk for obesity. The program will fund two types of research grants: studies to identify and/or evaluate promising school food environment and policy changes (12- to 18-month awards up to $100,000 each, and 18- to 36-month awards up to $400,000 each); and analyses of macro-level policy or system determinants of school food environments and policies (12- to 18-month awards up to $75,000 each). Applications must be submitted through the online process. Calls for Proposals and guidelines are available online.
Requirements: Preference will be given to those applicants that may be either public entities or nonprofit organizations that are tax-exempt under Section 501(c)3 of the Internal Revenue Code. Applicant organizations must be based in the United States or U.S. Territories. The focus of this program is the United States; studies of policies in other countries will be considered only to the extent that they may directly inform U.S. policy. The experience and qualifications of the research team is one of the primary criteria for proposal review. A doctorate or terminal degree is preferred for the principal investigator. Deadline for receipt of brief proposals - February 13. Select applicants will be invited to submit full proposals. Deadline for receipt of full proposals - May 23.
Restrictions: This program does not fund demonstration projects. The foundation does not award grants to private individuals or applicants from outside the U.S. or U.S. Territories. No faxed, emailed or mailed proposals will be accepted.
Amount of Grant: Up to 400,000 U.S.D
Contact: Kathy Kosiak, Research Coordinator; (800) 578-8636; fax (612) 624-9328; healthyeating@umn.edu
Internet: http://www.healthyeatingresearch.org/
Sponsor: Robert Wood Johnson Foundation
College Road E and Route 1, P.O. Box 2316
Princeton, NJ 08543-2316

RWJF Interdisciplinary Nursing Quality Research Initiative 4175
The initiative awards two-year grants to support the generation, dissemination, and translation of research to improve the contributions of nursing to the quality of hospital care. It will support interdisciplinary teams of nurse scholars and scholars from other disciplines to address gaps in knowledge about the relationship between nursing and health care quality. Topic areas include: identifying, developing, and testing new measures to advance quality nursing research; and testing the effects of implementing the nursing-sensitive performance measures endorsed by the National Quality Forum. Investigator-initiated ideas that are consistent with the broader set of program goals also will be considered. The two-phase application process begins with online submittal of a brief proposal dur by December 4; full proposals are by invitation, and are due by April 9. Guidelines are available online.
Requirements: Teams of researchers who are U.S. citizens or permanent residents are eligible. Each team must have at least one nursing researcher and at least one researcher from another discipline. Preference will be given to 501(c)3 applicants that are not 509(a) private foundations.
Date(s) Application is Due: Apr 9; Dec 4
Amount of Grant: Up to 300,000 U.S.D
Contact: Lori A. Melichar, Grants Administrator; (301) 571-8161 or (800) 978-8309; INQRIhelpdesk@constellagroup.com
Internet: http://www.inqri.org
Sponsor: Robert Wood Johnson Foundation
College Road E and Route 1, P.O. Box 2316
Princeton, NJ 08543-2316

RWJF Investigator Awards in Health Policy Research 4176
The program supports researchers whose cross-cutting and innovative ideas promise to contribute meaningfully to improving health and health policy in America. Each year, the program awards 24- to 36-month grants of up to $335,000 to highly qualified individuals to fund approximately 10 research projects that have national policy relevance. The program funds projects with a broader scope than traditional NIH or NSF grants. Data-driven health services research focusing on a narrowly defined research question is not appropriate for this program, nor are pilot projects that are designed to evaluate clinical care in a specific geographic area or to demonstrate service delivery programs or financing strategies.
Requirements: Individuals in the United States who are affiliated either with a U.S.-based educational institution or a 501(c)3 tax-exempt organization are eligible. Applications are welcomed from investigators in fields such as anthropology, business, demography, economics, engineering, ethics, genetics, health and social policy, history, journalism, law, medicine, nursing, political science, psychology, public health, science policy, social work, sociology and others. The program seeks a diverse group of applicants including minorities, researchers early in their careers, and individuals in nonacademic settings. Key Dates: Deadline for receipt of letters of intent - March 28; Applicants will be notified by mail if they have been selected to submit a full proposal. Deadline for receipt of full proposals - July 27.
Restrictions: Past awardees and teams of more than two researchers are not eligible to apply.
Date(s) Application is Due: Mar 28; Jul 27
Amount of Grant: Up to 335,000 U.S.D
Samples: Michael D. Cohen, Ph.D. - Handoffs in Hospitals: Research for the Design of Better Practices; Julie Fairman, Ph.D., FAAN, R.N. - Practice Politics: History of Nursing 1975 to the Present; Joseph J. Fins, M.D. - Minds Apart: Severe Brain Injury and Health Policy; Dominick L. Frosch, Ph.D., Jose A. Pagan, Ph.D. - Direct-to-Consumer Advertising: Do Television Pharmaceutical Ads Prompt More Than Just Prescription Requests?
Contact: Cynthia Church, Program Administrator; (732) 932-3817; fax (732) 932-3819; depdir@ifh.rutgers.edu
Internet: http://www.investigatorawards.org
Sponsor: Robert Wood Johnson Foundation
College Road E and Route 1, P.O. Box 2316
Princeton, NJ 08543-2316

RWJF Jobs to Careers: Promoting Work-Based Learning for Quality Care - Research Grants 4177
The program will support partnerships to advance and reward the skill and career development of incumbent (currently employed) workers providing care and services on the front lines of our health and health care systems. The program seeks to develop and redesign systems that support and institutionalize learning and career advancement and test new models of work-based learning. The research grants will support applied research to inform these efforts. Research funded through this program will complement and supplement learnings from the Jobs to Careers implementation grants, which will fund partnerships of health and health care employers and educational institutions to promote the program's objectives. Grant funds may be used for project staff salaries, consultant fees, data collection and analysis, meetings, supplies, project-related travel, and other direct project expenses, including a limited amount of equipment essential to the project.
Requirements: Eligible applicants for research grants through Jobs to Careers are scholars associated with educational institutions, research organizations, health care providers or other public or nonprofit organizations. Preference will be given to public entities or nonprofit organizations tax exempt under Section 501(c)3 of the Internal Revenue Code. Applicants may be independent of or associated with organizations currently receiving Jobs to Careers implementation grants. All proposals must be submitted only through the RWJF Grantmaking Online system.
Restrictions: Grant funds may not be used to subsidize individuals for the costs of their health care, to support clinical trials of unapproved drugs or devices, to construct or renovate facilities, for lobbying, or as a substitute for funds currently being used to support similar activities.
Date(s) Application is Due: Jun 19
Amount of Grant: Up to 200,000 U.S.D
Contact: Maria Kniesler Flynn, Program Director; (617) 728-4446; mflynn@jff.org
Internet: http://www.jobs2careers.org
Sponsor: Robert Wood Johnson Foundation
College Road E and Route 1, P.O. Box 2316
Princeton, NJ 08543-2316

RWJF Ladder to Leadership: Developing the Next Generation of Community Health Leaders Fellowships 4178
Ladder to Leadership: Developing the Next Generation of Community Health Leaders is a collaborative initiative of the Robert Wood Johnson Foundation (RWJF) and the Center for Creative Leadership. The initiative aims to enhance the leadership capacity of community-based nonprofit health organizations serving vulnerable populations. RWJF is committed to leadership development and improving the health of all Americans, particularly those in the most vulnerable communities where people are often economically disadvantaged and have limited educational opportunities and inadequate access to health services. The program will be delivered in nine priority communities on a staggered schedule over the next four years. Each phase of the three-part curriculum includes a mixture of face-to-face training sessions, individualized executive coaching and mentoring, and team project work all anchored within the context of the communities in which these leaders live and work to maximize the application and impact of the program concepts.
Requirements: Only one applicant from each organization will be considered. Qualified applicants must: be an early- to mid-career professional with one to five years of supervisory experience; be an employee of a community-based, health or health-related nonprofit or government organization that serves vulnerable populations; work in one of the priority communities targeted by the initiative; commit to a 16-month leadership development initiative and curriculum; have the support of their organization's executive director or senior leader and their board of directors; and be a citizen or permanent resident of the United States or its territories at the time of application.

Restrictions: Up to 30 fellows will be selected to participate in the program in each of nine targeted communities across the United States: Central New York; Cleveland, Ohio; Birmingham, Alabama; Albuquerque, New Mexico; Eastern North Carolina; Portland, Oregon; Mid-South Region (Western Tennessee, Eastern Arkansas and Northern Mississippi); New Jersey (specific site/region to be determined); and Starr County, Texas.
Contact: Christy Schaefer, Project Manager; (336) 286-4595; schaeferc@ccl.org or info@laddertoleadership.org
Internet: http://www.rwjf.org/applications/solicited/cfp.jsp?ID=20281
Sponsor: Robert Wood Johnson Foundation
College Road E and Route 1, P.O. Box 2316
Princeton, NJ 08543-2316

RWJF Local Funding Partnerships Program Grants **4179**
The foundation awards matching grants to establish funding partnerships with local grantmakers in support of innovative, community-based projects that focus on the health and health care of underserved and at-risk populations. Priority will be given to projects that involve cooperation among public- and private-sector service providers; have other financial sponsors; include continuation of the proposed innovation after support expires; and have local government support. Grants are awarded for 36 to 48 months and must be matched one-to-one. Grant funds may be used for project staff salaries, consultant fees, data collection and analysis, meetings, supplies, project-related travel, and other direct project expenses, including a limited amount of equipment deemed essential to the project. The full RFP is available online.
Requirements: Applicants may be either 501(c)3 public entities or nonprofit organizations and not classified as a private foundation under Section 509(a).
Restrictions: In-kind services and money to pay for capital costs or renovation may not be used to match funds.
Amount of Grant: 50,000 - 200,000 U.S.D
Samples: Northern Virginia Community College Educational Foundation (VA)—for community-based efforts to reduce health disparities, as part of the Local Initiative Funding Partners Program, $500,000.
Contact: Leticia Peguero; (609) 275-4128; lpeguero@localfundingpartnerships.org
Internet: http://www.rwjf.org/applications/solicited/cfp.jsp?ID=20781
Sponsor: Robert Wood Johnson Foundation
College Road E and Route 1, P.O. Box 2316
Princeton, NJ 08543-2316

RWJF New Careers in Nursing Scholarships **4180**
The Robert Wood Johnson Foundation New Careers in Nursing is a scholarship program to help alleviate the nursing shortage and increase the diversity of nursing professionals. Through grants to schools of nursing, the program will provide scholarships to college graduates with degrees in other fields who are enrolled in accelerated baccalaureate and master's degree nursing programs. A school of nursing may apply for between five and 30 scholarships per year to be awarded to students from underrepresented groups in nursing or disadvantaged backgrounds. Up to 400 scholarship awards will be made during this funding cycle. The deadline for receipt of applications and supporting documents submitted online is December 15.
Requirements: To be eligible, applicant institutions must: offer an entry-level accelerated baccalaureate nursing program or master's nursing program for non-nursing college graduates; and be accredited by a nursing accrediting agency recognized by the U.S. Department of Education.
Contact: Vernell DeWitty, (202) 463-6930; vdewitty@aacn.nche.edu
Internet: http://www.rwjf.org/applications/solicited/cfp.jsp?ID=20905
Sponsor: Robert Wood Johnson Foundation
College Road E and Route 1, P.O. Box 2316
Princeton, NJ 08543-2316

RWJF New Jersey Health Initiatives Grants **4181**
The program funds grants that promote the development of innovative, community-based health services in New Jersey. Special emphasis is placed on projects that improve access to primary health care, address the complex service needs of people with chronic health conditions, and reduce substance abuse. Grant funds may be used for staff salaries, project-related travel, data processing, supplies, a limited amount of equipment, and other expenses essential to the project. The primary emphasis of the program is the demonstration of new service delivery approaches, although a limited amount of funding will be available for planning, feasibility, and program development. The program will also support a small number of health services research or evaluation projects in the goal areas.
Requirements: Public entities or IRS 501(c)3 organizations located in New Jersey are eligible for funding. Applicants are asked to propose projects that respond to one of the following RWJF interest areas: Addiction Prevention & Treatment; Childhood Obesity; or Vulnerable Populations. Projects must address all of the following: Reach people who are not served by traditional health and social services; Focus on meeting the needs of individuals and families, touching their lives directly; Address the lack of policy, financing and service integration among state and local agencies that creates barriers to helping people.
Restrictions: Private 509(a) foundations are not eligible. Grant funds may not be used for capital costs (including construction, renovation, and most equipment purchases), to support ongoing general operating expenses or existing deficits, or to substitute for funds currently supporting similar services. No hard-copy brief or full proposals will be accepted.
Geographic Focus: New Jersey
Date(s) Application is Due: Jan 15
Amount of Grant: 50,000 - 300,000 U.S.D
Contact: Gretchen Hartling, Program Co-Director; (856) 225-6733; fax (856) 225-6736; info@njhi.org
Internet: http://www.njhi.org
Sponsor: Robert Wood Johnson Foundation
College Road E and Route 1, P.O. Box 2316
Princeton, NJ 08543-2316

RWJF Partners Investing in Nursing's Future Grants **4182**
The program is a collaborative initiative of the Robert Wood Johnson Foundation (RWJF) and the Northwest Health Foundation (NWHF), with the intention of addressing nursing issues at the community level through funding partnerships with local and regional foundations. Up to 10 awards of up to $250,000 each will be made for projects lasting up to 24 months. Local foundations, with other partners (such as state workforce investment boards, hospital and long-term-care associations, foundations and other funding sources) will match these awards with at least $1 for every $2 provided by the program. Funding will be commensurate with the size and scope of the proposed activity and the foundation's experience with nursing issues. Funding is not intended to support the continuation of ongoing projects, but it may support new programs led by foundations that are already working in the field of nursing. October 29 is the deadline for receipt of brief proposals, and January 5 is the deadline for receipt of invited full proposals.
Requirements: Eligible institutions include local or regional private, family or community foundations and public charities. Eligible foundations are those classified as tax-exempt under Section 501(c)3 as a public charity or private foundation, a nonexempt charitable trust treated as a private foundation under Section 4947(a)(1), or organizations that claim status as private operating foundations under Section 4942(j)(3) or (5) of the Internal Revenue Code. Government entities, corporate grantmakers and others may participate as part of funding collaboratives; however, they may not serve as the applicant institution. All proposals to this program must address one or more of these topics: (1) Diversity; (2) Educational Infrastructure and Faculty Development; (3) Public Health; (4) Long-Term Care; and/or (5) Mid-Level Management. Proposals for planning as well as implementation projects will be considered. Deadline for receipt of brief proposals - November 9. Deadline for receipt of invited full proposals - February 1.
Restrictions: This program accepts only proposals submitted through the RWJF Grantmaking Online system at http://grantmaking.rwjf.org/pin2.
Date(s) Application is Due: Jan 5; Oct 29
Amount of Grant: Up to 250,000 U.S.D
Contact: Judith L. Woodruff, Program Director; (971) 230-0093; judith@nwhf.org
Internet: http://www.rwjf.org/applications/solicited/cfp.jsp?ID=20903
Sponsor: Robert Wood Johnson Foundation
College Road E and Route 1, P.O. Box 2316
Princeton, NJ 08543-2316

RWJF Pioneer Portfolio Grants **4183**
The program aims to support novel, high-return ideas that may have far-reaching impact on people's health, the quality of care they receive and the systems through which that care is provided. It seeks ideas not only from the mainstream of health and health care but also looks to sources outside of these fields for innovations that might have transformative impact. Several projects apply approaches from diverse sectors such as finance, design and entertainment to forge new solutions in health and health care arenas.
Requirements: Preference is given to public agencies, public charities or are tax-exempt under section 501(c)3 of the Internal Revenue Code. Organizations must be located in the United States or one of its territories.
Restrictions: The organization does not fund any of the following: Ongoing general operating expenses or existing deficits; Endowment or capital costs; Basic biomedical research; Research on drug therapies or devices; International programs; Direct support of individuals; Lobbying of any kind. Additionally, RWJF no longer funds projects in the following areas: End-of-life care; Long-term care; Specific chronic conditions (that are not part of the strategies of other Program Areas); Physical activity for adults age 50 or older.
Samples: Ramanan Laxminaryan, Extending the Cure: Laxminaryan is reframing how we fight drug-resistant diseases by suggesting we manage antibiotics as a shared societal resource. Working with regulatory, health, pharmaceutical and economic experts, he is building evidence and consensus around incentive-based policy alternatives to ensure a vital pipeline of effective antibiotics and combat 'superbugs' that cannot be eliminated with existing drugs; Project HealthDesign: A national program that supports interdisciplinary teams to look beyond stand-alone personal health records (PHR) to design next-generation PHR systems that empower patients to better manage their health and health care; Influenza Prediction Markets: University of Iowa business school professors teamed up with an infectious disease expert to test whether the Iowa Electronic Markets - prediction markets that aggregate information and are best known for predicting election and film box office results - can target where influenza may strike in advance; Myelin Repair Foundation (MRF): MRF's innovative research model provides the infrastructure for five scientific teams spread across different university labs in the U.S. and Canada to work virtually toward treatment breakthroughs for patients suffering from multiple sclerosis.
Contact: Chinwe Onyekere; (609) 627-7504 or (888) 631-9989; conyeke@rwjf.org
Internet: http://www.rwjf.org/pioneer/submission.jsp
Sponsor: Robert Wood Johnson Foundation
College Road E and Route 1, P.O. Box 2316
Princeton, NJ 08543-2316

RWJF Pipeline, Profession and Practice: Community-Based Dental Education Grants 4184

The three national foundations provide grants to dental schools in the United States under the Pipeline, Profession & Practice: Community-Based Dental Education (Dental Pipeline) program, including: the Robert Wood Johnson Foundation (RWJF); the California Endowment; and W.K. Kellogg Foundation. The primary purposes of the Program is to assist dental schools in: increase services provided to vulnerable populations through dental school-community-based collaborations; train graduates with the cultural knowledge and communication skills they need to treat racially and ethically diverse patients; increase student body diversity; and graduate more dentists who choose to practice in communities-in-need.
Contact: Judith S. Stavisky, Senior Program Director; (888) 631-9989; fax (609) 627-6401; JStavis@rwjf.org
Internet: http://www.dentalpipeline.org/
Sponsor: Robert Wood Johnson Foundation
College Road E and Route 1, P.O. Box 2316
Princeton, NJ 08543-2316

RWJF Scholars in Health Policy Research Fellowships 4185

The program is a two-year fellowship targeted to outstanding new PhDs in economics, political science, and sociology to advance their involvement in health policy research. Each year, the program enables up to 12 individuals to spend two years at one of three nationally prominent universities—Harvard University, the University of California at Berkeley (in collaboration with the University of California at San Francisco), and the University of Michigan. Those selected as scholars will have access to the full range of university resources, including libraries, computers, databases and research support, and will receive stipends from their university of $80,000 for the first year and $83,000 for the second year. Application and guidelines are available online.
Requirements: Applicants must be U.S. citizens or permanent residents at the time of application and have received a doctoral degree after January 2002 but no later than July 15, 2007. Applicants must have received a doctoral degree in economics, political science, or sociology. Preference will be given to applicants who have not previously worked in the areas of health or health policy research.
Date(s) Application is Due: Oct 21
Amount of Grant: 89,000 U.S.D
Samples: Elizabeth Armstrong, Fetal Personhood: the Raw Edge of Obstetrical Practice and Ethics; Michael Greenstone,a project on the association between particulates in air pollution and infant mortality rates; Cathy Cohen, Race, Politics, and Adolescent Health: Understanding the Health Attitudes and Behaviors of African American Youth; Daniel Carpenter, Reputation and Regulation: A Study of Pharmaceutical Policy at the FDA
Contact: Program Coordinator; (617) 353-9220; fax (617) 353-9227; rwjf@bu.edu
Internet: http://www.rwjf.org/applications/solicited/cfp.jsp?ID=20841
Sponsor: Robert Wood Johnson Foundation
College Road E and Route 1, P.O. Box 2316
Princeton, NJ 08543-2316

RWJF State Coverage Initiatives Grants 4186

These grants are designed to support states in the early stages of planning coverage expansions and those states that have decided on a new coverage expansion mechanism and seek assistance designing and implementing the program. Both policy planning and demonstration grants are available. Specific new grant opportunities are occasionally announced on the sponsor's web site, however, states may submit project ideas at any time. The program offers primarily two types of grants: policy planning and demonstration grants. Policy planning grants are small, fast-track grants for up to $150,000 to support states in the early stages of the planning process toward expanding coverage. Grantees have used the funds to conduct health insurance surveys, analyze private market options, model specific strategies, hold policy summits, and bring together a consortium of stakeholders. Large demonstration grants of up to $1.5 million are awarded to states that have selected a coverage expansion mechanism and seek assistance in designing and implementing the coverage expansion programs. These grants have helped states to make substantial progress toward a sizable coverage objective, such as expanding eligibility to all children in a state or implementing public-private initiatives for the working uninsured. In certain circumstances, the foundation will provide special grants, such as implementation grants to assist in the successful implementation of a new coverage program.
Requirements: State organizations are eligible to apply. Contact sponsor for specific guidelines and application information.
Amount of Grant: Up to 15,000,000 U.S.D
Contact: Isabel Friedenzohn, (202) 292-6726; isabel.friedenzohn@academyhealth.org
Internet: http://www.statecoverage.net
Sponsor: Robert Wood Johnson Foundation
College Road E and Route 1, P.O. Box 2316
Princeton, NJ 08543-2316

RWJF Summer Medical and Dental Education Program (SMDEP) 4187

The program was originally established as the Minority Medical Education Program (MMEP) with the aim of increasing the number of highly qualified medical school applicants from minority groups that were underrepresented in medicine - primarily African Americans, Hispanics, and American Indians. In 2003, the program changed its name to the Summer Medical Education Program (SMEP), reflecting the inclusion of students representing a wide range of economic, cultural, racial, and ethnic diversity, also expanding to include pre-dental students who face challenges similar to those of pre-medical students. The program is a free (full tuition, housing, and meals) six-week summer medical and dental school preparatory program that offers eligible students intensive and personalized medical and dental school preparation. Stipends are designed to support scholars throughout the six weeks at SMDEP. Each program site determines the amount given based on local factors such as cost of living, whether or not meals or travel are included, etc. Even though amounts vary among program sites, the stipend allowance given to each scholar is adequate to support them with their day-to-day living expenses during the six weeks attending the program.
Requirements: To be eligible for the program, an applicant must be currently enrolled as a freshman or sophomore in college; have a minimum overall GPA of 3.0, with 2.75 in the sciences; and be a U.S. citizen or hold a permanent resident visa. Other factors considered in the admissions process include whether an applicant has a combined verbal and math SAT score of at least 950 or ACT score of at least 20; comes from a group that is racially/ethnically underrepresented in medicine and dentistry (as defined independently by each program site); comes from an economically or educationally disadvantaged background; has demonstrated interest in issues affecting underserved populations; has local and/or regional residency; and submits a strong personal statement and strong letters of recommendation.
Restrictions: Former participants are not eligible to apply.
Date(s) Application is Due: Mar 1
Contact: Charles Terrell, Program Director; (202) 828-0584; cterrell@aamc.org
Internet: http://www.smdep.org
Sponsor: Robert Wood Johnson Foundation
College Road E and Route 1, P.O. Box 2316
Princeton, NJ 08543-2316

RWJF Vulnerable Populations Portfolio Grants 4188

Programs within the Vulnerable Populations Portfolio have four elements in common: they offer an opportunity to improve health by taking a fresh approach to a long standing problem; they address poor health status in the context of other factors like housing, education and poverty; they make fundamental changes in how services are organized and delivered; and they address the lack of policy, financing, or service integration among local service providers and state and federal agencies. While RWJF awards most of its grants in response to calls for proposals, it also award grants in response to unsolicited proposals in the Vulnerable Populations Portfolio.
Requirements: Preference is given to public agencies, public charities or are tax-exempt under section 501(c)3 of the Internal Revenue Code. Organizations must be located in the United States or one of its territories. Guidelines and application information is available at the website.
Restrictions: This program will not provide funding for efforts that do not incorporate the social factors that drive health status as part of their proposed model. Additionally, the program does not fund documentaries, research, programs that address a single medical condition, or provide core support for free or safety net clinics, disease management models, or well-tested models whose effectiveness have already been established and that have been widely disseminated.
Contact: Thomas Andruszewski, (888) 631-9989; tandrus@rwjf.org
Internet: http://www.rwjf.org/programareas/approach.jsp?pid=1144
Sponsor: Robert Wood Johnson Foundation
College Road E and Route 1, P.O. Box 2316
Princeton, NJ 08543-2316

RWJF Young Epidemiology Scholars (YES) Competition 4189

The competition for original student research is designed to inspire talented students to investigate the many behavioral, biological, environmental and social factors that affect health and, based upon this knowledge, to identify ways to improve the health of the public. The competition awards up to 120 college scholarships each year to high school juniors and seniors who conduct outstanding research projects that apply epidemiological methods of analysis to a health-related issue.
Requirements: The student competition is open to high-school juniors and seniors who are citizens or permanent residents of the United States and enrolled in a high school in the United States, Puerto Rico, Guam, U.S. Virgin Islands American Samoa, Wake and, Midway Islands or the Marianas. Home-schooled students are also eligible. To enter the competition, you must register online and upload a written report of your research project. Reports sent via email, fax or mail will not be accepted.
Restrictions: Only one project per student will be accepted. Team projects are not eligible. If you are a previous national finalist or scholarship winner, you may not submit the same topic as a previous year unless you have significantly advanced your research (e.g., adding substantive new data, analysis or other material).
Date(s) Application is Due: Feb 1
Amount of Grant: Up to 50,000 U.S.D
Contact: YES Program Coordinator; (800) 626-9795, ext. 5849; fax (703) 707-5599; yes@collegeboard.org
Internet: http://www.collegeboard.com/yes
Sponsor: Robert Wood Johnson Foundation
College Road E and Route 1, P.O. Box 2316
Princeton, NJ 08543-2316

Ryan Gibson Foundation Grants 4190

The Foundation welcomes any and all research grant requests from around the country. The main focus is on blood related cancers. Applicants should submit a one to two page proposal with the following information: description of the procedure and or drug that is to be tested/developed, and what the desired/expected result will be; some background on the researcher and the facility that will be used (past projects success and or failures should be included); and a budget of anticipated costs. Mail all requests to the following address: The Ryan Gibson Foundation, Attn: Research Proposal, 18111 Preston Road, Suite 650, Dallas, TX 75252

Amount of Grant: 50,000 U.S.D
Contact: National Director; (214) 987-0008; fax (214) 365-9293; mail@trgf.org
Internet: http://www.trgf.org/grants.shtml
Sponsor: Ryan Gibson Foundation
5910 North Central Expressway, Suite 770
Dallas, TX 75206

S. D. Bechtel, Jr. Foundation / Stephen Bechtel Fund Preventive Healthcare and Selected Research Grants **4191**

The Foundation?fund believe that access to quality care, preventive health care and research programs can help individuals of all ages increase life expectancy and improve their quality of life. The directors have a particular interest in promoting nutritional health, fitness, and wellness in order to increase positive health outcomes. The Foundation/Fund support organizations working to achieve the following objectives: enabling San Francisco Bay Area community clinics and school-based health centers to improve the quality and delivery of medical care; enhancing and expand school-based efforts in the San Francisco Bay Area that improve the nutritional health and fitness of children; and advancing research related to Alzheimer's, aging, nutrition or other Board-determined specific research initiatives.
Requirements: The primary geographic focus of the Foundation/Fund is the San Francisco Bay area. The Foundation/Fund: support non-profit organizations providing quality programs in science, technology, engineering and math (STEM) education, environment, environmental education, character and citizenship development and preventive healthcare and selected research; provide capital support; provide operational support; and provide project support.
Restrictions: The Foundation/Fund do not provide endowment funding, international grants, or grants for individuals.
Geographic Focus: California
Date(s) Application is Due: Oct 1
Amount of Grant: 5,000 - 25,000 U.S.D
Contact: Program Coordinator; (415) 284-8675; fax (415) 284-8571; sdbjr@sdbjrfoundation.org
Internet: http://www.sdbjrfoundation.org/program_areas.htm#health
Sponsor: S. D. Bechtel, Jr. Foundation / Stephen Bechtel Fund
P.O. Box 193809
San Francisco, CA 94119-3809

S.H. Cowell Foundation Grants **4192**

The foundation awards grants for a wide variety of causes to nonprofit organizations in northern California. Projects of interest include those that support children, youth, and families; education; housing; alcohol abuse prevention; religious organizations, education, and welfare; school-to-work employment training; population and environment; family planning; and conventional arms control. Applicants are encouraged to obtain most of their operating and project funding from other sources. Matching and challenge grants will be awarded under appropriate circumstances. The foundation prefers to award grants for one-time capital needs or for specific projects that are time-definite in nature and likely to become self-sufficient within several years. The application process should begin with a phone call. If interested, the foundation will request a short letter and then a formal proposal.
Requirements: Grants are made only to 501(c)3 tax-exempt organizations primarily in northern California.
Restrictions: The foundation does not normally make grants to individuals, for start-up of new organizations, for academic or other research, for general support, for annual fund-raising, to governmental agencies, to churches for religious support, to hospitals for medical research or treatment, for conferences, for media projects, or for political lobbying.
Geographic Focus: California
Amount of Grant: Up to 500,000 U.S.D
Samples: Adventure Risk Challenge, Truckee, CA, $25,000 - to launch year-round expansion of the Adventure Risk Challenge (ARC) summer program; Boys & Girls Clubs of the North Valley Chico, CA, $500,000 - to launch a Boys & Girls Club in South Oroville; East Bay Asian Local Development Corporation Oakland, CA, $375,000 - to complete the budget to install a park at Lion Creek Crossings;
Contact: Susan Vandiver, (415) 397-0285; fax (415) 986-6786; info@shcowell.org
Internet: http://www.shcowell.org/grant/grant.php
Sponsor: S.H. Cowell Foundation
120 Montgomery Street, Suite 2570
San Francisco, CA 94104

S. Livingston Mather Charitable Trust Grants **4193**

The S. Livingston Mather Charitable Trust was established in 1953 in Ohio by Cleveland-Cliffs vice-president Samuel Livingston Mather. The Trust's primary areas of interest include cultural programs, education, child welfare, and social services. Support is also available for youth programs, the environment, and natural resources. Giving is primarily restricted to the northeastern Ohio area. Applicants should contact the Trust prior to submitting an application.
Requirements: Unsolicited requests for funds not accepted, contact the Trust before sending a proposal.
Restrictions: No support is available for: endowments; science; medical research programs; in areas appropriately supported by the government and/or the United Way; individuals; or deficit financing.
Geographic Focus: Ohio
Amount of Grant: 100 - 100,000 U.S.D
Samples: Cleveland Museum of Art, Cleveland, OH, $20,000; Cuyahoga Valley Countryside Conservancy, Peninsula, OH, $7,500.
Contact: Janet W. Havener, Director, c/o Glenmede Trust Company; (215) 419-6000
Sponsor: S. Livingston Mather Charitable Trust
1 Corporate Exchange, 25825 Science Park Drive, Suite 110
Beachwood, OH 44122

S. Mark Taper Foundation Grants **4194**

The S. Mark Taper Foundation, founded in 1989, is a private family foundation dedicated to enhancing the quality of people's lives by supporting nonprofit organizations and their work in the Los Angeles, Long Beach, Santa Ana, California communities. Areas of interest are broad and include but are not limited to education, the environment, independent living for the disabled, abused women, immigrant health care, children, hunger, housing, AIDS, teenage pregnancy prevention, job creation and economic revitalization, individuals with visual impairments, and the arts. Types of support include capital grants, challenge/matching grants, general operating grants, research grants, program/project grants, seed money grants, and scholarships. Grants are made generally for one year.
Requirements: 501(c)3 nonprofit organizations in California are eligible for grant support. Application forms are required therefore your initial approach should be a Letter of Inquiry containing one copy of the proposal and the following: brief history of organization and description of its mission; copy of most recent annual report/audited financial statement/990; detailed description of project and amount of funding requested; list of source(s) of last three years of funding for the specific project (if any) and, the organization.
Geographic Focus: California
Date(s) Application is Due: Dec 1
Amount of Grant: 5,000 - 500,000 U.S.D
Samples: Korean Health Education Information and Research Center, Los Angeles, CA - $500,000; Santa Clarita Child and Family Development Center, Santa Clarita, CA - $100,000; Los Angeles Unified School District Education Foundation, Los Angeles, CA - $200,000;
Contact: Raymond Reister, Executive Director; (310) 476-5413; fax (310) 471-4993; rreisler@smtfoundation.org or info@smtfoundation.org
Internet: http://www.smtfoundation.org/
Sponsor: S. Mark Taper Foundation
12011 San Vicente Boulavard, Suite 400
Los Angeles, CA 90049

Saginaw Community Foundation Senior Citizen Enrichment Fund **4195**

The Senior Citizens Enrichment Fund assists programs supporting charitable purposes that benefit the elderly in Saginaw County. The priority of this grant is to assist programs and projects that enhance the quality of life for seniors maintaining an independent residence.
Requirements: Applicants must first call the Foundation to discuss the project with the staff. To be eligible for funding, organizations must have a nonprofit status and directly benefit Saginaw County. Grants are ordinarily only made for one year. Applications can be downloaded at the website but must be received by mail or dropped off at the agency.
Restrictions: The Saginaw Community Foundation does not support: operating budgets; basic municipal services; basic educational functions; endowment campaigns; previously incurred debt; or sectarian religious programs.
Geographic Focus: Michigan
Date(s) Application is Due: Feb 1; May 1; Aug 1; Nov 1
Amount of Grant: Up to 10,000 U.S.D
Contact: Kendra Kempf, Program Associate/FORCE Coordinator; (989) 755-0545; fax (989) 755-6524; kendra@saginawfoundation.org
Internet: http://www.saginawfoundation.org/grants_and_scholarships/grants/
Sponsor: Saginaw Community Foundation
100 South Jefferson, Suite 201
Saginaw, MI 48602

SAG Motion Picture Players Welfare Fund Grants **4196**

The Motion Picture Players Welfare Fund (MPPWF) is an emergency financial assistance fund of the Screen Actors Guild New York region designed to assist eligible members who are struggling with a financial, personal or medical crisis. The Screen Actors Guild contracts with The Actors Fund to administer the MPPWF as well as provide an array of services to meet each members needs. Services include counseling and referrals for personal, family or work-related problems. Linkage to community resources for legal services, elder care, entitlement benefits or childcare. Workshops are offered on locating affordable housing, financial education as well as health insurance options. Financial assistance is available for rent, utilities, mental health and medical care as well as other basic living expenses. Grants are approved on a case-by-case basis based on need.
Requirements: All SAG members should feel free to call The Actors Fund for information, referrals and access to our social services. To be eligible for financial assistance from MPPWF you must be: living in the New York Tri-State area; a SAG member for at least six dues periods; a member in good standing at the time of application or be vested for pension purposes; and, able to document a need for emergency financial assistance.
Geographic Focus: New York
Contact: Program Director; (212) 221-7300, ext. 119
Internet: http://new.actorsfund.org/services/Partner_Programs/Motion_Picture_Players/index_html
Sponsor: Screen Actors Guild
360 Madison Avenue, 12th Floor
New York, NY 10017

Saigh Foundation Grants 4197

Fred M. Saigh was born in Springfield, Illinois on June 27, 1905, the son of Lebanese immigrants who owned a chain of grocery stores. He passed away in 1999. Although he is best known as the owner of the St. Louis Cardinals (1948 - 1953), many of his friends and acquaintances will always remember him as a perceptive and caring benefactor - "a one-man charity fund," in the words of sportswriter Mike Eisenbath. The Saigh Foundation continues the important work begun by Mr. Saigh by assisting St. Louis-area organizations that benefit children and youth, particularly in the areas of education and healthcare. The foundation is particularly interested in stimulating the development of new ventures, as well as in supporting organizations that feature innovative approaches or programs. Like Mr. Saigh, the foundation is especially dedicated to aiding those who might not otherwise receive assistance. Proposals are requested at least three months before any quarterly meeting. These are normally held in January, April, July, and October. The foundation uses a customized version of the Missouri Common Grant Application (CGA). It is available at the grant website along with guidelines and budget templates and should be submitted via mail or fax to the address given.

Requirements: Saint Louis-area nonprofit organizations are eligible.

Restrictions: The foundation does not participate in annual appeals, dinner functions, and fundraising events; capital campaigns; loans and deficits; grants for films and travel; or nonprofit organizations outside of the metropolitan Saint Louis area.

Geographic Focus: Missouri

Samples: Juvenile Diabetes Research Foundation (2011); Junior Achievement of Mississippi (2011); Salvation Army (2011).

Contact: JoAnn Hejna, Executive Director; (314) 862-3055; fax (314) 862-9288; saigh@thesaighfoundation.org

Internet: http://www.thesaighfoundation.org/grant_guide.html

Sponsor: Saigh Foundation
7777 Bonhomme Avenue, Suite 2007
Saint Louis, MO 63105

Saint Louis Rams Foundation Community Donations 4198

The foundation supports nonprofits that help inspire positive change for youth in the Saint Louis area. Programs that impact youth in the general fields of education, literacy, health, and recreation will be considered. Annually, the Rams provide to charitable groups more than 3,500 items, helping recipient organizations raise thousands of dollars through raffles, auctions and other fundraising endeavors. Other types of financial support include program development grants and general operating grants. The foundation does not accept unsolicited requests, but initial information may be sent for the office to keep on file for future opportunities.

Requirements: Nonprofits in the metropolitan Saint Louis, Missouri, area, including southern Illinois and eastern Missouri, are eligible. Preference is given to organizations that partner with other local nonprofits and offer creative approaches for more than grants (i.e., personnel involvement or in-kind support) and ways the Rams can participate.

Restrictions: The Rams do not provide monetary contributions or merchandise donations for the following: businesses, retail and otherwise; capital campaigns/start-up funding for new businesses; on-line auctions; chamber of commerce/city/neighborhood festivals such as homecoming celebrations and carnivals that do not directly benefit a charitable organization; class reunions; family reunions; pageant contestants (beauty and otherwise); student ambassador/exchange programs; or non-charity events and organizations such as company picnics, employee golf tournaments, employee recognition/incentive programs, card clubs, car shows, "poker runs", and organized adult leisure sports teams.

Geographic Focus: Missouri

Date(s) Application is Due: Jan 1; Jul 1

Contact: Donations Coordinator; (314) 516-8788 or (314) 982-7267; fax (314) 770-0392

Internet: http://www.stlouisrams.com/community/donations.html

Sponsor: Saint Louis Rams Foundation
1 Rams Way
Saint Louis, MO 63045

Saint Luke's Foundation Grants 4199

The foundation exists to improve the health and well being of individuals and families living in Greater Cleveland. Proposals are invited for activities intended to enhance community involvement and ownership; promote healthy behaviors and life styles by education and outreach programs aimed at increasing the capacity of individuals and families to protect and improve their own health; increase and improve health care by improving access to affordable, high-quality, comprehensive, culturally competent and appropriate health care; educate health care professionals to increase their capacity in serving the health and health care needs of inner-city residents; and increase knowledge by furthering research on: the determinants of health; innovative systems, strategies, and collaborations designed to improve health and well being in the community; and the underlying causes of diseases and disabilities particularly prevalent in the communities.

Requirements: 501(c)3 tax-exempt organizations and governmental units/agencies are eligible. The grant application is a 3 step process: step 1: Letter of Inquiry; step 2: site visit; step 3: proposal submission. See website for detailed description of each step. The deadlines posted are for the Letter of Inquiry.

Restrictions: The Foundation does not fund: individuals; religious organizations for religious or evangelical purposes; projects outside Greater Cleveland that do not directly benefit Cleveland residents; fundraising events; endowment funds; biomedical research; debt retirement; lobbying.

Geographic Focus: Ohio

Date(s) Application is Due: Jan 5; Apr 1; Jul 1; Aug 1; Oct 1

Contact: Peg Butler, Grants Manager; (216) 431-8010; fax (216) 431-8015; pbutler@saintlukesfoundation.org

Internet: http://www.saintlukesfoundation.org/grants/community-grants.html

Sponsor: Saint Luke's Foundation of Cleveland, Ohio
4208 Prospect Avenue
Cleveland, OH 44103

Saint Luke's Health Initiatives Grants 4200

The trust was established for the purpose of addressing critical health needs in Maricopa County, AZ. Community grants support innovative approaches that address community health from a broad perspective, including funding for such projects as professional education and training, public education/advocacy demonstration projects, program planning and expansion, and equipment to address the health needs of populations that are considered underserved and vulnerable. Community grants are awarded for up to two years. Bridge grants support small, emerging organizations or new programs of larger organizations that have infrastructure or developmental needs. Grants support strategic planning, technical training, purchasing equipment, or convening discussion groups on community health issues. Bridge grants are awarded for one year. Inquiries are encouraged prior to proposal submission.

Requirements: To be eligible for a community partnership grant, a prospective organization must: demonstrate tax-exempt status under Section 501(c)3 of the Internal Revenue Service Code or be recognized as an instrumentality of state/local government; be located in Arizona. Preference is given to organizations within Maricopa County, although a statewide program with impact on Maricopa County will be considered.

Restrictions: The trust does not fund individuals, participate in large capital campaigns or endowment drives, provide grants for ongoing annual operating support, or fund projects of sectarian or religious organizations.

Geographic Focus: Arizona

Date(s) Application is Due: Feb 12; Aug 14

Amount of Grant: 10,000 - 100,000 U.S.D

Contact: Jane Pearson, Associate Director; (602) 385-6500; fax (602) 385-6510; Jane.Pearson@slhi.org or info@slhi.org

Internet: http://www.slhi.org/community_grants/how_to_apply.shtml

Sponsor: Saint Luke's Health Initiatives
2929 North Central Avenue, Suite 1550
Phoenix, AZ 85012

Salem Foundation Charitable Trust Grants 4201

The Foundation provides funding for a large variety of community based programs that serve the greater Salem area. In partnership with donors, the Foundation supports nonprofit and other community organizations with funds for health and human services, affordable housing, early childhood development, community arts, culture and other important areas of need. Grant making decisions are made by the Distribution Committee, comprised of five Salem area residents.

Requirements: Must be a community-based organization that serves the greater Salem region. Grants may be made only to organizations which have received appropriate determination letters from the Internal Revenue Service that they are exempt under the Internal Revenue Code.

Geographic Focus: Oregon

Date(s) Application is Due: May 1; Dec 1

Samples: Camerata Musica—to support 2007-2008 concerts at the Salem Public Library; Family Building Blocks—for staff members to attend National Conference on Child Abuse and Neglect; Lord & Schryver Conservancy—Garden rehabilitation at Historic Deepwood Estate.

Contact: Trustee of the Salem Foundation; (503) 363-3136; salemfoundation@pioneertrustbank.com

Internet: http://www.pioneertrustbank.com/salemfoundation/index.html

Sponsor: Salem Foundation Charitable Trust
109 Commercial Street NE, P.O. Box 2305
Salem, OR 97308

Salem Foundation Grants 4202

The primary purpose of the Foundation is support of higher education, cultural activities, health services, hospital building funds, and Christian giving. Its fields of interest include: arts; Christian agencies and churches; the disabled; education (secondary and higher); the environment; family services; federated giving programs; health care (particularly heart and circulatory disease research); hospitals; human services; museums; and Protestant federated giving programs. Types of support include: annual campaigns, building/renovation projects, capital campaigns, general operating support, and program development. There is no formal application process and no deadlines. Please contact via letter.

Restrictions: Giving primarily in Florida, Maine and Minnesota. No grants to individuals.

Geographic Focus: Florida, Maine, Minnesota

Amount of Grant: Up to 10,000 U.S.D

Contact: Robert S. Parish, President; (952) 476-6292

Sponsor: Salem Foundation
2181 Springwood Road
Wayzata, MN 55391-2254

Salisbury Community Foundation Grants 4203

The foundation was created to address the needs of the Salisbury, NC, community and its residents. Nonprofit organizations may apply for project support in the categories of arts, education, environment, historic preservation, human services, health and medical research, public and civic affairs, religion, the elderly, and youth services. Applications are accepted throughout the year.

Requirements: Tax-exempt, charitable organizations in Salisbury, NC, and Rowan county are eligible.
Geographic Focus: North Carolina
Date(s) Application is Due: Jan 31; Apr 30; Jul 31; Oct 31
Amount of Grant: 10,000 U.S.D
Contact: McCray Benson, Assistant Secretary; (704) 973-4559; fax (704) 973-4946; mvbenson@fftc.org
Internet: http://www.fftc.org/affiliates/community/nc/salisbury
Sponsor: Salisbury Community Foundation
217 S Tryon Street
Charlotte, NC 28202

Salmon Foundation Grants **4204**
Established in New York in 1991, the Salmon Foundation provides funding to nonprofit organizations for programs providing for children and youth programs, family services, and education. Grants are awarded for projects, scholarship support, operating support, and capital funding. Funding ranges from $2,000 to $40,000.
Requirements: Funding provided to nonprofit organizations. Organizations tend to be on the east coast and include Connecticut, New Hampshire, Vermont, Ohio, Pennsylvania, Alabama, Tennessee, Virginia, California, Maryland, Colorado and District of Columbia.
Restrictions: No individual scholarships are made.
Geographic Focus: Alabama, California, Colorado, Connecticut, District of Columbia, Maryland, New Hampshire, Pennsylvania, Tennessee, Virginia
Amount of Grant: 2,000 - 40,000 U.S.D
Samples: Aspen Education Foundation, Aspen, Colorado, $7,500 - elementary math and literacy program support (2010); Baltimore Symphony Orchestra, Baltimore, Maryland, $15,000 - support for the OrchKids program (2010); The Family Place, Norwich, Vermont, $30,000 - for the Families Learning Together Program (2010).
Contact: Emily Grand, Administrator; (212) 708-9316 or (212) 812-4362
Sponsor: Salmon Foundation, Inc.
6 West 48th Street, 10th Floor
New York, NY 10036-1802

Salt River Project Health and Human Services Grants **4205**
Salt River Project (SRP) is an energy/utilities company serving electric customers and water shareholders in the Phoenix metropolitan area. SRP provides funding to nonprofit organizations that address critical needs within its service communities. SRP is committed to safe and healthy communities. Health and Human Services Grants support programs that reach out to underserved communities to promote the individual's ability to overcome barriers and be self-sufficient; increase the community's ability to care for individuals who are in need of food, shelter, and safety from violent or crisis situations; increase the ability of children to participate in youth programs which promote personal development and positive life choices; support increasing underserved communities' access to hospitals and medical care as an integral part of a thriving community; and sponsor programs that seek to highlight the never ceasing need for water and electric safety.
Requirements: Eligible applicants must be 501(c)3 nonprofit, organizations within SRP's service area. SRP's service area is central Arizona and includes the following cities and towns: Phoenix, Mesa, Tempe, Paradise Valley, Fountain Hills, Scottsdale, Apache Junction, Peoria, Queen Creek, Avondale, Chandler, Gilbert, Glendale, Guadalupe, and Tolleson. There are no specific grant deadlines. Requests are reviewed in an on-going process which typically takes eight weeks.
Restrictions: The following are ineligible: individuals, including support for specific students, researchers, travel expenses, conference fees; organizations that discriminate on the basis of race, creed, color, sex, or national origin; endowment programs; medical research projects or medical procedures for individuals; professional schools of art, academic art programs, individual high school or college performing groups; political or lobbying groups or campaigns; fraternal organizations, veterans' organizations, professional associations, and similar membership groups; public or commercial broadcasting programs; religious activities or church-sponsored programs limited to church membership; and debt-reduction campaigns. SRP does not donate services, including water or electricity, or equipment for which a fee is normally charged.
Geographic Focus: Arizona
Contact: Corporate Contributions Administrator; (602) 236-5900; webmstr@srpnet.com
Internet: http://www.srpnet.com/community/contributions/guidelines.aspx
Sponsor: Salt River Project
1521 North Project Drive
Tempe, AZ 85281-1298

SAMHSA Campus Suicide Prevention Grants **4206**
The purpose of this program is to facilitate a comprehensive approach to preventing suicide in institutions of higher education. This program is designed to assist colleges and universities in their efforts to prevent suicide attempts and completions and to enhance services for students with mental and behavioral health problems, such as depression and substance abuse, which put them at risk for suicide and suicide attempts. Applications are due by November 25.
Requirements: Eligibility for SAMHSA's Campus Suicide Prevention Grant program is limited to institutions of higher education. Applicants from both public and private institutions may apply, including State universities, private four-year colleges and universities (including those with religious affiliations), Minority Serving Institutions of higher learning, and community colleges.
Date(s) Application is Due: Nov 25
Amount of Grant: 100,000 U.S.D
Contact: Scott J. Salvatore, Center for Mental Health Services; (240) 276-1866; Scott.Salvatore@samhsa.hhs.gov
Internet: http://www.samhsa.gov/Grants/2009/sm_09_001.aspx
Sponsor: Substance Abuse and Mental Health Services Administration
1 Choke Cherry Road, Room 7-1085
Rockville, MD 20857

SAMHSA Child Mental Health Initiative (CMHI) Grants **4207**
The purpose of this program is to support States, political subdivisions within States, the District of Columbia, Territories, Native American Tribes and tribal organizations, in developing integrated home and community-based services and supports for children and youth with serious emotional disturbances and their families by encouraging the development and expansion of effective and enduring systems of care. Applications are due by January 15.
Requirements: Eligibility for this program is statutorily limited to public entities such as: state governments; Indian or tribal organizations (as defined in Section 4[b] and Section 4[c] of the Indian Self-Determination and Education Assistance Act); governmental units within political subdivisions of a State, such as a county, city or town; District of Columbia government; and Commonwealth of Puerto Rico, Northern Mariana Islands, Virgin Islands, American Samoa and Trust Territory of the Pacific Islands (now Palau, Micronesia and the Marshall Islands).
Restrictions: Proposed budgets cannot exceed the allowable amount in total costs (direct and indirect) in any year of the proposed project. Annual continuation awards will depend on the availability of funds, grantee progress in meeting project goals and objectives, timely submission of required data and reports and compliance with all terms and conditions of award.
Date(s) Application is Due: Jan 15
Amount of Grant: Up to 1,000,000 U.S.D
Contact: Gwendolyn Simpson, Grants Management; (240) 276-1408; gwendolyn.simpson@samhsa.hhs.gov
Internet: http://www.samhsa.gov/Grants/2009/sm_09_002.aspx
Sponsor: Substance Abuse and Mental Health Services Administration
1 Choke Cherry Road, Room 7-1085
Rockville, MD 20857

SAMHSA Conference Grants **4208**
The purpose of the Conference Grant program is to disseminate knowledge about practices within the mental health services and substance abuse prevention and treatment fields and to integrate that knowledge into real-world practice as effectively and efficiently as possible. The maximum Conference Grant award is $50,000 for a 12-month project period. Only direct costs will be funded under this program. Applications are due on the recurring dates of March 31 and September 30 each year.
Requirements: Applications may be submitted by public and domestic private nonprofit and for-profit entities.
Date(s) Application is Due: Mar 31; Sep 30
Amount of Grant: Up to 50,000 U.S.D
Contact: Gwendolyn Simpson, Grants Management; (240) 276-1408; gwendolyn.simpson@samhsa.hhs.gov
Internet: http://www.samhsa.gov/Grants/2008/OA_08_002cmhs.aspx
Sponsor: Substance Abuse and Mental Health Services Administration
1 Choke Cherry Road, Room 7-1085
Rockville, MD 20857

SAMHSA Drug Free Communities Support Program Grants **4209**
This Program is a collaborative initiative sponsored by ONDCP in partnership with SAMHSA in order to achieve two major goals: establish and strengthen collaboration among communities, private nonprofit agencies, and Federal, State, local, and tribal governments to support the efforts of community coalitions to prevent and reduce substance abuse among youth; and reduce substance abuse among youth and, over time, among adults by addressing the factors in a community that increase the risk of substance abuse and promoting the factors that minimize the risk of substance abuse. Substances include, but are not limited to, narcotics, depressants, stimulants, hallucinogens, cannabis, inhalants, alcohol, and tobacco, where their use is prohibited by Federal, State, or local law. Approximately $17 million for 130 FY 2009 DFC grants will be awarded through this RFA. DFC grants will be available to eligible coalitions in amounts of up to $125,000 per year over a five-year period. Applications are due by March 20.
Requirements: DFC grant funds are intended to support eligible community-based coalitions.
Date(s) Application is Due: Mar 30
Amount of Grant: Up to 125,000 U.S.D
Contact: Dan Fletcher, (240) 276-1270; dfcnew2009@samhsa.hhs.gov
Internet: http://www.samhsa.gov/grants/2009/sp_09_002.aspx
Sponsor: Substance Abuse and Mental Health Services Administration
1 Choke Cherry Road, Room 7-1085
Rockville, MD 20857

SAMHSA Strategic Prevention Framework State Incentive Grants **4210**
The purpose of the SPF SIG program is to provide funding to States, Federally recognized Tribes and U.S. Territories in order to: prevent the onset and reduce the progression of substance abuse, including childhood and underage drinking; reduce substance abuse-related problems; and build prevention capacity and infrastructure at the State, tribal, territorial and community-levels. Applications are due by November 7.
Requirements: Eligible applicants are the immediate Office of the Chief Executive (e.g., Governor) in the States, U.S. Territories or District of Columbia; and Federally

recognized Tribes. The Chief Executive of the State, Territory, or District of Columbia; or of the Federally recognized Tribe must sign the application.
Date(s) Application is Due: Nov 7
Amount of Grant: Up to 23,000,000 U.S.D
Contact: Mike Lowther, (240) 276-2581; mike.lowther@samhsa.hhs.gov
Internet: http://www.samhsa.gov/Grants/2009/sp_09_001.aspx
Sponsor: Substance Abuse and Mental Health Services Administration
1 Choke Cherry Road, Room 7-1085
Rockville, MD 20857

Samuel Freeman Charitable Trust Grants 4211

The trust has broad funding interests in supporting U.S. charitable, scientific, literary, and education programs. The trust awards grants to support cancer research and treatment; the preservation, exhibition, and operation of historical railway equipment; and secondary schools and universities. Types of support include general operating support, continuing support, annual campaigns, and seed money. There are no application deadlines; proposals are reviewed throughout the year.
Requirements: U.S. nonprofit organizations may apply.
Amount of Grant: 1,000 - 150,000 U.S.D
Contact: Linda Franciscovich, c/o U.S. Trust Company, (212) 852-3683; fax (212) 852-3377
Sponsor: Samuel Freeman Charitable Trust
114 W 47th Street
New York, NY 10036-1532

Samueli Foundation Health Grants 4212

The Foundation considers grants to agencies, primarily in Orange County, California, that serve the community, and whose programs meet the guidelines listed. Grants are usually approved for a defined period of time, but may be paid over a multi-year period. In the area of Health, its goals are to: support programs promoting alternative and integrative medicine; and utilize creative and innovative approaches to education, prevention, treatment and research of disease and illness.
Requirements: 501(c)3 tax-exempt organizations are eligible.
Restrictions: The foundation does not fund umbrella fund raising organizations, political campaigns, or grants to individuals.
Date(s) Application is Due: Nov 24
Amount of Grant: 1,000 - 50,000 U.S.D
Contact: Maggie Sherer, Grants Administrator; (949) 760-4400; fax (949) 759-5707; Info@samueli.org
Internet: http://www.samueli.org/health.aspx
Sponsor: Samueli Foundation
2101 East Coast Highway, 3rd Floor
Corona del Mar, CA 92625

Samueli Institute Scientific Research Grants 4213

The Institute operates in conjunction with the Uniformed Services University of Health Sciences and Universitatsklinikum Freiburg Institute. Its mission is is to transform health care through the scientific exploration of healing. The Institute is one of an elite group of organizations in the nation with a track record in complementary and integrative medicine, healing relationships and military medical research. The major purpose of funding is to investigate healing processes and their application in promoting health and wellness, preventing illness, and treating disease. Researchers should respond to Requests for Applications or Program Announcements.
Amount of Grant: 2,000 - 600,000 U.S.D
Contact: Matt Fritts, (703) 299-4800; fax (703) 299-4800; mfritts@SamueliInstitute.org
Internet: http://www.siib.org/funding/investigator-home/218-SIIB.html
Sponsor: Samueli Institute
1737 King Street, Suite 600
Alexandria, VA 22314

Samuel N. and Mary Castle Foundation Grants 4214

The Foundation is committed to providing resources to improve the life of Hawaii's children and families by improving the quality and quantity of early education. Efforts are concentrated on creating greater social equality and opportunity through improving access to high quality pre-K education. Secondarily, the Foundation provides limited support for independent K-12 education, the arts, health, and historical and cultural projects. Grants generally range from $5,000 to $25,000. Grants for major capital improvements typically range from $10,000 to $100,000.
Requirements: Eligible organizations must be tax exempt, publicly supported and charitable as determined by the Internal Revenue Service. Grants are primarily awarded to organizations located within the state of Hawaii, for programs and projects benefiting the people of Hawaii. Proposed programs or projects must be in response to a documented community need, and not solely an organizational need. Grants may be awarded for innovative programs, demonstration projects and "start-up" funding. Program and project support does not generally exceed three years, and funding must be applied for on a yearly basis. Applicants must contact the Foundation's Executive Director by letter, email, phone or a personal visit before making fund application. A site visit may be required if the organization has not applied before or in many years.
Restrictions: The following are ineligible: charter schools; endowments; regular operating costs such as salaries, rents, or maintenance; more than forty to fifty percent of total project costs; programs not open to all racial and ethnic groups; projects in which parents and the community have not been appropriately involved in planning and funding; publication projects; general student scholarships for tuition, travel, or conferences; video projects; and annual fund drives or sponsorships.
Geographic Focus: Hawaii
Date(s) Application is Due: Feb 1; Jun 1; Oct 1
Samples: Chaminade University of Honolulu, Honolulu, Hawaii, $83,000, Castle Colleagues Pre-School Directors Management Training Program (2011); Good Beginnings Alliance, Honolulu, Hawaii, $25,000, project support for Race to the Top - Early Education Challenge Fund Application and Project Management (2011); and The Storybook Theater of Hawaii, Hanapepe, Hawaii, $9,000, travelling program for pre-school children (2011).
Contact: Alfred L. Castle, Executive Director and Treasurer; (808) 522-1101; fax (808) 522-1103; acastle@aloha.net
Internet: http://foundationcenter.org/grantmaker/castle/prior.html
Sponsor: Samuel N. and Mary Castle Foundation
733 Bishop Street, Suite 1275
Honolulu, HI 96813

Samuel Roberts Noble Foundation Grants 4215

The foundation is engaged in basic plant biology research as well as agricultural research, consultation, and development. Grants are made in a number of areas, with primary focus on capital funding for higher education and on health research and delivery systems. Requests are initiated with a letter describing the organization and the project for which funds are requested. Applications are provided if the request is in a current area of interest. Public affairs requests are initiated with a two-page summary of the organization and the project. Applications must be submitted a month preceding a Trustee meeting. Requests may not be submitted by fax.
Requirements: Generally, grants are geographically limited to the southwestern United States, with primary emphasis in Oklahoma. Matching funds are required.
Date(s) Application is Due: Mar 1; Jun 1; Sep 1; Dec 1
Amount of Grant: 5,000,000 - 12,000,000 U.S.D
Samples: Heritage Foundation (Washington, DC)—for its research and educational programs and to finance the purchase of its building in Washington's Capitol Hill neighborhood, $5 million.
Contact: Donna Windel, (580) 223-5810, ext. 6245; fax (580) 221-7362; dkwindel@nobel.org
Internet: http://www.noble.org/Admin/Granting/grGuides.html
Sponsor: Samuel Roberts Noble Foundation
P.O. Box 2180
Ardmore, OK 73402

Samuel S. Johnson Foundation Grants 4216

The Samuel S. Johnson Foundation was incorporated in 1948 and supports organizations primarily in Oregon and Clark County, Washington region. The Foundation gives to: formal education programs leading to an R.N. status or baccalaureate or higher college/university degree in nursing; vocational education programs targeting high school drop-outs and high school grads who are not able to pursue junior college or higher formal education and which offer them job-specific technical training, mentoring or coaching; emergency food assistance programs; rural mobile health screening/care projects benefiting the uninsured medically needy; environmental programs, coastal & marine ecosystems, sustainable agriculture and communities.
Requirements: Grants are awarded to non-profit organizations in Oregon and Clark County, Washington. Contact the Foundation for current focus and guidelines with a phone call before submitting a proposal. No Application form is required, however you must include the following in your proposal: copy of IRS Determination Letter; brief history of organization and description of its mission; copy of most recent annual report/audited financial statement/990; listing of board of directors, trustees, officers and other key people and their affiliations; detailed description of project and amount of funding requested; contact person; copy of current year's organizational budget and/or project budget; listing of additional sources and amount of support. Include one copy of the proposal. The board meets twice a year, in May and October, with no deadline date for the submitting of proposals. If your proposal is accepted, you will receive notification within 2 - 3 weeks after the board meets.
Restrictions: No support for foreign organizations. No grants or scholarships to individuals, or for leadership training or staff development, campaigns to retire debt, annual campaigns, deficit financing, construction, sole underwriting of major proposals or projects, demolition or endowments.
Geographic Focus: Oregon, Washington
Amount of Grant: 400 - 100,000 U.S.D
Contact: Mary A. Krenowicz; (541) 548-8104; fax (541) 548-2014; mary@tssjf.org
Sponsor: Samuel S. Johnson Foundation
P.O. Box 356
Redmond, OR 97756-0079

San Antonio Area Foundation Grants 4217

Grants are usually awarded in the areas of health care and biomedical research, community and social services, arts and culture, education (early childhood education, higher education, medical schools, nursing schools, adult basic education and literacy), and animal services. Types of support include operating budgets, continuing support, annual campaigns, seed grants, emergency funds, equipment acquisition, matching funds, scholarship funds, research, lectureships, professorships, and building renovations. Applicants must submit a letter of intent to apply; if the letter is approved, the foundation will send a request for a proposal to the applicant. The foundation only reviews full proposals from applicants whose letters of intent have been approved. If asked to submit a proposal, the applicant will be sent an application package with a letter of notification in early February.
Requirements: Grants are made to organizations in the San Antonio, Texas, area.

Geographic Focus: Texas
Date(s) Application is Due: Nov 15
Amount of Grant: 250 - 250,000 U.S.D
Samples: San Antonio Symphony (TX)—for the Young People's Concert series, $25,000; Southwest Foundation for Biomedical Research (San Antonio, TX)—to purchase equipment, $15,000.
Contact: Lydia R. Saldana, Program Officer; (210) 228-3753 or (210) 225-2243; fax (210) 225-1980; lsaldana@saafdn.org or info@saafdn.org
Internet: http://www.saafdn.org/NetCommunity/Page.aspx?pid=254
Sponsor: San Antonio Area Foundation
110 Broadway Street, Suite 230
San Antonio, TX 78205-1948

San Diego Foundation for Change Grants **4218**
The foundation seeks funding proposals from grassroots community organizations who work to achieve social, economic, and environmental justice in San Diego County. Foundation concerns cover the range of progressive social issues, including: ending discrimination in any form; halting environmental pollution; promoting peace; expanding immigrant rights; improving San Diego's neighborhoods and workplaces; working toward racial equality; fighting for women's rights; developing youth projects; addressing domestic violence; and preventing violence in the community. Priority goes to projects with limited access to other sources of funding. In addition to the general funds, the foundation has the following special grants: the Danzig Award; and the James Mitsuo Cua Award. Organizations submitting a general application will be considered for these special grants with no additional application necessary. Letters of Intent are due on February 5, while full proposals are due by April 2.
Requirements: California nonprofit organizations in San Diego are eligible.
Restrictions: Grants do not support social or human service organizations that do not have a strong community organizing component; individual efforts; national or statewide projects unless there is a strong local focus; direct union organizing; private businesses or profit-making organizations; or electoral campaigns or candidates.
Date(s) Application is Due: Feb 5; Apr 2
Contact: John Fanestil, Executive Director; (619) 692-0527; fax (619) 255-3640; john@foundation4change.org or info@foundation4change.org
Internet: http://www.foundation4change.org/forapplicants.php
Sponsor: San Diego Foundation for Change
3758 30th Street
San Diego, CA 92104

San Diego Foundation Health & Human Services Grants **4219**
The San Diego Foundation's Health and Human Services Working Group was established with the vision to improve the health, well-being and self-sufficiency of individuals, families and communities in the San Diego region. The Health and Human Services Working Group's research, investigation and grant awards to date indicate that families who are vulnerable to episodic homelessness benefit from financial education that leads to budget management, traditional banking relationships, reduction of debt improved credit, and increased savings. This initiative will support financial education and asset building programs of community-based organizations that serve the residents of the San Diego region.
Requirements: To be eligible, organizations must be providing services in San Diego County. The organizations must have a 501(c)3 IRS tax exempt status. An organization may serve as a fiscal sponsor for a charitable organization that does not have a 501(c)3 status if a cooperative relationship between the two can clearly be demonstrated. The fiscal sponsor must be willing to administer the grant if awarded.
Restrictions: Generally, the Foundation does not make grants for: organizations that have previously received funding from the Foundation but have not submitted required final reports; major building/capital campaigns; scholarships; endowments; for-profit organizations; projects that promote religious doctrine; individuals; organizations outside San Diego County; marketing and/or promotional materials (annual reports, brochures, video productions); re-granting dollars to other nonprofit organizations or individuals; short-term, annual or one time events, including festivals, performances and conferences; travel outside of the San Diego region, or; existing obligations or debt.
Geographic Focus: California
Date(s) Application is Due: Jan 16
Amount of Grant: 30,000 U.S.D
Contact: Shelley Lyford, Research & Development Manager; (619) 235-2300; Shelley@sdfoundation.org
Internet: http://www.sdfoundation.org/communityimpact/cycle2006.html#ac
Sponsor: San Diego Foundation
2508 Historic Decatur Road, Suite 200
San Diego, CA 92106

San Diego Foundation Paradise Valley Hospital Community Fund Grants **4220**
The Foundation has created a joint process managed by the San Diego Foundation and Alliance Healthcare Foundation for the distribution of funds to tax-exempt community-based health and social service organizations. Grant making decisions and oversight of funded projects will be provided by an advisory committee comprised of diverse individuals representing stakeholders in the PVH service area, as well as members of The San Diego Foundation's Health & Human Services Working Group.
Requirements: To be eligible, organizations must be providing services in San Diego County. The organizations must have a 501(c)3 IRS tax exempt status. An organization may serve as a fiscal sponsor for a charitable organization that does not have a 501(c)3 status if a cooperative relationship between the two can clearly be demonstrated. The fiscal sponsor must be willing to administer the grant if awarded.
Geographic Focus: California
Contact: Anahid Brakke, Director of Health & Human Services; fax (619) 235-2300; anahid@sdfoundation.org
Internet: http://www.sdfoundation.org/communityimpact/healthservices/pvhfund.html
Sponsor: San Diego Foundation
2508 Historic Decatur Road, Suite 200
San Diego, CA 92106

San Diego HIV Funding Collaborative (SDHFC) Grants **4221**
The mission of the Collaborative is to raise, leverage and collaboratively allocate private funds to reduce the impact of HIV in the San Diego region. Donations are pooled and re-granted to local service organizations to help people affected by HIV/AIDS. Funds from the local community are used to make grants to local organizations and assist grantees with technical support. All grant applicants must start by submitting a Letter of Intent.
Requirements: The Foundation does not fund projects and programs outside San Diego and Imperial Counties.
Restrictions: The Foundation does not fund: research; lobbying; underwriting of medical expenses; general operating expenses we deem to be excessive; construction or renovation; the purchase of costly equipment; development activities, such as fund raising events, capital campaigns or annual fund drives; projects or proposals from individuals; or organizations that do not have 501(c)3 status.
Geographic Focus: California
Contact: Hamse Warfa, Program Officer; (858) 614-4892 or (858) 874-3788; fax (858) 874-3656; hwarfa@alliancehf.org
Internet: http://www.alliancehf.org/sdhiv/who_we_are.html
Sponsor: Alliance Healthcare Foundation
9325 Sky Park Court, Suite 350
San Diego, CA 92123

San Diego Women's Foundation Grants **4222**
The San Diego Women's Foundation (SDWF) strengthens and improves women's capacities to engage in significant philanthropy in the San Diego region. The foundation chooses one focus area of grant making each year. Its primary focus areas include: education, arts and culture, environment, employment and economic development, civil society, and health and human services. Check the web site each September for that year's focus guidelines.
Requirements: Only 501(c)3 organization within San Diego County should apply.
Geographic Focus: California
Date(s) Application is Due: Dec 6
Amount of Grant: 25,000 U.S.D
Samples: AjA Project in Community Heights, $40,000; Eveoke Dance Theater: Young Artists Program, $28,000; La Jolla Playhouse: Enriching Children's Lives Through Theater, $17,000.
Contact: Tracy Johnson, Director; (619) 814-1374 or (619) 235-2300; fax (619) 239-1710; tracy@sdfoundation.org
Internet: http://www.sdwomensfoundation.org/
Sponsor: San Diego Women's Foundation
2508 Historic Decatur Road, Suite 200
San Diego, CA 92106

Sandler Program for Asthma Research Grants **4223**
The mission of the foundation is to develop important new pathways of investigation in basic research regarding asthma. The program, which is open to investigators from all fields, particularly encourages applications from investigators not currently studying asthma. Innovation and risk are strongly encouraged. The program especially (but not exclusively) seeks applications in the following areas: development of new animal models for asthma; studies in humans or of human materials (excluding clinical trials); genetics of asthma; and technology development relevant to asthma. Three-year senior investigator awards and three-year junior investigator awards will be made. Awards to junior investigators are intended to support investigators who have demonstrated exceptional early accomplishment as independent investigators, permitting the expansion of their work into the field of asthma; they are not start-up funds. The proposed work should be directed toward uncovering basic mechanisms in the pathogenesis of asthma.
Requirements: Senior investigators will have well-established research programs and an international reputation for their research They will usually hold a full-time academic appointment as professor, tenured associate professor, or the equivalent. Junior investigators will have already established an independent research program. They will usually hold a full-time academic appointment as an assistant professor, early associate professor, or the equivalent. Investigators may apply from any U.S. or Canadian nonprofit organization that can and will provide the necessary facilities and infrastructure for the research, and that will assure compliance with guidelines for animal and human studies established by the U.S. Department of Health and Human Services. Applicants from Canadian institutions must also provide evidence that the institution has been designated as a public charity by the U.S. Internal Revenue Service.
Date(s) Application is Due: Feb 8
Amount of Grant: 150,000 - 250,000 U.S.D
Contact: William E. Seaman, Program Director; (415) 514-0730; fax (415) 514-0734; seamanb@sandlerresearch.org
Internet: http://www.sandlerresearch.org
Sponsor: Sandler Foundation
Box 0509, UCSF
San Francisco, CA 94143-0509

Sands Foundation Grants 4224

The Venetian Foundation (now the Sands Foundation) was formed December 7, 2000, by the Venetian Casino Resort. Today, the Las Vegas Sands Corporation's primary philanthropic initiative is pursued through Sands Foundation, a non-profit 501(c)3 organization. Sands Foundation pursues a mission of supporting charitable organizations and endeavors that concentrate on assisting youth, promoting health, and expanding educational opportunities within the local communities. The Foundation also supports causes that empower minority communities and improve underprivileged areas, as well as other valuable charitable and philanthropic activities permitted under relevant tax-exempt laws. Sands Foundation pursues a mission of supporting charitable organizations and endeavors that concentrate on assisting youth, promoting health, and expanding educational opportunities within our local communities. The Foundation also supports causes that empower minority communities and improve underprivileged areas, as well as other valuable charitable and philanthropic activities permitted under relevant tax-exempt laws. Charitable requests along with supporting documents may either be faxed or mailed.

Requirements: All charitable requests must be submitted in writing. Written requests should include the following: agency/organization information (brochures, information packet, list of the board of directors, history, background, or other helpful information); 501(c)3 tax identification number; contact person; mailing address and telephone number; overview of project or event at hand; date, time, location for event requests; purpose of request; very specific information about the amount/item(s) requested; and target population which will benefit from support.

Geographic Focus: Nevada, Pennsylvania, Macau, Singapore

Amount of Grant: 5,000 - 100,000 U.S.D

Samples: Keep Memory Alive, Las Vegas, Nevada, $25,000; Landmark Media Enterprises, Norfolk, Virginia, $37,500; Emeril Lagasse Foundation, New Orleans, Louisiana, $5,000.

Contact: Community Development Department; (702) 607-1677; fax (702) 607-1044; foundation@venetian.com

Internet: http://www.lasvegassands.com/LasVegasSands/Sands_Foundation/Donation_Request.aspx

Sponsor: Sands Foundation
3355 Las Vegas Boulevard South
Las Vegas, NV 89109-8941

Sandy Hill Foundation Grants 4225

The Sandy Hill Foundation offers awards to eligible nonprofit organizations in the areas of education, health care, and social services. The foundation gives primarily to the arts and culture, higher education, hospitals, health associations, social services, and federated giving programs. Their areas of interest include the arts; child and youth services; community and economic development; health organizations and associations; higher education; hospitals; human services; Protestant agencies and churches; recreation camps; and United Way and Federated Giving Programs. It also offers college scholarships for designated local area schools.

Requirements: There is no application or specific deadline for nonprofit giving. See contact information for current scholarship application due April 1.

Restrictions: The foundation gives primarily to the greater Hudson Falls, NY area. No grants are given to individuals.

Geographic Focus: New York

Date(s) Application is Due: Apr 1

Amount of Grant: Up to U.S.D

Contact: Nancy Juckett Brown, Trustee; (518) 791-3490

Sponsor: Sandy Hill Foundation
P.O. Box 30
Hudson Falls, NY 12839-0030

San Francisco Foundation Community Health Grants 4226

The San Francisco Foundation welcomes unsolicited applications twice a year. The Community Health Program works to expand access to health services and promote health prevention in the five counties we serve. Community Health's efforts focus on improving health on a community level and providing the social supports for communities to improve health. The Bay Area counties included are: Martin, Contra Costa, San Francisco, Alameda and San Mateo.

Requirements: Grants are made to nonprofit organizations in Alameda, Contra Costa, Marin, San Francisco, and San Mateo Counties of California.

Restrictions: The foundation generally does not fund projects outside the Bay Area community; long-term operating support; medical, academic, or scientific research; religious activities (religious institutions may apply for nonsectarian programs); direct assistance to individuals; or conferences or one-time events.

Geographic Focus: California

Date(s) Application is Due: Jan 16; Jul 1

Amount of Grant: 15,000 - 80,000 U.S.D

Samples: California Pan-Ethnic Health Network, Oakland, CA—to provide core operating support for public policy work and building relationships with primary constituents, $50,000;Latina Breast Cancer Agency, San Francisco, CA—to perform a needs and community assessment around the California Every Women Counts Cancer Detection Program, $25,000.

Contact: Denise Martin; (415) 733-8500 or (415) 733-8569; dkm@sff.org

Internet: http://www.sff.org/programs/community-health

Sponsor: San Francisco Foundation
225 Bush Street, Suite 500
San Francisco, CA 94104-4224

San Francisco Foundation Disability Rights Advocate Fund Emergency Grants 4227

The program awards emergency grants to positively impact large populations of people with disabilities in the Bay Area and northern California. Grants are available to organizations to assist mobilizing people with disabilities to act upon time-sensitive and urgent matters affecting the disability community. For example, funds may be used to cover the transportation and/or personal assistant costs associated with a community mobilization effort. Requests are accepted throughout the year; decisions are made, based on availability of funds, within four to six weeks of submission. Requests should be submitted via electronic mail. Guidelines are available online.

Requirements: Organizations located in the Bay Area and northern California are eligible.

Geographic Focus: California

Amount of Grant: 10,000 U.S.D

Contact: Denise Martin; (415) 733-8500 or (415) 733-8569; dkm@sff.org

Internet: http://www.sff.org

Sponsor: San Francisco Foundation
225 Bush Street, Suite 500
San Francisco, CA 94104-4224

San Juan Island Community Foundation Grants 4228

The foundation's mission is to enhance the quality of life on San Juan Island and make a positive difference in the community. The foundation awards grants to tax-exempt organizations in its areas of interest, including youth/education, the arts, environment, human services, and health care. Highest priority for funding will be given to requests for specific capital expenditures, special projects, one-time non-recurring expenditures, and those requests not otherwise able to obtain funding. Grantees are expected to publicize receipt of the grant from the foundation, and are required to submit a summary report one year after the allocation of funds. Application and guidelines are available online.

Requirements: San Juan Island 501(c)3 tax-exempt organizations and organizations with 501(c)3 fiscal sponsors are eligible.

Restrictions: Grants do not support religious organizations where the funds would be used to further the organization's religious purposes; individuals; other endowments; political purposes; or any purpose that discriminates as to race, creed, ethnic group, or gender.

Geographic Focus: Washington

Date(s) Application is Due: Jun 15

Contact: Grants Administrator; (360) 378-1001; grants@sjicf.org

Internet: http://www.sjicf.org/grantmaking.html

Sponsor: San Juan Island Community Foundation
P.O. Box 1352
Friday Harbor, WA 98250

Sanofi-Aventis Pharmaceutical Grants 4229

The program supports select activities in corporate communities. Philanthropic efforts seek to establish sustainable investment in children for health and educational programs; partner with organizations that address the health needs of the underserved and uninsured through innovative ways; and engage its employees to serve the community through volunteerism.

Requirements: 501(c)3 tax-exempt organizations are eligible.

Restrictions: The corporate philanthropy program does not fund individuals seeking sponsorship or help; travel; scholarships or memorials; political, religious, or veterans' organizations; marathons, runs and walks (for groups or individuals); pooled funds by groups of individuals; capital campaigns or endowments; service clubs (i.e., Lions, Rotary, Kiwanis); professional associations; organizations that conflict with company goals and products; school ads; or symposia and conferences.

Contact: Community Involvement Coordinator; (800) 981-2491

Internet: http://www.sanofi-aventis.us

Sponsor: Sanofi-Aventis Pharmaceuticals
300 Somerset Corporation Boulevard
Bridgewater, NJ 08807

Santa Barbara Foundation Grants 4230

The foundation makes capital and program grants to eligible charitable organizations in the following areas: Culture, Education, Environment, Health, Human Services, Personal Development and Recreation. Grants are also made to nonprofit agencies to provide scholarships for summer camp, sports programs, and childcare. One special fund provides grants for making the arts accessible to those with physical, visual, and hearing impairments, while another fund makes grants for activities that support blood and blood-related products and services.

Requirements: Applicant organizations must provide copies of 501(c)3 letters and also establish that the funds applied for will be used only within Santa Barbara County.

Restrictions: Grants are not awarded to individuals or for government entities, although affiliate agencies may apply.

Geographic Focus: California

Amount of Grant: 500 - 75,000 U.S.D

Samples: Boys and Girls Club of Santa Barbara (Santa Barbara, CA)—eight computers for computer lab, $8936; Santa Ynez Valley People Helping People (Santa Ynez, CA)—Advocates for Domestic and Child Abuse Prevention Program, $17,000.

Contact: Senior Program Officer; (805) 963-1873; fax (805) 966-2345; info@sbfoundation.org

Internet: http://www.sbfoundation.org/page11794.cfm

Sponsor: Santa Barbara Foundation
15 E Carrillo Street
Santa Barbara, CA 93101

Santa Fe Community Foundation Grants 4231

The mission of the community foundation is to provide leadership in helping meet the needs of the people of Santa Fe and northern New Mexico. Grants are awarded in three categories. General Grants address community concerns and help generate solutions in the areas of arts, civic affairs, education, environment, and health and human services. Technical Assistance Connector grants assist organizations to obtain professional expertise in board/staff development, fundraising, strategic planning, program evaluation, budgeting and financial management, public relations and marketing, volunteer management, and mergers. Lesbian and Gay Community Funding Partnership grants assist organizations serving lesbian, gay, and bisexual populations. Types of support include multiyear grants, general operating support, matching grants, and technical assistance grants. The foundation is particularly interested in funding initiatives that are collaborative, inclusive, respond to major gaps in service, and are rooted in the community. Listed deadline is for LGBT grants. Annual deadline dates may vary; contact program staff for exact dates.
Requirements: IRS 501(c)3 tax-exempt organizations serving residents of Santa Fe, Los Alamos, Rio Arriba, Taos, San Miguel, and Moro Counties in New Mexico are eligible.
Restrictions: The foundation does not award grants to individuals or for religious purposes, political purposes, capital outlay or capital campaigns, endowment, or governmental agencies experiencing operational budget cuts.
Geographic Focus: New Mexico
Date(s) Application is Due: Jul 1
Amount of Grant: 2,000 - 10,000 U.S.D
Samples: Santa Fe Children's Museum (Santa Fe, NM)—for a comprehensive three-year summer program for children aged five to seven from low-income Santa Fe families, $130,000; El Bien Estar de Familias Daycare Ctr (Tierra Amarilla, NM)—operating support for this child care center that serves the rural Chama/Tierra Amarilla area of New Mexico, $10,000.
Contact: Dolores Roybal, Program Officer; (505) 988-9715; fax (505) 988-1829; droybal@santafecf.org
Internet: http://www.santafecf.org/grants/general.php
Sponsor: Santa Fe Community Foundation
P.O. Box 1827
Santa Fe, NM 87504-1827

Sara Elizabeth O'Brien Trust Grants 4232

The Sara Elizabeth O'Brien Trust was established in 1981 to support charitable organizations that provide treatment or care to those who are blind or have cancer. The O'Brien Trust also supports medical research in the areas of blindness or cancer. The majority of grants from the O'Brien Trust are one year in duration; on occasion, multi-year support is awarded. Applicants must apply online at the grant website. Applicants are strongly encouraged to do the following before applying: review the downloadable state application procedures for additional helpful information and clarifications; review the downloadable online-application guidelines at the grant website; review the trust's funding history (link is available from the grant website); review the online application questions in advance; and review the list of required attachments. These will generally include: a list of board members, financial statements (audited, reviewed, or compiled by independent auditor); an organization summary; a list of other funding sources; an IRS Determination letter; and other required documents. All attachments must be uploaded in the online application as PDF, Word, or Excel files. The application deadline for the Sara Elizabeth O'Brien Trust is 11:59 p.m. on October 15. Applicants will be notified of grant decisions before December 31.
Requirements: Applicants must have 501(c)3 tax-exempt status.
Restrictions: The trust does not support requests from individuals, organizations attempting to influence policy through direct lobbying, or any political campaigns.
Geographic Focus: Massachusetts
Date(s) Application is Due: Oct 15
Samples: Foundation of the Massachusetts Eye and Ear Infirmary, Boston, Massachusetts, $60,000, to support research aimed at discovering genes associated with diabetic retinopathy (2010); Dana-Farber Cancer Institute, Boston, Massachusetts, $33,000, fellowships in pediatric palliative care (2010); MAB Community Services, Brookline, Massachusetts, $47,000, Seeing is Not Falling Project (2010).
Contact: Michealle Larkins; (866) 778-6859; michealle.larkins@baml.com
Internet: https://www.bankofamerica.com/philanthropic/fn_search.action
Sponsor: Sara Elizabeth O'Brien Trust
225 Franklin Street, 4th Floor, MA1-225-04-02
Boston, MA 02110

Sarah Scaife Foundation Grants 4233

Approximately 80 percent of the annual grants are made to public policy programs that address major domestic and international issues. There are no geographic restrictions. Other grants are made in the fields of education and culture, health and medicine, scientific research, public affairs, and recreation and equipment in all geographic areas, but primarily in western Pennsylvania. Types of support include general operating support, continuing support, program development, conferences and seminars, publication, seed grants, fellowships, research, and matching funds. Applications are accepted at any time. Foundation staff considers requests in February, May, September, and November.
Restrictions: Grants are not made to individuals or to national organizations for general fund-raising purposes.
Amount of Grant: 25,000 - 200,000 U.S.D
Samples: U of Pittsburgh (Pittsburgh, PA)—to construct facilities for the planned Pittsburgh Institute for Neurodegenerative Diseases, $5.4 million.
Contact: Michael Gleba, Vice President of Programs; (412) 392-2900
Internet: http://www.scaife.com/sarah.html
Sponsor: Sarah Scaife Foundation
1 Oxford Center, 301 Grant Street, Suite 3900
Pittsburgh, PA 15219

Sarasota Memorial Healthcare Foundation Grants 4234

The Foundation is an organization with the mission to improve the delivery of health care in the Sarasota area through the acquisition and utilization of philanthropic funds. Although the Foundation primarily supports Sarasota Memorial Health Care System, it also awards grants to other area nonprofit organizations to meet the community's changing health care needs. Grants totaling more than $32,000,000 have been provided since the Foundation's establishment in 1976. Current fields of interest include: health care; medical care; and community health systems.
Requirements: Health care programs and organization in the Sarasota, Florida, region are eligible to apply.
Geographic Focus: Florida
Contact: Alexandra Quarles, President; (941) 917-1286; alex-quarles@smh.com
Internet: http://www.smhf.org
Sponsor: Sarasota Memorial Healthcare Foundation
1515 S Osprey Avenue, Suite B4
Sarasota, FL 34239

Sarkeys Foundation Grants 4235

Major areas of foundation support include education and higher education, health care and medical research, children and family services, and cultural and humanitarian programs of regional significance. Types of support include building construction/renovation, capital campaigns, endowments, equipment, matching/challenge support, professorships, program development, and research grants. Grant proposals are considered at the October and April meetings of the board of trustees. The trustees may consider proposals at other regular or special meetings as circumstances warrant. The board may determine to condition awarding of a grant upon the fulfillment of an institution's additional fund-raising obligation for the project within a specified time period and may withdraw the grant if the institution fails to meet the required time allotted.
Requirements: 501(c)3 institutions located in Oklahoma are eligible.
Restrictions: Needs normally outside the range of the foundation's support include direct mail solicitations, annual campaigns, hospitals, operating expenses, vehicle purchases, permanent financing of a program, start-up funds for new organizations, or feasibility studies. Sarkeys Foundation will not accept proposals for new construction and/or major renovation projects this year and until further notice. In addition, Sarkeys will not accept any requests over $50,000
Geographic Focus: Oklahoma
Date(s) Application is Due: Feb 1; Aug 1
Amount of Grant: Up to 50,000 U.S.D
Samples: Rogers State U (Claremore, OK)—to endow a chair in business information technology, $250,000 matching grant.
Contact: Susan Frantz; (405) 364-3703; fax (405) 364-8191; Sarkeys@sarkeys.org
Internet: http://www.sarkeys.org/grants.htm
Sponsor: Sarkeys Foundation
530 E Main Street
Norman, OK 73071

Sartain Lanier Family Foundation Grants 4236

The foundation awards grants to Georgia nonprofits in support of education, health and human services, arts, environment, and community development. Types of support include building and renovation; capital campaigns; endowments; general operating support; program development; and program-related investments and loans. The foundation's board meets in May and November of each year to consider grant requests. Prior to submitting a full proposal, applicants should submit a letter limited to two pages summarizing the request. The letter should include a brief description of the organization and its purpose, the project for which funding is requested, the total cost of the project, and the amount being requested.
Requirements: Nonprofit organizations in the southeastern United States are eligible to apply, however, the majority of recipients are located in Georgia and specifically the Atlanta metro area.
Restrictions: The foundation does not make grants for individuals; churches or religious organizations for projects that primarily benefit their own members; partisan political purposes; tickets to charitable events or dinners, or to sponsor special events or fundraisers.
Geographic Focus: Georgia
Samples: Vanderbilt University (Nashville, TN)—for operating support, $700,000; Westminster Schools (Atlanta, GA)—for operating support, $100,000; Whitefoord Community Program (Atlanta, GA)—for operating support, $50,000.
Contact: Patricia Lummus, Assistant Director; (404) 564-1259; fax (404) 564-1251; plummus@lanierfamilyfoundation.org
Internet: http://www.lanierfamilyfoundation.org/grant.html
Sponsor: Sartain Lanier Family Foundation
25 Puritan Mill, 950 Lowery Boulevard
Atlanta, GA 30318

Sasco Foundation Grants 4237

The Foundation focus of support is on the arts, education, the environment, health, and human services. Although the main geographic areas are limited to Connecticut, Maine, and New York, the Foundation also occasionally supports programs of national importance. There are no deadlines and no formal applications. Please contact the Foundation for further details.

Requirements: Must be a 501(c)3 organization. Grants are not given to individuals.
Geographic Focus: Connecticut, Maine, New York
Contact: Uwe Linder, J.P. Morgan Private Bank Trustee; (800) 576-6209
Sponsor: Sasco Foundation
345 Park Avenue, 6th Floor
New York, NY 10154-1002

Savoy Foundation Post-Doctoral and Clinical Research Fellowships 4238
Fellowships of the Savoy Foundation are available to Canadian researchers and to foreign nationals or for projects conducted in Canada. Post-doctoral and clinical research fellowships of $30,000 will be awarded to scientists or medical specialists (Ph.D. or M.D.) wishing to carry out a full time research project in the field of epilepsy. The grant will be for one (1) year, but will be renewable, exceptionally, once upon request. As a rule, the modalities of attribution and amounts awarded will be similar to those of governmental funding agencies (MRC or FRSQ). The fellow with the highest mark will received the prestigious Steriade-Savoy award of an additional $1,500.
Restrictions: Fellows funded by the Savoy Foundation must not cumulate two major awards.
Geographic Focus: All States, Canada
Date(s) Application is Due: Jan 15
Amount of Grant: 30,000 U.S.D
Contact: Fellowship Coordinator; (450) 358-9779; fax (450) 346-1045; epilepsy@savoy-foundation.ca
Internet: http://www.savoy-foundation.ca/eng/mission/defaut.htm
Sponsor: Savoy Foundation
230 Foch Street
St Jean Sur Richelieu, QC J3B 6B2 Canada

Savoy Foundation Research Grants 4239
A small number of research grants are available each year to clinicians and/or established scientists working on epilepsy or related subjects. These grants will be attributed for the following purposes: launching of a new research project, preliminary studies in preparation of a more substantial request to be addressed to another agency, continuation or completion of a project, sharing of expenses, or a limited project not requiring funding in excess of the limit. As a rule, these grants are not renewable. Contact the foundation for application forms.
Requirements: Research grants are available to Canadian citizens or for projects conducted in Canada.
Geographic Focus: All States, Canada
Date(s) Application is Due: Jan 15
Amount of Grant: 25,000 U.S.D
Contact: Grants Administrator; (450) 358-9779; fax (450) 346-1045; epilepsy@savoy-foundation.ca
Internet: http://www.savoy-foundation.ca/eng/mission/defaut.htm
Sponsor: Savoy Foundation
230 Foch Street
St Jean Sur Richelieu, QC J3B 6B2 Canada

Savoy Foundation Studentships 4240
Studentships of the Savoy Foundation are available to Canadian researchers and to foreign nationals or for projects conducted in Canada. Studentships will be awarded to meritorious applicants wishing to acquire training and pursue research in a biomedical discipline, the health sciences or social sciences related to epilepsy. These studentships last one year, but will be renewable three times, upon request, for a maximum duration of four years. The stipend will be $15,000 per year. The student with the highest mark will receive the prestigious Van Gelder-Savoy award of an additional $1,500. An annual sum of $1,000 will be allocated to the laboratory or institution as additional support for the research project.
Requirements: To be eligible, the candidate must have a good university record (B. Sc., M.D. or equivalent diploma) and have ensured that a qualified researcher affiliated to a university and/or hospital, will supervise his/her work. Concomitant registration in a graduate program (M. Sc. or Ph. D.) is encouraged.
Restrictions: Students funded by the Savoy Foundation must not cumulate two studentship awards.
Geographic Focus: All States, Canada
Date(s) Application is Due: Jan 15
Amount of Grant: 15,000 U.S.D
Contact: Caroline Savoy, (450) 358-9779; fax (450) 346-1045; epileps@savoy-foundation.ca
Internet: http://www.savoy-foundation.ca/eng/mission/defaut.htm
Sponsor: Savoy Foundation
230 Foch Street
St Jean Sur Richelieu, QC J3B 6B2 Canada

SBC Foundation Grants 4241
The foundation supports nonprofit organizations in its company-operating areas in its areas of interest, including education, civic and community development, culture and the arts, and health and welfare. In the area of education, programs are supported that work to strengthen K-12 education and the teaching profession, address adult literacy, and retrain seniors and the disadvantaged. Economic development organizations with innovative programs designed to enhance and broaden the economic stability of communities receive consideration. In the area of health and welfare, applied research, educational programs, and other innovative programs in preventive health care, particularly those targeted to youth and seniors, receive support. Under arts and culture, performing arts and cultural programs in major metropolitan serving areas with outreach potential receive consideration. The foundation prefers to fund project-specific proposals rather than requests for grants to underwrite operating or capital budgets.
Requirements: 501(c)3 organizations serving residents in company-operating communities are eligible.
Restrictions: The foundation does not make grants to/for individuals, political activities or organizations, religious and fraternal groups serving only their members, goodwill advertising, local school systems, or disease-specific organizations.
Amount of Grant: Up to 100,000,000 U.S.D
Samples: U of Illinois (Urbana, IL)—for nine programs at its campuses in Chicago, Springfield, and Urbana-Champaign, $336,000; California Fire Foundation (Sacramento, CA)—for the National Parks Junior Ranger Program and recreation, recycling, and volunteer programs, $22 million value in grants and noncash donations over four years; American National Red Cross (Washington, DC)—for relief efforts related to tsunamis in South Asia, and for it Annual Disaster Giving Program, $1 million.
Contact: Laura Sanford, President; (800) 591-9663; fax (312) 207-1098
Internet: http://www.sbc.com/gen/corporate-citizenship?pid=5372
Sponsor: SBC Foundation
130 E Travis, Suite 350
San Antonio, TX 78205

Scaife Family Foundation Grants 4242
The foundation's grant program is primarily directed toward public policy programs that address major domestic and international issues. There are no geographical restrictions. Initial inquiries should be in letter form signed by the organization's president, or authorized representative, and have the approval of the organization's board of directors. The letter should include a concise description of the specific program for which funds are requested, and a program budget and timetable. Accompanying information should include a copy of the latest audited financial statements and annual report; current budget; list of officers and directors and their major affiliations; and a copy of the current determination letter from the Internal Revenue Service. Additional information may be requested, if needed, for further evaluation. The foundation meets in February, May, September, and November to consider grants. However, requests may be submitted at any time and will be acted upon as expeditiously as possible.
Requirements: 501(c)3 tax-exempt organizations are eligible.
Restrictions: The foundation does not make grants to individuals for any purpose or to nationally organized fundraising groups. Proposals for the following are usually declined: event sponsorships, endowments, capital campaigns, renovations, or government agencies.
Samples: U of Pittsburgh (PA)—for the Pittsburgh Institute for Neurodegenerative Diseases, $5.4 million.
Contact: Michael Gleba, Executive Vice President; (412) 392-2900
Internet: http://www.scaife.com/sarah.html
Sponsor: Scaife Family Foundation
301 Grant Street, Suite 3900
Pittsburgh, PA 15219-6401

Schering-Plough Foundation Community Initiatives Grants 4243
The Schering-Plough Foundation, a non-profit membership corporation established in 1955, is dedicated to working with the citizens of our communities to help them realize their full potential and enhance their quality of life. Support for community development takes many forms and allows the Foundation to reach out to numerous, highly diverse groups within its communities. The Foundation continues to support organizations that promote culture and the arts, environmental issues, legal services, etc. and are always seeking out new and innovative ways to serve the citizens of its communities.
Requirements: The Foundation considers requests from tax-exempt, 501(c)3 non-profit organizations located in the United States, or its possessions, whose goals and activities fall within its stated objectives and areas of interest. All requests for funding must be made online. National organizations are eligible to apply.
Restrictions: Grants are not made to individuals.
Amount of Grant: 5,000 - 200,000 U.S.D
Contact: Christine Fahey, Assistant Secretary; (908) 298-7232; fax (908) 298-7349
Internet: http://www.schering-plough.com/company/foundation.aspx
Sponsor: Schering-Plough Foundation
2000 Galloping Hill Road
Kennilworth, NJ 07033-0530

Schering-Plough Foundation Health Grants 4244
The Schering-Plough Foundation, a non-profit membership corporation established in 1955, is dedicated to working with the citizens of our communities to help them realize their full potential and enhance their quality of life. The Foundation's interest in improving the health and well-being of its communities remains a top priority and is reflected in the numerous initiatives that have been developed to benefit these communities. The Foundation invests in the future of those communities by building meaningful partnerships that enhance the quality of life. Currently, the Foundation is involved in several partnerships that seek to expand the quality and availability of health care in its underserved communities.
Requirements: The Foundation considers requests from tax-exempt, 501(c)3 non-profit organizations located in the United States, or its possessions, whose goals and activities fall within its stated objectives and areas of interest. All requests for funding must be made online. National organizations are eligible to apply.
Restrictions: Grants are not made to individuals.
Amount of Grant: 5,000 - 200,000 U.S.D
Samples: New Jersey Institute of Technology (Newark, NJ)—to enhance its master's degree program in pharmaceutical engineering, $250,000.
Contact: Christine Fahey, Assistant Secretary; (908) 298-7232; fax (908) 298-7349

Internet: http://www.schering-plough.com/company/foundation.aspx
Sponsor: Schering-Plough Foundation
2000 Galloping Hill Road
Kennilworth, NJ 07033-0530

Schlessman Family Foundation Grants 4245
The foundation awards grants to Colorado nonprofits in its areas of interest, including higher and other education, cultural programs, youth, disabled, and social services. Types of support include general operating support, continuing support, annual campaigns, capital campaigns, equipment acquisition, endowment funds, program development, curriculum development, scholarship funds, and matching grants. Emphasis will be given to programs in Front Range, CO. Applicants should submit a letter of request.
Requirements: Colorado nonprofits in the greater Denver area are eligible to receive grant support.
Geographic Focus: Colorado
Date(s) Application is Due: Dec 31
Amount of Grant: Up to 21,000,000 U.S.D
Samples: Urban Ministries (CO)—for operating support, $6000; Adult Learning Source (CO)—for operating support, $7500; Greeley Meals on Wheels (CO)—for operating support, $2000; Hospice of Metro Denver (CO)—for operating support, $3000.
Contact: Patricia Middendorf, Treasurer; (303) 831-5683; fax (303) 831-5676; pamiddendorf@qwest.net
Internet: http://www.schlessmanfoundation.org
Sponsor: Schlessman Family Foundation
1301 Pennsylvania Street, Suite 800
Denver, CO 80203-5015

Schnurmacher Foundations Grants 4246
The foundations award grants nationally, with emphasis on California, Connecticut, and New York. Areas of interest include arts, education, Jewish agencies and temples, human services, medical research, mental health, and religion. Types of support include capital campaigns, conferences and seminars, continuing support, curriculum development, endowments, equipment acquisition, general operating support, matching/challenge grants, research grants, and scholarship funds. There are no application forms or deadlines.
Requirements: Nonprofit organizations are eligible.
Amount of Grant: 1,000 - 200,000 U.S.D
Samples: Neurosciences Institute (La Jolla, CA)—for its campaign for endowment and research operations, $1 million; Beth Israel Medical Ctr (New York, NY)—for general operating support, $52,500.
Contact: Ira Weinstein, President; (212) 838-7766
Sponsor: Schnurmacher Foundations
155 E 55th Street, Suite 6J
New York, NY 10002

Schoenbaum Family Foundation Grants 4247
The foundation awards grants nationwide, with emphasis on Florida and West Virginia, in its areas of interest, including health, education, welfare, and Jewish federated giving programs. Types of support include program grants and scholarship funds. There are no application forms.
Date(s) Application is Due: Nov 1
Amount of Grant: Up to 13,000,000 U.S.D
Samples: Ohio State U, College of Human Ecology (Columbus, OH)—to create an early childhood-development center in the Weinland Park neighborhood east of campus, $2.5 million.
Contact: Raydel Walston, (941) 366-1708
Sponsor: Schoenbaum Family Foundation
P.O. Box 580
Sarasota, FL 34230-0580

Schrage Family Foundation Scholarships 4248
The Centier Bank-Schrage Family Foundation Scholarship rewards academic achievement, as well as school and community leadership. The $2,000 award is distributed for four years providing that the student achieves a GPA of 3.0 each semester. Centier Bank communities offering the Schrage Scholarship include Chesterton, Crown Point, Dyer, East Chicago, Gary, Griffith, Hammond, Hanover, Highland, Hobart, LaPorte, Lowell, Merrillville, Munster, Plymouth, Portage, Schererville, St. John, Valparaiso, and Whiting. Students should apply to their corresponding bank location.
Requirements: To be considered for the Schrage Scholarship, applicants must: be in the top 20% of their graduating class; obtain high level SAT or ACT scores; occupy leadership roles in school and/or community activities; and be willing to sign a publicity release form. Any college major is eligible for the scholarship, with preference given to Business majors. Students will be required to submit their grades to Centier Bank at the end of each school year and communicate openly to Centier Bank regarding any changes in their scholastic involvement (i.e. switching schools, internship programs, etc.).
Geographic Focus: Indiana
Amount of Grant: 2,000 U.S.D
Contact: Vicki Bukowski, Chief Learning Officer; (219) 755-6120, ext. 1815; vbukowsk@centier.com
Internet: https://www.centier.com/community/schrage.asp
Sponsor: Legacy Foundation
1000 East 80th Place, 302 South
Merrillville, IN 46410

Schramm Foundation Grants 4249
The foundation awards grants to Colorado nonprofit organizations in its areas of interest, including arts and culture, civic affairs, community development, education (elementary, secondary, and higher), healthcare, housing, humanities, medical research, science, social services delivery, technology, women's issues, and youth. Types of support include building construction/renovation, continuing support, equipment acquisition, general operating support, matching/challenge grants, program development, and scholarship funds. Applications are accepted from July 1 through the listed application deadline.
Requirements: Colorado 501(c)3 nonprofit organizations are eligible. Preference is given to requests from the Denver area.
Restrictions: Grants do not support advertising, advocacy organizations, individuals, international organizations, political organizations, religious organizations, school districts, special events, or veterans organizations.
Geographic Focus: Colorado
Date(s) Application is Due: Aug 31
Amount of Grant: 500 - 26,500 U.S.D
Samples: U of Denver (CO)—to renovate Spruce Hall of the Graduate School of Social Work, $100,000.
Contact: Gary Kring, President; (303) 797-6718
Sponsor: Schramm Foundation
800 Grant Street, Suite 404
Denver, CO 80203

Schurz Communications Foundation Grants 4250
The Schurz Communications Foundation, Inc. is a company sponsored foundation which was incorporated in 1940. The Foundations donors are: Schurz Communications, Incorporated; South Bend Tribune Corporation; WSBT, Incorporated. The Foundation supports community service clubs and organizations involved with arts and culture, higher education, youth development and disabled people.
Requirements: Giving is limited to the South Bend Indiana area.
Geographic Focus: Indiana
Amount of Grant: 1,000 - 40,000 U.S.D
Samples: Studebaker National Museum, South Bend, IN - $25,000; Boys and Girls Club of Saint Joseph County, South Bend, IN - $10,000; United Way of Saint Joseph County, South Bend, IN - $40,000;
Contact: David C. Ray, President; (574) 235-6241
Sponsor: Schurz Communications Foundation
225 West Colfax Avenue
South Bend, IN 46624

Scientific Communities Initiative Grants 4251
The coalition is awarding grants to digital ventures offering information in the areas of science, technology, and medicine (STM). The purpose of the grants is to stimulate and accelerate the creation of new university-based information communities serving users in key STM fields. Proposed projects should offer strategic alternatives to inefficiencies in the current, traditional scientific communication process. Projects should provide increased responsiveness to the needs of scientists—as authors and information users—and to libraries by offering an effective new electronic venue for publication of research; aggregating content/links needed by targeted users in sufficient breadth and mass; addressing important objectives of universities and colleges; delivering convenient, efficient access to this information in a Web environment optimized to the needs of scientists in the target discipline; and fostering a more open and competitive marketplace in which academe exercises an increased role in assuring broad dissemination of research. Partnerships and collaborations among universities and professional societies are encouraged. Letters of intent must meet the listed application deadline; full applications will be requested.
Amount of Grant: 100,000 - 500,000 U.S.D
Contact: Heather Joseph, (202) 296-2296; sparc@arl.org
Internet: http://www.arl.org/sparc/core/index.asp?page=f17
Sponsor: Scholarly Publishing and Academic Resources Coalition
21 Dupont Circle, Suite 800
Washington, DC 20036

SC Johnson Fund Grants 4252
The fund awards grants every year locally and globally to improve health, education, social progress, and environmental protection. Types of support include building construction/renovation, capital campaigns, employee matching gifts, equipment acquisition, and seed grants.
Geographic Focus: Wisconsin
Amount of Grant: Up to 95,000,000 U.S.D
Contact: Program Administrator; (800) 494-4855 or (262) 260-4855
Internet: http://www.scjohnson.com/community/global.asp
Sponsor: SC Johnson Fund
1525 Howe Street, MS 066
Racine, WI 53403

Scleroderma Foundation New Investigator Grants 4253
The foundation invites applications from promising new investigators who hold faculty or equivalent positions and who wish to pursue a career in research related to systemic sclerosis (SSc, scleroderma). This grant will support promising research that is likely to lead to individual research project grants. Before completion of the grant, investigators are encouraged to seek continuing support for research through other grants mechanisms. Applicants may request support for up to three years.

Requirements: Applications may be submitted by domestic nonprofit organizations, public and private, such as universities, colleges, hospitals, and laboratories. Applicants must have a doctorate degree in medicine, osteopathy, veterinary medicine, or one of the sciences, and must have completed a postdoctoral fellowship by the grant award date.
Restrictions: Foreign organizations and institutions are not eligible. New investigator grants may not be used to support thesis or dissertation research or fellowship training. Applicants who have been a principal investigator on foundation grants or other national, private, or government agencies other than fellowship grants are not eligible. Applicants may not simultaneously apply for this grant and the Physician Scientist Development Award, which is cofunded by the Arthritis Foundation.
Date(s) Application is Due: Sep 15
Amount of Grant: Up to 50,000 U.S.D
Samples: Dr. Constantin Bona, Mount Sinai School of Medicine (NY)—for the project Molecular Mechanisms IL-4 Induced Fibrosis in Scleroderma, $150,000,; Dr. Joseph Korn, Boston U School of Medicine (MA)—for the project Role of Integrin Alpha 11 in Fibroblast Function, $149,998.
Contact: Carolyn Weller, (800) 722-4673 or (978) 750-4499; fax (978) 750-9902; sfinfo@scleroderma.org
Internet: http://www.scleroderma.org/research/researchhome.htm
Sponsor: Scleroderma Foundation
12 Kent Way, Suite 101
Byfield, MA 01922

Scott County Community Foundation Grants **4254**
In order to fulfill most effectively its philanthropic responsibilities within the limits of its resources, the Scott County Community Foundation's grant program will emphasize focused and effective types of grants. The Scott County Community Foundation currently has one grants cycle per year. Applications are typically accepted in the summer and recipients are announced in early fall. The broad purpose of the Scott County Community Foundation is to improve the quality of life of the residents of Scott County, Indiana. Grant funding will focus on the encouragement of programs that enhance cooperating and collaboration among institutions within our community. Foundation funds ordinarily are not given to agencies for the purpose of meeting routine budgets. Funding will focus on effectiveness of special non-recurring projects which enrich health, education, cultural or recreational situations in Scott County. However, special consideration will be given to agencies that partner with other organizations to complete a unified project.
Requirements: The grant policies are: grants will be made only to organizations whose programs benefit the residents of Scott County; grants from the Scott County Community Foundation must meet legal and tax requirements as to the purpose and may be made only to charitable organizations and/or cause; grantees may be required to attend an interview by the Grants Committee; grantees must submit a final grant evaluation report; the Grantee will not be allowed to participate in any future Grant Cycles until the final report is on file in the Foundation office; grant money must be used for the specific purpose as stated in the line item budget provided in the grant application; any changes in purpose must be requested in writing and approved by the Foundation prior to any changes being made; any portion of the amount granted which is not used for the purpose of the grant will be repaid; grants will only be made to organizations, that as required by law make equal opportunities available to all persons without regard to race, age, color, religion, national origin, disability, citizenship status, or any other category protected under state or local law; after evaluation, the Grants Committee makes recommendations to the Board of Directors, which makes all funding decisions; applicants must submit 10 hole-punched copies of the application. Applications are available at the Foundation's office or through the Foundation's website.
Restrictions: The Foundation does not typically fund: on going operational expenses; existing obligations; services primarily supported by tax dollars or responsibility of a public agency; individuals or travel expenses; multi-year grants or repeat funding; advocacy or political purposes; religious or sectarian purposes; loans or endowments.
Geographic Focus: Indiana
Contact: L.L. Lowry, Executive Director; (812) 752-2057; fax (812) 752-9257; info@scottcountyfoundation.org
Internet: http://www.scottcountyfoundation.org/grants.htm
Sponsor: Scott County Community Foundation
60 North Main Street, P.O. Box 25
Scottsburg, IN 47170

Scranton Area Foundation Grants **4255**
The foundation awards grants to nonprofit organizations in the Scranton-Lackawanna area of Pennsylvania. Grants are made in broad general areas where needs always exist, including education (child development education, vocational education, and higher education), the environment, community health, human and social services, and the arts and humanities. The community foundation's board looks for projects and programs characterized by new approaches to addressing significant community problems or implementing projects, efficient use of community resources and applicants' resources, a well-planned approach to timely public issues, coordination and cooperation among charitable organizations and avoidance of program duplication, a base of other support, and citizen involvement. The board acts on grant applications quarterly. A preapplication conference with the executive director or a letter of intent regarding the project is strongly encouraged.
Requirements: IRS 501(c)3 tax-exempt organizations and publicly supported organizations such as government agencies, located in the Scranton-Lackawanna area of Pennsylvania are eligible.
Restrictions: Grants are generally not made for/to endowments, annual campaigns or special events, individuals, existing obligations, routine operating expenses, equipment acquisition, or building campaigns.
Geographic Focus: Pennsylvania
Amount of Grant: Up to 600,000 U.S.D
Samples: Housing Services Collaborative (PA)—for a cooperatively funded multi-organization housing initiative in Lackawanna County, $60,000; Needs Assessment: Clients of the Family Court of Lackawanna Court (PA)—for an assessment to identify the needs of families in distress and recommend appropriate programs, services, and delivery systems, $5250.
Contact: Jeanne Bovard, Executive Director; (570) 347-6203; fax (570) 347-7587; safinfo@safdn.org
Internet: http://www.safdn.org
Sponsor: Scranton Area Foundation
321 Spruce Street, Bank Towers, Suite 608
Scranton, PA 18503

Seabury Foundation Grants **4256**
In general, The Seabury Foundation makes one-time, special project grants, rather than general operating support. Areas of interest of the Foundation include: Arts, Culture and Humanities; Human Capacity-building; Health; Education; Conservation and Environment; Children, Youth and Families. On occasion, the Foundation may invite grant renewal requests for up to two consecutive succeeding years. After that period, the Foundation will not consider a request for funding from that organization until a minimum of two full years have passed since the final grant was awarded. Please note that all grants awarded are within the City of Chicago with the exception of projects personally sponsored by family members.
Requirements: Chicago, IL, nonprofit organizations are eligible. Initial contact should be through a Letter of Inquiry, letters need only be 1-2 pages in length. Deadline dates for Letter of Inquiry are: September 15; January 15; June 15. If after reviewing the Letter of Inquiry, if your organization is invited to submit a full proposal, the deadline dates are: November 1; March 1; August 1.
Restrictions: The Seabury Foundation do not entertain grants in support of gala events, annual fundraising events, or annual fundraising drives.
Geographic Focus: Illinois
Date(s) Application is Due: Jan 15; Mar 1; Jun 15; Aug 1; Sep 15; Nov 1
Samples: Infant Welfare Clinic of Oak Park, IL, $10,000 - Dental Clinic expansion; Chicago Jazz Philharmonic, Chicago, IL, $5,000 - Jazz Alive Education Program; Mary Rigg Neighborhood Center, IL $15,000 - Summer Day Camp Scholarships for 75 youth;
Contact: Boyd McDowell III, Director; (312) 587-7146; fax (312) 587-7332; bmcdowell@seaburyfoundation.org
Internet: http://www.seaburyfoundation.org/apply.htm
Sponsor: Seabury Foundation
1111 North Wells Street, Suite 503
Chicago, IL 60610-7633

Seagate Technology Grants **4257**
Seagate focuses its giving on K-12 science and technology education. Innovative programs that enhance creativity, encourage hands-on learning, and reach all populations (specifically low-income and minority communities where fewer opportunities exist) are sought. Programs that support and assist people in extreme need of shelter, food, healthcare, and support services are given preference. Arts and culture, civic, environmental, and diversity-focused programs are valuable to building a well-balanced, healthy community. Volunteerism is an important component of Seagate's philanthropic program, and consideration will be given to proposals that offer employee involvement opportunities. The company offers support through donations of cash; excess furniture, computer and office equipment; in-kind donations of disc drives, tape drives and software; and by encouraging employee volunteerism. There are no application deadlines. Submit requests on organization letterhead.
Requirements: Nonprofit organizations must: maintain program expenses at or below 20 percent; demonstrate a clear objective and have tools in place to measure, track, and report the impact of a contribution; operate within 50 miles of those communities where Seagate maintains a significant business presence (Silicon Valley, CA; Pittsburgh, PA; Longmont, CO; Bloomington and Shakopee, MN; Oklahoma City, OK; Springtown, Northern Ireland; Singapore; China; Indonesia; Malaysia; Mexico; Singapore; and Thailand); and hold tax-exempt status under IRC section 501(c)3.
Restrictions: Grants do not support organizations that raise funds for redistribution to other organizations; organizations, or events sponsored by organizations, that discriminate by race, creed, gender, disability, sexual orientation, age, religion or nationality; political parties, religious, or labor organizations (exceptions occasionally made for programs sponsored for the direct benefit of the community); luncheons, banquets or other forms of indirect support; private foundations; endowment funds or capital fund drives; fundraising organizations; special interest groups; athletic teams or events; individuals; travel funds; or courtesy advertising-related projects.
Geographic Focus: California, Colorado, Minnesota, Oklahoma, Pennsylvania, China, Mexico
Contact: Elena Sexton, (831) 438-6550; fax (831) 438-4127; elena.sexton@seagate.com
Internet: http://www.seagate.com/newsinfo/community/contribute/index.html
Sponsor: Seagate Corporation
920 Disc Drive
Scotts Valley, CA 95066

Searle Scholars Program Grants **4258**
The program's aim is to identify junior faculty members who have demonstrated a potential for doing innovative research in the biomedical sciences. Approximately 15 scholars are selected each year to receive the award. Scholars will receive a grant to

support their research over the next three years. Application forms are distributed through invited institutions.
Requirements: Applicants will be expected to be pursuing independent research careers in biochemistry, cell biology, genetics, immunology, neuroscience, pharmacology, related areas in chemistry, medicine, and the biological sciences. Candidates should be in their first or second year at the assistant professor level. This appointment must be a tenure-track position and must be in an academic department of an invited, degree-granting institution. Potential applicants whose institutions do not have tenure-track appointments should consult with the scientific director of the program prior to preparing an application.
Restrictions: Applications from any institution not invited to participate will not be considered. No more than two applications will be accepted from any one institution.
Date(s) Application is Due: Sep 30
Amount of Grant: 80,000 U.S.D
Contact: Mary Tarpey, Program Contact; (847) 714-1702; fax (847) 714-1716; mary.tarpey@kinshipcorp.com
Internet: http://www.searlescholars.net/apply
Sponsor: Searle Scholars Program
400 Skokie Boulevard, Suite 300
Northbrook, IL 60062

Seattle Foundation Benjamin N. Phillips Memorial Fund Grants **4259**
The Benjamin N. Phillips Memorial Fund was established by the estate of Joy Phillips to honor her late husband in 2006 as an area of interest fund of The Seattle Foundation. The goal of the Fund is to make grants to organizations improving the lives of Clallam County, Washington, residents. The Benjamin N. Phillips Memorial Fund is interested in supporting organizations that have: a mission statement that clearly defines the organization's purpose and reflects its understanding of the communities they serve; a clear articulation of why it believed what it is doing is important and that it will be effective and produce desired results; clearly defined priorities, goals and measurable outcomes; experienced and highly qualified staff and volunteer leadership; a skilled governing board whose knowledge includes management, fundraising and the community served; a funding plan appropriate to agency size and developmental state-guiding development efforts; sound financial management practices; support in the community and constituent involvement; and proven ability to mobilize financial and in-kind support, including volunteers. Grants are predominately made for one year, with no implied renewal funding. However, a two-year grant will be considered if a case is made for why funding is required for a longer period. An example of this exception is a planning or capacity-building process occurring over a two-year period of time. Approximately $250,000 will be distributed annually, with grants ranging in size from $1,000 to $25,000; the average grant size is $11,000.
Requirements: To qualify for a grant from the Foundation, an organization must: be a 501(c)3 tax-exempt nonprofit organization serving residents of Clallam County, Washington.
Geographic Focus: Washington
Date(s) Application is Due: Jul 1
Amount of Grant: 1,000 - 25,000 U.S.D
Samples: Compassion and Choices of Washington, Seattle, Washington, $10,000 - support two years of added outreach, partnership development and fund development activities in Clallam County (2011); First Book of Clallam County, Seattle, Washington, $1,000 - to support the purchase of books for low-income children (2011); Juan de Fuca Festival of the Arts, Port Angeles, Washington, $8,000 - support sponsorship of Baka Beyond workshops and concerts (2011).
Contact: Ceil Erickson, Grantmaking Director; (206) 515-2131 or (206) 515-2109; fax (206) 622-7673; c.erickson@seattlefoundation.org or phillips@seattlefoundation.org
Internet: http://www.seattlefoundation.org/nonprofits/phillips/Pages/benjaminphillipsmemorialfund.aspx
Sponsor: Seattle Foundation
1200 Fifth Avenue, Suite 1300
Seattle, WA 98101-3151

Seattle Foundation Doyne M. Green Scholarships **4260**
Doyne M. Green Scholarships are intended to support and honor female graduate students of medicine, law or social and public services. This scholarship is available to students who have completed at least one year of graduate work. The scholarship amount is $4,000, and two awards are given each year. The annual application deadline is March 1.
Geographic Focus: Washington
Date(s) Application is Due: Mar 1
Amount of Grant: 4,000 U.S.D
Contact: Ceil Erickson, Grantmaking Director; (206) 515-2131 or (206) 515-2109; fax (206) 622-7673; c.erickson@seattlefoundation.org
Internet: https://fortress.wa.gov/hecb/thewashboard/ScholarshipDetails/The+Seattle+Foundation/2012-2013/Doyne+M+Green+Scholarship
Sponsor: Seattle Foundation
1200 Fifth Avenue, Suite 1300
Seattle, WA 98101-3151

Seattle Foundation Hawkins Memorial Scholarship **4261**
he purpose of the Hawkins Memorial Scholarship is to honor and to give financial support to graduating seniors currently enrolled at Lake Stevens High School who are pursuing a course of study in life sciences, medicine or pharmacy. Preference will be given to students attending the University of Washington. The scholarship amount is $1,500, and one award is given each year. The annual application deadline is March 16.
Geographic Focus: Washington
Date(s) Application is Due: Mar 16
Amount of Grant: 1,500 U.S.D
Contact: Ceil Erickson, Grantmaking Director; (206) 515-2131 or (206) 515-2109; fax (206) 622-7673; c.erickson@seattlefoundation.org
Internet: https://fortress.wa.gov/hecb/thewashboard/ScholarshipDetails/The+Seattle+Foundation/2012-2013/Hawkins+Memorial+Scholarship
Sponsor: Seattle Foundation
1200 Fifth Avenue, Suite 1300
Seattle, WA 98101-3151

Seattle Foundation Health and Wellness Grants **4262**
The Seattle Foundation is committed to fostering health and wellness for people throughout the King County, Washington, region. The Foundation awards grants to nonprofits in all fields that improve the quality of life for county residents, and is focused on making sure that everyone in the county has access to quality care, including physical and dental health, cognitive, emotional, and mental health. Goals within the Health and Wellness element are to: improve access to: basic healthcare, especially services and treatment for those who are low-income, uninsured and/or underinsured; support efforts designed to reduce and/or eliminate disparities in health status due to poverty and/or race; foster efforts to strengthen the ability and capacity of providers to deliver quality services; and support efforts that protect the safety net for the vulnerable in our community through case management, treatment, and counseling services. The fields of interest and populations captured in this element include: healthcare, dental care, mental health, domestic violence, developmentally disabled, physically disabled, seniors, and birth to three programs, substance abuse programs and child welfare programs. The deadline for Health and Wellness is May 1.
Requirements: To qualify for a grant from the Seattle Foundation's Grantmaking Program, an organization must: be a 501(c)3 tax-exempt nonprofit organization; serve residents of King County; and provide programming in Health and Wellness that aligns with one of our three strategies.
Restrictions: The Foundation does not support disease-specific organizations and does not support health research projects. Multi-year grants are not considered.
Geographic Focus: Washington
Date(s) Application is Due: May 1
Amount of Grant: 15,000 - 50,000 U.S.D
Contact: Ceil Erickson, Grantmaking Director; (206) 515-2131 or (206) 515-2109; fax (206) 622-7673; grantmaking@seattlefoundation.org or c.erickson@seattlefoundation.org
Internet: http://www.seattlefoundation.org/nonprofits/grantmaking/healthwellness/Pages/HealthWellness.aspx
Sponsor: Seattle Foundation
1200 Fifth Avenue, Suite 1300
Seattle, WA 98101-3151

Seattle Foundation Medical Funds Grants **4263**
The Seattle Foundation administers the Medical Funds program to support medical research of potential benefit to the community and to address specific healthcare needs. In the area of medical research, grants are available in the fields of cancer, cardio-pulmonary disease, multiple sclerosis and diabetes. Special consideration will be given to research projects related to immunology, oncology, neurology, molecular biology and genetics. Grants will also be given to requests for specific equipment. Grants are also available to organizations that are administering projects addressing the healthcare needs of low-income children. In this area, grants will be given to requests for specific equipment and support for capital campaigns or facility renovation projects. A total amount of $200,000 is available for grants annually in this entire program. Typically no more than $50,000 is disbursed to any one organization.
Requirements: Preference will be given to organizations/institutions with the capacity to disseminate research findings and who receive regular support from other recognized funding sources (including the federal government).
Restrictions: These funds are not to be used for patient care. Research activities can include patient care, but the primary purpose of these grants is to support the purchase of equipment used in medical research.
Geographic Focus: Washington
Samples: Benaroya Research Institute at Virginia Mason, Seattle, Washington, $50,000 - to support the purchase of BD FACSVerse flow cytometer with 3-laser and 8-color capability (2011); Fred Hutchinson Cancer Research Center, Seattle, Washington, $18,879 - to purchase a QuantaMaster 40 Spectrofluorometer suite (2011); Puget Sound Blood Center, Seattle, Washington, $27,558 - to purchase a Perkin Elmer Micro Beta 2 Plate Reader to detect the biological profile of samples in small wells known as micro titer plates (2011).
Contact: Ceil Erickson, Grantmaking Director; (206) 515-2131 or (206) 515-2109; fax (206) 622-7673; grantmaking@seattlefoundation.org or c.erickson@seattlefoundation.org
Internet: http://www.seattlefoundation.org/nonprofits/medicalfunds/Pages/MedicalFunds.aspx
Sponsor: Seattle Foundation
1200 Fifth Avenue, Suite 1300
Seattle, WA 98101-3151

Seaver Institute Grants **4264**
The institute supports a variety of areas including the arts, private education, youth, health concerns, public affairs, the disabled, literacy, and science. Seed grants are awarded to nonprofit organizations for projects that offer the potential for a significant advancement in their fields. The institute seeks proposals that incorporate state-of-the-art scientific research, exceptional educational development, unique approaches to the creative arts, and the ramifications of shifting societal realities. There are no application deadlines.

Interested applicants should submit an inquiry letter of two or three pages and include the goals of the project, a preliminary budget, and the qualifications of the organization and principal investigator. If interested, the institute will invite a full proposal.
Requirements: U.S. 501(c)3 nonprofits are eligible.
Restrictions: Grants are not made for construction, endowments, ongoing projects, operating budgets, deficits, or individual scholarships.
Date(s) Application is Due: Apr 1; Nov 1
Amount of Grant: 1,000 - 35,000 U.S.D
Contact: Victoria Seaver Dean, President; (310) 979-0298; vsd@theseaverinstitute.org
Sponsor: Seaver Institute
11611 San Vicente Boulevard, Suite 545
Los Angeles, CA 90049

Self Foundation Grants 4265
The Self Family Foundation primarily serves the Greenwood, South Carolina and surrounding counties. The Self Family Foundation will consider requests from other regions of South Carolina, if they have statewide impact. The Foundation supports non-profit organizations in the following areas of interest: Education - enhancing children's school readiness and ability to achieve educational goals with success; working together with schools, families, youth-serving organizations and other community organizations to support the intellectual and social development of young people; Arts, Culture & History - increasing awareness of and access to activities that broaden the horizons and knowledge of all residents, within Greenwood, South Carolina; Civic & Community - improving neighborhood resources and community attitudes and services, emphasizing the utilization of existing facilities, skills and talents to create a stronger, more unified community through collaboration with organizations, businesses, public institutions and residents; Health and Human Services - promoting individual and community wellness, prevention and literacy.
Requirements: IRS 501(c)3 tax-exempt or 509(a) nonprofit organizations in South Carolina are eligible. Letters of Inquiry or requests for information may be emailed to: application@selffoundation.org or mailed to Foundation. Contact Foundation Program Officer by email or phone prior to submitting proposal. All applicants must have the financial potential to sustain the project on a continuing basis after any funding is approved.
Restrictions: No grants to individuals and no loans available.
Geographic Focus: South Carolina
Date(s) Application is Due: Feb 15; May 15; Aug 15; Nov 15
Amount of Grant: 1,500 - 50,000 U.S.D
Contact: Grants Administrator; (864) 953-2441 or (864) 941-4036; fax (864) 941- 4091; info@selffoundation.org
Internet: http://www.selffoundation.org/grantmaking4.htm
Sponsor: Self Family Foundation
P.O. Box 1017
Greenwood, SC 29648-1017

Sempra Energy Corporate Contributions 4266
Sempra Energy recognizes its role as a corporate citizen in its operating communities as well as within the global community. Corporate giving supports the following areas. Education—programs that motivate students to be prepared for higher education; that promote technology-based learning; and that encourage success in the fields of engineering, technology, business or international relations. Environment—programs that protect natural resources and promote healthy, safe communities. Business and economic development—programs that foster business retention and expansion. Community and civic leadership—programs that heighten leadership abilities and promote pride in diverse communities. Health and human services—programs that encourage life-long fitness, healthy lifestyles and human wellness, and help the homeless, hungry, and those in need. Arts and culture—programs that enhance the quality of life in communities where Sempra does business, create cultural dialog, and promote appreciation for diversity. Guidelines and application are available online.
Requirements: Tax-exempt organizations are eligible.
Restrictions: Grants generally are not awarded to benefit individuals, with the exception of company-sponsored scholarship programs; private foundations and endowment funds; organizations that discriminate based on age, gender, sexual orientation, marital status, physical or mental disabilities, race, color, religion, or national origin or ancestry; sectarian or denominational religious organizations, except for secular programs; or political parties, candidates, or partisan political organizations. Grants generally do not fund general operating expenses; travel expenses; loans or loan guarantees; debt reduction or past operating deficits; liquidation of an organization; capital improvement or building funds; or reducing or donating the cost of any gas or electric service that other customers of utility companies purchase.
Amount of Grant: Up to 8,000,000 U.S.D
Contact: Contribution Program Manager; (619) 696-4297; fax (619) 696-1868
Internet: http://www.sempra.com/community_1a1a.htm
Sponsor: Sempra Energy Corporation
101 Ash Street
San Diego, CA 92101

Seneca Foods Foundation Grants 4267
The Seneca Foods Foundation was established in 1988 in New York and is sponsored by the Seneca Foods Corporation. The foundation supports fire departments and organizations involved with education, health, Down Syndrome, agriculture and food, recreation, and youth. Funding is primarily in areas of company processing facilities, with emphasis on New York.
Requirements: Nonprofit organizations in Seneca Foods operating locations are eligible. Contact foundation for application form.
Restrictions: Grants are not made to individuals.
Contact: Kraig H. Kayser, President; (315) 926-8100; foundation@senecafoods.com
Sponsor: Seneca Foods Corporation
3736 South Main Street
Marion, NY 14505-9751

Sengupta Family Scholarship 4268
The Sengupta Family Scholarship is available to the valedictorian of Lake Central High School in St. John, Indiana. The award is $1,000 per year for four years. The applicant should contact the Lake Central High School Guidance Office for the current application, current deadline, and the Dollars for Scholars application. All college majors are eligible.
Geographic Focus: Indiana
Date(s) Application is Due: Jan 1
Amount of Grant: 1,000 U.S.D
Contact: Sengupta Scholarship Contact; (219) 365-8551; fax (219) 365-7156
Internet: http://www.legacyfoundationlakeco.org/scholarships/index.htm
Sponsor: Legacy Foundation
1000 East 80th Place, 302 South
Merrillville, IN 46410

Sensient Technologies Foundation Grants 4269
The foundation supports organizations involved with arts and culture; children/youth, services; community/economic development; education; education, fund raising/fund distribution; education, research; family services; food services; general charitable giving; health care; higher education; homeless, human services; hospitals (general); human services; medical research, institute; mental health/crisis services; nutrition; performing arts; residential/custodial care, hospices; United Ways and Federated Giving Programs; Urban/community development; and voluntarism promotion. The foundation's types of support include: annual campaigns; capital campaigns; emergency funds; endowments; general operating support; matching/challenge support; program development; research; and scholarship funds.
Requirements: An application form is not required. Submit a letter of inquiry as an initial approach. The advisory board meets in January and June/July or as needed, with deadline for proposal review one month prior to board meetings.
Restrictions: The foundation gives primarily to its areas of interest in Indianapolis, Indiana; St. Louis, Missouri; and Milwaukee, Wisconsin. It does not support sectarian religious, fraternal, or partisan political organizations. It does not give grants to individuals.
Geographic Focus: Indiana, Missouri, Wisconsin
Contact: Douglas L. Arnold, (414) 271-6755; fax (414) 347-4783
Sponsor: Sensient Technologies Foundation
777 E Wisconsin Avenue
Milwaukee, WI 53202-5304

Sequoyah Fellowship Program for Native American and Alaska Native Graduate Students 4270
These fellowships provide one-year, one-time unrestricted stipends to Native Americans and Alaska Natives. In past years, fellowships have been awarded to eligible recipients pursuing graduate degrees in such diverse fields as anthropology, microbiology, education, law, mechanical engineering, psychology, and business administration at universities throughout the United States. Approximately 10 fellowships are awarded each year around October 1. Student should provide the association with a biography describing background; tribal affiliation and a certificate showing tribal membership; university, year, and subject of study; career goals; interests; financial situation; two letters of recommendation; social security number; and the name, address, and phone number of the financial aid officer at the school/university he or she attends. Applications will be accepted from July 1 through October 1 each year.
Requirements: To be eligible, a graduate student must be an enrolled member of a Native American tribe or Alaskan Native corporation. Applicant must be minimally one-fourth degree Indian blood. A certificate of tribe enrollment and blood quantum from the tribe or BIA should be included.
Date(s) Application is Due: Aug 1
Amount of Grant: 1,500 U.S.D
Contact: Lisa Wyzlic, Scholarship Coordinator; (240) 314-7155; fax (240) 314-7159; lw.aaia@verizon.net
Internet: http://www.indian-affairs.org
Sponsor: Association on American Indian Affairs
966 Hungerford Drive, Suite 12-B
Rockville, MD 20850

Seva Foundation Community Grants 4271
The program supports local grassroots partnerships with Native American Peoples who have devised their own solutions in revitalizing/renewing spiritual and cultural beliefs and practices, health and wellness renewal, environmental protection and restoration, agricultural and sustainable community economic development, educational development, social renewal, and treaty rights protection. The foundation also provides technical assistance. Application and guidelines are available online.
Requirements: The foundation assists rural and urban Native communities in the United States.
Amount of Grant: 500 - 5,000 U.S.D
Contact: Cherie LaPointe, Program Director; (402) 878-2392; fax (402) 878-2092; clapointe@seva.org
Internet: http://www.seva.org/communitygrants.php
Sponsor: Seva Foundation
P.O. Box 225
Winnebago, NE 68071

Seventh Generation Fund Grants 4272

The fund works directly with grassroots Native American communities and traditional indigenous societies throughout the Americas in support of Native peoples' community organizing, issue advocacy, economic renewal, cultural revitalization, environmental justice, and related restorative development efforts. Grants support projects in the areas of arts and cultural expression, environmental health and justice, indigenous people of the Americas, sacred Earth, and sustainable communities. Types of support include small grants—seed money, general operating support; technical assistance grants—training and capacity building; and mini grants—community-based projects. Application and guidelines are available online.

Requirements: 501(c)3 (and organizations with 501(c)3 fiscal sponsors) focused in the Native American community are eligible. Mini-grant applicants need not have 501(c)3 status.
Date(s) Application is Due: Mar 1; Jun 1; Sep 1; Dec 1
Amount of Grant: 50 - 10,000 U.S.D
Contact: Elaine Quitiquit, Program Associate; (707) 825-7640; fax (707) 825-7639; eq7gen@pacbell.net
Internet: http://www.7genfund.org/we_help.html
Sponsor: Seventh Generation Fund
P.O. Box 4569
Arcata, CA 95518

Sexually Transmitted Diseases Postdoctoral Fellowships 4273

The ASHA Research Fund fellowships program is intended to expand the number of trained scientists working in the field of sexually transmissible diseases research by providing postdoctoral candidates with the opportunity to spend two years in a challenging, stimulating, and instructive STD research environment under the sponsorship of an established researcher. Applications for research training relative to any aspect of the sexually transmissible diseases are welcome. ASHA has supported projects in chancroid, human papilloma virus (HPV), group B streptococcal infection, herpes, gonorrhea, chlamydia, HIV, pelvic inflammatory disease (PID), and syphilis. In funding HIV-related research, ASHA gives priority to researchers exploring the relationship between HIV and other STDs. Although applications covering fellowship training in basic investigations are not discouraged, it is the intent of this program to focus on research of short- or medium-term duration. Requests for applications are usually issued in May of odd-numbered years, with an application deadline of October 15, and the award being given the following spring for a fellowship period commencing on July 1.

Requirements: Applications are requested from MDs, PhDs, DSc who will be conducting two-year research projects under the sponsorship of established STD investigators. Applicants are required to study at established U.S. institutions.
Restrictions: ASHA fellowships may not be used in support of faculty positions unless approved by ASHA's management. Current ASHA fellows are not eligible to apply for other ASHA fellowships.
Date(s) Application is Due: Nov 15
Amount of Grant: 35,500 - 36,000 U.S.D
Contact: Grant Administrator; (919) 361-8400; fax (919) 361-8425
Internet: http://www.ashastd.org/programs/arf.html
Sponsor: American Social Health Association
P.O. Box 13827
Research Triangle Park, NC 27709

Share Your Vision Art Contest 4274

This is a national art contest and exhibition, with funding from Roche. The program was created to help raise awareness of the impact of cytomegalovirus (CMV) retinitis on the lives of people with HIV. CMV retinitis is an AIDS-related opportunistic infection, which, if left untreated, can lead to blindness. Selected works will be displayed in an exhibition at Artists Space in New York and included in an accompanying exhibition catalogue. Winning artists and a guest will be invited to attend the opening reception, in October. Prizes will be awarded to winning artists, with a matching gift to an HIV/AIDS charitable 501(c)3 research or support organization of the winner's choice.

Requirements: The contest is open to HIV-positive artists who have been affected by or touched by CMV retinitis. Eligible work includes two- and three-dimensional work in any medium, including paintings, sculptures, and photography.
Date(s) Application is Due: Jun 16
Amount of Grant: 500 - 7,500 U.S.D
Contact: Amy Sadao, Executive Director; (212) 627-9855; fax (212) 627-9815; asadao@visualaids.org
Internet: http://www.thebody.com/visualaids/share_vision.html
Sponsor: Visual AIDS
526 W 26th Street, Suite 510
New York, NY 10001

Shaw's Supermarkets Donations 4275

Shaw's focuses its support on not-for-profit 501(c)3 organizations that help create healthy, thriving communities. Shaw's provides financial support, volunteer support and in-kind product donations to organizations that benefit communities in areas where they operate, and meet the following focus areas: (1) Hunger Relief - Ending hunger in the stores' local communities via both product and financial support to hunger-relief organizations. (2) Nutrition Education - support for organizations that educate and promote healthy lifestyles and nutrition by emphasizing disease management and prevention through diet. (3) Environmental Stewardship - environmental stewardship and sustainable operations and that will continually work to use energy more efficiently and reduce waste. Shaw's supports local efforts towards sustainability. (4) Community Connections - select local events that build strong communities by broadly engaging community pride and spirit.

Requirements: 501(c)3 tax-exempt organizations in Connecticut, Maine, Massachusetts, New Hampshire, Rhode Island, and Vermont are eligible. If the grant request involves an event, the request should be sent at least 8 weeks prior to event date. Download the application form from the website.
Restrictions: Shaw's will generally not support: individuals; travel or research expenses; fees for participation in competitive programs; veteran, fraternal or labor organizations; lobbying, political or religious programs; or, organizations that are not tax-exempt.
Geographic Focus: Connecticut, Maine, Massachusetts, New Hampshire, Rhode Island, Vermont
Contact: Donation Committee; shaws.donations@shaws.com
Internet: http://www.shaws.com/pages/community/donationRequests.php
Sponsor: Shaw's Supermarkets
P.O. Box 600
East Bridgewater, MA 02333

Shelley and Donald Rubin Foundation Grants 4276

The foundation's mission is to support innovative efforts to transform society's institutions and to empower individuals to develop their full potential. The foundation is primarily interested in supporting the inclusion of art from non-Western European cultures into the mainstream of scholarship and display. In addition, the foundation supports research, action, and other projects designed to reveal and understand barriers to the full access of all people to American society and the larger international community. Areas of particular interest include, but are not limited to: access to health care; AIDS and its effects on society's institutions; the celebration of ethnic and cultural diversity that simultaneously encourages intergroup understanding, and cultural and arts programs that encourage individual and community identity. Projects in other areas that are most often funded are those that in themselves serve as catalysts for social change, addressing emerging problems as well as better known, long-standing problems in new ways. On a limited basis, programs eligible for funding may also include evaluation, technical assistance, and demonstration projects leading to the development of innovative models. Some examples of technical assistance of interest are the establishment of Web sites, the development of Web/TV productions, and related outreach activities.

Requirements: Art-related projects eligible for funding most often fall within the areas of enhancing the foundation's web site, tibetart.org; preserving Himalayan art; supporting educational activities reaching a broad constituency; supporting Himalayan art and architecture restoration projects as well as traveling exhibits bringing Himalayan art to new audiences.
Restrictions: Grants do not support individuals, capital funds, operating funds, or funding for the delivery of direct services. The foundation does not accept letters of intent/proposals via email.
Amount of Grant: 5,000 - 60,000 U.S.D
Samples: Tibetan Children's Education Foundation (Helena, MT)—for a series of exhibitions on Tibetan tangka paintings in the larger context of raising awareness of Tibetan art and culture in the Midwestern United States, $6000; National Coalition Against Censorship (New York, NY)—to support the development of a national youth Free Speech and Democracy Film Contest, $3000; Kidsave International (Los Angeles, CA)—to implement Weekend Miracles, a program designed to match African-American foster parents with difficult-to-place African American children, $10,000; Commission on the Public's Health (New York, NY)—to develop and maintain an information network about the health care safety net in local neighborhoods in New York City, $2000.
Contact: Evelyn Jones Rich, Executive Director; (212) 780-2035; erich@sdrubin.org
Internet: http://www.sdrubin.org/guidelines.htm
Sponsor: Shelley and Donald Rubin Foundation
115 Fifth Avenue
New York, NY 10003

Shell Oil Company Foundation Grants 4277

Shell makes monetary grants to nonprofit organizations within the U.S. Priority consideration will be given to organizations in or near communities where Shell Oil Company or its affiliates in the United States have a major presence. Shell Oil Company and the Shell Oil Company Foundation will consider charitable contributions to eligible nonprofit organizations for select programs that fall within our focused areas of contributions: health and welfare; community outreach; workforce development / education and; environment. There are no deadline dates for submitting a proposal, proposals are reviewed on a regular basis.

Requirements: Nonprofit organizations in company operating locations are eligible.
Restrictions: Shell Oil Company and the Shell Oil Company Foundation will not consider contributions for the following purposes: individuals; private Foundations; non-profit organizations without a current 501(c)3 exempt status; conferences or symposia; endowment funds; fraternal and labor organizations; capital campaigns; conferences, workshops, or seminars not directly related to Shell business interests; religious organizations that do not serve the general public on a non-denominational basis; organizations located in or benefiting nations other than U.S. and its territories; organizational operating expenses.
Amount of Grant: 10,000 - 50,000 U.S.D
Contact: A.H. Myres, Vice President; (713) 241-3616; fax (713) 241-3329; socfoundation@shellus.com
Internet: http://www.shell.us/home/content/usa/responsible_energy/shell_in_the_society/giving_back/grant/grant_request_060208.html
Sponsor: Shell Oil Company Foundation
One Shell Plaza, 910 Louisiana, Suite 4478A
Houston, TX 77002

Shell Oil Products U.S. Corporate Contributions 4278

The company supports organizations that work to improve the quality of life and the general welfare of its employees and the citizens of their home communities. Contributions are made in the areas of social betterment and welfare, education, health, culture, civics, and certain unclassified causes. In the area of social betterment and welfare, in addition to local community funds, support may be given to institutions for the handicapped, associations for the blind, or other well-established organizations that enjoy wide community support and receive no support from local United Funds. Financial aid to education is directed primarily to accredited colleges and universities where the company has significant operations, where it recruits, or where research of interest or applicability to its business is being conducted. Approximately one-half of Pennzoil's corporate giving is directed toward educational support. Corporate contributions to hospitals, clinics, rehabilitation centers, and other health-related organizations are restricted to building programs, equipment additions, and unusual research programs. Contributions to campaigns conducted for the more prevalent diseases or disabilities may also be considered. Contributions are made to organizations engaged in the performing or visual arts, museums, zoos, libraries, and other cultural endeavors. Support will be given to selected civic organizations working to improve the quality of life in local Pennzoil communities. There may be causes outside the preceding classifications that are deserving of support and to which consideration will be given on an individual basis.

Requirements: Contributions are made to tax-exempt educational, charitable, health, and social betterment organizations. Those interested in applying for funding from Shell should be advised that Shell's contributions budget is completed in the fall of each year. Organizations making proposals following the completion of the budgeting process are not likely to be considered for funding during the following year.

Restrictions: Shell does not generally support strictly sectarian or denominational religious activities, secondary schools, veterans' organizations, fraternal organizations, charitable advertising, donations of products, or donations that are not tax deductible.

Contact: Constituency & Community Relations Manager; (713) 546-8590

Internet: http://www.shelloilproductsus.com

Sponsor: Shell Oil Products U.S.

1100 Louisiana, Suite 2346

Houston, TX 77002

Sheltering Arms Fund Grants 4279

The purpose of the fund is to complement and extend Sheltering Arms Hospital's mission by funding health related initiatives that serve uninsured and underinsured adults who are experiencing physical or cognitive disabilities or who are at risk for developing functional limitations. Emphasis will be placed on responding to the needs of individuals with a Rehab Impairment Category diagnosis as defined by the Centers for Medicaid and Medicare (CMS)

Requirements: Proposals will be accepted from charitable organizations, which serve the residents of the metropolitan Richmond and Central Virginia. Proposed programs should align with the goals of the fund: Improve access to health and related support services such as Primary Care, Specialty clinics (Diabetes, Podiatry, Dental, Women's Health, etc.), and Nutritional Support; Increase functional independence (Home care, Day care, Transportation, Use of assistive devices and technology, Facility renovations to improve handicapped access); Promote disability prevention for at risk populations through education and/or public health initiatives.

Geographic Focus: Virginia

Date(s) Application is Due: May 5

Amount of Grant: Up to 100,000 U.S.D

Contact: Susan Hallett, Program Officer; (804) 330-7400; fax (804) 330-5992; shallett@tcfrichmond.org

Internet: http://www.tcfrichmond.org/page2954.cfm#SAF

Sponsor: Community Foundation Serving Richmond and Central Virginia

7501 Boulders View Drive, Suite 110

Richmond, VA 23225

Sidgmore Family Foundation Grants 4280

The foundation supports programs that ensure the health, education, and general well-being of children, youth, and young adults. The foundation is interested in funding organizations with programs, projects, and research that utilize sports to promote the physical, mental, and spiritual development of young people; recognize that an impoverished environment limits the possibilities for children to develop and thrive; improve the quality of education available to youth, with special emphasis on programs designed to improve teacher training and teaching methods; and further advance knowledge and research in the medical fields of vision, hearing, and allergies. There are no application deadlines. Guidelines are available online.

Requirements: Nonprofit organizations are eligible. Preference is given to organizations that serve residents in the Washington metropolitan area, Maryland, and Virginia.

Restrictions: Grants do not support government agencies, national health organizations, political activities, or public policy advocacy groups.

Contact: Administrator; (516) 541-2713; fax (516) 541-2803; SidgmoreFound@aol.com

Internet: http://fdncenter.org/grantmaker/sidgmore/prior.html

Sponsor: Sidgmore Family Foundation

71 Leewater Avenue

Massapequa, NY 11758

Sidney Stern Memorial Trust Grants 4281

The foundation awards grants to nonprofit organizations in its areas of interest, including civil rights, children and youth, community development, disabled, education, health and science, social services, and Native Americans. Types of support include annual campaigns, building construction/renovation, deficit financing, emergency funds, endowment funds, equipment acquisition, land acquisition, scholarship funds, seed grants, special projects, and research. Requests from California receive preference. There are no application deadlines. The board meets monthly, except August.

Amount of Grant: 2,500 - 8,000 U.S.D

Contact: Program Contact; (310) 459-2117

Sponsor: Sidney Suitern Memorial Trust

P.O. Box 893

Pacific Palisades, CA 90272

Sid W. Richardson Foundation Grants 4282

Grants are provided to tax-exempt organizations in Texas in the areas of education (museums, learning centers, day schools, K-12 schools, higher education institutions, and business and economic education), health (medical schools, organ donor registries, hospitals, disease prevention, health science centers, and nursing associations), arts (arts councils, visual and performing arts festivals, museums, ballet, symphony orchestra, and arts education programs), and human services (boys' and girls' clubs, united funds, the elderly, crime prevention, the disabled, housing opportunities, food programs, and drug and alcohol abuse prevention); specific goals within these larger areas change periodically. Types of support include operating budgets, seed grants, building construction funds, equipment acquisition, endowment funds, research, publications, conferences and seminars, matching funds, continuing support, and projects/programs. Award amounts vary depending on proposed projects.

Requirements: Proposals must cover programs and projects within Texas.

Restrictions: Grants are not made to individuals.

Geographic Focus: Texas

Date(s) Application is Due: Jan 15

Contact: Valleau Wilkie Jr., Executive Director; (817) 336-0494; fax (817) 332-2176

Internet: http://www.sidrichardson.org/GrantInfo.html

Sponsor: Sid W. Richardson Foundation

309 Main Street

Fort Worth, TX 76102

Siebert Lutheran Foundation Grants 4283

Specific areas of interest vary from time to time. At present, the Foundation is supportive of the following: Clergy and lay education and training, community development and outreach, health ministry, evangelism and youth. Grants are occasionally made to provide seed money or start-up costs for a program or project. In such instances, the participation of other donors is desired. Grants are made only to Lutheran organizations exempt under Section 501(c)3 of the Internal Revenue Code and are generally awarded for a one-year period.

Requirements: The Foundation utilizes an on-line grant application. Potential grant applicants, who represent Wisconsin congregations or Wisconsin recognized service organizations in the Lutheran church, may complete the Foundation's formal, online, letter of inquiry. A telephone call is not necessary prior to completing the letter of inquiry for Lutheran organizations in Wisconsin. If your organization does not meet the above criteria and you would like to discuss possible funding, contact the Foundation office (262-754-9160).

Restrictions: The Foundation does not approve grants for the following: endowment funds, fellowships and scholarships, trusts, and other grant-making foundations. Grants are generally not made to churches for capital or operating expenses. No grants are made outside the United States.

Geographic Focus: Wisconsin

Date(s) Application is Due: Mar 1; Jun 1; Sep 1; Dec 1

Contact: Deborah Engel; (262) 754-9160; contactus@Siebertfoundation.org

Internet: http://www.siebertfoundation.org/grants.htm

Sponsor: Siebert Lutheran Foundation

300 North Corporate Drive, Suite 200

Brookfield, WI 53045

Sierra Health Foundation Responsive Grants 4284

The Sierra Health Foundation Responsive Grants are designed to promote health and well-being in Northern California communities. Fields of interest include: AIDS; alcoholism; biomedicine; child development, education; child development, services; children/youth, services; community/economic development; crime/violence/ prevention, youth; family services; health care; health organizations, association; human services; leadership development; medical care, rehabilitation; mental health/crisis services; nutrition; substance abuse, services; and youth development. The Foundation funds employee matching gifts, in-kind gifts, program development, program evaluation; and technical assistance. Examples of Foundation funding include food banks, homeless shelters, senior citizen agencies, youth and family centers, job readiness programs, and faith based organizations.

Requirements: Organizations may contact the Foundation for current application procedures and deadlines.

Restrictions: The Foundation funds the following northern California counties: Alpine, Amador, Butte, Calaveras, Colusa, El Dorado, Glenn, Lassen, Modoc, Mono, Nevada, Placer, Plumas, Sacramento, San Joaquin, Shasta, Sierra, Siskiyou, Solano (eastern), Stanislaus, Sutter, Tehama, Trinity, Tuolumne, Yolo and Yuba. The Foundation does not fund individuals or endowments.

Geographic Focus: California

Amount of Grant: Up to 25,000 U.S.D

Contact: Kathy Mathews, Grants Administrator; (916) 922-4755; fax (916) 922-4024; kmathews@sierrahealth.org or grants@sierrahealth.org

Internet: http://www.sierrahealth.org/doc.aspx?129

Sponsor: Sierra Health Foundation

1321 Garden Highway

Sacramento, CA 95833-9754

Sigma Theta Tau International /American Nurses Foundation Grant 4285

Sigma Theta Tau International (STTI), the only international nursing honor society worldwide, is a global community of nurse leaders dedicated to using knowledge, scholarship, service, and learning to improve the health of the world's people. STTI members belong to 475 chapters in 90 countries. STTI and the American Nurses' Foundation (ANF) have joined forces to provide funding for the STTI/ANF research grant. The purpose of the grant is to encourage the research career development of nurses through support of research conducted by beginning nurse researchers or experienced nurse researchers who are entering a new field of study. Applications must be submitted directly to STTI via their online submission system by the May 1 deadline date. Funding will begin in October. Links to an FAQ, application-review criteria, and help with grant-writing are available at the website.

Requirements: Eligible applicants must meet the following criteria: be a registered nurse with a current license; have a master's or doctoral degree or be enrolled in a doctoral program; have a clinical nursing research topic; submit a completed research application packagbe and signed research agreement; and be ready to implement the research project when funding is received. Preference will be given to Sigma Theta Tau members, other qualifications being equal. Allocation of funds is based on the quality of the proposed research, the future promise of the applicant, and the applicant's research budget.

Restrictions: Funds for this grant do not cover expenses incurred prior to the funding date.

Geographic Focus: All States, All Countries

Date(s) Application is Due: May 1

Amount of Grant: Up to 7,500 U.S.D

Contact: Research Services; (888) 634-7575 or (317) 634-8171; fax (317) 634-8188; research@stti.iupui.edu

Internet: http://www.nursingsociety.org/Research/Grants/Pages/grant_anf.aspx

Sponsor: Sigma Theta Tau International
550 W. North Street
Indianapolis, IN 46202-2156

Sigma Theta Tau International / Council for the Advancement of Nursing Science Grants 4286

Sigma Theta Tau International (STTI), the only international nursing honor society worldwide, is a global community of nurse leaders dedicated to using knowledge, scholarship, service, and learning to improve the health of the world's people. STTI members belong to 475 chapters in 90 countries. STTI and the Council for the Advancement of Nursing Science (CANS) have joined forces to provide funding for the STTI/CANS research grant. The purpose of the grant is to encourage qualified nurses to improve the health of the world's people through research. Proposals for clinical, educational or historical research, including plans for broadly disseminating the research findings, may be submitted for the grant. Applications must be submitted directly to STTI via their online submission system by the July 1 deadline date. Funding will begin in November. Links to an FAQ, application-review criteria, and help with grant-writing are available at the website.

Requirements: Eligible applicants must meet the following criteria: be a member in good standing of both STTI and CANS; hold at minimum a Master's degree or equivalent; submit a completed research application via STTI's on-line submission system; be ready to implement research project when funding is received; submit a final report; submit a completed abstract to Virginia Henderson International Nursing Library; submit the completed project abstract for presentation at the Sigma Theta Tau or Council for Advancement of Nursing Science research meeting; and credit research-grant partners in all publications and presentations of the research.

Restrictions: Funds for this grant do not cover expenses incurred prior to the funding date.

Geographic Focus: All States, All Countries

Date(s) Application is Due: Jul 1

Amount of Grant: Up to 5,000 U.S.D

Contact: Research Services; (888) 634-7575 or (317) 634-8171; fax (317) 634-8188; research@stti.iupui.edu

Internet: http://www.nursingsociety.org/Research/SmallGrants/Pages/CANS.aspx

Sponsor: Sigma Theta Tau International
550 W. North Street
Indianapolis, IN 46202-2156

Sigma Theta Tau International / Oncology Nursing Society Grant 4287

Sigma Theta Tau International (STTI), the only international nursing honor society worldwide, is a global community of nurse leaders dedicated to using knowledge, scholarship, service, and learning to improve the health of the world's people. STTI members belong to 475 chapters in 90 countries. STTI and the Oncology Nurses Society (ONS) have joined forces to provide funding through ONS's Small Research Grants program for this grant. From the perspectives of both organizations, the purpose of the grant is to stimulate clinically-related oncology-nursing research, increase the knowledge base for oncology-nursing practice, and to prepare future oncology-nurse researchers. Application guidelines are available through ONS; applications must be submitted directly to ONS. Contact information and a link to the ONS website is available at the grant website.

Requirements: Eligible applicants must meet the following criteria: be a registered nurse actively involved in some aspect of cancer-patient care, education or research and have earned a master's degree. Other qualifications being equal, preference will be given to Sigma Theta Tau International members.

Geographic Focus: All States, All Countries

Date(s) Application is Due: Oct 1

Amount of Grant: 10,000 U.S.D

Samples: Kathleen Ruccione, M.P.H., R.N., C.P.O.N.®, F.A.A.N. (U.S.A), $10,000 - "Iron, Anthracyclines and Cardiac Outcomes among Childhood Cancer Survivors" (2012); Shu-Feng Sung, Ph.D., R.N. (U.S.A), $10,000 - " Translaton Cardiovascular Risk and Breast Cancer Outcomes" (2010); Sharon Kozachik, R.N., Ph.D. (U.S.A), $10,000 - "Sleep, HPA Axis Activity and Paclitaxel-Induced Neuropathic Pain".

Contact: Debbie Kubiak, Grants Coordinator; (412) 859-6224 or (866) 357-4667; fax (877) 369-5497; dkubiak@ons.org

Kathi Dandracs, Grants Specialist; (412) 859-6233 or (866) 357-4667; fax (877) 369-5497; kkandracs@ons.org

Carlton G. Brown, R.N., Ph.D., A.O.C.N., President; (866) 357-4667 or (412) 859-6298; fax (877) 369-5497

Internet: http://www.nursingsociety.org/Research/Grants/Pages/grant_ons.aspx

Sponsor: Sigma Theta Tau International
550 W. North Street
Indianapolis, IN 46202-2156

Sigma Theta Tau International Doris Bloch Research Award 4288

Sigma Theta Tau International (STTI), the only international nursing honor society worldwide, is a global community of nurse leaders dedicated to using knowledge, scholarship, service, and learning to improve the health of the world's people. STTI members belong to 475 chapters in 90 countries. The purpose of STTI's Doris Bloch Research Award is to encourage nurses to contribute to the advancement of nursing through research. Allocation of funds is based on the quality of the proposed research, the future promise of the applicant, and the applicant's research budget. Applications from novice researchers who have received no other national research funds are encouraged and will receive preference for funding, other aspects being equal. Preference will be given to Sigma Theta Tau International members, other qualifications being equal. Applications must be submitted via STTI's online submission system. A link to the submission system will be available in July. Funding will begin in June. Links to an FAQ, application-review criteria, and help with grant-writing are available at the website. Funding is provided by a gift to the honor society's Research Endowment from the Doris Bloch estate.

Requirements: Eligible applicants must meet the following criteria: be a registered nurse with a current license; hold a master's or doctoral degree or be enrolled in a doctoral program; submit a completed research application package and signed research agreement; be ready to implement the research project when funding is received; complete the project within one (1) year of funding; submit to STTI a final report; and submit a completed abstract to STTI's Virginia Henderson International Nursing Research Library.

Geographic Focus: All States, All Countries

Date(s) Application is Due: Dec 1

Amount of Grant: Up to 5,000 U.S.D

Samples: Carina Katigbak, M.S., B.Sc.N. (U.S.A) - "Exploring the role of Community Health Workers upon Hypertension Related Behaviors with Filipino Americans in New York City" (2012); Hsiu-Chin Chen, R.N., Ph.D., Ed.D. (U.S.A) - "The significance of student input to nursing program evaluation: Part II" (2010); Jennifer Hobbs, R.N., Ph.D. (U.S.A) - "Standardizing the Particular in Clinical Information Systems: Nursing's Role in the Development of Clinical Standards during the Late 20th Century".

Contact: Research Services; (888) 634-7575 or (317) 634-8171; fax (317) 634-8188; research@stti.iupui.edu

Internet: http://www.nursingsociety.org/Research/Grants/Pages/grant_bloch.aspx

Sponsor: Sigma Theta Tau International
550 W. North Street
Indianapolis, IN 46202-2156

Sigma Theta Tau Small Grants 4289

Sigma Theta Tau International (STTI), the only international nursing honor society worldwide, is a global community of nurse leaders dedicated to using knowledge, scholarship, service, and learning to improve the health of the world's people. STTI members belong to 475 chapters in 90 countries. The purpose of STTI's Small Grants Program is to encourage nurses to contribute to the advancement of nursing through research. Ten to fifteen grants are awarded annually. Allocation of funds is based on the quality of the proposed research, the future promise of the applicant, and the applicant's research budget. Applications from novice researchers who have received no other national research funds are encouraged and will receive preference for funding, other aspects being equal. Preference will be given to Sigma Theta Tau members, other qualifications being equal. Applications must be submitted via STTI's online submission system. A link to the submission system will be available in July. Funding will begin in June. Links to an FAQ, application-review criteria, and help with grant-writing are available at the website.

Requirements: Eligible applicants must meet the following criteria: be a registered nurse with a current license; have a master's or doctoral degree or be enrolled in a doctoral program; have a clinical-nursing research topic; submit a completed research application package and signed research agreement; be ready to implement the research project when funding is received; complete the project within one (1) year of funding; submit a final report; and submit a completed abstract to STTI's Virginia Henderson International Nursing Research Library.

Restrictions: Funds for this grant do not cover expenses incurred prior to the funding date.

Geographic Focus: All States, All Countries

Date(s) Application is Due: Dec 1

Amount of Grant: Up to 5,000 U.S.D

Samples: Tara Albrecht, Ph.D., M.S.N., B.S.N. (U.S.A) - "Symptom Management and Psychosocial needs in Patients with Acute Myeloid Leukemia" (2012); Kathleen J. Finlayson, Ph.D., M.S.N. (Australia) - "Identifying Effective New Interventions to Prevent Recurrence of Venous Leg Ulcers in Adults with Chronic Venous Insufficiency" (2012); Orla Marie Smith, M.N. (U.S.A) - "Prevalence and Predictors of Anxiety, Depression and Risk for Post-Traumatic Stress in Family Members of Patients in the Intensive Care Unit: The ICU-Adapts Survey" (2012).

Contact: Research Services; (888) 634-7575 or (317) 634-8171; fax (317) 634-8188; research@stti.iupui.edu
Internet: http://www.nursingsociety.org/Research/Grants/Pages/small_grants.aspx
Sponsor: Sigma Theta Tau International
550 W. North Street
Indianapolis, IN 46202-2156

Sigrid Juselius Foundation Grants **4290**
The Sigrid Juselius Foundation, offers funding for domestic medical research. To enable established foreign medical researchers to work in a Finnish institution for 1-12 months, the Foundation also offers grants to nationals of other countries. Resources are allocated to senior and post doc researchers. The grants cover all disciplines of medical science. The grants can cover personal expenses (living costs) and travel.
Requirements: The Foundation does not award direct grants to foreign medical research. A foreign scientist interested in coming over to Finland has to get in touch with a Finnish scientist of his particular field, and the application should be made on the part of the Finnish host together with the proposed visiting researcher. A electronic application is available on the Sigrid Juselius Foundation website.
Restrictions: The funds are not available for studies or doctoral theses.
Date(s) Application is Due: Apr 15; Sep 15
Contact: Funding Officer; 358 20 710 9082; fax 358 20 710 9089
Internet: http://www.sigridjuselius.fi/index.php?page_id=48
Sponsor: Sigrid Juselius Foundation
Aleksanterinkatu 48 B
Helsinki, FI-00100 Finland

Simeon J. Fortin Charitable Foundation Grants **4291**
The Simeon J. Fortin Charitable Foundation was established in 1986 to promote the advancement of cancer research. The Fortin Charitable Foundation supports cancer research organizations dedicated to discovering a cure. To better support the capacity of nonprofit organizations, multi-year funding requests are strongly encouraged. Applicants must apply online at the grant website. Applicants are strongly encouraged to do the following before applying: review the downloadable state application procedures for additional helpful information and clarifications; review the downloadable online-application guidelines at the grant website; review the foundation's funding history (link is available from the grant website); review the online application questions in advance; and review the list of required attachments. These will generally include: a list of board members, financial statements (audited, reviewed, or compiled by independent auditor); an organization summary; a list of other funding sources; an IRS Determination letter; and other required documents. All attachments must be uploaded in the online application as PDF, Word, or Excel files. The application deadline for the Simeon J. Fortin Charitable Foundation is 11:59 p.m. on October 15. Applicants will be notified of grant decisions before December 31.
Geographic Focus: Massachusetts
Date(s) Application is Due: Oct 15
Samples: Childrens Hospital Corporation, Boston, Massachusetts, $90,000 (2010); Children's Hospital Corporation, Boston, Massachusetts, $75,000; UMass Memorial Foundation, Worcester, Massachusetts, $50,000.
Contact: Michealle Larkins; (866) 778-6859; michealle.larkins@baml.com
Internet: https://www.bankofamerica.com/philanthropic/fn_search.action
Sponsor: Simeon J. Fortin Charitable Foundation
225 Franklin Street, 4th Floor, MA1-225-04-02
Boston, MA 02110

Simmons Foundation Grants **4292**
The foundation awards grants to Maine nonprofits in its areas of interest, including arts, higher education, family services, health care, food services, health organizations, human services, children and youth, services and women. Types of support include general operating support, building construction/renovation, equipment acquisition, and scholarship funds.
Requirements: There are no application forms or deadlines however your initial approach should be in the form of a letter. Applicants should submit the following: detailed description of project and amount of funding requested; brief history of organization and description of its mission; descriptive literature about organization; copy of IRS Determination Letter.
Geographic Focus: Maine
Contact: Suzanne McGuffey, Treasurer; (207) 774-2635
Sponsor: Simmons Foundation
1 Canal Plaza
Portland, ME 04101-4098

Simone and Cino del Duca Grand Prix Awards **4293**
The Simone and Cino del Duca Foundation of the Institut de France awards each year a scientific Grand Prix based on the recommendations of a jury of members of the Institut de France. The amount of the Grand Prix will be 364,000 euros: 250,000 euros to finance the work of the award-winning group, 50,000 euros awarded to the director of the research group, and an additional 64,000 euros towards hiring one or more French or foreign postdoctoral researchers to be selected by the group director.
Requirements: The Grand Prix will reward an internationally recognized French or European mathematician proposing an ambitious research project based on the collective work of a group under his/her direction. The proposal should describe the scientific objectives and the means that will be used to implement the project (duration of the project, invitations, meetings, seminars, etc.).
Geographic Focus: All States, Albania, Andorra, Armenia, Austria, Azerbaijan, Belarus, Belgium, Bosnia & Herzegovina, Bulgaria, Croatia, Cyprus, Czech Republic, Denmark, Estonia, Finland, France, Georgia, Germany, Greece, Hungary, Iceland, Ireland, Italy, Kosovo, Latvia, Liechtenstein, Lithuania, Luxembourg, Macedonia, Malta, Moldova, Monaco, Montenegro, Norway, Poland, Portugal, Romania, Russia, San Marino, Serbia, Slovakia, Slovenia, Spain, Sweden, Switzerland, The Netherlands, Turkey, Ukraine, United Kingdom, Vatican City
Date(s) Application is Due: Jan 25
Amount of Grant: Up to 364,000 EUR
Contact: Grants Administrator; 01 47 66 01 21; fax 01 46 22 45 02; fondation-del-duca@institut-de-france.fr
Internet: http://www.enseignementsup-recherche.gouv.fr/cid23192/call-for-applications-for-the-grand-prix-scientifique-simone-and-cino-del-duca-2009.html
Sponsor: Institut de France
23, quai Conti
Paris, 75270 France

Simple Advise Education Center Grants **4294**
The Simple Advise Education Center supports organizations, centers & services, primarily in North Carolina. Giving to improve the quality of life of citizens with developmental disabilities.
Requirements: There are no specific deadlines with which to adhere. Contact the Foundation for further application information and guidelines.
Geographic Focus: North Carolina
Contact: Barbara J. Spigner, Director; (800) 677-7306
Sponsor: Simple Advise Education Center c/o BJ Hills and Assoc.
P.O. Box 758
Fayetteville, NC 28302-0758

Simpson Lumber Charitable Contributions **4295**
Simpson has been in the forest products business since 1890 and currently has two operating subsidiaries: Simpson Door Company and Simpson Tacoma Kraft Company, LLC. The mission of Simpson's contributions program is to improve the quality of life in communities where the company has a significant number of employees living and working; and to serve as a catalyst for employees to become involved and to provide leadership in their communities. The company's broad areas of interest include education, health and human services, and efforts to enhance its operating communities. Contributions are generally made in locations where the company has operations. To the extent possible, contributions will support organizations of interest to, or recommended by, Simpson employees. Simpson prefers to make capital contributions that will benefit the operating communities for the long term as opposed to contributing operating funds. Generally, support is committed for one year at a time and in amounts less than $5,000. Interested applicants should write to Simpson's Public Affairs department to request application materials. Simpson's review committees meet once per year to consider funding applications. The application deadline is May 15. For requests less than $1,000, applicants should contact Simpson's Public Affairs department.
Requirements: 501(c)3 organizations that serve Pierce, Thurston, Lewis, Mason, Grays Harbor, and Cowlitz counties in Washington are eligible to apply. Criteria taken into account in determining the amount of any contributions are as follows: degree of support from company employees; relative size and importance of company operations in the community (balance among Simpson communities); needs of organization or program for which funding is requested; amount of previous company contributions to the organization; amount committed by other companies, foundations, and/or governments (projects should demonstrate broad-based community support); and proximity of the requesting organization to Simpson operations or administrative offices.
Geographic Focus: Washington
Date(s) Application is Due: May 15
Amount of Grant: Up to 5,000 U.S.D
Contact: Raymond P. Tennison, President; (253) 779-6400
Beverly Holland, Public Affairs Manager; (253) 779-6400
Internet: http://www.simpson.com/communitycontribute.cfm
Sponsor: Simpson Lumber Company, LLC
917 East 11th Street
Tacoma, WA 98421

Singing for Change Foundation Grants **4296**
Grants are made to organizations serving children and families, the environment, and disenfranchised groups. Children and families grants support health, education, and programs concerned with the protection of children and families. Projects that foster self-esteem and self-sufficiency and that teach nonviolence and creative problem-solving are preferred. Environmental efforts, including awareness and programs to teach people methods of conservation, protection, and the responsible use of natural resources, receive support. Support also goes to organizations serving disenfranchised groups, including minorities; people with AIDS, disabilities or terminal illnesses; people with low levels of skill, education, or income; and homeless people. The foundation gives priority to organizations that keep their overhead low and that collaborate with other groups in their communities to find innovative ways of solving common problems.
Requirements: U.S. nonprofits with 501(c)3 status, or who have a sponsoring agency with such status, may submit applications for grant support.
Restrictions: The foundation does not make grants to individuals or to fund research, basic-needs programs, or single-service programs; governmental agencies; artistic, political or fund-raising organizations or events; programs that promote religious doctrine; or efforts to protect individual species or animals.

Amount of Grant: 500 - 10,000 U.S.D
Samples: Whale's Tale (Pittsburgh, PA)—to support two shelters for runaway and homeless youth under 21 years old; Transportation Alternatives (New York, NY)—for advocacy work to create better conditions for cyclists and walkers, improve public transit, and reduce dependence on cars.
Contact: Judith Ranger Smith, Executive Director; (816) 363-8132; fax (816) 363-1290; info@singingforchange.com
Internet: http://margaritaville.com/sfc/application.htm
Sponsor: Singing for Change
P.O. Box 7210
Kansas City, MO 64113

Sioux Falls Area Community Foundation Grants 4297
The purpose of the foundation's grantmaking program is to provide support primarily in the Sioux Falls area to a broad range of activities, such as charitable and educational programs, scientific and social research, care of the needy, improvement of social services and educational facilities, prevention and alleviation of social ills, encouragement of the arts, and protection of the environment. The board will consider grants to programs that address the underlying causes of specific problems of the community; programs that work to combat racism, sexism, and other forms of discrimination; new, innovative programs that require start-up funds; community studies, programmatic research, and original artistic works; established organizations that continue to be relevant to the needs of the community; and educational programs. Grant applications are due in February, April, June, August, October, and December; contact the office for specific grant deadlines. Initial inquiries may be made by phone or letter.
Requirements: Grants are made to 501(c)3 nonprofit organizations within the Sioux Falls, SD, area.
Restrictions: The board will not consider requests for capital or construction drives, ongoing operational support, elimination of an organization's deficit, computer hardware or software, payment of staff salaries, sectarian religious purposes, endowments, political advocacy projects, telephone solicitations, national fund-raising efforts, or travel for groups or individuals.
Geographic Focus: South Dakota
Amount of Grant: 3,000,000 U.S.D
Samples: Hope Haven International Ministries (SD)—to produce 500 CDs to recruit and train volunteers and offsets transportation costs for elderly volunteers who are unable to drive themselves to the distribution center, $4100; Lennox Area Historical Society (SD)—to purchase supplies and display cases for an exhibit commemorating the community's 125 years of main street business, $2000; Augustana College/Ctr for Western Studies (SD)—to support production of an updated history of the state of South Dakota and distribute it to libraries and schools throughout the state, $3500; Dakota Academy of Performing Arts (SD)—for middle school students to attend a performance of Live Wire, a play focused on AIDS education, $2600.
Contact: Sue Brown, President; (605) 336-7055; fax (605) 336-0038; sbrown@sfacf.org
Internet: http://www.sfacf.org/sfacf/aboutgrants.aspx
Sponsor: Sioux Falls Area Community Foundation
300 N Philips Avenue, Suite 102
Sioux Falls, SD 57104-6006

Siragusa Foundation Health Services & Medical Research Grants 4298
The Siragusa Foundation, established in 1950 by Ross D. Siragusa, is a private family foundation that is committed to honoring its founder by sustaining and developing Chicago's extraordinary nonprofit resources. The Foundation's Health Services and Medical Research Program is balanced between medical research organizations, which are concerned with scientific inquiry and the exploration of treatment methods, and health service organizations, which provide direct social and clinical care. The Foundation primarily funds fellowships, single-disease research programs and clinical treatment services. The Foundation believes it is critical to fund research efforts, health support services, patient care services and treatment, and advocacy programs for both adults and youth, all of which are highly complementary to each other.
Requirements: Nonprofit 501(c)3 organization in Chicago are eligible to apply. Before submitting a formal proposal to the Foundation, prospective applicants should submit a two-page letter of inquiry outlining the background of the organization and the proposed project. The letter of inquiry should describe succinctly the project for which funding is requested, including how it relates to the Foundation's interests, the target audience, the estimated budget and request amount. After reviewing the letter of inquiry, the Foundation may or may not request a formal proposal.
Restrictions: The Foundation limits potential grantees from applying for funds to once per calendar year. The Foundation does not accept unsolicited proposals from outside the Chicago area and does not support individuals or political advocacy. The foundation typically does not support capital expenditures or endowments.
Geographic Focus: Illinois
Samples: Arthritis Foundation, Greater Chicago Chapter, Chicago, Illinois, $5,000; Children's Memorial Foundation, Chicago, Illinois, $215,000; Rehabilitation Institute of Chicago, Chicago, Illinois, $16,666.
Contact: Kristen M. Buerster; (312) 755-0064; fax (312) 755-0069
Internet: http://www.siragusa.org/pages/program_areas/13.php
Sponsor: Siragusa Foundation
1 East Wacker Drive, Suite 2910
Chicago, IL 60601

Sir Dorabji Tata Trust Grants for NGOs or Voluntary Organizations 4299
Thoughtful and committed programs initiated by organizations in priority areas are considered for financial support. The Trust draws on the expertise of its staff and specialist consultants who travel widely, study project proposals and keep abreast of developments in each of the sectors (see below).
Requirements: The trust limits its funding to projects within India. This program focuses on specific sectors for its funding: Management of Natural Resources; Livelihood; Education; Health; and Social Development Initiatives. While no specific format is prescribed, applications should preferably include: (a) description of the proposed project with clearly articulated objectives, plans/programs or activities, and mechanisms for impact assessment; (b) profile of the project holder (qualifications, achievements, experience) and a brief background of the core team; (c) Registration Certificates of the organization and Income-tax Exemption Certificates; (d) The financial audited statements of the last three years and Annual Reports.
Geographic Focus: All States, Albania, Andorra, Armenia, Austria, Azerbaijan, Belarus, Belgium, Bosnia & Herzegovina, Bulgaria, Croatia, Cyprus, Czech Republic, Denmark, Estonia, Finland, France, Georgia, Germany, Greece, Hungary, Iceland, Ireland, Italy, Kosovo, Latvia, Liechtenstein, Lithuania, Luxembourg, Macedonia, Malta, Moldova, Monaco, Montenegro, Norway, Poland, Portugal, Romania, Russia, San Marino, Serbia, Slovakia, Slovenia, Spain, Sweden, Switzerland, The Netherlands, Turkey, Ukraine, United Kingdom, Vatican City
Contact: R.M. Lala; 91 22 6665 8282; fax 91 22 2204 5427; sdtt@sdtatatrust.com
Internet: http://www.dorabjitatatrust.org/ngo/ngo_grants.asp
Sponsor: Sir Dorabji Tata Trust
Bombay House, 24 Homi Mody Street
Mumbai, 400 001 India

Sir Dorabji Tata Trust Individual Medical Grants 4300
The Sir Dorabji Tata Trust provides individual medical grants to needy and deserving patients for medical and surgical care. The maximum disbursement for an individual patient is Rs 100,000, covering expenses incurred towards costs of major surgeries like neurosurgery, cardiac surgery, skin graft surgery for burns, cancer, kidney transplant and others. The grants are also available to patients who are physically handicapped. Though the quantum offered is modest and small, the grants bring much-needed financial relief to the patients and their families. The thrust is to help poor patients from government and municipal hospitals. Individual medical grants are one-time grants.
Requirements: The Trust gives priority to cases within Mumbai in view of the difficulties in assessing cases outside Mumbai. To apply for a medical grant from the Sir Dorabji Tata Trust, you are required to visit the Trust Office at Mulla House, 3rd Floor, 51 M.G. Road, Mumbai 400001 between 3:00 pm to 5:00 pm from Monday to Friday, and pick up the application form. Please ensure that you have all the following documents: (a) Doctor's Certificate; (b) Discharge Certificate; (c) Income Certificate; (d) Copies of Hospital/Medical Bills; (e) Proof of Residence i.e. Ration Card or any other document; (e) Recommendation Letter from Medical Social Worker from a general, municipal or government hospital; (f) Employers Reimbursement; (g) Medical Insurance. The completed forms should be submitted in person to the Trust Office along with the required documents within a month of collecting the form.
Restrictions: Limited to India.
Amount of Grant: Up to 100,000 INR
Contact: R.M. Lala; 91 22 6665 8282; fax 91 22 2204 5427; sdtt@sdtatatrust.com
Internet: http://www.dorabjitatatrust.org/ind_grants/ind_img.asp
Sponsor: Sir Dorabji Tata Trust
Bombay House, 24 Homi Mody Street
Mumbai, 400 001 India

Sir Halley Stewart Trust Grants 4301
The trust awards grants worldwide to organizations and individuals in the areas of medical research, social and education, and religion. Medical research grants focus on problems of the elderly, preventing disability in children, tropical diseases, and Crohn's disease. In the education and social category, grants support research and innovative development projects that attempt to prevent and resolve conflicts, promote reconciliation, and increase understanding within families and across racial, cultural, class, religious, and professional divides. Religion grants advance Christianity and include theological training, teaching in the UK about Christianity, encouragement of specific groups of people (e.g., the elderly, people with disability, students in higher education, etc.) to explore their spiritual needs and strengths, communicating the Christian message through the media, projects that attempt to help disadvantaged people, etc. Applications will not be accepted by fax or email.
Requirements: Nonprofit organizations worldwide are eligible.
Restrictions: Grants do not support general appeals of any kind; personal education fees or fees for taught courses—unless connected with research that falls within current priority areas; educational or 'GAP' year travel projects for young people; purchase, erection, or conversion of buildings; capital costs; running costs of established organizations; or university overhead charges.
Amount of Grant: Up to 875,000 GBP
Contact: Sue West, Administrator; 44-1954-260707
Internet: http://www.sirhalleystewart.org
Sponsor: Sir Halley Stewart Trust
22 Earith Road
Willingham, CB4 5LS England

Sisters of Charity Foundation of Cleveland Reducing Health and Educational Disparities in the Central Neighborhood Grants 4302

The Sisters of Charity Foundation of Cleveland is committed to helping families and individuals overcome the challenges of poverty. In this grant opportunity the Foundation focuses on health, housing and education as key components to building stronger families and stable neighborhoods. The Foundation seeks to reduce both health and educational disparities in Cuyahoga County. Grant funds may be used for program support, operating support, program-related capital support, planning, or capacity building. Grant periods will vary depending upon the particular grant. Most grants are for one year. Grant periods for planning may be shorter, and multi-year grants are possible.
Requirements: Applicants must be 501(c)3 organizations or governmental units or agencies (such as schools) that primarily serve the Cuyahoga County region.
Restrictions: The Foundation does not make grants to support endowments, fundraising campaigns (including annual appeals), membership drives, debt retirement, and individual scholarships. Further, the Foundation does not make grants to individuals.
Geographic Focus: Ohio
Date(s) Application is Due: Apr 7
Amount of Grant: 40,000 - 150,000 U.S.D
Contact: Ursula Craig; (216) 241-9300; fax (216) 241-9345; ucraig@socfdncleveland.org
Internet: http://www.socfdncleveland.org/sistersofcharity/OurFocusAreas/HealthDisparities/HealthintheCentralNeighborhood/tabid/344/Default.aspx
Sponsor: Sisters of Charity Foundation of Cleveland
The Halle Building, 1228 Euclid Avenue, Suite 330
Cleveland, OH 44115-1834

Sisters of Mercy of North Carolina Foundation Grants 4303

The foundation provides grants to tax-exempt health care, educational, and social service organizations that assist women, children, the elderly, and the poor to improve the quality of their lives. Types of support include start-up grants for new organizations or programs, ongoing operating expenses for individual organizations, program or project expenses, building renovation, and equipment acquisition. Preference will be given to organizations whose efforts are collaborative, ecumenical, and multicultural. Particular attention will be given to organizations that serve the unserved or underserved. Annual deadline dates may vary; contact program staff for exact dates.
Requirements: Tax-exempt health care, education, and social service organizations in North and South Carolina are eligible to apply.
Restrictions: The foundation does not ordinarily support projects, programs, or organizations that serve a limited audience; biomedical or clinical research; units of the federal government; political activities; publication of newsletters, magazines, books and the production of videos; conferences and travel; endowment funds; capital fundraising campaigns; annual giving campaigns; or social events or similar fundraising activities.
Geographic Focus: North Carolina, South Carolina
Date(s) Application is Due: Apr 1; Aug 1; Dec 1
Amount of Grant: 25,000 - 150,000 U.S.D
Samples: Greater Carolinas Chapter of the American Red Cross (Charlotte, NC)—to provide free transportation to and from medical appointments, $50,000.
Contact: Grants Administrator; (704) 366-0087; fax (704) 366-8850; contact@somncfdn.org
Internet: http://www.somncfdn.org/grantseekers.html
Sponsor: Sisters of Mercy of North Carolina Foundation
2115 Rexford Road, Suite 401
Charlotte, NC 28211

Sisters of Saint Joseph Charitable Fund Grants 4304

The charitable fund awards community support grants to nonprofits in West Virginia and Ohio counties in three areas: Healthy Communities grants support efforts to mobilize resources and citizen energy for communitywide health promotion, environmental protection, and local development; Healthy Families funding goes to efforts to provide children and youth with sustained, caring adult relationships that will encourage responsible behavior and positive self-identity; Healthy Senior Citizens grants support efforts to engage the elderly in the community by promoting wellness and preventive health care, independent living, voluntary service, and growth through learning. Community Initiative grants support grassroots community-based groups that are addressing health and human service needs directly affecting their local areas. The listed application deadline dates are for one-page letters of inquiry. If interested, the fund will send an application packet. Annual deadline dates may vary; contact program staff for exact dates.
Requirements: 501(c)3 tax-exempt organizations in Calhoun, Jackson, Pleasants, Ritchie, Roane, Tyler, Wirt, and Wood counties of West Virginia; and Athens, Meigs, and Washington counties of Ohio may apply for grant support. Eligible volunteer groups, such as neighborhood associations or church outreach communities, must be sponsored by a 501(c)3 organization.
Restrictions: The fund is unable to make grants to organizations without a 501(c)3 designation, whose projects do not benefit residents of the SSJCF service area, or whose projects do not promote health and wellness in the community it serves. Individuals are not eligible. Organizations that promote proselytizing or religious conversion as a goal of their programming or proposal; or Endowment funds.
Geographic Focus: Ohio, West Virginia
Date(s) Application is Due: Jan 9; Jul 12
Amount of Grant: 18,000 U.S.D
Contact: Grants Manager; (304) 424-6080; fax (304) 424-6081; info@ssjcharitablefund.org
Internet: http://www.ssjcharitablefund.org/grant_program.htm
Sponsor: Sisters of Saint Joseph Charitable Fund
P.O. Box 4440
Parkersburg, WV 26104-4440

Sisters of St. Joseph Healthcare Foundation Grants 4305

The Sisters of St. Joseph Healthcare Foundation is a non-profit, public benefit corporation, which addresses the needs of the working and indigent poor in: Southern California; San Francisco Bay Area; Humboldt County; Fresno County. The Sisters of St. Joseph Healthcare Foundation funds programs which directly serve the needs of the underserved, especially families and children at risk. The Foundation sponsors or supports long-term efforts which are closely identified with the Sisters of St. Joseph of Orange and their mission of bringing unity and healing where divisiveness and oppression exist. The foundation supports the concept of Healthy Communities and desire to fund programs and organizations that: provide direct health-related services; support and transform the individual, social, economic, institutional, and cultural aspects of communities; provide change within larger societal systems to benefit low-income and at-risk populations; and develop the leadership and capacity for self-determination of those served by our funding. The Sisters of St. Joseph Healthcare Foundation is particularly interested in proposals which fall into these funding categories: mental health services; health services; homeless services; violence prevention.
Requirements: Southern California, Humbolt County, Fresno County and San Francisco bay area nonprofits are eligible.
Restrictions: The Foundation does not fund direct support to individuals, annual fund drives, or capital campaigns.
Geographic Focus: California
Amount of Grant: 1,000 - 50,000 U.S.D
Samples: Access OC, Irvine, CA $50,000 - to improve access to specialty care for the safety net population; Boys & Girls Clubs, Garden Grove, CA $30,000 - to fund transportation to improve access to healthcare for low-income families; Casa Teresa, Orange, CA $40,000 - to assist homeless pregnant women with services to help them move from a low-income situation to self-sufficiency;
Contact: Sister Regina Fox; (714) 633-8121, ext. 7109; rfox@csjorange.org
Sponsor: Sisters of St. Joseph Healthcare Foundation
440 South Batavia Street
Orange, CA 92868-3998

Skillman Foundation Good Neighborhoods Grants 4306

The Good Neighborhoods program encourages the creation of safe, healthy, and vibrant neighborhoods where children with the support of caring adults, programs, and experiences can develop fully. Launched in January 2006, the program provides full-scale support to six Detroit neighborhoods where more than 65,000 children live, roughly 30% of the city's child population. Half of the children in these neighborhoods live in poverty. This program encourages the creation of safe, healthy and vibrant neighborhoods where children, with support of caring adults, programs and experiences, can develop fully. The six neighborhoods are: Brightmoor; Cody/Rouge; Northend (also known as Central); Osborn; Chadsey/Condon, in Southwest Detroit; and Vernor, in Southwest Detroit. The goal of the Good Neighborhoods program is to ensure that children experience safe, healthy and high-quality environments where they can thrive: neighborhoods with resources, opportunities and assets for children to develop fully and pursue prosperity.
Requirements: The applicant organizations must be tax exempt and may not be a 509(a) private foundation. The foundation's primary geographic area of focus is the Detroit metropolitan area.
Restrictions: The foundation does not award grants directly to individuals or provide loans of any kind; nor does it support sectarian religious activities, political lobbying, political advocacy, legislative activities, endowments, annual fund drives, basic research, or support of past operating deficits.
Geographic Focus: Michigan
Amount of Grant: 10,000 - 2,000,000 U.S.D
Contact: Tonya Allen, Vice President of Programs; (313) 393-1185; fax (313) 393-1187; info@skillman.org
Internet: http://www.skillman.org/good-neighborhoods/
Sponsor: Skillman Foundation
100 Talon Centre Drive, Suite 100
Detroit, MI 48207

Skin Cancer Foundation Research Grants 4307

Funding is provided annually for basic research, clinical studies, and educational programs related to skin cancer. The foundation's grants program is intended for pilot projects, which, if successful, can be further developed in order to be eligible for larger grants from other sources. Grants include: The Dr. Patricia Wexler Research Grant Awards (one-year, $10,000 grant for a research project); The Theodore S. Tromovitch Memorial Award (one-year, $5,000 grant for a research project); The Melissa K. Bambino Memorial Award (one-year, $8,000 grant for a research project); The Robert N. Cooke Memorial Award (one-year, $3,000 grant for a research project); Pevonia Botantica Research Award (one-year, $10,000 grant for a research project)
Requirements: Candidates must comply with all directions on the application form (available at the website). All projects are selected on the basis of their applicability to the foundation's mission of reducing the incidence, morbidity and mortality of skin cancer. Research must be conducted at institutions within the U.S. Preference is given to projects that address, at the basic science and clinical level, improved methods of prevention, detection, and treatment of skin cancers.
Restrictions: Overhead or indirect costs are not eligible.
Date(s) Application is Due: Oct 1
Amount of Grant: 3,000 - 10,000 U.S.D
Samples: Galina V. Shurin, PhD, University of Pittsburgh - Phosphatidylserine-based ointment for the topical therapy of skin cancer; Paul Nghiem, MD, PhD, University of Washington - Merkel cell carcinoma: Building a tumor and cell line toolbox; David

Polsky, MD, PhD, NYU Medical Center - The use of chloroquine to reduce ultraviolet light-induced DNA damage in the skin
Contact: Grants Administrator; (800) 754-6490; info@skincancer.org
Internet: http://www.skincancer.org/research-grants-2010.html
Sponsor: Skin Cancer Foundation
149 Madison Avenue, Suite 901
New York, NY 10016

SMH Heart Failure Research Grants 4308
The program is intended to foster clinical research that seeks to understand methods for improving the quality of care for patients hospitalized with a primary (or secondary) diagnosis of heart failure. A long-term goal is to foster the development of collaborations of hospitalists and other personnel within and across sites, particularly research collaborations with a high likelihood of leading to subsequent research applications to governmental agencies (e.g., NIH). Funds may only be used to support work directed at evaluation of the program, and cannot be applied to program implementation, support for clinical personnel, rental of space, or to support administrative personnel. Nonbinding letters of intent are requested by June 15; applications are due August 31 for applications. Guidelines are available online.
Requirements: Eligible institutions are for-profit or nonprofit acute-care hospitals or acute-care healthcare systems; or public or private institutions, such as universities, colleges, and physician practice groups. Individuals eligible to become principal investors are SHM members in good standing, as of July 1; and have the skills, knowledge, and resources necessary to carry out the proposed research, with an invitation to work with his or her institution to develop an application for support.
Date(s) Application is Due: Jun 15; Aug 31
Amount of Grant: 25,009 U.S.D
Contact: Charisse Braithwaite, (800) 843-3360 or (215) 351-2473; fax (215) 843-2536; Cbraithwaite@hospitalmedicine.org
Internet: http://www.hospitalmedicine.org//AM/Template.cfm?Section=Home
Sponsor: Society of Hospital Medicine
190 N Independence Mall W
Philadelphia, PA 19106-1572

Smith Family New Investigator Awards 4309
These awards are made to promote basic research in the areas of AIDS, cancer, cardiovascular disease, diabetes, neuroscience, autism, and related disorders. The program favors innovative research, clearly presented proposals, realistic objectives, qualified personnel, and financial viability. The program will make six to eight two-year awards.
Requirements: Massachusetts medical groups may submit proposals.
Geographic Focus: Massachusetts
Amount of Grant: 100,000 U.S.D
Contact: Grants Coordinator; (617) 451-0049, ext. 702; fax (617) 423-4619; research@tmfnet.org
Internet: http://www.tmfnet.org/grantmake.html
Sponsor: Medical Foundation, a Division of Health Resources in Action
95 Berkeley Street, Suite 208
Boston, MA 02116

Smithsonian Graduate Student Research Fellowships 4310
Students actively engaged in graduate study at any level may apply to study and conduct research for 10-week periods under the guidance of Smithsonian staff members. Applicants must propose to conduct research in disciplines that are pursued at the Smithsonian (American history, art, and culture; history of science and technology; history of art, design, crafts, and decorative arts; anthropology, archaeology, linguistics, and ethnic studies; evolutionary, systematic, behavioral, and environmental biology and conservation; earth, mineral, and planetary sciences; and materials characterization and conservation) and must plan to spend their entire tenure in residence. Applications are posted to the Web site in September.
Requirements: Students must be formally enrolled in graduate programs, have completed at least one semester, and have not yet been advanced to candidacy if in PhD programs.
Date(s) Application is Due: Jan 15
Amount of Grant: 4,500 U.S.D
Contact: Program Contact; (202) 275-0655; fax (202) 275-0489; siofg@si.edu
Internet: http://www.si.edu/ofg/fell.htm
Sponsor: Smithsonian Institution
750 Ninth Street NW, Suite 9300
Washington, DC 20560

Smithsonian Postdoctoral Research Fellowships 4311
Fellowships support independently conceived and designed research study by persons at advanced academic levels working with Smithsonian staff members and making use of Smithsonian collections, facilities, and other special resources. Proposals for research may be offered in fields of study conducted at the institute (American history, art, and culture; history of science and technology; history of art, design, crafts, and decorative arts; anthropology, archaeology, linguistics, and ethnic studies; evolutionary, systematic, behavioral, and environmental biology and conservation; earth, mineral, and planetary sciences; and materials characterization and conservation). The fellowship term is three to 12 months. Applications are posted on the Web site each September.
Requirements: Applicants must have a PhD or equivalent degree for less than seven years.
Date(s) Application is Due: Jan 15
Amount of Grant: 35,000 U.S.D
Contact: Program Contact; (202) 275-0655; fax (202) 275-0489; siofg@si.edu
Internet: http://www.si.edu/ofg/fell.htm
Sponsor: Smithsonian Institution
750 Ninth Street NW, Suite 9300
Washington, DC 20560

Smithsonian Predoctoral Research Fellowships 4312
Fellowships support independently conceived and designed research study by predoctoral students to work with Smithsonian staff members and use Smithsonian collections, facilities, and other special resources. Proposals are accepted for research in the areas of study conducted at the institute (American history, art, and culture; history of science and technology; history of art, design, crafts, and decorative arts; anthropology, archaeology, linguistics, and ethnic studies; evolutionary, systematic, behavioral, and environmental biology and conservation; earth, mineral, and planetary sciences; and materials characterization and conservation). The fellowship term is three to 12 months. Applications are posted to the Web site in September.
Requirements: Candidates must have completed all coursework and examinations and be ready to engage in dissertation research.
Date(s) Application is Due: Jan 15
Amount of Grant: 20,000 U.S.D
Contact: Program Contact; (202) 275-0655; fax (202) 275-0489; siofg@si.edu
Internet: http://www.si.edu/ofg/fell.htm
Sponsor: Smithsonian Institution
750 Ninth Street NW, Suite 9300
Washington, DC 20560

Smithsonian Senior Fellowships 4313
Each year, an occasional appointment is made to distinguished senior scientists for research in the fields of anthropology, natural history, environmental science and ecological processes, tropical biology, African art and culture, oriental art, astrophysics and related topics, history of science and technology, American cultural and social history, or art history. The fellowship term is three to 12 months. Prior to submitting applications, interested senior scientists should consult with Smithsonian staff in their chosen fields. Applications are posted to the Web site in September.
Requirements: Applicants must have held the PhD or equivalent degree for at least seven years.
Date(s) Application is Due: Jan 15
Amount of Grant: 35,000 U.S.D
Contact: Office of Fellowships; (202) 275-0655; siofg@si.edu
Internet: http://www.si.edu/ofg/fell.htm
Sponsor: Smithsonian Institution
P.O. Box 37012, Victor Building, Suite 9300, MRC 902
Washington, DC 20013-7012

Snee Reinhardt Charitable Foundation Grants 4314
The foundation awards a large variety of grants that have an impact on the social issues that Americans face every day. Recent grantees have included children's museums, universities, health agencies, theaters, schools, hospitals, youth issues, and environmental causes. Precedence is given to organizations situated within or whose activities are concentrated in, first, southwestern Pennsylvania, then northern West Virginia, northern Maryland, the Commonwealth of Pennsylvania and, lastly, the United States. Application and guidelines are available online.
Requirements: 501(c)3 nonprofit organizations are eligible.
Restrictions: Organizations, activities, and/or programs excluded from support include general capital improvement; general operating expenses, including salaries and fringe benefits; chairs or professorships; endowment funding; grants to individuals; political contributions; highly specialized health or medical programs that do not have specific impact on the community; organizations and programs designed to influence legislation or elect candidates to public office; organizations that discriminate by race, color, creed, gender, or national origin; or programs that promote, research, or support the prevention of life, abortion, the practice of euthanasia, or cruelty to animals.
Date(s) Application is Due: Apr 15; Aug 15
Amount of Grant: Up to 50,000 U.S.D
Contact: Grants Administrator; (412) 390-2690; fax (412) 390-2686
Internet: http://www.snee-reinhardt.org/index.html
Sponsor: Snee Reinhardt Charitable Foundation
River Park Commons Two, 2425 Sidney Street
Pittsburgh, PA 15203

SNM/ Covidien Seed Grant in Molecular Imaging/Nuclear Medicine Research 4315
The Society of Nuclear Medicine is pleased to offer the SNM/Covidien Seed Grant in Nuclear Medicine Research. This grant is designed to assist researchers in conducting new and innovative pilot projects that have potential for future support from foundations, corporations or government agencies.
Requirements: Applicant Eligibility: basic or clinical scientists employed by academic and research-oriented organizations; applicants from any country may apply; applicants must hold a full-time faculty position in an educational institution when the award period begins; applicants must have completed all advanced training and be no more than five years post training; applicants must not have served as a principal investigator on peer-reviewed grants totaling $50,000 or more in a single calendar year; preference will be given to individuals who have demonstrated great potential for a research career in the field of nuclear medicine/molecular imaging.
Restrictions: No grantee may receive more than one SNM research grant in any one year. The grant will not pay institutional overhead costs or indirect costs. No salary support

for principal investigator or co-principal investigators. Grants will be limited to a time span of one year; no extensions are allowed
Date(s) Application is Due: Feb 20
Amount of Grant: 25,000 U.S.D
Contact: Nicole Kern, SNM Program Manager; (703) 652-6795, ext. 1255; fax (703) 708-9015; nkern@snm.org
Internet: http://interactive.snm.org/index.cfm?PageID=2255
Sponsor: Society of Nuclear Medicine
1850 Samuel Morse Drive
Reston, VA 20190

SNM Molecular Imaging Research Grant For Junior Medical Faculty **4316**
The objective of this program is to provide salary support for one junior faculty member in an academic/research setting to enable them to engage in Molecular Imaging research related to diagnostic or therapeutic applications. The grant will support new, innovative pilot projects in any area of molecular imaging research. The grant may supplement other funds, but not duplicate other support.
Requirements: Applicant Eligibility: applicants must have MD degree (or equivalent) and have completed a residency and be certified in their specialty; applicants must be practicing in an academic/research setting as a faculty member in the United States; applicants must be within 5 years of their initial faculty appointment with an academic rank of instructor or assistant professor (or equivalent); applicants must provide a detailed proposal of the proposed research activity; applicants must not have received grant/contract amounts totaling $50,000.00 or more in a single calendar year as the principle investigator; preference will be given to individuals who have demonstrated great potential for a research career in the field of molecular imaging; multidisciplinary projects are encouraged; only one application per faculty member is accepted per grant; no grantee may receive more than one SNM research grant in any one year. Likewise no other research grant request or continuation will be considered until a satisfactory summary of an earlier grant is received; applicants are required to be a SNM member at the time of application.
Restrictions: The grant will not pay institutional overhead costs or indirect costs
Date(s) Application is Due: Jan 20
Amount of Grant: 100,000 U.S.D
Contact: Nicole Kern, SNM Program Manager; (703) 652-6795, ext. 1255; fax (703) 708-9015; nkern@snm.org
Internet: http://interactive.snm.org/index.cfm?PageID=7252
Sponsor: Society of Nuclear Medicine
1850 Samuel Morse Drive
Reston, VA 20190

SNM PDEF Mickey Williams Minority Student Scholarships **4317**
This scholarship supports minority students pursuing a two or four year degree in nuclear medicine. It honors the memory of Mickey Williams, a past SNMTS President who immigrated to the United States from Jamaica.
Requirements: To be eligible for the PDEF Mickey Williams Minority Student Scholarship, candidates must: be of African American, Native American, Hispanic American, Asian American or Pacific Islander decent [Eligible Native Americans include American Indian, Eskimo, Hawaiian and Samoan]; hold current U.S. residency status of U.S. citizen, U.S. national, or U.S. permanent resident; be currently enrolled in or accepted into an entry-level nuclear medicine technology program at the time of application; Award is open to students in associate and baccalaureate level programs only; demonstrate financial need; have maintained a cumulative grade point average of 2.5 or better on a four-point scale, with evidence of high academic performance. See SNM website for application.
Restrictions: Individuals with a visitor, student, or G-series visa are not eligible. Individuals with previous certificates or degrees in nuclear medicine sciences are ineligible.
Date(s) Application is Due: Jan 15
Amount of Grant: 5,000 U.S.D
Contact: Nicole Kern; (703) 652-6795, ext. 1255; fax (703) 708-9015; nkern@snm.org
Internet: http://interactive.snm.org/index.cfm?PageID=2682&RPID=2259
Sponsor: Society of Nuclear Medicine
1850 Samuel Morse Drive
Reston, VA 20190

SNM Pilot Research Grants **4318**
The pilot research grants are designed to help basic or clinical scientists in the early stages of their career conduct research that may lead to further funding. The Mitzi and William Blahd, MD Pilot Research Grant is awarded to the applicant with the highest-ranked proposal. Preference will be given to individuals who have demonstrated great potential for a research career in the field of nuclear medicine/molecular imaging.
Requirements: Applicant Eligibility: basic or clinical scientists employed by academic and research-oriented organizations in North America are eligible; applicants must not be no more than five years post training.
Date(s) Application is Due: Feb 20
Amount of Grant: 25,000 U.S.D
Contact: Nicole Mitchell, Program Manager; (703) 652-6795; nkern@snm.org
Internet: http://interactive.snm.org/index.cfm?PageID=2255
Sponsor: Society of Nuclear Medicine
1850 Samuel Morse Drive
Reston, VA 20190

SNM Pilot Research Grants in Nuclear Medicine/Molecular Imaging **4319**
The pilot research grants are designed to help basic or clinical scientists in the early stages of their career conduct research that may lead to further funding.
Requirements: Applicant Eligibility: basic or clinical scientists employed by academic and research-oriented organizations in North America are eligible; applicants must be no more than five years post training; applicant must not have served as the principal investigator of a peer-reviewed grant for more than $50,000 in a single calendar year; preference will be given to individuals who have demonstrated great potential for a research career in the field of nuclear medicine/molecular imaging.
Restrictions: The grant will not pay salaries, institutional overhead costs or indirect costs. Funds may supplement other funds, but not duplicate other support. Grants will be limited to a time span of one year. No grantee may receive more than one grant in any one year.
Date(s) Application is Due: Feb 20
Amount of Grant: 25,000 U.S.D
Contact: Nicole Kern, SNM Program Manager; (703) 652-6795, ext. 1255; fax (703) 708-9015; nkern@snm.org
Internet: http://interactive.snm.org/index.cfm?PageID=2255
Sponsor: Society of Nuclear Medicine
1850 Samuel Morse Drive
Reston, VA 20190

SNM Postdoctoral Molecular Imaging Scholar Grants **4320**
The postdoctoral program will support a two-year research endeavor that promotes integration of molecular imaging into the career of the trainee. The funds maybe used for new, innovative pilot projects in any area of molecular imaging research. The grant may supplement other funds, but not duplicate other support. A summary report and an accounting of funds will be requested at the halfway point and upon completion of the research project. Grantees are expected to submit results of the research as an abstract for presentation at the SNM Annual Meeting and/or as a scientific manuscript to the Journal of Nuclear Medicine. A final report is due within 60 days after completion of the research project.
Requirements: Applicant Eligibility: MD, MD/PhD, and PhD applicants are eligible; applicants from any country may apply; applicants must hold a full-time position in an educational institution when the award starts; applicants must have completed all basic training and be no more than 2 years post training; applicants must not have served as a principal investigator on peer-reviewed grants totaling $50,000 or more in a single calendar year; preference will be given to individuals who have demonstrated great potential for a research career in the field of molecular imaging; only one application per trainee is accepted; no grantee may receive more than one SNM research grant in any one year. Likewise no other research grant request or continuation will be considered until a satisfactory summary of an earlier grant is received.
Restrictions: Grant extensions are not often allowed, but will be reviewed and decided upon by the SNM Grants and Awards Committee if requested. The grant will not pay institutional overhead costs or indirect costs.
Date(s) Application is Due: Jan 20
Amount of Grant: 60,000 U.S.D
Contact: Nicole Kern, SNM Program Manager; (703) 652-6795, ext. 1255; fax (703) 708-9015; nkern@snm.org
Internet: http://interactive.snm.org/index.cfm?PageID=7252
Sponsor: Society of Nuclear Medicine
1850 Samuel Morse Drive
Reston, VA 20190

SNM Student Fellowship Awards **4321**
The SNM Student Fellowships are designed to stimulate interest in molecular imaging/nuclear medicine by supporting their full-time participation in clinical and basic research activities for three months (or less). These Fellowships are made possible by support of the Education and Research Foundation for SNM.
Requirements: Candidates from any country may apply. Students must be enrolled in a medical, pharmacy or graduate school, or be an undergraduate student demonstrating outstanding competence in nuclear medicine and/or molecular imaging research. Fellowship must be carried out during a three-month that starts no earlier then March and completed no later then December 31st.
Date(s) Application is Due: Jan 23
Amount of Grant: 3,000 U.S.D
Contact: Nicole Kern, SNM Program Manager; (703) 652-6795, ext. 1255; fax (703) 708-9015; nkern@snm.org
Internet: http://interactive.snm.org/index.cfm?PageID=2254
Sponsor: Society of Nuclear Medicine
1850 Samuel Morse Drive
Reston, VA 20190

SNM Tetalman Memorial Young Investigator Awards **4322**
Established by the family and friends of Dr. Tetalman, this award honors the work of a young investigator who is pursuing a career in nuclear medicine. It is named in memory of a highly respected and productive clinician and researcher, Mark Tetalman, M.D., whose career was cut tragically short. The award is based on a submitted paper supporting current research efforts and accomplishments, teaching, clinical service and administration. Service to the Society and a commitment to academic nuclear medicine will also be considered.
Requirements: Applicants must be 36 years old or younger as of July 1 of the application year and must have obtained certification in Nuclear Medicine or Nuclear Radiology or have completed a PhD program within the last seven years. Complete applications must

be must be received, not simply post-marked by November 14th. One original unstapled copy of the completed application must be sent to: SNM Development Office, Attention: SNM Committee on Awards
Date(s) Application is Due: Nov 14
Amount of Grant: 2,500 U.S.D
Contact: Nicole Kern, SNM Program Manager; (703) 652-6795, ext. 1255; fax (703) 708-9015; nkern@snm.org
Internet: http://interactive.snm.org/index.cfm?PageID=2253
Sponsor: Society of Nuclear Medicine
1850 Samuel Morse Drive
Reston, VA 20190

SNMTS Clinical Advancement Scholarships **4323**

The SNMTS Clinical Advancement Scholarships serve to support technologists who are pursuing clinical advancement through didactic educational programs. Ten (10) $500 scholarships will be awarded. Scholarship application form is available on the SNM website.
Requirements: To be eligible for the SNMTS Clinical Advancement Scholarship, candidates must: be a member of the SNMTS; demonstrate financial need; be currently enrolled in a didactic educational program(s) (ex. CT, DEXA, physics, statistics) which is/are college credit eligible; complete the said class or program. It is not required that class/program completion results in a degree.
Date(s) Application is Due: May 1
Amount of Grant: 500 U.S.D
Contact: Nicole Kern, SNM Program Manager; (703) 652-6795, ext. 1255; fax (703) 708-9015; nkern@snm.org
Internet: http://interactive.snm.org/index.cfm?PageID=7270&RPID=2259
Sponsor: Society of Nuclear Medicine Technologist Section
1850 Samuel Morse Drive
Reston, VA 20190

SNMTS Outstanding Educator Awards **4324**

The purpose of this award to is to recognize a SNMTS member who has significantly contributed to providing knowledge that advances and promotes the field of nuclear medicine technology through outstanding work in education. The winner will receive $750 along with a plaque to be presented at the SNM Annual Meeting.
Requirements: Eligible candidates include teachers but also educators in the non-traditional sense such as industry and clinical professionals. Evidence of their outstanding work must be shown in the following required supporting documentation: Curriculum Vitae, including listing of publications, presentations, programs, awards and honors; one letter of support from recent student/graduate (within 5 years) or participant in program; one letter of support from current program director or supervisor; one letter of support from a colleague; completed Nomination Form (see website).
Date(s) Application is Due: Mar 13
Amount of Grant: 750 U.S.D
Contact: Nicole Kern, SNM Program Manager; (703) 652-6795, ext. 1255; fax (703) 708-9015; nkern@snm.org
Internet: http://interactive.snm.org/index.cfm?PageID=2256
Sponsor: Society of Nuclear Medicine Technologist Section
1850 Samuel Morse Drive
Reston, VA 20190

SNMTS Outstanding Technologist Awards **4325**

The purpose of this award to is to recognize an SNMTS (SNM Technologist Section) member who has demonstrated outstanding service and dedication to the field of nuclear medicine technology. The winner will receive $750 along with a plaque to be presented at the SNM Annual Meeting.
Requirements: Eligible candidates include SNMTS members in any area of the field of nuclear medicine technology who have exhibited commitment to advancing the field in their workplace and through their involvement with the Society. Nominees must be involved with the Society at the local, regional and/or national level and have at least five years of experience in nuclear medicine technology. Evidence of their outstanding work must be shown in the following required supporting documentation: curriculum vitae- including listing of publications, presentations, programs, awards and honors; one letter of support from current supervisor; one letter of support from a colleague; one letter of support from a colleague who is a member of the SNMTS; completed Nomination Form (see website for application form).
Date(s) Application is Due: Mar 13
Amount of Grant: 750 U.S.D
Contact: Nicole Kern, SNM Program Manager; (703) 652-6795, ext. 1255; fax (703) 708-9015; nkern@snm.org
Internet: http://interactive.snm.org/index.cfm?PageID=2256
Sponsor: Society of Nuclear Medicine
1850 Samuel Morse Drive
Reston, VA 20190

SNMTS Paul Cole Scholarships **4326**

The Paul Cole Scholarship is named in memory of Paul Cole, CNMT, who served as President of the SNM Technologist Section (SNMTS) in 1986 and who was known as a champion of education for technologists. This scholarship is funded by the Education and Research Foundation for the SNM.
Requirements: To be eligible for the Paul Cole Scholarship, candidates must: demonstrate financial need; be enrolled in or accepted into an institution accredited through the Joint Review Committee on Educational Programs in Nuclear Medicine Technology (JRCNMT). To see if your institution is accredited with JRCNMT, visit their website at www.jrcnmt.org. Award is open to students in associate, baccalaureate, and certificate level programs; have maintained a minimum cumulative GPA of 2.5 or better (on a 4.0 scale) or B average in a nuclear medicine technology core curriculum.
Date(s) Application is Due: Jan 15
Amount of Grant: 1,000 U.S.D
Contact: Nicole Kern, SNM Program Manager; (703) 652-6795, ext. 1255; fax (703) 708-9015; nkern@snm.org
Internet: http://interactive.snm.org/index.cfm?PageID=2681&RPID=2259
Sponsor: Society of Nuclear Medicine Technologist Section
1850 Samuel Morse Drive
Reston, VA 20190

SOBP Domestic Travel Scholarships **4327**

The training of a new generation of psychiatric academicians is a major objective of the Society. Participation in professional meetings, where new information is exchanged and contacts are made, can have a critical impact on the career of a developing clinician-scientist. In recognition of the importance of academic exchange, the Society, through an educational grant from Eli Lilly and Company, Forest Laboratories and Elsevier Science, announces the continuation of annual Travel Fellowships. Each Fellowship will provide $1,500 to help defray the cost of attending the Society's Annual Meeting. In addition to the stipend and the opportunity to attend sessions of interest to them, Fellows also receive certificates of excellence and participate in a special ceremony held during the meeting in their honor. To receive the travel award, applicants are expected to attend the entire meeting and be present for the recognition events.
Requirements: For medical graduates, each University Department of Psychiatry in the United States, Canada and Mexico may nominate one candidate for this award. Medical graduates must be in their third, fourth, or fifth year of residency or in a bona fide research training program or sub-specialty training program within four years of leaving a general psychiatry residency training program.
Date(s) Application is Due: Jan 15
Amount of Grant: Up to 1,500 U.S.D
Contact: Maggie Peterson, (904) 953-2842; maggie@mayo.edu or sobp@sobp.org
Internet: http://www.sobp.org/i4a/pages/index.cfm?pageID=3372
Sponsor: Society of Biological Psychiatry
4500 San Pablo Road, Birdsall 310
Jacksonville, FL 32224

SOBP International Travel Fellowships **4328**

The Society offers International Travel Fellowships of up to U.S. $1,500 each to promising young investigators under the age of 35 from less-developed and less well-funded countries of the five IBRO Regions (Africa, Asian/Pacific, Central and Eastern Europe, Latin America, Western Europe) who would benefit from attending the annual meeting of the Society of Biological Psychiatry. This program will provide exposure of the fellowship recipients to the best science the Society has to offer and facilitate the professional contacts that may yield enduring collaborative and collegial relationships.
Requirements: Nominees must be: from a less developed and less well-funded country; and less than 35 years of age.
Geographic Focus: All Countries
Date(s) Application is Due: Jan 15
Amount of Grant: Up to 1,500 U.S.D
Contact: Maggie Peterson, (904) 953-2842; maggie@mayo.edu or sobp@sobp.org
Internet: http://www.sobp.org/i4a/pages/index.cfm?pageID=3375
Sponsor: Society of Biological Psychiatry
4500 San Pablo Road, Birdsall 310
Jacksonville, FL 32224

SOBP Ziskind-Somerfeld Research Awards **4329**

The Society offers an annual award of $2,500 in either basic or clinical research for the purpose of stimulating investigations in biological psychiatry by senior investigators who are members in good standing of the Society of Biological Psychiatry. The Ziskind-Somerfeld Research Foundation makes this award possible. No submission is required to be considered for this award. The research will be judged in comparison with the work of other senior investigators. Special consideration is given to the originality of the approach and independence of thought evident in the submission.
Requirements: Candidates will be at least thirty-five years of age at the time of submission. When more than one investigator is listed as an author on the submitted work, the principal investigators (first and last authors) must be over the age of 45.
Amount of Grant: 2,500 U.S.D
Contact: Maggie Peterson, (904) 953-2842; maggie@mayo.edu or sobp@sobp.org
Internet: http://www.sobp.org/i4a/pages/index.cfm?pageID=3380#top
Sponsor: Society of Biological Psychiatry
4500 San Pablo Road, Birdsall 310
Jacksonville, FL 32224

SOCFOC Catholic Ministries Grants **4330**

The Sisters of Charity Foundation of Cleveland belongs to a community of Catholic health and social service organizations inspired to service by the healing ministry of Jesus Christ. As a supporting foundation to the congregation of the Sisters of Charity of St. Augustine, the Foundation gives priority to the ministries of the Sisters of Charity Health System. A limited amount of remaining funds may be available to other Catholic ministries in health and human services in Cuyahoga County. Funding is limited to Invited Proposals.

Requirements: Each proposal to the Foundation must include a completed application form, proposal narrative, budget, and Budget Narrative. The application form and guidelines for the proposal narrative may be found on this website under the specific program area. Eligible organizations are 501(c)3 organizations or governmental agencies or units.
Restrictions: The Foundation does not make grants to individuals.
Geographic Focus: Ohio
Amount of Grant: 10,000 - 150,000 U.S.D
Samples: Secretariat for Parish Life & Development, OH, $10,000;2008; St. Vincent Charity Hospital, OH, $151,600;2008;
Contact: Lynn Berner, Program Officer; (216) 241-9300; fax (216) 241-9345; lberner@socfdncleveland.org
Internet: http://www.socfdncleveland.org/RespondingToOurCommunity/CatholicMinistries/tabid/313/Default.aspx
Sponsor: Sisters of Charity Foundation of Cleveland
The Halle Building, 1228 Euclid Avenue, Suite 330
Cleveland, OH 44115-1834

Society for Biomedical Screening Small Grants **4331**
The society's small grants program will provide start-up funding for research projects with potential to advance the disciplines, technologies, and skills involved in screening and molecular discovery, including the support of graduate or undergraduate training in these areas. The program is intended to support projects that may have an applied focus or those that may involve radically new concepts. Applications are considered without regard to race, religion, national origin, physical handicap, or age of the candidates.
Requirements: Applicants must be employed full time in a college, university, or nonprofit research institute in order to serve as principal investigators.
Date(s) Application is Due: Feb 18
Contact: David Roman, (203) 743-1336; fax (203) 748-7557; droman@sbsonline.org
Internet: http://www.sbsonline.org/awards_grants
Sponsor: Society for Biomedical Screening
36 Tamarack Avenue, #348
Danbury, CT 06811

Society for Imaging Informatics in Medicine (SIIM) Emeritus Mentor Grants **4332**
The mission of SIIM, The Society for Imaging Informatics in Medicine, is to advance computer applications and information technology in medical imaging through education and research. In support of this mission, the SIIM Research and Development Committee offers a $20,000 grant for a retired, or semi-retired, expert in imaging informatics to act as a mentor for a trainee (student, resident, fellow) or junior faculty member for hypothesis-driven research. Any area of research involving medical imaging informatics is eligible for support, as well as development of new hardware, software, or processes to support electronic imaging practice, education, or research.
Requirements: A senior, experienced investigator jointly applies with a resident, fellow, graduate student, or junior faculty for a SIIM emeritus mentor grant. Preference will be given to projects at SIIM institutions and to projects that contribute to the development of careers of new or prospective researchers. This grant may not be applied for in conjunction with another SIIM grant covering the same subject material.
Date(s) Application is Due: Aug 3
Amount of Grant: 20,000 U.S.D
Contact: Grant Manager; (703) 723-0432; fax (703) 723-0415; grants@siimweb.org or info@siimweb.org
Internet: http://www.scarnet.org/index.cfm?id=63
Sponsor: Society for Imaging Informatics in Medicine
19440 Golf Vista Plaza, Suite 330
Leesburg, VA 20176-8264

Society for Imaging Informatics in Medicine (SIIM) Micro Grants **4333**
The mission of SIIM, The Society for Imaging Informatics in Medicine, is to advance computer applications and information technology in medical imaging through education and research. In support of this mission, the SIIM Research and Development Committee offers a $5,000 grant for hypothesis-driven experiments of short duration and high risk. Prior preliminary data is not required. An imaging informatics project that needs resources falls between what a researcher is willing to fund personally or is more than his/her department provides in seed money and what is typically requested in grant applications. Examples include trying unconventional hardware or software in an imaging informatics experiment, trying a particular piece of hardware for a user-interface design, paying for short duration consulting assistance, or paying for an educational course that the researcher needs to enhance a particular study.
Requirements: There are no restrictions on who may apply for the grant. Trainees (residents, fellows, and graduate students) are required to include a letter from a faculty member, acting as an advisor, to show that the applicant's department supports the project. Preference will be given to projects at SIIM institutions and to projects that may yield high value results or provide enough preliminary data to justify additional work and application for traditional grant funding.
Date(s) Application is Due: Aug 3
Amount of Grant: 5,000 U.S.D
Contact: Grant Manager; (703) 723-0432; fax (703) 723-0415; grants@siimweb.org or info@siimweb.org
Internet: http://www.scarnet.org/index.cfm?id=63
Sponsor: Society for Imaging Informatics in Medicine
19440 Golf Vista Plaza, Suite 330
Leesburg, VA 20176-8264

Society for Imaging Informatics in Medicine (SIIM) Research Grants **4334**
The mission of SIIM, The Society for Imaging Informatics in Medicine, is to advance computer applications and information technology in medical imaging through education and research. In support of this mission, the SIIM Research and Development Committee offers two $50,000 grants for original research. Any area of research involving medical imaging informatics is eligible for support, as well as development of new hardware, software, or processes to support electronic imaging practice, education, or research.
Requirements: Residents, fellows, graduate students, or faculty may apply for SIIM research grants. Preference will be given to projects at SIIM institutions and to projects that contribute to the development of careers of new or prospective researchers. Residents, fellows, and students must conduct their projects under the guidance of experienced investigators. Grant recipients will not be eligible for concurrent support through other SIIM grants.
Date(s) Application is Due: Aug 3
Amount of Grant: 50,000 U.S.D
Contact: Grant Manager; (703) 723-0432; fax (703) 723-0415; grants@siimweb.org or info@siimweb.org
Internet: http://www.scarnet.org/index.cfm?id=63
Sponsor: Society for Imaging Informatics in Medicine
19440 Golf Vista Plaza, Suite 330
Leesburg, VA 20176-8264

Society for Imaging Informatics in Medicine (SIIM) Small Grant for Product Support Development **4335**
The mission of SIIM, The Society for Imaging Informatics in Medicine, is to advance computer applications and information technology in medical imaging through education and research. In support of this mission, the SIIM Research and Development Committee offers a $20,000 grant to support the development of a product or tool for use in the field of imaging informatics. Any area of product or tool development for use in the field of medical imaging informatics is eligible for support (e.g., open source application development, work flow measurement tools, replacement PACS change management tools).
Requirements: Residents, fellows, graduate students, or faculty may apply for product support development grants. Preference will be given to projects at SIIM institutions and to projects that contribute to the development of careers of new or prospective researchers. Residents, fellows, and students must conduct their projects under the guidance of experienced investigators.
Date(s) Application is Due: Aug 3
Amount of Grant: 20,000 U.S.D
Contact: Grant Manager; (703) 723-0432; fax (703) 723-0415; grants@siimweb.org or info@siimweb.org
Internet: http://www.scarnet.org/index.cfm?id=63
Sponsor: Society for Imaging Informatics in Medicine
19440 Golf Vista Plaza, Suite 330
Leesburg, VA 20176-8264

Society for Imaging Informatics in Medicine (SIIM) Small Training Grants **4336**
The mission of SIIM, The Society for Imaging Informatics in Medicine, is to advance computer applications and information technology in medical imaging through education and research. In support of this mission, the SIIM Research and Development Committee offers a $20,000 grant targeted at providing training opportunities in the field of imaging informatics for students and imaging informatics professionals. Any area of training within medical imaging informatics is eligible for support. The grant is designed to support hands-on training or experience with a designated mentor or sponsor in the imaging informatics arena. It is not intended to support tuition or other traditional education-related expenses. The period of training must last a minimum of three months and may extend no longer than twelve months.
Requirements: Residents, fellows, graduate students, or faculty members at recognized institutions may apply for SIIM training grants. Preference will be given to training experiences at SIIM institutions and to those that contribute to the careers of new or prospective imaging informatics personnel.
Date(s) Application is Due: Aug 3
Amount of Grant: 20,000 U.S.D
Contact: Grant Manager; (703) 723-0432; grants@siimweb.org or info@siimweb.org
Internet: http://www.scarnet.org/index.cfm?id=63
Sponsor: Society for Imaging Informatics in Medicine
19440 Golf Vista Plaza, Suite 330
Leesburg, VA 20176-8264

Society for Investigative Dermatology Career Development Award **4337**
The society makes one-year career development awards to encourage trainees in skin-related fields to become skin scientists. Applications will be considered in the following areas: skin biology, skin disease, skin research publishing, and presentation of skin research at a forum. Application forms may be obtained from the Dermatology Foundation, 1563 Maple Avenue, Evanston, IL 60201. Contact program staff to determine availability of awards.
Requirements: Eligible applicants include MDs and PhDs engaged in projects related to the skin and skin diseases.
Amount of Grant: 40,000 U.S.D
Contact: Department of Dermatology, Columbia University; (216) 579-9300; fax (216) 579-9333; sid@sidnet.org
Internet: http://www.sidnet.org
Sponsor: Society for Investigative Dermatology
820 W Superior Avenue, 7th Floor
Cleveland, OH 44113-1800

Society for the Arts in Healthcare Consulting Grants 4338
The Society for the Arts in Healthcare has partnered with the National Endowment for the Arts to provide hundreds of consulting grants to arts and/or health care organizations across the U.S.. Through a roster of over 20 trained arts in health care consultants, these grants assist organizations in utilizing a diverse array of arts media including performing arts, visual arts, literature, film, or design to promote the health and well-being of equally diverse populations. Organizations may apply for up to 20 hours of consulting with an experienced leader in the field. This grant pays for all consultant fees and grantees cover any additional costs such as long-distance phone calls or consultant travel and accommodation. Consultants are available to travel to organizations for site visits or to provide information and support via phone or email. The Consulting Service can assist in: program planning/strategic planning; artist training; fundraising and promotion; community collaborations and effective partnerships; evaluation; and improving health care facility design and environments. Projects can range from specific technical assistance such as designing a job description, creating a survey, feedback on a grant application, or developing guidelines for exhibitions to thinking through a strategic plan for a program. Prospective clients can request the number of hours that would be most appropriate for their needs.
Requirements: Applicant organizations must be based in the United States.
Date(s) Application is Due: Nov 16
Contact: Anita Boles, Executive Director; (202) 299-9770; fax (202) 299-9887; anita@thesah.org
Internet: http://www.thesah.org/template/page.cfm?page_id=466
Sponsor: Society for the Arts in Healthcare
2437 15th Street NW, B South
Washington, DC 20009

Society for the Arts in Healthcare Environmental Arts Research Grant 4339
The American Art Resources and Society for the Arts in Healthcare Environmental Arts Research Grant program offers a cash award sponsored by American Art Resources in Houston, Texas. The program is provided to a Principal Investigator to conduct or supplement Arts in Healthcare research. Selection is made on the merit, innovation and feasibility of the research proposal.
Requirements: Applicant organizations must be based in the United States.
Contact: Anita Boles; (202) 299-9770; fax (202) 299-9887; anita@thesah.org
Internet: http://www.thesah.org/template/page.cfm?page_id=570
Sponsor: Society for the Arts in Healthcare
2437 15th Street NW, B South
Washington, DC 20009

Society of Cosmetic Chemists Allan B. Black Award Sponsored by Presperse 4340
Dedicated to the advancement of cosmetic science, the Society strives to increase and disseminate scientific information through meetings and publications. By promoting research in cosmetic science and industry, and by setting high ethical, professional and educational standards, it strives to improve the qualifications of cosmetic scientists. Its mission is to further the interests and recognition of cosmetic scientists while maintaining the confidence of the public in the cosmetic and toiletries industry. With the Allan B. Black Award, a scroll and honorarium are awarded annually for the Best Paper on make-up technology either presented at an annual meeting, a seminar, or published in the Journal of Cosmetic Science, the Official Journal of the Society of Cosmetic Chemists.
Contact: Theresa Cesario, Executive Director; (212) 668-1500; fax (212) 668-1504; TCesario@SCConline.org
Internet: http://www.scconline.org/website/about_scc/awards_fellowships.shtml
Sponsor: Society of Cosmetic Chemists
120 Wall Street, Suite 2400
New York, NY 10005-4088

Society of Cosmetic Chemists Award Sponsored by McIntyre Group, Ltd. 4341
Dedicated to the advancement of cosmetic science, the Society strives to increase and disseminate scientific information through meetings and publications. By promoting research in cosmetic science and industry, and by setting high ethical, professional and educational standards, it strives to improve the qualifications of cosmetic scientists Its mission is to further the interests and recognition of cosmetic scientists while maintaining the confidence of the public in the cosmetic and toiletries industry. This award, given in the form of a scroll and honorarium, are awarded for the best paper presented at the Society's Annual Scientific Meeting.
Contact: Theresa Cesario, Executive Director; (212) 668-1500; fax (212) 668-1504; TCesario@SCConline.org
Internet: http://www.scconline.org/website/about_scc/awards_fellowships.shtml
Sponsor: Society of Cosmetic Chemists
120 Wall Street, Suite 2400
New York, NY 10005-4088

Society of Cosmetic Chemists Award Sponsored by The HallStar Company 4342
Dedicated to the advancement of cosmetic science, the Society strives to increase and disseminate scientific information through meetings and publications. By promoting research in cosmetic science and industry, and by setting high ethical, professional and educational standards, it strives to improve the qualifications of cosmetic scientists. Its mission is to further the interests and recognition of cosmetic scientists while maintaining the confidence of the public in the cosmetic and toiletries industry. Offered in the form of a scroll and honorarium, this honor is awarded for the Best Paper that makes the greatest scientific contribution to knowledge in the field of protecting against or ameliorating damage to human skin caused by exposure to UV radiation presented at the Society's Annual Scientific Seminar.
Contact: Theresa Cesario, Executive Director; (212) 668-1500; fax (212) 668-1504; TCesario@SCConline.org
Internet: http://www.scconline.org/website/about_scc/awards_fellowships.shtml
Sponsor: Society of Cosmetic Chemists
120 Wall Street, Suite 2400
New York, NY 10005-4088

Society of Cosmetic Chemists Chapter Best Speaker Award 4343
Dedicated to the advancement of cosmetic science, the Society strives to increase and disseminate scientific information through meetings and publications. By promoting research in cosmetic science and industry, and by setting high ethical, professional and educational standards, it strives to improve the qualifications of cosmetic scientists. Its mission is to further the interests and recognition of cosmetic scientists while maintaining the confidence of the public in the cosmetic and toiletries industry. The Chapter Best Speaker Award is given for the best scientific paper presented before each Chapter during a given calendar year. Recommendations for this award originates with the Chapters and are submitted to the National Office A scroll and honorarium ($250 for a manuscript submission and $150 for an abstract submission) is provided by National for presentation by the Chapter.
Contact: Theresa Cesario, Executive Director; (212) 668-1500; fax (212) 668-1504; TCesario@SCConline.org
Internet: http://www.scconline.org/website/about_scc/awards_fellowships.shtml
Sponsor: Society of Cosmetic Chemists
120 Wall Street, Suite 2400
New York, NY 10005-4088

Society of Cosmetic Chemists Chapter Merit Award 4344
Dedicated to the advancement of cosmetic science, the Society strives to increase and disseminate scientific information through meetings and publications. By promoting research in cosmetic science and industry, and by setting high ethical, professional and educational standards, it strives to improve the qualifications of cosmetic scientists. Its mission is to further the interests and recognition of cosmetic scientists while maintaining the confidence of the public in the cosmetic and toiletries industry. The Chapter Merit Award is awarded to a Chapter Member for outstanding service, dedication and voluntary services over a period of time. A scroll is provided by National for presentation by the Chapter. Recommendations for this award originates with each Chapter and are submitted to the National Office.
Contact: Theresa Cesario, Executive Director; (212) 668-1500; fax (212) 668-1504; TCesario@SCConline.org
Internet: http://www.scconline.org/website/about_scc/awards_fellowships.shtml
Sponsor: Society of Cosmetic Chemists
120 Wall Street, Suite 2400
New York, NY 10005-4088

Society of Cosmetic Chemists Frontier of Science Award 4345
Dedicated to the advancement of cosmetic science, the Society strives to increase and disseminate scientific information through meetings and publications. By promoting research in cosmetic science and industry, and by setting high ethical, professional and educational standards, it strives to improve the qualifications of cosmetic scientists. Its mission is to further the interests and recognition of cosmetic scientists while maintaining the confidence of the public in the cosmetic and toiletries industry. The Frontier of Science Award is an honorarium given to a speaker who has achieved exceptional national or international stature in the scientific community for delivering a lecture at the Annual Scientific Meeting.
Contact: Theresa Cesario, Executive Director; (212) 668-1500; fax (212) 668-1504; TCesario@SCConline.org
Internet: http://www.scconline.org/website/about_scc/awards_fellowships.shtml
Sponsor: Society of Cosmetic Chemists
120 Wall Street, Suite 2400
New York, NY 10005-4088

Society of Cosmetic Chemists Hans Schaeffer Award 4346
Dedicated to the advancement of cosmetic science, the Society strives to increase and disseminate scientific information through meetings and publications. By promoting research in cosmetic science and industry, and by setting high ethical, professional and educational standards, it strives to improve the qualifications of cosmetic scientists. Its mission is to further the interests and recognition of cosmetic scientists while maintaining the confidence of the public in the cosmetic and toiletries industry. The Hans Schaeffer Award, offered in the form of a scroll and honorarium, are awarded annually for the Most Innovative Paper presented at either an annual meeting or a seminar.
Contact: Theresa Cesario, Executive Director; (212) 668-1500; fax (212) 668-1504; TCesario@SCConline.org
Internet: http://www.scconline.org/website/about_scc/awards_fellowships.shtml
Sponsor: Society of Cosmetic Chemists
120 Wall Street, Suite 2400
New York, NY 10005-4088

Society of Cosmetic Chemists Joseph P. Ciaudelli Award 4347
Dedicated to the advancement of cosmetic science, the Society strives to increase and disseminate scientific information through meetings and publications. By promoting research in cosmetic science and industry, and by setting high ethical, professional and educational standards, it strives to improve the qualifications of cosmetic scientists. Its mission is to further the interests and recognition of cosmetic scientists while maintaining the confidence of the public in the cosmetic and toiletries industry. The

Joseph P. Ciaudelli Award, offered in the form of a scroll and honorarium, are bestowed annually for the best article appearing in the Journal of Cosmetic Science on the subject of hair care technology.
Contact: Theresa Cesario, Executive Director; (212) 668-1500; fax (212) 668-1504; TCesario@SCConline.org
Internet: http://www.scconline.org/website/about_scc/awards_fellowships.shtml
Sponsor: Society of Cosmetic Chemists
120 Wall Street, Suite 2400
New York, NY 10005-4088

Society of Cosmetic Chemists Keynote Award Lecture Sponsored by Ruger Chemical Corporation **4348**
Dedicated to the advancement of cosmetic science, the Society strives to increase and disseminate scientific information through meetings and publications. By promoting research in cosmetic science and industry, and by setting high ethical, professional and educational standards, it strives to improve the qualifications of cosmetic scientists. Its mission is to further the interests and recognition of cosmetic scientists while maintaining the confidence of the public in the cosmetic and toiletries industry. The Keynote Award is an honorarium given to a speaker who has achieved exceptional national or international stature in the scientific community for delivering a lecture at the Annual Scientific Meeting.
Contact: Theresa Cesario, Executive Director; (212) 668-1500; fax (212) 668-1504; TCesario@SCConline.org
Internet: http://www.scconline.org/website/about_scc/awards_fellowships.shtml
Sponsor: Society of Cosmetic Chemists
120 Wall Street, Suite 2400
New York, NY 10005-4088

Society of Cosmetic Chemists Literature Award **4349**
Dedicated to the advancement of cosmetic science, the Society strives to increase and disseminate scientific information through meetings and publications. By promoting research in cosmetic science and industry, and by setting high ethical, professional and educational standards, it strives to improve the qualifications of cosmetic scientists. Its mission is to further the interests and recognition of cosmetic scientists while maintaining the confidence of the public in the cosmetic and toiletries industry. Through the Literature Award, a scroll and honorarium are presented annually to the author or authors of scientific papers in basic research judged to be an outstanding contribution to cosmetic science and technology.
Contact: Theresa Cesario, Executive Director; (212) 668-1500; fax (212) 668-1504; TCesario@SCConline.org
Internet: http://www.scconline.org/website/about_scc/awards_fellowships.shtml
Sponsor: Society of Cosmetic Chemists
120 Wall Street, Suite 2400
New York, NY 10005-4088

Society of Cosmetic Chemists Maison G. de Navarre Medal **4350**
Dedicated to the advancement of cosmetic science, the Society strives to increase and disseminate scientific information through meetings and publications. By promoting research in cosmetic science and industry, and by setting high ethical, professional and educational standards, it strives to improve the qualifications of cosmetic scientists. Its mission is to further the interests and recognition of cosmetic scientists while maintaining the confidence of the public in the cosmetic and toiletries industry. The Society's highest honor, the Maison G. de Navarre Medal, recognizes individuals for their technical contributions to cosmetic science. The award is presented annually.
Contact: Theresa Cesario, Executive Director; (212) 668-1500; fax (212) 668-1504; TCesario@SCConline.org
Internet: http://www.scconline.org/website/about_scc/awards_fellowships.shtml
Sponsor: Society of Cosmetic Chemists
120 Wall Street, Suite 2400
New York, NY 10005-4088

Society of Cosmetic Chemists Merit Award **4351**
Dedicated to the advancement of cosmetic science, the Society strives to increase and disseminate scientific information through meetings and publications. By promoting research in cosmetic science and industry, and by setting high ethical, professional and educational standards, it strives to improve the qualifications of cosmetic scientists. Its mission is to further the interests and recognition of cosmetic scientists while maintaining the confidence of the public in the cosmetic and toiletries industry. The Merit Award is presented annually for outstanding service and distinguished leadership in Society activities.
Contact: Theresa Cesario, Executive Director; (212) 668-1500; fax (212) 668-1504; TCesario@SCConline.org
Internet: http://www.scconline.org/website/about_scc/awards_fellowships.shtml
Sponsor: Society of Cosmetic Chemists
120 Wall Street, Suite 2400
New York, NY 10005-4088

Society of Cosmetic Chemists Robert A. Kramer Lifetime Service Award **4352**
Dedicated to the advancement of cosmetic science, the Society strives to increase and disseminate scientific information through meetings and publications. By promoting research in cosmetic science and industry, and by setting high ethical, professional and educational standards, it strives to improve the qualifications of cosmetic scientists. Its mission is to further the interests and recognition of cosmetic scientists while maintaining the confidence of the public in the cosmetic and toiletries industry. The Robert A. Kramer Lifetime Service Award, given at the discretion of the Board of Directors, goes to an individual who has given extraordinary service over the course of his or her membership.
Contact: Theresa Cesario, Executive Director; (212) 668-1500; fax (212) 668-1504; TCesario@SCConline.org
Internet: http://www.scconline.org/website/about_scc/awards_fellowships.shtml
Sponsor: Society of Cosmetic Chemists
120 Wall Street, Suite 2400
New York, NY 10005-4088

Society of Cosmetic Chemists Shaw Mudge Award **4353**
Dedicated to the advancement of cosmetic science, the Society strives to increase and disseminate scientific information through meetings and publications. By promoting research in cosmetic science and industry, and by setting high ethical, professional and educational standards, it strives to improve the qualifications of cosmetic scientists. Its mission is to further the interests and recognition of cosmetic scientists while maintaining the confidence of the public in the cosmetic and toiletries industry. The Shaw Mudge Award, given in the form of a scroll and honorarium, are awarded for the best paper presented at the Society's Annual Scientific Seminar.
Contact: Theresa Cesario, Executive Director; (212) 668-1500; fax (212) 668-1504; TCesario@SCConline.org
Internet: http://www.scconline.org/website/about_scc/awards_fellowships.shtml
Sponsor: Society of Cosmetic Chemists
120 Wall Street, Suite 2400
New York, NY 10005-4088

Society of Cosmetic Chemists Student Poster Awards **4354**
Dedicated to the advancement of cosmetic science, the Society strives to increase and disseminate scientific information through meetings and publications. By promoting research in cosmetic science and industry, and by setting high ethical, professional and educational standards, it strives to improve the qualifications of cosmetic scientists. Its mission is to further the interests and recognition of cosmetic scientists while maintaining the confidence of the public in the cosmetic and toiletries industry. The Student Poster Award, sponsored by DD-Chemco, Inc., is given annually in conjunction with the Annual Scientific Seminar. Honorariums are given for first, second, third and fourth place Best Student Poster.
Contact: Theresa Cesario, Executive Director; (212) 668-1500; fax (212) 668-1504; TCesario@SCConline.org
Internet: http://www.scconline.org/website/about_scc/awards_fellowships.shtml
Sponsor: Society of Cosmetic Chemists
120 Wall Street, Suite 2400
New York, NY 10005-4088

Society of Cosmetic Chemists Young Scientist Awards **4355**
The Young Scientist Awards are presented each year to recognize the outstanding achievements of new cosmetic scientists throughout the industry. Each year every manufacturer, supplier or other organization in the personal care industry may select an individual in their employ as a recipient of this award. The criteria for selection is made by the individual company. Information is sent in March of each year and is posted on the SCC Website. The submission deadline is July 1st each year. The awards may be presented at an SCC Chapter Meeting or through an internal corporate presentation. Recipient names are listed in the SCC Winter Newsletter.
Requirements: Recipients be an active member of the Society for at least one year and working in the industry less than 10 years.
Date(s) Application is Due: Jul 1
Contact: Theresa Cesario, Executive Director; (212) 668-1500; fax (212) 668-1504; TCesario@SCConline.org
Internet: http://www.scconline.org/website/about_scc/awards_fellowships.shtml
Sponsor: Society of Cosmetic Chemists
120 Wall Street, Suite 2400
New York, NY 10005-4088

Society of Medical Friends of Wine Research Award **4356**
One award is given biennially for published original research identifying substances in wine; ascertaining the effects of wine components on living cells, tissues, or organs; or indicating appropriate clinical applications of wine in treatment or prevention of disease. Applications are accepted at any time.
Amount of Grant: 2,000 U.S.D
Contact: Susan Guerguy, Executive Secretary; (925) 933-9691; fax (925) 939-5224; marq@inreach.com
Internet: http://www.medicalfriendsofwine.org
Sponsor: Society of Medical Friends of Wine
511 Jones Place
Walnut Creek, CA 94597

Society of Neuroscience Minority Fellowships **4357**
The goal of the program is to increase the diversity of the pool of individuals participating in mental-health related neuroscience research and teaching programs. The program recruits and trains outstanding individuals of traditionally underrepresented racial and ethnic minorities. Training stipends are available for predoctoral (up to three years) and postdoctoral (up to two years) fellowships. The program includes training stipends in accordance with NRSA standard guidelines; travel assistance and registration to

attend the annual meeting, in addition to SFN member resources; enrichment programs including funds to participate in activities outside of the fellow's home laboratory; and mentoring opportunities, with a mentor from among SFN memberships. Details and application are available online.
Date(s) Application is Due: Jan 1; Apr 1; Sep 1
Contact: Jessica Pearce, Program Coordinator; (202) 462-6688; fax (202) 462-9740; jessica@sfn.org
Internet: http://www.sfn.org/content/Programs/DiversityinNeuroscience/mnfp/index.html
Sponsor: Society for Neuroscience
11 Dupont Circle NW, Suite 500
Washington, DC 20036

Solectron Social Responsibility Grants 4358
The company awards grants to nonprofit organizations in its areas of interest, including education at all levels, health and human services, environment and ecology, civic and community, and culture and the arts. There are no application deadlines. Complete guidelines are available online.
Requirements: 501(c)3 nonprofit organizations and exclusively public institutions (e.g., local school districts, universities, etc.) and comparable foreign organizations are eligible.
Contact: Social Investment Manager; (408) 956-6041 or (408) 957-8500; wwcommunityrelations@ca.slr.com
Internet: http://www.solectron.com/about/social.htm
Sponsor: Solectron Corporation
847 Gibraltar Drive
Milpitas, CA 95035

Sol Goldman Charitable Trust Grants 4359
The trust supports the arts, education, the environment, health organizations, human services, and Jewish and Protestant agencies. Preference is given to requests from the New York metropolitan area. There are no application deadlines or forms.
Samples: Johns Hopkins Medicine (Baltimore, MD)—to support a collaborative research initiative focused on developing novel means of earlier diagnosis and treatment of pancreatic cancer, $10 million; Yale U (New Haven, CT)—to endow the deanship of its law school, $2.5 million.
Contact: Grants Administrator; (212) 265-2280; fax (212) 582-0186
Sponsor: Sol Goldman Charitable Trust
640 Fifth Avenue
New York, NY 10019

Solo Cup Foundation Grants 4360
The foundation awards grants to Illinois nonprofit organizations in its areas of interest, including higher education, health, social services, and Christian organizations and churches. Types of support include capital campaigns, general operating support, and scholarship funds. There are no application deadlines or forms.
Requirements: Illinois nonprofit organizations are eligible.
Geographic Focus: Illinois
Amount of Grant: 5,000 - 85,000 U.S.D
Samples: La Rashida Children's Hospital (Chicago, IL)—for a capital campaign to modernize and upgrade its inpatient facility, $1 million; Immaculate Conception Church (Chicago, IL)—for restoration project, $85,000; Congregation of the Passion (Chicago, IL)—for general support, $50,000.
Contact: Robert M. Korzenski, President and CEO; (847) 831-4800; fax (847) 579-3245; info@solocup.com
Sponsor: Solo Cup Foundation
1700 Old Deerfield Road
Highland Park, IL 60035

Sonoco Foundation Grants 4361
The Sonoco Products Company is the company sponsor of the Sonoco Foundation. The foundation was established in 1983, in the state of South Carolina. The foundation supports organizations involved with education, crime and violence prevention, human services, youth, and community development. Giving primarily in areas of company operations, with emphasis on Hartsville, South Carolina.
Requirements: South Carolina nonprofit organizations are eligible to apply for these grants. There is no application form required. Applicants should submit a detailed description of project and amount of funding requested.
Restrictions: The foundation does not support religious, political, or lobbying organizations. Grants aren't available to individuals, or for educational capital funds programs, endowments, trips, or tours.
Geographic Focus: South Carolina
Amount of Grant: 1,000 - 600,000 U.S.D
Samples: University of South Carolina, Columbia, SC, $600,000; Boy Scouts of America, Florence, SC, $60,000; United Way of Florence County, Florence, SC, $14,000;
Contact: Joyce Beasley, (800) 377-2692 or (843) 383-7851
Internet: http://www.sonoco.com
Sponsor: Sonoco Foundation
1 North Second Street, MS A09
Hartsville, SC 29550-3300

Sonora Area Foundation Grants 4362
The community foundation supports nonprofit organizations in Tuolumne County, CA, in its fields of interest, including arts and culture; elementary, higher, and adult/continuing education; environment and animal welfare; hospitals and health care; social services; legal services and crime/law enforcement; federated giving; and public administration. Grants are awarded for equipment acquisition, programs and projects, conferences and seminars, seed money, scholarship funds, and matching funds. The Board of Directors reviews Letters of Inquiry at their scheduled board meetings (usually the second Tuesday) in February, April, June, August, October, December. Letters of Inquiry are due at the Foundation office by mid-month preceding these board meetings. Applications are available on the Web site.
Requirements: Nonprofits serving Tuolumne County, CA, may submit applications.
Restrictions: Grants are not awarded to/for sectarian purposes, private foundations, political purposes, general operating expenses, annual or capital campaigns, endowment funds, or debt retirement.
Geographic Focus: California
Amount of Grant: Up to 200,000 U.S.D
Contact: Grants Program Manager; (209) 533-2596; fax (209) 533-2412; acorn@sonora-area.org
Internet: http://www.sonora-area.org/grant-apps.html
Sponsor: Sonora Area Foundation
P.O. Box 577
Sonora, CA 95370

Sony Corporation of America Corporate Philanthropy Grants 4363
Sony Corporation of America consists of three operating companies as well as the corporate headquarters, which is based in New York City: Sony Electronics, Inc. (headquarters in San Diego, California); Sony Pictures Entertainment, Inc. (headquarters in Culver City, California); and Sony Music Entertainment, Inc. (headquarters in New York City). Each company, as well as the overall corporation, has its own philanthropic priorities and resources (e.g. grants, product donations, and recordings and screenings) that benefit a multitude of causes. Taken together the corporation's areas of interest cover arts and culture; health and human services; civic and community outreach, education; the environment; disaster response; and volunteerism. The core of Sony's various corporate philanthropy programs are their contributions to the communities in which Sony employees work and live; however, the corporation and subsidiaries contribute to national nonprofits as well. In the past, types of support have included general operating budgets, continuing support, annual campaigns, seed grants, building construction, equipment acquisition, endowment funds, employee matching gifts, internships, and employee-related scholarships. Sony Corporation of America and its subsidiaries welcome requests for support throughout the year. There is no grant application form. Requests must be submitted in writing to the corporation or its operating companies. Contact information is given on this page and at the website. Guidelines for what to include in the application as well as more information on types of programs the corporation and/or its subsidiaries have supported are available at the grant website. Notification of grant-request approval or rejection will be made in writing within one month of receipt of all proposed materials.
Requirements: U.S. nonprofits, including schools and school districts, are eligible.
Restrictions: The corporation does not consider multi-year requests for support. The following types of organizations will not be funded: organizations that discriminate on the basis of race, color, creed, gender, religion, age, national origin, or sexual orientation; partisan political organizations, committees, or candidates and public office holders; religious organizations in support of their sacramental or theological functions; labor unions; endowment or capital campaigns of national origin; organizations whose prime purpose is to influence legislation; testimonial dinners in general; for-profit publications or organizations seeking advertisements or promotional support; individuals seeking self-advancement; foreign or non-U.S.-based organizations; and organizations whose mission is outside of the U.S.
Amount of Grant: 1,000 - 100,000 U.S.D
Contact: Janice Pober, Senior Vice President; (310) 244-7737
Karen E. Kelso, Vice-President; (212) 833-8000
Michael T. Williams, Executive Vice President and General Counsel; (858) 942-2400
Internet: http://www.sony.com/SCA/philanthropy/guidelines.shtml
Sponsor: Sony Corporation of America
550 Madison Avenue, 33rd Floor
New York, NY 10022-3211

Sophia Romero Trust Grants 4364
The Sophia Romero Trust was established in 1948 to support and promote quality human-services and health-care programming for underserved elders living in Bristol County, Massachusetts. Special consideration will be given to organizations that serve older women living in Bristol County, Massachusetts. Grant requests for general operating support or program support are strongly encouraged. The majority of grants from the Romero Trust are one year in duration. Applicants must apply online at the grant website. Applicants are strongly encouraged to do the following before applying: review the downloadable state application procedures for additional helpful information and clarifications; review the downloadable online-application guidelines at the grant website; review the trust's funding history (link is available from the grant website); review the online application questions in advance; and review the list of required attachments. These will generally include: a list of board members, financial statements (audited, reviewed, or compiled by independent auditor); an organization summary; a list of other funding sources; an IRS Determination letter; and other required documents. All attachments must be uploaded in the online application as PDF, Word, or Excel files. The application deadline for the Sophia Romero Trust is 11:59 p.m. on February 1. Applicants will be notified of grant decisions before April 30.
Requirements: Applicants must have 501(c)3 tax-exempt status and serve the residents of Bristol County, Massachusetts.

Restrictions: The fund does not support requests from individuals, organizations attempting to influence policy through direct lobbying, or any political campaigns.
Geographic Focus: Massachusetts
Date(s) Application is Due: Feb 1
Samples: Family Services Association of Greater Fall River, Fall River, Massachusetts, $20,000, continuing support (2011); Southeast Massachusetts Home Health Aides, $4,000, program development (2011); YWCA of Southeastern Massachusetts, New Bedford, Massachusetts, $4,000, Widowed Person Program (WPP) (2011).
Contact: Emma Greene, Director; (617) 434-0329; emma.m.greene@baml.com
Internet: https://www.bankofamerica.com/philanthropic/fn_search.action
Sponsor: Sophia Romero Trust
225 Franklin Street, 4th Floor, MA1-225-04-02
Boston, MA 02110

Sosland Foundation Grants 4365
The foundation awards grants to nonprofit organizations in the metropolitan bistate Kansas City area. Areas of interest include arts and culture, higher education, health care, Jewish services, and social services. Types of support include program development, building construction/renovation, equipment acquisition, development grants, endowments, general operating grants, and matching gifts. There are no application forms or deadlines. Applicants should send a letter including a description of the program and the organization, a list of board and staff, and proof of tax-exempt status.
Requirements: Metropolitan bistate Kansas City 501(c)3 nonprofit organizations are eligible.
Restrictions: Individuals and publications or conferences are ineligible.
Geographic Focus: Kansas, Missouri
Amount of Grant: 1,000 - 5,000 U.S.D
Samples: William Jewell College (Liberty, MO)—to renovate Jewell Hall, $50,000.
Contact: Program Contact; (816) 756-1000; fax (816) 756-0494
Internet: http://www.soslandfoundation.org
Sponsor: Sosland Foundation
4800 Main Street, Suite 100
Kansas City, MO 64112

Sound Partners for Community Health Grants 4366
The program supports partnerships of public broadcasters and community organizations that focus on healthy living, public health challenges, quality of care, rural health care access, and vulnerable populations. The program is designed to combine the power of public-broadcast messages and the needs and resources of local and community organizations. The grant requires a 25 percent match of direct or in-kind donations from local organizations. The listed deadline is for preliminary proposals. Up to 60 public radio stations and up to 15 public television stations will be invited to submit full proposals.
Requirements: All 501(c)3 public radio and public television stations that receive Community Service Grants from the Corporation for Public Broadcasting are eligible. In addition, stations that are members of the National Federation of Community Broadcasters may apply.
Date(s) Application is Due: Apr 30
Amount of Grant: 7,500 - 10,000 U.S.D
Contact: Mark Sachs, (301) 565-0805; fax (301) 565-0808; ma.sachs@verizon.net
Internet: http://www.soundpartners.org
Sponsor: Sound Partners for Community Health
8730 Georgia Avenue, Suite 408
Silver Spring, MD 20910

Southbury Community Trust Fund 4367
The Fund awards grants in the areas of arts, community and economic development, education, the environment, and health and human services. High priority is given to program development, program implementation, organizational capacity building, and policy change addressing root causes of problems. Interested applicants should complete a letter of inquiry. Information for submission is available online.
Requirements: Organizations with IRS 501(c)3 status serving Southbury are eligible to submit letters of inquiry.
Restrictions: Funding is not available for: religious purposes; political campaigns; annual appeals; building fund/capital, endowment or memorial campaigns; loans or deficit financing; federal, state, or municipal agencies or departments supported by taxation; individuals.
Geographic Focus: Connecticut
Date(s) Application is Due: Jan 15
Contact: Anne Watkins, Director of Programs; (203) 334-7511; fax (203) 333-4652; awatkins@gbafoundation.org
Internet: http://www.gbafoundation.org/grants/apply.asp
Sponsor: Greater Bridgeport Area Foundation
211 State Street, 3rd Floor
Bridgeport, CT 06604

South Madison Community Foundation Grants 4368
The South Madison Community Foundation (SMCF) Grants serves the residents in Adams, Fall Creek, Green, and Stony Creek Townships of Madison County. The Foundation has developed a procedure for submission and evaluation of grant proposals. The Grants Committee makes its recommendation to the Board of Directors who give final approval for funding. Applicants are notified within one month of the submission deadline of approval or denial of their request for funding. Online applications are available at the Foundation website. Organizations are encouraged to read the Tips for Submitting Effective Proposals on the Foundation website.
Requirements: With limited funds, the Foundation must distribute grant funding that gives the greatest positive impact on the needs and growth of South Madison county. Applicants should focus on projects that: focus on the prevention of problems rather than the symptoms, and make a significant improvement to the community; maintain a proactive focus and an ability to respond to creative ideas; respond to the changing needs of the community; encourage programs that enhance cooperation and collaboration among institutions within the community; and leverage funds through the use of match and challenge grants.
Restrictions: Note that Cycle One of the funding deadlines is funded with a smaller budget than the other funding cycle dates, so that applicants should apply for no more than $1,000. Because of time constraints, all proposals to the Foundation must be submitted in writing, so that no personal presentations are accepted. The following items will not be funded: operational expenses of existing programs; endowments or deficit funding; funds for redistribution by the grantee; conferences, publication, films, television and radio programs unless integral to the project for which the grant is sought; religious or political purposes; travel for individuals or groups such as bands, sports teams, and classes; annual appeals and membership contributions; and major capital improvements.
Geographic Focus: Indiana
Date(s) Application is Due: Feb 1; Jun 1; Oct 1
Amount of Grant: 150 - 10,000 U.S.D
Samples: Indiana State Police Emergency Response Section, purchase of a thermal imaging camera, allowing police to operate during low and no light situations while conducting surveillance and/or pursuing suspects, $1,500 match (2011); Lapel High School, Character Counts Education program, help guide students in their decision-making process in the development of moral values, $5,000 (2011); Lapel Optimist Club, help with cost of equipment for participants of Little League Football Program, $1,500 (2011).
Contact: Barbara Switzer, Program Director; (765) 778-8444; fax (765) 778-9144; barbara@southmadisonfoundation.org
Internet: http://www.southmadisonfoundation.org/grants.html
Sponsor: South Madison Community Foundation
233 South Main Street
Pendleton, IN 46064

Southwest Florida Community Foundation Competitive Grants 4369
Program/project grants are primarily for new or expanded programs and pilot or demonstration projects, which meet a documented need. Capital grants primarily support the construction, acquisition and renovation of facilities. The foundation offers competitive grants for programs and projects in eight focus areas: animal welfare, arts and culture, community development, conservation and preservation, education, the environment, health, and human services.
Requirements: IRS 501(c)3 tax-exempt organizations undertaking programs to improve the quality of life in Florida's Lee County and contiguous counties, including Charlotte, Collier, Gledes, and Hendry, are eligible. (NOTE: Agencies located in Collier County must also serve one of the other four counties in order to be eligible.) Applicants must employ staff, elect a governing board, and conduct business without discrimination on the basis of race, religion, gender, sexual orientation, age, national origin or disability.
Restrictions: The foundation generally will not fund: Normal operating expenses or existing obligations; Annual campaigns; Endowments; Sectarian purposes (i.e. programs that promote or require a religious doctrine except where designated by a donor); Debt retirement Monetary awards; Professional conferences, sports team travel, class trips, etc; Individuals; Fraternal organizations, societies, etc; Political organizations or campaigns; Fund-raising or feasibility studies; Research; or, Re-granting.
Geographic Focus: Florida
Date(s) Application is Due: Jun 19
Samples: Calusa Nature Center & Planetarium - $26,000; Fellowship of Christian Athletes - $20,000; Harry Chapin Food Bank - $20,000; Literacy Council of Bonita Springs - $15,000; Take Stock in Children - $13,615; Vineyards Elementary School - $23,000
Contact: Carol McLaughlin, Chief Program Officer; (239) 274-5900, ext. 225; fax (239) 274-5930; cmclaughlin@floridacommunity.com
Internet: http://www.floridacommunity.com/grantseekers/
Sponsor: Southwest Florida Community Foundation
8260 College Parkway, Suite 101
Fort Myers, FL 33919

Southwest Gas Corporation Foundation Grants 4370
The corporate foundation supports nonprofit organizations in its service communities. General support grants and employee matching gifts are made in the areas of education (universities, colleges, and literacy programs) and social services (United Way organizations, youth groups, community service, and volunteer organizations). Additional areas of interest are arts/culture and health. Types of support include general operating grants, projects grants, conferences and seminars, building construction/renovation, capital campaigns, emergency funds, employee matching gifts,research grants, donated equipment, and in-kind services. In southern Arizona, contact Marty Looney, P.O. Box 26500, Tucson, AZ 85726-6500; (520) 794-6416.
Requirements: Nonprofit organizations in Arizona; San Bernardino County, CA; and Nevada may apply.
Geographic Focus: Arizona, California, Nevada
Amount of Grant: Up to 620,567 U.S.D
Contact: Suzanne Farinas, (702) 876-7247; fax (702) 876-7037
Internet: http://www.swgas.com
Sponsor: Southwest Gas Corporation
P.O. Box 98510
Las Vegas, NV 89193-8510

Southwire Company Grants 4371

The company awards grants to nonprofits located in company-operating communities in Cleburne and Randolph Counties, AL; Mississippi County, AR; Mohave County, AZ; Carroll, Clarke and Oconee, GA; Daviess and Hancock Counties, KY; Oktibeeha County, MS; and Clay County, IL. Corporate community relations supports local community health and development, families in need, at-risk youth, disadvantaged youth, youth education, and religious development. The one-year grants vary in size. Requests are reviewed monthly. Interested applicants should apply at least 45 days prior to the date funding is needed.

Requirements: Requests must be in writing. In Georgia and Alabama: Southwire Company, 1 Southwire Drive, Carrollton, GA 30119; in Arkansas: Southwire Company, Arkansas Plant, P.O. Box 685, Osceola, AR 72370; in Kentucky: Southwire Company, Kentucky Plant, P.O. Box 500, Hawesville, KY 42348; in Illinois: Southwire Company, Flora Plant, Eash Road, Flora, IL 62839; in Mississippi: Southwire Company, Starkville Plant, 103 Airport Road, P.O. Box 967, Starkville, MS 39759; in Texas: Integral Corporation, P.O. Box 151369, Dallas, TX 75223; and in Utah: Southwire Company, Utah Plant, 3295 West 8600 S, West Jordon, UT 84088.

Restrictions: The company does not make contributions to individuals; political, labor, athletic, or social groups; advertising campaigns; or fundraising benefits.

Geographic Focus: Alabama, Arkansas, Georgia, Illinois, Kentucky, Mississippi, Texas, Utah

Contact: Community Relations Manager; (800) 444-1700 or (770) 832-4242; fax (770) 832-4929; gift@southwire.com

Internet: http://www.southwire.com

Sponsor: Southwire Company

1 Southwire Drive

Carrollton, GA 30119

Sovereign Bank Grants 4372

The foundation is dedicated to improving the quality of life in communities served by Sovereign Bank. Charitable giving is focused in four major areas: community investment and economic development, youth and education, human service programs that improve the social needs of low- and moderate-income communities and individuals, and arts and culture. Proposals in the Mid-Atlantic region should be sent to the listed office; New England-region proposals should be sent to Sovereign Bank, 75 State Street, Boston, MA 02109.

Requirements: 501(c)3 nonprofit organizations within Sovereign Bank's principal assessment areas in Pennsylvania, New Jersey, Connecticut, Massachusetts, New Hampshire, and Rhode Island are eligible.

Geographic Focus: Connecticut, Massachusetts, New Hampshire, New Jersey, Pennsylvania, Rhode Island

Date(s) Application is Due: Feb 18; May 18; Aug 19; Oct 14

Contact: Joseph Schupp, Community Development Officer; (610) 320-8504

Internet: http://www.sovereignbank.com/companyinfo/foundation.asp

Sponsor: Sovereign Bank

601 Penn Street

Reading, PA 19601

Soy Health Research Program Incentive Awards 4373

This program supports the development of complete research proposals on the topic of soy and bone health, soy and prostate health, and the dose-response relationship between isoflavone consumption and health and efficacy.

Date(s) Application is Due: Mar 15

Amount of Grant: 10,000 U.S.D

Contact: Steve Veile, Project Manager; (573) 635-3265 or (888) 772-8449; stevev@communiqueinc.com or soy@communiqueinc.com

Internet: http://www.talksoy.com/Shrp/default.htm

Sponsor: United SoyBean Board

512 E Capitol Avenue, P.O. Box 237

Jefferson City, MO 65102

Special Olympics Health Professions Student Grants 4374

The purpose of this program is to engage health professions students to work with persons with intellectual disabilities as a way of filling in a gap that exists in most health program curricula. The program promotes short-term projects exploring issues that impact the health and well-being of all persons with intellectual disabilities, including, but not limited to, Special Olympics athletes. Projects may include: data collection and analysis on issues impacting persons with intellectual disabilities; measurement of attitudes, opinions and behaviors of health professionals, coaches, family/caregivers and athletes; follow-up assessments of existing programs; or health promotion projects. Projects that involve collaborations with Special Olympics Programs or other CDC grant recipients (e.g., state and local health departments) are encouraged. Special Olympics is not currently prioritizing specific research areas for funding, as the program's primary goal is to engage health professions students to work with the population of persons with intellectual disabilities. All awards will be made to the institution on behalf of the student grantee. Funding must include all expenses and no indirect costs may be covered by this grant. Applications are accepted year-round. Application and guidelines are available online.

Requirements: Health professions students at the bachelor, master, doctorate and post-doctorate levels from accredited institutions or programs in the United States or other countries are eligible. Fields or disciplines may include, audiology, dentistry, pre-med, medicine, nursing, optometry, physical therapy, podiatry, public health, social work, law, public administration, or other related health areas. All students??? projects will require a faculty advisor.

Amount of Grant: Up to 5,000 U.S.D

Contact: Darcie Mersereau, Manager, Research and Evaluation; (202) 628-3600 or (800) 700-8585; fax (202) 824-0200; 2006hastudentgrant@specialolympics.org

Internet: http://www.specialolympics.org/content.aspx?id=6336&terms=student

Sponsor: Special Olympics

1133 19th Street, NW

Washington, DC 20036-3604

Special People in Need Grants 4375

The program awards grants to individuals and organizations operating in the fields of education, health, and assistance to people who are disabled, ill, or economically disadvantaged. There are no application deadlines or forms. Applicants should submit a cover letter, which will be reviewed by the grant committee.

Requirements: Nonprofit organizations are eligible.

Amount of Grant: 500 - 5,000 U.S.D

Contact: Gary Kline, (312) 715-5235

Sponsor: Special People in Need Foundation

500 W Madison Street, Suite 3700

Chicago, IL 60661

Special People in Need Scholarships 4376

This scholarship is for undergraduate students with disabilities pursuing a postsecondary education. Applicants must demonstrate a strong financial need and a potential to benefit from an education, which might not otherwise be possible. The institution the student is attending must be willing to serve as a sponsor for disbursement.

Requirements: There are no formal deadlines. To be considered send an application with transcripts, letter of acceptance, and agreement from the institution to act as sponsor, letter of recommendation from an educator, financial statements, biographical data, and other relevant information.

Contact: Gary Kline, (312) 715-5235

Sponsor: Special People in Need Foundation

500 W Madison Street, Suite 3700

Chicago, IL 60661

Spencer County Community Foundation Grants 4377

The Spencer County Community Foundation is a nonprofit, public charity created by and for the people of Spencer County, Indiana. The Foundation helps nonprofits fulfill their missions by strengthening their ability to meet community needs through grants that assist charitable programs, address community issues, support community agencies, launch community initiatives, and support leadership development. Grant proposals are accepted once each year according to the grant cycle. Grants are normally given as one time support of a project but may be considered for additional support for expansions or outgrowths of an initial project. Proposals will be accepted from January through the March 2 deadline. Grant awards will be presented in June.

Requirements: The Foundation welcomes proposals from nonprofit organizations that are deemed tax exempt under sections 501(c)3 and 509(a) of the Internal Revenue Code and from governmental agencies serving the county. Proposals from nonprofit organizations not classified as a 501(c)3 public charity may be considered provided the project is charitable and supports a community need. Proposals submitted by an entity under the auspices of another agency must include a written statement signed by the agency's board president on behalf of the board of directors agreeing to act as the entity's fiscal sponsor, to receive grant monies if awarded, and to oversee the proposed project.

Restrictions: Project areas not considered for funding: religious organizations for religious purposes; political parties or campaigns; endowment creation or debt reduction; operating costs; capital campaigns; annual appeals or membership contributions; travel requests for groups or individuals such as bands, sports teams, or classes. A six month progress report and a final report at project completion are required by organizations whose proposals are approved for funding. Instructions and appropriate forms will be provided at the time the grant is awarded.

Geographic Focus: Indiana

Date(s) Application is Due: Mar 2

Amount of Grant: 400 - 10,000 U.S.D

Samples: Evansville Philharmonic Orchestral Corporation - for Young People's Concerts and classroom materials that help the teachers make the most of the cultural experience with their students, $500; Friends of Lincoln State Park, Inc. - for a nature center Lincoln Mural, $1000; Carter Fire District - to install a Weather Alert System for the west side of Dale, $5,108;

Contact: Barbara Hinton; (812) 649-5724; bhinton@spencercommunityfoundation.org

Internet: http://www.spencercommunityfoundation.org/disc-grants

Sponsor: Spencer County Community Foundation

321 Main Street, Suite A, P.O. Box 3

Rockport, IN 47635

Spina Bifida Young Investigator Awards 4378

The association (SBA) and the Spina Bifida Foundation award fellowships to support spina bifida research by promising young investigators. The purpose of the award is to provide initial or continuing financial support to a fellow in an established research program. The institution at which the established research program is located will be the awardee and will be expected to administer the award on behalf of the fellow, who will be the principal investigator. The grant will begin at the awardee institution's discretion between August 1 and December 1, and will be for a one-year period, unless prior written approval for a no-cost extension is given by SBA. Application and guidelines are available online.

Requirements: The fellow must possess an MD, PhD, DSW, DScN, or equivalent terminal degree prior to beginning the research under the award, but may not have held

a faculty title of assistant professor or higher for more than 24 months at the time the fellowship is to begin.
Date(s) Application is Due: May 9
Amount of Grant: 25,000 U.S.D
Contact: Young Investigator Awards Coordinator; (800) 621-3141 or (202) 944-3285; jgiffels@sbaa.org
Internet: http://www.sbaa.org/site/PageServer?pagename=research_young_investigator_2005_RFP
Sponsor: Spina Bifida Association
4590 MacArthur Boulevard NW, Suite 250
Washington, DC 20007-4226

Sprague Foundation Grants **4379**
Grants are made to hospitals, arts and cultural organizations, civic and community affairs, educational institutions, and health and human service organizations and charities located in New York and Massachusetts. Types of support include general operating support, matching and challenge funds, and program development. The initial approach should be a letter requesting the foundation's guidelines. Grants are awarded at the June and December board meetings.
Requirements: Only residents of New York and Massachusetts may apply.
Restrictions: No grants are provided to individuals, for building funds, or for loans.
Geographic Focus: Massachusetts, New York
Date(s) Application is Due: Apr 15; Oct 1
Amount of Grant: 1,000 - 55,000 U.S.D
Contact: Linda Franciscovich, c/o U.S. Trust Company of New York, (212) 852-3377
Sponsor: Seth Sprague Educational and Charitable Foundation
114 W 47th Street
New York, NY 10036-1532

Springs Close Foundation Grants **4380**
Since it was chartered in 1942, the Springs Close Foundation has contributed over $85 million to a wide variety of charitable and educational causes designed to improve the quality of life and well-being of the people in Chester, Lancaster and York Counties. Support includes, but is not limited to, food, shelter and medical assistance. This temporary change in focus is in response to high levels of unemployment and economic distress in the Foundation's service areas. The Foundation also makes occasional statewide grants in South Carolina. There are three major areas of program interest: Recreation and Environment, Public Education and Early Childhood Development, Community Service and Health
Requirements: Grants are made only to organizations that are tax-exempt under Section 501(c)3 of the Internal Revenue Code. No grants are made to individuals.
Restrictions: The Foundation will only consider grant requests from eligible nonprofit organizations that can effectively deliver emergency and basic support to citizens in Chester and Lancaster counties and a portion of York County.
Geographic Focus: South Carolina
Date(s) Application is Due: Mar 1; Oct 1
Amount of Grant: 5,000 - 50,000 U.S.D
Contact: Angela McCrae, President; (803) 548-2002; fax (803) 548-1797
Internet: http://www.thespringsclosefoundation.org/grants.htm
Sponsor: Springs Close Foundation
1826 Second Baxter Crossing
Fort Mill, SC 29708

Square D Foundation Grants **4381**
The foundation makes donations for operating support, capital development needs, and special projects to nonprofit organizations in the areas of education, social welfare, arts and cultural and civic affairs, and health. Each year the foundation supports United Way in communities where the company and its domestic subsidiaries have significant operations. Donations are not normally made to organizations already receiving support through United Way. Support of higher education is achieved through scholarships, endowments for faculty and acquisition or expansion of equipment or facilities, unrestricted operating support, and the matching gift program. Submit letters of application between June and August.
Requirements: Giving primarily to 501(c)3 tax-exempt organizations in areas of company operations, with emphasis on: Illinois; Indiana; Iowa; Kentucky; Missouri; Nebraska; North Carolina; Ohio; South Carolina; Tennessee; and Wisconsin.
Restrictions: Grants are not made to religious organizations for religious purposes, political groups and organizations, labor unions and organizations, organizations making requests by telephone, organizations listed by the U.S. attorney general as subversive, or to individuals.
Geographic Focus: Illinois, Indiana, Iowa, Kentucky, Nebraska, North Carolina, Tennessee, Texas, Wisconsin
Date(s) Application is Due: Aug 31
Amount of Grant: Up to 27,000,000 U.S.D
Contact: Harry Wilson, Secretary; (847) 397-2600
Internet: http://www.squared.com
Sponsor: Square D Foundation
1415 South Roselle Road
Palatine, IL 60067

SRC Medical Research Grants **4382**
The Scientific Council for Medicine supports research throughout the discipline of Medical Science, i.e. Medicine, Odontology, Pharmaceutics and Health Sciences. Medical research supported by the Scientific Council ranges from studies at molecular and cell level, and research involving laboratory animal science and patients, to epidemiological studies on groups of human subjects. The Program offers grants for research and junior research positions; grants for research periods for clinical researchers; grants for research stays (sabbaticals) abroad; project research grants (distributed, as a rule, for three years for salary, materials, travel, publication, etc); international postdoctoral fellowships and grants for postdoctoral positions in Sweden; networks; symposium; research equipment; collaboration grant and grant for inviting a visiting scientist.
Contact: Paola Norlin; +46 (0)8 546 44 311; fax 46 8 546 44 180; paola.norlin@vr.se
Internet: http://www.vr.se/responsibilities/researchfunding/medicine.4.69f66a93108e85f68d48000837.html
Sponsor: Swedish Research Council
Klarabersgviadukten 82
Stockholm, SE-103 78 Sweden

SSHRC Image, Text, Sound and Technology (ITST) Grants **4383**
The program will provide funding for researchers and their graduate students to develop expertise in applying advanced technology through summer institutes, workshops and conferences. The ITST program will help researchers refine their ability to apply digital multimedia technologies to their research on image, text and sound. The overall objectives of the ITST program are to: reflect on, interpret, and analyze new digital media, multimedia, and text-based computing technologies and integrate these into humanities and social sciences research; bring together theorists, experimentalists, and technologists from different disciplines, to share and nurture ideas and methods that challenge research to advance through the use of audio-visual and text-based technologies; and facilitate the creation of national and international networks of, and partnerships among, researchers, industries, governments, and individuals that will promote and sustain social sciences and humanities research and resources worldwide. Up to $50,000 is available for twelve months.
Requirements: Canadian postsecondary institutions are eligible to apply. The summer institute, workshop or conference must be held at the applying institution. Detailed instructions accompany the application form.
Geographic Focus: All States, Canada
Date(s) Application is Due: Sep 15
Amount of Grant: Up to 50,000 U.S.D
Contact: Garry Pinard; (613) 992-5129; fax (613) 947-0223; garry.pinard@sshrc.ca
Internet: http://www.sshrc.ca/web/apply/program_descriptions/itst/workshops_e.asp
Sponsor: Social Sciences and Humanities Research Council of Canada
350 Albert Street, P.O. Box 1610
Ottawa, ON K1P 6G4 Canada

SSHRC Research/Creation Grants in Fine Arts **4384**
The broad purpose of the pilot program is to support and develop excellence in research in artistic disciplines. The program invites artist-researchers affiliated with Canadian postsecondary institutions, within a broad range of disciplines, to apply for research grants of up to three years duration to support research/creation. Grants are available for programs of research/creation of up to three years. The value of these grants is up to a maximum of $100,000 per year, but totalling not more than $250,000 in a three-year period. SSHRC welcomes both small- and large-scale projects.
Requirements: Only those who fulfill the definition of artist-researcher are eligible to apply to this program. Humanities and social science scholars affiliated with Canadian postsecondary institutions who engage in research/creation are also eligible to apply.
Geographic Focus: All States, Canada
Date(s) Application is Due: Oct 20
Amount of Grant: 100,000 - 250,000 U.S.D
Contact: Mathieu Ravignat, Senior Program Officer; (613) 947-3724; fax (613) 947-0223; mathieu.ravignat@sshrc.ca
Internet: http://www.sshrc.ca/web/apply/program_descriptions/fine_arts_e.asp
Sponsor: Social Sciences and Humanities Research Council of Canada
350 Albert Street, P.O. Box 1610
Ottawa, ON K1P 6G4 Canada

SSMRF Julia's Hope Pilot Grants **4385**
The foundation supports research toward a cure for Sanfilippo Syndrome Type A, an inherited neurodegenerative disorder for which there is no treatment or cure, and related neurodegenerative disorders. Investigators are encouraged to utilize MPS-IIIA animal models. The grants are designed to bring sponsored research to a level at which additional, ongoing support may be secured through larger grants from NIH or other sources. Applications are available online.
Date(s) Application is Due: Mar 1
Amount of Grant: 50,000 U.S.D
Contact: Program Contact; (402) 330-8636; mail@juliashope.org
Internet: http://www.juliashope.org/grants.htm
Sponsor: Sanfilippo Syndrome Medical Research Foundation
P.O. Box 571
Boys Town, NE 68010

SSRC Collaborative Research Grants on Environment and Health in China **4386**
The collaborative grants program was started in 2008 to stimulate new research on environment and health in China that is directly linked to policy and practice, to encourage collaboration across disciplines, and to help develop a network of scholars working in this field. Each year 7-8 grants of $10-15,000 are given to Chinese institutes for projects that: integrate environment or health perspectives into research currently focused on a single domain by sharing existing data or conducting limited additional data collection;

develop a new focus on environment and health within the context of ongoing research on development; broaden the scope of analysis through collaboration with partners who bring additional disciplinary perspectives or international experience; or work with NGO or government partners to make existing research more accessible and useful.
Requirements: Each project must address an environmental health problem that affects a substantial number of people and it must be clearly useful in generating or disseminating knowledge that will inform policy or the work of government or civil society actors. In order to ensure feasibility, applicants must demonstrate the necessary access to data, or relationships with local communities or government agencies, to conduct the project successfully. If the project is being carried out at the local level, researchers must either live and work in the same province, or otherwise demonstrate how the project will build local capacity to work on environment and health issues.
Amount of Grant: 10,000 - 15,000 U.S.D
Contact: Jennifer Holdaway, Program Director; (718) 517-3659 or (212) 377-2700; fax (212) 377-2727; holdaway@ssrc.org
Internet: http://www.ssrc.org/fellowships/grants-for-collaborative-research-on-environment-and-health-in-china/
Sponsor: Social Science Research Council
One Pierrepont Plaza, 15th Floor
Brooklyn, NY 11201

SSSS Student Research Grants 4387
The society awards three grants annually for graduate research in areas of human sexuality. The purpose of the research is to advance the knowledge about sexuality. An applicant must send one copy each of a completed application fact sheet (available on the Web site) and a letter from the department chairperson stating student status within the institution and the educational purpose of the research. Also, an applicant must send seven copies each of a 10-page description of the proposed research, a proposed budget for the project, and a 150-word abstract.
Requirements: Students enrolled in a graduate degree-granting program working toward their master's degree or doctorate are eligible to apply, although the research need not be for the thesis.
Date(s) Application is Due: Feb 1; Sep 1
Amount of Grant: 1,000 U.S.D
Contact: Program Contact; (610) 530-2483; fax (610) 530-2485; thesociety@inetmail.att.net
Internet: http://www.sexscience.org/awards/index.php?category_id=427
Sponsor: Society for the Scientific Study of Sexuality
P.O. Box 416
Allentown, PA 18105

St. Joseph Community Health Foundation Burn Care and Prevention Grants 4388
The St. Joseph Community Health Foundation, sponsored by the Poor Handmaids of Jesus Christ, exists to steward resources that strengthen, improve, and sustain long-term community health in Allen County, Indiana. The Foundation mission is to invest, and collaborate with other community agencies to improve the physical, mental, and spiritual health of the poor and under served. The Burn Care and Prevention Grants are restricted funds to promote burn prevention education and to help educate health care professionals and first responders to emergency situations to treat severe burns with best care practices available. Grants typically range in size from $500 to $5,000.
Requirements: Applicants should have a demonstrated history of serving poor and powerless populations residing in Allen County. Applicants should be not-for-profit entities classified as 501(c)3 by the Internal Revenue Service. Some government agencies involved in related service areas may be eligible for support. The Foundation will consider requests for program, program-related equipment, staff continuing education, technical assistance and small capital grants. Applications are available at the Foundation's website.
Restrictions: Grant applications will not be accepted elimination of deficits, support of political activities, individuals, or projects all ready completed.
Geographic Focus: Indiana
Date(s) Application is Due: Mar 1; Jun 1; Sep 1; Dec 1
Amount of Grant: 500 - 5,000 U.S.D
Samples: Allen County Juvenile Firesetters Task Force, $5,000 - Funding for a seminar pertaining to the identification and treatment of juvenile fire setters; Public Safety Foundation of Northeast Indiana, Inc., $50,000 - Partial funding of construction costs for a new classroom/office building in the Safety Village.
Contact: Meg Distler, (260) 969-2001, ext. 201; fax (260) 969-2004; mdistler@sjchf.org
Internet: http://www.stjosephhealthfdn.org/index.php?option=com_content&view=article&id=72&Itemid=71
Sponsor: St. Joseph Community Health Foundation
2826 South Calhoun Street
Fort Wayne, IN 46807

St. Joseph Community Health Foundation Improving Healthcare Access Grants 4389
The St. Joseph Community Health Foundation was established in 1998 when the Poor Handmaids of Jesus Christ sold the St. Joseph Medical Center and reorganized the St. Joseph Community Health Foundation with a significant share of the proceeds. Proceeds from the Foundation's endowment are redistributed as grants typically ranging from $5,000 to $35,000 to advance programs, projects and partnerships that improve the access to quality health care and the health of the poor and powerless of Allen County, Indiana. Typically, only programs that can demonstrate that greater than 51 percent of their clients are very low income with health issues are considered for these grants. The Foundation also considers these values of the Poor Handmaids as a part of its grant review process: respecting and valuing each person; standing with the poor and powerless; using our talents and resources to respond to the emerging needs of society; nurturing leadership in our efforts to bring peace to the world. The Foundation will consider requests for program support, operations, seed monies, program-related equipment, staff continuing education, technical assistance, and matching funds. The Foundation will commit support on an annual basis only. Additional funding will be contingent upon program performance.
Requirements: Applicants should have a demonstrated history of serving poor and powerless populations with medical, dental, mental, and/or spiritual health care and wellness services. Applicants should be not-for-profit entities classified as 501(c)3 by the Internal Revenue Service. Partnerships between not-for-profits are encouraged. The proposed grant must be operated for the benefit of residents of Allen County, Indiana.
Restrictions: Grant applications will not be accepted for building projects, elimination of deficits, support of political activities, individuals, or projects already completed.
Geographic Focus: Indiana
Date(s) Application is Due: Mar 1; Sep 1
Amount of Grant: 5,000 - 35,000 U.S.D
Samples: AIDS Task Force, Inc., $9,000 - support for emergency medical assistance to people living with HIV and AIDS in Allen County; Boys & Girls Clubs of Fort Wayne, $7,000 - support for the Cavity-Free Zone program promoting dental health and hygiene in low-income children; Cedars H.O.P.E., Inc., $5,000 - matching grant to support the program staff who house mentally ill women suffering from complex psychiatric and medical conditions.
Contact: Meg Distler, Executive Director; (260) 969-2001, ext. 201; fax (260) 969-2004; mdistler@sjchf.org
Internet: http://www.stjosephhealthfdn.org/index.php?option=com_content&view=article&id=79&Itemid=70
Sponsor: St. Joseph Community Health Foundation
2826 South Calhoun Street
Fort Wayne, IN 46807

St. Joseph Community Health Foundation Pfeiffer Fund Grants 4390
The St. Joseph Community Health Foundation, sponsored by the Poor Handmaids of Jesus Christ, exists to steward resources that strengthen, improve, and sustain long-term community health in Allen County, Indiana. The Foundation mission is to invest, and collaborate with other community agencies to improve the physical, mental, and spiritual health of the poor and under served. The Pfeiffer grants are restricted to health serving organizations for capital improvements or acquiring medical & surgical equipment. These funds have primarily been directed to help local clinics acquire medical equipment. All grants will be made to non-profit agencies, not individuals.
Requirements: Applicants should have a demonstrated history of serving poor and powerless populations residing in Allen County, Indiana. Applicants should be not-for-profit entities classified as 501(c)3 by the Internal Revenue Service. Some government agencies involved in related service areas may be eligible for support. Applications are available on the Foundation's website.
Restrictions: Grant applications will not be accepted for elimination of deficits, support of political activities, individuals, or projects all ready completed.
Geographic Focus: Indiana
Date(s) Application is Due: Mar 1; Sep 1
Contact: Meg Distler, Executive Director; (260) 969-2001, ext. 201; fax (260) 969-2004; mdistler@sjchf.org
Internet: http://www.stjosephhealthfdn.org/index.php?option=com_content&view=article&id=106&Itemid=129
Sponsor: St. Joseph Community Health Foundation
2826 South Calhoun Street
Fort Wayne, IN 46807

St. Joseph Community Health Foundation Schneider Fellowships 4391
The late Dr. Louis and Mrs. Anne Schneider understood and valued the importance of well-trained health and wellness care providers in the community. As a result, they established these fellowships to assist local not-for-profit agencies build the capacity of their agencies and staffs to provide quality care to the people of Allen County, Indiana. A preference will be to provide Fellowships that broaden the base of community health knowledge with new skills, services, health and wellness care abilities of health professionals in the community. Agencies may apply to use the Fellowships to either send staff or volunteers to continuing education opportunities, or to bring an expert to the community to provide specialized health and wellness care training for a group of staff, volunteers, and/or community members. The Foundation will commit support on an annual basis only.
Requirements: Applicants should have a demonstrated history of primarily serving poor and powerless populations residing in Allen County. Applicants should be not-for-profit entities classified as 501(c)3 by the Internal Revenue Service. The Foundation will consider requests for program, program-related equipment, staff continuing education, and technical assistance. Grant applications are available at the Foundation's website.
Restrictions: The Foundation will not fund anything destructive of human life. Grant applications will not be accepted for building projects, elimination of deficits, support of political activities, individuals, or projects all ready completed.
Geographic Focus: Indiana
Date(s) Application is Due: Mar 1; Jun 1; Sep 1; Dec 1
Amount of Grant: 2,000 U.S.D
Contact: Meg Distler; (260) 969-2001, ext. 201; fax (260) 969-2004; mdistler@sjchf.org
Internet: http://www.stjosephhealthfdn.org/index.php?option=com_content&view=article&id=105&Itemid=123
Sponsor: St. Joseph Community Health Foundation
2826 South Calhoun Street
Fort Wayne, IN 46807

Stackner Family Foundation Grants 4392

The foundation awards grants concentrated in the Milwaukee metropolitan area for building programs and equipment acquisition, medical research, programs at all levels of education, social welfare, child welfare, and programs that assist the physically disabled. Additional categories of support include general operating support, continuing support, annual campaigns, capital campaigns, and program development. The board meets each year in January, April, July, and October. Application forms are not required.
Requirements: Nonprofit organizations in the greater Milwaukee, WI, area may request grant support.
Geographic Focus: Wisconsin
Date(s) Application is Due: Mar 15; Jun 15; Sep 15; Dec 15
Amount of Grant: 25 - 25,000 U.S.D
Samples: Horizon House (Milwaukee, WI)—for an outpatient program, $4000;Penfield Childrens Ctr (Milwaukee, WI)—for the capital campaign, $10,000.
Contact: Paul Tillman, (414) 646-5409; Stackner@MSH.com
Sponsor: Stackner Family Foundation
411 E Wisconsin Avenue
Milwaukee, WI 53202-4497

Stackpole-Hall Foundation Grants 4393

The foundation awards grants in Pennsylvania in support of higher/secondary education, and literacy and vocational education and projects; social services, including youth and child welfare agencies; the arts and cultural programs; health services, including mental health and drug abuse; community development, including civic affairs and leadership development, and volunteerism; and environmental conservation and protection. Grants are awarded in support of annual and capital campaigns, building construction/renovation, equipment acquisition, program and project development, matching funds, and seed money. Application forms are not required.
Requirements: Nonprofits in Elk County, PA, may submit grant proposals.
Restrictions: Grants are not awarded to individuals or for scholarships, fellowships, operating budgets, loans, or endowment funds.
Geographic Focus: Pennsylvania
Amount of Grant: 200 - 50,000 U.S.D
Contact: William Conrad, Executive Secretary; (814) 834-1845; fax (814) 834-1869; stackpolehall@alltel.net
Sponsor: Stackpole-Hall Foundation
44 S Saint Mary's Street
Saint Mary's, PA 15857

Stan and Sandy Checketts Foundation 4394

Established by Stan and Sandy Checketts in 1998, the Foundation's primary focus is on human services and helping individuals defray medical expenses. Primary fields of interest include: children and youth programs, health care, housing and shelter programs, human services, and recreation. Types of support are general operating funds and grants to individuals. Applicants should submit a detailed description of the project, along with the amount of funding requested. There are no deadlines, and the primary geographic focus is Utah.
Geographic Focus: Utah
Contact: Stan Checketts, President; (435) 752-1987; fax (435) 752-1948
Sponsor: Stan and Sandy Checketts Foundation
350 West 2500 North
Logan, UT 84341-1734

Staney Foundation Treatment Trials for Schizophrenia and Bipolar Disorder 4395

The foundation is interested in supporting studies of compounds not supported by the pharmaceutical industry in the treatment of schizophrenia and bipolar disorder, including novel agents and studies of off-label use of drugs approved for other conditions. Grants are awarded for a maximum of three years, depending on design. Applicant may apply online.
Date(s) Application is Due: Mar 1; Oct 1
Amount of Grant: 300,000 - 900,000 U.S.D
Contact: Xiomara Herrera, Grants Administrator; (301) 571-0760, ext. 124; fax (301) 571-0769; herrerax@stanleyresearch.org
Internet: http://www.stanleyresearch.org/application
Sponsor: Stanley Foundation
5430 Grosvenor Lane, Suite 200
Bethesda, MD 20814

Stanley Foundation Schizophrenia and Bipolar Disorder Research Grants 4396

The program supports researchers at all levels of development who are already in the field of schizophrenia and bipolar disorder, as well as researchers in other fields of medicine and biology (e.g., neurology, immunology, biochemistry, and epidemiology) who wish to initiate new research projects in schizophrenia and bipolar disorder. Up to two years of research will be supported.
Date(s) Application is Due: Mar 1
Amount of Grant: 75,000 U.S.D
Contact: Rhoda Marte, (301) 571-0760, ext. 118; fax (301) 571-0769; marter@stanleyresearch.org
Internet: http://www.stanleyresearch.org/application
Sponsor: Stanley Foundation
5430 Grosvenor Lane, Suite 200
Bethesda, MD 20814

Stark Community Foundation Women's Fund Grants 4397

The Women's Fund is a permanent endowment to benefit future generations and increase vital funding for programs which advance the economic, educational, physical, emotional, social, artistic, and personal growth of women and to educate and inspire women to become leaders in philanthropy. The interest area of this fund includes nonprofit agencies and their programs or projects that meet the needs of women and children in the Stark County area.
Requirements: The Women's Fund would like to fund grassroots nonprofit organizations, churches, and educational components that have unique projects serving women and/or women and children, and their basic needs for daily living.
Geographic Focus: Ohio
Date(s) Application is Due: Mar 5
Amount of Grant: 500 - 3,000 U.S.D
Contact: Cindy Lazor, Vice President for Program; (330) 454-3426; fax (330) 454-5855; cmlazor@starkcf.org
Internet: http://www.starkcf.org/the_womens_fund.asp
Sponsor: Stark Community Foundation
400 Market Avenue N, Suite 200
Canton, OH 44702

Starke County Community Foundation Grants 4398

The Starke Community Foundation offers grants to schools and teachers with the maximum amount of $500 per school. The following school systems are eligible to apply: Knox, North Judson-San Pierre and Oregon-Davis. Areas of interest include: community development; education; health and human services; youth; environment and recreation; and arts and culture. Applicants may fill out the online application, but are advised to contact the program coordinator to discuss their project before applying.
Requirements: The Foundation favors activities that: reach a broad segment of the community, especially those citizens whose needs are not being met by existing services that are normally expected to be provided by private rather than government sources; request seed money to realize innovative opportunities to meet needs in the community; stimulate and encourage additional funding; promote cooperation and avoid duplication of effort; help make a charitable organization more effective and efficient and better able to be self-sustaining; and one time projects or needs.
Restrictions: The Foundation will not consider grants for: religious organizations for the sole purpose of furthering that religion (this prohibition does not apply to funds created by donors who have specifically designated religious organizations as beneficiaries of the funds); political activities or those designated to influence legislation; national organizations (unless the monies are to be used solely to benefit citizens of Starke County); grants that directly benefit the donor or the donor's family; fundraising projects; and contributions to endowments.
Geographic Focus: Indiana
Date(s) Application is Due: Sep 30
Amount of Grant: 500 U.S.D
Contact: Grants Coordinator; (574) 223-2227 or (877) 432-6423
Internet: http://www.nicf.org/starke/grants.html
Sponsor: Starke County Community Foundation
1512 South Heaton Street
Knox, IN 46534

Starke County Community Foundation Lilly Endowment Community Scholarships 4399

The Lilly Endowment Scholarship provides funding for full time, four-year tuition for a field of study leading to a baccalaureate degree at any Indiana college or university. The scholarship also includes an $800 per year book stipend.
Requirements: Applicants must be U.S. citizens and residents of Starke County. Candidates can obtain the application from their high school guidance office. All college majors are eligible.
Geographic Focus: Indiana
Contact: Lilly Endowment Scholarship Contact; (574) 223-2227 or (877) 432-6423; fax (574) 224-3709
Internet: http://www.nicf.org/starke/availablescholarships.html
Sponsor: Starke County Community Foundation
1512 South Heaton Street
Knox, IN 46534

Starke County Community Foundation Scholarships 4400

Starke County offers a wide range of academic scholarships for students of various majors, with specific applications and deadlines. Applicants should refer to the website for application and deadline information or call the scholarship coordinator.
Geographic Focus: Indiana
Contact: Corinne Becknell Lucas, Scholarship Coordinator; (574) 223-2227 or (877) 432-6423
Internet: http://www.nicf.org/starke/scholarships.html
Sponsor: Starke County Community Foundation
1512 South Heaton Street
Knox, IN 46534

Starr Foundation Grants 4401

The foundation awards grants to nonprofit organizations in the areas of education, medicine and health care, human needs, public policy, and culture. Education grants support scholarships for deserving students, groups that offer need-based financial aid, and international-exchange programs. Human needs grants support food programs for the poor, job training, literacy, adequate housing, and programs for the disabled.

Medicine and health care grants include capital grants to hospitals, significant research grants, and grants to assist in the provision of health care to underserved communities. Public policy grants support international relations and democratic institutions around the world. Culture grants go to exchange programs and provide services to the elderly and/or disabled. The foundation rarely funds local charities outside of New York City but may fund national organizations that serve communities outside of New York. Types of support include multiyear grants, general operating grants, and capacity-building grants. There are no application deadlines. The board meets throughout the year.
Amount of Grant: Up to 70,000,000 U.S.D
Samples: World Trade Ctr Memorial Foundation (New York, NY)—to build the World Trade Center Memorial and a museum to remember the September 11, 2001, and February 26, 1993, terrorist attacks, $25 million; Johns Hopkins U (Baltimore, MD)—to endow a professorship and support other needs at the South Asia Studies program of the university's Nitze School of Advanced International Studies, $3 million; Dickinson College (Carlisle, PA)—to augment a scholarship endowment, $1 million; For relief efforts in South Asia and Africa—$2.5 million distributed.
Contact: Florence Davis, President; (212) 770-6882; florence.davis@starrfdn.org or grants@starrfoundation.org
Internet: http://fdncenter.org/grantmaker/starr
Sponsor: Starr Foundation
70 Pine Street, 14th Floor
New York, NY 10270

Staunton Farm Foundation Grants 4402
The Staunton Farm Foundation is a family foundation established in 1937 in accordance with the wishes of Matilda Staunton Craig, who wanted her estate to be used to benefit people with mental illness. The Foundation awards grants in the field of mental health in southwestern Pennsylvania. Projects that represent new and different approaches for organizations and ultimately affect patient care are encouraged. Support may be for more than one year, but the project must become self-sustaining following the grant period. Applicants should submit a letter of intent; full proposals are by invitation.
Requirements: Nonprofit organizations in the 10-county area in southwestern Pennsylvania including Washington, Greene, Fayette, Westmoreland, Armstrong, Butler, Lawrence, Beaver, Indiana, and Allegheny are eligible.
Geographic Focus: Pennsylvania
Date(s) Application is Due: Jun 1; Dec 1
Contact: Joni S. Schwager, Executive Director; (412) 281-8020; fax (412) 232-3115; jschwager@stauntonfarm.org
Internet: http://www.stauntonfarm.org
Sponsor: Staunton Farm Foundation
650 Smithfield Street, Suite 210
Pittsburgh, PA 15222

Steelcase Foundation Grants 4403
The foundation make grants to nonprofit organizations, projects, and programs in corporate communities. Grants are awarded in the areas of human service, health, education, community development, the arts, and the environment. Preference is given to programs designed to improve the quality life for disadvantaged, disabled, young, and elderly people. Matching gifts are made to educational, arts, and environment programs supported by Steelcase employees, retirees, and directors. To obtain a grant application, send a letter on letterhead, signed by the organization's chief executive officer. Include the following items in the letter: description of the organization or project; expected results of the project; amount of grant funds requested; and a copy of 501(c)3 nonprofit certification. Guidelines are available online.
Requirements: The foundation makes grants to IRS-certified nonprofit organizations in areas where Steelcase manufacturing plants are located (Grand Rapids, MI; Fletcher, NC; City of Industry, CA; Athens, AL; Toronto, Canada; and Tijuana, Mexico.)
Restrictions: The foundation does not provide grants to individuals or to organizations that discriminate on the basis of race, religion, sex, disability, or national origin; have received a Steelcase foundation grant within the past 12 months; request support for a conference or seminar; or request support for religious programs (nonsectarian programs for humanitarian purposes are eligible for consideration.)
Geographic Focus: Alabama, California, Michigan, Canada, Mexico
Amount of Grant: 2,000 - 673,000 U.S.D
Contact: Executive Director; (616) 475-2009; fax (616) 475-2200; sbroman@steelcase.com
Internet: http://www.steelcase.com/na/ourcompany.aspx?f=10042&c=10022
Sponsor: Steelcase Foundation
P.O. Box 1967, CH4E
Grand Rapids, MI 49501

Steele-Reese Foundation Grants 4404
The foundation's available income is divided equally for grants to operating charities in southern Appalachia (particularly Kentucky) and in the Northwest (particularly Idaho). In both areas the funds are devoted to education, health, welfare, and the humanities. Types of support include general operating support, equipment, endowment funds, matching funds, professorships, scholarship funds, and capital campaigns. The foundation gives preference to projects that have, among others, the following characteristics: rural, modest in ambition, narrow in function, unglamorous, based on experience, enjoying community financial support, and essential rather than merely desirable. While the foundation considers southern Kentucky and Idaho to be its territories of primary concern, it makes grants to organizations operating throughout southern Appalachia and in Oregon, Montana, and Wyoming.
Requirements: Applicant should submit a letter requesting guidelines. Only residents of Georgia, Idaho, Kentucky, Montana, North Carolina, Texas, and Wyoming are eligible to apply. Personal and telephone inquiries are not encouraged.
Restrictions: The foundation does not make grants to individuals, to community chest or similar drives, for conferences or workshops, for efforts to influence elections or legislation, for planning purposes or experimental projects, for emergencies, or for permanent support except for occasional endowment grants to organizations where stability is critically important.
Geographic Focus: Georgia, Idaho, Kentucky, Montana, North Carolina, Texas, Wyoming
Date(s) Application is Due: Mar 1
Amount of Grant: 10,000 - 50,000 U.S.D
Samples: Frontier Nursing Service, Wendover, KY, $40,000 - to purchase a new phone system for the Mary Breckenridge Hospital; Mountain Home Montana, Missoula, MT, $28,000 - salary support for the director of a teen-mother residency program; West Central Highlands Resource Conservation and Development Council, Emmett, ID, $10,000 - for upgrades and renovations of the Indian Valley Community Hall.
Contact: Charles U. Buice, (212) 505-2696; charlesbuice@hotmail.com
Internet: http://www.steele-reese.org/what.html
Sponsor: Steele-Reese Foundation
32 Washington Square West
New York, NY 10011

Stefan Batory Foundation Grants 4405
The foundation supports social, scientific, cultural, and educational activities that encourage the development of the market economy and democracy in Poland. It operates in the fields of the arts and humanities, culture, conservation and the environment, economic and international affairs, education, law and human rights, medicine and health, science and technology, and social welfare. Types of support include grants, scholarships, and fellowships.
Requirements: Only residents of Poland may apply.
Geographic Focus: All States, Poland
Contact: Anna Rozicka, Executive Director; 48-22-5360200; fax 48-22-5360220; batory@batory.org.pl
Internet: http://www.batory.org.pl/english
Sponsor: Stefan Batory Foundation
10a Sapiezynska Street
Warsaw, 00-215 Poland

Stella and Charles Guttman Foundation Grants 4406
The majority of grants are made to organizations providing services to people in the New York City metropolitan area. Contributions also are made to organizations performing scientific research or research relating to the delivery of social services without geographical limitations. The foundation supports programs that address the educational and social service needs of at-risk children and youth, including arts and cultural programs, drug prevention programs, family planning services, health and mental health services, intergenerational programs, programs for the physically disabled, and youth employment programs. The foundation also supports programs that promote the general welfare of the elderly and enhance their quality of life. Charities that conduct activities in Israel, including programs that foster improved relations between Jews and Arabs, also receive support. The foundation maintains a special grant category through which it can respond to special projects and address critical needs outside its principal program areas. Types of support include general operating support, continuing support, program development, and matching funds.
Requirements: Charitable 501(c)3 or 170(b)1 organizations are eligible with a strong emphasis on New York City and Israel.
Restrictions: The foundation does not make grants directly to individuals or to organizations not qualified as charitable, for foreign travel or study, to initiate or defend public interest litigation, to support anti-vivisectionist causes, or to religious organizations for religious observances.
Geographic Focus: New York
Amount of Grant: 5,000 - 30,000 U.S.D
Samples: Educational Alliance (New York, NY)—for general support, $25,000; Creative Ctr for Women with Cancer (New York, NY)—for general support, $25,000; Givat Haviva Educational Foundation (New York, NY)—for the Children Teaching Children program and Ibn Rouchd School in Jat, Israel, $16,000.
Contact: Elizabeth Olofson, Executive Director; (212) 371-7082; fax (212) 371-8936; info@guttmanfoundation.org
Internet: http://www.guttmanfoundation.org
Sponsor: Stella and Charles Guttman Foundation
122 E 42nd Street, Suite 2010
New York, NY 10168

Stephen S. Chang Award 4407
This award is given annually to recognize a scientist, technologist, or engineer who has made distinguished accomplishments in basic research, and the knowledge the person thus produced must have been utilized by industries for the improvement or development of products related to lipids. A prospective recipient must agree to be present for acceptance of the award and to deliver an award address at an AOCS annual meeting.
Requirements: Eligibility for nomination shall be demonstrated through high-quality scientific papers, patents, or the marketing of commercial products developed through the nominee's scientific or technological efforts. The award shall be made without regard for national origin, place of residence, race, color, creed, or gender.

Date(s) Application is Due: Oct 15
Amount of Grant: 6,000 U.S.D
Contact: Program Contact; (217) 359-2344; fax (217) 351-8091; membership@aocs.org
Internet: http://www.aocs.org/member/awards/awd-chang.asp
Sponsor: American Oil Chemists' Society
P.O. Box 3489
Champaign, IL 61826-3489

Sterling-Turner Charitable Foundation Grants **4408**
The charitable foundation awards grants to Texas organizations, primarily for higher and secondary education, adult basic education and literacy, social services, youth, the elderly, fine and performing arts groups and other cultural programs, church support and religious programs (Catholic, Jewish, and Protestant), hospitals and health services, hospices, research, conservation, and civic and urban affairs. Grants are awarded for general operating support, annual campaigns, capital campaigns, continuing support, building construction/renovation, equipment acquisition, endowment funds, program and project development, conferences and seminars, curriculum development, fellowships, scholarship funds, research, and matching funds. The board meets in April.
Requirements: Nonprofit Texas organizations in Tom Green, Kerr, Fort Bend, Harris and Travis County, may submit written requests for grants.
Restrictions: Individuals are ineligible.
Geographic Focus: Texas
Date(s) Application is Due: Mar 1
Amount of Grant: 5,000 - 25,000 U.S.D
Samples: Kinkaid School (Houston, TX)—for program support, $10,000; Houston Baptist U (Houston, TX)—for program support, $25,000; Museum of Fine Arts of Houston (Houston, TX)—for operating support, $166,667.
Contact: Patricia Moser, Executive Assistant; (713) 237-1117; fax (713) 223-4638; patricia@sterlingturnerfoundation.org
Internet: http://sterlingturnerfoundation.org
Sponsor: Sterling-Turner Charitable Foundation
815 Walker Street, Suite 1543
Houston, TX 77002-5724

Steuben County Community Foundation Grants **4409**
The Steuben County Community Foundation works to preserve and enhance the lifestyle and assets of Steuben County, Indiana, for current and future generations by providing ongoing assessment and financial support of identified needs through philanthropic giving and endowment building. The Foundation grants funding for arts and culture, education, health, environment, human services, recreation, and community development. Grants are not limited to these areas, and grant seekers are encouraged to respond to emerging community needs.
Requirements: Prospective applicants are strongly encouraged to discuss their grant requests with Community Foundation staff before beginning the application process. Applications can be obtained on the Foundation's website. The Steuben County Community Foundation encourages projects that: address priority community concerns; encourage more effective use of community resources; test or demonstrate new approaches and techniques in the solution of community problems; are intended to strengthens the management capabilities of agencies, and; promote volunteer participation and citizen involvement in community affairs.
Restrictions: If approved, each grant recipient must sign an agreement that includes the following obligations: public acknowledgment of the Foundation's support; expenditure of the monies as specified in the proposal; return of any unused portion of the grant; completion of an evaluation report; any special conditions as mutually agreed (failure to do so can adversely affect any subsequent requests).
Geographic Focus: Indiana
Date(s) Application is Due: Feb 28; Apr 29; Jun 30; Aug 31; Oct 31; Dec 30
Contact: Grant Contact; (260) 665-6656; fax (260) 665-8420
Internet: http://www.steubenfoundation.org/grants/index.html
Sponsor: Steuben County Community Foundation
1701 N Wayne Street
Angola, IN 46703

Steuben County Community Foundation Lilly Endowment Community Scholarships **4410**
The Lilly Endowment Community Scholarships help raise the level of education in Indiana. The scholarships cover tuition for a Steuben County resident for four years of full time study at an Indiana college or university, and includes an annual $800 stipend for books and equipment. All college majors are eligible.
Requirements: Candidates must be U.S. citizens and Steuben County residents. They must also be a high school senior graduating in the year of application and accepted to a full-time baccalaureate course of study at an Indiana public or private college or university. Applications are reviewed on the following criteria: leadership and community service; after school activities and work experience; essay; scholastic profile; and potential to succeed. Contact the Foundation office or high school guidance counselor for current application.
Geographic Focus: Indiana
Contact: Lilly Scholarship Contact; (260) 665-6656; fax (260) 665-8420
Internet: http://www.steubenfoundation.org/scholarships/availablescholarships.html
Sponsor: Steuben County Community Foundation
1701 N Wayne Street
Angola, IN 46703

Steuben County Community Foundation Scholarships **4411**
The Steuben County Community Foundation works to preserve and enhance the lifestyle and assets of Steuben County, Indiana, for current and future generations by providing ongoing assessment and financial support of identified needs through philanthropic giving and endowment building. The Foundation administers separate scholarship funds established by individuals, families, corporations, and organizations to assist local students in pursuit of advanced education. Together, these scholarships make up the Community Scholarship Program. Eligibility, criteria, and supplemental requirements vary for each scholarship. All scholarship applications are available at the local high school guidance departments, and at the Steuben County Community Foundation office. With few exceptions, most applications are due to the Steuben County Community Foundation Office by March 15th of the current year.
Geographic Focus: Indiana
Date(s) Application is Due: Mar 15
Contact: Bill Stockberger, Grant Administrator; (260) 665-6656; fax (260) 665-8420; bstockberger@steubenfoundation.org
Internet: http://www.steubenfoundation.org/scholarships/availablescholarships.html
Sponsor: Steuben County Community Foundation
1701 N Wayne Street
Angola, IN 46703

Stewart Huston Charitable Trust Grants **4412**
The purpose of the Trust is to provide funds, technical assistance and collaboration on behalf of non-profit organizations engaged exclusively in religious, charitable or educational work; to extend opportunities to deserving needy persons. Giving primarily in the Savannah, GA, area and Coatesville, PA.
Requirements: 501(c)3 nonprofit organizations are eligible.
Restrictions: Grants are not awarded for scholarship support to individuals, endowment purposes, purchases of tickets or advertising for benefit purposes, coverage of continuing operating deficits, or document publication costs. Support is not provided to intermediate or pass-through organizations (other than United Way) that in turn allocate funds to beneficiaries or to fraternal organizations, political parties or candidates, veterans, labor or local civic groups, volunteer fire companies, or groups engaged in influencing legislation.
Geographic Focus: Georgia, Pennsylvania
Date(s) Application is Due: Jan 15; Mar 1; Sep 1
Amount of Grant: 1,000 - 15,000 U.S.D
Samples: Graystone Society, Coatesville, PA, $325,000 - for building support; Grace United Methodist Church, Savannah, GA, $20,000 - for sponsorships; University of Pennsylvania Health System, Philadelphia, PA, $20,000 - for general support;
Contact: Scott Huston, Program Director; (610) 384-2666; fax (610) 384-3396; admin@stewarthuston.org
Internet: http://www.stewarthuston.org
Sponsor: Stewart Huston Charitable Trust
50 South First Avenue
Coatesville, PA 19320

Stocker Foundation Grants **4413**
The foundation supports creative initiatives that can be catalysts for constructive change. Preference is given to programs that have the promise of solution, with emphasis on education, the arts, women's issues, youth leadership, the disadvantaged, the physically challenged, and community needs. Specific areas of interest are arts education; community/grassroots partnerships; leadership; primary and secondary education enrichment programs; health care for the underserved; community involvement; and social services. Grants are awarded for general support, building funds, curriculum development, publications, equipment, endowments, projects, challenge/matching, and seed money. This award is considered three times throughout the year.
Requirements: Funds are made to organizations in Lorain County, Ohio; Pima County, Arizona; San Francisco County, California; Bernalillo and Dona Ana Counties, New Mexico; and King County, Washington.
Restrictions: Grants do not support annual campaigns, conferences, deficit financing, government services, public school services required by law, or research projects.
Geographic Focus: Arizona, California, New Mexico, Ohio, Washington
Date(s) Application is Due: May 15; Aug 15; Dec 15
Amount of Grant: 2,000 - 15,000 U.S.D
Samples: Douglas Ctr for Performing Arts (Douglas, AZ)—to support the youth theater program, $3500.
Contact: Patricia O'Brien, Executive Director; (440) 246-5719; fax (440) 246-5720; pobrien@stockerfoundation.org
Internet: http://www.stockerfoundation.org/fund.html
Sponsor: Stocker Foundation
401 Broadway Avenue, Suite 101
Lorain, OH 44052

Strake Foundation Grants **4414**
The purpose of the foundation's grant program is to fund primarily Roman Catholic-affiliated associations, including hospitals and higher and secondary educational institutions. Additional areas of interest include adult basic education and literacy, museums, and arts and culture. Support is provided for operating budgets, continuing support, annual campaigns, special projects, matching funds, scholarship funds, building construction/renovation, and research.
Requirements: Awards are made to organizations located only in the United States, primarily in Texas.

Restrictions: Awards are not made to support elementary schools or individuals, nor for deficit financing, consulting services, technical assistance, publications, or loans.
Geographic Focus: Texas
Date(s) Application is Due: Apr 1; Oct 1
Amount of Grant: 1,000 - 20,000 U.S.D
Samples: College of Saint Thomas More (Fort Worth, TX)—to support courses and materials related to the college's new bachelor of arts degree, $10,000.
Contact: George Strake Jr., President; (713) 216-2400; fax (713) 216-2401; foundation@strake.org
Sponsor: Strake Foundation
712 Main Street, Suite 3300
Houston, TX 77002

Stranahan Foundation Grants **4415**
Giving is largely for higher education but there is also support given for community foundations and community funds, youth, social service, health agencies, and cultural organizations, including a fine arts museum and performing arts groups.
Requirements: Nonprofit organizations in Ohio are eligible to apply.
Restrictions: Grants are not made to individuals.
Geographic Focus: Ohio
Date(s) Application is Due: Mar 1; Jul 1; Nov 1
Amount of Grant: 1,000 - 300,000 U.S.D
Contact: Pam Roberts, (419) 882-5575; proberts@stranahanfoundation.org
Internet: http://www.stranahanfoundation.org
Sponsor: Stranahan Foundation
4159 Holland-Sylvania Road, Suite 206
Toledo, OH 43623-2590

Streilein Foundation for Ocular Immunology Visiting Professorships **4416**
The SFOI Visiting Professorship is intended to stimulate the spread of knowledge concerning the immunological causes and approaches for managing ocular diseases. The SFOI Visiting Professorship should provide a clear scientific and educational benefit for the host institution. The Professorship will support a short visit of a senior scientist or clinician with a track record of ocular immunology research to a host institution. An important feature of the Visiting Professorship is to encourage and mentor young investigators. The visiting professor shall contribute one full day of teaching and interacting with students, residents, postdoctoral fellows, and faculty. The host institution will receive a $5,000 award to pay for coach airfare, and/or standard ground transportation, hotel (up to two nights), meals, honorarium, and expenses associated with this visit (e.g., refreshments, advertising). Transportation by personal car will be reimbursed at $0.51/mile. Meal allowance will be set by the host institution. Also included is an honorarium of $1,500 paid at the time expense reimbursement is submitted.
Requirements: Any institution of higher education or research institute is eligible to apply. The application for a SFOI Visiting Professorship must be submitted by a departmental Chairman of an educational institution, the Director of a research institute, or an official of similar stature from the host organization. The application must be received by the SFOI by July 1st of the calendar year in which the professorship is intended.
Date(s) Application is Due: Jul 1
Amount of Grant: 6,500 U.S.D
Contact: Joan Stein-Streilein, Ph.D., Treasurer; (617) 912-7494; fax (617) 912-0105; jstein@vision.eri.harvard.edu
Internet: http://www.streilein-foundation.org/fellowships.html
Sponsor: Streilein Foundation for Ocular Immunology
P.O. Box 6104
Boston, MA 02114

Streisand Foundation Grants **4417**
The foundation awards grants for national programs affecting women and at-risk youth. Areas of support include programs for children and youth, with a focus on the economically disadvantaged; civil rights, with emphasis on race relations between African Americans and Jews; AIDS research, advocacy, service, and litigation; nuclear disarmament; and environmental preservation. Types of support include general operating grants, technical assistance, and project-specific support. Applicants are asked to submit one- to two-page letters of inquiry along with a copy of the 501(c)3 determination letter. Applications are accepted between September 1 and the listed deadline. Guidelines are available online.
Requirements: 501(c)3 nonprofits and California projects in the Los Angeles area are eligible.
Restrictions: Grants are not made to individuals nor do they support start-up organizations, endowments, or capital campaigns.
Date(s) Application is Due: Dec 1
Amount of Grant: 10,000 - 20,000 U.S.D
Samples: Crossroads Community Foundation (Santa Monica, CA)—to support free quality arts and enrichment programming for children at Walgrove Elementary School in Venice, CA, $5000; Women Against Gun Violence (Culver City, CA)—to galvanize support for the reduction and prevention of gun violence, $5000; League of Conservation Voters Education Fund (Washington, DC)—to help grassroots environmental organizations motivate their members to be involved in the decision-making process, $20,000.
Contact: Program Contact; (310) 535-3767; fax (310) 314-8396; stfnd@aol.com
Internet: http://www.barbrastreisand.com/bio_streisand_foundation.html
Sponsor: Streisand Foundation
2800 28th Street, Suite 105
Santa Monica, CA 90405

Stroke Association Allied Health Professional Research Bursaries **4418**
The Research Bursary Awards are primarily intended for the purpose of training nurses and therapists, but consideration will be given to other health professionals. These bursaries will be awarded to UK departments that can demonstrate a track record and current participation in stroke research. The host department will be required to provide appropriate supervision and a research training program suitable for equipping the candidate for a career in stroke research, and to enable them to obtain a post graduate qualification. Bursaries may be taken up on a part-time, pro-rata basis.
Requirements: Research must take place in the United Kingdom.
Geographic Focus: All States, United Kingdom
Amount of Grant: 30,000 - 60,000 GBP
Samples: Trudi Pelton, University of Birmingham; Holly Robson, University of Manchester; Linda Hammett, St. George's, University of London;Andrea Atzori, University of Southampton; Emma Jones, University of Sheffield.
Contact: Research Department; (020) 7566-0345 or (020) 7490-2686; fax (020) 7251-9096; research@stroke.org.uk
Internet: http://www.stroke.org.uk/research/research_we_fund/allied_health_professional_bursaries/index.html
Sponsor: Stroke Association
Stroke House, 240 City Road
London, EC1V 2PR England

Stroke Association Clinical Fellowships **4419**
Fellowships are awarded to individuals for training in stroke medicine, and are intended to enable a specialist registrar to gain the appropriate clinical experience and accreditation required for a career in the prevention, treatment or rehabilitation of stroke.
Requirements: Applicants are required to nominate a supervisor and training program in a host institution that has been approved for stroke medicine training by the Stroke Medicine Sub-specialty Advisory Committee (SSAC). The funding is not to be taken as part of a one year taught course or for those who have completed the SpR training program and have certification.
Geographic Focus: All States, United Kingdom
Date(s) Application is Due: Jan 6
Amount of Grant: 40,000 GBP
Contact: Research Department; (020) 7566-0345 or (020) 7490-2686; fax (020) 7251-9096; research@stroke.org.uk
Internet: http://www.stroke.org.uk/research/research_we_fund/fellowships/index.html
Sponsor: Stroke Association
Stroke House, 240 City Road
London, EC1V 2PR England

Stroke Association Research Project Grants **4420**
Research project grants are awarded to doctors, professors, research fellows, therapists, psychologists and nurses at universities and hospitals across the UK. Project grants can last for up to three years. The Stroke Association's priorities for funding include research into the prevention, treatment, rehabilitation and long-term care of stroke patients. The majority of funded projects are patient orientated. The Stroke Association has recently undertaken a review of its activities and has developed a new Research and Development Strategy. As a result, applications for Research Project Grants have been revised and new application forms have been developed. All applications for funding by The Stroke Association will now undergo review by a panel of service users in addition to expert peer review. Therefore all applicants will now be required to complete a Project Grant Lay Summary Form, as well as the main Research Project Grant Application Form. Details of the new procedures and forms can be found on online.
Requirements: Research must take place in the United Kingdom.
Restrictions: Applications that are not submitted on the new forms will not be accepted for this application round.
Geographic Focus: All States, United Kingdom
Date(s) Application is Due: Feb 5
Amount of Grant: 70,000 - 210,000 GBP
Contact: Research Department; (020) 7566-0345 or (020) 7490-2686; fax (020) 7251-9096; research@stroke.org.uk
Internet: http://www.stroke.org.uk/research/apply_for_funding/index.html
Sponsor: Stroke Association
Stroke House, 240 City Road
London, EC1V 2PR England

Strowd Roses Grants **4421**
Strowd Roses, Inc. is a private charitable foundation which was established in 2001 under the will of Mrs. Irene Harrison Strowd of Chapel Hill, North Carolina. The Board of Directors of Strowd Roses, Inc. makes grants to qualified tax-exempt charitable, educational, religious and public organizations that are based in Chapel Hill or Carrboro, or are devoted primarily to benefiting the citizens of those communities. Grants may also be made to individuals engaged in projects designed to enhance the Chapel Hill/Carrboro community. Grants may, at the Board's discretion, include support for operating as well as capital expenditures, seed money and matching grants. Grants to individuals may be used for scholarships or fellowships; to produce a report or similar product; or to improve or enhance a literary, artistic, musical, scientific, teaching or similar capacity, skill or talent which will be used to benefit the life of the community. Particular consideration is given to those projects and purposes which further the interests of Mr. and Mrs. Strowd, including the welfare of children and youth, the enhancement of the environment, and the promotion of a sense of civic duty.

Requirements: Applicants should submit the appropriate application form (see foundations website), along with a proposed budget and evidence of their tax-exempt status (IRS determination letter or comparable documentation) to: Board of Directors, Strowd Roses, Inc., P.O. Box 3558, Chapel Hill, NC, 27515-3558.
Geographic Focus: North Carolina
Date(s) Application is Due: Jan 31; Apr 30; Jul 31; Oct 31
Amount of Grant: 10,000 U.S.D
Contact: Jennifer Boger, Grant Coordinator; (919) 929-1984; fax (919) 929-1990; jboger@strowdroses.org
Internet: http://www.strowdroses.org/grantApp.htm
Sponsor: Strowd Roses
P.O. Box 3558
Chapel Hill, NC 27515-3558

STTI/ATI Educational Assessment Nursing Research Grants **4422**
The purpose of this grant, co-sponsored by Assessment Technologies Institute (ATI), with the Honor Society of Nursing, Sigma Theta Tau International will support research that encourages the appropriate use of a standardized assessment program in a nursing education curriculum, including, but not limited to: use of formative assessment to drive remediation; use of summative assessment to guide curricular evaluation and revision; timing of assessment administration within a program of study; identification of key markers of at-risk student status.
Requirements: Eligibility Criteria: registered nurse with current license; have a master's or doctoral degree or be enrolled in a doctoral program; submit a completed research application package and signed research agreement via the online submission system; ready to implement research project when funding is received; submit a final report; submit completed abstract to the Virginia Henderson Research Library and credit research grant partners in all publications and presentations of the research.
Date(s) Application is Due: Jul 1
Amount of Grant: 4,500 U.S.D
Contact: Research Services; (888) 634-7575 or (317) 634-8171; fax (317) 634-8188; research@stti.iupui.edu
Internet: http://www.nursingsociety.org/Research/SmallGrants/Pages/STTIATI.aspx
Sponsor: Sigma Theta Tau International
550 W. North Street
Indianapolis, IN 46202-2156

STTI Emergency Nurses Association Foundation Grant **4423**
The ENA Foundation and Sigma Theta Tau International, Inc. (STTI) have combined resources to offer this annual research grant. The ENA Foundation/STTI research grant provides funding for research that will advance the specialized practice of emergency nursing. All relevant research topics will be considered. Priority will be given to research studies relating to the ENA/ENA Foundation research initiatives.
Requirements: Eligibility Criteria: principal investigator must be a registered nurse; team members may be from other disciplines; principal investigator must have a master's degree; submit a completed research application package and signed research agreement; be ready to start, or have already started, the research project. An electronic Research Grant Application is available on the ENA Foundation website. All applications MU.S.T be returned to ENA Foundation.
Restrictions: Application packets submitted to Sigma Theta Tau International will not be reviewed.
Date(s) Application is Due: Mar 1
Amount of Grant: 6,000 U.S.D
Contact: Department of Research; (847) 460-4119; fax (847) 460-4005; res@ena.org
Internet: http://www.nursingsociety.org/Research/SmallGrants/Pages/grant_ena.aspx
Sponsor: Emergency Nurses Association Foundation
915 Lee Street
Des Plaines, IL 60016 -6569

STTI Environment of Elder Care Nursing Research Grants **4424**
The Environment of Elder Care Nursing Research Grant co-sponsored by Hill-Rom Company will advance the science of nursing through research focused on critical aspects of elder care including clear lungs, no falls, safe skin, patient comfort and ease-of-use. Applications from novice researchers who have received no other national research funds are encouraged. Preference will be given to Sigma Theta Tau International members, other qualifications being equal. Allocation of funds is based on the quality of the proposed research, the future promise of the applicant, and the applicant's research
Requirements: Eligibility Criteria: registered nurse with a current license; hold at minimum a Master's degree or its equivalent, and/or be enrolled in a Doctoral program; submission of completed research application packet and a signed research agreement via our on-line submission system; ready to implement research project when funding is received; submit to STTI a final report; submit completed abstract to STTI's Virginia Henderson International Nursing Research Library; and submit an abstract for publication to Hill-Rom regarding the research done as a product of the grant, and credit research grant partners in all publications and presentations of the research.
Date(s) Application is Due: Jul 1
Amount of Grant: Up to 9,000 U.S.D
Contact: Research Services; (888) 634-7575 or (317) 634-8171; fax (317) 634-8188; research@stti.iupui.edu
Internet: http://www.nursingsociety.org/Research/SmallGrants/Pages/Environmentof Care.aspx
Sponsor: Sigma Theta Tau International
550 W. North Street
Indianapolis, IN 46202-2156

STTI Joan K. Stout RN Research Grants **4425**
The purpose of the Sigma Theta Tau International/Joan K. Stout, RN, Research Grant is to advance ongoing evidence-based study by nurse researchers on the impact of the practice of simulation education in schools of nursing and clinical care settings. The allocation of funds is based upon a research project that is ready for implementation. The proposed research project should be designed to ensure the ongoing practice of nurse-led simulation in improving quality of care in clinical and/or academic settings with the potential for further funding and ongoing research. Funds provided by interest from a research endowment established at Sigma Theta Tau International Foundation for Nursing, with a donation by the Hugoton Foundation.
Requirements: Eligibility Criteria: registered nurse with current license; have a master's degree, and/or be enrolled in a Doctoral program; submit a completed research application via online submission system; ready to implement research project when funding is received; complete project within one (1) year of funding; submit to STTI a final report; submit completed abstract to STTI's Virginia Henderson International Nursing Research Library.
Date(s) Application is Due: Jul 1
Amount of Grant: 5,000 U.S.D
Contact: Research Services; (888) 634-7575 or (317) 634-8171; research@stti.iupui.edu
Internet: http://www.nursingsociety.org/Research/SmallGrants/Pages/JoanKStout.aspx
Sponsor: Sigma Theta Tau International
550 W. North Street
Indianapolis, IN 46202-2156

STTI National League for Nursing Grants **4426**
Funds for this grant are made possible from a gift from Dr. Diane Billings to Sigma Theta Tau International Foundation for Nursing's research endowment. The purpose of this grant is to support research that advances the science of nursing education and learning through the use of technology in dissemination of knowledge.
Requirements: Eligibility Criteria: preference given to Sigma Theta Tau International and/or National League for Nursing members; grant proposals must advance the science of nursing education through the use of technology in the dissemination of knowledge; registered nurse with current license; hold a master's or doctoral degree or be enrolled in a doctoral program; submit a completed research application package and signed research agreement; ready to implement research project when funding is received. All applications must be submitted via the online submission system (see STTI website)
Date(s) Application is Due: Jun 1
Amount of Grant: 5,000 U.S.D
Contact: Research Services; (888) 634-7575 or (317) 634-8171; fax (317) 634-8188; research@stti.iupui.edu
Internet: http://www.nursingsociety.org/Research/SmallGrants/Pages/grant_nln.aspx
Sponsor: Sigma Theta Tau International
550 W. North Street
Indianapolis, IN 46202-2156

STTI Rosemary Berkel Crisp Research Awards **4427**
The purpose of the STTI Rosemary Berkel Crisp Research Awards is to support nursing research in the critical areas of women's health, oncology, and pediatrics. The allocation of funds is based upon a research project that is ready for implementation in the area of women's health, oncology, or pediatrics; the quality of the proposed research, future potential of the application, appropriateness of the research budget, and feasibility of time frame. Funds provided by interest from a gift from The Harry L. Crisp, II and Rosemary Berkel Crisp Foundation to the honor society's Research Endowment.
Requirements: Eligibility Criteria: registered nurse with current license; hold a master's or doctoral degree or be enrolled in a doctoral program; submit a completed research application package and signed research agreement via the online submission system; ready to implement research project when funding is received; member of Sigma Theta Tau International; Note: Some preference will be given to applicants residing in Illinois, Missouri, Arkansas, Kentucky, and Tennessee.
Date(s) Application is Due: Dec 1
Amount of Grant: Up to 5,000 U.S.D
Contact: Research Services; (888) 634-7575 or (317) 634-8171; fax (317) 634-8188; research@stti.iupui.edu
Internet: http://www.nursingsociety.org/Research/SmallGrants/Pages/grant_crisp.aspx
Sponsor: Sigma Theta Tau International
550 W. North Street
Indianapolis, IN 46202-2156

STTI Virginia Henderson Clinical Research Grants **4428**
The purpose of the Sigma Theta Tau International Virginia Henderson Clinical Research Grant is to encourage the research career development of clinically based nurses through support of clinically oriented research. Allocation of funds is based on a research project ready for implementation, the quality of the proposed research, the future potential of the applicant, appropriateness of the research budget and feasibility of the time frame. Funds for this biennial grant are provided by interest from the Virginia Henderson Clinical Research Endowment Fund.
Requirements: Eligibility Criteria: registered nurse actively involved in some aspect of health care delivery, education or research in a clinical setting; hold a master's or doctoral degree or be enrolled in a doctoral program; submit a completed research application package and signed research agreement via the online submission system; ready to implement research project when funding is received; member of Sigma Theta Tau International.
Date(s) Application is Due: Dec 1

Amount of Grant: Up to 5,000 U.S.D
Contact: Research Services; (888) 634-7575 or (317) 634-8171; fax (317) 634-8188; research@stti.iupui.edu
Internet: http://www.nursingsociety.org/Research/SmallGrants/Pages/grant_VHL.aspx
Sponsor: Sigma Theta Tau International
550 W. North Street
Indianapolis, IN 46202-2156

Subaru of Indiana Automotive Foundation Grants **4429**
The foundation is committed to making gifts to qualifying organizations, institutions, or entities within Indiana that will improve the quality of life and help meet the needs of the residents of the state. The foundation awards cash grants that are used to support the funding of specific capital projects in the areas of: arts and culture, education, and health and welfare. Grants must be used for investments in facilities, equipment, or real estate made by qualifying organizations. Application and guidelines are available online.
Requirements: Applying organizations must be 501(c)3 tax-exempt, with a chapter or office in Indiana; an education institution located in Indiana; or an Indiana governmental or quasi-governmental entity.
Restrictions: Support will not be provided for operating costs, routine expenses, or deficit reduction; endowments or memorials; fundraising events, conferences, meals, or travel; or annual fund drives. Support will not be provided to individuals; organizations located outside of Indiana or organizations that are not tax-exempt; for-profit businesses; organizations whose primary purpose is to influence legislation or support political candidates; religious institutions for religious purposes or fraternal organizations; or organizations that discriminate in the provision of services on the basis of race, sex, color, national origin, disability, age, religious affiliation, or any other unlawful basis.
Geographic Focus: Indiana
Date(s) Application is Due: Mar 31; Sep 30
Amount of Grant: 1,000 - 10,000 U.S.D
Samples: Christian Nursing Service Clinics (Montgomery County, IN)—funding for a photocopier with fax/scanner, television with video/DVD player, and office furniture to improve health services for the organization's free walk-in clinic, $1400; Town of Otterbein (Benton County, IN)—funding to replace Otterbein's drinking and waste water monitoring test equipment to benefit the more than 1300 citizens of Otterbein, $6480; North Newton After School Child Care (Newton County, IN)—funding for everyday equipment, supplies, toys, Lego's, games, jump ropes and balls for the after-school child care programs in three elementary schools in the North Newton School Corp, $1500.
Contact: Grants Administrator; (765) 449-6565; fax (765) 449-6952
Internet: http://www.siafoundation.org
Sponsor: Subaru of Indiana Automotive
P.O. Box 6479
Lafayette, IN 47903

Sulzberger Foundation Grants **4430**
The foundation awards grants to eligible nonprofit organizations in its areas of interest, including arts and culture, education, environmental protection and natural resource conservation, hospitals, and social services delivery. Types of support include annual campaigns, building construction/renovation, endowments, general operating support, multiyear grants, programs and projects, and scholarship funds. There are no application forms or deadlines. Preference is given to requests from New York City and Chatanooga.
Restrictions: Individuals are not eligible. Loans and matching grants are not awarded.
Amount of Grant: 100 - 255,000 U.S.D
Samples: New York Times Company Foundation (New York, NY)—for the Neediest Cases Fund, $100,000.
Contact: Marian Heiskell, President; (212) 556-1400
Sponsor: Sulzberger Foundation
229 W 43rd Street, Suite 1031
New York, NY 10036

Summit Foundation Grants **4431**
The community foundation awards grants to Summit County, CO, nonprofit organizations. Grants are awarded to support programs and projects in arts and culture, health and human services, early childhood and elementary education, the environment, sports, and scholarships to individuals. Types of support include continuing support, annual and capital campaigns, building/renovation, equipment acquisition, land acquisition, program and project development, conferences and seminars, seed money, curriculum development, scholarships, exchange programs, and matching funds. Application deadlines are in the spring and fall.
Requirements: Summit County, CO, nonprofit organizations may submit applications.
Restrictions: Grants are not awarded for support of religious organizations, political campaigns, individuals (except for designated scholarship programs), or operating budgets.
Geographic Focus: Colorado
Contact: Debra Edwards, Executive Director; (970) 453-5970; fax (970) 453-1423; sumfound@colorado.net
Internet: http://www.summitfoundation.org/gRequest.php
Sponsor: Summit Foundation
P.O. Box 4000
Breckenridge, CO 80424

Sunderland Foundation Grants **4432**
The foundation makes grants annually to publicly supported charitable organizations in its areas of interest, including higher education; churches; youth serving agencies; health facilities; community buildings; museums; civic projects; and and low maintenance, energy efficient housing projects sponsored by qualified tax-exempt organizations. The Foundation generally awards grants to larger, well-established nonprofit organizations. The Sunderland Foundation primarily makes grants in the geographic areas that have connections to the Ash Grove Cement Company. Grants are awarded in Western Missouri, Kansas, Nebraska, Arkansas, and Western Iowa, and to a lesser extent, in Idaho, Oregon, Washington, Utah and Montana.
Requirements: All grant requests should be submitted in writing, and proposals should include the following: a clear description of the project for which funds are being requested, including program goals and objectives, documentation of need and expected outcomes; a brief background on the proposing organization or agency; a detailed expense budget for the project indicating how the funds would be spent and over what time period; an income statement showing other sources of project support, public and/or private, which have been or will be solicited, including a statement of funds that have been received or pledged to date; a financial plan showing how the project will be supported beyond the grant period; the organization's current board of directors and their terms of office; a copy of the organization's most recent 501(c)3 tax-exempt ruling from the IRS; the organization's most recent certified audit or audited financial statement, where applicable. Proposals should be submitted by mail to the address listed in the "Sponsoring Organization" catagory. Proposals will also be accepted by email. Send a PDF file to the following address: sunderlandfoundation@ashgrove.com.
Restrictions: The Foundation does not award grants: for annual operating expenses; for programs or endowments; provide sponsorship for special events; to primary and secondary schools are generally not considered; to individuals. No scholarships are available from the Foundation.
Geographic Focus: Arkansas, Idaho, Iowa, Kansas, Missouri, Montana, Nebraska, Oregon, Utah, Washington
Samples: Pleasant Valley Baptist Church, Liberty, MO, $50,000 (2006%2Proxy-Connection: keep-alive Cache-Control: max-age=0); Salvation Army of Greater Kansas City, Kansas City, MO, $50,000; MD Anderson Cancer Center, Houston, TX, $100,000; Donnelly College, Kansas City, KS, $25,000;
Contact: Kent Sunderland; (913) 451-8900; sunderlandfoundation@ashgrove.com
Internet: http://www.sunderlandfoundation.org
Sponsor: Sunderland Foundation
P.O. Box 25900
Overland Park, KS 66225

Sunflower Foundation Bridge Grants **4433**
Bridge Grant funding is intended to provide core transitional financial support for new or expanded medical, mental, dental and/or public health services that provide primary care or primary prevention services to the uninsured and the underinsured. Funds are available for: salaries, benefits, allowable indirect expenses appropriate to the health care practitioners needed for the program; administrative services (e.g. financial & medical record clerks, clerical staff) resulting from the new or expanded services for which funding is being requested from the Foundation. Applicants must demonstrate a cost share in the project through a cash match of $1 to $1 of the total amount requested. Grants will not exceed $200,000 each and may be for projects up to three years in length.
Requirements: Proposals are invited from private nonprofit or public organizations that have a mission to improve health status or access to quality, affordable health care for Kansans.
Restrictions: The foundation will not fund ongoing general operating expenses or existing deficits; capital, endowment or specific fund-raising campaigns; fund raising events; routine continuing education; travel to conferences not directly related to the project; programs that are not Kansas-based; programs that require additional staff but demonstrate no clear means of sustainability after foundation funding; individual medical care or support; medical equipment; capital equipment (except for allowable technology); political purposes; or support of organizations that practice discrimination. Additional exclusions include: requests for new construction; and facility renovations. Any Safety Net clinic that qualifies for the Dental Hub initiative should contact the Sunflower Foundation program staff prior to application (there may be some restrictions for Dental Bridge Grant funding for organizations eligible for Dental Hub funding).
Geographic Focus: Kansas
Date(s) Application is Due: Oct 23
Amount of Grant: Up to 200,000 U.S.D
Contact: Cheryl Bean, Grants Manager; (785) 232-3000, ext. 104; cbean@sunflowerfoundation.org
Internet: http://www.sunflowerfoundation.org/areas_of_interest-health_care_access.php
Sponsor: Sunflower Foundation: Health Care for Kansans
1200 SW Executive Drive, Suite 100
Topeka, KS 66615-3850

Sunflower Foundation Capacity Building Grants **4434**
The Foundation funds two types of Capacity Building grants: Assessment (grant limit is based on scope of work) —the first step of any effort to build capacity within an organization is an assessment of present capacity to identify what is working well and what needs to be strengthened (the Foundation will fund technical assistance, consulting, and other expenses related to conducting a formal assessment of current organizational capacity); and Implementation ($20,000 grant limit)—the Foundation will fund the implementation of capacity building strategies. Allowable expenses typically include such items as technical assistance/consulting, technology (hardware and/or software), limited equipment (e.g. telephone systems), training and assessment/evaluation.
Requirements: Proposals are invited from private nonprofit or public organizations that have a mission to improve health status or access to quality, affordable health care for Kansans.
Restrictions: The foundation will not fund ongoing general operating expenses or existing deficits; capital, endowment or specific fund-raising campaigns; fund raising events; routine

continuing education; travel to conferences not directly related to the project; programs that are not Kansas-based; programs that require additional staff but demonstrate no clear means of sustainability after foundation funding; individual medical care or support; medical equipment; capital equipment (except for allowable technology); political purposes; or support of organizations that practice discrimination. Additional exclusions include: program expenses; existing staff salaries/benefits; Electronic Medical Records (EMR) technology; medical equipment; capital equipment (except for allowable technology); routine continuing education; and travel to conferences not directly related to the project.
Geographic Focus: Kansas
Date(s) Application is Due: Jun 19
Amount of Grant: Up to 20,000 U.S.D
Contact: Grants Administrator; (785) 232-3000 or (866) 232-3020; info@sunflowerfoundation.org
Internet: http://www.sunflowerfoundation.org/flash/request.html
Sponsor: Sunflower Foundation: Health Care for Kansans
1200 SW Executive Drive, Suite 100
Topeka, KS 66615-3850

Sunflower Foundation Walking Trails Grants **4435**
The Foundation is seeking projects that bring together public and private support to construct Walking Trails as a means to increase physical activity among Kansans. Grant funds may be used for Walking Trails building materials only. Walking Trails are generally expected to be at least ???? mile in length. Grants awards will be for one year and will not exceed $15,000. Organizations that are seeking matching funding to secure state or federal grants for larger Trail projects should contact the Foundation before submitting a proposal.
Requirements: Proposals are invited from private nonprofit or public organizations that have a mission to improve health status or access to quality, affordable health care for Kansans. Public access to the Walking Trail is required. Applicants must demonstrate a cost share in the project through a cash match of $1 to $1 of the total amount requested from the Sunflower Foundation.
Restrictions: The foundation will not fund ongoing general operating expenses or existing deficits; capital, endowment or specific fund-raising campaigns; fund raising events; routine continuing education; travel to conferences not directly related to the project; programs that are not Kansas-based; programs that require additional staff but demonstrate no clear means of sustainability after foundation funding; individual medical care or support; medical equipment; capital equipment (except for allowable technology); political purposes; or support of organizations that practice discrimination. Additional exclusions include: requests for physical fitness equipment; and fitness facility dues/membership fees.
Geographic Focus: Kansas
Date(s) Application is Due: Jun 29
Amount of Grant: Up to 15,000 U.S.D
Contact: Grants Administrator; (785) 232-3000 or (866) 232-3020; info@sunflowerfoundation.org
Internet: http://www.sunflowerfoundation.org/flash/request.html
Sponsor: Sunflower Foundation: Health Care for Kansans
1200 SW Executive Drive, Suite 100
Topeka, KS 66615-3850

Sunny and Abe Rosenberg Foundation Grants **4436**
The foundation awards grants to New York City area nonprofit organizations in its areas of interest, including health organizations—mental health, cancer research, and programs for the young, elderly, or physically impaired; education—excellence in public schools, libraries, and groups that fight discrimination; artistic and cultural—museums, performing arts, and arts education in public schools; recreation—city parks and youth camps; social services—food distribution to the needy and elderly, AIDS prevention, family planning, youth employment, and alcohol and drug rehabilitation; and international—orphanages and libraries. Grants generally are awarded for one year, but multiyear funding will be considered for exceptional projects.
Requirements: New York City-area nonprofit organizations are eligible.
Geographic Focus: New York
Amount of Grant: 10,000 - 50,000 U.S.D
Contact: Karen McQuade; (212) 660-1214; info@rosenbergfoundation.org
Internet: http://www.rosenbergfoundation.org
Sponsor: Sunny and Abe Rosenberg Foundation
888 7th Avenue, 24th Floor
New York, NY 10019

Sunoco Corporate Contributions **4437**
Fulfilling a civic responsibility as a good corporate citizen is the primary reason for Sunoco's corporate philanthropy. Funding areas include educational opportunities for diverse groups (schools, colleges, and universities are generally not included); health and human services—community organizations that address health and social problems (support for single diseases generally not included); arts and culture (generally not national or international institutions); civic and economic development—at-risk youth employment programs, adult job-training programs, economic development and improvement programs in specific communities and neighborhoods where Sunoco has a major presence; and environmental projects. Sunoco prefers to fund specific projects rather than operating budgets. All contributions made are on a one-time basis unless expressly noted otherwise. Requests must be signed and in writing.
Requirements: Proposals should be clear, concise, and include purpose, population served, time frame, budget, copy of 501(c)3 tax-exempt status, federal tax ID number, and amount requested from Sunoco.
Restrictions: Requests for deficit funding, individuals, benefit fundraisers, endowments, surveys, studies, religious groups, fraternal organizations, athletic groups, schools, single diseases, and non-tax exempt organizations are generally not considered.
Contact: Ed Hazzouri, Director; (215) 977-3401; fax (215) 246-8562
Internet: http://www.sunocoinc.com/aboutsunoco/contributions.htm
Sponsor: Sunoco
1801 Market Street, 10 Penn Center
Philadelphia, PA 19103-1699

Sunshine Natural Wellbeing Foundation Grants **4438**
The foundation awards grants to nonprofit organizations for existing or upstart programs relating to alternative treatments and research for asthma, immune disorders, and other similar illnesses, as well as children's educational programs relative to alternative health. Submit a proposal with specific goals, project budget, organization budget, audited financial reports, 501(c)3 letter, IRS 990 for previous year, list of current contributors, and list of board members.
Requirements: 501(c)3 nonprofits are eligible. Giving is primarily made in Florida.
Restrictions: Individuals are not eligible.
Geographic Focus: Florida
Date(s) Application is Due: Sep 1
Contact: Delores Wooldridge, (941) 346-5297
Sponsor: Sunshine Natural Wellbeing Foundation
6547 Midnight Pass Road, Suite 19
Sarasota, FL 34242

SunTrust Foundation MidAtlantic Grants **4439**
The foundation supports nonprofits in areas where SunTrust Banks, Inc., has a major presence. Priority areas include education—independent funds for private colleges and universities; health and human services—United Way, capital grants to hospitals, and groups such as the Salvation Army and Scouts; culture and arts—cultural enrichment, museums, performing arts, literary groups, arts councils, and historic preservation; and civic and community—economic vitality and free enterprise.
Requirements: Nonprofit organizations in areas where Crestar Bank has a major presence are eligible.
Geographic Focus: Virginia
Date(s) Application is Due: Oct 15
Amount of Grant: 1,000 - 20,000 U.S.D
Contact: Brenda Skidmore, President, c/o SunTrust Banks, Inc.; (804) 782-7907
Sponsor: SunTrust Foundation MidAtlantic
919 E Main Street
Richmond, VA 23219

Supelco AOCS Research Award **4440**
This award is given annually to a qualified scientist for outstanding original research on fats, oils, lipid chemistry, and/or biochemistry, the results of which have been published in technical papers of high quality. A travel and expense allowance and a plaque are provided. Application materials are available upon request.
Date(s) Application is Due: Nov 1
Amount of Grant: 8,000 U.S.D
Contact: Awards Officer; (217) 359-2344; fax (217) 351-8091; membership@aocs.org
Internet: http://www.aocs.org/member/awards/awd-research.asp
Sponsor: American Oil Chemists' Society
P.O. Box 3489
Champaign, IL 61826-3489

Support Our Aging Religious (SOAR) Grants **4441**
Support Our Aging Religious (SOAR) was established to raise national public awareness of the religious retirement crisis, to network concerned laity, and to provide financial resources to religious communities to assist them in the care of their elderly members. Each year the program awards grants to congregations across the country in its areas of interest, including the fields of education, social services, and health care. Grant categories include: building renovations to allow for handicapped accessibility, physical therapy, and medical equipment; and safety needs—fire alarms, sprinklers, hospital beds, and nurse call systems.
Requirements: SOAR welcomes applications from religious institutes that need assistance to provide critically needed items to care for their retired and infirm members.
Restrictions: SOAR is unable to provide: ongoing living expenses of retired religious in assisted living or skilled nursing facilities; major contributions to retirement funds; funds for construction of new buildings; funds for operational expenses; or cosmetic improvements that do not directly relate to the physical well-being of the retired. Proposals which request partial funding for large projects if additional funds have not been obtained cannot be supported.
Date(s) Application is Due: Nov 1
Amount of Grant: 500 - 30,000 U.S.D
Samples: Sisters, Servants of the IHM, Monroe, MI — handicapped accessible van; Society of the Sacred Heart, Atherton, CA — backup generator; Conception Abbey, Inc., Conception, Missouri — elevator upgrade.
Contact: Deborah Hudson Vornbrock, Grants Administrator; (202) 529-7627; fax (202) 529-7633; dvornbrock@soar-usa.org or info@soar-usa.org
Internet: http://www.soar-usa.org/published/index.html
Sponsor: Support Our Aging Religious
900 Varnum Street NE
Washington, DC 20017-2145

Susan G. Komen Breast Cancer Foundation Basic, Clinical, and Translational Research Grants 4442

The foundation awards funding for basic, clinical, and translational breast cancer research. Grants offer two years of support for a variety of research projects in such areas as complementary and alternative medicine; detection and diagnosis, prognosis; risk, prevention, and epidemiology; local treatment; psychosocial treatment; systemic treatment; and tumor cell biology. The foundation gives preference to projects that have potential for high impact in their field, may not be considered by other funders, and benefit the medically underserved in communities. Application deadlines are August 1 for electronic submittal of abstracts; August 8 for electronic submittal of applications; and August 15 for hard-copy applications. Information and application forms are posted on the foundation's Web site.
Requirements: A wide range of nonprofits, including research groups and health and human service agencies, are eligible to apply.
Date(s) Application is Due: Aug 1; Aug 8; Aug 15
Amount of Grant: Up to 250,000 U.S.D
Samples: Dr. Eliot Rosen, Long Island Jewish Medical Ctr (NY)—for the study BRCA1 as a Target for Chemoprevention Agents, $249,939; Dr. Margot Ip, Roswell Park Cancer Institute—for the study Stromal-Vascular Precursors as a Target for Breast Cancer Prevention, $249,987; Dr. Georgia Chenevix-Trench, Queensland Institute of Medical Research—for the study The role of the ATM Gene in Familial Breast Cancer, $249,300; Dr. Amanda Spurdle, Queensland Institute of Medical Research—for the study Evaluation of BRCA1 and BRCA2 Unclassified Sequence Variants, $248,250.
Contact: Grants Manager; (888) 300-5582; fax (972) 855-1640; grants@komen.org
Internet: http://www.komen.org/intradoc-cgi/idc_cgi_isapi.dll?IdcService=SS_GET_PAGE&nodeId=390
Sponsor: Susan G. Komen Breast Cancer Foundation
5005 LBJ Freeway, Suite 250
Dallas, TX 75244

Susan G. Komen Breast Cancer Foundation Breast Health, Breast Cancer Education, and Outreach Programs Grants 4443

Grants are made for breast health, breast cancer education, or outreach programs targeting services not otherwise available to the medically under served in Arizona. See website for grant deadline dates.
Requirements: Applicants must be 501(c)3 public charity, educational institute, and/or government agency based in Arizona.
Geographic Focus: Arizona
Contact: Administrator; (888) 300-5582; fax (972) 855-1640; email grants@komen.org
Internet: http://www.komen.org/grants
Sponsor: Susan G. Komen Breast Cancer Foundation
5005 LBJ Freeway, Suite 250
Dallas, TX 75244

Susan G. Komen Breast Cancer Foundation Brinker Award for Scientific Distinction 4444

The award recognizes outstanding work that has advanced basic research concepts, or affected clinical or social-behavioral applications, in the field of breast cancer research, screening, treatment, or prevention. The intent is to recognize scholars for a specific contribution, a consistent pattern of contributions, or leadership in the field that has had a substantial impact on breast cancer. Each award will include an honorarium, citation, and inscribed crystal statuette designed by Tiffany & Co. In the event that more than one person is selected for an award, the honorarium will be divided between the recipients and each will receive the citation and a statuette. It is strongly suggested that an individual scholar be nominated; however, researchers who have worked as a team or who have each made similarly outstanding contributions in the field may be nominated together in an award category. Awardees are expected to present their important findings in a lectureship at the annual San Antonio Breast Cancer Symposium. Nomination form and guidelines are available online.
Requirements: Breast cancer researchers who are affiliated with nonprofit institutions (academic, industry, government, or other) in the United States and abroad are eligible.
Date(s) Application is Due: Jun 30
Amount of Grant: 20,000 U.S.D
Contact: Award Administrator; (972) 855-1600; fax (972) 855-1605
Internet: http://www.komen.org/intradoc-cgi/idc_cgi_isapi.dll?IdcService=SS_GET_PAGE&nodeId=560
Sponsor: Susan G. Komen Breast Cancer Foundation
5005 LBJ Freeway, Suite 250
Dallas, TX 75244

Susan G. Komen Breast Cancer Foundation Hurricane Relief Grants 4445

The foundation awards grants to nonprofit institutions and providers of breast health and breast cancer care in areas affected by Hurricanes Katrina and Rita, and to organizations offering patient services in communities responding to the needs of displaced populations. Two types of grants are available. Rebuilding Grants are intended for nonprofit institutions and providers of breast cancer care in areas affected by the hurricanes for rebuilding efforts, including but not limited to, rebuilding facilities, replacing equipment, hiring staff, replacing/buying supplies, retrieving records, locating patients, and reassigning providers. Patient Services Grants are intended to enable nonprofit institutions and providers of breast cancer care in communities responding to displaced populations to maintain breast cancer services. Funding is designed to assist breast cancer care providers with patient treatment needs, including pharmaceutical and medical supplies; transportation of patients to and from treatment; and child care for patients undergoing treatment and patient navigation services. The foundation is unable to provide funds directly to individuals, but is collaborating with organizations like CancerCare to ensure patients' needs are being met. Individuals with breast health needs who were affected by the hurricane can contact the foundation's national toll-free Breast Care Helpline (800-462-9273). Application and guidelines are available online.
Requirements: Nonprofit organizations and providers of breast health and breast cancer care in areas affected by Hurricanes Katrina and Rita, and organizations offering patient services in communities responding to the needs of displaced populations are eligible.
Amount of Grant: 30,000 - 50,000 U.S.D
Contact: Grants Administrator; (972) 855-1600; fax (972) 855-1605
Internet: http://www.komen.org/intradoc-cgi/idc_cgi_isapi.dll?IdcService=SS_GET_PAGE&ssDocName=katrina
Sponsor: Susan G. Komen Breast Cancer Foundation
5005 LBJ Freeway, Suite 250
Dallas, TX 75244

Susan G. Komen Breast Cancer Foundation Interdisciplinary Fellowships 4446

The foundation serves as a catalyst to develop fellowship training for interdisciplinary programs in breast care that will prepare physicians for careers devoted to serving the multispecialty needs of the breast cancer patient; offer an interdisciplinary curriculum integrated into a comprehensive program; and provide special emphasis on enhancing the physician's understanding of the patient with benign and malignant breast disease while developing a better treatment environment for future patients. The program will fund two consecutive one-year fellowships. Guidelines are available online.
Requirements: The host institution must be an accredited medical institution in the United States or similarly recognized non-U.S. institution.
Restrictions: For-profit organizations are not eligible.
Date(s) Application is Due: Feb 7
Amount of Grant: 35,000 U.S.D
Contact: Dianne McDonald, Grants Manager; (888) 300-5582; fax (972) 855-1640; grants@komen.org
Internet: http://www.komen.org/intradoc-cgi/idc_cgi_isapi.dll?IdcService=SS_GET_PAGE&nodeId=395
Sponsor: Susan G. Komen Breast Cancer Foundation
5005 LBJ Freeway, Suite 250
Dallas, TX 75244

Susan G. Komen Breast Cancer Foundation Population Specific Research Grants 4447

The foundation supports research project grants in support of innovative efforts focusing on the epidemiology of breast cancer within specific populations at risk for the disease, such as African American, Asian/Pacific Islander, Hispanic, Native American, lesbian, and low-literacy populations. Grants provide two years of support. Annual deadline dates may vary; contact program staff for exact dates.
Requirements: U.S. nonprofit organizations and international institutions are eligible.
Amount of Grant: 250,000 U.S.D
Samples: Dr. Gail Tomlinson, U of Texas Southwestern Medical Ctr (TX)—for Decision Making and BRCA Testing in African Americans, $250,000; Dr. Deborah Erwin, U of Arkansas for Medical Sciences (AR)—for Development Of A Latina Witness Project Model, $248,480; Anna Wu, Ph.D. University of Southern California (CA)—for Soy, Tamoxifen, and Breast Cancer in Asian-Americans, $250,000.
Contact: Director, Director of Grants and Sponsored Programs; (972) 855-1600; fax (972) 855-1605; grants@komen.org
Internet: http://www.komen.org/grants/available.asp
Sponsor: Susan G. Komen Breast Cancer Foundation
5005 LBJ Freeway, Suite 250
Dallas, TX 75244

Susan G. Komen Breast Cancer Foundation Postdoctoral Fellowships 4448

Postdoctoral fellowships offer three years of support for basic, clinical, or translational cancer research and public health or epidemiology studies. Detailed program-specific eligibility requirements and application information are available directly from the foundation or its Web site. Abstracts submitted electronically are due August 1; applications submitted electronically are due August 8; hard-copy applications are due August 15.
Requirements: The fellowship candidate must be no more than three years post completion of PhD or equivalent or five years post completion of MD or comparable doctoral degree in the clinical health profession. U.S. citizenship or residency is not required.
Date(s) Application is Due: Aug 1; Aug 8; Aug 15
Amount of Grant: 45,000 U.S.D
Samples: Dr. Carolyn Bertozzi, U of California at Berkeley (CA)—fellowship recipient, for Disruption of 4-Epimerase: A Novel Anti-Metastasis Strategy for Breast Cancer, $135,000; Dr. Christopher Contag, Stanford U—fellowship recipient, for In Vivo Recruitment and Activation of NK-T Cells into Tumors, $135,000; Dr. Carol Lange, U of Minnesota Cancer Ctr (MN)—fellowship recipient, for Molecular Regulation of Breast Tumor Kinase, $135,000; Dr. Wayne Marasco, Dana-Farber Cancer Institute—fellowship recipient, for Isolation and Characterization of Human Anti-Human CXCR4 Antibodies for Immunotherapy of Breast Cancer Metastasis, $135,000.
Contact: Director, Director of Grants and Sponsored Programs; (888) 300-5582; fax (972) 855-1640; grants@komen.org
Internet: http://www.komen.org/intradoc-cgi/idc_cgi_isapi.dll?IdcService=SS_GET_PAGE&nodeId=391
Sponsor: Susan G. Komen Breast Cancer Foundation
5005 LBJ Freeway, Suite 250
Dallas, TX 75244

Susan G. Komen Breast Cancer Foundation STEP Grants **4449**
Screening, Treatment and Education Program (STEP) grants support community-based programs that provide the medically underserved with access to breast health/breast cancer education, screening, diagnostic, treatment, and support services. The funding period is for 12 months and based on our fiscal year of April 1st through March 31st. Application cover letter and guidelines are available online.
Requirements: The applicant organization must be nonprofit and federally tax-exempt; located in the Komen Greater NYC service area, which includes the five boroughs of New York City, Long Island, Westchester County, and Rockland County; and a current or planned provider of breast health services.
Geographic Focus: New York
Date(s) Application is Due: Sep 30
Amount of Grant: Up to 75,000 U.S.D
Contact: Kavita Das, (212) 560-9590, ext. 229; kdas@komennyc.org
Internet: http://www.komennyc.org/site/PageServer?pagename=grants_step
Sponsor: Susan G. Komen Breast Cancer Foundation
341 W 38th Street, 10th Floor
New York, NY 10018

Susan Mott Webb Charitable Trust Grants **4450**
The trust awards grants to eligible Alabama nonprofit organizations in its areas of interest, including the arts, civic and public affairs, animals/wildlife, community development, education, health care, religion, social services, youth programs and homelessness. Types of support include annual campaigns, building construction/renovation, capital campaigns, continuing support, curriculum development, emergency funds, endowments, equipment, general/operating support, internship funds, program development, publication, and technical assistance. Contact the Foundation for further application information and guidelines.
Requirements: Giving limited to the greater Birmingham, AL, area.
Restrictions: Grants do not support advertising, individuals, international organizations, or political organizations.
Geographic Focus: Alabama
Date(s) Application is Due: Apr 1; Oct 1
Amount of Grant: 2,000 - 100,000 U.S.D
Samples: Alabama Symphonic Association, Birmingham, AL, $40,000; Crisis Resource Center of Southeast Kansas, Pittsburg, KS, $25,000;
Contact: Laura Wainwright, Vice President, c/o Regions Bank; (205) 801-0380; fax (205) 581-7433; laura.wainwright@regions.com
Sponsor: Susan Mott Webb Charitable Trust
P.O. Box 11426
Birmingham, AL 35202-1426

Susan Vaughan Foundation Grants **4451**
The Susan Vaughan Foundation established in 1952, supports non-profits involved with: education, particularly to a library; arts; environment, natural resources; higher education; human services. The Foundation gives primarily in Houston and Austin Texas area with support in the form of: annual campaigns; building/renovation; capital campaigns; general/operating support; matching/challenge support.
Requirements: There is no formal application form required, applicant must be a non-profit in the Houston and Austin, Texas region. Initial contact should be made through a Letter of Inquiry.
Geographic Focus: Texas
Amount of Grant: 2,500 - 630,000 U.S.D
Samples: Clayton Library Friends, Houston, TX, $630,000; Houston Downtown Park Conservancy, Houston, TX, $50,000; Trees for Houston, Houston, TX, $30,000;
Contact: Jennifer Grosvenor, Grant Coordinator c/o Legacy Trust Co.; (713) 651-8980; jgrosvenor@legacytrust.com
Sponsor: Susan Vaughan Foundation
600 Jefferson Street, Suite 300
Houston, TX 77002-7377

Swindells Charitable Foundation **4452**
The Swindells Charitable Foundation was established in 1933 to support and promote quality health and human-services programming for underserved children and adults. The Swindells Charitable Foundation also makes grants to public charitable hospitals. Preference is given to organizations that serve sick or economically disadvantaged children or older adults. Special consideration is given to organizations that provide for the "basic human needs" of individuals. Grants from the Swindells Charitable Foundation are one year in duration. Applicants must apply online at the grant website. Applicants are strongly encouraged to do the following before applying: review the downloadable state application procedures for additional helpful information and clarifications; review the downloadable online-application guidelines at the grant website; review the foundation's funding history (link is available from the grant website); review the online application questions in advance; and review the list of required attachments. These will generally include: a list of board members, financial statements (audited, reviewed, or compiled by independent auditor); an organization summary; a list of other funding sources; an IRS Determination letter; and other required documents. All attachments must be uploaded in the online application as PDF, Word, or Excel files. The Swindells Charitable Foundation has biannual deadlines of February 1 and August 1. Applications are by 11:59 p.m. on the deadline dates. Applicants will be notified of grant decisions by letter within two to three months after each respective proposal deadline.
Requirements: Applicants must have 501(c)3 tax-exempt status.
Restrictions: Applicants will not be awarded a grant for more than three consecutive years. The trust does not support requests from individuals, organizations attempting to influence policy through direct lobbying, or any political campaigns. Capital requests will not be considered.
Geographic Focus: Connecticut
Date(s) Application is Due: Feb 1; Aug 1
Contact: Kate Kerchaert; (860) 657-7016; kate.kerchaert@baml.com
Internet: https://www.bankofamerica.com/philanthropic/fn_search.action
Sponsor: Swindells Charitable Foundation
200 Glastonbury Boulevard, Suite # 200, CT2-545-02-05
Glastonbury, CT 06033-4056

Sybil G. Jacobs Award for Outstanding Use of Tobacco Industry Documents **4453**
The award recognizes a person who has made a significant and well-recognized contribution to the health of the public in the recent past through use of tobacco documents. The award also honors innovation in the use and application of tobacco industry documents to improve the public's health and, where applicable, to further the goals of tobacco prevention and control in order to help build a world where young people reject tobacco and anyone can quit. Those nominated should be individuals who have made a notable impact through innovative use of tobacco industry documents as applied to research, policy, or advocacy.
Requirements: At least one of the following two criteria must be met by the nominated individual: Nominees must have made a remarkable research, policy, or advocacy contribution with the use of tobacco industry documents, and/or; Nominees must have employed innovative, creative approaches to the employment of tobacco industry documents that result in an improvement in the health or public awareness of a community or nation. Nominations may be made by individuals such as colleagues, peers, coworkers, instructors/professors, governments, or CBOs that are qualified to make such a nomination because of their familiarity with the nominee's work and contribution.
Restrictions: Nominees must not have any affiliation with the tobacco industry. Members of the awards committee may not nominate potential recipients, although they may provide written support as part of a nomination package. Employees and directors of the American Legacy Foundation and their relatives are not eligible.
Date(s) Application is Due: Feb 6
Amount of Grant: 7,500 U.S.D
Contact: Virginia Lockmuller, Grant Coordinator; (202) 454-5555; fax (202) 454-5599; awards@americanlegacy.org
Internet: http://www.legacyforhealth.org/awards/
Sponsor: American Legacy Foundation
1724 Massachusetts Avenue NW
Washington, DC 20036

Symantec Community Relations and Corporate Philanthropy Grants **4454**
Symantec's Community Relations and Corporate Philanthropy Program strives to have a positive impact on our local communities around the globe. Symantec supports four main philanthropic areas, all with the objective of creating a sustainable and diverse future for the technology industry; science, technology, engineering, and math education; equal access to education; diversity in engineering; environmental responsibility; and online safety.
Requirements: 501(c)3 entities and/or educational institutions that have philosophy programs that benefit the lives of children are encouraged to apply. Nonprofit organizations in Minneapolis, Minnesota; San Francisco Bay Area, California; Orlando, Florida; and Vienna, Virginia, are eligible.
Restrictions: Grants will not be made for political or religious purposes. Grants will not be made to organizations that discriminate on the basis of age, disability, religion, ethnic origin, sex or sexual orientation.
Geographic Focus: California, Florida, Minnesota, Virginia
Contact: Grants Director; (800) 327-2232 or (650) 527-8000; fax (650) 527-2908; community_relations@Symantec.com
Internet: http://www.symantec.com/about/profile/responsibility/community/index.jsp
Sponsor: Symantec Corporation
350 Ellis Street
Mountain View, CA 94043

Symbol Technologies Corporate Giving Grants **4455**
The corporate-giving program focuses on four specific areas—Partners in Education, Strength of Diversity, Symbol and the Arts, and A Helping Hand. Partners in Education supports educational institutions locally and nationally, with a special interest in assisting students pursuing technical careers on Long Island. Scientific innovations, particularly in university settings, are of interest. Strength of Diversity supports multicultural organizations. Symbol and the Arts supports local and global arts and cultural organizations, and efforts to offer art and music to wider audiences. A Helping Hand supports local and global community organizations, programs, and events that provide relief for the hungry and homeless, select health initiatives, and children's organizations. Requests are considered throughout the year. Allow two to four weeks for a response. If the matter is urgent, contact Symbol's Community Affairs Liaison immediately. Requests for equipment are entertained providing Symbol is in possession of such surplus equipment.
Requirements: Only solicitations submitted on letterhead by organizations qualified as 501(c)3 will be considered. Name, address, and contact information must be included, as well as a brief overview of the organization and how these efforts fall within Symbol's corporate contributions parameters. Documentation of tax exemption must be provided, as well as a summary of the organization's financial stability.

Restrictions: Symbol does not consider requests made by individuals, political, labor, religious or fraternal organizations, sports groups, and publications of books.
Contact: Director of Public Relations; (631) 738-4751 or (800)722-6234; abelsonb@symbol.com or info@symbol.com
Internet: http://www.symbol.com/about/overview/overview_community_affairs.html
Sponsor: Symbol Technologies
1 Symbol Plaza, MS A-12
Holtsville, NY 11742

T.L.L. Temple Foundation Grants **4456**
The foundation awards grants to Texas nonprofit organizations in its areas of interest, including health, community and social services, civic affairs, cultural activities, and education. Types of support include general operating support, capital campaigns, building construction and renovation, equipment acquisition, program development, scholarship funds, research grants, matching funds, and emergency funds. There are no application deadlines.
Requirements: Nonprofit organizations in Texas counties constituting the East Texas Pine Timber Belt are eligible.
Restrictions: Grants do not support private foundations or individuals, nor are grants made in support of deficit financing.
Geographic Focus: Texas
Amount of Grant: 5,000 - 60,000 U.S.D
Contact: A. Wayne Corely, Executive Director; (936) 639-5197
Sponsor: T.L.L. Temple Foundation
109 Temple Boulevard, Suite 300
Lufkin, TX 75901

Take Care to Learn: Teaching Clinical Care Management Grants **4457**
The program awards grants to primary-care residency and nurse-practitioner programs, as well as managed-care organizations. The goal is to support education programs leading to improving care for patients with asthma or diabetes and make sure trainees have the skills needed to deliver high-quality health care. Ten two-year grants will be awarded.
Requirements: Healthcare institutions are eligible.
Amount of Grant: Up to 300,000 U.S.D
Contact: Elizabeth March, Deputy Director; (617) 509-9811; fax (617) 859-8112; Elizabeth_March@harvardpilgrim.org
Internet: http://www.pqe.org/programs-takecare.html
Sponsor: Partnerships for Quality Education
133 Brookline Avenue, 6th Floor
Boston, MA 02215

Takemi Program in International Health Research and Advanced Training Fellowships **4458**
This midcareer program offers a limited number of fellowship positions for research and advanced training on critical issues of international health, especially those relating to developing countries. This interdisciplinary program has four major objectives: to create better methods for mobilizing and using health resources in both rich and poor countries; to promote cooperative research and comparative analysis of health policies and programs in different countries; to study transnational causes of health conditions such as population migration, disease transmission, and air and water pollution; and to bring together leading health professionals and scholars from many nations for research and training. Fellowship positions cover estimated costs for the program, which includes 10 months in residence at the Harvard School of Public Health. Contact program staff for deadlines.
Requirements: Applicants should have completed graduate degrees and demonstrated potential leadership capacity in their home countries. They are expected to show strong promise and appropriate preparation to enable them to benefit from a period at Harvard. Applicants may come from the disciplines of medicine, law, public health, economics, management, and social sciences. Applicants should submit a curriculum vita and publications list, a proposal of research and writing to be undertaken during the fellowship, and three letters of reference. The proposal should not exceed five typed pages and should include a clear statement of objectives, significance or relevance, data sources, and methodology.
Restrictions: Clinical research will not be funded. Individuals enrolled in degree-granting programs are ineligible.
Contact: Coordinator; (617) 432-0686; fax (617) 432-1251; takemi@hsph.harvard.edu
Internet: http://www.hsph.harvard.edu/takemi
Sponsor: Harvard University
665 Huntington Avenue, Building 1-1207
Boston, MA 02115-6021

Taproot Foundation Capacity-Building Service Grants **4459**
The program seeks to strengthen the capacity of strong charitable organizations that provide effective, high-quality programs. Applicant organizations must have a mission that fits in one of four categories: education—organizations that strengthen the public education (pre-K and K-12) system or enhance the educational success of students; environment—organizations that preserve or restore the environment; health—organizations that improve people's physical or emotional health; and social services—organizations that provide social services to those who suffer hardship or who have been unable to participate fully in the social and economic life of the community. The foundation makes grants of high-quality professional services that target capacity-building needs nonprofit organizations, including fundraising, marketing, information technology, and talent management. Each grant is staffed by professionals from the business community who donate their time and expertise to help the community. The grant also includes a project plan with estimated timing based on volunteers donating five hours per week. Guidelines and application are available online.
Requirements: 501(c)3 nonprofit organizations that provide direct benefits primarily or entirely to residents of the five boroughs of New York City (Manhattan, Bronx, Queens, Staten Island, and Brooklyn) and the six counties of the California Bay Area (Alameda, San Francisco, San Mateo, Santa Clara, Marin, and Contra Costa counties) are eligible. The organization must also have its headquarters in one of these areas.
Restrictions: The foundation does not support nonprofits with other than 501(c)3 status, for-profit corporations, or individuals. Service Grants are not expressly intended to support candidates for political office, or to influence legislation. Grants are not awarded to support religious purposes. (Agencies operated by religious organizations that serve secular needs are eligible.)
Geographic Focus: California, New York
Date(s) Application is Due: Mar 1; Jun 1; Sep 15; Dec 1
Contact: Grants Administrator; (415) 359-0226; national@taprootfoundation.org
Internet: http://taprootfoundation.org/npo/index.shtml
Sponsor: Taproot Foundation
466 Geary Street, Suite 300
San Francisco, CA 94102

Tauck Foundation Grants **4460**
The currently focuses its giving in five areas: destination grants—historical, cultural and environmental preservation projects in locales where Tauck World Discovery currently operates; youth development grants—after-school and summer programs for youth ages five and 18; community giving—community organizations in southwestern Connecticut; special community grants—one-time project needs of local (CT) organizations that have received past funding from the foundation; and discretionary giving—matches the charitable donations of family members. Letters of inquiry for destination grants are due by April 1 and August 1; for youth development grants January 15, April 1, and August 1. Full applications are by invitation. Guidelines are available on line.
Requirements: 501(c)3 tax-exempt organizations are eligible. Funding for youth development is distributed to organizations in southwestern Connecticut; Rochester, and Ithaca, NY; and Seattle, WA. Community giving grants are awarded in Fairfield County, CT, and nearby communities.
Geographic Focus: Connecticut, New York, Washington
Date(s) Application is Due: Jan 15; Apr 1; Aug 1
Amount of Grant: 1,000 - 75,000 U.S.D
Contact: Grants Administrator; (866) 828-2536, fax (203) 286-1340; info@tauckfoundation.org
Internet: http://www.tauckfoundation.org
Sponsor: Tauck Foundation
P.O. Box 5020
Norwalk, CT 06855-1445

Taylor S. Abernathy and Patti Harding Abernathy Charitable Trust Grants **4461**
The Taylor S. Abernathy and Patti Harding Abernathy Charitable Trust was established in 1988 to support religious, charitable, scientific, and educational purposes. To that end, the trust provides grants that support and promote quality educational, cultural, human services, and health care programming. Grant requests for general operating support and program support will be considered. The Trust generally supports organizations that serve residents of the Greater Kansas City Metropolitan area. There are no application deadlines for the Abernathy Charitable Trust. Proposals are reviewed on an ongoing basis. Grants are one year in duration. Grant requests for general operating support and program support will be considered.
Requirements: 501(c)3 organizations serving the residents of the Greater Kansas City Metropolitan area are welcome to apply.
Restrictions: Grant requests for capital support will not be considered.
Geographic Focus: Kansas, Missouri
Amount of Grant: Up to 25,000 U.S.D
Contact: Spence Heddens; (816) 292-4301; Spence.heddens@baml.com
Internet: https://www.bankofamerica.com/philanthropic/grantmaking.action
Sponsor: Taylor S. Abernathy and Patti Harding Abernathy Charitable Trust
1200 Main Street, 14th Floor, P.O. Box 219119
Kansas City, MO 64121-9119

Tech Museum of Innovation Awards **4462**
The awards honor the successful implementation of new technology or a novel implementation of technology when it is made accessible and useful. The program accepts nominations of individuals, companies, and organizations that are using technology to transform the way people live. Awards are made in five categories: health, education, environment, economic development, and equality. Up to five finalists in each category will be honored, and one finalist per category will receive the monetary prize. Nominations are due by the listed application deadline; application is by invitation.
Requirements: Eligible nominees worldwide include individuals, companies, and nonprofit or nongovernmental organizations. Individuals may nominate themselves.
Date(s) Application is Due: Apr 3
Amount of Grant: 50,000 U.S.D
Contact: Administrator; (408) 795-6338; fax (408) 918-0253; techawards@thetech.org
Internet: http://www.thetech.org/techawards
Sponsor: Tech Museum of Innovation
201 S Market Street
San Jose, CA 95113-2008

Tellabs Foundation Grants 4463

The foundation awards grants to eligible nonprofit organizations in the following areas: educational programs with a particular focus on local and national programs and curricula for engineering, science, mathematics and technology; health and human services programs for projects involving health and wellness-related, research, education and treatment in the United States (The primary focus is on projects involving hospitals and health care facilities); and environmental programs that encourage the understanding and protection of the environment. Generally, grants are awarded for specific programs rather than for general operating funds or capital projects. The primary focus of the foundation is to support programs in areas in which Tellabs employees live and work. Tellabs Inc and its affiliates also have direct giving programs, including an employee matching gift program, giving to the United Way, and limited direct corporate grants. The board meets quarterly (January, April, July and October); submit proposals at least four weeks before scheduled meetings.

Requirements: Unless invited by the foundation board to submit a full grant proposal, all applicants or programs must first submit a letter of inquiry. There is no set format, but it should be 1-2 pages in length. All grant recipients must be recognized by the IRS as tax exempt, not-for-profit organizations under Section 501(c)3 of the Internal Revenue Code. Contact program staff for company locations.

Restrictions: The foundation generally will not consider requests from: Organizations which do not have a 501(c)3 status; Political organizations or parties, candidates for political office, and organizations whose primary purpose is to influence legislation; Labor unions or organizations; Local athletic or sports programs; Service organizations raising money for community purposes; Individuals; Travel funds for tours, expeditions, or trips by individuals or groups; Dues or gifts to national or local alumni groups, clubs or fraternities; Institutional memberships or subscription fees for publications; Gifts to individual churches, synagogues or other entities organized exclusively for religious purposes; Donations for benefit events, raffle tickets or fundraising efforts that involve value returned to the donor; Organizations not operating for the benefit of the general public or that have discriminatory practices; Any grantee who receives funds of $50,000 or more shall not be entitled to submit another grant proposal for three years from the date of grant.

Amount of Grant: 10,000 U.S.D

Samples: Purdue U (West Lafayette, IN)—to construct a facility for the department of computer sciences, $30,000.

Contact: Meredith Hilt, Executive Director; (630) 798-2506; fax (630) 798-4778; meredith.hilt@tellabs.com

Internet: http://www.tellabs.com/about/foundation.shtml

Sponsor: Tellabs Foundation
1415 W Diehl Road, Mail Stop 10
Naperville, IL 60563

Temple-Inland Foundation Grants 4464

The foundation makes grants to IRS 501(c)3 tax-exempt organizations located in geographic operating areas of Temple-Inland Inc's subsidiary companies. Areas of interest include services for children and youth, Christian churches and organizations, education at all levels, health care and organizations, human services, and arts and culture. Types of support include general operating support, research, employee matching gifts, and employee-related scholarships. Requests should be received two weeks prior to quarterly board meetings. Submit a written request for guidelines and specific application deadlines. The listed deadline date is for scholarships.

Restrictions: Ineligible applicants include fraternal, veterans, political, local social, and service organizations.

Date(s) Application is Due: Mar 15

Amount of Grant: 1,000 - 30,000 U.S.D

Contact: Richard Warner, (936) 829-1721; fax (936) 829-7727

Internet: http://www.templeinland.com

Sponsor: Temple-Inland Foundation
1300 S Mopac
Austin, TX 78749

Templeton Advanced Research Grants 4465

The institute invites proposals to develop research projects in one of three topical areas: religion, spirituality, healing, and health outcomes—research that advances scientific understanding of the link, if any, between religion, spirituality, healing, and health outcomes; religion, spirituality, and human flourishing—research that scientifically explores how religion and spirituality contribute to the virtues and human strengths that reflect humanity's highest aspirations and noble qualities; and competitive dynamics and cultural evolution of religions and God concepts—inquiries that approach religion and spirituality as an important creative or causal factor in the formation of society. Successful proposals will demonstrate, among other qualities, the promise of taking these fields in new directions, the possibility of addressing a question or shortcoming proposed in the field analyses, methodological innovation, and the potential for significant impact in the research community. The listed application deadline is for letters of intent. Guidelines are available online.

Requirements: Scholars and researchers from fields such as anthropology, economics, history, political science, psychology, religious studies, and sociology are encouraged to apply.

Date(s) Application is Due: Aug 1

Amount of Grant: 150,000 - 1,000,000 U.S.D

Contact: Christopher Stawski, (215) 789-2200; fax (215) 789-2222; tarp@metanexus.net

Internet: http://www.metanexus.net/tarp

Sponsor: Metanexus Institute on Religion and Science
3624 Market Street, Suite 301
Philadelphia, PA 19104

Tenet Healthcare Foundation Grants 4466

The foundation awards grants to nonprofit organizations in Tenet communities helping meet unmet or underserved healthcare needs. Grants are awarded in the following categories: health and wellness, health education, human services, civic and community, arts and humanities, and education. The foundation has identified three national issues of foucs: the nursing shortage that is affecting the delivery of health care, the challenge of providing medical needs of the uninsured, and the ongoing effort to improve the quality of care and patient safety at U.S. hospitals. Guidelines and procedures are available online.

Requirements: 501(c)3 organizations are eligible.

Amount of Grant: Up to 11,000,000 U.S.D

Samples: Bea's Kids (Dallas, TX)—to provide education services for 150 underprivileged children, $15,000; Christian Health Ministry Clinic (Searcy, AR)—to treat uninsured and underserved people, $21,000.

Contact: Barbara Luton, Executive Director; (805) 563-6865; fax (805) 898-9104; foundation@tenethealth.com

Internet: http://www.tenethealth.com/TenetHealth/TenetFoundation/GrantsFunding

Sponsor: Tenet Healthcare Foundation
P.O. Box 31907
Santa Barbara, CA 93130

Tension Envelope Foundation Grants 4467

The foundation supports nonprofits in company-operating areas, with an emphasis on Jewish welfare funds, community funds, higher education, health, civic affairs, culture and the arts, and youth. There are no deadlines or guidelines, and the foundation does not distribute an annual report. To apply for a grant, send proposal, budget, and proof of 501(c)3 status. The board meets several times a year.

Requirements: 501(c)3 nonprofits in California, Iowa, Kansas, Minnesota, Missouri, North Carolina, Tennessee, and Texas are eligible.

Geographic Focus: California, Iowa, Kansas, Minnesota, Missouri, North Carolina, Tennessee, Texas

Amount of Grant: 100 - 70,000 U.S.D

Contact: Eliot Berkley, (816) 471-3800

Sponsor: Tension Envelope Foundation
819 E 19th Street, 5th Floor
Kansas City, MO 64108

Terrance Keenan Leadership Award in Health Philanthropy 4468

The award annually recognizes a grantmaker whose leadership and thoughtful application of philanthropic dollars has forged permanent improvements in health through creativity, risk-taking, and boldness. The award recipient's achievements exemplify responsiveness to people in communities, large or small, with a particular emphasis on reaching those in greatest need. Nominations are due by the listed deadline application date.

Requirements: The nominee must be, or have been within the last three years, an active grantmaker for at least five years with a U.S. grantmaking organization in the field of health, serving as a trustee, staff member, or donor. There must be evidence of significant contributions to the field of health philanthropy and leadership in the philanthropic community by the nominee. Guidelines and nomination form are available online.

Restrictions: Current Grantmakers in Health board members are not eligible for nomination.

Date(s) Application is Due: Sep 16

Amount of Grant: 5,000 U.S.D

Contact: Kate Treanor, (202) 452-8331; fax (202) 452-8340; ktreanor@gih.org

Internet: http://www.gih.org/calendar_url2665/calendar_url_show.htm?doc_id=261034

Sponsor: Grantmakers in Health
1100 Connecticut Avenue NW, Suite 1200
Washington, DC 20036

Texas Commission on the Arts Arts Respond Project Grants 4469

This program provides project assistance grants on a short-term basis and may include administrative costs directly related to the project. Projects must address one of the following priority areas: education, health and human services, economic development, public safety, criminal justice, natural resources, or agriculture. Organizations that are eligible are: Arts Organizations, Established Arts Organizations, Minority Arts Organizations, Rural Arts Providers.

Requirements: To be eligible for TCA grants, an organization must: be a tax-exempt nonprofit organization as designated by the Internal Revenue Service and/or must be an entity of government; be incorporated in Texas; have fulfilled all its outstanding contractual obligations to the State of Texas (i.e. student loans, child support, taxes, etc.); and comply with regulations pertaining to federal grant recipients including Title VI of the Civil Rights Act of 1964, Section 504 of the Rehabilitation Act of 1973, the Age Discrimination Act of 1975, the Education Amendments of 1972, the Americans with Disabilities Act of 1990, and the Drug Free Workplace Act of 1988.

Geographic Focus: Texas

Amount of Grant: Up to 45,000 U.S.D

Contact: Director of Programs; (512) 463-5535 or (800) 252-9415; fax (512) 475-2699; front.desk@arts.state.tx.us

Internet: http://www.arts.state.tx.us/index.php?option=com_wrapper&view=wrapper&Itemid=86

Sponsor: Texas Commission on the Arts
P.O. Box 13406
Austin, TX 78711-3406

Texas Instruments Corporation Health and Human Services Grants 4470

The Corporation builds healthy communities by funding health and human service programs that address the needs of the community. The Health and Human Services grants are primarily program specific, with limited support to United Way agencies. The Corporation board meets quarterly, and funding decisions are made within three weeks of each meeting.

Requirements: Applicants must be 501(c)3 tax-exempt nonprofit organizations. Grant applications are only accepted online. Applicants access the eligibility quiz at the website and if they are eligible, they gain access to the application.

Restrictions: Corporate grant funding is prioritized based on opportunities for TI employee/retiree involvement. Due to the TI Foundation's and TI employees' long-time support of United Way, only limited support of United Way agencies will be considered. The Corporation does not support the following: organizations that are not 501(c)3 charities; grants to individuals, including sponsorships; private foundations or endowment funds; sectarian or denominational religious organizations; political activities, parties or candidates; veteran organizations; hospitals; fraternal or labor organizations; courtesy advertising, including program books and yearbooks; entertainment events, scholarships or conferences; sporting events or teams; golf tournaments; unrestricted gifts to national or international organizations; travel or tours; or table sponsorships.

Geographic Focus: Texas

Contact: Andy Smith, Director of Corporate Philanthropy; (214) 480-3462; fax (214) 480-2920; wasmith@ti.com

Internet: http://www.ti.com/corp/docs/csr/giving.shtml

Sponsor: Texas Instruments Corporation
12500 TI Boulevard
Dallas, TX 75266-0199

Texas Instruments Foundation Community Services Grants 4471

The Foundation offers funding to enrich health and human services programs that meet the greatest community needs. This includes capital or civic campaigns. The Foundation considers supporting such campaigns based on community needs and economic impact. The Foundation primarily supports services through its annual United Way donations. Beyond United Way, it also provides limited support to organizations meeting the greatest community needs. The Foundation board meets quarterly (March, June, September, and November) and makes funding decisions within three weeks of each meeting.

Requirements: Applicants must be 501(c)3 nonprofit organizations. Grant applications are only accepted online. Applicants access the eligibility quiz at the website and if they are eligible, they gain access to the application. The Foundation primarily supports programs in the Dallas area.

Restrictions: The Foundation does not support the following: organizations that are not 501(c)3 charities; grants to individuals, including sponsorships; private foundations or endowment funds; sectarian or denominational religious organizations; political activities, parties or candidates; veteran organizations; hospitals; fraternal or labor organizations; courtesy advertising, including program books and yearbooks; entertainment events, scholarships or conferences; sporting events or teams; golf tournaments; unrestricted gifts to national or international organizations; travel or tours; or table sponsorships.

Geographic Focus: Texas

Contact: Ann Pomykal; (214) 480-6873; fax (214) 480-6820; giving@ti.com

Internet: http://www.ti.com/corp/docs/csr/giving.shtml

Sponsor: Texas Instruments Foundation
12500 TI Boulevard
Dallas, TX 75266-0199

Texas Instruments Foundation STEM Education Grants 4472

The Foundation funds programs that emphasize math and science education, including STEM (science, technology, engineering and math) teacher effectiveness. Grant decisions are made within three weeks of the Foundation board meetings (March, June, September, and November).

Requirements: Applicants must be 501(c)3 tax-exempt nonprofit organizations. Grant applications are only accepted online. Applicants access the eligibility quiz at the website and if they are eligible, they will then gain access to the application. The Foundation primarily supports programs in the Dallas area.

Restrictions: The Foundation does not support the following: organizations that are not 501(c)3 charities; grants to individuals, including sponsorships; private foundations or endowment funds; sectarian or denominational religious organizations; political activities, parties or candidates; veteran organizations; hospitals; fraternal or labor organizations; courtesy advertising, including program books and yearbooks; entertainment events, scholarships or conferences; sporting events or teams; golf tournaments; unrestricted gifts to national or international organizations; travel or tours; or table sponsorships.

Geographic Focus: Texas

Contact: Ann Pomykal; (214) 480-6873; fax (214) 480-6820; giving@ti.com

Internet: http://www.ti.com/corp/docs/csr/giving.shtml

Sponsor: Texas Instruments Foundation
12500 TI Boulevard
Dallas, TX 75266-0199

Textron Corporate Contributions Grants 4473

Textron Inc. was founded in 1923 as a small textile company and has since become one of the world's best known multi-industry companies. Textron focuses philanthropic giving in the following areas: workforce development and education; healthy families/vibrant communities; and sponsorships. In the area of workforce development and education, the company focuses on job-training and employment development (eg., school-to-work, welfare-to-work, job-training for underserved-audiences, literacy, and English-as a-Second-Language programs). In the area of healthy families/vibrant communities, the company focuses on arts and culture (with emphasis on outreach programs that enhance learning and target low- and moderate-income individuals), community revitalization (eg., affordable housing and economic development in low-income areas), and health and human-service organizations (eg., food pantries, homeless shelters, and services for low-income residents). In the area of sponsorships, the company encourages volunteerism and sponsors worthwhile events that benefit the communities where employees live and work. Textron's grant history has included funding of general-operating costs, capital campaigns, building construction/renovation, equipment acquisition, program development, conferences and seminars, publication, seed money, fellowships, scholarship funds, research, technical assistance, consulting services, and matching funds. Downloadable guidelines (PDF) and application (Word document) are available online. The completed application and required accompanying documentation must be received via mail by the deadline date.

Requirements: Textron targets its giving to nonprofit agencies located in its headquarters state of Rhode Island and those locations where the company has divisional operations. Organizations outside of Rhode Island should contact the Textron company in their area; a listing of Textron businesses along with their contact information can be accessed by clicking the "Contact Us" link at the Textron website.

Restrictions: Textron will review only one request per organization during a 12-month period. Contributions will not be made to the following types of organizations: organizations without 501(c)3 status as defined by the Internal Revenue Service; individuals; political, fraternal or veterans organizations; religious institutions when the grant would support sectarian activities; and organizations that discriminate by race, creed, gender, ethnicity, sexual orientation, disability, age or any other basis prohibited by law.

Geographic Focus: Georgia, Illinois, Kansas, Louisiana, Maryland, Massachusetts, New York, North Carolina, Pennsylvania, Rhode Island, Texas, Germany, Great Britain

Date(s) Application is Due: Mar 1; Sep 1

Samples: The Providence Center, Providence, Rhode Island, $10,000 - to provide training opportunities in receptionist, production, driver, and computer vocations (2011).

Contact: Karen Warfield; (401) 421-2800; fax (401) 457-2225

Internet: http://www.textron.com/about/commitment/corp-giving/

Sponsor: Textron Charitable Trust
40 Westminster Street
Providence, RI 02903

Thelma Braun and Bocklett Family Foundation Grants 4474

The Thelma Braun and Bocklett Family Foundation was established in 1980 to support and promote quality education, cultural, human-services, and health-care programming for underserved populations. Miss Braun was a gifted musician who spent many years playing the organ and piano for area organizations and churches. She never married and had no children of her own, but she truly loved the young people in the area. Special consideration is given to charitable organizations that serve the people of Grayson County, Texas, especially in the areas of arts and education. Applicants must apply online at the grant website. The Thelma Braun and Bocklett Family Foundation has rolling deadlines and decision dates. Applicants are strongly encouraged to do the following before applying: review the downloadable state application procedures at the grant website; review the downloadable online-application guidelines at the grant website; review the foundation's funding history (link is available from the grant website); review the online application questions in advance; and review the list of required attachments. These will generally include: a list of board members, financial statements (audited, reviewed, or compiled by independent auditor); an organization summary; a list of other funding sources; an IRS Determination letter; and other required documents. All attachments must be uploaded in the online application as PDF, Word, or Excel files.

Requirements: Applicants must have 501(c)3 tax-exempt status.

Restrictions: The foundation does not support requests from individuals, organizations attempting to influence policy through direct lobbying, or any political campaigns.

Geographic Focus: Texas

Samples: Denison Independent School District, Denison, Texas, $30,000 (2010); Grayson County Shelter, Denison, Texas, $25,000 (2010); Downtown Improvement Company of Denison, Denison, Texas, $15,000 (2010).

Contact: David Ross, Senior Vice President; tx.philanthropic@baml.com

Internet: https://www.bankofamerica.com/philanthropic/fn_search.action

Sponsor: Thelma Braun and Bocklett Family Foundation
901 Main Street, 19th Floor, TX1-492-19-11
Dallas, TX 75202-3714

Thelma Doelger Charitable Trust Grants 4475

The trust awards grants in the San Francisco Bay area to zoos, hospitals, and organizations supporting animal welfare and senior citizens. Grants are awarded for general purposes and to support operating budgets. There are no application deadlines. Application forms are required.

Requirements: Nonprofit organizations in the San Francisco Bay area of northern California may submit grant requests.

Restrictions: Individuals are not eligible.

Geographic Focus: California

Amount of Grant: 2,000 - 75,000 U.S.D

Samples: San Francisco Zoo (CA)—for operating support, $100,000; Marin Humane Society (Novato, CA)—for operating support, $70,000.

Contact: D. Eugene Richard, Trustee; (650) 755-2533

Sponsor: Thelma Doelger Charitable Trust
950 Daly Boulevard, Suite 300
Daly City, CA 94015-3004

Theodore Edson Parker Foundation Grants 4476

The Theodore Edson Parker Foundation's primary goal is to make effective grants that benefit the city of Lowell, Massachusetts, and its residents. Grants are made for a variety of purposes including social services, cultural programs, community development activities, education, community health needs, and urban environmental projects. The Foundation funds specific needs including special programs and projects, capital improvements and equipment purchases, and technical assistance. In his will, Mr. Parker suggested that his trustees give special consideration to the welfare of children, disadvantaged young women, and the elderly. The Foundation is especially committed to assisting these and other underserved individuals including refugees, immigrants, and people of color. Preference is given to applicants who have formulated creative approaches to societal problems and can provide leverage for Foundation funds. The Foundation favors projects that can demonstrate good prospects for continuation after the conclusion of Foundation funding. Interested applicants are welcome to contact staff prior to applying. All applications are accepted online.
Requirements: Eligible applicants are 501(c)3 organizations in Lowell, Massachusetts.
Restrictions: The Foundation does not usually fund the operating expenses of charitable organizations, endowments, or fund deficits. Applicants are limited to one per year. Grant recipients should expect to wait two years before submitting a new request.
Geographic Focus: Massachusetts
Date(s) Application is Due: Jan 15; May 15; Sep 15
Amount of Grant: 10,000 - 50,000 U.S.D
Samples: 911 Electronic Media Arts, Lowell, Massachusetts, $10,000 - support for paid coordinator for small, volunteer-led gallery (2011); Lowell Wish Project, Inc., Lowell, Massachusetts, $20,000 - expansion of the organizaion's Emergency Support Program (2011); and Girls Incorporated for Greater Lowell, Lowell, Massachusetts, $50,000 - renovations to create space for a program serving older girls (2011).
Contact: Philip Hall; (617) 391-3097 or (617) 426-7087; phall@grantsmanagement.com
Internet: http://parkerfoundation.gmafoundations.com/
Sponsor: Theodore Edson Parker Foundation
77 Summer Street, 8th Floor
Boston, MA 02110-1006

Theodore H. Barth Foundation Grants 4477

The foundation awards grants in its areas of interest, including education, environment and conservation, health, arts and culture, the disabled, and social services. There are no application deadlines. Submit a general letter of interest including a brief description of the problem, amount requested from the foundation, statement of project's main objectives, other planned sources of support, program description, anticipated results, plans to evaluate or publish project results, how project will be supported after the grant, qualifications of personnel, name and contact information, grant timetable, copy of IRS tax-exemption letter, total estimated project budget, and copy of most recent audited financial budget.
Requirements: Nonprofit organizations in Massachusetts and New York are eligible.
Restrictions: Student aid is not offered.
Geographic Focus: Maryland, New Jersey, Pennsylvania, Virginia
Amount of Grant: 5,000 - 10,000 U.S.D
Contact: Ellen Berelson, President; (212) 332-3466; barthfoundation@earthlink.net
Sponsor: Theodore H. Barth Foundation
45 Rockefeller Plaza, Suite 2006
New York, NY 10111

The Ray Charles Foundation Grants 4478

Although musician Ray Charles was blind since the age of seven, he always maintained that blindness was not a handicap, but rather that the inability to hear music constituted the truer loss. Because of this philosophy, he began making contributions to the field of hearing impairment and often anonymously contributed to cochlear implants for individuals who couldn't afford such surgeries. In 1986, Mr. Charles founded "The Robinson Foundation for Hearing Disorders" which was later renamed "The Ray Charles Foundation." Convinced of the tremendous need of education for youth, Mr. Charles also directed the foundation to make donations and support institutions and organizations for educational purposes. Over the past 23 years, the Ray Charles Foundation has provided financial donations to various institutions involved in the areas of hearing disorders as well as education. The foundation prefers to fund programs and projects that show promise of strengthening the community beyond the grant period and that offer maximum community impact to achieve long-term results. Types of funding the foundation provides are for capital, program, and core support. Capital funding is available for land, facility, and equipment purchases and renovations or new construction. Renovation and construction projects will be invited only for organizations that have already raised a substantial amount of their fundraising goal. The project must either be in construction or have a firm construction start date. Program funding is available for both new and expansion projects. While these must evidence a viable fundraising and sustainability plan, requests for program development or enhancement activiies will still be considered. Core support funding is available for established and well-managed organizations and programs. Communication with Foundation Directors is discouraged. Applicants to the Ray Charles Foundation should simply fill out and submit the foundation's Grant Application Packet via a mail carrier to the address given. Guidelines, instructions, and links to form are given at the website. Applications are accepted continually throughout the year.
Requirements: 501(c)3 organizations that have been in existence for at least three years and that are incorporated and delivering services to the United States are eligible to apply. Programs must advance the mission of the Foundation in the following program categories: Education, Hearing Disorders, and Culture and the Arts. Educational programs eligible for funding include, but are not limited to, after-school programs and academic activities aimed at college preparation for underprivileged youth, and academic/therapy services for youth who are blind and/or deaf. Hearing Disorder programs eligible for funding include research and treatment in the area of hearing disorders and educational programs and resources for youth with hearing disorders. Culture and Arts programs eligible for funding include institutions and museums whose mission it is to provide musical and cultural education and access to underprivileged and disadvantaged youth.
Restrictions: The Foundation does not fund organizations and programs outside of the United States. Most grants will be limited to one year in duration. Multi-year funding will be considered on an exceptional case by case basis. The Foundation generally does not approve grants to organizations on a continuing annual basis. If an agency applies for a 'one year grant' more than three years in a row, they will be declined and asked to wait at least two years before applying again. More obligations and responsibilities of grant recipients are listed at the website.
Contact: Grant Coordinator; (323) 737-8000; fax (323) 737-0148; info@theraycharlesfoundation.org
Internet: http://www.theraycharlesfoundation.org/GrantQualifications.html
Sponsor: Ray Charles Foundation
2107 W. Washington Boulevard
Los Angeles, CA 90018

Thomas and Agnes Carvel Foundation Grants 4479

This foundation awards support for general charitable purposes as well as for nutrition research and youth programs to nonprofit organizations in the tri-state area comprising New York, New Jersey, and Connecticut. Types of support include general operations grants, capital grants, grants in aid, program/project grants, matching gifts, and research grants. Application should be by letter and should outline the project, budget, and staffing of the project for which funding is being requested.
Requirements: 501(c)3 organizations in New York, New Jersey, and Connecticut are eligible.
Restrictions: Grants are not made to individuals.
Geographic Focus: Connecticut, New Jersey, New York
Date(s) Application is Due: Oct 1
Amount of Grant: 2,000 - 200,000 U.S.D
Samples: Seton Hall (South Orange, NJ)—to provide full scholarships to 10 high-school students from Newark, NJ, who commit to a year of volunteer service in Newark upon graduation, $50,000.
Contact: Ann McHugh, Vice President; (914) 793-7300
Sponsor: Thomas and Agnes Carvel Foundation
35 E Grassy Sprain Road
Yonkers, NY 10710

Thomas and Dorothy Leavey Foundation Grants 4480

Thomas Leavey and partner John C. Tyler began the Farmer's Insurance Company in 1928. The foundation was begun in 1952 by Leavey and his wife, Dorothy. Thomas Leavey died in 1980 and Dorothy actively led the foundation until her death in 1998. Giving primarily in southern California, the Foundation offers support for: hospitals, medical research, higher and secondary education, Catholic church groups; and provides scholarships to children of employees of Farmers Group.
Requirements: The foundation gives primarily in southern California.
Geographic Focus: California
Amount of Grant: 10,000 - 2,000,000 U.S.D
Samples: Childrens Museum of Los Angeles, Lake View Terrace, CA, $250,000; Santa Clara University, Santa Clara, CA, $2,000,000; Orthopaedic Hospital, Los Angeles, CA, $100,000.
Contact: Kathleen Leavey McCarthy, Chair; (310) 551-9936
Sponsor: Thomas and Dorothy Leavey Foundation
10100 Santa Monica Boulevard, Suite 610
Los Angeles, CA 90067-4110

Thomas Austin Finch, Sr. Foundation Grants 4481

The purpose of the Finch Foundation is the improvement of the mental, moral and physical well-being of the inhabitants of Thomasville, North Carolina, with emphasis on improving education, improving health related facilities and attracting new business to the community. However, grants are made for a wide variety of charitable causes throughout the greater Thomasville area. In general, grants are awarded for seed money, matching funds and general purposes. Primary consideration is given to projects of a non-recurring nature or to start up funding of limited duration. The foundation meets twice a year to review grant requests. Requests must be submitted by February 15th for the Spring Meeting and October 15th for the Fall Meeting. Application forms are available online. Applicants will receive notice acknowledging receipt of the grant request, and subsequently be notified of the grant declination or approval.
Requirements: Eligible applicants are IRS 501 (c)3 non-profit organizations located within the corporate limits of the City of Thomasville, North Carolina or, organizations where the benefits of the grant will inure primarily to the residents of Thomasville. Proposals should be submitted in the following format: completed Common Grant Application Form; an original Proposal Statement; an audited financial report and a current year operating budget; a copy of your official IRS Letter with your tax determination; a listing of your Board of Directors. Proposal Statements (second item in the above Format) should answer these questions: what are the objectives and expected outcomes of this program/project/request; what strategies will be used to accomplish your objective; what is the timeline for completion; if this is part of an on-going program, how long has it been in operation; what criteria will you use to measure success; if the request is not fully funded, what other sources can you engage; an Itemized budget should be included; please describe any collaborative ventures. Prior to the distribution of funds, all approved

grantees must sign and return a Grant Agreement Form, stating that the funds will be used for the purpose intended. Progress reports and Completion reports must also be filed as required for your specific grant. All current grantees must be in good standing with required documentation prior to submitting new proposals to any foundation.
Restrictions: Grants are generally not made for typical operational or maintenance-oriented purposes, political purposes, nor to organizations which discriminate on the basis of race, ethnic origin, sexual or religious preference, age or gender.
Geographic Focus: North Carolina
Date(s) Application is Due: Feb 15; Oct 15
Amount of Grant: 1,500 - 100,000 U.S.D
Samples: Memorial United Methodist Church, $30,000—general operating funds; Hospice of Davidson County, Inc., $60,000—construction of inpatient facility; Thomasville City Schools, $24,400—active classroom project.
Contact: Wachovia Bank, N.A., Trustee; grantinquiries6@wachovia.com
Sponsor: Thomas Austin Finch, Sr. Foundation
Wachovia Bank, NC6732, 100 North Main Street
Winston Salem, NC 27150

Thomas C. Ackerman Foundation Grants **4482**
The Thomas C. Ackerman Foundation was founded on August 6, 1991. By Spring, 1992, substantial assets had been received from the Thomas C. Ackerman Trust, funded as a result of Mr. Ackerman's death on February 13, 1991; the Board of Directors was fully constituted; and the guidelines and objectives for grant making were adopted. The Foundation's areas of interest include: arts and culture; education; and health and human services. To be considered for a grant from the Thomas C. Ackerman Foundation, an organization must initially submit a Letter of Intent online. Directions for completing the grant application will be provided to an organization, if it is invited to submit one. Awards generally range from $1,000 to $40,000.
Requirements: California nonprofits, primarily in San Diego County, are eligible.
Restrictions: It is a policy of the Foundation not to provide continuous support for any project to the extent that the project becomes dependent upon the Foundation for its continued existence. The Foundation may make multi-year commitments on occasion. Grants are not made to individuals. Generally, the Foundation does not support conferences or symposia. While occasional grant support is given to religious organizations, those grants are made for direct support of nonsectarian educational or service projects and not for projects which are of primary benefit to members of a particular religion or belief or which primarily promote a particular religion or belief. The Foundation will not consider grants relating to human medical or biomedical research.
Geographic Focus: California
Amount of Grant: 1,000 - 40,000 U.S.D
Samples: San Diego Museum of Art, San Diego, California, $5,000 - general operating support (2011); Nativity Prep Academy, San Diego, California, $34,000 - general operating support (2011); University of San Diego,San Diego, California, $40,000 - general operating support (2011).
Contact: Lynne Newman; (619) 741-0113; info@AckermanFoundation.org
Internet: http://www.ackermanfoundation.org/
Sponsor: Thomas C. Ackerman Foundation
600 W Broadway, Suite 2600
San Diego, CA 92101-3391

Thomas C. Burke Foundation Grants **4483**
The mission of The Thomas C. Burke Foundation is to support charitable organizations that ease human suffering and promote the health of people afflicted by disease, physical weakness, physical disability, and physical injury. The Foundation has a strong interest in programs focused on cancer prevention and treatment. Grants from The Thomas C. Burke Foundation are primarily one year in duration. On occasion, multi-year support is awarded. Applications should be submitted through the online application system by 11:59 p.m. on July 1. The link to the online application system is available at the grant website. Applicants are strongly encouraged to do the following before applying: review the downloadable state application procedures for additional helpful information and clarifications; review the downloadable online-application guidelines at the grant website; review the foundation's funding history (link is available from the grant website); review the online application questions in advance; and review the list of required attachments. These will generally include: a list of board members, financial statements (audited, reviewed, or compiled by independent auditor); an organization summary; a list of other funding sources; an IRS Determination letter; and other required documents. All attachments must be uploaded in the online application as PDF, Word, or Excel files. Applicants will be notified of grant decisions by letter within three to four months after the deadline.
Requirements: Charitable organizations must have 501(c)3 tax-exempt status and serve Bibb County and its surrounding communities. A breakdown of number/percentage of people served by specific counties is required on the online application.
Restrictions: Applicants will not be awarded a grant for more than three consecutive years. The foundation does not support requests from individuals, organizations attempting to influence policy through direct lobbying, or any political campaigns.
Geographic Focus: Georgia
Date(s) Application is Due: Jul 1
Contact: Quanda Allen, Vice President; (404) 264.1377; quanda.allen@baml.com
Internet: https://www.bankofamerica.com/philanthropic/fn_search.action
Sponsor: Thomas C. Burke Foundation
3414 Peachtree Road, N.E., Suite 1475 GA7-813-14-04
Atlanta, GA 30326-1113

Thomas F. and Kate Miller Jeffress Memorial Trust Grants **4484**
The Thomas F. and Kate Miller Jeffress Memorial Trust was established in 1981 to support basic research in chemical, medical, and other scientific research through grants to educational and research organizations in the Commonwealth of Virginia. The trust was established under the will of Mr. Robert M. Jeffress, a business executive and philanthropist of Richmond, Virginia. Grants are made to assist scientists in educational and research institutions to conduct investigations in the natural sciences, generally considered to include chemistry, biology (with the exception of field studies, classification, or other largely observational studies), and the basic medical sciences, such as biochemistry, microbiology and others. The principal evaluation criteria will be the scientific significance of the proposed work and the competence of the investigator. The trust is particularly interested in supporting fundamental research by scientists early in their careers and new areas of research or more speculative projects by established investigators. In general, grants will cover direct-expense items essential to the successful completion of the proposed project. Grants from the Jeffress Memorial Trust are one year in duration. Funding may be renewed at the discretion of the allocations committee. Application materials are available for download at the grant website. Applicants are strongly encouraged to review the state application guidelines for additional helpful information and clarifications before applying. Applicants are also encouraged to review the trust's funding history (link is available from the grant website). The Jeffress Memorial Trust has biannual deadlines of March 1 and September 1. Applicants for the March deadline will be notified of grant decisions by June 15. Applicants for the September deadline will be notified by December 31.
Requirements: Institutions requesting support for programs will be expected to give evidence of their interest in the programs by providing usual supplies, services, and indirect costs. Applications must be mailed.
Restrictions: Projects will be funded for a maximum of three years, including any renewals. The Jeffress Trust normally will not support clinical research (diagnosis or treatment of disease, abnormality, or malfunctions in people or animals or testing of drugs or procedures for their treatment). The trust normally will not provide: fringe benefits, indirect costs, common supplies and services, secretarial services, academic stipends or salaries, maintenance for institutionally-owned equipment; instrument or computer time, or travel to regularly scheduled meetings of societies or organizations.
Geographic Focus: Virginia
Date(s) Application is Due: Mar 1; Sep 1
Samples: Virginia Commonwealth University Foundation, Richmond, Virginia, $201,004 (2010); Edward Via Virginia College of Osteophathic Medicine, Blacksburg, Virginia, $25,000 (2010); University of Richmond, Richmond, Virginia, $50,000 (2010).
Contact: Sarah Kay, Vice President; (804) 788-2673; sarah.kay@baml.com
Internet: http://www.wm.edu/grants/OPPS/jeffress.htm
Sponsor: Thomas F. and Kate Miller Jeffress Memorial Trust
1111 E. Main Street, VA2-300-12-92
Richmond, VA 23219

Thomas Foundation Grants **4485**
The foundation awards grants to eligible nonprofit organizations in its areas of interest, including medical research and charitable giving. Grants are awarded for one year and may be renewed. Submit a written proposal and include tax-exemption documentation.
Requirements: 501(c)3 tax-exempt organizations serving southeastern Michigan and New York are eligible.
Geographic Focus: Michigan, New York
Amount of Grant: 1,000 - 115,000 U.S.D
Contact: Jay Brody, Director; (248) 528-1111
Sponsor: Thomas Foundation
201 W Big Beaver Road, Suite 600
Troy, MI 48084-4161

Thomas J. Atkins Memorial Trust Fund Grants **4486**
The Thomas J. Atkins Memorial Trust Fund was established in 1946 to support and promote quality educational, human services, and health care programming for underserved populations. The Trust Fund specifically serves the people of Middlesex County, Connecticut. Preference is given to organizations and programs that provide for the relief of human distress and suffering. The deadline for application is October 1. Applicants will be notified of grant decisions by letter within 2 to 3 months after the proposal deadline. Grants from the Atkins Memorial Trust Fund are 1 year in duration.
Requirements: 501(c)3 organizations serving the residents of Middlesex, Connecticut, are eligible to apply.
Geographic Focus: Connecticut
Date(s) Application is Due: Oct 1
Contact: Kate Kerchaert; (860) 657-7016; kate.kerchaert@baml.com
Internet: https://www.bankofamerica.com/philanthropic/grantmaking.action
Sponsor: Thomas J. Atkins Memorial Trust Fund
200 Glastonbury Boulevard, Suite #200, CT2-545-02-05
Glastonbury, CT O6033-4056

Thomas J. Long Foundation Grants **4487**
The foundation awards grants in California primarily for colleges and universities, libraries and library science, youth organizations, and health associations. Initial approach should be a letter requesting guidelines.
Requirements: California and Hawaii nonprofits are eligible.
Geographic Focus: California, Hawaii
Date(s) Application is Due: Apr 1; Oct 1
Amount of Grant: 5,000 - 50,000 U.S.D

Samples: U of the Pacific (Stockton, CA)—for the School of Pharmacy and Health Sciences, $13 million (2001); Santa Rosa Junior College (Santa Rosa, CA)—to equip the Community Health Education Ctr's computer classroom and to establish the pharmacy technician program, $100,000.
Contact: Marcia Sander, (925) 944-3800; fax (925) 944-3573
Sponsor: Thomas J. Long Foundation
2950 Buskirk Avenue, Suite 160
Walnut Creek, CA 94596

Thomas M. and Irene B. Kirbo Charitable Trust Grants 4488
The trust awards grants to nonprofit organizations in support of religion, hospitals, universities, and youth. The trust supports higher education, libraries, hospitals, youth organizations, and Christian agencies and churches. Application forms must be requested.
Requirements: Nonprofit organizations in Florida and Georgia are eligible.
Restrictions: Grants are not made to individuals.
Geographic Focus: Florida, Georgia
Amount of Grant: 1,000 - 125,000 U.S.D
Samples: Elon College (Elon College, NC)—for a new library, $150,000.
Contact: R. Murray Jenks, President; (904) 354-7212
Sponsor: Thomas M. and Irene B. Kirbo Charitable Trust
112 W Adams Street, Suite 1111
Jacksonville, FL 32202

Thomas Sill Foundation Grants 4489
The foundation provides encouragement and financial support to qualifying Manitoba organizations that strive to improve the quality of life in the province. Areas of interest include the performing and visual arts, historic preservation and historic sites, museums and libraries, adult and continuing education, literacy, universities, health, the environment, social sciences, community legal issues, special education, children and abused women, the visually and hearing disabled, physically disabled children, the mentally and learning disabled, sheltered workshops, and camps and day care programs. Funding priority is capital grants, pilot projects, and demonstration projects. Operating grants are rarely issued, and when awarded are usually for start-up purposes. Decisions are made monthly. Application forms and guidelines are available upon request.
Requirements: Charitable organizations may apply for grant support.
Geographic Focus: All States, Canada
Amount of Grant: 1,000 - 50,000 CAD
Samples: Misericordia General Hospital Foundation (Winnipeg, Manitoba)—for computer software, $C13,000; Gifts for Kids (Manitoba)—for Christmas gifts for needy children in remote locations, $C5000; Manitoba Theatre Centre (Manitoba)—for a new lighting system, $C10,000.
Contact: Program Contact; (204) 947-3782; fax (204) 956-4702
Internet: http://www.thomassillfoundation.com
Sponsor: Thomas Sill Foundation
600-175 Hargrave Street
Winnipeg, MB R3C 3R8 Canada

Thomas Thompson Trust Grants 4490
The Thompson Trust limits its distribution of funds to organizations located in Windham County, Vermont, (primarily the Town of Brattleboro) or in Dutchess County, New York (primarily the Town of Rhinebeck) which predominately serve residents located in those areas. The Will of Thomas Thompson was executed in 1867 and defined his charitable purposes rather narrowly as was the practice with 19th Century philanthropy. By successive court decrees in the 20th Century, the Trustees are now authorized to make grants to charitable organizations whose work and purposes promote health, education or the general social or civic betterment in the stated geographical areas; but the Trustees will continue to place particular emphasis on healthcare and other social services. The Trustees generally meet on the fourth Thursday in January, April, July and October. Applications are due by the first day of the month of the meeting.
Requirements: Organizations must have operated for three consecutive years before applying and be located in Windham County, VT, with preference given to Brattleboro; or Duchess County, New York, with preference given to Rhinebeck.
Restrictions: Grants are not made for general operating support, seed money, endowment purposes, or loans.
Geographic Focus: New York, Vermont
Date(s) Application is Due: Jan 1; Apr 1; Jul 1; Oct 1
Amount of Grant: 1,000 - 20,000 U.S.D
Contact: Susan T. Monahan, Program Contact; (617) 951-1108; fax (617) 542-7437; smonahan@rackemann.com
Internet: http://www.cybergrants.com/thompson/grant.html
Sponsor: Thomas Thompson Trust
160 Federal Street, 13th Floor
Boston, MA 02110-1700

Thomas W. Bradley Foundation Grants 4491
Established in 1976 in Maryland, the Thomas W. Bradley Foundation offer grants in the greater metropolitan Baltimore, Maryland, area. Funding is limited to organizations which work with or benefit mentally or physically handicapped children. Primary fields of interest include: children, children's services, and the developmentally disabled. Funding comes in the form of general operating support. Application forms are required, and applicants should submit the following: name, address and phone number of organization; copy of IRS Determination Letter; a brief history of the organization and description of its mission; a listing of board of directors, trustees, officers and other key people and their affiliations; a detailed description of project and amount of funding requested; and a copy of the current year's organizational budget and project budget. There are no specific deadlines with which to adhere.
Requirements: Only 501(c)3 organizations that serve the residents of Maryland should apply, and the Foundation gives preference in the Baltimore metropolitan area.
Restrictions: No grants, scholarships, fellowships, prizes, or similar benefits are made to individuals.
Geographic Focus: Maryland
Amount of Grant: 3,000 - 6,000 U.S.D
Samples: Muscular Dystrophy Association, Baltimore, Maryland, $5,000 - general operating support (2011); Sheppard Pratt Health Center, Baltimore, Maryland, $5,000 - for general operations (2011); Maryland School for the Blind, Baltimore, Maryland, $5,000 - general operating support (2011).
Contact: Robert L. Pierson, Trustee; (410) 821-3006; fax (410) 821-3007; info
Sponsor: Thomas W. Bradley Foundation
305 W. Chesapeake Avenue, Suite 308
Towson, MD 21204-4440

Thompson Charitable Foundation Grants 4492
Established in 1987, the Foundation supports organizations involved with education, Christian organizations, Autism research, health, human services, youth, including funding for capital and building improvements for human service organizations and educational institutions. Giving is limited to Bell, Clay, Laurel, and Leslie counties, Kentucky; Anderson, Blount, Knox, and Scott counties, Tennessee; and Buchanan and Tazewell counties, Virginia.
Requirements: Nonprofits operating in Bell, Clay, Laurel, and Leslie counties, Kentucky; Anderson, Blount, Knox, and Scott counties, Tennessee; and Buchanan and Tazewell counties, Virginia are eligible to apply. Initial contact should be a letter, no more then 2 pages long. The letter should include: a statement of the problem, the project will address; a detailed description of project, and amount of funding requested.
Restrictions: No support for religious or political organizations, budget deficits or endowments.
Geographic Focus: Kentucky, Tennessee, Virginia
Date(s) Application is Due: Mar 31; Sep 1
Amount of Grant: 2,000 - 500,000 U.S.D
Contact: Debbie Black, Executive Director; (865) 588-0491; fax (865) 588-4496; debbie@cf.org
Sponsor: Thompson Charitable Foundation
800 South Gay Street, Suite 2021
Knoxville, TN 37929-9710

Thomson Reuters / MLA Doctoral Fellowships 4493
The purpose of the Thomson Reuters/MLA Doctoral Fellowship is to foster and encourage superior students to conduct doctoral work in health-sciences librarianship or information science by providing support to individuals who have been admitted to candidacy. The award was originally established in 1986 by the Institute of Scientific Information (ISI, which is now a part of Thomson Reuters) and is administered by the Medical Library Association (MLA). The award is granted bi-annually in even-numbered years. The fellowship pays $2,000 in one payment toward the following types of expenses: project expenses; related travel expenses; or augmentation of another larger, separately-funded project relevant to health-science librarianship and which is part of the requirements of a doctoral degree. Guidelines and application materials are available at the website. All materials and letters of support must be received by MLA via email, fax, or mail by December 1. email applications and documents will be accepted as PDF or MS Word files only. Founded in 1898, the Medical Library Association (MLA) is a nonprofit, educational organization of more than 1,100 institutions and 3,600 individual members in the health sciences information field, committed to educating health-information professionals, supporting health-information research, promoting access to the world's health-sciences information, and working to ensure that the best health information is available to all.
Requirements: Applicants must satisfy the following eligibility criteria: be graduates of an American Library Association (ALA)-accredited school of library science; be candidates in a Ph.D. program with emphsis on biomedical and health-related information science; and be citizens of or have permanent residence status in the United States or Canada. Preference will be given to applicants who have at least 75% of their course work completed and dissertation prospectus either approved or in the approval process.
Restrictions: Past recipients of the Thomson Reuters / MLA Doctoral Fellowship are ineligible. The following types of expenses are not supported: tuition fees; tuition-related expenses; and living expenses.
Geographic Focus: All States, Canada
Date(s) Application is Due: Dec 1
Amount of Grant: Up to 2,000 U.S.D
Contact: Carla Funk, Program Contact; (312) 419-9094, ext. 14; fax (312) 419-8950; mlapd2@mlahq.org or grants@mlahq.org
Internet: http://www.mlanet.org/awards/grants/doctoral.html
Sponsor: Medical Library Association
65 East Wacker Place, Suite 1900
Chicago, IL 60601-7246

Thoracic Foundation Grants 4494
The major emphasis of the Thoracic Foundation of Boston is to offer support for medical research in thoracic diseases. Contributions are also made to hospitals providing service to thoracic and cardiac cases. Fields of interest include: asthma research; cancer research; children and youth services; heart and circulatory diseases; specialty hospitals; medical school education; and smoking cessation. In addition, funds are awarded to assist with

publications, conferences, and seminars in the same field. Grants are awarded thoughout Massachusetts, but primarily in the Boston metropolitan area. A letter of application may be submitted at any time, offering a detailed description of the project and amount of funding requested. Awards range from $5,000 to $75,000.
Geographic Focus: District of Columbia, Massachusetts
Samples: Action on Smoking and Health, Washington, DC, $5,000 - funding for non-smokers' rights (2011); Boston Biomedical Research Institution, Boston, Massachusetts, $75,000 - research and development related to thoracic disease (2011); Beth Israel Deaconess Medical Center, Boston, Massachusetts, $35,000 - experimental radiology program (2011).
Contact: Bradley R. Cook, Trustee, c/o Taylor, Ganson, and Perrin, LLP; (617) 951-2777; fax (617) 951-0989
Sponsor: Thoracic Foundation
160 Federal Street, 20th Floor
Boston, MA 02110-1700

Thrasher Research Fund Grants **4495**
The fund awards grants in support of research that seeks to improve children's health where insufficient research and investigation have been performed. In particular, the fund supports medical research that seeks to prevent or cure children's critical illnesses, injuries, and disabilities. Projects should be scientifically sound and culturally appropriate; document methods and practices that have sustainable benefits; have specific aims and a well-designed methodology; and evaluate the significance of impact on children's health. The fund normally supports projects no longer than three years. Financial support may be provided for supplies, minor equipment (items less than U.S.$1000), and technical personnel assistance related to a specific fund-sponsored project. The fund requires the principal investigator to submit semiannual project progress and financial reports. The initial application procedure is to submit a five-page prospectus, which should include a brief background statement justifying the proposed project; hypothesis to be tested; specific aims of the project; design and methodology of the proposed project, including evaluation component; time schedule and detailed budget; and investigator(s) curriculum vita (in addition to four-page prospectus). Applicants are encouraged to consult program staff prior to submitting a prospectus.
Restrictions: The fund does not award grants for general operations, construction or renovation of buildings or facilities, loans, scholarships, or the purchase of major equipment. Grants generally are not awarded for conferences, workshops, or symposia. Proposals in the areas of sexually transmitted diseases, abortion or reproductive physiology, and contraceptive technology will not be considered. In addition, proposals involving the use of therapeutically aborted human fetal tissue will not be considered.
Amount of Grant: 100,000 - 350,000 U.S.D
Samples: Dr. Rebecca Stoltzfus, Johns Hopkins School of Hygiene and Public Health (Baltimore, MD)—for a study to improve iron status and growth of preschool children in East Africa, $141,529 total; Paul Calvert, Intermediate Technology (Rugby, England)—for study on environmental health and sanitation for fishing communities, $117,250 total.
Contact: Julie Busse, (801) 240-4720; fax (801) 240-1625; BusseJ@thrasherresearch.org
Internet: http://www.thrasherresearch.org
Sponsor: Thrasher Research Fund
15 W South Temple Street, Suite 1650
Salt Lake City, UT 84101

Tides Foundation Syringe Access Fund **4496**
The Syringe Access fund is a collaborative grant making initiative that strives to reduce the risk of HIV infection and other blood-borne pathogens among injection drug users, their sexual partners and children through expanded access to sterile syringes. Grants are awarded to fund expanded access to clean syringes via syringe exchange programs and networking/capacity development among local syringe exchange programs; and coordinated, local/state-level education programs focusing on policies to expand access to sterile syringes, including the decriminalization of syringe exchange programs or the sale of syringes without a prescription. The program also seeks to broaden public support for syringe exchange programs as well as for policies designed to expand access to sterile syringes. The listed application deadline is for letters of intent, which must be submitted via email; proposals are by invitation. Guidelines and letter of intent form are available online.
Requirements: Applications are accepted from all 50 states, the District of Columbia, Native American reservations/tribal land, and Puerto Rico.
Amount of Grant: 25,000 - 50,000 U.S.D
Contact: Tierney Gleason, Syringe Access Fund Supervisor; (212) 509-4975 or (202) 408-4848, ext. 203; fax (212) 509-1059; syringeaccess@tides.org or saf@aidsfund.org
Internet: http://www.tidesfoundation.org/services-strategies/collective-giving/syringe-access-fund/index.html
Sponsor: Tides Foundation
55 Exchange Place, Suite 402
New York, NY 10005-1965

Tifa Foundation Grants **4497**
The Tifa Foundation has two grant making mechanism. The first, known as Pro-Active Grants, is through a proactive approach in which Tifa assumes an active role in working together with other organizations that correspond with Tifa's mission. The second, known as Open Grants, is through the endorsement of program proposals that conform to Tifa's strategic direction and priority/supporting programs.
Requirements: Grants are awarded to: organizations with proposed program activities that are in accordance with Tifa's vision and mission; organizations that possess unblemished an track record during their involvement with Tifa as well as with other donor institutions; organizations with legal entities; organizations that cooperate with government agencies; and institutions under the control of a particular university.
Restrictions: Grants are not eligible for: individuals, including proposals for scholarships, trips, conferences and others; all government agencies, legislative and judiciary bodies, political parties, military or those with affiliation/direct relationship with the aforementioned bodies; all organizations with profit-seeking interests; program proposals where the end product or outcome is sold to the public (except upon Tifa's approval); organizations that are still engaged in cooperation with Tifa for the ongoing period except upon prior approval.
Date(s) Application is Due: Jan 15; Jul 15
Contact: Yayasan Tifa, (62) 021 829 2776; fax (62) 021 837 83648; public@tifafoundation.org
Internet: http://www.tifafoundation.org/
Sponsor: Tifa Foundation
Jl. Jaya Mandala II No. 14 E
Menteng Dalam, South Jakarta, 12870
Indonesia

Times Mirror Foundation Grants **4498**
The foundation makes grants to nonprofit organizations in operating communities in its areas of interest, including journalism, education, literacy, community enrichment, and art and culture. The foundation focuses its support on programs that improve the quality of journalism, education and literacy, strengthen the fabric of the community, and enhance cultural appreciation and understanding. In addition, the foundation approves grants for unrestricted operating support, capital campaigns, seed grants, and special projects. Qualified organizations seeking grant support should send a two- to three-page letter briefly describing the organization, its purpose, programs, any special project to be considered, and its governing boards. Proof of tax-exempt status, list of current supporting organizations and amount of support, an organizational budget for current and upcoming fiscal years, and a copy of the most recent financial statements should be included.
Requirements: The foundation's geographical area is restricted to Southern California (Los Angeles, Orange, Ventura, San Bernardino and Riverside counties); Stamford, Greenwich and Hartford, Connecticut; and Allentown, Pennsylvania, and Queens and Long Island, New York. An applicant organization must include evidence of 501(c)3 status when requesting funds.
Restrictions: The foundation will not consider grants for production costs of films, videos, or television programs; religious or fraternal purposes; publications; conferences; or to individuals.
Geographic Focus: California, Connecticut, New York, Pennsylvania
Date(s) Application is Due: Apr 15; Oct 15
Amount of Grant: 1,000 - 25,000 U.S.D
Samples: Maryland African American Museum (Baltimore, MD)—for its capital campaign, $80,000; Archdiocese of Baltimore Chancery (Baltimore, MD)—for its Reading Scholars program, $25,000; Ctr for Nonprofit Management (Los Angeles, CA)—for its resource library of nonprofit-related materials, $15,000.
Contact: Michelle Williams, Executive Director; (213) 237-3945; fax (213) 237-2116
Internet: http://www.timesmirrorfoundation.org
Sponsor: Times Mirror Foundation
202 W First Street
Los Angeles, CA 90053

Tipton County Foundation Grants **4499**
Groups engaged in charitable projects or programs are invited to apply for a Tipton County Foundation Grant. Grant are made only to groups, organizations, or projects: nonprofit and charitable; that document their responsible fiscal management and adequate accounting procedures; proposed in writing by the organization's governing body. Proposals are reviewed quarterly on the first Monday of January, April, August, and October.
Requirements: Applicants must submit a letter of intent before applying for a grant with the Foundation. The LOI must include a detailed description of the organization, what the organization would like to do, who will benefit, and its cost, along with a timeline of the project, and a copy of the IRS tax exempt letter. If the organization submits a proposal, it must include its rationale, collaboration, and organizational strength in completing the project. The following attachments must be included with the proposal: a detailed list of the board of directors; evidence that the proposal has the board's approval; year-end and current financial statements; organizational budget; and a detailed project budget that describes revenue, expenses, in-kind revenue/expenses, and whether the revenue is pending, received, or approved.
Restrictions: The Foundation does not make grants to individuals or for the following: political purposes; programs or equipment that were committed to prior to the grant proposal period; debt reduction; annual appeals or membership contributions; building of endowments, unless located at the Tipton Foundation; projects that benefit only a few persons; travel expenses for individuals or groups.
Geographic Focus: Indiana
Contact: Lori Tragesser (765) 675-1941; fax (765) 675-8488; tcf@tipton.org
Internet: http://www.tiptoncf.org/grants.htm
Sponsor: Tipton County Foundation
1020 West Jefferson Street
Tipton, IN 46072

TJX Foundation Grants **4500**
The primary mission of the contribution program is to help needy families and children. The foundation also supports services in communities where the companies do business that promote sound mental and physical health, provide shelter, enhance education and job readiness, and build community ties. Contributions will be made in the areas of education, social services, health, civic and community affairs, and art and culture. Types of support include continuing support, general operating support, and program development.
Requirements: IRS 501(c)3 tax-exempt organizations are eligible.

Restrictions: The foundation will not fund individuals, political organizations, publications, cash reserves, seed money/start-up projects, endowments, religious groups, public policy research, travel grants, environmental issues, education loans, international organizations, film/photography, conferences/seminars, unrestricted grants, fellowships, or advocacy.
Amount of Grant: 5,000 - 25,000 U.S.D
Contact: Christine Strickland, Foundation Manager; (508) 390-3199; fax (508) 390-2091; Christy_Strickland@TJX.com
Internet: http://www.tjx.com/corprespons/commsupp.html
Sponsor: TJX Foundation
770 Cochituate Road, Route 1E
Framingham, MA 01701

Todd Brock Family Foundation Grants **4501**
Established in Texas in 2007, the Todd Brock Family Foundation offers support for educational programs, children, athletics, research, and community projects, as well Protestant agencies and churches. A formal application is required, and applicants should forward the entire proposal to the office. The Foundation rarely offers funding outside of Texas. There are no deadlines for submitting a completed proposal, and grants have most recently ranged from $300 to $250,000.
Geographic Focus: Texas
Samples: Memorial High School Boosters, Houston, Texas. $1,000 - funding for annual gold tournament (2010); Legacy Christian Academy, Beaumont, Texas, $250,000 - for general operations (2010); Yellowstone Academy, Houston, Texas, $25,000 - support for low-income children and families (2010).
Contact: Todd O. Brock, President; (409) 833-6226; fax (409) 832-3019
Sponsor: Todd Brock Family Foundation
1670 E Cardinal Drive, P.O. Box 306
Beaumont, TX 77704-0306

Tofias PC Charitable Foundation Grants **4502**
The foundation currently focuses on strengthening the organization skills of health-focused nonprofit organizations. Grants will be awarded to support organizational skills, including performance management, financial planning, fundraising and revenue generation, and building and management; and organizations working to improve people's health through education, research, awareness, advocacy, and/or direct services. The foundation is particularly interested in organizations with annual budgets up to $25 million; strong, financially stable organizations where a small grant could address a specific need; proposals with clearly defined infrastructure goals and achieveable implementation and evaluation plans; and organizations with strong leadership and an active volunteer network. One-year grants will be awarded. Applicants should submit a letter of intent (two-page maximum); full proposals are by invitation. Letters of intent are accepted on an ongoing basis.
Requirements: Organizations in the following Massachusetts counties are eligible: Bristol, Essex, Middlesex, Norfolk, Plymouth, Suffolk, and Worcester. Organizations in the following Rhode Island counties are eligible: Bristol, Kent, Providence, Newport, Washington.
Restrictions: The foundation will not support endowments, scholarships, or political offices.
Geographic Focus: Massachusetts, Rhode Island
Amount of Grant: 2,000 - 5,000 U.S.D
Samples: Boston Area Rape Crisis Ctr (Cambridge, MA)—to develop a system to evaluate its programs, $10,000; Visiting Nurse Assoc of Middlesex-East—to fund a development consultancy that will assess the agency's current fundraising program and its potential for growth, $5000; The Guidance Ctr—for a strategic planning consultant, $5000.
Contact: Grants Administrator; (617) 761-0600 or (888) 761-8834, ext. 637; foundation@tofias.com
Internet: http://www.tofias.com/foundation
Sponsor: Tofias PC Charitable Foundation
350 Massachusetts Avenue
Cambridge, MA 02139

Toledo Community Foundation Grants **4503**
The foundation, which serves northwest Ohio and southeast Michigan with a particular emphasis on the greater Toledo area, provides grants to support programs and services in the fields of education, social services, physical and mental health, neighborhood and urban affairs, youth programs, natural resources, and the arts. Preference is given to seed money requests for new projects and programs designed to meet community needs. Additional support mechanisms include program/project development and matching funds. Grant proposals are considered four times annually. Proposals should be five or six typewritten pages in length and include the project's purpose and need, implementation, evaluation, budget, and future support. Applicant organizations are notified in writing of approval/denial of funding requests.
Requirements: IRS 501(c)3 tax-exempt organizations serving residents in northwest Ohio and southeast Michigan are eligible.
Restrictions: The foundation usually does not make grants to support general operating budgets, annual campaigns, capital campaigns, or sectarian activities of religious organizations; to purchase equipment; or to produce films, videos, etc.
Geographic Focus: Ohio
Amount of Grant: Up to 54,000,000 U.S.D
Contact: Virginia Keller, Program Officer; (419) 241-5049; fax (419) 242-5549
Internet: http://www.pgdc.com/toledo
Sponsor: Toledo Community Foundation
608 Madison Avenue, Suite 1540
Toledo, OH 43604-1151

Tommy Hilfiger Corporate Foundation Grants **4504**
The foundation gives to children's programs nationally in areas of education, health, human services, and the arts, with the main focus being on education. Proposals for educational and cultural programs must address the following priorities: target K-12 and college students; expose students to career opportunities; develop skills in new technologies; leverage teacher, administrator, parental, and community involvement; include hands-on program activities; lead to comprehensive, systemic change on a regional and/or national basis; involve collaborative partnerships; demonstrate capacity to gain continuing support; result in dissemination and replication of lessons learned; have broad and positive impact on diverse populations with a special emphasis on women, minorities, and at-risk students; and develop evaluation component with measurable results Proposals that are health-related should be centered on the following: medical centers or hospitals; health education; and innovative therapy programs.
Requirements: 501(c)3 nonprofit organizations are eligible. Grant requests must be made in writing. Proposal seekers are advised to submit a three- to five-page proposal including a detailed budget (all pages must be numbered). A separate cover letter summarizing the proposal request is also required.
Restrictions: Grants are not made to/for individuals; political parties, campaigns, or causes; endowment funds or campaigns; film, video, television, and radio projects; or any organization that discriminates on the basis of race, creed, gender, or national origin.
Date(s) Application is Due: Apr 1; Oct 1
Samples: Dress for Success Worldwide, New York, New York, $30,000; Performing Arts Center Foundation, Clearwater, Florida, $25,000; Camp AmeriKids, Stamford, Connecticut, $20,000.
Contact: Program Officer; (212) 840-8888
Internet: http://companyinfo.tommy.com/#/tommy_foundation/grant_guidelines
Sponsor: Tommy Hilfiger Corporate Foundation
601 W 26th Street, 6th Floor
New York, NY 10001-1101

Topeka Community Foundation Kansas Blood Services Scholarships **4505**
The Kansas Blood Services Fund with the Topeka Community Foundation provides scholarships for individuals in the medical field of blood banking or a related field. The scholarship can be used to cover expenses related to attending a professional conference or meeting (including conference fees, hotel, meals and transportation).
Requirements: Residents of Nemaha, Riley, Pottawatomie, Jackson, Jefferson, Douglas, Shawnee, Wabaunsee, Osage and Lyon counties in Kansas are eligible to apply.
Geographic Focus: Kansas
Contact: Marsha Pope, Vice President; (785) 272-4804; fax (785) 272-4644; pope@topekacommunityfoundation.org
Internet: http://www.topekacommunityfoundation.org/newsarticle.cfm?articleid=10023450&ptsidebaroptid=0&returnto=page31525.cfm&returntoname=Grant Information&siteid=1897&pageid=31520&sidepageid=31525
Sponsor: Topeka Community Foundation
5431 SW 29th Street, Suite 300
Topeka, KS 66614

Topfer Family Foundation Grants **4506**
The mission of the foundation (TFF) is to fund programs and organizations that connect people to the tools and resources they need to build self-sufficient and fulfilling lives. Programs areas of interest include child abuse prevention and treatment; youth enrichment; job training and support services; children's health; and aging in place. The Topfer family is keenly interested in understanding the unique needs of the communities in which they invest foundation resources. Therefore, contributions are limited to the communities in which the family resides: the greater Austin, TX, and the greater Chicago, IL, metropolitan areas. In Illinois, preference is given to organizations serving Du Page and Cook Counties. Note to Chicago area applicants: The foundation is not accepting unsolicited applications in the Job Training or Children's Health program areas from the Chicago area at this time. Guidelines and application are available online.
Requirements: 501(c)3 nonprofit organizations in Illinois and Texas are eligible.
Restrictions: The foundation does not generally support: grants or loans to individuals; loans to charitable organizations; more than 20 percent of an organization's operating budget; administrative costs; organizations that exclude participants based on race or religion; the purchase of dinner, gala, or raffle tickets; school fundraisers or events (including sports and other extracurricular activities); tax-generating entities (municipalities, school districts, etc.) for services within their normal responsibilities; public or private educational institutions for recurrent operating, administrative, or capital expenses (i.e., acquisition, construction, improvement, and maintenance of buildings and equipment), or scholarships that subsidize existing funding base; academic or scientific research; advertising or marketing efforts; political campaigns or purposes; or any one program more than once a calendar year.
Geographic Focus: Illinois, Texas
Samples: Foundation Communities (Austin, TX)—to develop efficiency apartments for people with very low incomes, most of whom are homeless or in imminent danger of becoming homeless, $100,000; Bridge Communities (Du Page County, IL)—to support after school tutoring and educational enrichment for at-risk children, $20,000; Dress for Success (Travis County, TX)—to provide clothing for interviews for women in search of employment, $5000; Community Care Options (Berwyn, IL)—to support case management services for seniors, $50,000.
Contact: Grants Administrator; (866) 897-0298 or (512) 329-0009
Internet: http://www.topferfamilyfoundation.org
Sponsor: Topfer Family Foundation
5000 Plaza on the Lake, Suite 170
Austin, TX 78746

Toro Foundation Grants 4507

The foundation supports nonprofit organizations in corporate operating areas to improve the quality of life for employees and customers. The foundation's primary giving focus is to support organizations that preserve the outdoor environment. The foundation also contributes to programs that enhance education, health and welfare, culture, the arts, and civic projects. Preference is given to requests for program funds over endowment or capital fund requests and to organizations where Toro employees volunteer. Grants are evaluated based on criteria such as mission, employee involvement, geographical area, environmental component, industry relevance, and the presence of United Way funds. Proposals are accepted throughout the years. Guidelines are available online.

Requirements: 501(c)3 tax-exempt organizations in corporate communities are eligible, including Minneapolis/Saint Paul, and Windom, MN; El Cajon, Madera, and Riverside, CA; Oxford, MS; Beatrice, NE; Baraboo, Tomah, and Plymouth, WI; and Abilene, and El Paso, TX.

Restrictions: Grants do not support religious or political activities; general, capital, or endowment support for educational institutions; health treatment, scientific research, or substance abuse treatment programs; travel for individuals or groups; fundraising events or activities, social events, or goodwill advertising, except where Toro employees are participating; individuals or for-profit organizations; or organizations funded by the United Way.

Geographic Focus: California, Minnesota, Mississippi, Nebraska, Texas, Wisconsin

Contact: Grants Administrator; (800) 348-2424 or (952) 887-8870; fax (952) 887-8258

Internet: http://www.thetorocompany.com/community/foundation.html

Sponsor: Toro Foundation
8111 Lyndale Avenue S
Bloomington, MN 55420-1196

Tourette Syndrome Association Post-Doc Fellowships 4508

The Fellowship program is interested in supporting any research that has the potential to contribute to the understanding of Tourette Syndrome, including genetics, pathogenesis, pathophysiology, clinical treatments and animal model development. Post-doctoral research fellowships of up to $40,000 for one year are awarded. For preliminary screening, a Letter of Intent briefly describing the scientific basis and relevance of the proposed project should be sent to tsadevel@earthlink.net by October 1. The cover page to accompany each letter can be downloaded from the website. Applicants will be notified electronically as to whether or not to proceed with a full grant proposal. If invited to apply for a research grant, detailed instructions will be provided, and full proposals are due by December 9th.

Requirements: Candidates must have a M.D., Ph.D. or equivalent. Previous experience in the field of movement disorders is desirable, but not essential. Applicants may reside outside the United States.

Date(s) Application is Due: Oct 1; Dec 9

Amount of Grant: Up to 40,000 U.S.D

Contact: Fellowship Administrator; (718) 224-2999, ext. 222; fax (718) 279-9596; grantadministrator@tsa-usa.org

Internet: http://tsa-usa.org/research.html

Sponsor: Tourette Syndrome Association
42-40 Bell Boulevard
Bayside, NY 11361-2820

Tourette Syndrome Association Research Grants 4509

The Program is interested in receiving applications from any research study that has the potential to contribute to the understanding of Tourette Syndrome, including genetics, pathogenesis, pathophysiology, clinical treatments and animal model development. Research grants of up to $75,000 for one year are awarded. Applicants may apply for a second year of funding, which is reviewed competitively. For preliminary screening, a Letter of Intent briefly describing the scientific basis and relevance of the proposed project should be sent as an email attachment to tsadevel@earthlink.net by October 1st. The cover page to accompany each letter can be downloaded from the website. Applicants will be notified electronically as to whether or not to proceed with a full grant proposal. If invited to apply for a research grant, detailed instructions will be provided, and full proposals are due by December 9th.

Requirements: Candidates must have a M.D., Ph.D. or equivalent. Previous experience in the field of movement disorders is desirable, but not essential. Applicants may reside outside the United States.

Date(s) Application is Due: Oct 1; Dec 9

Amount of Grant: 75,000 U.S.D

Contact: Heather Cowley, Manager; (718) 224-2999, ext. 247; fax (718) 279-9596; Heather.Cowley@tsa-usa.org or grantadministrator@tsa-usa.org

Internet: http://tsa-usa.org/research.html

Sponsor: Tourette Syndrome Association
42-40 Bell Boulevard
Bayside, NY 11361-2820

Toys R Us Children's Fund Grants 4510

The Toys R Us Children's Fund is a public charity affiliated with Toys R Us, Inc. Since it was founded in 1992, the Children's Fund has contributed millions of dollars annually to qualified organizations that keep children safe and help them in times of need, including those providing disaster relief to children and families who are victims of large-scale crises. The Fund also provides grants to leading organizations that support children with special needs. Primary fund raising activities include the annual Toys R Us Children's Fund Gala, a private, invitation-only event that showcases the work of the Fund and raises money to support its many beneficiaries.

Requirements: U.S. 501(c)3 tax-exempt organizations are eligible.

Contact: Children's Fund Coordinator; (973) 617-3500 or (800) 869-7787; fax (973) 617-4040; TRUCF@toysrus.com

Internet: http://www9.toysrus.com/about/corpPhilanthropy.cfm

Sponsor: Toys R Us
1 Geoffrey Way
Wayne, NJ 07470-2030

TRDRP California Research Awards 4511

This program is focused on research projects that address questions specific to tobacco-related disease or tobacco control issues in California. The awards will provide funding for average annual direct costs of $140,000; $170,000 per year for research involving human subjects. Full indirect costs are paid to eligible institutions. Maximum duration is 3 years.

Requirements: Principal Investigators from California not-for-profit organizations are eligible for funding. Research undertaken with TRDRP funds must be conducted primarily in California. The Principal Investigator must conduct and supervise the research project directly and in person. Underrepresented racial and ethnic investigators are encouraged to submit applications. To be competitive, proposals must present a compelling justification, including evidence if available, supporting the California-specific criterion. The proposal should be fully developed, scientifically rigorous, and include sound background information, hypothesis, and promising preliminary studies or preliminary data.

Restrictions: All submitted applications will be screened for their direct relevance to tobacco use or tobacco-related disease. Proposed research that is not relevant will not undergo peer review for scientific merit. It is incumbent on the applicant to make a compelling case that the proposed work is directly relevant to the mission of TRDRP.

Geographic Focus: California

Date(s) Application is Due: Jan 15

Amount of Grant: 140,000 - 170,000 U.S.D

Contact: M.F. Bowen, (510) 987-9811; mf.bowen@ucop.edu

Internet: http://www.trdrp.org/fundingopps/callawardmechs.asp#calrchawd

Sponsor: Tobacco-Related Disease Research Program
Office of the President, 300 Lakeside Drive, 6th Floor
Oakland, CA 94612-3550

TRDRP Exploratory/Developmental Research Awards 4512

The purpose of these grants is to gather preliminary data or demonstrate proof-of-principle (i.e., pilot projects), or to conduct a research project within the specified limits of money and time. The ultimate goal of these awards is to provide the foundation for proposals for fully-developed research project awards from other funding programs. Total direct costs of $250,000 for the entire project duration (maximum of 2 years). Indirect costs are paid to eligible institutions.

Requirements: Principal Investigators from California not-for-profit organizations are eligible for funding. Research undertaken with TRDRP funds must be conducted primarily in California. The Principal Investigator must conduct and supervise the research project directly and in person. Underrepresented racial and ethnic investigators are encouraged to submit applications. Although Principal Investigators may submit as many proposals for these awards as they wish for separate projects, no more than one Exploratory/Developmental Research Award will be made to any individual PI in the same year. Only proposals that address one of TRDRP's Primary research area will be eligible and will undergo peer review. Exploratory/Developmental Research Awards are for new projects only; that is, competing renewal applications will not be accepted.

Restrictions: All submitted applications will be screened for their direct relevance to tobacco use or tobacco-related disease. Proposed research that is not relevant will not undergo peer review for scientific merit. It is incumbent on the applicant to make a compelling case that the proposed work is directly relevant to the mission of TRDRP.

Date(s) Application is Due: Jan 15

Contact: M.F. Bowen, (510) 987-9811; mf.bowen@ucop.edu

Internet: http://www.trdrp.org/fundingopps/callawardmechs.asp#explordev

Sponsor: Tobacco-Related Disease Research Program
Office of the President, 300 Lakeside Drive, 6th Floor
Oakland, CA 94612-3550

TRDRP New Investigator Awards 4513

This award mechanism is specifically designed to enable new investigators to initiate an independent research program in a tobacco-related field. These are awards for investigators who have a doctoral degree, have not yet served as a principal investigator of a research project grant (e.g., an NIH R01-type grant), and have had no more than five years experience as an independent investigator, i.e., no more than five years since completing formal postdoctoral training, or since the doctoral degree if no postdoctoral training. Direct costs of $90,000 per year can be averaged over the duration of the award. The indirect cost rate is capped at 8% for eligible institutions.

Requirements: New investigators must commit to a minimum of 50 percent time to the research project for the award duration.

Restrictions: Current or past new investigator awardees from TRDRP, other UC Special Research Programs, or other agencies (for example, current or past recipients of NIH K-series career development awards) are not eligible.

Geographic Focus: California

Date(s) Application is Due: Jan 15

Amount of Grant: 90,000 U.S.D

Contact: M.F. Bowen, (510) 987-9811; mf.bowen@ucop.edu

Internet: http://www.trdrp.org/fundingopps/callawardmechs.asp

Sponsor: Tobacco-Related Disease Research Program
Office of the President, 300 Lakeside Drive, 6th Floor
Oakland, CA 94612-3550

Tri-County Electric People Fund Grants 4514

The fund awards grants for charitable purposes to benefit people and organizations in the Tri-County Electric service area. Grants are awarded for food, shelter, clothing, health, and other humane needs for programs or services that benefit a significant segment of a community. An organization's financial statements, bylaws, and nonprofit status must be submitted with its application. Individuals or families must submit copies of previous federal income tax forms. The fund has the right to fully audit the use of donations at any time. Guidelines and application are available online.

Requirements: Individuals and organizations in mid-Michigan, including Ionia, Eaton, Ingham, Clinton, Gratiot, Montcalm, Isabella, and Mcosta Counties, are eligible.

Restrictions: No funds shall in any fashion be used to support any candidate for political office or any political purpose. An applicant may not reapply within six months from a previous application.

Geographic Focus: Michigan

Samples: Eaton County Clothing and Furniture Project (MI)—for their back-to-school program, $3000; EightCAP (Ionia and Montcalm Counties, MI)—for Camp Wah-Wah-Tay-See program expenses, $2000; Montcalm County family to help with medical expenses, $309.09; Women's Aid Service (Mount Pleasant, MI)—for replacement windows, $2000.

Contact: People Fund Supervisor; (877) 466-3957, ext. 1; tricoenergy@homeworks.org

Internet: http://www.tricoenergy.com/people_fund1/guidelines.htm

Sponsor: Tri-County Electric People Fund
P.O. Box 503
Portland, MI 48875

Tri-State Community Twenty-first Century Endowment Fund Grants 4515

The Foundation for the Tri-State Community makes grants in support of educational, cultural, scientific and other charitable purposes in Kentucky, Ohio, and West Virginia. Specifically, the Foundation encourages project proposals with the following goals: improve the quality of life in the Tri-State area; test or demonstrate new approaches and techniques to find solutions to important community problems; promote community volunteerism; strengthen non-profit agencies and institutions by reducing operating costs, increasing public financial support, and/or improving internal management; and use community resources to promote coordination, cooperation, and sharing among organizations to eliminate duplicate services.

Requirements: 501(c)3 tax-exempt organizations that serve Boyd and Greenup Counties, Kentucky; Lawrence County, Ohio; and Cabell and Wayne Counties, West Virginia are eligible. Applicants should contact the Foundation to obtain a copy of the grant application and to discuss the proposed project.

Restrictions: The foundation does not fund general operating support grants, endowment funds, individuals, operational deficits, or sectarian activities of religious organizations. Also, organizations may submit only one application per funding cycle, and past awardees should wait at least 12 months from the approval of their award to submit another request. The 21st Century Endowment grants also do not provide funding for components of projects begun before grant award decisions are made.

Geographic Focus: Kentucky, Ohio, West Virginia

Date(s) Application is Due: Jan 15; Apr 15; Jul 15; Oct 15

Samples: Helping Hands of Greenup County, Kentucky, $2,000 - for the purchase of a color printer (2010); Huntington Area Food Bank, Tri-State area, $4,000 - Primarius software program (2010); LEDC Symmes Creek Restoration Committee, Ohio, $2,121 - funding to develop and publish a Symmes Creek Canoe Trail Map (2010).

Contact: Mary Witten Wiseman, President; (304) 942-0046; fax (304) 942-0048; FTSC_MWWiseman@yahoo.com

Internet: http://www.tristatefoundation.org/html/grants.html

Sponsor: Foundation for the Tri-State Community
P.O. Box 2096
Ashland, KY 41105-2096

Triangle Community Foundation Donor-Advised Grants 4516

Established in 1983, the Triangle Community Foundation is a nonprofit organization that serves Chatham, Durham, Orange and Wake counties of North Carolina. There's no application process for the donor-advised grants, rather, fundholders advise Foundation staff as to where they would like their grants made. Often, fundholders seek guidance from staff about which nonprofits are doing work matching their charitable interests. For this reason, it is important nonprofits keep members of the Philanthropic Services Team informed about their organizations, including special projects and funding needs. For more information, nonprofit organizations should contact the following staff: Arts, Culture, & Humanites - Lori O'Keefe, lori@trianglecf.org; Community Building - Agnes Vishnevkin, agnes@trianglecf.org; Education & Youth Development - Robyn Fehrman, robyn@trianglecf.org; Health & Human Services - Sandra Rodriguez, sandra@trianglecf.org; Law, Justice, & Public Action - Robyn Fehrman, robyn@trianglecf.org; Natural Resources - Tracy Joseph, tracy@trianglecf.org; Volunteerism & Community Support - Libby Long, libby@trianglecf.org; Religious & Spiritual Development - Tracy Joseph, tracy@trianglecf.org.

Requirements: North Carolina 501(c)3 tax-exempt organizations, government units, and religious congregations planning projects that benefit residents of Chatham, Durham, Orange and Wake counties of North Carolina are eligible.

Geographic Focus: North Carolina

Contact: Lori O'Keefe, Director of Philanthropic Services; (919) 474-8370, ext. 144; fax (919) 941-9208; lori@trianglecf.org

Internet: http://www.trianglecf.org/page33699.cfm

Sponsor: Triangle Community Foundation
324 Blackwell Street, Suite 1220
Durham, NC 27701

Triangle Community Foundation Shaver-Hitchings Scholarship 4517

The Shaver-Hitchings Scholarship Fund was created in 1990 by Dr. George H. Hitchings and Dr. Joyce Shaver Hitchings to honor and reward individuals with a commitment to helping others in the area of drug and alcohol addiction. The Scholarship is available to any graduate student, physician's assistant, or medical student in the Triangle area who has worked (preferably as a volunteer) to help others with alcoholism, drug abuse and addictive disorder treatment or with preventive education on the subject of addiction. The fund is designed to recognize the recipient for his or her work or volunteer activity and to provide financial aid. Excellent candidates who have demonstrated a commitment to working with others in the area of addictive disorders, especially those who have done so as volunteers, are sought. One Scholarship of $1,500 will be awarded each year. The Scholarship may be applied toward expenses for tuition, required fees, and required books or materials not covered by other financial aid.

Requirements: Applicant must be a resident of Durham, Orange, Wake, or Chatham County.

Restrictions: The Scholarship does not cover living expenses such as room and board, transportation, or childcare.

Geographic Focus: North Carolina

Date(s) Application is Due: Mar 15

Contact: Libby Long Richards, Program Coordinator; (919) 474-8370, ext. 134; fax (919) 941-9208; libby@trianglecf.org

Internet: http://www.trianglecf.org/grants_support/view_scholarships/shaver-hitchings_scholarship/

Sponsor: Triangle Community Foundation
324 Blackwell Street, Suite 1220
Durham, NC 27701

Trinity Lutheran School of Nursing Alumnae Scholarships 4518

The Trinity Lutheran Hospital School of Nursing Alumnae Scholarship fund was established to assist persons who wish to pursue a career in nursing (such as undergraduate, advanced degree, etc.) and need financial assistance. Applications for the scholarship must be received by May 1st. Renewal of the scholarship is possible each year for those who reapply and meet the criteria.

Requirements: Applicants must: be a student currently enrolled in an accredited RN school of nursing or in an accredited higher education nursing program; have completed at least one semester at an accredited school; be a full or part-time student and the course work must be continuous; and have a cumulative GPA of 2.5.

Date(s) Application is Due: May 1

Contact: Becky Schaid, Executive Director; (816) 276-7515 or (816) 276-7555; fax (816) 926-2261; becky@btllf.org

Internet: http://www.btllf.org/funding.html

Sponsor: Baptist-Trinity Lutheran Legacy Foundation
6601 Rockhill Road
Kansas City, MO 64131

Triological Society Research Career Development Awards 4519

The purpose of this award is to provide support for the research career development of otolaryngologists (head and neck surgeons) who have made a commitment to focus their research endeavors on patient-oriented research, e.g., clinical trials, translational research. Research training supported by this award may be related to any research questions relevant to the specialty of otolaryngology, as long as it is demonstrated that the training will have a direct impact on the applicant's ability to pursue his/her long-range research objectives. Grants can be for one to two years, are non-renewable, and have a $40,000 maximum award limitation. There are approximately six grants awarded annually. Letters of Intent are due by December 15, with full proposals received by January 15 annually.

Requirements: Otolaryngologists-head and neck surgeons who hold full-time, part-time or contributed service medical school faculty appointments and who have made a commitment to focus their research endeavors on patient-oriented research may apply. Applicants must be sponsored by the Chair of his/her division or department and by an official representative of the institution which would administer the award and in whose name the application is formally submitted.

Date(s) Application is Due: Jan 15; Dec 15

Amount of Grant: Up to 40,000 U.S.D

Contact: Stephanie L. Jones; (703) 519-1586 or (703) 836-4444; sljones@entnet.org

Internet: http://www.triological.org/researchgrants.htm

Sponsor: Triological Society
555 North 30th Street
Omaha, NE 68131

Trull Foundation Grants 4520

The Trull Foundation was established to share God's bounty in Texas and beyond. The Foundation supports organizations involved with: health care, the coastal Texas environment, seniors, child abuse, neglect, farming, ranching, agriculture, birds, hunger, substance abuse, religion, charity, education, improving the quality of life, especially for those living in poor or oppressed conditions, with a special concern for the needs of Palacios and Matagorda County where the Foundation has its roots.

Requirements: 501(c)3 tax-exempt organizations and departments, agencies, and other services operated within federal, state, or local government agencies and institutions and agencies affiliated with organized religions and religious bodies are eligible. Proposals should be mailed directly to the foundation, faxed or emailed proposals will not be excepted. See foundation's website for additional proposal guidelines and application forms.

Restrictions: The foundation usually will not make long term commitments; make grants for buildings, endowments, or research; repeat grants to the same project longer than three years; or fund operational expenses except during initial years. Any scholarship

funds granted are made to institutions and are administered by those institutions. No scholarships or other funds are granted to individuals.
Amount of Grant: 250 - 20,000 U.S.D
Samples: El Buen Pastor Early Childhood Development Center, Austin, TX, $8,000 - Student Tuition Scholarships; Palmer Drug Abuse Program of Corpus Christi, Inc., Corpus Christi, TX, $7,000 - Reaching out to the Rural Areas of South Texas; Ducks Unlimited, Inc., Memphis, TN, $10,000 - Texas Prairie Wetlands; Louise Independent School District, Louise, TX, $3,000 - TI-83 calculators for students in Grades 9-12; Blessing Historical Foundation, Blessing, TX, $15,000 - Hotel Restoration;
Contact: E. Gail Purvis, Executive Director; (361) 972-5241; fax (361) 972-1109; gpurvis@trullfoundation.org
Internet: http://www.trullfoundation.org
Sponsor: Trull Foundation
404 Fourth Street
Palacios, TX 77465

Trust Family Foundation Grants 4521
The foundation awards program development grants to New England nonprofits in its areas of interest, including arts, education, Jewish philanthropy, and medical technology. Preference will be given to projects that leverage support, show collaborative efforts, can ultimately be self-funding, affirm the foundation's mission, and demonstrate high-quality development, management, and organization.The board meets in March, June, September, and December to consider requests.
Requirements: Nonprofits in New Hampshire and Massachusetts are eligible.
Restrictions: Grant requests will be denied for capital campaigns, city or state government, endowments, operating budgets, public schools, or unrestricted gifts.
Geographic Focus: Massachusetts, New Hampshire
Date(s) Application is Due: Feb 15; Apr 30; Jul 15; Oct 30
Amount of Grant: 250 - 275,000 U.S.D
Contact: Kathryn Wheeler, Foundation Administrator; (603) 898-6670
Sponsor: Trust Family Foundation
1 Stiles Road, Suite 202
Salem, NH 03079-4844

TSA Research Grant Awards 4522
The association accepts applications for grants in the following categories: proposals in basic neuroscience specifically relevant to Tourette syndrome (TS); clinical studies related to the etiology, pathophysiology, and treatment of TS; and one postdoctoral fellowship. Relevant scientific fields include animal models, basal ganglia physiology, biochemistry, epidemiology, genetics, molecular biology, neuroanatomy, neuroimaging, neurology, neuropathology, neuropsychology, neurophysiology, pharmacology, psychiatry, and psychology. As a means of preliminary screening to assess the relevance and quality of intended proposals, a letter of intent briefly describing the scientific basis of the proposed project is requested. An approximate level of project funding should be included. Letters of intent must meet the application deadline. All applicants will be promptly informed as to whether to proceed with a final proposal. Deadline listed is for letter of intent.
Requirements: Investigators already studying TS and those currently working in areas of science that can be applied to the problems of TS are invited to apply.
Date(s) Application is Due: Oct 17
Amount of Grant: 40,000 - 75,000 U.S.D
Contact: TSA Scientific Advisory Board; (718) 224-2999, ext. 222; fax (718) 279-9596; grantadministrator@tsa-usa.org
Internet: http://www.tsa-usa.org
Sponsor: Tourette Syndrome Association
42-40 Bell Boulevard
Bayside, NY 11361-2820

Tucson Electric Power Grants 4523
The company awards grants to Arizona nonprofit organizations for programs that directly enhance quality of life or essential programs for company service territory/operation sites. Focus areas for funding consideration include higher education, especially areas that promote careers related to a utility's workforce; environmental programs that further understanding and adoption of energy conservation, and protecting community; programs to support basic health, safety, and well-being; community economic development and civic activities; and child welfare and family-strengthening programs. Evaluation of requests is based on the potential impact on the community and on Tucson Electric Power (TEP); employee volunteer involvement; number of people served; programmatic strength and financial viability of applicant; and partnerships that promote collaborations and avoid duplication of services. Generally organizations will not be funded multiple times in a calendar year.
Requirements: Arizona 501(c)3 charitable nonprofits based in Pima County and serving at-risk youth or families in Pima County are eligible.
Restrictions: In general the company will not contribute to the following: religious, fraternal, social, or veteran's organizations; political or partisan organizations or candidates; individuals; individual schools K-12 (all schools in TEP service area are offered safety, environmental, and science related curricula and supplemental resources at no charge); organizations serving people suffering from a single disease or medical condition, or for medical research; endowments; travel funds of any kind; adult sports; programs outside TEP service territory/employee community; or agencies that contribute more than 10 percent of budget to outside service territory. TEP cannot donate electricity or electric service or waive fees where others would be charged.
Geographic Focus: Arizona
Date(s) Application is Due: Sep 12
Amount of Grant: 2,500 - 10,000 U.S.D
Contact: Sharon Foltz, (520) 884-3740; fax (520) 884-3606; sfoltz@tucsonelectric.com
Internet: http://www.tucsonelectric.com/Community/GrantsFunding/index.html
Sponsor: Tucson Electric Power Company
P.O. Box 711-UE102
Tucson, AZ 85702

Tull Charitable Foundation Grants 4524
Priority interest areas for awarding grants, in Georgia, are education, health and human services, youth development, and the arts. Grants support capital projects (e.g., buildings and endowments) in these areas. The amount of award varies depending on the project for which funding is requested.
Requirements: Grants are provided to organizations in Georgia.
Restrictions: Grants are not available for operating support, research, conferences and seminars, special events, or individuals.
Geographic Focus: Georgia
Date(s) Application is Due: Jan 1; Apr 1; Jul 1; Oct 1
Amount of Grant: 25,000 - 50,000 U.S.D
Contact: Program Contact; (404) 659-7079; jan@tullfoundation.org
Internet: http://www.tullfoundation.org/app_procedures.asp
Sponsor: Tull Charitable Foundation
50 Hurt Plaza, Suite 1245
Atlanta, GA 30303

Turner Foundation Grants 4525
The Foundation supports initiatives that enhance the quality of life in the greater Springfield/Clark County community through artistic, educational, environmental, recreational, family, healthcare, historic preservation, community beautification and revitalization programs.
Requirements: To be considered for a discretionary grant, each organization must serve the people of Clark County Ohio, be a tax-exempt non-profit with a 501(c)3 classification from the IRS, and have a governing board.
Restrictions: The Turner Foundation does not fund individuals, churches, legislative action groups, annual fundraising campaigns, scholarships, fraternal groups or political groups and issues.
Geographic Focus: Ohio
Date(s) Application is Due: Sep 18
Amount of Grant: 5,000 - 30,000 U.S.D
Contact: Grants Director; (937) 325-1300; email@hmturnerfoundation.org
Internet: http://www.hmturnerfoundation.org/html/grant.html
Sponsor: Turner Foundation
4 West Main Street, Suite 800
Springfield, OH 45502

TWS Foundation Grants 4526
The foundation awards grants to eligible nonprofit organizations in its areas of interest, including community development, arts, health care, public policy, research, and human services. Types of support include general operating support and scholarship funds. There are no application forms or deadlines.
Requirements: Nonprofit organizations, primarily in Connecticut, New York, Rhode Island, and Texas, are eligible.
Geographic Focus: Connecticut, New York, Rhode Island, Texas
Amount of Grant: 100 - 230,000 U.S.D
Samples: National Center for Policy Analysis (Dallas, TX)—$200,000; Westerly Hospital Foundation (Westerly, RI)—$100,000; Philharmonic-Symphony Society of New York (New York, NY)—$10,000.
Contact: Thomas Smith, Trustee
Sponsor: TWS Foundation
323 Railroad Avenue
Greenwich, CT 06830-6306

Tyco Electronics Foundation Grants 4527
The foundation provides grants to nonprofit organizations in geographic areas where Tyco Electronics Corp has a significant employee population and for specific projects or programs in broad categories, including education (with an emphasis on math and science), community impact, and arts and culture. In addition to a matching gifts program for employee contributions to accredited high schools, colleges, and universities, the foundation makes direct grants for programs that address a business or community concern of Tyco Electronics. Organizations that support precollege math and science education receive special attention. Agencies that promote personal growth, career opportunities, and economic self-sufficiency are encouraged to apply, as are local chapters of health- and civic-related organizations. Local United Way chapters benefit from donations that have a correlation to the level of employee giving. Organizations working to conserve or improve the environment through education may also receive funds. Special attention is given to communitywide arts organizations that solicit and allocate funds for a number of arts groups and institutions. Local public television and radio stations are encouraged to apply for funding of specific education initiatives. Capital campaigns of significant arts and cultural organizations serving communities in which the corporation has a major presence also will receive consideration. Grants also are awarded to support general operations, program development, and employee matching gifts. Applications are accepted throughout the year but are considered on the listed application deadlines.
Requirements: Organizations receiving funding must have IRS tax-exempt status and be located in company-operating areas (Roanoke, VA; the Triangle area of North Carolina; Rock Hill, SC; and central Pennsylvania).

Restrictions: The foundation generally will not support organizations in geographic areas where Tyco Electronics has few or no employees; individuals, private foundations, national organizations, or service clubs; fraternal, social, labor, or veterans organizations; organizations that discriminate on the basis of race, religion, color, national origin, physical or mental conditions, veteran or marital status, age, or sex; churches or religious organizations in general; political campaigns; general operating needs of United Way agencies; loans or investments; or programs that pose a potential conflict of interest.
Geographic Focus: California, Massachusetts, Michigan, North Carolina, Pennsylvania, South Carolina, Texas, Virginia
Date(s) Application is Due: Mar 15; Jun 15; Sep 15; Dec 15
Amount of Grant: 250 - 25,000 U.S.D
Contact: Mary Rakoczy, (717) 592-4869; fax (717) 592-4022; mjrakocz@tycoelectronics.com
Internet: http://www.tycoelectronics.com/about/foundation/vision.stm
Sponsor: Tyco Electronics Foundation
P.O. Box 3608, MS 140-10
Harrisburg, PA 17105-3608

Tylenol Scholarship 4528
The makers of the Tylenol Family of Products will award 170 scholarships in the amounts of $1000 and $5000 for higher education to students who demonstrate leadership in community activities and school activities and who intend to major in areas that will lead to health-related fields. Apply online at the sponsor's website.
Requirements: Application is open to students who will be attending an undergraduate or graduate course of study in the fall at an accredited two- or four-year college, university or vocational-technical school. This includes those students currently enrolled in an undergraduate or graduate course of study and have one or more years of school remaining. Applicants will be judged on leadership qualities and academic performance.
Date(s) Application is Due: May 15
Amount of Grant: 5,000 - 10,000 U.S.D
Contact: Administrators; (615) 320-3149; fax (615) 320-3151; comcast@spaprog.com
Internet: http://tylenol.com/page.jhtml?id=tylenol/news/subptyschol.inc&s_kwcid=ContentNetwork|1522864193
Sponsor: McNeil Consumer Healthcare
7050 Camp Hill Road
Fort Washington, PA 19034

Tyler Aaron Bookman Memorial Foundation Trust Grants 4529
The Tyler Aaron Bookman Memorial Foundation was established by Neil and Jill Bookman in memory of their eleven-year-old son, Tyler, who died of a brain tumor. His parents, based on Tyler's love of learning, started the Foundation with the money they had saved for his college education. To them, education is key to breaking the cycle of poverty. With funding centered in Camden, New Jersey, and adjoining states, the Foundation's primary fields of interest is support of health care and research, Catholic churches and agencies, and children and youth education. There are no specific deadlines with which to adhere, and applicants should initially approach the Foundation by way of letter requesting an application form.
Restrictions: No grants are given to individuals.
Geographic Focus: New Jersey, North Carolina, Pennsylvania
Amount of Grant: Up to 30,000 U.S.D
Contact: Neil S. Bookman; (215) 646-2192 or (267) 216-7718; fax (215) 654-6060
Sponsor: Tyler Aaron Bookman Memorial Foundation
426 Newbold Road
Jenkintown, PA 19046-2851

U.S. Cellular Corporation Grants 4530
The corporation awards grants to projects or programs that have significant relevance within its operating communities and that relate to the following strategic areas of concern: civic and community, education, health and human service, environment, and arts and culture. Guidelines and application are available online.
Requirements: 501(c)3 tax-exempt organizations are eligible.
Restrictions: The corporation will not make charitable contributions to the following entities: individuals; agencies that discriminate on the basis of race, color, creed, or national origin; political causes, candidates, legislative lobbying, or advocacy groups; endowments or memorials; construction or renovation projects (except Habitat for Humanity); religious organizations seeking to further a denominational or sectarian purpose; social, labor, alumni, or fraternal organizations serving a limited constituency; primary, secondary, or charter schools; special occasion, good-will, and single-interest magazines; local athletic or sports programs, walk-a-thons, and little leagues; or travel funds for tours, expeditions, or trips.
Contact: External Communications Department; (773) 399-8900; fax (773) 399-8937; externalcommunications@cellular.com
Internet: http://www.uscc.com/uscellular/SilverStream/Pages/a_charitable.html
Sponsor: U.S. Cellular Corporation
8410 W Bryn Mawr, Suite 700
Chicago, IL 60634

U.S. Department of Education 21st Century Community Learning Centers 4531
This program supports the creation of community learning centers that provide academic enrichment opportunities for children, particularly students who attend high-poverty and low-performing schools. The program helps students meet state and local student standards in core academic subjects, such as reading and math; offers students a broad array of enrichment activities that can complement their regular academic programs; and offers literacy and other educational services to the families of participating children.
Requirements: Awards are made to State Education Agencies (SEAs). Local education agencies (LEAs) and nonprofit organization may apply to states for subgrants. Formula grants are awarded to State educational agencies, which in turn manage statewide competitions and award grants to eligible entities. For this program, eligible entity means a local educational agency, community-based organization, another public or private entity, or a consortium of two or more of such agencies, organizations, or entities. States must give priority to applications that are jointly submitted by a local educational agency and a community-based organization or other public or private entity. Consistent with this definition of eligible entities, faith-based organizations are eligible to participate in the program. Each eligible entity that receives an award from the state may use the funds to carry out a broad array of before- and after-school activities (including those held during summer recess periods) to advance student achievement. Many states around the country are conducting competitions to award 21st Century Community Learning Center grants. The State Contact List (http://www.ed.gov/programs/21stcclc/contacts.html#state) now includes links to State websites and application due dates.
Contact: Peter Eldridge, Acting Program Contact; (202) 260-2514; fax (202) 260-8969; 21stCCLC@ed.gov or Peter.Eldridge@ed.gov
Internet: http://www.ed.gov/programs/21stcclc/index.html
Sponsor: U.S. Department of Education
400 Maryland Avenue SW, Room 3E246
Washington, DC 20202-6200

U.S. Department of Education Erma Byrd Scholarships 4532
The Erma Byrd Scholarship Program provides scholarships to individuals pursuing a course of study that will lead to a career in industrial health and safety occupations, including mine safety. This program is designed to increase the skilled workforce in these fields at both the fundamental skills level and the advanced skills level. The program has a service obligation component, requiring recipients of the scholarship to be employed in a career position directly related to industrial health and safety, including mine safety, for a period of one year upon completion of the degree program. The selected areas of study are: mining and mineral engineering, industrial engineering, occupational safety and health technology/technician, quality control technology/technician, industrial safety technology/technician, hazardous materials information systems technology/technician, mining technology/technician, and occupational health and industrial hygiene. Scholarship recipients must begin such employment no more than six months after the completion of their degree program. Average awards will be as follows: $2,500 associate's degree; $5,000 bachelor's degree; $10,000 graduate degree.
Requirements: Applicants must be: U.S. citizen, national, or permanent resident; pursuing a degree in mining and mineral engineering, industrial engineering, occupational safety and health technology/technician, quality control technology/technician, industrial safety technology/technician, hazardous materials information systems technology/technician, mining technology/technician, or occupational health and industrial hygiene; and, within two years of completing an associate's, bachelor's, or graduate degree at an accredited U.S. institution of higher education. As part of your application package, your Social Security Number must be submitted to the U.S. Department of Education via postal mail using the form on page 34 of the application package. For security reasons, DO NOT submit this form via email with the rest of your application.
Date(s) Application is Due: Jul 31
Amount of Grant: 2,500 - 10,000 U.S.D
Contact: Lauren Kennedy, (202) 502-7630; ermabyrdprogram@ed.gov
Internet: http://www.ed.gov/programs/ermabyrd/index.html
Sponsor: U.S. Department of Education
1990 K Street NW, Room 6121
Washington, DC 20006-8524

U.S. Department of Education Rehabilitation Engineering Research Centers Grants 4533
RERCs support activities that: (1) lead to the development of methods, procedures, and devices that will benefit individuals with disabilities, especially those with the most severe disabilities; or (2) involve technology for the purposes of enhancing opportunities for meeting the needs of and addressing the barriers confronted by individuals with disabilities in all aspects of their lives. Types of activities supported include: the development of technological systems for persons with disabilities; stimulation of the production and distribution of equipment in the private sector; and clinical evaluations of equipment. Awards are for five years, except that grants to new recipients or to support new or innovative research may be made for fewer than five years.
Requirements: States, Institutions of Higher Education (IHEs), Nonprofit Organizations, public or private agencies, including for-profit agencies, Indian tribes, and tribal organizations may apply. RERCs must be operated by or in collaboration with one or more institutions of higher education or nonprofit organizations.
Date(s) Application is Due: Apr 1; Aug 1
Amount of Grant: 850,000 - 1,000,000 U.S.D
Contact: Donna Nangle, (202) 245-7462; fax (202) 245-7323; donna.nangle@ed.gov
Internet: http://www.ed.gov/programs/rerc/index.html
Sponsor: U.S. Department of Education
400 Maryland Avenue SW, Room 5051, PCP
Washington, DC 20202-2700

U.S. Department of Education Rehabilitation Research Training Centers 4534
The RRTCs conduct coordinated and advanced programs of research, training, and information dissemination. Each RRTC has a major program of research in a particular area, such as mental illness, vocational rehabilitation (VR), or independent living, which is specified by the National Institute on Disability and Rehabilitation Research

(NIDRR). The RRTCs must serve as centers of national excellence and national or regional resources for providers and individuals with disabilities and their representatives. RRTC awards are for five years, except that grants to new recipients or to support new or innovative research may be made for fewer than five years.
Requirements: Note: Each year, competitions are held in specific areas that determine the types of projects. Eligible applicants include Institutions of Higher Education and Nonprofit Organizations. States, public or private agencies - including for-profit agencies, Indian tribes, and tribal organizations may also apply. Rehabilitation research and training centers must be operated by or in collaboration with: (1) one or more institutions of higher education or (2) one or more providers of rehabilitation or other appropriate services.
Date(s) Application is Due: Aug 21; Aug 22
Amount of Grant: Up to 675,000 U.S.D
Contact: Donna Nangle, (202) 245-7462; fax (202) 245-7323; donna.nangle@ed.gov
Internet: http://www.ed.gov/programs/rrtc/index.html
Sponsor: U.S. Department of Education
400 Maryland Avenue SW, Room 5051, PCP
Washington, DC 20202-2700

U.S. Department of Education Rehabilitation Training—Rehabilitation Continuing Education Programs—Institute on Rehabilitation Issues (IRI) 4535

The purpose of the Technical Assistance and Dissemination To Improve Services and Results For Children With Disabilities program is to promote academic achievement and to improve results for children with disabilities by providing technical assistance (TA), supporting model demonstration projects, disseminating useful information, and implementing activities that are supported by scientifically based research. The purpose of this priority is to support the establishment and operation of State Technical Assistance Projects To Improve Services and Results for Children Who Are Deaf-Blind (projects). Grants are available to support projects in the District of Columbia; Puerto Rico; and the Virgin Islands.
Requirements: Eligible applicants include: SEAs; LEAs, including public charter schools that are considered LEAs under State law; IHEs; other public agencies; private nonprofit organizations; outlying areas; FAS; Indian tribes or tribal organizations; and for-profit organizations. The projects funded under this competition must make positive efforts to employ and advance in employment qualified individuals with disabilities (see section 606 of IDEA). Applicants and grant recipients funded under this competition must involve individuals with disabilities or parents of individuals with disabilities ages birth through 26 in planning, implementing, and evaluating the projects (see section 682(a)(1)(A) of IDEA).
Restrictions: Funds awarded under this priority may not be used to provide direct early intervention services under Part C of IDEA, or direct special education and related services under Part B of IDEA. The Department will reject an application for a State project that proposes a budget exceeding the funding level for any single budget period of 12 months.
Geographic Focus: District of Columbia, Puerto Rico, U.S. Virgin Islands
Date(s) Application is Due: Jul 17
Amount of Grant: 30,000 - 65,000 U.S.D
Internet: http://www.ed.gov/programs/rsatrain/applicant.html#84264c
Sponsor: U.S. Department of Education
400 Maryland Avenue SW, Room 5038, PCP
Washington, DC 20202-2800

U.S. Department of Education Special Education—National Activities—Parent Information Centers 4536

The purpose of this program is to ensure that parents of children with disabilities receive training and information to help improve results for their children. Awards are made for parent information centers, community parent centers, and for technical assistance to such centers. This program includes the following grants competitions: Community Parent Resource Centers (CFDA 84.328C); Parent Training and Information Centers (CFDA 84.328M); and, Technical Assistance for the Parent Centers (CFDA 84.328R).
Requirements: Nonprofit organizations are eligible to apply. For Parent Training and Information (PTI) Centers (# 84.328M), parent organizations may apply. A parent organization is a private nonprofit organization (other than an institution of higher education [IHE]) that: (a) has a board of directors (1) the majority of whom are parents of children with disabilities ages birth through 26; (2) that includes (i) individuals working in the fields of special education, related services, and early intervention, (ii) individuals with disabilities, and (iii) the parent and professional members of which are broadly representative of the population to be served, including low-income parents of limited English proficient (LEP) children; and (b) has as its mission serving families of children with disabilities who are ages birth through 26, and have the full range of disabilities described in Sec. 602(3) of the Individuals with Disabilities Education Act(IDEA). For Community Parent Resource Centers (CPRC; # 84.328C), local parent organizations may apply (same requirements as listed above for PTI Centers). Additionally, *Community to be served* refers to a community whose members experience significant isolation from available sources of information and support as a result of cultural, economic, linguistic, or other circumstances deemed appropriate by the secretary of education. For Technical Assistance for Parent Training and Information Centers (# 84.328R), state education agencies (SEAs); local education agencies (LEAs); public charter schools that are LEAs under state law; IHEs; other public agencies; private nonprofit organizations; outlying areas (American Samoa, Guam, the Northern Mariana Islands, and the U.S. Virgin Islands); freely associated states; Indian tribes or tribal organizations; and for-profit organizations may apply.
Date(s) Application is Due: Feb 26; Apr 18
Amount of Grant: 100,000 - 500,000 U.S.D
Contact: Lisa Gorove, (202) 245-7357; fax (202) 245-7617; Lisa.Gorove@ed.gov
Internet: http://www.ed.gov/programs/oseppic/index.html
Sponsor: U.S. Department of Education
400 Maryland Avenue SW, Room 5051, PCP
Washington, DC 20202-2700

U.S. Department of Education United States-Russia Program: Improving Research and Educational Activities in Higher Education 4537

The program provides grants that demonstrate partnerships between Russian and American institutions of higher education that contribute to the development and promotion of educational opportunities between the two nations, particularly in the areas of mutual foreign language learning and the cooperative study of mathematics and science. Collaborative projects comprised of partnerships of Russian and American universities and colleges will be supported.
Requirements: Institutions of Higher Education (IHEs) or combinations of IHEs and other public and private nonprofit institutions and agencies are eligible to apply. Applicants must select one of three academic discipline as the subject area for their grant proposal: (A) Environmental Science Studies; (B) Biotechnology; or, (C) Any discipline, other than (A) and (B). Applications are invited from institutions of higher education with the capacity to contribute to a collaborative project in the areas listed with a Russian institution. The consortium partners, through promoting the study of and communication in foreign languages, are expected to increase awareness and understanding of the two cultures, and to strengthen the professional and scholarly ties between the two countries. Russian institutions will apply to The Russian Ministry of Education and Science for funding.
Restrictions: Any application that proposes a budget exceeding $150,000 for a single budget period of 12 months will be rejected. The Assistant Secretary for Postsecondary Education may change the maximum amount through a notice published in the Federal Register.
Geographic Focus: All States, Russia
Date(s) Application is Due: Jul 7
Amount of Grant: 189,000 U.S.D
Contact: Krish Mathur, Coordinator, U.S.-Russia Program; (202) 502-7512; fax (202) 502-7877; Krish.Mathur@ed.gov
Internet: http://www.ed.gov/programs/fipserussia/index.html
Sponsor: U.S. Department of Education
1990 K Street NW, Room 6155
Washington, DC 20006-8544

U.S. Department of Education Vocational Rehabilitation Services Projects for American Indians with Disabilities Grants 4538

The program provides financial assistance for the establishment and operations of vocational rehabilitation (VR) services programs for American Indians with disabilities living on or near a federal or state reservation. The purpose of this program is to assist tribal governments to develop or to increase their capacity to provide a program of vocational rehabilitation services, in a culturally relevant manner, to American Indians with disabilities residing on or near federal or state reservations. The program's goal is to enable these individuals, consistent with their individual strengths, resources, priorities, concerns, abilities, capabilities, and informed choice, to prepare for and engage in gainful employment. Program services are provided under an individualized plan for employment and may include native healing services.
Requirements: The governing body of an Indian tribe or consortia of such governing bodies located on federal and state reservations may apply.
Date(s) Application is Due: May 5; Jun 12
Amount of Grant: 350,000 - 450,000 U.S.D
Contact: Alfreda Reeves, (202) 245-7485; alfreda.reeves@ed.gov
Internet: http://www.ed.gov/programs/vramerind/index.html
Sponsor: U.S. Department of Education
400 Maryland Avenue SW, Room 5051, PCP
Washington, DC 20202-2700

U.S. Lacrosse EAD Grants 4539

U.S. Lacrosse is playing a leading role in ongoing research and is committed to educating the national lacrosse community about the life-saving value of having Automatic External Defibrillators (AEDs) available during lacrosse games and practices. In order to achieve that goal, U.S. Lacrosse, the national governing body of the sport, and Cardiac Science, a leading manufacturer of cardiology products, are offering an AED grant program which provides an AED and comprehensive management of AED and CPR training to lacrosse leagues or U.S. Lacrosse chapters.
Requirements: Applicant must be a current member of U.S. Lacrosse. Awarded group must budget $1295 towards the AED Grant Program.
Date(s) Application is Due: Nov 30
Contact: Sarah Newman, Manager of Programs and Grants; (410) 235-6882; fax (410) 366-6735; snewman@uslacrosse.org
Internet: http://www.uslacrosse.org/programs/AEDgrantprogram.phtml
Sponsor: U.S. Lacrosse
113 W University Parkway
Baltimore, MD 21210

UCLA / RAND Corporation Post-Doctoral Fellowships 4540

The post-doctoral training program is jointly housed in the RAND Health Sciences program and the UCLA School of Public Health. Post-doctoral fellows have the option of pursuing a Masters degree in the UCLA Department of Health Services. The grant covers all tuition and fees, as well as a stipend, which currently ranges from $35,568 to $51,036, depending on the number of years of post-doctoral experience. Travel money is also available for fellows to attend professional meetings.
Requirements: To be eligible, candidates must have obtained a Ph.D., medical degree, or other professional doctoral degree. Awards will last two years. Fellows will be required

to attend at least 8 courses at UCLA or RAND, including three courses in health services research methods, two in statistics, and three in health services issue areas. Those pursuing a Masters degree must complete a total of 12 courses. Upon arrival, fellows will align with an ongoing research project at RAND or UCLA. RAND and UCLA have numerous ongoing projects in the health services research area. These fall into several areas, including: access/utilization among the underserved; assessing quality, effectiveness, outcomes; health behaviors; alternative delivery systems; hospital and physician payment; health care cost containment; technology assessment; private health insurance; international comparisons. By the end of the first year, the fellow will identify a research idea and write a proposal. The proposal may be addressed to the principal investigator of an ongoing research project, or to an outside funding agency. By the end of the program, the fellow will prepare an article on research conducted during the fellowship, and submit it to a peer reviewed journal. This can either be done alone, or in collaboration with the research director or other member of the fellow's research team.
Amount of Grant: 35,568 - 51,036 U.S.D
Contact: RAND Health Communications Office; (310) 393-0411, ext. 7775; RAND_Health@rand.org
Internet: http://www.ph.ucla.edu/hs/post_doctoral.html
Sponsor: RAND Corporation
1776 Main Street, P.O. Box 2138
Santa Monica, CA 90407-2138

UCLA International Institute Global Fellowships **4541**
The institute awards postdoctoral global fellowships, tenable for one to two years, beginning September 1. Fellows will be drawn from the arts and humanities (including literary studies and philosophy), the natural sciences (including engineering, medicine, and public health), and the social sciences (including law, public policy, and management). Global fellows will pursue their own research and teach one undergraduate seminar. Fellows are expected to produce working papers, journal articles, and books, or works in other media, during their tenure at UCLA. Successful applicants will be of the highest intellectual caliber and will be doing pathbreaking work, preferably of multidisciplinary interest, on (or with obvious relevance to) the cultural, economic, social, and political dynamics of the contemporary world. Among the topics of particular interest to the institute are: new forms of governance and coordination, whether in the governmental, nongovernmental, or private spheres; security concerns that transcend traditional interstate conflict; the causes and consequences of globalization; transnational cultures and multiple identities (or resistance to these trends); and public health, the environment, and economic development. Applications and nominations should be submitted, on-line, between 1 October and 1 November. There are no citizenship requirements.
Requirements: Candidates will only be considered if they received their PhDs (or equivalent) between 1 January 1998 and 1 April 2005.
Date(s) Application is Due: Jan 14
Amount of Grant: 50,000 U.S.D
Contact: Program Coordinator; (310) 825-4921; globalfellows@international.ucla.edu
Internet: http://www.international.ucla.edu/globalfellows
Sponsor: UCLA International Institute
11349 Bunche Hall, UCLA
Los Angeles, CA 90095-1487

UCT Mentally Handicapped Citizens Teachers Scholarships **4542**
The purpose of these scholarships is to help teachers of the mentally handicapped in the continental United States and Canada obtain a degree and/or additional certification. In addition to requesting an application, the student must provide a resume of past work experience in the mentally handicapped field, indicate current college rank or standing, and state future educational and/or career goals. Application forms are available upon request. Applications are due prior to class registration and must be renewed annually.
Requirements: To be eligible, applicants must meet one or more of the following criteria: be a teacher of the mentally handicapped who needs additional coursework to be certified or to retain certification; be an experienced teacher who wishes to become certified to teach the mentally handicapped; have a bachelor's or master's degree and plan to pursue graduate work in the field of mental retardation; be a college junior or senior undergraduate whose course of study is mental retardation and who is planning to teach the mentally handicapped or have plans to become an instructor under a structured trade, vocational, or recreational program at a facility for the mentally retarded; and have a letter of recommendation from the instructor at the facility where the person intends to become a teacher in the field of mental retardation.
Restrictions: Applicants teaching in areas of special education other than mental retardation are ineligible.
Amount of Grant: 750 - 3,000 U.S.D
Contact: Ann Marshall; (800) 848-0123, ext. 126 or (614) 341-6994; amarshall@uct.org
Internet: http://www.uct.org/scholarships.html
Sponsor: Order of United Commercial Travelers of America
632 N Park Street
Columbus, OH 43215

UICC American Cancer Society International Fellowships for Beginning Investigators **4543**
The international fellowships fund 12-month basic or clinical research projects in the areas of epidemiology, prevention, detection, diagnosis, treatment, and psycho-oncology that foster a bidirectional flow of knowledge, experience, expertise, and innovation to and from the United States. Candidates should be in the early stages of their careers. Applications that are geared to the development of specific cancer control measures in developing and central and east European countries are particularly encouraged. Six to eight fellowships, with tenures beginning in April, will be awarded each year.
Requirements: Eligible candidates should hold assistant professorships or similar positions at their home institutes and have, ordinarily, a minimum of two and a maximum of ten years of postdoctoral experience after obtaining their MD or PhD degrees or equivalents.
Restrictions: Candidates who are physically present at the proposed host institute while their applications are under consideration are not eligible for UICC fellowships.
Date(s) Application is Due: Dec 1
Amount of Grant: 45,000 U.S.D
Samples: Paula Ximena Fernandez-Calotti, from Laboratory of Oncohematology Institute of Hematology Research, M.R. Castex, Argentina, to the Department of Biochemistry and Molecular Biology, University of Barcelona, Spain; Nikita A. Makretsov, from the Department of Pathology & Laboratory Medicine, Memorial University of Newfoundland, Canada, to the Department of Oncology Hutchinson / MRC Research Centre, United Kingdom.
Contact: Beate Vought; +41 22 809 1843; fax +41 22 809 1810; fellows@uicc.ch
Internet: http://www.uicc.org/index.php?option=com_content&task=view&id=14259&Itemid=182
Sponsor: International Union Against Cancer
62, route de Frontenex
Geneva, CH-1207 Switzerland

UICC International Cancer Technology Transfer (ICRETT) Fellowships **4544**
The program helps to facilitate rapid international transfer of cancer research techiques and technology and clinical management skills. It is hoped that funding will support the exchange of knowledge and enhance skills in basic, clinical, behavioral, and epidemiological areas of cancer research, in cancer control and prevention (including tobacco control). Participants should acquire appropriate clinical management, diagnostic and therapeutic skills for effective application and use in the home organization upon their return. There are 120 to 150 awards available annually, with applications accepted at any time and notification generally within 60 days of registration.
Requirements: Eligible applicants include appropriately qualified investigators, pathologists, epidemiologists, cancer registrars, behavioral scientists, tobacco control activists, laboratory technicians, clinicians and general practitioners, generally in the early stages of their careers.
Amount of Grant: 3,400 U.S.D
Contact: Beate Vought; +41 22 809 1843; fax +41 22 809 1810; fellows@uicc.ch
Internet: http://www.uicc.org/index.php?option=com_content&task=view&id=14262&Itemid=185
Sponsor: International Union Against Cancer
62, route de Frontenex
Geneva, CH-1207 Switzerland

UICC Raisa Gorbachev Memorial International Cancer Fellowships **4545**
This award is funded by UICC International Cancer Foundation and enables qualified beginning pathologists to carry out specific research or training project relating to the many fields of oncology, and should be centred on an appropriate training period in modern diagnostic tools in the field of pathology. The Fellowship is conditional on Fellows returning to the home institutes/country at the end of the fellowship period. The original Fellowships cannot be prolonged, or run concurrently with other awards, even those funded by other agencies. They can however, be extended subject to the written approval of home and host supervisors and at no additional cost to UICC i.e. up to a total duration of 3 months. Should return of the Fellow to their home country be delayed past these 3 months, 50 percent of the travel award must be reimbursed to UICC.
Requirements: The candidate must possess the appropriate professional qualifications and experience according to the specifications described on the website and must currently be engaged in cancer research, clinical oncology practice, or cancer society work. To permit effective communication at the host institute, the candidate must have adequate fluency in a common language. The candidate must also be on the staff payroll of a university, research laboratory or institute, hospital, oncology unit to where they will return at the end of a fellowship.
Restrictions: Candidates attached to commercial entities are not eligible. Fellowships cannot be granted to candidates who are already physically present at the proposed host institute whilst their applications are under consideration.
Contact: Beate Vought; +41 22 809 1843; fax +41 22 809 1810; fellows@uicc.ch
Internet: http://www.uicc.org/index.php?option=com_content&task=view&id=15956&Itemid=301
Sponsor: International Union Against Cancer
62, route de Frontenex
Geneva, CH-1207 Switzerland

UICC Yamagiwa-Yoshida Memorial International Cancer Study Grants **4546**
These study grants enable cancer investigators from any country to establish bilateral research projects abroad that exploit complementary materials or skills, including advanced training in experimental methods or special techniques. There are 14 to 16 three-to six-month grants awarded each year. Selection takes place twice annually, with recipients notified in mid-April and mid-October. Grants should be taken up within six months following notification. Travel awards contribute toward the least expensive international two-way airfare or other appropriate form of transport. Financial support is not provided for dependents. Application materials are available on the Web site.
Requirements: Candidates should have appropriate scientific or medical qualifications and be actively engaged in cancer research.
Date(s) Application is Due: Jan 15; Jul 1
Amount of Grant: 30,000 - 60,000 U.S.D

Samples: Dr. Hakan Akbulut, from Department of Medical Oncology, Cebeci Hospital, Turkey, to Sydney Kimmel Cancer Center, U.S.A; Dr. Gil Ast, from Department of Human Molecular Genetics & Biochemistry, Sackler School of Medicine, Israel, to Department of Biochemistry, School of Medicine, University of Washington, U.S.A.
Contact: Beate Vought, Fellowships Coordinator; +41 22 809 1843; fax +41 22 809 1810; fellows@uicc.ch
Internet: http://www.uicc.org/index.php?option=com_content&task=view&id=15958&Itemid=302
Sponsor: International Union Against Cancer
62, route de Frontenex
Geneva, CH-1207 Switzerland

UK Foundation for the Study of Infant Deaths Grants **4547**
The foundation awards grants in England, Wales, and Northern Ireland for research into sudden infant death syndrome. Specific areas of interest include epidemiology, immunology, pathology, statistics, and infection. Graduate researchers of all nationalities are considered.
Geographic Focus: All States, United Kingdom
Date(s) Application is Due: Mar 1
Amount of Grant: Up to 100,000 GBP
Contact: Secretary-General; 0870 787 0885; fax 0870 787 0725; research@sids.org.uk
Internet: http://www.sids.org.uk/fsid/grants.htm
Sponsor: UK Foundation for the Study of Infant Deaths
Artillery House, 11-19 Artillery Row
London, SW1P 1RT England

UniHealth Foundation Community Health Improvement Grants **4548**
UniHealth Foundation is an independent private health care foundation established in 1998 whose mission is to support and facilitate activities that significantly improve the health and well being of individuals and communities they serve. The objectives of the Community Health Improvement Grants are to: enable informed consumer health decisions, including appropriate use of healthcare services; promote improved individual health behaviors; identify and reduce risk factors for disease and injury; and ameliorate the impact of disease and enhance the quality of life. Grants supports community health improvement activities of hospitals by funding health education, prevention and treatment programs. Health education includes information dissemination on a broad range of topics, including health systems navigation and informed consumerism or the development of culturally sensitive educational materials. Prevention, in addition to clinical preventive services, might include injury and violence prevention or caregiver support services. Funding for treatment services targets activities aimed at improving health status and quality of life, including aspects of chronic condition management, dental care, end-of-life care and rehabilitation services. Grants range from $50,000 to $350,000. Throughout the year the Foundation announces other initiatives that focus on specific goals and objectives to promote foundation-identified priorities. New requests for proposals are listed on the Foundation's website.
Requirements: Most grants are made for the purpose of funding healthcare services and programs provided by or through qualified charitable hospitals in specified service areas in Los Angeles and northern Orange Counties. The service areas are: San Fernando and Santa Clarita Valley; Westside and Downtown Los Angeles; San Gabriel Valley; and Long Beach and Orange County. Applicants are required to submit a preliminary letter of inquiry. There are no deadlines. Those applicants invited to submit a full proposal will receive detailed information.
Restrictions: Individuals are not eligible for grants. Grants will not be given for propagandizing and/or influencing legislation or elections, political campaigns, programs that promote religious doctrine, events, endowments, annual fund drives, non-applied research, general operating expenses, or retirement of debt. Indirect costs for up to a maximum of ten percent of direct costs is allowable.
Geographic Focus: California
Contact: Caroline Chung, Grants Manager; (213) 630-6500; fax (213) 630-6509; Webadmin@unihealthfoundation.org
Internet: http://unihealthfoundation.org/program_areas.html
Sponsor: UniHealth Foundation
800 Wilshire Boulevard, Suite 1300
Los Angeles, CA 90017

Unihealth Foundation Grants for Non-Profit Oranizations **4549**
UniHealth Foundation is an independent private health care foundation established in 1998 whose mission is to support and facilitate activities that significantly improve the health and well being of individuals and communities they serve. The objectives of the Grants for Non-Profit Organizations are: to promote positive health behaviors; to overcome systemic barriers to healthy living; and to advance informed consumerism. Funding is provided for traditional health education and disease prevention strategies on a broad range of topics as well as health system navigation, caregiver support services or the development of culturally sensitive educational materials. Grants range from $10,000 to $350,000. Throughout the year the Foundation announces other initiatives that focus on specific goals and objectives to promote foundation-identified priorities. New requests for proposals are listed on the Foundation's website.
Requirements: Qualified nonprofit organizations providing health-related services consistent with the Foundation's mission are eligible. Applicants are required to submit a preliminary letter of inquiry. There are no deadlines. Those applicants invited to submit a full proposal will receive detailed information.
Restrictions: Individuals are not eligible for grants. Grants will not be given for propagandizing and/or influencing legislation or elections, political campaigns, programs that promote religious doctrine, events, endowments, annual fund drives, non-applied research, general operating expenses, or retirement of debt. Indirect costs for up to a maximum of ten percent of direct costs is allowable.
Geographic Focus: California
Contact: Caroline Chung, Grants Manager; (213) 630-6500; fax (213) 630-6509; webadmin@unihealthfoundation.org
Internet: http://unihealthfoundation.org/program_areas.html
Sponsor: UniHealth Foundation
800 Wilshire Boulevard, Suite 1300
Los Angeles, CA 90017

UniHealth Foundation Healthcare Systems Enhancements Grants **4550**
UniHealth Foundation is an independent private health care foundation established in 1998 whose mission is to support and facilitate activities that significantly improve the health and well being of individuals and communities they serve. The objectives of the Healthcare Systems Enhancements Grants are to: increase access to comprehensive, coordinated care at the appropriate point of service; improve the quality of healthcare services in terms of patient satisfaction and clinical outcomes; and promote the financial stability and viability of not-for-profit hospitals. Support is provided for programs which address access to health care, quality of clinical services, patient satisfaction and organizational capacity building. Such programs might include partnerships with other hospitals and/or community clinics and community based organizations to coordinate care. Projects that focus on administrative systems improvement, organizational development, change management and capital needs associated with program delivery will also be considered. Grants range from $50,000 to $350,000. Throughout the year the Foundation may announce other initiatives that focus on specific goals and objectives to promote Foundation-identified priorities. New requests for proposals are listed on the Foundation's website.
Requirements: Most grants are made for the purpose of funding healthcare services and programs provided by or through qualified charitable hospitals in specified service areas in Los Angeles and northern Orange Counties. The service areas are: San Fernando and Santa Clarita Valley; Westside and Downtown Los Angeles; San Gabriel Valley; and Long Beach and Orange County. Applicants are required to submit a preliminary letter of inquiry. There are no deadlines. Those applicants invited to submit a full proposal will receive detailed information.
Restrictions: Individuals are not eligible for grants. Grants will not be given for propagandizing and/or influencing legislation or elections, political campaigns, programs that promote religious doctrine, events, endowments, annual fund drives, non-applied research, general operating expenses, or retirement of debt. Indirect costs for up to a maximum of ten percent of direct costs is allowable.
Geographic Focus: California
Samples: Gateways Hospital and Mental Health Center, Los Angeles, California, $55,000, child and adolescent evidence based treatment program; Children's Hospital of Orange County, CHOC Children's positive patient identification pilot project, Orange, California, $350,000 (2010); and Children's Hospital of Orange County, Orange, California, $96,000, NICU extremely low birth-rate baby unit (2010).
Contact: Caroline Chung, Grants Manager; (213) 630-6500; fax (213) 630-6509; webadmin@unihealthfoundation.org
Internet: http://www.unihealthfoundation.org/program_areas.html
Sponsor: UniHealth Foundation
800 Wilshire Boulevard, Suite 1300
Los Angeles, CA 90017

UniHealth Foundation Innovation Fund Grants **4551**
UniHealth Foundation is an independent private health care foundation established in 1998 whose mission is to support and facilitate activities that significantly improve the health and well being of individuals and communities they serve. The objective of Innovation Fund Grants is to encourage new, creative approaches that address complex and enduring health challenges. The Foundation encourages the development of new ways of delivering and financing healthcare programs and services. Grant funding is designed to provide an opportunity for nonprofit organizations to test creative ideas and promising new practices through interventions which address demonstrated need. Grants range from $10,000 to $350,000. Throughout the year the Foundation announces other initiatives that focus on specific goals and objectives to promote foundation-identified priorities. New requests for proposals are listed on the Foundation's website.
Requirements: Qualified nonprofit organizations providing health-related services consistent with the Foundation's mission are eligible. Applicants are required to submit a preliminary letter of inquiry. There are no deadlines. Those applicants invited to submit a full proposal will receive detailed information.
Restrictions: Individuals are not eligible for grants. Grants will not be given for propagandizing and/or influencing legislation or elections, political campaigns, programs that promote religious doctrine, events, endowments, annual fund drives, non-applied research, general operating expenses, or retirement of debt. Indirect costs for up to a maximum of ten percent of direct costs is allowable.
Geographic Focus: California
Samples: The Alliance for Children's Rights, Los Angeles, California, $100,000, connecting to healthier futures (2010); Children's Bureau, Los Angeles, California, $50,000, health access (2010); and Los Angeles Mission Community Clinic, Los Angeles, California, $133,300, patient care team redesign (2010).
Contact: Caroline Chung, Grants Manager; (213) 630-6500; fax (213) 630-6509; webadmin@unihealthfoundation.org
Internet: http://unihealthfoundation.org/program_areas.html
Sponsor: UniHealth Foundation
800 Wilshire Boulevard, Suite 1300
Los Angeles, CA 90017

UniHealth Foundation Workforce Development Grants 4552

UniHealth Foundation is an independent private health care foundation established in 1998 whose mission is to support and facilitate activities that significantly improve the health and well being of individuals and communities they serve. The objectives of the Workforce Development Grants are to: increase the size and quality of the healthcare workforce; encourage and enable the practice of medicine in underserved communities; promote the training and retention of competent clinical staff; and foster skills improvement and knowledge enhancement through continuous education. Grants support programs to train and retain qualified clinical staff, foster ongoing professional development, including management and leadership skills, and promote allied health professions training. Throughout the year the Foundation announces other initiatives that focus on specific goals and objectives to promote foundation-identified priorities. New requests for proposals are listed on the Foundation's website.

Requirements: Most grants are made for the purpose of funding healthcare services and programs provided by or through qualified charitable hospitals in specified service areas in Los Angeles and northern Orange Counties. The service areas are: San Fernando and Santa Clarita Valley; Westside and Downtown Los Angeles; San Gabriel Valley; and Long Beach and Orange County. Applicants are required to submit a preliminary letter of inquiry. There are no deadlines. Those applicants invited to submit a full proposal will receive detailed information.

Restrictions: Individuals are not eligible for grants. Grants will not be given for propagandizing and/or influencing legislation or elections, political campaigns, programs that promote religious doctrine, events, endowments, annual fund drives, non-applied research, general operating expenses, or retirement of debt. Indirect costs for up to a maximum of ten percent of direct costs is allowable.

Geographic Focus: California

Contact: Caroline Chung, Grants Manager; (213) 630-6500; fax (213) 630-6509; webadmin@unihealthfoundation.org

Internet: http://unihealthfoundation.org/program_areas.html

Sponsor: UniHealth Foundation
800 Wilshire Boulevard, Suite 1300
Los Angeles, CA 90017

Unilever U.S. Grants 4553

The foundation awards grants to nonprofit organizations nationwide in company-operating locations in the areas of civic affairs, education, environment, health care, and social services. Capital grants are awarded only to hospitals in plant communities. Types of support include general operating grants and matching gifts. Applicants must submit a written proposal; proposals are accepted throughout the year.

Requirements: 501(c)3 nonprofits in company-operating areas are eligible to apply.

Restrictions: Grants do not support religious, labor, political, or veterans' organizations; individuals (except for employee-related scholarships); goodwill advertising, fundraising events, or capital campaigns; or loans.

Amount of Grant: 500 - 200,000 U.S.D

Contact: Grants Administrator; (212) 894-2236; fax (212) 318-3600

Internet: http://www.unilever.com

Sponsor: Unilever United States Foundation
700 Sylvan Avenue
Englewood Cliffs, NJ 07632

Union Bank, N.A. Corporate Sponsorships and Donations 4554

As part of its ten-year community committment, Union Bank has pledged to annually distribute at least two percent of its annual after-tax net profit to help meet the needs of the communities it serves, a commitment that has resulted in donations exceeding $72 million dollars during the first six years. This two-percent charitable commitment is achieved through contributions and sponsorships made directly by the bank (through the corporate contribution program) and through grants and investments made by the bank's foundation (Union Bank Foundation). The bank's corporate contributions program is intended to enhance the bank's reputation and visibility by supporting the charitable work of its employees and clients. The bank funds donations and sponsorships supporting a broad range of charitable categories, including community economic development, affordable housing, education, health and human services, culture and arts, emergency services, and the environment. The bank is particularly interested in donations and sponsorships that support low-income populations and promote and enhance diversity in all its forms. The bank's local-area contribution committees consider applications at their monthly meetings. Applications are accepted via an online application system accessible from the bank website. Please note that event-sponsorship applications should be submitted at least ninety days in advance of the event date. Prospective applicants should review the bank's application guidelines and instructions, which are available at the website. Questions may be directed to the foundation officers listed on this page.

Requirements: 501(c)3 nonprofits in company-operating areas in California and the Pacific Northwest are eligible (e.g., San Diego, San Francisco, Los Angeles, Anaheim, Berkeley, Del Mar, Fresno, Irvine, Mission Grove, Pasadena, Sacramento, Salinas, San Jose, Santa Ana, and Torrance, California). A branch locator is available at the bank's website.

Restrictions: The bank does not support the following requests from the following entities or for the following items: individuals; veterans, military, fraternal, or professional organizations; political organizations or programs; service club activities; other intermediary foundations (i.e., foundations which, in turn, make grants to other charities); churches or religious groups (except separately incorporated community development corporations); educational institution operating funds; and individual elementary or secondary schools.

Geographic Focus: California, Oregon, Washington

Contact: J.R. Raines, Assistant Vice President; (619) 230-3105; CSRGroup@unionbank.com or charitablegiving@unionbank.com

Internet: https://www.unionbank.com/global/about/corporate-social-responsibility/foundation/foundation-grants.jsp#products-tab-item-2

Sponsor: Union Bank, N.A.
350 California Street
San Francisco, CA 94104

Union Bank, N.A. Foundation Grants 4555

As part of its ten-year community committment, Union Bank has pledged to annually distribute at least two percent of its annual after-tax net profit to help meet the needs of the communities it serves, a commitment that has resulted in donations exceeding $72 million dollars during the first six years. The two-percent charitable commitment is achieved through contributions and sponsorships made directly by the bank and through grants and investments made by the Union Bank Foundation, a nonprofit public-benefit corporation which serves as an agent for the bank. Because of its belief that the long-term success of the Union Bank business-model is dependent upon the existence of healthy communities, Union Bank Foundation focuses its philanthropy on building innovative initiatives and partnerships to cultivate healthy communities, which it identifies as possessing the following characteristics: stable families with high rates of home ownership; availability of affordable housing; livable-wage job opportunities; accessible public transportation; convenient access to professional services (e.g., doctors, lawyers, and accountants); adequate public services (e.g., police, fire, and sanitation); safe public places to relax and recreate (e.g., parks, libraries, theaters); clean air and water supplies; a high level of civic engagement; a community constituency possessing diverse income levels; well-funded public schools; successful small business owners who live in the community; a variety of retail shops and restaurants; and traditional financial institutions providing access to capital in or adjacent to the community. With an eye toward being an agent of positive change in Union Bank communities, the foundation focuses on the following strategic program areas (targeting resources especially to benefit low- to moderate-income populations): Affordable Housing; Community Economic Development; Education; and Environment. In the area of Affordable Housing, the foundation focuses on for-sale housing, rental housing, special-needs housing, senior housing, transitional-living facilities, emergency/homeless shelters, youth housing, self-help housing, farm-worker housing, pre-development funding to nonprofit developers, and capacity building for nonprofit housing organizations. In the area of Community Economic Development area, the foundation focuses on small business development, individual development, and neighborhood development. Small business development includes micro-enterprise development and support, technical assistance/entrepreneurial training, organizations that promote access to capital for business or farms meeting Small Business Administration criteria, and job creation. Individual development includes job training/apprenticeship, welfare-to-work programs, wealth-accumulation/asset-building programs, life-skills training, financial-literacy/credit-counseling programs, mortgage credit counseling, business education, and intervention/prevention programs for at-risk youth. Neighborhood Development includes gang prevention/gang intervention programs, crime prevention, dispute resolution/mediation/violence prevention, reduction of liquor outlets, improved quality of food in local markets, childcare and daycare programs, drug- and alcohol-rehabilitation programs, independent living programs, organizational capacity building and funding for operating/administrative expenses, and community organizing to engage, inform and empower citizenry. In the area of Education, the foundation focuses on scholarship programs, tutoring programs, general education degree (GED) preparation, English as a second language (ESL) programs, computer education, support for the teaching profession, teacher training, literacy programs, parent education, visual- and performing-arts-organizations outreach programs, enrichment programs, and capacity-building. In the area of Environment the foundation focuses on brown-field remediation, science and education relevant to green building, energy upgrade and conservation, rehabilitation and cleanup, coastal/creek- and reserve-cleanup and preservation, urban green-space projects, environmental education, aquariums and museums, ecology and recycling centers, and state parks, nature centers, conservancy centers, botanical gardens, and wildlife centers. The Union Bank Foundation prefers program grants, but will consider requests for core operating support and/or capacity-building grants to support exceptional work within its strategic funding categories. The foundation considers applications at its bimonthly board meetings. Applications are accepted via an online application system accessible from the foundation website. Applicants must choose from three categories when they apply. These are requests for $1,000 or less, requests for $1001 to $25,000, and requests for over $25,000. Prospective applicants should review the foundation's application guidelines and instructions, which are available at the foundation website. Questions may be directed to the foundation officers listed on this page.

Requirements: 501(c)3 nonprofits in company-operating areas in California and the Pacific Northwest are eligible (e.g., San Diego, San Francisco, Los Angeles, Anaheim, Berkeley, Del Mar, Fresno, Irvine, Mission Grove, Pasadena, Sacramento, Salinas, San Jose, Santa Ana, and Torrance, California). A branch locator is available at the sponsor website.

Restrictions: The foundation does not support the following requests from the following entities or for the following items: individuals; veterans, military, fraternal, or professional organizations; political organizations or programs; service club activities; other intermediary foundations (i.e., foundations which, in turn, make grants to other charities); churches or religious groups (except separately incorporated community development corporations); educational institution operating funds; and individual elementary or secondary schools.

Geographic Focus: California, Oregon, Washington

Amount of Grant: 5,000 - 25,000 U.S.D

Contact: J.R. Raines; (619) 230-3105; charitablegiving@unionbank.com

Internet: https://www.unionbank.com/global/about/corporate-social-responsibility/foundation/foundation-grants.jsp

Sponsor: Union Bank Foundation
P.O. Box 45174
San Francisco, CA 94145-0174

Union Benevolent Association Grants 4556

The Union Benevolent Association attempts to distribute its limited resources as widely as is consistent with the needs of its recipients. It tends to select projects that aid large groups of individuals among the disadvantaged. Grants have frequently been given to help support summer camping and educational experiences for inner city youths, to aid programs for the elderly and handicapped, to assist those involved in legal, medical, and social difficulties, to improve the quality of life in deteriorating neighborhoods, and to provide deprived young people with employable skills and artistic and cultural experiences. The Board prefers to make grants for specific projects and, as a rule, does not contribute toward general operating revenues or major building programs. It limits its grants to projects serving the needs of residents of Philadelphia. Preference is given to organizations with budgets of 2 million and below, and which leverage the use of volunteers.

Requirements: The Association makes grants only to 501(c)3 non-profit organizations,that work with people in need, within the city of Philadelphia, PA. Submitted proposals must be on the Delaware Valley Grantmakers Common Application Form (available on the Union Benevolent Association website) and be received (not posted) by email AND regular mail on or before 5:00PM on the deadline date.

Restrictions: The Association does not make grants to support capital renovations or to individuals.

Geographic Focus: Pennsylvania

Date(s) Application is Due: Apr 30; Sep 30

Contact: Fernando Chang-Muy, Administrator; (215) 763-7670; fax (215) 731-9457; info@uba1831.org

Internet: http://www.uba1831.org/grant_guide.html

Sponsor: Union Benevolent Association

1616 Walnut Street, Suite 800

Philadelphia, PA 19103

Union Carbide Foundation Grants 4557

The foundation supports nonprofits in areas of company operations, including Danbury, CT; Kanawha Valley, WV; central New Jersey; Taft, LA; and Texas City and Seadrift, TX. Grants also are awarded outside of the United States. The majority of grant supports goes for restructuring elementary and secondary education: developing math and science programs, advancing technology through research, preparing women and minorities for technical careers, and environmental protection. Additional areas of consideration include health and social services, civic affairs, the environment, and higher education. Grants are awarded for general operations, emergency funds, in-kind gifts, and programs/projects. The board meets in the spring and fall to consider grant proposals.

Requirements: Nonprofits in company-operating areas in Connecticut, West Virginia, New Jersey, Louisiana, and Texas may apply.

Geographic Focus: Connecticut, Louisiana, New Jersey, Texas, West Virginia

Amount of Grant: 5,000 - 45,000 U.S.D

Samples: PhD Project (Montvale, NJ)—to attract successful African American, Hispanic American, and American Indian businesspeople to teaching careers at business schools where they can serve as role models for underrepresented students, $50,000.

Contact: Grants Administrator; (203) 794-2000

Internet: http://www.dow.com/facilities/namerica/texops/community/grants.htm

Sponsor: Union Carbide Foundation

39 Old Ridgebury Road, L-4

Danbury, CT 06817-0001

Union County Community Foundation Grants 4558

Established in 1989, the Union County Community Foundation is a regional affiliate of Foundation for the Carolinas. The Foundation assists donors in making charitable gifts to the community, provides services for nonprofit organizations to create new or manage existing endowments and makes grants for new projects. Preference will be given to programs that meet the following criteria: serve those who are mentally or physically challenged as a result of visual and/or hearing impairments; provide professional development for Union County public school teachers; or address community needs through innovative approaches (funding available only to programs that cannot be covered by normal budgets of existing charitable agencies). Giving limited for the benefit of Union County, NC.

Requirements: Non-profits in or benefiting the community of Union County, North Carolina are eligible to apply. Contact the Foundation, an application form is required to submit a proposal. Applicants should submit the following: signature and title of chief executive officer; copy of most recent annual report/audited financial statement/990; copy of current year's organizational budget and/or project budget; listing of board of directors, trustees, officers and other key people and their affiliations; copy of IRS Determination Letter.

Restrictions: No grants to individuals, or for capital campaigns, ongoing operating budgets beyond the seed level, publication of books, conferences, or endowment funds.

Geographic Focus: North Carolina

Contact: Karen Coppadge, Grants Specialist; (704) 973-4559; kcoppadge@fftc.org

Sponsor: Union County Community Foundation

217 South Tryon Street

Charlotte, NC 28202-3201

Union Pacific Foundation Health and Human Services Grants 4559

The foundation has a strong interest in promoting organizational effectiveness among nonprofits. To that end, this foundation dedicates the majority of their grants to help nonprofit organizations build their capacity, increase their impact, and operate more efficiently and effectively. Grants are made primarily to proposals that fall within three categories: community and civic, fine arts, and health and human services. The health and human services category assists organizations dedicated to improving the level of health care or providing human services in the community. Types of support include general operating support, continuing support, capital campaigns, building construction and renovation, curriculum development, equipment acquisition, program development, and matching funds.

Requirements: The foundation will accept only online applications; printed copies of the application are not available and will not be accepted. Grants are made to institutions located in communities served by Union Pacific Corporation and its operating company Union Pacific Railroad Company.

Restrictions: Grants will not be given to organizations that are not tax-exempt; specialized national health organizations; political organizations; organizations engaged in influencing legislation; religious organizations that are sectarian or denominational in purpose; veterans organizations; labor groups, social clubs, fraternal organizations, or individuals. United Way-affiliated organizations may apply only for capital projects that have received written approval from United Way. Unsolicited requests from formal educational institutions such as colleges and universities are not accepted.

Geographic Focus: Arizona, Arkansas, California, Colorado, Idaho, Iowa, Kansas, Louisiana, Minnesota, Missouri, Montana, Nebraska, Nevada, New Mexico, Oklahoma, Oregon, Tennessee, Texas, Utah, Washington, Wisconsin, Wyoming

Date(s) Application is Due: Aug 15

Amount of Grant: 50,000 - 10,000 U.S.D

Contact: Darlynn Myers, Director; (402) 271-5600; fax (402) 501-2291; upf@up.com

Internet: http://www.up.com/found/grants.shtml

Sponsor: Union Pacific Foundation

1400 Douglas Street, Stop 1560

Omaha, NE 68179

United Cerebral Palsy Research Grants 4560

The foundation is interested in stimulating and providing funding for pilot studies on research important to the prevention of cerebral palsy and to improvement in the quality of life of persons with disabilities due to cerebral palsy and related developmental disorders. Research grants generally are awarded in support of basic and clinical science pilot studies or for time-limited clinical studies or trials. High priority is given to prevention of prematurity, decrease of perinatal neurologic morbidity, prevention of central nervous system infections of the newborn and to medical, surgical, and bioengineering methods that may improve the functioning of persons with cerebral palsy addressing issues such as impaired mobility, impaired communication skills, swallowing, drooling, and effects of aging on further diminution of function. Funds are awarded initially for a period not to exceed two years and may be renewed for a one- or two-year period, to a total of four years.

Requirements: Grants are awarded to institutions or organizations on behalf of a principal investigator in support of biomedical and bioengineering research in areas that have a significant relationship to cerebral palsy.

Amount of Grant: Up to 50,000 U.S.D

Contact: Medical Director; (202) 973-7140; fax (202) 776-0414; lsmithslade@ucp.org

Internet: http://www.ucp.org/ucp_generaldoc.cfm/1/4/23/23-23/113

Sponsor: United Cerebral Palsy Research and Educational Foundation

1660 L Street NW, Suite 700

Washington, DC 20036-5602

United Hospital Fund of New York Health Care Improvement Grants 4561

United Hospital Fund awards provides funding to not-for-profit and public hospitals, and health care, academic, and public interest organizations, to improve health care in New York City. The Fund supports the development of model projects, sponsors research to analyze systemic problems, and fosters innovative solutions to health care issues. Health Care Improvement Grants are awarded to develop and evaluate innovative health care projects and conduct research and analysis of significant health systems issues that will shape the future of health care in New York City. Grants are intended to increase access to and use of appropriate, high quality, and efficient health care services, especially for low-income and other vulnerable persons. Applications should address one of the following issue areas: expanding health insurance coverage; strengthening health care finances; improving quality of care; or redesigning health care services. Projects should test innovative or creative responses to an important aspect of one of these issues. They also should yield information and lessons of significance to a broader community of health care providers and policymakers.

Requirements: Applicants must be not-for-profit organizations in New York City. Qualifying organizations may submit only one application per grant cycle. Applicants are available at the Fund's website.

Restrictions: Unallowable cost include: overhead or other indirect costs; general operating support; and capital costs (such as construction, acquisition, or renovation of facilities). Equipment expenses must be integral to the project and justified in the budget narrative and may not exceed five percent of the total request.

Geographic Focus: New York

Date(s) Application is Due: Mar 1; Jul 1; Nov 1

Contact: Hollis Holmes, Grants Manager; (212) 494-0700; fax (212) 494-0800; hholmes@uhfnyc.org

Internet: http://www.uhfnyc.org/grants

Sponsor: United Hospital Fund of New York

1411 Broadway, 12th Floor

New York, NY 10018

United Methodist Church Global AIDS Fund Grants 4562

This fund represents the commitment of every United Methodist to put a stop to HIV/AIDS in our world. It supports education, prevention, care, and treatment programs for people living with HIV/AIDS. The plan specifies that 25 percent of what each annual conference raises should be used in that conference for AIDS work, either locally or in global projects.

Requirements: Overseas programs fitting the mission of the Global AIDS Fund may apply to receive a grant. The application should propose work in one or more of the following types of activities: (1) Promoting HIV/AIDS awareness and prevention by education and information; (2) Behavior modification; (3) Voluntary testing and counseling; (4) Treatment of persons with AIDS; (5) Home Based Care; (6) Care of orphans and vulnerable children.
Restrictions: This program is only for grants covering primarily United Methodist or Methodist-related projects for up to a one-year period.
Samples: 2008: Mozambique Annual Conference- Communicating against HIV/AIDS newspaper ($10,000); University Church in Toledo, Ohio- HIV/AIDS education for pastors of sub-Saharan Africa, Zambia ($11,850); Helping Children Worldwide, Inc- Mercy Hospital HIV/AIDS Outreach, Sierra Leone ($10,000); Five Points Institlute for Pastoral Education, U.S.A- Pastoral Counseling Training, for pastors, health workers and social workers, Bedfordview Methodist Church ($3,000); Jamkhed Comprehensive Rural Health Project, India- Promote HIV/AIDS awareness and prevention education and information ($15,000).
Contact: Linda Bales, lbales@umc-gbcs.org
Internet: http://new.gbgm-umc.org/umcor/work/health/hivaids/aids-fund/
Sponsor: United Methodist Church
475 Riverside Drive
New York, NY 10115

United Methodist Church Malaria Control Grants 4563
United Methodists are joining forces with others to make malaria's grip on the lives of children and adults a part of history. These efforts include the United Methodist Community-Based program that provides education, medication, and mosquito nets to prevent and treat malaria.
Requirements: The application should propose work in one or more of the following types of activities: (1) Promoting Malaria awareness and prevention by education and information; (2) Distribution of Insecticide Treated Bed Nets; (3) Indoor spraying; (4) Testing; (5) Malaria case management; (6) Preventative treatment in pregnant women. Download the required application form at the sponsor's website.
Restrictions: This program is only for United Methodist and Methodist-related projects outside the U.S.
Contact: Dr. Cherian Thomas, (212) 870-3870; cthomas@gbgm-umc.org
Internet: http://new.gbgm-umc.org/umcor/work/health/malaria/
Sponsor: United Methodist Church
475 Riverside Drive
New York, NY 10115

United Methodist Health Ministry Fund Grants 4564
Grants are awarded to health care projects proposed by eligible organizations to respond to needs and build on assets of local, regional, and state situations. The Health Fund is targeting three areas for funding: Access to health care; Oral health; and, Healthy nutrition and exercise.
Requirements: Grants can be awarded only to 501(c)3 organizations or governmental entities with projects to benefit the health of Kansas residents. Grants have one or more of the following purposes: (a) Develop new or expanded, sustainable program resources to provide quality services; (b) Change the delivery system to meet demands, improve access/quality, or reduce cost; (c) Test innovative ideas for improved service delivery; (d) Offer public education for improvement of individual and community health care; (e) Provide group opportunities for health care providers to improve critical skills; (f) Develop technical expertise, collaborations, and similar supports for improvement and change in health care service delivery and education. Contact the sponsor before beginning a grant application - the Health Fund will not accept unsolicited and unauthorized applications, and will be automatically rejected them without consideration.
Restrictions: Grants are not awarded to individuals or for individual medical, dental, or other personal care treatment. Generally, capital projects and endowments are not funded. The fund does not provide regular operating expenses of on-going projects. The Health Fund does not fund projects in other U.S. states (outside of Kansas) or foreign nations.
Contact: Virginia Elliott, Vice President for Programs; (620) 662-8586; fax (620) 662-8597; velliot@healthfund.org
Internet: http://www.healthfund.org/funding.php
Sponsor: United Methodist Health Ministry Fund
100 East First Avenue, P.O. Box 1384
Hutchinson, KS 67504-1384

United Technologies Corporation Grants 4565
The corporation focuses its contributions programs in communities where UTC has major facilities with a substantial employee residence population. In Greater Hartford, UTC supports grants in arts and culture, math and science education, the environment, and human services (through the United Way). Nationally, UTC provides grants to arts organizations and groups engaged in environmental sustainability work. Funding proposals should include a brief letter, the organization's most recent financial statement, a list of its directors, and IRS tax-exemption documents. UTC has an employee matching gift program limited to degree-granting educational institutions, cultural, and environmental organizations. The corporation, on a limited basis, contributes to capital building funds in support of specific business objectives. General operating grants, program development grants, and work-study programs are additional types of support. UTC Corporate will accept online grant applications throughout the year. UTC business units accept grant applications between March 1 and June 1 of each year.
Requirements: U.S. 501(c)3 tax-exempt organizations and international tax-exempt organizations are eligible.
Restrictions: The corporation will not fund partisan political organizations, individuals, organizations, religious organizations with programs limited to or expressly for their membership, controversial social causes on which there are strong differences of opinion, organizations that are the recipients of federated drives to which the corporation contributes (capital campaigns excluded), and endowment campaigns.
Date(s) Application is Due: Jun 1
Contact: Krista Pilot; (860) 728-7000; fax (860) 728-7041; contribu@corphq.utc.com
Internet: http://www.utc.com/responsibility/social/index.htm
Sponsor: United Technologies Corporation
1 Financial Plaza, MS 503
Hartford, CT 06101

Unity Foundation Of LaPorte County Grants 4566
The Unity Foundation of LaPorte County makes discretionary and field of interest grants to charitable organizations in the area of the arts, education, health and human services, the environment, and the community for the benefit of the citizens of LaPorte County, Indiana. The Foundation is particularly interested in funding: projects that are not adequately being serviced by existing community resources; start-up costs for new programs; one-time projects or needs; projects that provide leverage for generating other funds and community resources; projects that facilitate cooperation and collaboration between organizations and the communities within LaPorte County.
Requirements: Only charitable organizations with a verifiable 501(c)3 IRS status operating or offering programs in LaPorte County will be considered for funding.
Restrictions: No grants will be made: to churches for sectarian religious programs; for operating budgets or for basic municipal or educational functions and services; for endowment campaigns or for previously incurred debts; to provide long-term funding. No grants will be made for post-event or after-the-fact situations.
Geographic Focus: Indiana
Date(s) Application is Due: Jul 18
Contact: Margaret Spartz, Primary Contact; (219) 879-0327 or (888) 898-6489; mspartz@uflc.net or unity@uflc.net
Internet: http://uflc.net/grants-scholarships/
Sponsor: Unity Foundation of LaPorte County
619 Franklin Street, P.O. Box 527
Michigan City, IN 46361

University of Calgary Nursing Alumni Scholarship 4567
One scholarship per year is given to a Canadian citizen or landed immigrant of high academic standing who is registered in or admitted to a full-time master's program, who intends to make a career in nursing, and who is a registered member of the University of Calgary Faculty of Nursing Alumni Association. The scholarship may be used for any area or discipline at the master's level related to nursing. Annual deadline dates may vary; contact program staff for exact dates.
Requirements: To be eligible, applicants must be Canadian citizens or landed immigrants; graduate students in a master's program who intend to make a career in nursing; and registered members of the University of Calgary Faculty of Nursing Alumni Association.
Geographic Focus: All States, Canada
Date(s) Application is Due: Dec 2
Amount of Grant: 1,000 CAD
Contact: Maria Lavoie, Graduate Scholarship Officer; (403) 220-8483; lavoiem@ucalgary.ca or ucawards@ucalgary.ca
Internet: http://www.grad.ucalgary.ca/funding/internal_scholarships/lev_4/nurs_alum_burs.htm
Sponsor: University of Calgary
2500 University Drive NW, Earth Sciences Building, Room 720
Calgary, AB T2N 1N4 Canada

University of California President's Postdoctoral Fellowships 4568
The fellowship, originally designed to encourage applications from minority and women scholars in areas where they were severely underrepresented, is now open to all qualified candidates. The program provides postdoctoral fellowships to qualified persons who intend to pursue academic careers and to enhance their prospects for appointment to faculty positions at the University of California and other institutions of higher education. In addition to financial support, fellowships include mentorship and guidance in areas critical to success in an academic career. Application materials are available upon request.
Requirements: Applicants must be U.S. citizens or permanent residents and must hold a PhD from an accredited university. Applicants are required to have earned their PhD degree by June 30 of the year for which they are applying.
Date(s) Application is Due: Nov 15
Amount of Grant: 44,000 U.S.D
Contact: Kim Adkinson, Fellowships Coordinator; (510) 987-9503; fax (510) 587-6077; kim.adkinson@ucop.edu
Internet: http://www.ucop.edu/acadadv/ppfp
Sponsor: University of California
1111 Franklin Street, 11th Floor
Oakland, CA 94607-5200

University of Georgia Counseling and Psychological Services Postdoctoral Fellowship 4569
The program is designed to develop advanced skills in the practice of clinical and counseling psychology. The postdoctoral program provides fellows an opportunity to refine and expand experience in the following areas: clinical practice with individuals, couples and groups, assessment, crisis intervention, workshops, and educational

programming. Postdoctoral fellows will work in close collaboration with senior staff and will function as professionals with many of the same responsibilities. Fellows attend all case conferences, professional development workshops, and staff meetings. In addition to the stipend, fellows are also eligible for staff benefits, including health and dental insurance, retirement, vacation, sick leave, professional development, exercise facility, library privileges, and licensing preparation time. Guidelines are available online. Annual deadline dates may vary; contact program staff for exact dates.
Requirements: Admission requirements include completion of all professional degree requirements from a regionally accredited institution of higher education or an APA/CPA–accredited program and completion of an APA approved predoctoral internship. Fellow must have received diploma before first day of fellowship or have a letter from Director of Graduate Studies verifying that all degree requirements have been met. Previous counseling center experience is preferred.
Date(s) Application is Due: Mar 11
Amount of Grant: 25,000 U.S.D
Contact: Dr. Al Clarke; (706) 542-2273; aclarke@uhs.uga.edu
Internet: http://www.uhs.uga.edu/CAPS/postdoc/index.html
Sponsor: University of Georgia
University of Georgia Health Center
Athens, GA 30602-1755

University of Oslo International Summer School (ISS) Scholarships 4570
A limited number of scholarships are available, mainly to applicants from developing countries, Eastern Europe, and the former Soviet Union. Applications must be made on a special scholarship application form that may be obtained from the ISS Head Office in Oslo. Students from Europe may apply for scholarships offered under the bilateral Cultural Agreements between Norway and their home country. Scholarship application procedures vary from country to country; inquire about them and request an application form at the Norwegian Embassy or Consulate General, Ministry of Education, or Ministry of Foreign Affairs in the home country. There are also some scholarships available to U.S. and Canadian students through several different private organizations, including Sons of Norway. Contact the ISS North American Branch Office for further details.
Requirements: Under the cultural agreements, scholarships are open to students with good academic records from Austria, Belgium, Bulgaria, Czech Republic, Egypt, Germany, Great Britain, Greece, Hungary, Ireland, Italy, Mexico, Netherlands, Poland, Portugal, Romania, Russia, Slovakia, Spain, and Turkey, typically. For scholarships to applicants from developing countries, preference is given to graduate level candidates from NORAD's main partner countries (i.e., Afghanistan, Angola, Bangladesh, Bhutan, China, Cuba, East Timor, Eritrea, Ethiopia, Guatemala, India, Indonesia, Iran, Laos, Malawi, Maldives, Mali, Mozambique, Nepal, Nicaragua, Nigeria, Pakistan, Palestinian Territories, South Africa, Sri Lanka, Tanzania, Uganda, Vietnam, Zambia and Zimbabwe).
Restrictions: Funding for transatlantic travel is not provided. Those ineligible to apply for scholarships include applicants to Norwegian language courses level I and II; medical doctors or nurses applying for Norwegian language courses; applicants to Norwegian language courses at any level residing in Norway; candidates from developing countries applying for undergraduate courses.
Geographic Focus: All States, Belgium, Bulgaria, Czech Republic, Germany, Hungary, Mexico, Poland, Romania, Russia, Slovakia, United Kingdom
Date(s) Application is Due: Feb 1
Contact: Charman Theresa Garibaldi, Administrative Assistant; 47-22-85-63-85; fax 47-22-85-41-99
Internet: http://www.uio.no/iss/iss.html
Sponsor: University of Oslo International Summer School
P.O. Box 1082 Blindern
Oslo, NO-0317 Norway

University of Queensland Travel Awards for International Collaborative Research 4571
The University of Queensland provides funds annually to facilitate visits to the university by scholars from other countries. Travel assistance is available to scholars actively engaged in research who will contribute substantially to research activity in the nominated school or center at the university. The assistance consists of: an air fare to Brisbane, inclusive of visa fees levied by the Australian government for the visitor only; and, subject to the visitor spending at least four weeks during semester at the university, a return air fare for the visitor only. In exceptional circumstances, and at the discretion of the selection committee, there is provision for a limited contribution to subsistence on a basis identified as genuine merit of proposal and genuine need, to be proportionate to the length of stay. Grants are normally reserved for suitably qualified staff of universities and internationally recognized research establishments. Other persons, however, may be considered where special circumstances exist and comparable standards are satisfied.
Requirements: Applicant must spend at least four weeks at the university during the semester. Applicants must be qualified members of a recognized university who are engaged in academic work that will contribute substantially to the department to which they are attached at the University of Queensland.
Restrictions: Applications will not be accepted from persons already in Australia.
Date(s) Application is Due: Aug 1
Contact: Office of Research, Postgraduate Studies; 61-7-336-53560 or 61-7-336-54584; fax 61-7-336-54455; oper@research.uq.edu.au
Internet: http://www.uq.edu.au/research/orps/?id=4793&pid=4547
Sponsor: University of Queensland
University of Queensland, St. Lucia
Brisbane, QLD 4072 Australia

Until There's a Cure Foundation Grants 4572
The foundation supports HIV/AIDS organizations throughout the United States. Grants focus on four priorities: care and services to meet the immediate needs of people living with HIV/AIDS, prevention education, vaccine development and research, and policy development and research. Types of support include capital campaigns, equipment acquisition, program development, and general operating support.
Requirements: 501(c)3 nonprofit U.S. organizations are eligible for grant support.
Restrictions: The foundation does not make grants for: for walk-a-thons, tournaments, or special events; to organizations with religious affiliations, unless the program is open to the entire community without regard to religious beliefs; for costs already incurred by an organization; or to fund general administration expenses of an organization rather than for specific projects and programs
Date(s) Application is Due: Apr 29
Amount of Grant: 1,000 - 40,000 U.S.D
Samples: AIDS Service Agency of North Carolina (NC)—$9000; AIDS Task Force Greater Cleveland (Cleveland, OH)—$41,818; AIDS Walk Atlanta (Atlanta, GA)—$15,000; Wisconsin AIDS Fund (WI)—$40,000.
Contact: Grants Administrator; (800) 888-6845 or (650) 332-3200; fax (650) 332-3210; grants@utac.org
Internet: http://www.utac.org
Sponsor: Until There's a Cure Foundation
560 Mountain Home Road
Redwood City, CA 94062

Upper Valley Community Foundation Grants 4573
The foundation awards grants in 61 Upper Valley communities to nonprofit organizations working in the arts and humanities, community development, education, environment, historic preservation, health, public affairs, and social/human services. Grantmaking priorities change each year. Types of support include project grants and strategic focus grants. Grant awards are generally limited to one year. Applicants may submit only one proposal in any grant cycle.
Requirements: 501(c)3 nonprofit organizations, community organizations, schools/school districts, and public agencies that serve the Upper Valley region (from Haverhill, NH, and Ryegate, VT, to Charlestown, NH, and Springfield, VT, and from Randolph, VT, to New London, NH) are eligible. Organizations without 501(c)3 status may apply through a 501(c)3 fiscal agent.
Restrictions: Grants do not support religious activities endowments, capital campaigns, annual appeals, capital improvements or major equipment purchases, special events, or deficit reduction.
Geographic Focus: New Hampshire, Vermont
Date(s) Application is Due: Apr 1; Oct 1
Amount of Grant: 5,000 - 10,000 U.S.D
Contact: Kevin Peterson, Executive Director; (603) 653-0387, ext. 102; kp@nhcf.org
Internet: http://www.nhcf.org/page17081.cfm
Sponsor: Upper Valley Community Foundation
P.O. Box 995
Hanover, NH 03755-0995

U.S./Israel Women to Women Grants 4574
The organization provides grants for grassroots efforts that advocate for the equal status and fair-treatment for all women in Israel. Preference is given to small, innovative programs run predominantly by women that have the potential to make a significant impact on the quality of life of women. A wide range of projects are supported, including the following initiatives: economic empowerment, antiviolence measures; health care; promotion of women's political leadership; and advocacy and policy reform. Each eligible organization may submit one proposal for one project. Proposals should be for projects that will be implemented during the upcoming year (January to December). Preference will be given to organizations in which women comprise 50 percent of their board of directors. Proposals can be sent by email attachment, fax, or postal mail.
Requirements: Applicant organization must be an Amuta or be affiliated with an Amuta that will receive the funding.
Restrictions: Individuals are not eligible. Applications are not considered from organizations whose operating budget is more than $1 million U.S. dollars). Colleges and universities are exempt from this criterion, but their department's budget will be taken into account.
Date(s) Application is Due: Jul 1
Amount of Grant: Up to 10,000 U.S.D
Contact: Debra Fattel, Director of Grants Management; (917) 351-0921; Debra@usisraelwomen.org or grants@usisraelwomen.org
Internet: http://www.usisraelwomen.org/grantmaking.guidelines.shtml
Sponsor: U.S./Israel Women to Women
45 W 36th Street, 10th Floor
New York, NY 10018

U.S.AID Developing Public Health Experts Grants 4575
U.S.AID-Bangladesh Developing Public Health Experts Grants will support the development of the next generation of public health experts. The objectives are: (1) expanding access to quality public health education among a diverse group of Bangladeshi students by addressing barriers to access and (2) improving and strengthening the quality of public health education in Bangladesh. These Grants will embrace new United States government initiatives and reforms, including the Global Health Initiative and the U.S.AID Forward procurement reforms, and will complement and support the

Government of Bangladesh health sector program. One Grant award covering a period of three years is anticipated. Applications are available on the website.
Requirements: Private institutions of higher education authorized under the laws of Bangladesh who have at least three years' experience offering public health degree programs of study are eligible.
Geographic Focus: Bangladesh
Date(s) Application is Due: Apr 16
Amount of Grant: 3,000,000 U.S.D
Contact: Kaiser Parvez Ali, (880-2) 885-5500, Ext: 2722; kali@usaid.gov
Internet: http://www.usaid.gov/bd/proposals.html
Sponsor: U.S. Agency for International Development
Ronald Raegan Building
Washington, DC 20523-1000

U.S.AID Disabilities Grants **4576**
U.S.AID-Georgia Disabilities Grants are to increase participation of people with disabilities in current U.S.AID programs and to strengthen the capacity of disabled people's organizations (DP.O.s). Program areas include, but are not limited to, inclusive education, policy and governance issues, civil society building, health care, and employment opportunities. Capacity building programs for DP.O.s may include organizational capacity, advocacy efforts, coordination, and leadership. Funding is expected to be over an 18-24 month period. Applications will be considered in two categories: (1) small fund programs ranging from $20,000 to $300,000; and (2) larger fund programs ranging from $300,000 to $600,000. Applicants may submit concept papers, and approved applicants will be invited to submit a full application.
Requirements: Eligibility applies only to Georgian non-government organizations.
Geographic Focus: All Countries
Amount of Grant: 20,000 - 600,000 U.S.D
Contact: Yana Adelber; (995-32) 544-144; yadelberg@usaid.gov
Internet: http://www07.grants.gov/search/search.do;jsessionid=HTdvPvkLdQLRQ60wvjXYG61gDyzRzyhBwpnhmSqfnqmv1JGXt2nT!-2099600874?oppId=142293&mode=VIEW
Sponsor: U.S. Agency for International Development
Ronald Raegan Building
Washington, DC 20523-1000

U.S.AID Empowerment of People Against Tuberculosis Grants **4577**
U.S.AID-Indonesia seeks to improve the quality of Indonesia's health system. The five-year Community Empowerment of People Against Tuberculosis Grants will be carried out through cooperative agreements awarded to up to three Indonesian non-governmental organizations (NGOs). The principal objective is to support the Indonesian National Tuberculosis (TB) Program in "achieving universal access to quality and early tuberculosis diagnosis and treatment through community empowerment." To do this, the project will use two components: (1) Mobilize communities to support improved TB care; and (2) Advocate for increased TB resources and improved TB services. Applicants must submit a concept paper by the published due date and, if approved, will be invited to submit a full proposal.
Requirements: Indonesian NGOs are eligible as prime partners. Applications may provide for partnering with experienced United States and Non-United States organizations in a direct partnership or joint venture to provide capacity building of Indonesian applicants.
Geographic Focus: All States, All Countries
Date(s) Application is Due: Jan 27
Amount of Grant: 4,000,000 U.S.D
Contact: Johanna Gardjito; +62-21-3435-9497; jgardjito@usaid.gov
Internet: http://www.grants.gov/search/search.do;jsessionid=LwLMPpvRkTgWm3GpRQCJ1Xs2rhttp://www.grants.gov/search/search.do;jsessionid=LwLMPpvRkTgWm3GpRQCJ1Xs2r1pTTMnLJQSdqMvJzs32fSv2HpTW!-1784104149?oppId=132873&mode=VIEW
Sponsor: U.S. Agency for International Development
Ronald Raegan Building
Washington, DC 20523-1000

U.S.AID Expanding Participation of People With Disabilities Grants **4578**
U.S.AID-Kazakhstan Expanding Participation of People With Disabilities Grants seek to increase participation of people with disability in U.S.AID activities and strengthen the capacity of disabled people's organizations (DP.O.s). Program areas could include, but are not limited to, education, health, government, civil society building, rule of law, HIV/AIDS, and employment opportunities. Capacity building programs for DP.O.s may include organizational capacity, advocacy efforts, coordination, and leadership. Concept papers for Kazakhstan and Uzbekistan will be considered. Approved applicants will be invited to submit a full application. Funding will be over an 18-24 month period. Concept paper instructions are available on the website.
Requirements: United States or non-United States entities, such as private, non-profit organizations (or for-profit companies willing to forgo profits), including private voluntary organizations, universities, research organizations, professional associations, and relevant special interest associations are eligible.
Geographic Focus: All States, All Countries
Date(s) Application is Due: Feb 24
Amount of Grant: 600,000 U.S.D
Contact: Natalie Portnagina, +7-2725-04-802; nportnyagina@usaid.gov
Internet: http://www.grants.gov/search/search.do;jsessionid=QglqPmxh0TngGZVC41Kf02QhnZV0Vw3cW1yX7n7Pk015sjzLanetZL!712472910?oppId=141453&mode=VIEW
Sponsor: U.S. Agency for International Development
Ronald Raegan Building
Washington, DC 20523-1000

U.S.AID Expanding Participation of People with Disabilities Grants **4579**
U.S.AID continues to work toward fuller inclusion of people with disabilities in foreign policy and development efforts, and towards that end, funds are made available to increase the participation of people with disabilities in U.S.AID activities and strengthen the capacities of Disabled People's Organizations (DP.O.s). Program areas could include, but are not limited to, education, health, government, civil society building, rule of law, HIV/AIDS, and employment opportunities. Capacity building programs for DP.O.s may include organizational capacity, advocacy efforts, coordination, and leadership. There is no deadline for submission of concept papers for the small fund programs (funding of $20,000 to $300,000). Papers will be reviewed in the order received, and funds will be allowed until depleted (approximately $2.8 million). The deadline for the large fund programs (funding of $300,000 to $600,000) is March 9.
Requirements: Applications should be from qualified United States or non-United States entities, such as private, non-profit organizations (or for-profit companies willing to forgo profits), including private voluntary organizations, universities, research organizations, professional associations, and relevant special interest associations.
Geographic Focus: All States, All Countries
Contact: Tyce Shideler, Grants Manager; armeniacontact@usaid.gov.
Internet: http://armenia.usaid.gov/en/node/48
Sponsor: U.S. Agency for International Development
Ronald Raegan Building
Washington, DC 20523-1000

U.S.AID Gender-Based Violence Access to Justice Grant **4580**
U.S.AID-Zambia's Stamping Out and Preventing Gender-Based Violence (GBV) Access to Justice Grant is to improve access to justice for adult and children survivors of GBV by building the capacity of GBV service providers as well as the policymakers, police, courts, and community leaders in GBV case management and implementation of laws. The program is also expected to provide legal aid to survivors to ensure success of cases in court. This five-year program to support the Government of the Republic of Zambia's goal of eliminating GBV in a holistic, systematic and comprehensive manner through a multi-sectoral and multi-dimensional approach. See the website for application.
Requirements: Any local non-governmental organization and for-profit organization is eligible.
Geographic Focus: Zambia
Date(s) Application is Due: Feb 20
Amount of Grant: 1,000,000 - 3,300,000 U.S.D
Contact: Shahid Pervaiz, Senior Acquisitions and Assistance Specialist; +260-211-357-195; spervaiz@usaid.gov
Internet: http://www07.grants.gov/search/search.do;jsessionid=5BF1PyRZL92D4vqQFTGQQ2C91y41G1TGTCNvMxVw1VtfBhgJmG8P!267572332?oppId=138993&mode=VIEW
Sponsor: U.S. Agency for International Development
Ronald Raegan Building
Washington, DC 20523-1000

U.S.AID Gender-Based Violence Prevention and Advocacy Grant **4581**
U.S.AID-Zambia's Stamping Out and Preventing Gender-Based Violence (GBV) Prevention and Advocacy Grant is to decrease societal acceptance of GBV, enhance protective factors, and improve the enabling environment to respond to GBV. Through this project, outreach efforts against GBV will be expanded, communities will be mobilized against GBV, and political and institutional commitment to GBV prevention and response with the Government of the Republic of Zambia (GRZ) will be strengthened. This is a five-year program to support the GRZ's goal of eliminating GBV in a holistic, systematic and comprehensive manner through a multi-sectoral and multi-dimensional approach. See the website for application.
Requirements: Any local non-governmental organization and for-profit organization is eligible.
Geographic Focus: Zambia
Date(s) Application is Due: Feb 13
Amount of Grant: 1,000,000 - 5,000,000 U.S.D
Contact: Shahid Pervaiz, Senior Acquisitions and Assistance Specialist; +260-211-357-195; spervaiz@usaid.gov
Internet: http://www07.grants.gov/search/search.do;jsessionid=5BF1PyRZL92D4vqQFTGQQ2C91y41G1TGTCNvMxVw1VtfBhgJmG8P!267572332?oppId=138973&mode=VIEW
Sponsor: U.S. Agency for International Development
Ronald Raegan Building
Washington, DC 20523-1000

U.S.AID Gender-Based Violence Survivor Suppport Grant **4582**
U.S.AID-Zambia's Stamping Out and Preventing Gender-Based Violence (GBV) Survivor Support Grant is to increase the availability and uptake of quality GBV services including clinical, psychological, and economic assistance, for adults and children survivors of GBV. The Grant is expected to build the capacity of local prime implementers of the Grants for GBV Prevention and Advocacy and GBV Access to Justice. This is a five-year program to support the Government of the Republic of Zambia's goal of eliminating GBV in a holistic, systematic and comprehensive manner through a multi-sectoral and multi-dimensional approach. See the website for application.
Requirements: Non-governmental organizations or for-profit companies willing to forego profit are eligible. U.S.AID encourages applications from potential new partners.
Geographic Focus: Zambia
Date(s) Application is Due: Feb 20
Amount of Grant: 1,000,000 - 7,900,000 U.S.D

Contact: Shahid Pervaiz, Senior Acquisitions and Assistance Specialist; +260-211-357-195; spervaiz@usaid.gov
Internet: http://www.grants.gov/search/synopsis.do;jsessionid=8LKQPyHLZDd4jGNP9mhJZZ8DJVLtKnSbb2LzQbwntYzwkb3sMDqP!267572332
Sponsor: U.S. Agency for International Development
Ronald Raegan Building
Washington, DC 20523-1000

U.S.AID Girls Empowerment Grants **4583**

U.S.AID-Tanzania recognizes the importance of a country-led platform to foster an enabling environment for sustainable impact. These Grants are for Girls Empowerment and Innovations in Youth, Reproductive Health, and Family Planning, and Service Outreach. The goal is to support the conditions that will enable girls and women to achieve positive social and economic outcomes. Four focus areas that support systemic cross-cutting approaches and strengthen catalytic change agents comprise the different application categories of grant opportunities. They are: (1) adolescent, youth and girls' empowerment; (2) innovations in reproductive health and family planning; (3) outreach services to increase access to and use of family planning services; and (4) capacity building for networks, coalitions, and advocacy groups supporting youth, girls' and women's empowerment. Applicants may submit a concept paper, and applicants with approved concept papers will be invited to submit full proposals.
Requirements: Locally registered Tanzanian Civil Society Organizations (CSOs) are eligible. CSOs include a wide range of non-state actors, such as non-government organizations including any non-profit or voluntary organizations, print and broadcast media, religious groups, labor unions, academic institutions, consulting firms and for-profit private companies. For profit organizations will not be eligible for fee or profit.
Geographic Focus: Tanzania
Date(s) Application is Due: Mar 31
Amount of Grant: 40,000 - 14,000,000 U.S.D
Contact: Samuel Kiranga; +255-22-266-8503; skiranga
Internet: http://www07.grants.gov/search/search.do;jsessionid=Z5jlPmQCysxnGHb5zTG1p5ZLpBLSdY7tK4Kp9GGhC8Tt4YNT4PRn!712472910?oppId=116933&mode=VIEW
Sponsor: U.S. Agency for International Development
Ronald Raegan Building
Washington, DC 20523-1000

U.S.AID Health Strategic Information Grants **4584**

U.S.AID-Mozambique is seeking concept papers from non-governmental organizations working in Mozambique (both registered local and international) in HIV-AIDS. The objectives are to: (1) Improve the strategic information capacity of the President's Emergency Plan (PEPFAR) for AIDS Relief Mozambique-funded implementing partners, local Mozambican community based organizations and selected Government of Mozambique ministries at both national and sub-national level; (2) Ensure effective PEPFAR Mozambique data systems, specifically in the areas of data quality, for greater program management, data use and decision making; and (3) Begin a process for the training, capacity development and placement of a cadre of Mozambican monitoring and evaluation specialists in Mozambican Government Institutions and selected international and local implementing organizations. Awards are expected to be made for periods up to three years. Applicants may submit a concept paper by March 1. Selected applicants will be invited to submit a full application.
Requirements: Eligible organizations include non-governmental organizations, faith-based organizations, community-based organizations, universities, multi-lateral organizations, foundations, private organizations affiliated with public academic institutions, professional associations, non-profit organizations and for-profit organizations willing to forego profit, and a consortia of the above. Organizations with no experience within the Mozambican health sector are strongly encouraged to include proposed collaboration with organizations with health sector experience.
Geographic Focus: All States, All Countries
Amount of Grant: 100,000 - 300,000 U.S.D
Contact: Albertina Cample; +258-21-352-000; fax +258-21-352-187; acample@usaid.gov
Internet: http://www07.grants.gov/search/search.do;jsessionid=jk6qPhMWTcyLzXsJLp0LhyhGp1p6515VhTRoadcnVJnCgvtkWQMMRB!1154909517?oppId=138473&mode=VIEW
Sponsor: U.S. Agency for International Development
Ronald Raegan Building
Washington, DC 20523-1000

U.S.AID HIV/AIDS Services Grants **4585**

U.S.AID-Tanzania seeks applications for Comprehensive, Integrated, and Sustainable HIV/AIDS Services Grants that will strengthen the capability of Tanzanian entities/institutions to: ensure quality service delivery using existing platforms to the extent possible; build linkages between clinic and community-based services and activities; integrate basic health and HIV/AIDS services within facilities; and help move HIV/AIDS services towards increased sustainability. Funding is expected to be from three to five years. Applicants may submit concept papers, and approved applicants will be invited to submit a full proposal.
Requirements: Tanzanian, international, or United States non-governmental organizations or private for-profit organizations are eligible. For-profit institutions will not be eligible for fee or profit.
Geographic Focus: All States, All Countries
Date(s) Application is Due: Aug 25
Amount of Grant: 30,000,000 - 100,000,000 U.S.D
Contact: Samuel Kiranga; +255-22-266-8503; skiranga@usaid.gov
Internet: http://www07.grants.gov/search/search.do;jsessionid=sBJhPm5JPzg7XTXTYzBCK6YLdKj2phXQHskwp41JCSFLmrXSbh0G!712472910?oppId=117653&mode=VIEW
Sponsor: U.S. Agency for International Development
Ronald Raegan Building
Washington, DC 20523-1000

U.S.AID HIV Innovations Grants **4586**

U.S.AID-South Africa seeks applications from local organizations interested in implementing five-year HIV Innovations for Improved Patient Outcomes for Priority Populations Grants. The purpose is to support South African organizations that identify, strengthen, and institutionalize better, cost-effective practices and innovations in HIV and AIDS care and treatment for priority and at-risk populations in accordance with South African government priorities and policies. Targeted priority and at-risk populations for this program include pregnant women; infants, children, and adolescents; men who have sex with men; commercial sex workers; hard to reach populations; and those co-infected with HIV/TB. Applications are available on the website.
Requirements: Any local South African partners, including non-governmental organizations, private voluntary organizations, faith-based and community organizations, and for-profit companies are eligible. Applications from potential new partners are encouraged.
Geographic Focus: South Africa
Date(s) Application is Due: Mar 15
Amount of Grant: 1,000,000 - 15,000,000 U.S.D
Contact: Beatrice Lumanade, Acquisitions and Assistance Specialist; +27 12 452 2000; applications4@usaid.gov
Internet: http://www07.grants.gov/search/search.do;jsessionid=GQQvPh8Mmtk6zmWM7qX06Jsy1Lane4znGW3LYTGgq7tFJxydv2f1L2!1154909517?oppId=135553&mode=VIEW
Sponsor: U.S. Agency for International Development
Ronald Raegan Building
Washington, DC 20523-1000

U.S.AID Improved Nutrition Grant **4587**

U.S.AID-Uganda anticipates awarding a five-year Therapeutic and Supplementary Products for Improved Nutrition Grant. The goal is to reduce the burden of under-nutrition in Uganda through strengthening the capacity of a local/regional companies to become a sustainable manufacturer and distributor of therapeutic and supplementary foods to meet national and/or regional demand. Expected results at the end of five years include: (1) Increased local production of therapeutic and supplementary foods to children, pregnant women, and people living with HIV with moderate or severe acute malnutrition; (2) An effective distribution system for therapeutic and supplementary foods within public and private sectors in Uganda established; and (3) Improved economic well-being of households of small scale farmers through livelihood empowerment. The application is available at the website.
Requirements: This is a limited regional competition, under which any type of organization, large or small, commercial (for profit) firms, faith-based, and non-profit organizations in partnerships or consortia from East African region, are eligible. U.S.AID encourages applications from potential new partners.
Geographic Focus: Burundi, Kenya, Rwanda, Tanzania, Uganda
Date(s) Application is Due: Mar 2
Amount of Grant: 22,000,000 U.S.D
Contact: Godfrey Kyagaba, Acquisitions and Assistance Specialist; +(256) 4143-060-01; gkyagaba@usaid.gov
Internet: http://www07.grants.gov/search/search.do;jsessionid=jGWvPnGMGqvq1LFQbnF3SbxpZBsj3vhTbnBmX9KtN2c2zJv2Qfby!712472910?oppId=140873&mode=VIEW
Sponsor: U.S. Agency for International Development
Ronald Raegan Building
Washington, DC 20523-1000

U.S.AID International Family Planning Organizations Grants **4588**

U.S.AID International Family Planning (FP) Organizations Grants are made to increase the use of FP services globally through strengthening selected international family planning organizations which have a global reach and an extensive, multi-country network of FP clinics in order to achieve maximum program impact and synergies. Applicant are expected to address four results to achieve project goals. The results are: (1) Srengthened organizational capacity to deliver quality FP services to target groups; (2) Internal quality assurance standards and results quantified and disseminated to strengthen FP performance at a global level; (3) Increased organizational sustainability of country level programs, including internal South to South support and technical assistance; and (4) Gender-sensitive FP services targeting youth strengthened at a global level. If an award is made, implementing partners will be supported to enhance their capacity to manage and scale up FP interventions worldwide and in specific targeted country settings. There is currently no closing date for concept papers. See the website for submission information.
Requirements: Eligible organizations must: (1) be United States organizations or non-United States organizations registered and working in U.S.AID priority countries with a worldwide network of FP clinics and a recognized track record in delivering FP services; (2) have documented experience in implementing FP programs in developing countries with FP being a central part of the organization's mission; (3) have the capacity to provide south to south training via networks of clinics in multiple countries and regions; (4) have links with community based support systems and non-government organizations and have a strong and a documented track record in training these organizations in FP; and (5) be able to expand U.S.AID's reach by operating in countries in which U.S.AID does not currently have programming, particularly in Sub-Saharan Africa and South Asia.
Geographic Focus: All States, All Countries
Amount of Grant: 40,000,000 U.S.D
Contact: Alisa Dunn, Contract Specialist; (212) 712-0908; adunn@usaid.gov

Internet: http://www07.grants.gov/search/search.do;jsessionid=pT8wPqGTTngC6tQHhh2yQmGGf1mtN8xJ52fjQnL76xz7C18M1Ysn!748281696?oppId=50154&mode=VIEW
Sponsor: U.S. Agency for International Development
Ronald Raegan Building
Washington, DC 20523-1000

U.S.AID Marketing Innovation for Health Grants 4589

U.S.AID-Bangladesh Marketing Innovation for Health Grants will support the implementation of dynamic new marketing innovation for health activity which builds on the successes of U.S.AID's long history of support for social marketing activities in Bangladesh. The new activity supports access to new and existing public health related products using new and innovative approaches to reach new beneficiaries using a private sector social marketing model. The program is expected to achieve the following results: (1) Increased distribution and sales of reproductive health products and a secured supply of contraceptive commodities; (2) Increased distribution and sale of ORS and Zinc to treat diarrhea and dehydration, safe delivery kit, and other MCH products for use in related services; (3) Increased distribution and sale of products for improving the nutritional status of children; (4) Increased distribution and sale of new and innovative products using social marketing techniques; (5) Improved health communications activities to reach target groups; (6) Increased training and referrals for long-term and permanent family planning methods, institutional delivery, management of sick newborn, and diagnosis and treatment of Tuberculosis; and (7) Strengthened linkages with other public and private sector partners. Grants are intended to cover a period of four years. Applications are available on the website.
Requirements: Bangladeshi Non-Government Organizations are eligible to apply.
Geographic Focus: Bangladesh
Date(s) Application is Due: Mar 12
Amount of Grant: 15,000,000 U.S.D
Contact: Kaiser Parvez Ali, Acquisitions and Assistance Specialist; (880-2) 885-5500, Ext: 2722; kali@usaid.gov
Internet: http://www.usaid.gov/bd/proposals.html
Sponsor: U.S. Agency for International Development
Ronald Raegan Building
Washington, DC 20523-1000

U.S.AID National Training Mechanism Grant 4590

The main objective of the National Training Mechanism Grant is to improve access to and the quality of training for the health workforce in Kenya. Four specific objectives include: (1) increasing the number of new workers graduating from a pre-service training institutions; (2) addressing existing health workforce gaps in skills and knowledge for professional development; (3) building training institutions' capacity; and (4) strengthening the regulatory bodies to effectively increase demand for training by health workers. A five-year award period is anticipated. The application is available on the website.
Requirements: Any United States or non-United States organization, non-profit, or for-profit, is eligible.
Geographic Focus: All States, All Countries
Date(s) Application is Due: Jan 30
Amount of Grant: 38,000,000 - 40,000,000 U.S.D
Contact: Camille Garcia, Agreement Officer; +254-20-862-2000; cgarcia@usaid.gov
Internet: http://www07.grants.gov/search/search.do;jsessionid=1KwQPhQf7HLmy4QCfMJPyLChnKGnBjwcW6Z6mVQTL2PzVfrhg57L!1154909517?oppId=133353&mode=VIEW
Sponsor: U.S. Agency for International Development
Ronald Raegan Building
Washington, DC 20523-1000

U.S.AID Northern Uganda Health Integration to Enhance Services Grants 4591

U.S.AID-Uganda is seeking applications to fund one or more organizations for a five-year health service integration program in Uganda. The purpose of the Northern Uganda Health Integration to Enhance Services (NU-HITES) Grants is to continue technical assistance and capacity building support similar to that provided under the Northern Uganda Malaria, AIDS and Tuberculosis project and to integrate additional support in Maternal, Neonatal and Child Health, Family Planning and nutrition while prioritizing the strengthening of district-level health systems. U.S.AID is transitioning its modes of support from humanitarian approaches that provide basic service provision to one of longer-term partnership with local structures and systems to ensure greater sustainability. This program will focus on supporting District governments to enable them to meet their primary responsibility for the provision of public sector health services. However, while public sector services and leadership are necessary, they are not sufficient to meet the population's health needs. NU-HITES will also support the growth and acceleration of a vibrant and sustainable private health sector that aims to serve the health needs of northern Uganda through innovations in service provision while complementing the public sector. The goal is the improved and sustained health and nutritional status of northern Uganda population through strengthening a district-based integrated package of quality health services. Program objectives are: (1) To increase use of quality health services at facility and community level by women, girls and children (Health Service Delivery); and (2) to strengthen systems for effective and sustainable delivery of quality health services (Health System Strengthening). Subject to the availability of funds, U.S.AID intends to provide approximately $50 million over the five-year period. Applications are available on the website.
Requirements: United States or non-United States organizations, non-profit, or for-profit entity is eligible. U.S.AID encourages applications from potential new partners.
Geographic Focus: All States, All Countries
Date(s) Application is Due: Mar 9
Contact: Godfrey Kyagaba, Acquisitions and Assistance Specialist; +(011) 2564-143-06001; gkyagaba@usaid.gov
Internet: http://www07.grants.gov/search/search.do;jsessionid=Hh1NPqMQy8LJCRsMQYTLQ9x63p5HCKcSGT64LzjWdVpvZZD0hcBv!748281696?oppId=141993&mode=VIEW
Sponsor: U.S. Agency for International Development
Ronald Raegan Building
Washington, DC 20523-1000

U.S.AID Support to People with Disabilities Grants 4592

U.S.AID-Vietnam Support to People with Disabilities Grants are a comprehensive and integrated disability program that will build on the accomplishments of the previous U.S.AID assistance to people with disabilities (PWD) living in communities in Danang, as well as include additional relevant public health activities. The geographic focus will be primarily Danang and to some extent other areas proposed by Grant recipients where there is a high disability burden, the need is greatest, and in regions where dioxin hot spots are located. Grants will also include health assessments for the two major hot spots (Bien Hoa and Phu Cat) and follow-on activities as a result of the assessments. Activities outside Danang should not exceed 30% of the total expected funding. Grants are intended to improve the quality and effectiveness of the disability service system. Grants are for a three-year period. Applications are available on the website.
Requirements: United States or non-United States non-governmental organizations (NGOs) and/or consortia are eligible.
Geographic Focus: All States, All Countries
Date(s) Application is Due: Feb 29
Contact: Aiyong (Paul) Seong; (66-2) 257-3065; aseong@usaid.gov
Internet: http://www07.grants.gov/search/search.do;jsessionid=yzNBPm6PGb1HLXC2YTvjLlQqk101KqLXPGvgFFnplMSyFfzNv19T!712472910?oppId=136253&mode=VIEW
Sponsor: U.S. Agency for International Development
Ronald Raegan Building
Washington, DC 20523-1000

U.S.AID Systems Strengthening for Better HIV/TB Patient Outcomes Grants 4593

U.S.AID-South Africa Systems Strengthening for Better HIV/TB Patient Outcomes Grants are intended to strengthen South African Government systems in order to improve patient outcomes and prevent HIV by supporting comprehensive clinic-based (hospitals, community health centers, and primary health care clinics) HIV-related services. Applicants may submit a concept paper, and approved applicants will be invited to submit a full application.
Requirements: Any local non-governmental organizations, private voluntary organizations, faith-based and community organizations, and for-profit companies willing to forego profit is eligible. Applications from potential new partners are encouraged.
Geographic Focus: South Africa
Date(s) Application is Due: Jan 6
Amount of Grant: 10,000,000 - 100,000,000 U.S.D
Contact: Beatrice Lumanade, Acquisitions and Assisstance Specialist; +27 12 452 2377; fax Applications4@usaid.gov
Internet: http://www07.grants.gov/search/search.do;jsessionid=GQQyPh8Mmtk6zmWM7qX06Jsy1Lane4znGW3LYTGgq7tFJxydv2f1L2!1154909517?oppId=133033&mode=VIEW
Sponsor: U.S. Agency for International Development
Ronald Raegan Building
Washington, DC 20523-1000

U.S. Airways Community Contributions 4594

U.S. Airways invests in community organizations and initiatives to enhance quality of life in markets served by the airline. The airline's funding priorities include arts and culture, health and human services, and education and environment in the following hub markets and focus cities: Boston, Charlotte, Las Vegas, New York, Philadelphia, Phoenix, Pittsburgh and Washington, DC.
Requirements: Tax exempt organizations in the airline's market areas are eligible to apply. All requests must be submitted in writing on the organization's letterhead 6-8 weeks prior to the event and any print deadlines. All completed proposals will receive a written response within 6-8 weeks from receipt. The Contributions Committee meets bi-weekly to review requests.
Restrictions: Funding will not be considered for: Pass through and start-up organizations; Individuals, including support for specific students, researchers or conference fees; Endowment programs; Foundations which are themselves grant-making entities; Professional associations; Debt-reduction campaigns; Fundraising activities related to individual sponsorship (i.e. walk-a-thons); Service clubs (i.e. Lions, Rotary, Kiwanis); Sports teams, political, labor or fraternal organizations, scouting groups and religious organizations; Organizations or programs funded 50% or more by government sources; Individual schools; district and system-wide education programs will be considered; Capital and building campaigns.
Geographic Focus: Arizona, District of Columbia, Massachusetts, Nevada, New York, North Carolina, Pennsylvania
Contact: Community Relations Manager; community.relations@usairways.com
Internet: http://www.usairways.com/awa/content/aboutus/corporategiving/default.aspx
Sponsor: U.S. Airways Community Foundation
4000 E Sky Harbor Boulevard
Phoenix, AZ 85034

U.S. Airways Community Foundation Grants 4595

The community foundation supports multi-year capital and building campaigns by 501(c)3 non-profit organizations operating in U.S. Airways' hub markets of Charlotte,

Philadelphia and Phoenix. Grant application deadlines are April 1 and October 1 of each year. Grants will be made in late May and November.
Requirements: Grantees must be 501(c)3 nonprofit organizations that operate in one of U.S. Airways' hub cities (Charlotte, Philadelphia and Phoenix) and improve the quality and availability of charitable health care; artistic and cultural organizations; education; and/or community services. Support includes multi-year capital and building campaigns only
Restrictions: The foundation will not consider funding for endowment programs; professional associations; debt-reduction campaigns; service clubs (i.e. Lions, Rotary and Kiwanis); political, labor/fraternal or sports organizations; individual schools; general operating support; staff/consultant fees; hardware, software; infrastructure; religious activities.
Geographic Focus: Arizona, North Carolina, Pennsylvania
Date(s) Application is Due: Apr 1; Oct 1
Contact: Julie Coleman, (480) 693-3652; community.relations@usairways.com
Internet: http://www.usairways.com/awa/content/aboutus/corporategiving/default.aspx
Sponsor: U.S. Airways Community Foundation
4000 E Sky Harbor Boulevard
Phoenix, AZ 85034

U.S.CM HIV/AIDS Prevention Grants **4596**
The U.S.CM HIV Prevention Grants Program, in cooperation with the Centers for Disease Control and Prevention, provides financial and technical assistance to local health departments and community-based organizations to implement HIV prevention programs. The goal of this effort is to strengthen the capacity of local service providers to carry out effective HIV/AIDS prevention activities.
Requirements: Grants are awarded on the recommendation of an independent external review panel composed of HIV/AIDS experts drawn from local and state health departments, national organizations, and local community-based organizations.
Contact: Crystal D. Swann, Assistant Executive Director; (202) 861-6707; fax (202) 293-2352; cswann@usmayors.org
Internet: http://www.usmayors.org/hivprevention/hiv_prevention_grant.asp
Sponsor: United States Conference of Mayors
1620 I Street NW
Washington, DC 20006

U.S. CRDF Leishmaniasis: Collaborative Research Opportunities in North Africa and the Middle East **4597**
The U.S. Civilian Research and Development Foundation (CRDF) is a nonprofit organization authorized by the U.S. Congress and established in 1995 by the National Science Foundation. This unique organization promotes international scientific and technical collaboration through grants, technical resources, and training. CRDF, with funding from the National Institute for Allergy and Infectious Diseases (NIAID), announces a targeted, collaborative grant competition on Leishmaniasis in the North Africa and the Middle East (MENA) Region. This grant competition is a follow-on activity to the June 2009 interdisciplinary research conference in Tunisia sponsored by NIAID that brought together regional and international scientists to discuss research opportunities on sand fly biology, leishmania parasites, immunology, leishmaniasis epidemiology, animal reservoirs, clinical manifestations, genetics, diagnostics, treatment, and prevention of leishmaniasis in North Africa and the Middle East. These follow-on grants will provide up to one- year's support to joint research teams of U.S. and MENA scientists, with the possibility of up to a one year no-cost extension, if justified. CRDF and NIAID anticipate that the research implemented through these awards will lay the foundation for future grant proposals to NIAID and other funding agencies. The average anticipated grant will provide up to $30,000 U.S. dollars for a period of one year. At least 80% of the funds awarded to each project will be used for project-related expenses of the MENA component of the research team, including institutional support. No more than 20% may be used for U.S. team expenses. As part of the grant, the CRDF strongly encourages a member of the MENA team to visit the U.S. team's laboratory, and members of the U.S. team are encouraged to visit the MENA laboratory.
Requirements: Each proposal submitted to the CRDF must have at least one MENA Principal Investigator and at least one U.S. Principal Investigator, who share overall responsibility for the project in their respective countries, coordinating all project participants and institutions. Each proposal must meet each of the following eligibility criteria: 1) Only workshop attendees and their institutions are eligible to apply. 2) Each Principal Investigator must work full-time in a civilian research environment. (U.S. Government Laboratories working on civilian-oriented R&D projects are eligible to apply to this program.) 3) The U.S. Principal Investigator and participants of the U.S. team may be foreign nationals, but must reside in the U.S during the course of the grant award. Graduate students on the U.S. team may be foreign nationals, but they must be enrolled in an accredited degree program at a U.S. institution. A PI from Navy Medical Research Unit 3 (NAMRU3) may serve as either a U.S. or MENA researcher. 4) All projects must be in one of the following subject areas: Sand fly biology; Leishmania Parasites; Immunology; Leishmaniasis Epidemiology; Animal reservoirs; Clinical manifestations, treatment, and prevention of leishmaniasis in North Africa and the Middle East (but not clinical trials involving interventions); Genetics; Diagnostics. 5) Only one proposal will be accepted from a U.S. or MENA PI. 6) All projects must be oriented toward non-military objectives and must be carried out in a civilian research environment. Use of human subjects requires appropriate approvals from all institutions involved including Local Ethics Committee (LEC) and U.S. IBR. Middle East countries: Afghanistan, Behrain, Iran, Iraq, Israel, Jordan, Kyrgyzstan, Kuwait, Lebanon, Oman, Pakistan, Palestine, Qatar, Saudi Arabia, Syria, Tajikistan, Turkey, Turkmenistan, United Arab Emirates, Uzbekistan, Yemen. North African countries: Algeria, Egypt, Libya, Morocco, Sudan, Tunisia, Western Sahara.
Restrictions: No clinical trials involving interventions will be accepted.
Date(s) Application is Due: Sep 30
Amount of Grant: Up to 30,000 U.S.D
Contact: Mr. Stuart Politi, Senior Program Manager; (703) 526-9720; fax (703) 526-9721; cgp@crdf.org
Internet: http://www.crdf.org/funding/funding_show.htm?doc_id=949719
Sponsor: U.S. Civilian Research and Development Foundation
1530 Wilson Boulevard, 3rd Floor
Arlington, VA 22209

U.S.DA Organic Agriculture Research and Extension Initiative **4598**
The U.S.DA Organic Agriculture Research and Extension Initiative (OREI) seeks to solve critical organic agriculture issues, priorities, or problems through the integration of research and extension activities. The purpose of this program is to fund projects that will enhance the ability of producers and processors who have already adopted organic standards to grow and market high quality organic agricultural products. Priority concerns include biological, physical, and social sciences, including economics. The OREI is particularly interested in projects that emphasize research and outreach that assist farmers and ranchers with whole farm planning Projects should plan to deliver applied production information to producers. Fieldwork must be done on certified organic land or on land in transition to organic certification, as appropriate to project goals and objectives. Funding amounts are based on specific project categories.
Requirements: The following are eligible to apply for the initiative: public and state controlled institutions of higher education; nonprofits having a 501(c)3 status with the IRS, other than institutions of higher education; nonprofits that do not have a 501(c)3 status with the IRS, other than institutions of higher education; private institutions of higher education; individuals; for profit organizations other than small businesses; and small businesses. The following entities are also eligible to apply: state agricultural experiment stations; colleges and universities; university research foundations; other research institutions and organizations; federal agencies; national laboratories; private organizations or corporations; individuals who are United States citizens or nationals; or a group consisting of two or more of the above entities.
Restrictions: Recipients are required to provide funds or in-kind support from non-federal sources in an amount at least equal to the amount provided by the funding agency.
Date(s) Application is Due: Mar 9
Amount of Grant: 50,000 - 2,000,000 U.S.D
Contact: Steven Smith; (202 401-6134; fax (202) 401-1782; sismith@nifa.usda.gov
Internet: http://www.nifa.usda.gov/funding/rfas/OREI.html
Sponsor: U.S. Department of Agriculture
1400 Independence Avenue SW
Washington, DC 20250

USDD Breast Cancer Clinical Translational Research Grants **4599**
Applications for the Breast Cancer Research Program (BCRP) are being solicited by the Assistant Secretary of Defense for Health Affairs, Defense Health Program. The BCRP was established in 1992 to support innovative research focused on ending breast cancer. The BCRP challenges the scientific community to design research that will address the urgency of the vision to end breast cancer. Specifically, the BCRP seeks to accelerate high-impact research, encourage innovation and stimulate creativity, bring new investigators into the breast cancer field, and facilitate multidisciplinary collaborations. The BCRP Clinical Translational Research (CTR) Grants are intended to promote significant improvements over current approaches to breast cancer prevention and therapy, such as studies that will prevent primary or recurrent breast cancer and studies that may result in a new treatment to prevent breast cancer progression to metastasis. CTR Grants support research projects that are likely to have a major impact on breast cancer by applying promising research findings to patients with, or populations at risk for, breast cancer. Principal Investigators (PIs) and their collaborators may have originated projects in their laboratories that will form the basis for 1-2 years of advanced translational research leading to a prospective clinical trial to be conducted during this award period. Alternatively, PIs may leverage partnerships with industry. The ability to conduct the required translational research and early phase clinical trial during the award period must be demonstrated. Investigators must demonstrate availability of and access to an appropriate patient population that will support meaningful clinical outcomes. Submission is a multi-step process requiring both pre-application submission through the CDMRP eReceipt System (https://cdmrp.org/) and application submission through Grants.gov (http://www.grants.gov/). The maximum period of performance is 5 years. The maximum allowable direct costs for the entire period is $12M plus indirect costs.
Requirements: PIs must be at or above the level of assistant professor (or equivalent).
Restrictions: Applicants are encouraged to sign up on Grants.gov for "send me change notification emails" by following the link on the Synopsis page for the Program Announcement/Funding Opportunity. If the application package is updated or changed, the original version of the application package may not be accepted by Grants.gov.
Date(s) Application is Due: Aug 15
Contact: Grants Officer; (301) 682-5507; help@cdmrp.org
Internet: http://www.grants.gov/search/search.do;jsessionid=G6z2PM6MT4vFzPGvmYl7TPJTc9lpLwG0CHlsTPC1jw2TJXJRC87n!488296553?oppId=146134&mode=VIEW
Sponsor: U.S. Department of Defense
1400 Defense Pentagon
Washington, DC 20301-1400

USDD Breast Cancer Era of Hope Scholar Grants **4600**
The Era of Hope Scholar Grants support individuals who are early in their careers and have high potential for innovation in breast cancer research. Since the intent is to recognize creative and innovative individuals rather than projects, the central feature of the award is the

innovative contribution that the Principal Investigator (PI) can make toward ending breast cancer. The PI should articulate a vision that challenges current dogma and demonstrates an ability to look beyond tradition and convention. Experience in breast cancer research is not required; however, the application must focus on breast cancer and the PI must commit at least 50% of his/her full-time professional effort during the award period to breast cancer research. Five awards are anticipated. The maximum period of performance is 5 years, and the maximum direct costs for the entire period is $2.5M plus indirect costs.
Requirements: Individuals should be exceptionally talented scientists who have demonstrated that they are the "best and brightest" in their field(s) through extraordinary creativity, vision, and productivity. They should have demonstrated experience in forming effective partnerships and collaborations and should exhibit strong potential for leadership in the breast cancer research community. The PI must be an independent investigator within 6 years of his/her last mentored position as of the deadline.
Restrictions: Postdoctoral fellows, clinical fellows (including residents and interns), and other researchers in training positions are not eligible for this award.
Date(s) Application is Due: Apr 19
Contact: Grants Officer; (301) 682-5507; help@cdmrp.org
Internet: http://www07.grants.gov/search/search.do;jsessionid=kvB0PNRhp6q1GBx0mTyPhVY6Lp0lyn05T431P0d9jvdVvysv7tSH!488296553?oppId=146114&mode=VIEW
Sponsor: U.S. Department of Defense
1400 Defense Pentagon
Washington, DC 20301-1400

USDD Breast Cancer Idea Grants **4601**
The Idea Award is designed to promote new ideas that are still in the early stages of development and have the potential to yield highly impactful data and new avenues of investigation. This mechanism supports conceptually innovative, high-risk/high-reward research that could lead to critical discoveries or major advancements that will accelerate progress toward ending breast cancer. Applications should include a well-formulated, testable hypothesis based on strong scientific rationale. Innovation and impact are the most important aspects of the Idea Award. Applications that demonstrate exceptional scientific merit but lack innovation and high potential impact do not meet the intent of the Idea Award. Presentation of preliminary data is not consistent with the intent of the Idea Award mechanism. While the inclusion of preliminary data is not prohibited, the strength of the application should not rely on preliminary data, but on the innovative idea and the potential impact it will provide. Submission is a multi-step process requiring both pre-application submission through the CDMRP eReceipt System (https://cdmrp.org/) and application submission through Grants.gov (http://www.grants.gov/). Approximately fifty awards are anticipated. The maximum period of performance is 3 years. The maximum allowable direct costs for the entire period is $375,000 plus indirect costs.
Requirements: Investigators at all academic levels (or equivalent) are eligible.
Restrictions: Applicants are encouraged to sign up on Grants.gov for "send me change notification emails" by following the link on the Synopsis page for the Program Announcement/Funding Opportunity. If the application package is updated or changed, the original version of the application package may not be accepted by Grants.gov.
Date(s) Application is Due: Aug 15
Contact: Grants Officer; (301) 682-5507; help@cdmrp.org
Internet: http://www07.grants.gov/search/search.do;jsessionid=xrZyPNZPBMTJ6V24yPlrLtPxyWHP8L5d3WzzXHTGqVwCSZdY5VyK!488296553?oppId=146113&mode=VIEW
Sponsor: U.S. Department of Defense
1400 Defense Pentagon
Washington, DC 20301-1400

USDD Breast Cancer Impact Grants **4602**
The Breast Cancer Research Program (BCRP) Impact Grants support research projects (from small to largescale) that specifically focus on scientific and clinical breast cancer issues, which, if successfully addressed, will revolutionize the understanding, prevention, and/or treatment of breast cancer and make major advances toward the goal of ending the disease. The BCRP particularly encourages applications that focus on poorly understood issues that are of critical significance in breast cancer, such as those related to prevention, susceptibility, recurrence, or metastasis. The proposed work must be based on sound overall research and fully supported by preliminary data and/or published reports. The most important aspect is the potential of the proposed research to have an unprecedented impact on reducing breast cancer incidence, recurrence, and/or mortality. The research project may be from any discipline or combination of disciplines, including basic, translational, clinical (clinical trials are allowed), behavioral, and/or epidemiological research. The potential impact may be near-term or long-term, but it must be significant and non-incremental. Applications are encouraged that include meaningful and productive collaborations between investigators. The Partnering PI Option is structured to accommodate two PIs, who will each receive a separate award. One partner is identified as the Initiating PI and the other partner is identified as the Partnering PI. The Initiating and Partnering PIs have different submission requirements; however, both PIs should contribute to the preparation of a single application. The collaborative partners may have expertise in similar or disparate scientific disciplines, but each partner is expected to bring different strengths to the application. New collaborations are encouraged but not required. It is the responsibility of the collaborating investigators to describe how their combined expertise in the collaboration will better address the research question and explain why the work should be done together rather than through separate efforts. Submission is a multi-step process requiring both pre-application submission through the CDMRP eReceipt System (https://cdmrp.org/) and application submission through Grants.gov (http://www.grants.gov/). Five awards are anticipated. The maximum period of performance is 5 years. The maximum allowable direct costs for the entire period is $2M plus indirect costs.
Requirements: PIs must be at or above the level of Assistant Professor (or equivalent).
Restrictions: Applicants are encouraged to sign up on Grants.gov for "send me change notification emails" by following the link on the Synopsis page for the Program Announcement/Funding Opportunity. If the application package is updated or changed, the original version of the application package may not be accepted by Grants.gov.
Date(s) Application is Due: Aug 15
Contact: Grants Officer; (301) 682-5507; help@cdmrp.org
Internet: http://www07.grants.gov/search/search.do;jsessionid=3JX5PNlGM67F6dpRvnM4TBpcZtQzdPD1pnJYwTGzv4FVp28xw3xg!488296553?oppId=146133&mode=VIEW
Sponsor: U.S. Department of Defense
1400 Defense Pentagon
Washington, DC 20301-1400

USDD Breast Cancer Innovation Grants **4603**
The Breast Cancer Research Program (BCRP) Innovator Grants support visionary individuals who have demonstrated creativity, innovative work, and leadership in any field including, but not limited to, breast cancer. The Innovator Grants will provide these individuals with the funding and freedom to pursue their most novel, visionary, high-risk ideas that could ultimately lead to ending breast cancer. Since the intent of the Innovator Grant mechanism is to recognize creative and innovative individuals rather than projects, the central feature of the award is the innovative contribution that the Principal Investigator (PI) can make toward ending breast cancer. Experience in breast cancer research is not required; however, the application must focus on breast cancer, and the PI must commit at least 50% of his/her full-time professional effort during the award period to breast cancer research. Submission is a multi-step process requiring both pre-application submission through the CDMRP eReceipt System (https://cdmrp.org/) and application submission through Grants.gov (http://www.grants.gov/). Three awards are anticipated. The maximum period of performance is 5 years. The maximum allowable direct costs for the entire period is $5M plus indirect costs.
Requirements: PIs must be at or above the level of associate professor (or equivalent). The PI should have a past record of creativity, promise for continued innovation in future work, and a vision that challenges current dogma and demonstrates an ability to look beyond tradition and convention. The PI is also expected to have demonstrated success at forming and leading effective partnerships and collaborations.
Restrictions: Previous recipients of the BCRP Innovator Grants and BCRP Integration Panel members are ineligible. Applicants are encouraged to sign up on Grants.gov for "send me change notification emails" by following the link on the Synopsis page for the Program Announcement/Funding Opportunity. If the application package is updated or changed, the original version of the application package may not be accepted by Grants.gov.
Date(s) Application is Due: Aug 2
Contact: Grants Officer; (301) 682-5507; help@cdmrp.org
Internet: http://www07.grants.gov/search/search.do;jsessionid=TR2GPNndhVCGQGnL6n9z32GJB9hkxn2NDRVlLLqDhVQGNF5Fhc2B!488296553?oppId=146115&mode=VIEW
Sponsor: U.S. Department of Defense
1400 Defense Pentagon
Washington, DC 20301-1400

USDD Breast Cancer Transformative Vision Grant **4604**
The Breast Cancer Research Program (BCRP) Transformative Vision Grant supports research projects that will fulfill an extraordinary vision for dramatically affecting the prevention or treatment of breast cancer. The Grant requires a plan that will test and achieve the vision as quickly as possible through the translation of the research ideas to individuals with, and/or those at risk for, breast cancer. The scope of the effort may include a broad spectrum of research spanning from basic to clinical studies. There are two critical components. The first is vision and impact. A vision for a new approach that will have a revolutionary impact on the prevention or treatment of breast cancer must be articulated. The time to the final impact may vary, but the success of the vision must be transformative and significantly advance the goal of ending breast cancer. The second is implementation. The vision must be supported by a detailed plan that identifies critical milestones, outlines the innovations and technical solutions that will be implemented to accomplish the milestones, and explains how these solutions will be translated to individuals with, and/or those at risk for, breast cancer. It is expected that the proposed plan will present an exceptional level of innovation and creativity and that the Principal Investigator (PI) will assemble the team necessary to realize the vision. Submission is a multi-step process requiring both pre-application submission through the CDMRP eReceipt System (https://cdmrp.org/) and application submission through Grants.gov (http://www.grants.gov/). One award is anticipated. The maximum period of performance is 5 years. The maximum allowable direct costs for the entire period is $12M plus indirect costs.
Requirements: PIs must be at or above the level of Assistant Professor (or equivalent).
Restrictions: Applicants are encouraged to sign up on Grants.gov for "send me change notification emails" by following the link on the Synopsis page for the Program Announcement/Funding Opportunity. If the application package is updated or changed, the original version of the application package may not be accepted by Grants.gov.
Date(s) Application is Due: Aug 15
Contact: Grants Officer; (301) 682-5507; help@cdmrp.org
Internet: http://www.grants.gov/search/search.do;jsessionid=vQJqPN4Q9PjYKC9JVyhJgwGfQj3dL3Sn40jyCLgQs42pN0X6wdsb!488296553?oppId=146116&mode=VIEW
Sponsor: U.S. Department of Defense
1400 Defense Pentagon
Washington, DC 20301-1400

USDD Broad Agency Announcement Grants 4605

The Department of Defense HIV/AIDS Prevention Program (DHAPP) has successfully engaged over 80 countries in efforts to combat HIV/AIDS among their respective military services. DHAPP is a partner collaborating with the United States State Department, Health and Human Services, United States Agency for International Development, and Centers for Disease Control and Prevention, in the President's Emergency Plan for AIDS Relief (PEPFAR). This BAA is intended to solicit existing partners and establish new partners in order to expand the DHAPP program. Proposals should focus on rapidly extending HIV/AIDS services. Applicants are encouraged to target specific needs with a practical business plan, using small grass-roots organizations to provide community-based services as a way to enhance organic capabilities and sustainability. Concept Papers may be submitted at any time during this period. Early submission is strongly encouraged due to the possible lack of funding for the later evaluation periods. The period of performance is generally one year, with up to two additional phases subject to availability of funding. Applicants are encouraged to propose additional phases if appropriate. This is a multiple year BAA which may be revised as needed. Changes will be posted at Grants.gov (http://grants.gov/applicants/find_grant_opportunities.jsp). Applicants are encouraged to sign up on Grants.gov for "send me change notification emails" by following the link on the synopsis page for the program announcement/funding opportunity.

Requirements: All responsible sources from academia, industry, and non-governmental organizations are eligible. Applicants respondents must demonstrate the active support of the in-country military and the Department of Defense representative in the corresponding United States Embassy in the planning and execution of proposals.

Restrictions: Funding is available for Angola, Botswana, Caribbean Regional, Central America Regional, Cote d' Ivoire, Democratic Republic of the Congo, Dominican Republic, Ethiopia, Ghana, Guyana, India, Indonesia, Lesotho, Malawi, Mozambique, Namibia, Rwanda, South Africa, Swaziland, Uganda, Ukraine, Vietnam and Zambia. A description of the work needed, along with the program areas and estimated funding is available in the BAA found at the website. All questions should be submitted to the points of contact by email.

Date(s) Application is Due: Sep 30

Amount of Grant: 300,000 - 700,000 U.S.D

Contact: Latrice Rubenstein, Contract and Grant Officer; (619) 532-4357; Latrice.Rubenstein@navy.mil

Internet: http://www.grants.gov/search/search.do;jsessionid=1GhLPGQSVtvhVkKQYJgRnyFGWgBT8SQvFXjYv31mW6Ky7klFQ1Vx!-757993493?oppId=67733&mode=VIEW

Sponsor: U.S. Department of Defense
5450 Carlistle Park, P.O. Box 2050
Mechanicsburg, PA 17055-0791

USDD Broad Agency Announcement HIV/AIDS Prevention Grants 4606

The Department of Defense HIV/AIDS Prevention Program (DHAPP) has successfully engaged over 80 countries in efforts to combat HIV/AIDS among their respective military services. DHAPP is a partner collaborating with the United States State Department, Health and Human Services, United States Agency for International Development, and Centers for Disease Control and Prevention, in the President's Emergency Plan for AIDS Relief (PEPFAR). This Broad Agency Announcement (BAA) allows DHAPP to assist countries where militaries are not funded under PEPFAR by soliciting existing partners and establishing new partners in furtherance of DHAPP and partner military program goals. Proposals should focus on rapidly extending HIV/AIDS services. Applicants are encouraged to target specific needs with a practical business plan, using small grass-roots organizations to provide community-based services as a way to enhance organic capabilities and sustainability. Concept papers are required and full proposals are by invitation only. Early submission is strongly encouraged. The period of performance is generally one year, with up to two additional phases subject to availability of funding. Applicants are encouraged to propose additional phases if appropriate. This is a multiple year BAA which may be revised as needed. Based on the availability of funds, up to three evaluation cycles will be conducted. Changes will be posted at Grants.gov (http://grants.gov/applicants/find_grant_opportunities.jsp). Applicants are encouraged to sign up on Grants.gov for "send me change notification emails" by following the link on the synopsis page for the program announcement/funding opportunity.

Requirements: All responsible sources from academia, industry, and non-governmental organizations are eligible. Applicants must demonstrate the active support of the in-country military and the USDD representative in the corresponding United States Embassy in the planning and execution of proposals.

Restrictions: Funding is available for Ghana and El Salvador. A description of the work needed, along with the program areas and estimated funding is available in the BAA found at the website. All questions should be submitted to the points of contact by email.

Amount of Grant: 60,000 - 500,000 U.S.D

Contact: Latrice Rubenstein; (619) 556-5276; Latrice.Rubenstein@navy.mil

Internet: http://www07.grants.gov/search/search.do;jsessionid=DBT6PFBSpD4JnP9CyYnwCqZjv7ZnwjK79s2tl1KPTPCTvZRwpQJn!1471941753?oppId=141233&mode=VIEW
xLjJrqQ0vdnQM3Wt08p5Y171y8Gj4qPxb!748281696?oppId=143433&mode=VIEW

Sponsor: U.S. Department of Defense
5450 Carlistle Park, P.O. Box 2050
Mechanicsburg, PA 17055-0791

USDD Care for the Critically Injured Burn Patient Grants 4607

The Assistant Secretary of Defense for Health Affairs, Defense Health Program Medical Research and Development Office is soliciting proposals for the Defense Medical Research and Development Program (DMRDP). The Telemedicine and Advanced Technology Research Center (TATRC) is administering the application process, and the United States Army Medical Research Acquisition Activity (U.S.AMRAA) is issuing this program announcement and will be negotiating all resulting awards. The goal is to advance the state of medical science in those areas of most pressing need and relevance to today's battlefield experience, in this case burn injuries. The objectives of the DMRDP are to discover and explore innovative approaches to protect, support, and advance the health and welfare of military personnel, families, and communities; to accelerate the transition of medical technologies into deployable products; and to accelerate the translation of advances in knowledge into new standards of care for injury prevention, treatment of casualties, rehabilitation, and training systems that can be applied in theater or in the clinical facilities of the Military Health System. The research results are expected to increase the body of knowledge available to professionals and practitioners in health, medical science and related fields. The research impact is expected to benefit the military community and civilian community, particularly as it relates to the following priority areas: (1) multicenter development and assessment of the impact of checklists on standardization of burn care and improvement of outcomes in burn patients; (2) multicenter studies of ICU-based rehabilitation outcomes, i.e., evaluation of physical and rehabilitation technologies/therapies directed at short-term ICU outcomes, understanding the effect of rehabilitation on pathophysiology in the ICU burn patient, and documenting the efficacy of these immediate ICU interventions on long-term quality-of-life outcomes such as the return-to-work/duty rate; (3) device/drug development and validation for inhalation lung injury; and (4) research into the formation of hypertrophic scarring and the development of therapies to manage the condition. Pre-applications are due by December 2, and applications are by invitation only. Three to six awards are anticipated.

Requirements: Principal investigators must be independent investigators at any academic level from academia, research institutions, industry, local, state or federal governments, and private foundations that possess the skills, knowledge, and resources necessary to carry out the proposed research. Applications from investigators within the military services and other federal agencies are encouraged, as are applications involving multidisciplinary collaborations among academia, industry, the military services, the Department of Veterans Affairs, and other federal government agencies.

Date(s) Application is Due: Mar 9

Amount of Grant: 1,500,000 U.S.D

Contact: Mary Rico, Grants Officer; (703) 518-4726; mary.rico@amedd.army.mil

Internet: http://www07.grants.gov/search/search.do;jsessionid=YYJpP9pQHVhR7BFjdtNMBm1grFHnpNT4H8rRHtKrSJtHcGVFsQnJ!712472910?oppId=129414&mode=VIEW

Sponsor: U.S. Department of Defense
1400 Defense Pentagon
Washington, DC 20301-1400

USDD Clinical Functional Assessment Investigator-Initiated Grants 4608

The United States Army Medical Research and Materiel Command (U.S.AMRMC) is seeking health information technology research that may lead to outcomes and products that may qualify as advanced development activities within the Department of Defense (USDD). The results of this research will support the Military Health System in the adoption and implementation of clinical functional disability assessments and standards, by leveraging the skills and talents of United States based businesses and teaching/research academic institutions for success. The intent is to create enduring initiatives and collaborations that will develop and support sustained business and enterprise solutions for the USDD as well as organizations such as the Veterans Benefits Administration, Centers for Medicare and Medicaid Services, and the general public. Proposals are invited that develop products and systems and generate new knowledge that supports medical protection and treatment for USDD personnel and beneficiaries. Military relevance will be a primary criterion for evaluation in all areas. Pre-proposals are due by November 30, and full proposals are by invitation only. One or two awards are anticipated.

Requirements: Eligible applicants include United States based businesses, teaching research academic institutions, and multi-center collaborations, to include international partners, that develop products and systems and generate new knowledge that supports medical protection and treatment for USDD personnel and beneficiaries. Principal investigators must be independent investigators at any academic level from academia, research institutions, government, industry, and private foundations that possess the skills, knowledge, and resources necessary to carry out the proposed research.

Date(s) Application is Due: Mar 15

Amount of Grant: 1,500,000 U.S.D

Contact: Mary Rico, Grants Officer; (703) 518-4726; mary.rico@amedd.army.mil

Internet: http://www07.grants.gov/search/search.do;jsessionid=YYJpP9pQHVhR7BFjdtNMBm1grFHnpNT4H8rRHtKrSJtHcGVFsQnJ!712472910?oppId=127813&mode=VIEW

Sponsor: U.S. Department of Defense
1400 Defense Pentagon
Washington, DC 20301-1400

USDD HIV/AIDS Prevention: Military Specific HIV/AIDS Prevention, Care, and Treatment for Non-PEPFAR Funded Countries Grants 4609

The Department of Defense HIV/AIDS Prevention Program (DHAPP) has successfully engaged over 80 countries in efforts to combat HIV/AIDS among their respective military services. DHAPP is a partner collaborating with the United States State Department, Health and Human Services, United States Agency for International Development, and Centers for Disease Control and Prevention, in the President's Emergency Plan for AIDS Relief (PEPFAR). This Broad Agency Announcement (BAA) allows DHAPP to assist countries where militaries are not funded under PEPFAR and to solicit existing partners and establish new partners in furtherance of DHAPP and partner military program goals. Proposals should focus on rapidly extending HIV/AIDS services. Applicants are encouraged to target specific needs with a practical business plan, using small grass-roots organizations to provide community-based services as a way to enhance organic capabilities and sustainability. Concept papers are required and full proposals are by invitation only. Submission within the first cycle, ending February 21, is strongly encouraged. The period of performance is generally one year, with up to two additional phases subject to availability of funding. Applicants are encouraged to propose additional phases if appropriate. This is a multiple year BAA which may be revised as needed. Based on the availability of funds, up to three evaluation cycles will be conducted. Changes will be posted at Grants.gov (http://grants.gov/applicants/find_grant_opportunities.jsp). Applicants are encouraged to sign up on Grants.gov for "send me change notification emails" by following the link on the synopsis page for the program announcement/funding opportunity.

Requirements: All responsible sources from academia, industry, and non-governmental organizations are eligible. Applicants respondents must demonstrate the active support of the in-country military and the Department of Defense representative in the corresponding United States Embassy in the planning and execution of proposals.

Restrictions: Funding is available for Chad, Niger, Mali, and Burkina Faso. A description of the work needed, along with the program areas and estimated funding is available in the BAA found at the website. All questions should be submitted to the points of contact by email.

Date(s) Application is Due: Sep 30

Amount of Grant: 30,000 - 290,000 U.S.D

Contact: Latrice Rubenstein, Contract and Grant Officer; (619) 556-5276; Latrice.Rubenstein@navy.mil

Cynthia Simon-ArndtCynthia.SimonArndt@med.navy.mil

Internet: http://www.grants.gov/search/search.do;jsessionid=2zGTPFTZnGnR2n0hFVQxLjJrqQ0vdnQM3Wt08p5Y171y8Gj4qPxb!748281696?oppId=141233&mode=VIEW

Sponsor: U.S. Department of Defense
5450 Carlistle Park, P.O. Box 2050
Mechanicsburg, PA 17055-0791

USDD HIV/AIDS Prevention Program Information Systems Development Grants 4610

The Department of Defense HIV/AIDS Prevention Program (DHAPP) has successfully engaged over 80 countries in efforts to combat HIV/AIDS among their respective military services. DHAPP is a partner collaborating with the United States State Department, Health and Human Services, United States Agency for International Development, and Centers for Disease Control and Prevention, in the President's Emergency Plan for AIDS Relief. This BAA is intended to solicit existing partners and establish new partners to support the development of a DHAPP.org web site. This site will serve as a source for program updates, with both a public side and a password protected intranet side. A secure online database is needed and needs to aggregate all individual partners' submissions into a single file for download by the DHAPP operations staff in a common format, such as XL and CSV and TXT. Other site capacities should include a file structure to upload and download files, internet and email messaging, and database functions to support information gathering and analysis. Proposed work should include an infrastructure plan and a plan for support and/or turnover of the site. The period of performance is generally one year, with additional phases at the discretion of the government. Applicants are encouraged to propose additional phases if appropriate. This is a multiple year BAA which may be revised as needed. Based on the availability of funds, up to three evaluation cycles will be conducted. Changes will be posted at Grants.gov (http://grants.gov/applicants/find_grant_opportunities.jsp). Applicants are encouraged to sign up on Grants.gov for "send me change notification emails" by following the link on the synopsis page for the program announcement/funding opportunity.

Requirements: All responsible sources from academia, industry, and non-governmental organizations are eligible. Applicants must demonstrate the active support of the in-country military and the USDD representative in the corresponding United States Embassy in the planning and execution of proposals. Applicants should have prior experience with development and hosting of websites preferably for funding multiple partners, from many countries and cultures and languages, and experience in training staff to use and maintain web based databases. Applicants should also have experience with development and hosting of USDD websites and be familiar with the most recent Information Assurance requirements.

Date(s) Application is Due: Sep 30

Amount of Grant: 80,000 - 100,000 U.S.D

Contact: Latrice Rubenstein, Grants Officer; (619) 556-5276; Latrice.Rubenstein@navy.mil

Cynthia Simon-Arndt, Technical/Progrommatic Officer; Cynthia.SimonArndt@med.navy.mil

Internet: http://www.afosr.af.mil/pdfs/BAA2005-1.pdf

Sponsor: U.S. Department of Defense
5450 Carlistle Park, P.O. Box 2050
Mechanicsburg, PA 17055-0791

USDD Medical Practice Initiative Breadth of Medical Practice and Disease Frequency Exposure Grants 4611

USDD Medical Practice Initiative Breadth of Medical Practice and Disease Frequency Exposure (MPI-BMP) Grants are being solicited by the Assistant Secretary of Defense for Health Affairs, Defense Health Program (DHP). The MPI is primarily focused on the development of medical training systems and competency assessment tools for the sustainment of military medical readiness. The preliminary focus will be on determining the viability of extracting relevant metrics from existing data for use in the determination of skills degradation and using this data to assist in identifying gaps. The MPI-BMP does not only concern itself with degradation from being removed from practice, but also looks at the breadth of clinical exposures and looks across each physician's expected clinical knowledge domain. The data collected in the course of this research will aid both public and private sector organizations in their efforts to improve the quality of care provided in the outpatient setting. The primary, but not exclusive, purpose is research that will: (1) identify when and why degradation of cognitive clinical skills occurs; (2) create validated analytical tools that can predict the probable onset of cognitive skills degradation and determine, with specificity, when skill or knowledge areas have degraded or will be likely to degrade; and (3) propose methods and tools which will enable physicians to preemptively refresh knowledge and maintain familiarity and fluency across the expected competencies for their specialty. Pre-proposals are due December 22, and proposals are by invitation only. Two awards are anticipated. The maximum period of performance is three years.

Requirements: Eligible organizations must be extramural and include nonprofit, public, and private organizations, such as universities, colleges, hospitals, laboratories, and commercial firms. Historically black colleges and universities/minority institutions are encouraged to apply.

Date(s) Application is Due: Apr 19

Amount of Grant: 4,000,000 U.S.D

Contact: Mary Rico, Grants Officer; (703) 518-4726; mary.rico@amedd.army.mil

Internet: http://www07.grants.gov/search/search.do;jsessionid=YYJpP9pQHVhR7BFjdtNMBm1grFHnpNT4H8rRHtKrSJtHcGVFsQnJ!712472910?oppId=127853&mode=VIEW

Sponsor: U.S. Department of Defense
1400 Defense Pentagon
Washington, DC 20301-1400

USDD Medical Practice Initiative Procedural Skill Decay and Maintenance Grants 4612

USDD Medical Practice Initiative Procedural Skill Decay and Maintenance (MPI-PSD) Grants are awarded by the United States Army Medical Research Acquisition Activity (U.S.AMRAA) and managed by the Army's Telemedicine and Advanced Technology Research Center. The MPI is primarily focused on the development of medical training systems and competency assessment tools for the sustainment of military medical readiness. The preliminary focus will be on determining the viability of extracting relevant metrics from existing data for use in the determination of skills degradation and using this data to assist in identifying gaps. The data collected in the course of this research will aid both public and private sector organizations in their efforts to improve the quality of care provided in the outpatient setting. The primary, but not exclusive, purpose is research that will: (1) identify when and why degradation of psychomotor and procedural clinical skills occurs; (2) create validated analytical tools that can predict the probable onset of skills2 degradation and determine, with specificity, when skills have degraded or will be likely to degrade; and (3) propose methods and tools which will enable physicians or surgeons to preemptively refresh the expected competencies for their specialty. The study must focus on one of more of the following physician practices: (1) surgical; (2) intensive care medicine; and (3) outpatient invasive procedures. Pre-proposals are due December 22, and proposals are by invitation only. Two awards are anticipated. The maximum period of performance is three years. The maximum period of performance is three years.

Requirements: Eligible organizations must be extramural and include nonprofit, public, and private organizations, such as universities, colleges, hospitals, laboratories, and commercial firms. Historically black colleges and universities/minority institutions are encouraged to apply.

Date(s) Application is Due: Apr 19

Amount of Grant: 2,000,000 U.S.D

Contact: Mary Rico, Grants Officer; (703) 518-4726; mary.rico@amedd.army.mil

Internet: http://www07.grants.gov/search/search.do;jsessionid=YYJpP9pQHVhR7BFjdtNMBm1grFHnpNT4H8rRHtKrSJtHcGVFsQnJ!712472910?oppId=127873&mode=VIEW

Sponsor: U.S. Department of Defense
1400 Defense Pentagon
Washington, DC 20301-1400

USDD President's Emergency Plan for AIDS Relief Military Specific HIV Seroprevalence and Behavioral Epidemiology Risk Survey Grants 4613

The Department of Defense HIV/AIDS Prevention Program (DHAPP) has successfully engaged over 80 countries in efforts to combat HIV/AIDS among their respective military services. DHAPP is a partner collaborating with the United States State Department, Health and Human Services, United States Agency for International Development, and Centers for Disease Control and Prevention, in the President's Emergency Plan for AIDS Relief (PEPFAR). This Broad Agency Announcement (BAA) is to solicit existing partners and establish new partners to assist select military troops with performing an independent HIV Seroprevalence and Behavioral Epidemiology Risk Survey. Proposals should focus on rapidly extending HIV/AIDS services. Applicants are encouraged to target specific needs with a practical business plan, using small grass-roots organizations to provide community-based services as a way to enhance organic capabilities and sustainability. Concept papers are required and full proposals are by invitation only. Concept papers may be submitted at any time. Early submission is strongly encouraged due to the possible lack of funding for the later evaluation periods. The period of performance is generally one year, with up to two additional phases subject to availability of funding. Applicants are encouraged to propose additional phases if appropriate. This is a multiple year BAA which may be revised as needed. Based on the availability of funds, up to three evaluation cycles will be conducted. Changes will be posted at Grants.gov (http://grants.gov/applicants/find_grant_opportunities.jsp). Applicants are encouraged to sign up on Grants.gov for "send me change notification emails" by following the link on the synopsis page for the program announcement/funding opportunity.

Requirements: All responsible sources from academia, industry, and non-governmental organizations are eligible. Applicants must demonstrate the active support of the in-country military and the USDD representative in the corresponding United States Embassy in the planning and execution of proposals.
Restrictions: Countries covered by this BAA are: Democratic Republic of Congo, Liberia, Malawi, Mali, Sierra Leone, South Africa, and Vietnam.
Date(s) Application is Due: Sep 30
Amount of Grant: 100,000 - 700,000 U.S.D
Contact: Latrice Rubenstein; (619) 556-5276; Latrice.Rubenstein@navy.mil
Internet: http://www.grants.gov/search/search.do;jsessionid=1GhLPGQSVtvhVkKQYJgRnyFGWgBT8SQvFXjYv31mW6Ky7klFQ1Vx!-757993493?oppId=65295&mode=VIEW
Sponsor: U.S. Department of Defense
5450 Carlistle Park, P.O. Box 2050
Mechanicsburg, PA 17055-0791

USDD U.S. Army Medical Research and Materiel Command Broad Agency Announcement for Extramural Medical Research Grants **4614**

The United States Army Medical Research and Materiel Command's mission is to provide solutions to medical problems of importance to the American warfighter at home and abroad. The scope of this effort and the priorities attached to specific projects are influenced by changes in military and civilian medical science and technology, operational requirements, military threat assessments, and national defense strategies. Research areas of interest for the Extramural Medical Research Grants include the following: military infectious diseases; combat casualty care; military operational medicine; clinical and rehabilitative medicine; medical biological defense; medical chemical defense; telemedicine and advanced technology; and special programs (including health care delivery; detection, diagnosis, control or eradication of specified diseases, conditions, or syndromies; and the advancement of military medical interests). This Broad Agency Announcement (BAA) is intended to solicit extramural research and development ideas. Funding has not been set aside specifically for this announcement and the number of awards has not been determined. Applicants are encouraged to submit proposals that span their entire research project up to five years. Pre-proposals and full proposals may be submitted at any time until the deadline.
Requirements: Eligible applicants include the following: private/public/state controlled institutions of higher education; Hispanic-serving institutions; historically black colleges and universities/minority institutions; tribally controlled colleges and universities; Alaska Native and Native Hawaiian serving institutions; nonprofits with 501(c)3 status (other than institutions of higher education; small businesses; for-profit organizations (other than small businesses); Indian/Native American tribal governments; Indian/Native American tribally designated organizations; non-domestic (non-United States) entities (foreign organizations); and federal, state, and local government agencies.
Date(s) Application is Due: Sep 30
Contact: Grants Officer; (703) 518-4726; (QA.BAA@amedd.army.mil),
Internet: http://www07.grants.gov/search/search.do;jsessionid=24YPPC4SdRy7mYT61phb5n46fQty11L2LySnpQGjn4qRpnSwgvG0!712472910?oppId=125873&mode=VIEW
Sponsor: U.S. Department of Defense
1400 Defense Pentagon
Washington, DC 20301-1400

USDD Workplace Violence in the Military Grants **4615**

The Defense Health Program is seeking applications for the Psychological Health and Traumatic Brain Injury (PH/TBI) Research Program for Workplace Violence in the Military. The PH/TBI Research Program was established to complement ongoing Department of Defense (USDD) efforts towards promoting optimal care for PH (including workplace violence across the USDD service areas) and TBI in the areas of prevention, detection and intervention. United States warfighters endure increasingly demanding and high tempo conditions both in garrison and in the combat field of operations in order to keep pace with ongoing wartime mission requirements. In addition to the loss of lives overseas, warfighters have lost their lives on an Army post stemming from acts of violence in the workplace. The tragic shooting of military personnel at Fort Hood in November 2009 highlighted the need for the USDD to thoroughly review its approach to force health protection and to broaden their historical focus on hostile external threats. The USDD force protection policies, programs, and procedures need to be improved and integrated into a cohesive and comprehensive approach. This Grant program stems directly from recommendations yielded from the Independent Review Related to Fort Hood, which determined that USDD programs, policies, process and procedures that address identification of indicators for violence and radicalization are outdated, incomplete, and fail to include key indicators of potentially violent behaviors. It was also determined that currently the USDD has no risk assessment system available to supervisors and commanders to assist them in the identification, mitigation, and tracking of internal threats. The Independent Review outlined actions needed to identify behavioral indicators of violence. Applications are sought to support a retrospective study and a prospective study across USDD service areas to develop a set of empirical behavioral-based predictors of potential violence. Applications may include a combined, two-tiered approach that would first involve the retrospective study with those results informing the design of the prospective study. Submitters may also choose to propose only a retrospective or a prospective study. Approximately $8M is available, and five awards are anticipated. Pe-applications should be submitted by February 27.
Requirements: Applications from investigators within the military services and other federal agencies are encouraged, as are applications involving multidisciplinary collaborations among academia, industry, the military services, the Department of Veterans Affairs, and other federal government agencies. A principal investigator must be an independent investigator at any academic level from government, academia, research institutions, industry, and private foundations that possess the skills, knowledge, and resources necessary to carry out the proposed research.
Date(s) Application is Due: Apr 26
Amount of Grant: 3,000,000 U.S.D
Contact: Grants Manager; (703) 674-2500; programannouncements@tatrc.org.
Internet: http://www.grants.gov/search/search.do;jsessionid=Q2lZP9lV4mvG5TgvVqhZ2YcTpzV1NQypcLBss0zn38199gQ1WYBM!-2099600874?oppId=140173&mode=VIEW
Sponsor: U.S. Department of Defense
1400 Defense Pentagon
Washington, DC 20301-1400

U.S.G Foundation Grants **4616**

The Foundation is committed to social responsibility and supports local and national charitable organizations that serve and educate the communities in which U.S.G Corporation operates. The Foundation originated in 1979 to enrich the lives of families, friends, colleagues, customers and citizens where U.S.G employees live and work. It provides financial assistance to non-profit organizations with solutions in mind for social, health and educational issues. For more information contact the Foundation.
Contact: Corporate Responsibility Administrator; (312) 606-4297; usgfoundation@usg.com
Internet: http://www.usg.com/company/corporate-responsibility.html
Sponsor: U.S.G Foundation
550 West Adams
Chicago, IL 60606

U.S. Trust Corporation Foundation Grants **4617**

The foundation supports worthy, nonprofit organizations in the communities in which U.S. Trust does business. Funded organizations include cultural and arts institutions and community services organizations that assist low-to moderate-income residents. The corporate contributions committee reviews proposals continuously throughout the year, and meets formally twice per year to award grants Types of support include general operating support, continuing support, annual campaigns, capital campaigns, building construction/renovation, endowment funds, program development, and matching funds. Preference is given to organizations that demonstrate an ability to solve problems and provide direct services, innovative programs with broad-based community support and community development impact, organizations in which U.S. Trust employees are active, privately supported rather than tax-supported organizations, and programs that encourage self-sufficiency of participants. Proposals are reviewed year round. Proposal deadlines are March 1 for culture and arts grants and September 1 for community services grants.
Requirements: Grant giving is concentrated in the communities in which U.S. Trust does business, and is directed to 501(c)3 nonprofit organizations.
Restrictions: Grants are not made to individuals for education or other purposes; educational, religious, veterans, fraternal, or labor organizations unless engaged in a significant project benefiting the entire community; organizations, projects, or programs outside the United States; political organizations or candidates; organizations requesting support for advertising; organizations supported by umbrella organizations; or organizations with operating budgets of $150,000 or less.
Geographic Focus: California, Connecticut, Delaware, District of Columbia, Florida, Minnesota, New Jersey, New York, North Carolina, Oregon, Pennsylvania, Texas, Virginia
Date(s) Application is Due: Mar 1; Sep 1
Amount of Grant: 10,000 - 30,000 U.S.D
Contact: Carol Strickland, Chairman of the Corporate Contributions Committee; (212) 852-1330; fax (212) 852-1341; cstrickland@ustrust.com or foundation@ustrust.com
Internet: http://www.ustrust.com/public/ustrust/our_firm/community_contributions?cmsid=P-564113&lvl1=our_firm&lvl2=community_contributions
Sponsor: U.S. Trust Corporation Foundation
114 W 47th Street
New York, NY 10036

Valley Foundation Grants **4618**

The foundation makes grants to nonprofit Santa Clara County, CA, organizations for programs and projects in the areas of medical services and health care for lower income households, the arts, seniors, and education. Types of support include research, general operating budgets, endowment, and matching funds. Applicants are required to submit a one- or two-page letter of intent describing the organization, need to be addressed, goals of the project, and total project budget. A full proposal will be requested if the project falls within the foundation's areas of interest. The board meets four times each year to consider requests. Letters of intent must be submitted electronically by the listed application deadlines.
Requirements: Nonprofit organizations in Santa Clara County, CA, may submit applications.
Restrictions: The foundation avoids grants that provide more than one-fourth of an organization's total budget in a 12-month period, usually expecting community and applicant commitment to a project through cost sharing. Applications cannot be accepted for the benefit of individuals, political purposes, or religious purposes.
Geographic Focus: California
Date(s) Application is Due: Feb 1; May 1; Jul 1; Nov 1
Amount of Grant: 10,000 - 100,000 U.S.D
Contact: Program Contact; (408) 358-4545; fax (408) 358-4548; admin@valley.org
Internet: http://www.valley.org
Sponsor: Valley Foundation
16450 Los Gatos Boulevard, Suite 210
Los Gatos, CA 95032-5594

VA Monthly Allowance for Children Born with Spina Bifida and to a Vietnam Veteran 4619
The objective of the program is to provide financial assistance to certain children born with spina bifida who are children of Vietnam veterans. Monthy payment range from $230 - $1,370 a month.
Requirements: Those children born with spina bifida, except spina bifida occulta, who are natural children of Vietnam veterans, regardless of age or marital status, conceived after the date on which the veteran first served in the Republic of Vietnam during the Vietnam era are eligible.
Amount of Grant: 230 - 1,370 U.S.D
Contact: Department of Veteran Affairs; (202) 273-7210; spina.inq@med.va.gov
Internet: http://www.va.gov/hac/spina/spina.html
Sponsor: U.S. Department of Veterans Affairs
P.O. Box 65025
Denver, CO 80206-9025

Vancouver Foundation Grants and Community Initiatives Program 4620
Vancouver Foundation's Grants and Community Initiatives program provides Community Impact Grants in nine fields of interest: Animal Welfare; Arts and Culture; Children, Youth and Families; Education; Environment; Health and Social Development; Health and Medical Research; Youth Homelessness; Youth Philanthropy. The Vancouver Foundation has a two-stage application process. The first stage is a letter of intent. This consists of a brief proposal to determine basic suitability. The second stage is submission of a full grant application by a specific date. The letter of intent may be submitted online. You will be notified, if you are invited to submit a full grant application. The following is a list of deadline dates for the letters of intent and applications. Animal Welfare, Arts and Culture, Children, Youth and Families, Education, Environment, and Health and Social Development grants have a deadline date of: Letter of Intent, June 18, Application, August 6. Youth Homelessness grants: Letter of Intent, August 6, Application, September 24. Youth Philanthropy grants: Letter of Intent, September 17, Application, October 1. Health and Medical Research grants deadline dates are undetermined, at this time.
Requirements: Under our Grants and Community Initiatives program, Vancouver Foundation funds eligible applicants, which include: registered charities and qualified donees under the Income Tax Act; First Nations that may be considered a public body performing a function of government.
Restrictions: The Foundation does not fund the following: 100% of a proposal's costs; an organization's ongoing operational or core expenses; retroactive funding, or for any expenses to be incurred prior to the Foundation's decision date; debt retirement or reserves; mortgage pay-downs; conferences, competitions, symposia, annual events, or travel to/attendance at such events; office equipment and furniture; activities of religious organizations that serve primarily their membership and/or their direct religious purposes, unless the community at large will benefit significantly; sabbatical leaves, student exchanges; medical facilities or equipment; league-based sports and recreation programs; library acquisitions and construction; school construction; publication of books; academic or dissertation research; research with human subjects; endowment matching grants; capital requests, with the exceptions noted above; activities previously supported through government funding; bursaries, scholarships and awards; school trips, annual events or equipment; grants to individuals or businesses; funding requests if reports on any previous grants are overdue.
Geographic Focus: Canada
Contact: Mark Gifford, Director; (604) 629-5362; markg@vancouverfoundation.ca
Internet: http://www.vancouverfoundation.ca/grants/communityinitiatives.htm
Sponsor: Vancouver Foundation
555 West Hastings Street, Suite 1200
Vancouver, BC V6B 4N6 Canada

Vancouver Foundation Public Health Bursary Package 4621
With funding from the BC Associated Boards of Health Dr. Ken Benson Memorial Bursary and the Isabel Loucks Foster Public Health Bursary Funds, bursaries are available to qualified students who are or wish to engage in post-graduate study in a field of Public or Community health practice within any recognized discipline. The funds are managed by the Vancouver Foundation and the Bursary funds will be released to a successful applicant(s) annually on the recommendation of the Health Officers' Council. The Health Officers' Council is comprised of the Medical Health Officers and public health physicians of British Columbia. Bursaries in an amount up to $3,000 are available to eligible applicants on a one- time-only basis. Successful candidates will be notified by October 31st.
Requirements: Eligibility for the Isabel Loucks Foster Public Health Bursary Fund: full-time students engaged in post-graduate study in a field of Public/Community Health Practice within an applicable discipline (e.g. nurses, nutritionists, environmental health officers, physicians, etc.); persons must be ordinarily resident in B.C; persons must have demonstrated excellence in performance and leadership ability; studies may be undertaken at any suitable educational institution either within or outside British Columbia.
Geographic Focus: Canada
Date(s) Application is Due: Sep 4
Amount of Grant: 3,000 U.S.D
Contact: Faye Wightman, President; (604) 688-2204; fax (604) 688-4170; info@vancouverfoundation.ca
Internet: http://www.vancouverfoundation.ca/grants/specialprograms.htm
Sponsor: Vancouver Foundation
555 West Hastings Street, Suite 1200
Vancouver, BC V6B 4N6 Canada

Vancouver Sun Children's Fund Grants 4622
In partnership with Vancouver Foundation, the Vancouver Sun Children's Fund provides grants twice yearly to registered charities in British Columbia that directly serve the needs of children and youth (up to18 years) with special challenges. Grants range from $100 to $5,000 and recommendations are made by Vancouver Foundation's Children, Youth and Families Advisory Committee whose members have knowledge and experience in the fields of children's physical, mental and emotional health.
Requirements: Eligible applicants are registered charities and other qualified organizations under the Income Tax Act. They must demonstrate fiscal responsibility and effective management. To apply, submit your Letter of Intent electronically.
Restrictions: Grants are not made to individuals or businesses.
Geographic Focus: Canada
Amount of Grant: 100 - 5,000 U.S.D
Contact: Mark Gifford Director, Grants and Community Initiatives; (604) 688-2204; fax (604) 688-4170; markg@vancouverfoundation.ca
Internet: http://www.vancouverfoundation.ca/grants/specialprograms.htm
Sponsor: Vancouver Sun Children's Fund c/o Vancouver Foundation
Harbour Centre, 555 West Hastings Street, Suite 1200, Box 12132
Vancouver, BC V6B 4N6 Canada

Vanderburgh Community Foundation Grants 4623
The Vanderburgh Community Foundation is a nonprofit, public charity created by and for the people of Vanderburgh County, Indiana. The Foundation assists nonprofits fulfill their missions by strengthening their ability to meet community needs through grants that assist charitable programs, address community issues, support community agencies, launch community initiatives, and support leadership development. The Foundation's grant cycle will be announced by the end of July.
Requirements: A Letter of Intent is the required first step for submitting funding requests to be considered by the Community Foundation. The Letter of Intent helps in determining whether a funding request qualifies for consideration before the applicant invests time and effort in completing a full grant proposal. The Foundation's office will provide additional information on proposal procedure or you may access the Foundation's website. The Foundation gives primarily to 501(c)3 tax-exempt organizations in Vanderburgh County, Indiana.
Geographic Focus: Indiana
Contact: Barbara Dicken, Director; (812) 422-1245; fax (812) 429-0840; barb@vcfoundation.org
Internet: http://www.vanderburghcommunityfoundation.org/disc-grants
Sponsor: Vanderburgh Community Foundation
18 NW Forth Street, Suite 220
Evansville, IN 47708

VanLobenSels/RembeRock Foundation Grants 4624
The foundation provides grants to organizations that serve vulnerable and underserved populations in Northern California. Proposals will be considered for project and program start-up funding for new ideas that demonstrate the potential for improvement and enhancement of life in our communities; projects and programs designed to provide fair treatment and equal access to the law for disadvantaged residents and newcomers; projects and programs targeted toward underserved and at-risk populations; and projects and programs that enhance access to services for underserved populations. Types of support include project development, conferences and seminars, publication, seed money, research, and matching funds. Proposals are considered every six to eight weeks, and grants are awarded throughout the year.
Requirements: Organizations must be located in California.
Restrictions: The foundation does not entertain applications that require medical, scientific, or other technical knowledge; request funding for an individual or family; or request funding for scholarships or stipends.
Geographic Focus: California
Samples: Black Women Organized for Educational Development (Oakland, CA)—for the Mentor Leadership program, $17,000; Ctr on Juvenile and Criminal Justice (San Francisco, CA)—for expansion of the entrepreneurship center, $10,000; Coalition for Alternatives in Mental Health (Berkeley, CA)—for trainee programs, $5000.
Contact: Toni Rembe, President; (415) 512-0500; fax (415) 371-0227; info@vlsrr.org
Internet: http://www.vlsrr.org/frameset_guidelines.html
Sponsor: VanLobenSels / RembeRock Foundation
131 Suiteuart Street, Suite 301
San Francisco, CA 94105

Varian Medical Systems Grants 4625
Varian Medical Systems makes grants, donations, and gifts to nonprofit organizations committed to supporting the provision of information and services involving the prevention, detection, and treatment of cancer. The company also supports education through matching gifts that employees make to nonprofit educational institutions. Applications must be received at least 60 days in advance of the time that the funding is needed. Application and guidelines are available online.
Requirements: Nonprofit organizations are eligible.
Restrictions: The following are ineligible: applications that are in any way associated with planned, current, or past purchases, or the recommendation to purchase, any Varian Medical Systems, Inc, products or services; local chapter programs already supported by Varian at the national level (i.e., National Survivor's Day, Relay for Life); capital campaigns for new buildings and/or major equipment purchases; sporting or entertainment events (i.e., golf); individual requests for personal needs/projects; a-thon fundraisers (i.e., walk-a-thon, telethon); religious or political organizations; or advertising programs (i.e., radio/television).

Contact: Program Contact; (650) 493-4000; fax (650) 424-6822; giving@varian.com
Internet: http://www.varian.com/com/ctz000.html
Sponsor: Varian Medical Systems
3100 Hansen Way, MS E-210
Palo Alto, CA 94304

VDH Commonwealth of Virginia Nurse Educator Scholarships 4626
Scholarships will be provided to full/part-time graduate nursing students who are accepted to or enrolled in a master's or doctoral level nursing program in the Commonwealth of Virginia. The purpose of this program is to increase the number of nursing faculty by providing master's/doctoral students with financial support. Additionally, this program will assist Virginia nursing schools to recruit and retain new nursing faculty, which enables the schools to increase enrollment.
Requirements: Full/part-time master's or doctoral nursing students enrolled or accepted in nursing program in Virginia will be eligible to apply. Students must complete their degree requirements in two years or less. Priority shall be given to master's degree candidates who will teach in community colleges. Applications will be accepted from June 1 through July 31 each calendar year. Complete and submit a Nursing Scholarship Application along with all required materials. Applications will be available online at: http://www.vdh.virginia.gov/healthpolicy/primarycare/incentives/nursing/index.htm
Restrictions: Recipients will be required to give two years of teaching service as a faculty member for every year they receive a scholarship. Recipients who do not take teaching positions at a Virginia nursing program will be required to pay back the scholarship with a 9% interest penalty. Repayment must be paid in full within two years of degree completion. Recipients must begin their teaching service within three months of completing their educational program.
Geographic Focus: Virginia
Date(s) Application is Due: Jul 31
Amount of Grant: 20,000 U.S.D
Contact: Aileen Harris, Health Care Workforce Manager; (804) 864-7436; fax (804) 864-7440; Aileen.Harris@vdh.virginia.gov
Internet: http://www.vdh.virginia.gov/healthpolicy/primarycare/incentives/nursing/va_nsp_guidelines.htm
Sponsor: Virginia Department of Health
P.O. Box 2448
Richmond, VA 23218

VDH Emergency Medical Services Training Fund (EMSTF) 4627
The EMS Training Fund program is designed to provide financial assistance for Virginia Certified EMS providers and Virginia Office of EMS approved Emergency Medical Services courses. These funds shall supplement local support for Emergency Medical Services Courses. The Emergency Medical Services Training Funds are monies available for student expenses related to attending EMS Certification programs, tuition reimbursement, auxiliary programs and continuing education programs whose lessons are based upon or resemble the learning objectives in the United States Department of Transportation's Intermediate-99 and Paramedic curricula and the Enhanced curricula as defined in 12VAC5-31. See website for application.
Requirements: These funds are designed for non-profit entities and individuals participating in Virginia's EMS System.
Geographic Focus: Virginia
Contact: Jackie Hunter, Program Support Technician; (804) 864-7585; Jacqueline.Hunter@vdh.virginia.gov
Internet: http://www.vdh.virginia.gov/OEMS/Training/EMSTF.htm
Sponsor: Virginia Department of Health
P.O. Box 2448
Richmond, VA 23218

VDH Rescue Squad Assistance Fund Grants 4628
The Financial Assistance for Emergency Medical Services Grants Program, known as the Rescue Squad Assistance Fund (RSAF) Grant Program is a multi-million dollar matching grant program for Virginia governmental, non-profit EMS agencies and organizations to provide financial assistance based on a demonstrated need. Items eligible for funding include EMS equipment and vehicles, computers, EMS management programs, courses/classes and projects benefiting the recruitment and retention of EMS members. RSAF is primarily a reimbursement grant that requires the grantee to make the purchase for the awarded item(s) and then submit an invoice for reimbursement.
Requirements: Applicant must be a Virginia non-profit agency/organization or governmental organization involved in emergency medical service (ems); applicant must submit verification of its FIN. Verification can be provided in the following formats: copy of the original letter from the IRS issuing FIN, copy of the latest tax returns (1st page only), statement from the County Administrator or City Manager of the municipality stating that the application is non-profit or a government agency and verifies their FIN; applicant must submit a copy (1st page only) of the most recent Federal Tax Return from the IRS (Form 990); applicant must submit the Virginia Office of EMS Affirmation Page in its entirety including the original signature of the Authorized Agent, the Fiscal Officer (Treasurer) and the Operational Medical Director (OMD). All Grant Applications must be submitted via the Internet. The Affirmation page is due by September 15 with original signatures. Faxed Applications will not be accepted. Submit all grant information to: Office of Emergency Medical Services, Attn: Amanda Davis - Grant Manager, 109 Governor Street, Suite UB-55, Richmond, VA 23219.
Restrictions: Applications submitted with line items less then $500.00 will be disqualified.
Geographic Focus: Virginia
Date(s) Application is Due: Sep 15
Contact: Amanda Davis, Grant Manager; (804) 864-7600; fax (804) 864-7580
Internet: http://www.vdh.virginia.gov/OEMS/Grants/index.htm
Sponsor: Virginia Department of Health
P.O. Box 2448
Richmond, VA 23218

Vectren Foundation Grants 4629
The foundation focuses its efforts in four major areas of giving: Education, Arts & Culture, Civic, and Health & Human Services. Priority will be given to those organizations and activities that serve to improve the quality of life in the communities Vectren serves.
Requirements: The only awards grants to organizations exempt from Federal Tax under Section 501(c)3 of the IRS code. The funding request application can be downloaded from the website. Applications should include a copy of your tax-exempt ruling from the IRS; a list of current board members and annual report, if available; and, a current operating and project budget. Applications for contributions to capital fund drives or to purchase equipment should also provide a clear description of the project, including its goals and objectives; and, a financial plan showing how the project will be funded. The plan should include other possible sources of support and funds that have been pledged or received to date.
Restrictions: The foundation will not: support programs, activities or campaigns that advance a specific religious agenda; make contributions to political, fraternal, labor or veteran's organizations, unless the activity is for direct community involvement; make contributions to issues-oriented organizations; will not award scholarships or grants that benefit individuals (This includes individuals raising funds for charitable purposes, student ambassador programs and sponsorship of individual sports or academic teams.); or, award grants involving travel outside of the Vectren Service Territory.
Geographic Focus: Indiana
Samples: Morgan County Public Library's Youth Art month; Cornerstone Center for the Arts, Arts Adventure program; Lakeview Volunteer Firefighters; United Way Campaigns
Contact: Mark H. Miller, Community Affairs Manager; (812) 491-4176; fax (812) 491-4078; mmiller@vectren.com
Internet: http://www.vectren.com/web/holding/discover/foundation/foundation_i.jsp
Sponsor: Vectren Foundation
One Vectren Square
Evansville, IN 47708

Ventura County Community Foundation Grants 4630
Each year, the foundation awards grants and scholarships, providing grantmaking opportunities for all areas of community need. Grant dollars are derived from designated funds, donor advised funds, endowment funds, field-of-interest funds, special funds, and greatest needs funds. Only certain funds are open to application. All requests for proposals (RFPs) are publicly announced and specifically describe the focus areas for which proposals are being sought. The foundation looks for projects where a moderate amount of grant money can effect a significant result, and where there is a well-planned and/or new approach to solving community problems. All competitive grants (except the SUN Fund) are made only in response to a formal RFP process that outlines specific areas of interest and information requested. Current grant focuses are posted five weeks prior to the due date. Applications are accepted once per year for each grant cycle. Applicants should not contact the foundation for funding outside of an RFP process. Application and guidelines are available online.
Requirements: California's Ventura County 501(c)3 nonprofit organizations are eligible.
Restrictions: Grants will not be made for political or partisan purposes, religious activities, general operating short falls, endowment funds, annual fund drives, or budget deficits. Individuals are ineligible, except through the foundation's scholarship program.
Geographic Focus: California
Contact: Tina Knight, (805) 988-0196, ext. 120; fax (805) 485-5537; tknight@vccf.org
Internet: http://www.vccf.org
Sponsor: Ventura County Community Foundation
1317 Del Norte Road, Suite 150
Camarillo, CA 93010

Verizon Foundation Health Care and Accessibility Grants 4631
The Verizon Foundation invests in projects that provide technology to help under served populations and people with disabilities access information on critical health issues. It also supports innovative technology that helps health care providers increase their efficiency, effectiveness and reach. The Foundation only accepts electronic proposals through its online process, and reviews unsolicited proposals on a continuous calendar year basis from January 1st through October 31st.
Requirements: Proposals will be considered from eligible tax-exempt organizations in certain 501(c)3 subsections as defined by the Internal Revenue Service (IRS). Proposals will also be considered from elementary and secondary schools (public and private) that are registered with the National Center for Education Statistics (NCES). Proposals may also be considered from eligible tax-exempt organizations in the subsection 170(B)(1)(a)(i)—Church, provided that the proposal will benefit a large portion of a community without regard to religious affiliation and does not duplicate the work of other agencies in the community.
Restrictions: The Verizon Foundation does not provide funding for any of the following: individuals; private charity or foundation; organizations not exempt under Section 501(c)3 of the Internal Revenue Code, and not eligible for tax-deductible support; religious organizations, unless the particular program will benefit a large portion of a community without regard to religious affiliation and does not duplicate the work of other agencies in the community; political causes, candidates, organizations or campaigns; organizations that discriminate on the basis of age, color, citizenship, disability, disabled veteran status, gender, race, religion, national origin, marital status, sexual orientation, military service or status or Vietnam-era veteran status; organizations whose primary purpose is to

influence legislation; endowments or capital campaigns; film, music, TV, video and media production projects or broadcast program underwriting; research studies, unless related to projects we are already supporting; sports sponsorships; performing arts tours; association memberships; or organizations that have received a grant from the Verizon Foundation in the last three consecutive years - organization may reapply after a one-year hiatus.
Amount of Grant: 5,000 - 10,000 U.S.D
Contact: Patrick Gaston, President; (800) 360-7955; fax (908) 630-2660; patrick.g.gaston@erizon.com
Internet: http://foundation.verizon.com/core/health.shtml
Sponsor: Verizon Foundation
1 Verizon Way
Basking Ridge, NJ 07920

Verizon Foundation Maryland Grants **4632**
The foundation invites applications, in Maryland, to receive financial grants that support education, literacy, health care access, workforce development, and the reduction of domestic violence. Consideration will be given to proposals for programs that provide job training for individuals, including welfare-to-work and school-to-work training programs; and assist people with disabilities in acquiring computer skills through adaptive technology that prepares them to enter the workforce. The Foundation only accepts electronic proposals through its online process, and reviews unsolicited proposals on a continuous calendar year basis from January 1st through October 31st.
Requirements: Proposals will be considered from eligible tax-exempt organizations in certain 501(c)3 subsections as defined by the Internal Revenue Service (IRS). Proposals will also be considered from elementary and secondary schools (public and private) that are registered with the National Center for Education Statistics (NCES). Proposals may also be considered from eligible tax-exempt organizations in the subsection 170(B)(1)(a)(i)—Church, provided that the proposal will benefit a large portion of a community without regard to religious affiliation and does not duplicate the work of other agencies in the community.
Restrictions: The Verizon Foundation does not provide funding for any of the following: individuals; private charity or foundation; organizations not exempt under Section 501(c)3 of the Internal Revenue Code, and not eligible for tax-deductible support; religious organizations, unless the particular program will benefit a large portion of a community without regard to religious affiliation and does not duplicate the work of other agencies in the community; political causes, candidates, organizations or campaigns; organizations that discriminate on the basis of age, color, citizenship, disability, disabled veteran status, gender, race, religion, national origin, marital status, sexual orientation, military service or status or Vietnam-era veteran status; organizations whose primary purpose is to influence legislation; endowments or capital campaigns; film, music, TV, video and media production projects or broadcast program underwriting; research studies, unless related to projects we are already supporting; sports sponsorships; performing arts tours; association memberships; or organizations that have received a grant from the Verizon Foundation in the last three consecutive years - organization may reapply after a one-year hiatus.
Geographic Focus: Maryland
Amount of Grant: 5,000 - 10,000 U.S.D
Contact: Diane Miles, Director; (410) 393-7450 or (800) 360-7955; fax (908) 630-2660; diane.f.miles@verizon.com
Internet: http://foundation.verizon.com/cybergrants/plsql/vznadmin.vz_www.iyc?x_zip=21202
Sponsor: Verizon Foundation
1 East Pratt Street, 10E
Baltimore, MD 21202

Verizon Foundation New York Grants **4633**
The Verizon Foundation—the philanthropic arm of Verizon Communications—supports a variety of programs and events in communities from Buffalo to Brooklyn. As one of the largest corporate contributors in New York, Verizon is proud to have directed more than 6,000 grants totaling more than $14 million to some 2,700 non-profits across the state. Those non-profits include school districts and universities, multicultural groups and hospitals, children's organizations and those assisting the disabled and victims of domestic violence. Through the Foundation, Verizon New York underscores the company's commitment to key funding priorities—combating literacy and domestic violence, empowering communities with technology, bridging the digital divide, creating a skilled work force, and galvanizing employees to volunteer. The Foundation only accepts electronic proposals through its online process, and reviews unsolicited proposals on a continuous calendar year basis from January 1st through October 31st.
Requirements: Proposals will be considered from eligible tax-exempt organizations in certain 501(c)3 subsections as defined by the Internal Revenue Service (IRS). Proposals will also be considered from elementary and secondary schools (public and private) that are registered with the National Center for Education Statistics (NCES). Proposals may also be considered from eligible tax-exempt organizations in the subsection 170(B)(1)(a)(i)—Church, provided that the proposal will benefit a large portion of a community without regard to religious affiliation and does not duplicate the work of other agencies in the community.
Restrictions: The Verizon Foundation does not provide funding for any of the following: individuals; private charity or foundation; organizations not exempt under Section 501(c)3 of the Internal Revenue Code, and not eligible for tax-deductible support; religious organizations, unless the particular program will benefit a large portion of a community without regard to religious affiliation and does not duplicate the work of other agencies in the community; political causes, candidates, organizations or campaigns; organizations that discriminate on the basis of age, color, citizenship, disability, disabled veteran status, gender, race, religion, national origin, marital status, sexual orientation, military service or status or Vietnam-era veteran status; organizations whose primary purpose is to influence legislation; endowments or capital campaigns; film, music, TV, video and media production projects or broadcast program underwriting; research studies, unless related to projects we are already supporting; sports sponsorships; performing arts tours; association memberships; or organizations that have received a grant from the Verizon Foundation in the last three consecutive years - organization may reapply after a one-year hiatus.
Geographic Focus: New York
Amount of Grant: 5,000 - 10,000 U.S.D
Contact: Sandy Wilson; (800) 360-7955; fax (908) 630-2660; sandra.c.wilson@verizon.com
Internet: http://foundation.verizon.com/cybergrants/plsql/vznadmin.vz_www.iyc?x_zip=10007
Sponsor: Verizon Foundation
140 West Street, 26th Floor, Room 2621
New York, NY 10007

Verizon Foundation Northeast Region Grants **4634**
The Verizon Foundation is committed to supporting programs and projects that leverage technology to improve basic computer literacy and domestic violence initiatives, seeking organizations that serve the needs of diverse communities, people with disabilities, and the economically and socially challenged. The Foundation only accepts electronic proposals through its online process, and reviews unsolicited proposals on a continuous calendar year basis from January 1st through October 31st.
Requirements: Proposals will be considered from eligible tax-exempt organizations in certain 501(c)3 subsections as defined by the Internal Revenue Service (IRS). Proposals will also be considered from elementary and secondary schools (public and private) that are registered with the National Center for Education Statistics (NCES). Proposals may also be considered from eligible tax-exempt organizations in the subsection 170(B)(1)(a)(i)—Church, provided that the proposal will benefit a large portion of a community without regard to religious affiliation and does not duplicate the work of other agencies in the community.
Restrictions: The Verizon Foundation does not provide funding for any of the following: individuals; private charity or foundation; organizations not exempt under Section 501(c)3 of the Internal Revenue Code, and not eligible for tax-deductible support; religious organizations, unless the particular program will benefit a large portion of a community without regard to religious affiliation and does not duplicate the work of other agencies in the community; political causes, candidates, organizations or campaigns; organizations that discriminate on the basis of age, color, citizenship, disability, disabled veteran status, gender, race, religion, national origin, marital status, sexual orientation, military service or status or Vietnam-era veteran status; organizations whose primary purpose is to influence legislation; endowments or capital campaigns; film, music, TV, video and media production projects or broadcast program underwriting; research studies, unless related to projects we are already supporting; sports sponsorships; performing arts tours; association memberships; or organizations that have received a grant from the Verizon Foundation in the last three consecutive years - organization may reapply after a one-year hiatus.
Geographic Focus: Massachusetts, Rhode Island
Amount of Grant: 5,000 - 10,000 U.S.D
Contact: Rick Colon, Regional; (781) 849-2046; richard.b.colon@verizon.com
Internet: http://foundation.verizon.com/cybergrants/plsql/vznadmin.vz_www.iyc?x_zip=02184
Sponsor: Verizon Foundation
125 Lundquist Drive, Room 1730
Braintree, MA 02184

Verizon Foundation Pennsylvania Grants **4635**
Verizon Foundation Pennsylvania is investing in online resources, community-based initiatives and state-wide programs that teach non readers to read and advance the complex skills necessary for educational achievement and job success in the 21st Century. The grants have enabled the recipient organizations to improve basic literacy skills, create new and diverse approaches to technology-based learning, and help domestic violence victims rebuild their lives. The Foundation only accepts electronic proposals through its online process, and reviews unsolicited proposals on a continuous calendar year basis from January 1st through October 31st.
Requirements: Proposals will be considered from eligible tax-exempt organizations in certain 501(c)3 subsections as defined by the Internal Revenue Service (IRS). Proposals will also be considered from elementary and secondary schools (public and private) that are registered with the National Center for Education Statistics (NCES). Proposals may also be considered from eligible tax-exempt organizations in the subsection 170(B)(1)(a)(i)—Church, provided that the proposal will benefit a large portion of a community without regard to religious affiliation and does not duplicate the work of other agencies in the community.
Restrictions: The Verizon Foundation does not provide funding for any of the following: individuals; private charity or foundation; organizations not exempt under Section 501(c)3 of the Internal Revenue Code, and not eligible for tax-deductible support; religious organizations, unless the particular program will benefit a large portion of a community without regard to religious affiliation and does not duplicate the work of other agencies in the community; political causes, candidates, organizations or campaigns; organizations that discriminate on the basis of age, color, citizenship, disability, disabled veteran status, gender, race, religion, national origin, marital status, sexual orientation, military service or status or Vietnam-era veteran status; organizations whose primary purpose is to influence legislation; endowments or capital campaigns; film, music, TV, video and media production projects or broadcast program underwriting; research studies, unless related to projects we are already supporting; sports sponsorships; performing arts tours; association memberships; or organizations that have received a grant from the Verizon Foundation in the last three consecutive years - organization may reapply after a one-year hiatus.
Geographic Focus: Pennsylvania
Amount of Grant: 5,000 - 10,000 U.S.D
Contact: Denise Loughlin; (215) 466-3351; denise.g.loughlin@verizon.com

Internet: http://www22.verizon.com/about/community/pa/community/comm_index.html
Sponsor: Verizon Foundation
3478 Kirkwood Road
Philadelphia, PA 19114

Verizon Foundation Vermont Grants 4636
The Foundation is committed to supporting programs and projects that leverage technology to improve basic computer literacy and domestic violence initiatives, seeking organizations that serve the needs of diverse communities, people with disabilities, and the economically and socially challenged. Last year, Verizon Foundation donated close to $250,000 to Vermont nonprofit organizations and our employees clocked more than 16,000 volunteer hours to community organizations. The Foundation only accepts electronic proposals through its online process, and reviews unsolicited proposals on a continuous calendar year basis from January 1st through October 31st.
Requirements: Proposals will be considered from eligible tax-exempt organizations in certain 501(c)3 subsections as defined by the Internal Revenue Service (IRS). Proposals will also be considered from elementary and secondary schools (public and private) that are registered with the National Center for Education Statistics (NCES). Proposals may also be considered from eligible tax-exempt organizations in the subsection 170(B)(1)(a)(i)—Church, provided that the proposal will benefit a large portion of a community without regard to religious affiliation and does not duplicate the work of other agencies in the community.
Restrictions: The Verizon Foundation does not provide funding for any of the following individuals; private charity or foundation; organizations not exempt under Section 501(c)3 of the Internal Revenue Code, and not eligible for tax-deductible support; religious organizations, unless the particular program will benefit a large portion of a community without regard to religious affiliation and does not duplicate the work of other agencies in the community; political causes, candidates, organizations or campaigns; organizations that discriminate on the basis of age, color, citizenship, disability, disabled veteran status, gender, race, religion, national origin, marital status, sexual orientation, military service or status or Vietnam-era veteran status; organizations whose primary purpose is to influence legislation; endowments or capital campaigns; film, music, TV, video and media production projects or broadcast program underwriting; research studies, unless related to projects we are already supporting; sports sponsorships; performing arts tours; association memberships; or organizations that have received a grant from the Verizon Foundation in the last three consecutive years - organization may reapply after a one-year hiatus.
Geographic Focus: Vermont
Amount of Grant: 5,000 - 10,000 U.S.D
Contact: Stephanie Lee, Regional Director; (617) 743-5440; stephanie.s.lee@verizon.com
Internet: http://foundation.verizon.com/cybergrants/plsql/vznadmin.vz_www.iyc?x_zip=05403
Sponsor: Verizon Foundation
185 Franklin Street, Room 1702
Boston, MA 02110

Verizon Foundation Virginia Grants 4637
Verizon Foundation grants have touched and will continue to affect thousands of lives in Virginia. Verizon is committed to investing in the technological, financial, and human resources necessary to build more literate communities that can prosper in the information era, and to foster domestic violence awareness, prevention, and recovery. The Foundation will continue to form strong partnerships with Virginia's non-profit community to achieve its funding priority goals and improve the quality of life for people across the Commonwealth. The Foundation only accepts electronic proposals through its online process, and reviews unsolicited proposals on a continuous calendar year basis from January 1st through October 31st.
Requirements: Proposals will be considered from eligible tax-exempt organizations in certain 501(c)3 subsections as defined by the Internal Revenue Service (IRS). Proposals will also be considered from elementary and secondary schools (public and private) that are registered with the National Center for Education Statistics (NCES). Proposals may also be considered from eligible tax-exempt organizations in the subsection 170(B)(1)(a)(i)—Church, provided that the proposal will benefit a large portion of a community without regard to religious affiliation and does not duplicate the work of other agencies in the community.
Restrictions: The Verizon Foundation does not provide funding for any of the following: individuals; private charity or foundation; organizations not exempt under Section 501(c)3 of the Internal Revenue Code, and not eligible for tax-deductible support; religious organizations, unless the particular program will benefit a large portion of a community without regard to religious affiliation and does not duplicate the work of other agencies in the community; political causes, candidates, organizations or campaigns; organizations that discriminate on the basis of age, color, citizenship, disability, disabled veteran status, gender, race, religion, national origin, marital status, sexual orientation, military service or status or Vietnam-era veteran status; organizations whose primary purpose is to influence legislation; endowments or capital campaigns; film, music, TV, video and media production projects or broadcast program underwriting; research studies, unless related to projects we are already supporting; sports sponsorships; performing arts tours; association memberships; or organizations that have received a grant from the Verizon Foundation in the last three consecutive years - organization may reapply after a one-year hiatus.
Geographic Focus: Virginia
Amount of Grant: 5,000 - 10,000 U.S.D
Contact: Stephan Clementi, (804) 772-1673 or (800) 360-7955; fax (908) 630-2660; steve.clementi@verizon.com
Internet: http://foundation.verizon.com/cybergrants/plsql/vznadmin.vz_www.iyc?x_zip=24012
Sponsor: Verizon Foundation
703 East Grace Street, 7th Floor
Richmond, VA 23219

Vermont Community Foundation Grants 4638
The foundation has broad program interests and will consider any project that meets a clearly defined community need in Vermont. Categories of support include, but are not limited to: the arts and humanities; education; the environment; historic resources; health; public affairs and community development; and social services.
Requirements: To be eligible, an organization must: have 501(c)3 tax-exempt status and not be classified as a private foundation; be located in or serve the people of Vermont; and employ staff and provide services without discrimination on the basis of race, religion, sex, age, national origin, disability, or sexual orientation. See Foundations website for additional guidelines, including rolling deadlines, and application forms.
Restrictions: The foundation does not make grants for endowments, annual operating or capital campaigns, religious purposes, individuals, debts, or equipment unless it is an integral part of an otherwise eligible project.
Geographic Focus: Vermont
Amount of Grant: Up to 25,000 U.S.D
Contact: Peter Espenshade, Vice President; (802) 388-3355, ext. 248; fax (802) 388-3398; pespenshade@vermontcf.org
Internet: http://www.vermontcf.org/apply/
Sponsor: Vermont Community Foundation
3 Court Street, P.O. Box 30
Middlebury, VT 05753

Vernon K. Krieble Foundation Grants 4639
The Foundation was established in 1984 by the family of Professor Vernon K. Krieble, scientist, educator, inventor, and entrepreneur. Recognizing that the Foundation's assets are the product of a free and democratic society, the founders considered it fitting that those assets be used "to further democratic capitalism and the preserve and promote a society of free, educated, healthy and creative individuals." The Foundation offers support to non-profit charitable and educational organizations that demonstrate leadership in furthering the original objectives, so that future generations can aspire to and achieve their full potential in a free society. Awards range from $2,500 to $50,000. There are no deadlines.
Requirements: Nonprofit 501(c)3 organizations are eligible. Funding is provided only for those organizations and projects which involve public policy research and education on issues supporting the preservation, and in some cases the restoration, or freedom and democracy in the United States, according to the principles of the Founding Fathers. Written proposals should include a summary of the project, the project budget, the amount requested, the qualifications of individuals involved, and a copy of the organization's Internal Revenue Service determination letter.
Samples: Foundation for West Hartford Schools, West Hartford, Connecticut, $2,500 - general support; Leadership Program of the Rockies, Denver, Colorado, $50,000 - general support; and Atlas Economic Research Foundation, Arlington, Virginia, $25,000 - general support.
Contact: Helen E. Krieble, President; (303) 758-3956; fax (303) 488-0068
Internet: http://www.krieble.org/grants
Sponsor: Vernon K. Krieble Foundation
1777 S Harrison Street, Suite 807
Denver, CO 80210

VFW Ladies Auxiliary Cancer Research Postdoctoral Fellowships 4640
The VFW Ladies Auxiliary offers three postdoctoral research fellowships annually so that recipient researchers can devote their full time for one year to the challenge of cancer research. The fellowship is not renewable. The grant is not to be used for lab equipment or supplies; it is a stipend for personal living expenses of the accredited researcher. The application must be filed on appropriate application form and must describe the proposed research project that the nominee will be conducting, including abstract and brief outline of methods, procedures, and significance. The proposed research must be the principal professional responsibility of the recipient during the stated period of the award. The application must be accompanied by a letter of recommendation.
Requirements: Nominees must be U.S. citizens who are qualified in a specific cancer-related specialty and must have earned a doctoral level degree by the time the fellowship is activated (July 1). Colleges and research institutions are invited to nominate researchers; institutions may nominate more than one person.
Date(s) Application is Due: Mar 1
Amount of Grant: 50,000 U.S.D
Contact: Judy Millick, Program Contact; (816) 561-3655; fax (816) 931-4753; jmillick@vfw.org or info@ladiesauxvfw.com
Internet: http://www.ladiesauxvfw.com/cancer%20fellowships.html
Sponsor: Veterans of Foreign Wars Ladies Auxiliary
406 W 34th Street
Kansas City, MO 64111

VHA Health Foundation Grants 4641
The Foundation has dedicated its resources to create a legacy strategy, which will build on its vision of the last decade by focusing on two critical issues that will significantly improve the delivery of health care: patient safety and disaster relief. Programs submitted for funding consideration must represent a new and innovative approach to hospital emergency preparedness and be beyond the concept stage - ready to test, implement, expand, or diffuse. Innovations should be either transformational (i.e. involve significant change in current operations) and/or disruptive (i.e., have profound impact on services and economics) in nature.
Requirements: Unsolicited inquires may be submitted at any time during the year. The Letter of Inquiry is the required first step in the process. Eligible programs should clearly incorporate four key elements: Innovation, Impact, Replicability and Sustainability. Grants are available to U.S. health care providers, including hospitals, health care

systems, clinics, and medical practices with IRS tax-exempt 501(c)3 status. Local partnerships are encouraged to apply, but, since the hospital is the foundation's primary audience, the health care provider must serve as fiduciary agent and play an integral role in the program. An organization may submit multiple inquiries for consideration. Previously funded organizations may apply. Applicants are required to financially invest in the program with cash from any source and/or through in-kind contributions specific to the grant period. The minimum level of the investment is equal to one-half of the funding requested from the foundation.
Restrictions: The foundation will not consider requests for seed money or planning grants; primary research and development of technology; construction or renovation; indirect costs or overhead; endowments; political activities or attempts to influence specific legislation; individual scholarships or tuition assistance; or programs outside the United States.
Amount of Grant: 100,000 - 250,000 U.S.D
Samples: MeritCare Health System, Fargo, ND and Medcenter One Health System, Bismarck, ND, $62,500—disaster relief, North Dakota flooding; Arkansas Methodist Medical Center, Paragould, AR, St. Bernard's Regional Medical Center, Jonesboro, AR, and Washington Regional Medical Center, Fayetteville, AR, $170,850—disaster relief, Arkansas winter storm.
Contact: Michael Regier, Senior Vice President; (877) 847-1450 or (972) 830-0422; fax (972) 830-0332; vhahealthfoundation@vha.com
Internet: https://www.vhafoundation.org/portal/server.pt?open=512&objID=1204&PageID=0&cached=true&mode=2&userID=504686
Sponsor: VHA Health Foundation
220 Easr Las Colinas Boulevard
Irving, TX 75039-5500

Victor E. Speas Foundation Grants **4642**

The Victor E. Speas Foundation was established in 1947 to provide medical care for the needy, to further medical research, and to support and promote quality educational, cultural, human-services, and health-care programming. In the area of arts, culture, and humanities, the foundation supports programming that: fosters the enjoyment and appreciation of the visual and performing arts; strengthens humanities and arts-related education programs; provides affordable access; enhances artistic elements in communities; and nurtures a new generation of artists. In the area of education, the foundation supports programming that: promotes effective teaching; improves the academic achievement of, or expands educational opportunities for, disadvantaged students; improves governance and management; strengthens nonprofit organizations, school leadership, and teaching; and bolsters strategic initiatives of area colleges and universities. In the area of health, the foundation supports programming that improves the delivery of health care to the indigent, uninsured, and other vulnerable populations and addresses health and health-care problems that intersect with social factors. In the area of human services, the foundation funds programming that: strengthens agencies that deliver critical human services and maintains the community's safety net and helps agencies respond to federal, state, and local public policy changes. In the area of community improvement, the foundation funds capacity-building and infrastructure-development projects including: assessments, planning, and implementation of technology for management and programmatic functions within an organization; technical assistance on wide-ranging topics, including grant writing, strategic planning, financial management services, business development, board and volunteer management, and marketing; and mergers, affiliations, or other restructuring efforts. Grant requests for general operating support and program support will be considered. Grants from the foundation are one year in duration. Application materials are available for download at the grant website. Applicants are strongly encouraged to review the state application guidelines for additional helpful information and clarifications before applying. Applicants are also encouraged to review the foundation's funding history (link is available from the grant website). There are no application deadlines for the Victor E. Speas Foundation. Proposals are reviewed on an ongoing basis.
Requirements: Applicants must have 501(c)3 tax-exempt status and serve the residents of Kansas City, Missouri. Applications must be mailed.
Restrictions: Grant requests for capital support will not be considered. The trust does not support requests from individuals, organizations attempting to influence policy through direct lobbying, or any political campaigns.
Geographic Focus: Missouri
Samples: Childrens Mercy Hospital Foundation, Kansas City, Missouri, $100,000, to fund research and clinical trials of reformulated adult drugs to benefit pediatric patients with rare and neglected blood and other cancers (2010); Open Options, Kansas City, Missouri, $25,000, United Cerebral Palsy of Greater Kansas City technology upgrade (2010); Metropolitan Community Colleges Foundation, Kansas City, Missouri, $50,000, for medical equipment for the Health Science Institute training facility (2010).
Contact: Spence Heddens; (816) 292-4301; Spence.heddens@baml.com
Internet: https://www.bankofamerica.com/philanthropic/fn_search.action
Sponsor: Victor E. Speas Foundation
1200 Main Street, 14th Floor, P.O. Box 219119
Kansas City, MO 64121-9119

Viets Medical Student Myasthenia Gravis Research Fellowships **4643**

This fellowship is for the purpose of initiating and furthering the interests of potential investigators from colleges and medical schools in the area of causative or treatment research in myasthenia gravis (MG), including clinical investigations as well as laboratory studies concerned with research on MG, underlying mechanisms of neuromuscular transmission, and the pathogenesis of MG. A qualified candidate desiring training in MG-related research should first find a scientist with an established research program who will agree to supervise her/his training and the research to be carried out. The research can be basic or clinical and conducted in the student's own medical school or in one close to the student's residence. The number of fellowships awarded each year is limited by the amount of funds available. Application guidelines are available online.
Requirements: Submit eight copies of the following: cover letter, summary of the research proposal and its association to Myasthenia Gravis or related neuromuscular conditions, curriculum vita of applicant and sponsoring preceptor, and letter of recommendation from preceptor that indicates acceptance of the candidate and outlines the proposed work plan for the research study.
Date(s) Application is Due: Mar 15
Amount of Grant: 3,000 U.S.D
Samples: Dan Chao, Stanford U, under the direction of Kang Shen, MD, PhD (CA)—for Elucidating Molecular Mechanisms of Synapse Specificity, $3000; Sidharth Ghosh, U of Texas Medical Branch at Galveston, under the direction of Premkumar Christadoss, MD (TX)—for Role of NK Cells in Secondary Immune Response to Acetylcholine Receptor in Experimental Autoimmune Myasthenia Gravis, $3000; Joshua Mozes, U of California at Davis, under the direction of David Richman, MD (CA)—for Efficient Production of Recombinant Anti-Acetylcholine Receptor Monoclonal Antibodies, $3000.
Contact: Fellowships Administrator; (800) 541-5454 or (651) 917-6256; fax (651) 917-1835; mgfa@myasthenia.org
Internet: http://www.myasthenia.org/research
Sponsor: Myasthenia Gravis Foundation of America
1821 University Avenue W, Suite S256
St. Paul, MN 55104

Vigneron Memorial Fund Grants **4644**

The Vigneron Memorial Fund was established in 1959 to support charitable organizations that work to improve the lives of physically disabled children and adults. Preference is given to charitable organizations that serve the people of the city of Providence or the town of Narragansett, Rhode Island. Capital requests that fund handicapped assistive devices (wheelchairs, walkers, etc.) or adaptive equipment (lift installation, ramp installation, etc.) are strongly encouraged. Grant requests for general operating or program support will also be considered. The majority of grants from the Vigneron Memorial Fund are one year in duration. Applicants must apply online at the grant website. Applicants are strongly encouraged to do the following before applying: review the downloadable state application procedures for additional helpful information and clarifications; review the downloadable online-application guidelines at the grant website; review the foundation's funding history (link is available from the grant website); review the online application questions in advance; and review the list of required attachments. These will generally include: a list of board members, financial statements (audited, reviewed, or compiled by independent auditor); an organization summary; a list of other funding sources; an IRS Determination letter; and other required documents. All attachments must be uploaded in the online application as PDF, Word, or Excel files. The Vigneron Memorial Fund shares a mission and grantmaking focus with the John D. & Katherine A. Johnston Foundation. Both foundations have the same proposal deadline date of 11:59 p.m. on April 1. Applicants will be notified of grant decisions before May 31.
Requirements: Applicants must have 501(c)3 tax-exempt status.
Restrictions: The fund does not support requests from individuals, organizations attempting to influence policy through direct lobbying, or any political campaigns.
Geographic Focus: Rhode Island
Date(s) Application is Due: Apr 1
Samples: Homefront Health Care, Providence, Rhode Island, $10,000, general operations (2010); Nickerson Community Center, Providence, Rhode Island, $5,000, for disability access ramp for the main entrance (2010); WaterFire Providence, Providence, Rhode Island, WaterFire Access Program (a highly successful outreach program for children and adults with disabilities and/or mobility restrictions) (2010).
Contact: Emma Greene, Director; (617) 434-0329; emma.m.greene@baml.com
Internet: https://www.bankofamerica.com/philanthropic/fn_search.action
Sponsor: Vigneron Memorial Fund
225 Franklin Street, 4th Floor, MA1-225-04-02
Boston, MA 02110

Viking Children's Fund Grants **4645**

The fund supports the needs of children in Minnesota and the upper Midwest. Grants support research at the University of Minnesota Department of Pediatrics into the major diseases and disorders of childhood; and nonprofits that focus efforts in the areas of health education and family services supporting the well-being of children in the upper Midwest.
Requirements: Nonprofit organizations in Minnesota and the upper Midwest, including Iowa, North Dakota, South Dakota, and Wisconsin, are eligible.
Geographic Focus: Iowa, Minnesota, North Dakota, South Dakota, Wisconsin
Date(s) Application is Due: Apr 1
Amount of Grant: 500 - 5,000 U.S.D
Contact: Patrick Leopold, Viking Children's Fund Coordinator; (952) 828-6500; info@vikings.com
Internet: http://www.vikings.com/commchildfund.html
Sponsor: Minnesota Vikings Football Club
9520 Viking Drive
Eden Prairie, MN 55344

Viola W. Bernard Foundation Grants **4646**

The foundation supports innovative programs that address the interplay between social conditions and the psychological health of children and families. Grants are generally limited to projects or programs for children and young people with a mental health

component. Recognizing that new social situations create new psychological needs, the foundation anticipates grant requests for mental health programs that evolve out of emerging issues and changing needs, requiring innovative approaches (i.e., the transition from welfare to work, from dependency to independence, from delinquency and rebellion to cooperation). Opportunities for grant support continuously emerge around foci of social conflict that often involve the police, school support, and social agencies. Only one proposal from an organizational entity will be funded at a given time. There is a preference for projects and programs in the New York City tri-state area, where site visits are practicable if deemed desirable. Applications may be submitted throughout the year and are reviewed biannually, in the fall and spring.
Restrictions: Applications from individuals will not generally be considered, although in rare cases an exception may be granted. Faxed and email applications will not be accepted.
Amount of Grant: Up to 200,000 U.S.D
Contact: Dr. Perry Ottenberg
Internet: http://www.violawbernardfoundation.org
Sponsor: Viola W. Bernard Foundation
210 W Washington Square, Suite 750
Philadelphia, PA 19106

Virginia Department of Health Mary Marshall Nursing Scholarships for Licensed Practical Nurses **4647**
The Mary Marshall Nursing Scholarships for Licensed Practical Nurses are funded by the Virginia General Assembly along with a donation from The Board of Nursing. The scholarship amount is dependent upon amount of scholarship source and the number of qualified applicants. It is important that all applicants fully understand the conditions of accepting a Mary Marshall Nursing Scholarship. These awards are not gifts. Scholarship recipients must agree to engage in full time nursing in Virginia for one month for every $100 received.
Requirements: To be considered for a Mary Marshall Nursing Scholarship, an applicant must meet the following criteria: residency in Virginia for at least one year; acceptance or enrollment as a full-time or part-time student in a practical school of nursing in the state of Virginia; and have submitted a completed application form and a recommendation from the Program Director regarding scholastic attainment and financial need prior to June 30. Applications and guidelines are available online from May 1 to June 30 every year. Applications must be typed, printed and mailed (with original signatures) to the: Virginia Department of Health, Office of Minority Health and Public Health Policy, ATTN: Nursing Scholarship, 109 Governor Street, Suite 1016, East Richmond, Virginia 23219.
Restrictions: The award recipient has 60 days from the date of graduation to obtain his/her license. Full time employment must begin within 90 days of the recipient's license date. Voluntary military service, even if stationed in Virginia, cannot be used to repay scholarship awards.
Geographic Focus: Virginia
Date(s) Application is Due: Jun 30
Contact: Aileen Harris, Health Care Workforce Manager; (804) 864-7436; fax (804) 864-7440; Aileen.Harris@vdh.virginia.gov
Internet: http://www.vdh.virginia.gov/healthpolicy/primarycare/incentives/nursing/lpn_guidelines.htm
Sponsor: Virginia Department of Health
P.O. Box 2448
Richmond, VA 23218

Virginia Department of Health Mary Marshall Nursing Scholarships for Registered Nurses **4648**
The Mary Marshall Nursing Scholarships are funded by the Virginia General Assembly along with a donation from The Board of Nursing. The scholarship amount is dependent upon amount of scholarship source and the number of qualified applicants. It is important that all applicants fully understand the conditions of accepting a Mary Marshall Nursing Scholarship. These awards are not gifts. Scholarship recipients must agree to engage in full time nursing in Virginia for one month for every $100 received.
Requirements: To be considered for a Mary Marshall Nursing Scholarship, an applicant must meet the following criteria: residency in Virginia for at least one year; acceptance or enrollment as a full time or part time student in a school of nursing in the state of Virginia; demonstration of a cumulative grade point average of at least 3.0 in required courses, not electives; demonstration of financial need, verified by the Financial Aid Office/authorized person at the applicant's nursing school; and have submitted a completed application form and an official grade transcript. Applications and guidelines are available online from May 1 to June 30 every year. Applications must be typed, printed and mailed (with original signatures) to the: Virginia Department of Health, Office of Minority Health and Public Health Policy, ATTN: Nursing Scholarship, 109 Governor Street, Suite 1016, East Richmond, Virginia 23219.
Restrictions: The award recipient has 60 days from the date of graduation to obtain his/her license. Full time employment must begin within 90 days of the recipient's license date. Voluntary military service, even if stationed in Virginia, cannot be used to repay scholarship awards.
Geographic Focus: Virginia
Date(s) Application is Due: Jun 30
Contact: Aileen Harris, Health Care Workforce Manager; (804) 864-7436; fax (804) 864-7440; Aileen.Harris@vdh.virginia.gov
Internet: http://www.vdh.virginia.gov/healthpolicy/primarycare/incentives/nursing/rn_guidelines.htm
Sponsor: Virginia Department of Health
P.O. Box 2448
Richmond, VA 23218

Virginia Department of Health Nurse Practitioner/Nurse Midwife Scholarships **4649**
The program awards nurse practitioner/nurse midwife scholarships on a competitive basis. Considerations for award selections include scholastic achievement; character; and stated commitment to postgraduate employment in a medically undeserved area of Virginia, in an employment setting that provides services to persons who are unable to pay for the service and participates in all government-sponsored insurance programs designed to assure access to medical care services for covered persons. Student recipients must agree to engage in full-time practice in a designated medically underserved area for a period of years equal to the number of annual scholarships received. Preference will be given to residents of the commonwealth; minority students; students enrolled in family practice, obstetrics and gynecology, pediatric, adult health, and geriatric nurse practitioner programs; and residents of medically undeserved areas of Virginia, as determined by the board of health, in accordance with the provisions of its regulations for that purpose. The application period for this scholarship is May 1st - June 30th.
Requirements: The applicant must meet the following criteria: residency in Virginia for at least one year; acceptance or enrollment as a full-time student in a nurse practitioner/nurse midwifery program in Virginia or a nurse midwifery program in a nearby state; demonstration of a cumulative grade- point average of at least 3.0 in graduate and/or undergraduate courses; have submitted a completed application form, official grade transcript of graduate and/or undergraduate courses, and a statement of intent to practice as a nurse practitioner/nurse midwife in an undeserved area of Virginia following graduation; submission of two reference letters; and submission of all materials to the Virginia Department of Health, Office of Minority Health and Public Health Policy, ATTN: Nursing Scholarship, 109 Governor Street, Suite 1016, East Richmond, Virginia 23219
Geographic Focus: Virginia
Date(s) Application is Due: Jun 30
Contact: Aileen Harris, Health Care Workforce Manager; (804) 864-7436; fax (804) 864-7440; Aileen.Harris@vdh.virginia.gov
Internet: http://www.vdh.virginia.gov/healthpolicy/primarycare/incentives/nursing/index.htm
Sponsor: Virginia Department of Health
P.O. Box 2448
Richmond, VA 23218

Virginia Environmental Endowment Kanawha and Ohio River Valleys Program **4650**
Currently limited to the Kanawha and Ohio River Valleys of Kentucky and West Virginia, this program supports research, education, and community action on water quality and the effects of water pollution on public health and the environment. Emphasis is given to collaboration by citizens, government, and industry to develop effective public policies and to increasing public participation in water quality monitoring and watershed management.
Requirements: Constructive, result-oriented projects that are conducted by existing nonprofit organizations are funded. Activities that unite business, government, and civic interests for environmental improvement are encouraged, as are projects that serve as models for other communities. Organizations must be in Kentucky or West Virginia.
Geographic Focus: Kentucky, West Virginia
Date(s) Application is Due: Jun 15
Samples: Kentucky Waterways Alliance (Munfordville, KY)—for a statewide citizens' group that works to monitor and protect water quality and to raise public awareness about related issues, $40,000 over two years; West Virginia Rivers Coalition (Elkins, WV)—to develop scientifically sound criteria for determining the level of nutrients in West Virginia rivers, $17,200.
Contact: Gerald McCarthy, (804) 644-5000; fax (804) 644-0603; info@vee.org
Internet: http://www.vee.org/programs.cfm
Sponsor: Virginia Environmental Endowment
1051 E Cary Street, 3 James Center, Suite 1400, P.O. Box 790
Richmond, VA 23218-0790

Virginia G. Piper Charitable Trust Grants **4651**
The trust awards grants to nonprofit organizations, primarily serving residents of Maricopa County, AZ, that have actively operated as a Section 501(c)3 organization for at least three years, in its areas of interest, including early childhood, youth, and arts and culture. The trust seeks proposals that link health, medical research and/or educational programs specifically to the needs of children, youth, and the elderly, as well as faith-based organizations that can use their considerable capacity and presence to address needs in these areas. Types of funding support include program grants, capital grants, endowment grants, challenge grants, and single- and multi-year grants. There are no application forms. Letters of inquiry can be submitted at any time.
Requirements: Arizona nonprofits, primarily in Maricopa County, are eligible.
Restrictions: Individuals are not eligible.
Geographic Focus: Arizona
Amount of Grant: 5,000 - 2,000,000 U.S.D
Samples: East Valley Children's Theatre (Mesa, AZ)—for marketing activities related to the theater's move to the new Mesa Arts Center, $12,000; Saint Mary's Food Bank (Phoenix, AZ)—to enable three local food banks to gather and distribute food donated by a chain of grocery stores, $225,000; Mount Claret Retreat Ctr (Phoenix, AZ)—to renovate and construct facilities at this small retreat and conference center owned by the Diocese of Phoenix, $250,000; Mesa Community College (AZ) and Scottsdale Community College (AZ)—to establish centers at which older adults can learn about health and physical fitness, as well as educational, paid, and volunteer opportunities, $533,946 over three years, and $347,010 over two years, respectively.
Contact: Karin Bishop, Grants Manager; (480) 948-5853; fax (408) 348-1316; info@pipertrust.org
Internet: http://www.pipertrust.org

Sponsor: Virginia G. Piper Charitable Trust
6720 N Scottsdale Road, Suite 350
Scottsdale, AZ 85253

Virginia Health Care Foundation Grants **4652**
The goal of the foundation is to increase access to primary care for Virginia's uninsured or for those in areas with no providers. Funded programs might include HIV/AIDS, substance abuse, and mental health, but programs must be connected to overall primary care. Proposals providing more than one service receive preference. The foundation seeks to match state funds from corporations. Listed deadlines are for concept papers; full proposals are by invitation. Annual deadline dates may vary; see web site for exact dates.
Requirements: Virginia health providers or health facilities may apply to one of four funding tracks.
Geographic Focus: Virginia
Date(s) Application is Due: Jan 29; Aug 9
Amount of Grant: 1,000 - 204,000 U.S.D
Contact: Elaine Summerfield, (804) 828-5804; fax (804) 828-4370; info@vhcf.org
Internet: http://www.vhcf.org
Sponsor: Virginia Health Care Foundation
1001 E Broad Street, Suite 445
Richmond, VA 23219

Virginia Tuition Assistance Grants **4653**
Grants are awarded to any full-time undergraduate or graduate (including medical or law) student, regardless of financial need, who is a domiciliary resident of Virginia enrolled at an eligible private college or university in Virginia. Completed applications must be submitted to the financial aid office of the institution on, or before, July 31 prior to the fall semester of enrollment. Applications submitted after July 31, but no later than December 1, will be considered for an award only if funds are available.
Requirements: Applicant must be a domiciliary resident of Virginia; enrolled full-time as an undergraduate, graduate, or professional student; and enrolled at an eligible institution pursuing an eligible degree that is not in religious training or theological education.
Geographic Focus: Virginia
Date(s) Application is Due: Jul 31; Dec 1
Amount of Grant: 1,900 - 2,500 U.S.D
Contact: Elizabeth Waddy, Student Aid Program Specialist; (804) 225-2632 or (804) 225-2600; fax (804) 225-2604; ElizabethWaddy@schev.edu
Internet: http://www.schev.edu/students/factsheetvtag.asp?from=schevs
Sponsor: State Council of Higher Education for Virginia
101 N 14th Street
Richmond, VA 23219

Virginia W. Kettering Foundation Grants **4654**
The Virginia W. Kettering Foundation, a private family foundation in Dayton, Ohio, was activated in 2003 when Virginia Kettering passed away at the age of 95. To continue her legacy of confident determination and passionate commitment to the Dayton area, Virginia Kettering directed the stewards of her foundation to support charitable organizations for charitable purposes within Montgomery, Greene, Clark, Miami, Darke, Preble, Butler and Warren Counties.
Requirements: 501(c)3 organizations located in, or that serve, the counties of Montgomery, Greene, Clark, Miami, Darke, Preble, Butler and Warren in Ohio.
Geographic Focus: Ohio
Date(s) Application is Due: Mar 1; Sep 1
Amount of Grant: 5,000 - 50,000 U.S.D
Contact: Judith M. Thompson, Executive Director; (973) 228-1021; fax (973) 228-2399; info@ketteringfund.org
Internet: http://vwketteringfoundation.org/main.html
Sponsor: Virginia W. Kettering Foundation
1480 Kettering Tower
Dayton, OH 45423

Visiting Nurse Foundation Grants **4655**
The foundation awards grants in support of home- and community-based health care for the medically underserved in Cook and the collar counties of Lake, McHenry, DuPage, Kane and Will, with a focus on Chicago. Priority areas are home health care services; prevention and health promotion; and early intervention. Preference is given to programs using nurses to provide the care. Types of support include program, capital, and general operating grants. Letters of intent should be submitted by deadline dates listed, and full proposals will be by invitation only.
Requirements: Nonprofit organizations in Cook, Lake, McHenry, DuPage, Kane, and Will counties of Illinois are eligible.
Restrictions: No grants are made to individuals.
Geographic Focus: Illinois
Date(s) Application is Due: Jan 19; Apr 19; Oct 20
Amount of Grant: 15,000 - 80,000 U.S.D
Contact: Robert N. DiLeonardi, Executive Director; (312) 214-1521; fax (312) 214-1529; info@vnafoundation.net
Internet: http://www.vnafoundation.net
Sponsor: Visiting Nurse Association Foundation
20 N Wacker Drive, Suite 3118
Chicago, IL 60606

Volkswagen of America Corporate Contributions Grants **4656**
The corporation accepts proposals broadly from programs servicing needs in healthcare, wellness, children, medical research, education and diversity.
Requirements: There are no specific deadlines, however the review committee does meet quarterly to go over proposals received. The committee considers the type of organization, its mission and the intended use of the funds.
Restrictions: Grants are available to 501(c)3 organizations only.
Amount of Grant: 1,000 - 2,500 U.S.D
Contact: Tara Jones, (248) 754-4693; tara.jones@vw.com
Sponsor: Volkswagen of America
3800 Hamlin Road
Auburn Hills, MI 48326

Vulcan Materials Company Foundation Grants **4657**
The company foundation supports nonprofit organizations in company-operating areas. Grants support a broad array of programs and projects in the areas of health and welfare, community and economic development, education (including an employee matching grants program), the environment, law and justice, women and children, arts and humanities, minorities, and literacy. Requests will be considered for support of general operating expenses, capital and annual campaigns, scholarship funds, research, continuing support, and seed money. There are no application deadlines.
Requirements: The foundation awards grants only to public charities and units of government, such as public schools and parks. A public charity is any Section 501(c)3 charitable organization that is not a private foundation because it meets one of the three internal Revenue Code definitions: 509(a)1, 509(a)2 or 509(a)3.
Restrictions: The foundation does not fund individuals; organizations outside the United States; telephone or mass-mail appeals, political organizations; testimonial dinners; sectarian religious activities; organizations which have discriminatory practices; or athletic, labor, fraternal and veterans associations. The Foundation generally will not consider requests from organizations located in communities where Vulcan has no operations, offices or employees.
Geographic Focus: Alabama, Arizona, Arkansas, California, Florida, Georgia, Illinois, Indiana, Iowa, Kansas, Kentucky, Louisiana, Maryland, Mississippi, Missouri, New Mexico, North Carolina, Pennsylvania, South Carolina, Tennessee, Texas, Virginia, Wisconsin
Amount of Grant: Up to 28,000,000 U.S.D
Contact: Program Contact; (205) 298-3222; giving@vmcmail.com
Internet: http://www.vulcanmaterials.com/social.asp?content=guidelines
Sponsor: Vulcan Materials Company Foundation
P.O. Box 385014
Birmingham, AL 35238-5014

W.C. Griffith Foundation Grants **4658**
The foundation awards grants, primarily in Indiana, to nonprofit organizations in its areas of interest, including arts and culture, cancer and medical research, children and youth services, Christian churches and organizations, community development, education, environmental programs, health organizations and hospitals, higher education, homeless services, human reproduction and fertility, library science and libraries, minorities, museums, music and performing arts, secondary education, and social services delivery. Types of support include building construction/renovation, capital campaigns, and continuing support. There are no application forms or deadlines. The board meets in June and November; proposals should be received in May and October.
Requirements: Nonprofit organizations are eligible. Preference is given to requests from Indianapolis, Indiana.
Amount of Grant: 500 - 50,000 U.S.D
Contact: Curt Farran, c/o National City Bank of Indiana; (317) 267-7262
Sponsor: W.C. Griffith Foundation
101 West Washington Street, Suite 600 East
Indianapolis, IN 46255

W. C. Griffith Foundation Grants **4659**
The W. C. Griffith Foundation was established in 1959 by a donation of William C. Griffith, and Ruth Perry Griffith. The Foundation gives primarily in Indianapolis-Carmel, Indiana area. The Foundation provides support primarily for hospitals, health associations, medical and cancer research, the arts, including music and museums, community funds and development, higher, secondary, and other education, family planning services, child welfare, the homeless, the environment, libraries, and Christian religious organizations.
Requirements: Applicants should submit the following: detailed description of project and amount of funding requested; 1 Copy of the proposal.
Geographic Focus: Indiana
Amount of Grant: 1,000 - 45,000 U.S.D
Samples: Eiteljorg Museum of American Indians and Western Art, Indianapolis, IN, $25,000; Providence Cristo Rey High School, Indianapolis, IN, $10,000;
Contact: Trust Coordinator; fax (317) 693-2504
Sponsor: W.C. Griffith Foundation
101 West Ohio Street, Suite 1450
Indianapolis, IN 46204-1998

W. Clarke Swanson, Jr. Foundation Grants **4660**
The Foundation, established in 1998, is the brain-child of W. Clarke Swanson, Jr., founder of Swanson Vineyards. The Foundation is primarily interested in supporting the arts, education, and health care in California, there is some national giving. The type of funding typically comes in the form of general operating support. Although there is no specific application form required, applicants should submit a detailed description of the project and amount of funding requested.

Amount of Grant: 3,000 - 250,000 U.S.D
Samples: Friends of Lincoln Theater, Yountville, California, $172,200; University of California at San Francisco Foundation, San Francisco, California, $50,000.
Contact: W. Clarke Swanson, Director; (707) 944-0905, ext. 20; clarke@swansonvineyards.com
Sponsor: W. Clarke Swanson, Jr. Foundation
1050 Oakville Cross Road, P.O. Box 148
Oakville, CA 94562

W. Glen Boyd Charitable Foundation Grants **4661**
The foundation supports programs that make a significant difference in empowering women and children to improve the quality of their lives, and/or that eliminate barriers that prevent success for women and children. Of special interest are organizations/programs that empower single mothers and women/children/youth living in poverty, violent, or at-risk situations; provide services for immediate and critical needs; empower women and/or children to build assets and promote resiliency; work on prevention; develop greater self-respect and self-sufficiency; develop personal, educational, professional, and leadership skills; provide access to/training or mentoring for greater educational and economic development; and provide supervised educational experiences for underserved children/youth. In addition, priority is given to organizations/programs that are grassroots-based and have difficulty accessing traditional funding; provide direct service; collaborate with other organizations to enhance services or advance progressive social change/systems change; value and actively include diversity and culturally specific programs and resources; and include voices of constituents (including youth) in the process of determining program needs, solutions, and evaluation. Letters of inquiry are accepted any time; full proposals are by invitation. Board meetings are generally held three months after the letter of inquiry deadline. Guidelines and application are available online.
Requirements: Minnesota, and Oregon 501(c)3 nonprofit organizations are eligible.
Geographic Focus: Minnesota, Oregon
Date(s) Application is Due: Jan 21; May 21; Sep 20
Amount of Grant: 5,000 - 20,000 U.S.D
Samples: Parrott Creek Child and Family Services (Oregon City, OR)—to expand the Community Safety Net Program, to remodel the dormitory of the residential-treatment program, to expand a transitional program for girls, and to support the early childhood program, $10,000.
Contact: Diana Dokos, Program Director; (877) 803-8830; WGBCF@ureach.com
Internet: http://wgbcf.org/gl.html
Sponsor: W. Glen Boyd Charitable Foundation
P.O. Box 8128
Portland, OR 97207-8128

W.H. and Mary Ellen Cobb Charitable Trust Grants **4662**
The Cobb's were originally from Kentucky, but moved to the Texas Panhandle where they owned and operated several clothing stores. Although they did not have children of their own, they were interested in the well-being of children and therefore established their charitable trust to benefit local charities in the Panhandle whose mission has a strong "emphasis on helping children." This foundation makes approximately five-seven awards each year and grants are typically between $1,000 and $25,000. Applicants must apply online at the grant website. Applicants are strongly encouraged to do the following before applying: review the downloadable state application procedures for additional helpful information and clarifications; review the downloadable online-application guidelines at the grant website; review the trust's funding history (link is available from the grant website); review the online application questions in advance; and review the list of required attachments. These will generally include: a list of board members, financial statements (audited, reviewed, or compiled by independent auditor); an organization summary; a list of other funding sources; an IRS Determination letter; and other required documents. All attachments must be uploaded in the online application as PDF, Word, or Excel files. The application deadline for this trust is 11:59 p.m. on September 30.
Requirements: The W.H. & Mary Ellen Cobb Charitable Trust considers requests from charitable organizations whose primary focus is the provision of services that benefit children. Organizations must: be geographically located within the Texas Panhandle; serve residents of the Panhandle, Amarillo, and vicinty; and have 501(c)3 tax-exempt status.
Restrictions: The trust does not support requests from individuals, organizations attempting to influence policy through direct lobbying, or any political campaigns.
Geographic Focus: Texas
Date(s) Application is Due: Sep 30
Amount of Grant: 1,000 - 25,000 U.S.D
Samples: St. Andrews Episcopal School, Amarillo, Texas, $10,000 (2010); High Plains Childrens Home and Family Services, Amarillo, Texas, $7,000 (2010); Maverick Boys and Girls Club of Amarillo, Amarillo, Texas, $5,000 (2010).
Contact: Mark J. Smith, Philanthropic Relationship Manager; (817) 390-6028; tx.philanthropic@baml.com
Internet: https://www.bankofamerica.com/philanthropic/fn_search.action
Sponsor: W.H. and Mary Ellen Cobb Charitable Trust
500 West 7th Street, 15th Floor, TX1-497-15-08
Fort Worth, TX 76102-4700

W. James Spicer Scholarship **4663**
The W. James Spicer Scholarship was established to recognize the teaching career of Mr. Spicer who taught in the Gary School Corporation for over 40 years. The one-time, $1,000 scholarship is awarded to a graduating senior at Wirt/Emerson School for the Visual and Performing Arts who will be pursuing higher education. All college majors are eligible.
Requirements: Applicants must be pursuing higher education, with a cumulative GPA of at least 2.0. Along with the completed application, candidates must submit a two page essay, the Scholastic Profile Form completed by their guidance counselor, an official high school transcript, a letter of recommendation from a teacher, a recommendation letter from someone other than a family member, and a letter of acceptance from the school they plan to attend.
Geographic Focus: Indiana
Date(s) Application is Due: Apr 15
Amount of Grant: 1,000 U.S.D
Contact: Spicer Scholarship Contact; (219) 736-1880; fax (219) 736-1940; legacy@legacyfoundationlakeco.org
Internet: http://www.legacyfoundationlakeco.org/scholarships/index.htm
Sponsor: Legacy Foundation
1000 East 80th Place, 302 South
Merrillville, IN 46410

W.K. Kellogg Foundation Healthy Kids Grants **4664**
The Foundation believes that children need nutrition, stimulation, healthy living conditions and access to quality health care. It helps many of them get all four, by funding organizations that improve birth outcomes and first food experiences, create access to healthy foods, improve health services, and educate families and communities about the inter-related factors that determine well-being. The Foundation especially focuses on children who are disadvantaged by multiple societal factors, a disproportionate percentage of whom are children of color.
Requirements: To be eligible for support, your organization or institution, as well as the purpose of the proposed project, must qualify under regulations of the United States Internal Revenue Service. The Kellogg Foundation does not have any submission deadlines. Grant applications are accepted throughout the year and are reviewed at its headquarters in Battle Creek, Michigan, or in its regional office in Mexico.
Restrictions: In general, the Foundation does not provide funding for: operational phases of established programs; capital requests (which includes the construction, purchase, renovation, and/or furnishing of facilities); equipment; conferences and workshops; films, television and/or radio programs; endowments; development campaigns; or research/studies (unless they are an integral part of a larger program budget being considered for funding).
Contact: Deborah A. Rey; (269) 969-2133 or (269) 968-1611; fax (269) 968-0413
Internet: http://www.wkkf.org/what-we-support/healthy-kids.aspx
Sponsor: W.K. Kellogg Foundation
1 Michigan Avenue East
Battle Creek, MI 49017-4012

W.K. Kellogg Foundation Secure Families Grants **4665**
The primary needs of the family must be addressed to create pathways out of poverty for children. The Foundation supports programs that increase family stability, foster quality jobs, careers and entrepreneurship, and promote secondary achievement and financial independence. It further supports strategies that increase income, assets, and aspirations of vulnerable children and their families and reduce disparities based on class, gender and race. The Foundation focuses on three areas, each with a direct impact on the ability of families to generate income and accumulate assets: family stability; career ladders; and financial independence.
Requirements: To be eligible for support, your organization or institution, as well as the purpose of the proposed project, must qualify under regulations of the United States Internal Revenue Service. The Kellogg Foundation does not have any submission deadlines. Grant applications are accepted throughout the year and are reviewed at its headquarters in Battle Creek, Michigan, or in its regional office in Mexico.
Restrictions: In general, the Foundation does not provide funding for: operational phases of established programs; capital requests (which includes the construction, purchase, renovation, and/or furnishing of facilities); equipment; conferences and workshops; films, television and/or radio programs; endowments; development campaigns; or research/studies (unless they are an integral part of a larger program budget being considered for funding).
Contact: Deborah A. Rey; (269) 969-2133 or (269) 968-1611; fax (269) 968-0413
Internet: http://www.wkkf.org/what-we-support/secure-families.aspx
Sponsor: W.K. Kellogg Foundation
1 Michigan Avenue East
Battle Creek, MI 49017-4012

W.M. Keck Foundation Medical Research Grants **4666**
The W.M. Keck Foundation concentrates on strengthening studies and programs in accredited universities and colleges and major, independent medical research institutions. Medical Research grants seek to advance the frontiers of the life sciences by supporting basic research that is high-risk and has the potential to transform its field. Successful projects are distinctive and novel in their approach to problems, push the edge of their field, or question the prevailing paradigm. Past grants have been awarded to major universities, medical schools, and independent research institutions to support pioneering biological research, including the development of new technologies, instrumentation, or methodologies. Historically, grants range from $500,000 to $5 million, and are typically $2 million or less. Applicants are strongly urged to contact Foundation staff well in advance of submitting a Phase I Application. The best times for these contacts are between January 1 and February 15 leading up to a May 1 submittal, or between July 1 and August 15 leading up to a November 1 submittal.
Requirements: Research universities, medical colleges and major private independent scientific and medical research institutes are eligible. Organizations must be exempt from federal taxation as defined by Section 501(c)3 of the Internal Revenue Code and be designated as a public charity (and not a private foundation) as defined by Section 509(a)1

or 509(a)2 or 170(b) of the Internal Revenue Code. If the institution is located in the State of California, the organization must also be exempt from California State Franchise or Income Tax under Section 23701(d) of the Revenue and Taxation Code.
Restrictions: Grants are not considered for: general operating expenses, endowments or deficit reduction; general and federated campaigns, including fundraising events, dinners or mass mailings; individuals; conference or seminar sponsorship; book publication and film or theater productions; public policy research; institutions that are located outside the United States; institutions that do not have at least three consecutive full, certified, audited financial statements; conduit organizations, unified funds or organizations that use grant funds from donors to support other organizations or individuals; or institutions that are subsidiaries or affiliates of larger entities that do not have a separate board of directors and independent audited financial statements.
Date(s) Application is Due: May 1; Nov 1
Amount of Grant: 500,000 - 5,000,000 U.S.D
Samples: Dirk Englund, Jonathan Owen, & Rafael Yuste, Columbia University, New York, New York, $1,000,000 - to develop a novel method for direct, real-time imaging of neuronal voltage signals (2011); Laura Ranum & Maurice Swanson, University of Florida, Gainesville, Florida, $1,000,000 - to pursue critical questions: How does RAN translation work? Is RAN translation a key, previously unrecognized cellular process? Are repetitive sequences throughout the genome translated into proteins and if so, what is their function? (2011).
Contact: Margie Antonetti, Grants and Database Specialist; (213) 680-3833; fax (213) 614-0934; mantonetti@wmkeck.org or MedRsch@wmkeck.org
Internet: http://www.wmkeck.org/grant-programs/medicalresearch.html
Sponsor: W.M. Keck Foundation
550 S Hope Street, Suite 2500
Los Angeles, CA 90071

W.M. Keck Foundation Science and Engineering Research Grants **4667**
The W.M. Keck Foundation concentrates on strengthening studies and programs in accredited universities and colleges and major, independent medical research institutions. Science and Engineering Research grants seek to benefit humanity by supporting projects that are distinctive and novel in their approach, question the prevailing paradigm, or have the potential to break open new territory in their field. Past grants have been awarded to major universities and independent research institutions to support pioneering science and engineering research, the development of new technologies, and to facilitate the purchase of advanced instruments where such instruments would further specific research ventures. Historically, grants range from $500,000 to $5 million, and are typically $2 million or less. Applicants are strongly urged to contact Foundation staff well in advance of submitting a Phase I Application. The best times for these contacts are between January 1 and February 15 leading up to a May 1 submittal, or between July 1 and August 15 leading up to a November 1 submittal.
Requirements: Research universities, medical colleges and major private independent scientific and medical research institutes are eligible. Organizations must be exempt from federal taxation as defined by Section 501(c)3 of the Internal Revenue Code and be designated as a public charity (and not a private foundation) as defined by Section 509(a)1 or 509(a)2 or 170(b) of the Internal Revenue Code. If the institution is located in the State of California, the organization must also be exempt from California State Franchise or Income Tax under Section 23701(d) of the Revenue and Taxation Code.
Restrictions: Grants are not considered for: general operating expenses, endowments or deficit reduction; general and federated campaigns, including fundraising events, dinners or mass mailings; individuals; conference or seminar sponsorship; book publication and film or theater productions; public policy research; institutions that are located outside the United States; institutions that do not have at least three consecutive full, certified, audited financial statements; conduit organizations, unified funds or organizations that use grant funds from donors to support other organizations or individuals; or institutions that are subsidiaries or affiliates of larger entities that do not have a separate board of directors and independent audited financial statements.
Date(s) Application is Due: May 1; Nov 1
Amount of Grant: 500,000 - 5,000,000 U.S.D
Samples: Alexander Fridman, Drexel University, Philadelphia, Pennsylvania, $1,000,000 - to develop a fundamental understanding of bio-molecular interaction pathways between strongly non-equilibrium electrical plasma and living cells (2011); Jules S. Jaffe, University of California, San Diego, California, $1,000,000 - to build the world's first high resolution, 3D, in situ, underwater microscope (2011).
Contact: Margie Antonetti, Grants and Database Specialist; (213) 680-3833; fax (213) 614-0934; mantonetti@wmkeck.org or SciEng@wmkeck.org
Internet: http://www.wmkeck.org/grant-programs/science-engineering.html
Sponsor: W.M. Keck Foundation
550 S Hope Street, Suite 2500
Los Angeles, CA 90071

W.M. Keck Foundation Southern California Grants **4668**
The Southern California program seeks to promote the education and healthy development of children and youth, strengthening families and enhancing the lives of people in the Greater Los Angeles area through its support of organizations that provide arts and cultural enrichment, civic and community services, early childhood and pre-college education, and health care. A special emphasis is placed on projects that focus on children and youth from low-income families, special needs populations, and safety-net services. Collaborative initiatives, as well as projects arising from the vision of one organization's strong leadership, are supported. Historically, grants range from $100,000 to $1 million, and typically are under $500,000. Applicants are strongly urged to contact Foundation staff well in advance of submitting a Phase I Application. The best times for these contacts are between January 1 and February 15 leading up to a May 1 submittal, or between July 1 and August 15 leading up to a November 1 submittal.
Requirements: Nonprofit organizations, including colleges and universities, pursuing relevant projects in Los Angeles County are eligible to apply.
Restrictions: Grants are not considered for: general operating expenses, endowments or deficit reduction; general and federated campaigns, including fundraising events, dinners or mass mailings; individuals; conference or seminar sponsorship; book publication and film or theater productions; public policy research; institutions that are located outside the United States; institutions that do not have at least three consecutive full, certified, audited financial statements; conduit organizations, unified funds or organizations that use grant funds from donors to support other organizations or individuals; or institutions that are subsidiaries or affiliates of larger entities that do not have a separate board of directors and independent audited financial statements.
Geographic Focus: California
Date(s) Application is Due: May 1; Nov 1
Amount of Grant: 100,000 - 1,000,000 U.S.D
Samples: Alliance for a Better Community, Los Angeles, California, $200,000 - to build the capacity of parents to participate in various forms of civic learning and engagement in order to support their student's achievement (2011); Alliance for Children's Rights, Los Angeles, California, $250,000 - o ensure that implementation of the California Fostering Connections to Success Act (2011).
Contact: Margie Antonetti, Grants and Database Specialist; (213) 680-3833; fax (213) 614-0934; mantonetti@wmkeck.org or SoCal@wmkeck.org
Internet: http://www.wmkeck.org/grant-programs/southern-california-program.html
Sponsor: W.M. Keck Foundation
550 S Hope Street, Suite 2500
Los Angeles, CA 90071

W.M. Keck Foundation Undergraduate Education Grants **4669**
The W.M. Keck Foundation concentrates on strengthening studies and programs in accredited universities and colleges and major, independent medical research institutions. The Foundation's Undergraduate Education program promotes distinctive learning and research experiences in science, engineering, and the liberal arts at four-year undergraduate colleges only in Foundation designated states, or through national organizations that address undergraduate needs. Historically, grants range from $200,000 to $1 million, and are typically under $500,000. On a case-by-case basis, the Foundation may consider the award of smaller seed grants for planning and piloting. Applicants are strongly urged to contact Foundation staff well in advance of submitting a Phase I Application. The best times for these contacts are between January 1 and February 15 leading up to a May 1 submittal, or between July 1 and August 15 leading up to a November 1 submittal.
Requirements: Public institutions and research universities located in the designated states of Alaska, Arizona, California, Colorado, Hawaii, Idaho, Kansas, Louisiana, Mississippi, Montana, Nevada, New Mexico, Oklahoma, Oregon, Texas, Utah, Washington, and Wyoming may apply.
Restrictions: Grants are not considered for: general operating expenses, endowments or deficit reduction; general and federated campaigns, including fundraising events, dinners or mass mailings; individuals; conference or seminar sponsorship; book publication and film or theater productions; public policy research; institutions that are located outside the United States; institutions that do not have at least three consecutive full, certified, audited financial statements; conduit organizations, unified funds or organizations that use grant funds from donors to support other organizations or individuals; or institutions that are subsidiaries or affiliates of larger entities that do not have a separate board of directors and independent audited financial statements.
Geographic Focus: Alaska, Arizona, California, Colorado, Hawaii, Idaho, Kansas, Louisiana, Mississippi, Montana, Nevada, New Mexico, Oklahoma, Oregon, Texas, Utah, Washington, Wyoming
Date(s) Application is Due: May 1; Nov 1
Amount of Grant: 200,000 - 1,000,000 U.S.D
Contact: Margie Antonetti, Grants and Database Specialist; (213) 680-3833; fax (213) 614-0934; mantonetti@wmkeck.org or UGProgram@wmkeck.org
Internet: http://www.wmkeck.org/grant-programs/undergraduate-program.html
Sponsor: W.M. Keck Foundation
550 S Hope Street, Suite 2500
Los Angeles, CA 90071

W.P. and Bulah Luse Foundation Grants **4670**
The W.P. and Bulah Luse Foundation was established in 1947. Mr. Luse, a self-made wildcatter in the early Texas oilfields, and his wife Bulah created this foundation to support and promote quality education, human-services, and health-care programming for underserved populations. Special consideration is given to charitable organizations that serve the people of Dallas, Texas, and its surrounding communities. The majority of grants from the Luse Foundation are one year in duration. Applicants must apply online at the grant website. Applicants are strongly encouraged to do the following before applying: review the downloadable state application procedures for additional helpful information and clarifications; review the downloadable online-application guidelines at the grant website; review the foundation's funding history (link is available from the grant website); review the online application questions in advance; and review the list of required attachments. These will generally include: a list of board members, financial statements (audited, reviewed, or compiled by independent auditor); an organization summary; a list of other funding sources; an IRS Determination letter; and other required documents. All attachments must be uploaded in the online application as PDF, Word, or Excel files. The W.P. & Bulah Luse Foundation has biannual deadlines of June 30 and December 31. Applications must be submitted by 11:59 p.m. on the deadline dates. Grant applicants for

the June deadline will be notified of grant decisions by October 31. Grant applicants for the December deadline will be notified of grant decisions by April 30 of the following year.
Requirements: Applicants must have 501(c)3 tax-exempt status.
Restrictions: If an organization receives a grant from the Luse Foundation, it must skip a year before submitting a subsequent application. The foundation does not support requests from individuals, organizations attempting to influence policy through direct lobbying, or any political campaigns.
Geographic Focus: Texas
Date(s) Application is Due: Jun 30; Dec 31
Samples: Dallas Lighthouse for the Blind, Dallas, Texas, $10,000 (2010); Texas A & M Foundation, College Station, Texas, $40,000 (2010); Methodist Health System Foundation, Dallas, Texas, $15,000 (2010).
Contact: David Ross, Senior Vice President; tx.philanthropic@baml.com
Internet: https://www.bankofamerica.com/philanthropic/fn_search.action
Sponsor: W.P. and Bulah Luse Foundation
901 Main Street, 19th Floor, TX1-492-19-11
Dallas, TX 75202-3714

W.P. and H.B. White Foundation Grants **4671**
The foundation awards grants to nonprofits in the metropolitan area of Chicago, IL, for social services; early-childhood, elementary, higher, and adult education; and health. Types of support include general operating budgets, continuing support, annual campaigns, building funds, capital campaigns, and special projects. The board meets in March, June, September, and December to consider requests.
Requirements: Only Illinois organizations are eligible to apply.
Geographic Focus: Illinois
Date(s) Application is Due: Feb 1; May 1; Aug 1; Nov 1
Amount of Grant: 10,000 - 30,000 U.S.D
Samples: Austin Career Education Ctr (Chicago, IL)—for operating support, $12,000; Greater Chicago Food Depository (Chicago, IL)—for operating support, $20,000; Roosevelt U (Chicago, IL)—for program support, $10,000.
Contact: M. Margaret Blandford, Executive Director; (847) 446-1441
Sponsor: W.P. and H.B. White Foundation
540 Frontage Road, Suite 3240
Northfield, IL 60093

W.W. Smith Charitable Trust Grants **4672**
The trust awards grants regionally in its areas of interest, including basic medical research dealing with cancer, AIDS, and heart disease; scholarship programs for undergraduate students at four-year universities and colleges; and proposals that address specifically food, clothing, and shelter for children and the aged. Resources are concentrated in areas where the most need can be discerned and where government or private assistance has not been available. Types of support include capital projects, challenge/matching grants, general operating grants, program grants, and research grants. Requests for grants for food, clothing, and shelter for children and the aged are considered throughout the year. Application deadlines are June 15 for cancer and AIDS research; and September 15 for heart research. Grants may be for more than one year but not for more than three years.
Requirements: Grant recipients are limited to organizations within the five-county area of Pennsylvania including Buck, Chester, Delaware, Montgomery, and Philadelphia Counties. Organizations must be tax-exempt and not classified as private foundations or private operating foundations.
Restrictions: After three consecutive years of funding, at least two years must elapse before further applications from the organization may be considered.
Geographic Focus: Pennsylvania
Date(s) Application is Due: Jun 15; Sep 15
Amount of Grant: 15,000 - 84,000 U.S.D
Contact: Frances Pemberton, Trust Administrator; (610) 397-1844; fax (610) 397-1680
Internet: http://www.wwsmithcharitabletrust.org
Sponsor: W.W. Smith Charitable Trust
200 Four Falls Corporate Center, Suite 300
West Conshohocken, PA 19428

W.W. Smith Charitable Trust Grants for Medical Research in Cancer, AIDS, and Heart Disease **4673**
Concentration of the grant program in this area is toward basic scientific research that can help uncover and solve some of the mysteries of cancer, heart disease, and AIDS. Types of support include challenge/matching grants, general operating grants, program development grants, research grants, and scholarships. Funding is directed to projects that are new or headed by little-known investigators who are not yet qualified for large government grants. The trust also supports promising research outside the conventional pathways. Grant applications should be submitted no later than June 15 for cancer and AIDS research and no later than September 15 for cardiovascular research. Each application should be accompanied by an audited financial statement for the institution of the researcher. Projects requested should be one to three years in length and should not request funds for equipment alone. A special application form is required.
Requirements: Institutions in the general geographic area of New Jersey, eastern Pennsylvania, and Maryland may submit proposals. Specialized application form and specific guidelines should be obtained from the trust prior to submitting a proposal.
Restrictions: Only 10 percent of indirect costs are allowed on all grant proposals. Funds to support students, fellows, and postdoctoral candidates should be supported by curriculum vita. Applications should be submitted for the work of individual investigators and not for general work of research centers.
Geographic Focus: New Jersey
Date(s) Application is Due: Jun 15; Sep 15
Contact: Frances Pemberton, Trust Administrator; (610) 397-1844; fax (610) 397-1680; info@wwsmithcharitabletrust.org
Internet: http://www.wwsmithcharitabletrust.org
Sponsor: W.W. Smith Charitable Trust
200 Four Falls Corporate Center, Suite 300
West Conshohocken, PA 19428

Wabash Valley Community Foundation Grants **4674**
The foundation's mission is to promote community investment for a better tomorrow. Giving primarily for arts and culture, education, human services, community development, and religion. Each year, the Foundation provides grants to charitable organizations for projects that meet the charitable needs in the Wabash Valley. These grants respond to specific community challenges and improve the vitality of the community. The Foundation teaches the area's youth leaders by engaging them in allocating limited resources for them to grant to worthy charitable projects. Deadlines for letters of intent are as follows: Clay County—May 1; Sullivan County—May 1; Vigo County—May 1 and November 15.
Restrictions: Giving is restricted to Clay, Sullivan, and Vigo counties in Indiana. No grants are given for endowments, deficit funding, annual appeals and membership contributions, travel for groups such as bands, sports teams and classes, conferences, publications, films, television, or radio programs (unless integral to the project for which the grant is sought).
Geographic Focus: Indiana
Date(s) Application is Due: May 1; Nov 15
Contact: Beth Tevlin, Executive Director; (812) 232-2234 or (877) 232-2230; fax (812) 234-4853; beth@wvcf.com
Internet: http://www.wvcf.com/wvcf_index3.htm
Sponsor: Wabash Valley Community Foundation
2901 Ohio Boulevard, Suite 153
Terre Haute, IN 47803

Wachovia Foundation Grants **4675**
The foundation awards grants to eligible tax-exempt organizations in four primary areas: education, community and economic development, arts and culture, and health and human services. Its mission is to build strong and vibrant communities, improve the quality of life, and make a positive difference in Wachovia communities. Contributions are coordinated through regional Wachovia Community Affairs offices (list available online), in Florida; Georgia; New Jersey, New York, and Connecticut; North and South Carolina (except for Charlotte); North Carolina—Charlotte); Pennsylvania and Delaware; Virginia, Maryland, and Washington, DC; Wachovia Securities; and national requests. Guidelines and application are available online.
Requirements: Applicant organizations must have a 501(c)3 nonprofit tax-exempt classification, be located or provide service in Wachovia's markets, have broad community support and address specific community needs; and demonstrate fiscal and administrative stability.
Restrictions: Grants are not made to support individuals; travel or conferences; political causes or candidates; organizations whose primary purpose is to influence legislation; religious, veteran, or fraternal organizations for programs limited to specific groups; retirement homes or communities; precollege-level private schools; or organizations already receiving support through United Way or a United Arts drive, except for approved capital campaigns.
Amount of Grant: 5,000 - 100,000 U.S.D
Samples: Virginia Commonwealth U (VA)—for the schools of business and engineering, $400,000; Rowan Regional Medical Ctr Foundation (Salisbury, NC)—for its capital campaign for a new emergency department, patient-care facility, and community-education center, $150,000.
Contact: Grants Administrator; (704) 374-4085; fax (704) 374-2484
Internet: http://www.wachovia.com/inside/page/0,,139_414_430,00.html
Sponsor: Wachovia Foundation
301 S College Street, Suite 4000, One Wachovia Center
Charlotte, NC 28288-0013

Waitt Family Foundation Grants **4676**
The Waitt Family Foundation is looking for programs with real potential to improve the lives of people in at-risk populations throughout the world These programs, supported by the Waitt Helping Hands Fund, will encourage collaboration across various disciplines or related organizations in order to change the way people see themselves, others, and the opportunities before them. The Foundation is interested in programs that foster independence, self-help, and self-reliance, and want people to obtain the skills necessary to better themselves and their communities. Most importantly, the Foundation is looking for projects that create long-term change to help the greatest number of people. There is a $100,000 minimum for all grant requests. Multi-year proposals will be considered.
Requirements: Organizations must have a U.S. tax identification number and be able to provide a current copy of their IRS 501(c)3 determination letter. Only organizations with federal non-profit agency tax status can receive financial support.
Restrictions: The Foundation does not fund the following categories: individuals; for-profit organizations; capital campaigns; endowments; debt reduction; arts; religious organizations; public education institutions; health care organizations or hospitals; or lobbying prohibited by the Internal Revenue Code.
Geographic Focus: California, Iowa, Nebraska, South Dakota
Date(s) Application is Due: Feb 1; May 1; Aug 1; Oct 1
Amount of Grant: 100,000 U.S.D
Contact: Cherie Jacobson, (858) 551-4400; fax (858) 551-6871; grants@waittfoundation.org

Internet: http://waittfoundation.org/grants/guidelines.html
Sponsor: Waitt Family Foundation
P.O. Box 1948
La Jolla, CA 92038-1948

Walgreen Corporate Relations Grants **4677**

The corporation awards grants to nonprofits in company-operating locations in its areas of interest, including health and human services, elementary education, civic and community, and arts and culture. Most funds support ongoing commitments to preselected national and Chicago-area organizations. There are no application deadlines. Applicants who do not receive a response within 90 days should resubmit their proposals. General information is available on the Web site.
Requirements: Nonprofits in the Chicago area and 38 states where Walgreen operates may apply.
Contact: Edward King, Director, Government and Community Relations; (847) 914-2856; fax (847) 914-3417
Internet: http://www.walgreens.com/about/community/guidelines.jhtml
Sponsor: Walgreen Company
200 Wilmot Road, MS 2255
Deerfield, IL 60015

Wallace Alexander Gerbode Foundation Grants **4678**

The Foundation is interested in programs and projects offering potential for significant impact. The primary focus is on the California San Francisco Bay area and Hawaii. The Foundation's interests generally fall under the following categories: arts and culture, environment, reproductive rights and health, citizen participation/building communities/inclusiveness, strength of the philanthropic process and the nonprofit sector, and special projects. Letter of inquiry are accepted on an ongoing basis. Awards range from $1,600 to $75,000. The Foundation also has a Special Awards program to commission new works of art. Special Awards program announcements, including guidelines, an application and deadlines, are posted on the Foundation's website. Generally Special Award programs are announced in May, applications are due in late August and awards are announced in January.
Requirements: Eligible organizations must be 501(c)3 nonprofits. Letters of inquiry should be sent to the Administrative Manager and include a brief description of the proposed work, a description of the organization, and a budget.
Restrictions: The Foundation generally does not support direct services, deficit budgets, general operating funds, building or equipment funds, general fundraising campaigns, religious purposes, private schools, publications, scholarships, or grants to individuals.
Geographic Focus: California, Hawaii
Samples: Alonzo Kings LINES Ballet, San Francisco, California, $75,000, support of commission; Aging In Place, San Francisco, California, $20,000, support of San Francisco Village-Northside; and Hastings College of the Law, San Francisco, California, support of the Center for WorkLife Law.
Contact: Olivia Malabuyo Tablante, Administrative Manager; (415) 391-0911; fax (415) 391-4587; olivia@gerbode.org
Internet: http://fdncenter.org/grantmaker/gerbode
Sponsor: Wallace Alexander Gerbode Foundation
111 Pine Street, Suite 1515
San Francisco, CA 94111

Wallace H. Coulter Foundation Grants **4679**

The foundation awards grants to support research and education in the fields of engineering, medicine, and science. Past grants also have supported humanitarian interests. There are no application deadlines or forms.
Restrictions: Grants do not support religious organizations or individuals.
Amount of Grant: 35,000 - 200,000 U.S.D
Samples: Boston U, Biomedical Engineering Dept (MA)—to help advance promising biomedical technologies to the clinical-application phase, $2.9 million.
Contact: Wayne Barlin; (305) 559-2991; fax (305) 559-5490; wb@whcf.org
Internet: http://www.whcf.org
Sponsor: Wallace H. Coulter Foundation
790 NW 107th Avenue, Suite 215
Miami, FL 33172

Walmart Foundation Facility Giving Grants **4680**

Walmart's founder, Sam Walton, introduced the philosophy, "operate globally, give back locally." The Walmart Foundation supports local, state and national organizations providing opportunities in the local communities served. Areas of focus include education, workforce development/economic opportunity, health and wellness, and environmental sustainability.
Requirements: Applicants must fit within one of the Foundation's four focus areas and be one of the following: hold a current tax-exempt status under Section 501(c)3, 4, 6 or 19 of the Internal Revenue Code; be a recognized government entity that is requesting funds exclusively for public purposes; be a K-12 public/private school, charter school, community/junior college, state/private college or university; or be a church or other faith based organization with proposed projects that address and benefit the needs of the community at large.
Restrictions: Applicants must offer programs that benefit the local community. Applicants receiving sponsorship at the national level are excluded from applying. Applications are online.
Date(s) Application is Due: Dec 1
Amount of Grant: 250 - 5,000 U.S.D
Contact: Grants Administrator; (800) 966-6546
Internet: http://walmartstores.com/CommunityGiving/238.aspx
Sponsor: Walmart Foundation
702 SW 8th Street
Bentonville, AR 72716

Walmart Foundation National Giving Grants **4681**

The Walmart Foundation's mission is to create opportunities so people can live better. The Grants support initiatives focused on enhancing opportunities in education, workforce development/economic opportunity, health and wellness, and environmental sustainability. The Foundation supports applicants implementing programs in multiple sites across the country or applications with innovative initiatives that are ready for replication nationally. Awards are $250,000 or higher.
Requirements: Applicants must be 501(c)3 organizations; recognized by the Internal Revenue Service as a "public charity" within the meaning of either Section 509(a)(1) or 509(a)(2) of the Internal Revenue Code; and must operate on a national scope through the existence of chapters or affiliates in a large number of states around the country; or possess a regional/local focus, but seek funding to replicate program activities nationally. In the case of proposals seeking funding for replication, organizations must demonstrate the capacity to support national expansion and be ready to begin the replication process.
Restrictions: The Foundation does not accept unsolicited proposals but does accept letters of inquiry providing a general understanding of the problem or issue, the need for the proposed solution and the applicant's capacity to carry out the work. If the applicant is selected to submit a full proposal, guidelines and additional information will be provided. The letter of intent is submitted online.
Contact: Grants Administrator; (479) 273-4000
Internet: http://walmartstores.com/CommunityGiving/8782.aspx
Sponsor: Walmart Foundation
702 SW 8th Street
Bentonville, AR 72716

Walmart Foundation State Giving Grants **4682**

The Walmart Foundation's mission is to create opportunities so people can live better. The Grants support initiatives aligned with their mission and having a long-lasting, positive impact within a state or region. Service areas include education, workforce development/economic opportunity, health and wellness, and environmental sustainability. The Foundation has a particular interest in supporting the following populations: veterans and military families, traditionally underserved groups, individuals with disabilities, and people impacted by natural disasters. Awards are $25,000 or higher.
Requirements: Applicants must be 501(c) 3 organizations. Applications accepted online only.
Restrictions: Funding awarded in a particular state must be fully allocated within that state. Applications accepted online only.
Date(s) Application is Due: Apr 22; Sep 28
Samples: Virginia Association of Free Clinics, Staunton, Virginia, $40,000 - healthcare equipment (2010).
Contact: Grants Administrator; (479) 273-4000
Internet: http://walmartstores.com/CommunityGiving/8168.aspx?p=8979
Sponsor: Walmart Foundation
702 SW 8th Street
Bentonville, AR 72716

Walt Disney Company Foundation Grants **4683**

The foundation funds medical and health services, health and welfare of children, youth activities, higher education, community funds, music, and cultural arts in Los Angeles and Orange Counties in California and Orange and Osceola Counties in Florida. Types of support include general operating support, continuing support, annual campaigns, capital campaigns, program development, scholarship funds, fellowships, and employee-related scholarships. Previous grants have supported downtown revitalization in both New York City, where Disney redeveloped an old Times Square theater, and Los Angeles, where support has been given for a new music hall that will be home to the Los Angeles Philharmonic Orchestra.
Requirements: 501(c)3 tax-exempt organizations in California and Florida may apply.
Restrictions: Disney will not make grants for scholarships, religious organizations, building campaigns, start-up campaigns, seed purposes, research, loans, conferences, general fund drives, annual charitable appeals, and political purposes.
Geographic Focus: California, Florida, New York
Amount of Grant: 1,000 U.S.D
Contact: Tillie Baptie, Executive Director; (818) 560-1006; fax (818) 563-5271
Internet: http://disney.go.com/disneyhand/contributions/wdcfoundation.html
Sponsor: Walt Disney Company Foundation
500 S Buena Vista Street
Burbank, CA 91521-0987

Walter L. Gross III Family Foundation Grants **4684**

The foundation awards grants to eligible Ohio and Kentucky nonprofit organizations in its areas of interest, including animals and wildlife, Christian churches and organizations, community outreach programs, education, environment, health care and medical, and social services. Grants are considered on a case-by-case basis and have been awarded to support building construction/renovation and general operating support. There are no application forms or deadlines with which to adhere, and applicants should begin by contacting the office directly.
Requirements: Ohio and Kentucky 501(c)3 nonprofit organizations are eligible to apply.
Geographic Focus: Kentucky, Ohio
Amount of Grant: 1,000 - 20,000 U.S.D

Samples: Habitat for Humanity, Cincinnati, Ohio—for general support, $10,000; Miami Uuniversity Foundation, Oxford, Ohio—for general support, $300,000; Our Daily Bread, Cincinnati, Ohio—for general support, $2,000.
Contact: Walter L. Gross III, President; (513) 785-6060 or (513) 785-6072; fax (513) 683-9467
Sponsor: Walter L. Gross III Family Foundation
9435 Waterstone Boulevard, Suite 390
Cincinnati, OH 45249-8227

Walter Nicolai Prize in Biomedical Gerontology **4685**
This award is given annually for meritorious research by a graduate or medical student in the area of biomedical gerontology. For purposes of the prize, biomedical gerontology is defined as basic biological or clinical research involving normal aging or age-associated diseases. The research should be oriented toward improving the quality of later life. There is one prize plus awards to the first two runners-up to help enable them to attend the annual meeting of the American Aging Association (AGE). Nomination should include four copies each of the following: six- to 10-page (double-spaced) report of the research with name and departmental affiliations in the format used by the AGE journal, sponsor's letter commenting on the significance of the work, and candidate's curriculum vita with current professional address. The sponsor's letter will serve as the nominating document. The recipient of the prize, as well as the two runners-up, will be notified and the prizes will be presented at the annual meeting.
Requirements: A nomination may be made by any member of a society associated with aging research. The sponsor may reside in the United States, Canada, or Mexico.
Amount of Grant: 500 U.S.D
Contact: Program Contact; (610) 627-2626; fax (610) 565-9747; ameraging@aol.com
Internet: http://www.americanaging.org
Sponsor: American Aging Association
110 Chesley Drive
Media, PA 19063

Warren Alpert Foundation Prize **4686**
The annual prize recognizes scientists and researchers from around the world for their contributions to humanity and their breakthroughs in the understanding and curing of major diseases. Many of the world's foremost scientific and medical professionals, working in a variety of disciplines, are nominated by their colleagues for this prestigious honor. Winners are presented with a monetary award at the foundation's annual luncheon. Recipients are selected by the foundation's scientific advisory board, which comprises internationally renowned biomedical scientists.
Amount of Grant: 100,000 U.S.D
Samples: David Goeddel, PhD; Sidney Pestka, MD; and Charles Weissmann, MD, PhD—award recipients, for their pioneering work on the purification, characterization, and cloning of human interferon-alpha, $150,000 each.
Contact: Patrick Gilligan, (617) 432-2094; patrick_gilligan@hms.harvard.edu
Internet: http://www.hms.harvard.edu/fa/alpertprize
Sponsor: Warren Alpert Foundation
25 Shattuck Street, Room 206
Boston, MA 02115

Warren County Community Foundation Grants **4687**
The Warren County Community Foundation Grants fund projects with focus on: innovative approaches or making significant improvements in solving Warren County problems; maintaining a proactive awareness with the ability to respond to creative ideas; and preventing problems rather than treating symptoms of problems. Grant categories include education, human services, recreation, arts and culture; citizenship, environment, and economic development.
Requirements: When applying for a Warren County Community Grant, organizations should first submit a pre-application letter, which will be reviewed by the Foundation's grant committee. The pre-application letter should be one page, provide a detailed description of the project, including the targeted group it will benefit, the amount of funding requested, and the projected time line, along with the applicant's contact information. Pre-application letters are accepted on the third Thursday in January for April funding, and the third Thursday in May for funding in August. If the project meets the guidelines and funds are likely to be available, the organization is asked to submit eleven copies of the Grant Application form and any attachments. Organizations should contact the Foundation office for a current application.
Restrictions: The Foundation will not fund any of the following: travel or expenses for individuals and groups such as clubs, sports and teams, and classes; projects outside of Warren County (unless it is significant to Warren County residents); organizations fully-funded by government; projects for day-to-day operations, endowments, or excessively long-term projects; religious activities or programs serving one denomination; political organizations or activities; nationwide and/or statewide fund raising efforts.
Geographic Focus: Indiana
Amount of Grant: 300 - 185,000 U.S.D
Samples: Warren County Summer Parks Program, $1,200 (2010); Iliana Antique Power Association, $2,200 (2010); Warren County Learning Center, $2,000 (2010).
Contact: Carol Clark, Executive Director; (765) 764-1501; fax (765) 764-1501; warrencountyfoundation@yahoo.com
Internet: http://www.warrencountyfoundation.com/Grants.html
Sponsor: Warren County Community Foundation
31 North Monroe Street
Williamsport, IN 47993

Warren County Community Foundation Mini-Grants **4688**
The Warren County Community Foundation funds community projects that have the greatest impact on meeting the needs of Warren County. Projects should focus on: innovative approaches or making significant improvements in solving Warren County problems; maintaining a proactive awareness with the ability to respond to creative ideas; and preventing problems rather than treating symptoms of problems. Grant priorities include: education, human services, recreation, arts and culture, citizenship, environment, and economic development. In order for the Foundation to react in a more timely matter to needs in the community, the Board has appropriated funds to the Mini-Grant Program. The Mini-Grants Program funds requests of $250 or less with a 30 day approval process. Established grant policies and exclusions will still apply.
Requirements: Applicants should submit a pre-application letter: a one page summary of the proposed project, the targeted group it will benefit, the amount of funding requested; and the projected time frame to complete the project. The applicant will be contacted to submit a grant application if the Foundation decides that the project is appropriate for funding.
Restrictions: Grants are made on a first-come, first-served basis, until the funds have been expended for that calendar year.
Geographic Focus: Indiana
Amount of Grant: Up to 250 U.S.D
Contact: Carol Clark, Executive Director; (765) 764-1501; fax (765) 764-1501; warrencountyfoundation@yahoo.com
Internet: http://www.warrencountyfoundation.com/Grants.html
Sponsor: Warren County Community Foundation
31 North Monroe Street
Williamsport, IN 47993

Warrick County Community Foundation Grants **4689**
The Warrick County Community Foundation is a nonprofit, public charity created by and for the people of Warrick County, Indiana. The Foundation helps nonprofits fulfill their missions by strengthening their ability to meet community needs through grants that assist charitable programs, address community issues, support community agencies, launch community initiatives, and support leadership development.
Requirements: The Foundation welcomes proposals from nonprofit organizations that are deemed tax-exempt under sections 501(c)3 and 509(a) of the Internal Revenue Code and from governmental agencies serving the Warrick County. Proposals from nonprofit organizations not classified as a 501(c)3 public charity may be considered provided the project is charitable and supports a community need.
Restrictions: Project areas not considered for funding: religious organizations for religious purposes; political parties or campaigns; endowment creation or debt reduction; operating costs; capital campaigns; annual appeals or membership contributions; travel requests for groups or individuals such as bands, sports teams, or classes.
Geographic Focus: Indiana
Date(s) Application is Due: Sep 1
Samples: Youth Resources of Southwestern Indiana $1,500 - for the Teen Advisory Council and Youth Make a Difference Grant Program; Lampion Center $1,937 - for the Nurturing Program for Abused Children; Evansville Philharmonic Orchestra $1,500 - for the 2008-2009 Warrick County Youth Concert Series.
Contact: Susan Sublett; (812) 897-2030; susan@warrickcommunityfoundation.org
Internet: http://www.warrickcommunityfoundation org/disc-grants
Sponsor: Warrick County Community Foundation
224 West Main Street, P.O. Box 215
Boonville, IN 47601

Washington Dental Service Foundation Grants **4690**
The foundation awards grants to dental-related nonprofit organizations and educational institutions in Washington State. Grants are awarded to provide equipment and supplies for community and low-income clients, train dental professionals, fund professional scholarships and continuing education seminars, and teach the public about good oral health practices. Grants are awarded four times each year.
Requirements: Dental-related nonprofits in Washington State are eligible.
Geographic Focus: Washington
Date(s) Application is Due: Apr 1; Oct 1
Samples: Puget Sound Neighborhood Health Ctrs (Seattle, WA)—for relocation of its Southeast Dental Clinic, $40,000; Children's Alliance (WA)—to host a conference on building oral health coalitions, $4000; Northwest Medical Teams International (WA)—to support the Mobile Health Care Clinic.
Contact: Heather Knaplund, Grant Administrator; (800) 572-7835, ext. 2337 or (206) 528-2337; fax (206) 528-7373; hknaplund@deltadentalwa.com
Internet: http://www.ddpwa.com/wdsfoundation/wdsfoundation.htm
Sponsor: Washington Dental Service Foundation
P.O. Box 75688
Seattle, WA 98125

Washington Gas Charitable Giving Contributions **4691**
For more than 160 years, Washington Gas has been an integral part of the growing metropolitan region of Washington D.C. Chartered by the 30th Congress and signed into law by President James Polk in 1848, the company has developed a Charitable Giving Program that is designed to make a meaningful and lasting impact on the communities it serves. Washington Gas focuses on three primary areas: education, the environment, and health. Types of support offered are grants, in-kind contributions, and volunteer resources. In the area of education, emphasis is placed on the development of math, science, technology, and business skills in K-12 youth. Consideration is also given to arts-related programs. In the area of the environment, emphasis is placed on programs that promote cleaner air and

water programs that protect and preserve the ecological system of the metropolitan area. In the area of health, consideration is given to health organizations that strive to improve the health and well-being of individuals within the community. Emphasis is also placed on energy assistance programs for low-income residents to heat and cool their homes, reducing ilness and casualties resulting from exposure to extreme temperatures. The company accepts applications on a rolling basis. Basic guidelines are listed at the website and provided in a downloadable PDF file. Applications must be mailed to the contact information given. Notification of acceptance or rejection generally takes fifteen business days or longer.
Requirements: 501(c)3 organizations are eligible to apply. Support is provided primarily in Washington, D.C., Maryland, and Virginia. The company prefers to support specific programs over general funding.
Restrictions: Support is not provided to religious organizations for sectarian purposes, political associations, organizations with strictly a sports focus, individuals, and requests for capital or endowment campaigns.
Geographic Focus: District of Columbia, Maryland, Virginia
Contact: Tracye Funn, (703) 750-1000; tfunn@washgas.com
Internet: http://www.washgas.com/pages/CharitableGiving
Sponsor: Washington Gas Company
101 Constitution Avenue, North West, 3rd Floor
Washington, DC 20080

Washington Post Contributions Program Grants 4692
The contributions program supports qualified nonprofit organizations of the Washington, DC, metropolitan community whose ongoing programs directly benefit those in the community who are in need and that focus on key issues in the areas of education, health and human services, the arts, civic betterment, and community service. Special interest is given to organizations whose programs serve the most needy and least able citizens; seek to fill unmet societal needs or solve long-standing community problems; and address key community issues and concerns in collaborative and imaginative ways. Grants are awarded to support specific programs or for general operating support. Funding rarely is available for capital campaigns. Applications for capital campaigns that meet the Post's general guidelines are held until December and will be funded only if resources are available at the time of review. There are no application forms. Proposals should include a one-page cover sheet; a three- to five-page application letter; supporting documents; list of board of directors and officers; a copy of the most recent annual report; and several relevant articles, reviews, brochures, etc., about the project or organization. Deadlines are the first Tuesday in March, June, September, and December.
Requirements: Organizations that are tax-exempt under section 501(c)3 of the Internal Revenue Code that operate in the Washington, DC, metropolitan area are eligible.
Restrictions: Funding is not available for individuals; national or other organizations whose programs do not directly serve Washington-area needs; religious organizations for religious purposes; fraternal, membership, or veterans organizations or athletic teams, volunteer fire/emergency services, or similar groups; political action or advocacy groups; multiyear grants; challenge grants; third-party fund-raising efforts; conferences; research; or start-up projects. Applications are not accepted.
Contact: Eric Grant, Director; (202) 334-6834; fax (202) 334-5609
Internet: http://washpost.com/community/charities/contributions.shtml
Sponsor: Washington Post Educational Foundation
1150 15th Street NW
Washington, DC 20071

Washington Women's Foundation Grants 4693
The grant committee of the Pooled Fund researches community needs in five predetermined issue areas: cultural arts, education, environment, health, and social services. The foundation funds projects that address at least one of the following criteria: responses to urgent and critical need; bold new ventures; and/or new approaches to time-worn problems. The foundation funds broadly; it does not limit funding to women's causes, and there are no geographic restrictions to eligibility. Beginning in January of each year, work groups identify giving priorities, compile data, and analyze information on organizations of interest. Agencies are selected to receive a formal request for proposal. Unsolicited proposals are not accepted. Organizations may submit an online letter of inquiry form to be considered.
Requirements: 501(c)3 nonprofit organizations are eligible.
Amount of Grant: 50,000 - 100,000 U.S.D
Samples: Academy of Children's Theatre of Richland (WA)—for a newly acquired building to house theater-art classes, outreach programs, and performance space for children in the Tri-Cities/Mid-Columbia Valley area, $50,000; Community for Youth (Seattle, WA)—to strengthen and expand its mentoring and after-school programs to help underachieving yet promising high-school students stay in school, $100,000; Jubilee Women's Ctr (Seattle, WA)—to provide safe, affordable, long-term housing in Seattle for women who are homeless or face an economic crisis, $85,000; Okanogan Family Planning (WA)—for medical equipment, furnishings, and other supplies at a new family-planning clinic, $85,000.
Contact: Administrator; (206) 340-1710; info@wawomensfoundation.org
Internet: http://www.wawomensfoundation.org/grants/guide.html
Sponsor: Washington Women's Foundation
2100 24th Avenue S, Suite 330
Seattle, WA 98144

Washington Women in Need Foundation Grants 4694
Grants are awarded in Washington to provide services to low-income women in the areas of health care, education, mental health counseling, and physical/dental/vision/hearing exams and treatment. There are no deadlines, but application materials must be obtained prior to applying. Opening dates are listed on the Web site.
Requirements: The foundation provides assistance to low-income women who are at least 18 and are residents of Washington. Our definition of low-income is based on the Federal Poverty Guideline. Eligibility is based on gross (pre-tax) income.
Geographic Focus: Washington
Contact: Sally Mackle, Executive Director; (425) 451-8838; fax (425) 451-8845; smackle@wawomeninneed.org
Internet: http://www.wawomeninneed.org/doc/program.htm
Sponsor: Washington Women in Need Foundation
1849 114th Avenue NE
Bellevue, WA 98004

Wasie Foundation Grants 4695
The foundation limits grantmaking activities to organizations in Minnesota and currently makes grants in the areas of higher education, specifically endowed scholarship programs, and the care and treatment of individuals afflicted with schizophrenia. Types of support include equipment acquisition, endowment funds, program development, conferences and seminars, scholarship funds, research, and matching funds. Call or write for application procedures and current deadlines; the foundation encourages phone contact.
Requirements: Organizations in Florida and Minnesota are eligible.
Restrictions: Grant making is limited to the Saint Paul/Minneapolis metropolitan area.
Geographic Focus: Florida, Minnesota
Amount of Grant: 15,000 - 200,000 U.S.D
Contact: Jan Preble, Program Associate; (612) 455-6880; fax (612) 455-6888
Sponsor: Wasie Foundation
4999 France Avenue S, Suite 250
Minneapolis, MN 55410

Wayne and Gladys Valley Foundation Grants 4696
The foundation awards grants to California nonprofits in its primary areas of interest, including education at all levels, health care and medical research, youth and families, human services, sciences and engineering, and religion. Types of support include general operating support, building construction/renovation, program development, professorships, research, and matching funds. There are no application forms or deadlines. The board meets in February, May, September, and November to consider requests.
Requirements: California nonprofits in Alameda, Contra Costa, and Santa Clara Counties are eligible.
Geographic Focus: California
Contact: Michael Desler, Executive Director; (510) 466-6060; fax (510) 466-6067; info@wgvalley.org
Sponsor: Wayne and Gladys Valley Foundation
1939 Harrison Street, Suite 510
Oakland, CA 94612-3532

Weatherwax Foundation Grants 4697
The foundation awards grants in Michigan to projects that promote and support education, civic and social programs, culture, science, and the arts, primarily in the greater Jackson area and, to a lesser extent, in the adjacent counties. Types of support include building construction and renovation; matching and challenge grants; general operating grants; program development; scholarships; annual and capital campaigns; and emergency funds. The initial approach is a two-page grant proposal that may be submitted at any time. Granting decisions are made in January, May, and August.
Requirements: Giving primarily in Hillsdale, Lenawee, and Jackson counties, MI.
Restrictions: No grants to individuals; or for computer purchases.
Geographic Focus: Michigan
Amount of Grant: 1,000 - 200,000 U.S.D
Samples: Albion College (MI)—for music facility renovation, $125,000; Michigan Theatre (MI)—for restoration, strategic planning, and executive search, $27,500; Family Service and Children's Aid (MI)—for general operating support and neighborhood outreach activities, $45,491.
Contact: Program Contact; (517) 787-2117; fax (517) 787-2118
Internet: http://www.lib.msu.edu/harris23/grants/wfbrochu.htm
Sponsor: Weatherwax Foundation
P.O. Box 1111
Jackson, MI 49204

Weber Family Foundation Grants 4698
The foundation is committed to supporting organizations and programs that improve the quality of life in the city of Atlanta and its surrounding communities. Areas of interest include: children and youth services, education, health associations, human services, and arts and cultural programs. Organizations that focus on any of the listed areas of interest are invited to submit a proposal. Guidelines are available online.
Requirements: Georgia tax-exempt organizations serving Atlanta and surrounding communities are eligible.
Date(s) Application is Due: Jan 1; Apr 1; Jul 1; Oct 1
Amount of Grant: 500 - 20,000 U.S.D
Samples: Bethany Christian Services (GA)—$5000; Charis Community Housing (GA)—$3000; United Negro College Fund (GA)—$1500; Zoo Atlanta (GA)— $1500.
Contact: Grants Administrator; (770) 804-1651, ext. 3; info@WeberFoundation.org
Internet: http://www.weberfoundation.org
Sponsor: Weber Family Foundation
P.O. Box 889063
Atlanta, GA 30356-9063

Webster Cornwell Memorial Scholarship 4699

The Webster Cornwell Memorial Scholarship is awarded each year to a graduating senior from Hammond High School. The $1,000 scholarship is designated for tuition, fees, and/or books, and is renewable for three years. A student will be selected based upon their cumulative GPA, high school and community activities, and an essay. All college majors are eligible.

Requirements: Applicants are required to have at least a 2.5 GPA and full time enrollment in a program of study leading to an associate or bachelor degree at an accredited college or university. In addition to the application, candidates must submit their high school transcript and the answer to an essay question posted in the application.

Geographic Focus: Indiana

Date(s) Application is Due: Apr 15

Amount of Grant: 1,000 U.S.D

Contact: Cornwell Scholarship Contact; (219) 736-1880; fax (219) 736-1940; legacy@legacyfoundationlakeco.org

Internet: http://www.legacyfoundationlakeco.org/scholarships/index.htm

Sponsor: Legacy Foundation
1000 East 80th Place, 302 South
Merrillville, IN 46410

Wege Foundation Grants 4700

The Wege Foundation awards grants in Michigan in five specific categories: education, environment, arts and culture, health care, and human services. Areas of interest include children and youth services; Christian agencies and churches; community development; elementary and secondary education; environmental resources; higher education; hospitals (general); human services; museums; and performing arts. The Foundation strives to be an inspiration for other communities. According to the Foundation's vision, "Grand Rapids inspires others to create communities that forge a balance in the environment, health, education, and arts to encourage healthier lives in mind, body, and spirit."

Requirements: Grants are awarded to nonprofit organizations in the greater Kent County, MI, area, with emphasis on the Grand Rapids area. The applicant may access the application after an online quiz.

Restrictions: The Foundation only funds organizations classified as tax-exempt under section 501(c)3 of the Internal Revenue Code. Grants are not awarded for operating budgets.

Geographic Focus: Michigan

Date(s) Application is Due: Feb 15; Sep 15

Contact: Jody Price, Corporate Financial Officer; (616) 957-0480; fax (616) 957-0616; jprice@wegefoundation.org

Internet: http://www.wegefoundation.com/seekingagrant/seekingagrant.html

Sponsor: Wege Foundation
P.O. Box 6388
Grand Rapids, MI 49516-6388

Weingart Foundation Grants 4701

Weingart Foundation makes grants to assist organizations that work in the areas of health, human services, and education. The Foundation gives highest priority to activities that provide greater access to people who are economically disadvantaged and underserved. Of particular interest to the Foundation are applications that specifically address the needs of low-income children and youth, older adults, and people affected by disabilities and homelessness. The Foundation also funds activities that benefit the general community and improve the quality of life for all individuals in Southern California. Weingart Foundation offers the following types of grants: core support; capital; capacity building; program; and a small grant program. See website for a description of each grant type: http://www.weingartfnd.org/default.asp?PID=5

Requirements: An organization that is certified as tax exempt under Section 501(c)3 of the U.S. Internal Revenue Code and is not a private foundation as defined in section 509(a) of that Code is eligible for consideration. Preference is given to organizations providing services in the following six Southern California counties: Los Angeles, Orange, Riverside, Santa Barbara, San Bernardino, and Ventura.

Restrictions: Grants are not made: for propagandizing, influencing legislation and/or elections, promoting voter registration; for political candidates, political campaigns; for litigation; to institutions limiting their services to persons of a single religious sect or denomination; for social or political issues outside the United States of America; to individuals; to federated appeals or for the collection of funds for redistribution to other nonprofit groups; for conferences, workshops, temporary exhibits, travel, surveys, films or publishing activities; for endowment funds; for contingencies, deficits or debt reduction; for fundraising dinners or events; for research. Grants generally are not approved for: national organizations that do not have local chapters operating in the geographic area of grant focus; projects or programs normally financed by government sources; refugee or religious programs, consumer interest or environmental advocacy; feasibility studies. The Foundation does not fund Section 509(a)(3) Type III non-functionally integrated supporting organizations.

Geographic Focus: California

Samples: National Urban Fellows, $25,000—program support for the Southern California Capacity Building Initiative (2010); Performing Arts Center of Los Angeles, $150,000—core support for Music Center education programs (2010); JWCH Medical Clinic, $900,000—capacity building support to strengthen administrative infrastructure (2010); University of Southern California, $10,000—support for the Los Angeles Leadership Group (2010); Church of Our Saviour, $150,000—core support (2010).

Contact: Administrator; (213) 688-7799; fax (213) 688-1515; info@weingartfnd.org

Internet: http://www.weingartfnd.org

Sponsor: Weingart Foundation
1055 W Seventh Street, Suite 3050
Los Angeles, CA 90017-2305

Welborn Baptist Foundation General Opportunity Grants 4702

The Welborn Baptist Foundation is seeking to fund lasting change in the communities it serves. Towards that end, the Foundation will assess grant proposals based on: demonstrated long-term results, or exceptional potential for such results based on well-established research; addressing root causes rather than symptoms; fit with the Foundation's target areas of emphasis; deep, established collaborations with (not just referrals from) other like-minded organizations; excellent prospects for long-term sustainability without Foundation resources; and a clear implementation plan that enables successes to be replicated elsewhere. To meet needs that do not naturally fit into the other priority target areas, but which are consistent with the Foundation's mission, the General Opportunity target area has been established. In the past, General Opportunity grants have focused on opportunities that enhance the general health and/or the health educational status of the community. Ideal opportunities for consideration in this category are those which address both health and health education. Proposals which have only a peripheral relationship to health or health education are not likely to be funded.

Requirements: Giving is limited to: Gallatin, Saline, Wabash, Wayne and White counties in Illinous; Dubois, Gibson, Perry, Pike, Posey, Spencer, Vanderburgh, and Warrick counties in Indiana; and Henderson County, Kentucky. Applicants must submit a Letter of Interest (LOI). All secular, church and other faith-based not-for-profit organizations that are tax exempt under section 501(c)3 of the IRS Code are eligible. Participation beyond the Letter of Interest stage is by invitation only.

Restrictions: The following program and project areas will not be considered for funding: scholarships, loans, grants or fellowship support directly to or for the benefit of specific and known individuals; establishment of, or contributions to, a permanent endowment, foundation, trust or permanent interest-bearing account; carrying on of propaganda or attempt to influence legislation or public elections; restricting the services, facilities or employment provided by the grant to individuals based on race, creed, color, sex, or national origin; any governmental agencies reporting to an elected or appointed official (except for schools governed by citizens boards); any requests for funding for deficits or retirement of debt; fundraising events; annual fund drives; venture capital for competitive profit making ventures; or basic scientific research. Additionally, it should be noted that the Foundation does not fund applications seeking replacement dollars (i.e., funding to substitute for dollars lost from another grantor).

Contact: Kevin Bain, Executive Director; (812) 437-8260 or (877) 437-8260; fax (812) 437-8269; info@welbornfdn.org

Internet: http://www.welbornfdn.org/grant-process/funding-targets

Sponsor: Welborn Baptist Foundation
Twenty-One Southeast Third Street, Suite 610
Evansville, IN 47708

Welborn Baptist Foundation Health Ministries Grants 4703

The Welborn Baptist Foundation is seeking to fund lasting change in the communities it serves. Towards that end, the Foundation will assess grant proposals based on: demonstrated long-term results, or exceptional potential for such results based on well-established research; addressing root causes rather than symptoms; fit with the Foundation's target areas of emphasis; deep, established collaborations with (not just referrals from) other like-minded organizations; excellent prospects for long-term sustainability without Foundation resources; and a clear implementation plan that enables successes to be replicated elsewhere. A subcategory of the Foundation's Community Health Status target is Health Ministries. A health ministry integrates faith and health within a church community for its members and the community it serves. It is an integral part of the overall ministry of the church, promoting physical, emotional, social and spiritual well-being. Health ministers are many – clergy, nurses and laity – coming together to share the compassionate love and grace of Jesus Christ. Furthermore, the Foundation will continue to offer technical assistance, training and seed funding to establish health ministries in congregations throughout the Foundation's Service Area.

Requirements: Giving is limited to: Gallatin, Saline, Wabash, Wayne and White counties in Illinous; Dubois, Gibson, Perry, Pike, Posey, Spencer, Vanderburgh, and Warrick counties in Indiana; and Henderson County, Kentucky. Applicants must submit a Letter of Interest (LOI). All secular, church and other faith-based not-for-profit organizations that are tax exempt under section 501(c)3 of the IRS Code are eligible. Participation beyond the Letter of Interest stage is by invitation only.

Restrictions: The following program and project areas will not be considered for funding: scholarships, loans, grants or fellowship support directly to or for the benefit of specific and known individuals; establishment of, or contributions to, a permanent endowment, foundation, trust or permanent interest-bearing account; carrying on of propaganda or attempt to influence legislation or public elections; restricting the services, facilities or employment provided by the grant to individuals based on race, creed, color, sex, or national origin; any governmental agencies reporting to an elected or appointed official (except for schools governed by citizens boards); any requests for funding for deficits or retirement of debt; fundraising events; annual fund drives; venture capital for competitive profit making ventures; or basic scientific research. Additionally, it should be noted that the Foundation does not fund applications seeking replacement dollars (i.e., funding to substitute for dollars lost from another grantor).

Geographic Focus: Illinois, Indiana, Kentucky

Contact: Kevin Bain, Executive Director; (812) 437-8260 or (877) 437-8260; fax (812) 437-8269; info@welbornfdn.org

Internet: http://www.welbornfdn.org/grant-process/funding-targets

Sponsor: Welborn Baptist Foundation
Twenty-One Southeast Third Street, Suite 610
Evansville, IN 47708

Welborn Baptist Foundation Improvements to Community Health Status Grants 4704

The Welborn Baptist Foundation is seeking to fund lasting change in the communities it serves. Towards that end, the Foundation will assess grant proposals based on: demonstrated long-term results, or exceptional potential for such results based on well-established research; addressing root causes rather than symptoms; fit with the Foundation's target areas of emphasis; deep, established collaborations with (not just referrals from) other like-minded organizations; excellent prospects for long-term sustainability without Foundation resources; and a clear implementation plan that enables successes to be replicated elsewhere. in the area of Improvements to Community Health Status, the Foundation has initiated a community-wide coalition designed to achieve long-term reduction in the proportion of residents that are either overweight or obese, through increased physical activity and healthier eating. Beyond this, the Foundation's funding objective in this target area is for all community residents to proactively take the appropriate steps that can lead to lower levels of chronic illness. The primary focus is therefore on: programs that increase the access of high-risk populations to health care services; and programs that address the prevention and successful management of chronic disease states.

Requirements: Giving is limited to: Gallatin, Saline, Wabash, Wayne and White counties in Illinous; Dubois, Gibson, Perry, Pike, Posey, Spencer, Vanderburgh, and Warrick counties in Indiana; and Henderson County, Kentucky. Applicants must submit a Letter of Interest (LOI). All secular, church and other faith-based not-for-profit organizations that are tax exempt under section 501(c)3 of the IRS Code are eligible. Participation beyond the Letter of Interest stage is by invitation only.

Restrictions: The following program and project areas will not be considered for funding: scholarships, loans, grants or fellowship support directly to or for the benefit of specific and known individuals; establishment of, or contributions to, a permanent endowment, foundation, trust or permanent interest-bearing account; carrying on of propaganda or attempt to influence legislation or public elections; restricting the services, facilities or employment provided by the grant to individuals based on race, creed, color, sex, or national origin; any governmental agencies reporting to an elected or appointed official (except for schools governed by citizens boards); any requests for funding for deficits or retirement of debt; fundraising events; annual fund drives; venture capital for competitive profit making ventures; or basic scientific research. Additionally, it should be noted that the Foundation does not fund applications seeking replacement dollars (i.e., funding to substitute for dollars lost from another grantor).

Geographic Focus: Illinois, Indiana, Kentucky

Contact: Kevin Bain, Executive Director; (812) 437-8260 or (877) 437-8260; fax (812) 437-8269; info@welbornfdn.org

Internet: http://www.welbornfdn.org/grant-process/funding-targets

Sponsor: Welborn Baptist Foundation
Twenty-One Southeast Third Street, Suite 610
Evansville, IN 47708

Welborn Baptist Foundation School Based Health Grants 4705

The Welborn Baptist Foundation is seeking to fund lasting change in the communities it serves. Towards that end, the Foundation will assess grant proposals based on: demonstrated long-term results, or exceptional potential for such results based on well-established research; addressing root causes rather than symptoms; fit with the Foundation's target areas of emphasis; deep, established collaborations with (not just referrals from) other like-minded organizations; excellent prospects for long-term sustainability without Foundation resources; and a clear implementation plan that enables successes to be replicated elsewhere. The School Based Health target area, known as HEROES, is a three-year commitment, during which a school will focus on implementing the Centers for Disease Control's Coordinated School Health Model (CSH) into their school environment and begin to change their school's culture. The funding from WBF is geared towards the Physical Education/Physical Activity and Nutrition components of CSH, but the schools are also expected to begin making changes in the other six components as well.

Requirements: Giving is limited to: Gallatin, Saline, Wabash, Wayne and White counties in Illinous; Dubois, Gibson, Perry, Pike, Posey, Spencer, Vanderburgh, and Warrick counties in Indiana; and Henderson County, Kentucky. Applicants must submit a Letter of Interest (LOI). All secular, church and other faith-based not-for-profit organizations that are tax exempt under section 501(c)3 of the IRS Code are eligible. Participation beyond the Letter of Interest stage is by invitation only.

Restrictions: The following program and project areas will not be considered for funding: scholarships, loans, grants or fellowship support directly to or for the benefit of specific and known individuals; establishment of, or contributions to, a permanent endowment, foundation, trust or permanent interest-bearing account; carrying on of propaganda or attempt to influence legislation or public elections; restricting the services, facilities or employment provided by the grant to individuals based on race, creed, color, sex, or national origin; any governmental agencies reporting to an elected or appointed official (except for schools governed by citizens boards); any requests for funding for deficits or retirement of debt; fundraising events; annual fund drives; venture capital for competitive profit making ventures; or basic scientific research. Additionally, it should be noted that the Foundation does not fund applications seeking replacement dollars (i.e., funding to substitute for dollars lost from another grantor).

Geographic Focus: Illinois, Indiana, Kentucky

Contact: Kevin Bain, Executive Director; (812) 437-8260 or (877) 437-8260; fax (812) 437-8269; info@welbornfdn.org

Internet: http://www.welbornfdn.org/grant-process/funding-targets

Sponsor: Welborn Baptist Foundation
Twenty-One Southeast Third Street, Suite 610
Evansville, IN 47708

Wellmark Foundation Grants 4706

The foundation encourages proposals that measurably decrease the number of community members suffering from an acute or chronic disease or improve the health status of community members afflicted with an acute or chronic disease. Priority areas include asthma, diabetes, heart failure and cardiovascular risk factors, major depression, and end-of-life care with an emphasis on pain control. Mini grants award smaller amounts of money available for projects that support one or more of the five priority areas. The listed application deadline is for mini grants.

Requirements: 501(c)3 tax exempt in Iowa or South Dakota are eligible.

Restrictions: 509(a) private foundations and primary contracting entities that have a contractual relationship with Wellmark are ineligible.

Geographic Focus: Iowa, South Dakota

Samples: Buena Vista Regional Medical Ctr Foundation (Storm Lake, IA)—to develop a palliative-care program in northwestern Iowa, $56,000; Front Porch Coalition (Rapid City, SC)—for a public-education campaign about depression that will also seek to reduce the stigma associated with the condition, $80,000; Mercy Medical Ctr Foundation-North Iowa (Mason City, IA)—for a pilot project designed to improve the diagnosis and treatment of major depression, $48,390; Primary Health Care (Des Moines, IA)—to provide diabetes education, including training on free blood-glucose meters and other self-management techniques, and to monitor low-income patients following the training, $33,663.

Contact: Grants Administrator; (515) 245-4819; fax (515) 235-4455; WMFoundation@wellmark.com

Internet: http://www.wellmark.com/community/wellmark_foundation/wellmark_foundation.htm

Sponsor: Wellmark Foundation
636 Grand Avenue, Station 150
Des Moines, IA 50309-2565

Wells County Foundation Grants 4707

The Wells County Foundation Grants give priority to programs having a positive effect on the Wells County community. Grant making fields of interest include arts and culture, education, civic affairs, youth, environment, community development, animal welfare, recreation, and health and human services. In reviewing grant proposals, the Foundation gives careful consideration to: the potential impact of the request and the number of people benefited; an innovative approach; the degree to which the applicant works with or complements the services of other community organizations; the extent of local involvement and support for the project; the organization's demonstrated fiscal responsibility and management qualification; and the organization's ability to obtain necessary additional and future funding.

Requirements: Applicants are encouraged to contact the Foundation to discuss their project. They should be prepared to give a brief discussion to enable the Foundation to determine whether the request falls within the grant-making guidelines. In addition, the staff will inform the applicant of the deadline for the most current grant cycle. Grant proposals for projects should include the following: a title page with the organization's contact information, the project's title and amount requested; a proposal narrative, summarizing what issues the project will address, its expectations, how many will benefit, the role of volunteers, its planned evaluation, and a signed endorsement by the Board of Directors; financial information, with the project budget, two pricing quotes for equipment requested, a list of other funding sources, and the project's funding for the future; and organizational information, including a brief history with mission and purpose, list of officers, financial statement or audit, and a copy of the tax exemption the IRS.

Restrictions: No grants will be made to any political organization or to support attempts to influence the legislation of any governing body other than through making available to the community at large the results of non-partisan analysis, study and/or research.

Geographic Focus: Indiana

Date(s) Application is Due: Feb 15; Jun 15; Oct 17

Samples: Bluffton Parks & Recreation Department, purchase of an ADA compliant pool lift chair for the Wells Community swimming pool so that patrons can enter and exit the pool independently, $7,000 (2011); Community Action of Northeast Indiana (CANI), provide emergency assistance to Wells County families through CANI's Homeles Prevention Rapid Re-housing Program so that families can work toward self-sufficiency, $3,000 (2011); Family Centered Services, Inc, to provide administrative funding for the Youth As Resources program so that young people can design and carry out service projects that address specific community needs, $7,000 (2011).

Contact: Tammy Slater, Chief Executive Officer; (260) 824-8620; fax (260) 824-3981; wellscountyfound@wellscountyfound.org

Internet: http://www.wellscountyfound.org/Grants.html

Sponsor: Wells County Foundation
360 North Main Street, Suite C
Bluffton, IN 46714

Western Indiana Community Foundation Grants 4708

The Western Indiana Community Foundation focuses its attention on local needs within the geographical boundaries as set by the Board of Directors. Primary fields of interest include: health, charitable service, education, cultural affairs, and community improvement. The Foundation is especially interested in learning of plans for: start-up costs for new programs; one-time projects or needs; and capital needs beyond an applicant's capabilities and needs. Grant applications may be submitted throughout the year. Applicants will be notified immediately following the Board of Directors decision.

Requirements: Organizations must fill out the online application and include the following information: their full contact information, description of the project, and grant request amount; other funding sources for the project, an itemized expenses list and project

timeline; the organization's IRS tax exempt status, how they plan to evaluate the project, and a description of public relations plans/foundation funding. The application packet should then be mailed to the Foundation,
Restrictions: The Foundation will not consider: grants for individuals; organizations for political or religious purposes; support for regular operating budgets; contributions to endowments; providing for long term funding; post-event situations; or apparel such as school/sport uniforms.
Geographic Focus: Indiana
Amount of Grant: Up to 10,000 U.S.D
Samples: City of Covington - $1,600 for mobile life guard chairs at the city swimming pool (2011); Friendship Circle Center - $3,373 annual payout for operations and parking lot resurfacing (2011); Fountain County Mentoring - $2,605 for youth field trip to Chicago's Shedd Aquarium and Museum of Science and Industry (2011); Covington High School - $500 - for SAFE-TALK training session for faculty and staff to help identify students who are high risk for suicide or other self-destructive behaviors (2011).
Contact: Dale White, Executive Director; (765)-793-0702; fax (765)-793-0703; dwhite@wicf-inc.org or info@wicf-inc.org
Internet: http://www.wicf-inc.org/grant_guidelines.asp
Sponsor: Western Indiana Community Foundation
135 South Stringtown Road
Covington, IN 47932-0175

Western New York Foundation Grants 4709

The Foundation supports sustainable organizations that improve the quality of life in Western New York. The Foundation makes investments that build on nonprofits' proven strengths in order to improve their effectiveness and their ability to fulfill their missions. Funding is provided in these categories: human services; education; urban and rural development; arts, culture, and humanities; and housing, park and land use.
Requirements: Western New York State 501(c)3 organizations located within one of the following counties are eligible: Allegany, Cattaraugus, Chautauqua, Erie, Genesee, Niagara, and Wyoming.
Restrictions: Only one application may be submitted at a time. Funded organizations may reapply two years following the final payment of an award. The following are ineligible: religious organizations for religious purposes; political organizations, campaigns, and candidates; municipal and government entities; grants or loans to individuals; fund-raising events, i.e. sponsorships, tables, dinners, and telethons; endowments; scholarships; operating expenses; hospital capital campaigns; and general capital campaigns.
Geographic Focus: New York
Date(s) Application is Due: Jun 30; Nov 30
Samples: Albright-Knox Art Gallery, Buffalo, New York, $25,000 - grand exterior staircase (2010); Child and Family Services, Buffalo, New York, $25,000 - Lockwood Mansion restoration (2010); and Compeer of Greater Buffalo, Buffalo, New York, $15,000 - high needs mentoring program (2010).
Contact: Beth Kinsman Gosch, Executive Director; (716) 839-4225; fax (716) 883-1107; bgosch@wnyfoundation.org
Internet: http://www.wnyfoundation.org
Sponsor: Western New York Foundation
11 Summer Street, Fourth Floor
Buffalo, NY 14209

Western Union Foundation Grants 4710

The Foundation generally awards grants in the United States and internationally in its areas of interest, including education; and health and human services addressing literacy, health care for the uninsured, nutrition, pre- and postnatal care, childhood immunizations, poverty, language barriers, and cultural adjustment. In order to maximize resources and create a greater impact in communities, the Foundation will not accept unsolicited proposals in 2010. The Foundation will work with its current NGO partners and solicit proposals from NGOs that closely align with the Foundation's education and economic development program areas.
Restrictions: In general grants do not support general operating support, individuals, endowments, special events, capital projects, early childhood education, other post-secondary scholarship programs, deficits or retirement of debt, re-granting agencies, and awards.
Geographic Focus: Colorado, Florida, Nebraska, New York, Texas
Date(s) Application is Due: Mar 1; Jun 1; Dec 1
Amount of Grant: 1,000 - 25,000 U.S.D
Samples: Southern Sudan Community Assoc (Omaha, NE)—to offer English classes to Sudanese refugees living in Omaha, $23,715; Open Door Mission (Omaha, NE)—for a program to provide emergency assistance to homeless and poor people, $15,000; CARE U.S.A (Atlanta, GA)—for its efforts to offset the food crisis in Niger, $80,000; International Federation of the Red Cross and Red Crescent Societies (Geneva)—for relief efforts in South Asia and Africa, $1 million.
Contact: Luella Chavez D'Angelo, President; (720) 332-4763 or (720) 332-6606; fax (720) 332-4772; luella.dangelo@westernunion.com
Internet: http://foundation.westernunion.com/ourProgramsDirect.html
Sponsor: Western Union Foundation
12500 East Belford Avenue, Suite 1-I
Englewood, CO 80112

West Foundation Grants 4711

The foundation awards grants nationwide to nonprofits in support of children and youth programs, especially in rural areas. Of primary consideration are programs for disadvantaged populations that focus on alcohol and drug abuse, abused children, battered women, mental health, education, and health. The foundation gives preference to requests from central Indiana, where it focuses on substance abuse programs, domestic violence prevention, and education. There is no application form. Proposals should include a cover letter including the amount requested, the nature of the program, and a description of the organization.
Requirements: Nonprofit organizations are eligible.
Geographic Focus: Indiana
Amount of Grant: Up to 14,000,000 U.S.D
Samples: Laubach Literacy (Syracuse, NY)—for the international component of its Women in Literacy: Women in Action campaign, $1 million over five years; Fellowship of Christian Athletes (Kansas City, KS)—for staff and a camping program for youth, $3000; National Indian Youth Leadership Project (Gallup, NM)—to continue a drug abuse prevention program for sixth-through eighth-grade students, $30,000.
Contact: Stephen West, President, c/o Bank One Center; (317) 283-5525
Sponsor: West Foundation
111 Monument Circle, Suite 220
Indianapolis, IN 46204-5168

Westinghouse Charitable Giving Grants 4712

The foundation makes charitable contributions to community priorities primarily where Westinghouse has a presence. Areas of emphasis include education—elementary, secondary, and high school programs that emphasize math and science (some non-fine arts programs); health and welfare—to improve quality of life in the community; and civic and social affairs—to support economic development, environmental quality, and the preservation of public safety. Within each area, the foundation actively encourages programs that meet the needs of special populations, such as the disadvantaged, the young, the elderly, the gifted, people with disabilities, minorities, and women. Because of the foundation's approach to funding in Westinghouse plant locations, all organizations, except colleges and universities, should submit their requests for support to the nearest Westinghouse location. Applications are accepted on an ongoing basis.
Requirements: 501(c)3 tax-exempt organizations are eligible.
Restrictions: The following organizations and activities are ineligible: individuals; political contributions; religious organizations; highly specialized health, medical, or welfare programs; fine arts organizations/programs; organizations that discriminate by race, color, creed, gender, or national origin; hospitals; colleges, universities, and two-year institutions; capital improvement or building projects; medical education; chairs or professorships; liberal arts, fine arts, or similar educational programs; educational research programs, regardless of academic discipline; graduate education, general endowment; equipment purchases at universities; United Way-affiliated agencies; tickets; or memberships.
Amount of Grant: Up to 5,000 U.S.D
Contact: Program Officer; (412) 374-6824; fax (412) 642-4874
Internet: http://www.westinghousenuclear.com/E2.asp
Sponsor: Westinghouse Electric Corporation
P.O. Box 355, ECE 575C
Pittsburgh, PA 15230-0355

Weyerhaeuser Company Foundation Grants 4713

The foundation's mission is to improve the quality of life in communities where Weyerhaeuser has a major presence and to provide leadership that increases public understanding of issues where society's needs intersect with the interests of the forest products industry. Community service grants include awards in the fields of education and youth, health and welfare, civic and community improvement, and culture and the arts. In addition to community grants, industry-related awards are made to educational institutions, environmental groups, and professional organizations that promote further understanding of how the forest products industry responds to a changing society. Types of support include seed money, building construction/renovation, equipment acquisition, employee-related scholarships, publication, conferences and seminars, fellowships, lectureships, operating budgets, research, program development, employee matching gifts, and technical assistance. Applicants should send a short letter that introduces the project and sponsoring organization and provide tax-exempt status evidence. If further consideration is warranted, the foundation may ask for additional information or a formal proposal. Proposals may be submitted at any time.
Requirements: Applying organizations must have nonprofit, tax-exempt status. To be considered for funding, an organization must: serve a community within a 50-mile radius of a major Weyerhaeuser facility; support a state-wide issue of interest to the Foundation and Weyerhaeuser in the key states of Alabama, Arkansas, Louisiana, Mississippi, North Carolina, Oklahoma, Oregon or Washington; or support a selected, high-priority national or international initiative directly related to the sustainability and importance of working forests. A limited number of smaller awards are also made to other locales where fewer employees are based.
Restrictions: Grants are not awarded to individuals or for political campaigns, activities that influence legislation, religious organizations seeking funds for theological purposes, or funds to purchase tickets or tables at fundraising benefits.
Geographic Focus: Alabama, Arizona, Arkansas, California, Colorado, District of Columbia, Georgia, Idaho, Illinois, Kentucky, Louisiana, Maryland, Michigan, Minnesota, Mississippi, Missouri, New Hampshire, New Jersey, New Mexico, North Carolina, Ohio, Oklahoma, Oregon, Pennsylvania, South Carolina, Texas, Utah, Virginia, Washington, West Virginia, Wisconsin
Date(s) Application is Due: Aug 31
Contact: Anne Levya, Team Coordinator; (253) 924-3159; fax (253) 924-3658; foundation@weyerhaeuser.com or anne.leyva@weyerhaeuser.com
Internet: http://www.weyerhaeuser.com/Sustainability/Foundation
Sponsor: Weyerhaeuser Company Foundation
P.O. Box 9777
Federal Way, WA 98063-9777

Weyerhaeuser Family Foundation Health Grants 4714
The Weyerhaeuser Family Foundation supports programs of national and international significance that promote the welfare of human and natural resources. These efforts will enhance the creativity, strengths and skills already possessed by those in need and reinforce the sustaining processes inherent in nature. The Foundation supports multi-site, national or international projects dealing with mental health, chemical dependency and population and family planning. Multi-site, national and international educational projects will also be considered. The Letter of Intent is the first step in the application process and should be no more than two pages, to which you must attach a one-page budget summary and an Application Cover Sheet-General Program.
Requirements: U.S. nonprofit, tax-exempt organizations are eligible.
Restrictions: The General Program does not fund projects serving only local or regional domestic areas; to be eligible, projects must be multi-site, national or international. In addition, the Foundation will not consider proposals in the following areas: books or media projects, unless the project is connected to other areas of Foundation interest; capital projects; individuals, scholarships or fellowships; land acquisitions or trades; lobbying activity; ongoing projects or general operating support for an organization; organizations located outside the United States; or research projects.
Date(s) Application is Due: Apr 1
Amount of Grant: 5,000 - 30,000 U.S.D
Contact: Peter A. Konrad, Program Consultant; (303) 993-5385
Internet: http://www.wfamilyfoundation.org/general_program.html
Sponsor: Weyerhaeuser Family Foundation
2000 Wells Fargo Place, 30 East Seventh Street
St. Paul, MN 55101-4930

WHAS Crusade for Children Grants 4715
The organization, established in Kentucky in 1954, is dedicated to helping the handicapped children of today and to expanding the ability to care for and prevent the handicaps of tomorrow. The crusade supports organizations which serve children suffering from a variety of mental and physical disabilities. WHAS Crusade for Children grants are made to non-profit agencies, schools and hospitals to help children (up to age 18) overcome physical, mental, emotional or medical challenges. Grants are for direct services only, and are made for specific programs or equipment that provide direct benefit to special needs children. Applications are mailed out each year in November and December. All agencies that received a grant this year receive an application for next year. To request an application, a non-profit agency, school or hospital should send a letter (on letterhead signed by the agency director or school superintendent) to the WHAS Crusade for Children at the address listed. The deadline for applications is late January each year.
Requirements: Agencies serving special needs children from Kentucky and southern Indiana are eligible to receive grants. The Crusade must have support from a county before grants can be made to a school, hospital or agency located in that county.
Restrictions: Grants are not made to individuals, and are not considered operating grants.
Geographic Focus: Indiana, Kentucky
Contact: Dawn Lee, President; (502) 582-7706; fax (502) 582-7712; dlee@whas11.com or contact-crusade@whascrusade.org
Internet: http://www.whascrusade.org/grants/
Sponsor: WHAS Crusade for Children
P.O. Box 1100
Louisville, KY 40201-1100

Whitaker International Fellowships and Scholarships 4716
The goal of the Whitaker funding is to assist in the development of outstanding professional engineers and scientists who will lead and serve the profession with an international outlook. The term of the Fellowship awards will be one academic year (as defined by the academic calendar of the host country). The term of the Scholarship awards can be for as little as one academic semester or as long as a full academic year. Benefits include maintenance allowance, airfare, accident and sickness insurance, partial tuition reimbursement (Fellows only), and access to grantee events, as well as membership in an elite alumni resource network. Applicants should consult the website for specific information about host institutions in Australia, Brazil, France, Germany, Ireland, Italy, The Netherlands, South Africa, Spain, Switzerland, and the United Kingdom.
Requirements: U.S. citizens and non-citizen permanent residents may apply. Fellows must have a BS or MS degree in biomedical engineering or bioengineering, and they will not have a doctorate at the time they start the grant. The Fellows applicant's most recent degree may not have been obtained more than two years prior to the start of the grant. Scholars must have either a degree in biomedical engineering and a doctorate which may not have been obtained more than two years prior to the start of the grant, or must be advanced Ph.D. candidates who have been admitted to candidacy and who will receive the Ph.D. prior to departure.
Date(s) Application is Due: Jan 23
Amount of Grant: Up to 35,000 U.S.D
Contact: Sabeen Altaf, Program Manager; (212) 984-5442; whitaker@iie.org
Internet: http://www.iie.org/Programs/Whitaker-International-Fellows-and-Scholars-Program
Sponsor: Institute of International Education
809 United Nations Plaza
New York, NY 10017-3580

White County Community Foundation - Adam Krintz Memorial Scholarship 4717
The White County Community Foundation - Adam Krintz Memorial Scholarship is available to students entering any field of study at a college, university or vocational school. Applicants must be graduates of Twin Lakes High School, with well-rounded activities in school, community, and church. They should also have a GPA of at least 3.0.
Requirements: Along with the application, students are required to submit a personal insight essay; an official high school transcript; a copy of a letter of acceptance from their college or university of choice; and two recommendation letters. The application packet can be returned to their guidance office or the Foundation.
Geographic Focus: Indiana
Date(s) Application is Due: Feb 27
Amount of Grant: 500 U.S.D
Contact: Lesley Wineland Goss, Executive Director; (574) 583-6911; fax (574) 583-8757; director@whitecf.org
Internet: http://www.whitecf.org/GrantScol.html
Sponsor: White County Community Foundation
1001 South Main Street
Monticello, IN 47960

White County Community Foundation - Annie Horton Scholarship 4718
The White County Community Foundation - Annie Horton Scholarship is available for a White County high school freshman, sophomore, or junior accepted to a summer program in the field of academics or leadership. The award will provide up to $500 for the cost of a summer program, which may include tuition, room and board. Preference will be given (but not limited) to a student accepted at a summer program in the field of journalism, art, music, science or leadership. Enthusiasm and interest in the summer program will be the most important criterion for acceptance, followed by academic standing and financial need.
Requirements: In addition to the application, students submit a personal essay and two letters of recommendation.
Geographic Focus: Indiana
Date(s) Application is Due: Apr 18
Amount of Grant: Up to 500 U.S.D
Contact: Lesley Wineland Goss, Executive Director; (574) 583-6911; fax (574) 583-8757; director@whitecf.org
Internet: http://www.whitecf.org/GrantScol.html
Sponsor: White County Community Foundation
1001 South Main Street
Monticello, IN 47960

White County Community Foundation - Landis Memorial Scholarship 4719
The White County Community Foundation - Landis Memorial Scholarship is available to Twin Lakes graduating seniors intending to pursue higher education at a college, university, or school of training. Applicants must be well-rounded students who are active in their school, community, or church, with a B average or better. Financial need will be a strong consideration. Any college major is eligible.
Requirements: In addition to the online application, student submit a personal insight essay; an official high school transcript; a copy of their college acceptance letter; and a written recommendation from a teacher. The application packet may be sent to the student's guidance office by February 7 or to the Foundation by February 29.
Geographic Focus: Indiana
Date(s) Application is Due: Feb 7; Feb 29
Amount of Grant: 10,000 U.S.D
Contact: Lesley Wineland Goss, Executive Director; (574) 583-6911; fax (574) 583-8757; director@whitecf.org
Internet: http://www.whitecf.org/GrantScol.html
Sponsor: White County Community Foundation
1001 South Main Street
Monticello, IN 47960

White County Community Foundation - Lilly Endowment Scholarships 4720
The Lilly Endowment Community Scholarship Program is designed to raise the level of educational attainment in Indiana and increase awareness of the potential of Indiana's community foundations to improve the quality of life of the state's residents. Scholarships pay for full academic tuition and required fees plus an annual $800 allocation for books and equipment for four years at any accredited Indiana college or university selected by the recipient. The scholarships are sponsored by Lilly Endowment Inc., and administered by community foundations throughout Indiana. Applicants must: be a U.S. citizen and have graduated from an accredited White County high school; have at least a 2.75 GPA; and have a combined SAT score of 800 or above (critical reading & math). Selection considerations include: financial need; community and school activities and evidence of leadership; the essay; grade point average and difficulty of high school courses taken; and the interview of finalists. Other scholarships, such as 21st Century Scholars, will be taken into consideration when calculating the applicant's financial need. Any college major may apply.
Requirements: In addition to the online application, students must submit a completed financial aid form, high school transcript, spring class schedule, an essay, and three letters of recommendation. The application packet should be submitted to the Foundation in early January.
Geographic Focus: Indiana
Date(s) Application is Due: Jan 17
Contact: Lesley Wineland Goss, Executive Director; (574) 583-6911; fax (574) 583-8757; director@whitecf.org
Internet: http://www.whitecf.org/GrantScol.html
Sponsor: White County Community Foundation
1001 South Main Street
Monticello, IN 47960

White County Community Foundation - Tri-County Educational Scholarships 4721
The White County Community Foundation - Tri-County Educational Scholarships offer funding for Tri-County seniors pursuing higher education at a college, university, or school of training. Applications will be reviewed using the following criteria: academic record; school-sponsored activities; community service and personal character; and a personal essay. Several memorial scholarships are available from $500 to $1,000. Any college major is eligible.
Requirements: Along with the online application, students must submit a personal insight essay, and an official high school transcript. The application packet is submitted to the student's guidance office by March 1.
Restrictions: Applicants must be graduating seniors from Tri-County High School.
Geographic Focus: Indiana
Date(s) Application is Due: Mar 1
Amount of Grant: 500 - 1,000 U.S.D
Contact: Lesley Wineland Goss, Executive Director; (574) 583-6911; fax (574) 583-8757; director@whitecf.org
Internet: http://www.whitecf.org/GrantScol.html
Sponsor: White County Community Foundation
1001 South Main Street
Monticello, IN 47960

White County Community Foundation - Women Giving Together Grants 4722
The White County Community Foundation - Women Giving Together Grants address White County's women and families, and inspire women to strengthen White County through charitable giving. Members of Women Giving Together believe that high priority issues should be addressed thoughtfully to inspire social change and improve the quality of life in White County. Grant funding focuses on community enhancement, education, and social/human services. Program priorities: must serve a charitable purpose; serve the needs of the women and families of the greater White County area; demonstrate an innovative and unique approach; and indicate other funding sources.
Requirements: Applicants are encouraged to contact the Foundation to be certain their project meets the Foundation's guidelines. Organizations submit the online application to include the following information: their organization's contact information and organization's mission statement; amount requested for the project with its timeline; a narrative that describes the organization, its need, project description, and how it will benefit the White County area; a list of those responsible, with timeline and funding sources other than the WGT; how the project will be financed in the future and how it will affect the community. Organizations should also submit a detailed budget; financial statements; the IRS determination letter; and a list of board members and their authorization for the project.
Restrictions: Projects not funded include: non-charitable projects and organizations; projects fully funded by local government; projects to fund an endowment, ongoing operating budgets, existing deficit, debt reduction, or multi-year, long-term funding; religious activities that do not serve the community as a whole; special events such as parades, festivals, or sporting activities; political organizations or campaigns; national and state fundraising efforts; project that indicate discrimination.
Geographic Focus: Indiana
Date(s) Application is Due: Oct 1
Contact: Lesley Wineland Goss, Executive Director; (574) 583-6911; fax (574) 583-8757; director@whitecf.org
Internet: http://www.whitecf.org/WGTGrant.html
Sponsor: White County Community Foundation
1001 South Main Street
Monticello, IN 47960

White County Community Foundation Grants 4723
The White County Community Foundation Grants fund projects in the following categories: education - to develop the untapped capacities of those of all ages, families, or communities; human services - to achieve a positive change in the conditions that adversely affect the elderly, disabled, or economically disadvantaged; recreation - to create or expand family-oriented leisure time opportunities; arts and culture - to encourage and stimulate the arts and historical preservation; citizenship - to generate an increased level of volunteerism, increased county-wide cooperation, and new leadership; environment - to help protect or enhance the environment that must support our life systems beyond the next century; and economic - to generate economic development activity within the area of direct job creation opportunity in White County. Priority is given to projects that: reach as many people as possible; identify and address an immediate community need; improve the ability of the organization to serve the community over the long-term; are run by non-profit organization; serve White County; and demonstrate fundraising from sources other that the Foundation as well as support from within the requesting organization.
Requirements: Before applying, organizations should contact the Foundation to verify deadlines and funding priorities, and to ensure they have the current application form. Once applicants reach this stage, they should include the following information with the online or hard copy application: a brief description of the organization; a narrative on the nature, purpose, and benefits of the project; a description of the project coordination; a project timeline with anticipated start and completion dates; a detailed budget; funding sources and future plan for funding; the impact of the project on the community; organization's most recent year-end financial statements and current operating budget; IRS determination letter; organization's board members; board member authorization for project funding.
Restrictions: The following projects are not eligible for funding: individuals; projects outside the White County area; programs normally funded by local government; projects to fund an endowment, ongoing operating budget, existing deficit, debt reduction; or multi-year, long-term funding; religious activities that predominantly serve one denomination; special events such as parades, festivals, and sporting activities; political organizations or campaigns; national and state fundraising efforts; and projects that indicate discrimination.
Geographic Focus: Indiana
Contact: Lesley Wineland Goss, Executive Director; (574) 583-6911; fax (574) 583-8757; director@whitecf.org
Internet: http://www.whitecf.org/CommGrant.html
Sponsor: White County Community Foundation
1001 South Main Street
Monticello, IN 47960

Whitehall Foundation Neurobiology Research Grants 4724
Through its program of grants and grants in aid, the foundation assists scholarly research in the life sciences. The foundation is currently interested in basic research in invertebrate and vertebrate (excluding clinical) neurobiology, specifically investigations of neural mechanisms involved in sensory, motor, and other complex functions of the whole organism as these relate to behavior. Research grants are available to established scientists of all ages working at accredited institutions and will be awarded for up to three years, with possible renewal on a competitive basis for a maximum of two years. Grants in aid are designed for researchers at the assistant professor level who experience difficulty in competing for research funds because they have not yet become firmly established. Grants in aid will also be made to senior scientists. The foundation's policy is to assist those dynamic areas of basic biological research that are not heavily supported by federal agencies or other foundations with specialized missions. Types of support include personnel (salary support for students and technicians); permanent equipment, materials, and supplies needed to complete the project; travel for research; publication costs; and overhead (no more than 25 percent or supplies). Applicant should initiate the proposal process by submitting a one-page letter written in technical language identifying the nature of the research proposal and indicating whether the request is for a grant or grant in aid. Grant application forms will be sent to those applicants whose proposed project matches the foundation's current funding interests. Deadlines listed are for letters of intent.
Requirements: Accredited U.S. institutions are eligible. The foundation prefers to support scientists at the beginning of their career and those senior scientists who have maintained productivity. The principal investigator must hold no less than the position of assistant professor or the equivalent.
Date(s) Application is Due: Jan 15; Apr 15; Oct 1
Amount of Grant: Up to 30,000 U.S.D
Samples: Daniel Pollen, U of Massachusetts (MA)—for research on the cerebral integration in the primate visual system, $40,000 of three-year grant; Richard Ivry, U of California (CA)—for research on the role of the cerebellum in temporal processing, $24,000 final payment on three-year grant; Paul Black, Boston U (Boston, MA)—to support research on brain-immune system interactions, $130,020 final payment on three-year grant.
Contact: Program Director; (561) 655-4474; fax (561) 655-4978; email@whitehall.org
Internet: http://www.whitehall.org
Sponsor: Whitehall Foundation
P.O. Box 3423
Palm Beach, FL 33480

Whitley County Community Foundation - Lilly Endowment Scholarship 4725
The Whitley County Lilly Endowment Scholarship covers full tuition and $800 per year for books and equipment for four years. Applicants must be Whitley County residents, high school graduates, pursuing a full-time baccalaurate course of study at a public or private college or university in Indiana. They must also have financial need and demonstrate pride in their community, responsibility, perseverance, self-esteem, and the desire to contribute through work or volunteerism. Any four year field of study is eligible.
Requirements: In addition to the application, students should include the following information: their high school transcript and SAT scores; two letters of recommendation; their parent or guardian's current tax forms; information regarding all financial resources available for funding their education, including other scholarships received or pending; and two essay discussions.
Geographic Focus: Indiana
Date(s) Application is Due: Jan 1
Contact: September McConnell, Executive Director; (260) 244-5224; fax (260) 244-5724; sepwccf@gmail.com
Internet: http://whitleycountycommunityfoundation.org/lilly.html
Sponsor: Whitley County Community Foundation
400 North Whitley Street
Columbia City, IN 46725

Whitley County Community Foundation Grants 4726
The Whitley County Community Foundation directs grants to charitable projects that will make a positive impact on Whitley County and its people, with particular interest in projects that shed new light on local needs and provide innovative, long-term solutions. Categories of support include: arts & culture, health, civic affairs, recreation, community development, welfare, and education. Grant reviews are scheduled on May 1 and December 1, but applicants are encouraged to submit proposals well in advance.
Requirements: The grant application must include a copy of the organization's IRS tax exempt letter, its annual budget and its projected budget. The cover letter should include: an introduction that establishes the organization's purpose and credibility, with background and accomplishments; a statement of need addressing why the project is necessary, documenting the need with statistics; what the organization hopes to accomplish with measurable objectives; what methods will be used to analyze results and refine the program; and how the project will be funded or maintained in the future. The packet may also include endorsement letters, a list of the board of directors, and a list of past support from other funders.

Restrictions: The Foundation is unlikely to support: annual campaigns, political activities, private schools, advertising, religious/sectarian causes, organizations outside the Whitley service area, and debt retirement.
Geographic Focus: Indiana
Date(s) Application is Due: May 1; Dec 1
Amount of Grant: 500 - 15,000 U.S.D
Contact: September McConnell, Executive Director; (260) 244-5224; fax (260) 244-5724; sepwccf@gmail.com
Internet: http://whitleycountycommunityfoundation.org/grantseekers.html
Sponsor: Whitley County Community Foundation
400 North Whitley Street
Columbia City, IN 46725

Whitley County Community Foundation Scholarships 4727
The Whitley County Community Foundation offers a variety of scholarships with criteria specific to the particular scholarships. The Foundation posts a list of scholarships categorized by high school, county, college major, and non-traditional student. Applicants should carefully read the information to see if they should submit a hardcopy of the application materials, submit the information as an email attachment to the community foundation, or if there is a specific business address to send the application materials.
Restrictions: Scholarships are available only to high school seniors from Whitko, Columbia City, and Churubusco.
Geographic Focus: Indiana
Contact: September McConnell, Program Manager; (260) 925-0311; fax (260) 244-5724; sepwccf@gmail.com
Internet: http://whitleycountycommunityfoundation.org/whitley.html
Sponsor: Whitley County Community Foundation
400 North Whitley Street
Columbia City, IN 46725

Whitney Foundation Grants 4728
The Foundation awards grants to nonprofit organizations primarily in Minnesota in its areas of interest, including AIDS prevention, arts, children and youth services, education, and human services. Types of support include program development, annual campaigns, and continuing support. Application forms are not required. Awards range from $100 to $5,000.
Requirements: Eligible applicants must be 501(c)3 organizations.
Geographic Focus: Minnesota
Samples: Marjorie McNeely Conservatory, St. Paul, Minnesota, $500 - general support (2010); Lake Forest Hospital Foundation, Lake Forest, Illinois, $250 - general support (2010); and Minnesota Zoo Foundation, Apple Valley, Minnesota, $250 - general support (2010).
Contact: Carol VanOrnum, (952) 835-2577
Sponsor: Whitney Foundation
601 Carlson Parkway
Minnetonka, MN 55305

WHO Foundation General Grants 4729
The foundation nationally supports grass-roots charities serving the overlooked needs of women and children. Grants are provided to organizations serving women and/or children in the United States and Puerto Rico. Specific projects and programs addressing health, education, and social service needs are the priority. The foundation recognizes the value of new programs created to respond to changing needs and will consider funding projects of an original or pioneering nature within an existing organization. Application and guidelines are available for download at the sponsor's website.
Requirements: 501(c)3 nonprofit organizations in the United States and Puerto Rico are eligible. Organizations must have been incorporated for a minimum of three years prior to application. Preference will be given to organizations with an operating budget of $3 million or less, those not dependent upon government grants, and those with greater organizational program costs than personnel costs. Funding requests must be made using the WHO Foundation application. Electronic and faxed submissions will not be accepted.
Restrictions: The following types of organizations, activities or purposes will not be considered: personal requests, loans or scholarships to individuals; educational institutions; endowment campaigns; international programs or projects; government agencies; fiscal agents; religious organizations (including young Men's and Women's Christian Association); political causes, candidates, organizations or campaigns; foundations that are grant making institutions; advertising in charitable publications; sports organizations; labor groups; research projects; travel for individuals or groups; conferences, galas, charity balls, sponsorships, seminars, or reunions; capital campaigns; salaries; or building campaigns (i.e. Habitat for Humanity).
Amount of Grant: 5,000 - 30,000 U.S.D
Contact: Cindy Turek; (800) 946-4663; fax (972) 386-8736; who@beauticontrol.com
Internet: http://www.whofoundation.org/WHO_FundingCriteria.htm
Sponsor: WHO (Women Helping Others) Foundation
P.O. Box 816029
Dallas, TX 75381-6029

WICHE Professional Student Exchange Program (PSEP) 4730
This program is open to students who are residents of Arizona, Colorado, Hawaii, Idaho, Montana, Nevada, New Mexico, North Dakota, Oregon, Utah, Washington, and Wyoming. The program has been developed to facilitate interstate resource sharing by providing means for states without specific professional schools to send residents to institutions in other WICHE states. Professional schools are those of medicine, dentistry, veterinary medicine, physical therapy, occupational therapy, optometry, podiatry, pharmacy, architecture, law, graduate library studies, graduate nursing education, public health, maritime technology, osteopathic medicine, and physician assistant. The student will pay (approximately) in-state or partial private tuition, and the sending state pays the receiving school an additional support fee. More information on the program, including eligibility requirements and restrictions, can be found on the Web site, or by contacting the program coordinator. In most states, the deadline for receiving completed applications for certification is October 15 of the year preceding admission. However, a few states have earlier deadlines or accept late applications. Contact program staff for exact dates.
Requirements: The program is limited to professional fields requested by member states (Alaska, Arizona, Colorado, Hawaii, Idaho, Montana, Nevada, New Mexico, North Dakota, Oregon, Utah, Washington, and Wyoming) and approved by WICHE. Students must meet state requirements. Support is limited to states that appropriate funds to assist specific numbers of students in specific fields.
Geographic Focus: Alaska, Arizona, Colorado, Hawaii, Idaho, Montana, Nevada, New Mexico, North Dakota, Oregon, Utah, Washington, Wyoming
Date(s) Application is Due: Oct 15
Contact: Coordinator; (303) 541-0214; sjackson@wiche.edu or info-sep@wiche.edu
Internet: http://www.wiche.edu/SEP/index.asp
Sponsor: Western Interstate Commission for Higher Education
P.O. Box 9752
Boulder, CO 80301-9752

Wichita Community Foundation Grants 4731
The foundation was established to serve the needs of the people of south-central Kansas. The foundation invites area nonprofit organizations to submit proposals that respond to community needs in the areas of arts and humanities, education, environment and conservation, and health and human social services. Emphasis will be placed on proposals that utilize innovative and creative approaches to the needs in each of these areas. Priority will be given for seed grants to initiate promising new projects, programs representing innovative and efficient approaches to serving community needs, and challenge grants. Proposals must include a cover sheet; a proposal, including the concept and objectives and the project's background, a time table, need, and benefits; a copy of the federal tax-exemption letter; the amount requested; a list of current members of the governing board; and a copy of the organization's latest audited financial statement. Application forms are available on the Web site.
Requirements: Grant requests will only be considered from nonprofit, tax-exempt agencies providing services to citizens of specified Kansas counties, including Harper, Kingman, Harvey, Reno, Butler, Cowley, Sedgwick, and Sumner.
Restrictions: Grants will not be made to or for general operating funds, in most cases; capital fund-raising campaigns, special events, or building projects; sectarian purposes; individual benefits; periods exceeding one year; endowment funds; debt elimination; fellowships or travel purposes; medical research; or charities affiliated with a political candidate or cause.
Geographic Focus: Kansas
Date(s) Application is Due: Aug 15; Aug 30
Amount of Grant: Up to 10,000 U.S.D
Samples: Community Mediation Task Force (KS)—to help train 200 citizens in the art and process of community mediation, $4000; Information Access Assoc (KS)—to purchase computer equipment including a Braille printer, $4700; Victory in the Valley (KS)—to establish an Emergency Medical Fund for travel, prescription drug, and medical supply costs for indigent patients, $5000.
Contact: Program Contact; (316) 264-4880; fax (316) 264-7592; wcf@wichitacf.org
Internet: http://www.wichitacf.org/new/grants.html
Sponsor: Wichita Community Foundation
200 W Douglas, Suite 250
Wichita, KS 67202

Widener Memorial Foundation in Aid of Handicapped Children Grants 4732
The foundation exists to support research into the causes, treatment, and prevention of diseases and conditions that handicap children orthopedically and to aid and assist public and private charitable organizations in the care, education, and rehabilitation of children with disabilities. Types of support include building/renovations, equipment, seed money, research, and program/project development. There are no application forms. The board meets each year in May and November to consider requests.
Requirements: Nonprofit organizations in Delaware Valley, Pennsylvania, may apply for grant support.
Restrictions: Grants are not awarded to individuals for scholarships or fellowships or for loans, endowment funds, or matching gifts.
Geographic Focus: Pennsylvania
Date(s) Application is Due: Apr 15; Oct 15
Amount of Grant: 15,000 - 100,000 U.S.D
Samples: West Chester U (PA)—to renovate an elevator in Mitchell Hall to improve access for disabled people, $100,000.
Contact: F. Eugene Dixon Jr., President; (215) 836-7500
Sponsor: Widener Memorial Foundation in Aid of Handicapped Children
665 Thomas Road, P.O. Box 178
Lafayette Hill, PA 19444-0178

Wiley Prize in Biomedical Sciences 4733
The prize is intended to recognize contributions that have opened new fields of research or advanced novel concepts or their applications in a particular biomedical discipline. The award may recognize a specific contribution or a series of contributions that demonstrate the nominee's significant leadership in the development of research concepts or their

clinical application. The award will consist of a grant, and each year's recipient will deliver a lecture at Rockefeller University.
Requirements: Wiley invites and encourages the nomination of exceptional PhD and MD scientists whose research has set the standard for excellence. More than one nomination can be made from the same organization. International nominations are eligible, and the number of years in a position is not relevant to the nomination.
Date(s) Application is Due: Jul 31
Amount of Grant: 25,000 U.S.D
Contact: Deborah Wiley, (212) 748-6783; fax (212) 748-6940; dwiley@wiley.com
Internet: http://www.wiley.com/legacy/wileyfoundation/nominations.html
Sponsor: Wiley Foundation
111 River Street
Hoboken, NJ 07030

Wilhelm Sander Foundation Grants **4734**
The purpose of the Foundations grants are to support research in the field of human medicine with a focus on cancer research.
Restrictions: The grant is limited to Germany and Switzerland.
Geographic Focus: All States, Germany, Sweden
Contact: Dr. J. Koppenhofer, (089) 544-1870; fax (089) 544-18720; info@sanst.de
Internet: http://www.wilhelm-sander-stiftung.de
Sponsor: Wilhelm Sander Foundation
Goethe route 74
Munich, 80336 Germany

Wilkins Family Foundation Grants **4735**
The foundation awards grants to U.S. nonprofit organizations in its areas of interest, including projects, community service organizations, economic development activities, and cultural organizations involved with education and health issues of children 17 years of age and younger. Letters of intent are accepted online; full proposals are by invitation.
Requirements: 501(c)3 tax-exempt organizations are eligible.
Restrictions: The foundation does not make multiyear commitments and prefers to fund causes, rather than management expenses.
Date(s) Application is Due: Feb 1
Contact: Grants Administrator; proposals@wilkinsfamilyfoundation.org
Internet: http://fdncenter.org/grantmaker/wilkinsfamily/procedure.html
Sponsor: Wilkins Family Foundation
302 N Rock Road, Suite 200
Wichita, KS 67206-2255

Willard and Pat Walker Charitable Foundation Grants **4736**
The foundation awards grants to Arkansas nonprofit organizations in its areas of interest, including arts, health organizations, higher education, and social services. Types of support include building construction/renovation, general operating grants, and program development. There are no application forms or deadlines.
Requirements: Arkansas nonprofit organizations are eligible.
Restrictions: Individuals are not eligible.
Geographic Focus: Arkansas
Samples: Ouachita Baptist (Arkadelphia, AR)—for its capital campaign, $2.5 million; U of Arkansas for Medical Sciences (Little Rock, AR)—to expand research and patient care in the areas of Alzheimer's disease, eye diseases, and mental illness, $21.5 million; U of Arkansas (Fayetteville, AR)—to construct a building as part of the College of Business, $8 million.
Contact: Tommy Karr, (501) 756-7031
Sponsor: Willard and Pat Walker Charitable Foundation
P.O. Box 5000
Springdale, AR 72765-5000

William A. Badger Foundation Grants **4737**
The Nabors to Neighbors Foundation was created in 2007 to assist charitable organizations focusing on need based projects for direct programming, capital and operating initiatives. This is a family foundation dedicated to organizations that deliver measurable results, seek partnerships and collaborations, utilize their resources within their respective communities while working to increase equity for those most in need. The Foundation encourages nonprofit organizations to apply who specialize in, though not limited to, improving the lives of children, education and medical initiatives. The Foundation makes no geographic restrictions on distributions. However, it has been the practice of the Trustees to make grants within Whitfield County, specifically Dalton, located in north Georgia. Requests must be postmarked by February 1, July 9 or October 1 in order to be considered.
Requirements: The Nabors to Neighbors Foundation makes grants to qualified 501 (c)3 organizations. All requests must include: background information on the organization, including a brief history, the organization's current address and phone number and the name and title or the primary contact; the goals, objectives, and budget for the one project or program for which funds are being requested; the amount of the grant requested; summary of how the funds will be used; supporting financial information on the organization, to include current financial status and listing of Board of Trustees; copy of organization's tax exemption letter from the Internal Revenue Service. Application form is available online and all requests must be postmarked by February 1, July 9 or October 1 in order to be considered.
Restrictions: Grants are not made to: individuals; an organization to be used as pass-through funds for an ineligible organization Faith-based organizations without a 501(c)3 exemption; organizations with political purposes, nor to organizations which discriminate on the basis of race, ethnic origin, sexual or religious preference, age or gender.
Date(s) Application is Due: Feb 1; Jul 9; Oct 1
Contact: Trustee, c/o Wachovia Bank; grantinquiries8@wachovia.com
Internet: https://www.wachovia.com/foundation/v/index.jsp?vgnextoid=108bf296ac21 2210VgnVCM100000617d6fa2RCRD&vgnextfmt=default
Sponsor: Nabors to Neighbors Foundation
3280 Peachtree Road NE, Suite 400, MC G0141-041
Atlanta, GA 30305

William Adlington Cadbury Charitable Trust Grants **4738**
The trust was established to support general charitable purposes in the West Midlands, the rest of the United Kingdom including projects in Northern Ireland, and UK charities working overseas. Areas of interest include social welfare, medical and health-care projects including medical research, education and training, The Religious Society of Friends (Quakers), places of religious worship and associated social projects, conservation of the environment including the preservation of listed buildings and monuments, arts, and penal affairs. Applications are considered on a continuing basis throughout the year. Application and guidelines are available online.
Restrictions: The trust does not fund individuals (whether for research, expeditions or educational purposes); projects concerned with travel, adventure, sports, or recreation; organizations that do not have UK charity registration (except those legally exempt from registration); or overseas charities not registered in the United Kingdom.
Geographic Focus: All States, United Kingdom
Contact: Carolyn Bettis, Trust Administrator; 44-121-472-1464
Internet: http://www.wa-cadbury.org.uk
Sponsor: William Adlington Cadbury Charitable Trust
Bristol Road, Selly Oak
Birmingham, B29 6QF England

William B. Dietrich Foundation Grants **4739**
The Foundation awards funding to nonprofits preferably for local needs. There are no submission deadlines. Areas of interest include children, the elderly, AIDS, museums, and libraries.
Requirements: Applicants may apply in writing, outlining the nature of the organization, the intended use of funds requested. A copy of the Internal Revenue Service determination letter should be submitted.
Restrictions: No grants are provided for individuals.
Samples: AIDS Project, Los Angeles, California, $25,000 - general purposes (2010); Fleisher Art Memorial, Philadelphia, Pennsylvania, $50,000 - general purposes (2010); and University of Pennsylvania, Philadelphia, Pennsylvania, $155,750 - Museum of Archeology and Anthropology (2010).
Contact: Frank G. Cooper, President; (215) 979-1919
Sponsor: William B. Dietrich Foundation
P.O. Box 58177
Philadelphia, PA 19102-8177

William B. Stokely Jr. Foundation Grants **4740**
The Foundation participates in scholarship funding at various colleges and universities mainly in eastern Tennessee. Grants are made to the educational institutions, which then distribute funds through their scholarship programs. Consideration also is given to requests from the arts, health service, civic organizations, and youth services. The Foundation does not require completion of a formal application, nor are there established deadline dates. All proposals must be submitted in writing.
Geographic Focus: Tennessee
Amount of Grant: 250 - 100,000 U.S.D
Samples: Joni and Friends Knoxville, Knoxville, Tennessee, $10,000 - disability ministry (2010); Mercy Health Partners, Knoxville, Tennessee, $100,000 - medical services (2010); and Friends of he Great Smoky Mountains, Kodak, Tennessee, $10,000 - land conservation (2010).
Contact: William Stokely III, President; (865) 966-4878
Sponsor: William B. Stokely Jr. Foundation
620 Campbell Station Road, Suite 27
Knoxville, TN 37922-1636

William Blair and Company Foundation Grants **4741**
Contributing to the community is an important part of the culture of William Blair and Company. The Foundation was officially established in Illinois in 1980, with giving primarily centered around metropolitan Chicago. All partners of the firm contribute part of their individual share of profits to the Foundation. Donation requests are made to the Foundation by partners and employees. The Foundation supports a broad range of causes including: civic affairs; public safety; arts and culture; higher education; youth-oriented activities; healthcare research; cultural affairs; and civic charities. Types of support include: annual campaigns; building and renovation; capital campaigns; general operating support; endowments; fellowships; internship programs; and scholarship funding. There are no specific deadlines or applications forms required. Funding typically ranges from $500 to $25,000.
Requirements: Requests can be made by sending a letter with a general description of the organization and its special purpose. Activities should have a significant impact on the Chicago metropolitan area.
Geographic Focus: Illinois
Contact: E. David Coolidge III, Vice President; (312) 236-1600
Sponsor: William Blair and Company Foundation
222 W Adams Street, 28th Floor
Chicago, IL 60606

William C. Woolf Foundation Grants 4742

The foundation exists to enhance the quality of life for citizens of Shreveport, LA, and awards one-year grants to eligible Louisiana nonprofit organizations for a variety of programs and activities. Proposals must meet the listed application deadline and include a copy of the IRS tax-determination letter.

Requirements: 501(c)3 tax-exempt organizations located in northwestern Louisiana are eligible.

Restrictions: Grants are not made to individuals.

Geographic Focus: Louisiana

Amount of Grant: 300 - 50,000 U.S.D

Contact: Barbara York, Chair

Sponsor: William C. Woolf Foundation
333 Texas Street, SH 2069
Shreveport, LA 71101

William G. and Helen C. Hoffman Foundation Grants 4743

Helen C. Hoffman resided in the Village of South Orange, New Jersey. Her foundation was established in 1998 in memory of herself and her husband after the death of their daughter Corinne Blair. Her testamentary wish was to establish this foundation for charitable, religious, scientific, literary, and educational purposes. Her preference was to support blindness and its cure. Approximately 90% of the grants will provide support to the blind and, to medical research for the prevention of blindness. The remaining 10% will fund annual grants in the following areas of interest: education, the arts, environment and, social/civic causes. Requests must be received by January 15 for the March meeting and, August 22 for the October meeting. Application forms are available online. Applicants will receive notice acknowledging receipt of the grant request, and subsequently be notified of the grant declination or approval.

Requirements: Any U.S. 501(c)3 non-profit organizations may apply. Proposals should be submitted in the following format: completed Common Grant Application Form; an original Proposal Statement; an audited financial report and a current year operating budget; a copy of your official IRS Letter with your tax determination; a listing of your Board of Directors. Proposal Statements (second item in the above Format) should answer these questions: what are the objectives and expected outcomes of this program/project/request; what strategies will be used to accomplish your objective; what is the timeline for completion; if this is part of an on-going program, how long has it been in operation; what criteria will you use to measure success; if the request is not fully funded, what other sources can you engage; an Itemized budget should be included; please describe any collaborative ventures. Prior to the distribution of funds, all approved grantees must sign and return a Grant Agreement Form, stating that the funds will be used for the purpose intended. Progress reports and Completion reports must also be filed as required for your specific grant. All current grantees must be in good standing with required documentation prior to submitting new proposals to any foundation.

Restrictions: Grants are not made for political purposes, nor to organizations which discriminate on the basis of race, ethnic origin, sexual or religious preference, age or gender.

Date(s) Application is Due: Jan 15; Aug 22

Amount of Grant: 5,000 - 15,000 U.S.D

Samples: Tri-County Scholarship Fund, $5,000—scholarships for economically disadvantaged students to receive a private elementary/secondary education; New Eyes for the Needy, Inc., $20,000— eyeglasses for poor Kentuckians; Christian Health Care Center Foundation, $10,000— support senior subsidies for adult day services.

Contact: Wachovia Bank, N.A., Trustee; grantinquiries2@wachovia.com

Internet: https://www.wachovia.com/foundation/v/index.jsp?vgnextoid=522852199c0aa110VgnVCM1000004b0d1872RCRD&vgnextfmt=default

Sponsor: William G. and Helen C. Hoffman Foundation
190 River Road, NJ3132
Summit, NJ 07901

William G. Gilmore Foundation Grants 4744

The foundation supports nonprofits primarily in the San Francisco Bay Area, CA; some funding also in Pueblo, CO and Portland, OR. Giving primarily for the arts, health, and children, youth, and social services.

Requirements: California, Colorado, and Oregon nonprofits are eligible to apply. Contact the Foundation for additional application information.

Restrictions: No grants to individuals.

Geographic Focus: California, Colorado, Oregon

Amount of Grant: 1,000 - 50,000 U.S.D

Samples: YMCA of Pueblo, Pueblo, CO, $50,000; Queen of the Valley Hospital Foundation, Napa, CA, $50,000; Pueblo Child Advocacy Center, Pueblo, CO, $5,000;

Contact: Faye Wilson, Executive Director; (415) 546-1400; fax (415) 391-8732

Sponsor: William G. Gilmore Foundation
120 Montgomery Street, Suite 1880
San Francisco, CA 94104-4317

William G. Kelley Foundation Grants 4745

The foundation's grants are available to help individuals and families overcome physical challenges. Grants are awarded based on physical and financial need. Priority will be given to applicants requesting assistance in the fields of assistive technology or universal design. Interested applicants are invited to send a letter of inquiry to the foundation, at the above address, describing the intended use of the funds, including estimates for materials and labor. Applicants should demonstrate financial need. One project grant will also be available to a nonprofit organization that provides support to individuals with physical challenges. Projects aiming to foster inclusion of people with challenges through accessibility and universal design will be given priority. Grant applications are accepted on a rolling basis. Application and guidelines are available online.

Requirements: Residents and nonprofit organizations of Essex and Middlesex Counties in Massachusetts may apply.

Geographic Focus: Massachusetts

Amount of Grant: Up to 2,000 U.S.D

Contact: Grants Administrator; info@williamgkelley.org

Internet: http://www.williamgkelley.org/news_events.htm

Sponsor: William G. Kelley Foundation
1 Marlboro Road
Georgetown, MA 01833

William G. McGowan Charitable Fund Grants 4746

The fund awards grants to eligible nonprofit organizations that are devoted to three specific philanthropic goals: developing the talents and gifts of the very young, especially those who have been disenfranchised by virtue of low income status, inner-city conditions or family situations; funding for selected areas of medical research directed toward finding cures for, or relieving the pain and suffering of, those afflicted with debilitating conditions or diseases; and the McGowan Scholars program, which provides financial assistance to selected college and university students interested in pursuing an undergraduate or graduate degree at a college or university business school accredited by the International Association for Management Education or Association of Collegiate Business Schools and Programs.

Requirements: Nonprofit organizations in the following states and cities are eligible: Chicago, IL; District of Columbia and its Virginia suburbs; metropolitan Kansas City, western New York; and northeast Pennsylvania.

Restrictions: The fund does not support multiyear grants; accept requests online; or fund art or theatrical activities, building funds, church renovation campaigns, or endowments.

Geographic Focus: California, District of Columbia, Illinois, Missouri, New York, Pennsylvania, Virginia

Date(s) Application is Due: Jan 2; May 1; Sep 1

Amount of Grant: 10,000 - 100,000 U.S.D

Samples: Rochester Institute of Technology, College of Applied Science (NY)—to construct the Center for Telecommunications, $2 million; U of the Pacific (Stockton, CA)—for a scholarship for a business student, $18,000; Foundation for the National Archives (Washington, DC)—for the National Archives Experience, which will make important American documents accessible to the public and teach visitors about the value of archives, $5 million.

Contact: Bernard Goodrich, Executive Director; (301) 320-8570; fax (301) 320-8627; goodric@aol.com or info@mcgowanfund.org

Internet: http://www.mcgowanfund.org/guidelines.html

Sponsor: William G. McGowan Charitable Fund
P.O. Box 40515
Washington, DC 20016-0515

William G. Rohrer Charitable Foundation Grants 4747

The foundation awards grants to New Jersey nonprofit organizations in its areas of interest, including higher education, health care, health associations, and human services. There are no application deadlines or forms.

Geographic Focus: New Jersey

Samples: Rowan U (Glassboro, NJ)—to improve the academic offerings at the College of Business and to form partnerships with local businesses, $10 million over 20 years.

Contact: Grants Administrator, c/o PNC Advisors; (856) 755-5735

Sponsor: William G. Rohrer Charitable Foundation
Route 38 at Eastgate Drive
Moorestown, NJ 08057

William H. Gates Foundation Grants 4748

The foundation awards grants to support projects dealing with education and with global health and population issues, and to finance charities that operate primarily in the Pacific Northwest. Grants are awarded to shrink the gap that increasingly prevents poor people from being able to take advantage of crucial developments in science and technology. Unsolicited proposals are strongly discouraged, but the foundation sometimes follows up on short inquiries it receives through the mail, as long as the projects fit within the foundation's grantmaking guidelines.

Restrictions: Grants are not awarded to individuals.

Contact: Program Contact; (206) 709-3140; info@gates.scholarships.cam.ac.uk

Internet: http://www.gatesfoundation.org/ForGrantSeekers

Sponsor: William H. Gates Foundation
P.O. Box 23350
Seattle, WA 98102

William H. Hannon Foundation Grants 4749

The foundation awards grants to eligible California nonprofit organizations in its areas of interest, including Roman Catholic education (elementary, secondary, and higher); churches; medical research; hospitals; health care and human services, especially for those least able to afford quality care; and needs of the disadvantaged, aged, sick, and the homeless. There are no application deadlines or forms. Applications should be sent at least three weeks before quarterly board meetings, in March, June, September, and December.

Requirements: California nonprofit organizations serving Los Angeles and southern California areas are eligible.

Geographic Focus: California

Samples: Loyola High School of Los Angeles (CA)—for the school's new science center, $1 million; Loyola Law School (Los Angeles, CA)—to endow a professorship, $3 million.

Contact: Kathleen Hannon Aikenhead, President; (310) 260-2470; fax (310) 260-9740; williamhannonfdn@yahoo.com
Internet: http://www.hannonfoundation.org
Sponsor: William H. Hannon Foundation
729 Montana Avenue, Suite 5
Santa Monica, CA 90403

William I.H. and Lula E. Pitts Foundation Grants **4750**
The foundation awards grants to Georgia charities that are affiliated with the Methodist Church. Specific areas of interest include education, children's homes, charitable hospitals, and organizations that provide care for the elderly. Grantees have included Andrew College, Emory University-Pitts Theological Library, Epworth by the Sea, LaGrange College, Magnolia Manor Inc, Superannuated Ministers, Open Door Community House, and United Methodist Children's Homes. Application and guidelines are available online.
Requirements: The foundation awards grants to Georgia charities that are affiliated with the Methodist Church.
Geographic Focus: Georgia
Date(s) Application is Due: Feb 15; Sep 15
Amount of Grant: 15,000 - 100,000 U.S.D
Contact: Grants Administrator; (404) 724-3773; rs.foundationinfo@suntrust.com
Internet: http://www.pittsfoundation.org/guidelines.html
Sponsor: William I.H. and Lula E. Pitts Foundation
P.O. Box 4655, MC 221
Atlanta, GA 30302

William J. Brace Charitable Trust **4751**
The William J. Brace Charitable Trust was established in 1958 to support and promote quality educational, cultural, human-services, and health-care programming, with a preference for the following three areas: the education and health of children; the health and care of older adults; and hospitals in Kansas City, Missouri. Grant requests for general operating support and program support will be considered. Grants from the Trust are one year in duration. There are no application deadlines for the Brace Charitable Trust. Proposals are reviewed on an ongoing basis. Applicants may download the application and Missouri state guidelines at the grant website. Applicants are strongly encouraged to review the state guidelines for additional helpful information and clarifications on the three areas of interest.
Restrictions: The Brace Charitable Trust generally supports organizations that serve the residents of Kansas City, Missouri. Grant requests for capital support will not be considered.
Geographic Focus: Missouri
Contact: Spence Heddens; (816) 292-4301; Spence.heddens@baml.com
Internet: https://www.bankofamerica.com/philanthropic/fn_search.action
Sponsor: William J. Brace Charitable Trust
1200 Main Street, 14th Floor, P.O. Box 219119
Kansas City, MO 64121-9119

William K. Warren Foundation Grants **4752**
The foundation awards grants to eligible nonprofit organizations in its areas of interest, including Catholic healthcare facilities, Catholic social services, education, and medical research. Types of support include building construction/renovation, endowment campaigns, general operating support, program development, and research grants. Grants are awarded primarily in Oklahoma. There are no application forms or deadlines. The board meets semiannually.
Geographic Focus: Oklahoma
Amount of Grant: 1,000 - 50,000 U.S.D
Contact: Grants Administrator; (918) 492-8100
Sponsor: William K. Warren Foundation
P.O. Box 470372
Tulsa, OK 74147-0372

Williams Companies Foundation Grants **4753**
The Foundation provides opportunities to improve the quality of life including but not limited to health and welfare, scholarships, and education. Applicants should submit proof of nonprofit status and a letter with an outline describing the organization, the amount and purpose of the request, and whether the request is for general operating support or for a specific program. There are no application deadlines.
Requirements: Nonprofits in Williams operating communities are eligible.
Restrictions: No grants are made to individuals.
Geographic Focus: Alabama, Louisiana, New Jersey, North Carolina, Pennsylvania, South Carolina, Texas, Virginia
Contact: Alison Anthony, Community Relations Coordinator; (918) 573-4903; communityrelationstulsa@williams.com
Internet: http://www.williams.com/community/foundation.asp
Sponsor: Williams Companies Foundation
One Williams Center, MD 50-5
Tulsa, OK 74172

Williams Companies Grants **4754**
Grants are awarded to 501(c)3 nonprofit organizations in areas with a large company presence in support of civic, arts and culture, education, and health and human services programs/projects. Preferred submission time for contribution requests is February through November. Upon receipt of completed information, please allow a minimum of six weeks for a response. Requests are reviewed on an ongoing basis.
Requirements: Nonprofit organizations located where Williams companies operate are eligible to apply.
Contact: Community Relations Office; (918) 573-4179; communityrelationstulsa@williams.com
Internet: http://www.williams.com/community
Sponsor: Williams Companies
One Williams Center, 47th Floor
Tulsa, OK 74172

Williamsport-Lycoming Foundation Grants **4755**
The foundation supports Lycoming County, PA, nonprofit organizations in the broad areas of arts and culture, economic and community development, education, health and human services, and recreation and the environment. A top priority is economic development and youth development. Mini grants of up to $5000 are reviewed quarterly. Proposed projects should address community needs, show evidence of qualified leadership, demonstrate a stable funding base, and use volunteers and in-kind support. All applicants must file a current application and required attachments. Potential applicants are encouraged to discuss funding requests with staff prior to submitting a proposal.
Requirements: 501(c)3 nonprofits in Lycoming County, PA, are eligible.
Restrictions: Generally no grants or loans are provided for endowments (unless created in the foundation), ongoing operational support, individuals, debt reduction, research grants of a highly technical or specialized nature, fellowships, travel grants, sectarian religious purposes, clubs, sports teams, fire companies, or services limited to a select membership.
Geographic Focus: Pennsylvania
Date(s) Application is Due: Mar 1; Sep 1
Amount of Grant: 1,000 - 300,000 U.S.D
Samples: Community Theatre League (PA)—to construct a new facility, $150,000; Williamsport-Lycoming Chamber of Commerce (PA)—for economic development marketing, $85,000 over two years.
Contact: Dawn Linn; (570) 321-1500; fax (570) 321-6434; dawnl@fcfpa.org
Internet: http://www.wlfoundation.org/Grants/foundation_guidelines.htm
Sponsor: Williamsport-Lycoming Foundation
330 Pine Street, Suite 401
Williamsport, PA 17701

William T. Grant Foundation Research Grants **4756**
The foundation supports research to improve the lives of youth ages 8 to 25 in the United States. This is done primarily by investing in high quality empirical studies. The foundation's Current Research Interests are understanding and improving social settings (i.e., families, schools, peer groups, organizations, programs, etc.), their effects on youth, and the use and influence of scientific evidence. Within these interests, the foundation is interested in descriptive and intervention studies. There are no geographical boundaries for its support of research projects.
Requirements: The grants are limited, without exception, to established tax-exempt, private organizations and institutions. To apply for a major grant, the Principal Investigator must submit a Letter of Inquiry to the grants coordinator briefly describing the project or program and the financial needs. Letters of Inquiry are due on the deadline dates listed. If it is determined that the project falls within the current program interests and priorities, a full proposal will be requested. Full proposal deadlines are four to six months prior to the quarterly board meetings in March and October.
Restrictions: The foundation does not support or make contributions to building funds, fundraising drives, endowment funds, general operating budgets, or scholarships. Grants are made to organizations or institutions, not individuals.
Date(s) Application is Due: Jan 3; Apr 1; Nov 1
Amount of Grant: 200,000 - 500,000 U.S.D
Contact: Nancy Rivera-Torres, Research Grants Coordinator; (212) 752-0071; fax (212) 752-1398; nrivera@wtgrantfdn.org
Internet: http://www.wtgrantfoundation.org/funding_opportunities/research_grants
Sponsor: William T. Grant Foundation
570 Lexington Avenue, 18th Floor
New York, NY 10022-6837

Wilson-Wood Foundation Grants **4757**
The Wilson-Wood Foundation established in 1983, supports organizations involved with: adult education; aging, centers/services; children/youth, services; education; health care; housing/shelter, development; human services; nutrition; women, centers/services. Giving is limited to the Manatee-Sarasota, Florida area, for for the underprivileged and the less fortunate in the community.
Requirements: Non-profits in the Manatee-Sarasota, Florida area are eligible to apply. Initial approach should be a phone call to the Foundation prior to submitting a the letter of inquiry (must be received by June 1). There is no application form required. Applicants must submit two copies of the proposal. Application must include the following information: timetable for implementation and evaluation of project; qualifications of key personnel; name, address and phone number of organization; copy of IRS Determination Letter, must be dated within the past 10 years; brief history of organization and description of its mission; copy of most recent annual report/audited financial statement/990; descriptive literature about organization; listing of board of directors, trustees, officers and other key people and their affiliations; detailed description of project and amount of funding requested; copy of current year's organizational budget and/or project budget; listing of additional sources and amount of support.
Restrictions: No support for foreign organizations, supporting organizations, or private foundations. No grants to individuals, or for endowment funds, deficit financing, travel projects, research, fundraising costs, multi-year projects, conferences, emergency funding or start up costs.

Geographic Focus: Florida
Date(s) Application is Due: Jun 1
Amount of Grant: 8,000 - 30,000 U.S.D
Contact: Susan Wood, Executive Director; (941) 966-3635
Sponsor: Wilson-Wood Foundation
930 Scherer Way
Osprey, FL 34229-6867

Windham Foundation Grants **4758**

The foundation has a particular interest in innovative programs and in furthering proven academic programs that are replicable in other Vermont locations. Types of support include general operating support, continuing support, building construction/renovation, equipment acquisition, program development, seed grants, curriculum development, technical assistance, scholarships to individuals, and technical assistance. Grants must be prepared according to specific guidelines. Applications are reviewed four times a year and consist of a letter of request, description of project, financial information including assets and liabilities of the organization, project budget, and a copy of the tax-exempt certification from the IRS. Annual deadlines may vary; contact program staff for exact dates.
Requirements: Grants are made to Vermont tax-exempt organizations in the areas of elementary and secondary education. Noneducation institutions may apply, but only for programs that relate directly to elementary an secondary education.
Restrictions: Grants are not made to individuals.
Geographic Focus: Vermont
Date(s) Application is Due: Feb 25; May 20; Aug 5; Nov 12
Amount of Grant: 5,000 - 10,000 U.S.D
Samples: Bellows Falls Union High School (VT)—for program support, $1000; Grafton Public Library (VT)—for operating support, $1000.
Contact: Administrator; (802) 843-2211; fax (802) 843-2205; winfound@sover.net
Internet: http://www.windham-foundation.org/foundation/grants/index.html
Sponsor: Windham Foundation
P.O. Box 70
Grafton, VT 05146

Windhover Foundation Grants **4759**

The foundation fuels innovative, entrepreneurial organizations — maverick in nature — that focus on meeting a pressing, unfilled social need. The foundation follows specific grant outlines when evaluating all applications. It often chooses to support charities in the communities where Quad/Graphics has a presence, preferring to give directly to charities rather than sponsoring events. The foundation has funded a variety of ventures including hospices, women's programs, art centers, libraries, playgrounds, parks, and arenas.
Requirements: Wisconsin 501(c)3 nonprofits are eligible.
Restrictions: Funds are not generally awarded to individuals or teams in any type of competition or contest, athletic or otherwise (including walks and runs); projects deemed to be purely religious or sectarian in nature; fund-raising events; or the children of Quad/Graphics employees.
Geographic Focus: Wisconsin
Contact: Eileen Graves, Quad Graphics Contact
Internet: http://www.qg.com/whoarewe/windhover.html
Sponsor: Windhover Foundation
N63 W23075 Main Street
Sussex, WI 53089-2827

Winifred & Harry B. Allen Foundation Grants **4760**

Established in 1963 in honor of Harry B. Allen, president of the Belvedere Land Company, and his wife, the Foundation has as its primary fields of interest: animal and wildlife preservation; education; the environmental programs; environmental education; protection of natural resources; health care; health organizations; human services; immigrants/refugees; performing arts; and visual arts. Though giving is centered around Marin County, California, the Foundation has also supported 501(c)3 organizations in Massachusetts, Georgia, New York, and the District of Columbia. Applicants should contact the Foundation in writing, stating the purpose and offering proof of tax exempt status. Though there are no specific deadlines, the Board meets four times annually, including April 15, June 15, September 15, and December 15.
Restrictions: No grants are given to individuals.
Geographic Focus: California, District of Columbia, Georgia, Massachusetts, New York
Amount of Grant: 500 - 5,000 U.S.D
Samples: Angel Island Association, Tiberton, California, $1,000—for general operating costs; Belvedere Public Library Foundation, Belvedere, California, $5,300—for general operations; Cornell Lab of Ornithology, Ithica, New York, $2,000—for research.
Contact: Howard B. Allen, (415) 435-2439; fax (415) 435-3166
Sponsor: Winifred and Harry B. Allen Foundation
83 Beach Road, P.O. Box 380
Belvedere, CA 94920-0380

Winn-Dixie Stores Foundation Grants **4761**

Winn-Dixie makes all donations through the foundation, focusing on the stores' trade area in the southern states. Winn-Dixie's giving is accomplished primarily through its matching gifts program. The majority of these funds have generally gone to United Way, with direct grants to United Way campaigns in many of its market communities. Grants for education are given to colleges, associated college funds, and elementary and secondary schools. Besides United Way, other health and human services donations include support for local branches of national disease organizations, hospices, mental health programs, and hospitals. Youth-related grantees include 4-H clubs, Scouts, YMCA/YWCA, and Junior Achievement. Additional types of support include continuing support, annual campaigns, building construction/renovation, equipment acquisition, program development, conferences and seminars, scholarship funds, research grants, and matching funds. There are no formal guidelines or application forms.
Requirements: IRS 501(c)3 tax-exempt organizations in Alabama, Florida, Georgia, southern Indiana, Kentucky, Louisianna, Mississippi, southern Ohio, Oklahoma, Texas, and Virginia are eligible, including national organizations.
Restrictions: Grants will not be made to individuals or to religious or political organizations. Matching contributions to fraternal, professional, social, or recreation groups will not be made. No grants except for matching contributions are made outside the trade territories served by Winn-Dixie: Alabama, Florida, Georgia, Kentucky, Louisiana, North Carolina, South Carolina, Texas, and Virginia.
Amount of Grant: 100 - 10,000 U.S.D
Samples: YMCA, Florida's First Coast Metropolitan (Jacksonville, FL)—$30,000; Urban League of Greater New Orleans (LA)—$25,000.
Contact: D.F. Henry, President; (904) 783-5000
Sponsor: Winn-Dixie Stores Foundation
5050 Edgewood Court
Jacksonville, FL 32254-3699

Winn Feline Foundation/AVMF Excellence In Feline Research Award **4762**

The Winn Foundation has partnered with the American Veterinary Medical Foundation (AVMF) to recognize the veterinary research scientist whose work mirrors Winn's efforts to improve the lives of "Every Cat, Every Day". The Winn/AVMF Excellence in Feline Research Award is accompanied by a $2,500 cash award, a crystal cat sculpture, plus up to $1,000 to offset round trip airfare and other travel expenses. The Research Award is paired with a matching scholarship award by the AVMF for a veterinary student interested in feline medicine. For nomination forms and additional information related to supporting materials, award criteria, and the selection process, visit www.avma.org/awards.
Requirements: Applicants should include a cover letter, a list of significant research achievements, and a concise curriculum vita (CV) including relevant publications. The cover letter must include the name of the award; the name, mailing address, telephone number(s), email address, and college and year of graduation of the nominee; the nature of the nominee's professional activity (type of practice or type of salaried work); organizational memberships (professional and scientific) of the nominee; a narrative sketch of the nominee's professional background; and a statement pertaining to the nominee's qualifications for the award. To be considered, nominees must have demonstrated Winn's mission to advance feline health and welfare through research; have a pattern of following good research study protocols; complete progress reports in a timely manner; and have their research published and acknowledge Winn as a donor of grants when appropriate.
Restrictions: No self-nominations will be considered.
Date(s) Application is Due: Feb 1
Contact: Janet Wolf, Executive Director; (856) 447-9798; winnfeline@aol.com
Eileen Hoblit, Award Contact; (800) 248-2862, ext. 6778; fax (847) 925-1329
Internet: http://www.winnfelinehealth.org/Pages/Researchers.html
Sponsor: Winn Feline Foundation
390 Amwell Road, Suite 402
Hillsborough, NJ 08844

Winn Feline Foundation Grants **4763**

The foundation supports research studies into medical problems affecting cats with the Winn Feline Foundation Grant and the Miller Trust Grant. The Foundation supports research on feline asthma, anemia, inflammatory bowel disease, hyperthyroidism, FIP, megacolon, polycystic kidney disease, transdermal drug delivery, liver disease, and vaccine-associated fibrosarcoma. Studies general in scope are primarily encouraged; however, the foundation is also interested in projects that address problems unique to individual breeds. Continuation of grants awarded will be considered. Application guidelines and current deadlines are available online.
Requirements: Applicants may be faculty veterinarians, postdoctoral fellows, practicing veterinarians, or veterinary students.
Restrictions: The foundation does not fund salaries of principal investigators, major equipment expenditures, travel or indirect costs.
Amount of Grant: Up to 35,000 U.S.D
Samples: DNA Array Analyses for Cat Diseases, University of California, Davis, $25,000 (2011); A Reproducible Protocol to Isolate a Characterized Population of Adult Feline Progenitor Cell, Louisiana State University School of Veterinary Medicine, $9,995 (2011); Evolution of Feline Infectious Peritonitis Virus Within FIP Cats and Tissue-specific Adaptation of the Virus to Activating Proteases, Cornell University, $23,986 (2011).
Contact: Janet Wolf, Grants Administrator; (908) 359-1184; fax (908) 359-7619; winn@winnfelinehealth.org
Internet: http://www.winnfelinehealth.org
Sponsor: Winn Feline Foundation
390 Amwell Road, Suite 402
Hillsborough, NJ 08844

Winston-Salem Foundation Competitive Grants **4764**

The foundation makes competitive grant awards to tax-exempt, nonprofit agencies in the greater Forsyth County, North Carolina, area. Its charitable purposes include: public interest, education and recreation, health, human services, arts and culture, youth, and older adults. Grants ordinarily are made for proposals that initiate, expand, or improve direct services to people and for endowment purposes under certain conditions. Additional types of support include program development, seed grants, scholarship funds, technical assistance, employee matching gifts, scholarships to individuals, and matching funds.

Except in unusual cases, grants will be approved for one year at a time. Potential applicants should contact the office for an appointment to introduce the proposed project.
Requirements: Organizations in Davidson, Davie, Forsyth, Surry, Stokes, Wilkes, and Yadkin counties in North Carolina are eligible to apply.
Restrictions: Grants for major equipment purposes and for capital campaigns in support of local community facilities are of low priority.
Geographic Focus: North Carolina
Amount of Grant: 15,000 - 20,000 U.S.D
Contact: Brittney Gaspari, Grants Director; (336) 725-2382; fax (336) 727-0581; bgaspari@wsfoundation.org or info@wsfoundation.org
Internet: http://www.wsfoundation.org/grant-seekers/types-of-grants/competitive-grants/
Sponsor: Winston-Salem Foundation
860 W Fifth Street
Winston-Salem, NC 27101-2506

Winston-Salem Foundation Elkin/Tri-County Grants **4765**
The Funds consist of three component trusts established by Richard T. Chatham, Lucy Hanes Chatham, and citizens of Elkin to benefit the community. Ordinarily, grants are made for projects that initiate, expand or improve direct service to people. Applicants should send a two or three page letter describing the need for the project, its goals and objectives, its cost and the portion of the cost requested.
Requirements: A list of board members is required, as well as organizational and project budgets, and evidence of non-profit 501(c)3 tax-exempt status. The chief board officer must sign the proposal or write an endorsement cover letter.
Restrictions: Grants will be made only to legally recognized non-profit organizations and educational institutions in Wilkes, Surry, and Yadkin Counties. Those that benefit residents of the Elkin-Jonesville-Dobson-Roaring River radius will be of priority interest to the advisory committee. Grants will not be made for on-going operating expenses. Foundation funds will not be granted to supplant tax or government funds for projects or institutions that would ordinarily receive public funding support.
Geographic Focus: North Carolina
Date(s) Application is Due: Jun 15
Amount of Grant: 500 - 7,000 U.S.D
Samples: Communities in Schools of Wilkes, $3,500; Hugh Chatham Memorial Hospital Foundation, $7,000; Mountain Valley Hospice and Palliative Care, $2,500.
Contact: Brittney Gaspari, Grants Director; (336) 725-2382; fax (336) 727-0581; bgaspari@wsfoundation.org or info@wsfoundation.org
Internet: http://www.wsfoundation.org/grant-seekers/types-of-grants/elkintri-county-grants/
Sponsor: Winston-Salem Foundation
860 W Fifth Street
Winston-Salem, NC 27101-2506

Winston-Salem Foundation Stokes County Grants **4766**
The Trust provides grants to non-profit organizations or informal groups in Stokes County for worthy public and charitable purposes with an emphasis on developing leaders and inspiring others. Leadership can be either formal and traditional, or demonstrated in less traditional ways that are creative and innovative, including leading by example. The trust is designed both to work with established organizations and to encourage new initiatives. Special attention will be given to encouraging leadership in certain focus areas: education; arts; environmental protection and recreational use; historical preservation; local government and community services; health; and organizations and issues affecting minorities and low-resource communities.
Requirements: Applicants can be informal groups of people with innovative ideas, including those that might grow into nonprofit organizations; or established organizations with 501(c)3 tax-exempt status. The Advisory Board will also consider letters from small groups of people with good ideas who need guidance about turning them into formal grant proposals in the future.
Restrictions: Grants will not be made for projects or institutions that would ordinarily receive public funding. However, public institutions such as schools or parks may submit proposals for innovative projects for which funds are not ordinarily provided.
Geographic Focus: North Carolina
Date(s) Application is Due: Sep 11
Amount of Grant: Up to 3,000 U.S.D
Samples: Walnut Cove Colored School, Inc., $1,120; King Senior Center, $1,000; Stokes Volunteer Center, $1,000.
Contact: Brittney Gaspari, Grants Director; (336) 725-2382; fax (336) 727-0581; bgaspari@wsfoundation.org or info@wsfoundation.org
Internet: http://www.wsfoundation.org/grant-seekers/types-of-grants/stokes-county-grants/
Sponsor: Winston-Salem Foundation
860 W Fifth Street
Winston-Salem, NC 27101-2506

Winston-Salem Foundation Victim Assistance Grants **4767**
The fund provides financial assistance to victims of violent crime to help minimize complications from the event. The fund also provides grants for statewide organizations that provide information in a supportive manner to victims of violent crime. Victims of violent crime such as rape or assault are eligible for assistance from the fund. The fund is also available to assist surviving family members of homicide victims and victims of violent crime resulting from domestic violence. The victim can receive monetary support for expenses such as the following: relocation expenses; locks, alarms, or repair of damages from break-ins; clothes, especially for rape victims when clothes are retained as evidence; transportation to counseling visits; lost wages that may be reimbursed to the Fund at a later date by North Carolina Crime Victims Compensation; and help with payment of bills that are late because of the consequences of the crime.
Requirements: Victims should be residents of Forsyth, Davie, or Stokes counties (or the area being served by Family Services, Inc.).
Restrictions: The fund is not for assistance to organizations serving victims (except, as noted in the purpose, for nonprofit statewide organizations that provide information in a supportive manner to victims of violent crime).
Geographic Focus: North Carolina
Contact: Betty Gray Davis, (336) 725-2382; fax (336) 727-0581; bgdavis@wsfoundation.org
Internet: http://www.wsfoundation.org/grant-seekers/types-of-grants/victim-assistance-grants/
Sponsor: Winston-Salem Foundation
860 W Fifth Street
Winston-Salem, NC 27101-2506

Wisconsin American Legion Auxiliary Scholarships **4768**
The American Legion Auxiliary in Wisconsin offers scholarships each year to high school students, college students, and graduate students to provide financial support for education to students with outstanding scholastic ability who are in financial need. Various named scholarships are available as well as scholarships for special areas such as nursing, health careers, and special education. When requesting an application form, the applicant should specify if he or she is a high school student, college student, or graduate student; there are three different forms. Some of the special area scholarships have additional requirements; ask for guidelines when requesting the forms.
Requirements: Students must be residents of Wisconsin; be the daughter, son, wife, or widow of a veteran; have a minimum 3.2 GPA on a 4.0 scale; and be in need of financial aid to continue education. The granddaughter or great-granddaughter may apply if she is an American Legion Auxiliary member.
Geographic Focus: Wisconsin
Date(s) Application is Due: Mar 15
Amount of Grant: 750 - 1,000 U.S.D
Contact: Grants Administrator; (608) 745-0124; fax (608) 745-1947; alawi@amlegionauxwi.org
Internet: http://www.amlegionauxwi.org
Sponsor: Wisconsin American Legion Auxiliary
P.O. Box 140
Portage, WI 53901-0140

Wolfe Associates Grants **4769**
Wolfe Associates Grants are made in the following fields: health and medicine; religion; education; culture, community service, youth skills development, and business. Awards range from $500 to $25,000.
Requirements: Applicants must be 501(c)3 organizations. Funding is primarily made to, but is not limited to, Ohio organizations. Application is made by sending a cover with a brief summary of the request, the amount requested, a copy of the organization's Internal Revenue Service determination letter and most recent form 990, and financial statements. There are no submission deadlines.
Geographic Focus: Ohio
Samples: Baldwin-Wallace College, Berea, Ohio, $5,000, education funding (2010); Columbus State Community College, Columbus, Ohio, $500, education funding (2010); Columbus Association for Performing Arts, Columbus, Ohio, $25,000, culture and environment funding (2010).
Contact: Rita J. Wolfe, Vice President; (614) 460-3782
Sponsor: Wolfe Associates
34 S Third Street
Columbus, OH 43215

Women's Foundation of California Community Action Fund **4770**
The foundation awards grants to support projects of organizations working to improve the lives of women and girls in California. Requests will only be considered in one of the following issue areas: economic rights and justice; environmental health and justice; and reproductive rights and justice. Types of support include general operating support, project/program-specific grants, and planning grants. Serious consideration will be given to cross-sector collaborations as well as to emerging organizations, especially in rural and non-urban areas. Funding for publications and one-time events will only be considered if they are integrated within a larger strategy for social change. Application and guidelines are available online.
Requirements: California organizations that work with low-income women and girls are eligible. Priority is given to projects serving women and girls of color; immigrant, undocumented, and refugees women and girls; lesbian, bisexual, transgender, intersex, and questioning women and girls; women and girls with disabilities; women over 50; indigenous women and girls; young and/or single mothers; and women and girls in rural and non-urban communities.
Restrictions: The program does not fund capital improvements, endowments, loans, individuals, debt reduction, expenses that accrued prior to the date of the grant award, or fundraising events.
Geographic Focus: California
Date(s) Application is Due: Dec 16
Amount of Grant: Up to 35,000 U.S.D
Contact: Grants Administrator; (415) 837-1113; fax (415) 837-1144
Internet: http://www.womensfoundca.org
Sponsor: Women's Foundation of California
340 Pine Street, Suite 302
San Francisco, CA 94104

Women's Foundation of California Grants 4771

The Women's Foundation of California envisions a future where communities in California are healthy, safe and economically prosperous. The Foundation achieve's this vision by focusing on women and girls as change agents through their central role in families and communities. The Foundation supports organizations that meet the following objectives: organizations mobilizing communities to change policies and systems; to improve the lives of women and girls; developing the leadership of women and girls working to change policies and systems and create stronger movements for social justice; promoting the human rights of women and girls in all sectors of society, including economic, political, social, environmental, sexual, cultural and civil rights.

Requirements: The foundation awards grants to: California 501(c)3 nonprofit corporations or California nonprofit groups with a 501(c)3 fiscal sponsor; organization/program prioritize low-income women and girls in California; health and safety: this includes reproductive health and justice, environmental health and justice and gender-based violence; economic prosperity: this includes efforts towards economic security and economic development. Organizations or programs that utilize one or more of the following strategies to bring about measurable and sustainable change: policy advocacy; community organizing; leadership development; media and/or public education; civic participation; participatory action research; youth-led programs and initiatives. Contact the Foundation for further application information and guidelines.

Restrictions: The Women's Foundation of California does not make grants to or for the following: individuals; for-profit organizations or small businesses; start-up loans; capital projects; direct services; fundraising events; shelters; scholarships; capital improvements; endowments; loans; debt reduction; expenses that accrued prior to the date of the grant award. Funding for publications and one-time events such as retreats and conferences will only be considered if they are integrated within a larger strategy for social change.

Geographic Focus: California

Amount of Grant: 10,000 - 40,000 U.S.D

Contact: Maya Thornell-Sandifor, Senior Program Officer; (415) 837-1113; fax (415) 837-1144; mayat@womensfoundca.org

Internet: http://www.womensfoundca.org/grantmaking.html

Sponsor: Women's Foundation of California
340 Pine Street, Suite 302
San Francisco, CA 94104

Women's Foundation of Southern Arizona Go Girls! Sports Grants 4772

The foundation is a social change organization committed to helping women and girls overcome economic, political, gender, and social barriers. Grant categories include health, safety, and well-being for women and girls; economic justice for women and girls; and innovative programs that improve the status of women and girls. Types of support include project specific support, general operating support, and planning or start-up support. Applicants must submit an electronic letter of intent by the listed application deadline. Guidelines are available online.

Requirements: Southern Arizona nonprofit organizations are eligible.

Restrictions: The foundation does not provide funds for individuals; campaigns to elect candidates to public office; capital fund drives; debt reduction; endowments; fundraising events or one-time conferences; scholarships, fellowships, and tuition reimbursement; programs inconsistent with federal, state, and local non-discrimination statutes regarding equal employment opportunity; programs that promote religious activities; programs outside of Pima, Cochise, and Santa Cruz counties in southern Arizona; or individual classrooms or schools.

Geographic Focus: Arizona

Date(s) Application is Due: Oct 1

Amount of Grant: Up to 10,000 U.S.D

Contact: Laura Penny, Program Officer; (520) 622-8886; fax (520) 770-1500; lpenny@womengiving.org

Internet: http://www.womengiving.org/programs/gogirl.php

Sponsor: Women's Foundation of Southern Arizona
2250 E Broadway Boulevard
Tucson, AZ 85719

Women's Fund of Rhode Island Grants 4773

The fund awards grants to eligible Rhode Island nonprofit organizations for projects designed to help level the playing field for women and girls. The fund is particularly interested in the following issues: leadership—including political participation and leadership, professional development opportunities, developing leadership in girls and women, and preteen prevention strategies to address self-esteem and self-advocacy issues; health, education, and well-being—including poverty, violence, teen pregnancy, self-esteem, health education and disease prevention, and K-16 education and adult literacy; economic autonomy—including women-owned enterprise, employment readiness and opportunity, and equal and livable wage; rights—including self-determination, public policy, and government action that assures the well-being of women and girls including immigrants. Types of support include project specific support, general operating support, planning, and start-up support. The fund considers funding on a year-to-year basis and does not make multiyear commitments. Application guidelines and forms are available online. The application deadline is for letters of intent.

Requirements: 501(c)3 nonprofits that serve women and girls who reside in Rhode Island and organizations applying through 501(c)3 fiscal agents that serve the target populations and area are eligible. The fund prefers to support organizations that have limited access to other donors.

Restrictions: The fund will not support projects that discriminate on the basis of ethnicity, race, color, creed, religion, gender, national origin, age, disability, marital status, sexual orientation, gender identity, or any veteran's status; or projects that present or incorporate religion in any manner. Grants do not support individuals or scholarships, capital or endowment, biomedical research or debt reduction, fundraising events or campaigns to elect candidates to public office.

Geographic Focus: Rhode Island

Date(s) Application is Due: Feb 18

Amount of Grant: 2,500 - 25,000 U.S.D

Contact: Marcia Conee-Tighe, (401) 274-4564; fax (401) 331-8085

Internet: http://www.rifoundation.org/wfri/html/index.htm

Sponsor: Rhode Island Foundation
1 Union Station
Providence, RI 02903

Women's Way Discretionary Fund Grants 4774

The organization awards grants for programs designed to meet the specific needs of women and girls. Grants support programs to end domestic violence, guarantee equal opportunity, challenge discrimination, foster economic self-determination, and affirm reproductive rights. Organizations must demonstrate that women are well represented at the managerial staff and board level.

Requirements: Nonprofit organizations that have an annual operating budget of $500,000 or less and that are located in Pennsylvania's Bucks, Chester, Delaware, Montgomery, or Philadelphia counties or New Jersey's Burlington, Camden, or Gloucester counties are eligible.

Geographic Focus: New Jersey

Amount of Grant: 1,000 - 3,000 U.S.D

Contact: Tamela Luce, Grants Administrator; (215) 985-3328, ext. 227 or (215) 985-3322; fax (215) 985-3369; tluce@womensway.org

Internet: http://www.womensway.org

Sponsor: Women's Way
1233 Locust Street, Suite 300
Philadelphia, PA 19107

Women of the ELCA Opportunity Scholarships for Lutheran Laywomen 4775

The purpose of these scholarships is to provide assistance to women studying for a career other than the ordained ministry. The following scholarships are offered: Amelia Kemp Scholarship— for ELCA women of color in undergraduate, graduate, professional, or vocational courses of study; Belmer, Flora Prince Scholarships—for ELCA women studying for ELCA service abroad; Kahler, Vickers/Raup, Emma Wettstein Scholarships—for ELCA women studying for service in health professions associated with ELCA projects abroad; Irene Drinkall Franke, Mary Seeley Knudstrup Scholarships—for ELCA women in graduate courses of study preparing for occupations in Christian service; and Cronk Memorial, First Triennium Board, General, Mehring, Paepke, Piero/Wade/Wade, Edwin, and Edna Robeck Scholarships—for ELCA women in undergraduate, graduate, professional, or vocational courses of study.

Requirements: The applicant must be at least 21 years old, be a U.S. citizen, hold membership in the Evangelical Lutheran Church in America, and have experienced an interruption in education of two or more years since completion of high school. Scholarship funds must be used within the year following the award.

Amount of Grant: 800 - 1,000 U.S.D

Contact: Valora Starr, Coordinator; (800) 638-3522, ext. 2741; fax (773) 380-2419; Valora.Starr@elca.org

Internet: http://www.elca.org/Growing-In-Faith/Ministry/Women-of-the-ELCA/Engage-in-action-and-support-one-another-in-our-callings/Scholarships/For-Lutheran-Laywomen.aspx

Sponsor: Evangelical Lutheran Church in America
8765 W Higgins Road
Chicago, IL 60631-4189

Wood-Claeyssens Foundation Grants 4776

The Foundation awards grants to eligible California nonprofit organizations. Types of support include annual campaigns, capital campaigns, continuing support, and general operating support. An application is available on the website.

Requirements: California 501(c)3 organizations serving Santa Barbara and Ventura Counties are eligible.

Restrictions: Funding is not available to individuals or to organizations that discriminate on the basis of age, gender, race, ethnicity, sexual orientation, disability, national origin, political affiliation or religious belief.

Geographic Focus: California

Date(s) Application is Due: Jun 30

Samples: Senior Concerns, Thousand Oaks, California, $15,000 (2010); Many Mansions, Thousand Oaks, California, $30,000 (2010); and Dream Foundation, Santa Barbara, California, $30,000 (2010).

Contact: Noelle Claeyssens Burkey, President; (805) 966-0543; fax (805) 966-1415; wcf0543@gmail.com

Internet: http://www.woodclaeyssensfoundation.com/Funding.htm

Sponsor: Wood-Claeyssens Foundation
P.O. Box 30586
Santa Barbara, CA 93130-0586

Wood Family Charitable Trust Grants 4777

The Wood Family Charitable Trust was established in 2007 to support and promote quality educational, human-services, and health-care programming for underserved populations. The Wood Charitable Trust specifically serves the people of Vale, Oregon and its surrounding communities. Grant requests for general operating support are strongly encouraged. Program support will also be considered. Small, program-related capital expenses may be included in general operating or program requests. The majority of grants from the Wood Charitable Trust are one year in duration; on occasion, multi-year support is awarded. Application materials are available for download at the grant

website. Applicants are strongly encouraged to review the state application guidelines for additional helpful information and clarifications before applying. Applicants are also encouraged to review the trust's funding history (link is available from the grant website). The application deadline for the Wood Family Charitable Trust is March 15. Applicants will be notified of grant decisions by May 15.
Requirements: The Wood Charitable Trust specifically serves the people of Vale, Oregon and its surrounding communities. Applicants must have 501(c)3 tax-exempt status.
Restrictions: The trust does not support requests from individuals, organizations attempting to influence policy through direct lobbying, or any political campaigns.
Geographic Focus: Oregon
Date(s) Application is Due: Mar 15
Samples: Vale High School, Vale, Oregon, $1,090, for digital cameras for photography class for students that cannot afford one (2010); Vale Elementary School, Vale, Oregon, $3,811, computer equipment for first grade teacher (2010); Vale Middle School, Vale, Oregon, $6,040, Wenger chairs and racks for music class (2010).
Contact: Cindy Keyser, Vice President; (800) 848-7177; cindy.s.keyser@baml.com
Internet: https://www.bankofamerica.com/philanthropic/fn_search.action
Sponsor: Wood Family Charitable Trust
800 5th Avenue, WA1-501-33-23
Seattle, WA 98104

Woodward Fund Grants **4778**
Grants support nonprofit institutions, corporations, and associations that are located in Georgia or one of its neighboring states and are organized and operated exclusively for religious, educational, and charitable and scientific purposes. Grantmaking focuses on capital projects. The distribution committee makes its decisions based on materials provided in the grant request. The trust does not require a formal application form, but requests a proposal outlining the project or program and containing the following information: project/program description; complete itemized project budget, including project schedule; need for the project; other funding sources, including the amount received from each source; names and qualifications of those conducting the project; objectives and how they will be achieved; a brief description of the applicant organization; method and criteria for project evaluation; list of officers, board of directors, or trustees; copy of the 501(c)3 designation letter; and plans for recognition of the trust in the project. Applicants should not contact committee members personally concerning a proposal or anticipated approach to the fund.
Requirements: 501(c)3 nonprofit organizations in Georgia and its neighboring states are eligible. Government agencies to which contributions by individuals are made deductible from income by IRS laws also are eligible.
Restrictions: Grants are not awarded to individuals or for scholarships or student loans.
Geographic Focus: Alabama, Florida, Georgia, North Carolina, South Carolina, Tennessee
Date(s) Application is Due: May 1; Nov 1
Amount of Grant: 5,000 - 300,000 U.S.D
Contact: Alice Sheets, c/o Wachovia Bank, Trustee; grantinquiries8@wachovia.com
Internet: https://www.wachovia.com/charitable_services/woodward_overview.asp
Sponsor: David, Helen, and Marian Woodward Fund
3280 Peachtree Road NE, Suite 400, MC G0141-041
Atlanta, GA 30305

Woodward Governor Company Charitable Trust Grants **4779**
The company-sponsored trust supports nonprofits in areas of company operations, including Rockford, IL; Fort Collins, CO; and Stevens Point, WI. Areas of interest include vocational education, adult basic education, and literacy programs; health care, health organizations, and hospices; delinquency prevention programs and services; services for minorities and immigrants; human services for the homeless, community and economic development; and arts/culture, including museums. Types of support include general operations, annual campaigns, capital campaigns, continuing support, equipment acquisition, seed funding, and emergency funding. There are no application deadline dates; receipt of proposals is preferred in March and July. Applications forms are not required.
Requirements: Only Colorado, Illinois, and Wisconsin organizations are eligible to apply.
Restrictions: Grants are not awarded to individuals or for endowment funds, research, scholarships, fellowships, special projects, publications, conferences, loans, or matching gifts.
Geographic Focus: Colorado, Illinois, Wisconsin
Amount of Grant: 500 - 200,000 U.S.D
Contact: Pam Johnson, Chair, Contributions Committee
Sponsor: Woodward Governor Company Charitable Trust
5001 N Second Street
Rockford, IL 61125-7001

World of Children Health Award **4780**
The World of Children Awards program was created to recognize and elevate those selfless individuals who make a difference in the lives of children here in the U S.A and across the globe, regardless of political, religious or geographical boundaries. These courageous leaders recognize that our children are the world's most important asset The Health Award, in the amount of up to $50,000, recognizes an individual who has made a significant contribution to children in the fields of health, medicine or the sciences.
Requirements: Nominations must be submitted in English. Nominees for the Award must have; created, managed or otherwise supported a sustainable program which has significantly contributed to the improved health of children; do this work over and above their normal employment, OR work for little or no pay; have been doing this for a minimum of 10 years; and have an existing non-profit organization in good standing, which can receive grant funds if awarded.
Restrictions: Organizations are not eligible; however a nominee or group of nominees may be part of an organization.
Date(s) Application is Due: May 1
Amount of Grant: 50,000 U.S.D
Contact: Lynn Wallace Naylor, Executive Director; (925) 399-6411; fax (925) 399-6001; contact@worldofchildren.org
Internet: http://www.worldofchildren.org/index.php/awards/awards-program
Sponsor: World of Children
6200 Stoneridge Mall Road, 3rd Floor
Pleasanton, CA 94588

World of Children Humanitarian Award **4781**
The World of Children Awards program was created to recognize and elevate those selfless individuals who make a difference in the lives of children here in the U.S.A and across the globe, regardless of political, religious or geographical boundaries. These courageous leaders recognize that our children are the world's most important asset. The Humanitarian Award, with a maximum of $50,000, recognizes an individual who has made a significant contribution to children in the areas of social services, education or humanitarian services.
Requirements: Nominations must be submitted in English. The nominee must: have created, managed or otherwise supported a sustainable program which has significantly contributed to children's opportunities to be safe, to learn, and to grow; do this work over and above their normal employment, OR work for little or no pay; have been doing this for a minimum of 10 years; and have an existing non-profit organization in good standing, which can receive grant funds if awarded.
Restrictions: Organizations are not eligible; however a nominee or group of nominees may be part of an organization.
Date(s) Application is Due: May 1
Amount of Grant: 50,000 U.S.D
Contact: Lynn Wallace Naylor, Executive Director; (925) 399-6411; fax (925) 399-6001; contact@worldofchildren.org
Internet: http://www.worldofchildren.org/nominations
Sponsor: World of Children
6200 Stoneridge Mall Road, 3rd Floor
Pleasanton, CA 94588

Wright Family Foundation Grants **4782**
The foundation awards grants to eligible nonprofit organizations in the areas of community, education, health, social needs, and the arts. Grants assist organizations in need of funding to support capital projects that will have a positive impact on the lives of people in Schenectady, NY (and surrounding counties), Brazoria County, TX, and Cocke County, TN. Application and guidelines are available online.
Restrictions: The foundation does not provide support for the following: organizations lacking proof of IRS 501(c)3 tax-exempt status; annual operating support or fund drives; religious or political organizations; athletic events or sponsorships; or social functions or advertising. Funding for underwriting or administrative support will not be considered. In addition, funding requests under $1000 will not be considered, however, there is no maximum limit.
Geographic Focus: New York, Tennessee, Texas
Date(s) Application is Due: Jan 20; May 20; Aug 20; Nov 20
Amount of Grant: 1,000 U.S.D
Contact: Program Contact; (518) 347-4530; fax (518) 347-6201; info@wrightfamilyfoundation.org
Internet: http://www.wrightfamilyfoundation.org
Sponsor: Wright Family Foundation
P.O. Box 1046
Schenectady, NY 12301

WWNFF Dissertation Grants in Women's Health **4783**
Dissertation grants encourage original and significant research on issues related to women's health. This program emphasizes the implications of research for the understanding of women's lives and significance for public policy or treatment. Previous grant recipients have explored such topics as postpartum depression; contraceptive law; sex hormones and ovarian cancer; dietary determinants of morbidity; sex and violence in everyday life; and sexuality on the Internet. Grant funds may be used for expenses connected with the dissertation, including but are not limited to, travel, books, microfilming, taping, and computer services. Ten awards will be made. Guidelines are available online. Annual deadline dates may vary; contact program staff for exact dates.
Requirements: Students in doctoral programs such as nursing, public health, anthropology, history, sociology, psychology, and social work, at graduate schools in the United States are eligible to apply. Candidates must have completed all pre-dissertation requirements by October 24 and expect to complete their dissertations by the summer of next year.
Restrictions: Candidates who are within a few months of completing their work should not apply.
Date(s) Application is Due: Oct 11
Amount of Grant: 6,000 U.S.D
Contact: Women's Health Dissertation Fellowship Director; (609) 452-7007; fax (609) 452-0066; contact@woodrow.org
Internet: http://www.woodrow.org/womens-studies/health/purpose.html
Sponsor: Woodrow Wilson National Fellowship Foundation
P.O. Box 5281
Princeton, NJ 08543-5281

Wyomissing Foundation Grants 4784

The foundation makes grants for charitable purposes and is primarily oriented to local giving within Berks County, PA, with emphasis on hospitals, higher education, and community agencies. Some grants are made that are nonregional in nature. In general, the foundation does not provide funds for operating budgets. Types of support include capital grants, challenge/matching grants, development grants, project grants, research grants, and seed funding. The trustees meet quarterly. Organizations will be made aware of the disposition of their application as soon as possible after the meeting.
Requirements: Giving primarily in Berks County, PA, and contiguous counties; limited support also in the mid-Atlantic area. Organizations wishing to apply should prepare an outline of the scope and need of the project on their own letterhead. The application should not exceed two pages and should include total budgeted costs and the funds anticipated from other sources.
Geographic Focus: Pennsylvania
Date(s) Application is Due: Feb 25; May 25; Aug 25; Oct 25
Amount of Grant: 1,000 - 25,000 U.S.D
Contact: Paul Roedel, President; (610) 376-7496; fax (610) 372-7626; wfbbec@nnl.com
Sponsor: Wyomissing Foundation
960 Old Mill Road
Wyomissing, PA 19610

Xoran Technologies Resident Research Grant 4785

The purpose of this grant is to stimulate original resident research in otolaryngology projects that are well-conceived and scientifically valid, with potential to clarify the role of head and neck computed tomography in diagnosis or treatment. Awards are for one year, are non renewable, and have a $10,000 maximum limit. Generally, a single award is made available annually. Letters of Intent are due on December 15, with full proposals due by January 15.
Requirements: Application is open to residents of an accredited otolaryngolgoy-head and neck surgery training program in the U.S. and Canada for projects that are well-conceived and scientifically valid, with potential to clarify the role of the head and neck computed tomography in diagnosis or treatment.
Date(s) Application is Due: Jan 15; Dec 15
Amount of Grant: Up to 10,000 U.S.D
Contact: Stephanie L. Jones; (703) 519-1586 or (703) 836-4444; sljones@entnet.org
Internet: http://www.entnet.org/EducationAndResearch/coreGrants.cfm
Sponsor: Xoran Technologies
One Prince Street
Alexandria, VA 22314-3357

Yampa Valley Community Foundation Cody St. John Scholarships 4786

The Foundation supports programs benefiting the Yampa Valley community. The Cody St. John Scholarships are administered by the Foundation to assist professional ski patrollers in furthering their medical education. This is a new Scholarship and details are still being worked out. Applicants may call the Foundation for further information.
Contact: Jennifer Shea, Program Manager; (907) 879-8632; jennifer@yvcf.org
Internet: http://www.yvcf.org/scholarship-apply.php
Sponsor: Yampa Valley Community Foundation
465 Anglers Drive, Suite 2-G
Steamboat Springs, CO 80488

Yampa Valley Community Foundation Grants 4787

The Foundation supports innovative programs benefiting the Yampa Valley community. Funding is provided for proposals that: promote the mission of the Foundation; serve either a broad or underserved population in the Yampa Valley; demonstrate the anticipated impact in the Yampa Valley; define the measurement and evaluation process to be used; effectively leverage the Foundation's resources; and exhibit sound business and financial practices. Grants may be awarded for innovative programs demonstrating progress toward achieving community goals in five focus areas: arts and culture, education, environment, health and human services, and recreation. Initially a letter of intent, providing an overview of the proposed project or program and the funding requested, should be submitted. Selected applicants will be invited to submit a full proposal.
Requirements: Eligible organizations must be a nonprofit 501(c)3 organization or fiscally sponsored by a qualifying organization in Routt or Moffat Counties that benefit the Yampa Valley. Only grant requests for charitable purposes that have a public benefit are considered.
Restrictions: The Foundation does not grant for debt reduction, endowments, political purposes or religious purposes. Proposals are typically declined for individual or professional development, team or travel expenses, retro-active grants for projects already completed or in process, or 100% of funding for a project. Funded programs and services may not include any political or religious intentions.
Geographic Focus: Colorado
Date(s) Application is Due: Jun 1
Samples: Advocates Building Peaceful Communities, Steamboat Springs, Colorado, $6,550 - general support; Healthcare Foundation for the Yampa Valley, Steamboat Springs, Colorado, $11,250 - CT scanner challenge match; and South Routt School District Oak Creek, Colorado, $10,000 - community greenhouse.
Contact: Jennifer Shea, Program Manager; (970) 879-8632; fax (970) 871-0431; jennifer@yvcf.org or nfo@yvcf.org
Internet: http://www.yvcf.org/grants.php
Sponsor: Yampa Valley Community Foundation
465 Anglers Drive, Suite 2-G
Steamboat Springs, CO 80488

Yampa Valley Community Foundation Volunteer Firemen Scholarships 4788

The Foundation supports programs benefiting the Yampa Valley community. The Steamboat Springs Volunteer Firemen sponsor two scholarships: the Rusty Chandler Memorial Scholarship and the Dave Linner Memorial Scholarship. The Rusty Chandler Memorial Scholarship is awarded to a high school senior and emphasizes community service. The Dave Linner Memorial Scholarship is awarded to a high school senior entering the EMS or medical related fields. Applications are available on the website.
Requirements: Applicants must be seniors graduating from Steamboat Springs High and have attended Steamboat Springs High School for at least two full semesters
Geographic Focus: Colorado
Date(s) Application is Due: Apr 15
Contact: Jennifer Shea, Program Manager; (907) 879-8632; jennifer@yvcf.org
Internet: http://www.yvcf.org/scholarship-apply.php
Sponsor: Yampa Valley Community Foundation
465 Anglers Drive, Suite 2-G
Steamboat Springs, CO 80488

Yawkey Foundation Grants 4789

The Yawkey Foundation is committed to continuing the legacy of Tom and Jean Yawkey by making significant and positive impacts on the quality of life of children, families, and the underserved in the areas of New England and Georgetown County, South Carolina. Funding supports the areas of health care, education, human services, youth and amateur athletics, arts and culture, and conservation and wildlife. Request should be limited to $25,000 unless otherwise directed. Applications are currently accepted only from organizations previously funded by the Foundation. Deadlines are: arts and culture, conservation, and health care March 1; human services June 15; education September 1; and youth and amateur athletics November 15.
Requirements: Eligible applicants must be tax-exempt 501(c)3 organizations. Proposals must provide significant benefits to a broad constituency either in New England or Georgetown County, South Carolina. The Foundation has a particular concern for organizations that serve disadvantaged children and families and also considers the following: relevance of the proposed project or program to the Foundation's areas of interest; need outlined in the proposal and how the organization has and will continue to address that need; the organization's fiscal health and ability to manage its resources effectively; ability of the project or program to leverage funding and support from other sources; ability of the organization and its staff to achieve the desired results; adequacy of proposed activities, budget, and timetable to achieve the desired results; and evidence of appropriate cooperation with other organizations.
Restrictions: All final reports for prior funding must be submitted before applying for additional funding. Only one request may be submitted during a twelve-month period. Organizations that have received three or more years of consecutive funding will not be eligible to reapply for funding for a one-year period. The Foundation does not make grants to: organizations that are not tax-exempt under section 501(c)3 of the Internal Revenue Service; individuals; private foundations; 509(a)3 Type III non-functionally integrated organizations; organizations or programs that provide benefits outside of the United States; legislative lobbying; foundations created by political or governmental or for-profit organizations; political campaigns and causes; government agencies, or agencies directly benefiting public entities; public school districts and public schools (including charter schools); community and economic development corporations or programs; advocacy groups; operating deficits or retirement of debt; general endowments; general capital campaigns; events, conferences, seminars, and group travel; awards, prizes, and monuments; fraternal, trade, civic, or labor organizations; music, video, or film production; feasibility or research studies; pass-through, intermediary organizations or foundations; religious organizations for sectarian purposes; and workforce development programs.
Geographic Focus: Connecticut, Maine, Massachusetts, New Hampshire, Rhode Island, South Carolina, Vermont
Contact: Nancy Brodnicki, Grants Program Administrator; (781) 329-7470
Internet: http://www.yawkeyfoundation.org/grant_guidelines.html
Sponsor: Yawkey Foundation
990 Washington Street
Dedham, MA 02026-6716

Young Ambassador Scholarship In Memory of Christopher Nordquist 4790

The annual $1,000 award to an eligible student is made possible through private contributions given to The Eye-Bank in memory of Christopher, who was two when he died and left the gift of sight. Through the years, his mother Andrea Nordquist has shared the story of their family's loss and subsequent decision to donate Christopher's corneas in hopes of increasing awareness of the good that eye donation can do. In this way, Christopher became one of The Eye-Bank's first Young Ambassadors. The Scholarship in his memory is intended to encourage young people to pursue learning and to help spread the message about the priceless gift almost anyone can leave after death – the gift of sight through eye donation. The scholarship award can be used for continuing education at an accredited university, college, trade or technical school certificate program.
Requirements: To apply, students must meet all of the following *Requirements:* Reside within The Eye-Bank's service area (New York City, Nassau, Suffolk, Westchester, Rockland, Duchess or Putnam Counties of New York State); and, be a student under 25 years of age who will be, or is presently enrolled, in a 2 or 4-year college, university, trade or technical school for the current academic year at the time of application deadline.
Geographic Focus: New York
Date(s) Application is Due: Jun 1
Contact: Andrea Nordquist
Internet: http://www.eyedonation.org/scholarship_program.html
Sponsor: Eye-Bank for Sight Restoration
120 Wall Street
New York, NY 10005-3902

YSA Get Ur Good On Grants **4791**

Get Ur Good On Grants are available for youth-led service initiatives on Global Youth Service Day, April 23-25. Available to children and youth around the world, YSA's Get Ur Good On Grants will award 100 $500 grants to support projects addressing critical community needs such as poverty, education, and environmental sustainability. Grant applications are due February 22. Additional guidelines are available at: www.YSA.org/grants.

Date(s) Application is Due: Feb 22
Amount of Grant: 500 U.S.D
Contact: Megan Brudney, Grants Manager; (202) 296-2992; fax (202) 296-4030; Brudney@ysa.org
Internet: http://www.ysa.org/grants/programs/get-ur-good-grants
Sponsor: Youth Service America
1101 15th Street NW, Suite 200
Washington, DC 20005

YSA UnitedHealth Heroes Service Learning Grants **4792**

The UnitedHealth HEROES program is a service-learning, health literacy initiative developed by UnitedHealthcare and YSA. The program awards grants to help youth, ages 5-25, create and implement local, hands-on programs to fight childhood obesity. Each grant engages participating youth in service-learning, an effective teaching and learning strategy that supports student learning, academic achievement, and workplace readiness. The grants encourage semester-long projects that launch on Martin Luther King, Jr. Day of Service and culminate on Global Youth Service Day.

Requirements: Schools, service-learning coordinators, non-profits, and students in the health professions located in all 50 states and the District of Columbia are eligible to apply for the $500-$1,000 grants.
Date(s) Application is Due: Oct 17
Amount of Grant: 500 - 1,000 U.S.D
Contact: Amanda McDonald, Grants Manager; (202) 296-2992; fax (202) 296-4030; McDonald@ysa.org or info@ysa.org
Internet: http://www.ysa.org/HEROES
Sponsor: Youth Service America
1101 15th Street NW, Suite 200
Washington, DC 20005

Z. Smith Reynolds Foundation Small Grants **4793**

The Z. Smith Reynolds Foundation now offers a Small Grants Process for grant requests of up to $35,000 per year for up to two years. Small grants are made in the areas of community building and economic development, the environment, governance, public policy and civic engagement, pre-collegiate education, and social justice or equity. In addition to funding projects that achieve the goals of each focus area, the foundation has an interest in building the capacity of organizations and in promoting organizational development. New programs, rather than those that are well-established and well-funded, receive priority consideration. Types of support include operating budgets, continuing support, annual campaigns, seed grants, matching funds, projects/programs, conferences and seminars, and technical assistance.

Requirements: The foundation makes grants only to nonprofit, tax-exempt, charitable organizations and institutions in North Carolina.
Restrictions: The foundation does not give priority to: the arts; capital campaigns; computer hardware or software purchases; conferences, seminars, or symposiums; crisis intervention programs; fund raising events; historic preservation; local food banks; or substance abuse treatment programs.
Geographic Focus: North Carolina
Date(s) Application is Due: Aug 3
Amount of Grant: Up to 35,000 U.S.D
Contact: Leslie Winner, Executive Director; (800) 443-8319, ext. 105 or (336) 725-7541, ext. 105; fax (336) 725-6069; lwinner@zsr.org or info@zsr.org
Internet: http://www.zsr.org/small_grants.htm
Sponsor: Z. Smith Reynolds Foundation
147 South Cherry Street, Suite 200
Winston-Salem, NC 27101-5287

Zellweger Baby Support Network Grants **4794**

This organization was started by a small group of parents whose lives have been affected in some way by a rare disorder. The Network provides support to families dealing with Zellweger syndrome. Its mission is to promote, advance, and improve awareness of Zellweger syndrome and other peroxisomal disorders, to assist, support, and aid, financially or otherwise, individuals and families affected by Zellweger syndrome. Grants are given to individuals. Grant amounts range from $100 to $1,000, and applications are available at the Foundation website.

Amount of Grant: 100 - 1,000 U.S.D
Contact: Pam Freeth, President; (605) 645-2983
Internet: http://www.zbsn.org/
Sponsor: Zellweger Baby Support Network
530 W. Jackson Boulevard
Spearfish, TX 57783

Subject Index

AIDS

Agribusiness

Agricultural Economics

Agricultural Extension

Agricultural Geography

Agricultural Management

Agricultural Planning/Policy

Agriculture

Behavioral/Experimental Psychology

Behavioral/Social Sciences

Beverages, Alcoholic

Bilingual/Bicultural Education

Bioassay

Biochemical Engineering

Biochemistry

Biomedical Research Resources

Biomedical Research Training

Cancer Prevention

Carbohydrates

Cardiology

Cardiomyocyte Mitochondria

Cardiovascular Diseases

Cardiovascular Health

Children's Museums

Children's Theater

China

Chiropractic

Chromatography

Chronic Illness

Civic Engagement

Civics

Civics Education

Civil Engineering

Civil Law

Civil Service

Civil Society

Civil/Human Rights

Classroom Instruction

Climatology

Clinical Medicine, General

Clinical Psychology

Clinical Research

Clinical Sociology

Clinical Trials

Clinics

Cloning of Cells

Coastal Processes

Cognitive Development/Processes

Colitis

Collaboration

College Students

College-Preparatory Education

Colombia

Communication Systems

Communications

Communicative Disorders, Hearing

Communicative Disorders, Speech

Community Colleges

Community Development

Community Services

Community and School Relations

Compensatory Education

Disaster Preparedness

Disaster Relief

Disasters

Discrimination

Disease, Chronic

Diseases

Distance Learning

Diversity

Documentaries

Domestic Violence

Economic Justice

Economic Opportunities

Economic Science

Economic Self-Sufficiency

Economic Stimulus

Economics

Economics Education

Ecosystems

Editing

Education

Education Reform

Education and Work

Educational Administration

Educational Evaluation/Assessment

Educational Finance

Educational Instruction

Electoral Systems

Electric Power

Electronic Media

Electronics/Electrical Engineering

Electrophysiology

Elementary Education

Environmental Issues

Environmental Law

Environmental Planning/Policy

Environmental Programs

Environmental Protection

Environmental Research

Environmental Services

Environmental Studies

Health Care

Health Care Access

Health Care Administration

Health Personnel/Professions

Health Planning/Policy

Health Records

Health Research

Health Sciences

Health Services Delivery

Higher Education Administration

Higher Education Studies

Higher Education, Private

Higher Education, Public

Hispanic Education

Hispanic Studies

Hispanics

Historical Preservation

History

Human Subjects Policy

Humanitarianism

Humanities

Humanities Education

Hunger

Jewish Studies

Job Training Programs

Journalism

Literary Arts

Literature

Liver Diseases and Disorders

Local History

Long-Term Care

Medical Ethics

Medical Informatics

Medical Physics

Medical Programs

Medical Research

Medical Sciences

Medical Technology

Medical/Diagnostics Imaging

Mental Health

Mental Retardation

Mentoring Programs

Metabolic Diseases

Metabolism

Metallurgy

Meteorology

Minorities, Ethnic

Minority Education

Nursing Education

Nursing Homes

Nutrition Education

Obstetrics-Gynecology

Occupational Health and Safety

Oceanography

Odontology

Olfaction

Oncology

Opera/Musical Theater

Patient/Physician Relationship

Peace/Disarmament

Pediatric Cancer

Pediatrics

Penology/Correctional Institutions and Procedures

Primary Care Services

Primatology

Print Media

Prion Disease

Prison Reform

Prisoners

Private and Parochial Education

Problem Solving

Process Development Engineering

Process Simulation and Control

Production/Operations Management

Professional Associations

Professional Development

Psychology Education

Psychology of Aging

Psychopathology

Psychosomatic Disorders

Psychotherapy

Public Administration

Public Affairs

Public Planning/Policy

Public Policy Systems Analysis

Public Relations

Religious Studies

Religious Welfare Programs

Research Resources (Health/Safety/Medical)

Respiratory Diseases

Respiratory System

Restoration and Preservation

Restoration and Preservation, Art Works/Artifacts

Restoration and Preservation, Structural/ Architectural

Science Education

Science Planning/Policy

Science and Technology

Science and Technology Centers

Science and Technology, History of

Scleroderma

Secondary Education

Social Services

Social Services Delivery

Social Stratification/Mobility

Social Work

Social Work Education

Sociology

Volunteers (Education)

Voter Educational Programs

Waste Management

Waste Management/Fossil Energy

Wastewater Treatment

Water Resources

Water Resources, Environmental Impacts

Water Resources, Management/Planning

Water Supply

Water Treatment

Waterways and Harbors

Weapons

Welding

Welfare Reform

Welfare-to-Work Programs

Wellness

Wetlands

Wildlife

Youth Services

Youth Violence

Zoology

Zoos

Program Type Index

NOTE: Numbers refer to entry numbers

Awards/Prizes

Basic Research

Building Construction and/or Renovation

Capital Campaigns

Centers: Research/Demonstration/Service

Citizenship Instruction

Community Development

Consulting/Visiting Personnel

Cultural Outreach

Curriculum Development/Teacher Training

Demonstration Grants

Development (Institutional/Departmental)

Dissertation/Thesis Research Support

Emergency Programs

Endowments

Environmental Programs

Fellowships

General Operating Support

Graduate Assistantships

Grants to Individuals

International Exchange Programs

Job Training/Adult Vocational Programs

Land Acquisition

Matching/Challenge Funds

Materials/Equipment Acquisition (Computers, Books, Videos, etc.)

Preservation/Restoration

Scholarships

Seed Grants

Service Delivery Programs

Symposiums, Conferences, Workshops, Seminars

Technical Assistance

Travel Grants

Vocational Education

Geographic Index

Note: This index lists grants for which applicants must be residents of or located in a specific geographic area. Numbers refer to entry numbers.

United States

Minnesota

North Carolina

North Dakota

Northern Mariana Islands

Ohio

Oklahoma

Oregon

Pennsylvania

Puerto Rico

Rhode Island

South Carolina

US Virgin Islands

Utah

Vermont

Canada

Other Foreign Countries

All Countries

www.ingramcontent.com/pod-product-compliance
Ingram Content Group UK Ltd.
Pitfield, Milton Keynes, MK11 3LW, UK
UKHW051421290926
15471UKWH00009B/1713

9 780983 762270